ENCYCLOPEDIA OF WORLD PROBLEMS AND HUMAN POTENTIAL

Union of International Associations, Brussels

München · New York · London · Paris

Selected Publications of UIA

Yearbook of International Organizations
27th edition, 3 volumes, 1990/1991, ISSN 0084-3814.

> **Vol.1 Organization Descriptions and Index**
> 27th edition, 1990/91, 1776 pages + Appendices (14). ISBN 3-598-22205-X.
>
> **Vol.2 International Organization Participation: Country Directory of Secretariats and Membership (Geographic Volume)**
> 8th edition, 1990/91, 1760 pages. ISBN 3-598-22206-1.
>
> **Vol.3 Global Action Networks : Classified Directory by Subject and Region (Subject Volume)**
> 8th edition, 1990/91, 1684 pages. ISBN 3-598-22203-3.

International Congress Calendar
31st edition, 1991, quarterly. ISSN 0538-6349.

Encyclopaedia of World Problems and Human Potential
3rd edition, 2 volumes, 1991. ISBN 3-598-10842-7.

> **Vol. 1 World Problems**
>
> **Vol. 2 Human Potential**

International Association Statutes Series
1st edition, 1988, 600 pages. ISSN 0933-2588.0. ISBN 3-598-21671-8

Who's Who in International Organizations
1st edition, 1991. ISBN 3-598-10908-3.

ENCYCLOPEDIA OF WORLD PROBLEMS AND HUMAN POTENTIAL

Volume 1: World problems

Edited by

Union of International Associations

3rd edition

K·G·Saur München·New York·London·Paris

ENCYCLOPEDIA OF WORLD PROBLEMS AND HUMAN POTENTIAL

The following people worked on the preparation of the current edition in different capacities and for different periods of time.

Editorial Staff
Marie Aeles *(Bibliographies)*
Nancy Carfrae
Anne Degimbe
Kristof Elst
Carine Faveere
Martine Gosse
Jon Jenkins *(World Problems)*
Maureen Jenkins
Jacqueline Nebel *(Human Development)*
Tarja Ryynänen *(World Problems)*
Rosemary Staniforth
Cecile Vanden Bloock

Computer support
Elisabeth Gale
Bernhard Knutsen
Paul Montgmery
Colin Mainoo
Stewart Woung

The programme through which this publication is produced is orchestrated by Anthony Judge.

Published Jan 1991 by
K.G. Saur Verlag KG
Ortlerstrasse 8
D-8000 München 70
Federal Republic of Germany

Information collected and edited by
Union of International Associations
40 rue Washington
B-1050 Bruxelles, Belgium

Computer typeset by
Computaprint Limited
39A Bowling Green Lane
London EC1R ONE, United Kingdom

Cover design by
Tim Casswell

Printed and bounded in Federal Republic of Germany

Deutsche Bibliothek Cataloguing-in-Publication Data

Encyclopedia of world problems and human potential / ed. by Union of International Associations. - München; New York; London; Paris: Saur.

ISBN 3-598-10842-7

NE: Union of International Associations; World problems and human potential

Vol. 1. World problems. - 3. ed. - 1991

Copyright 1991 by Union of International Associations. All rights reserved. No part of this work may be reproduced or copied in any form or by any means - graphic, electronic or mechanical, including photocopying, recording, taping, or information and retrieval systems - without written permission of the Secretary General, **Union of International Associations.**

ISSN 0304-0089 ISBN 3-598-10842-7

UAI Publication 299

Human life is driven forward by its dim apprehension of
notions too general for its existing language.
    Alfred North Whitehead

The names that can be named are not definitive names. Naming
engenders ten thousand things....Thirty spokes share the
wheel's hub. It is the centre hole that makes it
useful....Therefore profit comes from what is there;
usefulness from what is not there.
    Lao Tzu

In contrast with what is commonly assumed, a description, when
carefully inspected, reveals the properties of the observer. We,
observers, distinguish ourselves precisely by distinguishing what
we apparently are not, the world.
    Francisco Varela

Behind the misty wall of words, the diverse, even contradictory,
interpretations, motivations and utilisations are an indication of
fundamental divisions concerning values. In particular, the most basic
human rights are more frequently invoked as a weapon of attack or
defence against some party, rather than recognized as the royal
road to a positive relationship between individuals and groups in an
objective form of fraternity.
    René Maheu, Director-General, UNESCO

When men understand only one of a pair of opposites, or concentrate
only on a partial aspect of being, then clear expression also
becomes muddled by mere word play, affirming this one aspect and
denying all the rest....The wise man therefore sees that on both
sides of every argument there is both right and wrong.
    Chuang Tzu

Break the pattern which connects the items of learning and you
necessarily destroy all quality....The pattern which connects is a
metapattern. It is a pattern of patterns.
    Gregory Bateson

Neti Neti ("Not this; not that")
    Sanskrit aphorism

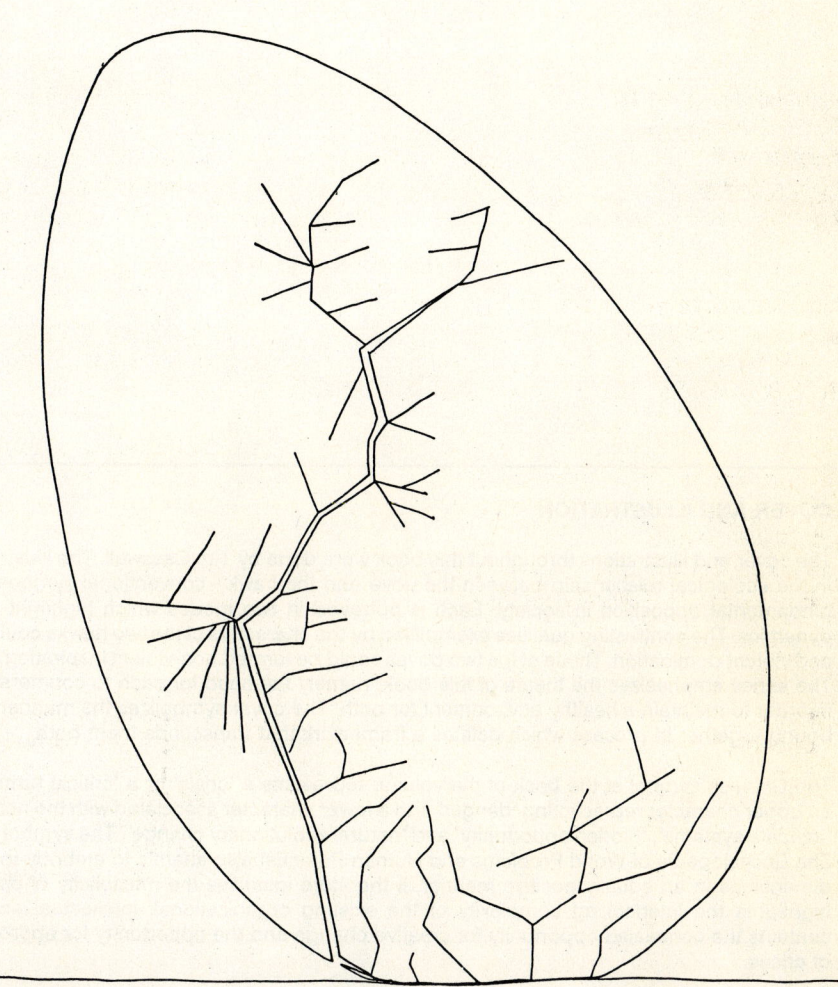

COVER AND ILLUSTRATION

The cover and illustrations throughout this book were done by Tim Casswell. The illustrations explore many phases in the ecological relationship between the dove and the hawk - conventional symbols for roles and processes in fundamental opposition in society. Each is potrayed in two modes which highlight the need for compensatory dynamics. The contrasting qualities exemplified by the characters of the two hawks could be termed heroic vigilance and violent domination. Those of the two doves could be termed non-violent inspiration and exploitative domination. The series emphasizes the theme of this book, namely the need for each to counteract the excesses of the other in order to maintain a healthy environment for both. The cover symbolizes the manner in which such extremes are bound together in process which defines a framework that transcends them both.

The Chinese symbol at the back of the volume represents a "crisis" or a "critical turning point". It is composed of an upper character representing "danger" and a lower character associated with the notions of "organic complexity", "intricate systems", "hidden opportunity" and "natural evolutionary change". The symbol is reproduced here because The Encyclopedia of World Problems and Human Potential also attemts to embody these seemingly incompatible dimensions in an equally positive manner. It therefore identifies the multiplicity of dangers to society, but it also highlights the interlocking complexity of the existing organizational, intellectual and personal resources. This contains the concealed opportunity for creative change and the opportunity for appropriate response to the crisis of crises.

Contents: Volume 1

INTRODUCTION

User guide	12
Overview	15
Content	29
Method	35
Assessment	43
Integrative insights	55

WORLD PROBLEMS

- **Section P: World Problems**

Section PA: Abstract fundamental problems	77
Section PB: Basic universal problems	109
Section PC: Cross-sectoral problems	137
Section PD: Detailed problems	235
Section PE: Emanations of other problems	425
Section PF: Exceptional problems	627
Section PG: Very specific problems	843
Section PJ: Problems under consideration	845

- **Section PX: VOLUME INDEX (for Volume 1)** — 847

- **Section PY: Bibliography** — 1085

- **Section PZ: Notes** — 1141

Contents: Volume 2

INTRODUCTION

    User guide   12
    Overview   15

HUMAN POTENTIAL

- **Section H: Human Development**

Section HH: Human development concepts	27	Section HX: Index to concept sets	315
Section HM: Modes of awareness	139	Section HY: Bibliography	335
		Section HZ: Notes	361

- **Section K: Integrative Knowledge**

Section KC: Integrative concepts	397	Section KX: Index	465
Section KD: Embodying discontinuity	433	Section KY: Bibliography	473
Section KP: Patterning disagreement	457	Section KZ: Notes	497

- **Section M: Metaphors and Patterns**

Section MM: Metaphors	519	Section MX: Index	573
Section MP: Patterns of concepts	533	Section MX: Bibliography	581
Section MS: Symbols	561	Section MZ: Notes	587

- **Section T: Transformative Approaches**

Section TC: Transformative conferencing	641	Section TX: Index	677
Section TP: Transformative policy cycles	667	Section TZ: Notes	685

- **Section V: Values**

Section VC: Constructive values	717	Section VX: Index	805
Section VD: Destructive values	743	Section VY: Bibliography	817
Section VP: Value polarities	769	Section VZ: Notes	825
Section VT: Value types	801		

VOLUME INDEX

- **Section X: Index to Volume 2**   841

APPENDICES (Section Z)

    Statistics   935
    Computers   939
    Union of International Associations: Profile   949

Introduction

User guide

    Access   12
    Warning   13
    Errata   14

Overview

    Contextual challenge   15
    Existential challenge   17
    Strategic assumptions   19
    Objectives   21
    Significance   23
    Intended uses of Encyclopedia   25
    Background and acknowledgements   27

Content

    General structure   29
    Sections and sub-sections   30
    Modifications, improvements and omissions   33

Method

    Logistical challenge   35
    Procedures   36
    Classification policy   37
    Language-determined distictions   40
    Response to diversity   42

Assessment

    International organizations as a source   43
    Biases   45
    Strengths and weaknesses   46
    Criticism   47
    Global modelling perspective   49
    Future possibilities   51
    Implications   52
    Processing system   54

Insights

    Section interrelationship   55
    Comprehension of sustainable integration   57
    Problem perception and deception   59
    Incommunicability of insights   60
    Problem perception and level of awareness   62
    Phases of human development through challenging problems   64
    Integration of perceived problems   66
    Barriers to transcendent insight and social transformation   68
    Interrelating possible viewpoints   69
    Human impotence and potential   73
    A new global organizational order?   74

***   Bibliographical references identified in abridged form in the following section refer to publications detailed, by author, in Section PR, which is the bibliography for Volume 1.

User guide: access

- **Volumes**

This Encyclopedia is divided into two volumes:

-- Volume 1: World problems
-- Volume 2: Human potential

- **Sections**

Each volume is divided into sections and sub-sections. Each section is denoted by one code letter (e.g. P= World problems; V= Human values). Each sub-section is denoted by two code letters (e.g. HH= Human development concepts; VC= Constructive values). Sections and sub-sections all appear in alphabetic order by code letter. The code letters also have some mnemonic significance. All sections and sub-sections are listed on the contents page.

- **Entries**

Each sub-section is composed of a series of entries. Each entry is numbered using the code letters of the sub-section (e.g. PE2370 = Abuse of tax havens). The entries appear in numeric order within the sub-section.

- **Volume indexes**

The easiest way to find an entry on a specific topic is by consulting the Volume Index, where names of all entries are listed together in alphabetic order by keyword. The index gives the sub-section and number where the entry is to be found. There is a separate Volume Index for Volume 1 (see Section PX) and for Volume 2 (see Section X).

- **Section indexes**

An alternative way to find an entry in Volume 2 is by using the mini-indexes located near the end of individual sections. They are more convenient for scanning the range of entries in a section. These provide an overview of entries within a sub-section. Section VX is the index for Section V, for example.

- **Explanations**

A brief introduction and commentary is provided at the beginning of each section and of each sub-section. More detailed comments are provided in the Notes at the end of each section. For example, the Notes for Section H are at HZ, the Notes for Section K are at KZ.

- **Cross-references**

Cross-references between entries are explained in the sub-section introductions where appropriate.

- **Bibliographical references**

Several sections have bibliographies. These are located near the end of each section. For example, the Bibliography for Section H is at HY, the Bibliography for Section K is at KY.

- **Classified index**

A classified index by subject (3000 categories) is provided to the World Problems section (Section P) in a companion series: *Yearbook of International Organizations* (Vol. 3). This also includes international organizations and treaties dealing with the same subject.

User guide: warning

1. Inconsistencies
The information collected in the Encyclopedia, and especially in the world problems section, is derived from a very wide range of sources. These reflect many levels of insight and expertise, as well as many cultures, ideologies, beliefs, priorities and biases. No attempt has been made to eliminate inconsistencies, although incompatible items have been treated as separate entries where appropriate. For example, both "capitalism" and "communism" are treated as world problems.

2. Juxtaposition
This Encyclopedia is deliberately organized in such a way as to juxtapose bodies of information which are normally kept apart. The hard reality of the "world problems" section is counterbalanced by various sections highlighting human values and development. Within the world problems section itself, for example, "counterarguments" are given (where such information is available) questioning or denying the facts presented in the problem description.

3. Perceptions
Wherever possible the information is compiled using extracts from documents of international bodies, whether governmental or non-governmental, formal or informal. In this sense the information may be viewed as factual. Given the different interpretations of these facts however, the information presented, especially in the case of world problems, can best be viewed as a collection of perceptions with which significant international constituencies identify strongly in advocating (or resisting) any social change. The Encyclopedia provides an overview of the world's hopes and worries, whether real or imaginary.

4. Editorial intervention
In honouring the biases active in the international community in this way, the editors have limited themselves to ensuring that the texts in the main sections, especially on world problems, make their point strongly and in as clear and concise a manner as the available material permits. In this period of imminent crisis, the editors have however accepted the need for a higher level of risk in exploring innovative possibilities. Some of the smaller sections are therefore the result of deliberate editorial experiments in gleaning and presenting information to highlight such possibilities, despite the risks of inadequacy and error.

5. Editorial bias
The basic bias of the editors is against limitation of information to reflect only a single viewpoint or paradigm, whether ideological, cultural, scientific or religious. Within any such paradigm, the information here also reflects different levels of ignorance, rather than attempting only to reflect a consensus prevailing amongst an elite group of authoritative experts (whose views may be poorly received outside their own circle). The bias is therefore to include information from some constituencies which may well be judged qualitatively inferior, misleading, irresponsible, or irrelevant by some other constituency.

6. Significance
The amount of information given on any problem, for example, does not reflect an editorial evaluation of its importance. Problems commonly accepted as important may be documented only briefly. This may be because of resource limitations, because of the profusion of relatively diffuse material available on them, or because they can be more effectively documented through their sub-problems. Little-known problems may be given relatively extensive coverage precisely because their existence is not well-recognized. Inclusion of information in this publication implies only that the editors considered the source from which it derived sensitive to and capable of reflecting the views of an international constituency, and therefore as being of significance to a wider audience.

7. Naivety
Information on phenomena such as world problems, values or modes of human development is widely assumed to be relevant to the design of any new broad-based initiatives in response to the global problematique. The editors have accepted the need for a certain naivety to break through the conceptual frameworks determining the general indifference of academic and governmental authorities to any questions concerning the actual number and variety of such phenomena. In identifying such phenomena within an open framework, some entries (on which whole libraries of books have been written) must necessarily appear naive. But despite the availability of such a wealth of detailed information, to the point of overload, there is a poverty of information on how to connect together this fragmented pattern. It is to this condition that this project responds by indicating possibilities, even if at times the result appears superficial or naive.

8. Pragmatism
The production of this book, within the constraints of modest resources, has been feasible only because of an extremely pragmatic approach to the collection and processing of information. Within these constraints the editors have deliberately set out to "open up", or highlight, neglected categories of information, fleshing out the content to the extent possible. Where there has been conflict between ability to locate and process adequate information (within a reasonable time period) and the elaboration of the pattern of categories, the latter has been given priority. The intention has been to provide as broad a coverage as was feasible. Hopefully, even where the information supplied is inadequate, readers will be oriented to new features of the global system which others view as meriting their attention.

9. Non-completion
This Encyclopedia (in its third edition) is the product of an ongoing project to explore ways of identifying and presenting categories of information relevant to the development process as perceived by international organizations. Major refinements will therefore continue to be made to many of the sections, and to the pattern of cross-references, especially in response to feedback on inadequacies. In this sense the book cannot be regarded as an unfinished product.

10. Solutions
This book in no way attempts to present an editorial view of "the answer" to the world's problems. Some sections in the past editions indicated the various kinds of answer, or bases for an answer, which are favoured within the international community. The editors have however endeavoured to respond to the challenge of how to interrelate inherently incompatible answers. The concern has been to respond to the possibilities formulating an appropriate meta-answer of practical signif' in such paradoxical circumstances.

User guide: errata

A publication of this scope, based on a multiplicity of sources of information, necessarily contains errors of the following kinds:

- Errors of content, due to the sources of information used;

- Errors of interpretation, due to the manner of selection and representation of the information used by the editors;

- Errors of typography and form;

- Errors arising from the process of selecting and registering cross-references;

- Errors arising from circumstances unforeseen in the design of the many computer programmes through which the data has been processed.

Considerable editorial effort has been made to reduce the number of trivial formal errors, but it has not been considered feasible to eliminate all of them within the resources and time available.

With regard to substantive errors, many of the entries on world problems, for example, contain information from one international group which some other international group would consider erroneous. In this sense this book documents the fallacies which are active in the international community by juxtaposing incompatible perceptions.

Through each successive edition, the editors have attempted to respond to error in the spirit advocated by Donald Michael:

"Changing towards long-range social planning requires that, instead of avoiding exposure to and acknowledgement of error, it is necessary to expect it, to seek out its manifestations, and to use information derived from the failure as the basis for learning through future societal experiment. More bluntly, future-responsive societal learning makes it necessary for individuals and organizations to embrace error. It is the only way to ensure a shared self-consciousness about limited theory as to the nature of social dynamics, about limited data for testing theory, and hence about our limited ability to control our situation well enough to expect to be successful more often than not." ("On the requirement for embracing error"; In: On Learning to Plan and Planning to Learn. 1973).

Overview: contextual challenge

Much has been written on the challenge of our times. It could be argued that further statements on the dimensions of the crisis are both repetitive and counter-productive. Specific problems are the topic of frequent media coverage and of reports by bodies of the highest authority. There are also many positive indications that nurture hope that major crises may be averted. The events in Eastern Europe are an example -- although it is not clear whether the excitement at such breakthroughs, and the new possibilities they offer, do not also serve to obscure other emerging crises to which we prefer not to give attention.

In such a context, what then is the value of a new edition of an Encyclopedia of this kind? Especially when information overload has itself become more than a minor problem, do we really need yet another book on the problems of the world?

The programme through which this Encyclopedia is produced is based on the assumption that our difficulties in responding to the challenge of the times lie as much in how we process information with a view to action as in the process of implementing solutions. There seems to be a prevailing confidence in the methods of the international community in response to the problems of the times. This is less than warranted by the very partial successes of the strategies implemented -- and the dimensions of the many problems that continue to grow. This confidence is sustained in part by the methods of the academic community, to the extent that their theoretical preoccupations are brought to bear on issues faced by society.

It would appear that a number of unquestioned assumptions are made in responding to the problems of society. The assumptions are implicit in difficulties such as the following:

1. Pseudo-objectivity of analyses of problems
The vast majority of descriptions of problems recognized by the international community are produced using methods which depend upon authoritative interpretations of the significance of data, whether quantitative or not. The manner by which the data is selected varies, as does the basis for the interpretation of any such data by different international organizations (or other constituencies). The importance, and even the existence, of many problems thus becomes questionable in the debates between constituencies. "Over-population" is the most striking example. And yet reports continue to be produced claiming objectivity in exhorting use of particular strategies, whilst implicitly or explicitly suggesting that other interpretations are suspect. The dynamic between such opposing perceptions and priorities is not captured. As a result any remedial programmes are undermined by the dynamics inherent in the relationship with any opposing perception. This can then be used as a convenient scapegoat in the event of failure.

2. Withholding of relevant information
The reports produced are those based on readily available, acceptable information. In the case of official reports, it is conveniently forgotten that standard procedures require that information embarrassing to particular governments or interests should be omitted or toned down. Much information is available only on a restricted basis, if at all. Information is only classified because of its importance, which suggests the conclusion that much that is made available for use in public reports is of little real consequence. Data significant to understanding of problems may simply be withheld, especially in the case of embarrassing incidents in which a cover-up policy is implemented. Issues relating to the incidence of leukaemia in people exposed to nuclear tests in the 1950s are an example. The non-disclosure of comparative international statistics on crime is another. This dimension is seldom reflected in authoritative reports.

3. Issue avoidance
Within a pattern of institutions mandated to deal with recognized problems, any indication of the emergence of problems that are inadequately handled is perceived as a threat to those whose budgets and careers depend upon positive evaluations of their incumbency. The tasks of organizations are complicated enough as they now stand. Further complication is therefore resisted. Reporting procedures, basic to budget and career assessments, therefore tend to avoid mention of programmatic weaknesses or the emergence of new problems (unspecified within the unit's mandate). The emphasis is on "upbeat" reporting in order to conceal deficiencies. Bad news is unwelcome at any level of an institutional hierarchy. The bearer, as was traditionally the case, may be severely penalized. Evaluations of institutional responses to problems tend to fail to reflect this dimension.

4. Misrepresentation of information
Information made available tends to be presented in such a way as to encourage the most favourable interpretations. Thus, aside from the process of issue avoidance, active steps may be taken to cast a positive light upon events or to support favoured arguments. Information may be "adjusted", especially in the case of statistics or the results of monitoring exercises. If necessary various forms of deception may be practised, even by official bodies. Disinformation is one such practice. The production of some reports can be seen as an effort to dissuade and to distract rather than to provide a basis for more appropriate action.

5. Biased expertise
Experts of any discipline survive through the fulfilment of contracts, as is increasingly the case for industry-funded university departments. Their continuing survival depends on their ability to meet the requirements of funding bodies. The work of eminent specialists may even be undertaken through organizations "fronting" for vested interests. Experts must therefore be sensitive to the kinds of conclusions that are acceptable. In such circumstances experts can, if necessary, be found to support any position -- if only that there is "no proven link" calling for politically sensitive action. The conclusions of some highly authoritative reports may therefore be pre-determined by ensuring the presence of appropriate experts on the investigatory body (a procedure known as "stacking" a committee). Any subsequent effort to question the report from other perspectives can be disparaged as quibbling by the unqualified serving other interests. The question of the degree to which some of the major reports have been biased in this way has not been explored.

6. Corruption
Many programmes have been carefully designed in response to problems and yet have failed or under-performed. Whatever the official explanation, there is much evidence to indicate that an important factor in such failure is the activity of involved individuals in attempting to profit to an unforeseen degree from the resources and influence that they control during the execution of the programme. Much of the evidence is anecdotal, although well-known to those with any experience in the field. Much is reported on a daily basis in the quality press. It often touches those at ministerial level. Such corruption, although possibly occurring in many forms, is not confined to any particular group of countries, as some in the industrialized countries would like to claim, although it does tend to be more discreetly organized there. Such semi-formal subversion of programmes is tacitly accepted, even at the highest level within international organizations. No international study of corruption has ever been made. Its potential for undermining new strategies is never officially mentioned when they are advocated.

7. Harassment
Individuals aware of any of the issues noted above are not free to report on them in written form -- or rather they do so at their own risk. Typically most official bodies require that employees sign non-disclosure agreements which may well apply after termination of their employment. Any attempt to provide hard evidence can severely affect career opportunities (such as through "blacklisting") in the case of official actions. In the case of unofficial actions, as with various systems of corruption, it can lead to severe peer group pressure from those who do not wish "the boat to be rocked". Harassment can also take physical form, especially against external activists and "whistleblowers", where cases of assassination have

been documented. The action of the French government against the Rainbow Warrior in New Zealand is an example. Naturally official reports tend to be discreet where such threats may be brought to bear.

8. Short-termism
In producing official surveys on the challenge of the times, the priority of many is naturally to ensure their survival through to the next budgetary cycle. In a political environment it is short-term issues which are the guarantee of survival. Longer-term issues can safely be given lower priority, even when they are exacerbated by short-term decision-making. A focus on short-term action creates the impression of effective action even though it may be counter-productive in the longer-term. This cannot be effectively addressed in reports addressed to bodies governed by short-term priorities.

9. Ambiguity
Consensus under the above constraints can be most easily achieved through use of ambiguity, each constituency projecting onto an agreement its own interpretation. "Development" is the most tragic example. According to a definition favoured by commercial interests, any degradation of the environment can be interpreted as a development achievement (as typified by "clearing" the land). "Sustainable development" can thus be widely approved through being understood as "sustainable competitive advantage". Clearly international reports run the risk of being shelved if they do not permit such ambiguity of interpretation. This effectively undermines what many are led to assume is the purpose of such strategies.

10. Unaccountability
The accountability of institutions and those responsible for them is limited. Many institutions cannot be effectively held accountable for their abuse of the social or natural environments. Thus the World Bank has been able to resist any sensitivity to environmental issues for at least a decade after these formally became a concern to the United Nations. Senior management can seldom be held accountable for the unethical actions of their employees, even when they are responsible for the pressure giving rise to such actions. The whole framework of "plausible deniability" runs throughout organizations. The the chief executive is thus well insulated and able to deny knowledge of any unethical action by the body for which he is responsible.

11. Violation of commitments
Commitments in response to problems are violated, whether they take the form of breaches of electoral promises, neglect of resolutions of international meetings, or failure to conform to the agreed provisions of intergovernmental treaties and conventions. Given the development of contract law and the regulation of advertising claims, it might be asked why similar standards are not applied to the promises and claims made by those seeking political power.

12. Loss of integrity
In the context defined by the above issues personal integrity is easily compromised or readily sacrificed. In any competition for resources, it becomes a luxury. In any particular case, it is unclear how compromised an eminent authority may be in supporting some position. This is also true of countries whose declared support for positive initiatives is often totally compromised by previous or parallel commitments to programmes having the reverse effect (typified by military aid to repressive regimes). In the case of international organizations, the most striking symbol of loss of integrity has been the Waldheim incident and the silence of the great powers aware of the facts (unless it is to be assumed that their intelligence agencies are unbelievably incompetent). It is extremely difficult to raise such questions in relation to new initiatives.

13. Loss of credibility
Much has been learnt as a result of the limited success of programmes over the past decades. People are increasingly aware of the issues indicated above. As a result there is a widespread erosion of the credibility of institutions and official expertise. Increasingly this extends to any organized activity. There is awareness of the ways in which the media are used to manipulate opinion and to spread disinformation. Much of this loss of credibility may indeed be unjustified and even paranoid. It nevertheless affects the ways in which reports and new programme initiatives are received.

14. Disparagement of complementary initiatives
The great majority of views on the problems of society, social directions, and possible alternatives are posited with little or no reference to any other views, past or present. When such reference is made, it tends to be made disparagingly or with condescension. Competition for scarce resources obliges organizations, and departments within organizations, to define an approach which establishes the irrelevance of other initiatives whose activities might under other circumstances have been considered complementary.

15. Dubious standards of proof
Within these constraints, the tragedy is that any truth about the challenge of the times has become something movable, an illusion to be marketed for the benefit of the few. There is no standard by which the pronouncements of collectivities can be assessed with any degree of confidence. The greater the resources controlled, the greater the pressure to deceive unless constrained to do otherwise. Standards of proof developed by science or judicial systems have been shown to lend themselves to abusive manipulation even in the most respected democracies. It has not been possible to prove, in the scientific sense of the term, that such abuse is exceptional rather than systemic. There appears to be no way that powerful institutions can prove their integrity or that of their representatives. To use the favoured phrase, there is "no proven link" between statements emanating from such collectivities and the reality with which the world is faced. It is clearer to state that their reports have higher or lower degrees of correspondence with that reality, according to the pressures to which they are subject.

It would be both naive and presumptuous to assume that any body could escape the above constraints. However, rather than seeking coherence in the presentation of information from a selected group of authorities -- the approach of many reports on the condition of the world -- a different approach is possible. This can open the door to more radical insights into the dilemma of the times precisely because it explicitly recognizes difficulties such as those indicated above.

Overview: existential challenge

Corresponding to the contextual challenge in responding to their environments, individuals ar faced with an existential challenge in redefining their self-image and the mind-set with which they respond to the world. The following are some of the features of this challenge.

1. Proliferation of explanations and injunctions
The hyper-development of the ability to explain and to label has fostered the pervasive illusion that this necessarily ensures that an environment so treated is somehow under control. Much effort is devoted to this process, whether by researchers, educators, legislators, administrators or managers. There is a resemblance to the enthusiastic reliance on pesticides by the agri-business. This process does appear to freeze portions of the environment, since readily comprehensible explanations tend to be in static terms. Not only does this render invisible any dynamic relationship to other aspects of the environment, but it also defines the explainer at the same reduced level of complexity -- at least in the relationship to what has been so explained.

2. Simplistic responses
Action on the environment, perceiving and responding to problems, is then viewed primarily as a question of reordering explained categories into a more appropriate pattern -- "sustainability" being the latest criterion. "Profitability" is a competing criterion. Irrespective of the criterion, there is a resemblance to the procedures by which radioactive products are handled in laboratories through "glove-box" manipulators. The person controlling the manipulators is of a much higher order of complexity than that aspect which is manifest through the possible movements of the manipulators. And yet problems are perceived and acted on at the level of complexity of the manipulators. The glove-box delimits the reality to which society is prepared to respond and constrains the manner of that response. But above all it protects the users of the glove-box from exposure to the invisible challenge of the products therein.

3. Paradigmatic entrapment
Action on problems thus becomes a matter of shuffling categories and institutional elements, combining and recombining them in an effort to increase the effectiveness of response. New categories and institutions are invented within the same pattern. Blame for problems is reallocated in a similar manner. In this way much change is apparent, together with many explanations as to why such change is sufficient to the challenge of the times. And yet this perception tends to remain unchallenged.

4. Failure to act on knowledge
Explanations do not respond to present (or future) suffering, although they may reduce anxiety about it. A physician, fully informed of the dangers of smoking, may continue to do so, irrespective of the recognized effects on his own health or the indirect consequences for others. Similarly a factory may continue to discharge pollutants, despite the manager being fully informed of the consequences for the environment. A walker may point with complaint to a piece of rubbish in a forest but not feel called to remove it. Such examples are indicative of the protection offered by the existential glove-box. It permits those using it to feel uninvolved. The pattern of explanations and injunctions has a numbing effect by which individuals are protected from any challenge to their own pattern of behaviour.

5. Unaccountability
The professionalism of international responses to the challenges of the times also protects individuals from any need to be personally concerned whether a programme succeeds or fails (provided explanations can be found to deflect any negative consequences for career advancement). But how to distinguish between the necessary detachment of a surgeon whose skills are unable to save a patient, and the indifference of a surgeon whose inappropriate action is aggravating the condition of the patient?

6. Disempowerment
The quantity of explanations and injunctions, and the eminence of those offering them, disempowers non-specialists. Those who are not mandated to provide authorized explanations are forced into a position of dependency for the construction of the reality in which they live. Imagination is crushed by the weight of explanations and by those who are empowered to impose them. Imagination itself is only acknowledged in those who have proven their commercial worth. As such it has become a product for consumption. In the glove-box, images are generated which trap the unwary into belief in their reality.

7. Unchallenged self-image
There has been much concern voiced about the need for a new paradigm and for non-linear approaches to the complexities of the environment. This seems to address the simplistic, even mechanistic, pattern of category shuffling -- a recognition that the glove-box only permits a restricted pattern of movements. But much of that discussion still seems to be calling for what amounts to a more complex set of manipulators for the same glove-box. The relationship of the user of the glove-box to the manipulations therein is not called into question. The dualistic relationship is not challenged. However rich the paradigm, to what extent will it call for a new self-image on the part of the user of the glove-box implying a new involvement in action? To what extent do perceived problems become existential challenges rather than merely a flavour-of-the-month?

8. Unchallenged relationship to the environment
The fashion for "holographic" paradigms, with the implication that everything is reflected in everything else, is an intellectual challenge calling for a broader and richer understanding. The Gaia model is of this kind. But whilst these call for a higher degree of responsibility and accountability on the part of individuals, they fail to render explicit the challenge to the articulation of the individual's relationship to the environment. A non-dualistic framework continually questions both what it means to be and act as an individual and what is the nature of the environment in which action is taken.

9. Need for a different mode of thought
It is indeed possible to avoid this challenge. It is possible to assume that one does not need to change and that the paradigm of "development decades" and "international organizations" is sufficient for the times. It may be assumed that "win-win" solutions are possible and that there need be no losers -- gain without pain. But there is increasing recognition that unless the life-style in industrialized countries is radically changed, the current system will become increasingly unsustainable. This calls for a different mode of thought.

10. Existential sacrifice
The prevailing mode of thought makes no explicit call for existential sacrifice, since sacrificing a category within the glove-box is not an existential operation. And yet in the larger reality people are indeed sacrificed through the imposition of austerity programmes -- structural adjustment with an "inhuman" face. A non-dualistic approach sees winning and losing as complementary phases in necessary learning cycles. Like inspiration and expiration, they are both necessary processes in a growing organism. To what sacrifice do administrators of programmes expose themselves in order to comprehend more subtle approaches to the environment?

11. Change vs pseudo-change
It is ironic that it is only those who are least appreciated who consciously expose themselves to being sacrificed in contemporary society. The dramatic examples are terrorists on suicide missions, self-immolating protesters, and soldiers in a jihad. However society also requires human sacrifices before legislative changes are considered necessary: children have to die before dangerous foodstuffs are prohibited by law, and demonstrators have to be willing to suffer, or lose their lives, before their cause receives attention. Real change is accomplished when people expose themselves to an existential challenge, thus becoming agents of change. Pseudo-change occurs when the initiators engage in manipulations within the glove-box which leave them totally untouched (other than through any loss of status or pride, as in losing a wager).

12. Distorted value of life
Again it is ironic that there is less and less in modern society that people are prepared to die for, or to allow others to die for. Whole societies can now be held to ransom for a single known hostage. Millions can be spent to maintain a comatose, brain-damaged patient on life-support for decades. Euthanasia is illegal, no matter what the desire of the person concerned. Exposure to risk is progressively designed out of society, to be replaced by vicarious experiences through videos or with the protection of required safety devices. The paradox is that unknown numbers are however sacrificed through carcinogenic products, abortion, structural violence, massacres, gang murders, cult rituals, "snuff" movies and associated perversions, or a failure of food and medical supplies.

13. Value sacrifice
The attitude to life has become as immature as that to death. Millions are spent on efforts to maintain youthfulness, whether through cosmetics, cosmetic surgery or attempts to reverse the ageing process. Every other value is sacrificed to save lives in industrialized societies, whilst allowing others to die elsewhere. Individuals in industrialized societies are prosecuted for life-endangering neglect. But these same societies fail to apply the same standards in their policies towards other societies. Reproduction is tacitly encouraged without any provision for the resulting population growth or the effects on the environment. Society evokes problems to provide solutions for its own irresponsibility -- a control mechanism for the immature lacking the insight for a healthy relationship to cycles.

14. Personal transformation
The challenge of the times would seem to involve a call for personal transformation through which social and conceptual frameworks can be viewed anew. Willingness to sacrifice inherited perspectives is an indication of the dimension of the challenge -- most dramatically illustrated by willingness to risk death. However physical death is not the issue, and may easily be a simplistic, deluded impulse lending itself to manipulation. Destruction of frameworks valued by others is equally suspect. Such dramatics provide rewards within the very frameworks whose nature the individual needs to question, but by which he may need to choose to be constrained.

15. Existential discipline
What are the existential disciplines by which an individual can progressively redefine what he or she is in relationship to the environment -- at present and in response to the emerging future? What does this imply for the organizations through which individuals may work or for the conceptual frameworks appropriate to that work? How does this understanding affect the individual's relationship to the piece of rubbish at his feet? The many spiritual traditions provide clues for further exploration. But their advocates are often dangerously enthusiastic about their own insights and disparaging of others. Insight is buried in dogma. The letter obscures the spirit and denies the awareness by which they may be distinguished. Instant faith is demanded to avoid the long-term challenge of disciplined acquisition of insight. This suggests that any such disciplines must also be applied in response to the purveyors of insights and to their products. But where there is no exposure to risk and the possibility of error, there is no learning or possibility of meaningful change.

16. Transformation of perspective
There is an irony in the call to ensure humanity's continued existence on the planet. It is a challenge to our existence, but it has not yet been recognized as an existential challenge. As with suppositions about the life hereafter, it is assumed that no change is implied for our understanding of ourselves. And yet it may be that such transformation of perspective is the key which the nature of the crisis will force us to recognize. As the Sufis suggest, the trick of insight required may be to remove the point from which we currently view. There is an illusion of who we are that needs to be sacrificed to give birth to a sustainable future.

17. Persistent personal egotism
There are many admirable advocates of change, whether involving social transformation or personal transformation. The most eminent, whether in the world of science, art, politics, or religion, have made striking contributions to this process and will continue to do so. And yet the level of egotism among those with most insight remains a major challenge, as illustrated by the following comment made by Richard Gardner and quoted by Michael Marien (*Societal Directions and Alternatives*, 1976): "We are afflicted not only by national but by personal egoism. That is what could eventually destroy us.. Many of these eminent people have such big egos that their principal preoccupation in life is to establish a piece of intellectual turf and preserve it against all comers, whatever the consequences. They're prepared to sacrifice truth -- prehaps not consciously, but subconsciously -- to the pursuit of idelgoy and the pursuit of ego."

Overview: strategic assumptions

Over the past 25 years, from the first International Development Decade, international groups and organizations have implemented or advocated every conceivable strategy offering some promise of counteracting the emergence of a crisis of crises. Whatever the successes, it is widely acknowledged that the basic trend has not been significantly affected. This recognition has itself been voiced so frequently through the Secretary-General of the United Nations that it has itself become an outworn generalization associated in the minds of many with the of loss of credibility of existing institutions, of democratic political processes and of academic research, all of which have proven incapable of more than token response to the global problematique. The series of special international commissions (Brandt, Palmer, Brundtland, South, etc) convened over the years to report on particular aspects of the emerging crisis have proven to be as much a symptom of collective impotence as capable of offering a foundation for new initiatives.

The same 25 years have seen the emergence of a widespread counter-culture which has offered the hope of alternative approaches. These have borne fruit in the form of new communities, personal growth movements, political activism, volunteer programmes, alternative technology, computer-supported networking and the green movement. These developments have been sustained in part by exciting breakthroughs in comprehension of the nature of self-organization, paradigm change, holism, implicate order, and the relationship between physics and consciousness. Nevertheless whilst these continue to offer the possibility of significant impact on the global problematique, this has not been forthcoming. And to a large extent such alternative approaches have appeared as luxuries irrelevant to the priorities of developing countries.

In envisaging the design of a project to respond to the challenges noted above, the following strategic constraints have been assumed:

1. Maintenance orientation
It would appear that collective ability to respond to the crisis of crises has been effectively paralyzed. The 1980s have seen the emergence of a sense of apathy, defeatism and despair in the international development community and in grass roots movements. This is largely disguised by public information programmes and media events designed to maintain confidence in projects and campaigns which do indeed have some measure of success. But as the food crisis in Ethiopia has demonstrated, although a magnificent one-time attempt can be made to remedy short-term problems in the spotlight of media coverage, the solutions to the underlying longer-term problems are not in sight. At this point in time programmes are deemed a success if they can slow the trend toward major crisis. An acceptable criterion is maintenance of the status quo, provided it lends itself to being described as innovation. Significant social innovation is seldom sought, however eloquently it is advocated.

2. Solution production
Many "answers", whether explanations, programmes, strategies, ideologies, paradigms or belief systems, are put forward in response to the current crisis, however it is perceived. The proponents of each such answer naturally attach special importance to their own as being of crucial relevance at this time, whether in the short-term for tactical reasons, or in the long-term as being the only appropriate basis for a viable world society in the future. This widespread focus on "answer production", is a vital moving force in society. However it obscures both the significance of the lack of fruitful integration between existing answers and the manner in which such answers undermine each other's significance. This mind-set also fails to recognize the positive significance of the continuing disruptive emergence of new "alternative" answers.

3. Questionable truths
Amongst this multitude of answers, explanations put forward as factual by scientific or government authorities are increasingly questionable because of peer group, religious, political, military, security and commercial pressures guiding objective evaluation and reporting. Recent examples include dubious evaluations by authorities of nuclear reactor and toxic waste hazards, official denial of the impact of acid rain on forests, and reassessment of the world population problem as non-critical. The situation has been epitomized in NASA, the model of western high-tech management, by the top executive pressures on engineers to withhold information on the gravity of problems associated with low-temperature effects on space shuttle launchings. Middle management in any bureaucracy is under considerable pressure to report positive achievement in the light of pre-defined policy objectives, rather than to indicate the dimensions of problems detected in the process. There is no assurance that such pressures do not affect the reporting of many other facts of social significance. Self-censorship is increasingly practised as in biology textbooks (to meet creationist objections) and in encyclopedias (to avoid raising unwelcome political questions concerning such social realities as corruption and institutionalized torture). Even in courts of justice, an expensive (astute) lawyer considerably increases the probability of a judgement favourable to his client. The truth of facts has become a question of interpretation, leaving authorities free to deny politically unacceptable conclusions by selecting experts prepared to declare that "there is no proven causal link" between the problems in question (even though such a link may be accepted by equivalent bodies in other countries).

4. Gladiatorial arena
Policy integration initiatives at this time are themselves fragmented and mutually hostile, to a degree usefully interpreted in terms of the metaphor of a "gladiatorial arena". The survival of any integrative answer must be bought at the price of the elimination of all other competitors. There is considerable confusion about the nature of integration and it is difficult to imagine that integrative processes favoured by one group would be considered to be of much significance by another. This phenomenon cannot be disguised by simply opting for consensual procedures, "networking" processes or by viewing it as a "healthy" feature of academic or political debate.

5. Irrelevance of alternatives
The most characteristic response to this confusion is to simplify the situation by establishing or affirming, explicitly or implicitly, the fundamental irrelevance of any other answers and perspectives that are viewed as incompatible, if their existence is recognized at all. The preoccupations of the other constituencies are thus defined as dangerously misguided or agonizingly irrelevant. As a consequence there is always a perfectly valid reason for not instigating any advocated course of action or for not considering any alternative perspective.

6. Projection of blame
Many would reject any such recognition of paralysis. But the basis for their rejection is that, "if only" some other portions of society would cease to block effective change then this would release the resources that would demonstrate the collective paralysis to be only momentary. Unfortunately it is precisely the number and variety of such "if only theys", which has ensured the spread of this paralysis and which guarantees that it will prevail for some time to come.

7. Assumption of innocence
Corresponding to this projection of blame onto other groups, as suitable scapegoats, is a widespread assumption of the unquestionable innocence of one's own group. This may well be perceived as making an untarnished significant contribution to the well-being of society. Whether it be academic disciplines or their corresponding professions, national or international organizations, public or private bodies, benevolent or alternative groups, each acts as though its contributions to society constituted an unmitigated good. However valuable these may be, the suspect consequences of these contributions can only be questioned at the risk of ridicule. Sanctions may be applied against those voicing such criticism, from within or without, whether in the case of the United Nations or of alternative groups. A perfect disguise is therefore provided for every possible systematic abuse.

8. Reinforcement of fragmentation

One major characteristic of the plethora of material documenting the ills of the global community is that it tends to reinforce the plaintive or angry plea, noted above, that "if only" some other group would act in some other way all could be well. Each such report focuses on one part of the network of problems, explicitly or implicitly denying the relevance of some other part with which others identify. It is understandable that any such other group would not be strongly motivated to respond to the concluding pleas of such a report. Furthermore it will probably associate itself with some other report denying, explicitly or implicitly, the relevance of the priorities laid out in the first. This process can be observed between the Specialized Agencies of the United Nations or their equivalents at the national level. It can be seen in the failure of the Brundtland Report to build on the Brandt Report, followed by the failre of the South Report to build on the Brundtland Report. It is however far from being limited to governmental bodies.

9. Narrow information base

The consequence of this process is that no group is motivated to recognize or document the full range of perceptions of the ills and opportunities of the world. Such information exists but has to be culled from documents in different locations, which very few are inclined or able to do. If such perceptions are not interrelated the chances of reducing the level of paralysis are handicapped. The argument is therefore that recognizing the full range of ills and opportunities by which groups are touched, and with which they identify, is a minimum requirement for exploring the ways in which they can be collectively empowered to release their contribution to the paralysis.

10. Single-focus dependence

Such fragmentation encourages, and is further reinforced by, dependence on single-factor explanations and single-policy initiatives. Each such initiative may necessarily formulated be in terms of a limited information base. This is usually discipline-oriented in the case of the academic community, but ideological, action-preference, "priority" and other filters may also be used. The integration of the approach is thus achieved artificially by deliberately avoiding the encouragement of a variety of complementary approaches capable of counteracting each other's weaknesses. When the opponents of such a unified approach can demonstrate its weaknesses, they then move to implement another simplistic approach of a countervailing nature in order to remedy them. Society thus moves spastically from policy to policy without any ability to acknowledge the merit of an ecology of policies and of alternation through a cycle of policies.

11. Initiative obsolescence

Single-focus dependence leads directly to the repetition of initiatives of a form which has failed in the past or whose success has been only marginal. Questioning strategies based on thinking of this kind, especially when they are defined with politically acceptable trigger words (population, energy, environment, food, health, education), may be considered tantamount to questioning the merits of motherhood. In the Club of Rome's terms, many such initiatives are maintenance-oriented and are incapable of innovative breakthroughs. The need to break through to new forms of initiative is not accepted by the international community. Even the eloquent pleas for a new order are made on the assumption that well-tried conceptual, policy, programme, organization and conference forms are appropriate to its conception and implementation, with perhaps some minor adaptation.

12. Disagreement phobia

Society has been unable to design any framework, whether conceptual or organizational, in which disagreement is an accepted, permanent integral feature. The frameworks now used are based on the assumption that consensus is the keystone on which any viable organization must depend. As a consequence, disagreement can never be accepted as an integral feature of society, except through structures or processes designed to eliminate it (conflict resolution, mediation, arbitration). These include competition and violent conflict, in which victory is sought, through the downfall of the opponent. Although disagreement is a daily and often creative reality, the fear of situations in which disagreement prevails is such that they are shunned, whether unconsciously or by well-rationalized processes. When they cannot be avoided, much effort is devoted to amplifying the significance of whatever minor items can be discovered on which agreement has been achieved. Agreement then becomes an essentially superficial pretence of little operational significance. Conceptual, organizational or legal structures based on such agreement are consequently totally inadequate to the innovative requirements of any dynamic development process in which disagreement is inherent. Stressing consensus as a key to development and social transformation comes dangerously close to destroying the basis of its dynamism. Development can only occur if there is disagreement with those maintaining the status quo.

13. Self-reflective paradox

Any attempt to reflect the widest possible range of perspectives on the ills and opportunities of the world is bedevilled by an interesting paradox. Given the prevalence of disagreement, whatever method is employed must necessarily engender disagreement. It cannot be expected to result in some ideal, objective approach that would engender universal consensus. Indeed the very attempt to reflect the fullest range of perspectives must naturally remain suspect to those with the vested interests necessary for any specific form of action. Any breakthrough into a more fruitful mode must therefore endeavour to give explicit recognition to this paradox and to the dynamics associated with it. In this light it would be unproductive to attempt to produce yet another "answer" to the condition of the world, however inadequate it might be.

14. Embodiment of discord

The widespread tendency to produce incompatible answers is a symptom of the underlying paralysis noted above. Any endeavour to break out of this paralysis must respond to this dynamic, if it is to be of any relevance to the current conditions. Under the prevailing linear approach, a particular position is taken up and defended, as required by the militaristic conventions of academic, religious, political or ideological debate. This could be contrasted with a complementary non-linear response, whereby such positions are identified both as perceived by those who hold them and by those who consider them nonexistent, irrelevant, misleading or downright evil.

A valid response is therefore to attempt to design a framework to internalize or embody discord, contradictions and logical discontinuity. The status within the framework of the perspective that the attempt itself represents must necessarily remain a paradox. A further step is therefore called for within such a framework to explore the adequacy of conceptual language to contain such incommensurable perspectives so characteristic of the dynamics of global society. The ultimate question is therefore how to interrelate inherently incompatible answers without producing yet another answer to compete with them in a process which has proved unable to transcend itself.

Overview: objectives

1. Organizational context

One of the original objectives, in initiating this project in 1972, was to endeavour to document how the network of international bodies focused on the network of world problems. Clearly some key problems attract much attention and many others attract very little, if any. But of greater interest is whether the organizations focusing on the same problem are in communication with each other, or whether the organizations dealing with one problem that closely affects another are in contact with the bodies dealing with the latter problem. How do problems escape the net of organizations? How does the network of organizations fail to encompass and contain the network of problems? What kinds of information would then be required to enhance transnational initiatives?

In addition to the major role that international organizations perform in identifying problems, many of them perform an important function in relation to human values. In fact the two functions are often intimately linked, as in the case of human rights issues. A significant group of organizations is also concerned with human development in its less material sense, as is the case of bodies concerned with religion and personal development. Again the question may be asked with which of the specific values is an organization associated, given that many of them carefully identify values in their statutes? Clearly there are very "popular" values, such as peace and justice, but are there values with which few, if any, bodies are associated? And to what extent are such values vital to the functioning of society? Is there also a mismatch between the network of organizations and the network of values?

2. Core concerns

(a) Clarification of "fuzzy" domains: The core concerns of the international community, whether problems, values or human development, remain conceptually "fuzzy". They are a continuing challenge to both scholars and practitioners. It is from this fuzziness that dilemmas, contradictions and paradoxes emerge. In considering the role of integrative approaches, and even the power of metaphor, in responding to this "fuzziness", it is apparent that here too there is considerable ambiguity and confusion. For this reason, despite the vital importance of these concerns, they are especially difficult to handle within information systems. Few information systems attempt to do so, preferring to deal with harder data. Experimenting with computer-based procedures to do so therefore constitutes a valid preoccupation.

(b) Recognition of "vectors of concern": The many international constituencies tend to disparage each others concerns, if they recognize the existence of concerns other than their own. Mapping the "vectors of concern" in relationship to one another provides a means of defining the nature and dimensions of the communication space within which the dynamics of the international community operate. Recognizing such vectors determines with what concerns constituencies identify, thus clarifying what moves them to act as well as the nature of the social reality within which they perceive themselves as functioning.

(c) Bridging between incommensurable domains: A major underlying concern is to create a framework within which it is possible to register links between specific world problems, human values and modes of human development. The intimate relationships between these domains call for more effective ways of processing information on them as a complex system.

(d) Anchoring transient insights: Insightful perceptions of subtle challenges and opportunities, of potentially major significance, appear in the literature. Because of their nature, and the categories and perspectives that they call into question, there are few places at which this kind of information can be collected. A suitable context is required to hold and "anchor" that information in relation to more conventional perspectives.

(e) Transformation enablement: There is a need for information in a
form which enables social transformation in the light of more appropriate values. This suggests as a valid objective the extension of information system design to incorporate the fuzzy conceptual dimensions which catalyze and motivate such transformation.

3. Objectives

The objective of the project through which this Encyclopedia is produced is threefold and may be described as follows:

(a) Collection and presentation of information
At this level the intention is to demonstrate the feasibility and value of assembling information reflecting the perspectives of a very wide range of international constituencies. In contrast to normal practice, this information should not be filtered by some particular criteria of "truth" or "importance". Every effort should be made to present it in terms of what is held to be true by the constituencies from which it originates, even if the information totally contradicts that from some other international constituency. It is a basic assumption of this project that it is the dynamics inherent in the interaction of such conflicting biases which reflect the reality of global society, as much as the fundamental insights emerging from any particular analysis of the global system in the light of criteria carefully selected by leading experts.

In organizing the information into the sections in this Encyclopedia, the intention has been to group material into classes corresponding to the terms conventionally used to describe and order any response to the global problematique and the possibilities of human development. Each of these tends in some way to be of fundamental concern to any international constituency, whatever the differences about the appropriate content of such classes.

In designing a framework "hospitable" to such a wide range of perspectives, whether mutually indifferent or inherently incompatible, a secondary objective has been to seek ways to juxtapose such perspectives in order to highlight the variety of relationships between them. The framework therefore contains the variety of incompatible perspectives by fragmenting the information into a very large number of descriptive entries. This deliberate disorganization is counterbalanced by a very extensive network of cross-references between such entries. When appropriate information has been obtained from appropriate sources, some form of counterargument is included in many entries, illustrating the limited or misleading nature of the perspective presented.

Metaphor: star mapping This objective can be usefully described in terms of the metaphor of an astronomical telescope. Whereas a limited number of astronomical objects are visible to the naked eye, their visibility from Earth is determined both by their intrinsic brightness and by their distance from the observer. The major problems cited by any international constituency are equivalent to the brightest of those objects. Others may be barely visible to them. By the use of a telescope the number of visible astronomical objects, whether stars or galaxies, increases enormously, depending on the resolving power of the telescope. The brightness of some of them, to an observer located elsewhere, may be very much greater than those visible from Earth. So for some other international constituency, a different, but possibly overlapping, set of problems appears to be of major importance. The challenge of this project is conceived as one of designing a telescope of sufficient resolving power to collect information from distant sources on the phenomena which are highly visible to them. This is achieved by using the whole array of international bodies as collectors, thus constituting (as with a radio telescope with a long base line) a much more powerful telescope than that based on dependence upon any one of them or upon any small group of them. As with recent discoveries concerning the dangers of exposure to low-level radiation, this may also help to demonstrate that long-term exposure to less visible problems can be as dangerous as short-term exposure to the more visible problems.

It is hoped that the collection and presentation of information in this reference book form will meet the information needs of many users.

(b) Clarification of conceptual challenge

At this level the intention is to clarify the challenge of interrelating perceived patterns of information with which people and constituencies can identify and by which they are empowered. In one sense this project is an endeavour to document the perceptions active in global society. For the resulting quantity of information to begin to become meaningful as a whole, this calls for new approaches to communication, with an emphasis on patterns of concepts. The perceptions documented are those with which different people identify and by which they are motivated. For such motivations to reinforce each other to achieve the required political will to change, greater understanding is required of how patterns of concepts may be nested together without doing violence to the particular perceptions with which people identify. For such social change to be fruitful, there is the even more challenging requirement of ensuring a comprehensible relationship between mutually incompatible patterns of concern that can correct each others' inadequacies and excesses.

Metaphor: electricity generation This second objective can to some extent be described in terms of the metaphor of electricity generation. The electrical current produced by some form of generator depends upon the degree to which opposite charges can be simultaneously generated within the same framework and conducted together (but insulated from each other) to the point where the difference between the charges can be used to do work. This project endeavours to accumulate and juxtapose within the same framework both extremely negatively charged information on world problems, and extremely positively charged information on human potential in various forms (values, subtler states of awareness, etc), rather as in the design of a battery. The hypothesis is that it is through an appropriate juxtaposition of the "bad news" and the "good news" that the generation of the will to change is effectively generated. This is in strong contrast to many other initiatives which endeavour to focus only on positive initiatives (solutions, values, etc), only on negative doom-mongering, or on a mixture from which the opposing charges cannot be effectively separated so as to empower people to act. In the light of this metaphor the latter efforts are as likely to succeed as attempts to design monopolar batteries or an electrical circuit with a single wire. When they do succeed in mobilizing people, their subsequent failures could be usefully compared to the dangerous discharges resulting from the generation of static electricity.

It is hoped that the information presented here will stimulate some users to contribute further to the clarification of this challenge.

(c) Enablement of paradigm alternation

At this level the intention is to explore indications of ways of moving beyond the sterile relationships between the existing paradigms within which the perceptions documented in this Encyclopedia are generated. For although the strengths and weaknesses of such paradigms continue to be demonstrated in many studies, the purpose of such studies tends to be that of proving the merits of some existing or alternative paradigm. The challenge then is to explore ways of moving beyond prevailing conceptual fragmentation whilst avoiding the opposite danger of simplistic holism under the guise of sterile relativism. The challenge is made more dramatic by the irresponsibility of experts. Whilst these may be qualified to justify some particular position, they are totally unable to offer any guidance to voters and decision-makers as to the manner by which their position can be reconciled with some totally contradictory position justified on other grounds.

In an isolated local context, or a simpler world, this difficulty may be avoided by establishing certain perceptions as true and others as false, misleading or totally irrelevant. Some people are then empowered by the acceptance of such a coherent pattern of truths and the challenge of articulating them. Others are empowered by the process of denying the corresponding falsehoods.

In the more complex modern world of interacting contexts, decision-makers are forced to recognize pragmatically that contradictory positions may both be true, possibly under different conditions, even though there is no coherent framework within which they may be reconciled. Some are even empowered by the opportunity this provides to "divide and rule" by "playing one side against the other". But there is also the recognition by others that neither position need be true, and they are then empowered by the process of rejecting the system constituted by both together.

Metaphor: fusion reactor This third objective can also be described in terms of the metaphor of the current technological challenge of designing a suitable magnetic container for plasma to enable nuclear fusion to take place. In order to generate energy in a fusion reactor, the problem is to discover the particular configuration of magnetic fields, values of plasma parameters, and means of protecting the plasma from contact with any material surface which would quench it. This can be achieved by "bouncing" the plasma around within the configuration of a magnetic cavity (or "bottle"). As in the case of plasma, any comprehensive understanding of the human condition (encompassing both the global problematique and the associated opportunities for human development) is "quenched" by any efforts to contain it within a particular conceptual framework. And as with plasma, transcending this difficulty seems to require the design of a container which ensures that such understanding can only emerge, exist and develop if it is continually "bounced" or alternated between an appropriate configuration of different conceptual perspectives. Although there are indications as to the possible design of such a container, the multi-perspective containers that have so far been designed reflect the lowest common denominator of the participating perspectives, rather than the highest common insight by which appropriate action in response to the global problematique could be empowered.

As this metaphor illustrates, this project is in many ways about the adequacy of the language used amongst international constituencies. To what extent are the challenges of society and the possibility of innovative response determined by the distinctions and connections permitted or forbidden by the language of the international community (and its various jargons)? Can the many distinct problems, values and strategies engendered by that language be meaningfully distinguished? Is it in some way fundamentally inadequate as a means of formulating distinctions and relationships that are required to respond appropriately to the global problematique?

It is hoped that this Encyclopedia may to some degree be used to explore the nature of the art of alternating between paradigms, languages or viewpoints as a way of enabling individuals and societies to be appropriately empowered in response to the conditions of the moment. The challenge appears to be to discover a comprehensible conceptual dynamic of sufficient complexity to permit an appropriate conscious alternation between the different combinations of acceptance and denial. This has been admirably illustrated in drawings by the artist Escher, especially as analyzed by Douglas Hofstadter in *Gödel, Escher, Bach*. As in the relationship between male and female, or between parent and child, it is the collective equivalent of the art of saying "yes" or "no" under changing conditions. This is at its most frustrating and enchanting as it explores the excluded middle ground forbidden by the boundaries of Aristotelian logic, however vital the latter may be in particular circumstances.

Overview: significance

The significance of each of the sections of this publication is treated separately in the introduction at the beginning of each section and in the Notes on each of them.

The significance of this publication as a whole can best be briefly illustrated by the following quotations which, taken together, indicate the importance of exploring the kind of approach attempted here.

1. Antiquated concepts and attitudes
"It is unforgivable that so many problems from the past are still with us, absorbing vast energies and resources desperately needed for nobler purposes: a horrid and futile armaments race instead of world development; remnants of colonialism, racism and violations of human rights instead of freedom and brotherhood; dreams of power and domination instead of fraternal coexistence; exclusion of great human communities from world co-operation instead of universality; extension of ideological domains instead of mutual enrichment in the art of governing men to make the world safe for diversity; local conflicts instead of neighbourly co-operation. While these antiquated concepts and attitudes persist, the rapid pace of change around us breeds new problems which cry for the world's collective attention and care: the increasing discrepancy between rich and poor nations; the scientific and technological gap; the population explosion; the deterioration of the environment; the urban proliferation; the drug problem; the alienation of youth; the excessive consumption of resources by insatiable societies and institutions. The very survival of a civilized and humane society seems to be at stake. The world is bursting out of its narrow political vestments. The behaviour of many nations is certainly inadequate to meet the new challenges of our small and rapidly changing planet. International co-operation is lagging considerably." (U Thant, Secretary-General of the United Nations on the occasion of United Nations Day, 1970).

2. Inadequacy of fractional approaches
"Many of the most serious conflicts facing mankind result from the interaction of social, economic, technological, political and psychological forces and can no longer be solved by fractional approaches from individual disciplines...Complexity and the large scale of problems are forcing decisions to be made at levels where individual participation of those affected is increasingly remote, producing a crisis in political and social development which threatens our whole future." (Bellagio, Declaration on Planning, 1968)

3. Crisis of crises
"What finally makes all of our crises still more dangerous is that they are now coming on top of each other. Most administrations...are not prepared to deal with...multiple crises, a crisis of crises all at one time... Every problem may escalate because those involved no longer have time to think straight." (John R Platt. What we must do Science, 1969)

4. Interwoven networks
"Society is not a crowd or cluster or clump of human beings; it is a set of networks of relations between human beings. Every human being is linked with others in a number of networks which are not mutually exclusive and are also not coextensive with each other." (Arnold Toynbee. Aspects of Psycho-history. Main Currents in Modern Thought, 1972)

5. Mismatch between organizations and problems
"The map of organizations or agencies that make up the society is, as it were, a sort of clear overlay against a page underneath it which represents the reality of the society. And the overlay is always out of phase in relation to what's underneath: at any given time there's always a mismatch between the organisational map and the reality of the problems that people think are worth solving... There's basically no social problem such that one can identify and control within a single system all the elements required in order to attack that problem. The result is that one is thrown back on the knitting together of elements in networks which are not controlled and where the network functions and the network roles become critical." (Donald Schon. What can we know about social change?, BBC Listener, 1970)

6. Entrapment by problems
"When anything becomes a problem we are caught in the solution of it, and then the problem becomes a cage, a barrier to further exploration and understanding." (J Krishnamurti. The Urgency of Change, 1971)

7. Arrogance of the disciplines
"...how is a practitioner of any one discipline to know in a particular case if another discipline is better equipped to handle the problem than is his? It would be rare indeed if a representative of any one of these disciplines did not feel that his approach to a particular organizational problem would be very fruitful, if not the most fruitful..." (R L Ackoff. Systems, organizations, and interdisciplinary research, General Systems, 1960)

8. Helplessness in the face of complexity
"Because our strength is derived from the fragmented mode of our knowledge and our action, we are relatively helpless when we try to deal intelligently with such unities as a city, an estuary's ecology, or the quality of life." (Editorial. Fortune, 1970)

9. Dependence on information
"Today, as we have seen, information is not primarily the triumphant standard of progress. It is the only means of maintaining sufficient control of evolution in order that humanity, strengthened by its knowledge and experiences and making appropriate use of all available information, can always maintain itself ahead of any threat which may lead to catastrophe." (Helmut Arntz, President, International Federation for Documentation, 1975)

10. Information overload
"The problem is that in most, if not all spheres of inquiry and choice, quantities of raw information overwhelm in magnitude the few comprehensive and trusted bodies or systems of knowledge that have been perceived and elaborated by man... Where, for example, does the novice urban mayor turn to comprehend the dynamic interrelationships between transportation, employment, technology, pollution, private investment, and the public budget; between housing, nutrition, health, and individual motivation and drive? Where does the concerned citizen or Congressman interested in educational change go for the best available understanding of the relationship between communications, including new technology, and learning?" (McGeorge Bundy. Managing knowledge to save the environment, US House of Representatives, 1970)

11. Category obsolescence
"The most probable assumption is that every single one of the old demarcations, disciplines, and faculties is going to become obsolete and a barrier to learning as well as to understanding. The fact that we are shifting from a Cartesian view of the universe, in which the accent has been on parts and elements, to a configuration view, with the emphasis on wholes and patterns, challenges every single dividing line between areas of study and knowledge." (P F Drucker, The Age of Discontinuity; guidelines to our changing society. 1968)

12. Evocation of fragmentation
"...the penalty for any principle which fails to express the whole is the necessity to co-exist with its opposite." (Lancelot Law Whyte. The Next Development in Man, 1950)

13. Challenge of synthesis
"...in face of the growing specialization of thought and action brought about by diversification in research and the division of labour, Unesco has a duty to promote interdisciplinary activities and contacts and to encourage broad views, in short, to emphasize the vital importance of the spirit of synthesis for the health of our civilization. I say vital advisedly since man - and I mean his essence, which is to say his judgment and his freedom of choice - is just as likely to be smothered by his knowledge as paralysed by the lack of it. Similarly, he is quite as likely to lose his identity in the confusion of competing social pressures as to atrophy in the condition known as under-developed."

(René Maheu, Director-General of Unesco, Address to a symposium on science and synthesis, 1967)

14. Integration of elements of thought
"*The synthesis we need involves a better integration of the elements of our thought, policies and institutions in order to solve national problems effectively, via the efficient achievement of national goals through the use, where appropriate, of such means as science and technology.*" (Robert W Lamson, Office of Exploratory Research and Problem Assessment, National Science Foundation, USA as read into the *Congressional Record* 93rd Congress, 2nd Session, 1974 by Charles S Gubser)

15. Facilitation of new forms of knowledge
"*Interdisciplinary knowledge can only develop through interdisciplinary education; it is a question of facilitating the emergence of a new form of knowledge... Whilst operating according to the norms of his specific dimension, the researcher must be able to encompass a mental space vaster than the epistemological cell within which his research runs the risk of confining him... The new understanding must be based on an affirmation of the functional unity of the human being as a focal point for all research intentions in the different domains of knowledge... This new understanding must be embodied in a new pedagogy, oriented to compensating for the deficiencies of specialization by stressing the combined unity of all domains of knowledge.*" (Georges Gusdorf. Interdisciplinaire (connaissance). In: Encyclopedia Universalis)

16. Need for unified personalities
"*The development of a world culture concerns mankind at large and each individual human being. Every community and society, every association and organization, has a part to play in this transformation; and no domain of life will be unaffected by it. This effort grows naturally out of the crisis of our time: the need to redress the dangerous overdevelopment of technical organization and physical energies by social and moral agencies equally far-reaching and even more commanding. In that sense, the rise of world culture comes as a measure to secure human survival. But the process would lose no small part of its meaning were it not also an effort to bring forth a more complete kind of man than history has yet disclosed. That we need leadership and participation by unified personalities is clear; but the human transformation would remain desirable and valid, even if the need were not so imperative. The kind of person called for by the present situation is one capable of breaking through the boundaries of culture and history, which have so far limited human growth. A person not indelibly marked by the tattooings of his tribe or restricted by the taboos of his totem: not sewed up for life in the stiff clothes of his caste and calling or encased in vocational armour he cannot remove even when it endangers his life. A person not kept by his religious dietary restrictions from sharing spiritual food that other men have found nourishing; and finally, not prevented by his ideological spectacles from ever getting more than a glimpse of the world as it shows itself to men with other ideological spectacles, or as it discloses itself to those who may, with increasing frequency, be able without glasses to achieve normal vision. The immediate object of world culture is to break through the premature closures, the corrosive conflicts, and the cyclical frustrations of history. This breakthrough would enable modern man to take advantage of the peculiar circumstances today that favor a universalism that earlier periods could only dream about.*" (Lewis Mumford. The Transformations of Man. 1956)

17. Mental defences
"*...That since wars begin in the minds of men, it is in the minds of men that the defences of peace must be constructed...a peace based exclusively upon the political and economic arrangements of governments would not be a peace which could secure the unanimous, lasting and sincere support of the peoples of the world...*" (Constitution of the United Nations Educational, Scientific and Cultural Organization)

18. Fostering self-knowledge
"*The relations between world culture and the unified self are reciprocal. The very possibility of achieving a world order by other means than totalitarian enslavement and automatism rests on the plentiful creation of unified personalities, at home with every part of themselves, and so equally at home with the whole family of man, in all its magnificent diversity... Without fostering such self-knowledge, balance, and creativity, a world culture might easily become a compulsive nightmare.*" (Lewis Mumford. The Transformations of Man, 1956)

19. Concealed neurotic processes
"*The fact which confronts us is that cultural change is limited by the restrictions imposed on change in individual human nature by concealed neurotic processes. At the same time there is continuous cybernetic interplay between culture and the individual, ie between the intra- psychic processes which make for fluidity or rigidity within the individual and the external processes which make for fluidity or rigidity in a culture. It would be naive to expect political and ideological liberty to give internal liberty to the individual citizen unless he had already won freedom from the internal tyranny of his own neurotic mechanisms... Therefore, insofar as man himself is neurotogenically restricted, he will restrict the freedom to change of the society in which he lives. This interplay is sometimes clearly evident, sometimes subtly concealed; but it is the heart of the solution of the problem of human progress.*" (Lawrence S Kubie. The nature of psychological change and its relation to cultural change. In: Ben Rothblatt (Ed) Changing Perspectives on Man, 1968)

20. New vision of selfhood
"*Man's principal task today is to create a new self, adequate to command the forces that now operate so aimlessly and yet so compulsively. This self will necessarily take as its province the entire world, known and knowable, and will seek, not to impose a mechanical uniformity, but to bring about an organic unity, based upon the fullest utilization of all the varied resources that both nature and history have revealed to modern man. Such a culture must be nourished, not only by a new vision of the whole, but a new vision of a self capable of understanding and co-operating with the whole. In short, the moment for another great historic transformation has come. If we shrink from that effort we tacitly elect the post-historic substitute. The political unification of mankind cannot be realistically conceived except as part of this effort at self-transformation; without that aim we might produce uneasy balances of power with a temporary easing of tensions, but no fullness of development.*" (Lewis Mumford. The Transformations of Man, 1956)

21. Choice
"*We can either involve ourselves in the recreative self and societal discovery of an image of humankind appropriate for our future, with attendant societal and personal consequences, or we can choose not to make any choice and, instead, adapt to whatever fate, and the choices of others, bring along.*" (Center for the Study of Social Policy of the Stanford Research Institute. Changing Images of Man, 1974)

In summary: It seems appropriate to attempt to bring together and interrelate within one framework information on: the problems with which humanity perceives itself to be faced; the organizational, human, and intellectual resources it believes it has at its disposal; the values by which it is believed any change should be guided; and the concepts of human development considered to be either the means or the end of any such social transformation.

Problems, organizations, concepts and human development are usually considered as though they were unrelated. But it is necessary to have a progressively more integrated conceptual structure in society before the interrelationships between the newer problems can be perceived. Both are needed before an attempt can be made to interrelate organizational units to handle the interlinked problems. Individual ability to tolerate and comprehend the complexity and dynamism of these interrelationships is directly related to the individual's own degree of personal development. Furthermore, a general increase in integration in any of these four domains will tend to increase integration in the other three. Equally, progressive fragmentation in any of the domains will provoke disintegrative tendencies in the others.

Even if the constraints make it impossible to achieve a satisfactory result through this particular exercise, it is to be hoped that through the process outlined here it will be possible to learn more about how information from very diverse sources can be concentrated and structured to the critical level required to provide the kind of integrative overview necessary for all to develop a sufficiently complex and strategically sound response to the world problem complex as it is now emerging.

Overview: intended uses of the Encyclopedia

A project of this kind evokes amongst some the response "Why bother, when we already know what ought to be done?" Who, after all, needs another book parading the range of problems with which the global community is confronted? Key people no longer have time to read more than one page summaries and each international body is acting as best it can to contain the problems to which it is sensitive.

In 1984 the Director of Political Affairs of one major intergovernmental body considered this project both presumptuous and ridiculous. He then went on to argue that problems did not "exist" in a way which allowed them to be identified and described in a book. For his institution they were agenda items which came and went according to the political currents of the moment, ceasing to "exist" once his organization was no longer obliged by political pressures to deal with them.

Others would argue that it is a grave mistake to focus on problems in any way because this "gives them energy", hindering the necessary "positive thinking" from which appropriate social transformation can emerge. There is widespread belief that the action required can be simply defined. Food aid is a topical example, although even major intergovernmental bodies are now acknowledging the counter-productive aspects of such generosity. A modicum of humility would require the recognition that most seemingly positive initiatives have at least minor counter-productive effects - omelettes cannot be made without breaking eggs.

There are however many who point out that international institutions are not containing the problems faced by the global community; rather they are being overwhelmed by them. To function at all, such bodies have to concentrate on very small portions of the pattern of problems, denying the relevance of other portions or even their very existence. This is especially the case when they are constrained to prove the value of their own initiatives even though they may aggravate such other problems. Many claim to know what needs to be focused on, or done, or avoided to resolve the crisis - if only everybody else would subscribe to their particular set of priorities. In such a context it is appropriate to present these many "action vectors" within a single framework, in effect bringing them collectively to consciousness rather them denying or repressing those which do not fall neatly within some favourite paradigm.

This Encyclopedia is therefore intended for those who question whether they are receiving information from a sufficiently broad range of perspectives. It is for those who believe that much might be learnt from the variety of perspectives on what constitute significant problems and significant responses to them. In particular it is for those who recognize the possible dangers and limitations of attempting to filter this variety down to a handful of "essential" problems which can be appropriately contained by a single policy, strategy or blueprint based on a single conceptual framework guided by a single set of values. The decision that any particular class of information in the Encyclopedia is irrelevant can be seen as raising valuable questions as to the nature of the assumptions on which each such judgement is made.

It is expected that the majority of readers will use this book to locate specific items or groups of information. Some users will respond to the challenge of ordering, comprehending and presenting such a range of information in new ways, because of the extent to which it reflects the variety of issues with which people and groups identify and by which they are motivated. It is hoped that some will also be further stimulated to explore the possibility of patterned dynamic relationships between incompatible conceptual languages, encompassing the discontinuity between them, in order to develop a dynamic conceptual foundation appropriate to the global order of the future.

The users of this volume will therefore include:

1. Commercial enterprises and entrepreneurs
Corporations concerned with navigating in a complex and turbulent social environment can detect problems which may affect them, whether directly or indirectly. For them, problems may also be looked upon as potential business opportunities, since some require heavy investment for remedial action.
Seemingly trivial problems, such as acne, may represent an important market. Each problem can be viewed as affecting or concerning a portion of the population. As such there is a potential market associated with each problem. Corporations may respond to this market with remedial products (as in the case of water pollution), with consultancy and other services (as in the case of environmental impact assessment), or with publication and information services (as with registers of pollutants).

2. Education
The information may be used in programmes with students in many fields who need to acquire an overview of the range of global issues, how they may relate to one another, and the difficulties of ordering such information within one conceptual framework. Of special interest is not only the information given here but also its weaknesses and the controversies associated with particular claims or counter-claims. It is an interesting challenge to students to attempt to detect problems which are not present here, especially in the light of their awareness of problems in their own environment. Such explorations can be extended to the international organizations supposedly concerned with the problem area.

3. Integrative studies
University departments (international relations, environments, law, social science) concerned with interdisciplinary issues can use it to stimulate discussion among students. It should be of particular value to departments responsible for designing general studies programmes for students.

4. Policy development
In many ways the Encyclopedia provides a form of checklist for policy-related issues. Ideally in attempting to elaborate a policy framework in a particular domain, the information here could be used to identify related issues which may need to be taken into consideration and which would otherwise be neglected until too late. It is especially valuable for relationships between problems across disciplinary and paradigm boundaries. For, whether concerns are a matter of established fact or deeply held opinion, they may need to be given serious consideration in any policy design.

5. Management sciences
The total set of problems suggests interesting lines of research into the modes of governance in a complex environment. This is especially the case where the issue is how to manage networks of problems using cycles of policies which encompass more than one budgetary or electoral cycle.

6. Foresight and futures research
Because the data included covers both currently fashionable and seemingly marginal or improbable concerns, it provides a much more appropriate source for use in anticipating new kinds of issues which may emerge to greater prominence than can currently be imagined. Many problems registered reflect the concerns of groups sensitive to issues that conventional bodies are unable to recognize.

7. Government
The information here may be used by government departments designing programmes which need to be sensitive to problems and possibilities in other sectors. Of special interest is the possibility of using the identified problems as a checklist to determine which government department, if any, is concerned with each problem, and thus evoking discussion about issues which are not the explicit responsibility of any department.

8. Briefings
The Encyclopedia is an interesting background document for briefings of diplomats or members of delegations, whether for

ministries of foreign affairs or for other bodies. As such it may also be used in training programmes. It provides a corrective to easy assumptions about mono-problem situations, based on single-factor explanations, leading to simplistic solutions.

9. Law and treaty preparation
Legal instruments are designed in part to regulate or contain problems. The range of problems included highlights the question of the degree of match between existing laws and the problems recognized in society. Does the existence of certain problems suggest the need for new laws, whether immediately or in a more distant future ?

10. International organizations
Governmental and nongovernmental bodies concerned with the potential range of problems should find it useful to explore this Encyclopedia when considering the design of new programmes. It could provoke useful discussion in the effort to locate counter-part organizations, focusing on related problems, with a view to collaboration or the exchange of information.

11. Information sciences
The Encyclopedia is basically an experiment in information collection and presentation. In order to handle the variety of fuzzily defined forms of data with which the Encyclopedia is concerned, methods have been used which raise interesting questions for further research in the information sciences, whether for classification theory or for the use of computers in database management, or in the graphic presentation and analysis of networks of concepts. The data may be used to test methods for handling such difficulties.

12. Expert systems
The data on the network of perceived problems continues to provide an interesting challenge to those working on expert systems and artificial intelligence because it is of a higher order of complexity than artificially constructed databases or those usually available. It should prove of even greater interest when available on CD-ROM.

13. Values and ethics research
Researchers grappling with the ill-defined fields of values, human development and states of consciousness, especially in their relationship to global problem-solving, will find an extensive range of information which is otherwise difficult to locate and assemble.

14. International relations
The Encyclopedia presents a much broader range than is usually available of information of potential interest to any research on international relations, especially that touching on international organizations. It raises many questions concerning the capacity of the network of international organizations to respond to the network of problems.

15. Security
For those bodies concerned with potential (and low probability) threats to national and international security, and with facts leading to the destabilization of societies, the information collected suggests leads for further investigation. Of special interest are the ways in which several minor threats may combine or interact so as to constitute a major threat.

16. Challenge to creativity
There are bodies and individuals who are specifically interested in having their perspectives and priorities challenged as one way of learning how to learn. The range of information, and the manner of its organization, highlights new linkages and evokes new levels of thinking. It reinforces recognition of the need for a paradigm shift. The information is a direct challenge to fixed patterns of thinking.

17. Foundations and funding agencies
A factor contributing to the difficulty in launching new initiatives is that funding agencies tend themseves to be locked into particular, and often outdated, patterns of priorities. The information collected here offers alternative perspectives which may suggest more frutiful approaches. It is also valuable in providing a sense of context for specific initiatives.

18. Media
The information gathered here constitutes a rich guide to possibilities for new investigatory reports and documentaries. It also offers perspectives from which established positions can be fruitfully challenged in any interview.

19. Human development programme design
The range of information on human development and modes of awareness provides a rich and unique source of insight into new possibilities for research in this area. It serves to demonstrate the scope of human development, as seen from many cultural perspectives. It provides reminders that there are many unexplored opportunities for human development whose existence has not been widely recognized. It is a direct challenge to the simplistic understanding of human development evident in official policies.

20. Imaginative responses to intractable problems
The information gathered here, and especially in Volume 2, suggests the possibility of new ways of thinking about the intractable problems (such as unemployment, substance abuse, poverty, violence, and the like). Much of that volume points to existing disciplines, and other possibilities, for using the imagination to reconfigure or reframe such problems into a more tractable form -- whilst simultaneously re-imagining the self that is exposed to such problems. It raises the possibility that problems such as drug abuse may, at least in part, be a consequence of imaginal deficiency reinforced by authoritative repression of imagination.

Overview: background and acknowledgements

1. Background

The first edition of this publication appeared in 1976 under the title *Yearbook of World Problems and Human Potential*. It was produced as an experiment arising from a joint project initiated in 1972 between the Union of International Associations (UIA) and the group Mankind 2000.

For the UIA it was then, and remains, a logical extension of its function as a clearinghouse for information on the networks of international agencies and associations, as documented annually in its 3-volume *Yearbook of International Organizations*.

For Mankind 2000, as catalyst of the international futures research movement, it was a means of bringing into focus its prime concern with the place and development of the human being in the emerging world society. The project was jointly financed by the two transnational non-profit bodies, with Mankind 2000 supporting the editorial costs, whilst the UIA funded the publishing and administrative costs.

Work on the second edition was initiated in 1983 and was completed in 1986. This third edition work was initiated in 1988 and completed in December 1990. The second edition was published under the present title. As with the present edition, the publication was jointly funded by the UIA and K G Saur Verlag, current publisher of the UIA's 3-volume *Yearbook*. The second edition was originally conceived as constituting a fourth volume within the *Yearbook* series, because of the degree of potential cross-referencing between the four volumes. It was subsequently decided to treat it as a separate publication under the current title rather than tie it to the established annual *Yearbook*.

Originally founded in Brussels in 1907, partly on the initiative of two Nobel Peace Laureates (Henri La Fontaine, 1913; Auguste Beernaert, 1909), the UIA as an international nongovernmental organization had activities prior to 1939 which illustrate its long-term interest in relation to the current project. These include publication of the *Annuaire de la Vie Internationale*, Vol I (1908-1909, 1370 pages), Vol II (1910-1911, 2652 pages) which included information on problems with which international organizations were concerned at that time. Also published was a *Code des Voeux Internationaux; codification générale des voeux et résolutions des organismes internationaux* (1923, 940 pages, under the auspices of the League of Nations). This listed those portions of the texts of international organization resolutions which covered substantive matters, including what are now regarded as world problems. It covered 1216 resolutions adopted at 151 international meetings. The subject index lists some 1200 items.

Paul Otlet, co-founder of the UIA, produced in 1916 a book entitled *Les Problèmes Internationaux et la Guerre* which identified many problems giving rise to and caused by war, and proposing the creation of a League of Nations. In 1935 he attempted a synthesis, *Monde*, which touched upon many problems and their solution within a society in transformation. The preface bore the title "*The Problem of Problems*", a topic he had first explored in 1918.

2. Acknowledgements

This publication is above all the fruit of continuing collaboration with a considerable number of the 25,000 international governmental and nongovernmental organizations listed in the *Yearbook of International Organizations* of the Union of International Associations. Such bodies make available a wide range of material on the areas of their concern. This is processed for all the publications of the UIA. Special requests were however made with regard to this particular publication.

Without this range of material from the complete spectrum of ideological and disciplinary perspectives, already structured to give a world focus rather than a national one, the editorial task would have been much more difficult, if not impossible, considering the resources available for a project of this kind.

Thanks are due to the United Nations and its associated bodies, especially the:
- Food and Agriculture Organization of the United Nations (FAO)
- International Labour Organisation (ILO)
- United Nations Children's Fund (UNICEF)
- United Nations Educational, Scientific and Cultural Organization (UNESCO)
- United Nations Industrial Development Organization (UNIDO)
- World Health Organization (WHO)

for the documents which they regularly send to the UIA and for those which were supplied in support of this project, either on a complimentary basis (regulations permitting) or at a discount on the normal sales price.

Thanks are equally due to other intergovernmental bodies which have supported this project in a similar manner:
- Commonwealth Secretariat
- Council of Europe
- Organization for Economic Cooperation and Development

Special thanks are due to the United Nations Library (Geneva) for facilitating access to other material, and especially to Theodor Dimitrov (Deputy Chief Librarian), for his support since 1972 in making such access possible.

This project benefited extensively from the participation of the Union of International Associations in the *Goals, Processes and Indicators of Development* project of the United Nations University's Human and Social Development Programme (1978-82), which was coordinated initially by Johan Galtung and subsequently by Carlos Mallman. A number of sections reflect the influence of that exercise, as well as from UIA participation in the subsequent UNU projects on *Information Overload and Information Underuse*, and on *Economic Aspects of Human Development*.

This edition would not have been possible without the continuing confidence of a unique phenomenon amongst publishers of international reference books. Where no international organization or funding agency has been prepared to recognize the opportunity of a project of this kind, Dr Klaus Saur has taken the risk, for the second time, in supporting this endeavour. Furthermore, late in the editorial cycle, he accepted that its scope be expanded into the current 2-volume format. The UIA is much beholden to him, and to Manfred Link, his Director of Production, for their creative flexibility in responding to the challenge of this publication. By the same token, the editors are grateful to Elizabeth Gale of Computaprint for her continuing proactive response to the complex challenge of computer typesetting an evolving publication.

Thanks are finally due to the staff of the UIA for adjusting to the disruption of this project, over and above their normal tasks, but especially to:
- Marie Aeles, for her patience in resolving the many difficulties of the bibliographies;
- Anne Degimbe, for responding to the many difficulties of converting to a new desktop publishing system;
- Jon Jenkins, for his effort to improve the quality of the world problems section under difficult circumstances.
- Jacqueline Nebel, for her care and insight in expanding the human development sections under unreasonable pressure;
- Tarja Ryynänen, for her work on the world problems;

And, not to be forgotten, are the five 108-megabyte disks of the UIA computer system which, over a three year period, have spun together the human problems and potentials they held. At 3,600 revolutions per minute, has any prayer wheel done more?

This project was originally conceived in 1972 by James Wellesley-Wesley with Anthony Judge. The first edition was made possible by the former's support through the foundation Mankind 2000.

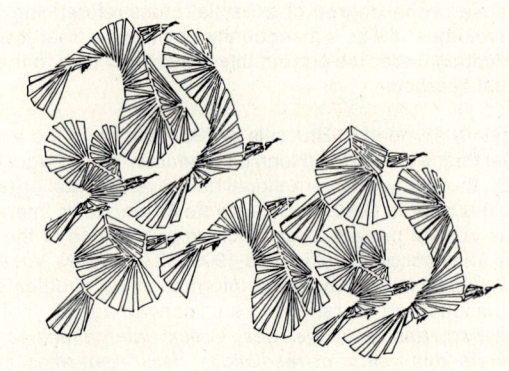
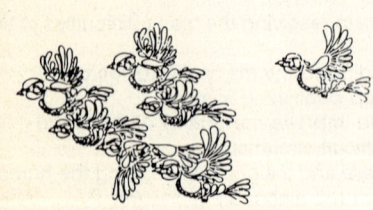

Content: general structure

1. Sections
This Encyclopedia is divided into two **volumes**.

- World Problems (Volume 1)
- Human Potential (Volume 2)

The 2 volumes together are organized into 6 main **sections**:

- World Problems (Volume 1)
    World problems (Section P)

- Human Potential (Volume 2)
    Human development (Section H)
    Integrative knowledge (Section K)
    Metaphors (Section M)
    Transformative approaches (Section T)
    Human values and wisdom (Section V)

Each of these main sections is composed of several **sub-sections**:

- World Problems (Volume 1)
    World problems (Section P)
        Abstract fundamental problems (Section PA)
        Basic universal problems (Section PB
        Cross-sectoral problems (Section PC)
        Detailed problems (Section PD)
        Emanations of other problems (Section PE)
        Fuzzy exceptional problems (Section PF)
        Very specific problems (Section PG)
        Problems under consideration (Section PJ)

- Human Potential (Volume 2)
    Human development (Section H)
        Human development concepts (Section HH)
        Modes of awareness (Section HM)
    Integrative knowledge (Section K)
        Integrative concepts (Section KC)
        Embodying discontinuity (Section KP)
        Patterning disagreement (Section KN)
    Metaphors and patterns (Section M)
        Metaphors (Section MM)
        Patterns of concepts (Section MP)
        Symbols (Section MS)
    Transformative approaches (Section T)
        Transformative conferencing (Section TC)
        Transformative policy cycles (Section TP)
    Human values and wisdom (Section V)
        Constructive values (Section VC)
        Destructive values (Section VD)
        Value polarities (Section VP)
        Value types (Section VT)

2. Additional sub-sections
In addition to the above sub-sections, in which the descriptive entries are located, each main section has three other sub-sections associated with it:

Bibliographical references
Index to the entries in the section
Notes (Comments and explanatory information)

3. Entries and entry number
Entries anywhere in the Encyclopedia are identified by a six-digit code. For example the entry on "Conflict" (denoted by PA0298) is located in Section P, in sub-section PA.

2. Classification of entries
Items within any sub-section are in most cases **not** grouped according to any classification scheme. This continues the policy adopted for the 1976 edition and is in accordance with that adopted for the *Yearbook of International Organizations*. Despite the strong arguments for classifying items, the fundamental reason for not doing so is that it avoids reinforcing the impression that such classification can be done in an unambiguous and satisfactory manner. One of the challenges of the times derives from the fact that there does not yet exist any classification scheme for the interdisciplinary, and often "fuzzy", topics characteristic of this Encyclopedia. What is therefore called for is a series of ongoing experiments with different classification schemes, some of which may eventually prove to be of value. To this end the data needs to be held in an arbitrary permanent order which facilitates such experiments without hindering the editorial tasks of maintaining the data on computer. This question is discussed in more detail on page ******. One such experiment in classifying the items in the world problems section by subject is published in *Global Action Networks* (Vol 3 of the *Yearbook of International Organizations*).

5. Cross-references between entries
There are cross-references between entries in the principal sections. If present, these are listed at the end of each entry. In some cases there are also cross-references between entries in different sections. Generally there are two main groups of cross-references:

(a) Cross-references indicating some form of logical relationship between entries in a section:
- **Broader, or more general, entries:** Other entries of which the entry may be considered a part;
- **Narrower, or more specific, entries:** Other entries which may be considered as part of that entry;
- **Related entries:** Other entries which may be considered as loosely related in some, as yet, unspecificied manner.

(b) Cross-references indicating some form of functional relationship between entries in a section:
- **Causally preceding entries:** Other entries that may be considered to precede this entry in any causal chain or process;
- **Causally following entries:** Other entries that may be considered to follow from this entry in any causal chain.

In the case of world problems, a further distinction may be made in each case between a **constructive** and a **destructive** causal chain.

6. Bibliographic references
The entries of some sections may contain abridged bibliographic references. This is also the case with the Notes (explanatory material concerning each section. Full bibliographic details in such cases may be obtained from the bibliographies associated with each section.

7. Indexes
There is **no** single index covering all entries in the Encylopedia. There are two types of index in the Encyclopedia, covering both titles and keywords of entries:

(a) Volume indexes: These provide index access to entries in all sections in a volume:
- Section PX: Index to world problems in Volume 1
- Section X: Index to all human potential entries in Volume 2

(b) Section indexes: For each section, these provide index access to entries in all sub-sections:
- Section HX: Index to human development (NB: Index to sets)
- Section KX: Index to integrative knowledge
- Section MX: Index to metaphors and patterns
- Section PX: Index to world problems (NB: also the Volume Index)
- Section TX: Index to transformative approaches
- Section VX: Index to human values and wisdom

8. Related international organizations
The number of international organizations makes it impractical to have a separate section on such bodies in this Encyclopedia, as was done for the 1976 edition. However the structure has been designed to interlink *in the future* with the 3-volume *Yearbook of International Organizations* through the system of cross-references. Entries on world problems (Section P) are specifically cross-referenced from the subject volume of that Yearbook where they are classified with international organizations concerned with that subject area.

Content: sections and sub-sections

In the Encyclopedia the **sections are positioned in an alphabetic order within each volume**. The position is determined by the initial mnemonic letter code. This enables the significance of cross-reference and index entries to be more easily remembered and understood during use. In Volume 2, for example, the Human Development (Section H) appears before the Integrative Knowledge (Section K).

In the following discussion of the contents, however, it is appropriate to review these sections in a particular logical sequence different from the mnemonic order. Other such sequences could also be usefully envisaged.

The Encyclopedia as a whole contains 20,958 entries. These are linked by 114,395 cross-references. The two major indexes for each volume contain a total of 91,385 entries. In addition there are 5 section indexes for Volume 2.

1. World problems (Section P)

(a) Intent and scope: The purpose of this section, filling Volume 1 as the largest in the Encyclopedia, is to identify the complete range of world problems perceived by international constituencies, whether as a focus for their programme activities, their research, their protest, their recommendations, or as part of their belief system. An entry has been established on each. This provides a context within which thenetwork of specific relationships perceived between these problems may also be identified.

As a whole this section endeavours to present all the phenomena in society that are perceived negatively by groups transcending national frontiers. These are the phenomena which engender fear and irrational responses as well as constituting a challenge to creative remedial action. Groups are very strongly motivated by the problems which infringe their values and arouse their indignation. As such they are a major stimulus driving the development of society.

(b) Content: The section contains entries on 13,167 "world problems" (28.7 percent more than in the 1986 edition). It is divided into 8 sub-sections. Of these, Sections PA through PF contain descriptive entries, whilst the entries of Sections PG and PJ are indexed and cross-referenced, but are not printed in this edition. Sections PG and PJ are used to register problems on which information is being sought, or which are inadequately distinguished from others already described, or which, as sub-problems, fall below a cut-off level of specificity presently documented in some hierarchy of problems appearing in Sections PA through PF.

With each entry may be associated up to 7 different types of cross-reference to other problems: more general, more specific, related, aggravating, aggravated, alleviating, alleviated. There are 80,394 cross-references of this kind (355 percent more than in the 1986 edition).

The index to the section (see Section PX) has 64,934 entries. There is a bibliography (see Section PY) of 4,745 items. Comments and explanations on the section are given in a set of Notes (see Section PZ)

(c) Constraints: Information on problems transcending national frontiers tends to be:

(a) widely available in excessive amounts in the case of macro-problems for which comprehensive strategies cannot be implemented effectively, or

(b) highly dispersed in modest amounts in the case of politically acceptable problems for which satisfactory programmes promising tangible results can be designed, or

(c) in the case of problems only recognized by experts, disguised or concealed within documents analyzing more acceptable problems or describing the range of detailed programmes in response to the latter, or

(d) reported infrequently in an unsystematic manner in the media and specialized press in the case of problems for which no organized response has yet emerged.

The majority of conventional responses to problems take the form of short-term budgetary commitments to politically acceptable short-term programmes, irrespective of the long-term nature of the problems which they are supposedly designed to contain. There is a need to group information on the network of perceived world problems to facilitate comprehension of their pattern as a whole, in all its variety and detail. There is also a need to facilitate comprehension of ways in which the constituent problems are interrelated, as a means of encouraging the emergence of more appropriate conceptual, strategic and organizational networks to contain them.

(d) Implications: The perceptions documented raise useful questions concerning the nature of problems and what is meant by the "existence" of a problem, especially when other groups consider that perception irrelevant, misleading or misinformed. There is great difficulty in obtaining and editing material on problems, rather than on incidents, remedial programme action, theories, or other frameworks through which perception of problems is filtered. So to that extent, it could be argued that this section assembles information on which people collectively have great difficulty in focusing, namely information whose significance, whether deliberately or inadvertently, is collectively repressed, displaced onto some less threatening problems, or projected in the form of blame onto some other social group.

2. Human values and wisdom (Section V)

(a) Intent and scope: The importance of values is frequently cited in relation to the global problematique, whether it be in debates in international assemblies, in studies criticizing "value-free" approaches to research, or in discussion of quality of life and individual fulfilment. Values are deemed especially important in questions of cultural development and are central to concern for the preservation of cultural heritage.

The purpose of this section is to register a complete range of values with which people identify, to which they are attracted or which they reject as abhorrent. The elusive notion of "wisdom" may be useful considered as the art of dealing with value dilemmas.

(b) Content: The section contains 2,270 entries. It is divided into four parts: Section VC, Section VD, Section VP, Section VT. Section VC contains 960 constructive value words (e.g. peace, harmony, beauty), whereas Section VD contains 1,040 destructive value words (e.g. conflict, depravity, ugliness). The entries in these two sections are linked by 6,000 cross-references to 225 entries in Section VP. These entries are value-polarities (e.g. agreement-disagreement, freedom-restraint, pleasure-displeasure) derived from the organization of *Roget's Thesaurus*. These in turn cross-reference 45 entries in Section VT in an attempt to identify major value categories. The section as a whole contains 14,463 cross-references.

None of the entries contain "descriptions" of the value(s) implied. In most cases this would be superfluous. The words in Section VC reflect values which tend to be accepted without questioning. Those in Section VD reflect values which tend to be rejected without questioning. The emphasis is placed on using the cross-references to indicate the range of connotations of particular value words. The entries on value polarities, Section VP, do however list proverbs, aphorisms or quotations selected to illustrate the dynamic counter-intuitive relationship between constructive and destructive values.

They endeavour to draw on popular wisdom or insight to demonstrate the negative consequences and limitations of blind adherence to constructive values or to demonstrate the positive consequences and creative opportunity of judicious action in the light of destructive values. They point to the existence of a more fundamental and challenging dynamic than that implied, for example, by peace-at-all-costs and total rejection of conflict.

The index also includes an index (see Section VX) of 2,549 entries and a bibliography (see Section VY) of 398 items. Comments and explanations on the section are given in a set of Notes (see Section VZ)

(c) **Constraints:** Whilst it had been hoped to develop lists of values from documents of international bodies, no adequate lists of values were located, even within the intergovernmental agencies (such as UNESCO) specifically concerned with human values, and despite numerous reports and meetings on "values" in recent years. The values referred to are very seldom named, although the commonest may be cited as examples. The list presented here has therefore been elaborated by the editors as an experiment based on the selection and interrelationship of constructive and destructive value words.

(d) **Implications:** This exploration of values is of special interest in relation to the world problems in Section P. Many problems are named in international debate using a destructive value word (e.g. insufficient, unrealistic, unjust, inappropriate). Problems defined in this way imply the existence of some corresponding value whose expression is infringed by the problem. Such values may or may not be noted in defining the purposes underlying remedial action in response to the problem, although often they form part of the wording of any rallying slogan in support of some international strategy (Section S, 1986). But the set of constructive and destructive value words does indicate a way of coming to grips with the range of problems which the existing language renders perceivable and nameable. They also indicate possible dimensions of human development. This section is of course limited at this stage by the biases inherent in *Roget's Thesaurus* and the English language. It does however create a framework which could enable these limitations to be transcended.

3. Human development (Section H)

(a) **Intent and scope:** The purpose of this section is to describe briefly the complete range of concepts of human development with which people identify, consider meaningful or reject in their search for growth and fulfilment in life. The scope of this section has been deliberately extended beyond the unrelated concepts accepted with great caution by intergovernmental agencies: the job-fulfilment orientation of ILO, the health-oriented concepts of WHO and the education-oriented concepts of UNESCO. It includes concepts legitimated by the psychological and psychoanalytical establishments as well as those promoted by the various contemporary growth movements. It also includes concepts from religions and from belief systems of different cultures. Entries are included on explicit concepts of human development and on therapies, activities or experiences in which a particular understanding of human development is implicit.

(b) **Content:** The section contains 4,051 entries (154 percent more than in the 1986 edition). It is divided into two parts: Section HH and Section HM. Section HH describes 1292 concepts of human development. Section HM endeavours to describe 2,759 modes of awareness, namely the experiential states associated with different stages in the process of human development as perceived by different groups (and preferably using wording with which such groups would identify).

The entries have been interlinked by 15,027 cross-references. These either indicate relationships between more general or more specific concepts, or, especially in Section HM, the relationship between succeeding modes of awareness in some process of human development (whether linear or cyclical).

The section includes a special index on numbered sets of concepts (see Section HX) of 3,356 items and a bibliography (see Section HY) of 2,488 items. Comments and explanations on the section are given in a set of Notes (see Section HZ)

(c) **Constraints:** The major constraint derives from the degree to which many entries in this section imply unconventional approaches to the use of language, challenging the rigidity of the boundary between subject and object.

(d) **Implications:** This section indicates ways in which people struggle within themselves for fulfilment and the experiences associated with that struggle which they find meaningful (whether or not such experiences are considered totally deluded or inappropriate by different scientific or religious establishments). That many of these experiences cannot be effectively "put into words" is indicated by the use of metaphors or symbols in naming them. These appear as strange to Western eyes as do others to Eastern cultures.

4. Integrative knowledge (Section K)

(a) **Intent and scope:** A principal characteristic of the global problematique is its inherent complexity. This calls for a complex response interrelating many different intellectual resources and insights and involving sensitivity to very different kinds of constraint. Integrative approaches of this kind have proved inadequate or exceedingly difficult to implement in a society characterized by specialization and fragmentation.

The purpose of this section is to assemble descriptions of the range of concepts or conceptual approaches which are, in some way, considered integrative and which are held by some international constituencies to provide the key to the organization of any effective strategic response to the global problematique. Many of the words used to label these concepts are those which are considered indicators of the power of an advocated approach. They frequently appear in project proposals to trigger favourable response, whether or not any content can be given to them in practice. Words like "global", "integrative", "networking" and "systematic" are the magical "words-of-power" in the modern organizational world.

(b) **Content:** The section contains 702 entries on integrative concepts. It is divided into three sub-sections. Section KC describes 632 integrative, interdisciplinary or unitary concepts in the broadest sense, namely it includes advocated methods of integrating awareness favoured by these who reject a purely conceptual approach.

The 70 entries in Section KD comment on recent efforts to interrelate incompatible conceptual approaches and the nature of the challenge that this implies. This material is derived from papers prepared by the editors during their participation in the Goals, Processes and Indicators of Development project of the United Nations University, especially on problems of methodology.

In Section KP, the final group of 20 entries is an exercise in designing a pattern of relationships between incompatible concepts in the light of insights in a wide range of different concept schemes that use sets of concepts of different sizes to contain qualitative complexity. Its merit lies in its deliberate attempt to internalize discontinuity and disagreement within the pattern.

The section includes an index (see Section KX) of 1,609 items and a bibliography (see Section KY) of 2,200 items. Comments and explanations on the section are given in a set of Notes (see Section KZ)

(c) **Constraints:** Iterest in interdisciplinarity in its own right, recent years have seen an emphasis on a project-by-project pragmatic approach, which avoids the need for any form of conceptual framework transcending individual disciplines, but begs the question as to the relationship between such projects.

(d) **Implications:** The section as a whole attempts to respond to the dramatic problem of how to interrelate vital conceptual insights which are essentially incommensurable and in practice often mutually antagonistic. A plurality of responses in not in itself an adequate response, especially since each fails to internalize the discontinuity, incompatibility and disagreement which its existence as an alternative engenders. It is for this reason that the second part explores the possibility, implicit or explicit in recent studies, that a more appropriate answer might emerge from a patterned alternation between alternatives. This calls for a focus on the models of alternation by which the pattern and timing of cyclic transformations can be ordered between mutually opposed alternatives. It highlights the possibility that the kind of integrative approach required may not be fully describable within the language of any single conceptual framework, however sophisticated.

5. Metaphors and patterns (Section M)

(a) Intent and scope: Any form of international "mobilization of public opinion" to engender the much sought "political will to change" is dependent upon communication, especially when the insights required to guide that change are complex, counter-intuitive or simply not clearly communicable within any one conceptual language.

The purpose of this section is therefore to review the complete range of communication possibilities and constraints. This is partly in response to the narrow focus of recent major intergovernmental initiatives under the extremely misleading titles of "International Commission for the Study of Communication Problems" (limited to the mass media) and the "International Communications Year" (telecommunications hardware) by UNESCO and ITU respectively. It is however a direct consequence of participation by the editors in the Forms of Presentation sub-project of the Goals, Processes and Indicators of Development project of the United Nations University.

(b) Content: The section consists of 444 entries. It is divided into three sub-sections: Section MM, Section MP, and Section MS. In different ways, these are each editorial experiments in the presentation of information. As a whole the section provides a framework within which to review alternative ways of interrelating items of information to facilitate comprehension and communication.

Section MM explores through 88 entries the possibility of designing metaphors that are appropriate to engendering a creative response to the global problematique. This section recognizes the unique importance of metaphor in politics, education, religion and scientific creativity as a means of communicating complex notions, especially in transdisciplinary contexts. The entries have been elaborated as an experiment to stimulate interest in this mode as one of the few means of rapidly stimulating innovative breakthroughs in development problems, since it is not dependent on lengthy, specialized education and can, for example, be intimately interwoven into pre-existing rural community experience.

Section MP explores "pattern language" of 253 patterns interlinked by 3,491 relationships. The pattern language was originally elaborated by a team led by the environmental designer Christopher Alexander as an aid to designing physical contexts in which quality of life is enhanced. Selected patterns have been used, according to the methods of the previous section, as substrates for metaphors such as to suggest ways in which social, conceptual and intra-personal contexts may also be "designed". Its special merit is the integration between the component patterns provided by relationships reflecting an understanding of the socio-physical environment which is both extremely realistic and exceptionally harmonious.

Section MS reviews in 103 entries the range of symbols used in modern and traditional cultures as a way of communicating multiple levels of significance in a compact and reproducible form. It emerges from the recognition of the special importance of symbols in embodying significance and giving focus to any campaign or programme and establishing its identity in relation to other initiatives. As a focus for public attention, their choice is far from being an arbitrary matter. It is a response to constraints which need to be better understood if human resources are to be more effectively mobilized. They give visual form to abstract concepts by which development processes are organized especially in traditional cultures which do not respond to conventional forms of presentation. The relationship between the symbols by which people are motivated (or alienated) is also of vital importance.

The section also includes an index (see Section MX) of 1,485 entries and a bibliography (see Section MY) of 299 items. Comments and explanations on the section are given in a set of Notes (see Section MZ)

6. Transformative Approaches (Section T)

(a) Intent and scope: The purpose of this section is to provide a context for the presentation of accessible techniques, which offer possibilities of making an immediate difference to the manner in which resources are mobilized in response to the global problematique.

(b) Content: The section contains 304 entries. It is divided into two sub-sections: Section TC and Section TP.

Section TC contains 207 entries with descriptions on new ways of conceiving meetings and meeting processes. Meetings, and especially international meetings, are a vital feature of social processes and the initiation of change. They are a principal means whereby different perspectives are "assembled". Through such occasions resources are brought to bear upon questions of common concern. They may also provide the environment in which supposedly unrelated topics can emerge and be juxtaposed. But despite the assistance of professionals and the increasing number of such events, there is rising concern that many do not fulfil the expectations of participants, nor of those whose future may depend upon the outcome. This is particularly true of events most concerned with social transformation. Current meeting procedures, despite efforts at innovation, on such questions tend to give rise to little of more than short-term public relations impact and in this form can themselves constitute an important obstacle to social change. In a very real sense meetings model collective (in)ability to act and the ineffectiveness of collective action. The challenge is therefore to provoke reflection on a new attitude or conceptual framework through which meeting dynamics may be perceived and organized in order that they may fulfil their potential role in response to the global problematique.

Section TP contains a network of 64 entries, linked by 384 relationships, based on the pattern of concepts implicit in the much-publicized Chinese classic, the *Book of Changes*. These are transposed into a language which highlights the significance of such a complex pattern of transformations in relation to sustainable policy cycles. Its special merit is the explicit recognition of the need to shift from condition to condition in order to ensure both healthy development and the ability to respond to a turbulent environment.

The section also includes an index (see Section TX) of 1,191 entries. Comments and explanations on the section are given in a set of Notes (see Section TZ)

Introduction \ Contents

Content: modifications, improvements and omissions

This note reviews the changes made to the structure and content of the volume since the 1976 edition, but especially the improvements made on the 1986 edition.

The 1976 edition was composed of 13 sections, interlinked by cross-references between items, both within a section and between sections. There was also a variety of introductory texts. Although this reflected the complexity of the material it made access to it more than necessarily difficult.

Although the general structure of the Encyclopedia now remains the same as the 1986 edition, it has been extensively restructured and redesigned. This is most evident in the division into two separate volumes, almost doubling the number of pages. The intention has been to focus on some priority sections and to simplify access. Editorial work has focused on the "world problems" section and on the "human development" section, as well as on developing the comments which appear as Notes..

The total number of entries in the 1976 edition was 18,563 linked by 30,455 relationships. This rose in the 1986 edition to 24,410 entries linked by 49,030 relationships. The current edition contains 20,958 entries linked by 114,395 relationships. This represents a decrease over the 1986 edition of 14.1 percent in the number of entries (due to the omission of a major section on strategies) and an increase of 133.3 percent in the number of relationships.

Five separate bibliographies totalling 10,130 items have been added. Separate indexes have been provieed for Volume 1 (64,934 items) and Volume 2 (26,451 items), together with 5 section indexes for Volume 2.

SECTIONS IN CURRENT EDITION

1. World problems (Volume 1: Section P)
(a) Increased scope: The number of entries in the 1986 edition was 10,233 (of which only 4,700 had descriptions, the remaining being title-only). In the current edition this has increased by 28.7 percent to 13,167 entries, of which 8,721 have descriptions (an increase of 85.6 percent).

Experience with the collection of information for the previous edition indicates that there are many categories of "problem" with which international organizations as such are not directly, or only by implication, concerned. These include a number of ways of looking at what can otherwise be called "problems". These are of interest because of the special sources of information on them.

- risks: as documented by the insurance industry

- criminal law: definitions and statistics on crime

- markets for products and services which in terms of their financial importance alone are indicative of problems which are perceived as important by the purchasers of such services. In the case of bodily conditions and ailments, for example, these are particularly interesting because such problems are normally considered of little significance in relation to social problems in general, and yet they are of much interest to industry because they represent a market for (new) products and services.

- religious/ethical/moral problems: such "problems" (including "sins") were avoided in the two previous editions in the process of documenting the core of "harder" problems. A deliberate attempt has been made to include them in this edition, especially when they reflect the sensibles of non-western cultures.

- abstract fundamental problems: as an experiment, derived from work on destructive values, many less tangible problems have been included because of their underlying importance in relation to more conventional problems. They are also closely related to moral and ethical problems.

- professional/disciplinary abuses: namely those with which the ethical committees of professions are increasingly concerned, often in response to consumer pressure.

(b) Ordering: Entries have been reordered by "level" to create subsections (A-G) to regroup problem entries by scope. This effectively moves more specific problems towards Section G and more general problems towards Section A (as is done in the companion 3-volume *Yearbook of International Organizations*). The criteria are described in the introduction to each sub-section. The purpose of this is to focus more clearly on the prominent problems and to isolate the very detailed problems (on which less work has been done) in sections which are not printed (although the names are indexed), as in the case of the many specific diseases. This reordering builds on the "qualifier" code used experimentally to distinguish such levels in the 1986 edition (especially in the classified index Section PX in that edition).

(c) Cross-references: The pattern of cross-references has been considerably extended and improved, aided by special software to facilitate the management of the networks of relationships. Much effort has been devoted to indicating references whereby problems are grouped into one (or more) hierarchies. The number of cross-references included increased from 13,574 in the 1976 edition to 17,636 in the 1986 edition (an increase of 29.9 percent), and to 80,394 in the current edition (an increase of 355.9 percent).

(d) Bibliographical references: Problem-specific bibliographical references have been added to many entries. A separate bibliography of 4745 items of problem literature is included for the first time.

(e) Indexing: There is an inherent difficulty in ordering the networks of problems into comprehensible patterns. The number of alternative titles given to problems has been significantly increased, leading as a result to a considerable increase in the number of index keywords through which problems can be located. In the current edition the index for Volume 1 contains 64,934 entries.

(f) Notes: The commentary has been completely restructured and extended.

2. Human development (Volume 2: Section H)
(a) Increased scope: This section has increased by 153.8 percent from 1596 entries in the 1986 edition to 4,051 entries in the current edition.

(b) Ordering: In the case of modes of awareness, a qualifier code has been introduced to distinguish them by level.

(c) Cross-references: The 4,461 cross-references in the 1986 edition have been increased by 236.9 percent to 15,027 linkages.

(d) Bibliographical references: Bibliographical references have been added to many entries, linking them to 2,488 items in a special bibliography.

(e) Indexing: A special index of 3,356 items in numbered concept sets has been introduced for the first time. The index to the section as a whole (incorporated into the Volume 2 index only) now contains 19,559 items.

(f) Notes: The commentary has been completely restructured and extended.

3. Human values and wisdom (Volume 2: Section V)
No changes have been made to this section although the explanatory Notes have been restructured and extended. The original link to "wisdom" through the material on value polarities has been explicitly recognized through the extension of the title of this section.

As in the previous edition there are 2,270 entries linked by 14,463 cross-references. A bilbiography of 398 items has been added. This covers values, as well as related literature on wisdom, proverbs and aphorisms.

4. Integrative knowledge (Volume 2: Section K)
No change to the entries on integrative concepts. The explanatory Notes have been restructured. A small section (Section KP) has been transferred to this location from its original position under Communication (Section C) in the 1986 edition.

A bibliography of 2,200 items has been added.

5. Metaphors and patterns (Volume 2: Section M)
This section, initially envisaged as focusing on "comprehension of information", is a completely restructured version of Section C (Communication) in the 1986 edition. It includes some of the 1986 information on patterns and symbols, although some of that has been transferred to Section K and some to Section T. It excludes the material on forms of presentation, omitted from this edition.

The main changes are a very considerable addition to the explanatory Notes in which the implications of metaphors for a more fruitful approach to problems are envisaged. A bibliography of 299 items has been included.

6. Transformative approaches (Section T)
The number of entries has been increased from 207 to 240. A section of 64 entries on transformative policy cycles has been adapted from a similar section included under Communication (Section C) of the 1986 edtion. The Notes have been extensively modified.

SECTIONS OMITTED FROM CURRENT EDITION

The following sections, present in either the 1976 or the 1986 edition (or both), have been omitted from this edition:

1. International organizations and associations (1976, Section A)
The function of this section is now achieved by grouping organizations, with world problems, by subject category in a complementary publication *Global Action Networks* (Vol 3 of the *Yearbook of International Organizations*).

2. Traded commodities and products (1976, Section C)
This material has been effectively incorporated into the World Problems section. It was originally designed to demonstrate how certain problems could be grouped in terms of the United Nations *Standard International Trade Classification.*

3. Intellectual disciplines and sciences (1976, Section D)
This section covering 1845 disciplines and sciences has been omitted. It was originally included to focus attention on the full range of conceptual resources available for more appropriate approaches to problems. The intention was to cross-reference disciplines to the problems to which they were relevant. A secondary concern was to recognize the specific ethical, and other, abuses associated with the misuse of particular disciplines.

4. Economic and industrial sectors (1976, Section E)
This material, like that on traded commodities, has been effectively incorporated into the World Problems section. It was originally designed to demonstrate how certain problems could be grouped in terms of the United Nations International *Standard Industrial Classification of All Economic Activities.*

5. Occupations, jobs and professions (1976, Section J)
This material has been omitted, although some of it is effectively incorporated into the World Problems section. The section was originally designed to demonstrate how certain problems could be grouped in terms of the *International Standard Classification of Occupations* of the International Labour Office.

6. Multinational corporations and enterprises (1976, Section M)
This section has been omitted. It dates from a period when it was considered appropriate to attempt to cross-reference world problems to the specific multinational corporations perceived as aggravating them in some way. A secondary objective was to open the possibility of cross-referencing such bodies to the problems for which they had constructive remedial programmes.

7. Human diseases (1976, Section Q)
This section is effectively incorporated into the World Problems section. It was originally designed to demonstrate how certain problems could be grouped in terms of the *International Classification of Diseases* of the World Health Organization.

8. International periodicals and serials (1976, Section S)
This section has been omitted. It was originally designed to demonstrate how certain problems could be usefully related to the periodical sources of information reporting on them.

9. Multilateral treaties, conventions and agreements (1976, Section T)
As with international organizations, the function of this section is now achieved by grouping such treaties, with world problems, by subject category in a complementary publication *Global Action Networks* (Vol 3 of the *Yearbook of International Organizations*).

10. Forms of presentation (1986, Section CF)
This omitted section was originally designed to focus attention on the range of ways in which information on world problems and the resources to deal with them could be usefully presented. The concern was to improve the handling of information through different media so that different treatments complemented rather than undermined each other. In the current edition it appeared more appropriate to focus attention on the value of metaphors and patterns, as a largely unexplored resource, rather than on forms of presentation in general. Metaphors in particular are deemed to offer much scope for the development of more appropriate policies and modes of action.

13. Strategies (1986, Section S)
This material has been excluded, in part because of lack of resources to focus on it, and in part because of the limited space available in the Encyclopedia. More fundamentally, however, it was considered that the Encyclopedia could perform a more useful function in focusing on issues underlying the formulation of strategies and remedial programmes. It can be argued that the current spectrum of strategies emerges from mind-sets governed by a relatively limited range of metaphors. To the extent that the range of metaphors is limited, and relatively simplistic, the resulting strategies are themselves simplistic and therefore inadequate to the complexity of the global problematique and its local manifestations.

Method: logistical challenge

1. Amount of information
The quantity of information available on "world problems" is more than daunting, as is the number of documents which could usefully be scanned to clarify the dimensions of "human potential". The major intergovernmental organizations claim that a large proportion of their huge document output is concerned with such problems and with aspects of the "development" process. There are extensive specialized libraries on particular clusters of problems, as there are on development. This quantitative challenge calls for search procedures capable of circumventing these obvious difficulties.

2. Unknown scope
Obtaining information on well-recognized topics, which can be communicated unambiguously in queries to appropriate information centres, is quite different from identifying those problems, or aspects of human development, known only to groups concerned with them and under labels of which others are unaware. This is especially demanding when different constituencies may use different sets of descriptors to denote such problems. This makes it quite difficult, if not impossible, to use conventional library search procedures.

3. Inaccessible library collections
(a) Major libraries obviously tend to be subject-oriented rather than problem-oriented. Collections tend to be organized and catalogued by subject. Even if a problem such as "illiteracy" is mentioned in the subject catalogue, it will tend to refer the reader to a large collection of material under "literacy". In most cases there is no such mention. The less obvious problems tend not to be mentioned on the title page of a document. It is therefore rare for them to be identifiable through the subject catalogue, particularly when discussion of the problem is incidental to the topic of the document. This applies to an even greater degree in the especially "fuzzy" area of human values and modes of awareness.

(b) Many libraries are not organized to permit scanning of books in their stacks by readers. Readers are often restricted to the number of books they can request at any one time. There are usually significant delays, often hours, before requested books are delivered. It is common for requested books to be "not on shelf".

(c) When undertaking massive document scanning operations in libraries or documentation centres, there is little point in gaining access to a document containing several pages of useful information if it is not possible to photocopy that document for later analysis in relation to other such materials on the same problem. Many libraries are not organized to facilitate such photocopying, whether because there are no such machines, or they are physically distant, or because of complex administrative procedures governing their use, or because of concern with copyright infringement, or because there are many people attempting to use them. Extracting information on problems requires extensive photocopying of odd pages in many documents. This is usually quite difficult in most libraries.

4. Impracticality of online database searches
In the light of some of the constraints noted above, it might be supposed that much could be gained by online searches. But although online searches can be efficient when precise descriptors are known, the situation is little different from libraries when this is not the case. Where descriptors are known, online searches can facilitate scanning of titles which may be of interest, but they do not help to determine which of them might be of value when the documents are eventually obtained. Even an abstract may fail to indicate that the document has very valuable paragraphs identifying problems. Dependence on online searching can also prove to be a major drain on limited resources.

5. Inaccessibility of expertise
There are no "qualified experts" on many of the entries in this Encyclopedia. This may be because a problem has only recently been recognized or is considered too marginal to merit such attention. In some cases it may be because the expertise in that area is focused on the scientific domain with which the problem is associated but that the problem itself is of marginal interest to the competent scholars. This is especially the case when a concrete problem has implications for several disciplines but is of limited interest to all of them. Only for certain types of problem is it possible for an individual to acquire recognized qualifications. Those with expertise tend also to have quite different interpretations of the "facts" relating to what some consider to be a problem. So, whereas it is usual to have encyclopedia entries specially written by experts on the topic, the difficulties of negotiating such arrangements and maintaining the associated correspondence for large numbers of entries was not considered cost effective. Given the current costs of expertise and consultants, this approach would have been quite impracticable. This is also the case with the human potential entries.

6. Bias of any particular pattern of sources
Because of the nature of the project, there was a strong desire to avoid dependence on any particular pattern of sources, even though this might have made certain aspects of the task much easier.

7. Severe resource limitations
In principle a project of this kind should be able to call upon a number of funding sources actively interested in supporting an international, interdisciplinary, multi-focus survey of the "world problems" perceived by significant international constituencies. In fact funding for such projects is increasingly limited. There is a preference for national (if not local), highly-focused, uni-disciplinary projects, even amongst international organizations -- and especially when they themselves have relatively narrow mandates. Much funding is ear-marked for currently fashionable topics because of its immediately apparent legitimacy. There is also a natural bias in favour of solution-oriented projects and a reluctance to increase the number of problems recognized. Interest in "human potential" is even more limited.

8. Appropriate compromise
The method developed (and described below) therefore called for a special compromise which could ensure a wide information collection and processing procedure without becoming paralyzed by the quantity of material.

Method: procedures

1. Scope
Details of the scope of each section are given in the introduction to each section and in the Notes on it. In general however every effort has been made to ensure coverage of perspectives from: industrialized and developing regions (North and South), socialist and capitalist economic systems (East and West), occidental and oriental cultures, and official and unofficial sources (governmental and nongovernmental). In doing so, attention was given to scientific and "unscientific" perspectives, whether well-documented or poorly-documented, fashionable or unfashionable, informed or "misinformed", and whether emanating from qualified elites or marginalized groups.

2. Method
Details of the method used for each section are given in the relevant Notes. In general method employed is an extension of that elaborated over many years to locate and process information on the 20,000 internationally-active organizations currently documented in the 3-volume *Yearbook of International Organizations*, with which this project is intimately linked at all levels.

The method of gathering and processing information may be outlined as follows:

(a) Inflow and collection of material: A constant flow of material is received, particularly from international organizations sensitive to the preoccupations of every sector of society in every region and culture of the world. This is mainly in response to: direct mail requests (partly in association with regular contacts involved in work on the *Yearbook of International Organizations*), mailing of proof pages from the previous edition, exchange agreements with international bodies, purchases or loans of publications (or microfiches) from intergovernmental bodies, and special requests. This material is received in many languages, although the text extracted from it (for the *Yearbook*) is presented in English.

(b) Special information gathering initiatives: Particular efforts, including library searches and bulk acquisition of documents, are made in the case of bodies such as the Specialized Agencies of the United Nations, OECD, the Commonwealth Secretariat, and the Council of Europe. As might be expected, such sources are supplemented by journal searches, reference books, press cuttings and unsolicited material from a wide variety of sources.

(c) Scanning and selection: Documents are scanned for relevant material and, if the document is complex, portions are photocopied for classification and filing in different locations.

(d) Provisional file allocation: During the process of scanning and classification, provisional decisions are taken as to whether the item represents a new category (of "problem", etc) or whether the document could be appropriately filed within an existing category.

(e) Editorial processing: Editors then work on files by item. Each file might itself contain many documents, including books, from very different sources. The editors attempt to elaborate the clearest and most succinct presentation of the item by combining information from different source documents as appropriate. Every effort is made to use existing texts supplied by international bodies. When this is not possible, adaptations of texts presented in a variety of other documents are made.

(f) File allocation review: During the editorial process the status of the item is reviewed. This may lead to its being further subdivided into separate items, or integrated with some other item, or simply rejected as low-priority material.

(g) Indexing: The editorial process is assisted by working indexes which are updated automatically as a result of that process. This includes automatic classification of items by subject code, where relevant.

(i) Computer-assisted editing: Special programmes are extensively used for error checking. For some sections of the Encyclopedia, very extensive use of computers has also been made to explore various ways of reordering and regrouping the items.

3. Comment

(a) Editing: The task of preparing the final text is therefore an editorial process of making the best use of any number of items touching on the nature of the world problem, or the human development concept, as the case may be. It should be stressed, particularly in the case of the world problems section, that the task is conceived as being an "editorial" one. It is not "research" in the narrower sense in that editors are not called upon to analyze material in order to formulate judgements or new hypotheses concerning the problems in any particular domain. This said, the task of determining from a mass of documents in a file what problems or sub-problems are being identified there, explicitly or implicitly, is necessarily a form of empirical research in the broader sense of the term. It is the role of the editors to clarify any presentation and to use supporting texts to reinforce any relevant opinion expressed, rather as in the formulation of a legal brief. It is not the role of the editors to impose their own opinion on the material. One clear exception to this, in the case of world problems (and discussed in that section), has been to clarify the names used to denote world problems, especially when these are confused, in conventional international jargon, with the names of associated values or remedial strategies.

(b) Editorial experiments: The Encyclopedia includes a number of smaller sections of a deliberately experimental nature, such as those on values and metaphors. As noted above, in each case the method used is discussed in the Notes for that section. Wherever possible it is an extension, or a variation, on the editorial procedure outlined above.

(c) Scope: The design and coverage of the Encyclopedia, namely the sections selected for inclusion in it, were partly determined by the experience of the previous editions and the possibility of updating or (temporarily) excluding certain of its sections. The existence and final form of some sections, especially that on values, was influenced by the opportunity of experimenting with various possibilities of manipulating and presenting information via the computer facility with which the editorial work is done.

(d) Pragmatism: It is appropriate to stress the strong pragmatic influence on working methods as they affected the design of the Encyclopedia in its present form. As in any design problem there were constraints on resources and in this case, due to the restricted level of editorial funding, they were very tight for a project of this scope. The detailed procedures were continually reviewed and modified to achieve a satisfactory final result with the most efficient use of resources. Since the page space was necessarily also limited, another concern was to "pack" information as efficiently as possibly. These factors influenced, and were influenced by, the manner in which the text database system could be used or modified to facilitate the procedures leading to the final product.

(e) Outside review: Despite the technical possibility of doing so, a decision was made not to use resources to submit edited texts in draft form to competent authorities for comment or improvement prior to publication. In the case of the world problems section, for example, the assumption was made that an adequate formulation could be adapted from the documents originally supplied by international organizations claiming some competence in the domain in question, particularly if these had been sent in response to proof texts from the previous edition. This procedure proved much more efficient than that of requesting such bodies to elaborate problem descriptions (as was done for the 1976 edition). With some minor exceptions, commissioning them to do so was beyond the resources of the project. As part of an ongoing project, the existing texts will be submitted as proofs to concerned bodies to trigger responses for the next edition, as is done for the *Yearbook of International Organizations*.

Method: classification policy

Strong arguments could be put forward for ordering the entries in each of the series in this Encyclopedia in some other way. The purpose of this note is to explain why the present method of ordering has been adopted.

1. Reasons for not using alphabetical order

(a) Many of the entries have names associated with them which are unsuitable as a basis for alphabetical ordering. The names may be derived from the description (of the problem, concept, *etc*) and may later be revised by substituting better alternatives. Some are long and contain many prepositions. Some names are artificial and not commonly accepted or recognized (*eg* those names in the world problems series starting with "inadequate"). Some entries have several possible names associated with them.

(b) Change of the name under which the entry is filed would result in the need to move the whole entry within the sequence of entries when an edition is revised. This is inconvenient for the user who does not have a permanent point of reference. It makes editorial control complicated, leading to cumbersome administrative procedures and to increased errors. It also changes and complicates the whole philosophy of file management when using computers, as well as increasing computer processing costs.

(c) If an edition (or an index) in another language is prepared, all the material must be reordered. This may necessitate the creation of a new set of indexes and cross-references. This leads to total lack of correspondence of the sequence of material between language editions. With the chosen policy, there no change of order between language editions, nor any change of indexes or cross-references. (Problem PA1234 in the English edition would be Problem PA1234 in a French edition).

(d) Use of alphabetical ordering artificially places entries next to one another on the basis of a chance superficial resemblance arising from the first words in the entry name. This may reinforce incorrect associations, or may imply to the reader that all related entries are together and that the problem of classification has been solved. With the chosen policy, readers are constantly reminded that the classification problem is an ongoing problem which is not solved by any particular classification scheme.

(e) Cross-references within entries, and indexes to entries, are more conveniently made to a number than to long alphabetical titles with many prepositions.

(f) Placing the emphasis on the name of an entry by the use of some alphabetical order prevents readers from being confronted with the fact that they are faced with a conceptual domain which may be ill-defined and labelled with many different names. By using a number, the same name may even be attached to different numbers used to denote different conceptual domains, or at least domains which are believed to be distinct.

2. Reasons for not using a classification scheme

It is interesting to note that a principal interest of academics in this project has been the choice of classification scheme (*eg* for the world problem series). A frequent argument is that it is impossible to collect information unless it is done through some classification scheme. The difficulty in this project (collecting information on world problems, for example) is that at no stage was it known what new variety of problem would emerge from the documents processed on the following day. Therefore at no stage was it appropriate to develop a definitive classification scheme. To do so prior to publication would deprive scholars of the pleasure of exposure to conceptually unordered data by supplying them with a classification scheme which it would be only too easy to criticize. Specific reasons for not developing any such scheme at this stage are as follows:

(a) The first problem in handling such varied material, is one of filing the entries and associated documents. This is a simple administrative task to which there are a limited number of viable solutions. The simpler the system, the simpler the administrative task. The classification of entries is a sophisticated intellectual exercise which ideally never ceases, since science is always in search of better categories and improved ways of structuring the relationship between elements of knowledge. There are many solutions to the problem of classifying any given series of entries. If a particular scheme of categories is chosen for a given series of entries, the filing problem is complicated and the administrative and intellectual activities interfere with each other. The conceptual exercise should follow the filing of the information, not accompany it, nor precede it (leading to the exclusion of unforeseen varieties).

(b) A particular classification scheme is always subject to criticism, pressure for revision, or the substitution of some alternative scheme. There is no consensus on a satisfactory scheme for classifying any of the series in this publication Such revision again represents an unnecessary interference between the administrative task and the intellectual task. If a computer system is based on a non-permanent code/number scheme, considerable difficulties arise.

(c) Many of the entries included (*eg* the integrative concepts) have been included precisely because they are "fuzzily" defined and cross conventional subject and disciplinary boundaries and are therefore difficult to classify. The same could be said for the more complex problems. Use of simplistic classification schemes to eliminate this difficulty would only serve to disguise this and would thus be a disservice to all concerned.

d) Structuring a publication in terms of a particular classification scheme only serves to alienate unnecessarily those who are intellectually frustrated by the idiosyncracies of such a scheme.

e) Even in the case of physically unambiguous entities like plants and animals, there is still considerable dispute about the allocation of particular entities to particular categories. There is even a school of thought which maintains that the classification enterprise itself is counter-productive. The situation is understandably more serious in the case of conceptual entities.

f) As stressed at a number of points in this volume, a conventional classification scheme would reinforce the mind-set that the entries here could be classified without doing violence to the relationships between them. For this reason the emphasis has been placed on exploring ways of portraying the network of relationships and using this as a basis for addressing the question of how to identify significant patterns of relationship.

g) By holding the entries in series on computer files, the subsequent intellectual operation of experimenting with a variety of different classification schemes is facilitated. The results of a number of such experiments can be incorporated as special indexes, regrouping the names of the entries, but without interfering with other experimental classifications or with the information as filed. Several experiments of this type have been included in this edition and are discussed below.

3. General principles

Since it may be argued that there is some resemblance between the task of classification of diseases of the human being and the task of classification of the problems of human society, it is useful to consider the general principles governing the former as printed in the introduction to the World Health Organization's *International Classification of Diseases* (Geneva, WHO, 1967). The section in question reads: "*Classification is fundamental to the quantitative study of any phenomenon. It is recognized as the basis of all scientific generalization and is therefore an essential element in statistical methodology. Uniform definitions and uniform systems of classification are prerequisites in the advancement of scientific knowledge. In the study of illness and death, therefore, a standard classification of disease and injury for statistical purposes is essential.*"

"*There are many approaches to the classification of disease. The anatomist, for example, may desire a classification based on the part of the body affected. The pathologist, on the other hand, is primarily interested in the nature of the disease process. The clinician must* consider disease from these two angles, but needs further knowledge of etiology. In other words, there are many axes of classification and the particular axis selected will be determined by the interests of the investigator. A statistical classification of disease and injury will depend, therefore, upon the use to be made of the statistics to be compiled.

The purpose of a statistical classification is often confused with that of a nomenclature. Basically a medical nomenclature is a list or catalogue of approved terms for describing and recording clinical and pathological observations. To serve its full function, it should be extensive, so that any pathological condition can be accurately recorded. As medical science advances, a nomenclature must expand to include new terms necessary to record new observations. Any morbid condition that can be specifically described will need a specific designation in a nomenclature.

This complete specificity of a nomemclature prevents it from serving satisfactorily as a statistical classification. When one speaks of statistics, it is at once inferred that the interest is in a group of cases and not in individual occurences. The purpose of a statistical compilation of disease data is primarily to furnish quantitative data that will answer questions about groups of cases.

This distinction between a statistical classification and a nomenclature has always been clear to medical statisticians. The aims of statistical classification of disease cannot be better summarized than in the following paragraphs written by William Farr a century ago:

"The causes of death were tabulated in the early bills of Mortality (Tables mortuaires) alphabetically; and this course has the advantage of not raising any of those nice questions in which it is vain to expect physicians and statisticians to agree unanimously. But statistics is eminently a science of classification; and it is evident, on glancing at the subject cursorily, that any classification that brings together in groups diseases that have considerable affinity, or that are liable to be confounded with each other, is likely to facilitate the deduction of general principles. Classification is a method of generalization. Several classifications may, therefore, be used with advantage; and the physician, the pathologist, or the jurist, each from his own point of view, may legitimately classify the diseases and the causes of death in the way that he thinks best adapted to facilitate his inquiries, and to yield general results. The medical practitioner may found his main divisions of diseases on their treatment as medical or surgical; the pathologist, on the nature of the morbid action or product; the anatomist or the physiologist on the tissues and organs involved; the medical jurist, on the suddenness or the slowness of the death; and all these points well deserve attention in a statistical classification. In the eyes of national statisticians the most important elements are, however, brought into account in the ancient subdivision of diseases into plagues, or epidemics and endemics, into diseases of common occurence (sporadic diseases), which may be conveniently divided into three classes, and into injuries, the immediate results of violence or of external causes."

A statistical classification of disease must be confined to a limited number of categories which will encompass the entire range of morbid conditions. The categories should be chosen so that they will facilitate the statistical study of disease phenomena. A specific disease entity should have a separate title in the classification only when its separation is warranted because the frequency of its occurence, or its importance as a morbid condition, justifies its isolation as a separate category. On the other hand, many titles in the classification will refer to groups of separate but usually related morbid conditions. Every disease or morbid condition, however, must have a definite and appropriate place as an inclusion in one of the categories of the statistical classification. A few items of the statistical list will be residual titles for other and miscellaneous conditions which cannot be classified under the more specific titles. These miscellaneous categories should be kept to a minimum.

Before a statistical classification can be put into actual use, it is necessary that a decision be reached as to the inclusions for each category. These terms should be arranged as a tabular list under each title, and an alphabetical index should be prepared. If medical nomenclature were uniform and standard, such a task would be simple and quite direct. Actually the doctors who practise and who will be making entries in medical records or writing medical certificates of death were educated at different medical schools and over a period of more than fifty years. As a result, the medical entries on sickness records, hospital records, and death certificates are certain to be of mixed terminology which cannot be modernized or standardized by the wave of any magician's wand. All these terms, good and bad, must be provided for as inclusions in a statistical classification.

The construction of a practical scheme of classification of disease and injury for general statistical use involves various compromises. Efforts to provide a statistical classification upon a strictly logical arrangement of morbid conditions have failed in the past. The various titles will represent a series of necessary compromises between classifications based on etiology, anatomical site, age, and circumstance of onset, as well as the quality of information available on medical reports. Adjustments must also be made to meet the varied requirements of vital statistics offices, hospitals of different types, medical services of the armed forces, social insurance organizations, sickness surveys, and numerous other agencies. While no single classification will fit the specialized needs for all these purposes, it should provide a common basis of classification for general statistical use.

4. Anti-developmental biases in classification scheme designs

In the conventional western approach to the design of classification schemes, a number of biases seem to emerge. These biases are inherently anti-developmental. The effects are particularly serious in the social science domain. The biases have been discussed elsewhere (A Judge, Anti-developmental biases in thesaurus design, 1981). They may be summarized as follows:

(a) Static bias associated with noun categories: Most classification schemes are concerned solely with ordering nouns or objects (called "subjects"). This emphasizes static, structural mind-sets, whereas the fundamental characteristic of development is change and movement. Thus, in the 1986 edition, to de-emphasize this static approach, strategies (Section S) were frequently named using a gerund form to stress their dynamic, action-oriented character.

(b) Low-context bias associated with western science: Western science is deeply preoccupied with specifics and classification systems are designed to handle them. Such systems are consequently not designed to reflect high sensitivity to context as is to be found in non-western cultures. And yet it is precisely such context sensitivity which is required to integrate the fragmented perspectives and engender more coherent approaches to seemingly unrelated issues.

(c) Pattern conservation bias: Classification schemes are usually designed on the assumption that they can grow by extension of a pre-defined pattern and not by transformation of that pattern. This is precisely the kind of thinking which reinforces the present non-transformative character of development.

(d) Dysfunctional bias: Most classification schemes are insensitive to the functional relationships between the phenomena classified. The fact that mercury may infiltrate food chains to affect seriously the survival of a bird species is not something a thesaurus is designed to highlight.

(e) Insensitivity to wider repercussions of classification schemes: Design of a classification scheme is seldom considered in terms of non-library users whose methods of organization will be reinforced by it, whether appropriately or inappropriately. This may be seen in: bookshop layout, agenda design, curricula design and organization charts. A mechanistic classification scheme reinforces mechanistic thinking in relating subject areas.

(f) Avoidance of top-of-hierarchy issues: Most of the effort in classification system design is directed to clarifying problems within particular domains. By contrast little effort is directed toward clarifying relationships between the major hierarchies within which this effort is made.

(g) Preference for adaptive "maintenance" thesauri: In the light of the Club of Rome distinction between "innovative learning" and "maintenance adaptive learning", most classification systems are designed after the fact in response to old issues. Every attempt is made to fit new issues (eg environment) into old frameworks. No provision can be made for the next interdisciplinary crisis which will call for information to be organized in a new way.

(h) Investment in rigid, anti-experimental systems: Classification systems are designed so that once implemented it is extremely costly to modify them. They are not designed to facilitate flexible experiments in reconfiguring patterns of categories.

(i) Concealment of contradictions: Every effort is made to render classification systems as impersonal as possible, de-emphasizing any personal or subjective elements. This both conceals the biases of the designers and avoids the need to classify biases more explicitly. And yet it is such biases which are a driving force behind developmental processes and problems.

(j) Concealment of values: The assumption is widely made that classification systems are value-free. A thesaurus which treats the real-world experience of "homelessness" as an aspect of the academic discipline of "sociology", and treats "war" as an aspect of the discipline of political science is taking a strong political position. A totally exploitative attitude towards the environment is suggested by the inter-agency institutional information system based on a *macrothesaurus* with categories of "fisheries", "fishing" and "fish processing", but not "fish" (as having a role in their own right within planetary ecosystems), nor over-fishing. It is no wonder that there has been so little institutional sensitivity to the ecological problems of deforestation.

5. Classification experiments

In this Encyclopedia several experiments in classification have been explored in this, and earlier, editions:

(a) Interfacing with international classification systems: For the 1976 edition four sections were organized on the basis of the adaptation of four specialized international schemes. Traded commodities and products (Section C) were organized according to the United Nations *Standard International Trade Classification*. Economic and industrial sectors (Section E) were organized according to the United Nations *Standard International Classification of all Economic Activities*. Occupations, jobs and professions (Section J) were organized according to the *Standard Classification of Occupations* of the International Labour Office. Human diseases (Section Q) were organized according to the *International Classification of Diseases* of the World Health Organization.

(b) "Pragmatic" approach: The range of international organizations in the *Yearbook of International Organizations* raises similar classificatory problems. For several editions, a simple pragmatic scheme has been used to distinguish international organizations by "levels" of internationality. The possibility of adapting this scheme to world problems was explored in the 1986 edition (Section PX) using qualifier codes. This system has been further developed in this edition to split the world problems into Sections PA through PJ.

(c) Patterns of concepts: In both the 1986 and current editions, small experiments have been made in relying on patterns of naumbers to order concepts (Section CP, 1986; Section KP, current). The use of taoist system of ordering has also been explored (Section CP, 1986; Section TP, current). Many of the entries on human development (Section H) form part of sets, or series, of concepts or modes of awareness. As stages they are numbered in their respective traditions. In the Buddhist tradition, for example, indexes based on such sets are occasionally made. This approach has been explored in this edition for the entries in Section H in the form of a special index (Section HX) ordered by number.

(d) Levels of awareness: In the current edition the possibility of grouping modes of awareness into levels has been explored by the use of a qualifier code (a through g), which is described in the relevant Notes (Section HZ). But, for lack of space, entries have not been extracted in this order.

(e) Human values: Both in the 1986 and current editions, the results of experimenting with classifying "value complexes" are presented in Section VT. This is based on clustering values, building on the organization of *Roget's Thesaurus*.

(f) Subject classification: Each year a more extensive classification experiment is undertaken with entries in this Encyclopedia, especially those in the World Problems section (Section P). In order to move towards establishing the links between international organizations and the world problems with which they are specifically concerned, both organizations and problems are classified by subject in the subject volume (Volume 3) of the *Yearbook of International Organizations*. This classification scheme, designed in detail in that volume, endeavours to highlight the interdisciplinary relationships so necessary to the cross-sectoral preoccupations of the international community.

Method: language-determined distinctions

This project is in many ways an exploration of the use of language and the effects of its use on the distinctions which are accepted in international discourse. The names under which the world problems described in Section P have been detected, for example, exploit many of the word variants available in the English language. The same is true of the entries on human development (Section H) and human values (Section V).

In reporting on the state of English as a language (C Ricks and L Michaels (Eds), The State of the Language, 1990), the South African author Njabulo Ndebele is quoted as follows:
"The problems of society will also be the problems of the predominant language of that society. It is the carrier of its perceptions, its attitudes, and its goals, for through it the speakers absorb entrenched attitudes. The guilt of the English then must be recognized and appreciated before its continued use can be advocated." Presumably this argument also applies to the potential for human development recognizable within a society.

1. Dilemmas
Whilst a reasonable amount of normalization and rationalization is possible, and has been undertaken, four questions remain (focused here on the situation with regard to problems):

(a) What degree of variance in the use of words reflects significant distinctions, rather than the simple use of alternative descriptors which are effectively synonymous?

(b) To the extent that different words do indeed reflect significant distinctions, to what extent does the availability of a set of words that could be used, for example, to construct a problem-type name, contribute to the recognition of problems which might not otherwise have been distinguished? And, conversely, in a culture (or discipline) with a more limited vocabulary, to what extent is the recognition of problems constrained?

(c) If problem-type names tend to conform to a limited set of patterns, to what extent is it appropriate to highlight the possible existence of problems whose names can be generated by combining available words in such patterns?

(d) Are there situations in which the repeated use of a powerful figure of speech to describe a problem situation can be erroneously interpreted as signifying the existence of a distinct problem? And, linked to this, do the frequent pronouncements by heads of government and judicial authorities that certain acts and people are "evil" constitute a recognition of "evil" as a problem?

In this edition several responses to these questions have been explored as indicated below.

2. World problems
In this case, as noted above, some degree of normalization and rationalization has been used. In part this was necessary to clarify the problematic nature of the many problems that have acquired conventional (possibly shorthand) labels with positive connotations (eg "youth", "development"). Thus "peace" would not be accepted as an appropriate problem name (except for those who perceive the state of peace to be a problem in its own right), especially since those who use the term to indicate a problem tend to subscribe to "peace" as a value or goal. Consequently an editorial rule was adopted that unless the name could be interpreted negatively (as in the case of "pacifism", for example), an appropriate negative qualifier had to be added to complete the problem name wherever possible. This was done using a prefix or suffix ("**mal**nutrition" rather than "nutrition", "**il**literacy" rather than "literacy", "meaningless**ness**" rather than "meaning"). Where this was not possible either quantitative qualifiers were added to the name (eg "shortage of...", "inadequate...", "lack of...") or qualitative qualifiers (eg "inappropriate...", "conflicting...", "deterioration of...", "instability of...").

Once the problematic dimension has been rendered explicit, patterns of problem names of types such as the following tend to emerge:

(a) **One key word:** Crime, unemployment. Hyphenated forms: child-marriage, anti-intellectualism.

(b) **Two key words:** Exploitation in employment, vested interests, soil pollution, racial discrimination. These illustrate combinations of an identifier (eg "pollution") with a delimiter (eg "soil", as opposed to "water").

(c) Multiple key words: The pattern may be further developed by refining the qualifiers and delimiters: eg Racial discrimination in public services.

(d) **Regional and group qualifiers:** The previous point illustrates the possibility of identifying more specific problems by the addition of qualifiers:
- Racial/Economic/Social/...
- discrimination/exploitation/...
- in the
- construction industry/agricultural industry/...
- against
- disabled/impoverished/...
- women/youth/...
- in
- tropical/semi-arid
- developing/industrialized
- countries/regions

It is clear from the last example that seemingly credible problem names can easily be "generated" by appropriate combinations of descriptors from descriptor sets (in a manner reminiscent of Ramon Llull's category generator). The availability of such descriptors sets encourages the recognition by the international community of problems which may or may not exist. Just as there are "letterhead organizations", there may be "descriptor-generated" problems (for which project funding may even by sought). Such problems may emerge into the literature as a result of efforts by delegations or the media to appear innovative.

Given the institutional resistance to the recognition of new problems however, systematic explorations of such a procedure suggests an approach to identifying potential problems rather than allowing them only to be registered in a haphazard manner as information is found on them. Perhaps category generators could usefully be developed to draw attention to categories of problem (strategy, value, etc) which may be neglected in conventional decision-making. In this project the procedure has only been adopted in the case of economic sectors and endangered species as a means of ordering problems at different levels of specificity, particularly in order to avoid isolated excessively-publicised entries (eg endangered panda, or endangered monkey-eating eagle) where more general entries would be more appropriate (eg endangered bears and birds of prey, respectively). The question remains however as to whether the distinctions between problems, generated using different verbal "operators", correspond to valuable nuances with which people identify, or whether such distinctions are purely contrived in order to arouse emotive effects through their novelty.

3. Human values
In this case the kind of information available is so diffuse and unstructured that it is fair to say that there are no lists of values with which the international community identifies, whether partially or completely. There are texts which reflect values, such as the Universal Declaration of Human Rights, but these do not identify values as such. There are no guidelines for naming values, and other than obvious values such as "peace", "justice" and "liberty", few appear to have been named.

For this reason the opposite approach was used, namely an effort was made to generate a complete set of potential values by identifying a comprehensive set of value-words which could be assumed to reflect the full range of values. Again the question arises as to whether these value words do effectively indicate distinct values. Here the difficulty is extreme because of the fluidity of language and the variety of connotations associated with any particular word. The procedure used was however designed to make this confusion explicit without attempting to resolve the issues which emerge. The procedure has the merit of not discriminating in favour or against any particular values as a result of emphasis on fashionable values.

4. Human development

In this case there is a great deal of information available and, even in the case of subtle modes of awareness, it is quite extensively articulated. The difficulty is that because of the subtlety of such modes and the essential subjectivity of the experience, the information frequently takes the form of metaphorical or symbolic allusions. These may convey only limited meaning to those from other cultures unfamiliar with the symbolic jargon used. This is especially true of western efforts to penetrate the extensive eastern literature on modes of awareness. Such cultures are quite explicit about the inadequacy of words to convey the nature of the essential experiences in personal development, although highly articulated terminologies may be used to distinguish such experiences. Such distinctions may be impossible to convey into a language such as English which lacks equivalent words. How, for example, is it possible to respect the distinction in Tibetan between 72 forms of what in English is simply indicated by the word "love"?

5. Patterns of concepts

In this case there is little information available on patterns of concepts. Concepts tend to be treated in isolation or set in a poorly articulated context. In order to elicit some indication of how a pattrn of concepts might appear, two experiments were undertaken using a metaphorical approach. The substrates selected for this approach were in both cases highly articulated patterns. In one case a set of 64 interconnected conditions central to traditional Chinese attitudes towards change was used. In the other, 253 interrelated patterns in environmental design were used. The result was necessarily artificial because, in seeking metaphoric parallels, the available language was stretched and distorted beyond normally acceptable usage. It raises questions as to whether the resulting verbal formulae indicate the existence of useful distinctions or whether they are totally contrived. In either case the challenge remains. For, as with values, there would appear to be important distinctions that need to be made in what is at present an extremely fuzzy and fluid non- material domain of significance. The available vocabulary, at least in non-literary English, would appear to be ill-suited to the task.

6. Strategies (1986, Section S)

In spite of the concrete action-oriented nature of strategies, there is relatively little structured information on the strategies advocated within the international community. Two complementary approaches were therefore used. The available information form international organization documentation and reference books was used to produce a series of entries (in Section SS) effectively presenting a "top- down" perspective. The second approach was based on the extraction of names of strategies from a series of 40 reports of community development dialogues undertaken by the Institute of Cultural Affairs, mainly in rural communities in developing countries (Section SP and SQ). To a large extent this presents a "bottom-up" perspective. The ICA material itself reflects the results of a struggle to use vocabulary creatively to make distinctions, especially using gerunds, which are motivating and empowering. One of the principal merits of this material is that it reflects a wide range of strategies, extending far beyond the limited set that is the concern of political economists.

Of potential interest in relation to further work on strategy terminology is the initiative of Henry G Burger (*The Wordtree; a transitive cladistic for solving physical and social problems*, 1984) to produce "*a transitive cladistic for solving physical and social problems*". "*The dictionary analyzes a quarter-million world-listings by their processes, branches them binarily to pinpoint the concepts, thus sequentially tracing causes to their effects.*" (Henry G Burger, The Wordtree: a transitive cladistic for solving physical and social problems. Merriam (Kansas), The Wordtree, 1984) It distinguishes the components of every action (transitive) verb in most dictionaries. Highly critical of the linearity of *Roget's Thesaurus*, Burger claims that by focusing on transitive verbs there is a built in emphasis on how action may be undertaken to solve problems. Solutions to problems are stored in language. He has developed a presentation to highlight this which contains a "quarter-million problems and solutions of behavior and goals"

The question remains however as to the degree to which the distinctions between similar strategies made with such linguistic devices are really meaningful, however much the subtle differences in emphasis are suggestive of potential significance which merits further investigation.

7. Human values

The words used to capture and label subtle distinctions between human values are remarkable for the degree of overlapping connotations. In Section V an attempt is made to chart this fuzziness. The question remains as to whether the distinctions made are to be considered meaningful. Furthermore, does the confusion of words constitute an ecosystem of necessary richness to ensure the survival of certain values that would otherwise become extinct.

8. Metaphor

The exploration in the Section M of the potential of metaphor in constructing realities is a direct response to the creative potential of language as a means of establishing more fruitful patterns of connections between seemingly unrelated concepts.

9. Comment

The different methods used all offer insights into the limitations and opportunities of language in increasing sensitivity to the range of phenomena that need to be borne in mind in responding to the global problematique. The acid test in reviewing the results can be performed as a thought experiment. For example, in the case of any given value word, can society afford to be insensitive to the value implied by that word, especially to the extent that the word indicates a value variant not adequately indicated by other words? Would the quality of life be diminished by ignoring that particular variant? Similarly, in the case of world problems, can society efford to neglect any one of the specific problems identified by a particular pattern of words on the assumption that, to the extent it is significant, it will be subsumed under some broader problem through which an effective response to it will be undertaken?

Method: response to diversity

1. Challenge of diversity
In contrast to many recent studies of world problems, or of the component units of the other sections of this publication, one intention of this project is to explore the consequences and difficulties of attempting to handle much larger numbers of units and their interrelationships.

Clearly the methods which are useful in the study of 5 to 10 units, the resulting conclusions, and the manner of their presentation, might have to be modified if it were considered admissable to focus on up to 100, or up to 1,000, or up to 10,000 units. Since the results and the value of studies of 5 to 10 major units have been, and continue to be, extensively explored (with questionable improvement in the ability of society to control itself), it may prove useful to give attention to a somewhat different approach.

2. Tolerance of diversity
The headings below are for different orders of magnitude. Under each heading a very rough attempt has been made to indicate what kinds of things are detected if the detail of a still higher order of magnitude is not to be tolerated. The point being that political debate would seem to limit itself to magnitudes 0 and 1. Social science investigations presumably venture into magnitude 2 (given that populations are seldom reduced to more than 100 classes) and some way beyond. In contrast with this relative intolerance of detail, biologists and chemists are obliged to work with a degree of variety corresponding to magnitude 6. (Biologists tolerate the existence of about 1,000,000 recognized plant species and a similar number of animal species; chemists tolerate the existence of about 1,000,000 recognized molecular compounds.)

3. Orders of magnitude

(a) Order of magnitude 0 (namely, 1 single unit only)
- **the** major problem (war, or pollution, or hunger, or unemployment *etc*)
- **the** major organization (United Nations, or NATO, or the Government, *etc*)
- **the** major value (peace, or happiness, or justice, *etc*)
- **the** major disciplines (science, or art, or physics, *etc*)

(b) Order of magnitude 1 (namely, approximately 10 units)
- the 5-10 major problems (war, pollution, hunger, *etc*)
- the 5-10 major organizations (UN, UNESCO, ILO, FAO, NATO, *etc*)
- the 5-10 fundamental values (peace, justice, health, *etc*)
- the 5-10 basic disciplines (physics, chemistry, biology, *etc*)

(c) Order of magnitude 2 (namely, approximately 100 units)
- continuous critical problems (of Ozbekhan or Battelle)
- major international (governmental) organizations
- disciplines taught at an average university
- major multinational corporations
- major multilateral treaties
- major recognized occupations
- sovereign nation-states

4. Identification with detail
By contrast, the sections on perceived world problems in this Encyclopedia cover in each case over 10,000 items (Order of magnitude 4). Those on values and modes of awareness cover over 1,000 items (Order of magnitude 3). The argument of this project is that people and organizations clearly identify with concepts distinguished at this level of detail. It is less easy for them to identify in any operational way with items distinguished at lower orders of magnitude, especially at level 0 or 1, although it is clearly in terms of concepts at level 0 or 1 that people can be temporarily swayed by the media.

5. Weakness of international information
Through the various international statistical series maintained by the United Nations Specialized Agencies (and published in their various Statistical Yearbooks), details are available on the number of cinemas, newspapers, radios, libraries, and so on, per capita in each country. But no systematic collection of information is **publicly** available on, for example, the number of local, regional, or national organizations in different areas, even in developed countries. The first type of data could be considered as covering methods of informing, instructing, or influencing individuals or groups - namely, the downward flow from centres of power and excellence. The second, and missing, type covers methods by which individuals and their minority groups express, protect, and further their particular interests and blend them into those of society as a whole.

Such devices are the channels through which individuals participate in society, refine definitions of values, and mould the direction of development. Bertram Gross notes: *"It is the intricate network of the subsystems that, more than anything else, establishes the framework of social structure." (The state of the nationals; social systems accounting. In: Social Indicators*, R A Bauer (Ed). Cambridge, MIT Press, 1966, p.194).

Serious questions are raised when the International Labour Office's *Yearbook of Labour Statistics* has data on industrial disputes and accidents, but no data on trade unions. Until data is available and attention is focused on the social devices (and their interrelationships) through which individuals initiate their own participation in society, it is questionable whether theories and policies based on aggregation of data on the social devices by which they can be influenced and controlled will result in policies which respond to (rather than create) social problems, or will be considered meaningful by individuals or the smaller groups which represent their special interests.

Assessment: international organizations as a source

The information received from these bodies is a prime source of material on the perceived world problems registered in this Encyclopedia.

1. Significance
This Encyclopedia is the product of a programme of the Union of International Associations whose context is the collection and processing of information on international nonprofit bodies, whether governmental or nongovernmental. This long-term activity, initiated in 1907, takes its most visible form in the companion series, the annual 3-volume *Yearbook of International Organizations*. In it over 25,000 bodies of these bodies are registered together with their specific relationships to each other.

International organizations, agencies, institutes and associations, whether governmental or nongovernmental, are, by their very nature, a focus for international action and concern of some kind. Many of these organizations are concerned specifically with direct or indirect action on world problems, and as such may be related to entries in the world problems section of this volume (Section P). Whether individually, or collectively as a community or a network, international bodies function as focal points for the mobilization of resources to be brought to bear on individual problems or groups of problems. In some cases, it is the creation of an international organization which signals the emergence and recognition of the new world problem on which it is focused.

2. Definition
Three categories of international organization are usually distinguished:

(a) intergovernmental organizations often known simply as international organizations since none of the other international organizations are inter-state organizations from a legal point of view;

(b) international nongovernmental nonprofit organizations, also known as international associations, international voluntary associations or agencies, and increasingly as transnational associations;

(c) international nongovernmental profit-making organizations, usually known as multinational corporations or enterprises, and increasingly as transnational enterprises.

The definitions given below apply to the selection of organizations for inclusion in the *Yearbook of International Organizations*:

(a) Intergovernmental organizations: The view of the Economic and Social Council of the United Nations concerning inter-governmental organizations is implicit in its Resolution 288 (X) of 27 February 1950: *"Any international organization which is not established by intergovernmental agreement shall be considered as a non-governmental organization for the purpose of these arrangements..."* The resolution was concerned with the implementation of Article 71 of the United Nations Charter on consultative status of non-governmental organizations, and it was amplified by Resolution 1296 (XLIV) of 25 June 1968: *"...including organizations which accept members designated by government authorities, provided that such membership does not interfere with the free expression of views of the organizations."*

Identifying eligible governmental organizations therefore presents no problem. All organizations established by agreements to which three States or more are parties have been included. Following the adoption of Resolution 334 (XI) of 20 July 1950 it was agreed with the UN Secretariat in New York that bodies arising out of bilateral agreements should not be included in the *Yearbook of International Organizations*.

(b) International associations: The problem of identifying eligible nongovernmental organizations is much more difficult. Ecosoc Resolution 288 (X) makes no attempt to explain what is meant by the term *"international organization"*. A number of aspects of organizational life have therefore been used to determine whether an organization is eligible for inclusion. Briefly these seek to ensure a reasonable balance in realtion to aountries with respect to the organization's: aims, membrship, structure, directorate, finance and activities.

3. Organizational relationships to this Encyclopedia
Although many organizations can be related directly to the world problems which are their main concern, others can be more closely associated with entries in other sections in this volume and only indirectly with specific world problems, if at all:

(a) Communication (Section C): A number of bodies are principally concerned with particular modes of communication and their use in furthering development processes.

(b) Human development (Section H): A number of bodies are principally concerned with promoting particular approaches to human development.

(c) Integrative knowledge (Section K): A limited number of bodies are primarily concerned with approaches to conceptual integration.

(d) Strategies (Section S, 1986): Many bodies are principally concerned with promoting and implementing strategies directly oriented towards alleviating particular world problems, whether or not the strategies are identified as such.

(e) Transformative approaches (Section T): In various ways a number of bodies are specifically interested in developing innovative techniques, especially in the domain of communication, in order to facilitate resolution of world problems.

(f) Human values (Section V): Whilst relatively few bodies focus specifically on the promotion of particular values, most indicate values in the aims specified in their constitution and make repeated references to such values in conference resolutions.

4. Inter-organiztional relationships
In addition to the relationships described above, international bodies are embedded in networks of formal and informal relationships amongst themselves. In the 1990-91 edition of the *Yearbook of International Organizations*, 65,174 relationships were specifically identified. Such relationships may include constitutionally defined participation of one organization in another's policy formulation; formal agreements to collaborate; membership of one in another; mutual consultation; and systematic or occasional exchange of information. Such networks are the international or transnational counterpart of even denser networks of relationships between organizations and groups within each country.

None of these networks of relationships has been mapped to any degree, and it is therefore impossible to determine systematically:

(a) To what extent the network of organizations matches the network of problems, namely the extent to which problems are matched by organizations or programs, and relationships between their counterpart organizations;

(b) In which areas organizations or inter-organizational relationships could be usefully recommended and encouraged at the international level;

(c) Where other entities (eg multilateral treaties, or international information systems) obviate the need for the establishment of organizations;

(d) Whether it is possible to predict with any precision the manner in which the network of organizations will develop, in order to foresee future structural problems;

(e) Which organizations merit additional funding by the nature of the focal role they perform in critical parts of the network;

(f) Where duplication of effort and competition for scarce resources occur, in order to determine whether or not such duplication is advantageous.

It is not even possible, with the information currently available, to determine whether such analyses, if performed, would be useful. Quite simply, the tools and information are not yet in a form to permit such networks to be understood and handled conceptually.

5. Relationships to world problems

In the 1976 edition of this volume, names and addresses of some 3,300 international organizations were included with cross-references to entries in other sections of the volume. Now that the number of bodies registered in the *Yearbook of International of Organizations* has reached over 25,000, alternative approaches to cross-referencing the bodies to problems, values and strategies need to be explored. One approach already adopted is to classify the problems with the organizations and multilateral treaties in *Global Action Networks* (Vol 3 of the *Yearbook of International Organizations*). This groups the entries within a pattern of some 3,000 subject categories. This gives an overview of the range of organizations and treaties associated with a particular problem domain.

One of the original aims in producing this Encyclopedia was to endeavour to document how the network of international bodies focused on the network of world problems. Clearly some key problems attract much attention and many others attract very little, if any. But of greater interest is whether the organizations focusing on the same problem are in communication with each other, or whether the organizations dealing with one problem that closely affects another are in contact with the bodies dealing with the latter problem. How do problems escape the net of organizations? How does the network of organizations fail to encompass and contain the network of problems?

In addition to the major role that international organizations perform in identifying problems, many of them perform a major function in relation to human values. In fact the two functions are often intimately linked, as in the case of human rights issues. An important group of organizations is also concerned with human development in its less material sense, as is the case of bodies concerned with religion and personal development. Again the question may be asked with which of the specific values is an organization associated, given that many of them carefully identify values in their statutes? Clearly their are very "popular" values, such as peace and justice, but are there values with which few, if any, bodies are associated? And to what extent are such values vital to the functioning of society? Is there also a mismatch between the network of organizations and the network of values?

6. Difficulties encountered

It has proved much more difficult to represent the degree of mismatch than was originally envisaged, despite extensive use of computers and sophisticated software. Progress and difficulties are briefly noted in the following:

(a) Fashionable information: Information on fashionable and uncontroversial problems is usually readily available. Here the difficulty is that many organizations produce information on the same fashionable problems. It is therefore of far less value to indicate the specific organization-to-problem link. The same is true of the fashionable values.

(b) Buried information: Whilst the documentation of international organizations is rich in relevant material, much of it is "buried" to the point of being inaccessible without the deployment of considerable resources. This is especially the case of new and controversial problems. It is also true of the specific problems which are articulations at a more detailed level of the fashionable problems (identified at a more general level). In these cases mention of the problems is made in chapter or paragraph headings, not in the title of the publication. In many cases they may only be identified in the body of a paragraph. Not only is information more difficult to obtain through normal bibliographic searches, but it becomes less appropriate to cite as an organization-to-problem link. A specific issue is whether mention at such a detailed level indicates that the organization is really concerned with the issue or is just mentioning it "in passing".

(c) "Solution language": It might be assumed, especially in the case of intergovernmental organizations, that conference resolutions would provide a rich source of information on organization-to-problem or organization-to-value links. This is not the case. Again fashionable issues may be specifically cited in resolutions but the difficulty is that this is often done ritualistically so that it is far from clear whether noting the link is meaningful. Where the resolutions concern action programmes, it is seldom clear from the language of the resolution exactly what problem is being addressed. The reason for this is that resolutions are presented in "solution language" and tend to focus on constructive actions such as institution building, without it necessarily being clear what problems an institution is intended to address -- other than at the most general level. With more effort it is possible to decode such language, but implicit problems may continue to remain elusive. The same may be said of the texts of multilateral treaties.

(d) Restricted information: Although much information collected by international bodies is intended for any interested parties, there are many practical difficulties in obtaining access to it. These are complicated by concern for certain classes of information which are collected and held on a confidential basis (possibly available only to organization members) or by the need to limit distribution of certain reports for cost reasons. There is of course always the suspicion that the dissemination of embarrassing information is restricted, whatever the justification. The fate of international statistics on crime is an example. Unfortunately the "expert committees" which articulate problems in many organizations produce their reports in a manner which renders them impossible to extract from the archives of those bodies, whether or not it was intended to restrict access to them.

(e) Failure to articulate concerns: It is easy to conclude, whether correctly or incorrectly, that few organizations endeavour to articulate clearly the problems or values with which they are concerned. For many it may indeed be sufficient to accept definitions articulated by others or to focus primarily on solutions. The currently accepted definitions may have been articulated in documents decades earlier. There may be no felt need to reproduce them in currently available documents. It is then less clear how to link such organizations to problems or values.

(f) Emphasis on articulating responses: In general the strengths of international organizations, whether governmental or nongovernmental, lie more in articulating responses to problems that have already been detected and defined rather than in the process of detecting and defining such problems. Presumably because of their relationship to their members, to those providing subventions, or to other bodies to whom they endeavour to establish their legitimacy, efforts must necessarily be the subject of upbeat reporting on programme achievements, rather than on a sharper articulation of existing problems or on the detection of new problems. Like any institution at the national level, they are not designed to be sensitive to problems that have not been anticipated in their programming mandate.

Although there are many exceptions to the above difficulties, it was decided once again not to attempt to indicate specific organization-to-problem or organization-to-value links. This absence is however compensated by the annual classification of problems from this Encyclopedia with organizations by subject in the *Yearbook of International Organizations* (Vol. 3). This has the advantage of being much more systematic in approach. It has the disadvantage of being insufficiently specific in many cases, since it based on keywords in the titles of organizations.

Assessment: biases

In the light of the scope and methods noted above, a further influence on the design of the publication was a number of specific biases, some of which strongly influenced the length of any description. Other biases are discussed in more detail in the methodological assumptions for World Problems (see Section PZ) and for Human Development (Section Section HZ), and in less detail for Human Values (see Section VZ).

1. Pluralism
As mentioned above, the whole editorial process was biased against any particular set of values, especially any particular concept of truth or falsehood, or of right or wrong, or of good or evil, or of strategic relevance or irrelevance, whether or not this resulted in texts which were acceptable or ridiculous in terms of the scientific, legal, religious, cultural, political or strategic priorities of others. The task was conceived as one of "telling things as they are" in the eyes of those who identify with a particular perspective, not of highlighting only what is important according to one such perspective.

2. Favouring the less well known
There is a definite bias towards giving more space to less well-publicized perspectives than they would normally warrant. Consequently less space has been given to the standard well-documented perspectives, for example the world problems of war, famine, pollution, etc.

3. Network presentation
The above bias is partly corrected by a bias in favour of presenting any conceptual complex (as in the case of a group of problems) as an interconnected set of many sub-items rather than as one long amalgamated description. In the case of the world problems, the sub-problem descriptions may in fact be longer than that of the parent problem.

4. "Opening up" categories
When information was inadequate or too much editorial work was required to process the available material into an appropriate form, there was a bias in favour of including the entry, even without a description, rather than excluding it so as to maintain an impression of entries of higher quality. There was therefore a bias in favour of "opening up" categories to which indexes and cross-references could refer in anticipation of work in future editions. This may be viewed as a bias in favour of lists.

5. Use of low-grade information
In contrast to other efforts to document world problems, for example, there is a definite bias against dependence on "high grade" information in which each "fact" must be substantiated by an approved authority. As pointed out earlier, such "facts" are quickly disputed, denied or ignored in counter-reports by those holding alternative views, whether "authoritative" or not. Where high grade information is available from international bodies it has been used. Where the information is too controversial to be approved by an international body, or where no concerned body, exists, "low grade" information circulating in the media has been used.

6. Avoidance of closure
This publication raises many questions about the use of language by the international community and the media. Whether, for example, a world problem (or a mode of awareness) denoted by a particular set of words "exists" in a manner distinct from that denoted by a related set of words (which appear to be partly synonymous) is a matter for continuing review. In this project there is a specific bias against premature resolution of such editorial/research difficulties. Obvious duplication has been avoided, but other cases have been allowed to co-exist especially in the human development section.

7. Avoidance of country-specific focus
In documenting world problems, assumed to affect a minimum of three countries, no direct effort has been made to focus on those problems as tied to, or arising from, specific countries. Obvious examples include apartheid (as a problem specific to South Africa) and the "Middle East problem" (as defined in relation to Palestine). In these cases, apartheid is treated as a more general problem (with examples from South Africa), and the "Middle East problem" is treated through many other detailed problems (occupied territories, repression, terrorism, military atrocities, and problems relating to political self-determination and living conditions) for which Palestine may provide examples. Problems are generalized across geographical regions wherever appropriate.

8. Classes of events
World problems are often associated with specific incidents and events, such as the "oil crisis" of the 1970s, a particular collapse of the stock market ("Black Monday"), or particular conflicts. In documenting world problems, such incidents are treated as examples of the appropriate class of problems. Particular incidents are not considered as problems in their own right. Problems are generalized across periods of time wherever appropriate.

9. Mono-lingual source material
As noted earlier, the limited resources imposed an unwelcome bias against material requiring translation into English (in marked contrast to the editorial practice for the *Yearbook of International Organizations*). The assumption was made that this was largely corrected by the extensive use of materials formulated in the multi-lingual environments of international organizations. Some exceptions were also made in the case of unique materials obtained in French. This bias has been partially corrected by a number of efforts, especially in the Human Potential sections, to present materials from other cultural perspectives, notably Buddhist and Islamic.

10. Avoidance of "definitive" classification
A final specific bias, associated with the previous point, is one against premature classification in this Encyclopedia. The task here is seen to be one of registering, describing and interrelating perspectives (in a non-linear manner, where necessary), not of classifying them in some framework which would eliminate significant inconsistencies. Hence the bias in favour of unstructured lists, complemented by indexing and cross-references. Classification, with all that it implies in terms of imposition of a particular conceptual (and often defensive) framework on data, is a separate matter. The same approach is adopted with regard to the international organizations and multilateral treaties in the *Yearbook of International Organizations* (vol 1). These are classified experimentally (in Volume 3) in an evolving integrated framework of some 3,000 categories, together with the world problems from this Encyclopedia.

11. Exploration of alternatives
There are many well-established schools of thought in relation to the materials documented in this Encyclopedia. Where appropriate, alternative approaches, in the form of "editorial experiments" have been used to collect, order and present information. The treatment of values is an example.

12. Enumeration
There are many other approaches to the materials of this Encyclopedia which are discursive in emphasis. It is then difficult to isolate conceptual entities such as "world problems", "values", "modes of awareness", and the like. For this project there is a strong bias in each case to endeavour to respond to the question of "how many" such entities there are in a given case. It is considered useful to insist on enumerating the elements of any universe of such entities. This bias is vital to the construction of a data base, although it is recognized that discursive approaches may be more appropriate under other circumstances.

Assessment: strengths and weaknesses

The strengths identified below may, from a different perspective, be understood as weaknesses. Similarly, the weaknesses identified may, under other circumstances, be considered strengths.

1. Strengths

(a) Range: The principal strength of this publication lies in the range of information presented, often derived from inaccessible documents, reflecting a broad spectrum of cultures, ideologies, disciplines and belief systems. Many of the topics are little-known, however vitally relevant they may appear to those who are especially sensitive to them.

A significant proportion of the information is of a kind which is normally avoided or ignored by institutions and academic disciplines, because there are no adequate procedures or frameworks for handling it. Many of the topics are therefore of a kind not to be found in available reference books, whether because they fall between conventionally recognized categories, or because they threaten them in some way (as with some types of problem).

(b) Juxtaposition of contrasts: A second strength lies in the juxtaposition of seemingly unrelated kinds of information (e.g. problems, values, human development) which emerge as complementary and call for the recognition of a pattern of relationships between them. The organization of the Encyclopedia is designed to permit very extensive cross-referencing of various types. It allows relationships, whether logical or functional, to be indicated in a much more precise manner than in other contexts.

(c) Juxtaposition of opposing perspectives: A third strength is the deliberate presentation of information so as to confront opposing iewpoints, whether through the arguments supporting or denying the existence of a particular problem, by matching constructive and destructive values, or by opposing strategies and counter-strategies. Wherever possible entries indicate the limitations of the perspective presented. The structure of the Encyclopedia therefore guards against dependence on any one particular perspective. Each may indeed be appropriate in particular circumstances, but it is more probable that it is only on the whole "gene-pool" of perspectives that humanity can safely depend in a turbulent social environment during a period of vulnerability to nuclear, ecological and food crises of an unpredictable nature.

(d) Exploration of the limits of language: A fourth strength is the exploration, both through the variety of information and through a number of editorial experiments, of the limitations of language in distinguishing both problems and responses to them (values, modes of awareness, etc). The approach used has made it possible to present sets of fuzzy categories, such as values, in a way which allows them to be usefully related to harder categories of information. Many neglected categories have been "opened up" in a manner which allows the significance of such distinctions to be explored. The approach usefully questions assumptions about the adequacy of language in responding to the global problematique and designing integrative strategies.

(e) Global, "top-down" approach: A final strength lies in the blending of the above strengths to stress the context within which problems and possibilities emerge. The result is a "high-context" initiative offering insights which are difficult to obtain from the conventional "low-context" approaches usually favoured. It represents a continuing effort to incorporate detail into a global context, without allowing specific concerns, however currently fashionable, to unbalance the whole.

2. Weaknesses

(a) Coverage of particular items: The principal weakness of the publication lies in the inadequacy of information on particular items. Whilst many of the entries are adequate, or more than adequate, there are exceptions where more appropriate information could usefully have been included. This is a direct consequence of the "top-down" method which was oriented to culling information from many sources. This did not permit (because of limitations on editorial resources) follow-up on particular items. This defect is also partly a consequence of the bias in favour of "opening up" neglected topics, as opposed to extending information on well-documented topics.

(b) Avoidance of classification: A second weakness for many is the absence of any scheme through which the large amount of information is ordered. To this extent it may appear as a "grab-bag" collection of disordered information of varying quality and significance. As is pointed out however, the absence of a classification scheme is deliberate because one of the fundamental challenges is the design of an adequate scheme which would be non-trivial and minimize distortion. The method used minimizes distortion and provides an information structure with which classification experiments can be undertaken, some of which are presented in this volume.

(c) Limited indication of sources: Despite the inclusion of a 10,000 item bibliography in this edition, a third weakness lies in the limited indication of sources, particularly since in recognizing the existence of a perspective in the international community it would be desirable to indicate what group or constituency holds a given view. The difficulty in including bibliographical references comes again from the method used. In the case of United Nations material, for example, literally tons of documents were scanned for the rare paragraphs defining a problem. In preparing the final entry, the file used might contain photocopies of many such paragraphs. It was not considered feasible to allocate scarce resources to time-consuming bibliographic work when the objective was to cross-reference the entry to the international body directly concerned with a topic, whether or not that body provided information on it. Indeed one of the basic difficulties in obtaining information on world problems, for example, lay in the fact that the bodies most concerned with an issue were frequently unable to supply a succinct description of it. More useful texts often came from other sources commenting in summary form on the issue.

(d) Incompletion: Although in its third edition, the Encyclopedia remains incomplete. With each edition the criteria are broadened, thus broadening the scope. Because of the emphasis on range and globality, gaps appear in the resulting pattern. Because of the "top-down" approach to information retrieval, these may remain unfilled. As an ongoing programme, the Encyclopedia reflects the defects of "work-in-progress" in response to evolving recognition of possibilities whereby it may be improved, rather than the strengths of a product completed in the light of a well-defined concept.

Assessment: criticism

1. Failure to advocate a position
A major criticism levelled at this project has been that it did not take a "position" or advocate a "stance". Such criticism fails to realize that the project is about the necessity of moving beyond the mind-set which engenders answer arenas where "stance-taking" is perceived as the only appropriate form of activity.

In a turbulent environment some more dynamic response is required than that of "drawing the line" somewhere - the conceptual equivalent of a "Maginot-line". A sailor on the deck of a ship in rough seas would fall over if he attempted to maintain a "stance" -- rather than shifting his weight from leg to leg in response to the movement of the ship. (cf Geoffrey Vickers: *Freedom in a Rocking Boat; changing values in an unstable society* (1970), or Donald Schon: *Beyond the Stable State; public and private learning in a changing society* (1971)). The problem of the sailor, if he is to achieve anything under such conditions, is to learn to "walk" rather than simply "standing".

Expressed differently, the criticism is that some central "point" is not being made. This is so. If anything, the "central point" here deals with the tangential strategies of "**not**-making" a central point, since **the overdefinition associated with any particular proposal seems to occupy and obstruct the necessarily undefined nature of the space through which transformative human and social development emerges "from the future".** This project focuses on the dynamics by which all points attempt to become the central point by denying the relevance of other points.

In the same geometric metaphor, this project does not favour a particular ideological "line" of argument, nor does it focus on a particular "area" of concern. The question is rather one of how such different "points", "lines" and "areas" fit together and interrelate to constitute a viable "container" for comprehension of the human and social development process. It is a question of tracking the "vectors of concern" with which people are identified. The peculiar feature of this container is that it must be able to contain the undefined.

The nature of the design problem has been compared to that of containing plasma as a source of fusion energy. Plasma also has unique global characteristics which call for a special configurative approach, especially since any contact with its container quenches it, draining away its energy, thus denaturing it.

Both in the last edition, and to an even greater extent in this one, the emphasis has shifted onto the manner in which our society is trapped by the inadequacies of the ways in which it uses language. To this extent, the "meta-stance" in effect taken is that new approaches to the use of language are required to understand the nature of the problems and possibilities which are engendered by the language used. In Geoffrey Vicker's terms: "A trap is a function of the nature of the trapped". The stress in this edition on new uses of metaphor is an effort to re-imagine both trap and trapped.

2. Excessive complexity
A second criticism has been that the project covered too many dimensions and was too complex. A response might be to question whether, except in the most specific settings, simple answers are productive or whether, at this time, they are not in fact downright dangerous.

There is widespread hope that a simple answer can be formulated to the challenge of the times. Many believe fervently that such answers exist in single-factor statements such as "peace", "love", "order", *etc.* Whilst a necessary feature of the psychosocial system, such belief obscures the richness and significance of the fundamental disagreement concerning the ways such conflicting answers can be implemented in practice.

Edgar Morin (1981) and Kenneth Boulding (1978) both note the dangers of single factor explanations at this time. In Boulding's words: "*The evolutionary vision sees human history as a vast interacting network of species and relationships of many different kinds, and there really is no "leading factor" always in the forefront. At times, changes in material technology are the major mutational developments and create niches for social changes of various kinds. At other times, however, intellectual or spiritual movements take the lead and create niches for new material artifacts and technologies; sometimes climatic changes dominate the scene, or sometimes biological mutations dominate, such as the disease bacteria that caused the great plagues.*" (1978, p.19-20)

To safeguard global society in the longer-term, the challenge would seem to be to find some comprehensible way (or set of ways) of interrelating the simple answers which must necessarily emerge as short-term local responses to such an environment. Hence the reason, in Section KD, for advocating **patterns of alternation between the necessary simplifications.**

The difficulty is illustrated by such admirable initiatives as those of the Brandt (1982) and Palme Commissions (1982) (formally titled the Independent Commission on International Development Issues and the Independent Commission on Disarmament and Security Issues), the Worldwatch Institute (1985), or the US Council on Environmental Quality (1982). Like their predecessors, these bodies have produced reports on the global situation with carefully thought out recommendations.

In the light of the arguments of this project it is difficult to escape the conclusion that such commendable **recommendations for global change are expressed in a language which is out-moded and incapable of engendering the credibility required to mobilize support to implement them.** This remains true even when considerable attention is devoted to visual presentation as in *The Gaia Atlas of Planet Management* (1985). Such weakness is disguised by the apparent success of the public relations exercises by which the reports are launched, the manner in which they are briefly taken up by parliaments, universities and the media, and the implementation of a few of their non-controversial recommendations from what was conceived as an integrated package. The limited effectiveness of such an approach is well-illustrated by a report evaluating the implementation of the Action Plan formulated by the 1972 United Nations Conference on the Human Environment (1982).

Such reports, in appealing to those who place great hope in simple answers (*eg* "cooperation" or "total disarmament"), fail to internalize the significance of other simplistic positions by which their implementation must necessarily be frustrated, as the historical record has repeatedly shown. The possibility of **transformative development emerges from the relationship between answers, not through the elimination of one or the other (or the constituencies to which they appeal).**

A different language is required to render such possibilities more credible and more fruitful. Such a language should not deny the simple answers, rather it should place them in a ("conceptual") context, which encodes the dynamics by which they need to challenge each other in order to separate the "essence" of each from the "dross" from which the dangerous abuses of any simple answer can emerge. In this way a form is given to the context in which each such answer has a function.

It is distressing that even in such an intellectually sophisticated country as France, for example, any individual capable of a leadership role or some degree of influence finds it necessary to align himself, right or left, and then engage in savage and often childishly unsympathetic misrepresentations of the difficulties of the other party, whilst disguising those of his own. Increasingly authorities of any tendency can only maintain credibility and dignity when those who disagree with them are absent or silenced.

What body or school of thought perceives the need for opposing tendencies in order to contain the complexities of the problematique? Presumably any such insight is confined to the much maligned "floating voters". No one of influence argues in public for the need to alternate continually between conflicting policies - and yet it is precisely through such alternation that organized society has developed. If everyone of influence is associated only with a **part**(y), who then speaks for (alternation between the parts within) the **whole? Can the whole be given more effective expression?**

The issue of "complexity" is now of increasing concern to the policy sciences, as illustrated by the initiative taken by the United Nations University in convening a meeting on the matter (United Nations University. *Science and Praxis of Complexity*. 1985). The challenge is once again that daily language does not normally provide a means for grasping that complexity in any useful way other than through over-simplification. The jargons of the specialists of whatever discipline, even when they can encompass that complexity, can only do so by becoming impenetrable to other disciplines, and especially to any wider public expected to approve actions based on such insights.

Whilst there may be ways through this "complexity barrier", as intimated by some of the initiatives reviewed in Section KD, it seems more probable that these will be rejected as incomprehensible. It is for this reason that greater stress has been placed on the potential of metaphors as a largely unexplored resource which appears to offer a more accessible way forward.

3. Complexity of language
A third criticism, linked to the preceding one, has been that the language used, for example in Section KD, was too complex. This was done deliberately to convey a better understanding of the very different conceptual languages used by authors with different backgrounds -- each offering new insights and shades of meaning on a central **undefined** concern -- using their own languages often in order to discuss language.

The essential argument of this project, as repeatedly emphasized, cannot be given explicitly because **there can be no one language appropriate to the meta-answer that appears to be required.** It can only be presented "tangentially" as a **configuration of distinct languages** - whether as the insights from different backgrounds or as an understanding of an N-fold set of distinct approaches, each from a particular background. This project is an exercise in presenting information in such a way.

None of the perspectives given in Section KD, as examples, is individually either necessary or sufficient, but some such set of contrasting perspectives is necessary to provide the requisite conceptual variety to contain the undefined. The problem is somewhat analogous to that of establishing a sufficiently long baseline in terrestrial or astronomical surveys, or to that of constructing a sufficiently large array of differently oriented receptors in radio-astronomy. Hopefully a **pattern of resonance** can be detected within the configuration of perspectives emerging from such very different languages, for it is only on the foundation of such resonance that a viable global approach can seemingly be designed.

Again it is through new uses of metaphor that it may be possible to comprehend and discuss the nature of the dynamic complexity it seems necessary to encompass.

4. Superficiality
A fourth criticism arose from those who have placed their hopes in global modelling as the key to understanding the global problematique. Indeed it was this group that first formulated the term "global problematique". From that perspective, the kind of information collected by this project is completely superficial since it does not analyze the relationship between economic and social factors and express them in mathematical terms so as to define a model which can be explored under different conditions. The modelling perspective is discussed in a following note.

5. Usability
A fifth criticism has concerned the usability of information of the type collected and presented here. Given the expectation that reference books should supply straightforward answers to questions, however complex, an Encyclopedia of this kind may seem to be inadequate.

The approach taken here has been to make individual entries relatively simple, but to increase considerably the complexity of the pattern of relationships between the entries. The Encyclopedia may therefore be used in a very simple manner, by avoiding the distractions of the pointers to other entries. As a "high-context" reference book, users derive most benefit from the Encyclopedia through the way in which it challenges expectations and habitual modes of thought.

In contrast to other reference books, this Encyclopedia is, in an important sense, of greater value in raising questions than in providing answers. It challenges the easy belief that reliance on particular answers is appropriate at this time.

6. Absence if images
Finally, in an Encyclopedia which stresses the value of visual imagery, the absence of such images represents a significant weakness. The point made here is that a new kind of imagery is required to give a deeper sense of the pattern of relationship between entries on problems, values and other concepts. Given the nature of this project, such network maps should be computer generated because of their degree of complexity. It has not proved possible to organize this within the resources of the programme as it now stands. However it is clear that a major obstacle lies in the lack of availability of the appropriate software (as discussed in Section Z).

The alternative route, chosen by many other reference books on world problems, is to use hand-drawn images. From the perspective of this Encyclopedia such imagery reinforces the impression that the pattrns of information are less complex than is in fact the case. They do not respond to the challenge of that complexity.

Assessment: global modelling perspective

1. Modelling based on equations
Global or world modelling may be understood as the attempt to represent rigorously the economic, political, social, demographic and/or ecological issues and their interdependencies on a global scale. The models maps these relationships as explicit mathematical equations which may be "run" forward in time to study their dynamic behaviour. They can thus be used to simulate future developments under a variety of conditions. Such modelling may be considered as the most sophisticated approach to dealing systematically with the nature of, and solution to, world problems.

Global modelling is seen by some as the key to understanding the global problematique. Indeed it was the group that initiated that approach that first formulated the term "global problematique". The accomplishments of this group through the 1970s have been reviewed, by some of those involved, in an exceptionally honest book: *Groping in the Dark; the first decade of global modelling* (Donella Meadow, et al, 1982).

The authors distinguish two types of model: (a) mental or verbal models; and (b) mathematical or computer models (which may be based on mental models). Mental models are complex, shifting and often unverbalized, and when they are verbalized can be understood as implying different mental models from those intended. Computer models express precise mathematical relationships such that conclusions can be calculated from initial assumptions, especially when the situation is so complex that it cannot be encompassed in words or simple equations. They perceive such computer models as providing the necessary guidance for policy decisions in response to the global problematique. Some may even perceive such guidance to be both necessary and sufficient.

Global modelling has continued to proliferate in the 1980s, both geographically and in terms of issues and methodologies. An overview of global modelling in 1985 was provided for UNESCO by Heinrich Siegmann of the Wissenschaftszentrum Berlin (UNESCO, 1987). Some 29 models were then considered relevant to the study.

Global modelling is considered to be still developing as a research field. According to Siegmann, hardly any model should be considered completed. The global models represent the confluence of modelling streams from three disciplines: political science, systems dynamics and econometrics. Global economic models have been put to use by policy-making institutions in order to aid in short-to-medium term projections. The modelling time horizon has in general become shorter. The issues addressed have become more specific.

Modellers regretted the cessation in 1981 of the Global Modelling Conference sponsored by the International Institute for Applied Systems Analysis. This provided a consolidating infrastructure for global modelling, a role which UNESCO considered providing prior to its own programming and budgetary difficulties from 1987.

2. Alternative approaches to modelling
Following the appropriation of the term "global modelling" by those designing models based on mathematical equations, it might be assumed that no other forms of "modelling" of the global problematique are possible. Mental and verbal models are less satisfactory for the reasons noted above, but interesting models of systems can also be explored using analog methods.

It is interesting to note that a number of disciplines use other kinds of models in order to grasp the nature of complex systems. In the case of chemistry, molecular structures made up of many thousands of atoms are displayed graphically under conditions where the real complexities of the system do not lend themselves to mathematical analysis.

This example points to a key issue, namely availability and adequacy of information in constructing equation-based models. For such an equation to be formulated, precise information must be available through which a mathematical function can be defined. If the information cannot be formulated in precise terms, then nothing can be incorporated into the model.

Unfortunately, very few non-economic problems can be formulated in precise terms. It is, however, possible to use a different kind of precise mathematical relationship to provide some means of reflecting fuzzy perceptions. A network of perceived relationships between problems can be progressively identified. Such a network model can be held and explored on computer (see Appendix ZF***). Such global modelling is based on graph theory (and related disciplines), a quite distinct branch of mathematics. This has, for example, been used in artificial intelligence research to model (and interrogate) systems of attitudes held by an individual.

3. Contrast to this Encyclopedia
Global modellers have not envisaged this possibility, which is the inspiration behind the project of which this Encyclopedia is a product. It is important to recognize the fundamental difference between these two approaches:

(a) assumptions concerning the objectivity of the "real world" vs. assumptions that "subjective" and "erroneous" assessments of "facts" attitudes may be as real to some constituencies and, as such, may have real consequences;

(b) dependence on quantitative ("hard") data vs. registration of qualitative ("soft") data;

(c) insights only emerge from mathematical analysis vs. insights emerge from relative positioning of nodes in a network;

(d) complexity can only be managed to the extent that it can be reflected in a set of equations vs. complexity can be navigated (in the same way that a complex road map is navigated);

(e) data sets must be reliable and complete to be of any value vs. recognition that data sets are necessarily partial and unreliable.

(f) the structure of the model does not involve an on-going dialogue with the constituencies whose views are represented there vs. the model is only useful to the extent that it reflects the current, and potential, biases of the views reflected therein.

The approach here allows the widest range of perceptions to be interrelated within the same framework, even as isolated perceptions. It treats the global problematique as a network of perceptions, some of which reflect a belief in the quantifiable reality of the perceptions.

4. Weaknesses of an equation-based perspective
From the perspective of conventional global modellers, the kind of information collected for this Encyclopedia is completely superficial since it does not analyze the relationship between economic and social factors and express them in mathematical terms so as to define a model which can be explored under different conditions.

There are a number of weaknesses in this position to which this project provides a partial response.

(a) Competing models: The early belief that it would be possible to produce a single satisfactory global model has been severely eroded by the emergence of some 30 competing models reflecting different assumptions, and producing conflicting conclusions. The authors of the above review are quite explicit about the disagreements between such competing approaches. What is strange is that there is no attempt to produce a model to model the disagreements between the models. The existing models are closed perceptual systems which are not designed to acknowledge alternative perceptual frameworks. They are scientific constructs representing only a portion of reality. This project is a step in the direction of modelling the conflicting perceptions about the many aspects of the global problematique. It attempts to encode that disagreement.

(b) Elitist: Whilst the Club of Rome's first model, described in *Limits to Growth*, attracted world-wide attention, the attention accorded to subsequent models has been limited to an increasingly smaller elite. Global models have been slow to integrate political factors in response to early criticism and failed to incorporate the many non-quantifiable issues to which the international community is obliged to respond (such as problems of human rights, corruption, alienation, drug abuse, etc.). Even many semi-quantifiable problems, such as threats to species and food chains, have not yet been incorporated into global models. There is an increasing gap between the world of the global modeller and the world in which many the international community perceive themselves as living. As such, modellers face increasing difficulty in legitimizing their efforts to the wider public which must approve any major policy decisions based on their models. Such a factor should be integrated into a realistic global model. For this reason this project is based on entries with which such constituencies identify, using their language wherever possible.

(c) Inability to encompass "surprises": Global models are unique because of the breadth of the theories they endeavour to embody. Whereas the human brain is well suited for finding complex relationships, it is only such computer simulations which are adept at tracing numerous dynamic relationships simultaneously over time. Unfortunately the theories and simulations are based on data series established and accepted in the past. New events, or configurations of crises, not envisaged by the theories and for which no theoretically acceptable data exists, cannot be rapidly incorporated into such models. The models may even have to be totally redesigned to handle such data which cross, interlink, and call into question, existing categories. As in warfare, there is a limit to the degree to which reliance can be placed on simulation of the evolution of a battle. Unforeseen situations require another form of data.

(d) Specialization: In response to the competition between global models, modellers have been encouraged to specialize. Indeed the authors of the review point out that many observers have argued that a policy-oriented model will be most useful when: it focuses on a specific, well-defined problem; the problem reflects the objectives and needs of a specific client; the sponsor and the client are the same (p.97). The authors also draw attention to the growing domination of economists in modelling initiatives. Seigmann stresses that world modelling continues to be scattered across a variety of fields. "*This affects both the consolidation of world modelling into a scholarly discipline, and the establishment of its usefulness to policy-makers and other possible sponsors of global modelling.*"
Models now tend to have boundaries that are more narrowly defined, and to have shorter term horizons and more disaggregation (p.100). Such models can only be said to be global in a very limited sense, reinforcing fragmentation of perspectives and begging the question of how such perspectives are to be interrelated to respond to a network of problems (the original inspiration of global modelling). This project is an effort to offer an alternative to this tendency.

(e) Scope: Following from the previous point, according to Seigmann, even world modellers often disagree on what comprises a world model and what objectives global modelling might have. Global models may create a false impression of adequacy in dealing with those issues for which they are designed, precisely because they are incapable of reflecting their inadequacies in dealing with other issues they may assume are irrelevant. This may completely undermine the value of the insights they offer. In this project, especially in relation to world problems, the emphasis is on registration of the issues which appear in the literature. The intention is to provide a framework within which different approaches to the data may be explored, enabling different experiments on its use for very different kinds of constituency.

(f) Uncertainties and scepticism: Seigmann argues that the future of global modelling, over twenty five years after Harold Guetzkow's Inter-Nation Simulations, and more than a decade after *Limits to Growth*, remains beset by uncertainties. Its utility to academics and policy-makers is not yet assured. He considers that scepticism regarding its usefulness may even have grown. He sees these problems and uncertainties as attributable to the cross-pressures involving world modellers themselves, the academic research environment, the potential clientele of world modelling in policy and decision-making, and the general public.

On the depth versus breadth of a model Seigmann's notes: "*This trade-off affects one's modelling aspirations as well as one's standing in academia and with potential clients in policy-making. Ordinarily, limited computer power, time constraints, limited resources and the task of coping successfully with complexity, require a compromise between modelling in sufficient levels of detail, yet having a broad scope. Global modelling thus becomes an exercise in brinkmanship. The specialist will tend to dismiss the model as irrelevant, if he finds it too simple -- which it usually is -- in his special area of expertise. The generalist, ie particularly in policy-making, will similarly reject the model as lacking in realism, if he finds certain sectors or regions not included.*" Global modellers may then be rated "*ineffective in reaching out to planners and civil servants due to the failure to be comprehensible and convincing.*"

(g) Comprehensibility to a wider public: With the exception of *Limits to Growth*, the results of global modelling have attracted little attention from the informed general public. On the question of popularization, Seigmann notes: "*Global modellers must therefore have an interest in delivering a "marketable" product. The broader the public discussion of model "results", the better the chances for continuing, extending and refining their work. That way, global modelling can create its own following (eg environmentalism) acting in its interest as a political pressure group. The popularization of global modelling, of course, entails considerable risks of two different kinds. First, in the frequent trade-off between scientific appropriateness and popularity, the latter might obtain to much weight. The modeller might tend to emphasize those aspects that are likely to spur the biggest headlines. The media might concentrate on those aspects of the model and its findings that they can sell best...Those models with the most spectacular "predictions"...will tend to be drawn into the limelight....A second kind of risk exists if the model, its "results", or the whole discipline, are being used -- deliberately or not -- to legitimize courses of action which might not be born out of "hard" facts.*"

Again it is interesting that there appears to be no recognition by such modellers that the status of a model needs to be reflected in the model itself. In the above quotations the model is treated as the product of an interest group seeking to impose its perspective. The model fails to encode the dynamics relating its perspective to others, whether other modellers, other disciplines, or the wider public.

By contrast, in the case of the project giving rise to this Encyclopedia, the dilemma created by conflicting perspectives is central to the whole approach. It is assumed that yet another formulation of "the truth" or "the methodology" would be as significant as the multiplicity of other such extant "answers" to the challenge of the times. The task is therefore defined as one of providing an appropriate framework through which conflicting and incommensurable answers and (perceptions of) "facts" can be related, at least to some degree.

At the data and analytic level, graph and network theory is seen as providing a way forward where more rigorous, and possibly over-demanding techniques cannot travel. In terms of comprehension of more complex dynamics and patterns by wider public, the role of metaphors has been emphasized. Given the current advances in graphic software techniques, there is no reason why these two approaches should not be "married" in some way that would be of immediate significance to policy-making. Organized in this way, specific equation-based models could be integrated as required.

4. Evaluating long-term developments using global models

In a follow up for UNESCO of the Seigmann study, Sam Cole (1987) concludes that conventional "*global modelling is relevant, because understanding long-term developments at the macro-level are important in defining a perspective for shaping the future. It is also important because the intellectual investigation of alternative policy options through scenario studies can reveal problems which have been overlooked but which are worth considering.*" Cole's report, intended to establish the relevance of such modelling for UNESCO, has "*no simple answer to the question of how global models may be integrated in the overall goals of UNESCO in the fields of education, science, culture and technology.*" The reason may be that the conventional modelling approach is quite inappropriate to the registration of most of the real, but only partially quantifiable, issues to which an organization such as UNESCO endeavours to respond.

Assessment: future possibilities

1. Further research possibilities
The information in this publication is maintained in computer files as summarized in Section Z. The project to date has been, and should continue to be, a data-collection and presentation exercise. The existence of an updated data base of this kind should also facilitate some types of research which have hitherto been almost impossible.

The data collected will not, for example, contribute directly to research using quantitative models, although it may suggest some problems and relationships for inclusion in such models. A precondition for conventional model building is a minimum of quantitative information on the dynamics of the relationship between two or more selected levels or quantities in the system. The "problems" become evident by interpretation of the results of the quantitative analysis. In the absence of quantitative information, or where the latter is vulnerable to criticism, no other systematic analysis has been possible.

One use of the data collected in this project will be to test whether results can emerge from analyzing the networks of relationships as networks in which qualitative rather than quantitative values are attached to the links between the nodes. The readily available tools of graph theory and topology, for example, have not yet been applied with computer assistance to such data (see Section Z).
For example, it should prove useful to conduct computer comparisons of the degree of isomorphism of a network of interrelated problems and the corresponding network of agencies (or treaties, or disciplines) which purport to focus upon them. If the functional interlinkages, particularly communication channels, of the latter do not correspond to the linkages (or degree of structural complexity) of the problems, then it is probable that problem complex is uncontained, and uncontainable, by the programmes of the agencies as they are currently implemented. It would seem that in a world system characterised by a number of relatively complex networks on which information is largely unavailable or inadequate for numerical analysis, such techniques could be used to identify and analyze clusters and critical points on for action might.

2. Improved relationship to international organizations
Since this volume is integrated with the *Yearbook of International Organizations*, any improvements will be closely related to improvements to that publication.

(a) Inclusion of new and previously undetected organizations (This is a standard procedure in connection with the regular publication of the *Yearbook of International Organizations*).

(b) Inclusion of other kinds of relationships between international organizations.

(c) Indication of relationships between international organizations and entries in sections of this volume, such as world problems, for example. This improvement could take the form of inclusion of a qualification on the nature of the relationship, namely whether it was associated with, for example: programme action in the field, research, information exchange, policy formulation, or the mobilization of public opinion.

(d) Inclusion, possibly in a separate section, of information on subsidiary units of international organizations which may have a more precise responsibility for a particular world problem. Relationships could then be indicated between such units and the parent body, and between the unit (rather than the parent body) and the world problem in question. This would give greater precision to the representation of the inter-organizational network.

(e) Inclusion of maps of the network of organizations centered on a particular world problem or centered around a particular key organization. Such maps could be produced by computer controlled plots of portions of the data held for production of this publication. Production of such maps, with errors and omissions, would be an excellent means of highlighting defects in the information already available in order to facilitate critical feedback. At the same time they provide a meaningful display of information with which those involved are already partially familiar. Such maps would become especially useful, at meetings or in documents, to the extent that they could be used to highlight any mismatch between the organizational network and the problem network on which it is focused.

(f) Inclusion, where appropriate, of claims made by an organization for the importance of its programmes with respect to particular world problems and the various obstacles to any increase in their effectiveness.

(g) Inclusion, where appropriate, of counter-claims by competent critics (such as other international organizations with the same focus) concerning the ineffectiveness of a particular organization's programme activity. Such information could of course only be included in situation where there is some degree of unanimity on the content of the counter-claim, and where information could be included from the organization in defence of its position.

3. Display possibilities
This project will succeed to the degree that it can render transparent the complexity it attempts to map. This poses the problem of developing a satisfactory form of display. An advantage of holding the relationships in computer files as components of directed graphs is that such graphs can be plotted (in colour) by computer with appropriate labelling and choice of projection.

Detailed problem "maps" can therefore be produced, printed and bound into "atlases" - the argument being that people (whether students, executives, researchers or policy makers) have at least as much need for such visual devices to orient themselves in the social system as they have for road and other maps (see Appendix ZF). Comprehension will be made easier by on-line computer graphics devices (discussed in Appendix ZF) with display screens to permit the user to interact with that part of the network he chooses progressively to explore, at the level of display complexity which he is prepared to tolerate, and with the ability to call up textual explanation, use computational power, or activate a parallel slide display whenever required.

Such exploration can be recorded on videotape for wider use (eg in a decision-making environment) or sectionalised for production as a series of printed maps. Although the hardware exists and some software has been developed to handle network structures in three dimensions and colour, this work has hitherto been confined to engineering design, architecture and chemistry, and its potential for handling the great complexity of social structures is poorly recognised.

Just as the structural analysis advocated above falls between the popular extremes of quantitative analysis and (case-oriented) "qualitative" studies, so the structural display falls between the extremes of tabular output (or the graph equivalent), text output (resulting from conventional information retrieval), and purely aesthetic displays (resulting from the increasing use of computers by artists).

Harold Lasswell's point with regard to policy makers could be made for all those not numerate, within and outside the research community: *"Why do we put so much emphasis on audio-visual means of portraying goal, trend, condition, projection, and alternative? Partly because so many valuable participants in decision-making have dramatizing imaginations... They are not enamoured of numbers or of analytic abstractions. They are at their best in deliberations that encourage contextuality by a varied repertory of means, and where an immediate sense of time, space, and figure is retained"*.

As a descriptive device for highly complex structures, apart form permitting relatively sophisticated analyses, a graph representation can be transformed into much more iconic forms than those conventionally used to describe psychosocial structures, and is thus more comprehensible.

Assessment: implications

1. Local vs. Global
This project may also be considered as an exploration of how the relationship between "local" and "global" may be comprehended in practice as a guide to action. The conventional geo-political interpretation is considered here to be merely one aspect of this problem, one which has the disadvantage of reinforcing nation-state oriented category schemes and single-factor approaches.

There is a dangerous trap in the belief that global thinking necessarily results from the interaction of (s)elected people from different nations and cultures. The very (s)election process ensures the specific, and consequently, non-global nature of such groups. Such elite groups, whether in the General Asssemblies of the United Nations or of the Fourth World "peoples groups", for example, may well be considered "local", as is evident from the fact that the participants usually have more in common with one another than with the masses whose interests they supposedly represent. In this sense the global characteristic which needs to be distinguished from what is conventionally called "global" is that in which the conventional global/local complementarity is embedded. Any definitions or institutionalizations of it are necessarily local phenomena.

2. Providing a context for the logical incompatible
This draws attention to another aspect of the global/local relationship associated with language in its most general sense and the "logical" problem of interrelating specific (local) conceptual or functional frameworks which have no "categories" through which to recognize each other's relevance. The conventional approach to this aspect has the disadvantage of reinforcing the fragmentation of the global into "local" disciplines and specializations with their associated institutions, curricula and mutually exclusive jargons and systems of categories. This leads to a fragmentation of whatever integrity is to be conceived as engaged in the process of human and social development.

This may be seen in a more positive light if global is related to holistic (right-hemisphere) thinking and local to linear (left-hemisphere) thinking. In which case the current slogan "Think globally; act locally" takes on a new significance.

3. Providing a context for temporal phases
A third aspect is that in which "local" denotes a specific period of time and "global" is the relationship between (all) such periods, however that is to be conceived. How is the succession of phases in any development process to be understood in terms of time? The conventional approach to this reinforces a distinctly linear and a-cyclic understanding which does not correspond to the richness of the human biological and psycho-social response to time (Carlos Mallmann and Oscar Nudler (Eds). *Time, Culture and Development*, 1982).

4. Underdefinition as the integrating function of global values
Perhaps even more difficult to clarify is the relationship between local and global in the case of values, especially when global values are subject to some localization process which obscures their nature, despite protests from local perspectives. Local values in their most explicit form determine the characteristics of behaviour patterns in particular socio-cultural settings.

Paradoxically, however, it would seem that the more global values are most effective when characterized by a considerable degree of ineffability and ambiguity, possibly associated with symbols allowing different levels of interpretation. It is their underdefined global nature which allows them to exert an integrative force on incompatible activities which have been overdefined locally or through any explicit programme.

Underdefinition in this sense is a characteristic of the "emptiness" given prominence in Eastern philosophies and of the "untouchability" of the sacred in both Eastern and Western religions. In this respect Bateson states: "*What now must be said is difficult, appears to be quite empty, and is of very great importance to you and to me. At this historic juncture. I believe it to be important to the survival of the whole biosphere...*" (1979, p. 8).

5. Sacredness and integration
It would seem that such underdefinition has the effect of "*pulling*" the human and social development process forward in a continuing attempt to "*fill the definitional vacuum*" - the nothingness of the "*semantic vacuum*" in Nalimov's terms (1982, p.75-94). As such it exerts a powerful integrative force which Kenneth Boulding notes in connection with sacredness: "*The whole question of the role of "sacredness" in human society has been inadequately explored. Sacredness is part of the integrative structure and its erosion may easily destroy those integrative structures that hold societies and organizations together. A good deal of human history indeed is written in terms of the substitution of one system of sacredness for another...But exactly what the dynamic processes are that create or destroy sacredness is a puzzling question.*" (1978, p.226-7)

Seen in this light there would seem to be merit in considering the vital role of (global) leadership in relation to the sacred conceived as the undefined. In effect, leaders have a special function as intermediaries processing, filtering and interpreting the inconceivable - a role many priesthoods have been happy to monopolize. The role is misused however when those led are completely deprived of the right to the undefined in an essentially overdefined society.

In this sense **access to the undefined is a catalyst for transformative human and social development**. It is in this respect that charismatic leaders function as a kind of integrative "keystone" in whom different groups, operating in a necessarily overdefined mode, can find whatever is needed to hold them together. Successful leaders therefore embody a certain degree of ambiguity in order to be "all things to all men". To what extent does the United Nations fulfil this leadership function and to what extent does it act to overdefine the domains in which it claims to lead?

6. Challenge of the multiplicity of unrelated answers
The current difficulty is then not so much with answers but with the lack of any operational perspective on the relationship between such answers. The impotence of the current approaches is unfortunately disguised by the plethora of unrelated studies on "motherhood" problems like "population", "energy", "environment", "food", and "health", whose limited global significance nobody dares to question.

In the Club of Rome's terms, the majority of such studies constitute maintenance (adaptive) learning by society, as opposed to the needed innovative (shock) learning required to anticipate new dimensions of the problematique (James Botkin. *No Limits to Learning*, 1979). Academic work does not seem able to move beyond its propensity to be satisfied with patterns of consistent categories within specialized (local) frameworks. Such a fragmented approach, and its inherent assumption of simple sectoral answers, is severely criticized as "developmentalism" by some (Herb Addo. *Development as Social Transformation*, 1985).

7. Development as societal learning
The emphasis in Section KD on development as learning introduces the challenge of a dynamic dimension which involves both the "observer" and the "developer" as participants in the transformation process rather than as manipulators. The learning process cannot be limited by the preoccupations of those who favour a single answer.

The learning perspective challenges the long-term global value of any "unified world model" or any corresponding "unified world government" with a "world action plan". Any monolithic over-arching structure, even if decentralized, can only fail to internalize the essentially discontinuous nature of transformative change, which must challenge pre-existing organization. Such a structure is therefore obliged, using a sexual metaphor, to take one of the two sex roles. If it takes the male role, at present it reinforces phallic authoritarian (alpha) structures which, when they are not paternalistic, will tend to "rape" the "peoples of the world" who are cast into the corresponding female role. If it takes the female role, at present it reinforces associative (beta) structures which, when they are not restrictively maternalistic, invite rape on the part of any group capable of adopting an authoritarian mode. Violence is discharged but not contained.

8. Alternation within learning cycles

Section KD attempts to clarify the learning cycles through which the essential dynamism of any more subtle ("dancing") relationship between these two modes might be embodied. "*The fixed idea is the enemy of all free thinking. It is far more difficult to accept that two opposing ideas may not be mutually exclusive than, in a desire for absolutes, to plump for one or the other.*" (1983, p.211)

It is in the dynamics of some kind of "androgynous" pattern of alternation or resonance between two or more such incommensurable modes that the possibilities for a planetary meta-answer appear to lie. But, as with the ideal of marriage, there are many well-recognized patterns of unfruitful organized relationship which are valuable to the non-transformative existence of both partners. Fruitful, transformative union, when it occurs, may involve shared ecstasy of long-term significance (on which ideals are focused), but the moment of union between opposites is temporary (although possibly recurrent). Permanent union is clearly impractical and sterile in the light of current understanding.

9. New global order

The new global order appropriate to the times may perhaps be best conceived as a **resonance hybrid composed of alternatives woven together by policy-learning cycles** rather than by structures. The medium of such cyclic action is partly the world-wide network of independent organizations which give form to world society and guarantee its "functional roundness" through the variety of their specific preoccupations (Union of International Associations. Global Action Networks, 1990/91).

Such bodies acquire and lose global significance according to the phases of the cycles. Within such a cyclic context, different local priorities are alternatively integrated together and then later displaced by others. **There is no ultimate integration or pattern of priorities** at the global level. This kind of global integration is not purely spatio-structural, it involves dynamics over time as expressed in multi-phase cyclic "structures". The required global answer can only be expressed dynamically, namely with an inherent degree of uncertainty, in contrast to the rigid conceptual, institutional and value structures by which answers are currently over- defined and localized, with the consequence that they can only attract limited support.

9. Reframing the context for action

This project attempts to reframe the context in which the immediate questions "what should **we** do?" or "how should **we** act?" can be usefully answered. Part of the difficulty lies in the **self-justificatory nature of the compulsion to act** which at present gives rise to a highly turbulent society. People are impelled to act by the perception of the discreteness of good and evil, beauty and ugliness, truth and falsity, and by the energy which such perception engenders. Such action is based on the decisions by which this discreteness emerges (Nalimov, 1982, p.17). And: "*We might say...that a person posing a question, on the unconscious level gets an answer as a probabilistically given preference function constructed on the semantic continuum. Then conceptualization takes place on the conscious, logically structured level: the continuum is cut into separate blocks corresponding to the maximum probability concentration. Clear-cut conceptualization oppositions create the polarization without which the passionate temperament of individuals...that* **provides society with its energy** *could not have been realized. But a person is never separated from his unconscious: the latter sooner or later liberates the person from the power of what it has generated on the conscious level.*" (Nalimov, 1982, p.294). For Nalimov **any such decision is perhaps absurd**, since it is an attempt to represent discretely a fuzzy situation which is by no means necessarily determined by a needle-shaped function of the distribution of probabilities.

Furthermore, discrete formulations of goal, success or failure are no less absurd. "*Goals emerge and spread in societies like infectious diseases.*" (Nalimov, 1982, p.10) There are "*many examples of a goal being too straight forwardly chosen, leading to wild perversions and turning from the coming blessing into an everyday burden.*" (Union of International Associations, 1986/1987, p.17) And yet **decisions to act, however misguided, are essential to the dynamic continuity of society** as Nalimov recognizes in quoting the Bhagavad- Gita: "*This world is linked by doing.*" (Nalimov, 1982, p.58)

10. Recgnition of an undefined common global focus

What Nalimov apparently fails to render explicit however, is the possibility that the set of all such discrete polarizations, of whatever quality, might be understood in terms of configurations, about an undefined **common global focus, offering a variety of local learning pathways**.

To the necessity of such intense **local "doing"** might then correspond some kind of **global "not-doing"** which Nalimov describes as follows: "*Perhaps the culture of the continuous vision of the world will become 'the culture of not-doing', where preference will be given to spontaneous development, and not to the unreserved and destructive activities in the name of a goal to which we are ascribing an unconditional value. But can we possibly imagine such a culture of 'not-doing'?...Contemporary technology tempts us to invent and realize grandiose projects. However, ecological forecasts, if possible at all, can only be made in a soft probabilistic form. Is it not safer to act more cautiously, by introducing into the projects beforehand ways of retreat...Is such a culture of soft doing possible at all?*" (Nalimov, 1982, p.17-18)

The danger in interpreting "not-doing" lies precisely in the fact that its significance lies in its undefined nature, tangentially described by sets of local "doing". In terms of development through alternation, **focus on not-doing (as a particular preference) must alternate with focus on doing**. This project is a contribution to understanding how this can be brought about - or better understood as already operating.

11. Transforming the "answer economy"

For there to be a viable response to the current condition in the immediate future, it would seem necessary to transform the present "answer economy" by reinterpreting it through a more seductive idea. There appears to be a need to embed "nation-state" thinking within a context of "alter-nation process" thinking. Hence the merit of propagating an essentially human, and inherently comprehensible, sexual metaphor to "contain" the dynamics of the discontinuity faced by humanity and to facilitate widespread understanding of the nature of the "pattern which connects". For, as Bateson warns: "*Break the pattern which connects the items of learning and you necessarily destroy all quality.*" (1979, p.8)

The question is not only whether we can find ways of rendering comprehensible the non-linear geometries which express parts of this pattern, and on which we have yet to learn collectively how to live in Atkin's terms (1981). For although configurations of metaphors are vital to collective comprehension of the **possibility** of "life on a different geometry", the immediate challenge is to learn from them how to catalyze the emergence of new organizations of values, concepts, information and people to reflect that understanding in **operational** programmes capable of managing our resources, material or otherwise.

Assessment: processing system

Although the obvious purpose has been to produce a physical product, the *Encyclopedia of World Problems and Human Potential*, any evaluation should also cover the computer software developed to that end and the working method it has made possible amongst a team of people whose activities are interwoven in relation to the data elements and their relationships.

(a) Product: The Encyclopedia is the visible result of a programme, initiated in 1972. In its third edition, every advantage has been taken of computer techniques to present the information in a comprehensible form, despite the inherent complexity of that information. It is however increasingly obvious, as discussed below, that there are limits to the ability to present such information in the traditional linear text mode, no matter how sophisticated the pattern of cross-references. To the average user, such reference books are decreasingly useful -- or rather they only respond to some of the user's needs.

(b) Software: The framework software used is Advanced Revelation (running on MS-DOS within a Novell Netware environment). The specially developed application programmes in this context are as much a product of this initiative as the Encyclopedia itself. At every stage in the evolution of this project, ways have been sought to increase the ability of the software to enable new styles of editorial and research work. In its current form, the software is an unusual hybrid between conventional text processing of entry descriptions and database processing of relationships between those entries.

It also permits a form of hypertext movement through the network of entries (as popularized recently by the Apple hypercard facility). In principle the software could be used for any similar project requiring continuing review and modification of the network of relationships between entities, including possible redefinition and regrouping of those entities.

(c) Working method: Much has been written about the change in writing methods with the advent of word processors. The relation between author and text is dramatically transformed, whether in the details of corrections and formatting, or in the creative implications of (re)structuring the pattern of headings within a document. Building on such working techniques, and others, has transformed the editorial approach to any given Encyclopedia entry.

Editors are decreasingly concerned with the task of editing the displayed text, and increasingly concerned with how that text can be meaningfully related to other texts. Editors make use of a range of software techniques to call up groups of entry titles, sort them, refine the list, and check details on particular entries, before editing the full description and linking it to other entries. It is the fluidity of this editorial technique which is in fundamental contrast to editorial approaches in more conventional databases.

(d) Process: The Encyclopedia project has always been seen as a long-term exercise, like its larger (but less complex) sister publication: the *Yearbook of International Organizations*. Work on these publications, and the related *International Congress Calendar* database, is a continuing process in which information for any part may have value for the whole. Any given publication elicits further source material from interested parties, especially international organizations. This leads to the continual improvement of the database as a whole, as well as ensuring appropriate updates. Through this process, defects and inadequacies in any one edition are gradually eliminated and a foundation is created for more challenging reorderings of the data.

(e) Groupware: It is obvious that sophisticated tools and complex databases are of little value without developing the skills of research and editorial personnel. As has been noted in other contexts, the whole approach to team work is transformed when people are linked together by a computer network such that what one person updates at 11.20 affects what others are doing at 11.21. The ways in which people think about what they are doing and how they relate their tasks to each other is totally changed, with many unexpected benefits.

Insights: section interrelationship

This Encyclopedia has been deliberately organized so as to juxtapose kinds of information that are usually kept apart. This is the case within sections where, as with the "world problems", information from seemingly unrelated subject areas are held together and cross-referenced. The justification for doing so has been presented in the comments on the individual sections. But it is also the case with regard to the sections themselves, which for many could be more appropriately presented in separate publications. The justification for establishing their relationship within a single publication is explored below.

This edition has three core parts based on information collected:

- World problems (Section P)
- Human development (Section H)
- Human values and wisdom (Section V)

There are an additional two parts concerned with ways of gaining a more integrative insight into the complexity the above represent:

- Integrative knowledge (Section K)
- Metaphors and Patterns (Section M)

These are discussed in the following note.

1. Relationship of world problems with...

(a) International organizations: The relationships between organizations focusing on problems is well-recognized in many international bodies. Not many would however want to think in terms of more than 10-50 organizations faced with 10-50 problems, and their perception of the network of organizations and the network of problems would be significantly distorted by their perception of the relative importance of the organization with which they themselves are associated and of the relative insignificance of other organizations and their preoccupations.

(b) Human values and wisdom: Problems tend to be recognized and dealt with as concrete, practical matters. They can be defined in terms which allow them to be the subject of well-managed programmes -- indeed they may be the integrative focus of those programmes, uniting together groups that would otherwise be competing. Even the term "problem" is recognizable in many languages and is an early part of any vocabulary concerned with practical matters. The term is applied to every level of obstacle or hindrance from the most personally intimate to the global level.

The case of human values contrasts sharply with that of world problems. Where it is common and meaningful to ask "do you have a problem", it is unusual and generally unacceptable to ask "do you have a value". The term is not common across languages and is not an early part of any vocabulary. It is far from being an immediate concern in any normal programme of action. And yet there is an intimate relationship between problems and values. Basically no problem is recognizable except in the light of a value. If "justice" is not a recognized "value", then "injustice" cannot be recognized as a problem.

Human values come to the fore as the driving force in many campaigns, where people's commitment is engaged through appeals to "freedom", "equality" and the like. As such they too can unite opposing groups under the same banner but with much less ability to focus on the concrete remedial action required. Much cultural endeavour is associated with articulating the interplay of values. Values are of increasing concern to the marketing of commercial products because of the way in which markets are segmented in terms of the value profiles of consumers. Values are of course an increasingly explicit question in the debate on "green" issues and options.

Problems tend to be explicit, whereas values tend to be implicit. But both are artefacts of the human mind. Despite being treated as concrete, problems as such (like values) cannot be photographed. People interpret certain (photographable) conditions as problematic. But the future will recognize other problems in photographs of conditions today which may now appear problem-free. It may be argued that awareness of a problem-value polarity is borne of exposure to certain conditions that cause some form of suffering. In different ways this suffering engenders learning through which sensitivity to a (new) value allows the suffered conditions to be constellated into a problem.

In summary, whilst problems tend to be concrete, relatively unambiguous, detailed features of normal organized activity, values are much more ambiguously defined and less easily related to specific programmatic steps. Problems provide focus through their concreteness and specificity in dealing with the present through established channels. Values provide focus through their inspirational value and their prescriptive potential in creating a more desirable future irrespective of established views.

By juxtaposing the section on world problems with that on human values it becomes possible to explore more systematically the relationships between them. Understanding of any system of values leads to greater understanding of the system of problems. In fact exercises in ordering the system of values may contribute to new ways of ordering the system of problems. Relating them may clarify the nature of the societal learning process through which problem-value polarities come to be recognized. A specific challenge is to identify more clearly the values associated with particular problems and to determine whether there are unrecognized problems following from acknowledgement of certain values.

(c) Human development: World problems can be seen as the obstacles encountered in the process of human development. But since there are very different understandings of what it means to be an individual in society, or of how the individual or the society develops, the development processes they engender may be opposed by quite different patterns of problems.

World problems can also be seen as constituting a learning crisis in the process of human development. In this sense problems are seen as a challenge to creativity, calling for new insights into the process of human development. Problems may then be understood as emerging through inadequate understanding of the complexities of that process.

Of special interest in the process of human development is the recognition of distinct phases and stages. In the individual these may be linked to distinct modes of awareness. In some cases these are thought of as a sequence of levels. The implication here is that, in responding effectively to the challenges of world problems, individuals may be called upon to shift to some new mode of awareness -- whether individually or collectively. This reflects the recognition that complex situations call for greater levels of understanding, requiring greater maturity in decision-making. In the absence of such maturity, inadequate understanding readily engenders new problems or sustains existing problems.

It is of course possible to perceive human development as requiring very little change on the part of individuals. The focus can then be placed on education or training, job satisfaction, and the fulfilment of basic needs, with occasional reference to ethical constraints. This view predominates in the international community which nevertheless continues to be faced with the mystery of how to generate "the political will to change" and with some recognition that individuals are going to have to "radically change their lifestyles" if sustainable development is to become a reality.

But despite the officially accepted view, individuals are investing very heavily in altering their modes of awareness, however these are to be understood. This is clearly seen in the funds allocated to drugs (currently of the same order as the international oil trade ????), to say nothing of those allocated to more legally acceptable stimulants. The extent of this investment, at all levels of society, is an indication that official views of human development are widely challenged.

Drugs are of course far from being the only method of achieving other modes of awareness. Many other approaches, often quite hostile to the use of drugs, have been advocated. Drug abuse is thus an example of a problem engendered by a frustrated approach to human development, however misguided it may appear. The much publicized difficulties with some cults provide another example, as does the increasing investment in magic and rituals dating from earlier periods of civilization, or the increasing alienation of employees from meaningless jobs.

The key question is whether there are processes of human development that offer access to more mature modes of awareness which could enable society to respond more appropriately to the challenges posed by world problems. Advocates of some processes are quite affirmative in their response. The issue then becomes whether it is possible to learn from them without becoming entrapped by the limitations associated with particular movements that are often culturally bound.

The notion that personal development leads to greater problem-perception and problem-solving ability is well-recognized by educators, but is not necessarily explicitly recognized outside the educational environments, particularly with regard to the decision and policy-making age group.

(d) Integrative knowledge: The notion that an interdisciplinary systems focus is important to any grasp of the current world problem situation is now widely acknowledged. The International Institute for Applied Systems Analysis, the United Nations University, and the United Nations Environment Programme all acknowledge this in developing their programmes.

2. Relationship of human development with....

(a) Human values and wisdom: Human development can be seen as the process of giving more effective expression to human values. Many of the advocated approaches to human development are quite explicit concerning the values in terms of which they are conceived or which they are desired to enhance. The more sophisticated approaches to policy-making and management are quite deliberate in their efforts to identify the values on which any action is to be grounded.

Through some processes of human development, providing access to more subtle modes of awareness, new value insights emerge. In such cases there may be a very intimate relationship between the state of awareness and comprehension of the value. Emerging awareness of certain states may even lead to the articulation of more subtle understanding of commonly identified values. Certain modes of awareness can be understood as the embodiment of specific values or configurations of values.

Perhaps of most importance is the manner in which certain processes of human development integrate together previously disparate insights. Values can easily decay into empty, "bloodless" categories unless they are sustained by appropriate levels of awareness. Human development may thus build a subtle connecting pattern between values. Such integration provides a new foundation from which action may be undertaken in a sustainable manner.

(b) International organizations: The notion that for an organization (or a network of organizations) to function more effectively, it may well be necessary to give attention both to the degree of personal development (maturity, *etc*) of staff members and to their further personal fulfilment through their job activity is increasingly recognized. Although both aspects may be explicitly recognized in selection and promotion of personnel at the higher executive levels of government and corporate administration, this may only be the case where political, seniority and other such factors are not of major importance. Personal fulfilment, in the form of job satisfaction, is increasingly a preoccupation of trade unions and employers. It is questionable, however, whether widespread acceptance is also accorded to the notion that both aspects are essential in ensuring that the organization responds creatively to its problem environment.

(c) Integrative knowledge: The notion that the integration of knowledge and the integration of the person are directly related is not widely recognized. It seems to be acknowledged in the work of Jean Piaget, Lancelot Law Whyte, and Georges Gusdorf and explicitly by some of those writing in the journal *Main Currents in Modern Thought* (Center for Integrative Education) and more recently, in the journal *Revision* (USA)

3. Interweaving world problems, human development and human values

(a) Negative commonality: Despite the apparent differences between the core parts, especially in the case of their more obvious entries, it is interesting to note the commonalities in the case of certain other entries. There are some which exhibit a considerable overlap between the three parts. Ironically the best examples of this cross-linking are associated with the traditional "seven deadly sins", typical of the Christian tradition, namely: gluttony, anger, greed, envy, pride, lust, apathy (and/or melancholy). The same would seem to be the case for the equivalents in other traditions (*****).

In each case these terms may be used as descriptors of negative values, of negative aspects of human development or of world problems. This is especially interesting because in those traditions such concepts are often considered to be at the "root" of the difficulties of the human condition. It is they which engender the more obvious difficulties in society and it is they which are the real barriers to human development. Perhaps even more ironically, the most fundamental entry common to the core parts is that of pain (or suffering), which again is a principal concern of the major spiritual traditions.

(b) Positive commonality: In contrast to such commonality through negativity, it is also possible to detect a positive commonality. To do this however it is necessary to accept the well-publicized Chinese insight (reflected in the ideogram for "crisis") that a problem of any kind can simultaneously be viewed as an opportunity. Although there is almost universal despondency in the face of the complex of world problems, there is a strong case for viewing them as a major historical challenge through which humanity will reach new levels of understanding. They provide a unique opportunity for collective learning.

In this light, as positive learning experiences, the descriptors associated with the traditional "seven virtues" (or their equivalents in other traditions) are to be found as common to positive modes of awareness, to positive human values, and to the most fundamental world problems. One such checklist gives: hope, will (courage), purpose (dedication), competence (discipline), fidelity, love, care and wisdom. It is through these (or their equivalents) that appropriate action can be conceived and undertaken. (Oriental traditions might however place greater stress on forms of abstinence and self-sacrifice, which ironically the Occident seeks to impose upon them through austerity programmes, whilst being unwilling to restrain its own life-style.) More fundamentally perhaps, common to the core parts is the learning dimension of new understanding or insight in response to suffering. This too, in one form or another, is common to the spiritual traditions.

(c) Basic incommensurability: Although there are many links between the core parts of this Encyclopedia, as the above paragraphs indicate, there is nevertheless a basic, even paradoxical, incommensurability that separates them in practice. One of the aims of this Encyclopedia is to create a context in which this tension can be recognized as a challenge to users in selecting the kinds of information that are considered relevant to the times. In a sense it is this tension which signals the instability of the present and impels towards the futures yet to be born.

(d) Juxtaposition of incommensurables: By assembling information on the extremes of "negativity" that the world problems represent, opposing it to the extremes of "positivity" that the human potential sections represent, conditions are created analogous to those in which electricity is generated. This may only facilitate some creative "sparks". But it would appear that there is a possibility, by suitably "wiring" the connections between the parts, that a more continuous stream of insight could be engendered in response to the challenges and opportunities of the times.

Insights: comprehension of sustainable integration

This Encyclopedia may be used like any other reference book. Using any of the indexes, entries on particular topics may be located and consulted. This may be sufficient for many. The organization of the book, with its many cross-references between entries, also permits users to explore "around" any particular entry or "from" it as a point of entry into a network of associated entries.

The deliberate lack of organization on any one page may also lead users to stumble onto entries totally unrelated to that they first consulted on that page. This can be a very fruitful and creative challenge as a boundary stretching exercise. Some users see this as the most useful feature of the book.

Such uses, whilst necessary for many purposes, are far from being sufficient in the face of the larger challenge of how to obtain some form of meaningful overview of such disparate kinds of information. Much has been written about the problems of information overload and information underuse, as well as on the problems of lack of information.

A first response with this material has been to group entries by subject, together with corresponding international organizations, in the companion volume Global Action Networks (Vol. 3 of the *Yearbook of International Organizations*). At one level this provides an equivalent to the "yellow page" telephone directory and may be used to locate organizations dealing with problems in a particular subject area. At another level the organization of the subjects is an ongoing experiment in the interrelationship of subject domains. This interweaving of subjects is designed to stress systemic relationships.

It is clear however that even more radical approaches are required to cultivate new levels of insight into complex patterns of information on "world problems", "human development", and "human values". This possibility is explored in three other parts of this Encyclopedia, of which one of them is implicit:

- Integrative concepts (Section K)
- Metaphors and patterns (Section M)
- Computer graphics (see Section Z)

Some of the implications of these approaches for an integrative understanding of the complexes of world problems, human development and human values are explored below.

1. Integrative Concepts

(a) Calls for integrative approaches: The past decades have stressed the importance of interdependence between problem areas and the disciplines capable of responding to them. There have been many calls for "integrative" programmes capable of handling the complexity of emerging networks of issues. "Interdisciplinarity" has been in vogue, especially through the work of systems specialists.

(b) Weakness of interdisciplinarity: But despite acknowledgement of the importance of this dimension, the integrative methodologies have tended to be weak or simplistic and cannot be said to have resulted in breakthroughs adequate to the challenge of the times. In fact such terms tend increasingly to be used as buzz words indicative of appropriate intentions but lacking significant content in practice.

(c) Integration through praxis: Currently the most integrative methodology is perhaps that which assembles a group of specialists to focus on a concrete problem. It is only the concrete case which ensures the integrative dimension. There is little methodology in the way that the disciplines are brought into play or in the way that their representatives interact. It has been remarked that interdisciplinary meetings tend to be integrative only in the binding of the book which holds the individual contributions together -- relegating the challenge of synthesis to the reader.

(d) Need for integrative overviews: The kinds of information in this Encyclopedia call strongly for an integrative overview, especially one which facilitates comprehension. It is for this reason that one section reviews the range of integrative concepts as a reminder of the approaches that have been considered and despite the limitations that each may have from some other perspective.

(e) Failure in response to competing perspectives: A major concern with existing integrative approaches is their essential failure in handling competing perspectives. Advocates of any one such approach are ill-equipped to respond proactively to another. It is as though each was an effort at some form of conceptual empire-building, with associated dynamics reminiscent of the geopolitical equivalent and its marginalization of certain underdeveloped zones.

(f) Challenge of incommensurability: An accompanying section (Section KD) therefore focuses on the challenge of forming comprehensible patterns of sets of essentially incommensurable insights. The authors reviewed there have in one way or another responded to the challenge of conceptual discontinuity and disagreement. They suggest lines of exploration which may help to move beyond the sterility of many current initiatives perceived as integrative.

2. Computer graphics

(a) Mapping networks: The kinds of complexity represented by the interrelated entries in this Encyclopedia strongly suggest the value of sophisticated use of computer graphics. There are many ways in which the networks of entries could be graphically represented on a computer screen or on paper. This would offer users a new way of approaching such complexity, explicitly highlighting patterns of connectedness.

(b) Beyond conceptual "tunnel vision": There is a desperate need to move beyond the conventional bias towards describing complexity using text or in tables of statistics. This reinforces what amounts to conceptual "tunnel vision" and fails to suggest new patterns which give clues to new ways of approaching the data.

(c) Non-linear sense of context: This opportunity is discussed in greater detail in Section Z. But the point can perhaps best be made by pointing to the role of the subway map as a vital guide to a complex transportation network. Such a map permits users to orient themselves in terms of where they are, where they want to get to, whilst allowing them to consider various options for getting there. It provides a non-linear sense of context and raises questions about other locations on the map and the ways of getting there.

(d) Towards an atlas of relationships: Equivalent maps could be produced for the network of problems, possibly bound in an "atlas" (or made available on computer via CD-ROM or via videodisc). They could also be related to other maps of the networks of international organizations. Such tools offer a new response to data that is already overwhelming in quantity.

3. Metaphors

(a) Inadequate dissemination of integrative insights: Whilst the two previous opportunities are indeed fruitful lines of exploration, there is a sense in which the urgency with which new insights are required makes such efforts of marginal value. Both are relatively specialized tools. It is not clear that they would be used creatively. Nor is it clear that any emerging insights could be effectively communicated to the non-specialists who would have to approve initiatives based upon them. The communication processes of the international community are not adequate to the task of disseminating integrative insight such as to prevent erosion of the richness that is a guarantee of its appropriateness.

(b) Metaphor as a widely accesible resource: Metaphor is widely used at all levels of society to communicate complex insights. It is used as much by the slum dweller endeavouring to empower himself in response to an essentially alienating environment as it is by the politician in communicating new programme proposals to the electorate. It is also used by hard-nosed managers in articulating business strategies and by nuclear physicists in endeavouring to comprehend mathematical abstractions. Metaphor is an important device in the creative process. It is fundamental to certain advertising techniques.

(c) Articulation of complex policy options: The use of metaphor as one of the major unexplored conceptual resources in responding to the challenge of the times is considered in Section MZ. The argument there is that metaphor provides a way of articulating complex policy options which could not otherwise be rendered credible. As such it provides the opportunity for exploring the kinds of policies which could be of sufficient complexity to respond to the dilemmas of the times -- several steps beyond the "nuclear umbrella".

Many of the metaphors considered specifically address the difficulty of interrelating essentially incompatible perspectives and of providing some understanding of the dynamics between them and how they can be appropriately managed.

(d) Communicability of metaphor: One of the principal merits of metaphors is that like humour and rumour they are readily understood, memorable and they travel well -- with relatively little loss of the richness which ensures their appropriateness. They call for relatively little investment and do not require expensive delivery systems. They do not threaten existing structures. They complement conventional educational techniques and media presentations. They are also extensively used in the political arena. The challenge is therefore more one of providing better metaphors and creating greater awareness of how to use them and of how to avoid becoming trapped by bad metaphors.

(e) Simplistic approaches to "sustainable development": The poverty of insight guiding current thinking may be illustrated by the concept of "sustainable development". The current credibility of that option is perhaps best illustrated by the metaphor of "having one's cake and eating it too". This illustrates the dilemma.

But the strategic complexity through which the dilemma might be resolved is not captured by that metaphor, or by any other. The metaphor of "stewardship" indicates an appropriate attitude but does not get to grips with the how -- and relies on imagery evoking benevolent paternalism (at its best). By contrast, a rich metaphor indicated in the text is that of "crop rotation", implying a cycle of distinct and complementary policies to ensure the sustainability of any long-term initiative. This honours the reality of opposing policies and indicates how they may function creatively in relation to one another.

(f) Metaphor as a traditional vehicle for insight: Exploration of metaphor as a response to current dilemmas is not as far-fetched as it may seem. In many traditions insight is carried from generation to generation through metaphor, often expressed in mythology. In crisis, cultures have drawn on such metaphor in articulating an appropriate response. Many of the insights concerning more profound states of awareness are expressed in metaphor, as are the spiritual journeys through which they are encountered.

The concern at this time is to understand what richer metaphors might be more appropriate, to clarify ways of designing and using metaphor in the policy process, and to empower people to use their own metaphors to redesign their own psycho-social environments in the light of their own insights.

Insights: problem perception and deception

1. Explicate versus Implicate

The way in which information on problems is handled, or denied, in society suggests that fruitful insights might be gained by contrasting explicit and implicit responses. It is obviously in the interests of institutions to accept the existence of only those problems which correspond to their mandates, and only then when they are under pressure to do so. This is an explicit situation, especially since such responses are determined and documented by written texts. Other problems may be recognized and discussed informally, or they may simply not be acknowledged at all. The ability to discuss problems whose existence cannot be explicitly acknowledged indicates what might be called an implicit situation.

The contrast between the two situations can be illustrated by two examples. Urban planners in developing countries tend to have maps indicating land use around a city. On the map certain areas may be indicated as "parkland" without any habitations. Decisions on that area will be based on its reality as parkland. Corresponding to this explicit administrative reality, detailed in technical and legalistic terms, there may well be an implicit reality in which that parkland is totally covered by semi-permanent shanty-town dwellings. These do not constitute an explicit problem (in terms of sanitation, etc) because they do not exist in the explicit reality.

2. "Honne" and "Tatemae"

Within the Japanese culture, special terms are used to distinguish these distinct realities. An example is the response of corporations competing for new university graduates. The "tatemae" of the situation, the explicitly stated reality, is that no potential employer starts recruiting before an agreed period of the year. But the "honne", the unspoken reality of the situation, is that the competitive hunt for graduates is undertaken as a free-for-all long before that date, to the point of threatening the survival of companies unable to compete with the largest. In another Japanese example, according to appearances, Toyota does not control the company Koito in which it has a 19 per cent stake, especially since it has no seats on the board. But the unspoken reality, recognized by all concerned, is that a number of the directors are former Toyota employees. Control is ensured through a range of friendly shareholders.

3. Official denials

Appearances ("tatemae") can thus be used as a more or less polite way of discouraging discussion of, or even obscuring, the underlying reality ("honne"). The distinction between the explicit and implicit realities is not sufficiently clear. "Official denials" concerning problems are made in the realm of explicit appearances and can thus never be construed as "lies" within that context. Such denials have totally undermined the credibility of institutions in the eyes of those exposed to the problems whose existence is being denied. The dissonance is perhaps greatest in the case of various forms of corruption. The existence of corruption may be officially denied, although it is a continuing lived reality for many.

4. The "shell game" as a metaphor of political deception

The history of the processes whereby problems are accepted into the explicit reality may well reveal phases in which advantage is deliberately taken of these two realities to manoeuvre public opinion. An interesting way of exploring this is by reflecting on the classical "shell game" as a basic metaphor for use of the interplay between the two realities for purposes of deception.

The shell game has a long tradition in many cultures. It is a basic trick of carnival magicians, although neglected by many of them as being unworthy of consideration because of its simplicity. Yet with three nut shells and a pea-sized pebble, the basic arts of deception can be explored -- in the process of guessing under which shell the pebble is currently located. Its special merit is that it relies on the close attention of those to be deceived. Its appeal lies in its evading qualities and the operator's mute challenge to the onlooker's perceptive faculty. The human mind, by instinct, does not admit defeat. The recurrent failure to designate the correct shell not only aggravates the' situation but improves the proceedings by creating greater interest.

The game is normally demonstrated with three split nut shells and a pebble. The pebble is covered with a shell and all three shells are then pushed around a flat surface with the ostensible purpose of confusing the victim.. Wen he is asked to pick the one covering the pebble, he fails to indicate the correct shell. This, in essence, is the effect of the Three Shell Game. In actual practice, the operator of course prevents the correct choice by controlling the secret passage of the pebble from one shell to another. The movement of the shells cannot by any means confuse an attentive spectator; it is an easy matter for anyone to follow visually the target shell around the table. The shells, however, are pushed around to satisfy two different aims. First, to provide a natural piece of misdirection. Secondly, to supply the operator with the logical and necessary excuse for touching the shells.

During the conveyance of the shells from one point to another, and undercover of which action, the operator accomplishes his secret designs. He withdraws the pebble, adroitly, from under the shell marked by the spectator and feeds it to another (using momentum transferred to the pebble from movement of the shell). On this account it stands to reason that so long as the spectator relies on his visual judgement, he will never succeed in making the right choice. In fact the closer the observation, the lesser the chance of making the right choice.

As a metaphor, the shells can be viewed as features of the explicit reality, whether as organizations, programmes, procedures or legal instruments. Any problem is handled through them. A new problem is thus "covered" by one of these organizational shells. Thus the observer is led to assume that it is being dealt with under that programme. But the dynamics of the political arena are complex -- an alphabet soup of organizations and programmes. Politics is the art of surviving and manipulating those dynamics. The concerned observer may continue to assume that the problem is being dealt with as before. But on investigation, when the "shell" is lifted, it becomes clear that it is not. The effective focus on the problem has shifted elsewhere, so it is claimed, to some organization to which the observer has not been paying attention.

5. Displacement of problem responsibility

The art of managing problems from the explicit reality is thus to constantly reassure the observer that the problem is being dealt with where he assumes, but to ensure that it is possible to claim, when challenged, that the real effort is taking place at some unsuspected location. Better still, the claim may be made that the problem was poorly formulated and is currently being more effectively tackled through some more fundamental problem handled by some other body. In a classical shell game, the observer must of course pay to see the location of the pebble. This is also true in the world of organizations where the observer must effectively pay for hard information, if only in loss of status through the failure of a challenge. He cannot afford to pay too often. Even elected representatives, members of parliament, can be continually led on by the confusion of information as to where, or whether, the problem is effectively being handled, if at all. As with shell game spectators, it is often easier to assume that you know where the pebble is rather than to pay to find out.

6. Emerging standards of subterfuge

It is useful to question appropriateness of the mind-set behind four fundamental arguments to avoid responding to issues:
(a) The "no proven link" argument based on the views of selected experts, when a "link" probabilty approach might be more appropriate;
(b) The "national security" justification, when introducing a "global security" perspective might be more appropriate;
(c) The "only obeying orders" justification, when recognition of the fallibility of all orders might avoid many crises;
(d) The "not our responsibility" response, when it is increasingly clear that everyone is increasingly responsible, if only indirectly.

Insights: social organization determined by incommunicability of insights

1. Constraints on communicability
It is readily assumed that new understanding of problems and opportunities can be communicated comprehensibly. This is not the case. Any new insight is understood to different degrees by different people. The resulting situation can be clarified using the work of Ron Atkin (*Multidimensional Man; can man live in 3-dimensional space?*, 1981) on q-analysis, namely the theory and application of mathematical relations between finite sets. He has applied this to the analysis of communication patterns within complex organizations.

2. Modelling the communication problem
The perceptual significance of this approach is well-illustrated by visual sensitivity to colours resulting from the three primary hues (red, green and blue). These may be represented on a simple triangle. Here the vertices (0-simplexes) represent the primary hues, the sides are twofold combinations (1-simplexes), and the combination of the three hues makes the central white (2-simplex).

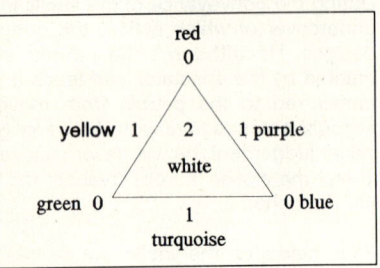

0-dimension vision:
    Red, Green or Blue
1-dimension vision:
    Yellow (=Red/Green);
    Purple (=Red/Blue); or
    Turquoise (=Blue/Green)
3-dimension vision:
    White
    (=Red/Green/Blue)

Now to be able to see all the colours, a person's vision needs to have the ability to function in the triangle as 2-dimensional "traffic" on that geometry, moving from location to location adjusting to the complexity of the geometrical structure which carries the visual traffic. If however the person's vision is limited to 1-dimensional traffic, then white could not be perceived because the visual traffic of seeing is then restricted to the edges and vertices only.

Similarly, if the person's colour vision is only 0-dimensional, then it is restricted to the vertices. It can only see one vertex colour at a time and never a combination (as represented by an edge). If vision was 3-dimensional, it would allow traffic throughout the geometry, but would perceive other colours as well, calling for a fourth vertex (forming a tetrahedron) in order to contain the full range of combinations.

3. Dimensions of comprehension
If the geometry represents problems or concepts (or modes of socio-economic organization) instead of colours, then it would be expected that some people, in relation to that set, would have 0-dimensional comprehension (i.e. sensitive to isolated primary problems only). In this sense there is an irony in the way that opposing political factions each tend to identify with a particular primary colour as a symbol. Others may have 1-dimensional comprehension (i.e. only sensitive to binary combinations of primary problems). The latter would be unable to maintain attention to three problems simultaneously in order to perceive the threefold combination (the central, integrated or underlying "white" problem). The threefold problem may then be termed a 2-hole in the pattern of communication connectivity amongst those involved. For 2-dimensional traffic however, the problem complex is coherent, comprehensible and well integrated. For the 1-dimensional traffic, it feels less secure as a whole, since the whole complex may only be experienced sequentially through a succession of experiences ("around the edges"). The shape of the whole may then be deduced but not experienced. For 0-dimensional traffic, the underlying problem does not exist, since experience is disconnected.

4. Social action as traffic in a geometry
Generally speaking it seems to be confirmed that action (of whatever kind) in the community can be seen as traffic in the abstract geometry. This traffic must naturally avoid the holes (because it is impossible for any such action to exist in a hole). The holes therefore appear strangely as objects in the structure, as far as the traffic is concerned. The difference is a logical one in that the word "q-hole" describes a static feature of the geometry, whilst the world "q-object" describes the experience of that hole by traffic which moves in that geometry.

5. Problems as comprehension inhibitors
This suggests **new ways of comprehending the nature of a problem**. As an "object" this phenomenon is an obstacle to communication and comprehension and obliges those confronted with it to go "around" in order to sense the higher dimensionality by which it is characterized. Communications "bounce off" such objects. As a "hole" this phenomenon engenders, or is engendered by, a pattern of communication. It appears to function both as "source" and "sink". Atkin suggests that, in some way, which is not yet fully understood, such object/holes act as sources of energy for the possible traffic around them. From the initial research it would appear that such objects/holes are characteristic of communication patterns in most complex organizational systems. It seems highly probable that they can also be detected in any partially ordered pattern of communication. "Societal problems", "human needs", and "human values" merit examination in this light from the perspective of different languages and modes of socio-economic organization.

6. Traffic in an organizational geometry
Very concretely, Atkin has investigated situations in which the "vertices" (which could themselves be n-simplexes in a multidimensional geometry) are individuals or offices linked together through various committees. They could also be governments or disciplines. There will then be a lot of 0-traffic and 1-traffic within and between offices due to the details of their intra-and inter-office (bilateral) operations.

This traffic will circulate around the holes/objects which they constitute. Any n-level traffic can only be encompassed, or be brought to rest, by an (n+1)-level body (e.g. an executive or a committee). If the latter does not exist, such traffic will continue to circulate around the q-objects in the structure and, according to Atkin, may be defined as noise. An "empire builder" (or any elite), for example, in such an organizational system will carefully create many q-holes underneath him (at the n-level), so that subordinate bodies answerable only to his appointees, are trapped in the flow of noise between them.

Atkin notes that even though the geometry may not have been rendered explicit, such structures generate the feeling throughout a community of some "power behind the scenes" acting to outwit the formal structure. The special value of q-analysis is that it can clarify why action/discussion in connection with (development) problems tends to be "circular" in the long-term, however energetic it may appear in the short-term. As such it shows how **social change is blocked by the way in which conceptual traffic patterns itself around any sensed core problem** which is never confronted as such because the connectivity pattern is inadequate to the dimensionality of that problem.

This would explain why so many problems go unresolved and why **the process of "solving" problems becomes institutionally of greater importance than the actual "elimination" of the problem.**

7. Representation of different modes of societal organization
The elements of the triangle may also be used to represent different modes of socio-economic organization comprehensible under different conditions. It is possible for a person or an organization to conduct all its communication in terms of one of these modes or frameworks. Communications in terms of other modes or frameworks would be incomprehensible and to some degree inconceivable.

It is possible to envisage a different paradigm, corresponding to the 1-dimensional traffic, which would permit movement between the primary modes via intermediate modes. This would correspond to the mind-set of a polyglot or a polymath, for example. Presumably more complex paradigms could also be envisaged.

Atkin analyzes much more complex situations in exploring information flows through the committee structure of a complex organization. He is especially concerned with how **information on**

committees without it being necessary to confront core issues or bring them into focus, namely the bureaucratic technique of handling information overload by avoiding use of that information.

8. Constraints on movement of communicable insights
Q-analysis gives precision to the recognition that traffic of different degrees of content connectivity finds (or creates) its appropriate level in any psycho-social communication complex, presumably including a language. **Communicable insights are level-bound**, especially where they are of high connectivity. In other words, **at the level within which it is possible to communicate, problems cannot necessarily be anchored unambiguously into terms and definitions which "travel well".** Precision introduces distortion which is only acceptable locally within any communicating society - although "locally" must be interpreted in the non-geographical sense in which all nuclear physicists are near neighbours, for example.

9. Compromise as warping communication geometry
The relation between two personal or institutional structures, conceived as a multidimensional backcloth, carries whatever traffic that constitutes the communication between them. If this backcloth changes by becoming dimensionally smaller, then its geometry loses vertices and the consequent connectivity properties. This is first indicated by the failure of higher dimensional traffic which the geometry can no longer carry. Such 4-traffic, for example, must then move through the structure to some new haven of 4-dimensionality or it must change its nature and become genuine 3-traffic.

This process of reducing communication expectations in order to continue to live within the new warped geometry is the classical problem of compromising. **The feeling of "having to compromise" is a painful one. It is the feeling of stress induced by the warping of the communication geometry,** namely the direct experience of a structurally induced force, in this case a 4-force. It is the feeling associated with the distortion of an unsatisfactory translation between languages. This approach clearly provides a very precise approach to understanding **more subtle forms of structural violence.** Atkin has applied it to an analysis of unemployment.

10. Human impoverishment and reduction of dimensionality
Such considerations suggest the power of q-analysis in clarifying approaches to human and social development in general. **Reducing the dimensionality of the geometry on which a person (or group) is able to live is an impoverishment associated with repressive forces.** Expanding the dimensionality induces positive, attractive forces through which a sense of development and enrichment is experienced. Q-analysis seems to be a valuable new language through which precision can be given to intuitive experiences and their communication, particularly since it provides an explicit measure of obstruction to change.

11. Disempowerment of response to problems
In the case of social development, it is probable that **most continuing societal problems should be seen as holes/objects**, especially given the well-established record of unfruitful action in response to them - however vigorous and dedicated. Typical examples are: peace/disarmament, development, human rights, environment, etc. Q-analysis could then provide understanding of why any **action tends to be drawn into a vortex of futility**, however much it satisfies short-term political needs for visible "positive" action. The participants in the action find themselves "circulating" around a central concern of which they are unable to obtain an overview due to the geometries of the overlapping conceptual and organizational structures through which they work (or which they somehow engender).

12. Dimensionality of human development
In the case of human development, Atkin shows how the individual can be defined in terms of a multidimensional geometry requiring a minimum of four levels. By relating this geometry to that of society, Atkin introduces an 8-level scheme within which the degree of integration or eccentricity of communication can be clarified in terms of developmental or anti-developmental forces.

13. Persistence of underdevelopment as low-connectivity
In such a multidimensional geometry it is clear that, **whether in the case of an individual, a group or society as a whole, it is not possible to eliminate "underdevelopment"** as associated with low dimensionality. Such a geometry will necessarily continue to have traffic of very low-level connectivity co-present with that of increasingly higher level connectivity. The simplest illustration arises from the continual birth of infants who will, when resources permit, continue to be educated through to the level of connectivity to which they can respond. But there will always be communication at both low and high-connectivity levels, especially about socio-political issues. The question is then how such learning communication between these different levels of connectivity can weave itself together within a social structure.

14. Designing better configurations of holes
It is the status of the holes/objects in relation to development which could provide an interesting point of departure for further investigation. As noted above, it is not a question of attempting vainly to eliminate such holes, especially when some of them may arise from alternative concepts of "development". Rather it is a question of how configurations of holes can be identified and/or designed. **It is such configurations of holes which provide the minimum structure (and communication dynamics) to stabilize and give form to the co-presence of the differing "answers" to the challenge of development.**

In effect such holes exist at a lower connectivity-level than the "macro-hole" of higher connectivity constituted by the world problematique at this time. This macro-crisis hole "absorbs" the development initiatives of society by engendering the immense volume of action/communication traffic around the hole so defined. This draws attention to the developmental implications of the probable presence of holes of yet higher dimensionality than can be readily sensed or made the subject of acceptable public (consensual) communication.

How then are "better" holes -- more appropriate problems -- to be engendered within such configurations? Now from one point of view it is necessary to avoid introducing an element of evaluation, because from each hole the perception of other holes will be distorted so that no communicable assessment can be usefully formulated. On the other hand, it may prove to be the case that, at the level of the configuration as a whole, more than one such configuration can be identified/designed in order to interrelate the perspectives associated with the set of holes. And at this level, without privileging any particular hole, more adequate interrelationships between the elements making up the holes can be identified.

15. Co-existence of multiple development strategies
Expressed differently, introducing evaluative judgements into the relationships between the holes within a particular configuration can only contribute to the dynamics between such holes in terms of perceived advantage/disadvantage. Excessive emphasis on this runs the risk of tearing the configuration apart. The identities associated with the holes can be respected in each of the configurations in a series constituting progressively more adequate or richer formulations of the relationships between "developments". **There is consequently a multiplicity of concepts of development operative in society. Individuals and groups may "progress" from one to another, possibly with a general tendency towards those of higher connectivity.** But other individuals and groups will emerge and find the concepts of lower connectivity more meaningful before moving on, if they do, to those of higher connectivity. (In this sense the "ontogenesis" of an individual tends to repeat the "phylogenesis" of his/her society). **Society in this sense is the arena within which individuals and groups refine their concept of development**

Insights: problem perception and levels of awareness

1. Degrees of immediacy of pain
Awareness of problems could be characterized in terms of the directness of the experience of that problem as a pain or a painful tension, with which it is possible to identify. Degrees of immediacy might then range as follows:

- *Experienced by oneself:* physically, emotionally, mentally, spiritually

- *Experienced by one's immediate family:* nuclear family, extended family and dependents (possibly including favoured animals)

- *Experienced by one's peers:* friends, neighbours, colleagues, associates

- *Experienced by known others:* acquaintances, friends of friends, public figures, role models

- *Experienced by unknown fellow citizens:* same tribe, race, religion or ideology

- *Experienced by any unkown human being:* ...however distantly located

- *Experienced by other entities:* in other realms of nature (animals, plants), possibly in their collective form (herds, species, forests)

- *Experienced by inanimate forms:* the pain of polluted rivers or degraded land

- *Experienced by the Earth as a whole:* the pain of Gaia.

2. Problem-need hierarchy (beyond Maslow)
The study of human needs has developed on Maslow's concept of a hierarchy of such needs. Implicit in any categorization of needs is a categorization of problems associated with fulfilling or failing to fulfil those needs. One categorization of human needs is that of Carlos Mallmann, during the course of work in the Goals, Processes and Indicators of Development project of the United Nations University ( ). He distinguishes the following:

- *Maintenance needs*: nutrition, rest, exercise, reproduction, shelter

- *Protection needs*: prevention, cure, restitution, defence

- *Affective needs*: self-esteem, friendship, sexual love, family love, attachment

- *Understanding needs*: psycholization, socialization, education, observation

- *Autonomous participation needs*: liberty, independence, autonomous decision-making

- *Recreation needs*: self- recreation, social recreation, recreation in the habitat

- *Creation needs*: creation by oneself, creation of social and habitational environments

- *Meaning needs*: self- realization, historic and prospective meaning, religious meaning and Weltanshauung

- *Synergy needs*: authenticity, solidarity, justice, altruism, responsibility, beauty, ecological equilibrium

It has been argued that, to the extent that such needs constitute a hierarchy, the earlier needs must be adequately fulfilled before attention can be devoted to the later ones. Some have questioned this as reinforcing bureaucratic tendencies to give priority to the earlier needs at the expense of any attempt to allocate resources to problems arising from inadequate fulfillment of the later ones. It is clear that the proportion of resources allocated to "higher" needs is low and easily neglected or discounted, as in the case of "peak" or self-realization experiences. There is even the irony that only through allocation of resources to the disputed "higher" needs is any solidarity felt with those having "lower" needs.

There would seem to be an inverted hierarchy complementing that above and grouping the attention accorded to corresponding problems. The two hierarchies, like two triangles, intersect somewhat as in the Star of David symbol. The apex of the inverted triangle is indicative of the relatively small amount of attention accorded to the concrete maintenance problems, especially the intractable problems like hunger (which are easily accepted by others as an unfortunate fact of life). Whereas the broad base of that triangle is indicative of the widespread attention given to (the rhetoric concerning) the problems relating to the higher needs like justice.

This may be another way of looking at a phenomenon described by analogy to the Peter Principle of career advancement. Namely problems tend to get abstracted to a level at which little can be effectively done about them, neglecting any focus at those levels amenable to action. Rhetoric can be used very elegantly to conceal the absence of effective action, precisely by focusing on the urgency of needs defined at a level where concrete action is necessarily difficult and slow to show results.

3. Complexity of response to problems
"Enemies", as dealt with by the approaches to problems detailed earlier, are essentially external, objective realities - - or at least are assumed to be so. But this need not necessarily be the case. As noted earlier, perception of problems is very much a product of a particularly Western mindset. It is possible that the following open up other levels of understanding and response to problems:

(a) Higher order problems: Just as there are basic maintenance problems (cf hunger, shelter), so there are logistical problems of responding to them (cf maldistribution of resources), and problems of principle in recognizing the more concrete problems (cf violation of rights to food, shelter). There may be problems of failure of conceptualization which undermine initiatives to recognize the problems of principle, or failure of the political will to act upon them. Beyond these may be fundamental weaknesses in the ideology or belief system in terms of which the preceding responses are undertaken. There is thus a case for more detailed exploration of higher order problems and the implications for dealing with those of a lower order. Of course the terms "higher" and "lower" imply a questionable bias, since those which are most existentially threatening can also be considered to be of the "highest" order.

(b) Contradictions: The convenient ways in which thinking has developed have tended to conceal the presence of essentially embarrassing "contradictions", which may constitute a special class of problems. These tend to be detected only when two or more problem- free perspectives are juxtaposed, although such juxtaposition may in itself be considered quite unreasonable.

Examples might include major discrepancies: between percentage salary increases accorded to employees and to executives; between mortality rates in developing and industrialized countries; between energy use per capita in developing and industrialized countries; between the rights and privileges accorded to different classes in any society; between the effective rights of men and those of women.

Thus whilst it may be considered reasonable to treat malnutrition in developing countries as a problem calling for action and the over-indulgence in foods in industrialized countries as another (at least to some), the coexistence of these problems may not be considered as an acceptable third problem in its own right. Rather any such perception may itself be considered as disruptive or subversive, and as such symptomatic of a problem in its own right.

(c) Problem complementarity: Much has been made of the inappropriateness of dealing with problems in isolation (cf ignorance and unhygienic food preparation) as has been the early practice of "specialized" international agencies. Integrated approaches are increasingly sought, implying some sense of a system of problems.

Perhaps beyond that is the recognition of a basic complementarity between certain problems, each evoking the other. Personal insecurity and excessive fertility may be a case in point, especially in developing countries. Psychological insecurity and substance abuse may constitute an equivalent complementary pair typical of industrialized countries. There may also be complementary triplets and quintuplets, whose detection is as yet beyond current facilities of conceptualization, given the emphasis on faculty-biased specialization.

(d) Tetra-lemmic problem articulation: Within the Western mindset favoured by the international community, a problem either exists or it does not. For other cultures, the situation may be more complex, as suggested by the work of Kinhide Mushakoji (*Scientific Revolutions and Inter-Paradigmatic Dialogue*, 1978). As noted in the Japanese examples cited in an earlier note, a problem may both not exist (in the explicit reality) whilst existing (in the implicit reality). An individual may move smoothly between both realities. Furthermore, there is a fourth condition in which a problem may be considered to neither exist nor to not-exist. This may be the most fruitful way of recognizing problems of "corruption".

(e) Paradoxical responses to problems: Within the Western mindset, a problem can always be solved if sufficient resources, of an appropriate kind, are allocated to the task. It is believed that there is always a technical solution, even though the "real" problem is the lack of political will to make available the required resources. But work in the field of psychotherapy suggests that there may be many kinds of problem (possibly including "lack of will") that respond much more readily to counter-intuitive strategies.

Paradoxical therapeutic strategies, ostensibly encouraging a client's negative or maladaptive behaviour, have been used with great success in working with difficult individuals, couples and families (Leon F Seltzer, *Paradoxical Strategies in Pyschotherapy*, 1986). Although seemingly irrational and sharply opposed to the kinds of expectation that clients bring to therapy, they appear to owe much of their effectiveness to the very unorthodoxy that that has so often rendered their use controversial.

(f) Zen of problem response: Intractable problems, like intractable enemies, can evoke unusual responses. The Eastern martial arts have developed through a philosophy in which it is the very energy of the opponent which is used to reorder the situation for mutual benefit. Such arts may of course be used to achieve a lower order response in which the opponent is neutralized. But such responses obscure the significance of those of higher order, beyond the "zero-sum game" of winners and losers. This may be more appropriate in dealing with the most challenging of opponents.

Of special interest is the expectation that there is something to be learnt, some insight to be gained, from the interaction with a really challenging problem. In a real sense the interaction, undertaken appropriately, dissolves or redefines the boundaries between the two parties and their environment, so that a new situation comes into being. This contrasts completely with lower order responses in which one of the parties is neutralized without effectively challenging or redefining their contextual situation. From a different perspective, it may be that problems can be usefully viewed as "koans", essentially insoluble within the mindset with which they are initially approached.

Insights: phases of human development through challenging problems

1. Experiential phases and modes
The contents of the core sections of this Encyclopedia might be understood as linked over time in terms of the problems and values encountered under different challenges to human development. There are many concepts of the phases of human development (Section H). The possibility of such an ordering might best be illustrated through one which links such phases to value dilemmas.

It is instructive, for some purposes, to view phases as succeeding each other in time, possibly over a life cycle. It can also be useful to view such phases as being possible at any stage of a life cycle, but with different probabilities. It may therefore be more fruitful to to consider that an individual of any physical age can be at different experiential ages with respect to each value dilemma.

Different people may thus be faced by different dilemmas at the same stage of life cycle, or by the same dilemmas when they are at different stages of a life cycle.

2. Value crises in a life cycle
In Erik Erikson's scheme (*Childhood and Society*, 1963), each individual goes through 8 stages in life. In each stage a value crisis is experienced which is crucial for continued development. The stages, with their corresponding crises are as follows:

- Infancy (basic trust vs. basic mistrust)
- Early childhood (autonomy vs. doubt)
- Play age (initiative vs. guilt)
- School age (industry vs. inferiority)
- Adolescence (identity vs. role confusion)
- Young adulthood (intimacy vs. isolation)
- Adulthood (generativity vs. stagnation or self-absorption)
- Mature adulthood (integrity vs despair)

Note that each value conflict is not resolved once and for all at the time the stage is traversed. It arises again at each subsequent stage of development. In transcending each crisis, it is neither necessary nor desirable to eliminate the negative portion of the value-polarity. The challenge is to ensure the emergence of an appropriate balance or dynamic between the two value extremes at each stage.

Resolution of any value dilemma cannot readily be based on any formula or argument. Whilst there may be logical arguments concerning the nature of the appropriate balance, these will be challenged by subtleties of experience that will highlight the existence of degrees of freedom other than those encompassed by any explicable pattern of concepts.

3. Moral and ethical dilemmas (virtues and sins)
An effort has been made by Donald Capps (*Deadly Sins and Saving Virtues*, 1987) to relate the stages in this life-cycle theory to the traditional basic sins and corresponding virtues of the Christian tradition (taking into account reservations concerning male bias noted by critics of Erikson's original theory). This is of interest because of the view that such root sins engender other problems by a sort of "domino effect". Analogous views can be found in other traditions, notably the Buddhist.

To make such an inquiry more topical, such root afflictions, or psycho-social traps, need to be recognized at a **group level** rather than solely at the individual level. In this way the link to **societal problems** is more firmly established.

Capps associates a "deadly sin" with each stage. Each such sin is appropriate to the corresponding stage as a prominent factor in the moral or spiritual life of that period, whose basic psychodynamics it reflects. The sins are not rigidly tied to particular stages but are linked to them through their common psychodynamics. Sins may thus emerge earlier or later than the stage with which they are primarily associated. Capps elaborates an 8-fold set of sins in the following sequence corresponding to the above stages: gluttony, anger, greed, envy, pride, lust, apathy, melancholy. There are striking resemblances to the Buddhist equivalents (see Section PZ).

4. Group sins or afflictions
With increasing reference in the 1980s to "corporate greed", it is interesting to explore the possible collective equivalents to these sins. In the light of Capps analysis, these might run as follows:

- Excessive consumption of resources, especially energy
- Collective anger, especially expressed in violence
- Collective greed, especially in the accumulation of resources
- Collective envy, especially for resources controlled by others
- Collective pride, typically as arrogance and triumphalism
- Collective lust for power, typically as expansionism
- Collective apathy, typically in response to emerging problems
- Collective despair, typically in acknowledging current impotence and in recollecting past failures

5. Appropriate responses and saving virtues
Traditionally, and as developed by Erikson and Capps, there are characteristic saving virtues through which people can most effectively respond to the above sins. Equivalents are to be found in the Buddhist and other traditions. These too tend to become particularly significant at different stages of the life-cycle. Using the same sequence, they are as follows, again expressed in terms of what might be their collective equivalents:

- Hope, which is expressed both individually and collectively
- Will (or Courage), especially in frequent appeals for the "generation of the political will to change"
- Purpose (or Dedication), increasingly evident in the formulation of "mission statements" and implicit in "resolutions"
- Competence (or Discipline), increasingly stressed as vital for effective management
- Fidelity or Loyalty, increasingly a concern of corporate human relations programmes and security procedures
- Love, increasingly explicit in "green" approaches to the environment and traditionally implicit in recognition of the "brotherhood of mankind"
- Care, especially evident in relief programmes
- Wisdom, occasionally acknowledged in calls for collective wisdom and statesmanship

In the Buddhist tradition, the equivalents might be considered to be the component elements of the Eightfold Noble Path:

- Right outlook;
- Right speech;
- Right acts;
- Right livelihood;
- Right endeavour;
- Right mindfulness;
- Right rapture of concentration.

6. Implications for sustainable development
From this perspective the challenge of sustainable development is one of both comprehending and giving form to balance. It is the imbalanced resolution of the value dilemmas which engenders problems. The difficulty is that whilst it may be easy to talk of "balance", it is quite another matter to comprehend its nature in practice (as is readily appreciated in learning to ride a bicycle). The dynamic balance, or Buddhist "middle way", involves eight degrees of freedom, when expressed in terms such as those above.

7. Proactive response to the challenge of appropriateness
It is ironic that understanding of any such scheme of sins and virtues in the West tends to be somewhat passive, in that any significance it has is determined by the slow development through a life-cycle. Any battles against "sin" remain private and personal matters, without any sense of strategy, as with the cultivation of "virtues". In this sense any form of personal improvement is considered to be largely an illusion within establishment institutions and disciplines (except under the guise of acquisition of marketable skills by training and experience).

By contrast, spiritual traditions in the East appear to challenge this passive determinism, rejecting the fatalistic subjection to the current life-cycle in favour of programmes of spiritual disciplines with acknowledged phases and insights through which the individual is transformed. The West has developed sciences of "development" designed to transform society, whilst assuming that human beings themselves only change through ageing and the acquisition of skills. The East has developed sciences of personal transformation, whilst assuming that any effects on society are lacking in lasting significance. The West has focused on the growth of society, neglecting the growth of the individual. The East has done the reverse. The West focuses on the life-cycle of the individual, whereas the East focuses on the spiritual cycle or journey (irrespective of how it may relate to the physical life-cycle).

8. Development of insight in learning cycles

Schemes such as the above suggest that people or groups at different learning stages generate different kinds of problems and can usefully cultivate corresponding strengths to counter them. It is unreasonable to expect any form of general consensus or shared understanding in such a dynamic context. This could only emerge through insights into interweaving cycles of development.

Whilst the management skills to organize such initiatives have been developed by the West, it is the East which appears to have a more profound articulation of the qualities of insight that need to be developed and how they need to be interwoven to reduce problem generation.

The situation is of course totally confused by the claims of both management "gurus" in the West and of spiritual "gurus" from the East, all with markets to cultivate and under competitive pressure to offer distinct products to potential customers. It remains to be discovered how their genuine insights can be effectively interwoven in response to the challenge of the times.

9. Disempowering injunctions

If there are eight things to be held in balance, as when learning to ride a bicycle, injunctions concerning any of them may be less than helpful. The difficulty is that, although the learner may have some knowledge of what is meant by any one injunction ("care", or "right mindfulness"), this knowledge is limited precisely because the person (or group) has not yet learnt its full significance in practice.

Efforts to ensure implementation of the injunction, through obedience to rules or procedures, do not guarantee achievement of the requisite level of insight. They may help to orient the learner, but they may also discourage and disempower. This is particularly the case when the learner has sufficient insight to recognize that the real challenge does not lie at the level of mechanical rules and procedures but in what amounts to the aesthetics of balance. At this level, it is less a question of whether the rules are obeyed to the letter and more a question of whether balance is maintained. Perfection may lie, as with an important principle of Japanese aesthetics, in the harmony of imperfections.

Exhortation and injunction may in many situations simply lead to what amounts to "learning fatigue" -- an appropriate complement to "compassion fatigue". In this sense they can be totally counter-productive.

In this light, the focus in the international community on elaborating declarations, rules and agreements may well orient usefully those adressed, but it fails to address the challenge of how they are to learn the secrets of balance. Worse, it reinforces the views of those focused on single-factor explanations and remedies, such as "market forces", "peace", "conservation", "equality", or "love". For them the answers are already self-evident and there is no collective learning challenge. Such approaches may be necessary, but they are not sufficient to obtain an understanding of the balance ultimately required for sustainable human development.

10. Intriguing dilemmas and developmental koans

In one sense the issue is the classic challenge of how the learning process can be made attractive, interesting or seductive. However the emphasis is not only learning things which can be taught mechanically or by rote. Rather it is the question of catalyzing the leap of imagination through which a new paradigm is grasped experientially enabling energies to be controlled in new ways. There are some classic responses to this challenge:

(a) Sufi tales of Nasruddin: The Sufis have deliberately cultivated an extensive set of teaching stories. They are brief, witty and call for a paradoxical switch in perspective. They may told purely for entertainment, thus ensuring their survival and wide dissemination, or they may be the basis for discussion and meditation. Such fables exist in other cultures. However it is those of the Sufis which are best designed to maintain the challenge to insight and to resist simplistic interpretation.

(b) Paradoxical aphorisms: All cultures have a store of paradoxical aphorisms which point to value dilemmas, holding their tension rather than indicating a simplistic way forward. Of course there are many other aphorisms which do the latter.

(c) Zen koans: These are deliberately designed as challenges to understanding, irritating the mind at the level at which it would like to respond to a dilemma so that finally it is forced to another level of understanding.

(d) Riddles and puzzles: In many traditions there are riddles and puzzles, often associated with magic. These point to the need to move beyond obvious modes of understanding to breakthrough to other forms of insight.

(e) Paradoxical strategies: In psychotherapy increasing attention is paid to the advantages of enjoining people to act in a manner contrary to that which they expect. Through encouraging them to act in a manner which, at one level, they know to be inappropriate, they achieve a more fruitful relationship to what they need to learn.

(g) Meditation: Given the attention of Buddhists to these issues, it is not surprising that they have developed very explicit meditation techniques concerned with the development of understanding of the appropriate attitude from which to response to the value dilemmas. These are designed specifically to avoid engendering the kinds of problems which result from imbalance. The techniques are not only considered with the imbalance associated with particular dilemmas, but also with the level of balance required to respond simultaneously to all the dilemmas. The mandala is one diagrammatic representation of this understanding although, as a mnemonic device, the issue is with what insight meditators can learn to "read" it.

(g) Computer graphics: New developments in computer-generated graphics are permitting imagery to be generated which does not conform to the rules of the physical universe. Viewing such imagery is a direct challenge to the imagination and calls for a basic shift in perspective. Such techniques could well be adapted to encourage insights into dilemmas and new forms of balance.

8. Encapsulating sets of dilemmas

The classical sets of eight value dilemmas represent a well-established approach to human development. It could be argued that the challenge of the times calls for a more powerful statement of the dilemmas of global society. If the issue is not one of learning facts and responding to injunctions, where is the set of learning "puzzles" enabling individuals to obtain their own unique insights into the kinds of balance required for physical and psychic survival?

Different traditions and cultures might be explored to locate the "riddles" to which we are called to respond. Others might be designed by different disciplines. In a period when education is increasingly problematic, sets of 3, 5, 8, 12, or more intriguing challenges could offer a powerful complement to factual learning. The possibility is seductive because the answers to the dilemmas cannot be effectively verbalized without denaturing them. The "right answer" is one which opens new vistas and feels "right" for the individual. They are a matter of personal (and possibly group) experience, difficult to share.

How might the challenge of sustainable human development be expressed in this way -- as a major step beyond the disempowering and ineffectual injunctions on which so much confidence is vainly placed by the well-intentioned?

Insights: integration of perceived problems

1. Shadow of humanity
In the terms of the analytical psychology developed by C G Jung and his followers, it may be asked whether humanity has a "shadow". In the case of an individual, the shadow is the sum of all the unpleasant, negative aspects of the personality -- those that one tends to hide from oneself and especially from others. In that context, it is argued that it is the shadow which makes the individual human. It cannot be eliminated, rather the aim should be to come to fruitful terms with it. In the light of this perspective, the system of world problems could be usefully viewed as the "shadow of humanity". This said, the question would then be what might be understood by the process of coming to terms with it.

For Jung, the shadow is *"a moral problem that challenges the whole ego-personality, for no one can become conscious of the shadow without considerable moral effort. To become conscious of it involves recognizing the dark aspects of the personality as present and real. This act is the essential condition for any kind of self-knowledge, and it therefore, as a rule, meets with considerable resistance."* (Jung, ***Aion) The inferiorities constituting the shadow have an emotional nature, a kind of autonomy, and accordingly an obsessive/possessive quality. While some traits peculiar to the shadow can be recognized without too much difficulty as one's own personal qualities, there are certain features which offer the most obstinate resistance to moral control and prove impossible to influence. These are usually associated with projections onto others as being undoubtedly at fault.

2. Transcending polarization
In Anthony Stevens' discussion of the question: *"Without some acknowledgement of the devil within us, individuation cannot proceed...True morality requires that the shadow achieve consciousness, because on that condition alone can an individual become responsible for the events of his life and render himself accountable for what he has projected onto others....It was Jung's contention that the two moral poles were capable of reconciliation: awareness of the shadow means suffering the tension between good and evil in full consciousness, and through that suffering they can be transcended. If one can bring oneself to bear the psychic tension that the opposites generate, the problem is raised to a higher plane, where the conflict is resolved: good is reconciled with evil, and a new synthesis follows between conscious and unconscious... The reconciliation is attained neither rationally nor intellectually, but symbolically, and it was to this symbolic process that Jung gave the term transcendent function. Through the transcendent function the conscious personality and inner adversary are both transformed: as new symbols arise from the unconscious...the opposites are reconciled and transcended; the personality becomes better balanced, more integrated... Phenomenologically, the experience is one of liberation combined with an awareness of the inner strength that comes of reaching harmony...with something greater than the mere ego."* (Stevens, pp 241-2)

3. Progressive integration of the shadow
This process of individuation is frequently depicted in Zen Buddhism by a traditional sequence of 10 ox-herding pictures, each with a brief commentary (Suzuki***). These are of special interest because of their indication of a person's progressive discovery and interplay with a shadowy element denoted by an ox. The following is an attempt to suggest how that classical sequence might be interpreted for clues to an unfolding relationship between humanity and its shadow (in the shape of the complex of world problems).

The phases in the sequence are:

(a) Undisciplined exploration of the problematique: Humanity, having violated its own inmost nature, loses track of the problematique and its significance. It is then led astray by the delusions to which it succumbs, such as desire for gain and fear of loss, and is confused by a multiplicity of views of right and wrong, appropriateness and inappropriateness. Although distracted by this confusion, and exhausted by its efforts, humanity continues its search for a sustainable solution. *At this time, it would appear that humanity, as represented by the international community, continues to be embroiled in the pre-systemic, single-factor perspectives of this first phase (ozone, acid rain, "health-for-all", substance abuse, illiteracy, terrorism, AIDS).*

(b) Recognizing traces of the problematique as an integrated system: Repeated (and basically unsuccessful) attempts to locate and contain the problematique through uncoordinated initiatives provide humanity with occasional insights into its nature, especially when more integrated approaches are used. Although recognizing that the problematique, by whatever means, is in some sense engendered by humanity as a whole, there remains a basic confusion between truth and falsehood, especially when it seems obvious to some that another particular group can be usefully blamed for specific problems. *Environmental and systems insights (tropical forests, global warming) are shifting the focus to this second phase.*

(c) Focusing on the problematic as a whole: Having cultivated a more intuitive insight, enabling it to integrate its complementary modes of perception, humanity focuses directly on the problematique, recognizing its many manifestations as consequences of different forms of inappropriate human intervention. *There are episodic exercises in focusing on the problematique as a whole (Brandt Report, Brundtland Report), although what they fail to take into account quickly condemns them as sub-systemic and inappropriate and encourages further initiatives of a similar nature.*

(d) Encompassing the problematique: Humanity grapples with the problematique directly for the first time. The momentum of the problematique, developed over the long periods during which it was uncontained, and the pressures and habitual opportunities of an undisciplined social environment, make it extremely difficult to control. Severe disciplinary measures are necessary. *The various development strategies, especially the current attempt at "sustainable development", correspond to this fourth phase, but only to the extent that efforts are made to implement them. On the national level, the structural adjustment required by the IMF is indicative of the political will required -- although typically such adjustment fails to take into account many facets of the problematique.*

(e) Orienting the problematique: Every insight concerning the problematique leads humanity to further insights in an endless pattern. With discernment these will all be of value. But when humanity deceives itself, confusion will prevail and the problematique will reassert itself in an inappropriate manner. Constant vigilance is required to discipline the problematique and orient its manifestations within appropriate bounds. *The seeds of this fifth phase may be seen in the increasing recognition of the need for a disciplined and radical change of life style, especially on the part of the industrialized countries.*

(f) Using the problematique as a vehicle for sustainable development: The struggle of humanity with the problematic is over. Humanity is no longer traumatized by gain or loss, which are assimilated as phases in a larger process that is now the focus of attention. Rhythms of action in harmony with nature are cultivated. The problematique is used as a vehicle moving in sympathy with those rhythms towards the re-enchantment of the Earth. The old modes of action are not considered viable and their advocates are no longer heeded. *Some indications of the nature of this phase are to be found in the writings of the "deep ecology" movement and in the preoccupations of some forms of sustainable agriculture -- although their obvious limitations lie in their inability to deal realistically with the conditions of industrialized, urban societies and the impoverishment of an overpopulated planet. The missing insight would seem to be how to achieve the transition to this stage by benefiting from the problematique itself.*

(g) Transcending the realm of the problematique: Having used the problematique as a vehicle to reach a sustainable condition, it is no longer required. However, the necessary disciplines for humanity to handle it remain available. Humanity can now act with serenity guided by insight that is no longer obscured by the dynamics of the problematique. There are writings on paradigm shifts into a new consciousness (in which the problematique no longer figures) and these do offer clues as to the nature of this phase. However, their neglect of the problematique would seem to be more a question of avoidance rather than transcendence, indicating that such perspectives lack vital insights.

(h) Disappearance of both humanity and the problematique: The dualistic mindset through which humanity is perceived, in opposition to the problematique and to other species, is itself transcended, as are the disciplines through which that relationship is articulated. Confusion disappears. But there is no question of being either entranced by more integrative insights or entrapped by lesser ones. The nature of this condition does not lend itself to definition. Typically, any desire for it renders it unattainable or unsustainable.

(i) Expression of essential humanity: Grounded in its essential nature, humanity stands untouched by inappropriateness. Processes of integration and disintegration are witnessed from a perspective that enfolds them. Neither formulation nor reformulation are necessary to ensure sustainability. Change, as perceived, is necessarily appropriate however paradoxical it may appear.

(j) Human intervention in the world: Human action is no longer associated with any particular mindset, nor does it follow any recognizable path. It cannot be assessed by any form of conventional wisdom, nor does it depend on any particular tools. No special effort is made to preserve forms of any kind -- including those of humanity itself. Insight into the emptiness underlying form enfolds any form of action in a more meaningful context, thus enabling greater appropriateness to emerge as required.

3. Comment

Some writings would also appear to imply a certain understanding of the later phases, but for humanity as a whole they lack the realism and credibility that will presumably emerge through painful societal learning in the future. Such understanding is in part reflected in the many religious visions of a saviour (a new Buddha, Christ, Imam or Messiah) scheduled to release humanity from the influence of evil forces. (The emergence of such saviours, or their precursors, is even the subject of occasional full-page advertisements in the quality press.) The weakness of such visions may lie in their tendency to remove responsibility and initiative from humanity in the belief that the burden could be more appropriately borne by such a saviour. It may be that a saviour should also be understood as a confused projection into the future of present intuitions as to the nature of that mode of humanity which could act to alleviate its own condition. Such projections are necessarily simplistic in their clarification of how redemption might successfully be brought about through some alchemical interweaving of the evil problematique, ordinary humanity and that insightful mode of the future. The many insights into individual development (some acknowledging a form of rebirth into the presence of a saviour "within"), are indicative of the accessibility of such levels of understanding (see Section H***), although their collective manifestation in groups, and humanity as a whole, is questionable as yet.

Although the notion of a sequence reflects a basic evolution of insight over long periods of historical time, it may also be fruitful to consider the different phases probabilistically, as conditions of comprehension by which humanity (or any part of it) can be determined, to different degrees, at any time. Thus some understanding of later phases may be achieved at any time, just as a child may occasionally exhibit extraordinarily mature insight. But such intimations can presumably only acquire their full meaning as a result of full experience and transcendence of the earlier learning phases. In a sense the perspectives of the earlier phases are always accessible from the later, and may even be appropriate under certain conditions -- just as indulgence in child-like behaviour may occasionally enrich the experience of an adult in maturity.

Insights: barriers to transcendent insight and social transformation

The kinds of individual and collective creativity through which innovative responses emerge to the problems of society depend largely on the ability to shift to perspectives of a significantly different order. The much sought "paradigm shifts" appear to call for a conceptual transcendence of some of the more obvious attitudes and behavioral processes by which we are collectively trapped.

Many of the spiritual traditions, as well as modern insights into spirituality, argue for the relevance of transcendent experience as providing a complementary integrative perspective through which the challenges of the times can be more appropriately encountered. It is therefore valuable to note some of the barriers to transcendent experience as a key to creative insight. A number of these points have been adapted from points made by Edwin Dowdy (Dowdy ****) and Alistair Kee (*** The Way of Transcendence).

Whilst the argument for transcendence needs to be stressed, some recognize that this can constitute another form of escapism which avoids essential issues. In this sense transcendence should not result in abandonment of the level transcended, rather it should involve experiencing it, and recreating it, in a more profound manner as a form of enlightened immanence (cf Morris Berman. *Coming to Our Senses*, 1989)

1. Individual development

(a) Opportunism: Numerous studies support the view that in the present competitive "rational" society, the individual must achieve an "autonomous ego", an "identity" of continuity and consistency, and "interactional competence". Adult status is acquired through learning experiences and crises in which the person must necessarily be egocentric. Attention is therefore fixed compulsively on felt needs, perceived threats and the opportunities offered by the environment.

(b) Egocentrism: Individual development through to young adulthood is characterized by the quite self-centered processes within the small child (expression of individuality, growth within the parental family), the struggle of the ego of the school-age child to gain control (development of a role-bound identity and detachment from internalized parental concepts), and development of the ability of the young adult to play the social game (whether in effective competition with others if successful, or otherwise by accepting a less desirable role). Identity is thus formed within a system of egocentric references with inbuilt barriers to transcendence or dissipation of the ego.

(c) Conformity: Conformity seems to prevail in identity-formation in which relatively few are at present able to transcend the developmental struggle and achieve some orientation to general principles such as the golden rule, the sanctity of life, the categorical imperative, or furthering the development of mankind. Such principles lack credibility in the presence of evident lack of normative regulation in many areas, with conflicts arising that cannot be overcome in the framework of normal role-behaviour.

(d) Role conflicts: Whilst there is evidence for subsequent stages in the development of moral maturity, involving contemplative experiences of a "non-egoistic and non-dualistic" type, relatively little importance is attached to them. Role conflicts, the stress of inter-generational distancing, and alienation erode any embryonic sense of being part of the whole of life or any sensitivity to a cosmic perspective. Ironically the need for such transcendent perspectives is expressed illegitimately through the seductive attraction of drug-induced experiences (whose value is "guaranteed" paradoxically by an alienating society's total repudiation of them).

2. Social structure

(a) Expediency: Government in a continuing period of chronic difficulties is reduced to a crisis-avoidance agency. It endeavours to maintain legitimacy by being seen to provide material guarantees such as defence, standard of living, and law and order. The prime issue becomes one of administrative expediency, irrespective of moral issues or the emancipation of individuals or groups. The organization of government is inimical to consideration of values because the steering mechanism interprets information according to its own systemic exigencies, with purposive-rational action justifying itself in terms of the criteria of its own ideology. The cognitive interest therefore dominates any emancipatory interest.

(b) Distorted priorities: The media, whether or not as direct tools of government, constantly portray the problems recognized by the state as being the most important problems, if not the only real problems. In initiating development policies, arbitrating conflicts of interest, and affecting immediate conditions of standard of living, government encourages recognition that its concerns are the core focus of any effective action. Concerns for inner quality of life and transcendent experience are seen as betraying collective action in response to social priorities. They can be readily portrayed as epitomizing the ineffectual.

(c) Marketable skills: Education, especially under increasing pressure from industry and government, has as its main priority the training of the young so that they may perform adequately the various roles which will maintain the operations of society. Curricula focus on skills that are marketable. Whilst there is some concern with the development of the individual's potential and with maturity of understanding, this tends to be vague and subject to the insights and whims of teachers. Sport emphasizes competitive instincts accompanied by occasional overtones of fair play. Religious education tends to be an alienating mix of doctrine, facts and stories. The only transcendent experience accessible to students is through extra-curricular sex experimentation and drugs. Those with creative talent wait until after their formal education to pursue their own self-development.

(d) Denigration of subjective experience: Psychology is primarily concerned with observable behaviour and thus reinforces the preoccupations of industry, education and the media. Mainstream psychology and psychiatry go further in actively denigrating subjective experience and emphasizing its delusional dimensions. Amongst the alternative psychologies and psychotherapies, the claims and counter-claims for special skills with respect to transcendent experience, often by practitioners of dubious reputation, can easily discourage exploration of that experience.

(e) Religious opposition and manipulation: Religion itself, at least in the Judeo-Christian tradition, is basically inimical to transcendent experience as a potential threat to orthodoxy. In general, much has been written concerning the alienating effects of institutionalized religion and on priesthoods of dubious spirituality with a questionable relationship to the transcendent. Amongst the alternative religions, the well-publicized exploitation and manipulation by cult leaders has made this route to transcendent experience less credible. However the Christian charismatic movements are offering new opportunities, as are fundamentalist initiatives in Hinduism and Islam. But it remains questionable whether such initiatives recognize that through some form of spiritual development, individuals might be better able to understand the challenges of the times and act on them more effectively.

(f) Exploitation: Recognition of the social barriers to transcendent experience has encouraged the emergence of many alternative initiatives of every colour and degree of insight. Their variety can itself create great difficulty for those attempting to locate some form of guidance in which they can have confidence. For even where there is great sincerity on the part of those offering such guidance, it is not necessarily matched by great insight. There are also situations in which great insight is offered by people whose aim is less than well-intentioned and possibly even malignant. Commitment to transcendent experience may then be distorted to other ends.

Insights: interrelating possible viewpoints

1. Dilemma of many possible views
This Encyclopedia seeks to respond to the dilemma of the many possible views concerning the nature of sustainable human development. It is the nature of any such view that it considers itself more appropriate than other competing views. Few views account for the existence of other views except as being predecessors or misguided. And yet it is the interaction of such views within a conceptual ecology which characterizes the dynamics of human society. Different views engender different styles of human development and give rise to different problems.

It therefore remains an interesting challenge to explore new ways of interrelating the network of views, especially if it is possible to embody in such an exercise features which give the whole a degree of complexity appropriate to the complexity it is intended to encompass. It is also desirable to build in features which depend not only on insights from western cultures.

2. Mapping the network of views
This is not a new challenge, although it may now appear more dramatic to some. An intriguing point of departure is a classic Buddhist text entitled the "*Brahmajala Sutta*" (The Discourse on the All-Embracing Net of Views). This appears to be unique in endeavouring to map out as a system the complete set of fundamental viewpoints. It is the first sutta in the entire collection of the Buddha's discourses in the Pali Tipitaka. Its importance stems from its primary objective namely the exposition of a scheme of 62 cases designed to include all possible views (past and future) on the central concern of speculative thought, the nature of the self in relation to the world. It is used as a way of establishing a context for the emergence of what amounts to an essentially more sustainable view characterized by a greater degree of insight. In systems terms, it is this higher level of insight which provides the necessary conceptual complexity to inhibit the continual generation of problems arising from less adequate perspectives (whatever vital function they may otherwise have).

3. Exploratory exercise
In using this point of departure, it should be clear that the following is merely an exploratory exercise. However, the intention here is not simply to point to an existing text born of a particular culture. What is intriguing about the text is that it is based on a relatively complex pattern with what seem to be many clues to a more systematic and fruitful way of representing that pattern. This is important because, read only as a checklist reinforcing linear patterns of thought, it is doubtful whether it could lead to any breakthroughs in response to the non-linear dilemmas of the times.

In what follows a translation of the text is first presented. Some clues to the patterning principles are then listed. The possibility of coding the pattern using a binary representation are then discussed, especially in relation to a similar endeavour of equal antiquity originating in another non-western culture, namely the Chinese *Book of Changes*. Then some implications for a more challenging understanding of sustainable human development are explored.

The text reflects insights formulated 2,500 years ago. It is not the intention to be strictly faithful to it. Rather the intention is to be guided by a method which is based on a fundamental patterning principle in the text itself. The logic of that pattern is to a high degree tetra-lemmic in the sense that it provides for four logical alternatives concerning any thesis: A, not-A, A and not-A, neither A nor not-A. It will be argued that these reflect a progressive complexification of understanding to whatever domain they are applied. Applied to this exercise, it is proposed to: start with the constraints of the text (A); to ignore those constraints, treating the text metaphorically, in search of a patterning principle (not-A); to endeavour to fit the pattern to the text with any necessary adjustments (A and not-A); and then to explore the insights implied by that result, reaffirming the pattern, but unconstrained by the limitations of the particular representation (neither A nor not-A).

Such an approach is characteristic of Madhyamaka Buddhism whose dialectic has been articulated by Nagarjuna (***). There the dialectic is designed to expose the inconsistencies in all sides of any issue as a preliminary to the experiential awareness of that which transcends such dualities. The Madhyamika recognizes the four alternatives as mutually exclusive and jointly exhaustive. The law of the excluded middle is accepted in logico-mathematical operations, where the alternatives are finite and mutually exclusive, but not in ontological contexts where this cannot be demonstrated. It should be stressed that the exercise here is exploratory, hopefully indicating possibilities which others can use with greater insight.

4. Summary of a classic net of views
The list of points presented below is an editorial adaptation of an appendix in the English translation by Bhikku Bodhi (1978) of the Buddhist text on *The All-Embracing Net of Views*. This provides a checklist of the views in the order in which they are discussed in the original text and its commentaries.

Despite the text's almost mechanical precision in classifying and distinguishing some views, all of them need to be considered connotatively rather than denotatively, especially as possible metaphors for more comprehensive levels of meaning than is apparent. This is especially the case given the antiquity of the text and the difficulties of translation from a non-western language.

5. Clues to patterning principles
(a) The text explicitly identifies 62 views as constituting a complete set of inappropriate or unsustainable views. Two other views are however implicit. One is that of the ultimately sustainable "correct" view (nirvana or nibbana). The other is the state of ignorance (samsara), possibly antedating the formation of any other view, and to which all other views are a response. These could bring the total elements in the set to 64.

(b) The majority of the sub-sets of views are based on patterns of 4 or 8 elements. One is 2, one is 5, one is 7, one is 16. Using the 2 omitted elements, and repositioning the sub-set of 2, a systematic patterning could be based on 4-fold and 8-fold sets, possibly within 16-fold and 32-fold sets.

(c) As noted above a number of 4-fold sub-sets (whether or not part of a larger sub-set) are explicitly structured in terms of: A, not-A, A and not-A, neither A nor not-A. As noted, these suggest a progressive complexification beyond the constraints of Aristotelian logic, which remain, however, a sub-set of the sequence. This implies a progressively increasing challenge to comprehension, suggesting that the more "nirvanic" views might be based on elements of the pattern governed by "neither A nor not-A" (as implied by the phrase "not this, not that", traditionally associated with that state). Correspondingly, the least nirvanic views might be based on elements of the pattern governed by "A".

(d) Wherever the 4-fold sub-sets occur, with textual variants of the above logic, the progression through the sequence is the same (see 2.1). This suggests that this sequence may hold in the global organization of the pattern as well as in the detail. If this were the case, then there would be 4 major 16-fold sets. These would be based on views about: past, future, past and future, neither past nor future. The first two are explicit. The last two may be valid interpretations.

(e) The views appear to be based on three conceptual domains or levels. These cannot be labelled satisfactorily since they are categories at the highest level of abstraction. The degree of abstraction is indicated by the dualistic form they take in the text. They might be tentatively labelled as follows: materiality/immateriality or form/formlessness; space/time; subject/object or knower/known. These are natural categories for any classification of philosophies.

(f) The overall pattern might then be represented as an interplay between dualism, the 3-fold organization of domains, and the 4-fold logic.

1. **Speculations about the past** (*Pubbantakappika*)
　1.1.1 Eternalism *(Sassatavada)*
　　1.1.1 Based on recollection of up to 100,000 past lives
　　1.1.2 Based on recollection of up to 10 aeons of world contraction and expansion
　　1.2.3 Based on recollection of up to 40 such aeons
　　1.2.4 Based on reasoning
　1.2 Partial-Eternalism *(Ekaccasassatavada)*
　　1.2.1 Theism
　　1.2.2 Polytheism held by beings who were gods corrupted by play
　　1.2.3 Polytheism held by being who were gods corrupted by mind
　　1.2.4 Rationalist dualism of an impermanent body and an eternal mind
　1.3 Extensionism: finitude and infinitude *(Antanantavada)*
　　1.3.1 View that the world is finite
　　1.3.2 View that the world is infinite
　　1.3.3 View that the world is finite in vertical direction but infinite across
　　1.3.4 View that the world is neither finite nor infinite
　1.4 Endless Equivocation *(Amaravikkhepavada)*
　　1.4.1 Held by one fearful of making a false statement
　　1.4.2 Held by one fearful of clinging
　　1.4.3 Held by one fearful of being cross-examined
　　1.4.4 Held by one who is dull and stupid
　1.5 Fortuitous Origination *(Adhiccasamuppannavada)*
　　1.5.1 Based on recollection of the arising of perception after passing away from the plane of non-percipient beings
　　1.5.2 Based on reasoning

2. **Speculations about the Future** (*Aparantakappika*)
　2.1 Percipient Immortality *(Sannivada)*, with the self immutable after death, percipient and:
　　2.1.1 Material
　　2.1.2 Immaterial
　　2.1.3 Both material and immaterial
　　2.1.4 Neither material nor immaterial
　　2.1.5 Finite
　　2.1.6 Infinite
　　2.1.7 Both finite and infinite
　　2.1.8 Neither finite nor infinite
　　2.1.9 Of uniform perception
　　2.1.10 Of diversified perception
　　2.1.11 Of limited perception
　　2.1.12 Of boundless perception
　　2.1.13 Exclusively happy
　　2.1.14 Exclusively miserable
　　2.1.15 Both happy and miserable
　　2.1.16 Neither happy nor miserable
　2.2 Non-percipient Immortality *(Asannivada)*, with the self immutable after death, non-percipient and:
　　2.2.1 Material
　　2.2.2 Immaterial
　　2.2.3 Both material and immaterial
　　2.2.4 Neither material nor immaterial
　　2.2.5 Finite
　　2.2.6 Infinite
　　2.2.7 Both finite and infinite
　　2.2.8 Neither finite nor infinite
　2.3 Neither Percipient nor Non-percipient Immortality *(N'evasanninasannivada)*, with the self immutable after death, neither percipient nor non-percipient:
　　2.3.1 Material
　　2.3.2 Immaterial
　　2.3.3 Both material and immaterial
　　2.3.4 Neither material nor immaterial
　　2.3.5 Finite
　　2.3.6 Infinite
　　2.3.7 Both finite and infinite
　　2.3.8 Neither finite nor infinite
　2.4 Annihilationism *(Ucchedavada)*
　　2.4.1 Annihilation of the self composed of the four elements
　　2.4.2 Annihilation of the divine sense-sphere self
　　2.4.3 Annihilation of the divine, fine-material-sphere self
　　2.4.4 Annihilation of the self belonging to the base of infinite space
　　2.4.5 Annihilation of the self belonging to the base of infinite consciousness
　　2.4.6 Annihilation of the self belonging to the base of thingness
　　2.4.7 Annihilation of the self belonging to the base of either perception nor non-perception
　2.5 Nibbana Here and Now *(Ditthadhammanibbanavada)*
　　2.5.1 Nibbana here and now in the enjoyment of the five trands of sense pleasure
　　2.5.2 Nibbana here and now in the first jhana
　　2.5.3 Nibbana here and now in the second jhana
　　2.5.4 Nibbana here and now in the third jhana
　　2.5.5 Nibbana here and now in the fourth jhana

6. Possible binary coding pattern

(a) Book of Changes

In the light of the above clues, the relationship to the 64-fold pattern of the Chinese *Book of Changes* calls for investigation, especially since the latter is similarly ambitious in scope. Of special interest is its early use in providing insights into the dilemmas of governance of Chinese society. The relevance of this pattern to understanding sustainable policy cycles is explored in Section TP of this Encyclopedia. The concern here is with the symbol system used to encode that pattern, not with its popular uses by those indifferent to its overall structure. It should be noted that two of the 64 elements there (denoting creativity and receptivity) have a primordial significance distinguishing them from the remaining 62. It is these two which can be suggestively associated with nirvana and samsara in the Buddhist pattern.

The *Book of Changes* originated as a set of linear signs for oracular pronouncements. At its simplest this took the form of an unbroken line for "Yes" and a broken line for "No", thus capturing the essence of the Aristotelian view and the excluded middle. Greater subtlety was required and the pattern was extended to a double line representation by combining the two basic possibilities, thus forming a set of four possible responses. It is these four which can be used to encode the 4-fold logic noted above.

The pattern of the *Book of Changes* was then further extended by adding a single broken (or unbroken) line to each of the four above. This gives the 8 possibilities, namely the 8 basic trigrams of that system. It is possible that these might prove appropriate to encoding the 8-fold sub-sets noted above.

The final extension of the pattern was by combining each of the trigrams with each other into hexagrams of six lines (broken or unbroken). It is these that are used to represent the 64 conditions of the Book of Changes.

(b) Genetic code and physical particles

Although the Book of Changes is an extremely interesting example of the use of a binary coding pattern, especially given its focus on the complex subtleties of psycho-social systems, another striking use of this same pattern is to encode the set of 64 codons of the genetic code. The binary code is of course also basic to digital computer operations, even in giving importance to sets of 64 elements. Another fundamental application of a binary system is the standard model mapping the entire range of physical particles in terms of 6 quarks in 3 pairs of 2 -- a first pair of up and down quarks, a second of charm and strange quarks and a third of bottom and top quarks (with each being harder to make than the previous pair). Each quark has an anti-matter counter-part. Mesons are two-quark particles (requiring a quark and an anti-quark, which in the case of a K-meson are an anti-strange quark and a down quark). Baryons are three-quark particles.

(c) Computer machine code

When used for computer purposes, the 64-possibilities build up from 6 "off" bit positions through a natural sequence to 6 "on" positions, thus encoding values ranging from 0 to 63. A simplistic first approximation to a pattern for the range of views would thus involve starting with a hexagram of six broken lines as representing primordial ignorance (samsara) and building up through the complete sequence to a hexagram of six unbroken lines as representing a final level of transcendental insight (corresponding to nirvana). By ignoring the first and last elements in the sequence, a correspondence could be obtained to the basic Buddhist pattern of views.

(d) Other possibilities for decoding
This is the crudest solution to mapping the views onto a pattern. It ignores difficulties created by exceptions in the above text, notably the single 2-fold set, the 5-fold set and the 7-fold set. Relocating the first of these to complete the last two, introducing there the two which were omitted, would lead to a second approximation.

Much more effort could however be devoted to thinking through the significance of the 4-fold logic and relating it to a representation using the 4 combinations of 2 lines (broken and unbroken). It is quite possible that insights from the Book of Changes might be helpful, especially in the case of the 4-fold Buddhist sets based on "material", "space", "perception" and "happiness" (see 2.1). Consider the following possible correspondences from that perspective:

- "Earth": material, finite, uniform perception, exclusively happy
- "Air": immaterial, infinite, diversified perception, exclusively miserable
- "Water": material and immaterial, finite and infinite, limited perception, happy and miserable
- "Fire": neither material nor immaterial, neither finite nor infinite, boundless perception, neither happy nor miserable.

Given the level of abstraction, it is appropriate to move beyond the particular instances, labels and metaphors, especially in order to capture meanings which are considered more active at this time. Consider the following:

- Space/Time: historical determinism ("past"), anticipation/vision ("future"), living in the present ("past and future"), proactive spontaneity ("neither past nor future")
- Subject/Object: objects without subjects, percipience without objects, subjects and objects, neither subjects nor objects.

Such an exploration could uncover ways of combining representations of the different views concerning the relationship between the three dualistic domains (materiality, objectivity, and space/time) as three pairs of two lines forming a single hexagram. These could be much more precisely linked to the views in the text. It would seem that the text contains sufficient indications to suggest that the final pattern might "lock" together in a totally unambiguous way, once the key was found. It might also provide a striking link to the insights and patterns of the Book of Changes such that each enhances the other.

As with any binary coding pattern, a finer pattern of distinctions can be obtained by adding further positions. Thus one extra would raise the number of distinctions to 128, and a second to 256.

7. Implications for sustainable human development

(a) Function of each view
It should be stressed that it is not the Buddhist approach that is being advocated here any more than that of the *Book of Changes*. As insights which have themselves stood the test of time, these are useful as an indication of directions to explore in identifying a pattern that can encompass the range of views of sustainable human development. The Buddhist approach has a strong bias in favour of a single view, one of the 64 in the derived pattern. It is important to understand the conditions under which the others may also appear politically desirable, whether or not importance is attached to different kinds and degrees of insight.

(b)
The text focuses strongly on the "self" and its perceptions. The implications would also seem to be valid for collectivities. The early mention of "lives" and the several references to "immortality" are of more immediate significance when understood metaphorically. A "life" may be understood as an unbroken period of attention or concentration. Attention to any matter may be broken by any distracting or disruptive influence onto which the attention is then shifted, whether individually or collectively. "Lives" can thus be understood as successive cycles of emergent focus and decay of attention, whether as a daily cycle of activity, a programmatic cycle of a group, or the life-cycle of an organization or of some intellectual or cultural fashion. In this sense the pattern aims to transcend the limitations of short-term concerns with a single programme, electoral cycle or business cycle. It is concerned with trans-cyclic sustainability. The original text could be rewritten to reflect such preoccupations.

(c) Traps on learning pathways
Degree of insight, area of insight and duration of insight are thus woven together to indicate the traps lying along the learning pathway from ignorant degradation (samsara) to insightful sustainability (nirvana).

(d) Forms of (un)sustainability
Perhaps most intriguing about the above pattern is the possibility that it represents a complete representation of forms of (un)sustainability, expressed at the most abstract level. It may be understood as both embodying and transcending the dualism which diminishes the significance of many systemic endeavours. In doing so it embodies increasing degrees of complexity through the later terms of the tetra-lemmic logic thus indicating the challenge to understanding and the learning process. It is as much a learning pathway (or "curriculum") as an explanation, thus denying superficial comprehension.

(e) Representation of the pattern
Having acquired a sense of the pattern, there is value in exploring ways of representing it so as to highlight features implicit in its structure which are of significance for sustainability:

- *'Mountain' Model*: Here the pattern is projected onto a tetrahedron so that the upper apex represents the sustainable condition (6 unbroken lines). The three edges leading up to it are used to represent the three dualistic domains (materiality/immateriality, knower/known, space/time). On each of those edges are four "islands of stability", indicated by the sequence of four values of the tetra-lemmic logic. This then gives the final lemma (neither A nor not-A) at the common apex and the first lemma (A) at the lower end of each edge (2 broken lines in each case). Each of the views is then defined by a triangular plane between those three edges. The initial view (ignorance) being given by the plane defining the bottom of the tetrahedron. The degree of unsustainability might then be represented by the slope of any plane (off which a coherent sphere of attention would eventually roll). Seemingly sustainable views would be those in which the plane was parallel to the base, corresponding to an equivalent degree of insight on all three dimensions. As a metaphorical mountain, the challenge is to ensure that planes are increasingly distant from the base and reduced in area, culminating in the apex position. Note that a form of sustainability may be achieved by repeatedly alternating between views (constantly correcting the tendency for a coherent sphere of attention to roll off any plane).

- *'Container (or Fortress)' Model*: The 6 (broken or unbroken) lines used to signify any particular view may be used to construct a tetrahedron, one line per edge. In this case there would be 64 ways in which the bordering edges could be defined. As a metaphorical container or fortress, the most vulnerable would be that in which the lines were all broken. The most sustainable (and least vulnerable) would be that in which the lines were all unbroken. Again a form of sustainability might be achieved by repeatedly alternating between views.

(f) Forms of intelligence
At the level of abstraction at which the 3-fold domains are defined, these may might well be understood as incorporating the distinctions currently made between three distinct and overarching intelligences (Gardner****): - object-related: ability to manipulate objects in the environment (bodily-kinaesthetic), recognize and imagine spatial relations among objects (spatial), and reason logically about things and their relations (logico-mathematical);
- object-free: competence in the sphere of language and music, which may not designate aspects of the physical environment at all;
- personal: intra-personal and inter-personal ability to assess emotional states, recognize strengths and weaknesses.

It is understood that these are developed somewhat independently by any individual. In the case of a group these might correspond to the functions of: human relations, **** savy ***

(g) Experiential stages
Similarly the progression of tetra-lemmic stages might also incorporate more experiential dimensions such as: sense of identity, encounter with otherness (opposing views or some "shadow" aspect in the Jungian sense), working relationship with otherness (toleration, etc), transcendence of the conflictual dynamics with otherness (proactive tolerance, creative detachment, etc). They may also be related to Erikson's life stages (see ***). Clearly the terms cannot make apparent the different levels of experiential significance associated with any such phases, which may be encountered and repeated at more profound levels of understanding. The corresponding insights at the group level merit exploration, for it is these that determine the emergence of more mature attitudes to effective cooperation between opposing factions.

(h) Requisite discipline to restrain problem generation
As far as achieving sustainable human development is concerned, the Buddhist pattern has direct practical implications. All effort in Buddhism is directed towards shifting understanding "upwards" in terms of the mountain model. Specific concern is given to clarification of understanding with regard to each of the four phases in the tetra-lemmic sequence. Furthermore there is specific recognition of what root problems are engendered by lack of clarification in each case. The Buddhist focus emphasizes individual meditation and discipline as the key to the successive phases of such clarification and identifies the problems in terms of personal weaknesses. But the same pattern may presumably be used to explore the collective implications, specifically to determine the kinds of discipline required to restrain tendencies to problem generation. Little attention has as yet been given to the stages of maturation of group insight through which longer-term, trans-cyclic programmes can be sustained.

(i) Bridge from the intellectual to the experiential
In contrast with western uses of the dialectic method, the tetra-lemmic approach provides a bridge from the intellectual operation of the dialectic to the experiential. It provides a way of questioning any view, including the approach itself. It distinguishes between the questionability of all views as products of involvement in a complex environment and levels of experience unmediated by such products. "Phenomenologically, one's world expands, not only incorporating the discursive as part and parcel of man's existence, but also including the nondiscursive vastness to which our categories do not and cannot come close to touching. The categories do not approximate reality, not because they are too simple, but because they are the products of particular perceptual-conceptual conditions." (MacDowell, p 9). From such a perspective it is valuable: to take account of all views, to consider all as inadequate, to accept both these positions together, and to consider that neither of these positions is adequate.

(j) Requisite variety and detachment from particulars
This 4-fold methodological perspective engenders a complex and subtle response both to the conditions of the world and to recommendations for human development. This would appear to embody the necessary complexity through which a sustainable dynamic can acquire meaningful credibility. It is therefore a subtle response to the absolutist viewpoints from which most approaches to human development derive. It both acknowledges their value and provides a freedom from their limitations, without necessarily becoming entrapped in nihilism or relativism that are themselves accepted as other views. It encourages and ensures full recognition of the value of choosing from various frameworks in order to deal with the phenomenal world. Relativism is thus only a possible characteristic of the realm of cognition but not of the experiential realm which transcends it and to which no categories apply. These two realms are held to be two aspects of the same reality, neither being more real than the other, since that distinction derives from human thought. Mundane reality is the real or transcendental reality on which concepts have been superimposed, whether through unconscious habit or, preferably, whenever pragmatically necessary. Wisdom is thus detachment from the need to cling to the elements of analysis, such as "mundane", "transcendental", "nirvana" or "wisdom".

(k) Sustainability through alternation between particulars
It may be that ultimately sustainability can only be achieved by appropriate alternation between all 64 views, weaving them into an appropriate ecology, without privileging any one view. For it is clear that in a complex psycho-social system, islands of what might be considered developmental regression and exaggeration will continually emerge and disappear in response to local conditions and environmental pressures (as is the case for any individual). But it is the pattern which connects these diverse perspectives. It is on the nature of that pattern, as a generator of viewpoints, that some focus is required. As with the genetic code or the periodic table of chemical elements, it is through that pattern that the diversity of perspectives emerges. Any new insight is thus subject to factionalization, with the possibility of 64 competing views, thus undermining the coherence of any response unless the connecting pattern is used to provide a context.

(l) Possibility of other patterns
Although the pattern explored here emphasizes 64 elements, there are undoubtedly other such patterns which can be usefully explored, whether with fewer or more elements and possibly lending themselves to other pattern logics. The point however is not the particular pattern used but much more the implications of any such pattern for sustainable development. To be of any significance it is necessary that the pattern be sufficiently complex to encode the diversity of concepts concerning sustainable development. To be of value it must facilitate the transcendence of the relatively sterile dynamics associated with polarized dialogues about sustainable development.

Insights: human impotence and potential

1. Organizational and intellectual impotence

Despite the many obvious successes of the past decades, there is much to suggest that humanity has lost its capacity to respond appropriately to the conditions which it engenders. Whilst its power to act continues to increase, its ability to act non-disruptively continues to decrease. As noted by Ervin Laszlo: *"We are applying piecemeal solutions to global problems and are fighting a holding action. We tend to forget that we are living in the midst of one of the most fundamental transformations in history."* (The Inner Lilits of Mankind, 1989)

The policies envisaged by institutions to improve the quality of their response are successful to the extent that they are isolated from the environments which their success impoverishes. Policies of wider scope and greater sensitivity are not viable under present circumstances nor under those that can be foreseen. They are undermined by political, resource and other constraints, but especially by the dead weight of an ever increasing population.

Corresponding to this bankruptcy of policies, there is an even more disappointing intellectual bankruptcy. The plethora of insightful recommendations for appropriate transformation fails to recognize the implications of such incoherence. The most powerful insights tend to acquire significant only to the extent that they ignore others which take into account different factors and dimensions. Interdisciplinarity has proved viable only in the most specific of cases.

Over the past decades hope has been placed in general systems, policy sciences, intergovernmental action, cooperation, development strategies, telecommunications, and other such approaches. They have accomplished much, but it is clear that the scope of such initiatives is inadequate to the complexity of the challenge.

Although there are many "answers" to the dilemma of the times, the very richness of proposals disguises humanity's essential impotence at this time (except ironically with respect to reproduction). Michael Marien, in a critical guide to 350 items of literature on social change proposals, noted that *"The great majority of the views on social change and alternatives are posited with little or no reference to any other views, past or present."* (Societal Directions and Alternatives, 1976). In the 1980s this is seen in the lack of substantive linkages between the Brandt Report (1980), the Brundtland Report (1987) and the South Report (1990). In 1990, Marien reviewed some 36 purportedly comprehensive agendas proposed for the USA alone. There is little interest in how people should comprehend or respond to such competing perspectives.

In this context it is to be expected that the way forward will emerge through the lowest common denominator amonst the extant proposals. The challenge of interrelating competing answers as necessary complements, to provide the requisite richness of an appropriate response, would seem to attract little interest.

2. Potential shift of perspective

Increasing frustration with the inappropriateness of intellectual and organizational responses has engendered many calls for "quantum leaps", paradigm shifts, radical changes of mind-set, and the like. Different levels of expectation are associated with such calls. For some the shift appears relatively straightforward -- merely a matter of acting with "common sense". For others it is a question of a "change of heart", leading to a more caring attitude. It may also be seen as a "shift in priorities", so that resources can be allocated more appropriately.

In response to the plethora of unrelated proposals, some see the way forward through the cultivation of consensus, especially on values. Marien (1976) argues that: *"Basically, our society lacks a comprehensive and inspiring vision, widely understood and supported, that can guide the formation of our basic policies."* Many would argue that their own group's vision should be accepted by others as the most appropriate guide to future development.

One of the core issues would seem to be not whether particular approaches, initiatives or visions are "right" or "wrong". Rather the question is how to interweave the initiatives and visions, and their advocates, such that the diverse insights into what can be understood as "right" can complement and enhance one another. But this process also needs to ensure that what is "wrong" in each initiative is counter-acted or constrained by others. The current notion of "checks and balances" is a very pale reflection of what is required. A new approach to complexity is required, although many assume that it will not be especially challenging.

Ervin Laszlo (1989) argues that: *"Regrettably, much current effort has been wasted: it has identified the wrong problems and identified them on the wrong scale...It is forgotten that not our world, but we human beings are the cause of our problems, and that only by redesigning our thinking and acting, not the world around us, can we solve them. The critical but as yet generally unrecognized issue confronting mankind is that its truly decisive limits are inner, not outer....On the personal level inner limits are constituted by the "hardening of the categories" of an age that is now behind us...On the cultural level inner limits reside in the atrophy of positive images and visions of the future of human society, and in the unwillingness to allow new realities to modify traditional conceptions of how to realize whatever ideals still hold sway today...And on the political level inner limits manifest themselves in the failure of the political will of the majority of the world's governments when it comes to actively implementing the objectively needed cooperative policies..."*

3. Addressing inner limits effectively

The challenge is to address these inner limits more effectively. It is they which constrain: the way in which strategies are envisaged; the way in which disciplines elaborate their patterns of categories and limit their integrative responses; and the possibilities for collaboration between organizations with supposedly complementary strategies.

These inner limits may be addressed: through novel approaches to human development (Section H); through exploring non-dualistic, pattern-oriented thinking (Section K); and through the pattern-exploring potential of interactive computer graphics (Section Z). But in all these cases, the strong suspicion remains that these approaches will remain inhibited by the pattern of categories through which such work is initiated -- "hardened categories" are indeed the issue. Such efforts may prove to be insufficient and too late.

4. Metaphor as an unexplored resource

It is for this reason that this Encyclopedia echoes others who perceive in metaphor (Section M) a major unexplored resource appropriate to the nature of the challenge. The argument here is that it may well provide a cognitive short-cut through which more powerful, complex and subtly appropriate initiatives can be envisaged. It may be used to by-pass the constraints of the conventional educational system, especially in developing countries and in the retrogressing areas of industrialized countries. People can learn to redesign, as they see fit, their perceptions of their environments so as to empower them to act and collaborate in new ways. Metaphors can provide the cognitive spectacles through which more appropriate uses of new technologies can be envisaged.

The desperate pleas for more encouraging social visions can be usefully seen as the search for new metaphors to provide a connecting pattern in an increasingly fragmented society. Metaphors offer an acceptable, but underdeveloped, way of operating cognitively on, and within, an increasingly oppressive reality. In this light many of the current initiatives can be challenged in terms of the adequacy of the metaphor that they currently imply.

What metaphors are implicit in the institutions and programmes of the United Nations? What metaphors are used in the effort to reconcile centralized planning and market economies? Is there a more powerful and appropriate metaphor for sustainable development than "having one's cake and eating it too"?

Insights: a new global organizational order?

The emergence of a more sustainable approach to the crises of the times might well be facilitated by exploring principles such as the following:

1. Recognition of disagreement on basic issues
People and groups (small or large) tend to disagree on basic issues when faced with complex problems and opportunities. Such disagreement often takes the form of unrestrained mutual hostility or perceived mutual irrelevance.

2. Counter-productive preoccupation with consensus
Preoccupation with achieving or imposing consensus absorbs considerable energy, alienates or represses many willing to contribute to a solution, and necessitates oversimplifications which are ultimately dangerous. Such consensus when achieved is usually of a token nature and can seldom be satisfactorily operationalized.

3. Cultivation of diversity and challenge
Social development is both the consequence and catalyst of individual human development which, although essentially undefinable, is characterized by increased ability to seek out and respond harmoniously to both diversity and challenging adversity.

4. Necessary variety of approaches
Complex problems may be understood and approached in different and seemingly contradictory ways - and the variety of such approaches tends to be essential to adequate containment and transmutation of the problem complex.

5. Appropriateness of coalitions
When coalitions can be formed on the basis of some degree of consensus this will and should be done. However, where there is resistance to such coalitions, or considerable resources are wasted on competition between coalitions, a "New Organizational Order" is vital to further success.

6. Limitation of focal centres
Conventional organizations, whether hierarchies or networks, achieve limited success by relying on performance at focal centres within domains over which consensus is maintained. The focalizing task at any such centre becomes virtually impossible, however, when the full range of harmonies and dissonances in the real world has to be encompassed. A "virtual centre" is called for.

7. New organizational order
A "New Organizational Order" may be brought into being by recognizing the fundamental distinction between local centres (focalizing local or specialized consensus) and the "unoccupied common centre" whose position is determined by the pattern of all local specialized centres constellated around it. It is the very pattern of harmonies and dissonances between the local centres which can engender the space of which the unoccupied centre is the focal reference point. This only occurs if the mutual rejection of those most strongly opposed is contained, by allowing them appropriate separation, and is thus itself used to maintain the form of the pattern.

8. Virtual common centre
The common centre can only exist and "function" by remaining free from the pattern by which it is defined. In a "New Organizational Order" communications cannot pass through such a centre or be mediated by it. They must travel along pathways through the pattern around the circumference (as is true on this planet). This permits many coalitions with profound differences of opinion to exist simultaneously (for example even as to whether it is "day" or "night" on the planet). However, it is their very complementarity within the unbounded overall pattern which maintains the stability of that pattern and contains its dynamism.

9. Higher order of consensus
Such dialectical freedom (the freedom to dialogue) can only be adequate operationally as an organizational response to the present challenges if the dialectical pattern is rendered explicit. The greater the diversity encompassed or tolerated within the pattern, the more explicit the structure of that pattern must necessarily be. Encompassing social reality in this way thus depends upon a higher order of consensus which does not itself depend upon universal consensus of a lower order at the verbal/conceptual level. However the unoccupiable central position can necessarily only be defined and understood to a very limited extent from any local centre within the pattern.

10. Appropriate complementarity
The "New Organizational Order" can only succeed by being open to the harmonious and conflictual redistribution of information and energy around the pattern as a whole. For the pattern to maintain its coherence and integrity, care must be taken to ensure the emergence of a complement to every portion of the pattern; such counter-patterns counteract and absorb each other's excesses and energize each other's evolution.

11. Alternative organizational patterns
The "New Organizational Order" cannot ultimately depend upon a single pattern to redistribute energies in response to present circumstances. Different patterns need to emerge according to the diversity to be interrelated and in response to the continuing pressures of human and social development. Familiarity with the range of patterning possibilities, and how their emergence may be facilitated, is an important factor in making this alternatives viable.

12. Alternation between patterns
The "New Organizational Order" will prove most significant when the transition between patterns can itself be made in harmony with a pattern of a yet higher order.

World Problems   P

Scope

The purpose of this section, the largest in the Encyclopedia, is to identify the complete range of world problems perceived by international constituencies, whether as a focus for their programme activities, their research, their protest, their recommendations, or as part of their belief system. An entry has been established on each. This provides a context within which the network of specific relationships perceived between these problems may also be identified.

Sub-sections

This section contains 8 sub-sections which are described on the following page.

Method

The entries are based on information obtained from international organizations, available in a wide variety of reference books, or reported in the international media. The procedures for identifying world problems are described in Section PZ. These were designed to detect both well-publicized problems as well as little-known problems, whether recognized by official bodies or not. The procedures include methods of handling hierachies of sub-problems which extend down to a level of specificity that it would be inappropriate to attempt to handle at this stage.

The procedures used in preparing this section are discussed in detail in Section PZ at the end of this volume.

Overview

Detailed comments are provided in Section PZ. As a whole, this section endeavours to present all the phenomena in society that are perceived negatively by groups transcending national frontiers. These are the phenomena which engender fear and irrational responses as well as those constituting a challenge to creative remedial action. Groups are very strongly motivated by the problems which infringe their values and arouse their indignation. As such problems are a major stimulus driving the development of society. The perceptions documented raise useful questions concerning the nature of problems, and what is meant by the "existence" of a problem, especially when other groups consider that perception irrelevant, misleading or misinformed. There is great difficulty in obtaining and editing material on problems, rather than on incidents, remedial programme action, theories, or other frameworks through which perception of problems is filtered. So to that extent, it could be argued that this section assembles information on which people collectively have great difficulty in focusing, namely information whose significance, whether deliberately or inadvertently, is collectively repressed, displaced onto some less threatening problems, or projected in the form of blame onto some other social group.

Index

A keyword index to entries is provided in Section PX.

Context

The contents of this section may be considered as complementing the other sections in ways such as the following:

Human development: By the manner in which human development is frustrated and impeded by world problems, and through the world problems engendered by the unbalanced pursuit of particular forms of human development (or by the conflict between different forms of human development).

Integrative knowledge: By the importance of integrative knowledge for comprehending the nature of the global problematique; by the manner in which that problematique calls for new kinds of integrative knowledge.

Metaphors and patterns: By the problems of communication in a global society and by the need to communicate the nature of world problems.

Transformative approaches: By the importance of such approaches in offering some new leverage in responding to the global problematique.

Human values: By the direct correspondence between disvalues and problems, and by they manner in which problems only become perceiptible in the light of the values upon which they infringe.

Reservations

The emphasis was placed on providing descriptions of less well-known problems, particularly when the extensive material available on the better known problems contained neither succinct descriptions of them nor descriptive material which could easily be reduced to succinct descriptions. The problem descriptions here represent a compilation of views from published documents (usually from international organizations) and are in no way intended as an accusation or a criticism of any particular group or country by the editors or publishers of this volume. By including or excluding particular world problems, the editors are in no way implying either approval or disapproval of the problem as conceived or as described. The same problems tend to be viewed differently by different groups in society. For one group a problem is of the utmost importance and urgency, for another the same problem is insignificant, does not exist, or is completely misconceived on the basis of available facts. Inclusion of a problem in this section is therefore not considered by the editors to mean that the problem "exists", but only that a functionally significant group of people in a number of countries believe, or claim to believe, that the problem exists on the basis of the facts available to them.

WORLD PROBLEMS

The section contains entries on 13,167 world problems. The entries are linked by 80,394 cross-references. As indicated below, it is divided into 6 sub-sections (PA through PF) containing descriptive entries. Two further sections (Sections PG and PJ) correspond to entries which are indexed (in Section PX) and cross-referenced (in Sections PA through PF), but are not printed. Entries may be located through the index (Section PX). Notes on the significance and methodology are given in Section PZ. Detailed statistics are available at the end of Volume 2 in Section Z.

Section PA:	Abstract fundamental problems	77
Section PB:	Basic universal problems	109
Section PC:	Cross-sectoral problems	137
Section PD:	Detailed problems	235
Section PE:	Emanations of other problems	425
Section PF:	Exceptional problems	627
Section PG:	Very specific problems	843
Section PJ:	Problems under consideration	845
• Section PX:	VOLUME INDEX (for Volume 1)	847
• Section PY:	Bibliography	1085
• Section PZ:	Notes	1141

Abstract fundamental problems

PA

Content

This section groups together world problems which are not usually the direct focus of international action. The problems here are characterized by their abstract or fundamental nature. For that reason, they are also especially characterized by terminological ambiguity. Despite their seemingly "fuzzy" nature, it is these problems which are frequently cited in debates and lead to recognition of the less ambiguous problems in subsequent sections.

Many of the problems here are best considered as problem complexes in that no single term adequately captures the underlying concern. Therefore, although each problem has a single main title, it also has many secondary synonymous titles. The situation is further complicated in that the words used in these titles frequently have other meanings. This results in their also being used to name other problems in this section (whether as main or secondary titles).

The terminological complexity is partially clarified by indicating cross-references from each problem to related problems in which the same words are used with somewhat different significance. An effort has also been made to show how many of these problems may be clustered within broader problems at an even higher level of abstraction. At this level of abstraction no words can be usefully used to name the problem. Such problems therefore have names with "*complex" appended to them.

This section groups 344 world problems for which there are 8,713 cross-references. Little attempt has been made to provide text descriptions to problems at this level.

Rationale

Although this information in this section is essentially "fuzzy", it is considered of value to that extent that it endeavours to map out the range of concerns which underly those with which international bodies are normally preoccupied. Because of its fuzziness, this section can only be considered as experimental. It is an experiment in handling the levels of ambiguity which bedevil communication concerning strategic issues.

Method

A general discussion of the approach used is provided in Section PZ. In contrast to the other entries in Section P, the entries here are based on work done on human values, and especially on negative values. This work is described in the section on Human Values (see Section VZ, Volume 2). Essentially it involved clustering negative value terms and determining ways to clarify, or map, the basic ambiguity by which such information is characterized. The intent was to "capture" the full spectrum of negative values because of their intimate relationship to fundamental problems.

Index

A keyword index to entries is provided in Section PX.

Comment

Detailed comments are given in the notes in Section PZ at the end of this volume.

Reservations

This sub-section is perhaps the most questionable in Section P. It can only be considered as experimental and tentative. It is designed to raise questions about the ways in which information of this degree of ambiguity and abstraction is handled. But although it deals with abstractions, it is precisely these abstractions which are most frequently used in ordinary speech and which are manipulated in many ways in international discourse.

Possible future improvements

Of special interest is the intimate relationship between the words used to name problems here and the negative values in the Human Values (Section V, Volume 2). The problems here could be explicity linked to those values, thus building a direct conceptual bridge between problems at values at the most fundamental level. There is also merit in considering ways of linking entries in this section to those in the section on Human Development (Section H, Volume 2), since many of the entries there are directly concerned with the problems at this level of abstraction.

ENTRY CONTENT AND ORGANIZATION

Ordering of entries
Entries are in **numeric order**. Entry numbers have been **allocated randomly**; they have no significance other than as a permanent point of reference to facilitate indexing, cross-referencing, and updating between editions.

Index access to entries
The location of an entry in this sub-section may be determined from the **Volume Index** (Section PX) on the basis of keywords in the name of the entry or its alternate titles.

Structure of entries
Entries may be composed of the following descriptive elements:

(a) **Entry number** This number has **no significance**, except as a convenient method of identifying the entry (particularly for indexing purposes), of filing information on it, and as an identifier to which cross-references from other entries (possibly in other Sections) may refer in this and future editions. The first letter of the entry number refers to the section of this volume in which the sub-section, denoted by the second letter, is located.

(b) **Problem name** This is printed in bold characters. It is the name selected as best indicating the nature of the problem. It may be followed by alternative problem names.

(c) **Nature** Description of the problem which attempts to identify the nature of the disruptive processes involved. The information included here, and in the following paragraphs, is compiled directly, to the extent possible, from available published documents. Where appropriate the text included may be reproduced, in a minimally edited form, from the publications of international organizations, such as those of the United Nations or its Specialized Agencies.

(d) **Incidence** Summary description of the extent of the problem which makes it of more than national significance.

(e) **Background** Describes briefly when and how the problem's importance was recognized initially, and how this recognition has evolved over time.

(f) **Claim** Stresses the special importance of this problem and why action is particularly urgent. This paragraph offers means of including statements which may deliberately exaggerate claims for the unique importance of the problem.

(g) **Counter-claim** Stresses, where appropriate, the relative insignificance or erroneous conception of the problem as described. This paragraph offers a means of including statements which may deliberately exaggerate the arguments refuting the evidence for the existence of the problem. Absence of such arguments from the text does not mean that they do not exist.

Cross-referencing of entries
At the end of any entry, there may be cross-references to other entries. These indicate the number and name of the cross-referenced entry, whether within this Section or in other Sections. There are 3 types of **hierarchical** cross-references between problems:
> **Broader** = Broader problem: more general problems of which the problem described may be considered a part. The described problem may be considered an aspect of several broader problems
> **Narrower** = Narrower problem: more specific problems which may be considered a part of the described problem
> **Related** = Related problem: problems that may be considered as associated in a hierarchically undefined way with the described problem.

There are 4 types of **functional** cross-references between problems:
> **Aggravates** = Problems aggravated by the described problem: a forward or subsequent negative causal link
> **Aggravated by** = Problems aggravating the described problem: a backward or prior negative causal link
> **Reduces** = Problems relieved, alleviated or reduced by the described problem: a forward or subsequent positive causal link
> **Reduced by** = Problems relieving or alleviating the described problem: a backward or prior positive causal link

♦ PA0005 Secrecy

Dependence on secrecy — Secretive people — Cover up — Intentional concealment — Inadequate public disclosure — Secrets

Nature Concealment of information with the intention of preventing others from learning of it, possessing it, making use of it or revealing it to others. The keeping of secrets acts like a psychic poison, alienating their possessor from the community. Secrecy can debilitate judgement by preventing criticism thus reinforcing erroneous beliefs, lowering resistance to the irrational and the pathological. It allows people to maintain facades that conceal negative traits. Whenever there is a tendency to negligence or abuse, people and especially institutions, seek to surround themselves with ever greater secrecy, often without any real justification. Long-term group practices of secrecy are especially likely to breed corruption and to spread its effects.

Incidence Secrecy is widely practised by governments, especially with respect to questions touching on national security, which may be interpreted in the very broadest sense. Anti-terrorist groups, by their nature, practice extreme secrecy. Their accountability is very limited. It is suspected that classified information contains many facts which would be very disturbing to the general public if they were made readily available. Secrecy is also practised by most institutions to prevent their strategies from becoming known to other bodies which would hinder or take advantage of them.

Claim Secrecy means impropriety. Everything secret degenerates, even the administration of justice; nothing is safe that does not show how it can bear discussion and publicity.

Counter-claim Secrecy is necessary to prevent knowledge of the strengths and weaknesses of a society from being exposed to its potential enemies. In the case of the individual it is necessary to protect privacy, intimacy and any understanding of the sacred. Every individual, group and society has secrets. It is only when the maintenance of the secret impairs the creative functioning of another is it necessary to bring them into the open.

Refs Bok, Sissela *Secrets* (1984).

Broader Unsociability (#PA6653) Uncommunicativeness (#PA7411).
Narrower Secret police (#PE6331) Media cover-up (#PD4383)
Military secrecy (#PC1144) Official secrecy (#PC1812)
Religious secrecy (#PF1106) Professional secrecy (#PD6576)
Abuse of banking secrecy (#PF5991) Hidden individual talents (#PG7887)
Corporation financial secrecy (#PE1571) Secrecy in scientific research (#PF1430)
Concealment of esoteric knowledge (#PF7077) Secrecy of medical facts and records (#PF5983)
Secrecy of national basic food stocks (#PF4763)
Suppression of information concerning environmental hazards (#PF4854).
Aggravates Ignorance (#PA5568) Abuse of power (#PB6918)
Unreported research (#PF9141) Unrecorded knowledge (#PF5728)
Social unaccountability (#PC1522) Unretrievable documents (#PF4690)
Manipulation of debates (#PD4060).
Aggravated by Avoidance of negative feedback (#PF5311).
Reduced by Informers (#PD8926).

♦ PA0072 Human death

Human mortality — Mortality

Nature Human death is the cessation of life, physical and mental, characterized by total and permanent cessation of the functions or vital actions of the human organism. In reality, two kinds of death are to be considered. First, there is somatic death, which implies the inability of the body to continue to function as an integrated entity. Without the use of artificial measures, somatic death is inevitably followed by cellular death, the cells of different organs and tissues of the body dying at different rates in accordance with their different oxygen requirements, those of the brain being the most delicate. For almost the entire history of man, therefore, death was synonymous with lack of oxygen due to cessation of cardiorespiratory function. But now those whose respiratory muscles have been paralysed by poliomyelitis can be maintained fully alive by mechanical respirators, as can victims of renal failure by renal dialysis. The function of the heart - which is but a sort of pump - can also be replace by mechanical means. But nothing can replace the functions of the brain. Once the fragile cells of the grey matter that forms its outer layer have been deprived of oxygen for more than a few minutes the person, if not the rest of his body, is irrevocably dead. He will never again think, feel, do, see, hear, smell, taste, touch, or be a person, although the action of his heart, lungs, and other vital organs may be maintained for years by artificial means and their cells continue to live. It is the technical possibility of using the organs of persons in this state to save another life that has made it urgent to arrive at a new and ethically acceptable definition of death. Increasingly, the conception of 'brain death' as constituting the essential criterion has gained ground. The activity of the brain may be measured electronically by the electroencephalograph, and when the brain has ceased to function this instrument records a tracing variously described as 'flat', 'linear', or 'isoelectric'. This conception of 'brain death' as signifying the end of the life in any meaningful sense has been so widely accepted that it has been incorporated into the legislation of 12 of the United States of America. The only controversial point is the time that should elapse before the flat tracing of the electroencephalograph should be considered irreversible.

Claim Death is an affront to the individual's right to life, liberty and the pursuit of happiness.

Counter-claim Death is the greatest psychical experience. Those reporting near-death experience, or who have been declared clinically dead but have nevertheless returned to life, report no great sorrow at leaving this life and a great reluctance or effort of will to return. In addition, death is sometimes welcomed as a liberation, because it puts an end to the strain of life or to suffering which for some people has become unbearable. It is also a 'vital' necessity to make room for new generations. Ultimately, death combats overpopulation.

Refs Becker, Ernest *The Denial of Death* (1973); Boerstler, Richard W *Letting Go* (1982); Charnaz, Kathlene C *The Social Reality of Death* (1980); Grey, Margot *Return from Death* (1985); Humphreys, S C and King, H *Mortality and Immortality* (1982); Kamerman, Jack B *Death in the Midst of Life* (1988); Kastenbaum, Robert (Ed) *Death and Dying* (1977); Krant, Melvin J *Dying and Dignity* (1974); Kubler-Ross, Elisabeth (Ed) *Death* (1975); Lonetto, Richard and Templer, Donald E *Death Anxiety* (1986); Lopez, Alan D and Ruzicka, Lado T (Eds) *Sex Differentials in Mortality*; Pontificia Accademia delle Scienze *The Artificial Prolongation of Life and The Determination of The Exact Moment of Death* (1985); Shapiro, Samuel et al *Sudden Death* (1983); Sheridan, Lilian B *Death* (1988); Woods, John *Engineered Death* (1978).

Broader Death of living creatures (#PF7043).
Narrower Suicide (#PC0417) Genocide (#PC1056) Homicide (#PD2341)
Massacres (#PD2483) War casualties (#PD4189) Infant mortality (#PC1287)
Bereaved children (#PE7691) Maternal mortality (#PD2422) Mind-induced death (#PF7918)
Inhumanity of capital punishment (#PF0399) Perinatal morbidity and mortality (#PD2387)
Inequality of life expectancy by gender (#PE1339)
Child mortality in developing countries (#PE5166)
Unequal regional distribution of deaths (#PC4312)
Unequal morbidity and mortality between countries (#PC6869).
Related Animal deaths (#PE7941) Insufficiency (#PA5473)
Exploitation of land for the burial of the dead (#PE9095)
Destruction of civilian populations and institutions (#PE8564).
Aggravates Intestacy (#PE5063) Fear of death (#PF0462)
Personal life crises (#PD4840) Indeterminacy of death (#PF0192)
Dehumanization of death (#PF2442) Imposed career interruptions (#PF4128)
Morbid preoccupation with death (#PD5086)
Uncertainty of death of missing persons (#PF0431).
Aggravated by War (#PB0593) Stroke (#PE1684) Sabotage (#PD0405)
Banditry (#PD2609) Asphyxia (#PE4104) Civil war (#PC1869)
Disasters (#PB3561) Witchcraft (#PF2099) Urban fires (#PD2211)
Human ageing (#PB0477) Aerial piracy (#PD0124) Mine disasters (#PD2278)
Food insecurity (#PB2846) Leeches as pests (#PE3660) Guerrilla warfare (#PC1738)
Terrorist bombing (#PE2368) Human cannibalism (#PF2513) Human contingency (#PF7054)
Psychotic violence (#PE7645) High mortality rate (#PJ6252) Religious sacrifice (#PD3373)
Childhood martyrdom (#PF8118) Inadequate riot control (#PD2207)
Human physical suffering (#PB5646) Hazardous remnants of war (#PF2613)
Inadequacy of civil defence (#PF0506) Human disease and disability (#PB1044)
Inadequate firearm regulation (#PD1970) Vulnerability of human organism (#PB5647)
Head hunting in tribal societies (#PF2666) Inhumane methods of riot control (#PD1156)
Inadequate safeguards against fire (#PD1631)
Denial of human rights in armed conflicts (#PC1454)
Vulnerability of women and children in emergencies (#PD1078)
Unethical experiments with drugs and medical devices (#PD2697)
Vulnerability of the elderly under states of emergency (#PD0096)
Vulnerability of the disabled during states of emergency (#PD0098).
Reduces Excessive longevity (#PJ5973) Denial of the right to die (#PF4813).
Reduced by Decrease in mortality rate (#PF0333).

♦ PA0224 Imbalance

Asymmetry

Broader Inequality (#PA6695) Distortion (#PA6790) Disagreement (#PA5982).
Narrower Fiscal and trade imbalances (#PC4879)
Imbalance between capital and technical assistance (#PE4866)
Imbalance of population growth between developed and developing countries (#PE4241).
Related Injustice of trials in absentia (#PE0424).

♦ PA0294 Human illness

Sickliness — Sickness — General poor health — General ill health — Physical and mental ill-health — Below par health — Lowered state of health

Nature Illness is the state that is perceived by the individual when he is suffering from disease or other disorders. It is an awareness of being unhealthy and is more abstract than the simple presence of symptoms, which is a manifestation. It includes the distress and discomfort associated with disease and the psychological response to the presence of disease.

Refs Bankowski, Z and Bryant, J H (Eds) *Health Policy, Ethics and Human Values* (1988); Brockington, I F and Kumar, R *Motherhood and Mental Illness* (1982); Guttentag *The Mental Health of Women* (1980); Watson, Douglas, et al *Mental Health, Substance Abuse and Deafness* (1983).

Broader Disease (#PA6799).
Narrower Shock (#PC8245) Hernia (#PE8961) Insanity (#PA7157)
Starvation (#PB1875) Health inequalities (#PC4844)
Human physical suffering (#PB5646) Human disease and disability (#PB1044).
Related Badness (#PA5454) Unkindness (#PA5643) Inexpedience (#PA7395).
Aggravates Multiple sclerosis (#PE1041).
Aggravated by Unwarranted pessimism (#PF2818)
Neglected health practices (#PD8607).

♦ PA0298 Conflict

Dependence on conflict

Nature Conflict may be defined as an incompatibility between aims or desires held by at least two parties within a social system. A party may be a person, a family, or a whole community; or it may be a class of ideas, an organization, an ethnic group, or a religion. Some of the causes of conflict are to be found in the aggressive behaviour that is almost universal among vertebrates.

Counter-claim 1. Conflict also has a positive function in helping to focus on what is really required in contrast to what is not really needed. Conflict can bring people closer together as they work to resolve their differences in a spirit of cooperation and positive intent. 2. The function of much conflict appears to be the control of food and reproduction through the control of territory. Conflict can be considered as one of the most motivating forces in our existence and as a necessary element to social life. The absence of conflict is contrary to basic human and social needs and is comparable to a state of death because there is no striving between need and need satisfaction, no oscillation between deprivation and gratification, no pendulum effect, no rhythm. It is an adynamic state, incompatible with life itself. This is why lack of conflict causes its own destruction, leading to conflict, just as conflict leads to actions aimed at attaining the state of conflictlessness.

Refs Brockner, J and Rubin, J Z *Entrapment in Escalating Conflicts* (1985); Choucri, Nazli *Multidisciplinary Perspectives on Population and Conflict* (1984).

Narrower War (#PB0593) Schism (#PF3534) Genocide (#PC1056)
Civil war (#PC1869) Imperialism (#PB0113) Social conflict (#PC0137)
Racial conflict (#PC3684) Ethnic conflict (#PC3685) Conflict of laws (#PF0216)
Social alienation (#PC2130) Economic conflict (#PC0840) Political conflict (#PC0368)
Lack of war relief (#PF0727) Religious conflict (#PC3292) Forced assimilation (#PC3293)
Ideological conflict (#PF3388) International conflict (#PB5057)
Psychological conflict (#PE5087) Conflict of information (#PF2002)
International insecurity (#PB0009) Psychological alienation (#PB0147)
Hazardous remnants of war (#PF2613) Intergovernmental disputes (#PJ5405)
Inadequacy of civil defence (#PF0506) Tribal conflict with states (#PG7758)
Pervasive fear of nuclear war (#PC3541) Organizational empire-building (#PF1232)
Conflict between minority groups (#PC3428) Destabilization of social systems (#PB5417)
Denial of human rights in armed conflicts (#PC1454)
Unsustainable development of forest lands (#PD4900)
Disruption of development by tribal warfare (#PD2191)
Conflict between government and the news media (#PE1643)
Private international trade and investment disputes (#PE8547)
Antagonism between government agencies and officials (#PE2719)
Denial of right of conscientious objection to military service (#PD1800)
Conflict of interest between governments or groups of governments (#PE8289)
Jurisdictional conflict and antagonism between international organizations (#PD0138)
Jurisdictional conflict and antagonism between government agencies within each country (#PE8308).
Related Enmity (#PA5446) Disaccord (#PA5532) Alienation (#PA3545)
Difference (#PA6698) Opposition (#PA6979) Revolution (#PA5901)
Aggression (#PA0587) Disagreement (#PA5982) Maladjustment (#PA6739)
Draft evasion (#PD0356) Class conflict (#PC1573)
Polarization of local conflicts (#PF1333).
Aggravates Chaos (#PF6836) Sabotage (#PD0405) Frustration (#PA2252)
Intolerance (#PF0860) Assassination (#PD1971) Psychic conflict (#PF4968)
Lack of cooperation (#PF2816) Undue political pressure (#PB3209)
Behavioural deterioration (#PB6321) Negative emotions and attitudes (#PA7090)
Inhumane methods of riot control (#PD1156) National insecurity and vulnerability (#PB1149)
Domination of government policy-making by short-term considerations (#PF0317).
Aggravated by Elitism (#PA1387) Distrust (#PA8653)
Injustice (#PA6486) Hero worship (#PF2650)
Fragmentation (#PA6233) Irrationality (#PA0466)
Human inequality (#PA0844) Lack of human unity (#PF2434)
Denial of human rights (#PB3121) International arms trade (#PC1358)

Discrimination in politics (#PC0934)
Lack of intersocietal resource channels (#PF2517)
Military manoeuvres in sensitive border areas (#PE3704).
Reduced by Restriction of arms supply (#PD1304).

♦ PA0429 Human violence

Human dependence on violence — Violent people — Himsa

Nature In its more limited sense, violence is an overt act of destruction, the exertion of physical force which is meant to affect another, or a type of behaviour that is designed to inflict personal injury to people or damage to property. When sanctioned by custom or tradition through the institutions of society it becomes institutionalized, especially in the most dramatic form as war. Human acts of violence may be defensive as well as offensive. When offensive they may be premeditated or not, and if not premeditated they may be provoked or unprovoked as in the case of so called senseless violence. Acts of violence may be perpetrated by individuals of any sex or age, or by groups in concerted and pre–planned acts, or spontaneously as in mob–violence. A major non–physical form is structural violence.

In a broader sense, as denoted by the Jain term himsa, violence also includes other harmful acts which do not involve physical assault. These may encompass violent thought, hurtful speech, deceit, greed, and pride or any forms of violation of personhood when applied to humans. The concept can also apply in relation to other life forms. In these broader senses, any act, whether intentional or unintentional, which depersonalizes can be an act of himsa through its transformation of the person into a mere object or be used or manipulated. Hoarding resources may not be an act of violence in its narrow sense but as an act of himsa it is a form of violence in the broader sense.

Background The Jains distinguish 432 types of himsa, some of which do not involve negative intent.

Incidence Society is most concerned about criminal violence to which it characteristically addresses more penal efforts than to anything else. To a lesser extent there is attention at national levels to juvenile violence, but here the efforts are usually seriously remedial. Civil violence statistics vary according to the presence of political confrontations, labour disputes, governmental repression of individual rights, unemployment and racial tensions.

Claim Societies may condition their population to accept violence as a legitimate solution to problems, and governments may manipulate groups of people to act violently. Violence has the characteristics of disease and the world suffers both from it and its proposed cure, the violent defence of peace by ideologies, who are prepared to annihilate each other.

Counter–claim Some forms of violence, in its broadest sense, are unavoidable. These include medical interventions, destruction of insects and habitats during harvesting. Violence may be the only form of effective self–defence.

Refs Archer, Dane and Gartner, Rosemary *Violence and Crime in Cross–National Perspectives* (1987); Commonwealth Secretariat (Ed) *Confronting Violence*; Kumar, Mahendra *Violence and Non–Violence in International Relations* (1974); Moonman, Eric (Ed) *The Violent Society*; Svalastoga, Kaare *On Deadly Violence* (1982); United Nations Educational, Scientific and Cultural Organization *Violence and Its Causes* (1980).

Broader Violence (#PA5414).
Narrower Feuds (#PE8210) Pogroms (#PJ2093) Revenge (#PF8562)
Banditry (#PD2609) Masochism (#PF3264) Brutality (#PC1987)
Harassment (#PC8558) Retaliation (#PF9181) Gang violence (#PG4444)
Violent crime (#PD4752) Civil violence (#PC4864) Video violence (#PE2224)
Rural violence (#PE3720) Youth violence (#PF7498) Violent deaths (#PD4666)
Urban violence (#PJ9519) Sexual violence (#PD3276) Family violence (#PD6881)
Blood vengeance (#PF7653) Battered husbands (#PG1898) Violent picketing (#PG5873)
Killing by humans (#PC8096) Violent revolution (#PC3229) Political violence (#PD4425)
Structural violence (#PB1935) Culture of violence (#PD6279) Violent sports fans (#PE6281)
Television violence (#PE4260) Personal covert himsa (#PF1978)
Violent demonstrations (#PG5872) Violence as a resource (#PF3994)
Violence against women (#PD0247) Indiscriminate violence (#PG1966)
Violence in comic books (#PG4262) Psychological aggression (#PG2578)
Violence as entertainment (#PD5081) Union monopoly and violence (#PG1735)
Harm to innocent bystanders (#PJ3378) Violence against prostitutes (#PE6209)
Abuses by private police forces (#PE4847) Jamming of satellite communications (#PD1244)
Criminally life endangering behaviour (#PD0437)
Violence by fanatical environmentalists (#PD5582)
Anti–social behaviour of university students (#PE7370)
Abusive treatment of patients in psychiatric hospitals (#PD0584).
Related Envy (#PA7253) Sadism (#PF3270) Unkindness (#PA5643)
Compulsion (#PA5740) Desecration (#PF9176) Inappropriateness (#PA6852).
Aggravates Mutiny (#PD2589) Victimization (#PF6987) Gunshot wounds (#PE9111)
Unreported violence (#PF4967) Inadequate firearm regulation (#PD1970)
Increase in anti–social behaviour in developing countries (#PD0329).
Aggravated by Racism (#PB1047) Dissent (#PA6838)
Militancy (#PC1090) Civil war (#PC1869)
Militarism (#PC2169) Aggression (#PA0587)
Culture shock (#PC2673) Dangerous toys (#PE1158)
Social conflict (#PC0137) Ethnic conflict (#PC3685)
Social injustice (#PC0797) Disruptive behaviour (#PD8544)
Retarded socialization (#PF2187) Inadequate riot control (#PD2207)
Violent interactive toys (#PE4297) Excessive television viewing (#PD1533)
Biologically determined aggression (#PF7490)
Environmental degradation of inner city areas (#PC2616)
Inadequate living and working conditions of immigrant labourers in industrialized countries (#PD3427).

♦ PA0430 Superstition

Superstitious people — Dependence on superstition

Nature Superstition is the belief in irrational or inexplicable phenomena as opposed to demonstrable facts and the reasoned theories of science and philosophy. Superstition has traditionally hindered progress, development and the effective use of technology. Superstition may also be a value judgement on religions or societies which have a different and possibly more 'backward' nature by technologically advanced cultures which fail to see the rites and customs performed by such societies as an integral part of their way of life, often well adapted to the environment. Superstition includes the belief in spirits, magic, taboos, witchcraft, and spiritual healing. Superstition as a value judgement applies to all primitive tribes and to less primitive groups where tradition still persists. Superstition of the same kind may exist in advanced cultures but less overtly and it may be a source of embarrassment to national governments which seek to eradicate it. Astrologists may be consulted by very eminent and cultured people in developed countries, and in Asia, but it is not a very open practice owing to the more general disbelief in it by sophisticated society, and it is frequently proscribed by law. Related are gypsy fortune–tellers at fairs and daily, weekly or monthly horoscopes which are widespread in journals. Superstition is an industry.

Claim Superstitious inclinations are evidenced in the 'mystique' and 'charisma' attributed to political leaders whose own rhetoric often includes the word 'destiny'. Political superstition is universal and has facilitated the rise of dictatorships. Religious and political superstitions have been combined in the Caribbean, Africa and Asia but also in Europe from the time of the Pythagorean Brotherhood, to the Knights Templars and the Freemasons.

Refs Daniels, Cora L and Stevans, C M *Encyclopedia of Superstitions, Folklore and the Occult Sciences of the World* (1971); Jahoda, Gustav *The Psychology of Superstition* (1970); Lasne, Sophie and Gaultier, Andre P *A Dictionary of Superstitions* (1984).

Broader Unbelief (#PA7392) Illusion (#PA6414).
Related Original sin (#PF8298) Hero worship (#PF2650) Human sacrifice (#PF2641)
Dependence on mysticism (#PF2590) Primitive secret societies (#PF2928)
Animal worship as a barrier to development (#PD2330)
Lack of understanding of spiritual healing (#PF0761)
Lack of appreciation of cultural differences (#PF2679)
Excessive medical intervention in childbirth (#PE7705).
Aggravates Bewitchment (#PF3956) Abuse of khat (#PE0912)
Underdevelopment (#PB0206) End of the world (#PF4528)
Fear of the unknown (#PF6188) Childhood martyrdom (#PF8118)
Execution of animals (#PF5415) Religious superstition (#PF1270)
Head hunting in tribal societies (#PF2666) Lack of family planning information (#PD1050)
Burdensome cost of religious ceremonies (#PF3313)
Undue religious influence on secular life (#PF3358)
Negative effects of claims of religious infallibility (#PF3376)
Irrational conscientious refusal of medical intervention (#PF0420).
Aggravated by Fear (#PA6030) Leprosy (#PE0721)
Despair (#PF4004) Tribalism (#PC1910)
Supernaturalism (#PF8433) Gypsy persecution (#PE1281)
Lack of education (#PB8645) Unexplained phenomena (#PF8352)
Religious discrimination (#PC1455).
Reduces Destruction of cultural heritage (#PC2114)
Lack of variety of social life forms (#PE8806).
Reduced by Agnosticism (#PF2333).

♦ PA0466 Irrationality

Incidence Underlying many conflicts are irrational; psychological currents. These add a powerful emotional charge to an otherwise minor problem or disagreement. These may include issues of jealousy, rivalry, attachment to past hurt.

Refs Brinton, Maurice *Irrational in Politics*.

Aggravates Conflict (#PA0298).
Aggravated by Irrationalism (#PF3399).

♦ PA0587 Aggression

Dependence on aggression — Aggressive people

Nature Aggression in human society is a highly complex phenomenon and covers a wide range of behaviour. It should be distinguished from aggressiveness. Aggressive behaviour can be considered to have a positive aspect in that it affords the natural drive to gain mastery over the external world, thus underlying the necessary attainment of individual independence and most of the greatest human achievements. It ordinarily fits the pattern of normal adjustment, while aggression tends towards maladjustment and abnormality, with behaviour that is essentially destruction oriented.

In a psychological sense, aggression refers to any manifestation of a self–assertive disposition. There is still considerable dispute as to whether it is simply a response to adverse external circumstances. It is nevertheless becoming increasingly apparent that particular styles of aggression relate to particular social contexts and roles and are thus culturally patterned.

In a political sense, aggression refers to any manifestation of an expansive policy; in a military sense to an unprovoked military attack; and in a legal sense to the use of armed force by a government in violation of an obligation under international law or treaty. In the last sense, the term has appeared in numerous treaties and official declarations since World War I, including the League of Nations covenant and the United Nations Charter. In 1933, the signatories of the 'Convention for the Definition of Aggression' agreed to define the aggressor in an international conflict as that state which is the first to commit any of the following actions: (1) declaration of war upon another state; (2) invasion by its armed forces, with or without a declaration of war, of the territory of another state; (3) attack by its land, naval, or air forces, with or without a declaration of war, on the territory, vessels, or aircraft of another state; (4) naval blockade of coasts or ports of another state; (5) provision of support to armed bands formed on its territory which have invaded the territory of another state, or refusal, notwithstanding the request of the invaded state, to take on its own territory all the measures in its power to deprive those bands of all assistance or protection. Furthermore, it was stipulated that no political, military, economic, or other considerations could serve as an excuse or justification for such acts of aggression.

Claim Man is the only vertebrate who destroys members of his own species. No other animal takes positive pleasure in the exercise of cruelty upon another of his own kind. The extremes of 'brutal' behaviour are confined to man and there is no parallel in nature to our savage treatment of each other.

The problem of war is more compelling than ever before in history and is compounded by the fear of destruction by nuclear weapons. The complexities of the circumstances which provoke war are such that no one man and no one viewpoint can possibly comprehend them all. One such complexity is the fact that instances of violations of pledged words are innumerable in the diplomatic and military history of the twentieth century. If stability in world affairs is ever to be achieved, the psychological point of view deserves equal consideration with the political, economic and other aspects. The study of human aggression and its control is, therefore, relevant to the problem of war although, alone, it cannot possibly provide a complete answer.

Counter–claim 1. Aggression is a necessary part of the biological inheritance of mammals birds and insects with which humans have to learn to co–exist. It has served and continues to serve to preserve us. 2. Except for a few pathological cases, aggression is not a basic motive for action; it is a tool of other more frightening motives, notably fanaticism. Aggression is merely a handmaiden that can be called into play once the heretics are identified and condemned, ostracized or silenced.

Refs Baron, R A *Human Aggression* (1977); Lauer, Hans E; Castelliz, K and Davies, Saunders *Aggression and Repression in the Individual and Society* (1981); Marsh, Peter and Campbell, Anne *Aggression and Violence* (1982).

Narrower Militancy (#PC1090) Militarism (#PC2169) Harassment (#PC8558)
Imperialism (#PB0113) Intimidation (#PB1992) Art vandalism (#PE5171)
Ethnic conflict (#PC3685) Physical aggression (#PG2579)
Violation of neutrality (#PC2659) International aggression (#PB0968)
Psychological aggression (#PG2578) International aggression (#PC7559)
Military and economic hegemony (#PB0318)
Aggression against nuclear power sources (#PE0403).
Related Defence (#PA5445) Tension (#PB6370) Conflict (#PA0298)
Disaccord (#PA5532) Hostility (#PB8538) Discourtesy (#PA7143)
Abuse of power (#PB6918) Loss of civility (#PC7013) Social discrimination (#PC1864)
Behavioural deterioration (#PB6321) Negative emotions and attitudes (#PA7090).
Aggravates Fear (#PA6030) Neutrality (#PF0473) Human violence (#PA0429)
Lack of human unity (#PF2434) Metal deficient diets (#PE1901)
Restriction of arms supply (#PD1304) Unjustified military defence policies (#PF1385).
Aggravated by Guilt (#PA6793) Frustration (#PA2252)
Inferiority complex (#PG2581) Retarded socialization (#PF2187)
Biologically determined aggression (#PF7490)
Military manoeuvres in sensitive border areas (#PE3704).
Reduced by Obscenity (#PF2634).

ABSTRACT FUNDAMENTAL PROBLEMS
PA0839

♦ PA0643 Pain

Dependence on pain — People in pain

Nature The ability to perceive pain constitutes a special sense which the body has evolved as a defence system, enabling it to preserve itself by avoiding damaging conditions. There are various types of pain. The most important is that caused by injury of the skin which contains special nerve–fibres for the conduction of painful impressions up the spinal cord. Internal parts of the body are less sensitive than the skin, and diseases in them usually give rise to a different sensation. These parts are not endowed with the capacity of feeling pain due to sudden injury, but inflammatory changes in these structures, and disturbances in their functions, are capable of influencing the brain so as to produce extremely severe pain. For example, any source of irritation on the course of a nerve is apt to produce the severe pain of neuralgia. Ordinary sensations of all sorts become painful when they are excessive, and thus liable to damage the organ in question.

Pain perception is not simply a function of the amount of physical damage alone. There is evidence that the intensity and quality of pain are also determined by past experience, expectation, anxiety, and the significance of the situation in which injury occurs. There is also physiological evidence that nerve impulses from the skin may be inhibited or modified during transmission through the spinal cord.

Incidence As an indication, the market for pain killers (including analgesics and headache tablets) in the UK in 1986 was £102 million.

Refs Cailliet, Rene *Hand Pain and Impairment* (1982); Kosterlitz, H W and Terenius, L Y *Pain and Society* (1980); Lewis, C S *The Problem of Pain* (1978); Sternbach, Richard A *Pain Patients* (1974); Swerdlow, M and Ventafridda, V *Cancer Pain* (1986).

Broader Disease (#PA6799).
Narrower Causalgia (#PG2694) Hyperalgesia (#PG2692) Chronic pain (#PE2694)
Abdominal pain (#PD8725) Phantom limb pain (#PG2695)
Joint pains due to torture (#PE3732)
Residual traumatic pains due to torture (#PE3798).
Related Insensibility (#PA5451) Unpleasantness (#PA7107) Human suffering (#PB5955)
Human physical suffering (#PB5646).
Aggravates Suicide (#PC0417) Fatigue (#PA0657) Attempted suicide (#PE4878)
Violence as a resource (#PF3994) Protein–energy malnutrition (#PD0339)
Death and disability from inhumane confinement (#PE5648).
Aggravated by Anxiety (#PA1635) Culture–induced fear (#PG2699)
Human disease and disability (#PB1044)
Excessive prolongation of the dying process (#PF4936)
Diseases of the skin and subcutaneous tissue (#PC8534)
Disruption of internal balance of the human body (#PE6603).
Reduces Injuries (#PB0855).
Reduced by Analgesia (#PG2701) Syringomyelia (#PG2700).

♦ PA0657 Fatigue

Nature There are basically two types of fatigue: physical (muscular) and general (mental). Physical fatigue is characterized by acute pain localized in the muscles, particularly those used for a long period in the exercise of a particular task. General fatigue is characterized by decreasing motivation and willingness to work, unusual sensitivity to situations and to other people, irrational stubbornness and disintegration of attention. Early researchers emphasized the importance of the accumulation of chemical products, such as lactic acid, in the production of fatigue. A more recent view is that fatigue is an outcome of frustration and conflict within the individual.

Incidence A survey of 200,000 workers in 10,000 private companies in Japan in 1982 showed that 72.8% of male workers complained of mental fatigue and 64.7% of physical fatigue caused by their job. The corresponding figures for female workers were 65.5% and 64.5%.

Refs Griffith, Linda L *Rattle Fatigue* (1986).

Broader Disease (#PA6799) Weakness (#PA5558) Inaction (#PA5806).
Narrower Job fatigue (#PD8052) Chronic fatigue (#PD4374)
Fatigue due to torture (#PE4229)
Human fatigue during control of complex equipment (#PE5572).
Related Apathy (#PA2360) Stress in human beings (#PC1648).
Aggravates Boredom (#PA7365) Neurosis (#PD0270)
Exhaustion (#PJ2732) Irritability (#PJ2736)
Sleep disorders (#PE2197) Multiple sclerosis (#PE1041)
Road traffic accidents (#PD0079) Psychological conflict (#PE5087)
Death and disability from inhumane confinement (#PE5648).
Aggravated by Pain (#PA0643) Fear (#PA6030)
Boring work (#PG2743) Malnutrition (#PB1498)
Climatic cold (#PD1404) Climatic heat (#PC2460)
Demoralization (#PF8446) Emotional strain (#PG2738)
Sleep deprivation (#PE2741) High altitude stress (#PD2322)
Human disease and disability (#PB1044) Health hazards of exposure to noise (#PC0268)
Excessive employment of married women (#PD3557)
International imbalance in the quality of life (#PB4993)
Health risks to workers in agricultural and livestock production (#PE0524).

♦ PA0831 Deprivation

Dependence on deprivation — Deprived people — Deprivation of basic necessities

Refs Misra, Girishwar and Tripathi, Lal Baccan *Psychological Consequences of Prolonged Deprivation* (1980).

Broader Poverty (#PA6434).
Narrower Homelessness (#PB2150) Childlessness (#PC3280)
Family breakdown (#PC2102) Enforced celibacy (#PD3371)
Cultural deprivation (#PC1351)
Deprivation of trade union funds and property (#PE8170).
Related Loss (#PA7382) Absence (#PA7270) Nonexistence (#PA5870)
Insufficiency (#PA5473) General obstacles to problem alleviation (#PF0631).
Aggravates Fear (#PA6030) Racial inequality (#PF1199)
Inequitable distribution of wealth (#PB7666) Criminally life endangering behaviour (#PD0437)
Political instability of developing countries (#PD8323)
Excessive social costs of structural adjustment in debtor developing countries (#PD8114).

♦ PA0832 Human destructiveness

Dependence on destruction — Destructive attitudes — Destructive people

Refs Fromm, Erich *Anatomy of Human Destructiveness* (1981).

Broader Destruction (#PA6542) General obstacles to problem alleviation (#PF0631).
Narrower Annihilation (#PF9169) Extermination (#PJ2839)
Property damage (#PD5859) Community damage (#PG9379)
Self–destruction (#PF8587) Amenity destruction (#PC0374)
Disruptive behaviour (#PD8544) Death of living creatures (#PF7043)
Environmental degradation (#PB6384) Homogenization of cultures (#PB1071)
Vibrations as a health hazard (#PE1145) Destruction of natural barriers (#PC1247)
Destruction of private property (#PG2842) Destruction of environmental oxygen (#PE5196)
Destruction inherent in development (#PF4829)
Inevitable destruction of natural environment by mankind (#PE2443).
Related Vandalism (#PD1350) Shallowness (#PA6993)
Human exceptionalism (#PF6730).
Aggravates Self–destructive excuses (#PF6044)
Fragmentation of the human personality (#PA0911).
Aggravated by Sabotage (#PD0405) Disasters (#PB3561)
War crimes (#PC0747) Aerial piracy (#PD0124)
Civil disorders (#PC2551) Guerrilla warfare (#PC1738)
Psychotic violence (#PE7645) Inadequate riot control (#PD2207)
International aggression (#PB0968) Hazardous remnants of war (#PF2613).

♦ PA0833 Discrimination

Dependence on discrimination — Denial of right to freedom from discrimination — Denial of right to equality

Nature Discrimination is the limitation or deprivation of the rights of certain categories of citizens based on their religion, race, ethnic origin, or sex. It is an antisocial problem that does not only affect those directly implicated and stems from hatred, false notions of racial superiority, antipathy, prejudice, ignorance, fear and intolerance.

Although discrimination is not usually persecution, in certain circumstances, especially when there is an accumulation of acts over time, or when there is an escalation in their intensity, it may then constitute persecution or lead to a well–founded fear of persecution.

Incidence Discrimination exists in employment, education, nutrition, health care, housing, immigration and migration laws, international relations, foreign aid, and in everyday social encounters.

Counter–claim Discrimination is drawing of distinctions and marking of differences. It is primarily the power or act of making a judgement, particularly in distinguishing right and wrong. It has both positive and negative aspects.

Refs Béteille, André (Ed) *Equality and Inequality*; Greenawalt, R Kent *Discrimination and Reverse Discrimination* (1983); Reardon, Betty *Discrimination* (1977); Vierdag, E W *Concept of Discrimination in International Law* (1973).

Broader Injustice (#PA6486) Negative emotions and attitudes (#PA7090).
Narrower Age discrimination (#PC2541) Ethnic discrimination (#PC3686)
Sexual discrimination (#PC2022) Social discrimination (#PC1864)
Racial discrimination (#PC0006) Dialect discrimination (#PF6016)
Cultural discrimination (#PC8344) Unequal property rights (#PJ2031)
Economic discrimination (#PC2157) Positive discrimination (#PF9539)
Legalized discrimination (#PC8949) Religious discrimination (#PC1455)
Discrimination in housing (#PD3469) Discrimination in politics (#PC0934)
Intellectual discrimination (#PF8590) Discrimination against women (#PC0308)
Discriminatory communication (#PD6804) Denial of rights to disabled (#PC3461)
Discrimination in employment (#PC0244) Discriminatory professionalism (#PC2178)
Flag discrimination in shipping (#PD0700) Discriminatory business practices (#PD8913)
Discrimination in public services (#PD8460) Discrimination against foreigners (#PD6361)
Discrimination in social services (#PC3433) Discrimination against minorities (#PC0582)
Discriminatory international order (#PB6021) Discrimination against trade unions (#PC4613)
Discrimination against adult students (#PE6258) Discriminatory imposition of standards (#PD5229)
Discrimination against foreign companies (#PD6417)
Discrimination against prisoners' families (#PE5043)
Violation of the rights of male homosexuals (#PE3882)
Discriminatory unwritten codes of behaviour (#PE7017)
Discrimination in family planning facilities (#PD1036)
Violation of the rights of female homosexuals (#PE5741)
Discrimination against indigenous populations (#PC0352)
Discrimination against use of accents of a language (#PE5141)
Discrimination against ex–prisoners and ex–detainees (#PE6929)
Tax discrimination against investment in a foreign country (#PD3047)
Denial of right to pursue spiritual well–being because of discrimination (#PE7310)
Distortion of international trade by discriminatory customs and administrative entry procedures (#PE2603).
Related Social neglect (#PB0883) Narrowmindedness (#PA7306)
Violation of human rights (#PB3860) Denial of right to justice (#PC6162)
Restrictions on freedom of movement between countries (#PC0935)
Discrimination against developing countries by the formation of regional groupings of developed countries (#PE1604).
Aggravates Traditionalism (#PF2675) Political apathy (#PC1917)
Public drunkenness (#PE2429) Cultural deprivation (#PC1351)
Cultural fragmentation (#PF0536) Impediments to marriage (#PF3343)
Lack of racial identity (#PF0684) Exploitation in employment (#PC3297)
Lack of community participation (#PF3307) Unequal distribution of social services (#PC3437)
Family poverty in industrialized countries (#PD1998)
Refusal to issue travel documents, passports, visas (#PE0325)
Domination of government policy–making by short–term considerations (#PF0317).
Aggravated by Acne (#PE3662) Taboo (#PF3310)
Hatred (#PA8487) Albinism (#PE2332)
Prejudice (#PA2173) Segregation (#PC0031)
Value erosion (#PA1782) Unreported births (#PF5381)
Incitement to hatred (#PE5952) Prejudice in children (#PD8973)
Infringement of privacy (#PB0284) Psychological alienation (#PB0147)
Denial of right to liberty (#PF0705) Refusal to grant nationality (#PF2657)
Victimization of workers' representatives (#PD1846)
Discrimination against foreigners in employment (#PD3529)
Inadequacy and insensitivity of intelligence testing (#PD1975)
Inequality of employment opportunity in developing countries (#PD1847)
Participation of transnational corporations in the apartheid system (#PE1996).

♦ PA0839 Domination

Hegemony — Dependence on domination — Dominating people — Dominating personal interests — Personal interest emphasis

Refs Knutsen, Torbjorn, L *Hegemony in the Modern International System.*

Narrower Elitism (#PA1387) Imperialism (#PB0113) Colonialism (#PC0798)
Paternalism (#PF2183) Intimidation (#PB1992) Exploitation (#PB3200)
Dictatorship (#PC1049) Power politics (#PB3202) Authoritarianism (#PB1638)
Civil disobedience (#PC0690) Cultural domination (#PG5055) Political inequality (#PC3425)
Political domination (#PC8512) Deception by government (#PD1893)
Organizational empire–building (#PF1232) Conspiracies for societal control (#PB7125)
Capture and use of wild animals as pets (#PD1179)
Deliberate lying by corporation officials (#PD4982)
Language domination by developed countries (#PD6029)
Denial of right to national self–determination (#PC1450)
National hegemony over United Nations agencies (#PF8946)
Abuse of dominant market position in international trade (#PD6002)
Restriction of outer space benefits to a limited number of developed countries (#PD0530).
Related Class conflict (#PC1573) National federalism (#PF0626)
Demeaning community self–image (#PF2093) Entrenchment of vested interests (#PD1231)
Untransposed community structures (#PF6450)
General obstacles to problem alleviation (#PF0631)
Limiting effect of individual survivalism (#PF2602)
Rivalry and disunity within developing regions (#PD0110)
Restrictions on foreign access to capital bond markets (#PD3135)
Domination of the world by territorially organized sovereign states (#PD0055)
Denial of effective national self–determination by capitalist exploitation (#PE3123)
Discrimination against developing countries by the formation of regional groupings of developed countries (#PE1604).

Aggravates Torture of animals (#PC3532) Excessive public debt (#PC2546)
Economic manipulation (#PC6875) Economic intimidation (#PC3011)
Inappropriate transplantation of industrialized country methods to developing countries (#PE1337).
Aggravated by Ethnocide (#PC1328) Freemasonry (#PF0695)
Pusillanimity (#PJ8038) Paternalistic lies (#PF7635)
Lack of self-confidence (#PF0879) Insect vectors of plant diseases (#PD7732)
Discrimination against dwarfs and midgets (#PE2635).
Reduced by Fragmentation (#PA6233) Racial discrimination in education (#PD3328).

♦ PA0844 Human inequality

Dependence on inequality

Nature Whereas within the advanced economies the trend has been towards greater economic and social equality, in the developing world development is tending to create new economic and social inequalities. In a general way, the developed countries are more egalitarian than the undeveloped because all classes of the population are better integrated into economic and social life, whereas in the less developed countries the obstacles to equality in education and the many forms of discrimination prevent certain groups from competing equitably with the others.

Claim A third of the world's population still produces and enjoys some 85 percent of the world's wealth; and the momentum of development is such that, in general, the more developed economies are still developing more rapidly than the less developed, thus further increasing disparities which are already grave. The most advanced economies still include undeveloped and even primitive sectors; virtually all the less developed economies now include highly advanced sectors: thus an increasing number of economies reproduce within themselves all the disparities and variants of the world picture.

Counter-claim Whereas the structural upheavals that occur at the outset of economic development may increase differences and inequalities, the subsequent course of economic growth, accompanied by the spread of education, the extension of social security, the diffusion of ownership of the capital of large undertakings and the increase of income from work relative to income from property, contributes to a reduction of the general inequality of income distribution.

Human beings are empirically unequal in intelligence, skills, moral qualities, physique and beauty. It is only some religious or humanistic ideal that suggests an equality at some ontological level that transcends these empirical differences.

Equality is closely related to justice and freedom and these have to be balanced against one another. Equality cannot be the sole, comprehensive social ideal. Some social inequalities are necessary in a given society and period of history but should be justified in light of individual freedom, a just society and the common good.

Refs Berthoud, Richard *Disadvantages of Inequality* (1976); Béteille, André (Ed) *Equality and Inequality*; Lewis, Michael and Beck, Bernard *The Culture of Inequality* (1978).

Broader Inequality (#PA6695).
Narrower Elitism (#PA1387) Class conflict (#PC1573) Racial inequality (#PF1199)
Social inequality (#PB0514) Economic inequality (#PC8541) Health inequalities (#PC4844)
Political inequality (#PC3425) Inequality in education (#PC3434)
International inequality (#PC9152) Distortionary tax systems (#PD3436)
Inequality of opportunity (#PC3435) Accumulation of privileges (#PF8025)
Unequal income distribution (#PD4962) Denial of right of assembly (#PC2383)
Discrimination in employment (#PC0244) Unequal property distribution (#PC3438)
Obsolete deliberative systems (#PD0975) Inequitable distribution of wealth (#PB7666)
Unequal distribution of social services (#PC3437)
Unequal distribution of fame and honours (#PF3439)
Unequal distribution of goods and services (#PE8603)
Inequitable tax systems in developing countries (#PD3046)
Area disparities in book production and distribution (#PF4385)
Unequal income distribution within developing countries (#PD7615)
Accentuated inequality between rural and urban development (#PE8569).
Related Chance (#PA6714) Injustice (#PA6486) Disagreement (#PA5982)
National federalism (#PF0626) Entrenchment of vested interests (#PD1231).
Aggravates Conflict (#PA0298) Ignorance (#PA5568) Prejudice (#PA2173)
Unjust peace (#PB7694) World anarchy (#PF2071) Human suffering (#PB5955).
Aggravated by Segregation (#PC0031) Economic intimidation (#PC3011)
Lack of models of equality (#PF8639) Underutilization of legal rights (#PF3464)
Denial of education to minorities (#PC3459)
Inequality inducing effects of television (#PD5833)
Inequality inducing effects of remote sensing systems (#PE9072).

♦ PA0857 Insecurity

Dependence on insecurity — Insecure people — Socio-economic insecurity

Nature Insecurity is a feeling of unprotectedness and helplessness against manifold anxieties arising from an all-encompassing uncertainty about one's self.

Insecurity is largely fostered by unpredictable fluctuations in which there is no cohesion. The resultant insecurities may be personal (insecurity regarding one's goals and ideals, one's abilities, one's relation to others, and the attitude one should take toward them) or societal (disintegration of society's values and mores, societal responsibilities, and societal cohesiveness).

Claim People's productivity is usually strongly influenced by the insecurity and unpredictability of their environment. Insecurity is thus exacerbated by social strife within countries and by political and military pressures or aggression of external origin.

Refs Park, James *Absurdity, Insecurity and Despair* (1975).

Narrower Social insecurity (#PC1867) Economic insecurity (#PC2020)
Emotional insecurity (#PD8262) Insecurity of property (#PC1784)
International insecurity (#PB0009) Inadequate security system (#PD6589)
Personal physical insecurity (#PD8657) Insecurity of Western marriages (#PE4985)
Prevailing community insecurity (#PD9044) Lack of airport and travel security (#PE8231)
National insecurity and vulnerability (#PB1149) European insecurity and vulnerability (#PD1863)
Military insecurity and vulnerability (#PC0541).
Related Fear (#PA6030) Danger (#PA6971) Uncertainty (#PA7309)
Psychological inertia (#PF0421) Retarded socialization (#PF2187)
Emotional dependency in marriage (#PD3244).
Aggravates Scepticism (#PF3417) Ideological apathy (#PF3392)
Lack of human unity (#PF2434) Fear of the unknown (#PF6188)
Unproductive use of resources (#PB8376).
Aggravated by Intimidation (#PB1992) Threat of war (#PF8874)
Enforced celibacy (#PD3371) Human contingency (#PF7054)
Denial of political rights (#PD8276) Psychological ungroundedness (#PF1185)
Exclusion of pre-adults from family decisions (#PE2268).

♦ PA0859 Instability

Refs Jahn, Egbert and Sakomoto, Yoshikazu (Ed) *Elements of World Instability* (1981).

Narrower Failure of materials (#PD2638) Political instability (#PC2677).
Related Danger (#PA6971) Change (#PF6605) Weakness (#PA5558)
Transience (#PA6425) Uncertainty (#PA7309) Irresolution (#PA7325)
Changeableness (#PA5490) General obstacles to problem alleviation (#PF0631).
Aggravates Chaos (#PF6836) Economic uncertainty (#PF5817)
Dangerous substances (#PC6913) Vulnerability of organisms (#PB5658)
Vulnerability of social systems (#PB2853).
Aggravated by Chance (#PA6714) Unfinished imperfect universe (#PF5716)
Unequal distribution of forces (#PJ7928) Competition in capitalist systems (#PC3125)
Excessive size of social institutions (#PF8798).

♦ PA0911 Fragmentation of the human personality

Dissociation of the human personality — Unintegrated individual personalities — Psychological fragmentation — Uncentered people — Inadequate personal integration — Fragmented selves

Nature Because of rapid economic and political innovations, changes in the nature of the family, and the decline of religion, the individual or essential self is suffering a breakdown, such that incidental or fragmented selves are being created, namely combinations of belief and emotional involvement, each of which could readily be abandoned for another. This process is exacerbated by the movement towards a mass capitalistic society in which individual selves have little input into the collective, in contrast to tribal societies where there is a more balanced ratio between communicator and receiver. This development is associated with alienation, powerlessness, meaninglessness, social isolation and self-estrangement. It is to be expected that people will increasingly search for experiences which validate personal existence by permitting the individual to feel fulfilled or at least alive. These experiences will tend to include a range of forms of violence, sexual behaviour, substance abuse, criminality and terrorism.

Claim 1. On all sides, individuals in modern societies are increasingly exposed to division, tension and discord. Social structures which defy all rules of justice and harmony have an impact on various levels of the human psyche. Such surroundings - the division of society into classes, alienation from work and its fragmented nature, the artificial opposition between manual and intellectual labour, the crises of ideologies, the disintegration of accepted myths, the dichotomies between body and mind or material and spiritual values, the manner in which education functions and is distributed to adolescents, the training given to the young, and unavoidable mass information - encourage dissociation of the elements of personality.

2. People who lack integration, wholeness and alignment, whether through guilt, destructive attitudes or wrong belief, are at war with themselves. Their physical and emotional health suffers as a consequence, as does the environment through which they endeavour to compensate for such inadequacies.

Broader Fragmentation (#PA6233).
Related Multiple personality disorder (#PE5048).
Aggravates General obstacles to problem alleviation (#PF0631).
Aggravated by Guilt (#PA6793) Negativism (#PF5950)
Wrong belief (#PJ1845) Mental illness (#PC0300)
Class consciousness (#PC3458) Alienation from work (#PD3076)
Human destructiveness (#PA0832) Disintegration of accepted myths (#PF8887)
Artificial opposition between manual and intellectual labour (#PE9114).

♦ PA1387 Elitism

Dependence on elitism — Elitist people

Refs Domhoff, G William and Dye, Thomas R (Eds) *Power Elites and Organizations* (1987); Klitgaard, Robert *Elitism and Meritocracy in Developing Countries* (1986); Pareto, Vilfredo *The Theory of the Economic and Political Elites in the Historical Scenario of the 20th Century* (1984); Walton, John *Elites and Economic Development* (1977).

Broader Injustice (#PA6486) Domination (#PA0839) Human inequality (#PA0844).
Narrower Sexism (#PC3432) Plutocracy (#PG3638) Urban bias (#PF9686)
Meritocracy (#PG3637) Racial elitism (#PG3643) Intellectualism (#PF2146)
Minority control (#PF2375) Cultural elitism (#PG3639) Military elitism (#PG3645)
Political elitism (#PE3647) Religious elitism (#PG3644) Scientific elitism (#PC1937)
Linguistic elitism (#PG3640) Elitist leadership (#PF1104) Educational elitism (#PC1527)
Ideological elitism (#PG3641) Elitist ruling classes (#PF4849)
Elitism in communist systems (#PC3170) Discriminatory professionalism (#PC2178)
Elitist intergovernmental groupings (#PD6896) Fraudulent nature of inherited titles (#PE5754)
Undisclosed control of national economies by limited number of individuals (#PF2344).
Related Prejudice (#PA2173) Segregation (#PC0031) Imperialism (#PB0113)
Freemasonry (#PF0695) Complacency (#PA1742) Social inequality (#PB0514)
Social discrimination (#PC1864) Political dictatorship (#PC0845)
Intra-state imperialism (#PC3197).
Aggravates Marxism (#PF2189) Conflict (#PA0298)
Eugenics (#PC2153) Nativism (#PF2186)
Ethnocide (#PC1328) Repression (#PB0871)
Puritanism (#PF2577) Scapegoats (#PF3332)
Intimidation (#PB1992) Anti-science (#PF2685)
Exploitation (#PB3200) Miscegenation (#PC1523)
Social conflict (#PC0137) Ethnic conflict (#PC3685)
Social injustice (#PC0797) Ethnic segregation (#PC3315)
Class consciousness (#PC3458) Lack of assimilation (#PF2132)
Arrogation of rights (#PD4680) Racial discrimination (#PC0006)
Ethnic discrimination (#PC3686) Denial of human rights (#PB3121)
Inflexible social structure (#PB1997) Denial of right of assembly (#PC2383)
Lack of community participation (#PF3307) Legalized racial discrimination (#PC3683)
Unequal political representation (#PC0655) Conflict between minority groups (#PC3428)
Underprivileged racial minorities (#PC0805) Underprivileged religious minorities (#PC2129)
Underprivileged linguistic minorities (#PC3324)
Discrepancies in human life evaluation (#PF1191)
Unequal distribution of goods and services (#PE8603)
Disruption of development by tribal warfare (#PD2191)
Use of undue influence to obstruct the administration of justice (#PE8829).
Aggravated by Destiny (#PF3111) Tribalism (#PC1910)
Feudalism (#PF2136) Massacres (#PD2483)
Obscurantism (#PF1357) Cultural invasion (#PC2548)
Political oligarchy (#PD3238) Political inequality (#PC3425)
Inequality before the law (#PC1268) Misuse of evolutionary theories (#PF3348)
Alienation in capitalist systems (#PD3112) Undemocratic political organization (#PC1015)
Race as a reinforcement of nationalism (#PF3352)
Unequal opportunities for media reception (#PD3039)
Inadequacy and insensitivity of intelligence testing (#PD1975)
Denial of the right to social security in capitalist systems (#PD3120).
Reduced by Populism (#PF3410) Anti-intellectualism (#PF1929)
Racial discrimination in education (#PD3328) Lack of participation in development (#PF3339)
Failure of individuals to participate in social processes (#PF0749).

♦ PA1635 Anxiety

Anxiety disorder — Anxiety neurosis — Dependence on anxiety — Anxious people

Nature An emotional state characterized by extreme apprehensiveness. Anxiety is a learned emotional response related to fear but differing from fear in that it tends to be of longer duration, of disproportionate intensity, more pervasive of the personality, and lacking a clear focus on an object. It also describes different degrees of susceptibility to fear. Highly anxious people are often considered to be neurotic introverts. It is more common in women than in men, and women are also more prone to psychiatric disturbances involving anxiety. While fear is a reaction to a real or threatening danger, anxiety is more typically a reaction to an unreal or imagined danger.

Soren Kierkegaard regarded anxiety as intimately related to freedom. When a person confronts all the undetermined choices yet to be made in their life, their freedom, they properly become dizzy, or anxious. This anxiety is painful but necessary. If they react only to its pain, they draw back, and thus never quite become human. But if they see the worth of the freedom despite the pain, anxiety becomes a kind of tutor as they become human, use their creative powers, and confront life as it is. Paul Tillich distinguished between ontological anxiety, like Kierkegaard, and pathological anxiety.

Counter–claim Creativity depends on anxiety as uncertainty and discomfort are the soul of creativity.

Refs Emmelkamp, P M G; Everaerd, W T A M and Kraaimaat, F (Eds) *Fresh Perspectives on Anxiety Disorders* (1989); Farrell, Alexander D *Anxiety Disorders* (1987); Kendall *Anxiety and Depression*; Last, Cynthia and Hersen, Michel (Eds) *Handbook of Anxiety Disorders* (1988); Lonetto, Richard and Templer, Donald E *Death Anxiety* (1986); Spielberger, Chas D, et al *Stress and Anxiety* (1988); Taeni, Rainer *Latent Anxiety* (1978).

Narrower Juvenile stress (#PC0877) Information anxiety (#PF7388) Anxiety resulting from torture (#PE0969).
Related Fear (#PA6030) Shame (#PF9991) Impatience (#PA6200) Difficulty (#PA5497) Unpleasantness (#PA7107) Psychic conflict (#PF4968) Stress in human beings (#PC1648) Negative emotions and attitudes (#PA7090).
Aggravates Pain (#PA0643) Phobia (#PE6354) Neurosis (#PD0270) Epilepsy (#PE0661) Headache (#PE1974) Hysteria (#PE6412) Alcohol abuse (#PD0153) Dehumanization (#PA1757) Sleep disorders (#PE2197) Psychosomatic disorders (#PD1967) Protein–energy malnutrition (#PD0339) Exclusion of pre–adults from family decisions (#PE2268).
Aggravated by Capitalism (#PC0564) Biochemical warfare (#PC1164) Inhibited grief process (#PD4918) False nuclear warfare alerts (#PF1236) Health hazards of exposure to noise (#PC0268) Unhealthy emotional responses to atomic energy (#PF0913).
Reduces Illusion of happiness (#PJ5224) Collapsed meaning of human creativity (#PF0936) Suppression of creativity and innovation (#PF0275).

♦ PA1742 Complacency

Dependence on complacency — Complacent people — Self–satisfaction — Citizen complacency

Nature Self–satisfaction may result in failure to cope with problems of various kinds. Complacency reinforces social attitudes against change, and may justify social inequalities and injustice by moralism or cynicism. It aggravates social conflict and may encourage violence as an only recourse.

Incidence Prime Minister Chamberlain's satisfaction after the Munich Accords with Chancellor Hitler was a notable example of one form of complacency. Victorian attitudes in England towards the British Empire were also complacent. There was complacency after the founding of the United Nations Organization, as there had been after that of the League of Nations; and there is complacency among the industrialized nations that they are doing all they can for development in the Third World.

Counter–claim Collective complacency has become rarer in an increasingly turbulent world, where governments and organizations of all kinds are faced with problems of a nature and scale never before known.

Narrower Moralism (#PF3379) Religious complacency (#PF1951) Government complacency (#PF6407) Complacency in science (#PD9848).
Related Apathy (#PA2360) Vanity (#PA6491) Elitism (#PA1387) Illusion (#PA6414) Decadence (#PB2542) Discontentment (#PA6011) Political apathy (#PC1917) Underdevelopment (#PB0206) Negative emotions and attitudes (#PA7090).
Aggravates Impaired vigilance (#PF6863) Fixation on partial solutions to problems (#PF9409).
Aggravated by Cynicism (#PF3418) Passivity (#PF6177) Lack of self–confidence (#PF0879).
Reduces Extremism (#PB3415).

♦ PA1757 Dehumanization

Dehumanized people

Nature Object–directed dehumanization is the tendency to view other individuals or groups as though they do not quite belong to the human race. Such a tendency can be socially dangerous. It leads to the 'rational' conclusion that it is not necessary to treat 'them' as if they were like 'us' or to be concerned with what they might be suffering if they were truly human beings.

Self–directed dehumanization includes diminution of the sense of personal responsibility for one's own actions, as well as feelings of powerlessness and of inability to question or to affect the course of events.

Claim There is an increasing tendency towards dehumanization as a response to many facets of modern life. Certain features of the preparations for nuclear war are particularly conducive to this reaction. In turn, dehumanization may increase the risk of nuclear warfare by inhibiting some of the psychological deterrents to it. Even under conditions of present conventional warfare, only a fraction of military personnel comes into face–to–face contact with the enemy. With increasing automation of weapon systems, there will be even less room for sympathy or empathy that might attenuate the suffering inflicted on others.

Broader Negative emotions and attitudes (#PA7090).
Narrower Immorality (#PA3369) Dehumanization of death (#PF2442) Dehumanization of man in the technological process (#PF5438).
Related Unkindness (#PA5643) Depersonalization (#PA6953) Ideological conflict (#PF3388) Double standards of sexual morality (#PF3259) Inadequate utilization of volunteer social service workers (#PF4892).
Aggravates Nuclear war (#PC0842) Loss of civility (#PC7013).
Aggravated by Fear (#PA6030) Anxiety (#PA1635) Torture schools (#PE2062) Double standards in morality (#PF5225) Competitive acquisition of arms (#PC1258) Proliferation of strategic nuclear arms (#PD0014) Excessive institutionalization of education (#PD0932).

♦ PA1766 Vulnerability

Nature Exposure to experiences of pain and grief. Vulnerability is not limited to or solely characteristic of oppressed groups, whether women, children or old people. People are only made vulnerable by the existence or perception of threat. The vulnerability of a form of life is roughly proportional to the weight of influences from afar, from outside the locality in which that form has achieved an ecological equilibrium.

Counter–claim A great deal of therapeutic work depends on the provision of enough safety to allow a person to be vulnerable, to feel anguish without having it denied, and to work through the pain towards healing. Choosing to be vulnerable is very different from enforced vulnerability. Appropriate self–protection is just as important as appropriate openness.

Narrower Vulnerability of organisms (#PB5658) Vulnerability of food chains (#PB2253) Vulnerability of social systems (#PB2853) Lack of protection for the vulnerable (#PB4353) National insecurity and vulnerability (#PB1149) Endangered species of plants and animals (#PB1395).
Aggravated by Crime (#PB0001).

♦ PA1782 Value erosion

Decay of traditional values

Nature In a society where a particular value is substantially realized, its status can be eroded; the value loses its attractiveness and comes to be downgraded by disenchantment and disillusionment.

Incidence Examples of value erosion in modern society are: efficiency in the era of automation, progress in an age of anxiety, economic security in a welfare state, and national independence for an emerging nation in socio–economic chaos.

Broader Negative emotions and attitudes (#PA7090).
Narrower Facile social concepts (#PF5242).
Related Erosion (#PC8193) Ideological conflict (#PF3388) Double standards of sexual morality (#PF3259).
Aggravates Discrimination (#PA0833) Social conflict (#PC0137) Family breakdown (#PC2102) Denial of human rights (#PB3121) Islamic fundamentalism (#PF6015).
Aggravated by Personal life crises (#PD4840) Double standards in morality (#PF5225) Untransposed significance of cultural tradition (#PF1373).
Reduced by Corruption (#PA1986).

♦ PA1986 Corruption

Dependence on corruption — Corrupt people — Defence against corruption

Nature Corruption is a debasement or subversion of integrity or purity and may occur in ideology or personal morality, culture, commerce, civil service (including police) or in politics. It constitutes a barrier to progress, and promotes conflict as a result of frustration and alienation; instability as a result of apathy; violence and crime. It may lead to social, ethnic, political, national and ideological disintegration and possibly revolution.

Corruption feeds on itself. People involved in the petty corruption become increasingly involved in more and larger amounts. Corruption has many victims. Public corruption offenses affect all citizens directly and personally, particularly those at th lower end of the income scale. Building inspectors paid small amounts of money to approve shoddy work results in dangerous construction, substandard electrical wiring, and inferior building materials, usually in the poorest areas of town. Judges and lawyers who line their pockets result in criminals set free to victimize citizens. Taxpayers pay for corruption. When officials are bribed to award construction contracts the cost of the bribe goes into the cost of the construction often 0ith interest. In a corrupt system, good people avoid public service. Honest contractors refuse to bid on government contracts; honest lawyers refuse to become prosecutors or judges; honest citizens avoid participating in politics either as voters or candidates.

Refs Alatas, Syed Hussein *Problem of Corruption*; Lal, H *Corruption* (1970); Lee, Rance P L *Corruption and Its Control in Hong Kong* (1981); Olusoga, S O *Management of Corruption*; Yari, Labo *Climate of Corruption* (1978).

Broader Improbity (#PA7363) Impairment (#PA6088).
Narrower Bribery (#PC2558) Corruptive crimes (#PD8679) Cultural corruption (#PC2913) Official corruption (#PC9533) Corruption of minors (#PD9481) Corruption in prisons (#PD9414) Corruption in politics (#PC0116) Ideological corruption (#PC2914) Corruption of documents (#PE7900) Malpractice in education (#PD4684) Unethical military practices (#PD7360) Institutionalized corruption (#PC9173) Unethical industrial practices (#PD2916) Unethical commercial practices (#PC2563) Unethical trade union practices (#PD4341) Unethical intellectual practices (#PC2915) Corruption in organized religion (#PC3359) Corruption of government leaders (#PC7587) Corruption in developing countries (#PD0348) Corruption of the good in human nature (#PE7917) Corruption in the entertainment industry (#PD3736) Corruption of sports and athletic competitions (#PE3754) Divergent national concepts of corrupt practices (#PF4210).
Related Vice (#PA5644) Lack of participation in development (#PF3339).
Aggravates Immorality (#PA3369) Uncleanness (#PA5459) Lack of leadership (#PF1254) Possession∗complex (#PA6686) Political disintegration (#PC3204) Loss of institutional credibility (#PF1963) Inequitable administration of justice (#PD0986) Unequal distribution of fame and honours (#PF3439) Inadequate disaster prevention and mitigation (#PF3566).
Aggravated by Fraud (#PD0486) Badness (#PA5454) Monarchy (#PF2170) Cronyism (#PF4549) Reversion (#PA5699) Complexity (#PA6468) Distortion (#PAo790) Embezzlement (#PD2688) Miseducation (#PA6393) Personality cults (#PC1123) Political appointees (#PF2031) Foreign exchange restrictions (#PF3070).
Reduces Value erosion (#PA1782).

♦ PA1999 Delay

Dependence on delay — Stalling — Delayed responses

Nature Accumulation of delay retards social processes and particularly social change processes. Action of any kind in a complex society involves a multiplicity of administrative, technical and conceptual operations. The more complex the society becomes, the greater possibility of inadvertent time lags and delays between such operations which, taken in isolation, may each be relatively efficient. These might be called delays of omission. Delays may also be deliberately created by judiciously postponing meetings or decisions. This technique of intentionally delaying decisions and actions is sometimes termed stalling. It is frequently used to gain monetary, military or diplomatic advantages. Since this forestalling of meetings, actions or decisions is deliberately obstructive, as it is intended to prevent action by others, it may be considered as delay of commission.

Incidence Delays and stalling on the international level have affected decolonization, for example, and repayments of loans. Arms reduction negotiations may be heavily influenced by delaying tactics.

Broader Limited access to social benefits (#PF1303).
Narrower Travel delays (#PE1977) Purpura in animals (#PG4342) Government inaction (#PC3950) Communication delays (#PF4453) Policy–making delays (#PF8989) Administrative delays (#PC2550) Non–recognition of problems (#PF8112) Delay in project implementation (#PF1470) Delay in administration of justice (#PF1487) Delays in community building programmes (#PF6502) Delays in delivery of goods and services (#PE3928) Delays in implementation of social change (#PC6989) Delays in centralized collective bargaining (#PF0685) Delays in elaboration of remedial legislation (#PC1613).
Related Non–payment of reparations by government (#PE4446).
Aggravates Wasted time (#PF8993) Wasted waiting time (#PF1761).
Aggravated by Strikes (#PD0694).

♦ PA2159 Human dependence

Nature Dependency in individuals relates to inability to make decisions, and the inclination to rely on others to initiate or lead in behaviour, while requiring them to be supportive, either materially, psychologically or emotionally. Dependency is normal for a child, and in some measure is normal for the infirm and elderly. In adults, dependent behaviour may prevent the development of full individuality. Dependent nations exhibit all of the foregoing traits. In addition, the well–known compensating psychology of dependence emerges among the pendant nations, that is, very demanding, very critical and accusative behaviour.

PA2159

Claim Dependence, like dominance, is a fact of life in world affairs. The absence of a "contract" prevents effective symbiosis, with responsiveness and sensitivity on the part of the dominant and responsibility and honour on the part of the subordinate. A muted tension exists between local and national governments; in the USA and the USSR between the governments of states and autonomous regions and Washington or Moscow; and childish reactions of weaker to stronger nations may be cited in the attitude of some Western European nations to the military might of the USA, of Commonwealth countries towards the UK, and the previous French colonies and territories towards France.

Narrower Paternalism (#PF2183) Dependency of women (#PC3426)
Parental over-protectiveness (#PF5255) Emotional dependency in marriage (#PD3244).
Aggravates National disintegration (#PB3384)
Vulnerability of the elderly under states of emergency (#PD0096)
Vulnerability of small nations to foreign intervention (#PD2374).
Aggravated by Slave trade (#PC0130) Bride-price (#PF3290)
Forced marriage (#PD1915) White slave trade (#PD3303)
Psychological inertia (#PF0421) Denial of rights to disabled (#PC3461)
Physically handicapped children (#PD0196) Capitalist political imperialism (#PC3193)
Expropriation of land from indigenous populations (#PC3304)
Vulnerability of the disabled during states of emergency (#PD0098).

♦ PA2173 Prejudice

Dependence on prejudice — Prejudice people

Nature Prejudice is the unsubstantiated prejudgement of an individual or group, favourable or unfavourable in character, tending to action in a consonant direction. Even favourable prejudices should be discouraged since they too represent unwarranted generalizations, mostly of an irrational nature. Although prejudice may apply to objects as disparate as trade-union leaders, women, or exotic foods, in practice it has been considered as dealing primarily with populations or ethnic groups distinguished by the possession of specific inherited physical characteristics (race or sex), or by differences in language, religion, culture, national origin, or any combination of these.

Counter-claim Prejudgements are normal and common because perception and cognition require the placing of particular items into more general categories. As long as the prejudgement is adjusted to correspond with reality no harm is done, and the process of thinking may be made more efficient.

Refs Keith, Arthur *The Place of Prejudice in Modern Civilization* (1982); Nisbet, Robert *Prejudices* (1982); Nugent, Ward J *Prejudice* (1988).

Narrower Taboo (#PF3310) Caste system (#PC1968) Racial prejudice (#PC8773)
National prejudice (#PG4368) Economic prejudice (#PG4371) Cultural prejudice (#PC8520)
Religious prejudice (#PD4365) Political prejudice (#PC8641)
Biased legal systems (#PF8065) Sexual discrimination (#PC2022)
Skin colour prejudice (#PC8774) Prejudice in children (#PD8973)
Intellectual prejudice (#PC3406) Prejudice against animals (#PC0507)
Discrimination in housing (#PD3469) Prejudice against minorities (#PC8494)
Disruptive personal prejudices (#PG9377) Prejudice against other languages (#PD8800).
Related Elitism (#PA1387) Badness (#PA5454) Misjudgement (#PA6607)
Inexpedience (#PA7395) Narrowmindedness (#PA7306)
Negative emotions and attitudes (#PA7090).
Aggravates Segregation (#PC0031) Antisemitism (#PE2131)
Social stigma (#PD0884) Culture shock (#PC2673)
Discrimination (#PA0833) Traditionalism (#PF2676)
Social conflict (#PC0137) Ethnic conflict (#PC3685)
Social injustice (#PC0797) Political apathy (#PC1917)
Linguistic purism (#PF1954) Social alienation (#PC2130)
Religious conflict (#PC3292) Ethnic segregation (#PC3315)
Racial stereotyping (#PF5452) Scientific censorship (#PD1709)
Social discrimination (#PC1864) Refusal to participate (#PF3226)
Lack of social mobility (#PF2195) Underprivileged minorities (#PC3424)
Inflexible social structure (#PB1997) Discrimination against women (#PC0308)
Religious conflict between sects (#PC3363) Underprivileged racial minorities (#PC0805)
Inadequate objectivity of institutions (#PF6691)
Lack of understanding of spiritual healing (#PF0761)
Inadequate cultural integration of immigrants (#PC1532)
Discrimination against foreigners in employment (#PD3529)
Incompatibility of traditional and new technologies (#PE3337)
Inadequacy and insensitivity of intelligence testing (#PD1975)
Inadequate and inaccurate textbooks and reference books (#PD2716).
Aggravated by Human inequality (#PA0844) Social inequality (#PB0514)
Racial segregation (#PC3688) Social fragmentation (#PF1324)
Fear of the abnormal (#PF7029) Psychological alienation (#PB0147)
Compulsory indoctrination (#PD3097) Denial of right to liberty (#PF0705)
Biased presentation of news (#PD1718) Bias in children's literature (#PD4773)
Lack of scientific investigation (#PF2720) Conflicting sense of sexual identity (#PF1246)
Biased and inaccurate history textbooks (#PD2082)
Biased and inaccurate geography textbooks (#PF1780).
Reduces Inadequate sense of personal identity (#PF1934).

♦ PA2252 Frustration

Dependence on frustration — Frustrated people

Nature Generally speaking, frustration may be seen as a limiting case of conflict, with only one goal state, and one actor, the goal state being blocked. From the psychoanalytic standpoint, frustration generally refers to the denial of gratification by reality.

Counter-claim Some degree of frustration seems to be necessary to activate the maturation process of the individual.

Refs Maier, Norman R *Frustration* (1982).

Related Defeat (#PA7289) Difficulty (#PA5497) Inexpectation (#PA5527)
Maladjustment (#PA6739) Psychological disorders (#PD8375)
Isolation of parent-child relationship (#PC0600).
Aggravates Neurosis (#PD0270) Obscenity (#PF2634)
Aggression (#PA0587) Withdrawal (#PF4402)
Parental permissiveness (#PD5344)
Exclusion of pre-adults from family decisions (#PE2268).
Aggravated by Conflict (#PA0298) Hypocrisy (#PF3377)
Breach of promise (#PF7150) Over-qualification (#PF3462)
Wasted woman power (#PF3690) Personal life crises (#PD4840)
Lack of racial identity (#PF0684) Denial of right to liberty (#PF0705)
Imbalance in the human sex ratio (#PF1128) Ill effects of educational failure (#PF2013)
Health hazards of exposure to noise (#PC0268) Inappropriate education of graduates (#PF1905)
Unequal distribution of fame and honours (#PF3439)
Ignorance of nonverbal communication skills (#PE0533)
Incompatibility of traditional and new technologies (#PE3337).

♦ PA2360 Apathy

Dependence on apathy — Apathetic people — Lack of motivation — Demotivation

Nature Apathy is a mental state in which the sufferer lacks the desire, will or energy to engage in any activity, whether intellectual or physical. It is variously called indifference, boredom, lassitude, languor, listlessness, laziness, lethargy or inertia. It may be symptom of mental disorder. Collectively, apathy may be expressed in social, economic, political or ideological paralysis, with all the available energy for change locked up in the institutions, systems and structures of society.

Narrower Lethargy (#PJ4782) Social apathy (#PC3412) Economic apathy (#PC3413)
Apathy of youth (#PF5949) Official apathy (#PF9459) Political apathy (#PC1917)
Religious apathy (#PC3414) Political inertia (#PC1907) Motivational death (#PF1948)
Ideological apathy (#PF3392) Neglect of the aged (#PD8945) Psychological inertia (#PF0421)
Inadequate hero images (#PF2834) Refusal to participate (#PF3226)
Interpersonal estrangement (#PB0034) Apathy in developing countries (#PD8047)
Lack of motivation in leadership development (#PF2208)
Apathy toward improvement of urban life styles (#PE8477)
Demoralizing constraints on housing rehabilitation (#PE2451)
External dependence and vulnerability of socialist countries (#PD1104)
International nongovernmental organization membership apathy (#PE7092)
Decline in communal spirit and village solidarity in developing countries (#PE0835)
Absence of convincing symbols connecting the individual's life to the cosmos (#PF1081).
Related Fatigue (#PA0657) Boredom (#PA7365) Avoidance (#PA6379)
Quiescence (#PA6444) Complacency (#PA1742) Incuriosity (#PA6598)
Indifference (#PA7604) Hopelessness (#PA6099) Unfeelingness (#PA7364)
Citizen apathy (#PF2421) Lack of community participation (#PF3307)
Inadequate sense of personal identity (#PF1934).
Aggravates Vice (#PB7743) Neglect (#PA5438) Extremism (#PB3415)
Mediocrity (#PF3900) Scepticism (#PF3417) Slave trade (#PC0130)
Absenteeism (#PD1634) Prostitution (#PD0693) Class conflict (#PC1573)
Student revolt (#PC2052) Social breakdown (#PB2496) Underdevelopment (#PB0206)
Social disintegration (#PC3309) Lack of political will (#PC5180)
Resignation to problems (#PF8781) Psychological alienation (#PB0147)
Rural poverty in developing countries (#PD4125).
Aggravated by Feudalism (#PF2136) Intimidation (#PB1992)
Mental depression (#PC0799) Culture of violence (#PD6279)
Totalitarian democracy (#PD3213) Overexposure of issues (#PG4785)
Inadequacy of religion (#PF2005) Nutritional deficiencies (#PC0382)
Abuse of coca and cocaine (#PD2363) Exploitation in employment (#PC3297)
Biased presentation of news (#PD1718) Inadequacy of social doctrine (#PF3398)
Radio and television censorship (#PD3029) Underutilization of legal rights (#PF3464)
Inadequate ideological frameworks (#PD0065) Loss of institutional credibility (#PF1963)
Malnutrition among indigenous peoples (#PC3319)
Ignorance of nonverbal communication skills (#PE0533)
Exploitation of indigenous populations in employment (#PD1092).
Reduced by Militancy (#PC1090).

♦ PA2658 Negligence

Dependence on negligence — Negligent people — Carelessness — Criminal negligence

Nature Negligence is a failure to give thought or to pay attention to the risks inherent in one's actions and to take the appropriate precautions against these risks. A negligent act may not result in harm but it, nevertheless, is an act of carelessness. Recklessness, involving the awareness of high risk, is contrasted with negligence in which there should have been awareness of the existence of such risk. Negligence refers to circumstances in which a reasonable person would have been aware of the risk.

Narrower Negligence in manslaughter (#PE0437)
Unethical professional practices (#PC8019)
Carelessness in dealing with infectious patients (#PE9105)
Increase in insurance claims for medical negligence (#PE4329)
Criminal negligence in performing socialist responsibilities (#PE5538).
Related Torts (#PD9022) Error (#PA6180) Chance (#PF7114)
Neglect (#PA5438) Disorder (#PA7361) Avoidance (#PA6379)
Inattention (#PA6247) Recklessness (#PD5349) Unpreparedness (#PA7341)
Irresponsibility (#PA8658).
Aggravates Inadequate protection and preservation of cultural property (#PF7542).
Aggravated by Lack of care (#PF4646) Unskillfulness (#PA7232)
Avoidance of work (#PC5528) Negative emotions and attitudes (#PA7090)
Human errors and miscalculations (#PF3702).

♦ PA3275 Sloth

Accidie — Acidy — Apathy

Nature Sloth is apathy and indifference, lack of interest and lack of concern, lack of engagement and lack of emotion. It is first alienation of the self from the world and then self-alienation. Finally it is withdrawal from all participation in the care of others and care of oneself. Sloth has both passive and active forms. In its passive form accidie results in lethargy, lifelessness and paralysis of the will. It believes in nothing. It cares and lives for nothing. It finds pleasure in nothing, and dislikes nothing. It finds meaning in nothing. It keeps itself alive because there is nothing for which to give its life. In its active forms sloth results in boredom and spiritual restlessness. Without hope, the slothful person makes life troublesome for others. His desires are distorted toward indifference, indifference about suffering of others or being the cause of distress. He lacks care for life in any of its manifestations. He is bored, impoverished and self-absorbed.

Broader Sin (#PF0641).
Related Envy (#PA7253) Lust (#PA4673) Anger (#PA7797)
Pride (#PA7599) Avarice (#PA6999) Idleness (#PA7710)
Gluttony (#PA9638) Original sin (#PF8298).

♦ PA3369 Immorality

Dependence on immorality — Immoral people

Nature Immorality - taken to include conduct or thinking contrary to established standards of morality, from which they may differ according to race, culture and creed - may include acts of violence, sexual misconduct, profanity, sacrilege or other acts considered subversive, perverse or harmful; it also encompasses ideological deviation, including heresy, subversive literature and pornography, or obscenity of any kind. It may even include such afflictions as alcoholism and drug addiction.

Broader Vice (#PA5644) Improbity (#PA7363) Dehumanization (#PA1757).
Narrower Cheating (#PJ7991) Obscenity (#PF2634) Scientific fraud (#PF1602)
Sexual deviation (#PD2198) Social deviation (#PC3452) Unmarried mothers (#PD0902)
Sexual immorality (#PF2687) Unmarried parents (#PD3257) Immoral literature (#PF1384)
Unwanted pregnancies (#PF2859) Illegitimate children (#PC1874)
Immoral public policy (#PF4753) Single parent families (#PD2681)
Ideological corruption (#PC2914) Abuse of confidentiality (#PF7633)
Abuse of scientific power (#PF2692) Interpersonal estrangement (#PB0034)
Inhumane scientific activity (#PC1449) Unethical commercial practices (#PC2563)
Nepotism in developing countries (#PD1672) Unethical intellectual practices (#PC2915)
Execution of inappropriate orders (#PF2418)
Moral offences in heterosexual pairing (#PG6445)
Circumvention of duties and assessments (#PD4882)
Vulnerability of marriage as an institution (#PF1870)
Distortion of corporation financial statements (#PE2032)
Irresponsible scientific and technological activity (#PC1153).
Related Moralism (#PF3379) Amoralism (#PF3349) Decadence (#PB2542)
Permissiveness (#PF1252) Abuse of credit (#PF2166) Mental pollution (#PB6248).
Aggravates Divorce (#PF2100) Banditry (#PD2609) War crimes (#PC0747)
Draft evasion (#PD0356) Human suffering (#PB5955) Social disintegration (#PC3309)

ABSTRACT FUNDAMENTAL PROBLEMS
PA5454

Behavioural deterioration (#PB6321) Film and cinema censorship (#PD3032)
Unethical personal relationships (#PF8759)
Denial of human rights in armed conflicts (#PC1454).
Aggravated by Ugliness (#PA7240) Corruption (#PA1986)
Indecent art (#PE5042) Impure thoughts (#PF5205)
Medical materialism (#PJp913) Retarded socialization (#PF2187)
Socio–economic poverty (#PB0388) Collapse of common values (#PF1118)
Corporation financial secrecy (#PE1571)
Denial of right to a people to freely dispose of natural wealth (#PE6955).

♦ PA3545 Alienation

Nature The term alienation has been used by philosophers, psychologists and sociologists to refer to a very wide range of psycho–social disorders including: loss of self, anxiety, anomie, despair, depersonalization, rootlessness, apathy, social disorganization, loneliness, atomization, meaninglessness, isolation, pessimism, and the loss of beliefs or values. Basically, alienation refers to an individual sense of estrangement from society; or to the state of isolation of those people who, because of their customs, allegiances, behaviour, race or other factors, are marked out by those with whom they mix as "different", strange and sometimes unacceptable. Thus, certain people, races, social classes or groups are alien for various reasons from the point of view of the surrounding society. The ensuing sense of estrangement or self–alienation involves the loss of a sense of identity, a feeling of powerlessness to affect social change and of depersonalization in a large and bureaucratic society.

Incidence Alienation has become a central notion of contemporary sociology.

Claim Alienation as a social process is inherent in antagonistic class society. Its main characteristic is the transformation of human work into a dominating, independent force that is hostile to the individual. The principal causes of alienation are the antagonistic division of labour and private property.

Refs Bier, William C (Eds) *Alienation* (1972); Geyer, Felix and Schweitzer, David (Eds) *Alienation* (1981); Geyer, R F and Schweitzer, D R (Eds) *Theories of Alienation* (1976); Hefner, Philip and Schroeder, W Widick (Eds) *Belonging and Alienation* (1976); Kanungo, Rabindra N *Alienation* (1982); Ludz, Peter C *Alienation as a Concept in the Social Sciences* (1975); Novack, George and Mandel, Ernest *Marxist Theory of Alienation* (1973).

Narrower Social alienation (#PC2130) Cultural alienation (#PC5088)
Political alienation (#PC3227) Psychological alienation (#PB0147)
Alienation in capitalist systems (#PD3112)
Inadequate sense of personal identity (#PF1934).
Related Enmity (#PA5446) Dissent (#PA6838) Duality (#PA7339)
Insanity (#PA7157) Conflict (#PA0298) Rhinitis (#PG5704)
Reversion (#PA5699) Opposition (#PA6979) Loneliness (#PF2386)
Dissatisfaction (#PA8886) Isolation of parent–child relationship (#PC0600)
Failure of individuals to participate in social processes (#PF0749).
Aggravates Youth gangs (#PD2682) Mental suffering (#PB5680)
Mental disorders (#PD9131) Underground press (#PD2366)
Motivational death (#PF1948) Social fragmentation (#PF1324)
Family disorganization (#PC2151).
Aggravated by Fear (#PA6030) Absurdity (#PF6991)
Family breakdown (#PC2102) Lack of human unity (#PF2434)
Personal life crises (#PD4840) Denial of human rights (#PB3121)
Ideological corruption (#PC2914) Discrimination in social services (#PC3433)
Loss of institutional credibility (#PF1963) Undemocratic political organization (#PC1015)
Inadequate welfare services for the deaf (#PD0601)
Inadequate facilities for children's play (#PD0549)
Vulnerability of marriage as an institution (#PF1870)
Excessive institutionalization of education (#PD0932).

♦ PA3917 Stagnation

Broader Inaction (#PA5806) Quiescence (#PA6444).
Narrower Political stagnation (#PC2494) Untransposed community structures (#PF6450).
Aggravated by Scientific censorship (#PD1709).

♦ PA3981 Unnaturalness

Narrower Perversion (#PB0869).

♦ PA4313 Accumulation

Acquisitiveness
Narrower Accumulation of titles (#PFp198) Accumulation of capital (#PC5225)
Accumulation of property (#PC8346) Accumulation of knowledge (#PF2376)
Accumulation of functions (#PF4174) Accumulation of privileges (#PF8025)
Accumulation of recognized merit (#PF7315).
Related Uncontrolled use of computer data (#PF4176).
Aggravated by Inappropriate ambitions (#PF9852)
Psychological ungroundedness (#PF1185).

♦ PA4565 Dependence

Dependency

Refs Geyer, R F and Zouwen, J van der *Dependence and Inequality* (1982).
Broader Restraint (#PA7296) Unconventionality (#PA7273).
Narrower Dependence on religion (#PF0150) Physically dependent people (#PD7238)
Lack of protection for the vulnerable (#PB4353).
Aggravated by Homelessness (#PB2150).

♦ PA4673 Lust

Nature Lust is using sexuality for the purposes of self enhancement and self satisfaction. Lust is not simply sexuality nor sexual drives; it is the perversion of these necessary and creative aspects of humanness. Lust is self destructive in the sense that it denies the ability to make choices about partners, it takes whoever is available. It denies the need and ability to give and receive. It only takes. Lust is destructive of the other. Its end is neither procreation nor an expression of love. Lust does not desire continuity of relationship but immediate gratification. At its root is pride and covetousness, pride in the desire to control and covetousness in the need to possess. Lust in its expression is cruel, at the least in the sense of being exploitative and a betrayal of the deepest levels of trust. It sometimes evokes passion in others only to reject them. At its worst sexual lust is rape. Lust is destructive of society. It refuses to accept responsibility for gratification of desires, sexual acts nor overwhelming desires, such as, the lust for power.
Broader Sin (#PF0641).
Related Envy (#PA7253) Anger (#PA7797) Pride (#PA7599)
Sloth (#PA3275) Avarice (#PA6999) Gluttony (#PA9638)
Melancholy (#PF7756) Promiscuity (#PC0745) Original sin (#PF8298)
Impure thoughts (#PF5205).
Aggravates Sexual immorality (#PF2687) Paternal negligence (#PD7297).

♦ PA5414 Violence

Inter–species violence — Intra–species violence

Claim It has not yet been determined why chimpanzees, genetically the closest relative of man, under some conditions live peacefully and at other times practice murderous genocidal and cannibalistic "warfare". Chimpanzees do engage in predation of other species, and it is therefore an open question how far this kind of violence is related to their intra–specific violence; particularly since they often eat the other chimpanzees that they kill. The exact interplay of innate violent predispositions, ecology, resource competition, mate competition, predation and territoriality are not yet understood.

Counter–claim Although fighting occurs widely throughout animals species, only a few cases of destructive intra–species fighting between organized groups of have ever been reported among naturally living species, and none of these involve the use of tools designed to be weapons. Normal predatory feeding upon other species cannot be equated with intra–species violence. Warfare is a peculiarly human phenomenon and does not occur in other animals.

Refs Dick, James C *Violence and Oppression* (1979); Hollon, W Eugene *Frontier Violence* (1974).
Narrower Human violence (#PA0429) Dangerous animals (#PC2321).

♦ PA5438 Neglect

Abandonment — Carelessness — Clumsiness — Default — Dereliction — Disregard — Forgetfulness — Haphazardness — Hastiness — Heedlessness — Inaccuracy — Inadvertence — Inattention — Inconsiderateness — Indifference — Insouciance — Lapse — Laxity — Laxness — Laziness — Looseness — Malingering — Messiness — Neglect — Negligence — Nonfeasance — Oblivion — Omission — Oversight — Permissiveness — Tactlessness — Thoughtlessness — Unpreparedness
Narrower Social neglect (#PB0883) Neglected health practices (#PD8607)
Neglect of sexual health of women (#PF5147)
Damage to infant brains from malnutrition and insufficient stimuli (#PE4874).
Related Envy (#PA7253) Fear (#PA6030) Vice (#PA5644)
Error (#PA6180) Chance (#PA6714) Refusal (#PA7321)
Absence (#PA7270) Poverty (#PA6434) Impiety (#PA6058)
Relapse (#PA5619) Badness (#PA5454) Idleness (#PA7710)
Disorder (#PA7361) Ugliness (#PA7240) Rashness (#PA7115)
Inaction (#PA5806) Improbity (#PA7363) Restraint (#PA7296)
Necessity (#PA6387) Avoidance (#PA6379) Disrespect (#PA6822)
Inaccuracy (#PF7905) Negligence (#PA2658) Opposition (#PA6979)
Quiescence (#PA6444) Regression (#PA6338) Inelegance (#PA6312)
Impatience (#PA6200) Impairment (#PA6088) Compulsion (#PA5740)
Unkindness (#PA5643) Unchastity (#PA5612) Malingering (#PE7701)
Abandonment (#PA7685) Uncertainty (#PA7309) Discourtesy (#PA7143)
Incuriosity (#PA6598) Inattention (#PA6247) Shortcoming (#PA6041)
Uncleanness (#PA5459) Lack of care (#PF4646) Irresolution (#PA7325)
Disobedience (#PA7250) Untimeliness (#PA7006) Imperfection (#PA6997)
Formlessness (#PA6900) Unimportance (#PA5942) Unfeelingness (#PA7364)
Forgetfulness (#PA6651) Insufficiency (#PA5473) Inhospitality (#PA5458)
Unpreparedness (#PF8176) Unskillfulness (#PA7232) Permissiveness (#PF1252)
Unintelligence (#PA7371) Unpreparedness (#PA7341) Disintegration (#PA6858)
Incompleteness (#PA6652) Thoughtlessness (#PA6940) Attitude*complex (#PA6983)
Indiscrimination (#PA6446) Nonaccomplishment (#PA6662) Misrepresentation (#PA6644)
Uncommunicativeness (#PA7411).
Aggravates Risk (#PF7580).
Aggravated by Apathy (#PA2360).

♦ PA5445 Defence

Aggression — Combativeness — Defensiveness — Invasion — Offence

Refs Vaillant, George E *Empirical Studies of Ego Mechanisms of Defense* (1986).
Broader Interaction*complex (#PA6429).
Narrower Invasion (#PD8779).
Related Envy (#PA7253) Vice (#PA5644) Reason (#PA5502)
Intrusion (#PA6862) Disaccord (#PA5532) Aggression (#PA0587)
Disrespect (#PA6822) Illegality (#PA5952) Discourtesy (#PA7143)
Disobedience (#PA7250).

♦ PA5446 Enmity

Acrimony — Alienation — Animosity — Antagonism — Antipathy — Belligerence — Bitterness — Coldness — Conflict — Disaccord — Disaffinity — Disunity — Enmity — Friction — Hatred — Hostility — Incompatibility — Inhospitability — Irreconcilability — Malevolence — Malice — Quarrelsomeness — Repugnance — Soreness — Sourness — Strain — Tension — Unfriendliness — Venom — Virulence
Broader Socialization*complex (#PA6373).
Narrower Malice (#PF5901) Friction (#PF1691) Hostility (#PB8538).
Related Hate (#PA7338) Envy (#PA7253) Cold (#PA6956)
Fear (#PA6030) Death (#PA7055) Reason (#PA5502)
Tension (#PB6370) Duality (#PA7339) Dissent (#PA6838)
Badness (#PA5454) Conflict (#PA0298) Insanity (#PA7157)
Inaction (#PA5806) Antipathy (#PA8810) Vengeance (#PA6606)
Avoidance (#PA6379) Repulsion (#PA5799) Reversion (#PA5699)
Disaccord (#PA5532) Alienation (#PA3545) Moderation (#PA7156)
Opposition (#PA6979) Difference (#PA6698) Impairment (#PA6088)
Unkindness (#PA5643) Separation (#PA7236) Discourtesy (#PA7143)
Malevolence (#PA7102) Irresolution (#PA7325) Disagreement (#PA5982)
Fragmentation (#PA6233) Unfeelingness (#PA7364) Maladjustment (#PA6739)
Unsociability (#PA6653) Unwillingness (#PA6509) Insufficiency (#PA5473)
Inhospitality (#PA5458) Insensibility (#PA5451) Unpleasantness (#PA7107)
Discontentment (#PA6011) Inexcitability (#PA5467) Incompatibility (#PF9047)
Unhealthfulness (#PA7226) Uncommunicativeness (#PA7411).

♦ PA5451 Insensibility

Anguish — Atrocity — Deadness — Discomfort — Distress — Dullness — Hurtfulness — Insensibility — Insensitivity — Irritability — Malaise — Misery — Pain — Soreness — Spasmodicness — Sufferance — Torment — Torture — Tortuousness — Wound
Broader Sense*complex (#PA6236).
Narrower Misery (#PB8167) Irritability (#PJ2736).
Related Pain (#PA0643) Envy (#PA7253) Fear (#PA6030)
Vice (#PA5644) Enmity (#PA5446) Torture (#PB3430)
Boredom (#PA7365) Disease (#PA6799) Poverty (#PA6434)
Badness (#PA5454) Disorder (#PA7361) Darkness (#PA6261)
Inaction (#PA5806) Weakness (#PA5558) Improbity (#PA7363)
Wrongness (#PA7280) Disrepute (#PA6839) Solemnity (#PA6731)
Agitation (#PA5838) Ignorance (#PA5568) Moderation (#PA7156)
Disrespect (#PA6822) Distortion (#PA6790) Impairment (#PA6088)
Unkindness (#PA5643) Difficulty (#PA5497) Discourtesy (#PA7143)
Imperfection (#PA6997) Unfeelingness (#PA7364) Unsavouriness (#PA7204)
Maladjustment (#PA6739) Discontinuity (#PA5828) Appropriation (#PA5688)
Unintelligence (#PA7371) Unpleasantness (#PA7107) Indiscrimination (#PA6446)
Inappropriateness (#PA6852) Unimaginativeness (#PA6738).

♦ PA5454 Badness

Abomination — Abuse — Affliction — Arrant — Atrocity — Awfulness — Badness — Banefulness — Baseness — Beastliness — Bestiality — Blame — Brutality — Calamity — Clumsiness —

PA5454

Corruption — Counterproductivity — Damage — Deleteriousness — Despicableness — Despoliation — Devilishness — Diabolic — Dirtiness — Disadvantage — Distress — Evil — Fiendish — Filth — Foulness — Fulsomeness — Grievance — Grimness — Grossness — Harmfulness — Horribleness — Hurtfulness — Illness — Inexpedience — Inferiority — Injury — Loathsomeness — Malevolence — Malignancy — Menace — Miasma — Mischievousness — Mischief — Noisome — Notoriety — Outrage — Poisonousness — Pollution — Prejudice — Repulsion — Rot — Sordidness — Squalor — Torment — Torture — Toxicity — Uncleanliness — Unpleasantness — Venom — Vexation — Viciousness — Vileness — Virulence — Woe — Worthlessness — Wound — Wrongness
Broader Adaptation*complex (#PA8178).
Narrower Evil (#PF7042) Torture (#PB3430).
Related Loss (#PA7382) Hate (#PA7338) Envy (#PA7253)
Fear (#PA6030) Vice (#PA5644) Death (#PA7055)
Error (#PA6180) Danger (#PA6971) Stench (#PA5981)
Enmity (#PA5446) Closure (#PA7391) Dissent (#PA6838)
Disease (#PA6799) Poverty (#PA6434) Neglect (#PA5438)
Disorder (#PA7361) Ugliness (#PA7240) Humility (#PA6659)
Grievance (#PF8029) Prejudice (#PA2173) Pollution (#PB6336)
Brutality (#PC1987) Improbity (#PA7363) Restraint (#PA7296)
Wrongness (#PA7280) Cheapness (#PA7193) Impotence (#PA6876)
Disrepute (#PA6839) Solemnity (#PA6731) Injustice (#PA6486)
Avoidance (#PA6379) Adversity (#PA6340) Vulgarity (#PA5821)
Repulsion (#PA5799) Reversion (#PA5699) Disaccord (#PA5532)
Dissuasion (#PA7343) Moderation (#PA7156) Opposition (#PA6979)
Disrespect (#PA6822) Distortion (#PA6790) Complexity (#PA6468)
Inelegance (#PA6312) Impairment (#PA6088) Illegality (#PA5952)
Unkindness (#PA5643) Unchastity (#PA5612) Difficulty (#PA5497)
Ungodliness (#PA7148) Destruction (#PA6542) Disapproval (#PA6191)
Shortcoming (#PA6041) Inferiority (#PA5652) Lamentation (#PA5479)
Uncleanness (#PA5459) Malevolence (#PA7102) Inexpedience (#PA7395)
Irresolution (#PA7325) Condemnation (#PA7237) Untimeliness (#PA7006)
Imperfection (#PA6997) Misjudgement (#PA6607) Misbehaviour (#PA6498)
Miseducation (#PA6393) Hopelessness (#PA6099) Unimportance (#PA5942)
Human illness (#PA0294) Unsavouriness (#PA7204) Nonconformity (#PA5878)
Appropriation (#PA5688) Insensibility (#PA5451) Unintelligence (#PA7371)
Unskillfulness (#PA7232) Unpleasantness (#PA7107) Unhealthfulness (#PA7226)
Meaninglessness (#PA6977) Narrowmindedness (#PA7306) Inappropriateness (#PA6852).
Aggravates Corruption (#PA1986).

♦ **PA5458 Inhospitality**
Abandonment — Inhospitability — Undesirableness — Ungraciousness — Unneighbourliness
Broader Socialization*complex (#PA6373).
Related Vice (#PA5644) Enmity (#PA5446) Absence (#PA7270)
Neglect (#PA5438) Restraint (#PA7296) Avoidance (#PA6379)
Unkindness (#PA5643) Unchastity (#PA5612) Abandonment (#PA7685)
Discourtesy (#PA7143) Inexpedience (#PA7395) Insufficiency (#PA5473)
Unpleasantness (#PA7107) Discontentment (#PA6011).

♦ **PA5459 Uncleanness**
Abomination — Beastliness — Bedraggled — Corruption — Dirtiness — Filth — Foulness — Impurity — Messiness — Pollution — Repulsion — Rot — Sordidness — Squalor — Stain — Uncleanliness — Vileness
Broader Integrity*complex (#PA7263).
Narrower Iron dust (#PE4048).
Related Hate (#PA7338) Vice (#PA5644) Stench (#PA5981)
Closure (#PA7391) Disease (#PA6799) Badness (#PA5454)
Neglect (#PA5438) Disorder (#PA7361) Ugliness (#PA7240)
Pollution (#PB6336) Improbity (#PA7363) Wrongness (#PA7280)
Cheapness (#PA7193) Disrepute (#PA6839) Injustice (#PA6486)
Avoidance (#PA6379) Vulgarity (#PA5821) Repulsion (#PA5799)
Reversion (#PA5699) Dissuasion (#PA7343) Opposition (#PA6979)
Distortion (#PA6790) Complexity (#PA6468) Impairment (#PA6088)
Unkindness (#PA5643) Unchastity (#PA5612) Destruction (#PA6542)
Disapproval (#PA6191) Imperfection (#PA6997) Formlessness (#PA6900)
Miseducation (#PA6393) Unimportance (#PA5942) Unsavouriness (#PA7204)
Unpleasantness (#PA7107) Unhealthfulness (#PA7226) Meaninglessness (#PA6977)
Inappropriateness (#PA6852) Uncommunicativeness (#PA7411).
Aggravated by Corruption (#PA1986).

♦ **PA5465 Loudness**
Discord — Loudness — Muteness
Broader Sense*complex (#PA6236).
Related Discord (#PA5565) Vulgarity (#PA5821) Disaccord (#PA5532)
Ineloquence (#PA6118) Disagreement (#PA5982) Colourlessness (#PA7301)
Uncommunicativeness (#PA7411).

♦ **PA5467 Inexcitability**
Agitation — Disturbance — Fear — Fierceness — Fuss — Nervousness — Panic — Strain — Trepidation — Uneasiness — Upheaval
Broader Feeling*complex (#PA6938).
Narrower Panic (#PF2633).
Related Envy (#PA7253) Fear (#PA6030) Defeat (#PA7289)
Vanity (#PA6491) Enmity (#PA5446) Disorder (#PA7361)
Inaction (#PA5806) Agitation (#PA5838) Disaccord (#PA5532)
Moderation (#PA7156) Impatience (#PA6200) Impairment (#PA6088)
Unkindness (#PA5643) Uncertainty (#PA7309) Destruction (#PA6542)
Inattention (#PA6247) Disapproval (#PA6191) Lamentation (#PA5479)
Irresolution (#PA7325) Insufficiency (#PA5473) Unpleasantness (#PA7107)
Discontentment (#PA6011) Unhealthfulness (#PA7226)
Uncommunicativeness (#PA7411).

♦ **PA5468 Motion*complex**
Narrower Slowness (#PA6166) Reaction (#PA6355) Repulsion (#PA5799)
Quiescence (#PA6444).

♦ **PA5471 Achievement*complex**
Narrower Defeat (#PA7289) Adversity (#PA6340) Difficulty (#PA5497)
Misbehaviour (#PA6498) Unskillfulness (#PA7232)
Nonaccomplishment (#PA6662).

♦ **PA5473 Insufficiency**
Abandonment — Absence — Austerity — Congestion — Defectiveness — Deficiency — Deprivation — Excess — Excessiveness — Exiguity — Extravagance — Extremism — Imperfection — Inadequacy — Incompetence — Incompleteness — Incontinence — Intemperance — Lack — Lavishness — Limitedness — Meanness — Nonfulfilment — Omission — Overabundance — Overdeveloped — Overemphasis — Overexpansion — Overextension — Overload — Overstrain — Oversupply — Overtax — Overweight — Overwork — Poverty — Prodigality — Prolixity —
Scantiness — Scarcity — Shortage — Smallness — Sparsity — Strain — Surfeit — Surplus — Tension — Thinness — Unconscionableness — Unreasonableness — Unsuitability — Verbosity
Broader Adaptation*complex (#PA8178).
Narrower Excess (#PB8952) Surplus (#PF4750) Shortage (#PB8238)
Scarcity (#PA5984) Overwork (#PD2778) Extremism (#PB3415)
Deficiency (#PJ5364) Socio-economic poverty (#PB0388).
Related Loss (#PA7382) Envy (#PA7253) Fear (#PA6030)
Vice (#PA5644) Error (#PA6180) Reason (#PA5502)
Enmity (#PA5446) Tension (#PB6370) Closure (#PA7391)
Boredom (#PA7365) Absence (#PA7270) Fewness (#PA7152)
Disease (#PA6799) Poverty (#PA6434) Neglect (#PA5438)
Insanity (#PA7157) Humility (#PA6659) Inaction (#PA5806)
Austerity (#PJ4983) Improbity (#PA7363) Restraint (#PA7296)
Wrongness (#PA7280) Cheapness (#PA7193) Undueness (#PA6921)
Impotence (#PA6876) Disrepute (#PA6839) Injustice (#PA6486)
Avoidance (#PA6379) Adversity (#PA6340) Vulgarity (#PA5821)
Disaccord (#PA5532) Lightness (#PA5491) Smallness (#PA7408)
Inadequacy (#PA8199) Littleness (#PA7285) Moderation (#PA7156)
Inequality (#PA6695) Impairment (#PA6088) Compulsion (#PA5740)
Unkindness (#PA5643) Unchastity (#PA5612) Human death (#PA0072)
Abandonment (#PA7685) Deprivation (#PA0831) Prodigality (#PA7269)
Selfishness (#PA7211) Environment (#PA7078) Affectation (#PA6400)
Shortcoming (#PA6041) Diffuseness (#PA5974) Inferiority (#PA5652)
Incompetence (#PA6416) Inexpedience (#PA7395) Irresolution (#PA7325)
Disobedience (#PA7250) Untimeliness (#PA7006) Imperfection (#PA6997)
Intemperance (#PA6466) Disagreement (#PA5982) Unimportance (#PA5942)
Nonexistence (#PA5870) Unsavouriness (#PA7204) Inhospitality (#PA5458)
Unintelligence (#PA7371) Unpreparedness (#PA7341) Unskillfulness (#PA7232)
Incompleteness (#PA6652) Discontentment (#PA6011) Inexcitability (#PA5467)
Narrowmindedness (#PA7306) Inappropriateness (#PA6852) Nonaccomplishment (#PA6662)
Uncommunicativeness (#PA7411) Political radicalism (#PF2177).

♦ **PA5479 Lamentation**
Callousness — Fuss — Grievance — Hardness — Insolence — Mournfulness — Petulance — Remorselessness — Shamelessness — Sorrow
Broader Feeling*complex (#PA6938).
Related Envy (#PA7253) Vice (#PA5644) Vanity (#PA6491)
Dissent (#PA6838) Badness (#PA5454) Disorder (#PA7361)
Rashness (#PA7115) Weakness (#PA5558) Grievance (#PF8029)
Improbity (#PA7363) Wrongness (#PA7280) Toughness (#PA6976)
Solemnity (#PA6731) Injustice (#PA6486) Agitation (#PA5838)
Disaccord (#PA5532) Moderation (#PA7156) Opposition (#PA6979)
Disrespect (#PA6822) Impatience (#PA6200) Unkindness (#PA5643)
Unchastity (#PA5612) Difficulty (#PA5497) Discourtesy (#PA7143)
Affectation (#PA6400) Inattention (#PA6247) Disapproval (#PA6191)
Irresolution (#PA7325) Pitilessness (#PA7023) Unfeelingness (#PA7364)
Intangibility (#PA5570) Unpleasantness (#PA7107) Discontentment (#PA6011)
Inexcitability (#PA5467) Unintelligibility (#PA7367).

♦ **PA5490 Changeableness**
Capriciousness — Changeableness — Eccentricity — Impetuousity — Impulsiveness — Inconsistency — Inconstancy — Instability — Irresponsibility — Mutability — Nonuniformity — Restlessness — Shapelessness — Transience — Uncertainty — Undependability — Unpredictability — Unreliability — Vacillation — Variation — Vicissitude
Broader Change*complex (#PA6721).
Narrower Uncertainty (#PA6438) Irresponsibility (#PA8658).
Related Danger (#PA6971) Reason (#PA5502) Unbelief (#PA7392)
Disorder (#PA7361) Ugliness (#PA7240) Rashness (#PA7115)
Inaction (#PA5806) Weakness (#PA5558) Improbity (#PA7363)
Adversity (#PA6340) Deviation (#PA6228) Agitation (#PA5838)
Difference (#PA6698) Inequality (#PA6695) Transience (#PA6425)
Impatience (#PA6200) Instability (#PA0859) Uncertainty (#PA7309)
Lawlessness (#PA5563) Eccentricity (#PF9063) Irresolution (#PA7325)
Invisibility (#PA6978) Formlessness (#PA6900) Irregularity (#PA6774)
Disagreement (#PA5982) Nonuniformity (#PA6890) Nonconformity (#PA5878)
Discontinuity (#PA5828) Unintelligence (#PA7371) Disintegration (#PA6858)
Discontentment (#PA6011) Unintelligibility (#PA7367)
Excessive consumption of specific foodstuffs (#PC3908).

♦ **PA5491 Lightness**
Burdensomeness — Congestion — Encumbrance — Overload — Overtax — Overweight
Broader Absolute properties*complex (#PA5939).
Related Closure (#PA7391) Inaction (#PA5806) Cheapness (#PA7193)
Adversity (#PA6340) Littleness (#PA7285) Difficulty (#PA5497)
Insufficiency (#PA5473) Unpleasantness (#PA7107).

♦ **PA5493 Depression**
Degradation — Overturn
Broader Relative motion*complex (#PA6752).
Related Vice (#PA5644) Defeat (#PA7289) Disrepute (#PA6839)
Solemnity (#PA6731) Impairment (#PA6088) Revolution (#PA5901)
Destruction (#PA6542) Unpleasantness (#PA7107) Disintegration (#PA6858).

♦ **PA5497 Difficulty**
Annoyance — Anxiety — Awkwardness — Bafflement — Barrier — Bewilderment — Bothersomeness — Burdensomeness — Complication — Counterproductivity — Defeat — Difficulty — Disadvantage — Disconcertion — Distress — Embarrassment — Encumbrance — Evil — Fixation — Frustration — Hardness — Impracticality — Inconvenience — Inhibition — Interference — Intricacy — Irksomeness — Obstacle — Obstruction — Opposition — Perplexity — Ponderousness — Problem — Prohibition — Repression — Restriction — Retardation — Ruin — Sabotage — Stalemate — Suppression — Toilsomeness — Torment — Trouble — Unmanageability — Upset — Worry
Refs Varma, Ved P *Anxiety in Children* (1984).
Broader Achievement*complex (#PA5471).
Related Loss (#PA7382) Envy (#PA7253) Fear (#PA6030)
Vice (#PA5644) Defeat (#PA7289) Anxiety (#PA1635)
Closure (#PA7391) Boredom (#PA7365) Refusal (#PA7321)
Dissent (#PA6838) Disease (#PA6799) Poverty (#PA6434)
Badness (#PA5454) Disorder (#PA7361) Insanity (#PA7157)
Humility (#PA6659) Inaction (#PA5806) Weakness (#PA5558)
Restraint (#PA7296) Wrongness (#PA7280) Toughness (#PA6976)
Undueness (#PA6921) Impotence (#PA6876) Intrusion (#PA6862)
Adversity (#PA6340) Exclusion (#PA5869) Agitation (#PA5838)
Ignorance (#PA5568) Disaccord (#PA5532) Lightness (#PA5491)
Repression (#PB0871) Moderation (#PA7156) Opposition (#PA6979)
Difference (#PA6698) Complexity (#PA6468) Inelegance (#PA6312)

ABSTRACT FUNDAMENTAL PROBLEMS

PA5568

Impatience (#PA6200) Impairment (#PA6088) Revolution (#PA5901)
Unkindness (#PA5643) Frustration (#PA2252) Aggravation (#PA7378)
Uncertainty (#PA7309) Environment (#PA7078) Destruction (#PA6542)
Inattention (#PA6247) Disapproval (#PA6191) Lamentation (#PA5479)
Inexpedience (#PA7395) Irresolution (#PA7325) Pitilessness (#PA7023)
Untimeliness (#PA7006) Imperfection (#PA6997) Disagreement (#PA5982)
Unfeelingness (#PA7364) Maladjustment (#PA6739) Forgetfulness (#PA6651)
Unwillingness (#PA6509) Impossibility (#PA6487) Appropriation (#PA5688)
Inexpectation (#PA5527) Insensibility (#PA5451) Unintelligence (#PA7371)
Unskillfulness (#PA7232) Unpleasantness (#PA7107) Unintelligibility (#PA7367)
Inappropriateness (#PA6852) Nonaccomplishment (#PA6662).

♦ PA5498 Existence*complex
Narrower Nonexistence (#PA5870) Extrinsicality (#PA7409)
Insubstantiality (#PA6959).

♦ PA5499 Unsexiness
Sexuality
Broader Life*complex (#PA6586).
Related Unchastity (#PA5612).

♦ PA5500 Rest
Broader Action*complex (#PA5553).
Related Crime (#PB0001).

♦ PA5502 Reason
Argumentativeness — Combativeness — Contentiousness — Contradiction — Controversy — Disingenuousness — Disputatiousness — Equivocation — Evasion — Fallacy — Faultiness — Feebleness — Flaw — Illogic — Incongruity — Inconsequence — Inconsistency — Insincerity — Invalidity — Irrationality — Misapplication — Mystification — Perversion — Prevarication — Quarrelsomeness — Senselessness — Sophistry — Speciousness — Subterfuge — Unauthenticity — Unreasonableness — Vainness — Wrangle
Broader Intellectual faculties*complex (#PA7197).
Narrower Fallacy (#PF4357) Sophistry (#PF8008).
Related Envy (#PA7253) Guilt (#PA6793) Error (#PA6180)
Vanity (#PA6491) Enmity (#PA5446) Oldness (#PA7131)
Dissent (#PA6838) Disease (#PA6799) Defence (#PA5445)
Disorder (#PA7361) Insanity (#PA7157) Weakness (#PA5558)
Improbity (#PA7363) Cheapness (#PA7193) Impotence (#PA6876)
Avoidance (#PA6379) Disaccord (#PA5532) Smallness (#PA7408)
Opposition (#PA6979) Distortion (#PA6790) Difference (#PA6698)
Impairment (#PA6088) Illegality (#PA5952) Aggravation (#PA7378)
Uncertainty (#PA7309) Affectation (#PA6400) Inexpedience (#PA7395)
Irresolution (#PA7325) Imperfection (#PA6997) Invisibility (#PA6978)
Miseducation (#PA6393) Hopelessness (#PA6099) Disagreement (#PA5982)
Unimportance (#PA5942) Disappearance (#PA6785) Unrelatedness (#PA5794)
Insufficiency (#PA5473) Contradictions (#PF3667) Unintelligence (#PA7371)
Unskillfulness (#PA7232) Disintegration (#PA6858) Changeableness (#PA5490)
Meaninglessness (#PA6977) Unintelligibility (#PA7367) Inappropriateness (#PA6852)
Misrepresentation (#PA6644) Uncommunicativeness (#PA7411)
Misinterpretability (#PA6741).

♦ PA5504 Ingratitude
Ingratitude
Broader Benevolence*complex (#PA7331).

♦ PA5527 Inexpectation
Bafflement — Blight — Defeat — Disappointment — Disillusionment — Dissatisfaction — Failure — Frustration
Related Envy (#PA7253) Vice (#PA5644) Error (#PA6180)
Defeat (#PA7289) Dissent (#PA6838) Impotence (#PA6876)
Impairment (#PA6088) Difficulty (#PA5497) Frustration (#PA2252)
Uncertainty (#PA7309) Destruction (#PA6542) Disapproval (#PA6191)
Shortcoming (#PA6041) Inferiority (#PA5652) Imperfection (#PA6997)
Hopelessness (#PA6099) Maladjustment (#PA6739) Unpleasantness (#PA7107)
Discontentment (#PA6011) Dissatisfaction (#PA8886) Disillusionment (#PA6453)
Nonaccomplishment (#PA6662).

♦ PA5532 Disaccord
Aggression — Antagonism — Argumentativeness — Battle — Belligerence — Bloody-mindedness — Chauvinism — Combativeness — Conflict — Contentiousness — Controversy — Defiance — Difficulty — Disaccord — Disaffection — Disaffinity — Discord — Disfavour — Disharmony — Disparity — Disputatiousness — Dissension — Dissidence — Divergence — Divisiveness — Division — Enmity — Ferocity — Fierceness — Friction — Fuss — Hostility — Incompatibility — Inharmonious — Irascibility — Militancy — Mischief — Misunderstanding — Quarrelsomeness — Rivalry — Tension — Truculence — Unfriendliness — Unpeacefulness — Unpleasantness — Variance — War — Warlike — Wrangle
Broader Interaction*complex (#PA6429).
Narrower Friction (#PF1691) Militancy (#PC1090) Ethnocentricity (#PB5765)
Misunderstanding (#PA8197).
Related War (#PB0593) Hate (#PA7338) Envy (#PA7253)
Fear (#PA6030) Death (#PA7055) Error (#PA6180)
Vanity (#PA6491) Reason (#PA5502) Enmity (#PA5446)
Tension (#PB6370) Dissent (#PA6838) Discord (#PA5565)
Badness (#PA5454) Defence (#PA5445) Conflict (#PA0298)
Disorder (#PA7361) Loudness (#PA5465) Hostility (#PB8538)
Improbity (#PA7363) Disrepute (#PA6839) Adversity (#PA6340)
Deviation (#PA6228) Exclusion (#PA5869) Agitation (#PA5838)
Repulsion (#PA5799) Aggression (#PA0587) Moderation (#PA7156)
Opposition (#PA6979) Difference (#PA6698) Inequality (#PA6695)
Impatience (#PA6200) Impairment (#PA6088) Unkindness (#PA5643)
Divergence (#PA5573) Difficulty (#PA5497) Separation (#PA7236)
Competition (#PB0848) Nationalism (#PB0534) Aggravation (#PA7378)
Discourtesy (#PA7143) Inattention (#PA6247) Disapproval (#PA6191)
Lamentation (#PA5479) Inexpedience (#PA7395) Irresolution (#PA7325)
Disobedience (#PA7250) Misbehaviour (#PA6498) Disagreement (#PA5982)
Unfeelingness (#PA7364) Nonuniformity (#PA6890) Maladjustment (#PA6739)
Unsociability (#PA6653) Nonconformity (#PA5878) Insufficiency (#PA5473)
Unpleasantness (#PA7107) Inexcitability (#PA5467) Incompatibility (#PF9047)
Narrowmindedness (#PA7306) Indiscrimination (#PA6446) Unintelligibility (#PA7367)
Misinterpretability (#PA6741).

♦ PA5553 Action*complex
Narrower Rest (#PA5500) Inaction (#PA5806) Unpreparedness (#PA7341).

♦ PA5558 Weakness
Changeableness — Cowardice — Debility — Decline — Decrepitness — Devitalization — Disability — Dullness — Effetism — Fatigue — Feebleness — Fragility — Hardness — Imperviousness — Impotence — Incapacity — Indecisiveness — Instability — Irresolution — Languor — Listlessness — Powerlessness — Softness — Tastelessness — Vapidity — Weakness — Weariness
Broader Power*complex (#PA7314).
Narrower Fatigue (#PA0657) Cowardice (#PJ4171).
Related Fear (#PA6030) Vice (#PA5644) Danger (#PA6971)
Reason (#PA5502) Boredom (#PA7365) Refusal (#PA7321)
Oldness (#PA7131) Disease (#PA6799) Unbelief (#PA7392)
Inaction (#PA5806) Cheapness (#PA7193) Toughness (#PA6976)
Impotence (#PA6876) Necessity (#PA6387) Avoidance (#PA6379)
Adversity (#PA6340) Vulgarity (#PA5821) Quiescence (#PA6444)
Transience (#PA6425) Impairment (#PA6088) Unkindness (#PA5643)
Difficulty (#PA5497) Incapacity (#PJ4312) Instability (#PA0859)
Uncertainty (#PA7309) Incuriosity (#PA6598) Shortcoming (#PA6041)
Lamentation (#PA5479) Irresolution (#PA7325) Pitilessness (#PA7023)
Imperfection (#PA6997) Invisibility (#PA6978) Formlessness (#PA6900)
Unimportance (#PA5942) Unfeelingness (#PA7364) Unsavouriness (#PA7204)
Nonuniformity (#PA6890) Insensibility (#PA5451) Unintelligence (#PA7371)
Unskillfulness (#PA7232) Unpleasantness (#PA7107) Changeableness (#PA5490)
Unproductiveness (#PA7208) Unintelligibility (#PA7367) Influencelessness (#PA6882)
Inappropriateness (#PA6852) Unimaginativeness (#PA6738).

♦ PA5559 Unlimitedness
Broader Credibility*complex (#PA5829).

♦ PA5563 Lawlessness
Anarchism — Anarchy — Authoritarianism — Chaos — Confusion — Despotism — Dictatorship — Disobedience — Disorder — Disorganization — Insubordination — Irresponsibility — Lawlessness — Licentiousness — Misrule — Revolution — Unruliness — Willfulness
Broader Compliance*complex (#PA5710).
Narrower Irresponsibility (#PA8658).
Related Chaos (#PF6836) Defeat (#PA7289) Vanity (#PA6491)
Refusal (#PA7321) Disease (#PA6799) Disorder (#PA7361)
Anarchism (#PC1972) Confusion (#PF7123) Improbity (#PA7363)
Restraint (#PA7296) Undueness (#PA6921) Agitation (#PA5838)
Moderation (#PA7156) Impairment (#PA6088) Illegality (#PA5952)
Revolution (#PA5901) Compulsion (#PA5740) Unchastity (#PA5612)
Separation (#PA7236) Uncertainty (#PA7309) Destruction (#PA6542)
Inattention (#PA6247) Irresolution (#PA7325) Formlessness (#PA6900)
Misbehaviour (#PA6498) Intemperance (#PA6466) Unwillingness (#PA6509)
Unskillfulness (#PA7232) Unpleasantness (#PA7107) Disintegration (#PA6858)
Changeableness (#PA5490) Meaninglessness (#PA6977) Narrowmindedness (#PA7306).
Aggravates Authoritarianism (#PB1638).
Aggravated by Disorganization (#PF4487) Political dictatorship (#PC0845).

♦ PA5564 Unorthodoxy
Misbelief
Broader Redemption*complex (#PAp259).
Related Error (#PA6180) Unbelief (#PA7392).

♦ PA5565 Discord
Discord — Disharmony — Dissonance — Harshness
Broader Sense*complex (#PA6236).
Related Disorder (#PA7361) Loudness (#PA5465) Disaccord (#PA5532)
Moderation (#PA7156) Difference (#PA6698) Unkindness (#PA5643)
Discourtesy (#PA7143) Pitilessness (#PA7023) Disagreement (#PA5982)
Unpleasantness (#PA7107).

♦ PA5568 Ignorance
Awkwardness — Barbarism — Darkness — Ignorance — Inanity — Incomprehensibility — Inexperience — Insensibility — Mindlessness — Misinformation — Naïvety — Paganism — Savagery — Shallowness — Superficiality — Unfamiliarity — Unintelligence — Witlessness — Dependence on ignorance — Uneducated people — Ignorant people — Lack of education — Education

Nature There is still no answer to many scientific questions and there exists an uncertain degree of knowledge with regard to major disciplines. The physical and behavioural sciences such as psychology and economics, which are so frequently looked to for certitude, do not always come up with clear cut answers; and unsubstantiated theories often disguise a lack of knowledge. While one category of ignorance applies to what is not known, another category applies to the inability to assemble all the facts needed. For example, although there may be considerable understanding of the effects of single policies on single objectives, where multiple and sometimes conflicting effects and objectives of programmes and policies exist the processes for evaluating all these may be too complex, so that decisions are made in ignorance of all the relevant data. Another type of ignorance allows for the knowledge to exist and also the capability and means to acquire it, but the will to do so is not present. Unwillingness to look at the facts constitutes intellectual bias. An emotional bias has the same effect, such as identification with a particular set of interests whether they are national, political, economic, ideological or religious. In the latter case, a tenet of Roman Catholic theology, for example, describes those who will never embrace Christianity due to their heredity, upbringing or environment, as 'invincibly ignorant', which aptly describes the 'closed' mind in any situation.

Counter-claim Surveys that claim to establish the relative ignorance of any population group, whether within countries or between countries, are methodologically suspect. The interpretation of any statistical results of such surveys in isolation is questionable in the absence of any accompanying discussion of the social, cultural and economic environment in which these supposedly ignorant individuals function as members of a productive society. It is impossible to make a connection between statistics of ignorance and the past or future status of a society. For example in recent comparisons of USA high school students with their foreign counterparts fails to take into account the continued elitism in of non-USA educational systems. The USA has many more people, with a greater range of socio-cultural backgrounds, in high school than in other countries.

Broader Intellectual faculties*complex (#PA7197).
Narrower Illiteracy (#PC0210) Innumeracy (#PC0143) Uncritical thinking (#PF5039)
Political ignorance (#PC1982) Deliberate ignorance (#PF8229) Scientific ignorance (#PD8003)
Ignorance of grammar (#PD7566) Ignorance of workers (#PD4506)
Ignorance of the law (#PG3516) Ignorance of history (#PD3774)
Bureaucratic ignorance (#PF8582) Mathematical ignorance (#PD6728)
Ignorance of drug users (#PJ2389) Ignorance of administration (#PE1234)
Ignorance concerning disease (#PD8821) Ignorance of cultural heritage (#PF1985)
Ignorance of reproductive processes (#PD7994)
Ignorance of lifelong human development (#PF5759).
Related Boredom (#PA7365) Oldness (#PA7131) Impiety (#PA6058)
Unbelief (#PA7392) Insanity (#PA7157) Darkness (#PA6261)
Impotence (#PA6876) Solemnity (#PA6731) Blindness (#PA6674)

Avoidance (#PA6379) Vulgarity (#PA5821) Moderation (#PA7156)
Inelegance (#PA6312) Unkindness (#PA5643) Difficulty (#PA5497)
Unbelievers (#PF8068) Inattention (#PA6247) Inexpedience (#PA7395)
Invisibility (#PA6978) Miseducation (#PA6393) Unimportance (#PA5942)
Unsavouriness (#PA7204) Disappearance (#PA6785) Maladjustment (#PA6739)
Insensibility (#PA5451) Unintelligence (#PA7371) Unskillfulness (#PA7232)
Unpleasantness (#PA7107) Meaninglessness (#PA6977) Thoughtlessness (#PA6940)
Indiscrimination (#PA6446) Unintelligibility (#PA7367) Inappropriateness (#PA6852)
Uncommunicativeness (#PA7411) Fragmentation of knowledge (#PF0944)
Ignorance of traditional herbal remedies (#PE3946).
Aggravates War (#PB0593) Distrust (#PA8653) Absurdity (#PF6991)
Confusion (#PF7123) Propaganda (#PF1878) Fragmentation (#PA6233)
Enterotoxaemia (#PJ7724) Social neglect (#PB0883) Human suffering (#PB5955)
Cultural suicide (#PF5957) Sexual mutilation (#PD5718) Ideological apathy (#PF3392)
Unexplained phenomena (#PF8352) Economic manipulation (#PC6875)
Loss of cultural identity (#PF9005) Unreported tax obligations (#PE9061)
Inconclusiveness of science (#PF6349) Inadequate personal hygiene (#PD2459)
Ritual slaughter of animals (#PF0319) Irrational religious beliefs (#PF6829)
Human errors and miscalculations (#PF3702) Limited individual attention span (#PF2384)
Unknown availability of subsidies (#PG9905) Abusive traffic in immigrant workers (#PD2722)
Exploitation of dependence on food aid (#PD7592)
Underestimation of the human potential (#PF7063)
Lack of intersocietal resource channels (#PF2517)
Endangered monuments and historic sites (#PD0253)
Inadequate disaster prevention and mitigation (#PF3566)
Inadequacy and insensitivity of intelligence testing (#PD1975)
Underdevelopment of industrial and economic activities (#PC0880)
Inadequate protection and preservation of cultural property (#PF7542)
Ineffectiveness and inefficiency of interdisciplinary meetings (#PF0409).
Aggravated by Secrecy (#PA0005) Illusion (#PA6414)
Toughness (#PA6976) Communism (#PC0369)
Solipsism (#PF5657) Distortion (#PA6790)
Regression (#PA6338) Censorship (#PC0067)
Rationalism (#PF3400) Forgetfulness (#PA6651)
Misinformation (#PD8523) Lost knowledge (#PF5420)
Ethnocentricity (#PB5765) Unreliable data (#PF6832)
News censorship (#PD3030) Human inequality (#PA0844)
Lack of education (#PB8645) Fear of knowledge (#PF9595)
Human contingency (#PF7054) Undomesticated men (#PF0551)
Educational wastage (#PC1716) Lack of information (#PF6337)
Defective reasoning (#PF5711) Over–specialization (#PF0256)
Inadequate education (#PF4984) Cognitive dissonance (#PF6638)
Unrecorded knowledge (#PF5728) Lack of communication (#PF0816)
Obstacles to education (#PF4852) Dependence on religion (#PF0150)
Refusal to participate (#PF3226) Elusiveness of reality (#PJ8032)
Obstruction of research (#PF9625) Reduced civic awareness (#PJ1835)
Lack of formal education (#PF6534) Disagreement among experts (#PF6012)
Biased presentation of news (#PD1718) Inaccessibility of knowledge (#PF1953)
Unlearned fundamental skills (#PG8901) Selective perception of facts (#PF2453)
Biased government information (#PF0157) Inflexible management patterns (#PF3091)
Shortage of funds for research (#PF5419) Irrelevant scientific activity (#PF1202)
Negative emotions and attitudes (#PA7090) Radio and television censorship (#PD3029)
Inadequate warning of disasters (#PF3565) Lack of scientific investigation (#PF2720)
Inadequate ideological frameworks (#PD0065) Irrelevance of educational curricula (#PF0443)
Irrelevance of science and technology (#PF0770) Inadequate objectivity of institutions (#PF6691)
Restrictions on freedom of information (#PC0185)
Mis–communications to societal learners (#PF7050)
Biased and inaccurate history textbooks (#PD2082)
Unpreparedness for surplus leisure time (#PF5044)
Unrecognized need for functional skills (#PF2995)
Unequal opportunities for media reception (#PD3039)
Biased and inaccurate geography textbooks (#PF1780)
Inadequate international map of the world (#PD0398)
Misdirection of human energies and desires (#PF8128)
Fragmentation of technological development (#PC1227)
Unequal global distribution of basic skills (#PF2880)
Uncritical acceptance of dogmas and standards (#PF2901)
Limited availability of learning opportunities (#PF3184)
Excessive childhood dependency in developing countries (#PD3491)
Absence of convincing symbols connecting the individual's life to the cosmos (#PF1081).
Reduced by Parascience (#PF9032)
Violation of sovereignty by trans–border broadcasting (#PE0261).

♦ PA5570 Intangibility

Callousness
Broader Sense*complex (#PA6236).
Related Vice (#PA5644) Unkindness (#PA5643) Lamentation (#PA5479)
Unfeelingness (#PA7364).

♦ PA5573 Divergence

Divergence — Division
Broader Relative motion*complex (#PA6752).
Narrower Divided cities (#PD7065).
Related Deviation (#PA6228) Exclusion (#PA5869) Disaccord (#PA5532)
Separation (#PA7236) Disagreement (#PA5982) Nonuniformity (#PA6890)
Nonconformity (#PA5878) Indiscrimination (#PA6446).

♦ PA5583 Punishment

Confinement — Retribution
Refs Odera, H O *Punishment and Terrorism in Africa* (1976).
Broader Retribution*complex (#PA7037).
Narrower Revenge (#PF8562) Eternal punishment (#PF7228)
Collective punishment (#PD6970) Unjust punishments for crimes (#PD4779)
Denial of right to freedom from cruel, inhumane or degrading punishment (#PC3768).
Related Restraint (#PA7296) Vengeance (#PA6606) Narrowness (#PA7256)
Environment (#PA7078).

♦ PA5589 Decrease

Proliferation
Broader Quantity*complex (#PA6108).

♦ PA5604 Femininity

Broader Life*complex (#PA6586).

♦ PA5611 Discovery

Inquisition
Refs Wilson, Robert A *The New Inquisition* (1987).
Broader Evaluation*complex (#PA7394).

♦ PA5612 Unchastity

Abandonment — Animality — Austerity — Bawdiness — Beastliness — Bestiality — Brutality — Coarseness — Concupiscence — Debauchery — Dirtiness — Filth — Foulness — Fulsomeness — Grossness — Illegality — Immodesty — Impropriety — Impurity — Inappropriateness — Incontinence — Indecency — Indecorum — Indelicacy — Indiscretion — Inelegance — Intemperence — Laxity — Licentiousness — Looseness — Lust — Mortification — Notoriety — Promiscuity — Prostitution — Ravishment — Seduction — Sensuality — Sexuality — Shamelessness — Unchastity — Uncleanliness — Vileness — Vulgarity — Wantonness — Lack of chastity
Broader Morality*complex (#PA6892).
Narrower Seduction (#PG5097) Impropriety (#PA6000).
Related Hate (#PA7338) Vice (#PA5644) Error (#PA6180)
Vanity (#PA6491) Stench (#PA5981) Closure (#PA7391)
Refusal (#PA7321) Absence (#PA7270) Disease (#PA6799)
Badness (#PA5454) Neglect (#PA5438) Disorder (#PA7361)
Ugliness (#PA7240) Rashness (#PA7115) Humility (#PA6659)
Austerity (#PJ4983) Brutality (#PC1987) Improbity (#PA7363)
Restraint (#PA7296) Wrongness (#PA7280) Undueness (#PA6921)
Disrepute (#PA6839) Injustice (#PA6486) Avoidance (#PA6379)
Vulgarity (#PA5821) Moderation (#PA7156) Inelegance (#PA6312)
Impairment (#PA6088) Illegality (#PA5952) Compulsion (#PA5740)
Unkindness (#PA5643) Unsexiness (#PA5499) Promiscuity (#PC0745)
Abandonment (#PA7685) Uncertainty (#PA7309) Discourtesy (#PA7143)
Destruction (#PA6542) Affectation (#PA6400) Disapproval (#PA6191)
Inferiority (#PA5652) Lawlessness (#PA5563) Lamentation (#PA5479)
Uncleanness (#PA5459) Prostitution (#PD0693) Inexpedience (#PA7395)
Irresolution (#PA7325) Untimeliness (#PA7006) Imperfection (#PA6997)
Misbehaviour (#PA6498) Intemperance (#PA6466) Disagreement (#PA5982)
Unimportance (#PA5942) Unsavouriness (#PA7204) Appropriation (#PA5688)
Insufficiency (#PA5473) Inhospitality (#PA5458) Unintelligence (#PA7371)
Unpreparedness (#PA7341) Unskillfulness (#PA7232) Unpleasantness (#PA7107)
Disintegration (#PA6858) Indiscrimination (#PA6446) Inappropriateness (#PA6852)
Uncommunicativeness (#PA7411).
Aggravated by Permissiveness (#PF1252).
Reduced by Chastity (#PG6123).

♦ PA5619 Relapse

Lapse — Regression — Reversion
Broader Integrity*complex (#PA7263).
Related Fear (#PA6030) Vice (#PA5644) Error (#PA6180)
Impiety (#PA6058) Neglect (#PA5438) Reversion (#PA5699)
Regression (#PA6338) Impairment (#PA6088).

♦ PA5643 Unkindness

Animality — Asperity — Atrocity — Austerity — Barbarity — Beastliness — Bestiality — Bitterness — Bloody–mindedness — Brutality — Callousness — Coldness — Cruelty — Cynicism — Damnation — Dehumanization — Devilishness — Diabolic — Disagreeableness — Ferocity — Fiendish — Fierceness — Forgetfulness — Grimness — Hardness — Harmfulness — Harshness — Heartlessness — Heedlessness — Illness — Inclemency — Inconsiderateness — Inhospitability — Inhumanity — Insensitivity — Malevolence — Malice — Malignancy — Meanness — Pitilessness — Roughness — Ruthlessness — Sadism — Savagery — Severity — Thoughtlessness — Truculence — Ungraciousness — Unkindness — Unnaturalness — Unthoughtfulness — Vandalism — Venom — Viciousness — Violence — Virulence — Withering
Nature Unkindness can be seen as subjective quality or as manifested objectively in outward behaviour, word, or act. It carries with it varying shades of bad will, which may be expressed in such terms as unrelenting anger, spirit of harshness, vengeance, callousness, etc.
Broader Benevolence*complex (#PA7331).
Narrower Malice (#PF5901) Cruelty (#PB2642) Cynicism (#PF3418)
Inhumanity (#PB8214).
Related Hate (#PA7338) Envy (#PA7253) Cold (#PA6956)
Fear (#PA6030) Vice (#PA5644) Death (#PA7055)
Sadism (#PF3270) Enmity (#PA5446) Disease (#PA6799)
Discord (#PA5565) Badness (#PA5454) Neglect (#PA5438)
Ugliness (#PA7240) Insanity (#PA7157) Humility (#PA6659)
Weakness (#PA5558) Austerity (#PJ4983) Vandalism (#PD1350)
Brutality (#PC1987) Wrongness (#PA7280) Cheapness (#PA7193)
Toughness (#PA6976) Disrepute (#PA6839) Solemnity (#PA6731)
Avoidance (#PA6379) Vulgarity (#PA5821) Ignorance (#PA5568)
Disaccord (#PA5532) Moderation (#PA7156) Disrespect (#PA6822)
Inelegance (#PA6312) Impairment (#PA6088) Compulsion (#PA5740)
Unchastity (#PA5612) Difficulty (#PA5497) Selfishness (#PA7211)
Ungodliness (#PA7148) Discourtesy (#PA7143) Incuriosity (#PA6598)
Destruction (#PA6542) Affectation (#PA6400) Inattention (#PA6247)
Disapproval (#PA6191) Inferiority (#PA5652) Lamentation (#PA5479)
Uncleanness (#PA5459) Malevolence (#PA7102) Inexpedience (#PA7395)
Irresolution (#PA7325) Condemnation (#PA7237) Pitilessness (#PA7023)
Imperfection (#PA6997) Misbehaviour (#PA6498) Hopelessness (#PA6099)
Disagreement (#PA5982) Unimportance (#PA5942) Human illness (#PA0294)
Unfeelingness (#PA7364) Unsociability (#PA6653) Forgetfulness (#PA6651)
Unwillingness (#PA6509) Nonconformity (#PA5878) Intangibility (#PA5570)
Insufficiency (#PA5473) Inhospitality (#PA5458) Insensibility (#PA5451)
Human violence (#PA0429) Dehumanization (#PA1757) Unintelligence (#PA7371)
Unpreparedness (#PA7341) Unskillfulness (#PA7232) Unpleasantness (#PA7107)
Inexcitability (#PA5467) Unhealthfulness (#PA7226) Narrowmindedness (#PA7306)
Indiscrimination (#PA6446) Unintelligibility (#PA7367) Inappropriateness (#PA6852)
Uncommunicativeness (#PA7411).

♦ PA5644 Vice

Abandonment — Abjection — Abomination — Amorality — Anger — Arrant — Atrocity — Avarice — Badness — Baseness — Blame — Callousness — Corruption — Debauchery — Decadence — Degeneration — Degradation — Depravation — Depravity — Dereliction — Devilishness — Diabolic — Dirtiness — Disgrace — Envy — Evil — Failure — Feloniousness — Fiendish — Foulness — Greed — Hardness — Heartlessness — Immorality — Imperfection — Impropriety — Impurity — Incorrigibility — Indiscretion — Infamy — Iniquity — Injury — Injustice — Irredeemability — Irreformability — Lapse — Lust — Malfeasance — Malpractice — Misconduct — Misdemeanour — Misfeasance — Nonfeasance — Offence — Omission — Outrage — Prodigality — Profligacy — Scandal — Shamelessness — Slothfulness — Transgression — Trespass — Turpitude — Unchastity — Uncleanliness — Ungodliness — Viciousness — Vileness — Villainy — Vitiation — Wantonness — Warpedness — Weakness — Wrongness — Dependence on vice
Nature Vice is the repetition of what is wrong, bad, or sinful.
Refs Shklar, Judith N *Ordinary Vices* (1985); Wallace, James D *Virtues and Vices* (1986).
Broader Morality*complex (#PA6892) Negative emotions and attitudes (#PA7090).
Narrower Evil (#PF7042) Pride (#PA7599) Anger (#PA7797)
Egoism (#PA6318) Infamy (#PB8172) Passion (#PA7030)
Depravity (#PC8974) Decadence (#PB2542) Immorality (#PA3369).
Related Loss (#PA7382) Hate (#PA7338) Envy (#PA7253)

ABSTRACT FUNDAMENTAL PROBLEMS PA5828

Fear (#PA6030) Error (#PA6180) Defeat (#PA7289)
Danger (#PA6971) Vanity (#PA6491) Stench (#PA5981)
Scandal (#PC8391) Avarice (#PA6999) Closure (#PA7391)
Absence (#PA7270) Disease (#PA6799) Impiety (#PA6058)
Relapse (#PA5619) Badness (#PA5454) Defence (#PA5445)
Neglect (#PA5438) Celibacy (#PA7410) Unbelief (#PA7392)
Ugliness (#PA7240) Rashness (#PA7115) Humility (#PA6659)
Inaction (#PA5806) Weakness (#PA5558) Hostility (#PB8538)
Improbity (#PA7363) Restraint (#PA7296) Wrongness (#PA7280)
Cheapness (#PA7193) Toughness (#PA6976) Undueness (#PA6921)
Impotence (#PA6876) Intrusion (#PA6862) Disrepute (#PA6839)
Solemnity (#PA6731) Injustice (#PA6486) Avoidance (#PA6379)
Vulgarity (#PA5821) Reversion (#PA5699) Corruption (#PA1986)
Dissuasion (#PA7343) Moderation (#PA7156) Disrespect (#PA6822)
Distortion (#PA6790) Inequality (#PA6695) Complexity (#PA6468)
Regression (#PA6338) Impairment (#PA6088) Illegality (#PA5952)
Unkindness (#PA5643) Unchastity (#PA5612) Difficulty (#PA5497)
Depression (#PA5493) Abandonment (#PA7685) Prodigality (#PA7269)
Selfishness (#PA7211) Ungodliness (#PA7148) Destruction (#PA6542)
Affectation (#PA6400) Disapproval (#PA6191) Shortcoming (#PA6041)
Inferiority (#PA5652) Lamentation (#PA5479) Uncleanness (#PA5459)
Inexpedience (#PA7395) Irresolution (#PA7325) Disobedience (#PA7250)
Condemnation (#PA7237) Pitilessness (#PA7023) Untimeliness (#PA7006)
Imperfection (#PA6997) Misbehaviour (#PA6498) Intemperance (#PA6466)
Miseducation (#PA6393) Hopelessness (#PA6099) Disagreement (#PA5982)
Unimportance (#PA5942) Unfeelingness (#PA7364) Unsavouriness (#PA7204)
Intangibility (#PA5570) Inexpectation (#PA5527) Insufficiency (#PA5473)
Inhospitality (#PA5458) Insensibility (#PA5451) Unintelligence (#PA7371)
Unskillfulness (#PA7232) Unpleasantness (#PA7107) Disintegration (#PA6858)
Incompleteness (#PA6652) Discontentment (#PA6011) Unhealthfulness (#PA7226)
Narrowmindedness (#PA7306) Indiscrimination (#PA6446) Unintelligibility (#PA7367)
Influencelessness (#PA6882) Inappropriateness (#PA6852) Nonaccomplishment (#PA6662)
Misrepresentation (#PA6644) Uncommunicativeness (#PA7411)
Bribery of public servants (#PD4541).
Aggravates Victimization (#PF6987) Human suffering (#PB5955)
Behavioural deterioration (#PB6321).
Aggravated by Malevolence (#PA7102).

♦ PA5652 Inferiority

Baseness — Deficiency — Demeaning — Failure — Imperfection — Inadequacy — Incompetence — Inferiority — Insufficiency — Littleness — Lowness — Maladroit — Meanness — Mediocrity — Pettiness — Servility — Smallness — Subjection — Subordination — Triviality — Vulgarity
Broader Quantity*complex (#PA6108).
Narrower Deficiency (#PJ5364).
Related Envy (#PA7253) Fear (#PA6030) Vice (#PA5644)
Error (#PA6180) Defeat (#PA7289) Fewness (#PA7152)
Lowness (#PA6798) Badness (#PA5454) Humility (#PA6659)
Improbity (#PA7363) Restraint (#PA7296) Cheapness (#PA7193)
Impotence (#PA6876) Disrepute (#PA6839) Solemnity (#PA6731)
Vulgarity (#PA5821) Smallness (#PA7408) Inadequacy (#PA8199)
Littleness (#PA7285) Disrespect (#PA6822) Inequality (#PA6695)
Impairment (#PA6088) Unkindness (#PA5643) Unchastity (#PA5612)
Selfishness (#PA7211) Discourtesy (#PA7143) Shortcoming (#PA6041)
Incompetence (#PA6416) Disobedience (#PA7250) Imperfection (#PA6997)
Unimportance (#PA5942) Nonconformity (#PA5878) Inexpectation (#PA5527)
Insufficiency (#PA5473) Unintelligence (#PA7371) Unpreparedness (#PA7341)
Unskillfulness (#PA7232) Unpleasantness (#PA7107) Incompleteness (#PA6652)
Discontentment (#PA6011) Narrowmindedness (#PA7306) Inappropriateness (#PA6852)
Nonaccomplishment (#PA6662) Uncommunicativeness (#PA7411).

♦ PA5663 Content

Broader Space*complex (#PA6713).

♦ PA5688 Appropriation

Bereavement — Depredation — Despoliation — Distress — Extortion — Fraudulence — Impoverishment — Rapacity — Ravishment — Spoilation — Subjugation — Usurpation
Broader Possession*complex (#PA6686).
Narrower Criminal coercion (#PD4469).
Related Loss (#PA7382) Fear (#PA6030) Death (#PA7055)
Defeat (#PA7289) Poverty (#PA6434) Badness (#PA5454)
Celibacy (#PA7410) Improbity (#PA7363) Restraint (#PA7296)
Cheapness (#PA7193) Undueness (#PA6921) Avoidance (#PA6379)
Unchastity (#PA5612) Difficulty (#PA5497) Bereavement (#PF3516)
Destruction (#PA6542) Intemperance (#PA6466) Insensibility (#PA5451)
Unpleasantness (#PA7107) Inappropriateness (#PA6852).
Uncommunicativeness (#PA7411).
Aggravates Grief (#PF5654).

♦ PA5699 Reversion

Alienation — Brainwash — Corruption — Desertion — Disenchantment — Regression — Retrogression — Reversion — Revulsion — Subversion — Traitorousness — Treason
Broader Change*complex (#PA6721).
Narrower Revulsion (#PG6146).
Related Vice (#PA5644) Error (#PA6180) Enmity (#PA5446)
Treason (#PD2615) Duality (#PA7339) Dissent (#PA6838)
Impiety (#PA6058) Relapse (#PA5619) Badness (#PA5454)
Insanity (#PA7157) Improbity (#PA7363) Avoidance (#PA6379)
Alienation (#PA3545) Dissuasion (#PAp343) Opposition (#PA6979)
Distortion (#PA6790) Complexity (#PA6468) Regression (#PA6338)
Impairment (#PA6088) Revolution (#PA5901) Destruction (#PA6542)
Disapproval (#PA6191) Uncleanness (#PA5459) Irresolution (#PA7325)
Disobedience (#PA7250) Miseducation (#PA6393) Unsociability (#PA6653).
Aggravates Corruption (#PA1986).

♦ PA5710 Compliance*complex

Narrower Refusal (#PA7321) Restraint (#PA7296) Compulsion (#PA5740)
Lawlessness (#PA5563) Nonobservance (#PA7362)
Obstacles to national development (#PF4842)
Inadequate local government financing (#PC6631)
Inequities in ship owner registration (#PE5875)
Energy deficient developing countries (#PE0379)
Instability in tourist dependent economies (#PF4112)
False assumptions on sustainable development (#PF2528)
Excessive anxiety on lending to developing countries (#PF4345)
Dependence of least developed countries on foreign aid (#PE8116)
Inadequate diversification of loans to developing countries (#PE4305)
Irrational fear of industrialization of developing countries (#PF4185)
Imbalance in economic and social planning in developing countries (#PF4837)
Uncertainty of development expenditures due to floating-rate loans (#PF4295)
Unavailability of land for agricultural purposes in developing countries (#PE5024).
Related Disobedience (#PA7250) Inequities in marine insurance (#PE5802)
Inviability of tropical island developing countries (#PE5808)
Vulnerability of small nations to foreign intervention (#PD2374).
Aggravated by Unequal global distribution of basic skills (#PF2880).

♦ PA5733 Antiquity

Broader Time*complex (#PA7222).

♦ PA5740 Compulsion

Austerity — Authoritarianism — Carelessness — Constraint — Duress — Enforcement — Inflexibility — Intimidation — Laxness — Looseness — Permissiveness — Relentlessness — Rigidity — Severity — Stiffness — Stubbornness — Violence
Broader Compliance*complex (#PA5710).
Narrower Intimidation (#PB1992) Inflexibility (#PA8555).
Related Envy (#PA7253) Fear (#PA6030) Error (#PA6180)
Vanity (#PA6491) Boredom (#PA7365) Refusal (#PA7321)
Neglect (#PA5438) Disorder (#PA7361) Rashness (#PA7115)
Softness (#PA5988) Austerity (#PJ4983) Restraint (#PA7296)
Toughness (#PA6976) Avoidance (#PA6379) Dissuasion (#PA7343)
Moderation (#PA7156) Permanence (#PA6802) Unkindness (#PA5643)
Unchastity (#PA5612) Uncertainty (#PA7309) Informality (#PA7170)
Discourtesy (#PA7143) Incuriosity (#PA6598) Affectation (#PA6400)
Inattention (#PA6247) Disapproval (#PA6191) Lawlessness (#PA5563)
Irresolution (#PA7325) Disobedience (#PA7250) Pitilessness (#PA7023)
Untimeliness (#PA7006) Unwillingness (#PA6509) Insufficiency (#PA5473)
Human violence (#PA0429) Permissiveness (#PF1252) Unskillfulness (#PA7232)
Disintegration (#PA6858) Narrowmindedness (#PA7306) Inappropriateness (#PA6852).
Aggravates Authoritarianism (#PB1638).

♦ PA5794 Unrelatedness

Misalliance — Misapplication — Separateness
Broader Relationship*complex (#PA6484).
Related Error (#PA6180) Reason (#PA5502) Duality (#PA7339)
Separation (#PA7236) Disagreement (#PA5982) Disintegration (#PA6858)
Inappropriateness (#PA6852) Misinterpretability (#PA6741).

♦ PA5799 Repulsion

Disaffinity — Repulsion
Broader Motion*complex (#PA5468).
Related Enmity (#PA5446) Badness (#PA5454) Disaccord (#PA5532)
Opposition (#PA6979) Uncleanness (#PA5459) Unfeelingness (#PA7364)
Unpleasantness (#PA7107).

♦ PA5806 Inaction

Boredom — Boringness — Burdensomeness — Debility — Dullness — Ennui — Exhaustion — Fatigue — Idleness — Impetuousity — Impulsiveness — Inactivity — Indifference — Indolence — Inertia — Intrusiveness — Languor — Laziness — Lethargy — Lifelessness — Listlessness — Overactivity — Overextension — Overstrain — Paralysis — Passivity — Shiftlessness — Slothfulness — Slowness — Stagnation — Strain — Stupor — Suddenness — Suspension — Tiresomeness — Toilsomeness — Torpor — Vagrancy — Weakness — Weariness — Wearisomeness
Broader Action*complex (#PA5553).
Narrower Fatigue (#PA0657) Lethargy (#PJ4782) Idleness (#PA7710)
Inactivity (#PB7991) Stagnation (#PA3917).
Related Loss (#PA7382) Fear (#PA6030) Vice (#PA5644)
Death (#PA7055) Danger (#PA6971) Enmity (#PA5446)
Boredom (#PA7365) Oldness (#PA7131) Disease (#PA6799)
Neglect (#PA5438) Unbelief (#PA7392) Rashness (#PA7115)
Darkness (#PA6261) Slowness (#PA6166) Weakness (#PA5558)
Paralysis (#PD2632) Impotence (#PA6876) Intrusion (#PA6862)
Solemnity (#PA6731) Avoidance (#PA6379) Exclusion (#PA5869)
Lightness (#PA5491) Exhaustion (#PJ2732) Quiescence (#PA6444)
Transience (#PA6425) Impatience (#PA6200) Impairment (#PA6088)
Difficulty (#PA5497) Incuriosity (#PA6598) Inattention (#PA6247)
Irresolution (#PA7325) Disobedience (#PA7250) Untimeliness (#PA7006)
Imperfection (#PA6997) Hopelessness (#PA6099) Unimportance (#PA5942)
Unfeelingness (#PA7364) Unsavouriness (#PA7204) Nonuniformity (#PA6890)
Maladjustment (#PA6739) Insufficiency (#PA5473) Insensibility (#PA5451)
Unintelligence (#PA7371) Unpreparedness (#PA7341) Unpleasantness (#PA7107)
Changeableness (#PA5490) Inexcitability (#PA5467) Influencelessness (#PA6882)
Inappropriateness (#PA6852) Unimaginativeness (#PA6738)
Uncommunicativeness (#PA7411).
Aggravated by Indifference (#PA7604).

♦ PA5811 Cessation

Broader Change*complex (#PA6721).

♦ PA5821 Vulgarity

Animality — Barbarism — Barbarity — Baseness — Bestiality — Brutality — Coarseness — Commonness — Exclusiveness — Gaudiness — Grossness — Ignobility — Impropriety — Inappropriateness — Incivility — Indecency — Indecorum — Indelicacy — Inelegance — Loudness — Lowness — Meanness — Overconscientiousness — Oversensitiveness — Profligacy — Roughness — Rudeness — Savagery — Snobbery — Tastelessness — Unsuitability — Vileness — Vulgarity
Broader Discriminative affection*complex (#PA7145).
Related Vice (#PA5644) Fear (#PA6030) Envy (#PA7253)
Vanity (#PA6491) Badness (#PA5454) Lowness (#PA6798)
Boredom (#PA7365) Loudness (#PA5465) Weakness (#PA5558)
Humility (#PA6659) Ugliness (#PA7240) Snobbery (#PJ3943)
Ignorance (#PA5568) Exclusion (#PA5869) Injustice (#PA6486)
Solemnity (#PA6731) Disrepute (#PA6839) Undueness (#PA6921)
Cheapness (#PA7193) Wrongness (#PA7280) Improbity (#PA7363)
Brutality (#PC1987) Unchastity (#PA5612) Unkindness (#PA5643)
Inelegance (#PA6312) Disrespect (#PA6822) Opposition (#PA6979)
Moderation (#PA7156) Uncleanness (#PA5459) Inferiority (#PA5652)
Disapproval (#PA6191) Discourtesy (#PA7143) Selfishness (#PA7211)
Unimportance (#PA5942) Disagreement (#PA5982) Misbehaviour (#PA6498)
Imperfection (#PA6997) Untimeliness (#PA7006) Inexpedience (#PA7395)
Insufficiency (#PA5473) Unsociability (#PA6653) Discontentment (#PA6011)
Unpleasantness (#PA7107) Unskillfulness (#PA7232) Colourlessness (#PA7301)
Unpreparedness (#PA7341) Unintelligence (#PA7371) Narrowmindedness (#PA7306)
Loss of civility (#PC7013) Inappropriateness (#PA6852).

♦ PA5828 Discontinuity

Brokenness — Discontinuity — Nonuniformity — Spasmodicness
Broader Order*complex (#PA7199).

Related Defeat (#PA7289) Disorder (#PA7361) Agitation (#PA5838)
Separation (#PA7236) Impairment (#PA6088) Inequality (#PA6695)
Moderation (#PA7156) Destruction (#PA6542) Dislocation (#PA7158)
Irregularity (#PA6774) Disobedience (#PA7250) Insensibility (#PA5451)
Nonuniformity (#PA6890) Changeableness (#PA5490) Disintegration (#PA6858).

♦ PA5829 Credibility*complex

Narrower Error (#PA6180) Unbelief (#PA7392) Uncertainty (#PA7309)
Unprovability (#PA7070) Unlimitedness (#PA5559) Impossibility (#PA6487).

♦ PA5838 Agitation

Agitation — Discomposure — Disorder — Disquiet — Disturbance — Fuss — Inquietude — Malaise — Nervousness — Perturbation — Restlessness — Spasmodicness — Spasticity — Trepidation — Turbidity — Turbulence — Unpeacefulness — Upset

Broader Relative motion*complex (#PA6752).
Narrower Nervousness (#PE4171).
Related Fear (#PA6030) Vanity (#PA6491) Disease (#PA6799)
Disorder (#PA7361) Disaccord (#PA5532) Solemnity (#PA6731)
Difficulty (#PA5497) Revolution (#PA5901) Impatience (#PA6200)
Moderation (#PA7156) Lamentation (#PA5479) Lawlessness (#PA5563)
Disapproval (#PA6191) Inattention (#PA6247) Destruction (#PA6542)
Uncertainty (#PA7309) Misbehaviour (#PA6498) Formlessness (#PA6900)
Insensibility (#PA5451) Discontinuity (#PA5828) Inexcitability (#PA5467)
Changeableness (#PA5490) Discontentment (#PA6011) Unpleasantness (#PA7107)
Unhealthfulness (#PA7226).
Aggravates Disorganization (#PF4487).

♦ PA5869 Exclusion

Blemish — Division — Exclusion — Expulsion — Insularity — Intrusiveness — Isolation — Nonconformity — Prohibition — Rejection — Repudiation — Restriction — Seclusion — Segregation — Severance — Snobbery — Suspension — Tightness — Withdrawal — Xenophobia

Broader Order*complex (#PA7199).
Narrower Isolation (#PB8685) Xenophobia (#PD4957) Loneliness (#PF2386)
Segregation (#PC0031).
Related Fear (#PA6030) Hate (#PA7338) Vanity (#PA6491)
Dissent (#PA6838) Fewness (#PA7152) Refusal (#PA7321)
Duality (#PA7339) Inaction (#PA5806) Ugliness (#PA7240)
Unbelief (#PA7392) Snobbery (#PJ3943) Disaccord (#PA5532)
Vulgarity (#PA5821) Avoidance (#PA6379) Necessity (#PA6387)
Intrusion (#PA6862) Cheapness (#PA7193) Restraint (#PA7296)
Separation (#PA7236) Difficulty (#PA5497) Divergence (#PA5573)
Impairment (#PA6088) Regression (#PA6338) Distortion (#PA6790)
Disrespect (#PA6822) Opposition (#PA6979) Withdrawal (#PF4402)
Disapproval (#PA6191) Incuriosity (#PA6598) Environment (#PA7078)
Disagreement (#PA5982) Imperfection (#PA6997) Untimeliness (#PA7006)
Disobedience (#PA7250) Irresolution (#PA7325) Nonconformity (#PA5878)
Counteraction (#PA6485) Unsociability (#PA6653) Unfeelingness (#PA7364)
Indiscrimination (#PA6446) Narrowmindedness (#PA7306) Inappropriateness (#PA6852).

♦ PA5870 Nonexistence

Annihilation — Deprivation — Nothingness — Nullity — Unreality

Broader Existence*complex (#PA5498).
Related Loss (#PA7382) Error (#PA6180) Death (#PA7055)
Poverty (#PA6434) Absence (#PA7270) Destruction (#PA6542)
Deprivation (#PA0831) Unimportance (#PA5942) Annihilation (#PF9169)
Insufficiency (#PA5473) Meaninglessness (#PA6977) Insubstantiality (#PA6959)
Unimaginativeness (#PA6738).

♦ PA5878 Nonconformity

Aberration — Deformation — Deviation — Divergence — Eccentricity — Inferiority — Malformation — Nonconformity — Shapelessness — Strangeness — Unnaturalness

Broader Order*complex (#PA7199).
Narrower Malformation (#PE4460).
Related Error (#PA6180) Badness (#PA5454) Dissent (#PA6838)
Insanity (#PA7157) Ugliness (#PA7240) Disorder (#PA7361)
Disaccord (#PA5532) Exclusion (#PA5869) Deviation (#PA6228)
Impotence (#PA6876) Restraint (#PA7296) Divergence (#PA5573)
Unkindness (#PA5643) Inelegance (#PA6312) Distortion (#PA6790)
Inferiority (#PA5652) Shortcoming (#PA6041) Affectation (#PA6400)
Uncertainty (#PA7309) Unimportance (#PA5942) Disagreement (#PA5982)
Formlessness (#PA6900) Imperfection (#PA6997) Disobedience (#PA7250)
Eccentricity (#PF9063) Counteraction (#PA6485) Nonuniformity (#PA6890)
Changeableness (#PA5490) Unintelligence (#PA7371) Unintelligibility (#PA7367)
Uncommunicativeness (#PA7411).

♦ PA5901 Revolution

Anarchism — Overturn — Revolution — Revulsion — Subversion — Upset — Rebellion

Nature A foundational change or reversal in direction of a social entity either in thinking such as a scientific revolution, in political systems as in the American, French or Russian Revolutions, in economic systems as in the Thatcher revolution or in culture as in the Cultural Revolution in China is designated as a revolution.

Claim Revolution is the moment of truth which exposes the falsehood in established structures. Revolutionaries are bearers of a righteousness not their own.

Refs Beilenson, Lawrence W *Power Through Subversion* (1972); Crew, Jennifer and Wright, D *Revolutions in the Modern World*; Fourquin, G *The Anatomy of Popular Rebellion in the Middle Ages* (1978); Jones, Howard M *Revolution and Romanticism* (1974); Midlarsky, Manus I *Inequality and Contemporary Revolutions* (1986); Russell, D E *Rebellion, Revolution, and Armed Force* (1974); Turok, Ben *Revolutionary Thought in the Twentieth Century* (1980).

Narrower Revulsion (#PG6146) Social revolution (#PC3236) Violent revolution (#PC3229)
Counter revolution (#PF3232) Economic revolution (#PC3233) Cultural revolution (#PF3235)
Political revolution (#PF3237) Ideological revolution (#PC3231)
Revolutionary communism (#PC3163) Technological revolution (#PC3234).
Related Fear (#PA6030) Defeat (#PA7289) Disorder (#PA7361)
Conflict (#PA0298) Reversion (#PA5699) Agitation (#PA5838)
Anarchism (#PC1972) Separation (#PA7236) Depression (#PA5493)
Difficulty (#PA5497) Illegality (#PA5952) Moderation (#PA7156)
Lawlessness (#PA5563) Inattention (#PA6247) Destruction (#PA6542)
Uncertainty (#PA7309) Disobedience (#PA7250) Fragmentation (#PA6233)
Change*complex (#PA6721) Seizure of power (#PC8270) Guerrilla warfare (#PC1738).
Aggravates Famine (#PB0315) Massacres (#PD2483) Secession (#PD2490)
Martial law (#PD2637) Nationalism (#PB0534) Dictatorship (#PC1049).
Aggravated by Armed insurrection (#PD8284) Ideological conflict (#PF3388).
Reduces Monarchy (#PF2170) Ethnic conflict (#PC3685)
Inaction on problems (#PB1423).

♦ PA5939 Absolute properties*complex

Narrower Cold (#PA6956) Darkness (#PA6261) Lightness (#PA5491).

♦ PA5942 Unimportance

Cheapness — Despicableness — Frivolity — Futility — Idleness — Inanity — Inconsequence — Indifference — Ineffectuality — Inferiority — Insignificance — Irrelevance — Levity — Littleness — Meanness — Mediocrity — Meritlessness — Nullity — Obscurity — Pettiness — Sadness — Shallowness — Smallness — Superficiality — Triviality — Unimportance — Vainness — Valueless — Vanity — Vapidity — Vileness — Worthlessness

Broader Adaptation*complex (#PA8178).
Narrower Indifference (#PA7604).
Related Vice (#PA5644) Envy (#PA7253) Reason (#PA5502)
Vanity (#PA6491) Neglect (#PA5438) Badness (#PA5454)
Fewness (#PA7152) Boredom (#PA7365) Weakness (#PA5558)
Inaction (#PA5806) Humility (#PA6659) Disorder (#PA7361)
Idleness (#PA7710) Smallness (#PA7408) Ignorance (#PA5568)
Vulgarity (#PA5821) Avoidance (#PA6379) Solemnity (#PA6731)
Disrepute (#PA6839) Impotence (#PA6876) Cheapness (#PA7193)
Restraint (#PA7296) Improbity (#PA7363) Unchastity (#PA5612)
Unkindness (#PA5643) Quiescence (#PA6444) Disrespect (#PA6822)
Littleness (#PA7285) Uncleanness (#PA5459) Inferiority (#PA5652)
Shortcoming (#PA6041) Disapproval (#PA6191) Inattention (#PA6247)
Incuriosity (#PA6598) Selfishness (#PA7211) Uncertainty (#PA7309)
Nonexistence (#PA5870) Disagreement (#PA5982) Hopelessness (#PA6099)
Formlessness (#PA6900) Invisibility (#PA6978) Imperfection (#PA6997)
Untimeliness (#PA7006) Irresolution (#PA7325) Inexpedience (#PA7395)
Insufficiency (#PA5473) Nonconformity (#PA5878) Disappearance (#PA6785)
Unsavouriness (#PA7204) Unfeelingness (#PA7364) Unpleasantness (#PA7107)
Unskillfulness (#PA7232) Unintelligence (#PA7371) Thoughtlessness (#PA6940)
Meaninglessness (#PA6977) Narrowmindedness (#PA7306) Nonaccomplishment (#PA6662)
Inappropriateness (#PA6852) Influencelessness (#PA6882) Unintelligibility (#PA7367).

♦ PA5952 Illegality

Anarchism — Anarchy — Bastardy — Feloniousness — Flaw — Illegality — Illegitimate — Lawlessness — Misdemeanour — Offence — Spurious — Transgression — Trespass — Unlawfulness — Violation — Wrongness

Refs Hamline University, Advanced Legal Education Staff *Issues of Illegality* (1985).

Broader Retribution*complex (#PA7037).
Narrower Crime (#PB0001) Illegitimate political regimes (#PC1461).
Related Vice (#PA5644) Envy (#PA7253) Error (#PA6180)
Reason (#PA5502) Defence (#PA5445) Badness (#PA5454)
Refusal (#PA7321) Disorder (#PA7361) Intrusion (#PA6862)
Undueness (#PA6921) Wrongness (#PA7280) Improbity (#PA7363)
Anarchism (#PC1972) Unchastity (#PA5612) Revolution (#PA5901)
Disrespect (#PA6822) Moderation (#PA7156) Lawlessness (#PA5563)
Affectation (#PA6400) Misbehaviour (#PA6498) Imperfection (#PA6997)
Untimeliness (#PA7006) Disobedience (#PA7250) Disintegration (#PA6858)
Meaninglessness (#PA6977) Inappropriateness (#PA6852)
Uncommunicativeness (#PA7411).
Aggravates Injustice (#PA6486).

♦ PA5974 Diffuseness

Extravagance — Indirection — Prolixity

Broader Communication*complex (#PA6732).
Related Boredom (#PA7365) Insanity (#PA7157) Deviation (#PA6228)
Cheapness (#PA7193) Improbity (#PA7363) Affectation (#PA6400)
Intemperance (#PA6466) Insufficiency (#PA5473)
Uncommunicativeness (#PA7411).

♦ PA5981 Stench

Badness — Foulness — Fulsomeness — Miasma — Mustiness — Stuffiness

Broader Sense*complex (#PA6236).
Related Vice (#PA5644) Danger (#PA6971) Badness (#PA5454)
Disease (#PA6799) Boredom (#PA7365) Closure (#PA7391)
Ugliness (#PA7240) Injustice (#PA6486) Disrepute (#PA6839)
Unchastity (#PA5612) Quiescence (#PA6444) Uncleanness (#PA5459)
Disapproval (#PA6191) Affectation (#PA6400) Destruction (#PA6542)
Misbehaviour (#PA6498) Irresolution (#PA7325) Unsavouriness (#PA7204)
Unpleasantness (#PA7107) Unintelligence (#PA7371) Unhealthfulness (#PA7226)
Narrowmindedness (#PA7306) Unimaginativeness (#PA6738).

♦ PA5982 Disagreement

Abnormality — Ambiguity — Ambivalence — Antagonism — Asymmetry — Conflict — Contradiction — Controversy — Disaccord — Disagreeableness — Discord — Disharmony — Disparity — Dissension — Dissidence — Dissonance — Disunity — Divergence — Heresy — Impropriety — Inapplicability — Inappropriateness — Inaptitude — Incoherence — Incompatibility — Incongruity — Inconsistency — Inequality — Infelicity — Inharmonious — Irreconcilability — Irrelevance — Maladjustment — Misalliance — Negation — Negativity — Nonconformity — Opposition — Repugnance — Self-contradiction — Unsuitability — Variance

Broader Relationship*complex (#PA6484).
Narrower Heresy (#PF3375) Imbalance (#PA0224) Impropriety (#PA6000)
Territorial disputes between states (#PC1888).
Related Vice (#PA5644) Envy (#PA7253) Hate (#PA7338)
Error (#PA6180) Enmity (#PA5446) Reason (#PA5502)
Discord (#PA5565) Disease (#PA6799) Dissent (#PA6838)
Refusal (#PA7321) Loudness (#PA5465) Insanity (#PA7157)
Disorder (#PA7361) Unbelief (#PA7392) Conflict (#PA0298)
Disaccord (#PA5532) Vulgarity (#PA5821) Exclusion (#PA5869)
Deviation (#PA6228) Injustice (#PA6486) Vengeance (#PA6606)
Solemnity (#PA6731) Undueness (#PA6921) Wrongness (#PA7280)
Ambiguity (#PF4193) Separation (#PA7236) Difficulty (#PA5497)
Divergence (#PA5573) Unchastity (#PA5612) Unkindness (#PA5643)
Inequality (#PA6695) Difference (#PA6698) Distortion (#PA6790)
Opposition (#PA6979) Disapproval (#PA6191) Destruction (#PA6542)
Uncertainty (#PA7309) Incoherence (#PF8094) Ambivalence (#PF1426)
Unimportance (#PA5942) Hopelessness (#PA6099) Misbehaviour (#PA6498)
Untimeliness (#PA7006) Disobedience (#PA7250) Irresolution (#PA7325)
Inexpedience (#PA7395) Insufficiency (#PA5473) Unrelatedness (#PA5794)
Nonconformity (#PA5878) Counteraction (#PA6485) Unwillingness (#PA6509)
Unsociability (#PA6653) Maladjustment (#PA6739) Nonuniformity (#PA6890)
Fragmentation (#PA6233) Changeableness (#PA5490) Discontentment (#PA6011)
Disintegration (#PA6858) Unpleasantness (#PA7107) Unskillfulness (#PA7232)
Unpreparedness (#PA7341) Contradictions (#PF3667) Incompatibility (#PF9047)
Human inequality (#PA0844) Inappropriateness (#PA6852) Unintelligibility (#PA7367).

♦ PA5984 Scarcity

Strain on world resources

Refs Ophuls, William *Ecology and the Politics of Scarcity* (1977); World Resources Institute Staff and International Institute for Environment *World Resources* (1987).

Broader Insufficiency (#PA5473).

Narrower Limited public land (#PJ0574) Scarcity of residential land (#PD8075)
Scarcity of appropriate transport (#PE8551)
Diminishing capital investment in small communities (#PF6477)
Inadequate supply of appropriate trained manpower in developing countries (#PE6243).
Related False image of scarcity (#PF3002).
Aggravated by Ineffective population control (#PF1020)
Conflicting roles of commodities in capitalism (#PF3115)
Excessive demand for goods in capitalist systems (#PC3116).

♦ PA5988 Softness

Inflexibility — Intractability — Stubbornness
Broader Relative properties*complex (#PA6091).
Narrower Inflexibility (#PA8555).
Related Compulsion (#PA5740) Disobedience (#PA7250) Irresolution (#PA7325)
Unwillingness (#PA6509).

♦ PA6000 Impropriety

Improper people
Broader Unchastity (#PA5612) Disagreement (#PA5982) Misbehaviour (#PA6498).

♦ PA6011 Discontentment

Complacency — Disappointment — Discontentment — Dissatisfaction — Envy — Inadequacy — Indefensibility — Insufficiency — Petulance — Resentment — Restlessness — Self-satisfaction — Sourness — Undesirableness — Uneasiness — Unhappiness — Unsuitability
Broader Feeling*complex (#PA6938).
Related Vice (#PA5644) Fear (#PA6030) Envy (#PA7253)
Error (#PA6180) Enmity (#PA5446) Vanity (#PA6491)
Dissent (#PA6838) Smallness (#PA7408) Vulgarity (#PA5821)
Agitation (#PA5838) Injustice (#PA6486) Solemnity (#PA6731)
Impotence (#PA6876) Wrongness (#PA7280) Impatience (#PA6200)
Inequality (#PA6695) Inadequacy (#PA8199) Lamentation (#PA5479)
Inferiority (#PA5652) Shortcoming (#PA6041) Disapproval (#PA6191)
Complacency (#PA1742) Unhappiness (#PA9191) Disagreement (#PA5982)
Hopelessness (#PA6099) Imperfection (#PA6997) Untimeliness (#PA7006)
Irresolution (#PA7325) Inexpedience (#PA7395) Inhospitality (#PA5458)
Insufficiency (#PA5473) Inexpectation (#PA5527) Inexcitability (#PA5467)
Changeableness (#PA5490) Incompleteness (#PA6652) Unpleasantness (#PA7107)
Unskillfulness (#PA7232) Unpreparedness (#PA7341) Dissatisfaction (#PA8886)
Inappropriateness (#PA6852).

♦ PA6030 Fear

Agitation — Alarmism — Anxiety — Apprehension — Awfulness — Baseness — Bothersomeness — Cowardice — Disconcertion — Discouragement — Dismay — Disquiet — Distress — Disturbance — Eeriness — Frightfulness — Grimness — Gruesomeness — Harassment — Heartlessness — Hideousness — Horribleness — Inquietude — Intimidation — Lapse — Malaise — Morbidity — Nervousness — Panic — Perturbation — Shock — Softness — Spiritlessness — Spunklessness — Strain — Suspense — Tension — Terror — Timidity — Torment — Tremulousness — Trepidation — Trouble — Uneasiness — Upset — Vexation — Weakness — Worry — Xenophobia — Dependence on fear — Fearful people — Fright — Scare — Horror
Nature A feeling of alarm caused by the expectation of danger, pain, disaster or other calamities, fear may lead to aggression and violence including war and conflict of all kinds. It may also lead to discrimination, conformism, inertia, alienation, delusion, dependency, lack of participation, corruption, injustice and inequality. Fear may exist in the form of general anxiety which has no object. Fear may also be phobic, that is, very specific to a particular object, condition, environment, etc, such as claustrophobia. Fear of grievous physical injury or death in the face of a real threat may cause tremor, sweating, heart palpitation, and involuntary passing of urine or faeces. Fear itself can be a cause of death, by shock or heart attack. Fear can grip whole armies or civilian populations.
Among the more peculiar phobias illustrating the range of affliction are taphephobia, pnigophobia, barophobia and batophobia; respectively the fears of being buried alive, of choking, of gravity, and of high objects. There is also phobophobia, the fear of fearing.
Refs Bamber, J H *The Fears of Adolescents* (1979); Campion, M G *Worry* (1986); Lewis, Michael and Rosenblum, Leonard A *The Origins of Fear*, Twitchell, James B *Dreadful Pleasures* (1985); Yardley, Stella S *Fear and Panic* (1988).
Narrower Panic (#PF2633) Terror (#PF8483) Phobia (#PE6354)
Hatred (#PA8487) Cowardice (#PJ4171) Morbidity (#PD4538)
Xenophobia (#PD4957) Nervousness (#PE4171) Hypochondria (#PE8322)
Fear of death (#PF0462) Fear of crime (#PF4682) Discouragement (#PF8948)
Fear of nature (#PF6803) Political fear (#PJ5666) Fear of failure (#PF4125)
Fear of intimacy (#PF8012) Fear of vandalism (#PJ9326) Fear of communism (#PF5233)
Fear of knowledge (#PF9595) Fear of ostracism (#PF2776) Excessive caution (#PF6389)
Fear of reprisals (#PF9078) Military reprisals (#PJ4986) Fear of officialdom (#PD9498)
Fear of the abnormal (#PF7029) Culture-induced fear (#PG2699)
Fear of new technology (#PF9127) Lack of self-confidence (#PF0879)
Fear of increased autonomy (#PE7706) Fear of sexual intercourse (#PF6910)
Fear of food contamination (#PF3904) Pervasive fear of nuclear war (#PC3541)
Fear of emotional sensitivity (#PF9209) Fear of contradicting popular views (#PF2040).
Related Vice (#PA5644) Envy (#PA7253) Hate (#PC7338)
Error (#PA6180) Death (#PA7055) Shock (#PC8245)
Enmity (#PA5446) Vanity (#PA6491) Danger (#PA6971)
Defeat (#PA7289) Neglect (#PA5438) Badness (#PA5454)
Relapse (#PA5619) Impiety (#PA6058) Poverty (#PA6434)
Disease (#PA6799) Boredom (#PA7365) Anxiety (#PA1635)
Tension (#PB6370) Weakness (#PA5558) Inaction (#PA5806)
Humility (#PA6659) Disorder (#PA7361) Unbelief (#PA7392)
Disaccord (#PA5532) Vulgarity (#PA5821) Agitation (#PA5838)
Exclusion (#PA5869) Adversity (#PA6340) Solemnity (#PA6731)
Disrepute (#PA6839) Impotence (#PA6876) Undueness (#PA6921)
Improbity (#PA7363) Difficulty (#PA5497) Unkindness (#PA5643)
Compulsion (#PA5740) Revolution (#PA5901) Impairment (#PA6088)
Impatience (#PA6200) Regression (#PA6338) Quiescence (#PA6444)
Moderation (#PA7156) Dissuasion (#PA7343) Insecurity (#PA0857)
Inferiority (#PA5652) Disapproval (#PA6191) Inattention (#PA6247)
Destruction (#PA6542) Uncertainty (#PA7309) Hopelessness (#PA6099)
Imperfection (#PA6997) Pitilessness (#PA7023) Irresolution (#PA7325)
Inexpedience (#PA7395) Insensibility (#PA5451) Insufficiency (#PA5473)
Appropriation (#PA5688) Unsavouriness (#PA7204) Unfeelingness (#PA7364)
Discontentment (#PA6011) Unpleasantness (#PA7107) Unintelligence (#PA7371)
Inexcitability (#PA5467) Unhealthfulness (#PA7226) Narrowmindedness (#PA7306)
Inappropriateness (#PA6852) Influencelessness (#PA6882)
Uncommunicativeness (#PA7411) Resistance to grace (#PF5266)
Anticipation*complex (#PA6562) Psychological inertia (#PF0421).
Aggravates Fatigue (#PA0657) Hysteria (#PE6412) Massacres (#PD2483)
Alienation (#PA3545) Superstition (#PA0430) Draft evasion (#PD0356)
Dehumanization (#PA1757) Mental disorders (#PD9131) Ancestor worship (#PD2315)
Lack of human unity (#PF2434) Emotional disorders (#PD9159)
(#PB0009) Human physical suffering (#PB5646)
Non-acceptance of reality (#PF1079) Inadequate firearm regulation (#PD1970)
Inadequate sense of personal identity (#PF1934)
Negative effects of claims of religious infallibility (#PF3376).
Aggravated by Sabotage (#PD0405) Alarmism (#PF4384)
Ugliness (#PA7240) Occultism (#PF3312)
Aggression (#PA0587) Witchcraft (#PF2099)
Deprivation (#PA0831) Intimidation (#PB1992)
Aerial piracy (#PD0124) Assassination (#PD1971)
Human sacrifice (#PF2641) Military secrecy (#PC1144)
Terrorist bombing (#PE2368) Human cannibalism (#PF2513)
Psychotic violence (#PF7645) Biochemical warfare (#PC1164)
National isolationism (#PF2141) Lack of understanding (#PJ4173)
Inadequate riot control (#PD2207) State sanctioned torture (#PD0181)
Inadequacy of civil defence (#PF0506) Biased presentation of news (#PD1718)
Head hunting in tribal societies (#PF2666) Inhumane methods of riot control (#PD1156)
Ineffective war crime prosecution (#PD1464) International non-military conflict (#PF3100)
Denial of human rights in armed conflicts (#PC1454)
Lack of understanding of spiritual healing (#PF0761)
Unhealthy emotional responses to atomic energy (#PF0913).

♦ PA6041 Shortcoming

Decline — Default — Encroachment — Failure — Imperfection — Inadequacy — Inferiority — Insufficiency — Shortage
Broader Relative motion*complex (#PA6752).
Narrower Shortage (#PB8238).
Related Vice (#PA5644) Error (#PA6180) Defeat (#PA7289)
Neglect (#PA5438) Badness (#PA5454) Poverty (#PA6434)
Absence (#PA7270) Refusal (#PA7321) Weakness (#PA5558)
Smallness (#PA7408) Adversity (#PA6340) Necessity (#PA6387)
Intrusion (#PA6862) Impotence (#PA6876) Undueness (#PA6921)
Cheapness (#PA7193) Restraint (#PA7296) Impairment (#PA6088)
Inequality (#PA6695) Inadequacy (#PA8199) Inferiority (#PA5652)
Unimportance (#PA5942) Imperfection (#PA6997) Insufficiency (#PA5473)
Inexpectation (#PA5527) Nonconformity (#PA5878) Discontentment (#PA6011)
Incompleteness (#PA6652) Unskillfulness (#PA7232) Unintelligence (#PA7371)
Nonaccomplishment (#PA6662).

♦ PA6048 Motivation*complex

Narrower Dissuasion (#PA7343) Informality (#PA7170)
Unconventionality (#PA7273) Criminal motivation (#PJ6406).

♦ PA6058 Impiety

Atheism — Blasphemy — Desecration — Desertion — Dishonour — Dubiousness — Godlessness — Impiety — Irreligiousness — Irreverence — Lapse — Paganism — Profanation — Sacrilege — Self-righteousness — Unctuousness
Broader Redemption*complex (#PA7259).
Narrower Blasphemy (#PF5630) Sacrilege (#PF0662)
Unbelievers (#PF8068).
Related Vice (#PA5644) Fear (#PA6030) Error (#PA6180)
Danger (#PA6971) Neglect (#PA5438) Relapse (#PA5619)
Atheism (#PF2409) Unbelief (#PA7392) Ignorance (#PA5568)
Reversion (#PA5699) Avoidance (#PA6379) Disrepute (#PA6839)
Wrongness (#PA7280) Improbity (#PA7363) Dishonour (#PF8485)
Impairment (#PA6088) Regression (#PA6338) Disrespect (#PA6822)
Disapproval (#PA6191) Inattention (#PA6247) Uncertainty (#PA7309)
Desecration (#PF9176) Irresolution (#PA7325) Unsociability (#PA6653)
Inappropriateness (#PA6852) Uncommunicativeness (#PA7411).
Aggravated by Repression (#PB0871).

♦ PA6073 Posterity

Broader Power*complex (#PA7314).
Aggravates Economic inflation (#PC0254) Economic stagnation (#PC0002).

♦ PA6088 Impairment

Blemish — Blight — Brokenness — Corruption — Damage — Decadence — Decline — Decomposition — Decrepitness — Degeneration — Degradation — Depravation — Depreciation — Depression — Derogation — Deterioration — Dilapidation — Disability — Disintegration — Disorganization — Dissolution — Effetism — Erosion — Failure — Injury — Involution — Lapse — Mischief — Misuse — Mortification — Mutilation — Perversion — Poisonousness — Pollution — Prostitution — Putrefaction — Regression — Retrogression — Rot — Ruin — Sabotage — Spoilage — Strain — Vitiation — Waste — Wear — Withering — Wound
Broader Integrity*complex (#PA7263).
Narrower Wear (#PB1701) Erosion (#PC8193) Spoilage (#PG3966)
Decadence (#PB2542) Corruption (#PA1986).
Related Vice (#PA5644) Fear (#PA6030) Envy (#PA7253)
Loss (#PA7382) Error (#PA6180) Death (#PA7055)
Enmity (#PA5446) Reason (#PA5502) Defeat (#PA7289)
Neglect (#PA5438) Badness (#PA5454) Relapse (#PA5619)
Impiety (#PA6058) Disease (#PA6799) Oldness (#PA7131)
Refusal (#PA7321) Boredom (#PA7365) Weakness (#PA5558)
Inaction (#PA5806) Humility (#PA6659) Ugliness (#PA7240)
Disorder (#PA7361) Disaccord (#PA5532) Reversion (#PA5699)
Exclusion (#PA5869) Adversity (#PA6340) Necessity (#PA6387)
Injustice (#PA6486) Solemnity (#PA6731) Disrepute (#PA6839)
Impotence (#PA6876) Cheapness (#PA7193) Improbity (#PA7363)
Pollution (#PB6336) Separation (#PA7236) Depression (#PA5493)
Difficulty (#PA5497) Unchastity (#PA5612) Unkindness (#PA5643)
Regression (#PA6338) Complexity (#PA6468) Distortion (#PA6790)
Disrespect (#PA6822) Dissuasion (#PA7343) Uncleanness (#PA5459)
Lawlessness (#PA5563) Inferiority (#PA5652) Shortcoming (#PA6041)
Disapproval (#PA6191) Inattention (#PA6247) Destruction (#PA6542)
Aggravation (#PA7378) Miseducation (#PA6393) Misbehaviour (#PA6498)
Misjudgement (#PA6607) Imperfection (#PA6997) Disobedience (#PA7250)
Irresolution (#PA7325) Inexpedience (#PA7395) Prostitution (#PD0693)
Insensibility (#PA5451) Insufficiency (#PA5473) Inexpectation (#PA5527)
Discontinuity (#PA5828) Disappearance (#PA6785) Fragmentation (#PA6233)
Inexcitability (#PA5467) Disintegration (#PA6858) Unpleasantness (#PA7107)
Unintelligence (#PA7371) Meaninglessness (#PA6977) Unhealthfulness (#PA7226)
Misrepresentation (#PA6644) Nonaccomplishment (#PA6662) Inappropriateness (#PA6852)
Misinterpretability (#PA6741) Uncommunicativeness (#PA7411).
Aggravates Disorganization (#PF4487).

♦ PA6090 Unastonishment

Broader Discriminative affection*complex (#PA7145).

♦ PA6091 Relative properties*complex

Narrower Softness (#PA5988) Toughness (#PA6976) Colourlessness (#PA7301).

♦ PA6099 Hopelessness

Apathy — Cheerlessness — Cynicism — Despair — Despondency — Disappointment — Futility — Grimness — Hopelessness — Impossibility — Incorrigibility — Irredeemability — Irreformability — Irremediability — Negativity — Pessimism — Slothfulness — Vainness

Broader Anticipation*complex (#PA6562).

Narrower Despair (#PF4004) Cynicism (#PF3418).

Related Vice (#PA5644) Fear (#PA6030) Envy (#PA7253) Error (#PA6180) Reason (#PA5502) Vanity (#PA6491) Chance (#PA6714) Apathy (#PA2360) Badness (#PA5454) Dissent (#PA6838) Refusal (#PA7321) Inaction (#PA5806) Ugliness (#PA7240) Avoidance (#PA6379) Solemnity (#PA6731) Impotence (#PA6876) Unkindness (#PA5643) Quiescence (#PA6444) Disrespect (#PA6822) Opposition (#PA6979) Disapproval (#PA6191) Destruction (#PA6542) Incuriosity (#PA6598) Unimportance (#PA5942) Disagreement (#PA5982) Irresolution (#PA7325) Inexpedience (#PA7395) Inexpectation (#PA5527) Impossibility (#PA6487) Unfeelingness (#PA7364) Discontentment (#PA6011) Unpleasantness (#PA7107) Meaninglessness (#PA6977) Nonaccomplishment (#PA6662) Inappropriateness (#PA6852).

♦ PA6108 Quantity*complex

Narrower Decrease (#PA5589) Smallness (#PA7408) Complexity (#PA6468) Separation (#PA7236) Inferiority (#PA5652) Incompleteness (#PA6652) Disintegration (#PA6858).

♦ PA6118 Ineloquence

Muteness

Broader Communication*complex (#PA6732).

Related Loudness (#PA5465) Uncommunicativeness (#PA7411).

♦ PA6164 Truth*complex

Narrower Denial (#PA7400) Dissent (#PA6838).

Related Disillusionment (#PA6453).

♦ PA6166 Slowness

Broader Motion*complex (#PA5468).

Related Boredom (#PA7365) Inaction (#PA5806) Untimeliness (#PA7006) Unintelligence (#PA7371).

♦ PA6176 Immateriality

Broader Life*complex (#PA6586).

♦ PA6180 Error

Aberration — Autism — Deception — Defectiveness — Delusion — Deviation — Disappointment — Disenchantment — Disillusionment — Distortion — Erroneousness — Failure — Fallacy — Falseness — Faultiness — Flaw — Hallucination — Heresy — Illogic — Imprecision — Inaccuracy — Inadvertence — Incorrectness — Indiscretion — Lapse — Laxity — Looseness — Misapplication — Misapprehension — Misbelief — Miscalculation — Miscarriage — Misconception — Misconduct — Misconstruction — Misfeasance — Misinterpretation — Misjudgement — Misstatement — Mistake — Misunderstanding — Misuse — Negligence — Omission — Oversight — Perversion — Self-contradiction — Spurious — Stupidity — Unauthenticity — Unreality — Wrongness

Refs Keeler, Leo W *The Problem of Error from Plato to Kant* (1977).

Broader Credibility*complex (#PA5829).

Narrower Heresy (#PF3375) Fallacy (#PF4357) Illusion (#PA6414) Misunderstanding (#PA8197).

Related Vice (#PA5644) Fear (#PA6030) Guilt (#PA6793) Reason (#PA5502) Defeat (#PA7289) Neglect (#PA5438) Badness (#PA5454) Relapse (#PA5619) Impiety (#PA6058) Disease (#PA6799) Rashness (#PA7115) Insanity (#PA7157) Disorder (#PA7361) Unbelief (#PA7392) Disaccord (#PA5532) Reversion (#PA5699) Deviation (#PA6228) Avoidance (#PA6379) Injustice (#PA6486) Impotence (#PA6876) Wrongness (#PA7280) Improbity (#PA7363) Deception (#PB4731) Unchastity (#PA5612) Compulsion (#PA5740) Illegality (#PA5952) Impairment (#PA6088) Regression (#PA6338) Difference (#PA6698) Distortion (#PA6790) Negligence (#PA2658) Inaccuracy (#PF7905) Unorthodoxy (#PA5564) Inferiority (#PA5652) Shortcoming (#PA6041) Disapproval (#PA6191) Inattention (#PA6247) Affectation (#PA6400) Selfishness (#PA7211) Uncertainty (#PA7309) Nonexistence (#PA5870) Disagreement (#PA5982) Hopelessness (#PA6099) Miseducation (#PA6393) Misbehaviour (#PA6498) Misjudgement (#PA6607) Imperfection (#PA6997) Untimeliness (#PA7006) Insufficiency (#PA5473) Inexpectation (#PA5527) Unrelatedness (#PA5794) Nonconformity (#PA5878) Unsociability (#PA6653) Maladjustment (#PA6739) Nonuniformity (#PA6890) Unfeelingness (#PA7364) Discontentment (#PA6011) Incompleteness (#PA6652) Disintegration (#PA6858) Unskillfulness (#PA7232) Unpreparedness (#PA7341) Unintelligence (#PA7371) Disillusionment (#PA6453) Indiscrimination (#PA6446) Insubstantiality (#PA6959) Misrepresentation (#PA6644) Nonaccomplishment (#PA6662) Unimaginativeness (#PA6738) Inappropriateness (#PA6852) Misinterpretability (#PA6741) Uncommunicativeness (#PA7411).

♦ PA6191 Disapproval

Abusiveness — Anathema — Blame — Censoriousness — Censure — Condemnation — Contemptuousness — Contumeliousness — Culpability — Damnation — Defamation — Denigration — Denunciation — Depreciation — Derogation — Dirtiness — Disappointment — Disapproval — Discontentment — Discredit — Disenchantment — Disesteem — Disfavour — Disgrace — Disillusionment — Disparagement — Displeasure — Disrespect — Dissatisfaction — Distaste — Exclusion — Execration — Filth — Foulness — Fuss — Indignation — Insinuation — Intimidation — Menace — Opposition — Opprobrium — Rejection — Reproach — Revilement — Ridicule — Satire — Scandal — Threat — Unctuousness — Unhappiness — Vileness — Vituperation

Broader Judgement*complex (#PA8528).

Narrower Satire (#PJ5950) Scandal (#PC8391) Anathema (#PG4069).

Related Vice (#PA5644) Fear (#PA6030) Envy (#PA7253) Hate (#PA7338) Loss (#PA7382) Error (#PA6180) Guilt (#PA6793) Stench (#PA5981) Vanity (#PA6491) Danger (#PA6971) Badness (#PA5454) Impiety (#PA6058) Dissent (#PA6838) Refusal (#PA7321) Closure (#PA7391) Humility (#PA6659) Ugliness (#PA7240) Disorder (#PA7361) Unbelief (#PA7392) Disaccord (#PA5532) Reversion (#PA5699) Vulgarity (#PA5821) Agitation (#PA5838) Exclusion (#PA5869) Necessity (#PA6387) Injustice (#PA6486) Solemnity (#PA6731) Disrepute (#PA6839) Intrusion (#PA6862) Cheapness (#PA7193) Wrongness (#PA7280) Improbity (#PA7363) Difficulty (#PA5497) Unchastity (#PA5612) Unkindness (#PA5643) Compulsion (#PA5740) Impairment (#PA6088) Impatience (#PA6200) Difference (#PA6698) Disrespect (#PA6822) Opposition (#PA6979) Moderation (#PA7156) Dissuasion (#PA7343) Uncleanness (#PA5459) Lamentation (#PA5479) Inattention (#PA6247) Affectation (#PA6400) Destruction (#PA6542) Displeasure (#PA6809) Unhappiness (#PA9191) Unimportance (#PA5942) Disagreement (#PA5982) Hopelessness (#PA6099) Misjudgement (#PA6607) Imperfection (#PA6997) Condemnation (#PA7237) Inexpectation (#PA5527) Unwillingness (#PA6509) Unprovability (#PA7070) Unsavouriness (#PA7204) Inexcitability (#PA5467) Discontentment (#PA6011) Unpleasantness (#PA7107) Disillusionment (#PA6453) Dissatisfaction (#PA8886) Inappropriateness (#PA6852) Uncommunicativeness (#PA7411).

♦ PA6200 Impatience

Anxiety — Fretfulness — Fuss — Haste — Hastiness — Impatience — Impetuousity — Intolerance — Passivity — Restlessness — Uneasiness

Refs Varma, Ved P *Anxiety in Children* (1984).

Broader Feeling*complex (#PA6938).

Related Fear (#PA6030) Envy (#PA7253) Vanity (#PA6491) Neglect (#PA5438) Anxiety (#PA1635) Inaction (#PA5806) Rashness (#PA7115) Disorder (#PA7361) Disaccord (#PA5532) Agitation (#PA5838) Difficulty (#PA5497) Transience (#PA6425) Quiescence (#PA6444) Moderation (#PA7156) Lamentation (#PA5479) Disapproval (#PA6191) Inattention (#PA6247) Intolerance (#PF0860) Untimeliness (#PA7006) Disobedience (#PA7250) Irresolution (#PA7325) Unfeelingness (#PA7364) Inexcitability (#PA5467) Changeableness (#PA5490) Discontentment (#PA6011) Unpleasantness (#PA7107) Narrowmindedness (#PA7306).

♦ PA6220 Spiritual void

Existential vacuum — Meaninglessness of life

Nature Individuals may arrive at the opinion that their life has no purpose or value, or that human life has no meaning, whether in spiritual or existential terms. Causes for this may lie in the body: middle-age crises, the onset of debilitating old-age, ill-health, clinical depression, reaction to medication, alcohol, drugs or other somatic agents; or they may lie in psychosomatic reactions from emotional shocks such as grief, and catastrophic upheavals in personal life. The same negative viewpoint may be reached by some schools of reasoning or philosophies of nihilistic tendency, or may result from being a victim of the disasters of war or other mass human destruction. Persons who are isolated, for whom no one appears to care, and in whose lives love has no place, are also vulnerable to the devastating impression of living in a void, which may culminate in suicide a in as its extreme result, or a listless and apathetic life at best, if not a mentally ill one.

Ontologically, persons experiencing a sense of void may be unsure of the reality of their own identity. They have a divided self, one part of which experiences something, the other part of which denies the experiences and the experiencer as being unreal or meaningless.

Narrower Cargo cults (#PF5375).

Related Theological collapse (#PF6358) Dehumanization of death (#PF2442).

Aggravates Deterioration of human environment (#PC8943).

♦ PA6228 Deviation

Aberration — Deviousness — Divergence — Indirection — Slant — Variation

Broader Relative motion*complex (#PA6752).

Narrower Violating taboos (#PF3976).

Relate Error (#PA6180) Insanity (#PA7157) Disaccord (#PA5532) Improbity (#PA7363) Divergence (#PA5573) Complexity (#PA6468) Difference (#PA6698) Distortion (#PA6790) Diffuseness (#PA5974) Disagreement (#PA5982) Nonconformity (#PA5878) Nonuniformity (#PA6890) Changeableness (#PA5490) Unintelligence (#PA7371) Misrepresentation (#PA6644) Uncommunicativeness (#PA7411).

♦ PA6233 Fragmentation

Disunity — Dependence on disunity — Lack of unity — Lack of integration

Claim Today's world is characterized by disunity. Disunity in politics, in thought in world undertakings, in freedom, in religion, in nations or races, and in language, allow man neither to understand nor to implement the organic oneness of humanity.

Counter-claim "Disunity" may actually only be man asserting his individuality which in and of itself propagates neither tension nor war, but is a very healthy and necessary avenue of expression and contrast.

Narrower Lack of human unity (#PF2434) Social fragmentation (#PF1324) Ideological conflict (#PF3388) Cultural fragmentation (#PF0536) Lack of national unity (#PF8107) Political fragmentation (#PF3216) Lack of a world religion (#PF3387) Political disintegration (#PC3204) Institutional fragmentation (#PC3915) Fragmentation of religious belief (#PF3404) Fragmentation of communist parties (#PD0923) Lack of world maritime integration (#PE5801) International economic fragmentation (#PC0025) Fragmentation of the human personality (#PA0911) Fragmentation of technological development (#PC1227) Lack of a sense of community and solidarity at the world level (#PF8704) Lack of autonomous world-level actor to identify and clarify world interests (#PF0053).

Related Enmity (#PA5446) Disorder (#PA7361) Separation (#PA7236) Impairment (#PA6088) Revolution (#PA5901) Destruction (#PA6542) Disagreement (#PA5982) Factionalism (#PF8454) Unnatural boundaries between states (#PF0090).

Aggravates Conflict (#PA0298) Disintegration (#PA6858) Human suffering (#PB5955) Ideological schism in communism (#PF3181) General obstacles to problem alleviation (#PF0631) Offences against the peace and security of mankind (#PC6239).

Aggravated by Ignorance (#PA5568) Regression (#PA6338) Liberalism (#PF0717) Deviant society (#PC2405) Lack of control (#PF7138) Ethnocentricity (#PB5765) Over-specialization (#PF0256) Lack of social mobility (#PF2195) Competition between states (#PC0114).

Reduces Domination (#PA0839) Social hierarchy (#PF4947) Over-centralization (#PF2711) Criminal solicitation (#PD7676) Military and economic hegemony (#PB0318) Domination of the world by territorially organized sovereign states (#PD0055).

♦ PA6236 Sense*complex

Narrower Stench (#PA5981) Discord (#PA5565) Loudness (#PA5465) Blindness (#PA6674) Invisibility (#PA6978) Insensibility (#PA5451) Intangibility (#PA5570) Unsavouriness (#PA7204) Disappearance (#PA6785).

♦ PA6247 Inattention

Atheism — Carelessness — Chaos — Confusion — Discomposure — Disconcertion — Disorder — Disorganization — Disorientation — Disregard — Distraction — Disturbance — Embarrassment — Frenzy — Frivolity — Fuss — Heedlessness — Inadvertence — Incuriosity — Indifference — Inobservance — Levity — Muddle — Negligence — Nonobservance — Perplexity — Perturbation — Shallowness — Superficiality — Thoughtlessness — Upset — Witlessness

Broader Attitude*complex (#PA6983).

Narrower Indifference (#PA7604).

Related Fear (#PA6030), Error (#PA6180), Chaos (#PF6836), Vanity (#PA6491), Defeat (#PA7289), Neglect (#PA5438), Impiety (#PA6058), Poverty (#PA6434), Disease (#PA6799), Refusal (#PA7321), Boredom (#PA7365), Atheism (#PF2409), Inaction (#PA5806), Humility (#PA6659), Rashness (#PA7115), Insanity (#PA7157), Disorder (#PA7361), Disaccord (#PA5532), Ignorance (#PA5568), Agitation (#PA5838), Avoidance (#PA6379), Necessity (#PA6387), Confusion (#PF7123), Difficulty (#PA5497), Unkindness (#PA5643), Compulsion (#PA5740), Revolution (#PA5901), Impairment (#PA6088), Impatience (#PA6200), Quiescence (#PA6444), Complexity (#PA6468), Disrespect (#PA6822), Opposition (#PA6979), Moderation (#PA7156), Negligence (#PA2658), Lamentation (#PA5479), Lawlessness (#PA5563), Disapproval (#PA6191), Destruction (#PA6542), Incuriosity (#PA6598), Uncertainty (#PA7309), Unimportance (#PA5942), Misbehaviour (#PA6498), Formlessness (#PA6900), Imperfection (#PA6997), Disobedience (#PA7250), Irresolution (#PA7325), Forgetfulness (#PA6651), Maladjustment (#PA6739), Disappearance (#PA6785), Unfeelingness (#PA7364), Inexcitability (#PA5467), Disintegration (#PA6858), Unpleasantness (#PA7107), Unskillfulness (#PA7232), Unpreparedness (#PA7341), Unintelligence (#PA7371), Indiscrimination (#PA6446), Nonaccomplishment (#PA6662), Unintelligibility (#PA7367).

♦ PA6261 Darkness
Darkness — Deadness — Lifelessness
Broader Absolute properties*complex (#PA5939).
Related Death (#PA7055), Boredom (#PA7365), Inaction (#PA5806), Ignorance (#PA5568), Blindness (#PA6674), Solemnity (#PA6731), Quiescence (#PA6444), Invisibility (#PA6978), Insensibility (#PA5451), Unsavouriness (#PA7204), Unintelligibility (#PA7367).

♦ PA6290 Imminence
Broader Causation*complex (#PA7406).

♦ PA6312 Inelegance
Awkwardness — Clumsiness — Gracelessness — Inelegance — Unnaturalness
Broader Communication*complex (#PA6732).
Related Neglect (#PA5438), Badness (#PA5454), Insanity (#PA7157), Ugliness (#PA7240), Ignorance (#PA5568), Vulgarity (#PA5821), Difficulty (#PA5497), Unchastity (#PA5612), Unkindness (#PA5643), Affectation (#PA6400), Inexpedience (#PA7395), Nonconformity (#PA5878), Unpleasantness (#PA7107), Unskillfulness (#PA7232), Uncommunicativeness (#PA7411).

♦ PA6318 Egoism
Dependence on egoism
Nature Egoism is an ethical stance in which private interests are viewed as the basic motive for all action and as the chief criterion of value applied to society and to an individual's immediate milieu. In both the life of an individual and the life of a nation, egoism can lead to a separation which sets individual or national goals as priorities, thus serving to polarize one being or one nation from the totality of universal cohesiveness.
Refs Nordau, Max *Egomania and the Psychology of Contemporary Man* (1986).
Broader Vice (#PA5644).
Narrower Pride (#PA7599), Solipsism (#PF5657).
Related Vanity (#PA6491), Avoidance of legal obligations by politicians (#PD4556).
Aggravates Banditry (#PD2609).

♦ PA6338 Regression
Lapse — Regression — Retrogression — Reversion — Withdrawal
Broader Relative motion*complex (#PA6752).
Related Vice (#PA5644), Fear (#PA6030), Error (#PA6180), Neglect (#PA5438), Relapse (#PA5619), Impiety (#PA6058), Dissent (#PA6838), Duality (#PA7339), Reversion (#PA5699), Exclusion (#PA5869), Avoidance (#PA6379), Impairment (#PA6088), Withdrawal (#PF4402), Incuriosity (#PA6598), Irresolution (#PA7325), Unfeelingness (#PA7364).
Aggravates Ignorance (#PA5568), Fragmentation (#PA6233), Negative emotions and attitudes (#PA7090).

♦ PA6340 Adversity
Accident — Affliction — Aggravation — Annoyance — Calamity — Decline — Destructiveness — Difficulty — Disaster — Grief — Lucklessness — Misfortune — Overload — Shock — Tribulation — Trouble — Vicissitude
Broader Achievement*complex (#PA5471).
Narrower Grief (#PF5654).
Related Fear (#PA6030), Envy (#PA7253), Death (#PA7055), Shock (#PC8245), Badness (#PA5454), Disease (#PA6799), Refusal (#PA7321), Weakness (#PA5558), Lightness (#PA5491), Disaccord (#PA5532), Necessity (#PA6387), Solemnity (#PA6731), Undueness (#PA6921), Cheapness (#PA7193), Difficulty (#PA5497), Impairment (#PA6088), Shortcoming (#PA6041), Destruction (#PA6542), Aggravation (#PA7378), Inexpedience (#PA7395), Insufficiency (#PA5473), Changeableness (#PA5490), Unpleasantness (#PA7107), Unintelligence (#PA7371), Unhealthfulness (#PA7226), Unintelligibility (#PA7367).

♦ PA6355 Reaction
Broader Motion*complex (#PA5468).
Aggravates General obstacles to problem alleviation (#PF0631)
Obstacles to the development of multidisciplinary approaches (#PF7923).

♦ PA6359 Manipulation
Dependence on manipulation — Manipulative people
Nature Dominating another person so that they act according to one's will. Manipulation implies that devious means are used and that the manipulated may well be acting against their own self-interest.
Refs Valorian Society *Human History Viewed As Sovereign Individuals vs Manipulated Masses* (1986).
Narrower Economic manipulation (#PC6875), Manipulative knowledge (#PF1609), Manipulation of students (#PE5777), Manipulation of civic education (#PF0996), Provocations of capitalist militarism (#PC0937).
Related Institutional lying (#PD2686).

♦ PA6373 Socialization*complex
Narrower Hate (#PA7338), Enmity (#PA5446), Celibacy (#PA7410), Unsociability (#PA6653), Inhospitality (#PA5458).

♦ PA6379 Avoidance
Abandonment — Acquisitiveness — Apathy — Avarice — Avoidance — Betrayal — Carelessness — Coldness — Defection — Dereliction — Desertion — Disappearance — Disinterest — Disregard — Disuse — Equivocation — Evasion — Fanaticism — Frenzy — Fury — Greed — Heedlessness — Inattention — Incontinence — Incuriosity — Indifference — Indiscrimination — Infatuation — Insouciance — Listlessness — Lovelessness — Lust — Malingering — Mindlessness — Negligence — Overzealousness — Passionlessness — Rapacity — Slothfulness — Sordidness — Voracity — Withdrawal — Zealotry
Broader Choice*complex (#PA7214).
Related Vice (#PA5644), Cold (#PA6956), Envy (#PA7253), Error (#PA6180), Enmity (#PA5446), Reason (#PA5502), Apathy (#PA2360), Neglect (#PA5438), Badness (#PA5454), Impiety (#PA6058), Disease (#PA6799), Dissent (#PA6838), Oldness (#PA7131), Absence (#PA7270), Duality (#PA7339), Avarice (#PA6999), Weakness (#PA5558), Inaction (#PA5806), Rashness (#PA7115), Insanity (#PA7157), Disorder (#PA7361), Unbelief (#PA7392), Ignorance (#PA5568), Reversion (#PA5699), Exclusion (#PA5869), Necessity (#PA6387), Disrepute (#PA6839), Cheapness (#PA7193), Restraint (#PA7296), Improbity (#PA7363), Unchastity (#PA5612), Unkindness (#PA5643), Compulsion (#PA5740), Regression (#PA6338), Quiescence (#PA6444), Difference (#PA6698), Disrespect (#PA6822), Opposition (#PA6979), Moderation (#PA7156), Withdrawal (#PF4402), Negligence (#PA2658), Uncleanness (#PA5459), Inattention (#PA6247), Incuriosity (#PA6598), Selfishness (#PA7211), Abandonment (#PA7685), Malingering (#PE7701), Unimportance (#PA5942), Hopelessness (#PA6099), Intemperance (#PA6466), Imperfection (#PA6997), Disobedience (#PA7250), Irresolution (#PA7325), Inhospitality (#PA5458), Insufficiency (#PA5473), Appropriation (#PA5688), Forgetfulness (#PA6651), Unsociability (#PA6653), Disappearance (#PA6785), Unfeelingness (#PA7364), Unpleasantness (#PA7107), Unskillfulness (#PA7232), Unpreparedness (#PA7341), Unintelligence (#PA7371), Indiscrimination (#PA6446), Narrowmindedness (#PA7306), Inappropriateness (#PA6852), Unintelligibility (#PA7367), Uncommunicativeness (#PA7411), Political radicalism (#PF2177).

♦ PA6387 Necessity
Contemptuousness — Decline — Denial — Disapproval — Disdain — Disregard — Exclusion — Indiscrimination — Nonacceptance — Rejection — Repudiation — Urgency
Broader Choice*complex (#PA7214).
Related Envy (#PA7253), Loss (#PA7382), Vanity (#PA6491), Denial (#PA7400), Neglect (#PA5438), Dissent (#PA6838), Refusal (#PA7321), Weakness (#PA5558), Unbelief (#PA7392), Exclusion (#PA5869), Adversity (#PA6340), Avoidance (#PA6379), Cheapness (#PA7193), Impairment (#PA6088), Disrespect (#PA6822), Opposition (#PA6979), Shortcoming (#PA6041), Disapproval (#PA6191), Inattention (#PA6247), Disobedience (#PA7250), Irresolution (#PA7325), Unprovability (#PA7070), Unpleasantness (#PA7107), Unintelligence (#PA7371), Indiscrimination (#PA6446), Inappropriateness (#PA6852).

♦ PA6393 Miseducation
Brainwash — Corruption — Misdirection — Misguidance — Misinformation — Mystification — Perversion — Sophistry
Broader Inadequate education (#PF4984), Communication*complex (#PA6732).
Narrower Sophistry (#PF8008), Misinformation (#PD8523).
Related Vice (#PA5644), Error (#PA6180), Reason (#PA5502), Badness (#PA5454), Ignorance (#PA5568), Reversion (#PA5699), Improbity (#PA7363), Impairment (#PA6088), Complexity (#PA6468), Distortion (#PA6790), Dissuasion (#PA7343), Uncleanness (#PA5459), Unskillfulness (#PA7232), Misrepresentation (#PA6644), Inappropriateness (#PA6852), Unintelligibility (#PA7367), Misinterpretability (#PA6741), Uncommunicativeness (#PA7411).
Aggravates Corruption (#PA1986).

♦ PA6400 Affectation
Artificiality — Blatancy — Censoriousness — Conceit — Elaborateness — Extravagance — Fakery — Hypocrisy — Insincerity — Ostentation — Pedantry — Pomposity — Pretentiousness — Sanctimony — Self-importance — Severity — Shamelessness — Spurious — Stuffiness — Sumptuousness — Unnaturalness
Broader Discriminative affection*complex (#PA7145).
Narrower Pedantry (#PG0181), Hypocrisy (#PF3377).
Related Vice (#PA5644), Error (#PA6180), Reason (#PA5502), Stench (#PA5981), Vanity (#PA6491), Boredom (#PA7365), Humility (#PA6659), Insanity (#PA7157), Ugliness (#PA7240), Cheapness (#PA7193), Wrongness (#PA7280), Improbity (#PA7363), Unchastity (#PA5612), Unkindness (#PA5643), Compulsion (#PA5740), Illegality (#PA5952), Inelegance (#PA6312), Quiescence (#PA6444), Moderation (#PA7156), Lamentation (#PA5479), Diffuseness (#PA5974), Disapproval (#PA6191), Discourtesy (#PA7143), Informality (#PA7170), Intemperance (#PA6466), Condemnation (#PA7237), Irresolution (#PA7325), Insufficiency (#PA5473), Nonconformity (#PA5878), Unintelligence (#PA7371), Narrowmindedness (#PA7306), Unimaginativeness (#PA6738), Uncommunicativeness (#PA7411), Inadequacy of formal logic (#PG7788).
Aggravates Defective reasoning (#PF5711), Inconclusiveness of science (#PF6349).

♦ PA6414 Illusion
Delusion — Misperception — Hallucination — Delusions
Nature Indefinite knowledge or incorrect interpretation can result from defective sensation. The defect may lie in the senses themselves, as a physiological fault in perception, or in the indefiniteness or blurring or contradiction in the sense stimuli, as found, for example, in optical illusions. The synthetic ability of the mind to piece together, or form a whole, from incomplete data often leads it to make errors. The mind sees what it is accustomed to seeing. Such errors in interpretation or imprecise knowledge may lead to incorrect judgements and as a result, to accidents. Unfortunately, when the habit of seeing what is desired to be seen instead of what is actually there is brought to bear on developmental and social issues, illusions may be maintained in official reports, particularly those showing satisfaction and giving positive evaluations for what is being done to address world problems.
Incidence An example in navigation is confused hearing of foghorns near high shores where echoes are thrown back, or in visual radar-control of fast-moving, apparently collision-bound aircraft, where rational after-images of blips can occur to the controllers.
Background The mind may substitute one identity for another, as in such cases as when an expected omnibus is seen although it is actually a truck. In other cases of illusion, the mind projects an identification where there is no object. For example the mind sees concrete images in abstract masses such as clouds, shadows, ink blots or tea leaves. In a third manner of illusion the mind compensates for what is there by seeing what ought to be there by custom, or by not seeing what is not there. Illusion by substitution, projection, and compensation is part of the mental process of perception.
Refs Wade, Nicholas *The Art and Science of Visual Illusions* (1983).
Broader Error (#PA6180), Maladjustment (#PA6739), Uncommunicativeness (#PA7411).

Narrower Superstition (#PA0430) Hallucinations (#PF2249)
Popular errors (#PF5627) Erotic delusions (#PG4542)
Illusion of consensus (#PF4327) Personal misperceptions (#PF4389)
Obsession with celebrities (#PF9617)
Symbols unrelated to human experience (#PF9070).
Related Complacency (#PA1742).
Aggravates Ignorance (#PA5568).
Aggravated by Schizophrenia (#PD0438).

♦ PA6416 Incompetence

Dependence on incompetence — Incompetent people

Nature Incompetence implies an interference with thinking which gives rise to defects in judgement and leads to behavioural abnormalities such as squandering, hoarding, or gullibility.

Incidence In the French 'Greenpeace Affair' of 1985, Frenchmen seemed most to condemn not their government's action of attacking the Rainbow Warrior, but their government's incompetence in the handling of the attack, as well as the subsequent botched cover-up attempt.

Refs Warren, Bennis G *The Unconscious Conspiracy*.

Narrower Citizen incompetence (#PE5742) Military incompetence (#PJ1069)
Governmental incompetence (#PF3953) Mediocrity of government leaders (#PF3962).
Related Impotence (#PA6876) Inferiority (#PA5652) Insufficiency (#PA5473)
Unskillfulness (#PA7232) Unpreparedness (#PA7341)
Behavioural deterioration (#PB6321)
General obstacles to problem alleviation (#PF0631).

♦ PA6425 Transience

Capriciousness — Changeableness — Impetuousity — Instability — Mutability — Transience — Volatility

Broader Time*complex (#PA7222).
Related Chaos (#PF6836) Danger (#PA6971) Weakness (#PA5558)
Inaction (#PA5806) Rashness (#PA7115) Disorder (#PA7361)
Impatience (#PA6200) Uncertainty (#PA7309) Instability (#PA0859)
Irregularity (#PA6774) Irresolution (#PA7325) Nonuniformity (#PA6890)
Changeableness (#PA5490) Unintelligence (#PA7371).

♦ PA6428 Dissimilarity

Broader Relationship*complex (#PA6484).

♦ PA6429 Interaction*complex

Narrower Defence (#PA5445) Disaccord (#PA5532) Compromise (#PA7218)
Opposition (#PA6979).

♦ PA6434 Poverty

Default — Deprivation — Distress — Embarrassment — Homelessness — Impoverishment — Lack — Liability — Mendicancy — Privation

Broader Possession*complex (#PA6686).
Narrower Deprivation (#PA0831) Socio-economic poverty (#PB0388).
Related Fear (#PA6030) Loss (#PA7382) Danger (#PA6971)
Neglect (#PA5438) Badness (#PA5454) Absence (#PA7270)
Duality (#PA7339) Humility (#PA6659) Difficulty (#PA5497)
Shortcoming (#PA6041) Inattention (#PA6247) Dislocation (#PA7158)
Uncertainty (#PA7309) Nonexistence (#PA5870) Imperfection (#PA6997)
Inexpedience (#PA7395) Insensibility (#PA5451) Insufficiency (#PA5473)
Appropriation (#PA5688) Unsociability (#PA6653) Incompleteness (#PA6652)
Unpleasantness (#PA7107) Inappropriateness (#PA6852).

♦ PA6436 Unnumbered

Broader Number*complex (#PA7412).

♦ PA6438 Uncertainty

Dependence on uncertainty

Nature Varying probabilities for events to happen, and ranges of error in human reasoned judgements, make outcomes uncertain. In science, all critical conditions must be fulfilled for certainty, presupposing exact and demonstrable knowledge of causes.

Background Uncertainty is one ot the basic psychological conditions obstructing decision-making at individual, organizational, governmental and intergovernmental levels. It is not the same as ignorance. While the human organism copes with uncertainty on an instinctual level, it has much greater difficulty in dealing with it in deliberative processes. The mathematical and statistical approaches to problem-solving involving probability theory are too recondite to find application in personal or official life. However, heuristic exploratory problem-solving with self-educating systems features such as feed-back, while familiar to science, can have a ready application to decision-making amid uncertainty. Unfortunately, human dialogue up to the level of intergovernmental deliberation and debate, faced with uncertainties, leads to 'leap into the dark' decisions or decisions to postpone decision-making, and little use is made of feed-back and other heuristic techniques, or of systems approaches to accelerate pragmatic, action-oriented proposals, resolutions and implementations in the world agenda.

Counter-claim The 'uncertainty principle' or 'principle of indeterminacy' discovered by W Heisenberg states that the position and velocity of an object cannot be measured accurately at the same time. Only for the exceedingly small masses of atoms and subatomic particles does the product of the uncertainties become significant for research purposes. Nevertheless, for practical purposes the uncertain coordinate of velocity with position for the atom does not prevent its splitting to release nuclear energy. The uncertainty principle in quantum mechanics has nothing to do with knowledge or action on the human level. However insofar as this principle illustrates the meaninglessness in nature of a concept of theoretical exact simultaneous knowledge of object position and location, it raises epistemological questions as to the basis for action which also were discussed by Plato, Aristotle and other philosophers. Chief amongst these are the idealists who fall into solipsistic and other subjectivist errors of divorcing philosophy from life with the view that all knowledge is uncertain, therefore the only logical action is to think and refrain from action, or more logically, to refrain from acting and thinking. The problem can perhaps be restated. Knowledge is certain in terms of its eventuality. Complete knowledge is unachievable. The level of knowledge must be determined that is appropriate to specific actions. For example, relief operations need not be delayed until a count of casualties is made; and justice need not require all the evidence, only a sufficiency. There are also signs which may not be certain but are indicative: for example, when dark clouds gather rain is expected; or when a country militarizes aggression may be anticipated. Thus action may be based on uncertain but probable knowledge; reasonable certainty.

Broader Danger (#PA6971) Unbelief (#PA7392) Uncertainty (#PA7309)
Invisibility (#PA6978) Irresolution (#PA7325) Changeableness (#PA5490)
Unintelligibility (#PA7367).
Narrower Economic uncertainty (#PF5817) Uncertainty of land zoning (#PG7739)
Inconclusiveness of science (#PF6349) Uncertainty of miners' future (#PG8265)
Uncertain toxicity thresholds (#PF5188) Inadequate negative capability (#PF7109)
Defence information uncertainty (#PE7679) Uncertain status of monetary gold (#PF2342)
Unpredictable governmental policy (#PF1559)
Uncertainty of death of missing persons (#PF0431)
Insecurity from future crop uncertainty (#PF8354)
Uncertainty of survival of the human race (#PE9085)
Unknowable future patterns of social choice (#PF9276)
Unexploited possibilities for local commerce (#PF2535)
Inconclusiveness of scientific and medical tests (#PD7415)
Uncertain environmental impact of current policy (#PF9450)
Non-equivalence of national educational qualifications (#PC1524)
Uncertainty of long-term health effects of radioactive fallout (#PE5324).
Aggravates Risk (#PF7580) Injurious accidents (#PB0731).
Aggravated by Chance (#PA6714).

♦ PA6444 Quiescence

Apathy — Closeness — Indifference — Indolence — Inertia — Languor — Lifelessness — Passivity — Stagnation — Stuffiness — Suspense — Torpor

Broader Motion*complex (#PA5468).
Narrower Stagnation (#PA3917).
Related Fear (#PA6030) Cold (#PA6956) Death (#PA7055)
Stench (#PA5981) Apathy (#PA2360) Neglect (#PA5438)
Boredom (#PA7365) Weakness (#PA5558) Inaction (#PA5806)
Darkness (#PA6261) Avoidance (#PA6379) Cheapness (#PA7193)
Impatience (#PA6200) Narrowness (#PA7256) Inattention (#PA6247)
Affectation (#PA6400) Incuriosity (#PA6598) Uncertainty (#PA7309)
Unimportance (#PA5942) Hopelessness (#PA6099) Imperfection (#PA6997)
Disobedience (#PA7250) Irresolution (#PA7325) Unfeelingness (#PA7364)
Unintelligence (#PA7371) Narrowmindedness (#PA7306) Unimaginativeness (#PA6738)
Uncommunicativeness (#PA7411).

♦ PA6446 Indiscrimination

Division — Imprudence — Indiscretion — Indiscrimination — Insensibility — Insensitivity — Muddle — Promiscuity — Segregation — Tactlessness

Broader Evaluation*complex (#PA7394).
Related Vice (#PA5644) Error (#PA6180) Neglect (#PA5438)
Rashness (#PA7115) Disorder (#PA7361) Disaccord (#PA5532)
Ignorance (#PA5568) Exclusion (#PA5869) Avoidance (#PA6379)
Necessity (#PA6387) Restraint (#PA7296) Separation (#PA7236)
Divergence (#PA5573) Unchastity (#PA5612) Unkindness (#PA5643)
Complexity (#PA6468) Inattention (#PA6247) Discourtesy (#PA7143)
Uncertainty (#PA7309) Promiscuity (#PC0745) Formlessness (#PA6900)
Insensibility (#PA5451) Unsociability (#PA6653) Maladjustment (#PA6739)
Unfeelingness (#PA7364) Unskillfulness (#PA7232) Unintelligence (#PA7371)
Narrowmindedness (#PA7306) Nonaccomplishment (#PA6662).

♦ PA6448 Obsession

Obsessive-compulsive disorder — Compulsion — Obsessed people — Psychological fixation — Impulse control disorder

Nature Obsessive-compulsive neuroses often result from a recurring repression of feeling or behaviour that is considered socially unacceptable, physically harmful, or morally wrong. When the obsession arises from repression of feelings, the type will be specified by the emotion repressed and frequently may also be related generally to the sexual instinct or the drive for self-assertion. Repressed emotions gradually react from the unconscious level, manifesting themselves by compulsive actions, some of which are not immediately evident as related to the repressed emotion. Obsessive-compulsive neuroses require treatment by psychotherapy; however, proper treatment is often not provided and individuals may spend an unhappy life-time with these conditions and cause additional unhappiness to those around them.

Broader Insanity (#PA7157) Maladjustment (#PA6739).
Narrower Neurosis (#PD0270) Pyromania (#PG5853) Kleptomania (#PG0515)
Hypochondria (#PE8322) Scrupulosity (#PF6404)
Pathological gambling (#PG4237) Obsession with people or things (#PG4164)
Intermittent explosive disorder (#PG6420).
Related Anorexia nervosa (#PE5758).

♦ PA6453 Disillusionment

Disillusionment

Narrower Waste paper (#PD1152) Compassion fatigue (#PF2819)
Imbalances between people's aspirations and the structure of opportunities and income available (#PF9008).
Related Error (#PA6180) Disapproval (#PA6191) Inexpectation (#PA5527)
Truth*complex (#PA6164).
Aggravated by Nativism (#PF2186) Oppressive prevalent images (#PF1365).

♦ PA6466 Intemperance

Avoidance — Debauchery — Dissipation — Excess — Excessiveness — Extravagance — Greed — Helplessness — Incontinence — Indulgence — Insobriety — Intemperence — Intoxication — Licentiousness — Prodigality — Rapacity — Self-indulgence — Voracity

Broader Morality*complex (#PA6892).
Narrower Obsolete methods of agricultural production (#PF1822).
Related Vice (#PA5644) Loss (#PA7382) Excess (#PB8952)
Disease (#PA6799) Refusal (#PA7321) Avarice (#PA6999)
Insanity (#PA7157) Avoidance (#PA6379) Impotence (#PA6876)
Undueness (#PA6921) Cheapness (#PA7193) Restraint (#PA7296)
Unchastity (#PA5612) Lawlessness (#PA5563) Diffuseness (#PA5974)
Affectation (#PA6400) Selfishness (#PA7211) Prodigality (#PA7269)
Insufficiency (#PA5473) Appropriation (#PA5688) Disappearance (#PA6785)
Uncommunicativeness (#PA7411) Over-intensive soil exploitation (#PC0052).

♦ PA6468 Complexity

Complication — Corruption — Deviousness — Involution — Muddle — Oversimplification — Pollution

Refs Aida, S et al *Science and Praxis of Complexity* (1985); Schoning, U *Complexity and Structure* (1986).

Broader Quantity*complex (#PA6108).
Related Vice (#PA5644) Badness (#PA5454) Disease (#PA6799)
Disorder (#PA7361) Reversion (#PA5699) Deviation (#PA6228)
Improbity (#PA7363) Pollution (#PB6336) Difficulty (#PA5497)
Impairment (#PA6088) Distortion (#PA6790) Dissuasion (#PA7343)
Uncleanness (#PA5459) Inattention (#PA6247) Uncertainty (#PA7309)
Miseducation (#PA6393) Formlessness (#PA6900) Unskillfulness (#PA7232)
Unpreparedness (#PA7341) Unintelligence (#PA7371) Unhealthfulness (#PA7226)
Indiscrimination (#PA6446) Nonaccomplishment (#PA6662) Inappropriateness (#PA6852)
Unintelligibility (#PA7367) Oversimplification (#PF8455).
Aggravates Corruption (#PA1986).

♦ PA6484 Relationship*complex

Refs Ickes, W (Ed) *Compatible and Incompatible Relationships*.

ABSTRACT FUNDAMENTAL PROBLEMS PA6607

Narrower Imitation (#PA6568)   Difference (#PA6698)   Inequality (#PA6695)
Disagreement (#PA5982)   Unrelatedness (#PA5794)   Nonuniformity (#PA6890)
Dissimilarity (#PA6428).

♦ PA6485 Counteraction
Nonconformity
Broader Power∗complex (#PA7314).
Related Dissent (#PA6838)   Exclusion (#PA5869)   Disagreement (#PA5982)
Disobedience (#PA7250)   Nonconformity (#PA5878).

♦ PA6486 Injustice
Bias — Discrimination — Foulness — Grievance — Impropriety — Indefensibility — Inequality — Inequity — Iniquity — Injury — Miscarriage — Outrage — Unconscionableness — Unlawfulness — Warpedness — Wrongness — Dependence on injustice — Unjust people — Denial of right to justice
Narrower Elitism (#PA1387)   Discrimination (#PA0833)   Secret societies (#PF2508)
Political injustice (#PC2181)   Physical intimidation (#PC2934)
Rejection of refugees (#PF3021)   Psychological intimidation (#PC2935)
Denial of right to justice (#PC6162)   Inaccessibility of justice (#PD8334)
Injustice of special courts (#PE0088)   Discrimination in employment (#PC0244)
Time lag in legal provisions (#PF6042)   Obsolete deliberative systems (#PD0975)
Injustice of religious courts (#PE0397)   Unjust punishments for crimes (#PD4779)
Injustice of trials in absentia (#PE0424)   Frivolous or vindictive litigation (#PF1542)
Denial of right to inherit property (#PF0886)   Deficiencies in civil justice systems (#PF4899)
Inequitable administration of justice (#PD0986)
Inadequate international judicial system (#PF2113)
Unequal distribution of fame and honours (#PF3439)
Deficiencies in military codes of justice (#PE8300)
Discrimination against prisoners' families (#PE5043)
Deficiencies in the criminal justice system (#PF4875)
Compulsory acquisition of land by government (#PC1005)
Inadequate legal counsel for political dissidents (#PF0732)
Unrestrained use of force in administration of justice (#PE8881)
Discrimination against juveniles in judicial proceedings due to protective legislation (#PE1295).
Related Envy (#PA7253)   Loss (#PA7382)   Vice (#PA5644)
Error (#PA6180)   Stench (#PA5981)   Badness (#PA5454)
Dissent (#PA6838)   Closure (#PA7391)   Ugliness (#PA7240)
Vulgarity (#PA5821)   Disrepute (#PA6839)   Undueness (#PA6921)
Cheapness (#PA7193)   Wrongness (#PA7280)   Improbity (#PA7363)
Grievance (#PF8029)   Unchastity (#PA5612)   Impairment (#PA6088)
Distortion (#PA6790)   Disrespect (#PA6822)   Moderation (#PA7156)
Inequality (#PA6695)   Oppression (#PB8656)   Uncleanness (#PA5459)
Lamentation (#PA5479)   Disapproval (#PA6191)   Destruction (#PA6542)
Disagreement (#PA5982)   Misbehaviour (#PA6498)   Imperfection (#PA6997)
Untimeliness (#PA7006)   Inexpedience (#PA7395)   Insufficiency (#PA5473)
Unsavouriness (#PA7204)   Discontentment (#PA6011)   Unpleasantness (#PA7107)
Unskillfulness (#PA7232)   Narrowmindedness (#PA7306)   Human inequality (#PA0844)
Misrepresentation (#PA6644)   Inappropriateness (#PA6852)
Judgement∗complex (#PA8528)   Uncommunicativeness (#PA7411)
Inadequacy of patent coverage (#PF3538)   Repressive detention of juveniles (#PD0634)
Repression of intellectual dissidents (#PD0434)   Violation of the integrity of creation (#PF5148)
General obstacles to problem alleviation (#PF0631)
Economic discrimination in the administration of justice (#PE1399).
Aggravates Conflict (#PA0298)   Human suffering (#PB5955)
Overthrow of government (#PD1964)   Socially unsustainable development (#PC0381)
Unequal distribution of social services (#PC3437)
Denial of human rights in armed conflicts (#PC1454)
Cultural discrimination in the administration of justice (#PE6529).
Aggravated by Illegality (#PA5952)   Secret laws (#PC6757)
Police state (#PD7910)   Victimization (#PF6987)
Abuse of confidentiality (#PF7633)   Underutilization of legal rights (#PF3464)
Ineffective war crime prosecution (#PD1464)   Inadequate national law enforcement (#PE4768)
Religious discrimination in the administration of justice (#PE0168)
Conflicts of national law in relation to international transactions (#PF9571).

♦ PA6487 Impossibility
Hopelessness — Impracticality
Broader Credibility∗complex (#PA5829).
Related Chance (#PA6714)   Solemnity (#PA6731)   Difficulty (#PA5497)
Hopelessness (#PA6099)   Unfeelingness (#PA7364).

♦ PA6491 Vanity
Arrogance — Boastfulness — Bumptiousness — Complacency — Conceit — Condescension — Confusion — Contemptuousness — Contumeliousness — Dictatorship — Disdain — Disrespect — Fuss — Immodesty — Impertinence — Impudence — Inarticulation — Insolence — Intimidation — Lordliness — Presumption — Pretentiousness — Ridicule — Rudeness — Self-admiration — Self-centredness — Self-esteem — Self-importance — Self-satisfaction — Selfishness — Shamelessness — Snobbery — Superciliousness — Vainness — Vanity — Dependence on vanity — Vain people — Vainglory
Narrower Selfishness (#PA7211)   Condescension (#PJ3277).
Related Sin (#PF0641)   Vice (#PA5644)   Fear (#PA6030)
Pride (#PA7599)   Reason (#PA5502)   Defeat (#PA7289)
Egoism (#PA6318)   Humility (#PA6659)   Rashness (#PA7115)
Disorder (#PA7361)   Snobbery (#PJ3943)   Disaccord (#PA5532)
Vulgarity (#PA5821)   Agitation (#PA5838)   Exclusion (#PA5869)
Necessity (#PA6387)   Intrusion (#PA6862)   Impotence (#PA6876)
Undueness (#PA6921)   Wrongness (#PA7280)   Improbity (#PA7363)
Confusion (#PF7123)   Arrogance (#PA7646)   Unchastity (#PA5612)
Compulsion (#PA5740)   Impatience (#PA6200)   Disrespect (#PA6822)
Opposition (#PA6979)   Moderation (#PA7156)   Dissuasion (#PA7343)
Lamentation (#PA5479)   Lawlessness (#PA5563)   Disapproval (#PA6191)
Inattention (#PA6247)   Affectation (#PA6400)   Discourtesy (#PA7143)
Uncertainty (#PA7309)   Complacency (#PA1742)   Hopelessness (#PA6099)
Misjudgement (#PA6607)   Formlessness (#PA6900)   Unimportance (#PA5942)
Unprovability (#PA7070)   Inexcitability (#PA5467)   Discontentment (#PA6011)
Disintegration (#PA6858)   Unpleasantness (#PA7107)   Unpreparedness (#PA7341)
Inappropriateness (#PA6852)   Unintelligibility (#PA7367)
Discriminative affection∗complex (#PA7145).
Aggravates Boasting (#PF4436)   Medical quackery (#PD1725).
Aggravated by Cosmetic and health problems related to human hair (#PE7111).

♦ PA6498 Misbehaviour
Badness — Discourtesy — Disorder — Impropriety — Mischievousness — Mischief — Misconduct — Misdemeanour — Roguery — Vandalism
Broader Achievement∗complex (#PA5471).
Narrower Floods (#PD0452)   Impropriety (#PA6000).
Related Vice (#PA5644)   Error (#PA6180)   Stench (#PA5981)
Danger (#PA6971)   Badness (#PA5454)   Disease (#PA6799)

Disorder (#PA7361)   Disaccord (#PA5532)   Vulgarity (#PA5821)
Agitation (#PA5838)   Injustice (#PA6486)   Undueness (#PA6921)
Wrongness (#PA7280)   Improbity (#PA7363)   Vandalism (#PD1350)
Unchastity (#PA5612)   Unkindness (#PA5643)   Illegality (#PA5952)
Impairment (#PA6088)   Disrespect (#PA6822)   Moderation (#PA7156)
Lawlessness (#PA5563)   Inattention (#PA6247)   Destruction (#PA6542)
Discourtesy (#PA7143)   Ungodliness (#PA7148)   Uncertainty (#PA7309)
Disagreement (#PA5982)   Formlessness (#PA6900)   Untimeliness (#PA7006)
Inexpedience (#PA7395)   Unskillfulness (#PA7232)   Unhealthfulness (#PA7226)
Loss of civility (#PC7013)   Inappropriateness (#PA6852)
Deterioration in product quality (#PD1435).
Aggravated by Disorganization (#PF4487).

♦ PA6509 Unwillingness
Antipathy — Aversion — Disagreeableness — Disobedience — Distaste — Indisposition — Indocility — Obstinacy — Opposition — Recalcitrance — Repugnance — Stubbornness
Broader Choice∗complex (#PA7214).
Related Envy (#PA7253)   Hate (#PA7338)   Enmity (#PA5446)
Disease (#PA6799)   Dissent (#PA6838)   Refusal (#PA7321)
Softness (#PA5988)   Antipathy (#PA8810)   Difficulty (#PA5497)
Unkindness (#PA5643)   Compulsion (#PA5740)   Difference (#PA6698)
Opposition (#PA6979)   Lawlessness (#PA5563)   Disapproval (#PA6191)
Disagreement (#PA5982)   Disobedience (#PA7250)   Irresolution (#PA7325)
Unpleasantness (#PA7107).

♦ PA6542 Destruction
Annihilation — Annulment — Banefulness — Blight — Brokenness — Calamity — Damnation — Death — Defeat — Depredation — Desolation — Despoliation — Destructiveness — Devastation — Disaster — Disintegration — Disorganization — Dissolution — Extermination — Extinction — Foulness — Irremediability — Negation — Overturn — Perdition — Ruin — Sabotage — Spoilation — Subversion — Suffocation — Suppression — Upheaval — Upset — Vandalism — Waste — Wastefulness — Withering
Broader Integrity∗complex (#PA7263).
Narrower Sabotage (#PD0405)   Suffocation (#PJ4103)   Annihilation (#PF9169)
Extermination (#PJ2839)   Human destructiveness (#PA0832).
Related Vice (#PA5644)   Fear (#PA6030)   Cold (#PA6956)
Loss (#PA7382)   Death (#PA7055)   Stench (#PA5981)
Defeat (#PA7289)   Badness (#PA5454)   Disease (#PA6799)
Dissent (#PA6838)   Refusal (#PA7321)   Closure (#PA7391)
Ugliness (#PA7240)   Disorder (#PA7361)   Celibacy (#PA7410)
Reversion (#PA5699)   Agitation (#PA5838)   Adversity (#PA6340)
Injustice (#PA6486)   Solemnity (#PA6731)   Disrepute (#PA6839)
Impotence (#PA6876)   Cheapness (#PA7193)   Restraint (#PA7296)
Vandalism (#PD1350)   Separation (#PA7236)   Depression (#PA5493)
Difficulty (#PA5497)   Unchastity (#PA5612)   Unkindness (#PA5643)
Revolution (#PA5901)   Impairment (#PA6088)   Disrespect (#PA6822)
Opposition (#PA6979)   Moderation (#PA7156)   Uncleanness (#PA5459)
Lawlessness (#PA5563)   Disapproval (#PA6191)   Inattention (#PA6247)
Ungodliness (#PA7148)   Uncertainty (#PA7309)   Nonexistence (#PA5870)
Disagreement (#PA5982)   Hopelessness (#PA6099)   Misbehaviour (#PA6498)
Condemnation (#PA7237)   Disobedience (#PA7250)   Inexpectation (#PA5527)
Appropriation (#PA5688)   Discontinuity (#PA5828)   Unsociability (#PA6653)
Disappearance (#PA6785)   Unsavouriness (#PA7204)   Fragmentation (#PA6233)
Inexcitability (#PA5467)   Disintegration (#PA6858)   Unpleasantness (#PA7107)
Unhealthfulness (#PA7226)   Unproductiveness (#PA7208)   Nonaccomplishment (#PA6662)
Inappropriateness (#PA6852).
Aggravated by Disorganization (#PF4487).

♦ PA6560 Instantaneousness
Broader Time∗complex (#PA7222).

♦ PA6562 Anticipation∗complex
Narrower Rashness (#PA7115)   Hopelessness (#PA6099).
Related Fear (#PA6030).

♦ PA6568 Imitation
Broader Relationship∗complex (#PA6484).

♦ PA6586 Life∗complex
Narrower Death (#PA7055)   Unsexiness (#PA5499)   Femininity (#PA5604)
Nonhumanity (#PA6926)   Immateriality (#PA6176).

♦ PA6598 Incuriosity
Apathy — Boredom — Boringness — Carelessness — Disinterest — Heedlessness — Indifference — Insouciance — Listlessness — Withdrawal
Broader Attitude∗complex (#PA6983).
Narrower Indifference (#PA7604).
Related Apathy (#PA2360)   Neglect (#PA5438)   Dissent (#PA6838)
Duality (#PA7339)   Boredom (#PA7365)   Weakness (#PA5558)
Inaction (#PA5806)   Rashness (#PA7115)   Disorder (#PA7361)
Exclusion (#PA5869)   Avoidance (#PA6379)   Unkindness (#PA5643)
Compulsion (#PA5740)   Regression (#PA6338)   Quiescence (#PA6444)
Withdrawal (#PF4402)   Inattention (#PA6247)   Unimportance (#PA5942)
Hopelessness (#PA6099)   Imperfection (#PA6997)   Irresolution (#PA7325)
Forgetfulness (#PA6651)   Nonuniformity (#PA6890)   Unfeelingness (#PA7364)
Unpleasantness (#PA7107)   Unskillfulness (#PA7232).

♦ PA6606 Vengeance
Irreconcilability — Retaliation — Retribution — Revenge — Vindictiveness
Refs Marongiu, Pietro and Newman, Graeme *Vengeance* (1987).
Broader Benevolence∗complex (#PA7331).
Narrower Revenge (#PF8562)   Retaliation (#PF9181).
Related Enmity (#PA5446)   Punishment (#PA5583)   Opposition (#PA6979)
Disagreement (#PA5982)   Irresolution (#PA7325).

♦ PA6607 Misjudgement
Censure — Depreciation — Disparagement — Exaggeration — Miscalculation — Misconstruction — Misinterpretation — Misjudgement — Overestimation — Prejudice — Presumption — Underestimation
Refs Murray, Allan P *Depreciation* (1971).
Broader Evaluation∗complex (#PA7394).
Related Loss (#PA7382)   Error (#PA6180)   Vanity (#PA6491)
Badness (#PA5454)   Rashness (#PA7115)   Disrepute (#PA6839)
Intrusion (#PA6862)   Undueness (#PA6921)   Cheapness (#PA7193)
Prejudice (#PA2173)   Impairment (#PA6088)   Distortion (#PA6790)
Disrespect (#PA6822)   Disapproval (#PA6191)   Condemnation (#PA7237)

-95-

PA6607

Inexpedience (#PA7395)　　Exaggeration (#PJ5960)　　Narrowmindedness (#PA7306)
Misrepresentation (#PA6644)　　Misinterpretability (#PA6741)
Uncommunicativeness (#PA7411).

♦ **PA6644 Misrepresentation**
Distortion — Exaggeration — Inaccuracy — Injustice — Misrepresentation — Misstatement — Perversion — Slant — Travesty
Broader Communication∗complex (#PA6732).
Narrower Exaggeration (#PJ5960).
Related Vice (#PA5644)　　Error (#PA6180)　　Reason (#PA5502)
Neglect (#PA5438)　　Deviation (#PA6228)　　Injustice (#PA6486)
Impairment (#PA6088)　　Inequality (#PA6695)　　Distortion (#PA6790)
Disrespect (#PA6822)　　Inaccuracy (#PF7905)　　Uncertainty (#PA7309)
Miseducation (#PA6393)　　Misjudgement (#PA6607)　　Imperfection (#PA6997)
Inappropriateness (#PA6852)　　Misinterpretability (#PA6741)
Uncommunicativeness (#PAp411).

♦ **PA6651 Forgetfulness**
Forgetfulness — Heedlessness — Oblivion — Repression
Nature Under normal circumstances, a thing is forgotten only by occupying the mind with something else, i.e. by losing interest in it through acquiring interest in something else. A thing cannot be forgotten by simply willing it to be gone. Loss of long term memory can result from damage to the brain, a faint original impression or agitation or excitement of a moment. A second type of forgetfulness is simply refusing to bring it to consciousness. For many a war or other tragic experience is dealt with this way.
Broader Attitude∗complex (#PA6983).
Related Loss (#PA7382)　　Neglect (#PA5438)　　Refusal (#PA7321)
Avoidance (#PA6379)　　Restraint (#PA7296)　　Difficulty (#PA5497)
Unkindness (#PA5643)　　Repression (#PB0871)　　Inattention (#PA6247)
Incuriosity (#PA6598)　　Unfeelingness (#PA7364)　　Thoughtlessness (#PA6940)
Structural amnesia in institutional systems (#PF7745).
Aggravates Ignorance (#PA5568)　　Cultural suicide (#PF5957)
Impaired vigilance (#PF6863).

♦ **PA6652 Incompleteness**
Defectiveness — Deficiency — Immaturity — Inadequacy — Incompleteness — Lack — Omission — Shortage — Underdevelopment
Broader Quantity∗complex (#PA6108).
Narrower Shortage (#PB8238)　　Deficiency (#PJ5364)　　Underdevelopment (#PB0206).
Related Vice (#PA5644)　　Error (#PA6180)　　Neglect (#PA5438)
Poverty (#PA6434)　　Disease (#PA6799)　　Absence (#PA7270)
Impotence (#PA6876)　　Inequality (#PA6695)　　Inadequacy (#PA8199)
Immaturity (#PF8413)　　Inferiority (#PA5652)　　Shortcoming (#PA6041)
Imperfection (#PA6997)　　Insufficiency (#PA5473)　　Discontentment (#PA6011)
Unskillfulness (#PA7232)　　Unpreparedness (#PA7341)　　Unintelligence (#PA7371)
Nonaccomplishment (#PA6662).

♦ **PA6653 Unsociability**
Autism — Coldness — Desertion — Desolation — Exclusiveness — Homelessness — Inaccessibility — Incompatibility — Loneliness — Seclusion — Secrecy — Segregation — Uncommunicativeness — Unfriendliness
Broader Socialization∗complex (#PA6373).
Narrower Secrecy (#PA0005).
Related Cold (#PA6956)　　Error (#PA6180)　　Enmity (#PA5446)
Impiety (#PA6058)　　Poverty (#PA6434)　　Duality (#PA7339)
Rashness (#PA7115)　　Disaccord (#PA5532)　　Reversion (#PA5699)
Vulgarity (#PA5821)　　Exclusion (#PA5869)　　Avoidance (#PA6379)
Solemnity (#PA6731)　　Restraint (#PA7296)　　Unkindness (#PA5643)
Disrespect (#PA6822)　　Opposition (#PA6979)　　Loneliness (#PF2386)
Destruction (#PA6542)　　Discourtesy (#PA7143)　　Dislocation (#PA7158)
Selfishness (#PA7211)　　Disagreement (#PA5982)　　Irresolution (#PA7325)
Unfeelingness (#PA7364)　　Unpleasantness (#PA7107)　　Incompatibility (#PF9047)
Indiscrimination (#PA6446)　　Unproductiveness (#PA7208)　　Narrowmindedness (#PA7306)
Unimaginativeness (#PA6738)　　Uncommunicativeness (#PA7411).
Aggravates Social isolation (#PC1707).

♦ **PA6659 Humility**
Abasement — Abjection — Arrogance — Baseness — Boastfulness — Chagrin — Conceit — Condescension — Disgrace — Embarrassment — Humiliation — Ingloriousness — Insinuation — Lordliness — Meanness — Mortification — Self-esteem — Servility — Smallness — Unimportance — Vanity
Broader Discriminative affection∗complex (#PA7145).
Narrower Condescension (#PJ3277).
Related Vice (#PA5644)　　Fear (#PA6030)　　Envy (#PA7253)
Vanity (#PA6491)　　Badness (#PA5454)　　Poverty (#PA6434)
Disease (#PA6799)　　Fewness (#PA7152)　　Smallness (#PA7408)
Vulgarity (#PA5821)　　Disrepute (#PA6839)　　Intrusion (#PA6862)
Cheapness (#PA7193)　　Wrongness (#PA7280)　　Restraint (#PA7296)
Improbity (#PA7363)　　Arrogance (#PA7646)　　Difficulty (#PA5497)
Unchastity (#PA5612)　　Unkindness (#PA5643)　　Impairment (#PA6088)
Disrespect (#PA6822)　　Unkindness (#PA5643)　　Littleness (#PA7285)
Inferiority (#PA5652)　　Disapproval (#PA6191)　　Inattention (#PA6247)
Affectation (#PA6400)　　Selfishness (#PA7211)　　Uncertainty (#PA7309)
Humiliation (#PF3856)　　Unimportance (#PA5942)　　Imperfection (#PA6997)
Condemnation (#PA7237)　　Disobedience (#PA7250)　　Insufficiency (#PA5473)
Unpleasantness (#PA7107)　　Narrowmindedness (#PA7306)　　Inappropriateness (#PA6852).

♦ **PA6662 Nonaccomplishment**
Conspiracy — Defeat — Failure — Futility — Mistake — Muddle — Neglect — Nonaccomplishment — Nonfeasance — Nonfulfilment — Omission — Uselessness
Broader Achievement∗complex (#PA5471).
Narrower Conspiracy (#PC2555)　　Electoral defeat (#PF4709).
Related Vice (#PA5644)　　Error (#PA6180)　　Defeat (#PA7289)
Neglect (#PA5438)　　Disorder (#PA7361)　　Impotence (#PA6876)
Difficulty (#PA5497)　　Impairment (#PA6088)　　Complexity (#PA6468)
Disrespect (#PA6822)　　Inferiority (#PA5652)　　Shortcoming (#PA6041)
Inattention (#PA6247)　　Destruction (#PA6542)　　Uncertainty (#PA7309)
Unimportance (#PA5942)　　Hopelessness (#PA6099)　　Formlessness (#PA6900)
Imperfection (#PA6997)　　Inexpedience (#PA7395)　　Insufficiency (#PA5473)
Inexpectation (#PA5527)　　Incompleteness (#PA6652)　　Unskillfulness (#PA7232)
Meaninglessness (#PA6977)　　Indiscrimination (#PA6446)　　Inappropriateness (#PA6852)
Misinterpretability (#PA6741)　　Uncommunicativeness (#PA7411).

♦ **PA6674 Blindness**
Darkness
Broader Sense∗complex (#PA6236).
Narrower Physical blindness (#PD0568).

Related Darkness (#PA6261)　　Ignorance (#PA5568)　　Solemnity (#PA6731)
Invisibility (#PA6978)　　Unintelligibility (#PA7367).

♦ **PA6686 Possession∗complex**
Narrower Loss (#PA7382)　　Poverty (#PA6434)　　Cheapness (#PA7193)
Prodigality (#PA7269)　　Appropriation (#PA5688).
Aggravates Crime (#PB0001).
Aggravated by Corruption (#PA1986)　　Social inequality (#PB0514).

♦ **PA6695 Inequality**
Asymmetry — Contrariety — Disparity — Inadequacy — Inequity — Injustice — Insufficiency — Nonuniformity — Unevenness
Broader Relationship∗complex (#PA6484).
Narrower Imbalance (#PA0224)　　Human inequality (#PA0844)
International inequality (#PC9152).
Related Vice (#PA5644)　　Dissent (#PA6838)　　Disorder (#PA7361)
Smallness (#PA7408)　　Disaccord (#PA5532)　　Injustice (#PA6486)
Impotence (#PA6876)　　Difference (#PA6698)　　Distortion (#PA6790)
Opposition (#PA6979)　　Inadequacy (#PA8199)　　Inferiority (#PA5652)
Shortcoming (#PA6041)　　Disagreement (#PA5982)　　Irregularity (#PA6774)
Imperfection (#PA6997)　　Insufficiency (#PA5473)　　Discontinuity (#PA5828)
Nonuniformity (#PA6890)　　Changeableness (#PA5490)　　Discontentment (#PA6011)
Incompleteness (#PA6652)　　Unpleasantness (#PA7107)　　Unskillfulness (#PA7232)
Misrepresentation (#PA6644).

♦ **PA6698 Difference**
Ambiguity — Ambivalence — Antagonism — Antipathy — Conflict — Contradiction — Disaccord — Disparity — Dissonance — Equivocation — Hostility — Inconsistency — Opposition — Perversity — Repugnance — Self-contradiction — Variation
Broader Relationship∗complex (#PA6484).
Related Envy (#PA7253)　　Hate (#PA7338)　　Error (#PA6180)
Enmity (#PA5446)　　Reason (#PA5502)　　Discord (#PA5565)
Dissent (#PA6838)　　Conflict (#PA0298)　　Disaccord (#PA5532)
Deviation (#PA6228)　　Avoidance (#PA6379)　　Hostility (#PB8538)
Antipathy (#PA8810)　　Ambiguity (#PF4193)　　Difficulty (#PA5497)
Inequality (#PA6695)　　Opposition (#PA6979)　　Disapproval (#PA6191)
Uncertainty (#PA7309)　　Ambivalence (#PF1426)　　Disagreement (#PA5982)
Irresolution (#PA7325)　　Unwillingness (#PA6509)　　Maladjustment (#PA6739)
Nonuniformity (#PA6890)　　Unfeelingness (#PA7364)　　Changeableness (#PA5490)
Disintegration (#PA6858)　　Unpleasantness (#PA7107)　　Contradictions (#PF3667)
Unintelligibility (#PA7367)　　Uncommunicativeness (#PA7411).

♦ **PA6706 Culmination**
Backlash
Broader Causation∗complex (#PA7406).

♦ **PA6713 Space∗complex**
Narrower Absence (#PA7270)　　Content (#PA5663)　　Dislocation (#PA7158).

♦ **PA6714 Chance**
Aimlessness — Haphazardness — Hopelessness — Impossibility — Purposelessness — Randomness — Dependence on chance
Broader Causation∗complex (#PA7406).
Narrower Social neglect (#PB0883)　　Mechanical failure (#PC1904)
Economic dependence (#PF0841)　　Excessive longevity (#PJ5973)
Cyclic business recessions (#PF1277)　　Human errors and miscalculations (#PF3702)
Maldistribution of world population (#PF0167)
Unreliability of equipment and machinery (#PC2297)
Inadequacy of governmental decision-making machinery (#PF2420).
Related Neglect (#PA5438)　　Disorder (#PA7361)　　Mutation (#PF2276)
Solemnity (#PA6731)　　Uncertainty (#PA7309)　　Hopelessness (#PA6099)
Impossibility (#PA6487)　　Unfeelingness (#PA7364)　　Meaninglessness (#PA6977)
Human inequality (#PA0844)　　Inappropriateness (#PA6852).
Aggravates Chaos (#PF6836)　　Ugliness (#PA7240)　　Instability (#PA0859)
Uncertainty (#PA6438)　　Economic uncertainty (#PF5817)
Genetic defects and diseases (#PD2389).

♦ **PA6721 Change∗complex**
Narrower Cessation (#PA5811)　　Reversion (#PA5699)　　Permanence (#PA6802)
Changeableness (#PA5490).
Related Revolution (#PA5901).

♦ **PA6731 Solemnity**
Anguish — Cheerlessness — Darkness — Dejection — Depression — Desolation — Despair — Despondency — Discontentment — Discouragement — Displeasure — Dolorousness — Grief — Grimness — Heartlessness — Hopelessness — Infelicity — Joylessness — Lowness — Malaise — Melancholy — Misery — Mournfulness — Oppression — Pessimism — Pleasurelessness — Sadness — Sorrow — Spiritlessness — Unhappiness — Wearisomeness — Woe
Broader Feeling∗complex (#PA6938).
Narrower Grief (#PF5654)　　Misery (#PB8167)　　Oppression (#PB8656)
Discouragement (#PF8948).
Related Vice (#PA5644)　　Fear (#PA6030)　　Envy (#PA7253)
Chance (#PA6714)　　Badness (#PA5454)　　Lowness (#PA6798)
Disease (#PA6799)　　Boredom (#PA7365)　　Inaction (#PA5806)
Darkness (#PA6261)　　Ugliness (#PA7240)　　Ignorance (#PA5568)
Vulgarity (#PA5821)　　Agitation (#PA5838)　　Adversity (#PA6340)
Blindness (#PA6674)　　Disrepute (#PA6839)　　Depression (#PA5493)
Unkindness (#PA5643)　　Impairment (#PA6088)　　Dissuasion (#PA7343)
Melancholy (#PF7756)　　Lamentation (#PA5479)　　Inferiority (#PA5652)
Disapproval (#PA6191)　　Destruction (#PA6542)　　Displeasure (#PA6809)
Unhappiness (#PA9191)　　Unimportance (#PA5942)　　Disagreement (#PA5982)
Hopelessness (#PA6099)　　Invisibility (#PA6978)　　Untimeliness (#PA7006)
Pitilessness (#PA7023)　　Irresolution (#PA7325)　　Inexpedience (#PA7395)
Insensibility (#PA5451)　　Impossibility (#PA6487)　　Unsociability (#PA6653)
Maladjustment (#PA6739)　　Unfeelingness (#PA7364)　　Discontentment (#PA6011)
Unpleasantness (#PA7107)　　Unproductiveness (#PA7208)　　Inappropriateness (#PA6852)
Unintelligibility (#PA7367).

♦ **PA6732 Communication∗complex**
Narrower Ineleganc (#PA6312)　　Diffuseness (#PA5974)
Ineloquence (#PA6118)　　Miseducation (#PA6393)
Misrepresentation (#PA6644)　　Uncommunicativeness (#PA7411).

♦ **PA6738 Unimaginativeness**
Autism — Barrenness — Dullness — Infertility — Stuffiness — Unimaginativeness — Unreality
Broader Attitude∗complex (#PA6983).

ABSTRACT FUNDAMENTAL PROBLEMS

PA6822

Related Error (#PA6180) Stench (#PA5981) Boredom (#PA7365)
Weakness (#PA5558) Inaction (#PA5806) Impotence (#PA6876)
Quiescence (#PA6444) Affectation (#PA6400) Selfishness (#PA7211)
Nonexistence (#PA5870) Imperfection (#PA6997) Irresolution (#PA7325)
Insensibility (#PA5451) Unsociability (#PA6653) Unfeelingness (#PA7364)
Unpleasantness (#PA7107) Unintelligence (#PA7371) Insubstantiality (#PA6959)
Unproductiveness (#PA7208) Narrowmindedness (#PA7306) Inappropriateness (#PA6852)
Human infertility (#PC6037).

♦ PA6739 Maladjustment
Ambivalence — Arrested development — Conflict — Dejection — Delirium — Delusion — Depersonalization — Disorientation — Dissociation — Fixation — Frustration — Hallucination — Insanity — Insensibility — Neurosis — Obsession — Overcompensation — Stupor
Broader Integrity*complex (#PA7263).
Narrower Illusion (#PA6414) Obsession (#PA6448).
Related Envy (#PA7253) Error (#PA6180) Enmity (#PA5446)
Defeat (#PA7289) Disease (#PA6799) Inaction (#PA5806)
Insanity (#PA7157) Conflict (#PA0298) Disaccord (#PA5532)
Ignorance (#PA5568) Solemnity (#PA6731) Difficulty (#PA5497)
Difference (#PA6698) Opposition (#PA6979) Inattention (#PA6247)
Uncertainty (#PA7309) Frustration (#PA2252) Ambivalence (#PF1426)
Disagreement (#PA5982) Irresolution (#PA7325) Insensibility (#PA5451)
Inexpectation (#PA5527) Unfeelingness (#PA7364) Disintegration (#PA6858)
Unskillfulness (#PA7232) Unintelligence (#PA7371) Mental illness (#PC0300)
Indiscrimination (#PA6446) Uncommunicativeness (#PA7411).

♦ PA6741 Misinterpretability
Distortion — Misapplication — Misapprehension — Misconception — Misconstruction — Misinterpretation — Misjudgement — Mistake — Misunderstanding — Perversion
Broader Meaning*complex (#PA7377).
Narrower Misunderstanding (#PA8197).
Related Error (#PA6180) Reason (#PA5502) Disaccord (#PA5532)
Impairment (#PA6088) Distortion (#PA6790) Miseducation (#PA6393)
Misjudgement (#PA6607) Imperfection (#PA6997) Unrelatedness (#PA5794)
Unskillfulness (#PA7232) Misrepresentation (#PA6644) Nonaccomplishment (#PA6662)
Inappropriateness (#PA6852) Uncommunicativeness (#PA7411).

♦ PA6752 Relative motion*complex
Narrower Following (#PA7241) Recession (#PA6917) Deviation (#PA6228)
Agitation (#PA5838) Depression (#PA5493) Regression (#PA6338)
Divergence (#PA5573) Shortcoming (#PA6041).

♦ PA6774 Irregularity
Capriciousness — Discontinuity — Nonuniformity
Broader Time*complex (#PA7222).
Related Disorder (#PA7361) Separation (#PA7236) Transience (#PA6425)
Inequality (#PA6695) Dislocation (#PA7158) Irresolution (#PA7325)
Discontinuity (#PA5828) Nonuniformity (#PA6890) Changeableness (#PA5490).

♦ PA6785 Disappearance
Disappearance — Dissipation — Dissolution — Extinction — Speciousness — Superficiality
Broader Sense*complex (#PA6236).
Related Cold (#PA6956) Loss (#PA7382) Death (#PA7055)
Reason (#PA5502) Absence (#PA7270) Boredom (#PA7365)
Ignorance (#PA5568) Avoidance (#PA6379) Separation (#PA7236)
Impairment (#PA6088) Inattention (#PA6247) Destruction (#PA6542)
Unimportance (#PA5942) Intemperance (#PA6466) Disintegration (#PA6858)
Unintelligence (#PA7371) Uncommunicativeness (#PA7411).

♦ PA6790 Distortion
Asymmetry — Bias — Blemish — Corruption — Deformation — Deviation — Disfigurement — Malformation — Misconstruction — Misdirection — Misinterpretation — Misrepresentation — Misuse — Mutilation — Perversion — Slant — Torture — Warpedness
Narrower Imbalance (#PA0224) Malformation (#PE4460)
Distorted media presentations (#PD6081) Personal physical disfigurement (#PD8076).
Related Vice (#PA5644) Error (#PA6180) Reason (#PA5502)
Badness (#PA5454) Torture (#PB3430) Ugliness (#PA7240)
Reversion (#PA5699) Exclusion (#PA5869) Deviation (#PA6228)
Injustice (#PA6486) Improbity (#PA7363) Impairment (#PA6088)
Complexity (#PA6468) Inequality (#PA6695) Dissuasion (#PA7343)
Uncleanness (#PA5459) Disagreement (#PA5982) Miseducation (#PA6393)
Misjudgement (#PA6607) Imperfection (#PA6997) Insensibility (#PA5451)
Nonconformity (#PA5878) Nonuniformity (#PA6890) Unpleasantness (#PA7107)
Unskillfulness (#PA7232) Narrowmindedness (#PA7306) Misrepresentation (#PA6644)
Inappropriateness (#PA6852) Structure*complex (#PA6944) Misinterpretability (#PA6741)
Uncommunicativeness (#PA7411).
Aggravates Ignorance (#PA5568) Corruption (#PA1986).
Aggravated by Inadequate objectivity of institutions (#PF6691).

♦ PA6793 Guilt
Complicity — Culpability — Faultiness — Guilt — Feelings of sinfulness — Guilt complex
Nature Guilt is the feeling an individual has of being personally culpable for some offence arising from an act or from a failure to act, behave or perform in some way. Associated with such a feeling typically are lowered self-esteem and a feeling that one should expiate or make retribution for the wrong that has been done. Guilt is often self-ascribed on an imaginary basis, deriving from an underlying life-uncertainty or feeling of inadequacy. Since such personality orientations are so frequently encountered, it is not surprising that there is almost a universal predisposition towards guilt, even towards imagined guilt. Guilt that arises with certitude from the breach of recognized standards or laws may often be terminated with the initiation of objective punishment. Guilt that arises from a supposed breach of obtusely evident standards may be more difficult for the personality to expurge. A particular case lies among more exalted religious ideals involving the practice of virtue, self-sacrifice and the performance of religious duties. Omission of such behaviour may easily give occasion, in those whose personalities are guilt-prone, for an imagined state of sinfulness. The sufferer may proclaim that he or she is estranged from God, a sinner who may be cast into the darkness. Imaginary sinfulness and real or imaginary guilt can cause serious depression and lead to nihilistic amorality, crime and suicide.
Incidence One classical form of guilt is that experienced by survivors of catastrophes. This was experienced by those who lived through the German concentration camps, and is frequently observed in survivors of terrorist attacks and disasters like the sinking of the Herald of Free Enterprise and the explosion of the Piper Alpha oil platform. People question their own right to survival, especially when they had to struggle with others for the few remaining chances of survival in a panic situation.
Claim Like cancer, guilt tends to take over all of the healthy responses and feelings in its path, and is very difficult to remove.
Counter-claim Martin Buber said, 'Man is the being that is capable of guilt, and capable of perceiving his guilt'. The capability of guilt, or moral responsibility, implies that an individual is capable of free self-determination, responsible behaviour and the assumption of responsibility.
The overemphasis on popular psychology has confused the distinction between guilt and the feeling of guilt. Feeling guilty may have nothing to do with the fact of guilt. A person responsible for a disaster may not feel guilty at all and a person not responsible for a disaster may feel guilty. There is tremendous potential for creativity and growth with the appropriate assumption of responsibility for an act, acknowledgement of actual guilt.
Broader Morality*complex (#PA6892) Irresponsibility (#PA8658)
Negative emotions and attitudes (#PA7090).
Narrower Sex guilt (#PJ2396) Masturbation guilt (#PF5609).
Related Sin (#PF0641) Error (#PA6180) Reason (#PA5502)
Opposition (#PA6979) Disapproval (#PA6191) Imperfection (#PA6997)
Original sin (#PF8298).
Aggravates Aggression (#PA0587)
Fragmentation of the human personality (#PA0911).
Aggravated by Pride (#PA7599) Complicity (#PF4983)
Impure thoughts (#PF5205) Inhibited grief process (#PD4918)
Inadequate sex education (#PD0759)
Discrimination against women in religion (#PD0127)
Limiting of responsibility to the personal (#PF1889).

♦ PA6798 Lowness
Broader Dimension*complex (#PA7319).
Related Vulgarity (#PA5821) Solemnity (#PA6731) Disrepute (#PA6839)
Inferiority (#PA5652).

♦ PA6799 Disease
Abnormality — Affliction — Badness — Complication — Debility — Defectiveness — Delirium — Disability — Disorder — Exhaustion — Fatigue — Feebleness — Fragility — Frenzy — Illness — Indisposition — Intoxication — Invalidity — Madness — Malaise — Morbidity — Mortification — Pain — Paralysis — Pestiferousness — Rot — Shock — Sickness — Spasticity — Unwholesomeness — Waste — Wound
Refs Brooksby, J B (Ed) *The Aerial Transmission of Disease* (1984); Sammen, Peter D *The Nails in Disease* (1986).
Broader Integrity*complex (#PA7263).
Narrower Pain (#PA0643) Fatigue (#PA0657) Morbidity (#PD4538)
Exhaustion (#PJ2732) Human illness (#PA0294) Plant diseases (#PC0555)
Animal diseases (#PC0952) Vector-borne diseases (#PD8385)
Human disease and disability (#PB1044).
Related Vice (#PA5644) Fear (#PA6030) Envy (#PA7253)
Loss (#PA7382) Error (#PA6180) Shock (#PC8245)
Reason (#PA5502) Stench (#PA5981) Danger (#PA6971)
Badness (#PA5454) Oldness (#PA7131) Weakness (#PA5558)
Inaction (#PA5806) Humility (#PA6659) Insanity (#PA7157)
Disorder (#PA7361) Agitation (#PA5838) Adversity (#PA6340)
Avoidance (#PA6379) Solemnity (#PA6731) Impotence (#PA6876)
Toughness (#PA6976) Cheapness (#PA7193) Wrongness (#PA7280)
Difficulty (#PA5497) Unchastity (#PA5612) Unkindness (#PA5643)
Impairment (#PA6088) Complexity (#PA6468) Moderation (#PA7156)
Uncleanness (#PA5459) Lawlessness (#PA5563) Inattention (#PA6247)
Destruction (#PA6542) Uncertainty (#PA7309) Disagreement (#PA5982)
Intemperance (#PA6466) Misbehaviour (#PA6498) Formlessness (#PA6900)
Invisibility (#PA6978) Imperfection (#PA6997) Irresolution (#PA7325)
Inexpedience (#PA7395) Insensibility (#PA5451) Insufficiency (#PA5473)
Unwillingness (#PA6509) Maladjustment (#PA6739) Incompleteness (#PA6652)
Unpleasantness (#PA7107) Unintelligence (#PA7371) Meaninglessness (#PA6977)
Unhealthfulness (#PA7226) Inappropriateness (#PA6852) Unintelligibility (#PA7367).
Aggravates Death of living creatures (#PF7043)
Economic and social losses due to disability (#PE4856)
Underdevelopment of industrial and economic activities (#PC0880).
Aggravated by War (#PB0593) Pests (#PC0728) Pollution (#PB6336)
Disease vectors (#PC3595).

♦ PA6802 Permanence
Rigidity
Broader Change*complex (#PA6721).
Related Compulsion (#PA5740) Irresolution (#PA7325).

♦ PA6809 Displeasure
Displeasure
Broader Feeling*complex (#PA6938).
Related Envy (#PA7253) Solemnity (#PA6731) Disapproval (#PA6191)
Unpleasantness (#PA7107).

♦ PA6822 Disrespect
Abusiveness — Affrontery — Arrogance — Atrocity — Contemptuousness — Contumeliousness — Cynicism — Discourtesy — Disdain — Disesteem — Dishonour — Disparagement — Disregard — Disrespect — Exclusiveness — Humiliation — Impudence — Injury — Insolence — Insult — Irreverence — Levity — Neglect — Offence — Outrage — Ridicule — Riskiness — Sarcasm — Satire — Servility — Snobbery — Superciliousness — Travesty — Withering — Lack of respect
Broader Appropriateness*complex (#PA7246).
Narrower Satire (#PJ5950) Cynicism (#PF3418) Verbal abuse (#PD5238).
Related Vice (#PA5644) Envy (#PA7253) Loss (#PA7382)
Vanity (#PA6491) Neglect (#PA5438) Defence (#PA5445)
Badness (#PA5454) Impiety (#PA6058) Humility (#PA6659)
Rashness (#PA7115) Snobbery (#PJ3943) Vulgarity (#PA5821)
Exclusion (#PA5869) Avoidance (#PA6379) Necessity (#PA6387)
Injustice (#PA6486) Disrepute (#PA6839) Wrongness (#PA7280)
Restraint (#PA7296) Improbity (#PA7363) Dishonour (#PF8485)
Arrogance (#PA7646) Unkindness (#PA5643) Illegality (#PA5952)
Impairment (#PA6088) Opposition (#PA6979) Moderation (#PA7156)
Lamentation (#PA5479) Inferiority (#PA5652) Disapproval (#PA6191)
Inattention (#PA6247) Destruction (#PA6542) Discourtesy (#PA7143)
Uncertainty (#PA7309) Humiliation (#PF3856) Unimportance (#PA5942)
Hopelessness (#PA6099) Misbehaviour (#PA6498) Misjudgement (#PA6607)
Disobedience (#PA7250) Irresolution (#PA7325) Inexpedience (#PA7395)
Insensibility (#PA5451) Unsociability (#PA6653) Unprovability (#PA7070)
Unpleasantness (#PA7107) Unskillfulness (#PA7232) Loss of civility (#PC7013)
Misrepresentation (#PA6644) Nonaccomplishment (#PA6662) Inappropriateness (#PA6852)
Uncommunicativeness (#PA7411).
Aggravates Shame (#PF9991) Disobedience of elders (#PF7149)
Disobedience of children (#PD5308).
Aggravated by Spiritual disobedience (#PF7467).

♦ PA6838 Dissent

Alienation — Annulment — Contradiction — Denial — Disaccord — Disapproval — Disparity — Dissatisfaction — Dissension — Dissidence — Grievance — Negation — Negativity — Nonconformity — Opposition — Rejection — Repudiation — Variance — Withdrawal

Broader Truth*complex (#PA6164).

Related Envy (#PA7253)
Reason (#PA5502)
Refusal (#PA7321)
Unbelief (#PA7392)
Reversion (#PA5699)
Necessity (#PA6387)
Difficulty (#PA5497)
Difference (#PA6698)
Withdrawal (#PF4402)
Destruction (#PA6542)
Hopelessness (#PA6099)
Inexpectation (#PA5527)
Unwillingness (#PA6509)
Discontentment (#PA6011)
Dissatisfaction (#PA8886)
Loss (#PA7382)
Denial (#PA7400)
Duality (#PA7339)
Celibacy (#PA7410)
Exclusion (#PA5869)
Injustice (#PA6486)
Regression (#PA6338)
Opposition (#PA6979)
Lamentation (#PA5479)
Incuriosity (#PA6598)
Disobedience (#PA7250)
Nonconformity (#PA5878)
Unprovability (#PA7070)
Unpleasantness (#PA7107)
Inappropriateness (#PA6852).
Enmity (#PA5446)
Badness (#PA5454)
Insanity (#PA7157)
Disaccord (#PA5532)
Avoidance (#PA6379)
Grievance (#PF8029)
Inequality (#PA6695)
Alienation (#PA3545)
Disapproval (#PA6191)
Disagreement (#PA5982)
Irresolution (#PA7325)
Counteraction (#PA6485)
Unfeelingness (#PA7364)
Contradictions (#PF3667)

Aggravates Human violence (#PA0429).

♦ PA6839 Disrepute

Abjection — Arrant — Atrocity — Baseness — Censure — Cheapness — Defamation — Degradation — Depravity — Despicableness — Discredit — Disesteem — Disfavour — Disgrace — Dishonour — Disparagement — Foulness — Fulsomeness — Grossness — Humiliation — Ignobility — Ignominiousness — Infamy — Ingloriousness — Littleness — Lowness — Meanness — Notoriety — Opprobrium — Pettiness — Reproach — Scandal — Smallness — Sordidness — Squalor — Stain — Vileness

Broader Discriminative affection*complex (#PA7145).

Narrower Infamy (#PB8172)
Scandal (#PC8391)
Depravity (#PC8974).

Related Vice (#PA5644)
Stench (#PA5981)
Lowness (#PA6798)
Humility (#PA6659)
Unbelief (#PA7392)
Vulgarity (#PA5821)
Solemnity (#PA6731)
Improbity (#PA7363)
Unchastity (#PA5612)
Disrespect (#PA6822)
Uncleanness (#PA5459)
Destruction (#PA6542)
Unimportance (#PA5942)
Condemnation (#PA7237)
Unprovability (#PA7070)
Unpleasantness (#PA7107)
Inappropriateness (#PA6852).
Fear (#PA6030)
Badness (#PA5454)
Fewness (#PA7152)
Ugliness (#PA7240)
Smallness (#PA7408)
Avoidance (#PA6379)
Cheapness (#PA7193)
Dishonour (#PF8485)
Unkindness (#PA5643)
Moderation (#PA7156)
Inferiority (#PA5652)
Selfishness (#PA7211)
Misjudgement (#PA6607)
Insensibility (#PA5451)
Unsavouriness (#PA7204)
Unintelligence (#PA7371)
Envy (#PA7253)
Impiety (#PA6058)
Closure (#PA7391)
Disorder (#PA7361)
Disaccord (#PA5532)
Injustice (#PA6486)
Wrongness (#PA7280)
Depression (#PA5493)
Impairment (#PA6088)
Littleness (#PA7285)
Disapproval (#PA6191)
Humiliation (#PF3856)
Imperfection (#PA6997)
Insufficiency (#PA5473)
Disintegration (#PA6858)
Narrowmindedness (#PA7306)

♦ PA6847 Contraction

Broader Dimension*complex (#PA7319).

♦ PA6852 Inappropriateness

Absurdity — Abuse — Aimlessness — Barrenness — Cheapness — Depletion — Desecration — Disuse — Effetism — Erosion — Exhaustion — Fatuity — Fecklessness — Futility — Harassment — Impotence — Impoverishment — Inanity — Inapplicability — Inappropriateness — Ineffectiveness — Ineffectuality — Injury — Malfeasance — Malpractice — Maltreatment — Misapplication — Misconduct — Misfeasance — Mismanagement — Misusage — Misuse — Obsolescence — Oppression — Outrage — Perversion — Pointlessness — Pollution — Profanation — Prostitution — Purposelessness — Rejection — Rubbish — Suspension — Torment — Triviality — Unproductiveness — Unsuitability — Uselessness — Vainness — Valueless — Vanity — Victimization — Violation — Violence — Waste — Worthlessness

Broader Adaptation*complex (#PA8178).

Narrower Erosion (#PC8193)
Absurdity (#PF6991)
Perversion (#PB0869)
Oppression (#PB8656)
Mismanagement (#PB8406).

Related Vice (#PA5644)
Loss (#PA7382)
Vanity (#PA6491)
Impiety (#PA6058)
Dissent (#PA6838)
Boredom (#PA7365)
Humility (#PA6659)
Celibacy (#PA7410)
Exclusion (#PA5869)
Injustice (#PA6486)
Impotence (#PA6876)
Wrongness (#PA7280)
Unchastity (#PA5612)
Illegality (#PA5952)
Distortion (#PA6790)
Moderation (#PA7156)
Inferiority (#PA5652)
Desecration (#PF9176)
Hopelessness (#PA6099)
Imperfection (#PA6997)
Inexpedience (#PA7395)
Insufficiency (#PA5473)
Unsavouriness (#PA7204)
Unpleasantness (#PA7107)
Unintelligence (#PA7371)
Meaninglessness (#PA6977)
Misrepresentation (#PA6644)
Influencelessness (#PA6882)
Uncommunicativeness (#PA7411).
Fear (#PA6030)
Error (#PA6180)
Chance (#PA6714)
Poverty (#PA6434)
Oldness (#PA7131)
Weakness (#PA5558)
Disorder (#PA7361)
Ignorance (#PA5568)
Avoidance (#PA6379)
Solemnity (#PA6731)
Undueness (#PA6921)
Pollution (#PB6336)
Unkindness (#PA5643)
Impairment (#PA6088)
Disrespect (#PA6822)
Exhaustion (#PJ2732)
Disapproval (#PA6191)
Unimportance (#PA5942)
Miseducation (#PA6393)
Untimeliness (#PA7006)
Prostitution (#PD0693)
Appropriation (#PA5688)
Discontentment (#PA6011)
Unskillfulness (#PA7232)
Human violence (#PA0429)
Unhealthfulness (#PA7226)
Nonaccomplishment (#PA6662)
Human infertility (#PC6037)
Envy (#PA7253)
Reason (#PA5502)
Badness (#PA5454)
Disease (#PA6799)
Refusal (#PA7321)
Inaction (#PA5806)
Unbelief (#PA7392)
Vulgarity (#PA5821)
Necessity (#PA6387)
Disrepute (#PA6839)
Cheapness (#PA7193)
Difficulty (#PA5497)
Compulsion (#PA5740)
Complexity (#PA6468)
Opposition (#PA6979)
Uncleanness (#PA5459)
Destruction (#PA6542)
Disagreement (#PA5982)
Misbehaviour (#PA6498)
Disobedience (#PA7250)
Insensibility (#PA5451)
Unrelatedness (#PA5794)
Disintegration (#PA6858)
Unpreparedness (#PA7341)
Thoughtlessness (#PA6940)
Unproductiveness (#PA7208)
Unimaginativeness (#PA6738)
Misinterpretability (#PA6741)

♦ PA6858 Disintegration

Anarchy — Brokenness — Chaos — Confusion — Decomposition — Degradation — Dilapidation — Disintegration — Disorganization — Dissociation — Dissolution — Erosion — Incoherence — Inconsistency — Looseness — Nonadherence — Separateness — Tenuousness — Wear — Abandonment of unity

Broader Quantity*complex (#PA6108).

Narrower Wear (#PB1701)
Erosion (#PC8193)
Disorganization (#PF4487)
Social disintegration (#PC3309)
Cultural disintegration (#PG3299)
National disintegration (#PB3384)
Political disintegration (#PC3204)
Territorial fragmentation (#PC2944)
Disintegration of organized religion (#PD3423)
Disintegration of technological capacity (#PD7719).

Related Vice (#PA5644)
Loss (#PA7382)
Error (#PA6180)
Death (#PA7055)
Vanity (#PA6491)
Duality (#PA7339)
Disorder (#PA7361)
Separation (#PA7236)
Compulsion (#PA5740)
Difference (#PA6698)
Lawlessness (#PA5563)
Uncertainty (#PA7309)
Formlessness (#PA6900)
Discontinuity (#PA5828)
Changeableness (#PA5490)
Unintelligibility (#PA7367).
Chaos (#PF6836)
Defeat (#PA7289)
Boredom (#PA7365)
Disrepute (#PA6839)
Depression (#PA5493)
Illegality (#PA5952)
Moderation (#PA7156)
Inattention (#PA6247)
Incoherence (#PF8094)
Disobedience (#PA7250)
Maladjustment (#PA6739)
Unpleasantness (#PA7107)
Reason (#PA5502)
Neglect (#PA5438)
Insanity (#PA7157)
Confusion (#PF7123)
Unchastity (#PA5612)
Impairment (#PA6088)
Littleness (#PA7285)
Destruction (#PA6542)
Disagreement (#PA5982)
Unrelatedness (#PA5794)
Disappearance (#PA6785)
Inappropriateness (#PA6852)

Aggravated by Fragmentation (#PA6233).

♦ PA6862 Intrusion

Encroachment — Impertinence — Insinuation — Interference — Intervention — Intrusiveness — Invasion — Presumption — Trespass

Refs Beazley, Kim and Clark, Ian *Politics of Intrusion* (1979).

Broader Contextuality*complex (#PA7213).

Narrower Invasion (#PD8779)
Intervention (#PG6007).

Related Vice (#PA5644)
Inaction (#PA5806)
Exclusion (#PA5869)
Illegality (#PA5952)
Disapproval (#PA6191)
Condemnation (#PA7237)
Vanity (#PA6491)
Humility (#PA6659)
Undueness (#PA6921)
Opposition (#PA6979)
Misjudgement (#PA6607)
Disobedience (#PA7250).
Defence (#PA5445)
Rashness (#PA7115)
Difficulty (#PA5497)
Shortcoming (#PA6041)
Untimeliness (#PA7006)

♦ PA6876 Impotence

Barrenness — Counterproductivity — Devitalization — Disability — Effetism — Failure — Fatuity — Fecklessness — Feebleness — Forcelessness — Futility — Helplessness — Imbecility — Impotence — Inability — Inadequacy — Inanity — Incapability — Incapacity — Incompetence — Incompleteness — Ineffectiveness — Ineffectuality — Inefficiency — Ineptitude — Inferiority — Insufficiency — Invalidity — Powerlessness — Sabotage — Softness — Uselessness — Vainness — Weakness

Broader Power*complex (#PA7314).

Narrower Inefficiency (#PB0843).

Related Vice (#PA5644)
Reason (#PA5502)
Defeat (#PA7289)
Oldness (#PA7131)
Inaction (#PA5806)
Ignorance (#PA5568)
Impairment (#PA6088)
Incapacity (#PJ4312)
Destruction (#PA6542)
Intemperance (#PA6466)
Irresolution (#PA7325)
Insufficiency (#PA5473)
Unsavouriness (#PA7204)
Unskillfulness (#PA7232)
Thoughtlessness (#PA6940)
Nonaccomplishment (#PA6662)
Influencelessness (#PA6882)
Fear (#PA6030)
Vanity (#PA6491)
Badness (#PA5454)
Boredom (#PA7365)
Unbelief (#PA7392)
Restraint (#PA7296)
Inequality (#PA6695)
Inferiority (#PA5652)
Unimportance (#PA5942)
Invisibility (#PA6978)
Inexpedience (#PA7395)
Inexpectation (#PA5527)
Discontentment (#PA6011)
Unpreparedness (#PA7341)
Meaninglessness (#PA6977)
Unimaginativeness (#PA6738)
Human infertility (#PC6037).
Error (#PA6180)
Danger (#PA6971)
Disease (#PA6799)
Weakness (#PA5558)
Smallness (#PA7408)
Difficulty (#PA5497)
Inadequacy (#PA8199)
Shortcoming (#PA6041)
Hopelessness (#PA6099)
Imperfection (#PA6997)
Incompetence (#PA6416)
Nonconformity (#PA5878)
Incompleteness (#PA6652)
Unintelligence (#PA7371)
Unproductiveness (#PA7208)
Inappropriateness (#PA6852)

Aggravated by Imbecility (#PE6314).

♦ PA6882 Influencelessness

Forcelessness — Impotence — Impressionability — Ineffectiveness — Ineffectuality — Powerlessness — Susceptibility — Weakness

Broader Power*complex (#PA7314).

Related Vice (#PA5644)
Weakness (#PA5558)
Impotence (#PA6876)
Irresolution (#PA7325)
Unintelligence (#PA7371)
Fear (#PA6030)
Inaction (#PA5806)
Unimportance (#PA5942)
Unsavouriness (#PA7204)
Unproductiveness (#PA7208)
Danger (#PA6971)
Unbelief (#PA7392)
Imperfection (#PA6997)
Unskillfulness (#PA7232)
Inappropriateness (#PA6852).

♦ PA6890 Nonuniformity

Boredom — Capriciousness — Changeableness — Deviation — Divergence — Mutability — Nonuniformity — Tedium — Variation

Broader Relationship*complex (#PA6484).

Related Error (#PA6180)
Inaction (#PA5806)
Deviation (#PA6228)
Inequality (#PA6695)
Incuriosity (#PA6598)
Irregularity (#PA6774)
Discontinuity (#PA5828)
Unpleasantness (#PA7107).
Boredom (#PA7365)
Disorder (#PA7361)
Divergence (#PA5573)
Difference (#PA6698)
Uncertainty (#PA7309)
Imperfection (#PA6997)
Nonconformity (#PA5878)
Weakness (#PA5558)
Disaccord (#PA5532)
Transience (#PA6425)
Distortion (#PA6790)
Disagreement (#PA5982)
Irresolution (#PA7325)
Changeableness (#PA5490)

♦ PA6892 Morality*complex

Narrower Vice (#PA5644)
Guilt (#PA6793)
Unchastity (#PA5612)
Selfishness (#PA7211)
Intemperance (#PA6466).

♦ PA6900 Formlessness

Chaos — Characterlessness — Confusion — Disorder — Featurelessness — Indecisiveness — Messiness — Muddle — Obscurity — Orderlessness — Shapelessness — Vagueness

Broader Structure*complex (#PA6944).

Related Chaos (#PF6836)
Neglect (#PA5438)
Boredom (#PA7365)
Disorder (#PA7361)
Complexity (#PA6468)
Lawlessness (#PA5563)
Unimportance (#PA5942)
Irresolution (#PA7325)
Disintegration (#PA6858)
Indiscrimination (#PA6446)
Unintelligibility (#PA7367).
Vanity (#PA6491)
Disease (#PA6799)
Weakness (#PA5558)
Agitation (#PA5838)
Moderation (#PA7156)
Inattention (#PA6247)
Misbehaviour (#PA6498)
Nonconformity (#PA5878)
Unpleasantness (#PA7107)
Insubstantiality (#PA6959)
Defeat (#PA7289)
Absence (#PA7270)
Ugliness (#PA7240)
Confusion (#PF7123)
Uncleanness (#PA5459)
Uncertainty (#PA7309)
Invisibility (#PA6978)
Changeableness (#PA5490)
Unskillfulness (#PA7232)
Nonaccomplishment (#PA6662)

Aggravates Injuries (#PB0855).

Aggravated by Disorganization (#PF4487).

♦ PA6917 Recession

Broader Relative motion*complex (#PA6752).

♦ PA6921 Undueness

Encroachment — Excess — Impropriety — Inappropriateness — Inconvenience — Lawlessness — Licentiousness — Presumption — Trespass — Trouble — Usurpation

Broader Appropriateness*complex (#PA7246).
Related Vice (#PA5644) Fear (#PA6030) Vanity (#PA6491)
Excess (#PB8952) Rashness (#PA7115) Vulgarity (#PA5821)
Adversity (#PA6340) Injustice (#PA6486) Intrusion (#PA6862)
Cheapness (#PA7193) Wrongness (#PA7280) Restraint (#PA7296)
Difficulty (#PA5497) Unchastity (#PA5612) Illegality (#PA5952)
Lawlessness (#PA5563) Shortcoming (#PA6041) Disagreement (#PA5982)
Intemperance (#PA6466) Misbehaviour (#PA6498) Misjudgement (#PA6607)
Untimeliness (#PA7006) Disobedience (#PA7250) Inexpedience (#PA7395)
Insufficiency (#PA5473) Appropriation (#PA5688) Unpleasantness (#PA7107)
Meaninglessness (#PA6977) Inappropriateness (#PA6852)
Uncommunicativeness (#PA7411).

♦ PA6926 Nonhumanity
Broader Life*complex (#PA6586).

♦ PA6938 Feeling*complex
Narrower Solemnity (#PA6731) Impatience (#PA6200) Lamentation (#PA5479)
Displeasure (#PA6809) Aggravation (#PA7378) Unfeelingness (#PA7364)
Inexcitability (#PA5467) Unpleasantness (#PA7107) Discontentment (#PA6011).
Related Boredom (#PA7365).

♦ PA6940 Thoughtlessness
Fatuity — Inanity — Oblivion — Unintelligence
Broader Intellectual faculties*complex (#PA7197).
Related Neglect (#PA5438) Boredom (#PA7365) Ignorance (#PA5568)
Impotence (#PA6876) Unimportance (#PA5942) Forgetfulness (#PA6651)
Unsavouriness (#PA7204) Unfeelingness (#PA7364) Unskillfulness (#PA7232)
Unintelligence (#PA7371) Meaninglessness (#PA6977) Inappropriateness (#PA6852).

♦ PA6944 Structure*complex
Narrower Closure (#PA7391) Formlessness (#PA6900).
Related Distortion (#PA6790).

♦ PA6950 Angst
Dread
Nature With conditions of sophistication and power every society paradoxically develops a morbidity or malaise among its members that can best be described as lying in the range of meanings of anxiety, despair and dread. An intimation of the nature of this malaise can be seen in the critical periods of history when there has been a turning away from traditional values. One period is the modern 'Age of Anxiety', originating after the industrial revolution with the arrival of the culturally dominating sciences and technologies including Darwinism, which helped erode authoritarian Christianity. Modern angst, as Kierkegaard wrote, is an overwhelming of the individual by world knowledge.
Background By various approaches many writers about angst come to the same conclusion, that it is a condition of deprivation, a result of man having lost something. It is pain and suffering, as Buddha taught; it is also strife or separation, called Eris in Greek religion. Thus, although it is emotional, it nevertheless arises from a real condition. What is separated or lost from the self is described differently by various thinkers. Some believe it to be meaning; others, security; others, power; and others, soul.
Aggravates Behavioural deterioration (#PB6321).
Aggravated by Human suffering (#PB5955).

♦ PA6953 Depersonalization
Dependence on depersonalization
Nature In this condition, the individual loses a sense of relationship with his own personality; the world becomes unreal and he is a stranger to himself. He imagines that the face he sees in the mirror has changed, it seems to be the face of someone else; his voice is not his own, his thoughts are bizarre, his actions are automatic. He may feel he has no body, that he is not real, that he does not even exist, or is a shadow or ghost.
Incidence Schizophrenia and serious memory disturbances, are frequently accompanied by symptoms of depersonalization. Mental patients are said to be estranged from themselves. Persons with certain nervous diseases, like allochiry (lacking self-perception of one half of the body), often suffer from a sense of partial or total depersonalization. Their limbs are not real limbs, or not their own limbs; their familiar surroundings become unreal; well-known faces assume a strange unreality and become unrecognizable. It can also result from long periods of insomnolence or sensory deprivation; can come as a result of epilepsy, hysteria, migraine, extreme torture or extreme pain, great emotional stress; psychedelic drugs can induce it as well, and ecstasy and visionary experience may result in temporary depersonalization.
Related Derealization (#PF4037) Dehumanization (#PA1757).
Aggravates Neurosis (#PD0270) Denial of human rights (#PB3121).

♦ PA6956 Cold
Closeness — Coldness — Extinction
Broader Absolute properties*complex (#PA5939).
Related Death (#PA7055) Enmity (#PA5446) Avoidance (#PA6379)
Cheapness (#PA7193) Unkindness (#PA5643) Quiescence (#PA6444)
Narrowness (#PA7256) Destruction (#PA6542) Unsociability (#PA6653)
Disappearance (#PA6785) Unfeelingness (#PA7364)
Uncommunicativeness (#PA7411).

♦ PA6959 Insubstantiality
Unreality — Vagueness
Broader Existence*complex (#PA5498).
Related Error (#PA6180) Uncertainty (#PA7309) Nonexistence (#PA5870)
Formlessness (#PA6900) Invisibility (#PA6978) Unimaginativeness (#PA6738)
Unintelligibility (#PA7367).

♦ PA6971 Danger
Alarmism — Badness — Crisis — Dangerousness — Doubtfulness — Dubiousness — Hazardousness — Insecurity — Instability — Liability — Menace — Risk — Susceptibility — Threat — Uncertainty — Undependability — Unpredictability — Unreliability — Untrustworthiness — Weakness
Refs Brecher, Michael, et al *Crises in the Twentieth Century* (1988).
Broader Integrity*complex (#PA7263).
Narrower Risk (#PF7580) Uncertainty (#PA6438).
Related Vice (#PA5644) Fear (#PA6030) Stench (#PA5981)
Badness (#PA5454) Impiety (#PA6058) Poverty (#PA6434)
Disease (#PA6799) Weakness (#PA5558) Inaction (#PA5806)
Unbelief (#PA7392) Impotence (#PA6876) Improbity (#PA7363)
Transience (#PA6425) Insecurity (#PA0857) Disapproval (#PA6191)
Uncertainty (#PA7309) Instability (#PA0859) Misbehaviour (#PA6498)
Invisibility (#PA6978) Imperfection (#PA6997) Irresolution (#PA7325)
Inexpedience (#PA7395) Unsavouriness (#PA7204) Changeableness (#PA5490)
Unintelligence (#PA7371) Unhealthfulness (#PA7226) Influencelessness (#PA6882)
Unintelligibility (#PA7367) Uncommunicativeness (#PA7411).

♦ PA6976 Toughness
Fragility — Hardness — Stiffness — Vulnerability
Broader Relative properties*complex (#PA6091).
Related Vice (#PA5644) Disease (#PA6799) Boredom (#PA7365)
Weakness (#PA5558) Difficulty (#PA5497) Unkindness (#PA5643)
Compulsion (#PA5740) Lamentation (#PA5479) Informality (#PA7170)
Pitilessness (#PA7023) Irresolution (#PA7325) Unfeelingness (#PA7364)
Unpreparedness (#PA7341) Unintelligibility (#PA7367).
Aggravates Ignorance (#PA5568) Mental pollution (#PB6248)
Exploitation of dependence on food aid (#PD7592).

♦ PA6977 Meaninglessness
Absurdity — Aimlessness — Futility — Inanity — Insignificance — Nonsense — Nullity — Purposelessness — Rot — Rubbish — Senselessness — Dependence on meaninglessness — Lack of purpose — Lack of ambition — Lack of law — Lawlessness
Nature There is a general sense of lack of meaning and purpose which dominates literature, art and philosophy. There is a feeling that the certainties provided by religion have been lost and can never be replaced. Science by solving our practical problems only makes this inner void more painfully obvious. Inevitably those who are most aware of this sense of meaninglessness feel alienated from society but although they respond to negative stimuli (pain, inconvenience and loss of freedom), freedom from these stimuli reveals their lack of purpose and incapacity for freedom.
Background The problem brought into existence the philosophy of existentialism which only confirmed the dimensions of the problem.
Claim Modern man lives amid an immense, complex civilization that he did little to create. It is not surprising if he feels passive and acted upon rather than an actor. His inclination to act is poisoned at the root by a feeling that anything he does takes place in a vacuum of meaninglessness. The sense of universal purpose offered by religion is a lie. Man believes himself important and unique because he has to, but in fact he is neither.
Counter-claim People who are most aware of the associated problems of contingency and absurdity are inclined to filter the world through their thought and rob it of the meaning which it naturally possesses.
Narrower Death instinct (#PF3849).
Related Reason (#PA5502) Chance (#PA6714) Badness (#PA5454)
Disease (#PA6799) Boredom (#PA7365) Insanity (#PA7157)
Disorder (#PA7361) Ignorance (#PA5568) Impotence (#PA6876)
Absurdity (#PF6991) Undueness (#PA6921) Impairment (#PA6088)
Illegality (#PA5952) Uncleanness (#PA5459) Lawlessness (#PA5563)
Nonexistence (#PA5870) Unimportance (#PA5942) Hopelessness (#PA6099)
Inexpedience (#PA7395) Disobedience (#PA7250) Unsavouriness (#PA7204)
Unintelligence (#PA7371) Thoughtlessness (#PA6940) Meaning*complex (#PA7377)
Economic apathy (#PC3413) Nonaccomplishment (#PA6662) Inappropriateness (#PA6852)
Human contingency (#PF7054)
Lack of central planning structures in small communities (#PF2540).
Aggravates Absenteeism (#PD1634) Proliferation of legislation (#PD5315).
Aggravated by Inadequate laws (#PC6848) Secret societies (#PF2508)
Police corruption (#PD2918) Political hostage-taking (#PD1886).
Reduced by Inappropriate ambitions (#PF9852).

♦ PA6978 Invisibility
Darkness — Feebleness — Obscurity — Uncertainty — Vagueness
Broader Sense*complex (#PA6236).
Narrower Uncertainty (#PA6438).
Related Reason (#PA5502) Danger (#PA6971) Disease (#PA6799)
Oldness (#PA7131) Weakness (#PA5558) Darkness (#PA6261)
Unbelief (#PA7392) Ignorance (#PA5568) Blindness (#PA6674)
Solemnity (#PA6731) Impotence (#PA6876) Uncertainty (#PA7309)
Unimportance (#PA5942) Formlessness (#PA6900) Irresolution (#PA7325)
Changeableness (#PA5490) Unintelligence (#PA7371) Insubstantiality (#PA6959)
Unintelligibility (#PA7367).

♦ PA6979 Opposition
Affrontery — Alienation — Antagonism — Antipathy — Arrogance — Bumptiousness — Complicity — Conflict — Contemptuousness — Contradiction — Contrariety — Defiance — Denial — Disaccord — Disdain — Disputatiousness — Disregard — Dissension — Enmity — Exclusiveness — Friction — Hostility — Impertinence — Impudence — Insolence — Intransigence — Irreconcilability — Negation — Negativity — Obstinacy — Opposition — Recalcitrance — Rejection — Repulsion — Rivalry
Broader Interaction*complex (#PA6429).
Narrower Friction (#PF1691).
Related Envy (#PA7253) Hate (#PA7338) Loss (#PA7382)
Guilt (#PA6793) Enmity (#PA5446) Reason (#PA5502)
Vanity (#PA6491) Denial (#PA7400) Neglect (#PA5438)
Badness (#PA5454) Dissent (#PA6838) Refusal (#PA7321)
Duality (#PA7339) Humility (#PA6659) Rashness (#PA7115)
Insanity (#PA7157) Unbelief (#PA7392) Conflict (#PA0298)
Disaccord (#PA5532) Reversion (#PA5699) Repulsion (#PA5799)
Vulgarity (#PA5821) Exclusion (#PA5869) Avoidance (#PA6379)
Necessity (#PA6387) Vengeance (#PA6606) Intrusion (#PA6862)
Hostility (#PB8538) Antipathy (#PA8810) Arrogance (#PA7646)
Difficulty (#PA5497) Inequality (#PA6695) Difference (#PA6698)
Disrespect (#PA6822) Alienation (#PA3545) Uncleanness (#PA5459)
Lamentation (#PA5479) Disapproval (#PA6191) Inattention (#PA6247)
Destruction (#PA6542) Discourtesy (#PA7143) Competition (#PB0848)
Disagreement (#PA5982) Hopelessness (#PA6099) Disobedience (#PA7250)
Irresolution (#PA7325) Unwillingness (#PA6509) Unsociability (#PA6653)
Maladjustment (#PA6739) Unprovability (#PA7070) Unfeelingness (#PA7364)
Unpleasantness (#PA7107) Contradictions (#PF3667) Inappropriateness (#PA6852).

♦ PA6983 Attitude*complex
Narrower Incuriosity (#PA6598) Inattention (#PA6247) Forgetfulness (#PA6651)
Narrowmindedness (#PA7306) Unimaginativeness (#PA6738).
Related Neglect (#PA5438) Excessive foreign public debt of developing countries (#PD2133).

♦ PA6993 Shallowness
Broader Dimension*complex (#PA7319).
Related Human destructiveness (#PA0832).
Aggravates Chaos (#PF6836) Vulnerability of organisms (#PB5658).
Aggravated by Disasters (#PB3561).

♦ PA6995 Unsanctity
Broader Redemption*complex (#PA7259).

♦ PA6997 Imperfection
Baseness — Blemish — Cheapness — Coarseness — Commonness — Defection — Defectiveness — Deficiency — Deformation — Dirtiness — Disfigurement — Distortion — Dullness — Erroneousness — Failure — Fallibility — Faultiness — Flaw — Immaturity — Impairment — Imperfection — Imprecision — Impurity — Inaccuracy — Inadequacy — Incompleteness — Indifference — Inferiority — Lack — Meanness — Mediocrity — Problem — Shortage — Stain — Tedium — Unevenness — Vapidity — Warpedness — Weakness
Broader Adaptation*complex (#PA8178).
Narrower Deficiency (#PJ5364) Personal physical disfigurement (#PD8076).
Related Vice (#PA5644) Fear (#PA6030) Envy (#PA7253)
Error (#PA6180) Guilt (#PA6793) Reason (#PA5502)
Danger (#PA6971) Defeat (#PA7289) Neglect (#PA5438)
Badness (#PA5454) Poverty (#PA6434) Disease (#PA6799)
Absence (#PA7270) Boredom (#PA7365) Weakness (#PA5558)
Inaction (#PA5806) Humility (#PA6659) Ugliness (#PA7240)
Unbelief (#PA7392) Shortage (#PB8238) Vulgarity (#PA5821)
Exclusion (#PA5869) Avoidance (#PA6379) Injustice (#PA6486)
Disrepute (#PA6839) Impotence (#PA6876) Cheapness (#PA7193)
Restraint (#PA7296) Improbity (#PA7363) Difficulty (#PA5497)
Unchastity (#PA5612) Unkindness (#PA5643) Illegality (#PA5952)
Impairment (#PA6088) Quiescence (#PA6444) Inequality (#PA6695)
Distortion (#PA6790) Inadequacy (#PA8199) Immaturity (#PF8413)
Inaccuracy (#PF7905) Uncleanness (#PA5459) Inferiority (#PA5652)
Shortcoming (#PA6041) Disapproval (#PA6191) Inattention (#PA6247)
Incuriosity (#PA6598) Discourtesy (#PA7143) Selfishness (#PA7211)
Uncertainty (#PA7309) Unimportance (#PA5942) Irresolution (#PA7325)
Inexpedience (#PA7395) Insensibility (#PA5451) Insufficiency (#PA5473)
Inexpectation (#PA5527) Nonconformity (#PA5878) Nonuniformity (#PA6890)
Unsavouriness (#PA7204) Unfeelingness (#PA7364) Discontentment (#PA6011)
Incompleteness (#PA6652) Unpleasantness (#PA7107) Unskillfulness (#PA7232)
Unpreparedness (#PA7341) Unintelligence (#PA7371) Narrowmindedness (#PA7306)
Misrepresentation (#PA6644) Nonaccomplishment (#PA6662) Unimaginativeness (#PA6738)
Inappropriateness (#PA6852) Influencelessness (#PA6882) Unintelligibility (#PA7367)
Misinterpretability (#PA6741) Uncommunicativeness (#PA7411).

♦ PA6999 Avarice
Dependence on avarice — Greed — Greedy people — Covetousness
Nature Avarice is the state of being in which an individual attaches such value to wealth and possessions that he makes the accumulation and retention of them the major goal of his life, to which he subordinates all else. It is not the normal human instinct to acquire, nor is it imposed by a materialistic society. It substitutes possessing for living. In the state of avarice the individual's life is so devoid of internal meaning that he creates personal significance with external things. It leads to a distorted view of reality and a narrowing of vision and imagination. Because it has no limit to its desire to do and to have, it disregards the property and rights of others. It involves a sense of ambition and purpose without limit. Such a perversion of values is totally disruptive of the moral life.
Claim The development of society cannot be sustained unless people can appreciate the need to abandon social processes based on greed.
Narrower Corporate greed (#PF7189).
Related Sin (#PF0641) Lust (#PA4673) Vice (#PA5644)
Envy (#PA7253) Anger (#PA7797) Sloth (#PA3275)
Gluttony (#PA9638) Avoidance (#PA6379) Cheapness (#PA7193)
Selfishness (#PA7211) Original sin (#PF8298) Intemperance (#PA6466).
Aggravates Careerism (#PF6353) Self-interest (#PA8760)
Lack of human unity (#PF2434) Conspiracy against the public (#PF4198)
Lack of relationship between wealth generation and the public good (#PF4730).

♦ PA7006 Untimeliness
Haste — Hastiness — Hesitation — Impropriety — Inconvenience — Inexpedience — Infelicity — Intrusiveness — Irrelevance — Lateness — Laxness — Laziness — Obstruction — Prematurity — Retardation — Slowness — Suspension — Tardiness — Unpreparedness — Unsuitability — Wrongness
Broader Time*complex (#PA7222).
Related Vice (#PA5644) Envy (#PA7253) Error (#PA6180)
Neglect (#PA5438) Badness (#PA5454) Boredom (#PA7365)
Closure (#PA7391) Inaction (#PA5806) Slowness (#PA6166)
Rashness (#PA7115) Idleness (#PA7710) Vulgarity (#PA5821)
Exclusion (#PA5869) Injustice (#PA6486) Solemnity (#PA6731)
Intrusion (#PA6862) Undueness (#PA6921) Wrongness (#PA7280)
Restraint (#PA7296) Difficulty (#PA5497) Unchastity (#PA5612)
Compulsion (#PA5740) Illegality (#PA5952) Impatience (#PA6200)
Uncertainty (#PA7309) Unimportance (#PA5942) Disagreement (#PA5982)
Misbehaviour (#PA6498) Irresolution (#PA7325) Inexpedience (#PA7395)
Insufficiency (#PA5473) Discontentment (#PA6011) Unpleasantness (#PA7107)
Unskillfulness (#PA7232) Unpreparedness (#PA7341) Unintelligence (#PA7371)
Unpreparedness (#PF8176) Inappropriateness (#PA6852) Oppressive reality (#PF7053).

♦ PA7023 Pitilessness
Cruelty — Hardness — Harshness — Heartlessness — Inclemency — Pitilessness — Relentlessness — Remorselessness — Ruthlessness
Broader Benevolence*complex (#PA7331).
Narrower Cruelty (#PB2642).
Related Vice (#PA5644) Fear (#PA6030) Discord (#PA5565)
Weakness (#PA5558) Solemnity (#PA6731) Toughness (#PA6976)
Difficulty (#PA5497) Unkindness (#PA5643) Compulsion (#PA5740)
Moderation (#PA7156) Lamentation (#PA5479) Discourtesy (#PA7143)
Irresolution (#PA7325) Unfeelingness (#PA7364) Unpleasantness (#PA7107)
Unintelligibility (#PA7367).

♦ PA7030 Passion
Dependence on passion — Insatiable desire
Claim The cause of human suffering is found in the desires of the physical body and sense-dominated mind whose passions are worldly illusions. Ardent affections arise out of the emotions where there lies a single enormous appetite or acquisitive instinct and power called desire. Much of the force of desire can sometimes be directed towards one objective; it then becomes an all-consuming passion. This passion may be for a person, a thing, or an idea. It may be for freedom or a quality of life or it may seek an artistic vision. It may seek spiritual reality. In all its forms, passion seeks consumption, and thus annihilation of itself. yet one passion may replace another, for this is the driving force for the personality; the appetite, desire or craving for life itself. Unexamined, it impels a person to chase one thing after another, often unsuccessfully, causing unhappiness and world-weariness. It creates the illusion that the satisfaction of passion by an acquisitive life will lead to contentment. The possibilities of seeing through this illusion, to detach one's self, particularly from objects, arises only after endless suffering and disappointments.
Counter-claim Only archaic, dualistic world views deny the positive nature of the life-force which expresses itself on the levels of the instincts, emotions and intellect. Human history would be impossible without, for example, the instinct for self-preservation, the reproductive and affective instincts associated with family life, and the acquisition of knowledge. All of these have their roots in desire, and the human race will have a history only as long as that desire is never sated.
Refs Marks, Joel (Ed) *The Ways of Desire* (1986).
Broader Vice (#PA5644) Excessive virtue (#PF7127).
Narrower Crimes of passion (#PG6526).
Aggravates Victimization (#PF6987).

♦ PA7037 Retribution*complex
Narrower Punishment (#PA5583) Illegality (#PA5952)
Condemnation (#PA7237).

♦ PA7055 Death
Annihilation — Banefulness — Bereavement — Bloody-mindedness — Calamity — Death — Destructiveness — Disaster — Dissolution — Eeriness — Extermination — Extinction — Gruesomeness — Lifelessness — Malignancy — Sacrifice — Self-destruction — Suffocation — Virulence
Broader Life*complex (#PA6586).
Narrower Bereavement (#PF3516) Suffocation (#PJ4103)
Annihilation (#PF9169) Extermination (#PJ2839).
Related Fear (#PA6030) Cold (#PA6956) Envy (#PA7253)
Loss (#PA7382) Enmity (#PA5446) Badness (#PA5454)
Boredom (#PA7365) Inaction (#PA5806) Darkness (#PA6261)
Ugliness (#PA7240) Celibacy (#PA7410) Disaccord (#PA5532)
Adversity (#PA6340) Separation (#PA7236) Unkindness (#PA5643)
Impairment (#PA6088) Quiescence (#PA6444) Moderation (#PA7156)
Destruction (#PA6542) Nonexistence (#PA5870) Appropriation (#PA5688)
Disappearance (#PA6785) Disintegration (#PA6858) Unpleasantness (#PA7107)
Unhealthfulness (#PA7226).

♦ PA7070 Unprovability
Contemptuousness — Denial — Discredit
Broader Credibility*complex (#PA5829).
Related Loss (#PA7382) Vanity (#PA6491) Denial (#PA7400)
Dissent (#PA6838) Refusal (#PA7321) Unbelief (#PA7392)
Necessity (#PA6387) Disrepute (#PA6839) Disrespect (#PA6822)
Opposition (#PA6979) Disapproval (#PA6191).

♦ PA7078 Environment
Confinement — Limitedness — Restriction
Refs Allaby, Michael (Ed) *Macmillan Dictionary of the Environment* (1988).
Broader Contextuality*complex (#PA7213).
Related Exclusion (#PA5869) Restraint (#PA7296) Difficulty (#PA5497)
Punishment (#PA5583) Narrowness (#PA7256) Insufficiency (#PA5473).

♦ PA7090 Negative emotions and attitudes
Dependence on negative emotions and attitudes
Broader General obstacles to problem alleviation (#PF0631).
Narrower Vice (#PA5644) Guilt (#PA6793) Obesity (#PE1177)
Fatalism (#PF6430) Nihilism (#PJ7927) Absurdity (#PF6991)
Competition (#PB0848) Humiliation (#PF3856) Technophobia (#PG6101)
Social stigma (#PD0884) Value erosion (#PA1782) Discrimination (#PA0833)
Dehumanization (#PA1757) Cosmopolitanism (#PF8331)
Unwanted pregnancies (#PF2859) Political alienation (#PC3227)
Unwarranted pessimism (#PF2818) Unaesthetic foodstuffs (#PD1126)
Lack of self-confidence (#PF0879) Increasing job monotony (#PD2656)
Non-acceptance of reality (#PF1079) Organizational empire-building (#PF1232)
Massive psychic traumatization (#PE6968) Ill effects of educational failure (#PF2013)
Irrational rejection of nuclear power (#PF8531)
Unhealthy emotional responses to atomic energy (#PF0913)
Alienation of skilled and committed personnel from international organizations and programmes (#PE1553).
Related Anxiety (#PA1635) Distrust (#PA8653) Monsters (#PF2516)
Confusion (#PF7123) Prejudice (#PA2173) Aggression (#PA0587)
Complacency (#PA1742) Intolerance (#PF0860) Harmful thought (#PF0441)
Lack of commitment (#PF1729) Unrecorded knowledge (#PF5728)
Promotion of negative images of opponents (#PF4133)
Misdirection of human energies and desires (#PF8128).
Aggravates Ignorance (#PA5568) Negligence (#PA2658)
Human suffering (#PB5955) Behavioural deterioration (#PB6321)
Inadequate personal hygiene (#PD2459).
Aggravated by Ugliness (#PA7240) Conflict (#PA0298)
Food fads (#PD1189) Regression (#PA6338)
Friendlessness (#PB5747) Ethnocentricity (#PB5765)
Maternal deprivation (#PC0981).

♦ PA7102 Malevolence
Malevolent people — Ill will — Animosity
Claim Vicious ill will, hatred and acts of destructiveness or spite, arise out of a propensity in the personality for this type of distortion. Malevolence is a pathological condition in the mind, a tumour-like complex of festering vengeance, grudge-holding, jealousy or self-justification. It can seize tribal or national leaderships and lead to bitter and bloody warfare.
Counter-claim Man does not live by bread alone, but also by the nourishment of animosities.
Narrower Intentional libel (#PG5898) Criminal motivation (#PJ6406)
Dishonourable grant acceptance (#PG8710) Intentional infecting with disease (#PD2651).
Related Hate (#PA7338) Enmity (#PA5446) Badness (#PA5454)
Unkindness (#PA5643).
Aggravates Vice (#PA5644) Demons (#PF6734) Enemies (#PF8404)
Human suffering (#PB5955) Loss of civility (#PC7013)
Affliction by malevolent spirits (#PF9043).

♦ PA7107 Unpleasantness
Abhorrence — Abomination — Affliction — Aggravation — Anguish — Annoyance — Antagonism — Antipathy — Anxiety — Atrocity — Aversion — Awfulness — Awkwardness — Banefulness — Baseness — Beastliness — Bedevilment — Bitterness — Boredom — Boringness — Bothersomeness — Burdensomeness — Chagrin — Cheerlessness — Confusion — Contrariety — Depression — Desolation — Despair — Despicableness — Difficulty — Disaffection — Disaffinity — Disagreeableness — Disapproval — Discomfort — Discomposure — Disconcertion — Discontentment — Disfavour — Disgust — Dislike — Dismay — Displeasure — Disquiet — Dissatisfaction — Distaste — Distress — Disturbance — Dolorousness — Dullness — Embarrassment — Encumbrance — Enmity — Ennui — Exasperation — Foulness — Fulsomeness — Grief — Grievance — Grimness — Grossness — Harassment — Harshness — Hatred — Hideousness — Horribleness — Hostility — Humiliation — Ignobility — Infelicity — Injury — Inquietude — Irksomeness — Joylessness — Loathsomeness — Lovelessness — Malaise — Melancholy —

Miasma — Misery — Mortification — Mournfulness — Noisome — Nuisance — Oppression — Pain — Pestiferousness — Pleasurelessness — Problem — Provocation — Repugnance — Repulsion — Sadism — Sadness — Shock — Sorrow — Sufferance — Tedium — Tiresomeness — Torment — Torture — Tortuousness — Tribulation — Trouble — Ugliness — Unbearableness — Uncomfortableness — Undesirableness — Uneasiness — Unhappiness — Unpleasantness — Vexation — Vileness — Wearisomeness — Woe — Worry — Wound

Broader Feeling*complex (#PA6938).
Narrower Grief (#PF5654) Misery (#PB8167) Ugliness (#PA7240)
Oppression (#PB8656) Provocation (#PJ2395).
Related Vice (#PA5644) Fear (#PA6030) Envy (#PA7253)
Hate (#PA7338) Loss (#PA7382) Pain (#PA0643)
Death (#PA7055) Shock (#PC8245) Enmity (#PA5446)
Stench (#PA5981) Vanity (#PA6491) Defeat (#PA7289)
Sadism (#PF3270) Badness (#PA5454) Discord (#PA5565)
Poverty (#PA6434) Disease (#PA6799) Dissent (#PA6838)
Boredom (#PA7365) Closure (#PA7391) Torture (#PB3430)
Anxiety (#PA1635) Weakness (#PA5558) Inaction (#PA5806)
Humility (#PA6659) Insanity (#PA7157) Disorder (#PA7361)
Lightness (#PA5491) Disaccord (#PA5532) Ignorance (#PA5568)
Repulsion (#PA5799) Vulgarity (#PA5821) Agitation (#PA5838)
Adversity (#PA6340) Avoidance (#PA6379) Necessity (#PA6387)
Injustice (#PA6486) Solemnity (#PA6731) Disrepute (#PA6839)
Undueness (#PA6921) Wrongness (#PA7280) Improbity (#PA7363)
Confusion (#PF7123) Hostility (#PB8538) Antipathy (#PA8810)
Grievance (#PF8029) Depression (#PA5493) Difficulty (#PA5497)
Unchastity (#PA5612) Unkindness (#PA5643) Impairment (#PA6088)
Impatience (#PA6200) Inelegance (#PA6312) Inequality (#PA6695)
Difference (#PA6698) Distortion (#PA6790) Disrespect (#PA6822)
Opposition (#PA6979) Moderation (#PA7156) Melancholy (#PF7756)
Uncleanness (#PA5459) Lamentation (#PA5479) Lawlessness (#PA5563)
Inferiority (#PA5652) Disapproval (#PA6191) Inattention (#PA6247)
Destruction (#PA6542) Incuriosity (#PA6598) Displeasure (#PA6809)
Discourtesy (#PA7143) Ungodliness (#PA7148) Uncertainty (#PA7309)
Aggravation (#PA7378) Humiliation (#PF3856) Unhappiness (#PA9191)
Unimportance (#PA5942) Disagreement (#PA5982) Hopelessness (#PA6099)
Formlessness (#PA6900) Imperfection (#PA6997) Untimeliness (#PA7006)
Pitilessness (#PA7023) Irresolution (#PA7325) Inexpedience (#PA7395)
Insensibility (#PA5451) Inhospitality (#PA5458) Inexpectation (#PA5527)
Appropriation (#PA5688) Unwillingness (#PA6509) Unsociability (#PA6653)
Nonuniformity (#PA6890) Unsavouriness (#PA7204) Unfeelingness (#PA7364)
Inexcitability (#PA5467) Discontentment (#PA6011) Disintegration (#PA6858)
Unskillfulness (#PA7232) Unintelligence (#PA7371) Unhealthfulness (#PA7226)
Dissatisfaction (#PA8886) Unproductiveness (#PA7208) Unimaginativeness (#PA6738)
Inappropriateness (#PA6852) Unintelligibility (#PA7367).

♦ PA7115 Rashness

Carelessness — Distrust — Haste — Hastiness — Hesitation — Impetuousity — Improvidence — Imprudence — Impudence — Indiscretion — Insolence — Mistrust — Overconfidence — Overzealousness — Presumption — Uncommunicativeness

Broader Anticipation*complex (#PA6562).
Related Vice (#PA5644) Envy (#PA7253) Error (#PA6180)
Vanity (#PA6491) Neglect (#PA5438) Inaction (#PA5806)
Insanity (#PA7157) Disorder (#PA7361) Unbelief (#PA7392)
Distrust (#PA8653) Avoidance (#PA6379) Intrusion (#PA6862)
Undueness (#PA6921) Unchastity (#PA5612) Compulsion (#PA5740)
Impatience (#PA6200) Transience (#PA6425) Disrespect (#PA6822)
Opposition (#PA6979) Lamentation (#PA5479) Inattention (#PA6247)
Incuriosity (#PA6598) Discourtesy (#PA7143) Uncertainty (#PA7309)
Misjudgement (#PA6607) Untimeliness (#PA7006) Irresolution (#PA7325)
Unsociability (#PA6653) Changeableness (#PA5490) Unskillfulness (#PA7232)
Unpreparedness (#PA7341) Unintelligence (#PA7371) Indiscrimination (#PA6446)
Uncommunicativeness (#PA7411).

♦ PA7131 Oldness

Age — Antediluvian — Archaism — Debility — Decrepitness — Disuse — Feebleness — Inexperience — Naïvety — Obsolescence — Puerility

Broader Time*complex (#PA7222).
Narrower Obsolete methods (#PF3713).
Related Age (#PA7333) Reason (#PA5502) Disease (#PA6799)
Weakness (#PA5558) Inaction (#PA5806) Unbelief (#PA7392)
Ignorance (#PA5568) Avoidance (#PA6379) Impotence (#PA6876)
Impairment (#PA6088) Invisibility (#PA6978) Irresolution (#PA7325)
Unskillfulness (#PA7232) Unintelligence (#PA7371) Inappropriateness (#PA6852).

♦ PA7143 Discourtesy

Aggression — Coarseness — Curtness — Discourtesy — Glibness — Harshness — Impoliteness — Incivility — Inconsiderateness — Insensitivity — Insolence — Rudeness — Severity — Surliness — Tactlessness — Truculence — Unfriendliness — Ungraciousness — Vulgarity

Broader Benevolence*complex (#PA7331).
Related Envy (#PA7253) Enmity (#PA5446) Vanity (#PA6491)
Neglect (#PA5438) Defence (#PA5445) Discord (#PA5565)
Rashness (#PA7115) Disaccord (#PA5532) Vulgarity (#PA5821)
Unchastity (#PA5612) Unkindness (#PA5643) Compulsion (#PA5740)
Disrespect (#PA6822) Opposition (#PA6979) Moderation (#PA7156)
Aggression (#PA0587) Lamentation (#PA5479) Inferiority (#PA5652)
Affectation (#PA6400) Misbehaviour (#PA6498) Imperfection (#PA6997)
Pitilessness (#PA7023) Irresolution (#PA7325) Insensibility (#PA5451)
Inhospitality (#PA5458) Unsociability (#PA6653) Unfeelingness (#PA7364)
Unpleasantness (#PA7107) Unpreparedness (#PA7341) Unintelligence (#PA7371)
Indiscrimination (#PA6446) Loss of civility (#PC7013)
Uncommunicativeness (#PA7411).

♦ PA7145 Discriminative affection*complex

Narrower Humility (#PA6659) Ugliness (#PA7240) Disrepute (#PA6839)
Vulgarity (#PA5821) Affectation (#PA6400) Unastonishment (#PA6090).
Related Vanity (#PA6491).

♦ PA7148 Ungodliness

Bedevilment — Fiendish — Mischievousness — Perdition

Broader Redemption*complex (#PA7259).
Related Vice (#PA5644) Badness (#PA5454) Insanity (#PA7157)
Unkindness (#PA5643) Destruction (#PA6542) Misbehaviour (#PA6498)
Unpleasantness (#PA7107).

♦ PA7152 Fewness

Smallness — Thinness — Tightness

Broader Number*complex (#PA7412).
Related Humility (#PA6659) Smallness (#PA7408) Exclusion (#PA5869)
Disrepute (#PA6839) Cheapness (#PA7193) Littleness (#PA7285)
Inferiority (#PA5652) Selfishness (#PA7211) Unimportance (#PA5942)
Insufficiency (#PA5473) Unsavouriness (#PA7204) Unintelligence (#PA7371)
Narrowmindedness (#PA7306).

♦ PA7156 Moderation

Acrimony — Agitation — Anarchism — Animality — Atrocity — Barbarity — Bitterness — Brutality — Chaos — Disturbance — Fanaticism — Fierceness — Frenzy — Fury — Fuss — Harshness — Inclemency — Inhumanity — Mercilessness — Mindlessness — Pitilessness — Roughness — Savagery — Severity — Spasmodicness — Spasticity — Stridency — Tempestuousness — Terror — Turbulence — Unconscionableness — Unruliness — Upheaval — Uproar — Upset — Vandalism — Vehemence — Venom — Viciousness — Violation — Violence — Virulence

Broader Power*complex (#PA7314).
Narrower Terror (#PF8483) Inhumanity (#PB8214).
Related Vice (#PA5644) Fear (#PA6030) Envy (#PA7253)
Hate (#PA7338) Death (#PA7055) Chaos (#PF6836)
Enmity (#PA5446) Vanity (#PA6491) Badness (#PA5454)
Discord (#PA5565) Disease (#PA6799) Insanity (#PA7157)
Disorder (#PA7361) Disaccord (#PA5532) Ignorance (#PA5568)
Vulgarity (#PA5821) Agitation (#PA5838) Avoidance (#PA6379)
Injustice (#PA6486) Disrepute (#PA6839) Cheapness (#PA7193)
Wrongness (#PA7280) Restraint (#PA7296) Improbity (#PA7363)
Brutality (#PC1987) Vandalism (#PD1350) Difficulty (#PA5497)
Unchastity (#PA5612) Unkindness (#PA5643) Compulsion (#PA5740)
Revolution (#PA5901) Illegality (#PA5952) Impatience (#PA6200)
Disrespect (#PA6822) Lamentation (#PA5479) Lawlessness (#PA5563)
Disapproval (#PA6191) Inattention (#PA6247) Affectation (#PA6400)
Destruction (#PA6542) Discourtesy (#PAp143) Uncertainty (#PA7309)
Misbehaviour (#PA6498) Formlessness (#PA6900) Pitilessness (#PA7023)
Disobedience (#PA7250) Irresolution (#PA7325) Insensibility (#PA5451)
Insufficiency (#PA5473) Discontinuity (#PA5828) Inexcitability (#PA5467)
Disintegration (#PA6858) Unpleasantness (#PA7107) Unpreparedness (#PA7341)
Unintelligence (#PA7371) Unhealthfulness (#PA7226) Narrowmindedness (#PA7306)
Inappropriateness (#PA6852).

♦ PA7157 Insanity

Aberration — Abnormality — Alienation — Bedevilment — Bigotry — Compulsiveness — Delirium — Disorientation — Distraction — Excessiveness — Extravagance — Extremism — Fanaticism — Fixation — Frenzy — Fury — Incoherence — Infatuation — Irrationality — Madness — Obsession — Overreligiousness — Overzealousness — Senselessness — Sickness — Strangeness — Unnaturalness — Witlessness — Zealotry

Refs Foucault, Michel *Madness and Civilization* (1988); MacSweeney, D *The Crazy Ape* (1982).
Broader Human illness (#PA0294) Intellectual faculties*complex (#PA7197).
Narrower Extremism (#PB3415) Obsession (#PA6448).
Related Envy (#PA7253) Hate (#PA7338) Error (#PA6180)
Enmity (#PA5446) Reason (#PA5502) Disease (#PA6799)
Dissent (#PA6838) Duality (#PA7339) Bigotry (#PC7652)
Rashness (#PA7115) Disorder (#PA7361) Unbelief (#PA7392)
Ignorance (#PA5568) Reversion (#PA5699) Deviation (#PA6228)
Avoidance (#PA6379) Cheapness (#PA7193) Wrongness (#PA7280)
Difficulty (#PA5497) Unkindness (#PA5643) Inelegance (#PA6312)
Opposition (#PA6979) Moderation (#PA7156) Alienation (#PA3545)
Diffuseness (#PA5974) Inattention (#PA6247) Affectation (#PA6400)
Ungodliness (#PA7148) Uncertainty (#PA7309) Incoherence (#PF8094)
Disagreement (#PA5982) Intemperance (#PA6466) Disobedience (#PA7250)
Irresolution (#PA7325) Insufficiency (#PA5473) Nonconformity (#PA5878)
Maladjustment (#PA6739) Disintegration (#PA6858) Unpleasantness (#PA7107)
Unintelligence (#PA7371) Mental illness (#PC0300) Meaninglessness (#PA6977)
Narrowmindedness (#PA7306) Unintelligibility (#PA7367)
Uncommunicativeness (#PA7411) Political radicalism (#PF2177).

♦ PA7158 Dislocation

Disarrangement — Discontinuity — Homelessness

Broader Space*complex (#PA6713).
Related Poverty (#PA6434) Duality (#PA7339) Disorder (#PA7361)
Separation (#PA7236) Irregularity (#PA6774) Discontinuity (#PA5828)
Unsociability (#PA6653).

♦ PA7170 Informality

Pedantry — Stiffness

Broader Motivation*complex (#PA6048).
Narrower Pedantry (#PG0181).
Related Boredom (#PA7365) Toughness (#PA6976) Compulsion (#PA5740)
Affectation (#PA6400) Irresolution (#PA7325).

♦ PA7193 Cheapness

Avarice — Closeness — Decline — Depreciation — Excess — Excessiveness — Exploitation — Extortion — Extravagance — Illiberalism — Meanness — Overtax — Prodigality — Profligacy — Sordidness — Tightness — Unconscionableness — Unreasonableness — Waste — Wastefulness

Broader Possession*complex (#PA6686).
Narrower Criminal coercion (#PD4469).
Related Vice (#PA5644) Cold (#PA6956) Envy (#PA7253)
Loss (#PA7382) Reason (#PA5502) Excess (#PB8952)
Badness (#PA5454) Disease (#PA6799) Fewness (#PA7152)
Refusal (#PA7321) Avarice (#PA6999) Weakness (#PA5558)
Humility (#PA6659) Insanity (#PA7157) Disorder (#PA7361)
Lightness (#PA5491) Vulgarity (#PA5821) Exclusion (#PA5869)
Adversity (#PA6340) Avoidance (#PA6379) Necessity (#PA6387)
Injustice (#PA6486) Disrepute (#PA6839) Undueness (#PA6921)
Improbity (#PA7363) Unkindness (#PA5643) Impairment (#PA6088)
Quiescence (#PA6444) Moderation (#PA7156) Narrowness (#PA7256)
Uncleanness (#PA5459) Inferiority (#PA5652) Diffuseness (#PA5974)
Shortcoming (#PA6041) Disapproval (#PA6191) Affectation (#PA6400)
Destruction (#PA6542) Selfishness (#PA7211) Prodigality (#PA7269)
Unimportance (#PA5942) Intemperance (#PA6466) Misjudgement (#PA6607)
Imperfection (#PA6997) Insufficiency (#PA5473) Appropriation (#PA5688)
Unintelligence (#PA7371) Narrowmindedness (#PA7306) Inappropriateness (#PA6852)
Uncommunicativeness (#PA7411).

♦ PA7197 Intellectual faculties*complex

Narrower Reason (#PA5502) Insanity (#PA7157) Ignorance (#PA5568)
Unintelligence (#PA7371) Thoughtlessness (#PA6940).

♦ **PA7199 Order*complex**
Narrower Disorder (#PA7361) Exclusion (#PA5869) Discontinuity (#PA5828) Nonconformity (#PA5878).

♦ **PA7204 Unsavouriness**
Deadness — Disgust — Foulness — Inanity — Noisome — Thinness — Vapidity — Weakness
Broader Sense*complex (#PA6236).
Related Vice (#PA5644) Fear (#PA6030) Stench (#PA5981) Danger (#PA6971) Badness (#PA5454) Fewness (#PA7152) Boredom (#PA7365) Closure (#PA7391) Weakness (#PA5558) Inaction (#PA5806) Darkness (#PA6261) Ugliness (#PA7240) Unbelief (#PA7392) Ignorance (#PA5568) Injustice (#PA6486) Disrepute (#PA6839) Impotence (#PA6876) Unchastity (#PA5612) Uncleanness (#PA5459) Disapproval (#PA6191) Destruction (#PA6542) Unimportance (#PA5942) Imperfection (#PA6997) Irresolution (#PA7325) Insensibility (#PA5451) Insufficiency (#PA5473) Unpleasantness (#PA7107) Unintelligence (#PA7371) Thoughtlessness (#PA6940) Meaninglessness (#PA6977) Unhealthfulness (#PA7226) Inappropriateness (#PA6852) Influencelessness (#PA6882).

♦ **PA7208 Unproductiveness**
Barrenness — Desolation — Impotence — Infertility — Sterility — Unproductiveness
Refs WHO Scientific Group, Geneva, 1975 *Epidemiology of Infertility* (1975).
Broader Power*complex (#PA7314).
Related Boredom (#PA7365) Weakness (#PA5558) Solemnity (#PA6731) Impotence (#PA6876) Destruction (#PA6542) Unsociability (#PA6653) Unpleasantness (#PA7107) Unimaginativeness (#PA6738) Inappropriateness (#PA6852) Influencelessness (#PA6882) Human infertility (#PC6037).

♦ **PA7211 Selfishness**
Acquisitiveness — Autism — Greed — Illiberalism — Littleness — Meanness — Pettiness — Self–admiration — Self–centredness — Self–esteem — Self–indulgence — Selfishness — Smallness
Broader Vanity (#PA6491) Morality*complex (#PA6892).
Narrower Cynicism (#PF3418).
Related Vice (#PA5644) Envy (#PA7253) Error (#PA6180) Fewness (#PA7152) Avarice (#PA6999) Humility (#PA6659) Smallness (#PA7408) Vulgarity (#PA5821) Avoidance (#PA6379) Disrepute (#PA6839) Cheapness (#PA7193) Unkindness (#PA5643) Littleness (#PA7285) Inferiority (#PA5652) Unimportance (#PA5942) Intemperance (#PA6466) Imperfection (#PA6997) Insufficiency (#PA5473) Unsociability (#PA6653) Unfeelingness (#PA7364) Narrowmindedness (#PA7306) Unimaginativeness (#PA6738).
Aggravates Family breakdown (#PC2102).
Aggravated by Hedonism (#PF2277).

♦ **PA7213 Contextuality*complex**
Narrower Intrusion (#PA6862) Environment (#PA7078).

♦ **PA7214 Choice*complex**
Narrower Avoidance (#PA6379) Necessity (#PA6387) Irresolution (#PA7325) Unwillingness (#PA6509).

♦ **PA7218 Compromise**
Broader Interaction*complex (#PA6429).

♦ **PA7222 Time*complex**
Narrower Age (#PA7333) Oldness (#PA7131) Antiquity (#PA5733) Transience (#PA6425) Infrequency (#PA7372) Untimeliness (#PA7006) Irregularity (#PA6774) Instantaneousness (#PA6560).

♦ **PA7226 Unhealthfulness**
Agitation — Badness — Destructiveness — Harmfulness — Noisome — Pestiferousness — Poisonousness — Pollution — Toxicity — Unwholesomeness — Venom — Virulence
Broader Integrity*complex (#PA7263).
Related Vice (#PA5644) Fear (#PA6030) Envy (#PA7253) Death (#PA7055) Enmity (#PA5446) Stench (#PA5981) Danger (#PA6971) Badness (#PA5454) Disease (#PA6799) Agitation (#PA5838) Adversity (#PA6340) Pollution (#PB6336) Unkindness (#PA5643) Impairment (#PA6088) Complexity (#PA6468) Moderation (#PA7156) Uncleanness (#PA5459) Destruction (#PA6542) Misbehaviour (#PA6498) Unsavouriness (#PA7204) Inexcitability (#PA5467) Unpleasantness (#PA7107) Inappropriateness (#PA6852).

♦ **PA7232 Unskillfulness**
Artifice — Awkwardness — Carelessness — Clumsiness — Cunning — Deceit — Gracelessness — Guile — Ignorance — Immaturity — Inability — Inadequacy — Inaptitude — Incapability — Incapacity — Incompetence — Ineffectiveness — Ineffectuality — Inefficiency — Inelegance — Inexpedience — Inexperience — Maladjustment — Maladroit — Malpractice — Mediocrity — Miscarriage — Misconduct — Misdirection — Misfeasance — Misguidance — Mismanagement — Misrule — Mistake — Muddle — Naïvety — Neglect — Negligence — Nonfeasance — Omission — Ponderousness — Sophistry — Subterfuge — Thoughtlessness — Trickery — Unfamiliarity — Unintelligence — Unmanageability — Unsophistication
Broader Achievement*complex (#PA5471).
Narrower Deception (#PB4731) Inefficiency (#PB0843) Mismanagement (#PB8406).
Related Vice (#PA5644) Error (#PA6180) Reason (#PA5502) Neglect (#PA5438) Badness (#PA5454) Oldness (#PA7131) Boredom (#PA7365) Weakness (#PA5558) Rashness (#PA7115) Ugliness (#PA7240) Disorder (#PA7361) Unbelief (#PA7392) Ignorance (#PA5568) Vulgarity (#PA5821) Avoidance (#PA6379) Injustice (#PA6486) Impotence (#PA6876) Improbity (#PA7363) Sophistry (#PF8008) Difficulty (#PA5497) Unchastity (#PA5612) Unkindness (#PA5643) Compulsion (#PA5740) Inelegance (#PA6312) Complexity (#PA6468) Inequality (#PA6695) Distortion (#PA6790) Disrespect (#PA6822) Inadequacy (#PA8199) Immaturity (#PF8413) Incapacity (#PJ4312) Lawlessness (#PA5563) Inferiority (#PA5652) Shortcoming (#PA6041) Inattention (#PA6247) Incuriosity (#PA6598) Uncertainty (#PA7309) Unimportance (#PA5942) Disagreement (#PA5982) Miseducation (#PA6393) Misbehaviour (#PA6498) Formlessness (#PA6900) Imperfection (#PA6997) Untimeliness (#PA7006) Irresolution (#PA7325) Inexpedience (#PA7395) Incompetence (#PA6416) Insufficiency (#PA5473) Maladjustment (#PA6739) Discontentment (#PA6011) Incompleteness (#PA6652) Unpleasantness (#PA7107) Unpreparedness (#PA7341) Unintelligence (#PA7371) Thoughtlessness (#PA6940) Indiscrimination (#PA6446) Nonaccomplishment (#PA6662) Inappropriateness (#PA6852) Influencelessness (#PA6882) Misinterpretability (#PA6741) Uncommunicativeness (#PA7411).
Aggravates Negligence (#PA2658).

♦ **PA7236 Separation**
Brokenness — Discontinuity — Disintegration — Dissolution — Disunity — Division — Revolution — Separateness — Severance
Broader Quantity*complex (#PA6108).
Related Death (#PA7055) Enmity (#PA5446) Defeat (#PA7289) Duality (#PA7339) Disorder (#PA7361) Disaccord (#PA5532) Exclusion (#PA5869) Divergence (#PA5573) Revolution (#PA5901) Impairment (#PA6088) Lawlessness (#PA5563) Destruction (#PA6542) Dislocation (#PA7158) Disagreement (#PA5982) Irregularity (#PA6774) Unrelatedness (#PA5794) Discontinuity (#PA5828) Disappearance (#PA6785) Fragmentation (#PA6233) Disintegration (#PA6858) Indiscrimination (#PA6446).

♦ **PA7237 Condemnation**
Blame — Censoriousness — Censure — Condemnation — Damnation — Denunciation — Insinuation — Recrimination — Reproach — Reprobation — Venial
Broader Retribution*complex (#PA7037).
Related Vice (#PA5644) Badness (#PA5454) Humility (#PA6659) Disrepute (#PA6839) Intrusion (#PA6862) Unkindness (#PA5643) Disapproval (#PA6191) Affectation (#PA6400) Destruction (#PA6542) Misjudgement (#PA6607).

♦ **PA7240 Ugliness**
Awfulness — Blemish — Clumsiness — Deformation — Desperation — Disfigurement — Elaborateness — Foulness — Frightfulness — Gracelessness — Grimness — Gruesomeness — Hideousness — Horribleness — Inelegance — Loathsomeness — Malformation — Shapelessness — Ugliness — Dependence on ugliness — Ugly people
Broader Unpleasantness (#PA7107) Discriminative affection*complex (#PA7145).
Narrower Malformation (#PE4460) Unaesthetic foodstuffs (#PD1126) Landscape disfigurement (#PC2122) Personal physical disfigurement (#PD8076) Landscape disfigurement from open–cast mining (#PD1637) Unaesthetic location of power transmission lines (#PD1665) Unaesthetic location of advertising hoardings and billboards (#PE2156).
Related Vice (#PA5644) Envy (#PA7253) Hate (#PA7338) Death (#PA7055) Stench (#PA5981) Neglect (#PA5438) Badness (#PA5454) Closure (#PA7391) Disorder (#PA7361) Vulgarity (#PA5821) Exclusion (#PA5869) Injustice (#PA6486) Solemnity (#PA6731) Disrepute (#PA6839) Unchastity (#PA5612) Unkindness (#PA5643) Impairment (#PA6088) Inelegance (#PA6312) Distortion (#PA6790) Uncleanness (#PA5459) Disapproval (#PA6191) Affectation (#PA6400) Destruction (#PA6542) Uncertainty (#PA7309) Hopelessness (#PA6099) Formlessness (#PA6900) Imperfection (#PA6997) Irresolution (#PA7325) Nonconformity (#PA5878) Unsavouriness (#PA7204) Changeableness (#PA5490) Unskillfulness (#PA7232) Unintelligibility (#PA7367) Monolithic architecture of high-rise buildings (#PE1925).
Aggravates Fear (#PA6030) Dogmatism (#PF6988) Absurdity (#PF6991) Hostility (#PB8538) Immorality (#PA3369) Friendlessness (#PB5747) Intolerance of imperfection (#PF7024) Negative emotions and attitudes (#PA7090) Personal physical unattractiveness (#PF4010).
Aggravated by Chaos (#PF6836) Chance (#PA6714) Disasters (#PB3561) Human ageing (#PB0477) Injurious accidents (#PB0731) Genetic defects and diseases (#PD2389) Human disease and disability (#PB1044) Human errors and miscalculations (#PF3702).

♦ **PA7241 Following**
Broader Relative motion*complex (#PA6752).

♦ **PA7246 Appropriateness*complex**
Refs Judge, A J N *Comprehension of Appropriateness* (1986).
Narrower Undueness (#PA6921) Wrongness (#PA7280) Disrespect (#PA6822).

♦ **PA7250 Disobedience**
Brokenness — Contumaciousness — Defiance — Disobedience — Disregard — Extremism — Indocility — Inobservance — Insubordination — Intractability — Lawlessness — Nonadherence — Nonconformity — Nonobservance — Obstinacy — Offence — Passivity — Recalcitrance — Revolution — Sedition — Servility — Stubbornness — Subjection — Traitorousness — Transgression — Trespass — Unruliness — Violation — Willfulness
Narrower Sedition (#PC2414) Disobedient wives (#PF4764) Citizen disobedience (#PD5707) Employee disobedience (#PD5244) Disobedience of elders (#PF7149) Spiritual disobedience (#PFp467) Disobedience of children (#PD5308) Disobedience of judicial order (#PD3879).
Aggravated by Lack of social discipline (#PF8078) Inadequate army discipline (#PD2543).

♦ **PA7253 Envy**
Acrimony — Affrontery — Aggravation — Anger — Animosity — Annoyance — Argumentativeness — Asperity — Belligerence — Bitterness — Contentiousness — Dejection — Disagreeableness — Disapproval — Discontentment — Displeasure — Disputatiousness — Dissatisfaction — Distrust — Exasperation — Fierceness — Fretfulness — Fury — Grimness — Hastiness — Indignation — Irascibility — Irritability — Jealousy — Meanness — Melancholy — Mercilessness — Mistrust — Offence — Outrage — Oversensitiveness — Perversity — Petulance — Provocation — Quarrelsomeness — Resentment — Rivalry — Soreness — Sourness — Surliness — Temper — Ugliness — Vehemence — Vexation — Violence — Virulence — Wound
Nature Envy in its most basic form is a state of discontent and ill will directed at another person because of their advantages or possessions. Envy is always directed at another person. Envy in the sense of being a sin is not desiring to better oneself or one's family. Its cause is not in inequities of people nor even in the injustices but in the envier's relationships to these realities. It is the state of being diminished or disgraced because of other person's advantage or possessions. These might be wealth or power but also humility, goodness or creativity. Envy brings two types of negative action. It seeks revenge through bringing down the other person by pointing out their failures or by actually causing them to fail. It seeks self destruction through acquiescing to the sense of diminished worth and refusing to work to one's own potential. Envy tells that life is found in hating the neighbour. By its very nature it destroys genuine, free, open, creative human relationships.
Background Envy has been recognized as a profound human problem from the time of the early Greeks. One of Aesop's fables tells of Zeus granting to an envious man any wish he desired with one stipulation. His neighbour receives twice as much as the man. There was no limit on what could be asked for. The man wished for the loss of one eye, for the terms of the promise from Zeus was that the neighbour would be totally blinded. Envy for the Greeks was a destroyer.
Refs Cohen, Betsy *The Snow White Syndrome* (1987).
Broader Sin (#PF0641).
Narrower Jealousy (#PF5013) Provocation (#PJ2395) Irritability (#PJ2736).
Related Vice (#PA5644) Fear (#PA6030) Hate (#PA7338) Lust (#PA4673) Death (#PA7055) Pride (#PA7599)

ABSTRACT FUNDAMENTAL PROBLEMS

PA7319

Anger (#PA7797)
Reason (#PA5502)
Badness (#PA5454)
Boredom (#PA7365)
Rashness (#PA7115)
Unbelief (#PA7392)
Disaccord (#PA5532)
Avoidance (#PA6379)
Solemnity (#PA6731)
Improbity (#PA7363)
Compulsion (#PA5740)
Impatience (#PA6200)
Opposition (#PA6979)
Lamentation (#PA5479)
Displeasure (#PA6809)
Aggravation (#PA7378)
Disagreement (#PA5982)
Untimeliness (#PA7006)
Original sin (#PF8298)
Inexpectation (#PA5527)
Unfeelingness (#PA7364)
Unpleasantness (#PA7107)
Dissatisfaction (#PA8886)
Sloth (#PA3275)
Neglect (#PA5438)
Disease (#PA6799)
Avarice (#PA6999)
Insanity (#PA7157)
Gluttony (#PA9638)
Vulgarity (#PA5821)
Necessity (#PA6387)
Disrepute (#PA6839)
Difficulty (#PA5497)
Illegality (#PA5952)
Difference (#PA6698)
Moderation (#PA7156)
Inferiority (#PA5652)
Discourtesy (#PA7143)
Competition (#PB0848)
Hopelessness (#PA6099)
Disobedience (#PA7250)
Insensibility (#PA5451)
Unwillingness (#PA6509)
Inexcitability (#PA5467)
Human violence (#PA0429)
Narrowmindedness (#PA7306)
Enmity (#PA5446)
Defence (#PA5445)
Dissent (#PA6838)
Humility (#PA6659)
Ugliness (#PA7240)
Distrust (#PA8653)
Adversity (#PA6340)
Injustice (#PA6486)
Cheapness (#PA7193)
Unkindness (#PA5643)
Impairment (#PA6088)
Disrespect (#PA6822)
Melancholy (#PF7756)
Disapproval (#PA6191)
Selfishness (#PA7211)
Unimportance (#PA5942)
Imperfection (#PA6997)
Irresolution (#PA7325)
Insufficiency (#PA5473)
Maladjustment (#PA6739)
Discontentment (#PA6011)
Unhealthfulness (#PA7226)
Inappropriateness (#PA6852).

♦ PA7256 Narrowness
Closeness — Confinement
   Broader Dimension*complex (#PA7319).
   Related Cold (#PA6956)   Cheapness (#PA7193)   Restraint (#PA7296)
   Punishment (#PA5583)   Quiescence (#PA6444)   Environment (#PA7078)
   Uncommunicativeness (#PA7411).

♦ PA7259 Redemption*complex
   Narrower Impiety (#PA6058)   Unsanctity (#PA6995)   Ungodliness (#PA7148)
   Unorthodoxy (#PA5564).

♦ PA7263 Integrity*complex
   Narrower Danger (#PA6971)   Relapse (#PA5619)   Disease (#PA6799)
   Impairment (#PA6088)   Uncleanness (#PA5459)   Destruction (#PA6542)
   Maladjustment (#PA6739)   Unhealthfulness (#PA7226).

♦ PA7269 Prodigality
Prodigality
   Broader Possession*complex (#PA6686).
   Related Vice (#PA5644)   Cheapness (#PA7193)   Intemperance (#PA6466)
   Insufficiency (#PA5473)   Uncommunicativeness (#PA7411).

♦ PA7270 Absence
Abandonment — Absence — Characterlessness — Default — Deprivation — Disappearance — Featurelessness — Lack — Nothingness
   Broader Space*complex (#PA6713).
   Related Vice (#PA5644)   Loss (#PA7382)   Neglect (#PA5438)
   Poverty (#PA6434)   Boredom (#PA7365)   Avoidance (#PA6379)
   Restraint (#PA7296)   Unchastity (#PA5612)   Shortcoming (#PA6041)
   Deprivation (#PA0831)   Abandonment (#PA7685)   Nonexistence (#PA5870)
   Formlessness (#PA6900)   Imperfection (#PA6997)   Inhospitality (#PA5458)
   Insufficiency (#PA5473)   Disappearance (#PA6785)   Incompleteness (#PA6652).

♦ PA7273 Unconventionality
Dependence
   Broader Motivation*complex (#PA6048).
   Narrower Dependence (#PA4565).
   Related Restraint (#PA7296).

♦ PA7280 Wrongness
Abnormality — Abomination — Amorality — Atrocity — Desecration — Disgrace — Evil — Ignominy — Impropriety — Inappropriateness — Incorrectness — Indecorum — Infamy — Profanation — Sacrilege — Scandal — Shamelessness — Unlawfulness — Unsuitability — Violation — Wrongness
   Broader Appropriateness*complex (#PA7246).
   Narrower Evil (#PF7042)   Infamy (#PB8172)   Scandal (#PC8391)
   Sacrilege (#PF0662).
   Related Vice (#PA5644)   Hate (#PA7338)   Error (#PA6180)
   Vanity (#PA6491)   Badness (#PA5454)   Impiety (#PA6058)
   Disease (#PA6799)   Humility (#PA6659)   Insanity (#PA7157)
   Vulgarity (#PA5821)   Injustice (#PA6486)   Disrepute (#PA6839)
   Undueness (#PA6921)   Improbity (#PA7363)   Difficulty (#PA5497)
   Unchastity (#PA5612)   Unkindness (#PA5643)   Illegality (#PA5952)
   Disrespect (#PA6822)   Moderation (#PA7156)   Uncleanness (#PA5459)
   Lamentation (#PA5479)   Disapproval (#PA6191)   Affectation (#PA6400)
   Desecration (#PF9176)   Disagreement (#PA5982)   Misbehaviour (#PA6498)
   Untimeliness (#PA7006)   Disobedience (#PA7250)   Inexpedience (#PA7395)
   Insensibility (#PA5451)   Insufficiency (#PA5473)   Discontentment (#PA6011)
   Unpleasantness (#PA7107)   Unpreparedness (#PA7341) Inappropriateness (#PA6852).

♦ PA7285 Littleness
Exiguity — Overdeveloped — Overweight — Scantiness — Smallness — Tenuousness
   Broader Dimension*complex (#PA7319).
   Related Fewness (#PA7152)   Humility (#PA6659)   Smallness (#PA7408)
   Lightness (#PA5491)   Disrepute (#PA6839)   Inferiority (#PA5652)
   Selfishness (#PA7211)   Unimportance (#PA5942)   Insufficiency (#PA5473)
   Disintegration (#PA6858)   Narrowmindedness (#PA7306).

♦ PA7289 Defeat
Bafflement — Brokenness — Confusion — Defeat — Failure — Frustration — Overturn — Panic — Ruin — Subjugation — Vanquishment
   Broader Achievement*complex (#PA5471).
   Narrower Panic (#PF2633)   Electoral defeat (#PF4709).
   Related Vice (#PA5644)   Fear (#PA6030)   Loss (#PA7382)
   Error (#PA6180)   Vanity (#PA6491)   Disorder (#PA7361)
   Impotence (#PA6876)   Restraint (#PA7296)   Confusion (#PF7123)
   Separation (#PA7236)   Depression (#PA5493)   Difficulty (#PA5497)
   Revolution (#PA5901)   Impairment (#PA6088)   Lawlessness (#PA5563)
   Inferiority (#PA5652)   Shortcoming (#PA6041)   Inattention (#PA6247)
   Destruction (#PA6542)   Uncertainty (#PA7309)   Frustration (#PA2252)
   Formlessness (#PA6900)   Imperfection (#PA6997)   Disobedience (#PA7250)
   Inexpectation (#PA5527)   Appropriation (#PA5688)   Discontinuity (#PA5828)
   Maladjustment (#PA6739)   Inexcitability (#PA5467) Disintegration (#PA6858)
   Unpleasantness (#PA7107)   Nonaccomplishment (#PA6662).

♦ PA7295 Distance
   Broader Dimension*complex (#PA7319).
   Aggravates Geographical isolation (#PF9023).

♦ PA7296 Restraint
Abandonment — Absolutism — Barrier — Confinement — Constraint — Dependence — Duress — Incontinence — Inferiority — Inhibition — Intemperance — Isolation — Laxness — Licentiousness — Prohibition — Repression — Restriction — Retardation — Seclusion — Segregation — Servility — Subjection — Subjugation — Subordination — Suppression — Tyranny — Unruliness
   Broader Compliance*complex (#PA5710).
   Narrower Isolation (#PB8685)   Absolutism (#PJ9995)   Dependence (#PA4565)
   Repression (#PB0871).
   Related Vice (#PA5644)   Loss (#PA7382)   Defeat (#PA7289)
   Neglect (#PA5438)   Badness (#PA5454)   Absence (#PA7270)
   Refusal (#PA7321)   Duality (#PA7339)   Closure (#PAp391)
   Humility (#PA6659)   Exclusion (#PA5869)   Avoidance (#PA6379)
   Impotence (#PA6876)   Undueness (#PA6921)   Difficulty (#PA5497)
   Punishment (#PA5583)   Unchastity (#PA5612)   Compulsion (#PA5740)
   Disrespect (#PA6822)   Moderation (#PA7156)   Narrowness (#PA7256)
   Lawlessness (#PA5563)   Inferiority (#PA5652)   Shortcoming (#PA6041)
   Destruction (#PA6542)   Environment (#PA7078)   Abandonment (#PA7685)
   Unimportance (#PA5942)   Intemperance (#PA6466)   Imperfection (#PA6997)
   Untimeliness (#PA7006)   Disobedience (#PA7250)   Irresolution (#PA7325)
   Inhospitality (#PA5458)   Insufficiency (#PA5473)   Appropriation (#PA5688)
   Nonconformity (#PA5878)   Forgetfulness (#PA6651)   Unsociability (#PA6653)
   Unintelligence (#PA7371)   Indiscrimination (#PA6446) Narrowmindedness (#PA7306)
   Uncommunicativeness (#PA7273).

♦ PA7301 Colourlessness
Gaudiness — Loudness
   Broader Relative properties*complex (#PA6091).
   Related Loudness (#PA5465)   Vulgarity (#PA5821).

♦ PA7306 Narrowmindedness
Apartheid — Authoritarianism — Bias — Bigotry — Chauvinism — Discrimination — Fanaticism — Illiberalism — Insularity — Intolerance — Littleness — Meanness — Pettiness — Prejudice — Segregation — Smallness — Stuffiness — Warpedness — Xenophobia
   Broader Attitude*complex (#PA6983).
   Narrower Bigotry (#PC7652)   Ethnocentricity (#PB5765).
   Related Vice (#PA5644)   Fear (#PA6030)   Envy (#PA7253)
   Hate (#PA7338)   Stench (#PA5981)   Badness (#PA5454)
   Fewness (#PA7152)   Boredom (#PA7365)   Humility (#PA6659)
   Insanity (#PA7157)   Smallness (#PA7408)   Disaccord (#PA5532)
   Vulgarity (#PA5821)   Exclusion (#PA5869)   Avoidance (#PA6379)
   Injustice (#PA6486)   Disrepute (#PA6839)   Cheapness (#PA7193)
   Restraint (#PA7296)   Prejudice (#PA2173)   Unkindness (#PA5643)
   Compulsion (#PA5740)   Impatience (#PA6200)   Quiescence (#PA6444)
   Distortion (#PA6790)   Moderation (#PA7156)   Littleness (#PA7285)
   Xenophobia (#PD4957)   Lawlessness (#PA5563)   Inferiority (#PA5652)
   Affectation (#PA6400)   Selfishness (#PA7211)   Uncertainty (#PA7309)
   Intolerance (#PF0860)   Nationalism (#PB0534)   Unimportance (#PA5942)
   Misjudgement (#PA6607)   Imperfection (#PA6997)   Irresolution (#PA7325)
   Inexpedience (#PA7395)   Insufficiency (#PA5473)   Unsociability (#PA6653)
   Unintelligence (#PA7371)   Discrimination (#PA0833) Indiscrimination (#PA6446)
   Unimaginativeness (#PA6738)   Uncommunicativeness (#PA7411).
   Aggravates Extremism (#PB3415) Authoritarianism (#PB1638)
   Cultural discrimination in the administration of justice (#PE6529).

♦ PA7309 Uncertainty
Ambiguity — Bafflement — Bewilderment — Bigotry — Changeableness — Confusion — Dangerousness — Disbelief — Discomposure — Disconcertion — Dismay — Disorder — Disorientation — Disturbance — Doubtfulness — Dubiousness — Embarrassment — Fallibility — Hazardousness — Hesitation — Imprecision — Inaccuracy — Incoherence — Indecisiveness — Insecurity — Instability — Insubstantiality — Irresolution — Laxity — Looseness — Muddle — Obscurity — Orderlessness — Perplexity — Perturbation — Problem — Randomness — Risk — Riskiness — Shapelessness — Suspense — Treachery — Unauthenticity — Uncertainty — Undependability — Unpredictability — Unreliability — Untrustworthiness — Upset — Vacillation — Vagueness
   Refs Bonatti, Luigi Uncertainty (1983); Keppel, David and Keppel, John Uncertainty (1982).
   Broader Credibility*complex (#PA5829).
   Narrower Risk (#PF7580)   Bigotry (#PC7652)   Uncertainty (#PA6438).
   Related Fear (#PA6030)   Hate (#PA7338)   Error (#PA6180)
   Reason (#PA5502)   Vanity (#PA6491)   Chance (#PA6714)
   Danger (#PA6971)   Defeat (#PA7289)   Neglect (#PA5438)
   Impiety (#PA6058)   Poverty (#PA6434)   Disease (#PA6799)
   Weakness (#PA5558)   Humility (#PA6659)   Rashness (#PA7115)
   Insanity (#PA7157)   Ugliness (#PA7240)   Disorder (#PA7361)
   Unbelief (#PA7392)   Agitation (#PA5838)   Improbity (#PA7363)
   Confusion (#PF7123)   Ambiguity (#PF4193)   Difficulty (#PA5497)
   Unchastity (#PA5612)   Compulsion (#PA5740)   Revolution (#PA5901)
   Transience (#PA6425)   Quiescence (#PA6444)   Complexity (#PA6468)
   Difference (#PA6698)   Disrespect (#PA6822)   Moderation (#PA7156)
   Insecurity (#PA0857)   Inaccuracy (#PF7905)   Lawlessness (#PA5563)
   Inattention (#PA6247)   Destruction (#PA6542)   Instability (#PA0859)
   Incoherence (#PF8094)   Unimportance (#PA5942)   Disagreement (#PA5982)
   Misbehaviour (#PA6498)   Formlessness (#PA6900)   Invisibility (#PA6978)
   Imperfection (#PA6997)   Untimeliness (#PA7006)   Irresolution (#PA7325)
   Inexpectation (#PA5527)   Nonconformity (#PA5878)   Maladjustment (#PA6739)
   Nonuniformity (#PA6890)   Inexcitability (#PA5467) Changeableness (#PA5490)
   Disintegration (#PA6858)   Unpleasantness (#PA7107) Unskillfulness (#PA7232)
   Indiscrimination (#PA6446)   Insubstantiality (#PA6959) Narrowmindedness (#PA7306)
   Misrepresentation (#PA6644)   Nonaccomplishment (#PA6662) Unintelligibility (#PA7367)
   Uncommunicativeness (#PA7411).
   Aggravated by Disorganization (#PF4487).

♦ PA7314 Power*complex
   Refs Bernholz, Peter International Game of Power (1985); Schott, Kerry Policy, Power and Order (1984).
   Narrower Weakness (#PA5558)   Posterity (#PA6073)
   Impotence (#PA6876)   Moderation (#PA7156)
   Reproduction (#PA7389)   Counteraction (#PA6485)
   Unproductiveness (#PA7208)   Influencelessness (#PA6882).

♦ PA7319 Dimension*complex
   Narrower Lowness (#PA6798)   Distance (#PA7295)   Narrowness (#PA7256)
   Littleness (#PA7285)   Contraction (#PA6847)   Shallowness (#PA6993).

♦ PA7321 Refusal
Decline — Denial — Disobedience — Exclusion — Illegality — Indulgence — Inhibition — Negativity — Nonacceptance — Nonobservance — Permissiveness — Prohibition — Rejection — Repression — Repudiation — Suppression

Refs Steward, D E *Contact Inhibition* (1985).

Broader Compliance*complex (#PA5710).
Narrower Repression (#PB0871).
Related Loss (#PA7382) Denial (#PA7400) Neglect (#PA5438)
Dissent (#PA6838) Weakness (#PA5558) Unbelief (#PA7392)
Exclusion (#PA5869) Adversity (#PA6340) Necessity (#PA6387)
Cheapness (#PA7193) Restraint (#PA7296) Difficulty (#PA5497)
Unchastity (#PA5612) Compulsion (#PA5740) Illegality (#PA5952)
Impairment (#PA6088) Opposition (#PA6979) Lawlessness (#PA5563)
Shortcoming (#PA6041) Disapproval (#PA6191) Inattention (#PA6247)
Destruction (#PA6542) Disagreement (#PA5982) Hopelessness (#PA6099)
Intemperance (#PA6466) Disobedience (#PA7250) Irresolution (#PA7325)
Unwillingness (#PA6509) Forgetfulness (#PA6651) Unprovability (#PA7070)
Unintelligence (#PA7371) Permissiveness (#PF1252) Inappropriateness (#PA6852).

♦ PA7325 Irresolution
Ambivalence — Betrayal — Bigotry — Capriciousness — Carelessness — Changeableness — Contumaciousness — Cowardice — Defection — Defiance — Desertion — Disloyalty — Dubiousness — Equivocation — Faithlessness — Fantasy — Fanaticism — Fear — Feebleness — Grimness — Hesitation — Impatience — Impetuousity — Impulsiveness — Inconsiderateness — Inconstancy — Incorrigibility — Indecisiveness — Indocility — Inflexibility — Instability — Intolerance — Intractability — Intransigence — Irascibility — Irreconcilability — Irresolution — Levity — Obstinacy — Overzealousness — Perversity — Petulance — Recalcitrance — Relentlessness — Repudiation — Rigidity — Stiffness — Strain — Stubbornness — Stuffiness — Suddenness — Traitorousness — Treason — Unadvisedness — Uncertainty — Undependability — Unmanageability — Unpredictability — Unreliability — Unruliness — Unthoughtfulness — Vacillation — Volatility — Wantonness — Weak-mindedness — Weakness — Willfulness — Withdrawal

Broader Choice*complex (#PA7214).
Narrower Bigotry (#PC7652) Cowardice (#PJ4171) Uncertainty (#PA6438)
Inflexibility (#PA8555).
Related Vice (#PA5644) Fear (#PA6030) Envy (#PA7253)
Hate (#PA7338) Enmity (#PA5446) Reason (#PA5502)
Stench (#PA5981) Danger (#PA6971) Neglect (#PA5438)
Badness (#PA5454) Impiety (#PA6058) Disease (#PA6799)
Dissent (#PA6838) Oldness (#PA7131) Refusal (#PA7321)
Duality (#PA7339) Boredom (#PA7365) Treason (#PD2615)
Weakness (#PA5558) Inaction (#PA5806) Softness (#PA5988)
Rashness (#PA7115) Insanity (#PA7157) Ugliness (#PA7240)
Disorder (#PA7361) Unbelief (#PA7392) Disaccord (#PA5532)
Reversion (#PA5699) Exclusion (#PA5869) Avoidance (#PA6379)
Necessity (#PA6387) Vengeance (#PA6606) Solemnity (#PA6731)
Impotence (#PA6876) Toughness (#PA6976) Restraint (#PA7296)
Improbity (#PA7363) Difficulty (#PA5497) Unchastity (#PA5612)
Unkindness (#PA5643) Compulsion (#PA5740) Impairment (#PA6088)
Impatience (#PA6200) Regression (#PA6338) Transience (#PA6425)
Quiescence (#PA6444) Difference (#PA6698) Permanence (#PA6802)
Disrespect (#PA6822) Opposition (#PA6979) Moderation (#PA7156)
Withdrawal (#PF4402) Lamentation (#PA5479) Lawlessness (#PA5563)
Inattention (#PA6247) Affectation (#PA6400) Incuriosity (#PA6598)
Discourtesy (#PA7143) Informality (#PA7170) Uncertainty (#PA7309)
Instability (#PA0859) Intolerance (#PF0860) Ambivalence (#PF1426)
Unimportance (#PA5942) Disagreement (#PA5982) Hopelessness (#PA6099)
Irregularity (#PA6774) Formlessness (#PA6900) Invisibility (#PA6978)
Imperfection (#PA6997) Untimeliness (#PA7006) Pitilessness (#PA7023)
Disobedience (#PA7250) Insufficiency (#PA5473) Unwillingness (#PA6509)
Unsociability (#PA6653) Maladjustment (#PA6739) Nonuniformity (#PA6890)
Unsavouriness (#PA7204) Unfeelingness (#PA7364) Inexcitability (#PA5467)
Changeableness (#PA5490) Discontentment (#PA6011) Unpleasantness (#PA7107)
Unskillfulness (#PA7232) Unintelligence (#PA7371) Narrowmindedness (#PA7306)
Unimaginativeness (#PA6738) Influencelessness (#PA6882) Unintelligibility (#PA7367)
Uncommunicativeness (#PA7411).

♦ PA7331 Benevolence*complex
Narrower Vengeance (#PA6606) Unkindness (#PA5643)
Ingratitude (#PA5504) Discourtesy (#PA7143)
Pitilessness (#PA7023).

♦ PA7333 Age
Broader Time*complex (#PA7222).
Related Oldness (#PA7131).

♦ PA7338 Hate
Abhorrence — Abomination — Animosity — Antipathy — Aversion — Bigotry — Bitterness — Dislike — Enmity — Execration — Hatred — Loathsomeness — Malevolence — Malice — Repugnance — Unchastity — Xenophobia

Refs Bychowski, Gustav *Evil in Man* (1968).

Broader Socialization*complex (#PA6373).
Narrower Malice (#PF5901).
Related Vice (#PA5644) Fear (#PA6030) Envy (#PA7253)
Enmity (#PA5446) Badness (#PA5454) Bigotry (#PC7652)
Insanity (#PA7157) Ugliness (#PA7240) Disaccord (#PA5532)
Exclusion (#PA5869) Wrongness (#PA7280) Antipathy (#PA8810)
Unchastity (#PA5612) Unkindness (#PA5643) Difference (#PA6698)
Opposition (#PA6979) Moderation (#PA7156) Xenophobia (#PD4957)
Uncleanness (#PA5459) Disapproval (#PA6191) Uncertainty (#PA7309)
Malevolence (#PA7102) Disagreement (#PA5982) Irresolution (#PA7325)
Unwillingness (#PA6509) Unfeelingness (#PA7364) Unpleasantness (#PA7107)
Narrowmindedness (#PA7306).

♦ PA7339 Duality
Alienation — Homelessness — Isolation — Loneliness — Seclusion — Separateness — Withdrawal

Broader Number*complex (#PA7412).
Narrower Isolation (#PB8685) Loneliness (#PF2386).
Related Enmity (#PA5446) Poverty (#PA6434) Dissent (#PA6838)
Insanity (#PA7157) Reversion (#PA5699) Exclusion (#PA5869)
Avoidance (#PA6379) Restraint (#PA7296) Separation (#PA7236)
Regression (#PA6338) Opposition (#PA6979) Alienation (#PA3545)
Incuriosity (#PA6598) Dislocation (#PA7158) Irresolution (#PA7325)
Unrelatedness (#PA5794) Unsociability (#PA6653) Unfeelingness (#PA7364)
Disintegration (#PA6858).

♦ PA7341 Unpreparedness
Coarseness — Fecklessness — Immaturity — Improvidence — Incapability — Incompetence — Negligence — Oversimplification — Roughness — Rudeness — Shiftlessness — Thriftlessness — Underdevelopment — Unpreparedness — Unsuitability — Vulnerability

Broader Action*complex (#PA5553).
Narrower Unpreparedness (#PF8176) Underdevelopment (#PB0206).
Related Error (#PA6180) Vanity (#PA6491) Neglect (#PA5438)
Inaction (#PA5806) Rashness (#PA7115) Disorder (#PA7361)
Vulgarity (#PA5821) Avoidance (#PA6379) Impotence (#PA6876)
Toughness (#PA6976) Wrongness (#PA7280) Unchastity (#PA5612)
Unkindness (#PA5643) Complexity (#PA6468) Moderation (#PA7156)
Negligence (#PA2658) Immaturity (#PF8413) Inferiority (#PA5652)
Inattention (#PA6247) Discourtesy (#PA7143) Disagreement (#PA5982)
Imperfection (#PA6997) Untimeliness (#PA7006) Inexpedience (#PA7395)
Incompetence (#PA6416) Insufficiency (#PA5473) Discontentment (#PA6011)
Incompleteness (#PA6652) Unskillfulness (#PA7232) Unintelligence (#PA7371)
Inappropriateness (#PA6852) Oversimplification (#PF8455).

♦ PA7343 Dissuasion
Corruption — Discouragement — Intimidation — Temptation

Broader Motivation*complex (#PA6048).
Narrower Temptation (#PA7736) Discouragement (#PF8948).
Related Vice (#PA5644) Fear (#PA6030) Vanity (#PA6491)
Badness (#PA5454) Reversion (#PA5699) Solemnity (#PA6731)
Improbity (#PA7363) Compulsion (#PA5740) Impairment (#PA6088)
Complexity (#PA6468) Distortion (#PA6790) Uncleanness (#PA5459)
Disapproval (#PA6191) Miseducation (#PA6393).

♦ PA7361 Disorder
Aimlessness — Anarchism — Anarchy — Bedraggled — Capriciousness — Carelessness — Chaos — Confusion — Disarrangement — Discomposure — Disharmony — Disintegration — Disorder — Disorganization — Disturbance — Frivolity — Fuss — Haphazardness — Looseness — Messiness — Misrule — Muddle — Negligence — Nonuniformity — Orderlessness — Perturbation — Promiscuity — Randomness — Senselessness — Shapelessness — Sordidness — Spasmodicness — Squalor — Turbulence — Uproar — Upset

Broader Order*complex (#PA7199).
Narrower Disorganization (#PF4487).
Related Fear (#PA6030) Error (#PA6180) Chaos (#PF6836)
Reason (#PA5502) Vanity (#PA6491) Chance (#PA6714)
Defeat (#PA7289) Neglect (#PA5438) Badness (#PA5454)
Discord (#PA5565) Disease (#PA6799) Rashness (#PA7115)
Insanity (#PA7157) Ugliness (#PA7240) Disaccord (#PA5532)
Agitation (#PA5838) Avoidance (#PA6379) Disrepute (#PA6839)
Cheapness (#PA7193) Confusion (#PF7123) Anarchism (#PC1972)
Separation (#PA7236) Difficulty (#PA5497) Unchastity (#PA5612)
Compulsion (#PA5740) Revolution (#PA5901) Illegality (#PA5952)
Impairment (#PA6088) Impatience (#PA6200) Transience (#PA6425)
Complexity (#PA6468) Inequality (#PA6695) Moderation (#PA7156)
Negligence (#PA2658) Uncleanness (#PA5459) Lamentation (#PA5479)
Lawlessness (#PA5563) Disapproval (#PA6191) Inattention (#PA6247)
Destruction (#PA6542) Incuriosity (#PA6598) Dislocation (#PA7158)
Uncertainty (#PA7309) Promiscuity (#PC0745) Unimportance (#PA5942)
Disagreement (#PA5982) Misbehaviour (#PA6498) Irregularity (#PA6774)
Formlessness (#PA6900) Irresolution (#PA7325) Insensibility (#PA5451)
Discontinuity (#PA5828) Nonconformity (#PA5878) Nonuniformity (#PA6890)
Fragmentation (#PA6233) Inexcitability (#PA5467) Changeableness (#PA5490)
Disintegration (#PA6858) Unpleasantness (#PA7107) Unskillfulness (#PA7232)
Unpreparedness (#PA7341) Unintelligence (#PA7371) Meaninglessness (#PA6977)
Indiscrimination (#PA6446) Nonaccomplishment (#PA6662) Inappropriateness (#PA6852)
Unintelligibility (#PA7367).

♦ PA7362 Nonobservance
Broader Compliance*complex (#PA5710).

♦ PA7363 Improbity
Baseness — Betrayal — Corruption — Deceit — Dereliction — Deviousness — Disaffection — Dishonesty — Dishonour — Disingenuousness — Disloyalty — Doubtfulness — Dubiousness — Duplicity — Evasion — Faithlessness — Falseness — Feloniousness — Fraudulence — Immorality — Improbity — Inconstancy — Indirection — Infidelity — Insincerity — Intrigue — Irresponsibility — Irritability — Notoriety — Perfidiousness — Roguery — Sedition — Self-absorption — Shamelessness — Traitorousness — Treachery — Treason — Turpitude — Unconscionableness — Undependability — Unfaithfulness — Unreliability — Untrustworthiness — Venality — Vileness — Villainy

Broader Judgement*complex (#PA8528).
Narrower Sedition (#PC2414) Deception (#PB4731) Corruption (#PA1986)
Immorality (#PA3369) Irritability (#PJ2736) Irresponsibility (#PA8658)
Lack of integrity (#PF7992) Unethical practices (#PC8247).
Related Vice (#PA5644) Fear (#PA6030) Envy (#PA7253)
Error (#PA6180) Reason (#PA5502) Vanity (#PA6491)
Danger (#PA6971) Neglect (#PA5438) Badness (#PA5454)
Impiety (#PA6058) Treason (#PD2615) Humility (#PA6659)
Unbelief (#PA7392) Disaccord (#PA5532) Reversion (#PA5699)
Vulgarity (#PA5821) Deviation (#PA6228) Avoidance (#PA6379)
Injustice (#PA6486) Disrepute (#PA6839) Cheapness (#PA7193)
Wrongness (#PA7280) Dishonour (#PF8485) Unchastity (#PA5612)
Illegality (#PA5952) Impairment (#PA6088) Complexity (#PA6468)
Distortion (#PA6790) Disrespect (#PA6822) Moderation (#PA7156)
Dissuasion (#PA7343) Uncleanness (#PA5459) Lamentation (#PA5479)
Lawlessness (#PA5563) Inferiority (#PA5652) Diffuseness (#PA5974)
Disapproval (#PA6191) Affectation (#PA6400) Uncertainty (#PA7309)
Unimportance (#PA5942) Miseducation (#PA6393) Misbehaviour (#PA6498)
Imperfection (#PA6997) Disobedience (#PA7250) Irresolution (#PA7325)
Insensibility (#PA5451) Insufficiency (#PA5473) Appropriation (#PA5688)
Unfeelingness (#PA7364) Changeableness (#PA5490) Unpleasantness (#PA7107)
Unskillfulness (#PA7232) Unintelligence (#PA7371)
Uncommunicativeness (#PA7411).

♦ PA7364 Unfeelingness
Animosity — Apathy — Autism — Callousness — Coldness — Disaffinity — Disinterest — Dullness — Hardness — Heartlessness — Hopelessness — Hostility — Imperviousness — Indifference — Insensitivity — Insouciance — Lethargy — Listlessness — Oblivion — Passionlessness — Passivity — Self-absorption — Sentimentality — Slothfulness — Soullessness — Spiritlessness — Spunklessness — Stupor — Torpor — Withdrawal

Broader Feeling*complex (#PA6938).
Narrower Lethargy (#PJ4782) Indifference (#PA7604).
Related Vice (#PA5644) Fear (#PA6030) Cold (#PA6956)

Envy (#PA7253)	Hate (#PA7338)	Error (#PA6180)
Enmity (#PA5446)	Chance (#PA6714)	Apathy (#PA2360)
Neglect (#PA5438)	Dissent (#PA6838)	Duality (#PA7339)
Boredom (#PA7365)	Weakness (#PA5558)	Inaction (#PA5806)
Disaccord (#PA5532)	Repulsion (#PA5799)	Exclusion (#PA5869)
Avoidance (#PA6379)	Solemnity (#PA6731)	Toughness (#PA6976)
Improbity (#PA7363)	Hostility (#PB8538)	Difficulty (#PA5497)
Unkindness (#PA5643)	Impatience (#PA6200)	Regression (#PA6338)
Quiescence (#PA6444)	Difference (#PA6698)	Opposition (#PA6979)
Withdrawal (#PF4402)	Lamentation (#PA5479)	Inattention (#PA6247)
Incuriosity (#PA6598)	Discourtesy (#PA7143)	Selfishness (#PA7211)
Unimportance (#PA5942)	Hopelessness (#PA6099)	Imperfection (#PA6997)
Pitilessness (#PA7023)	Disobedience (#PA7250)	Irresolution (#PA7325)
Insensibility (#PA5451)	Intangibility (#PA5570)	Impossibility (#PA6487)
Forgetfulness (#PA6651)	Unsociability (#PA6653)	Maladjustment (#PA6739)
Unpleasantness (#PA7107)	Unintelligence (#PA7371)	Thoughtlessness (#PA6940)
Indiscrimination (#PA6446)	Unimaginativeness (#PA6738)	Unintelligibility (#PA7367).

♦ **PA7365 Boredom**

Banality — Barrenness — Boringness — Characterlessness — Commonness — Deadness — Dullness — Effetism — Ennui — Inanity — Irksomeness — Lifelessness — Melancholy — Mustiness — Nuisance — Pointlessness — Ponderousness — Prolixity — Sameness — Slowness — Spiritlessness — Sterility — Stiffness — Stuffiness — Superficiality — Tastelessness — Tedium — Tiresomeness — Unimaginativeness — Vapidity — Wear — Weariness — Wearisomeness — Bored people — Boring people

Nature Modern industry, through progressive automation of tasks, has created numerous highly specialized, repetitive job situations which lead to a feeling of boredom in the worker. Workers who formerly actively participated in tasks have come to act as mere observers or checkers. Boredom sets in and sensory acuity is reduced, attention wanders, with serious consequences for efficiency. Employees who experience boredom frequently exhibit low morale.

Many people without regular jobs may also be bored, including the retired, the aged, teenagers, slum dwellers, the unemployed, and students. Boredom in the military is a particular problem and has lead to use of narcotics; thus presenting, among those who operate nuclear weapon systems, missiles, planes, and computers, the possibility of an accident under narcotic influence.

Claim Individuals and societies take ever greater risks to dispel boredom becoming the greatest threat to survival. Feeding the insatiable appetite to deal with boredom coupled with telling ourselves that disaster will never happen to us leads the individual to putting off visits to a doctor when obvious treatment is required, to not wearing seat belts or to taking greater chances. It leads societies to take greater risks, as did the Jews of Germany, the Americans in Vietnam, and the Europeans and Americans in World War I.

Cyril Northcote Parkinson, famous for his Parkinson's law that work expands to fill the time available, has created another law. The chief product of an automated society is a widespread and deepening sense of boredom.

Narrower Boredom of captive and domesticated animals (#PF7681).

Related Fear (#PA6030)	Envy (#PA7253)	Loss (#PA7382)
Death (#PA7055)	Stench (#PA5981)	Apathy (#PA2360)
Absence (#PA7270)	Weakness (#PA5558)	Slowness (#PA6166)
Darkness (#PA6261)	Inaction (#PA5806)	Ignorance (#PA5568)
Vulgarity (#PA5821)	Solemnity (#PA6731)	Impotence (#PA6876)
Toughness (#PA6976)	Difficulty (#PA5497)	Compulsion (#PA5740)
Impairment (#PA6088)	Quiescence (#PA6444)	Melancholy (#PF7756)
Diffuseness (#PA5974)	Inattention (#PA6247)	Affectation (#PA6400)
Incuriosity (#PA6598)	Informality (#PA7170)	Unimportance (#PA5942)
Formlessness (#PA6900)	Imperfection (#PA6997)	Untimeliness (#PA7006)
Irresolution (#PA7325)	Insensibility (#PA5451)	Insufficiency (#PA5473)
Disappearance (#PA6785)	Nonuniformity (#PA6890)	Unsavouriness (#PA7204)
Unfeelingness (#PA7364)	Disintegration (#PA6858)	Unpleasantness (#PA7107)
Unskillfulness (#PA7232)	Unintelligence (#PA7371)	Thoughtlessness (#PA6940)
Meaninglessness (#PA6977)	Feeling*complex (#PA6938)	Unproductiveness (#PA7208)
Narrowmindedness (#PA7306)	Unimaginativeness (#PA6738)	Inappropriateness (#PA6852)
Human infertility (#PC6037).		
Aggravates War (#PB0593)	Magic (#PF3311)	Pornography (#PD0132)
Shoplifting (#PE1113)	Job fatigue (#PD8052)	Underproductivity (#PF1107)

Student absenteeism (#PE4200) — Gambling and wagering (#PF2137)
Stress in human beings (#PC1648) — Excessive television viewing (#PD1533)
Human errors and miscalculations (#PF3702)
Human fatigue during control of complex equipment (#PE5572)
Substance abuse during control of complex equipment (#PE0680)
Damage to infant brains from malnutrition and insufficient stimuli (#PE4874).

Aggravated by Fatigue (#PA0657) — Loneliness (#PF2386)
Wasted waiting time (#PF1761) — Over-specialization (#PF0256)
Unstimulating entertainment (#PF8105) — Lack of community responsibility (#PJ3290)
Inadequate facilities for children's play (#PD0549)
Monotonous and unaesthetic architecture and design (#PF0867)
Inadequate coordination of international organizations and programmes (#PD0285).

♦ **PA7367 Unintelligibility**

Ambiguity — Complexity — Complication — Darkness — Difficulty — Equivocation — Hardness — Inarticulation — Incoherence — Incomprehensibility — Intricacy — Mystification — Obscurity — Opacity — Perplexity — Problem — Shapelessness — Uncertainty — Vagueness

Broader Meaning*complex (#PA7377).

Narrower Uncertainty (#PA6438).

Related Vice (#PA5644)	Reason (#PA5502)	Vanity (#PA6491)
Danger (#PA6971)	Disease (#PA6799)	Weakness (#PA5558)
Darkness (#PA6261)	Insanity (#PA7157)	Ugliness (#PA7240)
Disorder (#PA7361)	Unbelief (#PA7392)	Disaccord (#PA5532)
Ignorance (#PA5568)	Adversity (#PA6340)	Avoidance (#PA6379)
Blindness (#PA6674)	Solemnity (#PA6731)	Toughness (#PA6976)
Ambiguity (#PF4193)	Difficulty (#PA5497)	Unkindness (#PA5643)
Complexity (#PA6468)	Difference (#PA6698)	Lamentation (#PA5479)
Inattention (#PA6247)	Uncertainty (#PA7309)	Incoherence (#PF8094)
Unimportance (#PA5942)	Disagreement (#PA5982)	Miseducation (#PA6393)
Formlessness (#PA6900)	Invisibility (#PA6978)	Imperfection (#PA6997)
Pitilessness (#PA7023)	Irresolution (#PA7325)	Nonconformity (#PA5878)
Unfeelingness (#PA7364)	Changeableness (#PA5490)	Disintegration (#PA6858)
Unpleasantness (#PA7107)	Unintelligence (#PA7371)	Insubstantiality (#PA6959)
Uncommunicativeness (#PA7411).		

♦ **PA7371 Unintelligence**

Absurdity — Arrested development — Childishness — Credulousness — Decline — Decrepitness — Defectiveness — Deviousness — Dullness — Eccentricity — Fatuity — Feebleness — Frivolity — Grossness — Gullibility — Ignorance — Imbecility — Immaturity — Imprudence — Inadvisability — Inanity — Inattention — Incapacity — Incomprehensibility — Inconsiderateness — Indiscretion — Ineptitude — Inexpedience — Infatuation — Insanity — Insensibility — Irrationality — Lethargy — Madness — Mindlessness — Naïvety — Nonsense — Opacity — Puerility — Retardation — Senselessness — Sentimentality — Shallowness — Slowness — Softness — Stuffiness — Stupidity — Superficiality — Thinness — Thoughtlessness — Triviality — Unadvisedness — Unintelligence — Unreasonableness — Unthoughtfulness — Vapidity — Volatility — Weakmindedness — Weakness — Witlessness

Broader Intellectual faculties*complex (#PA7197).

Narrower Lethargy (#PJ4782)	Absurdity (#PF6991)	Imbecility (#PE6314).
Related Vice (#PA5644)	Fear (#PA6030)	Error (#PA6180)
Reason (#PA5502)	Stench (#PA5981)	Danger (#PA6971)
Neglect (#PA5438)	Badness (#PA5454)	Disease (#PA6799)
Oldness (#PA7131)	Fewness (#PA7152)	Refusal (#PA7321)
Boredom (#PA7365)	Weakness (#PA5558)	Inaction (#PA5806)
Slowness (#PA6166)	Rashness (#PA7115)	Insanity (#PA7157)
Disorder (#PA7361)	Unbelief (#PA7392)	Ignorance (#PA5568)
Vulgarity (#PA5821)	Deviation (#PA6228)	Adversity (#PA6340)
Avoidance (#PA6379)	Necessity (#PA6387)	Disrepute (#PA6839)
Impotence (#PA6876)	Cheapness (#PA7193)	Restraint (#PA7296)
Improbity (#PA7363)	Difficulty (#PA5497)	Unchastity (#PA5612)
Unkindness (#PA5643)	Impairment (#PA6088)	Transience (#PA6425)
Quiescence (#PA6444)	Complexity (#PA6468)	Moderation (#PA7156)
Immaturity (#PF8413)	Incapacity (#PJ4312)	Inferiority (#PA5652)
Shortcoming (#PA6041)	Inattention (#PA6247)	Affectation (#PA6400)
Discourtesy (#PA7143)	Unimportance (#PA5942)	Invisibility (#PA6978)
Imperfection (#PA6997)	Untimeliness (#PA7006)	Irresolution (#PA7325)
Inexpedience (#PA7395)	Eccentricity (#PF9063)	Insensibility (#PA5451)
Insufficiency (#PA5473)	Nonconformity (#PA5878)	Maladjustment (#PA6739)
Disappearance (#PA6785)	Unsavouriness (#PA7204)	Unfeelingness (#PA7364)
Changeableness (#PA5490)	Incompleteness (#PA6652)	Unpleasantness (#PA7107)
Unskillfulness (#PA7232)	Unpreparedness (#PA7341)	Mental illness (#PC0300)
Thoughtlessness (#PA6940)	Meaninglessness (#PA6977)	Indiscrimination (#PA6446)
Narrowmindedness (#PA7306)	Unimaginativeness (#PA6738)	Inappropriateness (#PA6852)
Influencelessness (#PA6882)	Unintelligibility (#PA7367).	

♦ **PA7372 Infrequency**

Broader Time*complex (#PA7222).

♦ **PA7377 Meaning*complex**

Narrower Unintelligibility (#PA7367) — Misinterpretability (#PA6741).

Related Meaninglessness (#PA6977).

♦ **PA7378 Aggravation**

Aggravation — Annoyance — Contentiousness — Deterioration — Exasperation — Provocation

Broader Feeling*complex (#PA6938).

Narrower Provocation (#PJ2395).

Related Envy (#PA7253)	Reason (#PA5502)	Disaccord (#PA5532)
Adversity (#PA6340)	Difficulty (#PA5497)	Impairment (#PA6088)
Unpleasantness (#PA7107).		

♦ **PA7382 Loss**

Bereavement — Damage — Denial — Depletion — Depreciation — Deprivation — Dissipation — Exhaustion — Injury — Privation — Repression — Ruin — Sacrifice — Spoilation — Suppression — Waste — Wear

Broader Possession*complex (#PA6686).

Narrower Wear (#PB1701)	Bereavement (#PF3516).	
Related Vice (#PA5644)	Death (#PA7055)	Defeat (#PA7289)
Denial (#PA7400)	Badness (#PA5454)	Poverty (#PA6434)
Disease (#PA6799)	Dissent (#PA6838)	Absence (#PA7270)
Refusal (#PA7321)	Boredom (#PA7365)	Inaction (#PA5806)
Unbelief (#PA7392)	Celibacy (#PA7410)	Necessity (#PA6387)
Injustice (#PA6486)	Cheapness (#PA7193)	Restraint (#PA7296)
Difficulty (#PA5497)	Impairment (#PA6088)	Disrespect (#PA6822)
Opposition (#PA6979)	Repression (#PB0871)	Exhaustion (#PJ2732)
Disapproval (#PA6191)	Destruction (#PA6542)	Deprivation (#PA0831)
Nonexistence (#PA5870)	Intemperance (#PA6466)	Misjudgement (#PA6607)
Inexpedience (#PA7395)	Insufficiency (#PA5473)	Appropriation (#PA5688)
Forgetfulness (#PA6651)	Disappearance (#PA6785)	Unprovability (#PA7070)
Disintegration (#PA6858)	Unpleasantness (#PA7107)	Inappropriateness (#PA6852).

♦ **PA7389 Reproduction**

Broader Power*complex (#PA7314).

♦ **PA7391 Closure**

Barrier — Congestion — Foulness — Obstacle — Obstruction

Broader Structure*complex (#PA6944).

Related Vice (#PA5644)	Stench (#PA5981)	Badness (#PA5454)
Ugliness (#PA7240)	Lightness (#PA5491)	Injustice (#PA6486)
Disrepute (#PA6839)	Restraint (#PA7296)	Difficulty (#PA5497)
Unchastity (#PA5612)	Uncleanness (#PA5459)	Disapproval (#PA6191)
Destruction (#PA6542)	Untimeliness (#PA7006)	Insufficiency (#PA5473)
Unsavouriness (#PA7204)	Unpleasantness (#PA7107).	

♦ **PA7392 Unbelief**

Apprehension — Denial — Disbelief — Discredit — Distrust — Doubtfulness — Dubiousness — Faithlessness — Gullibility — Heresy — Implausibility — Incredulity — Infatuation — Infidelity — Misbelief — Mistrust — Naïvety — Rejection — Softness — Superstition — Suspicion — Uncertainty — Unreliability — Unsophistication — Weakness

Broader Credibility*complex (#PA5829).

Narrower Heresy (#PF3375)	Uncertainty (#PA6438)	Unbelievers (#PF8068)
Superstition (#PA0430).		
Related Vice (#PA5644)	Fear (#PA6030)	Envy (#PA7253)
Loss (#PA7382)	Error (#PA6180)	Danger (#PA6971)
Denial (#PA7400)	Impiety (#PA6058)	Dissent (#PA6838)
Oldness (#PA7131)	Refusal (#PA7321)	Weakness (#PA5558)
Inaction (#PA5806)	Rashness (#PA7115)	Insanity (#PA7157)
Distrust (#PA8653)	Ignorance (#PA5568)	Exclusion (#PA5869)
Avoidance (#PA6379)	Necessity (#PA6387)	Disrepute (#PA6839)
Impotence (#PA6876)	Improbity (#PA7363)	Opposition (#PA6979)
Unorthodoxy (#PA5564)	Disapproval (#PA6191)	Uncertainty (#PA7309)
Disagreement (#PA5982)	Invisibility (#PA6978)	Imperfection (#PA6997)
Irresolution (#PA7325)	Unprovability (#PA7070)	Unsavouriness (#PA7204)
Changeableness (#PA5490)	Unskillfulness (#PA7232)	Unintelligence (#PA7371)
Inappropriateness (#PA6852)	Influencelessness (#PA6882)	Unintelligibility (#PA7367)
Uncommunicativeness (#PA7411).		

♦ **PA7394 Evaluation*complex**

Narrower Discovery (#PA5611) — Misjudgement (#PA6607) — Indiscrimination (#PA6446).

♦ **PA7395 Inexpedience**

Awkwardness — Damage — Deleteriousness — Disadvantage — Futility — Illness — Impairment

— Inadvisability — Inappropriateness — Inaptitude — Incongruity — Inconvenience — Ineptitude — Inexpedience — Infelicity — Injury — Liability — Mischief — Prejudice — Trouble — Undesirableness — Unsuitability — Uselessness — Worthlessness
Broader Adaptation*complex (#PA8178).
Related Vice (#PA5644) Fear (#PA6030) Loss (#PA7382)
Reason (#PA5502) Danger (#PA6971) Badness (#PA5454)
Poverty (#PA6434) Disease (#PA6799) Disaccord (#PA5532)
Ignorance (#PA5568) Vulgarity (#PA5821) Adversity (#PA6340)
Injustice (#PA6486) Solemnity (#PA6731) Impotence (#PA6876)
Undueness (#PA6921) Wrongness (#PA7280) Prejudice (#PA2173)
Difficulty (#PA5497) Unchastity (#PA5612) Unkindness (#PA5643)
Impairment (#PA6088) Inelegance (#PA6312) Disrespect (#PA6822)
Unimportance (#PA5942) Disagreement (#PA5982) Hopelessness (#PA6099)
Misbehaviour (#PA6498) Misjudgement (#PA6607) Imperfection (#PA6997)
Untimeliness (#PA7006) Inhospitality (#PA5458) Insufficiency (#PA5473)
Human illness (#PA0294) Discontentment (#PA6011) Unpleasantness (#PA7107)
Unskillfulness (#PA7232) Unpreparedness (#PA7341) Unintelligence (#PA7371)
Meaninglessness (#PA6977) Narrowmindedness (#PA7306) Nonaccomplishment (#PA6662)
Inappropriateness (#PA6852).

♦ PA7400 Denial
Denial
Broader Truth*complex (#PA6164).
Related Loss (#PA7382) Dissent (#PA6838) Refusal (#PA7321)
Unbelief (#PA7392) Necessity (#PA6387) Opposition (#PA6979)
Unprovability (#PA7070).

♦ PA7406 Causation*complex
Narrower Chance (#PA6714) Imminence (#PA6290) Culmination (#PA6706).

♦ PA7408 Smallness
Exiguity — Inconsequence — Insufficiency — Scantiness — Smallness — Unimportance
Broader Quantity*complex (#PA6108).
Related Reason (#PA5502) Fewness (#PA7152) Humility (#PA6659)
Disrepute (#PA6839) Impotence (#PA6876) Inequality (#PA6695)
Littleness (#PA7285) Inferiority (#PA5652) Shortcoming (#PA6041)
Selfishness (#PA7211) Unimportance (#PA5942) Insufficiency (#PA5473)
Discontentment (#PA6011) Narrowmindedness (#PA7306).

♦ PA7409 Extrinsicality
Broader Existence*complex (#PA5498).

♦ PA7410 Celibacy
Annulment — Bereavement — Malfeasance
Broader Socialization*complex (#PA6373).
Related Vice (#PA5644) Loss (#PA7382) Death (#PA7055)
Dissent (#PA6838) Destruction (#PA6542) Bereavement (#PF3516)
Appropriation (#PA5688) Inappropriateness (#PA6852).

♦ PA7411 Uncommunicativeness
Affectation — Artifice — Artificiality — Closeness — Collusion — Conspiracy — Cunning — Curtness — Deceit — Deception — Delusion — Demeaning — Dishonesty — Disingenuousness — Distortion — Dubiousness — Duplicity — Equivocation — Erroneousness — Evasion — Exaggeration — Excess — Extravagance — Fabrication — Faithlessness — Fakery — Fallacy — Falsehood — Falseness — Fraudulence — Furtiveness — Guile — Hallucination — Hypocrisy — Illegitimate — Improbity — Impurity — Indirection — Insincerity — Malingering — Mendacity — Misconstruction — Misdirection — Misguidance — Misinformation — Misrepresentation — Misstatement — Muteness — Mystification — Ostentation — Overemphasis — Overestimation — Perfidiousness — Perversion — Prevarication — Prodigality — Sanctimony — Secrecy — Slant — Sophistry — Speciousness — Spurious — Strain — Subterfuge — Surreptitiousness — Travesty — Treachery — Trickery — Truthlessness — Unauthenticity — Uncommunicativeness — Unctuousness — Unnaturalness — Victimization — Warpedness
Broader Communication*complex (#PA6732).
Narrower Fallacy (#PF4357) Secrecy (#PA0005) Illusion (#PA6414)
Hypocrisy (#PF3377) Sophistry (#PF8008) Conspiracy (#PC2555)
Misinformation (#PD8523) Political feuding (#PD4846) Unethical practices (#PC8247).
Related Vice (#PA5644) Fear (#PA6030) Cold (#PA6956)
Error (#PA6180) Enmity (#PA5446) Reason (#PA5502)
Danger (#PA6971) Excess (#PB8952) Neglect (#PA5438)
Impiety (#PA6058) Loudness (#PA5465) Inaction (#PA5806)
Rashness (#PA7115) Insanity (#PA7157) Unbelief (#PA7392)
Ignorance (#PA5568) Deviation (#PA6228) Avoidance (#PA6379)
Injustice (#PA6486) Undueness (#PA6921) Cheapness (#PA7193)
Improbity (#PA7363) Deception (#PB4731) Unchastity (#PA5612)
Unkindness (#PA5643) Illegality (#PA5952) Impairment (#PA6088)
Inelegance (#PA6312) Quiescence (#PA6444) Difference (#PA0698)
Distortion (#PA6790) Disrespect (#PA6822) Narrowness (#PA7256)
Uncleanness (#PA5459) Inferiority (#PA5652) Diffuseness (#PA5974)
Ineloquence (#PA6118) Disapproval (#PA6191) Affectation (#PA6400)
Discourtesy (#PA7143) Prodigality (#PA7269) Uncertainty (#PA7309)
Malingering (#PE7701) Miseducation (#PA6393) Intemperance (#PA6466)
Misjudgement (#PA6607) Imperfection (#PA6997) Irresolution (#PA7325)
Exaggeration (#PJ5960) Insufficiency (#PA5473) Appropriation (#PA5688)
Nonconformity (#PA5878) Unsociability (#PA6653) Maladjustment (#PA6739)
Disappearance (#PA6785) Inexcitability (#PA5467) Unskillfulness (#PA7232)
Narrowmindedness (#PA7306) Misrepresentation (#PA6644) Nonaccomplishment (#PA6662)
Inappropriateness (#PA6852) Unintelligibility (#PA7367) Misinterpretability (#PA6741).

♦ PA7412 Number*complex
Narrower Duality (#PA7339) Fewness (#PA7152) Unnumbered (#PA6436).

♦ PA7599 Pride
Dependence on pride
Nature Pride is in the first place rebellion against the limitations that existence places on an individual. It is the refusal to live without control over the world, the society and the self. Second, it is isolation. As a perverted form of self love, it denies the need for community of others. It drives people to ignore others, or hurt others, or patronize others. It refuses dependency on and help from others. Third, it is personal slavery. It alienates us from aspects of ourselves that we are ashamed to recognize or admit and thus prevents any personal growth. It is not self-respect but the perversion of self-respect. It is self-abasement because it is based on the lie that the self is the centre of the universe. It is not the normal human drive to personal significance but the trivialization of the rest of reality. Pride, one of the seven capital sins, destroys families, perverts relationships, divides communities, provokes wars, and corrupts the sources of human happiness.
Broader Sin (#PF0641) Vice (#PA5644) Egoism (#PA6318).
Related Lust (#PA4673) Envy (#PA7253) Anger (#PA7797)
Sloth (#PA3275) Vanity (#PA6491) Gluttony (#PA9638)
Excessive virtue (#PF7127).
Aggravates Guilt (#PA6793) Triumphalism (#PF8203) Original sin (#PF8298)
Excessive medical intervention in childbirth (#PE7705)
Loss of humility in relation to the environment (#PF2527).

♦ PA7604 Indifference
Dependence on indifference — Indifferent people
Incidence A study revealed that the New York Times ran 811 stories on the plight of European Jewry from November 1941 - November 1944; that these pieces ran 3 out of every 4 days; that on April 6, 1942 clear evidence of mass exterminations was reported; and that by June 30, 1942 the paper reported the death of one million Jews. But little play was given this news. There were no banner headlines, few front page articles, and few editorials. A 1983 UN study predicted a widespread famine in Ethiopia and Sudan if measures were not taken to rectify the situation; Pol Pot's regime massacred 2 million Cambodians as the news was reported to the world; and Amnesty International cites more than 50 countries where torture is a routine course of events.
Claim Of greater significance for humanity than war, pollution and other disasters is the widespread indifference to such problems and to the possibility of remedial action.
Refs Cizik, Richard *The High Cost of Indifference* (1984).
Broader Incuriosity (#PA6598) Inattention (#PA6247) Unimportance (#PA5942)
Unfeelingness (#PA7364).
Narrower Indifference to suffering (#PB5249) Indifference of students' families (#PG9324).
Related Apathy (#PA2360) Citizen apathy (#PF2421).
Aggravates Inaction (#PA5806) Lack of care (#PF4646) Lack of political will (#PC5180).
Aggravated by Passivity (#PF6177).

♦ PA7646 Arrogance
Dependence on arrogance — Arrogant people
Claim Arrogance develops out of fear and into aggression. It does not allow people, communities or nations to envisage the possibility of openness, expansion, acceptance and peace; but generates mistrust, prejudice, myopia and war. Arrogant people (and nations) are so involved with themselves and so consumed with competing with others that they cannot credit religions, viewpoints, philosophies and lifestyles which differ from their own. It is arrogance which creates and perpetuates the miseries prolific in today's world.
Narrower Cultural arrogance (#PF5178) Government arrogance (#PF8820)
Scientific arrogance (#PF7843) Intellectual arrogance (#PF7847)
Bureaucratic superiority (#PC1259) Arrogance of policy-makers (#PF2895)
Arrogance of intergovernmental agencies (#PF9561).
Related Vanity (#PA6491) Humility (#PA6659) Disrespect (#PA6822)
Opposition (#PA6979).
Aggravates Triumphalism (#PF8203) Arrogation of rights (#PD4680).

♦ PA7685 Abandonment
Desertion — Sense of abandonment
Nature Desertion of the aged, the sick, the deformed or crippled, the helpless, or of infants and children by parents, family groups, or communities is a time-rooted practice among many peoples. Technically, abandonment refers to the desertion of the aged and helpless, while the abandonment of infants left to perish is termed exposure. The basic causes of abandonment have always been economic, that is, the lack of food or the uselessness of the aged or handicapped to the group.
Incidence The growth of foundling homes in China, France, Cuba and other parts of the civilized world show that today most infants are abandoned in the hopes they will be picked up; leaving children in railway stations or railway station lockers still occurs. Some South African tribes take their old people into the wilderness and leave them in a small enclosure with a little food and water; Melanesians either burn or bury alive their elderly.
Background The Arabs either abandoned their old and helpless or buried them alive; the ancient Persians and Armenians left them in the deserts to be devoured by wild beasts; the early Romans threw everyone over 60 into the Tiber; many North American Indians left the old, the sick, and the weak behind when they moved camp, as did American pioneers crossing the plains in the 19th century.
Narrower Absence of God (#PF5764) Missing persons (#PD1380)
Abandoned wives (#PD1030) Juvenile desertion (#PD8340)
Abandonment of the dying (#PD4268) Desertion in marriage law (#PF3254).
Related Vice (#PA5644) Absence (#PA7270) Neglect (#PA5438)
Avoidance (#PA6379) Restraint (#PA7296) Unchastity (#PA5612)
Inhospitality (#PA5458) Insufficiency (#PA5473).
Aggravated by Soul murder (#PF4213).

♦ PA7690 Suffering
Incidence Suffering is almost universal. From the point where, in the evolutionary process, a brain is developed, suffering is a common experience through many forms. Evidence has also been presented that plants experience suffering.
Refs Slatoff, Walter J *The Look of Distance* (1985); Soelle, Dorothee and Kalin, Everett R *Suffering* (1975).
Narrower Human suffering (#PB5955) Animal suffering (#PD8812)
Suffering of plants (#PC7825).
Aggravates Struggle for existence (#PB4411)
Excessive foreign public debt of developing countries (#PD2133).

♦ PA7710 Idleness
Dependence on idleness — Laziness — Indolence
Refs Ward, S Alexander *How to Overcome Laziness and Achieve Your Goals* (1986).
Broader Inaction (#PA5806).
Narrower Unpreparedness for surplus leisure time (#PF5044).
Related Sloth (#PA3275) Neglect (#PA5438) Untimeliness (#PA7006)
Unimportance (#PA5942).
Aggravates Economic refugees (#PD4379) Religious backsliding (#PF6826)
Lack of community planning (#PF2605) Criminally life endangering behaviour (#PD0437).

♦ PA7736 Temptation
Dependence on temptation — Tempters
Nature Temptation, the inducement to sin, occurs when man is confronted by a temporal good (recognized as such) which conflicts with his eternal good.
Background In the Biblical sense, temptation is the situation in which man tempts his faith to God (having to choose between fidelity and infidelity to one's obligations to God) as well as tempting God Himself by testing Him to reward or punish.
Claim Temptation is rampant in today's world, accounting for the strife and desperation which abound. With so many choices and with such high regard placed on materialism, many figures both public and private have allowed themselves to fall prey to temptation, often at the sacrifice of fellow men.

ABSTRACT FUNDAMENTAL PROBLEMS PA9638

Counter–claim In one sense, temptation is man's lot; to escape all allurements of evil would require flight from this life.
Refs Bonhoeffer, Dietrich *Creation and Fall* (1965); Stanley, Charles *Temptation* (1988).
  Broader Dissuasion (#PA7343).
  Narrower Impure thoughts (#PF5205).
  Aggravates Religious backsliding (#PF6826).

♦ **PA7797 Anger**
Nature Anger resulting from an injury, mistreatment or opposition to an individual who seeks vengeance against the person or persons is destructive. It is destructive of the self. It is not not righteous anger; being enraged by the mistreatment of others. It is anger about how the angry person is being treated. It is not a result of compassion about the cruel treatment of others but because life is not treating him the way he thinks it should. At its centre it is egoism and as such self-destructive.
The angry person erects barriers against other people; it rejects genuine efforts of other persons to offer support and love. The constantly angry person becomes isolated and by this his quality of life diminished.
Counter–claim Anger is a healthy and normal emotional response to injustice, inhumanity, treachery, maltreatment of the weak and exploitation of the vulnerable. Looking on these things with calm detachment is to wither as a human being.
Refs Averill, James R *Anger and Aggression* (1982).
  Broader Sin (#PF0641). Vice (#PA5644).
  Related Lust (#PA4673) Envy (#PA7253) Sloth (#PA3275)
  Pride (#PA7599) Avarice (#PA6999) Gluttony (#PA9638)
  Melancholy (#PF7756) Over-eating (#PE5722) Original sin (#PF8298).
  Aggravated by Soul murder (#PF4213) Inhibited grief process (#PD4918).

♦ **PA8038 Spiritual hunger**
Spiritual desperation
Claim The hunger which drives many either to achieve the impossible or to acts of desperation is not a hunger for food. It may be suffered in the midst of plenty and with access to all material desires.
  Aggravated by Deterioration of human environment (#PC8943)
  Barriers to transcendent experience (#PF4371).

♦ **PA8178 Adaptation*complex**
  Narrower Badness (#PA5454) Inexpedience (#PA7395) Unimportance (#PA5942)
  Imperfection (#PA6997) Insufficiency (#PA5473) Inappropriateness (#PA6852).

♦ **PA8197 Misunderstanding**
  Broader Error (#PA6180) Disaccord (#PA5532) Misinterpretability (#PA6741).
  Narrower Misunderstanding of veterinary approach (#PG5402)
  Inappropriate understanding of progress (#PB8648).
  Aggravated by Cultural barriers (#PB2331).

♦ **PA8199 Inadequacy**
  Narrower Inadequate laws (#PC6848) Inadequate education (#PF4984)
  Inadequacy of doctrine (#PF3396) Inadequate infrastructure (#PC7693).
  Related Impotence (#PA6876) Inequality (#PA6695) Inferiority (#PA5652)
  Shortcoming (#PA6041) Imperfection (#PA6997) Insufficiency (#PA5473)
  Incompleteness (#PA6652) Unskillfulness (#PA7232) Discontentment (#PA6011)
  Insufficient financial resources (#PB4653).

♦ **PA8487 Hatred**
Hate
Claim Only the belief and execution of law of a free society determined to remain free can defeat hatred.
Refs Bychowski, Gustav *Evil in Man* (1968); Suttie, Ian *The Origins of Love and Hate*.
  Broader Fear (#PA6030).
  Narrower Misanthropy (#PJ1278).
  Aggravates Crime (#PB0001) Discrimination (#PA0833).
  Aggravated by Incitement to hatred (#PE5952).

♦ **PA8528 Judgement*complex**
  Narrower Improbity (#PA7363) Disapproval (#PA6191).
  Related Injustice (#PA6486).

♦ **PA8555 Inflexibility**
  Broader Softness (#PA5988) Compulsion (#PA5740) Irresolution (#PA7325).
  Narrower Inflexible social care structures in developing countries (#PF2493).

♦ **PA8653 Distrust**
Distrust blocking support — Distrust of loan structures — Lack of confidence — Suspicion — Mistrust
Nature The belief that a situation, person, action, organization, or even existence intends to do harm or, at the least, to do no good.
  Narrower Mistrust of police (#PF8559) Distrust of services (#PG8857)
  Mistrust of strangers (#PF8743) Distrust of storekeepers (#PG9249)
  Suspicion of bureaucracy (#PF8335) Mistrust of birth control (#PG8030)
  Suspicion of imposed change (#PG9094) Intergovernmental suspicion (#PC2089)
  Distrust of political dialogue (#PD2263)
  Distrust of business by the community (#PE8963)
  Distrust of interpersonal relationships (#PF4274)
  Distrust of professional service delivery (#PD0974).
  Related Envy (#PA7253) Rashness (#PA7115) Unbelief (#PA7392)
  Lack of work commitment (#PD2790) Negative emotions and attitudes (#PA7090)
  Loss of confidence in government leaders (#PF1097).
  Aggravates Crime (#PB0001) Doubt (#PF8474) Conflict (#PA0298)
  Demoralizing image of urban community identity (#PF1681).
  Aggravated by AIDS (#PD5111) Ignorance (#PA5568)
  Lack of leadership (#PF1254) Ideological conflict (#PF3388)
  International insecurity (#PB0009) Competition between states (#PC0114)
  Ineffective war crime prosecution (#PD1464)
  Deliberate misrepresentation in educational materials (#PF1183)
  Incomplete understanding of new societal service systems (#PF2212).

♦ **PA8658 Irresponsibility**
Abdication of responsibility — Irresponsible people — Avoidance of obligations — Evasion of obligations — Dereliction of duty
Nature Irresponsibility takes three forms: reducing one's context of responsibility, shifting or reducing to whom or to what one is responsible and denial of freedom of choice. Frequently in business, government or day to day situations the statement is made that it is not one's responsibility, one is not on duty or the phone ringing is not one's. This is reducing the context of responsibility. Shifting to whom one is responsible is illustrated by responses like the corporate officer telling customers that the business is responsible to shareholders. Another form of this is the suggestion that one is only responsible to one's self. Denial of freedom of choice is exemplified with the excuse of one is only following the law or doing what they were told.
From one perspective, both the free spirit acting on their own and the individual tied to a Kantian ethic of duty are equally irresponsible. The free spirit has reduced their obligations to their own feelings, intuitions or insights and thus enslaved their scope of choice to themselves while proclaiming themselves as ultimately free. The person of duty has reduced their obligations to a set of rules, authorities, or principles abdicating even the capacity of free choice. This perspective suggests that responsibility is found only in being both ultimately free and ultimately obligated.
  Broader Improbity (#PA7363) Lawlessness (#PA5563) Changeableness (#PA5490).
  Narrower Guilt (#PA6793) Indecision (#PF8808) Blame avoidance (#PF6382)
  Evasion of work (#PC5576) Evasion of issues (#PF7431) Avoidance of work (#PC5528)
  Evasion of the law (#PD4208) Nuclear irresponsibility (#PF6611)
  Irresponsible social science (#PF8032) Journalistic irresponsibility (#PD3071)
  Unethical commercial practices (#PC2563) Official evasion of complaints (#PF9157)
  Irresponsible transient occupants (#PG8879) Irresponsible international trade (#PC8930)
  Irresponsible genetic manipulation (#PC0776) Irresponsible international experts (#PF0221)
  Irresponsible pharmaceutical advertising (#PE2390)
  Irresponsible finders of personal property (#PF3859)
  Abdication of control by company directors (#PE9251)
  Irresponsible research using human subjects (#PC0080)
  Irresponsibility towards future generations (#PF9455)
  Irresponsibility of young people towards the family (#PE1832)
  Irresponsible scientific and technological activity (#PC1153)
  Irresponsible introduction of new species of animals (#PD1290)
  Social irresponsibility of transnational corporations (#PE5796)
  Irresponsible delimitation of policy responsibilities (#PF7823)
  Unaccountability of institutions degrading the environment (#PF3458)
  Irresponsible expression of emotions equated with free speech (#PF7798)
  Exploitation of regulatory loopholes in countries with underdeveloped legislation (#PE4339).
  Related Negligence (#PA2658) Draft evasion (#PD0356)
  Unethical medical practice (#PD5770)
  Socially irresponsible programmes of transnational banks in developing countries (#PE4360).
  Aggravates Induced abortion (#PD0158) Incorrect information (#PB3095)
  Alienation in capitalist systems (#PD3112)
  Excessive demand for goods in capitalist systems (#PC3116).
  Aggravated by Social unaccountability (#PC1522)
  Denial of the right to social security in capitalist systems (#PD3120).

♦ **PA8760 Self-interest**
  Broader Search for individualistic meaning (#PF6796).
  Narrower Compromise as a betrayal of principles (#PF3420).
  Related Avoidance of legal obligations by politicians (#PD4556).
  Aggravates Conspiracy against the public (#PF4198).
  Aggravated by Avarice (#PA6999).

♦ **PA8810 Antipathy**
  Related Hate (#PA7338) Enmity (#PA5446) Difference (#PA6698)
  Opposition (#PA6979) Unwillingness (#PA6509) Unpleasantness (#PA7107).

♦ **PA8886 Dissatisfaction**
  Narrower Lack of satisfaction (#PG6124).
  Related Envy (#PA7253) Dissent (#PA6838) Loneliness (#PF2386)
  Alienation (#PA3545) Disapproval (#PA6191) Inexpectation (#PA5527)
  Unpleasantness (#PA7107) Discontentment (#PA6011) Mental suffering (#PB5680).

♦ **PA9191 Unhappiness**
  Related Solemnity (#PA6731) Disapproval (#PA6191) Unpleasantness (#PA7107)
  Discontentment (#PA6011).
  Aggravated by Human sexual inadequacy (#PC1892).
  Reduced by Illusion of happiness (#PJ5224).

♦ **PA9226 Criminality**
Criminal state of mind
  Broader Inadequacy of prevailing mental structures to challenge of human survival (#PF7713).
  Narrower Criminal insanity (#PD9699).
  Related Inaccurate criminal stereotypes (#PF1244)
  Criminal negligence in performing socialist responsibilities (#PE5538).
  Aggravates Crime (#PB0001).

♦ **PA9638 Gluttony**
Nature Gluttony is the disease of spiritual emptiness. While it can manifest itself in overeating, excess drinking of alcohol or drug abuse, at its foundation is a malaise of the spirit; it cares–less about life. Human life like all energy seeks to consume. Gluttony is consumption in the immediate situation beyond the needs of the individual to the point of being life endangering, whether it is food and drink or at the level of society: goods, services and energy. It cares–less about beauty; gluttony is not concerned with the taste, smell or presentation of food, only the amount. It does not savour. It devours. It is destructive of companionship. It removes the individual from any caring for others and any caring of others for the individual.
Background Gluttony can become characteristic of society, and is often attributed to that of Ancient Rome and more recently to the consumer society of the West. In the Middle Ages, it was placed among the seven deadly sins.
  Broader Sin (#PF0641).
  Related Lust (#PA4673) Envy (#PA7253) Pride (#PA7599)
  Anger (#PA7797) Sloth (#PA3275) Avarice (#PA6999)
  Original sin (#PF8298).
  Aggravates Over-eating (#PE5722).

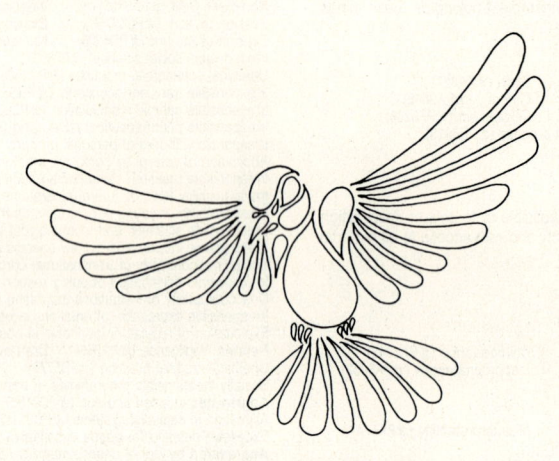

Basic universal problems   PB

Content

This section groups together the major multi-sectoral, world-wide problems which tend to be prominent on the agendas of the major international organizations and in the media. Such problems also tend to group, or focus, many of the more specialized problems which are described in subsequent sections. Indeed in many debates discussion of the more specialized problems may be subsumed under discussion of these major problems.

Note that further information relevant to an understanding of the problem may be present in other problems cross-referenced in the entry consulted.

Many of the problems in this section are of such proportions and complexity that no single organization or discipline can claim to encompass any one of them in all its aspects. The scope and implications of such problems tends to be a matter of continuing debate. They are are not sufficiently well-defined to respond to well-defined solutions. The nature of an appropriate solution to such problems is also a matter of continuing debate.

This section groups 141 problems for which there are 3,474 cross-references.

Rationale

Inclusion of such problems calls for no comment because of their widely recognized importance.

Method

The entries are based on information obtained from international organizations, from a wide variety of reference books, or as reported in the international media. The procedures for identifying world problems are described in Section PZ.

Index

A keyword index to entries is provided in Section PX.

Comment

Detailed comments are given in Section PZ at the end of this volume.

Reservations

The emphasis throughout this volume has been placed on providing descriptions of less well-known problems, particularly when the extensive material available on the better known problems contained neither succinct descriptions of them nor descriptive material which could easily be reduced to succinct descriptions. The problem descriptions here represent a compilation of views from published documents (usually from international organizations). A description does not necessarily constitute the best possible description of the problem since a compromise has had to be struck between availability of information, the resources to process it, and the space available in this volume.

In a number of cases a problem could have been allocated to another section. Inclusion of a problem in this section, rather than in a preceding or following section, has been based on a number of factors. The position of the problem in one or more hierarchies of cross-references was a major factor in determining its allocation to this section.

Possible future improvements

There is much scope for improving the quality of problem entries through feedback from interested bodies. More bibliographic references could be included where appropriate, as well as references to major resolutions concerning those problems recognized by the United Nations. There is also much scope for improving the pattern of cross-references, both between problems, to other sections of this volume (eg values) and to the 20,000 internationally-active bodies in the companion series (*Yearbook of International Organizations*).

ENTRY CONTENT AND ORGANIZATION

Ordering of entries
Entries are in **numeric order**. Entry numbers have been **allocated randomly**; they have no significance other than as a permanent point of reference to facilitate indexing, cross-referencing, and updating between editions.

Index access to entries
The location of an entry in this sub-section may be determined from
the **Volume Index** (Section PX) on the basis of keywords in the name of the entry or its alternate titles.

Structure of entries
Entries may be composed of the following descriptive elements:

(a) **Entry number** This number has **no significance**, except as a convenient method of identifying the entry (particularly for indexing purposes), of filing information on it, and as an identifier to which cross-references from other entries (possibly in other Sections) may refer in this and future editions. The first letter of the entry number refers to the section of this volume in which the sub-section, denoted by the second letter, is located.

(b) **Problem name** This is printed in bold characters. It is the name selected as best indicating the nature of the problem. It may be followed by alternative problem names.

(c) **Nature** Description of the problem which attempts to identify the nature of the disruptive processes involved. The information included here, and in the following paragraphs, is compiled directly, to the extent possible, from available published documents. Where appropriate the text included may be reproduced, in a minimally edited form, from the publications of international organizations, such as those of the United Nations or its Specialized Agencies.

(d) **Incidence** Summary description of the extent of the problem which makes it of more than national significance.

(e) **Background** Describes briefly when and how the problem's importance was recognized initially, and how this recognition has evolved over time.

(f) **Claim** Stresses the special importance of this problem and why action is particularly urgent. This paragraph offers means of including statements which may deliberately exaggerate claims for the unique importance of the problem.

(g) **Counter-claim** Stresses, where appropriate, the relative insignificance or erroneous conception of the problem as described. This paragraph offers a means of including statements which may deliberately exaggerate the arguments refuting the evidence for the existence of the problem. Absence of such arguments from the text does not mean that they do not exist.

Cross-referencing of entries
At the end of any entry, there may be cross-references to other entries. These indicate the number and name of the cross-referenced entry, whether within this Section or in other Sections. There are 3 types of **hierarchical** cross-references between problems:
> **Broader** = Broader problem: more general problems of which the problem described may be considered a part. The described problem may be considered an aspect of several broader problems
> **Narrower** = Narrower problem: more specific problems which may be considered a part of the described problem
> **Related** = Related problem: problems that may be considered as associated in a hierarchically undefined way with the described problem.

> There are 4 types of **functional** cross-references between problems:
> **Aggravates** = Problems aggravated by the described problem: a forward or subsequent negative causal link
> **Aggravated by** = Problems aggravating the described problem: a backward or prior negative causal link
> **Reduces** = Problems relieved, alleviated or reduced by the described problem: a forward or subsequent positive causal link
> **Reduced by** = Problems relieving or alleviating the described problem: a backward or prior positive causal link

♦ PB0001 Crime

Vulnerability of social defences — Dependence on crime — Criminals — Criminal liability

Nature Crime, on the increase throughout the world, includes both personal attacks (assaults, theft, muggings) and transnational offenses (diplomat kidnappings, airplane hijackings, drug trafficking), leading to a pervasive uneasiness amongst ordinary citizens who are increasingly resorting to active as well as passive self–protection measures.

Incidence In addition to criminal offenses often associated with low socio–economic groups (such as murder, burglary, rape, arson, manslaughter, and theft), there is now an acceleration in the dimensions of white collar crimes such as consumer fraud, illegal price fixing, tax evasion and contravention of anti–trust or anti–monopoly laws and labour legislation. Convictions of criminals by courts reflect biases inherent in police practice, as many criminal acts escape notice or their perpetrators cannot be identified. Organized crime, often impervious to police intensifications and court actions, is taken for granted as an aspect of modern industrial life in various countries, often putting legitimate businesses into bankruptcy as a result of unequal competition. Corrupt practices within official bureaucracies may divert funds intended for national development. Transnational crimes in their new political forms challenge social defence while disrupting world order and security.

About one in four American households was hit in 1989 by a violent or property crime including rape, robbery, assault, personal theft, burglary and motor vehicle theft. Same year the UK suffered 11,500 crimes a day of which 94 percent were crimes against property.

Background While studies show that in most countries urban areas have higher crime rates than the surrounding rural areas (due to unemployment, poverty, hopelessness of self–improvement, anonymity and overcrowding being more prevalent in cities), the corresponding economic strain has forced many countries to curtail their rehabilitative crime prevention programmes in favour of less effective short–term repressive policies. The overall general decline of family structure has led both to a mistrust of and intolerance towards the older and more established methods of prevention and control, and to impatience with the value systems upon which they have been based. As education spreads, higher expectations, if unfulfilled, appear to increase the vulnerability of young people to the temptation of illegal short cuts to wealth, power, and status. In any case, crime is predominantly an activity of young male members of society.

Claim Crime has effects which are as extensive as society itself. It can be as subversive to development and debasing to the quality of life as it is personally dangerous, socially damaging or politically embarrassing. Its quiet erosion of national achievement and the long–term influence on motivation can be far more detrimental to a society than is currently recognized. Also, crime is barbaric. Whether it is more so than war is arguable, and it is hard to distinguish between them as affronts to the thousands of years of civilization mankind has developed. One difference between war and crime is that in the former the adversaries have usually been combatants (until the advent of total wars). In crime there is only one combatant, the criminal, and the victim, the non–combatant, is frequently just a passive object. The victims of crime include people from all arenas of age, education, occupation, sex and nationality. The threat of possible crime, and the after effects of crimes perpetrated, violate the legitimate rights of the citizen and limit his freedom and opportunities, thwarting the aspirations of those who work for a happy community whether at the local or international level. Crime permeates all aspects of society and, as such, is a form of warfare waged on a world–wide scale, with the non–combatants on the losing side.

Counter–claim The distinction between crime and other forms of socially deviant behaviour is not clear–cut, and varies in time and place with the scope and limits of criminal law. Much criminal behaviour is irrational and may stem from a sense of guilt and the desire to be punished, or from a defect in the capacity to acquire conditioned avoidance responses to actions which elicit punishment (found in highly extroverted individuals). Other causes of criminal behaviour have been shown to be related to heredity, and to suspension of the socialization process inherent in family breakdown. Ability to make use of sophisticated language assists in mediating avoidance conditioning, and techniques are being developed using this and other factors which may reduce the inherent criminality in particular individuals.

Refs Archer, Dane and Gartner, Rosemary *Violence and Crime in Cross–National Perspectives* (1987); Braithwaite, John *Crime, Shame and Reintegration* (Date not set); Campbell, Duncan *That Was Business, This is Personal* (1990); Chang, Dae H and Blazicek, Donald L *An Introduction to Comparative and International Criminology* (1986); Council of Europe *Economic Crises and Crime* (1985); Eitzen, Stanley D and Timmer, Doug A (Eds) *Crime in the Streets and Crime in the Suites* (1989); Gross, F A *Crime* (1977); Hall, Richard *Dirorganized Crime*; Heidensohn, Frances *Women and Crime* (1985); Huggings, Martha K *From Slavery to Vagrancy in Brazil* (1985); Johnson, Elmer H (Ed) *International Handbook of Contemporary Developments in Criminology* (1983); Litan, Robert E and Winston, Clifford *Liability* (1988); Lopez–Rey, Manual *Guide to United Nations Criminal Policy* (1985); Meier, Robert F *Crime and Society* (1989); Messerschmidt, James *Capitalism, Patriarchy and Crime* (1986); Muslehuddin, Muhammad *Crime and the Islamic Doctrine of Preventive Measures* (1985); Nettler, Gwynn *Explaining Crime* (1984); Normandeau, André *International Bibliography on Criminal Statistics*; Reid, Sue T *Crime and Criminology* (1988); Shariff, M *Crime and Punishment in Islam* ; Shoham, S Giora and Rahau, Giora *The Mark of Cain* (1982); Von Hirsch, Andrew *Past or Future Crimes* (1985); Wilson, James Q and Herrnstein, Richard J *Crime and Human Nature* (1986).

Broader Illegality (#PA5952).

Narrower Torts (#PD9022) Criminals (#PC7373) Complicity (#PF4983)
Urban crime (#PD7399) Racketeering (#PE4914) Violent crime (#PD4752)
Economic crime (#PC5624) Statutory crime (#PC0277) Political crime (#PC0350)
Female criminals (#PE1837) Victimless crime (#PC5005) Unreported crimes (#PF1456)
Alcohol–related crime (#PE4131) Tools for criminal use (#PG6941)
Crimes against humanity (#PC1073) Violations of private law (#PD5727)
Periods of high crime rate (#PE4294) Locales of high crime rates (#PE7311)
Vicarious criminal liability (#PE6323) Crimes against national security (#PC0554)
Limited criminal liability of corporations (#PF7293).

Related Rest (#PA5500) Deviance (#PB1125) Low self esteem (#PF5354)
Lack of intersocietal resource channels (#PF2517).

Aggravates Vulnerability (#PA1766) Fear of crime (#PF4682)
Border controls (#PJ1718) Unsolved crimes (#PF6911)
Human suffering (#PB5955) Restrictions on freedom (#PC5075)
Illegally–obtained funds (#PD5433) Personal physical insecurity (#PD8657)
Inadequate prevention of crime (#PF4924)
Delays in elaboration of remedial legislation (#PC1613).

Aggravated by Hatred (#PA8487) Distrust (#PA8653)
Drug abuse (#PD0094) Urban slums (#PD3139)
Criminality (#PA9226) Abuse of law (#PC5280)
Prostitution (#PD0693) Culture shock (#PC2673)
Alcohol abuse (#PD0153) Mental illness (#PC0300)
Social conflict (#PC0137) Phenylketonuria (#PG3323)
Possession∗complex (#PA6686) Paternal negligence (#PD7297)
Criminal motivation (#PJ6406) Security risk people (#PD6818)
Retarded socialization (#PF2187) Bureaucratic corruption (#PC0279)
Criminal characteristics (#PF5544) Interpersonal estrangement (#PB0034)
National economic recession (#PD9436) Excessive television viewing (#PD1533)
Inadequate firearm regulation (#PD1970) Uncontrolled urban development (#PC0442)
Inequitable administration of justice (#PD0986) Inadequate provision of public safety (#PF2874)
Breakdown in community security systems (#PD1147)
Excessive portrayal of crime in the media (#PE7354)
Misuse of electronic surveillance by governments (#PD2930)
Disruption of family system in developing countries (#PD1482)
Increase in anti–social behaviour in developing countries (#PD0329)
Social environmental degradation from recreation and tourism (#PD0826)
Loss of traditional forms of social control in developing countries (#PD0144)
Inadequate living and working conditions of immigrant labourers in industrialized countries (#PD3427).

Reduced by Informers (#PD8926).

♦ PB0009 International insecurity

Dependence on international insecurity — Lack of international security — Violation of international peace — Violation of international security — Threats to international peace and security

Nature International insecurity results from the power relations of nation states. In the national sense 'insecurity' refers to the condition of a nation which is inadequately defended against dangers or aggression threatening its independence or territory from the outside. National and international insecurity are tied together and a contributory factor is the ever increasing destructiveness of wars. International insecurity may arise from the complexity of the strategic balance between superpowers and the arms race, from the persistence of armed conflicts, from the question of existing state borders and territorial claims. The attainment of national aims of a State and limitation of its freedom of action may also be effected by indirect economic and other non–military coercive action. The importance of non–military dangers in international life is growing rather than diminishing. Conflicts which would most likely have been expressed in the past by the application of military force, may now be implemented by the use of indirect coercive measures, economic or political. There are other potential factors of international insecurity, namely the threat of nuclear war, the economic imbalance between developed and developing countries, world food shortages, global pollution, and earth resources.

Broader Conflict (#PA0298) Insecurity (#PA0857).

Narrower Non–alignment (#PF0801) Imbalance of power (#PB1969)
Military influence (#PD3385) Imbalance of payments (#PC0998)
International aggression (#PB0968) Imbalance in strategic arms (#PC1606)
Competitive acquisition of arms (#PC1258) International non–military conflict (#PF3100)
Lack of a world food security system (#PF5137) Military insecurity and vulnerability (#PC0541)
European insecurity and vulnerability (#PD1863)
Vulnerability of nuclear power sources (#PD0365)
Environmental threats to national security (#PC4341)
Insecurity through unilateral structural disarmament (#PE7670)
Insecurity and vulnerability of nuclear weapon states (#PC4440)
Vulnerability of small nations to foreign intervention (#PD2374)
Insecurity and vulnerability of non–nuclear weapon states (#PD1521).

Related Imperialism (#PB0113) National insecurity and vulnerability (#PB1149).

Aggravates War (#PB0593) Distrust (#PA8653) Repression (#PB0871)
World federalism (#PF2088) Lack of intersocietal resource channels (#PF2517)
Denial of democracy in communist systems (#PC3176)
Misuse of satellite surveillance by governments (#PF3701).

Aggravated by Fear (#PA6030) Militarism (#PC2169)
Foreign control (#PC3187) Total disarmament (#PF8686)
Foreign dictatorship (#PC3186) Revolutionary communism (#PC3163)
Subversion of democracy (#PD3180) Disruptive foreign influence (#PC3188)
Communist economic imperialism (#PC3165) Unfulfilled treaty obligations (#PF2497)
Communist political imperialism (#PC3164) Capitalist economic imperialism (#PC3166)
Lack of international communication (#PG1851)
Monopoly of nuclear power techniques (#PD1741)
Subversion of international agreements (#PD5876)
Government opposition to population control (#PF1023)
Failure of disarmament and arms control efforts (#PF0013)
Military–industrial complex in capitalist systems (#PC3191)
Loss of credibility in international institutions (#PE8064)
Foreign intervention in internal affairs of states (#PC3185)
Deteriorated structures of essential corporateness (#PF1301)
Communist opposition to international organizations (#PF3162)
Military–industrial–governmental complex in communist systems (#PD1330)
Denial of right to national self–determination in communist systems (#PC3177)
Reinforcement of inappropriate development by privileged classes in developing countries (#PF6670).

Reduced by Pacifism (#PF0010).

♦ PB0034 Interpersonal estrangement

Dependence on interpersonal estrangement — Estrangement — Estranged relatives

Nature There is an increasing trend for bystanders to remain uninvolved in the event of some accident or emergency rather than help someone in trouble, particularly when other people are present. In part this is in order to avoid being drawn into complicated procedures whereby those assisting may be required to give evidence as witnesses (in the case of witnessed theft), in part in order to avoid physical danger (in the case of attack), and in part to avoid the risk of being sued for responsibility (in the case of assisting a wounded person prior to treatment in hospital).

Claim To some extent, this problem may be due to the increasing tendency for people to live 'independent', lives essentially estranged from their neighbours and all but very close friends. The media, television, newspapers and popular magazines play up, play upon and seek out the spectacular, the violent, the sensational, the sadistic, the horrible tragedies of modern life as a means of capturing and controlling the attention, using vastly overstimulating visual, auditory and verbal communications. Overstimulation tends to dull responsiveness, particularly when exposed to it hour after hour on a television set. One tends to be left in the position or role of vicarious spectator, without a sense of responsibility or humanitarian involvement. Indeed, this could even contribute to the sense of apathy related to the potential for nuclear war.

Broader Apathy (#PA2360) Immorality (#PA3369)
Psychological alienation (#PB0147).

Aggravates Crime (#PB0001).

Aggravated by Deteriorated structures of essential corporateness (#PF1301).

♦ PB0035 Unsustainable population levels

Unsustainable population growth — Over–population — Unconstrained social growth — Excessive population growth — Rapid population growth — Population disease — Demositis

Nature Human society is faced with an historically unique period of population growth, due to a dramatic decline in death rates coupled with constant fertility rates in much of the world. The momentum of population growth is so great that is difficult to slow down and it is unlikely to cease in forthcoming decades. Although there is some convergence of death rates between developed and developing countries, there remains nevertheless a substantial fertility differential Population growth thus results from the unbalanced acceleration of a long transitional process from the high birth and death rates that characterized most of human history to the low birth and death rates which characterize society today. The process is now in disequilibrium because the achievement of lower mortality has taken effect without any corresponding achievement of lowered fertility The situation has serious consequences for human values. Continued growth threatens both living standards and the environment in various ways. Population growth is a serious intensifier and multiplier of other social and economic problems, especially as it retards the prospects for

development of a better life in the poorer countries, and aggravates environmental pollution and resource depletion by the richer ones.
Incidence During the first millenia of human history, population growth was negligible (0.002 per cent per annum). By the middle of the eighteenth century, that rate had accelerated 150 times (to 0.3 per cent, namely 3,000 per million). By the 1950s, the latter rate of growth had accelerated threefold (to 1.0 per cent or 10,000 per million). At the present time, the rate of growth has doubled again (to 2.0 per cent per annum or 20,000 per million). The faster a population is growing, the shorter the time it takes to double in size. Whereas before 1650, world population took about 35,000 years (or about 1,400 generations) to double, at the present rate it will double in just one generation The world's population totalled approximately 2,982 million in 1960, 3,632 million in 1970, 4,763 million in 1984; it may reach 5,438 million by the end of 1990 and 6,494 million by end of the century. In these totals, the population of more developed regions may account for 976, 1,090, 1,210, 1,336, and 1,454 million respectively, rising at average annual rates of 1 per cent; somewhat less near the end of the century. The population of less developed regions at the same times may number 2,005, 2,542, 3,265, 4,102, and 5,040 million, growing at annual rates near 2.4 per cent, diminishing to 2 per cent by the end of the century. The world's combined population, therefore, would be increasing annually by 2 per cent until 1990, after which time the increase may fall to 1.8 per cent This seemingly very steady growth in global totals masks considerable differences among individual regions and countries, which are growing at different rates; accelerating in some instances and slowing down in others. On the whole, the projections imply acceleration followed by a slowing down as, sooner or later, the decreases in death rates are matched or overtaken by decreases in birth rates. In countries where birth rates have varied in the past, changes in age composition of the population would also affect their potential future growth. The projections referred to above implied slight increases in the rates of the population growth between 1970 and 1980 in Africa, Latin America, Northern America, Oceania and the USSR; and slight decreases in most of Asia and Europe. It is calculated that the general momentum of world population growth will not undergo much change until late in the century. When the projections are extended to the annum 2030, they anticipate that world population growth will diminish to the rate of 1.0 per cent per year Wars, famines and epidemics have occurred in various parts of the world in modern periods. Assessment of some of the effects of such calamities on past population trends has shown them to be considerable in the less developed regions during the second half of the nineteenth century, and in both the less developed and the more developed regions during the first half of the present one. Nevertheless it would be a fallacy to think that wars in modern times have significantly held population growth in check. It can be calculated schematically that all the national and regional disasters on record since 1850, terrible as they have been, have delayed the growth in world population by no more than about ten years. Had there been no wars, famines or epidemics since 1850, the world's population might have totalled more than 2.000 million in 1920 instead of in 1930, and more than 3,500 million in 1960 instead of in 1970.
Claim 1. Population growth is probably the greatest long–term threat to achieving ecological stability either locally or throughout the world. Each year 127 million children are born, each year 95 million come of school age, and each year 19 million people reach age 65. These totals are likely to rise steeply in the years ahead as more young adults swell the ranks of potential parents, and improved medical care advances life expectancy. At 2 per cent a year, the rate of world population growth is now more than double the rate in 1940. It may still rise. Each nation, each community, each family must assess in detail how these trends affect their hopes for higher living standards, a better education, and greater health and happiness 2. Excessive population growth diffuses the fruits of development over increasing numbers instead of improving living standards in many developing countries 3. Under certain conditions population growth may stimulate technological innovations or improvements. Such benefits may be outweighed by disadvantages when human institutions, market mechanisms and technology do not adapt quickly to changing conditions. Environmental circumstances may not favour innovations intended to respond to population growth, such as intensive agriculture in depleted soils which cannot support it **Counter–claim** (1) Population growth is a false issue deriving from a neo–colonial approach by the developed countries to hold down the developing countries for their own self–interest or, in extreme cases, as a subtle form of genocide. It is also used by developed countries as a diversionary, superficial form of reform to alleviate the present corrupt system, and to delay or defer the needed revolution. (2) Under correct economic conditions (such as public control of the means of production) population will take care of itself. The environmental consequences of population growth are due to the affluence and consumption of developed countries, which use up resources and produce pollution. (3) It is not the total numbers which are important, but rather their distribution over the land area. It is the concentration of people in urban conglomerations which is the real problem. (4) Population increase is advocated by some countries as a means of ensuring national sovereignty, and the minimum population requirements needed for a country to become a regional or world power. (5) The entire population problem and the solutions proposed for it pose fundamental questions in the moral and religious realm and may even introduce revolutionary upsets. Ideas such as freedom, happiness, justice and human dignity are put into question with no concern for what will replace them.
Refs Hirsch, F *Social Limits to Growth* (1977); Thronton, W T *Overpopulation and its remedy* (1971).
Broader Unsustainable development (#PB9419)
Unconstrained socio–economic growth (#PB9015).
Narrower Excessively large families (#PD7625).
Related Malthusianism (#PF4606).
Aggravates Agricide (#PE4045)
Illiteracy (#PC0210)
Malnutrition (#PB1498)
Unemployment (#PB0750)
Overcrowding (#PB0469)
Deforestation (#PC1366)
Human suffering (#PB5955)
Induced abortion (#PD0158)
Underdevelopment (#PB0206)
Denial of human rights (#PB3121)
Lack of community self–worth (#PF3512)
Inadequacy of postal services (#PF2717)
Obsolete deliberative systems (#PD0975)
Uncontrolled urban development (#PC0442)
Inadequate standards of living (#PF0344)
Unsustainable economic development (#PC0495)
Landlessness in developing countries (#PC0990)
Unemployment in developing countries (#PD0176)
Negative effects of family allowances (#PF0107)
Unsustainable agricultural development (#PC8419)
Ecologically unsustainable development (#PC0111)
Endangered species of plants and animals (#PB1395)
Delays in elaboration of remedial legislation (#PC1613)
Adjustment difficulties of new urban families (#PF1503)
Inappropriate education in developing countries (#PF1531)
Negative effects of over–crowding on mental health (#PF3850)
Increasing development lag against population growth (#PF3743)
Underdevelopment of industrial and economic activities (#PC0880)
Depletion of natural resources due to population growth (#PD4007)
Flooding of the urban labour market in developing countries (#PD0008)
Imbalance of population growth between developed and developing countries (#PE4241).
Aggravated by Nationalism (#PB0534)
Traditionalism (#PF2676)
Excessive longevity (#PJ5973)
Socio–economic poverty (#PB0388)
Lack of family planning (#PF0148)
Illegal induced abortion (#PD0159)
Decrease in mortality rate (#PF0333)
Ineffective population control (#PF1020)
Inadequacy of contraceptive methods (#PD0093)
Maldistribution of electrical energy (#PD3446)
Inadequate family planning education (#PD1039)
Inadequacy of male contraceptive methods (#PF1069)
Inadequate health care in family planning (#PD1038)
Religious opposition to population control (#PF1022)
Government opposition to population control (#PF1023)
High human fertility in developing countries (#PF0906)
Theological justification of population growth (#PF9671)
Inability of governments to regulate family size (#PF0401)
Opposition to population control and family planning (#PF1021).
Reduces Underpopulation (#PD5432).
Reduced by War (#PB0593)
Famine (#PB0315)
Human cannibalism (#PF2513)
Natural disasters (#PB1151)
Population decrease (#PF6441)
Declining birth rate (#PD2118)
Denial of the right to procreate (#PC6870)
Head hunting in tribal societies (#PF2666).

♦ PB0113 Imperialism

Dependence on imperialism
Nature The concept of imperialism basically designates the existence of relatively concentrated authority and rule and is diffused over broad territorial contours. In modern times, it has more specifically come to denote a type of political system through which one state has extended its rule over other states, mostly territorially noncontiguous ones, without entirely incorporating them into a framework of common political symbols and identity. It thus refers essentially to attempts to establish formal sovereignty over subordinate political societies, but is also often equated with the exercise of any form of political control or influence by one political community over another.
Incidence Imperialism has become part of a propaganda battle. In communist terminology, the word remains restricted to the policies of the West, in particular the United States; whereas Western authors have sought to identify communist policies with 'the new imperialism'. Writers in the developing countries have made the word interchangeable with 'neocolonialism'; others have extended the term to refer to the economic, political, and military policies of all industrialized states, including the Soviet Union, or of the white race as such, or even of any unsympathetic foreign state.
Claim 1. Imperialism continues to oppress many nations and presents a constant threat to peace and social progress. 2. Imperialism is the monopoly stage of capitalism, for it is the domination of the monopolies that constitutes its economic essence. Under imperialism the concentration of production and capital leads to the formation of monopolies, which take hold of the main resources of a society and take advantage of their supremacy to intensify their exploitation of the proletariat, enriching themselves at the expense of millions of common commodity producers and establishing their control over numerous small and middle entrepreneurs, thus reaping high monopoly profits.
Counter–claim The various political, financial, economic, technical and cultural activities of one state in another may aim only at the creation of sympathy, friendship, or influence. Also, the term 'imperialism' has lost its historical connotation and has become a theoretical concept, differently defined in the context of specific theoretical system. It has been blunted by over–frequent, emotional usage; but the resulting vagueness has not diminished its potency as one of the most powerful slogans of our time, used indiscriminately against any state, or even any group, regarded as inimical to a speaker's interest.
Refs Baumgart, Winfried and Mast, Ben V *Imperialism* (1982); Betts, Raymond F and Shafer, Boyd C *The False Dawn* (1975); Bukharin, Nikolai and Lenin, V I *Imperialism and World Economy*; Danmole, Masood *Heritage of Imperialism* (1974); Garnsey, P D and Whittaker, C R *Imperialism in the Ancient World* (1979); Kayyali, Abdul W (Ed) *Zionism, Imperialism and Racism* (1979); Lenin, V I *Imperialism*; Offiong, Daniel O *Imperialism and Dependency* (1982); Patnaik, Prabhat (Ed) *Lenin and Imperialism*; Warren, Bill *Imperialism* (1980).
Broader Conflict (#PA0298)
Domination (#PA0839)
Aggression (#PA0587).
Narrower Colonialism (#PC0798)
Foreign control (#PC3187)
Neo–colonialism (#PC1876)
Occupied nations (#PC1788)
Covert imperialism (#PF3199)
Economic imperialism (#PC3198)
Cultural imperialism (#PC3195)
Socialist colonialism (#PD4862)
Ecological imperialism (#PC5333)
Intra–state imperialism (#PC3197)
Lack of political independence (#PF0297)
Communist political imperialism (#PC3164)
Capitalist political imperialism (#PC3193)
Conflict of interests in imperialism (#PC6440)
Monolingualism in a multi-cultural setting (#PD2695)
Foreign intervention in internal affairs of states (#PC3185)
Discrimination against developing countries by the formation of regional groupings of developed countries (#PE1604).
Related Elitism (#PA1387)
Economic conflict (#PC0840)
Imbalance of power (#PB1969)
International insecurity (#PB0009).
Aggravates War (#PB0593)
Exploitation (#PB3200)
Miscegenation (#PC1523)
Non–alignment (#PF0801)
Guerrilla warfare (#PC1738)
Denial of right to national self–determination (#PC1450).
Aggravated by Destiny (#PF3111)
Freemasonry (#PF0695)
Nationalism (#PB0534)
Undue attachment to a social group (#PF1073)
Fraudulent nature of inherited titles (#PE5754).
Reduced by Failure of disarmament and arms control efforts (#PF0013).

♦ PB0147 Psychological alienation

Nature In Freudian psychology, alienation is the condition of civilized man. Life in a given society implies the acceptance of this society's rules and imposes the necessity to conform to certain roles and expectations. To do so, man represses and transforms vital instincts and impulses which would prove to be asocial. Thus, in becoming acceptable to others, he experiences self–estrangement.
Background Individuals may be at war with themselves, torn between different attitudes and values. This is particularly true where the value–system ingrained during the years of youth is later confronted by rapidly changing social conditions and demands. A person, in attempting to adjust to these, becomes alienated from his tradition, which gives rise to guilt, doubt and confusion and considerable inner suffering. Individuals may also lose their self–worth. In its intense form psychological alienation is pathological and sociopathic.
Broader Conflict (#PA0298)
Alienation (#PA3545).
Narrower Interpersonal estrangement (#PB0034).
Related Loneliness (#PF2386)
Social alienation (#PC2130).
Aggravates Prejudice (#PA2173)
Discrimination (#PA0833)
Psychotic violence (#PE7645)
Social fragmentation (#PF1324).
Aggravated by Apathy (#PA2360)
Impure thoughts (#PF5205)
Lack of communication (#PF0816)
Inadequate sense of personal identity (#PF1934).

♦ PB0205 Refugees

Asylum–seekers — Dependence on refugees
Nature A refugee, as defined by the 1951 UN Convention relating to the Status of Refugees, is any person who 'owing to well founded fear of being persecuted for reasons of race, religion, nationality, membership of a particular social group or political opinion is outside the country of his nationality and is unable or, owing to such fear, is unwilling to avail himself of the protection of that country; or who, not having a nationality and being outside the country of his former habitual

residence, is unable or owing to such fear and for reasons other than personal convenience, is unwilling to return to it'.

In addition to the problems encountered of settling in a host country and attempting to find acceptance by the community of that country, refugees encounter difficulties in finding employment and housing. They have to adapt themselves to new environments, lifestyles, ideas about values and customs, and often to a foreign language. They run the risk of losing their national culture, especially when the culture is oral, and are constantly confronted with a feeling of the inability to participate and belong, a feeling which often leads to cultural rootlessness. Their children have difficulty in getting admitted to schools and once admitted, experience vast problems in attempting to equate their previously obtained levels, degrees, and diplomas with the new standards. These children also have a higher incidence of nightmares, inability to concentrate in school, and of the various kinds of aggression, withdrawal, and mental breakdown observed in people who have seen horrors which cannot be eradicated from their minds.

For the receiving countries, waves of refugees may cause severe and abrupt changes in the environment. Population densities multiply overnight, traditional land systems are disrupted and markets are disorganized. The presence of refugees may create xenophobia among the permanent residents, stir up prejudice and lead to separate factions which disagree ideologically with one another.

To distinguish between refugees and immigrants is becoming increasingly difficult. People fleeing from war, civil disturbance and persecution are normally considered refugees while those seeking a better economic situation are not. What the difference is is frequently a matter of arbitrary choice.

Incidence It has been estimated that between 1900 and 1980 some 250 million people had fled their countries. The 1984 figures for unsettled refugees indicate that 10 million people per year are added to the list of refugees.

Background Refugees have existed since man began to raid and war. The Jews have been in a centuries old Diaspora; the Pilgrims fled to America in an attempt to flee religious persecution; and Protestant Europe in the 16th and 17th Centuries was plagued by refugees, most of whom belonged to the Protestant Church.

Claim While the idealistic notion of refugees is eventual return to their country of origin, the reality of today's refugee situation is quite grim. Refugees, the majority of whom are indefinitely settled in camps in the receiving country, are people with nothing. They have neither country nor home, culture nor possessions, rights nor much hope for improvement. Where and how they live is dictated to them and eventually even their dignity, self–esteem and individuality are damaged, often irreparably. They are victims and the road to recovery is a very long process.

Counter–claim Refugees are stereo–typed as hopeless victims, but it should be remembered that they usually add more to the host country than is perceived. They have contributed a large percentage of high achievers, writers, artists, and athletes. They have provided strong backs and a quiet willingness to do work shunned by nationals. They built America by muscle and brain and have opened up the myopic viewpoints of numerous countries by the addition of their culture, cuisine, art, music, and language.

Refs Bramwell, Anna C (Ed) *Refugees in the Age of Total War* (1988); Goodwin–Gill, Guy S *The Refugee in International Law* (1984); Hakovirta, H *Third World Conflicts and Refugeeism* (1986); ICIHI Staff *Refugees* (1986); Marrus, Michael R *The Unwanted* (1987); Miserez, D (Ed) *Refugees – The Trauma of Exile* (1989); Newland, Kathleen *Refugees* (1981); Rogge, John *Refugees* (1987); Scanlan, John A, et al *The Global Refugee Problem* (1983); Williams, Carolyn L and Westermeyer, Joseph *Refugee Mental Health in Resettlement Countries* (1986); Wittke, Carl F *Refugees of Revolution*.

Broader Man–made disasters (#PB2075).
Narrower Refugees by boat (#PD8034) Disabled refugees (#PC0768)
Economic refugees (#PD4379) Political refugees (#PD2549)
Displaced children (#PD5136) Non–settled refugees (#PC0519)
Environmental refugees (#PE3728) Socially handicapped refugees (#PD1507)
Exploitation of women refugees (#PD5025)
Physical insecurity of refugees and asylum–seekers (#PD6364).
Related Stowaways (#PJ1173) Displaced persons (#PD7822).
Aggravates Homelessness (#PB2150) Statelessness (#PE2485)
Rejection of refugees (#PF3021) Illegal movement across frontiers (#PC2367).
Aggravated by War (#PB0593) Civil war (#PC1869) Antisemitism (#PE2131)
Natural disasters (#PB1151) Regional tensions (#PC5917)
Deprivation of nationality (#PD3225) Refusal to grant nationality (#PF2657)
Expulsion of immigrants and aliens (#PC3207)
Denial of human rights in armed conflicts (#PC1454)
Non–equivalence of national educational qualifications (#PC1524).

♦ PB0206 Underdevelopment

Dependence on underdevelopment — Dependent development — Lack of development — Foreign dependence

Nature Underdevelopment occurs when some resources are not used to their full socio–economic potential, with the result that local or regional development is slower than it might be. As a system of self–reproducing hard–core poverty and stagnation, it is a complex system of mutually supporting internal and external factors that allow the less developed countries only a lop–sided development process. It hinges on the industrialized world's uneven economic conditions and the changes in the structure of the international division of labour since the Second World War; includes the division of the world into rich and poor countries as well as the disparities with in poor countries between their rich and poor inhabitants; and is convolutedly linked to the developing countries' deteriorating trade position.

The economic and social development of many developing countries is being held back by backward economies and social systems in which peasants and intermediate urban strata predominate. Almost all the developing countries suffer from large–scale hidden and partial unemployment exacerbated by an increase in population due to the decrease in child mortality. Their unequal trade situation stems from their dependence upon primary products (usually not more than three) for their export receipts. These commodities are often: in limited demand in the industrialized countries (for example: tea, coffee, sugar, cocoa, bananas); vulnerable to replacement by synthetic substitutes (jute, cotton, etc); or are experiencing shrinking demand with the evolution of new technologies that require smaller quantities of raw materials (as is the case with many metals). Prices cannot be raised as this simply hastens the use of replacement synthetics or alloys, nor can production be expanded as this rapidly depresses prices. Consequently, the primary commodities upon which most of the developing countries depend are subject to considerable short–term price fluctuation, rendering the foreign exchange receipts of the developing nations unstable and vulnerable. Development thus remains elusive.

Incidence Underdevelopment severely affects potential growth and stability for virtually all the world's developing countries (which include the majority of independent countries in Central and South America, Africa, the Middle East, and Asia; the main exceptions being South Africa, mainland China, Taiwan, and Israel). Most of these countries were colonies or semicolonies of the the imperialist powers or were dependent on them.

Claim As the industrialized countries continue to rely on their economic superiority and on their monopoly associations in the developing countries, they take advantage of the growing indebtedness of the underdeveloped nations (particularly as regards the scientific and technological advancements the underdeveloped countries are so eager to participate in) and try to keep them in a subordinate, underdeveloped position by strengthening the system of neo–colonialism.

Counter–claim The core problem is not the lack of development or stagnation, rather there has been too much development of an inappropriate kind.

Refs Bandyopadhyaya, J *Climate and World Order* (1983); Frank, Andre F *On Capitalist Underdevelopment* (1975); Kitching, G *Development and Underdevelopment in Historical Perspective* (1982); Stavenhagen, Rodolfo *Between Underdevelopment and Revolution* (1981).

Broader Incompleteness (#PA6652) Unpreparedness (#PA7341).
Narrower Traditionalism (#PF2676) Economic dependence (#PF0841)
Regional disparities (#PC2049) Lack of modernization (#PG5654)
Environmental poverty (#PD5261) Social underdevelopment (#PC0242)
Economic underdevelopment (#PC0281) Inadequate infrastructure (#PC7693)
Least developed countries (#PD8201) Lack of political development (#PD8673)
Urban slums in developing countries (#PD3489)
Economic and social underdevelopment (#PB0539)
Lack of participation in development (#PF3339)
Depressed regions in developed countries (#PD8183)
Inadequacy of aid to developing countries (#PF0392)
Administrative difficulties in new states (#PE1793)
Underdeveloped sources of income expansion (#PF1345)
Lack of economic and technical development (#PE8190)
Underdeveloped potential of basic resources (#PF3448)
Lack of appreciation of cultural differences (#PF2679)
Unattractive locale for economic development (#PF3499)
Inadequate weed control in developing countries (#PD3598)
Underprovision of basic services to rural areas (#PF2875)
Lack of means for local technological development (#PF6454)
Weakness of infrastructure in developing countries (#PC1228)
Restrictions imposed on aid to developing countries (#PF1492)
Structural rigidity in developing country economies (#PD2970)
Inadequate social discipline in developing countries (#PD0095)
Inequitable labour standards in developing countries (#PD0142)
Contradictions of capitalism in developing countries (#PF3126)
Inadequate political parties in developing countries (#PD0548)
Underdevelopment of industrial and economic activities (#PC0880)
Deterioration of terms of trade for developing countries (#PD2897)
Use of inappropriate technologies in developing countries (#PF0878)
Underutilization of facilities due to daily or seasonal peaks (#PF0827)
Decline in rural customs and traditions in developing countries (#PD1095)
Developed country limiting of trade between developing countries (#PD2961)
Ill–considered pressure to eliminate nakedness in developing countries (#PF3350)
Imbalance between urbanization and industrialization in developing countries (#PC1563).
Related Complacency (#PA1742) Desert nomadism (#PD2520)
Vulnerability of land–locked developing countries (#PD5788)
Insensitivity of transnational corporations to consumer needs in developing countries (#PE1011)
Inadequate relationship between transnational corporations and local industry in developing countries (#PE1511).
Aggravates Magic (#PF3311) Taboo (#PF3310) Cultural invasion (#PC2548)
Excessive public debt (#PC2546) Inadequate health services (#PD4790)
Malnutrition in developing countries (#PD8668)
Burdensome cost of religious ceremonies (#PF3313).
Aggravated by Apathy (#PA2360) Totemism (#PF3421)
Nomadism (#PF3700) Superstition (#PA0430)
Obscurantism (#PF1357) Caste system (#PC1968)
Lack of leadership (#PF1254) Economic imperialism (#PC3198)
Lack of social mobility (#PF2195) Inadequate land drainage (#PD2269)
Inflexible social structure (#PB1997) Ineffective population control (#PF1020)
Unsustainable population levels (#PB0035) Failure of development policies (#PF5658)
Reversal of development progress (#PF4718) Corruption in developing countries (#PD0348)
Deficiencies of developing countries (#PC4094) Underprivileged linguistic minorities (#PC3324)
Inadequate international judicial system (#PF2113)
Inadequate international map of the world (#PD0398)
Animal worship as a barrier to development (#PD2330)
Bilateralism in aid to developing countries (#PE9099)
Multiplicity of languages in a national setting (#PC1518)
Inappropriate education in developing countries (#PF1531)
Inadequate research and development capacity in developing countries (#PE4880)
Inadequate negotiation of entrance terms for transnational corporations in developing countries (#PE0853).
Reduces Destruction inherent in development (#PF4829).
Reduced by Communism (#PC0369) Socialism (#PC0115)
Capitalism (#PC0564).

♦ PB0262 Hunger

Dependence on hunger — Hungry people

Nature World hunger today is a paradox of scarcity amidst plenty. The reason so many millions in both the developing and the developed countries go to bed hungry is not due to a lack of supply but to a convoluted web of a myriad problems which stem from the common core of poverty. The world today produces enough food for each man, woman, and child to have a daily 3000 kilo–calories. 1983's world grain harvest totalled nearly 1.5 billion tons, and the European Community has stockpiled mountains of butter, beef and wheat. But it is the political and economic infrastructures which keep these foods from the hungry. Food aid is given and withheld according to political rather than physical needs.

Land is concentrated in the hands of the few. Four percent of the world's land owners control half of the world's cropland, while 58 percent of the world's landowners make do with 8 percent of the world's cropland. A far higher proportion of land owned by large land holders is underused than by small farmers. A vast number of rural populations have no land at all. The green revolution reinforced inequalities in food distribution by shifting food production from labour intensive agriculture to capital intensive agriculture requiring expensive inputs, such as, fertilizers, pesticides and irrigation, beyond the reach of the small farmer. It also tended to deplete non–renewable resources further endangering the long range viability of agriculture.

Since 1950, the growth rate of food production has been between 2.8 and 3.1 percent per year, both in developed and developing countries. However, there has been a steady decline in the growth rate in population in the developed countries and the populations of developing countries has grown almost at the rate as their food production.

The demand for edible calories in the form of animal proteins has increased further decreasing the availability of food among the low income cross–sections of populations in developing countries. Animal proteins require from twice (for poultry) to seven times (for red beef) cereal feed inputs as compared with the caloric yields of food cereals.

In both developed and developing countries, poverty is the greatest obstacle to good nutrition, and this can only be overcome by a broad process of equitable economic development. But this development process may take generations to complete, and the world's hungry cannot wait. Not only does hunger mean personal suffering and shortened, stunted lives for those personally afflicted by it, but it also means an enfeebled work force. It may be that the myopia of developed nations does not permit them to fully comprehend the impact of the toll of such a loss; but in developing nations, that toll is constantly visible. For many developing nations, hunger is an obstacle to the economic development that alone can free them from both hunger and poverty; and the non–realization of such economic freedom and stability creates danger. Throughout

history, food shortages and rising food prices have been a major cause of political instability, triggering riots and strikes, providing a unifying focus for opposition to established political systems, and ultimately causing governments to fall. It has also been a frequent source of international conflict, inciting wars of territorial expansion and compelling desperate people to migrate en masse across national borders.
Incidence Worldwide, 20 million people die each year from hunger or related diseases; at least 450 million go hungry every day. In the USA, 15% of American live below the poverty level; doctors in Boston (MA) report one out of every 10 poor children is physically stunted by malnutrition.
Claim Given the millions of dollars spent annually on reports and commissions to 'study the hunger problem', and the trillions of dollars spent on armaments and technology, it seems that the obvious solution to the world's precarious situation of peace lies not in those reports or technological breakthroughs, but very simply in feeding empty stomachs.
Refs Byron, William *The Causes of World Hunger* (1983); Collins, Joseph and Lappe, Frances M *World Hunger* (1986); Drèze, Jean and Sen, Amartya *Hunger and Public Action*; George, Susan *How the Other Half Dies* (1977); International Environment Liason Centre *International Environment–Development Facts March 1989* (1989); Le Magnen, Jacques *Hunger* (1986); United Nations *The Hunger Problematique and a Critique of Research* (1981); Woube, Mengistu *The Geography of Hunger* (1987).
Broader Starvation (#PB1875) Human physical suffering (#PB5646)
Vulnerability of human organism (#PB5647).
Narrower Maldistribution of food (#PC2801).
Aggravates Human cannibalism (#PF2513)
Malnutrition among indigenous peoples (#PC3319).
Aggravated by Famine (#PB0315) Plant diseases (#PC0555)
Poisonous algae (#PE2501) Food insecurity (#PB2846)
Racial discrimination (#PC0006) Vulnerability of food chains (#PB2253)
Shortage of fresh–water sources (#PC4815) Maldistribution of agricultural land (#PD9189)
Maldistribution of electrical energy (#PD3446)
Overexploitation of underground water resources (#PD4403)
Lack of purchasing power in developing countries (#PE8707)
Excessive social costs of structural adjustment in debtor developing countries (#PD8114).

♦ PB0284 Infringement of privacy

Dependence on infringement of privacy — Invasion of privacy — Denial of right to privacy — Arbitrary interference with privacy — Unlawful interference with privacy
Nature Infringement of privacy is a complex and difficult issue involving diverse meanings in different cultures. At the simplest level, unwanted intrusions on one's immediate situation of being alone or with family is a violation of privacy. The intrusion may be the physical presence of others or the noises one does not wish to hear. Degrees of privacy depend on the proximity of the intruder which is largely culturally defined. In some activities an important aspect of privacy is the freedom from observation, such as when praying, defecating or having sexual relations. In this vain privacy may be violated by being compelled to observe things which are offensive. While not every intrusion is a violation of privacy where the line is difficult to determine at any given moment because of shifting social expectations and the increasing interaction between people of different cultures.
A second dimension of infringements of privacy is the dispersion of private information about persons. Observations and physical intrusions may also be means of obtaining information that the individual want to keep secret. When information about a person is obtained against his will either by coercion or by force their right to privacy has been violated. When another person divulges information to a broader audience or it has been taken privacy has been violated. The right to privacy in some societies has been extended to include the freedom from inaccurate or misleading information being spread about an individual.
A third dimension of the infringement of privacy is lack of autonomy in making private decisions. While this is perhaps the most debatable dimension, it is, for example, acceptable by many legal and social systems for a married couple to choose whether to use contraceptives if it wishes. The right for a woman to choose to terminate a pregnancy is argued as a right to the autonomous making of private choices.
Different legal systems emphasize different aspects of the right to privacy; many claims to privacy are hard to distinguish from claims to respect for personal integrity, to personality, and to freedom from interference from government and other external agents. Litigations concerning violated rights of privacy may arise between celebrities or public figures and the media, particularly sensationalist tabloids. Under some countries' laws, public figures such as heads of state or royal families appear to have less rights to privacy than other people, as everything concerning them may be deemed legitimate news.
Claim Data available on individuals is divided into publically available data (such as the electoral register, birth registers, court proceedings and share registers) and data without public access: police files, tax files, medical files. Governments compile files on citizens' political activity, sexual deviations, legal offences, credit ratings, reputation and careers. No one controls the world of surveillance; not only do governments and the military have information, but large commercial computer banks have countless pieces of information on nearly all members of society. Technology provides the means for spying on people and many governments legalize its use.
Counter–claim In traditional societies, there is much less sense of individual separateness and so less concern for affording opportunities for development of individual attitudes and interests. Autonomy may be rejected as a matter of conscious choice by groups.
Many individuals do not know what is good for them and for society and therefore need to be guided by those who do perceive what is good. Plato suggests that the wise know what is best for society and they should determine what the rest of society should do. Marxists believe that desires of those living in bourgeois society are the product of false consciousness and that autonomy is unreal. Critics of modern society point out that belief in individuality is largely illusory because human beings are much more alike than different. The search for privacy is largely self–defeating sense too much solitude results in unhappiness. The social nature of people and the desirability of social planning need to be admitted.
Everything people do effects the other members of society and therefore, close regulation of human lives is justified to promote the general social good.
Refs Connors, S and Campbell, D *On the Record*.
Broader Collectivism (#PF2553) Arbitrariness (#PB5486).
Narrower Political surveillance (#PD8871) Defamation of character (#PD2569)
Unauthorized police search (#PD3544) Invasion of privacy by media (#PD9603)
Interception of communications (#PD7608) Uncontrolled use of computer data (#PF4176)
Denial of right to confidentiality (#PD6612) Invasion of privacy through testing (#PJ6946)
Denial of right to private home life (#PE6168)
Maintenance of political dossiers on individuals (#PD2929)
Unauthorized circulation of confidential information (#PG2189)
Invasion of privacy by compulsory telecommunications (#PE0223).
Related Harassment (#PC8558) Intrusive social science research (#PF0145)
Denial of political and civil rights (#PC0632) Unfair trials due to pre–trial publicity (#PE1692)
Misuse of electronic surveillance by governments (#PD2930).
Aggravates Anti–science (#PF2685) Discrimination (#PA0833)
Denial of right to liberty (#PF0705).
Aggravated by Gossip (#PE2192) Abuse of credit (#PF2166)
Uncontrolled media (#PD0040) Forced social intimacy (#PD4287)
Abuse of government power (#PC9104) Denial of right of complaint (#PD7609)
Irresponsible social science (#PF8032) Journalistic irresponsibility (#PD3071).
Reduced by Censorship (#PC0067) Erosion of journalistic immunity (#PD3035).

♦ PB0315 Famine

Dependence on famine — Massive starvation despite sufficient world food supply
Nature The problem of famine is manifold. 1. Once a famine has reached the proportions of a major disaster, it is too late to mount a fast and efficient relief operation. Supplies rushed to a country often get held up at the country's ports, unable to be distributed by the existing infrastructure. 2. Governmental organizations which issue relief aid are not set up to respond quickly or effectively, and volunteer agencies (which, not being bogged down in governmental bureaucracy, are quicker to respond) are neither designed nor equipped to cope with starving masses. 3. Information inadequacies exist. Although there are numerous statistics on crop failures or droughts, the study of more finely–tuned data such as the movements of local prices or mass migrations of people from their homes, is still in its infancy. 4. Attention must be paid to the governmental ideosyncracies, and to preferences of the donor countries; sometimes no action may be undertaken before an invitation is extended. 5. The governments of stricken countries may be unaware of or unconcerned with a rural famine; they may be unable to assemble, in enough time, the necessary technical case for aid; they may be unwilling to broadcast their problems to the world; and they may be hostile to western intervention.
Equally manifold are the problems confronting possible remedies. 1. Strategic stockpiles are needed in those countries usually most unlikely to be able to afford to keep such reserves, and they may well be prey to thievery or spoilage. 2. Too much developmental aid poured into an area can be impossible to absorb and/or can make farmers dependent upon such aid rather than their own resources. 3. Aid programs are often tied to restrictions and priorities that are either impossible or fruitless to achieve. 4. Too many helpers often cause confusion with their conflicting 'solutions' or their jealous rivalries.
Incidence The African continent in the mid 1980's suffered from famines on a scale never before witnessed. As of April 1985, 10 million people had abandoned their normal homes in search of food and water; 20 countries had been critically affected by drought; and 35 million lives were in danger.
Claim Famine is not a condition of lack of food but of inadequate planning, inadequate notification, slow response, government pride, misdirected aid, uncoordinated relief agency field work, politics, lethargic bureaucracy, ignorance, and incompetence. It is a grave problem which shakes the entire political, economic, and social foundations on which the stable and prosperous future for developing countries was to have been built.
Counter–claim Famine relief efforts are little more than a panacea until the next, and perhaps more profound, famine occurs; and indeed, may be a futile attempt to deep alive those people which nature is culling off in an attempt to remedy a surplus situation.
Refs Collins, Joseph and Lappe, Frances M *World Hunger* (1986); Lawrence, Peter (Ed) *World Recession and the Food Crisis in Africa* (1986).
Broader Starvation (#PB1875).
Narrower Man–made famine (#PD0571).
Aggravates Hunger (#PB0262) Beggars (#PD2500) Cholera (#PE0560)
Prostitution (#PD0693) Food hoarding (#PJ2225) Civil disorders (#PC2551)
Human cannibalism (#PF2513) Uncontrolled migration (#PD2229)
Protein-energy malnutrition in vulnerable groups (#PD0363).
Aggravated by War (#PB0593) Floods (#PD0452) Drought (#PC2430)
Tsunamis (#PD0033) Revolution (#PA5901) Hurricanes (#PD1590)
Insect pests (#PC1630) Climatic cold (#PD1404) Global cooling (#PF1744)
Plant diseases (#PC0555) Damage to crops (#PJ3949) Plant pathogens (#PD1866)
Natural disasters (#PB1151) Man–made disasters (#PB2075)
Distortionary tax systems (#PD3436) Long–term shortage of wheat (#PE2903)
Long–term shortage of food and live animals (#PE0976).
Reduces Decrease in mortality rate (#PF0333) Unsustainable population levels (#PB0035).

♦ PB0318 Military and economic hegemony

Dependence on military and economic hegemony
Nature Hegemony denotes dominance by one or more states over a group of states, or sometimes, when simply economic or military, within a regional area. Economic or ideological hegemony may be reinforced by the supply of weapons to the dominated nations by the dominating nation. Arms can be provided to enable local forces to perform military tasks which are in the interests of the supplying country. The arms may serve to strengthen the relationship between the supplying country and the recipient government (possibly through pre–emptive supply which forestalls any effort of another supplying country to enter into the same kind of relationship). The supply itself may provide an opportunity for influencing individuals in the recipient countries, especially those of the military establishment in countries where the military plays an important role in politics.
Incidence The use of arms trade as a means to secure the allegiance of a recipient regime has been used extensively by the USA, USSR and, to a lesser extent, France and the UK. The European countries tend to supply military aid to ex–colonies or to countries where there is a sizeable resident minority of the donor's nationals. Professional contacts between military personnel of supplying and recipient countries are therefore a useful way of maintaining sales channels. Through training, foreign servicemen become indoctrinated with the donor's defence thinking and weapons systems, and this encourages further purchase of such systems from the dominating country.
Claim Monopoly over military technology is beginning to play a cardinal role in a similar monopoly over new technology of immense importance in the economic life of nations and in the relations between powers in general. This leads to the emergence of a duopoly of the two superpowers in regard to modern technology, giving them a more and more dominating hegemony over world affairs. Only they can wield power over – and virtually only they have access to – new provinces of our planet (for example, sea–bed, outer–space, underground) which are being opened up due to the highly advanced technologies that originate in research and development for military purposes.
Counter–claim The hegemony of one or more states can be voluntarily delegated to them or acquiesced in, by the national member states in their league. In this sense US Leadership is democratic; the USSR's is not. Moreover, West European and Japanese initiatives do not allow for a 'duopoly' nor for world–conquest by any one power.
Refs Arrighi, G *A Crisis of Hegemony*.
Broader Aggression (#PA0587).
Narrower Preponderance of Western–style organizations (#PF8356)
Discrimination against developing countries by the formation of regional groupings of developed countries (#PE1604).
Related Domination of the world by territorially organized sovereign states (#PD0055).
Aggravates Neo–colonialism (#PC1876).
Aggravated by Competition between states (#PC0114)
Economic and social underdevelopment (#PB0539).
Reduced by Fragmentation (#PA6233).

♦ PB0388 Socio–economic poverty

Dependence on poverty — Poor people — Pauperism — Extreme poverty — Deterioration in

standard of living of the poor
Nature People are poor if they do not attain the minimum standard of living consistent with human dignity. The level of living below which a given family may be considered poor, the 'poverty line', is determined by both biological and socio–economic factors. For example, with development the diets of the population at large tends to change in such a way as to include less grain and more meat. It becomes a sign of poverty not to have meat, and since the real cost of meat is generally much higher than that of grain (in calorie terms, only part of the grain eaten by cattle is transformed into meat), the poverty line tends to rise with development.
Furthermore, with development, minimum consumption of other goods and services is set, legally or in the market, at increasingly higher standards. Minimum housing requirements tend to become higher and more stringent, the higher the general level of living. Consumer goods become of better quality and are packaged to include conveniences and services which were previously unobtainable. Since production is generally directed to classes and groups that possess the bulk of the purchasing power in the community, the poorer parts of the population become increasingly unable to obtain the lower quality and more elementary goods at reasonable prices. The traditional purchases of the poor (and their purchasing habits) may become expensive. In a modern urban community, a family prepared to buy a live chicken is not likely to find one and will instead pay for a chicken packaged via all the above–mentioned services. Similarly, cars (with bigger engines and push–button conveniences as standard features), buses and trains, replace the more traditional and lower cost means of transport such as the horse and carriage and the mule. Here again the general tendency of the poverty line is to rise with development.
An upward push to the poverty line arises from social and psychological factors which are referred to, perhaps not too adequately, as the 'demonstration effect'. Minimum socially acceptable standards of clothing rise generally with the rise of the average standard. Radios and television sets come to be regarded as essential items as more and more people purchase them. The poverty line in a society thus depends on the average level of consumption. In a closed community or nation, therefore, the problem of poverty would be solved if the poor were brought up to or closer to the level of living of the average. But this is not necessarily so in an open community or nation today. In most countries, many of the upward pressures on the poverty line – particularly those deriving from the demonstration effect – may be exercised from outside national boundaries. Standards of health, education and housing in rich nations affect expectations and demands in poor nations. Certain goods (refrigerators, for example) are not locally produced but imported from more developed countries; laws and regulations setting high standards are sometimes borrowed and put into effect with little actual change in conditions. Innovations and improvements in international transport and communications and technical assistance across borders have increasingly given the demonstration effect international significance. As a result, a country where levels of living are stationary may nevertheless experience a rising poverty line and may eventually find practically its entire population living under the line. The speed with which the poverty line rises in this instance relates to the degree to which a society is open to economic, social and cultural influence from the more advanced countries and to the difference between its level of living and the level of living of those countries.
Incidence It is estimated that two–thirds of the world population lives in a state of poverty. About 40 percent of the population of the developing countries lives in absolute poverty, according to the World Bank. The number of poor is forecasted at between 75 and 80 per cent of the world population by the year 2000, which means 5000 million persons, or more than the total world population in 1980.
Claim Poverty means death to many people. Those that cannot afford medical care die. The aged succumb to malnutrition and neglect. Expectant mothers receive inadequate pre–natal education and medical attention and infant mortality rates climb. Poverty means disease from unhygienic living conditions and the absence of adequate public sanitation and public health services. Poverty maims and cripples the young; poverty poisons family relationships; poverty corrodes human values. Poverty and hunger and their corollary, social injustice, are the major obstacles to the world's cultural and economic development, and to the universal improvement of the quality of life.
The definition of poverty is never relative while people suffer and perish from hunger, malnutrition, preventable diseases, and medical neglect. Yet to address the problem of poverty by inflating the figures still further by including the relativistic phenomena of the ever–increasing 'disadvantaged' is to blunt the thrust of the remedies. The disadvantaged are not poor in the absolute sense. In some countries of the North, they are those who do not have a second home in the country and can go only as far as the Atlantic or Mediterranean beach for their vacations. They are also those who can only afford a black–and–white television set and one automobile. They are people who collect unemployment benefits and yet do not know what real poverty is. The disadvantaged of the developed countries are a problem for those countries. The poor of the developing countries are a problem for the United Nations and have been on its agenda only to be crowded out by ideological cold wars, international discord and the arms race.
Counter–claim The traditional concept of poverty is limited and restricted, since it refers exclusively to the predicaments of people who may be classified below a certain income threshold. This concept is strictly economistic. The concern should be not with poverty but with poverties. In fact, any fundamental human need that is not adequately satisfied, reveals a human poverty. Some examples are: poverty of subsistence (due to insufficient income, food, shelter, etc), of protection (due to bad health systems, violence, arms, race, etc), of affection (due to authoritarianism, oppression, exploitative relations with the natural environment, etc), of understanding (due to poor quality of education), of participation (due to marginalization and discrimination of women, children and minorities), of identity (due to imposition of alien values upon local and regional cultures, forced migration, political exile, etc). But poverties are not only poverties. Each poverty generates pathologies.
Refs Amin, G A *Modernization of Poverty* (1980); Ayres, Robert L *Banking on the Poor* (1983); Beckerman, Wilfred, et al *Poverty and the Impact of Income Maintenance Programmes in Four Developed Countries* (1979); George, Vic *Wealth, Poverty and Starvation* (1988); Gordon, David M *Theories of Poverty and Underemployment* (1973); Haque, Serajul and Mohammed, Nur *Poverty, Inequality and Income Distribution* (1977); Mollat, Michel and Goldhammer, Arthur *The Poor in the Middle Ages* (1986); Pollitt, Ernesto *Poverty and Malnutrition in Latin America* (1980); Rodgers, Gerry *Poverty and Population. Approaches and Evidence* (1984); Scott, Hilda *Working Your Way to the Bottom* (1985); Srinivasan, T N and Bardhan, Pranab K *Rural Poverty in South Asia* (1988); Steidlmeier, Paul *Paradox of Poverty* (1987); United Nations *Apartheid, Poverty and Malnutrition* (1982).
Broader Poverty (#PA6434) Insufficiency (#PA5473).
Narrower Beggars (#PD2500) Unemployment (#PB0750) Rural poverty (#PC4992)
Urban poverty (#PC5052) Family poverty (#PC0999) Underemployment (#PB1860)
Economic hardship (#PD9180) Children in poverty (#PD4966) Economic insecurity (#PC2020)
Poverty and disability (#PD0723) Poverty in developed countries (#PC0444)
Inadequate standards of living (#PF0344) Poverty in developing countries (#PC0149)
Prohibitive cost of accommodation (#PD1842) Economically disadvantaged students (#PD2624)
Rural poverty in developing countries (#PD4125)
Destruction of rural subsistence economy (#PC2237)
Excessive burden on the poor due to legal delays (#PE1093)
Loss of traditional forms of social security in developing countries (#PD1543)
Decline of handicrafts and cottage industries in developing countries (#PD1250).
Related Temporal deprivation (#PF4644).

Aggravates Banditry (#PD2609) Socialism (#PC0115)
Immorality (#PA3369) Illiteracy (#PC0210)
Neo–fascism (#PF2636) Urban slums (#PD3139)
Youth gangs (#PD2682) Slave trade (#PC0130)
Homelessness (#PB2150) Malnutrition (#PB1498)
Prostitution (#PD0693) Grave robbing (#PF0491)
Human suffering (#PB5955) Civil disorders (#PC2551)
Abuse of credit (#PF2166) Social breakdown (#PB2496)
Psychotic violence (#PE7645) Public drunkenness (#PE2429)
Rural depopulation (#PC0056) Cruelty to children (#PC0838)
High mortality rate (#PJ6252) Illegal immigration (#PD1928)
Children's diseases (#PD0622) Socioeconomic stress (#PC6759)
Disadvantaged groups (#PB6320) Trafficking in women (#PC3298)
Environmental poverty (#PD5261) Gambling and wagering (#PF2137)
Lack of racial identity (#PF0684) Unsustainable development (#PB9419)
Inadequate diet of the poor (#PJ3529) Inadequate personal hygiene (#PD2459)
Dependence on social welfare (#PD1229) Mental disorders of the aged (#PD0919)
Sudden unexpected infant death (#PD1885) Unsustainable population levels (#PB0035)
Socially unsustainable development (#PC0381) Inadequate recreational facilities (#PF0202)
Trafficking in children for adoption (#PF3302) Abusive traffic in immigrant workers (#PD2722)
Criminally life endangering behaviour (#PD0437)
Excessive employment of married women (#PD3557)
Malnutrition among indigenous peoples (#PC3319)
Unequal regional distribution of deaths (#PC4312)
Inadequate welfare services for the blind (#PD0542)
Inadequate disaster prevention and mitigation (#PF3566)
Unequal morbidity and mortality between countries (#PC6869)
Restriction of indigenous populations to reservations (#PD3305).
Aggravated by Slavery (#PC0146) Debt slavery (#PD3301)
White slave trade (#PD3303) Underpayment for work (#PD8916)
Racial discrimination (#PC0006) Geographical isolation (#PF9023)
Intellectual terrorism (#PD6656) Inequality of opportunity (#PC3435)
Deteriorating quality of life (#PF7142) Ineffective population control (#PF1020)
International economic recession (#PF1172) Discrimination in social services (#PC3433)
Illiteracy among indigenous peoples (#PD3321) Unproductive subsistence agriculture (#PC0492)
Unequal distribution of social services (#PC3437)
Burdensome cost of religious ceremonies (#PF3313)
Exploitation of indigenous populations in employment (#PD1092).

♦ PB0469 Overcrowding

Excessively high population density
Nature There are some indications from animal experiments that population density in itself may affect mental health. Thus, under conditions of extreme overcrowding, an increase in male aggressiveness and an accompanying decline in the adequacy of maternal behaviour have been observed among rats. When the young were methodically removed from colonies of rats kept at high densities, it was noted in addition that bands of young males assaulted females and that there was an increase in homosexual behaviour. However, mice brought up under overcrowded conditions showed less stress behaviour than those transferred only later to such conditions. Some epidemiological studies have shown higher rates of schizophrenia, crime, suicide, alcoholism, and drug abuse in the central, more crowded areas of old established cities than in other areas of such cities, but correlations of this kind are by no means clear–cut.
Counter–claim While overcrowding may be a factor in some specific psychological and social problems it has never been proven as a sole cause with a human population. Animal populations of the densities used in these experiments are artificially created and maintained and therefore are of dubious value when extrapolated to human or even other animal situations.
Broader Social neglect (#PB0883).
Narrower Urban overcrowding (#PC3813)
Overcrowding of housing and accommodation (#PD0758).
Aggravates Cholera (#PE0560) Hepatitis (#PE0517) Common cold (#PE2412)
Infant mortality (#PC1287) Stress in human beings (#PC1648)
Lack of legal aid facilities (#PF8869).
Aggravated by Inadequate standards of living (#PF0344)
Uncontrolled urban development (#PC0442)
Unsustainable population levels (#PB0035)
Environmental stress on inhabitants of tall buildings (#PE4953).

♦ PB0477 Human ageing

Ageing — Old age — Senescence
Nature The tissues, as age advances, become more rigid and less elastic, the bones more brittle, the ligaments stiffer; deposition of fat in internal organs (for example, the heart) weakens their activity; the skin becomes thin so that cold is more acutely felt; the walls of the blood vessels become at first thicker, then more brittle, so that haemorrhage, more readily occurs. There is poorer blood supply to the brain, hence mental feebleness. Teeth and eyes weaken. Because people do not age at the same rate, a given individual may appear, physically or mentally, either younger or older than his true chronological age.
Incidence Within the next 45 years the total number of elderly in the world is expected to reach over 1.1 billion, with 75 percent of them residing in developing countries. By the year 2025, Asia will account for approximately 57 percent of the world's elderly, Africa and Latin America for about 20 percent, and the developed regions of Europe, North America and the USSR for approximately 25 percent.
Refs Baker, Scott R and Rogul, Marvin (Eds) *Environmental Toxicity and the Aging Process* (1987); Birren, J E and Danon, D *Aging*; Salk, Darrell, et al *The Werner's Syndrome and Human Aging* (1985); Terry, Robert D (Eds) *Aging and the Brain* (1988); United Nations *Report of the World Assembly on Aging* (1982).
Broader Human physical suffering (#PB5646) Human disease and disability (#PB1044).
Narrower Senility (#PE6402) Bereavement (#PF3516) Ageing women (#PE6784)
Ageing populations (#PD8561) Ageing war disabled (#PD0874)
Compulsory retirement (#PJ2411) Human sexual inadequacy (#PC1892)
Loss of capacity with age (#PC8310) Social withdrawal of aged (#PD3518)
Ageing of world population (#PC0027) Mental disorders of the aged (#PD0919)
Social disadvantage of the aged (#PD3517) Age discrimination in employment (#PD2318)
Rigidity and inadaptability in the aged (#PD3515)
Slowness of sensori-motor activities in the aged (#PD3514)
Susceptibility of the old to physical ill–health (#PD1043)
Retirement as a threat to psychological well–being (#PF1269).
Related Diabetes (#PE0102).
Aggravates Ulcers (#PE2308) Obesity (#PE1177) Suicide (#PC0417)
Ugliness (#PA7240) Deafness (#PD0659) Pneumonia (#PE2293)
Dysentery (#PE2259) Thrombosis (#PE5783) Loneliness (#PF2386)
Human death (#PA0072) Fear of death (#PF0462) Facial wrinkles (#PG9630)
Grey human hair (#PE6308) Sleep disorders (#PE2197) Teeth disorders (#PD1185)
Chronic bronchitis (#PE2248) Cirrhosis of the liver (#PE2446)
Dependency of the elderly (#PD8399) Inadequate income in old age (#PC1966)
Heat as an occupational hazard (#PE5720) Inadequate housing for the aged (#PD0276)
Personal physical unattractiveness (#PF4010) Religious theology disrelated to life (#PF5694)
Decline in cognitive ability with ageing (#PE9620)

(#PD0512)
Unsanitary environment for basic health in small rural villages (#PD2011).
Aggravated by Influenza (#PE0447) Cult of youth (#PF6766)
Alcohol abuse (#PD0153).

♦ PB0479 Unbalanced growth

Dependence on unbalanced growth

Nature Economic growth tends to proceed by surges and relapses, by over-investment booms and under-consumption crises. There can be unevenness both in the rate of expansion of the capacity to produce and in the degree of utilization of the productive capacity.

Unbalanced growth cannot be sustained long due to the emergence of bottlenecks. If, for example, the agricultural sector seriously lags behind the industrial sector, development may be interrupted through inflationary pressure or balance of payments disequilibria. Agriculture, on the other hand, unless it is totally export-oriented, cannot grow if the industrial sector is not able to develop sufficiently to create a steady increase in the demand for agricultural products. The heavy industry sector cannot grow irrespective of the wage-goods sector, for this may result in a general deficiency of demand.

Counter-claim Strictly speaking, only an economy which has an infinitely elastic supply schedule for every commodity and which is perfectly adjustable to changes in the demand pattern can aspire to balanced growth. It seems that disproportions are an essential dynamic element in the development process. These disproportions may take the form of a rapid advance in agricultural productivity which releases manpower and creates favourable conditions for a simultaneous or subsequent industrial upsurge. Within the industrial sector, the growth of heavy industry has a greater dynamic effect than the development of light industry. Hence a disproportion in favour of the former may be natural in certain stages of growth.

Broader Economic uncertainty (#PF5817).
Narrower Instability of economic and industrial production activities (#PC1217).
Aggravates Increasing development lag against socio-economic growth (#PC5879).
Aggravated by Economic inflation (#PC0254).

♦ PB0514 Social inequality

Dependence on social inequality

Nature Inequality appears to have always been an inherent feature in society and is no exclusive characteristic of a particular social pattern or period of history. The main forms of social inequality are those that result from disparities of wealth and income, those related to differential prestige or honour, and those derived from the distribution of power. These dimensions of inequality are related but not reducible to each other.

Social inequality can be based on natural differences of kind between people (sex, race, character traits, natural endowments) and on institutional variations (citizenship, religion, social position, etc); or on natural differences of rank (properties that are common to all but in varying amounts) in intelligence, age, strength, power, etc. It is the unequal treatment that results from these unequal characteristics that constitutes social inequality. It has a distributive aspect which refers to the ways in which different factors such as income, wealth, education, occupation, power, skill, etc, are distributed in the population and a relational aspect which refers to the ways in which individuals differentiated by these factors are related to each other within a system of groups and categories.

Incidence Throughout society, social inequality is still an extensive source of conflict. Although both types of industrial society - socialist and capitalist - have provided opportunities for social equality, and inequality has thus largely decreased during the last decades, it remains a fact that even in these affluent societies, people are still unequally placed. Nevertheless these inequalities seem small when compared to the far more obvious forms of social inequality in the Third World.

Refs Curtis *Inequality in American Communities* (1977); Dobkowski, Michael and Willimann, Isidor (Eds) *Research in Inequality and Social Conflict* (1988); Hamilton, M B and Hirszowicz, Maria *Class and Inequality* (1987); Leahy, Robert L *The Child's Construction of Social Inequality* (1983); Moscovitch, Allen and Drover, Glenn (Ed) *Inequality* (1981).

Broader Human inequality (#PA0844).
Narrower Feudalism (#PF2136) Segregation (#PC0031) Class conflict (#PC1573)
Minority control (#PF2375) Age discrimination (#PC2541) Racial exploitation (#PC3334)
Class consciousness (#PC3458) Social fragmentation (#PF1324)
Regional disparities (#PC2049) Social discrimination (#PC1864)
Ethnic discrimination (#PC3686) Unproductive dependents (#PC1420)
Lack of social mobility (#PF2195) Segregation in education (#PD3441)
Inequality before the law (#PC1268) Violation of human rights (#PB3860)
Discrimination against women (#PC0308) Social impediments to marriage (#PF3341)
Rural-urban income differential (#PE5022) Economic impediments to marriage (#PF3342)
Underprivileged linguistic minorities (#PC3324)
Excessive employment of married women (#PD3557)
Discrepancies in human life evaluation (#PF1191)
Separate and unequal development within different societies (#PE8416).
Related Elitism (#PA1387) Unmarried mothers (#PD0902) Political injustice (#PC2181)
Cultural deprivation (#PC1351) Denial of human rights (#PB3121)
Inadequacy of social doctrine (#PF3398).
Aggravates Slavery (#PC0146) Prejudice (#PA2173) Caste system (#PC1968)
Family poverty (#PC0999) Social conflict (#PC0137) Ethnic conflict (#PC3685)
Social injustice (#PC0797) Social revolution (#PC3236)
Possession*complex (#PA6686) Vulnerability of middle-class (#PC1002)
Poverty in capitalist systems (#PC3107) Conflict between minority groups (#PC3428)
Discrimination in social services (#PC3433) Racial discrimination in politics (#PD3329)
Prohibitive cost of farm machinery (#PF2457) Exploitation in capitalist systems (#PC3117)
Racial discrimination in education (#PD3328) Unequal coverage by social security (#PF0852)
Underprivileged religious minorities (#PC2129) Inequitable administration of justice (#PD0986)
Discrimination against women in public life (#PD3335)
Lack of individualism in capitalist systems (#PD3106)
Discrimination against men in social services (#PD3336)
Abuse of science and technology in capitalism (#PE3105)
Waste of human resources in capitalist systems (#PC3113)
Conflicting roles of commodities in capitalism (#PF3115)
Conflicting roles of money in capitalist systems (#PF3114)
Excessive demand for goods in capitalist systems (#PC3116)
Counterproductive capitalist investment financing (#PF3104)
Incompatibility of traditional and new technologies (#PE3337).
Aggravated by Destiny (#PF3111) Eugenics (#PC2153)
Tribalism (#PC1910) Illiteracy (#PC0210)
Exploitation (#PB3200) Sex segregation (#PC3383)
Ethnic segregation (#PC3315) Racial segregation (#PC3688)
Lack of human unity (#PF2434) Lack of assimilation (#PF2132)
Political inequality (#PC3425) Racial discrimination (#PC0006)
Plural society tensions (#PF2448) Exploitation in employment (#PC3297)
Denial of the right of association (#PD3224) Undemocratic political organization (#PC1015)
Structural rigidities in labour markets (#PD4011)
Racial discrimination in public services (#PD3326)
Confusion induced by rapid social change (#PF6712)
Inadequate education of indigenous peoples (#PC3322)
Inadequate cultural integration of immigrants (#PC1532).
Reduced by Restrictive legislation (#PD9012).

♦ PB0534 Nationalism

Dependence on nationalism — National chauvinism — Arab nationalism

Nature Nationalism is a political state of mind in which the supreme loyalty of the individual is felt to be due to the nation-state. The nation-state is regarded not only as the ideal, 'natural', or 'normal' form of political organization but also as the indispensable framework for all social, cultural, and economic activities. The main tenets of nationalism are: love of a common soil, race, language, or historical culture; desire for the political independence, security, and prestige of the nation; mystical devotion to a vague, sometimes even supernatural, social organism which, known as a nation or Volk, is more than the sum of its parts; the living of an individual exclusively for the nation, with the corollary that the nation is an end in itself; belief that the nation is or should be dominant among other nations and should take aggressive action to this end. Characteristics of nationalism are ideas of national superiority and national exclusiveness, which are developed to a greater or lesser degree, depending on the historical situation and the relations of a particular nation with other nations.

In addition to ethnic nationalism, there is statism where nationalist loyalties are focused on the state. Statism demands that all subjects obey and serve the state, as distinct from the nation or people, as the highest object of their allegiance. They are expected to pay taxes, serve in the armed forces, and whole-heartedly back the state's goals at home and abroad. These two forms of nationalism are frequently intermingled.

Background It was only at the end of the 18th century that, for the first time, civilization was considered to be determined by nationality and the principle was put forward that a man can be educated only in his own mother tongue, not in languages of other civilizations and other times. From the end of the 18th century on, the nationalization of education and public life went hand in hand with the nationalization of states and political loyalties. In many cases poets and scholars emphasized cultural nationalism first, preparing the foundations for the political claims for national statehood to be raised by people in whom they has kindled the spirit of nationalism.

National feeling was evident in certain groups at certain periods, especially periods of stress and conflict, before the 18th century. Its rise was prepared by a number of complex events: the creation of large, centralized states by the absolute monarchs, who destroyed the feudal allegiances and thus made possible the integration of all loyalties in one centre; the secularization of life and education which fostered the development of the vernacular languages and weakened the ties of religious or sectarian loyalties; the growing economic interdependence which demanded larger territorial units, which at the same time gave the necessary scope to the dynamic spirit of the rising middle classes and their capitalistic enterprise. Under the influence of the new theories of the sovereignty of the people and the rights of man, the people replaced the king as the centre of the nation. Nation and state became identified.

Nationalism is characterized by an extremely positive valuation of one's own nation, and is similar in form to that of a prejudice (that is, the development of positive reactions to symbols representative of one's own group and the rejection of alien groups). Some research suggests that nationalism forms the basis of further prejudices, such as anti-semitism and racism.

Claim Nationalism, and its extreme form, chauvinism, places the interests, of the nation-state above the interests of all other nation-states, promotes national arrogance, and exacerbates national hatreds and animosities. Nationalism is a quasi-religion that calls for a belief in the ultimate good of the nation and the ultimate sacrifice for its members.

Counter-claim Despite all the supra- or anti-nationalist trends, nationalism is still the most potent force for bringing about unity of action in the world. In the contemporary period, nationalism has a different character in countries fighting for political and economic independence against imperialism. Nationalism in these countries expresses in a limited way the idea of national liberation and national independence and frequently serves as a standard for the national liberation movement.

Nationalism acts as an accelerating factor because the policies of the newer or less developed nation-states have often been guided by the desire to catch up with the older and more highly developed nation-states. It can also act as a force preserving older forms of societal life and stressing the diversity within a world community that is based on the acceptance of the nation-state as the basic form of political and cultural organization.

Developing first in Western Europe with the consolidation of nation states, nationalism brought about the reorganization of Europe in the 19th and 20th centuries (unification of Germany and Italy; break-up of the Hapsburg and Ottoman empires); and has been the prime force in the political awakening of Asia and Africa. Nationalism has been a powerful source of inspiration in many of the arts and in the development of historical and language studies. This has proved to be as true of the 'new nations' of Africa and Asia as it was of 19th-century Europe.

Refs Burnell, Peter *Economic Nationalism in the Third World* (1985); Cameron, David M *Regionalism and Supranationalism* (1981); Fanon, Frantz *The Wretched of the Earth* (1966); Fayerweather, John *Host National Attitudes Toward Multinational Corporations* (1982); Hula, Erich and Thompson, Kenneth W *Nationalism and Internationalism* (1984); Marcu, E D *Nationalism in the Sixteenth Century* (1975); Ramadhan, S *Islam and Nationalism*; Sathyamurthy, T V *Nationalism in the Contemporary World* (1983); Stack, John F *The Primordial Challenge* (1986).

Broader Ethnocentricity (#PB5765).
Narrower Zionism (#PF0200) Anti-science (#PF2685) National communism (#PF3130)
National boundaries (#PF8235) Forced assimilation (#PC3293)
Parochial national interests (#PF2600) Nationalistic attitudes to currency (#PF6094)
Race as a reinforcement of nationalism (#PF3352)
Religion as a reinforcement of nationalism (#PF3351).
Related Racism (#PB1047) Disaccord (#PA5532) Intolerance (#PF0860)
Narrowmindedness (#PA7306) Ideological conflict (#PF3388)
Domination of the world by territorially organized sovereign states (#PD0055).
Aggravates War (#PB0593) Tribalism (#PC1910) Ethnocide (#PC1328)
Militarism (#PC2169) Imperialism (#PB0113) Neo-fascism (#PF2636)
Antisemitism (#PE2131) Hero worship (#PF2650) World anarchy (#PF2071)
Totalitarianism (#PF2190) Political repression (#PC1919) Racial discrimination (#PC0006)
Political dictatorship (#PC0845) Islamic fundamentalism (#PF6015)
Religious discrimination (#PC1455) Compulsory indoctrination (#PD3097)
Restrictions on emigration (#PC3208) Competition between states (#PC0114)
Unsustainable population levels (#PB0035) Radio and television censorship (#PD3029)
Newspaper and journal propaganda (#PD0184) Religious and political antagonism (#PC0030)
Threatened and vulnerable minorities (#PC3295) Protectionism in international trade (#PC5842)
Underprivileged ideological minorities (#PC3325) Inadequacy of international legislation (#PF0228)
Disruption of development by tribal warfare (#PD2191)
Destabilizing international telecommunications (#PD0187).
Aggravated by Revolution (#PA5901) Segregation (#PC0031)
National prejudice (#PG4368) Cultural prejudice (#PC8520)
National isolationism (#PF2141) National disintegration (#PB3384)
War and pre-war propaganda (#PD3092) Double standards in morality (#PF5225)
Biased government information (#PF0157)
Concentration of national governments activity on national affairs (#PE5132).
Reduces Social fragmentation (#PF1324) Political fragmentation (#PF3216)
Loss of institutional credibility (#PF1963)
Monolingualism in a multi-cultural setting (#PD2695).
Reduced by Foreign control (#PC3187) Occupied nations (#PC1788)
Foreign dictatorship (#PC3186).

♦ PB0539 Economic and social underdevelopment

Dependence on economic and social underdevelopment

Broader Underdevelopment (#PB0206).

Narrower Underemployment (#PB1860)
Inadequate savings (#PC0927)
Rural underdevelopment (#PC0306)
Social underdevelopment (#PC0242)
Economic underdevelopment (#PC0281)
Shortage of skilled labour (#PD0044)
Defaults on international loans (#PD3053)
Nakedness in developing countries (#PF8241)
Inappropriate employment incentives (#PD0024)
Weakness in trade among developing countries (#PC0933)
Inappropriate design of development projects (#PF4944)
Unequal income distribution between countries (#PC2815)
Inequitable tax systems in developing countries (#PD3046)
Domestic market restrictions in developing countries (#PD1873)
Dependence of developing countries on customs revenue (#PD2955)
Unrelated pioneer institutions in developing countries (#PF1724)
Restrictions on foreign access to capital bond markets (#PD3135)
Inadequate marketing of products of developing countries (#PE0523)
Contempt for agricultural labour in developing countries (#PD1965)
Outflow of financial resources from developing countries (#PC3134)
Dependence of developing countries on imported technology (#PF1489)
Disparity in social development within developing countries (#PD0266)
Restricted growth in export markets of developing countries (#PF1471)
Locational maladjustments of industry in developing countries (#PD1494)
Excessive dependence on export credits by developing countries (#PE0938)
Lack of response to monetary incentives in developing countries (#PD1432)
Burden of servicing foreign public debt by developing countries (#PD3051)
Reduction of the share of the developing countries in world exports (#PC2566)
Lack of processing industry for primary commodities in developing countries (#PD1554)
Processing in developed countries of commodities exported by developing countries (#PD0425)
Restrictions on industrial and economic development due to environmental policies (#PE4905)
International indebtedness arising from insurance transactions in developing countries (#PE0740)
Inappropriate transplantation of industrialized country methods to developing countries (#PE1337)
Inadequate adjustment assistance to industries and labour affected by developing country exports (#PE2844).

Related Underproductivity (#PF1107)
Capital shortage in developing countries (#PD3137)
Minimal export promotion by transnational corporations in developing countries (#PE1598).

Aggravates Unemployment (#PB0750)
Military and economic hegemony (#PB0318)
Social insecurity in developing countries (#PE4796).

Aggravated by Limited access to social benefits (#PF1303).

♦ PB0593 War

Dependence on war — Warfare — Armed conflict — Military conflict

Nature War generally denotes armed conflict. It can also mean sustained conflict, such as a cold war, in which the force of arms plays a highly significant part although armed forces may not come into direct confrontation. Methods of warfare include nuclear, chemical and biological, enhanced conventional as well as less sophisticated means of land, sea and air warfare, economic warfare and guerrilla warfare. Types of war include civil, international, nationalist, racial, religious and ideological. A steadily growing number of conflicts erupt within countries owing to economic, social, ethnic or religious differences and cause much damage. The repercussions of these conflicts have exacerbated the difficulties of the victims, who frequently are left unprotected since existing international legal instruments are not applicable to their situation.

Background The four Geneva Conventions define armed conflicts in three different ways: (a) An international armed conflict between two or more States which are party to the Geneva Conventions of 1949 and to Additional Protocol I of 1977, whether or not the conflict is a formally declared war and even if the state of war is not recognized by one of the parties; (b) Situations in which people are fighting against colonial domination, alien occupation and racist regimes, in the exercise of their right of self-determination as one of the fundamental principles of international law by the Charter of the United Nations; (c) A non-international conflict within the territorial limits of a State, when obvious hostilities break out between its armed forces and other organized armed groups, in situations where dissident forces are in conflict with the armed forces of the State, if the former are under responsible command and exercise such control over part of the territory as to enable them to carry out sustained and concerted operations.

The history of war shows evidence of a steady growth in the role of the economic factor in war. Until the 19th Century, wars had a comparatively narrow economic base and were waged by rather small professional armies. Since the second half of the 19th century, and particularly during the 20th century, wars have strained the economy of the belligerents and involved millions of people. More than 70 million people participated in World War 1 (1914–18) and more than 110 million in World War II (1939–45).

Incidence It has been estimated that in the last 5,500 years there have been 14,513 wars in which approximately 2,640 million people were killed. In the last global conflict, the Second World War, an estimated 30 million civilians and 30 million military personnel were killed. Since 1946, approximately 137 wars have been fought, with not a single day in which there was not a state of war somewhere. Over 85 nations have been involved, that is about one-half of all presently existing nations. Over 17 million civilians have been killed, namely 50 percent of the World War II total. In 1987 there were 25 wars going on that had taken some 3 million lives. Most of these conflicts have taken place in the poorer countries of Africa, Asia and Latin America, causing many casualties, especially among civilians. A 1983 study of world military and social expenditures found that in the first 348 years since the end of the Second World War, over 8 million civilians were estimated to have died as a result of 103 major armed conflicts worldwide.

Claim At an historical rate of about 3 armed conflicts a year, there will be over 40 wars of one size or another before 1999. Over 20 nations will be involved. Because nuclear weapons will be used, civilian casualties cannot be estimated, but may equal the 2,640 million killed in the last 5,500 years.

Counter-claim 1. Military action remains the ultimate sanction of the rule of law. In the absence of any impartial agency to uphold justice, nations may go to war in an effort to do so.

2. War is justified when certain conditions are met. St Augustine suggested that the reluctant and limited use of force may be one of the ways a Christian might be required in charity to serve the needs of an innocent neighbour under attack by an assailant. Thomas Aquinas listed right authority, just cause and right intention as the conditions of a just war. He also argued that force should be the last resort, should be proportionate to the evil remedied, should expect to succeed in its ends, and should contribute to a new state of peace.

3. Although 70 million people were killed during World War I and II, since then the major combatants have among them lost fewer than 200,000 in battle. They lose more in road accidents every three years.

4. A prince should have no other aim or thought nor take up any other thing for his study but war and its organization and discipline, for that is the only art necessary to one who commands and it is of such virtue that it not only maintains those who are born princes but often enables men of private fortune to attain to that rank. The chief cause of the loss of states is the contempt of this art, and the way to acquire them is to be well versed in the same.

5. War is a form of survival of the most fit at the national level. It eliminates, for the most part, those least needed and most undesirable in a nation. The moral fibre of the individuals who survive is tested and are better for it. It cripples or destroys nations least capable of functioning effectively. In the long run and on a global scale war benefits the nations of the world and the race.

6. War provides a machinery through which the motivational forces governing human behaviour are translated into binding social allegiance. No other institution has ever as successfully ensured social cohesion.

Refs Bartone, John C *War* (1984); Burton, John *Deviance, Terrorism and War* (1979); Butler, Smedley D; Crozier, Frank P and Thomson, Christopher B *Three Generals on War* (1973); California University Committee on *Problems of War and Peace in the Society of Nations* (1937); Dunnigan, James F and Bay, Austin *A Quick and Dirty Guide to War* (1985); Eagleton, Clyde (Ed) *Analysis of the Problem of War* (1972); Falk, Peter A *Law, Morality, and War in the Contemporary World* (1984); Glossop, Ronald J *Confronting War* (1987); Nettleship, Martin A; Givens, R Dale and Nettleship, Anderson (Eds) *War, Its Causes and Correlates* (1975); Novicow, Jacques and Cooper, S E *War and Its Alleged Benefits* (1972); O'Brien, William V *The Conduct of a Just and Limited War* (1981); Rice, Edward E *Wars of the Third Kind* (1988); Rollins, Leighton and Corrigan, Daniel *Disasters of War* (1981); Sturzo, Luigi and Gooch, G P *The International Community and the Right of War* (1971); Suter, Keith *Alternative to War*.

Broader Conflict (#PA0298).

Narrower Civil war (#PC1869)
Racial war (#PD8718)
Nuclear war (#PC0842)
Gang warfare (#PD4843)
Space warfare (#PD6439)
Undeclared war (#PG5529)
Ideological war (#PC3431)
Prisoners of war (#PC8848)
Unconventional war (#PC8836)
Biochemical warfare (#PC1164)
Conventional warfare (#PC4311)
Psychotronic warfare (#PD7986)
Psychological warfare (#PC2175)
Internal armed conflicts (#PD7661)
Mined international waterways (#PG2262)
Disastrous consequences of war (#PC4257)
Military expeditions against friendly powers (#PD7261)
Military conflict between communist countries (#PE5671)
War-time disruption of economies and production facilities (#PD8851).

Related Disaccord (#PA5532)
Intimidation (#PB1992)
Religious war (#PC2371)
Military secrecy (#PC1144)
Man-made disasters (#PB2075)
Psychological intimidation (#PC2935)
Violation of the integrity of creation (#PF5148).

Aggravates Panic (#PF2633)
Famine (#PB0315)
Disease (#PA6799)
Enemies (#PF8404)
Treason (#PD2615)
War debt (#PD3057)
Refugees (#PB0205)
Atrocities (#PD6945)
Human death (#PA0072)
Martial law (#PD2637)
Conscription (#PF6051)
Hero worship (#PF2650)
Profiteering (#PC2618)
War psychosis (#PE7867)
Combat trauma (#PE7912)
Art vandalism (#PE5171)
Human suffering (#PB5955)
Enjoyment of war (#PF4034)
Displaced persons (#PD7822)
Export credit risks (#PF3065)
Excessive public debt (#PC2546)
Crimes against humanity (#PC1073)
Human physical suffering (#PB5646)
Inadequate army discipline (#PD2543)
War damage in civilian areas (#PD8719)
Human disease and disability (#PB1044)
Social neglect of war veterans (#PD2077)
Wrecks and derelicts as hazards (#PE5340)
Imbalance in the human sex ratio (#PF1128)
Vulnerability of telephone system (#PE8254)
Civil crimes committed during war (#PG1112)
Inadequate emergency blood supply (#PE0366)
Socially unsustainable development (#PC0381)
Destruction of environmental oxygen (#PE5196)
Illegal international arms shipments (#PD4858)
Neglect of dependents of war victims (#PD2092)
Endangered monuments and historic sites (#PD0253)
Children engendered by occupying soldiers (#PD8825)
Denial of human rights in armed conflicts (#PC1454)
Inadequate welfare services for the blind (#PD0542)
Obstacles to international cultural exchange (#PF4857)
Destabilizing international telecommunications (#PD0187)
Destruction of cultural property during warfare (#PD7298)
Vulnerability of children during armed conflict (#PE8174)
Vulnerability of women and children in emergencies (#PD1078)
Disruption of food supply due to military activities (#PE8979)
Inadequate protection of civilians in armed conflict (#PE8361)
Restrictions on freedom of movement between countries (#PC0935)
Deterioration of the physical condition of art objects (#PD1955)
Vulnerability of the elderly under states of emergency (#PD0096)
Vulnerability of the disabled during states of emergency (#PD0098)
Domination of government policy-making by short-term considerations (#PF0317)
Obstructions to international personnel exchanges and cultural cooperation (#PE4785).

Aggravated by Racism (#PB1047)
Boredom (#PA7365)
Genocide (#PC1056)
Ignorance (#PA5568)
Militarism (#PC2169)
Colonialism (#PC0798)
Nationalism (#PB0534)
Imperialism (#PB0113)
Dangerous toys (#PE1158)
Economic rivalry (#PD8897)
Social injustice (#PC0797)
Imbalance of power (#PB1969)
Political repression (#PC1919)
Cognitive dissonance (#PF6638)
Political instability (#PC2677)
Religious intolerance (#PC1808)
International insecurity (#PB0009)
Intergovernmental disputes (#PJ5405)
War and pre-war propaganda (#PD3092)
Military-industrial complex (#PC1952)
Polarization of local conflicts (#PF1333)
Competitive acquisition of arms (#PC1258)
Shortage of fresh-water sources (#PC4815)
Private international arms dealers (#PD2107)
Fraudulent nature of inherited titles (#PE5754)
Inadequacy of international legislation (#PF0228)
Failure of disarmament and arms control efforts (#PF0013)
Economic and financial instability of the world economy (#PC8073)
Persistence of a technical state of war following cease-fire agreements (#PE2324).

Reduces Suicide (#PC0417)
Decrease in mortality rate (#PF0333)
Unsustainable population levels (#PB0035).

♦ PB0731 Injurious accidents

Risk of accident — Dependence on accidents

Nature Accidents cause tremendous suffering and countless disabilities, are a major cause of death at all ages, entail vast cost to society, and happen in all countries of the world.

Incidence Figures from many countries demonstrate that 20 per cent of hospitalizations on average are connected with accident injuries. In the USA – as in most industrialized countries – accidents are the main cause of death occurring during the first half of probable life expectancy, killing 150,000 people each year and injuring about 75 million. Of the injured, about 90,000 suffer head-injury in a car accident and require hospital treatment. Many are left with a permanent mental disability. Many others become juvenile disabled with some form of motor handicap – paraplegia or quadriplegia. Apart from the purely human aspects, the cost to society of road accidents in the USA in 1980 came to 2.6 per cent of the gross national product (GNP). In 50 developed and developing countries surveyed, motor vehicle accidents represented 40 per cent of all accidental deaths. A WHO study shows that deaths from motor vehicle accidents are increasing in the 15–24 age group. Between 1955–59 and 1970–74 they increased by over 600 per cent in Mexico, 250 per cent in Venezuela and 210 per cent in Chile, a much greater increase than in the developed countries. On the other hand, deaths from other kinds of accident are rising quite slowly, or even standing still. In industrialized and "newly industrialized" countries, accidents are now the main cause of death among women aged up to 34 years and among men aged up to 44. Women again become a high-risk group after the age of 65. Young women mostly suffer in traffic accidents, and men in work accidents. Older women are prey to domestic accidents, usually falls resulting in fracture of the thigh. Home accidents account for about 75 per cent of injuries to people over 65 in the industrialized countries. So far accidents account for only a small percentage of total deaths in the Third World, but as other causes of death and disability – such as malnutrition and communicable diseases – are gradually overcome, accidents will come to the fore. Many experts believe that for each death there are several hundred non-fatal accidents, and maybe ten or so

PB0731

permanent disabilities. Many seemingly trivial accidents, such as falls, turn our in fact to be more serious, particularly in the case of old people. In New Zealand, 28,000 people are injured each year in various types of fall which cost the insurance companies about $12 million.

Refs Hausen, Björn M *Woods Injurious to Human Health* (1981); ILO *The Cost of Occupational Accidents and Diseases* (1986).

Broader Risk (#PF7580).

Narrower Drowning (#PG2857)
Fractures (#PE7511)
Concussion (#PG3188)
Contusions (#PG5449)
Air accidents (#PD1582)
Oil pollution (#PE1839)
Mine disasters (#PD2278)
Marine accidents (#PD8982)
Accidental falls (#PE7113)
School accidents (#PE1990)
Firearm accidents (#PE2857)
Railway accidents (#PD0126)
Sprains of joints (#PG9214)
Superficial injury (#PG6807)
Radiation accidents (#PD1949)
Transport accidents (#PC8478)
Industrial accidents (#PC0646)
Pedestrian accidents (#PD0994)
Road traffic accidents (#PD0079)
Mechanical suffocation (#PG2860)
Electrical power failure (#PE1341)
Dislocation of the bones (#PG2523)
Accidents caused by fires (#PD4652)
Electric current accidents (#PG2862)
Accidents from falling objects (#PG2861)
Accidents involving astronauts (#PE6756)
Poisoning by solids and liquids (#PG2856)
Occupational domestic accidents (#PE4961)
Suffocation from ingested objects (#PG2855)
Accidents to agricultural workers (#PD5265)
Accidents to nuclear weapons systems (#PD3493)
Accidental large-scale contamination of the environment (#PD1386).

Related Unreported accidents (#PF2887).

Aggravates Lumbago (#PE1310)
Injuries (#PB0855)
Homicide (#PD2341)
Ugliness (#PA7240)
Pneumonia (#PE2293)
Disasters (#PB3561)
Ambivalence (#PF1426)
Teeth disorders (#PD1185)
Damage to goods (#PE4447)
Property damage (#PD5859)
Burns and scalds (#PE0394)
Human disability (#PC0699)
Underproductivity (#PF1107)
Physical blindness (#PD0568)
Rheumatic diseases (#PE0873)
Unavailability of first aid (#PJ1261)
Genetic defects and diseases (#PD2389)
Prohibitive medical expenses (#PE8261)
Inadequate cargo transportation (#PE0430)
Inadequate emergency blood supply (#PE0366)
Disorders of joints and ligaments (#PD2283)
Vulnerability of telephone system (#PE8254)
Inadequate prevention of disabilities (#PF0709)
Inadequate welfare services for the blind (#PD0542)
Inadequate assistance to victims of accidents (#PE4086).

Aggravated by Illiteracy (#PC0210)
Uncertainty (#PA6438)
Drunkenness (#PE8311)
Climatic cold (#PD1404)
Mental illness (#PC0300)
Poisonous gases (#PJ1939)
Evasion of work (#PC5576)
Sleep deprivation (#PE2741)
Traffic congestion (#PD0078)
Failure of materials (#PD2638)
Medical complications (#PE2863)
Environmental pollution (#PB1166)
Inadequate medical care (#PF4832)
Substance abuse at work (#PD9805)
Problems of migrant labour (#PC0180)
Transport of dangerous goods (#PD0971)
Unequal distribution of forces (#PJ7928)
Vulnerability of computer systems (#PE8542)
Threat of birds to aircraft safety (#PD1111)
Inadequate safeguards against fire (#PD1631)
Hazards of strong toxic substances (#PD0122)
Health hazards of exposure to noise (#PC0268)
Environmental hazards from chemicals (#PC1192)
Inadequate emergency medical services (#PD1428)
Proliferation of automobiles and motor vehicles (#PD2072).

♦ PB0750 Unemployment

Joblessness — Dependence on unemployment — Unemployed people — Excess labour — High unemployment rate

Nature In 1982 the International Conference of Labour Statisticians defined the unemployed as persons above a specified age who are without work, while at the same time are available for and are seeking work. Thus the unemployed include persons temporarily laid-off, unemployed students, homemakers and others mainly involved in non-economic activities. Other definitions are also used by governments and this definition may be interpreted very tightly. In the UK, for example, unemployment is defined as those looking for work within the past four weeks, available to start work within two weeks, and not having worked for any period whatsoever within the past week. Such definitions may exclude large numbers of people who have not actively sought work over the previous four months, for example.

Cyclical or deficiency of demand unemployment is one aspect of the problem, indicating that the total demand for goods and services is not sufficient to generate jobs for all those who want to work. Structural unemployment implies that there are particular sectors from which workers cannot quickly and easily move into other sectors in search of jobs. Inadequate mobility of workers ensures persistently high unemployment rates within certain sectors. This may arise because of technological change, because of a shift of economic activity out of a geographical region (leading to depressed areas), or because of a too rapid influx of workers of a particular type into a particular region. Immobility of workers tends to involve the least skilled and least educated and those at the extremes of age distribution; hence economic discrimination. In the developing countries the employment situation may also be characterized by chronic underemployment. The problem as a whole, however, is one of the results of economic development not keeping pace with the rapid rise of population.

Unemployment causes a sudden and drastic change in an individual's and a family's social habits. Those who are deprived of employment opportunities suffer a rapid deterioration in their skills and become increasingly unemployable as the duration of unemployment increases. The unemployed may find constructive ways of dealing with their frustration, such as collective work forms or self-employment. Many, however, turn to destructive ways such as participation in the black economy, organized crime or individual petty crime. Group action is unlikely, as the relative social isolation of the unemployed tends to increase individualistic concerns. Suicide rates have increased in areas with high unemployment, spotlighting the ill effects of joblessness on health.

Although unemployment has always existed in industrial civilization to a greater or lesser degree, everything seems to indicate that a new type of unemployment has now emerged as a structural component of the world economic system. A person suffering from extended unemployment goes through an emotional roller-coaster experience of at least four phases: shock, optimism, pessimism, fatalism. The last phase represents the transition from from frustration to stagnation, and from there to a final state of apathy, where the person reaches the lowest level of self-esteem. Extended unemployment totally upsets a person's fundamental needs system. Due to subsistence problems the person will feel increasingly unprotected, crisis in the family and guilt feelings may destroy affections, lack of participation will give way to feelings of isolation and marginalization, and declining self-esteem may very well generate an identity crisis. Extended unemployment generates collective pathologies of frustration.

Incidence At the beginning of 1984, the 24 member countries of the OECD recorded an unemployment rate of 9 per cent of the labour force, or 33 million people. Persons under 25 years of age are the greatest victims. In mid-1983, for example, youth unemployment reached nearly 20 per cent in the USA, over 20 per cent in Canada, France and the UK, about 30 per cent in Italy and about 40 per cent in Spain.

Background The personal insecurity and widespread hardship that resulted from market determination of the level of employment began to influence government policies of industrialized countries in the late nineteenth century. Intense concern with the welfare implications of widespread unemployment did not come until the great depression of the 1930s. From this followed economic planning and the almost universally espoused goal of full employment. Goals of development have come to be more socially oriented than in the past because of the recognition that the more limited approach of earlier periods has not successfully solved the problem of unemployment.

Claim Unemployment and under-employment, linked with an increasing disparity between the earnings of rich and poor, has certainly become the most disruptive social force, particularly in the developing nations. In many countries, employment creation and job distribution takes a subordinate rank to production and productivity as objectives of economic and social development.

Absence of jobs, grossly under-paid jobs, inequitable land distribution, and lack of self-employment skills and opportunities are already taking their toll in urban and rural areas amongst at least one-third of humanity. Unemployment and under-employment, with the resulting poverty, subsistence and below subsistence level living, affect everyone, including the rich, whose economic positions are threatened by the magnitude of income inequities. Population and employment projections indicate the possibility of several thousand million persons living in want in the near future.

Counter-claim Unemployment statistics measure only one dimension of the extent of utilization of labour, namely time used. The notion of clearly-defined, time-bound measurement of employment does not fit into a socio-economic structure in which traditional agriculture and the extended family are preponderant. The attempts to impose such a notion have resulted in the accumulation of a whole body of irrelevant data on unemployment in the rural economies of developing countries. An assessment of the phenomena of disguised unemployment in each of its settings is lacking. Proper and uninflated data may help towards the understanding that the pursuit of full employment tends to have inflationary effects on prices, which in turn negatively affects balance of payments. Social and economic development may in fact better proceed with a labour reserve whose needs are provided by governmental allocations and services.

Refs Council of Europe *The Psychological and Social Consequences of Unemployment* (1987); Crow, Iain, et al *Unemployment, Crime and Offenders* (1989); Gutkind, Peter C W *Bibliography on Unemployment* (1977); ILO *Year Book of Labour Statistics, 1988* (1988); ILO *Labour Force, Employment, Unemployment and Underemployment* (1982); ILO *World Employment Review* (1988); ILO *Employment, Growth and Basic Needs. A One-world Problem* (1978); Kelvin, Peter and Jarrett, Joanna *Unemployment* (1985); Newland, Kathleen *Global Employment and Economic Justice* (1979); OECD *Unemployment Compensation and Related Employment Policy Measures* (1979); OECD Staff *Job Losses in Major Industries* (1983); Schervish, Paul G *The Structural Determinants of Unemployment* (1983); WHO *Health Policy Implications of Unemployment* (1985).

Broader Socio-economic poverty (#PB0388).

Narrower Underemployment (#PB1860)
Rural unemployment (#PF2949)
Youth unemployment (#PC2035)
Educated unemployed (#PD8550)
Female unemployment (#PC5916)
Urban underemployment (#PC3490)
Seasonal unemployment (#PC1108)
Voluntary unemployment (#PC6720)
Structural unemployment (#PG7921)
Unproductive dependents (#PC1420)
Unemployed skilled labour (#PE1753)
Part-time farm employment (#PJ1074)
Imposed career interruptions (#PF4128)
Unemployment of older people (#PE5951)
Unemployment of married women (#PG9418)
Elimination of jobs by automation (#PD0528)
Unemployment in the wood industry (#PE5362)
Insufficient part-time employment (#PJ0108)
Non-productive members of society (#PF4000)
Underemployment of skilled workers (#PJ5489)
Short duration of athletic careers (#PG4186)
Unemployment in developed countries (#PC9718)
Inappropriate employment incentives (#PD0024)
Mass unemployment of human resources (#PD2046)
Unemployment in developing countries (#PD0176)
Graduate and post-graduate unemployment (#PD1162)
Unemployment as a perpetuator of failure (#PE8457)
Labour surpluses in developing countries (#PD0156)
Limited opportunities for significant work (#PF1403)
Artificial and arbitrary job qualifications (#PF2066)
Disparities in unemployment within countries (#PD1837)
Socio-economically inactive rural population (#PF4470)
Illegal private profit in socialist countries (#PC0939)
Disguised unemployment in socialist countries (#PC0940)
Unemployment caused by environmental conservation (#PD0467)
Antagonism between employment policy and technical advance (#PE5104)
Flooding of the urban labour market in developing countries (#PD0008)
Unemployment of premature school leavers in developing countries (#PE0015)
Inequality in distribution of natural resources between countries (#PF3043)
Decline of handicrafts and cottage industries in developing countries (#PD1250).

Related Ghost employees (#PE3993).

Aggravates Suicide (#PC0417)
Beggars (#PD2500)
Slavery (#PC0146)
Socialism (#PC0115)
Slave trade (#PC0130)
Imprisonment (#PD5142)
Urban poverty (#PC5052)
Human suffering (#PB5955)
Chattel slavery (#PC3300)
Social breakdown (#PB2496)
Underproductivity (#PF1107)
Illegal immigration (#PD1928)
Economic stagnation (#PC0002)
National economic recession (#PD9436)
Dependence on social welfare (#PD1229)
Obsolete deliberative systems (#PD0975)
Inadequate standards of living (#PF0344)
Denial of right to economic security (#PD0808)
Abusive traffic in immigrant workers (#PD2722)
Crimes committed during high unemployment (#PJ1139)
Denial of the right of trade union association (#PD0683)
Excessive claims for human development through sports (#PG4881)
Inequality of employment opportunity in developing countries (#PD1847)
Social environmental degradation from recreation and tourism (#PD0826)
Domination of government policy-making by short-term considerations (#PF0317)
Imbalances between people's aspirations and the structure of opportunities and income available (#PF9008).

Aggravated by Over-education (#PC6262)
Factory closures (#PE3537)
Natural disasters (#PB1151)
Limited job market (#PC7997)
Business bankruptcy (#PD2591)
Economic uncertainty (#PF5817)
Imbalance of payments (#PC0998)
Export of unemployment (#PD7466)
Segregation in employment (#PD3443)
Excess production capacity (#PD0779)
Limited employment options (#PF1658)
Cyclic business recessions (#PF1277)
Unjust dismissal of workers (#PD5965)
Inadequate manpower planning (#PJ2036)
Discrimination in employment (#PC0244)
Unsustainable population levels (#PB0035)
International economic recession (#PF1172)
Insufficient financial resources (#PB4653)
Exploitation in capitalist systems (#PC3117)
Economic and social underdevelopment (#PB0539)
Contradictions in capitalist systems (#PF3118)
Available jobs unrelated to education (#PJ9233)
Unfair distribution of land ownership (#PG2912)
Resource wastage in capitalist systems (#PC3108)
Structural rigidities in labour markets (#PD4011)
Financial destabilization of world trade (#PC7873)
Inadequate industrial retraining programmes (#PF4013)
General unproductivity of capitalist systems (#PF3103)
Limited means of marketing employable skills (#PE7344)
International imbalance in the quality of life (#PB4993)
Negative economic repercussions of disarmament (#PF0589)
Functional changes due to technological advance (#PE7943)

Decline in capital investment in productive capacity (#PE9265)
Contempt for agricultural labour in developing countries (#PD1965)
Denial of the right to social security in capitalist systems (#PD3120)
Imbalance between training and existence of openings in various professions (#PE5194).
Reduces Absenteeism (#PD1634) Economic inflation (#PC0254).
Reduced by Clandestine employment (#PC7607).

♦ PB0843 Inefficiency

Dependence on inefficiency — Inefficient people
Broader Impotence (#PA6876) Unskillfulness (#PA7232)
General obstacles to problem alleviation (#PF0631).
Narrower Educational wastage (#PC1716) Wasted waiting time (#PF1761)
Economic inefficiency (#PF7556) Unapplied scientific knowledge (#PF1468)
Uncontrolled industrialization (#PB1845) Inflexible management patterns (#PF3091)
Delay in administration of justice (#PF1487) Inadequate safeguards against fire (#PD1631)
Inappropriate education of graduates (#PF1905)
Delays in delivery of books and publications (#PF1538)
Delays in elaboration of remedial legislation (#PC1613)
Cost overruns in large-scale public programmes (#PD1644)
Exploitation of land for the burial of the dead (#PE9095)
Inadequacy of governmental decision-making machinery (#PF2420)
Ineffectiveness of international organizations and programmes (#PF1074)
Inefficiency due to mismatch of religious or national holidays (#PE4647)
Inadequate budgetary coordination within the United Nations systems (#PE2820)
Inadequate relationship between international governmental and nongovernmental organizations and programmes (#PE1973).
Related Behavioural deterioration (#PB6321).
Aggravates Inadequate personal hygiene (#PD2459).
Aggravated by Aircraft noise (#PE5799) Over-centralization (#PF2711)
Differing conceptions of time (#PF6665) Alcoholic intoxication at work (#PE2033)
Lack of professional standards (#PF3411) Nepotism in developing countries (#PD1672)
Health hazards of exposure to noise (#PC0268) Excessive size of social institutions (#PF8798)
Multiplicity of languages in a national setting (#PC1518)
Monopolistic activity by transnational enterprises (#PE0109)
Denial of right to union activity for special groups (#PE1355).

♦ PB0848 Competition

Dependence on competition — Competitiveness — Rivalry — Blocked action due to rivalries
Nature Competition results from the struggle for existence and for livelihood. The action of trying to gain what others are trying to gain at the same time may be either unconscious or conscious. As it becomes self-conscious it tends to pass over into social conflict.
Claim Hostility is a frequent result of competition in the workplace, the classroom, the home, the playing field; any place where one person's success depends on another's failure. This is what competition means: mutually exclusive goal attainment. Instead of labouring together toward a common end, people are obliged to work against one another. Since competition is a kind of aggression, it is not surprising that it often leads to physical violence.
Counter-claim Competition enters all major areas of man's life and generally connotes rivalry between two or more individuals or groups for a given prize. It is often an end in itself, as in the case of sporting events. In economic life competition is not a goal: it is a means of organizing economic activity to achieve a goal. The economic role of competition is to discipline the various participants in economic life to provide their goods and services skillfully and reasonably.
Refs Brenner, Reuven *Rivalry* (1990); Europa Institute, Leiden (Ed) *European Competition Policy* (1973); Semmler, W (Ed) *Competition, Instability, and Nonlinear Cycles* (1986).
Broader Negative emotions and attitudes (#PA7090).
Narrower Military rivalry (#PD9252) Economic rivalry (#PD8897)
Religious rivalry (#PC3355) Nuclear arms race (#PD5076)
Political rivalry (#PD8992) Athletic competition (#PE4266)
Ideological conflict (#PF3388) Interpersonal rivalry (#PD7617)
Intranational competition (#PE0580) Competition between states (#PC0114)
Misuse of international forums (#PF2216) Competition for scarce resources (#PC4412)
Ineffective worker organizations (#PF1262) Competition in capitalist systems (#PC3125)
Proliferation of strategic nuclear arms (#PD0014)
Competition between international organizations for scarce resources (#PC1463).
Related Envy (#PA7253) Disaccord (#PA5532) Opposition (#PA6979)
Unrecognized socio-economic interdependencies (#PF2969).
Aggravates Cheating (#PJ7991) Monopolization of knowledge (#PF5329)
Declining economic productivity (#PC8908).
Reduced by Mediocrity (#PF3900).

♦ PB0855 Injuries

Injured people — Wounded people — Wounds
Nature Wounds are mechanical injuries to organic tissues with disruption of the continuity of such structures as skin and mucous membranes. Wounds are characterized by three basic local symptoms: (1) separation of the edges of the wound, which varies with the extent, depth and location of the wound; (2) pain; and (3) bleeding. The last two result from injury to nerves and blood vessels. In addition to causing anatomical and functional disturbances to tissues and organs, some wounds are dangerous because they may lead to acute anaemia or shock as a result of heavy bleeding, or to wound infections, including such anaerobic infections as gangrene and tetanus.
Refs Durron, Daskin Rice *Injuries and Wounds I* (1985); Mendelson, George and Slovenko, Ralph *Psychiatric Aspects of Personal Injury Claims* (1988); Vaughn, Trudy W *Leg Injuries* (1987).
Broader Human disease and disability (#PB1044).
Narrower Gunshot wounds (#PE9111) Animal injuries (#PC2753)
Cuts and abrasions (#PG7877) Traumatic injuries (#PE6874)
Long-term injuries from sports (#PE5686).
Related Hazards to human health (#PB4885).
Aggravates Tetanus (#PE2530) Bursitis (#PE2320)
Peritonitis (#PE2663) Ageing war disabled (#PD0874)
Human physical suffering (#PB5646) Occupational risk to health (#PC0865)
Inadequate prevention of disabilities (#PF0709)
Death and disability from inhumane confinement (#PE5648)
Unsanitary environment for basic health in small rural villages (#PD2011).
Aggravated by Sabotage (#PD0405) Civil war (#PC1869)
Formlessness (#PA6900) Aerial piracy (#PD0124)
Victimization (#PF6987) Civil disorders (#PC2551)
Dangerous animals (#PC2321) Public drunkenness (#PE2429)
Psychotic violence (#PE7645) Injurious accidents (#PB0731)
Explosive substances (#PG7855) Industrial accidents (#PC0646)
Inadequate riot control (#PD2207) Hazardous remnants of war (#PF2613)
Inadequacy of civil defence (#PF0506) Vibrations as a health hazard (#PE1145)
Inhumane methods of riot control (#PD1156) Excessive speed of motor vehicles (#PE2147)
Inadequate safeguards against fire (#PD1631)
Inflammable and flammable substances (#PJ3667)
Deterioration in atmospheric visibility (#PE2593)
Denial of human rights in armed conflicts (#PC1454)
Continued operation of unsafe motor vehicles (#PE2240)
Vulnerability of women and children in emergencies (#PD1078)
Vulnerability of the elderly under states of emergency (#PD0096)
Vulnerability of the disabled during states of emergency (#PD0098).
Reduced by Pain (#PA0643).

♦ PB0869 Perversion

Unnatural acts — Dependence on perversion — Perverted people
Refs Stoller, Robert J and Stoller, Robert J *Perversion* (1986).
Broader Deviance (#PB1125) Unnaturalness (#PA3981) Inappropriateness (#PA6852).
Narrower Sadism (#PF3270) Masochism (#PF3264) Bestiality (#PE3274)
Exhibitionism (#PD4643) Sexual deviation (#PD2198).
Related Mental illness (#PC0300).
Aggravates Occultism (#PF3312).

♦ PB0871 Repression

Dependence on repression — Repressed people
Refs Lauer, Hans E; Castelliz, K and Davies, Saunders *Aggression and Repression in the Individual and Society* (1981).
Broader Refusal (#PA7321) Restraint (#PA7296).
Narrower Police state (#PD7910) Political police (#PD3542)
Racial repression (#PD8762) Economic repression (#PC8471)
Cultural repression (#PC8425) Political repression (#PC1919)
Religious repression (#PC0578) Ideological repression (#PC8083)
Lack of trans-frontier cooperation (#PF6855) Suppression of intellectual freedom (#PC5018)
Violent repression of demonstrations (#PD4811) Repression of intellectual dissidents (#PD0434)
Suppression of creativity and innovation (#PF0275)
Compulsory acquisition of land by government (#PC1005)
Denial of right to national self-determination (#PC1450).
Related Loss (#PA7382) Difficulty (#PA5497) Oppression (#PB8656)
Attachment (#PF6106) Forgetfulness (#PA6651).
Aggravates Impiety (#PA6058) Secret societies (#PF2508) Social injustice (#PC0797)
Political alienation (#PC3227) Student press weakness (#PE0628)
Communication with foreigners (#PF8565)
Denial of human rights in the administration of justice (#PD6927)
Misuse of nonprofit associations as front organizations by government (#PE0436).
Aggravated by Racism (#PB1047) Elitism (#PA1387)
Authoritarianism (#PB1638) Subversion of democracy (#PD3180)
International insecurity (#PB0009).

♦ PB0883 Social neglect

Dependence on social neglect
Nature Life expectancy at birth being a barometer of social well-being, it also depicts the hidden violence of social neglect. Life spans below the highest national averages may thus be seen as unnecessary loss of life primarily due to hunger and to illnesses related to unsafe water and poor sanitation.
Incidence In the world today 2,000,000,000 people live on incomes below $500 per year. At least one person in five is trapped in absolute poverty. In the US, militarily the most powerful nation in the world, the national poverty rate is now the highest in 17 years; there are 34 million people officially classified as poor under US poverty standards.
Worldwide, 600,000,000 have no jobs or are less than fully employed. In the Third World, one in three who wants to work cannot find a regular job. In all countries, it is the young people who are hardest hit by unemployment; in the US half of black teenagers are jobless.
11,000,000 babies around the world die before their first birthday. Relative to births, seven times as many infants die in the poorest countries as in the richest. As of 1980, less than 10 percent of children in the Third World were being immunized against the six common diseases of childhood; 5,000,000 died from them yearly.
Claim Structural violence (social neglect) is more destructive of human life than behavioural violence (war). Between 1945 and 1983 there were 15 million needless deaths per year from social neglect, almost as many as the 16 million war deaths over the entire period.
Social neglect, like military affluence, grows out of excessive militarization. The relation between the military assumption of political power and the official use of violence within countries, between war and a heavy death toll among civilians, are the already perceptible effects of an arms race out of control. Beyond these, there are far-reaching effects on society that are more hidden but no less destructive of human life. An intensive arms buildup kills whether or not the weapons are put to use.
Broader Chance (#PA6714) Neglect (#PA5438).
Narrower Overcrowding (#PB0469) Friendlessness (#PB5747)
Neglect of the aged (#PD8945) Non-settled refugees (#PC0519)
Maternal deprivation (#PC0981) Dehumanization of death (#PF2442)
Disaster unpreparedness (#PF3567) Neglect of victims of crime (#PD4823)
Social neglect of war veterans (#PD2077) Inadequate emergency blood supply (#PE0366)
Inadequate social welfare services (#PC0834) Inadequate rehabilitation facilities (#PD1089)
Institutionalized members of society (#PE4001) Inadequate disaster rescue and relief (#PF0286)
Inadequate protection of individual welfare (#PJ3312)
Inadequate disaster prevention and mitigation (#PF3566)
Lack of facilities for severely deformed people (#PD0211)
Increase in anti-social behaviour in developing countries (#PD0329)
Loss of traditional forms of social security in developing countries (#PD1543).
Related Discrimination (#PA0833) Denial of social rights (#PC0663)
Denial of rights to prisoners (#PD0520)
Delayed consequences of war-time imprisonment and deportation (#PF0726).
Aggravates Human suffering (#PB5955) Loneliness in old age (#PD0633)
Loneliness of children (#PC0239) Socially handicapped refugees (#PD1507)
Denial of rights of children and youth (#PD0513)
Inadequate facilities for children's play (#PD0549)
Economic and social losses due to disability (#PE4856)
International imbalance in the quality of life (#PB4993)
Vulnerability of women and children in emergencies (#PD1078)
Immigration barriers for handicapped family members (#PE4868)
Vulnerability of the elderly under states of emergency (#PD0096).
Aggravated by Ignorance (#PA5568) Certificate-based job market (#PJ8370)
Monolithic architecture of high-rise buildings (#PE1925).

♦ PB0968 International aggression

Dependence on international aggression
Nature Armed force may be used against the sovereignty and political independence of another state. It can take more shapes than declared or undeclared war. It may occur on the high seas in actions against merchant or naval vessels; it may occur in the air or in space, or on land in various ways. For example, the Nazis in World War II torpedoed merchant ships, invaded Poland, intimidated Austria into being annexed, rocket-bombed London, forced treaties, and allied themselves with three other aggressor-nations: Russia, Italy and Japan.
Refs Ferencz, Benjamin B *Defining International Aggression* (1975); Rifaat, Ahmed M *International Aggression* (1979).
Broader Aggression (#PA0587) International insecurity (#PB0009).
Narrower Violation of neutrality (#PC2659) Disruption of territorial integrity (#PC2945)
Aggressive uses of natural energy resources (#PD0408)

PB0968

Foreign intervention in internal affairs of states (#PC3185)
Unlawful interference with rights of innocent passage in territorial waters (#PE6755).
Aggravates Occupied nations (#PC1788)
Human destructiveness (#PA0832)
International crisis escalation (#PB6335)
Unsustainable development of energy use (#PC7517).
Aggravated by Economic rivalry (#PD8897)
Economic expansion (#PF8111)
Unfulfilled treaty obligations (#PF2497)
Inadequate international judicial system (#PF2113).

♦ PB1016 Maldistribution of resources

Maldistribution of resource utilization — International disparity in the consumption of resources — Breakdown of resource exchange

Claim Only very limited systems of interchange exist between those who hold the natural, human and technological resources and those who need them. This results in maldistribution throughout the globe.

Narrower Maldistribution of food (#PC2801)
Maldistribution of water (#PD8056)
Maldistribution of medical resources (#PD2705)
Maldistribution of electrical energy (#PD3446)
Maldistribution of science and technology (#PC8885)
Disparity between workers skills and job requirements (#PC1131)
Disparities in distribution of communication resources and facilities (#PD2762).
Related Monopoly of the economy by corporations (#PD3003).
Aggravates Lack of natural resources (#PC7928)
Under-utilized raw materials (#PF6590).
Aggravated by Unethical consumption practices (#PD2625)
Inequitable distribution of wealth (#PB7666)
Reluctant claims on external resources (#PF1226).

♦ PB1044 Human disease and disability

Dependence on human disease and disability — Human illness

Nature In the world as a whole there are three broad groups of diseases that account for a highly significant proportion of illness and death: communicable diseases; degenerative diseases; and neoplastic diseases (cancers). The relative frequency of each of these varies according to the socio–economic state of a country and the condition of its environment. Communicable diseases account for a large proportion of illness and death in developing countries, whereas infant mortality due to communicable diseases is extremely low in the developed world. On the other hand, degenerative diseases (such as those of the heart and circulatory system) and neoplastic diseases (cancer) account for a large proportion of illness and death in developed countries, where a significant proportion of the adult population becomes exposed to environmental factors that produce cancer and degenerative diseases. In the developing world, however, where infant mortality is relatively high, a much smaller proportion of people survive long enough to be exposed to agents capable of producing cancer and degenerative conditions. Even those who survive into adulthood may escape because the agents may not be present in the environment. It should be noted that these diseases are nevertheless widespread in the developing countries, but statistics on mortality and morbidity due to them are rarely available.

Incidence The health status of the majority of people in the disadvantaged areas of most countries of the world is low. This is shown by the high morbidity and mortality rates that exist in the rural and peri–urban populations that still constitute 80–85% of the population of the world, where some 550 million people suffer from absolute poverty. Although morbidity and mortality show a downward trend, problems such as malnutrition, communicable diseases, parasitic infestations, and others continue to take a heavy toll of people's lives, especially those of infants, children, and other vulnerable groups in the disadvantaged areas. The low health status of these people has not only manifested itself in terms of morbidity and mortality but has also affected human development and the capacity of individuals to develop their potentialities and lead a productive life. The health aspects of traffic accidents are of worldwide concern. It is estimated that more than 10 million people are injured on the world's roads each year; there are 250,000 deaths and the incidence of accidents is constantly increasing. The amount of disability that results from the associated morbidity is considerable, bearing in mind that in technically developed countries a substantial proportion of cases of cerebral injury in the community, as well as serious handicaps of a permanent nature, have been caused by road accidents. But the full extent of the morbidity cannot be estimated as not all injuries caused by road traffic accidents are officially recorded. The worldwide trends in smoking–related mortality and morbidity are alarming. Tobacco–smoking is a major cause of chronic bronchitis, emphysema, and lung cancer, as well as a major risk factor for myocardial infarction, certain pregnancy–related and neonatal disorders, and a number of other serious health problems. In the developed countries about half of all deaths are due to cardiovascular diseases, a fifth to cancer and a tenth to accidents. These problems are increasing in the developing countries too. Environmental health problems due to industrialization and urbanization are assuming growing importance; these same problems could affect developing countries as they build up their industries. Chronic disease increases as people grow older. In recent years there has been a steady increase in mental disorders and in social pathology such as alcohol and drug abuse. Lung cancer as well as other chronic lung diseases due to smoking, and obesity due to overeating, are common phenomena. The significant differences between the population pyramids of the developed and developing worlds are therefore partly explicable in terms of their differing mortality and morbidity patterns.

Claim Despite every effort and a heavy expenditure of material and human resources, the world health situation is grave and the present trends are developing into a major crisis which must be faced at once if costly reactions are to be averted.

Counter–claim Most people experience from time to time feelings that things are not as they should be; time is needed for change, rest or some other adaptation in their lifestyle. The medical profession has laid claim to these experiences as times when their expertise is required, because this sense of "dis–ease" is pathological and must be immediately dispelled.

Refs Aita, John A *Neurologic Manifestations of General Diseases* (1975); Cartwright, Frederick *Disease and History* (1972); Finegold *Anaerobic Bacteria in Human disease* (1977); Frank, Andrew (Ed) *Disabling Diseases* (1989); Henschen, Folke *The History and Geography of Disease* (1966); Howe, G Melvyn *A World Geography of Human Diseases* (1978); Prasad, Ananda S *Essential and Toxic Trace Elements in Human Health and Disease* (1988); Symons, Ronald C and Hughes, Charles C (Eds) *Culture–Bound Syndromes* (1985); Trowell, H C and Burkitt, D P *Western Diseases* (1981).

Broader Disease (#PA6799)
Human illness (#PA0294)
Human physical suffering (#PB5646).
Narrower Fever (#PD2255)
Stroke (#PE1684)
Ulcers (#PE2308)
Asthma (#PD2408)
Obesity (#PE1177)
Allergy (#PE1017)
Mycosis (#PE2455)
Q fever (#PE2534)
Bruxism (#PE5685)
Injuries (#PB0855)
Asphyxia (#PE4104)
Zoonoses (#PD1770)
Mutagens (#PD1368)
Mutation (#PF2276)
Headache (#PE1974)
Glanders (#PE2461)
Fractures (#PE7511)
Infection (#PC9025)
Hay fever (#PE6197)
Epidemics (#PC2514)
Paralysis (#PD2632)
Teratogens (#PE0697)
Laryngitis (#PE2653)
Ornithosis (#PE2578)
Gallstones (#PG3269)
Peritonitis (#PE2663)
Rickettsiae (#PE2572)
Human ageing (#PB0477)
Spina bifida (#PE1221)
Appendicitis (#PG2327)
Loss of blood (#PJ2230)
Alcohol abuse (#PD0153)
Cholecystitis (#PE2251)
Heart diseases (#PD0448)
Mental illness (#PC0300)
Liver diseases (#PE1028)
Heat disorders (#PE2398)
Minor ailments (#PE5953)
Sexual craving (#PE7031)
Hearing defects (#PD6306)
Teeth disorders (#PD1185)
Gland disorders (#PD8301)
Sleep disorders (#PE2197)
Motion sickness (#PE2611)
Speech disorders (#PE2265)
Hypersensitivity (#PE6898)
Human disability (#PC0699)
Toxic substances (#PD1115)
Poisonous plants (#PD2291)
Eating disorders (#PE5187)
Burns and scalds (#PE0394)
Kidney disorders (#PE2053)
Man–made diseases (#PD6663)
Human infertility (#PC6037)
Rift valley fever (#PE7552)
Gastric disorders (#PE1599)
Mental depression (#PC0799)
Learning disorders (#PD3865)
Physical disorders (#PG6226)
Iatrogenic disease (#PD6334)
Rheumatic diseases (#PE0873)
Chronic bronchitis (#PE2248)
Critical illnesses (#PE9038)
Lifestyle diseases (#PF4618)
Children's diseases (#PD0622)
Pregnancy disorders (#PD2289)
Neoplastic diseases (#PC3853)
Intestinal diseases (#PD9045)
Malignant neoplasms (#PC0092)
Hereditary diseases (#PG7966)
Childhood accidents (#PD6851)
Nutritional diseases (#PD0287)
Natural food poisons (#PD1472)
Unwanted pregnancies (#PF2859)
Periodontal diseases (#PE3503)
Adenovirus infections (#PE2355)
Medical complications (#PE2863)
Occupational diseases (#PD0215)
Diseases of the spine (#PD2626)
Degenerative diseases (#PD6216)
Diseases of the limbs (#PG3272)
Diseases of metabolism (#PC2270)
Human sexual disorders (#PD8016)
Natural human abortion (#PD0173)
Diseases of the spleen (#PE6155)
Psychomotor retardation (#PG2961)
Psychosomatic disorders (#PD1967)
Psychological disorders (#PD8375)
Thyroid gland disorders (#PE0652)
Occupational dermatoses (#PE5684)
Cardiovascular diseases (#PE6816)
Orthopaedic disabilities (#PG2956)
Nutritional deficiencies (#PC0382)
Diseases of the pancreas (#PE1132)
Diseases of the arteries (#PE2684)
Inadequate protein supply (#PC1916)
Pituitary gland disorders (#PE2286)
Unhealthy physical posture (#PD2838)
Complications of childbirth (#PC9042)
Lung disorders and diseases (#PD0637)
Diseases of the sense organs (#PC9623)
Mental disorders of the aged (#PD0919)
Environmental human diseases (#PD5669)
Genetic defects and diseases (#PD2389)
Diseases and injuries of bone (#PE3822)
Ill defined health conditions (#PC9067)
Diseases of connective tissue (#PD2565)
Foot diseases and disabilities (#PD2647)
Diseases of the nervous system (#PC8756)
Sudden unexpected infant death (#PD1885)
Uncontrolled tropical diseases (#PD4775)
Defective human immunity system (#PE3355)
Deterioration in physical health (#PC0716)
Diseases of the lymphatic system (#PD2654)
Environmentally induced diseases (#PD8200)
Infectious and parasitic diseases (#PD0982)
Disorders of joints and ligaments (#PD2283)
Perinatal morbidity and mortality (#PD2387)
Diseases of the circulation system (#PC8482)
Diseases of the respiratory system (#PD7924)
Facial or oral injury or deficiency (#PE8505)
Prohibitive cost of disease control (#PF2779)
Diseases of the reproductive organs (#PG3270)
Human physical genetic abnormalities (#PD1618)
Institutionalized members of society (#PE4001)
Diseases of the genito–urinary system (#PC4575)
Harmful effects of sensory deprivation (#PE5787)
Endemic disease in developing countries (#PD0103)
Biological agents as occupational hazards (#PE5696)
Trace element imbalance in the human body (#PE5328)
Diseases of blood and blood–forming organs (#PF8026)
Disease and injury from exposure to weather (#PF5739)
Somatic and psychosomatic effects of torture (#PE5294)
Diseases of the skin and subcutaneous tissue (#PC8534)
Disease and injury from physical confinement (#PE5763)
Diseases and deformities of the digestive system (#PC8866)
Disruption of internal balance of the human body (#PE6603)
Diseases of the oesophagus, stomach and duodenum (#PE8624)
Multiplicity of manual sign languages for the deaf (#PF2833)
Inadequate guardianship for mentally retarded adults (#PE3992)
Irrational conscientious refusal of medical intervention (#PF0420)
Harmful effects of ultrasonic radiation on the human body (#PD0748)
Desynchronization of bodily rhythm by international travel (#PE1904)
Unsanitary environment for basic health in small rural villages (#PD2011)
Excessive exposure to radiation from consumer goods and electronic devices (#PE1909).
Related Delayed consequences of war–time imprisonment and deportation (#PF0726).
Aggravates Pain (#PA0643)
Fatigue (#PA0657)
Ugliness (#PA7240)
Dystrophy (#PE3506)
Human death (#PA0072)
Human suffering (#PB5955)
Health inequalities (#PC4844)
High mortality rate (#PJ6252)
Dehumanization of death (#PF2442)
Behavioural deterioration (#PB6321)
Imposed career interruptions (#PF4128)
Intentional infecting with disease (#PD2651)
Inequality of life expectancy by gender (#PE1339)
Unequal regional distribution of deaths (#PC4312)
Child mortality in developing countries (#PE5166)
Economic and social losses due to disability (#PE4856)
Death and disability from inhumane confinement (#PE5648)
Unequal morbidity and mortality between countries (#PC6869)
Underdevelopment of industrial and economic activities (#PC0880).
Aggravated by War (#PB0593)
Pests (#PC0728)
Food fads (#PD1189)
Starvation (#PB1875)
Over–eating (#PE5722)
Absenteeism (#PD1634)
Insect pests (#PC1630)
Mental tension (#PB6302)
Water pollution (#PC0062)
Food insecurity (#PB2846)
Confined spaces (#PJ7974)
Forced exercise (#PD5628)
Tourist hazards (#PE8966)
Disease vectors (#PC3595)
Marine pollution (#PC1117)
Dangerous animals (#PC2321)
Lack of war relief (#PF0727)
Agricultural wastes (#PC2205)
Maternal deprivation (#PC0981)
Indoor air pollution (#PD6627)
Shortage of firewood (#PD4769)
Unhealthy environment (#PJ1680)
Unhygienic conditions (#PF8515)
Cobalt as a pollutant (#PE2339)
Rural underdevelopment (#PC0306)
Hazardous waste dumping (#PD1398)
Consanguineous marriage (#PC2379)
Inadequate medical care (#PF4832)
Hazards to human health (#PB4885)
Food spoilage in storage (#PD2243)
Domestic refuse disposal (#PD0807)
Insect vectors of disease (#PC3597)
Solid wastes as pollutants (#PD0177)
Pollution of inland waters (#PD1223)
Water system contamination (#PD8122)
Unethical medical practice (#PD5770)
Inadequate personal hygiene (#PD2459)
Understaffed health clinics (#PJ7980)
Inadequacy of civil defence (#PF0506)
Restrictive medical practices (#PD8831)
Denial of the right of health (#PJ2269)
Pest resistance to pesticides (#PD3696)
Unawareness of health benefits (#PG8511)
Inadequate hospital facilities (#PE5058)
Insect vectors of human disease (#PE3632)
Shortage of fresh–water sources (#PC4815)
Contamination of drinking water (#PD0235)
Hydrogen sulphide as a pollutant (#PE2329)
Industrial waste water pollutants (#PD0575)
Insect resistance to insecticides (#PD2109)
Politicization of health standards (#PD4519)
Inflammable and flammable substances (#PJ3667)
Limited psychiatric out–patient care (#PE0540)
Unsanitary disposal of human remains (#PE4725)
Airborne substances harmful to health (#PD2847)
Radio frequencies as a health hazards (#PE5099)
Water pollution in developing countries (#PD3675)
Prohibitive cost of hospital facilities (#PE4154)
Sonic boom generated by supersonic aircraft (#PE2435)
Violations of health and safety regulations (#PE4006)
Undesirable effects of animal feed additives (#PD1714)
Inadequate packaging of agricultural products (#PF3143)
Health hazards from water development schemes (#PE8692)
International imbalance in the quality of life (#PB4993)
Health hazards of exposure to cosmic radiation (#PF1686)
Health hazards of mining radioactive substances (#PE7419)
Environmental hazards of non–ionizing radiation (#PE7651)

Overexploitation of underground water resources (#PD4403)
Environmental hazards from food and live animals (#PC1411)
Inadequate teaching occupational health and safety (#PE8305)
Denial of the right to health for indigenous populations (#PE4459)
Increase in pests and diseases through perennial irrigation (#PE7832)
Inadequate working conditions in health and medical services (#PE7718)
Introduction of extraterrestrial infectious diseases and bacteria (#PF1312)
Inadequate supply of pharmaceutical products in developing countries (#PE4120).

♦ PB1047 Racism

Dependence on racism — Racialism

Nature Racism as a doctrine attributes the determination of human capacities to specific inherited physical traits that are considered to distinguish a race. Prejudice and discrimination on the ground of race, colour or ethnic origin occur in a number of societies, where physical appearance – notably skin colour – and ethnic origin are accorded prime importance. "Racism" has increasingly come to mean the hostility that one man feels for another because of his colour alone. These racist beliefs have been so widespread that although authoritatively and consistently proved to be erroneous, they still continue to be an important cause of prejudice.

Racism, which takes a number of forms, is a complex phenomenon involving a whole range of economic, political, historical, cultural, social and psychological factors. It is generally a tool used by certain groups to reinforce their political and economic power, the most serious cases being those involving apartheid and genocide. Racism exists in all parts of the world. Violence, even genocide against indigenous groups, has become endemic in many countries. Racism is often aggravated by international systems backed by powerful economic and military factors. Land rights claims of indigenous peoples are often rejected in the name of development and national security. Immigration policies and practices discriminate on the basis of race in many parts of Europe, Asia and North America. Education policies deny equality of opportunity and employment on the ground of race.

Incidence Under the dictatorship of Hitler in Germany, racism was made the official ideology of fascism and was used to justify the invasion of foreign territory, the physical annihilation of millions of people, and the incarceration, torture, and execution of German antifascists in concentration camps. Similar racist practices were carried out by Japanese militarists in China and other Asian countries and by Italian fascists in Ethiopia, Albania, and Greece. The practice of apartheid in South Africa is the most well known present-day example of a policy based on inequality of race.

Background The notion of race was familiar to the peoples of the ancient world. The physical characteristics of different populations were of keen interest to Greek, Latin, and Hebrew historians and geographers, who elaborated many theories as to their origin. The differentiating traits, however, were regarded as no more than accidental qualities of various types of human beings.

Racism as it exists in the contemporary attitude that holds certain racially different people to be inferior or superior, originated with the 16th-century expansion of Europe. As the Spanish and Portuguese conquerors came face to face with the people and cultures of the New World, they refused them a fully human status, as justification for denying them, through conquest and even slavery, the rights of other men. It has also been maintained that some passages from religious writings have reinforced racial prejudice. For example, when Christians tried to justify slavery, they claimed black skin was a punishment from God, and invoked the curse cast upon Cain. Making use of this symbolism, they invented causes for the malady, intended to justify in their own eyes a process of production based upon the exploitation of Negro labour.

The first "scholarly works" on racism appeared in the mid-19th century. J.A. de Gobineau declared in Essay on the Inequality of Human Races that the fair-haired, blue-eyed Aryans were the "higher" race, the creators of all the high civilizations and survived in the "purest" form among the aristocracy of the Germanic peoples. This theory was based on the scientifically unfounded identification of races with language groups and became the cornerstone of many racist conceptions. Later, racist ideas were frequently associated with social Darwinism, whose exponents applied Darwin's doctrine of natural selection and the struggle for survival to human society. The Darwinists made extensive use of Malthusianism to prove the superiority of the hereditary attributes of the ruling classes to those of the working people and to demonstrate the usefulness of artificial selection of conjugal pairs for improving the race. After World War I (1914–18) the "Nordic myth" became popular in reactionary circles, especially in Germany. The myth asserted the superiority of the northern (Nordic) race of tall, long-headed blonds, who were said to be genetically linked with the Germanic-speaking peoples. This myth was popularized in the works of many German pseudo scientists who openly supported Nazism.

A committee of experts on race problems, composed of anthropologists, psychologists and sociologists, convened in December 1949 in Paris under the auspices of UNESCO, agreed that the term "race" designates "a group or population characterized by some concentrations, relative as to frequency and distribution, of hereditary particles (genes) or physical characters, which appear, fluctuate, and often disappear in the course of time by reason of geographic and/or cultural isolation". The committee added that the biological fact of race and the myth of "race" should be distinguished. For all practical social purposes, "race" is not so much a biological phenomenon as a social myth.

Another approach to a definition of the term "race" is that appearing in paragraph 3 of the 1950 UNESCO statement, viz: A race, from the biological standpoint, may therefore be defined as one of the group of the populations constituting the species Homo sapiens. These populations are capable of inter-breeding with one another but, by virtue of the isolating barriers which in the past kept them more or less separated, exhibit certain physical differences as a result of their somewhat different biological histories. These represent variations, as it were, on a common theme.

A group of scientists, composed of physical anthropologists and geneticists, was convened by UNESCO from 4 to 9 June 1951. In concluding their *Statement on the Nature of Race and Racial Differences*, they set out what they considered to have been scientifically established concerning individual and group differences, as follows: (a) In matters of race, the only characteristics which anthropologists have so far been able to used effectively as a basis for classification are physical (anatomical and physiological). (b) Available scientific knowledge provides no basis for believing that the groups of mankind differ in their innate capacity for intellectual and emotional development. (c) Some biological differences between human beings within a single race may be as great as, or greater than, the same biological differences between races. (d) Vast social changes have occurred that have not been connected in any way with changes in racial type. Historical and sociological studies thus support the view that genetic differences are of little significance in determining the social and cultural differences between different groups of men. (e) There is no evidence that race mixture produces disadvantageous results from a biological point of view. The social results of race mixture, whether for good or ill, can generally be traced to social factors.

A conference of experts assembled in Moscow by UNESCO in August 1964 to give their views on the biological aspects of the race question also adopted a set of proposals on this subject. They stated, inter alia, that all men living today belong to a single species and are derived from a common stock; that pure races in the sense of genetically homogeneous populations do not exist in the human species; and that there is no national, religious, geographic, linguistic or cultural group which constitutes a race ipso facto. The proposals concluded: "The biological data given above stand in open contradiction to the tenets of racism. Racist theories can in no way pretend to have any scientific foundation".

Claim The unity of mankind from both the biological and the social points of view is the main thing. To recognize this and to act accordingly is the first requirement of modern man. The myth of "race" has created an enormous amount of human and social damage. In recent years it has taken a heavy toll of human lives and caused untold suffering. It still prevents the normal development of millions of human beings and deprives civilization of the effective cooperation of productive minds. The biological differences between ethnic groups should be disregarded from the standpoint of social acceptance and social action.

Popular notions of "race", however, have frequently disregarded the scientific evidence. They represent the totality of antiscientific conceptions concerning the physical and mental inequality of the races of man and the influence of these racial differences on the cultural development of human society throughout history. Racism propagates erroneous, misanthropic ideas regarding the original division of people into higher and lower races. The higher races are to be considered superior while lower races are to be considered, by nature, the objects of exploitation. As events in various parts of the world have shown recently, the concept of "race" is playing an increasingly important role in the affairs of modern life, a role characterized by deep-seated, often violent, passions, generated and existing irrespective of any scientific or rational truths.

Racial discrimination is the very negation of the principle of equality, and therefore an affront to human dignity. It is a negation, also, of the social nature of man, who can reach his fullest development only through interaction with his fellows.

Refs Frederickson, George M *The Arrogance of Race* (1988); Hodge, John L et al *Cultural Bases of Racism and Group Oppression*; Kayyali, Abdul W (Ed) *Zionism, Imperialism and Racism* (1979); Sadoux, Jean-Jacques *Racism in the Western Film from D W Griffith to John Ford* (1980); Smith, Joan, et al *Racism, Sexism, and the World-System* (1988); United Nations *Racism, Science and Pseudo-Science* (1983); United Nations Institute for Training and Research *Racism and Its Elimination* (1981).

Broader Ethnocentricity (#PB5765) Social injustice (#PC0797).
Narrower Nativism (#PF2186) Neo-fascism (#PF2636) Dispossession (#PG3275) Pauperization (#PG3276) Condescension (#PJ3277) Minority control (#PF2375) Religious racism (#PF4513) Racial inequality (#PF1199) Racial segregation (#PC3688) Racial discrimination (#PC0006) Ethnic discrimination (#PC3686) Restrictions on immigration (#PC0970) Bias in children's literature (#PD4773) Legalized racial discrimination (#PC3683) Threats against family or friends (#PE3308) Race as a reinforcement of nationalism (#PF3352) Excessive neutrality of intergovernmental official information (#PF3076).
Related Nationalism (#PB0534) Cultural discrimination (#PC8344).
Aggravates War (#PB0593) Eugenics (#PC2153) Genocide (#PC1056) Ethnocide (#PC1328) Repression (#PB0871) Racial war (#PD8718) Human violence (#PA0429) Ethnic conflict (#PC3685) Low self esteem (#PF5354) Guerrilla warfare (#PC1738) Racial exploitation (#PC3334) Political dictatorship (#PC0845) Consanguineous marriage (#PC2379) Underprivileged racial minorities (#PC0805) Excessively costly prestige projects (#PF3455) Endangered tribes and indigenous peoples (#PC0720) Denial of human rights in armed conflicts (#PC1454) Inadequacy and insensitivity of intelligence testing (#PD1975) Restriction of indigenous populations to reservations (#PD3305).
Aggravated by Destiny (#PF3111) Scapegoats (#PF3332) Hero worship (#PF2650) Exploitation (#PB3200) Racial conflict (#PC3684) Racist propaganda (#PD3093) Extremist ideologies (#PC6341) Plural society tensions (#PF2448) Undue attachment to a social group (#PF1073) Ideological impediments to marriage (#PF3345).
Reduced by Maldistribution of agricultural land (#PD9189).

♦ PB1071 Homogenization of cultures

Incidence Both minority cultures hitherto protected from external influence, and larger independent cultures, are threatened by a process of cultural homogenization.

Claim The imperative of technology to level all before it to a monotonous sameness constitutes the greatest menace of industrialized society. Cities, whether in their architecture, their road systems, their mass produced products, or their vehicles, suffer from an obsessive homogeneity. Sameness permeates the world like a blight. Under the guise of tourism, it inundates places heretofore relatively untouched by technological civilization. The blight of homogenization is part of the creeping madness engulfing the world. Every nation that succumbs to universal civilization reduces the potential for man to survive an unknown future.

Broader Human destructiveness (#PA0832).
Aggravates Marginalization (#PF4347) Gypsy persecution (#PE1281) Political monoculture (#PF4405) Discrimination against minorities (#PC0582) Threatened and vulnerable minorities (#PC3295).
Aggravated by Pseudo-culture (#PF5513) Cultural imperialism (#PC3195) Cultural fragmentation (#PF0536) Destruction of cultural heritage (#PC2114) Obsolete basis of cultural identity (#PF0836) Excessive use of foreign programmes for media (#PE9643) Vulnerability of socio-economic systems from globalization (#PF1245).
Reduces Racial discrimination (#PC0006).
Reduced by Underprivileged linguistic minorities (#PC3324).

♦ PB1125 Deviance

Nonconformity to social norms

Nature Deviance is a common phenomenon in the life of every human being, even in the so-called simple societies. It is always defined from the point of view of particular normative structure, and in a complex society where there are a multiplicity of groups and conflicting normative standards, each member of the society is at some time liable to be considered deviant by one standard or another. Often deviance simply involves conformity to the standards of a subgroup rather than those of the dominant social group. The consequences of deviance from normative standards are varied and may range from a frown to imprisonment or confinement for mental illness. However, people who deviate from social norms are not necessarily (or necessarily considered) mentally ill, nor does deviance necessary entail mental illness. The despised deviant from a particular society or social system may be regarded as a martyr or saint by another ethical philosophy or historical period. Deviance is not inherent in specific behaviour or attitudes but rather is a phenomenon of human interaction in a particular normative setting.

Refs Ben-Yehuda, Nachman *Deviance and Moral Boundaries* (1987); Burton, John *Deviance, Terrorism and War* (1979); Thio, Alex *Deviant Behavior* (1987).

Narrower Perversion (#PB0869) Moral offences (#PD9179) Deviant society (#PC2405) Social outcasts (#PD6017) Ideological offences (#PD6632) Juvenile delinquency (#PC0212) Insubordinate behaviour (#PJ6517).
Related Crime (#PB0001).

♦ PB1149 National insecurity and vulnerability

Internal insecurity — Political insecurity

Nature Many developing countries have neither a national identity inclusive enough to provide for the diverse sections of their populations, nor a sense of national will strong enough to surmount economic precariousness and strategic vulnerability. In developed countries, excessive will, as

exhibited by some superpowers, leads only to confrontations.
Refs Committee on Review of Switching, Synchronization and Network Control in National Security Telecommunications *Growing Vulnerability of the Public Switched Networks* (1989).
Broader Insecurity (#PA0857) Vulnerability (#PA1766).
Narrower Military insecurity and vulnerability (#PC0541)
National insecurity in developing countries (#PD3835)
Insecurity and vulnerability of non-nuclear weapon states (#PD1521).
Related International insecurity (#PB0009).
Aggravates Socio-economic burden of militarization (#PF1447)
Refusal to issue travel documents, passports, visas (#PE0325)
Restrictions on freedom of movement between countries (#PC0935)
Absorption of manpower resources by military activities (#PF0780).
Aggravated by Treason (#PD2615) Pacifism (#PF0010)
Conflict (#PA0298) Demonstrations (#PD8522)
Self-indulgent societies (#PF5466) Restriction of arms supply (#PD1304)
Unidentified submarine objects (#PE0712) Vulnerability of telephone system (#PE8254)
Environmental threats to national security (#PC4341)
Violation of sovereignty by trans-border broadcasting (#PE0261).
Reduced by Foreign military presence (#PD3496).

♦ PB1151 Natural disasters

Acts of God
Nature Relatively sudden and widespread disturbance of the social system and life of a community or region may be caused by one or more of the destructive forces of nature. Natural disasters are usually the result of geophysical or meteorological disturbances, the causes and mechanisms of which are now relatively well understood even though their occurrence and the detailed consequences cannot be predicted. The phenomena which mainly cause disasters are earthquakes and cyclonic storms, usually of tropical origin, but seismic activity under the sea can cause floods far from the centre of disturbance. Besides direct damage due to flooding, wind forces and earth movement, landslips and outbreaks of fire may occur to cause further damage and loss of life. Volcanic activity, besides being the cause of some earthquakes, can also cause damage from lava and ash.
Background Natural disasters have traditionally been viewed as acts of the gods. The term 'Acts of God' continues to be used in certain forms of legislation.
Incidence Over the past twenty years, natural disasters are estimated to have caused the deaths of 3 million people and to have affected the lives of 800 million others. For example, in 1988 there were serious floods in Nile, Ganges, and Brahmaputra, violent hurricanes in the Caribbean, fatal typhoons in the Philippines and the China Sea, earthquakes in Nepal, Yunnan and USSR Armenia. For one sole year, the death toll rose to more than 50,000; millions of people were left homeless.
Refs Taylor, James B et al *Tornado* (1970).
Broader Disasters (#PB3561).
Narrower Drought (#PC2430) Landslides (#PD1233) Forest fires (#PD0739)
Geological hazards (#PC6684) Topological disaster (#PC5010) Biological disasters (#PC5489)
Geomagnetic disasters (#PD0830) Meteorological disaster (#PD4065)
Evolutionary catastrophes (#PF1181) Disastrous failure of natural dams (#PE0715)
Destructive changes in ocean characteristics (#PC2087)
Natural disasters in least developed countries (#PE0299).
Related Man-made disasters (#PB2075)
Inadequate disaster prevention and mitigation (#PF3566)
Hazards to human health in the natural environment (#PC4777).
Aggravates Famine (#PB0315) Plague (#PE0987) Cholera (#PE0560)
Malaria (#PE0616) Scabies (#PE3359) Refugees (#PB0205)
Smallpox (#PE0097) Influenza (#PE0447) Hepatitis (#PE0517)
Diphtheria (#PE8601) Shigellosis (#PG3358) Unemployment (#PB0750)
Malnutrition (#PB1498) Tuberculosis (#PE0566) Encephalitis (#PE2348)
Homelessness (#PB2150) Typhoid fever (#PD1753) Leptospirosis (#PE2357)
Crop failures (#PG3363) Relapsing fever (#PE7787) Damage to crops (#PJ3949)
Water pollution (#PC0062) Schistosomiasis (#PE0921) Export credit risks (#PF3065)
Uncontrolled migration (#PD2229) Electrical power failure (#PE1341)
Declining economic productivity (#PC8908) Destruction of agricultural land (#PD9118)
Vulnerability of ecosystem niches (#PC5773) Vulnerability of telephone system (#PE8254)
Destruction of economy due to war (#PE8915)
Inadequate emergency medical services (#PD1428)
Economic dislocations in developing countries (#PD4063)
Vulnerability of women and children in emergencies (#PD1078)
Dependence of developing countries on food imports (#PE8086)
Vulnerability of the elderly under states of emergency (#PD0096)
Vulnerability of the disabled during states of emergency (#PD0098)
Instability in export trade of developing countries producing primary commodities (#PD2968).
Aggravated by Wrath of God (#PF8563) Instability of complex society (#PG3364)
Inadequate disaster rescue and relief (#PF0286).
Reduces Unsustainable population levels (#PB0035).

♦ PB1166 Environmental pollution

Environmental contamination — Dependence on environmental pollution
Nature Human activities inevitably and increasingly introduce material and energy into the environment; when that material or energy endangers or is liable to endanger man's health, his well-being or his resources, indirectly or directly, it is called a pollutant. A substance may be considered a pollutant simply because it is in the wrong place, at the wrong time, and in the wrong quantity. Pollutants can affect man with direct effects such as: acute effects from exposure to a toxic pollutant reaching man through air, water or food; long-term effects due to prolonged exposure to a pollutant at levels lower than those giving rise to overt toxic effects; synergistic interaction between pollutants or between a pollutant and malnutrition or disease; genetic effects that are manifested in future generations. Indirect effects on man may result from reduction of the food supply or deterioration of the environment. Such effects include: damage to plants and animals; disruption of ecological cycles such that a previously harmless species becomes a pest; damage to the human habitat (air pollutants that destroy forests and corrode buildings); water pollutants that destroy the recreational value of inland waters; alteration of the global climate (this is considered to be a future threat).
Incidence The volume of garbage and waste is constantly increasing, due in particular to increases in population, changes in living habits, increases in packaging material and increases in consumption. The capacity of waste and garbage disposal installations has long been left far behind. Water resources are to an increasing extent being polluted by the constantly growing volume of waste water from households and industry. The river Vistula, which flows through Warsaw before it reaches the Baltic, disgorges 5,000 tons of phosphorous, 90,000 tons of nitrogen, 130 tons of oil, three tons of phenol and lead, as well as unknown quantities of cadmium, mercury and zinc into the sea every year. Technical progress and mechanization has in recent decades led to a rapid increase in noise. about every fifth worker in the Federal Republic of Germany is subjected to a noise level of 90 db or over. The atmosphere is being polluted by dust, smoke and exhaust gases from industry, motor vehicles and domestic heating. In the Soviet Union fifty million people live in areas where the air pollution levels are 10 times the minimum health standard. In the Urals the city of Nizhny Tagil, for example, industrial enterprises spew nearly 700,000 tons of poisonous substances into the air every year **Background** Air, water and food always contain, and always have contained, varying amounts of 'foreign' matter, and in this sense the potential for pollution has always been present. Furthermore, one of the most widespread and oldest forms of pollution is that arising from contamination of the environment by pathogenic organisms. The present world-wide concern about pollution arises from the realization that today's problems originate essentially from human activity and are very much greater in magnitude and far more widespread than ever before.
Claim Three factors determine the magnitude and nature of the pollution problem, whether at the local or global level: the size of the human population; the rate of production and consumption; and the level and use of technology. But while the total stress resulting from these factors is increasing, the capacity of the environment to deal with their side effects is decreasing. It is for this reason that pollutants must be controlled. Pollution of one sort or another occurs throughout human societies and the effects of any given pollutant are frequently the same wherever they are felt. The present situation results principally from the unbridled application of technology in industrialized countries. Developing countries, however, are already encountering the same problems and are increasingly having to deal with the same pollutants. During this century both population growth and rapid industrialization have combined to poison the atmosphere; contaminate lakes, rivers and even oceans; erode the soil; and destroy many forms of life in the developed countries. Pollution cannot be contained within national boundaries. Wind and rain, ocean currents, migrating birds and fish carry pesticides, inorganic nitrogen fertilizer, oil, and atomic wastes to the far reaches of our planet.
Refs Goldstein, Eleanor C *Pollution* (1987); McCaffrey, Stephan C and Lutz, R E (Eds) *Environmental Pollution and individual rights* (1978); Novotny, Vladimir and Chesters, Gordon *Handbook of Nonpoint Pollution Sources and Management* (1981); Purves, David *Trace-Element Contamination of the Environment* (1985); Singer, S F (Ed) *Global Effects of Environmental Pollution* (1970); Sors, Andrew I and Coleman, David *Pollution Research Index*; United Nations *Airborne Sulphur Pollution Effects and Control*; Ware, G W (Ed) *Reviews of Environmental Contamination and Toxicology* (1987); Williams, John S Jr, et al *Environmental Pollution and Mental Health* (1973).
Broader Pollution (#PB6336).
Narrower Air pollution (#PC0119) Soil pollution (#PC0058)
Food pollution (#PD5605) Water pollution (#PC0062)
Polar pollution (#PE5993) Accumulated junk (#PD5510)
Trans-frontier pollution (#PF7945) Radioactive contamination (#PC0229)
Pollution of orbital space (#PD0089) Contamination of human body (#PF9150)
Havens for environmental pollution (#PE6172)
Environmental pollution in socialist countries (#PD9197)
Accumulation of pollutants in plants and animals (#PD5021)
Degradation of the environment through contamination (#PE4759).
Related Litter (#PD2541).
Aggravates Allergy (#PE1017) Human poisoning (#PD0105) Hazards to plants (#PD5706)
Health inequalities (#PC4844) Children's diseases (#PD0622) Injurious accidents (#PB0731)
Dangerous substances (#PC6913) Hazards to human health (#PB4885)
Wildlife pollution hazard (#PJ3387) Destruction of wildlife habitats (#PC0480)
Unsustainable economic development (#PC0495)
Long-term shortage of natural resources (#PC4824)
Health hazards of environmental pollution (#PC0936)
Economic and social losses due to disability (#PE4856)
Deterioration of the physical condition of art objects (#PD1955)
Degradation of semi-natural and natural habitats of flora and fauna (#PC3152).
Aggravated by Thermal pollutants (#PC1609) Agricultural wastes (#PC2205)
Sewage as a pollutant (#PD1414) Domestic refuse disposal (#PD0807)
Unethical industrial practices (#PD2916) Non-biodegradable plastic waste (#PD1180)
Waste of non-renewable resources (#PC8642) Industrial waste water pollutants (#PD0575)
Inadequate environmental education (#PD1370) Maldistribution of electrical energy (#PD3446)
Mismanagement of environmental demand (#PD5429)
Ecologically unsustainable development (#PC0111)
Methane gas emissions from landfill sites (#PE1256)
Environmental hazards of nuclear power production (#PD4977).

♦ PB1395 Endangered species of plants and animals

Vulnerability of populations of species
Nature Due primarily to the impact of man on the natural environment, whether directly or indirectly, a large number of species of plants and animals are endangered. In addition to the loss of entire species, many species are losing whole populations at a rate that reduces their genetic variability and thus their ability to adapt the kind of environmental adversity with which they will increasingly be confronted. Within species, there is also a loss of races and varieties.
There are three main causes for the plight of species: 1. Destruction of habitat, particularly rain forest, especially through urban expansion, logging, and agricultural and forestry development. 2. Direct exploitation for skins, ivory, feathers, shells, horns, oils and meats. 3. Pollution and the breakdown in natural food chains, the importation of new competitors, predators, parasites and diseases.
Incidence There are estimated 5 to 30 million species on planet Earth but only 1.7 have been scientifically investigated. Most live in the canopies of remote rainforest. A species probably becomes extinct about every 9 hours; it may be as frequent as every three hours; by the end of the century it may be as frequent as every 20 minutes. In 1981 the official endangered species list stood at 230. By 1988 it was 35,000.
Broader Vulnerability (#PA1766) Environmental degradation (#PB6384).
Narrower Endangered totemic species (#PE4184)
Endangered species of plants (#PC0238)
Endangered species of animals (#PC1713)
Endangered species of insects (#PC2326)
Loss of beneficial plants and animals (#PE8717)
Endangered species of medicinal plants (#PD4171)
Endangered animal and plant life due to radioactive contamination (#PD5157).
Related Degradation of the environment through the destruction of species (#PE5064).
Aggravates Abusive collection of specimens (#PE9417).
Aggravated by Deforestation (#PC1366) Inter-species warfare (#PF1925)
Unsustainable harvesting rates (#PD9578) Unsustainable population levels (#PB0035)
International trade in endangered species (#PC0380)
Inadequate plant genetic resources conservation (#PF3581)
Environmental hazards of new species introduction (#PC1617)
Degradation of semi-natural and natural habitats of flora and fauna (#PC3152).

♦ PB1423 Inaction on problems

Failure to act
Broader Lack of care (#PF4646)
General obstacles to problem alleviation (#PF0631).
Narrower Government inaction (#PC3950) Bureaucratic inaction (#PC0267).
Aggravates Credibility gap (#PB6314) Unsolved crimes (#PF6911)
Breach of promise (#PF7150) Oppressive reality (#PF7053)
Resignation to problems (#PF8781).
Aggravated by Passivity (#PF6177) Lack of political will (#PC5180)
Insufficient financial resources (#PB4653) Conceptual repression of problems (#PF5210)
Preoccupation with isolated problems (#PF6580)

(#PF9119)
Institutional preoccupation with obsolete problems (#PJ5014).
Reduced by Revolution (#PA5901).

♦ PB1498 Malnutrition

Dependence on malnutrition — Undernutrition — Badly nourished people — Malnutrition-based illness — Ill-fed

Nature The hunger problem of the world is no longer perceived to be primarily one of starvation or protein deficiency, but rather of chronic undernutrition, affecting a range of vulnerable groups.

Malnutrition can be defined as the lack of sufficient quantity or quality of nutrients to maintain the body system at some definable level of functioning. Its victims are mainly children under the age of 5, and it is visible in only about 2% of all actual cases - despite the image of the starving baby, visible and obvious malnutrition is rare, while invisible malnutrition touches approximately one quarter of the developing world's children. It may be invisible even to mothers, who see their children maintaining normal heights and weights for their age group; but nevertheless it saps a child's energy, lowers his resistance to infection, and holds back his growth. Malnutrition is not caused only by a lack of food (which may be seasonal) but by: infection (diarrhoea and intestinal parasites cause 50% of all malnutrition); by changing from breastmilk to commercial substitutes (the mother cannot read the directions for preparation, has neither time nor facilities for proper preparation, and cannot afford adequate amounts); and by improper weaning. Mild to moderate malnutrition, also called chronic undernutrition, is much more common than severe forms, while the severe (recognizable) forms are either due to insufficient protein and calories (marasmus) or to an acute protein loss or deprivation (kwashiorkor).

Malnutrition causes a child to withdraw into an inner world, because of the necessity to cease physical activity, and thus retards his mental as well as physical growth; it may cause a malnourished mother to give birth to an underweight baby and to exhaust herself in the vicious cycle of pregnancy, birth, and breastfeeding without adequate recuperation periods; and a malnourished adult man may be forced to reduce his activity at the expense of economic and community development. Ironically, malnutrition can also occur among the wealthy, whose over-nutritious diet, rich in fat and calories, leads to: reduced life expectancy, increased susceptibility to disease (obesity, heart disease, high blood pressure), and reduced productivity.

Incidence The majority of the 40,000 deaths every day among the developing world's infants and children are caused by infection as a result of malnutrition. A typical three-year-old in a developing country has one illness every three weeks. Various agencies have estimated the world's malnourished to number between 800 and 1100 million.

Refs Balderston, Judith, et al *Malnourished Children of the Rural Poor* (1981); Brozek, Josef *Malnutrition and Human Behavior* (1985); Gopalan, C *Nutrition - Problems and Programmes in South-East Asia* (1987); Reutlinger, Shlomo and Selowsky, Marcelo *Malnutrition and Poverty* (1976); Somogyi, J C *Malnutrition - A Problem of Industrial Societies?* (1988).

Broader Human physical suffering (#PB5646) Inadequate standards of living (#PF0344).
Narrower Urban malnutrition (#PD7473) Child malnutrition (#PD8941)
Nutritional diseases (#PD0287) Maternal malnutrition (#PE1085)
Nutritional deficiencies (#PC0382) Protein-energy malnutrition (#PD0339)
Protein deficiency in cereals (#PE3147) Subsistence-level malnutrition (#PJ1370)
Malnutrition in developing countries (#PD8668)
Malnutrition among indigenous peoples (#PC3319)
Trace element imbalance in the human body (#PE5328)
Inadequate immune responses in malnourished persons (#PE4883)
Physiological malnutrition arising from mental factors (#PE8925).
Related Starvation (#PB1875).
Aggravates Fatigue (#PA0657) Rickets (#PG2295) Dysentery (#PE2259)
Dystrophy (#PE3506) Meningitis (#PE2280) Tonsillitis (#PE2292)
Common cold (#PE2412) Tuberculosis (#PE0566) Liver diseases (#PE1028)
Hypersensitivity (#PE6898) Enteric infections (#PD0640) Cirrhosis of the liver (#PE2446)
Skeletal system disorders (#PE2298) Mental disorders of the aged (#PD0919)
Perinatal morbidity and mortality (#PD2387) Fragility of maintaining basic health (#PJ2524)
Inequality of life expectancy by gender (#PE1339)
Unequal regional distribution of deaths (#PC4312)
Inhibited growth of malnourished children (#PE4921)
Unequal morbidity and mortality between countries (#PC6869)
Unsanitary environment for basic health in small rural villages (#PD2011).
Aggravated by Slavery (#PC0146) Slave trade (#PC0130)
Family poverty (#PC0999) Underemployment (#PB1860)
Plant pathogens (#PD1866) Dietary torture (#PE4371)
Natural disasters (#PB1151) Dietary restrictions (#PJ1933)
Socio-economic poverty (#PB0388) Diseases of the pancreas (#PE1132)
Limited purchasing power (#PD8362) Export of nutritious food (#PJ1365)
Soil-transmitted diseases (#PD3699) Abuse of coca and cocaine (#PD2363)
Taboos against eating poultry (#PJ1326) Ineffective population control (#PF1020)
Rumination disorder of infancy (#PG9708) Poverty in developing countries (#PC0149)
Unsustainable population levels (#PB0035) Illiteracy among indigenous peoples (#PD3321)
Exploitation of dependence on food aid (#PD7592)
Inefficient use of proteins in factory farming (#PF2758)
Vulnerability of women and children in emergencies (#PD1078)
Vulnerability of the elderly under states of emergency (#PD0096)
Vulnerability of the disabled during states of emergency (#PD0098)
Use of agricultural resources for production of animal feed (#PD1283).
Reduces Diabetes (#PE0102) Decrease in mortality rate (#PF0333).

♦ PB1540 Secularization

Dependence on secularization — Decline of religion — Secularism — Militant secularism — Civil religion

Nature The influence of organized religion is being systematically eliminated in society. This encourages individualistic world views in which traditional, shared unifying concepts such as the existence of God may have no place.

Secularization affects medical-moral issues such as contraception, abortion and generally, the right to life. Although it might be evident first in public behaviour, it acquires a degree of officiality when it is embraced by government, and thereafter spreads quickly. As secularization permeates a society, government is eventually officially divorced from anything to do with religion, whether through a state church, subsidies for religious education, or any other institution of preferment that advances or establishes one creed above another. Acts of legislation may eliminate: tax advantages conceded to churches; religious tests or oaths for civil office; censorship and other religiously-inspired barriers to freedom of expression.

Secularization can be see as several types of concepts: 1) The decline of religion from an objective standpoint, such as, institutions, membership or participation in worship. 2) Institutions, practices and activities traditionally done by religion being assumed by non-religious social processes, such as, education being done by the state. 3) Norms from religion being transposed to the world, for example, the institutionalization of the norm of equality. 4) The world being desacralization, i.e. approaching the world through rational explanation and manipulation rather than through awe and a sense of mystery. 5) Religion conforming to the world.

Claim 1. Secularization destroys the one human, yet transcendent, unifying value that society has - the belief in God. In its place, modernism offers the state and its needs, whether they are nuclear arms or higher industrial production. Secularization means the death of inspiration for the arts; the denial of the right to life; and the elimination of hope that there exists, somewhere, true justice, and standards that do not change with the times.

2. All religious beliefs known in history help create and sustain the bond between man, other men and nature. But if faith weakens in the onslaught of science, technology and secularism, then man is truly isolated. In any increasingly secularized society, religious faith is less than ever a motivating force and an explanation of the surrounding world. Awareness of and concern with the fundamental problem of human existence has been rejected. People are no longer concerned with the meaning of life.

Counter-claim 1. The natural development of the human race involves elimination of institutions dedicated to primitive and superstitious ideas. The decline of religion has gone along with more social progress than the world has ever seen.

2. Secularization is not necessarily inimical to religion; it does not necessarily produce secularism, an ideology that opposes, or is indifferent to, religion. Many theologians argue that secularization is a part of the maturation of faith and that new demands on religion is giving birth to new creativity.

3. "The Church is bold in speaking of itself as the sign of the coming unity of mankind. However well founded the claim, the world hears it sceptically, and points to 'secular catholicities' of its own. For secular society has produced instruments of conciliation and unification which often seem more effective than the Church itself. To the outsider, the churches often seem remote and irrelevant, and busy to the point of tediousness with their own concerns."

Refs Dube, S C and Basilov, Vladimir N *Secularization in Multi-religious Societies* (1983); Mulder, D C (Ed) *Secularization in global Perspective* (1981).

Broader Ideological conflict (#PF3388).
Narrower Decline in religious broadcasting (#PF1433).
Related Proliferation of commercialism (#PF0815).
Aggravates Loss of faith in religion (#PF3863).
Aggravated by Compounding a crime (#PE1485)
Double standards in morality (#PF5225).
Reduces Ritualism (#PG2184) Official religion (#PF6091).

♦ PB1638 Authoritarianism

Dependence on authoritarianism — Authoritarian people — Authoritarian movements

Nature The demanding and enforcement of complete obedience to authority may be political, parental or ideological. It may be manifested in police brutality, governmental intimidation, or organizational indoctrination. Authoritarians claim their power to have been vested in them by virtue of supreme logic or science; by God; by the culmination of historical processes in their eventuality, or by destiny; or by hierarchical position in a structure of relationships. The authoritarian truly believes in a mystique that he or she somehow earned the right to such behaviour and, frequently, that he or she is the guardian of defenders against the forces of illogic, incompetence, sloth, immorality and in general, creeping entropy.

Claim The frequently friendless authoritarians are vectors of stress whether throughout the work unit, the enterprise, the community or the government, depending on their number or the reach of their influence. They can never be re-educated to be other than what they are and must simply be broken by time and events. Weak persons may revel in the idea that they are authoritarian, giving them an imaginary strength, while in reality they are petty, as is frequently exhibited in family roles **Counter-claim** Every strong leader or government is accused of being authoritarian. Leadership, however, entails the confidence of knowing that one is right. This is not a subject for negotiation, or the verbal wrongs of those who are in a subordinate place or of those whose many and diverse interests may have to be considered. The leader acts, not only speaks, for all. Great enterprises are launched and carried to success only by the strong, acting with dispatch and unfettered by consultations. Authoritarianism's emphasis on order and discipline has been a major factor in the economic success of many countries.

Refs Janos, Andrew C (Ed) *Authoritarian Politics in Communist Europe* (1976); Perlmutter, Amos *Modern Authoritarianism* (1984).

Broader Extremism (#PB3415) Domination (#PA0839).
Narrower Militarism (#PC2169) Paternalism (#PF2183) Dictatorship (#PC1049)
Educational authoritarianism (#PC1526) Authoritarian division of labour (#PC6089)
Authoritarian propagation of knowledge (#PF3706).
Aggravates Repression (#PB0871) Intolerance (#PF0860)
Social injustice (#PC0797) Underground press (#PD2366)
Civil disobedience (#PC0690) Police intimidation (#PD0736)
Restriction of freedom of expression (#PC2162).
Aggravated by Compulsion (#PA5740) Lawlessness (#PA5563)
Traditionalism (#PF2676) Narrowmindedness (#PA7306)
Excessive government control (#PF0304)
Inadequate political parties in developing countries (#PD0548)
Negative effects of claims of religious infallibility (#PF3376).
Reduces Lack of individualism in capitalist systems (#PD3106).
Reduced by Militancy (#PC1090) Permissiveness (#PF1252).

♦ PB1701 Wear

Wear and tear — Erosion of metals

Nature Wear is the removal of material from a solid surface as the result of mechanical action exerted by some other solid, possibly in the form of particles or as a result of the action of strong gas or liquid currents upon a surface, especially at high temperatures. Any sliding contact between solids leads to a measurable, if small, material loss. Such losses are particularly serious in modern precision mechanisms, for if they persist they lead rapidly to mechanical failure.

Broader Loss (#PA7382) Impairment (#PA6088) Disintegration (#PA6858).
Related Friction (#PF1691) Corrosion of iron and steel (#PE1945).
Aggravates Mechanical failure (#PC1904) Structural failure (#PD1230)
Failure of materials (#PD2638) Fatigue in materials (#PD1391).

♦ PB1845 Uncontrolled industrialization

Dependence on uncontrolled industrialization — Over-rapid industrialization — Inappropriate industrialization — Industrialism

Nature Because of its phenomenal recent growth and the revolutionary technological changes which have accompanied it, industry is the cause of many contemporary environmental problems. Where there has been no effective environmental planning, industries have been located in a haphazard way, often in conflict with residential and other areas. More recently, accelerating industrialization resulted in residential areas being increasingly located around industrial complexes. These have usually been concentrated in a few, large centres, resulting in excessive concentrations of population. Large, economically-active population centres provide both markets and labour, thus encouraging further industrial growth; so that urbanization further progresses in an uncontrolled manner.

Excessive concentrations of industry lead to a disparity in economic development between regions. In addition, over-concentration of industry causes serious environmental over-loading, particularly in terms of pollution and congestion, and a consequent threat to health and a reduction in the quality of life. In most countries the crux of the problem remains at the national and regional levels, where the development of reasonable industrial patterns within overall environmental development plans has still to be achieved.

PB1845

Emerging technologies offer the promise of higher productivity, increased efficiency, and decreased pollution, but many bring risks of new toxic chemicals and wastes and of major accidents of a type and scale beyond present coping mechanisms, especially in the case of transportation of hazardous materials and the disposal of toxic wastes.

Counter-claim Many so-called consequences of industrialization turn out not to be due to industrialization per se, but rather to the preservation or attempted preservation of pre-industrial ways of life in an alien and inappropriate environment. Thus, child labour in factories may be regarded as a continuation of the much less harmful rural customs of child labour on the farm; and urban slums often reveal a carry-over to the cities of rural methods of house construction, refuse disposal, use of water, and so on. In fact, many of the undesirable social consequences of industrialization are more properly regarded as results of failure to deal with the problems of social transition that inevitably arise from so basic a change in economic and social organization. Those that constitute social evils are not, generally speaking, necessary or inescapable consequences of industrialization itself. Given appropriate measures, they can be mitigated, if not avoided. Some of them, moreover, are actually related less to the growth of secondary industry itself than to one or another of the many political, cultural, legal or intellectual changes that tend to accompany that growth.

The social consequences of industrialization are, in many cases, little more than a transfer to the urban industrial environment (by population movement) of problems of destitution and need that had previously existed in the rural environment, where, being less concentrated, they were usually less noticeable. Where stagnant and depressed agricultural communities force into the industrial centres uprooted peasants and tribesmen in numbers far beyond available opportunities for gainful employment, urban growth tends to reflect not the expansion of industry but the wretchedness of agricultural conditions and the high incidence of under-employment in rural areas. In countries with rapidly increasing rural population, this disproportion between employment opportunities and labour supply in the industrial areas, increasing constantly through new influxes from the country, has exercised a depressing effect on urban levels of living, to the extent that in some cases the newcomer to the town has merely substituted urban misery for rural poverty.

Broader Inefficiency (#PB0843) Ineffective industry self-regulation (#PF5841).
Narrower Imbalance in city sizes within a country (#PF2120).
Aggravates Dust (#PD1245) Tuberculosis (#PE0566) Regional disparities (#PC2049)
Landscape disfigurement (#PC2122) Environmental degradation (#PB6384)
Uncontrolled urban development (#PC0442) Negative effects of the nuclear family (#PF0129)
Long-term change in atmospheric chemistry (#PF1234)
Disruption of family system in developing countries (#PD1482)
Increase in anti-social behaviour in developing countries (#PD0329)
Disparity in social development within developing countries (#PD0266)
Decline in rural customs and traditions in developing countries (#PD1095)
Loss of traditional forms of social control in developing countries (#PD0144)
Decline of handicrafts and cottage industries in developing countries (#PD1250)
Decline in communal spirit and village solidarity in developing countries (#PE0835).
Aggravated by Derelict industrial wastelands (#PE6005).

♦ PB1860 Underemployment

Dependence on underemployment

Nature In almost all the less developed countries, underemployment and consequent poverty is widespread in both urban and rural areas and is a central employment problem. In rural areas particularly, the majority of small-holders and landless agricultural labourers do not have adequate employment or income for the families dependent on them. Underemployment exists, visibly, when persons who are not working full time would be able and willing to do more work than they are actually performing, or, invisibly, when the income or productivity of persons in employment would be raised if they worked under improved conditions of production or transferred to another occupation, account being taken of their occupational skills.

Incidence India's example is typical and has statistical verification: one-quarter to one-third of the existing labour force in agriculture may be surplus to India's requirements. Pronounced seasonal underemployment also exists: where there are no irrigation facilities, available work is confined to 3–4 months in the year. In the urban sector, about 21 per cent of those gainfully employed work less than 28 hours per week. Japanese Household Economy Surveys yield estimates that approximately one-third to one-half of the agricultural labour force constituted surplus labour. The FAO has estimated that nearly one-third to two-thirds of agricultural workers in various areas of the Middle East, North Africa and Southern Europe are surplus. These and other examples show that one-quarter to two-thirds of workers may be affected in less developed countries depending on location and sector.

Claim Massive underemployment in a poor and stagnating economy means that low incomes are general and that a large proportion of human resources are not being applied to the task of human development.

Counter-claim In a dynamic economy, underemployment may well be preferable as a transitional arrangement to the harsh reality of full-time work and adequate incomes for some and no work and no income for others. In addition, in developing economies, sophisticated distinctions of unemployment and underemployment, in whatever terms, can have little relevance to the facts of poverty and may be used in a rhetoric that seeks to diminish the problem of unemployment.

Refs Ghosh, B N *Economics of Underemployment* (1983); ILO *Labour Force, Employment, Unemployment and Underemployment* (1982).

Broader Unemployment (#PB0750) Socio-economic poverty (#PB0388)
Economic and social underdevelopment (#PB0539).
Narrower Discrimination against part-time work (#PE6241).
Related Underproductivity (#PF1107).
Aggravates Malnutrition (#PB1498) Lack of job satisfaction (#PF0171)
Abuse of government employment (#PE4658)
Proliferation and duplication of international information systems (#PE0458).
Aggravated by Seasonal fluctuations (#PF8163) Inappropriate education (#PD8529)
Discrimination in employment (#PC0244) Inappropriate transfer of technology (#PE5820).

♦ PB1875 Starvation

Starving people

Nature Starvation is the state of an organism that is completely deprived of food, has an insufficient food intake, or suffers from disturbances in its assimilation of food. In man it may result from deprivation or insufficiency of food, the impairment of food intake, or specifically from diseases of the gastrointestinal tract. Starvation is considered complete if only water enters the organism; incomplete if food enters, but in an insufficient amount in relation to general energy output; and absolute when neither food nor water are taken in. How long life can be sustained under conditions of starvation depends upon varying factors: body weight (the greater the weight, the greater the energy reserve supply); age (children are most susceptible, the middle-aged most resistant); sex (women have a greater chance for survival); expenditure of energy (the less expenditure, the greater the endurance time); and peculiarities of each individual organism. The maximum length of time of starvation for man is considered to be 65–70 days (in absolute starvation, only a few days).

Incidence In East Africa and the Sahel, widespread starvation resulted in 1984, from long-term drought. The FAO lists fifteen countries which need exceptional or emergency assistance to prevent starvation: Angola, Bangladesh, Ethiopia, Haiti, Jamaica, Laos, Lebanon, Mozambique, Nicaragua, Peru, Sierra Leone, Somalia, Sri Lanka, Sudan and Vietnam.

Refs George, Susan *How the Other Half Dies* (1977); George, Vic *Wealth, Poverty and Starvation* (1988); Schwartz-Nobel, L *Starving in the Shadow of Plenty* (1982).

Broader Human illness (#PA0294).
Narrower Hunger (#PB0262) Famine (#PB0315) Thirst (#PE3818)
Nutritional deficiencies (#PC0382).
Related Malnutrition (#PB1498) Anorexia nervosa (#PE5758).
Aggravates Genocide (#PC1056) Ethnocide (#PC1328)
Human physical suffering (#PB5646) Human disease and disability (#PB1044).
Aggravated by Civil war (#PC1869) Plant pathogens (#PD1866)
Lack of war relief (#PF0727) Inadequacy of civil defence (#PF0506)
Vulnerability of food chains (#PB2253)
Unsustainable economic development (#PC0495).

♦ PB1935 Structural violence

Dependence on structural violence — Institutional violence — Institutionalized covert himsa — Structural injustice

Nature Societies are structured in ways that deny adequate standards of living and access to vital social services to the less privileged. Where institutions such as business, government, schools and prisons violate the personhood of society's members this is a form of violence (or, using Jain terms, covert himsa). Poor housing, racial discrimination, unemployment, disenfranchisement and repressive education are thus structural forms of violence. An example is medical care, for which key indices such as disease and early death rates per thousand are markedly disparate between the poorest and the richest.

Claim Personal violence for the amateur is dominance, structural violence is the tool of the professional. The amateur who wants to dominate uses guns, the professional uses social structure. The legal criminality of the social system and its institutions, of government, and of individuals at the interpersonal level is tacit violence. Structural violence is a structure of exploitation and social injustice. It seems to survive very well the changes from a slave society, via a feudal and capitalist order, to lodge in a socialist society.

Structural violence between countries is seen, for example, in the EEC, where there are colonial-type trade relations, elitist networks, highly varied bilateral interaction rates observed between hierarchically conscious states, and a process of recognizing some states as having a 'second class berth'in the EEC 'vehicle'. There are no statistics to illustrate the incidence of violence by consent – those products resulting from the social structure called apathy, ignorance, misinformation and resentment which allow for few opportunities to improve wages, working conditions, nutrition, living conditions and housing, and literacy and education. This is the violence all nations show to their underprivileged classes.

Broader Human violence (#PA0429).
Narrower Bureaucratic aggression (#PC2064).
Related Victimization (#PF6987) Violence as a resource (#PF3994).
Aggravates Human suffering (#PB5955).
Aggravated by Complicity with structural injustice (#PJ5026).

♦ PB1969 Imbalance of power

Dependence on imbalance of power

Broad · International insecurity (#PB0009).
Narrower Non-alignment (#PF0801) Monopoly of power (#PC8410)
Competitive acquisition of arms (#PC1258).
Related Imperialism (#PB0113) Intellectualism (#PF2146) Covert imperialism (#PF3199).
Aggravates War (#PB0593) Risk of war (#PF4215).
Aggravated by Imbalance in strategic arms (#PC1606).

♦ PB1992 Intimidation

Dependence on intimidation — Intimidating people — Threat — Danger

Nature Intimidation is the use of force or superior power to discourage people from certain activities or thoughts, or to enforce other activities or thoughts against their will. Intimidation may be physical or psychological and may relate to a wide range of situations. It induces fear, apathy and alienation, reinforces barriers to progress and facilitates exploitation.

Counter-claim A threat can be just or unjust, good or bad, depending upon whether the threatened retaliatory measure is morally justifiable. The sanction normally attached to positive law is, in effect, a threat of punishment to be inflicted upon its transgressors. To threaten punishment may therefore be reasonable and virtuous, and a parent, teacher, or a custodian of the law, would fail in his duty if he neglected in some circumstances to threaten punishment. Prudence and care for the society and the individual to be threatened must dictate the norms to be observed in making justifiable threats.

Refs Heurlin, Bertel *Threat as a Concept in International Politics* (1977); Milburn, Thomas W and Watman, Kenneth H *On the Nature of Threat* (1981).

Broader Domination (#PA0839) Aggression (#PA0587) Compulsion (#PA5740)
Behavioural deterioration (#PB6321).
Narrower Persecution (#PB7709) Death threats (#PD0337)
Racial intimidation (#PC2936) Social intimidation (#PC2940)
Aircraft harassment (#PE6006) Physical intimidation (#PC2934)
Economic intimidation (#PC3011) Unreported harassment (#PF4729)
Religious intimidation (#PC2937) Political intimidation (#PC2938)
Industrial intimidation (#PC2939) Threats against peasants (#PJ9069)
Intimidation of witnesses (#PJ5894) Psychological intimidation (#PC2935)
Harassment of public officials (#PD4915) Intimidation of public officials (#PD4734)
Intimidation of victims of crimes (#PJ1543)
Government intimidation of governments (#PE1622).
Related War (#PB0593) Threats against family or friends (#PE3308)
Inhumane interrogation techniques (#PD1362).
Aggravates Fear (#PA6030) Apathy (#PA2360) Insecurity (#PA0857)
Militarism (#PC2169) Prostitution (#PD0693) Exploitation (#PB3200)
Criminal coercion (#PD4469) Corruption of the judiciary (#PD4194)
Political barriers to effective legislation (#PC3201)
Enforced participation in community activity (#PD3386).
Aggravated by Elitism (#PA1387) Colonialism (#PC0798)
Terrorizing (#PE4466) Dictatorship (#PC1049)
Verbal abuse (#PD5238) Economic imperialism (#PC3198)
Lack of protection for victims of intimidation (#PE7793)
Use of undue influence to obstruct the administration of justice (#PE8829).

♦ PB1997 Inflexible social structure

Social stratification as an obstacle to development

Broader Lack of social mobility (#PF2195).
Narrower Traditionalism (#PF2676) Class consciousness (#PC3458)
Inflexible property laws (#PG4248) Inflexible management patterns (#PF3091)
Ineffective worker organizations (#PF1262)
Family structure as a barrier to progress (#PF1502).
Aggravates Bigamy (#PF3286) Caste system (#PC1968) Ancestor worship (#PD2315)
Underdevelopment (#PB0206) Social alienation (#PC2130)
Failure of development programmes (#PF8368) Destruction inherent in development (#PF4829)
Inadequate development of new social structures in developing countries (#PD0822).

BASIC UNIVERSAL PROBLEMS

PB2542

Aggravated by Sexism (#PC3432)
Prejudice (#PA2173)
Heterosexism (#PE0818)
Male domination (#PC3024)
Denial of right to leave any country (#PE3463).
Elitism (#PA1387)
Segregation (#PC0031)
Class conflict (#PC1573)
Lack of social planning (#PF8185)
Reduced by Counter culture (#PF0423).

♦ PB2075 Man-made disasters

Nature Relatively sudden and widespread disturbance of the social system and life of a community or region may be a result of the action or interaction of mechanisms or processes set up by man.

Claim Man-made disasters are on the increase. In the absence of a conflict between the Great Powers, the world might seem relatively at peace. But this is not the case as can be seen from the continuing struggles in in Chad, in countries of Central America, in Kampuchea, in Ethiopia, and Lebanon. The vulnerability of the world population to man-made disasters is growing. This is partly due to the steady population growth and partly due to the apparent neglect of the key element in the drive to reduce disaster-induced casualties: preparedness and prevention programmes.

Refs Cooper, M G (Ed) *Risk* (1985); Obeng, L E *Man-made Lakes and Their Problems* (1979).

Broader Disasters (#PB3561).
Narrower Refugees (#PB0205) Civil disturbances (#PD5372)
Disastrous accidents (#PC6034).
Related War (#PB0593) Natural disasters (#PB1151)
Pathologies of civilization (#PB3674)
Inadequate disaster prevention and mitigation (#PF3566).
Aggravates Famine (#PB0315) Homelessness (#PB2150) Damage to crops (#PJ3949)
Dangerous substances (#PC6913) Destruction of agricultural land (#PD9118)
Vulnerability of ecosystem niches (#PC5773) Vulnerability of plants and crops (#PD5730)
Inadequate disaster rescue and relief (#PF0286).
Aggravated by Human errors and miscalculations (#PF3702)
Long-term shortage of natural resources (#PC4824).

♦ PB2150 Homelessness

Dependence on homelessness — Homeless people — Destitution — Destitute people — Inadequate shelter

Nature Homelessness is the absence or attenuation of the affiliative bonds that link settled persons to a network of interconnected social structures, typified by the absence of any form of permanent accommodation. Such people or families may live in cars (in developed countries), or sleep in public places (such as railway stations) or in the streets. Homelessness is a matter of degree, ranging from a temporary condition (whilst establishing a new base), through a periodical one (in the case of migrant labourers, for example), to a permanent condition. In the latter case, the people may be self-supporting (gypsies, carnival performers) or dependent on society (beggars and vagrants).

Incidence There are 100 million homeless in the world today. In the USA there are an estimated 2 million, more than at any time since the Great Depression. According to the U.S. Conference of Mayors, 56 percent of the homeless in the U.S. are single men, 15 percent are single women, 28 percent are families, mostly single parent. One-third to one-half of the homeless men are military veterans. Most are under 40 years of age. Nearly one-third are severely mentally ill. Another one-third are chronic alcoholics. In the UK, there are 88,200 homeless families. One contributing factor is the shortage of cheap rental flats and single rooms. Natural disasters account for millions of homeless. In August 1988 25 million people were left homeless in Bangladesh due to flooding. More than a million people were left homeless in Khartoum, Sudan after flooding.

Refs Bingham, Richard D, et al *The Homeless in Contemporary Society* (1987); Quigley, Gary H *Homeless and Street People* (1988).

Broader Deprivation (#PA0831).
Narrower Abandoned children (#PD5734)
Homelessness in developing countries (#PD8856)
Homelessness in industrialized countries (#PD8285).
Related Statelessness (#PE2485).
Aggravates Contempt (#PF7697) Vagrancy (#PE5460)
Dependence (#PA4565) Humiliation (#PF3856)
Family breakdown (#PC2102) Underprivileged minorities (#PC3424)
Inadequately heated shelters (#PD5173) Inadequately cooled shelters (#PD9736)
Underprivileged home environment (#PE5199) Denial of right to private home life (#PE6168)
Vulnerability of women and children in emergencies (#PD1078)
Vulnerability of the elderly under states of emergency (#PD0096)
Vulnerability of the disabled during states of emergency (#PD0098).
Aggravated by Floods (#PD0452) Refugees (#PB0205)
Climatic cold (#PD1404) Climatic heat (#PC2460)
Social exclusion (#PC0193) Housing shortage (#PD8778)
Natural disasters (#PB1151) Inadequate housing (#PC0449)
Man-made disasters (#PB2075) Socio-economic poverty (#PB0388)
Lack of political will (#PC5180) Unrentable vacant housing (#PJ8683)
Insufficient housing funds (#PG8768) Unprepared adult leadership (#PF6462)
Prohibitive cost of accommodation (#PD1842) Denial of right to sufficient shelter (#PD5254)
Family poverty in industrialized countries (#PD1998).

♦ PB2253 Vulnerability of food chains

Food webs

Nature Food chains are the networks of feeding relationships which interlink different species within a biological community. Particular links in such networks may be threatened or eliminated by the action of man-made products (such as pollutants), thus placing all the members of the biological community in danger, even though the others may not be directly affected.

Broader Vulnerability (#PA1766).
Related Vulnerability of organisms (#PB5658) Shortage of fresh-water sources (#PC4815).
Aggravates Hunger (#PB0262) Starvation (#PB1875) Crop shortfalls (#PD5174)
Vulnerability of ecosystem niches (#PC5773).
Aggravated by Marine pollution (#PC1117).

♦ PB2331 Cultural barriers

Dependence on cultural barriers

Nature Cultural barriers, a result of a lack of appreciation of cultural differences and the reasons behind different customs, bolster misunderstanding and conflict. One of the most solid cultural barriers is language. Ideological conflicts due to different economic, political, religious or social systems also divide cultures. Customs and traditions can vary so widely between cultures, that, for example, a gesture which would elicit a favourable response in one place may be considered insulting in another; consequently, many cross-cultural development programmes, having a "one-sided" design, are doomed to failure.

Incidence In every country where there has been immigration, there are likely to be cultural barriers between the original and the new population. Immigrants often react to discrimination and isolation by clinging to their former culture as a source of unity and strength, thus feeding the misunderstandings between the two groups.

Broader Cultural fragmentation (#PF0536).
Narrower Untransferability of books between countries and cultures (#PF2126)
Barriers to the international flow of knowledge and educational materials (#PF0166).
Aggravates Culture shock (#PC2673) Misunderstanding (#PA8197)
Socially unintegrated expatriates (#PD2675).
Aggravated by Heterosexism (#PE0818) Lack of communication (#PF0816).
Reduces Destruction of cultural heritage (#PC2114)
Lack of variety of social life forms (#PE8806).
Reduced by Cultural invasion (#PC2548) Forced assimilation (#PC3293).

♦ PB2480 Ethical decay

Moral collapse — Moral relativism — Dependence on ethical decay — Moral pollution — Moral chaos — Decayed moral environment

Nature Moral relativism considers that something, such as abortion, can only be right or wrong in relation to one or another moral framework. Morality is then a human invention. The relevant moral principles are thus nothing but conventions that result from a process of tacit moral bargaining through which some groups can exert pressure on others in an effort to change the current moral conventions. This encourages the view that "anything goes" and that morality is just a matter of opinion. Relativism makes dialogue pointless, assuming that there is no binding truth or that partners in the dialogue are saying the same thing in different ways.

Claim Breakdown in traditional beliefs and a decline in respect for traditional authority, customs, and moral standards has led to a decay in conventional ethical values (often with nothing replacing them) and contributes to the emergence of new ideologies which are destructive and seek violent and simplistic solutions to problems. Concern for others, humility, hard work, and an unselfish outlook are lost as a sense of desperation fills people's minds and governments adopt militarist and hawkish ideologies.

Counter-claim Every one has and operates out of a system of ethical beliefs, even the absolute relativist or the most self-centered hedonist. It is not the decline of morality but whether or not the morality, or more accurately, moralities of today's society assists in the physical, social and spiritual growth of the individual and of society as a whole.

Broader Moral imperfection (#PB7712) Behavioural deterioration (#PB6321).
Narrower Deviant society (#PC2405) Misconduct in public office (#PD8227).
Related Conflict of duties (#PF0513).
Aggravates Loss of civility (#PC7013) Ineffective dialogue (#PF1654)
Collapse of common values (#PF1118) Inability to define moral standards (#PF7178)
Ineffective use of external relations relating to sportsmen (#PE6515).
Aggravated by Indecent art (#PE5042) Eclipse of reason (#PF7521)
Minimal church / school involvement (#PJ9011)
Biased and inaccurate biology textbooks (#PF9358)
Lack of meaningful educational context for ethical decisions (#PF0966).

♦ PB2496 Social breakdown

Social chaos

Nature Breakdown in the social structure or the smooth working of society may be temporary, or may lead to social change, dictatorship and ultimately social disintegration. Breakdown may take the form of strikes, violence, intimidation, delinquency and crime, anarchism, decadence including promiscuity, sexual deviation, marriage and family breakdown, alcoholism and drug addiction, or corruption.

Broader Social disintegration (#PC3309).
Narrower Social conflict (#PC0137) Family breakdown (#PC2102)
Failure of school systems (#PJ3896).
Aggravates Marital instability (#PD2103) National disintegration (#PB3384)
Criminally life endangering behaviour (#PD0437)
Inadequate development of new social structures in developing countries (#PD0822).
Aggravated by Apathy (#PA2360) Militancy (#PC1090)
Anarchism (#PC1972) Unemployment (#PB0750)
Class conflict (#PC1573) Counter culture (#PF0423)
Social deviation (#PC3452) Political inertia (#PC1907)
Underground press (#PD2366) Social fragmentation (#PF1324)
Socio-economic poverty (#PB0388) Lack of social mobility (#PF2195)
Inadequacy of social doctrine (#PF3398) Inadequacy of political doctrine (#PF3394)
Loss of institutional credibility (#PF1963)
Excessive employment of married women (#PD3557)
Family structure as a barrier to progress (#PF1502).

♦ PB2542 Decadence

Degeneracy — Dependence on decadence — Decadent people

Nature Decadence revels in an excess of sensual pleasure which may be obtained through alcohol, drugs, sex, risk or violence or foul language. It may result from a tolerant attitude in society, lack of religious or moral conviction, lack of legislation or law enforcement, corruption or general social stress. It may also result in disease, death and social disintegration. Decadence in art and in other aspects of culture has been theorized to be caused by a racial, genetic decay, hence the term 'degeneracy'.

Incidence Modern decadence has been attributed particularly to industrialized capitalist countries; but it also occurs in the sense of juvenile delinquency, alcoholism, drug addiction, sexual license and bureaucratic and party functionary indulgences in Eastern European countries and the USSR.

Background Decadence is said to have been a cause of the fall of the Roman Empire. Decadence in the sense of aestheticism, homosexuality, alcoholism and drug taking was noted in Western Europe at the end of the 19th century. In the first half of the 20th century, Nazism decreed the degeneracy of Cubism, Fauvism, Dada, Surrealism and other styles, also Jazz, Serial 12 tone music, and Bauhaus design.

Counter-claim Decadence and degeneracy are polemical terms used to denigrate changes and experimentation in behaviour and creative expression. No proof that a society can decay or a race degenerate has ever been offered, and the so-called decline and fall of the Roman Empire is considered by many scholars as merely a semantic invention.

Refs Camille, Paglia *Sexual Personae* (1990); Haworth, Lawrence *Decadence and Objectivity* (1977); Jameelah, Maryam *Western Civilization Condemned by Itself* (1979); Keppe, Norberto R *The Decay of the American People (and of the United States)* (1985); Talbot, Eugene S *Degeneracy* (1984).

Broader Vice (#PA5644) Impairment (#PA6088).
Narrower Debauchery (#PE8923) Promiscuity (#PC0745)
Aestheticism (#PG5029) Decadent clothing (#PE5607)
Decadent standard of living (#PD4037).
Related Excess (#PB8952) Immorality (#PA3369) Complacency (#PA1742)
Permissiveness (#PF1252) Gambling and wagering (#PF2137)
Behavioural deterioration (#PB6321).
Aggravates Obesity (#PE1177) Obscenity (#PF2634) Social conflict (#PC0137)
Hypersensitivity (#PE6898) Social alienation (#PC2130) Cultural corruption (#PC2913)
Social disintegration (#PC3309) Exaggerated tolerance (#PG5028)
Loss of faith in religion (#PF3863)
Decline in human genetic endowment (#PF7815).
Aggravated by Deterioration of industrialized countries (#PD9202).

♦ PB2619 Corruption of meaning

Verbicide — Dependence on corruption of meaning — Imprecise language

Nature Language that describes abstractions or ideas more readily loses its meaning than language used for concrete things.

Another form of corruption of a language is that of borrowing words from another language. In Japanese this is translated as loanwords. They come from several sources. First, some foreign words are introduced along with new things and new ideas from foreign cultures; many of these, such as the large number of technical terms, have no adequate equivalent in the local language. Second, even when there is a local equivalent, foreign words are used for their novelty or the sense of prestige they gave the speaker rather than for their usefulness in communications. An example of this in Japanese is the use of the word "young" in place of the word "wakamono", especially in the fashion world. Third, a foreign word is often substituted as a euphemism for a local word, as in the case in Japanese, the word "WC" or "toire" (from toilet).

Incidence Single words, compound terms or idiomatic expressions, once they are assigned to new concepts, enter on usage 'career paths', as exemplified by the term 'democracy'. For Aristotle this was hardly the most desirable form of government, but it is now the preferred term to describe optimum political and societal life by two opposed ideologies, free–market capitalism, and centrally–planned communism. The definition of democracy therefore is different in Hellenic terms from that in socialist or market ideologies. Other forms of democracy exist in some constitutional monarchies and tribal societies. The only criterion left for the word to have some semblance of meaning is when it appears that part of the population constitutes an electorate for some of the offices of government. A debate within the United Nations system on the meaning of 'democracy' has been abandoned. This example illustrates the corruption of language by a principle of diffusion of meaning. Terms become ideologically co–opted, endure numerous accretions and eventually become valueless except in the most cynical or ironic rhetoric. Another principle of corruption of meaning occurs when a term intended to be general and flexible is introduced into intergovernmental organizations or international diplomacy and thereupon is seized by different factions and given precise but conflicting definitions so that there no longer is a common term to base discussions on. On this shorter career path could be cited expressions such as development, appropriate technology, self–determination or new international economic order, for example.

Claim The use of foreign words aside from technical terms, tend to be rather emotional and vague and are often used with meanings and in grammatical contexts quite different from their original ones. It degrades the local culture and robs the young of the profundity of their own language.

Broader Human errors and miscalculations (#PF3702).
Narrower Excessive use of acronyms and abbreviations (#PF4286).
Related Corruption of documents (#PE7900)
Incomprehensibility of specialized jargon (#PF1748).
Aggravated by Misuse of language (#PF9598) Limited verbal skills (#PD8123)
Multiplicity of languages (#PC0178) Unethical intellectual practices (#PC2915).

♦ PB2642 Cruelty

Cruel people

Nature Cruelty is the wilful infliction of unnecessary pain and suffering, i.e., pain and suffering that are not necessary to achieve morally important ends and thus cannot be justified by those ends. Cruelty is the infliction of pain and suffering for their own sake on any sentient creature that can experience pain and undergo suffering. It can be physical, emotional or spiritual. It can be directed toward humans or other animals and some would suggest toward plants. It is destructive toward both the perpetrator and the victim.

Broader Masochism (#PF3264) Unkindness (#PA5643) Pitilessness (#PA7023).
Narrower Verbal abuse (#PD5238) Child cruelty (#PJ2150)
Mental cruelty (#PG2565) Family violence (#PD6881)
Disguised cruelty (#PG3022) Cruelty to children (#PC0838)
Violence against women (#PD0247) Maltreatment of animals (#PC0066)
Abusive treatment of the aged (#PD5501).
Aggravates Maltreatment of prisoners (#PD6005).
Aggravated by Inhumanity of capital punishment (#PF0399).

♦ PB2846 Food insecurity

Dependence on food shortage — Insufficient available food — Insufficient food supply — Household food insecurity — Food grain insecurity — Shortage of food supplies

Nature At the core of the world's food shortage problem, lies the inter–relationship of the world's undernourished with their respective national food systems and the linkages with broader international trade. The undernourished are profoundly affected by the movements and trends in the domestic and international trade of foods, as well as by the governmental policies designed to influence these movements and trends. Because food represents such a large portion of a poor family's expenditures, volatile or relatively high prices for food can have devastating consequences on the poor's chances for nutritional improvement.

In the decade between 1974 and 1984, there has been a major shift in philosophies and tactics behind eliminating the world's food problem. It was believed, in 1974, that there would be a period of tight food supplies brought about by population growth without a cushion of large reserves, and that mass starvation was a distinct possibility. Therefore, what was needed was the mobilization of the political will necessary to keep food issues in the forefront; then, with increased production in food–deficit areas, the worst aspects of hunger could be eradicated within a decade. What actually happened, however, was that the basic premise was not borne out; on the contrary, the years 1976–1978 brought a substantial recovery in cereal production and in the levels of reserves held by the major grain producers. 1981 and 1982 bumper world production levels brought in record surpluses and the lowest real market prices for cereals in 30 years. In the decade between 1974–1984, world population increased by one billion, but this has not led to a strain on the world food situation. Quite the contrary; by 1984 the world cereal situation was similar to that of the early 1970s – ample supplies at the global level, depressed grain prices, and unmarketable production in North America and the European Community.

Incidence Approximately 730 million people in different parts of the world, especially in the low–income countries of Asia, Africa, and Latin America, still go without the adequate food which would enable them to lead fully productive working lives. The apparent successes of global production of foodstuffs disguise substantial regional differences. For example, average African foodgrain productivity over 35 years has declined from 50 per cent to 20 per cent in relation to European productivity, with a drop in per capita food output of 1 per cent per year since the beginning of the 1970s. Estimates show that 64 developing countries (with a population of approximately 1.1 billion) will lack the resources to feed themselves in the year 2000.

Following major droughts, there was a 3 per cent drop in total world grain production in 1972. World food stocks dropped by more than 25 per cent after 1971 to 134 million tons in 1974. Although they recovered in the late 1970s to pre–crisis levels, the level of grain security at the end of the 1980s as a proportion of consumption remains below the 1969 level.

Claim The food shortage problem stems not from a lack of production, but from international market instability. This instability has arisen as a result of the insulation of domestic production from international market, conditions, creating distortions and fluctuations both in price and production levels, with particularly adverse effects on the low–income, commodity–export dependent countries. In the present international situation, most developing nations cannot shape their agricultural policies without reference to international market conditions and prices. Developed countries must begin to shape their agricultural policies not on protectionism, but on an understanding of their implications on international prices and their impact on the low–income countries. This is a central requirement for any meaningful long–term global strategy towards the equalizing of supply and demand, surplus and shortage.

Counter–claim The rate of cereal production is currently increasing at a rate in excess of the rate of population growth. More food is produced per capita at this time than at any time in human history. The agricultural resources are available, the issue is rather the maldistribution of food supplies and the inability of many people to pay for the food. In a survey of 117 developing countries, it has been estimated that they can collectively produce enough food to feed 50 per cent more than their projected population in the year 2000, even at low levels of technology.

Refs Brown, Lester R *The Changing World Food Prospect* (1988); FAO *Approaches to World Food Security* (1983); FAO *Food Aid and Food Security* (1985); FAO *World Food Security* (1985).

Broader Economic insecurity (#PC2020) Insufficient availability of goods (#PB8891).
Narrower Crop vulnerability (#PD0660) Limited food variety (#PF0479)
Food spoilage in storage (#PD2243) Inadequate protein supply (#PC1916)
Loss of micro–organic proteins (#PD5719) Excessive cost of animal protein (#PE4784)
Spoilage of agricultural products (#PC2027) Insecurity from future crop uncertainty (#PF8354)
Long–term shortage of food and live animals (#PE0976)
Inadequate packaging of agricultural products (#PF3143)
Inadequate staple food supply in developing countries (#PD4101)
Inadequate food supplies in least developed countries (#PE7954).
Related Food interdependence (#PJ0512) Consumer vulnerability (#PC0123).
Aggravates Hunger (#PB0262) Human death (#PA0072)
Nutritional deficiencies (#PC0382) Human disease and disability (#PB1044)
Unpreparedness for food emergencies (#PC5016)
Fragility of maintaining basic health (#PJ2524)
Long–term shortage of natural resources (#PC4824)
Dietary deficiencies in developed countries (#PD0800).
Aggravated by Food wastage (#PD8844) Food hoarding (#PJ2225)
Global cooling (#PF1744) Extinction of species (#PB9171)
Maldistribution of food (#PC2801) Vulnerability of farming (#PC4906)
Unethical food practices (#PD1045) Increased food consumption (#PJ1931)
Mismanagement of food resources (#PE6115) Imbalance in world food economy (#PD5046)
Biological contamination of food (#PD2594) Slowing growth in food production (#PC1960)
Vulnerability of plants and crops (#PD5730) Lack of a world food security system (#PF5137)
Seasonal fluctuations in agriculture (#PD5212)
Instability of food prices in developing countries (#PE4986)
Underdevelopment of food and live animal production (#PF2821)
Fluctuations in food production in developing countries (#PE8188)
Inadequate mechanisms for securing sufficient food supplies (#PF2857).
Reduced by Synthetic food products (#PG2222) Cyclic neuropenia in grey Collie dogs (#PG4509).

♦ PB2853 Vulnerability of social systems

Structural tensions within society

Nature Society is now characterized by a movement towards limits of what is feasible, the overloading of increasingly complex systems, and the lack of alternatives and safety fall–backs. Concurrently, there are deep structural changes occurring in the world economy, an enforced global reorganization of capitalist production by means of rationalization and relocation, and a crisis which impinges on the basic institutional and political structure of post–war capitalism.

Broader Vulnerability (#PA1766).
Narrower Geopolitical vulnerability (#PF5749) Vulnerability of developing countries (#PC6189)
Inadequacies of the international monetary system (#PF0048)
Vulnerability of socio–economic systems from globalization (#PF1245).
Aggravates Chaos (#PF6836) Economic uncertainty (#PF5817)
Growth of anti–systemic movements (#PJ0051) Criminally life endangering behaviour (#PD0437).
Aggravated by Instability (#PA0859) Unequal distribution of forces (#PJ7928)
Shortage of fresh–water sources (#PC4815) International economic injustice (#PC9112)
Limited access to social benefits (#PF1303)
Inadequate development of new social structures in developing countries (#PD0822).

♦ PB3095 Incorrect information

Dependence on incorrect information — False information

Nature Incorrect information may arise as a result of error or intentional distortion. Errors may occur from computer or other mechanical equipment that is insufficiently programmed or inaccurately used. They may occur as the result of insufficient research or lack of information. Intentionally false information may take the form of propaganda or indoctrination, false evidence, false confession, advertising abuse and other deceptions. The results may include confusion, ignorance, a sense of defeatism or apathy, unfair trial, imprisonment (political or for perjury and contempt of court). This may strengthen government control, political dictatorship, repression, organized crime, exploitation and corruption.

Broader Falsity (#PF5900).
Narrower Perjury (#PD2630) False alarms (#PF4298) Disinformation (#PB7606)
Unreliable data (#PF6832) False confessions (#PE7252)
Compulsory indoctrination (#PD3097) Contempt of judicial process (#PD9035).
Related Propaganda (#PF1878) Misleading information (#PF3096)
Suppression of information (#PD9146).
Aggravated by Irresponsibility (#PA8658) Corruption of documents (#PE7900)
Inoperative forums for public information (#PF7805).

♦ PB3121 Denial of human rights

Nature Active repression of human rights (including the right to work, education, social security, health, national self–determination, individual liberty, freedom of thought, expression, movement, privacy, religion, and ideology) or passive refusal to ensure human rights, usually on the part of governments, but also on the part of groups and individuals, occurs regardless of constitutions, legal provisions and bona fide statements. Human societies are so organized that in practice they tend to deny at least some of man's inalienable rights to some of its members on the grounds of race, colour, sex, language, religion, political opinion, national or social origin, property, birth or other status. The widespread violations of human rights over the globe relate to the insecurity of governments that do not have a broad popular support; to the need to maintain national security in times of real or perceived external threat; to the imposition of a form of organization of society on the minority or majority that do not accept it; to the maintenance of political stability seen as a sine qua non for economic and social progress; to, sometimes, the personal idiosyncracies or perversity of dictators; and, perhaps, to the conception of power seen and lived as limitless, by conviction or tactic.

As the notion of human rights has come to be understood in contemporary international usage, it means a set of justifiable or legitimate claims with at least six features: 1) they impose duties of performance or forbearance upon all appropriately situated human beings, including governments; 2) they are possessed equally by all human beings regardless of laws, customs, or agreements; 3) they are of basic importance to human life; 4) they are properly sanctionable and enforceable upon default by legal means; 5) they have special presumptive weight in constraining human action; and 6) they include a certain number that are considered inalienable, indefensible, and unforfeitable.

Claim Human rights are necessarily indivisible; their violation in the case of a single human being implies the flouting and denial of the very principle from which they spring. If only certain rights are recognized and guaranteed, the denial or disregard of others is a sufficient denunciation of the illusory character of such partial observance.
Human rights are still not effectively assured in most of the world's countries, and there is nothing to suggest that any steady, undeviating progress is being made towards improvement. Clashes of interest; overriding reasons of state; sudden, sharp changes in the economy and in social relationships; the vicissitudes of national and international policy; inter-group antagonisms; fluctuations in power relationships; the pressures of egoism, intolerance or obscurantism; the pretexts afforded by circumstances – all these are continually responsible for retreats. It would indeed be disastrous if, in the end, these retreats were to be met by the false wisdom or illusory realism of resigned acceptance.
Counter-claim Their is no agreed upon set of human rights. Some countries are interested in individual political rights, some in collective economic rights, some freedom from torture, some the rights of the child, and others freedom from racism and apartheid. In addition to the lack of agreement to what human rights are, so many rights have been added to the list that the list of rights has become impossibly complex, unenforceable and largely meaningless. The emphasis on human rights without the corresponding human responsibilities reinforces individualism and conveys a false and immoral understanding of human beings.
Human rights legislation can be as repressive as the criminal law and as inhibiting to free speech as the criminal law.
Refs Andrews, J A *Keyguide to Information Sources on the International Protection of Human Rights* (1986); Berth, Verstappen (Ed) *Human Rights Reports* (1987); Conway, Douglas R *Human Rights* (1988); Council of Europe *The European Convention on Human Rights* (1986); Council of Europe *Human Rights in Prisons* (1987); Donnelly, Jack and Howard, Rhoda E *International Handbook on Human Rights* (1987); Eze, Osita C *Human Rights in Africa* (1987); Organization of American States (Ed) *Basic Documents Pertaining to Human Rights in the Inter-American System* (1988); United Nations *The International Bill of Human Rights* (1978); Vincent-Daviss, Diana (Ed) *Bibliography of Human Rights*.
Broader Denial of rights (#PB5405).
Narrower Denial of right to work (#PC5281) Impediments to marriage (#PF3343) Denial of economic rights (#PD4150) Denial of cultural rights (#PD5907) Denial of the right to die (#PF4813) Denial of right to justice (#PC6162) Denial of freedom of thought (#PF3217) Denial of the right to procreate (#PC6870) Denial of political and civil rights (#PC0632) Denial of rights to vulnerable groups (#PC4405) Denial of human rights in armed conflicts (#PC1454) Denial of human rights in communist systems (#PC3178) Denial of human rights in capitalist systems (#PC3124) Denial of right to pursue spiritual well-being because of discrimination (#PE7310).
Related Social inequality (#PB0514) Violation of human rights (#PB3860) Deprivation of human rights (#PC7379) Infringement of human rights (#PC6003) Denial of rights of inanimate objects (#PF3710).
Aggravates Conflict (#PA0298) Alienation (#PA3545) Human suffering (#PB5955) Lack of social mobility (#PF2195) Unequal distribution of social services (#PC3437).
Aggravated by Elitism (#PA1387) Value erosion (#PA1782) Social injustice (#PC0797) Depersonalization (#PA6953) Military government (#PC0698) Police intimidation (#PD0736) Excessive government control (#PF0304) Unsustainable population levels (#PB0035) Undemocratic political organization (#PC1015) Limited acceptance of human rights treaties (#PE7300) Delays in elaboration of remedial legislation (#PC1613) Environmental degradation of inner city areas (#PC2616) Government harassment of human rights activists (#PE6934) Limited enforceability of international standards (#PF8927) Inadequacy of international human rights instruments (#PF6365) Official cover-up of government harassment of political activists (#PF3819) Psychological barriers to the judicial protection of individual rights (#PE1479).

♦ PB3200 Exploitation

Dependence on exploitation — Exploiters
Nature Exploitation is the process, condition or result of an individual, group or institution taking unfair advantage of another usually through coercion, deception or undue influence for their own ends.
In Marxist thought it is withholding from another person, through the market or production process, what is really his due.
The use of raw materials excessively without considering the long term ecological impacts of such use.
Refs Jenkins, R *Exploitations* (1970).
Broader Domination (#PA0839) Oppression (#PB8656).
Narrower Slavery (#PC0146) Class conflict (#PC1573) Sexual exploitation (#PC3261) Racial exploitation (#PC3334) Exploitation of trust (#PC4422) Economic exploitation (#PC8132) Political exploitation (#PC7356) Exploitation of children (#PD0635) Exploitative entertainment (#PD0606) Exploitation in employment (#PC3297) Exploitative personal services (#PC3299) Over-intensive soil exploitation (#PC0052) Lack of protection for the vulnerable (#PB4353) Endangered tribes and indigenous peoples (#PC0720).
Aggravates Racism (#PB1047) Marxism (#PF2189) Socialism (#PC0115) Tribalism (#PC1910) Social inequality (#PB0514) Cultural invasion (#PC2548) Restriction of indigenous populations to reservations (#PD3305).
Aggravated by Elitism (#PA1387) Ethnocide (#PC1328) Grey lies (#PF3098) Imperialism (#PB0113) Intimidation (#PB1992) Ethnic conflict (#PC3685) Racial segregation (#PC3688) Lack of human unity (#PF2434) Cultural fragmentation (#PF0536) National disintegration (#PB3384) Isolation of ethnic groups (#PC3316) Discrimination in employment (#PC0244) Discrimination against women (#PC0308) Short-term profit maximization (#PF2174) Discrimination against minorities (#PC0582) Illiteracy among indigenous peoples (#PD3321) Fraudulent nature of inherited titles (#PE5754) Discrepancies in human life evaluation (#PF1191) Enforced participation in community activity (#PD3386) Denial of right to national self-determination (#PC1450) Vulnerability of small nations to foreign intervention (#PD2374).

♦ PB3202 Power politics

Dependence on power politics
Nature The use of power politics implies the unscrupulous use of issues to make political capital for a particular party in order to obtain or maintain power. It differs from political opportunism in being a long-term activity, an undeclared part of party policy or a trend, and in being solely a group activity. The ultimate implication of power politics is political dictatorship, but it more often constitutes a barrier to political progress, inducing political apathy and alienation and serving to maintain existing social inequalities.
Broader Domination (#PA0839).
Narrower Totalitarianism (#PF2190) Undue political pressure (#PB3209) Political barriers to effective legislation (#PC3201).
Aggravates Grey lies (#PF3098) Propaganda (#PF1878) Dictatorship (#PC1049) Official secrecy (#PC1812) Political conflict (#PC0368) Political burglary (#PD1943) Political injustice (#PC2181) Defamation of character (#PD2569) Political disintegration (#PC3204) Unethical practices in politics (#PC5517) Unjust financing of political parties (#PE0752) Unequal distribution of fame and honours (#PF3439) Misuse of electronic surveillance by governments (#PD2930) Inadequate power of intergovernmental organizations (#PF9175).
Aggravated by Political pluralism (#PF2182).

♦ PB3209 Undue political pressure

Dependence on political pressure — Political influence and influencing politics — Excessive political influence by traditional vested interests
Nature Pressure on the governing or other political parties may be exerted by interest groups or opposing parties or factions, whether for political or non-political purposes. The most powerful pressure groups may be financial, highly organized and elitist in inclination. Where vested interest is concerned, much needed reforms may be effectively blocked. Where political representation is becoming more of a group activity than an individual one, the interest of minorities and individuals who are less well organized or less wealthy may be disregarded. Pressure from political groups or parties on non-political matters may provide a barrier to progress and reform by making an emotional issue of complex problems. This may occur on a national or international level. Methods of pressure include propaganda, intimidation, internment, and the use of secret police.
Political influence can be exerted in non-political affairs, and non-political entities can enter into political affairs (as exampled by transnational corporations and religious bodies). Political influence may be gained through the use of propaganda, intimidation or other pressure and may be a function of party or power politics or international conflict (including ideological conflict) for influence over other countries.
Incidence Political pressure groups are most noticeable in constitutional democracies. In dictatorships or single-party states they may occur more as factions than distinct groups.
Claim Political pressure groups are intrinsically unequal, either in economic strength or in the degree of organization; and their demands, if met, may discriminate against minorities or individuals. If opposing pressure groups are of equal or near-equal strength they may cause severe social and political conflict which may culminate in anarchy or political and national disintegration.
Vested interest groups have access to the means of decision-making and control it in order to perpetuate their own interests and preserve a static system. In the current era of rapid social change, such groups offer the comfort of familiar approaches and familiar solutions. Disestablished groups are largely disunited and lack access to the tools of the system, limiting effective action.
Broader Power politics (#PB3202) Political conflict (#PC0368).
Narrower Entrenchment of vested interests (#PD1231).
Related Political pluralism (#PF2182) Ideological conflict (#PF3388) Inflexible central government (#PD1061) Military-industrial malpractice (#PD4361).
Aggravates Social conflict (#PC0137) Political alienation (#PC3227) Undemocratic pressures (#PD3389) Political over-reaction (#PF4110) Political corruption of the judiciary (#PE0647) Vulnerability of government to lobbying (#PF5365).
Aggravated by Conflict (#PA0298) Demagoguery (#PC2372) Military influence (#PD3385) Double standards in morality (#PF5225) Political intervention by transnational corporations (#PE0032).

♦ PB3384 National disintegration

Nature National disintegration results in control, open or disguised, by another country. It may be caused by debt, poverty, cultural invasion, economic imperialism or social breakdown within the country. A country may be subjected to military occupation because of ideological deviancy or may be substantially altered by ideological 'support'. The results of disintegration may be ethnic conflict, violence or apathy, exploitation, and forced assimilation.
Incidence Although in the 20th century it has become more difficult for nations to be absorbed or annexed by others with the use of military force, political and economic influence of large powers is very strong. Small and developing countries are very susceptible.
Claim The present-day division of the world between the two super-powers has resulted in the larger nations taking the stance of gangsters while the smaller, less self-sufficient nations are forced into the role of prostitutes or "gangsters' molls", doing whatever a larger nation requires in order to receive its protection.
Broader Disintegration (#PA6858).
Narrower Destruction of cultural heritage (#PC2114).
Related Social disintegration (#PC3309) Ethnic disintegration (#PC3291).
Aggravates Martial law (#PD2637) Nationalism (#PB0534) Exploitation (#PB3200) Ethnic conflict (#PC3685) Forced assimilation (#PC3293).
Aggravated by Anarchism (#PC1972) Foreign control (#PC3187) Social breakdown (#PB2496) Human dependence (#PA2159) National federalism (#PF0626) Political disintegration (#PC3204) Inadequacy of political doctrine (#PF3394) Inadequate ideological frameworks (#PD0065) Inadequate sense of personal identity (#PF1934) Vulnerability of small nations to foreign intervention (#PD2374).
Reduced by Excessive government control (#PF0304).

♦ PB3407 Conformism

Dependence on conformism — Conformists
Nature Conformism is anti-individualist, reduces variety in living, and tends to follow established patterns. Conformism may occur within an ethnic or other social group, reinforcing prejudices and conflict with other groups. It may occur on a national level and may involve forced assimilation and lack of cultural integration of ethnic groups, leading to a loss of cultural heritage.
Broader Anti-intellectualism (#PF1929).
Narrower Excessive standardization (#PF2271) Dictatorship of the majority (#PD3239) Discriminatory unwritten codes of behaviour (#PE7017).
Aggravates Hypocrisy (#PF3377) Conservatism (#PF2160) Forced assimilation (#PC3293) Sexual discrimination (#PC2022) Religious discrimination (#PC1455) Discrimination against homosexuals (#PE1903) Threatened and vulnerable minorities (#PC3295) Inadequate sense of personal identity (#PF1934) Underprivileged ideological minorities (#PC3325) Threats to ideological movements and minorities (#PC3362).
Aggravated by Moralism (#PF3379) Ethnocide (#PC1328) Censorship (#PC0067) Class consciousness (#PC3458) Totalitarian democracy (#PD3213) Biased government information (#PF0157) Radio and television censorship (#PD3029) Forced participation in politics (#PD2910) Repression of intellectual dissidents (#PD0434).
Reduces Individualism (#PF8393) Ideological deviation (#PF3405).

♦ PB3415 Extremism

Dependence on extremism — Extremists — Zealotry — Fanaticism

PB3415

Nature The advocacy of extreme methods, particularly in the form of mass strikes and refusal to cooperate, may be practised by minority or special interest groups.
Fanaticism is a narrow-minded, passionate and combative attitude on the part of some people who are victims of propaganda and who then propagate exaggerated ideas which offer no compromise. Fanatics may be adherents of religious and political sects, health cranks or heralders of a Utopian world, and are usually completely intolerant of discussions which debate their ideas.
Incidence The anti-nuclear peace movement has its violent fringe who sabotage or illegally attempt to occupy or interfere with nuclear power plants and with military bases, supply-lines and personnel. Hunger strikes in the Soviet Union, Northern Ireland, and elsewhere direct the individual's extremism physically towards his own body and, morally, to those who he wishes to influence by the threat that those in power will be held responsible if he dies. Islamic extremism is a part of much of the Muslim world. The stability of Nigeria is threatened. Leaders in Indonesia and Malaysia are concerned. Extremism in the west often takes racial overtones. The Order, a white supremacy sect in the United States, plans to create an Aryan homeland, eliminating blacks, Jews and white traitors. El Rukn, a black Chicago street gang is suing for recognition as a religious organization. Move in Philadelphia in a confrontation with the police led to 11 deaths and the destruction of 61 tenement houses leaving 250 people homeless. The Klu Klux Klan is still active in much of the southern, midwestern and western parts of the United States.
Claim 1. The worst enemy of human society is fanaticism (whether xenophobic, ideological or both) and the human capacity for an intelligent routinization of fanaticism. 2. Religious and political extremists perpetrate kidnappings and assassinations; labour extremists disrupt national order by general strikes to support excessive demands; and governments adopt extremism in their reactions to domestic protest and unrest, with mass arrests, dismissals, censorship, and curtailment of civil liberties.
"Everybody hopes to become a martyr", states a Palestinian regarding his young nephew's death of wounds incurred by Israeli troops in the occupied territories. "That's our highest degree of death."
Broader Insanity (#PA7157) Insufficiency (#PA5473).
Narrower Authoritarianism (#PB1638) Religious extremism (#PF4954) Intrusive animal-rights campaigners (#PE4438).
Related Militancy (#PC1090).
Aggravates War crimes (#PC0747) Intolerance (#PF0860) Student press weakness (#PE0628) Political hostage-taking (#PD1886) Politicization of scholarship (#PF7220).
Aggravated by Apathy (#PA2360) Narrowmindedness (#PA7306) Political conflict (#PC0368) Ideological conflict (#PF3388) Compulsory indoctrination (#PD3097) Double standards in morality (#PF5225) Psychogenetic constraints on behaviour (#PF7076).
Reduced by Apartheid (#PE3681) Complacency (#PA1742) Legalized racial discrimination (#PC3683).

♦ PB3430 Torture

Dependence on torture
Nature Torture means any act by which severe pain or suffering, whether physical or mental, is intentionally inflicted on a person for such purposes as obtaining from him (or from a third person) information or confession, punishing him for an act he has committed, intimidating him or other persons, or as a source of pleasure to the torturer. The methods of torture include: electric shocks; beatings, especially in sensitive areas of the body; hanging prisoners upside down; threats against the families of prisoners; deprivation of food and sleep for days; singeing with cigarettes; administration of psychoactive, curare-like, sulphur and other drugs; etc. Instances of death under torture are known. For many victims surviving the torture has not necessarily meant a blessing. Physically maimed or psychologically shattered, these people are unable to lead a normal life. Deafness, loss of speech and brain damage are among the injuries that have been sustained.
Incidence Rarely before in history has torture been of such widespread use. Torture has become a common instrument of state policy practised against almost anyone that ruling cliques see as a threat to their power. Latin America is the continent where torture is still most systematically practised by officials for political reasons. Caning and flogging and in a few countries amputations are inflicted as court-ordered punishments on common criminals. In Turkey, political prisoners are only a small minority of those who suffer torture. Baljit Singh was blinded by the police in Bihar, India in 1980; 36 suspected criminals suffered the same fate. It is estimated that two thirds of the world's governments have recently tortured or cruelly treated their political or non-political detainees.
Torture may be of a political or quasi-legal nature or of a purely personal nature where the victims may be animals, plants, members of the torturer's family or friends or other individuals. Both categories of torture hold a personal element, but the former, being institutionalized, is more widespread and devastating in its effect.
Refs Amnesty International *Torture in the Eighties* (1984); Amnesty International Staff *Amnesty International Report*; Peters, Edward *Torture* (1985).
Broader Badness (#PA5454).
Narrower Human torture (#PC3429) Cruelty to plants (#PD4148) Torture of animals (#PC3532) Institutionalized torture (#PD6145).
Related Sadism (#PF3270) Distortion (#PA6790) Insensibility (#PA5451) Unpleasantness (#PA7107) Political crime (#PC0350) Police brutality (#PD3543) Crimes against humanity (#PC1073) Offences involving danger to the person (#PD5300) Inadequate assistance to victims of torture (#PE6936) Death and disability from inhumane confinement (#PE5648) Offences against the peace and security of mankind (#PC6239).
Aggravates Atrocities (#PD6945) Political repression (#PC1919).
Aggravated by Slavery (#PC0146) Slave trade (#PC0130) Television violence (#PE4260) Resistance movements (#PJ2066) Political mass murder (#PD5590) Political dictatorship (#PC0845) Unethical medical practice (#PD5770) Inhumane medical experimentation during war-time (#PE4781) Inhumane participation of the medical profession in torture (#PE4015).

♦ PB3561 Disasters

Disaster victim — Disaster
Nature In the event of a disaster, whether natural or occasioned by man, the day-to-day patterns of life are suddenly disrupted and people are plunged into helplessness and suffering, usually aggravated by a loss of protection, shelter, clothing, food, and medical care. A disaster disturbs the vital functioning of a society. It affects the system of biological survival (subsistence, shelter, health, reproduction), the system of order (division of labour, authority patterns, cultural norms, social roles), the system of meaning (values, shared definitions or reality, communication mechanism), and the motivation of the actors within all of these systems.
A disaster may result from natural phenomena such as: earthquakes; volcanic eruptions; storm surges; cyclones; tropical storms; floods; avalanches; landslides; forest fires; massive insect infestations; and drought. Equally, the activities of man may result in a disaster: armed conflict; industrial accidents; radiation accidents; factory fires; explosions or escape of toxic gases or chemical substances; pollution; mining or other structural collapses; transport accidents; and dam failures. The outbreak of infectious diseases may occur spontaneously or as a result of a disaster situation.
Claim With an ever-increasing population, combined with more industrialization and urbanization, the likelihood of disasters occurring appears to be on the increase. There is a pressing need for studying the causes and effects of disasters, their non-parochial nature, and the staggering impact they have, especially on developing nations.
Counter-claim Disasters produce many therapeutic effects on social systems. The sharing of a common threat to survival and the widespread suffering produced by disaster usually result in a dramatic increase in social solidarity and a temporary breakdown of social and economic distinctions. This resolves pre-existing trauma, and loss or privation motivates people to devote their energies to constructive purposes. It may also provide an opportunity for new innovative solutions to chronic problems of underdevelopment.
Refs American Health Research Institute Staff *Disasters and Disaster Planning* (1987); Banner, Hubert S and Tremayne, Errol E *Calamities of the World* (1971); Gist, *Psychosocial Aspects of Disaster* (1988); Gleser, Goldine C et al *Prolonged Psychological Effects of a Disaster* (1981); Kurzman, Dan *A Killing Wind* (1987); Nash, Jay R *Darkest Hours* (1976); Schmitz, Charles A *Disaster* (1987); Taylor, James B et al *Tornado* (1970).
Narrower Natural disasters (#PB1151) Man-made disasters (#PB2075) Unreported disasters (#PF7768) Unforeseen environmental crises (#PF9769) Disasters of extraterrestrial origin (#PF3562).
Aggravates Panic (#PF2633) Chaos (#PF6836) Ugliness (#PA7240) Shallowness (#PA6993) Human death (#PA0072) Annihilation (#PF9169) Human contingency (#PF7054) Human destructiveness (#PA0832) Food spoilage in storage (#PD2243) Inadequate prevention of disabilities (#PF0709) Protein-energy malnutrition in vulnerable groups (#PD0363) Vulnerability of women and children in emergencies (#PD1078).
Aggravated by Injurious accidents (#PB0731) Disaster unpreparedness (#PF3567) Inadequate warning of disasters (#PF3565) Inadequate disaster rescue and relief (#PF0286) Inadequate disaster prevention and mitigation (#PF3566).

♦ PB3674 Pathologies of civilization

Broader Behavioural deterioration (#PB6321).
Narrower Ignorance of cultural heritage (#PF1985).
Related Man-made disasters (#PB2075).
Aggravates Human suffering (#PB5955).

♦ PB3860 Violation of human rights

Refs Friedman, Julian R and Sherman, Marc I *Human Rights* (1985).
Broader Oppression (#PB8656) Social injustice (#PC0797) Social inequality (#PB0514).
Narrower Denial of social rights (#PC0663) Violation of civil rights (#PC5285) Denial of political rights (#PD8276) Deprivation of human rights (#PC7379) Infringement of human rights (#PC6003) Denial of right to develop as human beings (#PF2364).
Related Discrimination (#PA0833) Denial of human rights (#PB3121) Crimes against humanity (#PC1073).

♦ PB4353 Lack of protection for the vulnerable

Exploitation of the vulnerable — Dependency of vulnerable groups — Disempowerment of vulnerable groups
Broader Dependence (#PA4565) Exploitation (#PB3200) Vulnerability (#PA1766).
Narrower Social injustice (#PC0797) Dependency of women (#PC3426) Exploitation of women (#PC9733) Consumer vulnerability (#PC0123) Dependency of children (#PD2476) Dependency of the elderly (#PD8399) Exploitation of the elderly (#PD9343) Exploitation of the unemployed (#PD9347) Exploitation of dependence on food aid (#PD7592) Exploitation of the mentally handicapped (#PD9685) Lack of protection for victims of intimidation (#PE7793) Inadequate protection of civilians in armed conflict (#PE8361) Medical experimentation on socially vulnerable groups (#PD6760) Lack of commitment to the protection of vulnerable groups (#PF4662) Violation of rights of vulnerable groups during states of emergency (#PD3785).
Aggravated by Denial of rights to vulnerable groups (#PC4405).

♦ PB4411 Struggle for existence

Nature Response by living creatures, in the form of novel or intensified efforts, to their environmental difficulties and limitations. This process is characterized by continuing, and often painful or fatal, clashes both between the living creatures and with their environments. The struggle is aggravated by the tendency of any species towards over-population, by the dependency of members of the population on one another and on other species, and by unpredictable changes in the environment. The struggle is thus not limited to intraspecific competition. It takes place between both fellow-organisms of the same species and between organisms of very different natures.
Claim It is a mistake to consider that the struggle is necessarily directly competitive, sanguinary, and results in the immediate elimination of one of the parties. It may often be accurately described as an endeavour after well-being. In face of difficulties and limitations, one kind of organism may intensify competition, another may exhibit more elaborate parental care or greater mutual aid, another may adopt some form of parasitism and another may change its habitat. Half-understood technical concepts from biology have frequently encouraged inappropriate conclusions. The universal struggle in nature has been erroneously used to vindicate internecine competition and warfare among men.
Narrower Intraspecific competition (#PF9201) Interspecific competition (#PF9275).
Aggravated by Suffering (#PA7690).

♦ PB4653 Insufficient financial resources

Shortage of funds — Inadequate financial reserves — Lack of capital — Lack of capital reserves — Insufficient money — Lack of equity capital — Shortage of capital
Refs Giersch, Herbert (Ed) *Capital Shortage and Unemployment in the World Economy* (1978).
Broader Shortage (#PB8238) Long-term shortage of resources (#PB6112) Limited access to social benefits (#PF1303) Underdeveloped sources of income expansion (#PF1345).
Narrower Lack of venture capital (#PG7833) Limited purchasing power (#PD8362) Insufficient risk capital (#PJ1704) Lack of savings structures (#PF1348) Lack of funds for education (#PJ6410) Shortage of funds for research (#PF5419) Inadequate agricultural capital (#PJ1368) Insufficient capital investment (#PF2852) Inadequate resources for health (#PF9587) Lack of funding for infrastructure (#PJ8512) Inadequate recreational facilities (#PF0202) Chronic shortage of foreign exchange (#PC8182) Lack of finance in coastal communities (#PE2425) Capital shortage in developing countries (#PD3137) Incomplete access to development capital (#PF6517) Inadequate level of world monetary reserves (#PF3059) Strained capital resources in small communities (#PF3665) Foreign exchange shortage in developing countries (#PD3068) Inadequate level of developing country monetary reserves (#PF3060) Shortage of financial resources for action against problems (#PF0404).

Related Inadequacy (#PA8199) Unproductive subsistence agriculture (#PC0492)
Diminishing capital investment in small communities (#PF6477).
Aggravates Unemployment (#PB0750) Inaction on problems (#PB1423)
Lack of legal aid facilities (#PF8869)
Inadequate management of government finances (#PF9672).
Aggravated by Lack of political will (#PC5180) Limited funding expertise (#PJ0278)
Lack of funding structures (#PG9919) Lack of capital development (#PD8604)
Limited accumulation of capital (#PF3630)
Inappropriate use of financial resources (#PD9338)
Incomplete access to information resources (#PF2401)
Underdeveloped capacity for income farming (#PF1240)
Limited local availability of capital reserve (#PF2378)
Lack of channels for obtaining available local funding (#PF6544)
Disincentives for financial investment within developing countries (#PF3845).
Reduced by Overstated programme advantages (#PF8181).

♦ PB4731 Deception

Deceptive people — Dependence on deceit — Deceitful people — Dishonesty — Deceit

Nature Deception includes the range of means whereby people may be mislead. The most evident of these is lying. But it also includes withholding information which the person might find of immediate significance, as well as misleading the person into some alternative belief, or reinforcing such a belief.

Claim Dishonesty, including lying, is one of the major obstacles to appropriate relationships in society and is currently reaching epidemic proportions. It is not only practiced in personal relationships but as a routine at work, in religious life, and especially in politics. When people are unable to trust each other, the fabric of society is weakened and instability, chaos and violence often result.

Counter–claim It is widely recognized that there are cases in which considerable harm can be done by supplying a questioner with the information he requested, or which he would find significant. This justifies withholding such information and thus misleading the person.

Refs Hartshorne, Hugh and May, Mark *Studies in the Nature of Character* (1975); Henderson, M Allen *How Con Games Work* (1986); Witt, James G and Fischer, William E *Deadly Deceptions* (1987).

Broader Improbity (#PA7363) Unskillfulness (#PA7232)
Behavioural deterioration (#PB6321).
Narrower Fraud (#PD0486) White lies (#PF7631) Bogus firms (#PF0326)
Self-deception (#PF6362) Mutual deceits (#PJ8029) Denial of evidence (#PD7385)
Front organizations (#PE4358) Religious deception (#PF3495)
Misleading information (#PF3096) Official self-deception (#PF7702)
Deception by management (#PD3823) Withholding of information (#PF8536)
Bogus public interest groups (#PE7575) Fraudulent nature of inherited titles (#PE5754)
Non-destructible packaging and containers (#PD1754)
Unrepresentative international organizations (#PD4873).
Related Lying (#PB7600) Error (#PA6180) Propaganda (#PF1878)
Uncommunicativeness (#PA7411) Biased presentation of news (#PD1718)
Promotion of negative images of opponents (#PF4133).

♦ PB4788 Vulnerability of world genetic resources

Destruction of genetic diversity — Genetic erosion

Nature The availability of broadly based gene pools is an essential condition for adaptation to environmental change, both natural and man-made, such as: the replacement of pesticides by genetic defences; the adaptation of high-yield varieties to local conditions; the development of resistance to evolving parasites; and the correction of nutritional defects, such as low content of protein or specific amino-acids. Genetic diversity is required to counter the inadaptability to local conditions that sometimes follows the introduction of highly selected animal species. Continual selection for specific traits within a breed or type sometimes dangerously reduces genetic variability.

Man's impact on the biosphere is increasingly reducing the genetic resiliency of many species. Not only agricultural plant varieties, but also forest species, aquatic organisms, and certain types of animals and micro-organisms are affected. The full variety of microscopic organisms provides the indispensable link in the carbon and nitrogen cycles upon which all life depends: micro-organisms include bacteria, yeasts, molds, algae, protozoa, and viruses. The quality and flavour of man's food and drink often depends upon beneficial bacteria and fungi and industry uses micro-organisms to manufacture chemical products, including antibiotics. Micro-organisms help man to understand the underlying causes of many pathological conditions and pollutant organic wastes are rendered harmless by the use of bacteria.

The development, transforming and disrupting new areas for man's use is depleting or displacing valuable genetic resources. Wild species and primitive domesticates are lost. Areas in Asia, Latin America and Africa are threatened that have traditionally served as the 'centres of natural diversity' or the natural habitation of wild varieties and as the source of genetic resources for plant improvement. Indigenous crops are replaced by new higher yielding varieties of greater genetic uniformity and less adaptability to local conditions. Many plant characteristics, such as: protein quality, oils, unique growth habit, and dwarfness, may someday be required, but are being lost with the disappearance of wild species. The introduction by man of exotic diseases and insects poses a great risk to some of the world's gene resources. For example, the chestnut blight has wiped out all but scattered remnants of the American chestnut tree. Also threatened are the remnants of forest species whose populations, often critical for breeding, can be substantially reduced and sometimes eliminated.

The narrowing of the genetic base of many of the world's crops leaves them vulnerable to pests, diseases, and changes in soil and climate. Concurrent with this is the depletion of those genetic resources essential for both the reduction of that vulnerability as well as for the production of a large number of pharmaceutical and industrial products.

Incidence Mankind shares the Earth with 5 to 10 million other species, all of which have the right to survive. Almost entirely through loss of habitat, which in turn reflects the upsurge in human numbers and consumption, species are becoming extinct at a rate of hundreds and perhaps thousands each year – the majority of these extinctions occurring in the tropics. Indeed, the world's tropical rainforests are disappearing at the rate of about 7.3 million hectares a year. Unfortunately up to half the world's genetic diversity is concentrated on only 6 percent of its land surface, mostly in tropical forests. If present trends are not reversed, mankind may witness the elimination of one million of the planet's plant and animal species by the end of this century. This represents an irreversible loss of unique genetic materials. Such extinction means a loss of crucial ecological services such as the control of pests. Increasingly, species are contributing to agriculture, medicine, industry and energy. Because of intensive selection for high performance and uniformity the genetic base of much modern food production has grown dangerously narrow. Only four varieties of wheat produce 75 percent of the crop grown on the Canadian prairies; and more than half the prairie wheatlands are devoted to a single variety. Similarly, 72 percent of US potato production depends on only four varieties, and just two varieties supply US pea production. Almost every coffee tree in Brazil descends from a mere six plants from one place in Asia. These and other crops in a similar position are extremely vulnerable to outbreaks of pests and diseases and to sudden unfavourable changes in growing conditions. Unfortunately, while the genetic base of the world's crops and other living resources is narrowing rapidly, the means by which this dangerous situation could be corrected (the diversity of crop varieties and relatives) are being destroyed. Many wild and domesticated varieties of crop plants – such as wheat, rice, millet, beans, yams, tomatoes, potatoes, bananas, limes and oranges – are already extinct and many more are in danger of following them.

History has proved the dangers inherent in genetic uniformity: it was the unrecognized cause of the Irish potato famine, which left million dead in the mid-19th century.

Up to one half of the world's genetic diversity is concentrated on only 6 percent of its land surface, mostly in tropical forests, and the world's tropical rainforests are disappearing at the rate of 7.3 hectares a year. In addition, one species per day is being lost in these forests alone. Many species are losing sub-units such as races and populations, at a rate that greatly reduces their genetic variability. Even though these species are not being endangered in terms of their overall numbers, they are suffering a decline in their genetic stocks which leads to their having only a fraction of the genetic diversity they harboured only a few decades ago.

Claim Through their genetic resources, the 5–10 million species that inhabit this planet along with man provide essential materials for agriculture, medicine, industry, energy and other economic uses. The potential for further application has only begun to be explored but, considering the manifold benefits we already enjoy, the genetic reservoir is among the most valuable natural resources with which we can confront the unknown challenges of the future.

Counter–claim Maintaining natural genetic diversity will become less important over the very long term as advances in genetic engineering make possible substitution of genes among species.

Refs Committee on Criteria and Guidelines for the Evaluation of Projects Designed to Protect or Enhance Biodiversity, Board on Biology *Evaluation of Biodiversity Projects* (1989).

Broader Environmental degradation (#PB6384) Vulnerability of organisms (#PB5658)
Erosion of biological diversity (#PB9748).
Narrower Decline in human genetic endowment (#PF7815).
Aggravates Hereditary diseases (#PG7966).
Aggravated by Deforestation (#PC1366) Genetic inbreeding (#PD7465)
Monoculture of crops (#PC3606) Extinction of species (#PB9171)
Tropical deforestation (#PD6204) Genetic and ethnic weapons (#PC6664)
Irresponsible genetic manipulation (#PC0776) Excessive use of land for agriculture (#PD9534)
Decreasing genetic diversity of animals (#PC1408)
Degradation of the environment by trees (#PE7695)
Decreasing diversity of biological species (#PD7302)
Decreasing genetic diversity in cultivated plants (#PC2223).

♦ PB4885 Hazards to human health

Health hazards — Health hazard

Broader Risk (#PF7580).
Narrower Smoking (#PD0713) Road hazards (#PD0791)
Occupational hazards (#PC6716) Pesticide intoxication (#PE2349)
Motor vehicle emissions (#PD0414) Enhanced risks of disease (#PF1824)
Hazards of bottle-feeding (#PE4935) Health hazards of asbestos (#PE3001)
Health hazards of radiation (#PD8050) Occupational risk to health (#PC0865)
Vibrations as a health hazard (#PE1145) Circumcision as a health hazard (#PE6053)
Ultraviolet radiation as a hazard (#PE5672) Inadequate environmental monitoring (#PF4801)
Inadequate prevention of disabilities (#PF0709) Fragility of maintaining basic health (#PJ2524)
Environmental hazards from electricity (#PE1412)
Health hazards of environmental pollution (#PC0936)
Inadequate health care in family planning (#PD1038)
Undesirable effects of animal feed additives (#PD1714)
Health hazards of modern insulating materials (#PE1499)
Health hazards from water development schemes (#PE8692)
Hazards to human health in the natural environment (#PC4777)
Health hazards to tourists in developing countries (#PE7538)
Environmental hazards from economic and industrial products (#PC0328).
Related Injuries (#PB0855).
Aggravates Human disease and disability (#PB1044).
Aggravated by Wasted water (#PD3669) Hazardous wastes (#PC9053)
Agricultural wastes (#PC2205) Dangerous substances (#PC6913)
Sewage as a pollutant (#PD1414) Mercury as a pollutant (#PE1155)
Environmental pollution (#PB1166) Hazardous waste dumping (#PD1398)
Domestic refuse disposal (#PD0807) Herbicides as pollutants (#PD1143)
Fungicides as pollutants (#PD1612) Pesticides as pollutants (#PD0120)
Insecticides as pollutants (#PD0983) Rodenticides as pollutants (#PE3677)
Industrial waste water pollutants (#PD0575)
Organophosphorus insecticides as pollutants (#PG2104)
Environmental hazards of nuclear power production (#PD4977).

♦ PB4993 International imbalance in the quality of life

Nature A major objective of development is the growth of per capita income in the developing countries, as one of the means to improve living conditions and reduce international disparities in income and wealth. Further, the inadequacy of housing, and the lack of basic amenities such as drinking water and sanitation, not only affect the quality of life but also the rates of morbidity and mortality. An improvement in quality of life requires not only population policies, but also programmes to improve nutritional intake, provide new and renewable sources of energy, and to improve living conditions. While there is probably no single index of the level of living that can be applied in internationally, there are four recognized demographic variables associated with a better quality of life: longer life expectancy, lower mortality rates for all age groups of the population, lower morbidity rates, and lower fertility rates.

Incidence Income or output per capita is generally taken as the main indicator of the quality of life. In 1981 the per capita income of the richest country was nearly 220 times that of the poorest country and under present growth trends, international disparities will almost certainly widen. At present growth rates, it will take 70 to 90 years for the poorest countries to double their per capita income; even this will only slightly improve their standards of living. Meanwhile, their populations will double in 35 years or less.

Refs ILO *Improving Working Conditions and Environment* (1984).

Broader Denial of right to work (#PC5281) Deteriorating quality of life (#PF7142).
Narrower Inadequate working conditions for women (#PD3197)
Inadequate working conditions of teachers (#PE7165)
Inadequate working conditions for professionals (#PE3170)
International imbalance of quality of working life (#PD9170)
Inadequate conditions of work in the textile industry (#PE5823)
Inadequate conditions of work of agricultural workers (#PE4243)
Inadequate conditions of work in the construction industry (#PE6841)
Inadequate working conditions in health and medical services (#PE7718)
Inadequate conditions of work in the hotel and catering industries (#PE4493)
Inadequate conditions of work and employment in utility supply services (#PE8319)
Inadequate conditions of work and employment of public service personnel (#PE8728).
Related Poor living conditions (#PD9156) Denial of right to free choice of work (#PE3963).
Aggravates Fatigue (#PA0657) Unemployment (#PB0750) Civil disorders (#PC2551)
Social injustice (#PC0797) Political upheavals (#PC7660)
Occupational rheumatism (#PE0502) Lack of job satisfaction (#PF0171)
Human disease and disability (#PB1044).
Aggravated by Social neglect (#PB0883) Confined spaces (#PJ7974)
Discrimination against men in employment (#PD3338).

♦ **PB5057 International conflict**
Dependence on international conflict
Nature International conflict is the result of behaviour designed to destroy, injure, thwart or otherwise control another country or group of countries or their policies. It derives from the incompatibility of goals of at least two nations or groups of nations. When two groups have important but common goals which can be attained by one group only at the expense of the other, activities directed towards the attainment of such goals are likely to become hostile and aggressive. International relations are determined, apart from the psychological factors, by economic, political, military, technological and other related developments. The range of relations possible between two nations varies from mutual cooperation and support determined by fear of a common enemy, to cold war where national interests are perceived to be antagonistic; from peaceful coexistence when national interests are not conflicting, to local military conflicts where national interests clash strongly enough to lead to use of conventional weapons, or to nuclear or total war where national survival is perceived to be in jeopardy.
Incidence International conflict is traditionally expressed in anti–foreign demonstrations, negative sanctions, diplomatic protests or severance of relations, threats or acts of military presence, hostile military actions and declared wars. Conflict can also progress in ideological, cultural, linguistic, economic and other rivalries.
Claim Countries which refuse to settle their disputes by pacific means such as negotiation, arbitration, meditation, conciliation, or judicial settlement, often turn to the threat or the use of force, or other means of coercion for the settlement of their controversies.
Refs United Nations Institute for Training and Research *The United Nations and Collective Management of International Conflict* (1986).
Broader Conflict (#PA0298).
Narrower Offences against the peace and security of mankind (#PC6239).
Related Social conflict (#PC0137) Political conflict (#PC0368)
Psychological conflict (#PE5087).
Aggravates Human suffering (#PB5955).
Aggravated by Deteriorated structures of essential corporateness (#PF1301)
Foreign intervention in internal affairs of states (#PC3185).

♦ **PB5151 Misuse of resources**
Broader Abuse of government power (#PC9104).
Narrower Industrial processes geared to reduced social needs (#PE3939).
Related Unproductive use of resources (#PB8376)
Worldwide misallocation of resources (#PB6719).

♦ **PB5249 Indifference to suffering**
Failure to report suffering — Silence in response to suffering — Avoidance of reference to suffering — Indifference in response to injustice — Moral indifference — Moral tourism
Incidence The international community reports very selectively and to widely varying degrees on the suffering of peoples. Very extensive coverage may be given to accidents involving several hundreds of deaths. Little reference, or none at all, may be made to massacres of tens of thousands over a period of months. Whilst the media may successfully report on accidents and disasters, and occasionally on situations involving thousands of people over an extended period of time, international agencies are obliged to avoid referring to situations which are considered to be the internal concerns of particular countries or else limit themselves to emitting token resolutions protesting them. Examples include the massacres in Kampuchea and Uganda.
Claim Television has turned people into moral tourists: it is a trip to visit other people's suffering. Television has made it possible to witness existing horrors second–hand and to feel strong emotions, but the memory is short and the indignations are very brief.
Broader Indifference (#PA7604).
Narrower Silence about historical situations (#PF0608).
Aggravates Complicity with structural injustice (#PJ5026)
Failure to legally rehabilitate victims of miscarriage of justice (#PF7728).

♦ **PB5250 Natural environment degradation**
Disruption of ecosystems by human activity — Erosion of the human carrying capacity of the environment — Degradation of natural resources — Depletion of natural resources — Destruction of wilderness areas — Terracide
Nature Most degradation of natural resources results from the cumulative activities of farmers, households, and industries, all trying to improve their socio–economic well–being. These activities tend to be counter–productive for several reasons. People may not completely understand the long–term consequences of their activities on the natural resource base. Ill–defined or badly enforced property rights may result in environmental losses, as in the case of communal grazing lands, tree crops or water resources. Poverty can undermine the efficiency of market processes in accounting for long–term environmental concerns.
Environmental and ecological stress can be distinguished as the stress on humans experiencing such environmental degradation and resource depletion.
Incidence Throughout most of history, the interactions between human development and the environment have been relatively simple and localized. The complexity and scale of these interactions are now increasing, especially as resources became scarcer and competition for them increases. What were once local pollution incidents shared throughout a common watershed or air basin now involve several, or many, nations, as in the case of acid precipitation. Acute episodes of relatively reversible damage now affect multiple generations, as in the case of the disposal of radioactive wastes. Straightforward questions of ecological preservation versus economic growth how involve complex linkages, as in the case of the interaction between energy and crop production, deforestation and climate change.
The most important ways in which human activity is interfering with the global ecosystem are: (a) fossil fuel burning which may double the atmospheric carbon dioxide concentration by the middle of the next century, as well as further increasing the emissions of sulphur and nitrogen very significantly; (b) expanding agriculture and forestry and the associated use of fertilizers (nitrogen and phosphorous) are significantly altering the natural circulation of these nutrients; (c) increased exploitation of the fresh water system both for irrigation in agriculture and industry and for waste disposal.
According to present understanding the most important impacts of these changes in the long–term are: (a) a gradual change towards a warmer climate of which very little is known; (b) a decrease in the concentration of ozone in the stratosphere, due to the increased release of nitric oxides and chlorine compounds and increase in the troposphere, due to increased release of hydrocarbons and nitrate compounds; (c) an increase of the areas affected by lake and stream acidification in mid–latitudes and possibly also in the tropics, associated with the possibility of significant disturbance in the ion balance of soils (as is now being found in the case of aluminium); (d) a decrease of the extend of tropical forests, which will enhance the rate of increase in atmospheric carbon dioxide concentration and release other minor constituents to the atmosphere, which may also contribute to soil degradation; (e) due to loss of organic matter and nutrients, soil deterioration will occur and this implies a reduced possibility for the vegetation to return to pristine conditions; (f) a trend toward the eutrophication of estuarine and coastal marine areas; (g) more frequent development of anoxic conditions in freshwater and marine systems and sediments.
Counter–claim 1. Given the absorptive and regenerative capacity of the atmosphere, water and soil, numerous abuses to the environment may prove to have been transitory rather than irreversible.
2. A survey in 1989 indicates that at least one third of the land surface, more than 48 million square kilometres, is still wilderness. Most of the settled continents are between a quarter and a third wilderness, with the exception of Europe, where there is almost none. (These estimates exclude blocks of land under 4,000 square kilometres or those crossed by roads, tracks or other signs of human activity).
Refs Evernden, Neil *Natural Alien* (1985); Passmore, John *Man's Responsibility for Nature* (1980).
Broader Environmental degradation (#PB6384).
Narrower Soil erosion (#PD0949) Deforestation (#PC1366)
Forest decline (#PC7896) Endangered urban trees (#PD2025)
Landscape disfigurement (#PC2122) Anthropogenic climate change (#PC9717)
Destruction of the countryside (#PE3914) Long–term shortage of resources (#PB6112)
Disruption of seabed ecosystems (#PG7989) Cumulative environmental impacts (#PF1105)
Regional environmental degradation (#PD5845) Destruction of environmental oxygen (#PE5196)
Disruption of the hydrological cycle (#PD9670)
Unsustainable agricultural development (#PC8419)
Degradation of the environment by trees (#PE7695)
Criminal offences against the environment (#PC4584)
Environmental threats to national security (#PC4341)
Introduction of new species of insect pests (#PF3592)
Destructive changes in ocean characteristics (#PC2087)
Uncertain environmental impact of current policy (#PF9450)
Disruption of animal migration and movement patterns (#PC2279)
Degradation of the environment in developing countries (#PD3922)
Depletion of natural resources due to population growth (#PD4007)
Disruption of ecosystems in marginal agricultural lands (#PD6960)
Ecosystem modifications due to creation of dams and lakes (#PD0767)
Degradation of the environment through the destruction of species (#PE5064)
Degradation of semi–natural and natural habitats of flora and fauna (#PC3152).
Related Silting of water systems (#PD3654)
Environmental degradation of desert oases (#PD2285).
Aggravates End of nature (#PF9582) Environmental stress (#PC1282)
Irreversible problem emergence (#PF9790) Erosion of biological diversity (#PB9748)
Deterioration of human environment (#PC8943)
Lack of incentive for users to care for common property (#PF4516).
Aggravated by Maldevelopment (#PB6207) Undomesticated men (#PF0551)
Trade protectionism (#PC4275) Unreported accidents (#PF2887)
Human exceptionalism (#PF6730) Permafrost instability (#PD1165)
Environmental prodigality (#PF7318) Inappropriate development policy (#PF8757)
Vulnerability of ecosystem niches (#PC5773)
Unconstrained socio–economic growth (#PB9015)
Long–term shortage of natural resources (#PC4824)
Inappropriate design of development projects (#PF4944)
Disadvantageous terms for technology transfer (#PE4922)
Unconstrained exploitation of natural resources (#PF2855)
Outflow of financial resources from developing countries (#PC3134)
Incompatibility of environmental and economic decision–making (#PF9728)
Unsustainable short–term improvements in agricultural productivity (#PE4331)
Excessive social costs of structural adjustment in debtor developing countries (#PD8114).

♦ **PB5405 Denial of rights**
Violation of rights — Infringement of rights — Deprivation of rights — Contravention of rights
Narrower Denial of human rights (#PB3121) Denial of state's rights (#PD4814)
Violation of civil rights (#PC5285) Denial of rights of animals (#PC5456)
Denial of right of assembly (#PC2383) Denial of rights of businesses (#PD4728)
Denial of rights of minorities (#PC8999) Restriction of freedom of expression (#PC2162)
Denial of rights of inanimate objects (#PF3710) Denial to animals of the right to life (#PF8243)
Denial of right of a people to be self–determining (#PC6727).

♦ **PB5417 Destabilization of social systems**
Nature Destabilization is sabotage of the economy, political system or social forms of a rival in such a way as to destroy the ability to resist or retaliate. This is a popular form of government–sponsored conflict because it does not require as much accountability as more public actions.
Broader Conflict (#PA0298).
Narrower Destabilizing governments (#PD5693)
Military destabilization of developing countries (#PD7714)
Political destabilization of developing countries (#PD9792)
Aggressive economic destabilization of countries by external forces (#PE9420).
Related Sabotage (#PD0405).
Aggravated by Foreign intervention in internal affairs of states (#PC3185)
Political intervention by transnational corporations (#PE0032)
Vulnerability of socio–economic systems from globalization (#PF1245).

♦ **PB5486 Arbitrariness**
Caprice
Narrower Expulsion (#PC5313) Arbitrary exile (#PJ7268) Parole violation (#PE1121)
Legal inconsistency (#PF5356) Extra-legal executions (#PE6366)
Infringement of privacy (#PB0284) Arbitrary street search (#PG6632)
Inconsistent port charges (#PF5887) Abusive national leadership (#PD2710)
Inadequately worded agreements (#PF5421) Unnatural boundaries between states (#PF0090)
Arbitrary enforcement of regulations (#PD8697)
Arbitrary sequestration of real property (#PG1185)
Artificial and arbitrary job qualifications (#PF2066)
Government expropriation of private property (#PD3055)
Arbitrary external interference in family life (#PE4058)
Arbitrary evaluation of disability compensation (#PG7870).
Related Internment without trial (#PD1576).

♦ **PB5577 Social stratification**
Social status — Low status
Related Inadequate living and working conditions of immigrant labourers in industrialized countries (#PD3427).
Aggravates Hypergamy (#PF5430) Inferior status employment (#PD8996).
Aggravated by Certificate–based job market (#PJ8370).

♦ **PB5646 Human physical suffering**
Dependence on human physical suffering
Broader Human illness (#PA0294) Human suffering (#PB5955)
Vulnerability of human organism (#PB5647).
Narrower Hunger (#PB0262) Asphyxia (#PE4104) Human ageing (#PB0477)
Malnutrition (#PB1498) Stress in human beings (#PC1648)
Human disease and disability (#PB1044) Genetic defects and diseases (#PD2389)
Self–inflicted physical suffering (#PD7550)
Harmful effects of sensory deprivation (#PE5787)
Disease and injury from exposure to weather (#PF5739)
Disease and injury from physical confinement (#PE5763).
Related Pain (#PA0643) Mental suffering (#PB5680).

Aggravates Human death (#PA0072) Health inequalities (#PC4844)
Perinatal morbidity and mortality (#PD2387).
Aggravated by War (#PB0593) Fear (#PA6030) Injuries (#PB0855)
Starvation (#PB1875).

♦ PB5647 Vulnerability of human organism

Broader Vulnerability of organisms (#PB5658).
Narrower Hunger (#PB0262) Mutation (#PF2276) Mental suffering (#PB5680)
Learned helplessness (#PE6990) Stress in human beings (#PC1648)
Human physical suffering (#PB5646) Limited individual attention span (#PF2384)
Uncertainty of survival of the human race (#PE9085).
Aggravates Human death (#PA0072) Human suffering (#PB5955)
Human contingency (#PF7054) Indeterminacy of death (#PF0192).

♦ PB5658 Vulnerability of organisms

Broader Vulnerability (#PA1766).
Narrower Animal injuries (#PC2753) Vulnerability of human organism (#PB5647)
Vulnerability of plants and crops (#PD5730)
Vulnerability of world genetic resources (#PB4788).
Related Vulnerability of food chains (#PB2253).
Aggravated by Instability (#PA0859) Shallowness (#PA6993)
Annihilation (#PF9169) Genetic defects and diseases (#PD2389).

♦ PB5680 Mental suffering

Dependence on mental suffering
Broader Absence of God (#PF5764) Human suffering (#PB5955)
Vulnerability of human organism (#PB5647).
Narrower Male mid–life crisis (#PD5783) Female mid–life crisis (#PD5675).
Related Loneliness (#PF2386) Dissatisfaction (#PA8886)
Human physical suffering (#PB5646).
Aggravates Health inequalities (#PC4844).
Aggravated by Alienation (#PA3545) Humiliation (#PF3856)
Friendlessness (#PB5747) Oppressive reality (#PF7053).

♦ PB5747 Friendlessness

Dependence on friendlessness — Friendless people
Nature The nearly universal institution of close friendship is eroded by many forces in modern life, leading to loneliness, undue pressure on marriage to supply a partner's total needs for intimacy, accumulation of feelings of guilt and shame as friendship relationships are diminished and destroyed over the life of the individual, and resulting cynicism and decreasing possibilities for individual independence of judgement and action as the individual becomes more isolated. Though modern life supplies a person with a large number of acquaintances, colleagues and superficial friends, it tends to prevent those relationships from deepening into engagement and commitment because of the effects of personal, social, economic and physical mobility, among many other factors.
Claim Deep personal relationships are typical of human history. Their diminution increases feelings of alienation, anomie, and loneliness. Friendlessness contributes to mental and physical disease, to manipulation of the individual by governments or advertisers, and to the inability of individuals to develop, clarify and systematize their thinking about reality, independent of social conditioning.
Counter–claim The elimination of dyadic friendship opens the way for individuals to relate to broader groups, up to and including their identification with the whole of mankind. Friendship is a relationship typical of primitive social conditions. Increased wealth, political and physical security and various sorts of mobility allow individuals to free themselves from narrow bounds and to develop their potential in ever more various circumstances and, therefore, with ever more individual results.
Broader Social neglect (#PB0883).
Aggravates Mental suffering (#PB5680) Negative emotions and attitudes (#PA7090).
Aggravated by Ugliness (#PA7240).

♦ PB5765 Ethnocentricity

Dependence on ethnocentricity — Ethnic chauvinism — Ethnocentrism
Nature Ethnocentrism is the attitude and/or ideology concerning the relationship between an individual's own group and other groups. Positive aspects of the subject's group are strongly emphasized while features and members of other groups are judged in terms of standards applicable to the subject's group, and are often denigrated. An easy rejection of the unfamiliar is characteristic of ethnocentrism which therefore makes it a component of prejudice. In pluralistic societies, ethnocentricity can be destructive to patriotism and good citizenship, and lead to exaggerated demands for cultural and political autonomy.
Incidence North American Indian myths show various tribes claiming to have been created at 'the heart of the world'; anthropologists occasionally exhibit ethnocentric attitudes when they evaluate the culture or behaviour of members of another society by the light of their own culture; neo–colonialism is supported by ethnocentrics delineating 'modern civilization' from 'backwardness'; and in the Middle East, wars are being raged by various groups holding their cultures and ideologies to be superior to those of neighbouring groups.
Claim Racism, chauvinism, and other forms of ethnocentricity are by–products of exaggerated claims on behalf of the antecedents of current generations. Every conceivable virtue and gift may be claimed for a nation's forefathers or for the racial ancestors of a continent. Every achievement – artistic, humanistic, theological, or technological – may be cited as unique, and the combination of achievements presented as evidence of superiority. What is claimed for the forebears becomes the transmitted property – physically, culturally, genetically and psychologically – of the present wearers of an ethnic 'identity'. All the institutions and characteristics of the culture receive a sacred character by such excess. Change and improvement become obstructed and inter–ethnic dialogue ceases. Excessive cultural pride may lead to conceding to demands for the creating or maintaining of a separate church or single state religion, and of an official, state–approved, single (and sometimes unique) language.
Refs Reynolds, Vernon, et al (Eds) *The Sociobiology of Ethnocentrism* (1987).
Broader Disaccord (#PA5532) Narrowmindedness (#PA7306).
Narrower Racism (#PB1047) Nationalism (#PB0534).
Aggravates Ignorance (#PA5568) Fragmentation (#PA6233)
Multiplicity of languages (#PC0178) Negative emotions and attitudes (#PA7090).
Aggravated by Insufficient minority culture support (#PF5659).

♦ PB5858 Expansionism

Narrower Economic expansion (#PF8111) Territorial expansionism (#PC9547)
Destabilizing corporate expansion (#PD1220) Undirected technological expansion (#PC1730)
Undirected expansion of economic base (#PF0905).
Reduced by Denial of right to business growth (#PE2700).

♦ PB5955 Human suffering

Nature All human beings experience in varying degrees physical, mental and spiritual afflictions, which are the three facets of human suffering.
Counter–claim The voluntary acceptance of suffering, supernaturally motivated, has a definite place in authentic Christian asceticism, and there can be times and circumstances in which physical suffering is not only implied in the pursuit of Christian perfection but may even be demanded in adherence to basic Christian morality.
Refs Petrie, Asenath *Individuality in Pain and Suffering* (1978).
Broader Suffering (#PA7690).
Narrower Mental suffering (#PB5680) Human physical suffering (#PB5646).
Related Pain (#PA0643).
Aggravates Angst (#PA6950) Health inequalities (#PC4844)
Underestimation of the human potential (#PF7063).
Aggravated by War (#PB0593) Vice (#PA5644) Crime (#PB0001)
Absurdity (#PF6991) Ignorance (#PA5568) Injustice (#PA6486)
Immorality (#PA3369) Malevolence (#PA7102) Unemployment (#PB0750)
Original sin (#PF8298) Fragmentation (#PA6233) Global crisis (#PF6244)
Social neglect (#PB0883) Absence of God (#PF5764) Social conflict (#PC0137)
Lack of control (#PF7138) Human inequality (#PA0844) Family breakdown (#PC2102)
Human infertility (#PC6037) Human contingency (#PF7054) Moral imperfection (#PB7712)
Structural violence (#PB1935) Marital instability (#PD2103) Sexual unfulfilment (#PF3260)
Socio–economic poverty (#PB0388) Stress in human beings (#PC1648)
Denial of human rights (#PB3121) International conflict (#PB5057)
Restrictions on freedom (#PC5075) Lack of job satisfaction (#PF0171)
Inadequate child welfare (#PC0233) Behavioural deterioration (#PB6321)
Denial of right to liberty (#PF0705) Inadequate health services (#PD4790)
Pathologies of civilization (#PB3674) Human disease and disability (#PB1044)
Inadequate standards of living (#PF0344) Unsustainable population levels (#PB0035)
Vulnerability of human organism (#PB5647) Negative emotions and attitudes (#PA7090)
Denial of the right to procreate (#PC6870) Inequitable distribution of wealth (#PB7666)
Corruption of the good in human nature (#PE7917)
General obstacles to problem alleviation (#PF0631)
Disintegration of technological capacity (#PD7719)
Misdirection of human energies and desires (#PF8128)
Excessive medical intervention in childbirth (#PE7705).

♦ PB6021 Discriminatory international order

Dependence on discriminatory international order
Nature The tension between the developed and the developing countries inhibits developing countries from improving their position in the world community, because of the sacrifices such progress would inherently entail for the developed countries. Discrimination in the current international order may be outlined as follows: one quarter of the world's population currently lives in the developed countries of North America, Europe and Japan, with per capita incomes of 10 to 11 times those of the other 75 percent; the overall condition of life in developing countries is diminishing while in most developed countries it is improving. The WHO estimated in 1982 that 490 million people had less food than was necessary for basic survival; hundreds of millions suffer from debilitating diseases caused by malnutrition, insanitary conditions and pests; only one third of the people in developing countries have access to safe water supplies; the world's illiteracy rate of 50 percent is centred mostly on the developing countries where 40 percent of the people are under 15 years of age, compared with less than 25 percent for developed countries; and rural unemployment in developing countries is causing a massive drift to already overcrowded, besieged cities.
Scarcity of the resources vital to economic development and growth hits the developing countries the hardest: not only do they need to export most of their resources in order to develop their economies and to import those resources they lack, but many developed countries are stockpiling fossil fuels and minerals in order to secure their own future needs, thus diminishing prospects of the developing countries being able to secure supplies adequate for their present requirements. The maldistribution of industrial power and technology, and of the wealth they produce, exacerbates the discriminatory international order, as does the mounting pollution of soil, water and air by the wealthy countries, which are rapidly poisoning and extinguishing plants and animals necessary to the survival of the entire world.
Broader Discrimination (#PA0833)
Antiquated world socio–economic order (#PF0866).

♦ PB6112 Long–term shortage of resources

Resource depletion — Dependence on resource depletion
Nature The abundance of natural resources used to be generally assumed, but questions are beginning to be raised on a many fronts, including the availability of fuel and of nonfuel minerals, the potential productivity of agriculture, the supply of forest products and of water, and the increasing pressures of world population growth. In addition, industrial and agricultural processes now are capable of generating so much pollution that the life–support systems of the planet may be threatened. The interdependent systems of air, water, land and climate could be overloaded by waste by–products long before society is confronted with economic scarcities of other natural resources.
Counter–claim Resources will continue to be limited. But the crucial factor is that the ratio of resources to results can be vastly improved through new knowledge, new technology, and hard–won experience.
Refs Brown, Lester R *Resource Trends and Population Policy* (1979).
Broader Shortage (#PB8238) Natural environment degradation (#PB5250).
Narrower Loss of human resources (#PC7721) Insufficient financial resources (#PB4653)
Long-term shortage of natural resources (#PC4824)
Long–term shortage of uranium resources (#PG4417)
Depletion of natural resources due to population growth (#PD4007).
Aggravates Worldwide misallocation of resources (#PB6719).
Aggravated by Dependence on oil (#PJ4398) Inadequate mobilization of resources (#PF4979).

♦ PB6207 Maldevelopment

Dependence on maldevelopment — Imbalance of development — Inappropriate development — Uneven development — Exploitative development
Nature Maldevelopment in the newly liberated, erstwhile colonial, poor and backward countries is the result of many forces. At the base is the appalling problem of poverty, inherited inequality and the lack of equal opportunity as a legacy of the tribal and feudal past. This legacy was prolonged by the colonial system, which not only allowed tribal and feudal hierarchies to survive and perpetuate the unequal social relations that directly contribute to continuing poverty and keep economies at low levels of performance; but also introduced a new form and higher level of exploitation, by transforming the indigenous traditional political and economic elite into an intermediate stratum of collaborators in the colonial design of political domination and economic aggrandisement.
Counter–claim Very uneven development is acceptable as a means of initiating and maintaining the development process. Although those who benefit in the short–term are the rich and a small number of middle class and urban workers, it is they who are capable of buying products and services. Only through rapid development of the cities and the export economy can development get underway. National wealth thus grows at the fastest possible rate, providing more to trickle down to all. More goods become available and more poor people are able to get jobs as more factories open up. The new wages generate more demand for more goods and for more factories. Through this process the lower classes and those in rural areas will eventually be drawn more fully into the developing economy.

PB6207

Narrower Destruction inherent in development (#PF4829).
Aggravates Unsustainable development (#PB9419)
Natural environment degradation (#PB5250)
Ecologically unsustainable development (#PC0111)
Deterioration of industrialized countries (#PD9202).
Aggravated by Inappropriate loans (#PF4580) Uncontrolled markets (#PF7880)
Export-led development (#PE6237) Inappropriate development policy (#PF8757)
Inappropriate foreign investment (#PD8030) Maldistribution of productive capacity (#PC9785)
Inappropriate modernization of agriculture (#PF4799)
Incompatibility of environmental and economic decision-making (#PF9728)
Reinforcement of inappropriate development by privileged classes in developing countries (#PF6670).

♦ PB6248 Mental pollution
Dependence on mental pollution
Incidence Mental pollution occurs in all areas of life. It ranges from the violence and sex prolific in today's visual media, to South Africa's apartheid system, to the war in the middle East, to the cold war between the USA and the USSR.
Claim The pollution of man's mind by such pollutants as selfishness, jealousy, greed, anger and lust is responsible for the disequilibrium apparent in all aspects of today's world. The greatest danger lies in the fact that mankind does not recognize that mental pollution has gained dangerous proportions and that a polluted mind manifests itself by making the atmosphere tense and vicious. Mental pollution has spoiled the socio-economic atmosphere and political climate everywhere and has made man fight man, nation prepare itself militarily against nation, and allowed one person to exploit many others by subjecting them monetarily, politically, militarily, sexually and in other ways. Mental pollution is a global problem because every man is affected in one may or another.
Counter-claim Jealousy, selfishness, greed, anger and lust are, for better or for worse, inherent components of man's nature and it is unrealistic and idealistic to assume that these 'mental pollutants' will even be contained, much less disposed of.
Broader Pollution (#PB6336) Moral imperfection (#PB7712).
Narrower Psychological pollution by mass media (#PD1983).
Related Immorality (#PA3369) Psychological environment degradation (#PC3144).
Aggravated by Toughness (#PA6976) Misinformation (#PD8523)
Proliferation of information (#PC1298).

♦ PB6302 Mental tension
Nature Present-day reasons for increased stress are: rapid and disturbing changes due to modern science and technology; education that creates a spirit of competition and rivalry; the great expansion of cities; the change in values and lifestyles; the break-up of family life and loss of faith and love in mutual relations. This stress or build-up of mental tension does not only result in a lack of peacefulness but causes many physical and mental diseases such as: high blood pressure, insomnia, indigestion, asthmatic ailments, peptic ulcers, heart attacks and even cancer. It lowers man's immunity to diseases and accelerates the process of ageing.
Claim Although the problem of increased stress starts at the individual or group level, it acquires global dimensions. Wars, hot or cold, are an outcome of tension, as are communal riots or ethnic strife. Political tension may result in nuclear war which, in turn, could cause world catastrophe. Almost every individual suffers from tension and a man under tension often disturbs others and spreads waves of peacelessness. Sustained or frequent states of tension make a person physically and mentally weak so that he is unable to face even the slightest opposition or the smallest problem.
Broader Tension (#PB6370).
Related Premenstrual tension (#PE3761).
Aggravates Human disease and disability (#PB1044).
Aggravated by Stress (#PB9165).

♦ PB6314 Credibility gap
Dependence on credibility gap — Loss of credibility
Broader Obstacles to leadership (#PF7011).
Narrower Loss of institutional credibility (#PF1963)
Loss of confidence in government leaders (#PF1097)
Loss of credibility in international institutions (#PE8064)
Increasing scepticism about the accuracy of official information (#PF7649).
Aggravated by Institutional lying (#PD2686) Inaction on problems (#PB1423)
Broken government promises (#PF4558).

♦ PB6320 Disadvantaged groups
Nature Disadvantages of economically poor and/or minority groups typically include deprivations in housing, education, work opportunity and medical care (most often pre-natal); and are associated with family disruption, faulty identity formation or malignant identity diffusion, and excessively high rates of juvenile offences and of admissions to mental hospitals.
Aggravated by Socio-economic poverty (#PB0388).

♦ PB6321 Behavioural deterioration
Broader Underestimation of the human potential (#PF7063).
Narrower Deception (#PB4731) Consumerism (#PD5774) Eccentricity (#PF9063)
Intimidation (#PB1992) Ethical decay (#PB2480) Pusillanimity (#PJ8038)
Loss of civility (#PC7013) Political feuding (#PD4846) Institutional lying (#PD2686)
Disruptive behaviour (#PD8544) Pathologies of civilization (#PB3674).
Related Decadence (#PB2542) Aggression (#PA0587) Incompetence (#PA6416)
Inefficiency (#PB0843)
Criminal negligence in performing socialist responsibilities (#PE5538).
Aggravates Human suffering (#PB5955).
Aggravated by Vice (#PA5644) Angst (#PA6950)
Lying (#PB7600) Conflict (#PA0298)
Immorality (#PA3369) Mental illness (#PC0300)
Abuse of power (#PB6918) Salmonella infections (#PG5752)
Human disease and disability (#PB1044) Massive psychic traumatization (#PE6968)
Negative emotions and attitudes (#PA7090)
Corruption of the good in human nature (#PE7917)
Misdirection of human energies and desires (#PF8128).

♦ PB6332 Reactionary forces
Nature Reactionary forces not only resist change but seek to restore some earlier order of society, thus blocking change and annulling reforms already achieved.
Broader Conservatism (#PF2160) Traditionalism (#PF2676).
Aggravated by Lack of control (#PF7138) Excessive government control (#PF0304).

♦ PB6335 International crisis escalation
Nature Crisis escalation is the process whereby each side in turn increases the scope of an international crisis or the violence of an international conflict in the hope that its adversary's self-imposed limits will be reached before its own.
Incidence President Kennedy, during the 1962 Cuban missile crisis, deliberately escalated the crisis to the point where any further escalation exceeded the self-imposed limits of the Soviet Union; and his administration also developed an explicit doctrine of controlled escalation to give credibility to the US posture of extended deterrence in Western Europe. The Soviet Union now embraces a similar policy regarding the relation to a NATO attack on Warsaw Pact countries. President Reagan's Strategic Defence Initiative is an example of crisis escalation.
Aggravates Global crisis (#PF6244).
Aggravated by Tension (#PB6370) International aggression (#PB0968).

♦ PB6336 Pollution
Polluters
Background Until the 16th century, the verb 'to pollute' and the noun 'pollution' were used mainly in relation to morals and religion, thus pollution was defined as ceremonial impurity or defilement, or as profanation of some thing or place held to be sacred.
Claim Pollutions of varying type are the main causes of today's deteriorating quality of life. Not only is the physical environment contaminated with poisonous or unpleasant substances, but pornography may be considered literary pollution, and excessive noise may pollute places of work or entertainment.
Counter-claim The term pollution is generally used to describe the presence of chemical substances produced by man's activities but not to describe the natural existence of even higher levels of the same substance. Thus, if a factory discharges lead into a river, even if it is diluted down to harmless levels, this is considered to pollute the river; but the lead naturally leached from rocks, even yielding high levels which can be dangerous to fish, is not considered to be a pollutant. Such loose usage of the term 'pollution' is biased and the word should be employed only when damage occurs.
Refs Commission of the European Communities and Environmental Resources Ltd *The Law and Practice Relating to Pollution Control in the Member States of the European Communities* (1983); Goldstein, Eleanor C *Pollution*; Springer, Allen L *The International Law of Pollution* (1983).
Narrower Pollutants (#PC5690) Mental pollution (#PB6248) Domestic polluters (#PD4524)
Agricultural pollution (#PD0563) Governmental polluters (#PD5370)
Environmental pollution (#PB1166) Wildlife pollution hazard (#PJ3387)
Urban-Industrial pollution (#PC8745) Inhibition of crop growth by pollution (#PE8476).
Related Badness (#PA5454) Complexity (#PA6468) Impairment (#PA6088)
Uncleanness (#PA5459) Unhealthfulness (#PA7226) Inappropriateness (#PA6852).
Aggravates Disease (#PA6799)
Damage to cultural artefacts by environmental pollution (#PD2478).

♦ PB6370 Tension
Narrower Mental tension (#PB6302) International tension (#PB8287).
Related Fear (#PA6030) Enmity (#PA5446) Disaccord (#PA5532)
Aggression (#PA0587) Insufficiency (#PA5473).
Aggravates International crisis escalation (#PB6335).
Aggravated by Caffeine abuse (#PE0618) Lack of communication (#PF0816)
Lack of international cooperation (#PF0817).

♦ PB6384 Environmental degradation
Dependence on environmental degradation
Nature Increasing population density, urbanization, industrialization and other development schemes, are exerting ever increasing pressure on the carrying capacity of land and resources, leading in particular to a rapid rate of deforestation, expanding desertification in some countries, and the near extinction of some wildlife species. Physical factors such as earthquakes, hurricanes, monsoon rains and the immense problem of poverty in some countries accelerate the decline in environmental resources.
Broader Global crisis (#PF6244) Human destructiveness (#PA0832).
Narrower Oil pollution (#PE1839) Poisonous algae (#PE2501)
Soil degradation (#PD1052) Man-made diseases (#PD6663)
Radioactive fallout (#PC0314) Agricultural wastes (#PC2205)
Migrating sand dunes (#PD0493) Unhealthy environment (#PJ1680)
Evolutionary catastrophes (#PF1181) Destruction of coral reefs (#PD5769)
Destruction of land fertility (#PC1300) Destruction of natural barriers (#PC1247)
Natural environment degradation (#PB5250) Destruction of wildlife habitats (#PC0480)
Radio noise of industrial origin (#PE2473) Vulnerability of ecosystem niches (#PC5773)
Environmental consequences of war (#PC6675) Eutrophication of lakes and rivers (#PD2257)
Hazards of strong toxic substances (#PD0122) Vulnerability of marine ecosystems (#PC1647)
Deterioration of human environment (#PC8943) Destruction of environmental oxygen (#PE5196)
Hybridization of wild animal species (#PD2419) Deforestation of mountainous regions (#PD6282)
Psychological environment degradation (#PC3144)
Socio-cultural environment degradation (#PC4588)
Environmental degradation by pipelines (#PE6251)
Decreasing genetic diversity of animals (#PC1408)
Endangered species of plants and animals (#PB1395)
Environmental degradation by automobiles (#PE6142)
Vulnerability of world genetic resources (#PB4788)
Encouragement of drug resistant diseases (#PJ3767)
Environmental degradation of desert oases (#PD2285)
Environmental hazards from energy production (#PD6693)
Degradation of the environment through contamination (#PE4759)
Environmental degradation in industrialized countries (#PD7835)
Degradation of the environment in developing countries (#PD3922)
Accidental large-scale contamination of the environment (#PD1386)
Ecosystem modifications due to creation of dams and lakes (#PD0767)
Natural environmental degradation from recreation and tourism (#PE6920)
Environmental degradation by off-road and all-terrain vehicles (#PE1720)
Vulnerability of marine environment to catastrophic warfare damage (#PD5178)
Environmental degradation from recreational use of unsurfaced country roads and tracks (#PE7403).
Aggravates Restrictions on freedom (#PC5075) Human racial regression (#PF0411)
Long-term shortage of natural resources (#PC4824).
Aggravated by Vandalism (#PD1350) Biochemical warfare (#PC1164)
Environmental warfare (#PC2696) Uncontrolled industrialization (#PB1845)
Inadvertent modifications to climate (#PC1288)
Destruction of hedges and hedgerow trees (#PD1642)
Non-destructible packaging and containers (#PD1754)
Landscape disfigurement from open-cast mining (#PD1637)
Unaccountability of institutions degrading the environment (#PF3458).

♦ PB6719 Worldwide misallocation of resources
Claim Mankind has never before had such abundant financial and technical resources with which to overcome the global problems of mass hunger, starvation, disease and abject poverty; yet never before have so many people suffered and died unnecessarily. Even more will die in the future, as resources continue to accumulate in the stockpiles of weapons and overflowing granaries of the developed world.
The World Bank estimates suggest that over 20 percent of the world population lives 'below any rational definition of human decency'. High birth rates, malnutrition, disease, illiteracy and low income ensure that they will remain in this condition unless the developed world takes action to ensure a more equitable and just distribution of the world's resources. It is now widely recognized that the arms race presents a grave danger to the whole of mankind, and yet daily worldwide military expenditure exceeds the annual operating costs of the entire United Nations

system, a system designed to upgrade life. If only part of the money, manpower and research invested in military uses were diverted to development expenditure, the Third World could begin to look forward to a future where sufficient food and shelter were available for all and the basic rights of human dignity, justice and equity could become a reality.
Related Misuse of resources (#PB5151).
Aggravates Unproductive use of resources (#PB8376)
Mismanagement of food resources (#PE6115)
Inappropriate use of financial resources (#PD9338).
Aggravated by Local control of resources (#PF5539)
Unregulated global resources (#PF3183)
Conflicting use of resources (#PF6654)
Inefficient use of resources (#PE5001)
Long-term shortage of resources (#PB6112)
Waste of non-renewable resources (#PC8642)
Physically inaccessible resources (#PC4020)
Foreign control of natural resources (#PD3109)
Inadequate mobilization of resources (#PF4979)
Underutilization of natural resources (#PF1459)
Long-term shortage of natural resources (#PC4824)
Underutilization of renewable energy resources (#PE8971)
Excessive consumption of resources in developed countries (#PE5551).

♦ PB6918 Abuse of power

Nature One important difference between conventional crime and acts involving abuses of power is that many of the latter are committed under the guise of legitimacy. Concentration camps are justified by the need to protect internal security. Police abuse of power is legitimized as necessary to combat crime, to fight 'evil' with force. Extra-judicial executions are portrayed as acts of self-defence. The use of 'dirty tricks', undemocratic or outright illegal means, the overt or covert violations of civil liberties are thus rationalized and presented to an unsuspecting or an uninformed public as necessary and legitimate.
Claim The use of terror by individuals or groups is universally condemned, while its use by governments is defined as enforcement of the law. Definitions of terrorism are usually one-sided, referring to criminal acts directed against a state; but leaving out violent, repressive and oppressive actions of 'duly constituted' governments, including atrocities committed in violation such as the killing of civilians or the torture of prisoners.
Narrower Abuse of influence (#PC6307) Abuse of authority (#PC8689)
Abuse of economic power (#PC6873) Abuse of government power (#PC9104).
Related Aggression (#PA0587) Victimization (#PF6987).
Aggravates Oppression (#PB8656) Exploitation of trust (#PC4422)
Behavioural deterioration (#PB6321).
Aggravated by Secrecy (#PA0005) Martial law (#PD2637)
Monopoly of power (#PC8410) Inadequate political education (#PJ7906)
Lack of access to public archives (#PD1194)
Manipulation of the individual by mass media (#PE7448).

♦ PB7125 Conspiracies for societal control

Manipulative international plots
Broader Domination (#PA0839) Conspiracy (#PC2555).
Aggravated by Secret societies (#PF2508).

♦ PB7600 Lying

Dependence on lying
Nature Lying is an act contrary to truthfulness, or the virtue of veracity, consisting in the communication to another of a judgement that is not in accord with what the one who communicates thinks to be true. Lying not only has negative consequences for the liar and the deceived parties, but also gives rise to spreading deception and practices which may undermine entire communities. The veneer of social trust is often thin, and as lies spread - by imitation, or in retaliation, or to forestall suspected deception - trust is damaged. When trust is damaged, the community as a whole suffers; and when it is destroyed, societies falter and collapse.
Incidence Lying (whether in the form of outright untruth, exaggeration, understatement, omission, or 'white lies') is routinely used to manipulate the feelings and thoughts of others. It is also used in self-defence when people are confronted, and as a means of avoiding accountability. It may be essential to the process of selling goods and services, especially in encouraging people to purchase what they do not need.
Claim Lying creates emotional and physical tension that is detrimental to liars' health.
Counter-claim Lying may be a legitimate tool of defence, as in protection under repressive societies and regimes.
Lying is as much a part of normal growth as telling the truth. The ability to lie is a human achievement that tends to set them apart from all other species.
Refs Bok, Sissela *Lying; moral choice in public life* (1978); Bok, Sissela *Lying* (1979); Larson, John A, et al *Lying and Its Detection*; Shibles, Warren and Falkenberg, Gabriel *Lying* (1985).
Narrower Grey lies (#PF3098) White lies (#PF7631) Black lies (#PE4432)
Parental lying (#PD9145) Medical deception (#PD9836) Paternalistic lies (#PF7635)
Institutional lying (#PD2686) Political deception (#PF9583) Religious deception (#PF3495)
Deception in business (#PD4879) Deception by natural scientists (#PD9182)
Deception between sexual partners (#PE4890).
Related Deception (#PB4731) Denial of evidence (#PD7385)
Biased presentation of news (#PD1718).
Aggravates Behavioural deterioration (#PB6321).
Aggravated by Boasting (#PF4436) Evasion of work (#PC5576)
Educating people to lie (#PE3909).

♦ PB7606 Disinformation

Dependence on disinformation
Nature Disinformation is to communications what a diversion is to a military manoeuvre. It is information which has the semblance of truth, leaked or made available to espionage or to the media, in order to create attitudes or provoke actions favourable to the objectives of the disinformers. In some cases of social statistics, disinformation is provided to international collecting agencies solely to make programmes and policies of incumbent regimes appear progressive and productive. Disinformation is therefore static in the entire international communications network, civil or military, commercial or political.
Incidence Government, military and other bureaucracies are organized in such a way that reports from lower echelons must necessarily contain a certain percentage of disinformation in order to meet with the approval of the higher echelons to which they are submitted. Programme reports are written so as to give the best possible interpretation to the actions of those responsible for their implementation and thus further (or protect) their future careers. In intergovernmental organizations writing 'positively' is often encouraged.
Refs Deacon, Richard *The Truth Twisters* (1988).
Broader Incorrect information (#PB3095).
Narrower Covert smear campaigns by government (#PD7171).
Related Institutional lying (#PD2686).
Aggravates Misleading information (#PF3096) Lack of international cooperation (#PF0817).
Aggravated by Evasion of issues (#PF7431) Deception by government (#PD1893)
Competition between states (#PC0114)
Misrepresentation of geographical information (#PF9239).

♦ PB7666 Inequitable distribution of wealth

Dependence on inequitable distribution of wealth — Unequal distribution of wealth — Maldistribution of wealth
Broader Human inequality (#PA0844).
Narrower Personal wealth (#PC8222) Rural-urban income differential (#PE5022)
Maldistribution of wealth within developing countries (#PD5258)
Disparity between industrialized and developing countries (#PC8694).
Aggravates Human suffering (#PB5955) Maldistribution of resources (#PB1016)
International economic injustice (#PC9112).
Aggravated by Deprivation (#PA0831) Pursuit of affluence (#PF5864)
Economic dictatorship (#PC3240)
Excessive accumulation of wealth by government leaders (#PD9653)
Lack of relationship between wealth generation and the public good (#PF4730).

♦ PB7694 Unjust peace

Nature A condition of peace may obscure unresolved contradictions and injustices. These may result, on termination of a war, from the imposition of peace agreements which are deemed inequitable and unjust. They may also result, in the absence of war, from the progressive reinforcement of an international order which privileges some groups and countries at the expense of others.
Incidence The Treaty of Versailles terminating World War I was perceived as unjust and contributed to the buildup of resentment in Germany resulting eventually in World War II. The peace currently prevailing between industrialized and developing countries can be considered unjust in that many are deprived of a minimum standard of living, whilst a few are able to sustain a very high standard of living.
Broader Trivialization of peace (#PF5826).
Aggravates Unauthentic peace (#PF7643) Undemocratic political organization (#PC1015).
Aggravated by Human inequality (#PA0844) Unfairly negotiated treaties (#PF4787).

♦ PB7709 Persecution

Dependence on persecution — Persecuted people
Nature The suppression of liberty of belief and action, (or the punishment of those who adhere to a particular way of life or doctrine), whether due to religions, racial or national prejudices, may include attacks upon both individuals and groups, and may be executed by single tormentors, by organized gangs, or by governments.
Although discrimination is not usually persecution, in certain circumstances, especially when there is an accumulation of acts over time, or when there is an escalation in their intensity, it may then constitute persecution or lead to a well-founded fear of persecution.
Incidence To qualify as a refugee, a person must demonstrate a well-founded fear of persecution on account of race, religion, nationality, membership of a particular social group, or political opinion. Religions persecutions, for example, have been a part of history (e.g. the Diaspora of the Jews). Modern day immigration quotas which disfavour certain groups may be considered contemporary persecution.
Refs Arendt, Hannah *Eichmann in Jerusalem* (1977); Sheils, W J (Ed) *Persecution and Toleration* (1984).
Broader Intimidation (#PB1992).
Narrower Sexual harassment (#PD1116) Religious persecution (#PC5994)
Harassment of the media (#PD0160) Superstitious persecution of animals (#PD3453)
Government harassment of human rights activists (#PE6934).
Related Harassment (#PC8558).
Aggravates Underprivileged religious minorities (#PC2129).
Aggravated by Intolerance (#PF0860) Dependence on mysticism (#PF2590)
Denial of right of conscientious objection to military service (#PD1800).

♦ PB7712 Moral imperfection

Dependence on moral imperfection
Broader Underestimation of the human potential (#PF7063).
Narrower Ethical decay (#PB2480) Mental pollution (#PB6248)
Misconduct in public office (#PD8227).
Aggravates Human suffering (#PB5955) Intolerance of imperfection (#PF7024).
Aggravated by Lack of eugenic measures (#PD1091)
Inability to define moral standards (#PF7178).

♦ PB7991 Inactivity

Broader Inaction (#PA5806).
Narrower Minimal opportunities for corporate activities (#PF2316).
Aggravates Obesity (#PE1177).

♦ PB8031 Undemocratic social systems

Lack of democracy — Erosion of democratic principles
Narrower Undemocratic pressures (#PD3389) Undemocratic organizations (#PC8676)
Unethical catering practices (#PE6615) Undemocratic economic systems (#PJ1780)
Erosion of university autonomy (#PG6036)
Denial of democracy in communist systems (#PC3176)
Irresponsible expression of emotions equated with free speech (#PF7798)
Distortion of democratic procedures within international organizations (#PE5179).
Related Erosion (#PC8193) Undemocratic policy-making (#PF8703).
Aggravated by Political prisoners (#PC0562) Limitations of democracy (#PF6608)
Discrepancies between principles and practice (#PF4705).

♦ PB8167 Misery

Broader Solemnity (#PA6731) Insensibility (#PA5451) Unpleasantness (#PA7107).

♦ PB8172 Infamy

Nature Infamy is of two types: legal under canon law and the common loss of reputation which is the consequence of notorious conduct. The latter is a state of mind and the former is a punishment for those guilty of offenses against faith, such as, apostasy, heresy, schism, and desecration of Sacred Species; against ecclesiastical authority, such as, attacking the pope, a cardinal or a papal legate and against morals, such as, violation of a grave or body, bigamy, sexual relations with minors under 16, rape, sodomy, incest and pandering.
Broader Vice (#PA5644) Disrepute (#PA6839) Wrongness (#PA7280).
Aggravated by Rape (#PD3266) Incest (#PF2148)
Sodomy (#PE3273) Bigamy (#PF3286)
Heresy (#PF3375) Grave robbing (#PF0491)
Religious schism (#PF1939) Sexual abuse of children (#PE3265)
Desecration of cemeteries (#PD7258) Neglected community uniqueness (#PG9018)
Exploitation of the prostitution of others (#PE5303).

♦ PB8214 Inhumanity

Lack of humanity — Lack of humaneness

PB8214

Refs Lorenz, Konrad *The Waning of Humaneness* (1988).
Broader Moderation (#PA7156) Unkindness (#PA5643).
Narrower Growing size and impersonality of firms (#PE8706).
Aggravates Mutiny (#PD2589) Lack of war relief (#PF0727).

♦ PB8238 Shortage

Refs Kornai, J *Economics of Shortage* (1980).
Broader Shortcoming (#PA6041) Insufficiency (#PA5473) Incompleteness (#PA6652).
Narrower Labour shortage (#PC0592) Limited available land (#PC8160) Shortage of animal protein (#PC4998) Long-term shortage of resources (#PB6112) Insufficient financial resources (#PB4653) Long-term shortage of commodities (#PC1195) Insufficient educational material (#PE8438) Insufficient availability of goods (#PB8891).
Related Imperfection (#PA6997).
Aggravated by Capitalist speculation (#PC2194).

♦ PB8287 International tension

Broader Tension (#PB6370).
Aggravated by Competitive acquisition of arms (#PC1258) Biased and inaccurate history textbooks (#PD2082) Biased and inaccurate geography textbooks (#PF1780).

♦ PB8376 Unproductive use of resources

Waste of resources — Uncontrolled waste — Unrecycled waste
Claim The economy of the industrialized countries depends on producing all the unnecessary luxurious and wasteful things at an ever-increasing rate. To keep up the economy requires vast waste.
Refs Spooner, Henry J and Lord Leverhulme, *Wealth from Waste* (1973).
Narrower Waste paper (#PD1152) Mine wastes (#PG2548) Food wastage (#PD8844) Wasted water (#PD3669) Nuclear waste (#PD4396) Toxic substances (#PD1115) Hazardous wastes (#PC9053) Agricultural wastes (#PC2205) Waste of human resources (#PC8914) Chemical industry wastes (#PG2549) Waste of nuclear warheads (#PE4199) Hotel and restaurant waste (#PE1542) Unproductive labour resources (#PC6031) Waste of non-renewable resources (#PC8642) Restrictive use of available land (#PF6528) Marine pollution by plastic waste (#PE3741) Food manufacturing industry wastes (#PE8702) Dumping of consumer waste products (#PD8942) Unproductive subsistence agriculture (#PC0492) Competitive development of new weapons (#PC0012) Resource wastage in capitalist systems (#PC3108) Inappropriate aid to developing countries (#PF8120) General unproductivity of capitalist systems (#PF3103) Waste of resources invested in obsolete armaments (#PE9346) Wastage in governmental budgets and appropriations (#PD0183).
Related Radioactive wastes (#PC1242) Misuse of resources (#PB5151) Non-biodegradable plastic waste (#PD1180).
Aggravates Self-indulgent societies (#PF5466) Lack of natural resources (#PC7928) Prohibitive cost of basic services (#PF6527) Poor communications networks in rural areas (#PF6470).
Aggravated by Insecurity (#PA0857) Domestic refuse disposal (#PD0807) Unpredictable governmental policy (#PF1559) Insufficient recycling of materials (#PJ0525) Worldwide misallocation of resources (#PB6719) Excessive consumption of goods and services (#PC2518).

♦ PB8406 Mismanagement

Refs Persson, B (Ed) *Surviving Failures* (1979).
Broader Unskillfulness (#PA7232) Inappropriateness (#PA6852).
Narrower Policy-making errors (#PF0531) Agricultural mismanagement (#PD8625) Maltreatment of zoo animals (#PE4834) Unethical practices in forestry (#PD6701) Mismanagement of food resources (#PE6115) Incompetent financial management (#PF0760) Non-inclusive management decisions (#PF2754) Mismanagement in developing countries (#PD8549) Mismanagement of environmental demand (#PD5429) Long-term shortage of natural resources (#PC4824) Soil mismanagement in developing countries (#PD6482) Corruption and mismanagement of foreign aid (#PD0136) Mismanagement of aid to developing countries (#PF0175) Inadequate management of government finances (#PF9672) Intergovernmental organization mismanagement (#PD6628) Short range planning for long-term development (#PF5660) Inappropriate management of development projects (#PD3712) Agricultural mismanagement of housed farm animals (#PD2771) Inadequate economic policy-making in developing countries (#PF5964) Mismanagement by intergovernmental organization leadership (#PE6947).
Aggravates Deception by management (#PD3823).
Aggravated by Incompetent management (#PC4867) Difficult grant management (#PG9356) Absence of management training (#PD3789) Poor managerial communications (#PF1528) Inflexible management patterns (#PF3091) Limited accountability of public services (#PF6574) Inadequate management-employee communication (#PF3661) Unaccountable management of public information (#PF8074) Lack of management skills in developing countries (#PE0046) Inadequate management skills in rural communities (#PF1442) Lack of worker participation in business decision-making (#PF0574).

♦ PB8538 Hostility

Refs Bender, Lynn-Darrell *Politics of Hostility* (1975); Kaplan, Katheen R *Hostility, Characteristics and Behavior* (1988).
Broader Enmity (#PA5446).
Related Vice (#PA5644) Disaccord (#PA5532) Difference (#PA6698) Opposition (#PA6979) Aggression (#PA0587) Unfeelingness (#PA7364) Unpleasantness (#PA7107).
Aggravated by Ugliness (#PA7240).

♦ PB8613 Endangered cultures

Endangered languages — Dying cultures — Dying languages
Narrower Endangered indigenous cultures (#PC7203).
Aggravated by Political dictatorship (#PC0845) Denial of cultural rights (#PD5907) Racial discrimination in politics (#PD3329) Insufficient minority culture support (#PF5659) Migration of rural population to cities (#PE8768) Discrimination against minority languages (#PD5078) Denial of right of peoples to use their own language (#PE2142).

♦ PB8645 Lack of education

Broader Cultural deprivation (#PC1351).
Narrower Lack of training (#PD8388) Inadequate sex education (#PD0759) Lack of formal education (#PF6534) Inadequate preschool education (#PJ1962) Insufficient primary education (#PC6381) Inadequate education for nomadic children (#PE6206).
Aggravates Taboo (#PF3310) Ignorance (#PA5568) Illiteracy (#PC0210) Demagoguery (#PC2372) Superstition (#PA0430) Culture shock (#PC2673) Desert nomadism (#PD2520) Cultural invasion (#PC2548) Political ignorance (#PC1982) Lack of participation in development (#PF3339) Inadequate cultural integration of immigrants (#PC1532).
Aggravated by Street children (#PD5980).

♦ PB8656 Oppression

Nature The use of power or coercion whether violent or nonviolent to constrain another's freedom, to violate another's rights, to exploit another, or to deny another's claims to justice. Oppression is clearly at work whenever anyone believes himself to be better than another person by virtue of his membership in a self-styled superior group.
Incidence Obvious examples of oppression are the oppression of black people by whites, minorities by majorities, poor people by rich people and women by men. Less obvious examples include the oppression of physically handicapped people in an environment designed only for the able-bodied, the oppression of children in an adult-centered world, and the oppression of people who live on their own in a society where the family is the norm.
Counter-claim Individuals and groups are superior to other individuals and groups. Some people are more intelligent, wealthy, politically discerning, morally upright, sensitive to the needs of others, artistic, educated, disciplined and many many other things. The simple existence of these qualities does not make an oppressive situation and if it does, oppression has little meaning.
Refs Dick, James C *Violence and Oppression* (1979).
Broader Solemnity (#PA6731) Unpleasantness (#PA7107) Inappropriateness (#PA6852).
Narrower Coercion (#PC3796) Adultism (#PF6519) Exploitation (#PB3200) Violation of human rights (#PB3860).
Related Injustice (#PA6486) Repression (#PB0871).
Aggravates Civil disorders (#PC2551) Sense of powerlessness (#PF8618) Denial of right to liberty (#PF0705) Socially unsustainable development (#PC0381) Undemocratic political organization (#PC1015) Denial of right of conscientious objection to military service (#PD1800).
Aggravated by Abuse of power (#PB6918).

♦ PB8678 Insecurity of resources

Broader Economic insecurity (#PC2020).
Narrower Insecurity of property (#PC1784).
Related Cargo insecurity (#PE5103).
Aggravated by Fluctuations in real value of money (#PD9356).

♦ PB8685 Isolation

Remoteness — Inaccessibility
Broader Duality (#PA7339) Exclusion (#PA5869) Restraint (#PA7296).
Narrower Social isolation (#PC1707) Cultural isolation (#PC3943) Political isolation (#PC7569) Economic isolationism (#PC2791) National isolationism (#PF2141) Geographical isolation (#PF9023).
Aggravated by Social inaccessibility (#PC0237).

♦ PB8891 Insufficient availability of goods

Shortages of supply of goods — Limited commodity goods
Broader Shortage (#PB8238) Undirected expansion of economic base (#PF0905).
Narrower Food insecurity (#PB2846) Long-term shortage of chemicals (#PE1261) Shortage of industrial diamonds (#PG7754) Long-term shortage of manufactured goods (#PE0802) Long-term shortage of food and live animals (#PE0976) Long-term shortage of beverages and tobacco (#PE1253) Long-term shortage of mineral fuels and lubricants (#PE1712) Long-term shortage of inedible crude non-fuel materials (#PE0461) Long-term shortage of machinery and transport equipment (#PE1436) Long-term shortage of animal and vegetable oils and fats (#PE1188) Long-term shortage of miscellaneous manufactured articles (#PE0613).
Aggravates Rationing (#PF9026) Profiteering (#PC2618).
Aggravated by Prohibitive cost of goods and services (#PD1891) Excessive consumption of goods and services (#PC2518).

♦ PB8952 Excess

Broader Insufficiency (#PA5473).
Narrower Unnecessary personal consumption (#PF5931) Excessive size of metropolitan regions (#PD6120).
Related Cheapness (#PA7193) Undueness (#PA6921) Decadence (#PB2542) Intemperance (#PA6466) Uncommunicativeness (#PA7411).

♦ PB9015 Unconstrained socio-economic growth

Unwanted excessive growth
Narrower Unsustainable population levels (#PB0035) Unsustainable economic development (#PC0495).
Aggravates Natural environment degradation (#PB5250) Unconstrained exploitation of natural resources (#PF2855) Increasing development lag against socio-economic growth (#PC5879).
Aggravated by Anthropocentrism (#PF4096).

♦ PB9136 Restrictive practices

Narrower Restrictive trade practices (#PC0073) Restrictive legal practices (#PD8614) Restrictive social practices (#PC5537) Restrictive medical practices (#PD8831) Restrictive monetary practices (#PF8749) Restrictive religious practices (#PD8439) Restrictive scientific practices (#PD7875) Restrictive trade union practices (#PD8146) Restrictive professional practices (#PD8027) Restrictive patterns of traditional life (#PF3129) Restrictive channels of cultural interchange (#PF3037) Restrictive influence of religion on the masses (#PF3361) Restrictive pattern of business activities in small communities (#PD1415).
Aggravates Denial of the right of association (#PD3224).

♦ PB9165 Stress

Refs Golembiewski, Robert T, et al *Stress in Organizations* (1985); Hsu, Teng H *Stress and Strain Data Handbook* (1986); Murakami, Y *Stress Intensity Factors Handbook* (1987).
Narrower Strain (#PJ3545) Stress in materials (#PD7216) Stress in human beings (#PC1648) Animal stress in factory farming (#PD2760) Mental, physical and emotional strain (#PG2572).
Aggravates Mental tension (#PB6302).

♦ PB9171 Extinction of species

Species extinction — Lack of species preservation
Nature The number of species on earth is estimated to be anywhere between seven and 50 million, probably around 30 million species. Only about 1.4 million have been named and briefly described.Guessing the space of extinction is therefore complicated, but estimations have shown

a rate of one species a year. The present human-caused rate is hundreds of times higher, resulting perhaps to extinction of four species every hour. It is known that almost all past extinctions have occured by natural processes, but today human activities are overwhelmingly the main cause of extinction. The tropical deforestation alone will wipe out 5–15 percent of all species between 1990 and 2020.
Refs Ehrlich, Paul and Ehrlich, Anne *Extinction* (1981); Norton, Bryan G *The Preservation of Species* (1986); Ziswiler, V *Extinct and Vanishing Animals* (1967).
  Broader Death of living creatures (#PF7043)   Erosion of biological diversity (#PB9748).
  Narrower Animal extinction (#PD7989)   Wildlife extinction (#PC1445).
  Periodic mass extinctions of species (#PF4149)
  Degradation of the environment through the destruction of species (#PE5064).
  Aggravates Food insecurity (#PB2846)   Narrow range of food crops (#PD4100)
  Endangered species of medicinal plants (#PD4171)
  Vulnerability of world genetic resources (#PB4788)
  Decreasing diversity of biological species (#PD7302).
  Aggravated by Global cooling (#PF1744)   Geomagnetic reversal (#PF1588)
  Tropical deforestation (#PD6204)   Evolutionary catastrophes (#PF1181)
  Pesticide hazards to wildlife (#PD3680).

♦ **PB9419 Unsustainable development**
Incidence Unsustainable development now compromises the ability of future generations to meet their needs. Global warming, destruction of the ozone shield, acidification of land and water, desertification and soil loss, deforestation and forest decline, diminishing productivity of land and waters, and extinction of species and populations, demonstrate that human demand is exceeding environmental support capacities. The annual increase in industrial production in 1989 is as large as that of Europe's total production in the 1930s. The populations of 74 countries are doubling every 30 years or less. Population growth increases poverty and deprived people are forced to undermine the productivity of the land on which they live. It is extremely difficult for people, or other species, to adjust to change at this rate.

  Narrower Uncontrolled urban development (#PC0442)
  Unsustainable rural development (#PD4537)
  Unsustainable population levels (#PB0035)
  Unsustainable economic development (#PC0495)
  Socially unsustainable development (#PC0381)
  Ecologically unsustainable development (#PC0111)
  Unsustainable agricultural development (#PC8419)
  Unsustainable development of energy use (#PC7517)
  Unsustainable development of coast zones (#PD4671)
  Unsustainable development of fresh waters (#PD6923)
  Unsustainable development of forest lands (#PD4900).
  Aggravated by Maldevelopment (#PB6207)   Socio-economic poverty (#PB0388).

♦ **PB9748 Erosion of biological diversity**
Decreasing variety of life forms — Decreasing biodiversity
Nature The variety of natural life forms, whether eco-regions, habitats, species or gene pools, is being endangered by human activity. It is this variety which ensures the regeneration of harvested resources and the maintenance of ecological processes, whether as a vital part of world heritage or for its own sake. It also provides resources for the development and improvement of domesticated crops and livestock, for recreation and tourism, and for research and education.
Incidence The main levels of diversity of concern are: ecosystem diversity (the number and frequency of different communities of organisms and their environments); species diversity (the number and frequency of different species); and genetic diversity (meaning both genetic variability and the number and frequency of genetically distinct populations).
Refs Wolf, Edward C *On the Brink of Extinction* (1987).
  Narrower Extinction of species (#PB9171)   Vulnerability of ecosystem niches (#PC5773)
  Vulnerability of world genetic resources (#PB4788)
  Decreasing diversity of biological species (#PD7302)
  Decreasing diversity of biological habitats (#PD5386)
  Degradation of semi-natural and natural habitats of flora and fauna (#PC3152).
  Aggravated by Natural environment degradation (#PB5250).

Cross-sectoral problems   PC

Content

This section groups together the major cross-sectoral, world-wide problems which tend to be prominent on the agendas of international organizations with more specialized concerns, as well as in the media. Such problems also tend to group, or focus, many of the more specialized problems which are described in subsequent sections. Indeed in many debates discussion of the more specialized problems may be subsumed under discussion of these problems.

Note that further information relevant to an understanding of the problem may be present in other problems cross-referenced in the entry consulted.

The problems in this section are often sectoral variants on the broader or more basic problems described in the previous section (Section PB).

Many of the problems in this section are of such proportions and complexity that no single organization or discipline can claim to encompass any one of them in all its aspects. The scope and implications of such problems tends to be a matter of continuing debate. They are are not sufficiently well-defined to respond to well-defined solutions. The nature of an appropriate solution to such problems is also a matter of continuing debate.

This section groups 732 problems for which there are 10,910 cross-references.

Rationale

Inclusion of such problems calls for no comment because of their widely recognized importance. Where they are cross-sectoral variants of those in the previous section, their inclusion here prevents neglect of the sectoral specificity, as tends to be the case when such problems are subsumed under those of the broader problems in Section PB.

Method

The entries are based on information obtained from international organizations, from a wide variety of reference books, or as reported in the international media. The procedures for identifying world problems are described in Section PZ at the end of this volume.

Index

A keyword index to entries is provided in Section PX.

Comment

Detailed comments are given in Section PZ at the end of this volume.

Reservations

The emphasis throughout this volume has been placed on providing descriptions of less well-known problems, particularly when the extensive material available on the better known problems contained neither succinct descriptions of them nor descriptive material which could easily be reduced to succinct descriptions. The problem descriptions here represent a compilation of views from published documents (usually from international organizations). The text provided does not necessarily constitute the best possible description of the problem, since a compromise has had to be struck between availability of information, the resources to process it, and the space available in this volume.

In a number of cases a problem could have been allocated to another section. Inclusion of a problem in this section, rather than in a preceding or following section, has been based on a number of factors. The position of the problem in one or more hierarchies of cross-references was a major factor in determining its allocation to this section.

Possible future improvements

There is much scope for improving the quality of problem entries through feedback from interested bodies. More bibliographic references could be included where appropriate, as well as references to major resolutions concerning those problems recognized by the United Nations. There is also much scope for improving the pattern of cross-references, both between problems, to other sections of this volume (eg values) and to the 20,000 internationally-active bodies in the companion series (*Yearbook of International Organizations*)

ENTRY CONTENT AND ORGANIZATION

Ordering of entries
Entries are in **numeric order**. Entry numbers have been **allocated randomly**; they have no significance other than as a permanent point of reference to facilitate indexing, cross-referencing, and updating between editions.

Index access to entries
The location of an entry in this sub-section may be determined from the **Volume Index** (Section PX) on the basis of keywords in the name of the entry or its alternate titles.

Structure of entries
Entries may be composed of the following descriptive elements:

(a) **Entry number** This number has **no significance**, except as a convenient method of identifying the entry (particularly for indexing purposes), of filing information on it, and as an identifier to which cross-references from other entries (possibly in other Sections) may refer in this and future editions. The first letter of the entry number refers to the section of this volume in which the sub-section, denoted by the second letter, is located.

(b) **Problem name** This is printed in bold characters. It is the name selected as best indicating the nature of the problem. It may be followed by alternative problem names.

(c) **Nature** Description of the problem which attempts to identify the nature of the disruptive processes involved. The information included here, and in the following paragraphs, is compiled directly, to the extent possible, from available published documents. Where appropriate the text included may be reproduced, in a minimally edited form, from the publications of international organizations, such as those of the United Nations or its Specialized Agencies.

(d) **Incidence** Summary description of the extent of the problem which makes it of more than national significance.

(e) **Background** Describes briefly when and how the problem's importance was recognized initially, and how this recognition has evolved over time.

(f) **Claim** Stresses the special importance of this problem and why action is particularly urgent. This paragraph offers means of including statements which may deliberately exaggerate claims for the unique importance of the problem.

(g) **Counter-claim** Stresses, where appropriate, the relative insignificance or erroneous conception of the problem as described. This paragraph offers a means of including statements which may deliberately exaggerate the arguments refuting the evidence for the existence of the problem. Absence of such arguments from the text does not mean that they do not exist.

Cross-referencing of entries
At the end of any entry, there may be cross-references to other entries. These indicate the number and name of the cross-referenced entry, whether within this Section or in other Sections. There are 3 types of **hierarchical** cross-references between problems:
>
> **Broader** = Broader problem: more general problems of which the problem described may be considered a part. The described problem may be considered an aspect of several broader problems
> **Narrower** = Narrower problem: more specific problems which may be considered a part of the described problem
> **Related** = Related problem: problems that may be considered as associated in a hierarchically undefined way with the described problem.

> There are 4 types of **functional** cross-references between problems:
> **Aggravates** = Problems aggravated by the described problem: a forward or subsequent negative causal link
> **Aggravated by** = Problems aggravating the described problem: a backward or prior negative causal link
> **Reduces** = Problems relieved, alleviated or reduced by the described problem: a forward or subsequent positive causal link
> **Reduced by** = Problems relieving or alleviating the described problem: a backward or prior positive causal link

♦ PC0002 Economic stagnation

Declining economic growth — Unstable economic growth — Declining growth in world economy

Nature Economic stagnation exists when the total output (or output per capita) in goods and services is constant, falls slightly, or rises only sluggishly. It also exists when unemployment is chronic and increasing. Such conditions may exist in particular industries, in wider sectors of an economy, or in the economy as a whole.

Incidence In the 1950s and 1960s, the global economy grew by about 5 percent a year, sending living standards soaring in many parts of the world. In the 1970s, growth fell to slightly above 3 percent, and in the 1980s it has slipped to about 2.3 percent. Coupled with rising population, this means that the world economy, while growing, is very fragile indeed. A key factor in stagnant economies is the lack of incentives to invest accumulated capital. Since venture capital has become internationally mobile it may have nowhere to go in the cases of regional, hemispheric, or industrialized–nation stagnation. In times of stagnation, therefore, governments may feel that only their intervention in supplying money can stimulate growth. This can be successful in the short run but can exacerbate national debts and ultimately lead to higher taxes and interest rates. Elected officials who feel their tenure depends on their solving stagnation may turn to military spending or trade protectionism or both. Thus times of little or no economic growth may be periods of high international tension.

Claim Economic, social and environmental catastrophes can only be averted in many countries of the developing world, if global economic growth is revitalized. This means more rapid economic growth in both industrial and developing countries, freer market access for the products of developing countries, lower interest rates, greater technology transfer and significantly larger capital flows (both concessional and commercial).

Counter–claim A more rapid growth in the world economy would apply environmental pressures that are no more sustainable than the pressures presented by growing poverty. The resulting increased demand for energy and other non–renewable raw materials could significantly raise the cost of these items relative to other goods.

Refs Roberts, Brad *Slow and Uneven Global Economic Growth* (1986); Tarshis, Lorie *World Economy in Crisis*.

Narrower Decreasing agricultural growth per capita (#PF4326)
Declining economic growth in developing countries (#PD5326)
Declining economic growth in industrialized countries (#PF1737).
Aggravates Decline of philanthropy (#PF6221) Rural poverty in developing countries (#PD4125)
Domination of government policy-making by short–term considerations (#PF0317).
Aggravated by Posterity (#PA6073) Unemployment (#PB0750)
Economic apathy (#PC3413) Economic uncertainty (#PF5817)
Economic inefficiency (#PF7556) Restrictive monetary practices (#PF8749)
Denial of right to business growth (#PE2700)
Financial destabilization of world trade (#PC7873)
War–time disruption of economies and production facilities (#PD8851)
Mismatch of national macroeconomic policies among industrialized countries (#PF5000).
Reduces Unsustainable economic development (#PC0495)
Undirected expansion of economic base (#PF0905).

♦ PC0006 Racial discrimination

Dependence on racial discrimination — Denial of right to equality because of race — Unequal rights for different racial groups

Nature Racial discrimination includes any distinction, exclusion, restriction or preference based on race, colour, descent, or national or ethnic origin which has the purpose or effect of nullifying or impairing the recognition, enjoyment or exercise, on an equal footing, of human rights and fundamental freedoms in political, economic, social, cultural or any other fields of public life. Such discrimination may be based upon ideas or theories of superiority of one race or group of persons of one colour or ethnic origin, which may be used to justify or promote racial hatred and discrimination in a wide variety of forms. Such action may encourage acts of violence or incitement to such acts against groups of persons of another colour or ethnic origin.

The acceptance of such discrimination may lead to: unequal treatment before tribunals and other organs administering justice; unequal rights to security of person and protection by the State against violence or bodily harm (whether inflicted by government officials or by any individual, group or institution); unequal political rights (in particular the rights to participate in elections, to vote and to stand for election on the basis of universal and equal suffrage, to take part in the government as well as in the conduct of public affairs at any level, and to have equal access to public service); unequal enjoyment of other civil rights (right to freedom of movement and residence within the border of the State; right to leave any country, including the person's own, and to return to his country; right to nationality; right to marriage and choice of spouse; right to own property alone as well as in association with others; right to inherit; right to freedom of thought, conscience and religion; right to freedom of opinion and expression; right to freedom of peaceful assembly and association); unequal enjoyment of economic, social and cultural rights (right to work, free choice of employment, just and favourable conditions of work, protection against unemployment, equal pay for equal work; just and favourable remuneration; right to form and join trade unions; right to housing; right to public health, medical care and social security services; right to education and training; right to equal participation in cultural activities); unequal access to any place or service intended for use by the general public such as transport, hotels, restaurants, cafés, theatres and parks.

Although maintaining the conception of a socially unified society, racial divisions may be preserved but expressed in other terms, so that in reality there is racial differentiation. Some characteristics of society favour a racial differential, incorporated in polity, and expressed as cultural barriers, de–facto segregation, inequality and demographic recognition of racial and ethnic categories. Conflict may be acknowledged but the significance of racial and ethnic difference for the conflict may be denied.

Background Racial discrimination is historically rooted in colonialism and slavery. Colonialism in Africa (which was the most highly colonized continent) started out as an economic venture aimed at creating sources of cheap raw materials as well as captive markets for the manufactured goods of European industrial centres. It was an exploitative venture and later turned into a political game in which the desire for political power reflected a passion to ensure economic advantage. In instances where racial discrimination was already institutionalized, it was found necessary to maintain it to prevent the dominated racial groups from gaining power. Thus, the phenomenon of racial discrimination is inevitably linked with inequalities in power. It is obvious that political domination was one of the principal causes of racial discrimination, as may be seen from the fact that colonialism produced two distinct groups, one consisting of those discriminated against and the other comprising those practising discrimination. Colonization played a significant role in the development of racism and racial discrimination. With its bias towards enslavement, if not slavery, colonialism tended to enforce a spirit of dependence and helplessness on the indigenous population, and sought to justify its existence on the basis of the assumption that colonial peoples were 'a lower form of the species' or 'the white man's burden' to be borne as he saw fit.

Much has been done internationally to combat the problems caused by racism but the problem still remains despite, for example, the work of the United Nations Decade for Action to Combat Racism and Racial Discrimination. Part of the latter provides for governments to report every two years on action taken in this field. A recent report (1984) of the United Nations Economic and Social Council indicates progress in: legislative, judicial, administrative and other measures to prohibit manifestation of racism or racial discrimination (whether or not discriminatory practices actually prevail); guaranteeing the right of everyone to equality before the law without distinction as to race, colour, or national or ethnic origin; equality of economic, social and cultural rights; specific judicial and administrative procedures for effective recourse by individuals complaining of racial discrimination; measures to include questions relating to racism and racial discrimination in school curricula; measures to render illegal the dissemination of ideas based on racial superiority or hatred and the establishment of organizations based on racial prejudice; activities to protect migrant workers and their families and to prohibit racial discrimination with respect to laws and regulations on immigration; provision of assistance to the victims of racial discrimination; isolation of governments permitting the perpetuation of racialist policies; and publicity campaigns to mobilize national public opinion against racialism.

Claim The major adverse consequences of racially discriminatory practices include: failure to use human resources fully or efficiently, as a result of racial discrimination in education and training, employment and remuneration; high production costs entailed by payment of inflated wages in sectors of the economy where job competition is artificially eliminated by restrictive trade union practices and officially supported job reservations to protect the ruling racial group; failure to respond to the normal forces of supply and demand; reliance upon imported skilled labour while at the same time neglecting the training of local manpower; waste of land resources, and concentration of masses of the population in overpopulated and impoverished areas, where uncultivated tracts of land lie idle and reserved for the dominant ruling group; conversely, over–protection of land rights tending in turn to prevent the entry of private foreign capital and the economic exploitation of such resources under appropriate government control; combined crippling effects of restrictive labour and land policies on the geographical mobility of labour; perpetuation of migratory, unskilled inefficient labour; separate racial development, which tends to perpetuate dual economies, for example in African disintegrated economies in which the subsistence sector is predominantly African and the monetary sector is predominantly European; limiting effects on levels of production of discriminatory policies with regard to property ownership, credit policies, extension services and marketing facilities; waste of resources involved in administering the instruments of discrimination, particularly in the form of duplication of services and fixed capital; large disparities of incomes accruing to different racial groups, and the resulting wide gaps in standards of living; and narrow internal markets resulting significantly from the small purchasing power of the mass of the population, which in turn is associated with an inequitable distribution of national income within the complex of the racially restrictive policies and practices.

Broader Racism (#PB1047) Discrimination (#PA0833) Social injustice (#PC0797).
Narrower Miscegenation (#PC1523) Racial repression (#PD8762)
Racial segregation (#PC3688) Segregation in housing (#PD3442)
Legalized racial discrimination (#PC3683) Racial discrimination in politics (#PD3329)
Racial discrimination in education (#PD3328) Racial bias in sentencing offenders (#PE4907)
Refusal of sale because of buyer's race (#PE8823)
Racial discrimination in public services (#PD3326)
Discrimination against black working women (#PE6245)
Racial discrimination in sexual preferences (#PD4064)
Racial discrimination in according financial loans (#PE9054).
Related Religious persecution (#PC5994) Religious discrimination (#PC1455)
Denial of right to pursue spiritual well–being because of discrimination (#PE7310).
Aggravates Hunger (#PB0262) Slavery (#PC0146) Eugenics (#PC2153)
Genocide (#PC1056) Tribalism (#PC1910) Civil war (#PC1869)
Colonialism (#PC0798) Antisemitism (#PE2131) Forced labour (#PC0746)
Social conflict (#PC0137) Social inequality (#PB0514) Political conflict (#PC0368)
Racial exploitation (#PC3334) Illegal immigration (#PD1928)
Socio-economic poverty (#PB0388) Social impediments to marriage (#PF3341)
Denial of rights of minorities (#PC8999) Underprivileged racial minorities (#PC0805)
Threats against family or friends (#PE3308) Abusive traffic in immigrant workers (#PD2722)
Discrimination against indigenous populations (#PC0352)
Expropriation of land from indigenous populations (#PC3304)
Segregation of poor and minority population in urban ghettos (#PD1260)
Inadequate living and working conditions of immigrant labourers in industrialized countries (#PD3427).
Aggravated by Elitism (#PA1387) Nationalism (#PB0534)
Neo–fascism (#PF2636) Racial prejudice (#PC8773)
Racial inequality (#PF1199) Cultural prejudice (#PC8520)
Racial intimidation (#PC2936) Skin colour prejudice (#PC8774)
Inadequate cultural integration of immigrants (#PC1532)
Capital investments supporting racial discrimination (#PE4167)
Inadequate supply of pharmaceutical products in developing countries (#PE4120).
Reduces Dictatorship (#PC1049).
Reduced by Homogenization of cultures (#PB1071).

♦ PC0012 Competitive development of new weapons

Waste of resources on armaments research — Nuclear weapons research — Weapon experimentation

Nature Military research and development is concerned with stimulating the advancement of scientific knowledge and cultivating technical progress for military purposes. It is immediately directed, for the most part, at the creation of new and improved weapons, counter–measures and other military systems and equipment. Such research results in the replacement of existing offensive and defensive systems by generations of successively more complex, costly and lethal types; and is a direct incentive to the arms race, since each qualitative improvement in a weapon system by one country is a spur to further effort by the other. Military research and development has seriously distorted the whole pattern of world research and development away from the pressing needs of, for instance, agriculture, pollution and medicine. This includes the diversion of educated manpower from constructive civilian activities into the development of destructive weaponry, particularly of the more indiscriminate and inhumane variety.

Incidence Estimates of R and D expenditure during 1970–1980 are US $ 117 billion for the USA, and US $ 36 billion for the EEC countries. The world total may have exceeded US $ 250 billion in that decade (putting USA expenditure at approximately 46.8% and EEC countries at approximately 13.4% of total) compared with US $ 187 billion from 1961 to 1970. The USA, the UK, France, and the Federal Republic of Germany account for over 90 per cent of Western expenditure. It is estimated that the USSR outlay is 30–37% of the world total. In 1984, expenditure on military research and development was estimated at $70 to $80 billion world–wide, increasing at twice the rate of military expenditure as a whole.

Claim If even a fraction of military research and development funds were provided to attack some of the main economic and social problems of the world, much larger benefits of the peaceful uses of science could be expected, given a powerful sense of purpose and the same institutionalized techniques of organization and management which military research has stimulated. It has been estimated that at the beginning of the 1980s about one–quarter of the global expenditures for R and D was accounted for by military programmes. If countries are prepared to set the right priorities, they ought to be able to achieve even more rapid technological progress, without war or an arms race.

Counter–claim All countries want to be as secure as possible; they have no interest in pursuing policies that would weaken them in relation to potential enemies. Good will and mutual confidence are sorely lacking between superpowers of differing political ideologies; armaments provide a sense of heightened security. They are a necessary tool for negotiation and, as such, may be considered to be of a top priority.

Refs Betts, Richard K *Nuclear Blackmail and Nuclear Balance* (1987); Stockholm International Peace Research Institute *Yearbook 1990* (1990).

Broader Military secrecy (#PC1144) Unproductive use of resources (#PB8376)
Irresponsible scientific and technological activity (#PC1153).
Narrower Nuclear weapons testing (#PC2201).
Related Inadequate peace research support (#PF4848).
Aggravates Competitive acquisition of arms (#PC1258)
Inhumane and indiscriminate weapons (#PD1519)
Socio-economic burden of militarization (#PF1447)
Excessive consumption of goods and services (#PC2518).

♦ PC0018 Disruptions due to migration

Undesirable effects of migration — Disruptions due to emigration

Nature Due to the numbers involved and, more particularly, because the sex and age distribution of migrants often differs substantially from that of the rest of the population, migration can have pronounced effects on population composition, the rate of natural increase and the supply of human resources. It can also give rise to certain social and economic problems. Immigration can relieve manpower shortages, stimulate the economy and introduce desirable social changes; but it is frequently also a cause of sizeable problems of assimilation, housing and health, with related impacts on social and educational services. It can result in an immediate surplus of manpower for which there is no suitable economic opportunity.

Emigration, though capable of easing population and employment problems, has on occasion resulted in the reduction of needed manpower. This is particularly important in the developing countries, especially as it reduces already scarce supplies of highly trained personnel. Furthermore, emigration has sometimes resulted in markedly distorted sex and age structures.

Immigrant families frequently enter the socio-economic ladder at the bottom and tend to stay there longer than others. The process of discarding dimensions of their old culture and acquiring parts of a new one is slow and complex. It is psychologically difficult because personal identity and social orientation are tied to culture. They face issues of language, social customs and mores, economics, nutrition, housing, and government regulations. They are often discriminated against in housing and employment. They tend to be economically vulnerable, not to understand the society they are living in so they are taken advantage of and commit social errors. In education, their adopted society may place a different value on education and create stress. If a society expects high academic achievement and the immigrant has working as a higher value than education, it may take two or more generations for the family to adapt. In housing, immigrants arrive without a home and may find themselvesf needing an address before they can be employed and a job to rent accommodation. In employment, immigrants usually start in the lowest paying jobs, have difficulty with the language, may face legal restrictions in employment. They tend to be poor and live in conditions of poor housing, poor nutrition, ill health, mental difficulties and in an environment of high crime and physical danger. All of which discourages good social functioning. The considerable difference in role expectations within families and between families and the larger society may also cause stress. A family with strong extended family ties may be cut off from the normal social support and assistance when it migrates. A family that has been largely self-sufficient may find itself on welfare in its new society and the members experience themselves as social failures. Social services, educational institutions and other agencies with which migrants have contact may expect behaviour quite at odds with those of the migrant. For example, a migrant family from a strongly matriarchal culture moving to a patriarchal culture may find the head of the household ignored by members of the new society.

Incidence In what may well be the most massive demographic shift in human history – certainly in its scale and time frame – tens of millions of human beings are on the move, driven by war, oppression, natural disaster, and economic need to urban areas, open lands, and better hopes of survival. This vast resettlement of humanity is bound to increase society's tensions along the fault-lines of culture, religion, ethnicity, or race in the coming decades. Some of the areas where migration is posing problems of change are the Persian Gulf States which attract large numbers of Asian migrant workers, immigrants from Mexico and Central America into the United States; migrant workers from Turkey and Morocco in Northern Europe; equally pressed are "megacities" such as Mexico City, Sao Paulo and Lagos.

Narrower Failed migration (#PE3445) Resettlement stress (#PD7776)
Emigration of trained personnel from industrialized countries (#PE8093).
Aggravated by Excessive emigration from island developing countries (#PE5713).

♦ PC0025 International economic fragmentation

Distrusted economic co-operation — Non-cooperative economic relations — Lack of international economic coordination — Lack of economic cooperation

Nature The EEC, EFTA, the Nordic Council, CMEA and other regional organizations outside Europe; and the groupings of nations in the OECD, OPEC and various commodity cartels – all these fragment the world economic system into a drastic play of forces that are beyond any one group of nations to control. Rivalry forces the entire global network of trade and finance to drive upwards and downwards unpredictably with consequent hardships, usually falling more heavily on the smaller nations.

Claim For most developing countries except the largest, a new era of economic growth depends, in the short-term, on effective and coordinated management among major industrial countries in order to facilitate expansion, reduce real interest rates, and halt the tendency to protectionism. In the longer term, major changes are also required to make consumption and production patterns sustainable in a context of higher global growth. International economic cooperation to achieve the former is embryonic, and to achieve the latter is negligible.

Counter-claim Integration schemes tend to favour the more advanced members of a grouping, who benefit from the liberalization of trade and the expansion of their markets; while the less advanced members of a grouping tend to gain less, unless the scheme provides for structural changes in the latter countries. Special problems facing the less advanced countries include not only the smallness of their markets, but more particularly the absence of supplies exportable to the other partners of their grouping.

Broader Fragmentation (#PA6233) Lack of international cooperation (#PF0817).
Narrower Fragmentation of the international trading system (#PC9584)
Inadequate economic integration between regional groupings of developing countries (#PD9412).
Related World anarchy (#PF2071) World federalism (#PF2088).
Aggravated by Non-alignment (#PF0801) Institutional fragmentation (#PC3915)
Fragmented conduct of community operations (#PF1205)
Individualistic practices of local business (#PF1176)
Inadequate models of socio-economic development (#PF9576).

♦ PC0027 Ageing of world population

Elders' predomination in society

Nature Only when fertility levels begin to fall in conjunction with continued low mortality rates, does a population 'age', in the sense that population growth slows and the relative size of the adult population, especially the elderly age group, increases. The ageing of the world population raises critical economic, social and political questions, especially in the developing countries, where the transition from a currently youthful population to an aging one will occur much more rapidly than it has in more developed countries. In the market-economy countries this trend has reduced the potential long-term growth rate, lowered savings rates and increases the claim of pensions on private savings and budget revenues. It has therefore made productivity growth an increasingly important factor. The national capacity to look after old people in traditional ways is being seriously weakened, as the aged population rises. This makes the development of new methods of care necessary.

Incidence Due to the combined fluctuations in fertility, life expectancy at birth, and mortality, the average age of the world's population is expected to increase dramatically within the next forty years. While total world population is expected to treble in the 75 years from 1950 to 2025, the population of the over-60s will show a five-fold increase and the over-80s will increase to seven times their present number. This means that one person in every seven will be over 60 years of age in 2025 compared with just one in every 12 in 1950. In 1950 there were only an estimated 214 million people over the age of 60 in the world. In 1985 there were approximately 415.6 million. By 2025 that number is expected to reach 1,121 million. Consequently, their weight is expected to increase from 8.5 percent to 13.7 percent of the global population, while the weight of children aged 0–14 is expected to shift inversely from 13.4 percent to 8.4 percent – indicating a marked aging of the world population. The aging of populations will be most dramatic in the developing world where the over-60s are expected to increase nearly seven times between 1950 and 2025, when they will number 800 million. In 1985 the developing world holds 55.4 percent of the over-60s, very similar to the figure for 1950; but by 2025, over 70 percent of the over-60s will live there. The number of over-60s in the developing world will increase fastest between 2000 and 2025, when countries like Bangladesh, Brazil, Mexico and Nigeria will see their over-60s increase by nearly 15 times.

In the early 1980's the proportion of the population over 60 increased everywhere except in Africa and the USSR. The weight of the elderly in the very large populations of some developing regions is expected to increase rapidly, with implications for problems of age dependency and economic development already being encountered in 'aging' countries of the more developed regions. An expected 23% of the population of developed countries and 14% of the developing regions is expected to be over 60 by the year 2025. In Africa the number of persons over 60 is expected to grow by a factor of 4.4, from 22.9 million to 101.9 million, between 1980 and 2025 – a much larger increase than for the population as a whole. Indeed, people are growing old faster than children are being born to support them in their old age. In 1950 there were 19 people over 60 and 45 children under the age of 15 for every 100 adults aged 15–59. By 2025 there are expected to be 40 over-60s and only 35 children for every 100 active adults. There will be 270 million 'economically inactive' over 55-year olds in industrialized countries by 2020. That will mean 38 older dependents for every 100 workers; twice as many as in 1950. In Austria there is already one pensioner for every two workers. The 'dependency ratio' in East Asia is expected to double by 2025, when China will have one person over 60 for every three active adults.

In addition, the numbers of the very old are also rising. In 1980 there were an estimated 34.2 million people over the age of 80, that number having risen by almost 16% to 39.7 million in 1985. Again, although the numbers of the very old in the developed world at present outnumber those in the developing countries, by 2025 it is expected that about 60% of all those over 80 years will live in the Third World.

Background Only in the past few decades has the attention of national societies and the world community been drawn to the social, economic, political and scientific questions raised by the phenomenon of aging on a massive scale. Previously, while individuals may have lived into advanced stages of life, their numbers and proportion in the total population were not high. The twentieth century, however, has witnessed in many regions of the world the control of perinatal and infant mortality, a decline in birth rates, improvements in nutrition, basic health care and the control of many infectious diseases. This combination of factors has resulted in an increasing number and proportion of persons surviving into the later stages of life.

Refs OECD *Demographic Change and Public Policy* (1988).

Broader Human ageing (#PB0477)
Individualistic retaining of local tradition (#PF1705).
Narrower Ageing of the rural population (#PG3318).
Related Rigidity and inadaptability in the aged (#PD3515).
Aggravates Bereavement (#PF3516) Fear of death (#PF0462)
Mental disorders of the aged (#PD0919) Inadequate income in old age (#PC1966)
Criminalization of euthanasia (#PF2643) Inadequate housing for the aged (#PD0276)
Age discrimination in employment (#PD2318) Crisis in long-term pension funds (#PF5956)
Inadequate recreational facilities (#PF0202)
Inadequate welfare services for the aged (#PD0512)
Slowness of sensori-motor activities in the aged (#PD3514)
Retirement as a threat to psychological well-being (#PF1269).
Aggravated by Underpopulation (#PD5432) Excessive longevity (#PJ5973)
Social disadvantage of the aged (#PD3517)
Lameness caused by bone cyst in pedal bone (#PG1636).

♦ PC0030 Religious and political antagonism

Conflict between church and state — Conflict between religion and state

Nature Conflicts may arise over the amount of influence religion has in social and political affairs, or on direct ideological grounds (if the government is totally opposed to religion). Repression, trial and imprisonment of clerics and lay leaders may precede confiscation of sacred properties, or restrictive legislation. It may also lead to civil war, or international conflict. This conflict is most visible in developing countries where religion exerts powerful influence.

Background Every religion, with the growth of adherents, becomes a political force. Persecutions of the early Christians, and of the Jews in Europe, were undertaken from this perspective. Religious/political conflicts may take place from local levels (where sects attempt to control local government) right up to the international level of deliberations and resolutions of intergovernmental bodies which may be opposed by leaders of major faiths.

Broader Religious rivalry (#PC3355) Religious conflict (#PC3292)
Ideological conflict (#PF3388).
Narrower Anti-clericalism (#PF3360)
Undue religious influence on secular life (#PF3358)
Religion as a reinforcement of nationalism (#PF3351)
Accumulation and misuse of religious property (#PE3354).
Related Inadequacy of religion (#PF2005) Injustice of religious courts (#PE0397)
Discriminatory religious influence on the law (#PD3357).
Aggravates Religious war (#PC2371) Minimal church / school involvement (#PJ9011).
Aggravated by Nationalism (#PB0534) Religious repression (#PC0578)
Religious intolerance (#PC1808) Dependence on religion (#PF0150)
Religious indoctrination (#PD4890) Double standards in morality (#PF5225)
Religious discrimination in education (#PD8807)
Burdensome cost of religious ceremonies (#PF3313)
Conflict between government and the news media (#PE1643).
Reduced by Official religion (#PF6091).

♦ PC0031 Segregation

Nature Segregation is the establishment by law or custom of separate (and inferior) facilities for social, ethnic or religious groups providing separate educational, recreational, and other facilities. Segregation inevitably results in discrimination in favour of one group over the other or others. The word covers a whole range of discriminatory practices including the denial of employment and

voting rights and prohibition against intermarriage. More generally speaking, it also occurs in education, housing, public services and on age, sexual and class grounds.
Incidence Characteristic of societies with complex class systems, it manifests itself in areas as widely apart as India (caste system), Asia and the Middle East, as well as South Africa and North America. The implication that segregation was a natural phenomenon assisted its growth.
Broader Exclusion (#PA5869) Social injustice (#PC0797) Social inequality (#PB0514) Social fragmentation (#PF1324).
Narrower Caste system (#PC1968) Sex segregation (#PC3383)
Age segregation (#PD3444) Legal segregation (#PD3520)
Ethnic segregation (#PC3315) Racial segregation (#PC3688)
Spatial segregation (#PJ1728) Segregation in housing (#PD3442)
Segregation in marriage (#PD3347) Segregation in education (#PD3441)
Segregation in employment (#PD3443) Segregation in social services (#PD3440)
Segregation based on religious affiliation (#PC3365)
Segregation of poor and minority population in urban ghettos (#PD1260).
Related Elitism (#PA1387) Cultural fragmentation (#PF0536)
Inequality before the law (#PC1268).
Aggravates Slavery (#PC0146) Eugenics (#PC2153) Scapegoats (#PF3332)
Nationalism (#PB0534) Discrimination (#PA0833) Social conflict (#PC0137)
Human inequality (#PA0844) Social alienation (#PC2130) Lack of assimilation (#PF2132)
Lack of racial identity (#PF0684) Lack of social mobility (#PF2195)
Inequality in education (#PC3434) Unequal access to education (#PC2163)
Inflexible social structure (#PB1997) Inter–cultural misunderstanding (#PF3340)
Conflict between minority groups (#PC3428)
Discrimination against indigenous populations (#PC0352).
Aggravated by Taboo (#PF3310) Prejudice (#PA2173)
Antisemitism (#PE2131) Minority control (#PF2375)
Racial exploitation (#PC3334) Class consciousness (#PC3458)
Social discrimination (#PC1864) Plural society tensions (#PF2448)
Discrimination against minorities (#PC0582)
Inadequate cultural integration of immigrants (#PC1532)
Divisive effects of official cultural pluralism (#PF0152)
Inadequacy and insensitivity of intelligence testing (#PD1975).

♦ PC0052 Over–intensive soil exploitation

Soil mismanagement — Unimproved farm soils — Poor soil management — Unfertilized soil
Nature Soil erosion threatens serious and possibly irreversible damage across the world. In the Sahel and the Amazon, fragile ecosystems are threatened by climactic changes and massive development efforts. The soil in temperate countries is also endangered. Intensive monocrop cultivation with heavier and heavier equipment and ever–growing quantities of chemical pesticides and fertilisers is taking a serious toll. The side–effects of present agricultural practices are future loss of fertility of the soil and deep erosion. These practices continue because the prevailing system of grants and subsidies reward intensive, mechanized agro–industrial farming.
Incidence The United States Department of Agriculture found that about a quarter of US cropland exceeded its "tolerable" levels for erosion in 1980.
Refs FAO *Impact on Soils of Fast–Growing Species in Lowland Humid Tropics* (1980).
Broader Exploitation (#PB3200).
Narrower Soil compaction (#PD1416) Monoculture of crops (#PC3606)
Chemicalized farming (#PD7993)
Soil mismanagement in developing countries (#PD6482).
Related Intemperance (#PA6466)
Resource wastage in capitalist systems (#PC3108)
Individualistic retaining of local tradition (#PF1705).
Aggravates Soil pollution (#PC0058) Infertile land (#PD8585)
Expansive soils (#PE5036) Soil infertility (#PD0077)
Soil degradation (#PD1052) Vulnerability of plants and crops (#PD5730).
Aggravated by Agricultural mismanagement (#PD8625)
Shortage of cultivable land (#PC0219)
Unethical practice of soil sciences (#PD1110)
Underproductive methods of agricultural management (#PF6524).

♦ PC0056 Rural depopulation

Nature The urban population of the world has doubled since 1950 and is likely to do so again before 2000. Large shifts are occurring throughout the world from agricultural activities in rural areas to non–agricultural activities in urban areas, on a scale and at a rate hitherto unknown.
Background Steps that have been taken to stem the rising tide have been palliatives at best, mainly because they have often been formulated without an adequate knowledge of the causes and consequences of migration. Also, little is known on whether it is possible or desirable to control the flow. The issue is riddled with unanswered questions and, above all, with persistent myths that obscure the search for solutions.
Claim Fears have been expressed that migration is a major cause of rising urban unemployment, overcrowded housing, and relative shortage of public amenities. Migrants unable to find adequate employment or any work at all are forced to live in squatter settlements or inner–city slums lacking even the most basic facilities. The resulting pressure on residential land and housing causes speculation and excessive rents, and generally tends to depress living standards in the urban areas. Similarly migration to the cities affects rural areas; not only does it tend to draw away their more dynamic members, but it may also divert national investment resources towards the towns.
Counter–claim In the cities, the influx of migrants not only increases the labour supply but also generates new employment by stimulating industrial expansion and other economic activities. In rural areas, out–migration may lead to a reduction of the labour/land radio and provide a new environment conducive to changing rural production techniques. There may be a rising demand from the cities for rural output that stimulates agriculture and rural industrialization, thus helping to raise incomes of country dwellers. Similarly, remittances sent home by migrants may improve the distribution of income between the rural and urban population. Analysis of urban data suggests that migrants attain economic status comparable to that of urban natives in a remarkably brief period of time. There is no evidence that migrants are confined to marginal employment in the cities, or contribute disproportionately to urban underemployment.
Broader Internal migration (#PF4009).
Narrower Depopulation of mountainous regions (#PD1908)
Adjustment difficulties of new urban families (#PF1503)
Uncontrolled urbanization in developing countries (#PD0134).
Aggravates Urban poverty (#PC5052) Diplomatic errors (#PF1440)
Rural underdevelopment (#PC0306) Urban slums in developing countries (#PD3489)
Urban unemployment in developing countries (#PD1551)
Disruption of family system in developing countries (#PD1482)
Imbalance between urbanization and industrialization in developing countries (#PC1563).
Aggravated by Socio–economic poverty (#PB0388)
Migration of rural population to cities (#PE8768).
Reduces Rural unemployment in developing countries (#PD0295).

♦ PC0058 Soil pollution

Contamination of soil by toxic substances
Nature Many metals and organic substances have an affinity for organic matter or mineral particles. Soils thus provide a suitable medium for the accumulation of contaminants from aqueous sources and atmospheric deposition. Land pollution by toxic chemicals from agriculture and industry arises from the dumping on land of ever–increasing amounts of domestic and industrial solid wastes, and leads to contamination of soil, food and water. As world population and the degree of urbanization increase, this may prove to be a significant health hazard.
Incidence Major areas of concern in soil pollution include agricultural land pollution from chemical fertilizers, pesticides, or nutrients, solid wastes from industry and urban populations, and radioactive pollution from either nuclear explosions, release of liquid or solid radioactive wastes produced by industrial or research establishments, or radioactive fallout from the emissions of smoke–stacks of chemical works.
Compounding the problem of agricultural land pollution is the persistence of the degraded chemical products of soil additives, whose toxicity may, in fact, be more pronounced than that of the original additive. Slurry – the effluent of cattle and hogs – and the liquor from grass silage used as cattle feed are extremely polluting to land and water. Slurry can be 100 times and silage 200 times as polluting as untreated domestic sewage.
Regarding contamination from urban areas, the land available for waste disposal is diminishing, thus concentrating waste and its disagreeable effects, producing feeding grounds for insects and rodents and causing odours created from slowly smoldering fire or from organic decay. This creates a severe nuisance and public health hazard.
Industrialization has constantly increased the amount of wastes being dumped on the soil. It has been estimated that 50 per cent or more of the raw materials used by industry ultimately become waste products, of which about 15 per cent can be considered deleterious or toxic. In addition, inorganic contaminants such as fluorides, emitted from smoke stacks, can contaminate nearby farmland.
Levels in the northern hemisphere of radiation from fission products deposited in the soil by fallout are about 10 to 30 per cent of those due to natural radioactive substances in the soil. In time, this increased radioactivity could affect soil fauna or their predators.
Refs Assink, J W and Brink, W J van den (Eds) *Contaminated Soil* (1985).
Broader Environmental pollution (#PB1166).
Narrower Pollution of sediments (#PE5539) Soil–transmitted diseases (#PD3699)
Metal contamination of soil (#PD3668) Radioactive contamination of soil (#PE3383).
Aggravates Food pollution (#PD5605) Infertile land (#PD8585)
Water pollution (#PC0062) Shortage of cultivable land (#PC0219)
Environmental plant diseases (#PD2224) Destruction of land fertility (#PC1300)
Endangered species of insects (#PC2326).
Aggravated by Wasted water (#PD3669) Contradictions (#PF3667)
Soil degradation (#PD1052) Volcanic eruptions (#PD3568)
Agricultural wastes (#PC2205) Industrial emissions (#PE1869)
Mycoplasmal arthritis (#PG2302) Toxic metal pollutants (#PD0948)
Fluorides as pollutants (#PE1311) Hazardous waste dumping (#PD1398)
Pesticides as pollutants (#PD0120) Herbicides as pollutants (#PD1143)
Solid wastes as pollutants (#PD0177) Over–intensive soil exploitation (#PC0052)
Industrial waste water pollutants (#PD0575)
Damage by degradable organic matter (#PJ6128)
Environmental hazards from fertilizers (#PE1514).

♦ PC0062 Water pollution

Nature Water is considered polluted when it is altered in composition or condition so that it becomes less suitable for any or all of the functions and purposes for which it would be suitable in its natural state. This definition includes changes in the physical, chemical and biological properties of water, or such discharges of liquid, gaseous or solid substances into water as will or are likely to create nuisances or render such waters harmful to public health, safety or welfare, or to domestic, commercial, industrial, agricultural, fish or other aquatic life. It also includes changes in temperature, due to the discharge of hot water.
Pollution may be accidental (sometimes with grave consequences) but is most often caused by the uncontrolled disposal of sewage and other liquid wastes resulting from domestic uses of water, industrial wastes containing a variety of pollutants, agricultural effluents from animal husbandry and drainage of irrigation water, and urban run–off. The deliberate spreading of chemicals on the land to increase crop yields, or the addition of chemicals to water to control undesirable organisms, is another cause of pollution. Examples are the application of chemical fertilizers, and of pesticides for the control of aquatic weeds, insects and molluscs. Problems are compounded when national boundaries are involved, and cooperation in the management of transboundary waters is becoming essential.
Water as a part of the human environment occurs in four main forms: as groundwater, in freshwater surface masses, in the sea, and as vapour in the atmosphere. Human health may be affected by ingesting polluted water directly or in food, by using it in personal hygiene or for agriculture, industry or recreation, and by living near it. Two main categories of water–associated health hazards are from biological agents that may affect man following ingestion of water or other forms of water contact or through insect vectors, and from chemical and radioactive pollutants, usually resulting from discharges of industrial wastes. In addition, even well treated water sources may contain viruses from sewage which promote enteric disease.
Incidence In 1980 four out of five child deaths in the third world resulted from disease from dirty water supplies; 80% of people in developing countries have no sanitation facility. Water shortage and contamination kill 25,000 people a day. More than two thirds of India's water resources are polluted; 98% of China's sewage goes into rivers untreated, and in the Philippines domestic sewage makes up 70% of the Pasig River in Manila. About 2% of the groundwater supplies in the United States are polluted. Japan's Inland Sea has 200 red tides annually. Red tides are caused by the decaying of algae sapping large amounts of oxygen from the water, asphyxiating fish and other marine life. Commercial fishermen dump 22,000 metric tons of plastic packaging into the sea every year, along with 136,000 tons of plastic nets, lines and buoys.
Background During the 1970s, the pollution that caused most concern was due to sewage, agricultural chemicals, oil, and metals. Metal concentrations were clearly elevated in coastal waters, and in fish and shellfish living there. In some areas, mercury levels in species such as tuna were high enough to make these fish unsuitable as human food. Chemical contamination of the oceans appeared to be localized, with the worst conditions in estuaries and coastal areas in industrial regions, where ecological changes were apparent. Some of this pollution came via rivers: the amount of iron, manganese, copper, zinc, lead, tin and antimony that reached the sea by this route was far greater than would be supplied by natural geological processes. Other contaminants came through atmospheric deposition: the importance of this pathway for metals and synthetic chemicals was increasingly recognized during the decade. Offshore oil and gas exploration and dredging for sand and gravel in coastal areas also increased during the decade. Coastal zone development affected extensive estuarine areas, as well as mangrove swamps and coral reefs. Oil pollution killed sea–birds, fouled beaches and affected tourism. Although tanker accidents were the source of less than 5 per cent of all the oil entering the sea, accidents released large volumes in small areas, and were therefore especially damaging.
Claim The water crisis has become a major component of the environment crisis. Deterioration in the quality of water is due not only to the quantitative increase of consumption with the resulting increase of refuse; it is above all a consequence of the development of new products and new production technologies (use of detergents, development of the chemical industry, increasing use of chemical manures and pesticides in agriculture, etc), releasing into the environment substances that are all the more harmful for not always being biodegradable. The problem of

combating water pollution, at least in the industrialized countries, has become more difficult than that of ensuring an adequate quantitative supply of water. There is a risk of the same situation arising in the developing countries, where the factors making for pollution are tending to increase.

Refs Eckenfelder, W Wesley *Industrial Water Pollution Control* (1988); Jansson, Mats and Persson, Gunnar (Eds) *Phosphorus in Freshwater Ecosystems* (1988); United Nations Educational, Scientific and Cultural Organization *Effects of Urbanization and Industrialization on the Hydrological Regime and on Water Quality* (1977).

Broader Environmental pollution (#PB1166).
Narrower Marine pollution (#PC1117) Cyanide as a pollutant (#PG3653)
Water surface pollution (#PD5641) Pollution of groundwater (#PD2503)
Impurities in waste water (#PD0482) Pollution of inland waters (#PD1223)
Water system contamination (#PD8122) Transboundary water pollution (#PD1096)
Chemical contamination of water (#PE0535) Biological contamination of water (#PD1175)
Water pollution in developing countries (#PD3675).
Aggravates Air pollution (#PC0119) Aquatic weeds (#PD2232)
Animal diseases (#PC0952) Amenity destruction (#PC0374)
Water-borne diseases (#PE3401) Silting of water systems (#PD3654)
Insect vectors of disease (#PC3597) Rodent vectors of disease (#PE3629)
Pests and diseases of fish (#PD8567) Human disease and disability (#PB1044)
Environmental plant diseases (#PD2224) Pollution-induced fish diseases (#PE7584)
Eutrophication of lakes and rivers (#PD2257) Fragility of maintaining basic health (#PJ2524)
Inadequate water system infrastructure (#PD8517)
Unsustainable development of fresh waters (#PD6923)
Declining productivity of agricultural land (#PD7480).
Aggravated by Soil erosion (#PD0949) Wasted water (#PD3669)
Oil pollution (#PE1839) Soil pollution (#PC0058)
Natural disasters (#PB1151) Lead as a pollutant (#PE1161)
Acidic precipitation (#PD4904) Sewage as a pollutant (#PD1414)
Nickel as a pollutant (#PE1315) Arsenic as a pollutant (#PE1732)
Cadmium as a pollutant (#PE1160) Mercury as a pollutant (#PE1155)
Nitrates as pollutants (#PE1956) Hazardous waste dumping (#PD1398)
Chromium as a pollutant (#PE4072) Selenium as a pollutant (#PE1726)
Fluorides as pollutants (#PE1311) Detergents as pollutants (#PE1087)
Water pollution by fertilizers (#PE8729) Domestic waste water pollutants (#PD2800)
Industrial waste water pollutants (#PD0575) Radioactive contamination of water (#PE2441)
Damage by degradable organic matter (#PJ6128)
Polychlorinated biphenyls as a health hazard (#PE2432).

♦ PC0065 Dependence on external resources

Resource import dependence

Refs Sobhan, Rehman *Crisis of External Dependence.*

Narrower Dependence on oil (#PJ4398) Dependence on foreign labour (#PG1728)
Energy dependence and vulnerability (#PJ7735)
Dependence of developing countries on food imports (#PE8086)
Dependence of island developing countries on imports (#PD5677)
Dependence of developing countries on imported technology (#PF1489)
Dependence of industrialized countries on import of resources (#PE0537).
Aggravates Dependency on unpredictable sources of income (#PF3084).

♦ PC0066 Maltreatment of animals

Cruelty to animals

Nature Inhumane treatment of animals takes on various forms. Domestic and captive animals often live in inhumane or insanitary conditions. This is the case in many zoos. Unfit domestic animals can be abandoned or painfully put death. In some countries, the surplus of dogs is controlled by the use of strychnine. There is wanton abuse of animals for man's convenience, namely for the claims and interests of science, sport, entertainment and the production of food. Inhumane slaughter of food animals is still widespread in the meat industry and in the whaling and sealing industries. The cramming of geese is but another example of cruelty inflicted on livestock. Cruel transportation, whether by road, rail, sea or air, is one of the main causes of animal suffering.

Incidence A clear link has been established between the abuse of children, adult aggressiveness and cruelty to animals. Three quarters of the aggressive criminals interviewed in one study were abused and beaten as children (as compared to only ten percent of non-criminals). Sixty percent of this group committed cruelty to animals. One man put his girlfriend's cat into a microwave oven. Generally the motivation for these acts ranged from a desire to shock or to satisfy a grudge against another species to a displacement of hostility from a person whom the criminal did not attack to the animal which he could.

Background Contemporary philosophers question the traditional legitimizing of tyranny of human over nonhuman animals. Thomas Aquinas, in denying that animals have souls, laid the theological groundwork for post-Enlightenment philosophers to dismiss the non-human side of creation altogether.

Broader Cruelty (#PB2642) Inadequate animal welfare (#PC1167).
Narrower Zoosadism (#PG2204) Fishing for sport (#PE5286)
Cruelty to insects (#PE3468) Unwanted pet animals (#PD2094)
Livestock mutilation (#PF1849) Misuse of wild animals (#PD8904)
Maltreatment of livestock (#PE3394) Cruel animal transportation (#PD0390)
Inadequate feeding of animals (#PC2765)
Exploitation of animals for amusement (#PD2078)
Cruelty to animals in factory farming (#PD2768)
Cruelty to animals in food preparation (#PE0236)
Cruel treatment of animals for research (#PD0260)
Unsanitary and inhumane urban food animal conditions (#PE0395).
Related Denial of rights of animals (#PC5456)
Excessive commercial exploitation of farm animals by industrial concerns (#PD2772).
Aggravates Animal injuries (#PC2753) Broken-spirited animals (#PF5191)
Intrusive animal-rights campaigners (#PE4438).
Aggravated by Hunting of animals (#PC2024)
Unethical practices with domesticated animals (#PE4771)
Excessive animal sanitary regulations in international travel (#PF1555).
Reduced by Use of agricultural resources for production of animal feed (#PD1283).

♦ PC0067 Censorship

Dependence on censorship

Nature Censorship may involve restrictions on expression, or the public availability of books, newspapers and journals, films, plays, news, artwork, photography, broadcasts, and non-acceptance of new scientific thought. Censorship may lead to lack of information and subsequent development of apathy, ignorance, conformism and general stagnation. It may threaten democracy and encourage subversive activities. It may equally foster idealism through indoctrination and strengthen governmental control.

Incidence Censorship is universal, but particularly marked in political dictatorships or totalitarian regimes.

Claim Increasing censorship may be an indicator of national and international tension, or it may be a reaction against decadence. Censorship may be, in a distorted way, an expression of a search for values in fragmented societies. In a psychological sense censorship is a function exercised by the ego over drives and impulses which have an instinctual quality, and which the ego may seek to repress. Thus censorship in society also results in a bottling-up of forces; a repression that can lead to violent social outbursts, or societal dysfunctioning.

Refs Alderfer, Hannah, et al *Caught Looking* (1987); Boyle, Kevin (Ed) *Article Nineteen World Report 1988* (1988); Curry, Richard O (Ed) *Freedom at Risk* (1988).

Broader Restriction of freedom of expression (#PC2162).
Narrower Art censorship (#PD2337) Book censorship (#PD3026)
News censorship (#PD3030) Self censorship (#PF6080)
Video censorship (#PE5966) Postal censorship (#PD3033)
Theatre censorship (#PD3028) Military censorship (#PE5989)
Religious censorship (#PD5998) Scientific censorship (#PD1709)
Selective information (#PF6057) Film and cinema censorship (#PD3032)
Radio and television censorship (#PD3029) Censorship in communist systems (#PD3172)
Restrictions on freedom of information (#PC0185).
Related Political crime (#PC0350) Film propaganda (#PD3089)
Book propaganda (#PD3090) Official secrecy (#PC1812)
Art as propaganda (#PF3087) Racist propaganda (#PD3093)
Uncontrolled media (#PD0040) Religious propaganda (#PD3094)
Defamation of character (#PD2569) Photographic propaganda (#PD3086)
Radio and television propaganda (#PD3085) Inadequate government publications (#PF3075)
Restriction of access to news distribution media (#PF3082).
Aggravates Sedition (#PC2414) Ignorance (#PA5568)
Conformism (#PB3407) Student press weakness (#PE0628)
Suppression of information (#PD9146) Undemocratic political organization (#PC1015)
Repression of intellectual dissidents (#PD0434)
Suppression of creativity and innovation (#PF0275).
Aggravated by Unethical documentation practices (#PD2886).
Reduces Permissiveness (#PF1252) Immoral literature (#PF1384)
Infringement of privacy (#PB0284) Journalistic irresponsibility (#PD3071).

♦ PC0073 Restrictive trade practices

Dependence on restrictive trade practices — Restrictive business practices — Distortion of international trade by restrictive business practices

Nature "Restrictive business practice" means action or behaviour by an enterprise which, through abuse - or acquisition and abuse - of a dominant position of market power, limits access to markets or otherwise unduly restrains competition; this leading or being likely to lead to adverse effects on international trade, particularly that of developing countries, and on the economic development of these countries. It also refers to formal, informal, written or unwritten agreements or arrangements among enterprises which have the same impact. Although there is legislation in several countries to control restrictive business practices, in practice these apply to trade and transactions having effects on the domestic market and seldom to international trade when the adverse effects are felt abroad exclusively.

Refs Long, Frank *Restrictive Business Practices, Transnational Corporations and Development* (1981); United Nations Conference on Trade and Development *Problems of Protectionism and Structural Adjustment, Introduction and Part 1* (1987).

Broader Restrictive practices (#PB9136).
Narrower Cartels (#PC2512) Refusal to sell (#PF0468) Domestic cartel (#PD7963)
Transfer pricing (#PE1193) Unfair competition (#PC0099) Extraterritoriality (#PF2178)
Grant-back provisions (#PE5306) Challenges to validity (#PF1200)
Restrictions on research (#PF0725) Restrictions on publicity (#PF1575)
Restrictions on adaptations (#PF5248) Exclusive dealing arrangements (#PE0413)
Protection of company ownership (#PF4432) Restrictions on use of personnel (#PF3945)
Bilateralism in trade arrangements (#PF7642)
Restrictive transport insurance practices (#PD0881)
Patent pool or cross-licensing agreements (#PE4039)
Non-tariff barriers to international trade (#PC2725)
Collusive international trade arrangements (#PE0396)
Restrictive agreements on product standards (#PD0343)
Restrictive practices in trade in chemicals (#PE8600)
Restrictive practices in mineral fuels trade (#PE0141)
Exclusive sales and representation agreements (#PE4581)
Excessive concentration of business enterprises (#PD0071)
Restrictive practices in cargo airline services (#PE5910)
Preoccupation with reciprocity in trading relations (#PF3871)
Restrictive practices in trade in manufactured goods (#PD1797)
Payments after expiration of industrial property rights (#PF5292)
Restrictive practices in the food and live animals trade (#PE0342)
Ineffective regulation of restrictive business practices (#PF1596)
Restrictive practices in the beverages and tobacco trade (#PE7899)
Restrictive business practices in technology transactions (#PE1978)
Restrictive market divisions by transnational corporations (#PE3196)
Restrictions on export activity due to licensing arrangements (#PE0895)
Tying of supplies to subsidiaries by transnational enterprises (#PE0669)
Domestic bias in the regulation of restrictive business practices (#PE0789)
Disincentives for financial investment within developing countries (#PF3845)
Restrictive practices in trade in machinery and transport equipment (#PE7958)
Restrictive practices in trade in inedible crude non-fuel materials (#PE8351)
Restriction of free market competition by transnational corporations (#PE0051)
Restrictive business practices in relation to patents and trademarks (#PE0346)
Restrictive practices in trade in animal and vegetable oils and fats (#PE8880)
Extraterritorial application of restrictive business practices legislation (#PE8011)
Consequences of restrictive business practices of transnational enterprises (#PE1799)
Excessive injury to export interests developing countries due to export cartels (#PE2598)
Inadequate regulation of the restrictive business practices of state enterprises (#PE0225)
Direct foreign investment by transnational enterprises as a restrictive business practice (#PE0161)
Distortion of international trade by discriminatory customs and administrative entry procedures (#PE2603)
Restrictive business practices in the markets of developed countries against exports from developing countries (#PE5926).
Aggravates Economic conflict (#PC0840) Excessive external trade deficits (#PC1100)
Non-payment of compensation for damages to consumers (#PE0290).
Aggravated by Collusive tendering in international trade (#PE7072)
Non-viability of small states and territories (#PD0441)
Decline in competition due to entrance barriers (#PE2176)
Inadequate regulation of restrictive business practices in service industries (#PF0591)
Ineffective monitoring of restrictive business practices due to inadequate regulation (#PF2782).

♦ PC0080 Irresponsible research using human subjects

Nature Scientific researchers are able to obtain institutional facilities for research on humans who are not always in a position to give their free consent. Research has been conducted using mentally retarded children, prisoners, or military personnel without adequate regard for the social, moral and ethical implications. Because of the controversial nature of such research methods they tend to be used in an atmosphere of total or semi-secrecy.

Refs Spicker, Stuart F, et al *The Use of Human Beings in Research* (1988).

Broader Irresponsibility (#PA8658) Inhumane scientific activity (#PC1449)
Irresponsible scientific and technological activity (#PC1153).
Related Denial to experimental animals of the right to freedom from suffering (#PE8024).

♦ PC0092 Malignant neoplasms

Cancer — Household cancers

Nature Cancer, which is not a single disease but a spectrum of diseases that includes more than 100 kinds, is characterized by the unrestrained growth of cells. Cells naturally grow and multiply, but in a healthy body this growth is controlled by a complex series of regulatory mechanisms. Cancer occurs when this regulation fails and cell division goes haywire. In connective tissues (such as bone, cartilage, tendon, muscle) cancers are called sarcomas; in epithelial tissues (such as skin, bladder, lung, breast) they are called carcinomas; and in cells of the blood system they are named leukaemias. In most cancer cases unrestrained growth leads to the formation of tumours which spread into and often kill normal tissue. When not attended to, such a tumour (which is called malignant) ordinarily leads to death, but there can be a long delay between its onset and the appearance of obvious symptoms. Cancer is the second leading cause of death in the industrialized world and is rapidly becoming a major disease in the Third World as well. Though the actiology is still undetermined, certain cancers are linked to certain environmental and lifestyle factors such as carcinogenic substances in the work environment, cigarette smoking, excess exposure to the sun, alcohol, diets high in fat and low in fibre, and stress.

Incidence In 1980 it was estimated that there were 6,350,000 new cases of malignant neoplasms (cancers) worldwide and that neoplastic diseases resulted in 4.2 million deaths (20 per cent of mortality in industrialized countries; 5 per cent in developing countries). In 1984, WHO estimated 4.3 million deaths worldwide resulting from cancer, of which 2.3 million occurred in the developing regions and the remaining 2 million in developed areas. In addition, there are approximately 5.9 million new cancer cases annually, of which 2.9 million occur in the more developed countries and 3 million in the developing countries. Speculating about the origins of cancer, many respected oncologists describe a cancer personality – repressed and depressed – and draw attention to the loss or bereavement which commonly precedes the onset of the disease. Lung cancer patients are most often people unable to express strong emotions. Cancer of the cervix occurs most often among women with a tendency to helpless ness or a sense of hopeless frustration derived from some unresolved conflict in the preceding six months. Overall, the most important factors in the development of malignancy are: a loss of raison d'être; an inability to express anger or resentment; marked self dislike and distrust; and most significantly, loss of an important emotional relationship

Background The probable origin of the term 'cancer' is related to its growth pattern; that is, cancer often grows into surrounding tissue in cords that resemble the claws of a crustacean.

Refs American Cancer Society Staff *American Cancer Society's Complete Book of Cancer* (1986); Bammer, Kurt and Newberry, Benjamin H (Eds) *Stress and Cancer* (1981); Beck, L, et al (Eds) *The Cancer Patient* (1986); Bolognesi *Human Retroviruses, cancer and AIDS* (1988); Bracken, Jeanne M *Children with Cancer* (1986); Bresciani, Francesco, et al (Eds) *Hormones and Cancer Two* (1984); Chretien, et al *Current Therapy of Head and Neck Cancer* (1985); Crooke, Stanley T and Prestayko, Archie W *Introduction to Clinical Oncology* (1981); Engstrom, Paul F, et al (Eds) *Advances in Cancer Control* (1987); Hankins, W David and Puett, David (Eds) *Hormones, Call Biology and Cancer* (1988); Howe, Melvyn G (Ed) *Global Geocancerology* (1986); Kapoor, A S (Ed) *Cancer and the Heart* (1986); Maul–Mellott, Susan K and Adams, Jeanette, N *Childhood Cancer* (1987); Parkin, M D, et al (Eds) *International Incidence of Childhood Cancer* (1988); Preece, Paul, et al (Eds) *Cancer of the Stomach* (1986); Sax, N Irving *Cancer Causing Chemicals* (1981); Siegal, Mary–Ellen *The Cancer Patient's Handbook* (1986); Vaeth, J M and Meyer, J (Eds) *Cancer and the Elderly* (1986); Wagner, G, et al (Eds) *Cancer of the Liver, Esophagus and Nasopharynx* (1987); Waterhouse, J, et al (Eds) *Cancer Incidence in Five Continents* (1986); Williams, A Olufemi *Virus–Associated cancers in Africa* (1984).

Broader Human disease and disability (#PB1044).
Narrower Leukaemia (#PE0639) Liver cancer (#PE3233)
Breast cancer (#PE1175) Animal cancer (#PG1900)
Occupational cancer (#PE3509) Malignant neoplasm of bone (#PE9229)
Malignant neoplasm of skin (#PE5016)
Malignant neoplasm of mouth and throat (#PE9819)
Malignant neoplasm of digestive organs (#PE4303)
Malignant neoplasm of respiratory system (#PE7572)
Malignant neoplasm of genito–urinary organs (#PE5100)
Pleura and peritoneum cancers of the bronchi (#PE8228)
Neoplasms of lymphatic and haematopoietic tissue (#PE4637).
Related Neoplastic diseases (#PC3853) Diseases of the spine (#PD2626).
Aggravates Thrombosis (#PE5783) Tonsillitis (#PE2292)
Constipation (#PE3505) Diseases and injuries of the brain (#PD0992).
Aggravated by Ulcers (#PE2308) Eczema (#PE2465)
Smoking (#PD0713) Keratoses (#PG1906)
Air pollution (#PC0119) Benign tumours (#PD8347)
Biochemical warfare (#PC1164) Toxic food additives (#PD0487)
Stress in human beings (#PC1648) Motor vehicle emissions (#PD0414)
Thyroid gland disorders (#PE0652) Coitus as a cancer risk (#PD6033)
Pesticides as pollutants (#PD0120) Skeletal system disorders (#PE2298)
Lung disorders and diseases (#PD0637) Excessive consumption of fats (#PE4261)
Carcinogenic chemical and physical agents (#PD1239)
Environmental hazards of electrical power transmission lines (#PE9642).

♦ PC0099 Unfair competition

Refs Kitch, Edmund, et al *Selected Statutes and International Agreements on Unfair Competition, Trademarks, Copyrights and Patents* (1986); Shchetinin, V D U S *Monopolies and Developing Countries* (1985).

Broader Economic crime (#PC5624) Restrictive trade practices (#PC0073).
Narrower Business bribery (#PD8449) Misleading advertising (#PE3814)
Unfair competition from convict–made goods (#PE5506).
Aggravates Monopoly power due to advertising (#PE0081).

♦ PC0111 Ecologically unsustainable development

Environmental hazards due to economic development — Overdevelopment

Nature Environmental problems created by economic development can damage human welfare either directly or indirectly. Direct damage includes damage to health (from lead poisoning, for example, or lung disease aggravated by air pollution), social disruption (for example, displacement of people by mining operations or hydroelectric projects), and damage to the "quality of life" through congestion, noise, litter, etc. Indirect damage to human welfare occurs through interference with national biological systems. For example, the filling of estuaries and the pollution of coastal waters diminishes ocean productivity; and logging or overgrazing can accelerate erosion. The long–term consequences for human beings of chronic exposure to low concentrations of environmental contaminants may be more serious that those of acute pollution. A deteriorating relationship between human populations and the natural systems that sustain them is a major contributor to deepening poverty in many regions. The most serious threats of all, however, may well prove to be indirect and generated by mankind's disruption of the functioning of the natural environment.

Claim Pursuit of indiscriminate economic growth and affluent living standards has led to overdevelopment in the industrialized countries. Continued pursuit of these goals will produce accelerating problems of environmental destruction, resource and energy scarcity, conflict and social breakdown. The industrialized countries have an unsustainable way of life.

In some areas environmental degradation is due to large–scale commercial farming, ranching or forestry, in pursuit of short–term profit without regard to conservation – for example: destructive logging in South–East Asia; conversion of Latin American rainforest to pasture for beef cattle intended for the North American hamburger market. Any economy which has grown to the point where it cannot be sustained on ecological ground for a long future is overdeveloped and endangers all other economies.

Refs DeGaay Fortman, B (Ed) *Overdevelopment* (1979); Goudzwaard, Bob *Aid for the Overdeveloped West* (1975); Kapp, K William *Social Costs, Economic Development and Environmental Disruption* (1983).

Broader Unsustainable development (#PB9419) Destruction inherent in development (#PF4829).
Narrower Agricide (#PE4045) Hunting of animals (#PC2024)
Environmental hazards from mining (#PD2596)
Environmental hazards from electricity (#PE1412)
Environmental hazards from fishing industry (#PD0743)
Environmental hazards from forestry and logging (#PE1264)
Environmental hazards from manufacturing industries (#PD0454)
Environmental hazards from the construction industry (#PE8790)
Environmental hazards from economic and industrial products (#PC0328)
Environmental hazards from agricultural and livestock production (#PD0376)
Environmental hazards from mineral exploitation of seabed resources (#PE6682).
Aggravates Deforestation (#PC1366) Environmental poverty (#PD5261)
Environmental pollution (#PB1166) Vulnerability of plants and crops (#PD5730)
Excessive consumption of resources in developed countries (#PE5551).
Aggravated by Maldevelopment (#PB6207) Pursuit of affluence (#PF5864)
Environmental hazards (#PC5883) Uncontrolled urban development (#PC0442)
Unsustainable rural development (#PD4537) Unsustainable population levels (#PB0035)
Unsustainable economic development (#PC0495)
Mismanagement of environmental demand (#PD5429)
Unsustainable agricultural development (#PC8419)
Resource wastage in capitalist systems (#PC3108)
Unconstrained exploitation of natural resources (#PF2855)
Natural resource depletion due to high–level consumption (#PD4002).

♦ PC0114 Competition between states

National rivalry — Rivalry between nations — International rivalry

Nature The forces of competition between states, backed by military power, achieve unequal allocations of territory, influence, raw materials, or prestige goods. The fulfilment of national destiny is therefore dependent on the frustration of the goals of rival governments.

Claim The chauvinism of countries frequently exhibits the 'chosen people' syndrome backed up by an ideology, religious or otherwise, peculiar to each country. India, Sri Lanka, Israel, and the Arab States have their favoured religions as do a number of officially or historically Christian ones. Some countries claim unique cultures or histories which, as in China, Ethiopia, Iran and Japan, are partly based on racist notions. The Soviet Union claims itself to be the fountainhead of the true Marxist–Leninist heritage, and the US claims ownership of the 'Protestant ethic' (whereby the poor and miserable have only themselves to blame). France is unique, so is Italy, so is Great Britain and so is every other country, for to a unique ideology one can add the divisive value of linguistic chauvinism, and of paramount significance to ongoing rivalry, the historicity and 'hard–won' nature of national sovereignty.

Refs Brenner, Reuven *Rivalry* (1990).

Broader Competition (#PB0848).
Narrower Superpower rivalry (#PD9655) Second class states (#PD0579)
Unjustified military defence policies (#PF1385)
Declining international competitiveness (#PD8994)
Rivalry and disunity within developing regions (#PD0110)
Restrictions on foreign access to capital bond markets (#PD3135)
Reduction of the share of the developing countries in world exports (#PC2566)
Domination of the world by territorially organized sovereign states (#PD0055)
Differences in trading principles and practices between different economic systems (#PC2952)
Territories accorded a United Nations non–self–governing status disputed by the administering government (#PF2943).
Related Competition in capitalist systems (#PC3125)
Vulnerability of land–locked developing countries (#PD5788).
Aggravates Distrust (#PA8653) World anarchy (#PF2071) Fragmentation (#PA6233)
Disinformation (#PB7606) Nuclear irresponsibility (#PF6611)
Military and economic hegemony (#PB0318) Secrecy in scientific research (#PF1430)
Unfulfilled treaty obligations (#PF2497) Loss of international leadership (#PF8353)
Lack of trans–frontier cooperation (#PF6855)
Subversion of international agreements (#PD5876)
Lack of intersocietal resource channels (#PF2517)
Long–term shortage of natural resources (#PC4824)
Misuse of satellite surveillance by governments (#PF3701)
Politically unrealistic strategic warfare analysis (#PF1214)
Health hazards of hormone use in animal production (#PE2809)
Instability in relations between allies of superpowers (#PD7522)
Excessive expense of international athletic competitions (#PF4192)
Disparity between industrialized and developing countries (#PC8694)
Inefficient location of facilities of international organizations (#PE3538)
Imbalances in the distribution of the costs and benefits of economic integration (#PD0794).
Aggravated by Nationalism (#PB0534) Pursuit of national prestige (#PF8434)
Reallocation of aid funds to alternative priorities (#PF0648)
Contradictions of capitalism in developing countries (#PF3126).

♦ PC0115 Socialism

Nature A political philosophy comprising a vast range of ideological traits, socialism is expressed in three main ways: socialism which aims at revolution and the overthrowing of capitalist society (mainly equated with communism); socialism, including Christian socialism, which aims to gain improvements within the democratic constitutional framework (usually termed social democracy); fascism or state socialism which is violent, nationalistic and authoritarian. Socialism as an unqualified term may be used to describe all of these and also military and civilian dictatorships superimposed on nominally socialist regimes. Although socialism may be regarded as altruistically motivated, its aims and claims are very difficult to achieve in practice, giving rise to bitter factionalism, civil or general war, violence, subversive activities and general instability.

Incidence Although socialism as an idea for running societies is in disarray following the changes in Eastern Europe, the influence of socialism remains strong, especially in the advanced democracies held up as models superior to those of Marxist socialism. This influence takes the form of state–financed education systems of demonstrable inefficiency, but especially the state–financed welfare systems, especially in their more destructive forms through which dependency is encouraged.

Claim 1. The core of the socialist ideology is a negative position rather than an affirmation in that it is a reaction against the lack of government intervention in individual life, particularly economic life but also encompassing education, health, artistic expression, marital and family affairs and religious beliefs. The mentality of a socialist is characterized by a certain neurotic anxiety, which could be called the socialist anxiety, that if human affairs are not regulated in all ways, chaos is the inevitable outcome. In this respect socialism bears close resemblance to ancient Judaic legalism from the time of Moses to that of Jesus, and its historical origins seem to lie in this tradition as well. Such collectivist policy deprives individuals of moral responsibility and is an impediment to human development.

2. At the root of all socialist thought is supreme intellectual arrogance; a belief that the order of society should be made to conform to the order of a dominant mind.

Counter–claim 1. Socialism is the only political philosophy to have emerged which addresses, realistically and with human compassion, the systemic evils and structural inequities of a caste and class division of society that is inspired by primitive, aggressive behaviour. "Might makes right" has continued in the main to be the creed of the Western world. Economic and military super–power dominates the globe, paid for by the exploitation of people by business investors and their financial and managerial allies, by business men and women, and by presidents and prime–ministers and their millionaire backers. The socialist challenge to this system has had victories in the past: universal education, public health services, government social pension and unemployment pay schemes, trade unionism, emancipation of women, protection of minorities, public housing and so on. Its biggest victory, world peace, has proved elusive due to the statist imperatives continuing to be propagated by the exploiters who, sitting on their hoards in moral darkness, hypocritically wave their flags while the world arms itself for a nuclear confrontation.

2. Advocates of capitalism tend to confuse the triumph of markets over bureaucracy with the triumph of capitalism in its own right. Social democrats have long condemned the Soviet system as inconsistent with socialism, although that did not prevent people from accepting the idea that what existed in the USSR and its satellites was actually socialism. Both right–wing and left–wing dogmatists have found it convenient to discredit the socialist idea by this means. But Western capitalism has been extensively transformed by social democracy with pro–capitalist conservatives accepting many of the changes championed by the socialist movement such as unemployment insurance, old–age pensions, and health insurance.

Refs Liebich, André (Ed) *Future of Socialism in Europe* (1979).

Broader Ideological conflict (#PF3388).
Narrower Trade unionism (#PF8493) Socialist colonialism (#PD4862)
Bureaucratization of socialism (#PD2993) Nepotism in socialist countries (#PF6013)
Pseudo–socialism and state socialism (#PF4778)
Embourgeoisement in socialist countries (#PE5975)
Shortage of manpower in socialist countries (#PE8771)
Declining productivity in socialist countries (#PF7610)
Obstacles to economic reform in socialist countries (#PF3689)
Inadequate economic integration of socialist countries (#PF4884)
Obstacles to legal relations between socialist countries (#PF4886)
Criminal negligence in performing socialist responsibilities (#PE5538).
Related Fascism (#PF0248) Communism (#PC0369) Capitalism (#PC0564).
Aggravates Prohibitive labour costs (#PF8763) Dependence on social welfare (#PD1229)
Interception of communications (#PD7608) Politically emotive words and terms (#PF3128)
Nationalization of domestic enterprises (#PD1994).
Aggravated by Destiny (#PF3111) Unemployment (#PB0750)
Exploitation (#PB3200) Socio–economic poverty (#PB0388)
Double standards in morality (#PF5225)
Nationalization of foreign investments (#PC2172).
Reduces Underdevelopment (#PB0206)
Denial of the right to work in capitalist systems (#PC3119).
Reduced by Subversion of socialism (#PF9485).

♦ PC0116 Corruption in politics

Political corruption — Corruption of political parties

Nature Corrupt practices include tax evasion, nepotism, looting the national treasury, political bribery, unequal distribution of government contracts, unjust financing of political parties, unjust election administration, use of espionage (especially for domestic purposes), secret police and intimidation. This may lead to political blackmail and corruption in other spheres, to the encouragement of organized crime and violence, lack of credibility in institutions, alienation, apathy, subversive activities, revolution, disintegration, foreign intervention or stagnation. Although corruption is not the preserve of the public services nor solely a feature in developing countries, it is a major problem of public development administration and is a reflection of what happens in society generally. Its prevalence is linked with the power for development decisions being placed in the hands of public servants; and is important because of its impact on the way these decisions are used. The low level of remuneration of a majority of public sector employees, which very often rises at a rate well behind that of inflation and does not adequately compensate for changes in the cost of living, also encourages corruption.

Particularly striking cases of corruption relate to the wives and families of political leaders in countries with repressive governments. By amassing enormous wealth and transferring funds abroad, individuals may drain the economy by amassing so much wealth that their corrupt practices create a serious burden. Such individuals are protected from having to account publicly for their finances.

Incidence According to an American survey, 30 percent of people believed in 1990 that their own congressman was corrupt.

Refs Alatas, Syed H *Corruption* (1986); Bollens, John C and Schmandt, Henry J *Political Corruption* (1979); Clarke, Michael (Eds) *Corruption* (1984); Cox, Del *Corruption and Cover–Up* (1988); Dwivedi, S N *Political Corruption in India* (1967); Gibbons, Kenneth and Rowat, Donald C (Eds) *Political Corruption in Canada*; Gould, David J and Amaro–Reyes, Jose A *The effects of corruption on administrative performance* (1983); Heidenheimer, Arnold J et al (Eds) *Political corruption* (1988); LeVine, Victor *Political Corruption and the Informal Policy* (1971); Rose–Ackerman, Susan *Corruption* (1978); Williams, Robert *Political curruption in Africa* (1987).

Broader Corruption (#PA1986) Unethical practices in politics (#PC5517).
Narrower Nepotism (#PD7704) Political bribery (#PC2030) Political burglary (#PD1943)
Political blackmail (#PD2912) Unjust financing of political parties (#PE0752).
Related Political crime (#PC0350) Political apathy (#PC1917)
Political opportunism (#PC1897) Bureaucratic corruption (#PC0279)
Lack of political integrity (#PF0796) Institutionalized corruption (#PC9173)
Corruption in organized religion (#PC3359) Corruption in developing countries (#PD0348)
Political corruption of the judiciary (#PE0647)
Misuse of classified communications information (#PD5183).
Aggravates Political purges (#PC2933) Official secrecy (#PC1812)
Political deception (#PF9583) Tropical deforestation (#PD6204)
Disagreement among experts (#PF6012) Corruption of government leaders (#PC7587)
Frauds, forgeries and financial crime (#PE5516).
Aggravated by Corruptive crimes (#PD8679) Illicit drug trafficking (#PD0991)
Espionage in domestic politics (#PD1787) Unethical commercial practices (#PC2563)
Threat to parliamentary immunity (#PF0609)
Abusive distribution of political patronage (#PF8535)
Proliferation of public sector institutions (#PF4739).

♦ PC0119 Air pollution

Air pollutants — Atmospheric pollution — Noxious fumes — Airborne toxic and harmful agents

Nature A growing body of evidence seems to show a consistent association between air pollution and health impairment of varying degrees. Such associations are found between acute pollution exposure and morbidity and mortality; chronic lower–level exposure and morbidity and mortality; exposure and impairment of function and performance; exposure and symptoms of sensory irritation; and exposure and other effects on well–being.

A relatively new feature of the effects of air pollutants is the association with long term global problems. Considerable uncertainty surrounds scenarios relating to the effect of increased levels of carbon dioxide, but some predictions are of major impact on climate. The most significant polluting substances are sulphur dioxide and the oxides of nitrogen produced by power stations burning coal or oil and motor cars. The phenomenon of acid deposition is widespread throughout northern temperate regions, and researchers have associated 'acid rain' with damage to wildlife and wildlife habitats, including freshwaters and forests. Both wet and dry acid deposition is associated with damage to artifacts and materials. More locally, environmental contamination has resulted from long–term use of materials such as asbestos and lead, whose full impact as air pollutants has emerged only recently.

Airborne toxic agents, including micro–organisms and other harmful agents such as respirable dust, may have long–term effects such as genetic damage, carcinogenesis, and shortened life expectancy. The effect on an individual of foreign substances in the ambient environment depends on his health and on the degree and duration of exposure. Symptoms may not be readily distinguishable and their medical assessment is not always easy. Practical difficulties also arise from the inadequacy of long–term sampling procedures and analytical techniques.

The recommendation of maximum permissible limits for international adoption is a complicated procedure. Countries vary greatly in the amount of airborne known toxic and harmful agents workers may be exposed to so that one country may allow exposure to a toxic agent to be up to 90 times greater than the exposure permitted in another country. Permissible limits are normally based on an exposure of eight hours a day for five days a week. Different considerations must apply when people are exposed briefly to high concentrations, when working periods are longer than normal, or when workers are subjected to additional stresses such as high temperatures or poor nutrition.

Background Ambient air, when sampled close to ground level, contains gases, vapours and particulate matter derived either from natural sources, such as volcanoes, or from man's activities. Some components, such as spores, seeds, and pollen grains are not pollutants; they are natural constituents frequently found in the atmosphere. Although plant pollen may cause ill health in some people (hay fever and asthma), it is a natural and essential constituent of air and not one to which all individuals are equally sensitive. Air pollution is the result of the discharge into the atmosphere of foreign gases, vapours, droplets and particles, or of excessive amounts of normal constituents, such as the carbon dioxide and suspended particulate matter produced by the burning of fossil fuels.

Incidence In 1989 it was estimated that 20 per cent of the world's population, more than one billion people, is now breathing air contaminated above international safety limits and despite decades of effort to combat atmospheric pollution. The dimension of the problem is indicated by the fact that the US industry in 1989 spend $33 billion on clean air, more per capita than any of its trading partners. Further proposals suggest the need to increase that amount by $25 billion. Air pollution in the USA is estimated to cause up to 50,000 deaths per year and to cost $40 billion in health care and lost work days.

Refs Acid Rain Foundation Inc Staff *Air Pollutants* (1985); Amicarelli, V, et al *Identification of Air Quality and Environmental Problems in the European Community* (1987); Bevington, Ch F P *Identification and Quantification of Atmospheric Emission Sources of Heavy Metals and Dust from Metallurgical Processes and Waste Incineration* (1987); Bouscaren, R; Frank, R and Veldt, C *Hydrocarbons* (1987); Breuer, Georg and Fabian, P *Air in Danger* (1980); Dabberdt, Walter F *Atmospheric Dispersion of Hazardous–Toxic Materials from Transport Accidents* (1985); GESAMP *Atmospheric Transport on Contaminants Into the Mediterranean Region* (1985); Georgii, H W and Pankrath, J (Eds) *Deposition of Atmospheric Pollutants* (1982); Hutchinson, T C and Meema, K M *Effects of Atmospheric Pollutants on Forests, Wetlands, and Agricultural Ecosystems* (1987); Kennedy, Donald and Bates, Richard R (Eds) *Air Pollution, the Automobile and Public Health* (1988); Lidin, G D *Air Pollution in Mines* (1966); Loiy, Paul and Daisey, Joan M *Toxic Air Pollution* (1987); McGrath, James J and Barnes, Charles D *Air Pollution–Physiological Effects* (1982); Miller, E William and Miller, Ruby M *Environmental Hazards* (1988); Mudd *Responses of Plants to Air Pollution* (1975); Perkins, H C *Air Pollution* (1974); Rambo, A Terry *Primitive Polluters* (1986); Stern, Arthur *Air Pollution* (1977); Stern, Arthur *Air Pollution*; Stern, Arthur C *Air Pollution* (1986); Ulrich, B and Pankrath, J (Eds) *Effects of Accumulation of Air Pollutants in Forest Ecosystems* (1983); United Nations *Air Pollution Across Boundaries*; United Nations Economic Commission for Europe *Impact of Air–pollution Damage to Forests for Roundwood Supply and Forest Products Markets* (1987); WHO *Papers Presented at the WMO Technical Conference on Observations and Measurements of Atmospheric Contaminants (TECOMAC) (Vienna, Austria 1983)* (1985); WHO Collaborating Centre on Air Pollution Control and WHO Collaborating Centre on Clinical and Epidemiological Aspects of Air Pollution *Selected Methods of Measuring Air Pollutants* (1976); World Meteorological Organization *Dispersion and Forecasting of Air Pollution* (1972).

Broader Environmental pollution (#PB1166).
Narrower Soot (#PE1953) Aeroallergens (#PE2069) Irritant fumes (#PD3672)
Malodorous fumes (#PD1413) Indoor air pollution (#PD6627) Arctic air pollution (#PD6283)
Chemical air pollutants (#PD1271) Herbicides as pollutants (#PD1143)
Fungicides as pollutants (#PD1612) Biological air pollutants (#PD0450)
Urban–Industrial air pollution (#PJ5532) Plant–pathogenic air pollutants (#PE0155)
Airborne substances harmful to health (#PD2847)
Increasing atmospheric carbon dioxide (#PD4387)
Long–range transboundary air pollution (#PD3391)
Health hazards of air pollution for plants (#PE4744)
Health hazards of air pollution for animals (#PE9609)
Health hazards of air pollution for children (#PE8617)
Increase in atmospheric concentration of methane (#PE8815).
Aggravates Corrosion (#PD0508) Byssinosis (#PE2319)
Heart diseases (#PD0448) Global warming (#PC0918)
Marine pollution (#PC1117) Chronic bronchitis (#PE2248)
Malignant neoplasms (#PC0092) Occupational risk to health (#PC0865)
Lung disorders and diseases (#PD0637) Environmental plant diseases (#PD2224)
Endangered species of insects (#PC2326)
Endangered monuments and historic sites (#PD0253)
Deterioration in atmospheric visibility (#PE2593)
Carcinogenic chemical and physical agents (#PD1239)
Declining productivity of agricultural land (#PD7480)
Deterioration of stored documents and archives (#PE1669)
Deterioration of the physical condition of art objects (#PD1955).
Aggravated by Dust (#PD1245) Fires (#PD8054)
Smoking (#PD0713) Slashburning (#PE6264)
Water pollution (#PC0062) Ozone as a pollutant (#PE1359)
Industrial emissions (#PE1869) Temperature inversion (#PG1956)
Motor vehicle emissions (#PD0414) Detergents as pollutants (#PE1087)
Urban–Industrial pollution (#PC8745) Exploitation of fossil fuels (#PE4891)
Photochemical oxidant formation (#PD3663) Non–biodegradable plastic waste (#PD1180)
Industrial waste water pollutants (#PD0575)
Environmental hazards from petroleum (#PE1409)
Proliferation of automobiles and motor vehicles (#PD2072)
Environmental impacts of coal conversion plants (#PE8453).
Reduces Virus diseases (#PD0594) Pathogenic fungi (#PG1959)
Bacterial disease (#PD9094).

♦ PC0123 Consumer vulnerability

Inadequate consumer protection — Unprotected consumers

Nature Consumers are vulnerable to practices such as. inaccurate use of weights and measures; use of dangerous additives and preservatives in foodstuffs; lack of control of insecticides, pesticides and drug pre–testing; unsafe products (for example, unsafe toys, tyres, television sets and inflammable carpeting); fraudulent and misleading advertising; meaningless product guarantees and warranties; inadequate servicing and complaint handling facilities; inadequate indication of product quality, characteristics and degree of safety; lack of impartial testing services to evaluate and publicize the performance of competing products; uncontested price increases. The range of 'products' consumers can be abused on are: private goods such as cooking stoves; environmental goods like air and water systems; public services such as health or defence; government services in law making; tax–gathering and administration; and moral 'goods', for example the principle of protecting workers' health.

Refs Bourgoignie, Thierry, et al (Eds) *Unfair Terms in Consumer Contracts* (1983); Taperell, G Q; Vermeesch, R B and Harland, D J *Trade Practices and Consumer Protection* (1983); United Nations *Guidelines for Consumer Protection* (1986).

Broader Lack of protection for the vulnerable (#PB4353).
Narrower Dangerous toys (#PE1158) Unsafe aircraft (#PE1575)
Planned obsolescence (#PC2008) Violence against women (#PD0247)
Inappropriate labelling (#PD3521) Decrease in consumer choice (#PD6075)
Deterioration in product quality (#PD1435) Inadequate equipment maintenance (#PD1565)
Prohibitive cost of accommodation (#PD1842) Unsafe design of consumer products (#PF1379)
Continued operation of unsafe motor vehicles (#PE2240)
Non–payment of compensation for damages to consumers (#PE0290)
Lack of consumer choice in centrally–planned economies (#PD0515)
Insensitivity of transnational corporations to consumer needs in developing countries (#PE1011).
Related Food insecurity (#PB2846).
Aggravates Toxic food additives (#PD0487)
Misrepresentation of information to consumers (#PE6877).
Aggravated by Unethical commercial practices (#PC2563).

♦ PC0130 Slave trade

Trafficking in people

Nature There still exists a certain demand for exchanging slaves and for new slaves, hence a slave trade. Since it is prohibited by law in most countries, it is usually clandestine and not admitted by governments. The traditional type of slave trading, the buying and selling of both male and female slaves, is now limited by law. Other traffic in women, children and immigrants to Western European countries is still fairly widespread; the inheritance of new debt slaves, children and dependents of the original debtor who inherit his debt and his bondage, rather less so. Although traditional slave–trading is very difficult now owing to the abolition of slavery by most countries, other more subtle forms are still widely practised. Chief among these is traffic in women and children (in Africa, Asia and Latin America) and the illegal traffic in immigrants to Western Asia and Latin American and European countries. Other trading exists in the illegal kidnapping and and enticement of white people (mainly women). Insufficient public action is taken on this because it is felt that there must be compliance on the part of victims. In the case of immigrants, compliance is also an alleged factor and governments pay too little attention to misleading promises given by the exploiters, exploitative conditions of contract and poor working and housing conditions. Traffic in girls and women in the form of bride price and inheritance is accepted as tradition.

Incidence Swedish men pay 680 Pounds for three months of "service in bed" by women imported from Asia. The women arrive on three–month residence permits, which are renewable if their work proves satisfactory. A Swedish womens' protest group has called the trade "pure slavery".

Background The slave trade existed with self–sale and the sale of children in Egypt in 2600 BC; but it was not organized by slave traders until the 16th century, when Arab slave traders found a ready market among the English and Spanish colonists of the New World. The movement to abolish the slave trade and then slavery was led by William Wilberforce in the early 19th century in England.

Refs Blake, W O *History of Slavery and the Slave Trade* (1970); Eltis, David and Walvin, James *The Abolition of the Atlantic Slave Trade* (1981).

Broader Traffic in persons (#PC4442).
Narrower White slave trade (#PD3303) Trafficking in women (#PC3298)
Trafficking in children (#PD8405) Denial of right to liberty (#PF0705).
Related Unlawful business transactions (#PC4645).
Aggravates Torture (#PB3430) Tribalism (#PC1910) Brutality (#PC1987)
Prostitution (#PD0693) Malnutrition (#PB1498) Human dependence (#PA2159)
Disruption of development by tribal warfare (#PD2191).
Aggravated by Apathy (#PA2360) Slavery (#PC0146)
Unemployment (#PB0750) Human sacrifice (#PF2641)
Socio–economic poverty (#PB0388) Legalized racial discrimination (#PC3683).

♦ PC0137 Social conflict

Nature Conflict between different social groups may result in violence and death, or be the cause and result of a widespread variety of discriminatory practices. Philosophers disagree as to whether conflict in society is necessary or not and creative or not. Certainly when it results in violence it can be seen as harmful.

Incidence Social conflict occurs wherever social groups are in opposition to each other and blame each other for economic hardship or lack of political representation or intellectual freedom, or discrimination on the grounds of race, colour, religion, political ideology, sex, nationality, age, language and class. Conflict occurs particularly where different social groups lack contact with one another, either through physical segregation or through general lack of communication. The quality of leadership given to the opposing factions is crucial. Those leaders who are self–serving wish to appear strong, and so move directly from dissent to conflict without any attempt at persuasive negotiation or positive motivation of the other side with contingent rewards. In some instances of conflict there is a common transcending interest or problem, such as national economic recovery or national defence during war time, that effects a reconciliation. Usually, however, this is but a temporary state and the last four decades have seen numerous alliances of convenience degenerate into deadly rivalry and ensuing violent conflict. If prejudices run very high, conflict can result in genocide, or a campaign of mass extermination.

Refs Dobkowski, Michael and Willimann, Isidor (Eds) *Research in Inequality and Social Conflict* (1988); Kriesberg, Louis *Social Conflicts* (1982).

Broader Conflict (#PA0298) Social breakdown (#PB2496).
Narrower Scapegoats (#PF3332) Social deviation (#PC3452)
Lack of assimilation (#PF2132) Factionalism in developing countries (#PD1629)
Inadequate social development programmes (#PF4180)
Increase in anti–social behaviour in developing countries (#PD0329)
Loss of traditional forms of social control in developing countries (#PD0144).
Related Racial conflict (#PC3684) Ethnic conflict (#PC3685)
Social revolution (#PC3236) Social intimidation (#PC2940)
International conflict (#PB5057) Psychological conflict (#PE5087)
Inadequacy of social doctrine (#PF3398).
Aggravates Crime (#PB0001) Ethnocide (#PC1328) Martial law (#PD2637)
Human violence (#PA0429) Human suffering (#PB5955) Guerrilla warfare (#PC1738)
Public drunkenness (#PE2429) Political pluralism (#PF2182) Ideological conflict (#PF3388)
Social underdevelopment (#PC0242) Lack of political development (#PD8673)
Lack of community development (#PF7912) Undemocratic political organization (#PC1015)
Delays in elaboration of remedial legislation (#PC1613).
Aggravated by Elitism (#PA1387) Nativism (#PF2186)
Militancy (#PC1090) Prejudice (#PA2173)
Decadence (#PB2542) Segregation (#PC0031)
Conservatism (#PF2160) Miscegenation (#PC1523)
Value erosion (#PA1782) Statelessness (#PE2485)
Permissiveness (#PF1252) Traditionalism (#PF2676)
Age segregation (#PD3444) Social injustice (#PC0797)
Minority control (#PF2375) Social inequality (#PB0514)
Ethnic segregation (#PC3315) Racial segregation (#PC3688)
Social dictatorship (#PD3241) Racial exploitation (#PC3334)
Class consciousness (#PC3458) Religious repression (#PC0578)
Social fragmentation (#PF1324) Socioeconomic stress (#PC6759)
Racial discrimination (#PC0006) Social discrimination (#PC1864)
Metal deficient diets (#PE1901) Cultural fragmentation (#PF0536)
Unjust election timing (#PD2907) Segregation in housing (#PD3442)
Lack of social mobility (#PF2195) Plural society tensions (#PF2448)
Undue political pressure (#PB3209) Empty slogans and mottoes (#PF3212)
Segregation in employment (#PD3443) Exploitation in employment (#PC3297)
Underprivileged minorities (#PC3424) Segregation in social services (#PD3440)
Inter–cultural misunderstanding (#PF3340) Imbalance in the human sex ratio (#PF1128)
Discrimination against minorities (#PC0582) Racial discrimination in politics (#PD3329)
Negative effects of family allowances (#PF0107) Underprivileged ideological minorities (#PC3325)
Compromise as a betrayal of principles (#PF3420)
Vulnerability of marriage as an institution (#PF1870)
Lack of appreciation of cultural differences (#PF2679)
Denial of human rights in capitalist systems (#PC3124).

♦ PC0143 Innumeracy

Dependence on innumeracy — Number–blindness — Dyscalculia — Acalculalia — Developmental arithmetic disorder

Nature Even after receiving an education to an acceptable level of literacy, some people may continue to have great difficulty in understanding and manipulating numbers. In its extreme form, this revulsion against numbers has been termed number blindness. Acalculia, which is the inability to perform arithmetic operations, is seen most commonly with parietal lobe lesions.

Two aspects of innumeracy may be distinguished: the inability to think quantitatively and to realize the extent to which problems are problems of degree even when they appear as problems of kind; and the inability to make use of the scientific approach of observation, hypothesis, experiment and verification.

Incidence There have been few estimates of the extent of this problem, but it has been widely accepted that numbers do not mean much to most people although this is generally not the case where monetary questions are at issue. Otherwise well educated people fall victim all to often to the extreme misinterpretation of statistics which makes "pseudoscience" – phrenology, parapsychology, predictive dreams, astrology and television evangelism – so influential in daily life.

Broader Ignorance (#PA5568) Restrictions on the acquisition of knowledge (#PF1319).
Related Illiteracy (#PC0210) Mental illness (#PC0300) Learning disorders (#PD3865)
Reading disabilities (#PD1950) Developmental expressive writing disorder (#PE0330)
Developmental receptive language disorder (#PE9300).
Aggravated by Educational wastage (#PC1716) Mathematical ignorance (#PD6728).

♦ PC0146 Slavery

Slavery–like practices — Dependence on slavery — Denial of the right to freedom from slavery

Nature Slavery takes many forms, and although some of the more extreme and widespread of these forms have been virtually eradicated, more subtle forms still exist. The essence of slavery is ownership, its corollary is exploitation. Overt ownership still exists and also the sale of slaves, but more widespread is the trade in women, the paying of bride–price or the inheritance of a brother's widow or widows. The 'adoption' of children for a price is practised in Latin America and Asia. Personal services and debt slavery exist in Asia, Africa and South America. Forced labour occurs in South Africa and other African countries, and in Western Europe. Particularly in South Africa it forms part of a policy of racial segregation. In Europe it is the result of a traffic in immigrant workers and the restrictions placed on these by the governments and nationals of the countries in question.

Incidence Slavery–type practices remain very widespread throughout the world. From 1978 to 1981 among many instances reported to the ILO were those occurring in Tunisia, India, Italy, Taiwan, Columbia, Morocco, Palestine, Republic of Korea, United States of America (employment conditions of Mexican children), Spain and France. Slavery is illegal throughout the world except in the Sultanate of Muscat and Oman.

Background Self–sale or the sale of children to pay debts was recorded in Egypt in 2600 BC. Debt slavery and slavery from prisoners of war was widespread in all the near–Eastern cultures of that era. In the Greek and Roman civilizations slavery became a formal institution, but debt slavery ceased to be legal. After the fall of the Roman Empire slavery declined in western and central Europe, but persisted in southern Europe and the Middle East where it was given new impetus by Islam. With the colonization of the New World and the opening up of plantations, slave trade in negroes flourished and persisted until the advent of the anti–slavery movements in the 19th century. The liberation of these slaves did little to change attitudes in society towards them which still persist today.

Claim In 1990 it was suggested that there may be as many people in conditions properly denoted as slavery as there were 150 years before.

Refs Archer, Leonie *Slavery* (1988); Blake, W O *History of Slavery and the Slave Trade* (1970); Elliott, E N *Cotton Is King, and Pro–Slavery Arguments*; Fredericksson, George M *The Arrogance of Race* (1988); Miller, Joseph C *Slavery* (1985); Sawyer, Roger *Slavery in the Twentieth Century* (1986).

Broader Exploitation (#PB3200) Forced labour (#PC0746)
Social injustice (#PC0797) Denial of right to liberty (#PF0705).
Narrower Bondservice (#PE6342) Debt slavery (#PD3301)
Chattel slavery (#PC3300) Brothel slavery (#PD3888)
Dependence on breast feeding (#PE7627).
Related Colonialism (#PC0798) Forced marriage (#PD1915).
Aggravates Torture (#PB3430) Brutality (#PC1987) Slave trade (#PC0130)
Malnutrition (#PB1498) Miscegenation (#PC1523) Human sacrifice (#PF2641)
Ethnic conflict (#PC3685) Social discrimination (#PC1864)
Socio–economic poverty (#PB0388) Extra–economic constraints (#PE7784).
Aggravated by Masochism (#PF3264) Tribalism (#PC1910)
Feudalism (#PF2136) Segregation (#PC0031)
Unemployment (#PB0750) Social inequality (#PB0514)
Racial segregation (#PC3688) Racial discrimination (#PC0006)
Plantation agriculture (#PD7598) Segregation in employment (#PD3443)
Restrictions on immigration (#PC0970)
Discrepancies in human life evaluation (#PF1191)
Enforced participation in community activity (#PD3386)
Expropriation of land from indigenous populations (#PC3304).

♦ **PC0149 Poverty in developing countries**
Incidence The number of poor people in developing countries rose from 1,103 million in 1974 to 1,166 million in 1982. As a proportion of the population, however, those who could not afford to meet their basic needs fell, over this period, from 56 percent to 51 percent. In Asia there was an actual slight fall in the numbers estimated to be in poverty, from 759 million to 754 million, with a steep fall in the proportion of those in poverty (from 69 percent to 57 percent) thanks to rapid economic growth in the area which was less affected by recession than any other. There was a slight fall in the percentage in poverty in Latin America, among oil exporters of the Middle East and Africa, and in the drier African countries, but in all these regions the total number of poor increased. In tropical Africa, however, not only did the number in poverty increase dramatically, from 132 million to 187 million, but the proportion in poverty also rose, from 82 percent in 1974 to 91 percent in 1982.
Refs Garg, Usha and Vibhakar, Jagdish *Poverty of Nations and New Economic Order* (1985); Joshi, Nandini Umashankar *Challenge of Poverty* (1978); Sethuraman, S V *The Urban Informal Sector in Developing Countries* (1981).
Broader Socio–economic poverty (#PB0388).
Narrower Uncontrolled urbanization in developed countries (#PD3488).
Related Family poverty (#PC0999)
Family poverty in industrialized countries (#PD1998)
Inappropriate design of development projects (#PF4944).
Aggravates Infanticide (#PD3501) Malnutrition (#PB1498)
Family breakdown (#PC2102) Mental disorders of the aged (#PD0919)
Inappropriate development policy (#PF8757)
Unsustainable agricultural development (#PC8419)
Lack of purchasing power in developing countries (#PE8707).
Aggravated by Traditionalism (#PF2676) Lack of family planning (#PF0148)
Poor quality of domestic livestock (#PD2743)
Social insecurity in developing countries (#PE4796)
Unemployment in least developed countries (#PE9476)
Declining economic growth in developing countries (#PD5326)
Excessive social costs of structural adjustment in debtor developing countries (#PD8114).

♦ **PC0178 Multiplicity of languages**
Nature The multiplicity of languages is a major dividing factor in world society, reinforcing geographical, socio–economic (especially caste or class), political, ideological, professional and religious separatism. It prevents or hinders communication and the spread of education, and therefore aggravates international misunderstanding and mutual suspicion. Multilingualism within a country results in poor communications between members of different language communities and between those communities and the government. It can lead not only to mistrust and to political tension, but also to poor levels of literacy and problems in the judiciary when different languages may be used and transcripts required in order for a case to be heard at all.
Incidence The exact number of languages is not known because of difficulty in agreeing upon the distinction between a language and a dialect. The figure of 2,700 to 3,000 languages is frequently encountered. Of these approximately 150 (60 in developed and 90 in developing countries) have more than one million speakers. In Africa, for example, with approximately half the population of Europe, over 1,000 languages are spoken as against the 60 in Europe. In effect the average African language is spoken by less than 200,000 people. At the other extreme, languages such as English, French, Spanish, Arabic, Portuguese and German are each spoken internationally by a total each of from 100 to 350 million persons. Mandarin Chinese is spoken by some 650 million people internationally. Russian, Hindi, Bengali and Japanese languages have limited use outside their motherlands and thus their 660 million speakers do not have the advantage of belonging to an international linguistic community. On the other hand, the USSR and India each has a considerable number of languages spoken internally, which in India's case at least has caused a great deal of friction. Both India and Japan frequently resort to English for international communication, while Russia turns to French.
Background Different languages arose because of lack of contact between people of different regions, but especially because societies with distinct economic, moral and cultural traditions required specific vocabularies and linguistic structures. But at the same time, even within communities, distinctions between social groups – especially between a dominant elite and the mass of the population – came to be reflected in differences in idiom and vocabulary, in the meaning given to certain words, as well as in pronunciation. Millions of people today speak languages that are not understood by neighbouring groups, even though close economic and social links have been established and populations have intermingled. Thus, paradoxically, the very richness and diversity of languages can render communication difficult, just as its elaboration can perpetuate privilege.
Counter–claim The multiplicity of languages, each the incarnation of long traditions, is an expression of the world's cultural richness and diversity. The disappearance of a language is always a loss and its preservation the consequence of the struggle for a basic human right. Moreover, in the modern mass media as well as in traditional communication, the use of a variety of languages is an advantage, bringing a whole population on to equal terms of comprehension. The worldwide use of a small number of languages leads to a certain discrimination against other languages and the creation of a linguistic hierarchy. Because the small number of bilingual and multilingual people belong mainly to narrow local elites, the mass of the population is discriminated against since currently the spread of information tends to take place in the terms and the idiom of the linguistically powerful.
Narrower Linguistic purism (#PF1954) Semantic confusion (#PF5985)
Multiplicity of official languages (#PF6027)
Multiplicity of languages in a national setting (#PC1518)
Multiplicity of manual sign languages for the deaf (#PF2833)
Lack of terminological equivalents between languages (#PF0091)
Multiplicity of languages in international relations (#PC0410)
Prohibitive cost of linguistic interpretation legal proceedings (#PE1743)
Terminological confusion in weights, measures and numbering systems (#PF5670)
Bilingualism in national settings and regional linguistic controversies (#PE2488).
Related Differing conceptions of time (#PF6665)
Proliferation of national and international anniversaries and years (#PF2723).
Aggravates Language barriers (#PF6035) Corruption of meaning (#PB2619)
Cultural discrimination (#PC8344) Lack of legal aid facilities (#PF8869)
Excessive use of acronyms and abbreviations (#PF4286)
Inadequate labelling of dangerous substances (#PD2468)
Insufficient translation into minority languages (#PD0825)
Barriers to the international flow of knowledge and educational materials (#PF0166).
Aggravated by Ethnocentricity (#PB5765)
Excessive use of foreign programmes for media (#PE9643).
Reduced by Bilingualism (#PF7927)
Monolingualism in a multi–cultural setting (#PD2695).

♦ **PC0180 Problems of migrant labour**
Nature Many countries increasingly use large numbers of migrant workers, due to the inadequate supply of national labour, the difficulty encountered in filling certain arduous or low paid jobs, or due to the seasonal nature of the work. It is the conditions of work and life of the migrant workers that cause concern. Language difficulties may debar these persons from entering into harmonious human contact and working relationships, leading to social isolation prejudicial to the worker's mental health and safety problems resulting from inability to understand instructions. Provision of satisfactory accommodation is difficult. Families are broken up for the duration of absence or parents and children may keep entirely different working hours, as in the hotel industry. Young persons risk mental health problems and delinquency due to the promiscuous manner in which they are thrown together in the absence of parental authority. Workers run medical risks due to diet deficiencies, and they may bring infectious diseases from their country of origin and spread them to the resident population.
Incidence Pakistani workers abroad sent home more than $16 billion in the ten years to the end of 1986. In 1983 Bengalis working overseas sent back $610m, which is equal to 80 percent of Bangladesh's merchandise exports and one–quarter of its imports. Filipinos in the same year returned to the Philippines around $950m, which was 3 1/2% of the GDP. These figures do not include the flourishing black markets and informal arrangements which bypass banking procedures.
Background Rich countries have always imported labour from poorer ones. In the days of the British Empire, the British imported Tamils and Chinese to work on the plantations and mines of Malaya and Ceylon, Indians to build railways in East Africa and to cut sugar in South Africa. Chinese workers contributed massively to the railroads of the western United States. In the 1960, South Korea began to send coal miners and later nurses to West Germany. South Vietnam during the American occupation was another huge labour market for projects out of the firing line. The difference today is not of kind, but of scale and direction. The big Arab oil exporters have grand developmental ambitions but small populations – even Iraq has only 17m people. Singapore, Hong Kong and Taiwan need massive labour forces to continue their rapid economic growth.
Related Seasonal unemployment (#PC1108).
Aggravates Schistosomiasis (#PE0921) Injurious accidents (#PB0731)
Shortage of skilled labour (#PD0044).
Aggravated by Labour shortage (#PC0592) Insecurity of employment (#PD8211).

♦ **PC0185 Restrictions on freedom of information**
Denial of right to freedom of information — Prevention of the exchange of information
Nature Freedom of information is often restricted by political, economic and other barriers. In some totalitarian countries, there is a persistent dissemination of false or distorted information because the media are controlled by governments who tolerate no opposition or criticism. In some democratic countries, the main information media are in the hands of powerful press magnates who impose their political views within the press organs under their control and leave neither room nor opportunity for the expression of a different line of thought. Probably the most serious restrictions of a legal character arise from such concepts as "official secrets", "classified information" and "national security". The State obviously has a right to withhold information affecting national defence from the public domain, but such rights are abused when they extend to cover information of a political character, or in the technical and industrial spheres and – worst of all – public opinion. The vagueness of these restrictions make them all the more insidious. Access to news sources is a particularly thorny issue. Governments which do not restrict information in other ways may refuse to grant visas, restrict journalists' movements, place limitations on those whom newsmen may contact, withdraw accreditation of journalists or expel them from the country. There are very often discrepancies between the treatment of national journalists and foreign correspondents.
Incidence Pubiishers, editors, and reporters unpopular with totalitarian regimes may be assassinated, or confined as political prisoners, where they may be subjected to torture. In the last five years, such events have occurred in Latin America, Southeast Asia, Africa and the Middle East. In Western Europe, personnel associated with clandestine broadcasting of information to Eastern Europe have been murdered.
Broader Censorship (#PC0067).
Narrower Monopoly of the media (#PD3101) Suppression of information (#PD9146)
Withholding of information (#PF8536) Biased government information (#PF0157)
Denial of the right to inform (#PE7337) Erosion of journalistic immunity (#PD3035)
Refusal to grant licences to media (#PF3079) Prevention of the exchange of ideas (#PD8731)
Denial of the right to receive information (#PE4214)
Discriminatory design of information systems (#PD7450)
Restrictions on news coverage of legal affairs (#PF3073)
Restriction of access to news distribution media (#PF3082)
Restrictions on international freedom of information (#PC0931)
Conflict of laws on international restriction of information (#PD3080).
Related Denial of right to liberty (#PF0705) Denial of freedom of thought (#PF3217)
Denial of freedom of opinion (#PD7219) Restriction of freedom of expression (#PC2162)
Denial of right to correct misinformation (#PE7349).
Aggravates Ignorance (#PA5568) Compulsory indoctrination (#PD3097)
Ignorance of cultural heritage (#PF1985)
Social constraints on freedom of imagination (#PF4882).
Aggravated by Excessive government control (#PF0304)
Journalistic irresponsibility (#PD3071)
Deterioration of media standards (#PD5377)
Incomplete access to information resources (#PF2401).
Reduced by Misuse of classified communications information (#PD5183).

♦ **PC0193 Social exclusion**
Socially rejected people — Socially isolated groups
Narrower Disowned children (#PJ0827) Manipulative cults (#PE6336)
Exclusion of disabled persons from social and cultural life (#PD0784).
Aggravates Homelessness (#PB2150) Social isolation (#PC1707).

♦ **PC0195 Biological warfare**
Epidemic warfare — Bacteriological warfare — Germ warfare
Nature Bacteriological or biological agents of warfare are living organisms (or infective material derived from them) which are intended to cause disease or death in animals, plants, or man, and which depend for their effects on their ability to multiply in the person, animal or plant attacked. Various living organisms (for example, rickettsiae, viruses and fungi), as well as bacteria, can be used as weapons. The use of epidemic warfare on a strategic scale is liable to lead to the infection of a very high proportion of the population attacked, and even if the attacking country were protected (by immunization) from some specific strain, changes to more virulent forms might overwhelm the level of immunity. The chief types of bacterial and viral agents developed or considered have included: Anthrax (bacterial), which can cause death in 24 hours if lungs are attacked; Brucellosis or Undulant fever (bacterial), which is fatal in 5% of untreated cases; Encephalomyelitis (viral), which is up to 5% fatal if untreated and can cripple the nervous systems of survivors; Bubonic and Pneumonic Plague (bacterial); Psittacosis (viral) which can be fatal in up to 40% of cases ; Rocky Mountain spotted fever (rickettsial) which can kill in 3 days and Tularaemia (bacterial), for which the untreated rate of fatality is 5–8%.
Background Biological warfare technology was advanced in the Second World War by a unit of Japanese scientists working in occupied Manchuria using Allied prisoners for experimentation. Those responsible were exempted from prosecution as war criminals in return for handing over the results of their work to the United States. In 1945–72, the US Chemical Corps, with the participation of biologists from US universities under contract, conducted research at several places in the United States.
Incidence There is no military experience of the use of biological agents in warfare. One field trial

CROSS-SECTORAL PROBLEMS

PC0219

of the dissemination of such agents showed that 200 kg could be distributed from a ship travelling at a distance of 260 kilometers parallel to a coastline leading to a coverage of 75,000 square kilometers. The strength of such agents is illustrated by the fact that a similar degree of poisoning could be achieved with 0.5 kilo of Salmonella culture, with 5 kilos of botulinum toxin, 7 kilos of staphylococcal enterotoxin or 50 kilos of V–nerve agent, or for comparison, with 10 tons of potassium cyanide. In 1990 the USA was planning to introduce malumbia caterpillars, an indigenous pest of the coca plant, as part of a programme against producers of cocaine in Peru.

Counter–claim The course of an epidemic is not generally predictable and it would not be possible for a military commander to control its spread. Thus, although analogies are often drawn between the possible consequences of BW attack and the great pandemics of history, it is unlikely that any military commander in a rational state of mind would try to start an epidemic: once started (assuming that it could be started at will), it could well spread far beyond the confines of the target area, both in space and in time, and there would be little the commander could do to regulate it.

Refs Gander, T *Nuclear, Biological and Chemical Warfare* (1987); Geissler, Erhard (Ed) *Strenghtening the Biological Weapons Convention by Confidence–Building Measures* (1990); McDermott, Jeanne *The Killing Winds* (1987).
   Broader Biochemical warfare (#PC1164).
   Related Chemical warfare (#PC0872)
   Environmental hazards of new species introduction (#PC1617)
   Vulnerability of indigenous populations to introduction of diseases (#PE3721).
   Aggravates Epidemics (#PC2514)   Plant diseases (#PC0555)
   Animal diseases (#PC0952).
   Aggravated by International trade in biological weapons (#PD8621)
   Harmful synergistic interaction of biological agents (#PC1306).

♦ **PC0210 Illiteracy**
Dependence on illiteracy — Illiterate people — Wide spread illiteracy
Nature Illiteracy is the inability to read and write. A person is defined as illiterate if he cannot, with understanding, both read and write a short and simple statement on his everyday life; and as functionally illiterate if he cannot engage in all those activities in which literacy is required for the effective functioning of his group and community, and also for enabling him to continue to use reading, writing and calculation for his own and the community's development. Lack of such abilities prevents individuals from going about their daily activities in modern society, seeking suitable employment, or moving about normally with comprehension of the usual printed expressions and messages they encounter. Its consequences include inability to take up basic social services, fill in even simple forms, and understand traffic instructions or other danger signs.
The social causes of illiteracy are: lack of funds for education; poverty; isolation; hunger; and education systems imposed from outside. Millions of people speak non-transcribed languages. Many live in environments and conditions where written communications is not necessary or available.
Incidence Of the adult world population (aged 15 years and over) the number of illiterates in 1990 was 28 per cent of total, or 965 million; in 1980 was 28.6 percent of total, or 814 million; compared with 32.9 percent of total, or 760 million in 1970; and 39.3 percent, or 735 million in 1960. The burden of illiteracy falls hardest on the poorest and most disadvantaged groups, landless rural peasants, and slum dwellers. Nearly two thirds of those who are illiterate are women and the percentage is increasing. Most of the 965 million adults who cannot read or write are in developing countries, but recent statistics on illiteracy in developed countries include: one in three Americans will not be able to read this book (Jonathan Kozol on his book "Illiterate America"); 17 percent of soldiers entering the Israeli army cannot read or write (Israeli state comptroller Yitzhak Tunic). A recent study in Great Britain came to the conclusion that 7 million people are illiterate in the U.K. It is estimated that 1 in 4 Canadians are illiterate or functionally illiterate.
The absolute increase over the past decade has been of the order of 80 million, about the same figure as over the previous two decades taken together. But the tendency has been for the total of those who cannot read or write to grow much more slowly than the total of those who can. The proportion of illiterates in the total adult population has been shrinking gradually, even as their absolute number has increased. Four out of 10 adults were illiterate in the early 1950s; just over 3 out of 10 in 1970; fewer than 3 out of 10 in 1980. Illiteracy among younger adults is lower than among the adult population as a whole, the result of the recent expansion in primary schooling. Nevertheless, by 1980, 2 out of 10 young people were reaching the age of majority without having acquired even a rudimentary literacy. Past experience suggests that those who fail to learn the basic skills by the end of normal school age have limited prospects of acquiring them later as adults. They can be expected to form a sizeable, if diminishing, illiterate segment of the working-age population for the next 40 years or more, until at least the year 2020.
There is a distinct regional pattern to adult illiteracy in developing countries. The incidence is highest in Africa, where almost 6 out of 10 adults were unable to read or write according to estimates for 1980. In Asia and the Pacific, by far the most populous region, the proportion of illiterate adults is 4 out of 10. This is also the figure for the developing countries as a group. In Latin America only 1 in 5 adults remains illiterate. Over the past decade, the sharpest reduction in the illiteracy ratio has occurred in Africa. With the other two regions showing smaller (though still impressive) declines, regional disparities have become somewhat less pronounced. China's large population and relatively low level of illiteracy exerts a strong downward pull on the regional figure for Asia and the Pacific as well as on the over-all figure for developing countries. Brazil and Nigeria, the most populous countries in their respective regions, exert a less powerful influence in the opposite direction on the illiteracy ratio of their regions. The regional pattern for overall illiteracy coincides broadly with illiteracy across the income range. Latin America, the region with the highest average per capita income, has the lowest illiteracy ratio; Africa, with the lowest relative income, has the highest illiteracy. As just noted, within regions and among countries in similar economic circumstances, variations remain important. These tend to have deep cultural or historical roots and have diminished only slowly as efforts to raise the level of literacy in countries lagging in this respect continue to be checked by a limited ability to increase the number of those receiving instruction in reading and writing.

Refs Goldberg, Samuel *Army Training of Illiterates in World War Two*; Stevenson, Colin *Challenging Adult Illiteracy* (1986); United Nations Educational, Scientific and Cultural Organization *Statistics of Educational Attainment and Illiteracy, 1945–1974* (1977).
   Broader Ignorance (#PA5568)   Restrictions on the acquisition of knowledge (#PF1319).
   Narrower Economic ignorance (#PJ3349)   Cultural illiteracy (#PD2041)
   Computer illiteracy (#PG2575)   Functional illiteracy (#PD8723)
   Illiteracy among women (#PE4380)   Geographical illiteracy (#PD3984)
   Illiteracy in the fourth world (#PD6645)   Illiteracy in developed countries (#PC1383)
   Illiteracy in developing countries (#PD8329)   Illiteracy among indigenous peoples (#PD3321)
   Illiteracy as an impediment for leadership (#PE8177)
   Illiteracy as an obstacle to acquiring skills (#PE8246)
   Illiteracy as an inhibitor of business transactions (#PF7968).
   Related Innumeracy (#PC0143)   Non-productive members of society (#PF4000).
   Aggravates Social inequality (#PB0514)   Injurious accidents (#PB0731)
   Isolation of ethnic groups (#PC3316)   Low self image due to illiteracy (#PF9098)
   Language discrimination in politics (#PD3223)
   Political discrimination based on illiteracy (#PC3222)

Multiplicity of languages in a national setting (#PC1518)
Neglect of the role of women in rural development (#PF4959)
Underdevelopment of industrial and economic activities (#PC0880).
   Aggravated by Nomadism (#PF3700)   Desert nomadism (#PD2520)
   Phenylketonuria (#PG3323)   Lack of education (#PB8645)
   Educational wastage (#PC1716)   Reading disabilities (#PD1950)
   Diverse unilingualism (#PF3317)   Socio–economic poverty (#PB0388)
   Unsustainable population levels (#PB0035)   Underprivileged linguistic minorities (#PC3324)
   Inadequate education of indigenous peoples (#PC3322)
   Unavailability of scholarship funds for students (#PE5369)
   Increasing lag in education against population growth (#PE5369)
   Shortage of books and textbooks in developing countries (#PF0118).

♦ **PC0212 Juvenile delinquency**
Young criminals — Juvenile deviance
Nature Juvenile crime, as all crime, has been increasing. Brutal crime among young offenders also is increasingly evidenced in reports, particularly on urban areas. Some offenders are psychotic and their offences range from suicide to mass murder. Others are anti-social and given to minor acts of defiance. Ease of access to weapons; drug addiction; unemployment; and economic motives, are the more obvious circumstances leading to crime; but modern societal stress, breakdown of family life, threats of nuclear war and the confusion in values all contribute to aggravate violence among youth.
Incidence The extent of youthful crime is hard to judge. Since the Second World War, a substantial increase in juvenile convictions has been recorded in many countries. As offenders, boys outnumber girls in a ratio of about 10:1. Juvenile delinquency rates may rise with a higher general technological economic level and in situations of varied social change. Hence Western Europe, USA and Japan have high levels of juvenile delinquency. Youth gangs are noted also in Taiwan, South Africa, Australia, New Zealand, Poland, USSR and Yugoslavia. Juvenile delinquency has shown a sharp increase in such rapidly developing nations as Ghana and Kenya. Crimes against property are by far the most frequent type of offence. These include stealing from shops, houses, and cars; and the unauthorized taking of cars, usually for joy-riding. Theft seems to be associated more with the younger offender. Crimes against the person (assaults, fighting, robbery with violence), together with sex offences and, in industrially developed countries, traffic offences, come next and are more common among those aged from 17 to 21. Narcotic addiction and other types of drug dependence, though not always criminal offences, are a relatively new and disturbing form of deviance and seem to be increasing rapidly.
Background Despite the enormous amount of study devoted to it, a great many questions about juvenile delinquency still remain unanswered. The term covers a wide range of legally forbidden acts committed by young people who may be anything from 10 to 25 years of age. The highly varied misbehaviour of these young people, who differ greatly in personal background, development, experience, and situation, is no homogeneous phenomenon.
Refs Asher, Geoffrey *Custody and Control* (1986); Bortner, M A *Delinquency and Justice* (1988); Brusten, M, et al (Eds) *Youth Crime, Social Control and Prevention* (1986); Bynum, Jack E and Thompson, William E *Juvenile Delinquency* (1989); Denno, Deborah W and Schwarz, Ruth M *Biological, Psychological, and Environmental Factors in Delinquency and Mental Disorder* (1985); Friday, Paul C and Stewart, V Lorne *Youth Crime and Juvenile Justice* (1977); Gelstein, Sylvia S *Juvenile Delinquency* (1985); Hess, Albert G and Clement, Priscilla F (Eds) *History of Juvenile Delinquency* (1988); Leyton, Elliott *Myth of Delinquency*; Yablonsky, Lewis *Juvenile Delinquency* (1988).
   Broader Deviance (#PB1125)   Criminals (#PC7373).
   Narrower Youth gangs (#PD2682)   Infectious revenge (#PD5168).
   Related Hooliganism (#PD1109)   Limited individual attention span (#PF2384).
   Aggravates Vandalism (#PD1350)   Disowned children (#PJ0827)
   Repressive detention of juveniles (#PD0634)   Crimes committed in urban schools (#PJ1356)
   Excessive claims for human development through sports (#PG4881).
   Aggravated by Mental illness (#PC0300)   Juvenile stress (#PC0877)
   Street children (#PD5980)   Family breakdown (#PC2102)
   Police intimidation (#PD0736)   Criminal subculture (#PE5508)
   Lead as a pollutant (#PE1161)   Maladjusted children (#PD0586)
   Criminal association (#PE1178)   Corruption of minors (#PD9481)
   Illegitimate children (#PC1874)   Family disorganization (#PC2151)
   Retarded socialization (#PE2187)   Single parent families (#PD2681)
   Excessive television viewing (#PD1533)   Mental illness in adolescents (#PE0989)
   State custody of deprived children (#PD0550)   Dependence within extended families (#PD0850)
   Criminal investment in youth market (#PD5750)
   Excessive employment of married women (#PD3557)
   Negative effects of the nuclear family (#PF0129)
   Isolation of parent-child relationship (#PC0600)
   Inadequate care for children of prisoners (#PF0131)
   Inadequate facilities for children's play (#PD0549)
   Family poverty in industrialized countries (#PD1998)
   Discrimination against illegitimate children (#PD0943)
   Adjustment difficulties of new urban families (#PF1503)
   Inappropriate education in developing countries (#PF1531)
   Inadequate rehabilitation of juvenile offenders (#PE8803)
   Disruption of family system in developing countries (#PD1482)
   Increase in anti-social behaviour in developing countries (#PD0329)
   Narrow legal definition of the family in developing countries (#PD1501)
   Unemployment of premature school leavers in developing countries (#PE0015).

♦ **PC0219 Shortage of cultivable land**
Nature The amount of land used for crops world-wide is increasing but at slower rates than before. Millions of hectares are brought under cultivation for the first time each year, but almost as much is taken out of cultivation to be urbanized, returned to pasture or forest, or abandoned. With world population growing rapidly, and minimal increases in total arable land, arable land per capita is declining.
The quality of arable land is being impaired by a combination of urbanization, desertification, erosion and salinization; and in most countries the rate of soil loss from croplands is far in excess of the rate of soil formation. Expensive efforts to bring new lands under agriculture are partially offset by the loss of croplands to dehydration.
Incidence Of all potentially arable land (about 24 percent of the total surface of ice–free land) only about 44 percent is now cultivated. Almost 56 percent is not farmed because of inherent soil problems and man-induced problems. For example, in Mexico one half of the country's territory was suffering soil erosion by 1950 due to felling of trees for crops, pasture, fuelwood or intensive cattle raising. Actual arable land is almost fully utilized: 83 percent is already cultivated in Asia, 88 percent in Europe, 64 percent in the USSR, 51 percent in North America, and 22 percent in Africa. Per capita arable land declined at a rate of about 0.25 percent over the period 1968–80, and the rate of decline was faster in recent years. In the early 1970s, one hectare of arable land supported an average of 2.6 people, but by the year 2000, with present population projections, one hectare will have to support four people.
   Broader Limited available land (#PC8160)
   Long-term shortage of natural resources (#PC4824)
   Underdeveloped potential of basic resources (#PF3448).

–147–

PC0219

Narrower Restriction of wild animal range size (#PC0475)
Excessive land usage by transportation systems (#PE2525)
Defective land use planning in developing countries (#PD1141)
Alienation of land through acquisition by foreigners (#PE0896)
Unavailability of land for agricultural purposes in developing countries (#PE5024).
Aggravates Excessive land usage (#PE5059) Flood plain settlement (#PE0743)
Over-intensive soil exploitation (#PC0052) Slowing growth in food production (#PC1960)
Deforestation of mountainous regions (#PD6282)
Cultivation of marginal agricultural land (#PD4273).
Aggravated by Food wastage (#PD8844) Soil erosion (#PD0949)
Soil pollution (#PC0058) Inadequate land drainage (#PD2269)
Virus diseases in bacteria (#PD2562) Proliferation of second homes (#PF1286)
Spoilage of agricultural products (#PC2027) Unavailability of agricultural land (#PC7597)
Landscape disfigurement from open-cast mining (#PD1637)
Exploitation of land for the burial of the dead (#PE9095)
Increasing proportion of land surface devoted to urbanization (#PE5931)
Import-dependency in food staples in developing countries due to transnational corporations (#PE1806).

♦ PC0229 Radioactive contamination

Environmental hazards of atomic radiation

Nature Radioactive material may be suddenly or steadily introduced into the environment as a result of various human activities (industrial, medical, military, scientific) and is also naturally present throughout the earth and the atmosphere. Such material consists of unstable isotopes of various chemical elements (for example: carbon, hydrogen, iodine, strontium) called radio-nuclides, whose atoms undergo transformation into atoms of a different element, sometimes also unstable, at a known constant rate. Because they are chemically indistinguishable from their stable isotopes, radio-nuclides behave chemically like non-radioactive nuclides. They are thus similarly distributed among the various components of the environment and, through ingestion of food or inhalation of air containing them, may be deposited in various tissues of the human body, depending on the properties of the chemical compounds of which they are part.
The ionizing radiation that radio-nuclides emit during radioactive decay is what causes biological damage, whether the radiation reaches human tissues from inside or outside the body. The resulting effects depend on the amount of energy imparted by radiation per gramme of a specific tissue, a measurable quantity called the absorbed dose.

Incidence Radiation has always been a part of the natural environment, and the major part of the radiation dose received by the public at large is unavoidable. On the other hand, the uses of nuclear techniques in fields as diverse as industrial radiography (for quality assurance) and medicine (for diagnosis and treatment) is increasing; and nuclear energy is making an expanding contribution to electricity generation: currently ten percent of the world total. The radiation exposure of groups of workers and of the public at large is inevitably increased by the use of these nuclear techniques. Individual risks and 'detriment' to the population resulting from radiation exposure have been evaluated by national and international bodies of experts, and it is now widely recognized and accepted that practices conforming to the system of dose limitations recommended by the International Commission on Radiological Protection (ICRP) should have no significant radiological impact.

Counter-claim In the past few decades there has been increasing public and scientific discussion on the effects on man and his environment of exposure to low levels of ionizing radiation. Public attention has tended to focus on potential risks which might eventually arise from the projected expansion of nuclear-produced electricity, as a result of the increase in discharges of effluents contaminated with radioactive materials. It is for this reason that the nuclear industry has developed measures and techniques to prevent this radioactive 'pollution', which are generally considered more advanced and sophisticated than those used for other industrial pollutants. As a result, the nuclear power industry is a very minor contributor to population radiation exposure – accounting for a little over 0.1 percent of the average individual radiation dose in industrialized countries.

Refs Bertell, Rosalie *No Immediate Danger?*; Miller, E Willard and Miller, Ruby M *Environmental Hazards-Radioactive Materials and Wastes* (1985); SPC/SPEC/ESCAP/UNEP *Radioactivity in the South Pacific* (1984); United Nations *Sources, Effects and Risks of Ionizing Radiation*; United Nations *Environmental Contamination by Radioactive Materials* (1969); United Nations *Assessment of Radioactive Contamination in Man 1984* (1986).
Broader Environmental pollution (#PB1166).
Narrower Radioactive fallout (#PC0314) Plutonium pollution (#PE6285)
Health hazards of radiation (#PD8050) Radiation damage to materials (#PD1206)
Radioactive contamination of soil (#PE3383) Health hazards of irradiated food (#PD0361)
Ultraviolet radiation as a hazard (#PE5672) Radioactive contamination of water (#PE2441)
Contamination by natural radiation (#PC1299) Radioactive contamination of plants (#PD0710)
Microwave radiation as a health hazard (#PE6056)
Environmental hazards of solar radiation (#PE3883)
Long-term hazards of exposure to radiation (#PE4057)
Excessive occupational exposure to radiation (#PD1500)
Environmental hazards of non-ionizing radiation (#PE7651)
Harmful biological effects of ionizing radiation (#PE6294)
Radioactive contamination of animals and animal products (#PD1119)
Harmful effects of ultrasonic radiation on the human body (#PD0748)
Endangered animal and plant life due to radioactive contamination (#PD5157)
Excessive exposure to radiation from consumer goods and electronic devices (#PE1909)
Environmental hazards of extremely low frequency electromagnetic radiation (#PE7560)
Radioactive contamination of the marine environment and of fisheries products (#PE1431).
Related Chemical pollutants of the environment (#PD1670)
Excessive exposure of medical patients to radiation (#PE1704).
Aggravates Eczema (#PE2465) Animal abnormalities (#PD4031)
Genetic defects and diseases (#PD2389)
Environmental pollution by nuclear reactors (#PD1584).
Aggravated by Nuclear war (#PC0842) Radioactive wastes (#PC1242)
Radiation accidents (#PD1949) Industrial accidents (#PC0646)
Radiological warfare (#PC6666) Nuclear reactor accidents (#PD7579)
Transport of dangerous goods (#PD0971)
Accidents to nuclear weapons systems (#PD3493).

♦ PC0231 Inadequate drug control

Nature Inadequate drug control is a hugely complex set of local, national and international issues involving millions of people all over the world. They involve controlling the growing of raw materials, processing, smuggling, trafficking and using illegal drugs; preventing the corruption of businessmen, governments and their officials and enforcement officers; the murder and torture of competitors, judges, police officers, news reporters and concerned citizens. Money laundering operations need to be found and stopped. Potential users, their friends and families need to be educated and users need to be treated. Individuals, legitimate businesses, gangs and hugh multi-billion dollar crime syndicates are involved.
The raw materials for illegal drugs are grown in nearly every country in the world. Marijuana, for example, is raised in flower pots in Sydney, back yards in The Netherlands and mechanized farms in Northern California. Poppies, used to produce opium and its derivatives, including morphine and heroin are grown in the Middle East: principally in Iran, Afghanistan, Pakistan and Turkey; South and South East Asia: India, Burma, Laos and Thailand. Coca leaves are used to produce cocaine and its derivatives crack and basuco are grown in Bolivia, Peru, Ecuador and Columbia. The processing chemicals are manufactured, for the most part, in North America and Europe. Drugs are processed in a variety of ways and in a variety of locations. Marijuana is usually dried or used to produce hashish by the grower. Traditionally opium products were produced by middle men associated with big international smuggling and trafficking organizations but increasingly being processed by more local organizations which, in turn, are establishing their own smuggling networks. Cocaine products are, for the most part, manufactured by South American syndicates which control everything from growers to traffickers. Drugs used for medical and veterinary purposes are manufactured by legitimate businesses. Designer drugs such as ecstasy may be manufactured by bath tub chemists in any local neighbourhood.
Drug smuggling is equally complex. Individual carriers, including transportation personnel, tourists, businessmen, diplomats and plain criminals carry relative small amounts as they travel. Drugs are disguised as legitimate goods and shipped through legitimate means of transportation. Boat loads and plane loads of drugs are smuggled into little used landing spots.
Drug traffickers can be the occasional user selling to his friends, the small time hood or the highly organized gang.
Users range from the Peruvian farmer chewing coca leaves to the bank vice president sniffing cocaine. The street kid might sniff glue. The prostitute might smoke crack. The international athlete may be using performance enhancing drugs. The housewife might be abusing prescription drugs.
Drug control is also difficult because its wide spread use; the organization of producers; the number of smugglers and traffickers; and the amount of money involved. Police are frequently out armed and out spent. The judicial system is subject to bribes, intimidations and murders. Attempts to arouse public opinion by newspapers or concerned groups and individuals result in bombings, kidnappings and killings. Enforcement agencies are often at odds with each other because of different concerns and methods.
In addition different national control procedures may vary in the degree of restriction and supervision of legal drug production. The effects, especially the long-term effects of drugs used for medical purposes, both with and without prescription, are not necessarily fully explored before the marketing and use of the drug takes place (e.g. thalidomide, bromide hypnotics, barbiturates). Lack of public information, and even of medical information on the properties of drugs, may lead to unwittingly harmful use under certain circumstances or harmful combination with other drugs. Control of toxic substances such as solvents, used daily for domestic and industrial purposes, is very slight compared with their harmful potential. Such drugs constitute an occupational or medical hazard and may be abused in the same way as illicit drugs. Inadequate attention may be given to the intoxicating and addictive effects of alcohol, nicotine and caffeine.

Refs United Nations *National Laws and Regulations Relating to the Control of Narcotic Drugs Psychotropic Substances* (1983).
Broader Inadequate laws (#PC6848).
Narrower Inadequate drug quality control (#PD2392).
Related Substance abuse (#PC5536) Inadequate health control (#PF9401).
Aggravates Overdose (#PJ1995) Drug smuggling (#PE1880)
Drug dependence (#PD3825) Abuse of plant drugs (#PD0022)
Abuse of medical drugs (#PD0028) Abuse of sedatives and tranquillizers (#PE0139)
Inhaling of solvents and anaesthetic drugs (#PE1427).
Aggravated by Illicit drug trafficking (#PD0991) Law enforcement complexity (#PF2454)
Inadequate testing of drugs (#PD1190) Criminalization of drug use (#PF4735)
Inadequate information on drugs (#PF0603) Inadequate national law enforcement (#PE4768)
Excessive proliferation of medical drugs (#PD0644)
Economic dependence of some developing countries on the drug trade (#PE5296).

♦ PC0233 Inadequate child welfare

Broader Victimization of children (#PC5512) Inadequate social welfare services (#PC0834).
Narrower Untenable orphan care (#PJ9718) Neglected young children (#PE4245)
Lack of child welfare institutions (#PJ2673) Inadequate child day-care facilities (#PD2085)
Negative effects of family allowances (#PF0107)
Inadequate facilities for children's play (#PD0549)
Inadequate care for children of prisoners (#PF0131).
Related Neglect of adolescent health care (#PF6061)
State custody of deprived children (#PD0550)
Lack of self-development in the family (#PE8341)
Inadequate maternal and child health care (#PE8857).
Aggravates Human suffering (#PB5955) Displaced children (#PD5136)
Neglected children (#PD4522) Maladjusted children (#PD0586)
Loneliness of children (#PC0239) Repressive detention of juveniles (#PD0634).
Aggravated by Endangered children (#PD6065) Dependency of children (#PD2476)
Unproductive dependents (#PC1420)
Excessive employment of married women (#PD3557)
Denial of rights of children and youth (#PD0513)
Inadequate community care for transient urban populations (#PF1844).

♦ PC0237 Social inaccessibility

Narrower Absentee ownership (#PD2338) Distant representatives (#PJ0272)
Unrepresentative formal leaders (#PJ1823) Inaccessible government agencies (#PF5351)
Inaccessible administrative agencies (#PF2261)
Inaccessibility of decision-makers in multinational enterprises (#PE0573)
Restrictions on the distribution of confidential government information (#PD2926).
Aggravates Isolation (#PB8685).

♦ PC0238 Endangered species of plants

Lack of plant protection

Nature Plant species are under threat of extinction, whether due to by human activity (smuggling, digging up plants from the wild, pollution, ever-widening urban areas) or natural causes natural (droughts, floods, lightening storms and fires). Although endangered species of plant are protected by international convention, this convention is very hard to enforce and frequently updated. Many species are lost before or within a year or two after they are described by botanists. Also, those countries in which the endangered species are grown are often not signatories of the convention. Accidental removal of rare plants is also a problem. For example, neither the farmers uprooting plants and bulbs for sale as decorative house plants, nor the companies selling these plants in the mass market, may realize the endangered nature of the species with which they are dealing.

Incidence Of the 250,OOO flowering plant species currently estimated to exist, tens of thousands remain undiscovered and only some 5,000 have been tested for their pharmaceutical attributes. Some 25,000 of these species are currently under threat. The tropical, non-industrialized world contains almost all the areas rich in wild crop genetic resources. These areas are also those of high political and economic instability and in greatest need for development.

Claim Plants are our prime life-support system, having fed the world and cured its ills since time began. Loss in biological diversity due to shrinking of plant gene pools is one of the most pressing threats to human welfare. It is necessary, through the proper education of both governments and the public, to enact conservation and research activities supportive of reversing the threat of extinction.

Refs Lucas, G and Synge, H *IUCN Plant Red Data Book* (1978).
Broader Endangered species of plants and animals (#PB1395).

Narrower Endangered forests (#PC5165) Destruction of weeds (#PE3987)
Endangered species of flowering plants (#PE4314)
Endangered plantations of long-lifed trees (#PJ9565).
Related Endangered species of insects (#PC2326)
Declining breeds of cultivated plants (#PD5936).
Aggravates Suffering of plants (#PC7825).
Aggravated by Vulnerability of plants and crops (#PD5730).

♦ PC0239 Loneliness of children

Nature Lonely children are often 'unpopular' children, meaning children who are not readily accepted by their peers and who do not make friends. Children may be isolated or rejected because they are unaware of the modes of behaviour which contribute to peer acceptance, they lack insight into the adverse effects of their behaviour on peer relations and they lack the social skills necessary for making friends. The lonely child lacks a feeling of security and belonging. Major categories of isolated children are those with impairments; those who are new in a community; those whose parents are excessively conspicuous due to position, behaviour, history or unusual circumstances; those in linguistic, racial, ethnic or religious minorities; and children who are categorized, less scientifically, as introverted, intellectual or awkward. In the developmental cycle, the onset of puberty may create behavioural and psychological manifestations leading to separation from former friends. The adolescent may be the individual who feels loneliness most keenly. In some cases this may lead to leaving home prematurely under conditions as varied as premature marriage, military enlistment or running away.
Broader Loneliness (#PF2386).
Narrower Lack of ability (#PG2147).
Related Loneliness in old age (#PD0633) Lack of self-confidence (#PF0879).
Aggravates Loneliness in adults (#PG4829)
Lack of self-development in the family (#PE8341).
Aggravated by Child cruelty (#PJ2150) Social neglect (#PB0883)
Cruelty to children (#PC0838) Childhood aggression (#PD3907)
Maladjusted children (#PD0586) Inadequate child welfare (#PC0233)
Inadequate child day-care facilities (#PD2085)
Inadequate facilities for children's play (#PD0549)
Attitude manipulation of children through play (#PF2017).

♦ PC0242 Social underdevelopment

Dependence on social underdevelopment — Lack of social development
Nature Social development means the continuous improvement of the welfare of a population – taking place side by side with economic development, being supported by that development and, in turn, supporting it. But economic development which leads to social and political stability cannot progress very far with an illiterate, apathetic, undernourished, and disease-ridden population.
Incidence Virtually every country in the world, developed as well as developing, has a segment (or majority) of its populace suffering from social underdevelopment.
Claim Hopes for world peace will never be realized until the basic needs of people are met. This follows the Ghandian principle of not being able to fill an empty mind unless you first fill the empty stomach.
Broader Underdevelopment (#PB0206)
Economic and social underdevelopment (#PB0539).
Narrower Social apathy (#PC3412) Missing public signs (#PE4927)
Deficiencies in national and local legal systems (#PF4851).
Related Retarded socialization (#PF2187).
Aggravates Exploitation of child labour (#PD0164).
Aggravated by Child-marriage (#PF3285) Social conflict (#PC0137)
Cultural deprivation (#PC1351) Economic bias in development (#PF2997)
Conflict between minority groups (#PC3428)
Neglect of the role of women in rural development (#PF4959)
Negative effects of claims of religious infallibility (#PF3376)
Inadequate research and development capacity in developing countries (#PE4880).

♦ PC0244 Discrimination in employment

Dependence on discrimination in employment — Discriminatory employment practices — Prejudicial employment practices
Nature Discrimination in employment may be based on sex, religion, race, nationality, age, or other factors. While policies regarding such discrimination vary from country to country, even in those countries which outwardly forbid such practice, covert discrimination may be couched under such terms as 'distinctions', 'exclusions' or 'preferences'.
Incidence Women are notoriously discriminated against both in opportunity and pay; mandatory retirement is age discrimination; foreign workers often receive low-pay, low-status jobs; members of some religious sects may be discouraged from applying for certain positions; and in the USA, AIDS victims are currently battling what they view as discriminatory hiring and firing policies.
Refs ILO *Discrimination in Employment and Occupations. Standards and Policy Statements Adopted Under the Auspices of the ILO* (1967); Sullivan, Charles A and Zimmer, Michael J *Employment Discrimination* (1988).
Broader Injustice (#PA6486) Discrimination (#PA0833) Human inequality (#PA0844).
Narrower Closed professions (#PD8629) Discrimination against part-time work (#PE6241)
Discrimination against domestic servants (#PE4964)
Discrimination against non-union workers (#PD6019)
Discrimination against men in employment (#PD3338)
Discrimination against women in employment (#PD0086)
Employment discrimination against the elderly (#PD4916)
Discrimination in employment against immigrant workers (#PE4934).
Related Denial of right to work (#PC5281).
Aggravates Unemployment (#PB0750) Exploitation (#PB3200)
Underemployment (#PB1860) Waste of human resources (#PC8914)
Segregation in employment (#PD3443)
Maladjustment to disciplines of employment (#PD7650)
Limited availability of permanent employment in inner-cities (#PE1134).
Aggravated by Rural unemployment (#PF2949) Unethical personnel practices (#PD0862).

♦ PC0254 Economic inflation

Overheating of economy
Nature Inflation is an unacceptable decrease in the purchasing power of money, as measured by an accepted general price index. Not all increases in the general price index can be usefully labelled as inflationary, particularly when the increase over a period of months or years is relatively small. The term is usually reserved for larger increases, especially when they continue unchecked. Precise definitions are however lacking, although a general price increase of less than 1 per cent per quarter is often considered as 'creeping' inflation. An increase of 50 per cent per month is termed 'hyperinflation'. Although there is agreement on the general meaning of inflation, disagreement remains on how it is to be measured: which price level should be used to measure the purchasing power of money; what allowance should be made for innovations; whether both official and black market prices should be considered when the latter exist; whether prices should be gross or net of subsidies and sales taxes; and how price rises following disasters should be considered. Inflation is also a condition of generalized excess demand for stocks of goods and flows of real income, in which 'too much money chases too few goods'. It is also a rise of the money stock or money income in a society, either total or per capita. Other concepts of inflation also exist.
Background Historians have found traces of inflation at least as far back as the Roman Empire. From the time of Nero, the silver denarius and the gold aureus were tampered with, debasing its value and helping to wreck the economy. Later when other kings and emperors again sponsored metallic money, usually gold and silver, a common device was to clip the edges of coins so that a larger number could be minted from the same amount of metal. This worked until the populace caught on the and effect of more coins in circulation reduced the value of each in the marketplace. Paper money was introduced from China by Marco Polo to Europe and generally it was redeemable for coin. However during the 1920s, this was perceived as a hindrance to economic progress and governments received discretionary powers of control. Governments needing funds for reparations and rehabilitation simply printed the money required. Once started the inflationary spiral was accelerated by the general loss of confidence and the preoccupation with speculation rather than production. In Germany in 1923, the price level in terms of pre-war marks had risen thirteen hundred thousand million marks. The postage on an ordinary letter cost 100,000,000,000 marks. This hyperinflation is exceptional. Since World War II, the western industrial world has witnessed an unbroken rise in prices and money incomes, as yet without major, 1929-type, panics and without erosion of accepted business procedure. In some countries like Argentina and Israel, however, inflation in the mid-eighties reached the 400 percent level, bringing back the spectre of hyperinflation. The possibility of having a stable world-economic environment of low rate, creeping inflation, with islands of hyperinflation within it is unlikely, even though a recent ILO survey (comparing prices for December 1984 with those of December 1983) shows inflation to be on the decline worldwide.
Incidence For industrialized countries as a group inflation dropped from a peak of 9.4 per cent in 1980 to 4.8 percent in 1983 and declined further to 2.9 per cent in 1987. In the developed countries represented within the OECD, whenever the general price level is rising by 5 per cent or more, prices go up even in industries where productivity gains are well above average. Under these conditions, everyone is likely to become much more conscious of inflation. There is a general feeling of insecurity, and everyone tries to obtain priority for their income claims. An increase in inflationary expectations was clearly apparent in the acceleration in the collective bargaining cycle, the swelling demands for wage and other kinds of indexation, and the persistence of historically high interest rates. In the seventies and eighties 'double-digit' inflation appeared. Inflation feeds on itself, and unless and until people become convinced that over the longer run prices are going to be significantly more stable than they have been in the recent past, there is a clear danger that the whole process will gradually accelerate.
Contractual arrangements between borrowers and lenders, which can be modified to mitigate the adverse economic effects, pass the burden of price increases to others and make inflation increasingly difficult to control. It takes a long time before interest rates, contractual arrangements, pension schemes, etc, catch up with a given rate of inflation. The lags may be as long as 5, 10 or even 20 years, and thus there will always be distortions and inequities, the more so if the rate of inflation is accelerating. In particular, money rates of interest generally lag behind price rises thus stimulating excessive indebtedness and inefficient allocation of resources. Further, with high rates of inflation there must be much uncertainty as to what exactly the price rise is likely to be over a number of years, and this uncertainty will distort and inhibit long-run lending and investment decisions. It would therefore be quite wrong to conclude that, because the economic costs of inflation in the past have not been readily apparent, they could not quickly become considerable if the gradual but inexorable rise in inflationary expectations were to continue.
The social and political consequences of continuing inflation must also give great cause for concern. They are not easy to pin down because the impact of inflation on individuals and groups is extremely haphazard. Also, under inflationary conditions, income gains and the accumulation of wealth often appear to result not so much from work or sacrifice as from ingenuity and the exercise of economic and political power and influence. Resentment against inflation is incoherent and diffuses through the community. It therefore tends to strengthen other forces making, for disenchantment with government and existing political parties.
Further, the adverse consequences for the developing countries of inflation in the developed areas cannot be ignored. The impact of inflation and high interest rates on the net foreign exchange earnings of the developing countries, and on the value of aid and the cost of borrowing in real terms, is difficult to assess. But it is quite clear that successive periods of over-expansion followed by contraction, in the developed countries, greatly complicate the task of economic planning in developing countries. And it is evident that the volume and quality of development assistance has suffered from the fact that donors have so often been preoccupied by problems of domestic inflation, and the consequent balance of payments and budgetary difficulties.
Counter-claim The desire to continue to enjoy the benefits of economic growth, coupled with the difficulty of reconciling this with reasonable price stability, raises in an acute form the question of how much inflation really matters. Post-war experience shows that some economies may learn to live with moderate inflation and that in fact rising price levels provide a stimulus to the more productive segments of the economy, spurring profits, rewarding those factors most mobile and innovative, and penalizing the lethargic non-adaptive elements. Inflation is particularly necessary in the developing countries since there is no feasible alternative for financing the investment necessary for economic growth. Inflation provoking a redistribution of income in favour of profits and economic growth is often a necessary complement in developing countries. Increases in aggregate demand must result in increased prices because of structural bottlenecks in foreign trade, transportation, materials and food.
Disinflation, such as is presently occurring in the United States (October 1985), leads to drop in economic growth. The restructuring of businesses to allow for low inflation promotes fluctuations in prices which are independent of government economic policy and continue after pressure to reduce inflation is removed. Because return on tangible assets is depressed, investors sell such assets to buy high-yield securities and induce a wave of debt crises.
Refs Argy, Victor and Nevile, John (Eds) *Inflation and Unemployment* (1985); Brown, A J *The Great Inflation Nineteen Thirty-Nine to Nineteen Fifty-One* (1983); Brown, A J *World Inflation since Nineteen Fifty* (1985); Fuller, R *Inflation* (1980); Gilbert, Michael *Inflation and Social Conflict* (1986); Helpman, Elhanan and Leiderman, Leonardo *Stabilization in High Inflation Countries* (1987); ILO *Pensions and Inflation* (1980); OECD Staff *Measuring the Effects of Inflation on Income, Saving and Wealth* (1983); OECD Staff *International Aspects of Inflation – The Hidden Economy* (1982).
Narrower Prohibitive labour costs (#PF8763) Excessive cost of animal protein (#PE4784)
Repressed inflation in socialist countries (#PF0960)
Vulnerability of developing countries to inflation (#PD0367).
Aggravates Shoplifting (#PE1113) Unbalanced growth (#PB0479)
Imbalance of payments (#PC0998) Prohibitive cost of living (#PF1238)
Cyclic business recessions (#PF1277) Inadequate income in old age (#PC1966)
Crisis in long-term pension funds (#PF5956)
Loss of confidence in government leaders (#PF1097)
Cost overruns in large-scale public programmes (#PD1644)
Domestic market restrictions in developing countries (#PD1873)
Instability of economic and industrial production activities (#PC1217).
Aggravated by Strikes (#PD0694) Posterity (#PA6073)
Payment of interest (#PF5514) Export of inflation (#PD5351)
Excessive public debt (#PC2546) Economic inefficiency (#PF7556)

PC0254

Instability of prices (#PF8635) Restrictive monetary practices (#PF8749)
Over-reliance of government on money creation (#PF9560).
Reduces Obsolete deliberative systems (#PD0975).
Reduced by Deflation (#PD7727) Unemployment (#PB0750)
Dual exchange rate systems (#PE4476).

♦ PC0267 Bureaucratic inaction

Bureaucrat inertia

Nature Bureaucracies, or units within a bureaucracy, may use every possible tactic to avoid action, innovation, taking a position, taking responsibility or doing other than the absolute minimum consistent with fulfilling the letter of their mandate. Even assuming initial competence in bureaucracies, inaction in everything except paperwork leads to atrophy of the special knowledge and decision-making capabilities vital for governmental and intergovernmental intervention in the development process. Bureaucratic passivity may also lead to recruitment, retention and promotion of those who lack competence in energy and dedication, if not in knowledge, solely to ensure that a low level of practical contributions is maintained.

Incidence Recent statements have indicated problems in the USA (where bureaucratic inertia has been blamed for ineffectiveness in fighting terrorism), in the USSR (where preference for the status quo among cadres has led to a rejection of their economic plans for 1986–90); and in Egypt (where attempts at economic and industrial expansion are said to be thwarted by bureaucratic red-tape).

Broader Inaction on problems (#PB1423).
Narrower Governmental disregard for people as human beings (#PD8017).
Related Complex government regulations (#PF8053)
Bureaucracy as an organizational disease (#PD0460).
Aggravates Government inaction (#PC3950).
Aggravated by Official apathy (#PF9459) Formalism in developed countries (#PE5723)
Inefficient public administration (#PF2335).

♦ PC0268 Health hazards of exposure to noise

Noise pollution — High noise levels — Sonorous pollution — Lack of legislation restricting noise levels

Nature Excessive noise can affect human health by disturbing sleep, rest and communication. It can also damage hearing and evoke other psychological, physiological and pathological reactions. The effects of noise upon individuals, in order of decreasing noise magnitude, include: permanent hearing loss, speech communication disruption, individual annoyance, physiological stress reactions, psychological stress and sleep disturbance.

Background Noise, operationally defined as 'unwanted sound', has become an environmental contaminant of massive proportions. Nevertheless, major differences exist between noise and other forms of pollution. Noise is everywhere; it is not as easy to control as the sources of water and air pollution. Although certain effects of noise accumulate in the organism, if noise pollution were to cease there would be no noise residual in the environment, as there would be in the case of water and air pollutants. Unlike air and water pollution, the effects of noise are felt only close to the source. An essential awareness of noise and motivation to reduce the problem are not present: people are more likely to complain and demand political action about air or water pollution than about noise. Finally, noise is not likely to have genetic effects, while some forms of air and water pollution, such as radioactive pollution, can. However, the annoyance, frustration, impediment of learning and general stress caused by noise pollution may all have effects on future generations.

Incidence 15% of the population in the member countries of the OECD, that is over 100 million people, are exposed to an external noise (apart from the noise in the place of work) exceeding 55 dB. Continued daily exposure to noise over a period of years can cause hearing loss which may vary from partial to 'complete'. It has been estimated that there are about 500 professions and occupations which under the present conditions of industrial production involve the danger of impairment of learning due to noise.

Sound is measured in units called decibels (dB). Sound is generally considered to be dangerously loud when long-term exposure exceeds an 85 dB level. Due to the upward spiral in industrial technology and the increase of labour saving gadgets, noise sources have greatly proliferated during the past 20 years. Some countries including the USA have set the permissible noise exposure for workers at 90 dB for a duration of 8 hours a day. Even under such standards, it has been found that one-fifth of the exposed work force will suffer a disabling loss of hearing, and several countries have therefore lowered the limit to 80 dB (for example, the Netherlands). Present knowledge indicates that an upper limit of 75 dB would considerably reduce the risk for noise-induced hearing loss, as well as for other adverse effects of noise. Efforts to lower the decibel level have been intensified as a result of the remarkable increase in the number of industrial workers with hearing impairment. In Sweden, for example, 16,000 cases of hearing loss due to exposure to industrial noise were reported in 1977, compared to 5,000 in 1973.

Excessive noise costs the United States $4 billion dollars a year in compensation payments, accidents, inefficiency, and absenteeism. Noise depreciation of real estate values mounts into the billions, particularly in the vicinity of major airports. Of all present-day sources of noise, the noise from surface transportation – above all that from road vehicles – is the most diffused. In Europe and Japan it is the source that creates the greatest problems. Everywhere it is growing in intensity, spreading to areas until now unaffected and creating as much concern as any other type of pollution. Surveys carried out in the UK, France, Norway, Japan and Sweden show not only that traffic is considered to generate the most annoying kind of noise, but that it is often one of the most serious problems that town-dwellers must face. The world motor vehicle population (private cars and commercial vehicles) rose from 100 million units in 1960 to 200 million in 1970 and is thought to have exceeded 300 million units by 1980. In France, for example, it has been estimated that noise may have increased by 2–3 dB between 1970 and 1985. It can be concluded that as a result of the increase in road traffic, noise levels will increase unless preventive or corrective measures are taken. By the year 2000 the number of people exposed to noise exceeding 65 dB is expected to increase from 15% to 20%.

In addition to damaging humans physically and emotionally, noise generated by human activity is endangering fish and oceanic mammals, such as, seals and whales. It destroys the hair cells of the auditory organs of some fish, damages fish eggs and reduces the growth rate of fry.

Background Noise has been considered a nuisance since the beginning of civilization (a denunciation of 'the uproar of mankind' occurs in the five-thousand-year-old Epic of Gilgamesh), but it is only since the Industrial Revolution that noise was recognized as a real threat. The roaring machines of early industry probably caused premature deafness in thousands of people. It is now known that continuous noise over 85 decibels often causes deafness ('boilermaker's ear' was a hazard of nineteenth-century riveting factories). Deafness was one of the first recognized legal causes for workmen's compensation, and since the 1930s industry has been forced to lower the level of noise considerably. But the sheer volume of general noise has increased, because of both a rising population and a rising standard of living, which means more machines.

Claim The poor level of hearing of modern city dwellers is caused by noise and not by simple ageing. In the silent, desert environment in the Sudan, sixty-year-old men and women of the Mabban tribe have hearing essentially unchanged from their youth. In contrast, the average New York City dweller begins to suffer hearing loss at the age of twenty-five.

Refs American Health Research Institute Staff *Noise and Adverse Effects on Health* (1987); Industrial Health Foundation *Industrial Noise* (1973); Jones, Robert S *Noise and Vibration Control in Buildings* (1984); OECD Staff *Fighting Noise, Strengthening Noise Abatement Policies* (1986); Vergers, Charles A *Handbook of Electrical Noise* (1987).

Broader Pollutants (#PC5690) Inadequate laws (#PC6848)
Health hazards of environmental pollution (#PC0936).
Narrower Head noise (#PG5106) Traffic noise (#PD3664)
Aircraft noise (#PE5799) Neighbourhood noise (#PE5443)
Torture by continuous noise (#PE5691) Building and construction noise (#PG6070)
Noise in the working environment (#PD2831).
Related Vibrations as a health hazard (#PE1145)
Environmental hazards of vibration (#PJ2171).
Aggravates Fatigue (#PA0657) Anxiety (#PA1635) Deafness (#PD0659)
Migraine (#PE6357) Frustration (#PA2252) Hypertension (#PE0585)
Inefficiency (#PB0843) Mental illness (#PC0300) Sleep disorders (#PE2197)
Hearing defects (#PD6306) Domestic hazards (#PG2144) Injurious accidents (#PB0731)
Industrial accidents (#PC0646) Occupational deafness (#PD1361)
Stress in human beings (#PC1648) Cardiovascular diseases (#PE6816)
Occupational risk to health (#PC0865)
Health risks to workers in agricultural and livestock production (#PE0524).
Aggravated by False burglary alarms (#PG7982)
Proliferation of automobiles and motor vehicles (#PD2072)
Natural environmental degradation from recreation and tourism (#PE6920)
Environmental degradation by off-road and all-terrain vehicles (#PE1720).

♦ PC0277 Statutory crime

Broader Crime (#PB0001).
Narrower Tax evasion (#PD1466) Military offences (#PC0742)
Corruptive crimes (#PD8679) Criminal harm to property (#PD5511)
Misconduct in public office (#PD8227) Offences against public order (#PD7520)
Crimes against national security (#PC0554) Offences of general applicability (#PD4158)
Criminal violation of civil rights (#PD8709)
Offences involving danger to the person (#PD5300)
Crimes related to foreign relations and trade (#PE5331)
Crimes related to immigration, naturalization and passports (#PE3889)
Crimes against the integrity and effectiveness of government operations (#PD1163).
Related Criminals (#PC7373) Crimes against humanity (#PC1073)
Excessive community crime (#PJ7765).
Aggravates Inadequate local enforcement (#PF0336)
Instability of trade in inedible crude non-fuel materials (#PD0280).
Aggravated by Civil crimes committed during war (#PG1112)
Degradation of developing countries by tourism (#PF4115).

♦ PC0279 Bureaucratic corruption

Administrative corruption

Nature Integrity of behaviour is part of the tradition in well-established civil service systems. In some countries and cultures less importance is attached to such integrity with the result that laxity in ethics, bribery and corruption are often a major bureaucratic problem. This is particularly serious in some developing countries where the temptation is great and administrative control is lower – exports are priced out of world markets and the economy of the country is threatened; but the problem exists in the developed countries too.

Incidence Recent figures (May 1985) indicate that, for example, checking the customs office in Indonesia revealed it had received some US $200m a year in bribes, leading to inefficiency which left thousands of tons of goods waiting 11 years for customs clearance. Chinese officials in Hainan are said to have embezzled the equivalent of US $1.5 billion in the 14 month period up to March 1985. And the confusion resulting from 'official' tolerance of criminal business behaviour and unlicensed gambling in the USA is well documented.

Claim People who supply vices: gambling, prostitution, pornography, and drugs are members of the business, political and law enforcement communities, not simply members of a criminal society. Corruption of political-legal organizations is a critical part of the life-blood of the crime cabal. Crime cabals are intimately tied to, and in symbiosis with, the financial, political, commercial and judicial organizations of society. More important these organizations depend on crime cabals for their existence.

Refs Palmier, Leslie *Control of Bureaucratic Corruption* (1985).

Broader Official corruption (#PC9533).
Related Corruption in prisons (#PD9414) Corruption in politics (#PC0116)
Institutionalized corruption (#PC9173) Corruption in developing countries (#PD0348)
Frauds, forgeries and financial crime (#PE5516).
Aggravates Crime (#PB0001) Abuse of bureaucratic procedures (#PF2661)
Use of undue influence to obstruct the administration of justice (#PE8829).
Aggravated by Bureaucratic aggression (#PC2064)
Illicit drug trafficking (#PD0991)
Underpayment of government officials (#PD8422)
Proliferation of public sector institutions (#PF4739).

♦ PC0281 Economic underdevelopment

Nature The economic underdevelopment of a country can be measured on its balance sheet and by its gross national product; but in human terms it is the standard of living (as measured by comparative per capita figures) and the vital statistics for birth, health and death, that show what poverty brings. Economic underdevelopment not only means that the health of the people is affected, but also that local education and training and the importation of technologies for increased productively cannot be paid for. Thus there is no way out; and underdevelopment continues, maintained by expensive loans and inappropriate technology transfers.

Incidence Gross national product on a per capita basis in US dollars show that in 1980, the high-income oil exporting countries of the United Arab Emirates, Kuwait and Saudi Arabia ranged from 27,000 to 11,000; and the industrial economies, market and non-market, from 16,000 to 4,000. Over thirty economically underdeveloped countries had GNPs of less than 500; another twenty of less than 1000. Switzerland, with the highest figures for a non-oil exporter, had $16,440; while the lowest was Bhutan with $80 per person.

Refs Frank, André G *Dependent Accumulation and Underdevelopment* (1979).

Broader Underdevelopment (#PB0206)
Economic and social underdevelopment (#PB0539).
Narrower Underdeveloped approaches to local food production (#PF6493).
Aggravates Occupational diseases (#PD0215)
Inequality of employment opportunity in developing countries (#PD1847)
Emigration of trained personnel from developing to developed areas (#PD1291).
Aggravated by Disadvantageous terms for technology transfer (#PE4922).

♦ PC0293 Bad weather

Short-term variations in climate — Adverse weather conditions — Risk of bad weather — Winter weather

Nature Bad weather is of considerable economic significance. The construction industry (road building), manufacturing industry (sensitive processes and commodities), power industry (electricity and gas supply) are much affected by relatively minor weather changes. Road and rail communications are vulnerable to cold weather and particularly to snow. Accidents increase in bad weather. Bad weather, by keeping people at home, can cause a fall in r.ail sales. It has been estimated that savings resulting from response to weather reports was of the order of $1,000 million per year in the USA. Bad weather has a major impact upon agricultural activities, both during the plant growth phases and during crop storage.

Bad weather plays a factor in the physical and emotional health of people. Hot dry winds may be, in part, responsible for increases in respiratory, cardio–vascular problems, road accidents and crimes. Approaching high pressure systems apparently trigger increased blood pressure, fatigue and edginess which when combined can cause road accidents. The approach of a warm, humid front and a swift drop in temperature or atmospheric pressure can bring on a coronary. Winds in conjunction with pollution increase respiratory problems. A rise in temperature above the normal seasonal average (not just high temperatures) is responsible for cases of assault and rape during summer heat waves. Theft increases during the winter.

Incidence The regions of the world with the most variable climates are located near the tropics, namely those countries with less developed economies and seemingly less able to mobilize resources to compensate for the impact of adverse weather conditions.

Claim The most pressing climatic issue is that of the short–term variability in weather, not future medium– or long–term changes in the average weather. Furthermore it is probable that short–term impacts are the mechanism through which any longer term change becomes manifest.

Refs Hurst, G W and Rumney, R P *Protection of Plants Against Adverse Weather* (1971).

Broader Risk (#PF7580) Inhospitable climate (#PC0387).
Narrower Fog (#PE1655) Snow (#PG4193) Tornadoes (#PD1739)
Hurricanes (#PD1590) Hail storms (#PD0251) Climatic cold (#PD1404)
Air turbulence (#PD2127) Unreliable rainfall (#PD0489)
Wind damage to structures (#PE1334) Large–scale weather anomalies (#PC4987).
Related Long–term cyclic changes of climate (#PC6114).
Aggravates Hazards to plants (#PD5706) Crops pests and diseases (#PE7783)
Excessive wheat surpluses (#PE2902) Instability of wheat trade (#PE0385)
Instability of cocoa trade (#PE1549) Long–term shortage of wheat (#PE2903)
Pests and diseases of groundnut (#PJ1181) Vulnerability of plants and crops (#PD5730)
Slowing growth in food production (#PC1960) Vulnerability of crops to weather (#PE5682)
Weather as a factor of animal disease (#PD2740)
Vitamin E deficiency in domestic animals (#PE4760)
Health risks to workers in agricultural and livestock production (#PE0524)
Isolating effects of seasonal variations on undeveloped transportation (#PE3547)
Instability in export trade of developing countries producing primary commodities (#PD2968).
Aggravated by Inaccurate weather forecasting (#PF5118).
Reduces Instability of coffee trade (#PE0950).

♦ PC0300 Mental illness

Insanity — Non -psychotic mental disorders

Nature Mental diseases are characterized by disorders of the psyche, or mind. The view of objective reality is impaired, as are also the patient's self–correctness, his attitudes towards others and his behaviour. Some mental diseases are the result of primary affection of the brain, followed by a disturbance of the body as a whole. Others are caused by diseases of particular organs, with secondary disturbance of mental functions. They may be manifested in a variety of disorders: false sensory impressions, disturbances in thinking or mood, disturbances of consciousness and memory, and intellectual decline. Three types of mental illness are delineated; the psychoses (including schizophrenia and manic–depressive psychosis) constituting the most important group. The second group are nervous and mental disorders, including neuroses, psychopathies, and other nonpsychotic diseases. The third category is mental retardation.

The course of mental disease varies from single or rare attacks with complete remission, to severe, chronic psychoses with gross disorganization of mental activity and deterioration into feeblemindedness.

Mental illness is a non–organic, social–psychological disorder in which the individual is unable to protect his ego or social self sufficiently to participate in ordinary social life and obtain at least a minimal degree of social and psychological rewards.

Incidence In 1984, WHO estimated that at least 40 million people worldwide suffer from severe mental disorders, 20 million suffer from epilepsy, 200 million are incapacitated by less grave mental and neurological conditions, and countless millions are affected by alcohol and drug related problems. A study released that same year in the USA showed that 18.7% (or nearly one in five) of the adult population suffers from at least one mental disorder and a well–known British psychologist has said that a child born in Britain today stands a ten times greater chance of going to a mental hospital than to a university.

Background Since classical times Western thought has entertained three major explanations for behavioural disturbance: demonic possession, moral depravity and illness. Only since the latter half of the 19th century, with the ascendency of scientific medicine and psychiatry, has the illness paradigm achieved prominence. A recent trend, particularly in ethics and the law, to define mental illness as incompetence or irrationality.

Claim Mental illness has become one of the most important problems in medicine and public health in economically developed countries. The increased incidence of mental diseases is primarily due to nervous stress associated with the complex society that has developed as a result of the scientific and technological revolution. Modern life places growing demands on the individual, often bringing to the surface some forms of mental illness that would otherwise be counterbalanced; and certainly the higher number of mental diseases is also due to medicine's increased capability to detect such disorders.

Counter–claim Given the precarious balance of sanity, it is probably safe to say that we are all slightly mad at times. From infancy to senility man exhibits deviations from the normal standard; and, to some degree, dreams, drunkenness, hysteria, inspiration, sickness and ecstasy can all be described as aberrant conditions of mind. As Blaise Pascal wrote in the mid 1600's, 'Men are so necessarily mad, that not to be mad must be another form of madness'.

Any definition of mental illness is a complex and poorly delineated concept whose boundaries with normality (the sane), eccentricity (the odd) and moral culpability (the bad) are disputed.

Much of what is called mental illness by the oppressive society is healthy and at least semi–rational rebellion against conformity and against submission to and cooperation with oppression. The mental health paradigm and treatment system, including practitioners, patients, patients' family and friends, the mass media and pop understanding of mental health, are mechanism for controlling individual, in spite of good intentions. They are used to force submission, to enforce conformity and to imprison and destroy rebels and non–conformists. Women suffer the most, which is not unconnected from the vast majority of psychiatrists are men (86 percent in the U.S.). Male doctors will diagnose women as neurotic or psychotic twice as frequently as they do men with the same symptoms.

Refs American Health Research Institute Staff and Bartone, John C *Neurotic Disorders* (1984); Anchell, Melvin *Sex and Insanity* (1983); Black, Bertram J *Work and Mental Illness* (1988); Brenner, M Harvey *Mental Illness and the Economy* (1973); Carone, Pasquale, et al *Mental Health Problems of Workers and Their Families* (1985); Claridge, Gordon *Origins of Mental Illness* (1985); Egenter, Richard and Matussek, Paul *Moral Problems and Mental Health* (1967); Grunebaum, Henry, et al *Mentally Ill Mothers and Their Children* (1982); Hatfield, Agnes B and Lefley, Harriet P *Families of the Mentally Ill* (1987); Johnstone, Jay and Talley, Rick *Temporary Insanity* (1986); Menninger, W Walter and Hannah, Gerald (Eds) *The Chronic Mental Patient–II* (1987); Miles, Agnes *Women and Mental Illness* (1988); Nerozzi, Dina, et al (Eds) *Hypothalamic Dysfunction in Neuropsychiatric Disorders* (1987); Nunnally, Elam W, et al *Mental Illness, Delinquency, Addictions, and Neglect* (1988); Porter, Roy, et al (Eds) *The Anatomy of Madness* (1988); Smith, Robert J *The Psychopath in Society* (1978); Szasz, Thomas S *The Manufacture of Madness* (1971); Tsuang, Ming T and VanderMey, Randall *Genes and the Mind* (1980); Warr, Peter *Work, Unemployment and Mental Health* (1987).

Broader Human disease and disability (#PB1044).
Narrower Phobia (#PE6354) Mutism (#PE4526) Amnesia (#PD8297)
Neurosis (#PD0270) Monsters (#PF2516) Dyslexia (#PE3866)
Enuresis (#PE5431) Psychoses (#PD1722) Encopresis (#PG6903)
Tic disorders (#PE9344) Hallucinations (#PF2249) Panic disorder (#PE3575)
Drug dependence (#PD3825) Sleep disorders (#PE2197) Speech disorders (#PE2265)
Eating disorders (#PE5187) Conduct disorder (#PE3770) Mental deficiency (#PC1587)
Castration anxiety (#PG2400) Separation anxiety (#PE2401) Emotional disorders (#PD9159)
Fictitious disorders (#PG7686) Motor skills disorder (#PE7230)
Over–anxious disorder (#PE9580) Personality disorders (#PD9219)
Personal misperceptions (#PF4389) Psychosomatic disorders (#PD1967)
Non–acceptance of reality (#PF1079) Gender identity disorders (#PE6581)
Manic–depressive psychosis (#PD1318) Mental illness in adolescents (#PE0989)
Oppositional defiant disorder (#PE4634) Multiple personality disorder (#PE5048)
Massive psychic traumatization (#PE6968) Psychogenic physical disorders (#PE3974)
Avoidant disorder of childhood (#PE9256) Neurological effects of torture (#PD4755)
Anti–social personality disorders (#PF1721) Transient situational disturbances (#PG7846)
Developmental articulation disorder (#PE9712) Nutritionally induced mental illness (#PG1760)
Reactive attachment disorder of infancy (#PG5852)
Attention–deficit hyperactivity disorder (#PE9789)
Developmental expressive writing disorder (#PE0330)
Developmental receptive language disorder (#PE9300)
Developmental expressive language disorder (#PE5545)
Unhealthy emotional responses to atomic energy (#PF0913).
Related Insanity (#PA7157) Perversion (#PB0869) Innumeracy (#PC0143)
Maladjustment (#PA6739) Unintelligence (#PA7371)
Stigmatized diseases (#PD7279) Male mid–life crisis (#PD5783)
Female mid–life crisis (#PD5675).
Aggravates Crime (#PB0001) Alcohol abuse (#PD0153) Missing persons (#PD1380)
Criminal insanity (#PD9699) Mental impairment (#PF4945) Psychotic violence (#PE7645)
Public drunkenness (#PE2429) Injurious accidents (#PB0731) Juvenile delinquency (#PC0212)
Behavioural deterioration (#PB6321) Inadequate personal hygiene (#PD2459)
Fragmentation of the human personality (#PA0911).
Aggravated by Syphilis (#PE2300) Aircraft noise (#PE5799)
Oppressive reality (#PF7053) Maternal deprivation (#PC0981)
Personal life crises (#PD4840) Stress in human beings (#PC1648)
Psychological conflict (#PE5087) Inadequacy of psychiatry (#PE9172)
Diseases and injuries of the brain (#PD0992) Health hazards of exposure to noise (#PC0268)
Denial of rights of mental patients (#PD1148) Political prisoners in mental institutes (#PE4430)
Negative effects of over–crowding on mental health (#PF3850)
Denial of right to union activity for special groups (#PE1355)
Constraints on the development of mental health services (#PF4955).

♦ PC0306 Rural underdevelopment

Nature Rural underdevelopment is the absence of a series of quantitative and qualitative changes in a rural population that would effectively converge in raising the standard of living and improving the way of life of the people concerned. The most striking features may be: limited technical knowledge as characterized by relative technological stagnation, which is at the same time the cause and consequence of wrong land utilization; under–employment of the available rural manpower; a relatively low per capita income, and hence a chronic shortage of capital for financing further development; weak or inadequate socio–economic infrastructures for rural producers; and failing rural institutions, particularly the inadequate organization of the domestic market, resulting in a marked tendency to develop export crops (in fact, export monoculture) which are often the only profitable ones. The small peasant farmers, the share–croppers and the landless labourers continue to live in poverty using the same subsistence farming techniques as they have for centuries, and never receive the benefits of any food production techniques developed or profit realized from exported produce.

Refs Ghai Dharam P, et al *Overcoming Rural Underdevelopment* (1983); Lea, David A M and Chaudhri, D P (Eds) *Rural Development and the State* (1983).

Broader Economic and social underdevelopment (#PB0539).
Narrower Animal diseases (#PC0952) Attractive city jobs (#PD9869)
Long–term shortage of water (#PC1173)
Maldistribution of land in developing countries (#PD0050)
Lack of credit facilities for agricultural producers (#PE8516)
Inadequate commercial finance for rural development projects (#PF4340)
Inadequate systems of transport and communications in rural villages (#PF6496).
Aggravates Human disease and disability (#PB1044).
Aggravated by Urban bias (#PF9686) Rural depopulation (#PC0056)
Inadequate land drainage (#PD2269)
Instability of the primary commodities trade (#PC0463).

♦ PC0308 Discrimination against women

Denial of right to equality for women

Nature Discrimination against women on the basis of their sex occurs in religion, politics, education, employment, public life, social services, family and marital status, and before the law.

Incidence Women and girls are half of the world's population, do two–thirds of the world's work hours, receive a tenth of the world's income and own less than a hundredth of the world's property. 60 to 80 percent of all agricultural work in Asia and Africa is done by women. Also in Africa, women are responsible for 50 percent of all animal husbandry, and 100 percent of the food processing.

Background The causes of the inequality between women and men are directly linked with a complex historical process. The inequality also derives from political, economic, social and cultural factors. The form in which this inequality manifests itself is a varied as the economic, social and cultural conditions of the world community. Throughout history and in many societies women have been sharing similar experiences. One of the basic factors causing the unequal share of women in development relates to the division of labour between the sexes. This division of labour has been justified on the basis of the childbearing function of women, which is inherent in womanhood. Consequently, the distribution of tasks and responsibilities of women and men in society has mainly restricted women to the domestic sphere and has unduly burdened them. As a result, women have often been regarded and treated as men's inferior, and unequal in their activities outside the domestic sphere and have suffered violations of their human rights. They have been given only limited access to resources and to participation in every sphere of life, notably in decision–making; and in many instances institutionalized inequality in the status of women and men has also resulted.

Refs Blackstone, William T and Heslep, Robert D (Eds) *Social Justice and Preferential Treatment* (1977); Boulding, Elise *The Underside of History* (1976); Khushalani, Y *Dignity and Honour of Women as Basic and Fundamental Human Rights* (1982).

Broader Discrimination (#PA0833) Social injustice (#PC0797)
Social inequality (#PB0514).

PC0308

Narrower Dependency of women (#PC3426) Discrimination against rural women (#PE4947)
Discrimination against women in sports (#PE0197)
Discrimination against unmarried women (#PD8622)
Exclusion of women from decision making (#PE9009)
Discrimination against women in religion (#PD0127)
Discrimination against women in politics (#PC1001)
Discrimination against women in education (#PD0190)
Discrimination against women in employment (#PD0086)
Discrimination against women before the law (#PD0162)
Discrimination against women in public life (#PD3335)
Discrimination against women without children (#PE8788)
Sexual discrimination in contraceptive methods (#PF1035)
Discrimination against women in social services (#PD3691)
Discrimination against women in developing countries (#PC4898).
Related Sexual discrimination (#PC2022) Discrimination against men (#PC3258)
Denial of rights to vulnerable groups (#PC4405)
Denial of right to pursue spiritual well-being because of discrimination (#PE7310).
Aggravates Feminism (#PF3025) Prostitution (#PD0693)
Exploitation (#PB3200) Disobedient wives (#PF4764)
Wasted woman power (#PF3690) Trafficking in women (#PC3298)
Maternal malnutrition (#PE1085) Single parent families (#PD2681)
Denial of right to education (#PD8102) Discrimination against homosexuals (#PE1903)
Double standards of sexual morality (#PF3259) Insufficient leisure time for women (#PE8907)
Neglect of the role of women in rural development (#PF4959)
Vulnerability of women and children in emergencies (#PD1078).
Aggravated by Polygamy (#PD2184) Prejudice (#PA2173)
Disabled women (#PC0729) Forced marriage (#PD1915)
Sex segregation (#PC3383) Male domination (#PC3024)
Unmarried mothers (#PD0902)
Excessive employment of married women (#PD3557)
Vulnerability of marriage as an institution (#PF1870)
Denial to people of control over their own lives (#PC2381).

♦ PC0311 Stockpiles of nuclear warfare material

Excessive number of nuclear weapons

Nature A considerable amount of nuclear warfare material is stockpiled, the amount being in excess of that required to destroy the population of the planet. This stockpiled material might otherwise be available for use in nuclear power reactors to reduce the energy problem.

Incidence According to US Defence Secretary Brown, in 1981 the USA had 9,000 strategic nuclear warheads while the USSR had 7,000.

Broader Competitive acquisition of arms (#PC1258).
Aggravates Energy crisis (#PC6329).

♦ PC0314 Radioactive fallout

Radioactive dust

Nature Nuclear explosions, whether for military or peaceful purposes, and explosions involving nuclear material, like the one at the nuclear reactor at Chernobyl give rise to radioactive fission products which (in the case of the larger particles) may settle immediately to the ground or which may be transported by wind to other areas at distances well beyond the range of destruction and casualties caused by the blast, heat and initial radiation. Having reached these regions they may present a radiation hazard long after the explosion. This is particularly true of the smaller particles which are carried by the winds in the troposphere or stratosphere, generally in the directions of the prevailing westerlies. The debris in the troposphere circles the globe in about 2 weeks, while material injected into the stratosphere may travel much faster. North–South spreading is a much slower process. The radioactive debris is brought down to the surface of the planet mainly in precipitation, after remaining about 30 days in the troposphere. With high-yield explosions a single detonation may contaminate a huge area with radioactivity and make it uninhabitable for a long time.

Incidence Those substances that decay rapidly (that is, which have a short half-life) vanish very fast, but radioisotopes with a long half-life, such as strontium–90 and caesium–137 still fall in considerable quantities many years after the detonation of a nuclear explosion.

Radioactive fallout can affect man in two main ways: by radiation from fallout deposited on the ground; and by radiation from the active substances that man has taken into his body by ingestion (particularly of food) and by inhalation. The last comprehensive estimates by UNSCEAR of the dose commitments up to the year 2000 from fallout from nuclear tests carried out before 1968 indicate that the contributions to the gonadal dose from external radiation from caesium–137 and from shortlived fission products each amount to 36 mrem for the North Temperate Zone, and to 8 mrem for the South Temperate Zone. The contributions from internal radiation from caesium–137 and carbon–14 are 21 mrem and 13 mrem respectively for the North Temperate Zone, and 4 mrem and 13 mrem respectively for the South Temperate Zone. For this dose commitment, it is estimated that more than half the external dose had already been delivered by the end of 1967, and two thirds of that from internally deposited caesium–137. In contrast, only one tenth of the internal dose due to carbon–14 will have been delivered by the year 2000. It is not possible to quote an average exposure per year derived from this dose commitment. The genetic dose for the early seventies was less than 10 mrem per year for the northern hemisphere and 2 mrem per year for the southern hemisphere, with an average genetic dose to the world population of less than 7 mrem per year.

Exposure at this level (it was higher in previous years) will decrease over the subsequent years. This form of radiation differs, therefore, from other man-made radiation; the latter is tending to increase and will continue to exert its effects at least for decades.

Background Most of the knowledge about global fall–out is derived from the large scale series of atmospheric tests of nuclear weapons carried out mainly by the United States and the Soviet Union, before the Partial Test Ban Treaty of 1963. Since then 63 further atmospheric tests (and 52 underground tests) have been carried out by China and France. The other nuclear–weapon states have carried out about 670 tests since the Treaty but they were all underground explosions. By the end of 1980 the total number of tests of all kinds was estimated to be about 1,270. The number of tests which contributed to the global fall–out, that is to atmospheric, ground surface, water surface, and underground with venting, is believed to be about 550. The majority were fission bombs, many of a yield less than 1 kt, but the smaller number of thermonuclear bombs ranging up to 60 Mt contributed most of the explosive yield. The total yield was nearly 300 Mt, about half of which was due to fission. The majority of tests took place in the northern hemisphere and therefore most of the fall–out radioactivity remained in that hemisphere, predominantly in the temperate zone.

A report by the Lawrence Livermore National Laboratory in California in the United States concluded that the Chernobyl nuclear disaster emitted more long–term radiation into the world's air, water and topsoil than has been produced by all the atomic bombs and nuclear tests ever exploded.

Refs Izrael, Y A and Stukin, E D *The Gamma Emission of Radioactive Fallout* (1970); Klement, Alfred W *Radioactive Fallout from Nuclear Weapons Tests* (1965).

Broader Radioactive contamination (#PC0229) Environmental degradation (#PB6384).
Aggravates Human physical genetic abnormalities (#PD1618)
Uncertainty of long–term health effects of radioactive fallout (#PE5324).
Aggravated by Wind (#PE2223) Nuclear war (#PC0842)
Nuclear weapons testing (#PC2201).

♦ PC0328 Environmental hazards from economic and industrial products

Ecologically unfriendly products — Pollution–intensive goods — Hidden environmental costs of economic production

Nature The processing of certain raw materials may have significant negative side effects on the environment, as in the case of pulp and paper, oil, and alumina. The costs of such environmental damage, and of controlling or repairing it, tend to be hidden.

Incidence In industrialized countries, subject to more stringent environmental controls, the environmental costs are reflected to a greater degree in the prices of export products than in developing countries. Thus the costs tend to be paid by the importing nations, whether developed or developing. In the case of developing country production, the environmental costs tend to be borne domestically, in the form of damage to health, property and ecosystems, and are not reflected in the prices of the products exported. It has been estimated that in 1980 developing countries exporting to the OECD industrialized countries would have had to pay an extra $5.5 billion if their products had been processed according to the environmental standards prevailing in the USA, rising to $14.2 billion if the pollution control expenditures associated with the materials were also taken into account. Such estimates do not take account of the economic damages associated with resource depletion.

Counter–claim Developing countries are able to attract more investment to export manufactured goods because these costs are hidden and absorbed domestically. This would tend not to be the case if they were subjected to a more vigorous systems of global environmental control. Developing countries thus have a comparative advantage in pollution–intensive goods.

Refs Withers, John *Major Industrial Hazards* (1988).

Broader Hazards to human health (#PB4885)
Ecologically unsustainable development (#PC0111).
Narrower Large–scale industrial accidents (#PD2570)
Environmental hazards from beverages (#PE0849)
Environmental hazards from chemicals (#PC1192)
Environmentally unfriendly consumer products (#PD9310)
Vegetable oils and fats environmental hazard (#PE8136)
Environmental hazards from manufactured goods (#PE1344)
Environmental hazards of animal oils and fats (#PE8135)
Environmental hazards from food and live animals (#PC1411)
Environmental hazards from tobacco and tobacco manufactures (#PE0483)
Environmental hazards from inedible crude non–fuel materials (#PE0546)
Environmental hazards of miscellaneous manufactured articles (#PE8275)
Environmental hazards from mineral fuels, lubricants and related materials (#PE1346)
Environmental hazards from coffee, tea, cocoa, spices and their manufacture (#PE0481).
Aggravates Pollution in developing countries (#PC2023)
Inadequate legislation against environmental pollution in developing countries (#PE7141).
Aggravated by Unethical industrial practices (#PD2916).

♦ PC0350 Political crime

Crimes against the state — Crimes against the body politic

Nature Crimes committed against the state, its ideology, its representatives and its property are generally considered as political crimes. These may include acts of treason such as complicity with foreign espionage agents; and equally, corruption when it becomes too embarrassing to the government. Acts of terrorism are regarded as political crimes by the authorities and possibly also by other constitutional political parties or movements. Subversive activities ranging from violence to propaganda, but also strikes and peaceful opposition to the government may be considered as crimes by the administration, depending on the degree of political control exercised. Crimes committed by the state on political, religious or racial grounds on individuals and other states are also to be considered as political crimes, although they are not treated as such by the state – crimes against humanity. Political repression in various forms ranging from censorship to torture and equally elitism and exploitative practices reinforced by the political structure also constitute political crimes.

Pakistan's Zina Ordinance is an islamic law which defines adultery, fornication and rape as being crimes against the state.

Refs Ingraham, Barton *Political Crime in Europe* (1979); Proal, Louis *Political Crime* (1973).

Broader Crime (#PB0001).
Narrower Treason (#PD2615) Terrorism (#PD5574) Crimes by the state (#PG6455)
Political opposition (#PJ5891) Subversive activities (#PD0557) Political exploitation (#PC7356)
Avoidance of legal obligations by politicians (#PD4556).
Related Torture (#PB3430) Censorship (#PC0067) Political conflict (#PC0368)
Exploitation of trust (#PC4422) Corruption in politics (#PC0116).
Aggravates Strikes (#PD0694) Political prisoners (#PC0562)
Banned political parties (#PJ2274).
Aggravated by Official secrecy (#PC1812) Political repression (#PC1919)
Excessive government control (#PF0304) Undemocratic political organization (#PC1015)
Political corruption of the judiciary (#PE0647).

♦ PC0352 Discrimination against indigenous populations

Denial of rights of indigenous people — Internal colonialism

Nature Discriminatory practices used against indigenous populations in matters of education, employment, housing and social services involve their exploitation and an entrenchment of their dependency. Discrimination may also seek to destroy their cultural heritage by inadequate education and failure to incorporate a sense of their cultural identity with that of the national identity. Total disregard for them leaves the way open for ethnocide. Discrimination against indigenous populations may occur through inadequate government policy to provide for their welfare and their integration into national society without the loss of their cultural heritage.

Background Historically, indigenous populations have been regarded as a threat to the opening up of a new land and resources and have been dispossessed and killed. Those that remain have been exploited as cheap labour and, having none of the trappings of the developed society, are felt to be no good for anything better. Inadequate allocation of educational facilities and the difficulties involved in attempting to educate nomadic and working children perpetuate dependency and despondency. Companies sell inferior goods to indigenous populations. Legal restrictions may segregate them onto reservation land or deny political rights and the right to strike.

Among the many basic rights indigenous peoples are denied, are: to call themselves by their proper name and behaviorally express their own identity; to have official status and to be able to form their own representative organizations; to engage in foreign relations and trade with some degree of autonomy; to control their own local economies; to maintain the use of their own language and to have it recognized as officially acceptable in all transactions and for legal and governmental purposes; to be free to practice indigenous, or any other, religions or philosophies; to own land, preferably their original territorial base; to refuse removal to any reservation or any other dispersion; to control their own educational systems in every way.

Incidence Indigenous peoples are not racial, ethnic, religious or linguistic minorities. In certain countries the indigenous peoples constitute the majority of the population, and in certain countries indigenous peoples constitute the majority in their own territories.

Claim Indigenous peoples should have all these rights but also the right to participate freely on an equal basis in the affairs and development of their countries. For indigenous peoples, most of these rights of internal self-determination are denied; they are denied the full rights and obligations of external self-determination, namely, independence.

Broader Discrimination (#PA0833).
Narrower Disparagement of indigenous cultures (#PE1817)
Malnutrition among indigenous peoples (#PC3319)
Inadequate housing among indigenous peoples (#PC3320)
Discrimination against traditional economies (#PD4252)
Denial of social rights to indigenous peoples (#PE7765)
Expropriation of land from indigenous populations (#PC3304)
Exploitation of indigenous populations in employment (#PD1092)
Denial of rights to development for indigenous peoples (#PE4972)
Denial of the right to health for indigenous populations (#PE4459)
Denial of right to freedom of religion of indigenous peoples (#PE4332)
Denial of the right to legal services for indigenous populations (#PE2317)
Denial of right to social welfare services for indigenous peoples (#PE1506)
Denial of right of indigenous peoples to participate in political processes (#PE7312)
Denial of the right to an adequate standard of living for indigenous peoples (#PE5484)
Inadequate access of indigenous peoples to international decision-making processes (#PJ7596).
Related Ethnic discrimination (#PC3686) Denial of rights of minorities (#PC8999)
Legalized racial discrimination (#PC3683) Denial of rights to vulnerable groups (#PC4405)
Violation of treaties with indigenous populations (#PE7573).
Aggravates Ethnic conflict (#PC3685) Endangered indigenous cultures (#PC7203)
Destruction of cultural heritage (#PC2114) Lack of variety of social life forms (#PE8806)
Endangered tribes and indigenous peoples (#PC0720)
Inadequate education of indigenous peoples (#PC3322)
Restriction of indigenous populations to reservations (#PD3305)
Denial of rights of indigenous people to be self-governing (#PE1024).
Aggravated by Segregation (#PC0031) Colonialism (#PC0798)
Religious conflict (#PC3292) Diverse unilingualism (#PF3317)
Racial discrimination (#PC0006) Isolation of ethnic groups (#PC3316)
Unpredictable governmental policy (#PF1559) Illiteracy among indigenous peoples (#PD3321)
Inadequacy of international human rights instruments (#PF6365).

♦ PC0368 Political conflict

Nature Conflict between political parties and political beliefs, and for political influence in other countries, may lead to political disintegration, revolution, anarchy or dictatorship, or to social breakdown and civil war. It may also lead to international conflict, international war and foreign intervention.

Broader Conflict (#PA0298) Political revolution (#PF3237).
Narrower Marxism (#PF2189) Demagoguery (#PC2372) Political schism (#PC2361)
National federalism (#PF0626) Political radicalism (#PF2177) Political opportunism (#PC1897)
Political fragmentation (#PF3216) Single party democracies (#PD2001)
Undue political pressure (#PB3209) Inadequate political integration (#PF3215)
Inadequacy of political doctrine (#PF3394).
Related Racial conflict (#PC3684) Political crime (#PC0350)
Political feuding (#PD4846) Ideological conflict (#PF3388)
International conflict (#PB5057) Psychological conflict (#PE5087)
Political intimidation (#PC2938).
Aggravates Extremism (#PB3415) Aerial piracy (#PD0124)
Draft evasion (#PD0356) Guerrilla warfare (#PC1738)
Political prisoners (#PC0562) Subversive activities (#PD0557)
Inadequate riot control (#PD2207) Political disintegration (#PC3204)
Forced repatriation of prisoners of war (#PD0218).
Aggravated by Destiny (#PF3111) Propaganda (#PF1878)
Power politics (#PB3202) Ethnic conflict (#PC3685)
Political injustice (#PC2181) Political appointees (#PF2031)
Racial discrimination (#PC0006) Double standards in morality (#PF5225)
Illegitimate political regimes (#PC1461) Unequal political representation (#PC0655)
Underprivileged racial minorities (#PC0805) Racial discrimination in politics (#PD3329)
Undemocratic political organization (#PC1015) Unequal parliamentary constituencies (#PD2167)
Unjust financing of political parties (#PE0752) Underprivileged ideological minorities (#PC3325)
Political barriers to effective legislation (#PC3201).

♦ PC0369 Communism

Nature An economic, social and political system theoretically based on common property and an equal distribution of income and wealth, communism may be seen as temporary or part of an evolutionary process, or as inherently repressive and a menace which can only be avoided if the system is prevented democratically or otherwise from obtaining power. Many opponents of the system see no alternative to the latter interpretation.

Background Throughout history a variety of communist movements and societies have existed sporadically. Modern communism is exemplified by Marxism or Marxism-Leninism but the ideology of communism is still evolving and new forms and compromises with communist utopian theory are emerging in Western Europe and Asia.

Incidence It has been estimated that, at a minimum, the social cost of communism appears to have involved 50 million fatalities. Summary executions in the process of taking power involved at least 1 million people in the USSR, several million in China, about 100,000 in Eastern Europe and at least 150,000 in Viet Nam. If execution of political opponents after the acquisition of power is included, the combined total has been about 5 million. The extermination of people belonging to certain social categories deemed to be hostile, increases this by some 3 to 5 million. The liquidation of the independent peasantry adds a further 10 million. Fatalities associated with mass deportation and forced resettlement put the number of victims in the USSR at between 7 and 10 million, and in China at about 27 million. The execution or death of purged communists in labour camps resulted in the liquidation of over 1 million between 1936 and 1938. In Eastern Europe in the late 1940s and early 1950s tens of thousands of communists were killed or imprisoned. In China, particularly during the Cultural Revolution, it is estimated that several million must have suffered a similar fate.

Despite the overthrow of communist monopolies in Eastern Europe, local communist parties are using the new democracies to create a new role for themselves. In contrast with the crowds in the streets and their fledgling political competitors, they have the political habit, know-how and organization to encourage their ambitious and able members to seek a new role through which to influence the evolution of their societies. Given the weakness and fragmentation of their competitors, possibly following the immediate changes, there is some probability that their influence could predominate once again.

Claim Marx and Engels were only a movement in the history of the delusion that the State (or party) can solve all problems. The materialism of the communist philosophy and political science accords well with seventeenth and eighteenth century mechanistic conceptions of the universe and the belief that the well-ordered machine of humanity and the world could be planned, monitored, controlled or subjugated by edicts and dogmas.

Because of the iron discipline and utter ruthlessness, especially of Leninist parties, communists are more skilful at creating chaos and seizing power. And once in power their ideology enables them to eliminate opposition and hold power successfully. It is in this sense that the communist bloc has been described by the President of the USA as an "evil empire". It is not only oppressive and bereft of popular support, but it regards any deviation from the "party line" of the moment as "heresy". The sovereignty of such a political religion is different from, and infinitely more depraved than, any traditional political tyranny. It constitutes what has come to be known as totalitarianism.

Communism is what it professes to be, soulless. Its tyranny has enslaved nations. Most of the obstacles to peace today result from communist expansionist policies. Since 1974 over 100 million people have come under communist domination or have been alienated from the West by communist lies, cheating, government subversion and disrupted elections. Communists subsidize terrorism and wage wars by proxy. They stand behind the aggressors in virtually every one of the world's conflict areas and have instigated every postwar confrontation between the superpowers. As a result, since the end of World War II millions of refugees have been on the move, from communism to freedom. They flee from war, from communist coercion, and from extreme poverty.

In communist countries the growth rate is plummeting. Productivity is dropping, and worker and public morale are crumbling. The standard of living is sinking so much that the life expectancy is actually decreasing. The minimum wage in Brazil, a developing country, is higher than the average wage of most communist workers, even in so-called developed countries. Communists have never won a majority in a free election anywhere in the world. Their ideology is losing its appeal because their system has failed in practice to produce the results they predicted. Communism is inefficient, rigid, inhumane and irreversible except by revolution.

Counter-claim 1. Communism is a vital social force that arises spontaneously in every country when the excesses and deficiencies of the private ownership and privileged classes systems inevitably come to light. Communism offers the only rational solutions to the world problems which begin with personal selfishness and end in international aggression.

2. It is much too easy to assume that capitalism automatically safeguards democracy and human rights, by contrast with Communism. The horrors of apartheid were committed in the name of maintaining western values and defeating Communism, and were committed with the support of capitalist countries. There have been many successful capitalist systems, including that of Nazi Germany and Panama under Noriega, which have suppressed human rights in the name of anti-Communism. Many business interests, whether foreign or domestic, prefer to deal with autocratic systems because they can be more decisive and reliable than changeable democracies.

3. Communism has not yet existed anywhere. It remains a high ideal. What is condemned as communism is the perversion of socialism in the form of Stalinism and not socialism itself.

Refs Budenz, Louis F *The Techniques of Communism* (1977); Bukharin, Nikolai and Preobrazhensky, Evgeny *The ABC of Communism* (1988); Dallin, Alexander (Ed) *Diversity in International Communism* (1963); Daniels, Robert V *Documentary History of Communism* (1984); Roman, Stephen and Eugen, Loebl *Alternative to Communism and Capitalism* (1979).

Broader Ideological conflict (#PF3388).
Narrower Eurocommunism (#PF3876) National communism (#PF3130)
Communist repression (#PD8785) Socialist colonialism (#PD4862)
Revolutionary communism (#PC3163) Subversion of democracy (#PD3180)
Communist closed society (#PD3169) Elitism in communist systems (#PC3170)
Communist economic imperialism (#PC3165) Communist political imperialism (#PC3164)
Censorship in communist systems (#PD3172) Ideological schism in communism (#PF3181)
Contradictions in communist systems (#PF3179) Revisionism and anti-marxist crimes (#PE5546)
Denial of religion in communist systems (#PC3175)
Political prisoners in communist systems (#PD3171)
Denial of democracy in communist systems (#PC3176)
Economic competition in communist systems (#PC3167)
Denial of human rights in communist systems (#PC3178)
Denial of freedom of movement in communist systems (#PC3173)
Communist opposition to international organizations (#PF3162)
Suppression of private enterprise in socialist countries (#PD2048)
Denial of freedom of expression and thought in communist systems (#PC3174)
Denial of right to national self-determination in communist systems (#PC3177).
Related Fascism (#PF0248) Marxism (#PF2189) Socialism (#PC0115)
Capitalism (#PC0564)
Lack of integration of transport systems between neighbouring developing countries (#PD0664).
Aggravates Ignorance (#PA5568) Politically emotive words and terms (#PF3128)
Nationalization of domestic enterprises (#PD1994)
Enforced collectivization of agriculture (#PD7443).
Aggravated by Destiny (#PF3111) Anti-communism (#PF1826)
Economic dependence (#PF0841) Political instability (#PC2677)
Double standards in morality (#PF5225)
Nationalization of foreign investments (#PC2172).
Reduces Bourgeoisie (#PE7774) Underdevelopment (#PB0206)
Denial of the right to work in capitalist systems (#PC3119).
Reduced by Subversion of socialism (#PF9485).

♦ PC0374 Amenity destruction

Dependence on amenity destruction

Nature Urban life is based on amenities. They include sanitation systems and other hygienic services and facilities that range from inside plumbing to public enforcement of hygienic food-handling laws. Essential amenities are: food distribution to a wide number of outlet points; water for drinking, cooking, washing and for commercial and industrial use as well as for fire-fighting and recreation; law enforcement; hospitals and all health services; well-maintaining roadways and pavements; public transportation; parks and recreation areas; and breathable, fresh air. While urban life is based on amenities, its maintenance involves their deterioration and destruction. Natural resources such as city air and water are polluted; sunlight is blocked by high-rise buildings or smog; population pressure destroys parks and recreation facilities, brings excessive automobile and vehicle traffic and engine-exhausts, wears out public transportation equipment, and generally causes the deterioration of every city service including education, law enforcement, and health care.

Broader Human destructiveness (#PA0832).
Aggravated by Waste paper (#PD1152) Wasted water (#PD3669)
Water pollution (#PC0062) Marine pollution (#PC1117)
Malodorous fumes (#PD1413) Thermal pollutants (#PC1609)
Smoke as a pollutant (#PD2267) Mycoplasmal arthritis (#PG2302)
Hazardous waste dumping (#PD1398) Domestic refuse disposal (#PD0807)
Solid wastes as pollutants (#PD0177) Pollution of inland waters (#PD1223)
Non-biodegradable plastic waste (#PD1180) Photochemical oxidant formation (#PD3663)
Industrial waste water pollutants (#PD0575)
Damage by degradable organic matter (#PJ6128).

♦ PC0380 International trade in endangered species

Smuggling protected wildlife — Wildlife traffic — Trade in exotic species

Nature Export of live wild animals for zoos, private collections, medical research and as pets, seriously depletes populations of wildlife species, many of which are rare or in danger of extinction. The capture of animals may, for reasons of economic effectiveness, lead to the use of cruel methods and the death of the captured animals in transit.

Incidence There are no records of total numbers of wild animals imported into any of the major developed importing countries. It is known that India exports several million birds each year; and that the developed countries import over 100,000 monkeys annually for scientific research, including the production and testing of vaccines. It has been estimated that 80 per cent of wild

animals and birds die during capture or in transit. Fewer than 2 per cent of wild-caught birds survive a year in captivity.
Claim Despite the Convention on International Trade in Endangered Species of Wild Fauna and Flora (CITES), the World Wildlife Fund (WWF), and the Trade Records Analysis of Fauna and Flora in Commerce (TRAFFIC), the problem persists wherever jaded tastes for the exotic create a market that can be supplied by the less affluent, poaching on the diminishing preserves of nature.
Counter-claim Trade is necessary for breeding the endangered wildlife in Western countries which helps to preserve the species.
Refs Favre, David S *International Trade in Endangered Species* (1989); IUCN (Eds) *Convention on International Trade in Endangered Species of Wild Fauna and Flora - CITES*.
  Broader Irresponsible international trade (#PC8930).
  Narrower Instability of the fur trade (#PE1474)
   Trade in animal products of endangered species (#PD0389).
  Aggravates Fowlpest (#PE1400)   Zoonoses (#PD1770)
   Ornithosis (#PE2578)
   Endangered species of plants and animals (#PB1395).
  Aggravated by Unethical practices in transportation (#PD1012)
   Corruption of customs and excise officials (#PE4033)
   Unethical practices with domesticated animals (#PE4771).

◆ PC0381 Socially unsustainable development
Culturally unsustainable development
Nature Development is socially unsustainable as long as people have inadequate control over their own lives and experience their community identity as being continually eroded. Development is culturally unsustainable as long as it is incompatible with the culture and values of the people affected by it as determined by their own participation in the development process.
  Broader Unsustainable development (#PB9419)   Destruction inherent in development (#PF4829).
  Aggravated by War (#PB0593)   Terrorism (#PD5574)   Injustice (#PA6486)
   Oppression (#PB8656)   Socio-economic poverty (#PB0388)
   Denial to people of control over their own lives (#PC2381).

◆ PC0382 Nutritional deficiencies
Undernutrition — Undernourishment — Food deficiencies — Unbalanced diet — Unnutritious food consumption — Unnutritious eating habits — Low energy diet — Deficient nutritional practices — Inadequate nutrition — Dietary deficiencies — Improper dietary habits
Nature Observations from different parts of the world show that low protein intakes, often accompanied by low calorie intakes, occur in adults in many developing countries. In the industrialized world, excessive consumption of lipids, carbohydrates, sugars and alcohol make dietary considerations the major factor in most deadly diseases. The malnourished adult usually shows seasonal deficiencies, which are transient and do not cause obvious ill-health unless accompanied by acute or chronic disease. Dietary imbalance shows in reduced physical output in work among adults and developmental problems among children. An additional effect is a deterioration in the use of mental faculties — namely listlessness, lowering of initiative, lack of awareness, poor judgement, etc. At any age, poor nutrition means greatly decreased resistance to disease.
Background Undernutrition occurs almost exclusively in the context of poverty with its concomitants, which include: lack of opportunity to earn money; lack of land, or access only to land which yields little in relation to the labour and capital invested in it; lack of clean water; poor nutritional education; traditional food habits; low food quality; poor sanitation; limited access to medical and social services; and limited educational opportunities. All these problems are so closely linked that it is practically impossible to isolate the problem of undernutrition for an analysis of its causes and consequences and for the design of interventions which might improve the situation. In those cases where undernutrition is not a factor of poverty depletion of vitamins is caused by drug taking like the Pill and antibiotics, smoking cigarettes, drinking alcohol and eating faddish or vegetarian foods.
Refs FAO *Handbook on Human Nutritional Requirements* (1981); Golikere, R K *Vegetarian vs Non-Vegetarian* (1960); Saint Gompert, Eva *Nutrition Disorders* (1987).
  Broader Starvation (#PB1875)   Malnutrition (#PB1498)
   Human disease and disability (#PB1044).
  Narrower Rickets (#PG2295)   Metal deficient diets (#PE1901)
   Unbalanced family diets (#PJ0953)   Unbalanced infant diets (#PE0691)
   Inadequate dietary fibre (#PE4950)   Protein-energy malnutrition (#PD0339)
   Iodine deficiency disorders (#PD2726)   Inadequate diet of the poor (#PJ3529)
   Vitamin deficiencies in diet (#PD0715).
  Aggravates Acne (#PE3662)   Apathy (#PA2360)   Gastritis (#PG2250)
   Over-eating (#PE5722)   Constipation (#PE3505)   Eating disorders (#PE5187)
   Underproductivity (#PF1107)   Mental impairment (#PF4945)   Nutritional diseases (#PD0287)
   Periodontal diseases (#PE3503)   Diseases of metabolism (#PC2270)
   Heat as an occupational hazard (#PE5720).
  Aggravated by Food fads (#PD1189)   Plant pathogens (#PD1866)
   Food insecurity (#PB2846)   Nutritional ignorance (#PE5773)
   Predominance of fast food (#PF5940)   Inadequate health services (#PD4790)
   Inappropriate school lunches (#PJ0335)   Unavailability of dietary care (#PJ1332)
   Outdated forms of community health (#PF1608)   Prohibitive cost of nutritious food (#PF1212)
   Seasonal fluctuations in agriculture (#PD5212)   Decline in nutritional quality of food (#PE8938)
   Inadequate maintenance of physical health (#PF1773)
   Disrupted mechanisms for community health (#PF2971)
   Long-term shortage of food and live animals (#PE0976)
   Dietary deficiencies in developed countries (#PD0800)
   Excessive consumption of specific foodstuffs (#PC3908)
   Ignorance of women concerning primary health care (#PD9021)
   Neglect of the role of women in rural development (#PF4959)
   Debilitating deterioration of physical environment (#PD2672)
   Modification of environmentally adapted nutritional habits through food aid (#PF6078).
  Reduced by Increased food consumption (#PJ1931).

◆ PC0387 Inhospitable climate
Poor climate — Climate — Unfavourable climatic conditions — Climatic extremes — Extreme climatic conditions
Incidence Two types of handicap can be postulated. The first is of climatic features that are disadvantageous to productive human behaviour. These can include prevailing wind directions bringing sultry, humid or chilled air; natural radiation intensity, notably that of sunlight; windborne endemic microorganisms which give rise to local diseases and dysfunctions; unstable or unfavourable air pressures; and possibly, adverse geomagnetic interactions with the atmosphere. The second type of climatic feature that presents a handicap to development is that which interferes with sustainable agricultural and industrial production. Notably this includes recurring typhoons, hurricanes and electrical discharges, as well as excessive, flood-causing rains short growing seasons. High levels of rainfall in conjunction with heat, and persistent heat giving rise to drought or perpetual aridity, are also handicaps.
Claim Climate constitutes a major difference between developing and developed countries. Almost all the former are situated in the tropical, sub-tropical, or arctic zones. It is a fact that all successful industrialization in modern times has taken place in the temperate zones. This cannot be entirely an accident of history but must arise from special handicaps, directly or indirectly related to climate.
Counter-claim There are few countries whose climate in all regions at all times is inhospitable to development. As for the temperate zone being 'best', it would be interesting to ascertain the climates of Summer, Dilmun, Harappa and Jericho when they were the ancient centers of civilization. The only reason technology was able to advance in the North was because the Europeans, from the tenth century up through the Crusades, were able to plunder the learning of the Islamic world and the Byzantine empire. India and China also gave up their technological secrets on the eve of the so-called Industrial Revolution in the North. All these borrowings are never acknowledged. Moreover, industrialization is not synonymous with development. Many tropical nations such as the Mayas, Incas; and those of the 'Indies' had great wealth before the Industrial Revolution. All this was plundered as well. Today's less developed countries are not eager to imitate the societal aberrations that the temperate zone's industrializations have caused; and many have found that their climates have good seasons for tourism and are adequate for development in other suitable ways.
  Broader Deficiencies of developing countries (#PC4094).
  Narrower Humidity (#PD2474)   Bad weather (#PC0293)   Sand storms (#PD3650)
   Dust storms (#PD3655)   Ice accretion (#PD1393)   Climatic heat (#PC2460)
   Short-term climatic change (#PF1984)   Radio noise of natural origin (#PF1676)
   Long-term cyclic changes of climate (#PC6114)   Seasonal fluctuations in agriculture (#PD5212)
   Deterioration in atmospheric visibility (#PE2593)
   Long-term change in atmospheric chemistry (#PF1234).
  Aggravates Byssinosis (#PE2319)   Common cold (#PE2412)
   Tuberculosis (#PE0566)   Heat disorders (#PE2398)
   Heart diseases (#PD0448)   Occupational diseases (#PD0215)
   Food spoilage in storage (#PD2243)   Protein-energy malnutrition (#PD0339)
   Corrosion in tropical climates (#PE1811)   Inadvertent modifications to climate (#PC1288)
   Endangered monuments and historic sites (#PD0253)
   Destructive action of mould in tropical climates (#PE1265)
   Underutilization of facilities due to daily or seasonal peaks (#PF0827).

◆ PC0410 Multiplicity of languages in international relations
Nature When several languages are in use in international relations (and typically in an international organization), their structural differences often give rise to difficulties in translation. This is particularly important in the case of the texts of treaties and resolutions. Granting official status to several languages causes many difficulties and seriously threatens efficiency. There are considerable delays in the translation of some documents into the less-demanded languages. Use of an official language by non-mother-tongue speakers place them at a disadvantage unless they constitute the majority. The nations whose languages are the official languages are placed at a considerable advantage. The use of several languages is time-consuming and extremely costly in interpretation and translation.
Incidence The number of languages used in international relations is increasing, particularly in the largest international organizations. Approximately one-third of international organizations have one official language. Two-thirds have more than one. A 1960 survey indicated that out of 1,206 international organizations, 381 have a single official language, 346 have 2, 248 have 3, 147 have 4, 58 have 5, 15 have 6, and 11 have 7 or more. In a study carried out by the Union of International Associations, a survey of 315 international congresses showed that 18 written and 19 spoken languages had been used: in 298, English was used; in 266 – French; in German – 157; Spanish – 61; Russian – 16; Italian – 13; Portuguese – 9; Swedish – 8; Esperanto and Dutch – 6, etc. In 22.7 percent of the cases, consecutive interpretation was used; for 71.4 percent there was simultaneous interpretation, and both systems were used in the reamining cases. For two languages two interpreters are required so that they can alternate, with each interpreting into his dominant language. Three conference languages require six interpreters; four languages, twelve; five languages 20; and so on.
The following conference languages statistics are interesting as applied to intergovernmental organizations: the United Nations uses six, including Arabic, in the General Assembly, the Security Council and the Economic and Social Council. The European Communities presently use seven, with expectation of at least two more. The Council for Mutual Economic Assistance uses ten languages. Using the formula: x = n (n – 1), where "x" is the number of interpreters required and "n" the number of languages to be used, the number of interpreters required for a conference if all languages are used is: at the UN, 30; at the EEC, 42 (in the future, 56 to 72); and at CMEA, at present, 90. Also, staffing of interpreters is higher than a single conference requirement, so that although for a UN conference 30 interpreters are required, in practice there are more than double that number of staff. Document translation for all intergovernmental organizations also requires a fully staffed department for each language. UN translators may number 300 to 400 at any one time.
UN costs for all aspects of the translation and interpretation burden, for document production and distribution, cannot be measured accurately but the total cost for the entire UN family exceeds US $ 100 million annually. If the UN's six languages are compared to the EEC's seven and to CMEA's ten languages, the annual cost, for these three organizations alone, is probably US $ 500 million. The annual cost of the multiplicity of languages to all intergovernmental organizations may be in the vicinity of US $ 3 billion. The annual cost of document translation world-wide may reach US$ 20 billion, according to some sources. Japan alone, it is estimated, spends the equivalent of $ 2 billion. The cost of the multiplicity of languages is scarcely imaginable for everyone together: IGOs, INGOs, governments, commercial and industrial firms; and even individuals who must learn, and schools and universities that must teach, second, third and fourth or more languages to be used commonly. This world-wide cost is greater than the national budgets of most of the nations in the world and indicates how the multiplicity of languages is causing the greatest resource of all, human energy, to be depleted.
Counter-claim The costs of the multiplicity of languages are an investment. The payback is world understanding, international cooperation and cultural richness and diversity. Within the field of strictly official relations in multilateral and intergovernmental dealings, the restriction on the use of the number of official languages limits the choice of competent specialists and representatives. Computer-assisted translation can reduce time and cost expenditures.
  Broader Multiplicity of languages (#PC0178).
  Related Multiplicity of official languages (#PF6027).

◆ PC0417 Suicide
Nature Suicide is the act of intentional self-destruction. While most suicidal acts are self-performed and are clearly recognized by the means employed (gun, rope, knife, etc) other acts may be judged as apparent suicides, as in some cases of drowning and falling. Undetected suicide also occurs and may be a factor, for example, in automobile accident fatalities. Another kind of suicidal act involves using other people. For example, some murderers have committed their crimes in order to receive the death penalty; some criminals have asked the police to shoot them during apprehension; and many military personnel, during combat, have made solitary attacks to earn their death. Beyond all these forms, psychologists indicate that there are those who subconsciously wish to die, with implications for everyone they come into contact with, and physicians attest to psychic suicides among the critically ill or very old - people who will themselves to die and actually do so. Suicidal acts may be committed during insanity, temporary insanity, or gross dysfunctioning of the mind due to disease, drugs, drug or alcohol addiction, brain

damage, shock or neural depression. As such, they may be neither voluntary nor intended.

Background Suicide can occur in any psychiatric disorder. Alcoholics also contribute a disproportionately high number of suicides. Men outnumber women by three or four times. Suicide increases with increasing age and in both sexes there is a rise in the fifth and sixth decades. The larger the family, the smaller the risk of suicide. Suicides are more frequent among those who are divorced. Most suicides occur in the early morning hours and more occur on Monday and Tuesday than on any other day of the week. The spring is the season with highest incidence of suicide. Common motives include ill-health, domestic difficulties, and unhappy love affairs. Suicide often occurs when the patient seems to be recovering from an emotional crisis; approximately half occur within 90 days of such crisis. Thomas Aquinas views suicide as immoral because it deprives the group of resources and is an affront to its sensibilities. The obligations to other persons of a potential suicide make the act of voluntary self–destruction immoral. By killing oneself the groups one belongs to are deprived of the contributions: economic, psychological, political or religious; one could make. Members of the group may be offended because they view suicide as immoral on other grounds. Suicide is not only an offence against self and society it is a violation of God's sovereignty. It is an inappropriate response to God's gift of live.

Hindu and Jain traditions view a kind of passive ascetic self destruction in which a person accepts death from hunger or starvation is often commended as heroic. Generally Buddhist traditions view passive self–destruction as a misguided way to find release, yet it is highly respected in many Mahayana traditions as compatible with the life of a saint. In East Asian traditions, self–destruction for honour or vengeance has been viewed with approval.

Incidence Suicide statistics are subject to problems of faulty and incomplete record keeping, attempts at concealment and differences of definition. However, since the early 19th century it has been recognized that differences in rate between countries are too great to be explained away in terms of differential accuracy of registration. Examples are: Ireland, less than 3 per 100,000; Denmark and Hungary, over 20 per 100,000; France, Germany FR and Sweden, between 15–20; USA, England and Wales, 10–12; Spain, Italy and Norway between 5–10; Chile, less than 3; Uruguay over 10.

There is an increasing disproportion between male and female suicide rates. Over the period 1974–86, the percentage increase in the male/female suicide ratio is Scotland 75 per cent, England/Wales 63, Australia 58, Greece 56, Northern Ireland 46, Japan 41, USA 37, Canada 35, New Zealand 26, and Italy 24. Figures for men for 1986 give, per million population: Finland 430, Austria 421, Denmark 337, Switzerland 330, France 329. For women: Denmark 199, Austria 158, Japan 149, Belgium 141, Switzerland 132.

Claim Ambulatory potential suicides may be vectors of nihilism, to whom nothing matters and by whom nothing is considered worth doing. They may be like black holes in society in which acts of interest, love and help can often disappear without a trace. Social apathy, political indifference and, in general, a torpid personality and will, may be characteristics of what Freud called the mortido, a kind of death–wish or destructive force in man's nature.

Counter–claim Suicide is a rare act which, in order to qualify as voluntary and consciously intended, must be free of any constraint or force. In this sense the suicide, who is supposed to be morally responsible for his act, must not only be in perfectly sound mind and body, but must be free from any motive. From this point of view, the world's figures concerning so called suicides are in reality statistics of victims. They are in part victims of disease and in part victims of their fellow man. Despair, loneliness and selfish, callous societies are their murderers.

Refs Anderson, Olive *Suicide in Victorian and Edwardian England* (1987); Brody, Baruch A (Ed) *Suicide and Euthanasia* (1989); De Catanzaro, Denys *Suicide and Self-Damaging Behavior* (1981); Evans, Glen and Farberow, Norman L *The Encyclopedia of Suicide* (1988); Farberow *The Many Faces of Suicide*; Hopkins, Sidney J *Suicide* (1987); Maestri, William *Choose Life and Not Death*; McIntosh, John L *Research on Suicide* (1985); Osgood, Nancy and McIntosh, John L *Suicide and the Elderly* (1986); Seward, Jack *Hara-Kiri* (1968); Soubrier, J P and Vedrinne, J *Depression and Suicide* (1983).

Broader Human death (#PA0072) Self–destruction (#PF8587).
Narrower Ritual suicide (#PG0459) Prison suicides (#PD8680)
Juvenile suicide (#PE5771) Cognitive suicide (#PF5240).
Related Homicide (#PD2341) Death instinct (#PF3849).
Aggravates Dying a bad death (#PF1421).
Aggravated by Pain (#PA0643) AIDS (#PD5111)
Anomie (#PF6316) Loneliness (#PF2386)
Bereavement (#PF3516) Human ageing (#PB0477)
Unemployment (#PB0750) Culture shock (#PC2673)
Alcohol abuse (#PD0153) Being a burden (#PF9608)
Panic disorder (#PE3575) Human poisoning (#PD0105)
Mental depression (#PC0799) Cocaine withdrawal (#PG4267)
Marital instability (#PD2103) Social fragmentation (#PF1324)
Amphetamine withdrawal (#PG9778) Urban–Industrial pollution (#PC8745)
Mental disorders of the aged (#PD0919) Criminalization of euthanasia (#PF2643)
Uncontrolled urban development (#PC0442) Excessive parental drunkenness (#PE7700)
Dependency of women in marriage (#PD3694) Impairment of physicians' ability (#PE5746)
Abuse of sedatives and tranquillizers (#PE0139)
Post-hallucinogen perception disorder (#PG7457)
Decline in real wages in developing countries (#PD2769)
Retirement as a threat to psychological well–being (#PF1269)
Environmental hazards of electrical power transmission lines (#PE9642).
Reduced by War (#PB0593).

♦ PC0418 Uncontrolled application of technology

Indiscriminate application and distribution of technology — Bulldozer technology

Nature Those benefiting from technologies developed to meet specific needs often apply them as widely and quickly as is feasible, tending to ignore any feedback which questions such application. The desire to allow a technology to continue to supply needs hampers the ability to see negative consequences and overrides any decision to take the necessary steps to ensure that such technology operates in a global context relative to environmental protection. This often leads to environmental imbalances.

Claim Technology has served the interests of aggression, genocide, saturation bombings, and economic exploitation, making some classes and nations richer at the expense of others. The root of the problem is not in technology as such, but in its generation, its management, its use, and in the difficulty of controlling it. The need is not for the slower development of technology, either in advanced or in developing countries. Such a slowdown would cruelly sacrifice the interests of millions of underprivileged people whose hopes and expectations cannot begin to be met without more technology. What is absent is more thoughtful and careful application of all technologies to prevent long–range damage to the earth and violence to human values, and to foster social, economic and cultural development.

Broader Human errors and miscalculations (#PF3702).
Narrower Unnecessary gadgets (#PE3745)
Reduced images of environmental protection (#PF7614)
Indecisive response to technological changes (#PF1336)
Inadequate models of socio–economic development (#PF9576)
Unbalanced application of communications technology (#PE7637)
Fragmented social structures for environmental protection (#PE3977).
Related Limited image of employability (#PF2896)
Conflicting social service ideologies (#PD3190)
Breakdown in community security systems (#PD1147)
Unsystematic allocation of market facilities (#PD3507).
Aggravates Inadequate environmental education (#PD1370)
Inadaptation of technology to man in the industrialized societies (#PE5023).
Aggravated by Inappropriate transfer of technology (#PE5820)
Abuse of science and technology in capitalism (#PE3105)
Irresponsible scientific and technological activity (#PC1153).

♦ PC0432 Excessive specialization in education

Nature Children in secondary schools in some countries are required to select subjects in which they wish to specialize. The grades they obtain in those subjects may be a major factor determining the courses open to them at university. The movement is therefore from early specialization into continued study of the same subject. A change of subject leads to lost time required to gain the necessary grounding. Teaching of particular subjects is largely in the hands of specialists, whose success is often measured by the number of specialists they can produce. A vicious circle therefore exists whereby specialists create more specialists, with little attempt to supply any broad grounding from which the student could develop into other interests. The greater the degree of specialization, the greater the risk that the knowledge gained at university will date rapidly, so that the graduate may well find himself at a disadvantage in his career advancement when competing with graduates who left university at a later date. The specialization itself may then not correspond to the needs of society to the same extent as when the person left university.

Broader Anti–holism (#PF5745) Over–specialization (#PF0256)
Inadequate education (#PF4984).
Narrower Break–down in communications due to difference in training (#PE6650).
Aggravates Irresponsible international experts (#PF0221)
Inappropriate education of graduates (#PF1905)
Inadequate control over government administrative process (#PC1818).

♦ PC0442 Uncontrolled urban development

Urban growth — Urbanization — Overurbanization — Unplanned urbanization — Haphazard urban structure — Unsustainable urban development

Nature Under present conditions and levels of technology, the continued expansion of large urban centres creates risks of physical, economic and social breakdowns with the most serious political consequences. In both developing and developed countries, urban growth has been accompanied by severe social and economic problems, some of which appear likely to worsen as overall population growth is accompanied by the trend toward greater urban growth. In developed countries, problems of environmental deterioration (especially air and water pollution), traffic congestion, and other disamenities are encountered. In the developing countries, it is difficult to provide the minimum social services such as housing, water supply, sanitation, education, and medical facilities in the rapidly growing urban areas, or to absorb an ever expanding labour force into struggling urban economies. This results in a deterioration of environmental quality. In some countries, the growth of the cities is reducing land available for food production. In an average city there is no clearly recognisable structure or satisfactory layout. Most cities are built haphazardly resulting in a random character that confuses the identity of city communities, creates chaos in the pattern of land use, wastes resources and prohibits coherent patterns of any kind. Cities are not capable of providing neither intense activity in high density areas, nor intense quiet in low density areas.

Incidence The world's urban population increased by about 30 percent from 1970 to 1980, from 1,350 million in 1970 (37.5 percent of world population) to 1,800 in 1980 (41.3 percent), but the annual rate of urban growth remained at 2.9 percent, as in the 1960s. Regional differences are significant, the percentages of urban populations in 1970 and 1980 being: Africa (22.9 and 28.9); East Asia (28.6 and 33.1); South Asia (20.5 and 24.8); Latin America (57.4 and 64.7); North America (70.4 and 73.7); Europe (63.9 and 68.8); Oceania (70.8 and 75.9); USSR (56.7 and 64.8). The rural population, although decreasing in percentage terms, also increased in absolute numbers, from 2,310 million in 1970 to 2,600 million in 1980 Large and increasing proportions of the world's urban population are concentrated in big cities, some of them of historically unprecedented size. In 1960 there were eight cities in developing countries that had reached or exceeded a population size over 4 million compared to 10 cities of the developed regions. By 1980, there were 22 cities in the developing regions with more than 4 million population each, whereas in the developed regions there were only 16 such cities. So rapid was the urban growth rate in developing countries that it seems certain that if this trend continues the number of people living in cities will double by the year 2000. Projections suggest that the developing countries by that time will have about 61 cities of more than 4 million each, compared with about 25 in the developed regions. Eighteen cities in developing countries are expected to have more than 10 million inhabitants each by that year. Because this growth took place against a background of low incomes, it outstripped these countries' abilities to provide both accommodation and services, and the result was a mushrooming of squatter settlements around the perimeters of vast cities. From 20 to 80 percent of the urban populations of various cities lived in these shanty towns.

Counter–claim The process of urbanization must be understood as a basic condition for and as a functional consequence of economic, social and technological development. Indiscriminate efforts to avoid urbanization may only serve to delay development. The crucial issue is not the presence of the process by itself, but its quantity in relation to time and economic factors and its quality expressed in social and physical terms.

Refs Brown, Lester R and Jacobson, Jodi L *The Future of Urbanizations* (1987); Cheshire, Paul C and Hay, Dennis G *Urban Problems in Western Europe* (1988).

Broader Lack of control (#PF7138) Unsustainable development (#PB9419)
Ineffectiveness of individual participation in large communities (#PF6127).
Narrower Urban underemployment (#PC3490) Inadequate urban political machinery (#PC1833)
Boundary constraints on land planning (#PF0954)
Environmental degradation of suburbia (#PD2345)
Excessive dispersion of community facilities (#PF6141)
Unbalanced urban population density gradients (#PD6131)
Proliferation of automobiles and motor vehicles (#PD2072)
Uncontrolled urbanization in developed countries (#PD3488)
Uncontrolled urbanization in developing countries (#PD0134)
Unattractive pedestrian environments in urban areas (#PE6151)
Environmental stress on inhabitants of tall buildings (#PE4953)
Increasing proportion of land surface devoted to urbanization (#PE5931)
Imbalance between urbanization and industrialization in developing countries (#PC1563).
Related Inaccessibility of quiet zones in an urban environment (#PF6160).
Aggravates Crime (#PB0001) Suicide (#PC0417) Beggars (#PD2500)
Urban fires (#PD2211) Urban slums (#PD3139) Overcrowding (#PB0469)
Tuberculosis (#PE0566) Schizophrenia (#PD0438) Leishmaniasis (#PE2281)
Art vandalism (#PE5171) Ethnic conflict (#PC3685) Traffic congestion (#PD0078)
Inadequate housing (#PC0449) Pedestrian accidents (#PD0994)
Unhygienic conditions (#PF8515) Landscape disfigurement (#PC2122)
Motor vehicle emissions (#PD0414) Vibrations as a health hazard (#PE1145)
Neglect of property maintenance (#PD8894) Deterioration of human environment (#PC8943)
Destruction inherent in development (#PF4829) Negative effects of the nuclear family (#PF0129)
Ecologically unsustainable development (#PC0111)
Unsustainable development of coast zones (#PD4671)

Inadequate facilities for children's play (#PD0549)
Environmental degradation of inner city areas (#PC2616).
Aggravated by Urban overcrowding (#PC3813) Attractive city jobs (#PD9869)
Decrease in mortality rate (#PF0333) Uncontrolled industrialization (#PB1845)
Unsustainable population levels (#PB0035)
Urban unemployment in developing countries (#PD1551).
Reduces Tribalism (#PC1910).

♦ PC0444 Poverty in developed countries
Poverty in industrialized countries
Nature Despite nearly four decades of unparalleled economic growth in the industrialized world; despite anti-poverty programmes and admittedly improved social welfare schemes, the developed countries are not really any nearer to conquering poverty.
Certainly the poor in the West do not suffer the strict physical deprivation of their Third World counterparts. The safety net of social welfare has effectively reduced the worst effects of starvation and want. But poverty is much more than just a shortage of cash. It is fundamentally an absence of power and influence. That means dependency, helplessness and a lack of political clout. The fact is that the majority of the rich world's poor work for a living. They are poor because their wages are insufficient to maintain a decent living. Most of those who are born into poverty remain there, caught by the prevailing social attitudes towards poverty: that all those at the bottom are there because they are unintelligent and, in general, inferior.
Incidence About 32.5 million Americans, or 13.5 percent of the population, were living in poverty in 1987, the U.S. Census Bureau reports. The 1987 threshold for a family of four was an annual income of $11,612. During the same period, experts estimated that more than 20 percent of the Soviet population live in poverty. Some 6 million French now live below the Common Market's $5.00 a day poverty line. In 1990 it was estimated that 6 per cent of the Belgian population lived in extreme poverty with a further 20 per cent in and out of precarious circumstances. In 1990 it was estimated that 44 million people within the 12 countries of the European Community were below the poverty line with one million of these described as homeless. In those countries, to count as poor an individual's income must be less than half the average disposable income in the country in which they live. The poor in Europe are made up of rather different groups than in 1980. The unemployed are a growing proportion to the total number, while the elderly are a diminishing, but still very large, proportion. Single parent families have a very high chance of suffering from poverty.
Refs Garg, Usha and Vibhakar, Jagdish *Poverty of Nations and New Economic Order* (1985).
Broader Socio-economic poverty (#PB0388).
Narrower Family poverty in industrialized countries (#PD1998).
Aggravates Youth violence (#PF7498).
Aggravated by Deterioration of industrialized countries (#PD9202)
Lack of economic and technical development (#PE8190).

♦ PC0449 Inadequate housing
Dependence on inadequate housing
Nature The residential environment probably has the most profound impact on human health, behaviour and satisfaction, since this is where people spend the greatest part of their lives, rear children and develop social habits. Although decent shelter is a major human need, the current housing picture contains enormous deficiencies all over the world. More than a million people live in appalling housing conditions and there is a formidable global shortage of desperately needed dwellings. This situation is likely to worsen over the period ahead.
Mere statistics fail to capture the true dimension of the urban residential crisis. At the community level, the crisis is aggravated in many countries by a growing polarization of the population according to the location and quality of their houses. Although overall living standards have risen in most countries over the past decade, the supply of housing to low-income families remains far too small. The urban poor also bear the greatest burden of the mismanagement of the urban environment, as it is in the poorer areas that essential services are of the lowest standard. The residential crisis looms largest in the metropolitan areas of the less industrialized countries, resulting in scattered housing developments, mixed land uses, high rents, overcrowding and clandestine land occupancy.
Broader Inadequate infrastructure (#PC7693).
Narrower Urban slums (#PD3139) Housing shortage (#PD8778)
Haunted buildings (#PF0201) Insufficient rural housing (#PF6511)
Inadequately heated shelters (#PD5173) Inadequately cooled shelters (#PD9736)
Inadequate housing for the aged (#PD0276)
Substandard housing and accommodation (#PD1251)
Socially sterile rental accommodation (#PF6195)
Inadequate housing in developing countries (#PE0269)
Inadequate housing among indigenous peoples (#PC3320)
Monolithic architecture of high-rise buildings (#PE1925)
Inappropriate accommodations for single people (#PE6187)
Lack of housing for teachers in rural communities (#PE8103).
Related Denial of right to sufficient shelter (#PD5254).
Aggravates Homelessness (#PB2150) Chronic bronchitis (#PE2248)
Inadequate maintenance of physical health (#PF1773)
Overcrowding of housing and accommodation (#PD0758)
Illegal occupation of unoccupied property (#PD0820)
Depressing effect of poor housing construction (#PF1213).
Aggravated by Absentee ownership (#PD2338) Temporary residence (#PJ3760)
Housing destruction in war (#PE2592) Inaccessible housing loans (#PG8648)
Scarcity of residential land (#PD8075) Proliferation of second homes (#PF1286)
Uncontrolled urban development (#PC0442) Traditionally determined housing (#PJ0486)
Prohibitive cost of accommodation (#PD1842) Illiteracy among indigenous peoples (#PD3321)
Prohibitively expensive housing for the poor (#PE8698)
Demoralizing constraints on housing rehabilitation (#PE2451).

♦ PC0463 Instability of the primary commodities trade
Nature International trade in commodities is characterized by excessive price fluctuations, declining export earnings for producer countries, vulnerability to fluctuations in export earnings, restrictions on access to markets, reductions in the variety of commodities produced, limitations on the degree of processing of commodities by the producers, declining competitiveness of natural products, inadequate market structures, marketing and distribution systems.
Incidence Trade in primary commodities has long been characterized by market instability. In post-war years, the value of trade in individual commodities has been subject to year-to-year fluctuations. Instability indices for price and export earnings of non-fuel primary commodities of developing countries are about 12 and 14 percent respectively for 1961 - 1980. Agricultural products experienced higher instability in both price and value. Since the early 1970s, instability has been intensified drastically. Gross shortfalls in export earnings for the basket of primary commodities exported from developing countries amounted to about $ 45 billion during 1973 - 1982.
The instability of world markets for primary commodities is a general phenomenon; although there are significant differences in its incidence and magnitude, it is not confined to commodities of a particular nature nor to the trade of particular countries. Thus, among the commodities whose value on world markets has been least stable in recent years are many which only come to a relatively small extent from the developing countries: tallow, lard, wool, zinc, lead, soya beans. Nevertheless, many of the commodities of which the developing countries are the sole or chief suppliers (natural rubber, fibres, cocoa, coffee) have been among the least stable in terms of export value movements. The degree of instability is appreciably greater than is evident in markets for other products.
The instability in commodity export earnings is higher for individual countries as compared with developing countries as a group, and is caused mainly by volume instability. The instability may cause variations in incomes and, hence, in domestic expenditure on consumption or investment. It may also give rise to fluctuations in external purchasing power and, hence, in supplies of imports available for consumption or investment. These consequences pose numerous difficulties for national policy formulation extending beyond the immediate impact of the fluctuations in export proceeds upon public revenue. The fluctuations may result in immediate hardship to the small producers characteristic of many of these commodities, creating socially unacceptable inequities in distribution of income. Importation of supplies of capital equipment for integrated development programmes may be interrupted. The incentive to invest may be weakened or switched from long-term industrial ventures to short-term commercial investments. The credit-worthiness of the country, in the eyes of foreign governments and international banks, tends to be weakened as instability in the commodity market could lead to debt default and thus to destabilization of the world financial market.
Refs Sandizzo, Pasquale L and Diakosawas, Dimitris *Instability in the Terms of Trade of Primary Commodities 1900-1982* (1987).
Broader Instability of economic and industrial production activities (#PC1217).
Narrower Instability of chemicals trade (#PD0619)
Obstacles to commodity futures trading (#PF4870)
Instability of trade in manufactured goods (#PE0882)
Instability of trade in food and live animals (#PD1434)
Instability of trade in beverages and tobacco (#PE1641)
Instability of trade in inedible crude non-fuel materials (#PD0280)
Instability of trade in machinery and transport equipment (#PD0620)
Instability of trade in animal and vegetable oils and fats (#PE0735)
Instability of trade in miscellaneous manufactured articles (#PE0814)
Instability of trade in mineral fuels, lubricants and related materials (#PD0877).
Related Control of industries and sectors by transnational corporations (#PE5831).
Aggravates Rural underdevelopment (#PC0306)
Inadequate trade in agricultural commodities between developing countries (#PE4523).

♦ PC0475 Restriction of wild animal range size
Forcible displacement of wildlife
Incidence Human control of grazing lands has been achieved by the forcible displacement of wild animals, especially herbivores, and their restriction to smaller ranges. The large herds of gazelle, antelope, zebras and other animals in sub-Saharan Africa have been reduced in size and displaced onto reserves. The lands so acquired are now populated by cattle, sheep, goats and humans.
Broader Inadequate animal welfare (#PC1167) Shortage of cultivable land (#PC0219).
Related Denial to animals of the right to conditions of life and liberty proper to their species (#PE6270).
Aggravated by Crop damage by wildlife (#PC3150)
Destruction of wildlife habitats (#PC0480)
Disruption of arid zone ecosystems (#PD7096).

♦ PC0480 Destruction of wildlife habitats
Dependence on destruction of wildlife habitats — Unfavourable wildlife decisions
Nature Very few natural and cultural areas remain in the world and even these are threatened by increasing pressure for land by fast-growing human populations. Man destroys wildlife habitats by physical destruction, pollution, and over-exploitation.
Claim The UN Charter for Nature States states: 'Civilization is rooted in nature which has shaped human culture and influenced all artistic and scientific achievement, and living in harmony with nature gives man his best opportunities for creativity, rest, and recreation'. It goes on to assert that the excessive consumption and misuse of natural resources leads to the breakdown of the economic, social, and political framework of civilization.
Broader Environmental degradation (#PB6384)
Degradation of semi-natural and natural habitats of flora and fauna (#PC3152).
Narrower Pesticide hazards to wildlife (#PD3680)
Eutrophication of lakes and rivers (#PD2257)
Defoliation of insect breeding areas (#PJ4038)
Destruction of hedges and hedgerow trees (#PD1642)
Environmental degradation of desert oases (#PD2285)
Inundation of wildlife habitats through dams (#PE7794)
Environmental destruction in least developed countries (#PE8401)
Degradation of mountain environment by leisure activities (#PE6256).
Related Environmental warfare (#PC2696)
Degradation of the environment through the destruction of species (#PE5064).
Aggravates Endangered species of marsupials (#PD1762)
Restriction of wild animal range size (#PC0475).
Aggravated by Thermal pollutants (#PC1609) Environmental pollution (#PB1166)
Pollution of inland waters (#PD1223)
Mismanagement of environmental demand (#PD5429)
Over-use of designated wilderness areas (#PD7585).

♦ PC0492 Unproductive subsistence agriculture
Subsistence farming patterns — Peasant farming — Traditional agricultural methodology — Lack of mechanized agriculture
Nature Subsistence agriculture is characterized by extremely limited capital resources, constancy in the use of traditional methods of production and in the commodities produced, and low productivity of land and labour. These characteristics tend to perpetuate the existing situation whereby agriculture produces barely enough for survival, and cannot therefore make a substantial contribution to economic growth. As a result, countries in which the majority of the population is engaged in subsistence agriculture, and which have no other important natural resources, are inevitably poor and their economies remain stagnant. The low productivity of subsistence agriculture is perpetuated by a vicious circle of problems: from low productivity of resources to underemployment to low income to low savings to low investment in farm to low yields, back to low productivity. In order to acquire the necessary agricultural inputs to produce, the traditional farmer uses brokers and obtains funds at a very high price. He buys the inputs from merchants who charge high prices and impose harsh payment conditions. The farmer's production is acquired by middlemen who pay the lowest possible price. This situation is widespread and always to the disadvantage of the farmer who is unable to extricate himself from the cycle. Subsistence farming can be held to be responsible (at least in part) for deficient diets and subsequent malnutrition and disease, for keeping peasants largely isolated from the rest of society, and for an outlook which, through its sociocultural context, keeps the peasant at a subsistence level even though he could possibly improve his lot. In addition to resulting in poor economic production and participation, subsistence farming can be a causative factor in peasant revolts.
Incidence In 1980, approximately 70 million Latin American subsistence farmers had, after providing their own food requirements, less than $20 per year left over to spend on manufactured articles.

CROSS-SECTORAL PROBLEMS

PC0554

Claim Subsistence farmers are the most numerous class of mankind and industrial growth within individual nations, particularly developing ones, depends on a general increase in peasant spending power. Subsistence agriculture is characterized by extremely limited capital resources, constancy in the use of traditional methods of production and in the commodities produced, and low productivity of land and labour. And the critical characteristic of subsistence agriculture is low productivity. The result is grinding poverty, massive unemployment, drift to the cities, and a pervading atmosphere of unrest and irritation conducive to peasant risings, religious millennialism, and the empty–eyed apathy of those whose social circumstances make a mockery of hope.
Counter–claim Subsistence is far more than marginal existence. It is a way of living and often the basis of traditional economies and cultures. It provides food, shelter and clothing, albeit at a very low level, to all members of the farm family, thereby avoiding their becoming a charge to society as a whole. Subsistence farming develops skills and and understanding of the local environment enabling people to live directly off the land. The environment is protected because of the direct relationship between the farmer and nature. It involves cultural values and attitude different and often alien to those of the urbanized world, mutual respect, sharing, resourcefulness and understanding of the intricate interrelationships between the environment, animals and humans. It implies protecting the land from over exploitation.
Broader Unproductive use of resources (#PB8376).
Narrower Short–term gain (#PF8675) Undercutting vendor price (#PG9659)
Limited improvement capital (#PG9704) Misjudged borrowing practices (#PG9694)
Disoriented business practices (#PG9646) Individualistic economic practices (#PG9718)
Subsistence approach to capital resources (#PF6530)
Rural unemployment in developing countries (#PD0295)
Declining productivity of agricultural land (#PD7480)
Inhibiting effects of traditional life–styles (#PF3211)
Unconvincing alternatives to existing societies (#PF3826)
Inadequate domestic savings in developing countries (#PD0465)
Low productivity of agricultural workers in developing countries (#PE5883).
Related Narrow range of practical skills (#PF2477)
Insufficient financial resources (#PB4653)
Underdeveloped capacity for income farming (#PF1240)
Underdeveloped use of agricultural resources (#PF2164).
Aggravates Rural unemployment (#PF2949) Socio–economic poverty (#PB0388)
Subsistence–level malnutrition (#PJ1370) Demeaning community self–image (#PF2093)
Unmotivating subsistence employment (#PJ1555) Rural poverty in developing countries (#PD4125)
Lack of economic and technical development (#PE8190)
Subsistence agricultural income level in rural communities (#PE8171).
Aggravated by Infertile land (#PD8585) Subsistence life style (#PF1078)
Inadequate irrigation system (#PD8839)
Unrecognized benefits from cooperatives (#PF9729)
Maldistribution of land in developing countries (#PD0050).

♦ PC0495 Unsustainable economic development

Uncontrolled economic growth — Dependence on uncontrolled economic growth — Indiscriminate economic growth — Development as indiscriminate economic growth — Unsustainable development of energy and industry
Nature Economic growth is inconsistent with sustainable development when it results in the net reduction of the portfolio of assets which includes: natural assets, comprising resource assets (biological resources, agricultural land, geological resources) and environmental assets (ecological processes and biological diversity, including species and places that are valued aesthetically or for their own sake); manufactured assets (technology, buildings, equipment, infrastructure); and human assets (knowledge, skills).
Background Sustainable development requires economic efficiency and equity within and between generations. Economic efficiency is the production of the optimal combination of outputs by means of the the most efficient combination of inputs. Equity is the expansion of opportunities for the disadvantaged and passing on to future generations a portfolio of assets of equal or greater value than the existing one. If one group increases its welfare at the expense of another, it should compensate the other by transfer of assets of equal or greater value, both within generations and between generations. Such compensations should be made now rather than in the future, because the welfare of the deprived group has actually been reduced, and equity is not served by compensation that is merely hypothetical.
In order to ensure that the total value of the portfolio of assets is not diminished: the assets must be valued to reflect all their current and future contributions to future welfare; incentives must be provided so that individuals and organizations manage the assets according to these values; depletion or degradation of one asset must be compensated by an increase in the value of another.
Incidence The combined effects of increasing world population and of per capita consumption are putting pressure on the limited resources of the planet and on the limited capacity of man's natural ecosystems for self–regulation and self–regeneration. This pressure will necessarily bring about, in a non–remote foreseeable future, a general readjustment of the relationship between man and his natural ecosystems, at the cost of a catastrophic decrease in world population, due to massive mortality, together with a major degradation of the material and cultural standards of mankind.
Economic development is unsustainable when it increases vulnerability to crises. For example: a drought may force farmers to slaughter animals needed to sustain production in future years; a drop in prices may cause farmers or other producers to over–exploit natural resources to maintain incomes.
Claim To allow maximization of economic growth to be the overwhelming determinant of development is to guarantee that mostly inappropriate development will result. To conceive of development as indiscriminate economic growth is to favour the wealthy, since their fundamental interest lies in maximizing the amount of return on investment without having to be concerned whether capital ought to be invested more appropriately;, whether or not it is profitable.
Broader Unsustainable development (#PB9419) Destruction inherent in development (#PF4829)
Unconstrained socio-economic growth (#PB9015).
Narrower Undirected expansion of economic base (#PF0905)
Lagging transformation of agriculture in developing countries (#PD0946).
Aggravates Starvation (#PB1875) Cultural invasion (#PC2548)
Uncontrolled markets (#PF7880) Inadequate standards of living (#PF0344)
Excessive size of social institutions (#PF8798)
Ecologically unsustainable development (#PC0111)
Endangered tribes and indigenous peoples (#PC0720)
Restriction of indigenous populations to reservations (#PD3305)
Increasing development lag against socio–economic growth (#PC5879).
Aggravated by Traditionalism (#PF2676) Economic uncertainty (#PF5817)
Environmental pollution (#PB1166) Unsustainable population levels (#PB0035)
Long–term shortage of natural resources (#PC4824)
Negative economic repercussions of disarmament (#PF0589).
Reduced by Economic stagnation (#PC0002) Proliferation of technology (#PD2420).

♦ PC0507 Prejudice against animals

Speciesism
Nature Homo sapiens has a tendency to assume itself to be overwhelmingly superior to all other species, having the right to kill, mutilate and enslave the others at will. Speciesism has been defined as the high barrier placed by man between the human species and all the rest of the animal species.
Refs Ryder, Richard D *Animal Revolution* (1990).
Broader Prejudice (#PA2173).
Related Dangerous animals (#PC2321) Poisonous animals (#PE0175).
Aggravates Denial of rights of animals (#PC5456).

♦ PC0519 Non-settled refugees

Refugees in transit
Nature Despite good resettlement opportunities in various countries, the number of non–settled refugees wishing to emigrate is maintained at a high level by the continued influx of new refugees. In addition, it is often difficult to resettle severely handicapped or aged refugees. There are countries where no regular emigration procedure exists and where the number of refugees involved is too small to be handled by selection missions.
Incidence 200,000 refugees from Viet Nam, Laos and Kampuchea remain confined to camps stretching from Indonesia to Japan. In 1982, 49,000 new Indochinese refugees obtained temporary asylum somewhere in South East Asia, only half as many as the preceding year and the lowest number of arrivals since 1977. But the total for whom a durable solution has been found declined even further in 1982: 76,000 were resettled overseas against 168,500 in 1981. After five years of intense activity, resettlement of refugees in the United States is slowing down. In 1983 a total of 60,600 refugees from 20 countries were admitted to the USA, well below the 90,000 ceiling set by the President in consultation with Congress and the lowest number since 1978.
Broader Refugees (#PB0205) Social neglect (#PB0883).
Narrower Non–settled refugees living outside camps (#PE1508).

♦ PC0521 Monopolies

Captive markets — Duopoly
Nature Monopolies are large economic associations, such as cartels, syndicates, trusts, or concerns, which are owned by individuals, groups, or shareholders and which control industries, markets, or entire economies through a high concentration of capital and production. Their aim is to deny other producers the opportunity to compete. Their domination of the economy is the basis for their influence on all spheres of life. A monopoly may be granted by the state for certain purposes or may be acquired through the normal processes of business competition. In present day economics, a monopoly is presumed to exist when one firm has one third of the market, but the situation is considered to be more serious if three or four big firms have over half the market.
Background The increase in the amount and concentration of capital necessary for the formation of a monopoly was in part assured by greater centralization through the merger of independent companies. In the USA the first great wave of monopoly mergers occurred in the 1890's and early 20th century, when giant companies were formed bringing entire industries under their control, such as metallurgy, the petroleum industry, and the auto industry. The second great wave of monopoly mergers in the USA came on the eve of the 1929–33 depression, when monopolies were formed in the aluminium, glass, and other industries. Other forms of monopolization developed in the European capitalist countries, notably, syndicates and cartels. Cartels were also created internationally as a form of international monopoly.
After World War II new forms of monopoly association arose: the conglomerates, which became particularly widespread in the USA. The conglomerates brought together the most diverse industries, having no production connection and not even linked by common raw material or marketing conditions. The formation of conglomerates resulted from the increased concentration in the mid–20th century of scientific research and management. In the conglomerates, capital can flow from one branch to another, bypassing the traditional capital market.
Incidence In Europe big monopolies dictate air fares, international telephone charges and postal tariffs. Because of a lack of competition these, usually state–owned, companies can gat away with unreliable service and poor value for money.
Claim Monopolies can retard progress, in the technical sphere and in living standards, if it threatens profits. A monopoly economy is 'the exact opposite of free competition' (Lenin).
Counter–claim Because of the high concentration of economic resources at their disposal, capitalist monopolies have the potential to accelerate technological progress. A state monopoly of foreign trade is absolutely necessary as a defence against foreign economic and trade expansion. It promotes maximally affective export and import operations and guarantees the independent development of the national economy and the planned character of foreign trade.
Refs Baran, P and Sweezy, P *Monopoly Capital* (1966); Kierzkowski, Henryk *Monopolistic Competition and International Trade* (1984); Nukhovich, E *International Monopolies and the Developing Countries* (1980).
Broader Economic dictatorship (#PC3240).
Narrower State monopoly (#PJ4242) Food monopolies (#PE8018)
Aerospace monopolies (#PE7747) Aluminium monopolies (#PG5191)
Monopoly of the media (#PD3101) Liner shipping cartels (#PE3829)
Industrial gas monopolies (#PE1813) State–monopoly capitalism (#PE7947)
Unfair air transport practices (#PD9163) International monopoly of the media (#PD3040)
Indiscriminate anti–trust prosecution (#PF6386)
Monopoly of the economy by corporations (#PD3003)
Insurance monopolies in capitalist countries (#PE8006)
Legal profession's monopoly of court proceedings (#PE0405)
Monopolistic activity by transnational enterprises (#PE0109).
Related Cartels (#PC2512) Domestic cartel (#PD7963)
Frauds, forgeries and financial crime (#PE5516)
Excessive concentration of business enterprises (#PD0071).
Aggravates Grey lies (#PF3098).
Aggravated by Oligopolies (#PC3825) Collusive tendering (#PE4301)
Usury in developing countries (#PE2524)
Collusive tendering in international trade (#PE7072).

♦ PC0541 Military insecurity and vulnerability

Claim Given today's technological advances which have both made the world smaller and also increased military might for even the smallest nations, it is no longer possible for a nation to "bolt its doors" and keep out foreign trouble makers. Security is no longer a question of larger security forces or deadlier weapons; arms alone cannot provide security. What is needed is a political vision of a world in which nations can live in peace; in which political leaders exchange meaningful dialogue; a world where moral and legal principles govern the conduct of states; and where political ego is not a driving force.
Broader Insecurity (#PA0857) International insecurity (#PB0009)
National insecurity and vulnerability (#PB1149).
Narrower Pacifism (#PF0010)
Disparity in facilities for military mobilization and reinforcement (#PE3995).
Aggravates Unjustified military defence policies (#PF1385).
Aggravated by Total disarmament (#PF8686).

♦ PC0554 Crimes against national security

Offences against government
Broader Crime (#PB0001) Statutory crime (#PC0277).

Narrower Treason (#PD2615) Illegal exports (#PD4116)
Impairing military effectiveness (#PD4448) Aiding national security criminals (#PD7407)
Revisionism and anti–marxist crimes (#PE5546)
Armed crimes against national security (#PD8153)
Crimes related to military service obligations (#PD5941)
Crimes related to national security information (#PE3997).
Related Corruptive crimes (#PD8679).

♦ PC0555 Plant diseases

Nature Plant disease can be simply defined as any deviation from normal; and more scientifically as an injurious physiological process caused by the continued irritation of a primary causal factor, exhibited through abnormal cellular activity and expressed in characteristic pathological conditions called symptoms.

Some diseases can be attributed to inanimate and nonparasitic factors such as adverse environmental conditions. Physical and mechanical damage brought about by violent storms, improper cultivation practices, etc, besides being disastrous in themselves, may prepare the way for widespread infection by other disease agents. Living agents such as insects not only interfere directly with plant metabolism, but also frequently carry other disease agents from plant. A number of economically important disease–causing factors are plants themselves: bacteria and fungi. Algae are responsible for some relatively unimportant tropical plant diseases. Among flowering plants only a few forms are of occasional importance. Several slime moulds, numerous viruses and nematodes are also important agents of plant disease.

The symptoms of plant diseases may be death (necrosis) of all or any part of the plant, loss of turgor (wilt), overgrowths, stunting, or various other changes in the structure of the plant. A rapid death of foliage is often called blight, whereas localized necrosis results in leaf spots. Necrosis of stems or bark results in cankers. Overgrowths composed primarily of undifferentiated cells are called galls, or, less commonly, tumours. Chlorosis (lack of chlorophyll in varying degree) is the most common nonstructural evidence of disease. In leaves it may occur in stripes or in irregular spots (mosaic). These symptoms grade into one another and overlap. Often two or more bacteria, fungi, viruses or a combination of them together attack a plant to produce much greater damage than that resulting from a single agent alone. Furthermore, a plant which is already suffering from a deficient environment is more susceptible to attack by such agents.

Incidence Plant diseases have always caused substantial reductions in the yield of many economic crops and are a continuing threat to all of them. For example, in 1958, 70 per cent of the production costs of a large fruit company in Central America were expended in controlling banana diseases. In addition to the actual losses in production caused by plant diseases, there also have been violent fluctuations in crop yields. When crops were in short supply, prices went up; when large plantings were made to allow for losses that never occurred, there was a superabundance of crop with a corresponding fall in prices and disturbance in economic balance. Worldwide losses from plant diseases have not been estimated but in the USA alone the cost of crop losses due to disease as long ago as 1951–60 has been estimated to be more than 3,250 million dollars annually, or 7 per cent of the potential production.

Background Plant diseases are referred to in some of man's earliest writings, such as Aristotle and the Bible. An epidemic of late blight on potatoes caused the Irish famine of 1845–46. The epidemic continued for several years and starvation occurred in many places in western Europe. Other famines have occurred from time to time because of failures of wheat, rice and other food crops brought about by plant diseases. The extensive plantations of larch trees in Western Europe were destroyed about 1865 by larch canker. Powdery mildew in 1851 and downy mildew in 1878 began devastating epidemics in European vineyards. The epidemic of coffee rust in Ceylon, beginning in the 1870s is sometimes credited with helping to keep the English confirmed tea drinkers; after the rust destroyed the thriving coffee industry new plantations were grown to tea, and South America became the coffee–growing centre. In the United States, the American chestnut was practically wiped out following the introduction of the chestnut blight fungus in 1904, and beach yellows curly top of sugar beets and flaw wilt have been disastrous from time to time.

Claim Plant diseases pre–date agriculture and occur in habitats untouched by man, but their spread is favoured by the growing of a crop. Most pathogens are specialized parasites and can infect only a limited range of plants. The flora in most natural habitats is varied, so adjacent plants are unlikely to be hosts to the same pathogens, but in a crop, individual plants uniformly susceptible to the same parasites are crowded together, creating ideal conditions for rapid spread. With the elimination of variability in crops and the extensive deployment of monocultures, the potential for widespread plant disease is high. The possibility exists and is sometimes occurs that a new uncontrollable pathogen may seriously affect a major food plant. As it is, even without devastating important crops, plant diseases take a heavy annual toll in lost production.

Refs Archer, S A, et al (Ed) *European Handbook of Plant Diseases* (1988); Ayres, P G *Effects of Disease on the Physiology of the Growing Plant* (1982); Beniwal, S P S, et al *An Annotated Bibliography of Pigeonpea Diseases 1906–81* (1985); Bird, Julio and Maramorosch, Karl *Tropical Diseases of Legumes* (1975); Burdon, Jeremy J *Diseases and Plant Population Biology* (1987); CAB International Mycological Institute *Diseases of Banana*; CAB International Mycological Institute *Diseases of Rape*; CAB International Mycological Institute *Diseases of Tropical Forage Legumes and Grassess*; CAB International Mycological Institute *Irrigation and Plant Disease*; CAB International Mycological Institute *Diseases and Cultivation of Clove*; Chattopadhyaya, S B *Diseases of Plants, Yieldings, Drugs, Dyes and Spices* (1969); Dickson, James G *Diseases of Field Crops* (1971); Durbin, R D *Toxins in Plant Disease* (1981); ICRISAT *Coordination of Research on Peanut Stripe Virus* (1988); ICRISAT *Proceedings of the Consultants' Group Discussion on the Resistance to Soilborne Diseases of Legumes, 8–11 Jan 1979, ICRISAT Center, India* (1980); Khera, S S and Sharma, G L *Important Exotic Diseases of Live–Stock Including Poultry* (1967); Maramorosch, Karl and Harris, Kerry *Plant Diseases and Vectors* (1981); Mengistu, A, et al *An Annotated Bibliography of Chickpea Diseases 1915–1976* (1978); Meredith, D S *Banana Leaf Spot Disease (Sigatoka) Caused by Mycosphaerella Musicola Leach* (1970); Merino–Rodriguez, Manuel (Ed) *Lexicon of Plant Pests and Diseases* (1966); Singh *Diseases of Vegetable Crops*; Sivanesan, A and Waller, J M *Sugercane Diseases* (1986); Stevens, Neil E and Stevens, Russell B *Disease in Plants*; Stover, R H *Banana, Plantain and Abaca Diseases* (1972); Stover, R H *Fusarial Wilt (Panama Disease) of Bananas and Other Musa Species* (1962); Van Der Plank, J E *Plant Diseases* (1964).

Broader Disease (#PA6799).
Narrower Plant tumours (#PE5457) Plant pathogens (#PD1866)
Nematoid plant diseases (#PD2228) Uncontrolled plant diseases (#PJ1016).
Related Pests of plants (#PC1627) Hazards to plants (#PD5706)
Crops pests and diseases (#PE7783) Vulnerability of crops to weather (#PE5682).
Aggravates Hunger (#PB0262) Famine (#PB0315)
Destruction of weeds (#PE3987) Pests and diseases of wheat (#PE2222).
Aggravated by Soil infertility (#PD0077) Biological warfare (#PC0195)
Crop vulnerability (#PD0660) Plant disease vectors (#PD3596)
Insect pests of plants (#PD3634) Hazardous waste dumping (#PD1398)
Inadequate crop rotation (#PF3698) Inadequate plant quarantine (#PE0714)
Bird vectors of plant disease (#PD3601) Pest resistance to pesticides (#PD3696)
Plant vectors of plant disease (#PD3599)
Health hazards of air pollution for plants (#PE4744)
Decreasing genetic diversity in cultivated plants (#PC2223).

♦ PC0562 Political prisoners

Political imprisonment

Nature Individuals, whether military or civilian, with political or religious opinions considered undesirable by a government, or individuals engaged in activities considered undesirable, whether they express these violently or peacefully, may be interned without trial, or with a secret trial, or with a public trial distorted to suit propaganda purposes. Evidence may be fabricated and confessions extorted with the use of torture. Political prisoners may have their property confiscated. They may be deprived of nationality but be unable to obtain political asylum from another country. They may be sentenced to exile in an isolated part of the country, forced labour, life or long–term imprisonment, confinement among common criminals, confinement to a psychiatric prison–hospital and subsequent brainwashing or execution. The status and even the existence of a political prisoner may be denied.

Incidence In 1984 Amnesty International reported that political prisoners are being held in the following countries: **Af** Algeria, Angola, Benin, Burkina Faso, Burundi, Cameroon, Cape Verde, Central African Rep, Chad, Comoros, Congo, Djibouti, Egypt, Equatorial Guinea, Ethiopia, Gabon, Gambia, Ghana, Guinea, Guinea–Bissau, Kenya, Lesotho, Libya, Madagascar, Malawi, Mali, Mauritania, Morocco, Mozambique, Namibia, Niger, Nigeria, Rwanda, Sao Tome–Principe, Sierra Leone, Somalia, South Africa, Sudan, Swaziland, Togo, Tunisia, Uganda, Zaire, Zambia, Zimbabwe. **Am** Bolivia, Brazil, Chile, Colombia, Costa Rica, Cuba, El Salvador, Grenada, Guatemala, Haiti, Honduras, Mexico, Nicaragua, Paraguay, Peru, Suriname, USA, Uruguay. **As** Afghanistan, Bahrain, Bangladesh, Brunei, Burma, China, India, Indonesia, Iran, Iraq, Israel, Kampuchea, Korea PDR, Korea Rep, Kuwait, Laos, Lebanon, Malaysia, Nepal, Oman, Pakistan, Philippines, Saudi Arabia, Singapore, Sri Lanka, Syria, Taiwan, Thailand, Viet Nam, Yemen DR. **Eu** Albania, Bulgaria, Czechoslovakia, France, German DR, Greece, Hungary, Ireland, Italy, Poland, Romania, Spain, Switzerland, Turkey, USSR, UK, Yugoslavia.

Publications *Amnesty International Report 1984* Amnesty International (1984).
Broader Political injustice (#PC2181) Political repression (#PC1919).
Narrower Prisoners of conscience (#PC6935) State sanctioned torture (#PD0181)
Political prisoners in mental institutes (#PE4430)
Abusive detention in psychiatric institutions (#PE2932).
Related Internment without trial (#PD1576) Forced disappearance of persons (#PD4259)
Political confiscation of property (#PD3012)
Military political prisoners and detainees (#PD3014)
Inadequate protection of war correspondents (#PE3034).
Aggravates Undemocratic social systems (#PB8031).
Aggravated by Political crime (#PC0350) Political conflict (#PC0368)
Political corruption of the judiciary (#PE0647).
Reduced by Qualified amnesty (#PF3019) Political hostage–taking (#PD1886).

♦ PC0564 Capitalism

Nature Capitalism is an economic and social system, the main attributes of which are: dominance of commodity–money relations and private property in the means of production; developed social division of labour; increasing socialization of production; and transformation of labour power into a commodity. The shaping of the globe into a single, coherent system built on exploitation is the direct consequence of the pursuit of the valorization of capital. The current catastrophic state of the world system is a product of both the exigencies of the valorization of capital and the degrees of resistance to it encountered or engendered in the course of the history of the modern period. Since its inception, the capitalist system has combined the pursuit of valorization with unrestrained geographical expansion. It is this physical, and where necessary militarily aggressive, conquest which has given capitalism its world–wide systemic character.

In its process of development, monopoly capitalism evolves into state monopoly capitalism, which is characterized by the interlocking of the financial oligarchy with the upper echelons of the bureaucracy, the increased role of the state in all spheres of social life, the growth of the state sector in the economy, and increasingly active policies aimed at mitigating the socioeconomic contradictions of capitalism. When associated with imperialism, this engenders a profound crisis in bourgeois democracy, an intensification of reactionary tendencies, and the increased role of force in domestic and foreign policy. It is inseparable from the growth of militarism and military expenditures, the arms race, and tendencies towards the unleashing of aggressive wars.

Advanced capitalism is characterized by contradictions, namely the emergence of conditions, through the very success of capitalism, which are fundamentally antagonistic to capitalism itself, intensify with time, and cannot be resolved within the capitalist framework. The development of capitalism produces changes that call into question the social desirability of the drive for profits. It becomes incompatible with the further development of human potential and capacities. Such contradictions include: (a) Capitalism promises to meet basic needs but is increasingly unable to meet that promise, especially in developing countries. By its very nature is creates unequal development and is unable to institute economic reforms that would coopt burgeoning anti–imperialist struggles for liberation. (b) With continued economic growth, especially in the industrialized world, consumption fades in comparison to other dimensions of well–being, such as the availability of creative and socially useful work and meaningful individual development. But because capitalism must continually expand, the realization of these needs is incompatible with capitalist relations of production. (c) Capitalist economic growth becomes increasingly predicated on irrationality and production of waste (e.g. military expenditure and planned obsolescence of consumer goods), thus exhausting natural resources and threatening the ecological balance. (d) The expansion of capitalist production draws an ever–increasing share of the population within a country into alienating wage and salary work, sensitizing them to the oppressive conditions under which they are being called upon to function. (e) The internationalization of capitalism creates a corresponding world–wide proletariat which becomes progressively sensitized to the way in which the capitalist mode reinforces exploitative relations between countries. (f) Through its increasing need for a more educated labour force, workers become increasingly capable of grasping the essential irrationality of the system, the inequitable distribution of power within it and the associated social division of labour.

Incidence The capitalist world–system is not confined to those parts of the globe which can formally be designated as the capitalist countries. Some observers see the existing socialist countries as part of the capitalist world–system to the extent either that their internal organization follows capitalist criteria of efficiency and/or they participate in the world economy on terms set by capitalist competition and the law of value. In this sense they are not only victims of those ominous military, ecological and social developments and threats propelled by the dictates of capitalist accumulation, but they also bear their due proportion of responsibility for hampering social transformation. The deterioration of the conditions under which the mass of the population has to live in many areas of the world, the increase in conflicts between and within countries, the intensification of the world economic crisis, are demonstrating that unequal and uneven development, with all the forms of immiseration and alienation which it creates, cannot be surmounted within the capitalist world–system. The extent of the problem has been highlighted within the USA itself by a report from a conference of Catholic bishops stating that the distribution of income and wealth in the USA is so inequitable that it violates minimum standards of distributive justice. In 1982 the richest 20% of Americans received more income than the bottom 70% combined.

Claim Capitalism emphasizes limitless increases in consumption and waste, disregard for the poor, and indifference to any concept of appropriate development. The capitalist system is under severe attack from victims of inequality, alienation, racism, sexism, irrationality, and imperialism, who are struggling to free themselves from oppression and are learning that capitalism is one of their main enemies. The very existence of such challenges proves that capitalism is neither a smoothly operating system nor a system unsusceptible to change. The histories of capitalist systems give numerous instances of resistance from those whom capitalism has sought to subordinate. Often this resistance has been overcome only through the use of violence and coercion. The apparent vitality of capitalist economies is illusory in that it obscures recognition of the progressive devitalization of those sectors and countries on which that vitality depends for its resources.

Counter-claim 1. Capitalism is intended to serve the unique goals and needs of individuals. Throughout the history of capitalism, that individual goal has usually been upward social mobility. Essential to the achievement of all the different personal goals is the individual freedom that capitalism provides in greater measure than any other system of economic organization. 2. Capitalism is an exceedingly broad and somewhat vague term, covering societies as variously organized as Sweden, UK, Japan, France and the USA, in each of which the mixture of public and private enterprise, the legal rules governing the pursuit of profit, the approved market structure, the permitted accumulation of income and wealth, differs significantly from all the others. 3. The unjust nature of the global economy and the inappropriateness of development can be explained by the exploitative nature of exchange, market forces and distributive effects between nations and classes. However, these effects are only incidentally due to capitalism. They occurred under other imperial systems before capitalism emerged, and even if all nations were suddenly to become socialist they would still participate in a global market system with the very same distributive effects that currently produce underdevelopment.

Refs Amin, Samir *Unequal Development* (1977); Beaud, Michel *History of Capitalism Fifteen Hundred to Nineteen Eighty*; Brailsford, Henry N and Strauss, S *Why Capitalism Means War* (1972); Daems, H and Wee, M van der (Eds) *Rise of Managerial Capitalism* (1974); Deleuze, Gilles and Guattari, Felix *A Thousand Plateaus* (1987); Gilbert, Neil *Capitalism and the Welfare State* (1985); Gorizontov, B *Capitalism and the Ecological Crisis*; Hobson, John A *The Origins, Growth and Potential Survival of Modern Capitalism* (1985); Itoh, Makoto *The Basic Theory of Capitalism* (1987); Kostopoulos, Tryphon *The Decline of the Market* (1987); Kostopoulos, Tryphon *Decline of the Market* (1987); Kunio, Yoshihara *Rise of Ersatz Capitalism in South-East Asia* (1988); Lash, Scott and Urry, John *The End of Organized Capitalism* (1987); Messerschmidt, James *Capitalism, Patriarchy and Crime* (1986); Miles, Robert *Capitalism and Unfree Labour* (1987); Nevile, John W *Root of All Evil*; Preobrazhensky, E A and Day, Richard B *The Decline of Capitalism* (1985); Roman, Stephen and Eugen, Loebl *Alternative to Communism and Capitalism* (1979); Sixel, Friedrich W *Crisis and Critique* (1987); Wallerstein, I *The Capitalist World Economy* (1979); Wallich, Henry C *The Cost of Freedom* (1979).

Broader Ideological conflict (#PF3388).
Narrower Managerism (#PF8087)
Bourgeoisie (#PE7774)
Inadequate savings (#PC0927)
Crisis of capitalism (#PF7999)
Investment capitalism (#PJ5060)
Capitalist speculation (#PC2194)
State–monopoly capitalism (#PE7947)
Military–industrial complex (#PC1952)
Speculative flight of capital (#PC1453)
Poverty in capitalist systems (#PC3107)
Short–term profit maximization (#PF2174)
Capitalist economic imperialism (#PC3166)
Alienation in capitalist systems (#PD3112)
Capitalist political imperialism (#PC3193)
Non–productive capitalist elites (#PE3816)
Privatization of public services (#PE3391)
Uncertain status of monetary gold (#PF2342)
Competition in capitalist systems (#PC3125)
Imperialistic distribution system (#PD7374)
Exploitation in capitalist systems (#PC3117)
Contradictions in capitalist systems (#PF3118)
Boundary constraints on land planning (#PF0954)
Resource wastage in capitalist systems (#PC3108)
Tax obstacles to international investment (#PD0673)
Lack of individualism in capitalist systems (#PD3106)
General unproductivity of capitalist systems (#PF3103)
Denial of human rights in capitalist systems (#PC3124)
Abuse of science and technology in capitalism (#PE3105)
Internationalization of capitalist production (#PE7957)
Negative economic repercussions of disarmament (#PF0589)
Waste of human resources in capitalist systems (#PC3113)
Conflicting roles of commodities in capitalism (#PF3115)
Conflicting roles of money in capitalist systems (#PF3114)
Excessive demand for goods in capitalist systems (#PC3116)
Denial of the right to work in capitalist systems (#PC3119)
Abusive technological development under capitalism (#PD7463)
Inadequate domestic savings in developing countries (#PD0465)
Contradictions of capitalism in developing countries (#PF3126)
Dangers of private control of communications mass media (#PF2573)
Capitalist subversion in communist and neutral countries (#PE8349)
Denial of the right to social security in capitalist systems (#PD3120)
Restriction of educational opportunities in capitalist systems (#PD3122)
Inadequate incentives for increased productivity in developing countries (#PE8506)
Denial of effective national self–determination by capitalist exploitation (#PE3123)
Excessive foreign investment in traditional industries of developing countries (#PD0765).
Related Fascism (#PF0248) Socialism (#PC0115) Communism (#PC0369)
Capital shortage in developing countries (#PD3137).
Aggravates Anxiety (#PA1635) Banditry (#PD2609) Grey lies (#PF3098)
Anti–capitalism (#PF3110) Social injustice (#PC0797) Absentee ownership (#PD2338)
Interpersonal rivalry (#PD7617) Prohibitive labour costs (#PF8763)
Extra–economic constraints (#PE7784) Vulnerability of middle–class (#PC1002)
Usury in developing countries (#PE2524) Interception of communications (#PD7608)
Prohibitive cost of accommodation (#PD1842) Politically emotive words and terms (#PF3128)
Financial destabilization of world trade (#PC7873)
Dependence of least developed countries on foreign aid (#PE8116)
Abuse of dominant market position in international trade (#PD6002).
Aggravated by Destiny (#PF3111) Double standards in morality (#PF5225).
Reduces Underdevelopment (#PB0206).

♦ PC0569 Tariff barriers to international trade

Nature Virtually all of the developed countries, while permitting raw materials to enter free or at low rates of tariff, apply tariffs on processed products which as a rule are progressively higher the more highly processed is the imported product. This naturally tends to inhibit exports of processed products from developing countries, and favours the expansion of trade in raw or relatively less processed forms.

The matter of tariff barriers does not, however, end there. In the first place, as is being increasingly recognized, the nominal duty rates do not in themselves express the degree of protection accorded. That protection is expressed by what is called the effective or implicit rate of tariff, which takes account not only of the tariffs paid on the final product, but also of the value added in processing and of the duties that may be paid on materials used in the process of production. The effective tariff rate, which thus takes account of the whole tariff structure of the importing country, rises with increasing duties levied on the product of the manufacturing process and with decreasing values added. It declines with increasing duties levied on the materials used in the manufacturing process. It is negative if those duties exceed the duties levied on the product itself. It follows that the differences between the duty rates applied to imports of raw and processed products do not necessarily give a full picture of the tariff protection actually accorded to agricultural (or other) processing industries.

Further, even the level of the effective tariff rate is not a complete measure of the degree of protection it affords. Another factor is the supply and demand elasticities for the product in the exporting and importing countries. If both are high, and assuming that the change in tariff results in price changes, even a relatively small decrease in the effective tariff rate may lead to substantial increases in imports, and vice versa. Clearly these elasticities will vary not only from product to product but also from country to country, so that even identical changes in tariff structures will not always lead to the same changes in trade.

Finally, the effect of tariff changes on trade will also depend on the comparative advantage of the exporting and importing countries in producing the goods in question. The effect of tariffs is restrictive if they offset the comparative advantage of the exporting country. Conversely they are of little or no effect when levied on products for which the importing country in any case has a comparative advantage.

The last twenty years has seen a continuation of the general liberalization of tariff barriers undertaken in the context of the GATT. As a result of the Tokyo Round, the cuts in tariff rates on industrial products will be around 30 percent by 1987. While the major cuts were concentrated on industrial items of interest to the countries taking an active part in the negotiations, the principle of unilateral concessions to the developing countries was accepted. After years of negotiation, starting from the first session of the United Nations Conference on Trade and Development, a number of preferential schemes in favour of imports of manufactures on semi–manufactures from the developing countries was adopted by the end of 1972. The schemes have not, however, become as "generalized" as was initially hoped. Moreover, although most of the developing countries are covered by the preferential schemes, the lists of beneficiaries under the schemes are by no means uniform. For the European Economic Community the list varies according to product groups. In connection with its proposed scheme, the United States has pressed for the phasing out of the reverse preferences accorded by some developing countries.

As far as product coverage is concerned, while industrial products are included (apart from certain specifically defined exceptions or "negative lists"), agricultural commodities are included only if they appear on an explicitly selected positive list. The lists of exceptions for industrial products cover mostly "sensitive items", notably textiles, hydrocarbons, leather and leather products. In some cases, that of the European Economic Community for example, the product exceptions apply to particular country groups: in the case of cotton textiles, preference is limited to signatories of the Long–Term Textile Agreement; and, in the case of textiles and foot–wear, dependent territories are excluded.

Largely as a result of differences in the coverage of the negative lists, the share of dutiable industrial imports receiving preferences ranges from 100 per cent in the case of Denmark to 20 per cent in the case of Japan. For a group of 13 developed market economies, the average share is about one third. For agricultural products, the coverage under preferential schemes is generally smaller than for industrial products. For the same group of 13 countries, less than 5 per cent of dutiable agricultural imports is included, and wide country variations are observable.

In addition to the specific exclusion of sensitive commodities, there are safeguard mechanisms and escape clauses. The European Economic Community and Japan impose ceilings on all preferential entries. For several "sensitive" products, the EEC has established tariff quotas, while for 57 products, Japan has reduced tariffs by 50 per cent. The annual ceilings are determined by the volume of imports from beneficiary countries in a reference year (basic quotas), plus a fixed percentage (5 per cent in the case of EEC and 10 per cent in the case of Japan) of the value of imports from non–beneficiary sources (supplementing variable quotas). Preferential treatment for any group of products can be suspended for all countries if the preferential entry ceiling is reached and for any individual developing country if imports from it exceed 50 per cent (or sometimes less) of the ceiling. Community tariff quotas for according to a formula with no provision for reallocation. Preferences are automatically suspended for imports exceeding either the community quota or member country subquotas. Although ceilings and tariff quotas provide for moderate growth, a substantial proportion of imports from beneficiary countries may be subject to non preferential treatment if import flows exceed past performance.

Both the EEC and Japan also retain an escape clause with respect to agricultural products. The escape clauses retained by other developed countries would allow them to reintroduce tariffs if preferential imports caused or threatened to cause market disruption or serious injury. The United Kingdom would except Commonwealth countries from its safeguard measures, while the Nordic countries would resort to bilateral consultations with the developing country concerned before imposing them. Austria has adopted an escape clause which prohibits any safeguard acting before annual import growth surpasses a fixed percentage. Apart from the element of uncertainty introduced by the escape clauses and safeguard measures, the effectiveness of the preferential schemes is likely to be gradually diluted unless the schemes are progressively liberalized. It is precisely those lines in which the developed market economies find their competitive power weakening and the developing countries have gained a comparative advantage that constitute the "sensitive" items. Furthermore, ceilings calculated on the basis of past performance usually underestimate future capabilities especially for late–comers.

Claim It remains true that, even though in most cases the effective protection of agricultural processing in developed countries is not known, a reduction in the nominal tariff rate would of necessity result in a decrease in the effective protection. Given the urgent need of the developing countries to expand their exports as well as to develop their industries, the case therefore remains strong for a speedy reduction or elimination of the nominal duty rates on processed agricultural and other products imported from these countries. In the longer run, however, the greatest economic efficiency would be achieved by concentrating the effort to liberalize tariffs on products for which the demand and supply elasticities are such that the lowering or elimination of tariff protection is most likely to result in increased export earnings; and the competitive advantage lies on the side of the developing countries. With regard to the latter two points, and particularly competitive advantage, it is important to take a dynamic view, by making allowance for changes both in supply elasticities and in comparative advantage as the exporting countries develop their infrastructure and their industrial base.

Broader Distortion in international trade (#PC6761).
Narrower International trade barriers for primary commodities (#PD0057)
Tariff barriers to trade between developing and developed countries (#PC2369).
Aggravates Excessive external trade deficits (#PC1100).

♦ PC0578 Religious repression

Dependence on religious repression

Nature Religious repression includes confiscation of church property, banning and exile of religious orders, trial, imprisonment and execution of the clergy, and the banning and burning of religious books. It also includes persecution of adherents to the religion in question, general religious discrimination, religious war and mass murder.

Incidence Religious repression exists overtly in countries where ideology is contrary to religion (for example, in communist countries) and less overtly in countries where one religion strongly predominates and, by influence over the law and other social and political institutions, excludes the place in society of others.

Background Religious repression has existed from very early history. The Spanish inquisition is the best known. The religious inquisition of the Edo period (1600–1868) was designed to eradicate Christianity throughout Japan. Under it, the population was screened for the presence of the

PC0578

religion's missionaries and believers. Persons discovered to be Christian were forced to apostatize, the recalcitrant were subjected to psychological and physical tortures until they recanted, and those who refused to abandon their faith were executed. One of the most horrifying examples in the 20th century was the Nazi treatment of the Jews.

Broader Repression (#PB0871).
Narrower Iconoclasm (#PF4923) Religious war (#PC2371)
Religious intimidation (#PC2937) Persecution of religious sects (#PF3353)
Denial of religion in communist systems (#PC3175)
Segregation based on religious affiliation (#PC3365)
Restrictive influence of religion on the masses (#PF3361).
Related Racial repression (#PD8762) Denial of religious liberty (#PD8445).
Aggravates Social conflict (#PC0137) Anti–clericalism (#PF3360)
Religious conflict (#PC3292) Religious extremism (#PF4954)
Ideological conflict (#PF3388) Social fragmentation (#PF1324)
Religious intolerance (#PC1808) Religious and political antagonism (#PC0030).
Aggravated by Monasticism (#PF2188) Religious rivalry (#PC3355)
Religious discrimination (#PC1455) Fragmentation of religious belief (#PF3404)
Religious discrimination in the administration of justice (#PE0168).

♦ PC0580 Instability of manufacturing industries

Broader Instability of economic and industrial production activities (#PC1217).
Narrower Instability of woodworking industries (#PE0681)
Instability of basic metal industries (#PE2601)
Instability of paper and printing industries (#PE1927)
Instability of textile and clothing industries (#PE1008)
Underdevelopment of food processing industries (#PD0908)
Instability of machinery and equipment industries (#PE1852)
Instability of chemical and petrochemical industry (#PE0538)
Instability of non–metallic mineral products industry (#PE2599)
Instability of trade in miscellaneous manufactured articles (#PE0814)
Instability of manufacturing industries in developing countries (#PD9117).
Related Instability of trade in manufactured goods (#PE0882).

♦ PC0582 Discrimination against minorities

Underprivileged ethnic minorities

Nature Discrimination against minorities exists on racial, religious, linguistic, ideological, political or economic grounds and may take place in education, employment, housing and public services. Minority groups may be barred from certain schools or segregated in their own, by practice or by choice, which may be less adapted for conditions in the society at large. Their educational level, prejudice and fear of certain ideologies work against minority groups in recruitment for jobs, thus barring them from obtaining adequate housing, nourishment, clothing, etc. Minorities may be prevented from taking part in certain activities by law; in the case of religious sects, their services may be banned. Small nations may suffer from foreign debt problems and be discriminated against by donor or investing countries on the strength of their political ideology and economic system. Small island states and territories may be dominated by an outside power militarily, politically or economically.

Refs Ekstrand, L H (Ed) *Ethnic Minorities and Immigrants in a Cross–Cultural Perspective* (1986); Minority Rights Group Ltd *World Minorities*; Rothermund, Dietmar and Simon, John *Education and the Integration of Ethnic Minorities* (1986); Sen, Dhirani *Problem of Minorities* (1940); Simpson, George E and Yinger, Milton *Racial and Cultural Minorities*.

Broader Discrimination (#PA0833) Social injustice (#PC0797).
Narrower Discrimination against giants (#PE5578)
Denial of education to minorities (#PC3459)
Discrimination against dwarfs and midgets (#PE2635)
Discrimination against minority languages (#PD5078)
Violation of the rights of sexual minorities (#PD1914).
Related Economic discrimination (#PC2157) Dictatorship of the majority (#PD3239)
Threats to ideological movements and minorities (#PC3362).
Aggravates Genocide (#PC1056) Ethnocide (#PC1328)
Segregation (#PC0031) Exploitation (#PB3200)
Reservations (#PJ2562) Social conflict (#PC0137)
Ethnic conflict (#PC3685) Gypsy persecution (#PE1281)
Police intimidation (#PD0736) Forced assimilation (#PC3293)
Social fragmentation (#PF1324) Lack of social mobility (#PF2195)
Inequality before the law (#PC1268) Underprivileged minorities (#PC3424)
Persecution of religious sects (#PF3353) Underprivileged racial minorities (#PC0805)
Underprivileged religious minorities (#PC2129) Threatened and vulnerable minorities (#PC3295)
Underprivileged linguistic minorities (#PC3324)
Neglect of remote regions and islands (#PE5760)
Underprivileged ideological minorities (#PC3325)
Fragmentation within organized religions (#PF3364)
Segregation of poor and minority population in urban ghettos (#PD1260).
Aggravated by Colonialism (#PC0798) Dictatorship (#PC1049)
Homogenization of cultures (#PB1071) Refusal to grant nationality (#PF2657)
Prejudice against minorities (#PC8494) Religious conflict between sects (#PC3363)
Undemocratic political organization (#PC1015)
Lack of appreciation of cultural differences (#PF2679).

♦ PC0588 Non–participation

Decline in civic participation — Abstention from social processes — Disinterest in social processes

Nature Declining participation of citizens in civic activities has been attributed to: rising anti–governmental populism, long–term weakening of political parties, wholesale failures of government over an extended period of time, segmentation of the population into demographic ghettos by marketing and technology forces; loss of faith in the future and general loss of confidence in national goals; or to a compensating resurgence of activity at the local and community levels.

Incidence Voting, census returns and taxation are among the few venues for citizens to participate in public life at the national level. In the USA in 1989–90, 110–120 million Americans failed to vote, namely two–thirds of the electorate. 33 million failed to return their census forms, namely 37 per cent. For every $5 of federal tax, $1 is evaded, leading to a tax gap of $100 billion.

Narrower Lack of community participation (#PF3307)
Lack of participation in local welfare programmes (#PE8503)
Lack of participation in attempting to solve intergroup conflicts (#PE8357)
Inadequate international participation by island developing countries (#PE5761).
Related Political apathy (#PC1917)
Structural failure of citizen participation (#PF2347).
Aggravates Ineffectiveness of individual participation in large communities (#PF6127).
Aggravated by Blocked parental participation (#PJ5308)
Social disaffection of the young (#PD1544)
Sociological ignorance of citizen participation (#PF2440)
Ineffective structures for community participation (#PF2437).

♦ PC0592 Labour shortage

Manpower shortage — Inadequate labour force — Daytime manpower shortage

Broader Shortage (#PB8238) Underproductivity (#PF1107)
Unproductive utilization of plantation space (#PF6455).
Narrower Shortage of skilled labour (#PD0044) Shortage of technical skills (#PF6500)
Shortage of women instructors (#PG9482) Shortage of domestic servants (#PJ9711)
Shortage of military manpower (#PE4920) Understaffing of basic facilities (#PD9306)
Shortage of resident professionals (#PG8812)
Labour shortage in developing countries (#PD5045)
Lack of skilled workers in island developing countries (#PF5737)
Shortage of adequately trained personnel to act against problems (#PF0559)
Inadequate supply of appropriate trained manpower in developing countries (#PE6243)
Lack of skilled workers in the transport sectors of land–locked developing countries (#PE5884).
Related Lack of skilled manpower in rural areas of developing countries (#PE5170).
Aggravates Problems of migrant labour (#PC0180)
Exploitation of child labour (#PD0164).
Aggravated by Labour hoarding (#PE6333) Inadequate manpower planning (#PJ2036)
Unemployment in developing countries (#PD0176)
Lack of local leadership role models (#PF6479).
Reduced by Economic and social losses due to disability (#PE4856).

♦ PC0600 Isolation of parent–child relationship

Nature In the modern family, especially in industrialized countries, children are set apart from the world of adults; their parents are virtually their only adult contacts. Children are thus isolated from the mainstream of life and as a result there is uneasiness among young people. They may vent their frustrations on their parents or on society, or they may suffer from a lack of self–confidence.

Broader Social isolation (#PC1707) Negative effects of the nuclear family (#PF0129).
Narrower Generation communication gap (#PF0756).
Related Alienation (#PA3545) Loneliness (#PF2386) Frustration (#PA2252)
Dependency of children (#PD2476).
Aggravates Juvenile delinquency (#PC0212) Stress in human beings (#PC1648)
Lack of self–confidence (#PF0879)
Inadequate sense of personal identity (#PF1934).
Aggravated by Compulsory education (#PJ2615).

♦ PC0604 Endangered species of reptiles

Reptilia

Nature Due primarily to the impact of man on the natural environment, whether directly or indirectly, many of the 5,000 species of reptile are in danger of extinction.

Broader Endangered species of animals (#PC1713).
Related Endangered species of insects (#PC2326).

♦ PC0632 Denial of political and civil rights

Broader Denial of human rights (#PB3121) Violation of civil rights (#PC5285).
Narrower Denial of right to life (#PD4234) Denial of right to liberty (#PF0705)
Denial of the right to die (#PF4813) Denial of right to dignity (#PE6623)
Denial of right to security (#PD7212) Denial of the right of association (#PD3224)
Denial of right to resist oppression (#PE6949)
Denial of right to develop as human beings (#PF2364)
Restrictions on recognition of nationality (#PE4912)
Denial of right to participate in government (#PE6086)
Denial to animals of legal protection of their rights (#PE8643).
Related Travel restrictions (#PC8452) Infringement of privacy (#PB0284)
Denial of right to justice (#PC6162) Denial of right of assembly (#PC2383).
Aggravates Political repression (#PC1919).

♦ PC0646 Industrial accidents

Occupational accidents — Employment injuries — Work injuries — Accidents at work — Occupational injuries

Nature An industrial accident is one which takes place while work is being carried out; there are millions of industrial accidents throughout the world every year. Some of them are fatal, others result in permanent disablement, either complete or partial. The majority cause only temporary disablement, which, however, may last for several months. The contraction of certain types of disease, the presence of which may not be detected until long after the causative occupation has been left, is another form of industrial "accident". In addition to a loss of time and money, all accidents cause suffering to the victim and the family.

Incidence The statistics which exist on this matter in many industrialized countries are limited to a single category of accident, namely those accidents serious enough to require more than first–aid treatment and which therefore generally result in absence from work, prolonged invalidity or death. In many countries, these health risks are covered by social insurance schemes whose operation generates as a by–product the existing statistics. Industrial accidents which do not result in time off work are therefore virtually always missing from the statistical picture, even though safety experts and theoreticians consider such "minor" accidents to be at least as important as more serious accidents for understanding and knowledge of unsafe situations and behaviour.
The extent of industrial accidents and injuries varies widely over countries, industries and time periods, as also do the legislative provisions for the regulation of industrial processes and the installation of safety devices.
Every year in the United Kingdom about 1,000 people are killed at their work. Half a million workers suffer various injuries, and 23 million working days are lost annually because of industrial injury and disease. In the United States the frequency rate for disabling injuries (the number of reportable cases per million work–hours) rose from a low of 5.99 for 1961 to 10.87 for 1976, representing a huge increase of 81 percent. It was also estimated that accidents cost the nation US$ 51,100 million in lost wages, medical expenses, damage to property and administration costs. In 1976 alone, a million productive work–years were lost through accidents at work. These figures mean that injuries to workers would have had the same economic effect on the United States as if the whole of the nation's industry had been closed for one full week.
Today some countries (Japan, United States) regularly report over 2 million occupational accidents a year, and others (France, Federal Republic of Germany, Italy) over a million. The rate of fatal accidents in the developing countries has doubled or even trebled. Many countries, including some of the largest or most highly industrialized, still do not publish any figures, but it is reasonable to assume that over 15 million occupational accidents occur throughout the world every year.

Claim Every year more than 180,00 workers die and 110 million are injured as a result of occupational accidents.

Refs Branch, Turner W *Construction Accident Pleading and Practice* (1988); Glendon, A, et al *Bibliographical Review of Data Sources on Occupational Accidents and Diseases* (1986); ILO *The Cost of Occupational Accidents and Diseases* (1986); ILO *Occupational Injuries* (1982); ILO *Accident Prevention* (1986); International Labour Conference, 67th Session, *Safety and Health and the Working Environment, Report VI* (1981); Kletz, T *Learning from Accidents in Industry* (1988); Perron, Charles *Normal Accidents* (1985); Wang, Charleston C *How to Manage Workplace Derived Hazards and Avoid Liability* (1988).

Broader Injurious accidents (#PB0731) Occupational risk to health (#PC0865).
Related Occupational diseases (#PD0215).
Aggravates Injuries (#PB0855) Absenteeism (#PD1634) Disabled workers (#PD4673)
Traumatic injuries (#PE6874) Radioactive contamination (#PC0229).

Aggravated by Job fatigue (#PD8052)
Dangerous substances (#PC6913)
Seasonal unemployment (#PC1108)
Excessive hours of work (#PD0140)
Inadequate personal hygiene (#PD2459)
Inadequate safety legislation (#PJ2716)
Large-scale industrial accidents (#PD2570)
Drunkenness (#PE8311)
Dangerous occupations (#PC1640)
Unsafe port facilities (#PE4897)
Industries in difficulty (#PD5350)
Suppression of safety records (#PF2714)
Alcoholic intoxication at work (#PE2033)
Health hazards of exposure to noise (#PC0268)
Health risks to workers in commerce (#PE0688)
Inadequate training in decision-making (#PD2036)
Health risks to workers in service industries (#PE0875)
Health risks to workers in construction industry (#PE0526)
Health risks to workers in manufacturing industries (#PE1605)
Contempt for agricultural labour in developing countries (#PD1965)
Health risks to workers in agricultural and livestock production (#PE0524)
Health risks to workers in electricity, gas, water and sanitary services (#PE1159)
Health risks to workers in transport, storage and communication industries (#PE1581).
Reduced by Declining economic productivity (#PC8908).

♦ PC0655 Unequal political representation
Dependence on unequal political representation
Nature Unjust methods of achieving political representation lead to unequal advantage in politics. Such methods include: voting qualification requirements; unequal distribution of legislative seats; unfair choice of political candidates; the banning of opposition parties or certain parties held to be subversive; unjust polling procedure; unjust electoral campaigns; and the influencing of voters, including intimidation, bribery and corruption.
Broader Lack of representation (#PF3468)
Obstacles to community achievement (#PF7118).
Narrower Unjust electoral campaigns (#PD2919)
Unequal parliamentary constituencies (#PD2167)
Bias in selection of political candidates (#PD2931).
Aggravates Political conflict (#PC0368).
Aggravated by Elitism (#PA1387)
Political injustice (#PC2181)
Political inequality (#PC3425)
Limitations on right to vote (#PF2904).

♦ PC0663 Denial of social rights
Lack of social liberty
Broader Violation of human rights (#PB3860).
Narrower Denial of right to a family (#PE7267)
Denial of the right of health (#PJ2269)
Denial of right to extended family (#PE5241)
Deterioration of human environment (#PC8943)
Denial of right to social security (#PD7251).
Related Social neglect (#PB0883).

♦ PC0666 Inadequate standardization of procedures and equipment
Nature Behind all efforts made to remove trade barriers and obstacles and behind every action aimed at the creation of large markets and the achievement of economic integration, stands the basic factor of standardization. The standardization of terms, definitions and units of measurement, apart from its importance for the exchange of culture and scientific knowledge, plays an important role in facilitating trade, for it is impossible to conclude commercial transactions unless the terms used are standardized.
The standardization of technical symbols facilitates the understanding of industrial blue prints and designs. The standardization of sampling, testing and experimentation procedure further facilitates the process of delivery, receipt and arbitration in the exchange of commodities and products. The standardization of materials, ingredients and products assists in overcoming technical barriers and facilitates free movement across frontiers.
Advance towards international standardization is normally slow, although in some industries, such as chemicals, rapid progress may be expected. In other industries, such as those producing capital goods, international standardization may take longer than twenty or thirty years. The situation varies from industry to industry and from country to country.
The task of adjusting national standards to conform with international recommendations is a complicated one. First, adoption of international standards sometimes implies a large capital outlay for the necessary change-over in equipment, and intensifies the natural reluctance of producers who have built their plants and equipment on the basis of national standards. Second, the advantages of international standardization are found largely in the field of international trade, and the incentive to change is therefore often weak in industries and in countries whose domestic market constitutes the predominant element in total production.
One of the most important factors inhibiting the effort towards international standardization is the presence of two major systems of measurement: the metric system or Système International (SI) and the foot-pound system. Although metrication with its base-ten has proven its utility, the base-twelve foot-pound system is not alone as a relic of another age. The twelve month year, the seven day week, the twenty-four hour day, the sixty-minute hour and the sixty-second minute remain to be metricized into a scientific standard of time measurement.
Incidence The International Organization for Standardization (ISO) and the International Telegraph and Telephone Consultative Committee (CCITT) conflict in authority for standards applying to computer telecommunications equipment. At the same time, standards may also be proposed by the Institute of Electrical and Electronic Engineers, or by computer manufacturing trade associations in the US and Europe. The standards debate in the computer-telecommunications area has delayed the development of electronic mail, local area and metropolitan area networks.
Broader Inadequacy of international standards (#PF5072)
General obstacles to problem alleviation (#PF0631).
Narrower Incompatibility of transport modes (#PF2403)
Parochial telecommunications standards (#PE1840)
Non-uniformity in marking navigable waters (#PE6749).
Related Discriminatory imposition of standards (#PD5229)
Terminological confusion in weights, measures and numbering systems (#PF5670).
Aggravates Telephone delays (#PF1698)
Fragmentation of technological development (#PC1227).
Aggravated by Over-diversification of manufactured goods (#PD4907).

♦ PC0690 Civil disobedience
Resistance to government — Mass protests — Civil resistance
Nature Civil disobedience is any act of public defiance of a law, practice or policy of authorities, if that act is premeditated, known by the actor to be illegal, and done for public reasons. Civil disobedience may be direct as was the case of Henry David Thoreau's refusal to pay his poll tax. It may be indirect, such as, burning draft cards in protest against the war in Vietnam.
Refs Boyle, Francis A *Defending Civil Resistance under International Law* (1987); Russell, D E *Rebellion, Revolution, and Armed Force* (1974); Woodcock, George *Civil Disobedience.*
Broader Domination (#PA0839) Social injustice (#PC0797).
Narrower Draft evasion (#PD0356)
Demonstrations (#PD8522)
Passive resistance (#PF2788).
Related Citizen disobedience (#PD5707).
Aggravates Inadequate social reform (#PF0677).
Aggravated by Pacifism (#PF0010)
Nuclear war (#PC0842)
Unjust laws (#PC7112)
Authoritarianism (#PB1638)
Denial of right to liberty (#PF0705)
Illegitimate political regimes (#PC1461)
Undemocratic political organization (#PC1015).

♦ PC0698 Military government
Military dictatorship — Military regimes
Nature Military governments tend to have little respect for freedom nor for many human rights. Military strength involves laying plans for possible wars, and this necessarily leads to viewing other countries or internal factions as possible enemies. The result may be an oversimplified, black-and-white view of the world, divided into friends and enemies. In a country in which the ultimate control of policy rests with a civil government, these trends can be balanced, although an intense military preparedness tends to give the military men more influence on policy; a military government might not be subject to this restraining influence. When its position is insecure, it may attempt to strengthen it by exaggerating the threat from possible enemies, and if carried far enough, this trend may turn a country into an armed camp.
Incidence Typically the military seizes power in the midst of a crisis and attempts to restore order and rationalize the economic system. Whatever promises are made to the population, there is a tendency for the military to retain their hold over the political institutions. Examples include: Argentina (1966 and 1976), Bolivia (1971), Brazil (1964), Chile (mid-1973), Indonesia (1966), Turkey (1971 and 1980), and Uruguay (mid 1970s). Of the 45 black African nations, in 1990 23 were military dictatorships where no political parties were permitted.
Refs Wolpin, Miles D *Military Radicalism in the Middle-East and in the Mediterranean.*
Broader Authoritarian regimes (#PC9585)
Denial of right to national self-determination (#PC1450).
Aggravates Denial of human rights (#PB3121).
Aggravated by Militarism (#PC2169)
Military aid (#PE6052)
Competitive acquisition of arms (#PC1258).

♦ PC0699 Human disability
Disabled persons — Physically and mentally handicapped persons — Health impairment and handicap — Disablement
Nature Impairment is any loss or abnormality of psychological, physiological, or anatomical structure or function. Disability is any restriction or lack (resulting from an impairment) of ability to perform an activity in the manner or within the range considered normal for a human being. Handicap is a disadvantage for a given individual, resulting from an impairment or disability, that limits or prevents the fulfilment of a role that is normal, depending on age, sex, social and cultural factors, for that individual.
Disabled people do not form a homogeneous group. For example, the mentally ill and the mentally retarded, the visually, hearing and speech impaired, those with restricted mobility or with so-called 'medical disabilities': all encounter different barriers, of different kinds, which have to be overcome in different ways. Handicap is therefore a function of the relationship between disabled persons and their environment. It occurs when they encounter cultural, physical or social barriers which prevent their access to the various systems of society that are available to other citizens. Thus, handicap is the loss or limitation of opportunities to take part in the life of the community on an equal level with others.
The causes of handicaps vary throughout the world, as do the prevalence and consequences of disability. These variations are the result of different socio-economic circumstances and of the different provisions that each society makes for the well-being of its members. A survey carried out by experts has produced the estimate of at least 350 million disabled persons living in areas where the services needed to assist them in overcoming their limitations are not available. To a large extent, disabled persons are exposed to physical, cultural and social barriers which handicap their lives even if rehabilitation assistance is available.
The relationship between disability and poverty has been clearly established. While the risk of impairment is much greater for the poverty-stricken, the corollary is also true: the birth of an impaired child places heavy demands on the limited resources of the family, thus thrusting it deeper into poverty.
Many disabled people are denied employment or are given only menial and poorly paid jobs. In times of unemployment and economic distress, disabled persons are usually the first to discharged and the last to be hired. In some industrialized countries the rate of unemployment among disabled job-seekers is double that of able-bodied applicants.
Incidence There is a large and growing number of persons with disabilities in the world today. The estimated figure of 500 million is confirmed by the results of surveys of segments of population, coupled with the observations of experienced investigators. In most countries, at least one person out of 10 is disabled by physical, mental or sensory impairment, and at least 25 per cent of any population is adversely affected by the presence of disability.
Counter-claim Disabled is one of the distinctions that normal people make which easily leads to oppression, though it becomes clear that disability is a continuum of experience which nearly everybody experiences at some time in their life.
Refs Darnbrough, Ann and Kinrade, Derek *Directory for Disabled People* (1985); OECD *Disabled Youth* (1988); Rehabilitation International *The Economics of Disability* (1981); Rood, Lois S and Faison, Karen *Beyond Severe Disability* (1985); World Congress of Rehabilitation International *Participation of People with Disabilities* (1981).
Broader Human disease and disability (#PB1044).
Narrower Anosmia (#PE1066)
Disabled women (#PC0729)
Disabled workers (#PD4673)
Disabled refugees (#PC0768)
Mental deficiency (#PC1587)
Skill disabilities (#PG3271)
Ageing war disabled (#PD0874)
Poverty and disability (#PD0723)
Locomotor disabilities (#PE6769)
Dexterity disabilities (#PE6767)
Retarded socialization (#PF2187)
Disabled elderly persons (#PE3847)
Disabled victims of crimes (#PD0762)
Personal care disabilities (#PE6770)
Disabled victims of torture (#PD0764)
Neuro-muscular disabilities (#PG2957)
Socially handicapped refugees (#PD1507)
Body disposition disabilities (#PE6768)
Physically handicapped persons (#PD6020)
Increasing number of disabled persons (#PF0719)
Structural barriers for disabled persons (#PE0707)
Disabled persons in developing countries (#PD0724)
Unequal opportunities for disabled persons (#PE0706).
Aggravates Dependence of the disabled (#PF4296)
Physically dependent people (#PD7238)
Denial of rights to disabled (#PC3461)
Discrimination against the disabled (#PD9757)
Death and disability from inhumane confinement (#PE5648)
Arbitrary evaluation of disability compensation (#PG7870)
Unequal employment opportunities for disabled persons (#PE0783)
Inadequate educational facilities for disabled persons (#PF0775)
Inadequate recreational facilities for disabled people (#PE8833)
Exclusion of disabled persons from social and cultural life (#PD0784).
Aggravated by Injurious accidents (#PB0731)
Malicious physical disablement (#PE3733)
Inadequate community care for handicapped persons (#PE8924).

♦ PC0716 Deterioration in physical health
Lack of physical activity — Hypokinesia
Nature Working conditions in modern industrial societies are generally characterized by a gradual elimination of physical effort and a corresponding decline in physical fitness. This tendency is causing a deterioration in health, particularly among the middle-aged in developed countries. Many complaints about ill-health, pains, and the malfunctioning of various organs may be related to subnormally functioning muscular system. Weak muscles often cause more passive structures

of the body (joints, ligaments, and connective tissue) to be overloaded. Increasing mortality rates are associated with a higher incidence of cardiovascular disease and with a lower level of physical fitness.
Broader Human disease and disability (#PB1044).
Narrower Physical unfitness (#PD4475).
Related Stigmatized diseases (#PD7279) Unhealthy physical posture (#PD2838).
Aggravates Heart diseases (#PD0448).
Aggravated by Increasing job monotony (#PD2656)
Inadequate maintenance of physical health (#PF1773).

♦ PC0720 Endangered tribes and indigenous peoples

Exploitation of aboriginals — Abuse of native peoples — Violence against indigenous minorities — Victimization of indigenous peoples

Nature Native races are threatened by the cultural invasion of civilized outsiders who expropriate land, practise ethnocide and destroy existing social and religious structures, leaving the way open for prostitution, alcoholism, dependency and general social disintegration. Assimilation programmes are often inadequate and do not take enough account of the existing cultural base. Semi-slavery may ensue. Native races are also highly susceptible to disease brought by outsiders, for which they have no resistance and which are often fatal to them. Where political power is held by an ethnic minority, certain indigenous peoples are often denied proper political, cultural and economic well being through ethnic oppression and even ethnocide.

Incidence Programmes to repress indigenous peoples are actively supported by governments and businesses in many parts of North and Latin America, Europe, Asia, South-east Asia, and Africa. Repressive, discriminatory practices can be found worldwide.

Claim All the 'pacified' natives slowly lose their characteristics and authenticity and their culture is corrupted through contact with civilized outsiders.

Refs Bodley, John H *Tribal People and Development Issues* (1988); Burger, Julian *Report from the Frontier* (1987); ICIHI *Indigenous Peoples*; Moody, Roger *Indigenous Peoples* (1987); Morse, Bradford *Aboriginal Self-Government in Australia and Canada.*

Broader Exploitation (#PB3200) Threatened and vulnerable minorities (#PC3295).
Narrower Expropriation of land from indigenous populations (#PC3304)
Exploitation of indigenous populations in employment (#PD1092)
Restriction of indigenous populations to reservations (#PD3305).
Related Tribalism (#PC1910) Threatened sects (#PC1995) Ethnic disintegration (#PC3291)
Political exploitation (#PC7356) Isolation of ethnic groups (#PC3316)
Destruction of cultural heritage (#PC2114)
Threats to ideological movements and minorities (#PC3362)
Vulnerability of small nations to foreign intervention (#PD2374).
Aggravates Endangered indigenous cultures (#PC7203)
Lack of variety of social life forms (#PE8806)
Segregation of poor and minority population in urban ghettos (#PD1260).
Aggravated by Racism (#PB1047) Genocide (#PC1056)
Ethnocide (#PC1328) Colonialism (#PC0798)
Ethnic conflict (#PC3685) Cultural invasion (#PC2548)
Unsustainable economic development (#PC0495) Corporation-sanctioned assassination (#PE6356)
Underprivileged linguistic minorities (#PC3324)
Discrimination against indigenous populations (#PC0352)
Unconstrained exploitation of natural resources (#PF2855)
Vulnerability of indigenous populations to introduction of diseases (#PE3721).
Reduced by Segregation in marriage (#PD3347).

♦ PC0728 Pests

Harmful animals and plants

Nature Pests are nonpathogenic organisms that compete with man for food or space, damage his possessions or attack him personally. The term "pest" is subjective: an ecologist would not necessarily consider several leaf-eating larvae as pests, but the gardener probably would. Generally, though, to be considered a pest, an animal, plant, or insect must occur in such abundance as to distinctly threaten man or his interests.

Most pests are invertebrate animals, especially insects, mites, nematodes, snails and slugs. Vertebrate pests include many kinds of rodents, birds and deer. Weeds are also usually considered pests.

Refs Kitching, R L and Jones, R E *Ecology of Pests* (1981).

Narrower Weeds (#PD1574) Insect pests (#PC1630) Vampire bats (#PE1890)
Animal pests (#PD8426) Mollusc pests (#PG2849) Locust plagues (#PE0725)
Birds as pests (#PD1689) Pests of plants (#PC1627) Poisonous algae (#PE2501)
Arachnida pests (#PE3986) Household pests (#PD3522) Harmful wildlife (#PC3151)
Crustacean pests (#PG2848) Microbial pests in industry (#PD1607)
Pests and diseases of trees (#PD3585) Flea resistance to insecticides (#PE3572)
Inadequate weed control in developing countries (#PD3598).
Aggravates Disease (#PA6799) Instability of cocoa trade (#PE1549)
Human disease and disability (#PB1044) Vulnerability of plants and crops (#PD5730).
Aggravated by Inadequate quarantine (#PE2850)
Pest resistance to pesticides (#PD3696)
Diseases of beneficial insects (#PD2284)
Excessive use of chemicals to control pests (#PD1207)
Environmental hazards of new species introduction (#PC1617)
Irresponsible introduction of new species of fish (#PF3602)
Increase in pests and diseases through perennial irrigation (#PE7832).
Reduced by Pesticides as pollutants (#PD0120).

♦ PC0729 Disabled women

Nature The consequences of deficiencies and disablement are particularly serious for women. There are a great many countries where women are subjected to social, cultural and economic disadvantages which impede their access to, for example, health care, education, vocational training and employment. If, in addition, they are physically or mentally disabled their chances of overcoming their disablement are diminished, which makes it all the more difficult for them to take part in community life. In families, the responsibility for caring for a disabled parent often lies with daughters, which considerably limits their freedom and their possibilities of taking part in other activities.

Refs Matthews, Gwyneth Ferguson *Voices from the Shadows.*

Broader Human disability (#PC0699).
Aggravates Discrimination against women (#PC0308)
Denial of rights to disabled (#PC3461).

♦ PC0742 Military offences

Military crimes — Khaki-collar crime

Nature Both in times of peace and in times of war, individuals or groups from military forces may use their privileged position, particularly as foreigners based on allied territory, to act with immunity against civilians or civilian property. This may include deliberate destruction of property, such as field crops, by armoured vehicles on manoeuvre.

Refs Bryant, Clifton D *Khaki-Collar Crime* (1979).

Broader Statutory crime (#PC0277).
Related Military atrocities (#PD1881).
Aggravated by Foreign military presence (#PD3496).

♦ PC0745 Promiscuity

Claim The term promiscuity usually refers to indiscriminate, casual sexual encounters or a high frequency of sexual relationships with a large number of partners, such as is found in nymphomania, satyriasis, so called "Don Juans", and many homosexuals. It has come to include the careless attitude to sexual relations engendered in a society which allows even adolescents the knowledge and the opportunity to engage in sexual experimentation without the moral discipline or the understanding to control their emotions. It involves casual and exploitive sexual relations, limited communication and irresponsible parenthood.

Broader Decadence (#PB2542) Sexual immorality (#PF2687).
Narrower Group marriage (#PF3288).
Related Lust (#PA4673) Disorder (#PA7361) Unchastity (#PA5612)
Indiscrimination (#PA6446) Lack of family planning (#PF0148)
Inadequate sex education (#PD0759)
Discrimination in family planning facilities (#PD1036).
Aggravates Divorce (#PF2100) Adultery (#PF2314) Cohabitation (#PF3278)
Homosexuality (#PF3242) Family breakdown (#PC2102) Unmarried mothers (#PD0902)
Unmarried parents (#PD3257) Marital instability (#PD2103) Sexual unfulfilment (#PF3260)
Illegitimate children (#PC1874) Single parent families (#PD2681)
Sexually transmitted diseases (#PD0061)
Vulnerability of marriage as an institution (#PF1870).
Aggravated by Polyandry (#PF3289) Indecent art (#PE5042)
Permissiveness (#PF1252) Induced abortion (#PD0158)
Family disorganization (#PC2151) Double standards of sexual morality (#PF3259).
Reduces Prostitution (#PD0693).

♦ PC0746 Forced labour

Dependence on forced labour — Impressment of labour

Nature Forced labour covers a wide range of practices, from slavery to compulsory national service of a military or civil kind. Its effect varies from the total subjugation of prisoners, particularly political prisoners or prisoners-of-war, and of slaves, to the hardships endured by people recruited for national service. It may have severe adverse physical effects especially in the case of the two former. Forced labour associated with malnutrition and crippling diseases has led to the death of millions.

Incidence Forced labour in the racially segregated system of South Africa is legal, involving the exploitation of cheap African labour which is usually housed, fed and paid inadequately, segregated by sex and restricted in movement. The practice also occurs illegally in Latin America where Indians are enslaved. Forced labour of Mexican migrants in the South-Western United States also occurs illegally. Immigrants into western European countries recruited from Africa, the Middle East and Asia for menial work in factories find themselves excluded from unions and therefore unable to bargain for better conditions. Inadequately housed and unable to bring their families to live with them, they have to continue working under poor conditions because of the need to support their families.

Compulsory military service prevails in many industrialized countries, whereas other compulsory services are often exacted by law in developing countries. Prison labour is closely connected with forced labour but is given special consideration under the UN Forced Labour Convention. Labour may be imposed on persons awaiting trials, or as a means of punishment, or prisoners may be contracted out to private employers. Forced labour of political prisoners is known to be practised, but governments are reluctant to admit this or give information. The forced labour in Nazi concentration camps and its horrible conditions were not realized by the world until they were in existence almost ten years. However the tragedy of forced labour in the Soviet Union, particularly in the Gulag, has been exposed by a number of persons who have been interned there.

Refs Lasker, Bruno *Human Bondage in Southeast Asia* (1972); Novak, Daniel A *The Wheel of Servitude* (1978); Sherman, William L *Forced Native Labor in Sixteenth Century Central America* (1979).

Broader Social injustice (#PC0797).
Narrower Slavery (#PC0146) Conscription (#PF6051)
Abuse of prison labour (#PD0165).
Related Feudalism (#PF2136) Unpaid labour (#PD3056) Chattel slavery (#PC3300)
Political repression (#PC1919) Exploitative personal services (#PC3299)
Legalized racial discrimination (#PC3683)
Enforced participation in community activity (#PD3386).
Aggravates Restrictions on freedom of movement between countries (#PC0935).
Aggravated by Antisemitism (#PE2131) Racial discrimination (#PC0006)
Exploitation in employment (#PC3297) Unethical personnel practices (#PD0862)
Lack of protective legislation (#PJ2889) Inadequate national law enforcement (#PE4768)
Political discrimination in politics (#PC3221).

♦ PC0747 War crimes

Nature Violations in times of war as applied to civilians include: murder, rape, torture, illegal internment, enslavement, deportation, and harsh treatment or restrictions that affect physical and mental health. As applied to property, war crimes include plunder and wanton destruction or devastation of natural resources, productive land and communities, not justified by military objectives. Many military codes of justice provide for punishment of their occupying forces in cases of individual criminal activity against civilians; but collective military action against civilians, ordered or condoned, escapes justice.

Background The armed forces of every state have legal experts (judge advocates) whose primary responsibility is to administer military justice, maintain discipline and monitor the observance of the laws and customs of war.

Incidence War crimes in recent history include Nazi Germany's atrocities, involving the killing of millions of Jews, gypsies, and others considered as "Untermenschen" or sub-human. Whereas most war crimes result from the excesses of individuals, these were official crimes, as were certain massacres carried out by the USSR's NKVD. During World War II, Germany itself investigated some 10,000 documented war crimes against Germans, of which the files of 4,000 have survived. Of the latter, 50 per cent cover crimes perpetrated in the USSR (concerning massacres of thousands of people, although many of the accused were acquitted for lack of evidence or because of mistaken identity). The German records also contain investigations of war crimes allegedly committed by nationals of the USA (including many air attacks on German Red Cross installations and low-level gunning of agricultural workers), of the UK (including shootings of German shipwrecked crews), of France (including lynching of air crews and ill-treatment of prisoners of war), of Poland (including the massacre of 5,000 members of the German minority in Poland), of Yugoslavia and of other allies. The British-American decision on area bombing (especially the fire-bombing of Dresden in 1945) made no distinction between military and civil targets and resulted in the deaths of some 600,000 German civilians (in comparison with some 60,000 British victims of German air raids). The use of the atomic bomb in Japan by the USA has also been considered a war crime.

More recently such crimes include USA defoliation and village napalming in Vietnam, as well as incidents such as My Lai. A number of genocidal conflicts in Africa and South-East Asia have occurred since the Second World War. The civilian casualties in Lebanon are due in part to war crimes of deliberate murder. The USSR's occupation of Afghanistan has resulted in the slaughter of thousands in the name of pacification or reprisals, but the illegality of that occupation may be said to make all Soviet acts there criminal. Inadequate mechanisms for the identification and punishment of war-crimes are responsible in large degree for their partly hidden nature and

perpetuation.
Refs De Zayas, Alfred *The Wehrmacht War Crimes Bureau*; Tutorow, Norman E and Winnovich, Karen *War Crimes, War Criminals, and War Crimes Trials* (1986).
Broader Crimes against humanity (#PC1073).
Narrower Military atrocities (#PD1881)
Inhumane and indiscriminate weapons (#PD1519)
Persistence of a technical state of war following cease–fire agreements (#PE2324).
Aggravates Looting (#PE4152) Mental disorders (#PD9131)
Human destructiveness (#PA0832) Physically handicapped persons (#PD6020).
Aggravated by Sadism (#PF3270) Amoralism (#PF3349)
Extremism (#PB3415) Immorality (#PA3369)
Ideological war (#PC3431) Ineffective war crime prosecution (#PD1464)
Denial of human rights in armed conflicts (#PC1454).

♦ PC0768 Disabled refugees

Nature There are over 10 million refugees and displaced persons in the world today as a result of man–made disasters. Many of them are disabled physically and psychologically as a result of their sufferings from persecution, violence and hazards. Most are in third–world countries, where services and facilities are extremely limited. Being a refugee is in itself a handicap, and a disabled refugee is doubly handicapped.
Broader Refugees (#PB0205) Human disability (#PC0699).

♦ PC0776 Irresponsible genetic manipulation

Creation of transgenic plants and animals — Hazards of biotechnology — Irresponsible creation of new species — Hazards in genetic engineering accidents — Uncontrollable new species — Genetically modified feral species
Nature Genetic recombination consists in isolating and then splicing together DNA molecules from unrelated organisms to produce new hybrid organisms which may contain the genetic properties of both of the original organisms. Risks include the creation of new micro–organisms alien to human experience, and the possibility that one of these types of micro–organisms might escape the confines of the laboratory and cause human disease. Micro–organism changes with potential hazards include the emergence and spread of chloramphenicol–resistant S typhi and more virulent strains of shigella. Some of the artificial recombinant DNA molecules could prove biologically hazardous. One potential hazard in current experiments derives from the need to use a bacterium like E coli to clone the recombinant DNA molecules and to amplify their number. Strains of E coli commonly reside in the human intestinal tract, and they are capable of exchanging genetic information with other types of bacteria, some of which are pathogenic to man. Thus, new DNA elements introduced into E coli might possibly become widely disseminated among human, bacterial, plant, or animal populations with unpredictable effects.
Incidence There are studies underway in the USA to breed larger–than–normal animal species by the use of the injection of human genes. Early experiments on mice have proved successful, with mice growing to twice their normal size at twice their normal rate, but it is as yet unknown what consequences could result from this interference with nature. In the UK there are now reports of a feral tomato, a genetically modified species which has gone wild and can no longer be contained.
Genetically engineered soya bean, cotton, rice, corn, oilseed rape, sugarbeet, tomato and alfalfa crops are expected to enter the American market between 1993 and 2000.
Claim In recent years genetic engineers have perfected techniques for altering the detailed biology of plants and organisms. The prospect of new vaccines and pesticides made this way has attracted billions of dollars of investment. Environmentalists claim that these manipulated bacteria could have devastating side–effects when released.
In the case of crossing nature's boundaries by creating animal species that would normally never be able to mate and reproduce, science is acting unethically and environmentally inappropriately. The scientists involved are trying to redefine animal and human life. Genetic engineers are trying to transfer new genes to a person suffering from genetic disease, playing with a person's DNA and designing his destiny.
Counter–claim Successful genetic engineering could breed species of plants and animals which, in size and number, could have a considerable impact on world hunger.
The breeding of plants or animals and induced mutation in microbes have long been used to generate genetic variants with useful characteristics. The new recombinant DNA methods now being used to generate variants are no more inherently risky than the older methods. While the species line can be crossed with these methods the changes made in an organism's genetic make–up can be more precise and more limited than that of breeding methods. The accuracy of the methods diminishes the risk of inadvertently producing a dangerous variant. While there is no such thing as a new venture without risk, experiments can be done to test factors, such as persistence and spread of an organism, along with its effectiveness in doing what it was designed to do.
Refs McNally, Ruth and Wheale, Peter *Genetic Engineering* (1988).
Broader Irresponsibility (#PA8658)
Irresponsible scientific and technological activity (#PC1153).
Narrower Techno–mercenaries (#PE7153)
Release of genetically engineered micro–organisms (#PD7183).
Aggravates Laboratory waste water pollutants (#PE9813)
Decline in human genetic endowment (#PF7815)
Monopolistic control of new life forms (#PD7840)
Vulnerability of world genetic resources (#PB4788)
Introduction of high–yield crop varieties (#PF3146)
Accidental weed creation by genetic engineering (#PE5404)
Irresponsible introduction of new species of animals (#PD1290).
Aggravated by Mutation (#PF2276) Unethical practice of the biosciences (#PD7731)
Denial of the right of unborn children (#PF6616).

♦ PC0797 Social injustice

Dependence on social injustice
Nature Social injustice impedes growth and development, hampering or even halting improvement in living standards, fair distribution of income, creation of opportunities, and the elimination of inequalities. The inadequacy of economic growth, imbalances in economic structures, and imperfections in education and training systems contribute to, and are aggravated by, unjust conditions in the world.
Broader Lack of protection for the vulnerable (#PB4353).
Narrower Racism (#PB1047) Slavery (#PC0146) Scapegoats (#PF3332)
Segregation (#PC0031) Unjust laws (#PC7112) Forced labour (#PC0746)
Urban poverty (#PC5052) Ethnic conflict (#PC3685) Minority control (#PF2375)
Racial inequality (#PF1199) Civil disobedience (#PC0690) Age discrimination (#PC2541)
Social dictatorship (#PD3241) Racial exploitation (#PC3334) Social fragmentation (#PF1324)
Racial discrimination (#PC0006) Social discrimination (#PC1864)
Sexual discrimination (#PC2022) Religious persecution (#PC5994)
Unfit legal defendants (#PE4863) Religious discrimination (#PC1455)
Inequality before the law (#PC1268) Inequality of opportunity (#PC3435)
Violation of human rights (#PB3860) Discrimination against men (#PC3258)
Exploitation in employment (#PC3297) Underprivileged minorities (#PC3424)
Unequal income distribution (#PD4962) Discrimination against women (#PC0308)
Unequal property distribution (#PC3438) Discrimination against minorities (#PC0582)
Underprivileged racial minorities (#PC0805) Non–productive athletic activities (#PF4202)
Threatened and vulnerable minorities (#PC3295) Underprivileged linguistic minorities (#PC3324)
Discrepancies in human life evaluation (#PF1191)
Underprivileged ideological minorities (#PC3325).
Related Unmarried mothers (#PD0902) Political injustice (#PC2181)
Political inequality (#PC3425) Trafficking in women (#PC3298)
Ethnic discrimination (#PC3686) Prohibitive cost of farm machinery (#PF2457)
Discrimination against men in social services (#PD3336)
Delayed consequences of war–time imprisonment and deportation (#PF0726).
Aggravates War (#PB0593) Banditry (#PD2609) Human violence (#PA0429)
Class conflict (#PC1573) Social conflict (#PC0137) Unreported births (#PF5381)
Gypsy persecution (#PE1281) Forced mass expulsion (#PD0531)
Denial of human rights (#PB3121) Impediments to marriage (#PF3343)
Lack of social mobility (#PF2195) Distortionary tax systems (#PD3436)
Abuse of coca and cocaine (#PD2363) Prohibitive cost of living (#PF1238)
Deprivation of nationality (#PD3225) Exploitation of women refugees (#PD5025)
Conflict between minority groups (#PC3428) Abuse of bureaucratic procedures (#PF2661)
Racial discrimination in politics (#PD3329) Threats against family or friends (#PE3308)
Exploitation in capitalist systems (#PC3117) Denial of the right of association (#PD3224)
Racial discrimination in education (#PD3328) Denial of right to inherit property (#PF0886)
Double standards of sexual morality (#PF3259) Underprivileged religious minorities (#PC2129)
Excessive employment of married women (#PD3557)
Uncertainty of death of missing persons (#PF0431)
Unequal distribution of social services (#PC3437)
Discrimination against women in politics (#PC1001)
Racial discrimination in public services (#PD3326)
Discrimination against women in religion (#PD0127)
Discrimination against women in education (#PD0190)
Discrimination against women in employment (#PD0086)
Discrimination against women before the law (#PD0162)
Discrimination against women in public life (#PD3335)
Professional discrimination between educators (#PE0810)
Arbitrary evaluation of disability compensation (#PG7870)
Discrimination against women in social services (#PD3691)
Incompatibility of traditional and new technologies (#PE3337)
Inadequacy and insensitivity of intelligence testing (#PD1975)
Exploitation of indigenous populations in employment (#PD1092)
Inadequate living and working conditions of immigrant labourers in industrialized countries (#PD3427).
Aggravated by Strikes (#PD0694) Elitism (#PA1387)
Eugenics (#PC2153) Prejudice (#PA2173)
Apartheid (#PE3681) Capitalism (#PC0564)
Repression (#PB0871) Social apathy (#PC3412)
Deviant society (#PC2405) Racial conflict (#PC3684)
Authoritarianism (#PB1638) Social inequality (#PB0514)
Racial segregation (#PC3688) Lack of human unity (#PF2434)
Extradition refusal (#PF2645) Class consciousness (#PC3458)
Lack of assimilation (#PF2132) Prisoners of conscience (#PC6935)
Inequality in education (#PC3434) Plural society tensions (#PF2448)
Inadequacy of social doctrine (#PF3398) Limited access to social benefits (#PF1303)
Political discrimination based on illiteracy (#PC3222)
International imbalance in the quality of life (#PB4993).

♦ PC0798 Colonialism

Colonization — Slavery–like practices of colonialism
Nature Colonialism is the establishment and maintenance, for an extended time, of rule over a people that is separate from and subordinate to the ruling power. It imposes alien, authoritarian and more or less repressive regimes on materially inferior societies. Colonialism, because it moulds geographically, culturally, politically, socially and economically the life of subordinate societies to meet the needs of the colonial powers, not only violates, in the present, the elementary rights to self–determination but jeopardizes future possibilities of self–centered development.
Incidence Colonialism as described above, no longer occurs; the last colonial empire (Portugal) collapsed in 1975. If the term is still widely used, it is to denote (abusively) situations similar in some but not all aspects to colonial domination (apartheid, neo–colonialism) or that have risen with the process of decolonization (Commonwealth nations, dependent territories, associated States, dependencies, condominiums, UN Trust Territories, territories, unincorporated territories, overseas territories, overseas departments, self–governing territories, self–governing associations). Often colonial situations have been endorsed by the international community with the consequence that, although colonies as such have disappeared, the peoples have become minorities: more or less integrated, more or less discriminated against (Sahahuris, Basques, American Indians, Sikhs, Armenians, etc).
Background Colonialism was practised by ancient Tyre, Carthage, and the Greek city–states. A new phase of colonialism originated with the discovery of new continents during the fifteenth and sixteenth centuries. But whereas Spain and Portugal, to the detriment of their own productivity, were primarily interested in importing gold and other foods, modern colonialism as practised by other Western nations became a means to lighten the overproduction of material at home, to relieve overpopulation, and to secure bases for international commerce. The primary considerations were now economic.
Today, colonial imperialism seems to belong to the past, having vanished with the self–destruction of Old Europe during the world wars and with the concomitant rise of new national states on the continents of Asia and Africa. However, the tangle of paternalism, militarism, and superiority complexes survives in the struggle of the bigger powers for the dominance of smaller and often helpless territories. Western states claim to protect the emerging states against communist invasion, whereas the communists claim to liberate them from the evils of capitalism.
Claim Self–determination and independence are not the exclusive prerogatives of the powerful, but fundamental and inalienable rights of all peoples everywhere. Inadequacy of political, economic, social or educational preparedness cannot serve as a pretext for dictation by a minority different from the local majority in culture, history, beliefs and often race, of the conditions of social, economic and political life. The military or economic subordination is always accompanied by bestial exploitation and often by extermination of the indigenous population. After independence, domination and exploitation subsist under a new form, namely neo–colonialism.
Counter–claim Countries characterized by social, cultural, economic and political backwardness, require guardianship of a stronger state. The colonial mode of government is as legitimate as any other and has accomplished the advancement of dependent peoples in every sphere. For example, the lack of infrastructure capable of carrying relief supplies to famine victims in Ethiopia has been blamed on the fact that the country was never colonized. The colonial powers and their missionaries have forced certain countries to abolish at least the most inhuman customs and superstitions. Colonizing nations developed the resources of the overseas territories; without their organizing zeal, the formerly subjected countries would be even less competent to rule themselves than several of them are today. In some new nations many of the older inhabitants now ask themselves whether they felt more secure under foreign rule or under the present continual turmoil of revolutions now they are free to govern themselves.
Refs Engels, Friedrich and Marx *On Colonialism* (1980); Fanon, Frantz *The Wretched of the Earth*

(1966); Fanon, Frantz; Farrington, Constance and Sartre, Jean-Paul *Wretched of the Earth* (1965); Memmi, Albert *The Colonizer and The Colonized* (1990).
Broader Domination (#PA0839) Imperialism (#PB0113)
Denial of right of a people to be self-determining (#PC6727).
Narrower Neo-colonialism (#PC1876) Foreign control (#PC3187)
Occupied nations (#PC1788) Colonization of information (#PF4894)
Alien domination of peoples (#PC7384) Lack of political independence (#PF0297)
Foreign intervention in internal affairs of states (#PC3185).
Related Slavery (#PC0146) Occupied territories (#PD8021)
Violation of sovereignty by trans-border broadcasting (#PE0261).
Aggravates War (#PB0593) Tribalism (#PC1910) Feudalism (#PF2136)
Ethnocide (#PC1328) Freemasonry (#PF0695) Intimidation (#PB1992)
Miscegenation (#PC1523) Ethnic conflict (#PC3685) Minority control (#PF2375)
Cultural invasion (#PC2548) Covert imperialism (#PF3199) Religious conflict (#PC3292)
Forced assimilation (#PC3293) Cultural fragmentation (#PF0536)
Lack of racial identity (#PF0684) Plural society tensions (#PF2448)
National political dependence (#PF1452) Conflict between minority groups (#PC3428)
Discrimination against minorities (#PC0582)
Endangered tribes and indigenous peoples (#PC0720)
Administrative difficulties in new states (#PE1793)
Discrimination against indigenous populations (#PC0352)
Denial of right to national self-determination (#PC1450)
Unnatural boundaries between developing countries (#PD2544)
Exploitation of indigenous populations in employment (#PD1092)
Restriction of indigenous populations to reservations (#PD3305)
Inappropriate transplantation of industrialized country methods to developing countries (#PE1337).
Aggravated by Racial discrimination (#PC0006)
Territories accorded a United Nations non-self-governing status disputed by the administering government (#PF2943).
Reduced by Guerrilla warfare (#PC1738).

♦ PC0799 Mental depression

Mentally depressed people — Mental breakdown — Dejection — Depressive psychosis — Depressive neuroses

Nature Depression is the cause of unreasonable and unnecessary suffering for millions of people, often to the point of disabling the sufferer. Depressive disorders can be found throughout the world. Depressive patients account for a significant proportion of all those requiring mental heath care and, as the majority of them remain untreated, their suffering continues to disable them and to cause losses to their families and communities. The situation is especially severe in developing countries. Lack of adequate detection and treatment is due to poorly-trained health workers, scarce resources, and insufficient knowledge.

In psychiatry, depression refers to a clinical syndrome consisting of lowering of mood-tone (feelings of painful dejection), difficulty in thinking, and psychomotor retardation. It is a pathological state of conscious psychic suffering and guilt, accompanied by a marked reduction in the sense of personal values, and a diminution of mental, psychomotor, and even organic activity, unrelated to actual deficiency. As used by the layman, the word depression refers to the mood element, which in psychiatry would more appropriately be labelled dejection, sadness, gloominess, despair or despondency.

Incidence Depression is widespread; an estimated 100 million people develop clinically recognizable depression every year. Of 175,000 admissions to mental illness hospitals and psychiatric units in the UK in 1969, 66,000 were suffering from depression. Of an estimated 50,000-70,000 suicides in the USA each year, up to 60 per cent occur among persons suffering from depression. Approximately 125,000 people in the USA are hospitalized each year with depression, a further 200,000 are treated by psychiatrists, and 4 to 8 million more are in need of help but do not realize it. More women than men are treated for depression, and the largest occupational group is homemakers.

Counter-claim There is growing scientific evidence that depression is to some degree controlled genetically and that the same gene is responsible for creative genius. In fact, a much higher than average percentage of poets and other artists are manic-depressives and suffer from other depression disorders. Rather than finding ways of channelling this potential creativity, society treats the disease.

Refs American Health Research Institute Staff *Depression* (1982); Brown, George W and Harris, Tirril *Social Origins of Depression* (1978); French, Alfred and Berlin, Irving *Depression in Children and Adolescents* (1979); Kendall *Anxiety and Depression*; Kleinman, Arthur *Social Origins of Distress and Disease* (1986); Kleinman, Arthur and Good, Byron *Culture and Depression* (1985); Mahendra, B *Depression* (1987); Petti, Theodore A (Ed) *Childhood Depression* (1983); Seligman, Martin E *Helplessness* (1975).
Broader Human disease and disability (#PB1044).
Narrower Manic depression (#PG4434) Nervous breakdown (#PE6322)
Anaclitic depression (#PE1883) Post-natal depression (#PG9602)
Involutional depression (#PE0655) Depression due to torture (#PE0885)
Seasonal affective disorder (#PE0258) Mental depression in children (#PD3784).
Related Psychoses (#PD1722) Self disorders (#PF4843)
Manic-depressive psychosis (#PD1318) Ill defined health conditions (#PC9067).
Aggravates Apathy (#PA2360) Suicide (#PC0417) Caffeine abuse (#PE0618)
Sleep disorders (#PE2197) Bulimia nervosa (#PE4538) Mental impairment (#PF4945)
Attempted suicide (#PE4878) Abuse of amphetamines (#PE1558)
Forced repatriation of prisoners of war (#PD0218).
Aggravated by Grief (#PF5654) Stroke (#PE1684)
Alcohol abuse (#PD0153) Psychological inertia (#PF0421)
Personal unpopularity (#PF4641) Inhibited grief process (#PD4918)
Abuse of coca and cocaine (#PD2363) Biased presentation of news (#PD1718)
Ill treatment of prisoners of war (#PD2617) Abuse of sedatives and tranquillizers (#PE0139)
Post-hallucinogen perception disorder (#PG7457)
Solvent and methylated spirits drinking (#PE1349)
Inhaling of solvents and anaesthetic drugs (#PE1427)
Ignorance of nonverbal communication skills (#PE0533)
Monotonous and unaesthetic architecture and design (#PF0867).
Reduced by Unrealistically positive self-assessment (#PF4377).

♦ PC0805 Underprivileged racial minorities

Nature Racial minorities may be dominated by other racial and national groups, who discriminate against them, exploit them while keeping them in subjection and who try to destroy their cultural, social, religious and linguistic patterns.

Incidence Underprivileged racial minorities occur particularly where there has been colonization or mass immigration. The imposition of a more developed culture subjects primitive people to a state of slavery or complete dependence which shatters their existing social structure. Through discrimination they are not accorded equal rights or adequate special treatment to integrate them successfully into the new order. Racist or intensely nationalist policies may consciously attempt to obliterate minority culture for political reasons or may exclude racial minorities from citizenship rendering them stateless. They may be deported and their property expropriated, as in the case of the Ugandan Asians in the early 1970s. Underprivileged racial minorities may be used for the pretext of war by another country where the same or a similar racial group predominates.

Refs Blackstone, William T and Heslep, Robert D (Eds) *Social Justice and Preferential Treatment* (1977).
Broader Social injustice (#PC0797) Underprivileged minorities (#PC3424).
Narrower Nomadism (#PF3700) Desert nomadism (#PD2520)
Gypsy persecution (#PE1281) Isolation of ethnic groups (#PC3316).
Related Political repression (#PC1919) Exploitative personal services (#PC3299)
Underprivileged religious minorities (#PC2129) Underprivileged linguistic minorities (#PC3324)
Underprivileged ideological minorities (#PC3325)
Threats to ideological movements and minorities (#PC3362)
Expropriation of land from indigenous populations (#PC3304)
Restriction of indigenous populations to reservations (#PD3305).
Aggravates Secession (#PD2490) Racial conflict (#PC3684)
Political conflict (#PC0368) Abuse of coca and cocaine (#PD2363)
Unequal distribution of social services (#PC3437).
Aggravated by Racism (#PB1047) Elitism (#PA1387)
Prejudice (#PA2173) Cultural invasion (#PC2548)
Social fragmentation (#PF1324) Racial discrimination (#PC0006)
Ethnic discrimination (#PC3686) Lack of social mobility (#PF2195)
Discrimination against minorities (#PC0582)
Inadequate education of indigenous peoples (#PC3322).

♦ PC0834 Inadequate social welfare services

Denial of right to adequate welfare services
Broader Social neglect (#PB0883) Isolation of ethnic groups (#PC3316)
Limited access to social benefits (#PF1303).
Narrower Ageing populations (#PD8561) Inadequate child welfare (#PC0233)
Inadequate animal welfare (#PC1167) Regressive welfare system (#PG9346)
Inadequate health services (#PD4790) Inappropriate school lunches (#PJ0335)
Inadequate welfare services for the aged (#PD0512)
Inadequate welfare services for the deaf (#PD0601)
Inadequate welfare services for the blind (#PD0542)
Inadequate protection of individual welfare (#PJ3312)
Inadequate community care for handicapped persons (#PE8924)
Insufficient social security in the agricultural sector (#PD9155)
Inadequate community care for transient urban populations (#PF1844)
Denial of right to social welfare services for indigenous peoples (#PE1506).
Related Inadequate standards of living (#PF0344)
Insufficient emergency services (#PF9007)
Denial of right to sufficient food (#PE0324)
Denial of right to social security (#PD7251)
Denial of right to sufficient shelter (#PD5254)
Denial of right to sufficient clothing (#PE7616)
Denial of right to adequate medical care (#PD2028)
Inflexible social care structures in developing countries (#PF2493)
Denial of the right to social security in capitalist systems (#PD3120).
Aggravates Beggars (#PD2500) Exploitation of child labour (#PD0164)
Socially handicapped refugees (#PD1507) Unequal coverage by social security (#PF0852)
Denial of right to economic security (#PD0808) Unequal distribution of social services (#PC3437)
Economic and social losses due to disability (#PE4856)
Inadequate educational facilities for gifted children (#PD2051)
Inadequate educational facilities for disabled persons (#PF0775).
Aggravated by Nomadism (#PF3700) Gypsy persecution (#PE1281)
Neglect of the aged (#PD8945) Uninformed care techniques (#PG9655)
Denial of freedom of conscience (#PD7612) Governmental bias in statistics (#PF0019)
Insufficient special care vehicles (#PJ9008) Increasing cost of social security (#PF7911)
Individualistic welfare responsibility (#PF2560)
Non-adaptive local structure of social care (#PE5246)
Fluctuations in government social programmes (#PF0170)
Collapsed tension between care and responsibility (#PF5555)
Lack of participation in local welfare programmes (#PE8503)
Imbalance between capital and technical assistance (#PE4866).
Reduced by Inadequate utilization of volunteer social service workers (#PF4892).

♦ PC0838 Cruelty to children

Child abuse

Nature Child abuse is a range of maltreatment. In addition to physical harm and sexual abuse, it also includes serious neglect of a child's emotional and physical needs and forms of emotional abuse such as incessant berating of a child. Severe emotional abuse, such as being extremely cold during infancy, results in worse emotional and learning problems in children than emotionally responsive parents who are physically abusive. The earlier the maltreatment the more severe the consequences.

Refs Gelles, Richard J and Lancaster, Jane B (Eds) *Child Abuse and Neglect* (1988); Korbin, Jill E, et al *Child Abuse and Neglect* (1981); Leavitt, Jerome E (Ed) *Child Abuse and Neglect* (1983).
Broader Cruelty (#PB2642) Victimization of children (#PC5512).
Narrower Neglected children (#PD4522) Sexual abuse of children (#PE3265)
Emotional abuse of children (#PD7330) Exploitation of child labour (#PD0164)
Physical maltreatment of children (#PC2584) Trafficking in children for adoption (#PF3302).
Related Denial of rights of children and youth (#PD0513)
Inadequate care for children of prisoners (#PF0131)
Discrimination and harassment of children in public life (#PE6922).
Aggravates Maladjusted children (#PD0586) Loneliness of children (#PC0239)
Physically handicapped children (#PD0196) State custody of deprived children (#PD0550).
Aggravated by Drunkenness (#PE8311) Chaotic households (#PG0499)
Adolescent pregnancy (#PD0614) Trafficking in women (#PC3298)
Socio-economic poverty (#PB0388) Single parent families (#PD2681)
Dependency of children (#PD2476) Exploitation of children (#PD0635)
Physical intimidation by children (#PE2876).

♦ PC0840 Economic conflict

Economic invasion — Economic aggression — Economic infiltration — Trade war — Economic warfare

Nature Economic warfare is interference in international economic relations by a state, or group of states, for the purpose of improving their relative economic, political or military position. While conventional means of warfare are becoming increasingly less effective during long conflicts, economic and other less obvious means of warfare (guerrilla warfare) may be more effective or disruptive to the enemy.

Trade wars have resulted from attempts by governments to regulate their economies in such a way as to increase their power at the expense of rival countries. This primarily involves attempts to create surpluses of exports over imports. The belief that such surpluses signify national strength and security has led governments to follow policies that have come to be described as 'beggar thy neighbour' policies. By means of tariffs, quotas and devaluation of currencies, governments seek to reduce the entry of foreign goods into their own markets while trying to sell as much as possible to other countries.

Incidence The most disastrous trade wars occurred during the 1930s in Europe. They created such instability that trade declined and finally collapsed, bringing on the period of the 'great depression'.

Refs Conybeare, John A C *Trade Wars* (1987); Gordon, David R and Dangerfield, Royden *The Hidden Weapon* (1976).
Broader Conflict (#PA0298) International aggression (#PC7559).
Narrower Price warfare (#PJ4045) Economic unrest (#PD4012)
Economic civil war (#PD3765) Computer development trade war (#PG4302)
Economic sanctions against governments (#PF4260)
Distortion of international trade by embargoes and similar restrictions (#PE0522).
Related Imperialism (#PB0113) Economic influence (#PG4965)
Economic intimidation (#PC3011).
Aggravates Enemies (#PF8404) Cultural invasion (#PC2548)
Trade protectionism (#PC4275) Aggressive foreign policy (#PC4667).
Aggravated by Economic retaliation (#PD9389) Restrictive trade practices (#PC0073)
Excessive external trade deficits (#PC1100) Competition in capitalist systems (#PC3125)
Exploitation in capitalist systems (#PC3117) Contradictions in capitalist systems (#PF3118)
Vulnerability of intellectual property (#PF8854)
Distortion of international trade by dumping (#PD2144)
Waste of human resources in capitalist systems (#PC3113)
Conflicting roles of commodities in capitalism (#PF3115)
Conflicting roles of money in capitalist systems (#PF3114)
Excessive demand for goods in capitalist systems (#PC3116)
Denial of the right to work in capitalist systems (#PC3119)
Over-subsidized agriculture in industrialized countries (#PD9802)
Denial of the right to social security in capitalist systems (#PD3120).

♦ PC0842 Nuclear war

Effects of nuclear war

Nature Nuclear weapons add a completely new dimension to man's powers of destruction. Published estimates of the effects of nuclear weapons range from the concept of the total destruction of humanity to the belief that a nuclear war would differ from a conventional conflict only in scale. The situation, however, is not as arbitrary as opposing generalizations such as these might suggest. There is one inescapable and basic fact: the nuclear armouries in existence already contain large megaton weapons, every one of which has a destructive power greater than that of all the conventional explosives ever used in warfare since the day gunpowder was discovered. Were such weapons ever to be used in numbers, hundreds of millions of people might be killed; and civilization as we know it, as well as organized community life, would inevitably come to an end in the countries involved in the conflict. Many of those who survived the immediate destruction, as well as others in countries outside the area of conflict, would be exposed to widely-spread radio-active contamination, would suffer from long-term effects of irradiation and would transmit to their offspring a genetic burden which would become manifest in the disabilities of later generations.

Incidence A study has been made of the likely results of a nuclear attack on a hypothetical industrial region, consisting of nine cities each with populations of over 50,000 inhabitants (some well over), and also containing 140 smaller towns of fewer than 50,000 inhabitants (about sixty of which containing elements of key industry). Assuming that a one-megaton bomb burst at ground level in each of the nine cities, the study showed that simple cumulative estimates of casualties provide a very inadequate measure of the over-all effects of the attack. Such estimates showed that 20 per cent of the total population (or 30 per cent of the urban population, 35 per cent of the key-industrial population) would be killed. The houses destroyed would be 30 per cent of total (or 40 per cent of urban, 50 per cent of those occupied by key-industrial population). But cities are not isolated entities; they are linked in a variety of functional ways, being dependent on each other for raw materials of different kinds, as well as for semi-finished and finished manufactured goods. Taking the interaction of effects into account, the study showed that the percentage of key industry in the whole region (that is, industry with more than local significance) which would be brought to a stop would be between 70 per cent and 90 per cent of the whole. The lower figure of 70 per cent takes account of everything directly destroyed or completely disrupted inside the target cities; the higher figure of 90 per cent includes the areas surrounding the city which would also be indirectly 'knocked out' through, for example, failures of communications or supplies of raw materials and food. The more interdependent they are, the larger is the multiplying factor one has to bear in mind when estimating the cumulative effects of the destruction of single cities.

In hypothetical studies of this kind it has also been estimated that in the absence of special protection, blast-induced deaths alone resulting from 400 high level ten-megaton bombs aimed at United States metropolitan areas, would eliminate more than half of the total American population of over 200 million people. Even if they were all in substantial fall-out shelters the same proportion would be killed if the weapons were burst at ground level.

Background In 1945 atomic bombs were dropped on the Japanese cities of Hiroshima and Nagasaki.

Claim The effects of all-out nuclear war, regardless of where it started, could not be confined to the powers engaged in that war. They themselves would have to suffer the immediate kind of destruction and the immediate and more enduring lethal fall-out whose effects have already been described. But neighbouring countries, and even countries in parts of the world remote from the actual conflict, could soon become exposed to the hazards of radioactive fall-out, in precipitation at great distances from the explosion, of matter which had moved through the atmosphere as a vast cloud. Thus, at least within the same hemisphere, an enduring radio-active hazard could exist for distant as well as close human populations, through the ingestion of foods derived from contaminated vegetation, and the external irradiation due to fall-out particles deposited on the ground. The extent and nature of the hazard would depend upon the numbers and type of bombs exploded. Given a sufficient number, no part of the world would escape exposure to biologically significant levels of radiation. To a greater or lesser degree, a legacy of genetic damage could be incurred by the world's population.

Counter-claim 1. It is to be expected that no major nuclear power could attack another without provoking a nuclear counter-attack. It is even possible that an aggressor could suffer more in retaliation than the nuclear power it first attacked. In this lies the concept of deterrence by the threat of nuclear destruction. Far from an all-out nuclear exchange being a rational action which could ever be justified by any set of conceivable political gains, it may be that no country would, in the pursuit of its political objectives, deliberately risk the total destruction of its own capital city, the destruction of all its major centres of population and the resultant chaos which would leave in doubt a government's ability to remain in control of its people.

2. In the course of World War I and II some 70 million people were killed. Of these only 100,000 were killed by nuclear weapons. The remainder were victims of conventional weapons.

Refs Adams, Ruth and Cullen, Susan (Eds) *The Final Epidemic*; American Health Research Institute Staff *Nuclear Warfare* (1987); Catrina, Christian and Frei, Daniel *Risks of Unintentional Nuclear War*; Fox, Michael Allen and Groarke, Leo (Eds) *Nuclear War* (1987); Griffiths, Franklyn and Polanyi, John C (Ed) *Dangers of Nuclear War* (1979); Harwell, Mark A and Hutchinson, Thomas C *The Environmental Consequences of Nuclear War (SCOPE 28)* (1986); Malcolmson, Robert W *Nuclear Fallacies* (1985); Riordan, Michael (Ed) *The Day After Midnight*; United Nations Institute for Training and Research *Prevention of Nuclear War* (1984).
Broader War (#PB0593).
Narrower Sunlight inhibition by nuclear warfare soot (#PE6350)
Inadequate health services following nuclear war (#PD6265).
Related Unconventional war (#PC8836) Biochemical warfare (#PC1164)
Radiological warfare (#PC6666).
Aggravates Civil disobedience (#PC0690) Radioactive fallout (#PC0314)
Ecological imbalance (#PG3026) Radioactive contamination (#PC0229)
War damage in civilian areas (#PD8719) Disastrous consequences of war (#PC4257)
Insufficient nuclear weapon free zones (#PJ5335).
Aggravated by Dehumanization (#PA1757) Military secrecy (#PC1144)
Military-industrial complex (#PC1952) Competitive acquisition of arms (#PC1258)
Accidents to nuclear weapons systems (#PD3493).

♦ PC0845 Political dictatorship

Despotism — Autocratic rule — Tyranny

Nature Government by a single person, as opposed to a ruling clique. Rule tends to be maintained by intimidation, use of secret police and the armed forces, and by economic control. These encourage the persistence of social inequalities and elitism, and halt political and social development.

Some dictatorships, particularly in African nations, are based on traditional values where the structure of oppression already exists. The dictator merely diversifies the instruments of control, strongest of which may be the emphasis on traditional values and institutions. Most dictators do not appoint delegates based on their intelligence, but rather based on their lack of intelligence, lack of character, lack of constitution and courage, as these are the people who will neither challenge nor defy a dictator's rule. In its extreme form, dictatorship may become tyrannical, resulting in many excesses.

Background Despotism has been recognized as a form of government since the era of classical Greek philosophy when it was applied particularly to practices in Persia and the East. Despotism has been recorded through history to modern times and found an especially vivid expression in the absolutist monarchies in France and elsewhere in Europe during the 17th and 18th centuries.

Refs Friedrich, C J von and Brazezinski, Z K *Totalitarian Dictatorship and Autocracy* (1969); Jackson, Robert H and Rosberg, Carl G *Personal Rule in Black Africa* (1982); Wittfogel, Karl A *Oriental Despotism* (1981).
Broader Dictatorship (#PC1049) Totalitarianism (#PF2190)
Authoritarian regimes (#PC9585) Illegitimate political regimes (#PC1461).
Narrower Foreign dictatorship (#PC3186) Abusive national leadership (#PD2710).
Related Elitism (#PA1387) Political oligarchy (#PD3238)
Dictatorship of the majority (#PD3239).
Aggravates Torture (#PB3430) Lawlessness (#PA5563) Endangered cultures (#PB8613)
Lack of political development (#PD8673)
Monopoly of competence for intervention in the future (#PF0980).
Aggravated by Racism (#PB1047) Tribalism (#PC1910)
Massacres (#PD2483) Militarism (#PC2169)
Megalomania (#PF2108) Martial law (#PD2637)
Nationalism (#PB0534) Political apathy (#PC1917)
Social dictatorship (#PD3241)
Inadequate political parties in developing countries (#PD0548).
Reduces Disruption of development by tribal warfare (#PD2191).
Reduced by Assassination (#PD1971) State sanctioned torture (#PD0181)
Failure of individuals to participate in social processes (#PF0749).

♦ PC0865 Occupational risk to health

Occupational health risks

Refs Forssman, S and Coppee, G H *Occupational Health Problems of Young Workers* (1975); ILO *Ad Hoc Meeting on Civil Aviation, Geneva, 1977*; Parmeggiani, Luigi (Ed) *Encyclopedia of Occupational Health and Safety* (1983); Stern, R M, et al *Health Hazards and Biological Effects of Welding Fumes and Gases* (1986).
Broader Risk (#PF7580) Hazards to human health (#PB4885).
Narrower Industrial accidents (#PC0646) Occupational diseases (#PD0215)
Health risks to workers in commerce (#PE0688) Health hazards in the steel industry (#PE8263)
Health hazards in the glass industry (#PE8274)
Health risks to workers in service industries (#PE0875)
Health risks to workers in construction industry (#PE0526)
Health risks to workers in manufacturing industries (#PE1605)
Health risks to workers in agricultural and livestock production (#PE0524)
Health risks to workers in electricity, gas, water and sanitary services (#PE1159)
Health risks to workers in transport, storage and communication industries (#PE1581).
Related Occupational hazards (#PC6716) Ultraviolet radiation as a hazard (#PE5672).
Aggravates Pneumoconiosis (#PD2034) Dangerous occupations (#PC1640).
Aggravated by Injuries (#PB0855) Aeroallergens (#PE2069)
Air pollution (#PC0119) Lead as a pollutant (#PE1161)
Nickel as a pollutant (#PE1315) Cobalt as a pollutant (#PE2339)
Toxic metal pollutants (#PD0948) Mercury as a pollutant (#PE1155)
Arsenic as a pollutant (#PE1732) Lead as a health hazard (#PE5650)
Asbestos as a pollutant (#PE1127) Fluorides as pollutants (#PE1311)
Vanadium as a pollutant (#PE2668) Health hazards of asbestos (#PE3001)
Insecticides as pollutants (#PD0983) Manganese as a health hazard (#PE1364)
Heat as an occupational hazard (#PE5720) Cold as an occupational hazard (#PF5744)
Sulphur dioxide as a pollutant (#PE1210) Occupational hazards of benzene (#PE1849)
Solvents as an occupational hazard (#PE5708) Carbon monoxide as a health hazard (#PE1657)
Health hazards of exposure to noise (#PC0268)
Combined stresses as occupational hazards (#PE5656)
Biological agents as occupational hazards (#PE5696)
Shift work stress as an occupational hazard (#PE5768)
Improper lighting as an occupational hazard (#PE5780)
Excessive occupational exposure to radiation (#PD1500)
Polychlorinated biphenyls as a health hazard (#PE2432)
Conflicting standards for protection against chemical occupational hazards (#PE5651).

♦ PC0872 Chemical warfare

Chemical weapons — Binary chemical weapons

Nature Chemical agents of warfare include all gaseous, liquid or solid chemical substances which might be employed because of their direct toxic effects on man, animals or plants. (This excludes substances whose effect is primarily physical, such as incendiary weapons, high explosives and smoke). Nerve agents (tasteless chemicals of the same family as organophosphorus insecticides) are the most lethal, killing by poisoning the nervous system and disrupting bodily functions. Blister agents or vesicants (such as mustard gas) burn and blister the skin (causing more casualties than any other agent in the first world war). Choking agents (such as phosgene) are highly volatile liquids. Toxins (such as mycotoxins and botulinum toxin) are biological substances produced chemically which are very highly toxic. Tear gas and harassing agents are sensory irritants causing skin irritation, nausea and vomiting (and have been widely used as riot control weapons). Psycho-chemicals are drug-like chemicals intended to cause temporary mental derangement. Herbicides may also be used to poison or defoliate plants.

Incidence The intent to use chemicals in warfare is indicated by stockpiles of weapons and other military preparedness. For example, the Soviet Scud missile has a gas warhead and the chemical war corps of the USSR totals 80,000 men and 30,000 decontamination and reconnaissance vehicles. Soviet stockpiles of chemical weapons are extensive and include long-lasting VX nerve gas. Similar preparations by the USA include a 150,000 ton stockpile, and a multi-billion dollar budget to produce binary nerve gas weapons. France and the United Kingdom are also believed to be manufacturing and stocking chemical weapons. Over twenty nations have chemical arms besides the four powers. They include Czechoslovakia and at least two other Warsaw Pact countries, Egypt, Iraq, Vietnam, and North Korea.

Background Chemical weapons were first used on a large scale in the First World War, when there were 100,000 fatalities and over one million other poison–gas casualties. Some 100 million kilograms of chemical, lethal and non–lethal gases were used by the seven parties. Other major instances of use were in the Italian invasion of Ethiopia (1935–36), and the Japanese invasion of China (1937–45). Chemicals were used by the Nazis during the Second World War to gas concentration camp civilians; and, although chemical weapons were not employed in combat, almost all the belligerents maintained stocks, most of which were destroyed after the war. Defoliant chemicals were originally developed by Britain and used in the Malayan anti–terrorist campaign in 1948–60. During the Second Indochinese War, the United States used Agent Orange, a defoliating herbicide, and the riot–control gas CS. The USA used the same quantity of chemicals as was used by all the belligerents in the First World War, about 100 million kilograms (1965–1970). The Soviet Union has been charged with using Yellow Rain and other chemical weapons in Afghanistan since 1979, and Iraq has been charged with using mustard gas against Iran since 1983. Chemical attacks were reported in 1984 in the Gulf (Iran), Southeast Asia (Afghanistan, Laos, Kampuchea) and the Horn of Africa.

Claim The threat of chemical weapons is so real that computer studies have been made to demonstrate possible effects. For example, if 2,000 tons per day of nerve gas were used in a war in Europe by all combatants, the ratio of civilian casualties to military could reach 20 to 1, and total millions. The 1925 Geneva Protocol prohibiting chemical weapons is deficient and not regarded as a deterrent to their use, since a number of signatories reserve the right to retaliate in kind with such weapons and it is easy to claim or fabricate a claim of chemical attack.

Counter–claim Chemical weapons, toxic and non–toxic, are a humane alternative to nuclear war. A gas, for example, may still be developed that merely temporarily incapacitates.

Refs Aspen Strategy Group and European Strategy Group *Chemical Weapons and Western Security Policy* (1987); Gander, T *Nuclear, Biological and Chemical Warfare* (1987); Lundin, S J (Ed) *Non–Production by Industry of Chemical–Warfare Agents* (1989); Robinson, J P *Chemical and Biological Warfare Developments 1986–1987* (1990); Robinson, Julian (Ed) *The Chemical Industry and the Projected Chemical Weapons Convention* (1986); Stock, Thomas and Sutherland, Ronald (Eds) *National Implementation of the Future Chemical Weapons Convention* (1990); WHO *Health Aspects of Chemical and Biological Weapons* (1970).

Broader Biochemical warfare (#PC1164).
Narrower Nerve gases (#PG3050)
Blister agents (#PE7103)
Toxic substances (#PD1115)
Lung–damaging agents (#PE4555)
Psychochemical agents (#PD4248)
Indiscriminate use of tear gas (#PE4640)
Inhumane methods of riot control (#PD1156)
Anti-personnel use of toxic substances in peacetime (#PE9294).
Related Biological warfare (#PC0195).
Aggravates Defoliation (#PD1135)
Herbicide damage to crops (#PD1224).
Aggravated by Surface to surface missiles (#PE4515)
Trade in products for chemical warfare (#PE3808).
Reduces Inadequate riot control (#PD2207).

♦ PC0877 Juvenile stress

Adolescent disturbance — Youth anxiety

Nature The main characteristics of juvenile distress are: (a) a tendency towards anti–social behaviour, particularly as expressed in acts of unprecedented violence; (b) a revolution in sexual mores encouraging a tendency to promiscuity and perversion; (c) a wave of contagion that makes an obsession out of every new kind of stimulus (hot jazz, hot dancing, hot cars); (d) a leaning towards over–conformity with family and community or with a peer group; (e) an associated trend towards static–mindedness, a loss of adventure and creative spark; (f) a tendency to withdrawal, toward a loss of hope and faith, towards disillusionment and despair with progressive destruction of ideals; (g) a failure on the part of the adolescent to harmonize his goals with those of his family or society; (h) a trend toward disorientation, confusion, fragmentation of personal identity; (i) an increasing vulnerability of the adolescent to mental illness, as a result of aggravated orders of social adaptation.

Claim Adolescent disturbance needs is partly a result of the turmoil and instability of marital partnerships, the insecurity of parents and the disintegrative trends in family life as a whole. There is something deeply wrong, but it is not just with adolescents. It is with the whole mode of family life. Social health shows signs of failing and the effects of this failure cast a long shadow on our mental health. Family and community today fail to provide a receptive climate for the adolescent's needs.

Refs Garbarino, James, et al *Troubled Youth, Troubled Families* (1986); Irwin, Edna M *Growing Pains* (1978); Spielberger, Chas D, et al *Stress and Anxiety* (1988).

Broader Anxiety (#PA1635)
Stress in human beings (#PC1648).
Narrower Stress among children (#PE4421).
Aggravates Youth gangs (#PD2682)
Youth violence (#PF7498)
Apathy of youth (#PF5949)
Juvenile suicide (#PE5771)
Juvenile desertion (#PD8340)
Age discrimination (#PC2541)
Juvenile alcoholism (#PD1611)
Juvenile delinquency (#PC0212)
Idle youth lifestyle (#PG7766)
Juvenile prostitution (#PD6213)
Disobedience of elders (#PF7149)
Drug abuse by adolescents (#PD5987)
Exploitation of the elderly (#PD9343)
Mental illness in adolescents (#PE0989)
Adolescent sexual intercourse (#PD7439)
Abusive treatment of the aged (#PD5501)
Social disaffection of the young (#PD1544).
Aggravated by Lovesickness (#PF3385)
Family stress (#PD8130)
Marital stress (#PD0518)
Resettlement stress (#PD7776)
Competitiveness in education (#PD4178)
Psychological stress of urban environment (#PE6299)
Environmental stress on inhabitants of tall buildings (#PE4953).

♦ PC0880 Underdevelopment of industrial and economic activities

Broader Underdevelopment (#PB0206).
Narrower Lack of agricultural machinery (#PF4108)
Underdevelopment of fishing industry (#PE2138)
Rural poverty in developing countries (#PD4125)
Underdevelopment of forestry industry (#PG5324)
Political nature of development issues (#PF4175)
Underprovision of basic urban services (#PF2583)
Arrested development of labour potential (#PF6532)
Underdevelopment of manufacturing industries (#PF0854)
Inadequate trade–related structural adjustments (#PF4165)
Underprovision of basic services to rural areas (#PF2875)
Underdevelopment of mining and quarrying industries (#PE1858)
Underdevelopment of food and live animal production (#PF2821)
Ineffective economic structures in industrial nations (#PE4818)
Underdevelopment of agricultural and livestock production (#PD0629)
Inadequacy of the domestic market in developing countries (#PD0928)
Underdevelopment of hunting, trapping and game propagation (#PE8252)
Underdevelopment of the power industry in developing countries (#PF4135)
Underparticipation of developing countries in the airline industry (#PE4145)
Disparate development of economic sectors within developing countries (#PD1534)
Economic disadvantages of excessive food production in developing countries (#PF4130).
Related Economic apathy (#PC3413)
Lack of economic and technical development (#PE8190).
Aggravates Trachoma (#PE1946)
Permafrost instability (#PD1165).
Aggravated by Disease (#PA6799)
Ignorance (#PA5568)
Illiteracy (#PC0210)
Human disease and disability (#PB1044)
Unsustainable population levels (#PB0035)
Nepotism in developing countries (#PD1672)
Factionalism in developing countries (#PD1629)
Capital shortage in developing countries (#PD3137)
High labour turnover in developing countries (#PF0907)
Low occupational mobility in developing countries (#PD1493)
Inadequate staple food supply in developing countries (#PD4101)
Unrelated pioneer institutions in developing countries (#PF1724)
Underdevelopment of chemical and petrochemicals industry (#PE1483)
Dependence of developing countries on imported technology (#PF1489)
Inadequate electrical power supply in developing countries (#PE1900)
Inadequacy of the commercial sector in developing countries (#PD1865)
Inadequate transportation facilities in developing countries (#PD1388)
Lack of response to monetary incentives in developing countries (#PD1432)
Loss of traditional forms of social security in developing countries (#PD1543)
Shortage of industrial leadership and entrepreneurial ability in developing countries (#PC1820).
Reduced by Expropriation of land from indigenous populations (#PC3304).

♦ PC0918 Global warming

Overheating of the planet — Greenhouse effect — Deterioration of the atmospheric radiation balance — Increasing air temperature

Nature The climate of the earth is primarily conditioned by solar radiation. However, the amount of energy used by man (particularly through the burning of fossil fuels) over large areas of industrially developed countries is almost equal to the amount of solar energy that reaches these areas. At the same time, many processes lead to the production of carbon dioxide, chlorofluorocarbons, methane and water vapour. This is almost transparent to the incoming heat radiation from the sun but strongly absorbs and reflects back infrared radiation from the Earth's surface. The heat can therefore progressively build up. This raise the average temperature and considerably affect climatic patterns. It has been suggested that a fairly small increase of temperature could lead to changes in rainfall patterns, rises in sea levels, melting of the ice caps and widespread flooding. Evidence continues to accumulate that increases in atmospheric carbon dioxide (CO2) and other "greenhouse" gases are substantially raising the global temperature. Global warming causes more evaporation from the tropical oceans, putting more water vapour into the atmosphere. This leads to increased precipitation over cooler parts of the globe, including the polar regions where, in the short term at least, it is still cold enough for that precipitation to fall as snow. This increases the size of the polar ice sheets. The build up of carbon dioxide may also explain the increase in plankton blooms to an extent that two to three times more organic matter is being produced by photosynthesis than has been measured before. While considerable uncertainty exists concerning the rate and ultimate magnitude of such a temperature rise, current estimates range from a 1.5 to 4.5 degree increase will occur in the next 20 to 30 years to a 1 degree rise in the next 50 years. A warming of 0.4 degrees Centigrade in a decade is thought to be faster than most planets are able to accommodate.

Incidence Trends in global mean surface air temperature for the period 1900–1987 indicate an increase of 0.3 to 0.7 degrees Centigrade for that period. A warming trend o 0.09 degrees Centigrade per decade is found to be significant (at the 95 per cent confidence level) in the tropospheric 850–300 mb layer over the 30 year period 1958–1987, with a cooling of 0.62 degrees per decade in the stratospheric 100–50 mb layer during the interval 1973–1987. It is unlikely that the warmer temperatures predicted for the middle of the next century (warmer than previously recorded in history) can be avoided. Fossil fuel production and consequent carbon dioxide release is going to continue; it will take at least hundreds of years for it to be reabsorbed into the ocean. The rise in summertime temperatures will make certain areas more difficult to live in, although the changes in winter temperatures and snowfall patterns could make currently uninhabitable land more favoured. As the ocean warms, sea level will rise 25–140 centimetres due to thermal expansion and to the melting of ice in Antarctica and Greenland. This will have serious effects on coastal areas. A warmer ocean would also have other potential impacts such as altering hurricane frequency, diminishing sea ice concentration, and affecting fish and ocean biota. The exact nature of these results cannot be predicted with any certainty, but it is definite that mankind will need to adjust in a variety of ways. for many of the earth's plants and animals, a few degrees make the difference between survival and extinction.

Claim Such increases in the span of only a few decades represent an unprecedented rate of atmospheric warming which is leading to climatic changes of sufficient magnitude to produce major physical, economic and social dislocations on a global scale. Temperature increases are likely to be accompanied by dramatic changes in precipitation and storm patterns and a thermal expansion of the oceans giving raising the global average sea level. As a result, agricultural conditions will be significantly altered, environmental and economic systems potentially disrupted, and political institutions under stress. A one metre rise could displace 15 million people in Bangladesh, as may as 10 million in Egypt and large numbers elsewhere. Major river deltas in the United States, France, Italy, and Spain are among the vulnerable areas.

Counter–claim The greenhouse effect is not controversial, nor is it synonymous with global warming. Evidence from the ice ages shows that the carbon dioxide greenhouse has always played a crucial role in shaping the climate, but only as one element in a set of phenomena to which few pay attention. Owing to a large natural variability of the climate system, it has not been possible to ascribe the observed warming to changing greenhouse concentrations in a statistically rigorous manner. One recent analysis has concluded that there is no danger at present of a buildup of carbon dioxide in the atmosphere sufficient to cause a significant heating of the Earth. It is possible that the increase in the number of dust particles in the atmosphere will reflect back more of the sun's rays and produce a compensating reduction in the heating effect. Or global warming might have been caused by solar activity that will decrease next century and so a slight cooling will offset some of the greenhouse effect. Some scientists feel that as the planet warms, increased evaporation from the sea will produce more thick clouds, which will cool the Earth. There is also evidence that global warming is due to the warming cycle of natural rhythms of glaciation.

There are three ways of making predictions about the future climatic conditions: theory, computer models and existing data. All existing theoretical models about the climate are crude and very limited, as illustrated by the accuracy of weather forecasts of periods longer than 5 days. Computer models of the climate are based on theory and are limited. The use of existing data, that is past data, is a poor guide to the future. While there is a trend over the past century of a rise in global temperature, to conclude that this is due to the greenhouse effect is to believe that this is the sole cause.

Refs Flavin, Christopher *Slowing Global Warming* (1989); Forsdyke, A G *Meteorological Factors of Air Pollution* (1970); Kuo–Nan Liou, *An Introduction to Atmospheric Radiation* (1980); McCuen, Gary E *Our Endangered Atmosphere* (1987); Smith, I M *CO2 and Climate Changes* (1988); United Nations Environment Programme *The Greenhouse Gases* (1987); WHO *Report of the International Conference on the Assessment of the Role of Carbon Dioxide and of Other*

Greenhouse Gases in Climate Variations and Associated Impacts (1985) (1986).
Broader Long-term cyclic changes of climate (#PC6114)
Environmental threats to national security (#PC4341).
Narrower Sunlight inhibition by nuclear warfare soot (#PE6350).
Aggravates Hurricanes (#PD1590) Climatic heat (#PC2460)
Household pests (#PD3522) Evolutionary catastrophes (#PF1181)
Reduction of glacier size (#PE4256) Excessive environmental heat (#PD7977)
Slowing growth in food production (#PC1960)
Long-term changes in precipitation patterns (#PE4263)
Ecological disruption of animal breeding grounds (#PJ3994).
Aggravated by Soot (#PE1953) Air pollution (#PC0119)
Tropical deforestation (#PD6204) Meteorological disaster (#PD4065)
Anthropogenic climate change (#PC9717) Stratospheric ozone depletion (#PD6113)
Increasing atmospheric carbon dioxide (#PD4387)
Increased reflection of solar radiation (#PF5069)
Medium-term cyclic variations in solar radiant energy (#PE9528)
Increase in concentration of atmospheric water vapour (#PE9446).
Reduces Global cooling (#PF1744).

♦ PC0927 Inadequate savings
Uninvested personal savings — Spendthrift economies
Nature In most developing countries, total investment exceeds national savings; and the gap is filled by the net inflow of financial resources from abroad, on which the countries are therefore dependent.
Incidence In the USA net savings have fallen from more than 7 per cent of the economy in the 1960s to about 3 per cent in the 1980s at a time when increased domestic saving is of major importance for promoting higher rates of growth and an improved standard of living. In 1990 it was estimated that the global pool of savings was decreasing markedly. The very high Japanese savings rate that helped finance growth in the 1980s is slowly shrinking and other countries are seeing a decline in savings as well. The savings rate in virtually every country, other than the USA, is lower than it was in 1985. In the USA the savings rate remained at a historic low point in 1990. At the same time, the demands on savings are growing steadily.
Claim Among the factors determining the rate of economic growth, the capacity to save and to invest has long been considered one of the most important. The surplus of production over consumption creates the basis for the augmentation of future output. The lack of resources for investment has thus been identified as a principal constraint to economic and social growth, and policy prescriptions have accordingly been concerned with means of raising the rate of saving.
Counter-claim Research findings and practical experience accumulated over the years have caused many writers to question the central role assigned to capital accumulation. Referring to the historical experience of developed countries of widely varying institutional conditions and political systems in the nineteenth and twentieth centuries, they have observed that development has been the result of a combination of political, social and economic changes which have interacted to produce further changes. Such factors as institutional changes, technological progress, population growth and migration, a heightened spirit of enterprise, improvements in the quantity and quality of the labour force, and the widening of markets, have all played a part, along with capital formation, in promoting development. Some writers have even seen capital accumulation as more a consequence than a cause of development.
Broader Capitalism (#PC0564) Economic and social underdevelopment (#PB0539)
Undirected expansion of economic base (#PF0905).
Aggravated by Low bank interest rates (#PE3903).

♦ PC0931 Restrictions on international freedom of information
Nature Restrictions on the availability of national information to foreign media include: government secrecy; government control of information (both official data and propaganda); the expulsion of foreign correspondents and refusal of entry to others; and confiscation of articles, film, photographs, etc. Such restrictions hide existing injustices, inequality, exploitation and repression by keeping them closed to international scrutiny. They may encourage espionage and subversive activities or lack of cooperation and international tension.
Most governments feel that they should be able to refuse imported programmes. Some use this position to censor TV and radio broadcasting.
Broader Restrictions on freedom of information (#PC0185).
Narrower Suppression of information (#PD9146).
Related Denial of freedom of thought (#PF3217) Restriction of freedom of expression (#PC2162).
Aggravates Conspiracy (#PC2555) Lack of international cooperation (#PF0817).
Aggravated by Official secrecy (#PC1812)
Conflict of laws on international restriction of information (#PD3080).

♦ PC0933 Weakness in trade among developing countries
Decline in mutual trade between developing countries — Weakness in South - South interregional trade — Inadequate economic cooperation among developing countries
Nature Trade among developing countries is still at a much lower level, both absolutely and comparatively, than is generally realized. That which does occur takes place largely between countries of the same region or continent and is considerably inferior, in value terms, to trade with developed countries. As industrial requirements alter, intra-trade presents supply problems in the exporting countries as well as in the composition of exports. There is also an inability of developing country exporters to compete with the terms on which some commodities were being made available under aid-in-kind programmes of developed market economies. Another obstacle is the fact that the less advanced developing countries are less able to derive rapid benefits from, and bear the short-term burden of, regional cooperation than the more advanced ones, no matter how great the long-term benefits may be. This is frequently accompanied by insufficient linkages between trade liberalization measures, cooperation in production, insufficient channels of communication, restricted sources of finance, inadequate monetary arrangements and links, etc.
Incidence In 1981, the value of exports to the developing countries amounted to US $488,000 million, or about 25 percent of world exports. South-South exports in the same year made up 41 percent of the exports to the South. Although the bulk of developing country exports still consists of primary commodities, the main lag in intra-trade was in foodstuffs and raw materials (except petroleum, over 80 percent of developing country imports of which still comes from developing countries).
Background The low level and slow growth of trade among developing countries is largely determined by their past economic and political relationships with the industrialized part of the world. Historical links such as the flow of trade information, transport and communication systems, and export credit facilities all tend to encourage existing North-South trading patterns. The type of products that dominate developing countries' exports (primary products) find their most important markets in the developed countries. The imports of the developing countries come mainly from developed countries because they are very often more efficient sources of supply or the only sources for many goods vital for the development process; the goods produced in other developing countries, on the other hand, are very frequently similar to those produced domestically and are, therefore, kept out by protective measures.
Refs Bracho, Frank *Utopia and Reality of South-South Economic Cooperation* (1986).
Broader Fragmented regional cooperation (#PF9129)
Economic and social underdevelopment (#PB0539)
Imbalance in international trade patterns (#PC8415).
Narrower Inadequate trade between developing countries (#PD5176)
Weakness in intra-regional trade of developing countries (#PD0169)
Weakness in primary commodity trade amongst developing countries (#PE2951)
Weakness in trade in manufactured goods among developing countries (#PE2966)
Non-uniformity of tariff and non-tariff trade preferences of developing countries (#PJ8895).
Aggravated by Disparities between developing countries (#PD2963)
Weakness of infrastructure in developing countries (#PC1228)
Trade barriers and protectionism between developing countries (#PD2958)
Developed country limiting of trade between developing countries (#PD2961)
Bipolarization of trade between developed and developing countries (#PE4190)
Differences in trading principles and practices between developing countries (#PF2960).
Reduces Social insecurity in developing countries (#PE4796).

♦ PC0934 Discrimination in politics
Dependence on discrimination in politics — Political discrimination
Nature Discrimination in politics concerns voting rights, right to form and join political associations and the right to hold public office. Discrimination may occur on the grounds of race, colour, religion, language, nationality, sex, social status and possessions, political or ideological belief, or educational standing. Discrimination in politics causes conflicts and segregation, encourages exploitation and repression and may lead to violence, subversive activities, revolution and even war.
Broader Discrimination (#PA0833).
Narrower Racial discrimination in politics (#PD3329)
Language discrimination in politics (#PD3223)
Religious discrimination in politics (#PC3220)
Political discrimination in politics (#PC3221)
Ideological discrimination in politics (#PC3219)
Discrimination against women in politics (#PC1001)
Political discrimination based on illiteracy (#PC3222)
Property and occupational discrimination in politics (#PD3218)
Political discrimination in the administration of justice (#PE1828).
Aggravates Conflict (#PA0298) Social fragmentation (#PF1324).
Aggravated by Age discrimination (#PC2541) Political intimidation (#PC2938)
Deprivation of nationality (#PD3225) Limitations on right to vote (#PF2904).

♦ PC0935 Restrictions on freedom of movement between countries
Denial of right to freedom of international movement
Nature Restrictions of a legal and administrative nature to block the free movement of people between countries may result in statelessness for the persons concerned, or a reduction or annulment of their legal rights either in the country of origin or in the country of residence or transit.
Broader Travel restrictions (#PC8452) Violation of civil rights (#PC5285).
Narrower Harassment (#PC8558) Gypsy persecution (#PE1281)
Forced repatriation (#PD8099) Rejection of refugees (#PF3021)
Restrictions on emigration (#PC3208) Restrictions on immigration (#PC0970)
Foreign exchange restrictions (#PF3070) Denial of right to leave any country (#PE3463)
Harassment of travellers by immigration officials (#PE7780)
Refusal to issue travel documents, passports, visas (#PE0325)
Restrictions on nationals leaving their own country (#PE8414)
Restrictions on nationals returning to their country (#PE8817)
Restrictions on foreigners leaving country of sojourn (#PE8747)
Denial of the right to return to country of residence (#PJ7425)
Impediments to internationally mobile professionals and experts (#PF1068)
Violation of the right to the international freedom of movement of shipping (#PE8899).
Related Discrimination (#PA0833) Legalized racial discrimination (#PC3683)
Restrictions on freedom of worship (#PD5105) Repression of intellectual dissidents (#PD0434).
Aggravates Apartheid (#PE3681) Statelessness (#PE2485)
Disease transmission by international travel (#PD6421)
Discrimination against foreigners in employment (#PD3529)
Distortion of international trade by restrictive controls on movement of labour (#PE8882).
Aggravated by War (#PB0593) Forced labour (#PC0746) Illegal immigration (#PD1928)
Deprivation of nationality (#PD3225) Refusal to grant nationality (#PF2657)
Delay in issue of travel documents (#PE9123) National insecurity and vulnerability (#PB1149)
Excessive frontier formalities in international travel (#PE0208)
Denial of social security to nationals who have lived abroad (#PE8746)
Excessive animal sanitary regulations in international travel (#PF1555).
Reduces Emigration of trained personnel from developing to developed areas (#PD1291)
Inadequate living and working conditions of immigrant labourers in industrialized countries (#PD3427).

♦ PC0936 Health hazards of environmental pollution
Human exposure to hazardous environmental pollutants
Incidence Estimates of human exposure are made at two levels. First level, based on the measurement of pollutant levels in environmental media (e.g. drinking water, food) and an assessment of intake from all exposure routes. Second level, based on measurement of chemicals or their metabolites in body tissues (e.g. adipose tissue, blood), secreta (e.g. breast milk) and excreta (e.g. urine, faeces), providing an indication of the integrated exposure to a pollutant. More indirect approaches involve monitoring associated indicators as opposed to the pollutants themselves (e.g. cigarette consumption).
Refs Calabrese, Edward J *Pollutants and High Risk Groups* (1978).
Broader Hazards to human health (#PB4885).
Narrower Lead as a health hazard (#PE5650) Health hazards of radiation (#PD8050)
Chemical contaminants of food (#PD1694) Health hazards of exposure to noise (#PC0268)
Long-term hazards of exposure to chemicals (#PE4717).
Related Smoking (#PD0713).
Aggravates Human physical genetic abnormalities (#PD1618).
Aggravated by Environmental pollution (#PB1166).

♦ PC0937 Provocations of capitalist militarism
Incidence The militarist provocations of the USA, for example, include the extension of direct and indirect support for counter-revolutionary armies for civil war, as in Central America (Nicaragua, El Salvador, Guatemala, Honduras); the further expansion of the Rapid Deployment Force, which can assemble and put into the field 230,000 members of the Army, Navy, and Airforce in a crisis situation; the extension of military bases world-wide - the USA is currently equipping its strategic air forces in the Far East with new strategic bombers able to carry both nuclear weapons and the Short Range Attack Missile (SRAM); and the enormous weaponry of the United States itself, as well as that in the Third World. Military expenditure in the USA is planned to rise from an envisaged 1984 total of US $274 billion to US $465 billion by 1989. According to current planning, approximately US $2,000 billion will be spent on the US arms programme between 1985 and 1989. Finally, the continuing and expanding diversions of resources to military purposes is occurring in all market countries. (The growth rate of military spending of developing countries over the period 1972-1981 was about twice as high as that of their Gross Domestic Product).
Broader Manipulation (#PA6359) Aggressive foreign policy (#PC4667).

♦ **PC0939 Illegal private profit in socialist countries**
Undeclared employment in socialist countries — Second, shadow economy in socialist countries — Double employment in socialist countries — Second undeclared employment in socialist countries — Black market economies in socialist countries
Nature As its name implies, the second economy refers to all unlawful activities taking place outside the state–controlled economic system. It is centred primarily around the illegal production and distribution of goods and services for private profit. This illegal system has persevered and has even played a useful role in filling the gap resulting from the inefficient performance of the 'first' economy. However, one of its disadvantages is that it has given birth to a new privileged class that enjoys many amenities and luxuries not available to the majority of the population.
Counter–claim Yugoslavia, China, and Hungary now allow for private enterprise, albeit on a limited – and guarded – scale.
Refs Kornai, János *Contradictions and Dilemmas.*
Broader Unemployment (#PB0750) Underground economy (#PC6641)
Unlawful business transactions (#PC4645).
Narrower Currency black market in socialist countries (#PD2413).
Related Disguised unemployment in socialist countries (#PC0940).
Aggravated by Failure of centrally planned economies (#PC3894).

♦ **PC0940 Disguised unemployment in socialist countries**
Nature The high employment figures in socialist countries result from the fact that practically all blue and white collar workers are guaranteed employment by the state, usually in their current jobs. In fact, for managerial reasons, jobs do not fully utilize workers' skills and training, so that labour productivity is low.
Counter–claim This form of disguised unemployment is politically preferable to actual unemployment of workers.
Broader Unemployment (#PB0750).
Related Illegal private profit in socialist countries (#PC0939).

♦ **PC0952 Animal diseases**
Animal deficiency diseases
Nature Even in the developed countries, at the present time, losses through animal diseases and parasitic infestations represent a wastage of a considerable proportion of the national economic effort in livestock production, although efficient animal health services are operating and there is a high standard of livestock management. In the developing countries, such losses primarily concern not economics, but human existence, since the populations of nearly all these countries already suffer from a lack of animal protein. Moreover, many farmers rely on the strength of healthy animals for the power to till their soil.
The toll taken by disease is most obvious when sudden outbreaks are accompanied by heavy mortality, either naturally or as the outcome of a slaughter policy of control. This, however, is not the worst aspect. Less apparent, but much greater, are the continuing losses which, year in, year out, are caused by lowered productivity – smaller and slower liveweight gains, depressed milk yields, poorer work output, diminished fertility, and mortality among young stock. (One estimate gives an annual loss of US $300 million in the United States due to animal parasites alone, representing about 15 per cent of the value of the average annual consumption of livestock products). In the developing countries, a high animal–disease rate is the normal partner of low productivity, usually associated with poor management and inadequate feeding. As disease depresses production still further, a vicious cycle develops, with the result that the livestock becomes of such little value that investment either in disease control or in improved breeding or management appears very difficult to justify.
Diseases include those of a fungal, viral, bacterial, parasitic, infectious and communicable nature, and zoonoses. Contagious diseases of animals of particular concern include cattle plague, foot–and–mouth disease, contagious peri–pneumonia, anthrax fever, sheep–pox, rabies, glanders, dourine, swine fever and fowl plague.
Claim Throughout history, man has had to fight hard to make sure of his supplies of animal protein. Viruses, bacteria, fungi and parasites have repeatedly robbed him of these. Animal plagues ravaged Europe and Asia up to the early years of the present century, and throughout the world today disease limits the use of many areas that are otherwise suitable for livestock production. Despite advances in veterinary science, it is most unlikely that there will be any easy solution: this struggle for animal protein will continue, for as numbers of animals increase the problems of disease multiply.
Refs Cooper, J E and Jackson, O F *Diseases of the Reptilia* (1982); Curtis *Poultry Diseases* (1989); Dietz, O and Wiesner, E (Eds) *Diseases of the Horse* (1984); FAO *Animal Health Yearbooks* ; Gibson, T E (Ed) *Weather and Parasitic Animal Disease* (1978); Hayes, M Horace *Veterinary Notes for Horse Owners* (1988); Hofstad, M S et al *Diseases of Poultry* (1984); Holzworth, Jean *Diseases of the Cat* (1987); Houlton, John E and Taylor, Polly *Trauma Management in the Dog and Cat* (1987); Hámori, D *Constitutional Disorders and Hereditary Diseases in Domestic Animals* (1983); Karstad, L, et al (Ed) *Wildlife Disease Research and Economic Development* (1981); Kirkbride, C A *Control of Livestock Diseases* (1986); Mawdesley–Thomas, Lionel E, et al *Diseases of Fish* (1974); Mayer, E *International Congress on Diseases of Cattle (1980)* (1981); Naviaux, James L *Horses in Health and Disease* (1985); Sachrieder, Jürgen *Animal Diseases in Tropical and Subtropical Regions* (1970); Scholl, Erwin, et al *Diseases of Swine* (1986); Stohl, Leslee L *Cat Diseases and Veterinary Medicine* (1987); Taylor, David *Pig Diseases* (1983); United Nations *Emerging Diseases of Animals* (1968); Walton *A Handbook of Pig Diseases* (1987); World Meteorological Organization *Weather and Animal Diseases* (1970).
Broader Disease (#PA6799) Rural underdevelopment (#PC0306).
Narrower Zoonoses (#PD1770) Animal cancer (#PG1900) Bird diseases (#PD3323)
Fungal diseases (#PD2728) Cattle diseases (#PD0752) Enzootic diseases (#PD2733)
Epizootic diseases (#PD2734) Animal abnormalities (#PD4031) Poisoning in animals (#PD5228)
Diseases of wild animals (#PD2776) Skin diseases in animals (#PD9667)
Viral diseases in animals (#PD2730) Vectors of animal diseases (#PD2751)
Unrestrained animal damage (#PJ9217) Pests and diseases of fish (#PD8567)
Eye diseases and disorders (#PD8786) Bacterial diseases in animals (#PD2731)
Parasitic diseases in animals (#PD2735) Endocrine diseases in animals (#PD9654)
Diseases of senses in animals (#PD5535) Metabolic diseases in animals (#PD7420)
Infectious diseases in animals (#PD2732) Immune system diseases in animals (#PD4068)
Urinary system diseases in animals (#PD9293)
Inadequate control of animal diseases (#PD2781)
Diseases of nervous system in animals (#PD7841)
Circulatory system diseases in animals (#PD5453)
Reproductive system diseases in animals (#PD7799)
Vitamin E deficiency in domestic animals (#PE4760)
Ill–defined health conditions in animals (#PD9366)
Diseases of respiratory system in animals (#PD7307)
Diseases of the digestive system in animals (#PD3978)
Diseases of musculoskeletal system in animals (#PD7424).
Related Inadequate animal welfare (#PC1167) Deficiency diseases in plants (#PD3653).
Aggravates Animal deaths (#PE7941) Animal infertility (#PC1803)
Occupational diseases (#PD0215)
Underdeveloped use of agricultural resources (#PF2164)
Infected animal, meat and animal product shipments (#PE7064)
Economic loss through slaughter of diseased animals (#PE8109).
Aggravated by Water pollution (#PC0062) Factory farming (#PD1562)
Marine pollution (#PC1117) Poisonous plants (#PD2291)
Leeches as pests (#PE3660) Biological warfare (#PC0195)
Hazardous waste dumping (#PD1398) Inadequate feeding of animals (#PC2765)
Pesticide hazards to wildlife (#PD3680) Industrial waste water pollutants (#PD0575)
Water pollution in developing countries (#PD3675)
Health hazards of air pollution for animals (#PE9609)
Agricultural mismanagement of housed farm animals (#PD2771)
Inadequacy of agricultural education in developing countries (#PE9096).

♦ **PC0970 Restrictions on immigration**
Nature Quota, nationality or other restrictions on immigrants may be imposed by national governments. Restrictions may be based on employment opportunities and also sometimes on housing availability. Immigrants may effectively or explicitly be discriminated against on the grounds of colour or race or nationality if the country has preferential agreements with certain other countries.
Incidence The four traditional immigration countries, the United States, Canada, Australia and New Zealand, all have systems of restrictions. However, Israel, the Republic of South Africa, the USSR, and numerous Latin American countries also have highly selective immigration policies. The number of persons who would like to immigrate and whose applications are pending for long periods, in addition to those who are barred by restrictions, possibly numbers in the millions world–wide.
Broader Racism (#PB1047)
Restrictions on freedom of movement between countries (#PC0935).
Narrower Illegal immigration (#PD1928)
Refusal of entry to foreign workers' families (#PE8423)
Immigration barriers for handicapped family members (#PE4868).
Aggravates Slavery (#PC0146) Expulsion of immigrants and aliens (#PC3207)
Abusive traffic in immigrant workers (#PD2722).
Aggravated by Second–class citizenship (#PG3152)
Discrimination against immigrants and aliens (#PD0973).

♦ **PC0981 Maternal deprivation**
Infant emotional deprivation — Children deprived of affection — Maternal rejection — Maternal negligence — Hospitalism
Nature One of the essential ingredients for mental health is that the infant and young child should experience a warm, intimate, and continuing relationship with his mother or permanent mother substitute, such that both find satisfaction and enjoyment in the relationship. Maternal deprivation can thus be considered as an insufficiency of interaction between the child and a mother–figure, to the degree that identification with the maternal figure is not made and with the result that personality development is impaired.
Deprivation occurs when an infant or young child lives in an institution or hospital where he has no major substitute mother and where he receives insufficient maternal care, or when a young child lives with his mother or permanent substitute mother, from whom he receives insufficient care and with whom he has insufficient interaction. Deprivation may also come about through the child's own inability to interact with a mother–figure despite the fact that one is present and ready to give sufficient care – this inability to interact being consequent on and presumably caused by previous deprivation experiences.
Claim Maternal deprivation can result in the growth pattern of a failure to thrive, a decline from a previously established growth pattern, despite an adequate caloric intake.
Refs Grizzle, Anne F and Proctor, William *Mother Love, Mother Hate* (1988).
Broader Social neglect (#PB0883) Family rejection of children (#PC8127).
Related Anaclitic depression (#PE1883) Lack of intimate relationships (#PF4416).
Aggravates Soul murder (#PF4213) Mental illness (#PC0300)
Infant growth failure (#PE6909) Starving for attention (#PF1113)
Human disease and disability (#PB1044) Sudden unexpected infant death (#PD1885)
Negative emotions and attitudes (#PA7090).
Aggravated by Family breakdown (#PC2102).

♦ **PC0990 Landlessness in developing countries**
Nature Tenure reforms and parcelization of the erstwhile big landed estates of Asia and the Near East have resulted in wider diffusion of ownership; this has not only encouraged greater self–realization but also conferred the status of citizenship on those who were previously tenants, landless labourers or small farmers. Unfortunately, the individualization of tenure and the gradual transformation of land into a marketable commodity have been accompanied by a divorce between ownership of land and use of land (as witnessed in the case of tenancy), and divorce between management of land and labour on land. Many of these countries have therefore witnessed the phenomenon of a rising class of landless labourers and small farmers who are either forced to work for wages on the relatively big farms owned by rural, resident, non–cultivating landowners, or, in the absence of adequate opportunities for earning their livelihood, such resource–poor households must necessarily remain poor and are forced to overuse the environmental resource base in order to survive.
Incidence During the last decade, high concentration of land holdings continued in many developing countries, especially in Latin America, and was associated with a high degree of rural poverty despite a high level of per capita income. Some African countries, (Kenya and Malawi) which opted for freehold titles in land, experienced emergence of absentee ownership and landlessness. In the Near East, in Latin America, and in some countries of South East Asia, land settlements continue to be more prominent than redistribution of privately owned land. However, the land settlement programmes have been able to benefit only a small portion of tenants and landless workers. Also, the pattern of growth within agriculture in Latin America adversely affected the landless and marginal farmers.
Rough estimates for 1981 indicate that out of the agricultural population of 1.3 billion in the developing countries (excluding China), 745 million were small farmers, another 167 million were landless labourers. Between 1970 and 1981 there was an estimated addition of 124 million to these two categories, under the assumption of unchanged proportions of small holders and landless to total agricultural population. The bulk of these additions were in the Far East (75 million). This has been due to the increasing scarcity of agricultural land, as a consequence of higher growth rates of agricultural population than that of area under arable and permanent crops. Between 1970 and 1980, land per person in agriculture declined by 12 percent in Africa, 11 percent in the Near East and 9 percent in the Far East; it increased in Latin America. It is in the context of increasing land scarcity and landlessness that effective implementation of agrarian reform measures, coupled with measures to increase land productivity of small holders and commitment of resources for meeting the employment needs of the landless and mini–farmers, assumes even more importance in the 1980s than in the 1970s.
With existing patterns of land distribution, the number of smallholders and landless households in developing countries is expected to increase by some 50 million to 220 million by the year 2000. These groups represent about 75 per cent of the agricultural households in developing countries.
Refs Sinha, Radha *Landlessness* (1984).
Broader Deprivation of peasantry (#PC8862)
Maldistribution of land in developing countries (#PD0050).

CROSS-SECTORAL PROBLEMS PC1073

Aggravates Urban slums in developing countries (#PD3489).
Aggravated by Unsustainable population levels (#PB0035)
Inappropriate modernization of agriculture (#PF4799).

♦ PC0998 Imbalance of payments

Dependence on imbalance of payments — Balance of payments deficits

Nature Fluctuations in export earnings and a variety of other factors, both external and internal, may lead a country into a state of payments imbalance on transactions with other countries, whether in the form of a deficit or a surplus. In the absence of an adequate balance of payments adjustment mechanism or an adequate mechanism for the generation of international liquidity, imbalance of payments is regarded as an indicator of a need for a change in economic policy, and as an indicator of an actual or potential threat to the existing exchange system.

Balance of payments deficits are not always a reflection of the pressure of demand on productive capacity; often the contrary is true. Countries which face balance of payments difficulties are naturally anxious not to accentuate them even though they have unutilized capacity, while those which do not have these difficulties are worried about the heavy pressure of demand on capacity and the possibility of inflation.

Claim The competitive struggle between developed countries to avoid balance of payments deficits is not conducive to the reduction of protective or other barriers to trade, nor to the expansion of development assistance. Each developed country feels under pressure to keep down the amount of its development finance. This kind of restrictive pressure acts with particular force in a world economy in which the aggregate of the balance of payments positions of the developed countries amounts to little more than zero, so that some countries achieve balance of payments surpluses only at the expense of others suffering deficits. The balance of payments position is affected by foreign bond issues which could be used to finance projects in developing countries. The borrowing countries can of course be expected to spend the proceeds from bond issues on needed imports; but these expenditures will not necessarily be in the lending country, so a balance of payments effect remains. Countries in payment deficit may be tempted to lessen the impact of capital outflows on their payments position by requiring the proceeds of new lending to be spent domestically. This practice of aid-tying has become widespread, and is inherent in the system of export credits, the major forms of private credit to many developing countries.

Broader International insecurity (#PB0009).
Narrower Budget deficits in developing countries (#PD3131).
Aggravates Unemployment (#PB0750) Trade protectionism (#PC4275)
Cyclic business recessions (#PF1277)
Financial destabilization of world trade (#PC7873)
Restrictions imposed on aid to developing countries (#PF1492)
Restrictions on foreign access to capital bond markets (#PD3135)
Social environmental degradation from recreation and tourism (#PD0826).
Aggravated by Economic inflation (#PC0254) Fiscal and trade imbalances (#PC4879)
Speculative flight of capital (#PC1453) Disproportionate external public debt (#PC3056)
Socio-economic burden of militarization (#PF1447)
Decline in export credits to developing countries (#PE3066)
Excessive foreign public debt of developing countries (#PD2133)
Inadequate mechanism for balance of payments adjustment (#PF3062)
Inadequate mechanism for the creation of liquid reserves (#PF3061)
War-time disruption of economies and production facilities (#PD8851).
Reduced by Foreign exchange restrictions (#PF3070).

♦ PC0999 Family poverty

Poor families

Claim Family poverty results in high rates of population growth. Poor families, whether in terms of income, employment or social security benefits, need children, initially to work and later to sustain the elderly parents.

Broader Socio-economic poverty (#PB0388).
Narrower Abandoned children (#PD5734)
Family poverty in industrialized countries (#PD1998).
Related Poverty in developing countries (#PC0149).
Aggravates Widowhood (#PD0488) Urban slums (#PD3139)
Malnutrition (#PB1498) Street children (#PD5980)
Family breakdown (#PC2102) Desertion in marriage law (#PF3254).
Aggravated by Social inequality (#PB0514) Lack of family planning (#PF0148)
Unequal coverage by social security (#PF0852)
Social insecurity in developing countries (#PE4796).

♦ PC1001 Discrimination against women in politics

Denial of right to participate in government for women

Claim Everyone has the right to take part in the government of his or her country, directly or indirectly, through freely chosen representatives; and the right to equal access to public service in his or her country, as part of the essential human rights of every individual and the principle of equality of rights for men and women. Yet, in many countries, women are denied their political rights, including the right to vote in all elections, and the right to hold public office and to exercise all public functions, on equal terms with men and without any discrimination.

Refs Inter-Parliamentary Union *Participating of Women in Political Life and the Decision-making Process* (1988); Randall, Vicky *Women in Politics* (1987).

Broader Discrimination in politics (#PC0934) Discrimination against women (#PC0308).
Narrower Unequal franchise for women (#PU3201).
Related Denial of right to hold public office (#PE5608)
Exclusion of women from decision making (#PE9009)
Denial of right to participate in government (#PE6086).
Aggravates Dependency of women (#PC3426) Denial of right of family planning (#PE5226).
Aggravated by Social injustice (#PC0797) Restrictions on property rights (#PD8937)
Underutilization of legal rights (#PF3464)
Discrimination against women in education (#PD0190).

♦ PC1002 Vulnerability of middle-class

Elimination of bourgeois power

Claim Class antagonisms have been exacerbated to such a degree in market economies that even the bourgeois and petit bourgeois middle-classes, frequently willing aides and auxiliaries of the bourgeois-military state, are deprived of political power and threatened in terms of their economic existence.

Broader Social insecurity (#PC1867) Class consciousness (#PC3458).
Aggravated by Capitalism (#PC0564) Bourgeoisie (#PE7774)
Social inequality (#PB0514).

♦ PC1005 Compulsory acquisition of land by government

Broader Injustice (#PA6486) Repression (#PB0871)
Excessive government control (#PF0304).
Narrower Excessive military usage of land (#PE3402).
Aggravates Non-payment of compensation for forced relocation (#PE8898).

♦ PC1015 Undemocratic political organization

Dependence on undemocratic political organization — Absence of a democratic political process

Broader Undemocratic organizations (#PC8676).
Narrower Monarchy (#PF2170) Militarism (#PC2169) Totalitarianism (#PF2190)
Official secrecy (#PC1812) Lack of representation (#PF3468)
Secrecy in scientific research (#PF1430) Illegitimate political regimes (#PC1461)
Secret international agreements (#PF0419)
Failure of individuals to participate in social processes (#PF0749).
Related Military secrecy (#PC1144).
Aggravates Elitism (#PA1387) Apartheid (#PE3681) Alienation (#PA3545)
Political crime (#PC0350) Social inequality (#PB0514) Political conflict (#PC0368)
Civil disobedience (#PC0690) Political injustice (#PC2181) Political oligarchy (#PD3238)
Political repression (#PC1919) Political inequality (#PC3425)
Denial of human rights (#PB3121) Internment without trial (#PD1576)
Single party democracies (#PD2001) Discrimination against minorities (#PC0582)
Contempt for democratic processes (#PF0639)
Denial of human rights in the administration of justice (#PD6927).
Aggravated by Tribalism (#PC1910) Censorship (#PC0067)
Conspiracy (#PC2555) Attachment (#PF6106)
Oppression (#PB8656) Technocracy (#PF6330)
Unjust peace (#PB7694) Class conflict (#PC1573)
Social conflict (#PC0137) Threatened sects (#PC1995)
Political apathy (#PC1917) Political bribery (#PC2030)
Political ignorance (#PC1982) Political intimidation (#PC2938)
Political infiltration (#PD4798) Restriction of freedom of expression (#PC2162)
Threatened and vulnerable minorities (#PC3295) Lack of participation in development (#PF3339)
Maintenance of political dossiers on individuals (#PD2929).
Reduces Dictatorship of the majority (#PD3239).
Reduced by Compromise as a betrayal of principles (#PF3420).

♦ PC1049 Dictatorship

Nature Dictatorship arises from unrestricted domination by an individual, a clique or small group, a foreign power or by a majority, to the exclusion of minority rights and interests. Dictatorship may be economic, political or social.

Broader Domination (#PA0839) Authoritarianism (#PB1638).
Narrower Political oligarchy (#PD3238) Social dictatorship (#PD3241)
Economic dictatorship (#PC3240) Political dictatorship (#PC0845)
Dictatorship of the majority (#PD3239) Transnational corporation imperialism (#PD5891).
Related Paternalism (#PF2183).
Aggravates Informers (#PD8926) Intimidation (#PB1992)
Discrimination against minorities (#PC0582).
Aggravated by Revolution (#PA5901) Power politics (#PB3202)
Excessive government control (#PF0304)
Government support for repressive regimes (#PF4821).
Reduced by Racial discrimination (#PC0006).

♦ PC1056 Genocide

Dependence on genocide — Minority eradication — Genocidal massacres — Holocaust

Nature Genocide is a crime under international law and condemned by the civilized world. It includes any of the following acts committed with intent to destroy, in whole or in part, a national, ethnical, racial or religious group: 1. Killing members of the group; 2. Causing serious bodily or mental harm to members of the group; 3. Deliberately inflicting on the group conditions of life calculated to bring about its physical destruction in whole or in part; 4. Imposing measures intended to prevent births within the group; 5. Forcibly transferring children of the group to another group.

Incidence In all periods of history genocide has inflicted great losses on humanity. It has been practised against Armenians during the Ottoman Empire; against Hebrews, Slavs, and other racial groups during the Hitler regime (1933–1945); by the Khmer Rouge (under Pol Pot) in Kampuchea; against the Baha'is in Iran; by the Tutsi against the Hutu in Burundi in 1965, 1972 and 1988; by the Iraqis against the Kurds; by Paraguayans against the Ache indians before 1974 and a number of others.

The Holocaust was the systematic murder of 6 million Jews, gypsies, homosexuals, Jehovah's Witnesses, and "mental defectives" in Europe by the Nazis during and prior to World War II. Most were killed in gas chambers; but also infants were bayoneted for fun, workers thought to be slacking were casually shot in mid-conversation, pregnant women were kicked into parturition, and children's skulls were smashed against the wheels of railroad cars.

Refs Charny, Israel W (Ed) *Genocide* (1988); Charny, Israel W *Genocide* (1988); Kuper, Leo *Genocide* (1982).

Broader Conflict (#PA0298) Human death (#PA0072).
Narrower Massacres (#PD2483) Ethnocide (#PC1328).
Related Homicide (#PD2341) Infanticide (#PD3501) Mass-murder (#PJ0457)
Political mass murder (#PD5590)
Offences against the peace and security of mankind (#PC6239).
Aggravates War (#PB0593) Racial repression (#PD8762) Political repression (#PC1919)
Endangered tribes and indigenous peoples (#PC0720).
Aggravated by Racism (#PB1047) Tribalism (#PC1910)
Scapegoats (#PF3332) Starvation (#PB1875)
Antisemitism (#PE2131) Racial conflict (#PC3684)
Ethnic conflict (#PC3685) Racial discrimination (#PC0006)
Plural society tensions (#PF2448) Discrimination against minorities (#PC0582)
Vulnerability of indigenous populations to introduction of diseases (#PE3721).

♦ PC1059 Weakness of socio-economic infrastructure

Broader Antiquated world socio-economic order (#PF0866).
Narrower Underdevelopment of legal infrastructure (#PF4836)
Deficiencies in national and local legal systems (#PF4851)
Limited availability of public services in the small towns of developed countries (#PF6539).
Aggravated by Criminalization of abortion (#PF6169).
Reduces Undirected expansion of economic base (#PF0905).

♦ PC1073 Crimes against humanity

Incidence In addition to war crimes, crimes against humanity have occurred as a result of the treatment of POWs and internees at the end of war. The Allied decision at the Conferences of Teheran, Yalta and Potsdam to transfer more than 14 million Germans from pre-war Poland, East Prussia, Pomerania, East Brandenburg, Silesia, Sudetenland, Czechoslovakia and Hungary resulted in the deaths of 2.2 million people, mostly women and children (since the men were prisoners of war). The Yalta agreement to use the labour of German POWs as "reparations in kind" and Allied decisions to use "surrendered enemy personnel" and "disarmed enemy forces" constituted a form of slavery achieved by removing POWs from the protection of the Geneva Convention through their reclassification. This procedure was terminated by the USA in July 1945, by the UK in July 1948 and by the USSR in 1956. Official estimates of POW deaths, exceed 1.2 million, mostly in the USSR.

Broader Crime (#PB0001).
Narrower War crimes (#PC0747) Concentration camps (#PD0702)
Victimization of children (#PC5512).
Related Torture (#PB3430) Criminals (#PC7373) Statutory crime (#PC0277)
Violation of human rights (#PB3860) Illegality of nuclear weapons (#PF4727)

Government sanctioned killing (#PD7221) Legalized racial discrimination (#PC3683).
Aggravated by War (#PB0593) Denial of human rights in armed conflicts (#PC1454)
Remission of sentences for crimes against humanity (#PF1098).

♦ PC1090 Militancy

Nature An aggressive and activist attitude towards the furthering of a cause may be social, economic, political or ideological. Militancy includes industrial action, as well as social and ethnic action, non-cooperation, refusal to participate, and non-violent demonstration. Non-violent action may ultimately lead to violence either on the part of the militants or as repression and backlash. Militant action may cause social and economic breakdown.
Broader Disaccord (#PA5532) Aggression (#PA0587).
Narrower Non-cooperation (#PF8195).
Related Extremism (#PB3415).
Aggravates Strikes (#PD0694) Human violence (#PA0429) Social conflict (#PC0137)
Social breakdown (#PB2496).
Reduces Apathy (#PA2360) Authoritarianism (#PB1638).

♦ PC1100 Excessive external trade deficits

Incidence The United States's trade deficit in 1986 was over US$170 billion making it the largest debtor nation in the world. In June 1989 the United Kingdom's trade deficit widened to US$3.1 billion. In 1989 the Soviet Union recorded a trade deficit of £3.35 billion.
Broader Fiscal and trade imbalances (#PC4879)
Imbalance in international trade patterns (#PC8415).
Narrower Excessive external trade deficits of developing countries (#PE1496).
Related Budget deficit (#PD5492).
Aggravates Economic conflict (#PC0840) Disproportionate external public debt (#PC3056).
Aggravated by Trade protectionism (#PC4275) Obstacles to world trade (#PC4890)
Restrictive trade practices (#PC0073) Tariff barriers to international trade (#PC0569)
Collusive international trade arrangements (#PE0396)
Instability of mining and quarrying industry (#PE0993)
Restrictive practices in trade in manufactured goods (#PD1797)
Weakness in primary commodity trade amongst developing countries (#PE2951)
Tariff barriers to trade between developing and developed countries (#PC2369).
Reduced by Large trade surpluses (#PJ0207)
Preoccupation with reciprocity in trading relations (#PF3871).

♦ PC1108 Seasonal unemployment

Seasonal fluctuations in work — Seasonal labour — Seasonal work patterns — Seasonal employment — Limited off-season jobs
Nature A seasonal worker generally works in one to four month periods, most often in the building, construction, sugar, food preserves or hotel industries, or in agriculture, where, in the temperate zones, such workers are employed in grape, sugar-beet, potato, fruit and cereals harvests. In the tropical zones the work is mainly on cotton, coffee, cacao, rice and groundnut plantations. In the off-season, workers resident in the country may be faced with unemployment. Migrant workers present for the season seek other employment elsewhere.
Refs ILO *Labour and Social Problems Arising out of Seasonal Fluctuations of the Food Products and Drink Industries* (1978).
Broader Unemployment (#PB0750) Seasonal fluctuations (#PF8163)
Limited opportunities for significant work (#PF1403).
Related Problems of migrant labour (#PC0180).
Aggravates Strikes (#PD0694) Industrial accidents (#PC0646)
Insecurity of employment (#PD8211) Limited employment options (#PF1658)
Discontinuity of employment (#PD4461)
Diminishing capital investment in small communities (#PF6477).
Aggravated by Seasonal fluctuations in agriculture (#PD5212)
Underutilization of facilities due to daily or seasonal peaks (#PF0827).

♦ PC1117 Marine pollution

Nature The world's oceans can be divided into two zones: coastal and open. The coastal zone constitutes about 10 percent of the total ocean area and is the receptacle for wastes from rivers, direct run-off from the land, oil spills, and ocean dumping. When pollution occurs in marine waters, it is most likely to be observed first in the coastal waters, and its effects are likely to be serious there. The deep waters of the open ocean, usually 300 feet (91 meters) or more in depth, are mostly out of contact with coastal water and surface waters. The mixing of the deep and surface waters may take hundreds, even thousands of years, depending on the basin involved. Much of the manmade wastes dispersed to the open ocean: litter, plastics and oil slicks, are still in surface waters; but DDT and other chlorinated hydrocarbons are found in open ocean organisms; and radioactive isotopes of strontium and caesium from nuclear bomb detonations can now be found at depths of 1,000 meters. The open ocean changes slowly, but once changes do occur, they are not likely to be reversed quickly nor modified easily by humans.
Incidence Marine contamination occurs mainly in the Northern Hemisphere, in estuaries, bays and land-locked seas, such as the Mediterranean, Baltic and North Seas and certain estuaries in North America. In the Southern Hemisphere, Port Philip Bay in Australia and various coastal waters around southern Africa and South America have been reported to be contaminated.
Most contamination reaches the sea through rivers, direct coastal out-falls, drainage from human settlements and agricultural land, and deposition from the atmosphere. Some potential pollutants are, however, discharged from shipping and offshore structures such as oil rigs. Acute damage has occurred in areas around oil refineries and in industrialized estuaries, bays and coastal zones, where fish numbers have been reduced and many species eliminated. Although oil pollution is a nuisance, a bird-killer and a threat to coastal shellfish and tourism, it is still on the increase but cannot be proved to have had yet any serious impacts on a wide scale. Much of the suspended particulate matter in the sea originates from the land, and the particle-rich plumes of great rivers like the Amazon can be traced for as much as 2000 km. Particles of biological origin generally disintegrate in the top 500 m, but some detritus and faecal pellets along with part of the atmospheric fall-out reach the deep ocean floor. Such sediments are important transporters of pollutants, especially metals. The amount of iron, manganese, copper, zinc, lead, tin and antimony reaching the sea today through river discharges is of an order of magnitude greater than would be supplied by natural geological processes. Smelting and other industries may also contribute substantial quantities via the atmospheric pathways. Persistent solid wastes also enter the sea from urban areas, and 6.4 million tons are dumped annually from shipping: of this the proportion of plastic is currently low (under one percent) but is expected to increase unless controls are applied.
Claim Many marine scientists argue that even if hydrocarbon concentrations are low, the contamination of the sea is increasing, that chronic effects could appear slowly but then be virtually irreversible, and that the most stringent precautions are therefore essential.
Refs CPPS/UNEP *Sources, Levels and Effects of Marine Pollution in the South-East Pacific* (1983); Calimari, D *Selected Bibliography on Studies and Research Relevant to Pollution in the Mediterranean* (1977); Champ, Michael A and Park, P K *Global Marine Pollution Bibliography* (1982); FAO *Pollution* (1978); Ferguson Wood, E J and Johannes, R E (Eds) *Tropical Marine Pollution* (1975); Gerlach, S A *Marine Pollution* (1981); Geyer, R A *Marine Environmental Pollution* (1980); Okidi, C O *Regional Control of Ocean Pollution* (1978); Qing-Nan, Meng *Land-Based Marine Pollution* (1987); Soni, R *Control of Marine Pollution in International Law* (1985); Timagenis, G J *International Control of Marine Pollution* (1980); UNEP/FAO/UNESCO/WHO/WMO/IOC/IAEA *Selected Bibliography on the Pollution of the Mediterranean Sea* (1981); United Nations *Report of the FAO-UNEP Meeting on Toxity and Bioaccumulation of Selected Substances in Marine Organisms* (1985).
Broader Water pollution (#PC0062)
Obstacles to the utilization of coastal and deep sea water resources (#PF4767).
Narrower Coastal water pollution (#PD1356) Marine oxygen deficiency (#PE6289)
Marine pollution by plastic waste (#PE3741)
Vulnerability of marine animal communication (#PF7554)
Pollution of semi-enclosed bodies of seawater (#PE8175)
Vulnerability of marine environment to catastrophic warfare damage (#PD5178).
Aggravates Bird diseases (#PD3323) Animal diseases (#PC0952)
Poisonous algae (#PE2501) Amenity destruction (#PC0374)
Pollution of sediments (#PE5539) Pests and diseases of fish (#PD8567)
Endangered species of seals (#PE1656) Human disease and disability (#PB1044)
Vulnerability of food chains (#PB2253) Vulnerability of marine ecosystems (#PC1647)
Eutrophication of lakes and rivers (#PD2257)
Endangered species of echinodermata (#PD3158)
Endangered species of aschelminthes (#PD3159).
Aggravated by Air pollution (#PC0119) Oil pollution (#PE1839)
River pollution (#PD7636) Oil as a pollutant (#PE2134)
Agricultural wastes (#PC2205) Sewage as a pollutant (#PD1414)
Agricultural effluent (#PE8504) Marine dumping of wastes (#PD3666)
Industrial waste water pollutants (#PD0575)
Environmental hazards from petroleum (#PE1409)
Unethical practice of marine sciences (#PD4277)
Polychlorinated biphenyls as a health hazard (#PE2432).

♦ PC1123 Personality cults

Personality cults in politics — Media personality cults
Broader Human errors and miscalculations (#PF3702).
Aggravates Corruption (#PA1986) Insufficient role models (#PF8451)
Obsession with celebrities (#PF9617)
Uncritical acceptance of another person (#PF5973)
Unethical practices of philanthropic organizations (#PE8742).
Aggravated by Monarchy (#PF2170) Gullibility (#PJ2417)
Unethical practices in politics (#PC5517)
Media theatricalization of public life and politics (#PF9631).

♦ PC1131 Disparity between workers skills and job requirements

Nature There is a growing disparity between the skill-structure of those seeking work and available vacancies, with a consequent growth in the share of 'structural' or 'frictional' unemployment.
Broader Maldistribution of resources (#PB1016).
Narrower Obsolete vocational skills (#PD3548) Insufficient trained labour (#PD9113)
Unavailability of trainee employment (#PJ0336) Available jobs unrelated to education (#PJ9233)
Imbalance between training and existence of openings in various professions (#PE5194).
Related Disparity in social development within developing countries (#PD0266).
Aggravated by Limited means of marketing employable skills (#PE7344).

♦ PC1142 Abuse of police power

Uncontrolled police operations
Nature Police authority can be misused by individual officers or entire departments, or the national police force as a body may be the instrument for political repression.
Incidence Individual abuse of law enforcement authority includes extortion, accepting bribes to cover-up crime, and conspiracy or participation in criminal acts. Individual police officers may unjustly use excessive violence or unnecessarily 'shoot to kill' to apprehend or restrain criminals or suspects. Police departments, local station-houses or constabularies may have policies of protecting vice and gambling; of 'shaking-down' businesses; of harassing minorities and youths; and of maintaining officers on the payroll who are not regularly present for duty. As an instrument of the state, police may be used for surveillance, for maintaining dossiers on civilians, and for unjustly imprisoning civilians.
Broader Abuse of authority (#PC8689).
Narrower False evidence (#PF5127) Police brutality (#PD3543)
Police intimidation (#PD0736) Military police abuse (#PG6627)
Internment without trial (#PD1576) Unauthorized police search (#PD3544)
Abusive police interrogations (#PG3615) Abuses by private police forces (#PE4847)
Unethical practices by police forces (#PD9193)
Police crimes during narcotic investigations (#PE5037)
Military and police personnel participation in torture (#PE4119).
Related Harassment (#PC8558) Martial law (#PD2637)
Denial of right to life (#PD4234) Abuse of government power (#PC9104).
Aggravates Organized crime (#PC2343) Civil disorders (#PC2551)
Subversive activities (#PD0557) Public assaults on police (#PE7659)
Proliferation of weapons in civilian hands (#PE2449).
Aggravated by Police state (#PD7910) Political police (#PD3542)
Political repression (#PC1919).

♦ PC1144 Military secrecy

Nature Exact information concerning a nation's offensive and defensive capabilities is known only to a limited number of persons, both civilian and military, within a country. Details and aggregate statistics, military weapon technology and military planning are all classified according to a hierarchical system of increasing degrees of required secrecy. Thus the public and its elected representatives have no knowledge of military matters other than what they are told by their government. Ignorance of military affairs prevents the public from voicing approval or disapproval. Under the cloak of military secrecy: imprisonment of political 'criminals' may be conducted; operations may be undertaken against labour or civil unrest; and inhumane warfare and war crimes engaged upon.
Incidence In the USA, there was great secrecy concerning the Vietnamese War, the Bay of Pigs landing in Cuba, and the abortive rescue mission in Iran.
Claim National defence depends on military secrecy. The vital nature of military information is indicated by frequent espionage activities reported in the press. However, military secrecy can be self-defeating by, for example, delaying the flow of information and decisions needed for defence.
Broader Secrecy (#PA0005) Official secrecy (#PC1812).
Narrower Secret military operations (#PF7669)
Competitive development of new weapons (#PC0012)
Government secrecy concerning nuclear weapons testing (#PF4450)
Non-verifiability of compliance with nuclear power safeguards (#PF4455).
Related War (#PB0593) Espionage (#PC2140) Political injustice (#PC2181)
Undemocratic political organization (#PC1015)
Socio-economic burden of militarization (#PF1447).
Aggravates Fear (#PA6030) Militarism (#PC2169) Nuclear war (#PC0842)
Abuse of science (#PC9188).
Reduced by War resistance movements (#PU3343).

♦ PC1153 Irresponsible scientific and technological activity

Obstacles to proper use of science and technology — Unethical use of science and technology

Nature The natural pace of development of science and the continuing emergence of new techniques, open up new possibilities for research. In the absence of any control mechanism, some of these experiments (although scientifically interesting in isolation) may have unforeseen multiplier effects which disrupt the existing natural or social systems.

Incidence Examples of such experiments (proposed or implemented) are: warming the ionosphere to gain information on how it functions, underground nuclear tests in earthquake areas, creation of free-flowing interoceanic canals, construction of dams across interoceanic straits, melting polar ice-caps, release of artificial substances into the ionosphere and magnetosphere, and alteration of the atmosphere of Venus.

Claim Scientists are often inseparably involved with decisions on the use of discoveries. For example, many of the scientists working on the first US atomic bomb unhesitatingly recommended its employment, and some argued successfully for its use without any governmental warning to Japan.

Counter-claim Scientists should not be held responsible for the nature of the systems they study, nor for the nature of the social systems that exploit those studies. In these matters, their responsibility is no greater than that of other citizens.

Broader Immorality (#PA3369) Irresponsibility (#PA8658)
Adverse consequences of scientific and technological progress (#PF3931).
Narrower Eugenics (#PC2153) Techno-mercenaries (#PE7153)
Irresponsible social science (#PF8032) Abusive behaviour modification (#PE2690)
Abusive experimentation on humans (#PC6912) Deceptive social science research (#PF7634)
Irresponsible genetic manipulation (#PC0776) Inadvertent modifications to climate (#PC1288)
Unethical practice of the biosciences (#PD7731)
Competitive development of new weapons (#PC0012)
Irresponsible research using human subjects (#PC0080)
Non-acceptance of embryo transfer technology (#PE6039)
Abuse of science and technology in capitalism (#PE3105)
Abusive technological development under capitalism (#PD7463)
Dependence on sophisticated technology for development (#PD6571).
Related Inhumane scientific activity (#PC1449) Irrelevant scientific activity (#PF1202).
Aggravates Radiological warfare (#PC6666)
Uncontrolled application of technology (#PC0418).
Aggravated by Scientism (#PF3366) Unprofessional science (#PF6697)
Abuse of scientific power (#PF2692) Denial of rights of medical patients (#PD1662)
Fragmentation of academic disciplines (#PF8868)
Unaccountability of institutions degrading the environment (#PF3458).

♦ PC1164 Biochemical warfare

Dependence on biochemical warfare — Chemical and biological warfare

Nature All weapons of war are destructive of human life, but chemical and bacteriological (biological) weapons stand in a class of their own as armaments which exercise their effects solely on living matter. The fact that certain chemical and biological agents are potentially unlimited in their effects, both in distance and durability, and that their large-scale use could have irreversible effects on the balance of nature adds to the sense of insecurity and tension which the existence of this class of weapons engenders.

Chemical and biological weapons pose a special threat to civilians because the high concentrations in which they would be used in military operations could lead to significant unintended involvement of the civilian population within the target area and for considerable distances downwind. The large-scale or, with some agents, even limited use of chemical and biological weapons could cause illness to a degree that would overwhelm existing health resources and facilities.

Large-scale use of chemical and biological weapons could also cause lasting changes of an unpredictable nature in man's environment. The possible effects of chemical and biological weapons are subject to a high degree of uncertainty and unpredictability, owing to the involvement of complex and extremely variable meteorological, physiological, epidemiological, ecological and other factors. Although advanced delivery systems are required for the employment of chemical and biological agents on a militarily significant scale against large civilian targets, isolated and sabotage attacks requiring only simple delivery could be very destructive in certain circumstances with some of these agents.

The potential for developing an armoury of chemical and biological weapons has grown considerably in recent years, not only in terms of the number of agents but in their toxicity and in the diversity of their effects. Possible long-term effects of such warfare include: chronic illness caused by exposure to chemical and biological agents; delayed effects in persons directly exposed (causation of cancer, severe damage to the human foetus, and detrimental alterations in the human gene). Certain agents, such as the nerve gases, cause anoxia which can produce cerebral damage, leaving neurological sequelae even though the person's life is saved. Biological agents spreading from person to person can give rise to secondary attack waves which, in diseases like plague, may continue for months or years after the event. Massive attacks may alter the ecological relationship between man and lower animal and insect vectors of diseases previously latent. Decontamination would be required for persistent chemical and biological agents.

Claim Were these weapons ever to be used on a large scale in war, no one could predict how enduring the effects would be and how they could affect the structure of society and the environment in which we live. This overriding danger would apply as much to the country which initiated the use of these weapons as to the one which had been attacked, regardless of what protective measures it might have taken in parallel with its development of an offensive capability. A particular danger also derives from the fact that any country could develop or acquire, in one way or another, a capability in this type of warfare, despite the fact that this could prove costly. The danger of the proliferation of this class of weapons applies as much to the developing as it does to developed countries. Preparing for chemical and biological warfare increases the danger that stockpiled supplies of lethal materials may be disseminated accidentally or deliberately. Furthermore, such preparations in any one country stimulate preparations in other countries by giving credibility to those who react to the fear of annihilation with a justification of chemical and biological warfare. Thus, reciprocal fears between nations, and the projection and rationalization to which they give rise in defence, contribute to the spiralling of a chemical and biological weapons race that imperils all mankind.

Refs Gander, T *Nuclear, Biological and Chemical Warfare* (1987); Robinson, J P *Chemical and Biological Warfare Developments 1986–1987* (1990).

Broader War (#PB0593) Unconventional war (#PC8836).
Narrower Chemical warfare (#PC0872) Biological warfare (#PC0195)
Environmental warfare (#PC2696).
Related Nuclear war (#PC0842) State sanctioned torture (#PD0181)
Pervasive fear of nuclear war (#PC3541)
Inhumane and indiscriminate weapons (#PD1519).
Aggravates Fear (#PA6030) Panic (#PF2633) Anxiety (#PA1635)
Mutagens (#PD1368) Teratogens (#PE0697)
Malignant neoplasms (#PC0092) Hazardous waste dumping (#PD1398)
Environmental degradation (#PB6384) Inadequacy of civil defence (#PF0506)
Neglect of dependents of war victims (#PD2092).
Aggravated by Wind (#PE2223) Surface to surface missiles (#PE4515)
Competitive acquisition of arms (#PC1258).

♦ PC1167 Inadequate animal welfare

Claim Inadequate action is taken against undue suffering caused to animals. Lack of animal welfare encompasses abusive methods in factory farming, cruel methods of slaughtering for meat, inhumane conditions during transport, inhumane killing of stray and unwanted animals, inhumane treatment of laboratory animals, abusive commercial exploitation of non-food animals, cruel sports, abuse of animals used for transport, inadequate zoo facilities, and abuse in control of wild animal populations.

Refs Hood, D E and Tarrant, P V (Eds) *Problem of Dark Cutting in Beef* (1981).

Broader Inadequate social welfare services (#PC0834).
Narrower Poaching (#PD2664) Factory farming (#PD1562) Torture of animals (#PC3532)
Maltreatment of animals (#PC0066) Lack of care for animals (#PD8837)
Unjustified game cropping (#PE1327) Maltreatment of zoo animals (#PE4834)
Inhumane killing of stray animals (#PE2759) Restriction of wild animal range size (#PC0475)
Undesirable effects of animal feed additives (#PD1714)
Inadequate housing and penning of domestic animals (#PE2763)
Insanitary penning conditions as factor in animal diseases (#PE2764)
Excessive commercial exploitation of farm animals by industrial concerns (#PD2772).
Related Animal diseases (#PC0952) Wildlife extinction (#PC1445)
Denial to animals of the right to life (#PF8243)
Agricultural mismanagement of housed farm animals (#PD2771)
Inadequate community care for transient urban populations (#PF1844).
Aggravates Animal injuries (#PC2753) Animal extinction (#PD7989)
Animal road deaths (#PE1690)
Death and disability from inhumane confinement (#PE5648).
Aggravated by Poisonous plants (#PD2291) Stray dog populations (#PE0359)
Decreasing genetic diversity of animals (#PC1408)
Inadequate legislation for animal welfare (#PE5794).

♦ PC1173 Long-term shortage of water

Nature Great pressure is being exerted on the water resources of many countries, particularly those developing countries located in arid or semi-arid regions. Acute water shortage exists in many areas of the world and is likely to become more severe in future years. In some areas further economic growth will not be possible until adequate additional water supplies become available. In other areas failure to increase the water supply may well result in standards of living being reduced below present levels.

Although water statistics are still in their infancy, it is obvious that there is a persistent increase in the demand because of the growth in population and the rise in the standard of living, and the growth in commodity production and in service industries. World population is rising at present at an average rate of about 2 per cent per annum, which means that, other factors being equal, world water demand will double about every thirty-five years. This demand, however, is further raised by the rise in the standard of living, leading to greater requirements for water for human consumption. This is brought about by the movement of population from primitive to modern housing, with water supply and bathrooms; by the rapid urbanization evident in all countries; and by numerous other factors.

Practically every increase in the production of goods – from rice to electricity and chemicals – requires increased quantities of water. The same applies to service industries, from commercial laundries to hotels. The increased demand for water resulting from the growth in the production of goods and the expansion in service industries is difficult to estimate because much will depend on the type of commodities to be produced and the type of service industries to be established. It should be noted, however, that the increased demand for water for the production of goods and services is independent of the increased demand for water for the population and is therefore additional to the latter.

Incidence Demand for water is growing several times faster than population, as agriculture, industrial and domestic uses increase. Global water use doubled between 1940 and 1980, and it is expected to double again by the year 2000. Some 80 countries, with 40 per cent of the world's population, already suffer serious water shortage. According to one estimate, by the year 2000 half of all the earth's annually renewed water – precipitated on land – will be used by man. Unfortunately, because the actual availability of water bears little relationship to the distribution of population and demand, local shortages will become increasingly frequent as population and related consumption increase. This will cause special problems for some rapidly growing cities. In addition to problems of supply, the reliability of water flow is being disrupted in many areas as watersheds are deforested. More than half the population of developing countries, excluding China, lacks convenient access to safe water supplies. The resulting poor sanitation, in combination with undernutrition, accounts for the daily deaths of 40,000 infants and small children. In the face of population growth, many developing countries are unable to reduce the numbers who are lacking adequate water supplies.

Counter-claim Water shortage is not always the result of lack of water resources; not infrequently it is the result of lack of water resources development.

Refs FAO *Systematic Index of International Water Resources Treaties, Declarations, Acts and Cases by Basin* (1978).

Broader Rural underdevelopment (#PC0306)
Long-term shortage of natural resources (#PC4824)
Environmental threats to national security (#PC4341).
Narrower Instability of water supply (#PD0722) Shortage of industrial water (#PE5464)
Shortage of fresh-water sources (#PC4815).
Related Environmental stress on inhabitants of tall buildings (#PE4953).
Aggravates Environmental stress (#PC1282) Inadvertent modifications to climate (#PC1288)
Infection of industrial water supplies (#PJ6198)
Inadequate water system infrastructure (#PD8517).
Aggravated by Wasted water (#PD3669) Agricultural wastes (#PC2205)
Maldistribution of water (#PD8056) Lack of water conservation (#PJ3480)
Lack of accord on water use (#PF4839) Loss of water to industrial uses (#PE7433)
Overexploitation of underground water resources (#PD4403).

♦ PC1174 Uncontrolled environmental impact of technology

Nature The natural environment – air, water and soil – is not only essential for supporting life, but is also the medium through which the technological waste products of society are consumed and restored. At lower levels of industrialization and lower population densities than are now prevalent, the capacity of the natural environment to absorb and recycle waste appeared limitless.

Increasingly, however, nature's capacity to assimilate waste is being threatened because of the rise in industrial production which has resulted in a much larger volume of emissions of organic and inorganic compounds from mining, refining, chemical production and manufacturing activities. In addition, certain industries such as aluminium smelting plants and electricity generating stations require large amounts of cooling water and discharge large quantities of heat into streams and rivers. Technology is also responsible for production of discharges of fine particulates, sulphur dioxide and nitrogen oxides, into the atmosphere, and for the great number of undegradable waste products from the plastics, metallurgical, glass and petrochemicals industries. Further impacts are the increased number of motor vehicles and rising volume of traffic, leading to higher level

PC1174

emissions of carbon monoxide, nitrogen oxides and lead; and the increased use of incineration, both by municipalities and industries, as a means of disposal of solid wastes, leading to the emission of a complex mixture of pollutants.

Claim The rapid and uncontrolled introduction of new technology is not an unmitigated blessing. Technology creates negative as well as positive impacts. It displaces workers as well as increasing their productivity; it impinges on privacy as well as improving personal exchanges and contacts. In the case of the environment, in particular, the positive contribution of technology is increasingly challenged by its undesirable external effects.

There is evidence that there are limits, at least locally, to the demands we can make on the natural environment to assimilate this multitude of waste products. For example, the natural water flow conditions from the hydrological cycle are often insufficient to maintain adequate water supply to the polluted areas. There are urban areas where adverse climatic conditions do not permit self-cleaning of the atmosphere through wind currents and natural convection. In agriculture, there is a growing body of evidence that the widespread dispersion of chemicals leads to their reappearance in other parts of the ecological system. The increased use of plastic and glass containers leaves a large residue of undegradable waste products which accumulate indefinitely. The imbalance between the rate at which man produces waste products and nature assimilates them drastically changes the nature of the responsibilities of governments. Today a 'laissez-faire' policy of allowing free use of natural resources is no longer possible. Welfare considerations require that governments adopt a policy of deliberate control of human activities in order to preserve the quality of the environment.

Counter-claim The development and diffusion of new technology, with its attendant effect of raising the productivity level, is an essential source of economic growth. It is not possible to assign a precise figure to the contribution of technology to growth because its effects cannot be separated from those of capital investment, education and management. Nevertheless, technological change is widely accepted as an essential element of economic development. Improvements in productivity have ultimate limits which cannot be transcended except through infusion of new technology. No amount of investment or improvements in organization, management or skills, for example, could make the horse-drawn carriage as efficient as a railway. At the same time, technology has enormously increased the range of opportunities open to individual consumers. Advances in communication technology have led to more rapid and more accurate information, advances in transport technology to greater mobility and accessibility to distant places, advances in health technology to less disease, less suffering and greater life expectancy.

Broader Lack of control (#PF7138) Human errors and miscalculations (#PF3702).
Related Unbalanced application of communications technology (#PE7637).
Aggravated by Abusive technological development under capitalism (#PD7463).

♦ PC1192 Environmental hazards from chemicals

Chemical agents — Environmental hazards of chemical materials and products — Environmental hazards from chemicals and petrochemicals industries

Nature Many chemicals found in industrial, commercial, and household environments are toxic under certain conditions. Among the most common are asbestos, benzene, vinyl chloride, acrylonitrile, and phthalates. All are important industrial materials widely distributed in various forms, and all have serious health effects. Polychlorinated biphenyls (PCBs) were important industrial chemicals until their effects were demonstrated and production was stopped. The toxic hazards of certain pollutants, particularly in the petrochemical industry, may be increased by the fact that these pollutants are themselves often contaminated by impurities.

The flow of industrial chemicals into the environment varies with the chemical. Asbestos (including natural sources) and benzene are present in varying concentrations throughout the environment. Most asbestos fibers that are released directly into the air come from mining and milling; but once an asbestos product has been manufactured, release of fibers into the environment is highest in the use of brake linings and other friction products and in the deterioration of thermal insulation and construction materials. Demolition of buildings is an important source of asbestos contamination, and asbestos is also leached into drinking water from concrete water pipes. Industrial workers and miners are usually the most affected although, as materials made with asbestos deteriorate, the general public can be exposed. Benzene, an organic compound derived from petroleum, is an intermediate in the production of plastics, dyes, nylon, food additives, detergents, drugs, and fungicides. It is also used as a gasoline additive. Vinyl chloride gas is released directly at the workplace, although small amounts are released as consumer products deteriorate and are disposed of. Phthalates, a class of intermediate synthetic organic chemicals, are in widespread and growing use as resin in the production of plastics, models, cement, paints, and finishes. Acrylonitrile is used as a resin in the production of plastic bottles, acrylic fibers and textiles. Combined with butadiene and styrene, acrylonitrile forms a polymer, ABS, widely used in appliances, automobiles, luggage, telephones and many other common industrial and household products. Vinyl chloride is a gas used in the production of plastics. PCBs, liquids previously used in ink solvents, adhesives, textile coatings, and pesticides, are now used only in electrical transformers and other closed systems, but many products containing PCBs remain in use. The major source of PCBs is through improper waste disposal, especially into bodies of water. PCBs contaminate wildlife, fish in particular. The threat to human health is through the food chain – in fish, poultry, and meat. PCBs are widely distributed in the environment very much like DDT, which PCBs resemble structurally; and, like DDT, they are highly persistent.

Incidence An estimated 70,000 to 80,000 chemicals are now produced and traded, representing some 10 per cent of total world trade. A further 1,000 to 2,000 new chemicals enter the commercial market each year. Of a sample of 65,735 chemicals in common use in the USA in 1984, data for complete health hazard assessment was available for only 10 per cent of pesticides ad 18 per cent of drugs. No toxicity data was available for nearly 80 per cent of the chemicals used in commercial products and processes. Where chemicals are discovered to have toxic properties according to the criteria of one country, they may be withdrawn from testing (or not submitted), and then manufactured only for export to countries with less stringent regulations.

WHO estimates that 75 to 80 percent of all cancers are triggered by environmental pollutants, foremost among them industrial chemicals. In 1980 the chemical industry in the industrialized countries of Europe and North America had a total turn-over of US$ 550 billion, employing a full six percent of the entire workforce of the OECD member countries. Inadequate chemical waste disposal remains a serious threat to the health and welfare of Europeans and North Americans.

Unlike the industrially advanced countries, most developing countries do not have toxic chemical control laws, nor the technical or institutional capability for implementing such laws. Catastrophic results have resulted from escape of toxic chemicals into crowded neighbourhoods near factories as a result of industrial accidents. During the last decade, several cases have come to light where products banned or severely restricted in the industrialized countries were sold to, or 'dumped' on, the developing countries.

Production of benzene, phthalates, acrylonitrile and vinyl chloride has increased rapidly during the past 20 years. Almost 11 billion pounds of benzene were produced in 1978, making it one of the high-volume organic compounds in domestic production. Vinyl chloride production continues to grow, along with the use of plastics in industrial and consumer products. Asbestos production, on the other hand, has not increased appreciably. PCB production in the USA has declined since 1970 as a result of self-imposed restrictions by the major manufacturer. By law, the manufacture of PCBs was prohibited after July 1979, except when specifically exempted by EPA.

Refs Homburger, F *Safety Evaluation and Regulation of Chemicals* (1983); Marshall, Victor C *Major Chemical Hazards* (1987); OECD *OECD Guidelines for Testing of Chemicals* (1987); OECD Staff *Economic Aspects of International Chemicals Control* (1983); Paustenbach, Dennis J *The Risk Assessment of Environmental Hazards* (1988); Roe, J C *The Chemical Industry and the Health of the Community* (1985); Sax, N Irving and Lewis, Richard J *Hazardous Chemicals Desk Reference* (1987).

Broader Environmental hazards from manufacturing industries (#PD0454)
Environmental hazards from economic and industrial products (#PC0328).
Narrower Halogen compounds as pollutants (#PE4483)
Environmental hazards from petroleum (#PE1409)
Environmental hazards from fertilizers (#PE1514)
Chemical pollutants of the environment (#PD1670)
Environmental hazards of essential oils (#PE9056)
Long-term hazards of exposure to chemicals (#PE4717)
Environmental hazards from plastic materials (#PD8566)
Vulnerability of marine animal communication (#PF7554)
Environmental hazards from petroleum refineries (#PG5858)
Environmental hazards of tanning and dyeing industries (#PE8571)
Environmental hazards from manufacture of plastic products (#PE8651)
Environmental hazards of explosives and pyrotechnic products (#PE9177).
Aggravates Eczema (#PE2465) Liver diseases (#PE1028)
Injurious accidents (#PB0731).
Aggravated by Malodorous fumes (#PD1413).

♦ PC1195 Long-term shortage of commodities

Broader Shortage (#PB8238) Long-term shortage of natural resources (#PC4824).
Narrower Long-term shortage of chemicals (#PE1261)
Shortage of industrial diamonds (#PG7754)
Long-term shortage of manufactured goods (#PE0802)
Long-term shortage of food and live animals (#PE0976)
Long-term shortage of beverages and tobacco (#PE1253)
Long-term shortage of mineral fuels and lubricants (#PE1712)
Long-term shortage of inedible crude non-fuel materials (#PE0461)
Long-term shortage of machinery and transport equipment (#PE1436)
Long-term shortage of animal and vegetable oils and fats (#PE1188)
Long-term shortage of miscellaneous manufactured articles (#PE0613).

♦ PC1217 Instability of economic and industrial production activities

Uncoordinated production planning

Nature Industrial restructuring – the process of continuous change in the world pattern of production – is a logical consequence of growth and necessary for continued development. It is one aspect of the trend towards world interdependence and is marked by rapid growth and the shift of industrial activity away from the old industrial centres and towards the newly developed countries. Focus of technological development on a few industries and the commercial uncertainty of developing new processes has made government assistance (financial, technical, scientific and material) a necessity in these science-intensive industries. This has led to increased efficiency of such industries in developed countries, where government support is available, and a relative lagging behind of the traditional industries (such as textiles, clothing, leather, pulp and paper) in these countries. These latter industries have in consequence expanded rapidly in the developing countries, resulting in global specialization.

The magnitude and rate of these changes is not only enormous but largely unplanned, generating a momentum and direction of their own. The internal dynamics of national growth and the international spread of technology are governed by uncoordinated forces, including ad hoc governmental decisions and the dictates of the market; they encourage the perpetuation of the inequitable distribution of the benefits accruing internationally from the present rapid growth of industry and it is unclear whether they are the most efficient means of promoting such growth per se. As stated in the draft of the Lima Plan of Action on Industrial Development and Cooperation, "the unrestricted play of market forces is not the most suitable means of promoting industrialization on a world scale nor of achieving effective international cooperation in the field of industry".

Claim Instability is the essence of capitalist, laissez-faire markets, the results of which are unemployment, poverty, and disease or hunger for millions. Lack of central planning allows market forces to war on each other. Discord and confusion prevail in the economic and industrial system to the detriment of society and labour, but to the benefit of bankers and the privileged classes who move their capital in response to market forces.

Counter-claim Economic activity is an expression of human behaviour. As such it is inherently unstable. Were it less so, the fact still remains that food production and many essential commodities (such as wood, wood products and natural fibres) depend on the environment as well as on man's efforts, so that their availability is never certain as to specific quantities at specific times.

Broader Unbalanced growth (#PB0479) Economic uncertainty (#PF5817)
Economic and financial instability of the world economy (#PC8073).
Narrower Bank failure (#PE0964) Agricultural surpluses (#PC2062)
Instability of tea trade (#PE2054) Vulnerability of farming (#PC4906)
Excess production capacity (#PD0779) Prohibitive cost of living (#PF1238)
Instability of fishing industry (#PE1424) Instability of forestry and logging (#PE0459)
Instability of construction industry (#PE0509) Instability of manufacturing industries (#PC0580)
Introduction of high-yield crop varieties (#PF3146)
Vulnerability of world cable communications (#PD0407)
Instability of the primary commodities trade (#PC0463)
High labour turnover in developing countries (#PF0907)
Instability of mining and quarrying industry (#PE0993)
Limited number of available radio frequencies (#PF0734)
Instability of the maritime shipping industry (#PE5791)
Instability of production of food and live animals (#PD2894)
Instability of agricultural and livestock production (#PE8998)
Instability of hunting, trapping and game propagation (#PE9059)
Detrimental international repercussions of domestic agricultural policies (#PF2889)
Inadequate demand for primary commodities because of rising living standards (#PD2898).
Related Chance (#PF7114) Lack of control (#PF7138)
Underutilization of facilities due to daily or seasonal peaks (#PF0827)
Instability in export trade of developing countries producing primary commodities (#PD2968).
Aggravates US dollar dominance of world economy (#PD2463)
Obstacles for international ocean shipping (#PD5885).
Aggravated by Economic inflation (#PC0254) Hoarding of primary commodities (#PD0651)
Underdevelopment of food and live animal production (#PF2821).

♦ PC1227 Fragmentation of technological development

Incompatible technologies

Broader Fragmentation (#PA6233) Human errors and miscalculations (#PF3702)
Lack of international cooperation (#PF0817).
Aggravates Ignorance (#PA5568).
Aggravated by Institutional fragmentation (#PC3915)
Disintegration of technological capacity (#PD7719)
Inadequate standardization of procedures and equipment (#PC0666).

♦ **PC1228 Weakness of infrastructure in developing countries**

Nature Developing countries frequently lack adequate physical and social infrastructure of all kinds and their substantial improvement is essential for rapid economic development. In these countries, except in a few cities and towns, most areas are not served by modern transport and communications, and electric power is non-existent. Water supplies for domestic, agricultural and industrial use is limited. Sanitation facilities are non-existent for most people. Health and medical services are lacking.

The educational infrastructure in developing countries is weakest in the sciences and technologies, resulting in a lack of technological expertise to bring to bear on agricultural and other key areas of production.

Broader Underdevelopment (#PB0206)
Inappropriate design of development projects (#PF4944).
Narrower Disaster unpreparedness (#PF3567) Domestic refuse disposal (#PD0807)
Inadequate health services (#PD4790) Inadequate medical facilities (#PD4004)
Insufficient transportation infrastructure (#PF1495)
Inadequate housing in developing countries (#PE0269)
Inadequate disaster prevention and mitigation (#PF3566)
Lack of management skills in developing countries (#PE0046)
Structural rigidity in developing country economies (#PD2970)
Unrelated pioneer institutions in developing countries (#PF1724)
Inefficient public administration in developing countries (#PF0903)
Lack of sanitation in rural areas of developing countries (#PD1225)
Weakness of infrastructure in island developing countries (#PE5772)
Lagging transformation of agriculture in developing countries (#PD0946)
Underdevelopment of the power industry in developing countries (#PF4135)
Inadequacy of telecommunication facilities in developing countries (#PE0004)
Inadequate infrastructure and services in least developed countries (#PE0289)
Inadequate water supply in the rural communities of developing countries (#PD1204)
Lack of integration of transport systems between neighbouring developing countries (#PD0664).
Related Inappropriate sanitation systems (#PD0876).
Aggravates Inappropriate development policy (#PF8757)
Weakness in trade among developing countries (#PC0933)
Inadequate production capacity in developing countries (#PD4219)
Inadequate development of enterprises in developing countries (#PE8572).
Aggravated by Imbalance between capital and technical assistance (#PE4866).
Reduces Social insecurity in developing countries (#PE4796).

♦ **PC1242 Radioactive wastes**

Radioactive waste disposal — Radioactive effluents

Nature Radioactive wastes present special public health problems that are not common to other wastes and are unfamiliar to many health authorities. Difficulties arise from the fact that there is no method of neutralizing or modifying the radioactivity of the wastes. Their decay rate is fixed, being a specific invariable property of each radionuclide. Furthermore, radioactive wastes containing very low concentrations of radionuclides are usually disposed of by dilution into the environment. Once they have been discharged into the environment, in accordance with statutory limitations, no significant further control can be practised; they undergo dilution, retention, and reconcentration through the operation of natural processes, and under special circumstances may constitute a potential hazard to man.

The liquid wastes which constitute 99.9 per cent of the radioactivity in wastes create, because of their concentration, special problems of shielding, cooling and containment. In the case of solid wastes, the major problem is the sheer quantity which has to be disposed of, since its radioactivity is normally relatively low. Gaseous wastes create further complications.

High-level radioactive waste contains uranium-235 and plutonium-239. It is highly toxic, can be used to make nuclear weapons, and is difficult to store because radioactivity keeps it hot and corrosive for hundreds of years. For example, plutonium has a half-life of 24,300 years, which means that it will remain a danger to human health for up to half a million years or 16,666 generations. Some studies conclude that high-level waste, isolation from the biosphere for 10 million years may have to be considered for spent-fuel; and for reprocessed high-level waste, isolation time of 1 million to 10 million years are to be used for guidelines in developing appropriate disposal techniques. The least powerful but most plentiful form of high-level waste is uranium mine tailings. A ton of uranium ore yields only 4 pounds (1.8 kilograms) of usable uranium: the rest is tailings, which emit radioactive radon gas and will remain radioactive for practically an indefinite period. Most of the high-level waste existing today is a product of nuclear weapons production. However, the nuclear waste from peaceful uses is a hundred times more concentrated, so that civilian waste in the United States already contains more radioactivity than military waste.

Low-level radioactive waste - which includes any material that has picked up induced radiation, such as uranium mining tools, uniforms of workers in reprocessing plants, medical wastes, and cooling water from nuclear reactors - is the next most plentiful form of nuclear waste.

Incidence More than a million gallons (3.8 million litres) of low-level waste in steel drums was dumped in the ocean near San Francisco between 1946 and 1962, and about 25 percent of these drums are now leaking. The US Environmental Protection Agency (EPA) estimates that there could be as much as 400 million cubic feet (11.3 million cubic meters) of low-level waste in the United States alone by the year 2000.

United States military research and weapons production have produced 10 million cubic feet (280,000 cubic meters) of high-level waste and 50 million cubic feet (1.4 million cubic meters) of low-level waste; the Soviet Union has probably produced an equivalent amount, and the other nuclear powers - France, UK, China, and India - proportionately smaller quantities. Each year these totals are growing. During 1979, the global operation of nuclear reactors (assuming an average capacity of 500 megawatts) resulted in 530 million cubic feet (15 million cubic meters) of mine tailings, 380,000 cubic feet (11,000 cubic meters) of low-level waste, and 140,000 cubic feet (4,000 cubic meters) of high-level waste containing almost 50,000 pounds (23,000 kilograms) of plutonium.

It has been estimated that by the year 2000 high-level liquid wastes will be produced at the rate of 250,000 to 2,500,000 gallons per day. And by that time the total accumulated volume of high-level wastes will be of the order of 150,000 to 250,000 million gallons, and the total accumulated radioactivity will be more than 500,000 million curies. It is also estimated that by 1990 there will be 20,000 cubic metres of high-activity wastes arising from the reprocessing of spent fuel. Such wastes contain over 99 percent of the fission products present in the fuel, together with smaller quantities of actinides.

Because a considerable part of this accumulated activity will be due to strontium-90 and other long-life radionuclides, methods for ultimate waste disposal of these wastes must provide containment and control for at least several hundred years. It is doubtful whether any man-made structure could be guaranteed to provide permanent containment, so that the use of deep geological formations (salt deposits, antarctic ice, ocean deeps, etc) seems more suitable. High-level wastes are at present stored mainly in liquid form, and some constituents will remain dangerously radioactive for several hundreds of thousands of years. There is at present no generally accepted means whereby high-level waste can be permanently isolated from the environment and remain safe for very long periods.

Claim 1. A large backyard swimming pool 40 feet by 20 feet by 6 feet (12.2 meters by 6 meters by 2 meters) has a capacity of 4,800 cubic feet (135.8 cubic meters). High-level military waste from the United States alone would fill more than 2,000 of these pools. 2. No matter which method of disposal, radioactive waste storage facilities will have to be built to last tens of thousands of years longer than any building now existing. They will have to be our most permanent creations. The Egyptians are remembered for their pyramids, the Greeks for their temples and stadiums, the mediaeval Europeans for their cathedrals. Our legacy might be our poisonous radioactive garbage dumps.

Refs Brawner, Carroll O *First International Conference on Uranium Mine Waste Disposal* (1980); DOE Technical Information Center Staff *Radioactive Waste Management* (1984); DOE Technical Information Center Staff and McLaren, Lynda H *Radioactive Waste Management* (1984); Park, Kilho P, et al *Radioactive Wastes and the Ocean* (1983); Pottier, P E and Glasser (Ed) *Characterization of Low and Medium-Level Radioactive Waste Forms* (1986); Radioactive Waste Campaign *Deadly Defense* (1988); United Nations *Management of Low and Intermediate Level Radioactive Wastes* (1971); United Nations *Disposal of Radioactive Wastes into Rivers, Lakes and Estuaries* (1971).

Broader Hazardous wastes (#PC9053).
Narrower Nuclear waste (#PD4396).
Related Unproductive use of resources (#PB8376)
Transport and storage of radioactive material (#PE6884).
Aggravates Radioactive contamination (#PC0229).
Aggravated by Corrosion (#PD0508) Hazardous waste dumping (#PD1398)
Marine dumping of wastes (#PD3666) Inadequate waste treatment (#PD6795)
Vulnerability of nuclear power sources (#PD0365).

♦ **PC1247 Destruction of natural barriers**

Nature Destruction of natural barriers between ecosystems, such as by the creation of sea-level canals between previously separated oceans, may result in ecological disasters. Such destruction may result from atomic-blasting of channels as is proposed for the Panama link, or it may be progressive as a result of continued dredging operations as in the case of the Suez Canal which has been gradually deepened over the years, permitting more species to travel from the Red Sea to the Mediterranean.

Broader Human destructiveness (#PA0832) Environmental degradation (#PB6384).
Aggravates Environmental hazards of new species introduction (#PC1617).
Reduces Insufficient transportation infrastructure (#PF1495).

♦ **PC1258 Competitive acquisition of arms**

Competitive militarization — Arms race — Dependence on arms race — Surplus armaments — Myth of national security through increase in military capacity

Nature The arms race has two characteristic features. One is the multiplication and proliferation of primarily non-nuclear, tactical armaments. The other takes the form of a very rapid rate of product innovation and improvement and a constant search for new environments in which weapons can be used. At first sight it would seem that the effort to improve the quality of armaments, or to defend against them, follows a logical series of steps in which a new weapon or weapon-system is devised, then a counter-weapon to neutralize the new weapon, and then a counter-counter-weapon. But these steps neither usually nor necessarily occur in a rational time sequence. Those who design improvements in weapons are as a rule the same people who envisage the further steps to be taken. They do not wait for a potential enemy to react before they themselves react against their own creations. Before a new weapon is brought into service, the military designer is, as a rule, already designing a new model which - he hopes - will not only be more effective in performance, but also less vulnerable to defences which the other side might introduce in response to the new threat. Thus obsolescence becomes characteristic of the technological arms race. These features of the arms race show up very clearly in the field of long-range nuclear weapons. First there was a rapid change in the means of delivery, starting with the switch from manned bombers to liquid-fuelled ballistic missiles, beginning with intermediate and moving on to rockets of intercontinental range. Solid-fuelled missiles soon followed, deployed in concrete silos, in order to protect them from attack. In parallel, submarine-launched ballistic missiles were developed and deployed.

It does not necessarily follow that the process of action and reaction which characterizes the arms race, certainly the arms race in sophisticated weapons, means that security is increased as more is spent on armaments. Indeed in the field of nuclear weaponry the reverse appears to be the case. Each new step in the elaboration of such armaments usually ushers in a more perilous stage of uncertainty and insecurity. Furthermore, every new generation of weapons and weapon systems inevitably demands more and more resources which could be used for different economic and social purposes. By encouraging the development of certain areas of technology, and by providing resources for basic fields of science which might bear upon the development of sophisticated weapons, the arms race also inevitably affects the direction and tempo of a country's scientific and technological development. Its effect has been to encourage work in certain fields of knowledge and to retard progress in others. It stimulates a demand for certain classes of specialist and for certain kinds of specialized information, without which desired military projects could not be achieved. Short of powerful political decision in a contrary direction, this process, particularly so far as it concerns sophisticated modern weapons, could go on indefinitely.

The arms race has in fact become noticeably a technological race, the achievements of one side spurring the other to improve on the technological advances which it might have made itself. Sometimes the spur comes not from some clearly defined threat but from an imagined technical advance made by the other side. Secrecy in military affairs makes it inevitable that a potential enemy will usually be suspected of being stronger than he actually is. Consequently both sides strive continuously to improve the quantity and quality of their arms. So it is that the arms race becomes based on the 'hypothesis of the worst case', that is to say, one of two sides designs its programme of development on the assumption that its rival could, if it so decided, be the stronger.

Military expenditures not only divert resources from other uses, but also tend to disturb and destabilize the economy in general. Increased taxation or borrowing needed to raise money for arms (in developed market economies) slows the growth in personal consumption or private investment. If taxes are not raised, spending on such programmes as welfare services or education may be reduced, thus dislocating long-term social policies. Inflationary processes may be generated. In centrally planned economies, military expenditures limit the flexibility with which the economy can be planned, and the problem of preserving a proper balance between supply and demand for various industries and sectors becomes more difficult. In developing countries where the tax-base is limited, the pay of civil servants and the cost of military forces often take up much of the government's revenue. Revenues that might go into development are used instead for military purposes. In addition, military spending often puts a heavy burden on the balance of payments due to the purchase of arms from abroad.

The arms race is an important factor in limiting the expansion of international exchanges. Military considerations limit trade in so-called strategic commodities and products of advanced technology, and have led to creation of rival trade groupings. Strategic considerations inhibit technological and scientific exchanges between countries. Also, protectionist policies to favour domestic industry or agriculture are often defended on the grounds of maintaining the supply of vital commodities in time of war. This argument could not be advanced to justify trade barriers in a disarmed world. Trade between centrally planned economic and developed market countries has clearly been affected by the arms race and by the tensions between the two systems. This trade accounts for only 5 per cent of world trade. It would rise significantly the faster the arms race came

to a halt. As for the developing countries, the scarce foreign exchange resources used to obtain armaments could be applied to growth-producing purposes. In a world progressively disarmed, the level of trade could well be higher simply because of a higher level of world output.
Incidence Numerical counts of weapons alone are an inadequate measure because of the major trend to product improvement which makes, for example, a new combat fighter a very different weapons system from one built 10 years previously. In addition, national inventories of stocks of armaments are never published, but some figures are available which reflect these various qualitative changes:
At the outset of the 1960s no intercontinental ballistic missiles (ICBMs) had yet been deployed. By the end of the decade the estimated numbers were 2,150. In the late 1960s and early 1970s the USA tripled the number of its nuclear weapons with the introduction of multiple independently targeted reentry vehicles (MIRVs), and the USSR "MIRVed" their missiles during the next decade.
In 1960 the deployment of submarine-launched ballistic missiles was negligible. By the end of the decade, some 55 nuclear-missile submarines were operational, comprising about 800 missiles, capable of delivering about 1,800 warheads. The number has since doubled. From 1960 to 1968 the world stock of fighting vessels is estimated to have increased from 4,550 to 4,900. This relatively small increase in numbers masks the much larger increase in the value of this stock (at 1968 prices, the value of the stock in 1960 was about $34,000 million, as compared with $60,000 million in 1968, a 75 per cent rise). In 1960 the world stock of supersonic fighters was estimated at 6,000. By 1970 it had doubled. In 1960 there were 15 production programmes for supersonic aircraft; by 1970 these too had doubled. Doubling is a feature that continues and weapons proliferate in geometrical progression by number or power.
In July 1983 the superpowers were said to have reached a point of rough equivalency, the USA usually leading the way in new quantitative and qualitative technological changes, most recently with the Cruise missile, with the USSR never far behind and occasionally leading the way (as with very large missiles). Between 1983 and 1987 the USA is expected to spend more than US $1,600 billion, including a planned 3,900 aircraft (fighters, bombers and transport planes) and 8,800 tanks and cannon-carrying vehicles; and between 1983 and 1993, to produce 14,000 more strategic and tactical nuclear bombs and missiles. A future rate of real increase in military spending worldwide was estimated at a minimum of 2 to 3 percent annually. The latest escalation has been to extend the arms race beyond the earth's atmosphere in the so-called "Star Wars" defence programme.
Claim The threat of ultimate disaster the arms race has generated is by far the most dangerous single peril the world faces today - far more dangerous than poverty or disease, far more dangerous than either the population explosion or pollution - and it far outweighs whatever short-term advantage armaments may have achieved in providing peoples with a sense of national security. The worlds atomic arsenals contain the explosive power of 5 tons of TNT for every man, woman and child on earth. Their total explosive power equals that of one and a half million Hiroshima bombs. The arms race has resulted (1985) in the expenditure of the equivalent of nearly US $1,000 billion annually. George Kennan, former US ambassador to Moscow has said (May 1981): "We have achieved, we and the Russians together, in the creation of these devices and their means of delivery, levels of redundancy of such grotesque dimensions as to defy rational understanding".
Counter-claim In a situation of armed conflict, a balance of arms supplies to both sides in the conflict will have a deterrent effect and help to restrain the outbreak of war.
Refs Stockholm International Peace Research Institute *Arms Race and Arms Control* (1982); United Nations *Economic and Social Consequences of the Arms Race and of Expenditures* (1983).
Broader Imbalance of power (#PB1969) International insecurity (#PB0009)
International non-military conflict (#PF3100).
Narrower Naval arms race (#PD8412) Nuclear arms race (#PD5076)
Restriction of arms supply (#PD1304) Surface to surface missiles (#PE4515)
Monopoly of nuclear power techniques (#PD1741)
Stockpiles of nuclear warfare material (#PC0311)
Imbalance of conventional armed forces (#PC5230)
Insufficient nuclear weapon free zones (#PJ5335)
Proliferation of strategic nuclear arms (#PD0014)
Militarization of the deep ocean and sea-bed (#PD1241)
Proliferation of nuclear weapons and technology (#PD0837)
Military-industrial complex in capitalist systems (#PC3191)
Absorption of manpower resources by military activities (#PF0780)
Military-industrial-governmental complex in communist systems (#PD1330).
Related Plutonium overproduction (#PD2539).
Aggravates War (#PB0593) Nuclear war (#PC0842) Dehumanization (#PA1757)
Military government (#PC0698) Biochemical warfare (#PC1164)
Excessive public debt (#PC2546) International tension (#PB8287)
Nuclear weapons testing (#PC2201) Pervasive fear of nuclear war (#PC3541)
Irresponsible international trade (#PC8930) Private international arms dealers (#PD2107)
Socio-economic burden of militarization (#PF1447)
Lack of business opposition to the arms race (#PF7088)
Foreign intervention in internal affairs of states (#PC3185)
Hindrances to international spread of new technologies (#PE8758).
Aggravated by Personal wealth (#PC8222) Economic rivalry (#PD8897)
International arms trade (#PC1358) International status race (#PC5348)
Imbalance in strategic arms (#PC1606) Decadent standard of living (#PD4037)
Maldistribution of electrical energy (#PD3446)
Competitive development of new weapons (#PC0012)
Failure of disarmament and arms control efforts (#PF0013)
Politically unrealistic strategic warfare analysis (#PF1214).
Reduced by Negative economic repercussions of disarmament (#PF0589).

♦ PC1259 Bureaucratic superiority
Bureaucratic arrogance
Nature Public office has always been associated with the established privileges of a ruling class and an element of this consciousness remains as an attribute of high public office, even in the absence of such traditions. In most situations in which members of the general public interact with officials, the representatives of bureaucracy have many advantages: the public must pass through a single channel to further a given project; often there is no means of appealing against the decision of a particular official. It is therefore difficult for officials to avoid developing a sense of superiority and disdain towards the public in spite of being called "civil servants".
Broader Arrogance (#PA7646) Over-centralization (#PF2711).
Aggravates Government arrogance (#PF8820) Government complacency (#PF6407)
Arrogance of policy-makers (#PF2895) Lack of cooperation with officialdom (#PF8500)
Bureaucracy as an organizational disease (#PD0460).
Aggravated by Unchecked power of government bureaucracy (#PD8890).

♦ PC1268 Inequality before the law
Denial of right to equal protection by the law
Nature Unequal access to legal advice, representation and other legal facilities, which may or may not be written into the law, acts as a special restriction on the rights of certain sections of the community.
Refs Maltsev, G *Illusion of Equal Rights* (1984).
Broader Social injustice (#PC0797) Social inequality (#PB0514)
Denial of right to justice (#PC6162).
Narrower Discrimination before the law (#PC8726)
Denial of legal representation (#PF3517)
Lack of impartiality of the judiciary (#PE7665)
Denial of right to redress for rights violations (#PE4173)
Denial of right to recognition as a person before the law (#PE4716).
Related Segregation (#PC0031)
Narrow legal definition of the family in developing countries (#PD1501).
Aggravates Elitism (#PA1387) Distortionary tax systems (#PD3436)
Denial of right to liberty (#PF0705) Underutilization of legal rights (#PF3464)
Underprivileged religious minorities (#PC2129).
Aggravated by Legal segregation (#PD3520) Ignorance of the law (#PG3516)
Social discrimination (#PC1864) Discrimination against minorities (#PC0582)
Deficiencies in national and local legal systems (#PF4851)
Psychological barriers to the judicial protection of individual rights (#PE1479).

♦ PC1282 Environmental stress
Ecological stress
Nature Environmental degradation causes stress in individuals and groups, especially through the increasing scarcity of resources. Environmental stress may also result as a consequence of military conflict, although the existence of such stress through the degradation of resources may also be a cause of such conflict. Thus countries may engage in conflict to assert or resist control over raw materials, energy supplies, land, river basins, sea passages, and other environmental resources. When political processes are unable to handle the effects of environmental stress, such as those resulting from erosion and desertification which lead to the marginalization of sectors of the population, this can lead directly to violence.
Claim There can no longer be the slightest doubt that resource scarcities and ecological stresses constitute real and imminent threats to the future well-being of all people and nations.
Refs Hennssey, T C, et al *Stress Physiology and Forest Productivity* (1986); Mussell, Harry and Staples, Richard C *Stress Physiology in Crop Plants* (1979).
Aggravates Refugees by boat (#PD8034) Environmental refugees (#PE3728)
Environmental threats to national security (#PC4341).
Aggravated by Long-term shortage of water (#PC1173)
Natural environment degradation (#PB5250)
Competition for scarce resources (#PC4412)
Environmental consequences of war (#PC6675)
Unsustainable exploitation of fish resources (#PD9082)
Inequality in distribution of natural resources between countries (#PF3043).

♦ PC1287 Infant mortality
Child deaths
Nature Children under 5 years of age are the population group at most risk from adverse environmental conditions. Increased mortality is principally due to poor sanitation, lack of safe water, overcrowding, reduced intervals between pregnancies, lack of maternal education, malnutrition, and inadequate health care. Inadequate spacing, number, and timing of births can have a revolutionary impact on the growth and survival of children. Indeed, infant mortality increases steeply after too short an interval between births and after the third child. Also, children born to women under the age of 20 are approximately twice as likely to die in infancy as children born to women in their mid-20s.
Background The widespread acceptance of the loss of many infant lives has sometimes been attributed to a fatalistic outlook on life; it may also have been a cause of such fatalistic attitudes. During the seventeenth century, for example, the child mortality rate in Europe is estimated to have been in the region of 500 per thousand; before the introduction of new methods and the development of corresponding attitudes, almost everywhere at least 200 or more out of 1,000 liveborn infants died in their first year of life. In the history of every human society, the loss of many new-born infants has probably been regarded as unavoidable, and had to be accepted as such. In some of the less developed regions this may still be the case today.
Incidence The world infant mortality rate for 1980-1985 was 81 per 1000 live births, with more developed regions having a rate of 17 per 1000 and less developed regions' rates being 92 per 1000. The infant mortality rate for Africa was 114; Asia 87; Europe 16; Oceania 39; Latin America 63; and North America 12. In developing countries 50 per cent of total deaths are in children under 5 years of age, and for this age group mortality rates in excess of 300 per 1,000 live births occur in some regions. Infant mortality rates vary from less than 15 in most developing countries to more than 200 in the least developed countries.
Refs Honig, Alice S *Risk Factors in Infancy* (1986); Shapiro, Sam, et al *Infant, Perinatal, Maternal, and Childhood Mortality in the United States*; United Nations *Mortality of Children Under Age 5* (1989).
Broader Human death (#PA0072).
Narrower Neonatal mortality (#PD9750) Sudden unexpected infant death (#PD1885)
Perinatal morbidity and mortality (#PD2387)
Child mortality in developing countries (#PE5166).
Aggravated by Infanticide (#PD3501) Overcrowding (#PB0469)
Multiple pregnancy (#PG2047) Unhygienic conditions (#PF8515)
Lack of family planning (#PF0148) Unbalanced infant diets (#PE0691)
Neglected young children (#PE4245) Water system contamination (#PD8122)
Complications of childbirth (#PC9042) Human physical genetic abnormalities (#PD1618)
Insufficient birth spacing in families (#PF5968)
Inadequate maternal and child health care (#PE8857)
Ignorance of women concerning primary health care (#PD9021).

♦ PC1288 Inadvertent modifications to climate
Nature Although there is great scientific potential for the bettering of climate through control of the natural weather systems, there is also great danger to the world ecological system arising from uncontrolled and indiscriminate use of weather modification techniques to increase or decrease precipitation; increase or suppress hail, lightning or fog; and direct or divert storm systems. This is especially true if such modification leads to long-term changes in the atmospheric conditions. In particular the development of weather modification for military purposes creates a threat to peace and world order, because of their relative covertness and widespread potential for devastating impact on a target area and possibly on global weather patterns.
Claim In several countries, politicians and entrepreneurs, ignorant or impatient of the scientific facts and problems, are initiating and conducting major weather modification projects without the benefit of proper scientific direction, advice or criticism.
Broader Anthropogenic climate change (#PC9717)
Long-term change in atmospheric chemistry (#PF1234)
Irresponsible scientific and technological activity (#PC1153).
Related Hostile environmental modification (#PD7941).
Aggravates Environmental degradation (#PB6384).
Aggravated by Dust (#PD1245) Inhospitable climate (#PC0387)
Environmental warfare (#PC2696) Long-term shortage of water (#PC1173)
Adverse effects of urbanization on climate (#PE9020).

♦ PC1298 Proliferation of information
Information explosion — Information overload

Nature The amount of information produced is such that even within a very specialized field, it is difficult for people to keep informed of latest developments. It is even more difficult to determine developments in other fields which have implications for a given field.
Incidence World book titles increased 132 percent between 1950 and 1979. At any given time 3,000 million titles may be in print. Estimates suggest that there are about 80,000 regular scientific and technical journals out of a total of about 150,000 journals with valid information content. As an example, it took 32 years from 1907 to 1938 before Chemical Abstracts reached its millionth abstract; the fifth million was reached in 3 years and 4 months. According to the area of science chosen and the method used, an annual growth rate of 4 to 8 per cent is encountered with a doubling period of from 10 to 15 years. A spectacular example of the ability of information to reach people is the over 3,500 percent increase in television receivers since 1950, totalling some 500 million receivers in present use; this indicates an appetite for information that will be increasingly served by 24 hour programming, hundreds of television broadcasting stations, and thousands of dependent companies producing informational, educational or documentary programmes and films. In the decade from 1979 to 1989 the National Space Science Data Centre of the USA has accumulated some 6,000,000,000,000, or six trillion, bytes of information. That is about double the amount of information contained in all of the Library of Congress's 19 million books. The space probe of Venus, Magellan will increase this by an additional three trillion bytes. The Hubble Space Telescope scheduled to be launched in 1990 will generate several trillion bytes every year. If the Earth Observing System is launched, it will generate a trillions bytes of information every day. A host of other space probes are planned.
Refs Bellak, Leopold *Overload* (1975).
Broader General obstacles to problem alleviation (#PF0631).
Narrower Proliferation of advertising (#PD5034)
Proliferation of printed matter (#PD4552)
Proliferation of unprocessed scientific data (#PF1065)
Information overload during control of complex equipment (#PF6411).
Aggravates Paper shortage (#PE1616)
Mental pollution (#PB6248)
Over–specialization (#PF0256)
Fragmentation of knowledge (#PF0944)
Proliferation of computers (#PE3959)
Suppression of information (#PD9146)
Inaccessibility of knowledge (#PF1953)
Imbalanced distribution of knowledge (#PF0204)
Incomplete access to information resources (#PF2401)
Insufficient translation into minority languages (#PD0825)
Increasing development lag against information growth (#PE2000)
Proliferation and duplication of international information systems (#PE0458).
Aggravated by Adverse consequences of scientific and technological progress (#PF3931).

♦ PC1299 Contamination by natural radiation
Ill effects of radon on health
Nature Natural background radiation is high in certain parts of the world, and there is a risk of food contamination from this source. Certain fish can concentrate heavy metals, and will concentrate radioisotopes of these metals in the same way. Some marine animals (for example, molluscs) have dangerously highly levels, concentrated mainly in the shells or bones of the animals but also in their edible portions. For example, radioisotopes have been accumulated by oysters at levels sufficient to present a hazard from their consumption as food. Fish such as salmon, tuna, and anchovy also accumulate sufficient heavy metals for high levels of consumption to result in an increase in the body–burden of radioactivity. Frequent air travellers like airline pilots are exposed to cosmic radiation. Medical exposure through X–rays and barium enemas account for other forms of exposure. Radon is the biggest hazard. It seeps up through the soil from radioactive rocks, especially granite. Because the air pressure in houses is slightly lower than outside, they tend to suck in radon from the ground. Radon is the biggest single cause of lung cancer after smoking.
Incidence The radioactive gas radon seeps from the ground in Devon and Cornwall in the west of England, because of the high concentrations of uranium in the underlying rocks. Several thousands of houses are reported as having abnormally high radiation levels, some as much as eight times the level permitted for workers in the nuclear industry. Up to 12 percent of America's 75 million houses may have enough radon to warrant remedial action.
Refs Bodansky, David et al *Indoor Radon and Its Hazards* (1987); Brookins, Douglas G *The Indoor Radon Problem* (1990); International Commission on Radiological Protection (Ed) *Lung Cancer Risk From Indoor Exposures to Radon Daughters* (1987); International Commission on Radiological and Sowby, F D *Limits for Inhalation of Radon Daughters by Workers* (1981); Muller Associates Inc Staff and SYSCON Corporation Staff and Brookhaven National Laboratory Staff et al *Handbook of Radon in Buildings* (1988); National Research Council *Health Risks of Radon and Other Internally Deposited Alpha–Emitters* (1988); National Water Well Association Staff *Radon in Ground Water* (1987); Neiderhaus, Lee B *Radon* (1988); OECD *Metrology and Monitoring of Radon, Thoron and Their Daughter Products* (1985).
Broader Radioactive contamination (#PC0229).
Narrower Health hazards of radiation (#PD8050)
Health hazards of radiation in aircraft (#PE0962)
Health hazards of exposure to cosmic radiation (#PF1686).
Aggravates Indoor air pollution (#PD6627).

♦ PC1300 Destruction of land fertility
Nature The loss of arable land through various forms of soil degradation, particularly soil erosion, is the result of a complex set of often interacting economic, social and demographic forces and conflicting policy objectives, as well as of topographical and climatic factors. Relentless desertification has caused a continuing shrinkage of arable land in areas such as the Sudano–Sahel. Deterioration or loss of the productive capacity of the soil is often connected with salinization, including that caused by irrigation. Slat deposits, which are left in the soil after water runs off, can significantly reduce yields; and, in the long run, can make the land unsuitable for cultivation. At the same time, measures normally undertaken to prevent salinization are not always effective. In some cases they may also jeopardize the crop itself, especially when the effort to leach the excess salt from the surface of the soil leads to waterlogging which is itself damaging to crops. In many areas of the world, especially where irrigation is critical for crop production, the twin problems of waterlogging and salinity have become acute and their solution difficult and controversial.
A number of human activities so reduce the fertility of land, or degrade the soil, that large tracts of land must eventually be abandoned as being totally unproductive. This is the case in several agricultural methods. The failure to use the various forms of agricultural rotation – grass to crop to fallow, for instance – which were previously employed to remedy the 'soil fatigue' which uninterrupted growth of one crop often brings, has resulted in a decline in the organic humus content of the soil. Loss of fertility is usually quite irreversible, though there are examples where the process of intensive land–use, abandonment, and regeneration, have been incorporated into agricultural practices, as in the cyclical movements of 'slash–and–burn' forest tribes and of nomadic herdsmen on brush–land.
In many developing countries a heavy pressure of population, leading to increased demand for food, has been a major factor behind the cultivation of marginal lands, such as hill–sides and river banks, which directly leads to soil erosion. The increased irrigation needed for food production may also lead to problems such as salinization. Agricultural and animal husbandry practices, such as the shortening of the fallow period under shifting cultivation due to demographic pressure, and overgrazing, have often contributed to soil degradation. Rising costs of alternative energy as well as poverty and increase in population have led to increased demand for fuelwood and hence to accelerated deforestation. Crops that have high economic yield or commercial profitability (such as, in recent years, maize and soyabeans) have also been found on occasion to aid soil degradation.
Incidence By the year 2000, with population growth pushing ahead of the increase in potential production, the situation will have deteriorated for most developing countries: 64 countries (more than half the total) might then be in a critical position. Indeed, 38 of these would be able to support less than half their projected populations. It is estimated that if no long–term conservation measures are taken, land degradation in developing countries may in the long run depress food production by an average of 19 percent. The area of rain–fed cropland could shrink by as much as 544 million hectares – an area greater than the entire potential cropland of Southeast Asia. 30 percent of Central America's rain–fed cropland could be lost – and 36 percent of Southeast Asia's. The entire potentially cultivable lands of the 117 developing countries would be sufficient to support only 1.6 times the expected population of 2000 even if such land were used only for food crops or as grassland supporting livestock. This potential area is at least three times greater than the present cultivated area.
Broader Environmental degradation (#PB6384)
Long–term shortage of natural resources (#PC4824).
Narrower Destruction of agricultural land (#PD9118).
Aggravates Acidic soils (#PD3658)
Infertile land (#PD8585)
Soil infertility (#PD0077)
Deficiency diseases in plants (#PD3653).
Aggravated by Defoliation (#PD1135)
Sand storms (#PD3650)
Dust storms (#PD3655)
Forest fires (#PD0739)
Soil erosion (#PD0949)
Deforestation (#PC1366)
Alkaline soil (#PD3647)
Soil pollution (#PC0058)
Locust plagues (#PE0725)
Desert advance (#PC2506)
Soil compaction (#PD1416)
Soil degradation (#PD1052)
Diplomatic errors (#PF1440)
Soil salinization (#PE1727)
Migrating sand dunes (#PD0493)
Soil erosion by wind (#PE3656)
Soil erosion by water (#PD2290)
Inadequate land drainage (#PD2269)
Inȧdequate crop rotation (#PF3698).

♦ PC1306 Harmful synergistic interaction of biological agents
Compound disease
Nature Occasionally two biological agents combine to produce a more severe disease than either of the agents alone would produce. This synergistic interaction may be responsible for severe epidemics. It increases the unpredictability which accompanies the introduction of any new biological species into a new environment. Multiple–factor agents represent a new possibility in biological warfare, since two relatively harmless agents can be kept separate until released.
Incidence It is thought that the pandemic of 1918 in which influenza led to the deaths of many civilians and soldiers (more American lives were lost due to this disease than to gunfire) may have been such a compound disease.
Aggravates Biological warfare (#PC0195).

♦ PC1326 Endangered species of mammals
Nature Due primarily to the impact of man on the natural environment, whether directly or indirectly, many of the 4,500 species of mammal are in danger of extinction.
Incidence In 1600 (when reliable zoological records began), there were approximately 4,226 living species of mammals. By 1970, at least 36 (or 0.85 per cent) of these had become extinct and at least 120 more species (or 2.84 per cent) endangered. At the sub–species or geographical level, the situation is more serious; 64 sub–species have died out since 1600 and another 223 have been listed as rare or endangered. It is estimated that 83 per cent of the threatened mammals are in their present state because of man's activities.
Broader Endangered species of animals (#PC1713).
Narrower Endangered species of seals (#PE1656)
Endangered black rhinoceros (#PE6010)
Endangered species of llamas (#PE0904)
Endangered species of rabbits (#PD3480)
Endangered species of rodents (#PD3481)
Endangered species of elephant (#PD3771)
Endangered species of edentates (#PD3603)
Endangered species of Pangolins (#PG3566)
Endangered species of marsupials (#PD1762)
Endangered species of carnivores (#PD3482)
Endangered species of chiroptera (#PE3604)
Endangered species of monotremes (#PG3564)
Endangered species of Hyracoidea (#PG3568)
Endangered species of insectivores (#PD3479)
Endangered species of Flying lemurs (#PG3565)
Endangered species of Tubulidentata (#PG3567)
Endangered species of marine mammals (#PD3673)
Endangered species of non–human primates (#PE1570)
Endangered species of Odd–toed ungulates (#PG3569)
Endangered species of Even–toed ungulates (#PG3570).
Related Endangered species of insects (#PC2326).

♦ PC1328 Ethnocide
Cultural genocide — Cultural ethnocide — Dependence on ethnocide
Nature A culture may be suppressed by the prohibition of the use of its language, and the destruction or prevention of use of libraries, museums, ethnic schools, historic monuments, places of worship and other cultural institutions and objects. In its extreme form it may involve the prevention of births among an ethnic group and transference of children to another group. Cultural genocide may refer to ethnic groups but also to intellectual schools of thought and nations under foreign domination.
The systematic extermination of native races and indigenous tribes by outsiders, either 'nationals' of the country in question, or foreigners, is undertaken in order to expropriate land and safeguard implanted workers from attack. Ethnocide may be by killing or by destruction of the social structure and pauperization of the people or their enslavement, leading to disease, death, and lack of reproduction.
Incidence Ethnocide has been practised by colonists especially in Africa, the Americas and in Australia. It is alleged that ethnocide continues to be practised in some countries, with or without the tacit approval of the government. Recent examples are in Romania against the ethnic Hungarians and in Bulgaria against ethnic Turks.
Counter–claim "The Indians are of no interest, they are total savages, they have no law, no religion and live like animals", (former senior civil servant in Brazil: "The Primitive Tribes of South America" by Conrad Gorinsky).
Broader Genocide (#PC1056).
Related Ethnic disintegration (#PC3291).
Aggravates Domination (#PA0839)
Conformism (#PB3407)
Exploitation (#PB3200)
Loss of cultural identity (#PF9005)
Destruction of cultural heritage (#PC2114)
Lack of variety of social life forms (#PE8806)
Endangered tribes and indigenous peoples (#PC0720).
Aggravated by Racism (#PB1047)
Elitism (#PA1387)
Tribalism (#PC1910)
Starvation (#PB1875)
Nationalism (#PB0534)
Colonialism (#PC0798)

Ethnic conflict (#PC3685)
Forced assimilation (#PC3293)
Ideological deviation (#PF3405)
Cultural fragmentation (#PF0536)
Compulsory indoctrination (#PD3097)
Isolation of ethnic groups (#PC3316)
Discrimination against minorities (#PC0582)
Social conflict (#PC0137)
Anti-intellectualism (#PF1929)
Tropical deforestation (#PD6204)
Plural society tensions (#PF2448)
Denial of cultural rights (#PD5907)
Denial of freedom of thought (#PF3217)
Repression of intellectual dissidents (#PD0434).

♦ PC1351 Cultural deprivation
Dependence on cultural deprivation
Nature Cultural deprivation may be equated with a lack of education or the loss of cultural heritage. The adverse effects of the former include inequality of opportunity and other inequalities stemming from this, apart from a general lack of social development and lack of individual development. The latter may lead to social and ethnic disintegration.
Cultural deprivation in the educative sense may be caused by poverty and a general underprivileged environment which is not conducive to a child's development or to his or her gaining the most from the educational system (which may in any case, be deficient for these groups). It may also be caused by handicaps such as reading disabilities. In the purely cultural sense deprivation may be caused by a lack of emphasis on culture in educational policy which may direct studies away from the arts or philosophy to purely technical training. Cultural deprivation in the sense of loss of cultural heritage may be caused by a deliberate policy of forced assimilation to suppress a minority culture or by an educational policy which gives inadequate attention to the cultural heritage of minorities, not seeking to integrate them fully into society.
Broader Deprivation (#PA0831) Destruction of cultural heritage (#PC2114).
Narrower Lack of education (#PB8645) Anti-intellectualism (#PF1929)
Inadequate appreciation of culture (#PF3408)
Scarce options for involvement in culture (#PF6535).
Related Social inequality (#PB0514) Cultural fragmentation (#PF0536).
Aggravates Mental deficiency (#PC1587) Cultural revolution (#PF3235)
Retarded socialization (#PF2187) Social underdevelopment (#PC0242)
Inequality of opportunity (#PC3435) Lack of individual development (#PG3595)
Lack of variety of social life forms (#PE8806)
Retardation of psychomotor development in children (#PD1307).
Aggravated by Discrimination (#PA0833) Racial inequality (#PF1199)
Forced assimilation (#PC3293) Cultural illiteracy (#PD2041)
Reading disabilities (#PD1950) Ethnic disintegration (#PC3291)
Denial of freedom of thought (#PF3217) Inadequate educational facilities (#PD0847)
Lack of appreciation of cultural differences (#PF2679).

♦ PC1358 International arms trade
Nature The international traffic in arms consists mainly of arms sales between governments or between government controlled firms and other governments. The traffic facilitates arms races in certain regions or between blocs. The supply of arms to an area of potential conflict is a destabilizing factor which increases the risk of conflict. The trade in nuclear weapons is subject to treaty control (although many 'nonnuclear' weapons could be used as part of a nuclear delivery system), but there is very little coordination or agreement concerning the types of non-nuclear weapons to be supplied and the types of clients that may receive them.
Incidence As an indication, the value of imports by third world countries of major weapons (ships, aircraft, armoured fighting vehicles and missiles) in 1982 was estimated as $ 8,448 million (1975 constant dollars). In 1972, imports were valued at $ 3,473 million on the same basis. Information about the transfer of other weapons (rifles, automatic weapons, mortars, artillery pieces) is fragmentary and unreliable.
Claim Arms purchases by governments encourage producers to use resources in the production of arms instead of alternative types of production which might employ more people and strengthen the capacity of the economy in the producing country to produce items that would fulfil human needs. In addition, the arms trade often enhances the capacity of receiving governments to control their populations, hence arms trade may strengthen military regimes.
Counter-claim In the existing international structure certain arms transactions are clearly inevitable and, within the limitations of current international philosophy, admissible. So long as military alliances exist, it would be unrealistic to criticize the exchange of weapons among allies. The important area for examination is that which concerns the less developed countries.
Broader Unlawful business transactions (#PC4645)
Irresponsible international trade (#PC8930).
Narrower Arms trade with developing countries (#PD3497)
Illegal international arms shipments (#PD4858)
International trade in chemical weapons (#PD9692)
International trade in biological weapons (#PD8621)
Lack of business opposition to the arms race (#PF7088).
Aggravates Conflict (#PA0298) Competitive acquisition of arms (#PC1258).
Aggravated by Weapons (#PD0658) Private international arms dealers (#PD2107)
Socio-economic burden of militarization (#PF1447)
Government complicity in illegal activities (#PF7730).

♦ PC1366 Deforestation
Destruction of forests — Destruction of woodlands
Nature In many countries, and especially in the developing countries of the Southern hemisphere, systematic burning, grazing and cutting of forest-land is carried out in order to provide new land for agricultural or livestock purposes. It is often done without factors such as climate and topography having been sufficiently studied and on lands where slope, nature of the soil or other physiographic characteristics clearly indicate that the land involved is suitable only for forest. Although this practice may lead to a temporary increase in productivity, there are also many indications that in the long run there is usually a decrease in productivity per unit of surface and that erosion and irreversible soil deterioration often accompany this process.
Incidence Forests on which humans and other animals depend heavily take in carbon dioxide gas, provide oxygen and clean the air. Their water holding capacity maintains soil and water levels, preventing disasters such as landslides, floods and droughts. Tropical forests, the most important surviving woodlands, contain about two thirds of all plant and animal species. Tropical plants are the basis for several useful drugs, but vast numbers have not been tested for their medical properties. Tropical forests are also stores of genetic material to feed back into cultivated plants susceptible to disease and pests. At the present rate of deforestation, an estimated 15 percent of all species could disappear within the next two decades. Many factors contribute to deforestation: timber production, clearance for agriculture, cutting for firewood and charcoal, fires, droughts, strip mining, pollution, urban development, population pressures and warfare.
The Himalayan watershed covering Northern India, Nepal, and Bangladesh had lost 40 percent of its forest by 1980. The United States cleared most of its forests in the 19th century and is still felling trees. In the last decade, logs equivalent to a 600,000 acre forest were shipped outside the US. By the year 2000, timber will be cut nearly twice as fast on national forests as new trees can replace it. Costa Rica has lost a third of its forests, loses 60,000 hectares a year and will have none by the 2000. One million Indonesian farmers still use slash and burn techniques. Thousands of hectares are being lost due to warfare in Honduras, Nicaragua, El Salvador and Guatemala. More than a third of Switzerland's forests are dead or dying from pollution. Over 50 percent of the trees in West Germany are dead or dying.
Japan, Europe and the United States consume 66 million cubic metres of tropical hardwood a year. That is a 1500 percent increase in the past 30 years.
Claim Where it has not yet been ruined by modern man – forests devastated to set up pastures, industrial coffee plantations or vast plantations of other luxury crops for the industrialized world – the exuberant tropical vegetation is rich in plants, grasses and trees which are or could be used for food.
Refs Postel, Sandra *Air Pollution, Acid Rain, and the Future of Forests* (1984); Richards, John F and Tucker, Richard P (Eds) *World Deforestation in the Twentieth Century* (1988); Sahabat Alam Malaysia and Asia Pacific Peoples' Environment Network *Forest Resources Crisis in the Third World*.
Broader Endangered forests (#PC5165) Natural environment degradation (#PB5250).
Narrower Tropical deforestation (#PD6204) Destruction of alluvial forests (#PD6850)
Deforestation of mountainous regions (#PD6282).
Related Environmental degradation of desert oases (#PD2285)
Ecosystem modifications due to creation of dams and lakes (#PD0767).
Aggravates Floods (#PD0452) Sand storms (#PD3650) Soil erosion (#PD0949)
Riverine floods (#PD4976) Killing of plants (#PD4217) Unreliable rainfall (#PD0489)
Eichornia crassipes (#PE3815) Migrating sand dunes (#PD0493)
Shortage of firewood (#PD4769) Soil erosion by wind (#PE3656)
Soil erosion by water (#PD2290) Landscape disfigurement (#PC2122)
Silting of water systems (#PD3654) Destruction of land fertility (#PC1300)
Endangered species of animals (#PC1713) Disruption of the hydrological cycle (#PD9670)
Increased reflection of solar radiation (#PF5069)
Vulnerability of world genetic resources (#PB4788)
Endangered species of plants and animals (#PB1395)
Overexploitation of underground water resources (#PD4403)
Environmental hazards of new species introduction (#PC1617).
Aggravated by Forest fires (#PD0739) Forest decline (#PC7896)
Acidic precipitation (#PD4904) Environmental warfare (#PC2696)
Involuntary mass resettlement (#PC6203) Over-rapid timber exploitation (#PD9235)
Insufficient environmental laws (#PG9964) Unsustainable population levels (#PB0035)
Inadequate environmental education (#PD1370) Inundation of forests through dams (#PE7855)
Instability of forestry and logging (#PE0459) Maldistribution of electrical energy (#PD3446)
Excessive use of land for agriculture (#PD9534)
Ecologically unsustainable development (#PC0111)
Long-term shortage of natural resources (#PC4824)
Environmental degradation from high-speed roads (#PD6124)
Environmental hazards from forestry and logging (#PE1264)
Depletion of natural resources due to population growth (#PD4007)
Environmental degradation by off-road and all-terrain vehicles (#PE1720).

♦ PC1383 Illiteracy in developed countries
Incidence Because of a wide difference between way of determining illiteracy the estimates vary greatly. In one estimate having completed primary school is sufficient to be counted as literate. In another filling out an employment form or completing a multiple choice test is required. In 1970 it was estimated that 3.5 per cent of the adults in developed countries over 15 years of age were illiterate. Another estimate gives 13 percent of the population of Britain as functionally illiterate, with more illiterate primary school leavers in 1970 than in 1964. In 1982 the US Department of Education estimated that 10 percent of native English speakers and 48 percent of those without English as their first language were illiterate, giving an overall average of 13 percent in the USA. Recent statistics on illiteracy in developed countries include: one in three Americans will not be able to read this book (Jonathan Kozol on his book "Illiterate America"); 17 percent of soldiers entering the Israeli army cannot read or write (Israeli state comptroller Yitzhak Tunic). Following major hurricane damage in the USA in 1989, efforts to provide compensation were severely hampered by the inability of applicants for relief to complete written applications or sign them. In the southern USA, 25 per cent of adults left school at 14, rising to over 35 per cent in the case of blacks.
Broader Illiteracy (#PC0210).
Narrower Functional illiteracy (#PD8723) Illiteracy in the fourth world (#PD6645).
Aggravates Inadequate labelling of dangerous substances (#PD2468).
Aggravated by Deterioration of industrialized countries (#PD9202).

♦ PC1408 Decreasing genetic diversity of animals
Decreasing genetic diversity of animal breeding stock — Excessive use of domestic animals for breeding purposes — Excessive inbreeding of domestic animals — Genetic impoverishment of animals through domestication
Nature Animal breeding programmes, designed to produce animals with hardy high-yield characteristics after careful selection, can be counter-productive. The more successful a variety is, the more likely farmers are to choose the hybrid in question. An entire society may come to depend on a small range of highly selected varieties. When a new or mutant form of a disease appears to which the favoured hybrid has no inbred resistance, an entire breeding stock may be destroyed.
Both male and female animals which are finely bred or which have particularly commendable points from a breeding point of view, may be overused for breeding purposes, giving rise to inferior progeny or infertility. Females of certain species may be mated consecutively when it may be better for them to have a resting period between pregnancies. Males may be overused for financial reasons, or for their prowess or virility, or because of their fine breeding or special qualities. Males may be used too often with the same herd or the same individual animals. They may be overused for artificial insemination purposes, where demand for the services of sires of high merit is extensive, and where one ejaculation may be extended to 100 or more services, almost on the level of mass production.
Inbreeding of domestic animals is a common practice to improve stock and to preserve qualities found in one or two exceptionally fine animals. However, the effect of inbreeding also serves to increase the proportion of defective, weak, slow-growing animals, and unless the method is used very selectively, especially with species that are normally cross-fertilizing and outbred, these characteristics will predominate. Reasons for excessive inbreeding include lack of due care, general inexperience, overzealousness, and financial rewards.
Refs FAO *Proceedings of the Joint FAO/UNEP Expert Panel Meeting, 1983* (1984); FAO *Animal Genetic Resources Conservation and Management* (1981).
Broader Environmental degradation (#PB6384).
Narrower Decreasing genetic diversity of fish (#PD0547).
Related Capture and use of wild animals as pets (#PD1179)
Agricultural mismanagement of housed farm animals (#PD2771).
Aggravates Animal infertility (#PC1803) Animal abnormalities (#PD4031)
Inadequate animal welfare (#PC1167) Poor quality of domestic livestock (#PD2743)
Vulnerability of world genetic resources (#PB4788).
Aggravated by Genetic inbreeding (#PD7465) Genetic defects and diseases (#PD2389)
Limited agricultural education (#PF8835) Inadequate agricultural capital (#PJ1368)
Abuse of agricultural techniques (#PG5746)
Lack of expertise in agricultural techniques (#PE8752).

♦ PC1411 Environmental hazards from food and live animals
Health hazards from food and live animals

Broader Environmental hazards from economic and industrial products (#PC0328).
Narrower Harmful natural foodstuffs (#PD4238) Excessive consumption of sugar (#PE1894)
Environmental hazards from live animals (#PD0788)
Carcinogenic consequences of food preparation (#PE6619)
Environmental hazards from fruit and vegetables (#PE9029)
Environmental hazards from dairy products and eggs (#PE0505)
Environmental hazards from meat and meat preparations (#PE0133)
Environmental hazards from food processing industries (#PE1280)
Environmental hazards from fish, crustacea and molluscs (#PD0372)
Environmental hazards from cereals and cereal preparations (#PE8732)
Environmental hazards from animal feedstuffs, excluding unmilled cereals (#PE1331).
Related Toxic food additives (#PD0487).
Aggravates Food pollution (#PD5605) Human disease and disability (#PB1044)
Chemical contaminants of food (#PD1694) Biological contamination of food (#PD2594).
Aggravated by Underdevelopment of food processing industries (#PD0908).

♦ PC1420 Unproductive dependents

Nature The broad age span from 15 to 64 years coincides approximately with that of full working capacity. A 'dependency ratio' can be calculated, relating numbers of the population with ages less than 15 or greater than 65 to 100 persons of these 'working ages'. The above figures signify a dependency ratio of 73 per cent for the world, 57 per cent for the more developed regions, and 81 per cent for the less developed regions.
A sharp difference exists between the population compositions by age group in the more developed and the less developed regions. Because of generally low birth rates over decades in the past, the more developed regions have a much more moderate proportion of children than do the less developed regions, but the proportion of older persons (aged 65 years and over) is markedly greater in more developed than in less developed regions. It follows that the burdens of child rearing and education in the less developed regions will be proportionately much heavier if standards are to be reached comparable to those in more developed regions. So far these standards are not reached; children from the age of 9–10 years may take on almost full working responsibility, and before this age they may be working partially and contributing valuably to the family income. On the other hand, greater and still growing needs exist in more developed regions in the care of older persons, although working life or the possibility of working life may extend beyond the age at which social and commercial conditions generally tend to terminate it.
Broader Unemployment (#PB0750) Social inequality (#PB0514)
Underproductivity (#PF1107).
Narrower Excessive childhood dependency in developing countries (#PD3491).
Aggravates Inadequate child welfare (#PC0233) Abusive treatment of the aged (#PD5501).

♦ PC1445 Wildlife extinction

Dependence on wildlife extinction — Extermination of wild animals
Nature Animals and their natural habitat exist in a relatively stable, if fluctuating, equilibrium and it is unlikely that primitive hunting much influenced this. But modern man, using the advanced technology of an industrial society, has had neither the time nor the desire to adapt to the ecosystem as anything but a destructive force. Much use has been made of fire, water and land development in ways which have often devastated local habitats and food cycles. The problems of conservation are complex, partly because of inadequate knowledge of the ecosystemic relationships, partly because of conflicts between groups with different interests.
Growing pressures from increasing human populations and the consequent greater need for agricultural land tend to reduce the areas of land available for wildlife. Once land is taken over for agricultural and pastoral practices, its wildlife will be largely exterminated and its natural vegetation disturbed, if not completely destroyed.
Since the area of land which can be set aside for nature reserves is limited, ecological influences on animal populations and their distribution are distorted. Natural fluctuations in animal numbers cannot be absorbed in a limited habitat. The danger of destruction of the habitat by population explosions within the reserves (often a result of the imbalance produced by restraining man from his traditional hunting activities) cannot be prevented by the overspill or dispersion of the surplus population into the neighbouring areas, which would occur in a completely natural state. The human need for farm activities, which are pressing ever more closely upon the borders of reserved areas, takes priority over the needs for unlimited space of uncontrolled game herds, and domestic herds need to be kept within tolerable limits.
Incidence In the USA up to 1889, the government made use of a policy of intentional extermination of the bison and buffalo herds as a device to starve out unpacified Indian tribes. As a result of that policy a population of some 60 million animals was reduced to less than 100 wild animals.
Claim There is general agreement now that, on moral, aesthetic and scientific grounds, no animal species should be allowed to become extinct. National parks and reserves of various sorts exist in many countries, partly to accomplish this museum goal of the preservation of at least a viable nucleus of every species in a natural habitat. In the developing countries pressures are particularly great to make use of wild tracts of land for agricultural purposes. Wildlife conservation is also important to the economy through the fur trade, tourism and leisure (hunting, fishing).
Refs Tisdell, C *On the Economics of Saving Wildlife from Extinction* (1979).
Broader Extinction of species (#PB9171).
Related Animal extinction (#PD7989) Inadequate animal welfare (#PC1167)
Death of living creatures (#PF7043) Periodic mass extinctions of species (#PF4149)
Long-term shortage of natural resources (#PC4824).
Aggravates Crop damage by wildlife (#PC3150) Diseases of wild animals (#PD2776)
Unavailability of agricultural land (#PC7597).
Aggravated by Forest fires (#PD0739) Vulnerability of wetlands (#PC3486)
Endangered species of animals (#PC1713) Vulnerability of ecosystem niches (#PC5773)
Natural environmental degradation from recreation and tourism (#PE6920)
Degradation of semi-natural and natural habitats of flora and fauna (#PC3152).
Reduces Liver fluke (#PE2785) Epizootic diseases (#PD2734)
Viral diseases in animals (#PD2730) Vectors of animal diseases (#PD2751)
Wild animals as carriers of animal diseases (#PD2729)
Difficulty in identifying carriers of animal diseases (#PF2775).

♦ PC1449 Inhumane scientific activity

Dependence on inhumane scientific activity — Inhumane scientific research
Broader Immorality (#PA3369).
Narrower Misuse of medicines (#PD8402) Abusive psychosurgery (#PE1951)
Forced confessions with drugs (#PE4888) Abusive behaviour modification (#PE2690)
Misuse of evolutionary theories (#PF3348) Abusive experimentation on humans (#PC6912)
Cruel treatment of animals for research (#PD0260)
Dehumanized individual scientific research (#PF2112)
Irresponsible research using human subjects (#PC0080)
Unethical medical experimentation on prisoners (#PE4889)
Unethical experimentation using aborted foetuses (#PE4805)
Inadequacy and insensitivity of intelligence testing (#PD1975).
Related Unethical medical practice (#PD5770)
Irresponsible scientific and technological activity (#PC1153).
Aggravates Abusive detention in psychiatric institutions (#PE2932).
Aggravated by Abuse of science (#PC9188).

♦ PC1450 Denial of right to national self-determination

Nature Outside interference in the form of political, economic or cultural control of a small or weak country by a larger or stronger nation may hamper: national independence or free association or integration with another state; the free determination of social, cultural, political and economic systems; or the permanent sovereignty of a people over their natural resources.
Incidence Among the most well known examples of denial of the right to national self-determination include: 25 million Kurds distributed among Turkey, Iran, Iraq and Syria; the German nation remaining divided; the Palestinian people; Black South Africans; 150,000 Navajo living in the Big Mountain area of Arizona; Guatemala's Indians who make up nearly 60 percent of the population; Chile's 1 million Mapuche; the tribespeople of the Chittagong Hills in southeastern Bangladesh; some 800,000 Melanesians of West Papua; 200,000 Aborigines of Australia; and some 15,000 Dene of Canada's Northwest Territories.
Broader Domination (#PA0839) Repression (#PB0871)
Denial of right of a people to be self-determining (#PC6727).
Narrower Occupied nations (#PC1788) Military government (#PC0698)
Occupied territories (#PD8021) Foreign dictatorship (#PC3186)
Foreign military presence (#PD3496) Foreign control of natural resources (#PD3109)
Violation of sovereignty by trans-border broadcasting (#PE0261)
Denial of effective national self-determination by capitalist exploitation (#PE3123).
Related Denial of right to life (#PD4234) Lack of political independence (#PF0297).
Aggravates Exploitation (#PB3200).
Aggravated by Imperialism (#PB0113) Colonialism (#PC0798)
Economic dependence (#PF0841)
Foreign intervention in internal affairs of states (#PC3185).

♦ PC1453 Speculative flight of capital

Massive international capital flows — Hot money — Volatile capital — Export of capital — Exodus of private capital
Nature During the course of a monetary crisis, or when some such crisis is suspected as imminent, massive movements of short-term capital occur between currencies and capital markets. Such movements can take many forms ranging from leads and lags in commercial payments on through outright purchases of foreign securities. Such movements are extremely unsettling to the economy of an affected country and to the confidence in the international monetary system.
It is usually impossible to be precise about the amount of money that has moved across frontiers in currency crises: both those who gain and those who lose have every reason to conceal the truth in order to dampen the enthusiasm of speculators. The lessons of the past decade of currency crises are that each new one brings with it a larger wave of hot money than the previous one.
Incidence All estimates of capital flight are highly uncertain, but it has been shown to be a significant factor in the debt accumulation process. It is estimated that wealthy Latin Americans have at least US$180 billion personally invested outside their continent. This amounts to one half of the region's foreign debt. The estimate of cumulative capital flight from Argentina during 1974–82 was $31.3 billion. During 1979 – 1984 Argentina's capital flight was 60% of gross capital inflow. Nearly half of Venezuela's external debt is estimated to be due to capital flight. In Venezuela some US$27 billion was expatriated by citizens, or 117 percent of their nation's new borrowing. Zaire's ruling clan is estimated to have US$4 to $6 billion invested in Swiss accounts and foreign real estate. Zaire's total foreign debt in 1982 was US$4.2 billion. Mexico has a foreign debt of over $98 billion and it is estimated that more than $50 billion is invested by Mexicans outside the country. Some 10 billion sterling is held illegally in foreign bank accounts by Indians. Other obvious examples are the Philippines and Haiti under Ferdinand Marcos and Baby Doc Duvalier. The total for seven highly indebted countries was $92 billion, compared with an aggregate debt of $307 billion.
Refs Gyöngyössy, István *International Money Flows and Currency Crises* (1984).
Broader Capitalism (#PC0564).
Narrower Private domestic capital outflow from developing to developed countries (#PD3132).
Aggravates Tax evasion (#PD1466) Imbalance of payments (#PC0998)
Lack of venture capital (#PG7833)
Foreign exchange shortage in developing countries (#PD3068)
Fragmentation of the international trading system (#PC9584).
Aggravated by Government deficits (#PD5984) Devaluation of money (#PE3700)
Capitalist speculation (#PC2194) Misalignment of currencies (#PF6102)
Speculation on money markets (#PD9489) Non-repatriation of export proceeds (#PG6005)
Lack of international coordination of interest rates (#PF3141)
Inadequate mechanism for balance of payments adjustment (#PF3062).
Reduced by Foreign exchange restrictions (#PF3070).

♦ PC1454 Denial of human rights in armed conflicts

Nature Fundamental human rights, as accepted in international law and laid down in international instruments, may be wholly or partially ignored in situations of armed conflict. In the conduct of military operations, distinctions may not be made at all times between persons actively taking part in the conflict and civilian populations. Every effort may not be made to spare civilian populations from the ravages of war, and all necessary precautions may not be taken to avoid injury, loss or damage to the civilian populations. Civilian populations may even be made the actual object of military operations such as air bombardments or th use use of asphyxiating, poisonous or other gases. Dwellings and other installations that are used only by civilian populations may be attacked, as well as places or areas designated for the sole protection of civilians (such as hospitals). Civilian populations may be made the object of reprisals, forcible transfers or other assaults on their integrity. In particular, participants in resistance movements and freedom-fighters in territories under colonial and alien domination and foreign occupation, struggling for their liberation and self-determination, may not be treated, in case of arrest, as prisoners of war in accordance with the principles of the relevant conventions.
Broader Conflict (#PA0298) Denial of human rights (#PB3121).
Narrower Military atrocities (#PD1881) Environmental warfare (#PC2696).
Aggravates Fear (#PA6030) Panic (#PF2633) Refugees (#PB0205)
Injuries (#PB0855) War crimes (#PC0747) Human death (#PA0072)
Crimes against humanity (#PC1073) Inadequate hospital facilities (#PE5058)
Destruction of economy due to war (#PE8915)
Destruction of civilian populations and institutions (#PE8564).
Aggravated by War (#PB0593) Racism (#PB1047) Injustice (#PA6486)
Immorality (#PA3369).

♦ PC1455 Religious discrimination

Dependence on religious discrimination — Discrimination based on belief — Negative attitude to people of other faiths — Confessionalism
Nature Religious discrimination is universal. It may be subtle and not condoned by the law, but practised nonetheless. It may involve discrimination by one religion or sect against another, by the religious against those without religion, or by those without religion against the religious. Discrimination against religions is more familiar than discrimination by religions against non-religiously identified interests deemed to be in opposition to their tenets or authority. On a collective level, discrimination may take the form of boycotts, sit-ins, letter campaigns and demonstrations against governmental, non-governmental, and commercial organizations, and may involve violence, crime or even war. It is, however, particularly vicious on an individual level where non-believers or non-conformers may be called witches, heretics, or satanic, and barred

from employment, denied shelter, harassed, expelled or imprisoned, even on the basis simply of their not being co–worshippers in the same church or creed.

Although religious controversy is probably less pronounced today than at any previous time in history, and although people today are in general more tolerant of the religious beliefs of others, religious differences can nevertheless result in serious conflict, and from time to time complaints are heard about oppressive action taken against religious minorities. Like racial prejudice, religious intolerance is a matter of sentiment rather than reason, which makes it all the more difficult to counteract. Particularly well known is the persecution of the Jews (reaching a peak in the Nazi regime) and the persecution of the Palestinians currently going on in the Gaza Strip and on the West Bank.

Broader Discrimination (#PA0833) Social injustice (#PC0797).
Narrower Antisemitism (#PE2131) Religious racism (#PF4513)
Supralapsarianism (#PF3354) Religious discrimination in politics (#PC3220)
Religious discrimination in education (#PD8807) Disruptive secular impact of holy days (#PE7735)
Religious discrimination in the administration of justice (#PE0168).
Related Racial discrimination (#PC0006).
Aggravates Freemasonry (#PF0695) Superstition (#PA0430)
Threatened sects (#PC1995) Religious conflict (#PC3292)
Religious extremism (#PF4954) Religious repression (#PC0578)
Ideological conflict (#PF3388) Religious intolerance (#PC1808)
Cultural fragmentation (#PF0536) Underprivileged religious minorities (#PC2129)
Segregation based on religious affiliation (#PC3365)
Denial of right of conscientious objection to military service (#PD1800).
Aggravated by Theism (#PF3422) Conformism (#PB3407)
Nationalism (#PB0534) Traditionalism (#PF2676)
Religious prejudice (#PD4365) Religious intimidation (#PC2937)
Religious vilification (#PD5534) Dependence on mysticism (#PF2590).
Reduces Destruction of cultural heritage (#PC2114).

♦ PC1461 Illegitimate political regimes

Illegal political regimes

Nature Political regimes can be created without constitutional approval or the concensus of the national majority. Such regimes may be the result of a coup d'Etat, which necessarily entails the practice of political repression and tends to be exploitive and elitist. In an ex–colonial situation, legalized racial discrimination may be enforced. Such regimes may well be threatened from outside, resulting in very tight government control. This may lead to subversive activities often with foreign assistance, or alternatively apathy, or foreign intervention.

Broader Illegality (#PA5952) Occupied nations (#PC1788)
Undemocratic political organization (#PC1015).
Narrower Political dictatorship (#PC0845) Overthrow of government (#PD1964)
Excessive government control (#PF0304).
Aggravates Conflict of laws (#PF0216) Political conflict (#PC0368)
Civil disobedience (#PC0690) Political alienation (#PC3227)
Subversive activities (#PD0557) Unlawful government action (#PF5332).
Aggravated by Political instability (#PC2677).

♦ PC1463 Competition between international organizations for scarce resources

Rivalry between international organizations

Broader Competition (#PB0848) Lack of international cooperation (#PF0817).
Narrower Competition between intergovernmental organizations for scarce resources (#PE0063)
Competition between international nongovernmental organizations for scarce resources (#PE0259).
Related Ineffectiveness of international organizations and programmes (#PF1074).
Aggravates Inadequate funding of international organizations and programmes (#PF0498).
Aggravated by Ideological conflict (#PF3388)
Proliferation and duplication of international information systems (#PE0458)
Inadequate coordination of international organizations and programmes (#PD0285)
Proliferation and duplication of international organizations and coordinating bodies (#PE1029).

♦ PC1497 Bureaucratic bias

Nature An administrative apparatus dependent upon a skilled staff is limited in the degree to which it can admit the educationally disadvantaged without jeopardizing the level of its own performance. In order to fulfil its mandate, a government bureaucracy should be staffed by individuals appointed on the basis of their merit. In practice this condition cannot always be met. It has been difficult to avoid a preponderance of individuals from the higher and better educated social classes. Many of these individuals have direct or indirect links with political parties or business interests; in some countries, non–merit considerations such as party affiliation may be considered of prime importance.

Broader Over–centralization (#PF2711) Biased government information (#PF0157).
Narrower Governmental bias in statistics (#PF0019)
Inadequate objectivity of institutions (#PF6691).
Related Bureaucracy as an organizational disease (#PD0460).
Aggravates Abuse of bureaucratic procedures (#PF2661).

♦ PC1518 Multiplicity of languages in a national setting

Nature Within some countries several languages are in use. This language diversity leads to a variety of difficulties: retardation of political and economic development; aggravation of political sectionalism; hindrance to inter–group cooperation, national unity and multinational cooperation; hindrance to political enculturation, political support for the authorities and the regime, and political participation. This reduces governmental effectiveness and political stability, reduces occupational mobility, decreases efficiency and prevents the diffusion of innovative techniques.

Broader Multiplicity of languages (#PC0178).
Aggravates Inefficiency (#PB0843) Underdevelopment (#PB0206)
Political instability (#PC2677) Inadequate government (#PJ6362).
Aggravated by Tribalism (#PC1910) Illiteracy (#PC0210).
Reduced by Monolingualism in a multi–cultural setting (#PD2695).

♦ PC1522 Social unaccountability

Social non–accountability — Lack of social accountability — Absence of accountability structures — Irregular task accountability — Lack of structural accountability

Nature Few communities have the means by which an individual is accountable to any group beyond the family. There is a corresponding lack of responsibility taken by the individual and the family for these larger social contexts, whether community, national, regional or world.

Broader Collapsed images of vocation (#PF6098)
Paralyzing complexity of urban structures (#PF1776)
Paralyzing patterns between villages and administrative structures (#PF1389).
Narrower Ineffective student accountability (#PG9043)
Limited accountability of public services (#PF6574)
Unaccountable management of public information (#PF8074)
Unaccountable government intelligence agencies (#PF9184)
Lack of accountability in the disposal of wealth (#PE0503)
Absence of accountability in construction planning (#PF2804)
Inadequate system of political checks and balances (#PE4997)
Denial to animals of legal protection of their rights (#PE8643)
Public non–accountability in control of production processes (#PE3780)
Public non–accountability of organizations developing technology (#PE4032).
Related Unorganized development of work forces (#PF2128).
Aggravates Irresponsibility (#PA8658) Social disaffection of the young (#PD1544).
Aggravated by Secrecy (#PA0005)
Unchanging legal precedent undermines accountability (#PF5295).

♦ PC1523 Miscegenation

Non–acceptance of mixed races — Interracial marriage — Racial impediments to marriage — Interracial sex

Nature Mixed races occupy a distinctive position in most societies and are often neither absorbed nor accepted. Miscegenation has produced caste–like groups in some societies, and in others no significant distinction is made between the mixed groups and the socially inferior parental groups. If exclusive racial attitudes characterize both parental groups, the hybrid may be an object of exclusion by both.

Refs Cohen, Steven M *Interethnic Marriage and Friendship* (1980); Shapiro, Harry L *Race Mixture* (1953).

Broader Ethnic conflict (#PC3685) Forced assimilation (#PC3293)
Racial discrimination (#PC0006).
Related Segregation in marriage (#PD3347).
Aggravates Prostitution (#PD0693) Class conflict (#PC1573)
Social conflict (#PC0137) Impediments to marriage (#PF3343)
Inadequate sense of personal identity (#PF1934)
Discrimination against mixed race children (#PE5183).
Aggravated by Slavery (#PC0146) Elitism (#PA1387)
Imperialism (#PB0113) Concubinage (#PF2554)
Colonialism (#PC0798) Racial segregation (#PC3688)
Racial exploitation (#PC3334) Lack of social mobility (#PF2195)
Sexual exploitation of women (#PD3262) Nonacceptance by social groups (#PG3743)
Children engendered by occupying soldiers (#PD8825).
Reduces Lack of assimilation (#PF2132) Plural society tensions (#PF2448).
Reduced by Legal impediments to marriage (#PF3346)
Social impediments to marriage (#PF3341).

♦ PC1524 Non–equivalence of national educational qualifications

Uncertainty in university degree equivalencies

Nature University exchanges, as well as the high degree of mobility which is a feature of certain professions, demand a system of international equivalences of qualifications. Such a system does not exist. In practice the training of persons from different countries possessing the same qualifications (often with the same name) exhibit the widest disparities. In the case of various professions, the qualifications for entry have often been developed haphazardly, so that it is a matter of chance and of definition whether practical experience is essential before membership is granted in another country. The movement of a person possessing a qualification is thus normally restricted to countries which will accept his qualification; if he is obliged to move to a country which does not accept his qualifications, he may be forced to take up a job below his educational level.

Incidence Within the EEC, some professionals (hairdressers, midwives, cemetery directors) have Community–wide recognition and mobility, while other professionals (architects, accountants, opticians) cannot move from one EEC country to another and expect their degrees to be recognized. Attempts to rectify this inequality have met with little success due to the inevitable high costs and bureaucratic tie–ups involved.

Counter–claim Requirements for diplomas are not the same from country to country, thus the quality of services are unequal as well. Within countries, citizens can be reasonably sure that practising professionals have at least met governmental standards and thus will perform their services to a regulated minimum quality; this would not be so readily enforceable if many different diploma sources were recognized. Attempts to make diplomas and degrees equivalent would be too costly, time consuming, difficult to administer, and might even produce little effect (as evidenced by the doctor's directive which agreed upon bilateral movement, but resulting in less than 1000 of the EEC's 600,000 doctors choosing to move abroad).

Broader Uncertainty (#PA6438) Obstacles to education (#PF4852)
Lack of international cooperation (#PF0817).
Aggravates Refugees (#PB0205) Closed professions (#PD8629)
Educational wastage (#PC1716).
Aggravated by Restrictive professional practices (#PD8027).

♦ PC1526 Educational authoritarianism

Nature There is sometimes an over–rigid hierarchical relationship between teacher and student. When this relationship becomes too strict, it suffocates the learner's awareness of his own responsibilities and excludes a potential positive free response. Students may be deprived of democratic rights in determining, in a participative manner, the manner and content of what they are taught.

Broader Authoritarianism (#PB1638) Obstacles to education (#PF4852).

♦ PC1527 Educational elitism

Nature Educational systems tend to remain the exclusive preserve of an intellectual elite, the product of the bourgeois class which built the system and continues to dictate its law and moral values. Such systems do not aim to exclude people on the grounds of their social background, but co–opt the best, as defined by the existing social elite. Hence the school acts as a sieve, starting in the elementary classes and operating through successive stages of filtering with an eye to selecting the future elite. And if social mechanisms inevitably favour the academic success of children from privileged social and cultural backgrounds, this is seen as a consequence and not as an aim of the system.

This conception of social advancement through education is typical of blocked societies whose sole purpose is their own perpetuation, but it also affects societies in evolution, both in developing and developed countries.

Broader Elitism (#PA1387) Obstacles to education (#PF4852).
Aggravated by Segregation in education (#PD3441)
Inequality of access to education within countries (#PC1896).

♦ PC1532 Inadequate cultural integration of immigrants

Inadequate social integration of foreigners

Nature Cultural integration of immigrants is inhibited by non–acceptance of such immigrants by the original community. Finding themselves in an unequal position and their opportunities limited by discrimination and the law, immigrants may cling to their separatism and original 'nationality' as their sole means of strength and unity. However, the maintenance of a separate cultural identity by immigrants causes friction or conflict with the national majority, aggravating discrimination, segregation, and the refusal by the latter to accept any vestiges of the new culture **Counter–claim** Every decade newer immigrants are thought difficult to assimilate in contrast to earlier immigrants. But in each decade the immigrants have adjusted quickly both economically and culturally. Within a decade or two, immigrants tend to earn more than nationals with similar educational characteristics.

Broader Lack of assimilation (#PF2132) Cultural fragmentation (#PF0536)
Destruction of cultural heritage (#PC2114).
Narrower Restrictions on freedom of movement within countries (#PE8408).
Aggravates Segregation (#PC0031) Racial conflict (#PC3684)
Social inequality (#PB0514) Racial segregation (#PC3688)
Racial exploitation (#PC3334) Resettlement stress (#PD7776)
Racial discrimination (#PC0006) Social discrimination (#PC1864)
Xenophobia with regard to migrant workers (#PC5017)
Social maladjustment of children of migrants (#PE4258)
Marginalization of second-generation immigrants (#PE4990)
Biased media-image of foreign groups and peoples (#PE8802).
Aggravated by Prejudice (#PA2173) Lack of education (#PB8645)
Cultural arrogance (#PF5178) Illegal immigration (#PD1928)
Temporary residence (#PJ3760) Mistrust of strangers (#PF8743)
Inequality of opportunity (#PC3435) Proliferation of immigrants (#PD4605)
Abusive traffic in immigrant workers (#PD2722)
Discrimination against immigrants and aliens (#PD0973).

♦ PC1535 Endangered species of fish

Nature Due primarily to the impact of man on the natural environment, whether directly of indirectly, many of the 23,000 species of fish are in danger of extinction. Technological 'advances' have enabled the total world catch to increase by a factor of 70 over the last 150 years. Of that which is caught only 71 percent is considered fit for human consumption. The rest is fed to pigs and chickens, made into pet food, or used as fertilizer. Sharp drops in fish stocks can be related to changes in ocean currents and other natural phenomena that fishing fleets often do not take into account. Intense exploitation during a period of current-caused low productivity can deplete a fishery.
Incidence With reference to particular species: (i) It is known that the Atlantic salmon has been hunted to the verge of extinction. In 1967, the catch in estuaries, spawning rivers, and oceanic home waters was almost 10,500 tons. In 1982, with many more sporting hunters and much more efficient commercial fishing, the catch declined by 40 percent, to 6,100 tons. In 1984, spawning runs in Canadian rivers were reported down more than 50 percent from 1982. The Newfoundland commercial catch dropped to barely half that of three years ago. (ii) The California sardine fishery, which hit its peak of 1,500 million pounds (670 million kilograms) in 1936, ceased to exist by 1962. Sardines also practically vanished from such hitherto productive areas as Brittany, Portugal, and Morocco. (iii) Anchovies caught by Peruvian fishing boats in the Humboldt current are processed into fishmeal for export. Livestock throughout the world, especially chickens and pigs, are raised on this meal. Shifts in this current, coupled with overfishing, have drastically reduced the Peruvian anchovy fishery, which once made Peru the world's largest fishing nation.
Broader Endangered species of animals (#PC1713).
Related Endangered species of insects (#PC2326).

♦ PC1547 Inadequate structures for political dialogue and review

Claim Forms of political dialogue and review in both elected and appointed forms of government are inadequate. This leaves those who oppose existing regimes few opportunities for respected, peaceful participation, and little means to gain experience, to learn or to change. The result is often violent and confused opposition groups.
Broader Ineffective dialogue (#PF1654) Exclusion of opposing views (#PF3720).
Aggravates Political over-reaction (#PF4110).

♦ PC1563 Imbalance between urbanization and industrialization in developing countries

Nature Many urban areas of developing countries are cities in a demographic sense, but not in terms of the activities which they house. The economies of these urban centres are deficient in their arrangement of interrelated and mutually reinforcing economic components. Urbanization is increasing at a faster rate than industrialization, resulting in the creation of a large service economy. Whereas in the developed countries, the sequence of employment has moved from an agricultural economy to an industrial economy and then to a service economy, the developing countries have moved directly to a service economy because of migration and the natural increase. Consequently, unemployment and under-employment are common features of cities. A large proportion of the urban population consists of odd-job men who live on the edge of starvation. Manufacturing has been unable to absorb the population increase and the size of the service sector is completely out of balance with the income and development level in cities of the developing world.
Broader Underdevelopment (#PB0206) Uncontrolled urban development (#PC0442).
Narrower Lack of rural industrialization in developing countries (#PE8180).
Related Urban unemployment in developing countries (#PD1551).
Aggravates Rural unemployment in developing countries (#PD0295)
Inappropriate education in developing countries (#PF1531).
Aggravated by Rural depopulation (#PC0056)
Uncontrolled urbanization in developing countries (#PD0134).

♦ PC1573 Class conflict

Class war — Class exploitation — Dependence on class domination — Creation of a dominant class — Class domination
Nature The interests of the owners (private capital or the State) of the means of production, and its operators (labour, employees, etc) coincide as far as their viability depends on mutual dependence, but they differ as far as expectations of share in reward (or return on capital), and of benefits for risk and effort contributed. The primary phase of ownership which includes labour (as slaves or indentured servants) still persists, if only in the attitude that human beings can be hired or fired, replaced by machines and otherwise disposed of as a commodity, or cheated or denied political rights. Thus, extreme points-of-view taken by capital or ownership and its management representations evoke their counterparts, who wish to abolish the capital-labour dichotomy at the expense of capital, business for profit, and the free-market system, or state-controlled systems who wish to return to private-enterprise ways. The class conflict is evidenced by characteristics far beyond the simplistic concepts of nineteenth century thinkers.
Claim 1. Modern societies are fragmented in numerous categories of classes, of which the economic are accompanied by such other groupings as: party member, non-party member; bureaucrat, citizen; ethnic majority, minority; young, old, men, women; etc, all of whom have a political voice. The class conflict has extended to diplomatic levels where three or four blocs of nations participating in international organizations cannot adequately address the worlds' problems because of their disarray.
2. Underlying many manifestations of sexism and of racism is class domination based on economic exploitation and profit-motive, cultural captivity, colonialism and neo-colonialism.
Counter-claim Class war always advocates violence because there is no way that the ruling class will give up its power without such conflict. Historically this has always been the case. Violence is necessary to overthrow the state controlled by the ruling classes.
Refs Crouch, Colin *Class Conflict and the Industrial Relations Crisis* (1977).
Broader Exploitation (#PB3200) Human inequality (#PA0844)
Social inequality (#PB0514).
Narrower Populism (#PF3410) Anti-intellectualism (#PF1929).
Related Snobbery (#PJ3943) Conflict (#PA0298) Domination (#PA0839)
Racial conflict (#PC3684) Ethnic conflict (#PC3685) Cultural conflict (#PG3790)
Political exploitation (#PC7356).
Aggravates Enemies (#PF8404) Stepfamilies (#PF6064)
Social breakdown (#PB2496) Inflexible social structure (#PB1997)
Undemocratic political organization (#PC1015).
Aggravated by Apathy (#PA2360) Marxism (#PF2189)
Destiny (#PF3111) Grey lies (#PF3098)
Miscegenation (#PC1523) Minority control (#PF2375)
Social injustice (#PC0797) Illegal immigration (#PD1928)
Forced assimilation (#PC3293) Class consciousness (#PC3458)
Accumulation of privileges (#PF8025) Obsolete deliberative systems (#PD0975)
Poverty in capitalist systems (#PC3107) Alienation in capitalist systems (#PD3112)
Undue attachment to a social group (#PF1073)
General unproductivity of capitalist systems (#PF3103)
Waste of human resources in capitalist systems (#PC3113)
Denial of the right to work in capitalist systems (#PC3119)
Property and occupational discrimination in politics (#PD3218)
Inadequate living and working conditions of immigrant labourers in industrialized countries (#PD3427).

♦ PC1587 Mental deficiency

Mental retardation — Mentally handicapped — Mental feebleness — Impaired intelligence — Impaired mental activity — Low intelligence — Lack of intelligence — Mental degeneracy
Nature Mental retardation is not a single disease and the causes may be multiple. It is a condition in an individual characterized by intellectual defect, and its consequences are social inadequacy and persistent dependency. In the majority of those affected the precise causes of the mental defect are unknown. The condition is often evident at birth or at an early age and is normally of lifetime duration. The criteria of mental retardation in different countries depend to some extent on the degree of tolerance for persons showing deviation from the normal, on the complexity of the demands society makes on the individual, and on the availability of services for the retarded. The numbers of people relying upon help or special services will depend, among other things, on the economy, traditions, and culture of the society in which they live and on the services available in the country.
The problems concerning mental retardation appear to be increasing, since the severely retarded now have a higher expectation of life where the impact of advance in medical and social care are felt, and problems of community integration of the retarded are complicated by accelerated rates of urbanization and industrialization.
Incidence It is estimated that from 1 to 4.5 per cent of the population is mentally deficient. Mental retardation shades imperceptibly into normality in the higher ranges of the intelligence quotient. Of the persons classified as mentally retarded, about 75-85 per cent are only mildly retarded. Those classed as moderately, severely or profoundly mentally retarded comprise about 4 per 1,000 of the age-group 10-14 years.
Refs Bergsma, Daniel *X-Linked Mental Retardation and Verbal Disability*; Horobin, Gordon and May, David (Eds) *Living with Handicap* (1987); Jordan, Shannon M *Decision Making for Incompetent Persons* (1985); MacMillan, Donald *Mental Retardation in School and Society* (1982); Monat, Rosalyn K *Sexuality and the Mentally Retarded* (1982); Sternlicht, Manny and Sternlicht, Manny *Social Behavior of the Mentally Retarded* (1983); Talbot, Eugene S *Degeneracy* (1984); Wortis, J *Mental Retardation and Development Disabilities* (1986).
Broader Mental illness (#PC0300) Human disability (#PC0699).
Narrower Imbecility (#PE6314) Mental dullness (#PE4960) Absolute idiocy (#PE6413)
Feeblemindedness (#PE4821) Moderate mental retardation (#PG5531)
Mental deficiency in children (#PD0914) Borderline mental retardation (#PE4791).
Aggravates Speech disorders (#PE2265) Criminal insanity (#PD9699)
Driving delinquency (#PE6119) Reading disabilities (#PD1950)
Limited individual attention span (#PF2384)
Discrimination against mentally disabled (#PD9183)
Economic and social losses due to disability (#PE4856).
Aggravated by Obesity (#PE1177) Albinism (#PE2332)
Encephalitis (#PE2348) Spina bifida (#PE1221)
Premature birth (#PD1947) Children's diseases (#PD0622)
Cultural deprivation (#PC1351) Diseases of metabolism (#PC2270)
Genetic defects and diseases (#PD2389)
Inadequate guardianship for mentally retarded adults (#PE3992)
Inadequate rehabilitation facilities for the mentally handicapped (#PE8151).
Reduced by Denial of the right to procreate to the severely mentally handicapped (#PE4544).

♦ PC1606 Imbalance in strategic arms

Unequal strategic arms capability — International strategic instability — Missile gap
Nature In the military confrontation between the two superpowers (USSR, USA), due to the ease with which information on the military capability of one power may be misrepresented by the other in an effort to obtain resources to develop its own capability. Any evidence of inequality of capability becomes a threat to international peace and security and may gravely disturb the balance of power.
Background Measures of inequality of strategic nuclear arms capabilities are hindered by uncertainty of intelligence information, lack of mutual inspection, and technical problems of comparability on yield of weapons, range, number, targets, basing systems and survivability against counter-measures or a hostile nuclear attack.
Claim There is no comprehensive theory of strategic stability. Attempts to develop concrete programmes for strategic stabilization have so far been limited to proposals to either prevent surprise attacks and accidental wars or reduce the advantage of a nuclear first strike.
Broader International insecurity (#PB0009).
Aggravates Imbalance of power (#PB1969) Competitive acquisition of arms (#PC1258)
Proliferation of strategic nuclear arms (#PD0014).
Aggravated by Obstacles to unilateral nuclear disarmament (#PF7052).

♦ PC1609 Thermal pollutants

Nature Thermal pollution is the unfavourable product of man's actions; the major sources are heated effluents and solar heating. The major cause is the extension of the thermal electrical power industry. Through this disposal of waste heat, the temperature of surface waters throughout the world is being changed. Uncontrolled heat releases may destroy, dislodge, or debilitate positions of aquatic biota. Oxygen requirements of most aquatic life increase with a temperature rise; metabolic activity rises, and a point is reached where survival rapidly drops. With temperature rise, toxicity of pollutants increases, chemical reactions speed up, flocculation of finely suspended particles is hastened, and salinity increases. The alteration in behaviour, distribution, and migration of anadromous or schooling fishes can occur, and organisms can be killed from shock, or their life cycles can be affected.
Over cities, thermal inversions occur when heat radiates upward on clear nights and the ground layer cools. Over valley cities, or when high-pressure air masses stagnate over the city, the cool air layer is trapped by the warm air above. Gaseous pollutants (nitrogen oxides, sulphur dioxide, smog) collect in the cool air. Thus, thermal air pollution can augment the adverse effects of other air contaminants. The most serious thermal pollution stems from the use of water to cool industrial installations, especially fossil fuel and nuclear generating plants. Water taken from lakes, rivers,

or estuaries to cool reactors is sprayed into the air in cooling towers or is returned to the water body. If this water is too hot, it becomes a pollutant.
Incidence Electric power, iron and steel, oil refining, paper and wood pulp, and synthetic rubber industries account for 89 per cent of industrial water use in the USA. A thermal electric plant converts heat energy from fuel into electricity, with only 32 per cent average efficiency – the rest is waste. Vast quantities of water are now used to cool power plants, and it has been estimated that at least 25 percent of all fresh water flow in the United States will be used for this purpose by the year 2000.
Background Energy flows only in one direction through ecosystems; each time it is transformed from one form to another, or passes from one organism to another, a portion of the energy is converted to heat and is dispersed into space. The biosphere must tolerate this heat. Too much heat produced in a local arena becomes a thermal pollutant.
Broader Pollutants (#PC5690).
Narrower Environmental pollution by nuclear reactors (#PD1584)
Industrial and domestic heating emissions as air pollutants (#PE2824).
Aggravates Energy crisis (#PC6329) Amenity destruction (#PC0374)
Environmental pollution (#PB1166) Excessive environmental heat (#PD7977)
Destruction of wildlife habitats (#PC0480).
Aggravated by Industrial emissions (#PE1869) Industrial waste water pollutants (#PD0575)
Lack of conservation of energy by the private sector (#PE8599).

♦ PC1613 Delays in elaboration of remedial legislation
Delay in approval of urgent regulations
Nature Due to the proliferation of social problems and the limited period in each year during which legislative bodies are in operation, there is insufficient time to formulate and debate new legislative proposals. There is therefore considerable delay in preparing legislation for new problems which do not receive immediate priority attention.
Broader Delay (#PA1999) Inefficiency (#PB0843).
Narrower Resistance to internationally agreed standards (#PC4591).
Aggravates Inadequate laws (#PC6848) Denial of human rights (#PB3121)
Lack of parliamentary time to approve needed legislation (#PF8876).
Aggravated by Crime (#PB0001) Social conflict (#PC0137)
Conflict of laws (#PF0216) Policy-making delays (#PF8989)
Unsustainable population levels (#PB0035)
Bureaucracy as an organizational disease (#PD0460).

♦ PC1617 Environmental hazards of new species introduction
Disruption of ecosystems by exotic organisms — Translocation of living organisms — Hostile introduction of species
Nature Introduced species are plants and animals which have been translocated by human agency into lands or waters where they have not lived previously, at least during historic times. Such translocation of species always involves an element of risk if not of serious danger. Newly arrived species, depending on their interspecific relationships and characteristics, may act as or carry parasites or diseases, prey upon native organisms, display toxic reactions, or be highly competitive with or otherwise adversely affect native species and communities. They may be scientifically or aesthetically undesirable, and this is of special importance in natural areas which have been reserved for study or recreation. Some may become a nuisance through sheer overabundance. They may become liable to rapid genetic changes in their new environment or take advantage of abilities which were not significant to them in their original habitat. New food habits, for example, may cause irreversible damage in areas into which the species have been introduced. Possibly the greatest danger lies in the uncertainty of prior appraisal of a species proposed for an introduction. Many harmful introductions have been made by persons unqualified to anticipate the often complex ecological interactions which may ensue.
Incidence There exist numerous examples of adverse consequences of introductions, ranging from minor economic, cultural or scientific losses, up to unmitigated disasters.
Translocated plants, animals and micro–organisms taken together have been causes of widespread damage. Foot–and–mouth disease (epizootic aphthae), for instance, has often been transmitted both to domestic animals and wild ungulates by introduced livestock. Introductions of the water hyacinth Eichhornia crassipes to Africa and Asia, the European rabbit Oryctolagus cuniculus to Australia, the 'possum' Trichosurus vulpecula to New Zealand, the starling Sturnus vulgaris and gypsy moth Porthetria dispar to America were each, in various ways, calamitous. This epithet certainly also applies to many accidental introductions, such the chestnut blight Endothia parasitica fungus which has exterminated that beautiful and valuable tree Castanea dentata from North American forests, and the beetle which carries a parasite fungus Ceratostomella ulmi now decimating the American elm Ulmus americana and several rodents which despoil both standing and stored food–crops. The connected series of introduced rat, its parasitic fleas and the plague–virus of those insects, emphasizes the human disease aspect of introduced organisms.
Counter–claim 1. Introduced species have in many cases been successfully used in attacking and controlling crop insects, disease vectors, weeds and other organisms harmful to man. Again, many plants introduced into modified or degraded environments may be more useful than native species in controlling erosion or in performing other positive functions. In short, human welfare has been vastly enhanced by introductions, more especially of domesticated or semi–domesticated plants and animals, which, after trial and perhaps further genetic selection in new lands, have proven to be beneficial.
2. In the UK in 1989 it was reported that of 1,058 documented invasions or introductions of alien species of plants, animals or micro–organisms into the UK, about 10 per cent became established. Of these approximately 10 per cent became pests, varying in severity from relatively minor to highly damaging. It was believed that these figures overemphasized the risks, since 90 per cent of those established had not become pests.
Refs Kernan, R P, et al (Eds) *Introduction of Exotic Species, Advantages and Problems* (1979).
Broader Environmental hazards (#PC5883).
Narrower Aggressive honey bees (#PE0793)
Irresponsible introduction of new plant species (#PE1444)
Irresponsible introduction of new species of animals (#PD1290).
Related Biological warfare (#PC0195)
Degradation of the environment through the destruction of species (#PE5064).
Aggravates Pests (#PC0728) Endangered species of plants and animals (#PB1395).
Aggravated by Deforestation (#PC1366) Inadequate plant quarantine (#PE0714)
Destruction of natural barriers (#PC1247) Unethical practice of the biosciences (#PD7731)
Unethical practices with domesticated animals (#PE4771).
Reduces Weeds (#PD1574) Insect pests (#PC1630).

♦ PC1627 Pests of plants
Nature The term pests is here employed in the widest possible sense to include animals, insects, red spider mites and nematodes, which feed on or otherwise damage any parts of a plant, and weeds which compete with other plants for air, light and nutrients. Thus the plant pest is any living, but non–pathogenic agent which injures or limits the growth of a plant, and is of importance to man when crop yields are seriously damaged, either qualitatively or quantitatively.
Of the animal pests, rodents may on occasion be so abundant as to seriously damage unharvested crops. Other animal pests which may be of local importance if their breeding is allowed to progress unchecked include certain birds, mammals and reptiles. Biting–insects, mainly at the caterpillar stage, may feed on the foliage, the shoots, the flowers and the fruits, causing extensive damage and reducing yields. The sucking insects such as the aphids and the plant bugs live on the sap of plants. This may cause direct injury, as in the feeding of a large population of mites which causes the foliage to turn brown and become ineffective, but sucking insects are of even greater importance as the vectors of disease. Nematodes are particularly destructive when they damage the roots of a plant, hindering normal development. Weeds cause a constantly heavy annual toll in lost crop production. They compete directly with crops for air, light, space and nutrients and thereby reduce absolute yields. By being present in the harvested crop they reduce its value and usefulness, perhaps even rendering it useless.
Broader Pests (#PC0728).
Narrower Weeds (#PD1574) Birds as pests (#PD1689) Rodents as pests (#PE2537)
Insect pests of plants (#PD3634).
Related Plant diseases (#PC0555) Hazards to plants (#PD5706)
Insect pests of wood (#PD3586) Crops pests and diseases (#PE7783)
Pests and diseases of groundnut (#PJ1181).
Aggravates Suffering of plants (#PC7825) Destruction of weeds (#PE3987)
Infestation of seeds (#PE6271) Plant disease vectors (#PD3596)
Pests and diseases of trees (#PD3585) Pests and diseases of potato (#PE2219)
Pests and diseases of cotton (#PE2220) Pests and diseases of sugar–beet (#PE2975)
Excessive use of chemicals to control pests (#PD1207).
Aggravated by Crop vulnerability (#PD0660) Inadequate crop rotation (#PF3698)
Inadequate plant quarantine (#PE0714) Disastrous insect invasions (#PD4751)
Pest resistance to pesticides (#PD3696)
Decreasing genetic diversity in cultivated plants (#PC2223).

♦ PC1630 Insect pests
Injurious insects
Nature Insects inflict an enormous amount of damage upon man and his possessions, domestic animals and crops. Virtually all human communities are affected by insects in several important ways, the most obvious of which stems from their role as vectors or reservoirs of disease, whether of man, animals and plants. Although the harm done by insects as merely annoying pests is hardly comparable with the loss they cause as vectors of disease, it is by no means negligible. Insects consume and destroy large quantities of food stuffs.
Incidence In the USA, about 500 species of insects are of primary economic importance and losses caused by them range from US $4,000 million to $8,000 million annually.
Refs Hill, D S *Agricultural Insect Pests of Temperate Regions and Their Control* (1987); Hill, D S *Agricultural Insect Pests of the Tropics and their Control* (1987); Hurst, G W *Meteorology and the Colorado Potato Beetle* (1975); Metcalf, et al *Destructive and Useful Insects* (1962).
Broader Pests (#PC0728).
Narrower Lice as insect pests (#PE1439) Insect pests of wood (#PD3586)
Aggressive honey bees (#PE0793) Flies as insect pests (#PE2254)
Insect pests of plants (#PD3634) Insect bites and stings (#PE3636)
Mosquito breeding ponds (#PG9038) Insect vectors of disease (#PC3597)
Hemiptera as insect pests (#PE3615) Thysanura as insect pests (#PE3620)
Dermaptera as insect pests (#PE3618) Orthoptera as insect pests (#PE3641)
Disastrous insect invasions (#PD4751) Thysanoptera as insect pests (#PE3619)
Siphonaptera as insect pests (#PE3643)
Insect damage to stored and manufactured goods (#PD3657).
Related Animal pests (#PD8426).
Aggravates Itch (#PE3940) Famine (#PB0315)
Insecticides as pollutants (#PD0983) Environmental plant diseases (#PD2224)
Human disease and disability (#PB1044) Spoilage of agricultural products (#PC2027)
Vitamin E deficiency in domestic animals (#PE4760).
Aggravated by Agricultural wastes (#PC2205) Solid wastes as pollutants (#PD0177)
Pest resistance to pesticides (#PD3696) Inappropriate sanitation systems (#PD0876)
Insect resistance to insecticides (#PD2109)
Excessive use of chemicals to control pests (#PD1207)
Introduction of new species of insect pests (#PF3592).
Reduced by Endangered species of insects (#PC2326)
Environmental hazards of new species introduction (#PC1617).

♦ PC1640 Dangerous occupations
Broader Risk (#PF7580).
Related Occupational hazards (#PC6716).
Aggravates Industrial accidents (#PC0646) Occupational diseases (#PD0215)
Discrimination against women in employment (#PD0086).
Aggravated by Occupational risk to health (#PC0865)
Inadequate training in decision–making (#PD2036)
Inadequate knowledge and reporting of man–made disease (#PF7939).

♦ PC1647 Vulnerability of marine ecosystems
Nature From the point of view of protection of living resources in the marine environment, the ecological effects of pollution are of the most vital concern. Perhaps the most important effect of pollutants in the marine environment is ecological disruption, such as the imbalance created between organisms and their environment, and between communities of organisms of different species. This is often an insidious, long–term effect which can lead to large changes in populations of commercially important fish. The net result of pollution in inland bodies of water occurs by complex processes in the ecosystem which are poorly understood.
So far, there has been no conclusive evidence that populations of marine fish have been seriously affected by pollution alone. However, this may be related in part to our inability to clearly identify cause and effect in certain fisheries' problems. A whole ecosystem may be modified by input of a particular pollutant, because certain species are reduced in number or eliminated. Other hardier species may fill an ecological niche vacated by a sensitive species eliminated by the pollutant.
Incidence Even though there often appears to be no actual mortality due to a pollutant, the chronic effects may eliminate a population of organisms in the long–term. Fish populations have been eliminated by acid precipitation in lakes of the Scandinavian countries and in Canada. Salmon have been wiped out or severely reduced in numbers running up certain streams on both sides of the Atlantic by a combination of over–fishing, pollution and other anthropogenic effects. Marine mammals (seals and porpoises) in the Baltic Sea have exhibited reproductive failure associated with high DDT and PCB concentrations in their tissues.
Background There is a host of ecological effects due to the discharge of pollutants into coastal waters. Habitats of marine organisms may be adversely affected by solids settling on the bottom and by materials leached from them. Sedimentation from coastal mining operations may alter tropical waters unfavourably, especially coral reefs, where the light that is vital for photosynthesis is reduced. Erosion from improper land management may affect coastal spawning grounds as well as those in rivers. Organic substances in both dissolved and solid form decompose and remove oxygen from the water. This can be a serious problem in partially confined areas, such as bays and fjords, in which the water is not frequently replaced. It may even occur in basins on the exposed continental shelf where there is little or no flushing action by bottom currents. In areas where the volume waste is very large compared to the amount of water available to dilute it, even the salt composition of the sea water may be significantly changed. Normally, this is not a problem

in most coastal waters if there is adequate flushing action. Such problems as pH change, due to the input of highly alkaline or acidic wastes, which can be a serious complication in fresh waters, do not commonly occur in the sea because of the buffering action of sea water. The inflow of fresh water over coastal sea water causes a stratification to take place in the absence of tidal and wind mixing. This in effect reduces vertical mixing, and such processes as aeration of deeper water tend to be minimized. Moreover, the comparatively fresh surface layer may have low buffering capacity and can be affected by a pollutant in much the same way as a river or a lake.

Pollutants all exhibit toxicity to aquatic organisms in various degrees. Some may be acutely toxic even in low concentrations and kill aquatic organisms over a short period of exposure. Others may have a debilitating effect. It is perhaps the latter which is most important from a long-term ecological point of view. The sub-lethal, chronic effects of pollutants may include retardation of growth, alteration of chemoreception in food-finding and mating, aberrant behaviour, physiological stresses affecting the vigour of organisms, and reproductive failure. There are pollutional effects of substances introduced into the environment which are not necessarily characterized by toxicity to aquatic organisms. For example, coloured and suspended particulate materials may retard the penetration of sunlight and thereby inhibit photosynthesis. An excess of nutrients may cause dense algal growths that adversely affect higher forms of life such as fish and shellfish. The input of heat with cooling waters may not necessarily destroy organisms in the water, but again, the conditions in which they live may be adversely altered. One species may be encouraged at the expense of another. The migrating behaviour of certain fish species may be altered by temperature gradients introduced by cooling waters.

Refs Costlow, J D and Tipper, R C (Eds) *Marine Biodeterioration* (1983).

Broader Environmental degradation (#PB6384) Vulnerability of ecosystem niches (#PC5773)
Long-term shortage of natural resources (#PC4824).
Narrower Disruption of seabed ecosystems (#PG7989).
Aggravated by Marine pollution (#PC1117)
Unsustainable exploitation of fish resources (#PD9082).

♦ PC1648 Stress in human beings

Nature Stress denotes a broad range of biological and psychological reactions to environmental influences; as such, it is an essential component of the equipment that enables man to survive in an hostile environment. However, stress as a traditional method of adaptation has become inadequate in the psychological, social, and economic circumstances of modern society; it is this inadequate adaptation to change that increases the risk of disease. Certain psychosocial situations are potentially pathogenic because they induce in some individuals inadequate adaptational reactions. Stress is induced when a situation is interpreted by the individual, consciously or not, as a threat to his goals, integrity or well-being. Thus noise, dirt, bad air-conditioning, quarrelsome colleagues, lack of personal contact, new responsibility may be stressful to one individual but not to another.

Incidence Numerous studies attest that stressful life events such as death of a spouse or parent, marriage, divorce, desertion, loss of employment, birth of a handicapped child, etc, often precipitate physical as well as emotional illnesses. For example, the death rate of widows and widowers during the first year of bereavement has been shown to be 10 times that of other persons of the same age; divorced persons, in the year following their divorce, have an illness rate approximately 12 times that of married persons of similar age and situation. Young married women have exceptionally high referral rates to medical services, particularly for depressive and neurotic conditions, which raises many questions regarding the demands made on the housewife with a young family in a mobile, urban society. More in keeping with expectations are the high referral rates of the elderly to the psychogeriatric services because of disorders directly attributable to recent social problems, often of a kind that, with a little support from services, relatives, or neighbours, could have been remedied. Moreover, there are grounds for supposing that psychosocial stresses precipitate all kinds of illness, not simply mental ones. It has also, however, been very clearly demonstrated that people with depression and with schizophrenia, in the months immediately preceding the onset of their illness, experienced more stressful events than did a control group followed up for the same period.

Counter-claim Stress is often good for people. It keeps them from degenerating into mindless blobs. Too little stress, namely prolonged boredom, stagnation and lack of stimulation, is more unhealthy than too much.

Refs Alcorn, Randy and Alcorn, Nancy *Women under Stress* (1986); Appley, Mortimer H and Trumbull, Richard A *Dynamics of Stress* (1986); Asterita, Mary F, et al *Physiology of Stress* (1985); Baltruch, H J and Waltz, Millard *Cancer and Stress* (1987); Cooper, Carey L *Psychosocial Stress and Cancer* (1985); Day, Stacey B *Cancer, Stress, and Death* (1986); De Vries, Jan *Stress and Nervous Disorders* (1988); Gerhardt, Uta E and Wadsworth, Michael E J (Eds) *Stress and Stigma* (1984); Levi, Lennart *Society, Stress, and Disease* (1978); McCulloch, J Wallace and Prins, Herschel A *Signs of Stress* (1978); Orlandi, Mario, et al *Encyclopedia of Health* (1988); Shiraki, Kiezo; Yousef, Mohamed and Wilber, Charles G *Man in Stressful Environments* (1988); Sloan, S J and Cooper, C L *Pilots under Stress* (1987); Spielberger, Chas D, et al *Stress and Anxiety* (1988); Sullivan, John J and Foster, Joyce C *Stress and Pregnancy* (1987); Vingerhoets, A *Psychosocial Stress* (1985); Wiegele, Thomas C, et al *Leaders under Stress* (1985); Wilson, J P et al *Human Adaptation to Extreme Stress* (1988).

Broader Stress (#PB9165) Human physical suffering (#PB5646)
Vulnerability of human organism (#PB5647).
Narrower Trauma (#PD4571) Family stress (#PD8130) Heat disorders (#PE2398)
Marital stress (#PD0518) Juvenile stress (#PC0877) Occupational stress (#PE6937)
Resettlement stress (#PD7776) High altitude stress (#PD2322)
Socioeconomic stress (#PC6759) Professional burnout (#PF4833)
Stress among children (#PE4421) Dietary energy stress (#PJ1003)
Stress in perfectionists (#PF9624) Competitiveness in education (#PD4178)
Disruptive effect of household moving (#PE5928)
Psychological stress of urban environment (#PE6299)
Stress and trauma in a context of civil violence (#PE0525).
Related Fatigue (#PA0657) Anxiety (#PA1635)
Denial of rights to prisoners (#PD0520).
Aggravates Smoking (#PD0713) Neurosis (#PD0270)
Psychoses (#PD1722) Alcohol abuse (#PD0153)
Mental illness (#PC0300) Human suffering (#PB5955)
Medical quackery (#PD1725) Kidney disorders (#PE2053)
Juvenile suicide (#PE5771) Bronchial asthma (#PE3860)
Malignant neoplasms (#PC0092) Reading disabilities (#PD1950)
Loneliness in old age (#PD0633) Cardiovascular diseases (#PE6816)
Gastrointestinal diseases (#PE3861) Protein-energy malnutrition (#PD0339).
Aggravated by Boredom (#PA7365) Loneliness (#PF2386)
Overcrowding (#PB0469) Traffic noise (#PD3664)
Aircraft noise (#PE5799) Family breakdown (#PC2102)
Oppressive reality (#PF7053) Marital instability (#PD2103)
Personal life crises (#PD4840) Indoor air pollution (#PD6627)
Increasing pace of life (#PF2304) Elimination of jobs by automation (#PD0528)
Health hazards of exposure to noise (#PC0268) Isolation of parent-child relationship (#PC0600)
Overcrowding of housing and accommodation (#PD0758)
Discrimination against illegitimate children (#PD0943)
Retirement as a threat to psychological well-being (#PF1269)

Denial of right to union activity for special groups (#PE1355)
Inappropriate selection and examination procedures in education (#PF1266).
Reduced by Stress addiction (#PE4951).

♦ PC1707 Social isolation

Social distance — Social gap — Self-created isolationism

Nature The present-day tendency is for the individual and the community each to operate primarily for their own respective interests; the individual rarely sees himself as responsible for the community and the community does not represent all its individuals. There is resultant lack of mutual concern and involvement and neither party benefits fully.

Broader Isolation (#PB8685).
Narrower Isolated family units (#PG5240) Socially isolated women (#PG9712)
Individual isolationism (#PD1749) Social isolation as torture (#PD6810)
Social integration handicap (#PE6779) Social isolation of the elderly (#PD1564)
Mothers' self-imposed isolation (#PJ9425) Limited spheres of relationship (#PD1941)
Social isolation of women at home (#PE8681) Isolation of parent-child relationship (#PC0600)
Personal isolation in communities of industrialized countries (#PD2495).
Related Cultural isolation (#PC3943) Political isolation (#PC7569).
Aggravates Loneliness (#PF2386) Lack of social contact (#PF8695)
Obsolete vocational skills (#PD3548) Social disaffection of the young (#PD1544)
Abusive detention in psychiatric institutions (#PE2932).
Aggravated by Unsociability (#PA6653) Social exclusion (#PC0193)
Geographical isolation (#PF9023) Dependence on the media (#PD7773)
Restrictive social practices (#PC5537)
Disadvantages for homeworking employees (#PE6240)
Denial of the right to social security in capitalist systems (#PD3120).

♦ PC1713 Endangered species of animals

Nature Due primarily to the impact of man on the natural environment, whether directly or indirectly, a large number of species of animals are in danger of extinction.

Incidence Since 1600 (when reliable zoological records began) about one per cent of the higher animal species alive at that time have become extinct and another 2.5 per cent are now in danger. These figures are for full species, but geographical races have fared even worse. About 25 per cent of the 130 lost species disappeared through natural causes (since extinction is a natural part of evolution). The remainder have died out directly or indirectly as a result of man's avoidable actions: hunting, habitat disruption, introduction of species predators and competitors. Of the 304 species under threat today, 66 per cent of the birds and 83 per cent of the mammals are in their present state because of man's activities. About 75 species were wiped out by humans during the nineteenth century. So far in the twentieth century, the pace has accelerated, with approximately one species vanishing each year, a rate experts predict will increase drastically by the year 2000. According to a recent environmental study, 20 percent of all species on earth could become extinct during the next two decades, a loss unparalleled in human history. Many ecologists have predicted half a million extinctions in tropical rain forests alone by the end of the century - an average loss of about one hundred species per day. The rate of impact of these activities on the natural environment is accelerating.

Refs IUCN Conservation Monitoring Center Staff *IUCN Red List of Threatened Animals, 1988* (1988).

Broader Endangered species of plants and animals (#PB1395).
Narrower Endangered species of fish (#PC1535)
Endangered species of birds (#PD0332)
Endangered species of mammals (#PC1326)
Endangered species of reptiles (#PC0604)
Endangered species of amphibia (#PD3156)
Endangered species of farm animals (#PD7506)
Endangered species of invertebrates (#PD7513).
Related Monsters (#PF2516) Endangered species of insects (#PC2326)
Declining breeds of domesticated animals (#PD6305).
Aggravates Wildlife extinction (#PC1445).
Aggravated by Poaching (#PD2664) Animal pests (#PD8426)
Deforestation (#PC1366) Hunting of animals (#PC2024)
Endangered forests (#PC5165) Hybridization of wild animal species (#PD2419)
Superstitious persecution of animals (#PD3453)
Capture and use of wild animals as pets (#PD1179)
Commercial exploitation of wild animals (#PD1481)
Destruction of hedges and hedgerow trees (#PD1642)
Extermination of wild animal natural prey (#PD3155)
Trade in animal products of endangered species (#PD0389)
Aiding escape of prisoners of war or enemy aliens (#PE1200).

♦ PC1716 Educational wastage

Failure in school — School dropouts — Premature school leaving — Continuing dropout pattern — Repetition of educational stages — Wastage from school examination failures — Underperformance in education — Educational underachievement

Nature Educational wastage exists in the following forms: failure of the system to provide a universal education, failure to recruit children into the system, failure to hold children within the system, failure of the system to set appropriate objectives, and inefficiency in the achievement of such objectives. The most frequently suggested reason why children and adolescents in developing countries do not go to school or leave school early is that there are no schools to go to, or that there are not enough places in them. In addition the cost of attending school during periods of manpower shortage may be a determining factor, particularly in agricultural areas. Marriage customs may encourage early marriage and childbirth. Grave illness is also an important factor. Many children leave school early because they, or their parents, do not find what is taught at school relevant to their needs in future employment. Others leave because they are needed as helpers at home or on the farm. Finally, many parents feel that it is more important for their children to receive traditional education and training on the job rather than spend their time in classroom.

A major aspect of educational wastage occurs when students leave the educational system prior to the termination of an educational cycle. Dropping-out in this sense is not related to the existence or duration of compulsory schooling and therefore leaving school before the minimum age is not regarded as dropping out. However, those who leave before the end of a cycle, but who have satisfied the compulsory education laws by staying at school until they have reached the minimum age, would be regarded as dropouts; and in countries which do not have compulsory education, a child who left school before completion of the stage in which he had registered would be regarded as a dropout.

This definition conflicts to some extent with more general notions of premature leaving, interpreted as leaving before the minimum age. The term may also be applied to students leaving at the end of the compulsory period when a further period is considered desirable even though not required.

A major aspect of educational wastage is the repetition by a student of a year of work in the same class or grade and doing the same work as in the previous year. This may occur at any level, from elementary to university.

Incidence In the Third World, more than half of the pupils drop out entirely after the second year of primary school. Of primary school completers, only one in four obtains a place in a secondary

school. Only a fraction graduate.
In the early 1980s in France, dropouts made up between 100,000 and 250,000 of the 800,000 people leaving school annually. In UK 60 per cent of children, in Germany FR and the USA 10 per cent drop–out at the age of 16.
Some countries throughout the first and second levels systematically operate repetition in all grades, using end of year examinations and other information on which to base a decision on promotion, with a limited number of years in a grade permitted. A second group of countries resembles the first, except that the number of years in a single grade is not limited. A third group of countries promotes without regard to examination performance, and rarely permits repetition of grades. A UNESCO statistical study of school wastage, in selected countries, suggests that first and last elementary school grades were the focal points for repetition in Africa, and first grade in Latin America. In both these regions, first grade repetition was noticeably higher than in second through fifth. In selected countries in Asia and Europe, surveys show that a high first grade repetition rate was also indicated. The first year of high school, in these statistics, also shows a higher rate, as do the terminal grades of the two cycles of the 'general second level'. Clearly there are unaddressed psychological, social and economic factors in this wastage. The problems are only partly pedagogical.
Claim Educational wastage is a deeply moral issue. It is one of the highly sinister policies of nations, and includes exploitation of women; racial and class prejudice; abuse of labour; and inadequate health and social services. Educational wastage exists by intent of the privileged classes. If there is one, there is the other. The corollary is that if the one is done–away with, the other will be as well.
High rates of student drop–out ae both a personal tragedy for young people and a waste of human potential a nation can ill afford.
Refs Conklin, Agnes M *Failure of Highly Intelligent Pupils*; McIlroy, Ken *School Failure and What to do About It* (1979); United Nations *Coping with Drop–Out* (1987); United Nations Educational, Scientific and Cultural Organization *Wastage in Primary and General Secondary Education* (1980).
Broader Inefficiency (#PB0843) Waste of human resources (#PC8914).
Narrower Declining school enrolment (#PJ7844).
Aggravates Ignorance (#PA5568) Innumeracy (#PC0143)
Illiteracy (#PC0210) Youth violence (#PF7498)
Reading disabilities (#PD1950) Ill effects of educational failure (#PF2013)
Underutilization of intellectual ability (#PF0100)
Inappropriate selection and examination procedures in education (#PF1266)
Unemployment of premature school leavers in developing countries (#PE0015)
Inadequate research and development capacity in developing countries (#PE4880).
Aggravated by School phobia (#PE4554) Student absenteeism (#PE4200)
Inadequate teaching (#PC9714) Adolescent pregnancy (#PD0614)
Failure of school systems (#PJ3896) Inadequacy of formal education (#PF4765)
Irrelevance of educational curricula (#PF0443) Inappropriate education of graduates (#PF1905)
Unrealized use of education structures (#PF2568)
Attention–deficit hyperactivity disorder (#PE9789)
Decline in government expenditure on education (#PF0674)
Inappropriate education in developing countries (#PF1531)
Barriers to transfer between educational facilities (#PD0084)
Non–equivalence of national educational qualifications (#PC1524)
Shortage of books and textbooks in developing countries (#PF0118).

♦ PC1730 Undirected technological expansion

Nature The natural and technical sciences expand in an unintentional and narrow framework unconnected to the larger needs of society. The development of technology is beyond ethical constrains except the most pragmatic. Increasingly the direction of science is disrelated from the aspirations or opinions of the masses of people. Most scientists are untrained and unwilling to communicate with the general public.
Refs Kupperman, Robert H *Technological Advances and Consequent Dangers* (1984).
Broader Expansionism (#PB5858).
Narrower Scientific ignorance (#PD8003) Parochial scientific view (#PF1418)
Unapplied scientific knowledge (#PF1468)
Structurally blocked scientific co–operation (#PD7470)
Over–emphasis on immediate solutions in resource development research (#PE4059).
Related Absence of tactical methods (#PF0327) Human wisdom unrelated to daily life (#PD1703)
Antiquated intellectual methods to appropriate human depths (#PF1094).
Aggravates Educational curricula based on content rather than method (#PF3549).

♦ PC1738 Guerrilla warfare

Guerrilla war
Nature Guerrilla characteristics are shared by private armies, mercenaries, terrorists, militants, revolutionaries, freedom fighters, irregulars, outlaw gangs, highwaymen and brigands. Hit–and–run tactics are their hallmark although, when in force, they are capable of pitched battles. Modern communications and weapons have transformed the lonely, independent bands of men, each armed with a rifle, into well–disciplined forces. They are armies that are mobile, frequently fighting a defensive war whose only objectives are harassment, destruction, and, if possible, attrition of the enemy. Given this, they strike at will with no overall military strategy that can be anticipated, unless they are on near–equal terms with their enemy, at which time they act more like regular military forces.
Background Guerrillas are legitimatized when reconstituted as government forces. This occurred in the Soviet Union after World War I, and after the Cuban Revolution under Fidel Castro. Otherwise they are illegitimate and their fighters come under no international treaty for protection of combatant human rights. Likewise, guerrilla fighters may not observe Geneva Conventions or other so–called humane conventions of warfare. Viet Cong guerrilla terrorism directed against villagers is a conspicuous example.
Counter–claim Guerrilla warriors are the fighting vanguard as the masses of people whose armed struggle against oppression is the only means of liberation.
Refs Anand, V K *Insurgency and Counter–Insurgency* (1981); Ben–Rafael, E and Lissak, M *Social Aspects of Guerilla and Anti–Guerilla Warfare*; Corbett, Robin *Guerilla Warfare* (1986); Janke, Peter and Sim, Richard *Guerilla and Terrorist Organizations* (1983); Janke, Peter and Sim, Richard *Guerrilla and Terrorist Organizations* (1983).
Broader Civil war (#PC1869) Subversive activities (#PD0557).
Narrower Terrorism (#PD5574) Urban guerrillas (#PD1988)
Counter–insurgency warfare (#PG5689).
Related Revolution (#PA5901) Disruption of development by tribal warfare (#PD2191).
Aggravates Massacres (#PD2483) Militarism (#PC2169)
Human death (#PA0072) Illegal roadblocks (#PE9605)
Human destructiveness (#PA0832) Low–intensity conflict (#PE3988).
Aggravated by Racism (#PB1047) Tribalism (#PC1910)
Imperialism (#PB0113) Social conflict (#PC0137)
Unstable regimes (#PG3948) Political conflict (#PC0368)
Economic imperialism (#PC3198) Secret military operations (#PF7669)
Foreign intervention in internal affairs of states (#PC3185).
Reduces Apartheid (#PE3681) Colonialism (#PC0798).

♦ PC1777 Repression of self–consciousness

Repression of elements of the personality
Nature Individuals prefer to be bound by habits and institutions, reduced images of security and well being; and they attempt to avoid all encounters which would wake them up. Social controls are used to deny the things which they would prefer not to know. Only life styles which seem to be successful in this attempt to remain asleep are pointed to as significant ways to participate in society: the modern experience of the struggle to be significant in society is often called the rat race.
Broader Lack of commitment (#PF1729).
Narrower Sexual repression (#PF2922) Collapsed images of vocation (#PF6098)
Media reinforcement of materialism (#PF1673) Overemphasis on institutional security (#PC1835)
Institutionalized callousness of public services (#PF2006)
Cultural diversity ignored in social service agencies (#PE1862)
Exclusion from decision making processes of those who question the context of the process (#PE7155).
Related Consumerism (#PD5774) Evasion of issues (#PF7431)
Lost family role in society (#PF7456) Social isolation of the elderly (#PD1564)
Inadequate means for upholding global concern (#PF1817).
Aggravates Killing of plants (#PD4217).

♦ PC1784 Insecurity of property

Vulnerability of property to theft and damage — Unsecured goods — Unlocked property — Insecure property assets — Lack of protection for property
Broader Insecurity (#PA0857) Insecurity of resources (#PB8678).
Narrower Cargo insecurity (#PE5103) Political confiscation of property (#PD3012)
Vulnerability of intellectual property (#PF8854).
Aggravates Risk (#PF7580) Theft (#PD5552) Arson (#PE5505)
Burglary (#PD2561) Vandalism (#PD1350) Emotional insecurity (#PD8262)
Criminal harm to property (#PD5511) Personal physical insecurity (#PD8657).
Aggravated by Inadequate security system (#PD6589)
Breakdown of police protection (#PF8652)
Breakdown in community security systems (#PD1147).

♦ PC1788 Occupied nations

Captive nations
Nature Occupied nations are those which have been incorporated into larger groupings without any opportunity being given to the people of the areas in question to indicate whether or not they desire any form of independence or self–determination.
Incidence The most frequently cited example is Palestine under Israel, and the nations and nationality groups incorporated within the Soviet Union and Eastern European states – the most recent of which is the attempt on Afghanistan. Historical examples are groups such as the Celts and the Basques which in the distant past were incorporated into several nations; and, more recently, the American Indian nations which were overwhelmed by the settlers in both North and South America.
Broader Imperialism (#PB0113) Colonialism (#PC0798)
Denial of right to national self–determination (#PC1450)
Denial of right of a people to be self–determining (#PC6727).
Narrower Foreign dictatorship (#PC3186) Foreign military presence (#PD3496)
Illegitimate political regimes (#PC1461)
Contamination of public water supplies by sabotage (#PE1458).
Related Occupied territories (#PD8021) Lack of political independence (#PF0297)
Forced repatriation of prisoners of war (#PD0218).
Aggravates Lack of political development (#PD8673).
Aggravated by International aggression (#PB0968).
Reduces Nationalism (#PB0534).

♦ PC1803 Animal infertility

Nature Inability to conceive or to reproduce successfully may be manifested in barrenness (male or female), abortion, or still–birth. Infertility may arise from inadequate feeding and a resulting poor condition; from adverse environmental conditions or mismanagement (disregard for seasonal cycles of sexual activity); from genital diseases in either male or female; from hereditary or congenital abnormalities in male or female; or from physical or psychical inability or disturbance (incompatibility of male and female blood, old age, discrepancies in size, etc).
Refs Seren, E and Mattiolo, M *Definition of the Summer Infertility Problem of the Pig* (1987).
Broader Reproductive system diseases in animals (#PD7799).
Related Infertile land (#PD8585) Natural human abortion (#PD0173).
Aggravates Economic loss (#PE9013).
Aggravated by Animal diseases (#PC0952) Animal injuries (#PC2753)
Animal abnormalities (#PD4031) Inadequate feeding of animals (#PC2765)
Weather as a factor of animal disease (#PD2740)
Decreasing genetic diversity of animals (#PC1408)
Vitamin E deficiency in domestic animals (#PE4760)
Agricultural mismanagement of housed farm animals (#PD2771)
Inadequate housing and penning of domestic animals (#PE2763)
Congenital anomalies of the reproductive system in animals (#PG7447).

♦ PC1808 Religious intolerance

Dependence on religious intolerance
Nature Religious intolerance involves acts denying the right of people of another religious faith to practice and express their beliefs freely. Religious intolerance is expressed in discrimination, repression and religious rivalry, and results in or results from persecution. It leads to war and persistent hatred between nations and between peoples within nations.
Counter–claim There are limits to religious tolerance as senseless and obviously fraudulent cults proliferate. Governments may begin, as in the USA, to establish tests (there are presently some 14 criteria to be met according to one US government agency) to establish bonafide religious movements of an independent nature. The excesses, also, of traditional faiths are not always tolerated with equanimity when single religious leaders or councils promulgate violence to human life in various forms, or socially injurious doctrines. Socially irresponsible religions, religions whose political activities are self–serving, and religions that hold and seek wealth, cannot claim respect. That religious intolerance can be a good thing is seen by the actions of the missionaries who put down human sacrifice and cannibalism in the indigenous creeds.
Broader Sacrilege (#PF0662) Intolerance (#PF0860).
Narrower Puritanism (#PF2577) Catholicism (#PF8071)
Religious extremism (#PF4954) Denial of religious liberty (#PD8445)
Disruptive secular impact of holy days (#PE7735).
Related Dependence on religion (#PF0150).
Aggravates War (#PB0593) Apostasy (#PE9018) Fundamentalism (#PF1338)
Anti–clericalism (#PF3360) Sexual immorality (#PF2687) Religious rivalry (#PC3355)
Religious conflict (#PC3292) Religious propaganda (#PD3094)
Sexual discrimination (#PC2022) Religious intimidation (#PC2937)
Inadequacy of religion (#PF2005) Impediments to marriage (#PF3343)
Inadequate sex education (#PD0759) Injustice of religious courts (#PE0397)
Persecution of religious sects (#PF3353) Religious conflict between sects (#PC3363)
Religious and political antagonism (#PC0030) Discrimination against homosexuals (#PE1903)

Ill-considered missionary activity (#PF3370) Religious discrimination in education (#PD8807)
Fragmentation within organized religions (#PF3364)
Undue religious influence on secular life (#PF3358)
Discrimination against illegitimate children (#PD0943)
Negative effects of claims of religious infallibility (#PF3376)
Religious discrimination in the administration of justice (#PE0168).
Aggravated by Heresy (#PF3375) Divorce (#PF2100)
Idolatry (#PF3374) Freemasonry (#PF0695)
Religious war (#PC2371) Manipulative cults (#PE6336)
Religious repression (#PC0578) Dependence on mysticism (#PF2590)
Religious discrimination (#PC1455) Multidenominational society (#PF3368)
Double standards in morality (#PF5225) Fragmentation of religious belief (#PF3404)
Religious discrimination in politics (#PC3220)
Segregation based on religious affiliation (#PC3365)
Holy places as a focus of religious friction (#PF1816)
Peoples perceiving themselves as specially chosen (#PF4548).

♦ PC1812 Official secrecy

Official cover-ups — Excessive government secrecy — Official silence

Nature Data of special government interest may be considered unsuitable for public disclosure. The range of information deemed as secret is from all government information to specific information in foreign affairs, about criminal, espionage or terrorist activities, and about military affairs. Official secrets may mask government or political activity detrimental to the public interest or which would not be supported by a concensus. By keeping certain facts secret a government may be able to indoctrinate the public more successfully in its favour or achieve greater control over the nation in general. Insofar as official secrets are limited for access to officials directly concerned with them, they create an elite which may also be a technocracy. Because official secrets are restricted to a minimal number of people, where there is a leakage (especially to a foreign power) constituted by action from a legitimate member of that group, discovery of the leakage may easily be delayed because there is not sufficient general knowledge of the situation.

Governments excessively classify information as secret or restricted, and have a strong unwillingness to divulge sensitive activities. This leads to policy failures, abuses and embarrassment; it may also mask corruption or the leaking of confidential data either to a foreign power or to business or private interests. Government secrecy is demonstrated by the use of secret police and by espionage and surveillance activity at home and abroad. It may also involve the use of censorship and propaganda. Government secrecy at all levels may lead to citizen alienation or apathy and may be used to reinforce political injustice and inequality and repression, leading to government control and to dictatorship.

Secrecy also exists in intergovernmental organizations wherever personnel are obliged to sign contracts stipulating that they will not publish anything concerning the inner workings of their organization. Such restrictions protect inefficiencies and abuses. These may continue and eventually cause some lack of confidence. On the national level, governmental bureaux and departments may routinely withhold information from the public. This should be distinguished from secrets of state or official secrets. On the local levels of governments the withholding of information may also be common, particularly surrounding financial activities such as municipal bond issues, land use and expenditures of public funds, and concerning wrong-doing by party functionaries in office.

Incidence The circumstances of the British sinking of the Argentinian ship, the 'Belgrano' illustrates secrecy in a parliamentary democracy. In Latin American and the Philippines for example, governments make little attempt to explain the disappearance of citizens. Attacks on a refugee camp in Lebanon remain unexplained by the government guarantor of its protection (Israel). Documents may remain classified for excessively long periods as is illustrated by the fact that the UK still keeps secret the identity of "Jack-the-Ripper".

Claim All governments, whatever their political complexion, are tempted at one time or another to suppress or distort the truth to avoid scandal, humiliation or electoral defeat.
Broader Secrecy (#PA0005) Undemocratic political organization (#PC1015).
Narrower Military secrecy (#PC1144) Classified public information (#PF9699)
Secret international agreements (#PF0419)
Misrepresentation of geographical information (#PF9239)
Misuse of classified communications information (#PD5183)
Government secrecy concerning nuclear weapons testing (#PF4450)
Suppression of information concerning social problems (#PF9828)
Unaccountability of international financial institutions (#PF1136)
Suppression of information concerning environmental hazards (#PF4854)
Restrictions on the distribution of confidential government information (#PD2926)
Failure to notify the imprisonment or death in prison of foreign nationals (#PE6424).
Related Censorship (#PC0067) Biased government information (#PF0157)
Inadequate government publications (#PF3075)
Propaganda by intergovernmental organizations (#PE3077).
Aggravates Political crime (#PC0350) Unreported disasters (#PF7768)
Denial of access to news (#PF3081) Biased presentation of news (#PD1718)
Pervasive fear of nuclear war (#PC3541) Erosion of journalistic immunity (#PD3035)
International political espionage (#PC1868) Blackmail by government officials (#PD9842)
Unjust allocation of government contracts (#PF2911)
Abdication of government ministerial control (#PF7342)
Social constraints on freedom of imagination (#PF4882)
Unaccountable government intelligence agencies (#PF9184)
Restrictions on international freedom of information (#PC0931)
Excessive neutrality of intergovernmental official information (#PF3076).
Aggravated by Espionage (#PC2140) Power politics (#PB3202)
Corruption in politics (#PC0116) Excessive government control (#PF0304)
Inadequate system of political checks and balances (#PE4997).
Reduces Diplomatic embarrassment (#PG4043).

♦ PC1818 Inadequate control over government administrative process

Nature Under the conditions of the modern state, legislatures increasingly delegate authority to administrative agencies. With the consequent proliferation of governmental functions and reliance on expertise, the difficulties not only of parliamentary but also of executive control over the administrative process mount. Elected representatives are too few in number in comparison with the officials under their authority, less expert than the latter, and their control necessarily restricted to spot checks on performance.
Broader Lack of control (#PF7138) Inefficient public administration (#PF2335).
Narrower Inefficient public administration in developing countries (#PC0903).
Aggravates Abuse of bureaucratic procedures (#PF2661).
Aggravated by Excessive government control (#PF0304)
Inadequate public finance statistics (#PF7842)
Excessive specialization in education (#PC0432).

♦ PC1820 Shortage of industrial leadership and entrepreneurial ability in developing countries

Unconfident potential entrepreneurs — Inexperienced village entrepreneurship — Missing entrepreneur skills

Nature The general lack of indigenous leadership in the initial stages of the industrialization process may be due in part to the very newness of the factory system and its various concomitants. In addition to such initial difficulties there may be major impediments rooted in the social structure itself, in the rigidity of the social system and in the values a particular society attaches to different kinds of economic activity. In pre-industrial societies, the greater part of the population usually consists of peasant groups whose traditional outlook, closed family economy and general cultural background are not the best training ground for industrial leadership. The source of potential industrial leadership, therefore, may be limited to the numerically small upper classes. Even in this group, however, entrepreneurial ability is not purely a function of the education and wealth that its members may have, for in most cases their standards of values and ways of life do not dispose them towards industrial undertakings. So long as the role of businessman (whether that of merchant or that of industrial entrepreneur) is defined, not as a goal that is legitimate in itself, but as a means of advancement to other classes enjoying higher prestige, business may be deprived of indispensable incentives towards permanent careers and long-range undertakings. A preference for short-term commercial operations or for speculative enterprises rather than long-term industrial undertakings is common among merchants in many under-developed areas.

Counter-claim It is a myth that developing countries lack entrepreneurs. The entrepreneurs are different from those in developed countries: unregistered taxi and mini-bus operators, street-market traders, water vendors, money changers, informal credit brokers, growers of illegal crops, smugglers, etc.
Broader Lack of leadership (#PF1254) Insufficient skills (#PC6445)
Obstacles to leadership (#PF7011).
Related Inadequate local expertise in business practices in developing countries (#PE7313).
Aggravates Underdevelopment of industrial and economic activities (#PC0880).

♦ PC1833 Inadequate urban political machinery

Political fragmentation in urban areas — Balkanization of metropolitan government decision-making

Nature The suburban areas in metropolitan regions are becoming increasingly independent of the central areas and offer little support to policies for ameliorating the problems of the cities. Suburban self-sufficiency has frequently been reinforced by national 'New Towns' policies and programmes to strengthen smaller cities as 'growth points'. Central cities are vulnerable to the loss of their previous fiscal and political importance. A rising tax burden with a declining economic base now confronts many. The urban crisis has prompted action, particularly in the United States, where federal and state governments in cooperation with private enterprise, have developed and implemented many inner city renewals. 'Enterprise zone' mechanisms to attract investment in 'downtown' renewal area include favourable tax rates, licensing and permit concessions and a number of creatively conceived incentives. However, even in the USA, 'urban blight' has not been contained and the 'flight to the suburbs' continues. Undesirable suburban density increase and environmental destruction results have not as yet led to integrated urban – suburban planning.
Broader Political disintegration (#PC3204) Institutional fragmentation (#PC3915)
Uncontrolled urban development (#PC0442).

♦ PC1835 Overemphasis on institutional security

Claim Social organizations and institutions – school, family, corporation, etc – are all restricted by being based on (and emphasizing) narrow images of personal and institutional security. This traps people into thinking that all relationships should avoid stress, conflict and risk.
Broader Repression of self-consciousness (#PC1777).

♦ PC1864 Social discrimination

Dependence on social discrimination
Broader Discrimination (#PA0833) Social injustice (#PC0797)
Social inequality (#PB0514).
Narrower Social bias in planning of training programmes (#PF2885).
Related Elitism (#PA1387) Aggression (#PA0587).
Aggravates Nomadism (#PF3700) Scapegoats (#PF3332)
Segregation (#PC0031) Social conflict (#PC0137)
Minority control (#PF2375) Social alienation (#PC2130)
Age discrimination (#PC2541) Religious conflict (#PC3292)
Ethnic segregation (#PC3315) Social dictatorship (#PD3241)
Social intimidation (#PC2940) Class consciousness (#PC3458)
Social fragmentation (#PF1324) Single parent families (#PD2681)
Plural society tensions (#PF2448) Lack of social mobility (#PF2195)
Inequality before the law (#PC1268) Inequality of opportunity (#PC3435)
Inadequacy of social doctrine (#PF3398) Social impediments to marriage (#PF3341)
Discrepancies in human life evaluation (#PF1191)
Discrimination against immigrants and aliens (#PD0973)
Failure of individuals to participate in social processes (#PF0749).
Aggravated by Slavery (#PC0146) Divorce (#PF2100)
Nativism (#PF2186) Prejudice (#PA2173)
Caste system (#PC1968) Fear of the abnormal (#PF7029)
Negative effects of family allowances (#PF0107)
Inadequate cultural integration of immigrants (#PC1532).
Reduces Inadequate sense of personal identity (#PF1934).
Reduced by Restrictive legislation (#PD9012) Compulsory indoctrination (#PD3097).

♦ PC1867 Social insecurity

Inadequate security systems — Inadequate security system

Refs ILO *Into the Twenty-First Century. The Development of Social Security* (1986); ILO *Introduction of Social Security* (1984); United Nations *Demographic Development and Social Security* (1987).
Broader Insecurity (#PA0857) Ineffective structures of local consensus (#PF6506).
Narrower Pusillanimity (#PJ8038) Culture of violence (#PD6279)
Disorienting urban shift (#PG7981) Urban dependence mind-set (#PG8028)
Declining sense of community (#PF2575) Vulnerability of middle-class (#PC1002)
Unreliable phone installations (#PG8016)
Inadequate social security for migrants (#PE7611)
Dependence on family for social security (#PE8670)
Social insecurity in developing countries (#PE4796)
Insufficient social security in the agricultural sector (#PD9155)
Denial of the right to social security in capitalist systems (#PD3120)
Denial of social security to nationals who have lived abroad (#PE8746)
Denial of right to social welfare services for indigenous peoples (#PE1506).
Related Inadequate security system (#PD6589).
Aggravates Theft of works of art (#PE0323) Inadequate road conditions (#PJ0860)
Personal physical insecurity (#PD8657).
Aggravated by Human contingency (#PF7054) Increasing cost of social security (#PF7911)
Unrecognized benefits from cooperatives (#PF9729)
Inadequacies of the international monetary system (#PF0048).

♦ PC1868 International political espionage

Nature Such espionage involves covert gathering of information about the capabilities and intentions of foreign governments, about foreign areas where they may have a strategic interest, or about their general position on international relations. Methods include wiretapping, bugging,

compiling of political dossiers, microphotography, theft, use of computers, blackmail, intimidation, kidnapping, defection and abuse of diplomatic privilege. It may lead to international tension and cold war and also an extensive domestic political espionage network to ensure the survival of the country's international intelligence unit.
Background Although Sun Tsu in The Art of War wrote of the necessity for espionage systems in 530 BC, it began to evolve as a widespread phenomenon with the rise of nation states in Europe after the Treaty of Westphalia in 1648. After the French Revolution and the Napoleonic Wars, the scope of espionage systems expanded from their mainly military basis and started to become more politically orientated. It was not until the 20th century, the rise of national dictatorships, and the increase in the power and autonomy of intelligence bureaux in certain countries that espionage in its modern form evolved.
Claim When military strength decreases, the relative threat tends to increase. This requires more intelligence gathering that focuses on intentions of the potential enemies.
Broader Espionage (#PC2140).
Related Military espionage (#PD2922)
Revealing national security information to a foreign power (#PE4343).
Aggravated by Official secrecy (#PC1812).

♦ PC1869 Civil war
Intrasocietal military conflict
Nature A society that has no effective channels for settling political and economic grievances among its classes and parties may erupt in civil war. The term may be a misnomer in so far as civil refers only to civilians taking part, since military units of various types may join in. Such wars are not likely to show civility either, often being characterized by intense fratricidal hatreds. Those opposed to the government may be termed insurgents, revolutionaries, counter-revolutionaries, resistance fighters, guerrillas, and other names, depending on circumstances. Civil wars may involve sabotage, terrorism and other crimes. Civil war is recognizable when the insurgents do not disperse but manage to hold a territory, or when the government in power invokes martial law. Not all civil wars have the overthrow of the government as their intention, although this is always the aim of revolutions; some have the intent of forcing legislative changes or of securing powers or privileges which are felt due.
Background In European history, civil war is traceable from the time of the Greeks and Romans. The Roman Civil Wars brought the Emperor Augustus to power. The English Civil War (1642–51) resulted in King Charles' execution. The War of Independence that the American colonies fought with England, the French Revolution, the American Civil War, the Revolution in the USSR, the Boxer Rebellion, and many others may be cited among other instances. The Spanish Civil War, the Greek Civil War and the Algerian Civil War are more recent examples. The term civil war may be indiscriminately applied as as synonym for the broad class of intrasocietal military conflicts that have occurred in different forms in Latin America, Africa and Asia in recent years.
Broader War (#PB0593) Conflict (#PA0298) Violent revolution (#PC3229)
Unconventional war (#PC8836).
Narrower Guerrilla warfare (#PC1738) Economic civil war (#PD3765)
Resistance movements (#PJ2066) Internal armed conflicts (#PD7661).
Aggravates Terror (#PF8483) Refugees (#PB0205) Injuries (#PB0855)
Massacres (#PD2483) Starvation (#PB1875) Martial law (#PD2637)
Human death (#PA0072) Human violence (#PA0429) Lack of war relief (#PF0727)
Counter revolution (#PF3232) Low-intensity conflict (#PE3988)
Inadequate hospital facilities (#PE5058).
Aggravated by Secession (#PD2490) Ethnic conflict (#PC3685)
Political aggression (#PD8877) Racial discrimination (#PC0006)
Intra-state imperialism (#PC3197)
Disruption of development by tribal warfare (#PD2191)
Foreign intervention in internal affairs of states (#PC3185).

♦ PC1874 Illegitimate children
Children born out of wedlock — Illegitimacy
Nature In spite of greater information on sexual matters and greater availability of contraceptives, and the relaxation of norms on sexuality and marriage in many societies, this is, in the same societies, paralleled by a growing incidence of illegitimate births. Also, children born out of wedlock are less frequently than previously abandoned to charitable or public institutions for adoption by foster parents. Single women and unmarried couples belonging to the middle classes of affluent societies, are increasingly content to have children, but unwanted 'illegitimate' births are still mainly found among the poorer social groups.
Incidence Illegitimate children represent between 10 and 17 percent of total annual births in the developed countries. In some of these countries, this proportion of illegitimate births doubled during the 1960s and 1970s. In all of them, the trend is on a steady increase since the beginning of this century. At present, the bulk of this increase is among young women in the 15–20 age group. Such births occur more often in poorer social groups. In the United States in 1979, 55 percent of all black children were born out of wedlock, as compared with less than 10 percent of all white children; in 1988, two out of three birth were to single mothers. In the USA in 1988, 25 percent of birth were to an unmarried mother, namely about 1 million births. This represented an 8 percent increase on the previous year and a 51 percent increase on 1980. In other less economically developed societies, illegitimacy rates are also very high. In some Latin American countries, they reach 50 to 60 percent of all births, because consensual unions are so frequent that the frontiers between such unions and legal marriages are blurred.
In Sweden 46 percent of children were born out of wedlock in 1987. In Denmark and Iceland nearly one half of newborns are non-marital. In Norway, Austria, France and Britain it is one in four or one in five.
Background Illegitimate children are children born out of wedlock in infringement of a society's regulations of procreation. They may be the result of poverty, inadequate education, family breakdown, cohabitation, the non-validity of divorce, inadequate contraception, prostitution, promiscuity, slavery or war. Definition of illegitimacy depends on the law. For example, children of a polygamous marriage will be legitimate in a society where polygamy is legal; but they will be illegitimate if monogamy is the law.
Counter-claim The notion or illegitimacy itself is rapidly losing its content, as legislation follows changes in attitudes and secures equal rights to all children born in or out of wedlock, and to their parents. The differences in legal, social and psychological terms, between marriage and free union, legitimate and illegitimate birth, are fading. The formation of the family is more and more a private event. The social stigma attached to mothers of illegitimate children, and to the children themselves, has practically disappeared in a number of industrialized societies.
Refs Laslett, Peter, et al (Eds) *Bastardy and Its Comparative History* (1980); Teichman, Jenny *Meaning of Illegitimacy*.
Broader Immorality (#PA3369).
Narrower Denial of parental affiliation (#PD3255)
Discrimination against illegitimate children (#PD0943).
Related Concubinage (#PF2554) Single parent families (#PD2681).
Aggravates Unknown relatives (#PF0782) Juvenile delinquency (#PC0212)
Social fragmentation (#PF1324)
Unequal distribution of social services (#PC3437).
Aggravated by Incest (#PF2148) Adultery (#PF2314)
Polygamy (#PD2184) Promiscuity (#PC0745)
Prostitution (#PD0693) Cohabitation (#PF3278)
Group marriage (#PF3288) Permissiveness (#PF1252)
Family breakdown (#PC2102) Sexual immorality (#PF2687)
Unmarried mothers (#PD0902) Marital instability (#PD2103)
Sexual exploitation (#PC3261) Female prostitution (#PD3380)
Inadequate sex education (#PD0759) Non-validity of marriage (#PF3283)
Inadequate laws of adoption (#PD0590)
Vulnerability of marriage as an institution (#PF1870)
Discrimination in family planning facilities (#PD1036).
Reduced by Induced abortion (#PD0158).

♦ PC1876 Neo-colonialism
Nature Neo-colonialism designates the dominant characteristics of the relation of dependence of the less-developed countries on the more developed world, following their national independence. Post-independence dependence results in drain of income and manpower, in a lack of economic and social integration, in the persistence of a dual and distorted socio-economic structure, and in a spontaneous reproduction of under-development, allowing the less developed countries only a lop-sided development process.
Dependence takes various forms: (1) National security dependence: most of the less-developed countries are dependent for their military hardware on one great power or a combination of great powers. Without supplies of arms and spare parts, their national security could be in jeopardy; thus, they tend to become the client of one or other of the great power blocs. As a consequence, their sphere of autonomous action is severely limited. (2) Direct economic dependence: this is the control of a sizeable or critical sector of the economy by dominant foreign monopoly capital. Direct economic dependence is a legacy of the colonial days that is being gradually modified into indirect forms. (3) Trade dependence: a sizeable portion of trade and consequently foreign exchange earnings are tied to the economy of the dollar area, sterling area, franc area, rouble area, etc. (4) Financial dependence: this type of dependence is exerted primarily through banks and is therefore strongest in those countries where the banking system and, through it, the internal money circulation and the credit system, are under foreign control. Tied loans and grants, and various forms of foreign exchange control, constitute other kinds of financial dependence. (5) Technical dependence: this consists of a whole variety of intellectual forms of dependence, some direct and others indirect. The direct forms consist of the acceptance of imported technology and foreign technical advisers. The indirect forms are even more pervasive – the adoption of a particular foreign educational system with its biases, the influence of foreign-educated natives with their intellectual and emotional ties, etc. (6) Cultural dependence: the most pervasive and enduring kind of dependence lies basically in the increasing dominance in the modern world of a Western secular and scientific culture. In its more pedestrian forms, this kind of dependence consists of a whole variety of cultural imports (films, television programmes, books, periodicals, records, clothing styles, forms of consumption).
Claim The colonial age has ended. But the wish to dominate persists. Neo-colonialism comes wrapped in all types of packages –in technology and communications, commerce and culture.
Counter-claim If neo-colonialism exists, it is the inevitable product of an inherent imbalance between the advanced and the developing economies, irrespective of the political factors involved. This one-sided relationship will disappear only when the new states reach the position already achieved by Japan and become as powerful economically as the ex-colonial powers on whom they at present depend.
Broader Imperialism (#PB0113) Colonialism (#PC0798).
Narrower Preferential tariffs (#PG4097) Environmental colonialism (#PE3447).
Related Overthrow of government (#PD1964).
Aggravates National political dependence (#PF1452).
Aggravated by Economic dependence (#PF0841)
Military and economic hegemony (#PB0318)
Unstable supply of raw materials (#PD4270).

♦ PC1888 Territorial disputes between states
Unsettled territorial claims
Nature Territorial disputes may take the form of disputes over precise boundary demarcations or over the claims of a neighbouring state and a distant, usually colonial, power to a dependent territory. Disputes may also occur in connection with territory contiguous with both states, usually as a consequence of annexation during war by one state of an adjoining region. Claims by states to territorial waters along their coastline also give rise to disputes.
Incidence Many grievances and claims were left unresolved following the 1914–18 and 1939–45 wars, as was the case with previous European wars. The opinion of the Polish, Czechoslovak and Baltic peoples was not considered in determining the frontiers of those countries, which were established by agreements between other powers or by annexation. Territorial disputes may disrupt regional cooperation for development or defence – as in the disagreements between Argentina and Chile, and between Turkey and Greece (by which NATO is impacted).
The Antarctic regions constitute a special case in which seven nations have put forward conflicting claims. (These have been set aside for the moment under an agreement which facilitates unrestricted cooperative scientific exploration). A number of islands have been contested, from the Kurile to the Falklands (Malvinas). The seabed, outside territorial waters, is another special case, as are potential disputes over surfaces on the Moon, the planets or their satellites.
Refs Kratochwil, Friedrich, et al *Peace and Disputed Sovereignty* (1985).
Broader Disagreement (#PA5982).
Narrower Boundary disputes between states (#PD2946)
Conflicting claims by states to territories (#PC2362)
Conflicting claims to non-terrestrial territory (#PG4114)
Conflicting claims concerning Antarctic territory (#PG4115)
Conflicting claims concerning off-shore territorial waters (#PD1628)
Restrictions on passage through straits and interoceanic canals (#PD2948)
Inequitable allocation of rights to exploit sea-bed and marine resources (#PE1597).
Related Intergovernmental disputes (#PJ5405).
Aggravates Domination of the world by territorially organized sovereign states (#PD0055).
Aggravated by Zionism (#PF0200) Enclaves and exclaves (#PD2154)
Territorial expansionism (#PC9547) Undue attachment to territory (#PF3390).

♦ PC1892 Human sexual inadequacy
Nature Human sexual inadequacy is usefully grouped under four general headings, orientation, appetite, arousal and orgasm. In men arousal problems are either caused by a variety of psychological factors: stress, fear of failure, disease or punishment. Physical factors play an important part. Alcohol is also widely recognized as a cause of impotence. Age and fatigue may be implicated, and so may certain diseases such as multiple sclerosis, diabetes, tumours and hormonal disturbances. Increasingly it is being recognized that some medications, especially drugs used to reduce blood pressure may cause potency problems as a side-effect. Arousal disorder in women is less public. Sometimes it is a straight forward lack of lubrication, but the main problem, afflicting women much more than men, is that of orgasm failure. While a few women seem capable of orgasm by fantasy alone, most require manual or oral stimulation in addition to intercourse, and some never achieve orgasm at all. The flip side of female orgasm difficulty is premature ejaculation, though it is not always clear whose problem this is. If a man ejaculates before getting his trousers off it is fair to say it is his problem. At the other extreme is the definition that a man is a premature ejaculator if he reaches orgasm before his partner on more than

half of all occasions of intercourse regardless of how many hours he may perform. Some intermediate definition is more helpful. Disorders of appetite are increasingly recognized as important, particularly in women. A high proportion of women complain that they are just not sufficiently interested in sex to keep their partners happy. In some periods of history this would have been considered normal. Today with the powerful myth of male/female identity, women who have little interest in sex are made to wonder what is wrong with them. If men had the same interest in sex there would be no problem. Because the libido cannot easily be reduced by other than chemical or surgical means and an interest in sex can be increased by use of erotica, fantasy, foreplay, role playing, etc., the partner with the lower level of sexual appetite is more often seen as the one to be treated. Another disorder of appetite that effects men more often than women is the resistance that arises from repeated sex with the same partner. A tendency for males to be sexually recharged by novel females is observed in most mammals. This may manifest itself as a specific loss of appetite for the wife, even though the man may love her in other ways. The last category is that of orientation or target of sexual interest and is more common in men than women. Homosexual orientation, as opposed to occasional behaviour, occurs in something like 5 percent of men and 1 percent of women. Fetishism, transvestism, sado–masochism, paedophilia, zoophilia and other unusual preferences are almost exclusively male.

Refs Masters, William H and Johnson, Virginia E *Human Sexual Inadequacy* (1981).

Broader Human ageing (#PB0477) Sexual harassment of women (#PF3271).
Narrower Voyeurism (#PE3272) Bestiality (#PE3274)
Nymphomania (#PE8213) Sexual sadism (#PE6748)
Sado–masochism (#PE6137) Sexual fetishism (#PF6406)
Sexual impotence of men (#PF6415) Sexual inadequacy in the elderly (#PG4118).
Related Sexual unfulfilment (#PF3260) Sexual unfulfilment of the disabled (#PE5197).
Aggravates Lesbianism (#PF2640) Unhappiness (#PA9191)
Male homosexuality (#PF1390) Marital instability (#PD2103)
Sexual discrimination (#PC2022) Lack of self–confidence (#PF0879)
Psychological disorders (#PD8375) Sexual abuse of children (#PE3265)
Deception between sexual partners (#PE4890).
Aggravated by Prudery (#PF5892) Fear of failure (#PF4125)
Incompatibility (#PF9047) Inadequate sex education (#PD0759).

♦ PC1896 Inequality of access to education within countries

Nature Certain sectors of education are highly selective, depending on the social status of the classes for which they cater. Children of the poor or those who belong to groups suffering racial or social discrimination are in a difficult position from the outset, whether from lack of due care for physical and mental requirements of early childhood, or lack of pre–school education. They are handicapped, sometimes irremediably so, in comparison to children of wealthier classes or those from backgrounds more favourable to proper growth and development. Where school vacancies are increasingly limited as pupils mount the promotion ladder, a more or less arbitrary selection process prevents many who are capable of continuing their studies from so doing. The inadequate development of literacy programmes and out–of–school vocational training means that those who missed their chance of entering the school network at the outset find it less and less possible to educate themselves as they grow older. Thus, the universal right to education is often denied, by a complete reversal of justice, to the most underprivileged. They are the first to be denied this right in poor societies; and those most likely to be deprived in rich societies.

Equal access to education is only a necessary (not a sufficient) condition for justice. Equal access is not equal opportunity. This must comprise equal chance of success. But chances of this kind, on the contrary, are very unequal. The negative correlation between the financial, social and cultural status of families and the opportunity of access to the varying types of education, and thereafter of success, is far from having the same weight in every country, although it remains a universal phenomenon. Children from poor backgrounds Children from poor backgrounds are compelled to go to work prematurely, students who work have to do so in addition to studying. Poor hygiene, inadequate diets, and overcrowded homes hinder education development, and Other equally important causes include cultural conditions (especially language), that determine the level and content of academically useful pre–school knowledge.

Incidence Regional differences can reach such considerable proportions that figures relating to the educational situation in two different geographic sectors, for example, may vary by more than 50 per cent in relation to the national average for the same item. In India, significant disparities may be seen in comparisons between the average school enrolment rate for the 6–11 year old age group, which is 80 per cent, and regional rates, such as 121 per cent in Nagaland, 77 per cent in Andhra Pradesh, 68 per cent in Punjab and 57 per cent in Madhya Pradesh. In one country in Latin America, 66 per cent of primary schools in urban areas offered the full five years of the elementary educational cycle, whereas only 6 per cent did so in rural areas, and 59 per cent of village schools provided only two years of schooling. Other examples of inequality include the concentration of educational facilities in the major cities and towns, to to rural zones, and their concentration near city centres to the detriment of shanty towns, favelas and other poor districts lacking the schools available to their richer neighbours. There are numerous cases of ethnic or racial inequality, even in countries with ample material means to remedy the situation.

Broader Unequal access to education (#PC2163).
Related Limited availability of education in rural areas (#PF3575).
Aggravates Educational elitism (#PC1527) Inequality in education (#PC3434)
Underutilization of intellectual ability (#PF0100)
Discrimination against women in education (#PD0190).

♦ PC1897 Political opportunism

Nature Political opportunism, the unscrupulous use of events to gain political prestige or power, constitutes a barrier to political development and general progress, particularly from a lack of cooperation with partners or colleagues. Political opportunism may occur on the national or international level.

Refs Lenin, V I *Against Right–Wing and Left–Wing Opportunism, Against Trotskyism* (1983).

Broader Political conflict (#PC0368) Lack of political integrity (#PF0796)
Political barriers to effective legislation (#PC3201).
Related Careerism (#PF6353) Corruption in politics (#PC0116)
Political over–reaction (#PF4110) Political corruption of the judiciary (#PE0647).
Aggravates Martial law (#PD2637) Lack of political development (#PD8673).
Aggravated by Cynicism (#PF3418) Political rivalry (#PD8992)
Political pluralism (#PF2182).

♦ PC1904 Mechanical failure

Engineering failure risk — Mechanical equipment failure

Incidence Over a wide range of industrial sectors, the costs of unreliability and mechanical failure are generally 12 to 20 percent of a company's turnover.

Broader Risk (#PF7580) Chance (#PA6714).
Narrower Structural failure (#PD1230) Electronic equipment failure (#PD1475).
Aggravates Damage to goods (#PE4447) Unsafe aircraft (#PE1575)
Interruption risk (#PF9106) Accidental explosions (#PE3153)
Inadequate cargo transportation (#PE0430)
Accidents to nuclear weapons systems (#PD3493).
Aggravated by Wear (#PB1701) Corrosion (#PD0508)
Stress in materials (#PD7216) Fatigue in materials (#PD1391)
Failure of materials (#PD2638) Corrosion of iron and steel (#PE1945)
Defects in machinery design (#PE2462) Vibrations as a health hazard (#PE1145)
Defective product manufacture (#PD3998) Deterioration in product quality (#PD1435)
Inadequate maintenance personnel (#PJ0088) Inadequate equipment maintenance (#PD1565).

♦ PC1907 Political inertia

Dependence on political inertia

Nature Lag in political attitudes and legislation with respect to the demands of economic and technological or social change reflects the unwillingness of certain powerful groups with a vested interest in the status quo to change or to relinquish current benefits. It may also be a function of an outmoded or cumbersome political structure or of inadequate traditionalist opinions. It produces a barrier to economic, social and political progress and serves to maintain existing inequalities and injustices, ranging from poverty to pollution. Political lag can occur on either the national or the international level.

Broader Apathy (#PA2360) Inadequacy of political doctrine (#PF3394).
Narrower Traditionalism (#PF2676) Pre–electoral political inertia (#PF4713).
Related Inadequate political structure (#PC9058).
Aggravates Social breakdown (#PB2496) Political apathy (#PC1917)
Political instability (#PC2677).
Aggravated by Grey lies (#PF3098)
Bureaucracy as an organizational disease (#PD0460).

♦ PC1910 Tribalism

Nature Tribalism concerns groups of people having a common race, character, occupation, or interest. Status and ethnic or linguistic identity become linked (for example in England when it used to be accepted that a person with a certain accent would automatically be labelled "lower class") thus possibly leading to discrimination. Tribalism could also possibly lead to policies of ethnic chauvinism that grant advantages and privileges to an ethnic group as a whole or to certain of its members.

Claim Tribalism is one manifestation of a force of collective assertion, aggression and defence that is also inspired by race, religion, communal identity, a communal commitment exalted in some nations as patriotism. It is the primal urge to belong to one group and hate the rest.

Counter–claim A tribe is a family which has grown as a result of procreation. It is a big family; it provides its members with material benefits and social advantages the family provides for its members. It is a natural social umbrella for social security, bringing up children and providing a sense of identity.

Refs Rao, V Venkata *Century of Tribal Politics* (1977).

Broader Racial conflict (#PC3684) Ethnic conflict (#PC3685)
Ethnic discrimination (#PC3686).
Narrower Totemism (#PF3421) National instability due to tribalism (#PG6329)
Weak national identity due to tribalism (#PE8513)
Disruption of development by tribal warfare (#PD2191)
Ill–considered pressure to eliminate nakedness in developing countries (#PF3350).
Related Paternalism (#PF2183) Endangered tribes and indigenous peoples (#PC0720).
Aggravates Magic (#PF3311) Slavery (#PC0146) Elitism (#PA1387)
Polygamy (#PD2184) Genocide (#PC1056) Polyandry (#PF3289)
Occultism (#PF3312) Ethnocide (#PC1328) Witchcraft (#PF2099)
Militarism (#PC2169) Bride–price (#PF3290) Superstition (#PA0430)
Caste system (#PC1968) Hero worship (#PF2650) Abuse of khat (#PE0912)
Child–marriage (#PF3285) Traditionalism (#PF2676) Forced marriage (#PD1915)
Human sacrifice (#PF2641) Chattel slavery (#PC3300) Ancestor worship (#PD2315)
Social inequality (#PB0514) Guerrilla warfare (#PC1738) Human cannibalism (#PF2513)
Religious sacrifice (#PD3373) Social fragmentation (#PF1324) Trafficking in women (#PC3298)
Political dictatorship (#PC0845) Lack of national unity (#PF8107)
Primitive secret societies (#PF2928) Lack of political development (#PD8673)
Head hunting in tribal societies (#PF2666) Dependence within extended families (#PD0850)
Undemocratic political organization (#PC1015)
Discrepancies in human life evaluation (#PF1191)
Family structure as a barrier to progress (#PF1502)
Animal worship as a barrier to development (#PD2330)
Multiplicity of languages in a national setting (#PC1518)
Inadequate political parties in developing countries (#PD0548).
Aggravated by Slave trade (#PC0130) Nationalism (#PB0534)
Colonialism (#PC0798) Exploitation (#PB3200)
National boundaries (#PF8235) Racial discrimination (#PC0006)
Double standards in morality (#PF5225) Language discrimination in politics (#PD3223).
Reduced by Uncontrolled urban development (#PC0442).

♦ PC1916 Inadequate protein supply

Incidence The gap between supplies of animal protein and the effective demand in developing countries was an estimated 3.6 million tons in 1984. Even when the supply of protein foods in a particular region appears to be satisfactory (20 per cent above the average national requirements, for example), it is unevenly distributed and certain vulnerable groups go short.

Claim The world supply of food proteins amounts to approximately 79.5 million tons a year, 70 per cent of which is consumed by people in the developed countries, representing only a third of the world's population. The world needs good–quality food protein urgently, and the need will intensify with the continued increase in world population. Already the protein gap (the shortfall in protein production) is estimated at almost 2.5 million tons of animal protein annually. By the year 2000, it is estimated that even 60 million tons of animal protein will be insufficient for the population of over 6,400 million.

Counter–claim In a headlong drive to solve the problem of protein deficiency, resources which should be directed at the problem of simple starvation are being squandered. Revised estimates of the necessary dietary intake of protein are progressively lowering the amount required and reducing the protein gap by some 10 million tons of protein per year.

Broader Food insecurity (#PB2846) Human disease and disability (#PB1044).
Narrower Loss of micro–organic proteins (#PD5719).
Aggravates Protein–energy malnutrition (#PD0339).

♦ PC1917 Political apathy

Dependence on political apathy — Underutilization of political rights — Lack of participation in local politics — Lack of participation in politics

Nature Lack of participation in politics may be caused directly by discriminatory regulations or indirectly by the workings of bureaucracy, which may tend to make individuals or individual groups feel that action is not worth while since it will be blocked in the bureaucratic process. State controlled systems, while encouraging economic and social cohesion, provide no scope for democratic protest, whereas in democracies apathy may result from the dilution of proposals in order to obtain consensus. Lack of communication contributes to general apathy. Liberal ideology thinks in terms of the citizen's lack of representation in decision centres or his non–use of it. Socialist theory emphasizes that lack of political participation can occur at all levels of society.

Existing political rights may not be used because of political ignorance, ignorance concerning legislation, fear of authority, the need to form pressure groups rather than act as an individual, lack of information, or disinterest. It may cause or be a result of dictatorship, political repression and political alienation, or derive from a high standard of living or the effects of corruption in

blocking political participation.
Political apathy or lack of participation in politics, may lead to anarchy, fanaticism, and political, social or national disintegration. The underutilization of political rights undermines the practice and principles of democracy and may in extreme cases lead to dictatorship.
Counter–claim Political apathy at the national level is due to a healthy resurgence of interest in political activity at the local and community levels, especially through volunteerism, charitable giving and neighbourhood associations.
Broader Apathy (#PA2360).
Narrower Lack of political parties (#PU4387).
Related Complacency (#PA1742)
Social apathy (#PC3412)
Non–participation (#PC0588)
Corruption in politics (#PC0116)
Underutilization of legal rights (#PF3464)
Bureaucracy as an organizational disease (#PD0460)
Lack of worker participation in business decision–making (#PF0574).
Aggravates Social alienation (#PC2130)
Military influence (#PD3385)
Political injustice (#PC2181)
Political repression (#PC1919)
Political appointees (#PF2031)
Political stagnation (#PC2494)
Political instability (#PC2677)
Political dictatorship (#PC0845)
Bureaucratic aggression (#PC2064)
Impoverishment of political debate (#PF4600)
Undemocratic political organization (#PC1015)
Lack of participation in development (#PF3339)
Unjust allocation of government contracts (#PF2911)
Failure of individuals to participate in social processes (#PF0749).
Aggravated by Cynicism (#PF3418)
Prejudice (#PA2173)
Discrimination (#PA0833)
Political inertia (#PC1907)
Political ignorance (#PC1982)
Political alienation (#PC3227)
Lack of communication (#PF0816)
Single party democracies (#PD2001)
Inequality of opportunity (#PC3435)
Limitations on right to vote (#PF2904)
Denial of freedom of thought (#PF3217)
Lack of political development (#PD8673)
Lack of community participation (#PF3307)
Inadequate political integration (#PF3215)
Inadequacy of political doctrine (#PF3394)
Repression of intellectual dissidents (#PD0434)
Ideological discrimination in politics (#PC3219)
Restrictions on direct news coverage of parliamentary affairs (#PF3072).

♦ PC1919 Political repression

Dependence on political repression — Political oppression
Nature Political freedom is relative to what state systems will tolerate. There are therefore degrees of repression at all times everywhere, which may range from deterrent persuasion to sanctions, to incarceration, to the murder of dissidents in society by the state. Repression of political views, organizations and actions felt to be detrimental or subversive to a nationally accepted or imposed norm includes: the banning of political parties, literature and activities; internment; torture; disenfranchisement; confiscation of property; and maltreatment of families of offenders. The results may be political apathy, ignorance, alienation and instability or foreign intervention, international conflict, and war. Political repression may take the form of foreign intervention and may be particularly intense during war.
Police states, and governments active in political repression, are aided by technology. Some of the equipment and techniques used in controlling and penalizing victims of the state include sensory deprivation, electronic surveillance, and a range of computer software for citizen–victim 'files', data bases and non–conformist behaviour detection. The supplying of such instruments of government repression is an industry having its own suppliers, much like the defence industry, and some manufacturers supply arms, ammunition and various devices both to state police apparatus and to the military. Countries which pride themselves on their political freedom may in fact have lucrative industries which supply such instruments of repression to the governments of other, less politically free, nations.
Incidence Notable instances of political repression are reported to occur in Iran, USSR, South Africa and Brazil, among others. Repression of student activists is a frequently witnessed phenomenon. In the past decades it has been associated with Berkeley, Columbia, and Kent State in the USA, and with universities in Paris, Louvain, Caracas and Valencia (Venezuela). Other notable acts of repression are the now famous cases of dissident Soviet writers, and the foreign intervention in Hungary, Czechoslovakia and Afghanistan by the USSR. That political repression can result in murder or long–distance assassination was seen in the case of Leon Trotsky. However, political repression of intellectuals may often be due to the fact that they may be acting as spokesmen (whether students, artists, or theorists) for the masses.
Claim The greatest bulk of the people who suffer under political repression are the workers, and in developing countries, the peasants who live off the land.
Broader Repression (#PB0871).
Narrower Exile (#PC2507)
Imprisonment (#PD5142)
Political purges (#PC2933)
Political trials (#PD3013)
Political police (#PD3542)
Qualified amnesty (#PF3019)
Political prisoners (#PC0562)
Concentration camps (#PD0702)
Violation of amnesty (#PD3018)
Political intimidation (#PC2938)
Injustice of mass trials (#PE0597)
Injustice of special courts (#PE0088)
Injustice of trials in absentia (#PE0424)
Illegal movement across frontiers (#PC2367)
Repression of intellectual dissidents (#PD0434)
Mis–classification of political prisoners (#PF3020)
Military political prisoners and detainees (#PD3014)
Civilian political prisoners and detainees (#PD3015)
Abusive detention in psychiatric institutions (#PE2932)
Suppression of opposition groups or individuals (#PC7662)
Maintenance of political dossiers on individuals (#PD2929)
Inadequate legal counsel for political dissidents (#PF0732).
Related Forced labour (#PC0746)
Abuse of government power (#PC9104)
Forced political confessions (#PE3016)
Injustice of military tribunals (#PE0494)
Underprivileged racial minorities (#PC0805).
Aggravates War (#PB0593)
Informers (#PD8926)
Conspiracy (#PC2555)
Political crime (#PC0350)
Postal censorship (#PD3033)
Immoral literature (#PF1384)
Abuse of police power (#PC1142)
Lack of social mobility (#PF2195)
State sanctioned torture (#PD0181)
Political hostage–taking (#PD1886)
Internment without trial (#PD1576)
Denial of right of assembly (#PC2383)
Denial of rights to prisoners (#PD0520)
Forced disappearance of persons (#PD4259)
Conflict between minority groups (#PC3428)
Denial of the right of association (#PD3224)
Lack of individual rights to political asylum (#PF1075).
Aggravated by Marxism (#PF2189)
Torture (#PB3430)
Genocide (#PC1056)
Nationalism (#PB0534)
Paternalism (#PF2183)
Totalitarianism (#PF2190)
Political apathy (#PC1917)
Minority control (#PF2375)
Political instability (#PC2677)
Harassment of the media (#PD0160)
Single party democracies (#PD2001)
Elitism in communist systems (#PC3170)
Denial of freedom of thought (#PF3217)
Undemocratic political organization (#PC1015)
Political discrimination in politics (#PC3221)
Denial of political and civil rights (#PC0632)
Deficiencies in national and local legal systems (#PF4851)
Misuse of electronic surveillance by governments (#PD2930)
Political discrimination in the administration of justice (#PE1828).

♦ PC1937 Scientific elitism

Nature Scientific activities are insufficiently productive in relationship to the time and resources invested in them, because of the hierarchical institutionalism of the scientific establishment and its elitist nature.
Claim Science has been institutionalized along the lines of the upper class society which supports it, so that its products will perpetuate and strengthen individual, family and corporate wealth. Scientists are bought with excessive financial and social rewards. Their concerns are often limited to the self–interest of perpetuating high incomes through tenured teaching or research positions. They require power in order to intimidate or suppress work that is disdained or ideologically opposed by the national or international establishment. Often, therefore, scientists through their professional organizations or university faculties, behave as a self–contained, self–elected oligarchy or an apostolically–ordained priesthood whose episcopal synods lay down the dogmas of research and scientific truth.
Broader Elitism (#PA1387).
Aggravates Denial of rights of medical patients (#PD1662)
Irrelevance of science and technology (#PF0770)
Fragmentation of academic disciplines (#PF8868).
Aggravated by Incomprehensibility of specialized jargon (#PF1748).

♦ PC1952 Military–industrial complex

Nature The arms race tends to change traditional relationships between the civilian and military sectors of the economy. The military sector covers more than the military forces themselves. It includes the firms and industries which serve them, the scientific institutions where their research is done, and the political establishments and ministries that owe their power to the arms race – a combination which has come to be called the 'military–industrial complex'. The military–industrial complexes everywhere become concerned to preserve themselves, and consequently to maintain the circumstances which gave birth to them. Fear of a potential enemy leads a country to set up a military establishment, and this establishment in turn acts to keep the fear alive. It will suspect and question the sincerity of any conciliatory moves from the other side, and in general act to preserve a political image of the world as one which will always require a nation to maintain a high state of military preparedness with an industrial capability to support this.
Incidence Global industrial production for military purposes has been estimated as having an annual value exceeding US $125,000 million. In some countries there are whole industries which make the greater proportion of their sales to the government for defence purposes, for example the aero–space and ship–building industries in the leading NATO countries.
Military activities and the manufacture of armaments employ about 60 million people throughout the world. Half of them in industry, the other half in the armed forces.
Broader Capitalism (#PC0564).
Narrower National defence procurement procedures (#PE4097)
Military–industrial complex in capitalist systems (#PC3191)
Military–industrial–governmental complex in communist systems (#PD1330).
Aggravates War (#PB0593)
Nuclear war (#PC0842)
Pervasive fear of nuclear war (#PC3541).

♦ PC1960 Slowing growth in food production

Nature After two decades of impressive gains, global food production has slowed. In 1987 it fell below consumption and in 1988 a further decline in food stocks are expected. In 1987, world grain stocks in storage at the time of harvest dropped from 457 million metric tons to 390 million. In 1988 stocks could well fall below 300 million metric tons, a reduction that could easily raise grain prices to twice the 1987 level. The growth of grain production has slowed in several populous countries, including China, India, Indonesia and Mexico. The loss of momentum in world food output is widespread. The remarkable increases in food production during the last 15 years have come in part at the expense of soil and water resources. Since the 1970s soil erosion has increased sharply. For example, in the United States in 1976 farmers were estimated to be losing six tons of soil for every ton of grain produced. Water tables are falling in the United States, China, the Soviet Union and India where wells are running dry and thousands of villages are relying on tank truck for drinking water. The world area in grain has declined steadily from a record high in 1981. In the Soviet Union and the United States erodible land is being converted to grasslands and woodlands. Across the southern fringe of the Sahara the agricultural frontier is retreating as a result of declining rainfall, land degradation, and dune formation. China and the United States have reduced the area of irrigated land. Climatic changes may further reduce land and water resources. There are no technologies waiting in the wings that will lead to the quantum jumps in food output of the sort associated with the hybridization of corn, the eightfold increase in fertilizer use between 1950 and 1980, the tripling of irrigated area in the same period or the rapid spread of the Green Revolution.
Counter–claim An increase in global food production is not needed. It is the current and potential food–deficit countries (and regions within countries) that need to increase their food production — preferably through small farmers, to enable them to increase their income and so take some pressure of the land.
Broader Underdevelopment of food and live animal production (#PF2821).
Related Long–term shortage of meat and meat preparations (#PE1490)
Long–term shortage of cereals and cereal preparations (#PE1218)
Long–term shortage of sugar and honey and preparations thereof (#PE1120)
Long–term shortage of fruit and vegetables and preparations thereof (#PE1013)
Long–term shortage of coffee, tea, cocoa, spices and manufactures thereof (#PE1197)
Long–term shortage of salt–water fish, crustacea and molluscs and preparations thereof (#PE1783).
Aggravates Food insecurity (#PB2846).
Aggravated by Bad weather (#PC0293)
Desert advance (#PC2506)
Global warming (#PC0918)
Soil degradation (#PD1052)
Soil infertility (#PD0077)
Soil erosion by wind (#PE3656)
Migrating sand dunes (#PD0493)
Soil erosion by water (#PD2290)
Shortage of cultivable land (#PC0219)
Shortage of fresh–water sources (#PC4815)
Declining area of irrigated land (#PE4585)
Vulnerability of nuclear defence control systems (#PD4049)
Decreasing rate of development of major agriculture technologies (#PE5336).

♦ PC1966 Inadequate income in old age

Reduced income in old age — Income maintenance after retirement — Inadequate pension income — Poverty of the elderly
Nature In a number of countries, projected state pension funds will be insufficient for the needs of a growing population of elderly retirees. The tying pensions to cost–of–living indexes is being undermined and there have been attempts to de–couple payments from index rates, on a supposedly temporary basis. Inflation, for some commodities and necessary expenditures, has exceeded the index adjustments, so that retired people have difficulty sustaining the diet, the amount of home heating and the automobile usage, for example, that they were accustomed to. Personal property and real estate local taxes may be too heavy for them and proper medical care, including hospitalization, treatments and appliances may be too costly to be borne. Sole reliance on government pensions means immediate poverty for many retirees who are forced to sell their homes, or even to move away entirely from the district in which they spent their lives and have their families and friends, in order to find cheaper shelter. As people age, the economic and social structures which they created for the earlier years of their life must give way to new ones. Many people have no replacement for the loss of family home, employment, spouse, and friends as these things are lost. Thrown into a situation of economic dependence, elderly people often must struggle to get the simplest necessities of living. Even those fortunate enough to find themselves receiving social benefits of some kind find it difficult to adjust to a new style of spending.
Incidence In order to maintain the same standard of living, pensioners require an amount equal

to their average gross salary (for previous years) less all withholdings, including the costs of going to work, restaurant meals, transportation, and miscellaneous expenses. On the other hand, added to the amount should be equivalents for benefits lost – such as group health insurance, credit union access and the like. ILO economists believe the indexed pension should be in the range of 65 percent of a person's average gross earnings. Many governments believe the indexed range should not exceed a maximum irrespective of earnings beyond national average income levels, so that pensions may be as little as 20 or 30 percent of average gross income in many cases.
Refs ILO *Pensions and Inflation* (1980).
Broader Crisis in long–term pension funds (#PF5956).
Narrower Unequal distribution of old age pensions between men and women (#PE7942).
Related Low general income (#PD8568) Abusive treatment of the aged (#PD5501)
Inadequate welfare services for the aged (#PD0512)
Dependency on unpredictable sources of income (#PF3084)
Misallocation of resources to protect the aged (#PE0648).
Aggravates Social withdrawal of aged (#PD3518)
Mental disorders of the aged (#PD0919)
Criminalization of euthanasia (#PF2643)
Inadequate housing for the aged (#PD0276)
Inadequate recreational facilities (#PF0202)
Denial of right to economic security (#PD0808)
Susceptibility of the old to physical ill–health (#PD1043)
Retirement as a threat to psychological well–being (#PF1269).
Aggravated by Human ageing (#PB0477) Economic inflation (#PC0254)
Age discrimination (#PC2541) Ageing war disabled (#PD0874)
Ageing of world population (#PC0027) Unemployment of older people (#PE5951)
Social disadvantage of the aged (#PD3517) Marginal level of family income (#PD6579).
Reduced by Separation under marriage law (#PF3251).

♦ PC1968 Caste system

Caste prejudice — Casteism — Outcastes
Nature Rigid social stratification, in which ultimate lineage or parental position or profession determines occupation, marriage partner, rank or title, communal responsibilities, and the like, can be characterized as the caste system. The caste system, unlike the class system, or the mediaeval European system of the three estates of nobility, clergy, and commons, does not allow for upward mobility, at least in a single lifetime. As practised on the Indian sub–continent, it has been buttressed by sacred writings which, like the Old Testament, emphasize the concepts of purity and pollution. The effect of the caste system is repression and enslavement of generations of individuals and an incalculable retardation of national development owing to the fragmentation of society.
Incidence The caste system is best known in India where concepts like 'untouchables' still exist. The outcastes are the menial, depressed tribes or castes or they have been expelled from their own tribe or caste permanently or temporarily. However, wherever a priestly class has been able to dominate a society, like the Levites among the Jews, ritual and secular rules of purity have emerged; and where minority ethnic groups have also been present, they have conveniently been termed unclean or 'beyond the Pale'. This is true also of minority non–believers, so that caste system ideas enter into religious conflicts, for example in Assam and Punjab. In the state of West Bengal there are 41 scheduled tribes and 63 scheduled castes, accounting for only 26 percent of the population. The balance, 74 percent, are 'outcastes'.
Refs Channa, V C *Caste* (1979); Desai, I P *Caste, Caste–Conflict and Reservation* (1985); Kamble, N D *Deprived Castes and Their Struggle for Equality* (1983); Klass, Morton *Caste* (1980); Von Furer–Haimendorf, Christoph (Ed) *Caste and Kin in Nepal, India and Ceylon* (1982).
Broader Prejudice (#PA2173) Segregation (#PC0031).
Narrower Neutrality (#PF0473).
Related Racial segregation (#PC3688) Racial exploitation (#PC3334)
Ethnic discrimination (#PC3686).
Aggravates Racial conflict (#PC3684) Social outcasts (#PD6017)
Underdevelopment (#PB0206) Arranged marriage (#PF3284)
Class consciousness (#PC3458) Social discrimination (#PC1864)
Exploitative personal services (#PC3299)
Burdensome cost of religious ceremonies (#PF3313)
Inhibiting effects of traditional life–styles (#PF3211).
Aggravated by Tribalism (#PC1910) Traditionalism (#PF2676)
Racial inequality (#PF1199) Social inequality (#PB0514)
Lack of social mobility (#PF2195) Exploitation in employment (#PC3297)
Inflexible social structure (#PB1997) Undue attachment to a social group (#PF1073)
Trafficking in children for adoption (#PF3302).

♦ PC1972 Anarchism

Nature The ideal of anarchism is an extension of democracy and the ultimate in individualism. The practice of anarchy is usually equated with violence and total disruption.
Incidence Anarchism exists among minority and intellectual political movements and is especially notable in 'western' industrialized countries. It tends to increase in situations of civil or guerrilla warfare and to spread internationally (for example, the Baader–Meinhof terrorist group in West Germany was supplied with arms and a car by a Zurich anarchist group), because of the need to be supplied from outside sources.
Background The anarchist believes that it is practicable and desirable to abolish all forms of organized government. Forerunners of anarchism include the Greek philosopher Zeno and some hussite and Anabaptist religious reformers. Anarchist ideas were expressed by the French writers Rabelais and Fénelon and were familiar to 18th century French intellectuals. The English writer William Godwin gave anarchism its first modern exposition in 1793, advocating a stateless and communist society. The term 'anarchist' was coined as a reproach by the moderate Girondists of the French Revolution. In the 1840s anarchism gained momentum with the writings of Proudhon and the Hegelians Hess and Grun and later Nietzsche. In the 1870s anarchist groups split from the Marxist International Workingmen's Association and formed their own international in 1881. Mikhail Bakunin was its outstanding leader. Anarchism as a political movement subsided by the mid 20th century but continued to have intellectual influence.
Refs Berkman, Alexander *ABC of Anarchism* ; Institut Français d'Histoire Sociale *Anarchism* (1982); Miller, David *Anarchism* (1984).
Broader Ideological conflict (#PF3388).
Related Disorder (#PA7361) Revolution (#PA5901) Illegality (#PA5952)
Lawlessness (#PA5563) World anarchy (#PF2071).
Aggravates Social breakdown (#PB2496) National disintegration (#PB3384).
Aggravated by Nihilism (#PJ7927) Double standards in morality (#PF5225)
Inflexible deliberative systems of government (#PF7059).
Reduces Social constraints on freedom of imagination (#PF4882).

♦ PC1982 Political ignorance

Nature Political ignorance of the activities or working of governmental structures, and lack of awareness concerning political issues, may be general or it may be confined to underprivileged social and ethnic groups. General ignorance concerning the activities and working of the government may enable decisions to be taken undemocratically, and repression, torture and other crimes against humanity to be carried out without censure. It may equally lead to violent, rather than constitutional, means of change. Ignorance concerning political issues and the political ignorance of ethnic and other underprivileged social groups helps to maintain social inequalities, social injustice, exploitation, discrimination and social conflict, and may result in social breakdown or ethnic disintegration.
Broader Ignorance (#PA5568).
Narrower Sociological ignorance of citizen participation (#PF2440).
Related Citizen incompetence (#PE5742).
Aggravates Political apathy (#PC1917) Undemocratic political organization (#PC1015)
Political corruption of the judiciary (#PE0647).
Aggravated by Lack of education (#PB8645) Political alienation (#PC3227).

♦ PC1987 Brutality

Broader Human violence (#PA0429).
Narrower Police brutality (#PD3543) Military brutality (#PD4945)
Limited applicability of monetary grants (#PF2564).
Related Badness (#PA5454) Vulgarity (#PA5821) Moderation (#PA7156)
Unkindness (#PA5643) Unchastity (#PA5612) Aggravated assault (#PD0583).
Aggravates Mutiny (#PD2589) Atrocities (#PD6945).
Aggravated by Slavery (#PC0146) Slave trade (#PC0130).

♦ PC1995 Threatened sects

Nature Religious sects which exist legally may be threatened by the encroachments of modern civilization. This is because, in very strict religious sects such as the Amish community in the USA, the way of life does not follow the pattern of national society. It may seem to be a very old–fashioned or simplistic way of life. The emphasis on segregated education or a different kind of education than that prescribed by the State may cause conflict, and the State may threaten the unity of the sect through the dissemination of contrary beliefs and compulsory education. Forced assimilation and intolerance also constitute a threat.
Broader Threatened and vulnerable minorities (#PC3295)
Segregation based on religious affiliation (#PC3365)
Threats to ideological movements and minorities (#PC3362).
Related Ethnic disintegration (#PC3291)
Endangered tribes and indigenous peoples (#PC0720)
Vulnerability of small nations to foreign intervention (#PD2374).
Aggravates Social disintegration (#PC3309) Undemocratic political organization (#PC1015)
Lack of variety of social life forms (#PE8806).
Aggravated by Heterosexism (#PE0818) Social fragmentation (#PF1324)
Refusal of medical care (#PF4244) Religious discrimination (#PC1455)
Denial of freedom of thought (#PF3217) Religious conflict between sects (#PC3363).
Reduced by Segregation in marriage (#PD3347).

♦ PC2008 Planned obsolescence

Manipulated product life cycles
Nature Since product 'death' is inevitable – that is, because its design will be superceded by a new fashion or technology, or because its materials will wear out – and since industry is geared to the need for product replacement, products are engineered to die by material or mechanical failure. This makes a science out of replacement production when it is known that the product will not last more than ten years (automobiles) or not more than several weeks (batteries). Such planned obsolescence is a characteristic of every variety of consumer product and in some cases has resulted in serious accidents due to inadequate quality. The costs to the economy of planned obsolescence are incalculable.
Claim The planned obsolescence of many items is notorious. Possibly toys might top the list and, in some countries, clothing articles. Automobiles, batteries and some home appliances are well–known for failure, although the electric light bulb might also be given first place. In addition, the arms race is justified by military procurement based on planned obsolescence of aircraft, ships and weapons, and of delivery and support systems.
Counter–claim Product death transpires through competition, substitution, technology, cost–economies or changes in market demand. Products have a life–cycle that is necessary to understand and to anticipate in order to serve end–user needs at the most efficient cost levels. Users will not pay for over–engineered items. Automobiles and personal computers for example, can only be purchased cheaply if one is not looking for life–time usage.
Broader Consumer vulnerability (#PC0123)
Short–term planning of product life cycles (#PF1740).
Related Planned degradation in product quality (#PF7741).
Aggravates Scrapped automobiles (#PE5339) Corrosion of iron and steel (#PE1945)
Deterioration in product quality (#PD1435) Inadequate equipment maintenance (#PD1565)
Underutilization of second–hand equipment (#PD1484)
Prohibitive cost of equipment maintenance (#PE1722).

♦ PC2020 Economic insecurity

Dependence on economic insecurity
Broader Insecurity (#PA0857) Socio–economic poverty (#PB0388).
Narrower Food insecurity (#PB2846) Insecurity of resources (#PB8678)
Insecurity of employment (#PD8211) Insecurity of work changes (#PG9450)
Inadequate social security for migrants (#PE7611)
Insecurity from future crop uncertainty (#PF8354)
Social insecurity in developing countries (#PE4796)
Insufficient social security in the agricultural sector (#PD9155).
Related Unequal coverage by social security (#PF0852).

♦ PC2022 Sexual discrimination

Sexual prejudice — Gender discrimination
Nature Sexual roles tend to be fixed by most societies, but in different ways. Most 'western' industrialized countries do not accept homosexuality, whereas it is tolerated in Asia and among certain primitive tribes. Discrimination between the roles of men and women exists everywhere but is somewhat more liberal in certain industrialized countries.
Claim The allocation of fixed sexual roles includes general discrimination against women, against men, and the non–acceptance of homosexuality. In some cultures celibacy and the unmarried state are made virtually impossible and marriages are forced, under the influence of a societal demand for procreation. Childlessness is considered a curse.
Counter–claim Owing to a lack of complete scientific investigation, a differential psychology for males versus females remains to be elaborated. Despite substantial prejudice to the contrary, however, there is already a considerable amount of empirical and theoretical observations which assist in a probabilistic approach to role performance optimization for the sexes. More knowledge will enhance the preferential sex for some roles, as well as obliterate the need for specification in others.
Refs Walsh, Mary R *Doctors Wanted – No Women Need Apply* (1977).
Broader Prejudice (#PA2173) Discrimination (#PA0833) Social injustice (#PC0797).
Narrower Misogyny (#PE7062) Sex segregation (#PC3383) Gender abortions (#PD3947)
Gender stereotyping (#PD5843) Discrimination against men (#PC3258)
Sexually segregated schools (#PG3650) Sexual discrimination in education (#PD1468)
Inequality of life expectancy by gender (#PE1339)
Gender discrimination in developing countries (#PD9563).
Related Ethnic discrimination (#PC3686) Discrimination against women (#PC0308).

Aggravates Sexism (#PC3432) Transsexualism (#PF3277) Social fragmentation (#PF1324)
Lack of social mobility (#PF2195) Conflicting sense of sexual identity (#PF1246)
Lack of variety of social life forms (#PE8806).
Aggravated by Conformism (#PB3407) Traditionalism (#PF2676)
Male domination (#PC3024) Religious intolerance (#PC1808)
Human sexual inadequacy (#PC1892) Fear of emotional sensitivity (#PF9209)
Double standards of sexual morality (#PF3259).

♦ PC2023 Pollution in developing countries

Nature Although air, water and noise pollution are not yet matters of primary concern in urban areas of developing countries, such problems will grow more severe as these countries move toward their goals of economic development. Generally the devices and regulations presently in force to control pollution in developed countries are not applied to industrial processes in developing countries with equal efficiency or stringency. In an effort to provide increased economic well-being, environmental safeguards are neglected. Water supplies are not only contaminated with human wastes, but grow increasingly toxic as they receive the effluence from expanding industries. Air pollution increases with the material well-being of the urban population and emanates from power plants, industry, space heating and the growing number of motor vehicles.
Narrower Environmental destruction in least developed countries (#PE8401).
Aggravated by Urban slums in developing countries (#PD3489)
Environmental hazards from economic and industrial products (#PC0328).

♦ PC2024 Hunting of animals

Uncontrolled hunting — Recreational hunting
Nature Hunting is the seeking, pursuance and capture or destruction of game and wild animals for subsistence, profit or sport. Most governments establish "open" and "closed" hunting seasons and require hunters to be licensed. However, in some parts of the world the number of individuals wishing to hunt as a sport and who take advantage of the open season is increasing to such an extent that problems are being created. Uncontrolled hunting is extremely destructive of wildlife and may severely endanger rare species. Depending on the animals hunted, certain methods of hunting may be either cruel or wantonly destructive. Examples are the use of automatic weapons, nets, hand grenades, dynamite (in rivers and lakes), certain types of poison, or use of dogs. Despite claims by hunters that shooting an animal is more humane than death by starvation, many game animals are left wounded by unskilled hunters and many others are claimed to have been killed by error. Untended traps are a possible source of injury to humans, as is the possibility of bites from trapped animals. A major reason for the overpopulation of certain animals is the destruction of their natural predators by man. This upsets a delicate ecological balance in nature.
Background Mankind has developed through hunting. Many native peoples around the world continue to survive through hunting and do so only in respond to their basic needs. Modern forms of hunting arose with the invention of firearms through which killing for pleasure became a primary motivation, as with the slaughter of game in Africa and the bison in North America.
Incidence It is estimated that there are some 15 million hunters in the USA alone. In North America it is estimated that some 175 million animals are killed annually by hunters and trappers, including: 24,000 bear, 55,000 caribou, 67,000 moose, 84,000 antelope, 100,000 elk, 2.6 million deer, 21 million waterfowl, 27 million rabbits, 32 million squirrels, and 94 million upland birds. Game herds in Botswana have been decimated by illegal hunters. In many Muslim countries where pork is taboo, warthog and wild boar are hunted and killed as a threat to crops but the meat is wasted (in some cases thousands of pigs are left to rot).
Claim Hunters devise a number of spurious rationalizations to justify their cruelty to animals. Their involvement in wildlife management is merely designed to ensure that there are adequate numbers of particular species to be killed, not to ensure an appropriate ecological balance. As a traditional sport, it is difficult to justify when one party is both defenceless and unaware of the nature of the contest. Recreational hunting facilities are supported to a greater extent by the general tax-payer than through hunting licences and fees.
Counter-claim Hunters are lovers of nature. As such they have ensured the allocation of major resources to nature conservation thus ensuring that many species are saved from extinction. Hunting is an appropriate means of maintaining viable population levels of species whose numbers otherwise escalate.
Refs FAO *Legislation on Wildlife, Hunting and Protected Areas in Some European Countries* (1980).
Broader Lack of control (#PF7138)
Ecologically unsustainable development (#PC0111).
Narrower Hunting tourism (#PE3008) Killing of animals (#PD8486)
Trapping of animals (#PE5735) Hunting of marine animals (#PE0439)
Slaughter of animals for pelts (#PD4575).
Related Denial to animals of the right to freedom from mass killing (#PE9650)
Health risks to workers in agricultural and livestock production (#PE0524).
Aggravates Food wastage (#PD8844) Animal injuries (#PC2753)
Maltreatment of animals (#PC0066) Endangered species of animals (#PC1713).
Aggravated by Dangerous animals (#PC2321) Inadequate firearm regulation (#PD1970).

♦ PC2027 Spoilage of agricultural products

Post-harvest losses
Nature Faulty harvesting, handling and especially storage lead to agricultural product losses. The loss in quality is accompanied by a less widely recognized loss in quantity. In many areas people have always been accustomed to products that are more or less spoiled by insects; often they do not understood that this is why products may only command low prices on world markets, or why in some cases they cannot be sold at all.
Food supplies could be augmented to an important extent through trying to minimize crop losses. A number of surveys have shown post-harvest losses of food grains reaching 20 or even as much as 40 per cent, depending on crop and country. Losses occur all along the line: through poor management in harvesting; insect, rodent, and fungus damage in storage; poor quality sacks and other containers; inexperienced handling in transport; inefficient or poorly maintained milling equipment; and faulty distribution of milled products. The use of extremely low extraction rates to meet the presumed tastes of urban consumers means less food for humans and more feed for animals as well as a less nutritive end-product. Avoidable losses also occur in the handling and processing of other crops, such as oilseeds, where in many small factories an unnecessarily high proportion of oil is left in the by-products or where, alternatively, a part of the protein contained in the oilseed meal could be incorporated into foods suitable for human consumption. In the area of perishable fruit and vegetables, wastage in distribution quite often reaches 30 to 40 per cent. These examples indicate the scope for increasing food supplies through reduction of losses. But they also suggest that no single programme would have a significant impact on the problem. Because the problems are various so too their solution must be sought through improvements in several directions. Some of the most important are: investment in modern storage for grains and other crops, both in villages and in larger centres; better packaging materials and transport facilities; modifications in the milling practices for cereals and oilseeds; and more effort to modernize the organization of wholesale and retail distribution.
Incidence Losses, which may reach 100 per cent when produce is entirely spoiled for lack of proper storage facilities, are estimated to average 25 per cent globally. In 1990 it was estimated that 75 per cent of the USSR potato crop (fruit and vegetable crop, 40 per cent) never reaches consumers because of spoilage, exacerbated by inadequate transportation and storage. In all 25 per cent of USSR farm production is lost in this way.
Refs CAB International Mycological Institute *Post-Harvest Diseases and Disorders of Tropical Root Crops*.
Broader Food insecurity (#PB2846).
Narrower Food grain spoilage (#PD0811) Food spoilage in storage (#PD2243).
Aggravates Shortage of cultivable land (#PC0219)
Food poisoning through negligence (#PE0561).
Aggravated by Insect pests (#PC1630) Inadequate storage facilities (#PG4268)
Insufficient transportation infrastructure (#PF1495)
Inadequate packaging of agricultural products (#PF3143).

♦ PC2030 Political bribery

Nature Money or other privileges may be given in return for political concessions, such as moderation of legislation in favour of a certain interest group or individual. The donation may be directed to a key political personality or to a political party.
Broader Bribery (#PC2558) Corruption in politics (#PC0116).
Narrower Political gifts (#PG2915) Abuse of influence (#PC6307)
Trading in public office (#PE6948) Unjust financing of political parties (#PE0752).
Aggravates Undemocratic political organization (#PC1015).

♦ PC2035 Youth unemployment

High cost of youth unemployment — Youth job vacuum — Unavailability of youth employment
Nature The human wastage caused by unemployment is intensified when those affected are young. Adolescents' characters are still impressionable and the shock of their finding that society has no place for them, and in fact condemns them to poverty and marginal existence, is embittering. Large-scale youth unemployment is a loss to the whole of society, both in terms of training opportunities missed and by alienation of an entire generation.
Youth unemployment is disproportionately high in a number of countries, indicating that the problem is structural. Employers preferences seem to be for the 25 to 35 or 40 year-olds, whose experience assures higher productivity. Young persons are often forced to take jobs which offer no training and no future and are dead-ends.
Incidence In the developed OECD countries, approximately 40 percent of the unemployed are 14 to 24 years of age. In the EEC countries the average is nearly the same. In North America, figures have reached 50 percent in Canada for the same age group, while in the USA nearly 20 percent of the unemployed were young men between 16 and 19. Youth unemployment in the less developed countries is better told as tens of millions, rather than a percentage, and global youth unemployment may be of the order of 35 million persons.
Refs Commonwealth *Jobs for Young People*; Corvalan-Vasquez, Oscar *Youth Employment and Training in Developing Countries* (1984); OECD Staff *Youth Unemployment* (1980).
Broader Unemployment (#PB0750).
Narrower Unemployed educated youth (#PE1379).
Related Mass unemployment of human resources (#PD2046)
Unattractive locale for economic development (#PF3499)
Disruptive effect of changing employment patterns (#PF2303).
Aggravates Maladjustment to disciplines of employment (#PD7650).

♦ PC2049 Regional disparities

Underprivileged regions — Overprivileged regions — Deprived regions — Favoured regions
Nature During the course of any social transformation process, disparities between regions may emerge, or a long standing situation may be accentuated. Such disparities are found both among different countries at the international level, and among different regions within the same country. The inter-regional differences may occur in regard to such important needs as food, health, habitation, education, social mobility, administrative services and political representation. Many of these disparities have adverse consequences for the further process of local and global social transformation.
Refs Châu, Ta Ngoc and Carron, Gabriel (Eds) *Regional Disparities in Educational Development* (1980); Phillips, Paul *Regional Disparities*.
Broader Underdevelopment (#PB0206) Social inequality (#PB0514).
Narrower Regional underdevelopment (#PD4424)
Regional imbalance of power (#PC4291)
Regional environmental degradation (#PD5845)
Disparities between developing countries (#PD2963)
Imbalanced regional development within countries (#PJ3981).
Related Separate and unequal development within different societies (#PE8416).
Aggravated by Uncontrolled industrialization (#PB1845)
Fragmented regional cooperation (#PF9129)
Discrimination against developing countries by the formation of regional groupings of developed countries (#PE1604).

♦ PC2052 Student revolt

Student unrest — Student dissent — Student riots
Nature Student dissent has led to widespread criticism of the educational establishment. Some of this criticism was evoked from the very nature of the authoritarian responses to student revolt in places as widely separated as Paris, Berkeley, New York, and Mexico City.
Claim When the educational system remains the exclusive preserve of an intellectual elite, the product of the bourgeois class which built the system and continues to dictate its laws and moral values, students become confused by the divorce between an outmoded education and the reality of the world around them. They become frustrated, dissipate their energies, grow bored or put their hopes in radical protest.
Related Religious riots (#PE1417).
Aggravates Civil violence (#PC4864) Civil disorders (#PC2551)
Inhumane methods of riot control (#PD1156).
Aggravated by Apathy (#PA2360) Student press weakness (#PE0628)
Generation communication gap (#PF0756) Political impotence of students (#PE5729).

♦ PC2062 Agricultural surpluses

Nature Grain production, trade and consumption trends, and the appraisal of factors underlying them, indicate that surpluses, or the persistence of production in excess of effective demand, may now be considered a chronic feature of the world grain economy. The heart of the problem lies in the level of price or income guarantees to producers of wheat and other grains in many exporting as well as importing countries. These guarantees, combined with other aspects of national agricultural policies (if maintained substantially unchanged) will continue, together with technological advance, to stimulate an output larger than can be absorbed by effective demand. Independent measures of surplus disposal may therefore assume a semi-permanent character and affect an increasing part of the international trade in grains, thus adding to the marketing difficulties being experienced by exporting countries.
Heavily subsidized surpluses, often exported as food aid, depress international market prices of commodities and thus creating severe problems for developing countries whose economies are based on agriculture. They also tend to reduce the incentives for domestic food production.

Incidence In 1985 it was estimated that the EEC spends $ 1,000 million per year on storage of agricultural surpluses.
Counter–claim World food security is threatened because declining food stocks, declining growth in food production and warming of the atmosphere which will cause increased droughts. In 1988 food stocks were the lowest since the years following World War II.
Broader Surplus (#PF4750) Unnecessary reserves of material (#PF0687)
Instability of economic and industrial production activities (#PC1217).
Narrower Excessive wheat surpluses (#PE2902) Surplus domestic animal production (#PJ5775).
Aggravates Distortion of international trade by dumping (#PD2144).
Aggravated by Instability of trade in cereals and cereal preparations (#PE1769)
Over–subsidized agriculture in industrialized countries (#PD9802)
Detrimental international repercussions of domestic agricultural policies (#PF2889).
Reduces Inadequacy of food aid (#PF3949).
Reduced by Excessive consumption of goods and services (#PC2518).

♦ PC2064 Bureaucratic aggression

Nature Bureaucracies may use their powers and privileges, whether deliberately or inadvertently, in such a way as to favour policies and projects to the disadvantage of the regions, communities or interest groups on whom such projects have some impact, or for whose supposed benefit they have been conceived. They may be particularly successful in such attempts to extend their power and jurisdiction if they can avoid drawing attention to plans and proposals at the early stages when commitments have not yet been made. They also have a special advantage in that, in the event of protest, it is often their function to adjudicate between the views of protesters and their own proposals.
Broader Structural violence (#PB1935) Abuse of government power (#PC9104).
Related Bureaucracy as an organizational disease (#PD0460).
Aggravates Bureaucratic corruption (#PC0279).
Aggravated by Political apathy (#PC1917)
Failure of individuals to participate in social processes (#PF0749).

♦ PC2087 Destructive changes in ocean characteristics

Ocean current shift — Ocean temperature change — Ocean salinity change
Nature Several complex physical and chemical processes in the sea are capable of resulting in the catastrophic destruction of sea life. Such processes could also be triggered off by man–made events such as nuclear explosions, generation of electricity by transference of deep ocean water to the surface, or by the large scale diversion of rivers. The effects can take the form of: lowering of temperature due to unusually cold weather, lowering of salinity from exceptional river discharge, displacement of ocean current (such as the El Nino off the coast of Peru), vertical mixing (bringing low oxygen content water to the surface as in Walvis Bay, Namibia, or high hydrogen sulphide content water as in the Black Sea). Related phenomena may result in explosive growth of micro–organisms which concentrate substances toxic to some sea life. Other threats are in the form of urban and industrial development and destruction of highly productive coastal wetlands and reef areas; chemical and radioactive pollutants washed from the land, discharged and dumped into the ocean, or deposited from the atmosphere; uncontrolled exploitation of ocean resources; and mounting pressure on the world's fisheries.
Incidence Such changes may be of disastrous significance to economies highly dependent on the fishing industry, or to bird populations dependent on fish (such as guano birds in Peru). The impact of such activity on climate depends on its extent. Ocean temperature anomalies of a degree or larger produce large local changes in the fluxes of sensible and latent heat to the atmosphere, modifying climates even at great differences. Proposed large–scale diversion of rivers flowing into the Arctic could well destabilize the Arctic ice pack.
Broader Natural disasters (#PB1151) Anthropogenic climate change (#PC9717)
Natural environment degradation (#PB5250).
Related Obstacles to the utilization of coastal and deep sea water resources (#PF4767).
Aggravates Environmental hazards from fishing industry (#PD0743).

♦ PC2089 Intergovernmental suspicion

Mistrust between governments — Lack of confidence between governments — Lack of international cooperation due to personal mistrust
Broader Distrust (#PA8653).
Aggravates Government inaction (#PC3950) Lack of a world government (#PF4937)
Inefficient location of facilities of international organizations (#PE3538).
Aggravated by Abuse of confidentiality (#PF7633).

♦ PC2102 Family breakdown

Dependence on family breakdown — Breakdown of the family — Dismembered families — Family disintegration — Declining family values
Nature Families may breakdown as a result of marriage breakdown, or a break in the traditional structure of the family, nuclear or extended; it may also be indicated by a break in traditional family roles, such as a dominant father: submissive mother and children. Families also breakdown when destitute parents sell their children into slavery; or in a social welfare system when children may be taken from destitute parents and put into institutions. Homelessness and natural disasters, including death, may also split up families. Family breakdown may cause a loss of identity and severe adjustment problems.
Incidence Breakdown of the family due to marriage breakdown appears to be more frequent in developed countries, whereas that due to a break in traditional family structure is more marked in developing countries, where there has been cultural invasion; traditional family roles are changing more rapidly in developing countries with fast technological and economic change. Children are sold into slavery in Latin America, Africa and Asia.
Broader Deprivation (#PA0831) Social breakdown (#PB2496) Social disintegration (#PC3309).
Narrower Marital instability (#PD2103) State custody of deprived children (#PD0550)
Voluntary dissolution of the family (#PF4930)
Conflict concerning legal custody of children (#PF3252)
Disruption of family system in developing countries (#PD1482).
Aggravates Alienation (#PA3545) Prostitution (#PD0693)
Alcohol abuse (#PD0153) Permissiveness (#PF1252)
Human suffering (#PB5955) Orphan children (#PD7046)
Street children (#PD5980) Attempted suicide (#PE4878)
Abandoned children (#PD5734) Juvenile delinquency (#PC0212)
Maladjusted children (#PD0586) Maternal deprivation (#PC0981)
Loneliness in old age (#PD0633) Illegitimate children (#PC1874)
Annulment of adoption (#PF3281) Stress in human beings (#PC1648)
Single parent families (#PD2681) Inadequate laws of adoption (#PD0590)
Mental disorders of the aged (#PD0919)
Inadequate sense of personal identity (#PF1934).
Aggravated by Crack (#PE2123) Incest (#PF2148)
Divorce (#PF2100) Adultery (#PF2314)
Hedonism (#PF2277) Lesbianism (#PF2640)
Promiscuity (#PC0745) Selfishness (#PA7211)
Homelessness (#PB2150) Cohabitation (#PF3278)
Value erosion (#PA1782) Homosexuality (#PF3242)
Family poverty (#PC0999) Marital stress (#PD0518)
Induced abortion (#PD0158) Domestic quarrels (#PE4021)
Male homosexuality (#PF1390) Chaotic households (#PG0499)
Displaced children (#PD5136) Sexual unfulfilment (#PF3260)
Dependency of women (#PC3426) Natural human abortion (#PD0173)
Family disorganization (#PC2151) Predominance of fast food (#PF5940)
Generation communication gap (#PF0756) Separation under marriage law (#PF3251)
Poverty in developing countries (#PC0149) Negative effects of the nuclear family (#PF0129)
Family poverty in industrialized countries (#PD1998)
Vulnerability of marriage as an institution (#PF1870)
Adjustment difficulties of new urban families (#PF1503)
Fragmented patterns of extended family relationships (#PF1509)
Narrow legal definition of the family in developing countries (#PD1501)
Inadequate living and working conditions of immigrant labourers in industrialized countries (#PD3427).
Reduced by Traditionalism (#PF2676).

♦ PC2114 Destruction of cultural heritage

Nature The illegal removal of art from indigenous areas of developing countries continues, and in some cases it is the legal but unconscionable removal by local and national governments themselves that cause such destruction. Objects are taken from urban sites, villages, jungles, and even from underwater in oceans, rivers, and lakes.
Incidence More than 3,600,000 books were destroyed by fire and water in February, 1988 when the library of the Academy of Sciences of the Soviet Union at Leningrad burned down. Many pre–revolutionary volumes cannot be replaced.
Claim Mankind's heritage of culture extends back to 'Lucy', the East African earliest ancestor of mankind, and thereby covers the anthropology if not the archaeology of several million years. The culture heritage is diverse. It includes monuments and artefacts in all their range; it is the technology of primitive fire and weapon making, up to today's computers and scientific wonders; it is language and literature, behaviour, mines, and civilization. It is the human spirit. It is all these, but it is perishable. The material cultural heritage is breaking up, eroding, crumbling. The immaterial is perpetually changing and what went before goes unrecorded. Cultural objects are traded in for cash value with little regard for preservation, and laws for preservation and cultural documentation are inadequate.
Broader National disintegration (#PB3384).
Narrower Unwritten language (#PF3470) Cultural deprivation (#PC1351)
Theft of works of art (#PE0323) Vulnerability of sacred sites (#PD6128)
Vibration damage to cultural artefacts (#PE8162)
Endangered monuments and historic sites (#PD0253)
Abusive exploitation of cultural heritage (#PC7605)
Archaeological and anthropological looting (#PD1823)
Inadequate cultural integration of immigrants (#PC1532)
Destruction of historic documents and public archives (#PD0172)
Damage to cultural artefacts by environmental pollution (#PD2478)
Inadequate protection and preservation of cultural property (#PF7542).
Related Landscape disfigurement (#PC2122) Lack of racial identity (#PF0684)
Inter–cultural misunderstanding (#PF3340) Restriction of freedom of expression (#PC2162)
Endangered tribes and indigenous peoples (#PC0720).
Aggravates Ethnic segregation (#PC3315) Homogenization of cultures (#PB1071)
Ignorance of cultural heritage (#PF1985).
Aggravated by Vandalism (#PD1350) Ethnocide (#PC1328)
Cosmopolitanism (#PF8331) Cultural invasion (#PC2548)
Cultural arrogance (#PF5178) Cultural corruption (#PC2913)
Cultural illiteracy (#PD2041) Forced assimilation (#PC3293)
Anti–intellectualism (#PF1929) Cultural fragmentation (#PF0536)
Ill–considered missionary activity (#PF3370) Inadequate appreciation of culture (#PF3408)
Disintegration of organized religion (#PD3423) Threatened and vulnerable minorities (#PC3295)
Underprivileged linguistic minorities (#PC3324)
Monolingualism in a multi–cultural setting (#PD2695)
Discrimination against indigenous populations (#PC0352)
Trafficking in children for sexual exploitation (#PE6613)
Untransposed significance of cultural tradition (#PF1373)
Westernization of traditional modes of life in developing countries (#PF6592).
Reduced by Superstition (#PA0430) Cultural barriers (#PB2331)
Lack of assimilation (#PF2132) Religious discrimination (#PC1455).

♦ PC2122 Landscape disfigurement

Dependence on landscape disfigurement — Untended landscape areas — Degradation of culturally important landscapes
Nature In past centuries, landscapes developed slowly. Natural forces tended to keep a balance between plants, men and other species. In recent decades the balance has been upset by the impact of man, his machines and technologies, and the growing population. Industrialized societies expand rapidly, making heavy demands on resources. Fragile geological forms and living plant and animal communities, evolved over many thousands of years, can be destroyed very quickly. Once disintegration of a landscape begins it is difficult to reverse, leading to ugliness, wastelands, and erosion of fertile land (as a result of over–grazing, unwise cropping, mining, tourist facilities construction, and ill–considered deforestation). Culturally important landscapes include natural, modified, cultivated or built environments (separately or in combination), that symbolize a particular relationship between a society and the natural world. They may range from sacred groves and other sacred sites (as in South East Asia) to tracts of moorland (as in Europe) and alpine landscapes. Such landscapes are vulnerable to inappropriate development projects.
Incidence Many of the most beautiful landscapes in the UK have been seriously damaged by farming and development. Natural pastureland rich in wildlife has been replaced by cereal crops. The degradation has been accelerated by the large number of visitors.
Broader Ugliness (#PA7240) Natural environment degradation (#PB5250).
Narrower Dereliction (#PE5715) Scrapped automobiles (#PE5339)
Derelict industrial wastelands (#PE6005) Destruction of the countryside (#PE3914)
Non–biodegradable plastic waste (#PD1180) Degradation of natural seascape (#PJ4677)
Destruction of hedges and hedgerow trees (#PD1642)
Landscape disfigurement from open–cast mining (#PD1637)
Unaesthetic location of power transmission lines (#PD1665)
Unaesthetic location of advertising hoardings and billboards (#PE2156)
Natural environmental degradation from recreation and tourism (#PE6920).
Related Lack of community planning (#PF2605) Destruction of cultural heritage (#PC2114)
Vulnerability of ecosystem niches (#PC5773) Social inadequacy of large buildings (#PF6194).
Aggravates Hazards to plants (#PD5706) Vulnerability of plants and crops (#PD5730).
Aggravated by Litter (#PD2541) Deforestation (#PC1366)
Uncontrolled urban development (#PC0442) Uncontrolled industrialization (#PB1845)
Hazardous locations for nuclear power plants (#PD2718).

♦ PC2129 Underprivileged religious minorities

Denial of rights to religious minorities — Discrimination against sects
Nature Religious minorities may be persecuted or discriminated against by the religious majority group, resulting in lack of equal opportunity and poor living conditions.
Such religious minorities include sects and religious splinter groups which may be ostracized or persecuted, as well as minority groupings of major religions which occur in countries where another religion predominates. Where religion has been closely linked with politics or economic power, or where the practices and beliefs of a religious community are felt to be profane by the

community at large, religious minorities have been forced to emigrate, (for example, the Jews, Huguenots, Pilgrim Fathers), or made stateless if no other country will accept them. Discrimination and segregation may take place as a form of ostracism. Certain religious groups may be banned from attending the schools of another group or may insist on separate education for their children, which may not be the best preparation for employment opportunities in the society where they are living. Discrimination may occur in employment and housing. Immigrants bring their religion with them, even if this was not the reason for leaving their native country, and as such set up further frictions with the indigenous community. The protection of religious minorities may be used as a pretext for war by a country where the religion in question predominates.
Broader Underprivileged minorities (#PC3424) Denial of rights of minorities (#PC8999).
Related Underprivileged racial minorities (#PC0805)
Underprivileged linguistic minorities (#PC3324)
Underprivileged ideological minorities (#PC3325).
Aggravates Racial conflict (#PC3684) Religious conflict (#PC3292)
Unequal distribution of social services (#PC3437)
Peoples perceiving themselves as specially chosen (#PF4548).
Aggravated by Elitism (#PA1387) Persecution (#PB7709)
Social injustice (#PC0797) Social inequality (#PB0514)
Religious prejudice (#PD4365) Social fragmentation (#PF1324)
Segregation in education (#PD3441) Religious discrimination (#PC1455)
Inequality before the law (#PC1268) Inequality of opportunity (#PC3435)
Discrimination against minorities (#PC0582)
Religious discrimination in the administration of justice (#PE0168).

♦ PC2130 Social alienation

Refs Schaff, Adam *Alienation as a Social Phenomenon* (1980).
Broader Conflict (#PA0298) Alienation (#PA3545) Forced assimilation (#PC3293).
Narrower Loneliness in single people (#PD4392).
Related Psychological alienation (#PB0147).
Aggravates Hooliganism (#PD1109) Social apathy (#PC3412)
Mental disorders (#PD9131).
Aggravated by Prejudice (#PA2173) Decadence (#PB2542)
Agnosticism (#PF2333) Segregation (#PC0031)
Political apathy (#PC1917) Social discrimination (#PC1864)
Personal unpopularity (#PF4641) Inflexible social structure (#PB1997)
Lack of community participation (#PF3307) Limited access to social benefits (#PF1303)
Failure of individuals to participate in social processes (#PF0749).

♦ PC2140 Espionage

Spying
Nature The covert collection of information about a rival or an ally may concern social, military, industrial or political data and may occur between different countries and ideologies or between rival firms or rival political groups in the same country. Highly sophisticated technological means are often used.
Background Some of the most dramatic spy stories since World War II concern: the Soviet Union's penetration of British secret services (Burgess, Maclean, Philby, Blunt and others), and in the USA, with the U-2 and Pueblo missions and the Rosenberg case; and Israel's Mossad intelligence operations. Jonathan Jay Pollard spied on the US for Israel in the early 1980's. Marine Sergeant Clayton Lonetree helped the Soviets steal classified materials from diplomatic offices in Vienna and Moscow. A leading West German counter-espionage agent, Hans Joachim Tiedge, and three other West Germans in sensitive positions defected to East Germany. Vitally S. Yurchenko, a soviet defector to the US, exposed a number of spies in Europe and the US before redefecting to the USSR. A former analyst with the CIA, Larry Wu-Tai Chin was arrested for spying for China.
Incidence In the USA in 1990, a record budget of $30 billion was called for classified intelligence, of which more than half the funds would be allocated to intelligence gathering in Eastern Europe despite the moves towards democracy. Same time the KGB and other East block intelligence services have increased their efforts to obtain Western technology for modernizing the Soviet economy and information about Western intentions.
Increasingly the reasons for spying for a foreign power are financial and revenge rather than ideological. Espionage, if not global due to its high costs, or the lack of interest in some areas, exists both in the North and the South, and is mounted by countries that may be called developed or socialist. Of a peculiar nature unto themselves are systems of espionage orchestrated in the Vatican and by some Protestant secıs based in the USA.
Counter-claim Espionage serves a peace-keeping function by letting antagonists know what each is up to. This eliminates guess-work, over-anxiety and over-reaction. In the US Rosenberg case, where atomic secrets were passed to the Soviet Union, it was the contention of the defendants that they were serving peace by making nuclear weaponry possible elsewhere than in the USA alone. Some supporters contended that the only other strategic alternative for the Soviets was a preemptive strike against the USA after developing enhanced chemical-biological weapons. In the market place, where intense competition has resulted in industrial espionage, consumers often benefit from 'copying', which affords them lower prices while at the same time it distributes technological advances made but selfishly wished to be retained as proprietary items.
Refs Blackstock, Paul W and Schaf, Frank *Intelligence, Espionage, Counterespionage and Covert Operations* (1978); Buranelli, Vincent and Buranelli, Nan *Spy-Counterspy* (1982); Merrick, Lamar *The Mysterious, Conflicting Operations Carried Out in Both Europe and the United States by Spies and Secret Service* (1986).
Broader Subversive activities (#PD0557) Aggressive foreign policy (#PC4667)
International non-military conflict (#PF3100).
Narrower Counter-espionage (#PD2923) Military espionage (#PD2922)
Industrial espionage (#PC2921) Sociological espionage (#PF2924)
Treachery by double agents (#PE1578) Espionage in domestic politics (#PD1787)
International political espionage (#PC1868) Covert intelligence agency operations (#PD4501)
Mishandling national security information (#PE3749)
Misuse of satellite surveillance by governments (#PF3701)
Misuse of electronic surveillance by governments (#PD2930)
Revealing national security information to a foreign power (#PE4343).
Related Military secrecy (#PC1144)
Unaccountable government intelligence agencies (#PF9184)
Crimes related to national security information (#PE3997).
Aggravates Official secrecy (#PC1812) Political indoctrination (#PD1624)
Vulnerability of computer systems (#PE8542)
Inadequate protection of war correspondents (#PE3034).
Aggravated by Errant nationals (#PE0812) Unethical practices by employees (#PD4334)
Abuse of international cultural, diplomatic and commercial exchanges (#PF3099).

♦ PC2151 Family disorganization

Nature Family disorganization implies maladjustment, malfunctioning, psychological decay, and the existence of family problems. Whether the family is taken to mean a nuclear, extended or single parent family, the maladjustment of family life to prevailing conditions may result in emotional stress, crime, juvenile delinquency, promiscuity, poverty, and (ultimately) family breakdown. It may be the result of cultural invasion (primitive tribes and developing countries) or too rapid technical and economic change without corresponding social change.
Broader Disorganization (#PF4487) Social disintegration (#PC3309).
Narrower Marital instability (#PD2103) Maladjusted children (#PD0586)
Single parent families (#PD2681).
Aggravates Promiscuity (#PC0745) Family breakdown (#PC2102)
Domestic quarrels (#PE4021) Juvenile delinquency (#PC0212)
Generation communication gap (#PF0756)
Inadequate sense of personal identity (#PF1934).
Aggravated by Alienation (#PA3545) Illegal immigration (#PD1928)
Adjustment difficulties of new urban families (#PF1503)
Narrow legal definition of the family in developing countries (#PD1501).
Reduced by Traditionalism (#PF2676).

♦ PC2153 Eugenics

Nature Practices leading to human racial improvement based on judicious mating, ensuring that some attributes (selected as superior) prevail, and taking measures to prevent the dilution of the improved stock by those carrying attributes identified as qualitatively inferior.
Background At least since the time of Plato, individuals whose natural endowments seemed to equip them less well for success in society, e.g., the chronically ill or the mental disabled, have been discouraged by custom and law from having children. The desire for both negative eugenics (discouraging parenthood among the inferior) and positive eugenics (rewarding child-bearing among the most healthy or intelligent) has been around for a long time. It was only in the late 19th century that programs based on systematic, current scientific principles were suggested. In 1883 the British physician Sir Francis Galton coined the term "eugenics" to describe the study aimed as improving the human race by judiciously matching parents with superior traits. Galton focused on intelligence and his followers proposed that criminality, poverty, alcoholism, prostitution and other undesirable trait could be eliminated through eugenics. In the U.S.A. from 1905 until the 1930's a variety of state laws were enacted to prevent the "feeble minded", epileptics, the mentally ill, the retarded and many other types of people with undesirable traits from reproducing. People of different racial backgrounds were forbidden to marry, certain ethnic groups were not allowed to immigrate. In 1933 Adolf Hitler promulgated the Eugenic Sterilization Law in Germany.
Incidence Governments continue to adopt policies with eugenic objectives. In Singapore, the government has actively encouraged the academically educated members of the population to increase the number of their children. In Romania under Ceausescu women were obliged to have extra children, of whom some were separated from their parents at a very early age for a special upbringing within state orphanages.
Claim Several objections have been raised to the principles of eugenics. Its too zealous application holds serious risks of abuse, which have been vividly illustrated in the racist ideology of the Nazis, who had their own criteria of what was genetically suitable. Even without a racial basis, eugenics is confronted with the problem of determining who is fit and who is not. The human genetic mechanism is extremely complex, and normal and perfectly healthy parents, no less than those who are ostensibly unfit, can carry defective transmissible genes. Any attempt to make biology serve as a basis for elitist theories is rooted in a fundamental misconception, whether the 'elite' consists of certain individuals within each group, or of certain groups in themselves.
Counter-claim 1. Eugenics implies race improvement as a means of helping the evolution of humanity. Eugenics may be used to maintain and improve the genetic potentialities of a particular race or class. To achieve this, eugenics may concern itself with the elimination of the unfit, that is, those who are genetically unsuitable; including their sterilization, so that they cease to propagate. This would ensure that serious mental and physical defects would not be transmitted to future generations, and would prevent the social burden, responsibility, expense and manpower that would otherwise be needed to care for their progeny. Eugenicists feel that the 'absurd' UN Declaration of Human Rights according to which every married couple has the basic right to have a child and 'found a family' can be carried to the point of endangering the genetic pool of humanity.
2. Eugenic is the application of scientific knowledge to the improvement of humanity through evolution. In fact, eugenics is already practised by most nations through a variety of medical and social practices. Parents with hereditary defects in their families go to genetic counsellors with the result of avoiding having children when there is a high risk of birth defects. Birth control method have allowed women to "select" the best man to have a child by. Sperm banks screen donors for appropriate socio-economic and traits. Many dating agencies match partners on a variety of criteria much of which can be traced to genetic similarities. In some countries there are rewards for appropriate parents to have more children. Many birth control programs are aimed at specific population segments. Singapore has actively encouraged university graduates to marry and have children and other nations passively do the same.
Refs Parrish, Ruth G *Eugenics* (1988); Ramsey, Paul *Fabricated Man* (1970); Rosenberg, Charles *Problems in Eugenics* (1985).
Broader Irresponsible scientific and technological activity (#PC1153).
Narrower Forced motherhood (#PE5919) Government-enforced maternity (#PE3601).
Related Anti-science (#PF2685).
Aggravates Extermination (#PJ2839) Social injustice (#PC0797)
Social inequality (#PB0514) Economic repression (#PC8471)
Genetic defects and diseases (#PD2389) Inhuman methods of conception (#PE8634)
Exploitative property development (#PD8492) Denial of right to benefits of science (#PF6077)
Unlimited practice of human embryo storage (#PE5623).
Aggravated by Racism (#PB1047) Elitism (#PA1387)
Segregation (#PC0031) Racial exploitation (#PC3334)
Racial discrimination (#PC0006)
Irresponsibility towards future generations (#PF9455)
Inadequacy and insensitivity of intelligence testing (#PD1975)
Wastage of highly skilled personnel in the routine maintenance of complex systems (#PE1396).
Reduced by Lack of eugenic measures (#PD1091).

♦ PC2157 Economic discrimination

Denial of right to material well-being because of discrimination
Nature A country, or group within a country, is discriminated against whenever earnings are shown to be lower than that warranted by abilities.
Broader Discrimination (#PA0833).
Narrower Blacklisting (#PE0189)
Economic discrimination in the administration of justice (#PE1399).
Related Discrimination against minorities (#PC0582)
Denial of right to pursue spiritual well-being because of discrimination (#PE7310).
Aggravated by Economic prejudice (#PG4371)
Denial of right of equal pay for equal work (#PD1977).

♦ PC2162 Restriction of freedom of expression

Dependence on restriction of freedom of expression — Denial of right to freedom of expression — Denial of freedom of speech — Denial of right to communicate — Suppression of public debate
Nature Limiting freedom of expression may lead to exploitation, indoctrination, apathy, alienation and general stagnation as a result of inequality and injustice. It may serve to strengthen political dictatorship and government control or moralistic repression. Methods include censorship; the

refusal of licence (where it is necessary); injunctions; damages; denial of distribution and news access; restrictive taxation, subsidies and importation laws; interference; copyright; monopoly; commercialism; scarcity of resources; curtailment by governments of access to newsprint; corruption; and public opinion. Restrictions may be exercised by the government, private firms and authorities, or by the public.
Refs Vaina, Lucia and Hintikka, J (Eds) *Cognitive Constraints on Communication* (1983).
Broader Denial of rights (#PB5405).
Narrower Censorship (#PC0067) Lack of freedom of the press (#PE8951) Misappropriation of cultural property (#PE6074) Denial of freedom of expression in clothing (#PE5409) Denial of freedom of expression and thought in communist systems (#PC3174).
Related Uncontrolled media (#PD0040) Denial of right to life (#PD4234) Violation of civil rights (#PC5285) Denial of right to liberty (#PF0705) Denial of freedom of opinion (#PD7219) Destruction of cultural heritage (#PC2114) Restrictions on freedom of worship (#PD5105) Restrictions on freedom of information (#PC0185) Denial of right to correct misinformation (#PE7349) Restrictions on international freedom of information (#PC0931).
Aggravates Cultural stagnation (#PC8269) Compulsory indoctrination (#PD3097) Suppression of information (#PD9146) Undemocratic political organization (#PC1015) Insufficient communications systems (#PF2350).
Aggravated by Conservatism (#PF2160) Traditionalism (#PF2676) Authoritarianism (#PB1638) Newspaper monopoly (#PE0246) Monopoly of the media (#PD3101) Harassment of the media (#PD0160) Harassment of journalists (#PD3036) Radio frequency interference (#PD2045) Erosion of journalistic immunity (#PD3035) Refusal to grant licences to media (#PF3079) Inadequate protection of war correspondents (#PE3034) Restriction of access to news distribution media (#PF3082).
Reduces Libel (#PD3022) Slander (#PD3023).

♦ PC2163 Unequal access to education

Nature Education is much less accessible to children in developing countries than it is to children in developed countries. It is also less accessible to rural than urban populations, and less accessible to females than to males.
Incidence With only about one–third the total population and only one–quarter of the young people in the world, industrialized countries spend ten times more money on education than do developing countries; and this difference is increasing. About half of the world's school enrolments are recorded in developed countries, where those of school age (5–14) form only one–quarter of the world total for their age group. Conversely, the developing countries, twice as populous and containing three times as many children and young people as the developed countries, have hardly more than half of the world's school pupils. In Europe, USSR and North America, the increase in primary and secondary school enrolments run parallel to the increase in the population of young people between 5 and 19 years of age. In developing countries, the population in that age group increases by millions more than the increases in school enrolments.
Figures on higher education show that in North America about one in two people who have reached post–secondary age are in fact enrolled at that level. The attendance rate is markedly lower in Europe but remains considerable at about one in seven, whereas in Latin America and Asia this proportion is greatly reduced (about ten times less), falls sharply again among Arab States (fourteen times less) and is infinitesimal by comparison in Africa (thirty–four times less). There are nearly two–and–a–half times as many students in the European and North American higher educational systems than in all the other regions of the world combined. In North America, one student in eight attends a higher educational establishment, and the ratio in Europe is 1 in 20. Corresponding ratios in developing regions are: 1 in 38 in Asia, 1 in 45 in the Arab States, 1 in 49 in Latin America, and 1 in 90 in Africa. The favoured and less favoured nations reveal a similar disparity in the number of teachers, even without taking into account differences between primary–school teachers' professional training levels. Developing countries have 50 or 60 million more pupils than Europe and North America but approximately the same number of teachers. The primary–school situation, by region, is as follows: Europe and USSR, one teacher for 25 pupils; North America, 1 for 26; Latin America 1 for 32; Asia 1 for 36; Arab States 1 for 38; Africa, 1 for 40. Again, differences are marked in education for girls and women. In North America, Europe and Latin America, school enrolments of boys and girls at primary and secondary levels are approximately equal. But grouping Africa, Asia and the Arab States together, we find 50 per cent more boys than girls in primary schools, and 100 per cent more in secondary schools. World illiteracy figures show further the extent to which women are at a disadvantage. 80 percent of the adult population in low income countries are illiterate; of these over–800 million people, nearly 500 million are female.
Broader Obstacles to education (#PF4852) Unexercised responsibility for external relations (#PF6505).
Narrower Unequal school distribution (#PJ9458) Lack of vocational teachers (#PJ9603) Lack of school transportation (#PJ7849) Lagging training in social skills (#PF8085) Inaccessible educational facilities (#PD9051) Restrictions on early apprenticeship (#PJ7946) Unperceived educational opportunities (#PJ9762) Inequality of access to education within countries (#PC1896).
Aggravates Inequality in education (#PC3434) Denial of right to education (#PD8102) Debilitating education images (#PF8126) Lack of skilled manpower in rural areas of developing countries (#PE5170).
Aggravated by Segregation (#PC0031) Unbudgeted child education (#PG1363) Unpublished training opportunities (#PJ9985) Unavailability of scholarship funds for students (#PE3569) Limited availability of education in rural areas (#PF3575).

♦ PC2169 Militarism

Nature Militarism is a policy or doctrine or a system of power relationships that values war and accords primacy in state and society to the armed forces. It exalts the application of violence and the authoritarian structure of the military establishment. Militarism ceases to be simply an attitude when actions such as threats by the military or their regimes are made. Military coups d'Etat, juntas and the like may impose martial law or illegally suspend civil liberties. Militarism in tribal society takes the form of warrior prowess and hero worship. In more sophisticated society it may take the form of military government or effective military control or the encouragement to spend lavishly on defence or to engage in war.
Counter–claim Military expertise may be inadequately represented in the foreign policy process, since threats or actions of force may sometimes be required, if only to indicate defensive capabilities.
Refs Art, Robert J and Waltz, Kenneth N *The Use of Force* (1988); Fidel, Kenneth *Militarism in Developing Countries* (1975); McClellan, George B *The Armies of Europe* (1976); Sallantin, Xavier *L'invariant des jeux militaires, économiques et politiques* (1976).
Broader Aggression (#PA0587) Authoritarianism (#PB1638) Undemocratic political organization (#PC1015).
Narrower Martial law (#PD2637) Military influence (#PD3385) Military–industrial complex in capitalist systems (#PC3191) Absorption of manpower resources by military activities (#PF0780) Military–industrial–governmental complex in communist systems (#PD1330).
Related Pacifism (#PF0010).
Aggravates War (#PB0593) Hero worship (#PF2650) Militarization (#PD1897) Human violence (#PA0429) Minority control (#PF2375) Military espionage (#PD2922) Military government (#PC0698) Political dictatorship (#PC0845) International insecurity (#PB0009).
Aggravated by Tribalism (#PC1910) Nationalism (#PB0534) Intimidation (#PB1992) Military secrecy (#PC1144) Guerrilla warfare (#PC1738) Ideological conflict (#PF3388) Overthrow of government (#PD1964) Violent political revolution (#PD3230) Double standards in morality (#PF5225) International non–military conflict (#PF3100) Disruption of development by tribal warfare (#PD2191).

♦ PC2172 Nationalization of foreign investments

Risk of nationalization of overseas investments
Nature Nationalization of foreign investments may take several forms: the assets of the nationalized companies may be transferred to the state; or only the share capital may be transferred, leaving the company to continue operations under state controls. The process may be undertaken in an arbitrary manner with little or no compensation of foreign investors, or through compensation in non–convertible currencies. Typically, nationalization is applied to a particular foreign–owned enterprise if domestically–owned enterprises in the same sector do not exist.
Claim Nationalization is usually associated with attempts to implement socialist or marxist theories of government.
Counter–claim Nationalization may be undertaken to ensure state control of enterprises or industries of major importance to the health of the economy, particularly where the control of such enterprises is of political and social importance. In many developing countries, such enterprises are concerned with the exploitation of irreplaceable resources whose value on the international market may be subject to wide price fluctuations. The policies of a foreign–owned corporation with regard to the profitable sale of such commodities may be based on different criteria than those of the national government, particularly concerned by their domestic repercussions.
Refs Monsen, R J and Walters, K D *Nationalized Companies* (1983).
Broader Risk (#PF7580) World anarchy (#PF2071) Excessive government control (#PF0304).
Aggravates Socialism (#PC0115) Communism (#PC0369) Risk of capital investment (#PF6572) Defaults on international loans (#PD3053) Discrimination against foreign companies (#PD6417).
Aggravated by Political upheavals (#PC7660).

♦ PC2175 Psychological warfare

Covert psychological warfare operations — Clandestine psychological warfare
Nature A war of nerves, of misinformation, of exaggeration, but also of communication of truth, accompanies a war of armies and weapons. The targets are not fortifications or lines of men, but minds, and these may be those of the enemy's soldiers, its military leadership, its civilian leaders, or its populace, including children.
Incidence The forms of psychological warfare are various. They include the instilling of fear by the demonstration or use of terrible weapons even though in some cases these are not in production or are experimental. Leaflets can be dropped behind enemy lines with misinformation or facts on how their side is losing the war on other fronts. Radio broadcasts to the enemy homeland or soldiers in the field may use traitors to urge pacifism or surrender and to demoralize the civilian population. The authorities may start rumours to weaken the opposition or strengthen the unity of supporters. They may compare the enemy's high standard of civilian comforts during the conflict to the soldiers' own privations. At home, governments make propaganda statements and the media are exploited to misinform and exaggerate information concerning the progress of the war. Psychologists are employed to develop propaganda and other techniques to influence mass behaviour.
Refs Linebarger, Paul M *Psychological Warfare.*
Broader War (#PB0593) Offences against the peace and security of mankind (#PC6239).
Narrower Brainwashing of prisoners of war (#PD1652) Promotion of negative images of opponents (#PF4133) Deliberate misrepresentation in educational materials (#PF1183).
Related Propaganda (#PF1878) Psychic warfare (#PF4866) Psychotronic warfare (#PD7986) Covert smear campaigns by government (#PD7171) Psychic interference in decision–making (#PF0508).

♦ PC2178 Discriminatory professionalism

Professional discrimination
Nature The development of professional status, methods, character and standards which conform to a pattern for a given profession in a given country is a form of elitism, excluding newcomers on the basis of accepted (implied high) standards, whereas the real reason may be discrimination against another social group. Professionalism tends towards complacency and apathy in the face of new ideas or methods. Reverence for 'professional opinion' and the closed–shop aspect of professional organizations discourages individual initiative and may exploit the public with high consultancy fees. In sports or pastimes, professionalism discourages amateur participation.
Incidence Professionalism is particularly a phenomenon of capitalist countries, but it may exist in a disguised form in socialist countries, such as in sports where national prestige is at stake).
Broader Elitism (#PA1387) Discrimination (#PA0833).
Narrower Professional discrimination between educators (#PE0810).
Related Lack of professional standards (#PF3411) Restrictive trade union practices (#PD8146).
Aggravates Amateurism (#PG4384) Ethnic discrimination (#PC3686) Prohibitive cost of living (#PF1238) Lack of individual initiative (#PG4385) Property and occupational discrimination in politics (#PD3218).
Aggravated by Grey lies (#PF3098) Inappropriate selection and examination procedures in education (#PF1266).

♦ PC2181 Political injustice

Dependence on political injustice
Nature Unjust practice in politics includes all kinds of political discrimination, unequal representation and unjust election administration. Crimes of a political nature include war crimes, corruption, subversive activities, treason and repression. Political injustice may lead to alienation, apathy and stagnation or political instability, political or national disintegration, revolution or foreign intervention.
Broader Injustice (#PA6486).
Narrower Unfair elections (#PC2649) Political prisoners (#PC0562) Refusal to grant amnesty (#PF0182) Unjust electoral campaigns (#PD2919) Politically motivated arrests (#PD9349) Political prisoners in mental institutes (#PE4430).
Related Social injustice (#PC0797) Military secrecy (#PC1144) Social inequality (#PB0514) Unjust election timing (#PD2907) Corruption of the judiciary (#PD4194) Political party manipulation of elections (#PD2906) Bias in selection of political candidates (#PD2931).
Aggravates Political conflict (#PC0368) Political inequality (#PC3425) Political instability (#PC2677) Unequal political representation (#PC0655).

PC2181

Aggravated by Power politics (#PB3202) Political apathy (#PC1917)
Undemocratic political organization (#PC1015)
Abusive distribution of political patronage (#PF8535).

♦ PC2194 Capitalist speculation

Speculation in capitalism

Nature Speculative investment within the capitalist system may enrich the investors and the financial community, but in so doing it creates economic and political instability with negative repercussions on the welfare of the majority. The evolution of capitalism, in first creating an artificial and non–productive elite class with regard to the ownership of the means of production, tends towards an even greater abstraction of this class from the production process. Speculation may involve overseas and colonial holdings. The double role of commodities and money as exchange value as well as real value reaches its peak in speculation which may alter the exchange value drastically, according to artificially–created supply and demand. A direct result of speculation may be inflation.

Counter–claim In a world characterized by uncertainty, speculation is essential to the allocation of economic resources over time. There is no question of whether or not speculation should be permitted; the only economic issue is who will perform the service most effectively. The charge of 'over–speculation' is incorrectly framed. The issue is not one of amount but rather whether it is done well or poorly.

Broader Capitalism (#PC0564) Risk of capital investment (#PF6572)
Counterproductive capitalist investment financing (#PF3104).
Narrower Abuse of tax havens (#PE2370) Property speculation (#PD8202)
Commodity speculation (#PD9637) Speculation on money markets (#PD9489)
Vulnerability of stock markets (#PD5676) Speculation in developing countries (#PD1614)
Excessive concentration of business enterprises (#PD0071).
Related Protection of company ownership (#PF4432).
Aggravates Shortage (#PB8238) Leveraged buy–outs (#PE4963)
Cyclic business recessions (#PF1277) Speculative flight of capital (#PC1453).
Aggravated by Inflated art values (#PF7870) Uncertain status of monetary gold (#PF2342)
Lack of individualism in capitalist systems (#PD3106)
Abuse of science and technology in capitalism (#PE3105)
Disparity between share prices and underlying asset values (#PE3827).

♦ PC2201 Nuclear weapons testing

Thermonuclear weapons testing — Nuclear explosions in peacetime

Nature Nuclear explosions used to test new types of nuclear weapons cause pollution of the environment by radioactive materials, cause leukaemia, stimulate further rounds in the nuclear armaments race, encourage emulation of such developments by previously non–nuclear countries, and increase international tension.

Incidence The Threshold Treaty or USA–USSR Treaty on the Limitation of Underground Nuclear Weapon Tests limited testing to below the 150 kiloton yield level. This accord went into effect in 1976. The USA announced 14 underground tests in 1983 alone, but officially admitted through its Department of Energy, that it only announced the larger tests. Tests with 5 kiloton yield or less are kept secret. The figures for tests in 1980, 1981 and 1982 respectively are 14, 16 and 18 for the USA and 21, 21 and 21 for the USSR. USSR figures are obtained from seismic detectors outside the country, as all testing is secret.

Refs Cox, David and Goldblat, Jozef (Eds) *Nuclear Weapon Tests* (1988).

Broader Competitive development of new weapons (#PC0012)
Offences against the peace and security of mankind (#PC6239).
Narrower Nuclear explosions underground (#PE2095)
Government secrecy concerning nuclear weapons testing (#PF4450).
Related Environmental hazards constraining scientific research (#PF1789)
Environmental degradation through military activity during peace–time (#PE0736).
Aggravates Earthquakes (#PD0201) Radioactive fallout (#PC0314)
Pollution of groundwater (#PD2503).
Aggravated by Competitive acquisition of arms (#PC1258).

♦ PC2205 Agricultural wastes

Nature Agricultural practices may lead to pollution in the form of animal wastes, material eroded from land, plant nutrients, inorganic salts and minerals resulting from irrigation, herbicides and pesticides; to these may be added various infectious agents contained in wastes.

Incidence The total quantity of such wastes is large. In the USA, for instance, the production of animal wastes exceeds that of human wastes by a factor of at least five on a biochemical oxygen demand (BOD) basis, seven on a total nitrogen basis, and ten on a total solids basis. Most developing countries have economies based predominantly on agriculture. With agro–industrialization increasing during the past three decades, many countries find themselves with a growing surplus of agricultural wastes such as rice hull, jute stalk, groundnut shell, bagasse and coconut husk and pith. These materials are available in large quantities and are presenting serious problems of disposal.

Broader Environmental degradation (#PB6384) Unproductive use of resources (#PB8376).
Narrower Agricultural effluent (#PE8504) Unused livestock waste (#PG5411)
Misuse of grassland and rangeland (#PD5133).
Aggravates Dust (#PD1245) Zoonoses (#PD1770) Insect pests (#PC1630)
Soil pollution (#PC0058) Marine pollution (#PC1117) Malodorous fumes (#PD1413)
Sewage as a pollutant (#PD1414) Mycoplasmal arthritis (#PG2302)
Agricultural pollution (#PD0563) Environmental pollution (#PB1166)
Hazardous waste dumping (#PD1398) Hazards to human health (#PB4885)
Soil–transmitted diseases (#PD3699) Long–term shortage of water (#PC1173)
Human disease and disability (#PB1044) Eutrophication of lakes and rivers (#PD2257)
Damage by degradable organic matter (#PJ6128).

♦ PC2223 Decreasing genetic diversity in cultivated plants

Limited crop diversity — Limited crop variety

Nature Resistant, high–yield varieties of major food crops carefully crossfed and highly selected, are the success story of modern genetics. But the more successful a variety is, the more likely farmers and breeders are to choose its hybrid seed, and an entire society may come to depend on a few highly selected varieties. When a new or mutant form of pest or disease arrives to which the favoured crop has no inbred resistance, then entire crops may be decimated. Crop diseases often occur because crops are too narrowly based genetically to carry adequate resistance. In fact, disease losses have been shown to directly correlate with the degree of genetic homogeneity in some crops.

Incidence Attempts to combat disease with resistant varieties have resulted in an increase in the incidence of disease. This occurs because the pathogen is at least as variable as the host crop. The newly released resistant variety can actually select a strain of the pathogen virulent on it, thus removing itself from usefulness and necessitating a replacement. For example, pure–line varieties of oats have an average useful life expectancy of only about 5 years. Other crops also have a narrow genetic basis. A mere handful of parent wheat lines have contributed to the gene content of all the Australian wheats. A very substantial percentage of the intermediate wheat grass acreage in the USA is planted with strains that trace to a single introduction. The entire acreage of the grass, Digitarta decumbens, in Central America and the West Indies originated from a single clone introduced in 1940, and already it is necessary to spray pastures of this grass in Central America as a protection against aphids and fungal diseases. The situation has worsened considerably since hybridization involving single–gene resistance began, because now a single resistance gene can be quickly incorporated into crop varieties right across a continent. For example, various pure–line varieties of small grains were released in North America because they were highly resistant to prevalent races of rust fungi. Though these varieties resisted most rust races, they were susceptible to rare, virulent races which, though occurring in extremely small amounts in the rust population, could increase with sufficient rapidity in the absence of competition on a susceptible host to soon become devastating. Yet pure–line varieties of small grains continue to be released for commercial production.

Background For about 60 years attempts have been made to breed, systematically, new varieties resistant to disease, especially for cereal crops. But this has been done by minimizing, if not eliminating, diversity, and in some cases these efforts have worsened the situation and built up particular pathogens as more serious threats than they were originally.

Claim Varietal improvement programmes have so reduced the stabilizing genetic diversity of small grains that these crops are a ready substrate for one or more variants of the pathogen population.

Counter–claim The ideal way of controlling all diseases would be by means of resistant varieties. The more man can get crop plants themselves to resist diseases, the less work and worry he will have in controlling them. Moreover there are now no feasible methods of controlling some of the most destructive diseases except by means of resistant varieties.

Narrower Monoculture of crops (#PC3606).
Related Genetic inbreeding (#PD7465)
Individualistic retaining of local tradition (#PF1705)
Unproductive utilization of plantation space (#PF6455).
Aggravates Plant diseases (#PC0555) Pests of plants (#PC1627)
Vulnerability of plants and crops (#PD5730)
Vulnerability of world genetic resources (#PB4788).
Aggravated by Genetic defects and diseases (#PD2389)
Banned cultivation of plant species (#PE9773)
Inadequate plant genetic resources conservation (#PF3581)
Monopolization of agricultural genetic resources (#PE6788)
Unsustainable short–term improvements in agricultural productivity (#PE4331).

♦ PC2237 Destruction of rural subsistence economy

Nature The rural subsistence economy, the part of the economy from which until now the bulk of the rural population of developing countries produced its requirements, is being subjected to continual encroachments and destruction by capital–intensive export–agriculture, industrialization, and depredation of the environment, It is the destruction of this sector (a sector only partly integrated into the world economy and still partly autonomous) which is the root cause of the emigration of large parts of the rural population.

Claim Despite declarations by institutions such as the World Bank that the destruction of this sector should not run out of control, and the knowledge that it constitutes a source of cheap labour and the basis for political stability, nothing concrete has been done.

Broader Socio–economic poverty (#PB0388).
Related Declining breeds of domesticated animals (#PD6305).
Aggravated by Unsustainable rural development (#PD4537).

♦ PC2270 Diseases of metabolism

Metabolism disorders — Inborn errors of metabolism — Metabolic disease — Metabolic disorders

Nature In the healthy individual the many metabolic activities of the body are maintained in an optimum state by a multitude of regulating mechanisms. A disease is the morbid state resulting from a disturbance of this state, so any disease can be regarded as a metabolic disease. The term, however, is conventionally restricted to those disorders that are ascribable to primary disturbances in metabolic processes. The number of these is considerable. They can be classified as nutritional deficiency diseases, inborn errors of metabolism, endocrine disturbances, liver and renal disease, and miscellaneous metabolic diseases.

Refs Cardoso, Joel A *Metabolism with Inborn Errors* (1987); Cockburn, Forrester and Gitzelmann, Richard *Inborn Errors of Metabolism in Humans* (1982); Cornblath, Marvin and Schwartz, Robert *Disorders of Carbohydrate Metabolism in Infancy*; Holton, *Inherited Metabolic Disorders* (1987); Moore, M R, et al *Disorders of Porphyrin Metabolism* (1987); Plum, F *Brain Dysfunction in Metabolic Disorders* (1974); Randle, P J et al *Carbohydrate Metabolism and Its Disorders* (1981).

Broader Human disease and disability (#PB1044).
Narrower Gout (#PG3061) Diabetes (#PE0102) Acidosis (#PG4564)
Porphyria (#PG1855) Amyloidosis (#PG5192) Hypertension (#PE0585)
Hyperlipaemia (#PJ5192) Kidney disorders (#PE2053) Hyperalimentation (#PG5307)
Calcification of bone (#PG5433) Protein–energy malnutrition (#PD0339)
Familial periodic paralysis (#PG4558) Vitamin deficiencies in diet (#PD0715)
Disparities in calorie intake (#PD0446) Hepatolenticular degeneration (#PG4562)
Arteriosclerotic vascular disease (#PG4555)
Congenital disorders of lipid metabolism (#PE9816)
Congenital disorders of amino–acid metabolism (#PE9291)
Congenital disorders of carbohydrate metabolism (#PE4293).
Related Fibrosis (#PG4560) Gland disorders (#PD8301) Hemochromatosis (#PG4559)
Rheumatic diseases (#PE0873) Nutritional diseases (#PD0287).
Aggravates Ulcers (#PE2308) Scurvy (#PE2380) Rickets (#PG2295)
Obesity (#PE1177) Headache (#PE1974) Beriberi (#PE2185)
Pellagra (#PE2287) Cataract (#PE6817) Inanition (#PG4563)
Osteomalacia (#PG4566) Liver diseases (#PE1028) Night blindness (#PG4565)
Sleep disorders (#PE2197) Mental deficiency (#PC1587) Nutritional anaemia (#PD0321)
Human physical genetic abnormalities (#PD1618).
Aggravated by Trauma (#PD4571) Rubella (#PE0785)
Antimetabolites (#PG4574) Bacterial disease (#PD9094)
Nutritional deficiencies (#PC0382) Water–borne viral disease (#PG3399)
Genetic defects and diseases (#PD2389).

♦ PC2279 Disruption of animal migration and movement patterns

Nature Many animals migrate to different and often distant environments, usually on an annual cycle, and often for breeding purposes or to secure food supplies. A number of activities of man tend to disrupt such movement; examples are construction of fences, denial of use of island breeding grounds, and destruction of wetlands.

Broader Natural environment degradation (#PB5250).
Related Denial to animals of the right to conditions of life and liberty proper to their species (#PE6270).

♦ PC2297 Unreliability of equipment and machinery

Nature Reliability in equipment depends upon the quality of the initial research, its design, the quality of the materials used, the quality of the parts used in its construction, the external influences upon its operation, and the frequency and quality of maintenance. The continuing increase in the complexity of mechanical and electromechanical systems increases the probability of equipment failure.

Claim Quality control is uneven among the industrialized nations with some countries lagging well behind. In addition, there is an inadequate use of both the automatic, computerized safety features in installations, and the systems reliability measurements of machinery or equipment configurations.

Counter-claim Equipment and machinery are more reliable than their human operators. Industrial accident figures indicate this. The continuing increase in the complexity of industrial systems increases the probability of human failure.

Broader Chance (#PA6714) Inadequate research on problems (#PF1077).
Narrower Unreliability of weapons systems (#PF7801).
Aggravates Risk (#PF7580).
Aggravated by Industrial espionage (#PC2921) Failure of materials (#PD2638) Counterfeit machine parts (#PE5319) Defects in machinery design (#PE2462) Electronic equipment failure (#PD1475) Human errors and miscalculations (#PF3702) Deterioration in product quality (#PD1435) Inadequate equipment maintenance (#PD1565).

♦ PC2321 Dangerous animals

Aggressive animals — Violent animals

Nature Almost any animal can be dangerous at times, and an attack on man can result in death, severe shock, mutilation, or infection with an animal disease.

Large animals are especially dangerous when they are surprised or frightened. Domestic animals are no safer than wild ones: the bull is generally considered the most dangerous of domestic animals. Milder animals may inadvertently cause dangerous situations by, for example, straying onto roads. Animals trained to guard property pose a threat to those whose work obliges them to enter private property; even household pets can be dangerous if excited or frightened.

Smaller animals are equally dangerous. Parasites include insects, ticks, worms and protozoa. Snakes, spiders, scorpions, and lizards are often deadly, as are centipedes, insects, and some fish.

Broader Risk (#PF7580) Violence (#PA5414) Hazards to human health in the natural environment (#PC4777).
Narrower Sharks (#PJ3738) Harmful wildlife (#PC3151) Dangerous pet animals (#PE7175).
Related Prejudice against animals (#PC0507).
Aggravates Injuries (#PB0855) Hunting of animals (#PC2024) Human disease and disability (#PB1044).

♦ PC2326 Endangered species of insects

Nature The special function of many species in the stability of the ecosystem is not well known, nor is it known how many insect species are endangered by man's impact on the natural environment. Estimates of the number of species of insects vary from 700,000 to 2 million. They represent about 80 per cent of known animal life and are very widespread and adaptable. In the temperate zone for example, an average acre of soil may have over 4 million insects. Only a few thousand species are such as to affect man's activities in a manner which requires some form of control measures.

Incidence As an example of the extent of the threat, 25% of about 200 species butterfly are threatened, and in the last 20 years many populations have declined by at least 50%.

Claim Increasing habitat destruction is undoubtedly the most significant threat to many species of insects and results from forestry, agriculture, and industrial, urban and recreational development. The point has been reached where a code for collecting should be considered in the interests of conservation of the insect fauna.

Broader Endangered species of plants and animals (#PB1395).
Narrower Endangered species of butterfly (#PE3762) Endangered species of arthropoda (#PD3471).
Related Endangered species of fish (#PC1535) Endangered species of birds (#PD0332) Endangered species of plants (#PC0238) Endangered species of mammals (#PC1326) Endangered species of animals (#PC1713) Endangered species of reptiles (#PC0604).
Aggravated by Air pollution (#PC0119) Soil pollution (#PC0058).
Reduces Insect pests (#PC1630).

♦ PC2343 Organized crime

Yakuza — Mafia — Triads — Organized crime families and territories

Nature Organized crime can be segmented into four types: 1) the criminal gang, usually predatory mobile gangs involved in armed robbery, kidnapping and some kinds of drug trafficking; 2) the criminal syndicates cater to a specific segment of the population providing forbidden goods or services like drugs, sex and gambling; 3) criminal rackets extort money or business concessions from legitimate or illegitimate organizations including business and trade unions; and 4) criminal political machines. These types can overlap and any one criminal organization might be involved in one or all of these types Organized crime is difficult to eliminate for three main reasons: (1) Although individuals can be arrested and convicted, they can also be quickly replaced in the organization. (2) Although the criminal code penalizes individual acts or persons, is generally powerless to tackle the conspiracy of a permanent, sometimes international, well-structured organization. (3) As long as the public continues to demand certain services (such as gambling, drugs and prostitution) which are illegal or semi-legal, criminal entrepreneurs will find it profitable to employ the same techniques as industry in order to be profitable and efficient, which includes being efficient at avoiding prosecution. Techniques used include both the sound organization (usually on military lines with a chain of command up to the general staff) and the communications systems (international, national, local and field based) that effective operations require.

Incidence Crime syndicates and gangs have been associated with many cities, territories, and countries. Chicago, Marseilles, Corsica, New York, Sicily, London, and Hong Kong share the characteristics of dense population. Some organizations, especially the Italian-American Mafia and Cosa Nostra, have perhaps received more publicity than others because of American motion pictures and novels. However, organized crime is not restricted to these countries: examples are the Chinese Tong societies, the Japanese Yakuza, the Hell's Angels, the Cotronis of Canada, and the Medellin Cartel. Organized crime in the Soviet Union is growing; in 1988, an estimated 1200 gangs were operating Organized crime is responsible for the illegal trade in drugs, for an endless number of assassinations and abductions including that of USA labour leader James Hoffa, and for the corruption and attempted corruption of political life in many countries and at all levels, from local inspectors and mayors up to national leaders. Four of the largest trade unions in the USA, the International Brotherhood of Teamsters, Labourers International Union, Hotel Employees and Restaurant Employees International Union, and the International Longshoreman's Association, were cited in a report on Mafia controlled organizations. Members of the Mafia sub-culture usually live seemingly average, middle-class, respectable lives within the community; but this masquerade of respectability is a cover for some of the most heinous criminal activities – drug dealing, prostitution, extortion and gambling, leading in turn to murderous gang wars between so-called "families". This dangerous and insidious sub-culture will be eliminated only when the larger community no longer accepts its presence and ceases to give it protection by looking the other way. Regardless of its power and ability to generate fear, the Mafia could not exist without the complacency of the society at large It is estimated that each year in the USA, organized crime reduces the gross national product by $18.2 billion, reduces employment by more than 400,000, raises consumer prices about 0.3 per cent and reduces per capita income $77.22.

Refs Alexander, Shana *The Pizza Connection* (1988); Kelly, Robert J *Organized Crime* (1986); Posner, Gerald L *Warlords of Crime* (1988); Sifakis, Carl *The Mafia Encyclopedia* (1988); Sterling, Claire *The Mafia* (1990).

Broader Criminals (#PC7373).
Narrower Criminal gangs (#PD3837) International crime rings (#PG5512) Criminal investment in youth market (#PD5750).
Related Criminal subculture (#PE5508).
Aggravates Homicide (#PD2341) Prostitution (#PD0693) Gang warfare (#PD4843) Racketeering (#PE4914) Criminal threat (#PE4661) Corruptive crimes (#PD8679) Gambling and wagering (#PF2137) Animal fighting sports (#PE4893) Illicit drug trafficking (#PD0991) Transvestite prostitution (#PD4525) Unethical trade union practices (#PD4341) Unethical practices by employees (#PD4334) Corporation-sanctioned assassination (#PE6356) Unethical practices in transportation (#PD1012) Use of undue influence to obstruct the administration of justice (#PE8829).
Aggravated by Criminal conspiracy (#PD1767) Criminal association (#PE1178) Abuse of police power (#PC1142) Ineffective legislation against organized crime (#PE6699).

♦ PC2361 Political schism

Political dissent

Nature Fragmentation of political beliefs within a party, leading to factions and the development of new small political parties, brings political instability. It may also lead to political disintegration, repression, revolution, dictatorship or possible anarchy, foreign intervention, or social breakdown. Equally, such schism may result in a multi-party government with one party in power for a long period, such as has occurred in France.

Refs Degenhardt, Henry W *Political Dissent* (1983); Miller, Albert J *Confrontation, Conflict and Dissent* (1972); Moore, R I *The Origins of European Dissent* (1985).

Broader Schism (#PF3534) Political conflict (#PC0368).
Related Religious schism (#PF1939).
Aggravates Intolerance (#PF0860) World anarchy (#PF2071) Political instability (#PC2677) Political fragmentation (#PF3216) Political disintegration (#PC3204) Inadequate political integration (#PF3215) Political discrimination in politics (#PC3221) Weakness of multi-party parliamentary systems (#PF3214).
Aggravated by Ideological conflict (#PF3388).

♦ PC2362 Conflicting claims by states to territories

Disputed territories — Disputed islands

Nature Conflicting claims exist over certain territories. Should such territories become independent and apply for membership of international organizations, such as the United Nations, difficulties may arise if such claims have not been settled.

Incidence Previous examples of such claims are: Iraq's claim to Kuwait's territory; Morocco's claim against Spain over Ifni; Morocco and Mauritania against Spain over Spanish Sahara; Argentina against the United Kingdom over the Falkland (Malvina) Islands; Guatemala against the United Kingdom over British Honduras (Belize); Spain against the United Kingdom over Gibraltar; Ethiopia against Somalia over the French Territory of the Afars and Issas; Mauritius against the United Kingdom over Diego Garcia; Japan against the USSR over the islands of Etorofu, Kunashiri, and Shikotan; China, Vietnam, Taiwan, Malaysia and Philippines over the Spratly Islands in the South China Sea, and Turkey and Greece over Crete.

Broader Territorial disputes between states (#PC1888).
Narrower Boundary disputes between states (#PD2946).
Aggravates Non-viability of small states and territories (#PD0441) Domination of the world by territorially organized sovereign states (#PD0055).

♦ PC2367 Illegal movement across frontiers

Illegal political exit

Nature Movement by persons across internal boundaries or international frontiers may take place without legal process. The reasons may include the need for secrecy, which may apply to criminals; the need to flee another country with no time to meet legal requirements; and the desire to enter a country although permission will not be granted, either to leave or to enter. The intentions of movement may allow return, or they may be one way in the case of illegal refugees or migrants. Legal obstruction to movement across frontiers may cause serious crimes, from forgery, to stolen vehicles or aerial hijackings, to murder. Those in movement may also become helpless victims of mistreatment or murder by border guards, or victims of bandits and other criminals.

Incidence Large scale or recurring illegal movements have been from Eastern Europe, East Berlin, Kampuchea, Vietnam, Cuba, mainland China, Mexico, India and Pakistan, among others. Countries with an illegal immigrant problem include the USA, the UK, Germany FR, Netherlands, France and Belgium. In the UK, for example, some 1,600 illegal immigrants were deported in 1983. At the USA-Mexico border, thousands of Mexicans commute daily to jobs in the USA, although such "commuting" is for the most part illegal.

Broader Political repression (#PC1919).
Related Exile (#PC2507) Fortified frontiers (#PD5972) Lack of individual rights to political asylum (#PF1075) Crimes related to immigration, naturalization and passports (#PE3889).
Aggravates Illegal immigration (#PD1928).
Aggravated by Refugees (#PB0205) Aerial piracy (#PD0124) Restrictions on emigration (#PC3208) Foreign exchange restrictions (#PF3070) Denial of freedom of movement in communist systems (#PC3173) Refusal to issue travel documents, passports, visas (#PE0325).

♦ PC2369 Tariff barriers to trade between developing and developed countries

Nature A major obstacle to the efforts of the developing countries to expand their exports of manufactures is the tariff protection in developed market economy countries. Relatively high and escalating tariff rates applied by these countries tend to discourage the establishment of export-oriented industries in developing countries. Although negotiations have significantly reduced the overall level of tariff protection by these countries, the benefits of the tariff concessions, on average, have been far greater for the developed countries themselves than for the developing countries. Substantial tariff cuts have been made on those products in which developed countries dominate world trade (chemicals, machinery, transport equipment, etc), while in general only small tariff cuts have been made on products of current export interest to developing countries (foods, textiles, leather and leather goods and other labour intensive products).

The disadvantage faced by producers in developing countries is, however, much greater than is apparent from a cursory glance at the post-Kennedy nominal tariff rates. It is a familiar feature of the tariff systems of developed countries that tariffs on manufactures and semi-manufactures tend to be appreciably higher than tariffs on the raw materials used in their production. In fact, on raw materials not produced in the importing country, tariffs have frequently been zero. This combination of policies to ensure supplies of cheap imported raw materials and to protect domestic industries processing these materials heightens the effects of the nominal tariffs confronting the developing countries. Manufacturing costs in any case in these countries tend to be high relative to costs of primary production; and an important task in these countries is to bring down costs of manufacturing production relative to costs of primary production in order to render their manufactures internationally competitive at prevailing exchange rates. Not only do tariffs tend to be higher on manufactures than on raw materials, but as a general rule they tend to increase

PC2369

with the degree of processing, with the result that effective protection of manufactures and semi-manufactures in developed countries tends to be much higher than would be indicated by the nominal tariffs.

Claim Protection imposes economic losses on both developed and developing countries. In the developed countries, resources are used in protected activities in which they have a comparative disadvantage in relation to alternative uses. As a result, total real product is lower than it would otherwise be (there may, however, be an offsetting gain, though probably a relatively small one, to the extent that import restrictions turn the terms of trade in favour of the developed countries). At the same time, the developing countries concerned suffer a real income loss by having to channel their resources into less economic activities in the primary sector. For those countries which continue to export the commodities in question, there will also be a terms-of-trade loss arising from the developed countries' import restrictions.

For the developed countries a reduction in protection would allow the deployment of resources in more economic activities and, to this extent, would result in a gain in real income. For the developing countries, easier access to the markets of developed countries, resulting from a reduction in protection of commodities of particular interest to developing countries, would allow them to expand their export earnings. Moreover, in so far as the reduction in protection led to a reduction in world supply of particular commodities, prices could also tend to rise, thus adding to the expansion in export earnings resulting from the increased volume of exports to the developed countries. For certain primary products, the insulation of national markets in the developed countries has reduced the size of the residual world market. Since, for these commodities, it is the residual market which bears the burden of adjustment to changes in world supply and demand, the insulation of national markets tends to accentuate fluctuations in the world price, as measured in the residual market. To the extent that the reduction in protection results in a more unified market, operating on a wider base of transactions, fluctuations in world supplies will tend to result in smaller price fluctuations than under the present system and, for many commodities, in greater stability in the export earnings of the developing countries concerned.

Counter-claim For most developed countries, reduction of protection would conflict with the policy aim of ensuring a desired level of income for domestic producers of primary commodities.

Broader Tariff barriers to international trade (#PC0569).
Aggravates Excessive external trade deficits (#PC1100).

♦ PC2371 Religious war

Crusade — Religious justification of militarism

Nature Religion has often been a useful pretext for internal or international conflicts. On the one hand, the protection of religious minorities and communities provides an excuse for intervention, often foreign intervention. On the other hand, religious discrimination and intolerance are used to encourage violence, although they may be masked by other aims.

Background The most notable religious war in western civilization were the Crusades and the wars following the Reformation. The idea of crusade is one of three Christian attitudes toward war, alongside pacifism and the idea of just war. The crusade had four characteristics: holy cause, belief in divine guidance and aid, godly crusaders and ungodly enemies, and unsparing prosecution. The laws of Islam provided for Jihad, or "Holy War", and thus assisted the spread of that faith.

Incidence Although most of the major faiths stand for peace, religious justifications are put forward for militarism. Thus attempts are made to equate the defence of a particular economic and social system with that of "Western/Christian civilization". Certain repressive governments define themselves as "Christian" and profess to be defending "Christian values". Governments or groups in Islamic countries continue to declare Jihad.

Broader Religious conflict (#PC3292) Religious repression (#PC0578).
Narrower Jihad (#PF5681).
Related War (#PB0593) Racial war (#PD8718) Ideological conflict (#PF3388).
Aggravates Religious intolerance (#PC1808)
Benign neoplasm of bone and cartilage (#PG0819).
Aggravated by Religious extremism (#PF4954) Religious indoctrination (#PD4890)
Double standards in morality (#PF5225) Religious and political antagonism (#PC0030).

♦ PC2372 Demagoguery

Nature Political leadership on the basis of a popular appeal from impassioned speeches invoking popular emotion and prejudice, demagoguery contains elements of hero worship and megalomania and makes free use of propaganda.

Broader Propaganda (#PF1878) Political conflict (#PC0368).
Narrower Megalomania (#PF2108).
Aggravates Undue political pressure (#PB3209) Politically emotive words and terms (#PF3128).
Aggravated by Lack of education (#PB8645) Political alienation (#PC3227).

♦ PC2379 Consanguineous marriage

Inbreeding

Nature Marriages of blood relatives give rise to certain increased risks in the offspring. These risks arise both for traits controlled by recessive genes and those determined by polygenes. In either case, the result is to expose a proportion of the otherwise largely hidden component in human genetic variability.

Incidence Studies have shown little difference between the outcome of consanguineous and control marriages up to and including birth. From birth onwards, however, the findings are different; in a certain Japanese city a death-rate of 116 per 1000 was found during the first 8 years of life amongst the offspring of first cousins, against 55 amongst the controls. The proportion of major congenital abnormalities for children born of consanguineous parents was almost twice that of the others. In an American city, with a lower total death-rate, amongst the offspring of consanguineous unions the death-rate by the age of 10 years was 81 per 1000 compared with 24 per 1000 in the controls. The problem is not an urgent one in communities in which the rate of first-cousin marriages has already fallen to a low figure (below 3 per 1000), but in many parts of the world consanguineous marriages represent a rather high proportion of the total marriages. The likelihood of such marriages occurring increases in societies where the family size is large, in physically or culturally isolated communities (including small islands, inaccessible mountain valleys, sectarian communities).

Refs Arner, George B *Consanguineous marriages in the American population* ; Shields, William M *Philopatry, Inbreeding and the Evolution of Sex* (1983).

Related Incest (#PF2148) Illegal marriage (#PE7935).
Aggravates Chorea (#PG3096) Diabetes (#PE0102) Epilepsy (#PE0661)
Aniridia (#PG4813) Schizophrenia (#PD0438) Feeblemindedness (#PE4821)
Neurofibromatosis (#PE4814) Genetic inbreeding (#PDp465) Friedreichs ataxia (#PE8605)
Hereditary opticatrophy (#PG4816) Manic-depressive psychosis (#PD1318)
Human disease and disability (#PB1044) Genetic defects and diseases (#PD2389)
Diseases of connective tissue (#PD2565)
Human physical genetic abnormalities (#PD1618).
Aggravated by Racism (#PB1047) Class consciousness (#PC3458)
Religious prejudice (#PD4365) Geographical isolation (#PF9023)
Compulsory sterilization (#PF3240) Excessively large families (#PD7625).

♦ PC2381 Denial to people of control over their own lives

Nature People have been denied freedom from violence, oppression, and want; full participation in the decisions that affect them; and education and information.

Narrower Forced marriage (#PD1915) Excessive job control (#PE6836)
Denial of right to life (#PD4234) Undemocratic policy-making (#PF8703)
Criminalization of euthanasia (#PF2643) Denial of right of family planning (#PE5226)
Denial of right to educational choice (#PE4700) Denial of right to free choice of work (#PE3963).
Aggravates Discrimination against women (#PC0308)
Socially unsustainable development (#PC0381)
Expropriation of land from indigenous populations (#PC3304).

♦ PC2383 Denial of right of assembly

Nature The right of people to assemble in large numbers and in public places, although one of the most basic of human rights, is frequently and systematically denied in many parts of the world. Even in otherwise free societies, right of assembly may be limited.

Broader Human inequality (#PA0844) Denial of rights (#PB5405).
Narrower Violent repression of demonstrations (#PD4811).
Related Unlawful assembly (#PG5067) Denial of political and civil rights (#PC0632).
Aggravates Subversive activities (#PD0557)
Urban unemployment in developing countries (#PD1551)
Denial of the right of trade union association (#PD0683).
Aggravated by Elitism (#PA1387) Political repression (#PC1919)
Denial of the right of association (#PD3224).

♦ PC2405 Deviant society

Nature A deviant society is a society whose structure is contrary to human nature and does not allow the satisfaction of basic human needs. Developed societies are to a certain extent responsive to material needs, but non-material needs are largely unmet. It is ironic that in this sense, the so called underdeveloped countries are actually more 'developed'. Schizophrenia, for instance, is much less common in the Third World than in industrialized countries, perhaps owing to their still viable natural communities such as strong families and villages. In poor countries people cry for bread; in rich countries they a hunger for meaning and identity. The unprecedented material progress during the post-war period has not necessarily made people happier.

Background Human beings have basic material needs such as physiological needs for food, shelter, warmth, rest, activity, etc; as well as basic non-material needs such as the needs for love and belongingness, self-esteem and recognition, orientation and meaning, self-actualization and self-transcendence. These needs are fundamental to the survival of the individual.

Claim Numerous people in the industrialized societies long to get away from achievement-oriented society, away from soulless jobs on the factory floor, away from the grim climb up the career ladder. There is a longing for a softer society, a secret desire to drop out, to be free, to begin to live. Industrial and economic growth-oriented society thus suppresses basic non-material needs, especially those for love and belonging, because it is incompatible with viable natural communities. The family and the local community started to disintegrate with the advent of industrialism. The process of industrialization means that more and more people become organized in factories and offices which function according to impersonal bureaucratic rules. Large-scale enterprises may lead to increased productivity but they also take their toll in terms of depersonalization and anonymity.

Refs Thio, Alex *Deviant Behavior* (1987).

Broader Deviance (#PB1125) Ethical decay (#PB2480).
Aggravates Fragmentation (#PA6233) Social injustice (#PC0797).
Aggravated by Massive psychic traumatization (#PE6968).

♦ PC2414 Sedition

Broader Improbity (#PA7363) Disobedience (#PA7250).
Narrower Seditious writings (#PG4887).
Related Treason (#PD2615).
Aggravated by Libel (#PD3022) Censorship (#PC0067)
Disloyalty (#PJ1895) Lack of freedom of the press (#PE8951).

♦ PC2416 Natural pollutants

Nature Some substances are created by nature in sufficiently large quantities to temporarily throw an ecosystem out of balance. These include volcanic dust and ash, sea salt, sulphur dioxide from volcanoes, smoke from forest fires and the like. Some say that these cannot strictly be called pollutants because their impact is always short-term, and can be seen as a part of natural systemic regeneration. In the short term, however, they are definitely experienced as pollutants.

Broader Pollutants (#PC5690).
Related Biological pollutants (#PC5276) Silicates as pollutants (#PG6267)
Biological air pollutants (#PD0450).

♦ PC2430 Drought

Vulnerability to drought

Nature Severe drought is a natural event that results from disruptions of the water cycle, and that, in turn, interferes with the cycle of nutrients and flow of energy in ecosystems. During periodic onslaughts of drought, glacier melting is intensified, lakes and other water bodies dry up, and the groundwater level sinks, crops are destroyed and societies disrupted. Drought also greatly influences the development of ecosystems. Severe droughts cause widescale alterations in species composition, and major disruptions of ecological communities. As plants die, their roots are no longer able to hold the soil. This results in massive erosion of topsoil rich in nutrients and organic matter. During this period, succession is often set back to bare subsoil. Drought has very different hydrological causes and characteristics, and its effects are less spectacular than those of floods, but they are more insidious and lasting, since they are closely connected with endemic famine in the world, and may in addition undermine for many years the prosperity of an entire region (the case of the Sahel).

Incidence The drought in Africa lasted for much of the 1970's and 1980's and created a disaster out of an economic and social situation which was already serious. One of its worst effects is the increasing water shortage for human and animal consumption. As many as 21 out of the 24 food-aid dependent countries are also affected by water scarcity. In Mauritania, sand now covers 6 million extra hectares every year. Another 21 million hectares pass into zero productivity each year. In the last two decades Mauritania has lost 80 percent of its pasture to the sand. Chile lost 400,000 or 75 percent of its sheep population, experienced power shortages since electrical generation capacity was reduced by 20 percent, and faced increased unemployment as a result of the devastating drought in 1967-69. The 1971 Okinawan drought, the worst in history, destroyed sugar-cane and pineapple plantations and caused economic losses amounting to $10 million. During the whole 150 year period 1830-1980 the European and Asian parts of the USSR suffered severe drought damage (81 of those years were dry). Crop failures of 19 percent and 25 percent respectively were reported during the 1972 and 1975 droughts. During the last 200 years some 30 great droughts have scourged the USA.

Refs WHO *Drought* (1975); WHO *Drought and Agriculture* (1975).

Broader Natural disasters (#PB1151).
Related Meteorological disaster (#PD4065).
Aggravates Fires (#PD8054) Famine (#PB0315) Desert nomadism (#PD2520)

Endangered species of amphibia (#PD3156) Seasonal fluctuations in agriculture (#PD5212)
Vitamin E deficiency in domestic animals (#PE4760).
Aggravated by Overstocking (#PC3153) Global cooling (#PF1744)
Tropical deforestation (#PD6204) Inadequate land drainage (#PD2269)
Short-term climatic change (#PF1984).

♦ PC2460 Climatic heat

Heat wave

Nature There are regions, including many in developing countries, where high ambient temperatures combined with high humidity of the air and intensive solar radiation make it difficult for the human body to get rid of its surplus heat. The physiological rhythm of work, fatigue and recovery is disturbed by high body temperatures. Fatigue accumulates and efficiency in the performance of mental and physical tasks declines. Neither acclimatization nor adaptation can completely overcome the disadvantages of an unfavourable climate. Continuous exposure to heat for many hours may result in heatstroke.

High water temperatures in the Caribbean and adjacent waters has caused coral animals to loose the brown algae than normally live inside their cells in a symbiotic relationship. If the temperatures continue to be high the survival of coral in the Caribbean could be threatened.

Broader Inhospitable climate (#PC0387).
Related Meteorological disaster (#PD4065) Excessive environmental heat (#PD7977).
Aggravates Fatigue (#PA0657) Lightning (#PD1292) Homelessness (#PB2150)
Heat disorders (#PE2398) Mental impairment (#PF4945)
Food spoilage in storage (#PD2243).
Aggravated by Global warming (#PC0918).
Reduced by Climatic cold (#PD1404).

♦ PC2494 Political stagnation

Inadequate choice for voters — Voter fatigue — Electoral apathy — Decreasing number of election voters — Voter apathy — Unpopular voting patterns — Refusal to vote

Nature Qualified voters may boycott the polls because they have no faith in the integrity of the polling or election system. Alternatively, they may not vote for reasons of apathy, lack of communication or lack of education. Refusal to vote indicates a certain political instability and the possibility of unrest. It may also lead to dictatorship and extremism, or control by an elite.

Claim Lack of variety, lack of choice of doctrine and lack of progress in politics leads to political stagnation, which is manifested in apathy, alienation, cynicism, scepticism, materialism, conformism and political lag. It maintains a state of inequality and injustice and leads to violence, conflict, disintegration, revolution or dictatorship.

Broader Stagnation (#PA3917) Inadequacy of political doctrine (#PF3394)
Unrecognized socio-economic interdependencies (#PF2969).
Aggravates Citizen incompetence (#PE5742) Political instability (#PC2677).
Aggravated by Unfair elections (#PC2649) Political apathy (#PC1917)
Unrepresentative electoral systems (#PD9641) Impoverishment of political debate (#PF4600)
Inner-tensions within political parties (#PG5360)
Political party manipulation of elections (#PD2906).
Reduced by Compulsory voting (#PG5850) Forced participation in politics (#PD2910).

♦ PC2506 Desert advance

Desertification — Arid zone enlargement — Degradation of drylands

Nature Certain climatic changes, over a very long time scale, can by themselves result in the formation of deserts. But the actions and activity of man can accelerate this process, producing man-made deserts. Overgrazing, overcultivation, deforestation, bad irrigation and soil erosion all tend, where the climate is favourable, to produce desert and semi-desert conditions.

The spread of deserts is contagious. Sand not fixed in position by plants will spread in sandstorms; sand dunes, in a short period, will advance – engulfing roads, villages, crops, and previously fertile land. The dynamics of erosion are such that land already eroded by wind and rain is more susceptible to further erosion. Successive removals eventually create a soil condition wherein plant growth is minimized and erodibility greatly increased. Control becomes more and more difficult. In the extreme, the sands begin to drift and form unstable dunes which encroach on better surrounding lands. Moreover, there are adverse climatic consequences of erosion, for much rain is lost to the sea or to underground deposits instead of being retained by the soil and evaporated. Ruined land is thus a source of hot, dry air and so increases the aridity of conditions on neighbouring land.

The human costs of desertification often include malnutrition, threat of famine, and dislocation of people who must abandon their lands to seek employment elsewhere.

Incidence Attempts at reforestation in Spain, Italy and Greece would certainly have been more successful had the opposite shores of the Mediterranean still been covered with a wide belt of fertile land, as they once were. But the desert has already reached the shore of the Mediterranean on a wide front and sends out its drying winds to the European countries. Although approximately 100 countries are affected by desertification, the process is most serious in sub-Saharan Africa (particularly the Sahel), northwestern Asia, and the Middle East. Every year an additional 200,000 square kilometers – an area larger than Senegal – are reduced by desertification to the point of yielding nothing. The process is accelerating: some 3.5 billion hectares of cropland and 230 million people are now threatened.

Although it is probable that existing deserts will remain where they are over the next 100 years, it is not impossible that human misuse of the arid lands, whether contiguous or otherwise, could accelerate natural processes of desertification to the point where most arid lands could be converted into deserts over that period. This would have significant implications for regional and global climate change.

Refs Grainger, Alan *Desertification* (1986); Grainger, Alan *The Threatening Desert*; Hare, F K *Climatic Variations, Drought and Desertification* (1985); Heathcote, R L *Perception of Desertification* (1980); Mabbutt, J and Wilson, A (Eds) *Social and Environmental Aspects of Desertification* (1980).

Aggravates Infertile land (#PD8585) Unreliable rainfall (#PD0489)
Destruction of land fertility (#PC1300) Slowing growth in food production (#PC1960)
Unavailability of agricultural land (#PC7597)
Unsustainable agricultural development (#PC8419)
Increased reflection of solar radiation (#PF5069)
Environmental degradation of desert oases (#PD2285)
Declining productivity of agricultural land (#PD7480)
Endangered lifestyles of nomads and pastoralists (#PE8077)
Constraints to increased agricultural output in developing countries (#PD5114).
Aggravated by Defoliation (#PD1135) Sand storms (#PD3650)
Dust storms (#PD3655) Soil erosion (#PD0949)
Global cooling (#PF1744) Soil degradation (#PD1052)
Termites as pests (#PE1747) Migrating sand dunes (#PD0493)
Soil erosion by wind (#PE3656) Tropical deforestation (#PD6204)
Short-term climatic change (#PF1984) Misuse of grassland and rangeland (#PD5133)
Disruption of arid zone ecosystems (#PD7096)
Mismanagement of irrigation schemes (#PE8233).

♦ PC2507 Exile

Nature Prolonged absence from one's country of origin, either enforced by government authorities or by voluntary action in the fear of persecution for reasons of race, religion, nationality or political opinion, constitutes exile. Transportation of prisoners, political or otherwise, to isolated areas is another forms of banishment. Individuals forced into exile are made outcasts and are deprived of the comfort and protection of their group. Often they have to live in a different climate, learn a new language, new social customs and start their professional and family life anew. They may be resentful of, or resented by, society in their host country. Where exiles live freely in another country they may cause international conflict and may put out propaganda against the oppressive regime in their country of origin. Exile for political reasons often deprives a minority or opposition group of its leaders and spokesmen. If the exiles are allowed to return to the country after an extended period of time, they may face serious political, economic and social disadvantages.

Background Exile and banishment probably originated in tribal custom as a means of punishment. Transportation was a common practice in Europe from the 15th century to the 19th century; in the 20th century political reasons has become the major cause of exile.

Broader Political repression (#PC1919) Ideological repression (#PC8083).
Narrower Internal exile (#PJ9588) Arbitrary exile (#PJ7268).
Related Expulsion (#PC5313) Illegal movement across frontiers (#PC2367).
Aggravates Naturalization (#PG4974) Ineffective war crime prosecution (#PD1464).

♦ PC2510 Economically controlled political power

Nature Economic power, concentrated in large corporations, national governments and special interest groups, largely directs and sets guidelines for the agencies built to control the world's resources. This power is used to restrict efforts to create more equitable resource policies. Often any attempt to set priorities that recognize needs beyond resources' interests are muted or silenced. This is seen in such sectors as: production and extraction of natural resources; grain and sugar exports; oil exploration; and information access. These activities often are continued in the knowledge that their impact is detrimental to the long term economic priority of some parts of the world may eventually have implications related to the whole population or the entire planet.

Related Legal contract system reduced to individual needs (#PE5397).

♦ PC2512 Cartels

Nature Cartels are unions of sellers whose aims are to raise, support or fix prices or affect conditions of sale. They control production, imports or sales by allocating quotas for member's activities. Cartel agreements among firms in developed market-economy countries affect imports into those countries. The restrictive business practices resulting from these agreements directly hamper the interests of other countries, including developing countries, whose normal export development is impeded. The types of cartels distinguished include import cartels, rebate cartels, and agreements on standards. Cartel activities, when they affect the domestic market or foreign trade interests of the developed market-economy countries, are usually subject to the laws providing for controls, which range from prohibition to abuse control. This would seem to be one of the reasons why the known instances of such cartel activities are not considerable in number.

Refs Hobson, John A *Cartels, Trusts and the Economic Power of Bankers, Financiers and Money-Moguls* (1985); Stocking, George W and Watkins, Myron W *Cartels or Competition?* (1986).

Broader Economic crime (#PC5624) Restrictive trade practices (#PC0073).
Narrower Import cartels (#PD0336) Export cartels (#PD0470)
Rebate cartels (#PD0668) Domestic cartel (#PD7963)
Liner shipping cartels (#PE3829) Industrial gas monopolies (#PE1813)
Collusive international trade arrangements (#PE0396).
Related Monopolies (#PC0521) Violations against economic regulations (#PD7438).
Aggravates Collusive tendering (#PE4301)
Excessive injury to export interests developing countries due to export cartels (#PE2598).
Aggravated by Unethical practices of employers (#PD2879).

♦ PC2514 Epidemics

Nature The extent to which contagious and infectious diseases spread through human populations is highly variable. In a developed country a few hundred cases of typhoid or plague may produce anxiety or even mild panic, but the outbreak can usually be controlled by routine public health measures. Such an epidemic is not a threat to the survival of the community. In developing countries, diseases may be much more widespread and have a regular endemic character; in these cases, massive spread to the whole community is limited by a high degree of naturally acquired immunity, and public health measures may help. On the other hand, diseases (particularly new strains) like influenza or measles may attack a large proportion of a community, especially if it has acquired little immunity from previous exposure or by immunization. That some diseases (for example, influenza, measles and plague) are in fact capable of affecting a large part of a population in a highly disastrous fashion is a matter of historical fact. A major question is whether epidemics are rare accidents or whether they could be started deliberately. It is possible that, by a deliberate selection of mutants or through recombinants, highly virulent and spreading strains could be obtained.

Refs Bollet, Alfred J *Plagues and Poxes* (1987); Greenwood, M *Epidemics and Crowd Diseases* (1935); Hecker, Justus F *Dancing Mania of the Middle Ages* (1970); Prinzing, Friedrich *Epidemics Resulting from Wars* (1977); Webster, Noah *Brief History of Epidemics and Pestilential Diseases* (1970).

Broader Biological disasters (#PC5489) Human disease and disability (#PB1044).
Narrower AIDS (#PD5111) Plague (#PE0987) Anthrax (#PE2736)
Smallpox (#PE0097) Influenza (#PE0447) Hepatitis (#PE0517)
Tularaemia (#PE6872) Chikungunya (#PG4984) Coccidiosis (#PE2738)
Typhoid fever (#PD1753) O'nyong-nyong (#PG4985).
Aggravates Brucellosis (#PE0924) Yellow fever (#PE0985).
Aggravated by Biological warfare (#PC0195) . Unreported illness (#PF8090)
Medium-term cyclic variations in solar radiant energy (#PE9528)
Irrational conscientious refusal of medical intervention (#PF0420).

♦ PC2518 Excessive consumption of goods and services

Nature Societies that consume more goods and services than they produce or can purchase from which they produce are morally and financially in difficulty.

Incidence The United States at the present time must borrow over $150 billion a year externally to finance a 15 year long binge of household, social and defence spending. The United States has become the largest debtor in the world from the largest creditor in the last two decades.

Narrower Over-eating (#PE5722) Substance abuse (#PC5536)
Increased food consumption (#PJ1931)
Excessive consumption of specific foodstuffs (#PC3908)
Excessive consumption of resources in developed countries (#PE5551).
Related Food hoarding (#PJ2225) Proliferation of computers (#PE3959)
Decadent standard of living (#PD4037).
Aggravates Unproductive use of resources (#PB8376)
Waste of non-renewable resources (#PC8642)
Insufficient availability of goods (#PB8891)
Long-term shortage of manufactured goods (#PE0802)
Natural resource depletion due to high-level consumption (#PD4002)
Restrictive effects of traditional community decision-making (#PF3454).

PC2518

Aggravated by Competitive development of new weapons (#PC0012)
Production serving false consumption needs (#PF2639)
Use of agricultural resources for production of animal feed (#PD1283).
Reduces Surplus (#PF4750) Agricultural surpluses (#PC2062).

♦ PC2536 Stagflation

Nature National growth as measured by domestic output may show little gain from year to year (for example, less than 1 percent) yet wages and other costs may keep rising. This generates an inflation several multiples higher than the stagnant rate of growth, and leads to economic crises.

Broader Economic uncertainty (#PF5817).
Aggravated by Declining economic growth in developing countries (#PD5326)
Declining economic growth in industrialized countries (#PF1737).

♦ PC2541 Age discrimination

Age prejudice — Ageism — Discrimination against elders — Denial of rights to elders — Denial of rights to elderly — Age discrimination before the law

Nature The systematic oppression of older people by younger ones. People of advancing age are silenced by retirement and the cult of youth, and isolated from contact with other age groups. The result is a peculiarly suicidal vision of life as a journey to nowhere. Health, welfare and social programmes for the aged have to deal with this pervasive attitude not only among younger people, but often as strongly among the elderly themselves.

Refs Montague, Meg *Ageing and Autonomy* (1982).

Broader Discrimination (#PA0833) Social injustice (#PC0797)
Social inequality (#PB0514).
Narrower Ageing women (#PE6784) Age segregation (#PD3444)
Inadequate welfare services for the aged (#PD0512)
Denial of equal benefits to elderly workers (#PE1625)
Employment discrimination against the elderly (#PD4916).
Related Discrimination in housing (#PD3469) Neglected health practices (#PD8607)
Discrimination before the law (#PC8726) Denial of rights to vulnerable groups (#PC4405).
Aggravates Disrespect for elders (#PF3979) Discrimination in politics (#PC0934)
Inadequate income in old age (#PC1966) Generation communication gap (#PF0756)
Inadequate housing for the aged (#PD0276) Inadequate recreational facilities (#PF0202).
Aggravated by Senility (#PE6402) Cult of youth (#PF6766)
Juvenile stress (#PC0877) Social discrimination (#PC1864)
Youth oriented society (#PG9131) Legalized discrimination (#PC8949)
Social disadvantage of the aged (#PD3517) Age discrimination in employment (#PD2318)
Undue attachment to a social group (#PF1073) Rigidity and inadaptability in the aged (#PD3515)
Structural rigidities in labour markets (#PD4011)
Vulnerability of the elderly under states of emergency (#PD0096)
Inadequate rehabilitation facilities for the mentally handicapped (#PE8151).

♦ PC2546 Excessive public debt

National debt burden

Nature Public debt is the obligation on the part of the government, one of its agencies, or of local government, to pay specific monetary sums to holders of legally designated claims at particular points in time. Public debt is created by the act of public borrowing or sale of government securities. Such a debt is amortized or retired by a reverse transfer in which government gives up money for the bonds, Treasury bills or other debt instruments. Under present conditions, the net effect of such debts is to further inflation on the one hand and hamper incentives (through the taxes necessary for interest payments) on the other. It is also argued that the financing of government activities by borrowing results in the transfer of costs from the present to future generations, who must pay higher taxes to meet principal and interest obligations. An alternative or supplement to increasing revenue by higher taxes, is the government's selling of some of its assets in order to repay debt. Assets may be sold internally (privatized, in a sense) or to foreign governments or investors. Included may be off shore land, gold, art treasures, historic jewellery, oil, valuable minerals, weapons, historic manuscripts or documents, and anything else a government possesses or can appropriate. Whatever method is used, it results in national impoverishment, as does the alternative of drastically reduced public expenditures.

Incidence There is a current debt explosion which is not limited to a few countries, but rather is a worldwide phenomenon. Between 1974 and 1983 the ratio of central government debt to gross national or domestic product rose sharply in most industrial countries. For the seven largest industrial countries, that ratio rose from an unweighted average of 22 percent in 1974 to an average of 41 percent in 1983. In some of these countries the increase was quite sharp. In Japan, for example, the ratio rose from 12.2 percent of GNP in 1974 to almost 53 percent in 1983; in Italy it rose from 45.3 percent to almost 79 percent; and in Canada it rose from 16.3 percent to 35.5 percent. In the United States the ratio remained almost unchanged at around 28 percent up to 1981, but then it began to increase sharply, reaching almost 36 percent in 1983. In some of the smaller industrial countries the increases were even larger. From 1974 to 1983 the ratio increased by 50 percentage points in Belgium; by 70 percentage points in Denmark; by 54 percentage in Ireland; and by 34 percentage points in Sweden.

By 1988, the net public debt as a percentage of Gross Domestic Product was as follows for the seven largest industrial countries: United States 30 percent; Japan 24.6 percent; West Germany 23.8 percent; France 26.6 percent; Britain 39.1 percent; Italy 92.0 percent and Canada 36.7 percent.

Claim National debt is a government's debt to its own citizens. Unlike foreign debt, which has to be serviced out of export earnings, it is not a real burden on the economy, but a transfer between present and future generations. When borrowing is done for productive investment then the means of repaying the debt is also transferred to future generations. When borrowing is to pay bureaucrat's salaries then only the debt is transferred.

Counter-claim National debt is not a transfer between generations; it is an intra-generational transfer. If our grand-children, as a generation, inherit our national debt they also inherit dollar for dollar the assets (Treasury bond, etc.) represented by that debt. If our grandchildren pay extra taxes to service their inherited debt, they also, as a generation, get those same taxes back, dollar for dollar, as debt service paid on that debt. The net inter-generational transfer is exactly zero.

In periods of depression, the expansionary effects of government borrowing tend to bring about an increase in output. When governmental activities require capital outlays far in excess of usual expenditures, borrowing is not only virtually imperative if the outlays are to be made, but is entirely justifiable. The absolute figures of growth in government debt tend to exaggerate the actual growth in the debt relative to the economy as a whole.

Refs Arrow, Kenneth J and Boskin, Michael J (Eds) *The Economics of Public Debt* (1988); International Bank for Reconstruction and Development *World Debt Tables, 1989–1990 Edition* (1990); International Bank for Reconstruction and Development *World Debt Tables, 1987–1988 Edition*; International Environment Liason Centre *International Environment–Development Facts March 1989* (1989); Morris, Dirk *Government Debt in International Financial Markets* (1988).

Broader Global economic crisis (#PC5876).
Narrower War debt (#PD3057) Disproportionate external public debt (#PC3056)
Excessive debt of socialist countries to the West (#PD2502).
Aggravates Economic inflation (#PC0254) Government limitations (#PF4668)
Global financial crisis (#PF3612).
Aggravated by War (#PB0593) Domination (#PA0839) Underdevelopment (#PB0206)
Uncontrolled growth of debt (#PC8316) Competitive acquisition of arms (#PC1258)
Maldistribution of electrical energy (#PD3446).

♦ PC2548 Cultural invasion

Nature A foreign culture and technology can impinge on indigenous populations who may be backward and unable to compete against the invaders; they are therefore rendered dependent on the new system. Where cultural invasion occurs in developed countries it also has adverse dependency consequences which equally result in debt and a disruption of the existing society.

Incidence Cultural invasion formerly occurred in the sense of colonization and political domination. Now it is less overt but nonetheless widespread. It occurs in international trade and industrial development overseas when promoted by a powerful country. The economic invasion is closely connected with cultural invasion and with political influence. Countries which fall prey to excessive foreign investment find themselves heavily in debt (Philippines). Cultural invasion is particularly marked in developing countries where certain cultural aspects of the donor of aid pervade the receiver, disregarding the latter's own culture. The most notable effect of cultural invasion can be seen with indigenous tribes whose homelands have been seized for economic development. Unable to adjust, they become totally dependent, easily exploited and die from poverty, malnutrition and disease. Wherever cultural invasion occurs the loss in diversity of cultural heritage is marked.

Background Cultural invasion has occurred from the earliest recorded history with movements of population and tribal war.

Broader Cultural imperialism (#PC3195) Threatened and vulnerable minorities (#PC3295).
Narrower Technocracy (#PF6330) Social disintegration (#PC3309).
Related Invasion (#PD8779) Inter-cultural misunderstanding (#PF3340)
Ill-considered missionary activity (#PF3370).
Aggravates Elitism (#PA1387) Culture shock (#PC2673) Ethnic conflict (#PC3685)
Cultural domination (#PG5055) Political domination (#PC8512) Ethnic disintegration (#PC3291)
Cultural fragmentation (#PF0536) Isolation of ethnic groups (#PC3316)
Destruction of cultural heritage (#PC2114) Underprivileged racial minorities (#PC0805)
Endangered tribes and indigenous peoples (#PC0720).
Aggravated by Colonialism (#PC0798) Exploitation (#PB3200)
Underdevelopment (#PB0206) Lack of education (#PB8645)
Economic conflict (#PC0840) Cultural arrogance (#PF5178)
Economic imperialism (#PC3198) Investment capitalism (#PJ5060)
Unsustainable economic development (#PC0495) International monopoly of the media (#PD3040)
Inadequacy of aid to developing countries (#PF0392)
Foreign intervention in internal affairs of states (#PC3185).
Reduces Totemism (#PF3421) Traditionalism (#PF2676) Cultural barriers (#PB2331)
Human cannibalism (#PF2513) Head hunting in tribal societies (#PF2666).

♦ PC2550 Administrative delays

Bureaucratic delays — Red tape delays

Nature Large organizations, and particularly government bureaucracies, tend to develop a multitude of procedures and levels of approval through which proposals or applications for change must pass. Frequently decisions must be taken in committees which meet relatively infrequently. This introduces considerable delay in any projects which are dependent on the approval of such bodies.

Broader Delay (#PA1999) Inaccessible administrative agencies (#PF2261).
Narrower Delay in obtaining property titles (#PG1129)
Informational and procedural obstacles to world trade (#PE9107).
Related Delay in administration of justice (#PF1487).
Aggravates Unrealized use of education structures (#PF2568)
Delays in delivery of books and publications (#PF1538).
Aggravated by Bureaucratic fragmentation (#PC2662)
Government delaying tactics (#PF6119)
Complex government regulations (#PF8053)
Abuse of bureaucratic procedures (#PF2661)
Corruption in developing countries (#PD0348)
Unfamiliar bureaucratic procedures (#PJ9912)
Unchecked power of government bureaucracy (#PD8890)
Proliferation of public sector institutions (#PF4739).

♦ PC2551 Civil disorders

Riots

Refs Fogelson, Robert M *Violence As Protest* (1980).

Broader Civil violence (#PC4864).
Narrower Mutiny (#PD2589) Prison riots (#PE1675) Inciting riot (#PD6392)
Arming rioters (#PE5327) Religious riots (#PE1417) Engaging in riot (#PD4091)
Civil disturbances (#PD5372).
Related Unlawful assembly (#PG5067) Overthrow of government (#PD1964)
Offences against public order (#PD7520).
Aggravates Injuries (#PB0855) Stone throwing (#PJ7019) Political upheavals (#PC7660)
Human destructiveness (#PA0832).
Aggravated by Famine (#PB0315) Oppression (#PB8656)
Student revolt (#PC2052) Police brutality (#PD3543)
Armed insurrection (#PD8284) Abuse of police power (#PC1142)
Socio-economic poverty (#PB0388) Inadequate riot control (#PD2207)
Inhumane methods of riot control (#PD1156)
Environmental degradation of inner city areas (#PC2616)
International imbalance in the quality of life (#PB4993)
Monotonous and unaesthetic architecture and design (#PF0867).
Reduced by Excessive government control (#PF0304).

♦ PC2555 Conspiracy

Nature Conspiracy under Anglo-American law is a wide concept, usually described as an agreement between two or more persons to commit an unlawful act or accomplish a lawful end with unlawful means. Under this law an act which is not a criminal offence if committed by one person may become so if committed by more than one. Therefore the formation of labour unions or political opposition parties may be classified as conspiracy where these are banned. Other kinds of conspiracy include group and organized crime, smuggling, treason, corruption and fraud. Under Anglo-American law two kinds of conspiracy are recognized, 'chain conspiracy' as in organized crime and smuggling where agreement occurs at various levels though individuals may not know their contacts on a certain level; and 'wheel conspiracy' where a number of different people committing the same crime with the same person are held to have conspired, even if they do not know one another. The extent to which charges of conspiracy under this law can be brought, may be unjust and cause confusion. Continental European law (East and West) regarding conspiracy is much more narrowly defined. In most cases conspiracy constitutes a political crime against the state.

Background Conspiracy under old English law (Edward I, 1305) consisted of an abuse of the processes of criminal justice, malicious and false prosecution. The modern concept of agreement to commit an unlawful act derives from the activities of the Star Chamber in 17th century England, and during the 18th century was applied mainly to labour associations. The American law derives from the 17th century English concept and was also closely connected with labour disputes.

Refs Davis, David B *The Fear of Conspiracy* (1972); Goodspeed, D J *Conspirators* (1984); Graumann C F and Moscovici, S (Eds) *Changing conceptions of conspiracy* (1987); Shirley, Andrew *Plots and conspiracies* (1975).
Broader Nonaccomplishment (#PA6662) Uncommunicativeness (#PA7411).
Narrower Zionism (#PF0200) Catholicism (#PF8071) Freemasonry (#PF0695)
Jewish conspiracy (#PF8838) Criminal conspiracy (#PD1767)
Conspiracy against the public (#PF4198) Secret international agreements (#PF0419)
Conspiracies for societal control (#PB7125)
Secrecy concerning existence of extraterrestrials (#PF4331).
Related Torts (#PD9022).
Aggravates Banned trade unions (#PD3535) Banned associations (#PD3536)
Ineffective war crime prosecution (#PD1464) Undemocratic political organization (#PC1015).
Aggravated by Totalitarianism (#PF2190) Secret societies (#PF2508)
Political repression (#PC1919) Denial of the right of association (#PD3224)
Restrictions on international freedom of information (#PC0931).

♦ PC2558 Bribery

Kickbacks — Sweeteners
Nature Bribery includes the offering, demanding, giving and receiving of any payments or gifts to a person in connection with his official functions, with the intention of improperly influencing a decision. Bribery may be used in political, government, business, sports or any other field where the corruption of an individual can benefit the person doing the bribing.
Refs Jacoby, Neil H, et al *Bribery and Extortion in World Business* (1977).
Broader Corruption (#PA1986) Unethical personal relationships (#PF8759).
Narrower Giving bribes (#PC4631) Receiving bribes (#PC4701)
Business bribery (#PD8449) Political bribery (#PC2030)
Political appointees (#PF2031) Concealed government subsidies (#PD4532)
Corruption of sports and athletic competitions (#PE3754).
Related Financial scandal (#PD2458) Corruptive crimes (#PD8679).
Aggravates Police corruption (#PD2918) Corruption of the judiciary (#PD4194)
Abuse of bureaucratic procedures (#PF2661).
Aggravated by Criminal motivation (#PJ6406) Resignation towards bribery (#PF8611).

♦ PC2563 Unethical commercial practices

Irresponsible business practices — Unethical business practices — Irresponsible commercial practices — Negligence in commerce — Corruption in commerce
Nature Corrupt practices in business include embezzlement of company funds, tax evasion (use of company name to avoid taxation), and the use of bribery, intimidation and fraud. High company officials involved in corruption or seeking to extend the power of the company may offer bribes to public officials in return for favours. Commercial pressures on companies to improve performance increase susceptibility to unethical behaviour. Establishing criminal liability in such cases is a matter of law and evidence, generally with the requirement that the manager should know of wrongdoing or be willfully blind to actions by a subordinate.
A distinction can be made between unethical behaviour undertaken solely for personal interest and that undertaken with the tacit encouragement of superiors. Examples include cases where pollutants are dumped illegally to save disposal costs. It is considered impractical for top executives to know all the decisions taken by subordinates. In some cases such executives make it clear that they do not care how a job is done, provided it gets done. In other cases they deliberately turn a blind eye, or implicitly agree to it. And without in any way condoning unethical activity, executives may reward those subordinates who do get a job done under difficult circumstances. In all such cases the executives remain shielded from the blame by plausible deniability.
Incidence The retail industry in the USA has recently noted an increase in cases of vendors bribing buyers.
Broader Immorality (#PA3369) Corruption (#PA1986) Irresponsibility (#PA8658).
Narrower Contract fraud (#PD7876) Insider dealing (#PD3841)
Asset stripping (#PE9224) Corporate crime (#PD3528)
Commercial fraud (#PD2057) Business bribery (#PD8449)
Avoidance of copyright (#PD0188) Inappropriate labelling (#PD3521)
Unethical food practices (#PD1045) Commodities trading fraud (#PD3917)
Manufacture of illicit drugs (#PE2512) Defective product manufacture (#PD3998)
Restrictive shipping practices (#PD0312) Unethical real estate practice (#PD5422)
Irresponsible international trade (#PC8930) Exploitative property development (#PD8492)
Illicit production of alcoholic beverages (#PE7188)
Unethical practices in the apparel industry (#PD8001)
Unauthorized pharmaceutical manufacture and distribution (#PE0564)
Trading in products containing toxic substances with developing countries (#PE2061)
Coercive use of economic power by transnational enterprises against labour (#PE0207).
Related Institutionalized corruption (#PC9173).
Aggravates Tax evasion (#PD1466) Financial scandal (#PD2458)
Business bankruptcy (#PD2591) Inappropriate loans (#PF4580)
Misuse of advertising (#PE4225) Misleading advertising (#PE3814)
Consumer vulnerability (#PC0123) Corruption in politics (#PC0116)
Abuse of expense accounts (#PE4645) Unfair transport practices (#PD1367)
Usury in developing countries (#PE2524) Unethical industrial practices (#PD2916)
Deterioration in product quality (#PD1435) Inadequate equipment maintenance (#PD1565)
Industrial waste water pollutants (#PD0575) Ineffective industry self-regulation (#PF5841)
Evasion of customs and excise duties (#PD2620) Proliferation of direct mail advertising (#PE1810)
Victimization of workers' representatives (#PD1846)
Environmental degradation of inner city areas (#PC2616)
Excessive expense of athletic training programmes (#PF4196)
Alienation of land through acquisition by foreigners (#PE0896)
Excessive expense of international athletic competitions (#PF4192)
Harmful effects of advertising by transnational corporations in developing countries (#PE2004).
Aggravated by Unaccountability of institutions degrading the environment (#PF3458).

♦ PC2566 Reduction of the share of the developing countries in world exports

Nature A rapid rate of growth in export earnings of the developing countries is of strategic importance for their economic development; but most appear to be far from reaching this goal. The developing countries' declining share in world trade reflects their heavy dependence on exports of agricultural and certain primary commodities for which demand has grown only slowly. Moreover, the prospects for increasing exports of these products appear limited in view of the low elasticity of demand, the decrease in the raw material content of industrial products which is the result of technological progress, and the rising production of both natural and synthetic materials in the developed countries. Since the exports of manufactured products are generally not subject to such limitations, the main emphasis will increasingly need to be placed by developing countries on expanding and diversifying exports of manufactures and semi-manufactures. An important question, however, is whether such transformation of the export structure will take place at a pace sufficient to ensure an adequate rate of economic growth in the developing countries.
Economic growth and the expansion of exports of manufactures are, to some extent, interrelated. An adequate rate of expansion in exports of manufactures depends partly on the pace of economic growth in developing countries; for exports can be continuously expanded only if industrial capacity is steadily enlarged. On the other hand, the pace of industrial growth depends partly on the dynamism achieved in export earnings; for the ability to expand industrial capacity is related to the ability to import technology and capital equipment. But this interrelationship is flexible and, at many points, it may be considerably modified by the nature of the policies pursued in both developing and developed countries. Of particular concern are the restrictive business practices of transnational corporations whose cartel activities, and subsidiaries exports limitations are obstacles to the growth of markets for the developing countries.
Broader Competition between states (#PC0114)
Economic and social underdevelopment (#PB0539).
Narrower Minimal exports in least developed countries (#PE8306).
Related Declining international competitiveness (#PD8994).
Aggravated by Restricted growth in export markets of developing countries (#PF1471)
Over-production of primary commodities in developing countries (#PD2967)
Underproduction of primary commodities in developing countries (#PD3042)
Inequality in distribution of natural resources between countries (#PF3043)
Inadequate demand for primary commodities because of rising living standards (#PD2898)
Instability in export trade of developing countries producing primary commodities (#PD2968).

♦ PC2576 Self-interest driven investment

Nature Investment is fragmented and directed toward rapid accumulation of capital returns resulting in factoring out both the development of human resources and comprehensive and long term plans for a global economy.
Broader Dominance of economic motives (#PF1913).
Narrower Limited market development (#PF1086)
Inadequate credit policies (#PF0245)
Variations in national forms of currency (#PF2574)
Overemphasis on rapid returns on investment (#PF1275)
Proscriptive controls favouring the investor (#PF2607)
Inadequate access to negotiation on employment and reward (#PD1958).
Related Production of non-essentials (#PC3651) Belittling of grant recipients (#PF2708)
Non-inclusive management decisions (#PF2754)
Inadequate models of socio-economic development (#PF9576).
Aggravates Underdeveloped technological skill (#PF8552).

♦ PC2582 Intellectual dissent

Broader Ideological conflict (#PF3388).
Narrower Anti-science (#PF2685).
Aggravated by Double standards in morality (#PF5225).

♦ PC2584 Physical maltreatment of children

Battered children — Child beating — Child abuse
Nature Excessive physical punishment and abuse of young children, unable to protect themselves or seek outside protection, may result in physical or mental handicap, or both, or death. Often abused children are not living with both natural parents and martial problems are frequent causes of stress leading to abuse.
Physically abused children's traumas range from very mild to cases with injury to the skeleton and soft tissues. Children's health is often below par and frequently show evidence of neglect, including poor skin hygiene and multiple soft-tissue minor injuries. Bruises from blows and bites; head injuries from blows to the head or shaking; burns from scalding water, heated surfaces or open flames; and fractures are common; poisons and lacerations are relatively uncommon. Injuries to eyes can be caused by blows to the head or by shaking. The most common cause of death from child abuse is multiple rib fractures and lacerations of the lungs, liver, spleen, mesentery, and pancreas.
Incidence Accurate figures about the extent of child abuse world wide are difficult to gather because in perhaps most cases the whole family is involved. When a social atmosphere is created where people can freely talk about their experiences the figures are much higher than estimated. In the State of Florida one year the number of child abuse cases was seventeen the following year after a child abuse hot-line was established the number rose to over nineteen thousand. An estimated 10 percent of hospitalized, abused children die, another 10 percent suffer from permanent mental or physical handicap, or both (USA). Further statistics from the USA put the instances of known child abuse or neglect cases at 1.6 million in 1986. Another study suggests that there are approximately 6.5 million cases per year of children between the ages of 3 and 17 based on violent acts carried out toward children, rather than on injuries received. In West Germany in 1987 a thousand children are reported to died from beating from their parents. Some fifteen to eighteen thousand children a year are severely abused. In France recent surveys have put the number of abused children at about fifty thousand a year. Death from ill treatment is estimated at five hundred a year. In Britain over seven thousand children in 1983 were so badly hurt they require medical treatment.
Refs Bankowski, Z and Carballo, M (Eds) *Battered Children and Child Abuse* (1986); Bentovin, Arnon, et al *Child Sexual Abuse Within the Family* (1988); De Mause, Lloyd (Ed) *The History of Childhood* (1988); Maher, Peter *Child Abuse* (1987); Moorehead, Caroline (Ed) *Betrayal* (1989); Rao, Rama K *Classified and Annotated Bibliography on Child Abuse and Neglects* (1986); Scheper-Hughes, Nancy (Ed) *Child Survival* (1987).
Broader Cruelty to children (#PC0838) Victimization of children (#PC5512).
Narrower Parental punishment (#PD7187) Corporal punishment in schools (#PE0192).
Related Torture of children (#PD2851) Exploitation of child labour (#PD0164)
Discrimination and harassment of children in public life (#PE6922).
Aggravates Soul murder (#PF4213) Street children (#PD5980)
Corruption of minors (#PD9481) Physically handicapped children (#PD0196).
Aggravated by Family violence (#PD6881) Dependency of children (#PD2476).
Reduced by State custody of deprived children (#PD0550).

♦ PC2616 Environmental degradation of inner city areas

Inner city decay — Urban neglect — Neglect of urban housing — Deteriorating inner-city neighbourhoods — Urban decay
Nature Central areas, especially in the major cities of the west, are those parts of urban settlements where the concentration of the entire range of economic, financial, political and cultural activities is the greatest and the most complex. Traditionally they have been the scenes of co-existing housing, artisanal industries, business and many other economic and socio-cultural functions, in which they acted as clearing-houses for the exchange of goods, ideas and knowledge. This co-existence is now threatened. Although industries which in the past caused disruption and damage to the environment are gradually moving out, living and working in central areas can still be a traumatic experience. This is particularly true of the city centres in many industrialized countries, which exhibit environmental degradation. The numerous and highly competitive activities entailing land use overwhelm the limited space and create a situation of overcrowding, functional incompatibility and cultural degradation. Inner city areas have a high level of commercial specialization, a large number of offices and a sizeable daytime population. At the same time, city centres generally remain a sort of ghetto for a permanent, low-income population living in run-down housing and enjoying little in the way of public services and civic amenities. The concentration of the service industries inevitably entails the replacement of traditional housing and shops by office blocks, the provision of basic utilities at the expense of civic amenities and the provision of major access roads which eat up urban space. The ensuing over-concentration of traffic has the inevitable consequences of air pollution, visual obstruction by masses of vehicles and a disproportionate consumption of central spaces for transport needs. Structures of historic origin are often unable to meet modern requirements and, notwithstanding their value, frequently

face demolition. Some of the future–oriented activities hitherto dominating central areas tend to abandon the stifled centre and look for more favourable locations. It also uproots the rightful inhabitants, who, finding themselves jobless in the city centres, gradually move to the outskirts and thus hasten the decay of the inner city areas while greatly increasing the demand for utilities in the new locations.

The social disparities existing in central urban areas have a cumulative effect; underemployment and unemployment increase the inequalities and mean that the already under–privileged fall still further behind. Inter–related negative factors begin to operate and they often lead to alienation and violence with, as a consequence, the exodus of those who can afford to leave and their replacement by new arrivals fascinated by the big city. Decaying buildings with no settled occupants become haunts of crime, drug use, violence and fear. As businesses, good schools, and educated people leave, city centre's have the poor, uneducated, undesirable minorities and sick. Housing deteriorates. Schools become less capable of educating as violence increases, funds decrease and the best staff leave. Teachers see themselves as baby sitters or policemen. Social services deteriorate; medical services become more centralized into public hospitals and private practices move to better neighbourhoods; unemployment and welfare agencies become part of the means of oppression; and police services shift emphasis from prevention to restricting crime to defined geographical areas. Private housing becomes substandard as returns on investment drop below zero. Arson for the sake of insurance claims is the most profitable way of recovering investments from property. Residents see no need to maintain rental property as any improvement will not be paid for by the owner and is likely to attract criminals. Public housing is often the location of high rates of crime: women are raped in elevators and drugs pushed in hallways. Access to the political processes becomes more difficult until indifference sets in for the voting population and corruption and cynicism become mechanisms for survival for politicians. The private sector invests less and less. Jobs move further and further away. Transportation is by old junk cars for those who can afford them, luxury automobiles for the criminal and increasingly bad public services for the poor. Shopping for necessities means either paying higher prices for lower quality goods or travelling to distant and frequently alien centres. Criminal investments increase with gambling, prostitution and drugs. The self image of residents and employers becomes either one of a victim of circumstance or one of an outlaw. The absence of controls governing land use is unfortunate in the city centres of many market–economy countries where, in most cases, the incremental value of land flowing from community growth is not made available to the community to finance its further development, but is largely accrued to land speculators.

Claim The dilapidation of the inner city is actually the superficial manifestation of a much more general running down of traditional urban functions and institutions. Economic and social development hits hardest where it originates and changing cities are a reflection of changing societies just as the problems that have to be faced in inner city areas are the instant, physical materialization of the problems thrown up by modern society.

Broader Obstacles to community achievement (#PF7118).
Narrower Litter (#PD2541).
Aggravates Rats as pests (#PE3177) Human violence (#PA0429) Civil disorders (#PC2551) Denial of human rights (#PB3121) Massive urban emigration (#PG9773) Insensitive urban renewal (#PD7320) Urban slums in industrialized countries (#PE1887).
Aggravated by Urban poverty (#PC5052) Traffic congestion (#PD0078) National economic recession (#PD9436) Unethical commercial practices (#PC2563) Uncontrolled urban development (#PC0442) Urban–Industrial air pollution (#PJ5532) Environmental degradation by automobiles (#PE6142) Inadequate maintenance of infrastructure (#PD0645) Overcrowding of housing and accommodation (#PD0758) Segregation of poor and minority population in urban ghettos (#PD1260) Underutilization of facilities due to daily or seasonal peaks (#PF0827).

♦ PC2618 Profiteering
Excess profits — Windfall profits — Excessive profits

Nature Profiteering is the pursuit of gain at public expense. It may be by cheating the government, or it may be by robbing developing countries, or by over–charging consumers for medical, food and other necessities.

Incidence In 1990, as a result of the Gulf crisis, oil companies were accused of profiteering through the manner in which they increased prices in order to maintain profitability.

Claim The more notorious profiteers are those who benefit financially from wars, black marketeers and armaments manufacturers. In some views, to these can be added the major oil and pharmaceutical companies, medical, legal, and accounting malpractitioners, some private hospitals, funeral directors and undertakers in rich countries, corrupt government officials, some banks, lending or investing institutions (local or international), landlords and capitalists of all kinds, centrally planned economies that exploit their citizenry, and labour unions who force excessively high wages or profit for workers.

Counter–claim Profiteering is a polemical term that reveals an ignorance of the economics of national and private wealth and the concept of value. The existence, movement, expansion, consumption and replenishment of wealth are essential to profit economies. It is not subject to artificial levellings or controls, as it depends more on nature and human behaviour, the greatest variables of all, than on rational programmes or plans. The excess profits of one period are offset by capital or research investments, or by losses in another period. Only high liquidity provides risk and venture capital and the opportunity for an expanded employment base.

Broader Unethical practices (#PC8247) Short–term profit maximization (#PF2174).
Narrower Fraud by government agents (#PD8392).
Aggravates Price regulation (#PG5537) Dependency on middlemen (#PD4632).
Aggravated by War (#PB0593) Short–term gain (#PF8675) Corporate greed (#PF7189) Manufacture of munitions (#PG5539) Conspiracy against the public (#PF4198) Insufficient availability of goods (#PB8891) Distortion of international trade by embargoes and similar restrictions (#PE0522).

♦ PC2649 Unfair elections
Unjust election administration — Dishonest elections

Nature Administrative election procedure may not safeguard the rights and the convenience of voters, nor keep election expenses to a minimum, nor prevent fraud. The system of registration of voters, the assignment of party colours and symbols, ballot regulations, or regulations concerning the casting and counting of votes, may be at fault. The timing of the election, the hours which the booths are open and the location of polling stations may be unfairly organized. The organization of electoral campaigns and the choice of candidates may be unfair, and corruption may play a part in this; intimidation and undue influence over voters may go unchecked. Legislative seats may be unfairly distributed.

Broader Political injustice (#PC2181).
Narrower Unjust election timing (#PD2907) Obstruction of elections (#PD3982) Intimidation of electors (#PD2044) Political smear campaigns (#PD9384) Unrepresentative electoral systems (#PD9641) Bias in selection of political candidates (#PD2931) Electoral organization favouring political incumbents (#PF5153).
Related Unjust electoral campaigns (#PD2919).
Aggravates Political stagnation (#PC2494).
Aggravated by Lack of political integrity (#PF0796) Political party manipulation of elections (#PD2906).

♦ PC2659 Violation of neutrality
Broader Neutrality (#PF0473) Aggression (#PA0587) International aggression (#PB0968).
Aggravated by Unfulfilled treaty obligations (#PF2497).

♦ PC2662 Bureaucratic fragmentation
Nature In order to respond to the proliferation of claimants, constituents and contending groups, and the complexity of issues, bureaucracies become fragmented and specialized; they then tend to compete with one another for information and resources. Lines of organization become lines of secrecy and loyalty: each department restricts information that might advance the competing interests of the others. Such fragmentation may smother initiative. In the case of the extensive bureaucracy of central government, this process may go so far that the whole mechanism becoming too ponderous to be capable of anything other than token change.

Broader Institutional fragmentation (#PC3915).
Related Bureaucracy as an organizational disease (#PD0460).
Aggravates Lack of leadership (#PF1254) Administrative delays (#PC2550) Bureaucratic factionalism (#PF7979).

♦ PC2673 Culture shock
Nature Cultural invasion may be followed by the inability to adapt to a different cultural experience. The result is apathy and social disintegration, and in the case of certain primitive tribes it may be lead to their extinction. Culture shock may occur among immigrants, rendering them susceptible to exploitation and creating divisions and conflicts. Culture shock may also affect nations as a result of economic and cultural invasion leaving them dependent, in debt, and susceptible to foreign political influence.

Reverse culture shock is the experience of many expatriate workers. These returning business people, international agency workers and missionaries find it difficult to work with fellow country-men. Their children often know nothing of their parent's home country, school is difficult, friendships are hard to make, and they are treated differently by teachers, classmates and neighbours. Wives must adjust to smaller houses, no servants, and different systems of shopping. Working expatriates find different working conditions.

Incidence Culture shock is a widespread phenomenon. The social disintegration which it entails may cause alcoholism, drug dependence, suicide, promiscuity, crime and violence. In the inability of social groups to adapt economically it may cause debt, slavery, exploitation, general dependence, and as a result poverty, malnutrition and disease. It occurs in this way particularly with indigenous populations. Underprivileged immigrants into industrialized countries may similarly suffer. Other immigrants and sophisticated societies suffering from cultural invasion show more complicated psychological problems as a result, and the situation with nations is perhaps the most complex of all.

Counter–claim The negative social effects attributed to culture shock as a self–induced phenomenon removes the responsibility from the host or dominant culture. It is the dominant culture with its unfair discrimination at every conceivable level that causes sub–culture disintegration or apathy. On the other hand, just as giving birth involves a sustainable shock to the organism of the mother, almost all great civilizations were based on cultural integration and absorption. Thus, while sub–cultures may be in shock they have the chance to be re–born, metamorphosed as part of a broader cultural stream. Their isolation is broken down and they enter universal history with fuller opportunities for development.

Broader Socioeconomic stress (#PC6759).
Related Obsolete basis of cultural identity (#PF0836).
Aggravates Crime (#PB0001) Suicide (#PC0417) Alcohol abuse (#PD0153) Human violence (#PA0429).
Aggravated by Prejudice (#PA2173) Misconceptions (#PG5650) Cultural barriers (#PB2331) Cultural invasion (#PC2548) Lack of education (#PB8645) Lack of communication (#PF0816) Inter–cultural misunderstanding (#PF3340).

♦ PC2677 Political instability
Dependence on political instability — Unstable electoral system

Nature Lack of political continuity in the application of government tends towards sudden violent or disruptive change and fluctuation. Political instability is a function of a lack of consensus, which may arise as a result of colonial influence, a failure to materialize the sense of 'nationhood' in the citizenry, dictatorship, economic instability or foreign influence. Political instability may result in social, ethnic or national disintegration and provide a barrier to progress through repression and negative change. It may also cause foreign intervention or an increase in foreign influence.

Broader Instability (#PA0859) Inadequacy of political doctrine (#PF3394).
Narrower World anarchy (#PF2071) Political revolution (#PF3237) Weakness of multi–party parliamentary systems (#PF3214) Political instability of developing countries (#PD8323).
Related Political immaturity (#PJ5657).
Aggravates War (#PB0593) Communism (#PC0369) Political oligarchy (#PD3238) Political repression (#PC1919) Social disintegration (#PC3309) Human racial regression (#PF0411) Political disintegration (#PC3204) Illegitimate political regimes (#PC1461) Non–violent political revolution (#PD3228) International non–military conflict (#PF3100).
Aggravated by Political apathy (#PC1917) Political schism (#PC2361) Political inertia (#PC1907) Lack of leadership (#PF1254) Political injustice (#PC2181) Ideological conflict (#PF3388) Political stagnation (#PC2494) Subversive activities (#PD0557) Corruption in developing countries (#PD0348) Contradictions in capitalist systems (#PF3118) Lack of participation in development (#PF3339) Multiplicity of languages in a national setting (#PC1518).
Reduced by Conservatism (#PF2160) Foreign military presence (#PD3496).

♦ PC2696 Environmental warfare
Ecocide — Geophysical weapons

Nature In warfare, recourse to deliberate destruction of the environment is frequently an integral part of military strategy. Such warfare involves the defoliation or destruction of forest trees, the pollution or craterization of cultivated fields, and destruction or diversion of water sources. By these means it is hoped to deny the enemy cover, food, and the life–support of the countryside, thus making it more difficult for him to mass for effective attack.

Incidence Environmental warfare techniques were used by the USA in Vietnam. In the period 1962–1968, 4,560,600 acres of forest land, representing 10 per cent of the entire area of Vietnam, were sprayed with herbicides. These herbicides, such as Agent Orange, contaminate crops, have teratogenic effects on unborn children, and poison humans and animals. Plant cover can also be destroyed by the use of very large, bulldozer–driven Rome ploughs, and bombardment and artillery fire may also be used, deliberately or incidentally, to destroy the environment. Munitions create craters, which prevent the use of arable and timber land indefinitely, create breeding grounds for mosquitoes, accentuate soil run–off and erosion, and cause laterization of the land. Weather modification cloud–seeding has been used to increase rainfall to make roadways muddy and unusable; and land may be indiscriminately devastated by using electronic battlefield or systematic bombing techniques.

Background Geophysical weapons and weather modification techniques of warfare are in the experimental or research stages. Objectives include storm and rain production, stimulation of seismic earth shocks, ozone layer interference, production of tsunamis, creation of lightning, and other weather and environmental disturbances.

Refs Stockholm International Peace Research Institute *Warfare in a Fragile World* (1980); Stockholm International Peace Research Institute *Weapons of Mass Destruction and the Environment* (1977); Stockholm International Peace Research Institute *Ecological Consequences of the Second Indochina War* (1979); Westing, Arthur H (Ed) *Global Resources and International Conflict* (1986); Westing, Arthur H (Ed) *Environmental Warfare* (1984).

Broader Biochemical warfare (#PC1164)
Denial of human rights in armed conflicts (#PC1454).
Narrower Defoliation (#PD1135) Craterization (#PG5684).
Related Weapons (#PD0658) Destruction of wildlife habitats (#PC0480).
Aggravates Malaria (#PE0616) Teratogens (#PE0697) Soil erosion (#PD0949)
Deforestation (#PC1366) Laterization of soil (#PG5688)
Environmental degradation (#PB6384) Inadvertent modifications to climate (#PC1288).
Aggravated by Aggressive foreign policy (#PC4667)
Counter-insurgency warfare (#PG5689)
Mismanagement of environmental demand (#PD5429).
Reduces Plant pathogens (#PD1866).

♦ PC2724 Weakness in trade between different economic systems

Nature In terms of trading practices, countries may be allocated to one of three categories: developed market economies, centrally-planned (or socialist) economies, and developing countries. Trade between countries of different systems is weaker than that between countries of the same system.

Counter-claim While it is relatively weak, trade between COMECON countries and OECD countries is growing, after the 1982 recession. However, this trade flow is characterized by an unsatisfactory commodity structure, persistent imbalances, limitations and restrictions of different kinds and the level of cooperation is below what is feasible.

Refs Perry, C M and Pfaltzgraff, R L *Selling the Rope to Hang Capitalism?* (1987).

Broader Ideological conflict (#PF3388) Lack of international cooperation (#PF0817).
Narrower Limited tripartite cooperation (#PF0919)
Weakness in trade between socialist and developing economies (#PE2953)
Weakness in trade between socialist and developed market economies (#PE2954).
Aggravated by Double standards in morality (#PF5225)
Bilateralism in aid to developing countries (#PE9099)
Lack of consumer choice in centrally-planned economies (#PD0515)
Prohibitive cost of maintaining comprehensive document collections (#PE1122)
Differences in trading principles and practices between different economic systems (#PC2952).

♦ PC2725 Non-tariff barriers to international trade

Nature Tariff concessions may be rendered partly or wholly ineffective by non-tariff restrictions. The most conspicuous form of such restrictions is outright quantitative limitation. Less conspicuous but often more insidious forms include: discriminatory state trading, national procurement practices, content regulations, anti-dumping legislation, marketing obstacles, packaging and labelling restrictions, safety and health requirements, customs procedures, surcharges and miscellaneous fees.

The effect of these restrictions on the value of the trade of the developing countries is that an above-average proportion of imports from these countries into the developed market economies is subject to non-tariff barriers. It has been estimated that 28 per cent of the value of imports from developing countries is subject to these restrictions, as compared with 11 per cent for imports from the rest of the world. Although the striking difference may have been somewhat exaggerated by the fact that only products of export importance to the developing countries have been considered, the over-all picture is unmistakable.

Incidence In a recent study estimates that trade and technical barriers within the European Community were costing community industries Ecu 120 billion per year.

Broader Restrictive trade practices (#PC0073) Distortion in international trade (#PC6761).
Narrower Technical barriers to trade (#PC4382) Inter-cultural trade barriers (#PD9651)
Discriminatory exchange rate policies (#PE8583)
Distortion of international trade by quantitative restrictions (#PE9027)
Distortion of international trade by embargoes and similar restrictions (#PE0522)
Distortion of international trade through obstacles to patent protection (#PD0455)
Distortion of international trade by discriminatory preference agreements (#PD0340)
Distortion of international trade as a result of government participation (#PD2029)
Distortion of international trade by restrictive controls on movement of labour (#PE8882)
Distortion of international trade by selective indirect taxes and import charges (#PE8867)
Distortion of international trade by restrictive controls over foreign investment (#PE8525)
Distortion of international trade by discriminatory customs and administrative entry procedures (#PE2603)
Distortion of international trade by minimum pricing regulations and other measures to regulate domestic prices (#PE1182).
Related International trade barriers for primary commodities (#PD0057).

♦ PC2753 Animal injuries

Nature Damage to tissue may be caused accidentally or deliberately. Injuries include bone fractures, damage to muscles, tendons, ligaments, skin, or internal or external damage to organs, and may involve bleeding or bruising. Injuries can cause death or necessitate the destruction of an animal, or they may lead to disease if wounds become infected.

Incidence Injuries in domestic animals occur most frequently with dogs, cats and horses which are more likely to come into contact with traffic, or be used for activities which incur more risk than for animals such as cattle, sheep, pigs and poultry, or for caged domestic pets. Horses may sustain injuries from hunting, show jumping or other activities of this kind, or from traffic accidents. Dogs and cats may be injured from road accidents or from fighting one another. Rabies is spread in dogs by biting; and cattle may be injured by others horning them, by breaking through fences or falling into ditches; sheep may also be susceptible to the latter two causes. Injuries may be slight, such as punctures from barbed wire or thorns, but may provide the opportunity for infection from diseases such as blackleg. Domestic animals may be injured through cruelty, beating or mishandling, or may be the subject of or involved in cruel sports such as bullfighting, where horses are often very badly injured by the bull, as well as the bull being injured with darts long before it is killed. Wild animals may be injured by one another either in fights over territory or during mating or for leadership of a herd or group, or by predators. They may be caught in traps, shot without being killed, or may be injured while being hunted, either by falling or by dogs.

Broader Injuries (#PB0855) Vulnerability of organisms (#PB5658).
Aggravates Fungal diseases (#PD2728) Animal infertility (#PC1803)
Economic loss through slaughter of diseased animals (#PE8109).
Aggravated by Cruel sports (#PD1323) Bullfighting (#PG4834)
Hunting of animals (#PC2024) Trapping of animals (#PE5735)
Stray dog populations (#PE0359) Road traffic accidents (#PD0079)
Maltreatment of animals (#PC0066) Inadequate animal welfare (#PC1167)
Aiding escape of prisoners of war or enemy aliens (#PE1200).

♦ PC2765 Inadequate feeding of animals

Inadequate animal nutrition — Inadequate feeding of farm animals

Nature Insufficient nutrition of animals or an inadequate balance in their diet causes emaciation, death, anaemia, stress, and contributes to the severity of diseases such as mange.

Incidence The problem may occur as a result of drought or insufficient pastureland. It may be because of negligence or ignorance, or lack of financial means to provide adequate feeding. Alternatively, it may be deliberate, as with the production of white veal, where calves are deprived of all roughage and fed only on milk substitutes which induces anaemia. They suffer severe stress because of this, and may try to obtain fibre from the wooden crates in which they are housed.

Broader Maltreatment of animals (#PC0066).
Narrower Vitamin E deficiency in domestic animals (#PE4760).
Related Agricultural mismanagement of housed farm animals (#PD2771)
Denial to working animals of restorative nourishment and rest (#PE4793)
Denial to animals of the right to conditions of life and liberty proper to their species (#PE6270).
Aggravates Mange (#PE2727) Animal deaths (#PE7941) Animal diseases (#PC0952)
Animal infertility (#PC1803) Animal stress in factory farming (#PD2760)
Poor quality of domestic livestock (#PD2743)
Inferior meat quality from intensive animal farming units (#PE2770).
Aggravated by Insufficient pastureland (#PG5805)
Limited agricultural education (#PF8835)
Weather as a factor of animal disease (#PD2740)
Commercial exploitation of wild animals (#PD1481)
Inadequate legislation for animal welfare (#PE5794)
Underdeveloped sources of income expansion (#PF1345).

♦ PC2791 Economic isolationism

Economic isolation — Destructive economic isolation

Broader Isolation (#PB8685) Demeaning community self-image (#PF2093)
Detrimental international repercussions of domestic agricultural policies (#PF2889).
Narrower Contained village economy (#PJ0594).
Related Political isolation (#PC7569) National isolationism (#PF2141).
Aggravates Ineffective economic structures in industrial nations (#PE4818).

♦ PC2801 Maldistribution of food

Dependence on maldistribution of food — Unequal access to food

Nature The unequal distribution of food between and within countries causes hunger for many millions of the world's people. Those in developed as well as developing countries who go to bed hungry do so not necessarily because food is unavailable but because they can neither afford to buy it nor own the land necessary to produce it. A less noticeable food distribution problem exists within the family itself. Pregnant women, nursing mothers and children have the greatest food needs but are at the end of the distribution chain, with men assuming first claim to the available food.

Maldistribution of food is a technical and a political problem. Technically it is difficult to transfer food from an area of surplus to an area of need. The physical problems of moving the food from place to place include the lack of transportation infrastructure where the need for food is the greatest. The institutional problems of moving food from cities and towns to the rural poor include the lack of administrators to manage food transfers. On the political side it is unclear to what extent developed countries are committed to meeting the food and resource needs of poor countries. Not infrequently countries in need are not interested in their citizens most requiring food to receive it.

Incidence According to a 1984 FAO report, the total food production of the world, minus all the major losses (both inedible and avoidable wastages), when divided by the world's population, yields the equivalent of 2,743 kilocalories available to each person. However, breakdown by geographical areas shows wide disparities. The following figures include average daily energy requirements per person per region based on the age and sex structure of the population, the average body weight, the climate, and other relevant factors. The USA and Canada produce the equivalent of 5501 kilocalories per person per day, while 2650 is the average daily requirement; Western Europe produces 2893 per person, the average daily requirement being 2574; Australia and New Zealand, 7705 and 2650, respectively; Africa 1771 and 2332; Latin America, 3464 and 2416; the Near East, 2153 and 2426; the Far East, 2334 and 2230; Asian centrally planned economies (China, Kampuchea, North Korea, Mongolia, and Vietnam), 2336 and 2230; USSR and Eastern Europe, 3230 and 2574.

Claim The world needs a more equitable food distribution and crop production more than it needs more output. Per capita world grain output per person is approximately 325 kilograms. In some countries annual grain availability per person averages only 150 kilograms, requiring that it all be consumed directly; while other countries exceed 700 kilograms, which is largely converted into meat, milk, and eggs. A more equitable distribution of this bounty could greatly improve the nutritional status of millions of people and thereby greatly improve political tensions between both countries and regions.

Counter-claim That different regions have large populations as well as naturally limited resources is a fact at life. Food distribution on a world-wide basis is totally unrealistic. Cereal stocks have to be paid for and most developing countries are unable to afford it. The provision of free food aid (which, in times of economic recession in developed countries or in pursuit of particular foreign policies, may be rendered impractical) is merely a temporary expedient, making less developed countries dependent upon others and also distorting their own agricultural markets.

Broader Hunger (#PB0262) Maldistribution of resources (#PB1016).
Narrower Inadequacy of food aid (#PF3949) Export of nutritious food (#PJ1365).
Aggravates Food insecurity (#PB2846)
Unsustainable agricultural development (#PC8419)
Economic disadvantages of excessive food production in developing countries (#PF4130).
Aggravated by Inadequate mechanisms for securing sufficient food supplies (#PF2857).

♦ PC2815 Unequal income distribution between countries

Dependence on unequal income distribution between countries — International income gap — Income inequality between countries

Incidence Comparative per capita income in developed market countries is ten to eleven times higher than in developing market economies. On a regional basis for all countries: Africa is just below the average for developing market economies; Asia above the average and 30 to 40 percent above Africa; the Middle-East is about $300 higher than Latin America, but less than a third of the developed market economy's average; all European countries average over ten times Africa, and seven times Asia, but are 15 percent lower than the developed market economies; Australia-Oceania and North America are the highest – both are above the average for developed market countries, the former 5 percent above average, the latter 35 percent above; North American per capita income exceeds African by over 1,500 percent.

Claim The vast inequalities between the poor and the rich nations create one of the world's most serious and urgent problems. Failure to reduce the international income gap, or even to prevent it from widening further, is a major source of international tension and a threat to peace.

Counter-claim It is quite unlikely, with the best will in the world, that international inequalities can be reduced, or even prevented from increasing, for the rest of this century. But if one looks at realities rather than appearances, this does not seem an insoluble problem. The immediately urgent objectives are to achieve a substantial absolute improvement in the levels of the poorest countries (which is by no means identical with reducing inequality) and to bring about profound qualitative changes in all countries, which are needed for their survival in a world of limited resources and delicately poised ecological equilibria. To make a major issue of international

inequalities only distracts attention from these more serious and urgent problems. It plays into the hands of those who are interested not in achieving real economic progress but in making political capital out of alleged economic failure.
Broader Economic and social underdevelopment (#PB0539).
Related Unequal income distribution within developing countries (#PD7615).

♦ PC2913 Cultural corruption
Nature Debasement of culture in art forms, language, history, and ideology may arise during, and hasten the process of, ethnic disintegration; or it may constitute a decadent or sterile epoch in the course of cultural evolution. Cultural assets may be altered to suit a certain line of propaganda.
Broader Corruption (#PA1986).
Narrower Cultural stagnation (#PC8269).
Related Proverbial lore errors (#PF5653).
Aggravates Ethnic disintegration (#PC3291) Destruction of cultural heritage (#PC2114).
Aggravated by Decadence (#PB2542) Propaganda (#PF1878)
Forced assimilation (#PC3293) Cultural illiteracy (#PD2041)
Untransposed significance of cultural tradition (#PF1373).

♦ PC2914 Ideological corruption
Nature The debasement of religion and non–theistic ideology may take the form of deviation and schism, hypocrisy, heresy, decadence and excess, a dilettante dabbling in the occult and certain kinds of immoral experimentation. Ideological corruption actually serves to sharpen the need for ideology, and encourages apathy, alienation, cynicism, brutality and fear.
Counter–claim The accusation of corruption in the context of ideology is very often judgement based on outmoded traditional attitudes.
Broader Corruption (#PA1986) Immorality (#PA3369).
Related Amoralism (#PF3349).
Aggravates Alienation (#PA3545).
Aggravated by Inadequacy of doctrine (#PF3396).

♦ PC2915 Unethical intellectual practices
Intellectual irresponsibility — Intellectual corruption
Nature Intellectual capacity can be misused, for commercial or immoral purposes, to: achieve indoctrination; obtain high profits without due regard to the consequences; devise inhumane uses of scientific technique; develop brainwashing and torture techniques; develop weapons and sophisticated warfare techniques; develop information against the public interest in favour of vested interest.
Broader Corruption (#PA1986) Immorality (#PA3369).
Narrower Corruption of documents (#PE7900).
Related Unethical professional practices (#PC8019).
Aggravates Corruption of meaning (#PB2619).

♦ PC2920 Misappropriation of public funds
Embezzlement of public funds
Nature Public funds can be misused to the benefit of partisan, group or individual interests, and they may used unjustly to finance political parties, electoral campaigns, secret police, or espionage. Government contracts may be unfairly allocated; unbacked promissory notes may be issued and the funds used; appointed positions may be given to favourites and appointees may enjoy unequal salaries.
Broader Embezzlement (#PD2688) Corruption of government leaders (#PC7587).
Aggravated by Inappropriate public spending by government (#PF6377).

♦ PC2921 Industrial espionage
Clandestine market research — Organizational intelligence
Nature Industrial espionage aims at the covert gathering of information on rival enterprises' technological developments and secret processes. It may also include stealing marketing plans, data bases containing distributors names and distribution logistics, financial statements, overall corporate long–range planning including business models and forecasts, and computer programmes and software for management systems. Industrial spies usually operate within the jurisdiction of their employers marketing or market research departments, except technical specialists whose liaison is with industrial research.
Incidence In the UK in 1989 it was estimated that industrial espionage cost companies 5 million pounds per year. A higher estimate suggests 5 million pounds per incident, with some 2,500 UK companies spying on their competitors. As an indication over 100,000 bugging devices are sold annually in the UK.
Claim Where the survival of a company in highly competitive markets is concerned it is inevitable that traditional marketing intelligence/research will tend to move towards covert industrial espionage.
Counter–claim As with political and military espionage a large part of industrial espionage simply consists in analysing publicly available, or at any rate non–secret, information. Computer enhanced evaluation of competitors' finances, marketing and engineering or design style evolution, plus technical forecasting of industry and market share data, allows a great deal to be described that appears as inside information, and hence the results of espionage, when in fact it is highly developed comparative technical and marketing research.
Refs Bottom, Norman R and Gallati, Robert R *Industrial Espionage* (1984); Sable, Martin H *Industrial Espionage and Trade Secrets* (1985).
Broader Espionage (#PC2140).
Narrower Infringement of trade secrets (#PD3537).
Related Theft of property (#PD4691).
Aggravates Sabotage (#PD0405) Informers (#PD8926)
Counter–espionage (#PD2923)
Unreliability of equipment and machinery (#PC2297).
Aggravated by Economic rivalry (#PD8897)
Unaccountable government intelligence agencies (#PF9184).

♦ PC2933 Political purges
Nature Political purges check governmental development, tend to tighten government control, and cause alienation and apathy. Persons in political office whose policies and opinions are felt to be corrupt, impure or immoral may be eliminated by demotion, arrest, imprisonment, exile or execution. Purges are usually the result of ideological conflict and represent a hardening of attitudes against any liberalization which has evolved.
Incidence It is estimate that Stalin was responsible for the death of over 20 million during the 1930s. Few of any prominence over the age of 40 survived such purges, other than members of the Politburo.
Refs Conquest, Robert H *The Great Terror.*
Broader Political repression (#PC1919) Ideological conflict (#PF3388).
Aggravates Internment without trial (#PD1576).
Aggravated by Secret police (#PE6331) Ideological deviation (#PF3405)
Corruption in politics (#PC0116) Excessive government control (#PF0304)
Double standards in morality (#PF5225).

♦ PC2934 Physical intimidation
Bullying
Nature The threat of physical force to exploit others may be in the form of: police brutality; demonstrations; thuggery and juvenile delinquence; 'protection' tactics used by organized crime; terrorism and guerrilla tactics; various kinds of torture; or the maintenance of a large standing army, police corps or other group based on the use of force. Physical intimidation may be used in a variety of situations for a wide variety of reasons.
Refs Munthe, E and Roland, E (Eds) *Bullying* (1989).
Broader Injustice (#PA6486) Intimidation (#PB1992)
Lack of central planning structures in small communities (#PF2540).
Narrower Hazings (#PF5392) Harassment in playgrounds (#PE7768)
Physical intimidation by children (#PE2876)
Intimidation of pedestrians by vehicles (#PE6139).
Related Violence as a resource (#PF3994).
Aggravates Teasing (#PE4187) Juvenile suicide (#PE5771).
Aggravated by Personal unpopularity (#PF4641).

♦ PC2935 Psychological intimidation
Nature The use of irrational fear or uncertainty to exploit others may be on an individual, group, national or international level. Psychological intimidation may be accomplished by means of propaganda and other forms of indoctrination: advertising by bureaucracy and other methods of alienation; moralism and authoritarianism; the inducement of an inferiority complex by means of elitism and class distinction, including technocracy and meritocracy and the communication of information in a way that is hard to understand or misleading; the use of terror as a subversive tactic; secret police, show trials, etc; the use of threat and the creation of confusion. Psychological intimidation may be combined with physical intimidation, in conjunction with large standing army, military parades, heavy studded leather clothing and boots on teenage youths, carrying of knives, guns, keeping of fierce guard dogs, and so on.
Broader Injustice (#PA6486) Intimidation (#PB1992).
Related War (#PB0593).

♦ PC2936 Racial intimidation
Racial violence
Nature Force or superior power may be used to exploit a different racial group through fear. This may be achieved by physical or psychological intimidation, police brutality, inequality before the law, or indoctrination. Racial intimidation serves to maintain racial inequalities and exploitation and may lead to apathy or to subversive activities.
Claim The United States continues and even is escalating policies and practices of racist violence against women, children and families of an estimated 60 million racially oppressed people, including 32 million African Americans, 20 million Latino people and 8 million Asians and native Americans.
Broader Intimidation (#PB1992).
Related Racial conflict (#PC3684).
Aggravates Racial segregation (#PC3688) Racial discrimination (#PC0006)
Harassment of public officials (#PD4915).
Aggravated by Segregation in housing (#PD3442)
Legalized racial discrimination (#PC3683).

♦ PC2937 Religious intimidation
Nature Force or superior power may be used to exploit a different religious group through fear. This may be achieved by physical or psychological intimidation, terrorism, indoctrination and occultism. Religious intimidation may be part of a social, political or economic conflict between two religious groups which may conduct terrorist activities against each other. Indoctrination and moralism may produce a guilt complex. Fear of the occult and of being cursed may cause mental disorder, physical disease and even death (as is recorded in tribal societies).
Broader Intimidation (#PB1992) Religious repression (#PC0578).
Related Religious conflict (#PC3292).
Aggravates Religious extremism (#PF4954) Religious discrimination (#PC1455).
Aggravated by Religious intolerance (#PC1808)
Religious discrimination in the administration of justice (#PE0168).

♦ PC2938 Political intimidation
Nature The political exploitation or control of others may be achieved by means of physical or psychological intimidation, through the use of force or superior power used nationally or internationally. Methods of intimidation include indoctrination, diplomatic and military pressures, economic blockade, espionage, terrorism and the maintenance of large standing armies; and on a national level, the use of secret police, political purges, show trials, censorship, corruption, bureaucracy, strong government control over national activities, party control over individuals, inequality before the law, and elitism. Political intimidation may result in general apathy and alienation or, on the contrary, in subversive activities; it may reinforce a totalitarian or dictatorial regime, exploitation and political inequalities.
Broader Intimidation (#PB1992) Political repression (#PC1919).
Narrower Police intimidation (#PD0736) Intimidation of electors (#PD2044).
Related Political conflict (#PC0368).
Aggravates Discrimination in politics (#PC0934) Harassment of public officials (#PD4915)
Undemocratic political organization (#PC1015) Political corruption of the judiciary (#PE0647).

♦ PC2939 Industrial intimidation
Nature Force or superior power may be used to exploit employees or employers. Management may try to intimidate its employees by threatening dismissal, redundancy, short–time or court action (which may result in fines). Employees may try to intimidate management with the use of wild cat strikes, sympathetic strikes vertically within each production process of the industry, or threat of general strike and other restrictive practices. They may seek to intimidate each other with violent picketing or restrictive practices. Industrial intimidation techniques lead to a loss of production amd may result in general economic crisis and political conflict. They aggravate class conflict and division.
Broader Intimidation (#PB1992).
Related Economic intimidation (#PC3011).
Aggravates Strikes (#PD0694) Lock–out (#PD6808) Underproductivity (#PF1107)
Violent picketing (#PG5873) Global economic crisis (#PC5876).
Reduced by Arbitration (#PS2125).

♦ PC2940 Social intimidation
Dependence on social intimidation
Nature Force or superior power may be used to exploit other social groups or individuals through fear. Intimidation may arise out of class, racial, religious, age, sexual or other conflict, or stress. It may take the form of crime, violent or otherwise, or of indoctrination, in the sense of moralism and censorship or advertising and other social pressures, or alienation and isolation before the law. Social intimidation constitutes a barrier to social progress and serves to maintain existing segregation, exploitation and inequalities. It may induce conformism, apathy or deviation.
Broader Intimidation (#PB1992).

CROSS-SECTORAL PROBLEMS PC3113

Related Social conflict (#PC0137).
Aggravates Social fragmentation (#PF1324).
Aggravated by Social discrimination (#PC1864).

♦ PC2944 Territorial fragmentation

Balkanization

Nature Under certain conditions territories are recognized as having the right to self-determination. This gives rise to accepted difficulties in the case of existing dependent areas, but raises more serious problems when the principle is considered equally applicable to component parts of existing independent countries in which the majority of people of those areas express the desire for self-determination. There is no recognized limit to the application of this principle, although excessive fragmentation is undesirable.

Counter-claim The dangers of fragmentation should not be exaggerated. Often measures of decentralization and local autonomy satisfy the small groups involved. Furthermore, it is important to distinguish between self-determination of small entities already in existence and self-determination of sub-units which do not yet have any recognized status. Unlike the latter, the former do not have to prove that they have the right to self-determination.

Broader Disintegration (#PA6858).
Narrower Secession (#PD2490).
Related Non-viability of small states and territories (#PD0441).
Aggravates Disruption of territorial integrity (#PC2945).
Aggravated by Institutional fragmentation (#PC3915).

♦ PC2945 Disruption of territorial integrity

Denial of rights to territorial integrity

Nature The integrity of territory, particularly a small one, may be disrupted when the territory is separated legally, administratively and politically from the larger territory of which it was a part. Equally disruptive is the partial or complete replacement of the original population of a territory by immigrants, the demands of whose descendents must be considered along with those of the legitimate indigenous inhabitants. In another sense, a downstream riparian state may also consider that its territorial integrity has been disrupted when an upstream state reduces the quantity of the river flow or changes its quality.

Incidence In the case of small territories, the problem is mainly a heritage from the past and is likely to be replaced by the problem of territorial fragmentation. Examples include: claim of Spain against the United Kingdom concerning the status of Gibraltar; claim of Morocco against Spain concerning Ifni; claim of Guatemala against the United Kingdom concerning British Honduras; claim of Argentina against the United Kingdom concerning the Falkland Islands.

Broader International aggression (#PB0968).
Narrower Enclaves and exclaves (#PD2154).
Related Denial of state's rights (#PD4814).
Aggravates Non-viability of small states and territories (#PD0441).
Aggravated by Divided countries (#PD1263) Territorial expansionism (#PC9547) Territorial fragmentation (#PC2944).

♦ PC2952 Differences in trading principles and practices between different economic systems

Nature States having different economic systems have correspondingly different national trading policies and practices; these differences aggravate the problems of foreign trade organization and impede the growth of the international market. The accepted principles and the national organization of foreign trade in developed market economies have important characteristics in common, as do the principles and organization of centrally planned economies; but the actual trading practices of a country in either group may vary according to whether the trading partner is a member of the same group or of a different one. In the case of the third main trading group, the developing countries, there is not the same general similarity of trading organization and practices. They have some structural characteristics in common, but the types of economic and social organization vary from almost pure market economies to almost full state ownership, so the institutional arrangements in the field of foreign trade also differ widely.

Incidence Trade among countries having different economic and social systems accounts for 4 per cent of world trade, 2.5 per cent being trade between socialist countries of Eastern Europe and developed market-economy countries, and about 1.5 per cent between socialist countries of Eastern Europe and developing countries.

Broader Competition between states (#PC0114) Lack of international cooperation (#PF0817).
Narrower Differences in trading principles and practices between developing countries (#PF2960).
Related Ideological conflict (#PF3388) Discrepancies between principles and practice (#PF4705).
Aggravates Weakness in trade between different economic systems (#PC2724).
Aggravated by Double standards in morality (#PF5225).

♦ PC3011 Economic intimidation

Nature The use of force or superior power to economically exploit other groups may be on an individual, national or international level. On an individual level it may take the form of class conflict and elitism. On a national level as monopoly or oligopoly or government control, it may take the form of restrictive business practices such as price fixing and restriction of entry of other firms into the market, or under competition it may take the form of unfair competitive practice. On an international level, as economic imperialism, it may impose effective foreign control and political domination. It serves to maintain or widen economic and technological gaps, dependency and alienation. Corruption and espionage may also be used as methods of economic intimidation.

Broader Intimidation (#PB1992).
Narrower Domination of developing countries by transnational corporations (#PE0163).
Related Economic conflict (#PC0840) Economic imperialism (#PC3198) Industrial intimidation (#PC2939).
Aggravates Economic gap (#PD8834) Human inequality (#PA0844).
Aggravated by Domination (#PA0839).

♦ PC3024 Male domination

Machismo — Male supremacy — Traditional male dominance — Male chauvinism

Nature Masculine ideology of the dominant he-man male, particularly dominating women, embodies a feeling of the inherent superiority of the male of the species. He must always appear to be morally upright, strong, ambitious, energetic, virile, brave and able to direct, guide and support his family. He should not appear effete, or have a preference for homosexuality, or do domestic work or other kinds of work which are traditionally considered to be the women's domain. These very rigid role ideas cause social inflexibility and constitute a barrier to progress. They incidentally induce a certain amount of discrimination against men as a direct result of the broad discrimination against women that is generated.

Incidence Male chauvinism relegates women to subordinate roles in all spheres, while it does not actually deny them entry. Consequences for women may be alienation, apathy, stress or impoverishment.

Claim Macho attitudes have been linked to maltreatment of animals, insensitivity to the environment, and suppression of instincts of compassion.

Broader Ideological conflict (#PF3388).
Narrower Gender stereotyping of employment (#PJ6290).
Related Androcentrism (#PF6648).
Aggravates Feminism (#PF3025) Family violence (#PD6881) Emotional immaturity (#PJ5907) Sexual discrimination (#PC2022) Violence against women (#PD0247) Inflexible social structure (#PB1997) Discrimination against women (#PC0308) Insufficient leadership training (#PF3605) Gender discrimination in developing countries (#PD9563) Sexual discrimination in contraceptive methods (#PF1035).
Aggravated by Grey lies (#PF3098) Traditionalism (#PF2676) Double standards in morality (#PF5225).

♦ PC3056 Disproportionate external public debt

Foreign debt — Heavily indebted countries

Nature Governments may borrow funds on foreign money markets to finance developments within their country. Foreign-held debt gives a claim by foreigners against national output in the sense that payment of interest and principal by the capital-poor nations requires export of goods to earn national income. To the extent that foreign debt continues to rise, taxation may rise, and economy in terms of public services and the standard of living in general tend to fall. Increasing foreign debt may open the way to greater transnational corporation ownership of production facilities or greater penetration of the financial infra-structure. Excessive foreign debt leads to impoverishment until obligations are paid or until they are defaulted upon or until they are repudiated. Unpaid debt leads to being barred from the world capital markets, and can lead to peaceful or violent change in government or to war.

Incidence The percentage of export earnings required to pay interest alone (and not to retire the principal) exceeds 50 percent in four Latin American countries. In two others, it is 40 percent. In Turkey and the Philippines it is also about 50 percent. The 17 countries deemed in 1988 to have encountered severe debt servicing difficulties are: Argentina, Bolivia, Brazil, Chile, Colombia, Costa Rica, Côte d'Ivoire, Ecuador, Jamaica, Mexico, Morocco, Nigeria, Peru, Philippines, Uruguay, Venezuela, Yugoslavia. In 1990 it was estimated that 16 of the largest debtor countries owed $22 billion at the end of August, more than triple the amount owed in early 1989.

Refs Abbot, George C *International Indebtedness and the Developing Countries* (1980); Res, Zannis and Motamen, Sima *International Debt and Central Banking in the 1980s* (1987).

Broader Excessive public debt (#PC2546) Uncontrolled growth of debt (#PC8316).
Narrower Excessive foreign public debt of developing countries (#PD2133) Excessive foreign public debt of industrialized countries (#PD9168).
Aggravates Imbalance of payments (#PC0998).
Aggravated by Excessive external trade deficits (#PC1100).

♦ PC3074 Government propaganda

Nature Official information services provide news for national and international consumption. Information media used for government propaganda include radio, television, newspapers and journals, art and photographs, films and theatre. Government propaganda may be in favour of nationalism, war, racism, religious attitudes or any other policy or attitude needing popular support. More commonly, propaganda simply supports the economic, political and cultural status quo.

Broader Propaganda (#PF1878) Compulsory indoctrination (#PD3097).
Related Radio and television propaganda (#PD3085) Inadequate government publications (#PF3075) Restrictions on direct news coverage of parliamentary affairs (#PF3072).
Aggravates Conflict of information (#PF2002).
Aggravated by Excessive government control (#PF0304) Unaccountable government intelligence agencies (#PF9184).
Reduced by Uncontrolled media (#PD0040).

♦ PC3107 Poverty in capitalist systems

Nature Because the distribution of wealth under capitalism is uneven, accruing to an artificially created class (the bourgeoisie), the increasing number of unemployed and low wage-earners live under conditions of poverty.

Incidence In 1987 in the U.S. 13.6 percent of the population lives in poverty throughout the year, an a quarter experience poverty at some time during the year. For white people the poverty rate is 11 percent, for Hispanics 27.3 and for Blacks 31.1 percent. Recently in the United Kingdom some 9.4 million people live on or below the poverty line.

Broader Capitalism (#PC0564).
Aggravates Class conflict (#PC1573).
Aggravated by Social inequality (#PB0514).

♦ PC3108 Resource wastage in capitalist systems

Nature The capitalist profit motive necessitates an excess demand bolstered by advertising. The excess demand will only be promoted for certain goods (those which return the fastest profit) so that some goods which are necessary may be in short supply. The profit motive does not take account of the long-term availability of resources, which may be ruthlessly exploited while they remain at a low price, later creating scarcity and economic instability. Speculation by investors and financiers may create artificial shortages to increase profits, thereby causing inflation.

In a society in which a high standard of living demands particularly rapid consumption of new goods, many perfectly usable goods may be scrapped or goods may be made with a built-in obsolescence. The materials consumed in this kind of production are unlikely to be recovered in full, and are often not even partially salvaged; pollution is created by this overproduction of factory waste, scrap, and refuse, but fast profit-making industries tend to be less concerned with anti-pollution precautions. As machinery is developed to replace manpower (since it can produce more at a faster rate), so human resources are also wasted. The democratic capitalist ideal of providing universal education also wastes resources, by educating technicians or professionals for whom there may be no work either currently or in the future.

Broader Capitalism (#PC0564) Unproductive use of resources (#PB8376) General unproductivity of capitalist systems (#PF3103).
Narrower Excessive exploitation of raw material reserves by transnational enterprises (#PE0060).
Related Over-intensive soil exploitation (#PC0052) Waste of human resources in capitalist systems (#PC3113).
Aggravates Unemployment (#PB0750) Exploitation in capitalist systems (#PC3117) Ecologically unsustainable development (#PC0111).

♦ PC3113 Waste of human resources in capitalist systems

Nature Human resources are wasted under the capitalist system because of its basic motivation for profit rather than social welfare. The major profits of production return to the non-productive section of society – the artificially created property-owner class, the bourgeoisie. the proletariat, unable to acquire property, are denied self-development and therefore remain dependent wage-earners. With developments in science and technology, machines begin to take over the role of human labour, which is found to be too expensive in the quest for ever higher profits.

Broader Capitalism (#PC0564) Waste of human resources (#PC8914).
Narrower Underutilization of intellectual ability (#PF0100) Counterproductive capitalist investment financing (#PF3104).
Related Resource wastage in capitalist systems (#PC3108).
Aggravates Class conflict (#PC1573) Economic conflict (#PC0840).
Aggravated by Social inequality (#PB0514) Unrecorded knowledge (#PF5728) Lack of social mobility (#PF2195) Misdirection of human energies and desires (#PF8128).

♦ **PC3116 Excessive demand for goods in capitalist systems**
Excess demand for goods in capitalism
Broader Capitalism (#PC0564).
Narrower Competition in capitalist systems (#PC3125).
Aggravates Scarcity (#PA5984) Economic conflict (#PC0840)
Distortion of international trade by dumping (#PD2144)
General unproductivity of capitalist systems (#PF3103)
Abuse of science and technology in capitalism (#PE3105).
Aggravated by Irresponsibility (#PA8658) Social inequality (#PB0514).

♦ **PC3117 Exploitation in capitalist systems**
Claim Exploitation is seen as an inevitable result of the profit– motivation of production under the capitalist system and of the private ownership of property. These two factors create an artificial non–productive class, the bourgeoisie, whose wealth is created or sustained by the work of wage–earners. The absolute disparity between the income of the wage–earners and that of the property–owners increases as the owners engage in investment and speculative ventures, particularly abroad; and as machines are developed to replace manual production, thereby reducing labour's per–capita income, particularly by causing the loss of jobs. Exploitation is conspicuous in those techniques of speculative activity which create artificial scarcity or which, by other means, raise prices in order to secure profit, as this undermines the economic stability of the community or nation and may cause widespread unemployment and hardship. While the propertied class exploits the proletarian or working class on a national level, it also expands to exploit other countries, particularly the developing ones, in its quest for access to raw materials and labour at the lowest cost. Investments by rich countries impose heavy obligations on the poor ones, and much of the benefit and revenue is returned to the bourgeois in the 'donor' country.
Counter–claim In 19th century Marxist terms, the first phase of capitalism after the Industrial Revolution could be described as feudal, with wages, working hours and conditions subject to the whims of ownership. In the second half of the 20th century, free market economies exist with totally different social conditions. In addition to having an improved standard of living, well beyond that achieved under socialist systems, wage earners and their labour unions are now investors and consequently share in the benefits and risks of ownership. Labour union funds may be one of the important sources of investment capital needed to create or sustain jobs. Foreign investment means that as much as 80 percent of the cost of goods produced is expended in the host country, creating hundreds or thousands of jobs and stimulating the local economy, in addition to transferring technology.
Refs Waitakin, Howard and Waterman, Barbara *The Exploitation of Illness in Capitalist Society* (1974).
Broader Capitalism (#PC0564).
Narrower Denial of effective national self-determination by capitalist exploitation (#PE3123).
Related Economic exploitation (#PC8132).
Aggravates Unemployment (#PB0750) Economic conflict (#PC0840)
Denial of the right to work in capitalist systems (#PC3119).
Aggravated by Cheap labour (#PG5988) Social injustice (#PC0797)
Social inequality (#PB0514)
Resource wastage in capitalist systems (#PC3108).

♦ **PC3119 Denial of the right to work in capitalist systems**
Nature Since production in the capitalist system is based on profit–motive and not on necessity, the right to work is not recognized. At the same time no adequate provision is made for social welfare. The denial of the right to work is effectively a denial to bring together individual and national development.
Broader Capitalism (#PC0564) Denial of right to work (#PC5281).
Related Denial of right to free choice of work (#PE3963).
Aggravates Strikes (#PD0694) Class conflict (#PC1573) Economic conflict (#PC0840).
Aggravated by Exploitation in capitalist systems (#PC3117).
Reduced by Socialism (#PC0115) Communism (#PC0369).

♦ **PC3124 Denial of human rights in capitalist systems**
Nature Inherent from the contradictions in capitalism is the denial of certain human rights to some groups of society, notably the right to education, work, social security and national self–determination. This denial may be active or passive. Other rights such as freedom of thought, expression and movement and the right of privacy may also be abrogated. Human rights may be denied on a national or international basis; the latter particularly takes the form of colonialism, economic imperialism, foreign control and intervention. Poverty and alienation are heightened, causing conflict and eventual revolution, which under colonial conditions may take the form of guerrilla warfare and terrorism.
Broader Capitalism (#PC0564) Denial of human rights (#PB3121).
Related Denial of human rights in communist systems (#PC3178).
Aggravates Social conflict (#PC0137).

♦ **PC3125 Competition in capitalist systems**
Nature The free enterprise ethic of capitalism encourages the production–for–profit motive by private individuals whose enterprises compete with one another. The argument for this is essentially one of the survival of the fittest, and the predominance of the strongest and best. However, the conditions of competition in the capitalist world are infinitely more sophisticated and complex than this basic argument suggests. They involve artificial and unstable exchange values expressed in money (currency) and in commodities. Intense competition on the home and world markets induces a concentration of capital to withstand the pressure, which leads to inefficiencies, wastage of resources, unfair practices in restraint of trade (such as restriction of the entry of new firms into the market), and speculation and general difficulties in regulating the market mechanism. The net results include poverty and unemployment (especially through the development of intensive plant, machinery and capital, but also through bankruptcies of small firms), which sharpen alienation and class conflict and the likelihood of revolution. Industrial competition between developing countries mars their development progress. International industrial competition intensifies regional or strategic alliances and its aggressiveness can be the outcome of policies of economic, if not territorial, expansion.
Broader Capitalism (#PC0564) Competition (#PB0848)
Excessive demand for goods in capitalist systems (#PC3116).
Related Competition between states (#PC0114) Contradictions in capitalist systems (#PF3118)
Economic competition in communist systems (#PC3167).
Aggravates Instability (#PA0859) Economic conflict (#PC0840)
Cyclic business recessions (#PF1277).
Aggravated by Technology gap between developed countries (#PD0338).

♦ **PC3134 Outflow of financial resources from developing countries**
Excessive repatriation of profits by foreign investors in developing countries — Negative net resource transfers from developing countries — Transfer of surplus wealth from developing to industrialized countries — Loss of capital of developing countries
Nature Investment of foreign private capital in a developing country is purchased by payments from income, which results in an effective outflow of capital from the developing to the developed countries.
Incidence No estimates of the outflow of financial resources from developing countries are available on a comprehensive basis; but it has been estimated for Latin American that foreign investors transfer one third of the profits within the first 5 years, between one half and two thirds in the second 5 years, and thereafter continue taking about 90 per cent, converting the investment into a drain on, instead of contributing to, the monetary reserves of the developing countries. In 1982 and 1983 nearly $28,000 million more than was lent was received by the developed world's banks from the developing countries. At the same time bank lending decreased, as did direct investment by multinational enterprises.
Broader Economic and social underdevelopment (#PB0539)
Faltering structural adjustment in the world economy (#PF9664)
Economic and financial instability of the world economy (#PC8073).
Narrower Burden of servicing foreign public debt by developing countries (#PD3051)
Private domestic capital outflow from developing to developed countries (#PD3132)
Foreign private investment income outflow from developing to developed countries (#PE8957).
Aggravates Natural environment degradation (#PB5250)
Deterioration in external financial position of developing countries (#PE9567).
Aggravated by Transfer pricing (#PE1193)
Mismatch of national macroeconomic policies among industrialized countries (#PF5000).

♦ **PC3144 Psychological environment degradation**
Broader Environmental degradation (#PB6384).
Related Mental pollution (#PB6248).
Aggravates Deterioration of human environment (#PC8943).

♦ **PC3150 Crop damage by wildlife**
Ravaging wild animals
Nature Wild animals may cause damage to crops when agricultural land borders on virgin territory or game reserves.
Incidence In Africa, for example, the main animal which causes damage is the baboon. In some African regions, other animals such as queleas, starlings, bush pigs, monkeys, cane rats or elephants, become local problems; and wild dogs, hyenas, jackals, lions and leopards also cause problems. Control of these animals is sometimes difficult and always delicate; destruction of predators, for example, upsets the predator/prey cycle and often leads to great increases of other pest species. Destruction of leopards leads to overpopulation of baboons and an increase in rodent populations. In Botswana, increases in numbers of plague–carrying jerbils in desert areas may have resulted from the overhunting of small carnivores.
Broader Harmful wildlife (#PC3151) Debilitating content of village story (#PF2168).
Narrower Quelea (#PE1429).
Related Forest damage by wildlife (#PD0500).
Aggravates Endangered species of bear (#PD3483)
Restriction of wild animal range size (#PC0475).
Aggravated by Birds as pests (#PD1689) Wildlife extinction (#PC1445).
Reduced by Endangered species of non–human primates (#PE1570).

♦ **PC3151 Harmful wildlife**
Dangerous wild animals — Aggressive wild animals — Violent wildlife
Nature Wild animals compete with cattle for food and require land which could otherwise be used for agricultural purposes. When large areas of hitherto virgin territory are divided and developed into an intricate pattern of human use, or when protection of animals in parks or reserves results in great population increases amongst such animal species, then wild animals are likely to conflict with human activities. Most often wildlife competes with either crops or cattle. In addition, wild animals may bring disease into valuable domestic herds.
Incidence In order to keep wild animals from domestic animals, programmes for shooting out wildlife or fencing off huge areas have taken place in many parts of Africa. Fences in particular are very expensive to erect, they decimate animal populations whose migration routes they cross, and they are fairly quickly damaged.
Counter–claim Cattle overstocking has caused serious degradation of habitat, and cattle raising is thus, to some extent, counterproductive. There is disagreement as to the exact nature and importance of wildlife to cattle disease transmission. In such circumstances, it seems very unwise to shoot out or fence out wildlife, when it has been amply demonstrated that wildlife can itself be a most profitable form of land utilization.
Broader Pests (#PC0728) Dangerous animals (#PC2321).
Narrower Venomous animals (#PD6823) Crop damage by wildlife (#PC3150)
Forest damage by wildlife (#PD0500) Hazardous aquatic animals (#PE6815).
Aggravates Degradation of semi–natural and natural habitats of flora and fauna (#PC3152).

♦ **PC3152 Degradation of semi–natural and natural habitats of flora and fauna**
Modification of animal habitats — Destruction of biological niches
Nature Habitat degradation results primarily from overgrazing by wild or domestic animals, excessive burning to clear land for agricultural purposes, or misguided provision of waterpoints. The introduction of domestic animals leads to overstocking; this overstocking, particularly in areas which are often burned, inevitably leads to progressively less perennial grass surviving to the end of the dry season, and an increase in shrubs and trees. In Africa, tsetse flies spread into the area as the habitat becomes more suitable for them, and cattle are removed from the area either because of the tsetse fly or because of the degraded habitat. Usually at the stage of bush encroachment, animals which are favoured by this development, such as elephant and buffalo, increase; and as these animals overpopulate the area they themselves become an important factor acting on the environment to further degrade it.
Background Prior to about 40,000 BC humans were probably confined to the 'great world island' of Africa and Eurasia, together with Australia and parts of the Indonesian archipelago. Migration into the Americas began about that time, across a land link on the site of the present Bering Strait. The crossing of wider ocean passages to remote islands took place last of all. Over the centuries, the impact of man on terrestrial biota – the natural living resources of the continents – changed not only as a result of this progressive spread, but also as man's tools and technologies advanced and his needs escalated.
Incidence Mankind has increasingly modified the assemblages of plants and animals (biomes) found in different regions of the continents, and changed the distribution of species and the nature of the ecological interactions between them. Many of these changes have been an essential component of development; but, as a consequence, natural vegetation has disappeared from great stretches of the continents – especially those that, like Europe, are densely populated and highly developed. Where development has been hampered by poverty or driven by great urgency, the fertility of the land and stability of the soil have been placed in jeopardy. Whole species or populations of plants and animals have disappeared, causing a loss of genetic resources that is not only regrettable from an aesthetic or philosophical point of view but also threatens man's food supply.
Claim All creatures are susceptible to stress and panic when their natural habitat is threatened.
Refs FAO *Report of the Symposium on Habitat Modification and Freshwater Fisheries, Aarhus, Denmark, 1984* (1984).
Broader Erosion of biological diversity (#PB9748)

Natural environment degradation (#PB5250)
Unconstrained exploitation of natural resources (#PF2855).
Narrower Bush encroachment (#PG6021) Destruction of wildlife habitats (#PC0480)
Overgrazing in developing countries (#PD4812)
Ecological disruption of animal breeding grounds (#PJ3994)
Natural environmental degradation from recreation and tourism (#PE6920).
Related Overstocking (#PC3153)
Environmental degradation of desert oases (#PD2285).
Aggravates Erosion (#PC8193) Hazards to plants (#PD5706) Wildlife extinction (#PC1445)
Endangered species of bear (#PD3483) Vulnerability of plants and crops (#PD5730)
Vulnerability of ecosystem niches (#PC5773)
Endangered species of plants and animals (#PB1395)
Endangered species of non-human primates (#PE1570).
Aggravated by Fires (#PD8054) Harmful wildlife (#PC3151)
Environmental pollution (#PB1166) Unjustified game cropping (#PE1327)
Denaturalization of fauna (#PG6025) Destruction of the countryside (#PE3914)
Over-population of wild animals (#PE6024) Deforestation of mountainous regions (#PD6282)
Over-use of designated wilderness areas (#PD7585)
Hazardous locations for nuclear power plants (#PD2718).

♦ PC3153 Overstocking
Overgrazing
Nature The pressure of a high animal population on limited grazing areas has led to progressive deterioration of native grasslands in developed and developing countries. As grazing pressures rise, the better forage plants lose vigour, give up the struggle and disappear, being replaced by inferior species (often poisonous or unpalatable) resulting in the production of less and less food for man. As grazing pressure increases, the reserve forage needed to carry animals through severe droughts and floods disappears. Overgrazing also disrupts cycles that assure relatively even distribution of water and nutrients, which prevent erosion, especially on semi-arid grasslands that easily degrade into deserts.
Incidence Overgrazing is a principal factor in the deterioration of rangeland. In general, overgrazing reduces grazing duration during the growing season as well as during the dry season, often forcing settled farmers to buy food supplements. For nomads, overgrazing causes a decrease in the milk production of their animals, which forces them to increase the number of animals, further increasing the pressures on the range. In the long run, both farmers and nomads may decide to leave the region or settle around deep boreholes. A special disease may appear among cattle – botulism in the Sahelian areas is one example. Finally, in areas where water erosion occurs and where wells exist, sand dunes may fill up the wells.
Broader Unsustainable agricultural development (#PC8419).
Related Misuse of grassland and rangeland (#PD5133)
Degradation of semi-natural and natural habitats of flora and fauna (#PC3152).
Aggravates Floods (#PD0452) Drought (#PC2430) Poisonous plants (#PD2291)
Long-term shortage of food and live animals (#PE0976).

♦ PC3163 Revolutionary communism
Revolution in communism
Nature A revolutionary system must expand if it is to maintain its momentum; communist systems are born of revolution. Expansion and projection of a national communist revolution is often used as a tactic to increase national solidarity by diverting attention away from national issues over which there are in fact deep divisions. In this respect, an outward projection of communist ideology may be as much a manifestation of weakness as of the efficacy and unity of the proletarian revolution.
Refs Wild, Victor *The Science of Revolution* (1980).
Broader Communism (#PC0369) Revolution (#PA5901).
Related Contradictions in communist systems (#PF3179).
Aggravates Minority control (#PF2375) International insecurity (#PB0009)
Ideological schism in communism (#PF3181).
Aggravated by Ideological conflict (#PF3388).

♦ PC3164 Communist political imperialism
Nature Although communism professes itself to be anti-imperialist, certain communist states have practised imperialism in the form of military occupation, coercive economic commitments, exploitation, and foreign subversion. Revolutionary communism is of necessity expansionist partly because of domestic weakness (political dissent and economic underdevelopment in most communist countries) and partly because of its international weakness in economic terms. Acquired territory may be of strategic military importance, or important from the point of view of industry or raw materials. Influence is sought over the Third World, in competition with capitalist offers of development aid; but rival communist ideologies also combat one other in that sphere. Communist political imperialism usually takes the form of severe repression of nationalist tendencies and the implantation of foreign social and economic systems.
Incidence The dramatic increase in the Soviet hegemony over socialist Europe since World War II illustrates one aspect of communist imperialism while recent changes in Eastern Europe only demonstrate the levels of sophistication being used. The exportation of revolution by Cuba illustrates another.
Broader Communism (#PC0369) Imperialism (#PB0113)
Denial of right to national self-determination in communist systems (#PC3177).
Related Communist economic imperialism (#PC3165)
Capitalist political imperialism (#PC3193).
Aggravates International insecurity (#PB0009).
Aggravated by National communism (#PF3130) Economic dependence (#PF0841)
Economic and financial instability of the world economy (#PC8073).

♦ PC3165 Communist economic imperialism
Nature Communist countries, like capitalist countries, undertake aid programmes with the specific aim of gaining political influence and economic benefit. Economic aid and investment is not made with the aim of building up national communist parties in Third World countries, but rather of making sure that parties already in power conform to the policies of the donor country. Communist economic aid may be a manifestation of the conflict between different communist ideologies angling for support and dominance in the communist world. It is also undertaken in rivalry to capitalist economic imperialism, particularly in the form of the military, which raises the risk of war.
Incidence Soviet Russian purchases of Eastern Europe manufactured goods at above world market price levels is a form of political subsidy for the communist regimes. These implicit subsidies reached a total of over $20,000 million in 1980.
Broader Communism (#PC0369) Economic imperialism (#PC3198)
Denial of right to national self-determination in communist systems (#PC3177).
Related Communist political imperialism (#PC3164)
Capitalist economic imperialism (#PC3166)
Capitalist political imperialism (#PC3193).
Aggravates Economic dependence (#PF0841) International insecurity (#PB0009).

♦ PC3166 Capitalist economic imperialism
Claim Capitalist economic imperialism aims at the establishment of economic and hence political and social domination over other countries, particularly in the Third World, with capitalism as the economic base. In particular, over-sophisticated technology and machinery are exported to these countries. Because of the complex financial network of investment for development, the donor country can put extensive political pressure on the other to comply with its policies. The donor country may exact a high return revenue for its investment, which enriches the already wealthy instead of contributing to development. It may also gain control of a particular commodity market or its distribution. Political pressure may be exerted through military or other direct aid, or through effective control over international development funds. Unofficial pressure may be exerted by transnational companies whose revenue is often greater than that of the oppressed country.
Refs Addo, Herb *Imperialism* (1986); Berberoglu, Berch *The Internationalization of Capital* (1987); Sau, Ranjit *Unequal Exchange, Imperialism and Underdevelopment*; Willoughby, John *Capitalist Imperialism, Crisis and the State* (1986).
Broader Capitalism (#PC0564) Economic imperialism (#PC3198).
Narrower Imperialistic distribution system (#PD7374)
Control over international organizations (#PG6045)
Internationalization of capitalist production (#PE7957).
Related Communist economic imperialism (#PC3165)
Capitalist political imperialism (#PC3193)
Denial of effective national self-determination by capitalist exploitation (#PE3123).
Aggravates Economic dependence (#PF0841) International insecurity (#PB0009).

♦ PC3167 Economic competition in communist systems
Nature Despite the rejection in communist doctrine of the profit-motive base for an economic system (capitalism) and hence of competition, communist countries compete externally to gain trading agreements and influence in other countries. Prestige production items are exported and are not necessarily available on the domestic market. Internationally exchangeable foreign currency is sought in order to conduct foreign trade despite the outcry raised against capitalist finance and the false exchange value of money and commodities under the capitalist system. Communist economic competition may lead to conflict between different communist ideologies or may be used as a means of propaganda. A number of communist countries have introduced capitalist-style bonuses and other production incentives for workers. This creates confusion concerning the authority of Marxist-Leninist teachings and weakens allegiance to the state.
Broader Communism (#PC0369).
Related Competition in capitalist systems (#PC3125).
Aggravates Benign neoplasm of bone and cartilage (#PG0819).
Aggravated by Economic dependence (#PF0841).

♦ PC3170 Elitism in communist systems
Nature Despite the aim of communism to create a classless society elitism still exists in most communist states. The elite is usually composed of high-up party members, and 'heroes' of sport, industry, science and the arts; that is, those who are the best in their field and particularly those who have acquired international renown. The latter serve as propaganda abroad, the former provide propaganda for domestic consumption. Party members may rise in the hierarchy by denouncing their colleagues, but they themselves may be denounced and removed at a later date. 'Heroes' may also join the elite through conformism, particularly those in the arts and sciences, where innovative theories may be rejected by the party. The elite have greater wealth than the majority, and other benefits which include the possibility of more freedom and flexibility regarding restrictive laws. Although the idea of the worker is exalted in communist society, there are differences in pay for different occupations, and some countries even provide industrial incentives to management. Industrial 'heroes' may be the spokesmen or leading managers of a successful enterprise, and they may benefit more from the acclaim than the workers who should have equal share in it, but who are not credited individually.
Broader Elitism (#PA1387) Communism (#PC0369).
Related Single party democracies (#PD2001).
Aggravates Political repression (#PC1919).

♦ PC3173 Denial of freedom of movement in communist systems
Nature Although communist revolution is intended to liberate the masses from oppression and to be carried out with their full approval, many communist countries close their frontiers to those who wish to leave and often to foreigners who wish to enter; movement between communist countries may be difficult and expensive; and movement within the country itself may be restricted owing to the difficulty of obtaining housing. Holidays to non-communist countries are restricted by the refusal to issue foreign currency (which means that the people who take such holidays must have friends or relatives abroad) or refusal to issue travel documents. The denial of freedom of movement leads some people to make clandestine and very dangerous attempts to leave their country, which sometimes ends in their death if they are not successful. The denial of foreign entry to certain countries creates a lack of international understanding and contact which may cause conflict, while restriction of movement within the country retards social development and may make economic readjustment difficult.
Broader Communism (#PC0369) Travel restrictions (#PC8452).
Narrower Restrictions on socialist citizens working abroad (#PE1659).
Related Communist closed society (#PD3169) Restrictions on freedom of worship (#PD5105).
Aggravates Illegal movement across frontiers (#PC2367).

♦ PC3174 Denial of freedom of expression and thought in communist systems
Nature Communist regimes do not permit ideological or political dissent. The present practice of openness, or 'glasnost', underlines the dependence of the citizenry on governmental permission to voice their opinions. All 'deviant' themes are censored and their authors may be arrested, tortured, imprisoned or executed. To maintain conformity to party propaganda and indoctrination, a domestic espionage system is needed, and contact with non-communist countries cut off or reduced to a minimum. The right to choose is denied, all planning being made by the political elite. This causes widespread conformity, apathy and ignorance of alternatives to the imposed system. It encourages foreign pressure to reform and foreign support for dissidents, which may cause international conflict. Internally, it creates an inflexibility in the system which is anti-innovative and tends towards stagnation.
Broader Communism (#PC0369) Restriction of freedom of expression (#PC2162).
Narrower Censorship in communist systems (#PD3172).
Related Restrictions on freedom of worship (#PD5105).

♦ PC3175 Denial of religion in communist systems
Nature Communist doctrine is anti-religious in the established denominational sense. The ideal is the practical brotherhood of man through the international communist movement. Measures to superimpose the communist ideal on the traditional cultural heritage in the form of forced assimilation to the newly-created ideal may be brutal and do nothing to bridge the social and cultural gap. The repression of religion has not been successful in many communist countries, where it derives an increased fervour from the fact of persecution.
Incidence Roman Catholicism flourishes in Poland although it is in deep conflict with the communist regime, and the recent murder of a Catholic priest by the military set off a world-wide protest. In other communist countries of Eastern Europe (notably Russia), Jews, Evangelical Protestants, and Roman Catholics appear to suffer more than the Russian Orthodox, but all who profess their religious beliefs openly may lose their jobs, be harassed, imprisoned or tortured,

especially if they attempt to proselytize or to criticize the regime. In Albania and Ethiopia, all religions are suppressed.
Broader Communism (#PC0369) Religious repression (#PC0578).
Aggravates Religious extremism (#PF4954) Forced assimilation (#PC3293).

♦ **PC3176 Denial of democracy in communist systems**
Nature Ideological, political and religious freedoms are denied in communist systems, censorship is intense and ideological conformity strictly imposed, freedom of movement is denied, political representatives are chosen from one party and directives are given by the political elite, often with a centralized bureaucracy. In most communist countries there is little local autonomy or decision-making. Manifestations of denial of democracy under communism encourage foreign pressure and support for dissenters, which may lead to international conflict. Internally, denial of democracy induces apathy, cynicism, conformism and alienation.
Broader Communism (#PC0369) Undemocratic social systems (#PB8031).
Aggravated by International insecurity (#PB0009).

♦ **PC3177 Denial of right to national self-determination in communist systems**
Nature Because of the non-inevitability of communist revolution, communism has been imposed on certain countries where the opportunity for violent revolution arose. Nationalism is denied in strict communist doctrine, but is applied imperialistically by the dominant power under the guise of 'international communism'; that is, it imposes its national system on the countries over which it has control. This control may be politically or economically based. Attempts at national self-determination may be suppressed militarily (invasion, maintenance of military bases, control over secret police, the army) or they may be frustrated by economic dependence.
Broader Communism (#PC0369).
Narrower Communist economic imperialism (#PC3165)
Communist political imperialism (#PC3164).
Related Violation of sovereignty by trans-border broadcasting (#PE0261)
Denial of effective national self-determination by capitalist exploitation (#PE3123).
Aggravates International insecurity (#PB0009).

♦ **PC3178 Denial of human rights in communist systems**
Nature Fear of internal instability and of outside threats to communist regimes has led to dictatorship, the instigation of a single party, and rigid policy concerning all matters. The freedom to dissent is expressly denied and repressed. National self-determination may be denied, despite communism's anti-imperialistic doctrines. Movement is restricted. Individual and social liberties, such as freedom of association, freedom of movement and freedom of expression, are firmly repressed when perceived as a threat to the dominant ideology and its representatives.
Broader Communism (#PC0369) Denial of human rights (#PB3121).
Related Denial of human rights in capitalist systems (#PC3124).

♦ **PC3185 Foreign intervention in internal affairs of states**
Foreign interference in internal affairs of states — Super-power intervention in other countries — Intervention by major powers to protect investments by their citizens in foreign countries — Foreign government interference
Nature Intervention is the interference of one country in the internal affairs of another or in its relations with other countries. As contrasted with simple influence (intercession), intervention is aimed at deciding the domestic or foreign affairs of another country in the interests of the intervening country. Intervention may be overt (armed intervention) or covert, such as: the imposition of an alien political, economic or social system; the organization of conspiracies coups d'Etat, and civil wars to achieve such aims; the dispatching of spies, terrorists, and saboteurs; financing and supplying armaments; making loans with strings attached; and the use of radio, television, and press to conduct hostile propaganda. Although intervention is outlawed in numerous international treaties and agreements, including the UN Charter, it is employed by the major industrial powers leading to continual mistrust and covert activities.
Large countries may threaten or enact military, economic or political action against smaller nations with the primary justification either that the smaller state was unable to guarantee the safety of its foreign residents (in other words, the citizens of the intervening power), or that the smaller state was unable to govern itself and possibly posed a threat to the region. Such situations may arise from civil war or other disturbances of the peace, economic collapse, or other domestic disorder; and the circumstances for intervention may be engineered, or construed from a very distorted perspective in order to justify aggression and colonization or economic dominance through forced treaties and agreements.
Incidence The presence of the USSR in Afghanistan is the latest of Soviet interventions, which have also included presence in Hungary and Czechoslovakia; the presence of Israel in Lebanon and the Nicaraguan role in El Salvador are among other recent examples.
The USA criticizes the USSR for intervention in the other Warsaw Pact countries, Afghanistan and Vietnam, while the USSR denounces the USA for her prior intervention in Vietnam and Libya and her current intervention in Central America. Following the 1990 intervention in Panama, the USA indicated that it reserved the right to intervene in other countries to facilitate the democratic outcome of an electoral process.
Claim It is unrealistic to expect governments of capital exporting countries to remain passive when the property of their citizens is subject to discriminatory or confiscatory treatment by other countries.
Counter-claim The super-powers' influence is declining in the third world and with it their capacity to intervene. The super-powers are less welcome, their smaller nations are better armed, and smaller states are less fragile. There are fewer revolutions and coups taking place now when compared to the 1960s and 1970s. Outside intervention costs more and involves greater military risks. Domestic opposition to such intervention is stronger in both countries.
Refs Bull, Hedley *Intervention in World Politics* (1984).
Broader Imperialism (#PB0113) Colonialism (#PC0798)
International aggression (#PB0968).
Narrower Invasion (#PD8779) Military aid (#PE6052)
Disruptive foreign influence (#PC3188) Foreign military intervention (#PD9331)
Covert intelligence agency operations (#PD4501)
Government support for repressive regimes (#PF4821)
Vulnerability of small nations to foreign intervention (#PD2374)
Government action against regimes with alternative policies (#PF2199).
Related Government intimidation of governments (#PE1622).
Aggravates Civil war (#PC1869) Foreign control (#PC3187)
Guerrilla warfare (#PC1738) Cultural invasion (#PC2548)
Foreign dictatorship (#PC3186) International conflict (#PB5057)
International insecurity (#PB0009) Destabilization of social systems (#PB5417)
Denial of right to national self-determination (#PC1450)
Decline in foreign direct investment in developing countries (#PD3138)
Government seizure of foreign nationals in foreign countries (#PE6564).
Aggravated by Freemasonry (#PF0695) Political disintegration (#PC3204)
Competitive acquisition of arms (#PC1258) Threatened and vulnerable minorities (#PC3295).
Reduces Inappropriate foreign investment (#PD8030).
Reduced by Non-alignment (#PF0801) National isolationism (#PF2141).

♦ **PC3186 Foreign dictatorship**
Puppet governments
Broader Foreign control (#PC3187) Occupied nations (#PC1788)
Political dictatorship (#PC0845)
Denial of right to national self-determination (#PC1450).
Aggravates Subversive activities (#PD0557) International insecurity (#PB0009).
Aggravated by Invasion (#PD8779)
Foreign intervention in internal affairs of states (#PC3185).
Reduces Nationalism (#PB0534).

♦ **PC3187 Foreign control**
Broader Imperialism (#PB0113) Colonialism (#PC0798).
Narrower Foreign dictatorship (#PC3186) Disruptive foreign influence (#PC3188).
Related Denial of effective national self-determination by capitalist exploitation (#PE3123).
Aggravates National disintegration (#PB3384) International insecurity (#PB0009).
Aggravated by Invasion (#PD8779) Military influence (#PD3385)
International monopoly of the media (#PD3040)
Foreign intervention in internal affairs of states (#PC3185)
Foreign controls of newspaper and journal propaganda (#PD3041).
Reduces Nationalism (#PB0534).

♦ **PC3188 Disruptive foreign influence**
Foreign pressure
Nature The vitality of a particular society can impinge, regionally and globally, on many other nation-states in a number of ways, acting through a variety of mechanisms. Most notable of the influences of one state on another is that upon human behaviour, which includes language, customs, dress and consumer preferences. The foreign influences may also affect social relations, societal structures from the nuclear family upwards, values, norms, religious beliefs and cultural, ethnic or national identities. All foreign influences are disruptive to the extent that they usually introduce change in an abrupt manner. This is particularly true with regard to the artificially stimulated demand for free-market, developed countries' consumer products in less developed countries. Countries with significant aboriginal populations, or extremely low standards of living, are market targets for numerous sophisticated products that by any criteria would be called luxuries, or for products that are health hazards.
Incidence The intensive marketing of milk powders for bottle feeding of infants in countries where mothers are illiterate and do not understand the need to sterilize bottles or to avoid over-dilution or exposure of the formula to pests, and where mothers find too late that they can no longer pay for more milk powder or breast-feed their babies, has resulted in higher rates of infant mortality.
Broader Foreign control (#PC3187)
Foreign intervention in internal affairs of states (#PC3185).
Narrower Foreign controls of newspaper and journal propaganda (#PD3041).
Related Monolingualism in a multi-cultural setting (#PD2695).
Aggravates Non-alignment (#PF0801) International insecurity (#PB0009).
Aggravated by Lack of political independence (#PF0297)
Administrative difficulties in new states (#PE1793).

♦ **PC3191 Military-industrial complex in capitalist systems**
Broader Militarism (#PC2169) Military-industrial complex (#PC1952)
Competitive acquisition of arms (#PC1258).
Related Military-industrial-governmental complex in communist systems (#PD1330).
Aggravates International insecurity (#PB0009) Environmental consequences of war (#PC6675).
Aggravated by Cyclic business recessions (#PF1277).

♦ **PC3193 Capitalist political imperialism**
Nature Domination of the decision-making process of one people by that of another. The capitalist system of economics requires a political environment in which market forces are permitted to vie openly with one another. This environment is imposed upon a people that they may better participate in international capitalism as consumers and producers.
Broader Capitalism (#PC0564) Imperialism (#PB0113).
Related Communist economic imperialism (#PC3165)
Communist political imperialism (#PC3164)
Capitalist economic imperialism (#PC3166).
Aggravates Human dependence (#PA2159).

♦ **PC3195 Cultural imperialism**
Dependence on cultural imperialism
Nature One culture can dominate others by its commerce; by its superior products and technologies which create a demand; by its cultural achievements whether they are scientific, literary, artistic, intellectual or social; and negatively, by intimidation of size and nearness, and by forced political and military agreements. When cultural dominance is perpetuated without sensitivity to, and respect for, indigenous ways of life, it is imperialistic and expansionist, feeding on its own success. Where a population is susceptible it may experience cultural invasion from more than one source. For example, it may adopt as an additional language one that is foreign; it may follow consumer patterns from another model; and it may accept ideologies from still a third source. Populations may become culturally dependent on foreign importations, stifling their own development in literature, science, education, mass media, behaviour and language, and in economic growth.
Incidence Cultural invasion is particularly serious in its effects where the invading culture overwhelms and destroys the integrity of the receiving culture. This is best seen in the case of indigenous groups whose homelands have been appropriated for economic development. Deprived of their ancestral lands, and with their way of life disrupted, they lose their autonomy and self-respect; unable to adjust to changed economic conditions, they fall into poverty and suffer malnutrition and disease. Cultural invasion tends to have negative effects on cultural diversity.
Claim Cultural invasion formerly occurred in the sense of colonization and political domination. Now it is less overt but nonetheless widespread, even in the absence of direct colonization, via the mass media and the new information and communications technology. It occurs in international trade and in excessive and uncontrolled foreign investment overseas by powerful countries and transnational corporations. The economic invasion is attended by cultural invasion and with political influence.
Refs Sandbacka, Carola (Ed) *Cultural Imperialism and Cultural Identity* (1977).
Broader Imperialism (#PB0113).
Narrower Cultural invasion (#PC2548)
Monolingualism in a multi-cultural setting (#PD2695)
Excessive use of foreign programmes for media (#PE9643)
Excessive portrayal of perspectives of industrialized cultures in media (#PE3831).
Related Forced assimilation (#PC3293) International paternalism (#PF1871).
Aggravates Deculturation (#PJ1034) Cultural illiteracy (#PD2041)
Temporal imperialism (#PF1432) Homogenization of cultures (#PB1071)
National political dependence (#PF1452)
Administrative difficulties in new states (#PE1793).
Aggravated by Cultural arrogance (#PF5178) Economic imperialism (#PC3198).

♦ **PC3197 Intra-state imperialism**
Broader Imperialism (#PB0113).
Narrower National federalism (#PF0626).
Related Elitism (#PA1387).
Aggravates Civil war (#PC1869).

♦ **PC3198 Economic imperialism**
Dependence on economic imperialism
Nature Abuse of economic power by one or more states in order to place other nations in a subordinate or client position constitutes economic imperialism. This abuse can take the form of effective foreign control of a nation's economy leaving the appearance but not the reality of sovereignty, for the dependent country.
Incidence Economic imperialism, in varying degrees, describes the relation of North to South, but it is also seen in some North–North and South–South contexts.
Broader Imperialism (#PB0113).
Narrower Communist economic imperialism (#PC3165)
Capitalist economic imperialism (#PC3166)
Financial and industrial oligarchy (#PD5193)
Transnational corporation imperialism (#PD5891)
Concentration of power by transnational corporations (#PE0766).
Related Economic intimidation (#PC3011).
Aggravates Intimidation (#PB1992) Underdevelopment (#PB0206)
Guerrilla warfare (#PC1738) Cultural invasion (#PC2548)
Economic repression (#PC8471) Cultural imperialism (#PC3195)
Cultural fragmentation (#PF0536) National political dependence (#PF1452)
Inadequate political parties in developing countries (#PD0548)
Foreign controls of newspaper and journal propaganda (#PD3041)
Vulnerability of small nations to foreign intervention (#PD2374)
Economic dependence of some developing countries on the drug trade (#PE5296).
Aggravated by Economic dependence (#PF0841)
Dependence of industrialized countries on import of resources (#PE0537).

♦ **PC3201 Political barriers to effective legislation**
Nature Barriers to national development in the passing of legislation for the general improvement of social, labour, economic and other conditions, include the use of party politics in the fostering or opposing of measures, thus creating polarization and making agreement difficult. The suppression of political opposition and the effective expression of the interests of minority groups, as well as general political apathy and alienation, are also obstructive to the introduction of remedial or progressive legislation.
Broader Power politics (#PB3202).
Narrower Political opportunism (#PC1897).
Aggravates Inadequate laws (#PC6848) Political conflict (#PC0368)
Proliferation of legislation (#PD5315)
Lack of parliamentary time to approve needed legislation (#PF8876)
Limited local respect for regional and global legislation (#PF2499).
Aggravated by Intimidation (#PB1992) Class consciousness (#PC3458).

♦ **PC3204 Political disintegration**
Breakdown of political unity
Nature Breakdown and fragmentation of the political system may reach the point where it no longer functions adequately. Political disintegration may be initially characterized by factions and political schism, and the formation of many new parties with a small or short-lived following. This may lead to the necessity for coalition or qualified majority government, resulting in severe difficulty in effecting legislation. If society continues to be very divided in its political views and aspirations, the situation may degenerate into anarchy and ultimately revolution and dictatorship. Political disintegration, if it does not culminate in revolution and dictatorship, may lead to national collapse and foreign control.
Broader Fragmentation (#PA6233) Disintegration (#PA6858).
Narrower Martial law (#PD2637) World anarchy (#PF2071)
Inadequate urban political machinery (#PC1833).
Related Ethnic disintegration (#PC3291) Social disintegration (#PC3309)
Unequal parliamentary constituencies (#PD2167).
Aggravates Political revolution (#PF3237) National disintegration (#PB3384)
Foreign intervention in internal affairs of states (#PC3185).
Aggravated by Corruption (#PA1986) Power politics (#PB3202)
Political schism (#PC2361) Political conflict (#PC0368)
Political alienation (#PC3227) Subversive activities (#PD0557)
Political instability (#PC2677) Lack of representation (#PF3468)
Political fragmentation (#PF3216).

♦ **PC3205 Expulsion of ethnic minorities**
Nature The expulsion of unpopular ethnic groups from a country where they are either indigenous, or else immigrants of long-standing with a marked degree of assimilation or integration, renders such people stateless (particularly if they hold a passport of the country in question). The property of such people may be confiscated. They may be used as scapegoats in a period of social, economic and political unrest, or be the victims of racism.
Broader Expulsion (#PC5313).
Related Expulsion of immigrants and aliens (#PC3207).
Aggravates Ethnic conflict (#PC3685).

♦ **PC3207 Expulsion of immigrants and aliens**
Nature Expulsion is an act of a public authority by which an alien is requested under threat of penalty or, if necessary, compelled to leave the territory of the country of his residence or stay. The expression 'deportation' is used in the legislation of English-speaking countries; it has an entirely different meaning in French-speaking countries, where it denotes the punishment settlement outside the metropolitan territory of the country. Expulsion should be distinguished from exclusion (rejection), as expulsion affects entry by aliens into the country. The distinction between expulsion and exclusion is not clearly defined in the legislation of some states and sometimes both result from the same administrative procedures. In practically all states a close link exists between the two measures: expulsion may be utilized to enforce exclusion, and aliens may be expelled without any time limitation on the grounds that they were liable to exclusion at the time of their entry. Another measure which should be distinguished from expulsion is extradition – expulsion results from a unilateral decision of the state of residence or stay of the alien, and it is a measure taken in the interest of that state and does not prevent the expelled person, after leaving the territory of the state, from moving freely; extradition results from a bilateral arrangement between two interested states, and it is a measure taken in the interest of the state of enforced destination in cases where the person to be extradited is accused of having committed a criminal offence (other than a political offence or desertion), and consists in handing over that person to the authorities of the state requesting extradition Expulsion is carried out when the continued presence of an alien is considered to be undesirable by the authorities of the country of his residence or stay. This measure therefore ensures the enforcement of the country's immigration policy; it removes from its territory aliens who, since they have committed criminal offences, are deemed capable of committing such offences again, and aliens whose political views are not in accord with those of the government; expulsion diminishes public expenditure for relief to persons afflicted by indigency, illness and invalidism; and it is assumed to help the country economically by eliminating competition of foreigners with nationals and by removing manpower which is believed to be superfluous For the country of destination of the expelled person (generally his country of birth), his arrival is obviously considered a liability in the majority of cases. Even the family of the returning alien, if he has one, may not open its arms too widely; and the returning sick, criminal, aged, and dependent are often additional problems to plague a land usually more plentifully supplied with such than the land which has expelled them. As regards the impact of expulsion on immigrants, their removal from the country where they have been established seriously affects their material, social and moral position and that of their families. They find themselves deprived (sometimes in connection with circumstances of a temporary character, such as slackening of local production) of the positive acquisitions resulting from their residence in the country – relationships of personal, economic and social value; housing; fulfilment of periods of residence; and personal plans. Often such action causes a breaking-up of the person's family. Since the emigrant has severed his ties with his place of origin, his return there may mean greater hardship and increased difficulties. Particular difficulties and hardships occur in the case of expelled persons who are unable to meet the costs of their return to their country of origin, which may be distant, or who are not acceptable to any country, or who are refugees whose return to their country would endanger their life or freedom The effect on resident immigrants of the threat of removal through expulsion or exclusion, particularly if this may occur through no particular fault of their own but rather in connection with unemployment, illness or invalidism, retards the immigrants' adjustment to the environment and their assimilation **Background** The expulsion of immigrants became a particularly important social problem after World War I, when laws restricting the admission of immigrants and tightening controls on those already admitted were issued in the principal countries of immigration. The depression which began in 1929 initiated in many countries a tendency to curtail and render the employment of aliens difficult, to encourage their departure from the country, and to enforce such departure by frequently resorting to the use of expulsion both in direct and indirect forms. The extensive use of the power to expel and, occasionally, the extension of legislative authority for such action, have continued to reflect the frequent fear of aliens and the growing tendency towards more severe treatment and greater discrimination between alien and citizen. It is not possible to obtain reliable statistical data referring to the expulsion of aliens; the fact that in some countries the departure of aliens is often enforced without following the formal expulsion procedure affects the significance of such data.
Claim The deportation of an undesired alien may rid the country of immigration of that alien, but it does not necessarily solve the problems which he typifies; indeed, under some circumstances it may aggravate those problems. Deportation may divert attention from the removal of social conditions which, rather than the aliens themselves, are the real roots of the problems of crime, dependency, or mental disease. Deportation almost always affects others than just the deportees and in some cases it causes deep resentment both amongst the deported and amongst their friends and families who remain behind. Quite possibly total cessation of deportation would serve the national interest better than its continued indiscriminate use. The problem becomes still more complex, and the policy more questionable, if one permits considerations of the interest of other countries, of the immigrant himself and of those associated with him, or of international relations, to enter in. Deportation rarely solves the problem; it simply passes it on in an aggravated form.
Broader Expulsion (#PC5313).
Narrower Undesirable aliens (#PG6069).
Related Displaced persons (#PD7822) Expulsion of ethnic minorities (#PC3205).
Aggravates Refugees (#PB0205) Statelessness (#PE2485)
Resettlement stress (#PD7776)
Discrimination against immigrants and aliens (#PD0973).
Aggravated by Illegal immigration (#PD1928) Restrictions on immigration (#PC0970)
Declining sense of community (#PF2575).

♦ **PC3208 Restrictions on emigration**
Emigration restrictions
Nature Nationals may be refused the right to emigrate or even to leave the country. Government policy may make it difficult or impossible to leave by refusing to issue travel documents or foreign currency, by refusing to allow its own currency to be sold abroad, and by strict police control over frontiers and other departure points. Restrictions on emigration may lead to illegal exit and the collaboration on other subversive activities needed to achieve it.
Broader Restrictions on freedom of movement between countries (#PC0935).
Narrower Refusal to issue travel documents, passports, visas (#PE0325)
Excessive frontier formalities in international travel (#PE0208).
Related Lack of war relief (#PF0727)
Restrictions on international freedom of movement for national advantage (#PD0351).
Aggravates Internment without trial (#PD1576) Illegal movement across frontiers (#PC2367).
Aggravated by Nationalism (#PB0534).

♦ **PC3219 Ideological discrimination in politics**
Nature Ideological discrimination may be on the grounds of political, religious or other ideology, and may include unequal voting rights or the banning of political parties and religious sects. Discriminatory actions may be taken when a particular political party comes to power and wishes to consolidate its position, by denying political power to its opponents and at the same time rewarding its own members and supporters. Discrimination reinforces existing social prejudices, inequalities and injustice, and may lead to violence or even war.
Broader Ideological conflict (#PF3388) Discrimination in politics (#PC0934).
Narrower Underprivileged ideological minorities (#PC3325).
Aggravates Political apathy (#PC1917) Ideological repression (#PC8083)
Denial of political rights (#PD8276) Unethical practices in politics (#PC5517).
Aggravated by Intellectual prejudice (#PC3406) Double standards in morality (#PF5225).

♦ **PC3220 Religious discrimination in politics**
Nature Discrimination in politics on the grounds of religion, including the official banning of certain religious sects and the denial of the right to vote to adherents of a religion other than that officially recognized by the state, hardens religious intolerance and conflict and may lead to war in extreme cases.
Incidence The denial of the right to vote on grounds of religion occurs mainly where there is an established state religion.
Broader Religious discrimination (#PC1455) Discrimination in politics (#PC0934).
Related Religious discrimination in education (#PD8807)
Religious discrimination in the administration of justice (#PE0168).
Aggravates Religious conflict (#PC3292) Religious intolerance (#PC1808)
Denial of political rights (#PD8276).
Aggravated by Official religion (#PF6091) Persecution of religious sects (#PF3353).
Reduced by Anti-clericalism (#PF3360).

♦ **PC3221 Political discrimination in politics**
Nature Discrimination by the dominant or governing party may be practised against the adherents of political parties which are considered to be subversive or against the interests of the state, and may also take the form of banning political parties and intimidating their members and supporters,

manipulating the ballot in a 'free' election, or refusing the right of elections (such as in an absolute monarchy). It serves to maintain existing inequality and injustice and may lead to subversive activities, violence or revolution.
Broader Discrimination in politics (#PC0934).
Related Denial of political rights (#PD8276).
Aggravates Forced labour (#PC0746) Political repression (#PC1919)
Unethical practices in politics (#PC5517).
Aggravated by Political schism (#PC2361) Political prejudice (#PC8641).

♦ PC3222 Political discrimination based on illiteracy
Nature In certain countries, literacy tests must be passed before a person is allowed to vote. As modern methods of mass communication enable illiterate people to understand what they are doing in voting for one candidate or another, literacy tests are often maintained as a discouragement to certain sectors of the population. The most affected are women, low wage–earners, and indigenous people.
Broader Discrimination in politics (#PC0934).
Aggravates Social injustice (#PC0797) Restricted franchise (#PG6094)
Denial of political rights (#PD8276) Language discrimination in politics (#PD3223).
Aggravated by Illiteracy (#PC0210).

♦ PC3227 Political alienation
Citizen alienation
Nature Alienation of the public from the political executive, and alienation or polarization of political doctrines, may result in political instability, revolution, political, social or national disintegration, apathy, or lack of participation.
Broader Alienation (#PA3545) Negative emotions and attitudes (#PA7090).
Related Lack of political development (#PD8673).
Aggravates Demagoguery (#PC2372) Political apathy (#PC1917)
Political ignorance (#PC1982) Political disintegration (#PC3204).
Aggravated by Repression (#PB0871) Political appointees (#PF2031)
Political inequality (#PC3425) Undue political pressure (#PB3209)
Limitations on right to vote (#PF2904) Illegitimate political regimes (#PC1461)
Lack of community participation (#PF3307)
Bias in selection of political candidates (#PD2931)
Failure of individuals to participate in social processes (#PF0749).

♦ PC3229 Violent revolution
Nature The use of force, whether by the regular armed forces or by guerrilla or other subversive tactics, may be used initially to achieve revolution and may continue to be used in order to suppress the populace and certain key institutions or industries. Such violent means, when used to achieve social, political, economic, cultural or ideological revolution, may result in death, mutilation, massacre, internment without trial, torture, civil or guerrilla war, or social and national disintegration culminating in foreign intervention.
Claim It has been demonstrated again and again that not only are the ends of revolution corrupted by the tent that noble ends justify all means, but the dynamic of revolution brings bad government. The violence and authoritarianism needed to bring revolt to success reproduces itself in the new regime, sometime much worse than before. The kind of people who emerge as compelling leaders are seldom the kind of people who are willing and able to manage a decent government and practice the inspiring visions that they preach. Real stability comes from suppleness and flexibility, the capacity to perceive shifting needs and bend to them. That is not the legacy of romantic revolution but tedious reform.
Broader Revolution (#PA5901) Human violence (#PA0429)
Unethical practices in politics (#PC5517).
Narrower Civil war (#PC1869) Overthrow of government (#PD1964)
Violent political revolution (#PD3230).
Related Non–violent political revolution (#PD3228).
Aggravates Counter revolution (#PF3232).

♦ PC3231 Ideological revolution
Nature Radical change in ideological thinking, including cultural revolution, may be a direct result of political revolution or technological and economic revolution. It may give rise to the enforcement of a new social order if it is adopted by government, and may cause social conflict, repression and a lack of integration in society. Experimental religions may be banned or lead to decadence. Minority political ideologies may increase their strength and hold over society with the use of subversive activities.
Broader Revolution (#PA5901) Ideological conflict (#PF3388).
Aggravated by Double standards in morality (#PF5225).
Reduces Traditionalism (#PF2676).

♦ PC3233 Economic revolution
Nature The orientation of economic systems is influenced by such outside considerations as availability of resources, international economic independence, and by ideology. Economic revolution may follow from political revolution or be a cause of it; the change from one economic system to another may cause economic instability and inflation which, if it persists, may give rise to counter revolution. Economic revolution may occur as a result of foreign economic influence or control and may encourage foreign retaliation.
Broader Revolution (#PA5901).
Narrower Technological revolution (#PC3234).
Aggravates Economic and financial instability of the world economy (#PC8073).
Aggravated by International economic interdependence (#PF6105).

♦ PC3234 Technological revolution
Nature Rapid change in technological achievement may lead to lack of adaptation of social and economic systems and encourage economic imperialism and the widening of technological, social, and cultural gaps.
Broader Revolution (#PA5901) Economic revolution (#PC3233).
Narrower Technocracy (#PF6330) Social ill–effects of automation (#PE5134)
Antagonism between employment policy and technical advance (#PE5104)
Inadaptation of technology to man in the industrialized societies (#PE5023).
Related Abusive technological development under capitalism (#PD7463)
Unbalanced application of communications technology (#PE7637).
Aggravates Mental stress due to automation (#PE5164).

♦ PC3236 Social revolution
Nature As a radical change in social systems and structure, social revolution may result in or be aided by revolution, or may occur over time within the existing social and political framework. But legal and other institutions may not adapt fast enough to cope with social trends, and the persistence of traditional attitudes may result in lack of identity and social adaptation.
Incidence In communist countries, social change has been synonymous with economic and political revolution. In 'Western' industrialized countries, permissive society and the breakdown of the extended family is largely a result of technological change and a liberalizing of attitudes towards morality.
Broader Revolution (#PA5901).
Related Social conflict (#PC0137).
Aggravated by Social inequality (#PB0514).

♦ PC3240 Economic dictatorship
Nature Economic dictatorship may be domestic or international, comprising economic imperialism, oligopoly and monopoly under a market system, and government economic intervention or control. Economic imperialism and the activities of multinational companies may lead to foreign debt, foreign control and influence, and possibly to national disintegration. Monopoly, oligopoly and government intervention may lead to a high cost of living, inefficiency, unequal distribution of wealth and the bankruptcy of small and medium–sized firms. State controlled economies are often synonymous with political dictatorship.
Broader Dictatorship (#PC1049).
Narrower Monopolies (#PC0521) Nationalized industries (#PG6114)
Political intervention by transnational corporations (#PE0032)
Control of national economic sectors by transnational enterprises (#PE0042).
Related Disproportionate control of global economy by limited number of corporations (#PE0135).
Aggravates Inequitable distribution of wealth (#PB7666).
Aggravated by Grey lies (#PF3098).

♦ PC3258 Discrimination against men
Nature Discrimination against men arises in certain circumstances in employment, education, public services, and before the law (for example, as regards homosexuals). In the event of divorce the woman is more likely to be granted custody of the children than the man. The segregation of men's and women's work may make it difficult for men to obtain certain kinds of work.
Broader Social injustice (#PC0797) Sexual discrimination (#PC2022).
Narrower Social barriers to male pregnancy (#PG5358)
Discrimination against men in sports (#PE4232)
Discrimination against men in education (#PD8909)
Discrimination against unmarried fathers (#PD3256)
Discrimination against men in employment (#PD3338)
Discrimination against men before the law (#PD3692)
Discrimination against men in social services (#PD3336)
Discrimination against men in parental rights (#PE4010)
Discrimination against men in public services (#PE8507).
Related Discrimination against women (#PC0308).
Aggravates Sex segregation (#PC3383) Non–parental custody of children (#PF3253).
Aggravated by Feminism (#PF3025) Social inadequacy of men (#PF3613).

♦ PC3261 Sexual exploitation
Nature Exploitation of men, women and children for sexual purposes may lead to sexual violence, extortion, blackmail and intimidation, prostitution and slavery. It is encouraged by double standards of sexual morality, religious intolerance, and lack of sex education.
Refs Burgess, Ann W and Hartman, Carol R *Sexual Exploitation of Patients by Health Professionals* (1986); Russell, Diane E *Sexual Exploitation* (1984).
Broader Sexism (#PC3432) Exploitation (#PB3200).
Narrower Sexual exploitation of men (#PD3263) Sexual exploitation of women (#PD3262)
Sexual exploitation of children (#PD3267).
Related Sexual harassment (#PD1116) Exploitation in employment (#PC3297).
Aggravates Illegitimate children (#PC1874) Illegal induced abortion (#PD0159)
Sexually transmitted diseases (#PD0061).
Aggravated by Homosexuality (#PF3242) Sexual immorality (#PF2687)
Sexual unfulfilment (#PF3260) Inadequate sex education (#PD0759)
Double standards in morality (#PF5225) Double standards of sexual morality (#PF3259).

♦ PC3280 Childlessness
Infertility
Nature The frustration and sorrow of a couple who want but are unable to produce children may cause the marriage to break down, or may cause emotional disturbance (possibly manifested by baby–snatching). If the couple is not considered suitable as adoptive or foster parents by adoption authorities, they could remain childless. Traditionally childless couples are considered afflicted and may be even sinful, and the fear of childlessness in tribal society is the basis for polygamy; where polygamy has been erased in such communities by 'Western' culture, childlessness causes reversion to tradition; couples without a male child or with only one male child may be considered childless in such a setting. Although the inability to have a desired child is very painful for men as well as for women, the situation has graver consequences for the woman. In traditional societies, a woman who cannot bear children is denied the social identity of her sex, and in some cases may become an outcast.
Refs Behrman, S J, et al *Progress in Infertility* (1987); Campbell, Elaine *The Childless Marriage* (1986); Rowe, Patrick J and Vikhlyaeva, E M (Eds) *Diagnosis and Treatment of Infertility* (1988); WHO Scientific Group, Geneva, 1975 *Epidemiology of Infertility* (1975).
Broader Deprivation (#PA0831).
Narrower Paternal negligence (#PD7297) Lack of parental fulfilment (#PG6150).
Related Sexual unfulfilment (#PF3260) Non–parental custody of children (#PF3253).
Aggravates Child abduction (#PE6154) Underpopulation (#PD5432)
Marital instability (#PD2103) Inhuman methods of conception (#PE8634)
Trafficking in children for adoption (#PF3302).
Aggravated by Human infertility (#PC6037) Refusal of adoption (#PF3282)
Annulment of adoption (#PF3281) Sexual impotence of men (#PF6415).
Reduces Complications of childbirth (#PC9042).
Reduced by Polygamy (#PD2184) Inadequate laws of adoption (#PD0590).

♦ PC3291 Ethnic disintegration
Nature The falling apart or destruction of ethnic society may result from cultural invasion, compulsory education, and other national policies which do not take ethnic considerations into account.
Incidence Ethnic disintegration is particularly marked among primitive tribes, but can also be seen among sects and ideological communities in industrialized countries (such as Jewish Americans) with their increasing nationalism and uniform national social policies.
Broader Social disintegration (#PC3309).
Narrower Lack of racial identity (#PF0684).
Related Ethnocide (#PC1328) Threatened sects (#PC1995)
National disintegration (#PB3384) Political disintegration (#PC3204)
Endangered tribes and indigenous peoples (#PC0720).
Aggravates Cultural deprivation (#PC1351) Exploitative personal services (#PC3299)
Ignorance of cultural heritage (#PF1985)
Adjustment difficulties of new urban families (#PF1503).
Aggravated by Heterosexism (#PE0818) Cultural invasion (#PC2548)
Religious conflict (#PC3292) Cultural corruption (#PC2913)
Forced assimilation (#PC3293) Ill–considered missionary activity (#PF3370)
Inadequate sense of personal identity (#PF1934).
Reduces Ethnic segregation (#PC3315) Primitive secret societies (#PF2928).
Reduced by Traditionalism (#PF2676).

♦ PC3292 Religious conflict

Nature Religious conflict includes intolerance of other religions and discrimination against members of other religions, religious war, intellectual conflict and conflict between church and state. Such conflict is harmful to the overall credibility of religion and may cause religious apathy or disintegration. It may arise in the attempt to religiously convert tribal society and may result in ethnic disintegration and loss of cultural heritage. Religious conflict can ensue from political conflict.

Incidence Religious intolerance and discrimination exist on a worldwide scale. Conflict between church and state or other conflict occurs in the Middle East and in Northern Ireland and Cyprus. Islam and Christianity are competing for converts in parts of black Africa. Christian missionaries have made inroads into indigenous cultures, particularly in Latin America where native populations have been reduced to a state of poverty and dependency and in many cases are dying out. On the Indian sub–continent, where the Sikhs, Hindus and Moslems are involved in fratricidal blood–shed, the governments have not been able to mount a bulwark between rival religions and sects to prevent such conflict. In Southeast Asia and the Pacific tensions of greater or lesser religious nature are increasing in New Caledonia Fiji, Malaysia, Indonesia, Singapore, Sri Lanka, Australia and New Zealand.

Refs Kaniki, M *Religious Conflict and Cultural Accomodation* (1976).

Broader Conflict (#PA0298) Ideological conflict (#PF3388) Multidenominational society (#PF3368).

Narrower Jihad (#PF5681) Atheism (#PF2409) Religious war (#PC2371) Dependence on mysticism (#PF2590) Religious conflict between sects (#PC3363) Religious and political antagonism (#PC0030) Ill–considered missionary activity (#PF3370) Holy places as a focus of religious friction (#PF1816).

Related Ethnic conflict (#PC3685) Religious schism (#PF1939) Religious intimidation (#PC2937).

Aggravates Scapegoats (#PF3332) Aerial piracy (#PD0124) Lack of assimilation (#PF2132) Ethnic disintegration (#PC3291) Fragmentation of religious belief (#PF3404) Discrimination against indigenous populations (#PC0352).

Aggravated by Heresy (#PF3375) Sacrilege (#PF0662) Prejudice (#PA2173) Colonialism (#PC0798) Traditionalism (#PF2676) Minority control (#PF2375) Ethnic segregation (#PC3315) Religious propaganda (#PD3094) Religious repression (#PC0578) Experimental religion (#PF3367) Social discrimination (#PC1864) Religious intolerance (#PC1808) Inadequacy of religion (#PF2005) Religious discrimination (#PC1455) Religious indoctrination (#PD4890) Double standards in morality (#PF5225) Underprivileged religious minorities (#PC2129) Religious discrimination in politics (#PC3220) Religious discrimination in education (#PD8807) Underprivileged ideological minorities (#PC3325) Disruptive secular impact of holy days (#PE7735) Fragmentation within organized religions (#PF3364) Undue religious influence on secular life (#PF3358) Religion as a reinforcement of nationalism (#PF3351) Segregation based on religious affiliation (#PC3365) Lack of appreciation of cultural differences (#PF2679) Divisive effects of official cultural pluralism (#PF0152) Inadequate integration of religions into society (#PF3403).

♦ PC3293 Forced assimilation

Dependence on forced assimilation — Forced integration of peoples

Nature An enforced policy of assimilation may be used to suppress subcultures, as an act of discrimination against minorities, or as a political instrument to create national unity and suppress possible sources of subversive activity. Forced assimilation is often synonymous with a policy of nationalism, described as 'Americanization' or 'Russification', depending on the country. Forced assimilation has been exercised under colonial rule in a qualified form: the selection of an elite from the non–Western people.

Minorities may be dispersed among the majority so that in no area are they a majority. Leaders and potential leaders may be forced to work far from their homes. Minority presence in political processes may be restricted or eliminated all together. Minorities may not be allowed educational opportunities in their own language, culture or religious tradition. Teachers from minority groups may be discriminated against. Newspapers, magazines, periodicals, radio, television, books, theatrical presentations or exhibits and archives referring to minorities may be restricted or forbidden.

Tension and friction between population groups may be the consequence of a policy of forced assimilation. This occurs when the concept of a multi–ethnic or multi–cultural society has not been accepted by the politically dominant ethnic group of a country, which feels convinced of the need for the assimilation of other groups.

Incidence In 1985, ethnic Turks living in Bulgaria were forced, at the cost of many lives, to exchange their ancestral Turkish names for Bulgarian ones, in a governmental attempt to denationalize the ethnic and cultural distinctions of those Turks. A similar policy was used by Indonesia against Chinese Indonesians. The Hungarian minority in Romania may be assimilated.

Broader Conflict (#PA0298) Nationalism (#PB0534).

Narrower Miscegenation (#PC1523) Social alienation (#PC2130) Loss of faith in religion (#PF3863) Loss of linguistic tradition (#PG6175) Inadequate sense of personal identity (#PF1934).

Related National federalism (#PF0626) Cultural imperialism (#PC3195) Forced participation in politics (#PD2910).

Aggravates Nativism (#PF2186) Ethnocide (#PC1328) Class conflict (#PC1573) Cultural corruption (#PC2913) Cultural repression (#PC8425) Cultural deprivation (#PC135?) Ethnic disintegration (#PC3291) Lack of racial identity (#PF0684) Denial of cultural rights (#PD5907) Dictatorship of the majority (#PD3239) Destruction of cultural heritage (#PC2114) Lack of variety of social life forms (#PE8806) Monolingualism in a multi–cultural setting (#PD2695).

Aggravated by Conformism (#PB3407) Colonialism (#PC0798) Social fragmentation (#PF1324) Political domination (#PC8512) Cultural fragmentation (#PF0536) National disintegration (#PB3384) Discrimination against minorities (#PC0582) Ill–considered missionary activity (#PF3370) Denial of religion in communist systems (#PC3175).

Reduces Taboo (#PF3310) Magic (#PF3311) Totemism (#PF3421) Traditionalism (#PF2676) National unity (#PG6178) Ethnic conflict (#PC3685) Cultural barriers (#PB2331) Human cannibalism (#PF2513) Ethnic segregation (#PC3315) Subversive activities (#PD0557) Ethnic discrimination (#PC3686) Plural society tensions (#PF2448) Inter–cultural misunderstanding (#PF3340) Head hunting in tribal societies (#PF2666).

♦ PC3295 Threatened and vulnerable minorities

Threatened minorities

Nature Minorities are threatened with extinction, whether by ethnocide, by voluntary assimilation into a dominant culture, or by gradual suppression of the minority culture.

Incidence Threatened minorities include tribes, religious sects, ideological groups or movements and powerless small nations, which latter may be threatened with economic or political domination from outside and may even be absorbed by a powerful neighbour. Ideological groups are mainly threatened by police activity or internment, while religious sects and tribes may be as much threatened by the influence of the dominant culture as by any of its actions. In organizations, dissenters (collectively or individually) may be removed, relegated to marginal functions, downgraded, harassed, or made the subject of brain–washing, peer–pressure or other techniques. The organization may attempt to silence conscience in the case of those members with cultural and moral values differing from those accepted as the norm.

Broader Social injustice (#PC0797).

Narrower Threatened sects (#PC1995) Cultural invasion (#PC2548) Endangered tribes and indigenous peoples (#PC0720) Threats to ideological movements and minorities (#PC3362) Vulnerability of small nations to foreign intervention (#PD2374).

Aggravates Destruction of cultural heritage (#PC2114) Undemocratic political organization (#PC1015) Denial of the right to national sovereignty (#PE7906) Foreign intervention in internal affairs of states (#PC3185) Inequality of employment opportunity in developing countries (#PD1847).

Aggravated by Conformism (#PB3407) Nationalism (#PB0534) Economic expansion (#PF8111) Over–centralization (#PF2711) Social fragmentation (#PF1324) Homogenization of cultures (#PB1071) Discrimination against minorities (#PC0582) Insufficient minority culture support (#PF5659).

♦ PC3297 Exploitation in employment

Dependence on exploitation in employment

Nature The most extreme form of exploitation in employment is slavery. Most groups which are exploited in employment are underprivileged groups, such as indigenous populations, women, children, immigrants, the illiterate, the lowest levels of national society, the aged, and the disabled. Such groups may not have the knowledge or other means to combat exploitation and may sink into a state of apathy and resignation. Discrimination and segregation are two tools on which exploitation thrives. Existing trade unions may bar minority groups from becoming members, thus leaving them defenceless against unscrupulous employers. Where legislation exists against unjust employment conditions and rates of pay, it may be inadequate or inadequately enforced.

Broader Exploitation (#PB3200) Social injustice (#PC0797).

Narrower Excessive hours of work (#PD0140) Lack of minimum wage fixing (#PE6726) Denial of right to retirement (#PD4458) Exploitation of casual workers (#PD6930) Inadequate working conditions in developing countries (#PD1476) Inadequate living and working conditions of immigrant labourers in industrialized countries (#PD3427).

Related Feudalism (#PF2136) Sexual exploitation (#PC3261) Racial exploitation (#PC3334) Economic exploitation (#PC8132) Political exploitation (#PC7356) Exploitation in housing (#PD3465) Exploitation of children (#PD0635) Exploitation of indigenous populations in employment (#PD1092).

Aggravates Apathy (#PA2360) Caste system (#PC1968) Forced labour (#PC0746) Social conflict (#PC0137) Social inequality (#PB0514) Social disintegration (#PC3309) Abuse of prison labour (#PD0165) Exploitation of child labour (#PD0164) Illiteracy among indigenous peoples (#PD3321) Abusive traffic in immigrant workers (#PD2722) Inadequate education of indigenous peoples (#PC3322) Inadequate housing among indigenous peoples (#PC3320).

Aggravated by Discrimination (#PA0833) Denial of right of complaint (#PD7609) Unethical personnel practices (#PD0862) Unethical practices of employers (#PD2879) Restrictive trade union practices (#PD8146) Inadequate national law enforcement (#PE4768) Maladjustment to disciplines of employment (#PD7650) Inequitable labour standards in developing countries (#PD0142).

♦ PC3298 Trafficking in women

Enslavement and exploitation of females — Abduction of women — Sale of women — Bride selling — Mail–order brides

Nature The trade in women and female children as wives, concubines and prostitutes is still widespread. Girls may be promised in marriage (without their knowledge or consent), at a very early age to a man who is much older; or a man may either sell his wives and children or his heir inherit them upon his death. In Asia, concubines can still be bought by wealthy men; in Latin America and Asia, children are 'adopted' by wealthy families for a price, and they may become domestic servants or subjects of sexual exploitation. Wives, concubines and prostitutes may also be abducted or lured by false advertisements. While much of the foregoing illicit activity is to gratify the passions of the male and sometimes female 'buyers' (the 'white slave trade' may come into this category) some of it is to provide cheap slave labour.

Incidence End–markets for women have been reported in Puerto Rico, the Middle East (particularly Lebanon and Kuwait), Ivory Coast, Senegal, Australia and Japan.

In China in 1989 a series of police actions in one month uncovered 3,000 cases of abduction of women and children by some 900 gangs. In one relatively prosperous region, 4,810 women from other parts of the country were bought and sold in the three years from 1986. The women (some only 13 or 14 years old) are bought, taken by force, or duped by traders in the poor provinces and are then transported hundreds of miles to be displayed in markets and sold at auctions, especially in the north–west and coastal regions. The traditional practice of bride–selling has re–emerged as a result of agricultural reforms in the late 1970s and early 1980s. Where successful, this had the consequence of considerably increasing the cost of weddings and match–makers as well as increasing the pressure on farmers for male progeny. The limited social environment of Chinese villages made it much cheaper to purchase a wife than to pay for the cost of a traditional wedding.

Broader Slave trade (#PC0130) Traffic in persons (#PC4442) Exploitation of women (#PC9733).

Related Social injustice (#PC0797) Violence against women (#PD0247) Discrimination against women in employment (#PD0086).

Aggravates Forced marriage (#PD1915) Female prostitution (#PD3380) Cruelty to children (#PC0838) Sexual exploitation of women (#PD3262) Exploitation of child labour (#PD0164) Sexual exploitation of children (#PD3267).

Aggravated by Tribalism (#PC1910) Concubinage (#PF2554) Bride–price (#PF3290) Debt slavery (#PD3301) Prostitution (#PD0693) Child–marriage (#PF3285) Socio–economic poverty (#PB0388) Inadequate laws of adoption (#PD0590) Discrimination against women (#PC0308) Unethical personnel practices (#PD0862) Inadequate national law enforcement (#PE4768).

♦ PC3299 Exploitative personal services

Nature Personal services are services demanded of underprivileged people by public and religious authorities, landlords, recruiting agents, contractors and creditors, for no payment or nominal payment, or for food or the use of land, as in the feudal system of peonage of the Middle Ages. Personal services may include administrative, domestic, or agricultural work and usually involve the entire family, women and children included. In giving service to religious authorities, people may also be required to contribute cash.

Incidence Personal services are mainly performed by indigenous populations in Latin America and India. The expropriation of their land results in their dependency on landowners, authorities and other 'invaders' who exploit their lack of adaptation to modern market economy systems and distort their customs of community service. Since services are also required of children, their education ends at a very early age, thus denying them the possibility of escape from the system

of servitude. Trade in slaves and serfs has been noted by the ILO in Colombia and Venezuela, where Indians of one tribe sell Indians of another tribe either to whites or to other Indians. Although legislation has been passed in many countries abolishing slavery and the more blatantly exploitative forms of personal service, it has proved almost impossible to get rid of the system of peonage without a radical land reform policy. In India, the prevalence of the caste system keeps peonage in practice despite laws against it.
Broader Exploitation (#PB3200).
Related Forced labour (#PC0746) Economic exploitation (#PC8132) Inadequate laws of adoption (#PD0590) Underprivileged racial minorities (#PC0805).
Aggravated by Caste system (#PC1968) Ethnic disintegration (#PC3291) Exploitation of indigenous populations in employment (#PD1092).

♦ PC3300 Chattel slavery

Nature Chattel slavery is full slavery in its traditional form whereby slaves are the complete property of their master, can be bought and sold by him and treated in any way that he wishes, which may include torture and other brutality, excessively bad working conditions, and sexual exploitation. Chattel slavery includes the buying, selling and ownership of women and girls as concubines, wives or prostitutes, and of the children of slaves.
Incidence Chattel slavery, apart from the ownership of women and girls, is fairly limited in extent since the decree of King Faisal in Saudi Arabia in 1962 making slavery illegal. Until this time, slave owning and the slave trade had flourished particularly in Arabia, with Saudi Arabia as its centre. Slavery is still alleged to exist in Muscat and Oman, Aden and the Yemen, though in some cases slaves are legally free to leave their masters if they wish. In Africa, Zanzibar, Cameroon and Mauritania are centres of slavery used to supply the Arabian market. Chattel slavery of women survives in many African countries among tribespeople.
Broader Slavery (#PC0146).
Related Forced labour (#PC0746) White slave trade (#PD3303).
Aggravates Kidnapping (#PD8744) Concubinage (#PF2554) Forced marriage (#PD1915).
Aggravated by Tribalism (#PC1910) Bride–price (#PF3290) Unemployment (#PB0750) Child–marriage (#PF3285) Unethical personnel practices (#PD0862) Inadequate national law enforcement (#PE4768).

♦ PC3304 Expropriation of land from indigenous populations

Unsettled indigenous land claims — Invasion of indigenous lands — Denial of indigenous rights to lands and resources
Nature The lands of native tribes may be seized by 'civilized' outsiders for the development and exploitation of natural resources. Either the tribes are simply dispossessed and become dependent on the invaders or else they are relocated on land (reservations) which is usually of poor quality. If mineral or other resources are found on reservation territory, it may be expropriated without compensation to the tribes.
Incidence The problem has existed in all colonized countries and still exists where the indigenous population does not have control over the government and political independence. Even in such circumstances minority tribes may have their land expropriated (such as in Indonesia). The fact that many such tribes are nomadic, facilitates the seizure of their land and increases the risk of their extinction since the remaining less fertile land is insufficient for their way of life. They become dependent on the new landowners as a slave class and find it difficult to adapt to the new way of living so that it is difficult for them to improve their lot, thus being exploited in employment as they were exploited in the uncompensated seizure of their land. Expropriation denies the right of ownership to indigenous people. The Australian Supreme Court stated 'the Aborigenes belong to the land not the land to the Aborigenes' – re seizure of reserved territory in 1972 by a Swiss–led mining project. In New Zealand, the Waitangi Tribunal is considering claims by Maori tribes covering 60 percent of the land area and much of the offshore fisheries out to the 200 mile limit. Other examples include: North American Indians in the U.S. and Canada; Indians of Brazil, Guyana, Guatemala, Nicaragua, Ecuador, Peru, Bolivia and other Latin American countries; Igorot tribespeople of the Philippines; tribal people of India and Bangladesh; Melanesians of West Papua in Indonesia; Nighur people of China; Orang Ulu of Sarawak, Malaysia; and Lapps of Scandinavia. **Background** Expropriation began with early colonization in the 16th and 17th centuries, reached a peak with the European settlement of North America and Africa, and continues today.
Claim Indigenous people see their lands as gifts from their creator which they hold in trust for future generations. Expropriation of land is a violation of this belief. Native spirituality is intimately linked to the landforms and the natural world. The land is a source of food, medicine and clothing. It is a part of the human soul: the earth and its inhabitants are infused with the same spirit. Private ownership is an assault on the relationship between the indigenous people and the land.
Broader Endangered tribes and indigenous peoples (#PC0720) Discrimination against indigenous populations (#PC0352).
Narrower Violation of land rights of a people (#PD5218).
Related Underprivileged racial minorities (#PC0805) Government expropriation of private property (#PD3055).
Aggravates Slavery (#PC0146) Debt slavery (#PD3301) Human dependence (#PA2159) Social disintegration (#PC3309) Cultivation of marginal agricultural land (#PD4273) Exploitation of indigenous populations in employment (#PD1092) Restriction of indigenous populations to reservations (#PD3305) Denial of rights of indigenous people to be self–governing (#PE1024).
Aggravated by Nomadism (#PF3700) Racial discrimination (#PC0006) Ethnic discrimination (#PC3686) Outmoded legal systems (#PF2580) Unfairly negotiated treaties (#PF4787) Denial of right of complaint (#PD7609) Fraudulent mineral exploitation claims (#PE6975) Denial to people of control over their own lives (#PC2381) Violation of treaties with indigenous populations (#PE7573).
Reduces Underdevelopment of industrial and economic activities (#PC0880).

♦ PC3309 Social disintegration

Nature The falling apart or destruction of the social fabric, resulting in a widespread loss of identity, apathy and social conflict, may ultimately result in national disintegration. Social disintegration may involve ethnic disintegration and loss of cultural heritage, poverty, unemployment, hunger, homelessness, anarchy and immorality. It is aggravated by corruption, cultural invasion, war, natural disasters and debt.
Incidence The most marked social disintegration is to be seen in certain Third World countries. It may also exist to a less noticeable extent in industrialized countries.
Broader Disintegration (#PA6858) Cultural invasion (#PC2548).
Narrower Family breakdown (#PC2102) Social breakdown (#PB2496) Ethnic disintegration (#PC3291) Family disorganization (#PC2151) Lack of racial identity (#PF0684) Imbalance in the human sex ratio (#PF1128) Disintegration of organized religion (#PD3423) Vulnerability of marriage as an institution (#PF1870).
Related Counter culture (#PF0423) National disintegration (#PB3384) Political disintegration (#PC3204).
Aggravates Divorce (#PF2100) Martial law (#PD2637) Social apathy (#PC3412) Permissiveness (#PF1252) Marital instability (#PD2103) Conflicting sense of sexual identity (#PF1246) Inadequate sense of personal identity (#PF1934).
Aggravated by Apathy (#PA2360) Nativism (#PF2186) Decadence (#PB2542) Immorality (#PA3369) Agnosticism (#PF2333) Threatened sects (#PC1995) Social fragmentation (#PF1324) Political instability (#PC2677) Cultural fragmentation (#PF0536) Exploitation in employment (#PC3297) Inadequate ideological frameworks (#PD0065) Inadequate integration of ideology into society (#PF3402) Expropriation of land from indigenous populations (#PC3304).
Reduced by Traditionalism (#PF2676).

♦ PC3315 Ethnic segregation

Nature Segregation of ethnic groups into isolated or close–knit communities may be enforced by law or discriminatory practice or it may be voluntary. It promotes tension between ethnic groups and facilitates the continued subjugation of underprivileged communities.
Incidence Ethnic segregation may take the form of the confinement of indigenous populations to reservation land (including the system of apartheid in South Africa). It may also take the form of segregation among immigrants or different ethnic groups in urban areas (ghettos) or in ethnic distribution at the rural level. Individual communities, like the Amish Community in the USA, completely segregate themselves from outside influences in order to preserve their unique way of life.
Refs Peach, Ceri, et al *Ethnic Segregation in Cities* (1982).
Broader Segregation (#PC0031).
Narrower Restriction of indigenous populations to reservations (#PD3305).
Related Sex segregation (#PC3383) Age segregation (#PD3444).
Aggravates Social conflict (#PC0137) Ethnic conflict (#PC3685) Social inequality (#PB0514) Religious conflict (#PC3292) Lack of assimilation (#PF2132) Ethnic discrimination (#PC3686) Segregation of poor and minority population in urban ghettos (#PD1260).
Aggravated by Elitism (#PA1387) Prejudice (#PA2173) Social discrimination (#PC1864) Destruction of cultural heritage (#PC2114).
Reduced by Forced assimilation (#PC3293) Ethnic disintegration (#PC3291).

♦ PC3316 Isolation of ethnic groups

Nature The geographic isolation of certain ethnic groups creates difficulties in the administration of social welfare and education and facilitates the persistence of prejudice, discrimination and exploitation.
Incidence This problem occurs particularly with primitive and nomadic tribes who live and move in remote and inaccessible terrains. These people are under–privileged regarding social services because of the difficulty of reaching them; but they do not, in most cases, lack contact with civilized outsiders, who may kill them or exploit them. The introduction of a new way of life shatters their existing social patterns, reducing them to dependency, pauperization; and in many instances they resort to alcoholism and drug dependency to alleviate their situation. They contract new diseases from the outsiders to which they have no resistance and these often prove fatal. They are in need of social services, but these are not sufficiently available.
Broader Cultural isolation (#PC3943) Underprivileged racial minorities (#PC0805).
Narrower Nomadism (#PF3700) Inadequate social welfare services (#PC0834).
Related Endangered tribes and indigenous peoples (#PC0720) Geographic discrimination in the administration of justice (#PE1347).
Aggravates Ethnocide (#PC1328) Exploitation (#PB3200) Desert nomadism (#PD2520) Gypsy persecution (#PE1281) Human cannibalism (#PF2513) Diverse unilingualism (#PF3317) Inequality in education (#PC3434) Racial discrimination in education (#PD3328) Underprivileged linguistic minorities (#PC3324) Inadequate education of indigenous peoples (#PC3322) Discrimination against indigenous populations (#PC0352).
Aggravated by Illiteracy (#PC0210) Land enclosure (#PJ3523) Cultural invasion (#PC2548) Uncontrolled migration (#PD2229).

♦ PC3319 Malnutrition among indigenous peoples

Denial of the right to adequate food for indigenous populations
Nature Inadequate quantities of food and an unbalanced diet among indigenous peoples leads to disease, loss of productive capacity and the resorting to alcohol and drugs.
Incidence Malnutrition is widespread among the majority of indigenous populations, mainly due to poverty and nutritional ignorance. The basis for most diets is starchy vegetable or grain, no fruit, no milk, very little meat. Food therefore lacks variety and particularly vitamins. The result is a wide range of deficiency disease, gastro–intestinal infections and lack of resistance to other diseases or to fatigue. Indigenous people may resort to alcoholism or drug taking to dull the effects of their hunger. The poor quality of the soil of the land onto which indigenous people have been pushed accounts in many cases for the lack of variety in the crops grown and their poor quality. Food taboos and tradition also create obstacles to the nutritional education of primitive tribes.
Broader Malnutrition (#PB1498) Discrimination against indigenous populations (#PC0352).
Narrower Endemic goitre (#PE1924) Excessive cost of animal protein (#PE4784).
Related Denial of right to sufficient food (#PE0324) Inadequate housing among indigenous peoples (#PC3320).
Aggravates Apathy (#PA2360) Alcohol abuse (#PD0153) Gastrointestinal diseases (#PE3861) Abuse of coca and cocaine (#PD2363) Perinatal morbidity and mortality (#PD2387).
Aggravated by Hunger (#PB0262) Land enclosure (#PJ3523) Soil infertility (#PD0077) Dietary restrictions (#PJ1933) Nutritional ignorance (#PE5773) Socio–economic poverty (#PB0388) Vitamin deficiencies in diet (#PD0715).

♦ PC3320 Inadequate housing among indigenous peoples

Discrimination against indigenous populations in housing – Denial of the right to adequate housing for indigenous peoples
Nature Housing which does not provide enough protection against weather conditions, enough living space, sanitation or light, raises the risk of disease and early death.
Incidence The housing of indigenous people may be provided by their employers, as in the case of the reservations in North America and Australia and the mining camps in South America. It may be built by the indigenous people themselves but subject to restriction by their employers, such as agricultural labourers' accommodation on South American haciendas over which they have no right of ownership. Mineworkers in Bolivia and Peru suffer from serious overcrowding in large camps with no sanitation or water supply, no light or ventilation in the houses. Agricultural labourers build flimsy dwellings since they are temporary and cannot be owned. Their animals share the living space. These people may be less well–housed than the pedigree animals on the estate for which they are working. Indians in the USA are deficiently housed on reservations or tend to occupy slums in urban areas owing to low wages and discrimination in employment. Many indigenous people in mountain areas suffer from respiratory diseases owing to inadequate protection against the cold. In Chihuahua in Northern Mexico, Indians live in caves.
Broader Inadequate housing (#PC0449) Discrimination against indigenous populations (#PC0352).
Related Substandard housing and accommodation (#PD1251) Malnutrition among indigenous peoples (#PC3319).

Aggravates High mortality rate (#PJ6252) Inappropriate sanitation systems (#PD0876)
Overcrowding of housing and accommodation (#PD0758)
Illegal occupation of unoccupied property (#PD0820).
Aggravated by Waste paper (#PD1152) Mining camps (#PG6247)
Hacienda camps (#PG6248) Uncontrolled migration (#PD2229)
Inadequate ventilation (#PG6250) Exploitation in housing (#PD3465)
Inadequate land drainage (#PD2269) Exploitation in employment (#PC3297)
Inadequate housing construction (#PG0561) Denial of the right to ownership (#PE8411)
Insufficient storage space in homes (#PJ0221)
Inadequate education of indigenous peoples (#PC3322)
Restriction of indigenous populations to reservations (#PD3305).

♦ PC3322 Inadequate education of indigenous peoples

Discrimination against indigenous populations in education — Denial of the right of indigenous peoples to education

Nature Inadequate educational facilities result in inadequate educational achievement of indigenous peoples and lead to lack of assimilation into the national culture and persistent social and economic subjugation. Such inadequate facilities may arise from discrimination against indigenous peoples at the national level or inadequate policies to cope with the linguistic problems caused by a wide variety of unilingualist groups. Also, inadequate provision is made for the fact that children often have to work from the age of nine onwards, and that in so doing they may also become migratory. Isolation and wide geographic distribution also contributes to the difficulties in providing adequate education. Without sufficient education, indigenous people cannot be assimilated into the national culture and remain in an underprivileged position.

Refs Liegeois, P *La Scolarisation des Enfants Tziganes et Voyageurs* (1986).
Broader Inadequate education (#PF4984) Inequality in education (#PC3434).
Narrower Illiteracy among indigenous peoples (#PD3321).
Aggravates Illiteracy (#PC0210) Social inequality (#PB0514)
Lack of assimilation (#PF2132) Underpayment for work (#PD8916)
Lack of social mobility (#PF2195) Underprivileged racial minorities (#PC0805)
Inadequate housing among indigenous peoples (#PC3320).
Aggravated by Unwritten language (#PF3470) Diverse unilingualism (#PF3317)
Uncontrolled migration (#PD2229) Exploitation in employment (#PC3297)
Isolation of ethnic groups (#PC3316) Exploitation of child labour (#PD0164)
Inadequate educational facilities (#PD0847) Racial discrimination in education (#PD3328)
Discrimination against indigenous populations (#PC0352)
Inadequate adaptation of policy to educational difficulties (#PE8700).

♦ PC3324 Underprivileged linguistic minorities

Nature Linguistic minorities are segregated from the rest of the community by their language, which limits their opportunities in terms of education, employment and in general economic and political life. They have access only to a very limited proportion of the information generated in society, as media and governments are not able or willing to provide more than the minimum in minority languages. Their only recourse may be to give up their language and become assimilated into the culture of the dominant language, thereby losing the unifying basis of their culture.

Incidence In the case of pluralistic societies more than one language may be official, but one language may predominate over the others, giving greater opportunities to one group. In societies where only one language is official other language groups may consist of immigrants or colonized indigenous people. If immigrants learn the official language imperfectly or simply prefer their own, a ghetto situation arises, creating a very closed–in community, perpetuating its own education and community life apart from the community at large and often at a lower level, leaving the way open for social conflict. Indigenous people may be officially deprived of their language in order to 'nationalize' them; and poor linguistic communication may make it difficult for national authorities to render adequate social services.

Refs Coulmas, Florian *Linguistic Minorities and Literacy* (1984).
Broader Social injustice (#PC0797) Social inequality (#PB0514)
Underprivileged minorities (#PC3424).
Related Underprivileged racial minorities (#PC0805)
Underprivileged religious minorities (#PC2129)
Underprivileged ideological minorities (#PC3325).
Aggravates Illiteracy (#PC0210) Ethnic conflict (#PC3685)
Underdevelopment (#PB0206) Lack of social mobility (#PF2195)
Inequality of opportunity (#PC3435) Destruction of cultural heritage (#PC2114)
Unequal distribution of social services (#PC3437)
Endangered tribes and indigenous peoples (#PC0720)
Discrimination against minority languages (#PD5078).
Aggravated by Elitism (#PA1387) Social fragmentation (#PF1324)
Isolation of ethnic groups (#PC3316) Discrimination against minorities (#PC0582).
Reduces Homogenization of cultures (#PB1071).
Reduced by Lack of assimilation (#PF2132).

♦ PC3325 Underprivileged ideological minorities

Nature Religious, political, social or economic ideological minorities are discriminated against and harassed. Discrimination may take the form of police or individual harassment, arrest, or discrimination in employment, education and housing. Ideological minorities involved in political, social or economic matters are likely to be in confrontation with the national administration, and take this as a pretext for the involvement of an outside power favourable to their cause, or for war. Discrimination against such minorities (religious minorities included) may give rise to terrorism and a state of civil war.

Broader Social injustice (#PC0797) Underprivileged minorities (#PC3424)
Ideological discrimination in politics (#PC3219).
Related Underprivileged racial minorities (#PC0805)
Underprivileged religious minorities (#PC2129)
Underprivileged linguistic minorities (#PC3324).
Aggravates Social conflict (#PC0137) Political conflict (#PC0368)
Religious conflict (#PC3292) Ideological conflict (#PF3388)
Denial of education to minorities (#PC3459).
Aggravated by Conformism (#PB3407) Nationalism (#PB0534)
Compulsory indoctrination (#PD3097) Discrimination against minorities (#PC0582).

♦ PC3334 Racial exploitation

Dependence on racial exploitation

Nature Exploitation of other races on a class basis (usually lower than that of the exploiter) and as a source of high profit for little expenditure, may occur in the spheres of housing, employment and sexual relations. It may be legalized (as in the South African system of apartheid), or it may be the result of non–legalized discrimination and the lack of protection under the law. It may be illegal (such as traffic in immigrant workers to developed countries) but practised with the compliance of the immigrants who have been misled by exaggerated promises.

Broader Exploitation (#PB3200) Social injustice (#PC0797)
Social inequality (#PB0514).
Related Caste system (#PC1968) Ethnic conflict (#PC3685)
Political exploitation (#PC7356) Exploitation of children (#PD0635)
Exploitation in employment (#PC3297).
Aggravates Eugenics (#PC2153) Segregation (#PC0031)
Miscegenation (#PC1523) Social conflict (#PC0137)
Racial conflict (#PC3684) Social fragmentation (#PF1324)
Lack of social mobility (#PF2195).
Aggravated by Racism (#PB1047) Racial inequality (#PF1199)
Racial discrimination (#PC0006) Legalized racial discrimination (#PC3683)
Threats against family or friends (#PE3308) Denial of education to minorities (#PC3459)
Inadequate cultural integration of immigrants (#PC1532).

♦ PC3355 Religious rivalry

Competition between missionary bodies

Nature The rivalry for religious converts and influence increases the lack of religious unity and the level of religious intolerance. This may cause conflict between Church and state. One religion may lose strength to another because it supported the losing political faction in a power struggle.

Incidence This rivalry is not only among world religions, but among denominations, sects and cults and between all these and non–religious ideologies.

Background Missionary activity has traditionally been the cause of noticeable rivalries. In the 18th and 19th centuries the Christian missions fought against Hinduism in India, and in the 20th century Hindu missions fight against Christianity. Beginning with more traditional Hindu missions, this century has seen the emergence of a more entrepreneurial and extravagant style in the Transcendental Meditation, Hare Krishna and Divine Light organizations, with their hundreds of thousands of followers. Friction between the devotees of these cults and the Christian community has been high.

Broader Competition (#PB0848) Fundamentalism (#PF1338)
Fragmentation within organized religions (#PF3364).
Narrower Religious and political antagonism (#PC0030).
Aggravates Religious repression (#PC0578) Dependence on religion (#PF0150)
Fragmentation of religious belief (#PF3404)
Undue religious influence on secular life (#PF3358).
Aggravated by Evangelism (#PF6325) Religious intolerance (#PC1808)
Religious discrimination in education (#PD8807).

♦ PC3359 Corruption in organized religion

Nature Corruption of church officials may concern property or the manipulation of religious teachings to fit national or political aspirations. Accusations of corruption in the former category need not necessarily involve embezzlement or other forms of civil corruption. The possession and holding of wealth and the management of it in a similar manner to that of a commercial enterprise may be considered as corruption. Corruption may cause religious schism or anticlericalism and religious conflict. Religious influence may also be regarded as corruption or extending religious doctrines too far.

Background Accusations of corruption have been one of the main causes for religious schism ever since the Reformation.

Broader Corruption (#PA1986) Institutionalized corruption (#PC9173).
Related Ideological conflict (#PF3388) Corruption in politics (#PC0116)
Accumulation and misuse of religious property (#PE3354).
Aggravates Anti–clericalism (#PF3360) Superficial religion (#PJ5252)
Injustice of religious courts (#PE0397)
Religious discrimination in the administration of justice (#PE0168).
Aggravated by Dependence on religion (#PF0150)
Double standards in morality (#PF5225).

♦ PC3362 Threats to ideological movements and minorities

Nature Ideological minorities, whether religious or nontheistic ideological groups or political movements, may be threatened by conformism or ideological repression. They may be considered subversive and be severely repressed; they may find difficulties in coming to terms with the majority way of life; and they may be attacked by public authorities for not complying with public regulations on health or education.

Broader Ideological conflict (#PF3388) Threatened and vulnerable minorities (#PC3295).
Narrower Threatened sects (#PC1995).
Related Discrimination against minorities (#PC0582)
Underprivileged racial minorities (#PC0805)
Endangered tribes and indigenous peoples (#PC0720).
Aggravates Social fragmentation (#PF1324) Cultural fragmentation (#PF0536).
Aggravated by Conformism (#PB3407) Double standards in morality (#PF5225)
Religious conflict between sects (#PC3363) Repression of intellectual dissidents (#PD0434).

♦ PC3363 Religious conflict between sects

Nature Conflict between religious sects on ethnic or doctrinal grounds results in discrimination, segregation, intolerance, prejudice and inequality. The existence of minority sects may be threatened by larger sects which have wider recognition and are better integrated into national life. Under such conditions, compulsory education becomes a weapon of religious conflict.

Broader Religious conflict (#PC3292).
Narrower Multidenominational society (#PF3368)
Fragmentation within organized religions (#PF3364).
Related Religious discrimination in the administration of justice (#PE0168).
Aggravates Threatened sects (#PC1995) Discrimination against minorities (#PC0582)
Segregation based on religious affiliation (#PC3365)
Threats to ideological movements and minorities (#PC3362).
Aggravated by Prejudice (#PA2173) Religious extremism (#PF4954)
Religious intolerance (#PC1808) Injustice of religious courts (#PE0397).

♦ PC3365 Segregation based on religious affiliation

Nature Religious segregation may take the form of segregation in education, in employment, in class, in housing and in marriage. In public services, political discrimination can cause the apportioning of social benefits according to religion. Segregation results in inequality, religious intolerance, conflict, and sometimes civil war. Religious segregation may be inflicted by force, as with the segregation of the Jews in Nazi Germany and elsewhere in Europe under Nazi occupation during the 2nd World War.

Incidence Religious segregation is universal and particularly marked in multidenominational societies. For some fundamentalist or very traditional sects, segregation is almost complete.

Broader Segregation (#PC0031) Religious repression (#PC0578).
Narrower Threatened sects (#PC1995) Segregation in marriage (#PD3347).
Aggravates Religious conflict (#PC3292) Social fragmentation (#PF1324)
Religious intolerance (#PC1808).
Aggravated by Religious schism (#PF1939) Religious discrimination (#PC1455)
Religious conflict between sects (#PC3363) Religious discrimination in education (#PD8807).

♦ PC3383 Sex segregation

Nature Since no aspect of life is purely masculine or purely feminine, a world in which the separation of the sexes is extreme, distorts reality and perpetuates distortions. Segregation of the sexes in education, family life, employment, housing and public services leads to inequality and discrimination.

Incidence The world of a town is often split along sexual lines. Suburbs are for women, workplaces are for men; kindergartens are for women, professional schools for men; supermarkets are for women, hardware stores for men. Segregation in education fixes the roles for family life and later employment. Science is dominated by a masculine, and often mechanical, mentality; foreign diplomacy is governed by war, again the product of the masculine ego. Schools for young children are swayed by the world of women, as are homes. Religious schools are traditionally segregated according to sex, except perhaps for the very early years; this is most notable in Roman Catholic Schools in developed countries and their missionary schools in the developing world. Segregation in family life exists in its most complete form in traditional Muslim culture where boys have almost no contact with adult males until they have reached maturity, and girls are totally excluded from public life. Segregation in employment occurs universally, with a resulting segregation in public services based on the dependent status of women. Segregation in housing usually occurs with regard to institutions or hostels.
Counter–claim The institution of purdah, meaning segregation of the sexes, is indispensible in an Islamic society.
Refs National Research Council *Women's Work, Men's Work* (1985).
Broader Segregation (#PC0031) Sexual discrimination (#PC2022)
Neglect of the role of women in rural development (#PF4959).
Narrower Sexually segregated schools (#PG3650)
Social segregation of eunuchs (#PJ5599).
Related Ethnic segregation (#PC3315).
Aggravates Social inequality (#PB0514) Discrimination against women (#PC0308).
Aggravated by Discrimination against men (#PC3258).

♦ **PC3390 Intellectual conflict**
Nature Conflict of ideas in theory or philosophy may be manifested in propaganda; political, religious or social campaigns; and intolerance of other views, or refusal to consider other views, resulting in stagnation, alienation or schism.
Broader Irrationalism (#PF3399) Ideological conflict (#PF3388).
Related Racial conflict (#PC3684) Compulsory indoctrination (#PD3097).
Aggravates Propaganda (#PF1878).
Aggravated by Inadequacy of doctrine (#PF3396)
Intellectual prejudice (#PC3406)
Double standards in morality (#PF5225).

♦ **PC3406 Intellectual prejudice**
Nature The holding of biased intellectual views concerning social, political, ideological, or economic theories or philosophy may lead to stagnation and lack of flexibility in ideas, to intellectual conflict, and to intellectual repression where authorities are prejudiced against opinions which they feel could be damaging socially, economically, or politically.
Broader Prejudice (#PA2173) Intellectualism (#PF2146).
Related Repression of intellectual dissidents (#PD0434).
Aggravates Racial inequality (#PF1199) Self–righteousness (#PG6373)
Intellectual conflict (#PC3390) Ideological discrimination in politics (#PC3219).
Reduced by Anti–intellectualism (#PF1929).

♦ **PC3412 Social apathy**
Dependence on social apathy
Nature Lack of concern for social issues and the suffering of others may occur on an individual or national level. It may also be reflected in hostility or indifference shown towards other social groups by a group formed to protect its own self–interest. Social apathy reinforces social injustice, inequality and conflict and may lead to social breakdown, anarchy or dictatorship.
Broader Apathy (#PA2360) Social underdevelopment (#PC0242).
Related Political apathy (#PC1917)
Failure of individuals to participate in social processes (#PF0749).
Aggravates Social injustice (#PC0797).
Aggravated by Cynicism (#PF3418) Social alienation (#PC2130)
Social disintegration (#PC3309).

♦ **PC3413 Economic apathy**
Dependence on economic apathy — Industrial apathy
Nature Lack of desire to improve the standard of living, or inertia in the face of economic problems, may occur on a personal, group, or national level. On a personal level it may result in a low standard of living, lack of self–confidence and material insecurity. On a group level it tends to sharpen intergroup divisions along the lines of class. On a national level it may lead to economic stagnation and foreign intervention.
Broader Apathy (#PA2360).
Related Amoralism (#PF3349) Meaninglessness (#PA6977)
Underdevelopment of industrial and economic activities (#PC0880).
Aggravates Economic stagnation (#PC0002) Inadequate standards of living (#PF0344).
Aggravated by Inadequacy of economic doctrine (#PF3395).
Reduced by Strikes (#PD0694).

♦ **PC3414 Religious apathy**
Dependence on religious apathy
Nature Lack of religious conviction or activity may be marked by a lack of church attendance or other observance of religious customs, but does not necessarily mean that people leave their nominal church; they may remain within it but simply not practice religion in any particular form. Religious apathy may also be taken to mean an unquestioning attitude towards religious doctrine.
Incidence Lack of outward religious conviction but nominal adherence to a religion is widespread in Western industrialized countries. An unquestioning acceptance of religious doctrine is found particularly among poor people in developing countries.
Broader Apathy (#PA2360) Agnosticism (#PF2333).
Narrower Decreasing participation in collective religious worship (#PF8905).
Related Dependence on religion (#PF0150).
Aggravates Puritanism (#PF2577) Disintegration of organized religion (#PD3423).
Aggravated by Retarded socialization (#PF2187)
Inadequacy of religion (#PF2005).
Reduces Religious extremism (#PF4954).

♦ **PC3424 Underprivileged minorities**
Dependence on underprivileged minorities
Nature Minorities which are exploited or discriminated against by a dominant group include racial, religious, linguistic and ideological minorities, and vulnerable groups in the social strata, such as women, children, the aged, the disabled, the mentally retarded or deranged, the physically handicapped, and the socially handicapped (in the sense of deprivation in family upbringing, education, nutrition, or housing). Such minorities are usually inadequately protected under the law and may be outcasts in society.
Incidence Ethnic and cultural minorities existing outside the mainstream of activity and development are found in rich and poor countries alike. Relative poverty, malnutrition, high levels of mortality, ill health, limited educational opportunity, inadequate housing and inferior legal, social and cultural status characterize these groups. International minorities and vulnerable small states may also be exploited and inadequately protected.
Broader Social injustice (#PC0797) Plural society tensions (#PF2448)
Divisive effects of official cultural pluralism (#PF0152).
Narrower Underprivileged racial minorities (#PC0805)
Underprivileged religious minorities (#PC2129)
Underprivileged linguistic minorities (#PC3324)
Underprivileged ideological minorities (#PC3325).
Aggravates Social conflict (#PC0137) Inequality of opportunity (#PC3435).
Aggravated by Prejudice (#PA2173) Homelessness (#PB2150)
Social fragmentation (#PF1324) Lack of protective legislation (#PJ2889)
Discrimination against minorities (#PC0582).

♦ **PC3425 Political inequality**
Dependence on political inequality
Nature The unequal distribution of political benefits, such as effective political rights as opposed to nominal ones, results in political instability. Political inequality may be expressed as dictatorships' elitism, the superior political power of large pressure groups, or vested interest. It may equally be expressed as colonialism, foreign influence, and control or economic imperialism. Political inequality leads to social inequality and international inequality, alienation, apathy, stagnation and exploitation, and may result in political or national disintegration and revolution or possible foreign intervention.
Broader Domination (#PA0839) Human inequality (#PA0844).
Narrower Unfairly negotiated treaties (#PF4787) Unequal parliamentary constituencies (#PD2167).
Related Social injustice (#PC0797).
Aggravates Elitism (#PA1387) Social inequality (#PB0514) Political alienation (#PC3227)
Unequal political representation (#PC0655) Political corruption of the judiciary (#PE0647).
Aggravated by Political injustice (#PC2181) Undemocratic political organization (#PC1015).

♦ **PC3426 Dependency of women**
Nature Although women's situations are improving in some developed countries, in most countries throughout the world women may become a burden on society if they lose the economic support of a man. They may be left with children and this increases the burden on social welfare services or on other members of the family. Because the status of women is one of dependency, women may be ill–prepared to support themselves or a family; and even if willing to do so may not be able to in practice because of unequal pay and inadequate child–care facilities. Women are retired from work earlier than men despite a longer average life expectancy.
Incidence The problem is universal, though the burden is better absorbed in tribal and primitive society where most social categories are better integrated than in industrialized society.
Background Girl babies have always been unpopular. The Jewish Talmud says 'When a girl is born, the walls are crying; in the Book of Islam it is written 'When an Arab hears that a daughter has been born to him his face becomes saddened; and the Migures of Central Asia say 'It is better for a girl not to be born, or to die soon after birth. This indicates how women were felt to be a burden in traditional civilization, either from the point of view of maintenance, protection, and marriage settlement, or lack of dynamic contribution towards feeding or safeguarding the community.
Counter–claim There would be no society without women.
Broader Human dependence (#PA2159) Discrimination against women (#PC0308)
Lack of protection for the vulnerable (#PB4353).
Narrower Dependency of women in marriage (#PD3694).
Aggravates Family breakdown (#PC2102) Lack of social mobility (#PF2195)
Discrimination against men before the law (#PD3692)
Incompatibility of traditional and new technologies (#PE3337).
Aggravated by Imbalance in the human sex ratio (#PF1128)
Inadequate child day–care facilities (#PD2085)
Discrimination against women in politics (#PC1001)
Discrimination against women in employment (#PD0086)
Discrimination against women before the law (#PD0162)
Discrimination in family planning facilities (#PD1036)
Discrimination against women in social services (#PD3691).

♦ **PC3428 Conflict between minority groups**
Nature Conflict between minority groups may be fostered by the dominant group in order to maintain its domination. A group which is aspiring to join or come closer to the dominant group may despise its origins and suppress other groups which are still in a subordinate position. In the competition for available privileges, one subordinate minority might feel another is detracting from its chances, or it might see a privileged minority as the cause of its ills.
Broader Conflict (#PA0298) Inequality of opportunity (#PC3435).
Narrower Scapegoats (#PF3332) Self–hatred (#PG6396)
Inter–faith friction (#PG6397).
Related Fragmentation of communist parties (#PD0923).
Aggravates Ethnic conflict (#PC3685) Social fragmentation (#PF1324)
Social underdevelopment (#PC0242).
Aggravated by Elitism (#PA1387) Segregation (#PC0031)
Colonialism (#PC0798) Stereotypes (#PF8508)
Social injustice (#PC0797) Social inequality (#PB0514)
Political repression (#PC1919) Lack of social mobility (#PF2195)
Compulsory indoctrination (#PD3097).
Reduces Dictatorship of the majority (#PD3239).

♦ **PC3429 Human torture**
Denial of right to freedom from torture
Nature In many countries torture is used as a judicial instrument for extracting evidence and confessions from the accused. Individuals may inflict pain on other individuals for reasons of pleasure or vengeance. Individual torture includes acts of juvenile delinquency, such as the terrorizing and beating up of old or infirm people by gangs or individuals, or, for instance, the burning or humiliation of other young people who are not members of the group. Torture may be inflicted by sadists in an advanced state of mental disorder, or by 'protection' gangs and other organized crime units. Mental torture and less sophisticated means of physical torture may be practised within families or within a group of friends; this kind of torture may have a sexual basis. Individual torture may also be entered into under the effect of hallucinogenic drugs.
Incidence Human torture has been reported in the following countries: **Af** Algeria, Angola, Burundi, Cameroon, Chad, Comoros, Congo, Côte d'Ivoire, Djibouti, Egypt, Ethiopia, Gabon, Gambia, Ghana, Guinea, Guinea–Bissau, Kenya, Lesotho, Liberia, Mali, Madagascar, Mauritania, Morocco, Mozambique, Namibia, Niger, Nigeria, Rwanda, Somalia, South Africa, Sudan, Tanzania UR, Tunisia, Uganda, Zaire, Zambia, Zimbabwe. **Am** Argentina, Bolivia, Brazil, Canada, Chile, Colombia, Costa Rica, Cuba, Dominica, El Salvador, Grenada, Guatemala, Guyana, Haiti, Honduras, Mexico, Paraguay, Peru, USA, Uruguay. **As** Afghanistan, Bahrain, China, Hong Kong, India, Indonesia, Iraq, Iraq, Israel, Jordan, Korea Rep, Kuwait, Lao PDR, Lebanon, Malaysia, Nepal, Oman, Philippines, Singapore, Sri Lanka, Syrian AR, Taiwan (Rep of China), United Arab Emirates, Viet Nam. **Eu** Albania, Bulgaria, Czechoslovakia, Germany FR, Greece, Italy, Poland, Romania, Spain, Turkey, USSR, UK.
Claim The majority of torture victims, even in countries beset by widespread civil conflict, have

no security information about violent opposition groups to give away. They are tortured either to force confessions from them or as an acute message not to oppose the government. Even it torture could be shown to be efficient in some cases, it could simply never be permissible. From the point of view of the individual, torture, for whatever purpose, is a calculated assault on human dignity and for that reason alone is to be condemned absolutely. From the point of view of society, once justified and allowed for the narrower purpose of combating political violence, torture will almost inevitably be used for a wider range of purposes against an increasing proportion of the population. Just using torture once, almost invariably leads to its institutionalization and will erode the moral and legal principles that stand against a form of violence that could affect all of society. For the state, torture subverts a basic tenet of just punishment, a prescribed penalty for a proven offence. It destroys any amount of trust between citizens and rulers.
Counter–claim Though a disagreeable thing, torture has to be applied in select cases because there are hardened criminal who would not otherwise come out with the truth.
Refs Fanon, Frantz *The Wretched of the Earth* (1966).
Broader Torture (#PB3430)
Denial of right to freedom from cruel, inhumane or degrading punishment (#PC3768).
Narrower Physical torture (#PD8734) Religious torture (#PC7101)
Torture of children (#PD2851) Psychological torture (#PD4559)
Pharmacological torture (#PE4696).
Related Sadism (#PF3270) Cruelty to plants (#PD4148) Torture of animals (#PC3532)
Denial of right to life (#PD4234).
Aggravates Maltreatment of prisoners (#PD6005)
Suspicious deaths during detention (#PE6367).
Aggravated by Torture schools (#PE2062) Television violence (#PE4260)
Denial of right of complaint (#PD7609) Lack of access for prisoners' defence (#PE8637)
Military and police personnel participation in torture (#PE4119)
Dependence on excitement and danger among young people (#PE8933).

♦ PC3431 Ideological war
Dependence on ideological war
Nature War for ideological reasons includes religious war and nationalist and racial war where an ideology is involved, as well as cold war and political conflict between ideologies.
Incidence Religious wars are currently taking place in Ulster and the Middle East. The war over Bangladesh had a religious basis and Indian/Pakistani border clashes also contain a religious war element. Ideological war between Capitalism and Communism took place in Korea and Vietnam. The Middle East war contains elements of racism and nationalism as well as religion.
Broader War (#PB0593).
Narrower War between socialist states (#PD4952).
Related Racial war (#PD8718) Ideological repression (#PC8083)
Maldistribution of electrical energy (#PD3446).
Aggravates Enemies (#PF8404) War crimes (#PC0747)
Jamming of satellite communications (#PD1244).
Aggravated by Aggressive foreign policy (#PC4667)
Compulsory indoctrination (#PD3097).

♦ PC3432 Sexism
Nature The stereotyping of sexual roles in society is part of a widely accepted rigid ideology and social structure. In one aspect it manifests as a traditional division of labour, claiming descent from supposed pre–historic patterns where males were typically migrant and aggressive foragers and hunters, while females reared children, prepared food and did not go far from tribal encampments and dwellings. In another aspect, sexism manifests as the custom of disproportionate rewards, the greater share being demanded by, and given to, the male.
Claim Sexism is a pre–scientific attempted philosophy or psychology of behaviour, and the earliest form of social engineering, which attempted to establish some pre–conceived rationalization in the assignments of roles in society. It is thus inherent in all political sciences, so that the foundation documents and authorities of modern societies are sexist, mainly by unconsciously omitting provisions for equality of women. There is a considerable connection between masculine sexism and militarism, belligerency and war.
Refs Clark, Lorenne M G and Lange, Lynda (Eds) *Sexism of Social and Political Theory* (1979); Smith, Joan, et al *Racism, Sexism, and the World–System* (1988).
Broader Elitism (#PA1387) Extremist ideologies (#PC6341).
Narrower Sexual exploitation (#PC3261) Sexist education of children (#PF5967)
Sexually discriminating job terminology (#PF6014).
Aggravates Inflexible social structure (#PB1997) Discrimination in social services (#PC3433).
Aggravated by Sexual discrimination (#PC2022) Undue attachment to a social group (#PF1073).

♦ PC3433 Discrimination in social services
Inequality in social services
Broader Discrimination (#PA0833).
Narrower Segregation in social services (#PD3440)
Discrimination against men in social services (#PD3336)
Discrimination against women in social services (#PD3691)
Narrow legal definition of the family in developing countries (#PD1501).
Aggravates Alienation (#PA3545) Social fragmentation (#PF1324)
Socio–economic poverty (#PB0388).
Aggravated by Sexism (#PC3432) Social inequality (#PB0514)
Negative effects of family allowances (#PF0107).

♦ PC3434 Inequality in education
Discrimination in education — Dependence on inequality in education
Nature Unlimited educational opportunity is only available to the rich, and to a few among those of genius and talent who are poor, if they are discovered. In developed countries it is the universities and colleges which are out of reach for millions; in the developing countries it is education after age thirteen or fourteen. Even where free education is available, the poor student lacks equal nutrition, equal probability of good health, and equal reinforcement in his family environment for his scholastic endeavours. In addition, minority students are discriminated against in a number of ways, up to and including the practice of apartheid or racial segregation and enrolment in inferior schools.
Refs Quay, Richard H *In Pursuit of Equality of Educational Opportunity* (1977); Reynolds, Cecil R and Mann, Lester (Eds) *Encyclopedia of Special Education* (1987).
Broader Human inequality (#PA0844) Obstacles to education (#PF4852).
Narrower Denial of education to minorities (#PC3459)
Class discrimination in education (#PE0779)
Racial discrimination in education (#PD3328)
Sexual discrimination in education (#PD1468)
Religious discrimination in education (#PD8807)
Unequal opportunities for foreign students (#PE7726)
Inadequate education of indigenous peoples (#PC3322).
Related Segregation in education (#PD3441) Denial of right to education (#PD8102).
Aggravates Social injustice (#PC0797) Inequality of opportunity (#PC3435).
Aggravated by Segregation (#PC0031) Racial inequality (#PF1199)
Isolation of ethnic groups (#PC3316) Unequal access to education (#PC2163)
Lack of funds for education (#PJ6410) Inadequate educational facilities (#PD0847)
Economically disadvantaged students (#PD2624)
Inequality of access to education within countries (#PC1896)
Inadequacy and insensitivity of intelligence testing (#PD1975)
Inadequate educational facilities for gifted children (#PD2051).

♦ PC3435 Inequality of opportunity
Dependence on inequality of opportunity
Broader Social injustice (#PC0797) Human inequality (#PA0844).
Narrower Inequality in employment (#PD8903) Unequal property distribution (#PC3438)
Conflict between minority groups (#PC3428)
Unequal opportunities for media reception (#PD3039)
Unequal opportunities for disabled persons (#PE0706)
Undeveloped channels for commercial initiative (#PF6471)
Inequality of employment opportunity in developing countries (#PD1847)
Risk of unintentional nuclear war generated by the strategy of deterrence (#PF4162).
Related Unequal distribution of fame and honours (#PF3439)
Divisive effects of official cultural pluralism (#PF0152)
Incompatibility of traditional and new technologies (#PE3337).
Aggravates Age segregation (#PD3444) Political apathy (#PC1917)
Over–qualification (#PF3462) Socio–economic poverty (#PB0388)
Denial of right to liberty (#PF0705) Obsolete deliberative systems (#PD0975)
Underprivileged religious minorities (#PC2129)
Inadequate cultural integration of immigrants (#PC1532)
Failure of individuals to participate in social processes (#PF0749).
Aggravated by Racial inequality (#PF1199) Illegal immigration (#PD1928)
Cultural deprivation (#PC1351) Social discrimination (#PC1864)
Inequality in education (#PC3434) Segregation in education (#PD3441)
Underprivileged minorities (#PC3424) Lack of community participation (#PF3307)
Denial of education to minorities (#PC3459) Underprivileged linguistic minorities (#PC3324)
Racial discrimination in public services (#PD3326)
Discrimination against dwarfs and midgets (#PE2635)
Inadequate living and working conditions of immigrant labourers in industrialized countries (#PD3427).

♦ PC3437 Unequal distribution of social services
Deficiencies in the welfare state
Nature The modern welfare state is not able to deliver its services equitably, partly due to the inherent problem that the provision of social services creates the demand for them. Therefore social services to the aged, the disabled, to parentless children, to the sick, to the handicapped, to the unemployed and others, are overwhelmed by applicants' numbers. In addition, there are the inevitable bureaucratic inefficiencies that create distortions and slowdowns in the distribution of services. Other factors of a sinister, if not criminal, nature are corruption among civil servants, and discrimination against applicants based on ethnicity, language, age, religion, political party, gender or other bases. Government policy may also be to erode public welfare in order to save money for militarization.
Refs Hardy, G *Doom of the Welfare Society* (1975); OECD *The Welfare State in Crisis* (1981); Van Driel, G J; Hartog, J A and Van Ravenzwaaij, C *Limits to the Welfare State* (1980).
Broader Human inequality (#PA0844).
Narrower Limited access to social benefits (#PF1303).
Related Distortionary tax systems (#PD3436).
Aggravates Socio–economic poverty (#PB0388)
Restriction of indigenous populations to reservations (#PD3305)
Failure of individuals to participate in social processes (#PF0749).
Aggravated by Injustice (#PA6486) Discrimination (#PA0833)
Social injustice (#PC0797) Illegal immigration (#PD1928)
Illegitimate children (#PC1874) Denial of human rights (#PB3121)
Obsolete deliberative systems (#PD0975) Segregation in social services (#PD3440)
Economic impediments to marriage (#PF3342) Underprivileged racial minorities (#PC0805)
Inadequate social welfare services (#PC0834) Underprivileged religious minorities (#PC2129)
Underprivileged linguistic minorities (#PC3324)
Discrimination against unmarried fathers (#PD3256)
Racial discrimination in public services (#PD3326)
Inadequate development of new social structures in developing countries (#PD0822)
Inadequate living and working conditions of immigrant labourers in industrialized countries (#PD3427).

♦ PC3438 Unequal property distribution
Unequal distribution of land and assets
Nature Poverty has been exacerbated by the unequal distribution of land and other assets associated with the rapid rise in population, which has compromised the ability to raise living standards. Together with growing demands for the commercial use of good land, often to grow crops for export, these have forced many subsistence farmers onto poor land, thus depriving them of any hope of participating in the economic development of their country. Similarly, traditional shifting cultivators, who maintained a stable relationship with the forests on which they depended, now have neither land enough nor time to let the forests re–establish. Extending cultivation onto steep, and especially deforested, slopes is increasing soil erosion. Extending such cultivation into river valleys often increases vulnerability to frequent flooding.
Incidence In developed market economies, possession of property used to be a principal factor in income concentration, but its importance has declined. The top 5 percent of income recipients receive less than one third of their income from property where once the figure was between 50 and 100 percent. Property income – rent, dividends, profits – in total income had declined to between 10 percent and 20 percent, or even less; it was typically above 20 percent in the 1950s and much higher still in the pre–war period. Private ownership of property still remains concentrated, although the trend has been towards less concentration, due largely to weak equity prices and surge in house prices. At the same time, social ownership of the means of production has increased sharply, mainly through acquisition – often of faltering private enterprises – and sometimes through the establishment of new public industrial, commercial, or non–profit ventures. There has also been a pronounced trend to the separation of management from ownership. Management has become concentrated in fewer institutions, both through business concentration and growth of government. But equally, there has been a trend towards greater regulation over the use of property, both through private action and administrative and judicial regulation. An important effect of the parallel increases in concentration of management or control, and social regulation, has been that governments have been called on more frequently to adjudicate between the different and detailed claims of social groups and interests, and in the process the traditional machinery for decision making in the democratic state has come under severe strain.
Broader Social injustice (#PC0797) Human inequality (#PA0844)
Inequality of opportunity (#PC3435).
Narrower Maldistribution of agricultural land (#PD9189).
Related Obsolete deliberative systems (#PD0975)
Maldistribution of land in developing countries (#PD0050).
Aggravates Segregation in housing (#PD3442) Flood plain settlement (#PE0743)
Deforestation of mountainous regions (#PD6282).
Aggravated by Accumulation of property (#PC8346)
Distortionary tax systems (#PD3436).

♦ PC3452 Social deviation
Broader Immorality (#PA3369)   Social conflict (#PC0137)
Retarded socialization (#PF2187).
Narrower Hooliganism (#PD1109)   Social outcasts (#PD6017)
Counter culture (#PF0423)   Criminal subculture (#PE5508).
Related Sexual deviation (#PD2198).
Aggravates Social breakdown (#PB2496).

♦ PC3458 Class consciousness
Class distinction — Class bigotry — Class disparity resentment — Class division — Class system — Class consciousness as a social barrier — Class segregation — Class discrimination — Myth of classlessness
Nature Differences between acceptable behaviour, possessions, use of language and other habits form barriers between classes, which may be difficult or impossible to surmount, and cause considerable social prejudice. Individuals who make a partial transition from one class to another may be accepted by neither. Class consciousness as a social barrier can range from elitism (European nobility, American nouveau-riche) to self-degradation, the former possibly limiting association with people 'beneath one', the latter possibly limiting the social and educational opportunities which could lead to success. Both of these forms of class consciousness may limit practitioners from realizing their full potential, on both an individual and a global basis.
Incidence The classless society, envisaged by Marx and Lenin, has not yet materialized. In the Soviet Union where, officially, social class – whether defined by job, income, family or attitude – has been abolished, there are 'strata' of society: workers, peasants and intellectuals, distinctions which are supposed to disappear when full communism is reached. But years after the Bolshevik takeover, class consciousness still exists. There is a large and powerful state bureaucracy founded on privilege.
Counter-claim A healthy society needs both custodians and innovators. It needs custodians who feel obligated to pass things on to the next generation, without which society falls apart, loses all its savour, all its beauty, all its charm, all its virtue. Those who are best equipped to be custodians are the moneyed hereditary class. Ancestral connections of this class enrich schools, colleges, and regiments; they also enrich trade unions, businesses and indeed all human organizations.
Broader Bigotry (#PC7652)   Social inequality (#PB0514)
Inflexible social structure (#PB1997).
Narrower Inferior classes (#PF7428)   Second-class citizenship (#PG3152)
Vulnerability of middle-class (#PC1002)   Discredited moneyed hereditary class (#PE5341).
Related Schizmogenesis (#PE4593).
Aggravates Snobbery (#PJ3943)   Conformism (#PB3407)
Segregation (#PC0031)   Class conflict (#PC1573)
Ethnic conflict (#PC3685)   Social conflict (#PC0137)
Social injustice (#PC0797)   Minority control (#PF2375)
Lack of social mobility (#PF2195)   Consanguineous marriage (#PC2379)
Pursuit of personal prestige (#PF8145)   Social impediments to marriage (#PF3341)
Fragmentation of the human personality (#PA0911)
Political barriers to effective legislation (#PC3201)
Stifled potential for social interaction between different age groups (#PF6570)
Inadequate living and working conditions of immigrant labourers in industrialized countries (#PD3427).
Aggravated by Elitism (#PA1387)   Caste system (#PC1968)
Traditionalism (#PF2676)   Intellectualism (#PF2146)
Social discrimination (#PC1864)   Acceptance of hierarchy (#PJ3602)
Discrepancies in human life evaluation (#PF1191).

♦ PC3459 Denial of education to minorities
Discrimination against minorities in education — Violation of right to education to minorities
Nature Discrimination against minorities in education is effected by denial of all or some type or level of education, such as by limiting a person or class of persons to inferior quality education; by maintaining separate educational systems or institutions for a class of people; and by inflicting undignified conditions of education on students.
Broader Inequality in education (#PC3434)   Discrimination against minorities (#PC0582).
Related Segregation in education (#PD3441)   Denial of right to education (#PD8102)
Denial of rights of minorities (#PC8999).
Aggravates Human inequality (#PA0844)   Racial exploitation (#PC3334)
Inequality of opportunity (#PC3435).
Aggravated by Social fragmentation (#PF1324)   Inadequate educational facilities (#PD0847)
Racial discrimination in education (#PD3328)   Religious discrimination in education (#PD8807)
Underprivileged ideological minorities (#PC3325)
Discrimination against illegitimate children (#PD0943)
Inadequacy and insensitivity of intelligence testing (#PD1975)
Property and occupational discrimination in politics (#PD3218)
Restriction of educational opportunities in capitalist systems (#PD3122).

♦ PC3461 Denial of rights to disabled
Nature Although entitled to the same rights as all other human beings and to equal opportunities, disabled persons' lives are often handicapped by physical and social barriers, thus hampering their full participation in society. They are victims of stereotyped attitudes that have labelled disabled people as being incapable of any kind of worth, value or benefit to the family or the community; attitudes that maintain that they are only burden. These attitudes have led to and maintained inequities, discrimination, and the continued dependency of disabled people. As a result, disabled children and adults in all parts of the world often face a life that is segregated and debased. It is largely the environment which determines the effect of disability on a person's daily life: disabled persons are still denied the factors generally available in the community that are necessary for the fundamental elements of living, including family life, education, employment, housing, financial and personal security, participation in social and political groups, religious activity, intimate and sexual relationships, access to public facilities, freedom of movement and the general style of daily living. People with permanent disabilities who are in need of community support services, aids and equipment to enable them to live as normally as possible both at home and in the community do not always have access to adequate services due to a lack of knowledge and sensitivity to their problems. Disabled people are often forced into economic dependency, because employers feel they would be sufficiently productive. If they find jobs, they are usually low-paid and with no upward mobility.
Claim It is not a person's disability that handicaps them but society does by not providing access to building, education, transportation and all the other necessities granted to the non-disabled.
Broader Discrimination (#PA0833)   Denial of rights to vulnerable groups (#PC4405).
Narrower Discrimination against the disabled (#PD9757)
Inadequate welfare services for the deaf (#PD0601)
Discrimination against mentally disabled (#PD9183)
Inadequate welfare services for the blind (#PD0542)
Discrimination against physically disabled (#PD8627).
Related Discrimination against giants (#PE5578)
Discrimination against dwarfs and midgets (#PE2635).
Aggravates Human dependence (#PA2159)   Lack of self-confidence (#PF0879)
Institutionalization of the disabled (#PF4681).
Aggravated by Disabled women (#PC0729)   Human disability (#PC0699)
Inadequate rehabilitation facilities (#PD1089).

♦ PC3486 Vulnerability of wetlands
Destruction of wetland environments — Destruction of peat land
Nature Wetlands include marshes, bogs, fens, and all stretches of water, whether fresh or salt, static or flowing, temporary or permanent; these may be estuaries and marine shallows, brackish and saline lagoons, natural and artificial lakes, small ponds or pot-holes, reservoirs, flooded gravel pits, rivers, streams, flood-meadows, and swamps.
Wetlands are shrinking through drainage and reclamation projects, thus destroying a vital resource. There is also a tendency for water quality deterioration in water bodies due to human development efforts, particularly in agriculture and industry, such as evaporation ponds for drainage and industrial waste water disposal. There is evidence of these poor quality water bodies affecting wild life, notably migratory birds and in particular their reproductive capacities. There is an increasingly harmful effect of sediments and toxic chemical pollutants on fish and other types of aquatic life. Greenhouse warming effect could destroy most of the wetlands of the world by causing sea-levels to rise.
Background Wetland habitats support a vast range of plant and animal life, and serve a variety of important functions, the full value of which are even now only beginning to be recognized. (Functions include: water regime regulation, flood control, erosion control, nursery areas for food fishes, fish production, waterfowl production, recreation, plant production, aesthetic enjoyment, and wildlife habitat).
Claim Wetlands account for about 6 per cent of the global land area and are among the most valuable environmental resources. If they are conserved, they improve water quality by cycling and storing nutrients, recharge groundwater stores, delay floodwaters, store greenhouse gases, and provide habitats for many wild plants and animals, including valuable resource species. Wetlands are also among the most undervalued and threatened environments, and are widely degraded or lost completely to agriculture or urbanization.
Refs Brown, David E and Morehouse, Bonnie S *Arizona Wetlands and Waterfowl* (1985); Carp, E (Comp) *Directory of Western Palearctic Wetlands* (1980); International Institute for Environment and *Wetland Drainage in Europe* (1984); Matlby, Edward *Waterlogged Earth* (1986).
Broader Vulnerability of ecosystem niches (#PC5773).
Related Endangered parklands (#PE9282)
Disruption of ecosystems in marginal agricultural lands (#PD6960).
Aggravates Wildlife extinction (#PC1445)   Excessive use of land for agriculture (#PD9534)
Unsustainable development of fresh waters (#PD6923).
Aggravated by Inadequate empolderment of wetlands (#PD5110).
Reduced by Inadequate land drainage (#PD2269)
Underutilization of peat as an energy source (#PE8194).

♦ PC3490 Urban underemployment
Nature There is a good deal of underemployment – both urban and, especially, rural – in most countries; but as a general rule the criteria used for detecting underemployment, and especially the average standards of individual output of work, vary from one region to another and pertain to a particular type of society at a particular stage of its development. Although incomplete, the available estimates for Latin America reveal very high rates of urban underemployment, ranging from 20 percent and upwards in most countries. Estimates of urban underemployment in other regions are even scantier, although it is clear that urban underemployment exists everywhere. A broader but perhaps more significant indicator of the extent of urban underemployment than such non-comparable statistics is to be seen in the swelling of tertiary employment activities that characterize the great majority of developing countries.
Counter-claim Gradually, as a result of a reduction in the daily number of hours of work and an extension of the holiday periods which replace the numerous feastdays of traditional societies, there has been a return in the economies of the developed countries to an annual number of working days similar to, or even lower than, the number that had been customary prior to the Industrial Revolution. Yet, in assessments of the level of underemployment in developing societies, there is a tendency to regard the reduced number of hours of work in the developed countries as an unprecedented situation due to the high levels of productivity which modern technology has made feasible. In the South, another factor, which is generally overlooked, consists in the coercive influence of climate. Without taking any narrowly deterministic view of geography, there can be no doubt that the tropical, or semi-tropical, climate which reigns in the greater part of the Third World is less well suited, especially in certain seasons, to long hours of work than is the climate of the temperate regions in which the majority of developed countries are situated.
Refs Kahnert, Friedrich *Improving Urban Employment and Labor Productivity* (1987).
Broader Unemployment (#PB0750)   Uncontrolled urban development (#PC0442).
Related Urban unemployment in developing countries (#PD1551).

♦ PC3532 Torture of animals
Nature Deliberate torture of animals takes place under the pretence of scientific experimentation, religious ritual, economic production and sport.
Broader Torture (#PB3430)   Inadequate animal welfare (#PC1167).
Related Human torture (#PC3429)   Physical torture (#PD8734)
Cruelty to plants (#PD4148)   Torture of children (#PD2851)
State sanctioned torture (#PD0181)   Criminal killing of animals (#PJ1158)
Inhumane killing of stray animals (#PE2759).
Aggravates Animal deaths (#PE7941).
Aggravated by Domination (#PA0839).

♦ PC3541 Pervasive fear of nuclear war
Nature An atmosphere of fear is the natural result of living under the threat of nuclear war. Fear makes the maintenance of objectivity more difficult, and can lead to a rejection of reality and a retreat to false solutions based on distorted facts. When facts are distorted, freedom of expression may be misapplied, leading to pathological group action even on a national scale. Although concern about nuclear war and the arms race is reasonable and natural, it can lead either to a feeling of individual futility and impotence or to a rejection of social roles; such reactions contribute to increases in drug dependence and in the suicide rate.
Incidence The present arms race is a threat to the survival of mankind on this earth, including the people concerned with inventing and producing arms. Man may divert his aggressive instinct into remote control destruction which may ultimately destroy him because he does not have enough balance and insight to realize the result of his actions. Yet the recognition of the destructive capacity of modern weapons has created a mental stress which is undeniable, though not easy to assess in precise terms. Mass demonstrations against war increasingly reflect the anxiety of the world population.
Background The fears of nuclear destruction that have become pervasive since World War II were in evidence since near the turn of the Century. A French writer, Gustave Le Bon and a British chemist, Frederick Soddy warned of the use of atomic energy in 1903. Between these warnings and the actual development of the atomic bomb many of the mythical and visionary themes associated of nuclear energy were being used: transmutation, fiery destruction, utopia, transform-

ing rays, genetic monsters, and earth as a wasteland. With the bombing of Hiroshima and Nagasaki these images were given concrete expression and since have only be added to.
Claim An armed world, stocked with enough lethal power to wipe out all human life, always adding to its potential for conventional and mass destruction; a world spanned by modern surveillance systems; a world aware that no part can be protected from direct attack by nuclear missiles; is a fearful place for hundreds and hundreds of millions of peoples who strive to better their lot. The fear and tension which this induces is a factor which inflames conflicts between groups and between nations.
Refs Weart, Spencer R *Nuclear Fear* (1988).
Broader Fear (#PA6030) Conflict (#PA0298).
Related Biochemical warfare (#PC1164) Low-intensity conflict (#PE3988).
Aggravates Unhealthy emotional responses to atomic energy (#PF0913).
Aggravated by Exterminism (#PF7401) Official secrecy (#PC1812) Military-industrial complex (#PC1952) Competitive acquisition of arms (#PC1258) Unreliability of computer software (#PE4428) Failure of disarmament and arms control efforts (#PF0013).

♦ PC3595 Disease vectors
Agents causing infectious diseases
Nature A vector is a transmitter of disease from one animal to another or to man. Most of the causal agents of disease, microorganisms, have no means of locomotion and are dependent upon other agencies for their spread and propagation. Some are adapted for transport by such inanimate agencies as wind and water; others depend upon living vectors. Vectors may serve one of two functions in the propagation of disease: they may disseminate, which means to disperse or spread abroad; or they may transmit, which includes the inoculation of a new host, as well as the transport of inoculum. There are various types of animal vectors: arthropods (mosquitoes, flies, lice, fleas, bedbugs, ticks), rodents and snails, which are responsible for the transmission of a tremendous range of diseases, many of them of considerable importance from a public health point of view.
Incidence Most of the vector-borne diseases are in general found in the tropical areas of the world. However, some of them have a distribution extending up to the Arctic. The most important and the most widespread of the vector-borne diseases is malaria, of which there are an estimated 400 million cases. The vector is the Anopheles mosquito, sole carrier of the Plasmodium malaria-causing parasite. Many of the most important communicable diseases, in terms of morbidity and mortality, are vector-borne. Where the methods of combating the vector are still experimental and unproved, little progress seems to have been made: no great success has been achieved against the snail that helps to transmit bilharzia (schistosomiasis, with an estimated 200 million sufferers) or the gnat that transmits river blindness (onchocerciasis, with an estimated 20 million sufferers) or the fly that transmits sleeping sickness (trypanosomiasis, which is a constant threat in tropical Africa). These diseases as well as yaws (framboesia) and a variety of intestinal worms and skin infestations still debilitate workers and cause great suffering and premature death in many developing countries, especially in Africa. The obstacles to progress in this field are in part financial: high costs may be involved since the problems extend far beyond the range of biological and medical research or even of conventional public health. At issue is control over the territory occupied by the vector. What this may mean in terms of logistics and engineering effort is exemplified by the large-scale bush clearing experiments that have been used in campaigns against the shade-loving tsetse fly. Even the conventional public health measures tend to be extremely costly in developing countries in which basis physical infrastructure is weak or absent. To deal effectively with sewage and to ensure the potability of water supplies may require heavy investment and this may be the only certain way of reducing the incidence of many fly-borne and ingested diseases.
Refs McKelvey, John J, et al *Vectors of Disease Agents* (1981).
Narrower Ticks as pests (#PE1766) Microbial diseases (#PC7492) Plant disease vectors (#PD3596) Human disease vectors (#PD6651) Insect vectors of disease (#PC3597) Hemiptera as insect pests (#PE3615) Rodent vectors of disease (#PE3629) Man as vector of disease (#PD8371) Vectors of animal diseases (#PD2751) Flies as vectors of diseases (#PE4514) Animals as vectors of disease (#PD8360) Mosquitoes as vectors of disease (#PE1923) Introduction of extraterrestrial infectious diseases and bacteria (#PF1312).
Related Mites as pests (#PE3639).
Aggravates Disease (#PA6799) Vector-borne diseases (#PD8385) Human disease and disability (#PB1044) Infectious and parasitic diseases (#PD0982) Weather as a factor of animal disease (#PD2740).
Aggravated by Disease reservoirs (#PG6656) Inadequate quarantine (#PE2850) Irresponsible introduction of new species of animals (#PD1290).

♦ PC3597 Insect vectors of disease
Entomoses
Nature Insects are the most important vectors of disease, being prominent in the transfer of human, animal and plant diseases.
Incidence Examples of insect vector-borne diseases are sleeping sickness transmitted by the tsetse fly; Chagas disease, transmitted by a triatomid bug; onchocerciasis, a filarial disease carried by a black fly, the Simulium; Bancroftian and Brugian Filariases, transmitted by various species of mosquitoes in urban and rural areas; dengue and dengue haemorrhagic fever, transmitted by Aedes aegypti, a mosquito; yellow fever, also transmitted by A aegypti and other Aedes.
Refs WHO *Vector Control in International Health* (1972); WHO Scientific Group, Geneva 1972 *Vector Ecology* (1972).
Broader Insect pests (#PC1630) Disease vectors (#PC3595).
Narrower Ticks as pests (#PE1766) Lice as insect pests (#PE1439) Tsetse flies as pests (#PE1335) Flies as vectors of diseases (#PE4514) Insect vectors of human disease (#PE3632) Insect vectors of plant diseases (#PD7732) Insect vectors of animal diseases (#PD2748).
Related Blowflies as pests (#PE3627) Flies as insect pests (#PE2254) Rodent vectors of disease (#PE3629).
Aggravates Rickettsiae (#PE2572) Dutch elm disease (#PE1154) Botflies as pests (#PE3635) Maize pests and diseases (#PE3589) Pests and diseases of oak (#PE2984) Pests and diseases of rice (#PE2221) Human disease and disability (#PB1044).
Aggravated by Water pollution (#PC0062) Insect bites and stings (#PE3636) Insect resistance to insecticides (#PD2109) Water pollution in developing countries (#PD3675) Introduction of new species of insect pests (#PF3592) Insect damage to stored and manufactured goods (#PD3657).

♦ PC3606 Monoculture of crops
Single crop farming
Nature The short-term cost reductions of planting the same crop on the same land year after year is offset by a tendency for the build-up of pests and plant diseases, and for deterioration of the soil.
Broader Over-intensive soil exploitation (#PC0052) Decreasing genetic diversity in cultivated plants (#PC2223).
Related Stagnated development of agricultural production (#PD1285).
Aggravates Crop shortfalls (#PD5174) Vulnerability of world genetic resources (#PB4788) Degradation of agricultural land by cash crops (#PE8324).
Aggravated by Narrow range of food crops (#PD4100) Prohibitive cost of farm machinery (#PF2457).

♦ PC3651 Production of non-essentials
Nature Systems of production tend to be based on a media-created mode of consumption. Needs are invented to meet what a particular economy produces, thus neglecting real human needs and also productive efficiency.
Broader Dominance of economic motives (#PF1913).
Narrower False image of scarcity (#PF3002) Inflexible management patterns (#PF3091) Foreign control of natural resources (#PD3109) Accountability based solely on profit (#PF3551) Over-specialized supervisory personnel (#PF3588) Production serving false consumption needs (#PF2639) Market indicators exclusion of human requirements (#PE1843) Industrial processes geared to reduced social needs (#PE3939) Overemphasis on effective use of technical resources (#PF2959).
Related Belittling of grant recipients (#PF2708) Self-interest driven investment (#PC2576) Non-inclusive management decisions (#PF2754).

♦ PC3683 Legalized racial discrimination
Broader Racism (#PB1047) Racial discrimination (#PC0006) Legalized discrimination (#PC8949).
Narrower Apartheid (#PE3681).
Related Forced labour (#PC0746) Racial segregation (#PC3688) Segregation in housing (#PD3442) Crimes against humanity (#PC1073) Legal impediments to marriage (#PF3346) Racial discrimination in politics (#PD3329) Denial of the right of association (#PD3224) Racial discrimination in education (#PD3328) Racial bias in sentencing offenders (#PE4907) Racial discrimination in public services (#PD3326) Discrimination against indigenous populations (#PC0352) Restrictions on freedom of movement between countries (#PC0935).
Aggravates Slave trade (#PC0130) Racist propaganda (#PD3093) Racial intimidation (#PC2936) Racial exploitation (#PC3334) Restriction of indigenous populations to reservations (#PD3305).
Aggravated by Elitism (#PA1387) Denial of right of complaint (#PD7609).
Reduces Extremism (#PB3415).

♦ PC3684 Racial conflict
Broader Conflict (#PA0298) Scapegoats (#PF3332) Ethnic conflict (#PC3685).
Narrower Tribalism (#PC1910) Apartheid (#PE3681).
Related Class conflict (#PC1573) Social conflict (#PC0137) Political conflict (#PC0368) Racial intimidation (#PC2936) Intellectual conflict (#PC3390).
Aggravates Racism (#PB1047) Genocide (#PC1056) Aerial piracy (#PD0124) Social injustice (#PC0797) Social fragmentation (#PF1324) Lack of assimilation (#PF2132).
Aggravated by Neo-fascism (#PF2636) Caste system (#PC1968) Minority control (#PF2375) Racial inequality (#PF1199) Racist propaganda (#PD3093) Racial segregation (#PC3688) Illegal immigration (#PD1928) Racial exploitation (#PC3334) Segregation in housing (#PD3442) Lack of racial identity (#PF0684) Underprivileged racial minorities (#PC0805) Threats against family or friends (#PE3308) Racial discrimination in politics (#PD3329) Underprivileged religious minorities (#PC2129) Abusive traffic in immigrant workers (#PD2722) Lack of appreciation of cultural differences (#PF2679) Inadequate cultural integration of immigrants (#PC1532) Divisive effects of official cultural pluralism (#PF0152) Inadequate living and working conditions of immigrant labourers in industrialized countries (#PD3427).

♦ PC3685 Ethnic conflict
Ethnic difficulties — Minority turmoil — Ethnic unrest — Ethnic tensions — Inter-communal ethnic violence — Masked racism — Ethnic killings — Ethnic feuding
Nature Conflict of a physical nature or in the form of overt discrimination on ethnic (racial, religious, linguistic or national) grounds, ethnic conflict may be caused by cultural invasion and a lack of assimilation, maintaining ethnic and social differences and discrimination, by social inequality and by exploitation. Ethnic conflict is constantly exacerbated by mass poverty, limited access to resources, denial of human rights, lack of national integration and issues of international peace and security. It is at once the instrument of national integration and the darkhorse of internal disharmony and discord. Strikes, boycotts and other forms of disruption by minority groups in nations where there are a large number of different ethnic or minority groups threatens the stability of the nation. The national government is faced with the choice of ignoring the unrest which can result in the escalation of disruption and the discrediting of the party in power. It could use force to suppress the turmoil. In this case the source of the turmoil may go underground. In any case the disruption of the larger society by minorities is a detriment to the whole nation.
Incidence Ethnic battlefields are found in the Soviet Union, the United States, Sri Lanka, Burma, Northern Ireland, Spain, Yugoslavia, South Africa and elsewhere. Fewer than 10 of the 165 nation states of the world are ethnically homogenous, the rest are potential ethnic powder kegs.
Refs Horowitz, Donald *Ethnic Groups in Conflict* (1985).
Broader Conflict (#PA0298) Aggression (#PA0587) Social injustice (#PC0797).
Narrower Tribalism (#PC1910) Miscegenation (#PC1523) Racial conflict (#PC3684) National federalism (#PF0626).
Related Class conflict (#PC1573) Social conflict (#PC0137) Religious conflict (#PC3292) Racial exploitation (#PC3334) Plural society tensions (#PF2448).
Aggravates Genocide (#PC1056) Civil war (#PC1869) Ethnocide (#PC1328) Scapegoats (#PF3332) Exploitation (#PB3200) Human violence (#PA0429) Political conflict (#PC0368) Lack of assimilation (#PF2132) Ethnic discrimination (#PC3686) Occupational diseases (#PD0215) Lack of social mobility (#PF2195) Endangered tribes and indigenous peoples (#PC0720) Disruption of development by tribal warfare (#PD2191).
Aggravated by Racism (#PB1047) Slavery (#PC0146) Elitism (#PA1387) Prejudice (#PA2173) Colonialism (#PC0798) Hero worship (#PF2650) Cultural invasion (#PC2548) Social inequality (#PB0514) Ethnic segregation (#PC3315) Racial segregation (#PC3688) Class consciousness (#PC3458) Social fragmentation (#PF1324) Socioeconomic stress (#PC6759) Diverse unilingualism (#PF3317) Skin colour prejudice (#PC8774) Lack of racial identity (#PF0684) National disintegration (#PB3384) Blocked minority opinion (#PD1140) Prejudice against minorities (#PC8494) Expulsion of ethnic minorities (#PC3205) Uncontrolled urban development (#PC0442) Denial of rights of minorities (#PC8999) Conflict between minority groups (#PC3428) Discrimination against minorities (#PC0582) Underprivileged linguistic minorities (#PC3324) Insufficient minority culture support (#PF5659)

Lack of appreciation of cultural differences (#PF2679)
Discrimination against indigenous populations (#PC0352)
Divisive effects of official cultural pluralism (#PF0152)
Innate expectation of suppression of minority opinion (#PD2108).
Reduces Loss of cultural identity (#PF9005).
Reduced by Revolution (#PA5901) Forced assimilation (#PC3293).

♦ PC3686 Ethnic discrimination
Ethnic racism
Nature Ethnic discrimination, legal or unofficial, against culturally distinguishable groups, on the grounds of race, religion, language or nationality, may be subtle or overt. It may be for economic exploitation and gain, or for social status or ego.
Incidence The ethnic factor is often the primary motivating force in international politics and usually the major divisive issue in policies for national development and integration. The unique feature raised by ethnic tension is that the universalist ideologies of liberalism, socialism and communism have always played–down the importance of ethnicity as a remnant from an ancient, more feudal era. As a result, these ideologies per se no longer appear to contain easy answers or exhaust all policy approaches. No single nation, rich or poor, capitalist or socialist has a monopoly of the problem – or of the solution.
Refs Van Dijk, Teun A *Communicating Racism* (1987).
Broader Racism (#PB1047) Discrimination (#PA0833) Social inequality (#PB0514).
Narrower Tribalism (#PC1910) Antisemitism (#PE2131)
Ethnic and social discrimination in foreign language teaching (#PF5929).
Related Caste system (#PC1968) Social injustice (#PC0797)
Sexual discrimination (#PC2022) Cultural discrimination (#PC8344)
Collapse of common values (#PF1118)
Discrimination against indigenous populations (#PC0352).
Aggravates Lack of assimilation (#PF2132) Plural society tensions (#PF2448)
Lack of social mobility (#PF2195) Underprivileged racial minorities (#PC0805)
Illiteracy among indigenous peoples (#PD3321)
Expropriation of land from indigenous populations (#PC3304)
Exploitation of indigenous populations in employment (#PD1092).
Aggravated by Elitism (#PA1387) Ethnic conflict (#PC3685)
Ethnic segregation (#PC3315) Social fragmentation (#PF1324)
Compulsory indoctrination (#PD3097) Discriminatory professionalism (#PC2178).
Reduced by Forced assimilation (#PC3293).

♦ PC3688 Racial segregation
Nature Segregation on racial grounds in education, employment, housing, before the law or in public services, may be a matter of government policy (South Africa) or of custom and general practice. Racial segregation preserves racial conflict, prejudices and discrimination, lack of integration, lack of social mobility, and inequalities.
Incidence The phenomenon of racial segregation has appeared in all parts of the world where there are bi–racial communities, except where racial amalgamation has occurred on a large scale, as in Hawaii and Brazil. The problem of racial segregation is most acute in South Africa.
Background Segregation usually is a means of maintaining the economic advantages and the superior social status of the politically dominant racial group. In an urban industrial society it is difficult to maintain the physical segregation of races, and more dependence must be placed upon institutions to maintain it. Consequently, the races live in different social worlds and communication between them is restricted, regardless of physical proximity. As an ecological process, racial segregation is in a way a natural one since it results from a relatively impersonal competition between races for space or land. As far as the economic relations of races are determined only by competition, a racial division of labour may emerge, which results in the 'segregation' of racial groups in different occupations. The racial division of labour generally reflects the distribution of power in a community. Thus, the system of Negro slavery and the economic organization of life in the Southern USA as well as in the colonial areas in Africa represented a racial division of labour but with a clear relation between that division and the distribution of power.
Broader Racism (#PB1047) Segregation (#PC0031)
Racial discrimination (#PC0006).
Narrower Apartheid (#PE3681) Legal segregation (#PD3520).
Related Caste system (#PC1968) Legalized racial discrimination (#PC3683).
Aggravates Slavery (#PC0146) Prejudice (#PA2173) Exploitation (#PB3200)
Miscegenation (#PC1523) Social conflict (#PC0137) Racial conflict (#PC3684)
Ethnic conflict (#PC3685) Social injustice (#PC0797) Social inequality (#PB0514)
Lack of social mobility (#PF2195)
Segregation of poor and minority population in urban ghettos (#PD1260).
Aggravated by Racial inequality (#PF1199) Racial intimidation (#PC2936)
Plural society tensions (#PF2448) Racial discrimination in politics (#PD3329)
Racial discrimination in public services (#PD3326)
Inadequate cultural integration of immigrants (#PC1532).
Reduced by Desegregation (#PG6824).

♦ PC3768 Denial of right to freedom from cruel, inhumane or degrading punishment
Cruel punishment — Unusual punishment — Degrading punishment
Nature Article 5 of the Universal Declaration of Human Rights states: "No one shall be subjected to torture or to cruel, inhuman or degrading treatment or punishment." Article 7 of the International Covenant on Civil and Political Rights asserts: "No one shall be subjected to torture or to cruel, inhuman or degrading treatment or punishment. In particular, no one shall be subjected without his free consent to medical or scientific experimentation. In spite of these and national laws with similar guarantees, men, women and children are tortured, imprisoned, murdered and humiliated at the whim of officials in the majority of nations across the world.
Broader Punishment (#PA5583).
Narrower Expulsion (#PC5313) Human torture (#PC3429) Arbitrary exile (#PJ7268)
Institutionalized torture (#PD6145) Unjust punishments for crimes (#PD4779)
Denial of rights to prisoners (#PD0520) Inhumanity of capital punishment (#PF0399)
Disproportionately long prison sentences (#PE4602).
Related Denial of human rights in the administration of justice (#PD6927).

♦ PC3778 Elitist control of global economy
Unrepresentative control of international monetary system
Nature The major decisions about the global economy are made by a very small number of people of major industrialized countries. Other countries accept this control of their existence, and see as alternatives working within the system or giving up on having any impact on the economy.
Narrower Monetary bloc (#PE5247)
Disproportionate control of global economy by limited number of corporations (#PE0135).
Aggravates Economic inefficiency (#PF7556) Elitist control of production (#PD0154)
Over-centralization of global decision-making (#PF5472).
Aggravated by Undemocratic policy-making (#PF8703)
Elitist intergovernmental groupings (#PD6896)
Interlocking corporate directorates (#PF5522)
Unrepresentative international organizations (#PD4873)
Undisclosed control of national economies by limited number of individuals (#PF2344)
Disproportionate influence on national economies of limited number of corporations (#PE1922).

♦ PC3796 Coercion
Duress — Extortion — Blackmail
Nature Coercion is the threat or use of force to constrain another agent's freedom of action. Coercion and deception are two major ways to control the actions of others without their fully voluntary cooperation. The justification of coercion is often a major part of the justification of both violent and non-violent resistance.
Claim Both the state and opposition groups use coercion to achieve political ends. Gandhi used non–violent methods of coercion, a moral force on the British. Systems of law are based on coercion.
Refs Hodson, John D *Ethics of Legal Coercion* (1983); Mersky, Roy M (Ed) *Conference on Transnational Economic Boycotts and Coercion* (1978); Rothstein, Stanley W *The Power to Punish* (1984).
Broader Oppression (#PB8656).
Narrower Criminal coercion (#PD4469) Political blackmail (#PD2912)
Emotional manipulation (#PE9599).

♦ PC3813 Urban overcrowding
Broader Overcrowding (#PB0469).
Aggravates Uncontrolled urban development (#PC0442)
Impersonality of high density accommodation (#PF6156)
Unbalanced urban population density gradients (#PD6131)
Poor condition of open spaces in urban communities (#PF1815)
Segregation of poor and minority population in urban ghettos (#PD1260).
Aggravated by Urban slums (#PD3139) Urban poverty (#PC5052)
Migration of rural population to cities (#PE8768)
Uncontrolled urbanization in developing countries (#PD0134).

♦ PC3825 Oligopolies
Market domination
Refs Geroski, Paul A, et al (Eds) *Oligopoly, Competition and Welfare* (1985); Hay, Donald A and Vickers, John S *The Economics of Market Dominance* (1987).
Narrower Disproportionate control of global economy by limited number of corporations (#PE0135)
Disproportionate influence on national economies of limited number of corporations (#PE1922).
Aggravates Monopolies (#PC0521).

♦ PC3853 Neoplastic diseases
Neoplasms — Tumours — Growths — Sarcoma
Nature A tumour or neoplasm is the excessive, pathological growth of plant, animal, or human tissues in which the cells qualitatively change and lose their capacity for differentiation. They enlarge as a result of multiplication of their own cells. The first sign of tumour growth in a tissue is the appearance of a small number of cells which multiply with uncontrolled division. Tumour growth proceeds through stages of disorderly increase in the number of cells, focal growth, benign growth and malignant growth; the stages immediately preceding malignancy are called precancerous.
Refs Anderson, C K, et al *Germ Cell Tumours*; Pattillo, Ronald A and Hussa, Robert O *Human Trophoblast Neoplasms* (1984).
Broader Human disease and disability (#PB1044).
Narrower Benign tumours (#PD8347) Carcinoma in sita (#PE5308).
Related Malignant neoplasms (#PC0092).
Aggravated by Health hazards of radiation (#PD8050).

♦ PC3894 Failure of centrally planned economies
Nature Centrally planned economies have failed. Centralized national economies structurally lead to shortages in every sector. The combination of detailed central production and resource allocation plans result in permanent and chronic imbalances between sectors. In fact, the task of detailed central planning is to large to do and results in no effective coordination between sectors or even products. This imbalance causes factories to produce inputs they cannot get from the outside making them less efficient. Massive numbers of personal are required to do repairs rather than production. In some cases managers are know to send perfectly good machines to be repaired because repairs are part of the central plan. Unneeded goods produced as a result of unrealistically high production goals pile up in warehouses. In retail shops inventories for one item may exceed quarterly turnover and for another item may always be short.
Aggravates Over–centralization (#PF2711)
Illegal private profit in socialist countries (#PC0939)
Lack of consumer choice in centrally–planned economies (#PD0515).
Reduced by Absence of long–term economic planning agencies (#PF3610).

♦ PC3908 Excessive consumption of specific foodstuffs
Nature In industrialized countries, the major nutritional problems arise from dietary practices begun in early childhood that are based on an excessive consumption of, for example, animal fats, refined sugar and salt. These practices, continued for a lifetime, may lead to a number of disorders such as heart disease, hypertension, obesity, and possibly some types of cancer. Many developing countries are increasingly experiencing the same type of disorders.
Broader Excessive consumption of goods and services (#PC2518).
Narrower Excessive consumption of fats (#PE4261)
Excessive consumption of salt (#PE4231)
Excessive consumption of sugar (#PE1894)
Excessive consumption of spices (#PG5060)
Excessive consumption of protein (#PD7089)
Excessive consumption of vitamins (#PE4665)
Excessive consumption of animal flesh (#PD4518)
Excessive consumption of carbohydrates (#PG0605).
Related Food fads (#PD1189) Alcohol abuse (#PD0153) Changeableness (#PA5490)
Substance abuse (#PC5536) Dietary restrictions (#PJ1933) Toxic food additives (#PD0487)
Unbalanced food usage (#PJ7868) Reliance on canned food (#PJ8409)
Decline in nutritional quality of food (#PE8938).
Aggravates Over–eating (#PE5722) Nutritional deficiencies (#PC0382).

♦ PC3915 Institutional fragmentation
Sectoral fragmentation of institutional responsibility
Nature Most of the institutions with mandates to deal with the challenges of society at this time tend to be independent, fragmented and working to relatively narrow mandates with closed decision processes. The mandates of central economic and sectoral ministries of government are also often too narrow and too concerned with quantities of production or growth. They deal with one sector or industry in isolation, failing to recognize the importance of intersectoral linkages. These intersectoral connections create patterns of economic and ecological interdependence rarely reflected in the ways in which policy is made. Those responsible for managing natural resources and protecting the environment are institutionally separated from those responsible for managing the economy. Such institutions, and the policies which they engender, are inadequate in the face of the interlocked economic and ecological systems. Sectoral organizations tend to pursue sectoral objectives and to treat their impacts on other sectors as side effects, to be taken into account only if compelled to do so.

Claim Many of the environment and development problems confronting society have their roots in sectoral fragmentation of responsibility. This reinforces the difficulties of achieving sustainable development.
Broader Fragmentation (#PA6233).
Narrower Bureaucratic fragmentation (#PC2662) Fragmentation of health service (#PE5721)
Inadequate urban political machinery (#PC1833)
Fragmented social structures for environmental protection (#PE3977)
Inadequate coordination of the intergovernmental system of organizations (#PE0730).
Aggravates Political fragmentation (#PF3216) Territorial fragmentation (#PC2944)
Uncoordinated policy–making (#PF9166) International economic fragmentation (#PC0025)
Fragmentation of technological development (#PC1227)
Fragmented conduct of community operations (#PF1205).
Aggravated by Fragmented decision–making (#PF8448)
Fragmentation of academic disciplines (#PF8868)
Government resistance to institutional change (#PF0845).

♦ PC3943 Cultural isolation
Broader Isolation (#PB8685).
Narrower Isolation of ethnic groups (#PC3316) Local traditions of cultural isolation (#PF1696).
Related Social isolation (#PC1707) Political isolation (#PC7569).
Aggravates Untransferability of books between countries and cultures (#PF2126).
Aggravated by Cultural arrogance (#PF5178).

♦ PC3950 Government inaction
Governmental resistance and inertia in response to problems
Nature The inability or unwillingness for a government to act on perceived and acknowledged problems frequently exacerbates suffering and increases the extent of damage done by the problems.
Claim In some political systems inaction is made safe and attractive because the potential costs of action are much higher than penalties for inaction.
Broader Delay (#PA1999) Inaction on problems (#PB1423).
Narrower Inadequate response to societal needs (#PD1080)
Non–payment of reparations by government (#PE4446)
Governmental inaction concerning trade in services (#PF0041)
Progressive reduction in government action commitment (#PF5502)
Government inaction on alleged human rights violations (#PE1407).
Related Government delay in response to symptoms of problems (#PF6707).
Aggravates Indecisive multilateralism (#PF9564) Unfulfilled treaty obligations (#PF2497)
Delay in project implementation (#PF1470) Inadequate national law enforcement (#PE4768)
Inadequate international law enforcement (#PF8421).
Aggravated by Official apathy (#PF9459) Bureaucratic inaction (#PC0267)
Lack of political will (#PC5180) Government complacency (#PF6407)
Government limitations (#PF4668) Governmental incompetence (#PF3953)
Intergovernmental suspicion (#PC2089) Non–recognition of problems (#PF8112)
Official evasion of complaints (#PF9157)
Government resistance to institutional change (#PF0845)
Inadequate power of intergovernmental organizations (#PF9175)
Inadequate application of available knowledge to solve problems (#PF8191).

♦ PC4020 Physically inaccessible resources
Geographically remote resources
Broader Geographical isolation (#PF9023).
Narrower Inaccessible recreation areas (#PF6503)
Inaccessible natural resources (#PG7875).
Aggravates Worldwide misallocation of resources (#PB6719)
Inadequate water supply in the rural communities of developing countries (#PD1204).

♦ PC4094 Deficiencies of developing countries
Nature The causes of underdevelopment lie in deficiencies in developing countries which lack the internal conditions necessary to initiate and maintain a healthy development process. Such conditions include: inadequate capital, lack of skills, lack of work motivation. The are exacerbated by corruption in government, difficult geographical and climatic conditions and lack of western organizational habits and work values.
Counter–claim Although such deficiencies contribute to the difficulties of developing countries, they are relatively unimportant compared to the ways in which external market forces prevent poor countries from gaining access to an equitable share of world resources and ensuring the most appropriate forms of development.
Refs O'Flynn, Gráinne *World Survival*.
Narrower Inhospitable climate (#PC0387) Apathy in developing countries (#PD8047)
Capital shortage in developing countries (#PD3137)
Unattractive locale for economic development (#PF3499)
Lack of management skills in developing countries (#PE0046)
Lack of skilled manpower in rural areas of developing countries (#PE5170)
Inadequate local expertise in business practices in developing countries (#PE7313)
Inadequate supply of appropriate trained manpower in developing countries (#PE6243).
Aggravates Underdevelopment (#PB0206) Corruption of government leaders (#PC7587).

♦ PC4257 Disastrous consequences of war
Broader War (#PB0593).
Narrower War casualties (#PD4189) Prisoners of war (#PC8848)
Long–term effects of war (#PD7918) Hazardous remnants of war (#PF2613)
Housing destruction in war (#PE2592) War damage in civilian areas (#PD8719)
Industrial destruction by war (#PD8359) Poverty as a consequence of war (#PE5252)
War–time conditions and pressure (#PD9090) Environmental consequences of war (#PC6675)
Destruction of economy due to war (#PE8915).
Aggravated by Nuclear war (#PC0842).

♦ PC4275 Trade protectionism
Trade barriers — Protectionism — Unfair trade restrictions
Nature Trade protectionism is the result of governments acting to save certain sectors of their economies from foreign competition. The impact of this on the fledgling industries is considerable but they are in no position to retaliate. Protectionist policies have also been directed by developed countries against each other, but this is usually carefully negotiated to avoid retaliation.
Incidence In recessionary environments, the manufactured exports of developing countries have encountered heightened protectionist barriers, especially of the non–tariff kind, which have further exacerbated the ability of these countries to service and repay debt incurred at a time of brighter prospects for growth in world trade. There has been a pronounced movement away from the development consensus embodied, for example, in the Generalized System of Preferences, which was pioneered at UNCTAD II in New Delhi in 1968. The non–reciprocal non–discriminatory concessions extended under the GSP are being gradually withdrawn and to a large extent offset by discriminatory measures against developing countries' exports. Hence developing countries which in the period 1973–1981 were the most dynamic partners in the world exchange of goods, both as exporters and importers, are unable to make an effective contribution to the recovery of world trade, precisely at a time when such a contribution is most required. By September 1988 it was estimated that about 50 percent of world trade is affected by protectionist measures which assume the forms of voluntary export constraints, anti–dumping measures and all kinds of administrative procedures and obstacles.
Claim The cost of maintaining trade barriers is paid by the consumer through taxes which are used for subsidies to domestic industries, higher prices of domestic and uncompetitive goods and services and additional taxes to pay for expenditures by government to administer the barriers. The cost of protecting agriculture can be very high: in the United States 3 per cent of total farm input and 16 per cent in the EC. Non–tariff barriers such as Voluntary Export Restraints (VER) can cost to the importing country three times as much as equivalent tariff protection. Countries that do not protect home industries prosper because their more open markets draw in more and cheaper goods, keeping down inflation and interest rates. Their industries are toughened by having to meet global competition.
Counter–claim Protectionist policies serve to protect the national interest of the countries using them. They allow domestic industries vital to national defence to continue to be viable in spite of being uncompetitive in the world market. They guard against being cut off by foreign suppliers by ensuring domestic production, e.g. the arms industry of South Africa. They protect the domestic economy against a negative balance of payments.
Refs Commonwealth Secretariat *Protectionism* (1982); Schechter, Roger E *Unfair Trade Practices and Intellectual Property* (1986).
Broader Distortion in international trade (#PC6761).
Narrower Unpredictable barriers to trade (#PD5033)
Protectionism in international trade (#PC5842)
Protectionism in developing countries (#PD3714)
Protectionism in the services industries (#PD7135)
Government protection and national commodity price support (#PE8309)
Import substitution as a barrier to subsequent economic growth (#PE8792)
Conditional observance of multilaterally agreed trade commitments (#PF7838).
Related Anti–consumerism (#PF3511).
Aggravates Flag discrimination in shipping (#PD0700)
Natural environment degradation (#PB5250)
Excessive external trade deficits (#PC1100)
Fragmentation of the international trading system (#PC9584)
Ineffective economic structures in industrial nations (#PE4818).
Aggravated by Deflation (#PD7727) Economic conflict (#PC0840)
Imbalance of payments (#PC0998) Exchange rate volatility (#PE5930)
International economic recession (#PF1172)
Declining international competitiveness (#PD8994)
Vulnerability of economies to import penetration (#PD7486)
Over–reliance on economic interest groups by policy agencies (#PF1070).

♦ PC4291 Regional imbalance of power
Broader Regional disparities (#PC2049).

♦ PC4311 Conventional warfare
Broader War (#PB0593) Offences against the peace and security of mankind (#PC6239).
Narrower Siege (#PJ4917) Bombardment (#PE2306) Military blockade (#PJ5234)
Low–intensity conflict (#PE3988).

♦ PC4312 Unequal regional distribution of deaths
Unequal life expectancy
Incidence World statistics indicate that the number of deaths per 1000 population for 1970–1975 was 12 in the world, 19 in Africa, 13 in Asia, 10 in Europe, 10 in Oceania, 9 in North America, 9 in Latin America and 8 in the USSR.
Broader Human death (#PA0072).
Related Inequality of life expectancy by gender (#PE1339)
Unequal morbidity and mortality between countries (#PC6869).
Aggravated by Malnutrition (#PB1498) Socio–economic poverty (#PB0388)
Inadequate medical care (#PF4832) Human disease and disability (#PB1044)
Inappropriate sanitation systems (#PD0876).

♦ PC4321 Contempt for traditional modes of behaviour
Loss of reverence for life
Claim During most of human history, people have been guided in their actions by traditional customs and beliefs, not by consideration of short–term self interest. At present there is a growing contempt for such blind faith as people believe they are more rational. But old customs, which evolved over long periods, often (although not always) had deeper reasons behind them, even if these reasons were not understood by most of the people adhering to such customs. For example, the prohibition by certain religions against the consumption of pork makes sense when it is realized that pork is a frequent transmitter of parasitic diseases. And why did putting a horse in the field where cows were grazing inhibit the development of foot–and–mouth disease ? Traditional modes of behaviour often served the function of letting people live in peace with nature and with one another. As such traditions lose their power over people's ways of conducting themselves and are evaluated critically, they must be replaced by a new awareness of necessities, based on a more scientific understanding of the world which will lead people to behave in ways consistent with long–term survival. Otherwise loss of reverence for nature and for life may lead to disaster.
Broader Contempt (#PF7697).
Related Cargo cults (#PF5375)
Decline in rural customs and traditions in developing countries (#PD1095).
Aggravated by Discriminatory imposition of standards (#PD5229)
Untransposed significance of cultural tradition (#PF1373).

♦ PC4341 Environmental threats to national security
Environmental insecurity — Water insecurity
Nature Conflicts and tensions may arise not only because of political and military threats to national sovereignty. They may also result from environmental degradation and the pre–emption of development options which do not lend themselves to conventional military solutions to threats to national sovereignty.
Incidence The problem of water supply provides the most acute example, especially in the case of semi–arid countries. Some Arab countries consider that their access to three of the region's vital rivers (the Euphrates, the Jordan and the Nile) is being threatened by the non–Arab countries upstream. Thus Turkey is now controlling the flow of the Euphrates as part of a major dam project. Israel diverts some of the Jordan river flow for irrigation, leaving little for Jordan itself. Egypt is concerned at the implications of irrigation projects of Ethiopia and Uganda which could affect the flow of the Nile.
Broader International insecurity (#PB0009) Natural environment degradation (#PB5250).
Narrower Global warming (#PC0918) Decreasing land mass (#PF7435)
Long–term shortage of water (#PC1173) Transboundary water pollution (#PD1096)
Long–range transboundary air pollution (#PD3391).
Aggravates National insecurity and vulnerability (#PB1149).
Aggravated by Environmental stress (#PC1282).

♦ **PC4382 Technical barriers to trade**
Nature Increasingly in international trade, it is recognized that national standards and technical regulations which ostensibly are designed to protect the consumer and the environment turn out to be of a protectionist character.
Broader Non-tariff barriers to international trade (#PC2725).

♦ **PC4405 Denial of rights to vulnerable groups**
Nature Members of groups that are particularly vulnerable to arbitrary deprivation of their human rights and fundamental freedoms because of characteristics for which they are not responsible and which they are not in a position to change, such as children, mentally retarded persons, disable persons, persons belonging to ethnic, religious or linguistic minorities, persons born out of wedlock, non-citizens, and members of indigenous populations, are usually considered to be entitled to special measures to ensure their enjoyment of human rights and fundamental freedoms and to protect their welfare and, yet, these rights are frequently denied.
Broader Denial of human rights (#PB3121).
Narrower Maltreatment of prisoners (#PD6005) Denial of rights to disabled (#PC3461)
Denial of rights to soldiers (#PD4089) Denial of rights to students (#PD6346)
Denial of rights to prisoners (#PD0520) Denial of rights of medical patients (#PD1662)
Denial of rights of children and youth (#PD0513)
Denial of economic and social rights to refugees (#PE6375).
Related Age discrimination (#PC2541) Discrimination against women (#PC0308)
Denial of rights of minorities (#PC8999)
Discrimination against indigenous populations (#PC0352)
Inadequate community care for transient urban populations (#PF1844).
Aggravates Lack of protection for the vulnerable (#PB4353).

♦ **PC4412 Competition for scarce resources**
Competition for non-renewable resources — Competition for sources of energy
Refs Chandler, William U *Energy Productivity* (1985); Khan, A M and Hoelzl, A *Evolution of Future Energy Demands Till 2030 in Different World Regions* (1982).
Broader Competition (#PB0848).
Aggravates Environmental stress (#PC1282) Intranational competition (#PE0580).
Aggravated by Economic rivalry (#PD8897)
Inequality in distribution of natural resources between countries (#PF3043).

♦ **PC4422 Exploitation of trust**
Abuse of trust
Nature Human beings instinctively need to trust both individually and collectively. We seek to repose trust in family, friends, confessor or psychoanalyst. The need to trust is extended to institutions as well; institutions to which we are accountable and which have power, sometimes life and death power over our lives. The heads of institutions must be able to accommodate the trust of the many. Institutions are repositories of trust but usually trust in a specific sphere of society or culture. Trust is both given to an individual or institution and at the same time taken. Giving trust is an action and is not a passive process. When trust is given by the donor it becomes power in the hands of the recipient. Trust can be genuinely earned or acquired by deception or extortion. Trust can be limited to a very narrow field of activity, a lawyer can be trusted to represent one in a court but not to repair cars. When an individual acquires many kinds of trust: political, religious, etc., a demagogue like Hitler can result. The recipient of trust can use the resulting power to benefit those who conferred it or exploit it for their own gain. Repeated disclosures of the abuse of trust by the head of an institution will result in the withdrawal of trust.
Broader Exploitation (#PB3200) Political exploitation (#PC7356).
Narrower Abuse of law (#PC5280) Economic crime (#PC5624)
Abuse of science (#PC9188) Deception by natural scientists (#PD9182)
Abuse of commercial confidentiality (#PE6786) Frauds, forgeries and financial crime (#PE5516).
Related Political crime (#PC0350).
Aggravated by Black lies (#PE4432) Abuse of power (#PB6918).

♦ **PC4440 Insecurity and vulnerability of nuclear weapon states**
Nature Fear is a major proponent of the mutual, self-perpetuating distrust between nuclear weapon states. Their insecurity promotes the construction of "worst-case scenarios" wherein each nation imagines the enemy's most ingenious and devastating schemes and then prepares to be capable of retaliating with enough severity to deter anyone from ever putting such plans into action. Because the process is reciprocal, it risks turning the arms race into a mutually self-fulfilling prophecy. Adversaries which view one another as warlike and treacherous find their belief verified by the response each other makes in the face of such threats. Each notes the other's military build-up, aggressive propaganda, efforts at subversion, massive arms sales and donations abroad, and its violations of international law.
Claim Insecurity among nations undermines joint action to reduce the risks of nuclear catastrophe at every step. Keeping pace with the escalating powers of destruction, distrust also spurs them on, while blocking all moves to curb them. Out of a degree of insecurity and distrust that once would have seemed pathological, the nuclear weapon states now compete with one another in perfecting the means to global catastrophe.
Broader International insecurity (#PB0009).
Related Insecurity and vulnerability of non-nuclear weapon states (#PD1521).
Aggravates Domination of the world by territorially organized sovereign states (#PD0055).

♦ **PC4442 Traffic in persons**
Nature Men, women and children are bought and sold for the purposes of slavery, marriage, adoption, scientific experimentation, medical transplants, sexual exploitation including prostitution and pornography, and forced labour.
Broader Denial of right to liberty (#PF0705).
Narrower Slave trade (#PC0130) Forced marriage (#PD1915)
Marriage markets (#PD7282) Trafficking in women (#PC3298)
Trafficking in children (#PD8405) Vice and sex traffic offences (#PD8910)
Abusive traffic in immigrant workers (#PD2722)
Exploitation of the prostitution of others (#PE5303)
Trafficking in children for sexual exploitation (#PE6613).
Aggravates Abuse of brothel legislation (#PE6735).

♦ **PC4498 Lack of time**
Shortage of time
Claim In the industrialized societies, committed to saving time by every possible means, people are increasingly confronted with the lack of time. The time available is increasingly segmented and committed by prior engagements and projects. Tangential or discretionary time, previously an amenity of life, especially in traditional cultures, has become an expensive luxury. Despite a preoccupation with efficiency, people have less time for themselves or for each other.
Broader Constraint of time on individual and social development (#PF5692).
Narrower Limited shared time (#PJ9126) Limited leisure time (#PF9062)
Insufficient personal time (#PJ9534)
Lack of parliamentary time to approve needed legislation (#PF8876).
Related Time consuming procedures (#PJ8206).
Aggravates Sleep deprivation (#PE2741).
Aggravated by Wasted time (#PF8993) Hyperefficiency (#PF1706)
Temporal deprivation (#PF4644) Increasing pace of life (#PF2304).

♦ **PC4575 Diseases of the genito-urinary system**
Refs Bondi, A et al *Urogenital Infections* (1988); Mitchell, J *Urinary Tract Trauma* (1984); Neu, H C and Williams, J D *New Trends in Urinary Tract Infections* (1988).
Broader Human disease and disability (#PB1044).
Narrower Bright's disease (#PE2272) Diseases of breast (#PD9742)
Urinary bladder disorders (#PE2307) Diseases of male genital organs (#PD9154)
Diseases of female genital organs (#PD8775)
Malignant neoplasm of genito-urinary organs (#PE5100)
Symptoms referable to genito-urinary system (#PE9369).

♦ **PC4584 Criminal offences against the environment**
Environmental crime — Crimes against nature
Broader Economic crime (#PC5624) Natural environment degradation (#PB5250).

♦ **PC4588 Socio-cultural environment degradation**
Broader Environmental degradation (#PB6384).
Narrower Poor social environment (#PJ8742)
Untransposed significance of cultural tradition (#PF1373).
Related Unhealthy environment (#PJ1680).
Aggravated by Closure of social institutions (#PF3831)
Abuse of traditional cultural expressions of peoples (#PE4054).

♦ **PC4591 Resistance to internationally agreed standards**
Failure to conform to internationally agreed standards — Inadequate international harmonization of standards
Nature Differing standards constitute a significant barrier to the free movement of goods between countries.
Incidence In 1988 it was estimated that the costs of such standards in impeding trade throughout the European Community was ECU 20 billion.
Refs ILO *Measures to Overcome Obstacles to the Observance in the Construction Industry of ILO standards* (1984); ILO *Position of African Countries Regarding the Ratification and Implementation of International Labour Standards* (1983).
Broader Ineffective legislation (#PC9513)
Delays in elaboration of remedial legislation (#PC1613).
Narrower Violations of health and safety regulations (#PE4006)
Failure to conform to international health standards (#PE5239).
Aggravates Inadequacy of international standards (#PF5072).
Aggravated by Deficiencies in international law (#PF4816)
Discriminatory imposition of standards (#PD5229)
Inadequate legislation against environmental pollution (#PF9299).

♦ **PC4608 Inadequate enforcement of human rights**
Inadequate enforcement of civil rights
Refs Muravchik, Joshua and Kirkpatrick, Jeane *The Uncertain Crusade* (1986).
Broader Inadequate international law enforcement (#PF8421).
Aggravates Government failure to prosecute offenders effectively (#PE9545).
Aggravated by Inadequate national law enforcement (#PE4768)
Impunity of violators of human rights (#PF3474)
Connivance of authorities in human rights abuses (#PF9288)
Government inaction on alleged human rights violations (#PE1407).

♦ **PC4613 Discrimination against trade unions**
Discrimination against labour unions
Refs ILO *Protection Against Anti-Union Discrimination* (1976).
Broader Discrimination (#PA0833).
Narrower Government discrimination against trade unions (#PE4860).
Related Violation of trade union rights (#PD4695).
Aggravated by Ineffective worker organizations (#PF1262).

♦ **PC4631 Giving bribes**
Nature Bribery is giving material benefit to an employee or agent without the employer knowing with the intention of influencing the conduct of the employee to the detriment of the employer.
Broader Bribery (#PC2558).
Narrower Bribery of public servants (#PD4541).
Related Receiving bribes (#PC4701).

♦ **PC4645 Unlawful business transactions**
Illegal international business — Unlawful commerce — Illegal trade
Nature Conducting an international business transaction prohibited or declared to be unlawful by a statute with the intent to conceal the transaction from a government agency authorized to administer the statute or with the knowledge that his unlawful conduct substantially obstructs, impairs or perverts the administration of the statute or government functioning.
The exportation of weapons, ammunition and explosives to organizations and countries expressly band from such trade are examples of unlawful business transactions.
Broader Abuse of law (#PC5280).
Narrower Illegal exports (#PD4116) Money laundering (#PE7803)
Illegal ivory trade (#PE4991) Illicit drug trafficking (#PD0991)
International arms trade (#PC1358) Illicit export of works of art (#PE9004)
Trafficking in illegal firearms (#PE7711)
Illegal private profit in socialist countries (#PC0939)
Trade in animal products of endangered species (#PD0389).
Related Slave trade (#PC0130) Tax evasion (#PD1466)
Crimes related to foreign relations and trade (#PE5331).
Aggravates Diversion of high technology to hostile countries (#PE7174).
Aggravated by Unlawful government action (#PF5332)
Attraction of the forbidden (#PF5413)
Ineffective monitoring of illegal activity (#PF7264).

♦ **PC4667 Aggressive foreign policy**
Gun-boat diplomacy — Gunboat diplomacy
Broader Inappropriate policies (#PF5645)
Self-destructive government policy-making (#PF5061).
Narrower Espionage (#PC2140) Neutrality (#PF0473)
Threat of war (#PF8874) Secret military operations (#PF7669)
Border incidents and violence (#PD2950) International non-military conflict (#PF3100)
Provocations of capitalist militarism (#PC0937)
Limited acceptance of international treaties (#PF0977).
Aggravates Ideological war (#PC3431) Environmental warfare (#PC2696).
Aggravated by Economic conflict (#PC0840).

♦ **PC4701 Receiving bribes**

Nature Receiving, soliciting or agreeing to accept material benefit for conduct detrimental to the benefit of one's employer is receiving bribes.

Broader Bribery (#PC2558).
Narrower Bribery of public servants (#PD4541).
Related Giving bribes (#PC4631).
Aggravated by Police crimes during narcotic investigations (#PE5037).

♦ **PC4710 Tolerated atrocities**

Passive atrocities — Preventable deaths — Preventable disabilities

Narrower Malaria (#PE0616) Cretinism (#PE7905) Diarrhoea (#PD5971) Blindness in developing countries (#PD5139).

♦ **PC4726 Anti-social behaviour**

Refs Olweus, Dan, et al *Development of Antisocial and Prosocial Behavior* (1985).

Narrower Vandalism (#PD1350) Excessive television viewing (#PD1533) Increase in anti-social behaviour in developing countries (#PD0329).
Aggravated by Teasing (#PE4187).

♦ **PC4764 Vulnerability of protected natural areas**

Endangered protected sites — Endangered protected habitats

Nature Protected areas include: large tracts of land set aside for the protection of wildlife and its habitat; areas of great natural beauty or unique interest; areas containing rare forms of plant and animal life; areas representing unusual geologic formations; places of historic and prehistoric interest; areas containing ecosystems of special importance for scientific investigation and study; and areas which safeguard the needs of the biosphere. These areas serve a number of specific purposes: some provide recreation for large numbers of people without seriously detracting from the natural values; some aim to retain their more pristine beauty through greater restrictions; others (strict nature reserves) are reserved solely for scientific research as relatively undisturbed environments; and still others provide a reservoir of genetic materials in a spectrum of organisms adapted to a particular range of soil and climatic conditions.

Many national parks of high tourist value are being flooded by rising numbers of tourists and suffer from insufficient or inappropriate planning and management. Such parks, particularly in the developing countries, represent a major source of income which could be jeopardized if they deteriorate. Valuable wildlands are threatened by pressure detrimental to their protection and use. Damage frequently arises from a lack of understanding or interest, particularly in some developing countries, of the value of such wildlands. Deterioration often results from a lack of knowledge, or political or economic considerations generally inhibiting or delaying the required action, until the parks exist on paper only. The possibilities of such deterioration increase whenever a park or otherwise protected area is shared by two or more countries.

Broader Vulnerability of ecosystem niches (#PC5773).
Related Endangered monuments and historic sites (#PD0253).
Aggravated by Over-use of designated wilderness areas (#PD7585).

♦ **PC4777 Hazards to human health in the natural environment**

Refs United Nations Educational, Scientific and Cultural Organization *Select Bibliography of Unesco Publications, Reports and Documents Relating to Natural Hazards, 1961-1978* (1979).

Broader Hazards to human health (#PB4885).
Narrower Dangerous animals (#PC2321) Disease and injury from exposure to weather (#PF5739).
Related Dangerous paths (#PJ9888) Natural disasters (#PB1151) Environmentally induced diseases (#PD8200).
Aggravates Health hazards of mining radioactive substances (#PE7419).

♦ **PC4815 Shortage of fresh-water sources**

Shortage of drinking water — Shortage of water for food production — Scarce fresh water — Insufficient fresh water supply — Undeveloped water resources — Limited water supply

Nature That water is a finite resource to be conserved and protected is not yet universally perceived. Generally, fresh water is respected as a giver-of-life only in regions of chronic shortage. Of the world's water, five percent is fresh but four percent of the total is frozen in the polar regions. Thus all the water in lakes and rivers, all the moisture in the atmosphere, the soil and vegetation, and all the water underground amounts to only one percent of the total. Of this one percent that is liquid freshwater, 98 percent is groundwater: one percent is in lakes. Only half of the groundwater is within a half-mile of the surface and therefore within the reach of man. Groundwater often is found where surface water is insufficient or negligible, as in the Sahara and the southwestern United States. Reservoirs of such fossil water are a nonrenewable resource, deposited thousands of years ago. They are not being replenished and will be exhausted sooner rather than later. Although underground water should be reserved for drinking and household needs, it is used by industry even when surface water is available.

Of all environmental ills, contaminated water is the most devastating in its consequences. Each year 10 million deaths are directly attributable to waterborne intestinal diseases. One-third of humanity labours in a perpetual state of illness or debility as a result of impure water; another third is threatened by the release into water of chemical substances whose long-term effects are unknown. In the quest for greater food production, the management of water resources is central. Irrigation is as much a problem as a solution to improved agriculture. About 10 million hectares of arable land are being abandoned every year because of saturation, salinity, or alkalinization. The problem occurs in every country with substantial irrigation.

The securing of an adequate water supply has become one of the most critical problems facing many societies today. Fresh water, needed by human beings to sustain life, health and productive activities, constitutes only 0.8 percent of the world's total water supply, and it is not known just what portion of this amount is contaminated. Lack of water and ensuing poor sanitation are responsible for disability, disease and death, especially among infants and young children. Women, traditionally the world's water-bearers, suffer particular hardships. The time and energy they consume in fetching water from long distances might otherwise be devoted to caring for their families – or to educational or income earning pursuits of benefit to the entire community.

Incidence Global water use doubled between 1940 and 1980, and is expected to double again by 2000. Yet 80 countries, with 40 per cent of the world's population, already suffer serious water shortages. There will be growing competition for water for irrigation, industry and domestic use. River water disputes have already occurred in North America (the Rio Grande), South America (the Rio del la Plata and Parana), South and Southeast Asia (the Mekong and Ganges), Africa (the Nile), and the Middle East (the Jordan, Litani, Orontes, and the Euphrates).

Global use of energy has tripled during the past three decades. In some places the availability of water is becoming more of a constraint on hydroelectric energy production than the availability of primary fuel. A single coal gasification plant producing 7 million cubic meters of burnable gas per day would require as much as 25 million cubic meters of water a year. Obtaining and transporting the necessary water would be more damaging to the environment than strip-mining the coal itself. In strict economic terms, the advantages of cleaning up rivers and lakes are impressive. For example, the World Bank calculated a 30 percent rate of return on an $80-million investment aimed at cleaning up the Tiete River, which flows through the city of Sao Paulo, Brazil. The stench from the river is so strong that land along its banks has become almost valueless. Pollution abatement will make a large section of Sao Paulo healthier and more livable, and it will restore highly desirable land in the heart of the city.

Five industries – primary metals, chemicals, petroleum, pulp and paper and food processing – account for two-thirds of water's industrial use, yet water rarely represents more than one percent of their manufacturing costs. For a given unit of output some industrial plants withdraw 5 to 20 times as much water as other plants manufacturing the same product, and the largest amount of this is discharged with varying degrees of thermal, biological or chemical pollution. The key ecosystems of wetlands, watersheds and arid zones all depend on available water. The natural entity around which water management should be organized is the river basin, in effect a succession of watersheds. This is inconvenient in a system of state governments. Where several nations share the same river basin, the high degree of international cooperation required for successful water management on a regional scale has been lacking. There are incipient conflicts involving water resources everywhere. Hydrology is the oldest application of science to human welfare. The earliest exercise of government arose from the need to manage water. If clean and ample water is to be ensured for the future new applications of science, more effective organizations and governmental actions are essential. Without a fundamental change of approach to the interaction between water and land, it could well be impossible to support the population projected for the next century. Northern Africa and the Middle East would be particularly affected by severe constraints on water availability. By the year 2100, Africa is projected to have a population five times that of today; the population of Asia is expected to double. During the next century, the amount of water available on a per capita basis will be reduced in Asia from the 1980 figure of 5100 cubic meters, to 2600; and in Africa from 9000 cubic meters per person, to 1600. These continental averages conceal striking regional differences. Even if sophisticated computer-based conservation and management and heavy irrigation schemes were introduced, water availability would not allow for more than a 50 percent increase in population in North Africa and the Middle East.

In many countries, such as Israel, supplies from groundwater are decreasing because of rapid withdrawals. In other countries, such as Bangladesh, most wells are polluted because of lack of sanitation; and in yet other countries, such as Egypt and Libya, overpumping from the ancient aquifers has allowed sea water from the Mediterranean to seep into the wells. This is also a problem in certain parts of Australia where, because of the increasing amounts of groundwater being withdrawn for irrigation, the water table is dropping and seawater is entering the aquifers, affecting water supplies for farms.

Claim Water is essential for life. It makes up nine-tenths of the human body's volume and two-thirds of its weight. This vital element covers about three-quarters of the earth's surface. But some 97.4 percent is salt water in oceans, and 1.8 percent is frozen in polar regions.

Refs Engelbert, Ernest A *Water Scarcity* (1985); United Nations *Management of International Water Resources* (1975).

Broader Long-term shortage of water (#PC1173) Long-term shortage of natural resources (#PC4824).
Narrower Water salinization (#PE7837) Overexploitation of underground water resources (#PD4403).
Related Vulnerability of food chains (#PB2253) Destruction of environmental oxygen (#PE5196) Obstacles to the utilization of coastal and deep sea water resources (#PF4767).
Aggravates War (#PB0593) Hunger (#PB0262) Inadequate health services (#PD4790) Human disease and disability (#PB1044) Vulnerability of social systems (#PB2853) Slowing growth in food production (#PC1960) Prohibitive cost of basic services (#PF6527) Inadequate water system infrastructure (#PD8517) Unsustainable development of fresh waters (#PD6923) Declining productivity of agricultural land (#PD7480) Underprovision of basic services to rural areas (#PF2875) Underdeveloped provision of basic services in developing countries (#PF6473).
Aggravated by Impurities in waste water (#PD0482).

♦ **PC4824 Long-term shortage of natural resources**

Decreasing natural resources — Excessive demands for natural resources — Unintegrated biosphere and ecosystem management

Nature Unprecedented rates of population growth, swiftly rising incomes and per capita demand, and technological advances, impose requirements on natural systems which may exceed their capacity to respond. As a result the life-support system of the planet may eventually be damaged beyond repair. The fact that perturbations in remote and seemingly unimportant parts of the biosphere can trigger off a chain of cause-effect reactions that ultimately provoke profound changes in the entire system, underlines the absence of world-wide integrated resource management.

Claim Certain of the earth's resources, such as oil and coal, are non-renewable and sooner or later will be completely exhausted. Although other resources, such as corn, rice and other cereals, cattle, fish and timber, renew themselves and can be regularly cropped to provide the food, clothing and shelter essential to human survival, it is not so clearly realized that these resources are renewable only to the extent that their use is rationally planned and managed. There are limits to the extent to which we can draw on these resources; if these limits are are exceeded, this will destroy the capacity of resource renewal. Unfortunately, most current utilization of aquatic animals, of the wild plants and animals of the land, of forests and of grazing lands is not sustainable.

Counter-claim A recent compendium of studies covering almost the full spectrum of economic trends including: population, agriculture, energy and the environment conclude that if present trends continue the world in 2000 will be less crowded, though more populated; more ecologically stable; and less vulnerable to disruption of supply of resources than now. The people of the world will be richer, have more access to food and other necessities. Life for more people will be less precarious. Doom and gloom predictions neglect basic fundamentals of supply and demand. As demand for resources goes up and supplies diminish prices will rise forcing people to use other resources in more plentiful supply.

Refs Dasmann, R F *Planet in Peril?* (1972).

Broader Mismanagement (#PB8406) Long-term shortage of resources (#PB6112) Prohibitive cost of necessities in rural communities (#PF2385).
Narrower Waste paper (#PD1152) Wasted water (#PD3669) Electrical power failure (#PE1341) Unjustified game cropping (#PE1327) Scarcity of oil resources (#PG7957) Shortage of cultivable land (#PC0219) Long-term shortage of water (#PC1173) Destruction of land fertility (#PC1300) Exhaustion of mineral resources (#PD9357) Shortage of fresh-water sources (#PC4815) Long-term shortage of commodities (#PC1195) Vulnerability of marine ecosystems (#PC1647) Long-term shortage of electric energy (#PE1216) Long-term shortage of energy resources (#PF0334) Hazardous locations for nuclear power plants (#PD2718) Inadequate governmental energy conservation policies (#PF0037) Inadequate governmental resource conservation policies (#PF0038) Natural resource depletion due to high-level consumption (#PD4002).
Related Wildlife extinction (#PC1445) Instability of water supply (#PD0722) Long-term shortage of food and live animals (#PE0976) Depletion of natural resources due to population growth (#PD4007) Inadequate water supply in the rural communities of developing countries (#PD1204).
Aggravates Energy crisis (#PC6329) Deforestation (#PC1366)

PC4824

Man-made disasters (#PB2075) Lack of natural resources (#PC7928)
Natural environment degradation (#PB5250) Unsustainable economic development (#PC0495)
Worldwide misallocation of resources (#PB6719)
Natural environmental degradation from recreation and tourism (#PE6920).
Aggravated by Food insecurity (#PB2846) Excessive land usage (#PE5059)
Environmental pollution (#PB1166) Environmental degradation (#PB6384)
Competition between states (#PC0114) Conflicting use of resources (#PF6654)
Proliferation of second homes (#PF1286) Non-biodegradable plastic waste (#PD1180)
Irrational rejection of nuclear power (#PF8531)
Underdeveloped potential of basic resources (#PF3448)
Unconstrained exploitation of natural resources (#PF2855)
Degradation of the environment in developing countries (#PD3922).

♦ PC4844 Health inequalities

Unequal development in the promotion of health — Disparity in human healthiness

Nature The picture of current global health is marked by large regional disparities. Patterns of death and illness in the Third World today are similar to those found in the First World during the 19th century. Third World ill-health is principally related to malnutrition, poverty, and lack of access to basic needs; while for the First World, the critical issue is that of the health problems of an ageing population. Given projected trends for the next half-century, this contrast is unlikely to change significantly. Diseases of poverty will continue to be the hallmark of the Third World as geriatric care and the management of chronic diseases will persevere as the largest challenge in the developed countries. However, as socioeconomic conditions improve in the developing world, the importance of chronic diseases will also increase.

Claim Huge inequalities characterize the current picture of global health. In the developing world, health problems are related to malnutrition, poverty and lack of basic needs, while in the industrialized countries the health problems of the ageing populations present the greatest challenge. Impediments to the attainment of the desired level of mental and social well-being (as defined by the World Health Organization) are manifold. Health services in the developing countries have often been based on European or North American models, centering on highly technological, cost-intensive urban hospitals focused on curative rather than preventive health care. The absence of social justice and equity and, more particularly, the basic assumptions of superiority and inferiority of racial groups, indeed any form of racial discrimination in the field of health, all represent obstacles to progress because they undermine the most fundamental human right of all, the right to life.

Broader Human illness (#PA0294) Human inequality (#PA0844)
Underestimation of the human potential (#PF7063).
Narrower Unequal health benefits for women (#PE6835).
Aggravates Economic and social losses due to disability (#PE4856).
Aggravated by Human suffering (#PB5955) Mental suffering (#PB5680)
Psychosomatic disorders (#PD1967) Environmental pollution (#PB1166)
Lack of eugenic measures (#PD1091) Human physical suffering (#PB5646)
Inadequate health services (#PD4790) Protein-energy malnutrition (#PD0339)
Human disease and disability (#PB1044) Genetic defects and diseases (#PD2389)
Disregard for self-healing potential (#PG7709) Inequitable use of medical resources (#PJ5160)
Ignorance of lifelong human development (#PF5759).

♦ PC4864 Civil violence

Nature Groups of citizens, armed or unarmed, may initiate violent actions with the intent to demonstrate their opinion or position, to destroy or sequester property, to prevent the actions of certain persons, or to injure or kill others. Included in civil violence, therefore are: civil disobedience of certain types such as illegal sit-ins; certain kinds of actions which are illegal in labour disputes (on both sides), as well as in labour disputes with government; riots; organized terrorism; insurrection; and (taking citizenry not to exclude illegal militia) and vigilantism. The objectives of civil violence are frequently economic and political change, but may also be focused on religious, minority or other issues. The term 'civil violence' may be loosely applied to wartime conditions in a country not officially at war, where both legal and illegal militia are engaged.

Broader Human violence (#PA0429).
Narrower Terrorism (#PD5574) Hooliganism (#PD1109) Civil disorders (#PC2551)
Popular uprisings (#PJ5237) Armed insurrection (#PD8284) Civil disturbances (#PD5372).
Aggravates Vigilantism (#PD0527) Stone throwing (#PJ7019)
Stress and trauma in a context of civil violence (#PE0525).
Aggravated by Demonstrations (#PD8522) Student revolt (#PC2052).

♦ PC4867 Incompetent management

Inept directors — Inexperienced business management — Disordered local management — Distracted local management — Low-level management skills — Insufficient trained managers — Underused management skills — Absence of management training — Undiscovered managerial skills — Unexplored business training — Unstructured business training — Minimal business training — Insufficient management skills

Narrower Incompetent workers (#PD4535) Limited finance training (#PG5384)
Agricultural mismanagement (#PD8625) Limited commercial experience (#PG7798)
Poor managerial communications (#PF1528) Incompetent financial management (#PF0760)
Unprofessional building management (#PG8860)
Mismanagement in developing countries (#PD8549)
Undeveloped business skills in urban areas (#PE8048)
Lack of management skills in developing countries (#PE0046)
Underproductive methods of agricultural management (#PF6524).
Related Shortage of technical skills (#PF6500)
Unrecognized need for functional skills (#PF2995).
Aggravates Mismanagement (#PB8406) Incomplete cost projections (#PJ8109)
Unfocused style of community operations (#PF6559)
Discouraging conditions for small business (#PD5603)
Inefficiency of state-controlled enterprises (#PD5642)
Underutilization of locally available skills (#PF6538)
Lack of support for local commercial services (#PF6510)
Inadequate management skills in rural communities (#PF1442)
Inadequate road and highway transport facilities in developing countries (#PD0543).
Aggravated by Lack of training (#PD8388) Absence of management training (#PD3789)
Narrow range of business skills (#PE6554) Narrow range of practical skills (#PF2477)
Abdication of control by company directors (#PE9251)
Inadequate practical training in rural areas (#PF6472)
Unattractive locale for economic development (#PF3499)
Ineffective mechanisms for functional training (#PF1352)
Underutilization of potential in local communities (#PF6513)
Lack of local services for community leadership training (#PF2451)
Lack of opportunities for practical training in communities (#PF2837)
Unperceived relevance of formal education in rural communities (#PF1944)
Lack of skilled manpower in rural areas of developing countries (#PE5170)
Limited availability of technical agricultural and business training (#PF2698)
Inadequate local expertise in business practices in developing countries (#PE7313).

♦ PC4879 Fiscal and trade imbalances

Nature Fiscal and trade imbalances threaten to provoke recession, reduce opportunities for sustaining and increasing world economic growth and hamper the growth of developing nations. Because of fiscal and trade imbalances developing nations face the risk of prolonged stagnation in real per capita income, greater poverty and social unrest.

Broader Imbalance (#PA0224) Inadequate fiscal policies (#PF4850).
Narrower Large trade surpluses (#PJ0207) Excessive external trade deficits (#PC1100)
Imbalance in international trade patterns (#PC8415)
Imbalance between agricultural exports and imports in developing countries (#PE4956).
Aggravates Imbalance of payments (#PC0998).

♦ PC4890 Obstacles to world trade

Nature There are various types of obstacle to the development of world trade: (1) Obstacles in the conduct of negotiations, such as visa formalities, lack of information about decision-making procedures, complexity of the system, contacts with officials, contacts with and decision-making power of end-users, and slowness in conducting business. (2) Concerning business representation and servicing facilities, the main obstacles are lack of information, difficulties of an administrative character, utilization of local firms, servicing, and marketing. (3) The obstacles relating to standards and technical regulations are lack of information, difference in national standards, testing and certification procedures, and inspection of goods on production sites. (4) Concerning licensing procedures and related practices, obstacles vary according to whether exports are westbound or eastbound. Those relating to westbound exports are special requirements for granting permission for imports from eastern countries, absence of information about allocation of quotas, delays in publication of quotas, late delivery of licences, sub-division of quotas by periods, and sub-division of quotas between importing firms. Those relating to eastbound exports are lack of information about import possibilities, difficulties in obtaining information on available foreign exchange, and time-limits. (5) Obstacles related to priorities in imports are: preferences for domestic producers, preferences for specific countries, and compensatory transactions. (6) Related to duties, customs procedures and related practices are import duties (particularities of clearance procedures), facilities of a technical character, valuation, classification, and specific problems. (7) Obstacles induced by problems related to payment are impact of bilateral systems of payment, advance-deposit requirements, letters of credit, delay in transfer of payment, and other financial problems. (8) Other obstacles include packaging and labelling regulations, including mark-of-origin rules, transport facilities, and legal matters.

Broader Obstacles to national development (#PF4842).
Narrower Export credit risks (#PF3065) Excessive standardization (#PF2271)
Export credit competition (#PD3067) Discriminatory nuclear trade (#PD8124)
Foreign exchange restrictions (#PF3070) Non-convertibility of currencies (#PF3069)
Capital shortage in developing countries (#PD3137)
Extraterritorial intrusion of jurisdiction (#PE3140)
Inadequate barter system in international trade (#PE0117)
Protectionism in steel and basic metal industries (#PE5866)
International trade barriers for primary commodities (#PD0057)
Area disparities in book production and distribution (#PF4385)
Informational and procedural obstacles to world trade (#PE9107)
Decline in foreign direct investment in developing countries (#PD3138)
Imbalance in trade of cultural products between capitalist and socialist countries (#PE5678).
Aggravates Excessive external trade deficits (#PC1100).
Aggravated by Disincentives for financial investment within developing countries (#PF3845).

♦ PC4898 Discrimination against women in developing countries

Nature In addition to the problems that women in developing countries share with all human beings in the poor world - undernourishment, sickness, lack of resources, illiteracy - they also have to face those problems that are specific to women. Because of their status as second-class citizens, it is women who usually account for the highest figures of illiteracy and lack of material and other resources on international statistical charts. Women generally bear the heaviest burden of marginalization and exploitation by society as a whole. Moreover, they also often suffer oppression and exploitation by men. The specific problems of women in certain cultural contexts - strict norms which limit their possibilities of social, intellectual or emotional fulfilment - are added to the general problems of underdevelopment. Very often they find themselves occupying the lowest levels of society in general, and they often lack the knowledge and consciousness to struggle for their own liberation. Or, as is typical of a social group which is exploited and discriminated against, whether for ethnic, economic or sexual reasons, they discriminate against themselves and consequently consider that the discrimination imposed upon them is justified and lose interest in the struggle to free themselves from it.

Broader Discrimination against women (#PC0308).
Narrower Discrimination against women workers in multinational enterprises in developing countries (#PE4102).

♦ PC4906 Vulnerability of farming

Agricultural risks

Nature Farming is so subject to gluts and shortages and so vulnerable to weather that few countries can leave planning of agriculture to pure market forces.

Broader Risk (#PF7580)
Instability of economic and industrial production activities (#PC1217).
Narrower Crop vulnerability (#PD0660) Endangered family farms (#PD5962).
Aggravates Food insecurity (#PB2846).

♦ PC4987 Large-scale weather anomalies

Nature Present investigations of ocean-atmosphere interactions involve two concepts. The first is derived from experiments which suggest that regions in the ocean may significantly affect large-scale atmospheric processes over the continents, with a time-lag of 4-8 months. The second concept is a hypothesis that may help explain climate variability and large-scale weather anomalies. It assumes that water masses with abnormal temperatures may persist for long periods, are able to reach deep waters and migrate for long distances, and under certain conditions may reappear at the surface and induce large-scale anomalies of atmospheric circulation several months, or more probably, years later. These interactions of the ocean and the atmosphere determine the world's weather and are responsible for droughts, floods, storms, and other extreme weather conditions which take an annual toll in lives, crops, and property damage.

Incidence Much of these interactions begins in the Pacific Ocean, although they can affect climates as far away as North America and Africa. Among phenomena with significant influence on climate are those known as the El Niño/Southern Oscillation (ENSO). El Niño is the warming of South American coastal waters; Southern Oscillation is the movement of large areas of air pressure between the Pacific and Indian Oceans. These ocean-atmosphere events have occurred 9 times in the past 40 years.

During the 1982-83 El Niño, severe storms lashed the west coast of the United States, the first typhoon in 75 years hit French Polynesia, Australia suffered the worst drought in 200 years and China faced floods in the south and drought in the north. El Niño was blamed for 1,300 to 1,500 deaths worldwide, with damages of $2 billion to $8 billion.

Broader Bad weather (#PC0293).
Related Inaccurate weather forecasting (#PF5118).
Aggravates Hurricanes (#PD1590) Excessive rainfall (#PE4103)
Unreliable rainfall (#PD0489).

♦ **PC4992 Rural poverty**
Nature On average, 40 percent of rural people live below the poverty line; that is, they earn an income less than sufficient to supply their basic needs of food, health, water, housing and education. Behind these stark facts there is a mass of people condemned to hunger, malnutrition and ignorance. But something far worse is that in many countries poverty is on the increase. The post war 'Green Revolution' of capital–intensive industrialized agriculture with the introduction of HYVs (high yielding varieties) of wheat, maize and rice, has led to a growing polarization of rural rich and poor, despite increased output. Consolidation of landholdings results in the eviction of tenants and sharecroppers, while increasing mechanization means less demand for the growing number of landless labourers.
Why poverty has increased has more to do with the structure of the economy than its rate of growth. In a society characterized by extreme inequality of income and hence spending power, the very fact of inequality has a number of important consequences. The counterpart to the compression of the income of the poor is the concentration of the economic surplus in the hands of a minority. They way in which this surplus is used in turn largely determines the pace and nature of economic growth. Where the distribution of land is highly unequal the role of large land–owners is particularly crucial in determining the wages and incomes of the other members of rural society. Another reason for the persistence of poverty in rural areas is the pattern of investment in the country as a whole. The level of investment in developing countries is often extremely low; where it does occur it favours the towns. There is an 'urban bias' in the allocation of investment that deprives the rural areas of much–needed capital.
Incidence Despite the rise in average incomes over the past two decades, the incidence of rural poverty has shown little tendency to diminish and, in many cases, the standard of living of some socio–economic groups, notably the landless, has actually declined. The reasons for this may have less to do with aggregate or sectoral rates of growth than with such factors as the distribution of productive assets, the pattern of government investment, and the non–neutrality of technological advance. The experience of growth in the last quarter of a century has not succeeded in mitigating the problem of rural poverty in Africa, Latin America or Asia.
Over half the world's population is still rural and depends primarily on agriculture, forestry and fishing for a livelihood: more than 2,000 million people, the majority of whom live in developing countries. Of an estimated 750 million people in the developing world classified as living in poverty, more than 80 percent live in rural areas, and about 85 percent of all absolute poverty is in these rural areas. Among the rural poor, there are over 80 million small land holdings (less than 2 hectares), some 30 million or more tenant farmers, sharecroppers and squatters, and a growing numbers of landless or near landless workers (especially in Asia). The number of landless farm workers in the developing countries is increasing steadily. There are an estimated 47 million in India alone – about 1/3 of the active population in agriculture – and 10 million in Latin America.
Background The rural poor depend largely on agriculture for their livelihood, and have adapted their way of life to relative isolation, with little access to national resources and very little influence over their future. Their standard of living is low and often declining; their quality of life leaves them severely disadvantaged and less able to change their role without outside help. Their cultural traditions are strong and their societies are distinguished by marked divisions. The causes of rural poverty can be traced to low agricultural yields and low productivity of labour but they are rooted in a complex web of economic, social, political and geographical factors. One of the basic characteristics of poverty is lack of access to land and other rural assets. In addition, there is increasing population pressure on natural resources, with high rates of absolute and disguised unemployment. There are poor institutional mechanisms and extremely limited physical infrastructure and services in rural areas, so that access to the available resources and to decision–makers is severely curtailed. The result is that the circle of poverty is all–embracing and self–perpetuating; the chief challenge facing many of these rural people is merely to survive. Indeed, poor nutrition, bad shelter, low health standards, and inadequate and irrelevant education, combined with primitive farm and household technology, are the main components of rural poverty. These fundamental aspects of poverty are often compounded by seasonal and cyclical factors such as droughts or periodical fluctuations in commodity prices. Local wars or border conflicts also cause widespread misery for millions of rural people uprooted from their homes and land.
Refs Chambers, Robert; Longhurst, Richard and Pacey, Arnold (Eds) *Seasonal Dimensions to Rural Poverty* (1983); FAO *Development Strategies for the Rural Poor* (1984); ILO *Group-based Savings and Credit for the Rural Poor* (1984).
Broader Socio–economic poverty (#PB0388).
Related Urban poverty (#PC5052).
Aggravates Adjustment difficulties of new urban families (#PF1503).
Aggravated by Unsustainable rural development (#PD4537).

♦ **PC4998 Shortage of animal protein**
Nature Food supplies are being threatened not only by population growth but by the demand for animal protein inspired by rising affluence. This demand is now a major claimant on scarce supplies of both grains and feed–stock proteins. The latter are on the whole exported from poor to rich countries, where they are converted into animal protein providing one tenth to one fifth as much food value at a higher price. Grains – more than half the world's direct food supply – are consumed directly at a rate of about one pound per capita per day in most poor countries. Some rich countries consume less grain directly, but they consume grain at over twice that rate indirectly in the form of meat and beverages.
Rising demand for animal protein is now conflicting with three main limits on its production. (a) Marine protein production, having risen by 5 percent per year since about 1950, is now very close to its sustainable limit. This was formerly thought to be perhaps 100 to 120 million tons per year but, in the wake of the catastrophic drop in the Peruvian anchovy catch in 1972, it is now widely estimated to be about 70 million tons per year, and prices are higher. (b) Beef production, already limited by the fertility of cattle, has now encountered the sustainable limit of grazing and in many areas exceeded it. Further increases, therefore, depend on intensive feed–stock agriculture – at an energy cost of about 12 pounds of coal equivalent per pound of beef protein. (c) The increase in soybean production in the past few decades has in most cases been due to increased planting; no significant intensification of yields is in sight, and demand continues to rise much faster than supply.
Broader Shortage (#PB8238).
Aggravated by Unsustainable harvesting rates (#PD9578)
Unsustainable agricultural development (#PC8419)
Unsustainable exploitation of fish resources (#PD9082).

♦ **PC5005 Victimless crime**
Crimes without victims — Crimes against public morality
Refs Lincoln, Alan J *Crime in the Library* (1984).
Broader Crime (#PB0001).
Narrower Sodomy (#PE3273) Obscenity (#PF2634) Blasphemy (#PF5630)
Drug abuse (#PD0094) Prostitution (#PD0693) Birth prevention (#PE3286)
Public drunkenness (#PE2429) Gambling and wagering (#PF2137).
Related Moral offences (#PD9179).

♦ **PC5010 Topological disaster**
Broader Natural disasters (#PB1151).
Narrower Floods (#PD0452) Avalanches (#PD1146) Coastal erosion (#PE6734)
Ground failures (#PE5066) Migrating sand dunes (#PD0493).

♦ **PC5016 Unpreparedness for food emergencies**
Inadequacy of emergency food reserves — Inadequate emergency food supplies — Inadequate food stocks
Nature Nutritional emergencies arise from situations of mass starvation caused by the interruption of food supplies to the population over a long period. Unusual food shortages may be caused by major crop failures, war and civil conflicts, or natural disasters. According to the World Health Organization, the emergency subsistence level is the level below which large–scale starvation and death should be expected if the people are is of normal body size and are required to perform some work. Measurement of nutritional status in emergencies relies mainly upon taking body measurements, monitoring clinic records, or measuring the prevalence of oedema (swelling).
Incidence Food emergencies have generally been on the rise in recent years. WFP assistance approved for emergency projects increased from $ 91 million in 1978 to $ 92 million in 1980. In 1981, emergency assistance fell slightly to $ 178 million, but rose to $ 191.5 million in 1982. In 1987 the emergency resources of the WFP were not sufficient to cover the emergency needs for which assistance was requested. The increasing scale of food emergencies emphasizes the need for more resources to be devoted to emergency relief programmes.
Broader Disaster unpreparedness (#PF3567).
Related Suspension of rights during states of emergency (#PD6380).
Aggravates Failure to assist in emergencies (#PF5306).
Aggravated by Food insecurity (#PB2846) Mismanagement of food resources (#PE6115).

♦ **PC5017 Xenophobia with regard to migrant workers**
Nature Xenophobic attitudes and movements in various countries sometimes take the form of acts of violence with dramatic consequences. These prejudices are often exacerbated by the rise in unemployment which most industrialized countries are facing, and are supported by allegations of an economic nature (foreign workers are said to be taking jobs which could go to nationals); of a social nature (foreign workers are accused of enjoying social benefits and taking housing which could be allocated to nationals); of a moral nature (foreign workers are said to be the reason for increased violence and delinquency); and of an educational nature (the children of foreign workers are accused of slowing the progress of the classes they attend).
Claim The contributions of foreign workers are often unheeded. They have played an important role in the economic development of host countries (who initially invited them during periods of economic boom), and also usually accept those jobs which nationals do not want. It is also socially advantageous for nations to become less parochial by means of the introduction of varying cultural, artistic, linguistic and social backgrounds.
Broader Xenophobia (#PD4957).
Aggravated by Inadequate cultural integration of immigrants (#PC1532).

♦ **PC5018 Suppression of intellectual freedom**
Nature Culture is often a field of political struggle. Totalitarian governments tend to control cultural activities so as to avoid public expression of hostile attitudes or criticism of the regime they represent. They do so by curtailing intellectual freedom in various ways: by tightening control over intellectuals and universities, by restricting publication of works by dissident authors, by the dissolution of writers', actors', journalists' and artists' unions; and by any other move liable to still the expression of any opposition.
Incidence A recent example was in Spring 1984, when Poland was reported to be tightening control over intellectuals amid concern about the continued reluctance of academics, writers and others in professional and cultural fields to go along with government policies.
Broader Repression (#PB0871) Denial of freedom of thought (#PF3217).
Aggravates Underutilization of intellectual ability (#PF0100).

♦ **PC5038 Maldistribution of energy consumption**
Inequality in energy use — Disparities in energy consumption
Claim The rich industrialized countries are consuming over 75 per cent of world energy production, although they only include 25 per cent of the world population.
Narrower Maldistribution of electrical energy (#PD3446)
Energy deficient developing countries (#PE0379).
Aggravates Energy crisis (#PC6329) Energy dependence and vulnerability (#PJ7735).
Aggravated by Unethical consumption practices (#PD2625)
Unsustainable development of energy use (#PC7517).

♦ **PC5052 Urban poverty**
Urban underclass — Inner city poverty
Nature Urban poverty is, in a sense, an overflow of rural poverty. Because rural people in the low–income group find themselves 'unemployable' in the urban environment as a result of their deficient education and training, they continue to be poor.
In some developed countries the urban poverty has changed from being the result of migration to being among children living with one parent, in households headed principally by young women. Several factors, resulting from changes in the social and psychological situations of the poor are contributing to the persistence and growth of poverty. Poverty is increasingly concentrated in urban neighbourhoods that are deficient in social institutions that control and mediate social, political and economic relations and that provide resources and avenues of economic advancement. Poverty and its consequences, in themselves re–enforce these conditions.
In one community in Chicago, nearly one–half of the housing stock has disappeared since 1960 and that which remains is rundown. In 1985 the murder rate was twice that of the city of Chicago and 6 times that of the USA. Five percent of the youth were referred to court in 1985 alone. Infant mortality increased to 25 death per 1000 live births in the community while declining in the city and nation as wholes. In 1985, 70 percent of all babies born were born out of wedlock, half of these to women 21 years old or younger. The percentage of those receiving welfare assistance rose between 1970 and 1980 from one third to one half of the community's population.
In these types of communities, the number of employment opportunities are declining as smoke–stack industries which provided low–skilled jobs and steady income have closed. As these industries leave, the associated support industries and service business, i.e. banks, restaurants and stores also close or leave. The increasing crime rates cause insurance rates to rise forcing small business to close or move. While the number of service jobs in the urban area are increasing, these provide few entry level employment opportunities for unskilled or semi–skilled workers. At the same time, more and more unskilled jobs are being offered as part–time or contractual work.
Schools become less capable of providing education necessary for residents to break out of poverty.
The family situation also contributes to maintenance of poverty. The welfare system encourages single parent families with women heads of households because these types of units are receive the most benefits. The inner city images of masculinity and feminity re–enforce this family situation. Making marriage an unlikely route out of poverty.
Increasing drop–out rates, high crime rates, child maltreatment and drug abuse are effects of declining economies and local institutions such as school and churches; changing patterns of

PC5052

interaction between local institutions and residents; and inadequate informal social support networks in high risk neighbourhoods.
At the individual level, low self-esteem, lack of family cohesion and family discord and the unavailability of external support systems that encourage and reinforce an individual's coping efforts contribute to poverty.
Incidence One of the consequences of continuing rural poverty has been massive migration to the towns and cities of the developing world. By 1975, 795 million people lived in developing country urban areas and this is expected by the United Nations Environment Programme (UNEP) to increase to over 2000 million by the year 2000. The shanty towns around the major cities of the developing world have grown at a frightening pace during the last twenty years, accumulating massive problems of over-crowding, poor sanitation, bad housing, inadequate public services and unemployment. The slums and squatter settlements, erected on vacant public and private land around the major towns, constitute between 25 and 90 percent of the urban population of the developing countries.
Refs Davis, Lenwood G *Ecology of Blacks in the Inner City* (1975).
Broader Social injustice (#PC0797) Socio-economic poverty (#PB0388).
Narrower Urban slums (#PD3139).
Related Rural poverty (#PC4992).
Aggravates Criminal usury (#PE1181) Urban overcrowding (#PC3813)
Environmental degradation of inner city areas (#PC2616).
Aggravated by Urban bias (#PF9686) Unemployment (#PB0750)
Rural depopulation (#PC0056) Unequal income distribution (#PD4962).

♦ PC5075 Restrictions on freedom
Broader General obstacles to problem alleviation (#PF0631).
Narrower Denial of right to liberty (#PF0705) Denial of academic freedom (#PD4282)
Long-term shortage of energy resources (#PF0334)
Overcomplicated implementation of citizen rights (#PG7763).
Related Dietary restrictions (#PJ1933).
Aggravates Human suffering (#PB5955).
Aggravated by Crime (#PB0001) Taboo (#PF3310)
Environmental degradation (#PB6384).

♦ PC5088 Cultural alienation
Nature As communications have proliferated in recent decades and brought the external world to millions of people previously living in isolated communities, so they have generated two major concerns: (1) The development of mediated communication is a technical and social need, but may also be a threat to the quality and values of culture. (2) The indiscriminate opening of doors to new experiences and impressions by the media sometimes alienates people from their own culture.
With the speed and impact of the media explosion, certain harmful effects have been observed. For many people, their conception of reality is obscured or distorted by messages conveyed by the media. The rapid increase in the volume of information and entertainment has brought about a certain degree of homogenization of different societies while, paradoxically, people can be more cut off from the society in which they live as a result of media penetration into their lives. The introduction of new media, particularly television into traditional societies has shaken centuries-old customs, cultural practices and simple life styles, social aspirations and economic patterns. Too often the benefits of modern communications - which disseminate unfamiliar, vivid, absorbing information and entertainment originating in urban centres and, more often than not, from foreign sources - have been accompanied by negative influences which can dramatically disturb established orders. At the extreme, modern media have trampled on traditions and distorted centuries-old socio-economic patterns.
Incidence An analysis of the cultural flows between countries shows a serious imbalance. The media in developing countries take a high percentage of their cultural and entertainment content from a few developed countries. The flow in the other direction is a mere trickle by comparison. The developed countries get the selected best of the culture (chiefly music and dance) from developing countries; the latter get a lot of what on any objective standard is the worst produced by the former. This unequal exchange is inevitably harmful to national culture in developing countries. Their writers, musicians, film-makers and other creative artists find themselves shouldered aside by imported products. Local imitations of imported culture and entertainment do not improve the situation; they too lead to the imposition of external values.
Claim The commercial approach to culture operates to the detriment of true values. Transnational companies are playing an ever more active role in the world-wide provision of communication infrastructures, news circulation, cultural products, educational software, books, films, equipment and training. Although their role in extending facilities for cultural development and communication has been considerable, they also promote alien attitudes across cultural frontiers. Since similar cultures predominate in the countries where the transnationals have their roots, they transmit models and influences which are broadly alike. When these influences become dominant in very different cultures, the effect is to impose uniformity of taste, style and content. This is considered as cultural invasion, the type of intrusion that represents one of the major problems to be faced by everyone dealing with international communication issues.
The socio-cultural tastes of foreign countries have been widely disseminated and are familiar to and often admired by many; people imitate them and they may become adopted norms of human behaviour in the countries exposed to them. But the imitations of alien cultures are not the same as the true development of a national culture, for they can in reality inhibit growth of national cultures by adapting to standardized international patterns of mass culture. Another negative factor is that creative artists in developing countries - authors, musicians, playwrights, script-writers, film-makers - often find it difficult to stand up to the competition of the industrialized products of the big conglomerates.
Modernization and change are inevitable, in many cases desirable. There is, however, value for the world to retain cultural diversity. The search for international unity does not require the homogenization of peoples or the obliteration of national and cultural differences that today's international media appear to promote.
Counter-claim The process of modernization rarely takes place without some disrupting influence and effects. Moreover, in most societies some vestiges of the past are woefully and harmfully archaic, or even inimical to accepted present-day social philosophy and practice, thus should be changed to advance human progress.
Broader Alienation (#PA3545).
Related Cultural illiteracy (#PD2041).

♦ PC5165 Endangered forests
Forest vulnerability — Limited forest resources
Broader Endangered species of plants (#PC0238)
Vulnerability of ecosystem niches (#PC5773).
Narrower Deforestation (#PC1366) Forest fragmentation (#PD9490)
Acidic precipitation (#PD4904) Forest damage by wildlife (#PD0500)
Pests and diseases of trees (#PD3585) Over-rapid timber exploitation (#PD9235)
Misuse of tropical rain forests for agricultural development (#PE5274).
Aggravates Endangered species of animals (#PC1713)
Destruction of alluvial forests (#PD6850)
Long-term shortage of wood, lumber and cork (#PE1372).
Aggravated by Forest fires (#PD0739) Forest decline (#PC7896)
Termites as pests (#PE1747) Shortage of firewood (#PD4769)
Unethical practices in forestry (#PD6701) Underutilization of natural resources (#PF1459)
Unsustainable development of forest lands (#PD4900).

♦ PC5180 Lack of political will
Lack of political will to cooperate — Lack of political will to act on problems — Lack of political will to respond to needs of developing countries — Political unwillingness to change
Broader Resistance to change (#PF0557).
Narrower Conditional observance of multilaterally agreed trade commitments (#PF7838).
Related Political nature of development issues (#PF4175).
Aggravates Homelessness (#PB2150) Government inaction (#PC3950)
Inaction on problems (#PB1423) Political over-reaction (#PF4110)
Insufficient financial resources (#PB4653) Lack of international cooperation (#PF0817)
Government resistance to institutional change (#PF0845)
Inadequate power of intergovernmental organizations (#PF9175).
Aggravated by Apathy (#PA2360) Indifference (#PA7604)
Political fear (#PJ5666) Fear of failure (#PF4125)
Fear of reprisals (#PF9078) Fear of the abnormal (#PF7029)
Fear of vocational change (#PJ1318) Individual fear of future change (#PD2670)
Fear of contradicting popular views (#PF2040)
Rivalry and disunity within developing regions (#PD0110)
Institutional preoccupation with obsolete problems (#PJ5014).

♦ PC5225 Accumulation of capital
Accumulation of wealth
Refs Hamberg, Daniel *Economic Growth and Instability* (1978).
Broader Accumulation (#PA4313) Personal wealth (#PC8222).
Narrower Bourgeoisie (#PE7774)
Excessive accumulation of wealth by government leaders (#PD9653).
Related Lack of relationship between wealth generation and the public good (#PF4730).
Aggravated by Developmentalism (#PF9512).

♦ PC5230 Imbalance of conventional armed forces
Imbalance of conventional weapons — Imbalance of conventional arms
Nature International conflict is avoided when two enemies have equal capacity to inflict damage. When one enemy acquires better equipment, better trained or more combatants or has some other clear advantage the situation can quickly degenerate into war.
Counter-claim As any military historian knows, fighting and winning a war is a nearly infinitely complex task. The motivation of troops, their training, the skills and integrity of the officers, the reliability of intelligence and how it is used, where battles are fought and in what weather even the time of day, logistics, propaganda and weapons all play important roles. To equate, weapons parity with military balance is naive.
Broader Competitive acquisition of arms (#PC1258).
Aggravates Illegal international arms shipments (#PD4858)
Socio-economic burden of militarization (#PF1447)
Proliferation of weapons in civilian hands (#PE2449)
Creeping modernization of military weaponry (#PE7678).
Aggravated by Weapons (#PD0658).

♦ PC5276 Biological pollutants
Broader Pollutants (#PC5690).
Narrower Poisonous algae (#PE2501) Sewage as a pollutant (#PD1414)
Domestic waste water pollutants (#PD2800).
Related Natural pollutants (#PC2416).

♦ PC5280 Abuse of law
Broader Exploitation of trust (#PC4422).
Narrower Lawsuits abuse (#PE6622) Abuse of right to appeal (#PG4693)
Abuse of brothel legislation (#PE6735) Unfulfilled treaty obligations (#PF2497)
Unlawful business transactions (#PC4645) Violation of trade union rights (#PD4695).
Aggravates Crime (#PB0001).

♦ PC5281 Denial of right to work
Broader Denial of human rights (#PB3121).
Narrower Denial of right to free choice of work (#PE3963)
International imbalance in the quality of life (#PB4993)
Denial of the right to work in capitalist systems (#PC3119)
Discrimination against working women in socialist countries (#PD2872).
Related Discrimination in employment (#PC0244).

♦ PC5285 Violation of civil rights
Broader Denial of rights (#PB5405) Violation of human rights (#PB3860).
Narrower Unjust trials (#PD4827) Wrongful detention (#PD6062)
Travel restrictions (#PC8452) Needless incarceration (#PE5112)
Internment without trial (#PD1576) Denial of access to news (#PF3081)
Denial of religious liberty (#PD8445) Limitations on right to vote (#PF2904)
Criminal violation of civil rights (#PD8709) Denial of right to change religion (#PE6397)
Denial of right to trial by a court (#PE4737) Denial of political and civil rights (#PC0632)
Denial of right to manifest religion (#PF2850) Denial of right to hold public office (#PE5608)
Denial of right to fair and public trial (#PE3964)
Excessive length of pre-trial internment (#PE4887)
Denial of right to time to prepare a trial defence (#PE7624)
Restrictions on freedom of movement within countries (#PE8408)
Restrictions on freedom of movement between countries (#PC0935).
Related Denial of right to life (#PD4234) Restriction of freedom of expression (#PC2162).
Aggravated by Police crimes during narcotic investigations (#PE5037)
Political instability of developing countries (#PD8323).

♦ PC5313 Expulsion
Denial of right to freedom from arbitrary expulsion
Nature 'Expulsion' is an act, or a failure to act, by a State with the intention and the effect of securing the departure of a person or persons against their will from the territory of that State. The concept of expulsion encompasses indirect measures including ill-treatment, racial and other forms of discriminatory practices, harassment and other means of coercion designed to force people to leave - as well as the direct exercice of State power. Forms of indirect measures or practices are many and are sometimes of a subtle kind. These can be of a psychological as well as of an economic or social nature. In some cases, the authorities of a State tolerate, or even aid and abet, acts by its citizens with the intended effect if driving persons out of the territory of that State. An example is 'panic flight' where, for the purpose of removing the persons concerned, the authorities create a climate of fear or do nothing that can be reasonably expected of them to assure those comtemplating flight that they would be protected.
Broader Arbitrariness (#PB5486)
Denial of right to freedom from cruel, inhumane or degrading punishment (#PC3768).
Narrower Forced mass expulsion (#PD0531) Rejection of refugees (#PF3021)

Expulsion of ethnic minorities (#PC3205) Expulsion of immigrants and aliens (#PC3207).
Related Exile (#PC2507) Arbitrary exile (#PJ7268).
Aggravates Inhibition of personality development in exiled children (#PE7931).

♦ PC5323 Concentration of investment power

Nature An excessively concentrated degree of investment power is in the hands of very few, very large institutions. By virtue of their size and by virtue of the pressure that they are under to provide performance over a very short time period these institution have adopted stock exchange trading policies which can disrupt markets, because they are not counterbalanced by other types of investors who may be operating under different performance criteria or have different perceptions of value. The traditional balance between large institutions and individual investors, between short-term speculators and long-term investors, between occasional investors and regulars, which has always produced a degree of market equilibrium, is gone.
The complexity of modern stock markets and the speed at which they interact are beginning to alienate the individual investor, who feels he is incapable of coping with that complexity on his own. He is thus at a considerable disadvantage to the professional investor who has all these tools at his disposal. The alienation of the individual investor from the market tends to be followed by a certain alienation. If the individual feels he no longer has a direct stake in the economy and is forced to delegate his ownership and decision-making power to a large faceless, unaccountable institution, that leads to disillusionment with the entire system.
Broader Abuse of economic power (#PC6873).
Narrower Concentration of capitalist banks (#PE7975).
Related Uneven concentrations of power (#PG8703)
Excessive concentration of business enterprises (#PD0071)
Concentration of power by transnational corporations (#PE0766).
Reduced by Protection of company ownership (#PF4432).

♦ PC5333 Ecological imperialism

Broader Imperialism (#PB0113).

♦ PC5348 International status race

International prestige race
Narrower Pursuit of personal prestige (#PF8145) Pursuit of national prestige (#PF8434)
Pursuit of corporate prestige (#PF7983).
Aggravates Competitive acquisition of arms (#PC1258)
Excessively costly prestige projects (#PF3455).
Aggravated by Boasting (#PF4436) Inappropriate policies (#PF5645).

♦ PC5387 Forced participation in social processes

Narrower Participation in torture (#PD4478) Forced participation in politics (#PD2910)
Reluctant personal participation (#PG9362)
Enforced participation in community activity (#PD3386).

♦ PC5456 Denial of rights of animals

Nature As partners with humankind on this earth, members of the animal kingdom obviously have the right to appropriate care, respect of their habits and protection from unnecessary suffering. But even these minimal expectations are completely unfounded. Animals are pumped full of hormones and antibiotics to fatten them up for market, they are cooped up in factory farms without exercise or contact with the outdoors, and they are slaughtered with little regard for their lives as sensate, if not human, beings. So pervasive is the unconsciousness of humankind toward animals that even the outrageous protests of violent activist groups have made very little social impression at all.
Refs Magel, Charles R *A Bibliography on Animal Rights and Related Matters* (1981).
Broader Denial of rights (#PB5405).
Narrower Use of animals in warfare (#PE6443)
Denial to working animals of restorative nourishment and rest (#PE4793)
Denial to animals of the right to the attention, care and protection of humankind (#PF5121).
Related Maltreatment of animals (#PC0066) Maltreatment of livestock (#PE3394)
Intrusive animal-rights campaigners (#PE4438) Cruelty to animals in factory farming (#PD2768)
Denial of rights of inanimate objects (#PF3710)
Cruelty to animals in food preparation (#PE0236).
Aggravated by Prejudice against animals (#PC0507).

♦ PC5489 Biological disasters

Broader Natural disasters (#PB1151).
Narrower Epidemics (#PC2514) Disastrous insect invasions (#PD4751).
Related Forest fires (#PD0739).

♦ PC5512 Victimization of children

Broader Crimes against humanity (#PC1073).
Narrower Child beggary (#PG1103) Child cruelty (#PJ2150)
Child-marriage (#PF3285) Orphan children (#PD7046)
Child abduction (#PE6154) Pancreatic fluke (#PG3985)
Missing children (#PD6009) Unwanted children (#PE1907)
Child mutilations (#PG3726) Neglected children (#PD4522)
Displaced children (#PD5136) Abandoned children (#PD5734)
Child prostitution (#PE7582) Child malnutrition (#PD8941)
Cruelty to children (#PC0838) Torture of children (#PD2851)
Children in poverty (#PD4966) Endangered children (#PD6065)
Detention of children (#PE6636) Sacrifice of children (#PJ5597)
Children of alcoholics (#PD4218) Inadequate child welfare (#PC0233)
Sexual abuse of children (#PE3265) Neglected young children (#PE4245)
Drug abuse by adolescents (#PD5987) Militarization of children (#PE5986)
Emotional abuse of children (#PD7330) Exploitation of child labour (#PD0164)
Family rejection of children (#PC8127) Sexual intercourse with minors (#PE6522)
Physical maltreatment of children (#PC2584) Physical intimidation by children (#PE2876)
Forced disappearances of children (#PD5129) Medical experimentation on children (#PE6764)
Inadequate care for children of prisoners (#PF0131)
Involuntary loss of nationality of children (#PE6676)
Inadequate system of child support enforcement (#PF6076)
Discrimination and harassment of children in public life (#PE6922)
Lack of freedom of movement of children of separated parents (#PE4669).
Related Corruption of minors (#PD9481).
Aggravated by Trafficking in children for adoption (#PF3302).

♦ PC5517 Unethical practices in politics

Irresponsible politics — Political malpractice — Negligence in politics — Dirty politics
Broader Unethical practices (#PC8247).
Narrower Violent revolution (#PC3229) Corruption in politics (#PC0116)
Political smear campaigns (#PD9384) Espionage in domestic politics (#PD1787)
Covert smear campaigns by government (#PD7171)
Abusive distribution of political patronage (#PF8535)
Property and occupational discrimination in politics (#PD3218).
Aggravates Personality cults (#PC1123) Forced participation in politics (#PD2910).
Aggravated by Power politics (#PB3202) Military influence (#PD3385)
Political discrimination in politics (#PC3221) Ideological discrimination in politics (#PC3219)
Undue religious influence on secular life (#PF3358).

♦ PC5528 Avoidance of work

Working to rule
Nature In contrast with evasion of work, which involves deliberate and even illegal actions, avoidance of work occurs within acceptable patterns of behaviour and may be partly unconscious. It can include such forms as working to rule (as a result of labour disputes), reducing the working rhythm, marginally extending rest periods, delaying starting and stopping early, extending conversations into non-work topics to an unreasonable degree, inventing non-essential tasks requiring unnecessary movement around the work site, and extending (rather than curtailing) absences from work to ensure appropriate recovery from illness. Since it occurs within acceptable patterns of behaviour, it cannot be easily questioned or criticized. Consequently it can increase in scope in an insidious manner, whether in the case of an individual working independently (such as a student) or within a work force in a large organization.
Broader Irresponsibility (#PA8658).
Narrower Strikes (#PD0694) Absenteeism (#PD1634).
Aggravates Negligence (#PA2658) Unproductive labour resources (#PC6031).

♦ PC5536 Substance abuse

Refs American Association for Counseling *Substance Abuse*; Baker, Timothy B and Cannon, Dale S *Assessment and Treatment of Addictive Disorders* (1988); Cohen, Sidney *The Substance Abuse Problems* (1985); Einstein, Stanley *Drug and Alcohol Use* (Date not set); Ettorre, Betty *Women and Substance Abuse* (1989); Gottheil, Edward, et al *Etiologic Aspects of Alcohol and Drug Abuse* (1983); Gottheil, Edward, et al *The Combined Problems of Alcoholism, Drug Addiction and Aging* (1985); Mello, Nancy K *Advances in Substance Abuse* (1988).
Broader Excessive consumption of goods and services (#PC2518).
Narrower Smoking (#PD0713) Drug abuse (#PD0094) Alcohol abuse (#PD0153)
Abuse of khat (#PE0912) Caffeine abuse (#PE0618) Cannabis abuse (#PE1186)
Abuse of opiates (#PE1329) Abuse of antibiotics (#PE6629)
Abuse of medical drugs (#PD0028) Substance intoxication (#PD4027)
Substance abuse at work (#PD9805) Substance abuse by role models (#PE0742)
Solvent and methylated spirits drinking (#PE1349)
Inhaling of solvents and anaesthetic drugs (#PE1427).
Related Obesity (#PE1177) Over-eating (#PE5722)
Inadequate drug control (#PC0231) Limited individual attention span (#PF2384)
Excessive consumption of specific foodstuffs (#PC3908).
Aggravates Addiction (#PD6324) Mental impairment (#PF4945)
Invasion of privacy through testing (#PJ6946).
Aggravated by Panic disorder (#PE3575) Ignorance of drug users (#PJ2389)
Unethical entertainment (#PF0374) Forced confessions with drugs (#PE4888)
Excessive portrayal of substance abuse in the media (#PE3980).

♦ PC5537 Restrictive social practices

Broader Restrictive practices (#PB9136).
Narrower Restrictive social groups (#PG8682) Restrictive community size (#PJ0123)
Restrictive patterns of traditional life (#PF3129)
Restrictive effects of traditional community decision-making (#PF3454).
Aggravates Social isolation (#PC1707).

♦ PC5576 Evasion of work

Rejection of job opportunities by the unemployed
Nature In contrast to avoidance of work, which occurs within acceptable patterns of behaviour and therefore cannot normally be questioned, evasion of work usually involves deliberate and often illegal actions. It can include presenting false evidence enabling illness to be feigned, new job offers to be rejected, or the impression created that work has been done. In the latter case it may be closely associated with criminal negligence.
Broader Irresponsibility (#PA8658).
Narrower Malingering (#PE7701).
Aggravates Lying (#PB7600) Injurious accidents (#PB0731)
Unproductive labour resources (#PC6031).

♦ PC5601 Unequal global distribution of economic growth

International imbalance of economic activity — Unsynchronized economic growth
Narrower Imbalance in world food economy (#PD5046).
Aggravated by War-time disruption of economies and production facilities (#PD8851).

♦ PC5624 Economic crime

Commercial crime
Nature Owing to the generally recognized difficulty of giving an exact definition of economic crime, the European Committee on Crime Problems of the Council of Europe found it necessary to delimit the concept by means of the following list of offences: (1) cartel offences; (2) fraudulent practices and abuse of economic situation by multinational companies; (3) fraudulent procurement or abuse of state or international organizations' grants (4) computer crime (theft of data, violation of secrets, manipulation of computerized data); (5) bogus firms; (6) faking of company balance sheets and book-keeping offences; (7) fraud concerning economic situation and corporate capital of companies; (8) violation by a company of standards of security and health concerning employees; (9) fraud to the detriment of creditors (bankruptcy, violation of intellectual and industrial property rights); (10) consumer fraud (in particular falsification of and misleading statements on goods, offences against public health, abuse of consumers' weakness or inexperience); (11) unfair competition (including bribery of an employee of a competing company) and misleading advertising; (12) fiscal offences and evasion of social costs by enterprises; (13) customs offences (evasion of customs duties, breach of quota restrictions); (14) offences concerning money and currency regulations; (15) stock exchange and bank offences (fraudulent stock exchange manipulation and abuse of the public's inexperience); (16) offences against the environment.
Economic crime causes loss to a large number of people (partners, shareholders, employees, competitors, creditors), to the community as a whole, and even to the state, which has to bear a heavy financial burden or suffers a considerable loss of revenue; it harms the national and/or international economy and causes a certain loss of confidence in the economic system itself.
Incidence Economic crimes constitute a large proportion of the dark figure of unreported or insufficiently reported crimes. These appear to cause economic and social institutions and the public much greater harm than is suggested by the small number of crimes which are successfully prosecuted, because they trigger off chain reactions. Sweden has found that the receipts of different kinds annually withheld from taxation, through tax offences of a systematic kind in various economic activities, run to very substantial amounts; and a recent calculation of the taxation revenue withheld annually shows that in Sweden some 10 percent of the GNP escapes tax. This figure includes tax evasion by individuals as well as by large business firms.
A comparative study of economic and ordinary crime further shows the former's potential victims to be much more numerous than those of the latter: partners or shareholders (in the case of offences against company law, such as non-existent, inaccurate or incomplete accounts), the state (tax evasion); employees (failure to comply with labour legislation especially in respect of health and safety regulations); competitors (interference with free competition, whether direct by

means of cartels, or indirect by means of misleading advertisement); consumers (sometimes the same offences as the preceding category, but also offences in connection with price regulations or the quality of the goods offered); people with small savings (stock exchange offences); and neighbours or local residents (pollution of the environment). Such fraudulent practices are familiar and economic crime continues to increase; so much so that the question has often been asked whether the essential factor does not lie in the very economic systems of societies of a liberal type, where free enterprise in a market economy fosters, perhaps even stimulates, the commission of offences. But whilst certain forms of economic crime seem clearly to be associated with a specific economic system (interference with free competition), their equivalent none the less exists elsewhere (non-observance of the rules laid down in the plan). On the other hand, offences which seem totally independent of the economic system are on the increase almost everywhere in the world: tax evasion, customs offences, trafficking in foreign currencies (where exchange controls exist), pollution of the environment, infringements of labour legislation, and so forth.
Refs Chandra, Mahesh *Socio–Economic Crimes* (1979); Council of Europe *Economic Crime* (1981); Liebl, Hildegard and Liebl, Karlhans *International Bibliography of Economic Crime* (1985).
Broader Crime (#PB0001) Exploitation of trust (#PC4422).
Narrower Cartels (#PC2512) Bogus firms (#PF0326) Grant frauds (#PD4543)
Theft of data (#PD2957) Abuse of credit (#PF2166) Financial frauds (#PE2414)
Unfair competition (#PC0099) Collusive tendering (#PE4301) High interest rates (#PF9014)
Computer-based crime (#PE4362) Crimes against trade (#PG6404)
Banking law violations (#PE1208) Defrauding of secured creditors (#PD4401)
Hoarding of primary commodities (#PD0651) Evasion of social costs by companies (#PE3149)
Evasion of customs and excise duties (#PD2620)
Criminal offences against the environment (#PC4584)
Violations of health and safety regulations (#PE4006)
Misrepresentation of information to consumers (#PE6877)
Distortion of corporation financial statements (#PE2032)
Securities and commodities exchange violations (#PD4500)
Fraud concerning economic situation and corporate capital of companies (#PE5021).

♦ PC5690 Pollutants
Refs United Nations *Microbial Technologies to Overcome Environmental Problems of Persistent Pollutants* (1986).
Broader Pollution (#PB6336).
Narrower Thermal pollutants (#PC1609) Natural pollutants (#PC2416)
Biological pollutants (#PC5276) Toxic metal pollutants (#PD0948)
Pesticides as pollutants (#PD0120) Fungicides as pollutants (#PD1612)
Electromagnetic pollution (#PD4172) Biological air pollutants (#PD0450)
Inorganic salts as pollutants (#PD5227) Sulphur trioxide as a pollutant (#PE2964)
Health hazards of exposure to noise (#PC0268)
Chemical pollutants of the environment (#PD1670)
Environmental hazards of non–ionizing radiation (#PE7651).
Aggravates Allergy (#PE1017) Hypersensitives (#PE5169)
Degradation of the environment through contamination (#PE4759).

♦ PC5773 Vulnerability of ecosystem niches
Erosion of ecosystem diversity — Decreasing ecological diversity — Decreasing diversity of biological habitats
Nature Degradation or destruction of large natural environments, as well as environments partially modified or cultivated by man (including forests, rangelands, wetlands and aquatic ecosystems), designated wilderness areas, and culturally important landscapes (whether natural, modified, cultivated or built environments). This diversity of environments provides habitats for a variety of species and supports a range of activities of value to man (including timber, agriculture, livestock, and fish production).
Background The natural environment contains a number of terrain–discrete habitats. They include forests, mountains, plains, marsh or wetlands, desert, tundra, reefs, islands and many more, some of which are subdivisions of the above. Each habitat has any or all of the following: humans and human artifacts, animals, insects, plants, minerals, specific climate and other characteristics. There are food chains and other elements of ecosystems characteristic to each habitat. When one of the ecosystems is under attack as a result of natural or man-made disaster it is extremely difficult to calculate the ripple effects throughout nature. When two or more ecosystems are being degraded the probabilities of synergistic destructiveness multiply.
Incidence Ecosystems in many regions are threatened, despite their biological richness and their promise of material benefits. This has been documented with examples from ecosystems including forests (temperate, tropical, mangrove), coral reefs, savannas, arid zones and grasslands.
Claim The scale of oil spills and other pollution in the oceans threatens the entire sea environment. Ocean vegetation dies, ocean life dies and human hunger arises. Pollution of the atmosphere makes people and animals ill, affects crops, and alters weather patterns in persistent ways so that humans may not go out of doors in some places. However, oil slicks affect the atmosphere too, and a clouded sun and polluted atmosphere affect ocean temperature, sea level, and organic ocean life. As yet mankind has not been able to build a model of the global ecosystem interrelationships because they are not sufficiently understood. Pollution and degradation continue by sectors in the ecosystem niches, until some at least will fail. There may be a domino effect causing the human habitat to ultimately perish.
Broader Environmental degradation (#PB6384) Erosion of biological diversity (#PB9748).
Narrower Endangered forests (#PC5165) Vulnerability of wetlands (#PC3486)
Destruction of coral reefs (#PD5769) Vulnerability of marine ecosystems (#PC1647)
Depopulation of mountainous regions (#PD1908) Deforestation of mountainous regions (#PD6282)
Over-use of designated wilderness areas (#PD7585)
Vulnerability of protected natural areas (#PC4764)
Environmental degradation of desert oases (#PD2285)
Disruption of ecosystems in marginal agricultural lands (#PD6960).
Related Landscape disfigurement (#PC2122).
Aggravates Wildlife extinction (#PC1445) Natural environment degradation (#PB5250)
Extermination of wild animal natural prey (#PD3155)
Decreasing diversity of biological species (#PD7302).
Aggravated by Natural disasters (#PB1151) Man-made disasters (#PB2075)
Lack of biocoenosis (#PE5290) Vulnerability of food chains (#PB2253)
Degradation of semi–natural and natural habitats of flora and fauna (#PC3152).

♦ PC5842 Protectionism in international trade
Unfair restrictions in international trade — Economic nationalism
Claim Rather than fighting protectionism nations have accommodated and institutionalized it, disguising new barriers in such euphemisms as "bilateralism" and "managed trade". Under such agreements countries agree to open their markets to limited quantities of one other's products and exclude those from all other countries.
Refs Burnell, Peter *Economic Nationalism in the Third World* (1985); Riddell, Barry *Economic Nationalism* (1972).
Broader Trade protectionism (#PC4275).
Narrower Trade harassment (#PD7441) Union protectionism (#PG6833)
Protectionism in the shipping industry (#PE5888) Protectionism in the mining industries (#PG6817)
Protectionism in the automobile industry (#PG7661)
Protectionism in the commodities sectors (#PG7925)
Protectionism in labour–intensive industries (#PG6626)
Obstacles to international trade in services (#PD6223)
Protectionism in the high–technology industries (#PE8458)
Imposition of trade quotas for political reasons (#PE9762)
Protectionism in the defence and arms industries (#PE8664)
Protectionism in steel and basic metal industries (#PE5866)
Protectionism in the alcoholic beverages industry (#PE8134)
Protectionism in the consumer products industries (#PE8795)
Protectionism in the textile and apparel industries (#PE5819)
Protectionism against imports of service–related goods (#PE7025)
Trade barriers and protectionism between developing countries (#PD2958)
Protectionism in agriculture and the food production industries (#PD5830)
Protectionism in international trade against exports from developing countries (#PD9679).
Aggravated by Nationalism (#PB0534) Misalignment of currencies (#PF6102)
Unilateral interpretations of multilateral principles (#PF9629)
Mismatch of national macroeconomic policies among industrialized countries (#PF5000).

♦ PC5876 Global economic crisis
Claim World economic crisis threatens the economic future of all countries. It places a heavy strain on the international trade, finance and monetary systems and raises the spectre of trade warfare, competitive devaluations and financial collapse. It has had a strong impact on developing countries, resulting in a severe curtailment of their growth and seriously impairing their prospects for the years to come.
Refs Stajner, Rikard *Crisis* (1976).
Broader Global crisis (#PF6244).
Narrower Excessive public debt (#PC2546)
Interaction of deficiencies in world economical systems (#PF1739).
Aggravated by Crisis of capitalism (#PF7999) Industrial intimidation (#PC2939)
International economic interdependence (#PF6105)
Inadequacies of the international monetary system (#PF0048)
Imbalance in economic relationships between industrialized countries (#PC7459).

♦ PC5879 Increasing development lag against socio–economic growth
Increasing unsustainability of development
Narrower Increasing development lag against population growth (#PF3743)
Increasing development lag against information growth (#PE2000)
Increasing development lag against technological growth (#PE3078).
Aggravated by Unbalanced growth (#PB0479) Unsustainable economic development (#PC0495)
Unconstrained socio–economic growth (#PB9015).

♦ PC5883 Environmental hazards
Broader Risk (#PF7580).
Narrower Carcinogenic chemical and physical agents (#PD1239)
Environmental hazards of new species introduction (#PC1617).
Aggravates Ecologically unsustainable development (#PC0111).
Aggravated by Health hazards of asbestos (#PE3001).

♦ PC5890 Disabled children
Childhood disability
Refs Chigier, E (Ed) *Youth and Disability* (1986); Garbarino, James, et al (Eds) *Special Children – Special Risks* (1987).
Narrower Maladjusted children (#PD0586) Mental deficiency in children (#PD0914)
Physically handicapped children (#PD0196).

♦ PC5916 Female unemployment
Nature Women are generally more likely to be victims of unemployment than men, due to factors as diverse as recession, lack of education, and discrimination.
Incidence In the OECD countries, female unemployment exceeds male unemployment by anything from 30 to 100 percent (except Japan, where the rate is slightly higher for the males). In the EEC countries it averages 50 percent higher.
Broader Unemployment (#PB0750).

♦ PC5917 Regional tensions
Regional conflicts — Regional warfare — Area tensions
Nature The same commonalities that allow for conceptual groupings of nations may also represent objects of contention. Thus, geographically contiguous nations are so as a result of common borders which are subject to dispute or to illegal crossings or other infringement of territorial sovereignty. Such conflicts may escalate, or run the risk of escalating, thus determining responses from distant powers who seek to take advantage of the situation or prevent loss of their existing advantages in the region. In some cases this may result in minimalistic responses, based on cynical calculation or relative indifference, despite violence of genocidal proportions. For fear of being accused of opportunistic intervention, such powers (and the international community as a whole), may therefore limit their initiatives to deploring the scale of such violence.
Incidence Three long and intractable Third World conflicts have dominated the international community since World War II, namely Middle East (Israel), South East Asia (Vietnam) – later replaced by Central America – and Southern Africa (South Africa). More generally, nations that share the oceans resources (around the Mediterranean Basin, in the Indian Ocean, or the South Pacific, for example) may be areas in which tension builds up. Nations that use or wish to use straits, canals, seaways and other linking waterways, including rivers, may be embroiled in conflict, for example at Gibraltar, the Suez Canal, the Panama Canal, the Dardanelles, the Shatt al–Arab and similar places. The commonest designation of an area of nations is defined by ethnicity, language, culture and sometimes religion. The Levant, for example, comprising countries bordering the Eastern Mediterranean, if seen also in relation to Greece, Turkey, Cyprus and North Africa, is an area with a variety of tensions that aggravate each other, but in which ethnicity and religion play an important part. Areas designated in press reports include the Balkans, the Caribbean, Sub–Saharan Africa, Southeast Asia, Central America, Southern Africa and the Middle East. Some areas may be designated after a defence treaty, such as WTO or NATO countries; or an economic or other intergovernmental organization, such as EFTA, EEC, COMECON, OAS, or OAU.
Claim Where regional conflicts escalate to a level of genocide, this should engage the responsibility of international society as being a threat to the international community as a whole and as a threat to moral nature, especially when a weaker state or society is being subject to the violence and indifference of a stronger one.
Aggravates Refugees (#PB0205).

♦ PC5994 Religious persecution
Incidence Religious persecution has existed through history and is still a present–day reality. In 1983, for example, more than 250,000 Muslims living in nearly 100 countries as minorities faced persecution and oppression.
Broader Persecution (#PB7709) Social injustice (#PC0797).
Narrower Persecution of religious sects (#PF3353).

Related Racial discrimination (#PC0006).
Aggravated by Ideological conflict (#PF3388)
Peoples perceiving themselves as specially chosen (#PF4548).

♦ **PC6003 Infringement of human rights**
Refs Meron, Theodor *Human Rights in International Law* (1986).
Broader Violation of human rights (#PB3860).
Related Denial of human rights (#PB3121).   Deprivation of human rights (#PC7379).

♦ **PC6031 Unproductive labour resources**
Unproductive workers — Declining worker productivity
Broader Unproductive use of resources (#PB8376).
Narrower Low productivity of agricultural workers in developing countries (#PE5883).
Related Loss of animal productivity (#PD8469)
Declining productivity of agricultural land (#PD7480).
Aggravates Declining economic productivity (#PC8908).
Aggravated by Evasion of work (#PC5576)   Avoidance of work (#PC5528).

♦ **PC6034 Disastrous accidents**
Broader Man-made disasters (#PB2075).
Narrower Fires (#PD8054)   Transport accidents (#PC8478)
Collapsing physical structures (#PD4143)   Occupational domestic accidents (#PE4961)
Large-scale industrial accidents (#PD2570)   Disastrous technological failures (#PD4426).
Aggravated by Suppression of information concerning environmental hazards (#PF4854).

♦ **PC6037 Human infertility**
Barrenness — Human sterility
Nature The inability of many married couples to have children may be due to the husband or the wife, or both, owing to a variety of conditions, among which many may be described under the term infertility. There may be genetic or pathological sterility, or in the case of the male, impotence and related conditions resulting in the inability to naturally introduce sperm into the female. Male sterility is not uncommon. When it is not responsive to treatment, the couple wishing for the wife to conceive may resort to the use of a surrogate father or artificial insemination. When the female is infertile, the problem of inducing conception is addressed by a number of remedial approaches ranging from pharmaceutical treatments and microsurgery, to transferring fertilized embryos from one woman to another or from test-tube or in-vitro cultivation. There is also the resort to the use of a surrogate mother. Human breeding problems thus give rise to commercial services such as sperm banks, surrogates and their agents, and embryo transfer services. Infertility is particularly hard on women, for whom it attaches as a stigma, and may result in their being ostracized, abandoned, or divorced. Stress on the infertile young married female may lead to mental illness and suicide.
Incidence Infertile couples, unable to have one child or unable to have additional children, number in the tens of millions world-wide. In the USA alone, the number is estimated at 3 million.
Refs Lipshulz, Larry I and Howards, Stuart S (Eds) *Infertility in the Male* (1983); Spark, R F *The Infertile Male* (1988).
Broader Human disease and disability (#PB1044).
Related Boredom (#PA7365)   Impotence (#PA6876)   Unproductiveness (#PA7208)
Unimaginativeness (#PA6738)   Inappropriateness (#PA6852).
Aggravates Childlessness (#PC3280)   Human suffering (#PB5955)
Underpopulation (#PD5432).
Aggravated by Gonorrhoea (#PE1717)   Illegal induced abortion (#PD0159).

♦ **PC6089 Authoritarian division of labour**
Broader Authoritarianism (#PB1638).
Aggravated by Ineffective worker organizations (#PF1262)
Inadequate sense of community and solidarity amongst workers (#PE4179).

♦ **PC6114 Long-term cyclic changes of climate**
Long-term variations in solar radiant energy
Nature Climate changes during the course of long-term cycles which appear to reflect an astronomical forcing of the climatic system by changes in the seasonal patterns of insulation. These are driven by changes in the orbital parameters of the Earth, specifically due to variations in the tilt of the rotation of the axis and due to secular precession of the rotation axis. A complete precessional cycle is described every 26,000 years, and with the slow rotation of the orbit effectively changes the position of the Earth relative to the sun for any particular season. These changes in insulation play a critical role in determining whether ice sheets advance or retreat, as well as affecting the biological productivity of the oceans and the level of carbon dioxide in the atmosphere.
Incidence Major changes in climate occur approximately every 100,000 years, with associated changes of sea level of approximately 100 metres. Other changes occur in cycles of 41,000 years and 23,000 years, with subsidiary changes occurring in cycles of 59,000 years and 11,000 years. Changes in insulation of lesser significance are also expected, at 100,000 and 413,000 years due to changes in the ellipticity of the Earth's orbit. At present the Earth is relatively close to the sun during the northern winter, but the opposite situation will prevail in 11,000 years.
The great climatic changes of the past 2 million years created stress among living things. Forest, grassland and animal populations were repeatedly forced to change and migrate as glaciers waxed and waned. The most significant changes occurred in the sub-tropical deserts of the northern hemisphere. The Sahara was wetter (rock paintings there depict hunters of big game in extensive Savannah grasslands). The Indus Valley civilization of Mohenjo-Daro and Harappa arose during this wetter period, but subsequently faded because of climatic change. In the past 4000 years, this desiccation has predominated in much of the sub-tropical belt.
Refs Jackson, M; Ford-Lloyd, B V and Parry, M I (Eds) *Climate Changes and Plant Genetic Resources* (1990); WHO *Proceedings of the WMO/IAMAP Symposium on Long-term Climate Fluctuations (Norwich, 18-23 August 1975)* (1975).
Broader Change (#PF6605)   Inhospitable climate (#PC0387).
Narrower Global cooling (#PF1744)   Global warming (#PC0918).
Related Bad weather (#PC0293)
Medium-term cyclic variations in solar radiant energy (#PE9528).

♦ **PC6154 Insolvency**
Bankruptcy
Narrower Bank failure (#PE0964)   Business bankruptcy (#PD2591)
Personal insolvency (#PD9376)   Insolvent institutions (#PE6431)
Insolvent local authorities (#PJ5571).

♦ **PC6162 Denial of right to justice**
Broader Injustice (#PA6486)   Denial of human rights (#PB3121).
Narrower Inequality before the law (#PC1268)
Denial of human rights in the administration of justice (#PD6927)
Denial of the right to legal services for indigenous populations (#PE2317).
Related Discrimination (#PA0833)   Denial of political and civil rights (#PC0632).

♦ **PC6189 Vulnerability of developing countries**
External dependence of developing countries — Insecurity of developing countries
Refs Bergauist, James and Manickam, P Kambar *Crisis of Dependency in Third World Ministries* (1976); Subrahmanyam, K (Ed) *Insecurity of Developing Nations* (1986).
Broader Vulnerability of social systems (#PB2853)
International economic injustice (#PC9112).
Narrower Budget deficits in developing countries (#PD3131)
Vulnerability of land-locked developing countries (#PD5788)
Dependence of developing countries on food imports (#PE8086)
Vulnerability of developing countries to inflation (#PD0367)
Dependence of developing countries on customs revenue (#PD2955)
Dependence of least developed countries on foreign aid (#PE8116)
Dependence of developing countries on foreign insurance (#PE4280)
Dependence of developing countries on imported technology (#PF1489)
Vulnerability of island developing countries and territories (#PE5700)
Excessive dependence on export credits by developing countries (#PE0938)
Dependence of island developing countries on official assistance (#PE5724)
Economic dependence of some developing countries on the drug trade (#PE5296)
Instability in export trade of developing countries producing primary commodities (#PD2968)
Dependence of developing countries on external financing for development programmes (#PE7195).
Aggravates Exploitation of dependence on food aid (#PD7592).

♦ **PC6203 Involuntary mass resettlement**
Forced relocation — Transmigration — Relocation of population — Deportation of villages — Forced migration — Involuntary movement of people — Forced deportation
Nature Governments may move communities and tribal groups against their wishes to exploit their land or to break their resistance to assimilation or political domination. Such programmes are usually carried out with little regard for the social impact on family groups or communities, for the health of those concerned or for their means of livelihood at their destination. Villagers may be tempted to accede to such relocation without adequate information having been communicated to them to make an adequate decision.
Incidence Forced resettlements affect indigenous peoples everywhere, as in the case of native Americans in the USA, blacks in South Africa, Palestinians in Israel and aborigines in Australia. In Ethiopia at least 1.5 million people were being moved from arid north under a resettlement programme claimed to be voluntary by the government. Observers estimate that some 100,000 lives have been lost in the process. In Iraq approximately 300,000 kurdis have been forcibly deported and their homes destroyed. Relocation may also be implemented on a smaller scale, as for example when villages are to be flooded as the result of the construction of a dam. Although, sometimes vast numbers of people have been resettled, for example Ghana's Volta Dam saw the evacuation of some 78,000 people from over 700 towns and villages. Forced resettlement also occurs on the occasion of war, as in the case of the 3.25 million Sudeten Germans who settled (or fled) following the World War II, with the agreement of the Allies. Fatalities associated with mass deportation and forced resettlement put the number of victims in the USSR at between 7 and 10 million, and in China at about 27 million.
Counter-claim Relocation may be the only means of providing villagers with agricultural land appropriate to their needs.
Refs Hansen, Art and Oliver-Smith, Anthony *Involuntary Migration and Resettlement* (1982).
Narrower Marriage markets (#PD7282).
Related Back to the land (#PF4181)   Forced mass expulsion (#PD0531).
Aggravates Deforestation (#PC1366)   Displaced persons (#PD7822)
Resettlement stress (#PD7776)   Socially inappropriate housing (#PD8638)
Cultivation of marginal agricultural land (#PD4273)
Non-payment of compensation for forced relocation (#PE8898)
Insensitivity to diversity of cultural traditions (#PF8156).
Aggravated by Fear of resettlement (#PF9030)   Government insensitivity (#PF2808)
Enforced collectivization of agriculture (#PD7443)
Environmentally harmful dam construction (#PD9515)
Ecosystem modifications due to creation of dams and lakes (#PD0767).
Reduces Maldistribution of population within countries (#PC8192).

♦ **PC6215 Excessive growth of social expenditure**
Nature Spending for social welfare, health, pensions and education takes an increasingly larger share of major market economies' Gross Domestic Product. The burden of these expenditures contributes to budget deficits; unmanaged and uncontrolled, they may become unaffordable with disastrous consequences.
Incidence Among the seven largest OECD countries between 1961 and 1980, social expenditures rose from 15 percent of GDP to 23 percent, and are projected to reach 25 percent by 1987. The most recent figures indicate that pensions account for two-fifths of expenditures, with health care and education accounting for one-fifth each. Unemployment payments represent only one-twentieth of these outlays. However, there is considerable variance within OECD countries at the total level of social expenditure. For example, social expenditures as a percentage of GDP are Belgium: 38.0; Netherlands: 36.1; Sweden: 33.5; and Germany FR: 31.5. Of the fifteen remaining countries, seven had expenditures above the 1981 average of 24 percent. The lowest, for comparison, were Greece (12.8), Switzerland (14.9), Japan (17.5), Australia (18.6), New Zealand (19.6), USA (21.0), and Canada (21.7).
Aggravates Decline in government expenditure (#PF9108).
Aggravated by Inappropriate use of financial resources (#PD9338).
Reduces Decline in government social expenditure (#PF0611).

♦ **PC6239 Offences against the peace and security of mankind**
Crimes against peace
Nature Inter-state disturbances of peace may occur from unilateral military and paramilitary initiatives, which include the use of regular or irregular national armed forces or threat of use of military mobilization, or the organization, support or encouragement of non-national armed forces or irregular armed groups or terrorists hostile to another state, whether nationals of that state or not, and wherever their headquarters and operational bases. Non-military unilateral offences include initiating or encouraging psychological warfare, propaganda, and other activities, wherever emanating, calculated to foment civil strife elsewhere. Breaches of international law that are offences against the peace and security of states, are: any unilateral actions in violation of the provisions and spirit of international treaties designed to ensure respect for national sovereignty and international justice; illegal annexation of territories; and coercive economic or political measures enacted or threatened in order to intervene in another state's domestic or foreign affairs by extorting the behaviour desired. Intra-state disturbances of human peace and security of concern to the entire world include genocide in any form; mass-murdering of groups, classes, or randomly selected victims; and any harm perpetrated as a policy against the citizenry in general, or any societal segments. These include physical and mental abuse, internment, torture, deprivation of food, clothing, shelter, medical care, and freedom of movement; enslavement; forced sterilizations, abortions or removal of children from the family; internal or external exile; and restrictions as to education and employment. The foregoing offences may be the culpability of governments and the authorities of which they are constituted. Guilt may also pertain to offences of private individuals acting at the instigation or with the toleration of authorities. Other offences against the peace and security of mankind include atmospheric and ocean testing of

PC6239

nuclear weapons; the development and testing of weapons of mass-destruction in general, and any other acts that affect the global environment.
Incidence Offences against the peace and security of mankind may be enacted or threatened multilaterally. In one case it may be by the alliance of two or more powers, overtly or covertly; in other case two or more states may disturb world security by independently aggressive actions, or two or more groupings of states may offend the peace. Other offences may be made intergovernmentally, involving common action through a military treaty organization, or through improper decisions made by the United Nations Organization in respect to military interventions or other actions such as sanctions.
Claim Intergovernmental organizations may offend against the peace and security of mankind by perpetuating themselves even though they may be ineffective against the fundamental problems of population growth and increasing militarization in the world. Temporary UN peace-keeping forces may be created as a result of manipulation of the major powers, and forced upon a smaller state. Transnational corporations and cartels may offend against the peace and security of mankind by distorting world trade, by impoverishing or financially controlling small, developing countries, and by exploiting food, health and other consumer needs with overpriced, inferior, or dangerous goods.
Broader International conflict (#PB5057) International aggression (#PC7559).
Narrower Annexation (#PE5210) Threat of war (#PF8874)
Incitement to war (#PD4714) Conventional warfare (#PC4311)
Psychological warfare (#PC2175) Political mass murder (#PD5590)
Nuclear weapons testing (#PC2201) Cross border military operations (#PD5272).
Related Torture (#PB3430) Genocide (#PC1056).
Aggravated by Fragmentation (#PA6233) Unfulfilled treaty obligations (#PF2497).
Reduced by Conflict resolution (#PS3582).

♦ PC6242 Conflicts of integrated rural development
Nature Excessive urbanization and the concentration of investment in urban-oriented, capital-intensive activities have produced a severe imbalance within developing countries between the urban elite groups that monopolize power and wealth, and the majority of the rural population that remains poor; this despite the fact that the cost of bringing a person to an urban area is often much greater than the cost of providing social infrastructure to keep that person in the rural area. Rural industrialization is not meeting its aim of bringing city comforts to the rural areas, thereby creating a rural-urban continuum; it is, instead, creating a rural-urban conflict.

♦ PC6262 Over-education
Nature In industrialized countries particularly, the education of unemployed youth is at unprecedentedly high levels, and there are immense problems in finding jobs for graduates. In the past decade educational systems have produced more persons with higher educational credentials than there are job opportunities that can utilize such training. Thus, the unemployment rates of college and university graduates is rising; equally prevalent is the shift of such persons to occupations that traditionally have not required a college education. Graduates themselves are likely to have greater expectations with regard to their occupational attainments than the labour market can fulfil, and thus their job dissatisfaction is greater, with its deleterious consequences for productivity.
Broader Ineffective systems of practical education (#PF3498).
Aggravates Unemployment (#PB0750).
Reduced by Obstacles to education (#PF4852).

♦ PC6307 Abuse of influence
Trading in special influence — Influence peddling
Nature Trading in special influence is bribing a relative or other person with influence over a public servant or party official with respect to his legal or official duties.
Broader Abuse of power (#PB6918) Political bribery (#PC2030).
Narrower Use of undue influence to obstruct the administration of justice (#PE8829).
Aggravates Abuse of government power (#PC9104).
Aggravated by Abusive distribution of political patronage (#PF8535).

♦ PC6329 Energy crisis
Energy — Energy shortage
Nature World energy consumption is expected to rise by 75 percent between 1979 and 2000. Although oil consumption may rise by 30 percent over this period, oil's share of the energy balance is expected to fall from 45 percent to one third; filling the resultant gap may well prove to be the real energy crisis of the end of this century. Although rates of increase in energy use have been declining, the industrialization, agricultural development and rapidly growing populations of developing countries will need much more energy. However to bring the developing countries' energy use up to that of industrialized countries by the year 2025 would require an increase in the current global energy use by a factor of five. The planetary ecosystem could not stand this, especially if the increases were based on non-renewable fossil fuels. The current threats of global warming and acidification of the environment also tend to preclude even a doubling of energy use based on the present pattern of energy supply.
Background The energy crisis dates from the early 1970's and when Libya successfully challenged the world oil production-pricing policies traditionally dictated by the international oil companies. By 1980 the international oil companies had lost control of the world oil markets. Explorations in such areas of Alaska, the Arctic, the North Sea, and the continental USA failed to procure great benefits, and OPEC and Third World proven oil reserves failed to grown; thus the possibility of oil running out within 50 years became statistically demonstrable.
Counter-claim The so-called energy crisis owes more to politics, prices and technological gaps than to physical shortages of crude oil and other fuels. There are no shortages in fuels, because when prices of oil bases fuels rise enough effective alternatives will be found.
Refs Gordon, Richard L *An Economic Analysis of World Energy Problems* (1981); Kydes, A S and Geraghty, D M *Energy Markets in the Longer Term* (1985); United Nations Conference on Trade and Development *Characterization and In-dept Analysis of the Origins of the Current Crisis of the Market and of the Primary Tungsten and Intermediate Products Industry* (1987).
Broader Global crisis (#PF6244).
Narrower Shortage of firewood (#PD4769) Underdeveloped rural energy sources (#PF0393)
Energy dependence and vulnerability (#PJ7735) Insufficient nuclear power stations (#PD7663)
Lack of acceptable sites for power plants (#PE7519)
Use of agricultural land for fuel production (#PD6103)
Insufficient diversification in energy research (#PG5347)
Underutilization of oil shale as an energy source (#PF0445)
Underdevelopment of the power industry in developing countries (#PF4135)
Underproductivity of draught animal power in developing countries (#PF0377).
Related Acaricides as pollutants (#PG1836) Erosion of university autonomy (#PG6036)
Environmental hazards of decommissioned nuclear power plants (#PE7539).
Aggravates Exploitation of fossil fuels (#PE4891)
Landscape disfigurement from open-cast mining (#PD1637)
Proliferation of nuclear weapons and technology (#PD0837).
Aggravated by Friction (#PF1691) Oil as a pollutant (#PE2134)
Thermal pollutants (#PC1609) Inadequate cooking stoves (#PE7904)
Underutilization of solar energy (#PF0370) Underutilization of ocean energy (#PG5350)
Maldistribution of energy consumption (#PC5038) Stockpiles of nuclear warfare material (#PC0311)
Long-term shortage of natural resources (#PC4824)
Unsustainable development of energy use (#PC7517)
Proliferation of automobiles and motor vehicles (#PD2072)
Insufficient and inappropriate energy equipment in developing countries (#PE8592).

♦ PC6341 Extremist ideologies
Discriminatory ideologies
Incidence Extremist ideologies exist in all parts of the world: in the USA there is the Ku Klux Klan and the American Nazi Party, the League of Brothers operates in South Africa, the Revival Party (Fehiya) in Israel is based on racist principles, and the governments of Central and South American countries operate from military control that is synonymous with exclusiveness and intolerance.
Claim Extremist ideologies such as racism, terrorism, neo-nazism, neo-fascism, are based on a systematic denial of human rights and fundamental freedoms.
Broader Ideological conflict (#PF3388).
Narrower Sexism (#PC3432) Fascism (#PF0248) Neo-fascism (#PF2636)
Fascist liberalism (#PF9710).
Related Aggressive ideologies (#PJ4778).
Aggravates Racism (#PB1047).
Aggravated by Double standards in morality (#PF5225).

♦ PC6381 Insufficient primary education
Lack of preparatory education — Denial of right to free primary education
Nature It seems unlikely that the majority of developing countries will be able to introduce universal primary education, much less adequate primary education, in the forseeable future. Worldwide, but especially in developing countries, millions of children who are in principle able to enroll in primary school nonetheless receive an incomplete basic education, due to the necessity of travelling long distances to school, of working at home, the inability to pay, or the lack of either motivation or materials.
Incidence Half the children in the developing countries today do not receive a full cycle of primary education.
Refs Venkatasubramanian, K *Wastage in Primary Education* (1978).
Broader Lack of education (#PB8645) Inadequate education (#PF4984).
Related Restricted higher education (#PJ0572) Denial of right to education (#PD8102)
Denial of right to educational choice (#PE4700).

♦ PC6440 Conflict of interests in imperialism
Claim Dominance relations between nations, hegemonies, blocs and other political collectivities is a function of the structures and interests of the collectivities, of which imperialism arising out of an industrialist need for expanding markets is one special case. This case, however, is illuminated by the dynamics of the several interests, particularly where the objectives or even their methods or methodological theories differ. In the case of objectives or goals, the incompatibility of these within the relationship may not be initially apparent. In fact, actors in such relationships may not even know, or fully know, where their end interests lie. Some may already have been subordinated in a relationship to the extent that their consciousness of autonomous, particular objectives may have been distorted or subverted to serve the ends of the dominant member. In other cases, autonomous goals may be hidden or suppressed as a supposed strategem.
Whatever the aetiology of conflict, its symptoms are clearly indicated by a condition of comparative disadvantage or limitation in the subordinate constituencies. These disadvantages may be political, economic, social or cultural. When the gap between the advantages of the dominant members and the disadvantages of the subordinate members is very large, or is constantly growing, or the disadvantages affect the life and health of individuals or the real sovereignty of nations, the dominant partner in the conflict is imperialist, whether by active intent or by omission, to correct the tendency in relationships for some to sink into subordination. Much of the continuing imperialism today is due to neglect of international responsibilities by the powers, reinforced by self-interest and aggravated by the disarray among the developing countries among whom there are also conflicts of interests.
Broader Imperialism (#PB0113) Conflicts of interest (#PF9610).

♦ PC6445 Insufficient skills
Narrower Unprepared adult leadership (#PF6462)
Shortage of technical skills (#PF6500)
Limited construction manpower (#PG9875)
Narrow range of practical skills (#PF2477)
Underdeveloped technological skill (#PF8552)
Inadequate management skills in rural communities (#PF1442)
Lack of skilled manpower in rural areas of developing countries (#PE5170)
Shortage of industrial leadership and entrepreneurial ability in developing countries (#PC1820).
Aggravated by Lack of training (#PD8388) Unorganized transfer of skills (#PJ8603)
Lack of sharing of community skills (#PF3393)
Unrecognized need for functional skills (#PF2995).

♦ PC6631 Inadequate local government financing
Excessive reliance of local governments on grants
Nature Inadequate local government financing is caused and compounded by several factors. Towns tend to be caught in the dilemma that, while central government (due to the worsening economic situation or to ideological reasons) is unable or unwilling to fulfil its obligations, there is also reduction of their own revenues, caused by the exodus from the inner city. Taxpayers who leave tend to be relatively more prosperous than those who stay, so that government yield of local taxes decreases while an increasing percentage of expenditure goes to the support of the needy inhabitants who remain in the city.
Excessive reliance by local government on grants, or on unexpected increases in them, can result in poor use of public finances. Grants can encourage recipients to be less efficient and can decrease the fiscal autonomy of local regions.
Broader Compliance*complex (#PA5710).
Aggravated by Bureaucracy as an organizational disease (#PD0460).

♦ PC6641 Underground economy
Unreported income and business activities — Development of parallel economies — Black market economy — Shadow economy — Black market trading — Unreported financial transactions — Underground and illicit business — Black-markets
Incidence Unreported business activities and cash transactions occur in most countries and are the continuing subject of investigation by tax authorities. These activities include drug trafficking, gambling, prostitution, loan-sharking, bribery, use of state property for private gain, employment of workers without a permit and without paying social costs and misappropriation of public funds. In India, for example, it has been estimated that the underground economy is financially as important as the official economy.
Refs De Soto, Hernando *The Other Path* (1989).
Broader Tax evasion (#PD1466).
Narrower Currency black market (#PD5905) Urban informal sector (#PG4719)
Clandestine employment (#PC7607)

lack market (#PD5905)
Urban informal sector (#PG4719)
Clandestine employment (#PC7607)
Illegal private profit in socialist countries (#PC0939)
Development of informal sector in developing countries (#PD7978).
Aggravates Corruptive crimes (#PD8679) Unrecorded knowledge (#PF5728).
Aggravated by Unreported businesses (#PJ1192)
Economic inefficiency (#PF7556)
Marginal level of family income (#PD6579)
Lack of economic and technical development (#PE8190).

♦ PC6664 Genetic and ethnic weapons

Dependence on genetic and ethnic weapons — Race weapons

Nature It is theoretically possible to develop ethnic chemical weapons which would exploit naturally occurring differences in vulnerability among specific population groups.

Incidence During the Vietnam War, an elite group of scientists working for the Pentagon was employed to carry out blood tests on select groups of Asians in order to 'prepare a map portraying the geographic distribution of human blood groups and other inherited blood characters'. The US Department of Defence has been involved in research on coccidioides immitis, commonly known as Valley Fever. The disease can result in a mortality rate of 50–60 percent, but only 1–11 percent of whites will develop the fatal form while 20–59 percent of blacks will do so. The SADF may currently be researching viruses, chemicals, and diseases which will affect only blacks.

Background In November 1970 an article under the title of 'Ethnic Weapons's appeared 'Military Review', the professional journal of the US Army. The author states; 'In brief, human populations can be characterized by frequencies of distinct genes. Sometimes, gene frequencies agree fairly well between widely dispersed populations, but more often there are great differences'. He cites an example, 'Recently a series of widely debated observations have revealed an enzyme deficiency, in Southeastern Asian populations, making them susceptible to a poison to which Caucasoids are largely adapted', and concludes that 'the prospect may tempt an aggressor who knows he can recruit from a population largely tolerant against an incapacitating agent to which the target population is susceptible'.

Broader Planned weapons (#PF9269).
Aggravates Decline in human genetic endowment (#PF7815)
Vulnerability of world genetic resources (#PB4788).

♦ PC6666 Radiological warfare

Use of radiation as a weapon

Nature The use of radiological warfare is becoming a viable possibility as more countries are gaining the technology and materials needed for radioactive weaponry. Radiological warfare, which is radioactivity (excluding nuclear bombs) as a means of warfare, could be used on the defensive (to make an area impassable for enemy troops, to halt an armoured attack, to inhibit a river crossing, or to close mountain pass); or on the offensive (to force evacuation from cities, industrial areas and communication centres, thus a wreaking havoc with the economy and fighting potential of the country attacked); or it could simply be used to kill a large number of people.

Related Nuclear war (#PC0842).
Aggravates Radioactive contamination (#PC0229).
Aggravated by Weapons (#PD0658)
Irresponsible scientific and technological activity (#PC1153).

♦ PC6675 Environmental consequences of war

Military disruption of the biosphere — Pollution from military activities — Degradation of the environment by armed conflict

Incidence Environmental consequences of current and past wars include: hazards from unexploded weapons; physical and biological effects of damage to soil and landscape; and human suffering resulting from the disruption of social systems. The effects would be most devastating in the case of nuclear warfare. The long-term effects of chemical and biological warfare on a large-scale remain to be determined.

Refs Westing, Arthur H (Ed) *Cultural Norms, War and the Environment* (1988); Westing, Arthur H (Ed) *Herbicides in War* (1984); Westing, Arthur H (Ed) *Explosive Remnants of War* (1985).

Broader Environmental degradation (#PB6384) Disastrous consequences of war (#PC4257).
Narrower Environmental hazards of nuclear weapons industry (#PE5698).
Related Housing destruction in war (#PE2592) War damage in civilian areas (#PD8719)
Industrial destruction by war (#PD8359)
Environmental degradation through military activity during peace-time (#PE0736).
Aggravates Environmental stress (#PC1282)
Inadequate results of formal schooling (#PF6467).
Aggravated by Military-industrial complex in capitalist systems (#PC3191).

♦ PC6684 Geological hazards

Telluric and tectonic disasters

Nature Most geological processes, including the formation mineral bodies, are extremely slow, and even major tectonic movements are measured on the order of only centimetres per year. Few of these processes provide changes readily detectable in a decade. From time to time, however, they generate extreme natural events such as earthquakes, volcanic eruptions and tsunamis. These, with landslides and snow avalanches, constitute geological hazards which emphasize the natural variability in environmental systems, remind human communities of the great power of natural forces, and test the abilities of people and nations to respond to them.

Incidence The Mexico City earthquake of September 1985, killed approximately 5000 people, while the volcanic eruption that decimated Armero, Colombia (November 1985) claimed 25,000 lives.

Refs Steinbrugge, Karl V and Busch, Charles U *Earthquakes, Volcanoes, and Tsunamis* (1982).

Broader Natural disasters (#PB1151).
Narrower Tsunamis (#PD0033) Earthquakes (#PD0201) Volcanic eruptions (#PD3568)
Earth surface faulting (#PE5096).

♦ PC6716 Occupational hazards

Hazardous occupation

Broader Hazards to human health (#PB4885).
Narrower Confined spaces (#PJ7974) Lead as a health hazard (#PE5650)
Health hazards of asbestos (#PE3001) Manganese as a health hazard (#PE1364)
Vibrations as a health hazard (#PE1145) Heat as an occupational hazard (#PE5720)
Cold as an occupational hazard (#PF5744) Dust as an occupational hazard (#PE5767)
Sulphur dioxide as a pollutant (#PE1210) Occupational hazards of benzene (#PE1849)
Mental stress due to automation (#PE5164) Occupational hazards in farming (#PE3760)
Hazards to young people at work (#PE6868) Occupational hazards of painters (#PE9746)
Solvents as an occupational hazard (#PE5708) Carbon monoxide as a health hazard (#PE1657)
Occupational hazards of male workers (#PE6904)
Occupational hazards of female workers (#PE6902)
Biological agents as occupational hazards (#PE5696)
Combined stresses as occupational hazards (#PE5656)
Improper lighting as an occupational hazard (#PE5780)
Shift work stress as an occupational hazard (#PE5768)
Occupational hazards in the mining industry (#PE8428)
Excessive occupational exposure to radiation (#PD1500)
Occupational dangers in developing countries (#PD6885)
Polychlorinated biphenyls as a health hazard (#PE2432)
Occupational hazards in commerce and offices (#PE3957)
Environmental hazards of non-ionizing radiation (#PE7651)
Health hazards of computer visual display units (#PE5083)
Hazardous combination of effects in the work place (#PE8972)
Occupational hazards to workers in small industries (#PE7528)
Occupational risks and hazards of the medical profession (#PE5355).
Related Dangerous occupations (#PC1640) Occupational risk to health (#PC0865)
Inhaling of solvents and anaesthetic drugs (#PE1427).
Aggravated by Explosive substances (#PG7855)
Toxic metal pollutants (#PD0948)
Badly laid out work premises (#PJ2468)
Human errors and miscalculations (#PF3702)
Hazardous industrial installations (#PE4304)
Inflammable and flammable substances (#PJ3667)
Conflicting standards for protection against chemical occupational hazards (#PE5651).

♦ PC6720 Voluntary unemployment

Nature Voluntary unemployment has been described as a 'bourgeois problem', as it usually involves members of the educated classes who prefer to remain unemployed rather than accept work below the type they feel they can expect to obtain or at a wage lower than that paid to other workers with comparable levels of education.

Counter-claim If such people did accept less qualified and less well paid work, it would lower their probability of getting the desired employment later and would therefore lower their expected lifetime income.

Broader Unemployment (#PB0750).

♦ PC6727 Denial of right of a people to be self-determining

Broader Denial of rights (#PB5405).
Narrower Annexation (#PE5210) Colonialism (#PC0798)
Occupied nations (#PC1788) Occupied territories (#PD8021)
Alien domination of peoples (#PC7384) Lack of political independence (#PF0297)
Denial of rights to territories (#PD6620)
Denial of right to a people to live in peace (#PD5253)
Denial of right to national self-determination (#PC1450)
Denial of right to a people to pursue development (#PE1536)
Denial of rights of indigenous people to be self-governing (#PE1024)
Denial of right to a people to their own means of subsistence (#PE4399)
Denial of right to a people to freely dispose of natural wealth (#PE6955).

♦ PC6757 Secret laws

Secret regulations — Secret government procedures

Nature It is inconsistent with fundamental notions of justice to convict or punish persons for violations of laws, decrees, or regulations which have been kept secret, or have not received sufficient publication to give notice to such persons, or whose contents is not available or accessible to the general public or persons affected by such laws.

Aggravates Injustice (#PA6486) Conflict of laws (#PF0216)
Proliferation of legislation (#PD5315) Secret international agreements (#PF0419).

♦ PC6759 Socioeconomic stress

Nature Income and occupation; standards of housing, sanitation and nutrition; and the level of provision of health, educational, recreational and other services, may all be used as measures of socioeconomic stress. While poverty causes illness by depriving man of his basic needs of shelter and adequate nutrition, the type of poverty plays a large role in determining the type and extent of stress. People born into poverty are less susceptible to mental disorders than those upon whom poverty has been thrust by misfortune; the sense of social rejection and injustice felt by those who have memories of something better is a potent medium for psychological disorders resultant from socioeconomic stress.

Incidence Studies and reports show that persons committing or attempting suicide are overrepresented among the more affluent. A high incidence is associated with loss of occupation and lack of work for reasons other than illness; and ischaemic heart disease is also more prevalent in upper socioeconomic groups.

Broader Stress in human beings (#PC1648).
Narrower Culture shock (#PC2673) Occupational stress (#PE6937)
Plural society tensions (#PF2448).
Aggravates Social conflict (#PC0137) Ethnic conflict (#PC3685).
Aggravated by Socio-economic poverty (#PB0388).

♦ PC6761 Distortion in international trade

Broader International economic injustice (#PC9112).
Narrower Trade protectionism (#PC4275) Tariff barriers to international trade (#PC0569)
Non-tariff barriers to international trade (#PC2725)
Collusive tendering in international trade (#PE7072)
Conditional observance of multilaterally agreed trade commitments (#PF7838).

♦ PC6848 Inadequate laws

Insufficient law

Broader Inadequacy (#PA8199)
Inadequacy of governmental decision-making machinery (#PF2420).
Narrower Extradition refusal (#PF2645) Compulsory education (#PJ2615)
Obsolete legislation (#PF5435) Ineffective tax systems (#PF1462)
Inadequate drug control (#PC0231) Restrictive legislation (#PD9012)
Illegal induced abortion (#PD0159) Compulsory sterilization (#PF3240)
Inadequate laws of the sea (#PF5923) Inadequate laws of adoption (#PD0590)
Criminalization of drug use (#PF4735) Inadequate safety legislation (#PJ2716)
Inadequate firearm regulation (#PD1970) Lack of protective legislation (#PJ2889)
Criminalization of prostitution (#PF6231) Inadequate narcotics legislation (#PF6787)
Deficiencies in international law (#PF4816) Health hazards of exposure to noise (#PC0268)
Inadequacy of international legislation (#PF0228) Lack of anti-discrimination legislation (#PF7972)
Inadequate legislation for animal welfare (#PE5794)
Lack of legislative control on advertising (#PE8467)
Ineffective legislation against organized crime (#PE6699)
Inadequate legislation relating to action against problems (#PF1645).
Related Inhumanity of capital punishment (#PF0399).
Aggravates Meaninglessness (#PA6977) Proliferation of legislation (#PD5315)
Inadequate prevention of crime (#PF4924).
Aggravated by Inadequately worded agreements (#PF5421)
Political barriers to effective legislation (#PC3201)
Delays in elaboration of remedial legislation (#PC1613).

♦ PC6869 Unequal morbidity and mortality between countries

Nature There are marked differences, closely linked to social and economic policy, in mortality

patterns between the developed and developing countries. In the former, there is a sharp drop in death rate from the first year, while in the developing countries high death rates persist throughout early childhood.
Incidence The crude death rate for the world as a whole declined from an annual average of 19.7 per 1000 population during 1950–1955 to 10.6 per 1000 during 1980–1985 and is projected to fall to 9.1 per 1000 by 1995–2000. For the developing countries, the average annual crude death rate declined from 24.4 per 1000 during 1950–1955 to 11.0 per 1000 in 1980–1985 and is projected to decline to 9 per 1000 by 1995–2000.
Broader Human death (#PA0072)   Inequality in mortality rates (#PC9586).
Narrower Unequal mortality of the elderly between countries (#PE4354).
Related Inequality of life expectancy by gender (#PE1339)
Unequal regional distribution of deaths (#PC4312).
Aggravated by Malnutrition (#PB1498)   Unhygienic conditions (#PF8515)
Socio-economic poverty (#PB0388)   Inadequate medical care (#PF4832)
Human disease and disability (#PB1044).

♦ **PC6870 Denial of the right to procreate**
Fertility rights — Denial of right to found a family
Nature By means of coercion such as loss of maternal and child health benefits or educational assistance, tax benefits, or the imposition of penalties for going beyond a given number of children, governments deny their citizens the right to procreate. Subtle tactics such as antinatalist propaganda campaigns and/or quotas assigned to family planning workers are also employed.
Incidence China's One Child Policy is an extreme example of governmental intervention into procreation. Conversely, there was the Reagan administration's decision to cut off funding for family planning in countries where part of that funding goes to family planning clinics where abortions are performed.
Claim The individual's right to procreate is based on the fundamental tenet underlying all human rights – freedom of choice. Men and women should have the right to freely decide the number and spacing of their children and the right to the information, education, and means to do so.
Counter-claim Governments have a responsibility to provide people with improved standards of living, and population size and growth obviously affect those standards. Thus, if a government feels that an increase in population will hinder basic living improvements, it should have the prerogative to inform and try to coerce its people against procreation.
Broader Denial of human rights (#PB3121).
Narrower Denial of the right to procreate to the severely mentally handicapped (#PE4544).
Related Denial of right to life (#PD4234)   Denial of right to liberty (#PF0705)
Denial of right to a family (#PE7267)
Denial to animals of the right to conditions of life and liberty proper to their species (#PE6270).
Aggravates Human suffering (#PB5955)   Sexual unfulfilment (#PF3260).
Reduces Excessively large families (#PD7625)   Unsustainable population levels (#PB0035).

♦ **PC6873 Abuse of economic power**
Dependence on abuse of economic power
Broader Abuse of power (#PB6918).
Narrower Consumerism (#PD5774)   Abuse of banking secrecy (#PF5991)
Abuse of in kind payments (#PF6723)   Abusive use of trademarks (#PG3572)
Abuse of individual property (#PJ8392)   Exploitation in rural pricing (#PG8423)
Concentration of investment power (#PC5323)   Financial and industrial oligarchy (#PD5193)
Economic sanctions against governments (#PF4260)
Misuse of free production zones and export enclaves (#PE2858)
Abuse of dominant market position in international trade (#PD6002)
Abuse of monopoly power of state–owned or state–controlled enterprises (#PE0988).
Related Economic repression (#PC8471).

♦ **PC6875 Economic manipulation**
Dependence on economic manipulation
Nature Economic manipulation includes: illicit manipulation of commodity prices; speculation; hoarding; violations of exchange regulations; breach of governmentally fixed prices, of profit margins or of price freezes; sabotage or impairing the smooth functioning of the national economy; charging excessive trade or professional fees; and improper procurement practices.
Broader Manipulation (#PA6359).
Narrower Financial manipulation by sects (#PE6642)
Manipulation of commodity markets (#PD8647)
Manipulation of transfer prices by transnational corporations (#PE0245).
Aggravated by Ignorance (#PA5568)   Domination (#PA0839).

♦ **PC6912 Abusive experimentation on humans**
Human experimentation — Experimental exposure of humans to radiation
Nature Human subjects for experimentation (in order to acquire knowledge rather than improve the subject's condition) may be coerced into participation or participating unknowingly. They are ill-informed of potential effects; they may be unable to end their participation once the experiment has begun, even though there is a chance of permanent damage; and they may be unable to receive compensation should they be injured.
Background Past examples of abusive experimentation on humans include experiments on twins, dwarves and pregnant women in Nazi concentration camps; and Unit 731, a Japanese biological warfare centre during World War II, where experiments were carried out on on Asian and allied prisoners. Moral codes to be followed when human subjects are used for experimentation, have been adopted by the Nuremberg Tribunal; the Helsinki Declaration; and USA, UK and French medical associations.
Refs Katz, Jay *Experimentation with Human Beings* (1972).
Broader Inhumane scientific activity (#PC1449)
Irresponsible scientific and technological activity (#PC1153).
Narrower Medical experimentation on children (#PE6764)
Unethical medical experimentation on prisoners (#PE4889)
Inhumane medical experimentation during war-time (#PE4781)
Unethical experiments with drugs and medical devices (#PD2697)
Medical experimentation on mentally impaired persons (#PE8677)
Medical experimentation on socially vulnerable groups (#PD6760)
Medical experimentation on institutionalized subjects (#PE6763)
Medical experimentation on pregnant women and foetuses (#PE8343).
Related Undignified treatment of corpses (#PF5857)
Denial to experimental animals of the right to freedom from suffering (#PE8024).
Aggravates Trafficking in children for medical exploitation (#PD4271).
Aggravated by Unethical practice of radiology (#PD8290).
Reduced by Obstacles to medical experimentation (#PF4865).

♦ **PC6913 Dangerous substances**
Dangerous materials
Refs Japan Container Association (Ed) *Index of Dangerous Substances Classified in Various Regulations* (1975).
Broader Risk (#PF7580).
Narrower Corrosive substances (#PE6887)   Storage of dangerous substances (#PG3713)
Dangerous materials in frontier areas (#PE6860)
Inadequate labelling of dangerous substances (#PD2468)
Over-use of formaldehyde in building materials and personal products (#PE6214).
Aggravates Industrial accidents (#PC0646)   Hazards to human health (#PB4885)
Aggravated by Instability (#PA0859)   Man-made disasters (#PB2075)
Environmental pollution (#PB1166).

♦ **PC6935 Prisoners of conscience**
Nature Men and women may be imprisoned because of their political or religious beliefs or because of their colour or ethnic origin, although they have never used nor advocated violence. Most prisoners of conscience are detained for trying to exercise their rights of freedom of expression, association, assembly or movement. Some are held simply because of the political activity of members of their family. However, few governments admit that they detain people in violation of internationally agreed norms, and consequently treat prisoners of conscience as criminal offenders.
Broader Excessive virtue (#PF7127)   Political prisoners (#PC0562).
Related Squeamishness (#PF5735).
Aggravates Social injustice (#PC0797).
Aggravated by Denial of right of conscientious objection to military service (#PD1800).

♦ **PC6989 Delays in implementation of social change**
Broader Delay (#PA1999).
Narrower Delay in project implementation (#PF1470)
Delayed development of regional plans (#PF2018)
Delay in societal impact of education (#PE8318)
Delay in implementation of commitments (#PF3975)
Delay in societal impact of innovation (#PF8870).
Aggravated by Resistance to change (#PF0557)
Government resistance to institutional change (#PF0845).

♦ **PC7013 Loss of civility**
Incivility — Discourtesy — Bad manners — Rudeness
Nature When the members of a society lose consideration for each other it may be exhibited by indifferent behaviour, a decline in politeness, increasing rudeness and, ultimately, a propensity for violence. A culture of incivility exists in localities in many major cities. This is part of a spectrum which includes rude language, disregard for the rights of others, open violation of the law, aggressiveness, violence and crime.
Incidence In New York City, for example, a high degree of aggressiveness, violence and crime is tolerated. People watch passively as crimes such as murder, robbery and assault are committed and cries for help go unheeded. In the United Kingdom, there has been a marked increase in discourtesy, including hooliganism at sports events, swearing in public, aggressive and even dangerous driving, and destruction of public property.
Broader Behavioural deterioration (#PB6321).
Related Vulgarity (#PA5821)   Aggression (#PA0587)   Disrespect (#PA6822)
Discourtesy (#PA7143)   Misbehaviour (#PA6498)   Inhospitability (#PJ7162)
Disruptive behaviour (#PD8544).
Aggravates Decline in deference (#PF9618).
Aggravated by Malevolence (#PA7102)   Ethical decay (#PB2480)
Dehumanization (#PA1757)   Institutional lying (#PD2686).

♦ **PC7041 Worker maladjustment to technology**
Nature When workers with traditional skills are faced with highly advanced machinery and equipment as working partners, they find difficulty, when required, in overriding machine operation, or decisions. The workers feel diminished, feel that responsibilities are removed from them, and that machines set an intolerable pace independent of human decision-making. Employers do not know what skills are required in automated and semi-automated operations for workers, often do not hire the right people, and usually provide training inadequate to instil a sense of security and competence in those that are involved in the man-machine interface. There is both a human and a financial cost to such worker maladjustment.
Broader Lack of job satisfaction (#PF0171).
Aggravates Underproductivity (#PF1107).
Aggravated by Training inappropriate to structural and technological changes (#PE4596).

♦ **PC7101 Religious torture**
Nature Religious practices may require the inflicting of pain on one's self or others, as a test, as an element of sacrificial worship, as self-mortification, as discipline or punishment, as a consequence of rituals whose purposes are purificatory or sacramental, or from customs that are binding in the community and connected with its welfare. The methods of inflicting pain may be so unusual, or the instances of its application so frequent or humiliating, as to constitute a form of torture.
Incidence Religious torture in modern times exists in some inhumane applications of Koranic law to its infringements such as amputations, tongue-cutting and flogging in Sudan, Iran and elsewhere, and in the rites of circumcision both in Islam and Judaism. In Christianity an ascetic strain persists, ranging from a small Orthodox sect that practices self-mutilation, and snake-bitten Fundamentalists who prove their faith by handling poisonous serpents, to cloistered orders whose Rule and daily Hours includes solitary confinement, silence, hunger, self-flagellation and sleeplessness. Religious torture occurs among tribes that tattoo and deform themselves, or slowly kill their enemies. Some new cults imprison members, brainwash and torment them. These latest forms of religious mental torture include incessant mutual criticism, mutual spying and reporting on behaviour, and sophisticated psychological conditioning. They seem to parallel ideological indoctrination techniques in general.
Constant attacks on the religious beliefs of a group or individual in an attempt to erode these beliefs can be considered religious torture. This may involve violation of dietary practices, such as forcing Hindus or Buddhists to eat meat or Jews or Muslims to eat pork. Victims may be forced to defile sacred objects, such as icons; or witness mockeries of sacred rituals. People who believe in the sacredness of human life may be forced to witness or participate in murder.
Broader Human torture (#PC3429).
Related Physical torture (#PD8734)   Ideological conflict (#PF3388)
Pharmacological torture (#PE4696)   State sanctioned torture (#PD0181)
Institutionalized torture (#PD6145).
Aggravated by Double standards in morality (#PF5225).

♦ **PC7112 Unjust laws**
Legislative injustice — Laws in violation of human decency — Unjust legislation
Incidence Since the time of the Conquest, alien laws and regulations had been imposed on the Guatemalan people – a legal system that protected the conquerors and destroyed the centuries-old organizational structures of the indigenous population. For nearly five centuries the system for the administration of justice has served those who held economic and political power.
Broader Social injustice (#PC0797).
Narrower Inadequate evidence to convict known offenders (#PF8661).
Aggravates Informers (#PD8926)   Civil disobedience (#PC0690)
Outdated procedures (#PF8793)   Punishment of criminals by mutilation (#PE3488).

♦ PC7203 Endangered indigenous cultures
Destruction of indigenous cultures
Broader Endangered cultures (#PB8613).
Narrower Abuse of traditional customs (#PJ0675)
Degradation of indigenous cultures (#PJ4963).
Aggravated by Vulnerability of sacred sites (#PD6128)
Endangered tribes and indigenous peoples (#PC0720)
Discrimination against indigenous populations (#PC0352)
Degradation of developing countries by tourism (#PF4115)
Abuse of traditional cultural expressions of peoples (#PE4054)
Denial of right to freedom of religion of indigenous peoples (#PE4332).

♦ PC7356 Political exploitation
Broader Exploitation (#PB3200) Political crime (#PC0350).
Narrower Exploitation of trust (#PC4422) Exploitation in housing (#PD3465)
Misuse of food as a political weapon (#PF6202)
Exploitation of athletic competition for commercial or political ends (#PE4833).
Related Class conflict (#PC1573) Racial exploitation (#PC3334)
Economic exploitation (#PC8132) Exploitation in employment (#PC3297)
Endangered tribes and indigenous peoples (#PC0720).

♦ PC7373 Criminals
Criminal violence and disorders
Nature Criminal disorders may stem from any or all of the following factors: criminals have usually not adequately internalized the conditioning standards of behaviour which restrain well-adjusted people from expressing anti-social or criminal impulses - they do not possess the ability to say 'no' to themselves, therefore are consumed by their own selfishness; they have chronically low self-esteem, in which is manifest the idea that they themselves are worth little and thus any effort at self-improvement is futile; and they are usually isolated, encapsulated people with the loveless, alienated insensitivity that allows them to operate at a considerable psychological distance from their victims, a state of separative consciousness which atrophies the capacity to experience the humanity of those it victimizes.
Refs Morgan, Lanier *Understanding and Modification of Delinquent Behavior* (1984); Shover, Neal *Aging Criminals* (1985).
Broader Crime (#PB0001).
Narrower Vigilantism (#PD0527) Recidivists (#PE5581)
Youth violence (#PF7498) Organized crime (#PC2343)
Underworld milieu (#PG1850) Criminal trespass (#PD3794)
Psychotic violence (#PE7645) Dishonest employees (#PD9397)
Juvenile delinquency (#PC0212) Illegal armed groups (#PG1351)
Unconvicted war criminals (#PD4067) Increasing female criminality (#PE5592)
Preponderance of male criminal offenders (#PG6485).
Related Statutory crime (#PC0277) Crimes against humanity (#PC1073).
Aggravates Neglect of victims of crime (#PD4823)
Inadequate facilities for mentally disabled criminals (#PE8900).

♦ PC7379 Deprivation of human rights
Broader Violation of human rights (#PB3860).
Related Denial of human rights (#PB3121) Infringement of human rights (#PC6003).

♦ PC7384 Alien domination of peoples
Foreign domination of countries
Broader Colonialism (#PC0798) Denial of right of a people to be self-determining (#PC6727).
Narrower Occupied territories (#PD8021).
Aggravated by Violation of treaties with indigenous populations (#PE7573).

♦ PC7459 Imbalance in economic relationships between industrialized countries
Failure to restructure economic relations between developed countries
Narrower Structural imbalances between and within the three largest market economies (#PJ5979).
Aggravates Global economic crisis (#PC5876)
Disparity between industrialized and developing countries (#PC8694).

♦ PC7492 Microbial diseases
Microbes
Refs Roberts *Microbial Diseases of Fish* (1983).
Broader Disease vectors (#PC3595).
Narrower Virus diseases (#PD0594) Bacterial disease (#PD9094)
Fungal plant diseases (#PD2225) Microbial contamination of food (#PD9669).
Related Fungal diseases (#PD2728).

♦ PC7517 Unsustainable development of energy use
Inefficient use of energy resources — Inadequate energy conservation — Aggression against natural energy sources — Use of energy-intensive products — High energy development — Wasted energy resources
Nature Many aspects of modern consumer-oriented society demand large amounts of energy to sustain. Many products are energy-intensive in the manufacturing process, require energy for their use, and are constructed such that they cost more to repair than to replace. The packaging used for such products also requires large amounts of energy, whether through the use of chemicals, wood or plastics, or through the cost of their subsequent disposal.
Broader Unsustainable development (#PB9419) Destruction inherent in development (#PF4829).
Narrower Waste of electricity (#PE0984) Inadequate cooking stoves (#PE7904)
Resource-intensive packaging (#PE0635)
Lack of conservation of energy by the private sector (#PE8599).
Aggravates Energy crisis (#PC6329) Underutilization of wind energy (#PF0373)
Underutilization of solar energy (#PF0370) Underutilization of ocean energy (#PG5350)
Underdeveloped rural energy sources (#PF0393) Underutilization of fuelwood energy (#PJ0031)
Maldistribution of energy consumption (#PC5038) Underutilization of geothermal energy (#PJ7988)
Long-term shortage of energy resources (#PF0334)
Inability to reduce petroleum consumption (#PJ4533)
Insufficient utilization of renewable biofuels (#PF0357)
Underutilization of renewable energy resources (#PE8971)
Underutilization of oil shale as an energy source (#PF0445).
Aggravated by Uninsulated buildings (#PE0242) International aggression (#PB0968)
Unexplored energy alternatives (#PF7960)
Aggressive uses of natural energy resources (#PD0408)
Insufficient diversification in energy research (#PG5347)
Inadequate governmental energy conservation policies (#PF0037).

♦ PC7559 International aggression
Broader Aggression (#PA0587).
Narrower Economic conflict (#PC0840) Political aggression (#PD8877)
Offences against the peace and security of mankind (#PC6239).

♦ PC7569 Political isolation
Political isolationism
Broader Isolation (#PB8685).
Narrower National isolationism (#PF2141)
Extreme detachment from represented constituency (#PF0889).
Related Social isolation (#PC1707) Cultural isolation (#PC3943)
Economic isolationism (#PC2791).
Aggravated by Geographical isolation (#PF9023).

♦ PC7587 Corruption of government leaders
Corruption of government rulers — Corruption of government ministers — Corruption in government
Nature Whether due to an actual increase in governmental corruption, or the increase in an investigative press and judicial system, increasing numbers of government leaders are being implicated for corruption. Some withstand the charges, proving themselves innocent, but others are obliged to leave office, accept lesser posts, or remain in office but with a much weakened credibility.
Incidence Publicly reported scandals in industrialized countries include incidents in Federal Germany (involving bribes to ministers, the Prime Minister and the President), in Australia (involving a state Prime Minister and a high court judge), in Japan (involving bribery of a Prime Minister). Many of these incidents reportedly involved multinational corporations. Financial fraud and corruption of party leaders have been brought to public in the Soviet Union, and in Romania and Germany DR after the ousting of monolithic power of the Communist party. Similar incidents have been reported in the developing countries, for example in Philippines, Mexico, Panama, Tanzania, South-Korea, Nigeria, Zimbabwe, Venezuela, India, Haiti, Pakistan, Syria, Iraq, Zaire, Bolivia, Colombia, Ethiopia, Ivory Coast, Thailand, Paraguay. Especially in Latin American countries corruption of government leaders is connected to drugs.
Broader Corruption (#PA1986) Official corruption (#PC9533)
Institutionalized corruption (#PC9173).
Narrower Political appointees (#PF2031) Fraud by government agents (#PD8392)
Misappropriation of public funds (#PC2920)
Unjust allocation of government contracts (#PF2911)
Unethical practices by public service employees (#PE6702)
Excessive accumulation of wealth by government leaders (#PD9653).
Related Misconduct in public office (#PD8227) Corruption of ruling classes (#PD8380)
Mediocrity of government leaders (#PF3962) Unethical practices of government (#PD0814)
Frauds, forgeries and financial crime (#PE5516).
Aggravates Undervaluation of public assets (#PF1001)
Loss of confidence in government leaders (#PF1097)
Government leaders associated with sex scandals (#PE7937)
Corruption amongst relatives of government leaders (#PE9140).
Aggravated by Gerontocracy (#PD3133) Corruptive crimes (#PD8679)
Corruption in politics (#PC0116) Illicit drug trafficking (#PD0991)
Deficiencies of developing countries (#PC4094)
Proliferation of public sector institutions (#PF4739)
Conflict of interest among parliamentarians (#PE3735).

♦ PC7597 Unavailability of agricultural land
Need for agricultural land — Loss of agricultural land — Unavailability of farming land
Nature Most land that is best suited for crop production is already being farmed. If the current rates of conversion of agricultural land to non-agricultural uses continue in industrialized countries, and if the current rates of land degradation continue in developing countries, within 20 years more than one-third of the world's arable land could be lost or destroyed. As population pressures continue to increase, more people are being forced onto land that is becoming less and less productive, including important watersheds which become degraded and in turn reduce the productivity of agricultural land in their catchment areas.
Incidence Only about 11 cent of the world's land area (excluding Antarctica) offers no serious limitation to agriculture. The rest suffers from drought, mineral stress (nutritional deficiencies or toxicities), shallow soil depth, excessive water or permafrost. The world's cropland currently occupies 14 million Km2 and, even using the most optimistic assumptions, it would appear that croplands world-wide could be no more than doubled. Although because of ecological and economic constraints, only moderate additions to global arable land may be feasible. As people move onto marginal lands to ensure their livelihood, the global losses to desertification (at present estimated at 6 million hectares per year) will accelerate. In some areas of Ethiopia, people trying to eke out a living on eroded land are eroding it even more, cutting down the remaining trees for fuel, and denuding the countryside. Approximately half of India is already experiencing some form of soil degradation. There pressures of population are so sever in parts of the country that it is difficult to maintain a balance between food production and land degradation. In Java, soil erosion is an ecological emergency, a result of overpopulation which has led to deforestation and misuse of hillside areas by land-hungry farmers. Loss of arable land to erosion and urbanization is not confined to the developing world. The problem is acknowledged in the USA and Japan. The United Nations Environment Programme forecasts that over one-third of the world's arable land may be lost or destroyed by the turn of the century.
Broader Limited available land (#PC8160).
Narrower Inadequate grazing land (#PJ0404)
Limited availability of land for low-income and disadvantaged groups (#PF5008)
Unavailability of land for agricultural purposes in developing countries (#PE5024).
Related Shortage of urban land (#PD0384).
Aggravates Risk of capital investment (#PF6572)
Shortage of cultivable land (#PC0219)
Excessive use of land for agriculture (#PD9534)
Stagnated development of agricultural production (#PD1285).
Aggravated by Soil erosion (#PD0949) Desert advance (#PC2506)
Soil compaction (#PD1416) Soil degradation (#PD1052)
Wildlife extinction (#PC1445) Agricultural mismanagement (#PD8625)
Destruction of agricultural land (#PD9118)
Use of agricultural land for fuel production (#PD6103)
Inundation of agricultural land through dams (#PE3786)
Increasing proportion of land surface devoted to urbanization (#PE5931).

♦ PC7605 Abusive exploitation of cultural heritage
Conceptual asset stripping — Misappropriation of cultural symbols
Nature Use of valued aspects of a cultural heritage, such as a parts of a well-known piece of classical music, to improve the image of a commercial product, erodes the significance of that heritage. This is especially the case with younger generations which have not been exposed to that heritage in its unadulterated form. This phenomenon may also be seen in the progressive commercialization of traditional and religious festivals and pilgrimages.
Incidence A classic example is the misuse of the German cultural heritage by the Nazi regime. Many people alive today, who were in concentration camps, can recall their families being taken away to the gas chambers to the sound of Wagner's Meistersinger overture.
Broader Destruction of cultural heritage (#PC2114).
Narrower Facile social concepts (#PF5242).
Related Misappropriation of cultural property (#PE6074).
Aggravates Ignorance of cultural heritage (#PF1985)
Inadequate appreciation of culture (#PF3408).

Aggravated by Misuse of advertising (#PE4225)
Vulgar combination of sacred and erotic in advertising (#PE5190)
Social environmental degradation from recreation and tourism (#PD0826).

♦ PC7607 Clandestine employment

Nature Unlike double–jobbing, clandestine employment is often the only job that many workers have. It is carried out on the fringes of the law or outside it altogether, and takes three main forms – the undeclared employment of workers, undeclared self–employment, and undeclared multiple jobholding.

Incidence Clandestine employment particularly involves the unemployed, migrant workers, pensioners, unregistered self–employed workers, housewives, houseworkers, temporary staff, students and children. Their numbers are hard to estimate.

Counter–claim Clandestine employment has always existed because both the employer and the employed benefit from it. The employer does not pay social security contributions or taxes so that the cost of labour is relatively low for him; the employed usually receives a higher take–home pay.

Refs ILO *Annotated Bibliography on Clandestine Employment* (1987); de Grazia, Raffaele *Clandestine Employments. The situation in industrialised economy countries* (1984).

Broader Underground economy (#PC6641).
Aggravates Abusive traffic in immigrant workers (#PD2722)
Limited availability of permanent employment in inner–cities (#PE1134).
Reduces Unemployment (#PB0750).

♦ PC7644 Human consumption of animals

Meat eating — Eating animal flesh

Nature Data strongly suggest that a major influence on cholesterol levels and disease rates is the high consumption of animal–derived foods, including dairy products.

Incidence In the USA, the numbers of animals slaughtered annually for food includes: 4.5 billion cattle, calves, sheep, lambs, higs, chickens, ducks and turkeys. In the UK in 1987 the number included: 4 million cattle, 60,000 calves, 13.5 million sheep, and over 15 million pigs.

Claim Meat is not necessary for human survival. Its production involves a long and cruel process of forced imprisonment, biological manipulation, transportation over long distances in overcrowded and unsanitary conditions, followed by a violent death in a slaughterhouse.

Counter–claim This animal centered view, while trying to take a higher moral ground, seldom takes into account the suffering of a carrot or potato being boiled alive. For animals, even human animals, to live, they must kill other living things. Plant, too, are dependent on the death and decomposition of living matter. Only in this larger recognition of interdependence of death and life can a meaningful decision about meat eating be made.

Refs Adam, Carol J *The Sexual Politics of Meat* (1990); Adam, Carol J *The Sexual Politics of Meat* (1990); Golikere,R K *Vegetarian vs Non–Vegetarian* (1960).

Related Human consumption of animal products (#PD7699)
Excessive consumption of animal flesh (#PD4518).
Aggravates Cruelty to animals in food preparation (#PE0236)
Denial to animals of the right to a natural death (#PE8339).
Aggravated by Unethical consumption practices (#PD2625).

♦ PC7652 Bigotry

Dependence on bigotry — Bigoted people

Incidence Religious bigotry has been evidenced throughout history; Jews have often been the target. Political bigotry exists most commonly under totalitarian regimes headed by people with extremist positions. Class bigotry, in addition to the obvious caste system, is often discernible in labour – management conflicts and in the attitude of one national group to another.

Claim Bigotry is an intolerant, obstinate, and unthinking attachment to one's views, party, or religion. It is heedless of truth, flouts the obligation to treat one's fellow man with sympathy and love, and is highly disruptive of social unity. Bigotry involves powerful emotions, thus is easy to incite. It stems from various causes, often begun at home, developed in school, and maintained by the media and governmental laws.

Broader Uncertainty (#PA7309) Irresolution (#PA7325) Narrowmindedness (#PA7306).
Narrower Class consciousness (#PC3458) Religious prejudice (#PD4365)
Sexual bigotry in organized religion (#PE0567).
Related Hate (#PA7338) Insanity (#PA7157).
Aggravated by Racial stereotyping (#PF5452).

♦ PC7660 Political upheavals

Risk of political upheaval

Nature Mass executions have frequently followed violent changes of governments. Due to the institutional and legal vacuum immediately following the fall of a regime and during the transition period, armed forces, revolutionary tribunals, or even mass public rallies, assume the role of imposing 'justice'. Many executions are carried out without any trial. Even when trials are held death sentences are often delivered after brief or summary trials without procedures for safeguarding the rights of the accused, and sentences are often delivered without any legal base. Many of the convicted are executed immediately or within an extremely short period of time after sentencing, in many cases having been given no opportunities for appeal for review of the sentence or for pardon. Killings by or after torture in prison or detention camps are also commonly reported. Those executed are people suspected of their collaboration with the enemy, former government officials, military officers, policemen, supporters and associates of the former regimes, and others suspected of their opposition to the new regime and to the new government's policies. It is not unusual that family members and friends of those accused or executed, including women and children, are also among the victims. Mass executions are often justified by characterizing the victims as traitors, foreign agents, counter–revolutionaries, enemies of the people, etc.

Broader Risk (#PF7580).
Aggravates Human contingency (#PF7054) Export credit risks (#PF3065)
Nationalization of foreign investments (#PC2172).
Aggravated by Civil disorders (#PC2551)
International imbalance in the quality of life (#PB4993).

♦ PC7662 Suppression of opposition groups or individuals

Nature Opposition groups or individuals are often killed in order to eliminate movements critical of the regime. The victims may be trade–unionists, professionals, or simply those suspected of opposition. They may be assassinated in the streets, or abducted and subsequently found dead – often showing signs of torture. Even though these actions may be taken by 'death squads' with whom the government claims no association, there is often evidence to suggest governmental complicity.

Broader Political repression (#PC1919).
Narrower Frustration of the role of opposition (#PF1594).
Aggravates Structural failure of opposition groups (#PF3821).
Aggravated by Political opposition (#PJ5891)
Promotion of negative images of opponents (#PF4133).

♦ PC7674 Physically inaccessible services

Geographically remote facilities — Absence of essential services

Nature Many rural communities are only on the outer fringes of the services which are now a necessity for participation in contemporary society. For example: electricity is costly and may be available only privately rather than on a general, domestic basis; water may be distributed, but it is usually untested, and may be parasitically contaminated; outdoor washing areas result in stagnant open pools of waste water; there is rarely any central means of garbage disposal and most homes rely on open pits for garbage. Inoperative health outposts, distant medical services and inaccessible dental care facilities drain the vitality of the people and reinforce the life style of backward isolation. In addition, the daily expenditure of energy required to function with such rudimentary services severely minimizes productive output.

Broader Geographical isolation (#PF9023) Lack of essential local infrastructure (#PF2115).
Narrower Inaccessible job market (#PE8916) Isolated mass cemeteries (#PE7172)
Insufficient emergency services (#PF9007) Inaccessible educational facilities (#PD9051)
Inaccessible administrative agencies (#PF2261)
Inaccessible market and supply centres (#PF8299)
Unavailability of appropriate expertise (#PF7916)
Inadequate and insufficient immunization (#PF5969)
Prohibitively expensive housing for the poor (#PE8698)
Inaccessible commercial and financial services (#PE0718)
Inadequate medical laboratory facilities in developing countries (#PG0932).
Related Inadequate medical care (#PF4832) Prohibitive cost of fuel (#PJ0346)
Inadequate irrigation system (#PD8839).
Aggravates Stagnant surface water (#PE2634) Water system contamination (#PD8122)
Underprovision of basic services to rural areas (#PF2875).
Aggravated by Insufficient transportation infrastructure (#PF1495).

♦ PC7693 Inadequate infrastructure

Broader Inadequacy (#PA8199) Underdevelopment (#PB0206).
Narrower Inadequate housing (#PC0449) Inadequate maintenance (#PD8984)
Inadequate health services (#PD4790) Inappropriate sanitation systems (#PD0876)
Inadequate agricultural facilities (#PF6499) Insufficient communications systems (#PF2350)
Inadequate electricity infrastructure (#PD9033) Lack of essential local infrastructure (#PF2115)
Underprovision of basic urban services (#PF2583)
Inadequate water system infrastructure (#PD8517)
Inadequate maintenance of infrastructure (#PD0645)
Underdevelopment of legal infrastructure (#PF4836)
Insufficient transportation infrastructure (#PF1495)
Underprovision of basic services to rural areas (#PF2875)
Ineffective means for goods supply and distribution (#PF6495)
Weakness of infrastructure in island developing countries (#PE5772)
Inadequate infrastructure and services in least developed countries (#PE0289)
Insufficient and inappropriate energy equipment in developing countries (#PE8592).
Related Lack of economic and technical development (#PE8190)
Lack of skilled workers in island developing countries (#PF5737).
Aggravated by Lack of funding for infrastructure (#PJ8512)
Health risks to workers in electricity, gas, water and sanitary services (#PE1159).

♦ PC7721 Loss of human resources

Broader Long–term shortage of resources (#PB6112).
Narrower Depleted expertise of the rural labour force (#PF2973)
Emigration of trained personnel from developing to developed areas (#PD1291).

♦ PC7825 Suffering of plants

Broader Suffering (#PA7690).
Aggravated by Pests of plants (#PC1627) Killing of plants (#PD4217)
Hazards to plants (#PD5706) Cruelty to plants (#PD4148)
Endangered species of plants (#PC0238) Deficiency diseases in plants (#PD3653)
Vulnerability of crops to weather (#PE5682) Radioactive contamination of plants (#PD0710)
Destruction of hedges and hedgerow trees (#PD1642).

♦ PC7859 Non–restitution of property

Non–return of property — Non–restitution of monetary gold

Narrower Misappropriation of cultural property (#PE6074).
Aggravates Uncompensated damages (#PD7179).

♦ PC7873 Financial destabilization of world trade

Financial instability — Monetary instability — Instability of monetary markets — Fluctuations in world monetary and financial conditions — Vulnerability of national economies to vagaries of international financial system

Nature The stable expansion and balanced growth of world trade is hampered by by instability of exchange rates, inadequate multilateral systems of payment, disequilibrium in balance of payments, and destructive national and international measures taken to correct maladjustments in balance of payments.

Refs Chinkin, Davidson and Ricquier (Eds) *Current Problems of International Trade Financing* (1983); Gowda, K G V *Eurodollar Flows and International Monetary Stability* (1978); Sargent, J R *Europe and the Dollar in the World–Wide Disequilibrium* (1981).

Broader Economic and financial instability of the world economy (#PC8073)
Lack of central planning structures in small communities (#PF2540).
Narrower Exchange rate volatility (#PE5930) Speculation on money markets (#PD9489)
Foreign exchange restrictions (#PF3070)
Inadequate mechanism for balance of payments adjustment (#PF3062)
Inability of developing countries to adopt appropriate exchange rate policies (#PE7563).
Related Discriminatory exchange rate policies (#PE8583).
Aggravates Unemployment (#PB0750) Economic stagnation (#PC0002)
Economic inefficiency (#PF7556) Global financial crisis (#PF3612)
Disruption of financial markets (#PD4511)
Imbalance in international trade patterns (#PC8415)
Lack of economic and technical development (#PE8190)
Fragmentation of the international trading system (#PC9584).
Aggravated by Capitalism (#PC0564) Bank failure (#PE0964)
Imbalance of payments (#PC0998).

♦ PC7896 Forest decline

Unhealthy forests — Waldsterben — Mass tree deaths — Tree blight

Nature Accumulating evidence now confirms that the many recorded instances of mass tree deaths throughout the temperate and boreal forests of the northern industrialized hemisphere must be linked to a common cause. Whatever the final trigger mechanism (pests, diseases, drought), the tree population in many of these forests has been reduced to a susceptible condition because it has passed a threshold of tolerance to sustained attack from chemical pollutants. Although different experts claim predominance for one mechanism or another, there is little in the explanations which is mutually contradictory. Deaths and decline in many different species of tree are now so widespread over such a variety of sites among stands of different ages and at different altitudes that no single hypothesis can suffice to explain every instance. The most probable explanation remains a combination of mechanisms in different local circumstances, but induced by growth stress from excess nitrogen. Together these mechanisms add up to an overwhelming

stress burden upon the tree population which is only just beginning to show its effects and will require considerable social adjustment to reverse.
Incidence Recent evidence indicates that forests throughout the Eastern United States are in decline, perhaps seriously. Man-made pollution is the chief suspect. Some of the symptoms are similar to those that eventually led to a dramatic decline in central European forests. Research shows that some species of softwood trees are losing their foliage, dying and failing to reproduce at high elevations in some parts of the Appalachian Mountains. Some studies show a large-scale, rapid and simultaneous drop in the growth rates of a half dozen species of coniferous trees in the East. This trend started around 1960 and has apparently accelerated over the past 10 years. Some hardwood trees are also showing these symptoms, although to a lesser degree.
In 1990 it was estimated that at least 7 million hectares of forest in 15 European countries were suffering from Waldsterben, death of the forest. One third of Swiss forests are suffering moderate damage, 10 per cent are severely affected, and 2 per cent are dead. Another example of the extent of the damage is Germany, where official studies show that up to half the country's trees, which cover a third of its area, are either diseased or dead because of airborne industrial pollution and vehicle emissions.
  Broader Natural environment degradation (#PB5250).
  Narrower Endangered urban trees (#PD2025).
  Aggravates Deforestation (#PC1366)   Endangered forests (#PC5165)
  Eichornia crassipes (#PE3815).
  Aggravated by Acidic soils (#PD3658)   Acidic precipitation (#PD4904)
  Ozone as a pollutant (#PE1359)   Pests and diseases of trees (#PD3585)
  Demineralization of the soil (#PD9227).

♦ **PC7928 Lack of natural resources**
  Narrower Lack of natural resources in developing countries (#PD3625).
  Aggravates Desert nomadism (#PD2520).
  Aggravated by Maldistribution of resources (#PB1016)
  Unproductive use of resources (#PB8376)
  Long-term shortage of natural resources (#PC4824)
  Aggressive uses of natural energy resources (#PD0408)
  Unconstrained exploitation of natural resources (#PF2855).

♦ **PC7997 Limited job market**
Unorganized job market
  Broader Limits (#PF4677).
  Aggravates Unemployment (#PB0750)   Rural unemployment (#PF2949)
  Limited development of functional abilities (#PF1332)
  Dependency on unpredictable sources of income (#PF3084).

♦ **PC8019 Unethical professional practices**
Irresponsible professional activities — Professional malpractice — Professional negligence — Corruption in professional practices — Professional fraud
  Refs Bowman, James S, et al *Professional Ethics* (1984); Rotunda, Ronald D *Professional Responsibility* (1988); Zimmerman, Roy R *Malpractice II* (1985).
  Broader Negligence (#PA2658)   Unethical practices (#PC8247).
  Narrower Health fraud (#PD9297)   Unprofessional science (#PF6697)
  Malpractice in education (#PD4684)   Unethical medical practice (#PD5770)
  Unethical ophthalmic practice (#PE1369)   Unethical real estate practice (#PD5422)
  Military-industrial malpractice (#PD4361)   Unethical practice of radiology (#PD8290)
  Irresponsible international experts (#PF0221)   Unethical practices of interpreters (#PE6740)
  Unethical practices by police forces (#PD9193)   Unethical practices in psychotherapy (#PD5267)
  Unethical practices in the social sciences (#PD6626)
  Unethical practices in the legal profession (#PD5380).
  Related Institutionalized corruption (#PC9173)   Unethical intellectual practices (#PC2915).
  Aggravates Bias in scientific research (#PF9693).
  Aggravated by Lack of professional standards (#PF3411)
  Misuse of societally-endorsed professions to conceal socially unacceptable initiatives (#PJ0880).

♦ **PC8073 Economic and financial instability of the world economy**
Deterioration in international economic environment
  Refs Lundberg, Erik *Instability and Economic Growth* .
  Narrower High interest rates (#PF9014)   Exchange rate volatility (#PE5930)
  Financial destabilization of world trade (#PC7873)
  Faltering structural adjustment in the world economy (#PF9664)
  Outflow of financial resources from developing countries (#PC3134)
  Deterioration of terms of trade for developing countries (#PD2897)
  Instability of economic and industrial production activities (#PC1217)
  Disincentives for financial investment within developing countries (#PF3845)
  Private domestic capital outflow from developing to developed countries (#PD3132)
  Instability in export trade of developing countries producing primary commodities (#PD2968).
  Aggravates War (#PB0593)   Global financial crisis (#PF3612)
  Communist political imperialism (#PC3164).
  Aggravated by Economic revolution (#PC3233)   Economic inefficiency (#PF7556)
  Excessive foreign public debt of developing countries (#PD2133).

♦ **PC8083 Ideological repression**
  Broader Repression (#PB0871)   Denial of freedom of thought (#PF3217).
  Narrower Exile (#PC2507).
  Related Ideological war (#PC3431).
  Aggravates Atheism (#PF2409).
  Aggravated by Marxism (#PF2189)   Monasticism (#PF2188)
  Ideological discrimination in politics (#PC3219).

♦ **PC8096 Killing by humans**
  Broader Human violence (#PA0429).
  Narrower Homicide (#PD2341)   Serial killings (#PE9447)   Induced abortion (#PD0158)
  Killing of animals (#PD8486)   Killing non-human life (#PF6359)
  Inhumanity of capital punishment (#PF0399).
  Aggravates Death of living creatures (#PF7043).
  Reduced by Criminalization of euthanasia (#PF2643).

♦ **PC8127 Family rejection of children**
Claim Family and community today fail to provide a receptive climate for the adolescent's needs. In human relationships, both inside and outside the family, there is little genuine loving. There is no easy, spontaneous show of warmth and tenderness; no cherished touch. Family members fear and mistrust closeness and any open show of emotion. The competitive strivings of the parents are communicated to the children and to the relationships with the children.
  Broader Victimization of children (#PC5512).
  Narrower Maternal deprivation (#PC0981)
  Family rejection of physically handicapped (#PE2087).
  Aggravates Corruption of minors (#PD9481)   Infant growth failure (#PE6909)
  Neglected young children (#PE4245).
  Aggravated by Generation communication gap (#PF0756)
  Failure to recognize uniqueness of family members (#PF1750).
  Reduces Excessively large families (#PD7625).

♦ **PC8132 Economic exploitation**
  Broader Exploitation (#PB3200).
  Narrower Increase of defence budget (#PF6060)
  Abuse of project-tied migration (#PE7613)
  Commercial exploitation of wild animals (#PD1481)
  Excessive commercial exploitation of farm animals by industrial concerns (#PD2772)
  Economic exploitation of developing countries by industrialized countries (#PE2427).
  Related Political exploitation (#PC7356)   Exploitation in employment (#PC3297)
  Exploitative personal services (#PC3299)   Exploitation in capitalist systems (#PC3117)
  Misuse of tropical rain forests for agricultural development (#PE5274)
  Abuse of monopoly power of state-owned or state-controlled enterprises (#PE0988).
  Aggravates Exploitation of dependence on food aid (#PD7592).

♦ **PC8160 Limited available land**
Land shortage — Minimal prime land — Unavailability of prime land — Lack of land space
  Broader Shortage (#PB8238).
  Narrower Limited public land (#PJ0574)   Limited family land (#PG9748)
  Shortage of urban land (#PD0384)   Shortage of cultivable land (#PC0219)
  Scarcity of residential land (#PD8075)   Unavailability of building sites (#PJ8549)
  Unavailability of agricultural land (#PC7597)
  Insufficient common land in urban environments (#PE6171)
  Exploitation of land for the burial of the dead (#PE9095)
  Limited availability of land for low-income and disadvantaged groups (#PF5008).
  Related Inadequate care of community space (#PF2346)
  Unimaginative vision of resource utilization (#PF1316)
  Unattractive locale for economic development (#PF3499).
  Aggravates Factory farming (#PD1562)   Endangered parklands (#PE9282)
  Endangered monuments and historic sites (#PD0253).
  Aggravated by Decreasing land mass (#PF7435)
  Excessive land usage (#PE5059).

♦ **PC8182 Chronic shortage of foreign exchange**
Foreign exchange shortages — Shortage of foreign exchange
  Broader Insufficient financial resources (#PB4653)
  Inadequacies of the international monetary system (#PF0048).
  Narrower Foreign exchange shortage in developing countries (#PD3068).
  Aggravates Export credit risks (#PF3065)
  Social environmental degradation from recreation and tourism (#PD0826).

♦ **PC8192 Maldistribution of population within countries**
Nature Almost all developing countries consider the current geographic distribution of their population partially or wholly unacceptable. Their major cities have already become too large, and are growing too rapidly compared with smaller cities, towns and rural areas. Since urbanization is proceeding faster than industrialization, unemployment and underemployment are rampant; crime rates are increasing; and traditional social forms are disappearing.
  Broader Maldistribution of world population (#PF0167).
  Narrower Declining community population (#PJ8746)
  Unbalanced urban population density gradients (#PD6131).
  Aggravated by Insufficient housing funds (#PG8768).
  Reduced by Involuntary mass resettlement (#PC6203).

♦ **PC8193 Erosion**
  Refs Goldman, S J, et al *Erosion and Sediment Control Handbook* (1986); Holy, M *Erosion and Environment* (1980).
  Broader Impairment (#PA6088)   Disintegration (#PA6858)   Inappropriateness (#PA6852).
  Narrower Soil erosion (#PD0949)   Pay system erosion (#PE6725)
  Erosion of sovereignty (#PE5015).
  Related Value erosion (#PA1782)   Undemocratic social systems (#PB8031).
  Aggravated by Degradation of semi-natural and natural habitats of flora and fauna (#PC3152).

♦ **PC8222 Personal wealth**
Richness — Excessive wealth — Unnecessary personal wealth
Incidence The cumulative wealth of 400 richest Americans equals the savings that rest of the Americans have in commercial banks. The richest 10 percent of Americans controlled 68 percent of the nation's wealth in 1983.
  Refs Lampman, Robert J *The Share of Top Wealth-Holders in National Wealth, 1922–56* (1984); Sreenivasan, K *Anatomy of Wealth*.
  Broader Inequitable distribution of wealth (#PB7666).
  Narrower Accumulation of capital (#PC5225)   Proliferation of second homes (#PF1286)
  Unnecessary personal consumption (#PF5931)
  Excessive accumulation of wealth by government leaders (#PD9653).
  Aggravates Competitive acquisition of arms (#PC1258)
  Leadership as symbolic of wealth (#PF2870)
  Non-destructible packaging and containers (#PD1754).
  Aggravated by Pursuit of affluence (#PF5864)
  Lack of relationship between wealth generation and the public good (#PF4730).

♦ **PC8245 Shock**
Nature The sudden life-threatening condition resulting from severe injury and characterized by progressive impairment of all physiological systems of the body. The main characteristic of shock is the failure of capillary circulation in the tissues owing to impairment of cardiac output and over contraction of arteries and veins, capillary dysfunction, and changes in the rheological properties of blood. Shock can be brought on by an injury, burns, surgery, incompatible blood transfusion, anaphylaxis, functional cardiac disorders, loss of blood to tissues and organs and excessive loss of blood.
Shock, used in a different sense, is applied to those persons in an unusual mental state or persons with sever emotional disturbances, mental shock and emotional shock.
Background The condition of physical shock was described by the French surgeon H. F. Ledran in 1737 and by the Russian physician N. I. Pirogov in 1870.
  Refs Altura, Burton M et al *Handbook of Shock and Trauma* (1983); Proctor, R A *Clinical Aspects of Endotoxin Shock* (1985).
  Broader Human illness (#PA0294).
  Related Fear (#PA6030)   Disease (#PA6799)   Adversity (#PA6340)
  Unpleasantness (#PA7107)
  Insufficient provision of public services for communication (#PF2694).
  Aggravates Lung disorders and diseases (#PD0637).

♦ **PC8247 Unethical practices**
Improbity — Unethical people
  Broader Improbity (#PA7363)   Uncommunicativeness (#PA7411).
  Narrower Profiteering (#PC2618)   Unethical medical practice (#PD5770)
  Unethical practices in politics (#PC5517)   Unethical professional practices (#PC8019)
  Execution of inappropriate orders (#PF2418)
  Unethical practices by housing tenants (#PE7169).
  Aggravated by Inability to define moral standards (#PF7178).

♦ PC8269 Cultural stagnation
Broader Cultural corruption (#PC2913).
Narrower Insufficient minority culture support (#PF5659).
Related Obsolete basis of cultural identity (#PF0836).
Aggravated by Art censorship (#PD2337)   Theatre censorship (#PD3028)
Cultural illiteracy (#PD2041)   Anti-intellectualism (#PF1929)
Radio and television censorship (#PD3029)   Restriction of freedom of expression (#PC2162)
Untransposed significance of cultural tradition (#PF1373).

♦ PC8270 Seizure of power
Narrower Mutiny (#PD2589)   Revolt (#PE5144)   Armed insurrection (#PD8284)
Overthrow of government (#PD1964).
Related Revolution (#PA5901).

♦ PC8310 Loss of capacity with age
Broader Human ageing (#PB0477).
Narrower Deterioration of the mind with age (#PE4649).
Aggravates Slowness of sensori-motor activities in the aged (#PD3514).
Aggravated by Inadequate maintenance of infrastructure (#PD0645).

♦ PC8316 Uncontrolled growth of debt
Excessive development of credit
Narrower Consumer debt (#PD3954)   Understated debts (#PE1729)
Disproportionate external public debt (#PC3056)
Excessive debt of socialist countries to the West (#PD2502)
Excessive foreign public debt of developing countries (#PD2133)
Burden of servicing foreign public debt by developing countries (#PD3051).
Aggravates Excessive public debt (#PC2546)   Global financial crisis (#PF3612)
Fluctuations in real value of money (#PD9356)
Inadequacies of the international monetary system (#PF0048).
Aggravated by Leveraged buy-outs (#PE4963)
Lack of economic and technical development (#PE8190).

♦ PC8337 Personal and social maladjustment
Nature A condition or process involving the inability or unwillingness to conform to prevailing standards of external system or of one's own.
Broader Retarded socialization (#PF2187).
Narrower Social maladjustment of children of migrants (#PE4258)
Adjustment difficulties of new urban families (#PF1503).
Aggravated by Speech disorders (#PE2265).

♦ PC8344 Cultural discrimination
Cultural racism — Cultural bias
Broader Discrimination (#PA0833).
Narrower Left-handedness (#PF1215).
Cultural discrimination in the administration of justice (#PE6529).
Related Racism (#PB1047)   Ethnic discrimination (#PC3686).
Aggravates Racial discrimination in education (#PD3328).
Aggravated by Cultural arrogance (#PF5178)   Multiplicity of languages (#PC0178).

♦ PC8346 Accumulation of property
Claim The accumulation of money and property by the wealthy is always at the expense of the poor. The rich have a superfluous store of things which they do not need and which are neglected and wasted, while millions are starved to death for want of sustenance. Whether a person owns few or many possessions is not the issue. The basic problem is that the quest for more and varied possessions takes the form of an addiction to wealth, rather than legitimate need for and enjoyment of material things.
Broader Accumulation (#PA4313).
Narrower Accumulation of cultural property (#PD6907)
Accumulation and misuse of religious property (#PE3354).
Aggravates Unequal property distribution (#PC3438).
Aggravated by Property speculation (#PD8202).

♦ PC8390 Preservation of obsolete systems
Broader Inappropriate assumptions (#PF6814).
Narrower Obsolete methods (#PF3713)   Obsolete military bases (#PG8043)
Obsolete defence planning (#PJ1877)   Functional obsolescence of roads (#PE6704)
Obsolescence of rituals and customs (#PF1309)
Obsolescence of suburban mode of human settlement (#PD6150)
Architectural obsolescence of building structures (#PG7775)
Vocational obsolescence in the face of overwhelming need (#PF3844)
Institutional obsolescence in modern industrialized societies (#PE2862).

♦ PC8391 Scandal
Nature Scandal arises when a member of a community, by actions or opinions goes against the commonly accepted standards of the community and causes distress to the other members, perhaps even bringing the whole community into disrepute. Scandalous behaviour may range from the involvement of political leaders with extra-marital sex, illegal drugs or unreported cash transactions to a member of a celibate religious community getting married.
Counter-claim A scandal, a term far overused, is not necessarily a crime, involve corruption or bribery. When it is shocking, this says more about those shocked than the person involved in the scandal. For example, to describe as scandalous the use of prostitutes by politicians in a society where this practice is not only tolerated but expected is simply to expose one's own naivety.
Refs Kohn, George *Encyclopedia of American scandal* (1989).
Broader Disrepute (#PA6839)   Wrongness (#PA7280)   Disapproval (#PA6191).
Narrower Sex scandal (#PD9398)   Political scandal (#PD4651)
Financial scandal (#PD2458)   Unreported scandals (#PF5340)
Military-industrial malpractice (#PD4361).
Related Vice (#PA5644)   Biased presentation of news (#PD1718).

♦ PC8410 Monopoly of power
Monopolization of power
Broader Imbalance of power (#PB1969).
Narrower Gerontocracy (#PD3133)   Monopoly power due to advertising (#PE0081)
Monopoly of nuclear power techniques (#PD1741)
Excessive power and independence of transnational corporations (#PD5807)
Abuse of monopoly power of state-owned or state-controlled enterprises (#PE0988).
Related Accumulation of functions (#PF4174).
Aggravates Abuse of power (#PB6918).

♦ PC8415 Imbalance in international trade patterns
Broader Fiscal and trade imbalances (#PC4879).
Narrower Excessive external trade deficits (#PC1100)
Weakness in trade among developing countries (#PC0933)
Bipolarization of trade between developed and developing countries (#PE4190)
Imbalance in trade of cultural products between developed and developing countries (#PE5702).
Aggravated by Financial destabilization of world trade (#PC7873).

♦ PC8419 Unsustainable agricultural development
Unsustainable development of farmlands — Unsustainable development of agricultural resources — Degradation of agricultural resources
Broader Unsustainable development (#PB9419)   Natural environment degradation (#PB5250)
Destruction inherent in development (#PF4829).
Narrower Overstocking (#PC3153).
Aggravates Agricide (#PE4045)   Shortage of animal protein (#PC4998)
Ecologically unsustainable development (#PC0111)
Unsustainable development of coast zones (#PD4671).
Aggravated by Acidic soils (#PD3658)   Desert advance (#PC2506)
Soil degradation (#PD1052)   Chemicalized farming (#PD7993)
Overproduction of food (#PD9448)   Tropical deforestation (#PD6204)
Maldistribution of food (#PC2801)   Pesticides as pollutants (#PD0120)
Unstable shifting agriculture (#PD7516)   Unsustainable population levels (#PB0035)
Poverty in developing countries (#PC0149)
Inappropriate government intervention in agriculture (#PE1170)
Increasing proportion of land surface devoted to urbanization (#PE5931)
Neglect of agricultural and rural life in developing countries (#PF7047).

♦ PC8425 Cultural repression
Broader Repression (#PB0871).
Related Denial of cultural rights (#PD5907).
Aggravated by Forced assimilation (#PC3293).

♦ PC8452 Travel restrictions
Denial of right to freedom of movement of persons
Broader Violation of civil rights (#PC5285).
Narrower Prohibited travel for women (#PG5287)
Denial of freedom of movement in communist systems (#PC3173)
Restrictions on freedom of movement within countries (#PE8408)
Restrictions on freedom of movement between countries (#PC0935)
Lack of freedom of movement of children of separated parents (#PE4669).
Related Infrequent travel opportunities (#PG8342)
Denial of political and civil rights (#PD0632)
Restrictions on socialist citizens working abroad (#PE1659).
Aggravated by Taxation on travel (#PJ3126).
Reduced by Fast and easy travel (#PS1917).

♦ PC8471 Economic repression
Claim The high living standard of industrialized countries require extensive violence on the part of regimes of developing countries to gear their capital, soil and labour to the interests of developed countries. The economic repression is exercised by privileged regimes, but it is caused by the excessive consumption of the developed countries.
Broader Repression (#PB0871).
Related Abuse of economic power (#PC6873)   Denial of economic rights (#PD4150).
Aggravated by Eugenics (#PC2153)   Economic imperialism (#PC3198).

♦ PC8478 Transport accidents
Vehicle collisions — Transportation calamities
Nature Accidents occur between all human-operated vehicles, vessels and transport with enormous loss of life and personal injury. Mid-air collisions of airplanes occur between military, commercial and private aircraft, most often near heavily used airports; marine collisions occur in busy harbours between vessels of all kinds and sizes; and road collisions, which involve the most fatalities, occur between passenger cars, trucks, buses and other vehicles, and between such vehicles and pedestrians. Another type of collision occurs between transport and stationary objects and includes airplanes that have flown into mountains, bridges, buildings and electrical wires, for example; ships which have struck icebergs, jetties, wharves, port facilities, reefs, sand bars and submerged rocks or wrecks; and roadway and off-road vehicles which have struck trees, poles or other objects. In addition there are collisions with animals: aircraft collide with birds; fast moving ships and boats collide with, or are struck by, whales and other sea-creatures; and road collisions occur with large animals such as horses, deer, sheep and cattle.
Incidence Transport accidents are the leading cause of accidental mortality and morbidity in both industrialized and developing countries.
Claim Collisions also occur wherever humans are moving – whether in vehicles, on skates or using other devices for accelerating movement, or even (especially in crowded cities) between people who are merely walking. The collision-prone nature of humans is an individual physical limitation which, as a form of inattention, seems also to characterize world development as nations collide with each other, partly also as a consequence of a failure to exchange adequate signals.
Refs Society of Automobile Engineers *Crash injury impairment and disability* (1986).
Broader Injurious accidents (#PD0731)   Disastrous accidents (#PC6034).
Narrower Air accidents (#PD1582)   Marine accidents (#PD8982)
Railway accidents (#PD0126)   Road traffic accidents (#PD0079)
Military aircraft accidents (#PD5373).
Aggravates Large-scale industrial accidents (#PD2570).
Aggravated by Overloaded vehicles (#PE4127).

♦ PC8482 Diseases of the circulation system
Diseases of the circulatory system — Circulatory disorders — Circulatory system diseases
Broader Human disease and disability (#PB1044).
Narrower Hypertension (#PE0585)   Heart diseases (#PD0448)
Diseases of the arteries (#PE2684)   Diseases and injuries of the brain (#PD0992)
Symptoms referable to cardiovascular system (#PE3933).
Aggravates Sleep disorders (#PE2197).

♦ PC8494 Prejudice against minorities
Refs Dijk, Tean A van *Prejudice in Discourse* (1984).
Broader Prejudice (#PA2173).
Aggravates Ethnic conflict (#PC3685)   Discrimination against minorities (#PC0582).

♦ PC8512 Political domination
Broader Domination (#PA0839).
Narrower Annexation (#PE5210).
Aggravates Forced assimilation (#PC3293)   Cultural fragmentation (#PF0536).
Aggravated by Cultural invasion (#PC2548).

♦ PC8520 Cultural prejudice
Broader Prejudice (#PA2173).
Aggravates Nationalism (#PB0534)   Racial prejudice (#PC8773)
Racial discrimination (#PC0006).

♦ **PC8534 Diseases of the skin and subcutaneous tissue**
Skin diseases — Skin infections — Skin disorders — Skin irritation — Skin rash — Cutaneous diseases — Injury of the skin
Refs Adams, R M (Ed) *Occupational Skin Diseases* (1986).
Broader Human disease and disability (#PB1044).
Narrower Acne (#PE3662) Itch (#PE3940) Boils (#PG9261)
Corns (#PG1201) Eczema (#PE2465) Ecthyma (#PE9418)
Pyoderma (#PG6528) Scarring (#PG6749) Dandruff (#PJ5412)
Impetigo (#PG9729) Erythema (#PG0692) Psoriasis (#PE6325)
Urticaria (#PG4480) Pemphigus (#PG6116) Diseases of nail (#PG6646)
Acute lymphadenitis (#PG9858) Acquired acanthosis (#PG4949)
Occupational dermatoses (#PE5684) Benign neoplasm of skin (#PG9649)
Diseases of sweat glands (#PG9635) Leishmaniasis of the skin (#PG4608)
Malignant neoplasm of skin (#PE5016)
Cosmetic and health problems related to human hair (#PE7111).
Aggravates Pain (#PA0643).
Aggravated by Obesity (#PE1177) Diabetes (#PE0102)
Cement dust (#PE2854) Onchocerciasis (#PE2388)
Irritant fumes (#PD3672) Fungicides as pollutants (#PD1612).

♦ **PC8541 Economic inequality**
Broader Human inequality (#PA0844).
Narrower Unequal distribution of production between countries (#PF4336).
Aggravates Use of agricultural resources for production of animal feed (#PD1283).

♦ **PC8558 Harassment**
Broader Aggression (#PA0587) Human violence (#PA0429)
Restrictions on freedom of movement between countries (#PC0935).
Narrower Criminal threat (#PE4661) Sexual harassment (#PD1116)
Criminal menacing (#PE4467) Aircraft harassment (#PE6006)
Criminal harassment (#PD2067) Unreported harassment (#PF4729)
Harassment of the media (#PD0160) Harassment of journalists (#PD3036)
Harassment of public officials (#PD4915) Harassment of human rights monitors (#PF1585)
Sexual harassment in the working place (#PE8466)
Harassment of travellers by immigration officials (#PE7780)
Discrimination and harassment of children in public life (#PE6922).
Related Retaliation (#PF9181) Persecution (#PB7709)
Abuse of police power (#PC1142) Infringement of privacy (#PB0284)
Lack of international cooperation (#PF0817)
Refusal to issue travel documents, passports, visas (#PE0325).

♦ **PC8641 Political prejudice**
Broader Prejudice (#PA2173).
Aggravates Political discrimination in politics (#PC3221)
Political discrimination in the administration of justice (#PE1828).

♦ **PC8642 Waste of non-renewable resources**
Unproductive use of non-renewable resources
Nature When non-renewable resources, materials contained in the biosphere and the earth's crust which cannot be re-created in a human time scale, such as fossil fuels, minerals, groundwater and animal species are used they often undergo irreversible chemical change and cannot be used in the same form. At the current rate of consumption oil will run out in about 30 years' time, tin, cadmium, lead and zinc in 40 years, copper, antimony and nickel in about 70 years.
Broader Unproductive use of resources (#PB8376).
Aggravates Environmental pollution (#PB1166) Worldwide misallocation of resources (#PB6719).
Aggravated by Environmental poverty (#PD5261)
Environmental prodigality (#PF7318)
Excessive consumption of goods and services (#PC2518)
Dependence of government revenues on exploitation of environmentally inappropriate products (#PD1018).

♦ **PC8676 Undemocratic organizations**
Broader Undemocratic social systems (#PB8031).
Narrower Undemocratic political organization (#PC1015)
Unrepresentative international organizations (#PD4873).
Aggravated by Poor communications networks in rural areas (#PF6470).

♦ **PC8689 Abuse of authority**
Illegitimate authority — Illegal exercise or abuse of authority — Illegal use and abuse of administrative authority
Nature Abuse of administrative authority may be divided into two types: involuntary abuse, where the administrator made a bona-fide use of his powers, which, however, departed from the intention of the legislation; and voluntary abuse, where the administrator consciously departed from the rules governing him.
Broader Abuse of power (#PB6918) Dependence on authority (#PF8995).
Narrower Political trials (#PD3013) Abuse of police power (#PC1142)
Abuse of government power (#PC9104)
Misuse of personal authority for political purposes (#PE4635).
Related Corruptive crimes (#PD8679) Corruption of the judiciary (#PD4194)
Criminal violation of civil rights (#PD8709).
Aggravates Abuse of science (#PC9188) Unlawful government action (#PF5332)
Administrative difficulties in new states (#PE1793).

♦ **PC8694 Disparity between industrialized and developing countries**
Economic imbalance between developed and developing countries — Inequality between North and South — North-South gap — Limited progress in North-South negotiations — Failure to restructure economic relations between North and South
Refs Helleiner, Gerald K *International Economic Disorder* (1981).
Broader International inequality (#PC9152) Lack of international cooperation (#PF0817)
Inequitable distribution of wealth (#PB7666).
Narrower Economic gap (#PD8834) Information gap (#PF3397)
Technology gap between developed and developing countries (#PE7985).
Aggravates Malignant visions (#PF5691)
Lack of progress in establishing a New International Economic Order (#PF4306).
Aggravated by Competition between states (#PC0114)
Inappropriate understanding of progress (#PF8648)
Lack of economic and technical development (#PE8190)
Reallocation of aid funds to alternative priorities (#PF0648)
Imbalance in economic relationships between industrialized countries (#PC7459)
Economic exploitation of developing countries by industrialized countries (#PE2427).

♦ **PC8726 Discrimination before the law**
Broader Inequality before the law (#PC1268).
Narrower Discrimination against men before the law (#PD3692)
Discrimination against women before the law (#PD0162)
Legal discrimination in favour of offenders (#PD9316)
Discrimination against women in divorce rights (#PE8879)
Discrimination against foreigners in legal proceedings (#PF1798)
Cultural discrimination in the administration of justice (#PE6529)
Economic discrimination in the administration of justice (#PE1399)
Religious discrimination in the administration of justice (#PE0168)
Political discrimination in the administration of justice (#PE1828)
Geographic discrimination in the administration of justice (#PE1347)
Discrimination against juveniles in judicial proceedings due to protective legislation (#PE1295).
Related Age discrimination (#PC2541).

♦ **PC8745 Urban-Industrial pollution**
Industrial polluters
Broader Pollution (#PB6336).
Narrower Dust (#PD1245) Irritant fumes (#PD3672) Malodorous fumes (#PD1413)
Smoke as a pollutant (#PD2267) Urban-Industrial air pollution (#PJ5532)
Environmental impacts of coal conversion plants (#PE8453).
Related Employment at risk through elimination of industrial pollution (#PF7836).
Aggravates Suicide (#PC0417) Air pollution (#PC0119).
Aggravated by Unethical industrial practices (#PD2916).

♦ **PC8756 Diseases of the nervous system**
Nervous disorders — Neurological illness
Nature Diseases of the human nervous system underlie a large number of symptoms of disorders, ranging from the smallest disturbances of personality through such major problems as crippling, blindness, violent behaviour and even death. Diseases of the nervous system may be divided into the following types: Neurological manifestations secondary to other diseases, neurochemical disorders, developmental defects, diseases of peripheral nerves, diseases of the spinal cord, diseases of the autonomic nervous system, diseases of the brainstem and cranial nerves, motor system disorders, diseases of the cerebellum, intracranial tumours, diseases of the cerebrum, infections of the nervous system, demyelinating disorders, craniocerebral trauma, epilepsy, toxic poisoning, neuromuscular disorders.
Refs Adachi, Masazumi *Neuromuscular Diseases* (1989); Bobath, Berta *Abnormal Postural Reflex Activity Caused by Brain Lesions* (1985); Ecobichon, Donald J and Joy, Robert M *Pesticides and Neurological Diseases* (1982); Marquardt, Thomas P *Acquired Neurogenic Disorders* (1982).
Broader Human disease and disability (#PB1044).
Narrower Paraesthesia (#PG3547) Diseases and injuries of the brain (#PD0992)
Symptoms referable to nervous system (#PE9468)
Diseases of the central nervous system (#PE9037)
Diseases of nerves and peripheral ganglia (#PE8932)
Hereditary disorders of the central nervous system (#PE7915).
Aggravates Hallucinations (#PF2249) Sleep disorders (#PE2197).
Aggravated by Adenovirus infections (#PE2355).

♦ **PC8773 Racial prejudice**
Broader Prejudice (#PA2173).
Aggravates Racial discrimination (#PC0006).
Aggravated by Cultural prejudice (#PC8520).

♦ **PC8774 Skin colour prejudice**
Discrimination based on skin colour
Refs Adhinarayan, S P *Case for colour* (1964); Dovidio, John D and Gaertner, Samuel L *Prejudice, Discrimination, and Racism* (1986).
Broader Prejudice (#PA2173).
Aggravates Ethnic conflict (#PC3685) Racial discrimination (#PC0006)
Dissatisfaction with skin colour (#PF1741).

♦ **PC8801 Intractable diseases**
Incurable diseases
Nature Includes both diseases of unknown aetiology, with no established therapeutic remedies (and with the possibility of sequelae), and chronic diseases, which put a heavy social and economic burden on both the patient and the relatives. The definition may be narrowed to include only serious mental and physical disorders (progressive muscular atrophy and dystropy, etc), but may be extended to include: Behcet's disease, multiple sclerosis, myasthenia gravis, and others.
Narrower Chronic illness (#PD8239) Diseases of unknown aetiology (#PD8588).
Aggravated by Lack of understanding of spiritual healing (#PF0761).

♦ **PC8836 Unconventional war**
Broader War (#PB0593).
Narrower Civil war (#PC1869) Space warfare (#PD6439) Biochemical warfare (#PC1164)
Inhumane and indiscriminate weapons (#PD1519).
Related Nuclear war (#PC0842).

♦ **PC8848 Prisoners of war**
Broader War (#PB0593) Disastrous consequences of war (#PC4257).
Narrower Non-repatriation of prisoners of war (#PE0948)
Forced repatriation of prisoners of war (#PD0218)
Discriminatory treatment of foreign prisoners (#PE6883)
Delayed consequences of war-time imprisonment and deportation (#PF0726).
Related Housing destruction in war (#PE2592) War damage in civilian areas (#PD8719)
Industrial destruction by war (#PD8359) Persons missing in military action (#PE1397).
Aggravates Ill treatment of prisoners of war (#PD2617).

♦ **PC8862 Deprivation of peasantry**
Landlessness — Landless peasants — Loss of peasant title to land — Dispossession of peasant landholdings
Refs Eckholm, Erik *The Dispossessed of the Earth* (1979); Misra, R P *Third World Peasantry* (1986).
Narrower Landlessness in developing countries (#PC0990).
Aggravates Cultivation of marginal agricultural land (#PD4273).
Aggravated by Prohibitive cost of land (#PE4162).

♦ **PC8866 Diseases and deformities of the digestive system**
Digestive disorders — Digestive troubles — Disturbance of digestion — Impaired digestion — Poor digestion — Diseases of the digestive system
Refs Chey, William Y *Functional Disorders of the Digestive Tract* (1983); Sorenson, Joyce and Murray, Nancy *Digestive Disorders* (1983); Wanatabe, Shaw et al *Digestive Disease Pathology* (1988); Wanatabe, Shaw, et al *Pathology* (1988).
Broader Human disease and disability (#PB1044).
Narrower Hernia (#PE8961) Gastritis (#PG2250) Dyspepsia (#PE4724)
Peritonitis (#PE2663) Appendicitis (#PG2327) Liver diseases (#PE1028)
Intestinal diseases (#PD9045) Periodontal diseases (#PE3503) Cirrhosis of the liver (#PE2446)
Diseases of gallbladder (#PE9829) Benign neoplasm of digestive system (#PG4117)

PC8866

Malignant neoplasm of digestive organs (#PE4303)
Symptoms referable to digestive system (#PE9604)
Diseases of the oesophagus, stomach and duodenum (#PE8624).
Aggravated by Diabetes (#PE0102) Ascariasis (#PE2395)
Cement dust (#PE2854) Constipation (#PE3505)
Teeth disorders (#PD1185) Diseases of the pancreas (#PE1132).

♦ PC8885 Maldistribution of science and technology
Gap in scientific and technological capacity — Concentration of science and technology
Broader Maldistribution of resources (#PB1016).
Narrower Technology gap between developed countries (#PD0338)
Technology gap between developed and developing countries (#PE7985).
Aggravates Lack of information (#PF6337)
Inadequate research and development capacity in developing countries (#PE4880).
Aggravated by Adverse consequences of scientific and technological progress (#PF3931).

♦ PC8908 Declining economic productivity
Lack of productivity — Reduction in productivity
Narrower Loss of animal productivity (#PD8469)
Declining productivity of agricultural land (#PD7480)
Declining productivity in socialist countries (#PF7610)
Declining productivity in industrialized countries (#PD5543).
Aggravates Declining international competitiveness (#PD8994).
Aggravated by Competition (#PB0848) Natural disasters (#PB1151)
Proliferation of computers (#PE3959) Unproductive labour resources (#PC6031)
General unproductivity of capitalist systems (#PF3103)
Ineffective economic structures in industrial nations (#PE4818).
Reduces Industrial accidents (#PC0646).

♦ PC8914 Waste of human resources
Unproductive use of human resources — Wastage of human resources
Broader Unproductive use of resources (#PB8376).
Narrower Over-qualification (#PF3462) Educational wastage (#PC1716)
Mass unemployment of human resources (#PD2046)
Underutilization of intellectual ability (#PF0100)
Waste of human resources in capitalist systems (#PC3113).
Aggravated by Age segregation (#PD3444) Discrimination in employment (#PC0244)
Burdensome cost of religious ceremonies (#PF3313).

♦ PC8930 Irresponsible international trade
Broader Irresponsibility (#PA8658) Unethical commercial practices (#PC2563).
Narrower International arms trade (#PC1358)
International trade in endangered species (#PC0380)
Distortion of international trade by dumping (#PD2144)
Trade in animal products of endangered species (#PD0389)
Abuse of dominant market position in international trade (#PD6002)
Marketing of banned pharmaceutical drugs in developing countries (#PE6036)
Trading in products containing toxic substances with developing countries (#PE2061).
Aggravated by Competitive acquisition of arms (#PC1258).

♦ PC8943 Deterioration of human environment
Denial of right to environmental quality
Broader Denial of social rights (#PC0663) Environmental degradation (#PB6384).
Narrower Debilitating deterioration of physical environment (#PD2672).
Aggravates Spiritual hunger (#PA8038)
Lack of commitment to common symbols (#PE8814)
Symbols unrelated to human experience (#PF9070).
Aggravated by Spiritual void (#PA6220) Symbol system failure (#PF3715)
Faded community symbols (#PG8964) Oppressive prevalent images (#PF1365)
Uncontrolled urban development (#PC0442) Natural environment degradation (#PB5250)
Psychological environment degradation (#PC3144).

♦ PC8949 Legalized discrimination
Broader Discrimination (#PA0833).
Narrower Legalized racial discrimination (#PC3683)
Discrimination against foreigners in legal proceedings (#PF1798).
Aggravates Age discrimination (#PC2541)
Cultural discrimination in the administration of justice (#PE6529).

♦ PC8974 Depravity
Total depravity
Nature Total depravity is not only that every part of a human being is affected by sin, but that every part of every human is entirely corrupt.
Claim Humans are wholly defiled in all the faculties and parts of soul and body. They are utterly indisposed, disabled and made opposite to all good, and are wholly inclined to all evil.
Broader Vice (#PA5644) Disrepute (#PA6839).
Narrower Immoral literature (#PF1384).

♦ PC8999 Denial of rights of minorities
Denial of rights to ethnic minorities — Inadequate protection for linguistic, national and religious minorities
Nature No system in history has been able to guarantee the rights of minorities. In many countries they are not allowed to enjoy their own culture, to profess and practice their own religion, or to use their own language.
Counter-claim The national minorities' claims for their culture, religion, and language conceal a secessionist tendency.
Broader Denial of rights (#PB5405).
Narrower Underprivileged religious minorities (#PC2129)
Discrimination against immigrants and aliens (#PD0973)
Denial of economic and social rights to refugees (#PE6375).
Related Denial of education to minorities (#PC3459)
Denial of rights to vulnerable groups (#PC4405)
Discrimination against indigenous populations (#PC0352).
Aggravates Secession (#PD2490) Ethnic conflict (#PC3685)
Discrimination against dwarfs and midgets (#PE2635).
Aggravated by Racial discrimination (#PC0006).

♦ PC9025 Infection
Sepsis
Incidence As an indication, the market for antiseptics (and antiseptic disinfectants) in the UK in 1986 was £33.3 million.
Refs Finegold, Sydney M and George, W Lance *Anaerobic Infections in Humans* (1989); Gottschalk, G; Pfenning, W and Werner, H (Eds) *Anaerobes and Anaerobic Infections* (1980); Keith, Louis (Ed) *Common Infections* (1985); Keith, Louis (Ed) *Uncommon Infections and Special Topics* (1985); Klastersky, J (Ed) *Infections in Cancer Patients* (1982); Reid, Daniel, et al *Infections in Current Medical Practice* (1986); Siemens, Heide H *Infection and Infections* (1985); Strober, Warren, et al (Eds) *Mucosal Immunity and Infections at Mucosal Surfaces* (1988).
Broader Human disease and disability (#PB1044).
Narrower Diseases of the ear (#PD2567) Streptococcus infection (#PE3098)
Infection of the throat (#PG4579) Infection of the tonsils (#PG4580)
Infection of the sinuses (#PG4577) Urinary bladder disorders (#PE2307)
Diseases and injuries of the brain (#PD0992).
Aggravates Asthma (#PD2408) Neonatal mortality (#PD9750)
Foetal infection and death (#PE2041)
Inadequate immune responses in malnourished persons (#PE4883).
Aggravated by Susceptibility to infection (#PG4694).

♦ PC9042 Complications of childbirth
Complications of labour in childbirth — Complications of the puerperium
Nature These may include prolonged labour, breech delivery, cord prolapse, aspiration, and birth injuries. They may be complicated by premature labour and multiple pregnancies.
Refs Mauricev, Francis and Chamberlain, Hugh *The Diseases of Women with Child and in Childbed* (1985).
Broader Human disease and disability (#PB1044).
Narrower Birth trauma (#PE8911) Abnormal births (#PG6522)
Premature birth (#PD1947) Childbirth fever (#PG4639)
Low birth-weights (#PF5970) Puerperal pyrexia (#PG3941)
Puerperal phlebitis (#PG5854) Anaemia of pregnancy (#PG9577)
Disorders of lactation (#PG7881) Physical malformation of foetus (#PE2042)
Alienating child-birth environments (#PF6161)
Haemorrhage of pregnancy and childbirth (#PE4894).
Related Maternity (#PJ1893) Medical complications (#PE2863).
Aggravates Still-birth (#PD4029) Infant mortality (#PC1287)
Maternal mortality (#PD2422) Nutritional anaemia (#PD0321)
Personal life crises (#PD4840) Mothers' self-imposed isolation (#PJ9425)
Career interruption due to pregnancy (#PD7692)
Inequality of life expectancy by gender (#PE1339)
Discrimination against women in public life (#PD3335).
Aggravated by Domestic quarrels (#PE4021) Adolescent pregnancy (#PD0614)
Foetal erythroblastosis (#PG1921) Unsupervised home births (#PG7996)
Defective oxygen supply to the foetus (#PG9126)
Precipitate or ill-judged forceps delivery (#PG2953)
Medical experimentation on pregnant women and foetuses (#PE8343).
Reduced by Childlessness (#PC3280) Birth prevention (#PE3286).

♦ PC9053 Hazardous wastes
Incidence Bearing in mind the spectrum of definitions of "hazardous waste", it has been estimated that in 1984 some 325 to 375 million tons were generated world-wide, of which 5 million tons were in the newly industrialized and developing countries. Some 90 per cent of hazardous wastes are produced in the industrialized countries.
Refs Andelman, Julian B and Underhill, Dwight W (Eds) *Health Effects from Hazardous Wastes Sites* (1987); Hazardous Materials Control Research Institute *Management of Hazardous Wastes and Environmental Emergencies* (1985); International Environment Liason Centre *International Environment-Development Facts March 1989* (1989); OECD and NEA Staff *Near-Field Assessment of Repositories for Low and Medium Level Radioactive Waste* (1988); Wynne, B *Risk Management and Hazardous Wastes* (1987).
Broader Unproductive use of resources (#PB8376).
Narrower Radioactive wastes (#PC1242)
Methane gas emissions from landfill sites (#PE1256).
Aggravates Toxic waste smuggling (#PD9765) Hazards to human health (#PB4885)
Trading in products containing toxic substances with developing countries (#PE2061).
Aggravated by Toxic substances (#PD1115) Hazardous waste dumping (#PD1398)
Uncertain toxicity thresholds (#PF5188)
Ineffective official inspection of regulated activities (#PE4146).

♦ PC9058 Inadequate political structure
Related Political inertia (#PC1907).

♦ PC9067 Ill defined health conditions
Ill defined symptoms of disease — Ill defined diseases
Refs Britt, Beverley A (Ed) *Malignant Hyperthermia* (1987).
Broader Human disease and disability (#PB1044).
Narrower Uraemia (#PG2241) Acidosis (#PG4564) Headache (#PE1974)
Senility (#PE6402) Dehydration (#PE8062) Nervousness (#PE4171)
Hyperpyrexia (#PG2631) General debility (#PG3424)
Ill defined causes of morbidity (#PE5463) Symptoms referable to sense organs (#PE2665)
Symptoms referable to nervous system (#PE9468)
Symptoms referable to digestive system (#PE9604)
Symptoms referable to respiratory system (#PE7864)
Symptoms referable to cardiovascular system (#PE3933)
Symptoms referable to genito-urinary system (#PE9369)
Symptoms referable to musculoskeletal system (#PE4566).
Related Mental depression (#PC0799).

♦ PC9083 Cultural decline
Narrower Decline in religious broadcasting (#PF1433).
Related Obsolete basis of cultural identity (#PF0836).
Aggravated by Cultural illiteracy (#PD2041) Anti-intellectualism (#PF1929)
Rapidly changing cultures (#PF8521) Denial of right to a cultural life (#PE6561)
Untransposed significance of cultural tradition (#PF1373).

♦ PC9104 Abuse of government power
Misuse of political power — Abuse of public power — Political coercion
Nature The abuse of power involves the deliberate use of power for specific aims that could not be legitimately justified since they are often for the exclusive benefit of power itself, for the maintenance of a political regime or of an unjust social and economical system. High public officials who, protected by their position, accumulate unjustifiably their wealth, undertake all manner of business, favour their protégés and receive donations or improper "commissions".
Refs Airaksinen, Timo *Ethics of Coercion and Authority* (1988); Boggs, Carl *Social Movements and Political Power* (1987).
Broader Abuse of power (#PB6918) Abuse of authority (#PC8689).
Narrower Parole violation (#PE1121) Forced confession (#PE8947)
Misuse of resources (#PB5151) Official corruption (#PC9533)
Misuse of statistics (#PF4564) Political aggression (#PD8877)
Search without warrant (#PG6631) Bureaucratic aggression (#PC2064)
State sanctioned torture (#PD0181) Extortionate bureaucracy (#PD8655)
Abuse in government policy (#PF8389) Excessive extralegal powers (#PD3246)
Border incidents and violence (#PD2950) Governmental abuse of extradition (#PE6004)
Misuse of food as a political weapon (#PF6202)
Unchecked power of government bureaucracy (#PD8890)
Inadequate assistance to victims of torture (#PE6936)
Official cover-up of government harassment of political activists (#PF3819)

Misuse of nonprofit associations as front organizations by government (#PE0436)
Persistence of a technical state of war following cease–fire agreements (#PE2324).
Related Political repression (#PC1919)   Abuse of police power (#PC1142)
Abuse of international cultural, diplomatic and commercial exchanges (#PF3099).
Aggravates Torture schools (#PE2062)   Infringement of privacy (#PB0284).
Aggravated by Abuse of influence (#PC6307)   Contempt for democratic processes (#PF0639)
Abusive distribution of political patronage (#PF8535)
Unaccountable government intelligence agencies (#PF9184).
Reduced by Government limitations (#PF4668).

♦ **PC9112 International economic injustice**
Unjust international economic order — Unjust global economic system — Inadequate world economic system — Asymmetry in economic interdependence — Asymmetric interdependence between developed and developing countries
Nature The developing countries, which constitute 70 percent of the world's population, account for only 30 percent of the world's income. It has proved impossible to achieve an even and balanced development of the international community under the existing international economic order. The gap between the developed and the developing countries continues to widen in a system which perpetuates inequality and which was established at a time when most of the developing countries did not even exist as independent states. The present international economic order is in direct conflict with current developments in international political and economic relations. Since 1970, the world economy has experienced a series of grave crises which have had severe repercussions, especially on the developing countries, because of their generally greater vulnerability to external economic impulses.
Incidence Changes in the last four decades have resulted in the post-war consensus on the rules and regulations governing the world economy being continuously eroded. Most developing countries which were not present at the creation of these rules and regulations now question their legitimacy and equity. Some among them have asserted control over a strategic resource – petroleum – with spectacular results; others continue similar efforts, but without signal success to date. The revival of economic strength in post–war Europe and Japan led not only to a questioning of the rules they found inconvenient, but to the collapse of these rules; the most significant case being the breakdown of the world monetary system which had been agreed at Bretton Woods in 1944.
Claim The loss of consensus and the actual breakdown of some systems of global economic management have left the world economy in precarious circumstances. Currency values are unstable, trade is suffering as governments seek to shelter domestic producers, growth rates have slowed, and unemployment is increasing. And all this not because the world is incapable of producing enough for its growing population or has run out of raw materials, but simply because countries have not been able to agree on a fair system of managing the world's economy.
Refs Simri, M *Inderdependence and Conflict in the World Economy* (1981).
Broader Antiquated world socio–economic order (#PF0866).
Narrower Distortion in international trade (#PC6761)
Vulnerability of developing countries (#PC6189)
Financial paralysis of developing countries (#PD9449)
Ineffective economic structures in industrial nations (#PE4818)
Deterioration of terms of trade for developing countries (#PD2897)
Technology gap between developed and developing countries (#PE7985)
Undermining of multilateral forums by industrialized countries (#PE4289)
Exploitative transformation of international division of labour (#PF9281)
Protectionism in international trade against exports from developing countries (#PD9679)
Development by industrialized countries of products substituting for commodities exported by developing countries (#PD7682).
Related Inadequacy of economic doctrine (#PF3395).
Aggravates Deflation (#PD7727)   Uncontrolled markets (#PF7880)
Vulnerability of social systems (#PB2853).
Aggravated by Economic inefficiency (#PF7556)   Inequitable distribution of wealth (#PB7666)
Lack of progress in establishing a New International Economic Order (#PF4306)
Disproportionate control of global economy by limited number of corporations (#PE0135).

♦ **PC9152 International inequality**
Refs Wallerstein, Immanuel (Ed) *World Inequality*.
Broader Inequality (#PA6695)   Human inequality (#PA0844).
Narrower Unequal opportunities for media reception (#PD3039)
Disparity between industrialized and developing countries (#PC8694).
Aggravates Unfairly negotiated treaties (#PF4787).

♦ **PC9173 Institutionalized corruption**
Broader Corruption (#PA1986).
Narrower Corruption in prisons (#PD9414)   Corruption of the judiciary (#PD4194)
Unethical military practices (#PD7360)   Corruption of ruling classes (#PD8380)
Unethical industrial practices (#PD2916)   Corruption of government leaders (#PC7587)
Corruption in organized religion (#PC3359)   Corruption in developing countries (#PD0348)
Political corruption of the judiciary (#PE0647)
Unethical practices in local government (#PD5948)
Corruption amongst relatives of government leaders (#PE9140).
Related Police corruption (#PD2918)   Corruption in politics (#PC0116)
Bureaucratic corruption (#PC0279)   Unethical commercial practices (#PC2563)
Unethical professional practices (#PC8019).
Aggravates Resignation towards bribery (#PF8611)
Unchecked power of government bureaucracy (#PD8890)
Corruption and mismanagement of foreign aid (#PD0136).

♦ **PC9188 Abuse of science**
Misuse of scientific research — Misuse of scientific theories
Incidence Theories and data have been misused throughout the history of science to justify racial discrimination, violence and war. The theory of evolution has, for example, been used not only to justify war, but also genocide, colonialism and the suppression of the weak.
Broader Exploitation of trust (#PC4422).
Narrower Gender abortions (#PD3947)   Deceptive misuse of research (#PD7231)
Suppression of scientific information (#PF1615).
Aggravates Inhumane scientific activity (#PC1449)
Politicization of scholarship (#PF7220)
Biologically determined aggression (#PF7490).
Aggravated by Military secrecy (#PC1144)   Abuse of authority (#PC8689)
Secrecy in scientific research (#PF1430)   Deception by natural scientists (#PD9182)
Deceptive social science research (#PF7634).
Reduced by Anti–science (#PF2685).

♦ **PC9513 Ineffective legislation**
Ineffective regulations — Regulatory loopholes — Ineffective agreements
Nature Normal social legislation, particularly that without any time limits, although often based on inadequate and suspect knowledge, manages to persist regardless of its success or failure. This is partly because, once instituted, it gathers to itself many forms of vested interest opposed to its reversal, but also because it places the burden of proof of ineffectiveness on its opponents – usually under conditions which makes it virtually impossible to do so.

Narrower Ineffective tax systems (#PF1462)
Resistance to internationally agreed standards (#PC4591)
Ineffective legislation against organized crime (#PE6699)
Inadequate legislation against environmental pollution (#PF9299)
Exploitation of regulatory loopholes in countries with underdeveloped legislation (#PE4339).
Aggravates Proliferation of legislation (#PD5315).
Aggravated by Ineffective international agreements (#PF6992)
Limited local respect for regional and global legislation (#PF2499).

♦ **PC9533 Official corruption**
Corruption of officials
Broader Corruption (#PA1986)   Abuse of government power (#PC9104).
Narrower Police corruption (#PD2918)   Corruption in prisons (#PD9414)
Bureaucratic corruption (#PC0279)   Corruption of the judiciary (#PD4194)
Unethical military practices (#PD7360)   Corruption of government leaders (#PC7587)
Political corruption of the judiciary (#PE0647)
Unethical practices in local government (#PD5948)
Corruption of customs and excise officials (#PE4033).
Aggravates Unethical practices of regulatory inspectors (#PF8046)
Mismanagement by intergovernmental organization leadership (#PE6947).
Aggravated by Limited accountability of public services (#PF6574).

♦ **PC9547 Territorial expansionism**
Broader Expansionism (#PB5858).
Aggravates Occupied territories (#PD8021)   Territorial disputes between states (#PC1888)
Disruption of territorial integrity (#PC2945)
Conflicting claims to non–terrestrial territory (#PG4114)
Conflicting claims concerning Antarctic territory (#PG4115)
Unlawful interference with rights of innocent passage in territorial waters (#PE6755).

♦ **PC9584 Fragmentation of the international trading system**
Circumvention of international trading agreements — Inadequate international trading system — Breakdown of international trading discipline — Deterioration in long–term trade policies — Uncertainty in the international trading system — Declining credibility of the international trading system
Nature Growth and development require an increase in imports and hence in exports. Even if imports are financed in part through borrowing, exports need to grow to allow the additional debt to be serviced. A defective trading system which limits exports from developing countries can therefore hold back development. Indeed, at a time of weak commodity prices and contraction of net financial inflows, it can make it impossible for some countries to avert a downward spiral triggered by import reduction and investment cuts. The contraction in imports into developing countries inevitably has adverse effects on the output of the export industries of the industrialized countries, tending to create a self-reinforcing contractionary spiral that constitutes a major challenge for the international community.
Background One of the main factors undermining the international trading community's resolve to deal in a concerted manner with the tendencies eroding the multilateral system has been the widely shared perception that other trading partners are giving less and deriving more from the system. Part of this problem can be attributed to historical factors present at the birth of the system itself, in that the failure to put the Havana Charter of 1948 fully into effect resulted in issues of crucial interest to certain countries being left outside the system of contractual rights and obligations. Other factors have arisen from developments that could not have been foreseen at that time.
Incidence The post–Bretton Woods international financial and monetary system has in the 1980s signally failed to provide the necessary finance in amounts and on terms which would enable development to be effectively pursued. The international trading system is in disarray. Far from the integration of trade practices into a comprehensive and universal set of rules accepted by all governments, the past decade has seen a fragmentation of the trading system, with an ever-increasing number of trade issues being treated outside the framework of GATT and a large and growing number of GATT rules themselves being circumvented. Such circumvention is often the reflection of the particular interests of domestic high–cost producers, thus attenuating the broader international benefits of a properly functioning trading system.
From the outset the GATT system had a clear aim, namely the negotiation of binding commitments and of the reduction of tariffs and of tighter disciplines on non–tariff measures to reduce to a minimum the possibilities of "nullification" and "impairment". This course of action was pursued with success from 1947 through the Tokyo Round, which concluded in 1979. However, at the same time as the commitments and disciplines were being negotiated, tendencies serving to frustrate the benefits of these accomplishments emerged and threatened to reverse the whole process. By 1982, the international trading community recognized the "erosion" of the system, which has been manifested by a variety of symptoms. These include: (a) a neglect of the unconditional most-favoured-nation principle, the fundamental rule of GATT, in particular in application of discriminatory trading measures against developing countries; (b) the introduction (or resurrection) of concepts such as "market disruption" and "conditional" treatment to provide justification for such discrimination and for trade policies which conflict with basic principles of the multilateral system; (c) the decline of the relevance of the tariff as an instrument of trade policy; (d) an increased resort to non-tariff measures of flexible application, especially those designed to "harass" trade; (e) the proliferation of mechanisms for the management of the quantities, prices, and often the sources of imports; (f) an unravelling of previous commitments and undertakings, including those with respect to particular products or in favour of the developing countries; (g) general dissatisfaction with, and lack of commitment to, multilateral dispute settlement mechanisms; (h) an inability to translate multilateral rules and principles into national laws an regulations in such a way as to effectively guarantee their respect in the national context; (i) an emphasis on bilateral trade flows and a drift away from multilateral reciprocity. Developing countries have been particularly penalized by these tendencies.
Most governments are quick to demonstrate how the trading system works primarily to the advantage of others. The net result, however, is an ambiance of indiscipline, which has impeded the ability of the international community to address effectively the recognized problems confronting the trading system, created frustrations, exacerbated tensions in international trade relations and led to "pragmatic" solutions or retaliation.
In addition to this breakdown of trade discipline, trade policy has increasingly been affected by financial disturbances, in particular, by sharp swings in real effective exchange rates and interruptions of capital flows to developing countries. As a result of their severe external payments problems, the trade policies of developing countries are now being determined by external pressures arising from adverse movements in the terms of trade and changing exchange rates, interest rates, foreign investment patterns and capital flows, rather than by national development objectives.
Broader International economic fragmentation (#PC0025).
Narrower Inadequate barter system in international trade (#PE0117).
Aggravates Short range planning for long–term development (#PF5660)
Imposition of trade quotas for political reasons (#PE9762).
Aggravated by Trade protectionism (#PC4275)   Exchange rate volatility (#PE5930)
Misalignment of currencies (#PF6102)   Speculative flight of capital (#PC1453)

PC9584

Lack of international cooperation (#PF0817)
Financial destabilization of world trade (#PC7873)
Lack of international coordination of interest rates (#PF3141)
Unilateral interpretations of multilateral principles (#PF9629)
Deterioration of terms of trade for developing countries (#PD2897)
Over-reliance on economic interest groups by policy agencies (#PF1070)
Inability to negotiate effective multilateral safeguard systems (#PF5287).

♦ **PC9585 Authoritarian regimes**
Authoritarian government — Undemocratic regimes — Non-representational government — Ruling cliques — Government autocracies
Nature Authoritarian governments are characterized by continuity in leadership, insulation from societal pressures, well-established and integrated interest groups and the power to enforce decisions without the need to respond to the interests of minorities or the disenfranchised. In weak authoritarian governments, where political authority is maintained through personalistic, patron-client relations, tend to be ineffective at economic and social reform. The maintenance of political power often depends on the discretionary use of public funds so that any rational reform of public finances becomes irrational. Such regimes are likely to have more difficulty imposing reform than consultative democracies.
Claim Autocracies depend on the cadre of officials, police and military officers who regularly obey orders. Sometimes the obedience grows out of ideological or religious convictions, but usually it is enough that the relevant officials and officers believe that they will be punished if they fail to carry out orders and rewarded if they do. When they stop believing in the invincibility of the authoritarian regime, the autocrats will lose their power.
Counter-claim Strong authoritarian governments tend to be relatively successful in imposing the short-term costs of economic reform.
Refs Neufeld, Maurice F *Poor Countries and Authoritarian Rule* (1965).
  Narrower Military government (#PC0698)   Political dictatorship (#PC0845).
  Aggravates Elitist ruling classes (#PF4849)   Deception by government (#PD1893).

♦ **PC9586 Inequality in mortality rates**
Unequal mortality rates within countries
Refs Preston, S H *Mortality Patterns in National Populations* (1976).
  Narrower Inequality of life expectancy by gender (#PE1339).
  Unequal morbidity and mortality between countries (#PC6869)
  Unequal mortality of the elderly between countries (#PE4354).
  Aggravated by Inadequate health services (#PD4790).

♦ **PC9623 Diseases of the sense organs**
Disorders of sense organs
  Broader Human disease and disability (#PB1044).
  Narrower Diseases of the ear (#PD2567)   Eye diseases and disorders (#PD8786)
  Symptoms referable to sense organs (#PE2665).

♦ **PC9695 Dissidents**
Dissidence
Refs Degenhardt, Henry W *Political Dissent* (1983).
  Aggravates Political instability of developing countries (#PD8323).
  Reduced by Repression of intellectual dissidents (#PD0434).

♦ **PC9714 Inadequate teaching**
Inadequate standards of teaching — Low quality teaching
Claim With the modern teaching methods children are left alone to find things out for themselves. Teachers' role is passive: remaining in the background, suggesting topics and supplying materials. As a result the the pupils are unsure of the lessons' purpose, bored and misbehaving.
  Broader Inadequate education (#PF4984).
  Aggravates Educational wastage (#PC1716).

♦ **PC9717 Anthropogenic climate change**
Degradation of climate by man
Nature Human activities induce climatic changes. Indeed, mankind continues to modify the surface of the earth in ways that affect local climate; the main effects are three-fold. Firstly, the ratio of solar radiation reflected from and absorbed by the earth's surface is changed. Secondly, the ratio of convective and evaporative heat released from the earth's surface is changed. Finally, the hydrologic cycle is modified.
With human progress the direct effects of energy and mass transfer upon atmosphere are acquiring more and more significance. Thermal pollution of the atmosphere is observable over the largest cities. There is the pollution of atmosphere by aerosols and man-made industrial gases, such as soot and oxides of carbon and nitrogen. Man can also change the climate intentionally: meteorologists are trying to modify the weather, to prevent hail storms or to produce rain.
  Broader Natural environment degradation (#PB5250).
  Narrower Hostile environmental modification (#PD7941)
  Inadvertent modifications to climate (#PC1288)
  Adverse effects of urbanization on climate (#PE9020)
  Destructive changes in ocean characteristics (#PC2087)
  Adverse effects of power production on weather (#PE9134).
  Aggravates Global warming (#PC0918)
  Long-term changes in precipitation patterns (#PE4263).

♦ **PC9718 Unemployment in developed countries**
Unemployment in industrialized countries
Nature Unemployment constitutes a heavy social cost and contributes to a resurgence of protectionism. Political tension associated with high unemployment may also account for the reluctance of industrialized countries to expand their aid programmes.
Incidence Unemployment shows no sign of declining in many industrial countries. In Europe the average rate of unemployment has remained about 10 per cent since 1983.
  Broader Unemployment (#PB0750).
  Narrower Unemployment in developed countries resulting from participation of developing countries in manufacturing (#PE0402).

♦ **PC9733 Exploitation of women**
Claim The so-called "emancipation" of women means all too often means their merciless commercial exploitation, for example in the movie industry.
  Broader Lack of protection for the vulnerable (#PB4353).
  Narrower Trafficking in women (#PC3298)   Social subjugation of women (#PD4633)
  Sexual exploitation of women (#PD3262)   Lack of payment for housework (#PE4789)
  Exploitation of women refugees (#PD5025)
  Dependence of developing countries on unpaid female labour (#PE4451).
  Related Discrimination against women in employment (#PD0086).
  Aggravates Violence against women (#PD0247).

♦ **PC9785 Maldistribution of productive capacity**
Claim Appropriate development is inconceivable without radical change and redistribution. It will not result from continued pursuit of economic growth or the further enrichment of the wealthy, and any hoped for trickle down effects. The need is not so much for redistribution of existing wealth from the wealthy to the poor, rather what is most required is a redistribution of the existing productive capacity, especially the land, so that people can produce for themselves the things that they perceive as necessary for modest but adequate standard of living.
  Related Maldistribution of land in developing countries (#PD0050).
  Aggravates Maldevelopment (#PB6207).

Detailed problems   PD

Content

This section groups together the detailed and sectorally-specialized problems which tend to be prominent on the agendas of international organizations with specialized concerns, as well as in the media. Such problems also tend to group, or focus, many of the even more specialized problems which are described in subsequent sections. Indeed in many debates discussion of those more specialized problems may be subsumed under discussion of these problems.

Note that further information relevant to an understanding of the problem may be present in other problems cross-referenced in the entry consulted.

The problems in this section are often sectoral variants on the broader or more basic problems described in the previous sections (Section PB or PC).

This section groups 1,928 problems for which there are 17,269 cross-references.

Rationale

Inclusion of such problems calls for no comment because of their widely recognized importance. Where they are cross-sectoral variants of those in the previous section, their inclusion here prevents neglect of the sectoral specificity, as tends to be the case when such problems are subsumed under those of the broader problems in Section PB.

Method

The entries are based on information obtained from international organizations, from a wide variety of reference books, or as reported in the international media. The procedures for identifying world problems are described in Section PZ at the end of this volume.

Index

A keyword index to entries is provided in Section PX.

Comment

Detailed comments are given in Section PZ at the end of this volume.

Reservations

The emphasis throughout this volume has been placed on providing descriptions of less well-known problems, particularly when the extensive material available on the better known problems contained neither succinct descriptions of them nor descriptive material which could easily be reduced to succinct descriptions. The problem descriptions here represent a compilation of views from published documents (usually from international organizations). The text provided does not necessarily constitute the best possible description of the problem, since a compromise has had to be struck between availability of information, the resources to process it, and the space available in this volume.

In a number of cases a problem could have been allocated to another section. Inclusion of a problem in this section, rather than in a preceding or following section, has been based on a number of factors. The position of the problem in one or more hierarchies of cross-references was a major factor in determining its allocation to this section.

Possible future improvements

There is much scope for improving the quality of problem entries through feedback from interested bodies. More bibliographic references could be included where appropriate, as well as references to major resolutions concerning those problems recognized by the United Nations. There is also much scope for improving the pattern of cross-references, both between problems, to other sections of this volume (eg values) and to the 20,000 internationally-active bodies in the companion series (*Yearbook of International Organizations*)

ENTRY CONTENT AND ORGANIZATION

Ordering of entries
Entries are in **numeric order**. Entry numbers have been **allocated randomly**; they have no significance other than as a permanent point of reference to facilitate indexing, cross-referencing, and updating between editions.

Index access to entries
The location of an entry in this sub-section may be determined from
the Volume Index (Section PX) on the basis of keywords in the name of the entry or its alternate titles.

Structure of entries
Entries may be composed of the following descriptive elements:

(a) **Entry number** This number has **no significance**, except as a convenient method of identifying the entry (particularly for indexing purposes), of filing information on it, and as an identifier to which cross-references from other entries (possibly in other Sections) may refer in this and future editions. The first letter of the entry number refers to the section of this volume in which the sub-section, denoted by the second letter, is located.

(b) **Problem name** This is printed in bold characters. It is the name selected as best indicating the nature of the problem. It may be followed by alternative problem names.

(c) **Nature** Description of the problem which attempts to identify the nature of the disruptive processes involved. The information included here, and in the following paragraphs, is compiled directly, to the extent possible, from available published documents. Where appropriate the text included may be reproduced, in a minimally edited form, from the publications of international organizations, such as those of the United Nations or its Specialized Agencies.

(d) **Incidence** Summary description of the extent of the problem which makes it of more than national significance.

(e) **Background** Describes briefly when and how the problem's importance was recognized initially, and how this recognition has evolved over time.

(f) **Claim** Stresses the special importance of this problem and why action is particularly urgent. This paragraph offers means of including statements which may deliberately exaggerate claims for the unique importance of the problem.

(g) **Counter-claim** Stresses, where appropriate, the relative insignificance or erroneous conception of the problem as described. This paragraph offers a means of including statements which may deliberately exaggerate the arguments refuting the evidence for the existence of the problem. Absence of such arguments from the text does not mean that they do not exist.

Cross-referencing of entries
At the end of any entry, there may be cross-references to other entries. These indicate the number and name of the cross-referenced entry, whether within this Section or in other Sections. There are 3 types of **hierarchical** cross-references between problems:

    Broader = Broader problem: more general problems of which the problem described may be considered a part. The described problem may be considered an aspect of several broader problems
    Narrower = Narrower problem: more specific problems which may be considered a part of the described problem
    Related = Related problem: problems that may be considered as associated in a hierarchically undefined way with the described problem.

There are 4 types of **functional** cross-references between problems:
    Aggravates = Problems aggravated by the described problem: a forward or subsequent negative causal link
    Aggravated by = Problems aggravating the described problem: a backward or prior negative causal link
    Reduces = Problems relieved, alleviated or reduced by the described problem: a forward or subsequent positive causal link
    Reduced by = Problems relieving or alleviating the described problem: a backward or prior positive causal link

♦ PD0008 Flooding of the urban labour market in developing countries

Nature The rapid growth of the urban population in developing countries is largely due to rural migrations which, given their social and economic origin, are less predictable than are the other two growth factors, fertility and mortality rates. Some migratory movements are encouraged by the reality of growing urban industries, but it is mainly the erratic migrations from destitute rural areas, motivated by despair rather than by definite objectives, which furnish the greater part of the displaced peasants flooding the urban labour market in the cities of Asia and Latin America. Long years of unemployment gradually undermine their moral and physical strength and further reduce their chance for future economic and social advancement.

Claim The assumption that the influx of population to the urban areas is the response to an increased demand for labour is not valid for the third world. Unlike the situation in the rich countries where the growth of the service sector is an indicator of a high degree of economic development, in the poor countries it is more often the symptom of continued poverty and stagnation. There, while urban labour potential increases, many workers fail to find employment not only in industry which is lacking or underdeveloped but also in the organized service sector. Forming rather what may be termed a 'sub-' or 'pseudotertiary' sector, this substantial part of the third world labour force makes up the petty retail trade of hawkers and peddlers, as well as the low-productivity field of poorly paid personal and domestic services engaged by the upper social strata. They are the hand-to-mouth sector.

Broader Unemployment (#PB0750)
Disparity in social development within developing countries (#PD0266).
Narrower Urban unemployment in developing countries (#PD1551).
Aggravated by Urban bias (#PF9686) Unsustainable population levels (#PB0035).

♦ PD0014 Proliferation of strategic nuclear arms

Strategic arms competition — Strategic arms rivalry

Nature The two superpowers (USA, USSR) and regional powers (India, Israel and perhaps Brazil, Pakistan, Syria) have developed and deployed nuclear weapons in a wide variety of forms. Strategic weapons include intercontinental ballistic missiles and other weapons, such as long range bombers and submarine launched ballistic missiles, with which the two powers can threaten each other without using the territory of their allies. In each nation, the groups which govern policy have a hostile perception of the other nation engaged in the development and deployment of such weapons systems and will exaggerate the number and capabilities of its strategic weapons systems in an attempt to justify the development by their own nation of more and better systems. It has not proved possible to control this arms race, particularly since agreement reached on the limitation of the total number of delivery vehicles, for example, can be by-passed by substitution of a qualitatively superior vehicle for an existing type, or introducing multiple warheads delivered by a single vehicle. Such agreements may also be effectively by-passed by the construction of defensive anti-ballistic missile systems resulting in an offence/defence race. Agreement to cease research and development along particular lines always leaves the opportunity for each side to divert resources into alternative lines of advance.

The long-term effect of the deployment of MIRVs (multiple independently targetable re-entry vehicle) and ABM (anti-ballistic missile) systems will be to erode the strategic balance which exists between the superpowers, creating an environment of uncertainty in which neither side can know the strength of the other's strategic forces, increasing the threat of general nuclear war, and making arms control and disarmament measures very difficult, if not impossible, to negotiate.

Incidence The Stockholm International Peace Research Institute's figures for the end of 1982, indicate that the strategic nuclear weapon delivery capability for the USA is 1,264 nuclear bombs and 2,224 nuclear missiles (land, sea and air to air). For the USSR, the figures are 290 nuclear bombs and 6,366 nuclear missiles of all strategic types. The delivery capability is based on a single mission for delivery vehicles such as bombers and submarines; and on current strategic missile deployment. Nuclear weapon stockpiles are much larger, however. Various estimates place these at about 30,400 for the USA; 940 for China; 720 for France and 680 for the UK (one source has 1,700 for the UK). These include strategic, theatre and tactical weapons.

Counter-claim The development and deployment of nuclear weapons has had a general dampening effect on East-West antagonism, particularly by inhibiting the resort to war and in promoting management of crises. Because it was basically a bilateral balance in nuclear capability it tended to be stable. The pace of nuclear build-ups in the context of a structure of opposing relationships permitted time and circumstances for peaceful adjustments. The special structure and magnitude of the arms competition underlying the nuclear balance provided assurance against, rather than provocation to, an initiation of nuclear conflict. At all times there was a mutual recognition that the great risks and costs of any direct military encounter would clearly offset any value of political or economic objectives that might be gained by such an encounter.

Broader Competition (#PB0848) Competitive acquisition of arms (#PC1258)
Proliferation of nuclear weapons and technology (#PD0837).
Narrower Proliferation of nuclear weapons in developing countries (#PE9052)
Super-power monopoly of advanced nuclear warfare technology (#PD4445).
Aggravates Dehumanization (#PA1757)
Accidents to nuclear weapons systems (#PD3493)
European insecurity and vulnerability (#PD1863)
Failure of disarmament and arms control efforts (#PF0013).
Aggravated by Imbalance in strategic arms (#PC1606)
Surface to surface missiles (#PE4515).

♦ PD0022 Abuse of plant drugs

Nature Since so many plants contain chemicals which affect the human organism, abuse of plant drugs may be unintentional, i.e. through the unwitting ingestion of natural toxic plant substances. Typical drug reactions may also occur as a result of ingesting chemical fungicides, pesticides or other chemical preparations applied to plants. Many of these chemicals interfere with important enzyme systems and induce neurometabolic disorder. However, some plants yield psychoactive substances with little or no processing and may be readily abused.

Incidence Drugs easily extracted from plants by chewing, infusion or simple extractive processes such as maceration include mescaline, khat, coca, and betel. In tropical countries millions of persons in rural areas may be addicted to these substances.

Broader Drug dependence (#PD3825).
Narrower Abuse of khat (#PE0912) Cannabis abuse (#PE1186)
Abuse of opiates (#PE1329) Abuse of coca and cocaine (#PD2363).
Related Smoking (#PD0713) Abuse of medical drugs (#PD0028).
Aggravates Hysteria (#PE6412) Porphyria (#PG1855)
Schizophrenia (#PD0438) Multi-drug abuse (#PD0213).
Aggravated by Inadequate drug control (#PC0231)
Inadequate testing of drugs (#PD1190)
Inadequate information on drugs (#PF0603).

♦ PD0024 Inappropriate employment incentives

Ill-conceived incentives for employment — Reinforcement of unemployment — Inadequate work incentives

Nature As the "baby-boom" generation ages, skilled labour is becoming more difficult to find and to keep. Unemployment benefits in many developed countries are too close to minimum wages to make "drudgery" style employment worth the effort. There is a unprecedented demand for a sense of individual and team significance and for a level of involvement in the whole working process itself, for which few production systems are prepared. A culture of unemployment is developing, in which many people would prefer to be unemployed than to work in undesirable circumstances. This culture is not without its social price. The pattern of wages and salaries in some countries sustains a perverse incentive structure which over-encourages searching, or waiting, for certain types of work. The result is a number of candidates far exceeding that which can be absorbed in the occupations concerned. The incentive structure, making a limited number of job categories more desirable than the rest, may arise as a result of market forces freely creating a demand, and hence offering a premium to certain occupations. It may also be artificially created by government policies that attract large numbers of workers to one or a few types of developmental sectors, to the neglect of the infra-structure viewed as a totality.

Refs Hauk, Warren C (Ed) *Motivating People to Work* (1984).

Broader Unemployment (#PB0750)
Economic and social underdevelopment (#PB0539)
Over-specialized supervisory personnel (#PF3588).
Aggravates Educated unemployed (#PD8550) Shortage of skilled labour (#PD0044)
Limited employment options (#PF1658) Alcoholic intoxication at work (#PE2033)
Maladjustment to disciplines of employment (#PD7650)
Limited availability of permanent employment in inner-cities (#PE1134)
Emigration of trained personnel from developing to developed areas (#PD1291).
Aggravated by Unsatisfied need for continuing education (#PF0021)
Inappropriate education in developing countries (#PF1531).

♦ PD0028 Abuse of medical drugs

Nature The abuse of readily available legal drugs may take the form of overwillingness by doctors to prescribe strong drugs for ailments calling for less addictive or harmful means; overdosage, or other misuse, by the patient; or experiments by young people with drugs taken initially from medicine cabinets (illicit drug traffic is greatly aided by legal production). Further abuse is sometimes found in the medical profession itself (cocaine, opium, morphine, codeine, LSD, barbiturates, hypnotics, etc, are currently used in medicine). Inadequate control, testing, and general information have disastrous effects.

Broader Substance abuse (#PC5536) Drug dependence (#PD3825).
Narrower Abuse of antibiotics (#PE6629) Abuse of amphetamines (#PE1558)
Dependence on minor tranquillizers (#PE1821) Abuse of sedatives and tranquillizers (#PE0139).
Related Abuse of plant drugs (#PD0022) Abuse of animal drugs (#PE0043)
Inadequate medical care (#PF4832)
Excessive medical intervention in childbirth (#PE7705).
Aggravates Drug resistance (#PF9659).
Aggravated by Inadequate drug control (#PC0231).

♦ PD0033 Tsunamis

Tidal waves — Seismic sea waves

Nature Tsunamis are often called tidal waves, but this term is a misnomer. Unlike regular ocean tides, tsunamis are not caused by the tidal action of the moon and sun; they are very large ocean waves caused by vertical displacements along earthquake faults on the sea bottom, by submarine landslides and by volcanic eruptions beneath the sea. The velocity of these waves in the open sea is high but dependent upon the depth of the sea (500ft, 75 knots; 5,000ft, 240 knots; 15,000ft, 420 knots). Although they may be several metres in height at the point of origin and may reach 30 metres, they quickly lose height and become exceedingly long (up to 500 miles). The waves may last for a period between 10 and 60 minutes. They do not break in the usual manner but strike the coastline in a manner which gives rise to extremely dangerous and turbulent conditions. At the mouths of large rivers, tidal bores may be formed which may travel many miles upstream as solitary waves. Tsunamis may cause more havoc than the earthquakes which created them.

Incidence The circum-Pacific belt is the earth's most active seismic feature. Every island and coastal settlement in the Pacific Ocean area is vulnerable to the onslaught of the great waves. Those of 1868 and 1877 devastated towns in northern Chile, and caused death and damage across the Pacific. A series of waves generated by the eruption and collapse of Krakatoa in 1883 killed more than 36,000 persons in the East Indies. Japan lost 27,000 lives to the wave of 1896, and 1,000 more to that of 1933. There have been hundreds more whose effects were less spectacular but which took many lives and did much damage. The 1946 waves which struck Hawaii are a case in point, as are the waves sent out by the 1960 Chilean earthquakes; and the great Alaskan earthquake's sea waves, in 1964, caused damage as far away as California, Hawaii, Chile, and Japan.

Broader Geological hazards (#PC6684).
Related Storm surges (#PD2788).
Aggravates Famine (#PB0315) Floods (#PD0452) Tidal water damage (#PJ9589).
Aggravated by Earthquakes (#PD0201) Volcanic eruptions (#PD3568).

♦ PD0036 Cancer-causing foods

Carcinogenic diet — Food-related cancers

Nature The view that dietary practices might be a causative factor in cancer is not new, and not all cancer-producing agents are the results of man's industrial or agricultural activity. Some of the substances responsible for cancer occur naturally and are ingested by man through the food and drink he consumes daily.

Incidence There is increasing evidence that composition of daily food may significantly influence cancer incidence. This may be partially due to the presence of naturally occurring carcinogens but presumably mainly through the equilibrium of as yet ill-defined cancer-promoting and cancer-inhibiting factors in food. Some potent carcinogens arise from natural processes, and others are result from the pesticides and herbicides — many of which have been implicated in cancer —with which food products have been treated. Mutagens are present in substantial quantities in fruits and vegetables, and carcinogens are formed in cooking as a result of reactions involving proteins or fats. Rancid fats are possible causative agents of colon and breast cancer in humans, and fat intake increases the chances of getting such cancer. Burnt and browned materials formed by heating proteins during cooking are highly mutagenic. Salt-cured or pickled foods may be linked with increased risk of cancers of the stomach and oesophagus, and traditional smoking processes for ham, fish and sausage cause the absorption into the food of cancer-causing tars similar to those contained in tobacco smoke. In addition, cancer risks are higher among people who are 40 percent or more overweight and heavy drinking has been linked with cancers of the mouth, larynx and oesophagus.

Claim It has been estimated that in the UK, food-related cancers constitute 35 per cent of all cancers. Results of current studies are beginning to delineate more sharply specific causative agents, and when more definitive information is available, it should be possible for prudent people to choose fruits and vegetables that present minimal hazards. Until then, it is recommended that dietary fat should be reduced, trimming the fat from fish, meat or poultry, and that consumption of fibre-containing foods should be increased, in particular whole-grain foods and firm fruit and vegetables.

Broader Carcinogenic chemical and physical agents (#PD1239).
Related Unsafe artificial sweeteners (#PE6390).
Aggravated by Toxic food additives (#PD0487) Unethical food practices (#PD1045)
Carcinogenic consequences of food preparation (#PE6619).

♦ **PD0040 Uncontrolled media**
Nature Development in mass media techniques has been so rapid and far-reaching that it is now uncontrollable. Owing to commercialization, subversive organizations may use the media for their own purposes. Contributors to surveys may restrict their own freedom. Despite censorship, alien propaganda may be spread by the use of sophisticated techniques, and where there is freedom of expression, monopoly of media outlets may restrict this freedom and issue unrepresentative information. Modern reproduction and dissemination techniques may facilitate copyright infringement.
Incidence The media was heavily criticized in the June 1985 hijacking of a TWA airplane to Beirut, due to the amount of coverage it gave to the hijackers and their demands. The leaking by the press of the pre-summit letter (in the Reagan/Gorbachev summit, November 1985) from Defence Secretary Casper Weinberger to President Reagan, necessitated the launching of an investigation, as this leak was accused of being an attempt to sabotage the peace efforts of that meeting.
Broader Ineffective industry self-regulation (#PF5841).
Narrower Newspaper monopoly (#PE0246) Unfair trials due to pre-trial publicity (#PE1692) Misuse of classified communications information (#PD5183).
Related Censorship (#PC0067) Restriction of freedom of expression (#PC2162).
Aggravates Infringement of privacy (#PB0284) Unethical media practices (#PD5251) Distorted media presentations (#PD6081) Deterioration of media standards (#PD5377) Foreign controls of newspaper and journal propaganda (#PD3041) Restrictions on direct news coverage of parliamentary affairs (#PF3072).
Aggravated by Avoidance of copyright (#PD0188) Compulsory indoctrination (#PD3097) Proliferation of commercialism (#PF0815).
Reduces Government propaganda (#PC3074) Biased government information (#PF0157) Inadequate government publications (#PF3075) Propaganda by intergovernmental organizations (#PE3077).

♦ **PD0044 Shortage of skilled labour**
Decrease in skilled labour — Insufficient skilled workmen — Shortage of skilled manpower — Shortage of skilled workers
Nature In all countries, but particularly the under-developed, a serious shortage of qualified and technical personnel and certain special categories of labour threatens economic advance and also political stability. Reports of unemployment in the wealthier nations are consistently higher in the unskilled sectors, and nonexistent in certain highly skilled industries. The dearth of highly qualified administrators, executives, entrepreneurs, doctors, nurses, teachers, firemen, training instructors, etc, in the developing countries occurs side by side with massive under-employment and unemployment. Major projects are delayed and employment opportunities for unskilled workers are thereby restricted. The problem is intensified by overproduction of certain kinds of staff. For example, in the Philippines (1963-70) there was an estimated surplus of 15,000 engineers. The presence of migratory labour forces, skilled and unskilled, encourages minimum training for domestic labour and high turnover. What is lacking, in many instances, is an effective private enterprise manpower policy applicable to the development of commodity extraction, conversion, manufacturing or service industries and to the agriculture and food-processing sectors. Also lacking is the provision for such a policy to be integrated into total manpower development and educational policy, in conjunction with national planning.
Broader Labour shortage (#PC0592) Economic and social underdevelopment (#PB0539) Depleted expertise of the rural labour force (#PF2973).
Narrower Lack of artisans and craftsmen in developed countries (#PE4804).
Aggravates Incompetent workers (#PD4535) Shortage of military manpower (#PE4920) Lack of skilled manpower in rural areas of developing countries (#PE5170).
Aggravated by Labour hoarding (#PE6333) Lack of training (#PD8388) Occupational diseases (#PD0215) Functional illiteracy (#PD8723) Problems of migrant labour (#PC0180) Inadequate manpower planning (#PJ2036) Inappropriate employment incentives (#PD0024) Emigration of trained personnel from developing to developed areas (#PD1291) Reduction in demand for primary commodities due to technological change (#PD1276).

♦ **PD0047 Jurisdictional conflict and antagonism within international organizations**
Narrower Jurisdictional conflict and antagonism within intergovernmental organizations (#PE9011) Jurisdictional conflict and antagonism within international nongovernmental organizations (#PE1169) Jurisdictional conflict and antagonism within the specialized agencies of the United Nations (#PE8799).
Aggravates Ineffectiveness of international organizations and programmes (#PF1074).
Aggravated by Organizational empire-building (#PF1232) Politicization of intergovernmental organizational debate (#PD0457).

♦ **PD0050 Maldistribution of land in developing countries**
Impossibility to redistribute land
Nature The inequitable distribution of land for cultivation in many developing countries is a major factor in perpetuating a subsistence peasant economy. There are instances where landlords claim from the share-cropper as much as three-fourths of the produce. Similar problems of land tenure exist in tribal communities where neither the member of the tribe as an individual nor his family as a group is entitled to the continuous possession of land. These systems provide few incentives for land conservation and improvement, and result in land deterioration. With the application of agrarian reform laws, new social and economic problems are encountered. The size of the distributed farm in some countries may not make for economic viability. Holdings are generally too small to offer adequate employment opportunities for growing families. Drains and canals, which were previously owned by one landlord who had the resources to maintain them, are now, following reform, owned by a large number of shareholders. Programmes of land distribution often have to be carried out at the price of a decrease in productivity. While bigger land holdings permit the use of extensive techniques such as mechanized farming, smaller holdings may run into difficulties in supporting investments of this nature.
Claim A significant and growing proportion of the rural labour force remains landless. Demographic pressures are turning millions of marginal farms into non-economic productive units, and the gap between the rich and the poor in the rural areas is widening. Inadequate access to productive land is one of the most critical barriers to rural development.
Refs FAO *Agriculture* (1981).
Broader Rural underdevelopment (#PC0306).
Narrower Landlessness in developing countries (#PC0990) Uneconomical small farms in developing countries (#PD1007) Maldistribution of land through customary tenure systems (#PD0813) Maldistribution of land associated with large traditional estates (#PD0406).
Related Unequal property distribution (#PC3438) Reluctance to join in community action (#PF1735) Maldistribution of productive capacity (#PC9785) Maldistribution of wealth within developing countries (#PD5258).
Aggravates Insecure land tenure (#PD9162) Unproductive subsistence agriculture (#PC0492).
Aggravated by Traditional land distribution (#PJ1289) Maldistribution of agricultural land (#PD9189) Defective land use planning in developing countries (#PD1141).

♦ **PD0055 Domination of the world by territorially organized sovereign states**
National sovereignty as an obstacle to peaceful socio-economic development — National governments as an obstacle to representative democratic political organization — Nation-states as obstacles to world order
Nature The present system of world order, despite the rise of transnational economics and politics and the multiplication of specialized international activities of a functional kind, remains state-oriented. The quality of world order at a particular time depends on the pattern of voluntary or coercive relations among governments representing states. The nature of the adjustments made and the persisting pattern of state imperatives imperil human survival and the quality of life on earth.
Incidence The focus of government planning and action everywhere lags behind the emerging appreciation that the primary world order crises include the dynamics of economic and ecological disequilibrium. The pattern of statist imperatives constituted by economic growth, competition, maximization, self-help, absence of empathy, and autonomy of reproductive dynamics, indicates that the path to equilibrium barely exists with the present structure of interstate rivalry. Governments do not have the disposition to exercise self-restraint in such a way as to facilitate the adjustments that must accompany a transition to equilibrium.
Claim The disequilibria in the global economy and in global ecology cannot be dissociated from the issue of global security. The statist imperatives, particularly associated with the superpowers, are bringing the world closer to nuclear war.
Broader Competition between states (#PC0114) Lack of international cooperation (#PF0817).
Narrower Super-power chauvinism (#PE7778) Unbankruptability of sovereign states (#PF0478) Inadequate national empathy for external hardship (#PD0428) Governmental disregard for international values and procedures (#PF0292) Restrictions on international freedom of movement for national advantage (#PD0351) Lack of autonomous world-level actor to identify and clarify world interests (#PF0053) Discrimination against developing countries by the formation of regional groupings of developed countries (#PE1604).
Related Domination (#PA0839) Nationalism (#PB0534) World anarchy (#PF2071) Military and economic hegemony (#PB0318) Inability of governments to regulate family size (#PF0401) Domination of government policy-making by short-term considerations (#PF0317).
Aggravates Undue attachment to territory (#PF3390).
Aggravated by Erosion of sovereignty (#PE5015) Parochial national interests (#PF2600) Boundary disputes between states (#PD2946) Territorial disputes between states (#PC1888) Denial of right to equality between states (#PE4712) Conflicting claims by states to territories (#PC2362) Insecurity and vulnerability of nuclear weapon states (#PC4440) Financial and economic disputes between states and nationals of other states (#PE1911).
Reduced by Fragmentation (#PA6233).

♦ **PD0057 International trade barriers for primary commodities**
Nature It is convenient to distinguish barriers to trade which protect domestic producers of primary commodities against foreign competition, and those which are imposed wholly or mainly for revenue purposes, and also those barriers protecting domestic industries which process primary commodities.
Protection of domestic primary production: Although the forms of protection in force in developed countries vary considerably, they all have one general objective in common: to provide domestic producers of primary commodities with a level of real income judged to be reasonable in relation to the corresponding levels in other sectors of the economy. In most developed market countries, this income objective is sought indirectly through the support of domestic prices at remunerative and stable levels. In some countries, in addition, direct subsidies to domestic producers are made in order to maintain or increase the incomes of primary producers. However, the concern of governments to limit the budgetary cost of such assistance has generally restricted the scope for direct subsidy payments.
In some developed countries, the main effect of such protectionist measures is often to increase land values and rents, rather than to increase wages for farm workers, who usually represent an important part of the population engaged in primary production. Thus, in practice, these policies may not succeed in achieving their stated objective of increasing real incomes in the primary sector. Moreover, the resulting higher domestic prices for food and other goods mean that protectionist policies bear heaviest on the level of living of the poorer sections of the community.
The markets for most of the major agricultural commodities produced in the developed countries, and for many of the minor ones, are subject to some form of official intervention, although the forms of such intervention vary widely. Broadly speaking, such market interventions can be classified into three categories, according to whether they act primarily to support producer prices, to reduce the costs of agricultural inputs, or to influence the volume of foreign trade.
A number of developed countries impose duties for revenue purposes on the import, or internal sale, of a range of agricultural products, principally coffee, cocoa, tea, sugar, oilseeds and tobacco, and of petroleum.
In addition to barriers to trade in certain primary commodities, arising either from protection of domestic production in developed countries or from the imposition of revenue duties, all developed countries impose tariffs or quantitative restrictions on the processed forms of these commodities. For many commodities, restrictions on imports of the processed forms are combined with duty-free entry for crude materials, thus providing the domestic processing industries of the developed countries with a substantial degree of protection against imports, including imports from developing countries. Moreover, the degree of protection generally increases with the stage of fabrication, from semi-processed to further-processed commodities, and from these to fully manufactured goods.
Modification of the present tariff structures of countries with a developed market economy, designed to reduce the escalation in tariff rates on processed commodities, together with the removal of quantitative restrictions, would substantially widen the potential market for processing industries of developing countries. This would help to diversify the pattern of exports from developing countries and enable them to gain the benefit of the value added at the processing stages. It would, at the same time, allow the developed countries to use their resources more profitably in the production of the more advanced and sophisticated types of manufactured goods in which they have a substantial comparative advantage.
Broader Obstacles to world trade (#PC4890) Tariff barriers to international trade (#PC0569).
Narrower Ineffective industry self-regulation (#PF5841) Trade barriers and protectionism between developing countries (#PD2958) Trade barriers to manufactured goods from developing countries (#PE4170).
Related Non-tariff barriers to international trade (#PC2725) Barriers to the international flow of knowledge and educational materials (#PF0166).
Aggravates Excessive foreign public debt of developing countries (#PD2133).

♦ **PD0061 Sexually transmitted diseases**
Venereal diseases
Nature Sexually transmitted diseases (STD) are a wide ranging group of diseases and can have severe repercussions on the affected person. The range of STD – apart from the more common ones like herpes, gonorrhoea and syphilis – includes chancroid, trichomoniasis, candidiasis and

non-specific urethritis. These are transmitted from person to person primarily through sexual contact.

Incidence Sexually transmitted diseases, particularly syphilis, gonococcal infections and non-gonococcal urethritis, are among the most frequently occurring in the world; they are exceeded only by influenza during epidemics, malaria and schistosomiasis. In some countries with well-established health services, the evolution of STD over the last 10 to 15 years follows a steadily rising curve. For instance, in the USA about 800,000 cases of gonococcal infections were notified in 1973; the present total exceeds 2.5 million out of a population of 250 million. There is good reason to believe that the same increase is true of other countries. Studies of the distribution by age-group in several countries bring out an even more important point. The prevalence of STD in the 15–19 age group is twice as high as in the whole population; in the age-group 20–24 it is four or five times as high, and in the 20–29 age-group it is twice as high (these figures relate to the USA). Since 1975 an increasing number of countries have reported a significant increase in syphilis transmission. This trend had been expected as a result of the universal tendency to replace penicillin by other antibiotics in the initial treatment of gonorrhoea. The epidemic trend of gonococcal infections has been declining in some countries after reaching a peak about 1974–1975, but this decline has often been more than compensated for by an increase in the incidence of nongonococcal genital infections, the latter now being more common in nearly all countries. Particular attention is being paid to certain serotypes of Chlamydia trachomatis, which are associated with 35–50 percent of all cases of nongonococcal urethritis and cervicitis; there is strong evidence that some of the sequelae attributed to gonococcal infections (for example, salpingitis, peritonitis, infertility) could also be caused by C trachomatis. The perinatal transfer of this agent may cause inclusion conjunctivitis (usually self-limited) in 30–40 percent and pneumonitis in about 15 percent of infants born to infected mothers.

Of grave importance is the increasing rise of genital wart virus infection, caused by papillomaviruses. There is strong evidence linking HPV, especially types 16 and 18 HPV DNA with cervical neoplasia. It is now realized tha human immuno deficiency virus infection (HIV) and AIDS is to a very large extent sexually transmitted, being seen heterosexually transmitted in Africa, Asia and Latin America, and in the majority of cases, homosexually transmitted in Europe, North America and Australia, but with the potential for heterosexual transmission throughout the world.

Refs Holmes, King K, et al *Sexually Transmitted Diseases* (1984); Margolis, Stephen *Sexually Transmitted Diseases* (1984); Ostrow, David G, et al *Sexually Transmitted Diseases in Homosexual Men* (1983); WHO *WHO Expert Committee on Venereal Diseases and Treponematoses* (1986).

Broader Stigmatized diseases (#PD7279) Infectious and parasitic diseases (#PD0982).
Narrower AIDS (#PD5111) Herpes (#PE8615) Syphilis (#PE2300) Chlamydia (#PE5476) Gonorrhoea (#PE1717) Trichomoniasis (#PE2310) Vaginal candidiasis (#PG1873) Gonococcal urethritis (#PG1872).
Related Health risks of teenage sex (#PE6969) Legal impediments to marriage (#PF3346).
Aggravates Itch (#PE3940) Divorce (#PF2100) Infectious revenge (#PD5168) Distrust of interpersonal relationships (#PF4274).
Aggravated by Sodomy (#PE3273) Unsafe sex (#PE9776) Promiscuity (#PC0745) Prostitution (#PD0693) Group marriage (#PF3288) Sexual immorality (#PF2687) Sexual exploitation (#PC3261) Female prostitution (#PD3380) Inadequate sex education (#PD0759) Inhibiting social attitudes (#PG1879) Dissociation of sex from love (#PG1878) Inadequate health care in family planning (#PD1038) Social environmental degradation from recreation and tourism (#PD0826).

♦ PD0065 Inadequate ideological frameworks

Ideological confusion — Absence of social methodologies for a global ideology

Nature The inability of an ideological doctrine to convince people of its validity or sincerity may arise from its inability to meet practical demands, inconsistency of principle with practice, indoctrination with an opposing belief, or general dissension. An ideology which proves adequate for one society may prove inadequate for another. The social methodologies required to create, promulgate and sustain a global ideology have not yet been developed, so that parochial mindsets go unchallenged, pervading society with their limited views.

Claim The Chinese 'Cultural Revolution' was an unsuccessful attempt to halt modernization. The peace movement, which is world-wide, is quasi-ideological and has not had sufficient success in the most heavily armed countries. Its failure to convince all peoples and leaders may lead to world disaster.

Broader Inadequacy of doctrine (#PF3396).
Narrower Hedonism (#PF2277).
Related Inadequacy of religion (#PF2005) Inadequacy of social doctrine (#PF3398) Inadequacy of economic doctrine (#PF3395) Inadequacy of political doctrine (#PF3394).
Aggravates Apathy (#PA2360) Cynicism (#PF3418) Ignorance (#PA5568) Confusion (#PF7123) Lack of human unity (#PF2434) Ideological conflict (#PF3388) Social fragmentation (#PF1324) Social disintegration (#PC3309) National disintegration (#PB3384) Absence of tactical methods (#PF0327).
Aggravated by Heterosexism (#PE0818) Compulsory indoctrination (#PD3097) Human errors and miscalculations (#PF3702).

♦ PD0068 Economic non-viability of small developing countries

Nature Modern production methods, particularly in industry, are usually characterized by the important economies of scale they yield. The larger the domestic market for goods, the more likely efficiencies of scale will be realized. Yet over 50 developing countries have a population of less than 5 million, and about 80 per cent of the developing countries have less than 15 million inhabitants. The level of purchasing power, even in countries with larger populations, is often below that of small European countries. Thus, with respect to a very large number of productive activities, the domestic markets of the vast majority of developing countries are not large enough to enable them to benefit from the cost-reducing advantages of large-scale production methods. As a result, numerous industries are operating below capacity. Costs of production are driven up, and the parallel development of similar industries in each country of a particular region leads to a waste of resources that can amount, in several technologically advanced sectors, to some thousand millions of dollars in a particular continent. Often the size of the market simply precludes developing countries from entering new lines of production, thus impeding introduction or continuation of the import substitution process.

Broader Non-viability of small states and territories (#PD0441).

♦ PD0071 Excessive concentration of business enterprises

Proliferation of mergers and takeovers — Inadequate control of mergers

Nature The merger movement has strengthened the position of the transnational corporations in world trade and resulted in the concentration of an increasingly important proportion of world trade in the hands of a relatively small number of corporations. Mergers involving corporations with subsidiaries in other countries are generally arranged without any form of discussion with the host countries in which such subsidiaries are located. The absence of such consultation may create problems for developing countries, particularly problems arising from the concentration of market power, if two previously unrelated subsidiaries are brought together (for example, the international merger of two independent tyre manufacturers which also carry out significant manufacturing activities in other unrelated industries).

The merger movement has involved both horizontal concentration and vertical integration in industry, as well as the creation of conglomerates, in particular in Western Europe. The vertical integration resulting from mergers is particularly apparent in such fields as the production of aluminium and its end products. Such vertical integration tends to preclude exports of raw materials and intermediate goods except through the structure of the multinational corporations in question. In the industrial sectors of computers, agricultural machinery and motor vehicles, transnational corporations produce components and parts in many different countries of the world, including developing countries. As a result, the finished product represents a culmination of an international co-operative effort involving extensive intra-company trading marked by cross-shipments of sub-assemblies of industrial intermediates and components.

Claim Mergers generally speaking do not lead to increased shareholder value. One study suggests that only 2 out of 15 mergers studied increased the value of the company resulting from a merger. Secondly, in the United States and increasingly in Japan and Western Europe companies are to big. They are not more cost effective, economies of scale is a falsehood. They are not more innovative, middle sized companies tend to take on more risky products. Mergers disrupt the main purpose of the company. They can cause hundreds of employees to spend thousands of hours over years integrating the companies. On the other hand, hugh companies that have been broken up over the past few years have resulted in more profitable smaller organizations.

Counter-claim Carefully selected mergers which are not simply ways of diversifying can result in enormous opportunities. Inefficient multiple product factories can be cleaned up. Factories producing similar or identical products can be merged. Products and components can be exchanged within the group. Distribution and sales channels can be combined. Research and development departments can be utilized more effectively. The breadth of in house technology can be used more efficiently.

Broader Capitalist speculation (#PC2194) Restrictive trade practices (#PC0073).
Related Monopolies (#PC0521) Industrial gas monopolies (#PE1813) Concentration of investment power (#PC5323).
Reduced by Protection of company ownership (#PF4432).

♦ PD0077 Soil infertility

Sterile soils — Deterioration in soil fertility — Soil exhaustion — Depleted soil nutrients — Soil deficiency

Nature Soil infertility implies lack of the qualities which enable it to provide nutrient elements and compounds in adequate amounts and in proper balance for the growth of specified plants. Nutrient-weak soil may be aggravated by poor soil mechanics, particularly poor binding properties which do not hold soil particles together in a porous, water-stable structure. Soil binding is the result of microorganisms exuding gummy substances during organic decomposition. Infertile soils lacking in decomposing organic matter such as manure, will lack nutrients and binding qualities as well.

Background For healthy growth, all plants require certain nutrients from the soil. There are 16 elements (carbon, hydrogen, oxygen, nitrogen, potassium, phosphorus, sulphur, calcium, iron, magnesium, boron, manganese, zinc, copper, chlorine and molybdenum) which, in varying quantities, are essential to plant growth. If any of these is absent from the soil or present in insufficient quantities, then the soil is said to be deficient.

Incidence Soils have been found to be naturally deficient in one or more of all the essential elements. Almost all soils that have been cropped without fertilizer for 100 years or more will have become deficient in one or more of the plant nutrients. Nitrogen is deficient in all of the agriculturally important soils of the world. Phosphorus and potassium deficiencies are also very common. Zinc, manganese, iron, and copper deficiencies are commonly associated with sands and alkaline soils, while boron and molybdenum deficiencies are usually associated with the older, highly weathered soils. Calcium and magnesium deficiencies, which sometimes occur in the humid and semihumid temperate regions, are less of a problem since such soils are acid and the deficiencies are automatically corrected in the required liming programme.

Refs FAO *Improving Soil Fertility in Africa* (1978).

Broader Soil degradation (#PD1052) Underproductive methods of agricultural management (#PF6524).
Narrower Sandy soils (#PE3911) Acidic soils (#PD3658) Alkaline soil (#PD3647) Soil compaction (#PD1416) Calcareous soils (#PE7698) Soil salinization (#PE1727) Rock infested soils (#PG7495) Demineralization of the soil (#PD9227) Trace element deficiencies in soils (#PE1936) Reduction of soil fertility downstream due to impoundment (#PE8782).
Aggravates Plant diseases (#PC0555) Infertile land (#PD8585) Crop vulnerability (#PD0660) Environmental plant diseases (#PD2224) Deficiency diseases in plants (#PD3653) Slowing growth in food production (#PC1960) Malnutrition in developing countries (#PD8668) Malnutrition among indigenous peoples (#PC3319) Declining productivity of agricultural land (#PD7480).
Aggravated by Defoliation (#PD1135) Forest fires (#PD0739) Soil erosion by wind (#PE3656) Chemicalized farming (#PD7993) Soil erosion by water (#PD2290) Inadequate land drainage (#PD2269) Inadequate crop rotation (#PF3698) Inadequate irrigation system (#PD8839) Destruction of land fertility (#PC1300) Over-intensive soil exploitation (#PC0052) Unethical practice of soil sciences (#PD1110) Shortening of fallow periods on agricultural land (#PE9407).
Reduced by Floods (#PD0452) Domestic refuse disposal (#PD0807).

♦ PD0078 Traffic congestion

Narrower Port congestion (#PE4766) Air traffic congestion (#PD0689) Sea traffic congestion (#PD1486) Rail traffic congestion (#PJ5720) Road and highway traffic congestion (#PD2106) Interference between communications satellites (#PF0054).
Related Traffic noise (#PD3664).
Aggravates Injurious accidents (#PB0731) Motor vehicle emissions (#PD0414) Environmental degradation of inner city areas (#PC2616).
Aggravated by Inadequate traffic control (#PE8266) Uncontrolled urban development (#PC0442) Insufficient transportation infrastructure (#PF1495).

♦ PD0079 Road traffic accidents

Risk of traffic accidents — Automobile accidents — Car accidents

Nature The mortality and morbidity rates from road traffic accidents are increasing annually in nearly all the developed countries. The total number of deaths and injuries continues to increase despite the reduction in road casualties relative to the total number of motor vehicles and to the estimated annual number of kilometres driven, and to traffic density. A most important aspect of the problem is the disproportionate mortality rate and morbidity in the 15–24 age group (accounting in some countries for 40–50 per cent of all deaths among males in this age group). In addition to the pain and suffering caused and the tragedy of death or permanent disability, a serious economic loss to the community arises from such accidents, due to the costs of medical treatment, to the loss of the services of the injured person, and to damage to property.

Incidence Road traffic accidents are no longer the monopoly of rich countries. India, with fewer

PD0079

than five million motor vehicles, has 40,000 traffic fatalities a year. Accidents are an epidemic which has killed ten million people this century and which causes almost ten million casualties, including 250,000 killed, every year. Spain and Britain have an average of 16 road fatalities a day, France has 29 and the United States has 120.
Claim Despite the shock reaction to aircraft disasters killing hundreds of people, the prospect of the daily slaughter on the roads is largely ignored by the general public. Since less harmful accidents occur when people are driving carefully in hazardous conditions than occur in so-called 'perfect' driving conditions, it appears that the principle of 'risk compensation' is involved. The answer is not necessarily, therefore, the imposition of speed limits and seat-belt regulations. Too many safety regulations may give a feeling of security, leading people to take more risks than they would if driving were less secure.
Refs Calabresi, Guido *Costs of Accidents* (1970); Nordentoft, E L; Wallin, Johan A and Nielsen, Hans Victor (Eds) *Traffic Speed and Casualties* (1975); OECD *Young Driver Accidents* (1975); Society of Automotive Engineers *Automotive Crash Avoidance Research* (1987).
Broader Travel risks (#PD7716) Injurious accidents (#PB0731)
Transport accidents (#PC8478).
Related Animal road deaths (#PE1690).
Aggravates Animal injuries (#PC2753).
Aggravated by Fatigue (#PA0657) Road hazards (#PD0791)
Drunken driving (#PE2149) Driving delinquency (#PE6119)
Pedestrian accidents (#PD0994) Motor vehicle emissions (#PD0414)
Inadequate traffic control (#PE8266) Defects in machinery design (#PE2462)
Inadequate road maintenance (#PD8557) Urban road traffic congestion (#PD0426)
Photochemical oxidant formation (#PD3663) Excessive speed of motor vehicles (#PE2147)
Road and highway traffic congestion (#PD2106)
Inadequate emergency medical services (#PD1428)
Continued operation of unsafe motor vehicles (#PE2240)
Proliferation of automobiles and motor vehicles (#PD2072).

♦ PD0084 Barriers to transfer between educational facilities
Nature Artificial or outmoded barriers exist between different educational disciplines, courses and levels, and between formal and non-formal education. Students are unable to move easily from one level to another within a particular establishment or between educational establishments. They are unable to enter freely at various stages nor to leave at many different points.
Broader Excessive institutionalization of education (#PD0932).
Aggravates Educational wastage (#PC1716)
Increasing development lag against information growth (#PE2000).

♦ PD0086 Discrimination against women in employment
Denial of right of employment for women
Nature Although the lives of women have greatly changed due to demographic, technological and educational factors, the traditional discriminatory ideas about employing women outside the home have hardly evolved at all.
Incidence Employers in many countries refuse to hire married women or mothers of young children. Women, not being highly unionized, have little voice in the matter. Among the EEC countries, a woman is supposedly entitled to 12 weeks maternity leave with free medical aid and a cash grant, and her job must be open upon her return; but not all countries satisfy these conventions. In others cases, employers tend to state that work done by women is not equal to that done by men, thereby avoiding the need for equal pay. In the developing countries, women represent only 23.6 percent of the employees in the major non-agricultural occupations, only 11.3 percent of such employees being women in North Africa and the Middle East. Again, particular occupations show a very low share of the work going to women; they are particularly badly represented in administrative, managerial and production sectors of the developing countries. Women tend to be concentrated in low-skill, lowly paid occupations. Overtime is restricted, due to laws made some time ago under different conditions which limit the number of hours they may work; and some laws prevent women from working at night. In addition, there are prohibitions in jobs regarded as dangerous or unhealthy; real equality does not exist.
Claim There is a need for active research in order to provide a better basis for planning and action that may improve the present situation. It is difficult to combine a job with caring for young children, and measures must be taken to provide help for the mother who wishes to work. Also, in the light of the increases in divorce, single parent situations and families headed by a woman as the sole bread-winner (30% in many countries), women need work for their survival and that of their children. Beliefs connected with sex-typing of particular jobs are not supported by facts.
Counter-claim A woman's task is twofold. Primarily it is that of a mother or housekeeper, including the bringing up of children for whom a mother is irreplaceable. Secondarily, in our current type of civilization, it is to leaven and help humanize life in society. Apart from other reasons, it is more expensive to employ women because of: discontinuity in work and the need to replace temporarily women taking maternity leave; regulations which may require female workers to be given time to nurse their babies; regulations requiring the provision of creches; high absentee rates due to illness or child-care responsibilities; high turnover due to changes in family circumstances; separate facilities for women required by local cultural norms. Women lack muscular strength and make less effective supervisors.
Refs Bickner, Mei L and Shaughnessey, Marlene *Women at Work* (1977); Bose, Christine, et al *Hidden Aspects of Women's Work* (1987); Feinberg, Renee *Women, Education, and Employment* (1982); Paukert, Liba *The Employment and Unemployment of Women in OECD Countries* (1984); Tanimoto, Helene S and Inaba, Gail T *Women's Work, Collective Bargaining, Comparable Worth-Pay Equity, Job Evaluation - and All That* (1987).
Broader Discrimination in employment (#PC0244)
Discrimination against women (#PC0308).
Narrower Unequal pay for women (#PD0309)
Excessive employment of married women (#PD3557)
Discrimination against working mothers (#PD6812)
Discrimination against women executives (#PD9628)
Unequal employment opportunities for women (#PD5115)
Discrimination against black working women (#PE6245)
Inadequate maternity protection in employment (#PD6733)
Discrimination against women at retirement age (#PE6069)
Underrepresentation of women in science and technology (#PJ8394)
Discrimination against working women in socialist countries (#PD2872)
Discrimination in the employment of women with family responsibilities (#PE7206)
Discrimination against women workers in multinational enterprises in developing countries (#PE4102).
Related Prostitution (#PD0693) Trafficking in women (#PC3298)
Exploitation of women (#PC9733)
Discrimination against men in employment (#PD3338)
Economic barriers to women's access to the judicial process (#PE1198).
Aggravates Wasted woman power (#PF3690) Dependency of women (#PC3426)
Lack of social mobility (#PF2195) Denial of right of family planning (#PE5226)
Career interruption due to pregnancy (#PD7692)
Discrimination against women in social services (#PD3691)
Neglect of the role of women in rural development (#PF4959)
Inequality of employment opportunity in developing countries (#PD1847).
Aggravated by Maternity (#PJ1893) Social injustice (#PC0797)
Dangerous occupations (#PC1640) Exploitation of casual workers (#PD6930)
Inadequate child day-care facilities (#PD2085)
Lack of unionization among working women (#PE8345)
Belief in emotional instability of women (#PF8751)
Discrimination against women in education (#PD0190)
Discrimination against women without children (#PE8788).

♦ PD0087 Space weapons arms race
Military targets in space — Anti-satellite weapons
Nature The superpowers have long employed satellites for surveillance, command and control, and for treaty verification. And both the United States and the Soviet Union have had anti-satellite capabilities since the 1960's though the deployment of these interceptors was halted in the early 1970's. The moves to develop an all-out anti-satellite weapon race is threatening the stability of space. They violate the 1972 ABM treaty. The deployment of hundreds of sophisticated weapons and electronic jamming equipment would make the commercial development of space uneconomical. Many of the weapons could be used offensively against land targets. Lasers capable of destroying ballistic missiles can also destroy the enemy's cities by fire storms. The warning time would be seconds rather than minutes. The result of fire storms could effect the climate as bad or worse than a nuclear warfare.
Refs Jasani, Bhupendra *Space Weapons and International Security* (1987); Kirby, Stephen and Robson, Gordon (Ed) *The Militarisation of Space* (1987); United Nations *Disarmement*; United Nations Institute for Disarmament Research *Prevention of the Arms Race in Outer Space* (1986).
Broader Space warfare (#PD6439).
Narrower Anti-satellite arms race (#PE7004).

♦ PD0089 Pollution of orbital space
Accumulation of aerospace hardware in earth orbits — Space pollution — Space junk — Orbit junk
Nature Since the commencement of space operations and the use of satellites for different purposes, a considerable number of items of space hardware have accumulated in earth orbit. Many of these have ceased to perform any useful function and can no longer be controlled in any way from the ground. The continuing accumulation of such objects is a risk to future satellite and space operations (ranging from actual collision to interference between transmissions or obstruction of data transmission).
Incidence More than 7,000 objects larger than 10cms are currently being tracked. Computers extrapolations suggest that there may be up to 3.5 million smaller objects, most of them measuring between 1 and 50 microns in diameter. Some of these are due to the breakup of larger objects. Travelling at speeds between 5 and 10 miler per second, even the smaller objects can cause considerable damage to a space vehicle. Junk, such as burned-out rocket casings and other debris, accounts for over 3,800 items. Earth-circling payloads such as communications satellites and a USSR space station comprise over 1,300 objects. This orbital or satellite belt is located 35,500 kilometres above the equator and has a circumference of 266,000 kilometres. Objects in its space are a hazard to exploration and development missions, and to satellites, particularly as orbits may deteriorate. In the latter case objects fall through the atmosphere. Over 7,000 pieces of space junk have already descended. The Soviet Union's ocean reconnaissance Satellite Cosmos 954 came down in sizeable chunks on a sparsely populated part of Northern Canada in 1978. Cosmos 1402 burned up over the Indian Ocean in 1983. Others, like the US Skylab, may scatter debris over a wide range including populated areas.
Refs Graaff, W de and Reijnen, G C M *Pollution of Outer Space, in Particular of the Geostationary Orbit* (1989).
Broader Accumulated junk (#PD5510) Environmental pollution (#PB1166).
Aggravates Damage caused by space objects (#PE0250)
Limited number of geostationary satellite orbits (#PF0545).

♦ PD0093 Inadequacy of contraceptive methods
Lack of ideal contraceptives — Inadequacy of contraceptives — Contraceptive failure
Nature Many birth control methods are difficult to deliver in service settings, especially in developing countries. Problems of storage exist in countries where extended or multiple families all live in one area, and also where high humidity and heat are constant. For contraceptive methods that require paraphernalia, it is often difficult for a user to get to a clinic or other purchase point for re-supply due to: unavailable or unreliable transportation and road networks; the length of time involved in getting to and from the supply point; and also the long wait inevitable in most clinics in developing countries. Most current methods of birth control do not involve male participation, thus rendering the husband less involved in family planning. Finally, both oral contraceptives and the IUD have been linked to such diseases as cervical and breast cancers, high blood pressure, and heart disease.
None of the currently available methods of contraception is free from drawbacks. Both short-term side effects and long-term hazards to health are associated with hormonal contraceptives and intrauterine devices, while the barrier methods and behavioural techniques are difficult to use and often only partially effective. Sterilization and abortion both entail surgical risks and, especially in the case of abortion, are the subject of considerable controversy because of ethical and religious objections to their use. Thus, while the revolution in family planning technology has made current family planning programmes possible, today's technology continues to be deficient in ways that lead to unwanted pregnancies, high discontinuation-of-use rates, and exposure to health hazards among people who seek to limit their fertility. Improvements in contraceptive technology remains a major goal of population research efforts.
Incidence The reported contraceptive failure rates per 100 users are: 10-30 for natural family planning, 3-15 for condoms, 4-25 for the diaphragm, 10-25 for spermicides, 1-8 for combined oral contraceptives, less than 1 for injectable and implanted contraceptives, 1-5 for IUDs and less than 1 for surgical contraception. The only method that is 100 percent effective with no health risks is total abstinence.
In the USA no new types of contraceptives have appeared for 30 years because of punitive lawsuits against the manufacturers of new products. It is estimated that half of the 1.5 million abortions there result from contraceptive failure.
Broader Lack of family planning (#PF0148).
Narrower Lack of abortion facilities (#PF8481)
Inadequacy of male contraceptive methods (#PF1069)
Discrimination in family planning facilities (#PD1036).
Related Sexual discrimination in contraceptive methods (#PF1035).
Aggravates Unwanted children (#PE1907) Marital instability (#PD2103)
Single parent families (#PD2681) Illegal induced abortion (#PD0159)
Ineffective population control (#PF1020) Unsustainable population levels (#PB0035)
Inadequate health care in family planning (#PD1038).
Reduces Infanticide (#PD3501) Birth prevention (#PE3286).

♦ PD0094 Drug abuse
Illicit drug use — Illicit drug users
Nature Societies in all parts of the world have discovered substances which can alleviate pain and cure various ailments and which also give pleasurable sensations when consumed. However,

all these substances are dangerous because they share the quality of giving temporary euphoria and contentment; one can develop a craving for them which, in a short time, leads to complete dependence. It is this dependence, both physical and psychic, which causes obvious harm to the user and to society. Modern–day society has witnessed a spectacular increase in the materia medica in general and in the use of psychoactive drugs in particular.

Psychoactive drugs are substances which affect the activity of the central nervous system or, in simpler terms, exert a strong influence on the human mind and behaviour. The risks of psychoactive drug–taking are not limited to the individual drug taker; the habit can affect his environment and the society in which he lives. People can (and do) become addicted to certain psychoactive drugs and lead an entirely drug–orientated life (for example, the 'street' heroin addict).

An individual may start drug–taking for one or several reasons. The simplest reason is true for only a minority of persons – that drug–taking starts with the administration of a narcotic or psychotropic drug for therapeutic purposes; the majority of drug abusers start from a search for pleasure, or to identify with a group or with a drug–orientated society. The pleasurable effect of these drugs (euphoria) constitutes the first step; the second one comes when, after repeated administration of opiates, barbiturates, amphetamines or LSD, the individual does not experience the same pleasure, but requires greater and greater doses to produce the same pleasurable drug effect. This phenomenon is the development of tolerance. Regular and repeated drug administration leads to a state where the organism has so adapted itself to the presence of the drug that an interruption in the continuity provokes abstinence (withdrawal) symptoms. These symptoms can be painful and severe, sometimes even fatal, as often happens in the case of barbiturates and opiates. This is the physical dependence on a drug. There are some drugs, such as cocaine, cannabis or LSD, which do not produce physical dependence. But the development of psychic dependence can create an even stronger compulsion for regular drug–taking than the 'craving' experienced by the physically dependent addict.

There are two categories of psychoactive (mind affecting) drugs which are subjected to international control: narcotics and psychotropic drugs. Narcotic drugs are included in the Single Convention of 1961, and psychotropic substances in the Convention of 1971. Both terms are used in a legal sense and do not necessarily reflect the pharmacological properties of the respective drugs. Some drugs slow down mental activity (depressants of the central nervous system) and are thus useful medicines for: relieving pain (pain killers, analgesics, such as opium and morphine); inducing and/or maintaining sleep (hypnotics, such as some of the barbiturates); suppressing nervous excitement or curbing nervous disorders (sedatives such as barbiturates or glutethemide); or relieving anxiety (tranquillizers, such as meprobromate). Other drugs have an opposite effect on mental activity (stimulants of the central nervous system, such as dexamphetamine); that is, they induce states of excitement. For therapeutic purposes, they are used only when mental activity has to be intensified or when the appetite has to be curtailed. There is a group of substances which, in spite of their strong effect on mental activity, have very limited or no therapeutic use at all. These are hallucinogenic substances, that is, substances used to produce hallucinations or delusions.

Society's defensive reaction against drug pollution has now reached the level of the international community. The World Health Organization and the United Nations are variously involved in the delicate balance–keeping policy which calls for decisions on the following problems. which kind of drugs have to be controlled; what kind of national control is required; and how to complete this control by international action and assure the cooperation of governments which have the responsibility of drug control.

Incidence Drug abuse and drug addiction have taken on increasingly dangerous proportions in many parts of the world. The situation is aggravated by the continuous introduction of new psychotropic substances which are liable to be misused. The problem is not confined to one social stratum or age–group; it has been shown recently to be rapidly increasing among high–level executives, with resulting danger to other employees from drug–induced behaviour and also encouragement of other evils such as blackmail. These trends have affected not only developed countries but also developing countries, and are especially dangerous in the latter because of their limited resources to deal with the situation. The reasons for the spread of drug abuse are complex and are different in different countries, and some of them are very difficult to remove. Methaqualone, LSD, heroin, marihuana and other drug seizures net authorities over one hundred tons annually.

Background Until the end of the nineteenth century, the question of narcotic drugs was not widely regarded as an international problem calling for concerted action on a world–wide scale. Developments in the latter part of the nineteenth century, however, gave a new dimension to the problem. First, through technological progress, laboratories began producing from opium and coca leaf an increasing number of alkaloids and their derivatives. Further expansion of transport and international trade reduced geographical distances and natural barriers between nations until what originally seemed to be a local problem of a few countries became a matter of concern to the world community as a whole. Moreover, the relationships between drugs, misery, and crime contributed to a growing conviction that the sale of drugs could no longer be viewed as a regular commercial transaction, free from government interference. In some countries, addicts have been given prescriptions for a minimum daily 'fix' in an attempt to control or eliminate the black market in narcotics.

In the twentieth century, society has become one of drug consumers. The development of health care and of the pharmaceutical industry has contributed greatly to this situation. The chemical age has produced its side–effects – pollution, for example – and chemicals have penetrated not only the air, water and soil, but our very organisms.

Refs Andrews, Theodora *A Bibliography of Drug Abuse* (1981); Atkinson, Roland M (Ed) *Alcohol and Drug Abuse in Old Age* (1987); Austin, Gregory A, et al *Drug and Abuse* (1984); Boström, Harry and Ljungstedt, Nils (Eds) *Detection and Prevention of Adverse Drug Reactions* (1984); Chatterjee, S K *Drug Abuse and Drug–Related Crimes* (1988); Chetley, Andrew and Gilbert, David *Problem Drugs* (1986); Cumming, Gordon and Bonsignore, Giovanni *Drugs and the Lung* (1984); Goode, Elizabeth W (Comp) *Drug Abuse Bibliography for 1980* (1984); Goode, Polly T (Comp) *Drug Abuse Bibliography*; Lingeman, Richard R *Drugs from A to Z* (1974); Lodge *Drug and Alcohol Abuse in the Workplace* (1987); Spencer, Christopher and Navaratham, V *Drug Abuse in East Asia* (1981); United Nations *Report of the International Conference on Drug Abuse and Illicit Trafficking, Vienna, 17–26 June 1987* (1987); United Nations *The United Nations and Drug Control* (1987).

Broader Substance abuse (#PC5536) Victimless crime (#PC5005).
Narrower Overdose (#PJ1995) Generic drugs (#PG8691) Caffeine abuse (#PE0618) Drug dependence (#PD3825) Multi–drug abuse (#PD0213) Rural drug abuse (#PE9688) Drug abuse at work (#PD4514) Abuse of animal drugs (#PE0043) Drug abuse in prisons (#PE0978) Drug abuse by students (#PE5507) Children of drug addicts (#PE4609) Drug abuse among athletes (#PE4250) Drug abuse by adolescents (#PD5987) Drug use in animal sports (#PJ1135) Overprescription of drugs (#PE9087) Drug abuse by young drivers (#PG4511) Substance abuse by physicians (#PE7568) Drug abuse by military personnel (#PE5579) Drug abuse by government officials (#PD8696) Inhaling of solvents and anaesthetic drugs (#PE1427).
Related Low self esteem (#PF5354) Political indoctrination (#PD1624).
Aggravates Crime (#PB0001) Deafness (#PD0659) Drug smuggling (#PE1880) Inadequate sense of time (#PF9980) Criminally life endangering behaviour (#PD0437) Encouragement of drug resistant diseases (#PJ3767).
Aggravated by Self–destruction (#PF8587) Lack of self–confidence (#PF0879) Prescription of inappropriate drugs (#PE3799) Excessive proliferation of medical drugs (#PD0644) Degradation of developing countries by tourism (#PF4115) Economic dependence of some developing countries on the drug trade (#PE5296).

♦ PD0095 Inadequate social discipline in developing countries

Nature In varying degrees, developing countries lack adequate social discipline. There are deficiencies in their legislation and in law observance and enforcement. Public officials disregard the rules and directives that they should follow. Often they act in collusion with powerful persons and groups whose conduct it is theoretically their function to regulate. These tendencies act as impediments to policy making and policy implementation. Due to such laxity, arbitrariness and licentiousness, there is a widespread resistance to public controls; exploitation of such weaknesses leads to unjust enrichment of persons who have economic, social and political power.
Broader Underdevelopment (#PB0206) Lack of social discipline (#PF8078).
Aggravates Corruption in developing countries (#PD0348).

♦ PD0096 Vulnerability of the elderly under states of emergency

Lack of protection of the aged during disasters
Broader Violation of rights of vulnerable groups during states of emergency (#PD3785).
Aggravates Injuries (#PB0855) Human death (#PA0072) Malnutrition (#PB1498) Age discrimination (#PC2541) Vulnerability of the disabled during states of emergency (#PD0098).
Aggravated by War (#PB0593) Homelessness (#PB2150) Social neglect (#PB0883) Human dependence (#PA2159) Natural disasters (#PB1151) Vulnerability of women and children in emergencies (#PD1078).

♦ PD0098 Vulnerability of the disabled during states of emergency

Lack of protection of the handicapped during disasters
Broader Violation of rights of vulnerable groups during states of emergency (#PD3785).
Related Inadequate community care for handicapped persons (#PE8924).
Aggravates Injuries (#PB0855) Human death (#PA0072) Malnutrition (#PB1498) Human dependence (#PA2159) Physically handicapped children (#PD0196) Immigration barriers for handicapped family members (#PE4868).
Aggravated by War (#PB0593) Homelessness (#PB2150) Natural disasters (#PB1151) Vulnerability of women and children in emergencies (#PD1078) Vulnerability of the elderly under states of emergency (#PD0096).

♦ PD0103 Endemic disease in developing countries

Nature Many diseases such as cholera, yellow fever, typhus, malaria and others continue to cause thousands of deaths annually in developing countries despite the existence of therapeutic or prophylactic means of combating them. This situation is aggravated by the fact that the ravages of many diseases are accelerated where hunger and malnutrition exist, and by the fact that adequate health services and medical personnel are often lacking.
Broader Human disease and disability (#PB1044).

♦ PD0104 Inadequate information systems for international governmental decision–making processes

Institutional arteriosclerosis

Nature Intergovernmental organizations are used as instruments in dealing with problems which, by their nature, transcend national boundaries and are in effect regional or world–wide in scope. But it can hardly be expected that they can avoid the errors in decision–making which have afflicted the governments which control them unless adequate information is available for the decision–making process. This is now not the case. Further, unless governments themselves possess such information and are able to improve the decision–making machinery which has led to the present situation, the errors of IGOs will be compounded. The fundamental requirement for dealing with this situation has not been recognized: a common basis of processed information available to all governments and organizations.

Incidence The technology upon which international telecommunications systems are based is developing faster than the ability of international bodies to respond. The result is a tendency for communications policies to become political rhetoric rather than practical guidelines.

Claim The problems to be dealt with arise not in a series of bilateral relationships between governments and individual organizations, but in relation to the whole array of organizations now in existence. A decision taken in one organization ought to be considered not simply in regard to that body, but in regard to all others whose work is either directly or indirectly relevant. This may be a very large number. In the present situation, however, each government or organization makes its own investigations of more or less adequacy, drawing conclusions from differing selections of facts, all selections being inevitably incomplete. A very large proportion of the difficulties encountered in establishing a satisfactory programme arise at this point. Where matters are dealt with which, although sometimes simple in themselves, may in their consequences and relations to other actions and programmes be of great complexity, the lack of a common basis of full information aggravates the difficulty of finding an agreed intergovernmental solution. The information required is often so abundant and complex that even if it were all available, the unaided human intellect would be unable to grasp all of the factors so as to see all of their consequences in the round. In such circumstances, it is inevitable that each government or, as is often the case, the one specialist individual responsible for national policy in a complex technical matter, will seize upon certain aspects of that matter to the exclusion of others. Thus the distortion of views resulting from lack of complete information is only part of the problem: information also requires organization in order to be accessible. In an intergovernmental organization lacking this there arise divergent views which do not have their origin in real differences of interest or of policy, but which result from partial, and therefore selectively different, national or organizational approaches to the subject matter. When it is pointed out that, apart from basic factual information, governments are not even possessed of a synoptic method of seeing what is being done in all international organizations in a particular matter which they are considering, it will be understood how blundering their approach is likely to be in any one organization.
Narrower Failure of government intelligence services (#PF8819).
Aggravates Inappropriate policies (#PF5645) Inadequate research on problems (#PF1077) Inadequacy of intergovernmental decision–making process (#PF2876).

♦ PD0105 Human poisoning

Nature Accidental poisonings, suicide, and homicidal poisonings result largely from toxic substances being readily available at work and at home. Poisoning also occurs from over–use of otherwise harmless substances. Poisons vary according to their source, e.g. animal, plant or mineral, and according to their chemical composition. They can also be corrosives, irritants, or narcotics. Corrosive poisons produce immediate pain and swelling of lips, mouth and throat; death is rapid if the dose is large. Irritant poisonings produce vomiting, purging and abdominal pain. Narcotics produce giddiness, headache, interference with sight, and stupor preceded occasionally by convulsions. Narcotic–irritants produce at first symptoms of irritant poisons; later, delirium or convulsions. All poisonings can and often do end in death.

Incidence In the US poisoning may run to over 150,000 reported cases a year; in France, to some 70,000; and in the UK the figure exceeds 50,000 with about 24,000 of these being accidentally poisoned children. The US figure include 90,000 children less than 5 years of age. US statistics also indicate that about half of all poisonings are due to medicines or medicaments. Thus the correlation between the hazards of the family medicine chest and child poisoning incidence is extremely high, indicating negligence of parents and packagers of drugs to be a contributing factor. Poisoning is more prevalent in the industrialized world where several million cases, reported and unreported, are estimated to occur annually.

Refs Brookes, Vincent J *Poisons* (1975); Muggler, Dixie M *Poisoning and Medicine* (1985).

Narrower Metal poisoning (#PJ6543) Alcohol poisoning (#PG7437)
Product extortion (#PD5426) Accidental poisonings (#PE1935)
Intentional fatal poisonings (#PG1934) Food poisoning through negligence (#PE0561).
Aggravates Suicide (#PC0417) Homicide (#PD2341).
Aggravated by Poisonous algae (#PE2501) Poisonous gases (#PJ1939)
Toxic substances (#PD1115) Medical quackery (#PD1725)
Poisonous plants (#PD2291) Poisonous animals (#PE0175)
Misuse of medicines (#PD8402) Pesticide intoxication (#PE2349)
Motor vehicle emissions (#PD0414) Environmental pollution (#PB1166).

♦ PD0110 Rivalry and disunity within developing regions

Lack of sense of community and solidarity within developing countries

Incidence In the Arab world, on the Indian subcontinent, in Africa and in Latin America, solidarity is lacking and developmental rates are slowed by intra-regional rivalries and misunderstandings. In South America rivalry between Argentina and Brazil is notable, and there is a sense of difference between Andean countries and other OAS members. In Southeast Asia, ASEAN members and non-members cannot unify on a regional basis to further development. In Southern Africa, the region is neglected while civil wars and international hostilities continue. The lack of, or a weak sense of regional identity afflicts the Caribbean, the less developed countries of the Mediterranean basin including insufficiently developed parts of Europe, and it afflicts central Africa.

Counter-claim Austria and Turkey, North Korea and Japan, the Netherlands and Belgium are all examples of near neighbours who have mutual differences. What is it about the "developing country" label that allows the most simplistic slurs to find their way into acceptance ? The perception that a developing country is part of a homogenous region is a convenience of international organizations who for developmental aid purposes are largely responsible for introducing the regional concept, at least in the developmental perspective. The regional concept has more reality in international funding budgets than anywhere on the ground.

Broader Competition between states (#PC0114) Obstacles to community achievement (#PF7118).
Related Domination (#PA0839) World anarchy (#PF2071)
Intranational competition (#PE0580)
Lack of a sense of community and solidarity at the world level (#PF8704).
Aggravates Lack of political will (#PC5180) Anxiety resulting from torture (#PE0969)
Inadequate inter-regional cooperation on problems within countries (#PE8307).
Aggravated by Apathy in developing countries (#PD8047)
Fragmented regional cooperation (#PF9129)
Differences in trading principles and practices between developing countries (#PF2960).

♦ PD0120 Pesticides as pollutants

Pesticides — Chemical pesticides — Use of chemical pesticides — Pollutant residues — Pesticide residues as pollutants — Pesticide contamination of food chains — Bio-accumulation of pesticides

Nature Pesticides are chemical compounds used to control plants and animals that are classified as pests. They are most widely used on crops, but they are also used in and around the home on insects, rodents, weeds, and plant diseases; in wood processing and preserving; in paint; in food storage; and in public health programs. Insect and weed control are the two most common uses. Some pesticides are applied directly to plants or soil. Soon after application, they are dispersed into the environment, so that applications are often repeated. As the pesticides accumulate in the soil and wash into streams and rivers, they can affect fish and birds. Because of the toxicity of pesticides in concentrated form and because of the frequency of exposure, the most serious human health effects are found among agricultural and production workers. Long-term and chronic health effects occur as the chemicals are ingested and inhaled. Bioaccumulation (the buildup of toxic materials in tissues) is evident in fish and birds as well as in humans. Two-thirds of the insecticides used in agriculture are applied by aircraft, but only between 25 and 50 percent of this reaches the crop. A large proportion remains airborne and drifts or is lost through volatilization, leaching, and surface transport. Less than 1 percent actually comes in contact with an insect. Herbicides are usually applied directly to plants or the soil.

In the last few decades man has improved his living conditions in many ways. By altering his natural environment and by developing chemical defence mechanisms against unwanted species he has been able to control disease, to produce more and better crops and meat, and to create recreational areas free from nuisances. It has, however, also been observed that certain useful species, particularly some types of birds (eagles, pheasants, robins) and aquatic organisms (lake trout, commercial crab) now appear in reduced numbers in their normal sites or are occasionally found dead.

Some adverse effects on wildlife have been brought about by certain chemical compounds, used as pesticides, which have spread from their point of application into the general environment, carried by air and water, washed out by rain or snow, accumulated by soils and disseminated by living organisms.

Two different sets of problems are posed by the occurrence of persistent pesticides in the environment: localized problems, tending to be acute, leading to recognizable effects with assignable origins, which can be dealt with if successfully sorted out and if the will and powers exist; and more widespread problems, tending to be inferential with postulated effects, due to the universal presence of the materials in question. There is, for example, growing evidence from the amount of pesticide residues found in specimens of affected species as well as in the animals, fish, invertebrates, or plankton they feed upon, that there is a process of pesticide concentration in the food chains. As small amounts of persistent pesticides become more and more widely spread throughout the entire natural environment, they are absorbed by low forms of life. Where large numbers of these species serve as food for higher animals, some of the total pesticide remains in the eater. This can lead to a certain concentration of pesticides in a form of life at the end of the food chain. Eventually lethal doses may be reached for certain populations, or the species may be reduced because of adverse effects on reproduction or behaviour. Accumulation of pesticides in soils and waters may eventually render crops unsuitable for consumption. Excessive application of pesticides is a sign of bad agricultural practice, and a contributory factor in the problem; but non-observance of the necessary waiting time between last application and harvest is the main cause of increased residues.

The importance of changes in the environment is that they may be irreversible; species may disappear altogether. This will cause ecological imbalance – certain forms of life whose numbers are controlled by the disappearing species will no longer be held in check. Thus an apparently harmless species may become a pest. Another result of reduction of species diversity may be the loss of the genetic possibilities that each organism ultimately carries. Impoverishment of the 'gene pool' together with disturbance of the selective pressures of the environment will reduce the chances of future development of desired plants and animals. Furthermore, as men are at the top of the food chains, they too tend to concentrate residues in their bodies with as yet unknown effects on them. While perhaps sufficient attention has been directed to acute toxicity problems, too little attention has been paid to the effect of long term ingestion by man of small amounts of these chemicals.

A serious dilemma is that the cheapest pesticides are the organochlorines, known to accumulate in human fat. The substitution of organophosphorus pesticides brings both higher cost and a need for more carefully trained operators, both of which pose problems.

Incidence Concern about pesticide residues is largely associated with the organochlorine insecticides – notably dieldrin and DDT – for they are exceptionally persistent, a quality which also makes them highly useful in many pest control situations. It has been argued that their residues in foods, though initially minute, could accumulate in the human body and build up levels sufficient to affect health. The organochlorines are fat-soluble, so that any retained in the human body are mainly to be found in fatty tissues. The existing food laws in many countries prohibit the inclusion of most of the trace elements into food, but despite these laws considerable quantities of trace elements are brought into food by insecticides. In the USA the tolerance levels include highly toxic lead in 16 fruit and vegetable products, arsenic in 26, fluorine in 28, and antimony in 2. The presence of the trace element copper has been admitted in more than a hundred agricultural or horticultural products in the USA; bromide in 138 products including cereals, zinc in 71, and tin in one. In other countries the so-called trace elements are also admitted, with the occasional exception of lead, arsenic, antimony and fluorine.

Each year 1 pound of pesticides is used on crops, parks, forests, and lawns for each person on earth. In the United States, 6 pounds per person are used. Half a million people are poisoned by pesticides each year, and five thousand die. Although developed nations like the United States have restricted the most dangerous chemicals, they do not forbid their production for export. More than 140 million pounds of restricted or unregistered pesticides were exported by the United states in 1976. Though its use is banned in the United States, DDT is still produced for export and traces of it can be found on most coffee beans imported into the country. More than 800 million pounds of pesticides are used in developing countries that lack the resources or expertise to evaluate their safety. As pesticide use increases, so does resistance among pests. Pesticide use is then increased or new poisons are found to fight these new pests.

Claim Pesticide residues travel long distances and build up in the food chain, and so can affect people and other creatures far from the places where they were applied. They kill or injure large numbers of non-target organisms, including fish, birds, insects that pollinate crops, and animals that prey on others attacking crops. They have left unsolved many pest problems and created new ones.

Counter-claim The use of pesticides has helped to reduce crop losses due to all kinds of pest organisms. Crop losses due to pests are so severe in some places that there are good arguments for increasing pesticide application. At the same time, the demand for increased testing of new pesticides for environmental effects is slowing their introduction by industry, and meanwhile, pests in ever-increasing numbers are developing resistance to the older types. A sample of 38 developing countries applied 162,000 tonnes of pesticides in 1973, and this represented a 23 percent increase over the preceding few years. But to reduce serious pest damage, pesticide use should have been five times greater in the developing countries by 1985 than it was in 1970/71.

Refs American Society for Testing and Materials *Pesticides, Resource Recovery* (1986); Bull, David *A Growing Problem*; Matsumura, Fumio and Murti, Krishna C (Eds) *Biodegradation of Pesticides* (1982); Perring, F H and Mellanby, K *Ecological Effects of Pesticide* (1978); Postel, Sandra *Defusing the Toxics Threat* (1987); United Nations *Pesticides* (1988).

Broader Pollutants (#PC5690).
Narrower Pesticides mist (#PG3921) Agricultural poisons (#PD5277)
Pesticide intoxication (#PE2349) Herbicides as pollutants (#PD1143)
Fungicides as pollutants (#PD1612) Acaricides as pollutants (#PG1836)
Nematicides as pollutants (#PJ1961) Insecticides as pollutants (#PD0983)
Rodenticides as pollutants (#PE3677) Pesticide residues in food (#PE6480)
Water pollution by fertilizers (#PE8729) Chemical contamination of water (#PE0535)
Pesticide destruction of soil fauna and micro-organisms (#PD3574).
Aggravates Soil pollution (#PC0058) Food pollution (#PD5605)
Lake pollution (#PD8628) Malignant neoplasms (#PC0092)
Lead as a pollutant (#PE1161) Nickel as a pollutant (#PE1315)
Arsenic as a pollutant (#PE1732) Mercury as a pollutant (#PE1155)
Cadmium as a pollutant (#PE1160) Hazards to human health (#PB4885)
Pesticide damage to crops (#PD2581) Metal contamination of soil (#PD3668)
Environmental plant diseases (#PD2224) Pesticide hazards to wildlife (#PD3680)
Pest resistance to pesticides (#PD3696)
Unsustainable agricultural development (#PC8419)
Health risks to workers in agricultural and livestock production (#PE0524).
Aggravated by Excessive use of chemicals to control pests (#PD1207)
Polychlorinated biphenyls as a health hazard (#PE2432)
Concentration of noxious substances in food chains (#PE8154).
Reduces Pests (#PC0728) Liver fluke (#PE2785)
Infectious and parasitic diseases (#PD0982).

♦ PD0122 Hazards of strong toxic substances

Nature Recent surveys suggest that, in the USA alone, as many as 50,000 units processing chemicals were not designed to prevent the leaking of hazardous substances should the unit get out of control, although it is admittedly unlikely that such runaway reactions would or could occur in any but a minority of cases. Most emergency relief systems in the chemical industry have serious design problems (with potentially devastating consequences) and virtually all are inadequate. Emergency and safety systems are not given priority in industry or in undergraduate curricula; none of the 145 undergraduate chemical engineering programs in the USA requires a course in safety. Although the chemical industry is thought to be 'high-tech', much of the equipment used is antiquated and imprecise. Changes to improve matters would be costly – but not as costly as dealing with an accident should it occur.

Incidence Recent industrial accidents have broken records for the release of strong toxic substances. One, at the nuclear power reactor in Harrisburg, Pennsylvania (US), was the first known case of radioactive materials escaping into the environment resulting from commercial use of atomic energy. The Three Mile Island plant remains hazardous and problems of detoxification have not all been solved. Cancer, birth defects and other radiation-related illnesses have now begun to show higher incidences in the Harrisburg area. The full extent of human damage is not yet known however. The greatest industrial disaster of all time occurred in July, 1984 at Bhopal, India, where a transnational corporation's chemical factory malfunctioned leaking methyl isocyanate. Casualties may have exceeded 20,000 persons whose injuries include total or partial blindness, vision impairments, lung damage and other disorders. Genetic damage is also feared.

Broader Environmental degradation (#PB6384).
Aggravates Injurious accidents (#PB0731).
Aggravated by Moisture (#PG1974) Acid mists (#PE1976)
Lack of ventilation (#PG1972) Excessive environmental heat (#PD7977).

♦ PD0124 Aerial piracy

Hijacking — Skyjacking — Offences aboard aircraft

Nature Hijacking is the unlawful seizure of a vehicle, plane, or boat, by force or threat. Recently, hijacking has become synonymous with skyjacking, and especially refers to those who skyjack to make a political point, to seek asylum, to gain revenge or the release of prisoners, or to obtain other concessions. The safety of persons and property is jeopardized, and the operation of air services is seriously affected. Often, under ransom terms, prisoners are released. Increased security measures on the ground and in the air, although apparently successful, may well result in increased cunning, resourcefulness and determination to thwart such measures.

Incidence The first known skyjacking was in 1930, the next in 1947, but from then on it became far more common. Between 1930 and 1960 (30 years) there were 33 skyjackings; between 1961 and 1965 (5 years) there were 22. In 1971 alone there were 87. It is clear that the two decades from 1960 to 1980 have seen the emergence of a worldwide threat to air travel, with great increases from 1968 onwards. By 1970, the occurrences of violent acts such as unlawful seizure of aircraft and other forms of unlawful interference with civil aviation had reached dangerous proportions, although there has been a downward trend since then: there were 17 "successful" skyjacks in 1984, and 17 failed, compared with 373 failed hijacks in 1973.

Background Fantasies of flying, flight, defying the force of gravity, etc, seem to be widespread and to predispose some individuals to carry out dramatic acts such as skyjacking; and certain physical conditions such as vestibular system (inner ear) anomalies may contribute to such predisposition.

Claim The problem could be alleviated in four possible ways: (1) *Prevention*, through society's unwillingness to contribute to the crime by paying of ransom, feeding suicidal appetites or pandering to urges for notoriety. (2) *Control*, through a collection of airport-related techniques to detect potential offenders or produce fear within them so they do not go through with a planned act (guards, personal searches, magnetometers). (3) *Management* – overt behaviour of crew members to manipulate environmental factors known to them. (4) *Disposition* – the legal and/or psychiatric treatment of the offender, and the discovery of any common factors leading to such aberrant behaviour.

Refs Jones, David Lloyd *Three International Conventions on Hijacking and Offences on Board Aircraft* (1985); Joyner, Nancy D *Aerial Hijacking As an International Crime* (1974); McWhinney, Edward *Aerial Piracy and International Terrorism* (1987).

Broader Piracy (#PD1877) Aircraft environmental hazards (#PD8328)
Offences involving danger to the person (#PD5300).
Related Air terrorism (#PE4089).
Aggravates Fear (#PA6030) Injuries (#PB0855) Massacres (#PD2483)
Kidnapping (#PD8744) Human death (#PA0072) Air accidents (#PD1582)
Travel delays (#PE1977) Extradition refusal (#PF2645)
Human destructiveness (#PA0832) Illegal movement across frontiers (#PC2367).
Aggravated by Despair (#PF4004) Racial conflict (#PC3684)
Political conflict (#PC0368) Religious conflict (#PC3292)
Security risk people (#PD6818) Lack of international cooperation (#PF0817)
Lack of airport and travel security (#PE8231).
Reduced by Excessive government control (#PF0304).

♦ PD0126 Railway accidents

Railroad collisions — Rail transport accidents — Train accidents

Incidence In most countries with extensive road systems and use of the private automobile, transport accidents include a only small percentage involving the railways. In the UK, for example, in 1986 there were 220 deaths and injuries to passengers, staff and third parties as a result of specifically railway operations. However, multiplied on a world-wide scale, rail casualties and railway accidents without reported physical casualties are estimated to involve over 700,000 passengers annually and to create hazardous conditions for tens of thousands of rail-side community residents during derailments of freight cars and tanks containing hazardous chemical and explosive substances. Rolling stock, bridges, rails and switches are deteriorating in many countries including the USA, UK and other industrialized nations, creating additional risks to life and property.

Refs Sethi, S P *Railway Accidents* (1965).

Broader Travel risks (#PD7716) Injurious accidents (#PB0731)
Transport accidents (#PC8478).
Aggravated by Rail traffic congestion (#PJ5720).

♦ PD0127 Discrimination against women in religion

Denial of right to freedom of religion for women

Nature Discrimination against women in religious doctrines and in the practice of religion (also in the refusal to admit women as priests) has had a pervasive effect throughout society of maintaining the belief in the inferiority of women, and has become ingrained into civil law over the centuries. It represents a particular distortion of so-called 'traditional' values, particularly sexual morality, which uses the female gender as symbolic of the human lower nature **Counter-claim** Differing roles for women and men in some traditional religions are incorrectly perceived as discrimination. In Roman Catholicism the status of men and women in religious orders is equal. The Blessed Virgin Mary is placed above all saints and angels and is said to embody the ideal of womanhood. In Judaism, it is the woman of the house who lights the Sabbath candles, and in popular tradition, while men were preoccupied with Torah study and other religious observances, Jewish mothers became the economic centre of power in the family. In the essentially religious state of modern Israel, the fact that Ms Golda Meir could become Prime Minister indicates the recognition women have obtained in the Jewish faith. Discrimination against women in religion has been expressed more clearly in continental European Protestantism against the background of class-oriented, hierarchical society, and in Hinduism. Islam, Buddhism and indigenous religions rather provide different societal roles for women, expressly admitting their equality with men in terms of personal religion.

Refs Siriwardena, R (Ed) *Equality and the Religious Traditions of Asia* (1987).

Broader Denial of religious liberty (#PD8445) Discrimination against women (#PC0308).
Narrower Religious discrimination against women in priesthood (#PF6326).
Aggravates Guilt (#PA6793) Suttee (#PF4819)
Lack of family planning (#PF0148) Dependency of women in marriage (#PD3694)
Double standards of sexual morality (#PF3259)
Discrimination against women in education (#PD0190)
Discrimination against women before the law (#PD0162)
Discrimination against women in social services (#PD3691).
Aggravated by Social injustice (#PC0797).

♦ PD0132 Pornography

Nature In this particular form of exploitation of sex, written, graphic or other forms of communication are aimed at arousing sexual desires of a particularly lewd nature. Obscenity is an intrinsic tendency of the work itself and not the reaction of a particular person to it, be he genius, moron, or pervert. Opinions vary on whether it is more harmful to suppress pornography, therefore adding the excitement of guilt and secrecy, or to allow it to be freely available, hence perhaps corrupting children or encouraging sexual deviation. Unbiased information is lacking on the effects of pornography. Definitions of pornography range from any depiction of uninhibited nakedness or sexual activity to depicting women as limited beings with restricted sexual presence subservient to apparently specific male desires.

Incidence Pornography can be found in books, magazines, films, television and increasingly in live-shows. Since what is considered pornography reflects a society's and individual's degree of permissiveness in sexual matters, from one country to another and from one individual to another, nude pin-ups, sexual intercourse on stage, sexual suggestions in advertisements, or blue jokes, may be regarded as an enticement to sexual depravity, as a display of eroticism, or as completely innocuous. In the last 20 years many Western countries (for example, Denmark, Germany, USA, the Netherlands) have authorized the production and marketing of sex, although prostitution, also a sex-act for money, may still be banned. The consumers of commercial pornography are predominantly male often respectable. In the USA approximately 40 percent of video owners admit having rented at least one pornographic film, of which the largest proportion are gay.

Background Pornography is first recorded in ancient Greece as the writings of prostitutes, but it was not until the 15th century that Pietro Aretino developed writings and drawings specifically intended to arouse. The word "pornography" first appeared in English in 1850, but was primarily limited to use among classical scholars until the growth of commercial publishing in the 20th century.

Claim Pornography is a systematic practice of exploitation and degradation based on sex. It dehumanises sex so that beings are treated as things. It is degrades men, women, children and animals. Pornography expresses the culture's value for life. One major daily newspaper in the United Kingdom runs a new photograph of a bare breasted woman every single morning. Substantial exposure to such material is likely to increase the extent to which those exposed will view rape or other forms of sexual violence as less serious than they otherwise would have. It also sets the level of social imagery about its subjects. Whether one approves of such material or not, the picture enters the imagination. The self-understanding and social expectations are recreated daily by such photographs. The relationships between parents and children, workers and employers, teachers and students, fellow passengers on the train, are all informed.

Counter-claim There is no causal relationship between pornography and sexual offences. On the contrary, pornography reduces sexual crimes and in particular crimes committed against minors since it offers an outlet for repressed instincts.

Refs Alderfer, Hannah, et al *Caught Looking* (1987); Goldstein, Michael J, et al *Pornography and Sexual Deviance* (1973); Kendrick, Walter *The Secret Museum* (1988); Malamuth, Neil M and Donnerstein, Edward *Pornography and Sexual Aggression* (1984); McCuen, Gary E *Pornography and Sexual Violence* (1985); Sellen, Betty-Carol and Young, Patricia A *Feminists, Pornography, and the Law* (1987); Soble, Alan *Pornography* (1986); Yaffe, Maurice and Nelson, Edward *The Influence of Pornography on Behaviour* (1983).

Broader Obscenity (#PF2634) Sexual immorality (#PF2687).
Narrower Child pornography (#PF1349) Electronic pornography (#PE5402)
Exposure of children to pornography (#PJ8730)
Excessive portrayal of sex in the media (#PE7930).
Related Voyeurism (#PE3272) Prostitution (#PD0693) Indecent art (#PE5042)
Immoral literature (#PF1384).
Aggravates Sexual violence (#PD3276) Film and cinema censorship (#PD3032).
Aggravated by Boredom (#PA7365) Sexual unfulfilment (#PF3260)
Criminal investment in youth market (#PD5750).
Reduced by Art censorship (#PD2337) Book censorship (#PD3026).

♦ PD0134 Uncontrolled urbanization in developing countries

Unplanned urbanization in developing countries — Uncontrolled physical expansion of cities in developing countries

Nature The problems of the large cities of the developing countries are due largely to the fact that they have materialized ahead of any systematic movement towards modernization. Many of these cities formed transmission points from which raw materials and food were sent to the metropolises of Europe and North America, and to which manufactured goods returned. Rapid population growth has increased the tendency of cities in developing countries to outgrow the resources of the economies they are supposed to nourish and support. The traditional range of public services, utilities and welfare services taken for granted in the cities of developed countries are not generally available to the inhabitants of most of the cities of developing countries; even less so in the rural areas. Lack of finance, infrastructure and skills at all levels contribute to this situation.

One of the overriding influences is at the policy-making level, where the application of priorities, based on the experiences of developed countries, has led to a misunderstanding of the urbanization process. Indeed, neither the historically conventional European city, nor the colonial city, which has been the main focus of large-scale urbanization in the rest of the world, is really well-adapted to developing countries at the present time. Nor can the largest industrial cities, for all their success in other spheres, be accepted as socially or culturally desirable models, most having conspicuously failed to adapt to modern conditions and frequently having become sprawling industrial centres of dreary anonymity. The city throughout the developing world is in one sense the sign and symbol of a development process that could break down completely in the near future. Life in the city is failing to make good deficiencies in literacy and job skills or to provide work which the illiterate and unskilled can do.

Incidence Between 1950 and 1975 the population of cities in developing countries grew by 400 million people. It is predicted that between 1975 and 2000 these same cities will have grown by nearly 1,000 million. By the year 2000 there will be approximately 40 cities with populations of over five million in the developing world; by contrast there will be just 12 cities of this size in the industrialized countries.

In most cities in developing countries, the pressure on shelter facilities and services has degraded the urban fabric. The housing used by the poor is decrepit and civic buildings are frequently in a state of disrepair and advanced decay. The same may be said of essential infrastructure services (transport, public conveniences, water supply, drainage, sewage).

Claim There is no precedent in history for an increase of this order of magnitude. It has led to a degree of urbanization which, relative to the level of development, was excessive, so that it may be properly referred to as 'hyperurbanization' or 'overurbanization'.

Refs Gugler, Josef *The Urbanization of the Third World* (1988).

Broader Rural depopulation (#PC0056) Uncontrolled urban development (#PC0442)
Contradictions of capitalism in developing countries (#PF3126).
Aggravates Urban overcrowding (#PC3813) Malnutrition in developing countries (#PD8668)
Imbalance between urbanization and industrialization in developing countries (#PC1563)
Inadequate institutional structures for local government in developing countries (#PE0365).

♦ PD0136 Corruption and mismanagement of foreign aid

Abuses in international assistance programmes — Wasted foreign aid

Claim The effectiveness of foreign aid and international grants and assistance programmes is hindered by corruption and mismanagement. Abuses include fraudulent accounting procedures to cover poor management or to conceal misappropriation of funds for private gain. Misapplications and inefficient and questionable use of money, goods and services occur within granting national governmental bureaux, granting international organizations (public and private) and among the recipient counterparts who may be foreign officials inside or outside government service. Indifferent and incompetent management of essential services such as road and rail communications, and corruption which gives priority to luxuries for the advantaged rather than necessities for the disadvantaged, lead to the breaking down of distribution networks for goods which have reached the country where help is required but not the people in need. Although the total amount of aid in grant form is increasing, the growth in export credit loans more than makes up for lower debts from official aid loans. In addition, loans can have long-term commercial returns and create long-term dependency.

Broader Mismanagement (#PB8406) Corruption in developing countries (#PD0348).
Aggravates Compassion fatigue (#PF2819).
Aggravated by Institutionalized corruption (#PC9173)
Collusion between administrators of funding agencies and programme formulators (#PF8711).

♦ PD0138 Jurisdictional conflict and antagonism between international organizations

Broader Conflict (#PA0298).
Narrower Jurisdictional conflict and antagonism between intergovernmental organizations (#PE7901)
Jurisdictional conflict and antagonism between international nongovernmental organizations (#PE0064)
Jurisdictional conflict and antagonism between the specialized agencies of the United Nations (#PE2486)
Jurisdictional conflict and antagonism between international organizations at the country level (#PE7973)
Jurisdictional conflict and antagonism between regional intergovernmental organizations with common membership (#PE1583).
Aggravates Unreported crimes (#PF1456) Lack of a world government (#PF4937)
Inadequate integration of international information systems (#PE8066)
Inadequate coordination of international organizations and programmes (#PD0285)
Inadequate relationship between international governmental and nongovernmental organizations and programmes (#PE1973).
Aggravated by Glossitis in animals (#PG0416) Organizational empire–building (#PF1232)
Preponderance of Western-style organizations (#PF8356)
Misuse of nonprofit associations as front organizations by government (#PE0436)
Jurisdictional conflict and antagonism between government agencies within each country (#PE8308).

♦ PD0140 Excessive hours of work

Denial of right to reasonable work hours
Nature Excessive hours of work impair the health, safety and well–being of the worker, act as an obstacle to improvements in productivity and hinder rather than promote the achievement of full employment.
Incidence For all non–agricultural sectors, the weekly working hours in the 13 developing countries for which data are available remained practically unchanged at an average of about 47 hours over the 1970s. For manufacturing industries, data are available for 21 developing countries. Again the average of the country figures remained virtually unchanged at around 46 hours. Looking at the extremes of the data by individual countries, the latest figures show Burma and Israel with the lowest average hours of work in manufacturing, both with less than 40 hours a week. At the other end, Brunei, Ecuador, Egypt and the Republic of Korea report a more than 50–hour work week in their manufacturing industries. In the developed market economies, data covering all non–agricultural sectors are available for 17 countries, while there is information on manufacturing industries for 23 countries. For both of these categories the average working period declined over the decade, finally reaching the goal of a 40–hour work week by 1979. Looking at individual countries, by the end of the period Australia, Belgium, Italy, New Zealand, Portugal, Sweden, United Kingdom and United States reported an average work week in all non–agricultural sectors of fewer than 40 hours (the lowest figure: 35.6 hours in the United States); for manufacturing industries alone, additional data for Austria, Canada, Denmark and Norway place these countries in the under 40 hours–a–week group as well. For both categories of workers, the top spots in terms of the longest working week were taken by the Federal Republic of Germany, Spain and Switzerland; Ireland and Greece also ranked high for workers in manufacturing.
Background It is estimated that in Europe in the Middle Ages that an average of about 54 hours a week was worked; a total of nearly 2,900 hours a year. However, during the industrialization of Europe the yearly hours rose by about 700 to 3,600. Modern annual normal hours worked are mainly well under 2,500 in Europe, and in the United States under 2,000.
Claim Dangerous or very heavy work, or work exposed to the elements, or work requiring fine detail or complex intellectual operations may not be sustainable on the level of national averages. In many lines of work, hours are excessive with regard to the demands made.
Refs White, Michael *Working Hours* (1987).
Broader Limited leisure time (#PF9062) Lack of job satisfaction (#PF0171)
Exploitation in employment (#PC3297).
Aggravates Overwork (#PD2778) Industrial accidents (#PC0646)
Reduction of working time (#PJ3708).
Aggravated by Absenteeism (#PD1634) Irregular working hours (#PJ5329)
Disruption of work schedule due to computerization (#PE5074)
Working hours inappropriate to structural and technological changes (#PE7185).

♦ PD0142 Inequitable labour standards in developing countries

Nature Without strong trade union organization and collective bargaining procedures, employers can lay down unfavourable working conditions for their workers. The worker's interests are sacrificed to the objective of speeding up the development of the national economy. Without the organization of the rural proletariat into effective unions, the situation is perpetuated as the countryside continues to supply a source of cheap labour, thereby making it impossible to secure economic wages in any sector. The unprotected workers particularly affected are females and those of an age within the child or early adolescent range; exploitation of these two classes is even more pronounced than general. Also prevalent in developing countries is discrimination against ethnic minorities and the handicapped.
Claim The lack of fair labour standards retards social development and causes political unrest. It also causes alienation and crime.
Broader Underdevelopment (#PB0206)
Imbalance in economic and social planning in developing countries (#PF4837).
Related Underproductivity (#PF1107) Inadequacy of international standards (#PF5072)
Discriminatory imposition of standards (#PD5229).
Aggravates Exploitation in employment (#PC3297)
Emigration of trained personnel from developing to developed areas (#PD1291).
Aggravated by Inadequate credit policies (#PF0245)
Inadequate vocational education (#PF0422).

♦ PD0144 Loss of traditional forms of social control in developing countries

Nature The heavy incidence of crime and delinquency in towns and cities of new industrial growth is related, in the first instance, to the disruption of the traditional family system and the consequent weakening of family authority and control over individual members. The individual tends to lose the older controls before he has acquired the new and more personal moral codes and controls under the impersonal sanction of the law, which characterize urban societies. First generation town dwellers who repudiate their parents as peasants and reject traditional familial authority before they have had an adequate opportunity to acquire the values and controls appropriate to the urban industrial environment are apt to show a particular propensity for anti–social and criminal behaviour. Similar and particularly acute forms of social disorganization may occur in areas where the effects of the rapid growth of industrial centres are reinforced and aggravated by the simultaneous disintegration of traditional tribal systems. In such a situation conflicts among the heterogeneous norms of different tribal groups, brought into new and close contact within factory and town, as well as conflicts between tribal norms and those belonging to the alien urban pattern, may result in a state of moral confusion and social anonymity.
Broader Social conflict (#PC0137).
Aggravates Crime (#PB0001).
Aggravated by Uncontrolled industrialization (#PB1845)
Untransposed significance of cultural tradition (#PF1373)
Decline in rural customs and traditions in developing countries (#PD1095).

♦ PD0153 Alcohol abuse

Alcoholic intoxication — Dysfunctional drinking — Chronic alcoholism — Alcoholic excess — Excessive intake of alcohol — Excessive alcohol usage — Uncurbed alcohol use — Excessive consumption of alcohol — Intemperance — Alcoholism as a disease — Ineffective treatment for alcoholism
Nature Alcohol abuse can be distinguished from chronic alcoholism (a morbid dependence on alcohol with an easily awakened craving for alcohol, as well as loss of control) in that the physical and mental complications of the latter only occur in a minority of alcoholics. The complications of alcohol abuse occur primarily in the social sphere. The alcoholic very early comes to rely more and more on defence mechanisms including, as well as outright denial and rationalizations, the mechanism of projection, namely the ascribing of his own defects to others; thus everything and everybody is wrong, particularly his family who may already suffer at the alcoholic's hands at a time when friends still think him to be a decent and reasonable person. Children's personality development may often be significantly affected by the alcoholic's unpredictable attitude and behaviour, leading to disturbed emotional relations with parents. Complications also occur at work. Alcoholism releases aggression which may be directed against others in anti–social and criminal acts, as well as in traffic accidents Alcoholics are excessive drinkers whose addiction to alcohol has attained such a degree that there is noticeable mental disturbance or interference with their bodily or mental health, their interpersonal relations, and their smooth social and economic functioning, or those who show the signs of such development. They are unable to recognize the deleterious effects of their habit or, recognizing them, are nonetheless unable to curtail their alcohol consumption and continue in an almost compulsive way to drink heavily. Such persons drink intoxicating amounts of alcoholic beverages several times a day on a regular basis
Background Alcohol may be used to dull the pangs of hunger, fatigue, disease and mental depression brought on by hard labour and inadequate wages. From ancient times, drinking has been connected with excess and derived troubles. Drunkenness seems to have occurred early, and for the most part people did not distinguish between drunkenness and alcoholism as it is understood today—that is a psychobiological disease (a drug addiction or drug dependency) marked by an inconsistent helplessness of the victim to refrain from resorting to alcohol, and inconsistent helplessness to refrain from drunkenness. It was natural to condemn all drunkenness as wrongdoing. Seneca, in the first century, however, distinguished between drunkenness and addiction, and many have followed this line of thinking. Alcohol causes loss of muscle control by reducing blood flow to the cerebellum. Mood shifts are linked to imbalances thus created in the rate of metabolic activity in various parts of the brain cortex. Certain kinds of alcohol consumed have potentially fatal properties.
Incidence The abuse of alcohol is a major cause of morbidity and mortality in the countries of the European region. With very few exceptions, alcoholic beverage consumption has grown markedly all over the industrial world since the Second World War, a trend which is no less prevalent in developing countries. In some countries, overall per capita consumption levels (in terms of 100 percent ethanol) have doubled or tripled within the last two decades. In many Asian and African countries, colonialism commercialized and expanded alcohol production and consumption, a process which often escalated after independence. In this connection, one should not forget the amount of home–brewed product. The size of this is unknown to the authorities in most countries but has to be added to the officially registered per capita consumption figures. Finally, several countries report on increasing multiple use of various drugs and alcohol. This is still more serious as mixing alcohol and narcotic drugs is extremely hazardous It has been estimated that there are approximately 4.5 million chronic alcoholics in the USA; that the incidence of chronic alcoholism of all degrees is 4,390 per 100,000; that the incidence of chronic alcoholism with complications is 1,097 per 100,000. While it has been said that 85 percent of all chronic alcoholics are male, it is generally agreed that females are beginning to account for a greater percentage of chronic alcoholism than was formerly the case. About 80 percent of patients fall within the 35 to 50 year age group; those who appear superficially to be chronic alcoholics at an earlier age are likely to be in the psychopathic or schizophrenic groups; and those who are older more commonly have involutional reactions. In the USSR, as many as 85 percent of the workers in ordinary factories regularly drink to intoxication. Getting drunk is a common habit among the working class men who consequently suffer from high mortality. According to recent statistics, the average age of people suffering from alcoholism has fallen by between 5 and 7 years in the last decade, and according to research on ill patients, 90 percent of them started drinking before the age of 15, and one–third of them started before the age of 10.
Claim Alcohol is evil: spiritually in terms of sin; physiologically in terms of poison; economically in terms of waste and reduced potential of men, money and materials; psychologically in terms of loss of rationality, loss of moral judgement leading to insanity; legally in terms of crime. The persons who use alcohol without having problems serve as a peculiarly insidious role model to youth **Counter–claim** The use and misuse of alcohol as a beverage is complicated by the particular significance which alcoholic beverages have in the social life of many people. Already in primitive societies anthropological research has shown that alcohol consumption played a special role, such as that of providing group or tribe relief from tension on particular occasions, for instance after the harvest had been gathered in. Such community drinking occasions were strictly controlled by the tribal leaders. The incorporation of alcoholic beverage use into religious ritual, social hospitality, ceremonial functions and the accompaniment of festive occasions has given the alcoholic beverage a symbolic function in social life today. To this should be added the social function performed by the alcoholic beverage–selling establishment, whether it be the French bistro, the English pub or the ubiquitous bar. Thus few people can escape making judgments on alcoholic beverages, that is to say, whether they wish to drink them, how much they wish to drink, how much is good for them, etc. Such judgments will usually be couched in some such term as 'moderation' but as to what this implies there is no very precise indication. Those who drink 'moderately' are much less likely to suffer coronary heart disease than are teetotallers and heavy drinkers, probably because alcohol can alleviate stress. Most people are aware that alcohol abuse does exist and that certain persons become addicted to alcohol; but an ambivalent attitude is nevertheless displayed towards excessive consumption which is often a combination of amused tolerance and condemnation of the addict for his so–called 'vice' and inability to control his drinking.
Refs Adkins, Virgil R *The Static Position of Classifying Alcoholism and Drug Addiction As Identical Illnesses* (1986); Armyr, Gunno; Elmér, ke and Herz, Ulrich *Alcohol in the World of the 80s* (1984); Atkinson, Roland M (Ed) *Alcohol and Drug Abuse in Old Age* (1987); Bruun, Kettil, et al *The Gentlemen's Club*; Davies, P *Alcohol–related Problems in the European Community* (1985); Forrest, Gary G *Alcoholism and Human Sexuality* (1983); Heimann, H (Ed) *Alcohol Abuse in Indians of North and South America* (1978); Kurtz, Norman R, et al (Eds) *Occupational Alcoholism* (1984); Maxwell, Ph D and Milton, A *The Alcoholics Anonymous Experience* (1984); Tarter, Ralph E and Van Thiel, David H *Alcohol and the Brain* (1985); WHO *Problems Related to Alcohol Consumption* (1980).
Broader Substance abuse (#PC5536) Human disease and disability (#PB1044)
Consumption of alcoholic beverages (#PD8286).
Narrower Drunkenness (#PE8311) Public drunkenness (#PE2429)

Juvenile alcoholism (#PD1611) Alcoholic psychosis (#PE9263)
Alcohol–related crime (#PE4131) Excessive parental drunkenness (#PE7700)
Alcoholic intoxication at work (#PE2033) Uncomplicated alcohol withdrawal (#PE0375)
Alcohol idiosyncratic intoxication (#PG9365) Dementia associated with alcoholism (#PG9268)
Alcoholism amongst indigenous peoples (#PE7242)
Solvent and methylated spirits drinking (#PE1349).
Related Low self esteem (#PF5354)
Excessive consumption of specific foodstuffs (#PC3908).
Aggravates Gout (#PG3061) Crime (#PB0001) Ulcers (#PE2308)
Dropsy (#PG3235) Phobia (#PE6354) Suicide (#PC0417)
Obesity (#PE1177) Divorce (#PF2100) Amnesia (#PD8297)
Anaemia (#PD7758) Epilepsy (#PE0661) Delirium (#PG4263)
Vagrancy (#PE5460) Diabetes (#PE0102) Gastritis (#PG2250)
Pneumonia (#PE2293) Hepatitis (#PE0517) Laryngitis (#PE2653)
Absenteeism (#PD1634) Human ageing (#PB0477) Pancreatitis (#PG3330)
Liver diseases (#PE1028) Hallucinations (#PF2249) Drunken driving (#PE2149)
Phenylketonuria (#PG3323) Chronic illness (#PD8239) Criminal insanity (#PD9699)
Psychogenic fugue (#PE7451) Mental depression (#PC0799) Korsakoff syndrome (#PE0333)
Delirium alcoholicum (#PG4121) Cirrhosis of the liver (#PE2446)
Children of alcoholics (#PD4218) Alcoholic hallucinosis (#PG5738)
Sexual impotence of men (#PF6415) Foetal alcohol syndrome (#PE3853)
Alcohol–related violence (#PE7084) Inadequate medical resources (#PD7254)
Impairment of physicians' ability (#PE5746) Diseases and injuries of the brain (#PD0992)
Draining of resources due to alcohol (#PE8865)
Malignant neoplasm of mouth and throat (#PE9819).
Aggravated by Anxiety (#PA1635) Culture shock (#PC2673)
Mental illness (#PC0300) Family breakdown (#PC2102)
Stress in human beings (#PC1648) Retarded socialization (#PF2187)
Ambiguous shape of social identity (#PF6516) Susceptibility of women to alcohol (#PE7161)
Criminal investment in youth market (#PD5750)
Malnutrition among indigenous peoples (#PC3319)
Retirement as a threat to psychological well–being (#PF1269).

♦ PD0154 Elitist control of production

Nature Local people have little control over the means of production even in their own communities. Decisions over the types of production, waste control, products, use of capital and natural resources and most of the other aspects of production are controlled by those far removed from the actual site of the production and from the impact of these decisions on local people.
Broader Foreign control of natural resources (#PD3109).
Related Excessive job control (#PE6836).
Aggravated by Elitist control of global economy (#PC3778).

♦ PD0156 Labour surpluses in developing countries

Labour market flooding in developing countries
Nature The overabundance of labour is, in most developing countries, as serious a problem as the shortage of skills. In nearly all countries, the supply of unskilled and untrained manpower in the urban areas exceeds the available employment opportunities. The reasons are not hard to find. First, large urban populations are likely to build up prior to, rather than as a consequence of, the expansion of industrial employment. Then, as industrialization gains momentum, the productivity of factory labour tends to rise sharply, and this limits the expansion of demand for general industrial labour. Indeed, modern industrialization may even displace labour from cottage and handicraft industries faster than it is absorbed in newly created factories. Again, government service is able to provide employment for relatively few people. And finally, unless development is extremely rapid, trade, commerce, and other services simply do not absorb those who cannot find jobs in other activities. But despite relatively limited employment opportunities and over-crowded conditions, the modernization process impels people to migrate from the rural areas to the cities; and, as progress is made toward universal primary education, nearly every modernizing country is faced with the problem of mounting unemployment of primary school leavers. In overpopulated countries (such as Egypt or India) the rural areas are also overcrowded, resulting in widespread underemployment and disguised unemployment of human resources. Indeed, in many countries it is evident that total agricultural output could be increased if fewer people were living on the land and the size of agricultural units was increased. Thus, surplus labour in rural areas in most cases is not an asset and in some cases is definitely a liability for increasing agricultural output.
Broader Unemployment (#PB0750).
Narrower Unemployment in developing countries (#PD0176)
Urban unemployment in developing countries (#PD1551)
Unemployed intellectuals in developing countries (#PD1273)
Lack of technical development and excess of manpower in developing countries (#PE4933).
Related Rural unemployment in developing countries (#PD0295).

♦ PD0158 Induced abortion

Abortion — Foeticide — Killing of foetus — Abortificants — Abortion as murder
Nature Induced abortion is the deliberate disruption of the natural pre–natal development of a foetus.
Incidence It is impossible to produce a reliable estimate of induced abortions throughout the world since, in most countries, the majority are performed clandestinely. In recent years, induced abortions seem to have exceeded live births in several countries with low birth rates. In China alone, 53 million abortions have been performed in the last five years in keeping with the government's campaign to limit population growth. In the Soviet Union 90 percent of first pregnancies end in abortion. In countries with high birth rates, abortion ratios per 1000 births are thought to be low (WHO report, 1970).
Claim Abortion may harm a woman's health, consequent complications including perforation of the uterus, peritonitis and septicaemia, sometimes resulting in death. Although legalizing abortion where illegal abortion was highly prevalent may decrease the incidence of these complications, the reverse is true where the initial abortion frequency was low. Abortion, because of coercion or over–persuasion by authorities or by those on whom the pregnant woman depends, offends most moral and religious codes. A more positive approach might be to provide moral, social and material assistance to the woman and to her baby when it arrives. Induced abortion can be described as murder (in particular, this is the stance taken by the Roman Catholic Church); it is also considered by many to be the result of immorality and irresponsibility on the part of those performing the abortion, and on the part of the woman on whom it is performed, since it may be argued that she should take the consequences of her immoral or adulterous conduct. Induced abortion, either legal or illegal, tends towards breakdown of society, of morality and of the family and marriage, and leads to promiscuity.
Counter–claim Every 24 hours there are approximately one million conceptions throughout the world. Of these, nature aborts about 500,000, roughly 150,000 are deliberately terminated and the remaining 350,000 conceptions result in live births. As these figures so powerfully indicate, for every three live births one pregnancy is deliberately terminated. The fact that so many women are prepared to face the mental anguish, physical pain, and danger which often accompany an induced abortion, many without even the benefit of anaesthetic, indicates how common, and how passionate, can be the wish to avoid an unwanted birth. In contrast, the risk to the woman's health inherent in carrying out a legal abortion is considerably less than if the pregnancy were allowed to proceed to term. Those who deny a place for programmes to control the number and spacing of children are denying the right of the mother to control her own fertility, the right of the father to restrict the number of children he begets, and the right of the child to be wanted.
Refs Batchelor, Edwardd Jr (Ed) *Abortion* (1982); Brennan, William *The Abortion Holocaust* (1983); Cole, George F and Frankowski, Stanislaw (Eds) *Abortion and Protection of the Human Fetus* (1988); Goldstein, Robert D *Mother–Love and Abortion* (1988); Hafez, E S (Ed) *Spontaneous Abortion* (1984); Maestri, William *Choose Life and Not Death*; Sachdev, Paul (Ed) *International Handbook on Abortion* (1988); Tietze, Christopher and Henshaw, Stanley K *Induced Abortion* (1986); Tribe, Laurence *Abortion* (1990); Tribe, Laurence *Abortion* (1990).
Broader Killing by humans (#PC8096) Denial of the right of unborn children (#PF6616).
Narrower Late abortions (#PG7962) Gender abortions (#PD3947)
Abortion–related deaths (#PE3580) Illegal induced abortion (#PD0159)
Adolescent induced abortions (#PD1302).
Related Homicide (#PD2341) Birth prevention (#PE3286)
Natural human abortion (#PD0173) Unethical medical practice (#PD5770)
Violation of the rights of foetuses (#PE6369) Denial of rights of children and youth (#PD0513).
Aggravates Promiscuity (#PC0745) Underpopulation (#PD5432)
Family breakdown (#PC2102) Maternal mortality (#PD2422)
Marital instability (#PD2103) Post abortion syndrome (#PE5846).
Aggravated by Permissiveness (#PF1252) Irresponsibility (#PA8658)
Unsustainable population levels (#PB0035)
Deliberate imbalancing of population sex ratio (#PF3382).
Reduces Infanticide (#PD3501) Illegitimate children (#PC1874)
Criminalization of abortion (#PF6169).
Reduced by Lack of abortion facilities (#PF8481).

♦ PD0159 Illegal induced abortion

Amateur abortion — Variable criteria for legal induced abortion
Nature Illegal induced abortion may be performed by a qualified or unqualified person or by the woman herself. Most illegal abortions are carried out by unqualified persons using inefficient methods. Mortality from illegal abortions is high and the incidence of post–abortion infection and other physical damage is significantly greater than that of legally–performed abortions. The most common complications are massive haemorrhaging, perforation of the uterus, laceration, sepsis, and renal failure. These serious conditions require the immediate and skilled medical attention which is often unavailable in poor countries. Those who survive frequently suffer ill–health; in some cases, the damage is permanent.
The need for legal, induced abortions has been expressed differently by the various countries whose laws allow for such action. The most liberal, as for example in Cyprus, permit pregnancy terminations on grounds of demonstrated socio–economic and psychological conditions detrimental to full–term pregnancy and birth, as well as in cases of pregnancies resulting from rape. The more conservative countries consider only medical risk to the mother; however this may include her psychological health and sanity. Rape and grossly malformed foetuses are also considered adequate reasons for legal abortions in some conservative jurisdictions.
Anti–abortion laws in most countries are not strictly enforced and so serve only to deter the most reliable and skilled practitioners.
Incidence It is possible that in some communities the mortality rate from illegal abortions reaches or exceeds 1,000 per 100,000 abortions but the total number of illegal induced abortions is impossible to calculate.
Despite provision for legal abortion, in some countries (such as France) where the medical profession may come from a background of strong, professed religious belief, doctors may refuse to perform pregnancy terminations. In other countries also, Church opposition, even where abortion may be technically legal, creates considerable obstacles to making abortion available, for example via publicly supported clinics. Thus the situation, globally, shows that where abortion has been legalized the justification and criteria vary considerably; and in some countries, though criteria may be met and the law considers them legal, abortions may, in practice, be very difficult to obtain. The fact that many countries have not yet even considered the issues involved also indicates the cruel circumstances that unwanted or defective pregnancies impose on women.
Claim In countries where abortion is illegal, millions of women die each year as a result of severe illness due to backdoor abortions, and babydumping is also a dramatic consequence of refused abortion. The drain on scarce medical resources, the hidden effect on the woman and her family, and the loss of a woman from her home as a mother and a worker, all add up to make illegal abortion one of the most costly and inhuman plagues of the twentieth century.
Women have a right to control their own bodies. Legal abortion is a system of birth control many people find acceptable, more so than permanent drug taking against the possibility of becoming pregnant. No contraceptive method is fail–safe, all require planning. True spontaneity, with both partners able fully to enjoy sex when they wish without the fear of producing an unwanted child, is best achieved by making legal abortion easily available.
Counter–claim Except for strict physical and medical reasons, no abortions should be granted. Human life is sacred, even if conceived in violence. Once the door is opened to this and other psychological arguments for legal abortion, there is no end for the murder of children while in the womb. Wealthy people, finding pregnancy inconvenient, may purchase instant abortion service using the latest technology. This is an abuse of responsible adulthood, and of the powers of the medical profession. It is a crime against the unborn child. The acceptance of such legal abortions with the countless inevitable instances of criminality that it entails, cheapens human life and makes it just another commodity. In addition, it opens the way for the advocates of euthanasia who may argue just as emotionally for termination of the suffering, the dying and the aged, with all the consequent abuses that this entails. That the two concepts are related is seen in the case of modern China where both foetuses and infants (usually female) were murdered in order to keep the population down. In other words, naturalists may consider even population control is sufficient justification for murder; thus legalized abortion may eventually be ordered by the State for population control purposes.
Broader Homicide (#PD2341) Inadequate laws (#PC6848) Induced abortion (#PD0158).
Narrower Prenatal wrongful death (#PD6967) Abortion–related deaths (#PE3580).
Related Psychiatrism (#PF6351) Natural human abortion (#PD0173).
Aggravates Ectopic pregnancy (#PG4643) Human infertility (#PC6037)
Peritoneal adhesions (#PG2025) Unwanted pregnancies (#PF2859)
Lack of family planning (#PF0148) Unsustainable population levels (#PB0035)
Chronic pelvic inflammatory disease (#PG2024).
Aggravated by Unmarried mothers (#PD0902) Sexual exploitation (#PC3261)
Inadequate sex education (#PD0759) Lack of abortion facilities (#PF8481)
Prohibitive cost of abortion (#PJ2020) Adolescent induced abortions (#PD1302)
Inadequacy of contraceptive methods (#PD0093) Inadequate family planning education (#PD1039)
Discrimination in family planning facilities (#PD1036).

♦ PD0160 Harassment of the media

Harassment of the press
Nature Methods of harassment of the press include confiscation of articles, closure of press offices, censorship, injunctions and damages, imprisonment, trial, threats to journalistic immunity, police brutality, the compilation of political dossiers, general intimidation and in the case of foreign correspondents, deportation or banning entry.. It constitutes a restriction on the freedom of information and expression, which may lead to ignorance, apathy, alienation and indoctrination, encourage subversive activities, or strengthen dictatorship and government control.

Incidence Caught in a crossfire of deceit, red tape, ambition, censorship and logistics, truth is an early casualty anywhere news is actually taking place. A 1983 survey shows that a minority of the globe respects free speech. The rest of the world is 'gagged'. Most of the Eastern European countries continue to imprison outspoken journalists, as do most countries in Africa, the Middle East, Latin America, and Asia.
Broader Harassment (#PC8558) Persecution (#PB7709)
Denial of access to news (#PF3081).
Narrower Harassment of journalists (#PD3036) Erosion of journalistic immunity (#PD3035)
Newspaper and periodical censorship (#PD3027).
Aggravates Political repression (#PC1919) Lack of freedom of the press (#PE8951)
Restriction of freedom of expression (#PC2162).
Aggravated by Excessive government control (#PF0304)
Journalistic irresponsibility (#PD3071).

♦ PD0162 Discrimination against women before the law

Denial of right to justice for women
Incidence In some Islam countries efforts to bring laws in conformity with a certain interpretation of the Quran and Sunnah has led to serious miscarriages of justice for women. For example, the onus of providing proof in a rape of a woman rests on the woman herself. If she is unable to convince the court, her allegation that she has been raped is in itself considered as a confession of having sex outside marriage and the rape victim effectively implicates herself and is liable to punishment. Furthermore, the woman can be categorized as the rapist herself since it is often assumed that she seduced the man.
Broader Discrimination against women (#PC0308)
Discrimination before the law (#PC8726).
Narrower Unequal pension rights (#PJ2030) Dependency of women in marriage (#PD3694)
Unequal property rights for women (#PE4018)
Loss of civil capacity for married women (#PE8720)
Discrimination against women in divorce rights (#PE8879)
Discrimination against women in parental rights (#PE4019).
Related Discrimination against men before the law (#PD3692).
Aggravates Dependency of women (#PC3426)
Neglect of the role of women in rural development (#PF4959).
Aggravated by Social injustice (#PC0797)
Exclusion of women from decision making (#PE9009)
Discrimination against women in religion (#PD0127)
Discrimination against women in education (#PD0190).

♦ PD0164 Exploitation of child labour

Economic exploitation of youth — Necessity of child labour — Denial of rights to employed children — Child work force — Employment of children
Nature The economic exploitation of children may damage their health, minds, morals and personalities. It can cause children to be crippled, maimed and killed. It may be associated with malnutrition, disease, and mental and physical impairments that can be genetically transferred. Abuse of child labour creates and perpetuates the miseries of poverty, illiteracy, and sickness and a class of persons whose later social assistance needs can only cause the imposition of greater taxes on their former employers. It also creates a class that is disaffected and a source of political instability. Child labour abuse is criminal in most developed countries, but it is not always reported or detected. In some developing countries exploitation of children has led to conditions of slavery and forced prostitution. Even as a large group, children possess little strength; in small numbers they are yet more highly exploitable. Their vulnerability is manipulated, not only where there is greed and callousness, but also as part of traditional social patterns.
Incidence The ILO estimates that 50–55 million children up to age fifteen are at work through out the world. But even ILO experts admit this is a conservative. Figures sighted other places go as high as 145 million. Some one million may be illegally at work in developed countries. The greatest number, over 40 million, are in Asia, with 16.5 million in India alone; 10 million are in Africa; and some 3 to 4 million are in Latin America. In all regions children are now rarely employed in the larger and more modern industrial undertakings. Changed management attitudes, the introduction of more sophisticated machinery and rationalized production methods, the increased importance of high productivity, the presence of trade unions, the enactment of minimum age laws and the strengthening of inspection services have all contributed to the virtual elimination of child labour in such undertakings. However, child labour in factories has not altogether disappeared. Appreciable numbers of children clearly below the legal minimum age are employed in small marginal factories that rely on keeping wages and other costs to a minimum. Such factories are most numerous in Asia and, to a somewhat lesser extent, in Latin America and the Middle East, but they also exist in parts of Southern Europe and even in depressed areas of more industrialized regions. They seem to be particularly concentrated in certain industries; textiles, clothing manufactures, food processing and canning. In India, recent statistics show that children earn on average half the adult wage, gross violations of regulations and exploitation are rife, conditions are unhealthy and accidents frequent.
Background Child labour is a very broad term and the employment of children does not have the same characteristics everywhere. Such considerations as: the formal status of the working child (that is, whether he is a full–fledged employed person as opposed to an approximation to an informal trainee, an unofficial helper to an adult worker, an unpaid family worker or an 'adopted child'); the nature, intensity and regularity of the work; the hours of work and other conditions of employment; and the effect of work upon schooling, are at least as important as figures in judging the seriousness of the problem in a given situation.
Child workers are frequently represented as apprentices or learners, and many of them undoubtedly are in a sense, but the training they get is often minimal, the work strenuous, the treatment that of servants and the pay far below standard. In some cases such as the hand–made carpet industry, the work is handed out to women by middlemen who have none of the responsibilities of employers, and is performed at home by the women with their daughters or girls from other families. The girls are often practically infants and their employment and conditions of work are subject to no control. Similar practices are found in many countries in handicraft work. Craftsmen are given jobs on contract and are paid strictly by results; the payment and conditions of work of the child 'helpers' they commonly employ are their responsibility. Child labour in construction is found throughout most of Asia, Latin America and the Middle East.
Another largely uncontrolled and widespread occupation for children, traditional in many of the developing countries, is domestic service. In some countries it is common for very young children – mainly girls in Central America, the Middle East and some parts of Asia – to be brought to cities from rural areas by their parents, or purported parents, and virtually sold into domestic service. The children are usually unpaid and the practice is often described as 'adoption'. It is generally rationalized by the argument that these children enjoy much better conditions than they had in their previous homes. But, while this may often be true, in the total absence of outside control there is always a potential danger of overwork, neglect, mistreatment and exploitation. Non–industrial child labour is also a problem in Southern Europe, though different in nature and degree. Children below the legal minimum age are employed fairly widely in shops, cafés and restaurants and to a much lesser extent in markets and street trades. Employment is often combined with school attendance, or at least school enrolment, and provides a supplement to family income. In the more developed countries this problem sometimes also arises, but again it is of a different nature – it is not usually the case here that large numbers of children below the minimum age are being employed under illegal conditions or without safeguard and supervision.
The sector in which the bulk of the working children in every region are employed is, of course, agriculture. In the traditional, mainly subsistence sector of agriculture in the developing countries, the direct regulation of the employment of children is generally not practical. Until adequate educational facilities become available and until it becomes possible for most families to dispense with the work of their children, there is little chance that labour by unpaid children in the family will be reduced to any significant extent. On plantations, employment often takes the form of work as part of a family group: the parents do the main field work and the children either assist them, for example in plucking tea leaves, picking coffee beans or collecting latex, or do secondary jobs such as weeding, spreading fertilizer or caring for plants. This is a fairly common pattern in some Asian countries, including India and Pakistan; it is also found, but to a much lesser extent, in some African countries, especially during harvest periods. As a rule, light work by children under certain conditions is legally permitted, yet it is not uncommon for children to do full–scale agricultural work, including such heavy jobs as ploughing. In Brazil, for example, the planting of cotton, rice and sugar–cane and the harvesting of these crops and of coffee and cocoa are frequently done by both children and adults. Children can also be employed as wage earners on small farms. This is notably true in Egypt where children have long been hired for cotton picking, weeding and other work. In the Federal Republic of Germany nearly 100,000 children worked illegally in 1970, the year that two 10–year olds were killed in mine accidents. The estimated total of child workers in all countries of the North (American, Socialist, and Free–Market European) is over 1 million.
Claim Child labour is based on exploitation. Child workers are unable to defend themselves and are at the mercy of adults, their parents, employers and supervisors. While children working in family undertakings are not paid they are less exploited than wage earning children. Children not only frequently work longer hours than is legal but often longer than adults doing the same work. In the Philippines, for example, in 1976, over 57 percent of the children between the ages of 10 and 14 working in the non–agriculture sector worked for 40 hours or more. In Brazil, nearly a half a million children between 10 and 14 years of age work more than 49 hours a week. Child labourers receive low wages relative to adults doing similar work and relative to the value added by their work. Sometimes they receive no wages at all, for example 'apprentices' who are in fact working rather than learning. The cost to the employer is one of the main reasons they are employed. Children face greater occupational risks than adults. They don't realize the risks they are taking. They are often overworked, physically weak and undernourished and often work in high stress situations, all of which increases the chance of accidents. They often do more hazardous task than adults, for example creeping under moving parts of machinery to clean. They do the 'dirty work' like using solvents and cleaning sewers. They often exposed to toxic substances like pesticides and fertilisers in agriculture and glues in footwear manufacturing. In agriculture child labourers are at risk from disease, toxic poisoning and fatal accidents. In construction they are at risk from falls, falling objects and lifting heavy loads. In manufacturing they are at risk from unsafe machinery, lack of safety equipment, dangerous working conditions, lack of lighting and ventilation and inadequate training in the use of tools. Sitting in awkward positions, for example, in carpet making industries, can lead to bone deformities.
Working children some situation are also exposed to physical and mental abuse. They may be separated from their parents, isolated, virtually imprisoned, beaten and starved. They are not protected by clear written contracts, health care facilities, insurance nor social security. They lack education opportunities and frequently spend their whole lives at the bottom of the socio–economic ladder.
Refs Bequele, A and Boyden, J *Combating Child Labour* (1988); ILO *The Emerging Response to Child Labour* (1988); ILO *Child Labour. A Briefing Manual* (1986); ILO *Annotated Bibliography on Child Labour* (1986); ILO *The World of Work and the Protection of the Child* (1981); International Confederation of Free Trade Unions *Breaking Down the Wall of Silence* (1986); Lee–Wright, Lee *Child Slaves* (1990); Mendelievich, Elias (Ed) *Child Labour. A Briefing Manual* (1980); Mendglievich, Elias *Children at Work* (1979); Moorehead, Caroline (Ed) *Betrayal* (1989); Rodgers, Gerry and Standing, Guy *Child Work, Poverty and Underdevelopment* (1982); Sawyer, Roger *Children Enslaved* (1988); United Nations *Exploitation of Child Labour* (1982).
Broader Cruelty to children (#PC0838) Exploitation of children (#PD0635)
Victimization of children (#PC5512).
Narrower Militarization of children (#PE5986).
Related Physical maltreatment of children (#PC2584).
Aggravates Denial of right to education (#PD8102)
Illiteracy among indigenous peoples (#PD3321)
Trafficking in children for adoption (#PF3302)
Inadequate education of indigenous peoples (#PC3322)
Minimal opportunities for corporate activities (#PF2316).
Aggravated by Labour shortage (#PC0592) Trafficking in women (#PC3298)
Dependency of children (#PD2476) Social underdevelopment (#PC0242)
Exploitation in employment (#PC3297) Unethical practices of employers (#PD2879)
Inadequate educational facilities (#PD0847) Inadequate social welfare services (#PC0834)
Dependence within extended families (#PD0850) Denial of right to minimum work age (#PE1693)
Restrictive patterns of traditional life (#PF3129)
Limited development of functional abilities (#PF1332)
Inappropriate education in developing countries (#PF1531)
Trafficking in children for economic exploitation (#PD7266)
Exploitation of indigenous populations in employment (#PD1092).
Reduced by Underutilization of human resources (#PF3523).

♦ PD0165 Abuse of prison labour

Incidence More than 23 million pounds were lost through negligence and mismanagement of prison enterprises in Britain in 1985. There was concern in that case to distinguish between "limited efficiency", which is seen as desirable, and "inefficiency and neglect", which is what took place.
Broader Forced labour (#PC0746) Maltreatment of prisoners (#PD6005).
Related Denial of rights to prisoners (#PD0520)
Unsanctioned maltreatment of prisoners (#PE0998).
Aggravated by Exploitation in employment (#PC3297).

♦ PD0169 Weakness in intra–regional trade of developing countries

Nature The volume of trade among developing countries represents only a small proportion of their total foreign trade. It is estimated at about 20 percent. Characteristics of that trade is that it is low in primary commodities (excluding petroleum); that it is limited to the same region or same continent, and that it covers a narrow range of items.
Trade among developing countries is low because of trade barriers in developing countries and loss of market penetration due to increased competition with developed country exporters. High transportation and communications costs and inadequate export financing have made it difficult to many developing countries to compete. Severe anti–export bias in bias within the developing countries has added to the problem.
Broader Fragmented regional cooperation (#PF9129)
Weakness in trade among developing countries (#PC0933).
Related Minimal exports in least developed countries (#PE8306).
Aggravated by Lack of integration of transport systems between neighbouring developing countries (#PD0664).

♦ PD0172 Destruction of historic documents and public archives

Nature Archives constitute the organized body of records produced or received by public or semipublic institutions in the course of the transaction of their affairs. Preservation of archives over centuries is important for its value to research and as a part of the cultural heritage. Over periods of decades such preservation, particularly through periods of warfare, is important as a means of resolving a wide variety of claims. In time of war the cost of protection and preservation is such that much valuable material has to be destroyed or deteriorates for lack of proper attention. Adequate archival facilities are needed throughout the developing world although, over the past five years, twelve new functional national archives buildings have been inaugurated (Malaysia, Mexico, Botswana, Tanzania, Brazil, Indonesia, Solomon Islands, etc). Conservation and rehabilitation of vast accumulations of archives dating back to the XVIth and XVIIth centuries in countries like Mexico, Indonesia and the Philippines would require exceptional financial efforts for developing more efficient techniques and the training of conservationists. It is in near–tropical and tropical countries where archival document deterioration is in most need of alleviation.

Broader Destruction of cultural heritage (#PC2114)
Unethical documentation practices (#PD2886)
Inadequate protection and preservation of cultural property (#PF7542).
Narrower Records destruction by transnational enterprises (#PE7061).
Aggravates Deterioration of stored documents and archives (#PE1669).
Aggravated by Prohibitive cost of maintaining comprehensive document collections (#PE1122).

♦ PD0173 Natural human abortion

Miscarriage — Spontaneous abortion

Nature Spontaneous human abortion may take the form of foetal infection and death, or uterine contraction and rejection of the partially formed foetus which may be malformed or have congenital anomalies. Spontaneous abortion may be habitual. It may occur as a result of disease, malnutrition, genetic defects including radiation, ageing, multiple pregnancies or emotional disorder. If resulting from disease this may be that of the prospective mother or father. Spontaneous abortion may result in death or the continuance of disease complications or emotional disorder. It may create marital and family problems resulting in breakdown. Although miscarriage may be more common than childbirth, women are not prepared for the possibility or for the callousness with which their feelings of loss and bereavement are dismissed by relatives and medical practitioners.

Incidence The incidence of spontaneous abortion cannot be estimated accurately because of: the unrepresentative nature of populations available for study in the world; understatement due to failure to recognize abortion, lack of recall and lack of rapport between the woman and the interrogator; over statement, misreporting of temporary amenorrhoea, etc. A reasonable estimate seems to be an overall ratio of 15–20 spontaneous abortions per 100 pregnancies (WHO report, 1970). Because of the high percentage of genetic aberrations, between 12 and 15 per cent of of all embryos die within a few days or weeks, without the mother being especially aware of it. Between the second and sixth month about 10 to 15 per cent more will die as spontaneous miscarriages.

Counter–claim Miscarriages are a perfectly normal occurrence and serve a useful eugenic function in eliminating the unfit.

Refs Stonehouse, A and Sutherland, B *After a Miscarriage* (1986).

Broader Pregnancy disorders (#PD2289) Human disease and disability (#PB1044).
Narrower Foetal infection and death (#PE2041) Physical malformation of foetus (#PE2042).
Related Induced abortion (#PD0158) Animal infertility (#PC1803)
Illegal induced abortion (#PD0159).
Aggravates Bereavement (#PF3516) Family breakdown (#PC2102)
Marital instability (#PD2103) Emotional disorders (#PD9159).
Aggravated by Scurvy (#PE2380) Rubella (#PE0785)
Smallpox (#PE0097) Gonorrhoea (#PE1717)
Multiple pregnancy (#PG2047) Placental anomalies (#PG2046)
Sickle cell disease (#PE3724) Iodine deficiency disorders (#PD2726)
Cytomegalic inclusion disease (#PG2044).
Reduces Individual unfitness for survival (#PF4946).

♦ PD0176 Unemployment in developing countries

Insufficient employment opportunities in developing countries

Nature Lack of employment opportunities involves both those without work and those who have jobs but want to work longer hours or more productively. To individuals already employed it means lack of a reliable way of increasing income, achieving promotion and acquiring self–esteem. For societies, unutilized or underutilized skills are potential productive resources that are wasted. In some cases even where incomes are sufficient, lack of upward mobility causes resentment by individuals who feel they are missing opportunities which they feel they deserve. For developing countries, insufficient paths of advancement within employment means an inability to build a technical, professional or managerial infra–structure.

Claim For young persons, whether educated or not, to enter the labour force with a frustrating round of job seeking or resentment at missed opportunities which they feel they deserve, is hardly a good way to acquire the experience and work attitudes for a productive life, nor does it augur well for social stability.

Counter–claim Most of the so–called unemployed in developing countries in fact work extremely long hours (10–15 per day, 6 to 7 days per week) in the informal sector. The issue is rather that they are paid extremely low wages.

Refs Sethuraman, S V *The Urban Informal Sector in Developing Countries* (1981).

Broader Unemployment (#PB0750)
Labour surpluses in developing countries (#PD0156).
Narrower Rural unemployment (#PF2949) Insufficient part–time employment (#PJ0108)
Unemployment in least developed countries (#PE9476)
Urban unemployment in developing countries (#PD1551)
Rural unemployment in developing countries (#PD0295).
Related Underproductivity (#PF1107) Underdeveloped road network (#PE1055)
Underpayment for work in developing countries (#PE9199).
Aggravates Labour shortage (#PC0592)
Decline in real wages in developing countries (#PD2769)
Emigration of trained personnel from developing to developed areas (#PD1291).
Aggravated by Geographical isolation (#PF9023)
Inadequate manpower planning (#PJ2036)
Incomplete skills information (#PJ0140)
Unsustainable population levels (#PB0035)
Socio–economically inactive rural population (#PF4470)
Inappropriate education in developing countries (#PF1531)
Use of inappropriate technologies in developing countries (#PF0878).

♦ PD0177 Solid wastes as pollutants

Garbage — Refuse

Nature Solid wastes include domestic refuse and other discarded solid materials, such as those from commercial, industrial, and agricultural operations.

Domestic refuse includes paper, cardboard, metals, glass, food matter, ashes, plastics, wood, and other substances discarded from homes. It also includes old furniture, household appliances, and other items. Increasing amounts of paper, cardboard, plastics, glass, and other types of packaging are concomitants of improved standards of living. At the same time, the increasing use of gas, oil, and electricity for heating and cooking has resulted in a decline in the ash content of solid wastes. Thus, the proportion of paper and paper products in domestic refuse has already reached more than 50 per cent in some countries, and this trend will continue.

Commercial refuse includes the wastes discarded by markets, shops, restaurants, offices, and similar businesses. Such refuse is growing in importance not only as a result of increasing business activity, but also because there are few opportunities to deal with it on–site.

Industrial refuse comprises a very wide variety of wastes ranging from comparatively inert materials such as calcium carbonate to highly toxic and explosive compounds. Other examples are trimmings and scrap from manufacture, sludges and slag from industrial processes, and wastes from the food processing industry. Wastes produced in large quantities by mining and some other operations are usually treated by the respective industries themselves. The natural growth and diversity of industry generally, together with rapid technological developments, have resulted in substantial increases not only in the volume of industrial wastes, but also in their complexity.

Agricultural wastes include wastes arising from the production and processing of food and other crops and from the raising and slaughtering of livestock. In many areas the practice of agriculture near urban areas is resulting in the need to consider this type of waste, which is growing in volume, as part of the general urban waste–disposal problem.

General community waste includes demolition and construction debris, street refuse, discarded motor vehicles, and refuse arising from community services such as hospitals, abattoirs, transport systems, parks, canals and harbours. Special handling is required for potentially dangerous wastes such as those from hospitals, international ports and airports, and firms using radioactive materials. As in the other classifications, the volume of material arising from these sources is increasing.

Incidence It is safe to say that everywhere the amounts of solid wastes produced each day per person are increasing as a result of social, economic and technological changes. In addition, better quantitative data are now becoming available as a result of improved surveys. For example, the initiation about 20 years ago of scientific surveys in a few cities, encompassing all types of solid wastes, revealed a surprising discrepancy between earlier 'guess estimates' and actual production of wastes. Prior to about 1945, the production of solid wastes was assumed to be 350 to 400 kg per capita per year. Subsequent careful samplings and actual weighings in representative areas revealed that the actual values for all urban solid wastes ranged up to about 600 kg per capita annually.

Improved standards of living, the building boom, the growth of packaging of consumer goods, and vast increases in the use of paper, paper products, and synthetics have all contributed to an increase in the amount of urban wastes, so that the present average in the industrialized countries is probably at least 700 kg per capita per year. The annual increase is currently between 1 and 2 per cent per year. Furthermore, the density of refuse has been dropping, resulting in even greater increases in volume, with annual values of up to 5 cubic metres per capita not uncommon.

The above data show representative waste contributions in highly developed urban centres in industrialized countries with relatively high standards of living. Although the figures for other areas are lower at present, they can be expected to approach these figures in time. In developing countries, the amounts of solid wastes may well be very small percentages of the figures given. However, the data do not adequately reflect the additional solid wastes produced by agricultural and large industrial operations, nor do they take into account the solid pollutants increasingly being separated from gaseous and liquid wastes by improved treatment processes developed to comply with stricter environmental standards.

Broader Chemical pollutants of the environment (#PD1670).
Aggravates Dust (#PD1245) Fires (#PD8054) Insect pests (#PC1630)
Soil pollution (#PC0058) Malodorous fumes (#PD1413) Amenity destruction (#PC0374)
Mycoplasmal arthritis (#PG2302) Parasites of the human body (#PE0596)
Human disease and disability (#PB1044)
Damage by degradable organic matter (#PJ6128).
Aggravated by Sewage as a pollutant (#PD1414)
Lack of resource conservation (#PG2054).
Reduces Land reclamation (#PF2055).

♦ PD0181 State sanctioned torture

Political torture — Government sanctioned torture

Nature State sanctioned torture means any act by which severe pain or suffering, whether physical or mental, is intentionally inflicted by or at the instigation of a public official on a person for such purposes as obtaining from him (or from a third person) information or confession, punishing him for an act he has committed, or intimidating him or other persons. It does not include pain or suffering arising only from, inherent in, or incidental to, lawful sanctions to the extent consistent with the Standard Minimum Rules for the Treatment of Prisoners.

Incidence Governments are producing progressively more sophisticated methods of torture, including mind–shattering and audio–visual techniques. Torture has become a science, with schools, research facilities and international exchanges of information. While governments universally and collectively condemn torture, more than a third of the world's governments have used or tolerated torture or ill–treatment of political prisoners in the 1980s. Recent incidents and allegations of political torture include psychiatric hospital internment in the USSR with varying levels of so–called 'treatments'; and Operation Demetrius in Ulster, Ireland, conducted by the UK, in which political detainees, according to the European Commission of Human Rights, were subject to five kinds of sensory deprivation and resulting trauma. Amnesty International reports from some countries cite singular instances of political torture, while, under certain regimes, it is so recurrent as to be institutional, with a paraphernalia of schools, instructors of interrogation by torture, and medical doctors acting in complicity. Abuses may also be committed by opposition groups.

In a 1980–1983 global survey, Amnesty International reports allegation of political torture in the following countries: **Af** Algeria, Angola, Benin, Burundi, Cameroon, Chad, Comoros, Congo, Djibouti, Egypt, Ethiopia, Gabon, Gambia, Ghana, Guinea, Guinea–Bissau, Ivory Coast, Kenya, Lesotho, Liberia, Libya, Madagascar, Mali, Mauritania, Morocco, Mozambique, Namibia, Niger, Nigeria, Rwanda, Somalia, South Africa, Sudan, Tunisia, Uganda, Zaire, Zambia, Zimbabwe. **Am** Argentina, Bolivia, Brazil, Canada, Chile, Colombia, Costa Rica, Cuba, Dominica, El Salvador, Grenada, Guatemala, Guyana, Haiti, Honduras, Mexico, Paraguay, Peru, Suriname, USA, Uruguay. **As** Afghanistan, Bahrain, Bangladesh, China, India, Indonesia, Iran, Iraq, Israel, Jordan, Korea Rep, Kuwait, Laos, Lebanon, Malaysia, Nepal, Oman, Pakistan, Philippines, Saudi Arabia, Singapore, Sri Lanka, Syria, Taiwan, United Arab Emir, Viet Nam. **Eu** Albania, Bulgaria, Czechoslovakia, Greece, Italy, Poland, Romania, Spain, Turkey, USSR, Yugoslavia.

Counter–claim Authorities are obliged to defeat terrorists or insurgents who have put innocent lives at risk and who endanger both civil society and the state itself; and are therefore forced to resort to torture to obtain vital information.

Broader Political prisoners (#PC0562) Abuse of government power (#PC9104).
Narrower Torture schools (#PE2062) Psychological torture (#PD4559)
Pharmacological torture (#PE4696)
Military and police personnel participation in torture (#PE4119).
Related Sadism (#PF3270) Physical torture (#PD8734) Cruelty to plants (#PD4148)
Religious torture (#PC7101) Torture of animals (#PC3532) Biochemical warfare (#PC1164)
Torture of children (#PD2851) Concentration camps (#PD0702)

Unfulfilled treaty obligations (#PF2497) Forced disappearance of persons (#PD4259)
Inhumane interrogation techniques (#PD1362).
Aggravates Fear (#PA6030).
Aggravated by Political repression (#PC1919) Denial of right of complaint (#PD7609)
Repression of intellectual dissidents (#PD0434)
Intergovernmental exchange on torture methods (#PG2061).
Reduces Resistance movements (#PJ2066) Political dictatorship (#PC0845).

♦ **PD0183 Wastage in governmental budgets and appropriations**
Excessive governmental spending — Wastage of resources by government agencies — Extravagant waste of resources by government agents — Bureaucratic waste
Nature A recent report of the Organization for Economic Co-operation and Development (OECD) indicates that between 1960 and 1982, average government expenditure in relation to national income rose by more than 20 percentage points to 47 percent, representing an average annual increase of 2.75 percent. This is associated with a growing tendency to make transfer payments (benefits and interest payments) as opposed to providing goods and services; the former now account for over 50 percent of all public spending. At the same time, prices in the public sector increased more rapidly than in the private sector, although this tendency is starting to decrease.
There is little supervision of how the budget money has been spent by the several governmental agencies, because everybody is more interested in getting money than managing it afterwards. Agencies may have different accounting systems, no legislation or common rules to follow, no annual reports and many are just impossible to audit.
Incidence Brazil spends around 25 percent of gross domestic product on social programmes for health, education and nutrition. The result, however, places Brazil among the world's most backward nations. This is because spending is targeted away from the poorest sectors of society, allowing malnutrition, infant mortality and illiteracy to grow even as wealthier children improve their standard of living. Access to social benefits is not a right, but rather a favour provided by influential politicians or bureaucrats, making social expenditure priorities very difficult to change.
Claim Public expenditure is invariably wasteful as it lacks the discipline imposed by the market on the private sector. A particular problem are the increases in public sector borrowing, the consequent interest payments causing further spiralling of government costs, and in taxation.
Counter-claim Many of the increases in public spending are related to the oil crisis of the 1970s and the reduction in economic growth, with the consequent burden of payments to the unemployed and increased costs in services. There is evidence to suggest that welfare payments do not, in fact, give rise to any significant tendency to cheat or abuse the system, nor to stay out of work longer than necessary. Privatization of services and the free play of market forces would bring little if any improvement in costs.
  Broader Unproductive use of resources (#PB8376).
  Aggravated by Government inefficiency (#PF8491)
  Governmental incompetence (#PF3953)
  Proliferation of public sector institutions (#PF4739).

♦ **PD0184 Newspaper and journal propaganda**
Nature The use of news media and the press to disseminate information with intent to influence public opinion may include: government and official information; unrelated and misleading information; war, racist or religious propaganda; and writing from the underground and student press. Propaganda in the press may also include the use of cartoons and photographs. Newspaper and journal propaganda heightens international and domestic political conflict and may induce counterpropaganda, the confiscation of material felt to be subversive, imprisonment of those found with it in their possession or distributing it, and import restrictions on propagandist publications. There may be a conflict of information. Foreign influence or other subversive influence may induce ideological deviation, political unrest or revolution. Domestic political propaganda may result in the apathy and ignorance of the population.
Incidence Newspaper and journal propaganda is one of the most widely used means of indoctrination.
  Broader Propaganda (#PF1878).
  Narrower Underground press (#PD2366)
  Foreign controls of newspaper and journal propaganda (#PD3041).
  Related Ideological conflict (#PF3388) Misuse of advertising (#PE4225)
  Misleading information (#PF3096) Compulsory indoctrination (#PD3097)
  State control of communications mass media (#PD4597)
  Destabilizing international telecommunications (#PD0187).
  Aggravates Political unrest (#PD8168) Incitement to violence (#PJ2068).
  Aggravated by Nationalism (#PB0534) Double standards in morality (#PF5225)
  Proliferation of commercialism (#PF0815).
  Reduced by Newspaper and periodical censorship (#PD3027).

♦ **PD0187 Destabilizing international telecommunications**
Nature Many governments especially those holding monopolies of information, complain about radio transmitters broadcasting propaganda to listeners in their country because they represent unwarranted interference in their internal affairs. To protect their societies against alien propaganda messages, some countries try to filter or even halt incoming messages. Where this is true to a certain extent in the press, films, books, and exhibitions, it is almost virtually impossible in the case of radio broadcasting due to its technical capability of covering almost the entire globe at low costs of programme production and reception.
Incidence The USA's two radio programmes, Radio Free Europe (broadcast to Eastern Europe) and Radio Liberty (broadcast to the USSR) are considered examples of international broadcasting propaganda. Until 1971, these programmes were financed by the CIA, and during the 1956 Hungarian Revolution, Radio Free Europe was accused of making appeals for revolution.
Claim External radio broadcasting is generally shown to be most concentrated in surrounding areas of international tension.
Counter-claim External radio broadcasting can be an integral part of friendly relations between nations, supplying access to varying, non-political art (theatre, music).
  Broader Subversive activities (#PD0557) Compulsory indoctrination (#PD3097)
  International non-military conflict (#PF3100).
  Narrower Radio frequency interference (#PD2045).
  Related Misuse of advertising (#PE4225) War and pre-war propaganda (#PD3092)
  Newspaper and journal propaganda (#PD0184)
  Inappropriate use of telecommunications services (#PE4450).
  Aggravates Benign neoplasm of bone and cartilage (#PG0819).
  Aggravated by War (#PB0593) Propaganda (#PF1878) Nationalism (#PB0534)
  Misleading information (#PF3096).

♦ **PD0188 Avoidance of copyright**
Copyright pirates — Vulnerability of performers' rights — Literary piracy — Copyright infringement — Bootlegging
Nature Copyright covers the expression of ideas. For the purposes of copyright, piracy is the unauthorized reproduction for commercial purposes of works protected by copyright or similar rights, together with all subsequent commercial dealings in such reproductions.
Background There are numerous problems connected with copyright protection. Problems in high technology are due to 1) the development of techniques for the dissemination of creative works (reprography, computers, satellites, television by cable, video-cassettes, etc), 2) the ease with which communications technology can be imitated and 3) the globalization of the world economy, which has intensified both the incentives and harm of such violations. The technology and audio-visual media are now so widely used that it is urgently necessary to find solutions which reconcile the rights of authors or their assignees with users' interests. Problems related to copyright protection in the area of logos, industrial designs and inventions include approximations of these items, for example, a single letter may be changed in a brand name.
Incidence Abuse of intellectual property rights has led to a worldwide business estimated to be worth as much as $60 billion a year. In the European Community the estimated loss from phonogram piracy alone in 1984 was about $27 million. Illicit versions of British textbooks are available through Southeast Asia. British publishers claim that in Taiwan alone they are losing £25 million a year because locally pirated editions are available at a fraction of their UK price. Losses in Singapore are estimated at £16 million a year, in Korea £10 million a year, in Nigeria £6 million a year, in Indonesia £5 million a year and in Pakistan and Malaysia £4 million a year each. In the recording industry, 90 million pirated tapes were sold in Saudi Arabia alone in 1985. British companies claim this as a £23 million loss. Espionage in high technology for weapons systems and spacecraft has saved the Soviet Union more than $50 billion in research costs, according to American government sources.
Counter-claim While it is important to protect the legitimate interests of copyright proprietors, the control that they exercise over the use of protected works must not be allowed to become an obstacle to the development and improvement of documentation and education systems. It is therefore essential to formulate strategies which make it possible to promote the circulation of such materials.
Refs Chesterman, John and Lipman, Andy *The Electronic Pirates*; Commonwealth Secretariat *The Copyright System* (1983); Gadbaw, R Michael and Richards, Timothy J (Eds) *Intellectual Property Rights* (1987).
  Broader Counterfeiting (#PD7981) Unethical commercial practices (#PC2563)
  Vulnerability of intellectual property (#PF8854).
  Narrower Plagiarism (#PD3996) Computer piracy (#PE6625)
  Commercial piracy (#PG2080) Inadequacy of patent coverage (#PF3538).
  Related Theft of works of art (#PE0323) Infringement of trade secrets (#PD3537)
  Crimes against intangible property (#PE6486).
  Aggravates Uncontrolled media (#PD0040).
  Aggravated by Copyright barriers to transfer of knowledge (#PE8403).

♦ **PD0190 Discrimination against women in education**
Unequal access of women to education — Denial of right to education for women — Lack of training for women — Neglect of education for women
Nature Though attitudes have shifted considerably with the recent growth of feminism, women's primary destiny is still seen as mother and housewife, and education still discriminates against girls. This sex discrimination takes place not only by the visible sexist messages still to be found in many school textbooks but also by the invisible and more subtle messages within the classroom which are inherent in the way teachers relate to students and students to each other. These messages are so important that even if sex discrimination were purged from textbooks, social relations within the classroom might open another door allowing bias to enter. Relationships inside the classroom are powerful but also largely unconscious, because they reflect the dominant social relations between men and women outside school. Since women are generally subordinate in the public world of work and politics and do not really count in the larger society, they do not really count in the schools. Despite this, research shows that girls are generally more successful than boys in primary school but are not rewarded for this early academic success. Boys receive most of the teachers' time and the lessons are geared to their interests. Teachers tend to know them better as individuals and spend a lot of time helping and encouraging them. By secondary school the silent lessons of earlier grades have taken effect. An Australian research shows that by this time boys receive on average 70 percent of teachers' time. The overall academic achievement of girls begins to lag behind boys. In Britain 60 percent of girls leave school at age 16 without any educational qualifications. The sexes also begin to divide in subject choice with technical and scientific subjects dominated overwhelmingly by males.
Refs Feinberg, Renee *Women, Education, and Employment* (1982).
  Broader Discrimination against women (#PC0308).
  Denial of right to education (#PD8102)
  Sexual discrimination in education (#PD1468).
  Narrower Lack of education for women immigrants (#PE8699)
  Discrimination against women in athletic training (#PE4246).
  Related Discrimination against men in education (#PD8909).
  Aggravates Narrow range of practical skills (#PF2477)
  Denial of right of family planning (#PE5226)
  Discrimination against women in politics (#PC1001)
  Discrimination against women in employment (#PD0086)
  Discrimination against women before the law (#PD0162)
  Discrimination against women in public life (#PD3335)
  Discrimination against women in social services (#PD3691)
  Inflexible attitudes toward community social services (#PF3083).
  Aggravated by Social injustice (#PC0797) Gender stereotyping (#PD5843)
  Segregation in education (#PD3441) Dependency of women in marriage (#PD3694)
  Inadequate educational facilities (#PD0847)
  Discrimination against women in religion (#PD0127)
  Illiteracy among women in developing countries (#PE8660)
  Neglect of the role of women in rural development (#PF4959)
  Inequality of access to education within countries (#PC1896)
  Incompatibility of traditional and new technologies (#PE3337).

♦ **PD0196 Physically handicapped children**
Maimed children — Crippled children — Disabled children
Nature For many children, the presence of an impairment leads to rejection or isolation from experiences that are part of normal development. This situation may be exacerbated by faulty family and community attitudes and behaviour during the critical years when children's personalities and self-images are developing.
Refs Batshaw, Mark L; Perret, Yvonne M and Kasmer, Elaine *Children with Handicaps* (1986); Curtis, W S and Donlon, E T *Observational Evaluation of Severely Multi-Handicapped Children* (1985); Freeman, Peggy MBE *The Deaf/Blind Baby* (1986); McCarthy, Gillian T *The Physically Handicapped Child* (1984); Singer, Peter and Kuhse, Helga *Should the Baby Live?*.
  Broader Disabled children (#PC5890) Physically handicapped persons (#PD6020).
  Narrower Deaf children (#PE2083) Mute children (#PG2084)
  Blind children (#PE2082).
  Aggravates Human dependence (#PA2159) Maladjusted children (#PD0586)
  Lack of self-confidence (#PF0879) State custody of deprived children (#PD0550)
  Lack of self-development in the family (#PE8341)
  Family rejection of physically handicapped (#PE2087)
  Segregation of handicapped children in education (#PE8424)
  Immigration barriers for handicapped family members (#PE4868)
  Stress on families of the physically or mentally handicapped (#PD1405).
  Aggravated by Cruelty to children (#PC0838) Physical maltreatment of children (#PC2584)
  Inadequate community care for handicapped persons (#PE8924)
  Vulnerability of the disabled during states of emergency (#PD0098).

♦ PD0201 Earthquakes

Geological faults

Nature Major earthquakes and the majority of smaller ones are generated by sudden decrease or release, in the volume of rock, of elastic strain previously accumulated over an interval of time varying from a minimum of about one year in regions of great activity to many centuries in others. The accumulation of strain is produced by differential movements of contiguous portions of the earth's crust down to about 700 km. The release of strain results in the generation of a series of elastic waves (which should be distinguished from the continuous vibrations known as microseisms). These waves may: create landslides in hilly regions; change the direction of flow of rivers; disrupt man–prepared land areas; cause severe damage to buildings; or give rise to tidal waves, tsunamis. They are the most severe of natural disasters.

Incidence Between 1980 and 1985, earthquakes killed 28,404 persons across scattered parts of the world and caused damage estimated at US $36,439 million. Probably no area of the earth is entirely free of earthquakes if a sufficiently long time interval is considered. At present, however, the great majority are concentrated in a belt around the Pacific Ocean and in a wedge running from SE Asia through the Mediterranean to Portugal. About 50,000 earthquakes occur annually which are of sufficient size to be noticed without the aid of instruments. Of these, about 100 are large enough to produce significant damage if they occur in built up areas. The largest earthquakes occur at a rate of approximately one per year. During the last three centuries it is estimated that earthquakes have resulted in the loss of 2.5 million lives.

Refs NATO Advanced Study Institute Staff and Mansinha, L and Smylie, D E and Beck, A E et al *Earthquake Displacement Fields and the Rotation of the Earth* (1970).

Broader Bad omens (#PF8577) Geological hazards (#PC6684).
Aggravates Fires (#PD8054) Tsunamis (#PD0033) Dam failures (#PE9517)
Rock avalanches (#PG0476) Disastrous failure of natural dams (#PE0715)
Ground failures due to liquefaction (#PE5126)
Endangered monuments and historic sites (#PD0253)
Hazardous locations for nuclear power plants (#PD2718).
Aggravated by Global cooling (#PF1744) Earth surface faulting (#PE5096)
Nuclear weapons testing (#PC2201) Waste disposal in deep wells (#PG2096)
Nuclear explosions underground (#PE2095) Unethical practice of earth sciences (#PD0708)
Reservoir induced increases in seismicity (#PG7404)
Ecosystem modifications due to creation of dams and lakes (#PD0767).

♦ PD0211 Lack of facilities for severely deformed people

Nature Individuals who remain alive despite severe congenital malformations (typically combining several defects such as harelip, spina bifida, hydrocephalus together with mental abnormalities), require special care. Suitable facilities are only rarely available, with the result that the care received is inferior and places a severe strain on those who take responsibility for such cases, particularly if it is the parents who continue to do so.

Broader Social neglect (#PB0883)
Lack of facilities for the physically disabled (#PD8314).
Aggravated by Criminalization of euthanasia (#PF2643)
Human physical genetic abnormalities (#PD1618).

♦ PD0213 Multi–drug abuse

Nature Multi–drug abuse may be entirely or partially intentional, or unintentional. The first derives mainly from experimenting and the desire to explore the different effects of different drugs. Multi–drug abuse may also occur through the lack of availability of the desired alternative. Unintentional multi–drug abuse occurs as a result of the adulteration of 'street' sold drugs and the ignorance of the buyer. Marijuana and mescaline have been known to be doctored with animal tranquillizers. Multi–drug abuse is potentially much more dangerous than single drug abuse, since complications arising from the effects of different drug types render treatment difficult and in many cases impossible. Already complex drugs such as opiates are inherently incompatible with many other substances. Some of the most dangerous combinations of drug types are stimulants (amphetamines, coca) and depressants (barbiturates, tranquillizers, hypnotics), and either of these with alcohol. The use of alcohol in combination with other drugs, which may be prescribed medically as well as obtained illicitly, is particularly hazardous since alcohol is socially and legally acceptable and is not generally considered as a drug.

Refs Wesson, Donald R and Adams, Kenneth *Polydrug Abuse* (1978).

Broader Drug abuse (#PD0094).
Aggravates Psychoses (#PD1722).
Aggravated by Abuse of plant drugs (#PD0022).

♦ PD0215 Occupational diseases

Industrial diseases

Nature The increased mechanization and complexity of modern work processes has led to a variety of threats to human health. The result is considerable human suffering and economic losses both to the individual and his employer. In developing countries, where skilled labour is scarce, occupational health is of particular importance.

Incidence At least 100,000 workers die in the USA each year – and three or four times that number are disabled – as a result of occupational disease attributed to new chemicals, many untested for safety, being introduced into industrial products and processes.

Refs Adams, R M (Ed) *Occupational Skin Diseases* (1986); Bartone, John C *Occupational Diseases* (1983); Donham, K *Occupational Disease of Agricultural Workers* (1985); Glendon, A, et al *Bibliographical Review of Data Sources on Occupational Accidents and Diseases* (1986); Hemminki, Kari, et al (Eds) *Occupational Hazards and Reproduction* (1985); ILO *The Cost of Occupational Accidents and Diseases* (1986).

Broader Occupational risk to health (#PC0865) Human disease and disability (#PB1044).
Narrower Cataract (#PE6817) Silicosis (#PE1314) Leptospirosis (#PE2357)
Pneumoconiosis (#PD2034) Vibration sickness (#PE7788) Occupational cancer (#PE3509)
Occupational stress (#PE6937) Occupational deafness (#PD1361)
Occupational rheumatism (#PE0502) Occupational dermatoses (#PE5684)
Urinary bladder disorders (#PE2307) Occupational blood diseases (#PE6806)
Occupational psychopathology (#PD6880) Impairment of physicians' ability (#PE5746)
Occupational diseases of the voice (#PE6866)
Cardiac conditions in work environment (#PE6811)
Occupational illness in developing countries (#PD8841)
Pleura and peritoneum cancers of the bronchi (#PE8228)
Occupational hazards in commerce and offices (#PE3957).
Related Anthrax (#PE2736) Brucellosis (#PE0924) Industrial accidents (#PC0646)
Ultraviolet radiation as a hazard (#PE5672).
Aggravates Absenteeism (#PD1634) Underproductivity (#PF1107)
Shortage of skilled labour (#PD0044).
Aggravated by Pneumonia (#PE2293) Animal diseases (#PC0952)
Ethnic conflict (#PC3685) Inhospitable climate (#PC0387)
Dangerous occupations (#PC1640) Pesticide intoxication (#PE2349)
Lead as a health hazard (#PE5650) Economic underdevelopment (#PC0281)
Health hazards of asbestos (#PE3001) Heat as an occupational hazard (#PE5720)
Cold as an occupational hazard (#PF5744) Sulphur dioxide as a pollutant (#PE1210)
Solvents as an occupational hazard (#PE5708) Carbon monoxide as a health hazard (#PE1657)
Health risks to workers in commerce (#PE0688)
Combined stresses as occupational hazards (#PE5656)
Biological agents as occupational hazards (#PE5696)
Health hazards from basic metal industries (#PD0243)
Shift work stress as an occupational hazard (#PE5768)
Improper lighting as an occupational hazard (#PE5780)
Unattractive locale for economic development (#PF3499)
Health risks to workers in service industries (#PE0875)
Health risks to workers in construction industry (#PE0526)
Inadequate teaching occupational health and safety (#PE8305)
Health risks to workers in manufacturing industries (#PE1605)
Health risks to workers in agricultural and livestock production (#PE0524)
Health risks to workers in electricity, gas, water and sanitary services (#PE1159)
Health risks to workers in transport, storage and communication industries (#PE1581)
Conflicting standards for protection against chemical occupational hazards (#PE5651).

♦ PD0218 Forced repatriation of prisoners of war

Nature After a truce agreement, there is a return of prisoners of war to their native countries. Considerable disagreement arises over whether a prisoner must return to his homeland. Sometimes he chooses not to. This may be for ideological reasons, or due to brain washing, or due to collaboration with the enemy during internment.

Broader Prisoners of war (#PC8848) Forced repatriation (#PD8099).
Narrower Inaccessibility of religious scriptures (#PF3869).
Related Defection (#PG5533) Occupied nations (#PC1788)
Internment without trial (#PD1576) Refusal to grant nationality (#PF2657).
Aggravates Statelessness (#PE2485) Political unrest (#PD8168)
Lack of patriotism (#PG2131).
Aggravated by Antisemitism (#PE2131) Draft evasion (#PD0356)
Mental depression (#PC0799) Political conflict (#PC0368)
Ideological conflict (#PF3388) Collaboration with the enemy (#PG2132)
Brainwashing of prisoners of war (#PD1652) Ill treatment of prisoners of war (#PD2617)
Discriminatory treatment of foreign prisoners (#PE6883).

♦ PD0235 Contamination of drinking water

Nature The present wellbeing of most industrialized countries and the hopes for a brighter future in the less–developed countries depend to a significant extent on the availability of sufficient quantities of ground water of adequate quality. With the continuing and increasing presence of contaminants in major surface–water resources such as the Rhine, Danube and Great Lakes of North America, with the discovery of toxic contaminants in numerous subsurface waste–disposal sites in the industrialized nations, and with the pollution of aquifers underlying cities in the less–developed countries, there is a clear need for effective ground–water quality management throughout the world.

The various activities generating water contaminants are agriculture, mining, and the production of household, commercial and industrial wastes.

While certain solid and liquid agricultural wastes occur in concentrated forms, low–level ground–water contamination in rural areas occurs as a result of the widespread leaching of excess nutrients applied to both arable and pastoral land as inorganic and organic fertilizers. The principal contaminant is nitrate, derived both from fertilizers and as a result of the transformation of organically bound nitrogen in the soil to inorganic forms by bacteria (mineralization) following the ploughing of established and temporary grasslands. The potential quantities of nitrate released for leaching by ploughing may be large, exceeding the total annual quantities of nitrogen normally applied as fertilizers. Increased sulphate and chloride concentrations derived from the use of ammonium sulphate and potassium chloride fertilizers may also occur in drainage water from farmland. The residues of pesticides and herbicides may be leached from the soil and examples of ground–water contamination have been reported. The risk of infection of ground–water supplies by faecal bacteria from free ranging livestock is generally slight when livestock densities are low, as a result of the complete natural degradation of faeces and urine by soil bacteria when wastes are deposited in a disseminated manner. However, the problem is more serious where animals congregate in large numbers, for example around water holes or in stock yards, in which case large volumes of liquid and semi–solid faecal matter may readily infiltrate to the water table. Cryptosporidium from livestock slurry is a potentially deadly parasite that can not be destroyed by fast filtration or chlorination of water supply. Acute ground–water quality and human health problems have been recorded from rural areas in all parts of the world where domestic water supplies are taken from shallow wells or boreholes situated adjacent to farmyards. In many cases the wells are found to lack any sanitary protection in the linings, or sills and impermeable aprons around the well head, consequently massive contamination by animal or human faecal bacteria occurs readily. Even in areas where wells have sanitary protection, the rate of production and the highly polluting nature of farmyard slurries and especially silage effluents, necessitate either their collection for treatment or confinement to avoid ground–water contamination. The extremely high BOD (biological oxygen demand) values of silage effluents, together with the organic contaminants resulting from the fermentation, could lead to intense ground–water contamination, including the onset of anoxic conditions.

The principal solid mining wastes that have a potential to pollute ground water come from coal and metal ore exploitation. In the former case, the waste rock, generally sandstone, causes oxidation of the disseminated sulphides leading to the formation of sulphate ions, a lowering of the pH of interstitial fluids, the mobilization of iron and manganese and the physical disintegration of the shale. This produces a leachate high in sulphate, iron and dissolved and suspended solids and low in pH. Drainage waters from collieries characteristically possess both high suspended and high dissolved solids, particularly iron and sulphate ions resulting from the oxidation of ferrous sulphides in the host rocks. Chloride concentrations may also be high because of connate water trapped within the sedimentary rocks. Discharge of such waters on to the ground surface may result in widespread contamination of previously potable ground–water resources.

Household wastes contain a high proportion of putrescible matter which is broken down by biodegradation, leading to an initial temperature rise within the waste mass and the generation of carbon dioxide and methane gases. The leachate contains high concentration of total organic carbon (TOC), of which often more than 80 percent is in the form of volatile fatty acids (acetic, butyric, etc) and give waste disposal sites their characteristic odour. As the wastes age, the organic carbon component of the leachate changes to higher molecular weight substances such as carbohydrates. The period of this change is between 5 to 10 years in humid, temperate climates, but probably a shorter period in warmer climates. Under arid conditions the rate of bacterial degradation may be limited by lack of moisture.

The composition of industrial wastes varies with the source, ranging from cyanide wastes from metallurgical operations, through sulphite–rich paper and pulp manufacturing wastes, mercury–rich materials from the electrical industry, to solid residues, such as polychlorinated biphenyls (PCBs), pesticide or herbicide residues and phenol–rich tar wastes, from the petrochemical industries. In many cases toxic substances are present which, if solubilized, may present major threats to ground–water quality. In some cases the total quantity of waste occurring is relatively small and co–disposal may afford a method of using the absorption and biodegradation processes in landfills to attenuate the leachate from the industrial materials.

When chlorine is added to the water it may reacts with pollution creating more dangerous chemicals, like chlorophenols. The water pipes in old buildings may contain lead, cadmium and mercury.

Refs Calabrese, E J, et al (Eds) *Inorganics in Drinking Water and Cardiovascular Disease*

(1985).
Broader Pollution of inland waters (#PD1223).
Narrower Fluoridation of drinking water (#PE2871).
Disease-causing microbes in drinking water (#PG3400).
Aggravates Cholera (#PE0560) Dysentery (#PE2259) Encephalitis (#PE2348)
Poliomyelitis (#PE0504) Typhoid fever (#PD1753) Leptospirosis (#PE2357)
Dracunculiasis (#PE3510) Leeches as pests (#PE3660) Domestic hazards (#PG2144)
Epidemic myalgias (#PG2146) Animal myocarditis (#PG2145)
Lead as a health hazard (#PE5650) Contamination of human body (#PF9150)
Human disease and disability (#PB1044).
Aggravated by Wasted water (#PD3669) Sewage as a pollutant (#PD1414)
Hazardous waste dumping (#PD1398) Impurities in waste water (#PD0482)
Water system contamination (#PD8122) Industrial waste water pollutants (#PD0575)
Water pollution in developing countries (#PD3675)
Methane gas emissions from landfill sites (#PE1256).

♦ PD0243 Health hazards from basic metal industries
Incidence Some forms of cancer are linked directly to metal contamination. Most at risk are the people who work in particular industries or live near plants where production safeguards are inadequate. Arsenic workers have respiratory cancer mortality almost three times the expected rate. That for cadmium smelter workers is more than twice the expected rate, and lead smelter workers' is a third higher than expected. (The excess cancer rate in lead smelter workers may be attributable to the arsenic in the ore rather than to the lead). Lead has been linked to cancers of the respiratory and digestive organs, arsenic to cancer of the skin, mouth, and nose. Other serious and chronic health effects can result from metal poisoning and contamination. Lead and mercury are historically associated with mental disability, lead and arsenic with digestive difficulties, cadmium with kidney disease, and lead with anaemia.
Refs United Nations *Occupational Safety and Health in the Iron and Steel Industry* (1983); United Nations Environment Programme *Environmental Aspects of the Aluminium Smelting* (1981).
Broader Environmental hazards from manufacturing industries (#PD0454).
Narrower Iron and steel basic industries environmental hazards (#PE8397).
Non-ferrous metals basic industries environmental hazards (#PE8248).
Aggravates Occupational diseases (#PD0215).

♦ PD0247 Violence against women
Cruelty to women — Sexist violence
Nature Because women are 'easy' victims, they experience a great deal of direct behavioural violence in every society. Rape is a common worldwide phenomenon, with no more than half of all reported rapes being the work of strangers. In societies where the dowry prevails, dowry demands by husbands and their families result in bride burning, drowning and poisoning. It is more effective to have the wife incinerate, drown or poison herself than to have to do it for her. Wife-beating is less prevalent than rape. In times of war and civil rebellion women endure capture and torture not because they themselves are active in the fighting, but because they are the wives, mothers, or daughters of activists.
Incidence One in three women in Belgium has suffered sexual mistreatment and physical violence is even more wide spread, according to a recent government report. Ninety-nine of the abusers are men and two-thirds came from the woman's personal circles; either relatives (fathers and brothers) or friends and acquaintances. Most of the women who had suffered sexual abuse had first been molested at puberty or adolescence. One in six had be sexually abused at the age of 12. In 40 percent of the cases, the abuse happened more than once and often on a regular basis.
Broader Cruelty (#PB2642) Human violence (#PA0429)
Consumer vulnerability (#PC0123).
Narrower Rape (#PD3266) Wife abuse (#PD6758) Bride burning (#PE4718)
Female sexual mutilations (#PE6055) Exploitation of women refugees (#PD5025)
Alcohol-related violence against women (#PE7672).
Related Trafficking in women (#PC3298).
Aggravates Personal physical insecurity of women (#PE7750).
Aggravated by Family violence (#PD6881) Male domination (#PC3024)
Misuse of advertising (#PE4225) Exploitation of women (#PC9733)
Irresponsible pharmaceutical advertising (#PE2390).

♦ PD0251 Hail storms
Nature Convective storm clouds can provide the conditions for growth of hailstones, although the larger hailstones grow only inside the largest and severest storms where the core winds have very high upward velocities. Hail storms cause considerable damage to crops and property since the stones range in size from 5mm in diameter to 50 mm or more. The degree of damage is determined by whether the hail is accompanied by rain or wind.
Incidence Hail storms mainly occur in temperate climates. The air is seldom unstable enough in the polar regions and the freezing level is too high in the tropics. In the USA it is estimated that annual crop losses from hail storms are $200–300 million (greater than that caused by tornadoes).
Broader Bad omens (#PF8577) Bad weather (#PC0293).

♦ PD0253 Endangered monuments and historic sites
Destruction of historic buildings — Endangered ancient sites
Nature The decay of historical buildings as they age is a universal phenomenon from which no civilization and no country is exempt. Such decay is particularly serious in regions suffering from severe climatic conditions. The rate of deterioration has been considerably increased in recent years by air pollution in urban and industrial concentrations. Atmospheric pollution lies at the origin of a whole chemical and bacteriological attack on stone which defaces buildings and makes stonework pliable and brittle. Such pollution also disfigures historic buildings with layers of ash and soot, which, removed by abrasive action, cause further damage to the surface.
The vibrations set up by the passage of vehicles in the narrow streets of historic towns, and increasingly by supersonic aircraft, lead to further damage to the architectural heritage, especially in the case of fragile structures.
War always constitutes a major threat to monuments because of the psychological value to the attacker of destroying major symbols of the culture being attacked. In the 1930s, Nicholas Roerich promoted a Pact aiming at protecting cultural monuments in time of war.
Monuments may also be deliberately destroyed because of ignorance and lack of appreciation of the value of the heritage, particularly as a result of changes of architectural fashion and concepts of beauty. In some cases the population may have a latent antipathy for the architectural witnesses of civilizations or peoples whose contributions are rejected, disputed, or simply undervalued. This rejection also occurs when a population has changed its faith to a form which favours a completely different architectural emphasis. Wrong use or reuse is equally an important cause of the deterioration of monuments.
Finally, the historically recent processes of urbanization, industrialization and increases in leisure make much greater demand on space while at the same time leading to the abandoning of rural zones and thus of their heritage. Monuments and sites consequently come under heavy pressure. Governments may try to relieve pressure on traditional tourist attractions by encouraging more visitors to less known or remote sites.

Broader Destruction of cultural heritage (#PC2114).
Inadequate protection and preservation of cultural property (#PF7542).
Narrower Desecration of monuments (#PD4348) Vulnerability of sacred sites (#PD6128)
Destruction of archaeological sites (#PD4502)
Submergence of historical sites through dams (#PJ9372)
Damage to cultural artefacts by environmental pollution (#PD2478).
Related Vulnerability of protected natural areas (#PC4764).
Aggravated by War (#PB0593) Fires (#PD8054) Floods (#PD0452)
Ignorance (#PA5568) Hurricanes (#PD1590) Earthquakes (#PD0201)
Air pollution (#PC0119) Soil salinization (#PE1727) Rising water level (#PD8888)
Inhospitable climate (#PC0387) Limited available land (#PC8160)
Vibrations as a health hazard (#PE1145)
Social environmental degradation from recreation and tourism (#PD0826)
Natural environmental degradation from recreation and tourism (#PE6920).

♦ PD0260 Cruel treatment of animals for research
Misuse of experimental animals — Vivisection — Experiments on live animals — Maltreatment of laboratory animals — Medical experimentation on animals — Maltreatment of animals for educational purposes
Nature Biomedical research makes use of considerable numbers of animals in experiments, the majority of which may be for commercial rather than scientific purposes. Most of these experiments are not carried out in the interest of the animal and result in direct or indirect interference with its normal health or comfort or give rise to unforeseen consequences. Animals may be poorly cared for prior to, during, or following experiments, without adequate use of anaesthetics or appropriate use of euthanasia. Such experiments may be conducted without the appropriate laboratory facilities and by persons without the appropriate qualifications (such as children in school laboratories). Such research is therefore often unnecessarily cruel to the animal, unnecessary in terms of the advance of knowledge, and undesirable in the insensitive attitudes which it cultivates in those who practise such research or witness audio-visual records of it.
Incidence It has been estimated that 100–200 million animals die in laboratories around the world each year. In the USA researchers are estimated to use each year approximately: 45 million rodents, 700,000 rabbits, 200,000 cats, 500,000 dogs, 46,000 pigs, 23,000 sheep, 1,725,000 birds, 15–20 million frogs, 190,000 turtles, 61,000 snakes, 51,000 lizards, and over 85,000 primates. Animal experiments in the UK have diminished from a peak of 5,607,000 per year in 1971 to 3,112,051 in 1986, most of which were for the testing of pharmaceutical products. Recent figures produced by the Agricultural Department in the USA indicate that, apart from painful experiments, one sample of research institutions revealed that: 24% had major, repeated violation; 22% some major violation; and 29% minor violation of minimum standards of care. Only 24% fully complied with regulations. Advocates of vivisection reform have estimated that perhaps 15 per cent of animal experimentation is necessary (meaning there is no alternative) and proper (meaning following every effort to minimize the suffering, wasteful loss of life, and appropriate choice of animal) for the prevention and cure of disease.
The remaining 70 per cent of animal experimentation can be grouped as follows: (a) regular environmental testing to determine acceptable levels of toxins and other pollutants; (b) military testing of the effects of products designed for chemical and biological warfare, of exposure to radiation, and of other related war hazards (notably those associated with stress); (c) tests concerning human habits, especially drug addiction, and nicotine and alcohol dependency; (d) psychological tests, involving prolonged isolation and exposure to physical and psychological pain, supposedly because of insights of benefit to humans; (e) testing of pharmaceutical products to determine their harmful effects on humans; (f) research undertaken to satisfy scientific curiosity; (g) repetition of earlier experiments to avoid the need to check their published results in the literature; (h) teaching experiments conducted regularly in schools and universities, involving surgery, amputations and use of electrodes.
Claim 1. Vivisection, the performance of operations on live animals for physiological or pathological investigation, is cruel and heartless. Animals are subjected to all types of experimentation, often without the use of anaesthesia, and the experiments may be repeated for demonstration purposes. Evidence indicates widespread and callous (if not sadistic) treatment of animals. Anti-vivisectionist leagues are active in Great Britain and the USA but meet with strong opposition from the pharmaceutical industry as well as the medical science community. There are basically two positions within anti-vivisectionist thought: the absolutists and the reformers.
The absolutist position is that the ends do not justify the means. To inflict pain and death on an innocent being is always wrong. Human beings used in experiments which result in their suffering and death is considered morally wrong; similarly, the inflection of suffering on animals cannot be justified by reference to future benefits for human or other animals.
The reformer position is that while some experimentation may be necessary but most are not. Most experiments bring suffering and death to animals with no likelihood of significant benefits. Alternative methods, not involving animals, could replace experiments on animals, such as, the use of tissue cultures. Other methods could be developed.
2. Recent reviews of 10 randomly chosen animal "models" of human disease found little, if any, contribution towards the treatment of patients. The difficulty for researchers is that artificially induced disease in animals is never identical to the naturally arising disorder in people, making animal research a logically flawed process. Although experimenters search for animals species which most closely mimic human responses, a more effective and humane approach would be to concentrate resources on methods of direct relevance to people, such as epidemiology, clinical investigation of patients suffering from the illness, and in vitro experiments using human tissues.
Counter-claim 1. Medical research would be impossible without experiments on animals and most researchers treat their animals well. Scientists currently observe voluntary codes. In-house committees monitor research, and institutions are subject to government inspection. Badly kept animals are rare (if distressing) exceptions.
2. There are millions of people lying sick, in pain or dying in hospital from diseases that have not yet been cured. There are thousands of millions of people worldwide that are alive because of antibiotics and other treatments developed and tested using laboratory animals. To doubt this shows a total lack of appreciation of the central and fundamental role laboratory animals play in human welfare. There is hardly a life-saving or pain-relieving measure today that has not been largely a result of work with animals. Syphilis, diabetes, and Addison's disease all have effective treatments due to vivisectionist experiments. The introduction of chloroform; the discovery of the circulation of the blood; and the prevention of yellow fever, diphtheria, and smallpox have all resulted from animal experimentation, as have antibiotics, corticosteroids, kidney and liver transplants, cardiac surgery, hip replacements, poliomyelitis vaccine, and cytotoxic drugs for cancer therapy. The hopes for further improvements in the prevention and treatment of conditions such as coronary disease, heart failure, strokes, dementia, arthritis, cancer, cot deaths and AIDS, which cause so much suffering in the world, depend on such work continuing.
3. Today's drug industry cannot continue to develop new medicines without animal experiments, nor can it meet legal requirements on testing without first demonstrating the drugs' safety in animals. Vaccines, for instance, are tested on animals both to gauge potency and to ensure that they will not induce they very disease they are meant to prevent. If animals were not used in such research, it would be necessary to experiment on humans.
Refs Freund, Paul (Ed) *Experimentation with Human Subjects* (1972); Hutchings, M and Carver, M *Man's Dominion* (1970); Smith, C V *Meteorological Observations in Animal Experiments* (1970);

Vyvyan, John *The Dark Face of Sciences* (1971).
Broader Maltreatment of animals (#PC0066) Inhumane scientific activity (#PC1449)
Unethical experiments with drugs and medical devices (#PD2697).
Narrower Military use of animals (#PE1666) Experimental surgery on animals (#PE0412)
Experimental battering of animals (#PE1171)
Experimental exposure of animals to pain (#PE1670)
Use of animals in toxicological experiments (#PE9611)
Experimental exposure of animals to radiation (#PE0689)
Inhumane use of non-human primates in research (#PE1621)
Maltreatment of animals in aggression experiments (#PE1174).
Related Animal fighting sports (#PE4893) Obstacles to medical experimentation (#PF4865)
Cruelty to animals in food preparation (#PE0236)
Environmental hazards constraining scientific research (#PF1789).
Aggravates Boredom of captive and domesticated animals (#PF7681)
Trade in animal products of endangered species (#PD0389).
Aggravated by Unethical practice of the zoosciences (#PD4721)
Denial to experimental animals of the right to freedom from suffering (#PE8024).

♦ PD0266 Disparity in social development within developing countries

Nature In contemporary under-developed areas where rapid industrial expansion is induced, parallel changes in other sectors and other aspects of national life may lag far behind and fail to provide a basis for an integrated process of social and economic development. The result is apt to be a situation – found in many areas of Asia, Latin America and Africa, as well as in parts of Europe – in which modern urban industrial societies exist side by side with traditional rural societies but show few signs of close integration with them. This contrast has been accentuated in some countries by the fact that the industrial sector was established by representatives of a foreign culture, or closely modelled upon a foreign culture. The existence of a gulf between the modern industrial society and the traditional agrarian society may have important social repercussions for both sides, particularly at point of contact, when elements of the rural population move towards the cities or when the products of the factory begin to reach into the countryside.
Broader Economic and social underdevelopment (#PB0539).
Narrower Feudalism (#PF2136)
Flooding of the urban labour market in developing countries (#PD0008).
Related Disparity between workers skills and job requirements (#PC1131).
Aggravated by Forced marriage (#PD1915) Uncontrolled industrialization (#PB1845)
Imbalance between capital and technical assistance (#PE4866)
Disparate development of economic sectors within developing countries (#PD1534).

♦ PD0270 Neurosis

Psycho-neuroses — Psychoneurosis
Nature Neurosis is a group of diseases characterized by functional disorders in which the patient maintains a critical attitude to the disease as well as the capacity to control his behaviour.
Background The term was first used in 1776 by the physician W Cullen. In contemporary usage, neuroses include only disorders arising from psychic traumas and prolonged nervous tension. The predisposing factors are bodily constitution and trauma, intoxication, infection, and other debilitating diseases. Neurosis may be caused by an overpowering acute psychic trauma, such as the sudden loss of a loved one. Persistent psychic traumas, especially if they give rise to internal conflicts, may also result in neurosis. Overstrain of the nervous system is the basis for neuroses and this is more likely to occur in persons whose nervous systems are weak, imbalanced, or insufficiently responsive. The most common symptoms are undue fatigability, easy excitability, rapid exhaustion, disturbed sleep habits, sweating, and unpleasant sensations in the chest near the heart.
Refs Howells, John G *Modern Perspectives in the Psychiatry of the Neuroses* (1988).
Broader Obsession (#PA6448) Mental illness (#PC0300)
Mental disorders of the aged (#PD0919).
Narrower Phobia (#PE6354) Hysteria (#PE6412) Neurasthenia (#PE3520)
Hypochondria (#PE8322) Success neurosis (#PG4450) Hyperventilation (#PJ4225)
Nervous breakdown (#PE6322) Psychogenic fugue (#PE7451) Childhood neurosis (#PE3717)
Infantile neurosis (#PE3571) Depressive neurosis (#PG1829) Conversion disorder (#PG7717)
Psychogenic amnesia (#PG3919) Occupational neurosis (#PG7865)
Somatization disorder (#PG7558) Body dysmorphic disorder (#PG3728)
Somatoform pain disorder (#PG7292) Multiple personality disorder (#PE5048).
Related Borderline personality disorders (#PE4396).
Aggravates Down's syndrome (#PE2125) Abuse of sedatives and tranquillizers (#PE0139).
Aggravated by Fatigue (#PA0657) Anxiety (#PA1635)
Frustration (#PA2252) Depersonalization (#PA6953)
Stress in human beings (#PC1648) Inhumane interrogation techniques (#PD1362).

♦ PD0276 Inadequate housing for the aged

Blocked elderly housing — Old age pensioners' homes
Nature Progress in making the dwelling environment more appropriate has not kept pace with progress in extending the average human life-span. There is a need for the environment in which old people live to be of qualitatively higher standard to compensate for their somewhat sudden exclusion from participation in the productive aspects of modern industrial society and their consequent segregation. This is exacerbated by the demise of the 3-generation family structure and also by the preponderance of small houses being built. Married women are increasingly going out to work, and cannot look after aged relatives. Modern houses are expensive and beyond the means of many old people. Housing facilities tend not to take account of old people's needs: accessibility to collective social and medical services and isolation from noise and the usual stress of city life, while at the same time keeping touch with their families and friends.
Incidence A recent EEC survey showed that, for the four conveniences: running hot water; indoor lavatory; bathtub or shower stall; and central heating, in houses where the head of household was aged 54 or older, only 52% had all four facilities in the Federal Republic of Germany and only 30% in France. In UK households with head aged 65 or older, only 33% had central heating.
Refs Regnier Victor A, Pynoos Jon (Eds) *Housing the Aged* (1987).
Broader Inadequate housing (#PC0449).
Narrower Neglect of elderly in institutional care (#PE5288).
Related State custody of deprived children (#PD0550).
Aggravates Social withdrawal of aged (#PD3518)
Mental disorders of the aged (#PD0919)
Criminalization of euthanasia (#PF2643)
Inadequate welfare services for the aged (#PD0512)
Excessive institutionalization of vulnerable groups (#PF8209).
Aggravated by Human ageing (#PB0477) Age segregation (#PD3444)
Age discrimination (#PC2541) Rapidly changing cultures (#PF8521)
Ageing of world population (#PC0027) Inadequate income in old age (#PC1966)
Socially inappropriate housing (#PD8638) Social disadvantage of the aged (#PD3517)
Unavailability of building sites (#PJ8549)
Excessive employment of married women (#PD3557)
Rigidity and inadaptability in the aged (#PD3515)
Demise of three-generation family structure (#PG2180)
Susceptibility of the old to physical ill-health (#PD1043)
Slowness of sensori-motor activities in the aged (#PD3514)
Retirement as a threat to psychological well-being (#PF1269)
Underdeveloped provision of basic services in developing countries (#PF6473).

♦ PD0280 Instability of trade in inedible crude non-fuel materials

Broader Instability of the primary commodities trade (#PC0463).
Narrower Instability of trade in pulp and waste paper (#PE7914)
Instability of trade in wood, lumber and cork (#PE2521)
Instability of trade in metalliferous ores and metal scrap (#PE0553)
Instability of trade in oil-seeds, oil nuts and oil kernels (#PE0386)
Instability of trade in crude synthetic and reclaimed rubber (#PE0701)
Instability of trade in undressed hides, skins and fur skins (#PE1235)
Instability of trade in unprocessed textile fibres and their waste (#PE1550)
Instability of trade in crude fertilizers and crude minerals, excluding coal, petroleum and precious stones (#PE0760).
Aggravated by Statutory crime (#PC0277)
Restrictive practices in trade in inedible crude non-fuel materials (#PE8351).

♦ PD0285 Inadequate coordination of international organizations and programmes

Broader Inadequate planning of action against problems (#PF1467).
Narrower Inadequate coordination of the intergovernmental system of organizations (#PE0730)
Inadequate coordination of action on intergovernmental programmes at national level (#PE1375)
Inadequate coordination of international nongovernmental organizations and programmes (#PE1209).
Aggravates Boredom (#PA7365)
Ineffectiveness of international organizations and programmes (#PF1074)
Competition between international organizations for scarce resources (#PC1463).
Aggravated by Lack of a world government (#PF4937)
Inadequate organizational mechanism for international action (#PE8776)
Jurisdictional conflict and antagonism between international organizations (#PD0138)
Inadequate relationship between international governmental and nongovernmental organizations and programmes (#PE1973).

♦ PD0287 Nutritional diseases

Deficiency diseases — Human deficiency diseases
Nature Inadequate nutrition is a self-evident cause of ill health, and a major contributor to the high death rate among infants and young children in developing countries.
Incidence The most serious conditions are: protein-calorie malnutrition (because of its high mortality rate, its wide prevalence and the irreversible physical and mental damage it may cause); xerophthalmia, vitamin A deficiency, (because of its contribution to mortality of malnourished children, its relatively wide prevalence and the dramatic irreversible damage it causes, namely blindness); nutritional anaemias, iron deficiency anaemia and megaloblastic anaemias (because of their wide distribution, their contribution to mortality from many other conditions and their effects on working capacity); endemic goitre, iodine deficiency, (because of its wide distribution). In some specific areas of the world, other nutritional problems such as beriberi, vitamin B1 deficiency; pellagra, nicotinic acid deficiency associated with protein deficiency; or rickets, vitamin D deficiency, may be of considerable importance.
Refs Exton-Smith, A N and Caird, F *Metabolic and Nutritional Disorders in the Elderly* (1980); Halsted, Charles H and Rucker, Robert B *Nutrition and the Origin of Disease* (1988).
Broader Malnutrition (#PB1498) Human disease and disability (#PB1044).
Narrower Scurvy (#PE2380) Rickets (#PG2295) Pellagra (#PE2287)
Beriberi (#PE2185) Trichinosis (#PE2311) Dehydration (#PE8062)
Xerophthalmia (#PE2538) Endemic goitre (#PE1924) Kwashiorkor disease (#PE2282)
Thyroid gland disorders (#PE0652) Protein-energy malnutrition (#PD0339)
Vitamin deficiencies in diet (#PD0715) Nutritionally induced mental illness (#PG1760)
Trace element imbalance in the human body (#PE5328).
Related Diseases of metabolism (#PC2270) Deficiency diseases in plants (#PD3653).
Aggravates Physical blindness (#PD0568) Malnutrition in developing countries (#PD8668).
Aggravated by Nutritional ignorance (#PE5773) Nutritional deficiencies (#PC0382).

♦ PD0291 Inadequate level of investment within developing countries

Decline in investment in developing countries
Nature Although there are other sources of growth, the significance of investment lies in the fact that it not only adds to productive capacity but also provides the means for the transmission of technical progress. A high rate of investment increases the flexibility of the economy and facilitates its structural adaptation to changes in the economic environment. It is estimated that developing countries need to invest 15 to 20 per cent of their annual gross income in order to sustain an annual growth rate of 5 percent. Many countries are unable to achieve this.
The implications of capital formation for the expansion of productive capacity depend not only on the volume of investment but also on the effectiveness with which it is used. Given the scarcity of investment funds in most developing countries, the maximization of the efficiency of investment assumes special importance in the context of development efforts. The efficiency of investment is influenced by its distribution among different types of assets and by the share in the total of renewals and replacements. New investment in construction will generally have a lower output-growth potential in the short and medium-term relative to investment in machinery and equipment, which is more closely tied with production and whose contribution to output is more condensed in time. Residential construction, which generally absorbs an important part of fixed investment, contributes directly very little to expanding the productive capacity of the economy. Investment in infrastructure tends to exert its influence on production over a long period of time, whereas its capital requirements are relatively large. Another factor which tends to reduce further the growth potential of investment in construction arises from the fact that much construction is undertaken by public authorities in connection with activities whose measured output is nominal or even non-existent. With respect to investment in renewals and replacement, the question revolves around the extent to which such investment provides an opportunity for taking advantage of technical progress. The distribution of investment for replacement among construction, machinery and equipment becomes irrelevant when replacement is of the pure and simple kind and is not associated with any changes in production methods. In practice, however, replacement investment is much more likely to act as a vehicle for technical progress when it consists mainly of machinery and equipment than when it is mostly construction. Thus, in the case of both new and replacement investment, the growth potential of investment will tend to be higher the greater the share of machinery and equipment in the total, and lower the more preponderant the share of construction.
A second major factor influencing the investment-output relationship is demand. This is particularly true of the industrial sector of many developing countries where the persistence of excess capacity is a common phenomenon and, consequently, output is highly responsive to changes in the level of demand. The same is generally true of the bulk of export commodities whose output can usually be expanded relatively rapidly in response to higher world demand with no, or only little, additional investment incurred.
The investment-output relationship is also influenced by the sectoral distribution of output and investment. It is worth mentioning that the importance of weather conditions in determining agricultural output in some countries, which means that output in one period might have little or no relation to investment incurred during the same or in a preceding period; and the overwhelming contribution that the oil sector makes to output in some countries – the high elasticity of output to investment in this sector will distort intercountry comparisons of overall levels of investment efficiency.
Besides the above enumerated influences, apparent intercountry differences in the efficiency of investment could have at their origin differences in the relative importance of increases in

employment levels and in the suitability of the technology embodied in the investment to the prevailing resource endowments. Moreover, mere availability as such will not guarantee effective use. Allocation problems are also involved, as well as the supply of complementary resources, particularly those that have to be obtained abroad, and there are many organizational and institutional factors which can enhance or inhibit productivity;

Broader Contradictions of capitalism in developing countries (#PF3126).
Narrower Lack of capital investment in developing countries (#PE5790).
Aggravates Reversal of development progress (#PF4718)
Structural rigidity in developing country economies (#PD2970).
Aggravated by Subsistence approach to capital resources (#PF6530)
Inadequate domestic savings in developing countries (#PD0465)
Decline in foreign direct investment in developing countries (#PD3138)
Disincentives for financial investment within developing countries (#PF3845).

♦ PD0295 Rural unemployment in developing countries

Rural underemployment in developing countries

Nature Idle or partially idle people are the greatest waste of resources in developing countries. Low productivity both causes and results from this situation; there are simply not enough jobs off farms to employ all who are looking for them, or who would leave farming and take up other work if it were available. Thus the farm population in the developing countries is over large; the excess people on farms stay there because, in the absence of work in urban areas, they are at least reasonably sure of some food, of housing, and of the protective care of the family, which takes the place of a wide range of social services. Constant underemployment persists, especially at certain times of the year.

Incidence In Asia, seven persons engaged in agriculture produce only enough food for 10 families, whereas each farmer in an industrialized country produces enough food for 20 or more people – and that food is more adequate in nutritional terms, being primarily composed of high–cost proteins rather than primarily starchy products such as the cereals, roots, and tubers produced by his Asian counterpart.

Broader Unproductive subsistence agriculture (#PC0492)
Unemployment in developing countries (#PD0176).
Related Underemployment in developing countries (#PD8141)
Labour surpluses in developing countries (#PD0156)
Disparities in unemployment within countries (#PD1837).
Aggravates Urban slums in developing countries (#PD3489)
Subsistence agricultural income level in rural communities (#PE8171).
Aggravated by Low productivity of agricultural workers in developing countries (#PE5883)
Imbalance between urbanization and industrialization in developing countries (#PC1563).
Reduced by Rural depopulation (#PC0056).

♦ PD0305 Limited developing country capacity to absorb aid

Nature The absorptive capacity of a country may be defined as the amount of resources both of domestic and foreign origin which can be invested to yield a return over and above a socially acceptable rate. The factors limiting absorptive capacity can vary from country to country and over periods of time. They are either economic or institutional or, more broadly, may have their origin in socio–cultural conditions. If they stem from socio–cultural factors, they are difficult to overcome and only the development process itself is likely to eliminate them ultimately. The most important type of factor limiting absorptive capacity stems from limitations on the supply of resources. For example, proponents or promoters of development projects frequently overlook the decline in marginal efficiency of capital due to the scarcity of skilled manpower. Institutional limitations are frequently encountered in the form of: inefficiencies in the administrative infrastructure concerned with the formulation and evaluation of investment projects in the public sector; inefficiencies in the management of public or private enterprises; delays in the making of decisions, in taking action, or in changing administrative regulations.

Broader Mismanagement of aid to developing countries (#PF0175).
Aggravates Compassion fatigue (#PF2819).

♦ PD0309 Unequal pay for women

Denial of right to equal pay for women

Nature The gap between male and female earnings, although in a few countries narrowing somewhat in the wake of legislative efforts, has generally remained distressingly large, even for workers with roughly similar qualifications. It is now evident that this gap is due mainly to the concentration of women in the lower–paying occupations and industries rather than to men and women being paid differently in the same or similar jobs. Women are concentrated in the lower–paying occupations because these are sex–segregated keeping most men out, while the higher paying occupations tend to keep women out. On–the–job training to improve skills is biased towards males as well. Unequal pay for women is very evidently the tradition in many developing countries.

Refs Parcel, Toby L and Mueller, Charles W *Ascription and Labor Markets* (1983); Stolz, Barbara A *Still Struggling* (1985).

Broader Discrimination against women in employment (#PD0086).
Narrower Discrimination against women in payment and prizes for athletic events (#PG4236).
Related Discrimination against women executives (#PD9628)
Unequal employment opportunities for women (#PD5115)
Denial of right of equal pay for equal work (#PD1977).

♦ PD0312 Restrictive shipping practices

Unfair shipping practices — Unfair practices in maritime commerce

Nature Maritime commerce has many conflicting parties whose self–interest leads to unfair practices. They include governmental and industrial shippers, shipowners, bankers, brokers, forwarders, port labour and merchant marines. Others having important roles are: governmental and intergovernmental regulatory or investigatory bodies; conference line shipowners; owner consortia; vertically integrated transnational corporations who may control production and inland as well as maritime transport up to the end markets; and dockers and seafarers unions. Shippers, shipowners and union interests are too often competitively opposed to each other, and competition within the shippers and shipowners industries itself are the factors that can lead to unfair business practices.

Background Shipping conferences, 'rate agreements', 'freight agreements' and 'freight associations', groups of shipping lines operating on routes with basic agreements for charging uniform rates, for allocating routes, berthing and sailing rights, and for pooling cargo and revenues, and intended to shut out non–conference competition, are among the earliest cartels in international trade. A particular feature of shipping conferences is the power which they exercise in regulating the conditions under which liner services can operate in a particular trade. They make unilateral decisions which vitally affect the interests of users of shipping services, and hence the national or public interest of the countries whose trades they serve. From the earliest days they have caused considerable discontent on the part of shippers, who complain that the monopoly power of these conferences has led to abuse, and that they require regulation in the public interest.

Claim The protection of the national economy against the possible harmful effects of combinations of firms for the purpose of regulating markets is usually embodied in restrictive business practice legislation. There are two main practices which separately or together are the targets of such legislation: price fixing and other agreements which adversely influence competitive conditions. Since it is one of the objectives of conferences to concentrate market power, to influence the conditions of the trades in which they operate, to decide on who can engage in the trade, and to fix by common agreement the prices in those trades, these practices might prima facie be considered to be in conflict with the spirit of such legislation. Restrictive business practices legislation has not, however, been applied to shipping conferences by most of the countries which have such legislation.

Despite the rules which conferences lay down to regulate the behaviour of member lines, competition among the members of a conference may appear in certain forms which are in breach of the conference agreement. This includes such malpractices as intentionally miscalculating freight charges (for example, by charging freight on the basis of weight when it should have been charged on the basis of volume and vice versa) accepting a wrong description of the cargo or ignoring certain physical or chemical properties of cargo so as to give a shipper the advantage of a lower freight rate, or giving secret rebates to shippers. Advantages of a non–pecuniary kind may be given to shippers by the wrong dating of the bill of lading or accepting cargo after the booking for a particular sailing has been officially closed. These constitute only a few of the various malpractices which may occur and which are contrary to the spirit of conference agreements, although they are not specifically forbidden in all agreements.

Counter–claim Although it is recognized that liner conferences place restraints on competition, it is justified by their ability to coordinate regular and more frequent scheduled services than would be possible otherwise, a benefit for which conference and shippers both want to avoid paying too high a price. For many developed countries this requires that they maintain competitive access to cargo in their liner trades and that they permit conferences to choose their own members. In that sense such conferences are closed, but since members of such conferences do not compete freely among themselves, they are required to face non–members competing with them for all the cargoes they carry. Since ships are mobile assets, and liner trades can never therefore be impermeable markets, such competition can develop quickly, so long as governments do not restrict the carriers permitted to offer service or the cargoes for which they may compete. On some liner trade routes, non–conference lines may carry 30% or more of the trade indicating that the conferences and non–members are battling for market–share. The workings of market forces therefore will normally be sufficient to safeguard consumer interests and to produce an efficient allocation of resources.

Broader Unfair transport practices (#PD1367)   Unethical commercial practices (#PC2563)
Obstacles for international ocean shipping (#PD5885).
Narrower Inequities in marine insurance (#PE5802)
Unfair surcharges in ocean freight (#PF5922)
Reduction of ocean shipping services (#PE5898)
Protectionism in the shipping industry (#PE5888)
Unfair shipping practices in bulk trades (#PE5849).
Related Transnational corporation control of bulk shipping (#PE5804)
Ineffective self–regulation in the shipping industry (#PF5840)
Evasion of shipping regulations and taxes by flags of convenience (#PE5873).
Aggravated by Transfer pricing (#PE1193).

♦ PD0321 Nutritional anaemia

Nature Normal human beings have stores of iron, folate, and vitamin B12. If these are slightly reduced, no clinical or biochemical abnormality may result, but the ability to meet increased demands for nutrients (for example, during pregnancy) is decreased. A further depletion of these stores may produce biochemical and/or clinical effects, but not necessarily anaemia, whereas yet a further reduction results in anaemia. Anaemia is defined as a condition in which the concentration of haemoglobin is below the level that is normal for a given individual. Nutritional anaemia results from a deficiency in one or more essential nutrients, regardless of the cause of the deficiency. In any community in which anaemia is prevalent, the distribution of haemoglobin concentrations in anaemic persons overlaps that for persons with normal haemoglobin concentration.

Incidence About half of all women aged from 15 to 49 in developing countries, that is 230 out of 464 million, are anaemic, suffering from a deficiency of one or more essential nutrients, chiefly of iron, and less frequently of folate.

A higher percentage of pregnant women are anaemic than non–pregnant ones. About 50% of non–pregnant women and nearly two–thirds of pregnant women have haemoglobin concentrations below those laid down by WHO as being indicative of anaemia. This is due to the dramatic increase in nutrient requirements of pregnancy that is needed not only to replace body losses, but also to provide for the needs of the foetus and placenta and the increased blood volume of the mother. In Africa, 63% of the 15.1 million pregnant women are anaemic, as against 40% of the 77.1 million non–pregnant women. In Asia, China excluded, the figures are 65% anaemic of the 43.2 million pregnant, and 57% anaemic of the 253.2 million non–pregnant women. In Latin America, figures are 30% anaemic of the 9.6 million pregnant, and 15% anaemic of the 65 million non–pregnant women.

An adult woman needs three times as much iron as is required by an adult man but in many countries women's diets are frequently more deficient than men's. In certain societies food taboos, specially those that apply during pregnancy, aggravate malnutrition.

Anaemia can also be caused by parasitic diseases, the two chief culprits being intestinal parasites and malaria.

Broader Anaemia (#PD7758)   Genetic defects and diseases (#PD2389).
Narrower Nutritional anaemia in women in developing countries (#PE9185).
Aggravates Ulcers (#PE2308)   Deafness (#PD0659)   Leukopenia (#PG5543)
Diseases and injuries of the brain (#PD0992).
Aggravated by Rickets (#PG2295)   Purpura (#PG2234)
Uraemia (#PG2241)   Hookworm (#PE3508)
Haemophilia (#PE1920)   Haemorrhage (#PE2239)
Menstruation (#PE4838)   Loss of blood (#PJ2230)
Teeth disorders (#PD1185)   Gland disorders (#PD8301)
Bright's disease (#PE2272)   Toxic substances (#PD1115)
Diseases of metabolism (#PC2270)   Diseases of the spleen (#PE6155)
Inadequate intake of iron (#PG2237)   Gastrointestinal diseases (#PE3861)
Parasites of the human body (#PE0596)   Complications of childbirth (#PC9042)
Inadequate absorption of iron (#PG2238)   Occupational hazards of benzene (#PE1849)
Unethical consumption practices (#PD2625)   Haemolytic disease of the new born (#PE2399)
Diseases of blood and blood–forming organs (#PF8026).

♦ PD0329 Increase in anti–social behaviour in developing countries

Nature The general spread of crime and delinquency and other symptoms of the breakdown of social mores is a frequent concomitant of urban growth under rapid industrialization. In the expanding cities more than any other stratum the city proletariat is subject to the phenomenon of social and personal disorganization, not merely because of deplorable conditions under which its members live, but also because they consist largely of rural masses who have been attracted by industrialization and who suffer all the consequences of a maladjustment brought about by a rapid change in the cultural environment.

Broader Social neglect (#PB0883)   Social conflict (#PC0137)
Anti–social behaviour (#PC4726).
Aggravates Crime (#PB0001)   Juvenile delinquency (#PC0212).

Aggravated by Human violence (#PA0429) Uncontrolled industrialization (#PB1845)
Disruption of family system in developing countries (#PD1482)
Inadequate development of new social structures in developing countries (#PD0822).

♦ PD0331 Protein-energy malnutrition in infants and early childhood

Nature Protein-energy deficiency occurs at all ages but its incidence is greatest in the weaning and immediate post-weaning periods; deprived of a high quality protein food, the child is not yet old enough to fend for himself in the family circle and is particularly subject to dietary taboos and prejudices. Milder forms of dietary deficiency, however, continue to occur among children and adolescents in low-income groups in developing countries. The growth rates of children in developing countries deviate sharply from the norm at the time of weaning and continue at a low level throughout the entire period of growth, resulting in stunted adult stature. Apart from the effect on growth, mild or moderate protein deficiency renders infants and young children particularly susceptible to respiratory and gastrointestinal infections. The incidence of such diseases is much higher in malnourished than in well-nourished children; and mortality in the age group 1 to 4 years is 20 to 50 times higher in the developing than in the developed countries; it is probable that this difference is due in large part to malnutrition. Besides this, among people with low incomes in developing countries there is a high prevalence of weaning diarrhoea, because of the combined effects of poor hygiene and protein deficiency.
Incidence Protein malnutrition in young children is the major nutritional problem of the world, and if both its direct and indirect effects are considered, is a major cause of ill health. Frank protein deficiency is common in the less developed countries, and latent or subclinical protein deficiency is probably even more prevalent. Although protein deficiency may predominate, often calorie deficiency contributes important effects and the simultaneous insufficiency of other nutrients complicates the picture in varying degrees.
Claim Protein malnutrition is the biggest single contributor to infant and young child mortality in developing countries. This continuing mortality is of staggering proportions and contrasts with the progress made in reducing mortality in the latter years of life. In most developing countries, 25 to 30 percent of the children die before their fifth birthday. Children under five years of age account for two thirds to four fifths of all deaths in developing countries although they only constitute one fifth of the population.
Refs Cravioto, Joaquin; Hambraeus, Leif and Vahlquist, Bo (Eds) *Early Malnutrition and Mental Development* (1974).
Broader Child malnutrition (#PD8941)
Protein-energy malnutrition in vulnerable groups (#PD0363).
Narrower Reduced activity of malnourished children (#PE4465)
Damage to infant brains from malnutrition and insufficient stimuli (#PE4874).
Aggravates Diarrhoea (#PD5971) Mental impairment (#PF4945)
Diseases of the respiratory system (#PD7924)
Retardation of psychomotor development in children (#PD1307).
Aggravated by Unbalanced infant diets (#PE0691)
Ignorance of women concerning primary health care (#PD9021)
Substitution of inappropriate foodstuffs for breast feeding (#PE8255).

♦ PD0332 Endangered species of birds

Endangered species of aves
Nature Due primarily to the impact of man on the natural environment, whether directly or indirectly, many of the 8,590 species of birds are in danger of extinction. Pesticides kill them through accumulation in the food chain, or affect their reproductive capacity by causing thin and fragile egg shells or poisoned embryos. They suffer destruction by hunters and trappers. They are killed by poison bait left for other animals. Nests and sources of food can be destroyed by intensive farming techniques and by destruction of natural habitats. Over grazing, construction, afforestation, deforestation, and draining swamps and marshes destroy natural habitats.
Incidence In 1600 (when reliable zoological records began) an estimated 8,684 species of bird were extant. Of these, 94 (or 1.09 per cent) have since become extinct. A further 187 species (or 2.16 per cent) are now in danger of extinction. 164 subspecies have become extinct and a further 287 are in danger. Some species are becoming more wide-spread; others are declining in number and range. In some areas, exotic birds and birds that can thrive in a disturbed habitat are becoming more common at the expense of native birds.
Refs Collar, N J; Stuart, S N and Arlott, Norman *Threatened Birds of Africa and Related Islands* (Date not set); Greenway, James C *Extinct and Vanishing Birds of the World*; Jackson, Jerome A (Eds) *Bird Conservation* (1987); King, Warren B *Endangered Birds of the World* (1981).
Broader Endangered species of animals (#PC1713).
Narrower Endangered species of water fowl (#PE5067)
Endangered migratory bird species (#PE4938).
Related Endangered species of insects (#PC2326).
Aggravated by Bird shooting (#PE2693).

♦ PD0335 Insufficient diversification

Nature No single energy source can continue to meet all energy needs over a long period, nor are there new sources capable of taking over fully from those which now exist. In addition, increased reliance on certain types of energy resources (for example, coal and nuclear energy) result in environmental problems.
Claim Diversification programs must be implemented now before it is too late to begin such programs.
Narrower Inappropriate cash crop policy (#PF9187)
Underdeveloped capacity for income farming (#PF1240)
Insufficient diversification of urban and industrial energy supply (#PJ8028).
Aggravates Undiversified economies of developing countries (#PD2892).
Reduced by Over-diversification of manufactured goods (#PD4907).

♦ PD0336 Import cartels

Nature Import cartels involve agreements concluded among competing firms in one or several countries. Such cartels may collectively limit the aggregate amount of specified imported goods, determine the sources of supply for such imports and/or fix the prices and terms of purchase the cartel members will pay for such imports. Such activities may be: a defensive measure to achieve lower purchase prices for imported products because their importers encounter aggressive export policies pursued by suppliers through 'natural' monopolies, or export cartels or centralized State selling agencies; an aggressive measure aimed at preventing or limiting imports in order to protect members of the cartel from import competition, or to ensure that imports take place before rather than after the processing of the goods in question, or to minimize the buying prices paid by cartel members; part of an exclusive arrangement between exporters and importers of a particular product with the purpose of excluding other firms from the business. From the point of view of the developing countries, an aggressive import cartel would appear to be a most harmful activity since it could restrict the volume of exports from those countries and the range of the prices paid to them for the imports permitted.
Background The known cases of import cartels in the developed market economy countries after World War II appear to be few in number, since in most of these countries they are either prohibited, or only authorized in certain exceptional cases, or are subject to a control of abuses under the relevant restrictive practices laws. At the end of 1970 In Japan there were three authorized import cartels with regard to the import of certain agricultural products from particular developing countries in Asia. In addition, Japan had two import-export cartels for trade in certain textile products with developing and other countries. In the case of the United Kingdom, one import cartel was approved with regard to imports of sulphuric acid. In the Federal Republic of Germany there were two authorized import cartels concerned with imports of molybdenum and of tungsten. In the case of the latter two countries, the cartels were authorized as defensive arrangements. In the case of the United States there are no legal import cartels. No information is available as to whether legal import cartels exist in other countries. From time to time illegal import cartels have been found to exist in the developed market economy countries which affected trade with the developing countries. In certain cases, these arrangements have been aggressive in nature. In addition, similar illegal arrangements have been found to exist in relation to trade amongst certain developed market economy countries.
Broader Cartels (#PC2512).

♦ PD0337 Death threats

Nature Death threats are common for individuals with a high public profile and for members of well known institutions. They are intended to intimidate an individual or institution to demand ransom payments or change behaviour. In some cases they are intended to disrupt normal operations such as air flights. A number of death threats are received at examination time in American secondary schools. A third reason for death threats is publicity.
The use of death threats by government institutions and government sponsored groups as a form of torture designed to break down the will of the prisoner by demonstrating the utter helplessness of the person being tortured and the ultimate power of the torturers. Death threats are frequently accompanied by mock executions of the prisoner, threats of mutilation such as castration and amputation and threats of sexual abuse. Death threats are also indirect, such as, being forced to watch executions and being held with other prisoners to be executed.
Incidence The use of death threats as a means of torture has been reported in the following countries: **Af** Egypt, Kenya, Libyan AJ, Rwanda, Somalia, Zambia, Zimbabwe. **Am** Argentina, Bolivia, Colombia, Suriname. **As** Bangladesh, Indonesia, Israel, Korea Rep, Pakistan, Saudi Arabia, Taiwan (Rep of China). **Eu** Italy, Romania, Yugoslavia.
Broader Intimidation (#PB1992) Psychological torture (#PD4559)
Denial of right to security (#PD7212).
Narrower Sham executions (#PE4407).
Aggravated by Torture through mutilation (#PD7576)
Use of undue influence to obstruct the administration of justice (#PE8829).

♦ PD0338 Technology gap between developed countries

Nature Many developed countries, even if they belong to the OECD, EEC, COMECON or EFTA, may themselves be slow followers of the technology leaders. Their economic position may owe almost everything to the past and very little to any contribution, or major adaptation, to modern applied science: in process engineering, in industrial automation, in electronics or in computerization, for example.
Claim These slow followers may rely on their membership in economically cooperative regional associations or in defence alliances to give them the technological benefits at second hand. However, these benefits will be long-delayed as compared to those arising from national efforts in technology research and development, and there remains a wide technology gap among countries of the North due to a lack of national technology development programmes.
Seen from an international perspective, nations that are behind in technological development are not so because of lack of research and development funding, or that the firms doing the development are too small or that these firms are not protected and adequately assisted by their governments. The returns on R D investment are to low. Low demand for innovation reduces incentives to innovate. Weak technical infrastructure produces fewer young people who can innovate, produces inadequate links between universities and industry which block information flow. Some countries expend heavily on government research institutions which have neither the incentives of the market place nor the demand for academic excellence of the university, thus reducing innovation. Regulatory costs for new inventions often discourage innovation.
Broader Lack of international cooperation (#PF0817)
Maldistribution of science and technology (#PC8885).
Related Disadvantageous terms for technology transfer (#PE4922)
Abusive technological development under capitalism (#PD7463)
Inadaptation of technology to man in the industrialized societies (#PE5023).
Aggravates Competition in capitalist systems (#PC3125)
International monopoly of the media (#PD3040)
Unequal opportunities for media reception (#PD3039)
Increasing development lag against technological growth (#PE3078)
Risk of unintentional nuclear war generated by the strategy of deterrence (#PF4162)
Lack of technical development and excess of manpower in developing countries (#PE4933).

♦ PD0339 Protein-energy malnutrition

Protein-calorie malnutrition — Deficiency in dietary protein — Lack of protein — Low protein intake
Nature Protein malnutrition is the result of many factors, including: a deficiency of calories in general as well as of dietary protein; the consequences of a high incidence of infectious diseases including intestinal parasitism; ignorance of the principles of good nutrition; and inadequate purchasing power especially when children are numerous and closely spaced. Protein malnutrition is an important cause of infant and young child mortality, stunted physical growth, low work output, premature ageing and reduced life span in the developing world. Recent research has also revealed a link between malnutrition in infancy and early childhood, and impaired learning and behaviour in later life. The widespread occurrence of protein malnutrition especially among infants, pre-school children, and expectant and nursing mothers in many developing nations spells grave danger to the full expression of the genetic potential of the population of large sections of the world community.
The impact of malnutrition on economic and social development is all-pervasive. In spite of the high rates of child mortality, fertility levels are so correspondingly high as to maintain a very rapid population growth. There is a general agreement that the persistence of high mortality among infants and children is a major obstacle to family planning, as parents will not reduce the numbers of their children deliberately without greater assurance of their survival to adulthood. (Reduction in malnutrition among infants and children thus emerges as a prerequisite for fertility reduction, without which the population explosion is assuming disastrous proportions). The direct and indirect costs of malnutrition to the economy are often far more than would be required for its prevention. To direct increase in expenditures for medical care must be added the costs to society of rearing children who do not survive to a productive age, who having survived are less responsive to education, or who become constitutionally deficient adults. There is also an economic loss from absenteeism and reduced working capacity. The combined effect is to retard economic as well as social development.
Claim Protein malnutrition, which is a problem of crisis proportions for developing countries, must be recognized by the entire world community as a threat to world peace and stability which it can ignore only at its own peril. It is imperative that each developing country understand now the magnitude and appalling implications of the problem. Adequate nutrition is a prerequisite of the development of a nation's human resources. It is a fundamental goal as well as a basis for all economic and social progress.

Counter-claim In laboratory experiments, it has been shown that animals raised on a diet consisting of all the necessary vitamins and other nutrients, but only 60 to 65 per cent of the calories of their normal diet, will live significantly longer.

Refs United Nations *Protein-Energy Requirements Under Conditions Prevailing in Developing Countries* (1979).

Broader Malnutrition (#PB1498) Nutritional diseases (#PD0287) Diseases of metabolism (#PC2270) Nutritional deficiencies (#PC0382).
Narrower Kwashiorkor disease (#PE2282) Amino-acid imbalances (#PG2265) Disparities in calorie intake (#PD0446) Protein-energy malnutrition in vulnerable groups (#PD0363).
Aggravates Pellagra (#PE2287) Xerophthalmia (#PE2538) Health inequalities (#PC4844) Infectious and parasitic diseases (#PD0982).
Aggravated by Pain (#PA0643) Anxiety (#PA1635) Parasitosis (#PG2267) Burns and scalds (#PE0394) Inhospitable climate (#PC0387) Stress in human beings (#PC1648) Inadequate protein supply (#PC1916) Inadequate health services (#PD4790) Dietary deficiencies in developed countries (#PD0800).
Reduced by Excessive consumption of protein (#PD7089).

♦ PD0340 Distortion of international trade by discriminatory preference agreements

Preferential trade arrangements

Nature Preferential trade agreements between developed and developing countries may favour one or more developing countries at the expense of others. Discriminatory practices and privileged treatment may apply to developing countries' imports or exports or to both. Preferential trade agreements relating to customs duties, quantitative restrictions, licences, technical and safety specifications or standards, inspections, packing, loading and unloading, and various other procedures, and their related costs are closely related to preferential trade financing agreements such as credits and other direct aid for import purchases, monetary cooperation, import-related technical assistance, and banking and financial services support. Preferential agreements may also apply to bartering transactions, and similar arrangements. All of the foregoing practices and their many variants discriminate against a number of developing countries, denying them markets or adequate prices for their exports and increasing the costs of their imports. Such economic discrimination may be part of the foreign policy of the superpowers and their allies, intended to support friendly developing countries while repressing or undermining others.

Trade preferences, and in particular tariff preferences, do not by themselves provide a full explanation of the geographical pattern of the trade flows in individual commodities or even of a country's total trade. Several effects ascribed to trade preferences may in fact be due to special links existing in fields other than trade.

Under the conditions of depressed world commodity markets and falling prices, special trade preferences have tended to produce changes in the geographical pattern of trade; under such conditions, however, tariff preferences alone do little to stimulate new investment in the expansion of production.

Background Since the attainment of political independence, the economic structure of many developing countries, and particularly of those which acceded to political independence relatively recently, has remained largely determined by their past economic ties. They have continued to export predominantly primary commodities, and the developed countries linked with them by preferential ties have remained their most important markets as well as suppliers.

With a view to preserving their position in their most important export markets, securing better access to those sheltered markets for their future processed products and obtaining special financial and other advantages important for their development, most developing countries decide continue their special economic ties, including preferential tariffs and other conditions of mutual trade with the developed countries concerned.

The attainment of political independence by a great number of developing countries, particularly in Africa, coincided in time with the formation of regional economic groupings by the developed countries in Europe. Both the European Economic Community (EEC) and the European Free Trade Association (EFTA) include key countries of the major existing preferential systems involving a considerable number of developing countries. The regional economic integration of the developed countries has an important impact on special preferential arrangements.

The establishment of EEC led to a geographical extension of the area in which goods enjoy preferential duty-free access. Originally the overseas territories (in particular French and to a certain extent also Italian, Belgian and Netherlands) enjoyed preference in their respective metropolitan markets. This preferential access was extended to all other markets of EEC, for instance, to the market of the Federal Republic of Germany which - while being an important dynamic market - had until then been accessible to all developing countries on equal terms. On the other hand, there was a general reduction in preferential margins and a gradual transition took place from special marketing arrangements, securing for some products of the developing countries guaranteed prices and outlets (in France and Italy), to a system of world market prices.

The associated countries extended reciprocal tariff and quota preferences to all EEC members and began to dismantle their customs duties and quantitative restrictions (with certain exceptions) vis-à-vis all EEC members. In the case of EFTA, the position of the developing countries enjoying Commonwealth preferences was affected by the formation of the free trade area, since their industries and products have to compete on equal terms in the United Kingdom market not only with the British and other developed Commonwealth countries' industries, but also with duty-free imports from other EFTA members, so far as industrial products are concerned. The extent to which they are affected depends, of course, on the commodity composition and geographical pattern of their export trade. While in the case of EEC there has been practically no effect on the trade of the associated developing countries, owing mainly to the low level of development of their manufacturing industries, in the case of EFTA the consequence was a significant dilution of the competitive advantages derived from Commonwealth preferences.

The extension of special preferences by all the EEC countries to the associated States has not so far brought about a substantial change in the export pattern of the latter, a fact which has prompted them to call for the strengthening of special advantages. On the other hand, the extension of preferential access to other developed country markets previously open to all countries on equal terms gave rise to fears on the part of the outside developing countries that important markets for their traditional export products might be threatened. The further extension of preferential treatment to additional developed country markets, and to additional developing countries, as well as the lack of progress towards general solutions favourable to all developing countries, induced the developing countries that were being discriminated against to seek special solutions to their problems, either by applying for preferential treatment or by exploring other defensive measures.

Grave concern about the preferential policy of EEC persisted in Latin America. The Latin American countries, while requesting EEC to eliminate special preferences, in particular within the framework of international commodity arrangements, gave serious consideration to the possibility of counterbalancing their disadvantages by the creation of preferential trading arrangements in the western hemisphere.

The chain reaction of the kind witnessed during recent years has shown that any geographical extension of special preferences such as occurred in connection with the formation of EEC creates tendencies towards further extensions of the same nature. A proliferation of special preferential arrangements places in a particularly unfavourable position those developing countries which do not enjoy preferential access to any developed market, as well as those whose exports consist of one or a few commodities for which the developed countries participating in the groupings are the main outlet.

The territorial extension of the special preferential system implies not only a greater scope of discrimination, but also a greater potential internal self-sufficiency of the enlarged grouping. For countries outside the preferential area, the potential loss of markets is the more serious the more countries are included, the greater their production potential and the complementarity of their resources and the wider the preferential margin of discrimination. Moreover, the creation of large sheltered trade areas favours a possible shift of capital flows to the protected area.

Since a number of preferred countries are actual or potential producers of commodities that are also traditionally supplied by outside developing countries, a further proliferation of preferential trading arrangements is likely to have serious repercussions on other developing countries' exports to the developed countries belonging to such areas. Another important aspect is the repercussion which the reciprocal preferences enjoyed by developed countries in the preferential area produce on the progress in trade expansion, economic cooperation and regional or sub-regional integration among the developing countries.

The existing special preferential arrangements, as well as the tendency of several developing countries to arrive at similar special solutions constitute important obstacles to the negotiation of general solutions in favour of all developing countries.

Claim While the continuance of special preferences in favour of those developing countries whose commodities cannot compete on the world markets appear at first sight to be the simplest course of action in many cases, such special trade preferences providing particular advantages to exports of some developing countries but discriminating against others cannot be considered as a solution for the problems of all or even most developing countries. Moreover, even from the standpoint of the original preferred countries, the extension of special preferences to additional developing countries might in a number of cases remove valuable special trade advantages previously enjoyed only by them.

Taking account of the vital need of all developing countries, including those receiving special preferences, for a general solution of their fundamental trade and development problems, as well as of the dangers of the proliferation of the special preferential arrangements, the replacement of special preferences by adequate world-wide solutions is urgently required. Simultaneously, the elimination of reciprocal preferences enjoyed by developed countries would reduce the losses resulting for some developing countries from the abolition of special preferences.

Counter-claim The general economic, trade and financial position of countries enjoying special preferences explains why the transition from special preferences to equal competition may give rise to considerable problems for most of them and, in particular, for those which could not compete efficiently with some other developing countries or which depend on commodities for which the outlook is not encouraging.

The position of such individual countries varies according to the commodity composition and geographical breakdown of their exports, as well as according to the nature and scope of their existing links with the developed countries and the economic potential of each of them. For several of them, however, an elimination of special trade preferences would require important adjustments of their production and marketing to wider competition, an expansion of their exports to new markets, both developed and developing, and a greater geographical spread and diversification of their commodity trade. The adjustment, even if facilitated by a stabilization of world commodity prices at remunerative levels, by improved access to the markets of the developed countries and by financial assistance, as well as by continuation of other than discriminatory special links with the individual developed countries, would not as far as most primary commodities are concerned, be accomplished without a certain transitional period. Irrespective of the disappearance of special preferences and other transitional arrangements, the special economic, social and political links that have traditionally strengthened the relations between certain countries should continue to support the development efforts of the individual developing countries concerned; the correct function of these links, however, should be to serve as instruments of economic cooperation rather than as means of discrimination against trade with outside countries.

Broader Non-tariff barriers to international trade (#PC2725).
Related Distortion of international trade by discriminatory customs and administrative entry procedures (#PE2603).

♦ PD0343 Restrictive agreements on product standards

Nature It is generally recognized that standards applied in one country may have the effect of hampering, making more costly, or even excluding exports from other countries. Such standards may be intended to protect plant, animal and human life as well as the environment (health and safety standards); to provide information to buyers on the characteristics of the product (quality standards, labelling and packaging requirements, marks of origin); or to rationalize production and product use (norms). In many cases, especially as regards health standards, such standards are 'obligatory' that is compliance is required by law or governmental regulation even though in some cases the standards have been developed by private organizations. There are, however, also 'voluntary' standards either set up by independent private institutions, sometimes of a quasi-public character (for example, the International Organization for Standardization), sometimes by trade associations of the manufacturers concerned or by direct agreement among suppliers or buyers. The last two cases (trade association activities and agreements between competitors) come within the scope of private restrictive business practices.

In principle, agreements among competitors (within or outside a trade association) to apply uniform standards or to refuse to buy products which do not meet such standards are regarded as cartel practices and as such come within the scope of the controls applying to cartels in the respective developed market-economy-countries. In practice, such agreements are generally exempted. In the United States there is no statutory exemption from the general prohibition of cartels in section 1 of the Sherman Act, but it appears that agreements which apply objective, non-discriminatory standards are considered as 'reasonable' and thus lawful. As regards those countries where laws prohibit cartels in principle (Canada, France, Federal Republic of Germany, Japan, Norway), and the EEC and ECSC, some make provision for special exemptions for agreements on standards (Federal Republic of Germany and Japan) while it would seem that in the others such agreements would normally come under the general exemptions. A number of laws requiring notification and registration of cartels have exempted standardization agreements from this procedure (for example, Australia, United Kingdom). Moreover it would be highly unlikely that action under relevant abuse laws would be taken in the case of an objective, non-discriminatory standards agreement.

Incidence It is often difficult in these cases to draw a clear distinction between public and private standards. Even where standards are set by private organizations not directly linked with business or even by trade associations of the manufacturers of the product concerned, they may nevertheless in fact be 'compulsory' either because they are gradually written into public safety regulations or specifications for public contracts, or receive some less mandatory public approval (for example, where the government actively participates in or finances standardization institutions), or else because consumer acceptance of products becomes dominated by the need for the customary 'seal of approval'. The problem of standards as a barrier to imports must therefore be seen as a whole, often incorporating both public and private trade barrier aspects.

Broader Restrictive trade practices (#PC0073) Inadequacy of international standards (#PF5072).

♦ **PD0348 Corruption in developing countries**
Nature Developing countries have difficulty in introducing rational and ethical profit motives and market behaviour into the business sector and in eliminating motives of private gain from the government sector. Time, and frequent political upheavals in such countries, have afforded greater and greater opportunities for corruption, particularly large-scale graft by politicians and higher officials, but spreading downward to petty bribery. Corruption has been highly detrimental for development. It introduces an element of irrationality in all planning and plan implementation by distorting the actual course of development plans. A common method of exploiting a position of public responsibility for private gain is the threat of obstruction and delay; hence corruption impedes the processes of decision-making and execution at all levels. It increases the need for controls to check the dishonest official. Thus it tends to make administration cumbersome and slow, and prohibits rational delegation of authority. Corruption and the widespread knowledge of corruption counteract the strivings for national consolidation and in particular, decrease respect for and allegiance to the government.
Incidence The relative level of corruption in developing countries is difficult to assess with any certainty. It is however much higher than in the developed countries. Among factoral examples it is known that grants in aid of development have often been dissipated in large-scale corruption.
Claim Corruption is rampant in most developing countries and is growing, particularly among higher officials and politicians, including legislators and ministers. It is present in government purchasing agencies, offices issuing import licences and other permits and among those responsible for the assessment and collection of taxes and customs duties. It has also spread to the courts of justice and to the universities. The business world and organized crime are particularly active in promoting commerce-related corrupt practices among politicians and higher officials.
Refs Kameir, El-Wahtig and Kursany, Ibrahim *Corruption as the Fifth Factor of Production in the Sudan* (1985); Komuli, Suresh (Ed) *Corruption in India* (1975); Odekunle, Femi (Ed) *Nigeria* (1986).
Broader Corruption (#PA1986) Institutionalized corruption (#PC9173).
Narrower Nepotism in developing countries (#PD1672)
Corruption and mismanagement of foreign aid (#PD0136).
Related Corruption in politics (#PC0116) Bureaucratic corruption (#PC0279).
Aggravates Underdevelopment (#PB0206) Administrative delays (#PC2550)
Political instability (#PC2677) Disloyalty to government (#PG2273).
Aggravated by Corruptive crimes (#PD8679)
Inadequate social discipline in developing countries (#PD0095)
Increasing drug addiction in drug producing countries (#PJ0680)
Bribery by transnational enterprises in developing countries (#PE0322)
Reinforcement of inappropriate development by privileged classes in developing countries (#PF6670).

♦ **PD0351 Restrictions on international freedom of movement for national advantage**
Nature Governments overly restrict the flow of people across international frontiers. Increased mobility could induce excessively rapid equalizing tendencies (for example, labour shortages would be filled by unemployed manpower; high-wage societies would attract low-wage labour). However, the domestic pressures are such that governments are non-responsive to appeals to liberalize immigration laws or to allow numerous migrant or guest-labourers to enter, if only temporarily. Such restrictive policies accentuate inequalities, inducing migration of skilled individuals from places of greatest need to those of greatest opportunity. The countries that restrict movement the most are the least prepared to meet changing employment needs on a flexible basis.
Broader Domination of the world by territorially organized sovereign states (#PD0055).
Related Restrictions on emigration (#PC3208) Restrictions on freedom of worship (#PD5105).
Reduces Emigration of trained personnel from developing to developed areas (#PD1291).

♦ **PD0355 Mixed marriage**
Religious impediments to marriage — Interdenominational marriage — Inter-ethnic marriage
Incidence The Jewish people are prohibited biblically of marrying out of the faith. The main worry concerns Jewish men who take wives outside the faith because tradition is perpetuated through the maternal line. His children are considered lost.
Refs Bambawale, Usha *Inter-religious Marriages* (1982); Crohn, Joel *Ethnic Identity and Marital Conflict* (1986); Mayer, Egon *Children of Intermarriage* (1983); Rosenberg, Roy A, et al *Happily intermarried* (1988).
Broader Impediments to marriage (#PF3343).

♦ **PD0356 Draft evasion**
Weaknesses in military conscription systems — Draft dodging — Avoiding military service
Nature Some governments conscript in peace-time on a regular basis, usually when young men attain a certain age. This is for a period ranging from several months up to four years. Few men are exempt as this is considered a universal military training defence necessity. Other countries draft only during international or civil crises and conflicts. Under both systems, among those who are obliged to serve there are numerous legal and illegal evasions. Legal evasions to serving in fighting and active support units include: the purchase of an exemption, as for example, recently in Turkey; to elect alternate service in civilian status, an alternative some countries provide for conscientious objectors. Others, while accepting conscription, may retain virtual civilian status, living in comfortable quarters and in no danger, such as some higher reserve officers and those who by political influence have only nominal duties. In both latter cases the intent, if not the letter, of the draft law has been evaded. Under military law these people could be court-martialed for non-performance of duties or related offences and this does occur on rare occasions. Illegal evasion is accomplished by non-registration, especially if there was an unrecorded birth, living under assumed names, flight and migration. Instances of illegal evasion are severely prosecuted as a discouragement to others. It can include failure to report for induction, refusing induction, or refusing or failing to perform required alternative civilian work. The problem of administering and enforcing massive conscription efforts is that draft evasion be so successfully accomplished due to inadequate records and enforcement resources.
Broader Civil disobedience (#PC0690) Inadequate army discipline (#PD2543)
Crimes related to military service obligations (#PD5941).
Related Conflict (#PA0298) Irresponsibility (#PA8658)
Causing insubordination in the armed forces (#PE5782)
Obstruction of recruiting or induction into armed forces (#PE3912).
Aggravates Shortage of military manpower (#PE4920)
Forced repatriation of prisoners of war (#PD0218).
Aggravated by Fear (#PA6030) Hazings (#PF5392)
Immorality (#PA3369) Legal havens (#PE0621)
Conscription (#PF6051) Political conflict (#PC0368)
Ideological conflict (#PF3388).

♦ **PD0361 Health hazards of irradiated food**
Irradiation as a method of food preservation — Opposition to irradiation technology
Nature The irradiation of food, undertaken to prolong its shelf-life, is controversial as it is still unproven that there are no adverse side effects. Irradiation of foods have many possible repercussions. Radiant-resistant micro-organisms that may contaminate the food may be developed. Vitamins A, C, E, and especially B are damaged which would render valuable foodstuffs useless. Flavour, texture, and colour are changed. Genetic and reproductive irregularities have been observed in association with the consumption of irradiated food. Cancer causing chemicals known as aflatoxins, that are created by fungi and occur naturally in some foods are produced more abundantly than normal. Chemicals called radiolytic products are produced in foods by irradiation. Irradiation of meat results in the flesh not giving off a warning putrid smell when the meat has gone bad.
Food may be irradiated twice further exacerbating these effects. For examples potatoes may be irradiated to prevent sprouting and then used in a prepared meal which is irradiated before sale.
Background Irradiation is the process of exposing foods to gamma rays, x-rays or electrons over a limited period of time. The process has been around since 1921 when an American scientist discovered it could kill the Trichinella Spiralis parasite which can contaminate pork. Typically gamma rays are used which are produced from cobalt-60 or caesium-137. Both are waste products from the nuclear industry.
Claim Irradiation causes cancer and foetal abnormalities. At doses of 100,000 rads fruit and vegetable cells are killed and most insect larvae will be destroyed. Fungi, bacteria and viruses will not be killed or inactivated and may mutate causing worse contamination. Government advisory committees have concluded that it is quite safe and acceptable to eat food that has been subjected to radioactivity to reduce high levels of pathogen bacteria and parasites. Pathogens (such as salmonella) are found in raw sewage. Parasites can include flies, maggots, worms, etc. Such food should be destroyed, not eaten. In an age when malnutrition is a prevalent danger, the rendering less nutritious of otherwise nutrition-packed foods is ridiculous. Even if the irradiation of vegetables means that a person in a developing country would be able to get food, what good does it do if that food provides him with no vitamins and minerals, but only empty calories ? Food irradiation seems more to be a rich man's toy, providing the wealthy with the means of getting, for example, insect-free papayas from India. The time and money spent on irradiation research would be better employed in areas of increasing food (and thus economic and political) strength and self-sufficiency in developing countries.
Counter-claim Irradiation, which extends food shelf-life from a few days to a few weeks, can greatly aid the distribution of perishable foods to developing countries. In addition, research in the UK and by WHO has shown that irradiated food is not a radiation hazard. Any induced activity immediately after radiation is at least one million times less than the naturally occurring radiation, and after three days, this is reduced to one hundred million times less than naturally occurring radioactivity.
Refs FAO *Wholesomeness of Irradiated Food* (1977); United Nations *International Acceptance of Irradiated Food* (1979); United Nations *Aspects of the Introduction of Food Irradiation in Developing Countries* (1974).
Broader Radioactive contamination (#PC0229) Health hazards of radiation (#PD8050).
Related Unsafe artificial sweeteners (#PE6390).
Aggravates Eczema (#PE2465).
Aggravated by Unethical food practices (#PD1045)
Inadequacy of civil defence (#PF0506)
Harmful biological effects of ionizing radiation (#PE6294)
Environmental hazards from food processing industries (#PE1280).
Reduces Food spoilage in storage (#PD2243).

♦ **PD0363 Protein-energy malnutrition in vulnerable groups**
Nature Early childhood, pregnancy and the nursing period constitute the vulnerable periods when nutritional needs in general and the requirements for protein in particular are greatest. Protein deficiency arises either from the lack of adequate food supplies within a country or as the result of social, economic and cultural factors that limit the consumption of protein by vulnerable groups of the population, even when the total supplies are adequate for the population as a whole.
Incidence The average figures for protein intake expressed on a per capita basis do not reflect the frequently encountered gross inadequacy of protein in the diet of the vulnerable groups in the population. Moreover, a low intake of calories and frequent episodes of infection often lead to considerable wastage of even the small quantity of protein consumed and are major factors in the widespread protein-energy deficiency. During famine and other emergency situations, this problem becomes particularly serious.
Broader Protein-energy malnutrition (#PD0339).
Narrower Maternal malnutrition (#PE1085)
Protein-energy malnutrition in infants and early childhood (#PD0331).
Aggravated by Famine (#PB0315) Disasters (#PB3561).

♦ **PD0365 Vulnerability of nuclear power sources**
Nature Nuclear power is developed in a society in the belief that it will provide a relatively cheap energy source and reduce dependency on external energy sources. This form of energy, however, presents grave risks not only from reactor accidents and the storage of waste products, but also from the release of radioactive material by terrorist sabotage or conventional warfare. Short of converting to a garrison state, societies cannot effectively protect the nuclear fuel cycle against sabotage; this is a problem which will only increase as nuclear reactors proliferate. The diversion of plutonium by terrorists for conversion to atomic or radiological weapons presents additional risks, since unidentified or unlocated terrorists cannot be deterred by threat of retaliation.
In any future war, electricity-generating power stations are likely to be primary targets of attack, because their destruction could paralyse the whole war effort of a country. The trend towards ever larger power stations, arising from economy of scale, makes each such station a highly attractive target. These installations will no doubt be strongly defended, but the greatly improved performance of modern missiles – in terms of (even non-nuclear) explosive power, range, payload and accuracy – ensures a high degree of success in such attacks. All this applies especially to nuclear power stations which contain reactors of very high output – sometimes two or more in one station – and which may provide a significant proportion of the energy needs of a country. Putting them out of action could have a devastating effect on the economy, particularly in countries which plan to obtain most of their electricity from nuclear reactors, quite apart from the huge material loss, since reactors represent a very large concentration of capital investment.
There may be another important reason for making a nuclear power station a primary target, namely the release into the biosphere of an immense quantity of radioactivity. Such a release, with the consequent contamination of a large area and the panic that this would cause among the population, might indeed be the main purpose of the attack.
Broader International insecurity (#PB0009)
Environmental pollution by nuclear reactors (#PD1584).
Related Refusal to grant nationality (#PF2657).
Aggravates Nuclear accidents (#PD0771) Radioactive wastes (#PC1242)
Insufficient nuclear power stations (#PD7663)
Proliferation of nuclear weapons and technology (#PD0837)
Environmental hazards of nuclear power production (#PD4977).
Aggravated by Sabotage (#PD0405) Cardiovascular syphilis (#PG0121)
Nuclear reactor accidents (#PD7579) Deterioration of nuclear power plants (#PE5260).

♦ **PD0366 Insufficient health personnel**
Insufficient health surveillance — Insufficient health technicians — Limited health personnel
Broader Inadequate health services (#PD4790) Understaffing of basic facilities (#PD9306).

Narrower Insufficient doctors (#PE8303) Understaffed health clinics (#PJ7980)
Maldistribution of health personnel (#PF4126).
Related Limited reservoir of technical skills in rural communities (#PF2848)
Lack of skilled manpower in rural areas of developing countries (#PE5170).
Aggravates Virus diseases (#PD0594) Infrequent doctor contact (#PJ0362)
Inadequate welfare services for the blind (#PD0542).
Aggravated by Outdated forms of community health (#PF1608)
Decline in government health expenditure (#PF4586).

♦ PD0367 Vulnerability of developing countries to inflation
Nature The process of development in the majority of countries is accompanied by continuous inflationary pressure, translated, more often than not, into open and protracted inflation. In part this is because the growth process is driven by the aspiration of the masses to improve their standard of consumption, thus prompting governments to assume entrepreneurial functions and to encourage private entrepreneurs to embark on development schemes that offer a promise of future increases in consumption. The consumption habits of richer economies are adopted by developing countries without corresponding improved production techniques, thus resulting in smaller saving potential.
Refs Cline, William R, et al *World Inflation and the Developing Countries* (1981); Kapur, Basant K *Studies in Inflationary Dynamics* (1986).
Broader Economic inflation (#PC0254) Vulnerability of developing countries (#PC6189).
Aggravates Polarized protest against problems (#PF9691).
Aggravated by Underproductivity (#PF1107).

♦ PD0372 Environmental hazards from fish, crustacea and molluscs
Fish poisoning
Nature Fish do not naturally carry a wide variety of pathogens, but some may contaminate the marketed product and these are normally derived from the environment – for example, from pollution of the water in which the fish live, or improper handling after they are removed from the water. Many outbreaks of infectious hepatitis and typhoid have been attributed to shellfish as vehicles. Bivalve molluscs, such as oysters, clams and mussels, are the usual offenders.
Broader Environmental hazards from food and live animals (#PC1411).

♦ PD0376 Environmental hazards from agricultural and livestock production
Nature With the rapid increase in the use of technologically–enhanced agricultural inputs have come a number of environmental and resource problems. The natural productivity of an estimated one half of the world's cropland is declining because of soil erosion, waterlogging, salinization, and other environmental problems. In certain regions, the misuse of pesticides has led to the development of pesticide–resistant strains of pests, destroyed natural predators, killed local wildlife, and contaminated human water supplies. Improper application of fertilizers has changed the types of vegetation and fish species inhabiting nearby waterways and rivers. The availability of water may become the single most important constraint to increasing yields in the developing countries.
Broader Ecologically unsustainable development (#PC0111).
Related Plantation agriculture (#PD7598).

♦ PD0381 Accumulation of pollutants in terrestrial plants
Accumulation of contaminant residues in terrestrial plants
Nature Environmental pollution is characterized by the accumulation of toxic metals, organochlorine residues and radionuclides in terrestrial plants. Radionuclides are accumulated as a result of fallout from nuclear weapons testing and from nuclear reactor accidents. Mosses and lichens have a high capacity for interception and retention of airborne and waterborne contaminants.
Broader Accumulation of pollutants in plants and animals (#PD5021).
Narrower Radioactive contamination of plants (#PD0710).

♦ PD0384 Shortage of urban land
Nature The demand for urban land is growing, yet the supply is both genuinely and artificially limited. This situation radically increases land costs and in turn consumes scarce investment capital which could be better used elsewhere. It also irrationally distorts patterns of urban growth and development. This latter fact leads directly to a third round of undesirable consequences; as the urban infrastructure becomes more costly and inefficient, and institutions and facilities fail to provide adequate services to their populations, urban social and economic imbalances and injustices are intensified, the quality of the total urban environment erodes and it becomes difficult to harmonize man's activities with the components of the natural environment. Thus pollution, noise and other hazards all increase. The issue now is no longer the economic value of the land as determined by market processes, but the social value as determined by the goals and needs of urban society.
Claim In most countries with a market economy, efforts to implement housing and urban development programmes are hindered by the skyrocketing costs of urban land. In countries with high rates of economic and/or urban population growth, the price of land in urban areas is rising faster than incomes. Other factors driving up the value of urban land are the public decisions which change these land uses, the general world–wide preference pattern that exists for single–family houses, and inflation. Given these demand schedules and its inelastic supply, urban land has become a speculative investment commodity which is relatively risk–free and which may offer very high rates of return on capital. As such, it provides security against inflation and an uncertain future, an important consideration in developing countries where opportunities to invest personal savings in stocks, bonds and other 'safe' economic ventures are extremely limited.
Nevertheless, because of the artificial scarcity of land created by the withdrawal of this commodity from the market for speculative purposes, land speculation is another major factor contributing to the increase in urban land prices. This is particularly true of countries where small, powerful groups own or acquire the best urban land and, through oligopolistic marketing and pricing practices, further interfere with the free operation of the economic laws of supply and demand. These excessive land costs, taken together with the wasteful disuse of infrastructure which passes through land speculatively withheld from the market, comprise a steadily rising share of total housing costs, costs which are also on the ascendancy because of increased building and materials costs and the sometimes inordinate effective demand that exists for scarce housing. Because of these costs and the ability of commerce and luxury housing to outbid government–sponsored housing for prime urban real estate, low–income housing projects, such as sites and services, are usually located on the outermost edges of cities, far from urban services and downtown places of work, where the occupants lead geographically and socio–economically segregated lives.
Counter–claim The amount of urban land available for city use is generally neither scarce nor monopolized. There may be the appearance of land scarcity in small countries where population is large in relation to available sites, or in cities hamstrung by poor transportation or boundary problems. There may also be concentration in some parts of Latin America where latifundia survive. Finally, traffic concentrations, poor or inadequate development and sectional over–congestion may also be erroneously ascribed to land scarcity or monopoly. With sound planning, constructive legislation and adequate transportation and utilities, however, sufficient land generally can be made available for housing and the requisite neighbourhood facilities.
Broader Limited available land (#PC8160).
Related Unavailability of agricultural land (#PC7597).
Aggravates Excessive land usage (#PE5059)
Insufficient common land in urban environments (#PE6171).
Aggravated by Excessive use of land by automobiles (#PF6152)
Increasing proportion of land surface devoted to urbanization (#PE5931).

♦ PD0389 Trade in animal products of endangered species
Trade in furs and skins of endangered species
Nature Some animal products of rare or endangered species and species with diminishing populations are in high demand and give rise to extensive hunting and poaching, which further endangers their status.
Incidence Pelts and skins in demand for the fashion trade derive from the cheetah, ocelot, jaguar, snow leopard, clouded leopard, lynx, vicuña, tiger, and giant otter. Although legal limits have been imposed on the number of pelts or skins of certain species which may be exported from a country, these limits are exceeded due to a considerable illicit fur trade. Other products in demand include black or white rhino horn, crocodile hide, walrus tusks, sea turtle meat and by–products and, of course, elephant ivory. Much of the middle–men activity is done in Singapore, Hong Kong, Bangkok and Tokyo, with Honolulu and Brussels being other transit points. Countries whose rare and embattled species are diminishing include India, Kenya, Thailand, and Ecuador, while many of the oceans' species, which belong to the world, are being pirated for private gain. Illegal trading in ivory, the skins of endangered species and small live wild animals and birds is a $1.5 billion business annually.
Refs Martin, Esmond Bradley *International Trade in Rhinoceros Products* (1980).
Broader Unlawful business transactions (#PC4645)
Irresponsible international trade (#PC8930)
International trade in endangered species (#PC0380).
Narrower Illegal ivory trade (#PE4991).
Aggravates Poaching (#PD2664) Instability of the fur trade (#PE1474)
Endangered species of animals (#PC1713) Slaughter of animals for pelts (#PD4575).
Aggravated by Cruel treatment of animals for research (#PD0260).

♦ PD0390 Cruel animal transportation
Nature Common abuses in the transport of animals include: poor facilities; overcrowding; insufficient trained attendants; lack of supervision of attendants; lack of coordination in providing opportunities for feeding and watering; no arrangements for the emergencies or delays which inevitably occur; inadequate measures for safe unloading.
Incidence In Europe, for example, it has been established that each year several million animals are moved from one country to another for slaughter at the end of their journey. In many cases facilities do not exist to deal with so many tired, frightened, hungry, thirsty animals, nor for the sick, injured and dead.
Broader Maltreatment of animals (#PC0066).
Related Factory farming (#PD1562)
Excessive commercial exploitation of farm animals by industrial concerns (#PD2772)
Denial to animals of the right to the attention, care and protection of humankind (#PF5121)
Denial to animals of the right to conditions of life and liberty proper to their species (#PE6270).

♦ PD0398 Inadequate international map of the world
Nature A variety of factors combine to make it difficult to draw up and maintain an international map of the world adequate to the needs of modern life. There remain areas of the world which have been inadequately surveyed. Geographical names are only standardized with difficulty, since each language group names major world features in its own language (or even local features in a multilingual country). Economic development leads to the construction of new roads and towns and there is considerable delay in incorporating these into international maps. Also, the addition to such maps of national borders and country names raises major political issues where the borders are under dispute, or the name is not accepted by the regime in power. The technical problem of projecting a three–dimensional spherical surface satisfactorily into a two–dimensional map surface has not yet been solved so as to eliminate distortion.
Incidence Large scale maps (such as on the one millionth to one scale) serve as a basis for more specialized maps. Adequate maps are essential for development studies and surveys.
Refs United Nations *World Cartography* (1985).
Narrower Non–standardization of geographical names (#PF8511).
Aggravates Ignorance (#PA5568) Underdevelopment (#PB0206)
Biased and inaccurate geography textbooks (#PF1780).
Aggravated by Border incidents and violence (#PD2950)
Misrepresentation of geographical information (#PF9239)
Inadequate cartographic skills in developing countries (#PJ8291).

♦ PD0405 Sabotage
Monkeywrenching
Nature Acts of sabotage are aimed at threatening or destroying life and property, or at hampering or hindering specific activities. Sabotage is used during war, and is also used by terrorists, guerrillas, resisters and criminals during peace–time. Deliberate sabotage of projects which harm the earth and living systems, sometimes called "monkeywrenching", is a hardline ecological activity. These activities would include sinking metal spikes into trees to prevent them being felled or incapacitating bulldozers. Sabotage is also widely used among the more radical animal welfare activists.
Claim Sabotage has become a worldwide hazard, employed by extremist groups either to effect a radical change, strike at the real or imagined oppressor, or bring a situation to attention. Sabotage attacks have included bombings on the ground and in commercial airliners, destruction of property and destruction of life.
Broader Destruction (#PA6542) Impairing military effectiveness (#PD4448).
Narrower Computer viruses (#PD3102) Product extortion (#PD5426)
Contamination of public water supplies by sabotage (#PE1458).
Related Destabilization of social systems (#PB5417).
Aggravates Fear (#PA6030) Injuries (#PB0855) Human death (#PA0072)
Dam failures (#PE9517) Property damage (#PD5859)
Human destructiveness (#PA0832)
Vulnerability of nuclear power sources (#PD0365).
Aggravated by Conflict (#PA0298) Industrial espionage (#PC2921)
Security risk people (#PD6818) Lack of international cooperation (#PF0817).

♦ PD0406 Maldistribution of land associated with large traditional estates
Feudalistic land tenure — Latifundia
Nature The term 'large traditional estates' is often used to describe the tenure systems (prevailing in many Latin American countries) which are dominated by large estates having many, though not all, of the characteristics of feudalistic tenures. Under the impact of industrialization, some large estates have undergone changes towards commercialized agriculture; many feudalistic characteristics, however, continue to prevail. Unlike the case of peasant proprietorship or other individualized tenure structures, the tenure, production and supporting services structures are all fused into one highly centralized hierarchical system practically controlled by the owners of the large traditional estates. It is this particular characteristic that distinguishes it from customary tenure on the one hand and the private land ownership tenure (peasant proprietorship) on the other. Land

is concentrated into the hands of a few owners of latifundia, a sizeable proportion of which are of a traditional type. Most of the farm population in the rural areas who work on these latifundia are tied to the numerous minifundia or sub–family scale farms which are often too small to provide enough full–time employment for the family labourers.
Minifundistas and landless labourers are completely dependent on the owners of large traditional estates not only for employment but also for credit, marketing, roads and other services normally included in the category of physical or institutional infrastructure. In many customary tenure areas with their subsistence agriculture, as for instance in many countries of Africa, the supporting services structure is either relatively underdeveloped or practically absent. In the individualized tenure areas, as, for example, in countries of Asia, the Near East and North Africa, the separation between the tenure and production structure on the one hand and the supporting services on the other is virtually complete, resulting in distinction between non–cultivating land owner, landless tenant and money–lender–cum–trader, each representing the three different structures. Though in many Latin American countries which have not undertaken land reforms there is growing evidence of incipient separation between the three structures, the supporting services structure is often dominated by owners of large traditional estates ; the credit agencies, which are distinct, are invariably dominated by the estate owners.
Broader Maldistribution of land in developing countries (#PD0050).
Aggravated by Inappropriate modernization of agriculture (#PF4799).

♦ PD0407 Vulnerability of world cable communications

Nature There is an ever present hazard of damage to the many thousands of miles of submarine cable resting on the floors of the oceans of the world. Damage may be caused by fishing trawl gear or ships' anchors, earthquakes, volcanoes, exploitation of sea–bed resources or deliberate sabotage. Such cables carry several thousand telephone conversations simultaneously, together with telex, telegraph and data signals. They are costly and difficult to repair, particularly in mid–ocean.
Broader Instability of economic and industrial production activities (#PC1217).
Related Limited number of geostationary satellite orbits (#PF0545).
Aggravates Insufficient communications systems (#PF2350).

♦ PD0408 Aggressive uses of natural energy resources

Nature Oil is a commodity of basic strategic importance. No modern defence system can be maintained and no wars fought without a great supply of oil. Thus, significant quantities of scarce oil resources are currently being diverted for military purposes, resulting in less constructive uses of natural energy resources for humanitarian purposes.
Background World War I provided the first practical demonstration of the revolutionary effect of the military use of oil–powered ships, aircraft and vehicles. In World War II, the strategic importance of oil was even more crucial and oil installations were rated among the top–priority targets of attack. The strategic importance of oil has not diminished since then, but rather has increased with the emphasis on the two main characteristics of modern warfare: extensive mechanization and high mobility.
Claim Oil is not renewable. It is a limited resource becoming scarcer, as are other natural energy resources. These resources could be used to develop and maintain higher standards of living throughout the world, but the achievement of a high–quality civilization is impeded by the use of these resources for aggressive purposes.
Counter–claim The current high standards of living (which include freedom) in the developed countries can only be assured of continuation if there is a substantial defence network in these countries.
Broader International aggression (#PB0968).
Narrower Aggression against nuclear power sources (#PE0403).
Aggravates Lack of natural resources (#PC7928)
Unsustainable development of energy use (#PC7517).

♦ PD0412 Delay in administration of criminal justice

Claim The death penalty appeals process, especially in the USA, consumes millions of dollars and has made a mockery of the judicial system, contributing significantly to the public's lack of faith in the criminal justice system.
Broader Delay in administration of justice (#PF1487).

♦ PD0414 Motor vehicle emissions

Automobile emissions as air pollutants — Health risks of fuel exhaust
Nature The internal combustion engine of petroleum–powered motor vehicles discharges carbon monoxide, lead, nitrogen oxides, aldehydes, and ethylene and other aliphatic hydrocarbons into the atmosphere only a few feet from the breathing zone of the population. Local concentrations of these substances reach appreciable levels, which are highest in urban centres where traffic density is greatest. Under conditions of poor natural ventilation and strong sunlight, a complex series of reactions takes place between the nitrogen oxides and hydrocarbons leading to the formation of ozone, peroxyacyl nitrates (PAN), and several other substances (usually grouped together as 'photochemical oxidants'). This more extensive type of motor vehicle pollution affects the entire air environment of a community (Los Angeles 'smog'). Emissions of aliphatic hydrocarbons other than ethylene are not considered of importance, except in the causation of photochemical pollution. This may be significant regarding damage to forests and crops.
There are four recognized sources of pollution from the ordinary automobile, namely, the exhaust pipe, the crank case, the carburettor and the fuel tank. Tyre and road dust and asbestos particles from brake linings are not generally included in any discussion of the problem; some pollution from these sources will certainly continue, even if all other emissions can be eliminated.
The distribution of pollutants according to sources within the vehicle are: a) Evaporation losses, tank and carburettor (20 percent of the hydrocarbons); b) Crank case blow–by (25 percent of the hydrocarbons); c) Exhaust (55 percent of hydrocarbons and almost all of the lead, carbon monoxide and nitrogen oxides).
Exhaust gases from diesel engines contain negligible amounts of carbon monoxide, no lead and somewhat lower amounts of the lighter hydrocarbons per unit of fuel consumed than are emitted by petrol–powered vehicles. They are nevertheless responsible for much public criticism because they are offensive and malodorous and have a high content of particulate matter. Because diesel vehicles are small in number compared to petrol–powered vehicles, their overall contribution to pollution levels is not great. Nevertheless, because they often discharge pollutants in close proximity to people, they must be regarded seriously, and in cities diesels make a significant contribution to soiling of buildings and materials.
Incidence In most parts of the world, the traditional problems associated with industrial production and the consumption of fuels are still of more importance than those due to pollution by motor vehicles. By contrast, in the USA, the motor vehicle is considered the main source of pollution. Some 60 percent of the total weight of pollutants discharged into the atmosphere of the USA originates from this source. It has been estimated that an uncontrolled vehicle in the USA covering 12,000 miles a year emits the following pollutants: Exhaust: hydrocarbons 300 lb; carbon monoxide 1,700 lb; nitrogen oxides 90 lb; Crank case: hydrocarbons 130 lb; Evaporation: hydrocarbons 90 lb.
In other countries, the number of motor vehicles is much smaller but owing to their concentration in cities it is possible that pollution comparable to that in the USA may occur in the future. Local concentrations of the characteristic pollutants are inevitable in any city and the likelihood of photochemical pollution in widely scattered parts of the world must be recognized.
Air pollution from motor vehicles in the developing countries does not yet present a problem of such magnitude as in the highly industrialized countries. The number of cars in use is relatively small and the pollution caused by them is much less than that from industrial complexes. A typical example of such a situation is found in India, where large installations, such as chemical or petrochemical complexes, fertilizer and power plants surround, or are scattered among, most of the large cities. The discharges from these installations are so great that, proportionally, pollution from automobiles is insignificant, except in a few large cities.
However, the effects from automobile exhausts in these cities have some similarity to those in industrial cities in advanced countries because many vehicles have a high weight–to–horsepower ratio and they are often old and poorly maintained. The horsepower of 85 percent of the cars in India is between 10 and 14, and 60 percent of all vehicles are more than ten years old. Also maintenance is poor because spare parts are expensive or unavailable and technical competence is low. Consequently pollution is out of all proportion to the number of cars in circulation. Carbon monoxide peaks of 100 ppm have been recorded at street level at important intersections. As the number of vehicles continues to increase, it is expected that oxidant pollution may become a problem in other cities if control measures are not introduced.
Refs Ishinishi, Noburu, et al (Eds) *Carcinogenic and Mutagenic Effects of Diesel Engine Exhaust* (1986); Okken, P A; Swart, R J and Zwerver, S (Eds) *Climate and Energy* (1989).
Broader Hazards to human health (#PB4885)
Chemical pollutants of the environment (#PD1670).
Related Traffic noise (#PD3664)
Exploitation of fossil fuels (#PE4891).
Aggravates Corrosion (#PD0508)
Ciliostasis (#PG2313)
Air pollution (#PC0119)
Food pollution (#PD5605)
Human poisoning (#PD0105)
Malodorous fumes (#PD1413)
Malignant neoplasms (#PC0092)
Lead as a pollutant (#PE1161)
Ozone as a pollutant (#PE1359)
Smoke as a pollutant (#PD2267)
Nickel as a pollutant (#PE1315)
Road traffic accidents (#PD0079)
Asbestos as a pollutant (#PE1127)
Vanadium as a pollutant (#PE2668)
Hydrocarbons as pollutants (#PE0754)
Manganese as a health hazard (#PE1364)
Environmental plant diseases (#PD2224)
Sulphur dioxide as a pollutant (#PE1210)
Photochemical oxidant formation (#PD3663)
Nitrogen compounds as pollutants (#PD2965)
Particulate atmospheric pollution (#PD2008)
Carbon monoxide as a health hazard (#PE1657)
Deterioration in atmospheric visibility (#PE2593).
Aggravated by Traffic congestion (#PD0078)
Uncontrolled urban development (#PC0442).

♦ PD0425 Processing in developed countries of commodities exported by developing countries

Nature The commodity trade problems giving rise to the greatest difficulty, and hence to the greatest need of being placed in international perspective, are those involving commodities which are produced in the developed importing countries as well as in the developing exporting countries. Modifications of import policy in respect of commodities which are not domestically produced are likely to be achieved with much less difficulty. Whatever the domestic justification for measures tailored to the importing nation's needs, the restraint they exercise on consumption and imports is much less defensible in the context of the development needs of the exporting countries.
Broader Economic and social underdevelopment (#PB0539).
Aggravates Inadequate production capacity in developing countries (#PD4219)
Inadequate trade in agricultural commodities between developing countries (#PE4523).

♦ PD0426 Urban road traffic congestion

Nature Traffic jams now dominate life in the world's great cities. In Los Angeles drivers inch along 12–lane expressways. In Mexico City children start school late to avoid the morning smog. Bangkok's office–workers suffocate on packed buses for an average of 2 1/2 hours a day. All of this is caused by more people crowding into big cities but more importantly because people are moving more. Most of the extra moving is done in cars. Places like New York and London have nearly one–quarter of their land area devoted to roads and have vast metro systems under them and are near their capacity to do more for the driver.
Claim Cities can no longer accommodate the number of passenger cars and commercial vehicles that daily congest the main roads, city centres, side–streets, alleys and lots. Congestion, despite all ingenuities of traffic engineering, slows automobile speeds to pedestrian rates and fills the air with unhealthy exhausts (which also deface buildings and monuments with their sooty deposits and kill trees and plants). Vehicular traffic congestion is responsible for accidents that take lives and destroy property, is a factor in 'downtown decay', drives shoppers to suburbia where they can park, and probably contributes to city budget deficits because of the public services consumed. The passenger car itself has done more to destroy the quality of urban life than anything else. The influence of the automobile industry, retail stores, restaurants and theatres are among those whose interests are threatened by vehicle limitation proposals.
Broader Road and highway traffic congestion (#PD2106).
Narrower Environmental degradation from high–speed roads (#PD6124)
Inhibition of exploration by children of urban environment (#PF6159).
Related Intimidation of pedestrians by vehicles (#PE6139).
Aggravates Road traffic accidents (#PD0079)
Excessive use of land by automobiles (#PF6152)
Stultifying homogeneity of modern cities (#PF6155)
Environmental degradation by automobiles (#PE6142)
Impersonality of mass market shopping facilities (#PE6153).
Aggravated by Unidentifiable urban neighbourhoods (#PF6147)
Inadequate urban transport facilities (#PG5801)
Inaccessibility of water for recreation (#PF6138)
Maldistribution of urban shopping facilities (#PE6144)
Insufficient separation between urban subcultures (#PF6137).

♦ PD0428 Inadequate national empathy for external hardship

Nature The organization of the state involves mainly the satisfaction of the interests of the ruling groups. External calamity or hardship is not often operative as an effective basis for domestic sacrifice. Substantially, empathy normally ceases at the territorial boundaries even to the limited degree that it extends to diverse groups within the national domain. As a result there is no realistic hope that equalizing pressures can be generated within the nationalistic framework of global relationships, and without such pressures there is no prospect for a peaceful transition toward ecological equilibrium. Even the role of foreign aid as a form of world philanthropy is very suspect, partly because it involves nominal help and partly because it involves characteristic nationalistic pressure to use national wealth to project influence in foreign societies.
Broader Domination of the world by territorially organized sovereign states (#PD0055).

♦ PD0433 Exploitative commercial television

Exploitation of viewers — Advertiser controlled television programming
Nature The commercial exploitation of viewers by television is the result of the content of programs being determined by their mass appeal, and by advertisers. Using mass appeal means that neither the best nor even the best liked programs are made available. Producers create programs that would be the second, third or even fifth choice of viewers because at this level of choice the most number of people would watch the programme. The most exploited audiences

are children.
Incidence In the United States children spend more time watching television than any other activity than sleeping. Most of the programs are designed to sell products and not serve in any way the educational or informational needs of children. In the future, advertisers are expected to become progressively more involved in programme making, with programmes being devised with the aid of market research to test their appeal to specific audiences.
Refs Palmer, Edward L *Children and the Faces of Television* (1980).
Broader Exploitative entertainment (#PD0606).
Aggravated by Excessive television viewing (#PD1533).
Reduced by Television viewer fatigue (#PE4395).

♦ PD0434 Repression of intellectual dissidents

Nature Repression of opinions felt to be dangerous or subversive to the social, economic, ideological and political system of the state may be by censorship, internment, indoctrination, restriction on movement between countries, torture, exile and defamation of character. Repression results in apathy and conformism, forms a barrier to progress and strengthens totalitarianism.
Broader Repression (#PB0871) Political repression (#PC1919)
Denial of freedom of thought (#PF3217).
Related Injustice (#PA6486) Intellectualism (#PF2146) Intellectual prejudice (#PC3406)
Persecution of religious sects (#PF3353)
Excessive institutionalization of education (#PD0932)
Restrictions on freedom of movement between countries (#PC0935).
Aggravates Ethnocide (#PC1328) Conformism (#PB3407)
Political apathy (#PC1917) Ideological apathy (#PF3392)
Ideological conflict (#PF3388) State sanctioned torture (#PD0181)
Threats to ideological movements and minorities (#PC3362).
Aggravated by Censorship (#PC0067) Imprisonment (#PD5142).
Reduces Dissidents (#PC9695).

♦ PD0437 Criminally life endangering behaviour

Criminal harm to persons
Refs Buikhuizen, Wouter and Mednick, Sarnoff A (Eds) *Explaining Criminal Behaviour* (1988).
Broader Human violence (#PA0429).
Narrower Assault (#PD5235) Recklessness (#PD5349) Attempted murder (#PG0411)
Attempted suicide (#PE4878) Road traffic violations (#PE0930)
Offences involving danger to the person (#PD5300).
Aggravated by Idleness (#PA7710) Drug abuse (#PD0094)
Deprivation (#PA0831) Gang warfare (#PD4843)
Social breakdown (#PB2496) Socio-economic poverty (#PB0388)
Lack of social discipline (#PF8078) Vulnerability of social systems (#PB2853).

♦ PD0438 Schizophrenia

Nature Schizophrenia cannot be precisely defined because of the diversity of manifestations, courses and outcomes of this mental illness. It has been described as a slowly progressive deterioration of the entire personality, which involves mainly the affective life, and expresses itself in disorders of feeling, thought and conduct, and a tendency to withdraw from reality. In general it constitutes a gross failure to achieve or maintain functioning of an integrated personality. The person becomes unable to cope with problems of living and human relationships; he or she withdraws from reality, becoming refractory to influence and suggestion, with shallow and inappropriate emotional responses, foolish or bizarre behaviour, false beliefs (delusions), and false perceptions (hallucinations). The condition creates many difficulties for the family of the individual, although the human relationships in that setting may be an important factor in the emergence of it.
Incidence It is estimated that in the USA at any one time there are approximately 1.4 million patients in hospital because of schizophrenia, namely 25 per cent of the hospital population. The average length of stay is 13 years (although the majority leave within one year, large numbers remain until death). In the UK, 60,000 hospital beds are occupied by schizophrenics, and it is estimated that 600,000 others suffer from the disease; approximately 1 person in 100.
Claim Schizophrenia is madness par excellence and its status as a disease is the major raison d'être for the entire psychiatric profession. If neurosis and depression might be better treated with a bit of psychotherapy and a change of lifestyle, schizophrenia, says the psychiatrist, is considered a real illness calling for a real treatment.
Counter-claim There are three major problems with the prevailing attitudes towards, and methods of treating, schizophrenia: they do not allow for accurate diagnosis; they are based on the unproven assumption that schizophrenia is a disease; and the usual treatments of this alleged disease are crude assaults on the brain that are both dangerous and probably ineffective at bringing a cure.
Refs Bleuler, Manfred and Clemens, Siegfried M *The Schizophrenic Disorders* (1978); Cantor, Sheila *Childhood Schizophrenia* (1988); Drummond, Harold P *Schizophrenia* (1987); Gottesman, Irving I and Shields, James *Schizophrenia* (1982); Gottesman, Irving I and Shields, James *Schizophrenia and Genetics* (1972); Herner, Torsten *Challenge of Schizophrenia* (1982); Iversen, Leslie L et al *Neuroleptics and Schizophrenia* (1978); Miller, Nancy E and Cohen, Gene D *Schizophrenia and Aging* (1987); Strauss, John S and Carpenter, William T *Schizophrenia* (1981); WHO *Schizophrenia* (1979); Warren, Carol *Madwives* (1987).
Broader Psychoses (#PD1722).
Narrower Autism (#PE1222) Paranoia (#PE0435)
Residual schizophrenia (#PG9382) Catatonic schizophrenia (#PG9530)
Schizoaffective disorder (#PG6962) Paraphrenic schizophrenia (#PG9535)
Hebephrenic schizophrenia (#PG4301) Schizophreniform disorder (#PG9744).
Related Mental disorders of the aged (#PD0919) Borderline personality disorders (#PE4396).
Aggravates Illusion (#PA6414) Anhedonia (#PG0585) Hallucinations (#PF2249)
Sleep disorders (#PE2197).
Aggravated by Sadism (#PF3270) Solipsism (#PF5657)
Abuse of plant drugs (#PD0022) Consanguineous marriage (#PC2379)
Uncontrolled urban development (#PC0442).

♦ PD0441 Non-viability of small states and territories

Nature There are a large number of states and territories which can be categorized as 'small'. Many of the difficulties of most small states and territories are directly related to underdevelopment in general; lack of resources, inadequate cadres, illiteracy, etc. In addition, many handicaps of the small territories originate specifically in aspects of their 'mini' condition: physical isolation, small area and population, etc. For instance, physical isolation and difficulty of communication will normally result in psychological isolation and a lack of knowledge or understanding of the outside world. A population may be so small that it becomes extremely difficult or exorbitantly expensive to establish institutions which would be indispensable in moulding a group into a viable nation. It may be impossible to organize a higher education system if there is not a minimum supply of students, or to set up a diplomatic service if enough people cannot be spared to fill the necessary posts.
An additional difficulty is that such states and territories present a wide variety in status from the point of view of international law. Such entities, when independent, have difficulty in establishing the necessary administrative structures to conduct foreign relations through multilateral diplomacy, particularly through international organizations. Defence and national security give rise to special problems because of the lack of resources. Such territories are often reluctant to encourage the withdrawal of foreign power military bases on which they may depend for security and foreign exchange.
Refs Lloyd, Peter J *International Trade Problems of Small Nations*; Rapaport, Jacques, et al *Small States and Territories*.
Broader Geographically disadvantaged countries (#PF9247).
Narrower Vulnerability of land-locked developing countries (#PD5788)
Economic non-viability of small developing countries (#PD0068)
Vulnerability of island developing countries and territories (#PE5700).
Related Lack of human unity (#PF2434) Territorial fragmentation (#PC2944)
Administrative difficulties in new states (#PE1793).
Aggravates Second class states (#PD0579) Foreign military presence (#PD3496)
Restrictive trade practices (#PC0073) National political dependence (#PF1452)
Territories accorded a United Nations non-self-governing status disputed by the administering government (#PF2943).
Aggravated by Isolated islands (#PD2941) Disruption of territorial integrity (#PC2945)
Conflicting claims by states to territories (#PC2362).
Reduced by Micro-state participation in international organizations (#PD2942).

♦ PD0446 Disparities in calorie intake

Nature Human beings must obtain enough energy from the food they eat if they are to be healthy and active. Diets may, of course, provide sufficient calories and yet be grossly unsatisfactory in other respects, particularly in terms of protein content. Inequalities among countries are much more pronounced for proteins than for calories; protein shortages are both more frequent and relatively greater, and such shortages are more serious in their consequences. However, a serious calorie deficiency may deteriorate the protein status of a person, because in that case protein may be utilized for energy, even in people whose diets are already low in protein.
Incidence Recently it was estimated that the national per capita calorie and protein supply fell below the recommended minimum in at least half of the developing countries. However, national averages conceal wide disparities among population groups within countries.
Broader Diseases of metabolism (#PC2270) . Protein-energy malnutrition (#PD0339).

♦ PD0448 Heart diseases

Damage to heart
Nature Many general diseases affect the heart and can be classified according to the part of the heart affected or the nature of the changes produced. Inflammatory infections are divided into pericarditis, myocarditis, and endocarditis. Valvular diseases are one of the most important groups in which any of the four valves may be stenosed; associated with this group are hypertrophy in which the heart is enlarged, and dilation, in which one or more of the cavities is dilated. Degeneration of the muscular tissue may take place in the direction either of a fatty or, less commonly, of a fibroid change. There is also a class of functional disorders in which palpitation, irregularity, rapidity, slowness or even severe attacks of pain appear. If the defect of the heart be so great that it must be remedied by compensation, the pumping power of the heart weakens and symptoms appear either to the heart itself (pain and palpitation) or to other organs – for example: breathlessness, faintness, dyspepsia, swelling of the abdomen, dropsy of the feet.
Incidence Coronary heart disease (CHD) kills about two million people every year in the West, more people than any other disease or illness. Death rates in most industrialized countries are stationary or rising. In developing countries such as Malaysia, Singapore and Sri Lanka, the incidence has reached very high levels. In some developing countries, CHD has already emerged as a prominent public health problem afflicting especially men in the prime of life when their productivity and social and family responsibilities are greatest. In others, CHD poses a serious potential threat to health, and the probability is that, unless this threat can be averted or contained, it will soon reach proportions approaching those of the industrial countries.
Refs Ball, Madeleine and Mann, Jim *Lipids and Heart Disease* (1989); Braunwald, Eugene *Heart Disease* (1988); Cohn, Lawrence H and Ionescu, Marian I (Eds) *Mitral Valve Disease* (1985); Dembroski, T M, et al *Biobehavioral Bases of Coronary Heart Disease* (1983); Durron, Daskin R *Heart Injuries* (1987); Eaker, Elaine D, et al *Coronary Heart Disease in Women* (1987); Julian, D G and Wenger, N K *Cardiac Problems of the Adolescent and Young Adult* (1985); Nieveen, J (Ed) *Arrhythmias of the Heart* (1981); Sandler, G (Eds) *Coronary Heart Disease* (1987); Symons, C, et al *Specific Heart Muscle Disease* (1983).
Broader Human disease and disability (#PB1044)
Diseases of the circulation system (#PC8482).
Narrower Hypertrophy heart (#PG2364) Acute myocarditis (#PG5635)
Acute pericarditis (#PG9590) Acute endocarditis (#PE3898)
Ischaemic heart disease (#PE8158) Pulmonary heart disease (#PE9415)
Symptomatic heart disease (#PE4208) Degenerative heart disease (#PG2089)
Functional disorders of the heart (#PG2368) Inflammatory affections of the heart (#PG2362)
Cardio-pulmonary disease due to torture (#PE1556).
Aggravates Still-birth (#PD4029) Gland disorders (#PD8301)
Sudden unexpected infant death (#PD1885).
Aggravated by Sex (#PF9109) Smoking (#PD0713) Rubella (#PE0785)
Obesity (#PE1177) Diabetes (#PE0102) Syphilis (#PE2300)
Pneumonia (#PE2293) Gonorrhoea (#PE1717) Tuberculosis (#PE0566)
Hypertension (#PE0585) Air pollution (#PC0119) Water softness (#PE0199)
Rheumatic fever (#PE0920) Carbon disulphide (#PE5889) Rheumatic diseases (#PE0873)
Chronic bronchitis (#PE2248) Pregnancy disorders (#PD2289)
Chromium deficiency (#PG2372) Vanadium deficiency (#PG2384)
Smoke as a pollutant (#PD2267) High altitude stress (#PD2322)
Magnesium deficiency (#PG2377) Manganese deficiency (#PG2378)
Inhospitable climate (#PC0387) Coagulation disorders (#PE2373)
Diseases of the arteries (#PE2684) Abuse of coca and cocaine (#PD2363)
Elevated blood cholesterol (#PE2371) Lung disorders and diseases (#PD0637)
Genetic defects and diseases (#PD2389) Aerosols as industrial hazards (#PE1504)
Excessive consumption of sugar (#PE1894) Dust as an occupational hazard (#PE5767)
Deterioration in physical health (#PC0716).

♦ PD0450 Biological air pollutants

Nature Moulds, bacteria and insect-derived matinal produce disease and hypersensitive and allergic responses in many individuals. Aeroallergens, including allergens in ambient air, have not been fully studied. Airborne micro-organismo such as bacteria in tropical developing countries may be responsible for such diseases as cerebrospinal meningitis, Q-fever, histoplasmosis, anthrax and coccidioidomycosis. In addition to bacteria, biological air pollutants include viruses which cause a great number of viral diseases including influenza.
Broader Pollutants (#PC5690) Air pollution (#PC0119).
Related Aeroallergens (#PE2069) Natural pollutants (#PC2416).

♦ PD0452 Floods

Flooding — Unchecked seasonal flooding — Inadequate flood control
Nature Floods occur either because heavy rainfall causes rivers to overflow their banks into flood plains (this may be augmented by melting snow) or because high tides, possibly combined with large freshwater flows, cause coastal flooding. The flood plains affected may be very narrow (as

in mountainous regions) or many miles wide. Floods may have fatal consequences for individuals in the affected area; in any case considerable damage is caused to property, particularly where industrial and community development has encroached on the natural channels of larger rivers. Loss of crops and livestock can have severe economic consequences. Much damage may be caused to dwellings, creating many homeless people, particularly in developing countries. Flooding in a given area may only recur infrequently which makes it difficult to justify the cost of extensive preventive measures, especially in the case of freak floods.
Flood waves of high pressure and velocity may disrupt telecommunication lines and traffic, carrying large quantities of mud, rubble and trees, and eroding watercourses and bridge structures. Their effect is short but devastating. The lack of warning makes them particularly dangerous.
Incidence Over the period 1980–85, worldwide floods accounted for 33,542 deaths and some US $22,170 million in property loss. Few regions are immune from flooding. In the USA it is estimated that flood damage is approximately $300 million annually. The South-East Asian region is particularly vulnerable, and half the population of India is considered to be continually menaced by flooding.
Claim The main methods of controlling floods are building embankments to contain flood water within rivers and constructing reservoirs to impound flood waters before being released at a slow rate. For over 3000 years both methods have consistently failed. Embankments do not decrease the volume of water they dramatically increase the rate of flow. They simply transfer the threat of floods further down stream. Silt, normally deposited on the flood plains deposited in the river bed and as the bed rises embankments must rise until the river is above the surrounding country side. China's Yellow River bed is now five to ten meters above ground level as it crosses the Yellow Plains. This, in turn, increases the magnitude of the floods when they come. Dams frequently are not solutions to flood control. Very few dams are built solely for the purpose of flood control; out of 1,554 dams listed in "The International Registry of Large Dams", only 17 had been built only for flood control. The water level in flood control dams should be kept as low as possible while the water level in dams built for irrigation or hydro-electrical purposes should be kept as high as possible. This conflict in requirements for multipurpose dams usually means irrigation and electricity generation have priority and flood can result. These structural solutions, coupled with massive deforestation and increased development of flood plains result in disastrous floods with increasing frequency, cost and lose of life. It is not flooding that causes these disasters but man's responses operating in the illusion that floods can be eliminated. Indeed, floods have been used as a part of a sustainable agriculture for irrigating and fertilizing fields.
Refs Brown, J P *The Economic Effects of Floods* (1972); Clark, Champ *Flood* (1982); Roche, M and Rodier, J A *World Catalogue of Maximum Observed Floods*.
Broader Misbehaviour (#PA6498) Topological disaster (#PC5010)
Seasonal fluctuations (#PF8163).
Narrower Flash floods (#PE5140) Riverine floods (#PD4976)
Water salinization (#PE7837).
Related Coastal erosion (#PE6734)
Inundation of agricultural land through dams (#PE3786).
Aggravates Famine (#PB0315) Homelessness (#PB2150) Household pests (#PD3522)
Endangered monuments and historic sites (#PD0253)
Vitamin E deficiency in domestic animals (#PE4760)
Declining productivity of agricultural land (#PD7480)
Deterioration of the physical condition of art objects (#PD1955).
Aggravated by Storms (#PD1150) Tsunamis (#PD0033)
River ice (#PD3142) Hurricanes (#PD1590)
Flood waves (#PG2385) Storm surges (#PD2788)
Overstocking (#PC3153) Deforestation (#PC1366)
Soil erosion by water (#PD2290) Flood plain settlement (#PE0743)
Tropical deforestation (#PD6204) Silting of water systems (#PD3654)
Inadequate land drainage (#PD2269) Short-term climatic change (#PF1984)
Disastrous failure of natural dams (#PE0715)
Uncoordinated international river basin development (#PD0516).
Reduces Soil infertility (#PD0077).

♦ PD0454 Environmental hazards from manufacturing industries
Broader Ecologically unsustainable development (#PC0111).
Narrower Environmental hazards from chemicals (#PC1192)
Health hazards from basic metal industries (#PD0243)
Environmental hazards from woodworking industries (#PE0864)
Environmental hazards from food processing industries (#PE1280)
Environmental hazards from paper and printing industries (#PE1425)
Environmental hazards from textile and clothing industries (#PE1103)
Environmental hazards from machinery and equipment industries (#PE1859)
Environmental hazards from non-metallic mineral products industries (#PE0890).

♦ PD0455 Distortion of international trade through obstacles to patent protection
Nature Under present conditions, separate applications must be made in every country where patent protection is desired, resulting in a very costly duplication of searching and examining efforts. In view of the fact that most patent offices are labouring under steadily increasing backlogs of applications because of the rapid development of new techniques, efforts to eliminate this duplication are very much needed. Such difficulties discourage trade with marginal countries.
Broader Non-tariff barriers to international trade (#PC2725).
Related Distortion of international trade by discriminatory customs and administrative entry procedures (#PE2603).
Aggravated by Delay in recognition of patents (#PF7779)
Vulnerability of intellectual property (#PF8854).

♦ PD0457 Politicization of intergovernmental organizational debate
Nature Intergovernmental organizations no longer serve the interests of the international community of nations, when they are captured by blocs of member states through the mechanisms of majority voting power or biased leadership.
Incidence Within the United Nations system, UNESCO, the ILO, UNCTAD and a few smaller agencies have been criticized.
Broader Ideological conflict (#PF3388) Obstacles to community achievement (#PF7118)
Politicization of technical debates (#PD2860).
Narrower Proliferation and duplication of international organizations and coordinating bodies (#PE1029).
Aggravates Endemic abortion of ewes (#PG1537)
Loss of credibility in international institutions (#PE8064)
Jurisdictional conflict and antagonism within international organizations (#PD0047).
Aggravated by Glossitis in animals (#PG0416) Double standards in morality (#PF5225)
Preponderance of Western-style organizations (#PF8356).

♦ PD0460 Bureaucracy as an organizational disease
Nature Large organizational units tend to develop a multiplicity of tortuous procedures, narrow outlooks, and a high-handed manner in dealing with individuals and external bodies. This organizational malaise may include failure to allocate responsibilities clearly, application of rigid rules and routines with little consideration of cases, elevation of status over function, blundering officials, diffusion of responsibility, overstaffing, administrative delays, conflicting directives, duplication of effort, departmental empire building, and concentration of real power in the hands of relatively few people.
Incidence Contrary to current belief, bureaucracy as a disease is not confined to governmental agencies, but can be found in all kinds of organizational units, particularly large ones.
Background The modern theory of bureaucracy derives largely from the German sociologist Max Weber, who saw it as the formal codification of the idea of rational organization. Yet from the start popular writers have seen bureaucracy as an irrational force, dominating the lives of people, while political theorists have seen it as an independent force tending to swallow all of society in its maw.
Contrary to the prevailing belief that universal rules govern bureaucracies, in day to day operations, rule can and must be selectively applied which in turn invites corruption and misuse within the bureaucracy. The application of rules requires a high degree of discretion because rules can only specify what should be done when an action falls clearly within unambiguously specifiable categories, about which there can be no disagreement or difference in interpretation. Such categories are impossible. Ambiguity and vagueness can be found in any rule; moreover, conflicting rules or implications of rules can generally found which can be used to justify any decision an office holder wishes to make. Organizations develop their own common practices independent of rules and guidelines which take on the status of rules. Ultimately, the office holder has license to apply rules derived from a nearly bottomless pit of choices. Individual self-interest then depends on the user's ability to ingratiate himself to office holders at all levels in order to ensure that those rule most favourable are applied. The goals of the bureaucracy are frequently displaced by objectives in conflict with the purpose of the organization.
Counter-claim When people can no longer communicate on a face-to-face basis, they need formal regulations at every level of the organization. The tendency to equate all forms of institutions as bureaucratic creates meaningless conceptual frameworks from which to deal with the excesses of institutional organizations.
Refs Chaudhurry, Muzaffar Ahmed *Examination of the Criticism Against Bureaucracy* (1965); Ram, N V R *Games Bureaucrats Play* (1978).
Broader Ineffectiveness of individual participation in large communities (#PF6127).
Narrower Extortionate bureaucracy (#PD8655) Bureaucracy in cooperatives (#PJ5598)
Bureaucratization of socialism (#PD2993) Complex government bureaucracy (#PF8539)
Abuse of bureaucratic procedures (#PF2661)
Governmental disregard for people as human beings (#PD8017)
Over-development of bureaucracy in ex-colonial countries (#PD2511).
Related Political apathy (#PC1917) Bureaucratic bias (#PC1497)
Bureaucratic inaction (#PC0267) Bureaucratic aggression (#PC2064)
Bureaucratic fragmentation (#PC2662) Lack of community participation (#PF3307)
Lack of participation in development (#PF3339)
Institutional domination of organizational systems (#PF2825)
Inadequacy of the committee system of decision making (#PF2843).
Aggravates Anti-science (#PF2685) Political inertia (#PC1907)
Bureaucratic opposition (#PD7966) Suspicion of bureaucracy (#PF8335)
Unpredictable governmental policy (#PF1559)
Inadequate local government financing (#PC6631)
Uncertainty of death of missing persons (#PF0431)
Growing size and impersonality of firms (#PE8706)
Domination of individuals by institutions (#PF4220)
Displacement of natural light in buildings (#PF6198)
Delays in elaboration of remedial legislation (#PC1613)
Cost overruns in large-scale public programmes (#PD1644)
Inadequacy of governmental decision-making machinery (#PF2420)
Inadequate budgetary coordination within the United Nations systems (#PE2820).
Aggravated by Bureaucratic superiority (#PC1259)
Excessive government control (#PF0304)
Political opposition to administrative action (#PF2628)
Institutionalized callousness of public services (#PF2006)
Antagonism between government agencies and officials (#PE2719)
Inhibition of communication between non-proximate offices (#PF6197).

♦ PD0465 Inadequate domestic savings in developing countries
Bad investment of savings in developing countries — Uninvested community savings — Low level of domestic resource mobilization in developing countries
Nature It is generally agreed that the pace of economic development is associated with the growth of savings. This is because an act of saving makes possible the release of productive resources from consumption which may then be utilized to add to the stock of productive capital, thereby promoting the expansion of output. It is estimated that a rate of domestic savings of at least 15 per cent of gross income would be a necessary condition for self-sustained growth in the developing countries. Many countries are unable to achieve this.
Local savings in developing countries are often badly invested in terms of the facilitating of further economic growth. The shortage of long-term financing not only undermines capital formation but also shortens the time horizon of investors, and tends to direct funds into short-term speculative investments with quick pay-offs. An excessive preference for purchasing land leads only to its over-valuation. Over-investment in buildings in metropolitan cities is a feature of some developing countries, especially during inflation. Larger enterprises often merely reinvest in the extension of their own concerns where more diffused investment through the economy would bring larger returns to the economy as a whole. Lax monetary policies, high inflation and negative interest rates have helped to perpetuate this situation in some developing countries. Besides, the capacity to save, particularly in the case of debtor countries, has been impaired by the adjustment policies adopted in response to external imbalances.
Refs Fenton, Thomas P and Heffron, Mary J *Third World Resource Directory* (1984); United Nations *Savings for Development* (1985).
Broader Capitalism (#PC0564) Obstacles to community achievement (#PF7118)
Unproductive subsistence agriculture (#PC0492).
Aggravates Subsistence approach to capital resources (#PF6530)
Diminishing capital investment in small communities (#PF6477)
Excessive foreign public debt of developing countries (#PD2133)
Inadequate level of investment within developing countries (#PD0291)
Inadequate development of enterprises in developing countries (#PE8572).
Aggravated by High interest rates (#PF9014)
Inequitable tax systems in developing countries (#PD3046)
Inadequate economic policy-making in developing countries (#PF5964)
Subsistence agricultural income level in rural communities (#PE8171)
Excessive social costs of structural adjustment in debtor developing countries (#PD8114).

♦ PD0467 Unemployment caused by environmental conservation
Nature Environmental groups and conservationists have hurt the traditional economies of native peoples, increasing unemployment. For example, the banning of seal hunting has ruined the Inuit people in Greenland. Conservation of whales, even those which are not in danger of becoming extinct, has driven nearly all whale hunters into other fields of work or unemployment.
Claim Environmental conservation groups, their methods and goals frequently are as ham fisted, as insensitive to the needs of people and as narrow minded as any government or resource exploiting individual, company or government agency. They create as many problems as the, so called, solve.
Broader Unemployment (#PB0750) Unjust dismissal of workers (#PD5965).
Aggravates Redundancy of workers (#PD8007).

♦ PD0470 Export cartels

Nature Export cartels are arrangements between competing firms relating primarily or exclusively to export activity. Cartels relating exclusively to exports are sometimes called pure export cartels, as distinct from mixed export cartels which entail ancillary restraints on domestic competition for exports of the parties concerned – for example, where domestic export merchants are restrained by an export cartel of domestic manufacturers in their dealings with foreign customers. Mixed export cartels have to be distinguished from mixed cartels which control competition on the domestic market not as an ancillary measure of export cartel activity but as a primary measure to control domestic competition; but these cartels may to a greater or lesser extent be involved with exports. To ascertain the full scope of the cartellization of exports by developed market–economy countries not only export cartels in the strict sense, whether pure or mixed, have to be considered, but also domestic cartels whose members export to other countries.

A further distinction is that between national and international export cartels, the second type comprising firms from several countries. The operations of export cartels may consist of the complete pooling of the export business of the parties in a central agency, thus excluding any competition between them on export markets; or it may result in the exclusion, entirely or partly, of such competition by agreement on export prices, by the establishment of export quotas, by the allocation of export markets, or by the submission of fixed bids for tender.

Export cartels of firms in developed market–economy countries may affect the export interests of developing countries in three different respects: by affecting the price and/or supply of inputs used by exporting industries of developing countries; by monopolistic practices of such cartels vis–à–vis exporters from developing countries; and by international market allocations including subsidiaries of members located in developing countries. The question of whether particular export cartels have any of these effects can only be answered on the basis of information concerning: the nature of the product involved (that is, whether it is used as an input for exporting industries in developing countries); the supply situation, including the existence of substitute products and other competitive suppliers outside the cartel; and, finally, the actual operation of the cartel, such as what restrictions, prices, and terms of trade are applied to covered exports.

Thus, an export cartel of developed market–economy country manufacturers in respect of an input used in exporting industries of developing countries, may decide not to supply the cartellized product in question at all, or to supply it only to particular industries or countries, or only at less favourable prices, terms of payment, quantities and/or qualities than those quoted to buyers on the home market of the members or in other developed market–economy countries. Such a refusal to supply and other discriminatory treatments are likely to have adverse effects on the export interests of developing countries. The possibly adverse impact of the activities of export cartels on the balance of payments of the developing countries, especially through high import prices, is also recognized.

Incidence In all developed market–economy countries where laws have been enacted to restrict or prohibit domestic cartel activity, the activities of export cartels are less strictly controlled than those of domestic cartels. In those countries where legislation in principle prohibits cartels (Canada, France, Federal Republic of Germany, Japan, Norway, United States, and the EEC and ECSC), or where the legality of cartels is dependent upon registration (Austria, UK), export cartels are either specially exempted (Fed Rep of Germany, Japan, United States) or are exempted by definition or by an explicit provision outside the scope of the law. Only four of these countries (Fed Rep of Germany, Japan, UK, United States) require the notification of export cartels to competent national authorities. The extent to which the activities of export cartels are exempted and the extent to which details of their activities are published vary considerably in these countries.

Export cartels are the most frequent type of cartel activity in developed market–economy countries which have relevance to the export interests of developing countries. In the three developed market–economy countries which publish data on export cartels (the Fed Rep of Germany, Japan and the USA), there are about 300 export cartels. The number in the UK is over 50. Firms in other developed market–economy countries participate in some of the international cartels registered in the Federal Republic of Germany and the United Kingdom.

Broader Cartels (#PC2512).

♦ PD0482 Impurities in waste water

Waste water contamination

Refs Hulby, David and Chappell, Willard et al *Risk Assessment of Wastewater Disinfection* (1985).

Broader Water pollution (#PC0062).
Narrower Suspended matter in water (#PE0579) Chemical contamination of water (#PE0535)
Domestic waste water pollutants (#PD2800) Fouling of water supply systems (#PJ6457)
Industrial waste water pollutants (#PD0575) Biological contamination of water (#PD1175)
Radioactive contamination of water (#PE2441) Pollution of water by infected faeces (#PE8545)
Toxic organic compounds as water pollutants (#PE0617).
Related Water salinization (#PE7837).
Aggravates Wasted water (#PD3669) Silting of water systems (#PD3654)
Water system contamination (#PD8122) Contamination of drinking water (#PD0235)
Shortage of fresh–water sources (#PC4815)
Inadequate water system infrastructure (#PD8517).
Aggravated by Contamination of public water supplies by sabotage (#PE1458).

♦ PD0486 Fraud

Swindling

Nature Misrepresentation of facts intended to gain for the falsifier tangible or intangible property constitutes fraud. Criminal fraud occurs with the loss of a valuable possession by the victim. This can be money, real estate, jewels, securities, collectors items of diverse kinds such as paintings, historical documents; but also copyrights, patents and other ownership intangibles. Omission or concealment of facts constituting fraud may be prosecuted as criminal under European and derivative legal codes, but under Anglo–American legal systems may be treated as deceit in civil, rather than criminal legal actions.

Incidence Common types of criminal fraud include the use of bank cheques for which there are no funds; the confidence game; filing false claims for insurance, government subsidies and compensation; under stating the value of taxable property, over charging clients by, for example, stating higher costs in billing clients; not paying clients the full proceeds of sales; kickbacks; manipulation of prices; taking an advanced fee without delivering services or goods; using credit cards and other financial instruments without authorization; and impersonation.

Refs Canadian Institute of Chartered Accountants *Fraud and Error* (1982); Comstock, Anthony *Frauds Exposed; or, How People Are Deceived and Robbed, and Youth Corrupted.*

Broader Deception (#PB4731) Criminal harm to property (#PD5511).
Narrower Forgery (#PD2557) Mail fraud (#PE1404) Bogus firms (#PF0326)
Maritime fraud (#PE4475) Contract fraud (#PD7876) Ghost employees (#PE3993)
Medical quackery (#PD1725) Financial frauds (#PE2414) Scientific fraud (#PF1602)
Commercial fraud (#PD2057) Credit card fraud (#PE3592) Documentary fraud (#PE1110)
Agricultural fraud (#PE5687) Front organizations (#PE4358)
Fraudulent commodities (#PD4191) Fraud by government agents (#PD8392)
Exploitation of the elderly (#PD9343) Unethical real estate practice (#PD5422)
Military–industrial malpractice (#PD4361) Unethical practices of priesthood (#PF8889)
Frauds, forgeries and financial crime (#PE5516) Obtaining property by false pretences (#PE5076)
Fraudulent mineral exploitation claims (#PE6975)
Unethical use of social welfare benefits (#PD8859)
Trafficking in government benefit coupons (#PE7271)
Fraudulent certificates of origin of goods (#PJ6717)
Misrepresentation of information to consumers (#PE6877)
Corruption of sports and athletic competitions (#PE3754).
Related Torts (#PD9022) Unreported scandals (#PF5340).
Aggravates Corruption (#PA1986).
Aggravated by Gullibility (#PJ2417) Drunkenness (#PE8311)
Unexplained phenomena (#PF8352) Conspiracies of silence (#PF9493)
Police crimes during narcotic investigations (#PE5037).

♦ PD0487 Toxic food additives

Carcinogenic food additives — Toxic food colourings — Toxic food preservatives

Nature These substances, widely used in the more industrialized countries, are normally defined as non–nutritive substances which are intentionally added to food, generally in small quantities, to improve its appearance, flavour, texture or storage properties. Although not contaminants in the strict sense, they may give rise to harmful chemical changes and should be regarded as potentially toxic materials. For example, nitrates or nitrites have been widely used as preservatives, but have been found to cause methaemoglobinaemia, especially in young children, and may give rise to carcinogenic nitrosamines. Certain non–nutritive sweetening agents (such as cyclamates) have been widely used in recent years, but it has been found that bladder tumours develop in animals to which they are fed in relatively high doses. As a result, the use of these agents has been restricted or completely prohibited in a number of countries.

Certain methods of processing foods may lead to the formation of toxic contaminants. The smoking of foods has been shown to produce significant levels of a carcinogen and the incidence of neoplasms seems to be increased in areas where the consumption of smoked fish reaches a high level. Trace amounts of similar polycyclic hydrocarbons occur in extraction solvents; as a result, concentrated residues of such carcinogens may be present in processed food when the solvent has been evaporated. Artificial growth stimulants used on livestock and crops may have harmful consequences for human beings who eat such food if these chemicals manifest stability after intake. The hormone–like diethylstilbestrol, which is used to fatten livestock before slaughter, has been found to leave residues in the prepared meat; when pregnant women ingest a quantity of this drug, there is the hazard that their offspring may develop a rare type of cancer.

Incidence 31 artificial colours and more than 2,000 artificial flavours are actually listed. A single strawberry flavour of ice cream can contain as many as 55 different additives. In all, 2,800 substances are intentionally added to food in the United States.

Claim Most additives are use to replace flavours and colour destroyed by processing, disguise second–rate (and sometimes bad) food, and turn basic and relatively cheap ingredients into costly 'convenience' foods. Some additives are harmless, but some are known to be health risks, among them the azo coal tar dyes used for colouring, particularly the yellow dye Tartrazine, E102 and the dyes E104–133, 142, 151 and 154–155; the benzoate preservatives, E210–213; and the purine flavour enhancers, E627–635.

Counter–claim Food scientist, toxicologists and control officials mention food problems in the following decreasing order: microbial contamination of food, natural toxicants in food, environmental contamination of food and food additives. Food additives serve the interests of both the consumer and the producer of foodstuffs since they inhibit the spoilage of food, thus reducing the losses and enabling greater production at a lower cost. They increase the viability of the diet, make preparation of food more convenient and help stabilize the quality of food. Many countries require a toxicological analysis of any food additive which determines, in most cases, the average amount of a substance expressed in mg/kg of body weight which can be taken daily in the diet over a lifetime without risk considering all known factors.

It should be noted that food flavouring substances are different from food additives as there are several thousand flavouring substances which occur naturally. Most flavouring substances are identical substances to the tens of thousands which have been detected in the flavouring component of natural foods such as fruits, spices, vegetables, milk and meat products. The natural flavouring component of foods is invariably an extremely complicated mixture comprising a few dozen to several thousand different chemicals of natural origin.

Refs FAO *Specifications for the Identity and Purity of Food Additives and their Toxicological Evaluation; Emulsifiers, Stabilizers, Bleaching and Maturing Agents* (1974); FAO *FAO Nutrition Meetings Report Series* ; Friedman, Mendel *Nutritional and Toxicological Aspects of Food Safety* (1984); Miller, K *Toxicological Aspects of Food* (1987); United Nations Economic Commission for Europe *Regulations and Legislation on Food Additives and Chemicals for Food Packaging* (1984); Walker, R and Knowles, M E *International Symposium on Food Toxicology* (1984).

Broader Food pollution (#PD5605) Chemical contaminants of food (#PD1694)
Carcinogenic chemical and physical agents (#PD1239).
Narrower Unsafe artificial sweeteners (#PE6390).
Related Excessive consumption of sugar (#PE1894)
Excessive consumption of specific foodstuffs (#PC3908)
Environmental hazards from food and live animals (#PC1411).
Aggravates Malignant neoplasms (#PC0092) Cancer–causing foods (#PD0036)
Carcinogenic consequences of food preparation (#PE6619)
Environmental hazards from food processing industries (#PE1280).
Aggravated by Consumer vulnerability (#PC0123).

♦ PD0488 Widowhood

Nature Widowhood is a condition common to all societies, civilizations and communities. Its circumstances and problems have ranged from the former Hindu custom of suttee, where the widow had to die on her husband's funeral pyre, to the boredom of the wealthy middle–aged, upper class widow in her beautiful home. However, for most of the world's population, widowhood means economic deprivation. Homelessness and starvation are possible in cultures where extended families or state welfare does not offer support. Even with such support widows may become second–class citizens, appended to brothers, brothers–in–law or uncles' families, and they may effectively lose their children to the dominating couple in the extended group. Grief and loneliness are frequently the result of the loss of the spouse.

The case of young widows introduces the need for her to be self–supporting, and if there are children, to maintain a home for them, while often working fulltime outside. Few communities or countries offer adequate counselling services to widows, and funeral aid is often totally lacking.

Broader Neglect of dependents of war victims (#PD2092).
Aggravated by Loneliness (#PF2386) Family poverty (#PC0999).
Reduced by Suttee (#PF4819).

♦ PD0489 Unreliable rainfall

Monsoon failure — Failure of rainy season — Reduction in rainfall

Broader Bad weather (#PC0293) Inaccurate weather forecasting (#PF5118).
Aggravates Vulnerability of crops to weather (#PE5682)
Declining productivity of agricultural land (#PD7480)
Inadequate water supply in the rural communities of developing countries (#PD1204).
Aggravated by Deforestation (#PC1366) Desert advance (#PC2506)
Large–scale weather anomalies (#PC4987)
Long-term changes in precipitation patterns (#PE4263).

♦ **PD0490 Inadequate road and highway transport facilities**
Nature Inadequate road facilities for motorists impede the safe and efficient use of motorways. Road facilities including garages and service stations, and also accommodation for rest and meals at suitable intervals on the road network and particularly in less developed areas, are essential to the comfort and safety of motorists. Standardized first aid posts, properly staffed and equipped, as well as telephone booths for emergency phone calls, are often provided at too infrequent intervals along roadways, particularly in isolated areas.
Broader Insufficient transportation infrastructure (#PF1495)
Restrictions on effective means of transport (#PF2798).
Narrower Inadequate road and highway transport facilities in developing countries (#PD0543).
Related Overloaded vehicles (#PE4127) Inadequate road conditions (#PJ0860)
Underdeveloped road network (#PE1055).
Aggravates Road and highway traffic congestion (#PD2106).
Aggravated by Inadequate road maintenance (#PD8557).

♦ **PD0493 Migrating sand dunes**
Sand drift — Desert advance — Sand dune encroachment
Nature Sand dunes migrate constantly unless the sand of which they are composed is prevented by vegetation from blowing away.
Incidence All over the world advancing sand dunes threaten roads, villages and crops. The rate at which dunes move varies, depending upon the velocity of the wind and the height of the dunes, the smaller dunes migrating faster. In France, on the Bay of Biscay, the sands have advanced at a rate of 15 to 105 ft per year, burying forests, farms, vineyards, churches and whole villages. Miles of towns and cities in central Asia are buried under dunes. In North Africa, moving dunes constantly threaten human communities and engulf previously fertile land. Dunes constantly encroach on the Suez canal.
Refs FAO *Sand Dune Stabilization, Shelterbelts and Afforestation in Dry Zones* (1985).
Broader Topological disaster (#PC5010) Environmental degradation (#PB6384).
Related Sand storms (#PD3650) Dust storms (#PD3655)
Coastal erosion (#PE6734).
Aggravates Desert advance (#PC2506) Destruction of land fertility (#PC1300)
Slowing growth in food production (#PC1960).
Aggravated by Defoliation (#PD1135) Deforestation (#PC1366)
Soil degradation (#PD1052) Soil erosion by wind (#PE3656)
Narrow legal definition of the family in developing countries (#PD1501).

♦ **PD0496 Inadequate rail transport facilities**
Nature In the developed countries, passenger carrying rail service has steadily deteriorated. Rolling stock is antiquated; the passengers experience discomfort; and schedules have been cut, increasing waiting times or eliminating service from some stations altogether. Energy and labour costs are blamed, as well as truck and airplane competition for freight. In the developing countries there are insufficient electrification, rail lines, rolling and tractive stock, and personnel to manage, operate and maintain rail systems.
Broader Insufficient transportation infrastructure (#PF1495)
Restrictions on effective means of transport (#PF2798).
Aggravates Rail traffic congestion (#PJ5720).

♦ **PD0500 Forest damage by wildlife**
Nature The wildlife compact the soil, trample, dig for roots, browse, debark and damage seeds.
Broader Harmful wildlife (#PC3151) Endangered forests (#PC5165).
Related Crop damage by wildlife (#PC3150).
Aggravates Destruction of alluvial forests (#PD6850).

♦ **PD0508 Corrosion**
Nature Most materials, particularly metals and alloys, when exposed to most environments, are affected on the surface by simple oxidation, or electrochemical (electrolytic) or chemical reactions with the acidic or alkaline agents which may be present. Corrosion may be either wet or dry according to whether the environment is primarily liquid or gaseous, and may take the forms of superficial scaling (in some cases resulting in the formation of a desirable protective layer) or pitting, galvanic attack between dissimilar metals, intergranular cracking due to localized attack at grain boundaries, stress corrosion, dezincification or selective leaching of alloys, erosion-corrosion in the presence of wear, high temperature corrosion, or microbial and bacterial corrosion. The rate of corrosion often depends upon atmospheric conditions: rainfall and relative humidity, as well as air pollution, all tend to accelerate the process. As a health factor, corrosion in water supply systems (pipes, tanks, wells, etc) causes impurities directly, and also indirectly by allowing foreign matter to enter into the system.
Incidence Estimates of corrosion costs per year based on a 1978 study for an industrialized nation found that the cost of corrosion was approximately 4.2 percent of the Gross National Product. Of this total, about 15 percent was estimated to be avoidable through the application of best practice corrosion control technology.
Refs American Society for Testing and Materials *Corrosion in natural environment* (1974); Craig, H L *Stress Corrosion – New Approaches* (1976); Crooker, T W and Leis, B N (Eds) *Corrosion fatigue* (1983); Escalante, E (Eds) *Underground corrosion* (1981); Gross, H (Eds) *Dictionary of Corrosion and corrosion control* (1985); Guttman, V and Merz, M *Corrosion and Mechanical Stress at High Temperatures* (1981); National Corrosion Engl *Corrosion Basics – An introduction* (1984); National Corrosion England *Automotive corrosion by de-icing salts* (1981); Rabald, E *Corrosion Guide* (1978); Schweitzer *Corrosion and corrosion protection handbook* (1988).
Broader Failure of materials (#PD2638).
Narrower Corrosion of stonework (#PE5470) Corrosion in tropical climates (#PE1811)
Atmospheric corrosion of materials (#PE9525).
Aggravates Radioactive wastes (#PC1242) Mechanical failure (#PC1904)
Transport of dangerous goods (#PD0971).
Aggravated by Air pollution (#PC0119) Ozone as a pollutant (#PE1359)
Corrosive substances (#PE6887) Motor vehicle emissions (#PD0414)
Fungicides as pollutants (#PD1612) Radiation damage to materials (#PD1206)
Non-biodegradable plastic waste (#PD1180).

♦ **PD0512 Inadequate welfare services for the aged**
Denial of right to welfare services for the aged — Unserviced older homes — Unorganized elderly aid — Distant elders' programmes — Inadequate provision of home care services for the elderly
Nature In many developed countries, elderly people complain of shortcomings in such fundamental aspects of the quality of their lives as living arrangements, housing standards, health care, provision of social and counselling services, and public transport. They see social assistance primarily as an opportunity to improve their financial, social and mental independence, and as a means to self-help. However, measures taken by the public authorities often fail to improve access to cultural and educational activities, and also fail to improve the physical and mental mobility of the aged.
Counter-claim The real concern of elderly people is that increased assistance will lead only to increased dependence and social control.
Broader Age discrimination (#PC2541) Inadequate social welfare services (#PC0834).
Related Inadequate income in old age (#PC1966)
Abusive treatment of the aged (#PD5501)
Inaccessible educational facilities (#PD9051)
Social insecurity in developing countries (#PE4796)
Inadequate procedures for community planning (#PF0963).
Aggravates Senilicide (#PJ7124) Being a burden (#PF9608)
Social isolation of the elderly (#PD1564) Rigidity and inadaptability in the aged (#PD3515)
Insufficient transportation infrastructure (#PF1495)
Fragmented forms of care at the neighbourhood level (#PE2274).
Aggravated by Human ageing (#PB0477) Ageing war disabled (#PD0874)
Neglect of the aged (#PD8945) Rapidly changing cultures (#PF8521)
Ageing of world population (#PC0027) Mental disorders of the aged (#PD0919)
Inadequate housing for the aged (#PD0276)
Misallocation of resources to protect the aged (#PE0648)
Susceptibility of the old to physical ill-health (#PD1043)
Lack of central planning structures in small communities (#PF2540).

♦ **PD0513 Denial of rights of children and youth**
Nature The rights of the child, although being among the most essential of human principles, are also the most frequently disregarded. One of the main reasons for this is that the ability of parents to nurture the child has been weakened. Social upheavals and economic crises have undermined family structure and brought under pressure the primary responsibilities of parents, causing the dilution of traditional sustenance and support for millions of children. There are many – orphans, refugees, abandoned children, the offspring of broken homes – who are not protected at all by the family or the community.
General human rights provisions of constitutional or statutory law are, as a rule, not specifically made applicable to children. Some legal disabilities and sanctions imposed by municipal law on children are not justifiable in terms of child protection, whatever the intent behind them, but amount to repressive and unfair treatment of children collectively. This differential treatment of children as a class is tantamount to discrimination against them on the ground of non-age. Perhaps more prevalent than discrimination against children as a whole is discrimination against particular groups of children. There is the practice of differentiating between the sexes for purposes of the minimum age for marriage and similar gender discrimination with regard to access to education opportunities. The law in some countries favours male children for inheritance purposes. Discrimination on account of birth status (whether born within or out of wedlock) is still widespread and covers such matters as legal rights to inheritance and maintenance, as well as pension, insurance, and welfare benefits. Adopted children still encounter similar problems.
Refs Freeman, M D *The Rights and the Wrongs of Children* (1983).
Broader Denial of rights to vulnerable groups (#PC4405).
Narrower Exploitation of children (#PD0635) Denial of right to a name (#PE4624)
Violation of the rights of foetuses (#PE6369) Denial of right to minimum work age (#PE1693)
Involuntary loss of nationality of children (#PE6676)
Discrimination against illegitimate children (#PD0943)
Discrimination and harassment of children in public life (#PE6922).
Related Induced abortion (#PD0158) Cruelty to children (#PC0838)
Political impotence of students (#PE5729).
Aggravates Maladjusted children (#PD0586) Inadequate child welfare (#PC0233)
Denial of right to education (#PD8102).
Aggravated by Social neglect (#PB0883) Dependency of children (#PD2476).

♦ **PD0515 Lack of consumer choice in centrally-planned economies**
Lack of consumer goods in socialist countries
Nature In a centrally-planned or centrally-administered economy, there is the risk that the planning authority may be out of touch with what consumers want and are prepared to buy. Some goods may consequently be over-produced so that unsold stocks pile up in the shops. In other cases, production may be insufficient to meet consumer demands, so that shortages occur. Furthermore, the emphasis of the planning authority can easily become too geared to quantity, with insufficient regard being given to the quality or suitability of the product to consumer requirements. Moreover, the size of the planning authority's task can lead to delays and mistakes, and the more centralized the planning the more serious the mistakes.
Incidence In Poland, for example, a family may have to wait eight to twelve years for a suitable apartment and four to five years for a car, even if they already have the purchase money. In Czechoslovakia, a bottle of whisky costs one-eighth of an average month's salary.
Claim In countries with a more or less free economy, the consumer can generally choose between the national product and the imported product, subject to the payment of customs duty. In a collectivist economy, exporters' efforts, crucial when it comes to trade with a free economy, play a minor role compared with the decisions taken by planning bodies of the government.
Counter-claim One of the advantages of an administered economy is that, when it wants to do so, the planning authority can give a high priority to meeting the basic needs of people for food, shelter, clothing and so on. It also enables an overall view to be taken of the way in which resources are being used, and helps to avoid socially undesirable duplication of production by a number of enterprises.
Broader Consumer vulnerability (#PC0123) Decrease in consumer choice (#PD6075).
Aggravates Weakness in trade between different economic systems (#PC2724).
Aggravated by Failure of centrally planned economies (#PC3894).

♦ **PD0516 Uncoordinated international river basin development**
Lack of cooperation by riparian states — Uncomprehensive river plan — River water disputes — Upstream diversion of international rivers
Nature Two or more countries may have jurisdiction over portions of a geographical area which constitutes a hydrogeological unit as the catchment area for a single river. In the absence of a coordinated approach to the development of the river basin, one or more of these countries may be placed at a disadvantage. Water control projects (flood protection, dams, soil conservation) in one country may adversely effect the territory of another through which the same river passes. Disposal of sewage or industrial waste into a river by an upstream country and diversion of water from rivers flowing into a country may adversely affect its domestic water supply, irrigation requirements, power generation, etc. The withdrawal of underground water in one country may reduce the available river water in a second country.
Incidence Of 200 international river basins in a recent review, 148 (70 per cent) are shared by 2 countries, 30 by 3 countries and 22 by from 4 to 10 countries. About 25 per cent of the countries of the world are situated entirely within international river basins and, except for island countries, almost all countries are involved in the problems of international river basins to a greater or lesser extent. The number of related bilateral and multilateral treaties is about 300. Yet over one third of the 200 major international river basins are not covered by any international agreement, and fewer than 30 have any cooperative institutional arrangements. River water disputes have already occurred in North America (the Rio Grande), South America (the Rio del la Plata and Parana), South and Southeast Asia (the Mekong and Ganges), Africa (the Nile), and the Middle East (the Jordan, Litani, Orontes, and the Euphrates). Pollution, impoundment and diversion of water by upstream nations is likely to be a growing source of international tension and insecurity.
Broader Lack of international cooperation (#PF0817).
Aggravates Floods (#PD0452) Desiccation of rivers (#PJ4867)

fresh waters (#PD6923)
Vulnerability of land-locked developing countries (#PD5788)
Dependence on sophisticated technology for development (#PD6571).

♦ PD0518 Marital stress

Conflicts in marriage — Marital problems — Fighting between husband and wife — Enmity between husband and wife

Incidence In parts of the United States 50 percent of marriages end in divorce. A study in 1987 concluded that 30 percent of marriages in the United Kingdom show some form of problem. About 20 percent of all marriages in the UK end in divorce.

Refs Ikwue, K *Inconvenient Marriage* (1982); James, Adrian L and Wilson, Kate *Couples, Conflict and Change* (1986); Khanna, P N *Problems of Love and Sex in Marriage* (1970); Manocha, B L *Marriage Conflicts* (1983).

Broader Family stress (#PD8130) Stress in human beings (#PC1648).
Related Stress among children (#PE4421)
Psychological stress of urban environment (#PE6299).
Aggravates Juvenile stress (#PC0877) Family breakdown (#PC2102)
Nervous breakdown (#PE6322) Marital instability (#PD2103)
Occupational stress (#PE6937) Discontented marriages (#PD0941).
Aggravated by Sexual unfulfilment (#PF3260) Deception between sexual partners (#PE4890)
Disruptive effect of household moving (#PE5928)
Psychological inconsistency of marriage partners (#PF9818).

♦ PD0520 Denial of rights to prisoners

Inadequate prison conditions and penal systems — Inadequate gaol conditions — Inadequate gaol conditions — Overcrowding in prisons — Unacceptable conditions of imprisonment — Inhumane jails — Denial of right to humane imprisonment — Denial of rights to detainees

Nature In most industrialized countries there is a discrepancy between the growing daily average prison population and the availability of suitable staff, accommodation and treatment. The consequent overcrowding, and other factors such as out-of-date prison rules, old and unsuitable buildings and lack of suitable work, contribute to the inefficiency of imprisonment.

Prisons are monopoly public services with very low and often negative productivity, and as such are neglected by public authorities. Instead of a correctional system whose productivity would be measured in terms of serving to deter people outside from committing crimes and to stop people inside from doing so again, some prisons create more recidivism than they cure or deter.

Another aggravating factor is the bad communications between staff and inmates, the latter being unwilling captives and the former being assigned to maintain the prisoners' state of captivity. These fundamentally different basic aims of the two groups are reinforced by two very different value systems. The inmate culture is both criminal and antiauthoritarian. Its values are so hostile to authority that punishments by those in charge may bestow status on defiant prisoners, and rewards may produce suspicion or worse among fellow inmates. By contrast, the staff culture is authoritarian and, sometimes without recognizing it, concerned with emphasizing the status superiority of staff over inmates. The two groups tend frequently to be in a state of undeclared warfare, which sometimes results in prisoners rioting.

Prison conditions vary from country to country. In some countries overcrowding is common and in others isolation is more typical. Some prisons are without adequate heating or air conditioning. Most prisoners have adequate amounts of calories but as a rule the food is monotonous, usually consisting of beans cooked in fatty liquid. Food may be unappetizing and unhygienic and often contaminated with insects or waste products. There may be little meat or roughage. Prisoners may also be plagued by lice, fleas, mosquitoes and other insects. There are frequent cases of illness: hepatitis, salmonella, toxin-induced gastro-intestinal infections, dysentery and insect-borne infectious illnesses.

Incidence Between 1951 and 1971, Britain's prison population rose more than 100 per cent. In 1973 the Netherlands had 25.4 prisoners per 100,000 population; Belgium had 64.2, France 66, Denmark 71.7, and the UK 72.5. In Britain, with this higher daily average prison population than in other Western European countries, guards sometimes had to work as many as 70 hours a week.

Background The prison system as it operates in nearly all parts of the world is largely an American development, having its main origins late in the 18th century. The evolution of the prison system was primarily the consequence of the growth of new philosophies of human conduct. Prominent in this connection was the decay of faith in earlier methods of retributive vengeance designed to eliminate the criminal offender, ostracize him, or permanently stigmatize him. Concurrent with the decline of vengeance there developed a conception of rational man which assumed that criminal human behaviour can be regulated by a system of lesser punishments administered in an objective and impersonal manner by specialized agencies of government. The official punishments now used with greatest frequency are fines and various terms of imprisonment. Most of the methods still used in present day penal systems have long histories and many of the police, court, prison and even probation methods have traditions which have not always changed with the times.

Claim In the words of a recent President of the United States, the whole prison system is 'a convincing case of failure'. Demonstration by prisoners in favour of better treatment and threats of strikes by prison officers in favour of harsher treatment are signs of a penal system gone wrong.

Counter-claim Reforms or half-moves towards more participatory democracy in prisons often will put more power into the hands of the 'barons' and the thugs who so often tend to dominate convict groups, and any policy of appeasing prisoners who indulge in so-called 'passive' demonstrations will lead to more violent demonstrations later.

Refs Paulus, P B *Prison Crowding* (1988).

Broader Imprisonment (#PD5142) Denial of rights to vulnerable groups (#PC4405)
Denial of right to freedom from cruel, inhumane or degrading punishment (#PC3768).
Narrower Beating of prisoners (#PD2484) Solitary confinement (#PE4056)
Maltreatment of prisoners (#PD6005) Repressive detention of juveniles (#PD0634)
Excessive length of pre-trial internment (#PE4887)
Disproportionately long prison sentences (#PE4602)
Inadequate care for children of prisoners (#PF0131)
Inadequate rehabilitation of juvenile offenders (#PE8803)
Detention of offenders after completion of sentence (#PJ0717)
Denial of right of accused to segregation from convicted criminals (#PE7345)
Denial of right of juvenile criminals to segregation from adult criminals (#PE7209).
Related Social neglect (#PB0883) Stress in human beings (#PC1648)
Abuse of prison labour (#PD0165) Unjust punishments for crimes (#PD4779)
Abuse of rights of criminal suspects (#PE4183)
Political prisoners in communist systems (#PD3171)
Discrimination against ex-prisoners and ex-detainees (#PE6929).
Aggravates Recidivists (#PE5581) Prison riots (#PE1675)
Prison suicides (#PD8680) Homosexuality in prisons (#PE1363)
Disabled victims of torture (#PD0764) Privatization of public services (#PE3391)
Unsanctioned maltreatment of prisoners (#PE0998)
Maltreatment of prisoners by fellow inmates (#PE0428)
Abusive detention in psychiatric institutions (#PE2932).
Aggravated by Political repression (#PC1919).

♦ PD0527 Vigilantism

Death squads — Private armies — Paramilitary gangs — Para-military activities

Nature In the special units of armed forces of a considerable number of countries, secret police and security forces which act outside ordinary legal procedures exist under the supervision of the authorities or with their approval or connivance. In a number of situations, paramilitary groups of civilians, police and armed forces personnel, sometimes called 'death squads', have operated in a similar manner. High-level involvement is not uncommon, though such practitioners of torture and murder may be recruited from all sectors of society. In some countries, there may be prestigious or financial rewards.

These special units or forces have carried out arrests and detention, and in many cases have killed suspects without any of the legal formalities required by law and without reference to the judiciary. Their activities in most cases have been kept secret and outside the control of the judiciary. Information on arrests or detention is not communicated even to the families of the arrested or detained. Motive may be political, economic, or criminal. Since this method of repression requires speed and discretion, the chances are that identity errors and false accusations increase the numbers of innocent victims.

Vigilantes may act individually or in an organized group to suppress and punish crime without adhering to the established processes of law. They can be normally law-abiding citizens who feel that legal recourse is inadequate or non-existant, or they may be fanatics seeking an excuse to take the law into their own hands.

Incidence Paramilitary gangs have long been well recognized in Latin America as a deadly means of political repression. In Argentina there were 15,000 assassinations in the 18 months following President Peron's death in 1974; and from 1976 to 1981 between 7,000 and 10,000 people were declared dead or 'disappeared'. In the Philippines, 'salvaging' operations were directed primarily against opponents of the government of President Marcos, their best-known victim being the opposition leader Benigno S Aquino Jr. From 1973 to 1983, 1,166 people were killed, half of them after martial law was lifted in 1981; over 300 political offenders are listed as missing. Since Mrs. Aquino has come to power there has been an increase of anti-communist groups. In Indonesia, 'petrus' or 'mysterious killers' have been described as 'guardian angels' because of their mopping-up operations against suspected or known criminals; there were up to 4,000 victims in 1983. In El Salvador, ORDEN and paramilitary forces have been used to implement agrarian reform and have served as a radical means to settle land disputes. Both left wing and right wing death squads are increasing murdering political foes. In Columbia vigilantes have used by right wing groups in conjunction with drug cartel members to attack left wing political groups, peasants, union leaders, journalists and lawyers who oppose drug barons or promote human rights, and rival gangs. In Soviet bloc countries there has been an increase in the number of reports of unexplained deaths among national-rights advocates, religious dissidents, and leaders of independent trade unions and other organizations with a potential for broad support. In Bulgaria several hundred Turks, protesting against the government's policy of obliterating the minority's ethnic and cultural identity, were killed in 1984.

The most famous recent case of vigilantism involved Bernard Goetz who shot and wounded 5 youths he accused of attempting to rob him in a New York subway. He was brought to trial for attempted homicide and both he and his case became the basis for worldwide heated debate both condemning and condoning his action.

Broader Criminals (#PC7373) Government sanctioned killing (#PD7221)
Armed crimes against national security (#PD8153).
Related Assassination (#PD1971) Political police (#PD3542).
Aggravates Corruptive crimes (#PD8679).
Aggravated by Civil violence (#PC4864) Criminal gangs (#PD3837)
Psychotic violence (#PE7645) Criminal association (#PE1178)
Ineffective war crime prosecution (#PD1464).
Reduces Banditry (#PD2609).

♦ PD0528 Elimination of jobs by automation

Unemployment due to mass production techniques — Endangered jobs from technological innovation — Unemployment through introduction of new technologies

Nature Throughout the industrial revolution and its subsequent evolution, automation has been criticized as the harbinger of mass unemployment and the dehumanizer of the working man. The increasing use of computers and numerical control technology has prompted similar concerns. This new technology, although vital for industrial competitiveness, has enormous potential for eliminating jobs; in the industries where microcomputers have been extensively applied the impact on jobs is considerable.

Incidence In the early 1970s in the cash registers industry, a major manufacturer reduced its employment from 37,000 to 18,000 with the changeover to microelectronics. One micro-computer replaced 350 mechanical sewing machine parts, and an electronics plant in West Berlin cut employment from 1,800 to 800 by switching to this new technology. At a car plant in Sweden an automated welding line reduced the workforce from 50 to 10.

Background The fear that automation causes job elimination is not new, and has led to violent reactions throughout history. A mob in Germany in the 1660's drowned the inventor of a mechanized loom; steam engines and power machinery were frequently smashed in 19th century England; and in the 1920's in the United States claims were made that mechanization and the use of electrical machinery would ruin the economy.

Counter-claim New technologies create new products, new industries and new jobs. The application of new technology raises standards of living and improving working conditions. It increases equal opportunities for women. Many goods, such as home video-tape recorders and personal computers, could not be produced without new technology, and service sector jobs will increase sufficiently to make up for the loss of employment in manufacturing. In addition, new technology has the potential of decentralization of employment, enabling people to work at home or closer to their homes.

Refs Bhalla, A S (Ed) *Technology and Employment in Industry* (1985).

Broader Unemployment (#PB0750).
Narrower Elimination of jobs due to automation in developing countries (#PE4391).
Related Abuse of science and technology in capitalism (#PE3105).
Aggravates Friction (#PF1691) Stress in human beings (#PC1648).

♦ PD0530 Restriction of outer space benefits to a limited number of developed countries

Nature Although the benefits arising from space exploration have so far been of greater significance scientifically than economically, it is expected that this will change as the technology develops. Some countries, especially the developing countries, will be unable to benefit from such resources, thus further increasing the gap between the richer and the poorer nations.

Broader Domination (#PA0839).
Aggravates Damage caused by space objects (#PE0250).

♦ PD0531 Forced mass expulsion

Mass exodus

Nature Mass expulsions are generally the result of a disordered state of affairs due to political, social and economic factors. In the case of nationals, mass expulsions are frequently the result of troubled conditions arising from such factors as economic and social inequalities, the violation of basic human rights, terrorism, foreign intervention in internal affairs and acts of aggression.

Problems of development constitute additional factors. In the cases of aliens, economic and social conditions are also determining factors. Mass expulsion of resident aliens generally occurs in situations where there is no integration of minorities or of migrant worker populations. In situations where there is no policy or intention on either part of integration, such expulsions can become a distinct danger when economic or political conditions deteriorate in the receiving country. The position of foreign students is also precarious.
Claim By their nature, mass expulsions, whether of a direct or indirect nature, are generally arbitrary and discriminatory, entailing the violation of basic human rights and humanitarian standards and causing unnecessary suffering to the human beings involved and frequently damaging relations among States.
Broader Expulsion (#PC5313) Displaced persons (#PD7822).
Related Involuntary mass resettlement (#PC6203).
Aggravated by Social injustice (#PC0797).

♦ PD0540 Threatening public servants
Nature Threatening a public servant can involve several types. They or their family may be threatened with harm. The threat may be to commit a crime, to accuse someone of a crime or to expose a secret. Any of these act done with the intention of influencing a public servant in their official duties is a crime.
Broader Intimidation of public officials (#PD4734).
Narrower Political blackmail (#PD2912).

♦ PD0542 Inadequate welfare services for the blind
Denial of the right to welfare services for the blind
Broader Denial of rights to disabled (#PC3461) Inadequate social welfare services (#PC0834).
Narrower Inadequate rehabilitation methods for the blind (#PE8983).
Related Inadequate welfare services for the deaf (#PD0601)
Social insecurity in developing countries (#PE4796)
Inadequate community care for handicapped persons (#PE8924).
Aggravates Loss of self-respect (#PG2484).
Aggravated by War (#PB0593) Physical blindness (#PD0568) Injurious accidents (#PB0731)
Socio-economic poverty (#PB0388) Poor living conditions (#PD9156)
Insufficient health personnel (#PD0366)
Lack of economic and technical development (#PE8190).

♦ PD0543 Inadequate road and highway transport facilities in developing countries
Nature Roads are the veins and arteries of development along which move the means of production, and in the opposite direction finished goods, as people begin to emerge above the subsistence level. They are, however, a controversial innovation because the full return on the original investment has to be awaited for several or many years. This problem is seen in its most striking form in countries such as Afghanistan, Bolivia, Lesotho and Nepal, where terrain reaches up to the highest mountains in the world and makes road construction extremely costly. In such places the high cost of construction is very slowly offset by the economic returns from villages which are set far apart, even though some may be situated in very fertile land, have fine stands of timber or possess other good natural resources. Another difficulty is that many developing countries cannot afford, and perhaps do not need, the sophisticated vehicles of Northern manufacture, although simpler, less expensive, locally designed and manufactured vehicles would also create a demand for adequate roads and highway networks. While developing nations have invested from 15 to 35 percent of their national budgets to transportation infrastructure, of which three-quarters was spent on roads the networks are only growing at a rate of 0.2 to 9.5 percent in length. The density of road networks in developing countries is only about 10 percent of developed countries.
To a large degree, road transportation networks are slow to develop because of unfavourable financial and taxing policies, undue favour to railways to carry government sector cargoes, inefficient operations of trucking companies, lack of qualified personnel, inadequate maintenance of physical infrastructure, malpractices and inefficiency of government administrations, rapid changes in technology in developed countries often means spare parts are not available, and high cost of fuels.
Refs United Nations *Main Issues in Transport for Developing Countries During the Third United Nations Development Decade (1981-1990)* (1982).
Broader Insufficient transportation infrastructure (#PF1495)
Inadequate road and highway transport facilities (#PD0490)
Inadequate transportation facilities in developing countries (#PD1388).
Aggravated by Incompetent management (#PC4867)
Distortionary tax systems (#PD3436)
Inadequate financial services (#PJ8366)
Inefficient public administration (#PF2335)
Lack of funding for infrastructure (#PJ8512)
Inadequate maintenance of infrastructure (#PD0645)
Technology gap between developed and developing countries (#PE7985).

♦ PD0547 Decreasing genetic diversity of fish
Refs FAO *Conservation of the Genetic Resources of Fish* (1981).
Broader Decreasing genetic diversity of animals (#PC1408).

♦ PD0548 Inadequate political parties in developing countries
Nature In many developing countries, especially African, political parties are of tribal origin and as such are deprived of any national character. They are not based on any political or economic doctrine. Also, as within tribal rivalry, personal ambitions are often the sole motivation for seeking political power. These factors constitute a hindrance to national unity and to economic development.
Broader Underdevelopment (#PB0206).
Aggravates Hero worship (#PF2650) Authoritarianism (#PB1638)
Political dictatorship (#PC0845) National political dependence (#PF1452).
Aggravated by Tribalism (#PC1910) Economic imperialism (#PC3198)
Inadequate education (#PF4984).

♦ PD0549 Inadequate facilities for children's play
Inadequate spatial facilities for youth recreation — Unsupervised road play — Unsupervised children's play — Unprotected play areas — Lack of school recreation
Nature Typical suburban development, with its private lots opening off streets, almost confines children to their houses. Many urban settings are much worse. Parents, afraid of traffic or of their neighbours, keep their small children indoors or in their own gardens. Thus the children have few chance meetings with other children of their own age which enable them to form the groups essential to a healthy emotional development. This tendency to isolation is increasingly the case in modern urban environments.
Claim If children don't play enough with other children during the first five years of life, the chance is greater that they will have some kind of mental illness later in their lives. Since the layout of the land between the houses in a neighbourhood virtually controls the formation of play groups, it therefore has a critical effect on people's mental health.
Broader Inadequate child welfare (#PC0233)
Impersonality of high density accommodation (#PF6156).
Related Inadequate educational facilities (#PD0847)
Inadequate child day-care facilities (#PD2085)
Attitude manipulation of children through play (#PF2017)
Inhibition of exploration by children of urban environment (#PF6159).
Aggravates Boredom (#PA7365) Alienation (#PA3545)
Youth gangs (#PD2682) Family stress (#PD8130)
Juvenile delinquency (#PC0212) Loneliness of children (#PC0239)
Inhibition of adult life in small households (#PF6167)
Excessive claims for human development through sports (#PG4881)
Limited exposure to outside influences in rural villages (#PF2296).
Aggravated by Social neglect (#PB0883) Closure of recreation areas (#PE6276)
Uncontrolled urban development (#PC0442) Household segregation by age group (#PF6136)
Overcrowding of housing and accommodation (#PD0758)
Insufficient common land in urban environments (#PE6171)
Excessive intensification of the parent-child relationship (#PF6186).

♦ PD0550 State custody of deprived children
Nature Although all countries consider parents to be responsible for the welfare of their children, they also have legislation enabling the State to remove children who appear to be in danger, physically, morally, or intellectually, due to the parents' conduct. Once it has become necessary for the child to live apart from his parents, the treatment which is offered varies widely: countries may rely entirely on residential care, almost entirely on fostering, or on mixture of both. Placement in foster homes and the running of residential institutions may be either in the hands mainly of voluntary bodies, almost entirely a matter for the state, or a mixture of both. Deficiencies of the foster care system include: unnecessary state intervention in family life; concentration of work on ensuring the fairness of proceedings, expanding the availability of services to unnecessary and lengthy foster care; inappropriate placements and abusive treatment; lack of accountability of public agencies responsible for children in out-of-home placements; lack of representation for parents in cases where the state attempts to take away their children based on allegations of abuse or neglect. Comprehensive emergency services for abused and neglected children exist in only a few communities, and services for families are rarely available to prevent foster care placement or to facilitate family reunification.
Persons or institutions which assume the guardianship or custody of minors in this way receive assistance but are also subject to the supervision of the child welfare and assistance services.
Incidence In 1970 there were 27,841 children in state custody in Austria (with 73% in foster homes and 27% in institutional care); approximately 25,000 in the Netherlands (52% in foster homes and 48% in institutions); and 62,206 in the UK (50% in foster homes and 50% in institutions).
Claim A number of investigators have studied in great detail the adverse effects of institutional care, especially on infants and young children. They clearly show that the continuity of a warm relationship with a mother or mother-substitute is essential for mental health. Prolonged deprivation of maternal care during a young child's first three years may leave permanent damage to his personality - the first year of life seems to be of critical importance. Observations of pre-school and school-age children reveal that their adjustment is very poor in many respects. In particular there is marked inability to form interpersonal relationships: they have no capacity to care for people or to make friends. They show no real feelings and no emotional response in normal conditions. They seem inaccessible and exasperating to those trying to help. They show a curious lack of concern. They are often aggressive and destructive. Running away, stealing and lying are also frequently encountered. They have very poor performance at school, lacking interest and concentration. Their capacity for abstract thinking seems to be affected. Institutionalization during the first year of life affects language and social development most severely, and the longer the institutionalization the greater the damage.
It is clear that most children who enter public care are underprivileged from birth. If the effects of institutional care are added to this, with its high rate of emotional disturbance and educational retardation, the picture is grim indeed. Children placed in foster homes seem to suffer less ill effects, but research on the rate and after-effects of foster-home breakdown is so sketchy that no firm statement can be made. Only adoption at an early age appears to reverse the unfavourable factors. Finally, the financial cost of caring for children apart from their parents is extremely high.
Broader Family breakdown (#PC2102) Non-parental custody of children (#PF3253).
Narrower Repressive detention of juveniles (#PD0634).
Related Inadequate child welfare (#PC0233) Inadequate housing for the aged (#PD0276)
Conflict concerning legal custody of children (#PF3252).
Aggravates Emotional disorders (#PD9159) Juvenile delinquency (#PC0212)
Unauthorized police search (#PD3544).
Aggravated by Orphan children (#PD7046) Neglected children (#PD4522)
Abandoned children (#PD5734) Cruelty to children (#PC0838)
Desertion in marriage law (#PF3254) Physically handicapped children (#PD0196).
Reduces Physical maltreatment of children (#PC2584).
Reduced by Trafficking in children for adoption (#PF3302).

♦ PD0556 Abuse of hallucinogens
Psychedelic drug abuse — Abuse of hallucinogenic drugs
Nature Hallucinogens constitute a heterogeneous class of psychotropic substances not under international control, belonging to a wide variety of chemical and pharmacological groups, and including such drugs as LSD, peyote, psilocybin, and more recently phencyclidine (PCP). Their use causes distortion of vision, hearing, touch, smell and/or taste. Principal problems created by the use and abuse of hallucinogens are very different from those posed by the abuse of depressants and stimulants. The known hallucinogens are not medicines. Since there is no lawful consumption of hallucinogenic substances, except for scientific experiments, no distinction can be made between their use and abuse; for the present, their consumption is considered an abuse. Because the production of, trade in and consumption of hallucinogens are, in the vast majority of cases, illegal, both the quantities produced and the number of consumers are unknown.
Among the large number of substances used at present, the main one is certainly LSD. It is an abbreviation of the German expression for lysergic acid diethylamide. LSD is a synthetic or more precisely a semi-synthetic compound, since it is not found in any natural substance; but lysergic acid, the raw material required for its production, is a natural product found in ergot of rye (Secale cornutum). For a psychedelic trip, 100-250 microgrammes of LSD is used in the form of a diluted solution or in capsul. Drops of the solution are usually taken on a lump of sugar. Some people use STP, or a mixture of LSD and STP. The letters 'STP' (Serenity, Tranquillity, Peace) designate a synthetic compound produced exclusively in underground laboratories.
Other hallucinogens, the abuse of which is widespread, are derivatives of tryptamine, in particular dimethyltryptamine (DMT) and diethyltryptamine (DET), but other substances are also used. In addition to these 'classic' substances, 'new' toxic substances are also often taken; such experiments frequently end in hospitalization.
The absorption of substances in order to produce hallucinations, visions and depersonalization constitutes a very different phenomenon from the abuse of narcotics. The feeling of well-being produced by a narcotic such as opium does not include fantasies, illusions or hallucinations. The idea of these phenomena comes from literary sources, and in the imagination of the general public they occupy a place which has no connection with reality. The effect of a narcotic like opium,

although it is sometimes used in groups, is a solitary experience which cannot be shared with others; the feeling of well-being so induced is devoid of any social or sexual content. During a psychedelic trip, on the contrary, the feeling of being at one with the universe and in communication with others by extra-sensory means requires the presence of other people. The use of hallucinogenic substances is therefore usually part of a group activity.

During the 1960s, the myth that hallucinogenic substances could perfect the personality and broaden consciousness was strengthened by members of such groups at a time when young people ready for new experiences were faced with numerous material and cultural challenges of society. Drugs such as heroin were considered too dangerous to experiment with because of the possibility of addiction and the risk of incurring serious legal penalties. The new substances did not cause physical dependence, and their use was not (then) illegal; the danger of using them was therefore considered to be moderate. In this context, it should be noted that at the basis of this new wave of drug addiction, there are metaphysical elements which only occur sporadically in the history of the use and abuse of opiates, and that there is a certain undeniable similarity between the use of the new hallucinogens and that of 'ritual' drugs.

Many people are attracted only by the hallucinogenic effects of the psychedelic experience, which above all consist of unusual hallucinations that are difficult to describe. Hallucinogens may promote reactions of panic anxiety, fury, alteration in sense of time, touch, memory, identity, reality, lack of concentration and, as an after effect, paranoia or general insanity. The repeated abuse of hallucinogens is generally the 'wilful choice' of the user, rather than deriving from compulsion induced by the drug. Tolerance builds up rapidly and disappears rapidly, but chronic habitual overdose is very difficult if not impossible to treat. Physical reactions such as vomiting are frequent. The use of hallucinogens continues in tribal societies (particularly in Central and South America) for religious and spiritual purposes. Unstable and unpredictable drugs (e.g. LSD) have been connected with subsequent homicide, suicide and psychosis.

The complex reaction produced by LSD (in particular the phenomenon of depersonalization) changes the mental image of the body, and a feeling of detachment and of unreality are not limited to the duration of a 'trip'; on the contrary, the reappearance of these reactions several weeks or months after the 'trip' indicates the lasting effect of LSD on the central nervous system. The study of the connection between these reactions and the paranoid state of many people in hospital may reveal the price to be paid for experimenting with mental equilibrium.

Incidence Hallucinogen abuse is limited compared to most other drug types. There are about 2,008,000 abusers, a global rate of 0.47 per 1,000, but most of these are from the United States. In 1980, only 15 countries reported statistical data on hallucinogen abusers, although another 32 countries gave verbal reports of at least some abusers. Until 1976, the numbers generally reported were much larger, but since that time there appears to have been a marked decrease in the number of abusers worldwide. Only 2 countries have in recent years reported extensive abuse: the United States with 1,990,000 (9.1 per 1,000) and the Cayman Islands in the Caribbean with 35 abusers (3.2 per 1,000). Only 3 countries are in the moderate abuse range: Australia (0.7 per 1,000), Canada (0.14 per 1,000) and Switzerland (0.1 per 1,000).

These trends suggest that the international proselytizing by hallucinogen users in the 1960s may have been a time-limited phenomenon. The number of medical treatments after the use of hallucinogens is increasing. In 1967, a review of work on harmful reactions to LSD was published containing summaries of twenty detailed reports in which 221 adverse reactions are described, including 19 attempted suicides, 11 suicides, 4 attempted murders and 1 murder. Since this review, the number of medical reports has increased. However the number of fatal cases and the number of people treated do not indicate either the frequency or the danger of the use of LSD, since the 'trip' does not always end in hospitalization. Nonetheless, it should also be emphasized that, according to the reports on these reactions, it is impossible to guarantee a safe dosage, or a 'safe' personality that might not react in an unfavourable manner to LSD; a bad 'trip' is always possible, even in the case of an experienced 'traveller'.

Chronic effects of LSD have begun to be noted. Chronic states have been observed in those who use high doses of LSD (from 250 to 750 microgrammes per day) both frequently and regularly. According to some reports, the treatment of these persons is extremely difficult if not impossible.

Background Hallucinogens have been used from ancient times, particularly with regard to religious and spiritual experience; 1,000 hymns to 'soma' (a mushroom-based drug), the god narcotic of the Aryan invaders of India 3,500 years ago, survive in the 'Rig veda'. Tribal cultures from Siberia to southern Africa to the Americas (particularly profusely in Central and South America) have traditionally used hallucinogens for religious experience and malevolence (such as inducing insanity in others). The ingestion of unpurified ergot (LSD fungus) caused abortion, gangrene and madness in mediaeval Europeans.

Refs Hoffer, Abram and Osmond, H *Hallucinogens* (1967).

Broader Drug dependence (#PD3825).
Narrower LSD abuse (#PG2510)
Post-hallucinogen perception disorder (#PG7457).
Aggravates Psychoses (#PD1722) — Drug smuggling (#PE1880)
Hallucinations (#PF2249).
Aggravated by Illicit drug trafficking (#PD0991) — Manufacture of illicit drugs (#PE2512).

♦ PD0557 Subversive activities

Dependence on subversive activities

Nature Subversion is closely associated with ideological conflicts in which persons acting secretly in concert on behalf of an organized party of persons within the State, or outside the State, or acting in behalf of foreign nations, attempt to overthrow or weaken the domestic government and its political institutions. Methods include: ideological indoctrination; blackmail; corruption of loyalties of nationals; and all acts and disinformation calculated to destroy confidence in national leadership and exploit differences of opinion and other opportunities for furthering confusion, dissatisfaction and dissension.

Incidence Communist activities are sometimes alleged to be subversive in free-market and even socialist countries. Subversion has also been charged against the Freemasons, Roman Catholics, multinational enterprises and capitalist governments. In some countries subversion may be treated as treason, and prosecution of those allegedly responsible for such acts used for repression of political dissent.

Refs Abaya, Hernando J *Making of a Subversive* (1984).

Broader Political crime (#PC0350) — Secret societies (#PF2508).
Narrower Espionage (#PC2140) — Underground press (#PD2366)
Guerrilla warfare (#PC1738)
Destabilizing international telecommunications (#PD0187).
Aggravates World anarchy (#PF2071) — Political instability (#PC2677)
Internment without trial (#PD1576) — Political disintegration (#PC3204)
Inadequate army discipline (#PD2543) — Inhumane interrogation techniques (#PD1362).
Aggravated by Totalitarianism (#PF2190) — Political conflict (#PC0368)
Police intimidation (#PD0736) — Lack of assimilation (#PF2132)
Foreign dictatorship (#PC3186) — Abuse of police power (#PC1142)
Student press weakness (#PE0628) — Denial of right of assembly (#PC2383)
Illegitimate political regimes (#PC1461).
Reduced by Informers (#PD8926) — Forced assimilation (#PC3293).

♦ PD0563 Agricultural pollution

Agricultural polluters

Nature The individuals and organizations pollute the water, soil, and air through the use of pesticides and fertilizers in food production. The waste from cattle, pigs, and other stock raise the level of nitrogen in water supplies to unacceptable levels. Where the stubble of harvested crops are burned high levels of air pollution result.

Broader Pollution (#PB6336).
Narrower Water pollution from animal production (#PE3934).
Aggravates Inhibition of crop growth by pollution (#PE8476).
Aggravated by Agricultural wastes (#PC2205).

♦ PD0568 Physical blindness

Nature Universal agreement on the definition of blindness has not been reached although in 1954 the World Council for the Welfare of the Blind urged the acceptance of the following definition: (a) total absence of sight; (b) visual acuity not exceeding 10/200 or 3/60 in the better eye with correcting lenses; (c) serious limitation in the field of vision (generally not greater than 20 degrees). The Council recognized that many persons with sight in the better eye equal to 20/200 (or 6/60) are still seriously handicapped visually and it strongly urged that whenever possible the definition of blindness be expanded to include all those with this degree of visual loss. In 1972 the World Health Organization proposed a uniform definition of blindness as visual acuity of less than 3/60 in the better eye, with best possible correction. A distinction is made between economic blindness which impedes the capacity to work and social blindness which limits educational possibilities and the ability to take care of personal needs and maintain social relations

Causes of blindness can roughly be divided two broad categories: accidents and disease. Accidents are likely to affect all age groups. Loss of vision caused by disease may be linked to degenerative changes usually occurring late in life (for example glaucoma, diabetic retinopathy and vascular disease) particularly in industrialized countries. In less developed countries, infections and malnutrition are the most important causes of loss of vision and they tend to manifest themselves during childhood or throughout the entire lifespan. The four leading diseases are Xerophthalmia, cataracts, onchocerciasis and trachoma. At least two-thirds of the cases of blindness in the world could have been prevented or could be cured by the application of existing knowledge.

Incidence About 28 million people in the world are blind, that is unable to count fingers at 3 metres distance, and 42 million have suffered such severe visual impairment that they cannot count fingers at 6 metres; of these 42 million, an estimated 80 per cent are in the developing countries. The Southeast Asia region accounts for almost 50 per cent of the world's blind. Of these, India alone has nine million. A total of about 500,000 pre-school age children develop corneal xerophthalmia due to malnutrition every year in Bangladesh, India, Indonesia and the Philippines. The incidence of milder forms of xerophthalmia is probably ten times higher.

Claim It is expected that the number of blind will double early in the next century unless preventive action is taken. In most countries facilities for the prevention and relief of blindness are totally inadequate, yet blindness is one of the most expensive disabilities in terms of economic loss, and the most economically remediable in terms of medical and rehabilitation resources.

Refs Amaura, Edward G (Eds) *Blindness* (1987); Norris, Mirriam and Spaulding, Patricia J *Blindness in Children* ; Wilson, John and International Agency for the Prevention of Blindness *World Blindness and its Prevention*.

Broader Blindness (#PA6674) — Stigmatized diseases (#PD7279)
Eye diseases and disorders (#PD8786) — Visually handicapped persons (#PD2542)
Human physical genetic abnormalities (#PD1618).
Narrower Blind workers (#PG3370) — Blind children (#PE2082)
Partial loss of vision (#PE4196) — Blindness in developing countries (#PD5139).
Related Night blindness (#PG4565) — Colour deficiencies (#PE6343)
Inflected loss of vision (#PE3377).
Aggravates Inadequate welfare services for the blind (#PD0542)
Inadequate rehabilitation methods for the blind (#PE8983).
Aggravated by Leprosy (#PE0721) — Smallpox (#PE0097)
Trachoma (#PE1946) — Glaucoma (#PE2264)
Syphilis (#PE2300) — Cataract (#PE6817)
Gonorrhoea (#PE1717) — Filariasis (#PE2391)
Xerophthalmia (#PE2538) — Toxoplasmosis (#PE3659)
Onchocerciasis (#PE2388) — Conjunctivitis (#PE7974)
Primary glaucoma (#PG4523) — Secondary glaucoma (#PG4524)
Diseases of retina (#PE4584) — Injurious accidents (#PB0731)
Congenital glaucoma (#PG4522) — Nutritional diseases (#PD0287)
Parasites of the human body (#PE0596) — Genetic defects and diseases (#PD2389)
Diseases and injuries of the brain (#PD0992).

♦ PD0571 Man-made famine

Nature While farmers have faced droughts for millennia, it is only recently that dry years result in widespread famine. Formerly, in good years, a farmer would fill grain stores with enough to survive for several consecutive seasons. Grazing was done considering available pastures and water holes so over grazing was avoided. Fields were left fallow for long periods to enable the soil to recover. This balance has been disrupted by a number of things. Veterinary medicine by treating animals has resulted in vast herds with little grass. Mechanical pumps increase the use of water creating the likelihood of drought. Demands for market crops because of pressures for exports and encourage short term thinking. Grain silos are emptied eliminating buffers against famine. The land is no longer left fallow and the soil becomes exhausted.

Incidence Major examples have been associated with experiments in centralized planning including the famine in the Ukraine in 1932-1933 (estimated 5 million died) and that in China following the Great Leap Forward programme in 1960 (conservative estimate that 20 million died).

Broader Famine (#PB0315).

♦ PD0575 Industrial waste water pollutants

Discharge of dangerous substances into industrial waste water

Nature Industrial wastes usually contain traces or larger quantities of the raw materials, intermediate products, final products, co-products and by-products, and any ancillary or processing chemicals used. The composition and amount of pollutants discharged by a specific industry can usually be determined only by detailed analysis of its effluents. The complete enumeration of the substances present in industrial waste waters as a whole would run into thousands. They include detergents, solvents, cyanides, heavy metals, mineral and organic acids, nitrogenous substances, fats, salts, bleaching agents, dyes and pigments, phenolic compounds, tanning agents, sulphides and ammonia. Of the compounds mentioned, many are biocidal and toxic. In spite of this variety, many industrial wastes can be measured by the same parameters as those applicable to municipal wastes, such as biological oxygen demand (BOD) and chemical oxygen demand (COD), turbidity, and suspended solids; however, the lack of information on the composition of industrial discharges has caused the greatest difficulties in water management.

Refs Bromley, J, et al (Eds) *Chemical Waste* (1986).

Broader Impurities in waste water (#PD0482) — Discharge of dangerous substances (#PD4542)
Chemical pollutants of the environment (#PD1670).
Narrower Mine wastes (#PG2548) — Chemical industry wastes (#PG2549)
Food manufacturing industry wastes (#PE8702).

Related Wasted water (#PD3669).
Aggravates Air pollution (#PC0119) Soil pollution (#PC0058)
Infertile land (#PD8585) Animal diseases (#PC0952)
Water pollution (#PC0062) Marine pollution (#PC1117)
Malodorous fumes (#PD1413) Thermal pollutants (#PC1609)
Amenity destruction (#PC0374) Nickel as a pollutant (#PE1315)
Sewage as a pollutant (#PD1414) Mycoplasmal arthritis (#PG2302)
Mercury as a pollutant (#PE1155) Cadmium as a pollutant (#PE1160)
Environmental pollution (#PB1166) Fluorides as pollutants (#PE1311)
Hazardous waste dumping (#PD1398) Hazards to human health (#PB4885)
Detergents as pollutants (#PE1087) Marine dumping of wastes (#PD3666)
Pollution of inland waters (#PD1223) Pests and diseases of fish (#PD8567)
Metal contamination of soil (#PD3668) Human disease and disability (#PB1044)
Lack of resource conservation (#PG2054) Contamination of drinking water (#PD0235)
Eutrophication of lakes and rivers (#PD2257)
Damage by degradable organic matter (#PJ6128)
Water pollution in developing countries (#PD3675)
Polychlorinated biphenyls as a health hazard (#PE2432).
Aggravated by Unethical commercial practices (#PC2563).

♦ PD0577 Theft of documents

Theft of books — Theft of public records — Theft of archives
Broader Theft of property (#PD4691) Unethical documentation practices (#PD2886).
Narrower Stolen archives (#PG0672).
Related Frauds, forgeries and financial crime (#PE5516).

♦ PD0579 Second class states

Government loss of leadership role in world affairs
Nature States may become preoccupied at the lack of consideration accorded to their country in world affairs or to the erosion of any leadership role they may have held over some period. The risk of acquiring the reputation of a second class state may drive the government to extremes to acquire new status. Such extremes may include wastage of resources on prestige projects or the development of a high technology weapons capability.
Broader Competition between states (#PC0114).
Aggravates Excessively costly prestige projects (#PF3455)
Vulnerability of small nations to foreign intervention (#PD2374)
Insecurity and vulnerability of non-nuclear weapon states (#PD1521).
Aggravated by Denial of right to equality between states (#PE4712)
Non-viability of small states and territories (#PD0441).

♦ PD0583 Aggravated assault

Nature Aggravated assault is willfully causing serious bodily harm to another human being or causing bodily harm to another person while attempting to harm anyone with or without the use of weapons.
Broader Assault (#PD5235).
Narrower Family violence (#PD6881) Inflected loss of vision (#PE3377).
Related Brutality (#PC1987) Simple assault (#PE1144).

♦ PD0584 Abusive treatment of patients in psychiatric hospitals

Physical brutality in psychiatric hospitals
Nature Patients in psychiatric hospitals, particularly those committed for long terms have been abused in many ways. Patient's money is pilfered. They are neglected; hospital managers don't manage; senior physicians abdicate from care; senior nurses are rigid and fail to adopt modern methods of care and government agencies fail to inspect hospitals. Patients are physically and sexually abused. The physical surrounds are unpleasant. Activities are boring. Many lack adequate medical treatment and even sufficient food.
Broader Human violence (#PA0429) Denial of rights of mental patients (#PD1148).
Narrower Mental cruelty (#PG2565).
Related Unethical practices in psychotherapy (#PD5267)
Unethical practices of health services (#PE3328).
Aggravates Mental disorders (#PD9131) Inadequate mental hospitals (#PF4925).
Aggravated by Inadequate rehabilitation facilities (#PD1089).

♦ PD0586 Maladjusted children

Problem children
Incidence In the US in 1971, 3 percent of US school children were hyperactive according to the US Department of Health, Education and Welfare. In 1974, official estimates put the number at 15 percent. In 1978, 1.7 percent – 1.8 percent of US school children were receiving drugs to control 'hyperactivity'. In the UK, there were 600 'maladjusted' children in 1950; 8,000 'maladjusted' children in 1966; 20,000 'maladjusted' children in 1976; and in 1978, 16,000 to 18,000 school children were receiving drugs to control 'hyperactivity'.
Claim Today's mental welfare activities have focused increasingly on children considered to be hyperactive or maladjusted. They have turned the raising of children into a technical problem that needs the constant monitoring of experts. When the parents fail to fulfil their role in the manner considered appropriate, then the experts are prepared to 'intervene' as surrogate parents. This intervention is sometimes blatantly repressive; there are many cases of children 'in care' being massively drugged to quell their protests. In the 1960s, it was suggested that hyperactivity stems from otherwise undetectable 'minimal brain damage', and that drugs can put it right. Within a few years, about a million American children were living under permanent sedation to control their 'hyperactivity'.
Broader Disabled children (#PC5890) Family disorganization (#PC2151)
Retarded socialization (#PF2187).
Narrower Childhood aggression (#PD3907).
Related Children's diseases (#PD0622).
Aggravates Prostitution (#PD0693) Juvenile delinquency (#PC0212)
Loneliness of children (#PC0239) Lack of self-confidence (#PF0879).
Aggravated by Family breakdown (#PC2102) Displaced children (#PD5136)
Cruelty to children (#PC0838) Marital instability (#PD2103)
Female prostitution (#PD3380) Inadequate child welfare (#PC0233)
Physically handicapped children (#PD0196) Repressive detention of juveniles (#PD0634)
Negative effects of the nuclear family (#PF0129) Denial of rights of children and youth (#PD0513)
Adjustment difficulties of new urban families (#PF1503).

♦ PD0590 Inadequate laws of adoption

Nature Although the institution of the adoption of children exists in most countries, there are differing views as to the principles which should govern adoption, both in procedure and in legal consequences. The legal status of an adoptive child may be different from that of a natural child, and he may have restricted rights of inheritance. Since the natural mother's or the parents' consent is usually necessary, the adoption process may be hindered, or the parents may reclaim the child later. Some foster parents discriminate when choosing children, often leaving the unintelligent, less attractive, deformed or racially unacceptable children. In rare cases foster parents prove to be inadequate. Others effect a sham 'adoption', acquiring children for a price from poor parents for use as servants or possibly for sexual exploitation.
Background Adoption first became institutionalized under the Greek and Roman civilizations, although the earliest records come from before those civilizations. Most adoptions were in order to perpetuate the male line in a family, and the adopted were therefore usually male and often adult. The adoption movement in its modern form began in the 19th century in the USA, Canada and New Zealand; the UK introduced adoption laws in 1926, and France in 1923.
Claim Arrangements for adoption are fraught with many unnecessary hazards. Instead of promoting the rights and interests of the child it can be instrumental in violating them. That inadequate adoption procedures do violate children's rights is evident in the fact that from time to time reports of alleged profiteering and racketeering through child-adoptions appear in the media of several nations. That such a situation should still exist is tragic for, if the adoption could take place without mishap, the child might gain a happy home life.
Refs Sachdev, Paul (Ed) *Adoption* (1983).
Broader Inadequate laws (#PC6848) Dependency of children (#PD2476)
Impediments to adoption of children (#PF7353).
Narrower Orphan children (#PD7046) International adoptions (#PE4296)
Decreasing number of adoptive parents (#PF4205)
Decreasing number of adoptable children (#PF4200).
Related Exploitative personal services (#PC3299)
Non-parental custody of children (#PF3253).
Aggravates Trafficking in women (#PC3298) Illegitimate children (#PC1874)
Exploitation of children (#PD0635) Restrictions on property rights (#PD8937)
Denial of right to inherit property (#PF0886) Trafficking in children for adoption (#PF3302)
Inadequate sense of personal identity (#PF1934).
Aggravated by Divorce (#PF2100) Family breakdown (#PC2102)
Unmarried mothers (#PD0902) Desertion in marriage law (#PF3254).
Reduces Childlessness (#PC3280).

♦ PD0594 Virus diseases

Viral infections — Viral diseases — Bacteriophages — Pathogenic viruses
Nature Virus diseases are widespread diseases in nature caused by viruses and found in humans, animals, birds, fish, insects, plants, protozoa, and even bacteria. In man, virus diseases include smallpox and chicken pox, measles, herpes, influenza, German measles, mumps, poliomyelitis, viral hepatitis, endemic encephalitis (transmitted by ticks, mosquitoes, and other insects), trachoma, and yellow fever. The incubation period of virus diseases is from two or three days (influenza, certain kinds of encephalitis, and others) to 30 days and longer (rabies, epidemic hepatitis, and others). Infection can occur via air, food, milk, water, or various objects, and through the bite of bloodsucking arthropoda (mosquitoes, sand flies, and ticks).
Refs Evans, A S *Viral Infections of Humans* (Date not set); Hotchin, J *Persistent and Slow Virus Infections* (1971); Kurstak, E, et al *Viruses, Immunity and Mental Disorders* (1987); Mackenzie, J S *Viral Diseases in South East Asia and the Western Pacific* (1983); Persley, G J (Ed) *Bacterial Wilt Disease in Asia and the South Pacific* (1986); Shipley, Elizabeth H *Bacterial Infections* (1987); Tyrell, D A J (Ed) *Aspects of Slow and Persistant Virus Infections* (1979).
Broader Microbial diseases (#PC7492) Infectious and parasitic diseases (#PD0982).
Narrower Mumps (#PE2356) Rabies (#PE1325) Cowpox (#PE6886)
Herpes (#PE8615) Rubella (#PE0785) Measles (#PE1603)
Smallpox (#PE0097) Trachoma (#PE1946) Fowlpest (#PE1400)
Influenza (#PE0447) Hepatitis (#PE0517) Chickenpox (#PG9663)
Common cold (#PE2412) Lassa fever (#PG9222) Yellow fever (#PE0985)
Dengue fever (#PE2260) Encephalitis (#PE2348) Poliomyelitis (#PE0504)
Coxsackie virus (#PE3852) West Nile fever (#PG6834) Viral infections (#PG3236)
Rift valley fever (#PE7552) Contagious ecthyma (#PE6831) Haemorrhagic fevers (#PE5272)
Nairobi sheep disease (#PE6268) Foot-and-mouth disease (#PE1589)
Infectious mononucleosis (#PE5550) Bat salivary gland fever (#PG4980)
Chronic fatigue syndrome (#PE1914) Water-borne viral disease (#PG3399)
Viral diseases in animals (#PD2730) Virus diseases in protozoa (#PG7376)
Virus diseases in bacteria (#PD2562) African green monkey disease (#PG7524)
Cytomegalic inclusion disease (#PG2044) Argentinian haemorrhagic fever (#PG7151)
Disease-causing viral combinations (#PE6403).
Aggravates Deafness (#PD0659) Paralysis (#PD2632)
Ornithosis (#PE2578) Chronic bronchitis (#PE2248)
Pregnancy disorders (#PD2289) Viral plant diseases (#PD2227)
Biological contamination of food (#PD2594) Perinatal morbidity and mortality (#PD2387)
Diseases of the respiratory system (#PD7924).
Aggravated by Viruses (#PE3680) Man as vectors of disease (#PD8371)
Insufficient health personnel (#PD0366) Animals as vectors of disease (#PD8360)
Insect vectors of human disease (#PE3632).
Reduces Bacterial diseases in animals (#PD2731).
Reduced by Air pollution (#PC0119).

♦ PD0601 Inadequate welfare services for the deaf

Denial of right to welfare services for the deaf
Nature Many hearing-impaired people are illiterate, unemployed and under-employed due to widespread prejudice by the general public and a lack, especially in developing countries, of the basic tools for rehabilitation (such as education in lipreading, and hearing aids). Special education, when it exists, is often disregarded or postponed. As a rule the deaf are taught a manual art or craft, regardless of their individual attitudes and preferences; vocational pre-counselling and counselling systems are neglected and often the choice is confined to the three or four trades taught in the school workshop.
Background The history of the education of the deaf begins properly in the 16th century. Before this time those born deaf were the subject of philosophic speculation but it was generally assumed that they were incapable of education. However, despite increasing interest, it was only in 1778 in Leipzig, Germany, that the first state school for deaf children was opened. One year later the state school for deaf children in Vienna was founded. The deaf were the first group of handicapped children to receive special education.
Claim One cannot over-emphasize the seriousness of a situation which deprives disabled people of one of the fundamental rights of man: the choice of employment. Its psychological consequences are self-evident, for in most cases the deaf are forced into a type of work they do not like, which makes them feel all the more ill at ease in society.
Broader Denial of rights to disabled (#PC3461) Inadequate social welfare services (#PC0834).
Related Inadequate welfare services for the blind (#PD0542)
Social insecurity in developing countries (#PE4796)
Inadequate community care for handicapped persons (#PE8924).
Aggravates Alienation (#PA3545).
Aggravated by Hearing defects (#PD6306).

♦ PD0606 Exploitative entertainment

Incorporation of advertising into entertainment — Subliminal advertising — Cinematic product placement
Incidence Although subliminal advertising is not currently practised, the deliberate insertion of commercial products as props in various forms of entertainment, especially film and television, is increasingly sought. Cinematic product placement became common in the 1980s and has now become rationalized as a specialized branch of advertising. Increasingly advertisers recognize that movies are an alternative advertising and promotional medium with an essentially captive audience. Corporations may gain control (or fund producers) of entertainment in order to be able to influence the content of the film, building in verbal or visual plugs for specific products. Such

plugs need not be overt since the effect may be achieved through positive associations with the star or with the emotions aroused. Such vehicles are also deliberately used to build negative associations around a competitor's product by ensuring its presence at a negative moment in the film.
Claim The rise of product placement has damaged movie narrative, not only through the shattering effect of individual plugs, but also more profoundly through the partial transfer of creative authority out of the hands of film-making professionals and into the purely quantitative universe of the company executives. It represents the usurpation by advertising of those authorial prerogatives once held by directors and screenwriters. The basic decisions of film-making are now often made indirectly by advertisers concerned with their value as vehicles for the presentation of products.
Counter-claim There is a long tradition of using entertainment to carry messages not explicitly sought by the audience. Morality plays are a prime example. Many films are financed by groups concerned to put over a particular message, possibly with religious, moralistic or human rights overtones. Soap opera may be deliberately used to carry moralistic messages of a certain kind.
Broader Exploitation (#PB3200).
Narrower Exploitative films (#PE6328) Exploitative commercial television (#PD0433)
Exploitation of athletic competition for commercial or political ends (#PE4833).
Related Exploitation of animals for amusement (#PD2078).

♦ PD0610 Sanctions against trade union workers
Denial of rights of trade unions to organizers — Harassment of organizers — Intimidation of worker's representatives
Broader Trade unionism (#PF8493).
Narrower Arrest of trade union leaders (#PD7630)
Threats against trade union leaders (#PD7471)
Assassination of trade union leaders (#PE0252)
Death threats against trade union leaders (#PE4869)
Victimization of workers' representatives (#PD1846)
Forced disappearances of trade union leaders (#PE5882).
Related Denial of the right to picket (#PE8712) Violation of the right to strike (#PE5070)
Denial of right to collective bargaining (#PE3970)
Denial of right to organize trade unions (#PE5398)
Violation of the right of trade unions to function freely (#PE1758).
Aggravated by Unethical practices of employers (#PD2879).

♦ PD0614 Adolescent pregnancy
Teenage pregnancy — Early childbearing — Adolescent motherhood — Adolescent fatherhood — Early teenage motherhood — Adolescent parenthood — Pregnant schoolgirls
Nature Adolescents in any country who become pregnant face severe health risks whether they are married or not. Mothers under 20 suffer more complications in pregnancy and delivery than women who bear children at age 20 or later. The infants of adolescent mother have a higher incidence of low birth weight, prematurity, stillbirth, and perinatal mortality. In many societies, pregnant teenage women are force to leave school before they are equipped with vocational, health and social skills necessary to secure an adequate live for themselves and their children. In those societies where having a child represents a rite of passage into adulthood, teenagers are under social pressure to have children as early as possible even when they are not socially or psychologically prepared for the responsibility. Teenage pregnancy has dire consequences for the economic future of the teenager and her family. Teenage mothers frequently fail to complete their education and this, in turn, leads to employment in low paying positions, unemployment or reliance on public support.
Incidence Although the extent of pregnancy among young women is not know with precision, it is believed to be quite widespread. Worldwide, pregnancy related complications are the main cause of death among 15–19 year old women. According to data from the World Fertility Survey, over half of all women aged 25–29 had their first birth before the age of 20 in such countries as: Benin, Ghana, Kenya, Senegal, Jordan, Sudan, Yemen Arab Republic, Indonesia, Bangladesh, Pakistan, Dominican Republic, and Jamaica. In the Caribbean, almost 60 percent of first babies are born to women under 19 and half of these are born to mothers under 17 years of age. In Indonesia, 41 percent of women have their first baby before they reach 17 years of age. In the United States, more than a million teenage girls become pregnant every year, about 30,000 of them under 15.
Refs Corbett, Margaret-Ann and Meyer, Jerrilyn H *The Adolescent and Pregnancy* (1987); Furstenberg, Frank F Jr *Unplanned Parenthood* (1976); Jones, Elise F, et al *Teenage Pregnancy in Industrialized Countries* (1987); Lancaster, Jane B and Hamburg, Beatrix A (Eds) *School-Age Pregnancy and Parenthood* (1986); Lindsay, Jeanne W and Monserrat, Catherine *Teenage Marriage* (1988); Rickel, Annette U *Teenage Pregnancy and Parenting* (1989).
Related Maternity (#PJ1893) Health risks of teenage sex (#PE6969).
Aggravates Unmarried mothers (#PD0902) Disowned children (#PJ0827)
Cruelty to children (#PC0838) Educational wastage (#PC1716)
Single parent families (#PD2681) Complications of childbirth (#PC9042)
Adolescent induced abortions (#PD1302) Mothers' self-imposed isolation (#PJ9425).
Aggravated by Early marriage (#PE7628) Sexual immorality (#PF2687)
Paternal negligence (#PD7297) Adolescent sexual intercourse (#PD7439)
Pre-marital sexual intercourse (#PD5107) Neglect of adolescent health care (#PF6061)
Ignorance of reproductive processes (#PD7994).

♦ PD0619 Instability of chemicals trade
Broader Instability of the primary commodities trade (#PC0463).
Narrower Instability in tanning and dyeing trade (#PE9089)
Instability of trade in manufactured fertilizers (#PE0806)
Instability of trade in chemical elements and compounds (#PE0500)
Medicinal and pharmaceutical products trade instability (#PE7996)
Instability of trade in essential oils and perfume materials (#PE8232)
Instability of trade in regenerated cellulose and artificial resins (#PE8769)
Instability of trade in mineral tar and crude chemicals derived from coal, petroleum and natural gas (#PE8464).

♦ PD0620 Instability of trade in machinery and transport equipment
Broader Instability of the primary commodities trade (#PC0463).
Narrower Instability of trade in non-electrical machinery (#PE8828)
Trade instability of electrical machinery, apparatus and appliances (#PE8875).

♦ PD0622 Children's diseases
Nature The intensive physical growth and development of children create the special anatomic and physiological nature of the body, and thus the special nature of the pathology of children's diseases. Even diseases usually affecting adults run a distinct course in children.
The infants considerable food requirements strain its as yet imperfect digestive system, giving rise to: gastrointestinal diseases (dyspepsia); gastrointestinal bacterial infections (dysentery, enterocolitis); chronic nutritional disturbances (infant dystrophy). Improper feeding and insufficient air and sunlight may lead to rickets. The vulnerability of the mucous membranes gives rise to pneumonias.
In early childhood incidence of acute infectious diseases increases: measles; whooping cough; chicken pox; scarlet fever; diphtheria. Susceptibility to tuberculosis is also greater. Allergies are more frequent and are a factor in the development of: bronchial asthma; rheumatism: eczema; nephritis.
Children of school age suffer endocrine disturbances, rheumatism, cardiovascular diseases, and psychoneuroses.
Refs Altman, Arnold and Schwartz, Allen D *Malignant Diseases of Infancy, Childhood and Adolescence* (1983); Jelliffe, D B and Stanfield, J P (Eds) *Diseases of Children in the Subtropics and Tropics* (1978); Koocher, Gerald P and O'Malley, John E *The Damocles Syndrome* (1981); Moll, H *Atlas of Pediatric Diseases* (1976); Rendle-Short, et al *Synopsis of Children's Diseases* (1985); Steihm, *Immunology Disorders in Infants and Children* (1988); Von Rosenstein, Nicholas R *The Diseases of Children and their Remedies* .
Broader Human disease and disability (#PB1044).
Narrower Dystrophy (#PE3506) Tonsillitis (#PE2292) Tuberculosis (#PE0566)
Appendicitis (#PG2327) Genetic defects and diseases (#PD2389)
Infectious and parasitic diseases (#PD0982).
Related Maladjusted children (#PD0586).
Aggravates Infanticide (#PD3501) Mental deficiency (#PC1587)
Mutilation and deformation of the human body (#PD2559).
Aggravated by Socio-economic poverty (#PB0388)
Poor living conditions (#PD9156)
Environmental pollution (#PB1166)
Parasites of the human body (#PE0596).

♦ PD0629 Underdevelopment of agricultural and livestock production
Broader Underdevelopment of industrial and economic activities (#PC0880).
Narrower Lack of meat and egg production (#PE9115)
Poor quality of domestic livestock (#PD2743)
Underproductive methods of agricultural management (#PF6524).
Related Stagnated development of agricultural production (#PD1285).
Aggravated by Epizootic diseases (#PD2734) Illiteracy among indigenous peoples (#PD3321)
International movement of animals as factor in animal diseases (#PD2755).

♦ PD0633 Loneliness in old age
Nature Elderly people can become socially isolated due to: mandatory retirement policies which cut them off from work relationships; mobility of children causing them to live further away; death of the spouse, relatives and friends; and loss of membership of organizations.
Incidence Loneliness, desolation and isolation characterize the social lives of many of the aged, particularly in developed countries where geographic or familial isolation increases with age, together with physical incapacity and dependence. In France, for example, more than 50 percent of the elderly live alone or with an elderly spouse. The most severe situations are those of elderly farmers living alone outside of any community or hamlet. Loneliness is also more frequent for older women due to the longevity gap. A recent study in New York City, for example, found that 30,000 older people, most of them women, were living in total isolation.
Background In traditional societies, old people have always enjoyed a privileged position based on respect, consideration, status and authority. The extended family involved the old, made demands on them, gave them a privileged place in society, something useful to do, allowed them to go on feeling worthwhile and abolished solitude. With the coming of the nuclear family, the elderly person has become isolated. No means of caring for the elderly has been found that is as effective as the extended family; neither the institutions of the welfare state, nor home-delivered hot meals, charity or private nursing homes, can deliver the same quality of care and basic human decency to the old and the infirm.
Broader Loneliness (#PF2386) Household segregation by age group (#PF6136)
Inhibition of individual psychological development through life cycle (#PF6148).
Narrower Self-imposed loneliness (#PU2665) Insufficient personal contact (#PU2666).
Related Loneliness of children (#PC0239).
Aggravates Senility (#PE6402) Emotional disorders (#PD9159).
Aggravated by Social neglect (#PB0883) Family breakdown (#PC2102)
Stress in human beings (#PC1648) Loneliness in single people (#PD4392).

♦ PD0634 Repressive detention of juveniles
Injustice in the detention of juveniles
Nature Juvenile detention, while often sharing the same characteristics as adult detention (solitary confinement, confiscation of personal possessions, wearing of a uniform), is not determined by the same criteria as adult detention. The majority of children in detention are held for conduct which would not be considered criminal in an adult, such as promiscuity, truancy, running away, incorrigibility and very minor offences. Children are denied trial by jury, individual access to a lawyer, and bail. Sentences may be indeterminate, which may mean that children serve longer than the adult maximum for a given crime, or may remain in detention when the crime was minor or nonexistent but there is no other institution to take them. In this way children without families often end up with longer sentences for minor crimes than those who have committed serious crimes (theft, assault, rape, murder) but whose families welcome them back. The effect of such confinement tends to make hardened criminals of people who might not have otherwise been so, especially since there are little or no rehabilitation facilities.
Broader Wrongful detention (#PD6062) Denial of rights to prisoners (#PD0520)
State custody of deprived children (#PD0550).
Narrower Inadequate segregation of different categories of juvenile offenders (#PE8215)
Denial of right of juvenile criminals to segregation from adult criminals (#PE7209).
Related Injustice (#PA6486) Inadequate rehabilitation of juvenile offenders (#PE8803).
Aggravates Maladjusted children (#PD0586).
Aggravated by Hooliganism (#PD1109) Abandoned children (#PD5734)
Corporal punishment (#PD8575) Juvenile delinquency (#PC0212)
Dependency of children (#PD2476) Inadequate child welfare (#PC0233)
Lack of child welfare institutions (#PJ2673)
Confinement for non-criminal reasons (#PG2670)
Discrimination against juveniles in judicial proceedings due to protective legislation (#PE1295).

♦ PD0635 Exploitation of children
Denial of right to freedom from exploitation of children
Nature The types of exploitation of children are not reducible to a common denominator; they differ in severity and in significance. The most repulsive include the sweat-shop system, bondservice, maids-of-all-work in a situation of virtual bondage, and prostitution. Children are highly exploitable, and in certain circumstances their vulnerability is manipulated to their lasting disadvantage. Children are sold into slavery, trained to be criminals, used for human organ transplants, forced into prostitution, abandoned, used for force labour, forced to work in inhuman and dangerous conditions and enlisted into armies. Human foetuses are sold for use in pharmaceutical and cosmetic industries.
Refs Moorehead, Caroline (Ed) *Betrayal* (1989).
Broader Exploitation (#PB3200) Denial of rights of children and youth (#PD0513).
Narrower Exploitation of child labour (#PD0164) Sexual exploitation of children (#PD3267).
Related Child-marriage (#PF3285) Racial exploitation (#PC3334)
Corruption of minors (#PD9481) Exploitation in employment (#PC3297)
Discrimination and harassment of children in public life (#PE6922).

Aggravates Cruelty to children (#PC0838).
Aggravated by Dependency of children (#PD2476)
Inadequate laws of adoption (#PD0590).

♦ PD0637 Lung disorders and diseases

Respiratory diseases — Pulmonary embolism — Pulmonary diseases — Occupational chronic pulmonary diseases — Chronic obstructive pulmonary disease — Pulmonary disease — Lung diseases

Nature Inflammatory processes resulting from infections of bronchi and lungs, scarring and tissue destruction, as well as impaired pulmonary circulation, are frequent lung disorders. Since the lungs literally filter the venous blood drainage from the entire body, they often become involved secondarily by blood-borne processes; many malignant tumours eventually metastasis to the lungs. More important still is the steadily rising incidence of primary malignancy of the lung and its relationship to cigarette smoking and environmental pollution.

Incidence In the last decade, communicable diseases of the respiratory system, as a group, were one of the principal causes of morbidity and mortality in many countries. Data reported by 88 countries with a total population of about 1200 million show that, in the year 1972 alone, more than 666,000 deaths were related to acute respiratory infections. This represents an average of 6.3 percent of all deaths reported, although there are considerable differences between continents and between countries, with an overall range of from 3.0 percent to 13.6 percent. Pneumonia, both viral and bacterial, accounted for 75.5 percent of the deaths related to acute respiratory infections Mortality rates for acute respiratory infections were highest in infants and, in some countries, exceeded 2000 for every 100,000 liveborn babies. The rate declined in childhood and early adult life, but increased progressively with age in the middle and old age groups. However, acute respiratory infections in infants and children below 15 years of age accounted for 20.3 percent of the total number of deaths from all causes, against the 4.2 percent represented by deaths from the same cause in persons belonging to the age group 55 years and over Some of the acute respiratory infections leave patients with sequelae and are known also to exacerbate already existing diseases of the respiratory tract; both situations may lead to the development of chronic lung conditions. As regards chronic respiratory diseases (including chronic bronchitis, emphysema, and bronchial asthma) a total of 304,298 deaths was reported from the above-mentioned 88 countries for 1972. This represents an average of 2.9 percent of all deaths reported by them for that year, the highest mean percentage being reported from Africa (6.3%) and the lowest from Asia (2.0%). Of all deaths from chronic respiratory diseases, 18.4% occurred in infants and children and 81.6% in adults (15 years and over) Chronic respiratory diseases are responsible also for widespread morbidity and invalidity in several parts of the world, in spite of the fact that some causative or aggravating factors (such as smoking, air pollution, socioeconomic conditions and respiratory infections in children) are already known and could be ameliorated or removed With due reservations concerning differences in criteria, data from the World Health Organization over recent years indicate cases of this problem in the following countries:

Af Mauritius. Am Brazil, Canada, El Salvador. As Hong Kong, Israel, Japan, Kuwait, Singapore, Sri Lanka, Syria, Thailand. Pa Australia, New Zealand, Papua New Guinea. Eu Austria, Bulgaria, Czechoslovakia, France, Germany FR, Greece, Hungary, Iceland, Ireland, Italy, Luxembourg, Malta, Netherlands, Poland, Portugal, Romania, UK, Yugoslavia.

Refs Burrows, B *Respiratory Disorders* (1983); Cotes, et al *Occupational Lung Disorders* (1986); Dosman, James A and Cotton, David J *Occupational Pulmonary Disease* (1980); Fishman, Alfred P *Pulmonary Diseases and Disorders* (1988); Goldhaber, Samuel Z *Pulmonary Embolism and Deep Venous Thrombosis* (1985); Gray, Frank D *Pulmonary Embolism*; Hume, Michael et al *Venous Thrombosis and Pulmonary Embolism* (1970); Merchant, James A *Occupational Respiratory Diseases* (1986); Sawicka, E and Branthwaite, M *Respiratory Emergencies* (1988); Smith, Barbara E *Digging Our Own Graves* (1987); Vladutiu, Adrian O *Pleural Effusion* (1986).

Broader Lack of community planning (#PF2605) Human disease and disability (#PB1044).
Narrower Pleurisy (#PG6121) Pneumonia (#PE2293) Byssinosis (#PE2319)
Berylliosis (#PG2678) Tuberculosis (#PE0566) Pneumoconiosis (#PD2034)
Bronchiectasis (#PE8579) Hernia of the lung (#PG4463)
Hypostatic pneumonia (#PG1124) Pulmonary tuberculosis (#PE2526)
Blood circulation disorders (#PE3830) Acute respiratory infections (#PE7591)
Cardio-pulmonary disease due to torture (#PE1556)
Inflammatory infections of the respiratory organs (#PE9151)
Inflammatory affections of the bronchial tubes and lungs (#PE8822).
Aggravates Heart diseases (#PD0448) Malignant neoplasms (#PC0092).
Aggravated by Shock (#PC8245) Ulcers (#PE2308)
Smoking (#PD0713) Obesity (#PE1177)
Ringworm (#PD2545) Influenza (#PE0447)
Air pollution (#PC0119) Chronic bronchitis (#PE2248)
Inhalation of dusts (#PG2683) Ozone as a pollutant (#PE1359)
Smoke as a pollutant (#PD2267) Diseases of the pancreas (#PE1132)
Diseases of the arteries (#PE2684) Hazards of bottle-feeding (#PE4935)
Health hazards of asbestos (#PE3001) Dust as an occupational hazard (#PE5767)
Photochemical oxidant formation (#PD3663)
Diseases of the central nervous system (#PE9037).

♦ PD0640 Enteric infections

Nature Due to slow progress in the improvement of water supply and excreta disposal, enteric diseases, including cholera, are continuing to spread in receptive areas and are becoming endemic.

Incidence Little is known about the real situation because of the difficulties of surveillance. Although there is a lack of reliable information, it has been estimated that in 1975 there were about 500 million cases of enteric infection (diarrhoea) in children under 5 years of age in Asia, Africa and Latin America.

Refs Hillary, Irene B and Hennessen, W (Eds) *Enteric Infections in Man and Animals* (1983).

Broader Infectious and parasitic diseases (#PD0982).
Narrower Typhus fever (#PG1685) Typhoid fever (#PD1753).
Related Cholera (#PE0560).
Aggravates Diarrhoea (#PD5971).
Aggravated by Malnutrition (#PB1498) Houseflies as pests (#PE3609)
Sewage as a pollutant (#PD1414) Inappropriate sanitation systems (#PD0876).

♦ PD0644 Excessive proliferation of medical drugs

Incidence The average physician has some 50,000 drugs from which to choose when writing out a prescription. Few are able to make effective use of more than 100.

Claim The proliferation in the use of medical drugs is a result of: over-prescription by doctors; the discovery of new drugs (a boom over the last decade); an increase in concern about and awareness of personal health on the part of the general public; and increasing belief in the efficacy of drugs for the majority of physical and psychological disorders on the part both of doctors and the public. Although the World Health Organization has drawn up a "Model list of Essential Drugs", with only some 220 drugs and vaccines, pharmaceutical companies continue to produce new variations.

In many countries the proliferation of drugs results in prescriptions that read like a shopping list, and in many unnecessary "medicines" being sold. In 1982 the Bangladesh government reckoned that one third of the money spent on drugs was wasted on totally useless tonics, vitamins and other dubious preparations.

Counter-claim The over-prescription by doctors is not a guiding force in the development of new drugs. It is a fact that new drugs coming on the market are efficient the previous ones, otherwise they would not be approved for market.

Broader Misuse of medicines (#PD8402).
Narrower Reverence for drugs in medicine (#PG2702).
Aggravates Drug abuse (#PD0094) Inadequate drug control (#PC0231)
Overprescription of drugs (#PE9087) Inadequate testing of drugs (#PD1190).
Aggravated by Inadequate information on drugs (#PF0603)
Irresponsible pharmaceutical advertising (#PE2390)
Lack of integration of traditional and Western medicine (#PF4871).
Reduced by Inadequate medical care (#PF4832).

♦ PD0645 Inadequate maintenance of infrastructure

Neglected maintenance of local infrastructure — Ageing infrastructure

Nature Highways, bridges, airports, and systems of mass transit, power and telecommunication networks, water supply and storage, waste water and solids treatment, and hazardous waste treatment not being adequately maintained, repaired and improved to meet current and future demands. When the national infrastructure is not maintained there may be loss of life, millions of dollars lost and a decrease in productivity for a region or the whole nation. The failure to maintain these services may be the result of delays in approving projects, of failure to invest in infrastructure, high cost in meeting national or local requirements, money being allocated in wasteful ways, inefficient forms of awarding infrastructure contracts or low political appeal for maintenance verses new projects. Projects may be delayed because of multiple layers of sometimes conflicting government agencies required to approve projects. Costs might increase faster than incomes for projects.

Incidence In the USA infrastructure expenditures declined from 2.3 per cent of GNP in the 1960s to a mere 0.4 per cent by the early 1980s.

Broader Inadequate maintenance (#PD8984) Inadequate infrastructure (#PC7693)
Inappropriate level of technological equipment (#PF2410).
Narrower Unmaintained bridges (#PE2471) Outdated basic utilities (#PG8871)
Inadequate road maintenance (#PD8557) Dysfunctional public utilities (#PE7647)
Inadequate maintenance equipment (#PD6520) Deterioration of nuclear power plants (#PE5260)
Ageing industrial plants and processes (#PE2866)
High cost of natural gas trade infrastructure (#PE9137)
Lack of technical infrastructure for maritime commerce in developing countries (#PE5814).
Related Inadequate port infrastructure (#PE5792)
Inadequate housing in developing countries (#PE0269)
Inadequate infrastructure and services in least developed countries (#PE0289).
Aggravates Loss of capacity with age (#PC8310)
Neglect of property maintenance (#PD8894)
Lack of economic and technical development (#PE8190)
Environmental degradation of inner city areas (#PC2616)
Inadequate road and highway transport facilities in developing countries (#PD0543).

♦ PD0651 Hoarding of primary commodities

Broader Economic crime (#PC5624)
Violations against economic regulations (#PD7438).
Narrower Food hoarding (#PJ2225) Hoarded monetary gold (#PD3045)
Hoarding in developing countries (#PD1751).
Aggravates Instability of economic and industrial production activities (#PC1217).

♦ PD0658 Weapons

Nature The purpose of weapons is to destroy, incapacitate or to impose the will of the user on his victim. Their very existence in an invitation to their use.

Refs Lifton, Robert Jay and Falk, Richard *Indefensible Weapons* (1984).

Narrower Napalm (#PG6516) Nerve gases (#PG3050) Home-made bombs (#PG4790)
Planned weapons (#PF9269) Fuel-air explosive (#PE4027)
Non-violent weapons (#PF9327)
Inhumane and indiscriminate weapons (#PD1519)
Lack of appreciation for nuclear weapons (#PE7476)
Incendiary weapons of massive destructiveness (#PD3492).
Related Environmental warfare (#PC2696).
Aggravates Radiological warfare (#PC6666) International arms trade (#PC1358)
Imbalance of conventional armed forces (#PC5230)
Government secrecy concerning nuclear weapons testing (#PF4450).

♦ PD0659 Deafness

Nature The term 'deaf' should be applied only to individuals whose hearing impairment is so severe that they are unable to benefit from any amplification. Permanent or temporary deafness may be the result of: exposure to sound (stimulation deafness); injury; disease; or developmental anomalies, either early in life or during aging. About half of all hearing-impaired children suffer from hereditary deafness. Acquired deafness has been reduced with the help of antibiotics and vaccinations.

Many who are born deaf or who became deaf early in life, experience language difficulties all through life; if no special education is available the child remains mute. Communication problems apart, there are indications that deafness has a pervasive effect on the total personality of the deaf person, including his social adjustment, perception and general motor activity. He is severely limited in understanding the world around him, in making himself understood, and in making the most of his learning experiences. He is deprived of the ability to enjoy not only music but also environmental sounds.

Myths and superstitious beliefs about causes of deafness reinforce negative attitudes and determine the way the deaf person is perceived and treated, even in his own family circle. In parts of East Africa, a congenitally deaf child is believed to be obeying an injunction by a god who has warned him not to divulge some secrets confided to him; to avoid possible risks, the child 'chooses' to be born deaf and mute. This, in turn, provides a reason for other people to avoid him.

Incidence Whereas the prevalence is practically the same in America, Asia, and Europe, there seem to be far fewer cases of deaf-mutism in Africa and Oceania although as yet. There exists no indication of an ethnically determined predisposition to deaf-mutism. There are more cases of deaf-mutism among males than among females, even in countries where women outnumber men.

Refs Austin, Gary *Bibliography on Deafness* (1976); Fellendorf, George W (Ed) *Bibliography on Deafness* (1977); Van Cleve, John V *Gallaudet Encyclopedia of Deaf People and Deafness* (1987).

Broader Hearing defects (#PD6306) Diseases of the ear (#PD2567)
Stigmatized diseases (#PD7279).
Narrower Anaconsis (#PG2749) Dysaconsis (#PG2751)
Diplacusis (#PG2752) Hypoaconsis (#PG2750)
Deaf children (#PE2083) Acoustic trauma (#PE4109)
Genetic deafness (#PG9360) Occupational deafness (#PD1361).
Related Mutism (#PE4526).
Aggravates Speech disorders (#PE2265)
Multiplicity of manual sign languages for the deaf (#PF2833).

PD0659

Aggravated by Mumps (#PE2356)
Measles (#PE1603)
Drug abuse (#PD0094)
Sore throat (#PE4651)
Otosclerosis (#PE2746)
Traffic noise (#PD3664)
Head injuries (#PG5105)
Aircraft noise (#PE5799)
Ménière's disease (#PG2760)
Pregnancy disorders (#PD2289)
Diseases and injuries of the brain (#PD0992)
Human physical genetic abnormalities (#PD1618)
Denial of right to union activity for special groups (#PE1355).
Rubella (#PE0785)
Syphilis (#PE2300)
Common cold (#PE2412)
Human ageing (#PB0477)
Typhoid fever (#PD1753)
Scarlet fever (#PG2757)
Virus diseases (#PD0594)
Nose inflammation (#PG2755)
Nutritional anaemia (#PD0321)
Adenovirus infections (#PE2355)
Health hazards of exposure to noise (#PC0268)

♦ PD0660 Crop vulnerability

Nature Crops are vulnerable to extensive damage from numerous sources: disease; pests; adverse environmental conditions; unfavourable weather; weeds. Lack of crop protection, therefore, seriously reduces potential yields.
The problems of crop protection have changed dramatically since 1945. There is now a whole arsenal of chemicals with which to combat agricultural pests and diseases, but this development has itself many drawbacks. Such sophisticated techniques are available only to a minority of farmers; in most parts of the world the standard of crop protection remains abysmally low. In addition, modern crop protection methods have been criticized for relying too heavily on chemical control. Biological controls, both natural and contrived, have been neglected. In some cases involving misuse of agricultural chemicals, crops must be protected from the very measures intended for their protection. Meanwhile previously localized pests and diseases continue to spread worldwide.
Incidence In 1970 insects, diseases and weeds caused pre–harvest losses worldwide of 34 percent ($ 70 million) of the total potential value (estimated at $ 208,000 million). This loss was borne largely by developing countries in Africa, Asia and South America. To these figures must be added the postharvest losses of about 20 percent, due mainly to insects and rodents.
The Food and Agriculture Organization of the United Nations has published world figures which indicate that 35 percent of the wheat which could theoretically be harvested is lost as a result of pests and diseases. The corresponding figure for potatoes is 40 percent, for sugarbeet 24 per cent, for apples 30 percent, for cotton 60 percent and for tobacco 62 percent. Even where modern methods of crop protection are employed, losses are still very heavy.
Background Recognition of the importance of insect pests dates back about 100 years, although from earliest times farmers must have noticed the damaging effects of pests, diseases and the environment. Early writers could do little more than describe the life cycles of insects as there were few, if any, means of dealing with them. Early twentieth century attempts to take direct action against pests were limited, in the USA to controlling scale insects on citrus crops with oil sprays, and in Europe to moth control with lead arsenate. Few other chemicals were available and the means of applying them in sprays to trees was, by modern standards, quite primitive. Fumigation of citrus trees was also practised in the USA and in the Mediterranean region by tenting trees and releasing hydrogen cyanide gas into the tents.
As knowledge about pests and diseases gradually accumulated, so did an awareness that this knowledge should be made available to the farming community. In the USA in 1887, the Hatch Act provided for Agricultural Experiment Stations in each state and for the information to be passed on to farmers and agricultural students.
Broader Food insecurity (#PB2846)
Vulnerability of farming (#PC4906)
Vulnerability of plants and crops (#PD5730).
Narrower Inadequate plant quarantine (#PE0714)
Vulnerability of crops to weather (#PE5682)
Excessive use of chemicals to control pests (#PD1207).
Related Hazards to plants (#PD5706).
Aggravates Weeds (#PD1574)
Plant diseases (#PC0555)
Pests of plants (#PC1627)
Business bankruptcy (#PD2591)
Plant disease vectors (#PD3596)
Insect pests of plants (#PD3634).
Aggravated by Soil infertility (#PD0077)
Inadequate crop rotation (#PF3698)
Pest resistance to pesticides (#PD3696)
Plant–pathogenic air pollutants (#PE0155)
Dissemination of plant diseases by man (#PD3593)
Introduction of new species of insect pests (#PF3592).

♦ PD0664 Lack of integration of transport systems between neighbouring developing countries

Nature Regional financing and design is necessary for efficient international road and rail links, commodity transport and distribution networks, and road and rail connections to harbours and airports. Many small developing countries attempting unilateral infrastructure improvement of transport facilities encounter obstacles resulting from neighbouring countries circumstances. For example, a land–locked country may build a road or rail line to its border where it is forced to stop, owing to lack of arrangements to join a transit line or road to the seaport.
Broader Weakness of infrastructure in developing countries (#PC1228).
Related Communism (#PC0369).
Aggravates Weakness in intra–regional trade of developing countries (#PD0169).

♦ PD0668 Rebate cartels

Nature A rebate is a return or a reduction by a seller to a buyer of some part of the purchase price. It may be given: when the buyer has purchased a certain quantity (quantity rebate); when he has purchased exclusively from the seller or from a group of which the seller is a member (loyalty rebate); or when he has performed a particular function in the distribution of the product, for example wholesaling (functional rebate).
The basis for the granting of a quantity rebate may be the quantity of a particular product purchased from a particular seller either at one time or over a period of time (simple quantity rebate). It may also be the aggregated quantity of several products purchased from a single seller or several sellers, or of a single product purchased from several sellers. Where quantity rebates depend upon the aggregated quantity purchased by the buyer from a group of sellers, they are usually called aggregated rebates or aggregated quantity rebates. However, this term is also used to denote rebates that depend on the quantity of several products bought from the same seller. Aggregated rebate systems based on purchases made from several sellers may be adopted by a seller with or without there being a specific agreement among the several sellers. Where there is an agreement, it is usually referred to as an aggregated rebate cartel.
Inasmuch as a rebate cartel involves an agreement that effects the pricing of a product, it can be considered as a type of price cartel. These formal rebate systems must be differentiated from the informal ad hoc preferential treatment frequently given by individual sellers to established customers, in such forms as payment terms, delivery dates, etc.
Incidence Rebate cartels of suppliers in developed market–economy countries, unless they relate exclusively to exports to foreign markets (in which case the rules concerning export cartels would apply), are in principle subject to the same rules as apply to cartels generally. In the USA rebate cartels are, therefore, prohibited per se as a form of illegal price–fixing. In a number of other countries (Canada, France, Federal Republic of Germany, Japan and Norway), and in the EEC and ECSC, cartels are prohibited in principle but may qualify for a general or special exemption. Only one country (Federal Republic of Germany) provides for a special exemption for rebate cartels; such cartels may be exempted in individual cases but only if the agreed rebates are such as to constitute a genuine compensation for services rendered and if the cartel does not seriously damage outsider suppliers or discriminate against customers. In principle, to meet these requirements, purchases from firms outside the cartel in any foreign countries must be included, but this principle has not always been enforced. In 1970 there were 33 officially registered exempted rebate cartels in the Federal Republic of Germany.
In the period 1957 to 1968, aggregated rebate cartel arrangements were adjudicated as violating French law by the French Technical Commission for Cartels and Dominant Positions. Other known cases have related to trade within the European Free Trade Association (EFTA) and the European Economic Community (EEC). Deferred rebate and dual rebate systems of loyalty rebates are applied by a number of international shipping conferences.
Broader Cartels (#PC2512).

♦ PD0673 Tax obstacles to international investment

Nature In an effort to correct temporary imbalances of payments, countries may introduce permanent fiscal measures which have the effect of inhibiting international capital movements by imposing a heavier burden of tax on both inward and outward movements of income over national borders. Although such measures may appear consistent with immediate national policies, they embody two extremely undesirable and inappropriate features: effects are not limited to new capital movements and are of far longer duration than the circumstances ordinarily require. Such measures run the risk of provoking retaliatory measures and a reversion to economic isolationism.
Broader Capitalism (#PC0564).
Narrower International double taxation (#PD0858)
Tax discrimination against non–residents of a country (#PD3048)
Tax barriers to the dissemination of technical knowledge (#PD3050)
Tax discrimination against investment in a foreign country (#PD3047).
Related Inequitable tax treaties between developed and developing countries (#PD1477).

♦ PD0678 Distortion of international trade by selective domestic subsidies

Overdependence on subsidies — Discriminatory subsidies
Nature A wide variety of techniques is used by governments to subsidize production costs. They may be grouped as subsidies to specific industries, as regional development programmes, or as subsidies for particular economic activities (which seek to raise growth rates in broad groups of industries, to bring about significant structural adjustments, and to facilitate the adjustment of enterprises to economic shocks). Such government aids to domestic industries result in some degree of distortion of trade patterns. Some of these forms of aid may represent efforts either to offset tariff and other trade concessions or to improve balance of payments positions.
Counter–claim Domestic subsidies and aids are prompted by greater sense of public responsibility toward improving economic conditions for those employed in depressed industries or regions, and raising the rate of growth both generally and in selected industries.
Broader Discriminatory business practices (#PD8913)
Distortion of international trade as a result of government participation (#PD2029).
Narrower Distortion of national economies from food subsidies (#PE7413)
Over–subsidized agriculture in industrialized countries (#PD9802)
Distortion of international trade by export subsidies and countervailing duties (#PE1961).
Related Unknown availability of subsidies (#PG9905)
Distortion of international trade by discriminatory customs and administrative entry procedures (#PE2603).
Aggravates Disincentives against farming (#PD7536)
Decline in government health expenditure (#PF4586).
Aggravated by Denial of right to business growth (#PE2700)
Underdeveloped approaches to local food production (#PF6493).
Reduced by Limited availability of financial credit (#PF2489).

♦ PD0683 Denial of the right of trade union association

Nature In many countries of the world, workers are denied the right to establish and join trade union organizations. Trade unions are liable to be dissolved or suspended by administrative action, and they are not free to establish and join federations or to affiliate to international trade union organizations. Workers are not adequately protected against acts of anti–union discrimination regarding their employment, nor are trade unions protected against acts of interference by employers in their establishment, functioning or administration. Collective bargaining and the right to strike are often severely limited.
Incidence In recent years in practically all Latin American countries, a large number of trade unionists have lost their lives under unexplained and unexamined circumstances. In Morocco, many workers were killed in 1981 when a general strike was violently suppressed by the authorities. In Iran in 1983 new measures of repression were used against several strikes involving thousands of workers who were beaten, arrested and in many cases dismissed. In Turkey the right to strike has been suspended since the military coup in 1980. The right to strike is also seriously restricted or completely denied in countries such as Algeria, Chad, Ethiopia, Gabon, Liberia, Mauritania, Nigeria, Sudan and Tanzania.
In several African countries the trade union movement continues to be dominated by the political party in power. In communist regimes the official trade unions are strictly subordinated to the Communist Party and act as instruments for the implementation of economic plans and for the maintenance of labour discipline. In Cuba in 1982, fifty workers of the State construction undertaking were arrested for attempting to establish an independent trade union to defend their interests. In Poland after a period of severe repression of unofficial trade union activities, many leaders and members are still in gaol either awaiting their trial or after being sentenced to heavy terms of imprisonment.
In the West, too, there are restrictions on trade union activities. In Belgium, collective bargaining on wage increases was blocked in 1982 for a period of several years. The Canadian government adopted legislation in 1982 by which it suspended collective bargaining rights and the right to strike in the federal public service for a period of 24 months. This example was followed by several provincial governments.
Claim Freedom of assembly for trade union purposes constitutes one of the fundamental human rights. Basic trade union rights are essential to the existence of free and democratic trade unions capable of defending and furthering the workers' interests
Broader Violation of trade union rights (#PD4695)
Denial of the right of association (#PD3224).
Narrower Banned trade unions (#PD3535)
Denial of the right to picket (#PE8712)
Lack of trade union recognition (#PG4050)
Violation of the right to strike (#PE5070)
Denial of right to organize trade unions (#PE5398)
Violation of right of workers to join trade unions (#PE5192)
Violation of the right of workers' organizations to establish confederations (#PE4071).
Related Infringement on the functioning of legitimate organizations (#PE5222).
Aggravates Segregation in employment (#PD3443).
Aggravated by Unemployment (#PB0750)
Minority control (#PF2375)
Denial of right of assembly (#PC2383).

♦ PD0689 Air traffic congestion

Overloaded airports — Air traffic paralysis — Congested airspace — Airport congestion
Nature Air traffic control continuously lags behind the increasing volumes of military and civilian

planes using air space. All major international airports and heavily used national airports are exposed to constant dangers of mid-air and ground collisions. In addition, major inter-city and inter-continental air-lanes are heavily congested. Aggravating factors are military plane manoeuvres which interfere with civil aviation, and airline economics which may result in less than optimal maintenance and equipment. The increased use of private airlines also poses a threat as inexperienced pilots inadvertently enter commercial air lanes. Justification for an expensive satellite-based plane tracking system over the Atlantic included the possibility of still greater air traffic density.
Air traffic congestion causes atmospheric pollution and noise pollution, especially near airports.
Incidence Every year hundreds of near-miss collisions are reported, many at the more than 25 major world airports which each serves between 5 and 30 million passengers annually. Congestion-caused waiting times often extend to hours at major airports, and fuel is wasted by airplanes maintaining circular patterns, stacked above the runways.
In Europe the poorly organized system costs governments $500 million annually, airlines $980 million a year for delays an economy $400 million indirectly a year. On top of that, the extra mileage, inefficient routings and altitudes and other problems resulting from complex airspace cost $1.8 billion annually. Total annual cost of inefficiencies are $4.2 billion and it is estimated to reach $31.5 billion by the year 2000.
Counter-claim Aircraft are funnelled into one-way corridors with safe distances between corridors and between aircraft flying along the same corridor. The airspace is not overcrowded, but the airspace management system is overloaded, because the air transport industry and the air traffic control community are not willing to embrace new technologies, such as satellite navigation systems, to reduce the separation between aircraft and between corridors.
Broader Traffic congestion (#PD0078).
Aggravates Air accidents (#PD1582) Air traffic delays (#PF9464)
Reservation overbooking (#PE8667).
Reduced by Underparticipation of developing countries in the airline industry (#PE4145).

♦ PD0693 Prostitution

Nature Prostitution is the promiscuous bartering of sex favours for money or gifts. Most prostitutes are females serving male customers, the next largest group are male prostitutes serving homosexual clients. There are some male prostitutes who serve female customers and a few female prostitutes who have lesbian clients. In some countries prostitution is a crime; in others it is legalized in so far as all prostitutes must be registered and submit to regular medical examinations to check for venereal disease. Where it is illegal, prostitutes are usually subject to intimidation and extortion from pimps or organized crime. Prostitutes often blackmail their clients or rob them. Social stigma may prevent prostitutes from becoming rehabilitated into normal life.
Prostitution is closely related to urban life and mobile populations. In primitive tribes it is virtually unknown, although promiscuity before marriage, and polygamy or polyandry, may be accepted. Prostitution is often very marked in poverty stricken areas where there is little employment for either men or women, or among indigenous tribes which have suffered cultural invasion and have not been able to adapt economically to a new way of life. Prostitutes may be very young and coerced into prostitution. They may also come from broken homes or socially or economically deprived backgrounds.
Incidence Prostitution continues to spread and, along with it, the exploitation of prostitution. In one country studied, about 10 per cent of women aged between 15 and 30 live from prostitution; in another, the proportion of prostitutes in the female population of the capital is over 13 percent. Prostitution is beginning at an increasingly younger age.
Background In Babylon, Cyprus and among the Phoenicians and in parts of Western Asia, women prostituted themselves as a religious duty at the sanctuary of a goddess. Prostitution was a means of earning a dowry in some ancient cultures. Prostitutes existed in the ancient Greek and Roman civilizations. In the Middle Ages they were tolerated as the lesser of evils, although efforts were made to check the practice; the Church made attempts to reform and rehabilitate prostitutes. Restrictions on prostitution increased after the Renaissance and became more formalized with the creation of police forces in the 19th century. Until after World War II, the sale of extramarital sex was highly institutionalized in Japan; government officials legally recognized and attempted to supervise brothels from the 12th century on. Prostitution was and still is unknown in many so-called 'primitive' societies, but is found today, to varying degrees, in most cultures and in all parts of the world, especially where the population is very dense and where money changes hands frequently.
One can approach prostitution from the angle of ethnology, sociology or cultural history. From the point of view of political economy, one can see the world of prostitution as a closed economic system; or, from the point of view of criminology, as a branch of the criminal world because of the procuring involved. Prostitution can also be judged by the standards of public health, religion or morality. From the human rights approach, prostitution can be considered as a form of slavery: like slavery in the usual sense, prostitution has an economic aspect. While being a cultural phenomenon rooted in the masculine and feminine images given currency by society, it is a very lucrative market. The merchandise involved is men's pleasure, or their image of pleasure, and is supplied by physical intimacy with women or children. Thus, the alienation of the person in prostitution is more far-reaching than in slavery in its usual sense, where what is alienated is working strength, not intimacy.
Refs Bullough, Vern L and Elcano, Barrett W *A Bibliography of Prostitution* (1977); Joardar, Biswanath *Prostitution in Historical and Modern Perspectives* (1984); Mukherji, S K *Prostitution in India* (1986); Sereny, Gitta and Wilson, Victoria *The Invisible Children* (1985); Weininger, Otto P *Motherhood and Prostitution* (1983).
Broader Victimless crime (#PC5005) Sexual immorality (#PF2687).
Narrower Brothel slavery (#PD3888) Male prostitution (#PD3381)
Child prostitution (#PE7582) Female prostitution (#PD3380)
Juvenile prostitution (#PD6213) Transvestite prostitution (#PD4525)
Abuse of tourism for sexual purposes (#PE4437)
Exploitation of the prostitution of others (#PE5303).
Related Adultery (#PF2314) Obscenity (#PF2634) Masochism (#PF3264)
Impairment (#PA6088) Unchastity (#PA5612) Pornography (#PD0132)
Concubinage (#PF2554) Forced marriage (#PD1915) Inappropriateness (#PA6852)
Male homosexuality (#PF1390) Therapeutic sex surrogacy (#PF4221)
Vice and sex traffic offences (#PD8910) Sexual exploitation of children (#PD3267)
Discrimination against women in employment (#PD0086).
Aggravates Crime (#PB0001) Sexual unfulfilment (#PF3260) Trafficking in women (#PC3298)
Illegitimate children (#PC1874) Sexually transmitted diseases (#PD0061).
Aggravated by Famine (#PB0315) Apathy (#PA2360)
Slave trade (#PC0130) Intimidation (#PB1992)
Homosexuality (#PF3242) Miscegenation (#PC1523)
Organized crime (#PC2343) Family breakdown (#PC2102)
Sexual deviation (#PD2198) Marriage markets (#PD7282)
White slave trade (#PD3303) Marital instability (#PD2103)
Maladjusted children (#PD0586) Socio-economic poverty (#PB0388)
Unethical entertainment (#PF0374) Foreign military presence (#PD3496)
Discrimination against women (#PC0308) Double standards in morality (#PF5225)
Double standards of sexual morality (#PF3259) Criminal investment in youth market (#PD5750)
Unemployment of premature school leavers in developing countries (#PE0015).
Reduced by Promiscuity (#PC0745) Permissiveness (#PF1252).

♦ PD0694 Strikes

Collective stoppages of work — Industrial action — Strikers
Nature Disputes involving the labour force that result in collective work stoppages are a grievous source of economic loss that can occasion irreparable harm. One form of collective work stoppage is strikes. They usually take the form of employer-employee or management-labour conflict. In this context, there are different methods: a sympathy or solidarity strike to support another branch of workers who are striking; a slowdown or work-to-rule (which results in slowdown) strike, which allows work to continue but at a reduced pace; a wildcat strike, called by local union branches but not endorsed officially by the union as a whole (and generally disapproved of by union headquarters); a work stoppage, when workers cease work but do not leave the place of work; and a rotating or checkerboard strike, affecting first one area of production then another so that the work schedule is disrupted.
There are other interests that can diverge enough to cause a strike. Organized labour can be its own adversary in territorial battles between rival unions, and strike actions may be initiated to force or to prevent elections by the work force as to which union they wish to represent them. Local or national government may intervene in industrial affairs by embargoes, general labour legislation, industry regulation, expropriation, subsidy, nationalization, and emergency powers. In socialist countries labour disputes can only be with the government.
Strikes may be illegal, as breaches of contract with employers, or in violation of public laws or governmental restrictions. Similarly employer action (whether private or public ownership) can illegally employ 'strikebreakers' and 'scabs' to physically intimidate or replace striking workers by non-union workers.
Strikes inevitably involve government, possibly commencing with governmental arbitration or proceeding to the courts of justice. In a general strike, the striking workers, and the political parties and individuals supporting them, may come into violent conflict with the police or military forces of a country. There have been a substantial number of deaths caused by strikes, particularly general strikes, and labour issues are perhaps the single most polarizing factor in the developed countries.
Background Labour is organized, generally in trade unions, at the site and regional level, and also at the national and international level. Moreover, the trade unions of the various crafts and industries are united nationally and internationally. For example the American AFL-CIO, which consists of hundreds of trade unions, is a member of the International Confederation of Free Trade Unions, whose numbers world-wide number some 90 million workers. This means that strikes correspond in their extent to these levels of organization. There are internationally coordinated strikes, national general strikes involving one or more industries, strikes involving one employer at all its locations, or one employer at one location, and strikes involving more than one union at every level.
Strikes became fairly common in the 16th to 18th centuries, though the right to strike was not recognized until the 19th century, and then only in industrialized countries. The first strike is believed to have been in Florence (Italy) in 1345.
Incidence At the beginning of the 20th century, one third of all strikes in the United States were by unorganized employees. The proportion decreased and now, since 1933, strikes are uniformly confined to unionized workers.
In 1986 the International Labour Organization says that 65.3 million working days were lost through strikes in 52 countries. There were 14,200 strikes in 1986 down from 15,500 during the previous year. The size of the strikes increased from an average from 810 workers in 1985 to 827 workers and lasted an average of 5.5 days an increase of one half day. Increasing automation has, however, reduced the effectiveness of strike actions. A 1959 strike by workers in the American oil industry failed because a relatively high rate of production was maintained by the automated system alone. Similarly a 1983 strike by communications workers failed because of the computerized telephone exchange systems.
Economic decline has also been a limiting factor. In the US auto industry, a strike failed because manufacturers were content to use up a large inventory before bargaining to resume production. Foreign competition has often encouraged employees and employers to agree on no-strike provisions to ensure stability of domestic production, such as in the steel industry in 1982. Increasingly, too, public opinion does not oppose the hiring of unemployed persons to replace strikers, particularly when the latter have used violent techniques. An example of this is the air traffic controllers strike in the US in 1982, when public opinion was against the strikers.
Refs Chamberlain, Neil W and Schilling, Jane M *Impact of Strikes* (1973); Hyman, Richard *Strikes* (1989); ILO *Conciliation in Industrial Disputes* (1985); ILO *Industrial Disputes* (1987); ILO *Prevention and Settlement of Labour Disputes in ASEAN* (1985); Jackson, Michael P *Strikes* (1987); Walsh, Kenneth *Strikes in Europe and the United States* (1984).
Broader Economic unrest (#PD4012) Avoidance of work (#PC5528)
Non-violent weapons (#PF9327).
Narrower Undeclared strikes (#PD5384)
Undeclared strikes in socialist countries (#PD1882)
Transnational strike action by trade unions (#PE1541).
Aggravates Delay (#PA1999) Social injustice (#PC0797) Factory closures (#PE3537)
Economic inflation (#PC0254) Inadequacy of postal services (#PF2717)
Denial of the right of association (#PD3224)
Instability in export trade of developing countries producing primary commodities (#PD2968).
Aggravated by Militancy (#PC1090) Trade unionism (#PF8493)
Political crime (#PC0350) Passive resistance (#PF2788)
Seasonal unemployment (#PC1108) Employee disobedience (#PD5244)
Underpayment for work (#PD8916) Poor living conditions (#PD9156)
Industrial intimidation (#PC2939) Increasing job monotony (#PD2656)
Lack of job satisfaction (#PF0171) Decline of craftsmanship (#PG2798)
Cyclic business recessions (#PF1277) Inadequate manpower planning (#PJ2036)
Fluctuations in real value of money (#PD9356)
Growing size and impersonality of firms (#PE8706)
Victimization of workers' representatives (#PD1846)
Denial of the right to work in capitalist systems (#PC3119)
Isolation of trade union members from their representatives (#PJ7969)
Collusion of trade union leaders with employers and government (#PE8367).
Reduces Economic apathy (#PC3413).
Reduced by Violation of the right to strike (#PE5070).

♦ PD0700 Flag discrimination in shipping

Nature Flag discrimination consists of a wide variety of acts and pressures exerted by governments designed to direct cargoes to ships of the national flag, regardless of the commercial considerations which normally govern the routing of cargoes. Restrictions are either imposed on all foreign vessels or discriminate against ships which are registered in certain countries. Flag discrimination thus places impediments in the path of the free flow of international trade, disturbing trade between all countries and all sectors of the economy. Developing countries see the growing use of 'flags of convenience' by shipowners in developed countries as a substantial impediment to their efforts to expand their own merchant fleets.
Broader Discrimination (#PA0833)

(#PD5885).
Aggravates Inequities in ship owner registration (#PE5875).
Aggravated by Trade protectionism (#PC4275).

♦ PD0702 Concentration camps

Mass detention of political prisoners — Gulags — Labour prisons

Nature The mass internment of civilians by military or police forces for indefinite periods of time and under inhumane conditions characterizes concentration camps and gulags. Many prisoners serve out their sentences in reform through labour camps, whose function is not fully explained in the legal code. Prisoners may even be forbidden to return home after they are released and must remain in the area, thus amounting to a form of internal exile.

Background The Spaniards detained Cuban civilians in the rebellion of 1895. The British employed civilian internment centres in the Boer War in 1901. The Russian government, after the 1917 revolution, sent tens of thousands of persons to the concentration camps of Tambor province, the Solvetsky Islands, and the Kolyma – the country of Gulag. In 1918, under Lenin, the waves of civilian detentions began, and according to Solzhenitsyn, millions were sent to the Gulag country under Stalin in 1929–1930, again in the infamous purges of 1934–38, and yet again in the 1950s. The Gulag archipelago held six million political prisoners at one time in the 1930s and 1940s; the total who died in the Gulag from 1918 to the present may be two to four times this number. In just 12 years, between 1933 and 1945. Nazi Germany's camps systematically killed an estimated 20 to 26 million, over half of all those who were detained. Total concentration camp or gulag deaths of political prisoners exceeds 50 million persons in this century.

Incidence Political prisoners are detained en masse in South Africa, Cuba and the USSR. Refugee camp conditions, as in Lebanon, assume the features of mass detention. In China in 1989 it was estimated that some of the labour camps held up to 40,000 inmates together with their families, juvenile offenders and sme freed workers who were confined to the region.

Claim The effects of Nazi concentration camps of 1933 to 1945 persist in the sufferings and anxieties of ex-prisoners. Effects of the camp stress are still present not only in the victims themselves, but are also evident in their offspring. The stigma of the concentration camp is probably the most important mark of war borne by present society.

Refs Mendelsohn, J *The Final Solution in the Extermination Camps and the Aftermath* (1982); Mendelsohn, John *Medical Experiments on Jewish Inmates of Concentration Camps* (1982); Weglyn, Michi *Years of Infamy* (1976).

Broader Wrongful detention (#PD6062) Political repression (#PC1919)
Crimes against humanity (#PC1073).
Narrower Military concentration camps (#PG2816)
Political concentration camps (#PG2815).
Related Internment without trial (#PD1576) State sanctioned torture (#PD0181)
Ill treatment of prisoners of war (#PD2617).
Aggravates Death and disability from inhumane confinement (#PE5648)
Inhumane medical experimentation during war-time (#PE4781).
Aggravated by Antisemitism (#PE2131).

♦ PD0708 Unethical practice of earth sciences

Irresponsible geologists — Negligence by geologists — Malpractice in geology — Corruption of geologists — Underreporting of earthquake risks — Underreporting of groundwater pollution hazards

Claim Geologists, under pressure from their employers, have adopted practices which lead to the underreporting of earthquake hazards in relation to the siting of nuclear tests and installations, to the underreporting of groundwater pollution hazards in relation to the siting of toxic waste dumps, to the underreporting of geological hazards in relation to the construction of large dams, and to the failure to investigate adequately the nature of such hazards. Geologists are intimately involved in those corrupt practices associated with oil and mineral exploration, especially with the establishment of fraudulent claims

Narrower Unethical practice of soil sciences (#PD1110).
Aggravates Earthquakes (#PD0201) Pollution of groundwater (#PD2503)
Unpredictability of earthquakes (#PF4928) Fraudulent mineral exploitation claims (#PE6975)
Inadequate earthquake resistant construction (#PE6257)
Hazardous locations for nuclear power plants (#PD2718).

♦ PD0710 Radioactive contamination of plants

Nature Plant material may become contaminated from the atmosphere by fission products that do or do not become diluted with soil substances. Direct contamination may occur on leaves, fruits and seeds. Depending on the nature of the plant, material not retained in this way may be partly absorbed by the basal parts and surface roots of the plant. Material also may enter the soil and then the plant via the roots. The soil route of entry, in addition to leading to dilution, allows material deposited before the plant developed to enter the plant. Soil entry is unimportant, however, with short-lived nuclides. If the soil is undisturbed, fission products will move downward only slowly. Surface rooted crops are therefore likely to absorb the largest amounts from soil deposits.

The extent to which different fission products are absorbed by plants depends on many factors. Radionuclides contaminating plants may reach man directly by way of foods of plant origin or indirectly through animal products. Sources of radioactive contamination are: fallout from nuclear explosion, movement of radioactive nuclides from the soil, radioactive soil dust borne on the wind, and irrigation contaminated by radioactive water.

Incidence The increase in the radioactivity of the soil due to artificial radioactive nuclides is small compared to the radioactivity from naturally occurring radioactive nuclides. The increase in radioactivity of crops is relatively greater than that in the soil and the concentration of radioactive isotopes is especially high in water plants.

Refs Aleksakhin, R M *Radioactive Contamination of Soil and Plants* (1965).

Broader Radioactive contamination (#PC0229)
Accumulation of pollutants in terrestrial plants (#PD0381)
Endangered animal and plant life due to radioactive contamination (#PD5157).
Narrower Radioactive contamination of soil (#PE3383).
Aggravates Suffering of plants (#PC7825) Deformation of plant life (#PD9480)
Radioactive contamination of animals and animal products (#PD1119).

♦ PD0713 Smoking

Cigarette smoking — Tobacco habit — Smoking epidemic — Tobacco smoking — Nicotine abuse — Chewing tobacco

Nature The ill effects of smoking become manifest only after a period of years and may thus not appear to be obviously linked with the habit. Smoking is a major etiological factor in a number of disabling and fatal diseases, notably ischaemic heart disease, lung cancer, chronic bronchitis and emphysema. The mortality of cigarette smokers from many other diseases is greater than that of non-smokers, particularly in the case of peptic ulcer, and of cancers of the larynx, oral cavity, oesophagus and bladder.

Since the economic benefits from tobacco-growing and the manufacture, marketing and taxation of tobacco products are enormous, many governments hesitate to take firm action against a habit whose dangers are now generally understood. The adverse health consequences of smoking and the higher work absenteeism of smokers bring about huge economic losses to countries.

Advertizing is an integral part of corporate expansion of the tobacco industry. The spending of large sums of money on advertizing serves to spread and maintain the idea of smoking, associating it with success, pleasure, relaxation, freedom, or, in its crudest form, with the attractive sexual attributes of feminity or virility. Promotion of tobacco conveys the message that smoking is socially acceptable. Through advertising the industry buys the silence of those media that should be actively pointing out the negative consequences of smoking'. Most importantly, advertising conveys the message, often in a most subtle way, that tobacco is a legal product which is quite properly on the market and should be treated like all other legal products.

Incidence In 1990 it was estimated that tobacco-caused diseases would kill nearly 10 per cent of the current world population, namely 500 million, unless present smoking patterns change. About 250 million of the predicted victims would die in middle age (from 35 to 69), losing approximately 20 years of life. Of the current population under 20 years of age, about 800 million are expected to become smokers.

The average decrease of lifespan for a person smoking 20 cigarettes a day is about five years. Among British physicians 35 years of age and older, studied over a long period of time, more than twice as many smokers as non-smokers died before reaching age 65. A study of adults in a specific community in the USA, published in 1990, indicated that men who smoke cigarettes throughout their lives die nearly 18 years earlier than men who never start. Thus a 30-year old man who smokes will reduce his life expectancy, on average, by about one fourth — dying at the age of 64 rather than 82.

On the basis of an exhaustive review of worldwide scientific evidence, a US report reached the conclusion that overall mortality for all cigarette smokers is about 70 percent higher than for non-smokers. Life expectancy for a 30-year old, two pack-a-day smoker is eight years less than for a non-smoker of the same age. Mortality rates are higher among those who have smoked for longer periods, those who started smoking at a younger age, and those who smoke cigarettes with a higher tar and nicotine content. Ex-smokers, however, experience a decline in mortality rates.

There is recent evidence of adverse effects of smoking by pregnant women on the foetus and the new born baby, and of increased risk of ill-health for very young children in smoking families. The synergistic effect of cigarette smoking and exposure to toxic industrial agents is well documented. Literature has also appeared on the effects of involuntary or 'passive' smoking (inhalation by non-smokers of sidestream smoke from the lighted tip of a cigarette between puffs as well as of exhaled mainstream smoke). In France it has been calculated that as many as 60,000 deaths a year - including 18,500 cancer deaths, or 15 percent of all cancer deaths - are related to smoking. Since France has approximately 500,000 deaths per year, this means that 11 to 12 percent of all deaths are linked to smoking. For 1984, the USA Surgeon General officially attributed 350,000 deaths to smoking.

A particularly pressing cause for concern is the increase in tobacco production and consumption in the developing countries. According to FAO, tobacco production in the developing countries rose by 28 percent between 1969–71 and 1977, while in the developed countries it rose by only 15 percent. World consumption of tobacco rose by about 3–4 percent annually during the decade 1965–1975; in 1975 and 1976 consumption slowed down in the developed countries, but continued to rise in the developing countries by about 5 percent per annum. Research conducted in India, Jamaica, Pakistan, Papua New Guinea, and Singapore has linked smoking to cancer of the lung, oral cavity, oesophagus, and to bronchitis and peptic ulcers, and points to it as a risk factor in cardiovascular diseases. If forceful government action is not taken promptly in developing countries, the smoking epidemic is likely to spread there within the next decade, affecting their populations with the numerous smoking related diseases before communicable diseases and malnutrition have been brought under adequate control.

The combined sales of the 3 largest tobacco companies in the USA came to more than $34,000 million in 1984; the industry earns about 17 per cent a year on its equity. The importance of advertizing and sales promotion to the tobacco companies is reflected in the amount of money they spend for this purpose. Global advertizing costs of tobacco transnational conglomerates (TTCs) amounted to some $1,800 million in 1978. In the United States, the cigarette industry spends more than $422,000,000 annually on advertizing in newspapers, magazines, and on billboards; in 1980, it was reported to spend more in one day than the principal government agency on smoking and health spent in a year. Recent information from the United Kingdom shows a substantial increase in regular tobacco advertizing expenditures. In Malaysia, the tobacco companies spent approximately $5,000,000 in 1977 to promote smoking.

Counter-claim The campaign against cigarette smoking fails to take account of the lack of evidence against pipe smoking.

Refs Balfour, D J *Nicotine and the Tobacco Smoking Habit* (1984); Chandler, William U *Banishing Tobacco* (1986); Iversen, L L *The Biology of Nicotine Dependence* (1990); Ramström, Lars M (Ed) *Smoking Epidemic* (1980); Reginald, Jorge S *Smoking* (1987); WHO *Tobacco Smoking* (1986).

Broader Substance abuse (#PC5536) Hazards to human health (#PB4885)
Inappropriate personal habits (#PD5494).
Narrower Nicotine withdrawal (#PE9253) Smoking in work places (#PG5425)
Smoking by adolescents (#PE6219) Smoking in developing countries (#PE4996)
Health hazards of passive smoking (#PE5146)
Health hazards of smoking for women (#PE4995).
Related Caffeine abuse (#PE0618) Smoke as a pollutant (#PD2267)
Abuse of plant drugs (#PD0022)
Health hazards of environmental pollution (#PC0936).
Aggravates Fires (#PD8054) Silicosis (#PE1314) Byssinosis (#PE2319)
Laryngitis (#PE2653) Lung cancer (#PE7085) Air pollution (#PC0119)
Heart diseases (#PD0448) Pneumoconiosis (#PD2034) Chronic bronchitis (#PE2248)
Malignant neoplasms (#PC0092) Urinary bladder disorders (#PE2307)
Health hazards of asbestos (#PE3001) Lung disorders and diseases (#PD0637)
Malignant neoplasms of female genital organs (#PE1905)
Instability of trade in tobacco and tobacco manufactures (#PE0572).
Aggravated by Immoral public policy (#PF4753) Stress in human beings (#PC1648)
Irresponsible tobacco and cigarette advertising (#PE9093)
Dependence of government revenues on exploitation of environmentally inappropriate products (#PD1018).
Reduces Asthma (#PD2408).

♦ PD0715 Vitamin deficiencies in diet

Avitaminoses

Nature Vitamin deficiency is a morbid condition caused by either a complete absence, an inadequate intake and/or production, or an increased destruction of vitamins in the organism. Vital functions require a certain amount of vitamins, some of which are synthesized by intestinal bacteria while others must be ingested. An inadequate diet (including improperly stored or improperly prepared foods) will gradually lead to vitamin deficiency, as will certain diseases which inhibit the natural synthesis of vitamins in the body.

Incidence As an indication, the market for vitamin supplements (excluding foods) in the UK in 1986 was £63 million.

Refs Kemm, J R *Vitamin Deficiency in the Elderly* (1985); Lichtenstein, Michael J *Vitamin Deficiencies* (1988).

Broader Nutritional diseases (#PD0287) Diseases of metabolism (#PC2270)
Nutritional deficiencies (#PC0382).
Narrower Scurvy (#PE2380) Rickets (#PG2295) Beriberi (#PE2185)

Pellagra (#PE2287) Xerophthalmia (#PE2538)
Vitamin B deficiency (#PG4635).
Aggravates Malnutrition among indigenous peoples (#PC3319).
Aggravated by Diseases of the pancreas (#PE1132).

♦ PD0722 Instability of water supply

Unpredictable water supply — Sporadic water supply — Erratic water system — Variable water supply — Insufficient water pressure — Inadequate water pressure

Nature Inland waters occur in strictly limited volume (about 0.01 percent of global flow in rivers and their associated lakes and swamps), they undergo both seasonal and yearly fluctuations, and they are subject to man-made changes in their physical and biological qualities. Inland water bodies also suffer over-enrichment (eutrophication) and pollution caused by discharges from industries, drainage from agricultural and domestic chemicals and wastes, and acidification from acid rain.

Incidence During the 1970s there was a world-wide increase in the absolute numbers of people without access to safe water or sanitary facilities. Increased eutrophication and pollution of inland water bodies have been only partially offset by the biological revival of some rivers and lakes through various remedial and control measures. The continuing growth of the world population and acceleration in water use had by 1970 already begun to strain the water resources of some areas, and the problems were aggravated by pollution and the continued prevalence of water-borne disease. Withdrawals of water for use in agriculture, industry and in the home continued to increase during the decade, although in many developed countries more slowly than in the preceding decade. Statistics remained uneven, but domestic water supplies barely kept pace with population growth in many developing regions, and waste water disposal services fell behind in many areas. Ground-water quality also deteriorated in many areas, and statistics for this section of the freshwater resource remained inadequate.

Whereas in developing countries the proportion of the urban population with access to safe water supply rose from 67 percent in 1970 to 77 percent in 1975 and then declined slightly to 75 percent in 1980, although the proportion of rural people served by safe water supply increased from 14 percent in 1970 it had still reached only 29 percent in 1980. The waste water treatment situation was even less heartening. While a high proportion of the developed urban populations had adequate services, the proportion of developing country urban population served by sewers and other sanitary facilities declined during the decade, from 71 percent to 53 percent. In rural areas the numbers served were 11 percent in 1970 and little better (13 percent) in 1980.

The total water use in 1980 was in the order of 2,600 to 3,000 km3/year; this is projected to have reached 3,750 km3 in 1985 – about 8 to 10 percent of the average run-off in all continental river basins. The three major uses of water are: irrigation (73 percent), industry (21 percent) and domestic and recreational uses (6 percent). The 1970s saw further extension of irrigation (and improved drainage) to newly reclaimed lands, especially in arid regions. Industrial uses increased during the decade, but savings were also made through increased efficiency. In Japan, for example, total industrial withdrawals increased from about 50 million cubic metres in 1974, but by the mid-1970s two-thirds of this was recycled water compared with one-third in 1965. Total water demand for the year 2000 is predicted to be between two and four times that for 1970.

Broader Long-term shortage of water (#PC1173).
Narrower Inadequate water supply in the rural communities of developing countries (#PD1204).
Related Long-term shortage of natural resources (#PC4824)
Undeveloped channels for public and private resources (#PF3526).
Aggravates Inadequate water system infrastructure (#PD8517)
Limited accountability of public services (#PF6574).
Aggravated by Inadequate maintenance equipment (#PD6520)
Underdeveloped provision of basic services in developing countries (#PF6473).

♦ PD0723 Poverty and disability

Impoverished and disabled persons

Nature The relationship between disability and poverty has been clearly established. While the risk of impairment is much greater for the poverty- stricken, the corollary is also true: the birth of an impaired child, or the occurrence of disability in the family, often places heavy demands on the limited resources of the family and strains on its morale, thus thrusting it deeper into poverty. The combined effect of these factors results in higher proportions of disabled persons among the poorest strata of society. For this reason, the number of affected families living at the poverty level steadily increases in absolute terms. The negative impact of these trends seriously hinders the development process. Existing knowledge and skills, if appropriately applied, could prevent the onset of many impairments and disabilities, could assist affected people in overcoming or minimizing their disabilities, and could enable nations to remove barriers which exclude disabled persons from everyday life.

Refs United Nations *Disability* (1986).
Broader Human disability (#PC0699) Socio-economic poverty (#PB0388).
Aggravates Personal care disabilities (#PE6770).

♦ PD0724 Disabled persons in developing countries

Nature The problems of disability in developing countries need to be specially highlighted. As many as 80 percent of all disabled persons live in isolated rural areas in the developing countries. In some of these countries, the disabled are estimated to represent as many as 20% of the population and thus, if families and relatives are included, 50% of the population could be adversely affected by disability. The problem is made more complex by the fact that, for the most part, disabled persons are also usually extremely poor people. They often live in areas where medical and other related services are scarce or even totally absent, and where disabilities are not and cannot be detected in time. When they do receive medical attention, if they receive it at all, the impairment may have become irreversible. In many countries, resources are not sufficient to detect and prevent disability and to meet the need for the rehabilitation and supportive services of the disabled population. Trained personnel, research into newer and more effective strategies and approaches to rehabilitation, and the manufacturing and provision of aids and equipment for disabled persons are quite inadequate.

In such countries, the disability problem is further compounded by the population explosion, which inexorably pushes up the number of disabled persons both in proportional and absolute terms. There is thus an urgent need, as the first priority, to help such countries to develop demographic policies to prevent an increase in the disabled population and to rehabilitate and provide services to the already disabled.

Broader Human disability (#PC0699).

♦ PD0736 Police intimidation

Police harassment — Police reprisals — Police retaliation — Threat of police reprisals

Broader Abuse of police power (#PC1142) Political intimidation (#PC2938)
Unethical practices by police forces (#PD9193).
Related Political police (#PD3542) Police brutality (#PD3543)
Harassment of public officials (#PD4915).
Aggravates Fear of police (#PF8378) Fear of reprisals (#PF9078)
Mistrust of police (#PF8559) Juvenile delinquency (#PC0212)
Subversive activities (#PD0557) Denial of human rights (#PB3121)
Public assaults on police (#PE7659) Forced political confessions (#PE3016)
Discrimination against prisoners' families (#PE5043).
Aggravated by Authoritarianism (#PB1638) Unauthorized police search (#PD3544)
Discrimination against minorities (#PC0582).

♦ PD0739 Forest fires

Wildfires — Wildland fires — Bush fires

Nature Throughout most of the world the worst enemy of the forest is considered to be fire. It has special uses, under expert control as a management tool (clearing fuel by control burning, removing debris from clearing operations, back-burning for fire suppression), but otherwise fire destroys wildlife, damages timber resources, weakens trees, paves the way for attack by insects and diseases, and increases soil deterioration and surface run-off of water. Carbon emissions from the burnings of Brazilian rain forests pollute air and contribute significantly to global warming. Forest fires also threaten local communities, destroying lives and property. More indirectly, they damage watersheds and recreational resources.

Incidence Climate and the nature of the forest are important determinants of the incidence of forest fires. Fires are common wherever there is a long dry season (such as in parts of Latin America and Africa) and especially where thunderstorms are accompanied by little rain (such as in parts of Asia and North America). Wildfires are virtually unknown in the humid tropical evergreen forests and are uncommon where forests are divided into small, isolated units as in central Europe. Wildfire is most common in many other types of drier tropical forests and may, under exceptional circumstances, destroy tropical evergreen forests (as happened in Kalimantan, Ivory Coast, Samoa). Coniferous forests burn much more readily than broad-leaved forests.

The US Forest Service has developed a fire danger index which combines all the major determinants (including moisture content and arrangement of the trees and other vegetation, atmospheric temperature, humidity, wind and topography) of fire susceptibility, and this provides a measure of the potential fire danger in a given area at a particular time. Nevertheless, a fire requires an igniting agent which may be man-caused, such as a match or cigarette, or natural, such as lightning. During 1966, 65 per cent of the 110,000 wildfires occurring in the United States were started accidentally by man, 26 per cent were of incendiary origin, and 9 per cent were started by lightning. These percentages vary widely by region, however. (In the 12 Rocky Mountain states, 52 per cent were started by lightning and 2 per cent were incendiary; in the 13 Southern states, 1 per cent were started by lightning and 38 per cent were of incendiary origin). Extensive information on the incidence, causes and costs of forest fires is only available for the United States. Private, state and federal agencies spend $300,000,000 annually for controlling and preventing fire on 1,000,000,000 acres of forest and watershed lands. In the period 1957–66, there were 1,120,000 forest fires covering more than 46,000,000 acres in the U.S. Annually 100,000 – 150,000 fires burn 4,000,000 – 8,000,000 acres. The total damage is inestimable but timber losses alone are about $450,000,000 annually. During the last 20 years the number of fires and the area burned have both been reduced with the development of more efficient fire prevention and control technology. In the period 1943–7, an annual average of 167,000 fires burned 22,500,000 acres. During 1963–7, these rates were reduced to an annual average 128,000 fires over 4,600,000 acres.

Counter-claim In the natural state, fire, started by lightning or other natural agents, has a beneficial effect and to some extent the forest is adapted to it. In the wild forest, fallen trees and branches may form a deposit many feet deep on the forest floor that provides shelter and nourishment for tree parasites, including insects, fungi and bacteria. This deposit, through which seedling trees have to push to reach light and air, also represents an accumulation of valuable minerals which are withheld from the living trees. Fire then plays a beneficient role as it sweeps through this tangled deposit, sterilizing, providing space in which new trees can thrive, and returning minerals to the soil. Indeed, fire is an essential agent in the regeneration of the forest in its natural state. Certain tree species have become so adapted to fire that it is unlikely that their cones will open to shed seeds without it. Fires which start naturally should therefore be allowed to run their course in the light of responsible conservation principles. In some cases, forest fires should even be started, to help forests to evolve back to natural state.

Refs Artsybashev, E S *Forest Fires and Their Control* (1984); Chandler, Craig C, et al *Fire in Forestry* (1983); Clarke, B and Show, S B *Forest Fire Control* (1978); FAO *Wildland Fire Management Terminology* (1986); Institute of Paper Chemistry *Forest Fires* (1974); WHO *Systems for Evaluating and Predicting the Effects of Weather and Climate on Wildland Fires* (1978).

Broader Fires (#PD8054) Natural disasters (#PB1151).
Related Defoliation (#PD1135) Biological disasters (#PC5489).
Aggravates Soil erosion (#PD0949) Deforestation (#PC1366)
Soil infertility (#PD0077) Endangered forests (#PC5165)
Wildlife extinction (#PC1445) Smoke as a pollutant (#PD2267)
Pests and diseases of trees (#PD3585) Destruction of land fertility (#PC1300)
Deforestation of mountainous regions (#PD6282).
Aggravated by Inadequate safeguards against fire (#PD1631).

♦ PD0743 Environmental hazards from fishing industry

Nature Effluent and waste from fish farms may damage wild fish, seals and shellfish. Fish farmers use tiny quantities of highly toxic chemicals to kill lice: one overdose could be devastating.

Broader Ecologically unsustainable development (#PC0111)
Obstacles to the utilization of coastal and deep sea water resources (#PF4767).
Narrower Unsustainable exploitation of fish resources (#PD9082).
Aggravated by Destructive changes in ocean characteristics (#PC2087).

♦ PD0748 Harmful effects of ultrasonic radiation on the human body

Nature Ultrasound is sound (a mechanical vibration phenomenon) having a frequency above the range of human hearing (typically above 16 kHz) which, unlike electromagnetic radiation, requires a medium through which to propagate. Exposure to ultrasound can be divided into two distinct categories: airborne and liquid-borne. Exposure to airborne ultrasound occurs in many industrial applications such as cleaning, emulsifying, welding and flaw detection; through the use of consumer devices such as dog whistles, bird and rodent controllers, and camera rangefinders; and commercial devices such as intrusion alarms. Liquid-borne exposure occurs predominantly through medical exposure in diagnosis, therapy, and surgery. As with any other physical agent, ultrasound has the potential to produce adverse effects at sufficiently high doses. In addition, biological effects of unknown significance have been reported at low exposure levels under laboratory conditions.

Incidence Exposure of human beings to low frequency ultrasound (16 – 100 kHz) can be divided into two distinct categories; one is via direct contact with a vibrating solid or through a liquid coupling medium, and the other is through airborne conduction. For airborne ultrasound exposure, at least one of the critical organs is the ear. Effects reported in human subjects exposed to airborne ultrasound include: temporary threshold shifts in sound perception, altered blood sugar levels, electrolyte imbalance, fatigue, headaches, nausea, tinnitus, and irritability. Studies on skeletal tissue indicate that bone growth may be retarded following exposure to ultrasound at high therapeutic intensities, even if the transducer is kept in motion during treatment. If the transducer is held stationary, bone and other tissue damage occurs at lower intensities. Both in vitro and in vivo exposures of muscle tissue have been reported to trigger contractions. Therapeutic intensities of ultrasound have also been reported to alter thyroid function in man. A number of reports indicate that lower foetal weight and increased foetal abnormalities occur following exposure to ultrasound in the low therapeutic intensity range. One study suggests that lower birthweights may

result from exposure to diagnostic ultrasound in utero. As the practice of ultrasound diagnosis becomes more widespread, it will be difficult to find adequate control populations and opportunities for satisfactory epidemiological studies may become increasingly rare.
Counter-claim Although the health implications from a number of effects already reported indicate the need for a prudent approach to the ultrasound exposure of human subjects, the benefits of this imaging modality far outweigh any presumed risks.
Broader Radioactive contamination (#PC0229) Human disease and disability (#PB1044).

♦ PD0752 Cattle diseases
Refs Ristic, Miodrag and McIntyre, Ian (Eds) *Diseases of Cattle in the Tropics* (1981).
Broader Animal diseases (#PC0952).
Narrower Fat cow syndrome (#PG4372) Texas cattle fever (#PG3990)
Double muscling in cattle (#PG4099) Abomasal disorders of cattle (#PE9364)
Lumpy skin disease of cattle (#PG7207) Respiratory diseases of cattle (#PE5524)
High-mountain disease of cattle (#PG3920) Farmer's lung disease in cattle (#PG5383)
Haemophilus septicaemia of cattle (#PG0354)
Sudden death syndrome of feeder cattle (#PG7866)
Tracheal oedema syndrome of feeder cattle (#PG7167).
Related Brucellosis (#PE0924).

♦ PD0758 Overcrowding of housing and accommodation
Nature There is widespread evidence to show that overcrowding in small dwellings causes psychological and social damage. In chronic cases, overcrowding increases the incidence of contagious disease and generally contributes to the deterioration of health, morale and safety. Overcrowding makes privacy for individuals or couples virtually impossible, which may lead to psychological stress.
Emphasis on physical conditions rather than on overcrowding has frequently been responsible for policies of indiscriminate slum demolition which tend to intensify the overcrowding rather than expand the housing supply.
Incidence In Bombay, for example, the crowding of 10 persons in a room 10 by 15 feet in size is not unusual. Occupancy in tenements ranges from 6 to 9 persons per room, with an overall average of more than 7 persons per room; the bulk of these tenements consist of only one room. In Calcutta, Bombay, Ahmedabad, Cawnpore and Nagpur in India, 60 to 90 per cent of the working-class families and 50 per cent of the middle-class families live in single rooms. Population density reaches 1,200 to 1,500 per acre. Only 10 to 14 per cent of the built-up areas in the bigger cities consist of open spaces in the form of streets. In socialist countries, where construction in cities is mainly directed towards basic equipment to the prejudice of housing, overcrowding of several households in a small flat is quite common. In Japan, the 1978 housing census shows that the average number of persons per room was 0.8 and the average number of rooms per dwelling was 4.5. The corresponding figures for the United States in 1970 were 0.6 and 5.1 respectively, according to the United Nations Statistical Yearbook, 1978.
Broader Overcrowding (#PB0469).
Narrower Impersonality of high density accommodation (#PF6156).
Aggravates Leprosy (#PE0721) Smallpox (#PE0097) Trachoma (#PE1946)
Dysentery (#PE2259) Tuberculosis (#PE0566)
Stress in human beings (#PC1648)
Substandard housing and accommodation (#PD1251)
Stultifying homogeneity of modern cities (#PF6155)
Inadequate facilities for children's play (#PD0549)
Illegal occupation of unoccupied property (#PD0820)
Environmental degradation of inner city areas (#PC2616)
Depressing effect of poor housing construction (#PF1213)
Limits on participation in community development (#PF3560)
Inadequate living and working conditions of immigrant labourers in industrialized countries (#PD3427).
Aggravated by Inadequate housing (#PC0449) Segregation in housing (#PD3442)
Inadequate housing in developing countries (#PE0269)
Inadequate housing among indigenous peoples (#PC3320).

♦ PD0759 Inadequate sex education
Nature Sexual matters are often inadequately explained to persons needing advice, whatever their age, because of religious beliefs or taboo. The inadequacy of education of children in practical matters of contraception and intercourse is often based on the belief that this information will encourage promiscuity. Older women and men may be given insufficient advice by doctors and hospitals. Contraception is banned in certain countries; abortion is illegal in a great many; masturbation is usually held to be deviant. Lack of sexual education reinforces existing taboos and leads to sex by trial and error, possible impotence or lack of satisfaction, guilt complexes and, where contraceptive advice is lacking, unwanted pregnancies and abortion because of ignorance. Sexual repression may lead to sexual violence and deviancy. Anti-authoritarianism among young people may make them shy of using advice facilities.
In countries where sex education is not part of the accepted curriculum and is being introduced and expanded, great sensitivity has to be exercised to allow for public opinion and cultural and religious mores. Sectors of the public may consider that, because of their own moral values, information given in a sex education programme could have a corrupting influence on the young and they will try to prevent its introduction. The term 'sex education' may be an obstacle in itself, as it is often seen in isolation as 'education in sexual intercourse' rather than as a preparation for adulthood.
Incidence Sex education for young people in the European region ranges from areas where it is a compulsory and accepted part of the school curriculum to areas where it is non-existent. In no country is it prohibited by law. However, even in countries where it has been compulsory since 1971 and integrated into the school curriculum, the specific aim of preventing sexual and personal problems due to ignorance of sexual matters has not been met completely.
It is recognized that knowledge alone will not prevent emotional and personal problems in this essential area of life, but findings reveal a lack of preparation even for immediate events such as puberty. A recent study showed that twice as many boys (25 percent) as girls (12 percent) had had no information prior to the event and were unprepared for the biosocial changes of menarche and spermarche. The median age when information on sex was introduced was 12-13 years for boys as well as girls, and only a minority had had any sex education before the age of 10 years. The primary source of sex education for boys was the teacher (53 percent), with mothers (5.5 percent) the next most frequent source, while for girls, mothers (41 percent) and teachers (32 percent) were quoted. Fathers played a very small part. As a group, girls were better informed at an earlier age and had knowledge of a wider variety of topics. Boys were found to be much more dependent on teaching at school. The fact that fathers were hardly ever mentioned may reflect the fact that in many societies they spend less time with their children and are less involved in their intimate care. It may, however, also be due to the fact that women relate sex to reproduction and babies, which provides an easier introduction to the topic, whereas men may related it more to pleasure or desire, which may be harder to discuss with their children.
Not all teachers are naturally adept at sex education, although their skills can be improved with training, which should be part of the basic teacher-training syllabus. Similarly, doctors and nurses may have the factual knowledge that would appear to make them suitable sex educators, but they may be unskilled in communication and would also benefit from further training.
Refs Lewin, B *Sex and Family Planning* (1984); McCarthy, Wendy and Fegan, Lydia *Sex Education and the Intellectually Handicapped* (1984).
Broader Lack of education (#PB8645) Inadequate education (#PF4984)
Inadequate secondary education (#PD5345).
Related Promiscuity (#PC0745).
Aggravates Guilt (#PA6793) Homosexuality (#PF3242) Sexual violence (#PD1276)
Sexual deviation (#PD2198) Sexual immorality (#PF2687) Unmarried parents (#PD3257)
Male homosexuality (#PF1390) Sexual unfulfilment (#PF3260) Sexual exploitation (#PC3261)
Illegitimate children (#PC1874) Human sexual inadequacy (#PC1892)
Illegal induced abortion (#PD0159) Sexual exploitation of women (#PD3262)
Sexually transmitted diseases (#PD0061)
Discrimination against illegitimate children (#PD0943).
Aggravated by Taboo (#PF3310) Religious intolerance (#PC1808)
Generation communication gap (#PF0756).
Reduced by Nudism (#PF2660).

♦ PD0762 Disabled victims of crimes
Nature With the emergence of 'victimology' as a branch of criminology, the true extent of injuries inflicted upon the victims of crime, causing permanent or temporary disablement, is only now becoming generally known.
Broader Human disability (#PC0699) Neglect of victims of crime (#PD4823).
Related Grief (#PF5654).

♦ PD0764 Disabled victims of torture
Nature Victims of torture are disabled physically or mentally not by accident of birth or normal activity but by the deliberate infliction of injury. Torture victims often experience the mental sequelae as the worst ones: impairment of memory and concentration, nightmares and other sleep disturbances, sexual disturbances, fear, depression, fatigue, sense of guilt, feeling of isolation, loss of identity and very low self-esteem. This may be due to the fact that family and friends tend to respond negatively to psychological symptoms and with sympathy toward physical ones. Frequent physical sequelae are pains in muscles, joints and bones, headaches, gastro-intestinal symptoms and specific problems related to specific tortures. In additions the conditions of their imprisonment aggravate the physical and mental problems they face. They may suffer from malnutrition, infections and diseases from over crowding.
Background While victims of torture have existed since ancient Egypt times, the tendency of the past few years has been for torture to be used not only as a means of repression of the individual, as was the case up to now, but as a means of mass political repression. At the same time the social situation changed. Those exercising torture today normally desire to leave as few obvious tracks as possible. Using the latest techniques and equipment torturers can be more systematic and goal oriented. Most victims of torture are released only after all objective effects have disappeared. Those who are injured beyond healing are often executed.
Broader Human disability (#PC0699) Torture of children (#PD2851)
Institutionalized torture (#PD6145).
Narrower Family of torture victims (#PE2119) Neurological effects of torture (#PD4755)
Somatic and psychosomatic effects of torture (#PE5294).
Aggravated by Torture through confinement (#PD4590)
Denial of rights to prisoners (#PD0520)
Depriving prisoners of medical treatment (#PD1480)
Inadequate assistance to victims of torture (#PE6936).

♦ PD0765 Excessive foreign investment in traditional industries of developing countries
Claim The process of takeover of traditional industries by foreign investors at best meets with resentment from the local business community and government, and at worst encounters solid opposition. By creating an unfavourable reaction, such action can lead to restrictive measures against all foreign private investors, including those offering scarce technology.
Counter-claim Foreign investment represents high levels of management skills and production, distribution and marketing technology. The profit motive of transnational corporations and its satisfaction assures countries of assistance in the development of traditional industries. Excessive government investment or nationalization of traditional industries assures a slow-down in development.
Broader Capitalism (#PC0564) Inappropriate foreign investment (#PD8030).

♦ PD0767 Ecosystem modifications due to creation of dams and lakes
Environmental degradation due to creation of dams and lakes — Dam construction — Adverse effects of damming — Impounded rivers
Nature The damming of rivers produces dramatic and far-reaching environmental changes: of the first order are hydrological changes, including water quality; the second order affects channel morphology and aquatic, riparian, and floodplain vegetation; and the third order includes vertebrates and fish. Thus there are important linkages between the physical, chemical, and biological components of river systems that are disordered when a river is impounded.
The most significant changes take place in the hydrological regime which, in turn, affects many other natural processes. Stream flow is most influenced by reservoirs with long-term and marked seasonal regulation; the upstream levels can change over a year by amounts ranging from a few tens of centimetres to tens of metres and in some reservoirs by over a hundred metres. Marked variations in level and water exchange rate affect the flow, temperature, and hydrochemical and hydrobiological regimes of reservoirs and downstream rivers as well as silting and changes in beds and banks. The flow of suspended and dissolved materials, including mineral biogenic substances and trace elements, also drops sharply.
Reservoirs cause substantial changes in ground water. Depending on the local hydrogeological conditions, the range over which the ground-water head spreads can vary from tens of metres to many kilometres. In some cases this is intended and considered favourable but where substantial vertical and horizontal spreading of the ground-water head swamps can be created and land flooded.
Reservoirs greatly influence the landscape of the river valleys, including the proportion of land to water area as well as the drainage network pattern. Changes in the landscape are also a result of changes in soils, vegetative cover and wildlife. The areas, volumes, widths and lengths of the reservoirs vary markedly over a year; the strips of land that are regularly drained may become partially covered with vegetation, or turn into sandy or silty shoals. Downstream flooding of the plains decreases or ceases altogether, which results in the former channels and flood lakes becoming dry, silted up and overgrown. Downstream channels and flood plains can change their shape, through erosion of the banks and bottom and drying up of flood-plain branches and mainstream channels. Transformation of the channels usually continues until equilibrium is reached between degradation and aggradation. The erosion zone gradually moves downstream. Channel erosion leads to a lowering of the water level. At flood time, because of the lower high-water level in the main stream, the current velocity in the tributaries rises and the erosion of their beds and banks increases. Some alterations also occur in the downstream flow and delta. The flow is redistributed between the branches in favour of bigger and deeper ones, which results in silting and drying up of the smaller branches and erosion of the bigger ones.
The creation of large reservoirs changes the climate, soils, vegetation and fauna. The area affected depends on local factors, the dimensions and operating regime of the reservoir and its

geographical position. The main alterations in the meteorological conditions are as follows: radiation and evaporation increase, the climate becomes less continental, wind speeds rise, and breeze-like winds appear. Downstream of the water installations, winter discharges increase, and their influence on the air temperature and humidity is stronger. The thermal conditions of the water in cold climates changes considerably: winter temperature are substantially higher, air humidity is higher and fogs may occur.
The mode and intensity of the reservoir's effects on the soil and vegetation depend upon the ground-water regime, the lithological composition of the soil, the mode of regulation and changes in the micro-climate. The ares of shore affected around large reservoirs may vary from hundreds of metres to several kilometres. Changes in the downstream soil and vegetation are not so marked as at the reservoirs. Less, if any, fertile mud is deposited on grasslands, the amounts of moisture stored in the ground of flood plains are reduced, and the ground-water level is lowered. In the majority of river valleys, the soil and vegetation zones shift down towards the river channels.
Reservoirs cause substantial changes in the wildlife over wide areas. During filling, the number of many species of animals and birds is considerably reduce, as their habitats and feeding areas are disturbed. In mountain regions, big reservoirs disturb the migration paths of hoofed animals. During the operation of reservoirs, the habitat conditions are greatly affected by the water level variations. Because of changes in the shores, numerous animals and birds are deprived of their habitats. On many reservoirs, the amount and, sometimes, the number of species of waterfowl increase. Over time a balanced biocoenosis gradually develops.
During the first years after the reservoir starts operation, the water quality deteriorates owing to organic and mineral matter being washed out from the bottom and shores, decay of vegetable residues and the negative influence of peat. The downstream quality of water depends directly on how much water is released and how often. As a rule in amount of nutrients entering the water is reduced. As compared with natural watercourses, reservoirs are, as a rule, less capable of self-purification in some important ways.
Conflicting with contact uses of water in reservoirs, particularly in the tropics, have been water-borne or water-organism-vectored diseases and invasion by aquatic nuisance plants. Urinary schistosomiasis should now be considered an expected impact of dams constructed in endemic areas. The explosive growth of aquatic weeds and the establishment of a wide range of invertebrates, among them the intermediate hosts of human parasites, lead to the spread of schistosomiasis and other public health problems in a dam area. Domestic water supply and waste disposal are seldom adequate in most resettlements and in such conditions, with the immigration of infected people to the dam area where there is constant contact with water containing vector snails, schistosomiasis soon becomes established and quite often exceeds onchocerciasis and malaria in severity. These health problems are among the most severe of the impacts directly affecting people associated with dams.
Counter-claim The question of the merit or otherwise of constructing a reservoir has no simple answer. Dams large and small are necessary for development.
Refs FAO *Selected Bibliography on Major African Reservoirs* (1983); Goldsmith, Edward and Hildyard, Nicholas (Eds) *The Social and Environmental Effects of Large Dams*; National Research Council *Safety of Dams* (1985); Serafim, J Laginha (Ed) *Safety of Dams* (1984).
Broader Environmental degradation (#PB6384) Natural environment degradation (#PB5250)
Environmentally harmful dam construction (#PD9515).
Narrower Impoldering risks (#PE6347) Silting of water systems (#PD3654)
Increase in pests and diseases through perennial irrigation (#PE7832).
Related Deforestation (#PC1366).
Aggravates Earthquakes (#PD0201) Onchocerciasis (#PE2388)
Water-borne animal diseases (#PE2787) Involuntary mass resettlement (#PC6203).

♦ PD0771 Nuclear accidents
Risk of atomic accidents
Nature Nuclear accidents can involve X-ray equipment, radioactive isotopes, particle accelerators, nuclear fuel production plants, nuclear power plants and atomic and hydrogen bombs. Some accidents threaten public health, contaminate crops and animals and could destroy the country side for hundreds of square kilometres. There have been more accidents involving nuclear weapons or reactors than the authorities have officially admitted.
It is extremely difficult to predict exactly what kinds of accidents will occur and thus build in safeguards. All major US nuclear accidents have involved events completely unforeseen in any detailed government study of reactor safety. Furthermore, operator error is likely to play a major role in future accidents.
Claim Nuclear accidents, like the Challenger disaster, have repeatedly demonstrated that fools of sufficient magnitude can be found to overcome any foolproof system.
Refs Beres, Louis R *Apocalypse* (1980); International Atomic Energy Agency *Severe Accidents in Nuclear Power Plants*; Lebow, Richard N *Nuclear Crisis Management* (1988); United Nations *International Conventions on Civil Liability for Nuclear Damage* (1976).
Broader Risk (#PF7580) Disastrous technological failures (#PD4426).
Narrower Nuclear reactor accidents (#PD7579) Nuclear accidents in space (#PE5080)
Accidents to nuclear weapons systems (#PD3493).
Related Large-scale industrial accidents (#PD2570).
Aggravates Risk of unintentional nuclear war due to accidents (#PF4346).
Aggravated by Job fatigue (#PD8052) Nuclear irresponsibility (#PF6611)
Theft of nuclear materials (#PD3495)
Ease of manufacture of nuclear bombs (#PF3494)
Deterioration of nuclear power plants (#PE5260)
Inadequate nuclear reactor safeguards (#PF6084)
Vulnerability of nuclear power sources (#PD0365)
Environmental hazards of nuclear power production (#PD4977)
Non-verificability of compliance with nuclear power safeguards (#PF4455).
Reduced by Restrictive regulation of nuclear power (#PF8654)
Unilateral structural disarmament of nuclear weapons (#PE4051)
Environmental hazards of decommissioned nuclear power plants (#PE7539).

♦ PD0779 Excess production capacity
Overcapacity — Overcapacity in industrial production — Idle factories — Low capacity utilization of manufacturing plant — Decline in production capacity utilization
Nature At the national, international and plant levels excess capacity may occur. At the national and international levels, overcapacity occurs either as a result of the emergence of too many competing enterprises (as in the case of airlines serving the North Atlantic route, for example) or as a result of reaching the limit for import substitution (as in the case of consumer goods and light industries in many Asian countries). In the latter case it is generally agreed that the high cost of production in relation to the income generated by the investment or its failure to reach an adequate level is the ultimate cause of the failure of demand to rise. In countries such as India, this explanation is perhaps less important than the shortage of foreign exchange (due to the sluggish growth of exports) which has prevented the import of spare parts and special materials. At the manufacturing plant level, overcapacity may result from the size of parts and product inventories, excessive number of designs to meet market needs, assembly and design processes or an excessive number of employees.
Broader Lack of central planning structures in small communities (#PF2540)
Instability of economic and industrial production activities (#PC1217).
Aggravates Unemployment (#PB0750) Factory closures (#PE3537)
Over-production of commodities (#PD1465)
Decline in capital investment in productive capacity (#PE9265)
Inadequate production capacity in developing countries (#PD4219).
Aggravated by Lack of purchasing power in developing countries (#PE8707).

♦ PD0784 Exclusion of disabled persons from social and cultural life
Nature Disabled persons are often denied the opportunities of full participation in the activities of the socio-cultural system of which they are a part. This deprivation comes about through physical and social barriers that have evolved from ignorance, indifference and fear. Such attitudes and behaviour often lead to the exclusion of disabled persons from social and cultural life. People tend to avoid contact and personal relationships with those who are disabled. The pervasiveness of the prejudice and discrimination affecting disabled persons and the degree to which they are excluded from normal social intercourse produce psychological and social problems for many of them. Too often, the professional and other service personnel with whom disabled persons come into contact fail to appreciate the potential for participation by disabled persons in normal social experiences and thus do not contribute to the integration of disabled individuals and other social groups.
Because of these barriers, it is often difficult or impossible for disabled persons to have close and intimate relationships with others. Marriage and parenthood are often unattainable for people who are identified as 'disabled' even when there is no functional limitation to preclude them. The needs of mentally handicapped people for personal and social relationships, including sexual partnership, are not adequately recognized.
Many persons with disabilities are not only excluded from the normal social life of their communities but are in fact confined in institutions. While the leper colonies of the past have been partly done away with and large institutions are not as numerous as they once were, far too many people are today institutionalized when there is nothing in their condition to justify it.
Refs Boswell, David M and Wingrove, Janet M *The Handicapped Person in the Community* (1974).
Broader Social exclusion (#PC0193) Denial of cultural rights (#PD5907).
Aggravated by Human disability (#PC0699) Institutionalization of the disabled (#PF4681).

♦ PD0788 Environmental hazards from live animals
Broader Environmental hazards from food and live animals (#PC1411).
Aggravates Vulnerability of plants and crops (#PD5730).

♦ PD0790 Soliciting obstruction of proceedings
Nature It is a crime to ask others to obstruct official judicial, legislative or administrative proceedings. Obstruction includes asking people not to testify, or to appear as a witness; or to disrupt the proceedings by noise or violent behaviour. The process of law is disrupted. The person being asked is in danger of being a criminal.
Broader Contempt of judicial process (#PD9035)
Crimes against the integrity and effectiveness of government operations (#PD1163).
Aggravated by Legal prevarication (#PF9756).

♦ PD0791 Road hazards
Hazardous road passages
Nature Within national road transportation systems, different sections have different characteristics and effects with respect to safety, speed and economy in the movement of people and goods. Uncorrected high-risk locations continue to be the scene of accidents, injuries and deaths.
Broader Hazards to human health (#PB4885) Human errors and miscalculations (#PF3702).
Narrower Unsafe school crossing (#PU0149).
Aggravates Road traffic accidents (#PD0079)
Isolating effects of seasonal variations on undeveloped transportation (#PE3547).
Aggravated by Fog (#PE1655) Inadequate road maintenance (#PD8557)
Deterioration in atmospheric visibility (#PE2593).

♦ PD0794 Imbalances in the distribution of the costs and benefits of economic integration
Nature Integration schemes have the tendency to favour the more advanced members of a grouping who benefit from the liberalization of trade and the expansion of their markets, while the less advanced members of a grouping tend to gain less, unless the scheme provides for structural changes in the latter countries. It is generally true that the special problems facing the less advanced countries include not only the smallness of their markets, but more particularly the absence of supplies exportable to the other partners of their grouping.
Counter-claim So as not to become self-defeating, measures taken with a view to correcting imbalances in the distribution of costs and benefits must not obstruct the development of the region as a whole. Some measures which are necessary to deal with short-term problems may not be sufficient to correct underlying imbalances, while other measures appropriate for the latter purpose would not be useful for resolving immediate problems. There is a danger that exclusive concern with long-term problems may allow short-term problems to assume such proportions that a crisis is provoked. In that case progress towards the solution of long-term problems will at best require far more time than would have been needed if emergency provisions had been available in the first place; at worst, the grouping dissolves without having had the opportunity of tackling the long-term problems at all.
Broader Lack of international cooperation (#PF0817).
Aggravated by Competition between states (#PC0114).

♦ PD0800 Dietary deficiencies in developed countries
Unnutritious eating habits — Poor nutritional habits — Unnutritious food consumption — Junk food consumption
Nature In developed countries, the intake of rich foods with little nutritional value tends to increase, whilst the intake of milk products, fruit and vegetables decreases. Though poor diets cannot be directly related to malnutrition, malnutrition might result from consumption of a poor diet over a period of months or years.
Incidence In the USA in 1965, for example, it was estimated that 20 per cent of the families existed on a nutritionally poor diet. This included both high and low income families. Only 50 per cent of USA families were consuming a good diet. (Previous figures for 1955 gave 15 and 60 per cent respectively). A usual mixed diet in Scandinavia, Britain, or North America provides 6 mg or less of iron per 1000 calories daily; a woman ingesting 1800 or 2000 calories per day therefore obtains 10-12 mg of dietary iron. This amount is insufficient to meet current standards of allowances for women, especially during pregnancy and lactation. Its adequacy may be further compromised if the dietary composition is not such as to favour iron absorption, that is, if it is devoid of or low in vitamin C or high in bulk or phytate.
Background Caloric needs determine the limits of food that can healthily be ingested. These needs are small because of the life-style of the affluent. In order to maintain desirable protective levels of nutrients in the diet consistent with the acceptable food habits of society, some 40-50 percent of the total diet must consist of high-nutrient density foodstuffs supplying needed quantities of the essential nutrients: amino acids, vitamins, minerals, and trace elements. This

means that 40–60 percent of the energy value of the diet must be derived from high nutrient density foodstuffs: products such as meat, milk, eggs, cheese, vegetables, fruits, and cereals. In order to further maintain nutrient density of the diet, selected enhancement of the nutrient content of some foodstuffs, such as cereals, is desirable.
Claim In UK between 1982 and 1986 the price of healthy food rose faster than unhealthy food. For the 4 million British adults and children who depended on income support, and others on low incomes, a healthy diet was beyond their reach.
Refs Somogyi, J C and Varela, G (Eds) *Nutritional Deficiencies in Industrialized Countries* (1981).
Broader Inappropriate personal habits (#PD5494)
Deterioration of industrialized countries (#PD9202).
Narrower Unbalanced food usage (#PJ7868) Reliance on canned food (#PJ8409)
Excessive consumption of fats (#PE4261).
Related Trace element imbalance in the human body (#PE5328).
Aggravates Obesity (#PE1177) Dietary restrictions (#PJ1933)
Nutritional deficiencies (#PC0382) Protein–energy malnutrition (#PD0339).
Aggravated by Food insecurity (#PB2846) Predominance of fast food (#PF5940).

♦ PD0807 Domestic refuse disposal
Garbage disposal — Rubbish disposal — Trash disposal — Unorganized trash collection — Sporadic trash removal — Infrequent rubbish collection — Unremoved heavy trash — Consumer waste disposal
Nature The difficulties in disposing of rubbish are: increasing shortage of labour while public demands for a better service grow more exacting; adapting machines and methods to suit the growing volume of light but bulky waste; the pressing problem of where to dump treated or untreated refuse as existing tips are filled. The passing years have seen a mounting volume of 'bulky waste' – such items as old washing machines, refrigerators, cars and other objects that will not fit into the standard dustbin and which are often dumped about the countryside because there are no lawful ways of getting rid of them.
Incidence Rubbish disposal ranks as one of the affluent society's major headaches. The United States spends more than $3000 million a year on storing, collecting and disposing of solid wastes, which are produced at a rate equivalent to 3.5lb every day for each one of the 200 million American citizens. French consumers produce about 1.7lbs of garbage per day, Japanese produce 1.9lbs and West German produce 2.3lbs. In England and Wales (with about 48 million inhabitants) local authorities collect about 14 million tons of house and trade refuse annually, at a cost of over £60 million.
Broader Unformed style of cooperative action (#PF6514)
Demoralizing images of rural community identity (#PF2358)
Weakness of infrastructure in developing countries (#PC1228).
Related Underdeveloped provision of basic services in developing countries (#PF6473).
Aggravates Dust (#PD1245) Fires (#PD8054) Waste paper (#PD1152)
Malodorous fumes (#PD1413) Amenity destruction (#PC0374)
Smoke as a pollutant (#PD2267) Mycoplasmal arthritis (#PG2302)
Vector–borne diseases (#PD8385) Environmental pollution (#PB1166)
Hazards to human health (#PB4885) Wood deterioration and decay (#PD2301)
Human disease and disability (#PB1044) Unproductive use of resources (#PB8376)
Non–biodegradable plastic waste (#PD1180)
Damage by degradable organic matter (#PJ6128)
Limited accountability of public services (#PF6574).
Aggravated by Insufficient recycling of materials (#PJ0525)
Non–destructible packaging and containers (#PD1754).
Reduces Soil infertility (#PD0077) Land reclamation (#PF2055)
Diplomatic errors (#PF1440).

♦ PD0808 Denial of right to economic security
Broader Inadequate standards of living (#PF0344).
Narrower Denial of right to benefits for invalids (#PE5211)
Denial of right to benefits to survivors (#PE4531)
Denial of right to economic security during periods of unemployment (#PE5406).
Related Denial of right to sufficient food (#PE0324)
Denial of right to social security (#PD7251)
Denial of right to sufficient shelter (#PD5254)
Denial of right to sufficient clothing (#PE7616)
Denial of right to adequate medical care (#PD2028).
Aggravated by Unemployment (#PB0750) Inadequate income in old age (#PC1966)
Inadequate social welfare services (#PC0834).

♦ PD0811 Food grain spoilage
Nature Large amounts of the world's food grain are wasted because of inadequate handling and storage at various stages between production and consumption. If grain is stored in unsuitable and primitive ways, the storage losses (caused by germination, microbes, rodents, birds and insects) can be as much as 20 to 40 per cent while at the same time the quality of the rest may be badly impaired. The problem is especially difficult in tropical developing countries where there the extremely high temperatures coupled with widely fluctuating humidity facilitate every type of wastage from insect infestation to mould growth. Even in developed countries, the growth of fungi in existing storage facilities gives rise to the production of substances toxic to man and to farm animals.
Incidence The variability of foodgrain losses from season to season, between different crops, from country to country and under different kinds of treatment, makes accurate measurement of their extent very difficult. Reasonable average estimate of wastage is at least 10 per cent of the total grain production of the world, amounting to 75 million tons per year, or a calorie loss sufficient to supply the needs of 250 million people. Some estimated losses per crop year are: India, 25 per cent; tropical Africa, 30 per cent; USA, $500 million.
The most significant losses frequently occur towards the end of the storage season (when small scale farmers, at least, will be short of food) and tend to arise at the farm level when the traditional capability to conserve grain is disturbed with the introduction of new improved varieties or multiple cropping. The traditional handling, drying and storage systems may be inadequate to cope with the increased yield. Furthermore, the characteristics of the crop may be changed so that the grain is more susceptible to insect or fungal attack, or the timing of harvest may be changed from the usual dry season to the less favourable wet season.
Broader Spoilage of agricultural products (#PC2027).
Narrower Food spoilage in storage (#PD2243).
Aggravates Food wastage (#PD8844).
Aggravated by Inadequate food storage facilities (#PE4877)
Instability of trade in food and live animals (#PD1434)
Inadequate packaging of agricultural products (#PF3143).

♦ PD0813 Maldistribution of land through customary tenure systems
Nature Customary tenures, whereby the pattern of use rights over land are recognized by the community for centuries without necessarily the sanction of statutory or recorded evidence, are prevalent on varying scales in many countries of the developing world. The problem is of crucial importance in Africa. Though this type of tenure system was eminently suitable for regulation of land use in the past, it has undergone such modifications under the impact of the modern economic systems that new problems have arisen largely as a result of growing trends towards individualization.
Monetization and commercialization of agriculture is widely recognized as an important prerequisite for developing agriculture but the existing tenure system in its present state of breakdown acts as a limiting factor. The progress in regard to remedial measures has been very uneven, and many new problems have arisen. For instance, a cardinal feature of the old customary tenure is the delineation of specific areas as belonging collectively to a specific tribe or community. The conflict between the interest of subsistence farmers operating within the customary tenure system, individual farmers cultivating cash crops on a small scale, and that of the big commercial plantations, organized on modern lines, has become acute; well–intentioned but ad hoc efforts to tinker with tenure systems have adversely affected the interest of the subsistence farmer.
While the problem of landlessness has not yet arisen in Africa on the same scale as in Asia and Latin America, there is great danger that unregulated individualization of land and introduction of the concept of land as marketable commodity may eventually result in a growing class of rural landless, unless adequate steps are taken at this stage to solve these basic problems.
Claim The problem of customary tenure is closely linked with the problem of subsistence economy, unequal population pressure between different regions reserved for different tribes, and the need for balanced development of different regions. As many countries have not formulated a comprehensive land policy, a large segment of the subsistence producers has often failed to participate in the benefits flowing from economic development.
Broader Maldistribution of land in developing countries (#PD0050).

♦ PD0814 Unethical practices of government
Government improbity
Narrower Misuse of statistics (#PF4564) Deception by government (#PD1893)
Historical misrepresentation (#PF4932) Government sanctioned killing (#PD7221)
Unreported government spending (#PF2990)
Unethical practices in local government (#PD5948)
Official cover–up of government harassment of political activists (#PF3819).
Related Corruption of government leaders (#PC7587).

♦ PD0820 Illegal occupation of unoccupied property
Squatting — Appropriation of unoccupied housing
Nature Squatting, the appropriation of private or public land or property for one's own use without title or right, occurs in rural and urban areas, and in developed and developing countries. Squatting may take the form of open or furtive mass movements or of individual operations. It is the illicit consequence of the struggle for shelter, the trespass of desperation; sometimes it is denounced, sometimes disesteemed; often it is tolerated for want of a practical alternative. Uncontrolled, it harbours a formidable threat to the structure of private rights established through the centuries. The buildings are substandard, they lack electricity, sanitation, water and access roads. They are dangerous to the inhabitants, embarrassing to governments and debilitating to society.
Incidence Rural migration to the cities for jobs, where absence of jobs or housing facilities exists, causes massive squatting in Asia, particularly in India, and in Africa and Latin America. Squatting in Europe exists, and although relative to the other continents it is small, it is shocking to Europeans to realize that cities such as Amsterdam and London have large numbers of homeless. The 1981–1982 recession produced considerable numbers of squatters in the United States.
Counter–claim Although uncontested squatting on property can lead, in some countries, to passing of its ownership title to the squatter after a number of years, this remote threat does not remove the moral responsibility of governments, organizations and individuals to give shelter to those who lack it. Harsh evictions have resulted in exposure to the weather for the homeless with consequent disease or death.
Broader Criminal trespass (#PD3794).
Related Insecure lease tenure (#PJ8344).
Aggravates Urban slums (#PD3139).
Aggravated by Housing shortage (#PD8778) Inadequate housing (#PC0449)
International migration (#PF4008) Unrentable vacant housing (#PJ8683)
Housing destruction in war (#PE2592) Deteriorated vacant houses (#PJ8678)
Scarcity of residential land (#PD8075) Prohibitive cost of accommodation (#PD1842)
Substandard housing and accommodation (#PD1251)
Overcrowding of housing and accommodation (#PD0758)
Inadequate housing among indigenous peoples (#PC3320)
Prohibitively expensive housing for the poor (#PE8698)
Unfeasible housing alternatives in urban areas (#PE8061).

♦ PD0822 Inadequate development of new social structures in developing countries
Lack of associations in developing countries
Nature In developed countries, the modern industrial worker is not merely a factor of production in the industrial system, but a participant in it. In contrast, the industrial worker in newly industrialized countries may fail to gain a comparable place in the social structure; newcomers to industry, divested of their traditional social roles, may be absorbed by the industrial system not as social persons, but largely as a market commodity. Industrialization, under these circumstances, leads to the formation of human aggregates which are no longer kept together by ties of family or community, but have not yet evolved new forms of social organizations fitting them for full participation in urban society.
In many of the less developed countries a major difficulty in such popular participation in social change arises from the fact that members of the pre–industrial society are poorly equipped with the mechanisms that play a major part in social change. For example, modern industrial society is to a considerable extent an 'associational' society; it involves an intricate framework of associations and groups, organized to foster (directly or through the government) special interests and purposes: professional, welfare, economic, political, artistic, religious and so on. Of particular significance in the this context are social reform movements organized for the deliberate purpose of introducing changes. There is a notable absence of such associations in most pre–industrial societies, where organized human relationships are limited largely to those defined by the structure of the family and local community.
Claim Where urban deterioration sets in or improvement is unduly slow, it is often a measure of the disparity between the pace of technological and industrial change and the pace of social change. The lag in social change and the resultant worsening of the problems of transition usually reflect the failure of a community to develop new institutions, organizations, habits and ways of life – in respect of the provision of security, personal status, social acceptance, moral controls of behaviour, leadership, forms of recreation – to replace those associated with the extended family or local community which are no longer appropriate to an industrial society.
Related Non–adaptive local structure of social care (#PE5246)
Inflexible social care structures in developing countries (#PF2493).
Aggravates Lack of social identity (#PJ4185) Vulnerability of social systems (#PB2853)
Unequal distribution of social services (#PC3437)
Increase in anti–social behaviour in developing countries (#PD0329).
Aggravated by Social breakdown (#PB2496) Inflexible social structure (#PB1997)
Economic bias in development (#PF2997).

♦ PD0825 Insufficient translation into minority languages
Nature Due to the considerable increase in published material in the major world languages, and

to the insistence by minority cultures on the use of local languages, publication of adequate translations (particularly of textbooks) has not kept pace.
Incidence Translations represent only 8 to 9 per cent of world publication output. Of this small volume, 75 per cent of translations are from 5 main languages widely spoken throughout the world and 75 per cent are made in 12 main producing countries. There is therefore an imbalance which affects the whole book market.
Claim A number of bottlenecks seriously impede book development in the world, particularly in regions where a book shortage prevails. One of the keys to the financing of intellectual production is the translation or adaptation of existing works in so far as this will provide a market for writers in countries with languages which are not widely spoken and will enable countries which cannot meet the demands of their inhabitants to supplement national production. Translation is, however, a medium of exchange from which the developing countries are least able to benefit although it is those countries which have the most urgent need of it.
Counter-claim As the Kenyan novelist, Ngugi wa Thiongo, argues, if authors would write in their indigenous languages, the problem of translations would be markedly reduced.
Broader General obstacles to problem alleviation (#PF0631).
Aggravates Inaccessibility of knowledge (#PF1953).
Aggravated by Multiplicity of languages (#PC0178)
Proliferation of information (#PC1298).

♦ PD0826 Social environmental degradation from recreation and tourism
Cultural degradation from recreation and tourism — Over-exploitation of tourism — Excessive tourism
Nature Excessive tourism, or the increase in the number of tourists towards a saturation level, is evidenced by: diversion of land to the accommodation of tourist facilities, thus preventing its use for houses, schools and open space; threat to the local employment structure by the growth of seasonal and low-paid employment needed to service the tourist industry; and pressure on the general urban infrastructure, particularly transportation, through the steadily increasing (although seasonal) tourist flow. These factors alienate local residents, particularly when the tourist facilities consist primarily of casinos and strip-shows which attract crime and prostitution. Deterioration of tourist areas is usually progressive, involving misuse of resources and spoilage of the very assets that brought the tourists in the first place. Overdevelopment drives tourists away to the next popular and unspoilt location. Social problems often accompany the growth of tourism, including: social friction arising from the importation of foreign workers; confrontation such that the better-off traveller gives orders to the less-well-off native; resentment of residents having to share services with tourists; disappearance or commercialization of local cultures and customs leading to a monotonous world non-culture of identical styles, behaviour patterns, entertainment, food, and language. Economic problems include: hindrance to the growth of a country's economy through the commitment of a large portion of the labour force to a service activity with poor productivity prospects; inflationary consequences of excessive tourist activity, in particular enormous increases in the price of basic foodstuffs and other commodities required by both the tourists and the local population; unfavourable impact on the balance of payments due to the initial investment, imports for maintenance and outward movement of profits; heavy infrastructure costs; loss of control over the economy due to the absence of fiscal regulators where taxes are low or non-existent in order to encourage foreign investment; and overdependence of an economy on one product vulnerable to changes of fashion.
Incidence In 1971, 181 million tourists spent $19,900 million (excluding payments for international transportation). Only 12 countries account for 75 per cent of all tourists, and two of these (USA and Fed Rep of Germany) account for 40 per cent of all tourist arrivals, although they represent less than 10 per cent of the world's population. The problem will considerably increase when incomes rise sufficiently in other areas of the world to permit international travel.
Claim The tourist, in his search for something different, inevitably erodes and destroys that difference by his very enjoyment of it. The growth in tourist numbers will tend to be provided by people who are unaware of the problem to which they give rise and who will therefore be unable to mitigate the ill effects which they, albeit unconsciously, may cause.
Counter-claim Tourism is a major source of foreign exchange, particularly for developing countries. It is a growth industry, often with much better prospects than other export industries. In some countries, the tourist potential may offer an inherent competitive advantage over other countries, and in some cases a complete monopoly. (Historic monuments, pleasant scenery, good beaches, sun or snow are as valuable to a country as more tangible resources). In addition, development of the tourist industry may enable a country to promote a better image of itself and thereby achieve other objectives. The industry is a major source of employment, often in areas of unemployment. It can be an instrument of regional policy aimed at establishing an equitable balance between industrial areas and the rest of the country.
Broader Inappropriate policies (#PF5645) Ineffective industry self-regulation (#PF5841).
Narrower Hunting tourism (#PE3008) Political tourism (#PD7276)
Abuse of traditional customs (#PJ0675) Abuse of tourism for sexual purposes (#PE4437)
Degradation of developing countries by tourism (#PF4115)
Abuse of traditional cultural expressions of peoples (#PE4054).
Related Tourist hazards (#PE8966)
Alienation of land through acquisition by foreigners (#PE0896)
Abuse of international cultural, diplomatic and commercial exchanges (#PF3099).
Aggravates Crime (#PB0001) Foreign ownership (#PE4738)
Sexually transmitted diseases (#PD0061) Disappearance of local culture (#PF3012)
Endangered monuments and historic sites (#PD0253)
Abusive exploitation of cultural heritage (#PC7605)
Insufficient transportation infrastructure (#PF1495)
Deterioration of cultural artefacts from tourism (#PE9825)
Abuse of monopoly power of state-owned or state-controlled enterprises (#PE0988).
Aggravated by Unemployment (#PB0750) Imbalance of payments (#PC0998)
Unethical entertainment (#PF0374)
Chronic shortage of foreign exchange (#PC8182).

♦ PD0830 Geomagnetic disasters
Broader Natural disasters (#PB1151).
Narrower Geomagnetic storms (#PD1661) Geomagnetic reversal (#PF1588)
Geomagnetic field anomalies (#PF2407).

♦ PD0837 Proliferation of nuclear weapons and technology
Diversion of nuclear materials
Nature Proliferation, in the sense of spread of independent ownership of nuclear weapons to more nations than those which currently possess them, is detrimental to international peace and security. The risks of using nuclear weapons are likely to increase the more there are independent possessor nations in the world. Nuclear source materials used for peaceful purposes cannot be safeguarded against diversion to military activities. There is danger in the lack of agreement by countries having nuclear reactors, to allow inspection by the International Atomic Energy Agency of all such facilities for monitoring of the nuclear fuel cycle in order to detect or prevent diversion to military purposes. IAEA nations without nuclear weapons have agreed that membership is conditional upon such inspections, but the nuclear weapon nations are on a voluntary basis to negotiate terms of inspections and other safeguards. Only the UK and the USA have done so, while the USSR has made a token agreement to allow inspection of a portion of its 40 operating reactors.
Incidence Nuclear technology is spreading rapidly around the globe, because of the need for alternative energy production. In 1986, 26 countries had about 380 operating commercial nuclear power stations. As a by-product of this power production, considerable quantities of plutonium are produced each year: over 100 tons in 1980. About one third of this plutonium is produced in the present non-nuclear-weapon countries. This in theory could correspond to the production of about 100 nuclear weapons of nominal size per week in such countries. In the US, 23 university research reactors using highly enriched, weapons-grade uranium (such as the University of Missouri with 45 kilograms and MIT with 29 kilograms) have insufficient security to prevent theft, but are unwilling to convert to low-enriched uranium, according to the Nuclear Regulatory Commission.
Claim The potential for the spread of nuclear weapons is one of the most serious threats to world peace.
Refs Finch, Ron *Exporting Danger* (1986); Fischer, David; Goldblat, Jozef and Szasz, Paul (Eds) *Safeguarding the Atom* (1985); Goldblat, Jozef *Non-Profileration* (1985); United Nations *Problems Associated with Export of Nuclear Power Plants* (1979).
Broader Competitive acquisition of arms (#PC1258).
Narrower Insufficient nuclear weapon free zones (#PJ5335)
Proliferation of strategic nuclear arms (#PD0014).
Aggravates Theft of nuclear materials (#PD3495)
Ease of manufacture of nuclear bombs (#PF3494)
Environmental hazards of nuclear weapons industry (#PE5698)
Insecurity and vulnerability of non-nuclear weapon states (#PD1521).
Aggravated by Energy crisis (#PC6329) Techno-mercenaries (#PE7153)
Unethical practice of physics (#PD1710) Unethical practices by employees (#PD4334)
Illegal exports of nuclear materials (#PE3968)
Vulnerability of nuclear power sources (#PD0365)
Diversion of high technology to hostile countries (#PE7174)
Adverse consequences of scientific and technological progress (#PF3931).
Reduced by Monopoly of nuclear power techniques (#PD1741)
Misuse of satellite surveillance by governments (#PF3701).

♦ PD0847 Inadequate educational facilities
Limited educational facilities — Undesignated school facilities — Lack of educational facilities — Minimal preschool facilities — Insufficient educational facilities — Expensive training facilities — Advanced training facility — Insufficient training facilities
Broader Obstacles to education (#PF4852).
Narrower Lack of secondary schools in the countryside (#PE8797)
Lack of facilities for educating older people (#PF2012)
Inadequate educational facilities for gifted children (#PD2051)
Inadequate educational facilities for disabled persons (#PF0775).
Related Inadequate facilities for children's play (#PD0549).
Aggravates Cultural deprivation (#PC1351) Lack of social mobility (#PF2195)
Inequality in education (#PC3434) Segregation in education (#PD3441)
Exploitation of child labour (#PD0164) Inadequate secondary education (#PD5345)
Denial of education to minorities (#PC3459) Restrictive use of available land (#PF6528)
Illiteracy among indigenous peoples (#PD3321)
Fragmented planning of community life (#PF2813)
Lack of essential local infrastructure (#PF2115)
Discrimination against women in education (#PD0190)
Unavailable education for effective living (#PF2313)
Inadequate education of indigenous peoples (#PC3322)
Inadequate practical training in rural areas (#PF6472)
Haphazard transmission of practical technology (#PF3409)
Neglect of the role of women in rural development (#PF4959)
Restriction of indigenous populations to reservations (#PD3305).
Aggravated by Ineffective methods of practical education (#PF2721)
Unperceived relevance of formal education in rural communities (#PF1944).

♦ PD0850 Dependence within extended families
Nature Negative effects of the extended family include: lack of family planning and the use of the family as a means of social security, which involves child labour; obligations to support members of the family in need – this may include many people and the burden may fall on one breadwinner; family inheritance, which in certain communities divides land equally between sons and thus reduces the economic viability of holdings. The extended family in its social context may not be recognized by the law. New urban families face severe adjustment problems as they move into a more nuclear family situation, but the extended family may also be transplanted into the urban setting with increased family tensions from overcrowding, and the threat to the traditional authoritarian hierarchy from wider social affiliations.
Broader Family structure as a barrier to progress (#PF1502)
Narrow legal definition of the family in developing countries (#PD1501).
Narrower Arranged marriage (#PF3284)
Dependence on family for social security (#PE8670).
Related Negative effects of the nuclear family (#PF0129)
Adjustment difficulties of new urban families (#PF1503).
Aggravates Abandoned children (#PD5734) Juvenile delinquency (#PC0212)
Lack of family planning (#PF0148) Exploitation of child labour (#PD0164)
Discrimination against illegitimate children (#PD0943).
Aggravated by Tribalism (#PC1910) Traditionalism (#PF2676).

♦ PD0858 International double taxation
Nature When an individual or a business enterprise is recognized by more than one country to be engaged in economic activities, the tax regulations of each country may be formulated in such a way that the individual or enterprise is taxed not only on the income in the country in question but on all income earned in other countries. This penalizes inequitably any economic activity with links outside a given country. It is a special difficulty for individuals with income in one country who are also resident and employed in a second country.
Incidence The extent of the problem is indicated by the fact that the accepted method of avoiding the difficulty is through bilateral double taxation agreements between the country pairs concerned. There are over 140 countries between each of which such agreements have to be negotiated. The problem is aggravated by the difficulty of determining in precisely which country the income was generated, particularly since corporate accounting systems may conceal rather than reveal the exact manner in which the income was generated.
Refs International Fiscal Association (Ed) *International Double Taxation of Inheritances and Gifts* (1985); International Fiscal Association (Ed) *Currency Fluctuations and International Double Taxation* (1986); Pires, Manuel *International Juridical Double Taxation of Income* (1989).
Broader Tax obstacles to international investment (#PD0673)
Tax discrimination against non-residents of a country (#PD3048)
Tax discrimination against investment in a foreign country (#PD3047).

♦ PD0862 Unethical personnel practices
Irresponsible personnel practices — Corruption in personnel practices
Claim Employers may adopt a range of unethical practices ranging from explicit or implicit discrimination against job applicants, exploitative hiring policies, misleading employees as to the

nature of work required or the conditions of work, and biased employee evaluations with a view to career advancement. Personnel may be hired illegally, as with the employment of minors or illegal immigrants. Hiring and promotion may be done on the basis of nepotism or in return for favours (whether sexual or otherwise). The form of salary payment may arranged to ensure tax avoidance or tax evasion.

Broader Unethical practices of employers (#PD2879).
Aggravates Tax evasion (#PD1466) Forced labour (#PC0746) Chattel slavery (#PC3300) Secret societies (#PF2508) White slave trade (#PD3303) Trafficking in women (#PC3298) Exploitation in employment (#PC3297) Discrimination in employment (#PC0244) Unethical practices by employees (#PD4334) Trafficking in children for adoption (#PF3302) Sexual harassment in the working place (#PE8466).

♦ PD0868 Parasites

Parasitism

Nature Parasites are organisms living in or on another of a different kind, deriving their subsistence form the living material, digested food, secretions or other products of their host, are inextricably bound up with their host or hosts in the continuance of their life, and is rather injurious than beneficial in their influences. Parasitism is a relation of dependence, always nutritive, often more, between the parasite and the host but the inter-relation takes so many forms that absolutely precise definition is impossible, and it is not easy to separate off parasitism from other vital associations. A living creature habitually growing on a plant is called an epiphyte; it is not a parasite unless it gets its food in whole or in part from its bearer, as mistletoe does. A living creature growing on an animal is called a epizoic, like a barnacle on a whale; it is not a parasite unless it gets its food in whole or in part from its bearer, such as many fish lice. Parasitism is not symbiosis, a mutually beneficial internal partnership between two organisms of different kinds, such as unicellular algae which live within Radiolarians, some polyps, and a few worms. Nor is it commensalism, a mutually beneficial external partnership between two organism of different kinds, such as some kinds of hermit crabs which are always accompanied by sea-anemones. The distinction between parasitism and these other forms of relations is difficult to make. A distinction between parasites and predators must be made because killing off the host is not in the interest of the parasite.

Parasitism is also a term used metaphorically to describe the poor and the rich, depending on who is using it, and other human relationships.

Refs Flynn, Robert J *Parasites of Laboratory Animals*; Gaafar, S M, et al (Eds) *Parasites, Pests and Predators* (1985); Mettrick, D F and Desser, S S (Eds) *Parasites* (1982); Read, Clark P *Parasitism and Symbiology*.

Broader Biological contamination of water (#PD1175).
Narrower Bedbugs (#PE3617) Myiasis (#PE3633) Mites as pests (#PE3639) Leeches as pests (#PE3660) Parasitic plants (#PD6284) Botflies as pests (#PE3635) Blowflies as pests (#PE3627) Protozoan parasites (#PE3676) Lice as insect pests (#PE1439) Black flies as pests (#PE3646) Tsetse flies as pests (#PE1335) Flies as insect pests (#PE2254) Hemiptera as insect pests (#PE3615) Parasites of the human body (#PE0596) Siphonaptera as insect pests (#PE3643) Parasitic diseases in animals (#PD2735) Mosquitoes as vectors of disease (#PE1923) Kissing bugs as vectors of disease (#PE3616).
Aggravated by Houseflies as pests (#PE3609).

♦ PD0874 Ageing war disabled

Broader Human ageing (#PB0477) Human disability (#PC0699).
Aggravates Social withdrawal of aged (#PD3518) Mental disorders of the aged (#PD0919) Inadequate income in old age (#PC1966) Social disadvantage of the aged (#PD3517) Inadequate welfare services for the aged (#PD0512) Susceptibility of the old to physical ill-health (#PD1043).
Aggravated by Injuries (#PB0855).

♦ PD0876 Inappropriate sanitation systems

Inadequate sewerage systems — Inadequate sanitation infrastructure — Undeveloped sanitation services — Improper sewage disposal — Overloaded sewage systems — Insanitary drainage facilities — — Unsanitary toilet facilities — Poor sanitation facilities

Incidence Rapid expansion of urban populations presents problems for the provision of basic services. Particularly where squatter settlements proliferate at the outskirts of cities, a common occurrence in developing countries, access to sanitation (or drinking facilities) may be entirely lacking or inadequate. As an indication, the market for toilet cleaners in the UK in 1986 was £88 million.

Claim Modern sewage systems are excessively expensive means of taking household wastes and disposing of them in such a way as to prevent nutrients from returning to the soil.

Broader Unhygienic conditions (#PF8515) Inadequate infrastructure (#PC7693).
Narrower Agricultural effluent (#PE8504) Lack of cesspool fencing (#PG9786) Inadequate waste treatment (#PD6795) Inadequate toilet facilities (#PG7986) Lack of sanitation in rural areas of developing countries (#PD1225) Inadequate facilities for the transport of sanitary wastes (#PE8475).
Related Dysfunctional public utilities (#PE7647) Weakness of infrastructure in developing countries (#PC1228).
Aggravates Cholera (#PE0560) Hepatitis (#PE0517) Insect pests (#PC1630) Typhoid fever (#PD1753) Malodorous fumes (#PD1413) Enteric infections (#PD0640) Industrial effluent (#PG6771) Sewage as a pollutant (#PD1414) Disease-prone housing (#PG7994) Tapeworm as a parasite (#PE3511) Excreting in public places (#PE1602) Sanitation system accidents (#PE4570) Faecal transmission of disease (#PG4438) Insect vectors of human disease (#PE3632) Pollution of water by infected faeces (#PE8545) Water pollution in developing countries (#PD3675) Unequal regional distribution of deaths (#PC4312) Methane gas emissions from landfill sites (#PE1256) Demoralizing images of rural community identity (#PF2358) Unsanitary environment for basic health in small rural villages (#PD2011).
Aggravated by Random defecation (#PG6578) Prohibitive cost of basic services (#PF6527) Unfocused design of community space (#PF1546) Substandard housing and accommodation (#PD1251) Fragmented planning of community life (#PF2813) Inadequate housing in developing countries (#PE0269) Inadequate housing among indigenous peoples (#PC3320) Underprovision of basic services to rural areas (#PF2875) Ineffective structures for community decision-making (#PF1781) Undeveloped channels for public and private resources (#PF3526) Dependence on sophisticated technology for development (#PD6571) Prohibitive costs of purification of effluents and emissions (#PF5743) Underdeveloped provision of basic services in developing countries (#PF6473) Shortage of sanitary plumbing, heating and lighting fixtures and fittings (#PE7940).
Reduced by Priorities to sanitation funding (#PS8523).

♦ PD0877 Instability of trade in mineral fuels, lubricants and related materials

Broader Instability of the primary commodities trade (#PC0463).
Narrower Instability of trade in gas (#PG5604) Instability of trade in electric energy (#PE7907) Instability of trade in coal, coke and briquettes (#PE8206) Instability of trade in petroleum and petroleum products (#PD0909).
Aggravates Health risks to workers in agricultural and livestock production (#PE0524).

♦ PD0881 Restrictive transport insurance practices

Nature Some governments that are anxious to expand all sectors of their economy attempt to defend their insurance market by restricting the trader's freedom of choice in the placing of transport insurance. This is done by requiring the buyer or seller to insure his imports in the country of importation or his exports in the country of exportation; by imposing discriminatory taxes on marine insurance placed with foreign companies; or by the operation of their import licensing and exchange control regulations. In so doing, governments overlook the fact that insurance, more than any other transaction, is based on confidence and is thus incompatible with any form of coercion. When insurance affects international transactions, then the need for confidence is all the greater.

Transport insurance of goods has all the elements of an international transaction. The goods themselves move from one country to another, frequently on the high seas. The supplier is in one country, the buyer in another and the carrier may belong to a third country. Consequently, restrictive measures affecting transport insurance inevitably have a direct effect on international trade. It is therefore international trade as a whole which suffers from restrictive measures imposed in relation to transport insurance; and the economy of the country which imposes them loses the benefits to be derived from a free choice of the transport insurance arrangements. If, furthermore, other countries adopt a similar attitude in retaliation, this leads to the paradoxical situation where no commercial transaction remains possible between two countries without violating the law of one or the other.

Claim It is not so much the relatively small amounts of money involved in relation to the business turnover as the burden of business complications and administrative vexation resulting from such restrictions in the field of transport insurance which constitute the undeniable harm done to international trade; the role of transport insurance should be to facilitate and not to impede the flow of trade.

Broader Restrictive trade practices (#PC0073).
Narrower Inequities in marine insurance (#PE5802).
Aggravates Inadequate insurance (#PF8827) Obstacles for international ocean shipping (#PD5885).

♦ PD0884 Social stigma

Stigmatization

Nature A social stigma is a mark of disgrace imposed on an individual by other individuals or a social group. In popular usage it often refers to any negative sanction or to disapproval for nonconformity.

An undesirable differentness of an individual that disqualifies him or her from full social acceptance. It is an attribute or stereotype that departs negatively from the expectations of others.

Refs Ainlay, Stephen C, et al *The Dilemma of Difference* (1986); Gerhardt, Uta E and Wadsworth, Michael E J (Eds) *Stress and Stigma* (1984); Goffman, Irving *Stigma* (1963); Page, Robert M *Stigma* (1984).

Broader Negative emotions and attitudes (#PA7090).
Narrower Stigmatized diseases (#PD7279) Dependence on breast feeding (#PE7627) Discrimination against illegitimate children (#PD0943).
Aggravates Teasing (#PE4187) Physical intimidation by children (#PE2876) Discrimination against homosexuals (#PE1903) Discrimination against HIV-infected persons (#PE4299).
Aggravated by Divorce (#PF2100) Prejudice (#PA2173) Bestiality (#PE3274) Mental disorders (#PD9131) Unmarried mothers (#PD0902) Marital instability (#PD2103) Prejudice in children (#PD8973) Sexual exploitation of men (#PD3263) Human pseudo hermaphroditism (#PE2246) Children engendered by occupying soldiers (#PD8825) Discrimination against mixed race children (#PE5183) Failure to legally rehabilitate victims of miscarriage of justice (#PF7728).

♦ PD0894 Lack of meaningful personal and social paradigms

Nature A lack of meaningful images prevents people from making decisions which are unrelated to the future of society. Most images do not portray the interrelatedness of people – the concepts of individualism and nationalism have become overruling forces. Social images and roles neglect to portray that each person has the possibility of creating his own future. Images that convey the escaping from situations have emerged.

Broader Tensionless image of free choice (#PF1675).

♦ PD0902 Unmarried mothers

Discrimination against unwed mothers

Nature The unmarried mother is considered to be the single woman who has given birth to a child out of wedlock, and not the mother who, because of widowhood, desertion, or some other reason, is the sole parent. Discrimination against unmarried mothers still exists, both in law and in fact, in many countries. Situations exist where the unmarried mother is denied the recognition in law of any status as a parent. But the consequences of births out of wedlock are not reflected exclusively in the legal field. The unmarried mother may also feel them in various other aspects of her life, especially in relation to the community in which she lives. The fact that a woman has given birth to a child out of wedlock affects the attitude of society towards her, one of the consequences being the ostracism to which she is often subjected. Also, her responsibilities as a mother are most often made heavier as a result of the lack of the moral and financial assistance of a husband. In certain countries unmarried mothers may have difficulty in establishing the legal paternity of the child and therefore be unable to obtain maintenance. Social discrimination and lack of child care facilities may be such that it is difficult for an unmarried mother to find employment and housing. In extreme cases in some societies an unmarried mother may be considered to have brought shame and degradation on her family, and be murdered by her brother or her father – while little or no punishment is given to the man involved.

Background Formerly, in some countries, unmarried mothers could be imprisoned for several months for a first pregnancy and for life for a second.

Claim The statistics indicate that the problems faced by unmarried mothers are affecting millions of children since, in the final analysis, whatever contributes to the neglect or the improvement of the position of unmarried mothers has a definite impact on their children, and the well-being and future participation of those children in the development of society.

Broader Immorality (#PA3369) Unmarried parents (#PD3257) Single parent families (#PD2681).
Related Social injustice (#PC0797) Social inequality (#PB0514) Discrimination against unmarried fathers (#PD3256).
Aggravates Social stigma (#PD0884) Illegitimate children (#PC1874) Illegal induced abortion (#PD0159) Inadequate laws of adoption (#PD0590) Discrimination against women (#PC0308) Denial of parental affiliation (#PD3255) Non-parental custody of children (#PF3253) Trafficking in children for adoption (#PF3302) Inadequate child day-care facilities (#PD2085) Inadequate medical care for pregnant women (#PE4820).

Aggravated by Promiscuity (#PC0745)
Paternal negligence (#PD7297)
Adolescent pregnancy (#PD0614)
Desertion in marriage law (#PF3254)
Sexual exploitation of women (#PD3262)
Lack of family planning information (#PD1050)
Discrimination in family planning facilities (#PD1036).

♦ PD0908 Underdevelopment of food processing industries

Broader Instability of manufacturing industries (#PC0580)
Underdevelopment of manufacturing industries (#PF0854)
Underdevelopment of food and live animal production (#PF2821).
Related Instability of trade in tobacco and tobacco manufactures (#PE0572).
Aggravates Environmental hazards from food and live animals (#PC1411)
Environmental hazards from food processing industries (#PE1280).

♦ PD0909 Instability of trade in petroleum and petroleum products

Instability of petroleum prices — Oil price fluctuations

Incidence Until 1970, the world oil prices never rose higher than US $2 a barrel, although oil products were sold at prices several hundred percent above cost. In the 1970s and 1980s, OPEC countries raised the per barrel reference prices from less than US $2.00 to $34.00. Since 1979, OPEC's market share has dropped from 63 to 38 percent of world oil sales. Oil production has grown enormously in China, Latin America and Northern Europe while it has halved in the Middle East. The role of the Arab oil-producing countries and the importance of the Straits of Hormuz to the world economy has dramatically declined not only because of the Iran-Iraq war, but also because of the rapid development of non-petroleum-based sources of energy. present oil glut has caused production or price roll backs in OPEC and in non-OPEC producers such as the Soviet Union, Norway and Britain. Oil prices may stabilize around $20-25 a barrel in the 1990's.

Refs Mabro, Robert (Ed) *The 1986 Oil Price Crisis* (1988).
Broader Instability of trade in mineral fuels, lubricants and related materials (#PD0877).
Aggravates Criminal contempt (#PD5705)
Distortion in international politics from fuel shortages (#PE8080).

♦ PD0914 Mental deficiency in children

Mentally retarded children — Mentally subnormal children

Nature The term 'mental subnormality' describes an incomplete or insufficient general development of the mental capacities. The term 'mental deficiency', more commonly used to describe the disability of persons suffering from incomplete development of the intelligence, is also often applied to conditions in which only the individual's emotional development is incomplete while intellectual development continues up to normal levels. These latter are also sometimes referred to in certain countries as 'moral defectives', especially in connection with certification or commitment procedures. Their disabilities, however, differ in kind from those of persons described as mentally subnormal; and it is probably preferable to deal with those who have normal intellectual abilities but incomplete or distorted emotional development which leads them into conflict with the law as 'anti-social persons with psychopathic personalities', rather than to equate them with the mentally subnormal.

The picture is further complicated by the fact that in the USA the term 'feeble-mindedness' is now coming to be replaced by the terms 'mental deficiency' or ' 'mental retardation'. In British usage, however the term 'retardation' has a developmental implication.

Although it is not possible to draw a sharp distinction between mental subnormality of varying degrees of severity, in many ways useful practice is to divide the condition into three grades: 'mild', 'moderate' and 'severe' degrees of subnormality.

Incidence Few attempts have been made to assess the prevalence of mental subnormality in different countries, and it is not possible to give definitive figures since prevalence rates depend on many factors which differ both with society and with social and economic conditions. The proportion of children regarded as educationally subnormal in different countries varies greatly according to the criteria employed. Dutch estimates based on eight large cities give a mean rate of 2.6 per cent; French estimates range from 1.5 per cent to 8.6 per cent, depending on age; English educational practice aims to make provision for 1 per cent of schoolchildren in special schools, while a further 8 per cent or 9 per cent are considered to require special educational provision within the ordinary school system. Varying estimates have been given in different states of the USA and in Switzerland. For adults the prevalence rates are lower, and the recognition and even the manifestation of mild subnormality in adulthood is dependent mainly on thresholds of community tolerance and the complexity of social life, both of which fluctuate widely.

Estimates are thus valid only for the time and place at which they are made. It is, however, agreed by all that the number of mildly subnormal far exceeds that of more severe cases; and English statistics which have been widely quoted suggest that among every 100 mentally subnormal persons, the following proportions will be found: 75 mild, 20 moderate, and 5 severe cases. In other words, the very great majority of the subnormal are of mild grade and potentially capable of being taught to make a fairly adequate social adaptation in appropriate circumstances.

Background It is probable that in all societies the grossest forms of mental subnormality have been recognized and given special names, but in communities in which little attention is paid to mental health, the severely defective alone are regarded as remarkable. It is only during the last century, and in particular within the last 40 or 50 years in parts of Europe and America, that some consideration has been given to the special problems presented by persons suffering from milder forms of mental subnormality or social incompetence. The need for such consideration has arisen partly due to the industrialization of society and the consequent growth of cities, partly because of compulsory education, and partly through the demands made by industrialism for rapid occupational adaptation. The increases in the social services which characterize modern society have also led to a growing recognition of this need. The mild and even moderate degrees of mental retardation frequently associated with environmental deprivation are often revisable by appropriate intervention applied at an early age, and are preventable.

Refs Forest, Marsha *Education – Integration* (1984); Manocha, Sohan L *Malnutrition and Retarded Human Development* (1972); Robinson, Halbert B and Robinson, Nancy M *The Mentally Retarded Child* (1976).
Broader Mental deficiency (#PC1587)
Disabled children (#PC5890).
Related Autism (#PE1222)
Mental illness in adolescents (#PE0989).
Aggravated by Inadequate guardianship for mentally retarded adults (#PE3992).

♦ PD0917 Parochial escapist media entertainment

Bias in the media against development and social transformation — Mass media bias towards entertainment and reassurance — Parochial news coverage

Claim The present and potential influence of the entertainment and information media is enormous. The escapist, diversion-oriented programming which today indicates one dimension of our paralysis, is being sustained by the managers of media as a service to the established structures. This contributes to the isolation, victimization and parochialism of local people and is a key factor in limiting their global possibilities.

Broader Escapism (#PF7523)
Distorted media presentations (#PD6081)
Thwarted technological communications (#PF0953).
Aggravates Meaningless recreation (#PF0386)
Dependence on the media (#PD7773)
Insufficient cultural media (#PJ8476)
Deterioration of media standards (#PD5377)
Media theatricalization of public life and politics (#PF9631).
Aggravated by Excessive television viewing (#PD1533)
Incomplete access to information resources (#PF2401).

♦ PD0919 Mental disorders of the aged

Incidence There is a rise in the prevalence of mental illness in persons above the age of 65, as well as tendency for the rate of mental illness to vary with marital status: widowed persons, in particular, have a higher rate than married ones. More striking differences are found in relation to social class: there is a marked trend for the frequency of mental illness to vary inversely with social status. There is also a strong positive association found between mental disorder and physical handicap, which holds good for both men and women, and for all age groups over 65.

Refs Butler, and Lewis, *Aging and Mental Health* (1983); Pitt, Brice (Ed) *Dementia* (1987).
Broader Human ageing (#PB0477)
Human disease and disability (#PB1044)
Susceptibility of the old to physical ill-health (#PD1043).
Narrower Neurosis (#PD0270)
Paranoia (#PE0435)
Senile dementia (#PE3083)
Reading disabilities (#PD1950)
Changes of personality (#PG3085)
Atrophic senile psychosis (#PG3080)
Arteriosclerotic psychosis (#PG3081)
Manic-depressive psychosis (#PD1318)
Rigidity and inadaptability in the aged (#PD3515)
Decline in cognitive ability with ageing (#PE9620).
Related Schizophrenia (#PD0438)
Slowness of sensori-motor activities in the aged (#PD3514).
Aggravates Suicide (#PC0417)
Fear of death (#PF0462)
Social withdrawal of aged (#PD3518)
Deterioration of the mind with age (#PE4649)
Inadequate welfare services for the aged (#PD0512).
Aggravated by Bereavement (#PF3516)
Living alone (#PF3089)
Malnutrition (#PB1498)
Family breakdown (#PC2102)
Ageing war disabled (#PD0874)
Socio-economic poverty (#PB0388)
Poor living conditions (#PD9156)
Rapidly changing cultures (#PF8521)
Ageing of world population (#PC0027)
Inadequate income in old age (#PC1966)
Excessive population mobility (#PF3806)
Poverty in developing countries (#PC0149)
Inadequate housing for the aged (#PD0276)
Age discrimination in employment (#PD2318)
Inadequate recreational facilities (#PF0202)
Diseases and injuries of the brain (#PD0992)
Negative effects of the nuclear family (#PF0129)
Retirement as a threat to psychological well-being (#PF1269)
Inadequate rehabilitation facilities for the mentally handicapped (#PE8151).

♦ PD0923 Fragmentation of communist parties

Nature There is considerable evidence that the communist parties of the world are no longer acting in harmony, and that the common ideology of Marxism-Leninism has lost its integrating and mobilizing force. The Soviet Communist Party has been instrumental in bringing this about by putting the interests of the Soviet state above the interests of the communist movement as a whole. Thus, attempts by satellites of the Soviet Union to bring about reform (as in Poland in the 60s and 80s and Czechoslovakia in the late 60s) were stifled, and this gave rise to disillusionment with any idea of reform in the countries of the Eastern bloc. The Soviet Union has been unable to establish 'good relations' with the countries in its Eastern European sphere of power, and has been seen to be unable to act as an equal partner with them. This has influenced the actions of the communist parties of other countries.

Incidence In recent years, communism in Western Europe has lost much of its original dynamism and fascination. One reason is that the various communist parties find it difficult to reconcile changes in economic and social structures in their various countries (and the resulting shifts in awareness and values) with their theories. They are also divided among themselves on such fundamental problems as the strategy of transformation to be pursued and on their assessment of Soviet policy. The Italian Communist Party, for example, which is the largest and most influential in the West, has publicly called into question the socialist nature of the Soviet system and called the construction of socialism in Eastern Europe 'a misguided development forced upon these countries from outside'. There are, however, many anomalies. The Chinese Communist Party has resumed relations with parties critical of Moscow (such as the League of Communists of Yugoslavia, the Italian Communist Party and the Spanish Communist Party) but also with the French Communist Party whose foreign relations are still oriented towards Moscow.

Broader Fragmentation (#PA6233).
Related Conflict between minority groups (#PC3428).

♦ PD0928 Inadequacy of the domestic market in developing countries

Small size domestic market in developing countries

Nature Low per capita national income is one of the principal attributes of less developed countries. Low personal incomes, in turn, are reflected in a pattern of expenditure which is very unfavourable to secondary industry; in general, the lower the income the higher is the proportion spent on food, much of which is unprocessed. It is only as income rises that a market for manufactured products is created. This is one of the ways in which the process of industrialization, when soundly conceived and executed, tends to be cumulative: the higher incomes which flow from new investment and employment are among the principal elements of an expanded demand for manufactures. Nevertheless, many under-developed countries are too small – quite apart from income levels – to be able to sustain factories of economic size in more than one or two segments of secondary industry. The attainable level and patterns of industrialization in such countries are obviously limited.

Incidence The following least developed countries are reported to have a negative annual growth rate per capita GDP, 1980-85: **Af** Burkina Faso, Central African Rep, Chad, Djibouti, Ethiopia, Gambia, Guinea, Lesotho, Malawi, Mali, Mauritania, Niger, Sao Tome-Principe, Sudan, Tanzania UR, Togo. **Am** Haiti. **As** Yemen Dem. **Pa** Kiribati, Samoa, Vanuatu.

Broader Domestic market restrictions in developing countries (#PD1873)
Underdevelopment of industrial and economic activities (#PC0880).
Aggravates Capital shortage in developing countries (#PD3137)
Inadequate development of enterprises in developing countries (#PE8572).

♦ PD0929 Inadequate facilities for international organization action

Broader Inadequate buildings, services and facilities for organized action against problems (#PD2669).
Narrower Inadequate facilities for international nongovernmental organization action (#PF2016).
Aggravates Ineffectiveness of international organizations and programmes (#PF1074).
Aggravated by Inadequate funding of international organizations and programmes (#PF0498).

♦ PD0930 Blocked global marketing

Nature The marketing of goods across the globe is obstructed in many ways. There is no detailed current analysis of present and future global needs; a constantly fluctuating monetary system; and, in many instances, a severe lack of necessary credit.

Broader Monopoly of the economy by corporations (#PD3003).

♦ PD0932 Excessive institutionalization of education

Claim Education constitutes an independent variable in every society and a direct factor in social contradictions. The school's position in society and the play of forces to which it is subjected make it incapable, however, of being the instrument of true education in the service of mankind or of promoting conviviality. On the contrary, it often serves the purposes of repressive, alienating and

PD0932

dehumanizing societies. Unless educational institutions are suppressed, individuals cannot regain their freedom in society, resume control of the institutions, and thereby recover their initiative in education.
Broader Inadequate education (#PF4984) Faceless social institutions (#PF2081)
Institutional domination of organizational systems (#PF2825).
Narrower Barriers to transfer between educational facilities (#PD0084).
Related Repression of intellectual dissidents (#PD0434)
General obstacles to problem alleviation (#PF0631).
Aggravates Alienation (#PA3545) Dehumanization (#PA1757).
Aggravated by Excessive size of social institutions (#PF8798)
Fostering of dependency by social institutions (#PF1755).
Reduced by Loss of institutional credibility (#PF1963).

♦ PD0941 Discontented marriages
Unhappy marriages
Claim A survey in France indicated that 50 per cent of French women would choose a different husband if given the choice, in contrast to 20 per cent of Swedish women.
Aggravated by Marital stress (#PD0518).

♦ PD0943 Discrimination against illegitimate children
Denial of rights to illegitimate children
Nature Illegitimate children are discriminated against under the law and in social attitudes. Acts of discrimination include: the denial of inheritance or of filiation; legal impediments to marriage; an unequal distribution of social services on account of inferior status or lack of filiation; discrimination in education. This results in lack of social mobility and may also result in juvenile delinquency and emotional stress.
Incidence In most countries children born out of wedlock do not have the same rights as legitimate children, despite the trend in favour of greater equality between them. As a rule, it is only when they have been legitimated that illegitimate children have the same rights and obligations as the children born of the marriage. If, on the other hand, they have merely been recognized by the father, although entitled to be fed and brought up at their father's expense, they cannot claim the same rights as legitimate children, particularly in the matter of inheritance. However, they may, in certain cases, apply to the putative father for a maintenance and education allowance. In Belgium, for example, illegitimate children can exercise this right up to eighteen years of age, in Finland up to seventeen and in Guyana up to sixteen.
Background In Roman times, bastards were classified as 'nothi' and were entitled to support from their father but not to inheritance. Under Germanic law, only children born of parents of the same social rank were legitimate. Under English common law, parental rights and duties did not apply to illegitimate children who were regarded as the 'sons of no one'. Until the 16th century in England the parish maintained such children but an act of 1576 laid the responsibility on the parents.
Broader Social stigma (#PD0884) Illegitimate children (#PC1874)
Denial of rights of children and youth (#PD0513).
Narrower Children engendered by occupying soldiers (#PD8825)
Discrimination against children of prostitutes (#PE0392).
Related Unmarried parents (#PD3257).
Aggravates Juvenile delinquency (#PC0212) Social fragmentation (#PF1324)
Stress in human beings (#PC1648) Legal impediments to marriage (#PF3346)
Denial of parental affiliation (#PD3255) Denial of education to minorities (#PC3459)
Denial of right to inherit property (#PF0886)
Discrimination and harassment of children in public life (#PE6922).
Aggravated by Religious intolerance (#PC1808) Inadequate sex education (#PD0759)
Dependence within extended families (#PD0850).

♦ PD0946 Lagging transformation of agriculture in developing countries
Nature The decisive changes in agricultural structure, productivity and food production which created a vital surplus above subsistence level, and preceded the growth of industrial cities in Europe and America and later in Japan, have begun to occur on a sufficient scale only very recently in the developing countries.
Because the agricultural sector is predominant and pervasive in most developing countries, its performance has a profound influence on growth and the quality of life. Because industrialization alters the methods of production and distribution in all economic sectors, a close relationship between agriculture and industry is necessary in these countries to significantly improve the standard of living. However, during the past two decades of development, the links between the two sectors have been weak and transient. At times, agricultural development may have suffered from a diversion of investment funds to industry. The world food crisis of 1972–1973 was followed by staggering price increases for certain basic commodities (cereals, sugar, oil seed), which were, in fact, alarm signals set off by the great imbalances between the world's food supplies and its food needs. While the imbalances were aggravated by poor harvests, in the long term there has been a disturbing tendency for agriculture to stagnate. For example, in 1965–1975 agricultural output in the developing countries expanded very slowly. Several developing countries experienced a decline in agricultural output in absolute terms.
The agricultural sector is also exposed to an extremely high degree of risk arising out of natural factors, principally weather conditions. The importance of this class of risks stems from the relatively large number of events that may cause agricultural output to drop (floods, drought, frost, hail, pests, diseases, etc), and also from the frequency with which these events actually occur and the magnitude of the crop losses they may cause. Moreover, most of these events resulting in losses are beyond the control or prevention of individuals or of societies as a whole.
Incidence Latin America remains divided between latifundia too large for the entrepreneurial competence of their owners and minifundia too small even for subsistence. Many parts of Africa still have to invent acceptable and workable post-tribal forms of food production for the market. In the Indian subcontinent, managerial skills are only beginning to be sufficiently developed for decisive increases in productivity. Food production in the developing countries in general is not able to do more than keep pace with the growth of population at the old, inadequate levels of diet; this modest increase is not sufficient to offer a comparable rise in employment for the new millions born on the land. In Latin America, for instance, the economically active population grew from 32.5 million to 86.2 million between 1925 and 1960; farming offered only 12 million new jobs. Urban areas had to absorb the balance. The recent dramatic advances in agricultural production through the use of hybrid grains, coupled with improved fertilizer and irrigation practices, are changing this discouraging record. But at best they can only relieve the food shortage for about two decades, a short respite for achieving a better balance between population and resources.
Broader Unsustainable economic development (#PC0495)
Weakness of infrastructure in developing countries (#PC1228).
Narrower Lack of agricultural machinery (#PF4108).
Aggravates Lack of economic and technical development (#PE8190).
Aggravated by Neglect of agricultural and rural life in developing countries (#PF7047).

♦ PD0947 Gerontocracy in developing country agriculture
Nature A tradition still fairly prevalent in tropical Africa, for example, is that of farming by the extended family, under the direction of its head. This involved, and still involves, the great disadvantage that the main economic power is given to the older people, who tend to be less amenable to progress. This likelihood is increased by the fact that the average age of village head-men, who are also entrusted with considerable economic power by the community, is even higher than that of the heads of large families.
Broader Gerontocracy (#PD3133).

♦ PD0948 Toxic metal pollutants
Toxic metals
Nature There are 35 metals of major concern with regard to occupational exposure. About 20 of these may constitute occupational health hazards if not properly handled and may lead to toxic effects in humans. Toxicity may be either acute or chronic.
Acute toxic effects from metals are usually the result of the inhalation of air or the ingestion of liquids containing the metals in very high concentrations. Dangerously high concentrations of metals in workroom air may occur after industrial accidents (such as explosions, sudden leakage of metal fumes from normally closed containers, sudden failure in ventilation systems). Inhalation of high concentrations of metals is irritating and may cause severe damage to the respiratory tract. Fatalities may occur either immediately after exposure or up to several weeks later. Ingestion of high concentrations of metals may take also place through their leakage from containers and pipes into beverages. For example, acute intoxication has often occurred from consumption of home-brewed alcoholic drinks distilled in lead radiators or copper pans. The symptoms following the ingestion of excessive amounts of metals are those of food poisoning, with nausea, vomiting, abdominal pains and, in certain cases, diarrhoea.
The effects of chronic occupational exposure to metals are the most frequent, and are mainly derived from industrial activity. The toxicity of some metals, such as lead, has been known to man for centuries, and may lead to lead paralysis and colic. The inhalation of mercury vapour in mining and in the felt hat industry frequently causes severe central nervous-system symptoms; and long-term exposure to cadmium in workroom air causes pulmonary emphysema and kidney damage with proteinurea.
Incidence Regarding the effects of exposure in the general environment, most problems stem from cadmium, mercury, arsenic and lead. Some worldwide examples are the following: In Japan, cadmium has given rise to the Itai-Itai (or ouch-ouch) disease in hundreds of people. Methylmercury poisoning resulting from the consumption of contaminated fish also caused the Minamata disease in Japan, when methylmercury accumulated in the brains of people who ate contaminated fish caused irreversible damage to the central nervous system with associated ataxia, paraesthesia and decreased visual field; furthermore, a number of children having severe cerebral palsy-like brain damage were delivered by mercury-exposed mothers. In Iraq poisoning was causing by eating seed treated with methylmercury; about 6,000 people were hospitalized, and there were some 500 fatalities.
Lead poisoning in children is a serious world-wide problem. In addition to lead in food and air, children are exposed to lead in contaminated dust and soil, and from the consumption of paints containing lead (pica). It has been estimated that hundreds of thousands of children in the United States are in peril of subclinical lead poisoning.
Arsenic in drinking water has caused severe health problems in Asia and in South and North America. Consumption of water containing arsenic may give rise to vascular changes, hyperpigmentation and keratosis of the skin, and skin cancer.
Broader Pollutants (#PC5690) Chemical pollutants of the environment (#PD1670).
Narrower Tin as a pollutant (#PE1438) Lead as a pollutant (#PE1161)
Nickel as a pollutant (#PE1315) Cobalt as a pollutant (#PE2339)
Mercury as a pollutant (#PE1155) Cadmium as a pollutant (#PE1160)
Arsenic as a pollutant (#PE1732) Vanadium as a pollutant (#PE2668)
Selenium as a pollutant (#PE1726) Plutonium overproduction (#PD2539)
Antimony as a health hazard (#PE1989) Metal contamination of soil (#PD3668)
Manganese as a health hazard (#PE1364) Beryllium as a health hazard (#PE2209).
Aggravates Soil pollution (#PC0058) Occupational hazards (#PC6716)
Occupational risk to health (#PC0865) Chemical contamination of water (#PE0535).
Aggravated by Marine dumping of wastes (#PD3666).

♦ PD0949 Soil erosion
Top soil erosion — Progressive loss of top soil — Soil depletion — Lack of soil conservation
Nature Erosion is the deterioration of soil by the physical movement of soil particles from a given site. Wind, water, glacial ice, animals, and the use of tools by man may be agents of erosion. The two most important agents are wind and water; but in most instances these are important only after man, animals, insects, diseases or fire have removed or depleted the natural vegetation. Exceptions exist, though, where the major causal factor of soil erosion is climatic change. Accelerated erosion is the most serious form of soil degradation, when surface soil may sometimes be blown or washed away right down to bedrock.
It is vegetation that keeps soil (in its natural state) from eroding. Undisturbed by man, soil is usually covered by a canopy of shrubs and trees, by dead and decaying leaves or by a thick mat of grass. Whatever the vegetation, it protects the soil when the rain falls or the wind blows. Root systems of plants hold the soil together. Even in drought, the roots of native grasses, which extend several metres into the ground, help tie down the soil and keep it from blowing away. With its covering of vegetation stripped away, soil is vulnerable to damage. Whether the plant cover is disturbed by cultivation, grazing, deforestation, burning, or bulldozing, once the soil is laid bare to the erosive action of wind and water, the slow rate of natural erosion is greatly accelerated. Losses of soil take place much faster than new soil can be created, and a kind of deficit spending begins with the topsoil.
Incidence Nearly every semi-arid area with cultivation or long-continued grazing, every hill land with moderate to dense settlement in humid, temperate and subtropical climates, and all cultivated or grazed hill lands in the Mediterranean climate areas, suffer to some degree from such erosion. Thus recognized problems of erosion are found in such culturally diverse areas as southern China, the Indian plateau, South Australia, South Africa, the Soviet Union, Spain, the southeastern and midwestern US, and Central America.
Wind erosion is most commonly a problem in (seasonally) dry, windy regions, with a smooth, flat terrain, whereas water erosion is more common in (seasonally) wet regions with a sloping or hilly or mountainous terrain. Both result in a loss of topsoil, rich in humus, and lead to a decline in long-term productivity. With the destruction in soil structure, eroded land is even more susceptible to erosion. In the United States, soil erodes at an average rate of 9 to 12 tons per acre per year. Almost half of America's cropland is losing soil faster than it can be replaced. In developing countries erosion rates per acre are twice as high, partly because population pressure forces land to be more intensively farmed. New topsoil is constantly being formed, of course, but at nature's leisurely pace of about 1.5 tons per acre per year (3.4 metric tons per hectare). The difference between creation and loss represents an annual loss of 7.5 to 10 tons per acre (16.8 to 23.5 metric tons per hectare) worldwide. Erosion is not a new phenomenon, but it has now reached unprecedented proportions, and, should it continue at the current rate, one-third of the world's arable land will be depleted within the next 20 years.
Background Man-induced soil erosion existed even before the development of agriculture, when man employed fire to clear forests. Land which long ago was very fertile, supporting ancient civilizations, is now barren desert, barely supporting a few nomadic tribes. The desert of North Africa was once 'the granary of Europe'. Accelerated soil erosion has been known throughout history wherever men have tilled or grazed slopes or semi-arid soils. There are many evidences

of the physical effects of accelerated erosion in the eastern and central parts of the Mediterranean basin, in Mesopotamia, in China, and elsewhere. The hill sections of Palestine, Syria, southern Italy, and Greece experienced serious soil losses from grazing and other land use mismanagement many centuries ago. Accelerated water erosion on the hills of southern China and wind erosion in northwestern China also date far back into history.
Claim Soil formation is extremely slow. Where the climate is moist and warm, it takes thousands of years to form just a few centimetres of soil. In cold or dry climates, it takes even longer, or soil may not form at all. While soil is technically a renewable resource, its slow rate of formation makes it practically irreplaceable. In the face of the continuously expanding world demand for grain and the associated relentless increase in pressures on land, soil erosion is accelerating. In effect, mounting economic pressures are degrading the resource base. In 1980 in the United States, soil erosion was described as epidemic in proportion. Although soil erosion is a physical process, it has numerous economic consequences, affecting productivity, growth, income distribution, food sufficiency, and long-term external debt. Ultimately, it affects people. When soils are depleted and crops are poorly nourished, people are often undernourished as well.
Refs Blaikie, Piers *The Political Economy of Soil Erosion in Developing Countries* (1986); Brown, Lester *Soil Erosion* (1984); Chisci, G and Morgan, R P *Soil Erosion in European Community—Impact of Changing Agriculture* (1986); El-Swaify, S A, et al *Soil Erosion and Conservation*; FAO *Keeping the Land Alive* (1983); Guérin, Laurent and Leblond, Bernard *Soil Conservation* (1988); Jacks, Graham V and Whyte, Robert O *Vanishing Lands*; Kirkby, M J and Morgan, R P *Soil Erosion* (1981); Morgan, R P *Soil Erosion and Its Control* (1985); Morgan, Royston P *Soil Erosion and Conservation* (1986); Zachar, D *Soil Erosion* (1982).
Broader Erosion (#PC8193) Soil degradation (#PD1052)
Natural environment degradation (#PB5250).
Narrower Soil erosion by wind (#PE3656) Soil erosion by water (#PD2290)
Land erosion brought about by site development (#PE7099).
Related Insufficient access to technology for agricultural upgrading (#PF3467).
Aggravates Land misuse (#PD8142) Desert advance (#PC2506)
Infertile land (#PD8585) Water pollution (#PC0062)
Silting of water systems (#PD3654) Pollution of inland waters (#PD1223)
Shortage of cultivable land (#PC0219) Demineralization of the soil (#PD9227)
Destruction of land fertility (#PC1300) Unavailability of agricultural land (#PC7597)
Declining productivity of agricultural land (#PD7480).
Aggravated by Defoliation (#PD1135) Forest fires (#PD0739)
Deforestation (#PC1366) Diplomatic errors (#PF1440)
Environmental warfare (#PC2696) Tropical deforestation (#PD6204)
Inadequate land drainage (#PD2269) Deforestation of mountainous regions (#PD6282)
Excessive use of land for agriculture (#PD9534)
Substitution of fast growing plant species (#PE6396).

♦ PD0971 Transport of dangerous goods

Risk of transporting dangerous goods
Nature Dangerous goods constitute a hazard when transported, due to the possibility that such goods may either cause accidents to persons or damage to the means of transport or to other goods. Some explosions occurring during the transport of such goods have caused considerable damage to entire cities. The likelihood of accidents and the amount of damage caused are increased by inadequate packaging and labelling of such substances. If a substance is improperly labelled, it is not possible to determine and rapidly apply the correct counteractant.
Such goods account for an increasing proportion of international trade and new substances of this type are constantly appearing on the market. They include: explosives, gases (compressed, liquefied or dissolved under pressure), inflammable liquids, inflammable solids and substances which on contact with water emit inflammable gases, oxidizing substances and organic peroxides, poisonous (toxic) and infectious substances, radioactive substances, and corrosives.
Refs Dangerous Goods Panel of Air Navigations *Technical Instructions for the Safe Transport of Dangerous Goods by Air 1989–90* (1988); Dangerous Goods Panel of Air Navigations Staff and Commission of ICAO Staff *Technical Instructions for the Safe Transport of Dangerous Goods by Air, 1987–1988* (1986); United Nations *European Agreement Concerning the International Carriage of Dangerous Goods by Road (ADR) and Protocol of Signature* (1985); United Nations *Transport of Dangerous Goods/Recommandations on the Transport of Dangerous Goods* (1985).
Broader Risk (#PF7580).
Narrower Toxic waste smuggling (#PD9765) Dangerous cargo handling (#PD9108).
Aggravates Fires (#PD8054) Urban fires (#PD2211) Poisonous gases (#PJ1939)
Injurious accidents (#PB0731) Accidental explosions (#PE3153)
Radioactive contamination (#PC0229).
Aggravated by Corrosion (#PD0508) Hazardous waste dumping (#PD1398)
Inadequate labelling of dangerous substances (#PD2468).

♦ PD0973 Discrimination against immigrants and aliens

Discrimination against migrant workers and their families — Denial of rights to migrant workers — Discrimination against non-nationals — Discrimination against non-citizens — Discrimination against foreigners
Nature Discrimination against immigrants and aliens may occur in housing, education, employment, public services, politics and before the law. It may also take the form of restricted entry or expulsion of immigrants and aliens, or exclusion from citizenship. Discrimination fosters segregation, lack of integration of foreigners and immigrants into society, and social and racial conflict when immigrants cannot find employment but are eligible for social security benefit; they are resented by the indigenous population which is unwilling to pay, through taxes, for the maintenance of foreigners or naturalized citizens.
Refs ILO *Migrant Workers - Occupational Safety and Health* (1977); Sundberg-Weitman, Brita *Discrimination on Grounds of Nationality* (1977); United Nations *The Social Situation of Migrant Workers and Their Families* (1986).
Broader Denial of rights of minorities (#PC8999) Discrimination against foreigners (#PD6361).
Narrower Refusal to grant citizenship (#PE5453)
Refusal of entry to foreign workers' families (#PE8423)
Discrimination in employment against immigrant workers (#PE4934).
Aggravates Nativism (#PF2186) Resettlement stress (#PD7776)
Restrictions on immigration (#PC0970)
Inadequate cultural integration of immigrants (#PC1532)
Inadequate living and working conditions of immigrant labourers in industrialized countries (#PD3427).
Aggravated by Social discrimination (#PC1864) Proliferation of immigrants (#PD4605)
Socially unintegrated expatriates (#PD2675) Expulsion of immigrants and aliens (#PC3207)
Language discrimination in politics (#PD3223) Abusive traffic in immigrant workers (#PD2722).

♦ PD0974 Distrust of professional service delivery

Nature The distrust by experts of the whims of the public, and the distrust of professionals by ordinary people, prevents services from being delivered to the vast majority who need them. Specialists have become increasingly isolated from the poorest and most needy. The public attitude toward those specialists has turned cynical. Specialized language has also contributed to a lack of communication of needs and of available skills.
Broader Distrust (#PA8653) Individualistic utilization of expertise (#PF5639).
Aggravates Lack of legal aid facilities (#PF8869).

♦ PD0975 Obsolete deliberative systems

Outmoded deliberative systems — Outmoded deliberative systems of government
Nature Questions, problems and debates put to existing deliberative systems are in conflict with the archaic concepts, the supply of wisdom, knowledge and data, and the analytical methods of those systems. Those wanting to participate in political power are frustrated. Priorities set by decision-making processes are unrelated to people's needs and expectations.
Claim Deliberative systems are being seriously threatened and rendered ineffective and non-comprehensive by the complexity and size of highly urbanized societies. Problems are developing which are of a very different nature from those out of which the systems were created. Ineffectiveness and parochialism are evident in the face of the demand for meaningful effective grassroots participation.
Broader Injustice (#PA6486) Human inequality (#PA0844)
Inflexible central government (#PD1061).
Related Unequal property distribution (#PC3438).
Aggravates Class conflict (#PC1573) Social fragmentation (#PF1324)
Obsolete legislation (#PF5435) Obsolete policy-making (#PF5009)
Unequal distribution of social services (#PC3437).
Aggravated by Unemployment (#PB0750) Distortionary tax systems (#PD3436)
Inequality of opportunity (#PC3435) Unsustainable population levels (#PB0035).
Reduced by Economic inflation (#PC0254).

♦ PD0982 Infectious and parasitic diseases

Communicable diseases
Incidence Infectious diseases remain great killers in the developing world, the six most serious taking the lives of millions every year; and parasitic diseases remain rampant. More parasites, bacteria and vectors are becoming resistant to more drugs and pesticides, creating additional problems for the control of several diseases. While the average life expectancy at birth improved steadily almost everywhere during the 1970s, in many of the least developed countries it is still less than 50 years.
Claim The greatest challenges for environmental medicine in the years to come lie in the less developed regions, especially in curbing parasitic infections and the diseases of squalid settlements, rather than in the developed countries, where the indications are that environmental factors are not major causes of premature mortality and where a major challenge to medicine is to overcome the so-called diseases of civilization (such as coronary heart diseases, cancer, hypertension, etc) and social and behavioural problems, and to adopt the pattern of health care more closely to the needs of people.
Refs Cheng, Thomas C *Parasitic and Related Diseases* (1986); Christie *Infectious Diseases* (1987); Cook, G C *Communicable and Tropical Diseases* (1988); Croll, Neil A and Cross, John H *Human Ecology and Infectious Diseases* (1983); Fiennes, T W *Infectious Cancers of Animals and Man* (1982); Krugman, et al *Infectious Diseases of Children* (1985); Pontificia Accademia delle Scienze *The Interaction of Parasitic Diseases and Nutrition* (1985); Robineault, Manfred J *Communicable Diseases* (1985); Velimirovic, B *Infectious Diseases in Europe* (1984).
Broader Children's diseases (#PD0622) Human disease and disability (#PB1044).
Narrower AIDS (#PD5111) Mumps (#PE2356) Plague (#PE0987)
Rabies (#PE1325) Cholera (#PE0560) Malaria (#PE0616)
Rubella (#PE0785) Measles (#PE1603) Mycosis (#PE2455)
Myiasis (#PE3633) Smallpox (#PE0097) Trachoma (#PE1946)
Influenza (#PE0447) Hepatitis (#PE0517) Acariasis (#PG5128)
Diarrhoea (#PD5971) Meningitis (#PE2280) Filariasis (#PE2391)
Brucellosis (#PE0924) Chicken pox (#PE7775) Sarcoidosis (#PE0264)
Pediculosis (#PG5542) Tuberculosis (#PE0566) Yellow fever (#PE0985)
Dengue fever (#PE2260) Poliomyelitis (#PE0504) Elephantiasis (#PE1601)
Typhoid fever (#PD1753) Leishmaniasis (#PE2281) Toxoplasmosis (#PE3659)
Helminthiasis (#PE6278) Trichomoniasis (#PE2310) Virus diseases (#PD0594)
Hydatid disease (#PE2354) Animal influenza (#PE3157) Bacterial disease (#PD9094)
Enteric infections (#PD0640) Spirochaetal diseases (#PE3254)
African trypanosomiasis (#PE1778) Arthropod-borne diseases (#PE7796)
Parasites of the human body (#PE0596) Zoonotic bacterial diseases (#PD6363)
Environmental human diseases (#PD5669) Acute respiratory infections (#PE7591)
Sexually transmitted diseases (#PD0061) Intestinal infectious diseases (#PE9526).
Related Fungal diseases (#PD2728).
Aggravates Gastritis (#PG2250) Encephalitis (#PE2348)
Malnutrition in developing countries (#PD8668)
Inhibited growth of malnourished children (#PE4921).
Aggravated by Disease vectors (#PC3595) Leeches as pests (#PE3660)
Marine dumping of wastes (#PD3666) Inadequate health services (#PD4790)
Protein-energy malnutrition (#PD0339) Failure in disease notification (#PG3162)
Lack of economic and technical development (#PE8190).
Reduced by Pesticides as pollutants (#PD0120).

♦ PD0983 Insecticides as pollutants

Biocides — Insecticides — Use of chemical insecticides
Nature Three major types of insecticides are used by farmers: organochlorines, organophosphates, and carbamates. The organochlorine compounds are being replaced as it becomes evident that they cause extensive, possibly irreparable, harm to wildlife. Because of the wide spectrum of life forms which they attack, some call these insecticides biocides.
Incidence Production of specific insecticides has been restricted and their use curtailed. DDT was banned for most uses by the EPA in 1972, but some is still exported to tropical countries for mosquito control. Aldrin and dieldrin were banned for most uses in 1974. Heptachlor and chlordane were banned for crop use in 1975. Toxaphene, now the most widely used organochlorine, is under review. The total amount of insecticides applied to major crops has not increased appreciably in recent years, although use, measured in terms of acres treated, grew 32 percent from 1971 to 1976, largely because of increased corn applications.
Refs Hutson, D H and Roberts, T R (Eds) *Insecticides* (1986); WHO *International Conference on Alternative Insecticides for Vector Control, Atlanta, Feb 1971* (1971); WHO Expert Committee on Insecticides, Geneva, 1970 *Application and Dispersal of Pesticides* (1971).
Broader Pesticides as pollutants (#PD0120).
Narrower DDT as a pollutant (#PE5028) Organochlorine pollution (#PG2007)
Carbamate insecticides as pollutants (#PE1282)
Organophosphorus insecticides as pollutants (#PG2104).
Aggravates Hazards to human health (#PB4885)
Occupational risk to health (#PC0865)
Insecticide damage to crops (#PD3695)
Pesticide hazards to wildlife (#PD3680)
Insect resistance to insecticides (#PD2109).
Aggravated by Insect pests (#PC1630)
Excessive use of chemicals to control pests (#PD1207).
Reduces Water-borne animal diseases (#PE2787)
Insect vectors of animal diseases (#PE2748)
Worms as vectors of animal diseases (#PE2750).
Reduced by Vectors of animal diseases (#PD2751).

♦ PD0986 Inequitable administration of justice

Discrimination in the administration of justice — Elitist justice — Elitist legal judgements
Nature Justice serves only the the rich and influential who have any control over the structures

of law when majority of people are not able to participate in articulating the need for new (and the revision of old) statutes to meet present-day needs. The poor, outcasts and minorities are discriminated against in the enforcement and the interpretation of law. for change offered by sub-groups are rejected by the majority body as not serving the needs of the whole.
Broader Injustice (#PA6486)
Exclusion of the masses from setting criteria in judicial judgements (#PD1060).
Narrower Denial of human rights in the administration of justice (#PD6927)
Economic discrimination in the administration of justice (#PE1399)
Cultural discrimination in the administration of justice (#PE6529)
Political discrimination in the administration of justice (#PE1828)
Geographic discrimination in the administration of justice (#PE1347).
Related Denial of right to life (#PD4234)
Religious discrimination in the administration of justice (#PE0168).
Aggravates Crime (#PB0001) Mistrust of system of justice (#PD8217).
Aggravated by Corruption (#PA1986) Social inequality (#PB0514)
Geographical isolation (#PF9023) Miscarriage of justice (#PF8479)
Delay in administration of justice (#PF1487) Inadequate national law enforcement (#PE4768)
Unethical practices in the legal profession (#PD5380)
Unrestrained use of force in administration of justice (#PE8881).

♦ PD0991 Illicit drug trafficking
Drug trade — Illegal drug trade — Illicit narcotics trade
Nature The production, transport, distribution and sale of drugs used for illicit purposes is greatly aided by the production and manufacture of legal drugs, such as alcohol, tobacco and caffeine. The distribution of opiates is mainly carried out by organized crime, whereas the distribution of LSD and cannabis is mainly done by amateurs. The reason for this is the greater profit obtained from the illicit traffic of opiates and the higher court penalties involved which are of less risk to a well-organized network than to individuals. The initial cost of traffic in opiates may be higher, which would also preclude amateur participation. Monopolies tend to develop in the illicit opiate trade. The smuggling of illicit drugs is presumed to be aided by some official corruption since the quantities apprehended by the authorities, despite tighter control measures, remain relatively small. Illicit heroin demand in USA (the largest known market) seems to be fully supplied.
Many drugs (for example: heroin, mescaline, and marijuana) are doctored with other substances before being sold on the streets. Usually these substances are inert (quinine and lactose), but when these are unobtainable (recent tighter control of quinine in US) other more dangerous substances may be added, for example, animal tranquillizer.
Drugs can be divided into three categories. Dangerous drugs include "hard" drugs, e.g., heroin, potent hallucinogens like LSD, injectable amphetamines and potent cannabis preparations like hashish. Abusable drugs include barbiturates, oral amphetamines, marihuana and peyote. Restricted drugs include non-prescriptive medications, such as cough syrups.
Incidence Opium production is carried on in the Golden Triangle (border area of Burma, Laos and Thailand), in the Golden Crescent (parts of Afghanistan, Iran, Pakistan), Turkey and Mexico. The difficulties in trafficking opium help account for the extremely high use of the drug in territories and lands adjacent to the poppy-growing areas. Opium is bulky and has a characteristic odour; for this reason it is usually converted to heroin for trafficking internationally. Coca, from which cocaine is processed, is grown in Western South America, chiefly in Peru and Bolivia, which also grow cannabis or marijuana. Forty-five tons of andean cocaine is trafficked to the USA, the ultimate destination for drugs from many countries and 10 tons go to Europe. Ganja or cannabis from Jamaica, Mexico and the Andes is trafficked to the US by the ton. Illicit drug trafficking, moving hundreds of tons of drugs around the world, besides using organized crime distribution rings, also depends on younger tourists, of whom tens of thousands illegally transport drugs each year.
Counter-claim Dangerous drugs should be not only banned but the laws strictly enforced for both the user and the trafficker. Abusable and restricted drugs should be decriminalized and effective treatment programs should be set up. In all cases honest, direct and effective mass media descriptions of the short and long term consequences of drug taking should be aired.
Refs United Nations *Report of the International Conference on Drug Abuse and Illicit Trafficking, Vienna, 17-26 June 1987* (1987); United Nations *The United Nations and Drug Control* (1987).
Broader Corruptive crimes (#PD8679) Offences against public order (#PD7520)
Unlawful business transactions (#PC4645).
Narrower Drug smuggling (#PE1880) Possession of drugs (#PE5556)
Adulteration of illicit drugs (#PE0456) Illicit trade in prescribed drugs (#PE4946)
Economic dependence of some developing countries on the drug trade (#PE5296).
Related Illegal international arms shipments (#PD4858).
Aggravates Cannabis abuse (#PE1186) Narco-terrorism (#PE9210)
Border controls (#PJ1718) Abuse of opiates (#PE1329)
Abuse of animal drugs (#PE0043) Abuse of hallucinogens (#PD0556)
Corruption in politics (#PC0116) Inadequate drug control (#PC0231)
Bureaucratic corruption (#PC0279) Abuse of coca and cocaine (#PD2363)
Corruption of the judiciary (#PD4194) Unethical military practices (#PD7360)
Corruption of government leaders (#PC7587) Abuse of sedatives and tranquillizers (#PE0139).
Aggravated by Gang warfare (#PD4843) Organized crime (#PC2343)
Criminal investment in youth market (#PD5750).

♦ PD0992 Diseases and injuries of the brain
Cerebrovascular disease — Tumour of the brain — Organic disease of the brain — Degenerative changes of the brain — Brain tumour — Brain damage — Brain derangement — Diseases of the brain — Brain disease — Brain infection
Refs Almli, C Robert and Finger, Stanley *Early Brain Damage* (1984); Black, Perry (Ed) *Brain Dysfunction in Children* (1981); Cervos-Navarro, J and Sarkander, H I (Eds) *Brain Aging* (1983); Gerosa, M, et al (Eds) *Brain Tumors - Biopathology and Therapy* (1986); Gotoh, F and Letchner, H (Eds) *Clinical Haemorheology - A New Approach to Cerebrovascular Diseases* (1986); Hetzel, B S and Smith, R M (Eds) *Fetal Brain Disorders* (1981); Kameyama, Masakuni and Tomonaga, Masanori *Cerebrovascular Disease* (1988); Katzman, Robert (Ed) *Biological Aspects of Alzheimer's Disease* (1983); Maurer, K and Wurtman, R J *Organic Brain Disorders* (1990).
Broader Infection (#PC9025) Diseases of the nervous system (#PC8756)
Diseases of the circulation system (#PC8482).
Narrower Stroke (#PE1684) Amnesia (#PD8297) Agraphia (#PE0280)
Paralysis (#PD2632) Concussion (#PG3188) Encephalitis (#PE2348)
Memory defects (#PD8484) Paralysis agitans (#PE2206) Language disorders (#PE3886)
Cerebral ischaemia (#PJ5087) Cerebral infarction (#PD9057) Alzheimer's disease (#PE7623)
Blood circulation disorders (#PE3830) Chemical imbalances in the brain (#PE4715)
Intracerebral and other intracranial haemorrhage (#PD8461).
Aggravates Deafness (#PD0659) Epilepsy (#PE0661)
Headache (#PE1974) Meningitis (#PE2280)
Hydrocephalus (#PJ3190) Motor aphasia (#PE0345)
Mental illness (#PC0300) Hallucinations (#PF2249)
Sleep disorders (#PE2197) Speech disorders (#PE2265)
Neurological rage (#PE0429) Physical blindness (#PD0568)
Cerebral paralysis (#PE0763) Mental disorders of the aged (#PD0919)
Diseases of connective tissue (#PD2565).
Aggravated by Syphilis (#PE2300) Ascariasis (#PE2395)
Mastoiditis (#PG3510) Head injuries (#PG5105)
Alcohol abuse (#PD0153) Rheumatic diseases (#PE0873)
Malignant neoplasms (#PC0092) Nutritional anaemia (#PD0321)
Diseases of the ear (#PD2567) Parasites of the human body (#PE0596)
Diseases of the lymphatic system (#PD2654).

♦ PD0994 Pedestrian accidents
Incidence If road users are classified conventionally into car occupants, motorcyclists, bicyclists, occupants of heavy goods vehicles, occupants of public service vehicles, and pedestrians, then in terms of traffic fatalities pedestrians constitute the largest single group. In the USA, they account for 18 per cent of fatalities, UK 41 per cent, Poland 46 per cent. No firm data are available from countries where the car is present only in relatively small numbers, but in those very countries it is pedestrians who predominate as the main victims of traffic accidents.
Broader Travel risks (#PD7716) Injurious accidents (#PB0731).
Aggravates Road traffic accidents (#PD0079).
Aggravated by Drunken pedestrians (#PG8377) Uncontrolled urban development (#PC0442).

♦ PD1007 Uneconomical small farms in developing countries
Nature In many countries, individual family farms are very small and cannot provide a decent standard of living for their owners. Smallholders with limited education and economic understanding still pursue the traditional system of land utilization. Primitive agricultural practices and over-grazing of pastures further contribute to the deterioration of the soil.
The substantial rise in the number of smallholders in many developing countries is a result of many diverse factors, such as growing population pressures, slowdown in the growth rate of industrialization, acquisition of land by the big owner-cultivators, acquisition of land by urban resident investors and subdivision and fragmentation of land due to the laws of inheritance.
Further, the small peasants and the tenants with uneconomic-sized holdings have perforce to work on the big farmer's holdings and thus have to compete with the landless labourer for the limited employment opportunities. The medium-sized farmers have justifiably tried to take land on tenancy from other farmers in order to increase the size of their holdings to make them more viable; this has, however, meant competition between the (small) pure tenant and the medium-sized owner-cultivator.
Claim The position of the small peasant, the tenant and the labourer has actually deteriorated with the increase in population; inadequate access to services offered by government and other institutional agencies only worsens his position vis-à-vis that of big farmers. In their anxiety to encourage improved techniques, many governments have adopted policies which in effect subsidize the big farmer to the virtual neglect of small farmers; such policies, though quick-yielding in the short run, may create grave long-term socio-economic problems.
Broader Maldistribution of land in developing countries (#PD0050).
Related Uneconomic size of farms (#PJ2079).

♦ PD1012 Unethical practices in transportation
Irresponsible transport practices — Negligence in the transport industry — Corruption in the transport industry
Claim Because of its highly competitive and lucrative nature, especially in the case of international transport between legal jurisdictions, the transport sector is vulnerable to a wide range of unethical practices. These include: abuse of safety standards (as in the case of ferry boats), improper maintenance (as in the case of aircraft, trucks and automobiles), use of personnel who are improperly qualified and/or working excessive hours (as in the case of pilots), price fixing (as with air passenger transport and cargo shipping conferences), diversion and/or relabelling of cargoes (as in sanction beating), smuggling (as in the case of arms, drugs and black market goods), transport of crew or passengers with improper travel documents (as in the case of illegal immigrants), avoidance of tax and other regulations (as with flags of convenience). In many such instances, and especially cargo handling, organized crime is a significant factor.
Narrower Maritime fraud (#PE4475).
Aggravates Oil pollution (#PE1839) Drug smuggling (#PE1880)
Port congestion (#PE4766) Cargo insecurity (#PE5103)
Toxic waste smuggling (#PD9765) Unsafe port facilities (#PE4897)
Marine dumping of wastes (#PD3666) Inconsistent port charges (#PF5887)
Unfair transport practices (#PD1367) Illicit export of works of art (#PE9004)
Unethical maintenance practices (#PD7964) Evasion of customs and excise duties (#PD2620)
Illegal international arms shipments (#PD4858)
International trade in endangered species (#PC0380)
Fraudulent acquisition or use of passports (#PE4496)
Inadequate enforcement of safety regulations (#PD5001)
Restrictive practices in cargo airline services (#PE5910)
Human fatigue during control of complex equipment (#PE5572)
Diversion of high technology to hostile countries (#PE7174)
Excessive costs of inefficient port cargo-handling (#PE5899)
Evasion of shipping regulations and taxes by flags of convenience (#PE5873).
Aggravated by Organized crime (#PC2343).

♦ PD1018 Dependence of government revenues on exploitation of environmentally inappropriate products
Dependence of government revenue on substance abuse — Dependence of government income on products dangerous to health — Dependence of government revenues on exploitation of non-renewable resources
Incidence In the case of developed-market economies, the three major categories, in terms of revenue collection, on which fiscal charges are levied on importation are mineral fuels (including petroleum), alcoholic beverages, and tobacco. For example revenues raised by developed market economy countries on tobacco alone were $22 billion in 1983. In the case of the UK, government revenue from alcohol consumption was 6,700 million pounds in 1987-8.
Aggravates Smoking (#PD0713) Waste of non-renewable resources (#PC8642)
Consumption of alcoholic beverages (#PD8286)
Underutilization of renewable energy resources (#PE8971).
Aggravated by Government hypocrisy (#PF9050).

♦ PD1026 Breakdown in covenants for life
Claim There has been a collapse of the covenantal nature of the social structures related to marriage and the family. Today's society reflects the fragmentation of youth and adult relationships: the non-covenantal nature of these relationships can be seen in the way individuals relate to families, the way families relate to the community and the way families relate to the world through the community.
Broader Lost family role in society (#PF7456).
Related Escapist family life styles (#PD4069).

♦ PD1030 Abandoned wives
Deserted wives
Refs Brandt, Lilian *Five Hundred Seventy Four Deserters and Their Families* (1972).
Broader Abandonment (#PA7685) Desertion in marriage law (#PF3254).
Related Abandoned children (#PD5734).

♦ **PD1036 Discrimination in family planning facilities**
Nature Access to family planning facilities may be barred or discouraged on the basis of age, marital status, number of children, income or race. The main areas where discrimination occurs are: access to family planning information; to all or certain kinds of contraceptives; and to abortion.
Incidence A specific form of family planning discrimination was the experimental – and failed – programme initiated by the Singapore government in 1984. Under the programme, women with university degrees were encouraged (via priority registration at pre–primary and primary school levels), to bear more children, while women without higher education were granted large sums of money if they agreed to be sterilized after their first or second child.
Broader Discrimination (#PA0833)
Inadequacy of contraceptive methods (#PD0093).
Narrower Sexual discrimination in contraceptive methods (#PF1035).
Related Promiscuity (#PC0745).
Aggravates Unmarried mothers (#PD0902) Unwanted children (#PE1907)
Dependency of women (#PC3426) Illegitimate children (#PC1874)
Illegal induced abortion (#PD0159)
Discrimination against unmarried fathers (#PD3256)
Inadequate health care in family planning (#PD1038).
Aggravated by Prohibitive cost of contraceptives (#PJ3244)
Inadequate family planning education (#PD1039)
Discrimination against unmarried women (#PD8622)
Religious opposition to population control (#PF1022)
Government opposition to population control (#PF1023)
Discrimination against women without children (#PE8788).

♦ **PD1038 Inadequate health care in family planning**
Health hazards of contraceptives — Increased use of birth control pills — Health hazards of birth control
Nature General health care may be underemphasized in promotion of family planning and family planning education, and this may discourage acceptance of the practice. Inadequacy of available contraceptive methods to overcome discomfort or other side–effects, and insufficient medical follow–up on this, may also lead to a rejection of the idea as a whole. The health risks related to pregnancy often go unstated. The fact that general health care is not emphasized in family planning makes the task of overcoming religious and societal taboos much harder, and provides no counterweight to the arguments in favour of large families as a means of insurance against sickness and old age.
Health hazards of various birth control devices include breast and cervical cancers, cardio–vascular complications, birth defects, reduced lactation, abnormal menstrual cycles, migraine headaches, hair loss, fluid retention, raised blood pressure, weight gain, emotional disturbances, and either temporary or permanent infertility.
Incidence Every year in Africa and Asia alone, half a million women die from pregnancy, childbirth and after–birth effects – leaving behind over 1 million motherless children. In Latin America, illegal abortion is now the number one killer of women between the ages of 15 and 39. Worldwide, 25 million women a year suffer serious illness or complications during pregnancy and childbirth. Fifteen million of the 125 million babies born every year will not reach their first birthday.
It was estimated, in 1980, that approximately 270 million people used some form of contraception, about 2/3 of them in developed countries, and 1/3 in developing countries.
Claim 1. Being pregnant, giving birth and breast feeding are exhausting processes for a woman's body. If the recovery time is too short, health pays the price. Infants are more likely to be malnourished. Mothers suffer from anaemia, toxaemia and plain exhaustion. Babies are prone to low birth weights and the consequent 20 times greater risk of death in infancy.
The age of the mother, as well as the frequency of birth, also links family planning to health. Outside the age band 20–35, there is a higher incidence of unwanted pregnancy, a higher risk to the mother, and a higher rate of mortality among the infants born. And roughly one–third of all births in the world are to mothers younger than 20 or older than 35.
2. Merging family planning with health services helps to avoid duplicating personnel and administration where resources are scarce; it helps in the many cases where contraceptives themselves have adverse effects on health; it helps that there is a relationship of trust between people and their health workers. But most important of all, family planning is "one of the numbers in the code which releases the combination lock of community health".
Broader Hazards to human health (#PB4885) Inadequate health services (#PD4790).
Narrower Inadequate supervision of contraceptive use (#PU3247).
Related Inadequate maternal and child health care (#PE8857)
Inadequate health care in least developed countries (#PE9242).
Aggravates Neglected young children (#PE4245)
Sexually transmitted diseases (#PD0061)
Unsustainable population levels (#PB0035)
Malignant neoplasms of female genital organs (#PE1905).
Aggravated by Desert nomadism (#PD2520) Inadequacy of contraceptive methods (#PD0093)
Inadequate family planning education (#PD1039)
Discrimination in family planning facilities (#PD1036).

♦ **PD1039 Inadequate family planning education**
Lack of family–planning education — Inadequate family planning research
Nature Family planning education may be inadequate insofar as it provides too little information or in that it is ill–suited to the needs of those who are being educated. The aspect of general health care may be underemphasized. Some social groups may be discriminated against, particularly the young and unmarried and certain racial and religious groups, or information may simply be less available for the less well–off, of for those in rural and inaccessible areas.
Incidence Some failures in family planning education policies and strategies are: the exclusion of natural methods of family planning; inadequate research in the psycho–social area; inability of policies to adequately respond to needs for a safe, reliable method; lack of coordination among family planning providers; and failure to target information to potential target groups in a meaningful or comprehensible way.
Broader Inadequate education (#PF4984) Obstacles to family life (#PF7094)
Limited medical knowledge (#PD9160).
Narrower Lack of family planning information (#PD1050)
Inadequate family planning education for men (#PE9060)
Negative effects of family planning education on children (#PE0341).
Aggravates Lack of family planning (#PF0148) Illegal induced abortion (#PD0159)
Unsustainable population levels (#PB0035) Insufficient birth spacing in families (#PF5968)
Inadequate health care in family planning (#PD1038)
Discrimination in family planning facilities (#PD1036)
Opposition to population control and family planning (#PF1021).
Aggravated by Inadequate secondary education (#PD5345)
Inadequacy of male contraceptive methods (#PF1069)
Limited availability of functional information (#PF3539).
Reduces Birth prevention (#PE3286).

♦ **PD1043 Susceptibility of the old to physical ill–health**
Illness of the elderly
Nature A large number of factors in the aging process contribute to the elderly's susceptibility to physical ill–health. The defence systems against disease decline by 20 percent of their peak. Between the ages of 30 to 70 the average person loses 30 to 40 percent of the body's mass. The heart's pumping efficiency drops 30 percent in the same period. By 40 years of age men and women begin to lose bone because of a decreased ability to absorb calcium.
Broader Human ageing (#PB0477).
Narrower Mental disorders of the aged (#PD0919)
Rigidity and inadaptability in the aged (#PD3515)
Decline in cognitive ability with ageing (#PE9620).
Aggravates Incontinence (#PE4619) Fear of death (#PF0462)
Teeth disorders (#PD1185) Social withdrawal of aged (#PD3518)
Spouses of torture victims (#PE0753) Physically dependent people (#PD7238)
Criminalization of euthanasia (#PF2643) Inadequate housing for the aged (#PD0276)
Negative effects of the nuclear family (#PF0129)
Inadequate welfare services for the aged (#PD0512)
Slowness of sensori–motor activities in the aged (#PD3514).
Aggravated by Ageing war disabled (#PD0874) Inadequate income in old age (#PC1966)
Social disadvantage of the aged (#PD3517) Inadequate recreational facilities (#PF0202)
Retirement as a threat to psychological well–being (#PF1269).

♦ **PD1045 Unethical food practices**
Irresponsible practices in the food industry — Negligence in the food industry — Corruption in the food industry
Broader Unethical commercial practices (#PC2563).
Narrower Unethical catering practices (#PE6615)
Infected animal, meat and animal product shipments (#PE7064).
Aggravates Food fads (#PD1189) Food wastage (#PD8844)
Food insecurity (#PB2846) Food monopolies (#PE8018)
Ritual pollution (#PF3960) Cancer–causing foods (#PD0036)
Inadequacy of food aid (#PF3949) Food spoilage in storage (#PD2243)
Chemical contaminants of food (#PD1694) Inappropriate cash crop policy (#PF9187)
Imbalance in world food economy (#PD5046) Biological contamination of food (#PD2594)
Health hazards of irradiated food (#PD0361) Food poisoning through negligence (#PE0561)
Exploitation of dependence on food aid (#PD7592)
Cruelty to animals in food preparation (#PE0236)
Dumping of food in developing countries (#PE0607)
Environmental hazards from food processing industries (#PE1280)
Denial to food animals of the right to freedom from suffering (#PE3899)
Domination by transnational corporations of the domestic name–brand food sector in developing countries (#PE1796).
Aggravated by Ineffective self–regulation in the food–processing industries (#PE8472).

♦ **PD1050 Lack of family planning information**
Nature Lack of family planning information may result from discrimination, doctors' reticence, lack of research, or communication difficulties (such as language problems, illiteracy, cultural gap). Governments may specifically forbid the dissemination of family planning information as part of a population expansion programme.
Discrimination in the dissemination of information may be practised against certain social groups, particularly the young and unmarried and certain racial and religious groups, or information may simply be less available for the less well–off, and in rural and inaccessible areas.
Incidence The birth control pills distributed by the American AID programme come packaged with very little information, and that information is written in English. Most women who receive these packages cannot read their own language, let alone English, so the pills go unused. There are not even diagrams to show what to do. On these same packages, the only picture is a profile of a woman; the profile is coloured blue. As blue denotes depression, many women think that perhaps the pills are anti–depressants and/or that they themselves might turn blue if they take the pills. Both of the mentioned problems could easily be alleviated by proper education of both the packager and supplier of these pills as well as the recipients.
Broader Inadequate family planning education (#PD1039).
Narrower Birth control misinformation (#PG5266).
Related Lack of medical information (#PJ8069).
Aggravates Unmarried mothers (#PD0902) Retarded socialization (#PF2187)
Opposition to population control and family planning (#PF1021).
Aggravated by Superstition (#PA0430).

♦ **PD1052 Soil degradation**
Soil deterioration — Destruction of soil structure — Land degradation — Diminution in biological productivity of land — Deterioration of soil resources — Decline in soil productivity
Nature Over the last 100 years, with the growth in consumption, world population and man's technical capacity, the cultivation of soils has increased, both extensively and intensively, but without a parallel increase in soil conservation measures. The result has been accelerated soil degradation. Soil deterioration has many causes, the main ones being inappropriate land–use, erosion, salinization and waterlogging, and chemical degradation. In general, the removal or depletion of vegetation or vegetable residues which protect the land is the basic cause of soil degradation. Excessive or untimely tillage, over–fertilization, improper implements, burning crop residues and excessive livestock grazing have all contributed to this process. Lack of earthworms, humus and the roots of grasses make soil incapable of forming and maintaining a crumb–like structure. In wet areas, inability to maintain a "crumb–like" structure causes the soil to become waterlogged and turn into an intractable sticky mass with the consistency of butter. Repeated passage by tractors and other farm machinery consolidates such a mass to considerable depths below the surface, making it difficult for roots of plants to penetrate and for oxygen, moisture and nutrients to diffuse freely. During dry weather this clay shrinks and sets to a hardness like concrete. In dry areas, the crumb structure of the soil becomes less pervious or even impervious by the formation of a crust, which reduces water infiltration and the availability of water to plants. Because of the resulting increased run–off water, erosion may occur. Reduced protection of the soil by vegetation also makes the area more liable to wind erosion.
In advanced stages of erosion, all soil, and therefore all capacity for production, may be removed. More frequently, lack of soil conservation results in the loss of the most productive layers of the soil – those having the highest capacity for retention of moisture, the highest soil nutrient content, and the most ready response to artificial fertilization. Moderate to slight soil deterioration cannot be regarded as having serious social consequences, except over many decades. As an income depressant, however, it does prevent a community from reaching full productive potentiality. More severe erosion has led to very damaging social dislocation, reducing dependent communities to subsistence level. This has been illustrated in the hill and mountain lands of the south–eastern United States and in Italy, Greece, Palestine, China, and elsewhere for many millions of peasant people. Illiteracy, short life spans, nutritional disease, poor communications and isolation from the rest of the world have been the marks of such communities. It has been estimated that about half the area of originally potentially productive land in the world has become useless because of the absence of soil conservation.
Background Mankind has lost a considerable part of its cultivable soil during the course of recorded history. We can distinguish three zones on the globe which were successively the homes of dominating civilizations and where the soil has become progressively devastated in proportion to the time that has elapsed since they were first settled. *First zone*: In the Sahara hundreds of archaeological finds and cave paintings indicate that this was once a flourishing country with many

PD1052

lakes and rivers. *Second zone*: From West China, Turkey, Afghanistan, Iran, Iraq, Jordan, Sinai up to North Africa, there is presently a continuous range of stone, salt and sand deserts. In ancient times, these latitudes were inhabited by Sumerians, Babylonians, Persians, Macedonians, and Phoenicians – names which are all connected with world power and wealth. *Third zone*: Palestine, Syria, Asia Minor, Greece, Italy and Spain. It is true that the Southern European States, (Greece, Italy and Spain), are neither steppes nor deserts, nevertheless their denuded mountains justify the statement once made by Henry C. Wallace, United States Minister of Agriculture: 'Nations live as long as their humus'. These countries, to which for centuries numerous foreigners came in search of fertile land, nowadays see their own inhabitants emigrating to all parts of the world to find better living conditions.

Incidence Statistics for 1970 to 1980 show that soil degradation – erosion, salinization and alkalinization and chemical degradation – occurred in all parts of the world to varying degrees and caused production losses. According to various estimates, between 30 and 80 percent of all lands under irrigation have been subject to salinization, alkalinization and waterlogging. Salinization and waterlogging are believed to be seriously affecting 200,000 – 300,000 hectares of the world's best land each year.

Claim The soils of our planet are known to have been formed over the lapse of thousands of years in conditions which have long since ceased to exist. The destruction of soils through man's mistakes, however, can take place in just a few years, and in most cases this process is irreversible or difficult to correct. The soil cover as the bearer of fertility and a component of the biosphere must be guarded from destruction and preserved for future generations.

More than three billion hectares, almost a quarter of the world's land surface, is at risk from desertification, salinization from bad irrigation or other degradation. Worldwide, an estimated 26 billion tons of topsoil are washed or blown off cropland each year. Every year 6 million hectares of productive dryland becomes desert. US$6 billion damage a year is done off-site by eroded soil deposited on coral reefs, in dams and fisheries, in the USA each year. 8 million metric tonnes of hazardous chemical wastes are leaking into Dutch soil. Cleaning up the 21,000 abandoned chemical hazard sites in West Germany will cost US$4 billion.

Soil degradation erodes the overall resource base for agriculture. Loss of croplands encourages farmers to overuse the remaining agricultural land and to move into forests and onto rangelands. Such approaches to agriculture deplete the soil and are not sustainable.

Refs Andel, J van, et al (Eds) *Disturbance in Grasslands* (1987); Boels, D, et al *Soil Degradation* (1982); Chisholm, Anthony and Dumsday, Robert *Land Degradation* (1988); FAO *Guidelines for the Control of Soil Degradation FAO/UNEP* (1983); FAO *Assessing Soil Degradation* (1977); FAO *Land Degradation* (1977); WHO *Meteorological Aspects of Certain Processes Affecting Soil Degradation – Especially Erosion* (1983).

Broader Environmental degradation (#PB6384).
Narrower Soil erosion (#PD0949) Infertile land (#PD8585)
Soil compaction (#PD1416) Soil infertility (#PD0077).
Aggravates Soil pollution (#PC0058) Desert advance (#PC2506)
Migrating sand dunes (#PD0493) Destruction of land fertility (#PC1300)
Vulnerability of plants and crops (#PD5730) Slowing growth in food production (#PC1960)
Unavailability of agricultural land (#PC7597)
Unsustainable agricultural development (#PC8419).
Aggravated by Land misuse (#PD8142) Soil salinization (#PE1727)
Metal contamination of soil (#PD3668) Destruction of alluvial forests (#PD6850)
Over-intensive soil exploitation (#PC0052) Radioactive contamination of soil (#PE3383)
Unethical practice of soil sciences (#PD1110)
Pesticide destruction of soil fauna and micro-organisms (#PD3574).

♦ PD1060 Exclusion of the masses from setting criteria in judicial judgements

Incidence The civil rights movement, born out of demonstrations initiated by Martin Luther King, is a classic example of a response to the closing off of the judicial process from the people, depriving them of the socially established channels for dealing with grievances and disputes.

Claim When the judicial system is experienced as closed, people have no recourse but to seek solutions through unconventional channels. Present-day covenants and codes do not reflect what the individual experiences as the social reality in which he or she participates. The processes of judicial review and creation of legislation have so far neither enabled people to articulate their understanding of the malfunctioning of the system nor to participate in the re-creation of legal principles. The social order is unable to transform the present situation into viable alternatives for the future.

Broader Reduced understanding of globality (#PF7071).
Narrower Outmoded legal systems (#PF2580) Inequitable administration of justice (#PD0986)
Closed channels of dialogue with the judiciary (#PE7725)
Unchanging legal precedent undermines accountability (#PF5295).
Related Narrow context for counsel (#PF0823) Undemocratic policy-making (#PF8703)
Inflexible central government (#PD1061).

♦ PD1061 Inflexible central government

Claim The more complicated government becomes, the less flexible and creative it is in relation to the demand placed upon it by changing needs. As societies grow in size and complexity, central governments themselves grow more complex in the attempt to deal with the ambiguities in these societies. There is a contradiction within deliberative systems in that, because of the unwillingness of the central governments to set up the structures necessary to allow local people a significant voice and to participate in the decision-making process, such government tyrannises rather than serves.

Broader Reduced understanding of globality (#PF7071).
Narrower Obsolete deliberative systems (#PD0975)
Inflexible deliberative systems of government (#PF7059).
Related Undue political pressure (#PB3209) Narrow context for counsel (#PF0823)
Undemocratic policy-making (#PF8703)
Exclusion of the masses from setting criteria in judicial judgements (#PD1060).
Aggravates Defensive life stance (#PF0979) Excessive government control (#PF0304).

♦ PD1078 Vulnerability of women and children in emergencies

Vulnerability of women during armed conflicts — Vulnerability of women during foreign military occupation — Lack of protection of women and children during disasters

Nature In the case of famine, flood, earthquake, civil conflict, epidemic disease, or war, women and children separated from the men of their families are exceptionally vulnerable. Armed conflicts affect men and women alike in terms of ecpnomic and social dislocation, displacement of people, physical disability, and emotional and psychological damage. Women, however, are more vulnerable to violation of their basic human rights, because they are less able to defend themselves. Several forms of physical violence and discrimination are endemic in most societies. In situations of armed conflict, women are especially vulnerable because they are fleeing persecution, because of the social disruption caused by flight, because they are sometimes separated from their families and the protection of their community, and finally because they are foreigners in an alien environment. Often it is the children and the aged, usually women, who succumb first, lacking the resources or abilities to overcome the hazards with which they are surrounded. Lasting traumatic damage often accompanies the harrowing experiences of man-made or natural disasters, and unprotected children thus become impaired adults.

Claim Mothers and children have the right to continue the propagation of their kind in peace and good health. No society which neglects their care during natural and human-made disasters and emergencies continues to grow.

Broader Violation of rights of vulnerable groups during states of emergency (#PD3785).
Aggravates Injuries (#PB0855) Human death (#PA0072) Malnutrition (#PB1498)
Women combatants (#PE8220) Displaced children (#PD5136)
Vulnerability of the elderly under states of emergency (#PD0096)
Vulnerability of the disabled during states of emergency (#PD0098).
Aggravated by War (#PB0593) Disasters (#PB3561) Homelessness (#PB2150)
Social neglect (#PB0883) Natural disasters (#PB1151) Refusal of adoption (#PF3282)
Militarization of children (#PE5986) Discrimination against women (#PC0308).

♦ PD1080 Inadequate response to societal needs

Claim Social service institutions are failing to deal with problems that are more complex and numerous than ever before. In many cases, their own bureaucratic organization may block the services they intend to deliver. In other instances, specialization narrows the field of services delivery. Efforts to demand accountability are generally only on a rather stop-gap basis.

Broader Government inaction (#PC3950) Absence of tactical methods (#PF0327).
Aggravated by Official evasion of complaints (#PF9157).

♦ PD1089 Inadequate rehabilitation facilities

Nature Rehabilitation is still often perceived by the public, and even by the medical profession, as a sophisticated and complex process aimed at helping the severely disabled to overcome functional limitations due to various impairments such as blindness or locomotor deficiencies; and is not perceived as including the amelioration of those deficiencies resulting from non-visible disabilities, such as mental retardation, or from other mental health problems, chronic diseases and chronic pain.

While rehabilitation has made progress as regards children and people of employable age, the challenge of rehabilitating the elderly has only recently been taken up. One of the main reasons for the slow development in this area might be the unduly pessimistic view of the value of treatment, held not only by the old people themselves, their relatives, friends and neighbours, but also, regrettably, by health personnel.

Claim Due attention is not always paid to all the contributing factors of the disability, whether medical, social, psychological, occupational or economic. It is not always realized that rehabilitation is not just a uni-directional process during which the patient is simply an object, but an active process involving the disabled person and the therapist or the rehabilitation team; nor that it is not only a process of the individual adapting to the environment, but also one of changing the latter to meet the needs of people with certain restricted abilities.

Broader Social neglect (#PB0883) Inadequate medical facilities (#PD4004)
General obstacles to problem alleviation (#PF0631).
Narrower Inadequate rehabilitation of juvenile offenders (#PE8803)
Inadequate rehabilitation methods for the blind (#PE8983)
Inadequate rehabilitation facilities for the mentally handicapped (#PE8151)
Inadequate vocational rehabilitation facilities for disabled persons (#PE7317).
Aggravates Displaced children (#PD5136) Denial of rights to disabled (#PC3461)
Abusive treatment of patients in psychiatric hospitals (#PD0584).

♦ PD1091 Lack of eugenic measures

Lack of eugenics — Insufficient intervention in human evolution — Discrimination against eugenics

Nature Eugenic measures may be taken to keep the overall burden of hereditary disease and disability as low as possible (negative eugenics) although in the future it may be possible to devise measures to improve genetic health (positive eugenics). In the absence of eugenic measures, persons suffering serious hereditary diseases, with a high risk of transmitting them to their children, are not discouraged from having children. In addition, genetic counselling procedures would not be effective in the case of persons suffering from high-grade mental retardation which is of special importance due to its frequency, and some form of enforcement may be required.

Counter-claim Eugenic measures will never cause hereditary diseases or disabilities to disappear completely. Spontaneous mutations, the coming together of heterozygous carriers of recessive conditions, and other factors will always bring about new cases.

Refs Ramsey, Paul *Fabricated Man* (1970).

Broader Lack of control (#PF7138) Inadequate medical care (#PF4832).
Related Mutation (#PF2276) Social evolutionism (#PF2196).
Aggravates Human contingency (#PF7054) Moral imperfection (#PB7712)
Health inequalities (#PC4844) Genetic defects and diseases (#PD2389)
Decline in human genetic endowment (#PF7815)
Underestimation of the human potential (#PF7063)
Foetal malformation in diabetic pregnancies (#PE4808).
Reduces Eugenics (#PC2153) Unlimited practice of human embryo storage (#PE5623).
Reduced by Government-enforced maternity (#PE3601).

♦ PD1092 Exploitation of indigenous populations in employment

Discrimination against indigenous populations in employment — Denial of the right to employment for indigenous populations

Nature The unfair treatment of indigenous populations concerning working conditions, payment, and legal restrictions on the right to combine and strike implies their exploitation in employment.

Claim The expropriation of the land of indigenous populations has reduced many to a state of dependency in which they are easily exploited by unscrupulous landowners and companies. When segregated and living on reservations low prices for produce are hard to combat. Legislation may exist (as in South Africa and other African countries) to deny the right to combine and strike in industry. Lack of education and general discrimination serve to keep other populations subjugated. Nomadism contributes to this, as does the seasonal nature of harvest time occupations for which excessively low wages are paid. The vicious circle is perpetuated by the need for children to work from an early age, thus denying them the benefits of education. Exploitation includes inadequate housing for workers and lack of welfare services in hazardous occupations.

Broader Endangered tribes and indigenous peoples (#PC0720)
Discrimination against indigenous populations (#PC0352).
Related Exploitation in employment (#PC3297).
Aggravates Apathy (#PA2360) Socio-economic poverty (#PB0388)
Exploitation of child labour (#PD0164) Exploitative personal services (#PC3299).
Aggravated by Nomadism (#PF3700) Colonialism (#PC0798)
Social injustice (#PC0797) Ethnic discrimination (#PC3686)
Inadequate protection of individual welfare (#PJ3312)
Expropriation of land from indigenous populations (#PC3304).

♦ PD1095 Decline in rural customs and traditions in developing countries

Nature In the less developed countries, traditional social structures and cultural values are being undermined by the process of economic development. Mass communication media are also bringing about changes in the social structures of the less developed countries by disseminating new ideas and values on a universal scale. There is an increasing availability of books, newspapers and radio and television receivers in these countries, and the ensuing loss of customary occupations, combined with the breakdown of mutual aid and communal cooperation, tends to disrupt rural society, bringing about a general decline in village custom and tradition and thereby reducing the satisfaction obtained by the individual from life in rural areas.

Claim This crumbling and shrinking of village life is of the utmost importance: village festivals and ceremonies gave value and colour to the lives of peasants and compensated for the lack of the numerous comforts possessed by city dwellers. They prevented the villager from feeling poor, strengthened his social ties and standing with his fellow villagers, filled his life, and gave him a fixed place in the world. With the weakening and falling-off of such customs, the village becomes poorer and emptier. The ties, support and contentment of individual life are removed and nothing of equal value put in their place.

Counter-claim Although social customs in the agrarian sector of many developing countries usually represent the result of lengthy adjustment to the environment and are often commendable from a cultural point of view, they often act as a brake on economic development. Even where there is social justification for traditional practices, as in the case of mutual aid, it may be outweighed by economic disadvantages. In some cases, peasants have been glad to renounce older forms of community cooperation, regarded as inefficient, expensive (in the outlay of food and drink required), and not dependable, however colourful and festive. Hierarchical social structures have had to open up to provide channels of social mobility. Wherever the traditional social system fails to do this, political and social pressures begin to mount. Wherever the social system is flexible and allows for individual mobility, more fluid class and status structures, based on education and achievement as well as lineage, have begun to replace the rigid traditional hierarchies of caste or colour.

Broader Underdevelopment (#PB0206).
Related Contempt for traditional modes of behaviour (#PC4321)
Decline in communal spirit and village solidarity in developing countries (#PE0835).
Aggravates Loss of traditional forms of social control in developing countries (#PD0144).
Aggravated by Uncontrolled industrialization (#PB1845)
Untransposed significance of cultural tradition (#PF1373).

♦ PD1096 Transboundary water pollution

Broader Water pollution (#PC0062)
Environmental threats to national security (#PC4341).

♦ PD1104 External dependence and vulnerability of socialist countries

Nature Growing reliance on Soviet raw materials and markets and on Western credit has driven the East European countries into a position of external dependence and vulnerability. They are increasingly dependent on imported energy and raw materials. Much of their meat, grain and animal feedstuffs already come from the United States and other hard-currency sources. In addition, they imported so much Western plant equipment during the 1970s that the continued importation of components and spare parts remains a necessity. This does not yet mean, however, that they are technically equipped to provide exports that would earn them hard currency. Rather, there is a growing dependence on Western international corporations, not only for technology, but also for marketing services. In addition, these Eastern European countries are increasingly vulnerable to competition from the rapidly-industrializing developing countries, where low wages are still paid.

Broader Apathy (#PA2360) Underproductivity (#PF1107).

♦ PD1106 Militant individualism

Claim When there are radical changes in the external world and familiar social environments disappear, people have no way of recognizing their personal significance in the universe nor of understanding their own importance. In order to shield themselves from the threat of being exposed to larger questions, they bury themselves in their jobs, families and leisure activities, and leave everything else out. They no longer function as self-aware persons.

Broader Individualism (#PF8393) Individual isolationism (#PD1749).
Narrower Individualistic welfare mindset (#PS9473).

♦ PD1109 Hooliganism

Yobbery

Nature Hooliganism can be defined as wanton destruction of property or injury to persons, sometimes involving theft, whether by a gang or a small group of (usually) young people. Injury to persons may vary from gang rape to assault on unsuspecting individuals (often old people) or inter-gang warfare. Hooliganism is characterized as a lack of self-control, love of malicious mischief, indifference to the comfort of suffering of others, idleness passing into dishonesty and crime, horseplay passing into violence. It may be expressed through petty rudeness. In areas characterized by hooliganism or yobbery, fear of attach prevents people from going out. When they do they are met by litter, graffiti and a brutal lack of consideration.

Background Between about 1914 and 1950, the level of violence and other forms of anti-social behaviour in industrialized countries was abnormally low. Though unproven, this lull may well have resulted from the terrible blood-letting of the world wars. When wars occur, they provide an excellent outlet for the aggressive of young men. Football hooligans would be praised as heroes on any battlefield. In peace it is harder to find any excuse for fighting of which society would approve.

Counter-claim Spirited youth craves risk, excitement, danger. It cannot find fulfilment by polishing shoes and passing accountancy exams, but few alternatives are available. The more violent sports have traditionally been one way to work out aggression, but respectable society dislikes such recreations almost as much as war. The idea that the rude, violent streak in any culture can be removed is however a delusion. Throughout history it finds better and worse expressions, but it does not disappear.

Refs Buford, Bill *Among the Thugs* (1990); Pearson, Geoffrey *Hooligan* (1983).

Broader Civil violence (#PC4864) Social deviation (#PC3452).
Related Vandalism (#PD1350) Juvenile delinquency (#PC0212)
Offences involving danger to the person (#PD5300).
Aggravates Destruction of private property (#PG2842)
Repressive detention of juveniles (#PD0634).
Aggravated by Youth gangs (#PD2682) Social alienation (#PC2130).

♦ PD1110 Unethical practice of soil sciences

Irresponsible pedologists — Negligence by pedologists — Malpractice in pedology — Corruption of pedologists — Underreporting of hazards to soils

Claim Soil scientists, under pressure from their employers and fertilizer interests, have adopted practices which lead to the underreporting of hazards to soils as a consequence of soil mismanagement and mechanized agriculture, especially phenomena such as top soil erosion, degradation of soil fertility and demineralization of soils.

Broader Unethical practice of earth sciences (#PD0708).
Aggravates Soil infertility (#PD0077) Soil degradation (#PD1052)
Over-intensive soil exploitation (#PC0052)
Environmental hazards from fertilizers (#PE1514).

♦ PD1111 Threat of birds to aircraft safety

Bird strikes against aircraft

Nature Birds threaten the safety of aircraft in flight when they may strike and destroy the windshield at a critical moment, possibly leading to sudden decompression within the aircraft. They may also be sucked into jet intakes. On the ground, resting birds may hinder the movement of aircraft landing or taking off.

Incidence Birds can fly at very high altitudes and are regularly detected at 20,000 feet. Many birds have collided with aircraft in mid-air with disastrous results. Particular species most often involved include starlings, gulls, condors, albatrosses, several species of eagle, and particularly the whistling swan. In Europe in 1989, some 3,000 bird strikes were recorded against aircraft., although only 30 caused damage requiring repairs.

Broader Aircraft environmental hazards (#PD8328).
Aggravates Air accidents (#PD1582) Injurious accidents (#PB0731).
Aggravated by Bird netting (#PG6093) Birds as pests (#PD1689)
Delayed development of regional plans (#PF2018).

♦ PD1115 Toxic substances

Toxins

Nature Toxic substances are chemicals or mixtures of chemicals, either synthetic or natural, that are poisonous to humans, plants, or animals under expected conditions of use and exposure. These substances include pesticides, some industrial chemicals, drugs, hazardous wastes, and radioactive materials. Toxic substances can cause immediate poisoning, behavioural and other nervous system disorders, and other illnesses. Over time, exposure can lead to chronic ailments and disability and can affect the growth of cells, sometimes causing cancer, genetic damage, or mutation Owing to their widespread distribution, use, and disposal, toxic substances pervade the environment.

Toxic effects depend not only on the composition and basic properties of a substance or mixture but also on the dosage, route and conditions of exposure, susceptibility of the organism exposed and other factors.

Incidence Technological and social developments have multiplied the hazards to which the population, especially in urban areas, is exposed, such as the harmful effects of chemicals on the central nervous system. More than one-third of the industrial chemicals listed in the American "table of threshold limit values" affect the nervous system at the threshold concentration, or at concentrations twice to ten times greater than the recommended level. Many industrial chemicals, such as carbon disulphide, mercury, manganese, tin, lead compounds, trichloroethylene, decaborane, and carbon monoxide have been shown to be selective neurotoxic agents producing neurological and behavioural disturbances. The critical period of vulnerability is during foetal and immediate postnatal life, and infancy; at these stages of cerebral maturation such compounds can produce serious and irreversible damage. The dosage and duration of exposure to chemical agents is also very important. For example, lead poisoning in children can produce irreparable brain damage with permanent mental retardation. Severe exposure often occurs in children from slum areas of industrialized cities and leads to chronic impairment of the nervous system.

Refs Faulstich, H, et al *Amanita Toxins and Poisoning* (1980); Kurata, Hiroshi and Ueno, Yoshio (Eds) *Toxigenic Fungi* (1984); WHO *Aquatic (marine and freshwater) Biotoxins* (1984).

Broader Chemical warfare (#PC0872) Human disease and disability (#PB1044)
Unproductive use of resources (#PB8376).
Narrower Mycotoxins (#PE7514) Toxic hazards of cassava (#PE6840)
Hazardous combinations of substances (#PD5256).
Aggravates Human poisoning (#PD0105) Hypersensitivity (#PE6898)
Hazardous wastes (#PC9053) Nutritional anaemia (#PD0321)
Bioaccumulation of toxic substances (#PD7205).

♦ PD1116 Sexual harassment

Sexual provocation — Sexual exhibitionism

Nature Sexual harassment varies widely from demanding sex in exchange for a promotion or a job to lewd language or photos in the work place. It includes repeated, offensive flirtations, advances and propositions; continual or repeated verbal abuse of a sexual nature; graphic verbal commentaries about an individual's body; sexually degrading words used to describe an individual; display in the workplace of pornographic or suggestive material; sexually explicit graffiti in bathrooms or other public places and opening meetings or other gatherings with "dirty" jokes. Teachers may harass students. Landlords may harass tenants. Law enforcement personnel may harass defendants or criminals. Wherever one person seeks power over another sexual harassment may be present. To a large degree the driving concern behind sexual harassment is not lust but demonstrating power over people.

Refs Herbert, Carrie M H *Talking in Silence* (1989); International Confederation of Free Trade Unions *Sexual Harassment at Work* (1987).

Broader Harassment (#PC8558) Persecution (#PB7709)
Unethical personal relationships (#PF8759).
Narrower Sexual harassment of men (#PE1293) Sexual harassment of women (#PF3271)
Environmental sexual harassment (#PD7707)
Sexual harassment in the working place (#PE8466).
Related Sexual exploitation (#PC3261).
Aggravates Teasing (#PE4187).
Aggravated by Exhibitionism (#PD4643) Sexual immorality (#PF2667)
Paternal negligence (#PD7297).

♦ PD1119 Radioactive contamination of animals and animal products

Nature Animal products are an significant link in the food chain by means of which radionuclides reach the human population, because the grazing animal is an effective collector of contamination via plants. There are many factors that influence the degree of exposure of farm animals to radionuclides, such as metabolic properties of the specific nuclide, and feeding and management practices. Exposure affects the wellbeing of animals which will generally acquire higher body burdens than will the human populations dependent upon them.

Incidence Animals can be a source of radioactivity when their meat and bones are used as food and through their products: for example, milk the and eggs. When 'fresh' radioactive materials enter the organism of the cow, iodine, molybdenum, strontium and barium can be found in the milk. Iodine is the critical substance and cows should be put on stored fodder if pastures have been subjected to radioactive fallout. One survey noted that hen's eggs accumulated a considerable amount of radioactive nuclides - up to 8 percent of the daily uptake of iodine-131. There was 20-50 times more radioactivity in the yolk than in the egg whites. It was noted that if eggs, laid during the early days of fallout, are immediately washed, much of the radioactivity is removed from the shell.

Broader Radioactive contamination (#PC0229)
Endangered animal and plant life due to radioactive contamination (#PD5157).
Aggravated by Radioactive contamination of plants (#PD0710).

♦ PD1126 Unaesthetic foodstuffs

Nature As the inadequacy of the food supply increases with the increase in population, a variety of synthetic (and less appealing) foodstuffs and protein concentrates (from green leaves and seeds, for example) will come to form the staple diet.

Broader Ugliness (#PA7240) Negative emotions and attitudes (#PA7090).
Aggravated by Neglected food resources (#PF7808)
Long-term shortage of food and live animals (#PE0976).

♦ PD1133 Non-comprehensive wage scales

Nature There is no standard universal income scale which guarantees all people the basic requirements necessary to sustain human life. There are also no adequate criteria that recognize the income needs of individual families.

Broader Inadequate access to negotiation on employment and reward (#PD1958).
Narrower High severance pay for top managers (#PE3872).

♦ PD1135 Defoliation

Nature Defoliation, a form a chemical warfare, is the use of herbicides to destroy plant or forest landcover in order to improve visibility and cut off food supplies to the enemy. The use of chemical herbicides cannot be limited to affecting only the enemy, however. It both drifts over borders into neutral neighbouring areas and countries and also affects innocent civilians and animals in the warring countries themselves. Chemical agents have been proved to produce birth defects in humans and animals; have caused the starvation of civilians due to the destruction of food crops (sometimes so completely as to render regeneration impossible); have caused the death – and possible extinction – of species of animals which depend on foliage for food and concealment; and leave toxic residual in humans and animals which can eventually accumulate to lethal levels.

Incidence In the Vietnam War, the USA used considerable amounts of herbicides against Vietnam; in 1967, 50 million pounds of herbicides were sprayed over 1 million acres of South Vietnam; in 1968 10 million gallons infected 4 million acres (one third of which was crop land); usage was discontinued in 1970. A 1971 report shows that by 1970, 1.2 million acres of mangrove forests in Vietnam had been totally destroyed and 600,000 people had been cut off from their food supplies. Records from a Saigon children's hospital indicate a dramatic increase (from 26 per 1000 to 64 per 1000) in cleft palate, spina bifida, and other birth defects after 1966, the year in which heavy antiplant spraying began. In 1984, Kampuchea complained to the United Nations Commission on Human Rights that the Vietnamese occupying forces in Kampuchea were using chemical weapons in water sources and food supplies, leading to severe diarrhoea, dysentery, fever, and possible death.

Background The first chemical to be used in warfare was chlorine gas in World War I, although during the course of the war other chemicals were used as well. Italy used chemical weapons against Ethiopia in 1935–1936 and both Ally and Axis powers produced and stockpiled chemical weapons in World War II. The Geneva Protocol of 1925 and the Bacteriological and Toxin Convention of 1972, the key documents prohibiting biochemical warfare, have not been successful in preventing its use. The first document outlawed first–use but failed to prohibit either possession or retaliatory use; the second bans the production, possession, transfer, and use of chemical weapons but lacks verification procedures. It has been more the fear of retaliation (thus leading to usage mainly against those countries unable to reciprocate) than legal or moral restrictions that have limited usage.

Claim The use of defoliants is even more inhumane than traditional warfare because of its unobservable and thus undetectable presence. It kills civilians as well as combatants and the consequences of its resultant ecological calamity remain long after the combat ceases.

Counter–claim Defoliants have been proved to be highly effective. They can kill personnel without great material destruction and the weapons used are relatively inexpensive. They are particularly useful in combating guerrilla warfare by exposing enemy positions usually concealed in foliage, thus making defensive positions possible. The use of defoliants on North Vietnamese supply routes was shown to have seriously impeded North Vietnamese movements.

Broader Environmental warfare (#PC2696).
Narrower Defoliation of insect breeding areas (#PJ4038).
Related Forest fires (#PD0739).
Aggravates Sand storms (#PD3650) — Soil erosion (#PD0949) — Desert advance (#PC2506) — Soil infertility (#PD0077) — Killing of plants (#PD4217) — Migrating sand dunes (#PD0493) — Herbicide damage to crops (#PD1224) — Destruction of land fertility (#PC1300).
Aggravated by Chemical warfare (#PC0872) — Herbicides as pollutants (#PD1143).
Reduces Vectors of animal diseases (#PD2751) — Wild animals as carriers of animal diseases (#PD2729).

♦ PD1140 Blocked minority opinion

Paralysis of minority voices — Political minority status

Nature Minority voices in a political system are often blocked from expressing their criticism of established practices because of their own sense of being a victim of the social system. They, therefore, do not participate in shaping society further perpetuating their victim image.

Claim In these times of rapid change, fears of social chaos render society brittle and inflexible relative to the articulation of unpopular opinions. No matter how "free", societies have to struggle to relate to their minority members. Unfortunately, excessive suppression of these minority voices heightens social rigidity and often results in precisely the chaos so feared.

Broader Exclusion of opposing views (#PF3720).
Narrower Majority rule mindset (#PF0851) — Fear of social anarchy (#PF1088) — Demeaning minority self-image (#PF1529) — Innate expectation of suppression of minority opinion (#PD2108).
Related Inadequate political networks (#PD2213) — Collapse of societal engagement (#PF2340) — Irresponsible expression of emotions equated with free speech (#PF7798) — Unsystematic use of powerful relationships by rural communities (#PE1101).
Aggravates Ethnic conflict (#PC3685).

♦ PD1141 Defective land use planning in developing countries

Nature Defective land tenure structure in many developing countries is a major hindrance to effective introduction of integrated land use planning. In the absence of effective social control over private land use, misuse of grazing and pasture land is continuing at an alarming rate in many countries; similarly, conservation and development of water resources have been made difficult due to a defective pattern of water rights. Deforestation is further evidence of ineffective land use planning and absence of adequate state control over the national resources. Coordination between land reform and land use planning is crucial to the integrated development of cultivated land, grazing areas and forests, and to solving the problems of shepherds and nomads, who have not been integrated into the general economy of the country.

Broader Land misuse (#PD8142) — Shortage of cultivable land (#PC0219) — Human errors and miscalculations (#PF3702).
Aggravates Maldistribution of land in developing countries (#PD0050).

♦ PD1143 Herbicides as pollutants

Herbicides

Nature Herbicides are chemicals used for weed control. The first chemical weed killer was common salt. Other inorganic substances like arsenates, copper sulphate and sulphuric acid have been tried, but unfortunately these materials also poison the soil and kill beneficial plants. Modern herbicides, although very effective in removing unwanted weeds, are potentially hazardous and the tons of weedkillers used each year in agriculture have deleterious side affects. Many of these chemicals persist in the soil – often affecting the subsequent year's crop – or find their way into domestic water sources. Some herbicides are suspect as carcinogens, mutagens, or they may interfere with embryo development. Older types, such as arsenic, have the added hazards of persistence.

As usually applied, herbicides can create serious problems with drift, killing plants not intended. Even on calm days, sprays of these materials can be detected surprising distances away, and damage may occur a mile or more from point of application. Danger can also arise from the use of lawn conditioners which contain herbicides such as 2,4–D, and 2,4,5–T in granular form. In hot weather these can volatilize and affect susceptible plants nearby.

Incidence There is a great number of commercial herbicides available. For example, there are 7,000 formulations of some 130 organic chemicals. Besides these, there are scores of inorganic preparations. Their dangerous properties are mainly those of injury to health and, in some instances, of inflammability. As an indication, the market for farden herbicides in the UK in 1986 was £14.5 million.

Broader Air pollution (#PC0119) — Agricultural poisons (#PD5277) — Pesticides as pollutants (#PD0120).
Narrower Vapours (#PS3339).
Aggravates Defoliation (#PD1135) — Soil pollution (#PC0058) — Hazards to human health (#PB4885) — Herbicide damage to crops (#PD1224) — Environmental plant diseases (#PD2224) — Carcinogenic chemical and physical agents (#PD1239).

♦ PD1146 Avalanches

Nature Avalanches are masses of snow moving rapidly down a mountain slope or cliff and are responsible for a large number of injuries each year, particularly on building and civil engineering sites in mountainous areas. There are two main types of avalanche: slab avalanches and avalanches of loose snow.

Various factors intervene in the release of avalanches. A heavy fall of snow may produce a layer of snow that moves as an immediate avalanche within 3 days of its deposition. The structure of the snow layer is one of the prime factors in avalanche causation; wind and temperature are other decisive factors. Avalanches usually occur during periods of high wind which cause overloading and wind slabs. There is usually a rise in temperature, a factor that normally accompanies heavy precipitations; this temperature rise reduces the cohesion between snow strata but at the same time promotes compaction and consolidation. Temperature rise therefore has a dangerous effect at first, which does not persist except when there is significant melting. On slopes with an incline of over 140% there is no danger of avalanche since the snow cannot accumulate over the long term. The lower limit of dangerous incline is quite low (30%), although avalanche defence construction is not built on slopes of less than 70% incline. Local conditions such as relatively smooth ground surfaces or the presence of convex slopes will increase the avalanche hazard.

Incidence Immediate avalanches account for 60–80% of all avalanches and, for example, the Alpine avalanche disasters of 1951 were the consequence of exceptionally heavy snowfalls. In France it is accepted that, at an altitude of 1,500 m, following snowfalls of 50 mm water equivalent (about 50 cm of snow), there is a serious avalanche hazard. When a level of 100 mm water equivalent has been exceeded, the danger becomes widespread and avalanches are numerous.

Refs FAO *Avalanche control* (1985); International Commission on Snow and Ice of the International Association of Hydrological Sciences *Avalanche Atlas* (1981).

Broader Topological disaster (#PC5010).
Narrower Rock avalanches (#PG0476) — Snow avalanches (#PE7838).
Related Coastal erosion (#PE6734) — Ground failures (#PE5066).
Aggravates Disastrous failure of natural dams (#PE0715).

♦ PD1147 Breakdown in community security systems

Inadequate community security — Individualistic images of security

Nature Physical, economic, political and cultural security is perceived as dependent upon the individual's ability to provide it. This self–centredness, while an effective refuge in the midst of changing times, is yet a very weak basis on which to build social cooperation.

Claim The anonymity of many inner city communities has created an atmosphere of isolation which enhances the possibility of crime, and at the same time increases the obstacles to crime prevention. For security measures, citizens have found it necessary to put multiple locks on their doors, conceal fire arms in their homes and places of business and be suspicious of even their own neighbours. Systems of community–wide security have collapsed and there is little intra–neighbourhood support for the few cooperative efforts to prevent property thefts and personal violations. The prevailing sense of hopelessness and fear perpetuates an atmosphere of ambivalence in which crime goes unreported, insecure situations which are not of immediate threat to personal welfare are ignored, and simple property watch between neighbours is no longer trusted. This condition is compounded by incidents in which individuals who report crimes are threatened with retaliation, while others are tempted to participate in pay–offs for personal protection. The tendency for crime to become acceptable behaviour increases.

Broader Obstacles to community achievement (#PF7118) — Overemphasis on self–sufficiency with respect to interdependence (#PF3460).
Narrower Risk of capital investment (#PF6572) — Necessity of business payoffs (#PG8854) — Restricted local participation (#PF3287) — Limited image of employability (#PF2896) — Increase in abandoned buildings (#PJ1665) — Conflicting social service ideologies (#PD3190) — Absence of images of social responsibility (#PF3553) — Unsystematic allocation of market facilities (#PD3507) — Fragmented individual decision–making process (#PF3559) — Absence of traditional patterns of community life (#PF3531).
Related Exclusion of opposing views (#PF3720) — Reduced understanding of globality (#PF7071) — Uncontrolled application of technology (#PC0418) — Legal contract system reduced to individual needs (#PE5397).
Aggravates Crime (#PB0001) — Insecurity of property (#PC1784).

♦ PD1148 Denial of rights of mental patients

Violation of the rights of mentally–ill persons

Nature Mental patients are denied the same legal rights as are accorded to ordinary citizens and even to criminals. A mental patient may be committed voluntarily or against his will to a psychiatric hospital from which, if he wishes to leave, he may have to prove before a tribunal that he is not insane (even in countries where a prisoner on trial is deemed innocent until proven guilty). He may be allowed no legal representation at the tribunal and once its decision is made, there may be no right of appeal. If he is not guilty of a criminal offence, habeas corpus cannot be invoked for his release. During internment there is no machinery for appeal against abuse by psychiatrists and psychiatric workers. The treatment alone may be brutal and disturbing, but other 'illegal' excesses have also been brought to light in recent years. Patients may be committed to hospital for political or family reasons. Diagnosis of mental disorder and hence treatment is widely disputed among different psychiatrists and yet in most cases the opinion of only one is needed to certify a person insane in the first instance and to prolong hospitalization thereafter. Relatives, psychiatric workers and the state (in the case of political prisoners) may confiscate the property of certified patients.

Incidence In the UK, for example, a mental patient may apply for a discharge 6 months after he has been committed, and again after 6 months, and also at the end of the 2nd year; from then on, only at 2–year intervals. Under Japanese law, a superintendent of a mental hospital can commit a patient indefinitely without he patient's consent and without the opinion of independent psychiatrists if the person responsible for the patient, usually a family member, gives consent.

Refs Heginbotham, Chris *The Rights of Mentally Ill People* (1987); United Nations *Principles, Guidelines and Guarantees for the Protection of Persons Detained on Grounds of Mental Ill-Health or Suffering from Mental Disorder* (1985).
Broader Denial of rights of medical patients (#PD1662)
Discrimination against mentally disabled (#PD9183).
Narrower Exploitation of the mentally handicapped (#PD9685)
Abusive treatment of patients in psychiatric hospitals (#PD0584).
Aggravates Mental illness (#PC0300) Inadequate mental hospitals (#PF4925)
Abusive detention in psychiatric institutions (#PE2932).
Aggravated by Inadequacy of psychiatry (#PE9172).

♦ PD1150 Storms

Nature An atmospheric disturbance involving perturbations of the prevailing pressure and wind fields on scales ranging from tornadoes (1 km across) to extratropical cyclones (2,000–3,000 km across) which may be accompanied by rain, blizzard, lightning and related phenomena, and preceded or accompanied by squalls. Storms may: cause flooding and damage to crops; uproot trees; damage roofs and chimneys; break windows, leading to rain damage; overturn trucks; affect transportation, communication and energy supplies; delay building construction; and destroy traditional landmarks. In their more violent form, storms may cause severe damage and loss of life.
Incidence It has been estimated that approximately 1,800 thunderstorms are active over the world at any one time. Over the period 1970–81, 353,832 persons were killed, and property worth 12,600 million dollars damaged by storms. Early in 1990 in Europe, over a period of 6 weeks, 5 storms killed nearly 200 people and caused over $11 billion damages. Large storms of increasing intensity and frequency are expected by some climatologists as a result of the greenhouse effect, since more heat provides greater energy for storms.
Refs Godschalk, David R, et al *Catastrophic Coastal Storms* (1989).
Broader Meteorological disaster (#PD4065).
Narrower Wind (#PE2223) Tornadoes (#PD1739) Hurricanes (#PD1590)
Snowstorms (#PG3352) Dust storms (#PD3655) Excessive rainfall (#PE4103).
Aggravates Floods (#PD0452) Lightning (#PD1292) Air turbulence (#PD2127).

♦ PD1152 Waste paper

Improper paper disposal — Newspaper waste
Incidence In West Germany some 5 million tonnes of waste paper is collected each year to be recycled, but dwindling prices for inferior household paper and low sales are discouraging paper manufactures from recycling paper. In 1989 in UK the supply of low-grade printed matter has escalated due to voluntary waste paper collection so much that many mills have large stockpiles and have to discourage further recycling projects.
Broader Disillusionment (#PA6453) Unproductive use of resources (#PB8376)
Long-term shortage of natural resources (#PC4824).
Related Paper shortage (#PE1616).
Aggravates Fires (#PD8054) Litter (#PD2541) Domestic emissions (#PG3368)
Amenity destruction (#PC0374) Smoke as a pollutant (#PD2267)
Industrial emissions (#PE1869)
Inadequate housing among indigenous peoples (#PC3320).
Aggravated by Domestic refuse disposal (#PD0807)
Non-biodegradable plastic waste (#PD1180).

♦ PD1156 Inhumane methods of riot control

Nature Many riot control methods are detrimental to health; although there are weapons that cause no longlasting distress to able-bodied people, the elderly, the sick and young children can be affected quite seriously.
Incidence Plastic and rubber bullets used by riot police as the first line of defence have killed 15 people in Northern Ireland between 1969 and 1986.
Claim The use of riot control weapons seems to be dictated by the wish to discourage people from exercising their democratic rights of peaceful demonstration. The use of many weapons is indiscriminate and although theoretically safe, when used during a riot – a time of pressure – they can be mishandled.
Broader Chemical warfare (#PC0872).
Related Police brutality (#PD3543) Civil disturbances (#PD5372).
Aggravates Fear (#PA6030) Panic (#PF2633) Injuries (#PB0855)
Human death (#PA0072) Demonstrations (#PD8522) Civil disorders (#PC2551)
Inflected loss of vision (#PE3377) Harm to innocent bystanders (#PJ3378).
Aggravated by Conflict (#PA0298) Student revolt (#PC2052)
Generation communication gap (#PF0756).

♦ PD1157 Irregular payments of international financial obligations

Irregular repayment of international debts — Irregular international payment of interest due
Nature Developing countries are often unable to repay interests or instalments on principal of international loans in uniform amounts with regular periodicity. Foreign exchange problems, or internal liquidity crises may intervene, but mechanisms to adjust payments due are of an ad hoc nature. In the United Nations system uniform payments of assessments suffer from these and from political variables.
Broader Lack of international cooperation (#PF0817).
Related Defaults on international loans (#PD3053).

♦ PD1162 Graduate and post-graduate unemployment

Unemployment of highschool graduates — Chronically unemployed graduates
Nature Universities in developed countries have organized themselves to increase considerably the number of post-graduate students. The demand for such students is now decreasing, with the result that there is a considerable rise in postgraduate unemployment. This decrease has in part been due to decreases in government expenditure on research as a result of national economic problems. In less developed countries, high school graduates in great numbers are among the jobless, particularly in Asia.
Broader Unemployment (#PB0750) Rural unemployment (#PF2949)
Educated unemployed (#PD8550).
Aggravated by Inappropriate education of graduates (#PF1905).

♦ PD1163 Crimes against the integrity and effectiveness of government operations

Broader Statutory crime (#PC0277).
Narrower Impersonating officials (#PE7687) Bribery of public servants (#PD4541)
Hindrance of law enforcement (#PD5515) Contempt of judicial process (#PD9035)
Falsification of public records (#PD4239) Intimidation of public officials (#PD4734)
Obstruction of government function (#PD6710) Soliciting obstruction of proceedings (#PD0790)
Crimes against government and public property (#PE5634)
Criminal negligence in performing socialist responsibilities (#PE5538).

♦ PD1165 Permafrost instability

Tundra ecosystem fragility — Thermokarst
Nature Permafrost is rock and soil material that has remained below zero centigrade for two or more years continuously. Much permafrost is composed of silty soil with a high frozen water content. Any disturbance to the thin vegetative cover which insulates the permafrost from warm air and sunlight, causes it to thaw into an unstable slurry of watery, oozing mud. This thawing can cause differential settlement of the ground surface, erosion, drainage problems and subsequent frost action. Once the equilibrium has been upset, the whole process continues irreversibly. It can be initiated simply by the passage of a tracked vehicle which destroys the vegetation mat and erodes into gullies many feet deep. This instability makes economic exploitation difficult (for example, road construction, oil pipelines with oil at temperatures of 160 fahrenheit) and is a threat to the fragile tundra ecosystems.
Incidence The distribution of the permafrost is not precisely known but more than one fourth of the Earth's surface has it in some degree. It is a phenomenon of high latitudes in both hemispheres and of high altitudes of major mountain systems.
Refs Harris, Stuart *The Permafrost Environment* (1985).
Broader Climatic cold (#PD1404).
Related Frost (#PD2244).
Aggravates Natural environment degradation (#PB5250).
Aggravated by Environmental degradation by pipelines (#PE6251)
Underdevelopment of industrial and economic activities (#PC0880)
Environmental degradation by off-road and all-terrain vehicles (#PE1720).

♦ PD1175 Biological contamination of water

Biological water pollutants
Broader Water pollution (#PC0062) Proliferation of pets (#PD2689)
Impurities in waste water (#PD0482).
Narrower Parasites (#PD0868) Water-borne diseases (#PE3401)
Water-borne viral disease (#PG3399)
Disease-causing microbes in drinking water (#PG3400)
Pollution of semi-enclosed bodies of seawater (#PE8175).
Aggravates Fungal diseases (#PD2728) Bacterial disease (#PD9094).

♦ PD1179 Capture and use of wild animals as pets

Domestication of wild animals — Taming of wild animals
Nature There are a number of species of wild animals which are found in reasonable abundance and are hardy and adapted to captivity. These make admirable pets it they are properly housed and fed. Parakeets or love-birds, hamsters and many species of tropical fish are examples. However, with the recent booming interest in wild animals, very large numbers of much less suitable species are being captured and shipped to meet the ever increasing demand. Many of the most sought after species are rare and their wild populations cannot withstand the rate of depletion brought about by the thriving market for the pet trade. The scarcer the species is, the higher the price it brings, and the search becomes more intense.
Most wild animals have very specific requirements for habitat, food, and social environments. In captivity they require special care, diets, and conditions of temperature and humidity which are beyond the ability of the casual buyer and most dealers to provide. Most animals that are taken from the wild and sold as pets are condemned to a short life, the torture of unnatural confinement, and inadequate care.
Incidence In the UK in 1988, of the 200,000 wild birds imported as pets, 13 percent died in transit. Investigations show flagrant disregard by suppliers, airlines and government officials worldwide of international regulations designed to protect animals during transportation.
Broader Domination (#PA0839) Misuse of wild animals (#PD8904).
Related Decreasing genetic diversity of animals (#PC1408).
Aggravates Broken-spirited animals (#PF5191) Maltreatment of pet animals (#PD1265)
Endangered species of animals (#PC1713).
Aggravated by Unethical practices with domesticated animals (#PE4771).

♦ PD1180 Non-biodegradable plastic waste

Nature Plastic wastes present formidable problems as they are at present not biodegradable, are bulky and can resist incineration. Incineration in fact may not be possible due to production of noxious or toxic fumes. Inadequate levels of recycling aggravate the waste problem. In addition, in a world where plastic production may, in a decade, reach over 60 million metric tons per year, plastic waste is a major loss of an important commodity.
Broader Hazardous waste dumping (#PD1398) Landscape disfigurement (#PC2122)
Environmental hazards from plastic materials (#PD8566).
Related Unproductive use of resources (#PB8376).
Aggravates Corrosion (#PD0508) Waste paper (#PD1152)
Air pollution (#PC0119) Amenity destruction (#PC0374)
Smoke as a pollutant (#PD2267) Environmental pollution (#PB1166)
Long-term shortage of natural resources (#PC4824).
Aggravated by Domestic refuse disposal (#PD0807).

♦ PD1185 Teeth disorders

Tooth decay — Dental caries — Diseases of the mouth and jaw
Nature Disturbances of the teeth and related structures often produce serious impairment of general health, since mastication and subsequent digestion may be reduced. Also bacteria or their products from a dental infection may invade other tissues, either by direct extension or by passing into the bloodstream. There is a strong positive relationship between general bodily health and that of the oral cavity and teeth.
Dental caries, or cavity formation, results from the gradual deterioration of the enamel, dentine, and finally the tooth pulp. It is due to many contributing factors. However, the action of microorganisms on fermentable carbohydrates contained within adherent accumulations on the tooth surface called plaques produce acids which appear to initiate the process. The ensuing bacterial invasion leads to continued destruction of the enamel, the dentine, and finally cavitation of the tooth. Dentoalveolar abscesses are acute or chronic inflammations from bacterial infections that occur in the root canals of the jaws. Periodontal disease, or pyorrhoea, is an inflammation of the gum margin and tooth sheath (periosteum) from local irritation or infection. Toothache, or ondontalgia, is pain arising from stimulation of the dental nerves by any process such as inflammation. It may occasionally occur as a reflex stimulation of the nerve roots and fibres from a lesion between the teeth and the brain. Diseases of the jaws are intimately related to tooth disorders since the upper and lower jaws form series of pockets for the teeth and carry their blood and nerve supply.
Incidence It is common for 95 percent of a population to have 10 to 15 teeth per person showing evidence of attack by the age of 20. Periodontal or gum diseases are the second most important cause of poor oral health after tooth decay. They affect nearly 9 out of 10 adults in many countries. According to the latest data on caries prevalence, the incidence of oral disease in the world falls into three general categories: very low to low in most countries in Africa and parts of Southeast Asia and the Pacific; moderate in many countries in the Eastern Mediterranean Region and some countries in Asia; and high to very high in most of the industrialized nations. Superimposed upon the caries problem is that of periodontal disease; broadly speaking, this disease tends to be worse in developing populations, especially in Africa and Asia, and somewhat less so in Europe and the Americas.
The number of dental schools in the world rose from 320 in 1958 to 371 in 1963 and 478 in 1974, but an analysis of the dental manpower situation shows that in general the number of dentists has increased less rapidly than the population. The population per dentist in the world was 7310 in 1963 and rose to 7566 in 1975. The dentist/population ratio differs widely from country to country,

ranging from 1:1,140 in Sweden to 1:1,150,000 in Niger. The overall figures for the world, even in 1975, were far from high and the increase in the number of dentists is tending to become less rapid than it was 20 years ago. The geographical distribution of dentists within countries varies considerably, rural areas almost everywhere having fewer dentists and a much higher population/dentist ratio. Although this is a universal problem, it is particularly serious in developing countries where the proportion of the population living in large cities is low. As an indication, the market for toothbrushes and toothpaste in the UK in 1986 was £121 million (plus a further £23 million for denture cleansers).

Refs Ivanyi, L *Immunological Aspects of Oral Diseases* (1986); Kelly, James E *Decayed, Missing and Filled Teeth Among Youths 12–17 Years* (1975); Lehner, T and Cimasoni, G *The Borderland Between Caries and Periodontal Disease II* (1980); Thyistrup, A (Ed) *Dentine and Dentine Reactions in the Oral Cavity* (1987).

Broader Human disease and disability (#PB1044).
Narrower Cleft palate (#PE5117) Dental caries (#PG3892) Periodontal diseases (#PE3503) Loosening of the teeth (#PG3416) Dento–facial anomalies (#PG9483) Diseases of the tongue (#PG5613) Irregularity of the teeth (#PG3415) Inflammation of the tooth fang (#PG3419).
Aggravates Headache (#PE1974) Diarrhoea (#PD5971) Nutritional anaemia (#PD0321) Unaesthetic appearance (#PG3421) Bodily ailments, lesions and malfunctions (#PG3423) Diseases and deformities of the digestive system (#PC8866).
Aggravated by Human ageing (#PB0477) General debility (#PG3424) Injurious accidents (#PB0731) Pregnancy disorders (#PD2289) Diseases of the jaw (#PG3427) Inadequate dental care (#PJ5478) Inadequate dietary fibre (#PE4950) Excessive consumption of sugar (#PE1894) Susceptibility of the old to physical ill–health (#PD1043).

♦ PD1189 Food fads

Nutritional quackery

Nature Since earliest recorded history, magic qualities have been attributed to certain foods and an aura of mystery has surrounded many of them. As knowledge of nutrition has increased, people generally have become more aware of the importance of food, but they are still susceptible to folklore and faddism in food choices. Most people as they grow older have deep desires to maintain perpetual youth and health. Few have any knowledge concerning the relationships between food and health, and many lack confidence in the medical community's assistance in maintaining their health. Thus, they fall prey to the promises and false premises of the sellers of pseudo–nutrition and food nonsense.

Incidence Nutritional quackery is the biggest fraud in the health field. Conservative estimates indicate that US $500 million to $2,000 million are spent annually in the USA alone on non–prescribed and unnecessary food supplements and vitamin and mineral products. In recent years new fads in foods have included 'health foods', 'organic foods' and 'natural foods'. In almost any city in the USA, foods described by this terminology are available for sale, usually at several times the cost of regular foods. At least 26 common herbal teas contain toxic ingredients. These teas have already caused a wide range of diseases of the digestive system, blood, the heart, the nervous system and skin. In America at least four people have died from herbal tea poisoning.

Claim Damaging statements have come from misguided enthusiasts and even from some in responsible positions in agriculture and in industry, who are promoting the thesis that poor soils are responsible for a decrease in the nutritive value of specific food crops and animal products, and thus result in serious malnutrition. These ideas continue to be set forth without adequate evidence for them, and in spite of specific evidence to the contrary. These enthusiasts are doing both agriculture and the cause of nutrition harm by undermining public confidence in the nutritive value of agricultural products.

There is no difference in nutritive value between organically grown food and food grown with the aid of chemical fertilizers and chemical pesticides. This has been known for many years by nutritional and soil scientists and is based on thorough research. Fertilization, regardless of the type, does not influence the nutrient composition of the plant with regard to its content of protein, fat, carbohydrate, or the various vitamins; these nutrients are influenced primarily by the genetic composition of the seed and the maturity of the plant at the time of harvest.

The following items of diet have all been found to cause "health problems" in various health surveys: tea, coffee, milk, water, orange juice, apple juice, sugar, butter, margarine, bread, bran, bacon, salt, wine, beer, eggs, beef, pork, lamb, chicken, sausages, potatoes and chocolate. It can be assumed that surveys will eventually find a health risk in all known comestibles.

Broader Health fraud (#PD9297).
Related Predominance of fast food (#PF5940) Unsafe artificial sweeteners (#PE6390) Excessive consumption of specific foodstuffs (#PC3908).
Aggravates Medical quackery (#PD1725) Nutritional deficiencies (#PC0382) Human disease and disability (#PB1044) Negative emotions and attitudes (#PA7090) Prohibitive cost of nutritious food (#PF1212).
Aggravated by Unethical food practices (#PD1045) Fear of food contamination (#PF3904).

♦ PD1190 Inadequate testing of drugs

Nature Medical, animal and plant drugs (pesticides especially) may not be fully tested for harmful effects before being put on the market. Animal and plant drugs may be well tested in so far as they relate to animals or plants (though the latter may upset the ecological balance), but not regarding their effect on humans. Medical drugs may be tested on animals but reactions in humans may be significantly different.

Adequate pharmaceutical research is complicated by high cost and long duration, inadequate national and international control, inadequate exchange of information on both the national and international level, and the attempt of legislation to break pharmaceutical monopolies, which results in the granting of licences to drug–copying firms which do no research of their own. Equally the lack of availability of human subjects for research may lead to unethical experiments without consent of the patient, or straightforward marketing of the product without further research. Lack of understanding of allergies and the effects of combination of substances is aggravated by a proliferation of new drugs and the increasing tendency of doctors to prescribe powerful drugs for ailments which were formally treated by less drastic means. New commercial drugs available without prescription are equally, if not more, dangerous (for example, thalidomide). The effects of animal and plant drugs may initially go unrecognized initially because of the indirectness of their entry into the human body.

Lack of coordination of knowledge regarding these three areas and effects noted in different countries contributes to delayed safety measures in the event of disaster (such as with and the fungicide infection in Turkey 1956). Irresponsibility and commercial interest also play a part. Inadequate general testing of drugs may occur, but there is also insufficient standardization of pharmaceutical manufacture and quality control of proven products.

Broader Inconclusiveness of scientific and medical tests (#PD7415).
Narrower Inadequate testing of animal drugs (#PU3433).
Related Lack of medical information (#PJ8069).
Aggravates Irritating drugs (#PJ4491) Medical deception (#PD9836) Abuse of plant drugs (#PD0022) Inadequate drug control (#PC0231) Medication side effects (#PD9807) Inadequate information on drugs (#PF0603) Irresponsible pharmaceutical advertising (#PE2390) Increasing drug experimentation in developing countries (#PE6201).
Aggravated by Excessive proliferation of medical drugs (#PD0644).

♦ PD1194 Lack of access to public archives

Nature A considerable amount of material which is preserved in public archives remains inaccessible either to scholars or to the general public. In most countries where access difficulties are encountered by users, these are due primarily to the lack of adequate archival budgets that would allow proper housing, equipping and staffing of the archival institutions. This impedes the transfer of inactive records from the agencies and their processing (arrangement and description) by qualified professionals. The most liberal and intelligent access regulations declassifying documents are useless if the archival institutions are not in a position to take over the records and make them accessible through employing appropriate finding aids. Given the information explosion, however, no archival budgets would ever be set to allow for the processing of the flood of hard–copy documents. Government production of microform documents allowing computerized description and retrieval is severely lagging.

Where facts are only available through such records, this lack of access helps to perpetuate misinterpretation of events.

Broader Ignorance of cultural heritage (#PF1985).
Aggravates Abuse of power (#PB6918) Historical misrepresentation (#PF4932).

♦ PD1204 Inadequate water supply in the rural communities of developing countries

Distant spring water — Arduous water collection — Inconvenient water supply — Limited water access — Difficult water access — Distant water sources

Incidence Over 1.5 billion people in developing countries are without safe drinking water, and even more do not have basic sanitary facilities. Everyday, over 30,000 people in the developing countries die as a result of inadequate water supply and sanitation. Lack of safe water is the start of a tragic disease cycle; millions of people are unable to work due to water–related diseases, and countless are crippled, blinded or maimed for life. Tens of millions of women and children, in particular, have to walk miles and spend half each day fetching water, often polluted, and carrying heavy buckets of water home. It has been estimated that 80 percent of all illness in the developing countries is associated with polluted water. Although the percentage of people without safe water is falling the total number is increasing.

Claim The financial resources needed to improve water supply and sanitation in developing countries have been estimated by the World Bank at US $30 billion per year during the period 1981–1990. This price is far less than what is now being paid in terms of human suffering and economic loss.

Broader Instability of water supply (#PD0722) Limited access to social benefits (#PF1303) Weakness of infrastructure in developing countries (#PC1228).
Narrower Limited catchment facilities (#PG9676).
Related Long–term shortage of natural resources (#PC4824).
Aggravates Prohibitive cost of basic services (#PF6527) Inadequate water system infrastructure (#PD8517) Unproductive utilization of plantation space (#PF6455) Inadequate community care for transient urban populations (#PF1844).
Aggravated by Unreliable rainfall (#PD0489) Maldistribution of water (#PD8056) Physically inaccessible resources (#PC4020).

♦ PD1206 Radiation damage to materials

Nature Harmful changes to the properties of liquids, gases and solids are caused by interaction with nuclear radiation; such damage to inanimate materials being limited at present to those used structurally or otherwise within the radiation field of a nuclear reactor.

Broader Radioactive contamination (#PC0229).
Aggravates Corrosion (#PD0508) Nuclear reactor accidents (#PD7579).

♦ PD1207 Excessive use of chemicals to control pests

Lack of biological control of pests

Nature The problems of controlling insects and other pests have reached an all–time high. Pest control researchers, extension specialists and farmers are faced with mounting pest problems while certain avenues to solutions are meeting constraints of a technical, economic or social nature. These problems include: the damaging effects of persistent pesticide residues on terrestrial fresh water and marine organisms; the destruction of beneficial controlling organisms, thereby causing an increase in the number of pest species and their abundance; and the development of pest strains resistant to the pesticides used to control them. This resistance has in turn led to use of a still greater variety of pesticides and their use in greater quantities. In some situations it has become uneconomic even to attempt to control certain crop pests, and large areas have suffered great economic and social stress.

Claim The effect of new technologies on living systems has been consistently overlooked. New technologies have been introduced often in almost total ignorance or disregard of the biological and ecological systems that they subsequently disturb, and of the dynamic and evolving nature of living systems.

Broader Crop vulnerability (#PD0660).
Aggravates Pests (#PC0728) Insect pests (#PC1630) Pesticides as pollutants (#PD0120) Insecticides as pollutants (#PD0983) Disastrous insect invasions (#PD4751) Pesticide hazards to wildlife (#PD3680).
Aggravated by Pests of plants (#PC1627) Diseases of beneficial insects (#PD2284) Loss of beneficial plants and animals (#PE8717).
Reduces Vectors of animal diseases (#PD2751) Snail vectors of animal diseases (#PE2747).

♦ PD1211 Non–political market allocation

Claim The refusal to acknowledge that allocation of market facilities is a political process has led to a seemingly unbridgeable gap between the grassroots and the dominating economic powers, thus preventing the equitable allocation. The current injustices in market facilities demand that the wealthy elite at the top of the "pyramid" participate with those at the base in planning and defining equal access.

Broader Unsystematic allocation of market facilities (#PD3507).

♦ PD1220 Destabilizing corporate expansion

Claim A particularly disturbing aspect of the present era of economic uncertainty is the incidence of unregulated corporate expansion in the public as well as in the private sector. The importance placed on expansion by large–scale economic forces means that human, environmental and even quality issues take second place. This heedless expansionism creates not only social but also economic chaos.

Broader Expansionism (#PB5858) Unregulated ownership of the means of production (#PF2014).

♦ PD1223 Pollution of inland waters

Freshwater pollution

Nature During recent decades, with continued economic growth, population increase, urbanization and technical development, more and more waste materials have been discharged into bodies of water. Toxic substances and heavy organic loads usually have a severe local impact. This is especially true of smaller bodies of water such as streams, where the composition of invertebrate communities may be used to derive practical indices of pollution severity.

A particularly insidious effect of pollution occurs when toxic materials accumulate within the tissues of a species used as food. Such materials include PCBs and mercury. Fish in the Lower Rhine, for example, accumulate sufficient quantities of phenol for this to be detectable by taste. Heavy metal (especially mercury) accumulation in fish is causing concern about its fitness for human consumption in some areas.

Thermal pollution due to discharges from the cooling systems of power plants affects the aquatic ecosystem to various degrees. Heat influences all biological activity, ranging from feeding habits and reproduction rates of fish via metabolism to changes in nutrient levels, photosynthesis, eutrophication and the degradation rate of organic material.

The possible environmental impact of the long-range transport of sulphur and nitrogen oxides deposited as strong mineral acids in precipitation has emerged as a new problem. Sulphur oxides and nitrogen oxides derived from fossil fuel combustion both contribute appreciably to the acidity of precipitation. Some lakes have shown a decline in pH values of as much as 1.8 units since the 1930s - that is, acidity as measured by hydrogen ion concentration has increased almost a hundred fold. There is evidence that acidification reduces the diversity of plant plankton and affects a number of other organisms in the aquatic food web. The growth of rooted plants is reduced, and the abundance of bog moss (Sphagnum), growing as benthic carpets, increased. The rate of decomposition of organic matter declines and fungal felts may develop over sediments, mineralizing nitrogen and phosphorous much more slowly than the normal bacteria. Productivity of the lake ecosystem is reduced.

Incidence Atmospheric transport of sulphur compounds and other acidifying components has led to extensive regional acidification of water courses in parts of eastern North America and southern Scandinavia that are not far from centres of industry and where there is little natural buffering. Elsewhere in Europe the acidification of surface waters has been reported in Belgium, the Netherlands, Denmark, Italy and the United Kingdom. Fish populations declined during the 1960s and 1970s in lakes over an area of 33,000 km2 in southern Norway, in the Adirondack Mountains in the United States and in several areas in Canada.

Refs FAO *Symposium on the Nature and Extent of Water Pollution Problems Affecting Inland Fisheries in Europe* (1972).

Broader Water pollution (#PC0062).
Narrower Lake pollution (#PD8628) Pond pollution (#PE5385)
River pollution (#PD7636) Stagnant surface water (#PE2634)
Pollution of groundwater (#PD2503) Water system contamination (#PD8122)
Contamination of drinking water (#PD0235).
Aggravates Aquatic weeds (#PD2232) Amenity destruction (#PC0374)
Pests and diseases of fish (#PD8567) Human disease and disability (#PB1044)
Destruction of wildlife habitats (#PC0480) Eutrophication of lakes and rivers (#PD2257).
Aggravated by Soil erosion (#PD0949) Wasted water (#PD3669)
Contradictions (#PF3667) Oil as a pollutant (#PE2134)
Sewage as a pollutant (#PD1414) Lack of water conservation (#PJ3480)
Industrial waste water pollutants (#PD0575).

♦ PD1224 Herbicide damage to crops

Misuse of herbicides — Inconsiderate use of herbicides

Broader Citizen apathy (#PF2421) Misuse of chemicals (#PD5904)
Pesticide damage to crops (#PD2581).
Aggravated by Defoliation (#PD1135) Chemical warfare (#PC0872)
Herbicides as pollutants (#PD1143).

♦ PD1225 Lack of sanitation in rural areas of developing countries

Nature Man-made pollution of the environment has become a main concern of the governments of industrialized nations, but environmental hazards are perhaps most pernicious in the rural regions of the tropics. Epidemics and endemic disease are a constant threat to millions of people, most of them helpless to overcome the danger. There has been a great deal of progress in public health and in international cooperation to fight disease, but much remains to be done. In the more advanced rural communities, sanitation is often of a high standard, with programmes covering nearly the complete range of man's environment. Such communities approach well-developed urban settlements in quality of service. However, in those scattered rural communities not yet collected into villages, and those areas where the rural village is underdeveloped, sanitation is usually minimal or non-existent.

Incidence Three out of four people in the Third World excluding China have no sanitary facilities. Although the percentages are following, the total number of people without proper sanitation facilities still increases every year.

Broader Inappropriate sanitation systems (#PD0876)
Weakness of infrastructure in developing countries (#PC1228).
Aggravates Insufficient rural housing (#PF6511)
Unsanitary environment for basic health in small rural villages (#PD2011).

♦ PD1229 Dependence on social welfare

Dependence on social security — Debilitating effects of social welfare — Unmotivating welfare dependence — Welfare weakened initiative — Dependence on government welfare benefits — Destructive welfarism

Nature Welfare systems tend to create poverty rather than to diminish it, by exceeding their role of providing temporary relief to deserving unfortunates. Such systems encourage dependency and permanent entrapment.

Background In a time of rapid economic expansion, some nations are finding it possible to guarantee a level of economic support for citizens who are disabled, unemployed or otherwise incapable of self-support. In addition, although many inner city residents receive some kind of public assistance for food, housing, medical care, child support or aid to the aged or disabled, there is little incentive for welfare recipients to leave welfare rolls. It may be risky and often a job results in less income than public assistance offers. Welfare systems also produce their particular forms of social immobility and social strata as the individual uses special currency and special clinics. This often results in aimless activity, fraud and indolence. A committed approach is necessary to counter the long-term patterns of social welfare programs.

Incidence In 1990 in the Netherlands, nearly one in seven members of the labour force claims to be incapacitated and unable to work. More than 100,000 extra people are so declared each year. A third of those claiming disability identify psychological reasons, including nervous exhaustion. Each receives 70 per cent of their previous salary.

Claim Existing welfare systems produce a large stratum of dependents, inclined to the drug culture and to casual relationships and therefore to illegitimacy, family breakdown and crime. They also engender an interested elite by which they are operated and which derive their income and power from them. Publicly financed welfare has meagre results and cannot create affluence.

Counter-claim The OECD study "The Role of the Public Sector" challenges the view that overly generous government welfare programs are eroding the will to work, save and invest.

Refs Manning, N P *Social Problems and Welfare Ideology* (1985).

Broader Overdependence on government (#PF9530)
Limited access to social benefits (#PF1303).
Narrower Insufficient health payments (#PG8867)
Over-acceptance of socio-economic dependency (#PF8855).
Related Low self esteem (#PF5354).
Aggravates Benefit overpayments (#PD7500) Ambiguous shape of social identity (#PF6516).
Aggravated by Socialism (#PC0115) Unemployment (#PB0750)
Socio-economic poverty (#PB0388).

♦ PD1230 Structural failure

Failure of engineering materials and structures — Building failures — Constructional failure — Building collapse — Material fractures

Nature The failure of engineering materials and structures may have detrimental effects on safety and costs. Maintaining the integrity of structures throughout their lifetime plays a key role for safety in many areas where a structural failure would entail health hazards or losses of life. Evident examples are nuclear reactors, chemical industries, transport, buildings, bridges, and off-shore platforms. Structural failure may be aggravated by the use of structures under traffic and loading conditions which were not originally taken into account in the design specifications.

Incidence Studies performed in the United States and in other countries have shown that the cost of the three main processes that durable materials must withstand, i.e. fracture, corrosion, and erosion and wear is probably about 10 percent of the Gross National Product in industrial countries. Fracture was estimated to cost about $120 billion (1982 dollars) per year in the USA and corrosion at least as much. In the USA more than 400,000 highway bridges were built prior to 1935, and many therefore no longer adequately or safely serve current traffic demands. In Mexico City subsidence in the District Federale has caused a very considerable amount of structural failure.

Refs American Society of Civil Engineers Staff *Structural Failures*; Ransom, W H *Building Failures* (1987).

Broader Mechanical failure (#PC1904).
Narrower Dam failures (#PE9517) Collapsing physical structures (#PD4143)
Uneven settling of building foundations (#PE8020)
Inadequate earthquake resistant construction (#PE6257).
Aggravated by Wear (#PB1701) Dry rot of wood (#PE1606)
Wet rot of wood (#PE4767) Concrete fatigue (#PE1626)
Rodents as pests (#PE2537) Termites as pests (#PE1747)
Wood-boring beetles (#PE0886) Unmaintained bridges (#PE2m71)
Insect pests of wood (#PD3586) Brittle fracture of metals (#PG5588)
Defects in machinery design (#PE2462) Corrosion of iron and steel (#PE1945)
Low-quality construction work (#PD7723) Neglect of property maintenance (#PD8894)
Human errors and miscalculations (#PF3702)
Malpractice in the construction industry (#PD9713).

♦ PD1231 Entrenchment of vested interests

Entrenched privileges — Parochial interests

Nature When any large scale activity has developed giving a livelihood to a group of people, they will use all kinds of arguments and propaganda to prove that it is in the public interest and morally right that that activity should continue (as in the case of: slave trade, tobacco, tranquillizers, nuclear power). International, national and local relationships create structural blocks to the full development of production potential of the global society. Parochial interests and national policy determine what will be traded, and what materials will be used in manufacturing. These same impediments create a military oriented or national security oriented mindset reenforcing cultural taboos about ownership and distribution priorities.

Claim Entrenched vested interests of both nations and corporations insure the continuation of laws which maintain closed markets, thus sustaining the hierarchical status quo and hindering the development of responsible and responsive political bodies that would open trade and create the possibility of common markets and grassroots cooperatives.

Broader Undue political pressure (#PB3209) Imperialistic distribution system (#PD7374)
Unsystematic allocation of market facilities (#PD3507).
Related Domination (#PA0839) Human inequality (#PA0844)
Proliferation of commercialism (#PF0815).
Aggravates Vulnerability of government to lobbying (#PF5365)
Reinforcement of inappropriate development by privileged classes in developing countries (#PF6670).

♦ PD1233 Landslides

Landslip disasters — Mudslides

Nature Landslides occur when a mass of soil or rock moves down an incline under the influence of gravity. The causes may be either natural or artificial. Natural landslides can be caused by the access of groundwater, the properties of the soil itself, the gradient of a slope, the height of a slope, or the erosion of a slope at its base. The causes of artificial landslides lie in the way soil has been cut or banked. The velocity of a landslide varies greatly, ranging from a slow of 1 cm a year to a speed of several metres a day. This velocity can change over a period of time. In the case of a landslide covering a wide area, several slides may take place concurrently, each one having its own rate of slippage. The movement occurs not only in the direction of the slope but also upwards and downwards.

Even though landslides are generally not so spectacular or costly as earthquakes, major floods, hurricanes and some other natural catastrophes, they are more widespread and, over a period of years may cause more property loss than any other geologic hazard. Moreover, much of the damage and sometimes a considerable proportion of the loss of life attributed to earthquakes and intense storms is really due to landslides. In many developing regions landslides constitute a continuing and serious impact on the social and economic structure, of which the true measure is not in monetary units but rather in disruption and attendant misery of human lives.

Incidence In the May 1970 earthquake in Peru which cost about 70,000 lives, about 20,000 people perished in the debris avalanche from the north peak of Nevado Huascaran. During the period 1971-75, of some 19,000 lives lost in earthquakes, tsunamis, volcanic eruptions, landslides, and snow or ice avalanches, about 84 percent of the casualties were attributed to earthquakes and 14 percent to landslides. Annual direct or indirect costs of landslides in the United States have been estimated to exceed US $1,000 million. Annual losses in Italy have been estimated at US $1,140 million.

Refs Crozier, Michael J *Landslides* (1986); Schuster, Robert L (Ed) *Landslide Dams* (1986).

Broader Natural disasters (#PB1151).
Related Ground failures (#PE5066) Coastal erosion (#PE6734).
Aggravates Disastrous failure of natural dams (#PE0715).
Aggravated by Deforestation of mountainous regions (#PD6282).

♦ PD1239 Carcinogenic chemical and physical agents

Environmental cancer — Environmental cancer hazards

Nature Some studies on epidemiological approaches to cancer aetiology suggest that 80 percent of all cancers can be attributed to environmental factors. This means that they are not an inevitable consequence of the ageing process and are therefore potentially preventable. Factors with which such environmental (exogenous, as distinct from endogenous) cancers are associated include, in order of importance: diet, tobacco and reproductive/sexual behaviour as major causes; and occupational exposure and exposure to pollutants as minor causes. While few biological agents are known to be associated with cancer in man (although, for example, the fluke infection, schistosomiasis, is known to cause bladder cancer), such an association has been established

for a number of physical and chemical agents. The term chemical carcinogen in its widely-accepted sense is used to indicate a substance that is known conclusively to induce or enhance the incidence of neoplasia.
Incidence There are 18 chemicals, groups of chemicals and industrial processes which are known to be carcinogenic in man, and a further 18 which are probably carcinogenic. Several hundred chemicals have been shown to be carcinogenic in experimental animals. All substances for which conclusive evidence of carcinogenicity exists in man have also been proved to be carcinogenic in animals, with the possible exception of arsenic. Whether an animal carcinogen causes cancer in man depends on many factors, including inter-species biological differences, and the duration and extent of exposure and the measures taken to reduce these.
Changes in cancer incidence patterns among immigrants is revealing – the patterns gradually change from those prevalent in the country of origin to approximately those of the host country (as happened with Japanese immigrants to California and Hawaii). The recent increase in malignant melanoma is the direct result of changes in clothing habits and the fashion of being tanned. Other increases result from medical practices – the increase in thyroid cancer as a consequence of childhood exposure of the head and neck to X-rays, and the increase in endometrial cancer as a consequence of the use of menopausal oestrogens. The most important environmental factor in carcinogenesis is tobacco smoking, which alone accounts for 25–35 percent of all cancer deaths in men and 5–10 percent in women.
Refs Sugimura, Takashi, et al (Eds) *Environmental Mutagens and Carcinogens* (1981).
Broader Environmental hazards (#PC5883).
Narrower Toxic food additives (#PD0487) Cancer-causing foods (#PD0036)
Medicinal use of occupational carcinogens (#PG6568)
Carcinogenic consequences of food preparation (#PE6619)
Incorporation of carcinogens into consumer goods (#PE8934).
Related Mutagens (#PD1368) Teratogens (#PE0697).
Aggravates Malignant neoplasms (#PC0092).
Aggravated by Air pollution (#PC0119) Oil as a pollutant (#PE2134)
Occupational cancer (#PE3509) Cobalt as a pollutant (#PE2339)
Herbicides as pollutants (#PD1143).

♦ PD1241 Militarization of the deep ocean and sea-bed
Nature The waters of the ocean under 700 metres comprise an immense volume of space which has hitherto been relatively unexploited for military purposes. Marine technology has now opened up this space and the under lying sea-bed, making it available for both mobile and fixed weapons systems to be developed for operation from such locations. Such a development is desirable for the superpowers (USA, USSR) as a means of securing complete invulnerability for their strategic deterrence forces. From a military point of view, the advantage of this environment is that it is opaque to most forms of electromagnetic radiation and the vast spaces provide considerable room for manoeuvre; its use would enable the superpowers to keep their 'second strike' or 'assured destruction' capability secure under most conceivable circumstances.
Incidence Current under-sea weaponry consists of nuclear submarine-launched ballistic missiles (SLBMs). Over 100 submarines, principally from the superpowers, were known to be in operation in the 1980s. They carried over 1,000 missiles, many of them of MIRV design.
Claim The advantages of deep-sea based nuclear missiles are shifting from deterrent towards potential first-strike weapons. The armament of the present US Polaris submarines, for example, is itself sufficient to destroy about 200 million people and the whole of the industry of any enemy.
Counter-claim The 1972 treaty banning emplacement of nuclear weapons or other weapons of mass destruction on or in the sea-bed beyond twelve miles from territorial sea limits was signed and ratified by the USA, USSR, UK and 67 other states.
Broader Militarization (#PD1897) Competitive acquisition of arms (#PC1258)
Obstacles to the utilization of coastal and deep sea water resources (#PF4767).

♦ PD1244 Jamming of satellite communications
Deliberate interference with satellite communications
Nature An increasing proportion of world voice, data and image communication is transmitted via communication satellites. These may easily be jammed; all that is needed is a powerful transmitter with which to beam a jamming message towards the satellite. This will overload its transmitter and thus effectively drown its signals in noise so long as the jamming transmission is maintained. Such quenching, effected from any one point on the earth's surface, could reach any satellite visible from that point. It is possible also to jam receivers in a region by transmitting noise on the channel used by satellites, a method that will not interfere with the satellite itself.
Broader Human violence (#PA0429) Radio and television propaganda (#PD3085)
Radio and television censorship (#PD3029).
Narrower Radio frequency interference (#PD2045).
Related Radio noise of natural origin (#PF1676) Radio noise of industrial origin (#PE2473).
Aggravates Insufficient communications systems (#PF2350).
Aggravated by Ideological war (#PC3431).
Reduces Violation of sovereignty by trans-border broadcasting (#PE0261).

♦ PD1245 Dust
Excessive wind-blown dirt
Nature Dust is a particulate contaminant suspended in the atmosphere. Like carbon dioxide it limits the amount of energy arriving and leaving the surface of the planet, with consequent changes in weather patterns. Industrial, transport and domestic equipment all release particles of dust into the atmosphere. Such man-made contaminants include: carbon and metal flakes, nuclear bomb debris, pesticides and other agricultural chemicals, and lead from combustion engines. The exposure of man to dusts can lead to a wide variety of respiratory diseases, including pulmonary fibrosis, obstructive lung disease, allergy and lung cancer.
Refs Pye, Kenneth *Aeolian Dust and Dust Deposits* (1987).
Broader Urban-Industrial pollution (#PC8745) Particulate atmospheric pollution (#PD2008).
Narrower Oil smoke (#PG0786) Acid mists (#PE1976)
Coal fly-ash (#PG5363) Carbon black (#PE3804)
Mineral dusts (#PE4679) Pesticides mist (#PG3921)
Interstellar dust (#PF9002) Metal dust and fumes (#PE5439)
Smoke as a pollutant (#PD2267) Dust as an occupational hazard (#PE5767).
Related Aeroallergens (#PE2069).
Aggravates Silicosis (#PE1314) Tuberculosis (#PE0566)
Air pollution (#PC0119) Typhoid fever (#PD1753)
Pneumoconiosis (#PD2034) Chronic bronchitis (#PE2248)
Asbestos as a pollutant (#PE1127) Inadvertent modifications to climate (#PC1288)
Deterioration in atmospheric visibility (#PE2593)
Long-term change in atmospheric chemistry (#PF1234)
Demoralizing images of rural community identity (#PF2358)
Obstruction of astronomical observation by environmental pollution (#PE7244).
Aggravated by Wind (#PE2223) Dust storms (#PD3655)
Volcanic eruptions (#PD3568) Agricultural wastes (#PC2205)
Soil erosion by wind (#PE3656) Domestic refuse disposal (#PD0807)
Solid wastes as pollutants (#PD0177) Uncontrolled industrialization (#PB1845).

♦ PD1250 Decline of handicrafts and cottage industries in developing countries
Incidence Artisan and cottage production provides employment for about 50 to 75 percent of the number employed in large-scale manufacturing in the developing countries, depending on the degree of industrialization. Craftsmen working in or out of small shops, or at home, as owners or employees, are thus outnumbered by up to two to one by factory workers.
Claim Until the advent of the disrupting forces of technology, the village was the basic economic and cultural unit. Now its self-sufficiency has disappeared and it is dependent on the city, the nation and the outside world. Village industries, such as spinning and weaving, pottery, brass-ware, oil pressing, vegetable dyes, hand paper making and lacquerwork have gone into abeyance as machine-made goods, such as aluminium ware, kerosene, textiles and synthetic dyes, take their place. A superfluidity of cheap manufactures has displaced the craftsman, and his hereditary skills are disappearing. The ecological balance has been disrupted and the sense of social solidarity, of esprit de corps, generated by the rapport between the farmer and his land, the craftsman and his craft, are disappearing too, leaving a mobile, restless and shiftless population. The family structure of the village has been transformed into a mob-like adjunct of the factory.
Counter-claim Non-productive tribal and village economic activity has not declined enough in the face of the need for developing countries to enter and compete in the industrialized world. Clinging to traditional ways is an admission of defeat, a denial that the country is able to improve the living standards, literacy, health and security of its people.
Broader Unemployment (#PB0750) Socio-economic poverty (#PB0388).
Aggravated by Uncontrolled industrialization (#PB1845).

♦ PD1251 Substandard housing and accommodation
Violation of house codes
Nature Substandard housing is associated with an absence of public sanitation measures such as: clean, uninfected drinking water; human waste disposal; and garbage removal. It is also associated with communities which have inadequate medical, educational and other services. An indicator of the sub-standard housing problem is that demand for accommodation in government-sponsored council houses and low-income-family apartment developments exceeds supply. Even in the more developed countries, housing stocks contain large numbers of dwellings that are unfit and even not repairable. In the developing countries, corresponding to the distribution of incomes, a certain duality has come to characterize the major urban areas: the city proper is often barely distinguishable from its counterparts in the high-income countries, while alongside it and linked to it in the economic nexus are the barricados, bustees, bidonvilles, favelas and shanty towns that house what has become known as the marginal or transitional population.
Incidence In the UK, for example, nearly 20 per cent of the housing stock is classified as unfit or in an unsatisfactory condition. Government statistics show that over 3.8 million dwellings require expenditure totalling £18.84 billion at an average cost of £4,900 per dwelling. Twenty percent of all primary schools are house unmodernized buildings which are over 80 years old. A quarter of all schools also have outdoor lavatories and one in 10 children are taught in temporary accommodation.
Claim Much of the world's population lives in substandard housing, even judged by local conditions. This may not necessarily indicate a total absence of plumbing, electricity or telephones, but it does imply: unhygienic and unsafe overcrowding both in dwellings and in the density of dwellings per given area; dwellings that are structurally unsound, or open to the elements; dwellings that are infested with vermin; and general living conditions that give rise to stress and a sense of despair, and which motivate crime.
Broader Inadequate housing (#PC0449) Poor living conditions (#PD9156)
Violations of private law (#PD5727).
Narrower Uninsulated buildings (#PE0242) Disease-prone housing (#PG7994)
Inadequate housing construction (#PG0561) Unhygienic dirt floors in houses (#PG9668)
Insufficient storage space in homes (#PJ0221).
Related Unknown building codes (#PJ8820) Inadequate maintenance (#PD8984)
Deteriorated vacant houses (#PJ8678) Low-quality construction work (#PD7723)
Inadequate hospital facilities (#PE5058)
Paralyzing complexity of urban structures (#PF1776)
Inadequate housing among indigenous peoples (#PC3320).
Aggravates Inappropriate sanitation systems (#PD0876)
Illegal occupation of unoccupied property (#PD0820).
Aggravated by Insecure lease tenure (#PJ8344) Inadequate building standards (#PF8829)
Inadequate electricity supply (#PJ0641) Socially sterile rental accommodation (#PF6195)
Alienating public housing assignments (#PJ9479)
Overcrowding of housing and accommodation (#PD0758)
Restrictive building codes in urban areas (#PE8443)
Ineffective self-regulation in the housing construction industry (#PE8265).

♦ PD1260 Segregation of poor and minority population in urban ghettos
Ghettoization
Nature The physical environment and social well-being of central cities depend largely on employment opportunities, a rising tax base, and income distribution. Unskilled immigrants are an increasing proportion of the urban population, and the decreasing number of jobs available to them means they rely on government for welfare, medical care, housing, food, and other essentials. At the same time, the ability of central cities to provide these services is hampered by the continuously declining tax base resulting from the suburbanization of industries and the movement of the middle class away from city centres.
Background The assimilation of newcomers into urban areas has to be assisted by the inhabitants already there, who must nevertheless compete with those they help for the same unskilled or semi-skilled jobs. The friction thus engendered is compounded in the current situation by the existence of prejudice against minority groups. Largely middle-class populations have felt their property and person to be threatened by peoples of a different colour or ethnic group, and have moved to the suburbs, where they now defy others to follow. What used to be a socially and ethnically diverse population living in districts or 'villages' within the central areas of cities has become a more homogeneous population consisting mainly of less educated minority groups often demoralized by their lack of occupational opportunities, by substandard housing conditions, and by the general deterioration of living conditions in the city centre. Even where actual prejudice does not exist, cultural and educational differences stimulate urban middle-income families to seek out suburban neighbours like themselves.
Broader Segregation (#PC0031) Discrimination in housing (#PD3469).
Aggravates Zionism (#PF0200) Locales of high crime rates (#PE7311)
Environmental degradation of inner city areas (#PC2616).
Aggravated by Urban bias (#PF9686) Ethnic segregation (#PC3315)
Racial segregation (#PC3688) Urban overcrowding (#PC3813)
Racial discrimination (#PC0006) Cultural fragmentation (#PF0536)
Segregation in housing (#PD3442) Discrimination against minorities (#PC0582)
Endangered tribes and indigenous peoples (#PC0720).

♦ PD1263 Divided countries
Partition
Nature Armistice agreements which terminate conflict may divide a larger political and cultural unit into two smaller units. This may give rise to tensions and border incidents which are a threat to the international peace and security in the area. Where interaction across the boundary is inhibited, it may also be source of social and economic difficulty, particularly for families; and this

is aggravated when the boundary cuts through a city (as in the case of Berlin). A different form of division may occur when two or more foreign powers take on joint responsibility for one territory as in the case of Andorra, the New Hebrides, and the Canton/Enderbury Islands.
Incidence Examples of this problem are: Federal Republic of Germany and the German Democratic Republic (reunited after 45 years); North and South Korea; India, and West and East Pakistan (now Bangladesh); Yemen; and Cyprus.
Refs Fraser, T G *Partition in Ireland, India, and Palestine* (1984); Huls, Mary E *Partitions* (1986).
Broader Unnatural boundaries between states (#PF0090).
Related Divided cities (#PD7065).
Aggravates Border incidents and violence (#PD2950)
Disruption of territorial integrity (#PC2945).

♦ PD1265 Maltreatment of pet animals
Maltreatment of companion animals
Claim The concept of an animal as being owned as personal property results in much animal suffering. Since such animals are believed to exist primarily for the pleasure of humans, they have become an object of human violence, vanity and neglect.
Broader Exploitation of animals for amusement (#PD2078).
Aggravates Unwanted pet animals (#PD2094) Domestic animal bites (#PE4931).
Aggravated by Proliferation of pets (#PD2689) Kidnapping of pet animals (#PE0805)
Capture and use of wild animals as pets (#PD1179).

♦ PD1271 Chemical air pollutants
Broader Air pollution (#PC0119)
Chemical pollutants of the environment (#PD1670).
Narrower Malodorous fumes (#PD1413) Lead as a pollutant (#PE1161)
Cadmium as a pollutant (#PE1160) Asbestos as a pollutant (#PE1127)
Fluorides as pollutants (#PE1311) Organochlorine pollution (#PG2007)
Gaseous organic compounds (#PJ5013) Hydrocarbons as pollutants (#PE0754)
Manganese as a health hazard (#PE1364) Beryllium as a health hazard (#PE2209)
Sulphur dioxide as a pollutant (#PE1210) Titanium dioxide as a pollutant (#PE5195)
Nitrogen compounds as pollutants (#PD2965) Particulate atmospheric pollution (#PD2008)
Carbon monoxide as a health hazard (#PE1657)
Chlorofluorocarbons as an environmental hazard (#PE4378).

♦ PD1273 Unemployed intellectuals in developing countries
Nature A university education creates very high employment expectations. In some countries, a university degree may be looked upon almost as a guarantee of a soft and secure job in the government service, and in most it is assumed to be a membership card of the elite class. But, even in rapidly modernizing countries, the purely administrative jobs in the government service become filled fairly rapidly: for example, the demand for lawyers is certainly not as great as the demand for technically trained personnel. And in those societies where large enterprises are owned and managed by members of family dynasties, even the opportunities for professionally trained engineers and technicians may be limited, at least in the early stages of development. In many developing countries there appear to be too many lawyers or too many arts graduates, and there may also be instances of unemployed or underemployed engineers, scientists, economists, and even agronomists. The unused intellectual, however, is sometimes unemployed because he is unwilling to accept work which he considers beneath his educational level. Rather than accept work beneath his status, or employment in remote rural areas, the university graduate, and sometimes the secondary school leaver as well, may prefer to join the ranks of the unemployed. A sizeable quantity of unused human capital of this kind reflects a wasteful investment in human resource development and poses a serious threat to a country's social and political stability.
Broader Underutilization of intellectual ability (#PF0100)
Labour surpluses in developing countries (#PD0156).
Related Inadequate research and development capacity in developing countries (#PE4880).
Aggravated by Inappropriate education in developing countries (#PF1531).

♦ PD1276 Reduction in demand for primary commodities due to technological change
Production of synthetic substitutes for primary commodities — Competition between synthetics and primary commodities
Nature The demand for many primary commodities, particularly those produced by developing countries, has been seriously eroded by the evolution of man-made substitutes. Half the world requirement of rubber, for example, is now met by synthetic products. Pulp-based rayon and acetate, and petroleum-based noncellulosic fibres are rapidly increasing their share of the total fibre market, thus restricting the rate at which cotton and wool sales can be expanded. In almost every field technical advances are resulting in economies in use. Leather produced from animal hides is replaced by plastic materials produced from natural gas; aluminium is substituted for wood; the products of the chemical industry for the output of the farm and the forest. The market share of natural products is falling steadily, which has the side effect of raising world price-elasticity of demand to at or near the level of synthetic prices.
Aggravates Instability of wool trade (#PE2056) Instability of jute trade (#PE1794)
Shortage of skilled labour (#PD0044) Instability of cotton trade (#PE1510)
Instability of trade in crude synthetic and reclaimed rubber (#PE0701)
Instability of trade in undressed hides, skins and fur skins (#PE1235)
Instability of vegetable fibre trade, excluding cotton and jute (#PE1513)
Instability in export trade of developing countries producing primary commodities (#PD2968).

♦ PD1283 Use of agricultural resources for production of animal feed
Nature Agricultural resources in developed countries (land, water, etc) are used primarily to raise crops to feed to farm animals for domestic consumption and production of meat and also to export feed-grains. This does not feed the poor and hungry in the developing countries but feeds farm animals for the more affluent to consume.
Absence of animal flesh and products in the diet - vegetarianism - is practised widely in some Asian countries, and to a much lesser extent in the western industrialized countries. Certain religions do not permit the killing of animals for food on the grounds that a human being should not inflict harm on any sentient creature.
Incidence One third of the world's grain is fed to livestock, but animal foods account for only one tenth of the world's caloric intake. In the United States, about two thirds of the harvested acreage is used to grow food for livestock. Western developed nations export 3 million metric tons of grain protein to underdeveloped nations and import 4 million metric tons of protein in oilseeds and fish from these countries. Most of the oilseeds and fish is fed to livestock. One third of Africa's protein-rich peanuts are fed to European livestock. To feed the entire world an average American diet would require twice the world's existing arable land and 80 percent of the world's energy.
Claim The consumption of animal flesh (whether meat, fowl or fish) or the use of animal products (leather goods, products of bone-processing) is unethical, unaesthetic, uneconomic, and without nutritional justification. It may be considered unethical because human beings should not needlessly kill sentient animals. The consumption of animal flesh is not the most economic method of obtaining the highest yield of nutritional products from agricultural land. It is also believed that animal products are less nutritious than other foodstuffs.
Counter-claim It is not proven that a vegetarian diet is more nutritious than a diet with flesh. Omission of animal products from diets in industrialized societies may be dangerous if adequate nutritional substitutes are not used. Refusal to consume animal products in some developing countries, such as India, aggravates the problem of providing the poor with adequate protein.
Aggravates Malnutrition (#PB1498)
Excessive consumption of animal flesh (#PD4518)
Excessive consumption of goods and services (#PC2518).
Aggravated by Taboo (#PF3310) Economic inequality (#PC8541).
Reduces Haemorrhoids (#PE3504) Constipation (#PE3505)
Maltreatment of animals (#PC0066) Inadequate diet of the poor (#PJ3529)
Cosmically inharmonious way of life (#PG3530).

♦ PD1284 Conceptual plagiarism
Theft of ideas — Unacknowledged copying of ideas
Incidence In innovative environments, including academic research, product development and creative design, in which new ideas form part of the normal subject of professional exchanges, informal conversation and memos, those originating ideas frequently find them taken up by others, possibly in a disguised for, without acknowledgement. This aggravates professional jealousies and tensions, reinforces secretism and generally acts as a break on the free exchange of ideas.
Broader Plagiarism (#PD3996) Vulnerability of intellectual property (#PF8854).
Aggravates Professional jealousy (#PD8488) Prevention of the exchange of ideas (#PD8731).

♦ PD1285 Stagnated development of agricultural production
Nature Agricultural production in third world villages is not meeting the needs of residents and in some cases is declining, although present knowledge of technology, hybrid crops and animal husbandry should make self-sufficiency in food production possible in virtually any community. Problems inhibiting such progress include: decrease of available water while populations increase; tenant farmers working on as short-term a basis as one year or even one crop; non-resident owners who may receive 50% of the crop but leave upgrading of tools and supplies to the tenant, thus leaving little means or incentive to improve seed varieties, fertilizers or equipment, or to experiment with the practical application of new farming methods. Unless a new emphasis is placed on the more effective use of land and livestock, communities will be able to depend less and less on local food production and economic self-sufficiency will not be possible.
Broader Decreasing agricultural growth per capita (#PF4326).
Narrower Off-farm employment (#PG9739) Limited family land (#PG9748)
Low-yield rice crops (#PG9724) Limited farm produce (#PG9716)
Untreated chicken diseases (#PG5454).
Related Insecure land tenure (#PD9162) Monoculture of crops (#PC3606)
Fragmented agricultural development (#PG8315)
Underdevelopment of agricultural and livestock production (#PD0629).
Aggravates Lack of care for animals (#PD8837)
Long-term shortage of food and live animals (#PE0976)
Unequal distribution of agricultural production (#PD4316).
Aggravated by Absentee ownership (#PD2338) Lack of agricultural machinery (#PF4108)
Limited agricultural education (#PF8835) Inadequate agricultural facilities (#PF6499)
Unavailability of agricultural land (#PC7597)
Obsolete methods of agricultural production (#PF1822)
Resistance to changing agricultural methods (#PF3010)
Underdeveloped use of agricultural resources (#PF2164)
Uncoordinated efforts in agricultural development (#PF1768)
Underproductive methods of agricultural management (#PF6524)
Insufficient access to technology for agricultural upgrading (#PF3467)
Unavailability of replacement parts for agricultural machinery (#PE7504)
Decreasing rate of development of major agriculture technologies (#PE5336)
Constraints to increased agricultural output in developing countries (#PD5114)
Unavailability of land for agricultural purposes in developing countries (#PE5024)
Insufficient fertilizers for agricultural development in developing countries (#PE4140).

♦ PD1290 Irresponsible introduction of new species of animals
Animal invasions — Reintroduction of animals
Nature Man has transported both wild and domesticated animals across the world. In their new environments, many species have become established, have multiplied and have become nuisances.
Incidence The pattern of European colonization resulted in the introduction of new species which either displaced or destroyed indigenous species. The Europeans brought cats, dogs, horses, pigs, goats, rabbits, donkeys, mongoose, rats and sheep to the Americas. Cats, rats and dogs decimated or extinguished (in over a 100 documented cases) birds and mammals. The mongoose eliminated ground-nesting birds, as well as other mammals and reptiles. Pigs, rabbits and goats destroyed the flora on lush tropic islands. Introduced bird species such as starlings and sparrows displaced many indigenous birds.
Some introductions that have been deliberate, and even well-intentioned, have still had serious consequences. Thus, the rabbit in Australia (which devastated the Australian grasslands), the sea lamprey in Canada (which resulted in the extinction of the most important commercial fish from the Great Lakes), the giant African snail in the Pacific and the edible snail in California were purposely introduced to serve as food but multiplied beyond expectations to become pests. Many other destructive species have been and still are being introduced unintentionally to new regions. The problem is acute with commensals, many of which have spread from their centres of evolution to become partially or totally coextensive with man.
The reintroduction of exterminated species is bound to fail if the chosen animal became extinct in the area too long ago and if the environment itself has undergone too many changes. Reintroduction needs years of careful planning - the approval of local population, technical conditions of the release, feeding system, protection and breeding control - and even then some unexpected problems may arise.
Broader Irresponsibility (#PA8658)
Environmental hazards of new species introduction (#PC1617).
Narrower Wasted woman power (#PF3690)
Introduction of new species of insect pests (#PF3592)
Irresponsible introduction of new species of fish (#PF3602).
Aggravates Animal pests (#PD8426) Disease vectors (#PC3595).
Aggravated by Inadequate animal quarantine (#PE2756)
Irresponsible genetic manipulation (#PC0776)
Unethical practice of the zoosciences (#PD4721).

♦ PD1291 Emigration of trained personnel from developing to developed areas
Brain drain from developing countries — Non-return of developing country students studying in foreign countries — Loss of developing country human resources to industrialized countries — Migration of skilled people from rural to urban areas — Reverse transfer of technology
Nature The migration of highly trained persons has caused concern among nations in recent years. This concern was first generated by the movement of engineers and scientists between a number of developed countries. However, in recent years the migration of trained persons from developing to developed countries has been attracting more attention. There are several reasons for this growing concern. First, the flow between developed countries has been reduced by a

number of factors, and there is a growing awareness that such migration is not of high and continuing concern among the nations involved. At the same time, the migration of trained persons from developing to developed countries raises important questions related to such matters as the equity of transfers of talent from developing to developed countries at high cost to the former and at no cost to the latter. Further, technological progress is a prerequisite to the economic development of developing countries, and trained persons are indispensable to the development and functioning of a society dependent upon such progress. The migration of these persons is therefore perceived by some countries as a serious threat to successful development efforts and to the successful execution of the economic and social changes which these countries desire. Yet, many developing nations see large numbers of their trained persons migrate to the advanced industrialized countries and contribute to the steadily growing total and per capita output of the developed countries at a time when they may be desperately needed at home.
Some developing countries, where the early stage of development of higher education necessitates the training of many students abroad, are faced with the problem of students who failed to return home after completing their courses. Such students remained abroad either to pursue their studies up to a level higher than that needed by their own country or to accept employment. Further difficulties created by training abroad are the frequent failure of foreign curricula to meet national needs, and the inappropriate attitudes which were all too easily engendered during a prolonged stay in a more developed country.
Incidence The migration of highly trained persons is to a relatively few advanced nations having market economies, including particularly USA, UK, Australia, Canada, France and the Federal Republic of Germany, and provides these nations with a valuable resource for which they pay virtually nothing. Their intellectual life and their research capacity are enriched. They augment at low cost their supply of trained personnel, particularly for critically important positions in the health services. The general availability to these advanced countries of highly trained immigrants has had a tendency to divert attention from the need to expand their own supplies of highly trained persons.
The financial loss suffered by a developing country due to the outflow of trained personnel is only the visible tip of the iceberg. The effects upon the process of development are as important, if not more so. The capacity of developing countries to achieve the progress associated with development depends upon the existence of structures within which this progress can take place and upon the existence of trained persons who can organize these structures and play key leadership roles within them. The structures include government, industries, agriculture and the social services; the trained persons include engineers, scientists, teachers, doctors and nurses. These persons can function only in the context of organized structures. Conversely, the structures are useless without the trained personnel to make them run. When, therefore, these trained personnel emigrate, the operation of the structures is seriously disrupted unless replacements are found from a continuing pool of similarly trained persons. The loss of such leaders therefore causes serious damage to programmes of development.
Counter-claim The brain drain is largely an imaginary malady. Countries with high emigration rates often have impressive rates of growth. In the 1970's, of the nine countries with the highest levels of emigration, had above average growth rates. Many countries benefit substantially through remittances from unskilled or semi-skilled workers abroad, which makes it difficult for them to put up barriers to movement. The principal motivation for emigration is lack of suitable jobs. Countries with a reasonable balance between the number of professional graduates and jobs available have had little problems with brain drain. Where brain drain does exist, it is primarily in those states where education has out paced infrastructure.
Refs Atal, Y and Dall'Oglio, L (Eds) *Migration of Talent* (1987); Beyer, G *Brain Drain* (1972); Chopra, S K (Ed) *Brain Drain* (1986); Cortes, Carlos E (Ed) *The Latin American Brain Drain to the United States* (1981); Kao, Charles H *Brain Drain* (1980); Oh, Tai K *The Asian Brain Drain* (1977).
Broader International migration (#PF4008) Loss of human resources (#PC7721).
Narrower Migration of rural population to cities (#PE8768).
Related Youth migration to cities (#PJ7859)
Emigration of trained personnel from industrialized countries (#PE8093).
Aggravates Shortage of skilled labour (#PD0044)
Inadequate industrial retraining programmes (#PF4013)
Inadequate research and development capacity in developing countries (#PE4880)
Unavailability of trained teachers in the rural areas of developing countries (#PE8429).
Aggravated by Attractive city jobs (#PD9869) Lack of job satisfaction (#PF0171)
Economic underdevelopment (#PC0281) Inadequate manpower planning (#PJ2036)
Social disaffection of the young (#PD1544) Inappropriate employment incentives (#PD0024)
Unemployment in developing countries (#PD0176)
Lack of economic and technical development (#PE8190)
Inequitable labour standards in developing countries (#PD0142)
Lack of technical development and excess of manpower in developing countries (#PE4933).
Reduces Limited exchange of skills among developing countries (#PF5006).
Reduced by Restrictions on freedom of movement between countries (#PC0935)
Restrictions on international freedom of movement for national advantage (#PD0351).

♦ PD1292 Lightning
Risk of lightning striking
Nature A large discontinuous discharge is produced through the air, generally under turbulent conditions of the atmosphere associated with thunderstorms. The electricity is generated in cumulonimbus clouds by separation of the electric charge associated with the upward movement of air and the freezing of water droplets. The main discharge of lightning runs from the earth upward (return stroke) along a channel prepared by a leader discharge. The peak value of the lightning current exceeds 35 kA (35,000 amps) in 50 percent of cases, with one percent exceeding 200 kA. The temperature in the lightning channel may reach 20,000 to 30,000 deg C and the rapid heating of air produces an explosion which is heard as thunder. When lightning strikes it may cause loss of life, destroy structures or cause fires. The high voltage or current produced in electrical and electronic equipment by lightning causes severe damage to them. Microelectronic devices are especially sensitive to the secondary effects of lightning (such as induced voltage) if they are not protected.
Incidence Electronic counters produce data which provide an estimate of about about two lightning strikes per sq km in temperate zones and between 10 and 15 strikes per sq km in tropical zones each year. Lightning discharges between clouds occur about twice as frequently. In forest areas in the USA it has been estimated that one sizeable fire results per annum from lightning for each 40 sq km of forest. Estimates for lightning-caused deaths are: USA, 160 per annum, Germany FR, 41 per annum (92 people struck each year). Despite assumptions to the contrary, lightning may cause fatal damage to aircraft. It has been estimated that over Europe one aircraft is hit by lightning for every 2400 flying hours, and once every 6,000 flying hours over oceans. The cost of damage to electrical and electronic devices due to secondary effects of lightning may be ten times higher than that produced directly by a stroke of lightning.
Refs Hart, William C, et al *Lightning and Lightning Protection* (1979).
Broader Bad omens (#PF8577) Meteorological disaster (#PD4065).
Narrower Ball lightning (#PG9472).
Related Thunderstorms (#PE3881).
Aggravates Fires (#PD8054) Air accidents (#PD1582) Damage to goods (#PE4447)
Interruption risk (#PF9106) Dying a bad death (#PF1421).
Aggravated by Storms (#PD1150) Climatic heat (#PC2460).

♦ PD1294 Inadequate facilities for the transport of water supplies
Broader Inadequate water system infrastructure (#PD8517)
Insufficient transportation infrastructure (#PF1495).

♦ PD1302 Adolescent induced abortions
Teenage abortions — Adolescent abortion — Teenage abortion
Nature Adolescent girls get pregnant out of ignorance about sexuality or a lack of access to contraception. Frightened at the prospect of being found out, either by parents who will disapprove of her sexual liaison or by a school administration that will expel her from classes, she resorts to legal or illegal abortion.
Incidence About one quarter of the 1.5 million abortions performed each year in the USA are on teenagers, and 40 per cent of adolescent pregnancies end in abortion. Teens aged 14 or under obtained one percent of the abortions. According to a study in Ibadan, Nigeria, of unmarried women aged 14-25 who became pregnant, 9 out of every 10 had an abortion.
Counter-claim Childbearing is more than 14 times more likely to result in death than is abortion for teens aged 15 - 19.
Broader Induced abortion (#PD0158).
Related Health risks of teenage sex (#PE6969).
Aggravates Illegal induced abortion (#PD0159).
Aggravated by Adolescent pregnancy (#PD0614)
Adolescent sexual intercourse (#PD7439).

♦ PD1304 Restriction of arms supply
Nature Governments may apply restrictions on the sale of arms to another country. Such restrictive measures may only come into force when the recipient country becomes involved in an international conflict, when it is feared that the supply of arms to countries in conflict will implicate the supplier in the dispute and compromise its neutrality. Such policies effectively prevent the supply of spare parts and replacements in times of conflict, thus endangering the military viability of the recipient country and its ability to defend itself against aggression.
Counter-claim Restrictive policies help to restrain or contain conflicts and prevent them from escalating or involving the supplying countries.
Broader Competitive acquisition of arms (#PC1258).
Aggravates National insecurity and vulnerability (#PB1149).
Aggravated by Neutrality (#PF0473) Aggression (#PA0587).
Reduces Conflict (#PA0298).

♦ PD1307 Retardation of psychomotor development in children
Nature Malnutrition and cultural deprivation in infancy and early childhood tend to lead to retardation of psychomotor development, impaired learning and behaviour. Cerebral development being of an irreversible character, deprivation in the course of its evolution, especially during the immediate post-natal period, has irreversible consequences. Undernourishment in the mother can make the new-born baby's nervous system more fragile, and may lead to serious damage later on. Premature weaning can compromise the process whereby the neurones acquire myeline. The critical period of the brain's physical growth being between the fifth and the tenth month following birth, malnutrition at this time may reduce the number of cells in the brain, which has virtually finished growing by the end of the second year. Studies in Central Africa and Central America show that malnutrition during the first four years of life lead to mediocre intellectual performance when children reach school age. If there are other deficiencies in the environment, (psychological conditions, family situation, irregular school attendance) the child runs greater risk of suffering from other handicaps at school. Educational deficiency, the wrong kind of education or lack of education, even among children living in comfortable material conditions, may have disastrous consequences for cerebral development.
Aggravated by Cultural deprivation (#PC1351) Inadequate education (#PF4984)
Protein-energy malnutrition in infants and early childhood (#PD0331).

♦ PD1308 Inadequate image of roles within marriages
Sex-role conditioning — Inflexible sexual roles
Nature Both men and women have an inadequate understanding of marriage. They are either unclear about how to act out their individual roles or have an inadequate image of what their roles should be. Such an image may be the illusion that the marriage covenant is simply to do with individuals working and living together; or it may derive from the illusion created by the mass media of roles that are either jokes, a soap opera, or so one dimensional that they are impossible in reality. Such images induce an individualistic approach such that each person attempts to fulfil his obligation to the covenant and to himself without reference to his partner, thus denying the whole meaning of covenantal relationship.
Broader Static and unrelated social roles (#PF1651)
Individually defined operating structure of marriage (#PD2294).
Aggravated by Attitude manipulation of children through play (#PF2017)
Modern disruption of traditional symbol systems (#PF6461).

♦ PD1317 Possessive attitude of parents
Claim Present day parents view their pre-adult children as a family "possession" and may thus look upon their "investment" in youth in much the same way that other means of economic security are fostered and nurtured. Pre-adults are placed in the untenable position of "meaning givers" without the possibility of full participation in the decision-making process. The relationship of the pre-adult to the entire set of family relationships is exemplified by the reaction of adults who attempt to hold on to such a possession by trying to buy back a runaway youth through emotional or material gifts.
Broader Parental domination (#PF4391).
Narrower Exclusion of pre-adults from family decisions (#PE2268).

♦ PD1318 Manic-depressive psychosis
Affective psychosis — Affective psychoses
Nature Manic-depressive psychosis is a severe disturbance of affect (feeling or emotion) characterized by extreme and pathological elation such as extreme excitability, volubility and foolhardiness, sometimes leading to physical exhaustion, alternating with severe dejection or depression; both of which may last for months or years. There may be normal intervals between these states, and cases are known in which only the manic or the depressive attack occurs once or repeatedly. Attacks usually alternate with intervals of complete health.
Due to the unpredictable nature of manic-depressives, it is impossible to rely on them for responsible action and judgement. While they may be very alert and mentally capable during their manic phase, they can experience extreme feelings of guilt and unworthiness and even lean toward suicidal tendencies in their depressive modes. As manic-depressive psychosis is not immediately recognizable, unknowing persons may get involved in the patient's manic schemes, only to be disappointed when the intended outcomes never materialize.
Incidence The condition is relatively common. Some 20 to 25 percent of all admissions to psychiatric hospitals consist of manic-depressive patients. Frequently the diagnoses of patients with manic-depressive psychoses are changed to schizophrenia after there have been several attacks of disturbed behaviour.
Refs Hershman, Jablow and Lieb, Julian *The Key to Genius* (1988); Jackson, Stanley W

Melancholia and Depression (1987).
Broader Psychoses (#PD1722) Mental illness (#PC0300)
Mental disorders of the aged (#PD0919).
Narrower Mania (#PE4650) Manic episode (#PG9342) Manic depression (#PG4434)
Involutional melancholia (#PG0378) Brief reactive psychosis (#PG4654).
Related Mental depression (#PC0799).
Aggravated by Consanguineous marriage (#PC2379).

♦ PD1323 Cruel sports
Blood sports
Incidence Cruel sports, which provide amusement for people but agonizing death for animals, include bullfighting (for example, in Spain, Portugal, France, Mexico, Colombia, Ecuador, Japan); fox hunting with hounds (for example, in the UK and the USA); hare coursing (England and Ireland); turkey shoots (USA); cockfighting (Cuba, Indonesia, Japan); deer hunting on horseback accompanied by hounds (France); and wild pig-sticking (Germany FR).
Broader Unethical entertainment (#PF0374)
Exploitation of animals for amusement (#PD2078).
Narrower Bird shooting (#PE2693) Prizefighting (#PE8766)
Trapping of animals (#PE5735) Animal fighting sports (#PE4893).
Aggravates Animal injuries (#PC2753).
Aggravated by Athletic competition (#PE4266).

♦ PD1330 Military-industrial-governmental complex in communist systems
Nature The production of State-owned farms and manufacturing in planned economies continues to be directed towards fulfilling military and strategic needs first, and consumer requirements last. Shortages of food and essential consumer products are frequent characterizations of East European life under Communism. Consumer products, as a result, are rarely exported due to inferior quality.
Background Communist development of heavy industry to support militarization was accelerated when, in the late 1940s, Germany, Poland and Czechoslovakia were stripped of much of their industrial machinery so that it could be shipped further east.
Broader Militarism (#PC2169) Military-industrial complex (#PC1952)
Competitive acquisition of arms (#PC1258).
Related Military-industrial complex in capitalist systems (#PC3191).
Aggravates International insecurity (#PB0009).

♦ PD1350 Vandalism
Recurring property vandalism — High vandalism rates — Criminal mischief
Nature Destructive actions against urban public property by minors is the most prevalent form of vandalism. Vandals can be senselessly destructive - breaking glass windows, tearing up paving stones, destroying parks, etc; their violence can be a by-product of street gang conflicts; or they can be reprisals against school and other public authorities. Modern vandalism in a number of cities has been abetted by the invention of the spray paint can and the resulting graffiti; covering over sprayed paint or removing it entails very high expenditure.
Incidence British Telecom loses £18 million a year in damaged public telephone equipment. All together vandalism cost Britain more than £1.8 billion a year in the end of 1980s.
Claim Vandalism is not only associated with urban poverty but often with urban density. Poorly designed public housing is believed, by some, to cause or aggravate this kind of socially deviant behaviour. Others blame images and role models projected by entertainment media. The abundance of theories as to the causes of vandalism, none of them conclusive, is either an indication that there is no one immediate, specific cause, or that the cause escapes detection because it is so diffused, such as social malaise, or the disintegration of a society or a culture and consequent moral bankruptcy.
Refs Lévy-Leboyer, Claude (Ed) *Vandalism* (1984).
Broader Anti-social behaviour (#PC4726).
Narrower Art vandalism (#PE5171) Uncontrolled local vandalism (#PJ0154)
Defacement of urban structures (#PD5305).
Related Unkindness (#PA5643) Moderation (#PA7156) Destruction (#PA6542)
Hooliganism (#PD1109) Misbehaviour (#PA6498)
Human destructiveness (#PA0832).
Aggravates Property damage (#PD5859) Fear of vandalism (#PJ9326)
Theft of works of art (#PE0323) Unhelpful vandalism image (#PG8732)
Environmental degradation (#PB6384) Vulnerability of sacred sites (#PD6128)
Destruction of cultural heritage (#PC2114).
Aggravated by Juvenile delinquency (#PC0212) Insecurity of property (#PC1784)
Disruptive behaviour in schools (#PE9092)
Stagnated images of community identity (#PF6537)
Insufficient care of community property (#PF1600).

♦ PD1354 Increase in trace gases in the atmosphere
Nature There are a number of gases, other than carbon dioxide, which are present in trace amounts in the atmosphere and which are known to aggravate the problems of global warming.
Incidence Only a few such gases are present in sufficient concentration and have sufficiently strong radiation absorption bands in the appropriate thermal radiation window that contains up to 30 per cent of the upward infrared radiation from the Earth's surface. However practically all these gases are increasing in the troposphere and their total effect may be comparable to that of increasing carbon dioxide. The gases include: carbon dioxide, methane, nitrous oxide, ozone, and chlorofluorocarbons..
Broader Degradation of the atmosphere (#PD9413).
Narrower Ozone as a pollutant (#PE1359) Stratospheric ozone depletion (#PD6113)
Increasing atmospheric carbon dioxide (#PD4387)
Increase in nitrous oxide in the atmosphere (#PE9443)
Chlorofluorocarbons as an environmental hazard (#PE4378)
Increase in atmospheric concentration of methane (#PE8815).

♦ PD1356 Coastal water pollution
Beach pollution
Nature Pollution of coastal waters may arise from various sources, such as: the discharge of sewage and industrial waste from coastal outfalls; the dumping of wastes at sea; the discharge of sewage and rubbish from ships; the handling of cargo; the exploration and exploitation of the sea bed and ocean floor; accidental pollution by oil and other substance of pollutants from the land by air and other routes. Undoubtedly the most frequent cause of coastal pollution problems is the discharge of municipal sewage and industrial wastes into coastal waters or into estuaries through unsatisfactory disposal facilities. If wastes contain persistent pollutants, discharge into rivers even at considerable distances upstream from the mouth can result in substantial quantities reaching the sea. The major classes of pollutant reaching coastal waters are decomposable organic materials, heavy metals and other toxic matter, dissolved and suspended non-toxic inorganic substances, and pathogenic organisms.
Many factors, such as dilution, temperature, adsorption, sedimentation and nutrient deficiencies influence self-purification of the sea. The marine environment is generally unfavourable to the survival of most pathogenic organisms. However, under special circumstances, particularly in temperate and warm coastal waters near large cities, pathogenic agents may be found in marine waters in the proximity of the coast-line and in estuaries.
Incidence In the Mediterranean an estimated 85 per cent of sewage flows into the sea without adequate treatment, creating risks of diseases such as viral hepatitis, dysentery, and poliomyelitis and typhoid, all of which are endemic in the region. The UN's Mediterranean Action Plan estimates that 20 per cent of the beaches are too polluted for safe swimming.
Broader Marine pollution (#PC1117).
Aggravates Recreational contact with sewage (#PE6685)
Unsustainable development of coast zones (#PD4671).
Aggravated by Poisonous algae (#PE2501)
Unsanitary disposal of human remains (#PE4725).

♦ PD1361 Occupational deafness
Nature Occupational deafness is hearing loss caused by specific working conditions. Exposure of a person with normal hearing to intense noise will cause a hearing loss reflected by an elevated threshold of audibility. Noise-induced hearing loss may be of the chronic type if it develops over a period of years, or an acute type if it occurs in a relatively limited time, produced by an acoustic stimulus which is intense but of short duration. In either instance, noise-induced hearing loss is an irreversible and incurable disease which can only be corrected to a small degree by hearing aids. If severe enough, it can cause permanent work loss, and severely limits a person's ability to function effectively even in normal daily activities.
Broader Deafness (#PD0659) Occupational diseases (#PD0215).
Aggravates Deterioration in product quality (#PD1435).
Aggravated by Noise in the working environment (#PD2831)
Growth in size of production unit (#PF8155)
Health hazards of exposure to noise (#PC0268).

♦ PD1362 Inhumane interrogation techniques
Nature Although in violation of the United Nations Universal Declaration of Human Rights, inhumane interrogation techniques are presently carried out in many countries, both developed and developing. Coercive techniques are often carried out under official orders, and may include: sleep and food deprivation; drug induced confessions; sexual abuse including rape; disorientation techniques; sensory deprivation; and the infliction of physical pain, often by advanced technological methods which leave no marks as evidence.
Incidence Inhumane interrogation has been reported in a number of countries.
Broader Induction of incongruent actions (#PF3790).
Narrower Interrogation while naked (#PE7233) Forced political confessions (#PE3016)
Forced confessions with drugs (#PE4888).
Related Intimidation (#PB1992) State sanctioned torture (#PD0181)
Abusive police interrogations (#PG3615) Unethical interrogation methods (#PE3553)
Death and disability from inhumane confinement (#PE5648).
Aggravates Neurosis (#PD0270) Psychoses (#PD1722)
Mental disorders (#PD9131).
Aggravated by Subversive activities (#PD0557) Internment without trial (#PD1576).

♦ PD1367 Unfair transport practices
Narrower Cabotage (#PE0670) Restrictive shipping practices (#PD0312)
Unfair air transport practices (#PD9163) Unfair rail transport practices (#PG3619)
Restrictive practices in cargo airline services (#PE5910).
Related Insufficient transportation infrastructure (#PF1495).
Aggravates Inadequate air transport service (#PJ0260).
Aggravated by Unethical commercial practices (#PC2563)
Unethical practices in transportation (#PD1012).

♦ PD1368 Mutagens
Nature A mutagen acts by changing the genetic material that is transferred to daughter cells when cell division occurs, with the result that the new cells have new inheritable characters. Such changes in the genetic material may consist of the alteration of one or more nucleotides, or of chromosomal alterations, resulting in an altered number of chromosomes or an altered chromosomal structure. If a mutagen acts on the germ cells (spermatozoa or ova) of man (or any other sexually reproducing organism) some of the offspring will carry the mutant genes in all their cells. The mutant may be so disadvantageous that death occurs before birth, and if this occurs at a very early stage of foetal growth, the pregnancy may not even be detected. If pregnancy goes to term, however, an abnormal offspring may be born, but the appearance of such an offspring is not in itself evidence that a new mutation has occurred, since abnormal offspring may be due to a mutation in previous generations or may be due to teratogenesis. A mutagen may also have an effect on somatic cells without necessarily affecting germ cells. In the latter case the mutated cell may die or it may be turned into a cancer cell. The resultant cancer cell does not necessarily develop into a clinically apparent tumour because other processes such as promotion and progression are also necessary. Some of these latter processes may also involve mutations. A gross similarity between mutagenesis and carcinogenesis can be said to exist, since both processes produce heritable changes in the phenotype. The current investigations on oncogenes, activated during tumorigenesis in humans, indicate a close connection between mutagenesis and (part of) carcinogenesis, although the precise relation is still unknown. Similar interference with the genetic material can apparently start up uncontrolled cell division. If the products of such division displace or invade normal tissues, the result is a cancer.
In both these instances, the mutagen responsible would have manifested activity as a carcinogen. A gross similarity between mutagenesis and carcinogenesis can be said to exist, since both these processes produce heritable changes in the phenotype.
Broader Human disease and disability (#PB1044).
Narrower Mutagenic effects of drugs (#PE4896).
Related Teratogens (#PE0697) Carcinogenic chemical and physical agents (#PD1239).
Aggravates Mutation (#PF2276).
Aggravated by Biochemical warfare (#PC1164) Ozone as a pollutant (#PE1359).

♦ PD1370 Inadequate environmental education
Lack of environmental education
Claim The growing awareness of the need for environmental protection could be turned to action through an appropriate educational approach. Yet environmental protection is underemphasized in educational systems. Individuals and communities should be given the opportunity to find out about environmental protection, what action is needed, and what it would be appropriate for them to do.
Broader Inadequate education (#PF4984).
Aggravates Deforestation (#PC1366) Environmental pollution (#PB1166).
Aggravated by Uncontrolled application of technology (#PC0418).

♦ PD1376 Instability of trade in live animals
Broader Instability of trade in food and live animals (#PD1434).
Narrower Instability of trade in sheep and goats (#PE8861)
Instability of trade in asses, mules and hinnies (#PE8593).

♦ **PD1380 Missing persons**
Refs Simpson, John and Bennett, Jana *The Disappeared and the Mothers of the Plaza* (1985).
  Broader Abandonment (#PA7685).
  Narrower Missing children (#PD6009)   Juvenile desertion (#PD8340)
  Forced disappearance of persons (#PD4259)   Persons missing in military action (#PE1397).
  Related Kidnapping (#PD8744).
  Aggravates Unknown relatives (#PF0792)
  Uncertainty of death of missing persons (#PF0431).
  Aggravated by Mental illness (#PC0300)   Weakening of family ties (#PG3632)
  Desertion in marriage law (#PF3254)
  Unexplained appearances and disappearances of persons and objects (#PE8631).

♦ **PD1386 Accidental large-scale contamination of the environment**
Claim With the growing use of technology, and the increasing use of the environment as a sink into which the waste products of technology are deposited, the margin of error for adding pollutants to the environment is constantly decreasing, thus increasing the probability of accidental large-scale contamination.
  Broader Injurious accidents (#PB0731)   Environmental degradation (#PB6384).
  Aggravated by Large-scale industrial accidents (#PD2570).

♦ **PD1388 Inadequate transportation facilities in developing countries**
Nature Secondary industry is particularly vulnerable to transportation difficulties, for the actual manufacturing process is often no more than a brief act of separation, synthesis or transformation preceded by the lengthy operation of assembling raw materials and fuel and followed by the distribution of the product to its multiple consumers. Economic secondary production consists very largely in the correct choice of a location which will minimize the combined costs of assembly and distribution, and is thus very dependent upon the availability, cost and effectiveness of the transportation system. In many of the less developed countries, transport difficulties have been a major force tending to inhibit industrialization altogether. In others, transport difficulties have tended to reinforce those centripetal influences making for industrial concentration, often in coast cities with ports that provide an important transport link with the outside world. In most of the developing countries that have been undergoing industrial expansion in recent years, transport has tended to become a bottleneck, retarding the pace of further development.
Refs Moavenzadeh, F and Geltner, D *Transportation, Energy and Economic Development* (1984).
  Broader Insufficient transportation infrastructure (#PF1495).
  Narrower Deficient transport in remote island developing countries (#PE5664)
  Inadequate road and highway transport facilities in developing countries (#PD0543)
  Inadequate transportation facilities for rural communities in developing countries (#PE6526).
  Aggravates Underdevelopment of industrial and economic activities (#PC0880).

♦ **PD1391 Fatigue in materials**
Nature Fatigue in materials is the progressive change of structure and mechanical properties produced by frequent loading and unloading of the material. Such repetition of stress cycles leads to progressive deterioration of the cohesion of the material, particularly in the case of metals. This deterioration is the cause of a high proportion of failure in structures and machine parts. Such failures can cause loss of life, as in the case of aircraft accidents of the collapse of bridges.
Incidence It is estimated that material fatigue accounts for 90 percent of the failure of parts or structures in service.
Refs American Society for Testing and Materials *Effects of Environment and Complex Load History on Fatigue Life – STP 462* (1970); Puskár, Anton and Golovin, Stanislav A *Fatigue in Materials* (1984); Taske, C E, et al *Thermal and Environmental Effects in Fatigue* (1983).
  Broader Failure of materials (#PD2638).
  Narrower Concrete fatigue (#PE1626)
  Deterioration of stored documents and archives (#PE1669)
  Deterioration of the physical condition of art objects (#PD1955).
  Aggravates Mechanical failure (#PC1904).
  Aggravated by Wear (#PB1701)   Friction (#PF1691)
  Stress in industry (#PE6996)   Corrosion of iron and steel (#PE1945).

♦ **PD1393 Ice accretion**
Nature A coating of ice builds up under appropriate conditions, on railway equipment, aircraft, and ships in such a way as to hinder or completely prevent normal operations
  Broader Inhospitable climate (#PC0387)   Meteorological disaster (#PD4065).
  Narrower River ice (#PD3142)   Hazardous glaciers (#PE6824)
  Ice-blocked seaways (#PD2498).
  Related Icebergs (#PE1289).

♦ **PD1398 Hazardous waste dumping**
Toxic waste dumping — Disposal of hazardous waste
Nature Waste is hazardous which because of its quantity, concentration or physical, chemical or infectious characteristics may cause or contribute to an increase in mortality or an increase in illness or pose a substantial present or future hazard to human health or the environment. Hazardous wastes result from manufacturing processes and from non-manufacturing sources such as waste oil from filling stations, ash from power plants, dredged materials, and waste from hospitals. The problem is two fold: the ongoing treatment of waste as it is continuously produced and remedying damage to the environment and human health cause by improper disposal of wastes in the past. Undisciplined disposal of these wastes can cause fires, explosions, air, water and land pollution, contamination of food and drinking water, damage to people who get them on their skins or inhale their vapours, and harm to plants and animals.
One of the most worrying features of the problem is that very little is known about the long-term consequences of exposure to the chemicals, although a good deal is known about their short-term effect. It is known, however, that consequences over long periods can include cancer, delayed nervous damage, malformations in unborn children, and mutogenic changes that could produce disability and disease in future generations. The situation is made even more difficult because, once they are in the environment, chemicals spread in a very complex way and may be converted into other substances which have different effects.
As the amount of hazardous waste produced increases annually, places and methods of disposal that are scientifically, technologically, politically and environmentally safe are becoming more difficult both to discover and to agree upon. Many alarming events in the past decade have resulted from indiscriminate dumping of wastes. Wastes have been considered to be worthless, and so there has been little economic incentive to recover them, and, instead, considerable desire to dump them as quickly and cheaply as possible. But improper disposal may render a relatively harmless substance dangerous. There are laws to control the use of the thousands of new chemicals put on the market each year, but there is little regulation over disposal. By-products and intermediate chemicals, created during the manufacturing process, usually end up in the wastes; these may be more chemically and biologically active than the finished products, and thus potentially more harmful in the environment. The notification schemes recently brought in by law do not cover wastes; their potential damage may go unrecognized. It is also physically difficult to screen wastes for toxicity because they are highly complicated mixtures of substances. If these wastes are exported for disposal or used for other purposes, they may do serious damage because people are unaware of the risks. Particularly vulnerable are the developing countries, because of relatively cheap prices, few environmental regulations, few people trained in necessary technology, corruption and poverty. In highly industrialized countries fragmented waste treatment laws, overlapping enforcement agencies and too few inspectors means that a significant amount of toxic waste is improperly treated.
Incidence Hazardous waste is about 10–20 percent of the world's manufacturing waste. Tens of millions of tons of toxic and otherwise hazardous substances enter the environment every year as unwanted wastes. The managing and disposing of this waste is a significant problem. Until recently, many hazardous wastes were disposed of without proper evaluation of the environmental consequences. There have been several hundred cases of contamination of wells by poisons from hazardous wastes – the most common of all the dangers to arise from improper waste disposal. These often occurred because the wastes were put into sand or gravel pits or old mine workings.
The USA generates 60 million tons of hazardous wastes a year; the EEC generates approximately 20 to 30 million tons.
Claim Businesses treating toxic wastes run from highly reputable firms dealing with complex problems as responsibly as possible to criminal organizations endangering millions of lives and destroying the environment. These questionable firms used a variety of quasi–legal and illegal practices in dealing with toxic substances. Bogus labels are placed on containers. Substances are mixed with non-toxic wastes and redefined. Waste is shipped to countries which have no idea what the substances are nor their impact on humans or the environment. They are burned at sea polluting it. In some cases waste is dumped overboard in full knowledge of the dangers.
Toxic waste dump sites must be cleaned up as they are a menace to society. In addition to the health hazards they pose, there are also intangible problems, such as loss of real estate value for homes neighbouring toxic waste dump sites. As these problems have largely resulted from industrial carelessness and shortsightedness, the bulk of the financing of the research, development, and implementation of clean-up efforts should come out of industrial taxes.
Counter-claim Due to the staggering magnitude of the problem, which includes prohibitive costs, numbers of sites, (in the USA alone, and estimated 10,000 sites need to be cleaned up), length of time involved, and inappropriate methodology, (again in the USA), methods to either move or cover over toxic dump sites have proved very costly and ineffective), there seem to be no definitive solutions to this problem. As controls on industrial toxic waste removal have become stricter, many large industries attempt to circumvent paying for cleanups by moving their operations or exporting their wastes abroad. The former situation results in valuable jobs and tax revenues leaving the home country; the latter situation renders developing countries (where waste removal laws are less strict) vulnerable to such pollution exports. Many smaller companies are being forced into bankruptcy when required clean up toxic waste dumps.
Refs Armour, Audrey (Ed) *Not-in-My-Backyard Syndrome* (1984); Bromley, J, et al (Eds) *Chemical Waste* (1986); European Communities *Safety Aspects of Hazardous Wastes* (1986); Forester, William S and Skinner, John *International Perspectives on Hazardous Waste Management* (1987); Gupta, Joyeeta *Toxic Terrorism* (1990); Huismans, J W and Suess, M J (Eds) *Management of Hazardous Waste* (1983); International Technical Information Institute (Ed) *Toxic and Hazardous*; Mickan, B *Parameters Characterizing Toxic and Hazardous Waste Disposal Sites* (1987); Miller, E Williard and Miller, Ruby M *Environmental Hazards–Industrial and Toxic Wastes* (1985); Postel, Sandra *Defusing the Toxics Threat* (1987).
  Broader Inadequate waste treatment (#PD6795)
  Dumping of consumer waste products (#PD8942).
  Narrower Marine dumping of wastes (#PD3666)   Non-biodegradable plastic waste (#PD1180)
  Discharge of dangerous substances (#PD4542)
  Unsanitary disposal of human remains (#PE4725)
  Trading in products containing toxic substances with developing countries (#PE2061).
  Aggravates Nuclear waste (#PD4396)   Soil pollution (#PC0058)
  Plant diseases (#PC0555)   Infertile land (#PD8585)
  Animal diseases (#PC0952)   Water pollution (#PC0062)
  Hazardous wastes (#PC9053)   Radioactive wastes (#PC1242)
  Amenity destruction (#PC0374)   Hazards to human health (#PB4885)
  Transport of dangerous goods (#PD0971)   Human disease and disability (#PB1044)
  Contamination of drinking water (#PD0235).
  Aggravated by Biochemical warfare (#PC1164)   Agricultural wastes (#PC2205)
  Toxic waste smuggling (#PD9765)   Cultivation of illegal drugs (#PD4563)
  Storage of dangerous substances (#PG3713)   Industrial waste water pollutants (#PD0575)
  Havens for environmental pollution (#PE6172).

♦ **PD1404 Climatic cold**
Cold spell — Cold wave — Cold climates
Nature There are regions in the world where low ambient temperatures hinder or prevent undertaking of normal physical tasks. Skilled jobs requiring manual dexterity are difficult. The likelihood of errors or accidents increases, particularly when the temperature of the hand falls below 15 deg. C. At 51 deg. C, metal becomes as brittle as glass, camera film snaps and rubber soles of shoes break.
Incidence Outdoor jobs in cold climates are found not only at high latitudes but also in temperate zones at high altitudes and in winter. Those at risk in the temperate zones are all outdoor workers exposed to environmental temperatures below about 10 deg C, especially when wind and rain supervene.
In 1987 a severe cold wave swept across Europe paralyzing transportation, closing schools, businesses and governmental offices, causing more than 264 deaths, raising oil prices and causing a large number of accidents.
Refs Kano, Eiichi (Ed) *Current Researches in Hyperthermia Oncology* (1988).
  Broader Bad weather (#PC0293)   Meteorological disaster (#PD4065).
  Narrower Frost (#PD2244)   Permafrost instability (#PD1165).
  Aggravates Famine (#PB0315)   Fatigue (#PA0657)   Lumbago (#PE1310)
  Homelessness (#PB2150)   Cold disorders (#PE0274)   Bright's disease (#PE2272)
  Chronic bronchitis (#PE2248)   Multiple sclerosis (#PE1041)   Injurious accidents (#PB0731)
  Ice-blocked seaways (#PD2498)   Sudden unexpected infant death (#PD1885)
  Non-viability of cold countries (#PF8663).
  Aggravated by Global cooling (#PF1744).
  Reduces Climatic heat (#PC2460).

♦ **PD1405 Stress on families of the physically or mentally handicapped**
Nature Stress on the family of handicapped people has often gone unnoticed as it is the handicapped person himself who usually receives all the attention.
Background For the families of handicapped persons living at home, the stress may come immediately, if the condition is recognizable at birth. This is probably less traumatic than the ensuing waiting period between birth and diagnosis, if the condition is not readily diagnosed. The greatest stress usually falls upon the mother. If she is not young, if the child is her first and she feels it may be her last, the stress is more acute. Even if she anticipates other pregnancies, there is the ever-present fear of delivering another handicapped child. As time goes by and the child does not progress at a normal rate, the mother may suffer acute physical strain because her child continues to behave as a baby for a long time. When the child reaches school age, the mother may be emotionally upset when she sees other children of the same age as her child going to

school whilst hers remains at home. As the child reaches adolescence, the mother has anxiety about the future for her child, and if the child does become self-sufficient enough to live in a workshop or group housing situation with other handicapped persons, the separation between mother and child may be unduly severe as the physical and emotional bond between them (and the level of dependency) had been unnaturally close for so many years.

Many of the stresses indicated for the mother apply to the father as well but this is in part dependent upon how much time he spends at home and how active an interest he takes in his child. With the mother spending so much time on caring for the handicapped child, the father may feel neglected. As the child reaches adolescence, the father has to accept the possibility that he may have to support him for the remainder of his life, rather than having the child become financially independent after an expected (and socially accepted) length of time. It the handicapped child is a boy, the father may severely miss the normal father–son relationship he may have dreamed of.

Broader Family stress (#PD8130).
Aggravates Inadequate community care for handicapped persons (#PE8924).
Aggravated by Physically dependent people (#PD7238)
Physically handicapped children (#PD0196)
Discrimination against the disabled (#PD9757).
Reduced by Institutionalization of the disabled (#PF4681).

♦ PD1410 Undercapitalized waste use schemes

Nature Effective use of waste products is blocked by undercapitalization of waste–use plans. The highly industrialized nations have the ability, because of technological advancement, to use much of the waste products of the world. The problem lies in the lack of provision by these nations of the necessary financial incentives and guarantees to industry for waste collection and centralization schemes. Industries are thereby hindered from investing in further development of ways to recycle unfinished products on a low–risk basis.
Broader Imperialistic distribution system (#PD7374).

♦ PD1413 Malodorous fumes

Nature Unpleasant odours may be produced as a result of industrial processes, fuel combustion, processing of animal products, and improper disposal of liquid and solid wastes. In streets carrying heavy traffic, and under certain meteorological conditions, bad odours may be given off by motor vehicles. Such malodorous substances, though not clearly defined, are principally products of incomplete combustion and various components of fuels. The concentration of such substances may be very low but they may greatly distress pedestrians and the occupants of buildings along the roadside. Despite the reassurance of toxicologists, the public tends to fear such odours not only because of the discomfort they cause but also because they are believed to be a threat to health. In certain individuals, such psychological reactions may provoke somatic responses.

There is no single statute aimed specifically at the control of odour emissions in the way that limits are set for the discharge of substances with known effects on people, vegetation, the fabric of buildings and other materials; so there are no targets for industry to meet This unsatisfactory situation over smells arises mainly because their environmental nuisance cannot be described in unequivocal, measurable terms. Moreover, most smelly substances are odorous at concentrations of 10 parts per million, some at less than 0.001 parts per million, or at a fraction of the quantities adopted to safeguard individuals against the toxic properties of compounds.

Broader Air pollution (#PC0119) Chemical air pollutants (#PD1271)
Urban–Industrial pollution (#PC8745).
Narrower Carbon disulphide (#PE5889) Hydrogen sulphide as a pollutant (#PE2329)
Thiols and mercaptans as pollutants (#PE5436).
Aggravates Headache (#PE1974) Amenity destruction (#PC0374)
Environmental hazards from chemicals (#PC1192)
Environmental hazards from food processing industries (#PE1280).
Aggravated by Agricultural wastes (#PC2205) Industrial emissions (#PE1869)
Sewage as a pollutant (#PD1414) Motor vehicle emissions (#PD0414)
Domestic refuse disposal (#PD0807) Solid wastes as pollutants (#PD0177)
Inappropriate sanitation systems (#PD0876) Industrial waste water pollutants (#PD0575).

♦ PD1414 Sewage as a pollutant

Discharge of dangerous substances into sewage systems

Nature Since human beings are the definitive host for many excreta–related pathogens responsible for communicable diseases, inadequate disposal or lack of sewage systems has an effect both at the individual and at the community level. The effect of faecal pollution and pathogens goes even further: as international tourism and business travel increase all people who visit endemic areas are at risk and, if infected, may return to their homes as carries of the disease. The common practices of burying excreta in the ground (latrines), dumping it on open ground, or disposing of it as sewage – raw or treated – invariably pollute water sources, contaminate soil and infect food and other objects, which become the main vehicles of transfer of the pathogens back to man.

In developing countries, sewage disposal poses a great problem. In inadequately equipped urban areas and in the suburban slum districts proper disposal is virtually unheard of – instead, residents resort to using open pit latrines, which create a direct public health risk. The carrying of wastewater in open drains is a characteristic feature of many developing countries. Except for a few exceptions, such as hospitals, educational institutions, government–reserved residential areas, some private housing developments and army barracks, most urban areas have no central sewerage system or sanitary excreta disposal system. Excreta, together with refuse, are indiscriminately discharged into open drains, and ultimately finds their way into storm water drains, streams and nearby rivers.

Also, during the past ten years, industry has developed hundreds of new chemicals that are known to pass undegraded through conventional systems for sewage treatment; these substances constitute a serious pollution problem. Special attention has been given to detergents, which until recently were a serious threat to both underground and surface waters, as they contained chemical substances that could not be broken down by biological means.

Refs Davis, R D; Hucker, G and L'Hermite, P (Eds) *Environmental Effects of Organic and Inorganic Contaminants in Sewage Sludge* (1983).

Broader Biological pollutants (#PC5276) Discharge of dangerous substances (#PD4542).
Aggravates Cholera (#PE0560) Salmonellosis (#PE7562) Lake pollution (#PD8628)
Water pollution (#PC0062) Poisonous algae (#PE2501) Marine pollution (#PC1117)
Malodorous fumes (#PD1413) Enteric infections (#PD0640) Protozoan parasites (#PE3676)
Mycoplasmal arthritis (#PG2302) Environmental pollution (#PB1166)
Hazards to human health (#PB4885) Marine dumping of wastes (#PD3666)
Soil–transmitted diseases (#PD3699) Wildlife pollution hazard (#PJ3387)
Solid wastes as pollutants (#PD0177) Pollution of inland waters (#PD1223)
Metal contamination of soil (#PD3668) Environmental human diseases (#PD5669)
Contamination of drinking water (#PD0235) Recreational contact with sewage (#PE6685)
Eutrophication of lakes and rivers (#PD2257)
Damage by degradable organic matter (#PJ6128)
Polychlorinated biphenyls as a health hazard (#PE2432).
Aggravated by Wasted water (#PD3669) Contradictions (#PF3667)
Agricultural wastes (#PC2205) Inappropriate sanitation systems (#PD0876)
Industrial waste water pollutants (#PD0575).

♦ PD1415 Restrictive pattern of business activities in small communities

Nature Present patterns of business activity in small communities tend to restrict rather than encourage development and increased money circulation. Until these activities are expanded in scope and range, the small amount of capital which the community has to finance its development will continue to be drained elsewhere. Typically, business activities in remote geographic locations have to deal with high freight and overhead costs; increases in such operating costs have caused decline in profits in, for example, the beef industry. Commercial development may be hampered as retail stores must obtain wholesale goods from outside the community and pay freight costs on these goods; expansion of fishing and agricultural industries may be hindered by limited marketing outlets whose fluctuating prices are controlled by other phases of food processing; local businesses tend to have a single product operation which does not utilize by–products and potential markets are largely undeveloped. In addition, most small village's shopping is done in other towns and cities because surfaced roads have made transportation convenient, so that potential customers for local businesses are only those residents who do not have access to transportation.

Broader Restrictive practices (#PB9136).
Narrower Single product operation (#PG9235) Limited traditional markets (#PG9299)
Unutilized fish by–products (#PG9331) Increase in operating costs (#PF7932)
Insufficient small businesses (#PG9196) Individualized marketing practices (#PG5334).
Related Unchanging business pattern (#PG5367)
Confined scope of business operations (#PF2439).
Aggravates Unexplored potential markets (#PF0581).
Aggravated by Geographical isolation (#PF9023)
Prohibitive cost of transportation (#PE8063).

♦ PD1416 Soil compaction

Lack of soil aeration

Nature Hardening of the soil, whether due to man–made or natural causes, frequently restricts root growth through mechanical resistance to root extension. As a result, it is common to find a normally deep–rooted plant having a shallow root system. When this happens, the plant cannot effectively use moisture and nutrients stored in the sub–soil, and drought hazard to the plant is greater Compacted subsurface layers, termed pans, are present almost without exception in cultivated land and are formed by the passage of tillage implements and heavy machinery over the soil. Subsequent tillage loosens the ploughed layer but leaves an abrupt boundary between the ploughed soil and the compact layer beneath it. The strength of compacted zones is influenced as much by the moisture content of soil as by its degree of compaction. The force required to penetrate wet soil is often increased fivefold or more after drying. While high–strength soil layers can stop root growth completely, it is more common to find root proliferation retarded by intermediate levels of resistance.

Broader Soil degradation (#PD1052) Soil infertility (#PD0077)
Over–intensive soil exploitation (#PC0052).
Related Expansive soils (#PE5036).
Aggravates Infertile land (#PD8585) Soil erosion by water (#PD2290)
Environmental plant diseases (#PD2224) Destruction of land fertility (#PC1300)
Deficiency diseases in plants (#PD3653) Unavailability of agricultural land (#PC7597)
Declining productivity of agricultural land (#PD7480).
Aggravated by Land misuse (#PD8142) Inadequate land drainage (#PD2269).

♦ PD1428 Inadequate emergency medical services

Nature Many victims of traffic and other accidents die on the scene or shortly thereafter, or suffer severe and possibly permanent disabilities, because they do not receive adequate emergency medical care in a prompt and efficient manner. Such accident victims die or are maimed despite the fact that within the medical community of almost all nations, the state–of–the–art in the delivery of emergency medical care in hospitals and medical centres is well developed.

Broader Inadequate medical care (#PF4832) Insufficient emergency services (#PF9007).
Narrower Inadequate emergency blood supply (#PE0366).
Aggravates Risk (#PF7580) Travel risks (#PD7716) Injurious accidents (#PB0731)
Road traffic accidents (#PD0079).
Aggravated by Natural disasters (#PB1151).

♦ PD1432 Lack of response to monetary incentives in developing countries

Nature The modern industrial system assumes a certain response by workers to wage incentives. However, this cannot be taken for granted among peasant populations or other traditional groups of the agrarian society from which industrial manpower has to be drawn in developing countries. Wage incentives may have relatively little appeal to people whose economic organization has been of a non–monetary nature and who find their satisfactions chiefly in traditional rewards, whether economic or via recognition and prestige. Customary types of consumption may lead to money being regarded as no more than a means of fulfilling certain limited requirements. Thus, in addition to the expected limitations of the potential labour supply due to ignorance and ill–health, the flow of workers into industry in developing countries may also be subject to powerful restraints arising from different expectations and cultural patterns of the population.

Broader Economic and social underdevelopment (#PB0539).
Aggravates Underdevelopment of industrial and economic activities (#PC0880).
Aggravated by High labour turnover in developing countries (#PF0907).

♦ PD1434 Instability of trade in food and live animals

Nature Instability in world agricultural markets, reflected in excessive fluctuations of prices and also in the uncertainty which surrounds the availability of agricultural products in world markets, is a recurrent problem. The current phase of instability is one of the worst experienced in recent history. Enhancing the stability of markets is a necessary step toward the establishment of an orderly trading system in which both importers and exporters can take advantage of trade for their internal development. Though some countries may temporarily gain from exceptional price increases, these are usually shortlived windfalls and do not provide an adequate basis for production planning. In the same way, low prices have attraction for consumers but they could lead to shortages and high prices at a later date if they weaken incentives for producers. There are several causes of instability: fluctuations in production; variations in the level of demand; the technical lags in production of many agricultural goods in response to price changes, which result in supply changes inappropriate to current market conditions; and monetary problems or fluctuations in currency rates.

Absolute price stability is neither possible nor of itself desirable in a situation where price has to play the role of reflecting or influencing relative demand and availability. Equally, undue price fluctuations indicate that the market is not operating satisfactorily. Individual countries can and do take action to ensure a greater degree of stability on domestic markets. Some countries hold stocks of foodstuffs which can be managed in such a way as to enhance domestic price stability. Others have to provide various forms of subsidies. Action is often taken also by means of flexible import restrictions or export aids, or price control in the market for food. Other stabilization instruments are available in the form of bilateral or regional trade agreements.

The problem for the international community is that stabilization of a limited part of a global market may lead to greater instability in those parts of the market excluded from the stabilization arrangement. To the extent that many developing countries depend on the 'free' part of otherwise managed markets, both for export outlets and for imported supplies, this presents a serious

problem. Hence a reasonable level of stability of the world or 'free' markets for all food commodities would be a desirable aim of international policy.
The world market for cereals holds the key to any measure designed to control fluctuations in the prices of basic foodstuffs. Of the markets for the different cereals, that for wheat is the most significant in this regard, although stability in the market for rice and feed grains is also of great importance. The inter-linking of these markets must be acknowledged in any stabilization policy. Any action taken on the storage and marketing of wheat, for instance, must take into account developments in the markets for other grains. This would in turn reduce the size of reserves needed for each particular grain.
The inevitable link between stocks and prices raises different problems for different governments. If national stocks become too 'large' in a particular year and prices begin to fall, governments of the main grain exporting countries might be forced to adopt adjustment measures to reduce production and to protect farm incomes from falling. But if such measures were to be adopted before the world as a whole had built up adequate security stocks, it could expose the main importing countries to dangers of serious shortages and very high prices in case of crop failures in subsequent years. At the same time, however, the exporting countries cannot be expected to carry the entire burden of security stocks and also suffer any depressing effect on prices which might be created by large stocks. The dilemma created by this inherent conflict between what could be regarded as legitimate national objectives of many governments and the international requirements for food security and price stability cannot be resolved without some agreed international approach for sharing the burden of security stocks on an equitable basis and for stabilizing grain prices within agreed limits.
Broader Instability of the primary commodities trade (#PC0463).
Narrower Instability of sugar trade (#PE0383) Instability of trade in legumes (#PG6527)
Instability of trade in live animals (#PD1376)
Instability of trade in dairy products and eggs (#PE0576)
Instability of trade in meat and meat preparations (#PE0755)
Instability of trade in vegetable-based foodstuffs (#PE9164)
Instability of trade in miscellaneous food preparations (#PE1683)
Instability of trade in cereals and cereal preparations (#PE1769)
Instability of trade in fruit and vegetables and preparations thereof (#PE0961)
Instability of trade in animal feedstuffs, excluding unmilled cereals (#PE8816)
Instability of trade in coffee, tea, cocoa, spices and manufactures thereof (#PE0915)
Instability of trade in fish, crustacea and molluscs and preparations thereof (#PE0972).
Related Instability of production of food and live animals (#PD2894).
Aggravates Food grain spoilage (#PD0811).
Aggravated by Food spoilage in storage (#PD2243)
Vulnerability of plants and crops (#PD5730)
Disruption of food supply due to military activities (#PE8979).

♦ PD1435 Deterioration in product quality

Manufacture of substandard products — Poor workmanship — Low quality merchandise — Inadequate quality control — Shoddy workmanship
Nature Competition for markets has led to a great increase in the rate of introduction of new designs for products, and increasing pressure is placed upon all concerned to produce at a lower price than competitors. It is therefore no longer possible to continue long-term production of a product once it has been developed to an adequate level of quality. In addition, considerations of economy, and the enormous increase in the scale of manufacturing, have led to heavy dilution of the skilled labour force formerly employed to produce quality products; and competition is further intensified as developing countries enter the market with competing products.
It is useful to distinguish between (a) poor quality from the perspective of the producer and (b) the user. Levit distinguishes between custom designed or built products and mass produced ones. Custom designed dams, software, ships, financial investments, consulting work, etc, are different because there is no clear difference between design, production and delivery. They all happen effectively at the same time.
One can also distinguish between (a) poor quality products (b) deterioration of time in the quality of products and (c) poor quality not just linked to products but also services, research, etc.
Broader Consumer vulnerability (#PC0123) Lack of qualitative excellence (#PF5703)
Ineffective means for goods supply and distribution (#PF6495).
Related Misbehaviour (#PA6498).
Aggravates Mechanical failure (#PC1904) Electronic equipment failure (#PD1475)
Unreliability of equipment and machinery (#PC2297).
Aggravated by Mediocrity (#PF3900) Planned obsolescence (#PC2008)
Occupational deafness (#PD1361) Counterfeit machine parts (#PE5319)
Unethical commercial practices (#PC2563) Unethical practices by employees (#PD4334)
Planned degradation in product quality (#PF7741)
Quantitative pressure on standards of quality (#PF7227).

♦ PD1464 Ineffective war crime prosecution

Ineffective deterrent against war crimes — Inadequate punishment of war criminals
Nature The measures of punishment that may be used against war criminals were set forth in the statutes of the International Military Tribunals (1945–1946 and 1946–1948) as well as in the law of the Control Council of Germany (1945). In 1968, the 23rd session of the United Nations General Assembly adopted a convention concerning the inapplicability of the statute of limitations to war crimes and crimes against humanity. In the four decades since the end of the Second World War, preliminary or criminal proceedings have been initiated against thousands of persons believed to have taken part in Nazi crimes, but in the majority of cases the proceedings did not end with convictions either because the suspects were not extradited back to Europe, or there was insufficient evidence against them.
War crimes cannot be deterred by the threat of prosecution against individual perpetrators. The trials and sentences of 24 Nazi defendants in Nuremberg in 1945–1946 were provided on ex post facto grounds, no firm legal precedent having been established, and are not universally regarded by international jurists as having certain legal effect. This is not to say that they cannot be invoked, although other instruments have been proposed – such as the International Law Commission's Code of War Offences, prepared in 1954, and the Convention on the non-applicabilities of Statutory Limitations to war crimes, in force since 1970. Crimes of declared and undeclared wars are known since 1954 but have mainly been unpunished.
Claim International treaties to prevent war crimes are unenforceable unless individuals can be prosecuted. If treaties are not effective in this sense, nations may indulge in indiscriminate reprisals during and after conflict. Personal responsibility for war crimes is therefore a doctrine in which every field combatant, in every service and in every grade and rank, should be instructed; and such responsibility extends further, through headquarters and chains of command. A government which does not prosecute its own perpetrators of crime during war may itself be criminal. The My Lai massacre in Korea led to the 1969 prosecution of an American officer by his superiors, and there may have been less publicized prosecutions elsewhere since the Second World War. Inadequate punishment, however, may leave superior officers liable in the eyes of the enemy, as well as in the view of world justice; and the invocation of the Nuremberg precedent remains possible in such cases.
Counter-claim Trials of enemy officers are a mask for summary execution. Their very possibility aggravates cruelty in the conduct of wars as no concept of decency and honour on the battlefield is left.
Broader Deficiencies in international law (#PF4816)
Deficiencies in the criminal justice system (#PF4875).
Narrower Remission of sentences for crimes against humanity (#PF1098).
Related Military atrocities (#PD1881) Citizen disobedience (#PD5707)
Releasing of repatriated prisoners (#PJ1505)
Government failure to prosecute offenders effectively (#PE9545).
Aggravates Fear (#PA6030) Distrust (#PA8653) Injustice (#PA6486)
War crimes (#PC0747) Vigilantism (#PD0527)
Unconvicted war criminals (#PD4067).
Aggravated by Exile (#PC2507) Conspiracy (#PC2555)
Extradition refusal (#PF2645) Lack of international cooperation (#PF0817)
Government approved employment of war criminals (#PE4697)
Connivance of authorities in human rights abuses (#PF9288).
Reduced by Assassination (#PD1971).

♦ PD1465 Over-production of commodities

Primary commodity surplus
Incidence When, as has occurred, a number of countries expand exports of the same commodities at the same time, the 'fallacy of composition' becomes evident. Increased demand for primary commodities does not seem to have been induced significantly by the drop in prices. The price elasticities of demand for these commodities are generally low, and in the case of raw materials their demand is derived from that for industrial products, the prices of which do not tend to respond commensurably to changes in material input prices.
Claim The production of commodities has gone beyond actual need, resulting in increased sales promotion to create a market for the over-abundant supply of goods. Such a market tends to be self-perpetuating rather than extending into areas where markets are unrealized.
Broader Limited market development (#PF1086).
Narrower Overproduction of food (#PD9448).
Aggravates Unnecessary reserves of material (#PF0687).
Aggravated by Excess production capacity (#PD0779)
Inflexibility of commodity supply (#PF9432)
Introduction of high-yield crop varieties (#PF3146).

♦ PD1466 Tax evasion

Fiscal fraud — Customs tax evasion — Lack of compliance with fiscal laws
Nature Tax evasion is the deliberate use of a variety of financial and fiscal devices of doubtful legality, including the deliberate concealment of relevant information, in order to avoid payment of tax. Of special concern is the technique of implying in each country through which funds move that they are being taxed in some other country. This takes advantage of the relatively poor exchange of information between national tax administrations, particularly in developing countries. Tax evasion may also be closely associated with illegal business transactions; it may be the result of collusion between taxpayers in different countries, and of claims for tax deductions in developed countries by taxpayers with business in developing countries.
Incidence In the USA it is estimated that for every $5 of federal taxes, $1 is being evaded, mostly by single proprietors and small businesses. The annual tax gap is expected to exceed $100 billion in 1990.
Refs Balter, Harry G *Tax Fraud and Evasion* (1983); Chopra, O P *Tax Ethics* (1985); International Bureau of Fiscal Documentation *International Tax Avoidance and Evasion* (1981); United Nations *International Cooperation in Tax Matters* (1987).
Broader Statutory crime (#PC0277)
Circumvention of duties and assessments (#PD4882).
Narrower Underground economy (#PC6641) Abuse of expense accounts (#PE4645)
Evasion of social costs by companies (#PE3149) Evasion of customs and excise duties (#PD2620)
Unlawful trafficking in taxable objects (#PD4221)
Illicit production of alcoholic beverages (#PE7188)
Distortion of corporation financial statements (#PE2032).
Related Corruptive crimes (#PD8679) Financial scandal (#PD2458)
Unlawful business transactions (#PC4645) Frauds, forgeries and financial crime (#PE5516).
Aggravates Distortionary tax systems (#PD3436).
Aggravated by Errant nationals (#PE0812) Abuse of tax havens (#PE2370)
Unreported businesses (#PJ1192) Ineffective tax systems (#PF1462)
Speculative flight of capital (#PC1453) Corporation financial secrecy (#PE1571)
Unethical personnel practices (#PD0862) Unethical commercial practices (#PC2563)
Unethical practices of employers (#PD2879)
Ineffective tax systems in developing countries (#PF2124)
Inequitable tax treaties between developed and developing countries (#PD1477).

♦ PD1468 Sexual discrimination in education

Refs Barton, Len and Walker, Stephen *Race, Class and Education* (1983); Klein, Susan S *Handbook for Achieving Sex Equity Through Education* (1985).
Broader Sexual discrimination (#PC2022) Inequality in education (#PC3434).
Narrower Discrimination against men in education (#PD8909)
Discrimination against women in education (#PD0190).

♦ PD1472 Natural food poisons

Naturally occurring poisonous substances in food-stuffs
Nature These include mycotoxins and marine biotoxins. Apart from being responsible for crop losses, certain mycotoxins such as aflatoxin, ergot and "yellow rice" result from growth or storage of crops under conditions leading to infestation by fungus, which produce highly toxic chemicals, frequently invisible, that can not be destroyed by normal processing and cooking. With the rapid expansion of fisheries as a means of providing a source of protein (particularly in developing countries), there is increasing concern over the presence of toxins in fish. The hazard of marine biotoxins is particularly insidious because species of fish that have been eaten for many years without danger may suddenly become poisonous, causing many deaths.
Refs Gontzea, I and Sutzescu, P *Natural Antinutritive Substances in Foodstuffs and Forages* (1968); Watson, D H (Ed) *Natural Toxicants in Food* (1987).
Broader Human disease and disability (#PB1044).
Related Harmful natural foodstuffs (#PD4238) Food poisoning through negligence (#PE0561).

♦ PD1475 Electronic equipment failure

Nature Although amounts spent by manufacturers on production testing are increasing rapidly, a rising percentage of faulty equipment is being delivered to customers. Every step in the manufacture of an electronic product, for example, produces defects of which a certain number inevitably remain undetected. There is a direct relationship between the percentage of undiscovered defects and the cost of the final product, since some defects must be accepted if costs are to be kept down. Such difficulties are likely to get worse rather than better with the introduction of more sophisticated techniques, such as integrated circuit technology, and with the increasing complexity of equipment which such techniques make possible.
Incidence Electronic equipment failure most often occurs in complicated radio-electronic equipment, such as computers and automatic control devices. The principle causes are imperfect soldering, poor contact in plug and socket connectors and internal defects in the elements.
Broader Mechanical failure (#PC1904).
Aggravates Unreliability of equipment and machinery (#PC2297).

Aggravated by Humidity (#PD2474)
Defects in machinery design (#PE2462)
Deterioration in product quality (#PD1435)
Inadequate electrical maintenance (#PG0584).
Failure of materials (#PD2638)
Defective product manufacture (#PD3998)
Inadequate equipment maintenance (#PD1565)

♦ PD1476 Inadequate working conditions in developing countries
Nature Levels of wages that are inadequate to maintain family life, long hours of work, women's night work, industrial home work, sweatshop methods and child labour have been common in the early stages of industrial expansion in nearly all countries. In spite of improvements introduced through protective social legislation, as well as through humanitarian policies adopted by some employers, these conditions are still found in many countries, particularly in industrializing areas where population density is high or rural conditions poor. As countries grow wealthier, the income of adult workers tends to increase, and it ceases to be a matter of economic necessity for wives to work in factory night shifts or for children to be sent out to earn supplementary wages. In many of the less developed countries, however, poverty is still a compelling force inducing these and other socially undesirable forms of industrial labour, notwithstanding laws and regulations to the contrary. These abuses are difficult to correct, especially where factories are small in size and large in number, public administration weak and the cost of adequate policing and inspection disproportionately high.
Broader Exploitation in employment (#PC3297).

♦ PD1477 Inequitable tax treaties between developed and developing countries
Nature In the past 50 years a network of over 500 tax treaties has been established which has greatly contributed to the dramatic progress achieved by industrialized countries in exchange of goods and investments. However, whilst essentially fair when used between developed countries, the patterns and methods established become inequitable when applied to relations between countries at different levels of development, when investment, and often certain trade components, follow a one-way direction. Developing countries therefore face the dilemma of whether to forgo the tax treaty instrument as an effective tool for stimulating international investment and trade, or to apply the traditional treaty pattern at severe loss of scarce revenue.
Broader Unfairly negotiated treaties (#PF4787).
Related Circumvention of duties and assessments (#PD4882)
Tax obstacles to international investment (#PD0673).
Aggravates Tax evasion (#PD1466)
Ineffective economic structures in industrial nations (#PE4818).
Aggravated by Unreported tax obligations (#PE9061).

♦ PD1480 Depriving prisoners of medical treatment
Nature Depriving prisoners of medical treatment is a part of the techniques to torture people. Some victims are left to die other's are provided with enough medical treatment to continue suffering. This increases the sense of isolation, and humiliation and prolongs the pain of physical tortures.
Incidence Denial of medical treatment has been reported in the following countries: **Af** Comoros, Mali, Morocco, Tanzania UR, Zambia. **Am** Cuba, Guyana, Haiti, Paraguay, Uruguay. **As** Afghanistan, Indonesia, Korea Rep. **Eu** USSR.
Broader Torture by deprivation (#PD3763) Inadequate medical care (#PF4832).
Narrower Underutilization of livestock in least developed countries (#PE8595).
Related Inhumane participation of the medical profession in torture (#PE4015).
Aggravates Disabled victims of torture (#PD0764).

♦ PD1481 Commercial exploitation of wild animals
Broader Economic exploitation (#PC8132) Misuse of wild animals (#PD8904).
Aggravates Endangered species of animals (#PC1713)
Inadequate feeding of animals (#PC2765)
Cruelty to animals in factory farming (#PD2768)
Agricultural mismanagement of housed farm animals (#PD2771)
Inadequate housing and penning of domestic animals (#PE2763).

♦ PD1482 Disruption of family system in developing countries
Nature Industrialization tends to break up the extended family and uproot workers from their traditional communities; it hastens the decay of those devices of self-help and mutual aid which, within the limits permitted by the generally low standards of living in the rural society, offer some protection to all members of the group, irrespective of their ability.
The migration of a large proportion of the able-bodied and younger men from rural, tribal areas to wage-earning occupations in urban industries frequently leaves the family system unbalanced and incapable of properly carrying out its conventional social and economic tasks, including the production of food. The balance of primitive agriculture is upset and no new markets appear to to give rise to a new equilibrium.
The subordination of the elementary needs of the family and the community to the manpower requirements and technical exigencies of the factory system finds physical expression in barracks systems, factory dormitories and compounds and workers tenements, workshop-sleeping and street-sleeping, and in numerous variants of shanty-towns, bidonvilles and native locations. Bad housing contributes to the disruption of the worker's family life and augments the losses which the traditional family system sustains in the course of adaptation to the industrial environment. Connected with bad housing, both as cause and effect, is the shiftlessness of a large proportion of many new urban populations. Where housing conditions are unsuited to family life, the worker is discouraged from bringing his family into the industrial area; he does not feel permanently settled in the area of his place of work and remains tied to his former community. Even when there are housing opportunities, industrial workers of peasant origin often fail to become integrated and assimilated into modern urban society and the stream of urban life; instead they tend to form a sub-culture of low status, living separately in special quarters and pursuing a mode of life that is half urban, half rural.
Claim In the new industrial society the urban worker who is crippled, unemployed, or too old to work may have no tie with any group that feels a direct responsibility for his welfare. At the same time, the instability of the urban family results in increasing numbers of deserted mothers and abandoned children. In many cases, begging becomes their only means of livelihood. In south and south-eastern Asia, for example, there are vast numbers of persons disabled by disease and malnutrition, many of them without the customary claims on family or village and for whose sustenance the modern economy fails to provide.
Broader Family breakdown (#PC2102).
Aggravates Crime (#PB0001) Juvenile delinquency (#PC0212)
Social insecurity in developing countries (#PE4796)
Increase in anti-social behaviour in developing countries (#PD0329).
Aggravated by Rural depopulation (#PC0056) Uncontrolled industrialization (#PB1845).

♦ PD1484 Underutilization of second-hand equipment
Nature Considerable and increasing quantities of second-hand equipment become available every year in advanced industrial countries. Although it may no longer be economic to employ such equipment in the conditions (such as a particular industrial sector in a developed country) under which it was previously used, it might still be used to economic advantage under different conditions (such as in a developing country). Nevertheless, such resources tend to be ignored in favour of the purchase of new equipment.
Broader Underproductivity (#PF1107).
Aggravated by Planned obsolescence (#PC2008).

♦ PD1486 Sea traffic congestion
Broader Traffic congestion (#PD0078)
Obstacles to the utilization of coastal and deep sea water resources (#PF4767).
Aggravates Oil pollution (#PE1839) Marine accidents (#PD8982).

♦ PD1493 Low occupational mobility in developing countries
Nature The actual flow of workers into industry in developing countries may be subject to powerful restraints arising from peculiarities of the social organization and culture pattern of the population. The modern industrial system assumes a certain degree of occupational mobility. But occupational mobility, facilitating the upward movement of workers to jobs on a higher economic level, may have little meaning in societies where production is organized on a family basis, and occupation is inseparable from the family status and kinship roles. In some societies, the immobilizing influences of a closely-knit family organization or well-integrated local community may be augmented by the checks and restraints of a wider social system. Caste systems and various forms of peonage, servitude or bondage, including indebtedness, segregation laws and colour bars are all detrimental to the manpower and mobility requirements of an expanding industrial economy. These handicaps are widespread among less developed countries, though there are significant local and regional differences in the degree to which they actually limit the size and effectiveness of the available labour force. In general, these social factors are more potent in Africa than in Latin America or Asia, where urban labour reservoirs tend to be much larger.
Broader Underproductivity (#PF1107).
Aggravates Absenteeism (#PD1634)
Underdevelopment of industrial and economic activities (#PC0880).
Aggravated by High labour turnover in developing countries (#PF0907).

♦ PD1494 Locational maladjustments of industry in developing countries
Regional imbalance
Nature Economic growth if not checked by appropriate measures has the tendency to limit itself to its original centres or regions. When a region becomes developed, it tends to make other areas appear backward and unpromising. Private enterprise easily overestimates the external economies accruing to these growing centres. They may have scavenged the opportunities that exist in the centres but are reluctant to venture out of the area to investigate possibilities elsewhere. Government policy tends to be influenced by this impressive picture of growth poles. Thus it is that in all developing countries the bulk of public investment in new plants and projects goes to centres that have already 'taken off'. Like private operators, the majority of governments in developing countries are either too timid or find it too difficult to break away from the growing centres to establish new basic industries in underdeveloped regions.
At some stage in the development of an industrial complex, maximum economies of scale are achieved and further expansion may result in 'dis-economies' which offset or more than offset the economies. There is, for example, the constant competition between various industries which increases the price of materials and other factors of production such as skilled labour, land, capital and transportation. There may be militant labour unions whose activities affect wages and attitudes towards labour. Strikes may be frequent and widespread. When the provision of services and amenities increases in cost, taxes are raised to pay for them. Costs rise as a result of congestion and the strain on existing services such as transportation, water supply and waste disposal. Rural population, attracted by employment opportunities in growing centres, migrate to these centres in quest of jobs ranging from work in trades, administration and services, to labouring jobs such as digging, carrying, loading and cleaning. The urban-industrial complexes are usually ill-prepared to receive the 'in-migrants'. The preparation of building sites, the provision of roads and services and the construction of houses cannot keep pace with the influx. Without money or possessions and willing to live at a lower standard than other residents, the newcomers crowd in with friends who have arrived earlier, become squatters and create shanty towns; in extreme cases some band together into gangs and go foraging.
The negative aspects of economic concentrations are not only localized within the concentrations themselves, but at the national level they have produced one of the most urgent problems facing the large developing countries, namely the problem of regional inequality and stranded or neglected areas. Undue emphasis has often been laid on the development of the privileged areas. Consequently, other regions, some of them very promising growth poles, may be left relatively inaccessible and virtually neglected.
Claim Governments in developing countries are faced with serious locational maladjustments as a result of unguided population movement and an unprecedented development of urban areas. The growth of the 'primate city' has become an inevitable concomitant of economic development. With these cities have come the social and economic problems connected with excessive urban centralization and severe congestion of both population and industrial enterprises. While these urban-industrial centres exist and flourish in each country, a vast interior awaits settlement and development.
Counter-claim The processes by which urban concentrations grow in the early stages of development are the same in all developing countries. Many industrial processes are generally attracted to the same location and the process is cumulative. The initial attraction, for example, may have been the availability of raw material, or the presence of some source of energy, or such favourable nodal situations as a major port or a national administrative capital. Whatever the initial reason for the establishment of industrial enterprise, the very existence of industry often makes the location attractive to other industries. Some may want to use a by-product, previously wasted, and therefore available at a low cost. New industrial undertakings may also be established to furnish existing firms with various goods and services. As the nucleus grows it gradually becomes a centre of concentrated earning power and, hence, of purchasing power. It becomes an increasingly better market for the consumer goods industries, which will in turn be attracted to the already growing centre. Service industries will also grow with the expanding centre, as will the labour force, both in quantity and in the variety of its skills. This adds further to the attractiveness of the concentration for yet other processes.
Through such geographic concentration, opportunities for economizing in certain areas will increase. Just as a single industry can, up to a point, achieve internal economies as it expands its output and capacity, so can the mass of industries in a concentrated area as the industrial capacity of the area grows. As the nucleus grows, banking and insurance facilities become available, maintenance and repair services are established, journals are produced and a host of tertiary activities develop.
Enormous economic advantages can thus be derived from the concentration of economic activities within a few large centres. Consequently, the development of these concentrations is inevitable in the early stages of industrial development and their existence in developing countries is, to a large extent, justified.
Broader Economic and social underdevelopment (#PB0539).
Related Imbalance in city sizes within a country (#PF2120).

♦ PD1500 Excessive occupational exposure to radiation
Nature Although the data may be incomplete, the population dose due to occupationally exposed

persons should not exceed 1 per cent of the natural background radiation, even after allowing for an increased use of nuclear energy.
Incidence The radiation exposure of the majority of persons professionally engaged in handling radiation sources and radioactive substances is less than about 1 rem per year, or less than one fifth of the maximum permissible dose allowed for radiation exposure of the gonads. A number of countries have reported that the genetic dose to the whole population arising from the dose received by occupationally exposed persons, who constitute only a small fraction of the total population, is about 0.1 – 0.2 mrem per year.
Broader Occupational hazards (#PC6716) Radioactive contamination (#PC0229).
Aggravates Occupational risk to health (#PC0865).
Aggravated by Unethical practice of radiology (#PD8290).

♦ PD1501 Narrow legal definition of the family in developing countries
Nature Although legal recognition of the family may be restricted to husband, wife, children and immediate dependents, in many communities the actual family structure is far more extended. It may cover cousins, uncles, aunts, nephews, nieces of any degree and also brothers, sisters, grandparents, grandchildren, great–grandchildren; and concubines, their children and other relations. In certain communities the 'family' is anyone with the same family name.
Claim Insofar as kinship patterns involve obligations which have remained despite transition from tribal to more modern society, national legal administrative procedure, often based on the western principle, is ill adapted to meet social needs. This results in continuing poverty, inadequate social security, abandoned and illegitimate children, delinquency, crime, unemployment and general family disorganization.
Broader Obstacles to extended families (#PF3127)
Discrimination in social services (#PC3433)
Family structure as a barrier to progress (#PF1502).
Narrower Cohabitation (#PF3278) Denial of parental affiliation (#PD3255)
Dependence within extended families (#PD0850)
Discrimination against unmarried women (#PD8622).
Related Inequality before the law (#PC1268)
Social insecurity in developing countries (#PE4796)
Jurisdictional conflict and antagonism within intergovernmental organizations (#PE9011).
Aggravates Orphan children (#PD7046) Family breakdown (#PC2102)
Abandoned children (#PD5734) Juvenile delinquency (#PC0212)
Migrating sand dunes (#PD0493) Family disorganization (#PC2151)
Denial of right to extended family (#PE5241)
Adjustment difficulties of new urban families (#PF1503).
Aggravated by Desertion in marriage law (#PF3254).

♦ PD1507 Socially handicapped refugees
Nature Socially handicapped refugees include: unaccompanied young people; mothers with several children; unmarried couples with children; families with four or more children, or with an aged or disabled dependent; households in which the main potential breadwinner is an unemployed or under–employed refugee over 45 years of age who does not have the requisite skills; chronic alcoholics, prisoners, ex–prisoners and refugees unadapted to work, and families in which such a person is the potential breadwinner. The social disabilities affecting a number of refugees of this type arise from mental disorders such as psychosis, psychoneurosis or mental deficiency, which reveal themselves through alcoholism, idling, isolation, and aggressive or promiscuous behaviour.
Broader Refugees (#PB0205) Social outcasts (#PD6017) Human disability (#PC0699).
Related Immigration barriers for handicapped family members (#PE4868).
Aggravated by Social neglect (#PB0883) Inadequate social welfare services (#PC0834).

♦ PD1519 Inhumane and indiscriminate weapons
Dependence on inhumane and indiscriminate weapons
Nature A fundamental principle of the laws of war is that choice of means of injuring an enemy is not unlimited, weapons should not cause superfluous injuries and they should not be employed indiscriminately against non–combatants and combatants. These provisions are being undermined by current means and strategies of warfare. The massive firepower of modern weapons systems and the use of chemical sprays, area weapons, delayed action fuses and a variety of means of environmental destruction tend to undermine those regulations intended as far as possible to protect civilian populations from the exigencies of armed conflict.
Incidence In addition to nuclear and bio–chemical weapons, those that appear likely to cause indiscriminate killing or injury if developed and employed, include high volume sound at ultra low frequencies (infrasonics); radiation – either electromagnetic (plasmic fire–balls, differential lightning bolts, huge standing waves, etc) or nuclear (neutron beams); and human biological energy (psychotronics).
Refs Stockholm International Peace Research Institute *The Law of War and Dubious Weapons* (1976).
Broader Weapons (#PD0658) War crimes (#PC0747) Unconventional war (#PC8836).
Narrower Illegality of nuclear weapons (#PF4727)
Incendiary weapons of massive destructiveness (#PD3492).
Related Biochemical warfare (#PC1164).
Aggravated by Competitive development of new weapons (#PC0012).

♦ PD1521 Insecurity and vulnerability of non–nuclear weapon states
Nature The proliferation and stockpiling of nuclear weapons by nuclear powers is a threat to the peace and security of those nations which do not possess nuclear weapons. Fear ranges from anxiety over possibilities of nuclear blackmail to dread of an exchange of missiles by any of the nuclear powers.
Claim Even a nuclear exchange detonating a megaton (1 million tonnes of TNT is the equivalent) would cause a global winter so that the southern hemisphere as much as the north would be affected in this regard. In addition, many military targets are located in countries which have no nuclear, and in some cases very little conventional, defences. For these reasons, and the disruption of all trade movements, all nations have a vital stake in maintaining peace.
Broader International insecurity (#PB0009) National insecurity and vulnerability (#PB1149).
Narrower Insufficient nuclear weapon free zones (#PJ5335).
Related Insecurity and vulnerability of nuclear weapon states (#PC4440).
Aggravated by Second class states (#PD0579)
Proliferation of nuclear weapons and technology (#PD0837).

♦ PD1533 Excessive television viewing
Increase in antisocial behaviour due to television — Habitual television watching — Late television watching
Nature Wherever television becomes available for a number of hours a day it dominates the leisure time of children, even if the programmes are not such as they would be expected to find interesting. Television may contribute to antisocial behaviour by triggering off acts of delinquency or reinforcing the importance of violent behaviour in solving human problems. It may teach a potential criminal a new skill.
Incidence It is estimated that, in any country where more than a few hours of television is available, an average child of between 6 and 16 years of age spends 500–1,000 hours a year watching television, namely 6,000–12,000 hours during 12 school years. The latter figure is not much different from the amount of time an average child spends at school during those same years. It is estimated that the average USA citizen watches the equivalent of 3,000 entire days (nearly 9 years) of television between the ages of 2 and 65: during this time an act of violence is screened every 14 minutes and a killing every 45 minutes. The 1983 daily average family viewing time in the USA was well over 7 hours. This is approximately a 55 percent increase from 1953. Comparable figures are emerging in Western Europe where colour, cable and international programming make this form of home entertainment increasingly popular.
Claim The values, knowledge, health and behaviour of television viewers are corrupted by the violence and pornography of which they are passive spectators. Television demeans and trivializes everything and everyone connected with it. As such it is contributing to the decay and downfall of civilization, brought about by the dazzling assault of information on the senses, rather than by any suppression of information. In modern society everything becomes a branch of entertainment since this provides a form of coherence which is not readily achieved in a world characterized by complexity and information overload. Such coherence is found in the endless pursuit of disconnected sights, sounds, images and interesting personalities that offer a pretence of coherence. Television encourages people to avoid dealing with reality directly, distorting ability to deal with complexity, and possibly altering irretrievably any desire to confront reality. It is particularly insidious because, rather than offering a trivial simulation of reality, it has become both a simulation of reality and reality itself. People addicted to TV have shortened attention spans and a lack of reflectiveness. Imaginations atrophy. TV addiction is associated with low achievement in school. As a technology, electronic scanning, effectively shuts down the functioning of those parts of the brain that reason and think actively and logically. One consequence of this shutting down is the emotional content of a programme becomes paramount.
Counter–claim Similar castigation to that earned by television in the 20th century was directed to playing cards in the 13th century. Card playing, chess, hobbies, sports, dancing, novel reading, theatre–going and a host of other recreations could, if carried to excess by enthusiasts, always be said to have a negative effect on something. Any means of family enjoyment could raise the same scruples as those focused on television, and their abuse considered a bad formative influence on children.
Infants as young as 10 months old watching certain kinds of television programmes are helped learn language skills, the ability to perform physical tasks and an understanding that what they are watching is related to the rest of life around them.
Refs Milavsky, J Ronald et al *Television and Aggression* (1982).
Broader Addiction (#PD6324) Anti–social behaviour (#PC4726)
Psychological pollution by mass media (#PD1983).
Narrower Television violence (#PE4260).
Related Inappropriate personal habits (#PD5494).
Aggravates Crime (#PB0001) Human violence (#PA0429) Juvenile delinquency (#PC0212)
Declining sense of community (#PF2575) Mental illness in adolescents (#PE0989)
Radio and television propaganda (#PD3085) Exploitative commercial television (#PD0433)
Parochial escapist media entertainment (#PD0917)
Collapse of distinctions between categories (#PF7938)
Media theatricalization of public life and politics (#PF9631)
Lack of responsible involvement in community affairs (#PF6536)
Ineffectiveness of international organizations and programmes (#PF1074).
Aggravated by Boredom (#PA7365) Deterioration of media standards (#PD5377)
Unpreparedness for surplus leisure time (#PF5044).

♦ PD1534 Disparate development of economic sectors within developing countries
Nature Disparate development occurs in those countries which have within their borders subsidiary subsistence economies or more or less self–contained village societies. This is the situation to a greater or lesser degree in almost every developing country, and it tends to inhibit the growth of manufacturing industry in a number of ways. Within the subsistence sectors–proper, the pre–commercial economic organization practises too rudimentary a degree of division of labour for any industrial specialization to occur. By the same token, a subsistence sector cannot form part of the general market of the country. This insulation of a portion of the population – a very large portion in most African countries, not much smaller in many Asian countries and significant in several Latin American countries – affects economic growth in general but is particularly inimical to the development of secondary industry. It inhibits the flow of labour into occupations requiring industrial skills and keeps off the market a significant proportion of the country's potential purchasers of industrial products.
Broader Underdevelopment of industrial and economic activities (#PC0880).
Aggravates Disparity in social development within developing countries (#PD0266).

♦ PD1543 Loss of traditional forms of social security in developing countries
Nature Loss of traditional forms of security may be an important impediment to the stabilization of the newcomer in industry and to his assimilation in the new social environment. The traditional society usually has established patterns of family reciprocity and mutual aid, which provide for the individual's economic and social security in times of need. In traditional societies, old people have always enjoyed a privileged position based on respect, consideration, status and authority; but this is starting to be upset under the influence of modern trends, and that privileged position is now being questioned. Attachment to the traditional forms of social security and confidence in their efficacy – as compared with the uncertainties of industrial employment – are important influences preventing the peasant migrant from identifying his future and his life interests with a career as industrial worker, and from settling permanently in an urban environment.
Broader Social neglect (#PB0883) Socio–economic poverty (#PB0388).
Aggravates Social insecurity in developing countries (#PE4796)
High labour turnover in developing countries (#PF0907)
Underdevelopment of industrial and economic activities (#PC0880).
Aggravated by Untransposed significance of cultural tradition (#PF1373).

♦ PD1544 Social disaffection of the young
Youth alienation — Frustration of youth — Grievances of youth — Youth protest — Youth unrest — Unengaged community youth — Community youth displacement — Isolated community youth — Lack of youth participation — Unplanned youth participation — Unreleased potential of youth participation — Unclear youth roles — Unclarity on youth roles — Inadequate youth roles — Insufficient youth responsibility — Limited youth accountability — Separation of youth / adult activities — Lost direction for youth engagement — Unclear youth needs — Undefined youth needs — Unsatisfied youth ambitions
Nature The young constitute a major force in the development of any community, yet often their resources remain untapped. Programmes and activities for young people often fail because of lack of support either form the adult members of the community, or from the young people themselves, who resist organization from 'outside'. In addition, lack of formal structuring of time outside school hours hinders participation even when the desire is present.
As times change, formal school education is being taken more seriously in third world communities. Youth are pulled away from both gainful work and the home. This shift in values derived from increased education and exposure produces acute conflicts between youth and adults. When rural youth come home from school they are expected to do a number of household chores. Some feel they need as much leisure time as youth in urban areas. As the conflict increases, the forms of discipline demand more attention. Discipline in the schools has sometimes been harsh or physically harmful. Communities waver between upholding old expectations and rapidly advancing

into a modern style. While providing education for its youth some communities lack local social forms to engage youth in community areas broader than formal schooling. The combination of a vacillating set of social expectations and the youth's feelings of being stifled by the generational differences contribute to departures of young people from their homes for training and work. Unless this trend toward insufficient youth engagement is curbed soon, the future development of local neighbourhoods and villages will be deprived of vigour and vision.
Claim The transient nature of everyday experience has led to an increasing lack of identification of young people with every aspect of society. In virtually all areas of knowledge, yesterday's truths are becoming today's fiction. This encourages a sense of mistrust and disassociation which prevents youth from making long-range plans.
Broader Generation communication gap (#PF0756)
Disorientation of the young due to lack of social forms (#PD2050)
Failure of individuals to participate in social processes (#PF0749).
Narrower Idle youth lifestyle (#PG7766) Limited youth activities (#PJ0106)
Blocked parental participation (#PJ5308) Lack of community responsibility (#PJ3290)
Unstructured afterschool engagement (#PG8575).
Related Limited leisure time (#PF9062).
Aggravates Non-participation (#PC0588) Undemocratic policy-making (#PF8703)
Emigration of trained personnel from developing to developed areas (#PD1291).
Aggravated by Grievance (#PF8029) Juvenile stress (#PC0877)
Social isolation (#PC1707) Rural unemployment (#PF2949)
Social unaccountability (#PC1522) Insufficient role models (#PF8451)
Collapse of societal engagement (#PF2340) Haphazard forms of social ethics (#PF1249)
Deteriorating community identity (#PF2241) Static and unrelated social roles (#PF1651)
Obstacles to community achievement (#PF7118) Unformed style of cooperative action (#PF6514)
Restricted scope of local employment (#PF2423)
Unexplored avenues of leadership potential (#PF2797)
Disorganized attempts of upgraded employment (#PF6458)
Undeveloped channels for commercial initiative (#PF6471)
Ineffective structures for community participation (#PF2437)
Fragmented patterns of extended family relationships (#PF1509).

♦ PD1551 Urban unemployment in developing countries

Nature Owing to the combination of a number of unprecedentedly powerful factors (heavy density of settlement in agricultural areas, wide and probable growing difference between rural and urban levels of income, very rapid growth in the number and proportion of young people who have been to school), the rural-urban drift has proceeded in most of the countries of the Third World at a very fast pace. This movement has resulted in a very serious disequilibrium in the job supply/demand situation in the urban areas, an imbalance which is reflected in a swelling of the tertiary sector, considerable underemployment in a number of sectors and, in particular, urban unemployment which has reached extremely high rates in a large number of developing countries.
Incidence Since World War II there has been a net emigration of over 150 million persons to urban areas. The scale of this movement, to which is added a considerable natural increase in the population of towns (due to a high birth rate and a reduced death rate resulting from the age composition of the population), has led to an extremely rapid growth of the urban population during this period, which increased at an average rate of over 4.5 percent a year. The 750 million jobless in the Third World in 1975 is expected to increase to 1,320 million by the year 2000. Meanwhile, the number of new jobs needed is expected to be 630 million in Asia, 250 million in black Africa, 190 million in Latin America (as compared with 360 million jobs needed in in Western countries). Figures for 1975 show that whereas unemployment as a whole in the developing countries was 5 percent of the total labour force, rural unemployment was 3.6 percent and in the towns 8 percent of the potential workforce was jobless.
Refs Anker, Richard and Hein, Catherine *Sex Inequalities in Urban Employment in the Third World* (1986); Bairoch, Paul *Urban Unemployment in Developing Countries. The Nature of the Problem and Proposals for its Solution* (1976).
Broader Unemployment in developing countries (#PD0176)
Labour surpluses in developing countries (#PD0156)
Flooding of the urban labour market in developing countries (#PD0008).
Related Urban underemployment (#PC3490)
Disparities in unemployment within countries (#PD1837)
Imbalance between urbanization and industrialization in developing countries (#PC1563).
Aggravates Uncontrolled urban development (#PC0442)
Decline in real wages in developing countries (#PD2769).
Aggravated by Rural depopulation (#PC0056) Denial of right of assembly (#PC2383).

♦ PD1554 Lack of processing industry for primary commodities in developing countries

Lack of vertical diversification in developing countries
Nature Since nominal tariff rates of the developed countries increase with the degree of processing, and since there is a long-standing vertical integration of industry (often with transnational corporation involvement) that already has control of all or most processing, the obstacles to achieving increased national participation in processing are high. In addition, absence of capital and technical knowledge leaves little for developing countries to attempt development in these areas on their own.
Broader Economic and social underdevelopment (#PB0539).
Aggravates Instability in export trade of developing countries producing primary commodities (#PD2968).

♦ PD1562 Factory farming

Negative effects of factory farming — Intensive animal husbandry
Nature Under modern systems of livestock production, traditional husbandry has been superseded by advances in technology and associated values of productivity and efficiency. Animals are no longer raised under free-range or extensive conditions. They are being confined and raised as if they were mindless and emotionless cogs in the complex machinery of factory farming. Often this is done at the expense and compromise of their needs and rights to humane consideration and treatment.
Factory farming is a very attractive commercial proposition for industrial investors and with a greater demand for meat and egg products and a comparatively higher standards of living in these countries, the trend towards intensive farming methods is likely to increase. The competition for land space also contributes to the trend. Factory farming is characterized by the widespread use of antibiotics in animal feeds, which could lead to drug resistance in bacteria pathogenic to both humans and animals. Hormones, which may cause cancer in humans, are used as implants to stimulate growth in beef cattle. Pesticides, nitrate fertilizers, herbicides and moulds (aflatoxins), contaminate grains and other crops used in the animals' feed. Some of these substances, which are hazardous to humans if ingested in sufficient amounts, may be stored in the animal's fat, muscles and internal organs, or be concentrated and excreted in milk, which is ingested by humans. Although some of these substances are identified animal carcinogens, their use is often justified on the grounds that they help reduce production costs.
Incidence Factory farming exists mainly in the developed Western countries. The highest incidence of factory farming is to be found in the USA; in Europe the highest incidence is found in the UK. Most cattle and sheep are still raised for part of their lives under extensive conditions ranging from feedlot corrals to half-open barns which afford varying degrees of freedom and protection from the elements. However, the free-range and semi-intensive operations require the most labour. Thus, there is a growing trend to raise cattle and sheep in total confinement buildings, especially in northern climates. Almost all broiler chickens and laying hens and some sixty to seventy percent of all hogs are now housed intensively in total confinement. The relatively high cost of installing factory farming units is a restricting factor on their incidence in developing countries.
Background Extensive, free-range conditions were first replaced by semi-intensive conditions of raising livestock in enclosed pastures or fields. Next, with increasing land costs dictating a more frugal use of arable land, animals were removed from the land to intensive confinement systems.
Claim Factory farming is an inadequate means of producing protein, leading to the spread of animal diseases and drug resistant animal and human diseases, animal stress and malformation; a threat to the farming community from industrial investors; maintenance of high food prices because of the cost of installation; and export of high protein grain crops from protein-deficient countries to developed countries.
Refs Mason, Jim and Singer, Peter *Animal Factories* (1982).
Broader Inadequate animal welfare (#PC1167)
Agricultural mismanagement of housed farm animals (#PD2771).
Narrower Animal stress in factory farming (#PD2760)
Animal malformation in factory farming (#PD2761)
Inefficient use of proteins in factory farming (#PF2758)
Spread of animal diseases through factory farming (#PD2752)
Abuse of antibiotics and vaccines in factory farming (#PE8383).
Related Cruel animal transportation (#PD0390).
Aggravates Fowlpest (#PE1400) Animal diseases (#PC0952)
Malnutrition in developing countries (#PD8668)
Encouragement of drug resistant diseases (#PJ3767)
Threat from industrial combines to farming communities (#PE8014).
Aggravated by Limited available land (#PC8160)
Increased demand in developed countries for meat and egg products (#PE8013).
Reduces Lack of meat and egg production (#PE9115).

♦ PD1564 Social isolation of the elderly

Isolated senior citizens — Isolated non-participating elders in urban environments — Isolation fostering structures of engagement for elders — Preclusion of elders' participation by image — Limited participation — Elders' self-images preclude participation
Nature The self-image of large numbers of elders is such that they cannot participate in the community or social processes. Elders many times see themselves as spectators, or as insignificant, as having made their contribution to society. A sense of impending death may leave many socially paralysed. Besides, current structures designed for the care of elders often isolate them from society, forcing them to cling to the past, with draw further from society, and lose any sense of having a meaningful role. There is some evidence that environments which are not stimulating encourage the onset of senility.
Claim In many nations the elderly members of the community are increasingly isolated from society, both physically and figuratively. They come to view themselves as insignificant, irrelevant spectators, severed from a society with which they cannot share their wisdom and experience, and which cannot therefore profit from their leadership. Young people thus lack the necessary perspective from which to view their future and adults are left with only a shallow context for their lifestyles and values.
Broader Social isolation (#PC1707) Obstacles to community achievement (#PF7118)
Restrictive effects of traditional community decision-making (#PF3454).
Narrower Erosion of elders' wisdom (#PF1664).
Related Lost family role in society (#PF7456) Repression of self-consciousness (#PC1777)
Inadequate means for upholding global concern (#PF1817).
Aggravates Lack of self-confidence (#PF0879) Dependence on the media (#PD7773)
Undemocratic policy-making (#PF8703) Obsolescence of rituals and customs (#PF1309)
Avoidance of a confrontation with death (#PF1586)
Limited community responsibility of adults (#PF1731)
Disorientation of the young due to lack of social forms (#PD2050).
Aggravated by Lack of commitment (#PF1729) Disobedience of elders (#PF7149)
Social withdrawal of aged (#PD3518)
Underprovision of basic urban services (#PF2583)
Inadequate welfare services for the aged (#PD0512)
Restrictive patterns of traditional life (#PF3129)
Fragmented patterns of community activity (#PF6504).

♦ PD1565 Inadequate equipment maintenance

Deterioration in the quality of equipment maintenance and servicing — Disrepair of equipment — Inadequate maintenance of machines
Nature Poor maintenance and repair generally cause economic losses through lower efficiency of the installed machinery, lower quality of products and higher costs of production. In addition, poor maintenance and repair lead to the deterioration and consequently the shortened physical lifetime of installed equipment, which therefore constitutes a significant waste of capital, particularly in developing countries. Industrial development in developing countries is therefore hampered and burdened with greater obstacles, which are aggravated by the financial limitations placed on the purchase of new equipment.
Incidence More than 3,600,000 books were destroyed by fire and water in February, 1988 when the library of the Academy of Sciences of the Soviet Union at Leningrad burned down. Many pre-revolutionary volumes cannot be replaced.
Broader Consumer vulnerability (#PC0123) Inadequate maintenance (#PD8984).
Narrower Inadequate vehicle servicing (#PJ0909).
Aggravates Underproductivity (#PF1107) Mechanical failure (#PC1904)
Electronic equipment failure (#PD1475)
Unreliability of equipment and machinery (#PC2297)
Health risks to workers in agricultural and livestock production (#PE0524).
Aggravated by Planned obsolescence (#PC2008)
Unethical commercial practices (#PC2563)
Inadequate maintenance equipment (#PD6520)
Prohibitive cost of equipment maintenance (#PE1722).

♦ PD1566 Fragmentation of social structures in depressed areas

Nature When a community loses its main source of income, social care structures become fragmented. The company which provided the majority of people with employment (such as the mining company in a village which has grown up around a mine), usually also provided every form of care and social life. The closing of such a business creates an atmosphere of uncertainty and diminishing services. The community seems to lose its sense of direction and forgets how to work together. Local festivals become less significant and lose their appeal because they are oriented to a life that no longer exists. Young people have to travel considerable distances to good schools, involving them in time and expense and leaving little time for extra-curricular activities, since disillusionment among teachers and adults in general means that local schools find it difficult to provide socially relevant education or after school activities. Previously active older people are restricted by the changes occurring and find their roles in the town are curtailed. General social life declines, leaving a number of empty halls whose use no longer justifies the cost of their

upkeep.
Broader Social fragmentation (#PF1324).
Narrower Undetermined social base (#PG8303) Lapsed volunteer services (#PG8298)
Inadequate public facilities (#PG9297) Fragmented community interaction (#PG8301)
Restricted extra-curricular activities (#PG8299)
Disjointed patterns of community identity (#PF2845).
Related Social withdrawal of aged (#PD3518).
Aggravated by Limited access to social benefits (#PF1303)
Insufficient community celebrations (#PJ0188).

♦ PD1574 Weeds

Nature Major weeds are plants which either through their abundance or persistence or physical properties, hinder or harm man's social or economic activities, and threaten his livelihood and well-being. In particular, weeds limit food production, by damaging crops; competing with crops for water, light and nutrients; hindering the tending and harvesting of crops; and by their invasion of grazing land. Aquatic weeds add another dimension to the problem: by inhibiting the flow of irrigation water, they too can limit crop production.
Incidence An average of more than 10 % of agricultural production is lost world-wide by weed competition. The world sale of herbicides in 1980 was 4.7 billion US dollars. From one particular study of 5,000 weeds that occur throughout the world, it was concluded that 200 are of concern to agriculture, based upon their geographical distribution and severity. About 30 of these weeds have very wide geographical distribution and may be important in 10 or more world crops. Another 50 species are widely distributed across the world but are more a nuisance than a threat. They are constant companions of man, require much labour to keep them at bay, but are not likely to destroy crops. The weeds of the next group, numbering about 100, are restricted to certain crops or geographical regions. They seem to find ecological barriers, which have not yet been studied. The remaining 20 weeds fall between the three groups and are very difficult to place or classify. For 75 per cent of these weeds, there is almost no background information.
Counter-claim Many of the world's crops are also weeds, including guavas, pomegranates, maize, rice, lettuce and asparagus. Many weeds also serve as crops. Of the world's 17 worst weeds, 13 are exploited as crops, whether as animal fodder, medicines or potherbs. Five of them are also cultivated. Thousands of other plants considered weeds are useful as foods or medicines.
Plants which are weeds because they are pioneer plants (for example, plants which can quickly occupy denuded areas) sometimes prove very useful in providing vegetation cover on abandoned or temporarily idle land. Such plants are generally very good at reducing water runoff and providing cover, and hence preventing water and wind erosion of the soil. Weeds are part of the complex web of plant life, and clearly should be respected as such, even if they are uninvited guests.
Refs FAO *Improving Weed Management* (1984); FAO *Weeds in Tropical Crops* (1981).
Broader Pests (#PC0728) Pests of plants (#PC1627).
Narrower Aquatic weeds (#PD2232)
Accidental weed creation by genetic engineering (#PE5404).
Related Insect pests of plants (#PD3634).
Aggravates Pests and diseases of rice (#PE2221)
Pests and diseases of wheat (#PE2222)
Pests and diseases of sugar cane (#PE2217).
Aggravated by Crop vulnerability (#PD0660) Inadequate weed control (#PG9029)
Inadequate weed control in developing countries (#PD3598).
Reduced by Destruction of weeds (#PE3987)
Environmental hazards of new species introduction (#PC1617).

♦ PD1576 Internment without trial

Arbitrary arrest — Denial of right to freedom from arbitrary detention — Administrative detention without charge — Arrest on dubious grounds
Nature Arrest and detention on grounds of suspicion rather than factual evidence may follow from a suspension of habeas corpus on the declaration of a state of national emergency. Internment without trial may be justified by the authorities as being necessary in a state of war or guerrilla or terrorist attacks or in an extreme strike situation, but may equally be used to suppress legitimate political opposition and human rights. If internment is on a large or sophisticated level, concentration camps may be set up and torture used in order to obtain confessions.
An individual may, in various circumstances, be arrested in certain countries without being given any grounds for such arrest. There have been cases of arrest on grounds which, in their very wording, seem to constitute violations of human rights; or on the basis of charges so vaguely framed that they could easily serve as a pretext for violations of human rights. Persons may be arrested for having expressed, in a non-violent manner, their political, philosophical or religious opinions; the purpose is to forestall any attempt to oppose a regime. In view of the policy of secrecy practised by various governments on the alleged basis of public security requirements, it is often difficult to obtain information on the identity and the number of persons thus detained, precisely because of the absence of charges and of reports of proceedings.
Incidence Internment without trial has been noted in the following countries: **Af**: Angola, Chad, Egypt, Ethiopia, Ghana, Kenya, Lesotho, Libya, Madagascar, Mali, Mozambique, Namibia, Nigeria, Senegal, Sierra Leone, Somalia, South Africa, Sudan, Tanzania, Togo, Tunisia, Uganda, Zambia, Zimbabwe. **Am**: Brazil, Paraguay. **As**: Bahrain, Brunei, India, Indonesia, Israel, Malaysia, Singapore, Sri Lanka, Vietnam, Yemen DR. **Eu**: Czechoslovakia, Greece, Italy, Portugal, Spain, UK.
Refs McGuffin, John *Internment*.
Broader Wrongful detention (#PD6062) Abuse of police power (#PC1142)
Violation of civil rights (#PC5285).
Narrower Unlawful imprisonment (#PD4489) Forced disappearance of persons (#PD4259)
Government seizure of foreign nationals in foreign countries (#PE6564).
Related Arbitrariness (#PB5486) Political trials (#PD3013)
Political prisoners (#PC0562) Concentration camps (#PD0702)
Politically motivated arrests (#PD9349)
Forced repatriation of prisoners of war (#PD0218).
Aggravates Inhumane interrogation techniques (#PD1362).
Aggravated by Apartheid (#PE3681) Martial law (#PD2637)
Unjust trials (#PD4827) Secret police (#PE6331)
Political purges (#PC2933) Vagueness of laws (#PF9849)
Political repression (#PC1919) Violation of amnesty (#PD3018)
Subversive activities (#PD0557) Refusal to grant amnesty (#PF0182)
Restrictions on emigration (#PC3208) Unauthorized police search (#PD3544)
Suspension of habeas corpus (#PG3794) Excessive government control (#PF0304)
Undemocratic political organization (#PC1015)
Political prisoners in communist systems (#PD3171).
Reduced by Qualified amnesty (#PF3019).

♦ PD1582 Air accidents

Aircraft collisions — Aircraft near-collisions
Incidence There were an average of about 20 fatal aircraft accidents per year in the early 1980s, among scheduled services with, on average, 20 to 40 fatalities per accident. In 1980 and 1981 combined, the actual figures were 39 accidents with 1176 fatalities. Per 100 million passenger-kilometres the range of fatalities is about 0.04; the number of fatal aircraft accidents per 100,000 aircraft hours flown is about 0.12; and the number of fatal aircraft accidents per 100,000 landings, about 0.17 (in 1981). Additionally, non-scheduled commercial operations reported to the ICAO showed 47 fatal accidents, 1980–81, with 520 fatalities. Complete statistical information on safety in general (private) aviation operations is not available on a worldwide basis, but estimates put the range at about 1,000 fatal accidents per year with about 2,000 deaths. The above figures, which exclude the USSR, indicate that aviation accidents cause over 2,000 deaths annually, excluding military aviation accidents. The fatalities for the scheduled commercial operations alone totalled some 16,000 over the twenty year period ending December, 1981. These were caused by 562 fatal accidents. This averages 28 fatal accidents per year with an average of about 30 fatalities. Total fatalities in the twenty year period for all aviation must be placed in the 30,000 range, conservatively, and in the post-war period since the 1940s, at least 50,000. The number of people who have been involved in some kind of air accident in the post-war period, and who were killed, maimed, injured, or merely shocked as a result, has to be placed in the range of several hundred thousand.
Broader Travel risks (#PD7716) Injurious accidents (#PB0731)
Transport accidents (#PC8478).
Narrower Military aircraft accidents (#PD5373).
Related Aircraft environmental hazards (#PD8328).
Aggravated by Lightning (#PD1292) Aerial piracy (#PD0124)
Air turbulence (#PD2127) Unsafe aircraft (#PE1575)
Air traffic congestion (#PD0689) Human errors and miscalculations (#PF3702)
Threat of birds to aircraft safety (#PD1111).

♦ PD1584 Environmental pollution by nuclear reactors

Radiation pollution from reactors — Heat pollution
Nature Nuclear reactors, and particularly those used for the generation of electricity, pollute the environment by their thermal and radioactive discharges. The radioactive discharges may be considerably magnified as a result of reactor accidents, faulty safety precautions, faulty construction or human error. These considerations make it very difficult to select suitable sites for reactors.
In the case of thermal discharges, nuclear plants may inject over 60 per cent more heat to the condenser cooling water than do the most modern fossil-fuelled plants of the same size. Radioactive pollution from a nuclear reactor is created by fission in the reactor core. Some of the radionuclides created by the process escape along various pathways to the environment. The actual emission level depends on the reactor type and the funds invested in containing it. There is disagreement concerning how much radiological pollution from such a source should be officially tolerated.
Refs Stockholm International Peace Research Institute *Nuclear Radiation in Warfare* (1981); United Nations *Impact of Nuclear Releases into the Aquatic Environment* (1976).
Broader Thermal pollutants (#PC1609).
Narrower Gamma ray disturbance in space (#PF1441)
Vulnerability of nuclear power sources (#PD0365)
Environmental hazards of nuclear power production (#PD4977)
Environmental hazards of decommissioned nuclear power plants (#PE7539).
Aggravated by Radioactive contamination (#PC0229)
Human errors and miscalculations (#PF3702)
Hazardous locations for nuclear power plants (#PD2718).

♦ PD1590 Hurricanes

Tropical cyclones — Typhoons — Lack of typhoon protection
Nature Hurricanes appear as a nearly circular but somewhat assymetrical vortex with winds spiralling in towards the low-pressure centre. Winds reaching speeds of more than 75 mph are said to be of hurricane force although speeds up to 200 mph have been measured. Enormous destruction results from both winds and tidal surges (10–20 ft waves), in coastal areas, and to a lesser extent from effects of heavy rains (10–20 in). Diameter of the cyclonic circulation is generally less than 1000 km, with hurricane force winds restricted to a 100–300 km central ring. Typhoons, the term used in the North Western Pacific, sometimes reach 2000 km in diameter.
Incidence On an average, 60 tropical storms, accompanied by wind speeds of more than 40 mph, form on the surface of the globe each year. About 20 of them form in the North Western Pacific, about 7 form in the Western North Atlantic and the Caribbean. Typhoons have, for centuries, ravaged Asia and the Far East. During 1961–1970 alone the damage sustained amounted to US $6,750 million, equivalent to about 0.5 per cent of the GNP of the countries concerned. Some 14 million people were affected, 12,000 lives were lost, 7.6 million hectares of land were flooded or otherwise damaged and 4 million buildings were partially or totally damaged. The hurricane which struck southern Britain on October 19, 1987 cost insurance companies in the UK a total of over £860 million, uprooted some 2–3 million trees and killed 19 people. Hurricane Gilbert struck the Caribbean in September 1988 killed over 250 people, left over 1 million homeless and caused at least $10 billion of property damage. An estimated 10,000 people died and 250,000 were left homeless by a cyclone in Bangladesh. Experts have suggested that global warming could give hurricanes 50 per cent more power.
Counter-claim Records suggest that hurricanes remove the 'big trees' of the eastern forests every 100 to 150 years; when this happens in mature climax forests, the light that reaches the forest floor allows light-demanding plants to grow on mounds of soil raised by the tipped roots of the toppled trees. In this way, hurricanes (and other wind storms) create a diversity of species by providing gaps for species that demand light.
Refs Nalivkin, D and Bhattacharya, B B *Hurricaines, Storms and Tornadoes* (1983); WHO *The Quantitative Evaluation of the Risk of Disaster from Tropical Cyclones* (1976).
Broader Storms (#PD1150) Bad weather (#PC0293).
Aggravates Famine (#PB0315) Floods (#PD0452) Storm surges (#PD2788)
Wind damage to structures (#PE1334)
Endangered monuments and historic sites (#PD0253).
Aggravated by Global warming (#PC0918) Large-scale weather anomalies (#PC4987).

♦ PD1591 Lack of individual historical context

Nature When an individual does not know or rejects his own family history he has no context in which to make decisions for the future. Such rootlessness has a paralyzing effect, so that the past no longer appears as a series of related events and the individual does not see his own life as part of history.
Broader Individual isolationism (#PD1749).

♦ PD1593 Endangered species of whale

Endangered species of cetacea
Nature Many of the world's whale populations have been over-exploited by the whaling industry to the point where it is believed that some of the remaining species are in danger of extinction. Eleven species of whale have been hunted, formerly for lamp oil and lubricants and presently as a source of oil for the manufacture of margarine, soap and glycerine, as meat for human and animal consumption, or for soups and flavourings. Sperm whale oil and spermaceti provide industrial scouring and hardening agents, lubricants and cosmetics. All of these products could be replaced by synthetic derivatives from vegetable and mineral sources, but it is still economically attractive to produce them from whales. Despite warnings that the catches were considerably larger than the stocks could withstand in the long term, whaling nations have allowed their industries to take almost as many whales as they could, arguing that they could not otherwise

survive economically.
The whaling industry throughout the world is currently holding a three year moratorium on hunting ending in 1990. Japan and Iceland both kill whales for "scientific" purposes.
Incidence In the Antarctic the fin whale probably numbered 480,000 originally and in 1971; 77,000 (the maximum sustainable yield level is around 222,000 when they yield 10,000). The sei whales in the Antarctic have been fished intensively only since the early 1960s, when the fin whale declined most rapidly in numbers. In this time the sei whale stocks have been reduced from about 150,000 to 75,000–83,000 (this is above the maximum sustainable yield level of 52,000 which yields about 5,000). In the North Pacific the original stock of fin whales probably numbered about 43,500 and now stand at about 15,000 (the maximum sustainable yield level is 27,000 which yields 1,200). Sei whales have been reduced from about 70,000 to 46,000 (the maximum sustainable yield level is around 40,000 with a yield of about 3,000).
In the case of the sperm whales, numbers are best considered for each sex separately, because of the difference in body size and behaviour in this polygynous species. In the North Pacific the initial population of males is estimated to have been 134,000 and there are now 64,000 remaining (with a sustainable yield of about 4,000 males). This stock is just above the population level required to maintain reproduction with the initial number of females, estimated to have been about 124,000 (the females are still well above the maximum sustainable yield level with a yield of about 5,000). In the Indian/Antarctic area, numbers are estimated at 20,000–49,000 (with sustainable yields of 1,000 of each sex).
A number of species which have been fished in the past are protected. These include the right whale, the grey whale, the blue whale (of major interest to the industry, depleted in the antarctic from an estimated 200,000 to some 2,000 animals by the time the species was given total protection in 1965), and humpback whales. Some protected species are deliberately caught in defiance of regulations or quotas which there are no means of enforcing.
Incidental kills of small cetaceans in tuna fisheries, for example, have risen considerably in recent years; 250,000 porpoises and dolphins are estimated to be killed each year in the USA tuna fishery alone. The overall quota limiting the world catch set for the 1981 season was 13,900 whales, a sharp decline from the 46,600 killed per year a decade earlier. But even with protection, there is little sign that whale stocks, except for the California grey whale (Eschrichtius robustus), are recovering.
Whales may be more threatened by human waste than weapons. One example is in the saltwater St Lawrence River estuary, home to about 400 beluga whales. During the 5 year period 1983–85 88 beluga carcasses laden with toxic chemicals have washed ashore. The whales were so polluted, said one biologist, that most were technically toxic waste.
Refs Klinowska, Margaret *Dolphins, Porpoises, and Whales of the World* (1988); Schevill, William E *The Whale Problem* (1974).
Broader Endangered species of marine mammals (#PD3673).
Narrower Endangered species of rorquals (#PG3808)
Endangered species of porpoises (#PE3806)
Endangered species of grey whale (#PG3807)
Endangered species of sperm whales (#PG3799)
Endangered species of right whales (#PG3809)
Endangered species of beaked whales (#PG3802)
Endangered species of river dolphins (#PG3801)
Endangered species of white whale and narwhal (#PE8187)
Endangered species of rough–toothed and white dolphins (#PE8302).
Aggravated by Polychlorinated biphenyls as a health hazard (#PE2432).

♦ PD1607 Microbial pests in industry
Nature Considerable problems are caused by the growth of microbes in rolling mill oils, cable ducts, aviation kerosene, metal plating solutions, and other unexpected but troublesome locations. Eliminating the slime in nuclear coolant ponds and dressing grinding wheels fouled by infected oils, are costly processes. The presence of such organisms reduces productivity, introduces delays and reduces product quality.
Broader Pests (#PC0728).
Narrower Infection of industrial water supplies (#PJ6198).

♦ PD1611 Juvenile alcoholism
Alcohol consumption by children and young people — Excessive consumption of alcohol by adolescents — Youth alcoholism — Unobserved drinking age laws — Under–age drinking
Nature Alcoholic toxicomania may develop out of the combination of a certain personality, the environment, and the influence of the drug. The personality of the minor who is in the developmental stage is qualitatively different from that of an adult person. The environment in which the minor lives is usually also different from that of the adult. This, then, forms the special complex of problems concerning alcohol abuse by minors and alcoholism of minors. In the literature dealing with alcohol toxicomania, papers concerning alcoholism in minors are quite exceptional. This is due to the fact that adult drinkers do not like to admit that they often used to consume alcohol long before reaching adult age. (In a test survey, of 400 adults, 87 per cent gave a mendacious answer). Another reason is that minors who drink very rarely come under professional anti-alcoholic care because their relatives often adopt a tolerant attitude towards them. Neither does the majority of the anti–alcoholic corps show any special interest in the work with under–age drunkenness, since such work requires not only experience with toxicomaniacs but also knowledge of psychology and psychiatry of children and adolescents.
Incidence Alcohol is a significant causal factor in young driver traffic accidents. It has been found that in one locality in the USA, 40 percent of the alcohol–related fatal accidents involved drivers aged 24 years or under whereas 20 percent of the licensed drivers were in this age group. A Norwegian road survey revealed that 2 percent of the drivers had a blood–alcohol level higher than the legal limit (0.05 percent) and 20 percent of these drivers were under 23 years of age. The accident vulnerability (involvement with exposure held constant) of the 18–19 year old driver group, but not of 20–65 year old drivers, is increased at relatively low blood–alcohol levels (0.01–0.04 percent).
Refs Armyr, Gunno; Elmér, ke and Herz, Ulrich *Alcohol in the World of the 80s* (1984).
Broader Alcohol abuse (#PD0153) Citizen apathy (#PF2421)
Parental permissiveness (#PD5344).
Aggravates Children of alcoholics (#PD4218).
Aggravated by Juvenile stress (#PC0877) Corruption of minors (#PD9481).

♦ PD1612 Fungicides as pollutants
Fungicides
Broader Pollutants (#PC5690) Air pollution (#PC0119)
Pesticides as pollutants (#PD0120).
Related Agricultural poisons (#PD5277).
Aggravates Eczema (#PE2465) Corrosion (#PD0508)
Hepatitis (#PE0517) Eye irritation (#PE6785)
Food pollution (#PD5605) Burns and scalds (#PE0394)
Hazards to human health (#PB4885) Fungicide damage to crops (#PD3577)
Blood circulation disorders (#PE3830) Pesticide hazards to wildlife (#PD3680)
Diseases of the skin and subcutaneous tissue (#PC8534).

♦ PD1614 Speculation in developing countries
Nature Capital that is not dissipated in conspicuous consumption or used for financing land transactions may not be available for industrial uses in developing countries because it is invested instead in speculative commercial undertakings.
The quick turnover usually involved in such activities affords opportunities for speedy liquidation which are greatly valued under conditions of uncertainty. In comparison with industrial activities, moreover, they are simpler both to start and to operate. And, in some countries, they are in a better position to escape taxation and various government controls.
Broader Capitalist speculation (#PC2194) Risk of capital investment (#PF6572).
Related Hoarding in developing countries (#PD1751).
Aggravates Capital shortage in developing countries (#PD3137).

♦ PD1618 Human physical genetic abnormalities
Congenital malformation — Congenital anomalies — Congenital disorders — Congenital deformities — Human birth defects — Congenital abnormalities — Human monstrosities
Nature Abnormalities of any part of the human body may develop during pregnancy. They may be relatively minor (such as excess toes), not affecting the health of the individual, or they may be very serious and lead to physical disability and even death. Cases include rudimentary or seal–like limbs, incomplete or undeveloped extremities, fusion of the lower extremities, absence of the brain, abnormally small brain and head, enlargement of brain and head, protrusion of the brain through the head, absence of eyes, abnormally small eyes, defective closure of lines of junction (harelip, cleft palate, spina bifida), partial or wholly double individuals (Siamese twins), or individuals with double heads or bodies Many such conditions may be regarded as resulting from the action of an unfavourable environment on a genetically susceptible individual. The relative importance of the environment varies with the condition.
Incidence The frequency of such traits, reported in all countries, has been estimated to be 1.5 per cent of the live–born, or higher if still–births are included. An additional 1 per cent of affected children can be detected by the age of 5 years. Many unsuspected abnormalities can only be detected by laboratory tests and x–ray With due reservations concerning differences in criteria, data from the World Health Organization over recent years indicate cases of this problem in the following countries:
Af Mauritius (57). Am Bahamas (22), Barbados (16), Belize (4), Brazil (8,544), Canada (1,469), Chile (1,287), Costa Rica (309), Dominican Rep (402), El Salvador (172), Guatemala (719), Haiti (5), Martinique (7), Neth Antilles (14), Panama (157), Paraguay (130), Puerto Rico (194), St Christopher–Nevis (2), St Lucia (2), Suriname (25), USA (13,938), Virgin UK (1). As Hong Kong (269), Israel (452), Japan (4,345), Kuwait (302), Singapore (177), Sri Lanka (434), Syria (177), Thailand (660). Au Australia (937), New Zealand (226), Papua New Guinea (22). Eu Austria (409), Belgium (488), Bulgaria (559), Czechoslovakia (1,111), France (2,599), Germany FR (2,458), Greece (777), Hungary (915), Iceland (19), Ireland (358), Italy (2,910), Luxembourg (12), Malta (33), Netherlands (770), Northern Ireland (135), Poland (3,948), Romania (1,956), Scotland (298), UK (England and Wales) (3,037), Yugoslavia (1,317).
Refs Barrett, Rowland P, et al *Advances in Developmental Disorders* (1987); Blauth, W and Schneider–Sickert, F *Congenital Deformities of the Hand* (1980); Chervenak, Frank A, et al *Anomalies of the Fetal Head, Neck and Neural Axis* (1988); Ciba Foundation Staff *Congenital Disorders of Erythropoiesis* ; Eriksson, A W et al *Population Structure and Genetic Disorders* (1981); Goodmann, Richard M and Gorlin, Robert J *The Malformed Infant and Child* (1983); Kalter, Harold (Ed) *Issues and Reviews in Teratology* (1984); Milunsky, Aubrey *Genetic Disorders and the Fetus* (1986); Mizejewski, Gerald and Porter, Ian S *Alpha–Fetoprotein and Congenital Disorders* (1985).
Broader Human disease and disability (#PB1044)
Genetic defects and diseases (#PD2389).
Narrower Dwarfism (#PE2715) Club foot (#PE3836) Microcephaly (#PG3838)
Cleft palate (#PE5117) Spina bifida (#PE1221) Anencephalus (#PG6828)
Physical blindness (#PD0568) Congenital glaucoma (#PG4522)
Foetal alcohol syndrome (#PE3853) Congenital heart disease (#PE2365)
Congenital toxoplasmosis (#PG6734) Congenital hydrocephalus (#PG6205)
Congenital anomalies of eye (#PE7549) Human pseudo hermaphroditism (#PE2246)
Inhibited human physical growth (#PD5177) Vulnerability of children to AIDS (#PE4276)
Congenital anomalies of nervous system (#PE9296)
Congenital anomalies of genital organs (#PE4249)
Congenital anomalies of alimentary tract (#PG9487)
Congenital anomalies of circulatory system (#PE6924)
Congenital anomalies of respiratory system (#PG3834)
Foetal malformation in diabetic pregnancies (#PE4808)
Congenital anomalies of musculoskeletal system (#PE8589)
Congenital syndromes affecting multiple systems (#PE9324).
Related Birth injuries (#PE4828).
Aggravates Deafness (#PD0659) Still–birth (#PD4029)
Hypertension (#PE0585) Infant mortality (#PC1287)
Neonatal mortality (#PD9750) Sudden unexpected infant death (#PD1885)
Perinatal morbidity and mortality (#PD2387) Decline in human genetic endowment (#PF7815)
Lack of facilities for severely deformed people (#PD0211)
Lack of adequate clothing supply for dwarfs, midgets and giants (#PE8496).
Aggravated by Teratogens (#PE0697) Changelings (#PE9453)
Radioactive fallout (#PC0314) Diseases of the spine (#PD2626)
Diseases of metabolism (#PC2270) Mercury as a pollutant (#PE1155)
Consanguineous marriage (#PC2379)
Health hazards of environmental pollution (#PC0936).

♦ PD1624 Political indoctrination
Political re–education — Thought reform — Brainwashing
Nature Brainwashing refers to an extreme form of indoctrination of individuals, or the reindoctrination of ideological deviants. It may involve police torture techniques to weaken resistance, including the the denial of sleep and food or use of drugs to create confusion and a blank, receptive mind for propaganda. In the case of political prisoners, brainwashing may be the preparation for a show trial where the victim confesses guilt, thus creating propaganda for use at home and abroad. In the case of prisoners of war, they may become collaborators during the war or be repatriated afterwards with subversive intent. Foreign espionage agents may be brainwashed to act as double agents.
Incidence Brainwashing evolved in the East and recent practices are largely confined to countries such as North Korea, China and the USSR, where its objective is to form or reform good communists. In the West its most notorious emergence has been among new religious cults, particularly in the USA.
Refs Goff, Kenneth (Ed) *Brain–Washing* (1983); Williams, John J *Electromagnetic Brainblaster* (1985); Winn, Kenneth *The Manipulated Mind; brainwashing, conditioning and indoctrination* (1983).
Broader Ideological conflict (#PF3388) Compulsory indoctrination (#PD3097).
Narrower Brainwashing of prisoners of war (#PD1652).
Related Drug abuse (#PD0094) Soul murder (#PF4213)
Forced political confessions (#PE3016)
Torture through behavioural regulation (#PE6914)
Psychic interference in decision–making (#PF0508).
Aggravates Mental disorders (#PD9131) Personal disempowerment (#PF0549)
Inadequate sense of time (#PF9980)

(#PD9131)
Personal disempowerment (#PF0549)
Inadequate sense of time (#PF9980)
Unethical practices of philanthropic organizations (#PE8742).
Aggravated by Espionage (#PC2140) Ideological deviation (#PF3405)
Double standards in morality (#PF5225) Inadequate political education (#PJ7906).

♦ PD1628 Conflicting claims concerning off-shore territorial waters

Unilateral claims to off-shore territorial waters

Nature Coastal states claim different widths of sea adjacent to their shores as being subject to their jurisdiction. Claims vary from 3 nautical miles to 200 miles. The purpose of such claims is to reserve for the state in question the right to exploit the economic resources of the area. The different nature of these resources gives rise to related problems of seabed exploitation (usually in relation to the adjacent continental shelf) and fishing rights. This problem is aggravated by the special difficulties of reaching agreement on and determining the exact shore line from which the territorial water limit should be measured (for example, low-tide line or high-tide line), particularly since these change over time. Further difficulties occur in the case of bays, estuaries and inlets, off-shore islands, and straits.

Land-locked countries are placed at a disadvantage if the larger claims are accepted, since this considerably reduces the area falling beyond national jurisdiction and thereby limits access to the resources in the areas encompassed by such claims

Incidence Vessels from the UK, the USA, the USSR, Peru, Japan and many other ocean fishing nations have been charged in recent years with fishing within 200 mile (exclusive economic zones) of other nations. Several of these incidents have included detention or dangerous harassment of vessels and some have been referred to as 'tuna wars'.

Broader Territorial disputes between states (#PC1888).
Aggravates Vulnerability of land-locked developing countries (#PD5788)
Restrictions on passage through straits and interoceanic canals (#PD2948).
Aggravated by Unsustainable exploitation of fish resources (#PD9082)
Unilateral declarations of independence by extra-territorial bases (#PF1066).

♦ PD1629 Factionalism in developing countries

Nature Factionalism, found in all subdivisions and at all levels of pre-industrial society, is the tendency of a society to be divided by caste or class cleavages, ethnic or religious distinctions, differences in cultural tradition or social pattern, kinship loyalties or regional identifications including linguistic nuances. These divisions tend to inhibit the development of a feeling of unity in the society and of identity among its members, and as a result the individual's sense of personal loyalty and duty may be limited to the members of a very narrow social environment, his family, clan, local community or other parochial circles and groupings. The normative pressures rooted in modern, factional environments may profoundly affect the conduct of the individual in external situations and relations. In particular, they may pose difficult problems for personnel recruitment and management in industrial or developmental undertakings.

Broader Factionalism (#PF8454) Social conflict (#PC0137)
Ideological conflict (#PF3388).
Aggravates Nepotism in developing countries (#PD1672)
Underdevelopment of industrial and economic activities (#PC0880).
Aggravated by Double standards in morality (#PF5225).

♦ PD1631 Inadequate safeguards against fire

Inadequate fire code — Continuing fire hazards — Unavailability of fire protection — Inadequate fire protection — Inadequate fire prevention

Nature Many features of building construction are closely related to fire safety. Although the modern types of construction generally contain much less combustible material than the old plank-on-heavy-timber or joisted brick-walled buildings, they collapse much sooner when exposed to fire temperature. Unprotected steelwork fails at relatively low temperatures.

In cases requiring fire-resistant construction such as hazardous occupancies, vital occupancies such as storage rooms, transformer rooms, electrical substations, laboratories, high-value occupancies where any serious fire would cause excessive property damage and production loss, and in multistorey buildings, fire-resistant construction is not always used.

Incidence In the UK in 1990 some 3.5 billion pounds of military equipment was held in stores which did not meet government fire protection standards.

Broader Inefficiency (#PB0843) Human errors and miscalculations (#PF3702)
Underprovision of basic services to rural areas (#PF2875).
Narrower Unreliability of self-extinguishing plastics (#PU3849).
Aggravates Injuries (#PB0855) Human death (#PA0072) Forest fires (#PD0739)
Property damage (#PD5859) Interruption risk (#PF9106) Injurious accidents (#PB0731)
Occupational domestic accidents (#PE4961) Prohibitive cost of basic services (#PF6527)
Inadequate prevention of disabilities (#PF0709)
Demoralizing images of rural community identity (#PF2358).
Aggravated by Standards setting procedure (#PS3851).

♦ PD1632 Misuse of telephone surveillance by governments

Wiretapping — Telephone bugging — Illegal phone tapping

Nature The interception of telephone conservations or telegraph messages may be legal or illegal and may be used to collect information and evidence on suspected criminals or as an espionage technique for political or business purposes. Legal wiretapping in some countries requires a court order and is used by local constabulary or state security police. If police corruption or political corruption is involved the information may be misused. The effect of wiretapping may be to reinforce government control or strengthen police power and there is risk of abuse. New forms of wiretapping apply to telecommunication data or code pulses, and hence also to computer-tapping. Illicit wiretapping is closely related to tape-recording of conversations without permission, which may be a civil, rather than a criminal offence against the rights of privacy. Wiretapping may also be an adjunct to electronic bugging.

Incidence In 1985, for example, it was reported that all but one of the member 21 countries of the Council of Europe made use of telephone surveillance. It is claimed that in 1978 the government of Romania was capable of monitoring 40 per cent of of the population's telephone communications and was tapping 90 per cent of government telephone communications and 80 per cent of army communications.

Refs French, S *The Big Brother Game* (1986).

Broader Unlawful government action (#PF5332) Interception of communications (#PD7608)
Misuse of government surveillance of communications (#PD9538).
Related Political surveillance (#PD8871)
Inappropriate use of telecommunications services (#PE4450).

♦ PD1634 Absenteeism

Labour absenteeism — Employee absenteeism

Nature By increasing the overall cost of labour, absenteeism accounts for part of the high prices of commodities and services. People may be absent from the workplace for a variety of reasons ranging from holidays or illness to attending football matches and so on. The word 'absenteeism' is used to describe absence when an employee is normally expected to attend for work and therefore excludes holidays and strikes. The main type of absenteeism is that attributed to incapacity (illness or injury), and this usually accounts for not less than three-quarters and often almost all industrial absenteeism. The attribution to incapacity may be supported by a medical certificate, depending upon local rules or social insurance regulations. Voluntary absenteeism rates are affected by working conditions and motivation. On an individual basis, emotional adjustment is a factor; those with problems, in or out of the workplace, tend to greater absence. There is also a variation with age and sex.

Incidence In most organizations about half the time lost is caused by 10 percent or less of the workforce. Beyond this, statistics are difficult to elaborate owing to enormous varieties of absences and definitions. However, if 1,600 hours represented the desired output of work of every employee annually, and 80 paid hours are lost in absent days and part-days (equivalent to 10 days a year), one could estimate the effect if this immediately translated into a 5 percent increase in prices (as a result of a 5 percent loss in productive output) or a 5 percent decrease in profit per employee (as a result of increased labour overhead). In a 1985 survey, by the Industrial Society concluded that absenteeism costs the British economy 200 million working days a year. Absentee rates among skilled manual workers in Britain runs about 10 percent, 9 percent for semi-skilled, 12 percent among 16–17 years olds and at least 8 percent in all blue collar industry groups except construction. In Sweden, on a typical day, 25 per cent of workers are absent (of these 10 per cent are sick).

Refs Blennerhassett, Evelyn and Gorman, Patricia *Absenteeism in the Public Service* (1986); Goodman, Paul S, et al *Absenteeism* (1984).

Broader Avoidance of work (#PC5528).
Narrower Student absenteeism (#PE4200).
Related Inadaptation of work to family needs (#PF5145).
Aggravates Excessive hours of work (#PD0140) Human disease and disability (#PB1044).
Aggravated by Apathy (#PA2360) Influenza (#PE0447)
Malingering (#PE7701) Common cold (#PE2412)
Alcohol abuse (#PD0153) Schistosomiasis (#PE0921)
Meaninglessness (#PA6977) Gastric disorders (#PE1599)
Rheumatic diseases (#PE0873) Industrial accidents (#PC0646)
Occupational diseases (#PD0215) Occupational rheumatism (#PE0502)
Lack of job satisfaction (#PF0171) Rigid family obligations (#PJ8340)
Alcoholic intoxication at work (#PE2033) Unethical practices by employees (#PD4334)
Age discrimination in employment (#PD2318)
Unsatisfied need for continuing education (#PF0021)
Economic and social losses due to disability (#PE4856)
Low occupational mobility in developing countries (#PD1493).
Reduced by Unemployment (#PB0750).

♦ PD1637 Landscape disfigurement from open-cast mining

Strip-mining

Nature Although open-cast mining is the most economic and safest method of exploiting reserves of coal and minerals which lie just below the surface, it presents serious disadvantages. Large earth-moving equipment strips the over-burden and stacks it in a bank parallel with the cut and the uncovered ore is then fragmented, loaded and transported from the pit. This process disfigures the surface of the land, and in the absence of reclamation, leads to permanent scars. The process buries the vital topsoil, disrupts drainage patterns, destroys the productive capacity of agricultural and forest land, as well as impairing their aesthetic and social value. It may also poison rivers and totally disrupt ecosystems.

Incidence The limiting depth at which underground mining becomes profitable has deepened constantly and opencast mining dominates in the recovery of many raw materials. Today 70–90% of the world's exploitable iron, copper, phosphate, asbestos, bauxite, manganese and lignite are extracted in opencast mines. Nearly all materials for construction are recovered from opencast mines and in the case of hard coal and uranium, a substantial share of approximately 25–30% is workable by opencast mining techniques. In the OECD countries, strip mining of coal is practised on a large scale in the Federal Republic of Germany, Australia and the USA and on a smaller scale in the UK, Canada, Ireland and Greece. Over 500,000 acres in the eastern USA have been stripped for coal, of which only half have been restored to close to their original state. About 50 per cent of the coal produced in the USA comes from surface mines. So far only 1 million of the 54 million acres of government owned land containing coal have been leased for mining. Reclamation is very expensive and mining deeper coal reserves would considerably increase the cost of energy.

Counter-claim Since opencast mining has many advantages compared with underground mining it will gain in significance in the future. The mining losses are low in opencast mining operation. In general, 90% or more of a mineral deposit can be recovered. Furthermore, the large quantities – in opencast mines these amount to far more than 100,000 tons per day – necessary for economical raw material treatment cannot be produced from underground operations even in the case of maximum mechanization. Finally, the mechanization of all opencast operations with high equipment capacity and the working conditions in the open pit decrease staff requirements. In times when it is more and more difficult to find suitable labour to work in underground mines, this is a great advantage of opencast mining technology.

Refs Munn, Robert F *Strip Mining* (1973).

Broader Ugliness (#PA7240) Landscape disfigurement (#PC2122).
Aggravates Environmental degradation (#PB6384)
Shortage of cultivable land (#PC0219).
Aggravated by Energy crisis (#PC6329).

♦ PD1642 Destruction of hedges and hedgerow trees

Nature With the introduction of modern farming methods and the pressure for higher crop yields, hedges and hedgerow trees are eliminated to create more rational field sizes. Such destruction reduces the amenity value of the countryside and eliminates valuable habitats for birds and insects which help destroy pests and pollinate crops.

Claim 1. Quickset hedges increase yield and foster early growth in certain sizes of fields, by reducing evapo-transpiration or increasing temperature. 2. Protection is offered against the damaging effects of wind and hail; hedges provide the animals with shelter, thus reducing veterinary costs. 3. Hedgerows act as water regulating reservoirs, limiting soil erosion and bleaching. 4. Land covered by quickset hedges can be considered ecologically balanced, if no one factor can predominate (pullulation of rodents, drought etc). The presence of a wide variety of species has an overall balancing effect (birds of prey as predators of the common mole, insectivorous birds, insects as pollenizing agents, parasites of crop ravagers).

Counter-claim Removal of hedges allows modern farming machinery to be used to best advantage, avoids the cost of maintaining hedges, prevents the propagation of pests and weeds associated with them, eliminates the shade which in the case of high hedges reduces crop yield, and increases the available land.

Broader Landscape disfigurement (#PC2122) Destruction of wildlife habitats (#PC0480).
Related Over-cultivated gardens (#PE7265) Destruction of the countryside (#PE3914).
Aggravates Suffering of plants (#PC7825) Environmental degradation (#PB6384)
Endangered species of animals (#PC1713).
Aggravated by Inappropriate modernization of agriculture (#PF4799).

♦ PD1644 Cost overruns in large-scale public programmes

Nature The estimated cost of large-scale projects when submitted to and approved by decision-

making bodies tends to be lower than the final cost. This may be due to deliberate underestimating by sub-contractors, increases in the cost of wages and materials during the life of the programme, or to unforeseen research and development difficulties.
The biggest contributor to cost overruns is delays in equipment or materials delivery. Very large and complex construction projects are usually unique. Since bringing all parts and equipment the site is not feasible, each phase of the building requires on time delivery of every necessary part and piece of construction equipment. If one item is delayed, the whole project may be delayed until it arrives.
Broader Inefficiency (#PB0843) Human errors and miscalculations (#PF3702).
Aggravates Socio-economic burden of militarization (#PF1447).
Aggravated by Economic inflation (#PC0254) Incomplete cost projections (#PJ8109) Overstated programme advantages (#PF8181) Underestimation of programme costs (#PF8499) Bureaucracy as an organizational disease (#PD0460).

♦ PD1652 Brainwashing of prisoners of war
Nature The war crime of POW brainwashing has come to be an increasing possibility in ideological conflicts.
Background During the Second World War, prisoners were subjected to propaganda and a very few were subverted. The stress of confinement with its hardships and privations, and the fear of death, were the influencing factors. Concerted psychological behaviour and thought modification techniques applied to POWs were not evident as a feature of modern conflict until the Korean War.
Broader Psychological warfare (#PC2175) Political indoctrination (#PD1624) Ill treatment of prisoners of war (#PD2617).
Aggravates Forced repatriation of prisoners of war (#PD0218).

♦ PD1661 Geomagnetic storms
Magnetic storms
Nature Solar flares produce bursts of ionizing radiation which result in geomagnetic storms in the earth's ionosphere. A magnetic storm is one of the geomagnetic anomalies, and consists of a period of rapid variation of the earth's magnetism, which can upset long-distance radio transmissions, cause compasses to give false readings. Magnetic storms are caused by sudden emissions of electrons and protons by the sun.
Incidence The storms, which may last several days or even weeks, produce widespread changes in the radioreflecting layers of the earth's atmosphere. This leads to poor radio reception (and even to complete black out on some frequencies for stations in polar regions), severe interference to telegraph communications, and even overloading of power lines and transformer-blowout. Geomagnetic storms may also affect weather patterns, to an extent which varies with the 11 year sunspot cycle. The storms may begin suddenly and simultaneously all over the earth within about one minute.
Broader Geomagnetic disasters (#PD0830) Geomagnetic field anomalies (#PF2407).
Aggravates Electrical storms (#PJ4133) Radio frequency interference (#PD2045).
Aggravated by Medium-term cyclic variations in solar radiant energy (#PE9528).

♦ PD1662 Denial of rights of medical patients
Nature The doctor-patient relationship is not a therapeutic alliance in which the doctor and the patient decide together the best treatment, but usually the doctor alone decides the medically best treatment.
Refs Peschel, Richard E and Peschel, Enid R *When a Doctor Hates a Patient* (1986).
Broader Denial of rights to vulnerable groups (#PC4405).
Narrower Institutionalized patients (#PG4705) Denial of rights of mental patients (#PD1148) Medical practitioners refusing to treat patients (#PE5027).
Related Excessive exposure of medical patients to radiation (#PE1704) Denial to experimental animals of the right to freedom from suffering (#PE8024).
Aggravates Irrelevant scientific activity (#PF1202) Irresponsible scientific and technological activity (#PC1153).
Aggravated by Scientific elitism (#PC1937).

♦ PD1665 Unaesthetic location of power transmission lines
Broader Ugliness (#PA7240) Landscape disfigurement (#PC2122).
Related Environmental hazards of electrical power transmission lines (#PE9642).

♦ PD1668 Parochial family responsibility
Nature The internal organization of the family tends to be haphazard and without regard to the objective restrictions and demands of its community and society. Structuring is limited to individual criteria for deciding what is needed to sustain the family, and responsibilities are family-centred and parochial in context. Accountability is reduced to tasks which are pertinent to the individual family needs.
Broader Reduced interior structure of families (#PF3783).

♦ PD1670 Chemical pollutants of the environment
Nature As modern man's dependence on chemicals for societal benefits has grown, the potential for widespread pollution or contamination has also increased. There are now some 70,000 chemicals on the commercial market, and many of these are currently used and released into the environment with little or no knowledge of their potential long-range effects. It is estimated that 1,000 new chemical enter the market each year. There are five potential sources of pollution from chemicals: 1) Chemical products themselves, such as CFCs, pesticides and nitrate fertilizers; 2) Hazardous waste and its treatment; 3) Chemical emissions, such as from factories, power plants and automobiles; 4) Accidents, such as the fire in 1986 that destroyed a chemical store in Switzerland belonging to Sandoz; and 5) Transport of chemicals.
Incidence Besides those on the market there are about 4 million chemical substances identified. Probably about one million of these are produced each year as intermediates, waste or laboratory chemicals that are not marketed but which may reach the public through contamination. Broad classes of pollutants are antibiotics and hormones use in the production of farm animals; chemical pesticides; nitrate and phosphate fertilizers; industrial liquid wastes; industrial gaseous emissions; industrial solid waste, scrap or process slag; (plastics, heavy metals and other dangerous substances can be released into the environment in gas, liquid or bulk); and some specially toxic substances worth mentioning as a class by themselves such as vinyl chlorides, PCBs, and acrylonitrile.
Refs International Maritime Organization *Manual on Chemical Pollution*.
Broader Pollutants (#PC5690) Environmental hazards from chemicals (#PC1192).
Narrower Dioxin poisoning (#PE7555) Ozone as a pollutant (#PE1359) Agricultural poisons (#PD5277) Nitrates as pollutants (#PE1956) Toxic metal pollutants (#PD0948) Asbestos as a pollutant (#PE1127) Motor vehicle emissions (#PD0414) Chemical air pollutants (#PD1271) Detergents as pollutants (#PE1087) Phosphates as pollutants (#PE1313) Solid wastes as pollutants (#PD0177) Hydrocarbons as pollutants (#PE0754) Sulphur dioxide as a pollutant (#PE1210) Titanium dioxide as a pollutant (#PE5195) Sulphur compounds as pollutants (#PG6442) Hydrogen sulphide as a pollutant (#PE2329) Nitrogen compounds as pollutants (#PD2965) Industrial waste water pollutants (#PD0575) Particulate atmospheric pollution (#PD2008) paper products (#PE7416).
Related Radioactive contamination (#PC0229) Photochemical oxidant formation (#PD3663).
Aggravated by Chemical trespass (#PE9363) Unethical practice of chemistry (#PD4265).

♦ PD1672 Nepotism in developing countries
Nature In an efficient industrial enterprise, the basic criteria for the recruitment of personnel and the assignment of tasks must be the ability to do the required work and a sense of responsibility in performance. The admission of exceptions may mean the difference between success and failure. Hence the particularist spirit which dominates segments of pre-industrial society, both in economic activities and in public life, and which tends to place personal loyalties and obligations to kin and friends above other considerations, may easily clash with the demands of industry. It may foster extreme practices of nepotism - such as putting relatives on the payroll even though they are incompetent or do not report for work - which may have a crippling effect on a small industrial undertaking and seriously reduce the efficiency of even the largest enterprise. Inimical to efficiency in most circumstances, such practices may prove a major deterrent to economic development in societies in which industrial enterprises are just beginning to emerge.
Counter-claim The family does play an important part in business in situations where there is economic immaturity of the population, the absence of a tradition of impersonal service in industry, and unreliability of employees who have no kinship ties to the firm.
Broader Nepotism (#PD7704) Immorality (#PA3369) Corruption in developing countries (#PD0348).
Aggravates Inefficiency (#PB0843) Underdevelopment of industrial and economic activities (#PC0880).
Aggravated by Factionalism in developing countries (#PD1629).

♦ PD1689 Birds as pests
Nature In most countries, most species of birds are considered to be a valuable part of the fauna but there are some species of bird that have feeding and roosting habits which conflict seriously with agricultural and urban life.
Incidence Those that most often draw complaints from farmers in the USA and Western Europe because of serious depredations on crops include a number of species of blackbirds and finches, crows, starlings, English sparrows, robins and several species of ducks and geese. The damage caused is widespread, and of no small economic significance. It includes crop damage to corn, sorghum, rice and fruits. In urban areas, the starling, English sparrow, and rock pigeon are the species that commit most of the objectionable roosting on buildings and in shade trees.
Economic losses caused by large roosts of starlings, pigeons and sparrows on buildings have not been fully estimated but they are of considerable economic significance in many cities. In Washington DC, for example, sizable expenditures are now being made to bird-proof many of the public buildings, such as the Treasury, Capitol, and Supreme Court. This is accomplished by installing a low-voltage electrical system on building ledges that serve as roosting sites.
Broader Pests (#PC0728) Animal pests (#PD8426) Pests of plants (#PC1627).
Narrower Quelea (#PE1429).
Related Insect pests of plants (#PD3634).
Aggravates Crop damage by wildlife (#PC3150) Threat of birds to aircraft safety (#PD1111).

♦ PD1694 Chemical contaminants of food
Chemical contamination of dietary intake
Incidence Contaminants of major concern include: aflatoxins; PCBs; toxic metals (lead, cadmium, mercury, tin); organochlorine pesticides (aldrin plus dieldrin, DDT complex, hexachlorocyclohexane, lindane, heptachlor and heptachlor epoxide, hexachlorobenzene, endrin and endosulfan); organophosphorous pesticides (diazinon, malathion, parathion methyl and fenitrothion).
Refs WHO *Assessment of Health Risks in Infants Associated with Exposure to PCBs, PCDDs and PCDFs in Breast Milk* (1988).
Broader Food pollution (#PD5605) Health hazards of environmental pollution (#PC0936).
Narrower Tin as a pollutant (#PE1438) Lead as a pollutant (#PE1161) Toxic food additives (#PD0487) Cobalt as a pollutant (#PE2339) Mercury as a pollutant (#PE1155) Cadmium as a pollutant (#PE1160) Selenium as a pollutant (#PE1726) Organochlorine pollution (#PG2007) Pesticide residues in food (#PE6480) Manganese as a health hazard (#PE1364) Organophosphorous insecticides (#PG3909) Carbamate insecticides as pollutants (#PE1282) Residues of veterinary drugs in foods (#PE4903) Polychlorinated biphenyls as a health hazard (#PE2432) Concentration of noxious substances in food chains (#PE8154).
Related Biological contamination of food (#PD2594).
Aggravated by Unethical food practices (#PD1045) Food manufacturing industry wastes (#PE8702) Environmental hazards from food and live animals (#PC1411).

♦ PD1703 Human wisdom unrelated to daily life
Nature An emphasis on vocational skills and technical expertise has completely overshadowed the need for "non-technical" learning. In fact, people have been so convinced of the irrelevance of intellectual disciplines (which have anyway never been seen as accessible) that, as a result, they do not see human wisdom as relevant and applicable to everyday life. They have no interest in learning or using the humanities. This leads to limited and weak participation in people's own culture and to their being cut off from the global cultural context. The resulting lack in comprehensive knowledge of the past and of the world as a whole means an inability to make informed decisions on social issues.
Narrower Fragmentation of knowledge (#PF0944) Avoidance of the irrational (#PF1610) Failure of methods to appropriate data (#PE0630) Incomprehensibility of specialized jargon (#PF1748) Failure to profit from patterns of history (#PF1746).
Related Absence of tactical methods (#PF0327) Undirected technological expansion (#PC1730) Antiquated intellectual methods to appropriate human depths (#PF1094).
Aggravates Educational curricula based on content rather than method (#PF3549).
Aggravated by Underutilization of popular wisdom (#PF2426).

♦ PD1709 Scientific censorship
Non-acceptance of new scientific theories
Nature Some systems of scientific thinking are too inflexible to allow for new thinking - which may even be branded as a kind of heresy if current scientific thinking also forms part of political ideology. Otherwise, new theories may simply be written off as eccentricities or not worth considering. Scientific rivalry may play a part in holding back new ideas.
Claim A principle mechanism in scientific censorship is the requirement by the fraternity that discoveries be published in refereed, scientific journals of irreproachable standing. The referees are appointed by the editor to review proposed publications for their 'worth'. Referees or official reviewers are frequently partisan when they are informed, and just as often, they are ill-informed in regards to sub-specializations in which they have no working experience. The referee gains a sense of power in his rejections.
Counter-claim All systems of thinking, scientific or otherwise, have points of inflexibility even if it is the assumption that the establishment is biased. All journals, except those who no one want

PD1709

to write for, have the problem of selection, i.e. discrimination and therefore can be accused of some form of censorship, given a loose and very abstract meaning of the word.
Broader Censorship (#PC0067) Intellectual discrimination (#PF8590).
Related Heresy (#PF3375) Self censorship (#PF6080).
Aggravates Stagnation (#PA3917) Secrecy in scientific research (#PF1430)
Suppression of scientific information (#PF1615)
Biased and inaccurate biology textbooks (#PF9358).
Aggravated by Prejudice (#PA2173) Grey lies (#PF3098)
Traditionalism (#PF2676) Scientific rivalry (#PG3918)
Politicization of scholarship (#PF7220).

♦ PD1710 Unethical practice of physics
Irresponsible physicists — Negligence by physicists — Malpractice in physics — Corruption of physicists
Claim The defence industry makes extensive use of physicists, who devote their skills to the design of weapons of destruction, notably nuclear weapons. Such practices contribute to the further proliferation of nuclear weapons in developing countries. Physicists employed to monitor nuclear power stations have been encouraged to downplay risks or design faults in installations and to cover-up major accidents. Given the especially competitive nature of the discipline, some practitioners make unethical use of the insights and results obtained by others. Others, in pursuit of personal ambition, lend their authority to disproportionate claims for the allocation of limited resources to prestige projects in high energy physics.
Aggravates Proliferation of nuclear weapons and technology (#PD0837)
Misappropriation of resources for high cost research projects (#PF0716).

♦ PD1714 Undesirable effects of animal feed additives
Hormonal growth promoters — Animal feed additives — Health hazards of drug use in meat production — Health hazards of hormone use in animal production
Nature A widespread practice in animal husbandry in recent years has been to add substances to animal feeds, or to inject hormone–like agents, in order to promote the growth of animals. These substances include a wide variety of antibiotics and other chemotherapeutic agents (such as antiprotozoal and antibacterial compounds). They are presumed to act mainly as suppressors of disease agents, although they may also exert as yet ill–defined non-specific effects that promote animal growth. Undesirable effects of these substances include the emergence of antibiotic-resistant strains of micro–organisms that, when transferred to man, may cause considerable medical and public health problems. Allergic reactions and toxic effects in man may also occur.
Incidence An example is that of stilbene oestrogens used in animal production as anabolic agents. These are orally active, persist in food, pose some environmental problems because of their low biodegradability. Diethyl stilbesterol (DES) is a known carcinogen. On reaching adolescence, the children of mother who had taken DES in pregnancy were developing cancer, the girls in the vaginal tract and the boys in the prostate, bladder or testicles. Sometimes the cancer showed as early as seven, sometimes as late as the mid–twenties. Abnormalities showed in 1978 in Milan in children who were fed veal which had been treated with DES. Knowledge of the toxicology of many other used substances is still rudimentary.
Claim Some synthetic hormones, such as diethyl stilbenes, are notorious cancer causing agents.
Refs FAO *Hormones in Animal Production* (1982).
Broader Hazards to human health (#PB4885) Inadequate animal welfare (#PC1167).
Related Denial to working animals of restorative nourishment and rest (#PE4793)
Denial to food animals of the right to freedom from suffering (#PE3899).
Aggravates Human disease and disability (#PB1044)
Residues of veterinary drugs in foods (#PE4903).
Reduces Vitamin E deficiency in domestic animals (#PE4760).

♦ PD1718 Biased presentation of news
Sensationalism
Nature News media may only present sensational items which give an exaggerated impression of a certain sector of social, economic or political life, and which will tend to strengthen prejudices and antagonisms and maintain inequalities and injustices, or alternatively they may mask these and lead to ignorance and apathy. News items may be scandalous and tend to defame a person's reputation. They may also be titillating or gossipy or simply of a low standard.
Newsworthy items, or those that are felt to be so by the directors of the various news media, tend to be negative items or exaggerated items which are not necessarily related to context. The latter may be good or bad, depending on the political views of the controllers.
Broader Distorted media presentations (#PD6081).
Narrower Gossip (#PE2192) Titillation (#PG3929) Sensationalism (#PG2304)
Lack of perspective (#PG3926) Junk food journalism (#PE1750).
Related Lying (#PB7600) Scandal (#PC8391) Deception (#PB4731)
Propaganda (#PF1878) News censorship (#PD3030) Political feuding (#PD4846)
Misleading information (#PF3096) Irrelevance of science and technology (#PF0770)
Promotion of negative images of opponents (#PF4133).
Aggravates Fear (#PA6030) Apathy (#PA2360) Prejudice (#PA2173)
Ignorance (#PA5568) Mental depression (#PC0799)
Defamation of character (#PD2569).
Aggravated by Grey lies (#PF3098) Official secrecy (#PC1812)
Proliferation of commercialism (#PF0815).

♦ PD1722 Psychoses
Psychosis
Nature Many so–called psychotics are likely to become incapacitated at one moment or another of their lives. This condition, which provokes immense human suffering, may lead to permanent reclusion in mental hospitals and at worst may induce criminal acts. Psychosis is a symptomatology in which the individual fails to discriminate between stimuli and information received from the external world and stimuli arising within himself. An imaginary process of reorganization of the world becomes the sole way to deal with conflict, accentuating the incapacity for corrective learning. This condition generates disorders of perception, of thinking, of consciousness, of affectivity and mood and of activity. Psychoses are of two types, based on the underlying physical disease: organic psychoses including brain tumours, senile dementia, and general panesis; and functional psychoses including schizophrenia, manic–depressive psychosis, and affective psychosis.
Incidence Psychosis accounts for 5 to 15 percent of the total admissions to mental hospitals; approximately half of these are people diagnosed as suffering from schizophrenia, the major so–called functional psychosis. The majority of the remaining cases are due to a variety of organic psychoses, predominantly aged people suffering from senile or arteriosclerotic psychoses.
Refs Buckley, Peter *Essential Papers on Psychosis* (1988); Massie, Henry N and Rosenthal, Judith *Childhood Psychosis in the First Four Years of Life.*
Broader Mental illness (#PC0300) Self disorders (#PF4843)
Mental disorders (#PD9131).
Narrower Paranoia (#PE0435) Schizophrenia (#PD0438) War psychosis (#PE7867)
Senile dementia (#PE3083) Socio–psychosis (#PE6965) Alcoholic psychosis (#PE9263)
Atrophic senile psychosis (#PG3080) Arteriosclerotic psychosis (#PG3081)
Postencephalitic psychosis (#PG7886) Manic–depressive psychosis (#PD1318)
Occupational psychopathology (#PD6880).
Related Mental depression (#PC0799) Borderline personality disorders (#PE4396).
Aggravates Infanticide (#PD3501) Hallucinations (#PF2249)
Psychotic violence (#PE7645).
Aggravated by Multi–drug abuse (#PD0213) Abuse of amphetamines (#PE1558)
Abuse of hallucinogens (#PD0556) Stress in human beings (#PC1648)
Inhumane interrogation techniques (#PD1362).

♦ PD1725 Medical quackery
Quack doctors
Nature Quackery is the use of primitive methods of medical treatment and healing by those who only pretend to have the necessary knowledge and qualifications to heal. Sick people may put all their hope and money into quack healing, hoping it will alleviate the diseases against which modern medicine is impotent, and those diseases which standard doctors refuse to treat. Such blind faith, usually encouraged by the quack healers, can lead to severe emotional and financial distress if the alternative methods – often viewed as the last resort – fail to produce the desired effects.
Background Quack healers have played a significant role in medicine throughout history. In Britain and Europe, treatment was long in the hands of old wives, lay healers, priests and herbalists, many of whom were genuine and who followed old–established folk methods, but few of whom could boast any formal training. Some techniques were considerably less reputable than others, but many practitioners would continue in practice despite the obvious ineffectiveness of the methods they prescribed. Further opportunity for charlatanism arose with advances in advertising techniques, and by 1843 medical advertising is quoted as taking a large proportion of advertising space in the press. Although legislation has been introduced to protect the gullible and the desperate from some of the more extravagant and misleading claims, quack remedies are still widely advertised in the press and, increasingly, on the radio and on television.
Counter–claim The derisive term "quack" has repeatedly been bestowed on those who step out of established order, who are ahead of their times, and whose methods may gain future respect and acceptance. The early advocates of antisepsis, anaesthesia, immunization and other innovations were thus branded. The mesmerism of the 18th century has become today's hypnosis or psychosomatic sleep, and the bone–setter of early days is today's physiotherapist. But other practices, such as homeopathy, acupuncture, naturopathy, and herbology are still regarded with scepticism, perhaps because they provide relatively inexpensive remedies for the patient to apply by himself and thus threaten the expensive – and extensive – monopoly today's doctors have over healthcare.
Refs Jarvis, William T, et al *Quackery and You* (1983).
Broader Fraud (#PD0486) Health fraud (#PD9297)
Unethical medical practice (#PD5770).
Aggravates Human poisoning (#PD0105).
Aggravated by Vanity (#PA6491) Food fads (#PD1189)
Occultism (#PF3312) Fear of death (#PF0462)
Lack of knowledge (#PF8381) Belief in miracles (#PG3935)
Unexplained phenomena (#PF8352) Stress in human beings (#PC1648)
Misleading advertising (#PE3814)
Lack of understanding of spiritual healing (#PF0761).

♦ PD1739 Tornadoes
Tropical cyclones — Water spouts — Twisters
Nature Tornadoes are rapidly rotating columns of air (160–800 km/hour) of small diameter (averaging 300–400 meters) which travel along a path up to 80 km (but occasionally up to 480 km) and cause considerable damage to buildings and crops encountered by the base of the column, which sucks dust, water and debris up into its rotating spiral. On the occasion when a violent tornado strikes a heavily populated area, it may cause millions of dollars of damage, and the loss of many lives.
Incidence The tornado occurs in its most violent form in the south–eastern USA, in the USSR and in Australia. Tornadoes are less violent and less common in other continents. They tend to occur in the vicinity of a severe thunderstorm, in the warm sector of a cyclone, and are most frequent in late spring and early summer. Damage and loss of the life caused by tropical cyclones vary widely from year to year. In an average year, about 80 tropical cyclones form over the warm ocean waters in certain parts of the tropics; tornadoes in these conditions are referred to as water spouts. Whilst these events are relatively few, each may have serious results and, on average, about 20,000 people lose their lives each year because of them; the damage caused may reach US $6,000 to US $7,000 million.
Broader Storms (#PD1150) Bad weather (#PC0293).
Aggravates Wind damage to structures (#PE1334).
Aggravated by Thunderstorms (#PE3881).

♦ PD1741 Monopoly of nuclear power techniques
Incidence Five countries (USA, UK, USSR, China, France) are nuclear powers, and one, India has exploded a nuclear device. Seven other countries (Argentina, Brazil, Israel, Pakistan, South Africa, and Spain) bring to the total twelve countries which have the technical capacity to enrich uranium, the capacity to manufacture it, and also access to it. While the number of countries with such abilities is expected to number 59 by the end of the century, the present monopoly by those twelve countries endangers self–protection measures for non–nuclear nations.
Counter–claim Nuclear weapons in the hands of governments led by fanatics is even more frightening a thought than now exists with nuclear weapons mainly in the hands of the USA and the USSR. Although there is always an impending threat of nuclear war, this threat is, to some degree, diminished by the limited number of countries currently capable of nuclear attack.
Broader Monopoly of power (#PC8410) Competitive acquisition of arms (#PC1258).
Related Super–power monopoly of advanced nuclear warfare technology (#PD4445).
Aggravates International insecurity (#PB0009) Insufficient nuclear power stations (#PD7663).
Reduces Proliferation of nuclear weapons and technology (#PD0837).

♦ PD1749 Individual isolationism
Claim The concept of self is particularly evident in this century, as oppressed people become independent; such awareness could give rise to historically–significant corporate action. But authentic awareness of self is aborted by individualism, which results in an isolation that cuts people off from their past and results in their clinging to self–created images. This in turn merely traps and confuses them, so they can no longer relate to what is happening around them.
Broader Social isolation (#PC1707) Defensive life stance (#PF0979).
Narrower Militant individualism (#PD1106) Lack of individual historical context (#PD1591)
Paralysis in individual decision-making (#PF1457)
Confusing decision–making methodologies (#PF4756).
Related Low self esteem (#PF5354) Tensionless image of free choice (#PF1675).

♦ PD1751 Hoarding in developing countries
Nature In many societies there is a reluctance to disclose the possession of monetary wealth to relatives and neighbours who might expect to benefit from it. In some communities a good deal of hoarding takes the form of investment in jewellery, usually for the adornment of women

DETAILED PROBLEMS
PD1800

and children. This custom had its origins at a time when the possession of a commodity that was indestructible, transportable and divisible, was sound practice as insurance against the consequences of political insecurity or periodical economic disasters, such as drought or famine. Though it has largely outlived its economic justification in some of the countries in which it was once a wise provision, it is still firmly rooted in existing family usages in some developing countries, where it diverts resources and tends to lessen industrial capacity.

Broader Hoarding of primary commodities (#PD0651).
Related Speculation in developing countries (#PD1614).
Aggravates Capital shortage in developing countries (#PD3137).

♦ PD1753 Typhoid fever

Enteric fever

Nature Typhoid fever is an acute infections disease, affecting only humans, and characterized by fever, septicaemia, and lesions of the cardiovascular, nervous, and digestive systems.

Incidence This is one of the diseases being explored because of its value in biological warfare.

Background It was first described in the early 1800's and is prevalent in countries with poor sanitary conditions. Infection occurs when bacteria enters the month from the contaminated hands of a sick person or bacteria carrier. The bacteria multiply in milk, water, vegetables, and fruit. Treatment of typhoid fever patients includes confinement, a special bland diet, antibiotics, and systemic restorative and symptomatic drugs. Vaccination of an entire population when there are indications of an impending epidemic is an auxiliary protective measure; the main protective measures being the availability of sanitary and hygienic public facilities (especially in restaurants, grocery stores, food industry enterprises), and the education of the public in the necessity for good personal hygiene.

Broader Epidemics (#PC2514) Enteric infections (#PD0640)
Stigmatized diseases (#PD7279) Infectious and parasitic diseases (#PD0982).
Narrower Epidemic typhus (#PG3894) Non–epidemic typhus (#PE3895).
Related Diseases of the lymphatic system (#PD2654).
Aggravates Fever (#PD2255) Deafness (#PD0659) Toxaemia (#PG3957)
Gangrene (#PG5675) Pneumonia (#PE2293) Meningitis (#PE2280)
Peritonitis (#PE2663) Haemorrhage (#PE2239) Encephalitis (#PE2348)
Cholecystitis (#PE2251) Heart failure (#PE3958)
Osteomyelitis, periostitis and other infections involving bones and acquired deformities of bone (#PE8912).
Aggravated by Dust (#PD1245) Rickettsiae (#PE2572)
Wasted water (#PD3669) Rats as pests (#PE3177)
Contradictions (#PF3667) Ticks as pests (#PE1766)
Mites as pests (#PE3639) Natural disasters (#PB1151)
Chronic bronchitis (#PE2248) Houseflies as pests (#PE3609)
Cockroaches as pests (#PE1633) Lice as insect pests (#PE1439)
Flies as insect pests (#PE2254) Mycoplasmal arthritis (#PG2302)
Unhygienic conditions (#PF8515) Siphonaptera as insect pests (#PE3643)
Contamination of drinking water (#PD0235) Inappropriate sanitation systems (#PD0876)
Damage by degradable organic matter (#PJ6128)
Pollution of water by infected faeces (#PE8545).

♦ PD1754 Non–destructible packaging and containers

Nature Paper and cardboard are being replaced for wrapping and packaging by plastic and other materials which are almost indestructible and which, when discarded, damage the environment, unless collected for disposal. It is also increasingly acceptable to supply certain goods in non–returnable containers and to use non–returnable plastic cups, plates, spoons, forks, etc, for serving food. Any saving to the consumer in the initial cost of the product is likely to be offset by the higher indirect charges for waste disposal and by the damage to the environment.

Claim Packaging is specifically designed to become instant waste. Packaging makes up more than a third of all household waste in Western countries. 42 percent of paper products used in Britain are used for packaging, and are thrown away.

Broader Deception (#PB4731).
Aggravates Litter (#PD2541) Domestic refuse disposal (#PD0807)
Environmental degradation (#PB6384) Resource–intensive packaging (#PE0635)
Unstable supply of raw materials (#PD4270).
Aggravated by Spoilage (#PG3966) Personal wealth (#PC8222)
Misleading advertising (#PE3814).

♦ PD1762 Endangered species of marsupials

Endangered species of macropods

Nature Many of the marsupial species (particularly the kangaroo) are perceived as being in competition with sheep for grazing land; and therefore considerable effort has been devoted to exterminating them. Others are threatened by the degradation of their habitat. Others, including rare species, are slaughtered as a cheap source of meat.

Incidence The marsupials are found only in Australasia. Seven species are extinct, twelve are endangered and 29 remaining are being slaughtered by millions every year.

Broader Endangered species of mammals (#PC1326).
Narrower Endangered species of koala (#PG3976)
Endangered species of wombats (#PG3977)
Endangered species of opossums (#PG3968)
Endangered species of thylacine (#PG3980)
Endangered species of bandicoots (#PG3978)
Endangered species of rat opossums (#PG3971)
Endangered species of honey possum (#PG3975)
Endangered species of pigmy possums (#PG3974)
Endangered species of marsupial moles (#PE3981)
Endangered species of ringtail gliders (#PG3973)
Endangered species of phalangers and cuscuses (#PE8537)
Endangered species of kangaroos and wallabies (#PE8708)
Endangered species of dasyures and marsupial mice (#PE8527).
Aggravated by Destruction of wildlife habitats (#PC0480).

♦ PD1763 Endangered species of cats

Felidae — Cats

Broader Endangered species of carnivores (#PD3482).
Narrower Endangered species of felis (#PG3982)
Endangered species of neofelis (#PG3983)
Endangered species of panthera (#PG3984)
Endangered species of acinonyx (#PG3981).

♦ PD1767 Criminal conspiracy

Nature A person agreeing with one or more persons to engage in or to cause criminal activity is guilty of the crime of criminal conspiracy. Agreement to commit conspiracy does not need by explicit but may be implicit in light of the circumstances of the agreement.

Broader Conspiracy (#PC2555) Offences of general applicability (#PD4158).
Narrower Criminal association (#PE1178).
Related Complicity (#PF4983) Criminal attempt (#PD5321) Criminal facilitation (#PD6845)
Criminal solicitation (#PD7676).
Aggravates Organized crime (#PC2343)
Vitamin E deficiency in domestic animals (#PE4760).

♦ PD1770 Zoonoses

Zooanthroponoses — Anthropozoonoses — Inadequate control of zoonoses

Nature Zoonoses are diseases and infections of animal origin which are communicable to man and may be viral, bacterial, parasitic or fungal. The frequency varies with the type and source, as well as with the geographical location and occupational exposure. In the majority of zoonoses the infection remains limited to the affected individual, and person–to–person transmission is rare or exceptional. Often, the infection causes observable disease only in man; the animal 'carrier' being symptomless or only mildly sick, as is the case, for example, in Q–fever.

Incidence There are over 150 zoonoses carried by a wide variety of animals. The highest incidence of zoonoses is noticed in persons who come in close contact with animals or animal products or those who share with animals environments containing suitable vectors Common zoonoses are anthrax, brucellosis, Chaga's disease, equine encephalomyelitis, equine infectious anaemia, foot–and–mouth disease, glanders, hydatid disease, leptospirosis, listeriosis, liver flukes, louping illness, lymphocytic choriomeningitis, Newcastle disease, psittacosis–ornithosis, Q fever, rabies, rat–bite fever, rift valley fever, ringworm, Rocky Mountain disease, Russian spring–summer virus, salmonellosis, scabies, schistomiasis influenza, echinoceccosis tapeworms, tick–bite fever, tick paralysis, toxocara, toxoplasmosis, trichinosis, tuberculosis, tularaemia, vesicular stomatitis, Warburg disease, Wesselsbron disease, typhoid, yellow fever, and plague. In spite of their varied nature these diseases have one common feature – in nature, they are transmissable from animals to man either directly or through animal products and sometimes through invertebrate vectors (insects, ticks, mites, molluscs.

Refs Fiennes, Richard N *Zoonoses and the Origins and Ecology of Human Disease* (1979); Hubbert, William T, et al *Diseases Transmitted from Animals to Man* (1975); Soulshy, E J *Parasitic Zoonoses* (1974); WHO *Bacterial and Viral Zoonoses* (1982).

Broader Animal diseases (#PC0952) Human disease and disability (#PB1044).
Narrower Rabies (#PE1325) Liver fluke (#PE2785) Rickettsiosis (#PE5530)
Hydatid disease (#PE2354) Fungal diseases (#PD2728)
Arthropod–borne diseases (#PE7796) Soil–transmitted diseases (#PD3699)
Zoonotic bacterial diseases (#PD6363).
Related Water–borne animal diseases (#PE2787)
Environmental human diseases (#PD5669)
Parasitic diseases in animals (#PD2735)
Human vectors of animal diseases (#PD2784)
Domestic animals as carriers of animal diseases (#PD2746).
Aggravates Animals as vectors of disease (#PD8360)
Health risks to workers in agricultural and livestock production (#PE0524).
Aggravated by Unclean food (#PJ2532) Enzootic diseases (#PD2733)
Epizootic diseases (#PD2734) Agricultural wastes (#PC2205)
Stray dog populations (#PE0359) Abuse of animal drugs (#PE0043)
Diseases of wild animals (#PD2776) Inadequate health control (#PF9401)
Viral diseases in animals (#PD2730) Vectors of animal diseases (#PD2751)
Bacterial diseases in animals (#PD2731)
Inadequate control of animal diseases (#PD2781)
International trade in endangered species (#PC0380)
Wild animals as carriers of animal diseases (#PD2729)
Inadequate carcass disposal of diseased animals (#PE2778)
Difficulty in identifying carriers of animal diseases (#PF2775)
Inadequate disinfection of pastureland after disease outbreak (#PD2774)
International movement of animals as factor in animal diseases (#PD2755)
Inadequate disinfection measures for animal housing and equipment (#PE2757)
Importation of infected carcass meats as factor in animal diseases (#PE2777).

♦ PD1787 Espionage in domestic politics

Nature Political espionage used for domestic purposes furthers political repression and may be used to incriminate opponents in either constitutional or totalitarian systems. It includes the building up of political dossiers, wire–tapping, burglary, censorship and intimidation; and may be aided by the political appointment of supporters to key positions. Espionage may be used during elections or to achieve indoctrination at other times. It serves particularly to tighten government control and to promote the interests of an elite. Acts of corruption and other injustices may be effectively covered up by intelligence activities.

Broader Espionage (#PC2140) Unethical practices in politics (#PC5517).
Narrower Political burglary (#PD1943).
Aggravates Secret police (#PE6331) Corruption in politics (#PC0116).
Aggravated by Lack of political integrity (#PF0796).

♦ PD1791 Disparity of national tax systems

Nature When countries which are trading partners make use of tax systems which are not harmonized, this may introduce considerable obstacles to effective trade, particularly when the trading partners are within an economic union or common market. Differences in fiscal systems may lie in their basic structures, in the way they make use of a certain tax, or in the proportion of direct to indirect taxation. One view argues that taxes should be uniform throughout such a common market in order to equalize opportunity for business enterprises in all countries of the market and to prevent distortions of free competition. Another view holds that complete uniformity may in fact impair the international flow of capital and that an optimum degree of non–uniformity is necessary.

Broader Lack of international cooperation (#PF0817).
Narrower Inappropriate taxation of not–for–profit and philanthropic charitable organizations (#PF3049).

♦ PD1797 Restrictive practices in trade in manufactured goods

Nature Technical specifications for communications consumer products may be so designed as to eliminate some potential imports, or to raise their cost of manufacture. Computer–compatible peripheral equipment may be limited by technical obstacles so that main–frame or original equipment manufacturers (OEMs) retain a near–monopoly position on these items.

Broader Restrictive trade practices (#PC0073).
Aggravates Excessive external trade deficits (#PC1100).

♦ PD1800 Denial of right of conscientious objection to military service

Incidence Liberality, if that is the word, is characteristic only of the Germanic countries (including Scandinavia, the UK, the Commonwealth, and the USA). Denial of the right of conscientious objection seems to be more clearly prevalent in Latin countries (including Latin America, Portugal, Italy, France), in Socialist countries, and in countries which are party to international political disputes (for example, Israel). Since the Koran and the Bhagavad Gita both justify war, orthodox Moslems and Hindus may not be able to be religious conscientious objectors. This situation is also reflected in Shinto, an influence on historic Japanese martiality.

Refs Eide, Asbjørn and Mubanga–Chipoya, Chama *Conscientious Objection to Military Service*.

Broader Pacifism (#PF0010) Conflict (#PA0298)
Denial of freedom of conscience (#PD7612).

Narrower Conscientious objection at the factory (#PE7007).
Aggravates Persecution (#PB7709) Prisoners of conscience (#PC6935)
Internment of conscientious objectors (#PG4022).
Aggravated by Oppression (#PB8656) Religious discrimination (#PC1455).
Reduces Conscientious objection (#PD4738).

♦ PD1802 Non-productive use of cattle and livestock

Nature In most parts of eastern and southern Africa, cattle-keeping has little relation to the traditional subsistence economy, except as a means of simultaneously producing wealth and 'banking' it. The animals are rarely slaughtered for commercial sale or domestic consumption; they are used chiefly for ceremonial feasts and religious celebrations associated with death rites or ancestor worship; for validating marriages and cementing kinship bonds; for paying fines; settling conflicts; or financing undertakings requiring hired labour. Ownership of cattle determines a man's position in the community; his social standing, influence and potential economic power are measured by the size of his herd.
Counter-claim Cattle keeping in Eastern and Southern Africa by the peasant farmers is an intrinsic part of the traditional subsistence economy. While cattle are a source of wealth and status and are not kept primarily for the provision or sale of beef, they have many other roles in society, particularly in societies which practise cropping. These roles are difficult to quantify in terms of value, but have a value, in monetary terms, that is probably greater than the value of the animal as a source of saleable beef.
Though not necessarily in order of priority, the role of livestock, particularly cattle, in traditional societies, is: 1) the provision of drought animals for land preparation; 2) the provision of milk for home consumption; 3) the provision of manure for fertilization of cropping lands; 4) as a source of wealth and status; 5) for religious, social and ceremonial purposes; and 6) as a source of cash from the sale of beef.
It is recognized that peasant farmers who own cattle enjoy a higher level of productivity and standard of living than those who do not, and with limited land resources there is therefore keen competition between peasant families for cattle ownership and grazing rights.
In more arid region, where cropping is not practised, priorities for cattle ownership change and off-take from herds for sale is higher, but cattle remain a valuable source of milk and in some cases blood for family nutrition.
Broader Underproductivity (#PF1107).
Narrower Underutilization of livestock in least developed countries (#PE8595).
Aggravates Trespassing livestock (#PE4898)
Capital shortage in developing countries (#PD3137).

♦ PD1809 Alienation of support for international organizations and programmes

Broader Ideological conflict (#PF3388) Lack of international cooperation (#PF0817).
Aggravated by Religious opposition to population control (#PF1022).

♦ PD1823 Archaeological and anthropological looting

Clandestine excavations
Nature Archaeological sites may be excavated clandestinely to obtain objects and relics for sale on the art market or to museums. Such looting of sites destroys their value as a source of information on the cultures that lived there and on the origins of the modern culture in the region. It may also destroy the importance or beauty of sites for visitors, whether foreigners or citizens of the country.
Incidence Recently clandestine or illicit excavations have increased considerably and in some cases are conducted on a large scale with advanced equipment. The Four Corners area in the southwest of the United States as estimated 60–95 percent of some 1.5 million sites including cliff dwellings, dry caves and buried villages have been looted. Under-water thieving of historical objects has grown to considerable proportions as well.
Background The pillaging of burials is believed to have been a reason for the construction of the pyramids. In more recent centuries, underprivileged nations have traditionally provided much of the raw material for anthropologists and archaeologists from more prosperous countries. As a result, many museums may own artifacts acquired under circumstances that were not legal or which resulted from scientific excavation. In the US indian artifacts are bringing high prices from private collectors; a single Mimbres Indian, a tribe extinct since the 12th century, bowl recently sold for $40,000.
Claim Willingness of private collectors, and private and public museums to purchase artifacts that lack documentation and proper export papers, provides a persisting market for thieves.
Refs Houdek, Frank G *Protection of Cultural Property and Archaeological Resources* (1988).
Broader Destruction of cultural heritage (#PC2114)
Unethical practice of anthropology (#PD2623).
Related Theft of property (#PD4691) Theft of works of art (#PE0323)
Abusive collection of specimens (#PE9417).

♦ PD1827 Educational curricula over emphasizing method rather than content

Nature Students with startling gaps in cultural, historical and literary knowledge is the result of an over emphasis on teaching method or process and ignoring content. Curriculum emphasizes skills over knowledge. Teachers are taught teaching methods rather than subject matter. Textbooks tend to be a compendium of disconnected facts.
Claim If children continue to be taught to think without teaching them something to think about there is the danger of unwittingly proscribing a society's own heritage.
Broader Irrelevance of educational curricula (#PF0443).
Related Lack of emphasis on basic education (#PF1548).
Aggravates Narrow scope of education (#PF3552)
Inadequate economic integration of socialist countries (#PF4884)
Lack of meaningful educational context for ethical decisions (#PF0966).
Reduces Educational curricula based on content rather than method (#PF3549).

♦ PD1837 Disparities in unemployment within countries

Regional unemployment within countries
Nature The greater the regional variations in unemployment within a country, the higher will be the national rate of unemployment consistent with stable inflation.
Incidence In 1987 unemployment rates in USA ranged from 2.5 per cent in New Hampshire to 11.8 per cent in Louisiana. In Italy jobless rate varied from 6.4 per cent in Lombardy to 22.2 in Campagnia.
Broader Unemployment (#PB0750).
Narrower Rural unemployment (#PF2949).
Related Urban unemployment in developing countries (#PD1551)
Rural unemployment in developing countries (#PD0295).

♦ PD1842 Prohibitive cost of accommodation

Inflated house values — Disparity between income and cost of housing — Limited low-cost housing — High housing costs — High cost of housing
Nature The proportion of income which a family can pay for shelter varies according to the country and the times. The disparity between what the lower income groups can pay and the rent required to acquire a building and amortize its cost, has generally widened with the years because of technological lag, rising costs of materials, labour shortages and building laws which impose more costly standards. When wages have risen, so have shelter costs, and the gap has continued unbridged.
Broader Consumer vulnerability (#PC0123) Socio-economic poverty (#PB0388)
Prohibitive cost of living (#PF1238).
Narrower Prohibitive cost of home maintenance (#PJ9022).
Related Prohibitive ownership costs (#PJ0033)
Prohibitive cost of necessities in rural communities (#PF2385).
Aggravates Homelessness (#PB2150) Inadequate housing (#PC0449)
Insufficient rural housing (#PF6511) Marginal level of family income (#PD6579)
Illegal occupation of unoccupied property (#PD0820).
Aggravated by Capitalism (#PC0564) Underpayment for work (#PD8916)
Prohibitive cost of land (#PE4162) Limited accumulation of capital (#PF3630)
Limited individual capital reserves (#PF2899).

♦ PD1846 Victimization of workers' representatives

Denial of right to protection of union representatives — Inadequate protection and facilities for workers' representatives
Broader Lack of job satisfaction (#PF0171)
Sanctions against trade union workers (#PD0610).
Aggravates Strikes (#PD0694) Discrimination (#PA0833).
Aggravated by Unethical commercial practices (#PC2563)
Lack of trade union recognition (#PG4050)
Ineffective worker organizations (#PF1262)
Denial of right to organize trade unions (#PE5398).

♦ PD1847 Inequality of employment opportunity in developing countries

Nature Inequalities of opportunity and treatment may arise either from deliberate acts of discrimination or from passive situations resulting from economic, social, cultural or geographic factors. Groups affected include aboriginal and tribal populations; those occupying specific backward regions; different ethnic groups; different religious and linguistic communities; and foreign workers. In some cases minorities are alleged to practise discrimination against the majority. Very often this discrimination is not so much a matter of fact as one of feeling.
Background Before independence, political but not economic unification was imposed in many developing countries. When the colonisers left, the more important jobs tended to be taken over by the educated middle classes and denied to poorer and minority groups.
Refs Blaug, Mark *Education and the Employment Problem in Developing Countries* (1981).
Broader Inequality in employment (#PD8903) Inequality of opportunity (#PC3435).
Aggravates Discrimination (#PA0833).
Aggravated by Unemployment (#PB0750) Lack of leadership (#PF1254)
Economic underdevelopment (#PC0281) Denial of right to education (#PD8102)
Threatened and vulnerable minorities (#PC3295)
Discrimination against women in employment (#PD0086)
Prohibitively expensive housing for the poor (#PE8698)
Inability of communities to take advantage of training for business, industry or public service (#PE8509).

♦ PD1863 European insecurity and vulnerability

Nature European security is jeopardized by the confrontation of the military forces of the NATO and Warsaw Pact alliances (and the threat of attack which they represent), and the absence of political, economic, scientific, technical, and cultural relations between the two blocs of countries. Symptomatic of the problem are the absence or near-absence of notification of military manoeuvres, the location of such manoeuvres in sensitive border areas, surveillance activities, construction of fortifications in border areas, flights of foreign planes carrying nuclear weapons, presence of foreign troops and military bases, and the actual size of military budgets and armaments.
Broader Insecurity (#PA0857) International insecurity (#PB0009).
Aggravated by Surprise attack (#PE3705) Foreign military presence (#PD3496)
Proliferation of strategic nuclear arms (#PD0014)
Military manoeuvres in sensitive border areas (#PE3704)
Misuse of satellite surveillance by governments (#PF3701)
Allocation of television frequency bands for satellite transmission (#PF3703).

♦ PD1865 Inadequacy of the commercial sector in developing countries

Nature The development of domestic industry depends very largely upon the size of the local market, but this in turn, though ultimately a function of the national income and its distribution, depends partly upon machinery for taking the product to the potential consumer. The effectiveness of this distributive organization is one measure of the adequacy of the economic framework. Where the commercial sector is poorly equipped to handle the output of local factories, the absorptive capacity of even the small domestic market is not fully realized. Inadequacy of the commercial sector not only reduces the size of the accessible market and throws the burden of carrying stocks of finished goods and organizing their distribution, at least partly, onto the factory; but it also magnifies problems of supply, making it necessary for the producer to maintain larger stocks of raw materials and consumable stores than would be required if ordinary trade channels were more effective.
Increased stocks – and the consequent increase in costs – are also a factor in the organization of equipment and machinery, for the lack of repair facilities and ancillary industries in the under-developed country may render it necessary to carry more spares and replacements than would otherwise be required and, in some cases, to instal standby plants in order to assure continuous production or at least avoid unduly long breakdown delays.
Narrower Lack of local commercial services (#PF2009).
Aggravates Underdevelopment of industrial and economic activities (#PC0880).
Aggravated by Domestic market restrictions in developing countries (#PD1873).

♦ PD1866 Plant pathogens

Refs Maramorosch, Karl and McKelvey, John J *Subviral Pathogens of Plants and Animals* (1985).
Broader Plant diseases (#PC0555).
Narrower Plant cancer (#PE1899) Parasites on plants (#PD4659)
Viral plant diseases (#PD2227) Fungal plant diseases (#PD2225)
Nematoid plant diseases (#PD2228) Bacterial plant diseases (#PD2226)
Environmental plant diseases (#PD2224) Deficiency diseases in plants (#PD3653).
Related Plant disease vectors (#PD3596).
Aggravates Famine (#PB0315) Starvation (#PB1875) Malnutrition (#PB1498)
Nutritional deficiencies (#PC0382) Pests and diseases of rice (#PE2221).
Aggravated by National political dependence (#PF1452).
Reduced by Environmental warfare (#PC2696).

♦ PD1873 Domestic market restrictions in developing countries

Broader Economic and social underdevelopment (#PB0539).
Narrower Bias against private enterprise (#PF1879)
Excessive protection of industries by customs duties (#PD1987)
Inadequacy of the domestic market in developing countries (#PD0928)
Excessive government participation in the economies of developing countries (#PD1902).
Aggravates Inadequacy of the commercial sector in developing countries (#PD1865).
Aggravated by Economic inflation (#PC0254).

DETAILED PROBLEMS

PD1915

♦ **PD1877 Piracy**
Nature Illegal acts of violence, detention or any act of depredation committed outside the jurisdiction of any State, for private ends, by the crew or the passengers of a private ship or a private aircraft is piracy if it is directed against their own or other ships or aircraft, or against persons or property onboard, overboard, or ashore (where there is no effective jurisdiction). Pirates may be abetted by persons performing acts of voluntary participation in the operation of a ship or of an aircraft with knowledge of facts, making it a pirate ship or aircraft, or by any act of inciting or of intentionally facilitating participation in piracy. Piracy can be committed with the use of a warship, government ship or government aircraft whose crew has mutinied and taken control. The terms ship and aircraft may also be specified as to mean 'any sea-going vessel' (submarines, etc) and 'any air- or spacecraft'. Hybrid vessels such as sea-based air- or spacecraft, and any unmanned vessels or craft controlled remotely, may be involved in future piratical acts.
Incidence The most recent widespread and publicized piracy concerns the piratical atrocities committed against the Indo-Chinese boat people in the aftermath of the Vietnam war.
Broader Theft (#PD5552).
Narrower Piracy at sea (#PD8438) Aerial piracy (#PD0124).

♦ **PD1881 Military atrocities**
Military excesses
Nature During military conflict or internal unrest, groups of civilians or defenceless prisoners of war may be systematically murdered or mutilated, women raped and business and homes looted. This may be done by soldiers either acting under orders from high command, or as isolated acts perpetrated without explicit orders. Such atrocities may be carried out as reprisals for attacks by partisan or resistance groups, as a means of maintaining control of the country or as rewards to soldiers.
Broader War crimes (#PC0747) Atrocities (#PD6945)
Denial of human rights in armed conflicts (#PC1454).
Narrower Massacres (#PD2483) Ill treatment of prisoners of war (#PD2617)
Physical insecurity of refugees and asylum-seekers (#PD6364).
Related Military offences (#PC0742) Government sanctioned killing (#PD7221)
Ineffective war crime prosecution (#PD1464).
Aggravated by Torture schools (#PE2062) Mercenary troops (#PD2592)
Military brutality (#PD4945) Military disobedience (#PD7225)
Inadequate army discipline (#PD2543) Execution of inappropriate orders (#PF2418)
Brutalization of military personnel (#PD7602)
Racial discrimination by security forces (#PD3519).

♦ **PD1882 Undeclared strikes in socialist countries**
Nature Strikes are not officially permitted in socialist countries, but they do occur. In some cases, industrial unrest takes the form of 'hidden strikes' which include widespread absenteeism, the slowing down of the production process, the increased production of defective goods, and the fluctuation of the work force.
Broader Strikes (#PD0694) Undeclared strikes (#PD5384).

♦ **PD1885 Sudden unexpected infant death**
Sudden infant death syndrome — Crib deaths — Cot deaths
Nature Cot deaths - the sudden unexpected death of infants during sleep - are now recognized as a disease syndrome not caused by suffocation or by parental neglect. This phenomenon concerns a young child, usually from two to six months of age, who was either quite well or had only trivial symptoms; the child is found dead in its cot or bed, almost invariably in the morning. This circumstance is so frequent that the universally accepted name is 'Cot Death' or, in the USA, 'Crib Death'.
Incidence The age range within which cot death occurs is from two weeks to two years. It is thus not a condition of new-born children, and death in the first couple of weeks of life almost always excludes this syndrome. The great majority of cases occur between two and six months, with a peak at four months. Few deaths occur after nine months, and those in the second year are very rare. Recent surveys have shown a significant preponderance of boys over girls; most of the evidence suggests that such deaths affect boys more than girls in the ratio of about 3:2. There is a striking seasonal variation in the deaths. Cot deaths occur in the cold, wet seasons when respiratory infections are at their height. The preponderance of cot deaths in the lower income groups also suggests that over-crowding, especially in the sleeping places, and possibly less prevention of cross infection within households are significant contributing factors. A number of recent studies have shown that the cot-death syndrome is very closely connected with smoking during pregnancy: an American study covering nearly 20,000 births showed that 70 percent of the mothers whose children had fallen victim to the syndrome had smoked during pregnancy.
Refs Golding, Jean, et al *Sudden Infant Death* (1985); Shapiro, Samuel et al *Sudden Death* (1983); Tildon, Tyson, et al *Sudden Infant Death Syndrome* (1983).
Broader Infant mortality (#PC1287) Human disease and disability (#PB1044).
Aggravated by Crack (#PE2123) Asphyxia (#PE4104)
Pneumonia (#PE2293) Suffocation (#PJ4103)
Climatic cold (#PD1404) Heart diseases (#PD0448)
Premature birth (#PD1947) Multiple births (#PE4107)
Maternal deprivation (#PC0981) Socio-economic poverty (#PB0388)
Neglected young children (#PE4245) Human physical genetic abnormalities (#PD1618)
Foetal malformation in diabetic pregnancies (#PE4808)
Smoking during pregnancy and breast-feeding (#PE5026).

♦ **PD1886 Political hostage-taking**
Political kidnapping — State-sanctioned hostage-taking
Nature The taking of a person or group of persons as hostage and threatening to kill or mutilate them if certain demands (such as the release of prisoners, the granting of political asylum, the publicity for the group or the donation of money) are not met, is often used by dissident minorities against governments. Of the techniques used by political terrorists and political dissenters, perhaps the most insidious is skyjacking. The randomness of political skyjackings, as well as the violence, makes them particularly abhorrent.
Incidence Political kidnappings have been widely used in recent years in many different countries, notably in Spain, Italy, South Africa, Central America and the Middle East.
Broader Kidnapping (#PD8744).
Related Hostage taking (#PE4108) Political burglary (#PD1943)
Non-repatriation of prisoners of war (#PE0948).
Aggravates Meaninglessness (#PA6977).
Aggravated by Extremism (#PB3415) Political repression (#PC1919).
Reduces Political prisoners (#PC0562).

♦ **PD1891 Prohibitive cost of goods and services**
Over-priced goods and services — Inflated prices of goods and services
Nature Because of high material and labour costs in countries producing goods, the base costs within price schemes are also high. This results in limited market access, since a number of countries do not have adequate buying power. Future development is geared towards the production of the goods which countries can afford to buy; and some parts of the world either do not get the goods that they need or else pay more than their buying power justifies.
Refs Philips, L *Predatory Pricing* (1987).
Broader Limited market development (#PF1086).
Narrower Inflated art values (#PF7870)
Prohibitively expensive housing for the poor (#PE8698).
Aggravates Prohibitive cost of living (#PF1238) Insufficient availability of goods (#PB8891).

♦ **PD1893 Deception by government**
Deliberate lying by government officials — Misrepresentation of facts by national leaders — Deliberate distortion of official news and information — Official over-reporting and under-reporting — Diplomatic lying — Perjury by government agents — Susceptibility of electorate to government repetition of untruths — Intellectual dishonesty in government
Nature The use or alteration of intelligence reports, the creation of misinformation, and the use of unconfirmed information for the purposed of beginning or continuing an unpopular or potentially unpopular policy, of maintaining power in government or of insuring a public image, such as being a great statesman.
Incidence Governmental lying is a continuous phenomena. Almost every month there is an exposure. It may be in South America (Chile, Argentina), Southeast Asia (Vietnam, the Philippines), South Africa, India, or anywhere that governments are under pressure. Some instances of governmental deception have achieved great notoriety. Hitler's dictum that "a lie thrice repeated becomes a truth" was put into practice during the regime of the Third Reich. The Soviet intelligence apparatus has a bureau of disinformation, and counterparts exists for domestic purposes. The Non-aggression Pact of Stalin and Hitler was one of many misrepresentations of the two leaders. In the USA there have been Watergate (concerning President Nixon's falsehoods), and earlier misrepresentations concerning the conflict in Vietnam. In the UK, the Belgrano affair and the 1985 national coal strike have occasioned allegations of misrepresentation or lying.
Refs Deacon, Richard *The Truth Twisters* (1988); Handel, Michael *Military Deception in Peace and War*.
Broader Domination (#PA0839) Institutional lying (#PD2686)
Unethical practices of government (#PD0814).
Narrower Official self-deception (#PF7702) Unreported government spending (#PF2990)
Concealed government subsidies (#PD4532).
Aggravates Disinformation (#PB7606) Political deception (#PF9583)
Terminological deception (#PF5383) Official fabrication of evidence (#PD8716)
Mediocrity of government leaders (#PF3962) Blackmail by government officials (#PD9842)
Loss of confidence in government leaders (#PF1097).
Aggravated by Government treachery (#PF4153)
Authoritarian regimes (#PC9585)
Governmental incompetence (#PF3953)
Misconduct in public office (#PD8227)
Avoidance of negative feedback (#PF5311).

♦ **PD1897 Militarization**
Narrower Militarization of children (#PE5986) Militarization in developing countries (#PD9495)
Militarization of the deep ocean and sea-bed (#PD1241).
Related Socio-economic burden of militarization (#PF1447).
Aggravated by Militarism (#PC2169).

♦ **PD1902 Excessive government participation in the economies of developing countries**
Nature In virtually all the developing countries, the acceleration of industrial growth has brought an equally accelerated increase in government intervention in economic activity. In sectors other than transportation, power and basic services (and in those fields of industrial activity whose very magnitude makes private participation difficult or impossible), such participation merely obstructs and increases the cost of development and generates ever greater exclusion of the private sector from productive activities in which private enterprise has proven to be more efficient.
Claim Those who aspire to replace private enterprise by government control forget that the spirit of enterprise, imagination, courage and creative genius are characteristic of the men who forged the machinery of progress in the continuous struggle that is private enterprise.
Broader Domestic market restrictions in developing countries (#PD1873)
Excessive government intervention in the private sector (#PD4800).

♦ **PD1908 Depopulation of mountainous regions**
Nature Due to the continually increasing industrialization of lowland areas; and to the relatively harder living conditions in mountainous regions and to the relatively low birth rate there, such regions are being depopulated, despite the need for maximum use of land suited to farming and lumber.
Broader Rural depopulation (#PC0056) Vulnerability of ecosystem niches (#PC5773).
Aggravates Restrictive use of available land (#PF6528)
Deforestation of mountainous regions (#PD6282).

♦ **PD1914 Violation of the rights of sexual minorities**
Discrimination against sexual minorities — Denial of the rights of sexual minorities
Nature Sexual minorities can be defined as persons who engages in homosexuality, incest, paedophilia and other forms of deviant sexual behaviour. All societies prohibit some or all of these activities either legally, morally or by social custom.
Broader Discrimination against minorities (#PC0582).
Narrower Discrimination against homosexuals (#PE1903)
Violation of rights of transsexuals (#PE8548).
Aggravated by Sexual deviation (#PD2198).

♦ **PD1915 Forced marriage**
Denial of right of choice in marriage — Marriage without consent — Levirate marriages
Nature Customs of forced marriage include: the paying of a bride-price; child marriage; concubinage; and abduction. Also included are arranged marriages where neither the man or the woman, or where just the woman, has no choice in the matter. In forced marriage it is usually the woman who is denied the freedom of choice.
Incidence Bride-price, child marriage and the inheritance of widows occur in traditional tribal society, mainly in Africa. Concubinage exists in Africa, but also in Asia and Arabia. Abduction, enticement and blackmail play an important part in the acquisition of concubines, apart from their being bought and sold. Abduction, the blackmail of parents, and forced marriage is a traditional pattern in Sicily. Arranged marriages still occur in Asia, where parental domination is strong. The French press carries regular reports of young girls and women taken out of school to be sent to their home countries as involuntary brides, particularly from North African communities in France.
Levirate marriages are the custom of compulsory marriage with a childless brother's widow, particular to the Hebrew culture and some others.
Claim An involuntary or forced marriage which lacks the free consent of both parties, particularly in the case of a young child, is a violation of essential human rights and a form of slavery with regard to the person who is forced to participate in the marriage.
Broader Traffic in persons (#PC4442)
Denial to people of control over their own lives (#PC2381).

Narrower Concubinage (#PF2554) Arranged marriage (#PF3284).
Related Slavery (#PC0146) Prostitution (#PD0693) Illegal marriage (#PE7935).
Aggravates Child–marriage (#PF3285) Early marriage (#PE7628)
Human dependence (#PA2159) Marriage markets (#PD7282)
Non–validity of marriage (#PF3283) Discrimination against women (#PC0308)
Disparity in social development within developing countries (#PD0266).
Aggravated by Tribalism (#PC1910) Bride–price (#PF3290)
Traditionalism (#PF2676) Chattel slavery (#PC3300)
Trafficking in women (#PC3298).

♦ PD1928 Illegal immigration

Clandestine immigration — Clandestine emigration — Unlawful entry

Nature Immigration requirements may involve restrictive quotas and permits. In most countries work permits are required – and where restrictions are in force, such permits may be unobtainable. This often leads to trafficking in and exploitation of immigrants. Immigration restrictions and quotas may be imposed for racial or colour reasons, often under the guise of employment policy. Despite this, the demand for immigrant workers by firms unwilling to apply legally for work permits may be quite high. They may apply for such a permit once the immigrant is working for them (having entered on a temporary tourist visa) or they may prefer to keep him on an illegal basis so that they can pay less in wages. The families of immigrant workers may be refused entry by law but may also be smuggled in, perhaps at a price. Illegal trafficking takes place at great cost to the immigrant and he may also be a prey to extortion afterwards, or imprisonment and deportation if he is discovered by the authorities.

Counter–claim Illegal immigration is an overblown scare designed to create public fear. In the USA ill–informed reports indicated that in 1976 there were from 4 to 12 million illegal aliens, whereas research subsequently showed that at its peak there were only 3 million.
Broader Restrictions on immigration (#PC0970).
Narrower Undocumented migrants (#PE6951).
Related Lack of individual rights to political asylum (#PF1075)
Denial of economic and social rights to refugees (#PE6375)
Crimes related to immigration, naturalization and passports (#PE3889).
Aggravates Class conflict (#PC1573) Racial conflict (#PC3684)
Criminal coercion (#PD4469) Resettlement stress (#PD7776)
Family disorganization (#PC2151) Inequality of opportunity (#PC3435)
Expulsion of immigrants and aliens (#PC3207) Abusive traffic in immigrant workers (#PD2722)
Unequal distribution of social services (#PC3437)
Inadequate cultural integration of immigrants (#PC1532)
Restrictions on freedom of movement between countries (#PC0935)
Inadequate living and working conditions of immigrant labourers in industrialized countries (#PD3427).
Aggravated by Unemployment (#PB0750) Racial discrimination (#PC0006)
Socio–economic poverty (#PB0388) Illegal movement across frontiers (#PC2367)
Discrimination against foreigners in employment (#PD3529)
Refusal to issue travel documents, passports, visas (#PE0325)
Demand for unskilled labour in industrialized countries (#PE8092).

♦ PD1941 Limited spheres of relationship

Nature As social complexity increases, individuals tend to limit drastically the number of people with whom they develop a relationship. For some it is only their family that is important; others may relate only to those speaking the same language, living in the same neighbourhood or working in the same concern. Relationships with humanity as a whole thus disappear.
Broader Social isolation (#PC1707).
Related Social integration handicap (#PE6779).
Aggravated by Media reinforcement of materialism (#PF1673).

♦ PD1942 Lack of essential freedom due to unawareness of actuality

Claim Both formal and informal educational systems leave a basic unawareness concerning the gift of human actuality, leading to a breakdown of the individual's basic freedom. This in turn limits his awareness of the possibilities inherent in a realistic understanding of his actual situation. It prevents a rational delineation of escape patterns. This can be seen in children who have not been taught to reappropriate their own failures, and to realistically relate those failures to the world rather than escaping from them.
Broader Tensionless image of free choice (#PF1675).

♦ PD1943 Political burglary

Nature Theft for political reasons, national or international, may be for the purposes of maintaining secrecy or of exposing misdemeanours of an opponent.
Broader Burglary (#PD2561) Corruption in politics (#PC0116)
Espionage in domestic politics (#PD1787).
Related Political hostage–taking (#PD1886).
Aggravated by Power politics (#PB3202).

♦ PD1947 Premature birth

Premature still–birth

Nature Premature birth is the result of the precipitous ending of pregnancy. The causes are various, some dependent on the mother (previous abortions, muscular insufficiency of the upper uterine cervix, inflammatory processes of the sexual organs), and on with the foetus (improper foetal position, anomalies of placental attachment). The smaller a baby is at birth, the less are its chances of survival (especially if the necessary medical attention is unavailable), the more specialized are its nutritional requirements (need for more calories, poorer fat absorption, more prone to iron–deficiency anaemia), and the more it is prone to mental deficiency.

Incidence Approximately 2.5 to 6 per cent of all births are premature.

Refs Escalona, Sibylle *Critical Issues in the Early Development of Premature Infants* (1988).
Broader Complications of childbirth (#PC9042).
Related Foetal malformation in diabetic pregnancies (#PE4808).
Aggravates Rickets (#PG2295) Still–birth (#PD4029) Mental deficiency (#PC1587)
Neonatal mortality (#PD9750) Sudden unexpected infant death (#PD1885)
Perinatal morbidity and mortality (#PD2387).
Aggravated by Inadequate medical care for pregnant women (#PE4820).

♦ PD1949 Radiation accidents

Risk of radiation accidents

Nature Radiation accidents are occasions on which irradiation or radioactive contamination occurs in accidental circumstances; and cases of irradiation or contamination which are likely to lead to overexposure are usually considered to be accidental. 'Overexposure' is to be understood as meaning any radioactive irradiation or contamination such that the maximum permissible levels laid down in the relevant regulations are likely to be exceeded. The consequences of such events depend on the circumstances and extent of the irradiation or contamination.

Incidence In addition to the accidents at Chernobyl, Three Mile Island and a number of other nuclear facilities, radiation accidents are increasingly common from non–nuclear facilities. Radium needles are leaking in a hospital in Khartoum. In Brazil, an unwanted radiocaesium source was stolen, the thieves cut it open, there was contamination, several people died. There has been a similar incident in Mexico. There are unwanted teletherapy sources in Jordan and Zaire, old radium needles in Afghanistan and Bolivia, gauging sources abandoned at a mine in Zambia, a neutron source in broken–down equipment in Uganda.

Refs International Atomic Energy Agency Staff and NEH (OECD) Staff *Optimization of Radiation Protection* (1987); OECD Staff and NEA *The Radiological Impact of the Chernobyl Accident in OECD Countries* (1988); United Nations *Radiation Doses, Effects, Risks* (1986).
Broader Injurious accidents (#PB0731).
Related Large–scale industrial accidents (#PD2570).
Aggravates Leukaemia (#PE0639) Damage to goods (#PE4447)
Interruption risk (#PF9106) Radioactive contamination (#PC0229).

♦ PD1950 Reading disabilities

Reading disorders

Nature Some children fail to learn to read as easily, or in the same manner, as their peers. This creates a persistent educational problem and considerable heartache for the children themselves. Three major groups of children are to be distinguished: (1) Those in whom the reading retardation is due to frank brain damage manifested by gross neurologic deficits. In these cases there are clearly demonstrable major aphasic difficulties, and they are similar to adult dyslexic syndromes. (2) Those with no history or gross clinical findings to suggest neurological disease, but in whom the reading retardation is viewed as primary. The defect appears to be in the basic capacity to integrate written material and to associate concepts with symbols. (3) Those cases demonstrating reading retardation on standard tests but in whom there appears to be no defect in basic reading learning capacity. These children have a normal potential for learning to read but this has not been utilized because of exogenous factors, common among which are anxiety, negativism, emotional blocking and limited schooling opportunities.

Failure to distinguish between these major types of reading disability leads to confusion and inadequate diagnosis; children of the third group may be (mistakenly) considered unintelligent, careless or even mentally retarded.

Incidence Estimates of the percentage of children handicapped in this way differ from country to country but range from 5 to 20 per cent: boys outnumber girls 3 to 1. Learning disabilities have been described as the most pervasive educational problem in the USA.

Refs Bakker, Dirk J; Satz, Paul and De Wit, Jan *Specific Reading Disability* (1970); Bartoli, Jill and Botel, Morton *Reading – Learning Disability* (1988); Doehring, Donald et al *Reading Disabilities* (1981); Leong, C K *Children With Specific Reading Disabilities* (1987); Lynn, Robert E *Chicorel Abstracts to Reading and Learning Disabilities* (1984); Malatesha, R N and Aaron, P G *Reading Disorders* (1982); Ross, R R *Reading Disability and Crime* (1977).
Broader Learning disorders (#PD3865) Mental disorders of the aged (#PD0919).
Narrower Dyslexia (#PE3866).
Related Innumeracy (#PC0143) Developmental expressive writing disorder (#PE0330)
Developmental receptive language disorder (#PE9300).
Aggravates Illiteracy (#PC0210) Cultural deprivation (#PC1351).
Aggravated by Mental deficiency (#PC1587) Sensory disorders (#PG4201)
Student absenteeism (#PE4200) Educational wastage (#PC1716)
Stress in human beings (#PC1648) Genetic defects and diseases (#PD2389)
Ineffective methods of practical education (#PF2721).

♦ PD1955 Deterioration of the physical condition of art objects

Nature Art objects: sculpture, architecture and paintings deteriorate naturally with the passage of time. This deterioration is considerably aggravated by air pollution, particularly in modern cities. Inappropriate air–conditioning and heating, prolonged dampness, mould, insects, rough handling and transportation (to exhibitions), together with the effects of natural disasters (such as fire or flood) and bombardment in time of war, all lead to a diminution in the quality of extant works of art.
Broader Fatigue in materials (#PD1391).
Related Debasement of works of art (#PE0558).
Aggravated by War (#PB0593) Fires (#PD8054) Floods (#PD0452)
Air pollution (#PC0119) Environmental pollution (#PB1166)
Damage to cultural artefacts by environmental pollution (#PD2478).

♦ PD1958 Inadequate access to negotiation on employment and reward

Claim People are a medium of exchange, investing their brain, muscle power or other talents. All people have the right to exchange their full abilities and potential in return for just compensation for their work. However, everyone does not have access to negotiations on the nature of his or her investment and the rate of return. Few people have any guarantee of a full and just compensation for their work.
Broader Self–interest driven investment (#PC2576).
Narrower Non–comprehensive wage scales (#PD1133)
Outdated labour negotiation procedures (#PF1786)
Artificial and arbitrary job qualifications (#PF2066)
Limited means of marketing employable skills (#PE7344).
Related Limited market development (#PF1086) Inadequate credit policies (#PF0245)
Variations in national forms of currency (#PF2574).
Aggravates Limited ways of matching talent and jobs (#PF2792).

♦ PD1964 Overthrow of government

Coups d'état — Incitement to rebellion — Rebellion — Advocating overthrow of government

Nature The take–over or alteration of government by force, or the show or threat of force, when carried out by a relatively small number of members of the state apparatus (usually backed by some key military unit) distinguishes the world's frequent coups d'état from broad–based popular or guerrilla managed revolutions. Very often the coup is a military one engineered by a group of officers.

Rebellion against government is defined as a crime in many criminal codes.

Incidence If unsuccessful coups associated with a nation's military officers' positioning themselves during major political disorders are included, then the post World War II count exceeds 100. Over 50 were in Latin America (which also had 50 in the first half of the century); over 30 in Asia; over 20 in Africa; and 5 in Europe. Between two–thirds and three–quarters of coups d'état are successful.

Claim Coups d'état, even by the military, are ideologically nourished by the civilian culture in which they take place. They always reflect society's loss of a sense of lawfulness, and not only the loss of a sense of lawfulness on the part of those doing the coup.

Refs Wiking, Staffan *Military Coups in Sub–Saharan Africa* (1983).
Broader Seizure of power (#PC8270) Violent revolution (#PC3229)
Illegitimate political regimes (#PC1461).
Narrower Assassination (#PD1971).
Related Mutiny (#PD2589) Treason (#PD2615) Neo–colonialism (#PC1876)
Civil disorders (#PC2551) Violent political revolution (#PD3230).
Aggravates Militarism (#PC2169) Unstable regimes (#PG3948)
Export credit risks (#PF3065).
Aggravated by Injustice (#PA6486)
Over–development of bureaucracy in ex–colonial countries (#PD2511)
Government action against regimes with alternative policies (#PF2199).

♦ PD1965 Contempt for agricultural labour in developing countries

Claim In some countries fledgling urban industrialization or natural modernization make rural living appear unattractive. Traditional society breaks down and there is little to hold large numbers of people to the agricultural life. Government perception that campaigns are needed to return people to the farms as well as to increase the productivity and rewards there is lacking. This may be due in part to political leaders who themselves disdain labour and do not even have the capacity to pose for a photograph with a pick or a sickle in their hands.

Broader Contempt (#PF7697) Underproductivity (#PF1107)
Economic and social underdevelopment (#PB0539).
Aggravates Unemployment (#PB0750) Industrial accidents (#PC0646)
Neglect of agricultural and rural life in developing countries (#PF7047)
Deterioration of domestic food production in developing countries (#PD5092).

♦ PD1967 Psychosomatic disorders

Nature A psychosomatic illness is one in which a bodily ailment is, at least in part, attributable to emotional factors. The role of anxiety and its emotional derivatives is a key concept of psychosomatic medicine and clinical studies have shown the temporal relation between the course of a patient's disease and the emotional vicissitudes of his life. In psychosomatic illnesses, the anxiety associated with a stressful situation is generally unconscious, but the physical repercussions of the emotion occur just the same.

Background The bodily reactions that accompany emotion serve a useful purpose when the emotion is temporary and appropriate. Fear or anger, for example, in the face of external danger or threat arouses a defensive response. Once appropriate action has been taken, the physiological emotional changes that have enabled the individual to reach into his reserves of strength usually subside. However, when the emotions are not discharged, due to being unconscious and part of a basic personality conflict, or because the person is involved in a repetitive and inescapable situation, the physical effects persist. Tissue change in an organ may be the end result. Thus, psychosomatic symptoms are the instruments that a patient has unconsciously adopted to handle some difficult situation. Although he consciously desires relief from his physical suffering, he will unconsciously resist relinquishing these safety mechanisms until some solution for the underlying problem seems imminent.

Refs Hill, O *Psychosomatic Medicine* (1970).

Broader Mental illness (#PC0300) Human disease and disability (#PB1044).
Aggravates Health inequalities (#PC4844).
Aggravated by Anxiety (#PA1635)
Lack of understanding of spiritual healing (#PF0761)
Somatic and psychosomatic effects of torture (#PE5294).

♦ PD1970 Inadequate firearm regulation

Lack of gun laws

Nature Offences of violence involving firearms are rising faster than violent crime rates themselves. The principal concern is with concealable hand-guns which may be carried illegally on the person, and with the uncontrolled purchase of such weapons, frequently on a one-time basis per individual. However, the same inadequate controls allow firearms to be sold in quantities, and in some countries, permit sales to unidentified persons of automatic weapons, explosives and other arms.

Incidence A typical year in the USA will see 52 percent of murders committeed with a handgun, with rifles or shotguns accounting for another 13 percent. Only 35 percent of homicides are committed without firearms. One in ten deaths of children aged 1 to 20 in the US are due to firearms. There are 25 million pistols and revolvers in the US. About 2 million new weapons of this type are placed in the US market each year.

Counter-claim Gun control is the thin-edge of the wedge to allow greater government interference and regulation of individual, constitution-granted personal liberties. Every male of fighting age in Switzerland has at least one firearm in his house and there is little abuse. It is not the weapons that need regulation, but people's behaviour and the economic inequities in the world system that give rise to crime.

Broader Inadequate laws (#PC6848).
Aggravates Crime (#PB0001) Homicide (#PD2341) Banditry (#PD2609)
Human death (#PA0072) Assassination (#PD1971) Gunshot wounds (#PE9111)
Youth violence (#PF7498) Hunting of animals (#PC2024) Psychotic violence (#PE7645)
Proliferation of weapons in civilian hands (#PE2449).
Aggravated by Fear (#PA6030) Dishonour (#PF8485)
Human violence (#PA0429) Police corruption (#PD2918)
Pursuit of personal prestige (#PF8145)
Discrepancies in human life evaluation (#PF1191).
Reduced by Excessive government control (#PF0304).

♦ PD1971 Assassination

Nature Assassination can be defined as murder for a political end, often by an non-interested agent. It has been justified, sometimes even urged, as a revolutionary means, especially when no other way of overthrowing the establishment seems viable. As assassinations are stealthily carried out, the intended victim can never fully protect himself.

Incidence The 'Golden Age' of political assassinations began in 1865 with the assassination of Abraham Lincoln and continued with an assassination virtually every year until 1901 when another US President, William McKinley, was killed. Recent assassinations include Anwar Sadat, president of Egypt, in 1981; Indian Prime Minister Indira Gandhi in 1984; and the attempts on Pope John Paul II and President Ronald Reagan (USA), both in 1981; and Prime Minister Margaret Thatcher (UK), 1984.

Background The original "assassins" were members of a fanatical oriental sect, which held the religious belief that the imprisoned soul was released from the body after death by executing orders given by the Universal Soul. This belief allowed devotees to perform any deed without question or fear. The assassinations for which the sect was famous were originally committed against its persecutors, but developed into being committed for anyone willing to pay for such services. Assassins come to be trained for assassination: they were taught foreign languages and religious customs; the adoption and maintenance of disguises; and how to win the confidence of their intended victims. The name of the sect is derived from the Arabic 'Hashshashin' – hash eaters – because it was suspected that they intoxicated themselves prior to carrying out their missions.

Refs Rapoport, David C *Assassination and Terrorism*; Wilcox, Laird *Terrorism, Assassination, Espionage and Propaganda* (1988).

Broader Homicide (#PD2341) Overthrow of government (#PD1964).
Narrower Political assassination (#PE5614) Government sanctioned killing (#PD7221)
Assassination of trade union leaders (#PE0252) Corporation-sanctioned assassination (#PE6356)
Assassination of environmental activists (#PE5005).
Related Terrorism (#PD5574) Vigilantism (#PD0527).
Aggravates Fear (#PA6030) Unstable government (#PG4223).
Aggravated by Conflict (#PA0298) Political unrest (#PD8168)
Inadequate firearm regulation (#PD1970).
Reduces Political dictatorship (#PC0845) Ineffective war crime prosecution (#PD1464).
Reduced by Excessive government control (#PF0304).

♦ PD1975 Inadequacy and insensitivity of intelligence testing

Abuse of IQ evaluations — Lack of definition of intelligence — Inadequate means of measuring intelligence

Nature The theory that there is one, over-all general kind of intelligence is not scientifically demonstrable; the profusion of intelligence testing and the assignment to an individual for most of his or her student life of an arbitrary intelligence quotient rating may be an infringement of civil liberties. Such ratings may be used to unfairly screen out students for educational programmes, or applicants for job or promotional opportunities. The tests themselves are often biased towards privileged backgrounds and the dominant ethnic culture. Thus gifted but disadvantaged individuals' scores will not reflect their capabilities.

Claim Aptitude, intelligence and general ability testing are inseparable phenomena illustrating the belief that human nature and human potential are fully understood by scientists; that they can be classified in 7, 13, 22, or 66 categories, and individuals pigeon-holed according to their predominant top scores. These premises underlie a large amount of educational thinking that seeks an assembly-line efficiency by putting students on the supposed right track in accordance with their abilities at the earliest age. Such educational perspectives accord very well with state politics, centrally-controlled social behaviour, and 'scientific' human breeding programmes. Intelligence testing has been turned into a tool for role modification if not for behaviour modification, and is one of the growing techniques for the factory-farming of human beings.

Counter-claim Identical twins separated at or soon after birth have IQs which correlate higher than either the IQs of ordinary siblings reared apart or the IQs of fraternal twins reared together. This suggests that, according to some studies, genetic factors account for nearly 60 percent of the variability in IQs. Most experts in the field, while agreeing that there is no adequate understanding of the nature of intelligence, believe IQ tests measure something that can be reasonably called general intelligence. IQ tests also predict reasonably well the academic and job performance of both blacks and whites suggesting that they are not culturally biased.

Broader Inhumane scientific activity (#PC1449).
Aggravates Elitism (#PA1387) Eugenics (#PC2153) Segregation (#PC0031)
Discrimination (#PA0833) Racial inequality (#PF1199)
Inequality in education (#PC3434) Lack of individual development (#PG3595)
Denial of education to minorities (#PC3459).
Aggravated by Racism (#PB1047) Prejudice (#PA2173)
Ignorance (#PA5568) Social injustice (#PC0797)
Acceptance of hierarchy (#PJ3602) Invasion of privacy through testing (#PJ6946).

♦ PD1977 Denial of right of equal pay for equal work

Broader Low general income (#PD8568) Denial of economic rights (#PD4150).
Narrower Disparity in remuneration between public and private sector employees (#PE7760).
Related Unequal pay for women (#PD0309).
Aggravates Economic discrimination (#PC2157).

♦ PD1983 Psychological pollution by mass media

Nature The development of mass media and public information techniques has increased the amount of information directed at the individual to the point at which he can no longer continue to absorb it. The information which is not absorbed is not harmless because it helps to neutralize the thinking of the individual, leading to indifference through familiarity. The overabundance and sensationalization of information disturbs the psyche, hinders the individual in distinguishing between real and unreal, true and false, and useful and useless. The critical faculty is eroded and the culture is distorted.

Broader Mental pollution (#PB6248)
General obstacles to problem alleviation (#PF0631).
Narrower Excessive television viewing (#PD1533).
Aggravates Inadequacy of psychiatry (#PE9172).
Aggravated by Exploitative films (#PE6328) Distorted media presentations (#PD6081)
Excessive portrayal of negative information by the media (#PE1478).

♦ PD1987 Excessive protection of industries by customs duties

Nature The inordinate protection given industry by customs duties in a given country acts as a deterrent to, and decisively affects, the domestic market for industrialization. Whilst such protection was undoubtedly considered essential to the development of industry in the early stages of industrialization, its persistence as a policy has created such a barrier of tariffs and restrictions that industries become isolated from world markets and thereby are deprived of the advantages offered by healthy competition from abroad. Because of the exclusive concern with the domestic market, together with the limited size of such a market, restrictive and monopolistic practices develop which weaken incentives to technical progress and the corresponding increase in productivity.

Background Some developing countries have been able to use moderate protection to expand manufacturing output steadily and efficiently, but many countries have wasted both capital and labour resources in the course of industrial development. Experience indicates that although protection was introduced in these countries to assist infant industries, in the belief that the dynamic benefits would exceed the short term costs, escalating levels of protection for manufacturing led to the neglect of, and biases against, agriculture and other non-manufacturing activities, and permitted a high degree of inefficiency with concomitantly low social returns on the human and capital resources invested in manufacturing itself. Excessive protection for manufacturing also tended to handicap overall economic growth by raising the costs of inputs into agriculture and other primary production and service industries instead of making such inputs widely and cheaply available. It has usually accentuated regional imbalances by the undue attraction of resources into large cities. It has led to the restriction of potential domestic markets, undue limitations on the scale of production, and further cost burdens to the economy. Where excessive protection led to balance of payments problems, it became a brake on development rather than an instrument for its acceleration. Balance of payments problems have not been solved with the progress of industrialization as the proponents of infant industry protection expected. Indeed, where protection has been excessive these problems have been exacerbated.

In the past, import restrictions did not cause serious social hardships when most manufactured goods were imported, because imports of such luxury goods as automobiles could be restricted in times of balance of payments stringency, and liberalized when export income permitted. With industrial progress, however, balance of payments difficulties have necessitated the restriction of imports of industrial inputs rather than of final goods and this has led to unemployment. 'Stop and go' policies have thus become very costly. Excessively protected infant industries often failed to grow up, though some have become competitive exporters. Where foreign investors were attracted by high levels of protection that enabled them to make monopoly profits, high volumes of remittances added to balance of payments problems.

Broader Domestic market restrictions in developing countries (#PD1873).

♦ PD1988 Urban guerrillas

Broader Guerrilla warfare (#PC1738).

♦ PD1990 Desiccation of lakes

Desiccation of inland seas

Aggravated by Desiccation of rivers (#PJ4867)
Mismanagement of irrigation schemes (#PE8233).

♦ PD1994 Nationalization of domestic enterprises
Nature Nationalization is the process by which government alters or terminates control or ownership of an private enterprise and transfers the assets or share capital to the state or to a state-controlled body. Nationalization, although similar to expropriation, differs from it in motive and degree. It is usually associated with the implementation of communist or socialist theories of government, and tends to be applied to major domestic industries whose control and policies influence social and political questions beyond the economic sphere. Compensation, if any, received by the original owners of the enterprise may be inadequate and unfair.
Broader Collectivism (#PF2553).
Aggravates State monopoly (#PJ4242)
Inadequate development of enterprises in developing countries (#PE8572).
Aggravated by Socialism (#PC0115) Communism (#PC0369).

♦ PD1998 Family poverty in industrialized countries
Nature Families may be poor for a large variety of reasons, ranging from a lack of education on the part of the breadwinner and handicaps of various kinds (physical or mental), to discrimination on the grounds of race, religion, colour, sex, age, political ideology, or language. A low family living standard may be inadequately compensated by national social security benefits, or the structure of these benefits may be such that it creates a disincentive to work, therefore bringing recriminations on the family and particularly the breadwinner, from better-off sections of society which contribute higher taxes.
Since methods of social security and assistance to underprivileged families, where they exist, are rarely fully adequate and are usually fraught with bureaucratic stipulations, it is difficult to eradicate family poverty. Family poverty is especially likely to occur among minority groups which are shunned or discriminated against by the community at large.
Incidence In the USA, an estimated ten percent of all families are poor. One-third of all US children live in families with incomes below the 'minimum but adequate' level. More than 40 percent of black children under 6 years of age are poor; and in the southern USA, more than 80 percent of black children whose the mother is the only bread-winner are poor.
In most industrialized countries the poorest fifth of the population still receives less than seven percent of the national income. And the distribution has become more unequal over time, not less.
Broader Family poverty (#PC0999) Poverty in developed countries (#PC0444).
Related Poverty in developing countries (#PC0149).
Aggravates Homelessness (#PB2150) Family breakdown (#PC2102)
Juvenile delinquency (#PC0212) Desertion in marriage law (#PF3254)
Urban slums in industrialized countries (#PE1887).
Aggravated by Discrimination (#PA0833).

♦ PD2001 Single party democracies
Nature Political rule by one party involves the refusal of the right to opposition; the banning both of existing parties and of the formation of others; imprisonment of leaders and followers of other parties; and general repression. The situation manifests as political instability and may cause social and political conflict or general apathy.
Incidence In 1990, out of 45 black African nations, only 4 (Botswana, the Gambia, Mauritius and Senegal) allow their people to vote, choose their leaders and express themselves freely. In sub-Saharan Africa, only 5 can be considered multi-party democracies. Military dictatorships prevail in 23 countries, where no political parties are permitted, and the remainder are one-party states ruled by dictators-for-life.
Claim Single party rule is justified as a means whereby certain elites can retain power accrued through corruption and patronage.
Counter-claim Single party rule provides unity where otherwise tribal and ethnic differences would prevail and undermine the stability of the country.
Broader Political conflict (#PC0368)
Bias in selection of political candidates (#PD2931).
Related Fascism (#PF0248) Elitism in communist systems (#PC3170).
Aggravates Political apathy (#PC1917) Political repression (#PC1919)
Political monoculture (#PF4405) Elitist ruling classes (#PF4849)
Reallocation of aid funds to alternative priorities (#PF0648).
Aggravated by Undemocratic political organization (#PC1015).

♦ PD2008 Particulate atmospheric pollution
Airborne particles
Nature Suspended particulate matter is one of the most prevalent atmospheric pollutants. Natural sources of particulates, including windblown dusts and sea-spray, pollen, forest fires and volcanoes are estimated to exceed man-made emissions possibly by as much as twenty times. Man-made particulate emissions are emitted directly from fossil fuel combustion, industrial and agricultural sources or are a result of gas-particle conversions.
Refs Pickett, E E (Ed) *Atmospheric Pollution* (1987).
Broader Chemical air pollutants (#PD1271)
Chemical pollutants of the environment (#PD1670).
Narrower Dust (#PD1245) Photochemical oxidant formation (#PD3663).
Related Acidic precipitation (#PD4904).
Aggravates Deterioration in atmospheric visibility (#PE2593).
Aggravated by Motor vehicle emissions (#PD0414).

♦ PD2011 Unsanitary environment for basic health in small rural villages
Debilitating conditions of health in rural communities
Nature In some small villages, although everyone desires to be healthy, there are certain traditional modes of behaviour that stand in the way. For example, animal waste is often collected in close proximity to humans both in the house and on the village streets and pathways. Raw sewage is deposited near the village, and constitutes a safety, as well as a health, menace to children and animals. Rubbish collection is too infrequent and there are no collection bins, with the result that some villagers simply throw their refuse over the side of a road into a public dump, thus perpetuating another threat to public health.
Residents of most rural villages have little access to the benefits of modern health care. Their vitality may be sapped by chronic infestations of common parasites and worms due to inadequate and contaminated water supplies. Overcrowded living accommodation, in buildings with dirt floors and thatched roofs, promotes the harbouring of dangerous insects. Diets lack adequate vegetables and proteins, either because fruit and vegetables are not grown or because they are not used to their full potential. Many villages have no local milk production or cold storage capabilities. Academic achievements of students may be hampered by chronic infections which are often due to low resistance and nutritional deficiencies, and yet villagers do not understand why they are frequently ill and what makes them sick. Despite growing international awareness that preventive medical procedures are urgently needed to upgrade the physical energy levels of the people of developing communities, until these sanitation problems are solved, the basic health of the people in small villages will remain at an unnecessarily low level; and the human energy required when a particular effort is necessary may simply not be available.
Broader Unhygienic conditions (#PF8515) Human disease and disability (#PB1044)
Obstacles to community achievement (#PF7118).
Narrower In-home animal shelter (#PJ9800) Lack of cesspool fencing (#PG9786).
Aggravated by Injuries (#PB0855) Human ageing (#PB0477)
Malnutrition (#PB1498) Inadequate waste treatment (#PD6795)
Inappropriate sanitation systems (#PD0876)
Lack of sanitation in rural areas of developing countries (#PD1225).

♦ PD2019 Educationally reinforced egocentric attitudes
Nature Education currently reinforces reliance upon external criteria of success or encourages individualism by using personal satisfaction as incentive. People are educated to be programmable pieces of equipment in the production-consumption cycle or permitted to develop creativity without regard for social mission. Society believes employment means security, and the more money one makes, the higher is that person's worth or status. Schools reflect this belief by valuing grades and college acceptance tests, both viewed as indicators of potential for success.
Broader Collapsed images of vocation (#PF6098).

♦ PD2025 Endangered urban trees
Urban hazards to trees — Endangered park trees
Nature The trees growing on urban streets are subject to a number of hazards, and they often appear sickly, as is noticeable to city residents. Disasters such as the epidemic of 'Dutch elm disease' and the death of trees from the use of de-icing salt have made an awareness of 'urban forestry' particularly pertinent. The most important problems for the vitality of city trees have been listed in the following order of significance: dryness, salt pollution, and nutrient deficiency. In smaller cities with a less pronounced urban climate, mechanical damage, through motorized traffic, construction projects, and vandalism, is predominant. Other factors considered to be responsible for the deteriorating condition and stunted growth of urban trees are natural gas, fumes from industry and traffic, over-shadowing by tall buildings, and reduction in the air content of soil by excessive compression.
Broader Forest decline (#PC7896) Environmental plant diseases (#PD2224)
Natural environment degradation (#PB5250).
Related Plant-pathogenic air pollutants (#PE0155).

♦ PD2028 Denial of right to adequate medical care
Broader Inadequate standards of living (#PF0344).
Narrower Refusal of medical care (#PF4244)
Inadequate community care for handicapped persons (#PE8924).
Related Denial of the right of health (#PJ2269) Denial of right to sufficient food (#PE0324)
Denial of right to social security (#PD7251) Inadequate social welfare services (#PC0834)
Denial of right to economic security (#PD0808) Denial of right to sufficient shelter (#PD5254)
Denial of right to sufficient clothing (#PE7616) Denial of right to benefits for invalids (#PE5211)
Denial of right to benefits to survivors (#PE4531)
Denial of right to economic security during periods of unemployment (#PE5406).
Aggravates Unhealthy emotional responses to atomic energy (#PF0913).
Aggravated by Inadequate medical care (#PF4832).

♦ PD2029 Distortion of international trade as a result of government participation
Broader Non-tariff barriers to international trade (#PC2725).
Narrower Distortion of international trade by selective domestic subsidies (#PD0678)
Distortion of international trade by export subsidies and countervailing duties (#PE1961)
Distortion of international trade by state-trading and government monopoly practices (#PE8267)
Distortion of international trade by discriminatory customs and administrative entry procedures (#PE2603)
Distortion of international trade by discriminatory government and private procurement policies (#PE0347).

♦ PD2034 Pneumoconiosis
Nature Pneumoconiosis is caused by the accumulation of dust in the lungs and the tissue reaction to its presence.
Incidence The appearance of pulmonary disability from pneumoconiosis is related to the amount of dust that has been inhaled. This amount will vary with the fibrogenicity of the dust. A dust which has a high fibrogenic potential is capable of incapacitating a larger amount of lung tissue following a shorter exposure than a dust having a low fibrogenic potential. In general, a very small percentage of workers exposed to a dust with a low fibrogenic potential (such a soft-coal miners) become incapacitated solely because they have developed simple pneumoconiosis. However, breathlessness and incapacitation may develop when simple pneumoconiosis becomes converted into the complicated variety and an excessive amount of functioning lung tissue is destroyed by progressive massive fibrosis. Of the pneumoconioses caused by fibrogenic dusts, silicosis and asbestosis are the most important. Silicosis is characterized by multifocal nodular fibrosis whereas asbestosis is typically a non-uniform diffuse pulmonary fibrosis that tends to be more pronounced in the basilar portions of the lungs. Silica and asbestos workers as well as coalminers and workers in non-dusty trades may develop breathlessness from chronic bronchitis or emphysema or both. The cause of this breathlessness is most often ascribable to the destruction of lung tissue or inflammation of the air passages (or both) by cigarette smoke. Cigarette smoke is also the main factor in the production of lung cancer in asbestos workers.
Refs ILO *Pneumoconiosis Conference* (1971).
Broader Occupational diseases (#PD0215) Lung disorders and diseases (#PD0637)
Diseases of the respiratory system (#PD7924).
Narrower Silicosis (#PE1314) Barytosis (#PG4272) Siderosis (#PE4274)
Byssinosis (#PE2319) Berylliosis (#PG2678) Anthracosis (#PG2676)
Farmer's lung (#PE6899).
Related Pleura and peritoneum cancers of the bronchi (#PE8228).
Aggravates Dyspnoea (#PG4278) Pneumonia (#PE2293)
Meningitis (#PE2280) Tuberculosis (#PE0566)
Chronic bronchitis (#PE2248).
Aggravated by Dust (#PD1245) Smoking (#PD0713)
Health hazards of asbestos (#PE3001) Occupational risk to health (#PC0865).

♦ PD2036 Inadequate training in decision-making
Lack of decision-making experience
Nature Today's society has inadequate training structures for making decisions. When faced with a great proliferation and variety of possibilities, intangible as well as tangible, many people do not have the skills to make an intelligent choice. The emerging generation is faced with "overchoice", where the advantages of diversity are cancelled by the complexity of choice. Being confronted with the need to make increasingly rapid decisions, these are frequently made on trivial criteria, often by default.
Broader Lack of training (#PD8388) Irrelevance of educational curricula (#PF0443).
Aggravates Indecision (#PF8808) Unmarried parents (#PD3257)
Industrial accidents (#PC0646) Dangerous occupations (#PC1640)
Disorientation of the young due to lack of social forms (#PD2050).
Aggravated by Individualistic retaining of local tradition (#PF1705).

♦ PD2041 Cultural illiteracy
Nature There are startling gaps in the knowledge of people in history, literature and other cultural subjects because of an over emphasis on teaching students how to think without giving them any content. Curriculum emphasizes skills over knowledge. Teachers are taught to stress methods

over subject matter. Text books are flea markets of disconnected facts. Even testing programs such as the Standard Aptitude Test avoid evaluating substantive knowledge.
Incidence A recent survey in the United States found 43 percent of 17 year old tested could not place World War I in the correct half century, 39 percent could not do the same for the writing of the U.S. Constitution and nearly a third placed the date of Columbus's first landing in the New World after 1750.
Broader Illiteracy (#PC0210).
Related Cultural alienation (#PC5088) Computer illiteracy (#PG2575) Geographical illiteracy (#PD3984).
Aggravates Cultural suicide (#PF5957) Cultural decline (#PC9083) Cultural corruption (#PC2913) Cultural stagnation (#PC8269) Cultural deprivation (#PC1351) Cultural fragmentation (#PF0536) Limited cultural context (#PF2504) Destruction of cultural heritage (#PC2114) Inadequate appreciation of culture (#PF3408).
Aggravated by Cultural imperialism (#PC3195) Narrow scope of education (#PF3552) Ignorance of cultural heritage (#PF1985) Narrow range of cultural exposure (#PF3628) Acculturation as a dilution of cultural heritage (#PD4272).

♦ PD2044 Intimidation of electors
Threats against voters — Electoral violence
Broader Unfair elections (#PC2649) Political intimidation (#PC2938).
Related Violence as a resource (#PF3994).

♦ PD2045 Radio frequency interference
Jamming radio frequencies — Interference in radio and television communications
Nature Clear reception of radio signals depends upon the ability to exclude unwanted radio frequency energy that interferes with desired transmissions. Interfering signals include those from another transmitting station using the same frequency in another part of the world, or else one whose assigned frequency is so close that the harmonic content of the unwanted signal overrides or beats with the wanted wave. A further type of interference may result from cross-modulation with a more powerful wave from another station. (This problem should be distinguished from that of radio noise or atmospherics which may also interfere with radio reception).
Such interference may also be caused by freak weather: atmospheric thunderstorms and tropospheric conditions (causing reception of normally distant and non-interfering stations).
Other sources of interference include industrial, medical scientific and radio control transmitters such as garage door openers, microwave ovens, radio controlled models, radio and television receiver oscillators, induction heating machinery, wireless intercom and diathermy and other medical apparatus. Radio frequency transmitters include power tools, automobile ignitions, appliances, fluorescent and neon lights, microcomputer, power transformers and touch control lamps.
The victims of radio frequency interference include TV sets, am and fm radio receivers, amateur radio receivers, medical equipment, motor vehicle anti-skidding devices, hi-fi audio systems, intercom systems, blasting devices, smoke detectors, video cassette recorders and automobile cruise controls.
Incidence It has been estimated that at one stage the USSR spent more resources on jamming the BBC World Service to the USSR than the BBC spent in producing it.
Claim Man-made intentional interference adds to the congestion in telecommunication frequencies. For political reasons, some countries are jamming frequencies on which foreign broadcasters are beaming messages to their homelands.
Broader Jamming of satellite communications (#PD1244) Destabilizing international telecommunications (#PD0187).
Aggravates Poor television reception (#PG9867) Restriction of freedom of expression (#PC2162).
Aggravated by Geomagnetic storms (#PD1661) Radio noise of natural origin (#PF1676) Radio noise of industrial origin (#PE2473) Limited number of available radio frequencies (#PF0734).

♦ PD2046 Mass unemployment of human resources
Nature Mass unemployment is a complex matter in transient, suburban, developing communities. Large numbers of people are tending to move to the city to seek employment. Many of these newcomers (and many of the original residents as well) are not trained in the skills and methods necessary for a rapidly expanding, technologically oriented economy. In addition, the young people find, when they finish school that they are vying with hundreds of others for every available job in the city. Likewise, the young educated women are not meaningfully employed in the over-crowded families but cannot find openings in the job market. Family care structures ensure the support of these people, often "carrying" them for years in the hope that an opportunity will arise. These circumstances have produced a large force of unemployed urban residents while an extensive amount of necessary work remains untackled in their own home villages. This mass of disengaged and unproductive people not only means that something positive that could be done is not done, but has the negative effect of a drain both economically and psychologically on the community: the support of the unemployed leads to even more reduced family resources; and the inactivity and unused energy of the unemployed, and especially the young, fosters a mood of hopelessness.
Refs Malinvaud, Edmond *Mass Unemployment* (1985).
Broader Unemployment (#PB0750) Waste of human resources (#PC8914).
Narrower Educated unemployed (#PD8550) Unemployed skilled labour (#PE1753) Lack of programmes for dropouts (#PG8047).
Related Youth unemployment (#PC2035).
Aggravates Marginal level of family income (#PD6579) Obstacles to learning due to hunger (#PE8442).
Aggravated by Prohibitive cost of education (#PF4375) Inequitable education selection (#PG8023).

♦ PD2048 Suppression of private enterprise in socialist countries
Dependence of socialist countries on private enterprise
Nature Generally speaking, almost all ruling communist parties regard private economic activity as undesirable. However, there has recently been a trend to the creation of small or private businesses in some socialist countries.
Incidence The most active country in recent years has been Hungary. In 1979, a Western observer noted that there were nearly 11,000 private traders there, assisted by about 500 family members and some 1500 registered assistants. In the same year, 91,000 private artisans and skilled workers supplied 46 percent of the services available. Shopkeepers are offered tax concessions to open stores, especially in the provinces, and since January 1982 more than 3000 private shops have been opened and also several thousand licences for private taxis issued. In addition, 13,000 so-called small enterprises have been created. The government has started to set up offices where requests for the private services of full- and part-time artisans and skilled workers are matched with offers to provide such services, and private construction teams can be found legally building private residences, especially in the countryside. Official statistics show that the private sector provides about one-third of agricultural output, and it is now thought that 8 percent of all Hungarian industrial exports are produced through some form of private enterprise.
Broader Communism (#PC0369) Denial of rights of businesses (#PD4728).
Aggravates Inadequate development of enterprises in developing countries (#PE8572).
Aggravated by Bias against private enterprise (#PF1879).

♦ PD2050 Disorientation of the young due to lack of social forms
Nature Young people are currently caught between the collapse of old social forms and the creation of new, with no guidelines on how to operate in the present. Their present impact on society is in questioning old values and demanding that social issues be responsibly dealt with.
Narrower Absence of rites of passage (#PF1674) Obsolete educational values (#PF8161) Social disaffection of the young (#PD1544) Young people's lack of context for the future (#PF2068).
Related Elder paralysis over the future (#PF3973) Limited community responsibility of adults (#PF1731) Denial by old people of the significance of the past (#PF2830).
Aggravated by Social isolation of the elderly (#PD1564) Inadequate training in decision-making (#PD2036).

♦ PD2051 Inadequate educational facilities for gifted children
Nature Children gifted with exceptional intellectual ability or other talents may not be recognized within conventional national or local governmental school systems. Even when they are identified, it is only rarely possible to provide suitable facilities in which they can develop at their own accelerated pace and undertake work commensurate with their abilities. In the absence of such facilities it is only by chance that the child can develop his talents into constructive channels.
Incidence In the USA for example, about one percent of the elementary school population may be intellectually gifted as measured by intelligence testing. The only programmes for such students exist in a small number of the very largest cities, yet the greatest proportion of the gifted students assumed to exist, live in small cities, towns and rural districts.
Counter-claim The segregation of exceptionally intelligent children into special facilities tends to create an intellectual aristocracy which is widely considered objectionable, particularly since they tend to be found amongst the privileged classes. In addition, such separation may make it difficult for these children to integrate themselves into society at a later stage in life. Specialized high schools for the performing or plastic arts, for vocational preparation, or for accelerated science programmes isolate students from their peer group, whose areas of excellence may lie in their character, their general development, their physical coordination, or simply, but importantly, in the diversity of their backgrounds. High schools need to develop special opportunities for all their exceptional students, but they should maintain a general character, in order to reflect the society which created them.
Refs Delp, Jeanne L and Martinson, Ruth A *A Handbook for Parents of Gifted and Talented (Also Helpful for Educators)*; Gearheart, Bill R, et al *The Exceptional Student in the Regular Classroom* (1988); Reynolds, Cecil R and Mann, Lester (Eds) *Encyclopedia of Special Education* (1987); Rivlin, Harry R, et al *Advantage - Disadvantaged Gifted*; Special Learning Corporation *Exceptional Children* (1984).
Broader Inadequate educational facilities (#PD0847).
Aggravates Inequality in education (#PC3434).
Aggravated by Inadequate social welfare services (#PC0834).

♦ PD2057 Commercial fraud
Nature The classic frauds remain the most common, sometimes aided and abetted by modern technology, such as high quality photocopying and printing. Tried and tested frauds involve false credit, sham securities, fake assets, fake documents to show shareholdings and property ownership, accounts fraud and bad housekeeping by the company.
Broader Fraud (#PD0486) Unethical commercial practices (#PC2563).
Aggravates Documentary fraud (#PE1110).
Aggravated by Conspiracy against the public (#PF4198).

♦ PD2067 Criminal harassment
Nature The crime of harassment is communicating by mail or phone with the intention of frightening or intimidating another person the threat of a violent crime, or making a anonymous phone call, or using offensively coarse language on the phone, or making repeated calls with or without speaking with no purpose of communicating.
Broader Harassment (#PC8558).
Narrower Obscene telephone calls (#PE5757).
Related Criminal threat (#PE4661).

♦ PD2072 Proliferation of automobiles and motor vehicles
Increase in road vehicles
Incidence There are in the world, over 550 million automobiles and other motorized vehicles (excluding motorbikes and cycles). About 455 million are registered for road use, the rest are off-the-road vehicles used by government, military, industry and agriculture, and by private persons for recreational use. Every 23 kilometers of a four-lane road uses 100 hectares of land, so that the price of a five-minute drive is the loss of about three farms and several tons of annual produce, or alternatively, a woodland of a thousand trees. In exchange for the lost oxygen and herbage, in the USA for example, cars emit over 75 million tons of pollutants. Proliferation rates for the auto have seen a growth of about 2 percent in the USA; 5 percent in Western Europe and well over 10 percent in Japan.
Claim The trend away from public transport to privately owned cars is universal. Major metropolitan areas will soon reach a point of inability to sustain traffic. Many cities already have inadequate parking. Inadequate public transport and restrictions on private automobile use will mean a loss of mobility and forced decentralization. Transportation development policies and investments in most developing countries are focused on encouraging motorization which increases the dependency on imported oil to fuel motorized transport.
Broader Uncontrolled urban development (#PC0442).
Aggravates Air pollution (#PC0119) Energy crisis (#PC6329) Injurious accidents (#PB0731) Road traffic accidents (#PD0079) Health hazards of exposure to noise (#PC0268) Road and highway traffic congestion (#PD2106).
Reduced by Maldistribution of private automobiles (#PF4480) Maldistribution of commercial vehicles (#PF4485).

♦ PD2077 Social neglect of war veterans
Nature War-time mobilization of citizens, followed by military training and active service for a number of years, causes an interruption in peoples' lives. This affects marriage, employment, education and, to a certain extent, behaviour patterns. The returning veteran may find that he has marriage or family problems, may face unemployment, or find that he is unable to study any longer or to finance further studies if his wishes. His military experience may also overqualify him for his civilian work opportunities. He may have sustained physical injuries that require ongoing medical attention. The processes of having to kill time and again and of frequently and unexpectedly having his life in danger may lead to emotional problems that require long term treatment. All of these veterans' needs, and others, are not met fully by society, and post-war periods often evidence a forgetfulness of the men and women who served their countries.

Broader Social neglect (#PB0883).
Aggravated by War (#PB0593).

♦ PD2078 Exploitation of animals for amusement

Maltreatment of animals for entertainment

Nature In addition to the exploitation of animals in cruel sports, zoos, menageries and circuses are increasing in number as more people with few qualifications enter the business of displaying exotic animals. Animals may be confined for long periods in small cages. They may have untreated wounds, exhibit neurotic symptoms, or be chained. Animals ranging from mice to dogs may be fed alive to carnivores. The animals may be improperly fed, mixed with others when they are naturally solitary, or deprived of any means of distraction (in the case of primates in particular).

Claim The use of animals for entertainment involves their forced captivity, degradation, depersonalization, stress and suffering. They are deprived of their natural habitats, isolated from their normal social environment, prevented from fulfilling their reproductive cycles, and required to adapt to human vagaries and abuses.

Broader Maltreatment of animals (#PC0066).
Narrower Cruel sports (#PD1323) Maltreatment of zoo animals (#PE4834)
Maltreatment of pet animals (#PD1265) Maltreatment of performing animals (#PE4810)
Maltreatment of animals in aquaria (#PE5461)
Exploitation of animals in spectator sports (#PE0891).
Related Exploitative entertainment (#PD0606)
Cruelty to animals in food preparation (#PE0236).

♦ PD2082 Biased and inaccurate history textbooks

Nature History textbooks tend to reflect the perspectives and stereotypes that one culture or nation has concerning another. Historical events are interpreted, either deliberately or inadvertently, to favour the country, culture or language–system for which such books are intended, often to the disadvantage of other cultures. Such bias or inaccuracy reinforces stereotypes and aggravates tensions between cultures, ethnic groups and nations.

Broader Misleading information (#PF3096)
Inadequate and inaccurate textbooks and reference books (#PD2716).
Related Biased and inaccurate geography textbooks (#PF1780).
Aggravates Prejudice (#PA2173) Ignorance (#PA5568)
International tension (#PB8287).

♦ PD2085 Inadequate child day-care facilities

Lack of child care facilities — Missing child care — Inadequate child–care facilities

Nature Child day–care facilities in most countries are not enough in quantity and many are deficient in the quality of the care given and the facilities which are provided. The lack or unsuitability of such facilities may aggravate the frustrations of a mother who wishes or needs to go out to work, and this may be reflected in her own treatment of the child or the family atmosphere in general. Children from deprived or unhappy homes may be denied the benefits of emotional and physical wellbeing that could be provided by adequate day–care. The inadequacy of day–care facilities aggravates discrimination against women in employment, as most mothers are expected to put their children before their job, and consequently firms will not give equal opportunities or equal pay to women.

Claim Great care may be given to promoting recognition of the right of women to the same treatment as men, and to ensuring that they receive the best and most appropriate training - but all will be of little real use to most women unless good child–care services are developed and made available to them.

Refs ILO *Work and Family* (1988).

Broader Inadequate child welfare (#PC0233).
Narrower Preoccupying child care (#PU0956) Limited child–care structures (#PU0850).
Related Inadequate facilities for children's play (#PD0549)
Fragmented forms of care at the neighbourhood level (#PE2274).
Aggravates Underpopulation (#PD5432) Wasted woman power (#PF3690)
Dependency of women (#PC3426) Loneliness of children (#PC0239)
Discrimination against women in employment (#PD0086)
Inadequate community care for transient urban populations (#PF1844)
Limited availability of public services in the small towns of developed countries (#PF6539).
Aggravated by Unmarried mothers (#PD0902) Single parent families (#PD2681)
Excessive employment of married women (#PD3557).

♦ PD2091 Sexual scientism

Nature Understanding of sexual behaviour is too often reduced to a body of clinical research. This tendency began as a creative response to the abundance of dehumanizing sexual images inherited from the 19th century. There is today, however, insecurity about expressing the more mysterious and subjective aspects of maleness and femaleness.

Broader Individualistic perception of sexual activity (#PF1682).

♦ PD2092 Neglect of dependents of war victims

Neglect of war widows and orphans

Nature Governments give only token support, if any at all, to elderly parents, dependent spouses, offspring or wards of deceased or severely handicapped military personnel or civilian casualties.

Broader Lack of war relief (#PF0727).
Narrower Widowhood (#PD0488) Orphan children (#PD7046).
Related Denial of right to benefits to survivors (#PE4531)
Inadequate assistance to victims of human rights violations (#PD5122).
Aggravated by War (#PB0593) Biochemical warfare (#PC1164)
Children engendered by occupying soldiers (#PD8825).

♦ PD2094 Unwanted pet animals

Abandoned animals and pets

Nature As the result of the domestication of animals, their birth rate has escalated in an uncontrolled manner without any adequate means to house and feed them. Such animals face much suffering and deprivation. Since they are unwanted, they face maltreatment by their owners.

Incidence Many unwanted animals are sent to animal shelters where, in the USA alone, some 25 million are destroyed annually. Millions are sold to laboratories where they are the subject of cruel experiments. The remainder fend for themselves as strays.

Claim Abandoned animals create health hazards and are a public menace. Wild dogs and cats breed in streets, gutters and abandoned shelters; they are often rabid and disease ridden; they may bite humans, infect humans with fleas, and steal food from gardens or marketplaces.

Broader Maltreatment of animals (#PC0066).
Aggravates Excessive stray animal populations (#PE5776).
Aggravated by Proliferation of pets (#PD2689) Maltreatment of pet animals (#PD1265)
Unethical practices with domesticated animals (#PE4771).

♦ PD2103 Marital instability

Marriage breakdown — Estrangement in marriage — Broken marriages — Broken homes

Nature Marital breakdown covers divorce, separation and desertion as well as estrangement and the continuation of living together though without marital relations. The result is severe emotional stress in both spouses and in children of the marriage. Divorce, desertion and separation may result in poverty and adjustment problems.

Incidence Marriage breakdown in the sense of estrangement and living together is not usually recorded, though it may be more extensive than the incidence of divorce and separation. Desertion, except where it is given as a reason for divorce, is also not usually recorded. The incidence of divorce is highest among industrialized and communist countries.

Refs Hussain, S Jaffer *Marriage Breakdown and Divorce Law Reform in Contemporary Society* (1983).

Broader Family breakdown (#PC2102) Family disorganization (#PC2151).
Narrower Single parent families (#PD2681) Separation under marriage law (#PF3251)
Insecurity of Western marriages (#PE4985) Emotional dependency in marriage (#PD3244)
Non–parental custody of children (#PF3253) Religious or civil refusal of divorce (#PF3248).
Aggravates Bigamy (#PF3286) Suicide (#PC0417) Adultery (#PF2314)
Prostitution (#PD0693) Cohabitation (#PF3278) Social stigma (#PD0884)
Human suffering (#PB5955) Attempted suicide (#PE4878) Maladjusted children (#PD0586)
Illegitimate children (#PC1874) Stress in human beings (#PC1648)
Increasing number of single person households (#PJ8818).
Aggravated by Divorce (#PF2100) Promiscuity (#PC0745)
Bisexuality (#PF3269) Childlessness (#PC3280)
Marital stress (#PD0518) Unrequited love (#PF6096)
Incompatibility (#PF9047) Induced abortion (#PD0158)
Social breakdown (#PB2496) Sexual unfulfilment (#PF3260)
Social disintegration (#PC3309) Natural human abortion (#PD0173)
Human sexual inadequacy (#PC1892) Desertion in marriage law (#PF3254)
Dependency of women in marriage (#PD3694) Deception between sexual partners (#PE4890)
Inadequacy of contraceptive methods (#PD0093)
Excessive employment of married women (#PD3557)
Negative effects of the nuclear family (#PF0129)
Adjustment difficulties of new urban families (#PF1503)
Psychological inconsistency of marriage partners (#PF9818).

♦ PD2106 Road and highway traffic congestion

Incidence The cost of traffic jams in Belgium is $375,000,000 and 25 million hours a year. Vehicles wear our faster, petrol is consumed faster and accidents increase because of traffic jams. In Los Angeles, motorists who battle congestion are exposed to between two and four times the levels of cancer–causing toxic chemicals found elsewhere outdoors.

Broader Traffic congestion (#PD0078).
Narrower Urban road traffic congestion (#PD0426).
Aggravates Urban fires (#PD2211) Road traffic accidents (#PD0079).
Aggravated by Proliferation of automobiles and motor vehicles (#PD2072)
Inadequate road and highway transport facilities (#PD0490)
Unjustified restrictions on the free movement of commercial vehicles (#PE0897).

♦ PD2107 Private international arms dealers

Nature Private arms dealers facilitate sale of arms to governments or other groups, either by acting as agents for manufacturers, or by obtaining their arms directly from manufacturers in the supplying country, or else by purchase of equipment declared surplus or obsolete by the government of that country. Governments experience difficulty in controlling private transactions, which may involve a number of middlemen and the movement of the arms through several countries before reaching their final destination, but this has not prevented the sale of considerable amounts of surplus armaments to such dealers.

Incidence Private arms dealers exist primarily in countries where the economy is not centrally controlled. Less than 5 per cent of the arms trade with the developing countries is in the hands of private dealers and only a very small proportion of these operate without government approval.

Background At the close of the 1914–18 war, and within the framework of the League of Nations, there were strong suspicions that the role of the arms trade in general, and private arms dealers in particular, were a major cause of war. Since the 1930s, the regulation of the arms trade has not been a major subject of international discussion. After the 1939–45 war, the arms trade was mainly carried on by governments, and the role of the private arms dealers was reduced considerably.

Narrower Illegal international arms shipments (#PD4858).
Aggravates War (#PB0593) International arms trade (#PC1358).
Aggravated by Competitive acquisition of arms (#PC1258).

♦ PD2108 Innate expectation of suppression of minority opinion

Nature Minority voices have a deep fear of suppression. Consequently, when this suppression is articulated through intimidation and harassment, minorities tend to withdraw altogether from the decision–making process or to rebel.

Broader Blocked minority opinion (#PD1140).
Aggravates Ethnic conflict (#PC3685).

♦ PD2109 Insect resistance to insecticides

Nature The misuse of pesticides has inadvertently created strains of monster bugs that can no longer be chemically killed. There are about 30 species that can not be killed by insecticides. These pests have developed enzymes that detoxify a pesticide or slow its penetration.

Incidence Malaria had been nearly eliminated from many parts of the world until the World Health Organization decided to eradicate it by using insecticides. WHO spent 21 years and $2 billion before giving up in 1976. The result was a race of malaria–carrying mosquitoes virtually immune to insecticides.

Cotton bollworms were not a big problems until farmers decided to spray them. Most of the bollworms were controlled by natural predators. With the use of pesticides the predators were killed.

Broader Drug resistance (#PF9659) Pest resistance to pesticides (#PD3696).
Narrower Louse resistance to insecticides (#PE3576)
Blackfly resistance to insecticides (#PE3578)
Mosquito resistance to insecticides (#PE3582)
Housefly resistance to insecticides (#PE3583)
Cockroach resistance to insecticides (#PE3579).
Related Resistant bacteria (#PE6007) Drug resistant viruses (#PG6399)
Encouragement of drug resistant diseases (#PJ3767).
Aggravates Insect pests (#PC1630) Insect pests of plants (#PD3634)
Insect vectors of disease (#PC3597) Human disease and disability (#PB1044).
Aggravated by Mutation (#PF2276) Insecticides as pollutants (#PD0983).

♦ PD2118 Declining birth rate

Nature Nearly every nation in the world is experiencing a decline in the rate of births. While this is beneficial to most nations, the pressure this places on nations with replacement birth rates or lower, like Singapore and much of Western Europe, means that these cultures will eventually be lost to the world.

Counter–claim Birth rates increase and decrease over long periods of time involving generations to be concerned about declining birth rates over a period of less than 50 years is being short sighted.

Aggravates Underpopulation (#PD5432) Population decrease (#PF6441)
Maldistribution of world population (#PF0167)
Socio-economically inactive rural population (#PF4470)
Imbalance of population growth between developed and developing countries (#PE4241).
Reduces Ineffective population control (#PF1020)
Unsustainable population levels (#PB0035)
Depletion of natural resources due to population growth (#PD4007).

♦ PD2127 Air turbulence

Clear air turbulence

Nature Turbulence is one of the major causes of aircraft accidents which are brought about by structural failure, forced gains or losses of altitude (up to thousands of feet), or loss of control. The term turbulence in meteorology normally refers to atmospheric motions smaller than the scale which is designated as the mean flow, and therefore encompasses a wide spectrum of motions. Only a relatively narrow band of turbulence is a significant problem for aircraft flight. There are several types of turbulence: convective (near cumulo-nimbus clouds and thunderstorms), low-level clear-air turbulence (caused either by rough terrain or vortices in a large aircraft's wake), violent air motion near mountains (mountain waves), and high-level clear-air turbulence (above 20,000 feet). The latter form of turbulence is particularly hazardous since it is difficult to detect by radar and could prove especially dangerous to supersonic aircraft.

Incidence High-level turbulence occurs in patches with horizontal dimensions between 80 and 500 km, and vertical dimensions from 20 up to 600 metres. Between 5 and 10 per cent of high-level turbulence is heavy, and 1 to 3 per cent of that which occurs is violent or extreme.

Refs Hopkins, R H *Forecasting Techniques of Clear Air Turbulence Including that Associated with Mountain Waves* (1977).

Broader Bad weather (#PC0293).
Narrower Mountain waves (#PS4322).
Aggravates Air accidents (#PD1582).
Aggravated by Storms (#PD1150) Wind shear as a hazard (#PE5085).

♦ PD2133 Excessive foreign public debt of developing countries

External debt crisis of developing countries

Nature The external public debts of developing countries arise as a result of funds borrowed under a variety of conditions in order to finance different aspects of the development process within the country. In the general case, the problems of large development debts are of a structural origin. When service on foreign debt grows faster than export earnings, foreign exchange budgets become constricted and vulnerable. New aid relieves the burden only partially, for debt service is payable in untied and convertible foreign exchange, whereas aid is usually tied. Moreover, the stimulation of the export earnings needed to meet foreign debt is not helped by the protectionism of creditors. Where there are also structural limits to the rate at which exports can grow, very large-scale lending may create unmanageable debt situations. Current high rates of interest tend to make the problem even more difficult. There is a widespread impression that the problem of indebtedness is highly concentrated in relatively few developing countries, facing difficulties either because of an excessive accumulation of debt (particularly short-term debt on hard terms) or because of a sudden reduction in the capacity to service debt due to unfavourable developments in foreign exchange receipts. However, there is growing evidence that experience to date can serve only as a warning of a much broader and more widely diffused problem resulting from the inappropriateness of the terms of aid, and from the pressure upon developing countries to seek suppliers' credits to make up for the shortfall in the volume of aid. Because of the large current account deficits of many developing countries in the early 1980s, substantial adjustments have been required to cover interest obligations. The best way to have adjusted would have been to combine cuts in spending with policies to switch production into exports and into efficient import substitution, which usually requires a real depreciation in the exchange rate. The process is less costly if it does not occur too swiftly, if a large proportion of domestic output is easily traded and if it is easy to expand exports rather than necessary to compress imports. Unfortunately, many of the principal debtor countries got into difficulties just because these conditions did not exist: their exchange rates had become seriously overhauled; their producers were heavily protected, often by import controls that reduced the incentive to sell abroad; export industries were relatively underdeveloped and, in addition, faced growing restraints in foreign markets.

The world debt crisis, caused both by the oil price shocks of 1974 and 1980-81 and by domestic policies which advocated high levels of spending, expansion and borrowing, has reached the point where many debtor nations are unable to pay even the interest rates on their loans and are threatening non-payment if their loans are not re-structured by the IMF and other banks.

Incidence In 1982 and 1983, official creditors and banks had to reschedule over US $100 billion in loan repayments owed, while in 1983 alone, non-oil exporting countries paid US $50 billion just for debt-servicing interest. In 1986 debt service payments of all developing countries reached $101 billion. At the end of 1983, the major Latin American debtor countries had interest payments which alone consumed 40 percent of all their export revenues; every percentage point of US interest rates costs Latin American countries $3 billion.

Claim When debtor countries have to borrow in order to pay interest, an unsustainable economic and political threshold has been reached, which could result in economic and political catastrophe. What is at stake in the debt crisis is the nature of the relationship of developed to developing countries and of democracy to communism. The world is interdependent, and creditors and debtors may serve only to ruin each other by their own tests of strength and obstinacy.

Refs Delamaide, Darrell *Debt Shock* (1984); Korner, Peter, et al *The IMF and the Debt Crisis* (1986); Lal, S N *Problems of Public Borrowing in Under-Developed Countries* (1978); Miller, Morris *Coping is Not Enough The international debt crisis and the roles of the World Bank and the International Monetary Fund* (1986); Naylor, R T *Dominion of Debt* (1985); Nunnenkamp, Peter *The International Debt Crisis of the Third World* (1986); OECD *External Debt* (1988); OECD *Financing and External Debt of Developing Countries - 1986 Survey* (1987); Reisen, Helmut and Van Trotsenburg, Axel *Developing Countries Debt* (1988).

Broader Uncontrolled growth of debt (#PC8316) Disproportionate external public debt (#PC3056).
Narrower Increasing foreign debt of island developing countries (#PE5748)
Inadequate diversification of loans to developing countries (#PE4305)
Uncertainty of development programmes due to short-term loans (#PF4300).
Related Attitude*complex (#PA6983)
Contradictions of capitalism in developing countries (#PF3126)
Deterioration in external financial position of developing countries (#PE9567).
Aggravates Global crisis (#PF6244) Imbalance of payments (#PC0998)
Unemployment in least developed countries (#PE9476)
Financial paralysis of developing countries (#PD9449)
Economic and financial instability of the world economy (#PC8073)
Decline in commercial bank lending to developing countries (#PE4655)
Burden of servicing foreign public debt by developing countries (#PD3051)
Excessive social costs of structural adjustment in debtor developing countries (#PD8114).
Aggravated by Suffering (#PA7690) Austerity (#PJ4983)
Inappropriate loans (#PF4580) Exchange rate volatility (#PE5930)
Mismanagement in developing countries (#PD8549)
Militarization in developing countries (#PD9495)
Budget deficits in developing countries (#PD3131)
Inadequate control of development projects (#PF9244)
Inappropriate design of development projects (#PF4944)
Economic dislocations in developing countries (#PD4063)
Military destabilization of developing countries (#PD7714)
Low return on investment in developing countries (#PE9811)
Inappropriate management of development projects (#PD3712)
Political destabilization of developing countries (#PD9792)
Dependence of developing countries on food imports (#PE8086)
Structural rigidity in developing country economies (#PD2970)
Low complementarity of developing country economies (#PE8184)
Inadequate domestic savings in developing countries (#PD0465)
Excessive borrowing by state-controlled enterprises (#PE7474)
Prohibitive cost of inadequate development policies (#PF9101)
International trade barriers for primary commodities (#PD0057)
Inadequate negotiating capacity of developing countries (#PE9646)
Excessive dependence on export credits by developing countries (#PE0938)
Deteriorating terms of financial loans to developing countries (#PE4603)
Aggressive economic destabilization of countries by external forces (#PE9420)
Private domestic capital outflow from developing to developed countries (#PD3132)
Inadequate external debt management capacity within developing countries (#PE7184)
Decline in concessional financial resources available to developing countries (#PE3812)
Inability of developing countries to adopt appropriate exchange rate policies (#PE7563)
Rescheduling of debts of developing countries at market-related interest rates (#PE8110)
Distortion of international trade by export subsidies and countervailing duties (#PE1961)
Instability in export trade of developing countries producing primary commodities (#PD2968)
Dependence of developing countries on external financing for development programmes (#PE7195)
Exploitative financial policies of transnational corporations in developing countries (#PE6952)
Restrictive conditions on loans to developing countries through intergovernmental facilities (#PE9116)
Restrictive business practices in the markets of developed countries against exports from developing countries (#PE5926).

♦ PD2144 Distortion of international trade by dumping

Nature International trade may be distorted when an enterprise effectively reduces the foreign price of its product below that of the domestic price, where any such difference is not due to actual differences in the cost of selling, production, or transportation. The price reduction may be disguised by quality differentials, spurious quantity discounts, or favourable credit terms. Governments may support such forms of export through special export subsidies, tax rebates, or other special concessions.

Claim Dumping is an unfair trade practice requiring regulation by national and international agencies.

Counter-claim Accusations of dumping are being used to impose discriminatory protection of local interests by erecting trade barriers and promoting local industry.

Broader Irresponsible international trade (#PC8930).
Narrower Food wastage (#PD8844)
Dumping of food in developing countries (#PE0607).
Related Distortion of international trade by discriminatory customs and administrative entry procedures (#PE2603).
Aggravates Economic conflict (#PC0840).
Aggravated by Agricultural surpluses (#PC2062)
Excessive demand for goods in capitalist systems (#PC3116).

♦ PD2154 Enclaves and exclaves

Nature Enclaves and exclaves are discontiguous territories of states which are located within the territory of other states. In the case of islands they may exist within the territorial waters of another state. They are a particular form of disrupting influence on territorial integrity. As such they may be a source of military insecurity, and administrative and communication problems.

Incidence The number of enclaves has declined considerably since 1900. About 15 remain in Europe. Some remain in Africa and Asia, but most have been eliminated with the independence of the countries in those areas. New enclaves have been created as a result of armistice agreements (for example, West Berlin). Other well-known enclaves are Gibraltar and Hong Kong (to 1997).

Background When a nation refers to an extra-territorial possession which is encapsulated in another state, it refers to its exclave. The term enclave is also used for a culturally distinct area enclosed by its host country, where another language is frequently spoken, for example the Russian enclave in Shanghai after the revolution.

Broader Disruption of territorial integrity (#PC2945).
Related Unilateral declarations of independence by extra-territorial bases (#PF1066).
Aggravates Secession (#PD2490) Boundary disputes between states (#PD2946)
Territorial disputes between states (#PC1888) Evasion of customs and excise duties (#PD2620)
Unnatural boundaries between developing countries (#PD2544).
Aggravated by Changing river courses (#PG4357).

♦ PD2167 Unequal parliamentary constituencies

Nature Legislative seats which are apportioned according to population and administrative districts may give unequal representation. Where candidates from different parties contest an election within the limits of a constituency, minority interests may suffer, never winning an election, but having a considerable proportion of votes overall. Unequal distribution of legislative seats may lead to political conflict as a result of exploitation, or to apathy and alienation. In certain countries this problem is countered by a system of proportional representation.

Incidence In the UK, for example, the Ulster Unionist party (whose members are concentrated in one region, Northern Ireland) have virtually the same number of seats in the House of Commons as have the Liberal and Social Democratic parties combined (with votes approaching those of the main Conservative and Socialist parties, but with a wide regional distribution of members).

Broader Political inequality (#PC3425) Unequal political representation (#PC0655).
Narrower Political party manipulation of elections (#PD2906).
Related Political disintegration (#PC3204).
Aggravates Political conflict (#PC0368).

♦ PD2184 Polygamy

Polygyny — Sororal polygyny

Nature Polygyny is the system under which a man is married to two or more wives at the same time; it includes bigamy which is the (usually illicit) marriage to two women; and where polygamy is forbidden under the law, it can take the form of cohabitation with many women. As well as being discriminatory against women, men who cannot afford the bride-price and therefore cannot wed, can also be perceived as victims of this nuptial practice. Although polygyny ensures the birth of many children, as well as an additional supply of labour (both of which contribute, in traditional societies, to the wealth and prestige of the family), in a changing society it causes a strain on economic resources and can be a barrier to population control.

Incidence Polygamy in its traditional form occurs mainly in Africa, although concubinage still exists in Asia. Harems exist in Arabia and the Middle East but the Muslim law of taking up to four legal wives is much less followed in current practice, particularly because of the cost. It has been abolished in certain countries (Tunisia, Turkey), and severely restricted elsewhere (Pakistan,

PD2184

Egypt). But the rich in both Asia and Arabia continue to keep harems and this encourages traffic in women and children, particularly to Arabia. Bigamy occurs in isolated cases throughout developed society. A feature of the Mormon religion used to be polygamy, but the practice was outlawed in 1896 as a condition for Utah to become a state in the United States.
Counter–claim Even with equal numbers of men and women in a society, there is always more marriageable women than men because women mature and marry earlier. Therefore polygamy provides full adult status in a traditional society for as many women as possible.
Serial polygamy is practised in most developed nations through the practice of divorce and remarriage or cohabitation and separation.
Refs Clignet, Remi *Many Wives, Many Powers*; Deer, John *Polygamy and Polyandry* (1986); Khan, Mazahar–ul Hag *Purdah and Polygamy* (1982).
Broader Group marriage (#PF3288) Sexual immorality (#PF2687).
Narrower Concubinage (#PF2554).
Related Adultery (#PF2314) Polyandry (#PF3289).
Aggravates Illegitimate children (#PC1874) Discrimination against women (#PC0308).
Aggravated by Tribalism (#PC1910).
Reduces Divorce (#PF2100) Childlessness (#PC3280).

♦ PD2191 Disruption of development by tribal warfare
Tribal conflicts
Nature Warfare between different tribal groups, usually within the same 'national' boundary, causes disruption and disunity in the nation. Tribal warfare and conflict have been known to disrupt rural development programmes. This factor comes into play particularly when an extensive development programme takes in several ethnic groups. Since rural development programmes attract strangers into otherwise closely–knit homogenous rural communities, conflicts in morals, manners and of personality often create potentially explosive situations which require the talents of competent social workers to resolve.
Incidence Tribal warfare has been recently reported in the Philippines, Burundi, Sudan, and South Africa.
Broader Conflict (#PA0298) Tribalism (#PC1910).
Related Guerrilla warfare (#PC1738) National instability due to tribalism (#PG6329)
Race as a reinforcement of nationalism (#PF3352).
Aggravates Civil war (#PC1869) Militarism (#PC2169)
Social fragmentation (#PF1324) Lack of national unity (#PF8107)
Lack of political development (#PD8673).
Aggravated by Elitism (#PA1387) Slave trade (#PC0130)
Nationalism (#PB0534) Ethnic conflict (#PC3685)
National boundaries (#PF8235).
Reduced by Political dictatorship (#PC0845).

♦ PD2198 Sexual deviation
Sexual abnormality — Sexual perversion — Infringement of sexual taboos — Abuse of sex
Nature Sexual perversions depart from the normal in the following respects: disregard for the barriers between species (man and animals); insensitivity to barriers imposed by disgust (faeces, urine, etc); transgression of the incest barrier (prohibition of sexual gratification with close blood relatives); homosexuality; and transference of the part played by the sexual organs to other organs and different parts of the body. Deviant sexual behaviour patterns may involve or result in sexual violence and may cause marriage breakdown or impotence. They encourage prostitution and may lead to crime and can result in disease, injury, death and murder.
Incidence Sexual deviations include incest, male homosexuality, lesbianism, trans–sexualism, sadism, masochism, sodomy, bestiality, voyeurism, sexual exhibitionism, paedophilia, necrophilia and masturbation; they may be manifested in pornography or in obscenities and violence.
Counter–claim Sexual perversion depends on the what is meant by normalcy and leads to a wide variety of meanings. If normal sex is whatever minimal erotic activity required to promote conception with an appropriate partner, all else is irrelevant or immoral. If sex is for pleasure to consenting adults then perversion is inflicting pain in the short or long term. If sex is the expression of mutual love, perversions are expressions of hate or being manipulative. If sex is a way of communicating, deceit is a perversion. Masturbation, while not exactly perverse, it is not sexual as talking to yourself is not a whole conversation.
Refs Allen, Clifford *The Sexual Perversions and Abnormalities* (1979); Docter, R F *Transvestites and Transsexuals* (1988); Masters, R E L *The Hidden World of Erotica* (1973); Mathis, James L *Clear Thinking about Sexual Deviations* (1972); Rosen, Ismond *Sexual Deviation* (1979).
Broader Perversion (#PB0869) Immorality (#PA3369) Violating taboos (#PF3976)
Mental disorders (#PD9131).
Narrower Incest (#PF2148) Sodomy (#PE3273) Obscenity (#PF2634)
Voyeurism (#PE3272) Narcissism (#PF7248) Bestiality (#PE3274)
Bisexuality (#PF3269) Frotteurism (#PE5844) Strip tease (#PG6142)
Necrophilia (#PF6957) Nymphomania (#PE8213) Masturbation (#PE4426)
Transvestism (#PE6348) Homosexuality (#PF3242) Exhibitionism (#PD4643)
Sexual sadism (#PE6748) Transsexualism (#PF3277) Sexual violence (#PD3276)
Sexual masochism (#PE3851) Sexual fetishism (#PF6406)
Transvestic fetishism (#PG7757) Sexual exploitation of children (#PD3267).
Related Sexual craving (#PE7031) Social deviation (#PC3452)
Sexual abuse by women (#PJ1767) Human pseudo hermaphroditism (#PE2246).
Aggravates Prostitution (#PD0693) Satanic rituals (#PF7887)
Spiritual impurity (#PF6657) Sexual unfulfilment (#PF3260)
Sexual impotence of men (#PF6415) Sexual abuse of children (#PE3265)
Public displays of sexuality (#PE4882)
Vulnerability of marriage as an institution (#PF1870)
Violation of the rights of sexual minorities (#PD1914).
Aggravated by Sex (#PF9109) Permissiveness (#PF1252) Hormonal disorder (#PE4431)
Male prostitution (#PD3381) Sexual repression (#PF2922) Immoral literature (#PF1384)
Retarded socialization (#PF2187) Human sexual disorders (#PD8016)
Inadequate sex education (#PD0759).

♦ PD2207 Inadequate riot control
Aggravates Rape (#PD3266) Fear (#PA6030) Panic (#PF2633)
Looting (#PE4152) Injuries (#PB0855) Human death (#PA0072)
Gang violence (#PG4444) Human violence (#PA0429) Demonstrations (#PD8522)
Civil disorders (#PC2551) Police brutality (#PD3543)
Human destructiveness (#PA0832) Harm to innocent bystanders (#PJ3378).
Aggravated by Political conflict (#PC0368).
Reduced by Chemical warfare (#PC0872).

♦ PD2211 Urban fires
Nature Particular areas of risk are: the older parts of cities, where buildings tend to be crowded together and it can be difficult to open up clear avenues as a precaution against the spread of fire; in the case of skyscrapers; and where additional storeys have superimposed on an old building. While the building materials may be fireproof, furnishings and other objects can feed a fire for hours and create high temperature levels. Many synthetic substances are readily combustible and, once set alight, are difficult to extinguish. An additional problem is that the insulation afforded by plastic on electrical leads drops sharply at temperatures above 500 deg C. In a fire, vapours and acids attack these leads and cause short–circuits which in turn start secondary fires.
Incidence Total annual fire losses in the USA, for example, are estimated at US $5,000 million, of which 80–85 per cent are due to urban fires. Buildings are sometimes set on fire deliberately by criminal acts or attacks in war. Towards the end of World War II, 60 per cent of the bombs dropped on towns were incendiary. In major city fires, the rate of damage is in direct proportion to the density of construction. When buildings occupy 20–30 per cent of the surface area, destruction may amount to 65 per cent; when the density is 40 percent, the rate of destruction may be 80 per cent.
Broader Fires (#PD8054).
Aggravates Human death (#PA0072) Burns and scalds (#PE0394)
Smoke as a pollutant (#PD2267).
Aggravated by Transport of dangerous goods (#PD0971)
Uncontrolled urban development (#PC0442)
Flammable construction materials (#PG4448)
Road and highway traffic congestion (#PD2106).

♦ PD2213 Inadequate political networks
Claim Traditional political forms tend to be inflexible to alternative perspectives and to lack creative tension. There are no adequate networks to act as intermediary between the local and the global and provide for rapid, effective and accountable involvement in the political process.
Broader Exclusion of opposing views (#PF3720).
Narrower Distrust of political dialogue (#PD2263)
Insufficient images of political involvement (#PF5298)
Diversity of limited local decision–making bodies (#PD2245)
Governmental disregard for people as human beings (#PD8017)
Inflexible intermediary political implementation networks (#PE0174).
Related Blocked minority opinion (#PD1140) Collapse of societal engagement (#PF2340)
Irresponsible expression of emotions equated with free speech (#PF7798).

♦ PD2224 Environmental plant diseases
Nonparasitic plant diseases — Physiological plant diseases — Physiogenic plant diseases
Nature Plants in an adverse or unfavourable environment or subject to injury by physical or chemical agents may vary so far from their normal growth habits that they are considered diseased. Accidents, poisons or defects in the environment often result in diminished growth or diseased tissues. Subsequently, the plant is more susceptible to attack by microorganisms which may cause further damage.
Incidence Crops may be seriously damaged by fire, water, frost, insects and poor agricultural techniques. Pollution of the air by noxious gases, dust and smoke may provoke injury and disease. A growing problem in developed countries is plant–pathogenic air pollutants. The most important are sulphur dioxide, ozone and peroxyacetyl nitrate. The last two are photochemical pollutants formed from the interaction of other pollutants. Agricultural chemicals, when improperly used, can cause serious physiological diseases. Damage is fairly common from the misuse of sprays, seed treatments, fumigation, fertilizers, herbicides, soil treatment and hormones. Too high a temperature may result in sun scald and death in the tips and margins of leaves. With insufficient light, plants may become chlorotic. Low temperatures may cause damage, such as net necrosis in potatoes, and frost injury is common. Too little water causes stunting, wilting and burning; but too much water may induce flooding of tissues, resulting in such diseases as water core of apple, celery heart rot, and tomato blossom end rot. For healthy growth, plants require various essential elements. If these elements are absent, or present in insufficient quantities, plants will display characteristic symptoms of deficiency diseases. Besides nitrogen, potash and phosphorus, which plants need in relatively large amounts, smaller quantities of sulphur, magnesium and calcium are required. Trace elements, which are necessary in minute amounts for healthy plant growth, include boron, iron, copper, manganese, molybdenum and zinc.
Broader Plant pathogens (#PD1866).
Narrower Endangered urban trees (#PD2025) Pesticide damage to crops (#PD2581).
Related Plant cancer (#PE1899) Parasites on plants (#PD4659)
Viral plant diseases (#PD2227) Fungal plant diseases (#PD2225)
Bacterial plant diseases (#PD2226) Environmental human diseases (#PD5669)
Deficiency diseases in plants (#PD3653).
Aggravated by Insect pests (#PC1630) Acidic soils (#PD3658)
Air pollution (#PC0119) Alkaline soil (#PD3647)
Soil pollution (#PC0058) Water pollution (#PC0062)
Soil compaction (#PD1416) Soil infertility (#PD0077)
Soil salinization (#PE1727) Ozone as a pollutant (#PE1359)
Motor vehicle emissions (#PD0414) Pesticides as pollutants (#PD0120)
Herbicides as pollutants (#PD1143) Hydrocarbons as pollutants (#PE0754)
Metal contamination of soil (#PD3668) Plant–pathogenic air pollutants (#PE0155)
Photochemical oxidant formation (#PD3663) Nitrogen compounds as pollutants (#PD2965)
Environmental hazards from fertilizers (#PE1514).

♦ PD2225 Fungal plant diseases
Nature Fungi are responsible for by far the greatest number and diversity of plant diseases. All crop plants are apparently attacked by some variety of fungus; often a dozen or more different fungi induce disease on a single species. The distance to which fungus spores may be carried by air currents, for example, is very great. Because of this and because of their astonishing productivity, fungi are responsible for a much larger number of the rapidly spreading, hence epidemic diseases, than are viruses or bacteria. It is the sporadic nature of this disease that brings about the greatest hardships on the individual farmer.
There is the correlation between the mode of dissemination of the fungus and its relation to the host. In general leaf, stem, and fruit diseases are caused by airborne or insect–carried fungi, root diseases by soil inhabiting species. Some vascular wilts are caused by soil fungi, some by fungi possessing insect vectors.
Incidence Among the most common and widespread diseases of plants caused by fungi are the various downy mildews (of grape, onion, tobacco, etc), the powdery mildews (of grape, cherry, apple, peach, rose, lilac); the smuts (of maize, wheat, onion); the rusts (of wheat, oats, beans, asparagus, snapdragon, hollyhock); apple scab; brown rot of stone fruits ; and various leaf spots, blights and wilts. These diseases are responsible for millions of dollars' worth of damage annually, to growers all over the world.
Certain fungi are strictly local in their effect, producing lesions on leaf, stem or root system, though their localism may at times be an expression of host resistance. Other fungi are selective with respect to particular tissues, as exemplified by the vascular wilt organisms, which are confined to the water–conducting tissues; or the chestnut blight fungus, injuring the cambium layer. Still others are indiscriminate, establishing themselves at various points, and at times destroying the entire plant.
Background Rust of cereals has always presented a major problem to man, but it was not until 1800 that it was understood that rust was caused by a fungus. A devastating epidemic of late blight fungus on potatoes beginning in Europe in 1845 brought about the famine, particularly in Ireland, that caused starvation, death and mass migration. In America, the chestnut was practically eliminated at the beginning of this century by a fungus imported from the orient.
Refs Colhoun, J *Club Root Disease of Crucifers Caused by Plasmodiophora Brassicae Woron* (1958); Holliday, Paul *Fungus Diseases of Tropical Crops* (1980).

DETAILED PROBLEMS
PD2228

Broader Plant pathogens (#PD1866)
Microbial diseases (#PC7492)
Pests and diseases of trees (#PD3585).
Narrower Oomycetes (#PE5465)
Rust fungi (#PE6255)
Zygomycetes (#PE0614)
Ascomycetes (#PE4586)
Basidiomycetes (#PE0364)
Deuteromycetes (#PE4346)
Moulds in plants (#PE8051)
Pathogenic fungi (#PG1959)
Dutch elm disease (#PE1154)
Club root of cabbage (#PG3396)
Black mildew in plants (#PE1607)
Slime moulds in plants (#PE1371).
Related Plant cancer (#PE1899)
Fungal diseases (#PD2728)
Parasites on plants (#PD4659)
Viral plant diseases (#PD2227)
Bacterial plant diseases (#PD2226)
Environmental plant diseases (#PD2224)
Deficiency diseases in plants (#PD3653).
Aggravates Infestation of seeds (#PE6271)
Maize pests and diseases (#PE3589)
Pests and diseases of tea (#PE2980)
Pests and diseases of elm (#PE2982)
Pests and diseases of oak (#PE2984)
Pests and diseases of vines (#PE2985)
Pests and diseases of cocoa (#PE2979)
Pests and diseases of palms (#PE2981)
Pests and diseases of cotton (#PE2220)
Pests and diseases of potato (#PE2219)
Pests and diseases of rubber (#PE2977)
Pests and diseases of olives (#PE2978)
Pests and diseases of coffee (#PE2218)
Pests and diseases of chestnut (#PE2983)
Pests and diseases of citrus fruit (#PE2976)
Pests and diseases of deciduous fruit (#PE3591).
Aggravated by Aphids as pests (#PE3613)
Homoptera as insect pests (#PE3614)
Fungal resistance to fungicides (#PE4456).

♦ PD2226 Bacterial plant diseases

Agrobacterium — Bacillus — Clavibacter — Erwinia — Pseudomonas — Xanthomonas — Mycoplasma-like organisms (MLOs)

Nature A number of different bacterial species incite diseases in plants. Some of the major bacterial diseases can seriously impair the economic value of such host crops as cotton, sugar cane, potato, apple and pear.

Incidence There are more than 170 different kinds of bacteria which cause diseases in flowering plants belonging to 150 genera or 50 families. Bacterial diseases of plants occur in almost anywhere that is reasonably moist or warm. Their destructiveness varies from year to year and place to place, according to the presence or absence of a critical environmental condition under which the bacteria operate on the host plant.

Plants respond in many ways to invasion by bacteria. Among the symptoms of bacterial infections are galls, wilts, slow growth, dwarfing, imperfect fruits or ears, rots, colour changes of various plant parts, retarded ripening, distortion of leaves, cankers, brooming, fasciation, and leaf spots. Among the many well-known bacterial galls are olive knot, cane gall, beet pocket rot, sweet pea fasciation, hairy root, and crown gall. All contain large swollen cells and small, rapidly dividing cells along with vascular cells in a relatively disorganized arrangement. Eventually these gall structures may interfere with the normal transmission of water and food supplies, and the plants may die.

Bacterial wilt may be quite destructive – for example, in sweetcorn, cucumber, tobacco and related plants. Such bacteria may produce a slime, which plugs the water-conducting tissue of the invaded plant. Closely related are such diseases as black rot of cabbage, ring rot of potatoes, and tomato canker, which may start in the water-conducting tissue but subsequently result in disintegration of surrounding tissue. Cankers develop from extensive tissue destruction – for example, that caused by the fire blight bacteria or from the lesions of the tomato canker organism.

Local spots occur most commonly on the leaves, but sometimes appear elsewhere, as on many fruits. Symptoms of black arm of cotton show when the angular leaf spot bacteria enter the stem and girdle it. Bacterial blight of beans, halo blight of oats, potato scab, and many others appear primarily as local spots. The bacteria causing halo blight of oats and wildfire of tobacco produce toxic substances that are responsible for the yellowish areas immediately around the dead spots where the bacteria have invaded the tissue.

Soft rots develop in relatively fleshy tissues when certain bacteria invade them extensively. Such bacteria produce an enzyme that dissolves the pectic substance that cements plants' cell walls together. The result is a slimy, often foul-smelling, mass. The soft rots often follow and extend invasion and damage by some other pathogen. For example, black rot of cabbage and late blight of potatoes would be much less serious except for the subsequent soft rot.

Symptoms of disease appear at varying lengths of time after bacteria attack and grow in a plant. Soft rots are sometimes evident within a day or so, angular leaf spot of cotton within 10 days, corn wilt within 1 to 2 months. Crown gall of orange may take 2 years.

Background The first plant disease proven to be of bacterial origin was fire blight on apple, pear and related plants. The discovery was made a century ago when the disease was devastating orchards in the USA. It has since driven certain high-quality apples and pears out of cultivation in the eastern US.

Refs Fahy, P C and Persley, G J (Eds) *Plant Bacterial Diseases* (1983); Gorlenko, M V *Bacterial Diseases of Plants* (1961); Hiruki, Chuji (Ed) *Tree Mycoplasma and Mycoplasma Diseases* (1987); Lozano, J Carlos (Ed) *Fifth International Conference on Plant Pathogenic Bacteria* (1981); Maramorosch, K and Raychaudhuri, S P (Eds) *Mycoplasma Diseases of Crops* (1988); Maramorosch, Karl and Raychaudhuri, S P *Mycoplasma Diseases of Trees and Shrubs* (1981).

Broader Plant pathogens (#PD1866)
Bacterial disease (#PD9094).
Narrower Crown gall (#PE2230)
Fire blight (#PE2229)
Bacterial bean blight (#PG4094)
Root nodules of legumes (#PG3343)
Rhizobium plant diseases (#PG5311)
Bacterial wilt of cucumber (#PG4650)
Bacterial soft rot in plants (#PE4706).
Related Plant cancer (#PE1899)
Parasites on plants (#PD4659)
Viral plant diseases (#PD2227)
Fungal plant diseases (#PD2225)
Environmental plant diseases (#PD2224)
Bacterial diseases in animals (#PD2731)
Deficiency diseases in plants (#PD3653).
Aggravates Infestation of seeds (#PE6271)
Pests and diseases of potato (#PE2219)
Pests and diseases of cotton (#PE2220)
Pests and diseases of olives (#PE2978)
Pests and diseases of chestnut (#PE2983)
Pests and diseases of sugar cane (#PE2217)
Pests and diseases of sugar-beet (#PE2975)
Pests and diseases of citrus fruit (#PE2976)
Pests and diseases of deciduous fruit (#PE3591).
Aggravated by Aphids as pests (#PE3613)
Resistant bacteria (#PE6007)
Abuse of antibiotics (#PE6629)
Homoptera as insect pests (#PE3614)
Substitution of fast growing plant species (#PE6396).

♦ PD2227 Viral plant diseases

Virus diseases in plants

Nature A number of infectious plant diseases result from viruses. Major crops that are seriously affected by virus diseases include: tobacco, potato, sugar beet and cane, peach, orange, cotton and wheat. Most of the plant viruses impair or destroy chlorophyll, causing the plant to wilt or die; some viruses, though, stunt or otherwise deform plants without seriously affecting the chlorophyll.

Incidence Virus diseases of plants are widespread, and many of them cause economic loss. The whole organization of the potato industry is conditioned by the necessity for minimizing virus diseases. For more than a century it has been known that if a farmer in England continued to use his own seed potatoes for successive crops, degeneration progressed gradually until the great majority of the plants were malformed and stunted. Among other important crop diseases caused by viruses are spotted wilt of tomato, tobacco mosaic, leaf curl of cotton in the Sudan, and swollen shoot disease of cacao. The tobacco mosaic virus besides affecting many members of the nightshade family (Solanaceae), which includes tobacco, tomato and potato, occurs on some 30 species of plants in 14 other families.

Viruses originate in local areas all over the world. Through long association native plants have developed a tolerance to the local viruses that enables infected individuals to survive with little injury. When crop plants are introduced into an area they frequently become subject to infection with the native viruses, against which they have had no opportunity to develop resistance. Such a virus may cause extensive losses to a crop plant, not only in the areas of original distribution of the virus, but also in the other areas to which it may spread on the recently attacked crop plant. Virus diseases produce a wide range of symptoms and types of injury on plants. Sometimes they kill the plant in a short time, as with spotted wilt and curly top on tomato. More often they cause lesser injuries that result in reduced yields and lower quality of product. With respect to the general symptoms produced, most viruses are of two rather clearly defined groups: those that cause mottling or spotting of leaves, and those that cause a yellowing leaf, curling, dwarfing, or excessive branching, but little or no mottling or spotting.

Background The importance of virus diseases on crop plants has increased tremendously in the past 50 years. Since 1900, more than 200 new plant viruses have been discovered. Many of them have done widespread damage to crop plants. Curly top caused almost complete abandonment of the sugar beet industry in parts of western United States from 1926 to 1932 and still causes severe injury to tomatoes, beans and a number of other crops. Sugarcane mosaic caused extensive losses to the sugarcane industry in the United States, Argentina, Brazil, and other countries beginning about 1917. Spotted wilt has become widespread and now causes losses to tomato and other crops in many parts of the world. Since 1940, swollen shoot has caused extensive damage to the cacao industry of West Africa. Virus diseases of citrus trees have become more destructive and from 1936 to 1946 tristeza caused the loss of 7 million orange trees in the state of Sao Paulo, Brazil, alone. It has attacked or now threatens millions of trees in various tropical and subtropical areas. This increase in destructiveness of virus diseases and in the number of known viruses has come about largely as a result of the expansion of agricultural enterprises and the increased movement of plants and plant products in recent years.

Refs Bos, L *Symptoms of Virus Diseases in Plants* (1978); Francki, R I, et al *The Plant Viruses*; Martyn, E B (Ed) *Plant Virus Names* (1968).

Broader Plant pathogens (#PD1866).
Narrower Aster yellows (#PG1059)
Tobacco mosaic (#PG4459)
Leaf curl of cotton (#PG4455)
Potato latent mosaic (#PG1964).
Related Plant cancer (#PE1899)
Parasites on plants (#PD4659)
Fungal plant diseases (#PD2225)
Bacterial plant diseases (#PD2226)
Viral diseases in animals (#PD2730)
Virus diseases in bacteria (#PD2562)
Virus diseases in protozoa (#PG7376)
Environmental plant diseases (#PD2224)
Thysanoptera as insect pests (#PE3619)
Deficiency diseases in plants (#PD3653).
Aggravates Maize pests and diseases (#PE3589)
Pests and diseases of elm (#PE2982)
Pests and diseases of oak (#PE2984)
Pests and diseases of cocoa (#PE2979)
Pests and diseases of palms (#PE2981)
Pests and diseases of potato (#PE2219)
Pests and diseases of cotton (#PE2220)
Pests and diseases of sugar cane (#PE2217)
Pests and diseases of sugar-beet (#PE2975)
Pests and diseases of citrus fruit (#PE2976)
Pests and diseases of deciduous fruit (#PE3591).
Aggravated by Virus diseases (#PD0594)
Aphids as pests (#PE3613)
Homoptera as insect pests (#PE3614)
Inadequate plant quarantine (#PE0714)
Disease-causing viral combinations (#PE6403)
Insect vectors of viral diseases of plants (#PD3600).

♦ PD2228 Nematoid plant diseases

Nematodes — Eelworms

Nature Nematodes, or eelworms, differ from most of the other organisms that cause plant diseases in that they themselves are animals, not plants. A great number live in the soil and, though most can be classed as harmless, several hundred species are known to feed on living plants as parasites and to be the causes of a variety of plant diseases. Information accumulated during the past century indicates that all of the crop and ornamental plants grown in the world can be attacked by plant parasitic nematodes.

Incidence The golden nematode of potatoes (Heterodera rostochiensis) is a leading menace to the potato industry of Europe. A great effort is being made to control it. A related species, the sugar beet nematode (H schachtii), is a critical pest that has restricted the acreage of sugar beets in Europe, Asia and America. The citrus nematode (Tylenchulus semipenetrans) occurs throughout the citrus-producing regions and exacts a heavy toll, symptoms of which are the common slow decline and dieback visible in many groves over 15 years old. The burrowing nematode (Radopholus similis) is a serious parasite in the tropics, where it attacks citrus, banana, pepper, abaca and other important crops, causing severe losses.

There are no statistics for the world wide damage caused by nematodes, but it has been estimated that in the USA alone the damage done to crops is at least several hundred million dollars per year. The use of soil fumigants for nematode control during the past several years has often produced dramatic proof that nematodes in the soil can make the difference between a good crop and one not worth harvesting. Yield increases of 25 per cent to 50 per cent after soil fumigation are common. Experiments with soil fumigation have also made it evident that severe nematode damage can occur on a great variety of crops, including tree crops. The underground parts of plants, roots, tubers, corns, and rhizomes are more apt to be infected than above ground parts.

Damage to plants attacked by nematodes is due primarily to the feeding of the nematodes on the plant tissues. The most common types of nematode damage are manifested as rotting of the attacked parts and adjacent tissue or the development of galls and other abnormal growths. Either can interfere with the orderly development of the plant and cause shortening of stems or roots, twisting, crinkling or death of parts of stems and leaves, and other abnormalities. The various kinds of nematode damage interfere with the growth of plants. Reduction in the size of the root system by rotting or galling restricts its efficiency in obtaining the food and water the plant must get from the soil. Root knot galls distort the tissue that has the function of conducting food and nutrients to the upper part of the plant. Damage to stems and leaves also interferes with normal growth. Consequently the yield of crop plants is reduced. Crippled plants cannot produce a high-quality crop. With some crops such as carrots and white potatoes, galls and rot caused by nematodes can make culls out of what would otherwise be saleable produce.

Refs Solov'eva, G I *Parasitic Nematodes of Woody and Herbaceous Plants* (1975).

Broader Plant diseases (#PC0555)
Plant pathogens (#PD1866)
Parasites on plants (#PD4659).
Related Parasitic plants (#PD6284).
Aggravates Pests and diseases of potato (#PE2219)
Pests and diseases of cotton (#PE2220)
Pests and diseases of sugar-beet (#PE2975)
Pests and diseases of citrus fruit (#PE2976).
Reduced by Nematicides as pollutants (#PJ1961).

♦ PD2229 Uncontrolled migration

Nature In a world where access to transportation has widened horizons, human migration in search for work, food, housing freedom from oppression or a better quality of life have radically affected population patterns across the world. Social tensions are created between newcomers and the existing population. The homelands lose valuable resources. Migration from poor areas to wealthier one increases patterns of inequality whether it is the brain drain from the south to the north, the migration of Asians from East Africa or the movement of desert nomads into squatter camps.

The mass movement of many species of mammals, birds and insects in order to take advantage of two different habitats. These animals are difficult to protect because of the lack of effective international agreement between countries along their migratory paths.

Narrower Internal migration (#PF4009) International migration (#PF4008).
Aggravates Underpopulation (#PD5432) Failed migration (#PE3445)
Isolation of ethnic groups (#PC3316)
Inadequate education of indigenous peoples (#PC3322)
Inadequate housing among indigenous peoples (#PC3320).
Aggravated by Famine (#PB0315) Natural disasters (#PB1151).

♦ PD2232 Aquatic weeds

Water weeds

Nature Aquatic weeds are plants which hinder or harm man's social or economic activities. Aquatic plants can limit the usefulness of irrigation systems. They can inhibit fishing and the passage of ships and they may harbour insect pests. Aquatic weeds pose a particular problem where there is a concern for water loss because they dramatically increase the rate of transpiration.

The problem is becoming greater with increased trade and development which tend to spread and encourage the growth of weeds. In particular, man-made lakes, reservoirs, canals and irrigation systems, with their different ecology from natural water courses, often provide an ideal environment for the spread of aquatic weed, especially where the water is shallow and rich in nutrients. The concentration of human settlements, the development of sewage systems and the increasing use of fertilizers ensure that large amounts of nutrients are entering water courses, there to stimulate weed growth.

Aquatic weeds are spread very quickly through the medium of the river. Boats distribute weeds up- and downstream. Currents carry them downstream and floods push them into pools and swamps and backwaters. Seeds may be carried by wildlife and the wind. Man-made water systems are particularly susceptible to weed growth, as man-made reservoirs and distributaries are often shallow and clear, as well as rich in nutrients, and weed growth follows. For example, submerged aquatic weeds have cut the flow in one large irrigation canal system in India by 80 per cent. The result is that the reduced flow encourages seepage from the canal and thus contributes to water-logging and salinity. Finally, water cannot be moved to crops, animals, and people in sufficient quantity, on schedule.

Incidence The most serious problems are caused by the water hyacinth (Eichhornia crassipes) which has become ubiquitous in warm waters. For example in Florida, more than 40,000 hectares of water are covered despite multi-million dollar control programs. In Bangladesh, water hyacinth rafts weighing up to 300t/ha are carried by floodwaters and can destroy rice fields. In the 90,000 hectare lake Laguna de Bay, Philippines, which provides much of the fish supply for Manila, water hyacinths impede access to fish pens and fish cages and frequently destroy fish culture installations when piled up by high winds, especially in typhoons. Another problem weed is the aquatic fern Salvinia, which covers about 12,000 ha of swamp and rice fields in Sri Lanka and 600 to 850 km2 (annual fluctuations) of the largest man-made lake in the world, Lake Kariba, Africa (40,000 km2). The edges of the lake are permanent mats of weed which cover hundreds of acres in river estuaries. Harbours are so blocked that ships can scarcely move. Life is uncertain on the lake because the wind and current moves the mats unpredictably. Fishing nets placed at night may become hopelessly entangled and shifted by morning. Fishing camps may be forced to move if the shore is blocked by weed mats to prevent the use of canoes.

Counter-claim Some aquatic weeds have high food potential. The best known example is water spinach (Ipomoea aquatica) which is widely consumed by Southeast Asian peoples and has a high protein content (19–34% dry matter). The aquatic fern Azolla Prinnata is widely used as livestock feed and as a nitrogenous fertilizer in Vietnam and the Peoples Republic of China. Water hyacinth and other aquatic weeds can be made into compost or silage for agricultural use. Aquatic weeds are also used to feed herbivorous fish, particularly the grass carp, in freshwater ponds. Chopped water hyacinth is being investigated as a fishpond fertilizer since it can provide nutrients which stimulate plankton growth and can also act as a matrix for the growth of detrital bacteria: important foods for some warm water cultured fish, particularly tilapias. Similarly some water weeds are used on a small scale as human food. Young flower buds of water hyacinth are eaten in the Philippines and the leaves of Ipomoea peptans are eaten in many parts of the world. It is the fibre in leaves that limits their widespread use as a source of food, but the protein can be mechanically extracted from water weed and is of good nutritional value.

Broader Weeds (#PD1574) Environmentally harmful dam construction (#PD9515).
Narrower Eichornia crassipes (#PE3815) Salvinia auriculata (#PE6254).
Aggravates Eutrophication of lakes and rivers (#PD2257)
Water losses from irrigation systems (#PE8796)
Destruction of fisheries through dams (#PE8148).
Aggravated by Water pollution (#PC0062) Soil erosion by water (#PD2290)
Detergents as pollutants (#PE1087) Silting of water systems (#PD3654)
Pollution of inland waters (#PD1223) Inadequate irrigation system (#PD8839)
Inadequate weed control in developing countries (#PD3598).

♦ PD2239 Parochial attitudes of organizations' members

Claim Intersocial organizations are often incapable of dealing effectively with one another because their grassroots supporters are often parochially minded, such that they are unable to envision a world role other than that of conflict and have, at best, primitive means of participating in any sort of global decision-making.

Broader Misuse of international forums (#PF2216).

♦ PD2243 Food spoilage in storage

Nature Foods, especially perishable foods, may become spoilt as a result of lack of adequate storage, preservation or packaging. This leads to abnormal flavour, colour, odour or consistency, which may or may not be harmful.

Incidence In 1986, about one half of the European refrigerated surplus food stocks were no longer fit for consumption by human or animal.

Refs FAO *Food Loss Prevention in Perishable Crops* (1983); FAO *The Prevention of Losses in Cured Fish* (1982); FAO *Manual of Pest Control for Food Security Reserve Grain Stocks* (1985); Hall, D W *Handling and Storage of Food Grains in Tropical and Subtropical Areas* (1980); MATCOM *Food Spoilage and Preservation. A Learning Element for Staff of Consumer Cooperatives* (1984); Zakladnoi, G A and Ratanova, V F *Stored-Grain Pests and Their Control* (1987).

Broader Food insecurity (#PB2846) Food grain spoilage (#PD0811)
Spoilage of agricultural products (#PC2027).
Related Fruit rot (#PE5373).
Aggravates Human disease and disability (#PB1044)
Biological contamination of food (#PD2594)
Instability of trade in food and live animals (#PD1434)
Environmental hazards from food processing industries (#PE1280).
Aggravated by Humidity (#PD2474) Disasters (#PB3561)
Climatic heat (#PC2460) Rodents as pests (#PE2537)
Inhospitable climate (#PC0387) Unethical food practices (#PD1045)
Inadequate food storage facilities (#PE4877).
Reduced by Health hazards of irradiated food (#PD0361).

♦ PD2244 Frost

Nature Frost includes the direct crystallization of water on exposed objects and, of more importance, the occurrence and effect of subfreezing temperatures on plants and crops. Damage to crops is caused by freezing of the aqueous solutions in plant, cells causing these to burst and destroy the plant partially or completely. The occurrence of frost determines the length of the growing season in many regions. This may reach critically short proportions for most usable crops in some regions. Citrus fruits are particularly sensitive. In colder regions, frost action is exceedingly important in weathering and erosion, in disturbing vegetation (destruction of root systems), and in engineering problems. In such regions, frost causes the ground to heave, disturbing foundations and landscaped surfaces.

Refs Bagdonas, A; Georg, J C and Gerber, J F *Techniques of Frost Prediction and Methods of Frost and Cold Protection* (1978); Sakai, A and Larcher, W *Frost Survival of Plants* (1987).

Broader Climatic cold (#PD1404).
Related Permafrost instability (#PD1165).

♦ PD2245 Diversity of limited local decision-making bodies

Nature Access to decision-making structures results in being able to influence every level of society. At present it is difficult for the individual to understand and relate to the decisions made on various levels, from the local to the global, within mass society. Emerging "grass-root" groupings for local power are too diverse and parochial to be effective, and the average person lacks the means of making his individual vision known at all levels.

Broader Inadequate political networks (#PD2213).

♦ PD2255 Fever

Feverishness

Refs Kluger, Matthew J *Fever* (1979).

Broader Human disease and disability (#PB1044).
Aggravates Hallucinations (#PF2249).
Aggravated by Q fever (#PE2534) Malaria (#PE0616)
Hay fever (#PE6197) Swine fever (#PE6266)
Dengue fever (#PE2260) Trench fever (#PG3689)
Yellow fever (#PE0985) Typhus fever (#PG1685)
Tuberculosis (#PE0566) Leptospirosis (#PE2357)
Typhoid fever (#PD1753) Scarlet fever (#PG2757)
Rat-bite fever (#PG6702) Pappataci fever (#PG4504)
Relapsing fever (#PE7787) Rheumatic fever (#PE0920)
Childbirth fever (#PG4639) East Coast fever (#PE7946)
Blackwater fever (#PG2628) Rift valley fever (#PE7552)
Texas cattle fever (#PG3990) African swine fever (#PE5207)
Haemorrhagic fevers (#PE5272) Malignant catarrhal fever (#PE6280)
Rocky Mountain spotted fever (#PG3899).

♦ PD2257 Eutrophication of lakes and rivers

Nature Eutrophication which may be natural or 'man-made', is the response in water to overenrichment by nutrients, particularly phosphorus and nitrogen. 'Man-made' eutrophication, in the absence of control measures, proceeds much faster than the natural phenomenon and is one of the major types of water pollution. The resultant increase in fertility in affected lakes, reservoirs, slow-flowing rivers and certain coastal waters causes symptoms such as algal blooms, heavy growth of certain rooted aquatic plants, algal mats, deoxygenation and, in some cases, unpleasant tastes and odours. These often adversely affect the vital uses of water such as water supply, fisheries, or recreation, and impair aesthetic qualities. In short, man-made eutrophication of inland bodies of water has become synonymous with the deterioration of water quality and is frequently the cause of considerable cost increases.

Eutrophication of water can produce conditions in lakes and slow moving water previously only associated with the stagnant village pond; namely water filled with plant life and little else, evil smelling from lack of oxygen to purify decaying products and generally unacceptable for any recreational or functional use except perhaps by ducks. Furthermore, it is undesirable, and perhaps dangerous, to dump unwanted products of industry, agriculture or sewage into such water since the oxygen deficiency prevents the natural purification processes from operating.

The exact cause of the change from clear water conditions to an explosive growth of algae is unknown. Certainly there must be an excess of nutrients of which nitrates and phosphates from sewage, chemical fertilisers and detergents account for some 90 per cent or more of the man-added nutrient burden. Other chemical or bio-chemical substances present in effluents and land drainage can act as micro-nutrients. Irrespective of the nature of the trigger action which initiates eutrophication, the development of this undesired state in water is certainly made possible by the existence in water of a large added nutrient burden.

Incidence Early in the 1960s, it became obvious that a large number of lakes and reservoirs, particularly those located in industrialized countries, were rapidly changing in character and becoming increasingly fertile (eutrophication) because of addition of plant nutrients originating largely from human activities. The main nutrient sources were municipal and industrial wastewater and agricultural and urban runoffs. The incidence of these conditions, lakes covered at certain periods of the year with algal blooms, the water full of microscopic algae and weeds, with a resultant deoxygenation of the water and incomplete decomposition processes, are difficult to predict because of varying conditions (limnological, ecological or climatological). The change, however, can take place in 10 years or less, thus preventing the use of the lake or river for swimming and many recreational purposes. The use of eutrophied water for drinking or industrial purposes requires expensive treatment processes which may not be really successful in today's state of knowledge (for example, the water taste may not be acceptable). Eutrophic water is poisonous to many species of fish, especially the more desired game fish, and can kill animals drinking it.

Claim Destruction of touristic attraction of a district and loss of water recreation facilities near industrial and urban areas are important consequences to governments under pressure for improved balance of payments and increased standards of living. Economically the penalties are in the clogging of navigational channels, in the need for purification for industrial use and in the elimination of nearby potable water. In general it has proved impossible to eliminate residual organic matter which promotes the growth of bacteria and other micro-organisms in the water distribution system.

Refs Brock, T D *Eutrophic Lake* (1985); Landner, Lars and Wahlgren, Ulf *Eutrophication of Lakes and Reservoirs in Warm Climates* (1988).

Broader Environmental degradation (#PB6384) Destruction of wildlife habitats (#PC0480)
Unsustainable development of fresh waters (#PD6923).
Related Marine oxygen deficiency (#PE6289).

Aggravates Inadequate water system infrastructure (#PD8517).
Aggravated by Wasted water (#PD3669)
Aquatic weeds (#PD2232)
Lake pollution (#PD8628)
Water pollution (#PC0062)
River pollution (#PD7636)
Marine pollution (#PC1117)
Agricultural wastes (#PC2205)
Sewage as a pollutant (#PD1414)
Soil erosion by water (#PD2290)
Detergents as pollutants (#PE1087)
Phosphates as pollutants (#PE1313)
Pollution of inland waters (#PD1223)
Industrial waste water pollutants (#PD0575)
Damage by degradable organic matter (#PJ6128)
Mismanagement of environmental demand (#PD5429)
Environmental hazards from fertilizers (#PE1514).

♦ PD2263 Distrust of political dialogue

Claim The positive aspects inherent in the creative tension of political dialogue has not been incorporated into the political process or seen as a priority of society. The give–and–take necessary for equitable interchange of perspectives and ideas is viewed with distrust. Both establishment and dissident groups see divergent points of view and comprehensiveness as a threat to their internal order.
Broader Distrust (#PA8653)
Inadequate political networks (#PD2213).
Aggravated by Ineffective dialogue (#PF1654).

♦ PD2267 Smoke as a pollutant

Smoke
Nature A cloud of fine particle suspended in a gas, usually as the result of the incomplete burning of biomass and fossil fuels. Smoke pollution cuts out sunlight, leaves a deposit of soot on surfaces, and can be a major lung irritant.
Broader Dust (#PD1245)
Urban–Industrial pollution (#PC8745)
Photochemical oxidant formation (#PD3663).
Narrower Oil smoke (#PG0786)
Health hazards of passive smoking (#PE5146).
Related Smoking (#PD0713).
Aggravates Morbidity (#PD4538)
Heart diseases (#PD0448)
Chronic bronchitis (#PE2248)
Amenity destruction (#PC0374)
Lung disorders and diseases (#PD0637)
Deterioration in atmospheric visibility (#PE2593).
Aggravated by Waste paper (#PD1152)
Urban fires (#PD2211)
Forest fires (#PD0739)
Slashburning (#PE6264)
Industrial emissions (#PE1869)
Motor vehicle emissions (#PD0414)
Domestic refuse disposal (#PD0807)
Non–biodegradable plastic waste (#PD1180).

♦ PD2269 Inadequate land drainage

Inadequate drainage system — Ill–considered land drainage — Inadequate gully maintenance — Inadequate maintenance of drainage ditches — Ineffective temporary drainage — Unstandardized drainage culverts
Nature Large tracts of land are agriculturally unproductive because of excess water in or on the soil, due to inadequate land drainage. Water in excess of that needed by plants may inhibit growth of the economically important parts of the plant. High water content also lubricates the soil particles and frequently leads to unstable conditions unsuitable to machine and other crop operations. In the extreme, when land is drowned, waterlogged or marshy, it may be of no value to crop production. Excess water creates problems in agricultural production over vast areas.
Incidence There are no figures to indicate the scale of the problem on a world wide basis, but it has been estimated that, in the USA alone, about a quarter of agricultural land would benefit from land drainage. Inadequate drainage on irrigated lands presents a different facet of the same problem. Badly drained irrigated lands can become marshy or salty and have to be abandoned. There are throughout the world 20–25 million hectares of now saline and barren lands which were once fertile. The loss of irrigated lands to agriculture is continuing, and every year between 200,000 and 300,000 hectares of irrigated lands fall victim to salinization or revert to marshland.
Counter–claim Land drainage can have negative side effects, such as: the increasing of drought hazard in rain–fed agriculture during dry spells; the decomposition of organic matter in the soil; the disappearance of peats through oxidation upon exposure to air and sunshine; the formation of acids in potentially acid sulphate soils; and the impoverishment of indigenous flora and fauna. The negative effects of land drainage usually owe their source to adverse soil conditions in the area being drained; pollution of drainage water is usually caused by other activities in the area, such as over–fertilization, the application of chemicals, or the use of poor–quality irrigation water. If the water being discharged through the drain is polluted by elements leached from the soil, this can have a harmful effect on the environment. The polluting elements may be fertilizers, toxic chemicals, salts, or acids. These may severely affect the quality of the water, or even render it unfit for further use, and cause the ecosystem to be gravely disturbed. Land drainage as such in areas prone to inundation is mostly an improper measure, as it creates new pathways for the water and may thus strengthen the inundations and cause additional erosion.
Drainage has long been regarded by many decision–makers as a step towards national prosperity because of the increase in agricultural land and the control of flooding. The examples of the unforeseen results of drainage are found in almost every river system of Europe, and in many cases the ill–effects are still accumulating, since the cure to one problem is often the cause of several more.
Broader Underprovision of basic services to rural areas (#PF2875)
Inefficient extraction and utilization of natural resources (#PF2204).
Related Insufficient land upkeep (#PJ8634).
Aggravates Floods (#PD0452)
Drought (#PC2430)
Soil erosion (#PD0949)
Acidic soils (#PD3658)
Alkaline soil (#PD3647)
Soil compaction (#PD1416)
Soil infertility (#PD0077)
Underdevelopment (#PB0206)
Underproductivity (#PF1107)
Soil salinization (#PE1727)
Rising water level (#PD8888)
Soil erosion by water (#PD2290)
Rural underdevelopment (#PC0306)
Shortage of cultivable land (#PC0219)
Destruction of land fertility (#PC1300)
Limited accountability of public services (#PF6574)
Inadequate housing among indigenous peoples (#PC3320).
Aggravated by Silting of water systems (#PD3654)
Unrecognized opportunities (#PF6925)
Fragmented planning of community life (#PF2813)
Undeveloped potential of informal leadership (#PF1196)
Demoralizing images of rural community identity (#PF2358)
Inadequate dissemination and use of available information (#PF1267)
Limited reservoir of technical skills in rural communities (#PF2848)
Insufficient access to technology for agricultural upgrading (#PF3467)
Detrimental effect of jungle environment in tropical villages (#PE2235)
Underdeveloped provision of basic services in developing countries (#PF6473).
Reduces Vulnerability of wetlands (#PC3486).

♦ PD2278 Mine disasters

Mining accidents
Nature Underground fires or explosions may lead to serious mine disasters and may be caused in many ways, in particular by the careless use of naked lights, by damage to electrical cables or other apparatus, or by spontaneous combustion of highly bituminous coals.
Incidence Despite many years of effort to diminish mining risks in general and large–scale accidents in particular, the accident rate in mines still remains high.
Broader Injurious accidents (#PB0731)
Collapsing physical structures (#PD4143).
Aggravates Human death (#PA0072).

♦ PD2283 Disorders of joints and ligaments

Refs Huskisson, E C and Hart, F Dudley *Joint Disease* (1987).
Broader Human disease and disability (#PB1044)
Diseases of connective tissue (#PD2565).
Narrower Bunion (#PG9308)
Bursitis (#PE2320)
Synovitis (#PG4616)
Club foot (#PE3836)
Arthrodesis (#PJ1191)
Tenosynovitis (#PE5606)
Disc disorders (#PG5560)
Tumours of joints (#PG4619)
Ankylosis of joint (#PG3782)
Infections of joints (#PG4617)
Inflammation of tendons (#PG5607)
Joint pains due to torture (#PE3732)
Vertebrogenic pain syndrome (#PE9461).
Related Rheumatic diseases (#PE0873).
Aggravated by Syphilis (#PE2300)
Hysteria (#PE6412)
Haemophilia (#PE1920)
Syringomyelia (#PG2700)
Rheumatic fever (#PE0920)
Arteriosclerosis (#PE2210)
Tuberculous milk (#PG4614)
Injurious accidents (#PB0731)
Overemphasized adult sports (#PG5243)
Diseases of the central nervous system (#PE9037).

♦ PD2284 Diseases of beneficial insects

Nature Diseases of insects are important when the insects concerned are beneficial to man. A diseased insect may be suffering from an infectious disease caused by a microorganism or a noninfectious disease, such as a metabolic disturbance, a genetic abnormality, a nutritional deficiency, a physical or chemical injury, or injury caused by predators or parasites.
Beneficial insects, for which diseases have important consequences for man, include those which produce useful products, such as the silkworm and the honeybee, and those which are predators or parasitic upon various pests. Until recently, concern has focused solely upon the infectious diseases of insects which produce useful products, such as pébrine, flacherie, polyhedroses and muscardine of the silk worm, and foulbrood and nosema of the honeybee. These and other diseases still cause losses to the respective industries. Recently increasing attention has been paid to noninfectious diseases of insects, especially those where injury is caused by chemical agents such as insecticides. Whole populations of honeybees have been wiped out by drifting insecticidal sprays. More disturbing are reports that efforts to kill pests have been more effective in killing their natural predators (either directly through poisoning with pesticides or indirectly through feeding on poisoned prey or through the disappearance of adequate prey), which may result in population explosions of pests, either when chemical control is relaxed or when the pest develops some type of immunity. Furthermore, pesticides applied to the soil indiscriminately destroy harmful and beneficial insects. The latter may be vital in the regeneration of the soil.
Refs Metcalf, et al *Destructive and Useful Insects* (1962).
Related Denial to animals of the right to a natural death (#PE8339).
Aggravates Pests (#PC0728)
Loss of beneficial plants and animals (#PE8717)
Excessive use of chemicals to control pests (#PD1207).
Aggravated by Misuse of pesticides (#PD4629)
Pesticide destruction of soil fauna and micro–organisms (#PD3574).

♦ PD2285 Environmental degradation of desert oases

Destruction of oasis ecosystems
Incidence Most of the world's dry lands are considered to be at 'high risk' or 'very high risk' of losing their ability to support useful species of plant and animal life. As much as 30 percent of the earth's land surface, covering part of one hundred nations and occupied by one sixth of the world's population, is designated as at least 'threatened'.
Broader Environmental degradation (#PB6384)
Destruction of wildlife habitats (#PC0480)
Vulnerability of ecosystem niches (#PC5773).
Related Deforestation (#PC1366)
Natural environment degradation (#PB5250)
Degradation of semi–natural and natural habitats of flora and fauna (#PC3152).
Aggravated by Desert advance (#PC2506)
Mismanagement of environmental demand (#PD5429).

♦ PD2289 Pregnancy disorders

Complications of pregnancy — Difficulty with child–bearing — Transplacental disorders
Nature Nausea, vomiting, fatigue, constipation, dental and skin problems are the common relatively minor complaints of pregnancy. Serious disorders include pernicious vomiting, toxaemias, and haemolytic diseases. Other diseases affect specific parts of the reproductive organs, foetus and embryo.
Refs Bellisario, R and Mizejewski, G J (Eds) *Transplacental Disorders* (1990); Faro, Sebastian and Gilstrap III, Larry C (Eds) *Infections in Pregnancy* (1990); Goldstein, Phillip *Neurological Disorders of Pregnancy* (1986); WHO *The Hypertensive Disorders of Pregnancy* (1987); WHO *Prenatal and Perinatal Infections* (1985).
Broader Human disease and disability (#PB1044).
Narrower Ectopic pregnancy (#PG4643)
Hydatidiform mole (#PG4640)
Anaemia of pregnancy (#PG9577)
Natural human abortion (#PD0173)
Haemorrhage of pregnancy and childbirth (#PE4894)
Toxaemias of pregnancy and the puerperium (#PE8022)
Foetal malformation in diabetic pregnancies (#PE4808).
Related Medical complications (#PE2863).
Aggravates Chorea (#PG3096)
Deafness (#PD0659)
Syphilis (#PE2300)
Haemorrhage (#PE2239)
Tuberculosis (#PE0566)
Hypertension (#PE0585)
Haemorrhoids (#PE3504)
Heart diseases (#PD0448)
Teeth disorders (#PD1185)
Childbirth fever (#PG4639)
Kidney disorders (#PE2053)
Neonatal mortality (#PD9750)
Haemolytic diseases (#PG4641)
Health risks of teenage sex (#PE6969)
Foot diseases and disabilities (#PD2647)
Perinatal morbidity and mortality (#PD2387).
Aggravated by Rickets (#PG2295)
Virus diseases (#PD0594)
Inadequate medical care for pregnant women (#PE4820).

♦ PD2290 Soil erosion by water

Nature Soil erosion occurs primarily when land is exposed to the action of rain. Unprotected by a cover of vegetation, and the binding action of roots, each raindrop hits the naked soil with the impact of a bullet. Soil particles are loosened, washed down the slope of the land, and either end up in the valley below or are washed out to sea by streams and rivers.
Incidence Soil erosion by rain and snow and by streams and rivulets occurs throughout the whole world, though its seriousness varies with environmental conditions. Natural causes result in the washing away of 9.3 billion tons of soil a year, but human intervention increases that figure to approximately 24 billion tons of soil a year, which ends up in rivers and finally in the oceans. The FAO estimates that 11.6 percent of Africa north of the Equator, and 17.1 percent of the Near East, are subject to water erosion, as are 90 million hectares (of a total of 297 million) in India. In Nepal, the removal of topsoil by the monsoon rains does double harm, first by denuding the hillsides, and second by filling the Himalayan rivers with silt. The beds of some rivers now rise a foot every year. The swollen rivers then overflow their banks to progressively wider levels, sometimes even changing course because of the floods, thus drowning the best farmland. The Malagasy Republic is being virtually washed away; in some areas as much as 250 metric tons of soil per hectare is lost every year. The causes are both man–made and natural. Madagascar's soils tend to be

erosion–prone, a condition aggravated by tropical cyclones which can deluge the island with up to 15 mm of rain in 15 minutes.
Refs FAO *Soil Erosion by Water* (1978).
Broader Soil erosion (#PD0949).
Narrower Coastal erosion (#PE6734) Uncontrolled river erosion (#PJ9518).
Aggravates Floods (#PD0452) Aquatic weeds (#PD2232) Rock avalanches (#PG0476) Soil infertility (#PD0077) Silting of water systems (#PD3654) Destruction of land fertility (#PC1300) Slowing growth in food production (#PC1960) Eutrophication of lakes and rivers (#PD2257).
Aggravated by Deforestation (#PC1366) Soil compaction (#PD1416) Diplomatic errors (#PF1440) Inadequate land drainage (#PD2269).

♦ PD2291 Poisonous plants
Nature Poisonous plants produce and accumulate poisons during vital activity; the plants are toxic to animals and humans. The principle active substances are alkaloids, glycosides (including saponins), essential oils, and organic acids. They are generally found in all parts of the plant, but frequently in different quantities.
Incidence Poisonous plants include many mushrooms, equisetics, club mosses, ferns, gymnosperms, angiosperms. In countries with temperate climates they are most widely represented in the families: Ranunculaceae (crawfoot), Papaveraciae (poppy), Euphorbiaceae (spurges), Asclepiadaceae (milkweeds), Apocinaceae (dogbane), Solanaceae (nightshade), Scrophulariaceae (figwort), and Araceae (arum). Many plant poisons are valuable medicinal substances, for example, morphine, strychnine, atropine, and physostigmine.
Refs Hails, Michael R *Plant Poisoning in Animals* (1986); Hirono, I *Naturally Occurring Carcinogens of Plant Origin* (1987); James, Lynn F, et al *The Ecology and Economic Impact of Poisonous Plants on Livestock Production* (1987); Keeler, Richard F, et al *Effects of Poisonous Plants on Livestock* (1978); Leeuwenberg, A J *Medical and Poisonous Plants of the Tropics* (1987); Vahrmeijer, J *Poisonous Plants of Southern Africa that Cause Stock Losses*.
Broader Human disease and disability (#PB1044).
Related Allopathy (#PF7703) Ignorance of traditional herbal remedies (#PE3946).
Aggravates Human poisoning (#PD0105) Animal diseases (#PC0952) Inadequate animal welfare (#PC1167) Lack of integration of traditional and Western medicine (#PF4871).
Aggravated by Overstocking (#PC3153).

♦ PD2294 Individually defined operating structure of marriage
Nature The operating structures of contemporary marriage are defined from an individualistic perspective. When the family becomes dysfunctional or a member of the family is in trouble it is assumed that there is an individual who is responsible rather than a family structure that is responsible. There are few if any symbols of family unity, few rites rehearsing the integrity of the family unit. In fact most of the symbols and images related to the family in society at large point to either its breakup, its incapacity to function effectively or place it in a unrealistic social setting.
Broader Lost family role in society (#PF7456).
Narrower Inadequate image of roles within marriages (#PD1308).
Related Reduced interior structure of families (#PF3783) Vulnerability of marriage as an institution (#PF1870) Individualistic perception of sexual activity (#PF1682) Exclusion of pre–adults from family decisions (#PE2268).
Aggravated by Pneumonia (#PE2293) Tuberculosis (#PE0566).

♦ PD2301 Wood deterioration and decay
Nature Wood in its natural state, whether living or felled, is subject to decay and deterioration which causes serious economic loss. High temperatures, exposure to salt water, and certain climatic conditions favour the development of harmful fungi and insects. All wood and wood products are subject to decay if the moisture content of the wood is at or above the fibre saturation point (approximately 30 per cent), although for significant decay, temperatures in the range of 50 to 90 deg F, a supply of oxygen and a moisture content in excess of 90 per cent are required. (Wood that is dry or immersed in water does not rot).
Mould and stem fungi, confined mainly to sapwoods, cause discoloration, increase the absorptiveness of the wood and reduce its strength. In the USA these fungi cause average annual losses of about US $10 million. Decay fungi reduce the specific gravity and the strength properties of wood. Brown–rot fungi attack the wood cellulose, and white–rot fungi attack both the lining and the cellulose.
Insects can damage the appearance of wood or seriously reduce its strength. Termites attack wood structures both above and below ground level. Bark, ambrosia and power–post beetles cause various kind of damage to wood. Carpenter ants damage partially rotted wood in buildings and poles. Various mollusc and crustacean groups, termed marine borers, damage pilings and boats.
Incidence In the USA alone losses due to wood decay amount to approximately US $300 million annually, not including the cost of protective measures.
Refs Nicholas, Darrel D *Wood Deterioration and Its Prevention by Preservative Treatments* (1984).
Broader Insect damage to stored and manufactured goods (#PD3657).
Narrower Teredos as pests (#PE3624).
Related Insect pests of wood (#PD3586) Plant diseases in storage and transit (#PD3587).
Aggravated by Termites as pests (#PE1747) Domestic refuse disposal (#PD0807).

♦ PD2315 Ancestor worship
Cult of the dead
Incidence Ancestor worship exists among the Bantu in Southern Africa, the Chinese, the Japanese and in the Melanesian and the Polynesian cultures. While it existed in the cultures of Greece, Rome, Babylonia, Ancient Europe and Asia, it mainly survives in folklore, although its influence is present for example in the sraddha rites in India, the imperial ancestor cult in Japan, and the European All Souls Day customs to honour the dead.
Claim Ancestor worship is always associated with tenacious clinging to the old ways, and in modern primitive societies is an obstacle to development. Worship of the dead, particularly of ancestors, stems from the belief that they can intervene in the affairs of the living; it is the fear that this intervention will be detrimental, or even destructive, that motivates the living to offer propitiation to the dead through prayer, ritual, and sacrifice.
Broader Idolatry (#PF3374) Superstition (#PA0430) Communion with the dead (#PF9195).
Related Hero worship (#PF2650).
Aggravates Human sacrifice (#PF2641) Vulnerability of sacred sites (#PD6128).
Aggravated by Fear (#PA6030) Tribalism (#PC1910) Traditionalism (#PF2676) Inflexible social structure (#PB1997).

♦ PD2318 Age discrimination in employment
Constraints against employment of older people — Discrimination in favour of youth in employment
Nature Some workers experience difficulties on account of their age in keeping or obtaining employment. Although the ageing processes advance at different speeds in different individuals, most people reach their peak, physically and mentally, between the ages of 25 and 30, and thereafter decline. Manual skills are at their maximum at an even younger age. Decreasing abilities may be at least partially offset by increasing stamina, experience, and common sense. However, the chance of illness also increases with age.
Broader Human ageing (#PB0477).
Aggravates Absenteeism (#PD1634) Age discrimination (#PC2541) Mental disorders of the aged (#PD0919) Heat as an occupational hazard (#PE5720) Employment discrimination against the elderly (#PD4916).
Aggravated by Grey human hair (#PE6308) Rheumatic diseases (#PE0873) Occupational rheumatism (#PE0502) Ageing of world population (#PC0027) Rigidity and inadaptability in the aged (#PD3515) Slowness of sensori–motor activities in the aged (#PD3514) Retirement as a threat to psychological well–being (#PF1269).

♦ PD2322 High altitude stress
Hypoxia — High altitude — High elevation conditions — Polycthemia — Mountain sickness
Nature High altitudes constitute the natural environment which is probably of greatest stress to man. The shortage of oxygen (hypoxia) and very low barometric pressure can give rise to many physiopathological problems. The condition is aggravated by the increased intensity of ultra-violet radiation from the sun. Acute exposure to high altitude can severely reduce working capacity. Above 3,500 metres, hypoxia may affect the nervous system on initial exposure and altitude sickness may occur, and although individual response as regards severity and duration may vary, working capacity is always affected. Persons transferred to high altitudes, even when acclimatized, remain susceptible to headache, insomnia, hypoxia and related conditions. Persons transferring to and from high altitudes are susceptible to attacks of mountain sickness.
Incidence Approximately 25 million people live and work at altitudes over 3,000 metres.
Refs Winslow, Robert M and C, Carlos Monge *Hypoxia, Polycythemia, and Chronic Mountain Sickness* (1987).
Broader Stress in human beings (#PC1648).
Aggravates Fatigue (#PA0657) Headache (#PE1974) Heart diseases (#PD0448) Sleep disorders (#PE2197).
Aggravated by Loss of humility in relation to the environment (#PF2527).

♦ PD2330 Animal worship as a barrier to development
Nature Animals are worshipped as gods themselves, as representatives of gods, as focal points for rituals, or as central figures in any ceremonial behaviour. Sacrifices often play a prominent role. The strong magical and taboo element in animal worship may form a barrier to development. Worship is inspired by a sense of identification, such as is exemplified in a belief in transformation from one form to another (eg werewolves), and thus may impede adjustment to a modern technical world.
Incidence The totemistic relationship between man and animals may be predominantly magical, concerned with the increase of the species, as it is among the tribes of central Australia. It may stress the taboo aspect, as in Africa, or a personal guardian spirit or animal alter–ego, as in the nagual of the Central American Indians or the panther or serpent of the African Fang. Ceremonial behaviour is characteristic of hunting cultures. Special societies often have animal names or symbols; this custom is still alive among modern societies and clubs. In developed countries the practice of witchcraft or pseudo–witchcraft uses animal worship as part of magic rites. Protection of an animal with a religious identification may strain food resources yet give little in return, as is the case with cows in India.
Background The best–known examples of animal worship in antiquity are to be found in the culture of Ancient Egypt. Animal ancestry and symbolism featured also in Greek civilization.
Broader Taboo (#PF3310) Totemism (#PF3421) Occultism (#PF3312).
Narrower Lycanthropy (#PE8450) Primitive secret societies (#PF2928).
Related Magic (#PF3311) Idolatry (#PF3374) Superstition (#PA0430) Superstitious persecution of animals (#PD3453) Symbols unrelated to human experience (#PF9070).
Aggravates Underdevelopment (#PB0206).
Aggravated by Tribalism (#PC1910) Traditionalism (#PF2676).

♦ PD2337 Art censorship
Censorship of photography
Nature Restriction on expression and content of classical art (such as painting, sculpture, etc) and topical art (such as political cartoons) may be censored as politically harmful or as obscene. Commissioned work may be specified in such a way as to propagandize an ideal. Censorship of art may lead to cultural stagnation and a lack of cultural appreciation, alternatively it may lead to political ignorance, apathy or idealism. For the artist him or herself, it may lead to arrest and imprisonment. Certain art styles may be proscribed because of traditional thinking and lack of appreciation of new ideas.
Incidence Restrictions may be placed on the display of works of art and photographic reproductions considered morally or politically damaging. Photographs may be censored from newspapers and journals, books, exhibitions, postal communications and from theatre and cinema publicity. Censored photographs may be pornographic or unacceptably nude; they may show scenes which would be damaging to government prestige, or they may be a breach of official secrecy. Photographs may be censored by confiscation; books or other publications containing them may be banned, confiscated and destroyed; and cinema publicity nudity may be blacked out at strategic points, rendering the photograph even more titillating than before. Photograph censorship may lead to moralism and guilt complexes regarding sexual matters, or to immorality. Concerning political matters, it may result in arrest and imprisonment.
Broader Censorship (#PC0067).
Related Self censorship (#PF6080).
Aggravates Cultural stagnation (#PC8269) Inadequate appreciation of culture (#PF3408).
Aggravated by Moralism (#PF3379) Traditionalism (#PF2676) Excessive government control (#PF0304).
Reduces Nudism (#PF2660) Obscenity (#PF2634) Pornography (#PD0132).

♦ PD2338 Absentee ownership
Unapproachable absentee landlords — Uninvolved absentee landlords — Obscure land ownership — Outside land ownership — Absentee land ownership — Non–resident home ownership — Absentee business ownership — Absence of land holders — Absentee property ownership — Absentee landlords of urban property
Nature Absentee owners of property tend to lack interest in the function which the property fulfils or could fulfil in the community within which it is located. They may also be indifferent to the problems of tenants or those dependent on the property in question. This may apply not only to owners of agricultural estates but also to owners of large industrial complexes, because they control prices and production although they are not themselves engaged in the work.
In many urban industrial communities, where only a small percentage of the population own their own homes, absentee land owners of residential properties assume little responsibility for property management and improvement, while residents do not see any means of assuming such responsibility. This is intensified by the occasions when, through tenant effort, improvements to property are made, only to have the landlords increase the rent beyond the tenants' capacity to pay. Complaints to management are foreseen as resulting in eviction proceedings. In many cases

the land holders hide behind a screen of agents and false corporations, making residents' access to owners a major research task.
Shifting plans of local governments leave residents with vague and contradictory ideas about zoning regulations; people are uncertain how to plan and, assuming malevolence, fear further encroachment by business, industry and other interests. Many of such economic ventures, although physically situated in an urban community, do not feel they have a responsibility to it. Often the seat of decision-making for such a firm is located in another city or state, and therefore local management sees no responsibility for its empty warehouses and unkempt, deserted buildings. The result is a community that sees no real way to project a local design and improvement plan. It sees itself as powerless, with the control of its destiny in someone else's hands.
Incidence In the USA two–thirds of land rented out is owned by absentee landlords. In Latin America up to 90 percent of large landowners are absentee. The practice is also prevalent in Asia and Africa.
Background Absentee ownership has been a political and social issue for centuries, well–known examples being the court mobility in pre–revolutionary France and English landowners in 19th century Ireland.
Broader Social inaccessibility (#PC0237) Lack of community participation (#PF3307)
Uncoordinated use of community property (#PF3005)
Ownership as a basis for land allocation (#PF6460).
Narrower Absentee owned natural resources (#PG8995)
Land controlled by business groups (#PG7736).
Related Uncertainty of land zoning (#PGp739).
Aggravates Fear of reprisals (#PF9078) Inadequate housing (#PC0449)
Deteriorated vacant houses (#PJ8678) Demeaning community self–image (#PF2093)
Unethical real estate practice (#PD5422) Inadequate care of community space (#PF2346)
Lack of finance in coastal communities (#PE2425)
Ineffective utilization of public environment (#PF6543)
Stagnated development of agricultural production (#PD1285)
Disorganized approach to land ownership in tropical villages (#PF2365).
Aggravated by Capitalism (#PC0564) Foreign ownership (#PE4738)
Proliferation of second homes (#PF1286)
Inhibiting effects of traditional life–styles (#PF3211).

♦ PD2341 Homicide
Murder — Manslaughter — Premeditated killing — Execution — Strangulation
Nature Homicide can be criminal (murder and manslaughter) or non–criminal (executions authorized by civil or military authorities and killing by police authorities to prevent commission of a felony); but nonetheless the fact that humans kill one another is unacceptable by any definition. Realistically, while the circumstances can be comprehended for some of these deaths, (excusable homicides may occur by misadventure, and in some cases as self–defence), the millions of homicides annually occurring world–wide indicate the inadequate nature of the attempts at mitigation. The most common method is strangulation of women and stabbing of men, and the most vulnerable group to be murdered are babies under the age of one year.
Incidence In the UK, for example there are approximately 750 murders per year, as against approximately 22,500 for the USA. In the UK, this constitutes 9 women per million population or 14 men per million. The murder rate has increased progressively since 1945, being less than 250 per year in the UK during the 1950s. One index of society's increasingly violent nature is the steady rise in the proportion of murders carried out by people unknown to the victim. In 1980 in the UK, about 20 per cent of murders were by strangers increasing to approximately 30 per cent in recent years (37 per cent for men). Of those killed by people they knew, only 6 per cent of men were killed by spouse or lover, compared 37 per cent in the case of female victims. Men and women are equally likely to die at the hands of their children who are responsible for 14 per cent of the victims. But 3 per cent of men and 6 per cent of women were murdered by a parent with 68 babies under one year murdered per million. In the USA in 1990 homicides increased by 8 per cent over the previous year to an estimated 23,220, namely 2,000 than in the previous year, and up from 9,110 in 1960.
Refs Allen, Nancy *Homicide* (1980); Daly, Martin and Wilson, Margo *Homicide* (1988); Gardiner, Muriel and Spender, Stephen *The Deadly Innocents* (1985); Jerath, Bal K *Homicide – A Bibliography* (1984); Jerath, Bal K, et al *Homicide – A Bibliography of over 4,500 Items* (1982).
Broader Human death (#PA0072) Killing by humans (#PC8096)
Denial of right to life (#PD4234).
Narrower Lynching (#PE6287) Senilicide (#PJ7124) Infanticide (#PD3501)
Mass–murder (#PJ0457) Parenticide (#PE0651) Manslaughter (#PD7952)
Assassination (#PD1971) Human sacrifice (#PF2641) Serial killings (#PE9447)
Hunting of humans (#PD5493) Justifiable homicide (#PE7533)
Illegal induced abortion (#PD0159) Negligence in manslaughter (#PE0437)
Intentional fatal poisonings (#PG1934) Government sanctioned killing (#PD7221).
Related Suicide (#PC0417) Genocide (#PC1056) Duelling (#PF5382)
Induced abortion (#PD0158) Killing non–human life (#PF6359)
Prenatal wrongful death (#PD6967) Criminal killing of animals (#PJ1158)
Criminalization of euthanasia (#PF2643)
Offences involving danger to the person (#PD5300).
Aggravates Inhumanity of capital punishment (#PF0399).
Aggravated by Drunkenness (#PE8311) Gang warfare (#PD4843)
Armed robbery (#PE4739) Human poisoning (#PD0105)
Organized crime (#PC2343) Unsafe transport (#PG4741)
Injurious accidents (#PB0731) Inadequate firearm regulation (#PD1970)
Police crimes during narcotic investigations (#PE5037)
Failure of disarmament and arms control efforts (#PF0013).

♦ PD2345 Environmental degradation of suburbia
Destruction of green inner suburbs
Nature The suburbs in developed countries are boring and visually depressing. The ecology of suburbs is destroyed by concretization and over–cultivation of gardens.
Broader Uncontrolled urban development (#PC0442).
Aggravates Over–cultivated gardens (#PE7265).

♦ PD2363 Abuse of coca and cocaine
Cocaism — Long–term effects of cocaine abuse
Nature The use, distribution and sale of cocaine and its by products are major social and health problems. The social problems include crime: murder, smuggling, selling illegal drugs and laundering money; prostitution; degeneration of the family; losses in productivity for businesses; increasing medical costs; and corruption of police and other government officials.
In addition to addiction, cocaine chief effects are on the nervous system and the blood vessels. Blood vessels constrict when cocaine is used causing a quick rise in blood pressure sometimes causing an angina. The flow of oxygenated blood to the heart is reduced causing damage to heart cells, irregularity or increase in heartbeat and heart attacks. Those who have not previously had heart problems can develop them over time. Cocaine attacks the brain and nervous system causing constricted arteries, broken arteries, stokes, seizures, tremors delirium and psychosis. Blood supplies to the intestines are restricted. Cells in the liver are destroyed. The lungs accumulate fluid. When cocaine is snorted the cells in the lining of the nose are damaged and the sense of smell is lost. With long term use, men may experience difficulty in maintaining an erection or in ejaculating and women may find it difficult to reach orgasm. Cocaine use during pregnancy can cut off oxygen to the foetus, fatally or with complications as premature delivery or detachment of the placenta from the womb. New born babies may be under weight or suffer from withdrawal symptoms: tremors, mood swings and irritability.
Coca dulls both mind and body, impairing the intelligence of the user's children, and increasing occupational risk when the user is working with a machete or other dangerous instruments. Excessive consumption of coca renders the individual more susceptible to disease and early death. Coca leaves (from an evergreen shrub native to Western South America and certain Far Eastern regions) may be chewed or used for the manufacture of cocaine, a powerful stimulant drug. Coca chewing has a debilitating effect, reducing both physical capacity and sensations of hunger. Cocaine intake produces a sense of euphoria and increased muscular strength, but also hallucinations, anxiety and fear; stimulation is followed by depression, dependence being mental rather than physical; overdose may be fatal.
Incidence There are over 6 million abusers of these substances. Three– quarters are cocaine abusers, mainly in the USA. Cocaine abuse is found more often among rich communities. Coca–leaf chewing is a traditional practice among South American Indians and although officially recognized to be harmful, it is encouraged by their present–day employers, some of whom even pay partly in coca–leaf. Coca–leaf chewing enhances their poverty, malnutrition and exploitation, keeping peasant populations in bondage through debts incurred, and therefore undernourished because of the need to pay debts in kind (food, animals). The excessive consumption of coca among South American Indians is most marked in the high plateau lands of Peru and Bolivia but is also noted in other South American countries. Coca is used mainly to dull the pangs of hunger, fatigue, disease and mental depression brought on by hard labour and inadequate wages. The traditional belief that coca is mainly beneficial is encouraged, and it is even declared a 'food' by certain Bolivian landowners.
In the USA the estimated street consumption of cocaine soared from 31 metric tons in 1982 to 72.3 metric tons in 1985. According to a National Survey on Drug Abuse, 22.2 million people said they had tried cocaine or crack at least once, while 5.8 million said they had used it in the last 30 days. In 1990 study estimates that nearly 2.2 million Americans are using cocaine at least once a week.
Background Coca was used for centuries by South American Indians and since 1884 in western medicine as a local anaesthetic. Sigmund Freud used it in the belief that it was a cure–all before he recognized the threat of addiction. In the 1880's cocaine for inhalation was easily purchased at the corner drugstore in the US. It was also available in forms to be smoked. In the 1890's Cocaine reached its peak per–capita consumption in America that probably never been equalled. As users became addicted, became involved in violent crime and became paranoid and as the drug became associated with prostitution and criminal gangs social concern grew. In 1906 the Food and Drug Act required any over–the–counter remedy containing cocaine list it as an ingredient. In 1914 the federal government enacted a law controlling the distribution of opiates and cocaine. The state of New York outlawed the use of cocaine for any reason including medical ones. Cocaine became the archetype of the dangerous drug. About 1970 the drug reappeared in the US.
Refs Allen, David F *The Cocaine Crisis* (1987); American Health Research Institute Staff *Medical Subject Research Index of International Bibliography Concerning Cocaine* (1982); Anglin, Lise *Cocaine* (1985); Brink, Carla J (Ed) *Cocaine* (1985); Washton, Arnold M *Cocaine Addiction* (1989).
Broader Abuse of plant drugs (#PD0022).
Narrower Crack (#PE2123) Basuco (#PE5245) Cocaine withdrawal (#PG4267)
Cocaine intoxication (#PG6889).
Related Smoking during pregnancy and breast–feeding (#PE5026).
Aggravates Apathy (#PA2360) Malnutrition (#PB1498) Drug smuggling (#PE1880)
Heart diseases (#PD0448) Liver diseases (#PE1028) Mental depression (#PC0799)
Drug abuse among athletes (#PE4250) Perinatal morbidity and mortality (#PD2387).
Aggravated by Social injustice (#PC0797) Illicit drug trafficking (#PD0991)
Manufacture of illicit drugs (#PE2512) Underprivileged racial minorities (#PC0805)
Malnutrition among indigenous peoples (#PC3319).

♦ PD2366 Underground press
Illicit literature — Clandestine press
Nature Illegal or semi–legal newspapers, news sheets and journals may be published in opposition to the traditionally established press. Such periodicals may contain information of a politically or morally subversive nature, or even be actively anti–governmental.
Counter–claim The underground press may be the only means whereby injustices, inequalities and official hypocrisy may be exposed regarding pornography, homosexuality, discrimination (racial, ethnic, women, etc), drug addiction, radicalism, war, sex, police brutality and similar issues.
Refs Miles, *The Official History of the Underground Press* (Date not set).
Broader Subversive activities (#PD0557) Subversion of democracy (#PD3180)
Newspaper and journal propaganda (#PD0184).
Related Student press weakness (#PE0628) Lack of freedom of the press (#PE8951).
Aggravates Social breakdown (#PB2496).
Aggravated by Moralism (#PF3379) Alienation (#PA3545)
Authoritarianism (#PB1638) Newspaper monopoly (#PE0246)
Newspaper and periodical censorship (#PD3027).
Reduces Propaganda (#PF1878) Compulsory indoctrination (#PD3097).

♦ PD2374 Vulnerability of small nations to foreign intervention
Vulnerability of small states — Plight of small states within the world–system — International insecurity of small states
Nature Small nations without the resources to withstand foreign influence are vulnerable to intervention, interference in their affairs, economic coercion, diplomatic and political pressure, external attack and possible annexation. Although they may play the superpowers off against each other to obtain favourable terms and the maximum degree of independence, and although most are likely to remain at least nominally independent, they tend to be very heavily influenced by one superpower in political, economic, social and ideological matters. Others may be coerced into obedience (such as Czechoslovakia after 1968). The international community does not always respect their independence nor feel an obligation to provide effectively for their territorial integrity. Four security problems generally exist for small states faced with either external attack or internal uprising: a lack of trained security forces; a lack of arms and training in their use; inadequate surveillance of exclusive economic zones and/or illegal trade such as drug–running; and an inability to gather and process information and intelligence.
Incidence Third world nations are at the greatest disadvantage regarding infringements of their independence. They may be heavily in debt to countries which invest in their natural resources and may be unable to maintain national unity. In extreme social and national disintegration recourse may be had to armed forces which will be supplied from a foreign power. If a faction wins power in this way it will owe allegiance to the foreign power and be dependent on it for its strength (such as South Korea).
Refs Ayoob, Mohammed *Conflict and Intervention in the Third World* (1980); Commonwealth Secretariat *Vulnerability* (1985).

PD2374

Broader International insecurity (#PB0009) Threatened and vulnerable minorities (#PC3295)
Foreign intervention in internal affairs of states (#PC3185).
Narrower Protectorates (#PG4803) Military dependency (#PG4802)
Economic dependence (#PF0841) Political dependency (#PG4801).
Related Threatened sects (#PC1995) Compliance*complex (#PA5710)
Endangered tribes and indigenous peoples (#PC0720).
Aggravates Annexation (#PE5210) Exploitation (#PB3200)
National disintegration (#PB3384) National political dependence (#PF1452).
Aggravated by Human dependence (#PA2159) Second class states (#PD0579)
Economic imperialism (#PC3198).
Reduced by Fragmented regional cooperation (#PF9129).

♦ PD2387 Perinatal morbidity and mortality

Perinatal morbidity

Nature Perinatal mortality comprises neonatal deaths and still-births. The perinatal mortality in an area is very much an indication of the quality of the maternal of maternal and child health services in general, and of the quality of antenatal and maternity care in particular. The causes of perinatal mortality are those of the component problems.

Incidence Perinatal mortality ratios (PMRs) reported by individual countries for 1965 varied from 18.3 to 82.0 per 1000 live births. In developing countries, where 72% of the world's population live, the risk of death in the perinatal period is between 40 and 80 per 1000, a figure several times higher than that of developed countries. Sample surveys suggest that even higher PMRs may occur. The influence of poor diet, disease, and impoverished social environment may result in a higher mortality rate after the first week and especially between 1 and 4 years; thus, for 1000 live births, 300 children might die before the age of 5 years. On the other hand, a number of countries, mostly highly developed, reported PMRs of under 25, less than half those recorded 30 years ago. Nonetheless, in developed countries perinatal mortality is a more important problem than is mortality after the first week of life. Some European countries have mortality rates of less than 10 per 1000 live births for the period one week to one year of age One of the targets of the 1974 World Population Plan of Action specified that "countries with the highest mortality levels should aim by 1985 to have an infant mortality rate of less than 120 per 1000 live births. Although infant mortality rates have dropped in virtually all regions of the developing world, this goal has not been met by the countries with the highest infant mortality levels. The gap between the average level of infant mortality between developed and developing countries is very wide - 17 infant deaths per 1000 live births in developed countries compared with 92 per 100O in the developing world.

Refs Milunsky, Aubrey, et al (Eds) *Advances in Perinatal Medicine* (1981); WHO Expert Committee, Geneva *Prevention of Perinatal Mortality and Morbidity* (1970); WHO Seminar, Tours *Prevention of Perinatal Mortality and Morbidity* (1972).
Broader Human death (#PA0072) Infant mortality (#PC1287)
Human disease and disability (#PB1044).
Narrower Still-birth (#PD4029) Neonatal mortality (#PD9750)
Foetal infection and death (#PE2041).
Aggravates Decrease in mortality rate (#PF0333).
Aggravated by Rubella (#PE0785) Obesity (#PE1177)
Tetanus (#PE2530) Diabetes (#PE0102)
Syphilis (#PE2300) Malnutrition (#PB1498)
Virus diseases (#PD0594) Birth injuries (#PE4828)
Premature birth (#PD1947) Maternal mortality (#PD2422)
Pregnancy disorders (#PD2289) Human physical suffering (#PB5646)
Abuse of coca and cocaine (#PD2363) Health risks of teenage sex (#PE6969)
Haemolytic disease of the new born (#PE2399) Human physical genetic abnormalities (#PD1618)
Malnutrition among indigenous peoples (#PC3319).
Reduced by Inadequate medical care for pregnant women (#PE4820)
Foetal malformation in diabetic pregnancies (#PE4808).

♦ PD2389 Genetic defects and diseases

Hereditary defects and diseases — Genetic defects — Genetic errors

Nature Genetic defects and diseases are a major contributing factor to human mortality and morbidity. The burden imposed upon the individual, the family, and society by genetic disease and genetically determined detrimental deviations from normal, although varying according to situation, is considerable. Genetic disorders can be either chromosomal or genic. Chromosomal disorders involve the lack, excess, or abnormal arrangement of chromosomes. Genic disorders are caused by either major mutant genes or by the interaction of many genes, each with a small effect. Environmental factors often cause modifications as well, and sometimes involve a developmental or physiological threshold. The origin of genetic disorders is mutation. Broadly defined, mutation refers to any stable change in the genetic material; it includes both numerical and structural chromosomal aberrations, such as extra or missing chromosomes, inversions, translocations, duplications, and deletions, as well as the whole range of single-gene alterations such as duplications, frame-shift insertions or deletions, inversions, and base-pair substitutions It is recognized today that between three and four percent of all individuals are born with genetic anomalies. One-third of these are attributable to gene-mutations, whose frequency is 1 per 1,000 individuals (per generation and per gene): the diseases in this group include haemophilia, albinism and various metabolic diseases (alkaptonuria, phenylketonuria, galactosemia, etc). Another third are due to chromosomal aberrations - one out of every hundred persons is born carrying an anomaly of this type. Trisomy 21 (Down's syndrome, formerly called mongolism) is the most frequent example. The rest are induced by untoward events during the intra-uterine life of the foetus such as X-rays or other forms of radiation, rubella (German measles) or certain other viruses, and chemical substances. The vast majority of the 'errors' caused by chromosome rearrangements or by gene mutations are of a non-beneficial character for the individual affected Each category of genetic disease presents different problems with respect to causes, prevention, diagnosis and treatment. There are 3,000 to 4,000 human genetic diseases, 500 of them linked to a defect in a single gene. Many of the disorders are extremely rare, but they also include: cystic fibrosis, sickle-cell anaemia, haemophilia, and Tay-Sachs disease.

Incidence It has been estimated that 10 per cent of all admissions to hospitals in industrialized countries are for genetic diseases. In the UK, about 15,000 children are born every year suffering from severe genetic physical and mental disease and genetic predispositions. Infants with defects are increasing at a rate of 12 per cent per year and 40 percent of paediatric deaths in the UK are more or less directly related to genetic disease.

Refs Evans, H J and Lloyd, D *Mutagen-Induced Chromosome Damage in Man* (1979); Fox, Richard G (Ed) *Extra Y Chromosome and Deviant Behaviour* (1970); Schulman, Joseph D and Simpson, Joe L *Genetic Diseases in Pregnancy* (1981); Valentine, G H *The Chromosomes and Their Disorders* (1986).
Broader Children's diseases (#PD0622) Human physical suffering (#PB5646)
Human disease and disability (#PB1044).
Narrower Diabetes (#PE0102) Genetic deafness (#PG9360) Friedreichs ataxia (#PE8605)
Nutritional anaemia (#PD0321) Hereditary diseases (#PG7966)
Hereditary regression (#PD8149) Hereditary opticatrophy (#PG4816)
Criminal characteristics (#PF5544) Mucopolysaccharide diseases (#PE4414)
Hereditary metabolic diseases (#PG4843) Genetic susceptibility to disease (#PG2834)
Human physical genetic abnormalities (#PD1618)
Compounding effect of treating genetic diseases (#PF7727)
Hereditary disorders of the central nervous system (#PE7915).
Aggravates Asthma (#PD2408) Albinism (#PE2332) Ugliness (#PA7240)
Hay fever (#PE6197) Haemophilia (#PE1920) Heart diseases (#PD0448)
Hypersensitivity (#PE6898) Mental deficiency (#PC1587) Physical blindness (#PD0568)
Health inequalities (#PC4844) Reading disabilities (#PD1950) Animal abnormalities (#PD4031)
Diseases of metabolism (#PC2270) Vulnerability of organisms (#PB5658)
Criminalization of euthanasia (#PF2643) Vulnerability of plants and crops (#PD5730)
Decline in human genetic endowment (#PF7815)
Decreasing genetic diversity of animals (#PC1408)
Decreasing genetic diversity in cultivated plants (#PC2223).
Aggravated by Incest (#PF2148) Chance (#PA6714)
Mutation (#PF2276) Eugenics (#PC2153)
Genetic inbreeding (#PD7465) Injurious accidents (#PB0731)
Consanguineous marriage (#PC2379) Lack of eugenic measures (#PD1091)
Environmental influences (#PG4844) Parochial scientific view (#PF1418)
Radioactive contamination (#PC0229) Mutagenic effects of drugs (#PE4896)
Lack of genetic counselling (#PG4845) Health hazards of radiation (#PD8050).

♦ PD2392 Inadequate drug quality control

Nature Manufactured drugs are not subject to the same quality controls as certain other commodities. Many contain the same active principle but vary in therapeutic effectiveness. The absence of standards and equivalences is a threat to public health and often results in unnecessary, costly purchases of ineffective medicines. If, at the national level, official supervision is inadequate, transfer to the international level by exporting the product magnifies the problem. Inadequate testing is compounded by a lack of internationally accepted standards for classifying and packaging drugs.

Incidence Aminophenazone, chloramphenicol clioquinol, diphenoxylate, phenylbutazone, practotol, tomotil and thalidomide are some of the drugs whose dangers were found out too late. In the USA, for example, it is thought that as many as 600 drugs are either ineffective or have minimal benefits which do not outweigh their side effects. As many as one in eight prescription drugs fall into this category, with a retail value of over $1,000 million.
Broader Inadequate drug control (#PC0231) Inadequate information on drugs (#PF0603).
Narrower Inadequate drug classification (#PG4852).
Related Irresponsible pharmaceutical advertising (#PE2390).

♦ PD2408 Asthma

Nature If asthma is caused by obstructed respiratory system it is said to be bronchial; if red blood cells fail to carry oxygen away from the lungs, it is cardiac. The three cardinal features of bronchial asthma are shortness of breath and other signs of oxygen deficiency (such as laboured breathing blueness of lips or a rapidly beating heart); wheezing (a unique sound made by air passing through narrowed, mucus-filled bronchi); and coughing, a reflex that produces expectoration. Asthma has specific causes (such as allergies and infections) but heredity, environment, hormones and stress are important factors determining its course. This disease has a high morbidity with a surprisingly low mortality, though its treatment may have negative side-effects.

Incidence Ten to 15 million Americans suffer from varying degrees of asthma and nearly 4,000 a year die from it. In the UK it causes 2,000 deaths a year. Asthma sends more children to the hospital than any other complaint. It is the leading reason that children miss school and adults miss work. Although asthma is a treatable disease, its incidence is on the increase.

Claim The drugs that treat the symptoms of asthma for long periods do not treat the disease and drugs that treat the disease offer no short term relief from the symptoms. Patients tend to stop treatment of the disease and continue treatment of symptoms. This leads to continued damage of the airways.
Broader Human disease and disability (#PB1044)
Diseases of the respiratory system (#PD7924).
Narrower Cardiac asthma (#PG4866) Potters' asthma (#PG3552)
Kidney disorders (#PE2053) Bronchial asthma (#PE3860).
Related Eczema (#PE2465) Allergy (#PE1017).
Aggravated by Sinusitis (#PG4873) Hay fever (#PE6197)
Infection (#PC9025) Tonsillitis (#PE2292)
Bright's disease (#PE2272) Bacterial disease (#PD9094)
Chronic bronchitis (#PE2248) Emotional disorders (#PD9159)
Allergy inducing pollens (#PG3219) Genetic defects and diseases (#PD2389)
Aerosols as industrial hazards (#PE1504) Symptoms referable to sense organs (#PE2665).
Reduced by Smoking (#PD0713).

♦ PD2413 Currency black market in socialist countries

Nature The currency black market in socialist countries reinforces a form of internal economic discrimination since one of the ways Eastern bloc countries earn hard currencies such as dollars is to sell scarce consumer goods in stores that only accept hard currency. Even when the shelves in state stores are empty, there is ample stock in hard-currency stores. So the greater the shortages in state stores, the higher the premium consumers will pay for dollars on the black market, which is the only place they can get them.

Incidence In Poland, at the beginning of 1983, the black-market premium was about six times the official rate. In Romania, it was 250 percent. In the other socialist countries, the black-market rate lies between 50 percent and 200 percent above the official rate. Only in Hungary, which enjoys an ample supply of consumer goods, is it under 25 percent.
Broader Currency black market (#PD5905)
Illegal private profit in socialist countries (#PC0939).
Aggravated by Declining productivity in socialist countries (#PF7610).

♦ PD2419 Hybridization of wild animal species

Nature Some formerly pure stocks of wild animals are being modified or even destroyed by interbreeding and hybridization with introduced stocks (as with dogs), or with animals better able to compete in an environment altered by human activities.
Broader Environmental degradation (#PB6384).
Related Hybrids (#PF3646)
Denial to animals of the right to the attention, care and protection of humankind (#PF5121).
Aggravates Endangered species of animals (#PC1713).

♦ PD2420 Proliferation of technology

Dependence on technology

Refs HÖll, Otmar *Austria's Technological Dependence* (1980).
Related Proliferation of commercialism (#PF0815).
Aggravates Scientism (#PF3366) Agnosticism (#PF2333)
Technological monoculture (#PF4741) Blind faith in technology (#PF4989)
Disintegration of organized religion (#PD3423)
Dehumanization of man in the technological process (#PF5438)
Dependence on sophisticated technology for development (#PD6571)
Dependence of developing countries on imported technology (#PF1489).
Reduces Unsustainable economic development (#PC0495).

♦ PD2422 Maternal mortality

Maternal deaths

Nature The term maternal mortality is used to refer to all deaths ascribed to childbirth and puerperium.

DETAILED PROBLEMS

PD2483

Incidence Half a million women a year die as the result of pregnancy or pregnancy related causes. All but 6,000 of these are in the developing world. Pregnancies which happen before a woman is biologically ready for childbearing; pregnancies which occur towards the end of her reproductive span and pregnancies which follow many previous birth carry additional biological and behavioural risks independently of a woman's setting or life circumstances.

In a number of developing countries, maternal mortality rates in excess of 500 per 100,000 live births are by no means exceptional, and rates of over 1000 per 100,000 have been reported in parts of Africa. In a survey in rural areas in Bangladesh, maternal mortality was found to be 570 per 100,000 live births, the mortality rate for the youngest group of mothers being as high as 1770 per 100,000. Maternal mortality accounted for 57% of deaths of women aged 15–19 years in the area and 43% of deaths of women aged 20–29. In Afghanistan, maternal mortality has been estimated to be around 700 per 100,000 live births. Variations within countries are considerable. In Afghanistan, for example, the urban rate was almost half the figure quoted above, and in Malaysia the highest rate for a district was 18 times the lowest rate.

It has become increasingly obvious that the high mortality of women in developing countries in their middle years is a cause for grave concern. Age-specific death rates for women rise sharply between the ages of 20–30 in many countries, where women often have less chance than men of surviving the years between 15 and 45. In a number of countries in Asia, life expectancy at birth is actually lower for women than for men. Despite the known underrecording, maternal causes are still among the leading causes of death for women in the child-bearing ages. In almost all developing countries, deaths from maternal causes are among the 5 leading causes of death for women aged 15–44; in one-third of these countries they come first or second. Because childbearing is spread over more years in developing than in developed countries, women in developing countries not only undergo a higher risk per pregnancy but they are at risk over a longer period of their lives. It has been estimated that, in the areas with the highest maternal mortality (such as most of Africa and West, South, and East Asia) about half a million women die from maternal causes every year, leaving behind at least one million motherless children. In Latin America, the maternal mortality rates are much lower, but several studies have shown serious underreporting of maternal causes of death; in some countries up to half of such deaths were not reported accurately.

Postpartum haemorrhage, often with anaemia as an underlying or associated cause, and sepsis, are the most frequent causes of maternal deaths and are directly related to the absence or inadequacy of prenatal and delivery care. In addition, hypertensive disorders of pregnancy – toxaemia – are important, not only in the developed countries where they account for 23–25% of all maternal deaths, but probably even more in the developing countries.

Refs International Bank for Reconstruction and Development *Preventing the Tragedy of Maternal Deaths*; United Nations *World Population Trends and Policies – 1987 Monitoring Report – Special Topics*.
Broader Human death (#PA0072).
Aggravates Perinatal morbidity and mortality (#PD2387).
Aggravated by Induced abortion (#PD0158)   Lack of family planning (#PF0148)
Complications of childbirth (#PC9042)
Haemorrhage of pregnancy and childbirth (#PE4894)
Toxaemias of pregnancy and the puerperium (#PE8022).

♦ **PD2438 Lack of means for achieving consensus**
Undetermined consensus procedures — Unclear consensus patterns — Incomplete consensus method — Inadequate consensus process — Inadequate consensus structure — Insufficient need for consensus — Collapsed consensus structure
Nature There is a lack of comprehensive models for organizing consensus and forming objective opinions. Groupings find it hard to form a strong consensus beyond a common dissatisfaction with the status quo.
Incidence These difficulties are evident in the Peace Movement, where highly diverse groups congregate without methods for building a comprehensive peace plan. Another example is the fragmentation of opposition parties in India.
Broader Ineffective structures of local consensus (#PF6506)
Inadequate procedures for community planning (#PF0963)
Ineffective structures for community decision-making (#PF1781).
Related Collapse of societal engagement (#PF2340)
Unstructured local decision-making (#PF6550)
Deteriorated structures of essential corporateness (#PF1301).
Aggravates Disagreement within alliances (#PD2629)
Lack of responsible involvement in community affairs (#PF6536).
Aggravated by Ineffective dialogue (#PF1654).

♦ **PD2458 Financial scandal**
Financial fraud
Broader Scandal (#PC8391).
Narrower Bank fraud (#PE1398)   Insider dealing (#PD3841)
Theft of property (#PD4691)
Inadequate management of government finances (#PF9672).
Related Bribery (#PC2558)   Tax evasion (#PD1466)   Embezzlement (#PD2688)
Political scandal (#PD4651)   Theft of works of art (#PE0323)
Unethical financial practices (#PE0682).
Aggravated by Unethical commercial practices (#PC2563).

♦ **PD2459 Inadequate personal hygiene**
Bodily uncleanliness — Lack of cleanliness
Broader Inappropriate basic hygiene (#PD8294)   Inappropriate personal habits (#PD5494).
Narrower Irregular cleanup campaigns (#PU8849)
Unchanging sanitation practices (#PU5422).
Related Unsociable human physiological processes (#PF4417).
Aggravates Acne (#PE3662)   Leprosy (#PE0721)   Dandruff (#PJ5412)
Hepatitis (#PE0517)   Ascariasis (#PE2395)   Bad breath (#PE6558)
Spiritual impurity (#PF6657)   Industrial accidents (#PC0646)
Human disease and disability (#PB1044)   Disagreeable human body odour (#PE4481)
Personal physical unattractiveness (#PF4010).
Aggravated by Ignorance (#PA5568)   Inefficiency (#PB0843)
Mental illness (#PC0300)   Unhygienic conditions (#PF8515)
Socio-economic poverty (#PB0388)   Inadequate standards of living (#PF0344)
Negative emotions and attitudes (#PA7090).

♦ **PD2463 US dollar dominance of world economy**
Overvalued dollar — Weakness of dollar
Incidence The rise in the cost of the dollar with respect to other currencies – about 70 percent since 1980 – resulted in American farmers and businessmen being unable to compete in overseas markets; domestic American industries being wiped out because foreign good cost much less than domestic products; developing countries' debts rising every time the dollar goes up; and the USA trade deficit reaching $150 billion in 1985, thus making the USA a debtor nation for the first time since World War I.
Claim The world economy is dominated by the United States, whose dollar has superseded the established gold standard for world economy. The danger in allowing one nation's economy to support the equilibrium of the global economy is twofold: fluctuating trust in the United States' economy leads to world-wide monetary instability; and national pride in the dollar has become unduly inflated. The risk of renewed weakness of the dollar in the 1990s with rapid depreciation could push up US inflation, raise domestic interest rates, and –if international investors become reluctant to hold dollar assets at prevailing yields – cause instability in financial markets.
Broader Misalignment of currencies (#PF6102)
Variations in national forms of currency (#PF2574).
Aggravated by Instability of economic and industrial production activities (#PC1217).

♦ **PD2468 Inadequate labelling of dangerous substances**
Nature After a dangerous substance has been packaged, it may be handled by many different people in many locations before it reaches its point of usage. The handling or use of such products may lead to accidents, yet the label does not always contain sufficient information for the correct counteractant or antidote to be determined rapidly. Furthermore, since such packages may pass through countries using different languages, the description on the label may be incorrectly interpreted or be completely incomprehensible.
Broader Dangerous substances (#PC6913)   Inappropriate labelling (#PD3521).
Aggravates Transport of dangerous goods (#PD0971).
Aggravated by Multiplicity of languages (#PC0178)
Illiteracy in developed countries (#PC1383).

♦ **PD2469 Parochial monetary agreements**
Nature An international monetary agreement would help provide the stabilization necessary for universal economic subsistence; but, because nations continue to fear loss of their individual power, national governments have failed to establish concrete global agreements which include the type of commitment and mutual trust necessary for a universal monetary system. By default, all countries in fact participate in the present reduced monetary systems in which emerging nations are tyrannized by powerful nations.
Broader Restrictive monetary practices (#PF8749)
Variations in national forms of currency (#PF2574).

♦ **PD2474 Humidity**
Hot humid climates
Nature High humidity combined with high temperature, as encountered in the tropical zones, provide an ideal incubation condition for fungi which can cause tremendous damage within a few weeks. Though not as severe in the temperate zones, humidity can still render improperly designed or unprotected equipment inoperative. Even properly designed equipment will encounter difficulties under some conditions.
Broader Inhospitable climate (#PC0387).
Aggravates Hookworm (#PE3508)   Household pests (#PD3522)
Food spoilage in storage (#PD2243)   Corrosion of iron and steel (#PE1945)
Electronic equipment failure (#PD1475)
Deterioration of stored documents and archives (#PE1669)
Destructive action of mould in tropical climates (#PE1265).

♦ **PD2476 Dependency of children**
Broader Annulment of adoption (#PF3281)   Lack of protection for the vulnerable (#PB4353).
Narrower Refusal of adoption (#PF3282)   Inadequate laws of adoption (#PD0590).
Related Dependency of women in marriage (#PD3694)
Isolation of parent-child relationship (#PC0600).
Aggravates Child-marriage (#PF3285)   Cruelty to children (#PC0838)
Emotional disorders (#PD9159)   Inadequate child welfare (#PC0233)
Exploitation of children (#PD0635)   Exploitation of child labour (#PD0164)
Emotional dependency in marriage (#PD3244)   Repressive detention of juveniles (#PD0634)
Physical maltreatment of children (#PC2584)   Denial of rights of children and youth (#PD0513).

♦ **PD2478 Damage to cultural artefacts by environmental pollution**
Pollution damage to paintings — Pollution damage to frescoes — Pollution damage to monuments — Pollution damage to sculpture
Incidence Acid rain produced by oxides of nitrogen and sulphur from oil refineries and tourist buses on the Yukatan peninsula is destroying the temples, murals and megaliths of ancient Mayan civilization. The acid rain eats into the stone surface at once, corroding crevices, ledges and moulding.
Broader Destruction of cultural heritage (#PC2114).
Endangered monuments and historic sites (#PD0253).
Narrower Deterioration of stained glass due to acid rain (#PE0082).
Related Deterioration of cultural artefacts from tourism (#PE9825).
Aggravates Deterioration of the physical condition of art objects (#PD1955).
Aggravated by Pollution (#PB6336).

♦ **PD2479 Inequitable distribution of skilled specialists**
Nature Highly skilled people in medicine, science, engineering and agriculture are required to reduce the world's suffering. But the concentration of technological training and development centres in the developed world has resulted in a disproportionate number of skilled specialists in the West. The more equitable distribution of these skills is inhibited in many ways: (1) The responsibility for training specialists is left in the hands of those most closely related to the field, who tend to follow the trend towards specialization; this tends to reduce the vision of the social role of those professionals. (2) In many fields, expertise has an associated social status which may create a dichotomy between the values and expectations of the expert and those he is serving; rather than confront their differences, the expert and the recipient may avoid dealing with each other. (3) The priority for distribution of skills is high in neither the government nor the private sector. (4) There are inadequate mechanisms for mediating between the public's need and available expertise. (5) Emphasis on individual values rather than social values mediates against professionals assuming a larger social responsibility.
Broader Limits on areas of research (#PF2529).
Related Thwarted technological communications (#PF0953)
Gap between the function of social techniques and the needs they address (#PF3608).
Aggravated by Unarticulated goal of educational methods (#PF2400).

♦ **PD2483 Massacres**
Mass killings
Incidence Since 1948, an average of 100,000 persons have been killed every year, frequently as a result of massacres, in acts of war. Over 200,000 East Timorese were killed by Indonesian military personal in the mid 1970's. A million Cambodians died in the 1970's under the regime of Pol Pot. In the early 1980's 100,000 people were massacred in Mozambique, mainly by the rebel group Renamo. In Burma, the army killed an estimated 3,000 people as they protested 26 years of authoritarian rule in September 1988. In the past 50 years more than a million Tibetans have died at the hands of the Chinese. In Brazil, on 28 March 1988, 14 Tikuna Indians were massacred and in April 1988, 20 Yanomanis in the State of Toraima were killed. On March 1987 more than 1,000 Dinkas of Sudan were massacred in the most abominable manner by the inhabitants of a town in the west of the country where they had taken refuge after repeated attacks on their villages by Arab militia. Between 1969 and 1988, as many as 300,000 Hutu people of Burundi were killed

by the Tutsi minority. It is estimated that the purges of Stalin cost as many as 40 million lives. Massacres have recently occurred in Uganda under Idi Amin and Milton Obote, the village of My Lai in Vietnam by American soldiers, South Korea, Zimbabwe, The Philippines, India, Sri Lanka and China.
Broader Genocide (#PC1056) Human death (#PA0072) Military atrocities (#PD1881).
Aggravates Elitism (#PA1387) Depopulation (#PG4958) Political dictatorship (#PC0845).
Aggravated by Fear (#PA6030) Civil war (#PC1869) Amoralism (#PF3349) Revolution (#PA5901) Aerial piracy (#PD0124) Guerrilla warfare (#PC1738).

♦ PD2484 Beating of prisoners
Flogging offenders — Torture by beating
Nature In some countries, prisoners are beaten (sometimes to death). This may either be done as a form of torture to obtain information, to frighten the citizenry, or as a method of execution. Prisoners may be beaten with blunt instruments, clubs, rifle butts, sticks or truncheons. They are slapped, kicked or punched. Falanga, blows to the soles of the feet, is a frequent form of torture. Whippings with rope, hide, wire or barbed wire are common. Canning and flogging are still practised as standard or alternative punishments in some countries. Victims have had blankets placed over them; motor tyres placed over the blankets restraining their arms and them beaten through the blankets. Often victims are hung upside down and beaten.
Incidence In a recent survey of victims of torture 92 percent were beaten while in prison. Beatings has been reported in the following countries: **Af** Angola, Benin, Burundi, Cameroon, Chad, Comoros, Congo, Côte d'Ivoire, Djibouti, Egypt, Ethiopia, Gabon, Ghana, Guinea–Bissau, Kenya, Lesotho, Liberia, Libyan AJ, Madagascar, Mali, Mauritania, Morocco, Mozambique, Namibia, Rwanda, Somalia, South Africa, Tanzania UR, Tunisia, Uganda, Zaire, Zambia, Zimbabwe. **Am** Argentina, Bolivia, Brazil, Canada, Chile, Colombia, Costa Rica, Cuba, Dominica, El Salvador, Grenada, Guyana, Haiti, Honduras, Mexico, Paraguay, Peru, Suriname, USA, Uruguay. **As** Afghanistan, Bangladesh, China, India, Indonesia, Iran Islamic Rep, Iraq, Israel, Korea Rep, Malaysia, Nepal, Pakistan, Philippines, Saudi Arabia, Singapore, Sri Lanka, Syrian AR, Taiwan (Rep of China). **Eu** Albania, Bulgaria, Czechoslovakia, Greece, Italy, Poland, Romania, Turkey, USSR, UK, Yugoslavia.
Broader Corporal punishment (#PD8575) Maltreatment of prisoners (#PD6005) Denial of rights to prisoners (#PD0520).
Related Corporal punishment in schools (#PE0192) Inhumanity of capital punishment (#PF0399).

♦ PD2487 Inadequate inland waterway transport facilities
Nature Although inland waterway transport is traditionally an extremely important economic activity in South America, Asia and Africa, it is frequently conducted under primitive conditions with modern facilities a rarity except at riverheads; even there they may be inadequate. In developed countries, the infrastructure of inland waterways is deteriorating.
Incidence The UK inland waterways are narrow and shallow and are being lost to recreational use only.
Counter–claim Major inland rivers have been important transport corridors since prehistoric times, and remain so in many countries today. Navigation on the Rhine, the Main, the Seine and its tributaries, and the Danube system, for example, forms an important component in the freight transport system of Europe. The great potential of the trans–European east–west waterway linking the Black Sea and the North Sea had been recognized as far back as 1920, and its construction, which involved 13 riparian countries, continues. Engineering works have opened the lower reaches of many river systems to small vessels capable of navigation in the coastal seas, while combined barge and ship systems, such as barge–aboard–catamarans or lighter–aboard–ships, have been developed to allow small crafts designed for river traffic to be conveyed across intervening stretches of open sea. The combination of inland waterways and coastal shipping remain an important means of freight transport in Europe, Japan and the United States. In Europe, although total traffic remains steady, international waterborne freight is increasing.
Broader Insufficient transportation infrastructure (#PF1495).

♦ PD2490 Secession
Separatism — Nationalist agitation — Minority group separatism — Separatism of indigenous peoples — Cultural separatism
Nature Secession is the process by which a part of a country separates, as a result of its own decision, from the country as a whole, so as to create a new country which may become internationally recognized as an indivisible and separate entity. Attempts at separation are disruptive of the political and social life of a country during the period when the secessionists attempt to build up sufficient pressure on the country as a whole to permit the secession to take place. There is some difficulty in distinguishing internationally between the 'illegality' of secession and a people's right to self–determination.
Refs Craven, Gregory *Secession* (1986); Kapur, Rajiv A *Sikh Separatism* (1986).
Broader Schism (#PF3534) Territorial fragmentation (#PC2944).
Aggravates Nativism (#PF2186) Civil war (#PC1869).
Aggravated by Revolution (#PA5901) Social fragmentation (#PF1324) Enclaves and exclaves (#PD2154) Cultural fragmentation (#PF0536) Denial of rights of minorities (#PC8999) Underprivileged racial minorities (#PC0805) Denial of rights of indigenous people to be self–governing (#PE1024).

♦ PD2495 Personal isolation in communities of industrialized countries
Nature In many communities of the developed world residents are isolated geographically, socially and structurally. The often rich cultural heritage is hidden from public view and is evident only in the homes and memories of individual families. Children may attend schools in different districts. Summer residents are often not integrated into the community. Newcomers in rural areas are resented, and attempts on their part to get involved are viewed as trying to impose urban values upon a rural setting. Feuds arise because of divergent opinions on land use, inadequate land agreements, and unobserved property rights.
Broader Social isolation (#PC1707) Deterioration of industrialized countries (#PD9202).
Narrower Mistrust of strangers (#PF8743) Disagreements over land use (#PG7933) Scarcity of residential land (#PD8075) Fragmenting district boundaries (#PG7968) Denial of the right to ownership (#PE8411) Fragmentation of resident relationships (#PG7949) Neglect of agricultural and rural life in developing countries (#PF7047).
Related Disrelationship from community history (#PJ0836).
Aggravates Dependence on the media (#PD7773) Boundary disputes between neighbours (#PE7903).
Aggravated by Demeaning community self–image (#PF2093) Insufficient communications systems (#PF2350).

♦ PD2498 Ice–blocked seaways
Nature Important seaways and waterways in the Northern Hemisphere are blocked or menaced by ice during many months of the year. Depending upon the climatic conditions, the period of difficulty may increase to the detriment of shipping movements upon which vital economic activities may depend.
Incidence Sea ice prevents effective exploitation of mineral and oil deposits along the northern shores of Canada, Alaska and the USSR. Recent climatic changes have resulted in the Icelandic fishing industry being paralyzed by drift ice which had been absent since the 1920s. Such ice causes damage to harbours and ships. Ice in the Great Lakes and St. Lawrence Seaway is hazardous to ships and requires aerial surveillance.
Broader Ice accretion (#PD1393) Obstacles to the utilization of coastal and deep sea water resources (#PF4767).
Related River ice (#PD3142).
Aggravates Insufficient transportation infrastructure (#PF1495).
Aggravated by Climatic cold (#PD1404).

♦ PD2500 Beggars
Begging
Nature Individuals may gain all or part of their living by begging, whether from door to door or in public places. In some cases attempts to increase such income may be made by deforming or mutilating the body. This may be done deliberately by the beggar or may be done by an adult, possibly the parent, dependent on the begging of children.
Incidence Begging occurs frequently, particularly in developing countries, where unemployment and shortage of food are problems. With the rise of homelessness and destitution in industrialized countries, it is increasingly evident there. In Washington, for example, where the number of blacks living below the poverty line has increased by 30 per cent during the 1980s, male beggars are to be encountered on many street corners. In 1990 federal court decided for the first time that panhandling is free–speech right protected by the First Amendment.
Claim Beggars are evidence that either society has failed or the beggar has failed. Where society has failed, beggars confront individuals with this failure. Where the beggar has failed, they erode city life, it is a form of blackmail, and a protection racket. Frequently, beggars make more money begging than if they work. A beggar in China is reported to have made an average daily income equal to a months salary for a worker. A well known beggar in Bombay is known to be delivered to the place where he operates by his engineer son in a new car. Community has a right to protect a minimally civilized ambience in public spaces, like in subways, and ban panhandling.
Refs Kumarappa, Jagadisan Mohandas *Our Begger Problems* (1945); Ribton–Turner, C J *A History of Vagrants and Vagrancy, and Beggars and Begging* (1972).
Broader Socio–economic poverty (#PB0388) Offences against public order (#PD7520).
Aggravates Compassion fatigue (#PF2819) Emotional manipulation (#PE9599) Mutilation and deformation of the human body (#PD2559).
Aggravated by Famine (#PB0315) Unemployment (#PB0750) Uncontrolled urban development (#PC0442) Inadequate social welfare services (#PC0834).

♦ PD2502 Excessive debt of socialist countries to the West
Incidence The combined debts of socialist countries to the West stood at over US $115,000 million in 1987. This was nearly 30 percent higher than three years before. The debts are principally to private commercial sources, to Western governments, and to the World Bank and the International Monetary Fund. They were largely incurred for the importation of food and agricultural products, and for Western technology, fuels and energy raw material. Consumer goods are also imported. The degree of indebtedness varies considerably from one socialist country to another. In 1983, the figures were: Albania $25 million; Bulgaria $3,300 million; Czechoslovakia $2,750 million; East Germany $10,750 million; Hungary $7,250 million; Poland $30,900 million; Yugoslavia $20,500 million, plus $3,400 million owed by Comecon banks. In 1990 Poland owed about $41 billion, Hungary $21 billion and Bulgaria $9 billion to foreign creditors.
Broader Excessive public debt (#PC2546) Uncontrolled growth of debt (#PC8316).

♦ PD2503 Pollution of groundwater
Groundwater contamination — Contaminated well water — Nitrate pollution of groundwater — Nitrate pollution of aquifers
Nature Untreated non–saline groundwater is normally much safer to drink than any untreated surface water, since the ground itself provides an effective purifying medium; such water constitutes a major source of drinking–water supplies. Nevertheless, groundwater can be polluted by domestic and industrial waste waters, by soluble materials leached out of tips for municipal refuse or industrial waste, by accidental spillage of other liquids, especially oil, and by intrusion of highly saline water.
Groundwater in its percolation through soil and rocks, leaches out soluble salts; it is thus typically mineralized, and sometimes heavily so. The vulnerability of groundwater to contamination is determined by the hydrological setting of the aquifer, the nature of the contaminant and the effectiveness of regulatory action. Of all the activities of man that influence the quality of groundwater, agriculture is probably the most important, as a diffuse source of pollution from fertilizers, pesticides and animal wastes. Of the main nutrients in nitrogen, phosphorus and potassium fertilizers, nitrogen in the form of nitrate is the most common cause of degradation of groundwater near agricultural lands. Industrial wastes include a wide spectrum of materials from all types of industry, and contain many organic and inorganic chemicals which are potential pollutants. Industrial wastes reach groundwater from impoundments or lagoons, spills, pipeline breaks and land disposal sites.
Septic tanks and cesspools contribute filtered sewage effluent directly to the ground, and are the most frequently reported sources of groundwater contamination, especially in rural, recreational and suburban areas. In many areas, the solid residual material known as sewage sludge – which contains a large number of potential contaminants – is spread on agricultural land. In some regions liquid sewage that has not been treated or that has undergone partial treatment is sprayed on the land surface. Such application of liquid sewage and sewage sludge to the land provides valuable nutrients such as nitrogen and phosphorus to the soil, with benefits to agriculture. However, the waste water or sludge can add to the contamination of groundwater. The soil profile shows a considerable ability to remove or detoxify several of the compounds found in the waste water, but some may nonetheless affect groundwater quality. The soil may also effectively eliminate the pathogenic bacteria through filtration and soil microbiological processes, but survival of viruses is still an open question.
Refs FAO *Groundwater Pollution* (1979); United Nations *Aquifer Contamination and Protection* (1980); Zwirnmann, K H (Ed) *Nonpoint Nitrate Pollution of Municipal Water Supply Sources* (1982).
Broader Water pollution (#PC0062) Pollution of inland waters (#PD1223) Water system contamination (#PD8122).
Aggravated by Corrosive groundwater (#PE4740) Nuclear weapons testing (#PC2201) Unethical practice of earth sciences (#PD0708) Unsustainable short–term improvements in agricultural productivity (#PE4331).

♦ PD2511 Over–development of bureaucracy in ex–colonial countries
Nature At the time of independence most ex–colonial countries have well–developed governmental institutions and a strong civil service; this facilitates the transition to self–government. But the continued professionalization and expansion of the civil service tends to outpace the growth of executive, legislative and judicial organs of government, and of parallel development of interest groups in the private sector. It also favours an apparent carry–over from the colonial period of attitudes of superiority and disdain toward the public, and a preoccupation with law–and–order

considerations rather than with programmes of economic development and social welfare. Under conditions of economic scarcity, government posts are much sought after at the expense of the development of the private sector.
Broader Bureaucracy as an organizational disease (#PD0460)
Unchecked power of government bureaucracy (#PD8890).
Related Administrative difficulties in new states (#PE1793).
Aggravates Overthrow of government (#PD1964).

♦ PD2520 Desert nomadism

Nature Nomads in developing countries may be the least privileged as regards education, since they are isolated from educational facilities and may find it impossible to attend school regularly. Inaccessibility also works against them in health matters; and in the event of drought, the administration of food aid may be difficult. Inefficient use of limited resources and grazing of nomadic flocks, in particular goats, cause soil erosion and strain existing limited pasture, so that development of the area may be hampered. With desert irrigation and reclamation programmes, it is important for the authorities to control both the animal population and the high birth rate among nomads. The incentives given to make nomads comply with these policies imply their settling in one place and relinquishing nomadism.
Broader Nomadism (#PF3700) Underprivileged racial minorities (#PC0805).
Narrower Land reclamation (#PF2055).
Related Underdevelopment (#PB0206) Proliferation of immigrants (#PD4605).
Aggravates Illiteracy (#PC0210)
Inadequate health care in family planning (#PD1038).
Aggravated by Drought (#PC2430) Infertile land (#PD8585)
Lack of education (#PB8645) Social fragmentation (#PF1324)
Lack of natural resources (#PC7928) Isolation of ethnic groups (#PC3316)
Inadequate irrigation system (#PD8839).

♦ PD2539 Plutonium overproduction

Nature Plutonium is a man-made element which is a necessary component of most modern warheads. Plutonium production increased in the '80s to heighten arms capability. Now, as arms limitations agreements limit the need for production of new warheads, there is an excess of plutonium. The plants which manufacture it are aging, and there is possibility that quantities of the element could escape should there be any accident.
Broader Toxic metal pollutants (#PD0948).
Related Plutonium pollution (#PE6285) Competitive acquisition of arms (#PC1258)
Failure of disarmament and arms control efforts (#PF0013)
Protectionism in the defence and arms industries (#PE8664).

♦ PD2541 Litter

Unremoved public litter — Damaging unremoved litter
Broader Environmental degradation of inner city areas (#PC2616).
Related Environmental pollution (#PB1166).
Aggravates Landscape disfigurement (#PC2122).
Aggravated by Waste paper (#PD1152) Unenforced littering laws (#PJ0380)
Non-destructible packaging and containers (#PD1754).

♦ PD2542 Visually handicapped persons

Refs Bonner, R et al *The Visually Limited Child* (1970); Meighan, Thomas *An Investigation of the Self-Concept of Blind and Visually Handicapped Adolescents* (1971).
Broader Physically handicapped persons (#PD6020).
Narrower Night blindness (#PG4565) Physical blindness (#PD0568)
Colour deficiencies (#PE6343).
Aggravated by Visual deficiencies (#PD8179).

♦ PD2543 Inadequate army discipline

Lack of army discipline — Strict army discipline codes — Military indiscipline
Nature Military personal who are disorganized, inefficient, slothful and disregard hard work are incapable of carrying out their roles in the military. They are a danger to their country because they are incapable of defending it. They are a danger to themselves because undisciplined troops have higher mortality rates than disciplined ones under similar command and tactical situations. They are a danger to civilians, particularly enemy civilians.
Army personnel may sometimes use arbitrary or disproportionate force against individuals or groups of individuals, without being punished for their acts. Often, no investigation of these wrongdoings or misconduct is carried out, and a code of conduct may be ignored or even non-existent. When codes of conduct do exist, the training of army personnel may not pay due attention to the rights of those suspected, arrested or detained.
Broader Military disobedience (#PD7225) Lack of social discipline (#PF8078).
Narrower Draft evasion (#PD0356).
Aggravates Hazings (#PF5392) Disobedience (#PA7250) Unpreparedness (#PF8176)
Military atrocities (#PD1881) Ill treatment of prisoners of war (#PD2617)
Drunkenness of military personnel and troops (#PE8495).
Aggravated by War (#PB0593) Grievance (#PF8029)
Subversive activities (#PD0557).

♦ PD2544 Unnatural boundaries between developing countries

Nature Many states which have only recently achieved independence have boundaries which cut across language, tribal and cultural groupings. The arbitrary and illogical position of the boundaries produces states which lack natural unity and are subject to the divisiveness resulting from the assertion by the component peoples of their right to their particular cultures and languages. It also sets the stage for the difficulties in producing larger regional groupings, which the colonial powers were unable or unwilling to pass to their successors.
Incidence This problem occurs mainly in Africa and to a lesser extent in South East Asia as a heritage of Western colonialism.
Background During the colonial period, Western powers designated boundaries to their own satisfaction and without much reference to the social and cultural situation in the areas through which the boundaries cut. Having split up a continent in this way, there was little further attempt to regroup the territories into larger political or administrative units.
Broader Unnatural boundaries between states (#PF0090).
Aggravated by Colonialism (#PC0798) Enclaves and exclaves (#PD2154).

♦ PD2545 Ringworm

Nature Ringworm is a common term used to describe the appearance of lesions caused by fungal infections of the skin. Fungal agents which have a propensity to infect skin and hair are called dermatophytes. Dermatophytes can be classified as either geophilic, zoophilic or anthropophilic, according to their primary nidi, which are soil, lower animals, or man, respectively. Several of the geophilic and zoophilic organisms are occupational hazards for agricultural workers. Agricultural workers may be exposed to dermatophytes by direct or indirect contact with infected animals or soil. Breaks in the skin enhance the ability of the organism to establish an infection and agricultural workers are very prone to lacerations or abrasions on their hands and arms. As hands and arms are commonly unprotected with clothing and are frequently in contact with potentially infected animals or fomites, these parts are the most common sites for infection.
Broader Fungal diseases (#PD2728).
Related Dermatophytoses (#PG4553).
Aggravates Itch (#PE3940) Mycetoma (#PG5054) Pneumonia (#PE2293)
Meningitis (#PE2280) Madura foot (#PG5044) Coccidiosis (#PE2738)
Phycomycosis (#PG5053) Actinomycosis (#PE2353) Blastomycosis (#PE4928)
Histoplasmosis (#PG3696) Cryptococcosis (#PE4932) Sporotrichosis (#PG4935)
Diseases of the ear (#PD2567) Pityriasis versicolour (#PG5047)
Lung disorders and diseases (#PD0637).
Aggravated by Mycosis (#PE2455).

♦ PD2547 Indecent advertising

Gratuitous sex in advertising — Nudity in advertising
Nature Advertisements may contain statements or visual presentations offensive to public decency. Complaints about naked women and gratuitous sex in advertising are however seldom upheld.
Incidence In the UK in 1989, 76 per cent of women and 62 per cent of men agreed that using sex as a selling device in advertising is offensive.
Broader Indecency (#PF8842).
Narrower Vulgar combination of sacred and erotic in advertising (#PE5190).
Related Nudism (#PF2660) Misuse of advertising (#PE4225).
Aggravates Youth violence (#PF7498).

♦ PD2549 Political refugees

Defectors
Broader Refugees (#PB0205).

♦ PD2552 Profit-oriented interest payments

Nature Profit-oriented credit exchange systems results in rigidity in credit availability and terms, so that someone needing a loan for hospital bills may pay the same rate of interest as someone obtaining a loan to take a vacation. With a view toward maximizing a secure return on funds, extensive loan restrictions are placed on applicants, excluding needy applicants who cannot meet rigid credit requirements.
Broader Inadequate credit policies (#PF0245).
Related Inconsiderate choice of payment times by creditors (#PE3926).
Aggravates High interest rates (#PF9014).

♦ PD2557 Forgery

Nature Forgery is knowingly making, completing or altering any writing with the intention of deceiving or harming a person, institution or government.
Incidence The US State Department has reported a number of forged government documents. Among them was a memorandum appearing to bear President Reagan's signature on forged White House stationary calling for the creation of a Latin American military force to contain Soviet expansion. Another document was in the form of a memorandum from the intelligence service of Zaire and implicated the US in training of guerrillas to undermine African countries.
Broader Fraud (#PD0486) Vulnerability of intellectual property (#PF8854).
Narrower Art forgery (#PE2382) Counterfeiting (#PD7981)
Historical forgery (#PE5051) Forged scientific data (#PF0255)
Forgery of wills and testaments (#PG1190)
Fraudulent certificates of origin of goods (#PJ6717).
Aggravates Economic loss (#PE9013)
Obtaining property by false pretences (#PE5076).

♦ PD2559 Mutilation and deformation of the human body

Nature Permanent or semipermanent modification of the human body may be undergone by removal of superficial features (such as hair, teeth) or breaking of the skin and by introducing a change of its shape. Such processes may be accidental or unintended. But frequently, they are undergone voluntarily or without strong social censure, as a result of custom, occupation, magical or medical rites, aesthetic considerations, religion (ascetic mortification), tribal initiation, or to increase the income from begging. When used as a form of punishment (including amputation of limbs, drug induced mutilation and torture) or when dictated, mutilation and deformation constitute a violation of the human rights and in any case can represent a health hazard for the person.
Incidence Intentional and unintentional mutilations and deformations occur in all societies: tribes over-fatten their women for aesthetic reasons; skin marking in the form of tattooing or cicatrization is frequently practised, particularly in tribal societies; depilation is practised by the less-hairy Eastern races and by Western women. Deformation of the head, neck, trunk, limbs or feet by special bindings has been practised in tribal societies on all continents; the nose is frequently pierced for the admission of various forms of decoration or modified under plastic surgery for aesthetic reasons; the lips may also be pierced and stretched, some teeth are removed, filed, encrusted or blackened in various societies. Removal of the epicanthic fold over the eyes is common in some Eastern societies to mimic the Caucasoid appearance. Perforation of the earlobe is common, and various techniques are used to increase the size or modify the shape of the breast. The genitalia are mutilated in a variety of ways, including circumcision, unilateral castration, castration, female circumcision, infibulation and artificial defloration, and blinding is used as an alternative to incarceration in some countries.
Counter-claim In many cultures various modifications of the human body are symbolic expressions of inclusion in the community or group. They marks of a change in status within the community. They are necessary to participate in adult roles. Individuals are clearly distinguished and confusion about the individual's roles on the part of the individual and the society in reduced.
Refs Camporesi, Piero and Murray, Tania C *The Incorruptible Flesh* (1988); Favazza, Armando R *Bodies under Siege* (1987); Walsh, Barent W and Rosen, Paul M *Self-Mutilation* (1988).
Narrower Tattooing (#PG5079) Child mutilations (#PG3726)
Male sexual mutilation (#PE6054) Female sexual mutilations (#PE6055)
Torture through mutilation (#PD7576) Punishment of criminals by mutilation (#PE3488)
Deliberate deformation of childrens' bodies (#PE1646).
Aggravated by Leprosy (#PE0721) Beggars (#PD2500)
Children's diseases (#PD0622) Malpractice in plastic surgery (#PE4429)
Sexual discrimination in contraceptive methods (#PF1035)
Unethical experiments with drugs and medical devices (#PD2697).

♦ PD2561 Burglary

Breaking and entry
Nature Burglary is surreptitiously entering or remaining in a building with the intent of committing a crime, whether theft, rape, robbery or kidnapping.
Incidence In the UK 900,000 burglaries are reported each year.
Refs Clarke, R V G and Hope, Tim *Coping with Burglary* (1984); Rengert, George and Wasilchick, John *Suburban Burglary* (1985); Walsh, Dermot *Heavy Business* (1986).
Broader Theft of property (#PD4691).
Narrower Safecracking (#PG1145) House-breaking (#PG1115)
Political burglary (#PD1943).

Related Theft (#PD5552) Criminal intrusion (#PE6771)
Surreptitious entry during surveillance (#PE3973).
Aggravated by Insecurity of property (#PC1784).

♦ PD2562 Virus diseases in bacteria
Bacteria as vectors of viral diseases
Broader Virus diseases (#PD0594).
Related Viral plant diseases (#PD2227) Viral diseases in animals (#PD2730)
Virus diseases in protozoa (#PG7376).
Aggravates Shortage of cultivable land (#PC0219).

♦ PD2565 Diseases of connective tissue
Collagen diseases — Diseases of the musculoskeletal system and connective tissues — Diseases of the muscoskeletal and connective tissues — Diseases of the musculo-skeletal system — Diseases of muscles
Refs Buckle, Peter *Musculo-Skeletal Disorders at Work* (1987); Fleischmajer, Raul, et al (Eds) *Biology, Chemistry and Pathology of Collagen* (1986).
Broader Human disease and disability (#PB1044).
Narrower Osteoarthritis (#PG9454) Sprains of joints (#PG9214)
Myasthenia gravis (#PE3638) Rheumatic diseases (#PE0873)
Diffuse scleroderma (#PG5083) Diseases of the spine (#PD2626)
Skeletal system disorders (#PE2298) Infantile muscular atrophy (#PG5514)
Malignant neoplasm of bone (#PE9229) Keratoconjunctivitis sicca (#PG4981)
Systemic lupus erythematosus (#PG5084) Diseases and injuries of bone (#PE3822)
Congenital anomalies of spine (#PG9817) Disorders of joints and ligaments (#PD2283)
Benign neoplasm of bone and cartilage (#PG0819)
Symptoms referable to musculoskeletal system (#PE4566).
Aggravates Rheumatic fever (#PE0920) Dermatomyositis (#PE5086)
Rheumatoid arthritis (#PE5081) Generalized necrotizing arthritis (#PG5081)
Generalized necrotizing arteritis (#PG5087).
Aggravated by Consanguineous marriage (#PC2379)
Diseases and injuries of the brain (#PD0992).

♦ PD2567 Diseases of the ear
Diseases of the mastoid process — Ear tumour — Infection of the ear — Earache — Eardrum injury
Refs Ballenger, John J (Ed) *Diseases of the Nose, Throat, Ear, Head and Neck* (1985); Becker, W, et al *Ear, Nose and Throat Diseases* (1988); Maran, A G *Logan Turner's Diseases of the Nose, Throat and Ear* (1988); Mawke, Michael and Jahn, Anthony F *Diseases of the Ear* (1987); Schuknecht, Harold F *Pathology of the Ear* (1974).
Broader Infection (#PC9025) Diseases of the sense organs (#PC9623).
Narrower Deafness (#PD0659) Dizziness (#PE5101) Head noise (#PG5106)
Mastoiditis (#PG3510) Deaf mutism (#PE5261) Otosclerosis (#PE2746)
Otitis media (#PG6581) Labyrinthitis (#PG5102) Acoustic trauma (#PE4109)
Ménière's disease (#PG2760).
Aggravates Hearing defects (#PD6306) Diseases and injuries of the brain (#PD0992).
Aggravated by Mycosis (#PE2455) Measles (#PE1603)
Ringworm (#PD2545) Tonsillitis (#PE2292)
Common cold (#PE2412) Diseases of the nose (#PE5122)
Diseases of the respiratory system (#PD7924) Diseases of upper respiratory tract (#PE7733).

♦ PD2569 Defamation of character
Denial of right to freedom from attacks on personal honour and reputation — Character assassination
Nature Injuring of a person's good name and reputation may take the form of libel or slander; and may be accomplished with malice or be unintentional. In either case, particularly in the instance of libel, which is a criminal offence, it may give rise to prosecution, the penalties for which may be imprisonment, fine, or the award of damages. Defamation of character may be used extensively in propaganda, which on an international level may go unchecked by the law. It may be used to a certain extent in domestic political power struggles and also in business or other power struggles. Defamation of character may be so phrased that it is no longer a clear-cut crime under the law and may be too difficult and costly to prove. If the harmful comments come near to the truth or other misconduct might be unearthed in the course of court proceedings, the victim may not wish to try to obtain full satisfaction and may be subsequently blackmailed.
Broader Torts (#PD9022) Infringement of privacy (#PB0284).
Narrower Libel (#PD3022) Slander (#PD3023).
Related Censorship (#PC0067) Propaganda (#PF1878).
Aggravates Dishonour (#PF8485).
Aggravated by Humiliation (#PF3856) Power politics (#PB3202)
Criminal coercion (#PD4469) Biased presentation of news (#PD1718).

♦ PD2570 Large-scale industrial accidents
Industrial disasters — Chemical accidents
Nature The term "industrial accidents" can also refer to plant failures which cause explosions, dangerous leaks, fires, floods, and other major disasters.
Incidence Continual industrial growth increases the difficulty of isolating plants from centres of population. In a world which is increasingly dependent on chemical products (many of them highly toxic) and highly complex large-scale technologies, industrial accidents with catastrophic consequences are likely to increase. The world-wide production of organic chemicals has risen from 63 million tonnes in 1970 to about 250 million tonnes by 1985. In OECD countries more than 200 serious chemical accidents occur annually.
Major industrial accidents this century include: Halifax (Canada) 1917, munitions and explosives (2,000 to 3,000 killed, 8,000 injured); Oppan (Germany) 1921, fertilizer extraction (560 killed, 3,000 injured); Hawk's Nest (USA) 1931-35, silicosis during tunnelling (2,000 to 3,000 killed); Texas City (USA) 1947, explosive-triggered fire (600 killed, 2,000 injured); Ludwigshafen (Germany) 1948, chemical factory fire (200 killed, 4,000 injured); Seveso (Italy) 1976, dioxin manufacturing plant explosion (40,000 domestic and farm animals killed; 400 children affected by chloracne, 400 abortions, significant rise in incidence of cancer); San Carlos de la Rapita (Spain) 1978, road tanker fire (200 killed); Salang Tunnel (Afghanistan) 1982, petrol tank truck fire (2,000 to 2,700 killed); Cubato (Brazil) 1984, explosion from leaking pipeline (600 killed, 3,000 injured); Ixhautepec (Mexico) 1984, explosion of stored petroleum gas (1,500 killed, 7,000 injured).
In December 1984, a chemical leak at the Union Carbide Corporation plant in Bhopal, India killed over 3,000 people and injured 200,000. A similar accident at the Sequoyah Fuels Corporation plant in Gore, Oklahoma, USA killed one worker and injured at least 30 others. At Institute, West Virginia, another leak injured at least 135 people. On October 31, 1986 a gigantic spill of mercury and pesticides from the Sandoz chemical plan in Basel, Switzerland poisoned some 200 miles of the Rhine river.
Refs Baxi, Uprendra and Paul, Thomas *Mass Disasters and Multinational Liability* (1986); Kurzman, Dan *A Killing Wind* (1987); Lagadec, P *Major Technological Risk* (1982); Long, Duncan *Surviving Major Chemical Accidents and Chemical-Biological Warfare* (1986).
Broader Disastrous accidents (#PC6034) Disastrous technological failures (#PD4426)
Environmental hazards from economic and industrial products (#PC0328).
Narrower Accidental explosions (#PE3153).
Related Nuclear accidents (#PD0771) Radiation accidents (#PD1949)
Electric current accidents (#PG2862).
Aggravates Industrial accidents (#PC0646)
Accidental large-scale contamination of the environment (#PD1386).
Aggravated by Transport accidents (#PC8478)
Unsafe industrial, laboratory and medical equipment (#PE4859).

♦ PD2579 Exclusive nationally-oriented language systems
Claim The deeply entrenched attachment to national language systems is a major hindrance to the globally-oriented language systems that are called for in the present time. Although they project the linguistic symbols that preserve the wisdom and heritage of a culture, national educational plans have universally failed to revise language systems to include the trans-social explosion of symbols and to recognize the need for adapting a secondary linguistic technique for use in global exchange.
Broader Discriminatory communication (#PD6804)
Thwarted technological communications (#PF0953).
Aggravates Semilinguism (#PF2789).

♦ PD2581 Pesticide damage to crops
Nature Among agricultural malpractices, the improper use of chemicals such as fungicides, insecticides and herbicides is the most common cause of plant injury and disease. The misapplication of pesticides to soils and plants may lead to plant injury or growth retardation. Many pesticides are toxic to some plant species, and a few, such as the fumigants, are toxic to most plants. Crop plants will be injured or killed if these chemicals have not decomposed or volatilized. Certain pesticides decompose to simple inorganic substances which may injure sensitive plants. Toxic decomposition residues of pesticides include compounds of arsenic, bromide, chlorine, copper, iodine and mercury. Several plant species are sensitive to inorganic bromides or chlorides: onions and citrus fruit may be damaged by bromide-containing pesticides; avocado by chloride-containing pesticides. Pesticides which decompose very slowly may, if used continuously, build up to such concentration in the soil that they seriously retard crop growth.
Treatment of the soil with any pesticide which kills soil microbes may result in plant injury. The killing of bacteria which oxidize ammonia results in the accumulation of toxic concentrations of ammonia from decomposing organic matter. If large numbers of microbes are killed, a temporary phytotoxicity may occur, manifested by reduced absorption of phosphorus, zinc and copper. The toxicity may last from a few weeks to a year; young citrus, peach, and certain other tree seedlings, are especially sensitive to this toxicity.
Chemicals used for seed treatment are frequently toxic, especially to some species of plants. For example, plants of the cabbage family are stunted by copper-containing seed treatment materials. Vegetative organs, such as potato tubers, are very susceptible to chemical injury, and strong poisons, such as mercuric chloride, can cause considerable damage. Materials applied to the soil to control fungi, bacteria and nematodes may injure plants grown in the soil too soon after treatment.
Broader Environmental plant diseases (#PD2224).
Narrower Fumigant damage to crops (#PE3584) Herbicide damage to crops (#PD1224)
Fungicide damage to crops (#PD3577) Insecticide damage to crops (#PD3695).
Related Pesticide intoxication (#PE2349) Plant-pathogenic air pollutants (#PE0155).
Aggravated by Pesticides as pollutants (#PD0120)
Metal contamination of soil (#PD3668)
Pesticide destruction of soil fauna and micro-organisms (#PD3574).

♦ PD2586 Unethical practice of hydrology
Irresponsible hydrologists — Negligence by hydrologists — Malpractice in hydrology — Corruption of hydrologists — Underreporting of disruptions in natural water cycles
Claim Hydrologists, under pressure from their employers, have adopted practices which lead to the underreporting of disruptions in natural water cycles to avoid public protest, especially in the case of major dam and irrigation systems, given the implications of silting and sedimentation for traditional patterns of agriculture.
Aggravates Maldistribution of water (#PD8056) Lack of accord on water use (#PF4839)
Mismanagement of irrigation schemes (#PE8233).

♦ PD2589 Mutiny
Nature Mutiny is usurping command of a vessel or organization through force or the threat of force.
Broader Civil disorders (#PC2551) Seizure of power (#PC8270)
Impairing military effectiveness (#PD4448).
Related Revolt (#PE5144) Overthrow of government (#PD1964).
Aggravates Military disobedience (#PD7225).
Aggravated by Brutality (#PC1987) Inhumanity (#PB8214)
Human violence (#PA0429) Lack of social discipline (#PF8078)
War-time conditions and pressure (#PD9090).

♦ PD2591 Business bankruptcy
Risk of insolvency — Business failure
Nature Business enterprises may be declared bankrupt, through a judicial process, when it is proved that they are unable to pay their debts. Adverse economic conditions or various forms of unnatural trade competition may force otherwise viable enterprises into bankruptcy. Bankruptcy provisions of the law are abused in many countries when officers of enterprises voluntarily file for bankruptcy after concealing assets or paying themselves gratuitous sums, or in the course of fraud.
Incidence Approximately 12,000 American companies filed for reorganization under chapter 11 of America's bankruptcy laws in 1988 and 20,000 were expected to do so in 1989. One year in the mid-1980's saw 25,000 French companies and 16,000 German companies fail. During that same year more British companies closed their doors than in year in history.
The last kind of business to fail during a recession are the banks. Bank failures in the USA for example, climbed from 6 in 1977 to 42 in 1982 and 48 in 1983. The amount of deposits in failed banks from 1977 through 1983 was $18,520 million, which would have been lost to depositors if the funds were not insured by the government. By 1984, 630 banks were considered to have serious problems, representing over 4 percent of all US banks.
Counter-claim 1. The concept of the elimination of inefficient firms is essential to capitalism. Bankruptcy is as necessary for capitalism as profit; together they make up the stick and carrot which persuade businessmen to work. 2. In the USA for example, bankruptcy laws have been amended to allow for reorganization of the enterprise and time to pay its debts in an orderly fashion. Therefore bankruptcy does not automatically mean failure or business liquidation but can be a safety-net or a last chance to save a business. In China bankruptcy is now being considered as an legitimate way of controlling inefficient loans.
Broader Risk (#PF7580) Insolvency (#PC6154).
Narrower Small business failures (#PE9405).
Related Personal insolvency (#PD9376).
Aggravates Unemployment (#PB0750) Abuse of credit (#PF2166)
Factory closures (#PE3537).
Aggravated by Embezzlement (#PD2688) Crop vulnerability (#PD0660)

Creative accounting (#PE6093) Cyclic business recessions (#PF1277)
Risk of capital investment (#PF6572) Usury in developing countries (#PE2524)
Unethical commercial practices (#PC2563)
Discouraging conditions for small business (#PD5603).

♦ PD2592 Mercenary troops

Mercenaries — Condottieri — Use of mercenaries as a means of impeding right to self-determination

Incidence Over 60 mercenary troops from Sri Lanka attacked the Maldives on November 3, 1988 in an attempt to overthrow the government. The South African government is known to use mercenaries against Angola, the South West Africa People's Organization and the African National Congress. Mercenaries have helped to train teams of assassins for Columbian cocaine traffickers and their allies. Thirty French and Belgian mercenaries had a virtual control of the Comoro Islands for 11 years and killing 2 presidents. The Gurkha troops engaged in Nepal by the British Government for service in British colonies and territories may be considered as mercenaries of one form.

Refs Mockler, Anthony *The New Mercenaries* (1987); United Nations *Thirty-Ninth Session supplement.*

Related Hunting of humans (#PD5493)
Violation of sovereignty by trans-border broadcasting (#PE0261).
Aggravates Military atrocities (#PD1881).
Aggravated by Errant nationals (#PE0812)
Unlawful recruiting for and enlistment in foreign armed forces (#PE4484).

♦ PD2594 Biological contamination of food

Nature Food may become contaminated with microbes at many points during its production, processing, transportation, storage, distribution, and preparation for consumption. The degree of hazard and points of maximum danger vary, depending on the types of contamination, and on the food and its method of production, including handling and processing procedures.

Broader Food pollution (#PD5605).
Narrower Botulism (#PE7766) Rickettsiosis (#PE5530)
Microbial contamination of food (#PD9669).
Related Paramphistomes (#PG5082) Chemical contaminants of food (#PD1694)
Infected animal, meat and animal product shipments (#PE7064).
Aggravates Food insecurity (#PB2846) Fungal diseases (#PD2728).
Aggravated by Unclean food (#PJ2532) Virus diseases (#PD0594)
Food spoilage in storage (#PD2243) Unethical food practices (#PD1045)
Food manufacturing industry wastes (#PE8702)
Environmental hazards from food and live animals (#PC1411)
Unsanitary and inhumane urban food animal conditions (#PE0395).

♦ PD2596 Environmental hazards from mining

Environmental destruction from open-pit mines and quarries

Nature Mining pollution is caused by land excavation; the discharge of mine pit water or waste water; the dumping of waste rock, tailings or slag; and the discharge of metallic smoke and dust into the atmosphere.

All forms of open-pit mining and quarrying destroy the habitat. Slag and water-table disturbances and stream pollution from underground mining are equally destructive. Mining and quarrying are noisy, and may use blasting techniques causing noise pollution and driving off wildlife. Dust from coal and other mines and quarries can pollute large areas.

Incidence In the USA for example, some 10,500 miles of stream have been affected by acid mine drainage. About half a million acres of land remain unrestored from coal mining operations alone.

Refs Sendlein, L V, et al *Surface Mining Environmental Monitoring and Reclamation Handbook* (1983).

Broader Ecologically unsustainable development (#PC0111).
Narrower Coal mining environmental hazards (#PE5160)
Natural gas production environmental hazards (#PG5161).

♦ PD2608 Inequitable distribution of construction expertise

Claim Inequitable distribution to all people of the practical application of construction expertise in the sectors of building trades, civil engineering and architectural design has resulted from parochialism on the part of those having construction skills, and from a lack of determination on the part of most of the world to respond to local construction needs. Illiteracy, ignorance, poverty and a lack of powerful support further weaken the position of those who wish an equitable share in present-day construction expertise.

Broader Profit motivated utilization of construction technology (#PF2464).

♦ PD2609 Banditry

Nature Banditry is robbery through direct attack or threat of attack on a traveller or isolated individual. Soviet law, for example, requires three conditions to be met before a person is considered a bandit: 1) the participation of two or more people (a band); 2) the possession of arms, even if by only one member of the band; and 3) the cohesion and organization of the participants. A crime is considered having been committed from the moment the armed band is organized, even if the band has committed no attack.

Broader Robbery (#PD5575) Human violence (#PA0429).
Aggravates Revenge (#PF8562) Bloodshed (#PG5183)
Human death (#PA0072) Illegal roadblocks (#PE9605).
Aggravated by Egoism (#PA6318) Outlaws (#PE4409)
Capitalism (#PC0564) Immorality (#PA3369)
Social injustice (#PC0797) Socio-economic poverty (#PB0388)
Inadequate firearm regulation (#PD1970).
Reduced by Vigilantism (#PD0527) Excessive government control (#PF0304).

♦ PD2614 Fear of losing cultural identity

Claim Individual cultural groups are resistant to creating comprehensive, engaging structures because they fear that a new culture will involve sacrificing the uniqueness of their own traditions. They tend to stress the uniqueness of their own culture to the exclusion of all others.

Broader Parochial national interests (#PF2600).
Related Loss of cultural identity (#PF9005).

♦ PD2615 Treason

Treason felony — Traitors — Petty treason — High treason

Nature When a citizen of a country which is engaged in an international war intentionally participates in or aids military activity with the intent of furthering the aims of the enemy or prevent the victory of his country is guilty of treason.

Refs Benda, Julien *Treason of the Intellectuals* (1969); Pincher, Chapman *Traitors* (1987).

Broader Political crime (#PC0350) Crimes against national security (#PC0554).
Related Sedition (#PC2414) Reversion (#PA5699) Improbity (#PA7363)
Irresolution (#PA7325) Lese majesty (#PF4341) Trade unionism (#PF8493)
Government treachery (#PF4153) Overthrow of government (#PD1964).
Aggravates National insecurity and vulnerability (#PB1149).
Aggravated by War (#PB0593) Informers (#PD8926) Unpopular regimes (#PG5527).

♦ PD2617 Ill treatment of prisoners of war

Nature Prisoners of war are detained for the purpose of preventing them from becoming combatants. They are not criminals, yet they may be abused and denied necessities and forced to live under inhumane conditions.

Incidence War prisoners may be denied essential medical care, forced to labour to aid the enemy, beaten, tortured, or starved. They may be obliged to serve in the enemy's forces, or they may be confined in the combat zone where they may be shelled, bombed, or fired upon. Mistreatment of POWs includes solitary confinement, inability to receive Red Cross transmitted communications or necessities, political brain-washing or indoctrination, and victimization by psychological, surgical or medical experimentation.

Broader Military atrocities (#PD1881) Maltreatment of prisoners (#PD6005)
Denial of rights to soldiers (#PD4089).
Narrower Brainwashing of prisoners of war (#PD1652).
Related Concentration camps (#PD0702)
Denial of rights to wounded military personnel (#PE4758).
Aggravates Defection (#PG5533) Demoralization (#PF8446)
Mental depression (#PC0799) Forced repatriation (#PD8099)
Poor living conditions (#PD9156) Collaboration with the enemy (#PG2132)
Inadequate hospital facilities (#PE5058)
Forced repatriation of prisoners of war (#PD0218)
Delayed consequences of war-time imprisonment and deportation (#PF0726).
Aggravated by Prisoners of war (#PC8848) Lack of war relief (#PF0727)
Inadequate army discipline (#PD2543).

♦ PD2620 Evasion of customs and excise duties

Smuggling — Customs fraud

Nature Smuggling is the clandestine movement of commodities or persons across borders including: false identification of products, their contents, their country of origin or other information. Its objectives may be financial (to avoid duties or import or export restrictions); criminal; political; or humane.

Incidence To indicate their inexhaustible range, a list of some smuggled consignments could include butter, cocaine, machine guns, political manuscripts, escaped criminals, stolen art masterpieces, refugees, alcoholic beverages, protected species of animals, tobacco, abducted women and children, and diamonds. Because of the difficulty involved in detection, some countries (such as the USA and the UK) may have very few smuggling convictions when compared with the thousands of minor and hundreds of major undetected smuggling operations annually.

The US Customs Service estimates that about 10 percent of all imports are fraudulent. In fiscal years 1984 and 1985, only 27 percent of the textile and apparel imports involved in customs fraud were even detected.

Refs Finn, M C *Complete Book of International Smuggling* (1986).

Broader Tax evasion (#PD1466) Economic crime (#PC5624).
Narrower Drug smuggling (#PE1880) Toxic waste smuggling (#PD9765)
Illicit export of works of art (#PE9004).
Related Frauds, forgeries and financial crime (#PE5516).
Aggravated by Documentary fraud (#PE1110) Enclaves and exclaves (#PD2154)
Unethical commercial practices (#PC2563) Unethical practices in transportation (#PD1012)
Corruption of customs and excise officials (#PE4033).

♦ PD2623 Unethical practice of anthropology

Irresponsible anthropology — Negligence by anthropologists — Malpractice in anthropology — Corruption of anthropologists — Underreporting of hazards to minority cultures

Claim Anthropologists, under pressure from their employers, have adopted practices which lead to the underreporting of hazards to minority cultures, especially as a consequence of intrusion by other cultural systems. Bioscientists have failed to investigate adequately the nature of such hazards in the process of further developing knowledge about such cultures. There is little peer control of irresponsible intervention in minority cultures ith the associated introduction of exotic artefacts and modes of behaviour. Anthropologists participate in acquisitive practices by universities, universities and art galleries which deprive cultures of their traditional artefacts, some of which may be vital symbols of their cultural heritage.

Broader Unethical practices in the social sciences (#PD6626).
Narrower Archaeological and anthropological looting (#PD1823).

♦ PD2624 Economically disadvantaged students

Broader Socio-economic poverty (#PB0388).
Aggravates Inequality in education (#PC3434).

♦ PD2625 Unethical consumption practices

Irresponsible consumption of resources — Corruption in the consumption of resources

Aggravates Leukopenia (#PG5543) Consumer debt (#PD3954)
Nutritional anaemia (#PD0321) Decadent standard of living (#PD4037)
Maldistribution of resources (#PB1016) Human consumption of animals (#PC7644)
Unnecessary personal consumption (#PF5931) Consumption of alcoholic beverages (#PD8286)
Human consumption of animal products (#PD7699)
Maldistribution of energy consumption (#PC5038)
Natural resource depletion due to high-level consumption (#PD4002)
Excessive consumption of resources in developed countries (#PE5551).
Aggravated by Tuberculosis (#PE0566) Gaucher's disease (#PG5547)
Rheumatic diseases (#PE0873) Cirrhosis of the liver (#PE2446)
Infectious mononucleosis (#PE5550)
Misrepresentation of information to consumers (#PE6877).

♦ PD2626 Diseases of the spine

Spinal paralysis

Refs Bland, John H *Disorders of the Cervical Spine* (1987); Cyriax, James *The Slipped Disc* (1980); Jeanmart, L *Tumors of the Spine* (1985).

Broader Human disease and disability (#PB1044)
Diseases of connective tissue (#PD2565).
Narrower Lumbago (#PE1310) Spina bifida (#PE1221) Disc disorders (#PG5560)
Whiplash injury (#PG5559) Rheumatic diseases (#PE0873).
Related Cerebral paralysis (#PE0763) Malignant neoplasms (#PC0092).
Aggravates Paralysis (#PD2632)
Human physical genetic abnormalities (#PD1618).

♦ PD2629 Disagreement within alliances

Nature Disagreements within treaty organizations increase tensions within the alliance and provide exploitable opportunities for enemies of the alliance.

Incidence During recent negotiations between the United States and the Soviet Union about reductions in nuclear arms members of NATO disagreed about which weapons should be reduced. There seems to be a lack of overall arms control plan on the part of NATO.

Broader Subversion of international agreements (#PD5876).
Aggravated by Conflicting priorities (#PF5766) Unfulfilled treaty obligations (#PF2497)
Lack of means for achieving consensus (#PD2438)
Limited acceptance of international treaties (#PF0977).

♦ **PD2630 Perjury**
False swearing
Nature Perjury is the deliberate concealment of facts or conscious distortion of the truth by a witness or victim, either during a preliminary investigation or in court. Perjury could result in the conviction of an innocent person, or the acquittal of a criminal, thereby rendering court decisions vulnerable.
Broader Incorrect information (#PB3095) Hindrance of law enforcement (#PD5515)
Falsification of public records (#PD4239).
Narrower False evidence (#PF5127)
Failure to appear as witness to produce information or to be sworn (#PE1756).
Aggravates False statements (#PF4583).
Aggravated by Use of undue influence to obstruct the administration of justice (#PE8829).

♦ **PD2632 Paralysis**
Palsy
Broader Human disease and disability (#PB1044)
Diseases and injuries of the brain (#PD0992).
Narrower Hemiplegia (#PG2943) Paraplegia (#PE2945)
Tetraplegia (#PE2946) Poliomyelitis (#PE0504)
Elephantiasis (#PE1601) Wasting palsy (#PG5567)
Facial paralysis (#PG5569) Paralysis agitans (#PE2206)
Cerebral paralysis (#PE0763) Functional paralysis (#PG5570)
Infantile monoplegia (#PG2944)
Paralysis of throat, eye, limb and respiratory muscles (#PE8173).
Related Inaction (#PA5806).
Aggravated by Stroke (#PE1684) Syphilis (#PE2300)
Hysteria (#PE6412) Neuritis (#PG4892)
Spina bifida (#PE1221) Encephalitis (#PE2348)
Virus diseases (#PD0594) Multiple sclerosis (#PE1041)
Diseases of the spine (#PD2626)
Diseases of nerves and peripheral ganglia (#PE8932).

♦ **PD2637 Martial law**
Broader Militarism (#PC2169) Political disintegration (#PC3204)
Deficiencies in national and local legal systems (#PF4851).
Narrower Curfew (#PG5582).
Related Military law (#PG5584) Military influence (#PD3385)
Abuse of police power (#PC1142).
Aggravates Abuse of power (#PB6918) Political dictatorship (#PC0845)
Internment without trial (#PD1576) Cessation of functions of civil courts (#PE8218).
Aggravated by War (#PB0593) Civil war (#PC1869) Revolution (#PA5901)
Social conflict (#PC0137) Political opportunism (#PC1897) Social disintegration (#PC3309)
National disintegration (#PB3384).

♦ **PD2638 Failure of materials**
Nature Industrial and engineering progress increasingly involves operating many kinds of mechanisms under more onerous conditions than they were previously operated, whether it be in terms of speed, size, load, temperature range, or environmental conditions. Structures and engineering components fail by instability, fatigue, creep, corrosion, or brittle fracture, possibly aggravated by excessive wear. Such failures lead to accidents, breakdown, or more usually to the need to maintain and replace individual parts in machinery and structures.
Incidence Steel and concrete, for example, are universally used materials. The failure of the former can cause immediately fatalities when occurring in airplanes, spacecraft and in vehicles operated at high speeds. The failure of concrete has caused structural collapse; excessive loading and stresses due to vibrations or earth tremors being critical factors. Failure of materials in containers or equipment used for hazardous toxic or explosive substances occurs with some regularity and safety specifications may be lacking or inadequate.
Broader Instability (#PA0859).
Narrower Corrosion (#PD0508) Metal failure (#PD7215) Creep in metals (#PG5589)
Creep of concrete (#PE6313) Fatigue in materials (#PD1391)
Brittle fracture of metals (#PG5588) Intolerance of imperfection (#PF7024).
Aggravates Mechanical failure (#PC1904) Injurious accidents (#PB0731)
Electronic equipment failure (#PD1475)
Unreliability of equipment and machinery (#PC2297).
Aggravated by Wear (#PB1701) Stress in industry (#PE6996)
Corrosion of iron and steel (#PE1945) Human errors and miscalculations (#PF3702)
Atmospheric corrosion of materials (#PE9525)
Vibration damage to cultural artefacts (#PE8162).

♦ **PD2647 Foot diseases and disabilities**
Refs Enna, C D *Peripheral Denervation of the Foot* (1989); Klenerman, L *The Foot and its Disorders* (1982).
Broader Human disease and disability (#PB1044).
Narrower Flat foot (#PG5603) Club foot (#PE3836) Metatarsalgia (#PG5605)
Tenosynovitis (#PE5606) Inflammation of tendons (#PG5607).
Aggravated by Pregnancy disorders (#PD2289) Pyogenic infections (#PG5610).

♦ **PD2651 Intentional infecting with disease**
Nature Prisoners are intentionally infected with diseases to increase suffering and as a means of torturing them. In some cases they are infected and released from prison so that they die after detention.
Broader Malevolence (#PA7102) Physical torture (#PD8734).
Narrower Infectious revenge (#PD5168).
Aggravated by Human disease and disability (#PB1044).

♦ **PD2654 Diseases of the lymphatic system**
Broader Human disease and disability (#PB1044)
Diseases of blood and blood-forming organs (#PF8026).
Narrower Leukaemia (#PE0639) Lymphosarcoma (#PE5548)
Chronic lymphadenitis (#PG7746) Tuberculous lymphadenitis (#PG5624).
Related Rubella (#PE0785) Typhoid fever (#PD1753) Scarlet fever (#PG2757).
Aggravates Kidney disorders (#PE2053) Diseases and injuries of the brain (#PD0992).
Aggravated by Filariasis (#PE2391) Hypertension (#PE0585)
Parasites of the human body (#PE0596)
Government action against regimes with alternative policies (#PF2199).

♦ **PD2656 Increasing job monotony**
Dehumanization of work — Monotonous and repetitious work
Nature A high degree of mechanization may increase psychosomatic disorders, reduce job satisfaction, and contribute to a higher rate of absenteeism. Factors such as inter-personal relations at work, work stability, shift work, speed, and safety are important. Workers engaged in repetitive tasks, and controlled by machines, derive less satisfaction from their work. Shift work to sustain factory output capacities creates a psychosocial working environment that may adversely influence the health of the worker. Night work, and the change of working hours from one shift to another, may subject the workers to certain stresses. Such stresses affect the nervous system, increasing the frequency of peptic ulcer and of nervous symptoms, such as fatigue, nervousness, irritation, and insomnia. These nervous symptoms are usually related to lack of sleep, which in turn may be related to housing conditions, and especially to disturbance of sleep by noise during the day, if the worker is on the night shift.
Claim An environment of machines creates ambient electromagnetism. Light quality or frequencies, sound vibrations and electrical and magnetic forces may disrupt the physical organism in subtle ways, for example acting through the nervous system, to affect attention, reaction times, or technical inspection or evaluation processes. Also where the ratio of work-space to worker is very high, or where workers are isolated for safety reasons or where strict regulations prevent human interactions on the job, stress may be induced. All these and other varieties of monotonous or dehumanized work are counter-productive.
Broader Lack of job satisfaction (#PF0171) Negative emotions and attitudes (#PA7090).
Narrower Aircraft pilot fatigue (#PE3870).
Related Social ill-effects of automation (#PE5134).
Aggravates Strikes (#PD0694) Alienation from work (#PD3076)
Substance abuse at work (#PD9805) Deterioration in physical health (#PC0716).

♦ **PD2664 Poaching**
Illegal hunting of protected and endangered species — Inadequate enforcement against game poaching
Nature Game or fish may be trapped or shot in areas where such rights are either privately owned or specially restricted for conservation reasons. Until the 20th century, most poaching took the form of subsistence poaching by impoverished peasants wishing to augment inadequate diets. The main incentive is now commercial profit which encourages a much more organized approach. It is not just bandit armies anymore that kill endangered species for their own profit; corrupt government officials, members of military forces and high-placed families have become rich through the slaughter of wildlife.
Incidence The world's poaching industry is estimated to generate $1.5 billion per year. In the developed countries, poaching is now limited to shooting deer and game birds (pheasant) and fishing (salmon). In Scotland a single small boat of salmon can bring in up to £7,000. In developing countries, poachers operate within game parks. It is almost impossible to police the very large areas, owing to limited manpower. In addition poachers may operate in large highly mobile groups. Continuing reports indicate the existence of poaching gangs in Kenya. These gangs may number as many as 50 persons or more with individual firearms, but also with gang-owned automatic weapons (Bren and Sten guns, for example) which are used to machine-gun protected game. No animals are immune to poachers; for example, ivory from massacred African elephants regularly finds it way to purchasers in Japan and Hong Kong where it brings £200 a kilogram. Rhinoceros horn retails for £2,000 in Taiwan.
Claim Since 1980 the conflict between poachers and game warden has become ever more deadly, claiming casualties on both sides, but that happens in every war.
Broader Theft of property (#PD4691) Inadequate animal welfare (#PC1167).
Aggravates Endangered species of animals (#PC1713)
Ineffective use of external relations relating to sportsmen (#PE6515).
Aggravated by Inadequate national law enforcement (#PE4768)
Trade in animal products of endangered species (#PD0389).

♦ **PD2669 Inadequate buildings, services and facilities for organized action against problems**
Broader General obstacles to problem alleviation (#PF0631).
Narrower Inadequate facilities for international organization action (#PD0929).

♦ **PD2670 Individual fear of future change**
Social change powerlessness — Paralyzing image of the impossibility of change
Claim Although many people realize that it is they who create the future, that there is nothing which will occur except that which is created, yet within people there is an innate fear of the future. This is intensified by present-day society's heavy emphasis on defence. Individuals are trapped and helpless in the conscious expectation of participating in creating a different future while not knowing how to appropriate this role.
Broader Obstacles to leadership (#PF7011)
Inhibited capacity to visualize a creative future (#PF2352).
Narrower Fear of vocational change (#PJ1318).
Related Unrecognized socio-economic interdependencies (#PF2969).
Aggravates Resistance to grace (#PF5266) Resistance to change (#PF0557)
Lack of political will (#PC5180) Inadequate social reform (#PF0677)
Oppressive prevalent images (#PF1365).
Aggravated by Temporal deprivation (#PF4644).

♦ **PD2671 Meaningless corporate engagement**
Nature Overall reductionistic images have resulted in a disparity between the goals of the corporate community and individual purpose. People lack a global context from which to see their own experience as significant, resulting in a gap between individual engagement and a corporate thrust. Viewed from a limited framework, present-day individual commitment is not experienced as as meaningful.
Aggravated by Unarticulated goal of educational methods (#PF2400).

♦ **PD2672 Debilitating deterioration of physical environment**
Blighted land use
Nature Overall care for buildings and open spaces in the urban environment is often obscured by uncontrolled litter in streets and vacant lots, and by abandoned automobiles and discarded appliances. Poor housing upkeep adds to the sense of an unkempt and uncared-for community. Ill-timed and insufficient garbage collection adds to the situation as dogs and rats strew refuse across the area. Sensing it to be uncared for, people from outside the community use the many empty spaces as dumping grounds for garbage. The cumulative effect is a resigned acceptance of being unable to control the environment and continuing careless destruction of the community. Residents themselves feel that the community is forgotten; and without a significant demonstration that with corporate effort it is possible to care for buildings and space, the cycle continues.
Broader Deterioration of human environment (#PC8943).
Narrower Unvalued diplomas (#PG7719) Overgrown vacant lots (#PG7718)
Negative social context (#PF9003) Restrictive social policies (#PF8282)
Low-quality construction work (#PD7723) Deteriorated building conditions (#PG7716)
Poor condition of open spaces in urban communities (#PF1815).
Related Inadequate care of community space (#PF2346).
Aggravates Nutritional deficiencies (#PC0382)
Unequal opportunities for foreign students (#PE7726).

♦ **PD2675 Socially unintegrated expatriates**
Nature Expatriates domiciled in a foreign country may have been exiled by their own country or may have exiled themselves. They may have sought political asylum. They may keep their original nationality and be working abroad for a multinational company, for their diplomatic service, for an international organization or more rarely for an indigenous company or organization. Expatri-

ates are often elitist, especially in an ex–colonial situation or where they are a technocracy; very often they remain segregated and isolated from the rest of society. This may cause racial and cultural conflict. If expatriates are exiles they may produce propaganda against their country of origin. Unless expatriates have taken the nationality of the country of residence, they may be excluded from political representation, which may leave them an underprivileged minority or an affluent elite with little concern about conditions for others in the country.
Broader Plural society tensions (#PF2448) Foreign exchange restrictions (#PF3070).
Related Administrative difficulties in new states (#PE1793).
Aggravates Social fragmentation (#PF1324) National political dependence (#PF1452) Discrimination against immigrants and aliens (#PD0973).
Aggravated by Cultural barriers (#PB2331) Transnational corporation imperialism (#PD5891).

♦ PD2681 Single parent families
Single parenthood — Single mothers — Lone–parent families — One–parent families
Nature Difficulties of single parents include inadequate social security or other income, lack of child–care facilities and debt, resulting in general poverty and stress, as well as emotional problems. Single parents may find themselves discriminated against in housing and may have to make do with inadequate facilities. Single parents may be widows or widowers, divorcees, or unmarried mothers or fathers.
Incidence The number of single parents rises substantially with an increase in the divorce rate. In this respect the incidence may be higher in developed countries, where traditional family patterns are changing. The number of unmarried mothers, and to a lesser extent unmarried fathers, is rising in so–called 'permissive society' where birth control is not fully effective. In other societies, single parents may occur as a result of widowhood or in a system of cohabitation such as that of the West Indies. Discrimination against women plays a large part in the difficulties of single parents, who are for the most part women.
In 1971, 8 per cent of families had a single parent in UK. By 1988, there were twice as many single parent families, 16 per cent. The biggest category of single parents are separated or divorced wives, 60 per cent.
Refs Renvoize, Jean *Going Solo* (1985); Roberts, Paula *Women, Poverty, and Child Support* (1986).
Broader Immorality (#PA3369) Marital instability (#PD2103) Family disorganization (#PC2151).
Narrower Unmarried mothers (#PD0902) Unmarried parents (#PD3257) Discrimination against unmarried fathers (#PD3256).
Related Illegitimate children (#PC1874) Conflict concerning legal custody of children (#PF3252).
Aggravates Youth violence (#PF7498) Unknown relatives (#PF0782) Cruelty to children (#PC0838) Juvenile delinquency (#PC0212) Social fragmentation (#PF1324) Lack of social mobility (#PF2195) Inadequate child day–care facilities (#PD2085).
Aggravated by Divorce (#PF2100) Promiscuity (#PC0745) Family breakdown (#PC2102) Adolescent pregnancy (#PD0614) Social discrimination (#PC1864) Lack of family planning (#PF0148) Desertion in marriage law (#PF3254) Discrimination against women (#PC0308) Separation under marriage law (#PF3251) Inadequate provision of alimony (#PE3247) Non–parental custody of children (#PF3253) Inadequacy of contraceptive methods (#PD0093) Inadequate system of child support enforcement (#PF6076).

♦ PD2682 Youth gangs
Street gangs — Wilding — Steaming
Nature Groups of adolescents, usually male, from urban working class or underprivileged districts, take part in aggressive and delinquent activities both within the gang and outside it, fighting other gangs, committing assault and theft and damage to property. Rarely are such gangs organized crime units, more often they are delinquent as a means for obtaining 'kicks'. Increasingly street gangs are involved in drug trafficking, intimidation and violence.
Incidence Youth gangs have developed in many countries, increasing (like the general level of juvenile delinquency) in countries with a higher economic level or with rapid social and economic change. In 1988, 622 wilding robberies were referred to New York City's family court. It is the second most common crime among youths in New York City, after crack dealing. In Los Angeles in 1990 there were some 750 gangs. Gang–related murders were put at 570 in 1989 and robberies at 1,800.
Counter–claim Gangs offer an identity and opportunity for self–assertion to youths under conditions where life holds out little else.
Broader Juvenile delinquency (#PC0212).
Related Criminal gangs (#PD3837).
Aggravates Gang rape (#PG5670) Hooliganism (#PD1109) Gang warfare (#PD4843).
Aggravated by Alienation (#PA3545) Urban slums (#PD3139) Juvenile stress (#PC0877) Corruption of minors (#PD9481) Socio–economic poverty (#PB0388) Inadequate facilities for children's play (#PD0549).

♦ PD2683 Misuse of postal surveillance by governments
Monitoring of mail
Nature Postal monitoring may be used by government agencies to keep track of the activities of individuals or groups considered to be acting against the interests of the government.
Refs French, S *The Big Brother Game* (1986).
Broader Political surveillance (#PD8871) Misuse of government surveillance of communications (#PD9538).
Related Inadequacy of postal services (#PF2717) Interception of correspondence (#PE5093) Inappropriate use of the mail service (#PE6754) Unaccountable government intelligence agencies (#PF9184).
Aggravated by Excessive government control (#PF0304).

♦ PD2686 Institutional lying
Incidence Confidence in public officials and in professionals has been seriously eroded. Incidences like Watergate, the covert American bombing of Cambodia, France's Greenpeace Affair, and numerous other political scandals have served to undermine public confidence, thereby affecting governments' images in the world, perhaps fostering further deception and cover–up.
Claim Lies can only persist as long as they are shielded from critical scrutiny that would result from the free flow of ideas and people. Therefore for their survival, lies require barriers to such free flow, usually achieved through the introduction of censorship and travelling restrictions. Such restrictions promote differences in the information available to different people and generate differing views of the world. These differences in perception breed prejudice and lead to the emergence of dogmas which stand in the way of mutual understanding between groups. All these trends reinforce the emergence and persistence of lies and half–truths. If the spread of ideas, people or products is too rapid for existing institutions to adjust to them, these institutions will try to put up barriers to preserve themselves.
Broader Lying (#PB7600) Behavioural deterioration (#PB6321).
Narrower Paternalistic lies (#PF7635) Deception by government (#PD1893) Historical misrepresentation (#PF4932) Deliberate lying by corporation officials (#PD4982).
Related Manipulation (#PA6359) Disinformation (#PB7606) Irrational religious beliefs (#PF6829).
Aggravates Credibility gap (#PB6314) Loss of civility (#PC7013) Loss of confidence in government leaders (#PF1097).
Aggravated by Religious deception (#PF3495).

♦ PD2688 Embezzlement
Nature The misappropriation, or fraudulent conversion, of property (often funds) by a person who already has it in his lawful possession. Unlike larceny, which is a crime against possession, embezzlement is a crime against ownership.
Incidence Notorious examples of embezzlers are Ferdinand and Imelda Marcos who siphoned more than $100 million from the Philippines.
Refs Snyder, Stanley G *Embezzler's Dirty Tricks* (1986).
Narrower Misappropriation of public funds (#PC2920).
Related Theft of property (#PD4691) Financial scandal (#PD2458) Frauds, forgeries and financial crime (#PE5516).
Aggravates Corruption (#PA1986) Business bankruptcy (#PD2591).
Aggravated by Unethical practices by employees (#PD4334).

♦ PD2689 Proliferation of pets
Nature Large numbers of pets and domestic animals are maintained in households, particularly in developed countries. These pets consume considerable amounts of food at a time of increasing food shortage. The situation has worsened because of the changing composition of diets for cats and dogs. Recent developments in pets' feeding habits, arising from the growth of the pet–food industry, puts them in competition for food with the human population. Pets are also an important factor in the transmission of disease to man and to other domestic animals. Sidewalks of crowded inner cities are littered with dog excrement which causes disease, particularly among young children. The acquisition of wildlife as pets from areas where rabies or rarer diseases are endemic constitutes a special public health problem. The novelty of exotic pets appeals to some people, but there seems to be a lack of public information on how to handle or keep such animals as ocelots, raccoons, tarantulas, and armadillos. Three–quarters of these animals die within six months or are deposited in a humane society shelter or roadside zoo. Research also indicates that what is assumed to be animal love is, in many instances, animal exploitation or an expression of the owner's vanity.
Incidence In the USA, households now shelter more animals than people, with pets outnumbering humans more than three to one. Estimates of the pet population are 40 million dogs, 40 million cats, 15 million birds, ten million other warm blooded creatures ranging from rodents to monkeys, and 600 million fish. In the UK, 54% of households are estimated to own pets. Pets are to be found in 55 per cent of French households (one dog in every 3 families, 1 cat in every 4), involving expenditure of $5.2 billion in 1988 to care for the 35 million animals.
Refs Szasz, Kathleen *Petishism* (1969).
Narrower Dangerous pet animals (#PE7175) Biological contamination of water (#PD1175).
Aggravates Unwanted pet animals (#PD2094) Domestic animal bites (#PE4931) Maltreatment of pet animals (#PD1265) Unethical practices with domesticated animals (#PE4771) Domestic animals as carriers of animal diseases (#PD2746) Disproportionate allocation of resources to pet animals (#PF1085) International movement of animals as factor in animal diseases (#PD2755).
Aggravated by Loneliness (#PF2386).

♦ PD2695 Monolingualism in a multi–cultural setting
Linguistic imperialism
Nature Use of a single language in a multi–cultural setting leads to difficulties and resentment. Speakers using a foreign language have their confidence undermined, and are relegated to using simplistic terms which may not necessarily convey the subtleties or power of their argument. This prevents their audience from understanding the point, which therefore does not receive its full attention. The foreign speaker is left with a feeling of frustration and inadequacy at finding himself unable to make the full contribution he would otherwise be capable of. As language conditions those who use it, speakers gradually identify themselves with the language and the culture it incarnates. Non–speakers of a foreign language may therefore experience it as a threat to their identity. Monolingualism renders speakers unexposed to the variety of alternative concepts, structures and perceptions that exist in other languages; creating a gap between the educated elite speaking the single language and the people in the many other cultures on which the monolingualism is effectively imposed.
Incidence An historical example of monolingualism in multi–cultural settings is the imposition of Latin by the Romans in order to administratively unify their empire. More recent examples include Spanish and French military and cultural imperialism prior to World War II; the imposition of Russian in Ukrainia Georgia, and in post–war Warsaw Pact relations; the dominance of official Hindi in India; official Mandarin and Pin–yin in China; and Swahili in East Africa. The most evident example is the increasing use of English in international relationships of varying types.
Broader Imperialism (#PB0113) Cultural imperialism (#PC3195).
Related Disruptive foreign influence (#PC3188).
Aggravates Linguistic purism (#PF1954) Destruction of cultural heritage (#PC2114).
Aggravated by Forced assimilation (#PC3293).
Reduces Multiplicity of languages (#PC0178) Multiplicity of languages in a national setting (#PC1518).
Reduced by Nationalism (#PB0534).

♦ PD2697 Unethical experiments with drugs and medical devices
Unregulated medical experiments on humans
Nature Experimental drugs and other treatments may be prescribed for patients who are not informed that the full effects of their treatment are not known and who are not asked for their consent to take part in an experiment. Such patients are generally those who can be most easily pressured into accepting unknown and in some cases unwanted treatments, such as the poor and uneducated. Reports of unethical experimentation have been made particularly with regard to contraceptive methods and methods of induced abortion.
Incidence Experiments, ethical and unethical, have been done on humans since the beginning of medicine. During World War II, the Nazis experimented on prisoners in concentrations camps and the Japanese experimented on prisoners of war. In the US, during the 1960s, cancer cells were injected subcutaneously into senile patients without their knowledge at the Jewish Chronic Disease Hospital; at the Willowbrook State Hospital, retarded children were deliberately infected with viral hepatitis; 300 rural black males were left untreated for diagnosed syphilis in the Tuskegee Syphilis Study.
Broader Unethical medical practice (#PD5770) Abusive experimentation on humans (#PC6912) Obstacles to medical experimentation (#PF4865).
Narrower Medical experimentation on children (#PE6764) Cruel treatment of animals for research (#PD0260) Inhumane medical experimentation during war–time (#PE4781) Medical experimentation on mentally impaired persons (#PE8677)

Medical experimentation on socially vulnerable groups (#PD6760)
Medical experimentation on institutionalized subjects (#PE6763)
Medical experimentation on pregnant women and foetuses (#PE8343)
Increasing drug experimentation in developing countries (#PE6201).
Related Inadequate medical care (#PF4832)
Unethical medical experimentation on prisoners (#PE4889).
Aggravates Human death (#PA0072)
Mutilation and deformation of the human body (#PD2559).
Aggravated by Inadequate information on drugs (#PF0603).

♦ PD2705 Maldistribution of medical resources

Nature The decisions about distribution of medical and health resources are made at the international, national, regional, clinical and individual practitioner levels, all of which, provide opportunities for maldistribution. At the microallocation end of the spectrum are the allocation of scarce lifesaving resources between competing claimants. At the macroallocation end are decisions about the allocation of resources between competing social needs.
Broader Inadequate medical resources (#PD7254)
Maldistribution of resources (#PB1016).
Narrower Maldistribution of health personnel (#PF4126).
Aggravates Inadequate medical care (#PF4832).

♦ PD2710 Abusive national leadership

Tyrannical dictatorship — Rule by fiat — Arbitrary leadership — Psychopathic national leaders — Self-aggrandizing national leaders — Undemocratic leadership
Nature Some despots, once secure in their hold over a country, feel free to indulge in a wide range of arbitrary acts. These may include grandiose public works at the expense of the poor, self-aggrandisement, adoption of unusual dress style, attribution of grandiloquent titles, blatant favouritism in appointments, unwarranted claims to expertise and wisdom, and imposition of an idiosyncratic ideology through the education system (possibly based on an extensive volume of unsubstantiated writing). Policies may result in extensive violence to major sections of the population, including forced resettlement, systematic violations of human rights and extra-judicial executions, and support for international terrorism.
Incidence Only 6 of the more than 150 heads of state in the history of post-colonial Africa have relinquished power voluntarily. The remainder were evicted or assassinated in military coups for economic incompetence, political tyranny and other failings. Even though many of them gained independence from colonial rule, this did not enable them to take on the leadership of their countries effectively and avoid economic ruin.
Claim Because of the likelihood of reprisals, it is standard practice within the international community to refrain from naming the countries or leaders with such characteristics, since "everyone knows who they are" from media reports. In the 1970s and 1980s it has been argued that many of the dictators of developing countries were actually psychologically disturbed. For example, Gaddafi of Libya, Amin of Uganda, Khomeini of Iran, and Bokassa of Central African Empire are all reported to have authorized and/or participated in atrocities ranging from the massacre of a classroom of school children, to the storming of foreign embassies and taking hostages, or even to cannibalism.
Refs MacMunn, George F *Leadership Through the Ages* (1935).
Broader Arbitrariness (#PB5486) Political dictatorship (#PC0845).
Aggravates Denial of right to state succession (#PE0241)
Excessively costly prestige projects (#PF3455)
Excessive accumulation of wealth by government leaders (#PD9653).
Aggravated by Boasting (#PF4436) Leadership impaired by illness (#PF8387).

♦ PD2712 Elitist control of construction technology

Nature The power to implement construction expertise belongs to an elite segment of the global community which is oriented toward growth of its own profit and capital assets, rather than toward comprehensive global benefit. Such power increases proportionately with continuing implementation, leading to a corresponding decrease in the power at grass roots level to implement and direct the course of construction planning. Without this local regulation, distribution and development are irregular and sporadic.
Broader Profit motivated utilization of construction technology (#PF2464).

♦ PD2716 Inadequate and inaccurate textbooks and reference books

Nature Provision of textbooks to students can be constrained by school budgets, professional qualifications and levels of achievement of teachers and school administrators, and ideological influences. Inadequate budgets can lead, for example, to retention of dated, inaccurate accounts of history or expositions of the sciences. Lack of procedures to assure acquisition of the best possible texts can lead to the use of those that are pedagogically inferior. This also results from indiscriminate preference for authors of the some institutions, school systems or regions on the 'native son' principal, or for financial reasons; or conversely, the blind following of the leading text-book publishers, or university presses because of their expensive textbook promotion campaigns, or 'safe' reputation. Thirdly, ideological influences account for textbooks with racial bias, excessive nationalism, bias against theories in biology (such as evolution), cosmogenesis ('Big Bang' versus 'Continuous Creation'), or in psychology. Bias against socialism, communism, capitalism, militarism and other 'isms' are sometimes demanded by their respective opposite environments. All of the foregoing result in generations of ill-informed and, frequently, prejudiced citizens.
Broader Ideological conflict (#PF3388) Human errors and miscalculations (#PF3702).
Narrower Biased and inaccurate history textbooks (#PD2082)
Biased and inaccurate biology textbooks (#PF9358)
Biased and inaccurate geography textbooks (#PF1780)
Deliberate misrepresentation in educational materials (#PF1183).
Aggravated by Prejudice (#PA2173) Double standards in morality (#PF5225).

♦ PD2718 Hazardous locations for nuclear power plants

Risk of ecoaccidents — Risk of ecocatastrophe
Nature Nuclear power plants and reactors may be projected and installed in locations where vulnerability to surface subsidence, seismic shock and earthquake faults were insufficiently studied. Other ignored factors may include hazards to the environment from radioactively contaminated steam or gases, or contaminated solid or liquid residue disposal.
Background Nuclear power plants proposed, under construction, or installed, may be subject to protest actions by the nearest community residents in whose view a hazard exists for the population. Opposition is not only specific to 'melt-down' and other nuclear disaster possibilities or slow radioactive contamination, but generally, against the size of the power plant, the clearing of landscape for it, etc. Frequently opposition is psychologically motivated by the fact that the community may have had no chance to participate in the location decisions made by private power companies or local government or both.
Refs Cairns, John *Ecoaccidents* (1985).
Broader Human errors and miscalculations (#PF3702)
Long-term shortage of natural resources (#PC4824)
Environmental hazards of nuclear power production (#PD4977).
Aggravates Landscape disfigurement (#PC2122) Nuclear reactor accidents (#PD7579)
Insufficient nuclear power stations (#PD7663)
Environmental pollution by nuclear reactors (#PD1584)
Unhealthy emotional responses to atomic energy (#PF0913)
Degradation of semi-natural and natural habitats of flora and fauna (#PC3152).
Aggravated by Earthquakes (#PD0201) Land subsidence (#PD5156)
Unethical practice of earth sciences (#PD0708).

♦ PD2722 Abusive traffic in immigrant workers

Traffic in immigrant workers — Illicit labour trafficking
Nature Workers from developing countries may be induced to migrate by companies and other entrepreneurs giving a false impression of working conditions and the value of remuneration. The traffic also includes illegal immigration for which a heavy price is paid in cash by the immigrant and in prison sentences or deportation if he is discovered and arrested. Agents for this trade often make promises of obtaining work permits and visas for the immigrants which they may or may not carry out.
Incidence Because of the demand for cheap unskilled labour in industrialized countries to carry out menial tasks that nationals no longer want to do, recruiting programs have been set up in certain underdeveloped countries giving false hopes to the impoverished and unemployed. Although if they are successfully transplanted they earn higher wages than at home, they are in no position to benefit from this owing to the higher cost of living in industrialized countries and the need to send money home to the family. Many immigrants recruited in this way enter illegally and pay a high price to an agent or middle man. They may or may not obtain permits later. They face deportation and prison sentences and a total loss of their cash outlay if they are caught by the authorities. They are usually transported and housed in appalling conditions and are very often paid below minimum wage. Exploiters are often their own countrymen. Governments and firms tacitly condone unjust working and living conditions. The exploited come mainly from Africa, Asia and the Middle East.
Refs Council of Europe *Temporary Employment Businesses* (1985); United Nation's *Exploitation of Labour Through Illicit and Clandestine Trafficking* (1986).
Broader Traffic in persons (#PC4442) Proliferation of immigrants (#PD4605).
Related Segregation in employment (#PD3443)
Crimes related to immigration, naturalization and passports (#PE3889).
Aggravates Racial conflict (#PC3684) Border controls (#PJ1718)
Labour shortage in developing countries (#PD5045)
Discrimination against immigrants and aliens (#PD0973)
Inadequate cultural integration of immigrants (#PC1532)
Lack of capital investment in developing countries (#PE5790)
Inadequate living and working conditions of immigrant labourers in industrialized countries (#PD3427).
Aggravated by Ignorance (#PA5568) Unemployment (#PB0750)
Illegal immigration (#PD1928) Racial discrimination (#PC0006)
Socio-economic poverty (#PB0388) Clandestine employment (#PC7607)
Exploitation in employment (#PC3297) Restrictions on immigration (#PC0970)
Unethical practices of employers (#PD2879)
Political instability of developing countries (#PD8323).

♦ PD2726 Iodine deficiency disorders

Nature Iodine deficiency causes goitre, hypothyroidism, retarded physical development and impaired mental function, increased rate of spontaneous abortion and stillbirth, neurological cretinism, including deaf mutism and myxoedematous cretinism, including dwarfism and severe mental retardation.
Claim Of the many disabling conditions that come massively in the way of human development, the endemic deficiency in iodine is second to none, in the severity of its consequences coupled with the spread of its prevalence. In some areas of the world 4 to 15 percent of the newborn infants are condemned to the destruction of their mental and physical health.
Broader Nutritional deficiencies (#PC0382).
Aggravates Cretinism (#PE7905) Simple goitre (#PG1279)
Natural human abortion (#PD0173) Thyroid gland disorders (#PE0652).

♦ PD2728 Fungal diseases

Fungal diseases of animals — Fungus infection — Fungus disease
Nature Parasitic fungi attack living organisms, penetrate their outer defences, invade them and obtain nourishment from living protoplasm, causing disease and sometimes death of the host.
Incidence More than 50 species of fungi cause disease in animals and man, forming two subsections: dermatophytoses, superficial infections of the skin, hair and nails; and mycoses, infections of the deeper tissues and organs. Dermatophytoses include ringworm and actinomycosis. Dermatophytoses are caused by fungi present on the surface of the skin; mycoses on the other hand live and grow independently of the host and infection is caused by inhaling or ingesting spores, or by the implantation of fungus under the skin through small punctures, cuts or bruises. Mycosis infections include coccidioidomycosis, histoplasmosis, cryptococcosis, and adiaspiromycosis.
Refs Wilson, J Walter and Plunkett, Orda A *The Fungus Diseases of Man* (1965).
Broader Zoonoses (#PD1770) Animal diseases (#PC0952)
Infectious diseases in animals (#PD2732).
Narrower Mycosis (#PE2455) Ringworm (#PD2545) Oomycosis (#PJ9289)
Candidosis (#PE4923) Coccidiosis (#PE2738) Geotrichosis (#PG0787)
Mucormycosis (#PG9715) Actinomycosis (#PE2353) Chromomycosis (#PG9572)
Histoplasmosis (#PG3696) Cryptococcosis (#PE4932) Paecilomycosis (#PG7192)
Sporotrichosis (#PG4935) Dermatophytoses (#PG4553) Adiaspiromycosis (#PG5703)
Rhinosporidiosis (#PG9835) Thrush in chickens (#PE4497)
Entomophthoromycosis (#PG6047) Nocardiosis in animals (#PG6309)
Epizootic lymphangitis (#PG9632) Mycotoxicosis in poultry (#PG6792)
Aspergillosis in animals (#PG9721) North American blastomycosis (#PG5620).
Related Moulds in plants (#PE8051) Microbial diseases (#PC7492)
Fungal plant diseases (#PD2225) Viral diseases in animals (#PD2730)
Infectious and parasitic diseases (#PD0982)
Denial to animals of the right to a natural death (#PE8339).
Aggravates Enzootic diseases (#PD2733) Epizootic diseases (#PD2734)
Skeletal system disorders (#PE2298)
Wild animals as carriers of animal diseases (#PD2729).
Aggravated by Animal injuries (#PC2753) Vectors of animal diseases (#PD2751)
Biological contamination of food (#PD2594) Biological contamination of water (#PD1175).

♦ PD2729 Wild animals as carriers of animal diseases

Nature Wild animals may help to spread epidemics that affect animals, especially domesticated ones, or may keep a disease enzootic when the epidemic has died out, thus causing serious difficulty in controlling animal diseases.
Incidence The responsibility of wild animals for the transmission of diseases to domestic animals is not known, but they are blamed for maintaining and spreading some of the most devastating diseases, such as anthrax, brucellosis, foot-and-mouth disease, Newcastle disease, liver fluke, rinderpest, trypanosomiasis, and rabies. Transmission may occur through direct contact, but is more likely to occur through pastureland and water infected by wild animals. Serious diseases of domestic animals, such as trypanosomiasis, may have a less serious effect on wild animals so that they remain alive and thus are able to transmit the disease more freely. The wild animal sector

DETAILED PROBLEMS

PD2735

of disease is very difficult to control without mass extermination, which is undesirable from an ecological and cultural point of view, as well as being impracticable. Wild animals imported for zoos have introduced diseases to countries where it was not known before, such as American liver fluke brought to Italy by the importation of North American elk.
Broader Vectors of animal diseases (#PD2751) Animals as vectors of disease (#PD8360).
Narrower Snail vectors of animal diseases (#PE2747)
Insect vectors of animal diseases (#PD2748)
Worms as vectors of animal diseases (#PD2750)
Wild birds as vectors of animal diseases (#PE2749).
Aggravates Zoonoses (#PD1770) Rinderpest (#PE2786)
Brucellosis (#PE0924) Enzootic diseases (#PD2733).
Aggravated by Fungal diseases (#PD2728) Diseases of wild animals (#PD2776)
Viral diseases in animals (#PD2730) Bacterial diseases in animals (#PD2731)
Parasitic diseases in animals (#PD2735).
Reduced by Defoliation (#PD1135) Wildlife extinction (#PC1445).

♦ PD2730 Viral diseases in animals
Viral animal infections — Virus diseases in animals
Nature Animal diseases can be caused by viruses, which are self-reproducing agents that multiply only within susceptible living cells and lead to a wide range of infections and to cancer. The four main categories of viruses: animal and human; insect; plant; and bacterial, may be interchangeable and transmissible. For viruses to survive, there must be a means by which virus units can reach susceptible cells, multiply within these cells, and liberate into the environment. Viral animal diseases include rinderpest, rabies, distemper, hog cholera, fowl pox, infectious laryngotracheitis, Newcastle disease, and some cancers.
Refs Appel, Max J (Ed) *Virus Infections of Carnivores* (1987); Kahrs, Robert F10 *Viral Diseases of Cattle* (1981); McFerran, J B and McNulty, M S (Eds) *Acute Virus Infections of Poultry* (1986); Wittmann, G (Ed) *Herpesvirus Diseases of Cattle, Horses, and Pigs* (1989).
Broader Virus diseases (#PD0594) Animal diseases (#PC0952)
Infectious diseases in animals (#PD2732).
Narrower Rabies (#PE1325) Cowpox (#PE6886) Fowlpest (#PE1400)
Fowl pox (#PG5711) Rinderpest (#PE2786) Bluetongue (#PE6297)
Lassa fever (#PG9222) Swine fever (#PE6266) Duck plague (#PG7496)
Border disease (#PG4244) Coxsackie virus (#PE3852) West Nile fever (#PG6834)
Bovine leukosis (#PE1243) Ephemeral fever (#PG3738) Marek's disease (#PE9689)
Canine distemper (#PG5710) Rift valley fever (#PE7552) Caprine arthritis (#PG5248)
Egg drop syndrome (#PG5400) Contagious ecthyma (#PE6831)
Wesselsbron disease (#PG3959) African swine fever (#PE5207)
Feline panleukopenia (#PG4070) Nairobi sheep disease (#PE6268)
Reticuloendotheliosis (#PG6856) Foot-and-mouth disease (#PE1589)
African horse sickness (#PE1805) Equine viral arteritis (#PG4283)
Swine vesicular disease (#PG3897) Bat salivary gland fever (#PG4980)
Viral hepatitis of ducks (#PG9368) Viral hepatitis of geese (#PG4030)
Equine infectious anaemia (#PG7882) Malignant catarrhal fever (#PE6280)
Peste des petits ruminants (#PG7639) Akabane disease of animals (#PG7858)
Viral hepatitis of turkeys (#PG9326) Canine parvoviral infection (#PG7845)
Infectious canine hepatitis (#PG7668) Infectious laryngotracheitis (#PG5712)
African green monkey disease (#PG7524) Canine herpesviral infection (#PG0962)
Vesicular exanthema of swine (#PG6600) Feline infectious peritonitis (#PG9245)
Lymphoid leukosis of chickens (#PG3875) Airborne viral animal diseases (#PD2741)
Argentinian haemorrhagic fever (#PG7151) Vesicular stomatitis in animals (#PG3759)
Crimean-Congo haemorrhagic fever (#PG6985) Coronaviral enteritis of turkeys (#PG9619)
Rotaviral infections in chickens (#PG9408) Haemorrhagic enteritis of turkeys (#PG5040)
Malabsorption syndrome of chickens (#PG9749) Marble spleen disease of pheasants (#PG4464)
Infectious bursal disease of chickens (#PG3991)
Coronaviral encephalomyelitis of swine (#PG3788)
Lymphoproliferative disease in turkeys (#PG6583)
Haemorrhagic anaemia syndrome in chickens (#PG5506)
Encephalomyocarditis virus disease of pigs (#PG7729).
Related Hepatitis (#PE0517) Yellow fever (#PE0985) Encephalitis (#PE2348)
Fungal diseases (#PD2728) Viral plant diseases (#PD2227)
Virus diseases in protozoa (#PG7376) Virus diseases in bacteria (#PD2562)
Bacterial diseases in animals (#PD2731)
Denial to animals of the right to a natural death (#PE8339).
Aggravates Zoonoses (#PD1770) Enzootic diseases (#PD2733)
Epizootic diseases (#PD2734)
Wild animals as carriers of animal diseases (#PD2729).
Aggravated by Stray dog populations (#PE0359)
Vectors of animal diseases (#PD2751)
Water-borne animal diseases (#PE2787)
Disease-causing viral combinations (#PE6403)
Inadequate carcass disposal of diseased animals (#PE2778)
International movement of animals as factor in animal diseases (#PD2755)
Inadequate disinfection measures for animal housing and equipment (#PE2757)
Importation of infected carcass meats as factor in animal diseases (#PE2777).
Reduced by Wildlife extinction (#PC1445).

♦ PD2731 Bacterial diseases in animals
Nature Animal diseases can be caused by unicellular micro organisms (bacteria), leading to infections and toxaemia. The virulence of bacterial animal diseases depends on the degree of immunity of the host. There are four main types of bacteria: the spherical or coccus form; the rod or bacillus type; the spirally twisted spirillum; and a long filamentous type. The extent of the disease is dependent on the ability of the bacteria to multiply in the host. Transmission may be by contagion, inhalation, infection from food, via insects, or via wounds. Bacterial animal diseases include plague, glanders pseudotuberculosis, enteritis, mastitis, pneumonia, infectious anaemia, anthrax, food poisoning, tuberculosis, Johne's disease, meningitis, and haemorrhagic septicaemia.
Broader Animal diseases (#PC0952) Bacterial disease (#PD9094)
Infectious diseases in animals (#PD2732).
Narrower Anthrax (#PE2736) Glanders (#PE2461) Blackleg (#PE2737)
Shigella (#PG4520) Pneumonia (#PE2293) Enteritis (#PE4973)
Strangles (#PJ9367) Tularaemia (#PE6872) Listeriosis (#PE3779)
Paratuberculosis (#PG7828) Actinobacillosis (#PG9855) Swine erysipelas (#PG5031)
Campylobacteriosis (#PG4274) Septicaemia in foals (#PG7429)
Legionnaire's disease (#PE6783) Meningitis in animals (#PG7785)
Omphalitis in poultry (#PG4443) Streptococcus infection (#PE3098)
Haemorrhagic septicaemia (#PG4458) Mastitis in milking cows (#PG5698)
Toxaemic animal diseases (#PG5714) Spirochetosis in poultry (#PG9204)
Allergy inducing bacteria (#PG3224) Colibacillosis of poultry (#PG9827)
Streptococcosis in poultry (#PG7658) Alimentary anaemia in stock (#PG5189)
Staphylococcosis in poultry (#PG9371) Post-dipping lameness in sheep (#PG6897)
Bovine pneumonic pasteurellosis (#PG9381) Porcine streptococcal infections (#PG3795)
Food poisoning through negligence (#PE0561) Nonsuppurative polyarthritis in lambs (#PG9230).
Related Plague (#PE0987) Tetanus (#PE2530) Brucellosis (#PE0924)
Tuberculosis (#PE0566) Bacterial plant diseases (#PD2226)
Viral diseases in animals (#PD2730) Parasitic diseases in animals (#PD2735).
Aggravates Zoonoses (#PD1770) Enzootic diseases (#PD2733)
Epizootic diseases (#PD2734)
Wild animals as carriers of animal diseases (#PD2729).
Aggravated by Resistant bacteria (#PE6007) Stray dog populations (#PE0359)
Vectors of animal diseases (#PD2751) Water-borne animal diseases (#PE2787)
Soil-borne diseases in animals (#PE2739)
Inadequate carcass disposal of diseased animals (#PE2778)
International movement of animals as factor in animal diseases (#PD2755).
Reduced by Virus diseases (#PD0594).

♦ PD2732 Infectious diseases in animals
Infectious animal diseases
Nature Diseases can be passed from one animal to another, via direct contact, contagion, or via inhalation. Highly infectious diseases may cause epidemics, with substantial losses. Infectious animal diseases include the four main categories of animal diseases: viral, bacterial, fungal and parasitic.
Refs Bryans, J T and Gerber, H (Eds) *Equine Infectious Diseases 3* (1973); Cunha, Burke A *Infectious Diseases in the Elderly* (1988); Davis, John W, et al *Infectious Diseases of Wild Mammals* (1981); Merchant, I A and Barner, Ralph D *Outline of Infectious Diseases of Domestic Animals* (1964); National Research Council *Guide to Infectious Diseases of Mice and Rats* (1988).
Broader Animal diseases (#PC0952).
Narrower Q fever (#PE2534) Glanders (#PE2461) Fowl cholera (#PG6565)
Fungal diseases (#PD2728) Glässer's disease (#PG6669) Sweating sickness (#PG9662)
Equine ehrlichiosis (#PG9732) Canine ehrlichiosis (#PG2402) Elokomin fluke fever (#PG4029)
Lyme disease in dogs (#PG6391) African horse sickness (#PE1805)
Bovine petechial fever (#PG9270) Clostridial infections (#PE7769)
Salmon poisoning disease (#PG9553) Viral diseases in animals (#PD2730)
Bacterial diseases in animals (#PD2731) Septic pasteurellosis of sheep (#PG5206)
Insect vectors of animal diseases (#PD2748) Sporadic bovine encephalomyelitis (#PG7381)
Contagious caprine pleuropneumonia (#PE6293) Pasteurella antipestifer infection (#PG7534)
Erysipelothrix infection in poultry (#PG0792)
Mycoplasma synoviae infection of poultry (#PG5618)
Canine infectious cyclic thrombocytopenia (#PG9616)
Mycoplasma meleagridis infection of turkeys (#PG7510)
Mycoplasma gallisepticum infection in poultry (#PG9387)
Congenital anomalies due to infectious diseases in animals (#PE6419).
Related Parasitic diseases in animals (#PD2735)
Denial to animals of the right to conditions of life and liberty proper to their species (#PE6270).
Aggravates Enzootic diseases (#PD2733) Epizootic diseases (#PD2734).
Aggravated by Unclean food (#PJ2532)
Inadequate control of animal diseases (#PD2781)
Inadequate carcass disposal of diseased animals (#PE2778)
Spread of animal diseases through factory farming (#PD2752)
Difficulty in identifying carriers of animal diseases (#PF2775)
Insanitary penning conditions as factor in animal diseases (#PE2764)
Inadequate disinfection of pastureland after disease outbreak (#PD2774).

♦ PD2733 Enzootic diseases
Nature Enzootic animal diseases are indigenous to or always present in particular areas or among particular species of animals. Enzootic diseases are liable to become epizootic under certain circumstances, in which case many animals are attacked and the disease may spread over a large area or over an entire continent. Enzootic animal diseases raise the risk of infection in man if they are zoonotic, that is to say communicable and common to man and animals.
Incidence Rabies is enzootic in Europe; braxy and louping ill are enzootic among sheep in Scotland and the north of England; anthrax is enzootic among horses in Asia Minor; foot-and-mouth disease is enzootic in many tropical countries; and rinderpest is currently enzootic or epizootic in most tropical African countries.
Broader Animal diseases (#PC0952).
Related Epizootic diseases (#PD2734) Parasitic diseases in animals (#PD2735).
Aggravates Zoonoses (#PD1770).
Aggravated by Fungal diseases (#PD2728) Viral diseases in animals (#PD2730)
Bacterial diseases in animals (#PD2731) Infectious diseases in animals (#PD2732)
Airborne viral animal diseases (#PD2741)
Wild animals as carriers of animal diseases (#PD2729)
International movement of animals as factor in animal diseases (#PD2755).

♦ PD2734 Epizootic diseases
Nature Epizootic animal diseases affect many animals in a large area at the same time, and spread with great rapidity. Epizootic diseases are generally contagious from one animal to another but may be spread by intermediate agents such as forage, straw, insects, fertilizers, contaminated water, movements of game or wild animals and bird migration, or sea seasonal factors. Insect-borne diseases have a higher propensity to become epizootic, especially when the insects are numerous, such as with flies and trypanosomiasis in Africa.
Broader Animal diseases (#PC0952).
Related Enzootic diseases (#PD2733).
Aggravates Zoonoses (#PD1770)
Underdeveloped use of agricultural resources (#PF2164)
Underdevelopment of agricultural and livestock production (#PD0629).
Aggravated by Fungal diseases (#PD2728) Diseases of wild animals (#PD2776)
Viral diseases in animals (#PD2730) Bacterial diseases in animals (#PD2731)
Parasitic diseases in animals (#PD2735) Infectious diseases in animals (#PD2732)
Airborne viral animal diseases (#PD2741)
Confusion of symptoms in animal diseases (#PF2780)
Spread of animal diseases through factory farming (#PD2752)
Difficulty in identifying carriers of animal diseases (#PF2775)
Insanitary penning conditions as factor in animal diseases (#PE2764)
Inadequate disinfection of pastureland after disease outbreak (#PD2774)
International movement of animals as factor in animal diseases (#PD2755)
Inadequate disinfection measures for animal housing and equipment (#PE2757).
Reduced by Wildlife extinction (#PC1445)
Economic loss through slaughter of diseased animals (#PE8109).

♦ PD2735 Parasitic diseases in animals
Nature Parasitic animal diseases cause severe economic losses, and some may also be transmitted to human beings. They inflict traumatic, lytic, obstructive, intoxicative, allergenic and proliferative damage on the host, or cause loss of nutriment. In addition, the damage done by parasites may enable bacteria and other pathogens to enter the tissues and set up diseases. There are two main varieties of parasites: protozoa and metazoa. The main classes of protozoa are rhizopods, ciliates, flagellates and sporozoans. Metazoa comprise helminths (parasitic worms) of five separate species: nematoda (roundworms), nematomorpha (gordian worms), platyhelminthes (flukes and tapeworms), acanthocephala (spiny-headed worms) and annelida (leeches); and arthropods, which can be pathogenic 'per se' or may transmit diseases (by virus, bacteria, protozoa and helminths). Main parasitic animal diseases include trypanosomiasis, liver fluke, blood fluke, trichinosis, toxoplasmosis, babesiosis, theileriosis, besnoitiosis, myiasis, leishmaniasis, mange, filariasis. Bacteria, viruses and fungi may also be parasitic.
Broader Parasites (#PD0868) Animal diseases (#PC0952).
Narrower Mange (#PE2727) Myaisis (#PG5725) Filariasis (#PE2391)

Liver fluke (#PE2785) Coccidiosis (#PE2738) Theileriosis (#PG5723)
Hexamitiasis (#PG9766) Leishmaniasis (#PE2281) Trypanosomiasis (#PE5725)
Heartwater disease (#PE2680) Coccidiosis in poultry (#PG9814)
African trypanosomiasis (#PE1778) Blood sporozoa of birds (#PE4284)
Histomoniasis in poultry (#PG9380) Trichomoniasis in poultry (#PG4423)
Fluke infections in birds (#PG4324)
Manson's eyeworm infection in poultry (#PG7126).
Related Zoonoses (#PD1770) Ascariasis (#PE2395) Trichinosis (#PE2311)
Toxoplasmosis (#PE3659) Chagas' disease (#PE0653) Sarcosporidiosis (#PG9770)
Enzootic diseases (#PD2733) Trematode diseases (#PE6461) Parasites on plants (#PD4659)
Tapeworm as a parasite (#PE3511) Bacterial diseases in animals (#PD2731)
Infectious diseases in animals (#PD2732).
Aggravates Epizootic diseases (#PD2734)
Wild animals as carriers of animal diseases (#PD2729).
Aggravated by Vectors of animal diseases (#PD2751)
Water-borne animal diseases (#PE2787)
Insect vectors of animal diseases (#PD2748).

♦ PD2740 Weather as a factor of animal disease

Nature Weather conditions may contribute to the incidence of animal diseases via their effect on fodder and therefore on nutrition and the general resistance of animals to disease. They may also contribute directly to produce a favourable environment for the growth of spores and parasites which cause disease, or to the spread of viruses and bacterial spores. In addition weather factors affect the proliferation of the intermediate hosts, such as flies and mosquitoes. Seasonal considerations can account for animal infertility.

Incidence Extreme heat or extreme cold adversely affect animal resistance to disease and may lead to undernourishment through lack of fodder. Extreme cold is the more harmful. Weather factors affecting the growth of spores and parasites and the incidence of viral or bacterial infection include wet weather, wind, rainfall followed by hot weather, hot dry weather, and drought. Mosquitoes and flies tend to proliferate during the rainy season where intermittent rainfall is followed by summer temperatures. Diseases affected by the weather include anthrax, African horse sickness, foot-and-mouth disease, fowlpest, and coccidiosis.

Narrower Water-borne animal diseases (#PE2787)
Airborne viral animal diseases (#PD2741).
Related Denial to animals of the right to conditions of life and liberty proper to their species (#PE6270).
Aggravates Animal infertility (#PC1803) Inadequate feeding of animals (#PC2765)
Soil-borne diseases in animals (#PE2739).
Aggravated by Bad weather (#PC0293) Disease vectors (#PC3595).

♦ PD2741 Airborne viral animal diseases

Nature Climatic conditions are significant in the spread of airborne viral animal diseases. The virus is commonly attached to particles only a few microns in size, such as dust or dried saliva. These particles may be carried by the wind and then dispersed on the ground by rainfall. For this reason, diseases such as foot-and-mouth are of higher incidence during wet weather. Buildings act as a filter and trap for airborne material. Modern intensive animal rearing units therefore contribute to the risk of disease, especially with the use of mechanical ventilation which increases the throughput of air. Airborne viruses may travel distances of 30 or even 150 km. Animal fodder may be contaminated by infected airborne particles, but the importance of this in comparison with inhalation, depends on the dosage.

Broader Viral diseases in animals (#PD2730)
Airborne substances harmful to health (#PD2847)
Weather as a factor of animal disease (#PD2740).
Related Wind-borne animal diseases (#PG5736).
Aggravates Enzootic diseases (#PD2733) Epizootic diseases (#PD2734)
Contaminated pastureland (#PJ5737)
Inadequate disinfection of pastureland after disease outbreak (#PD2774).
Aggravated by Inadequate control of animal diseases (#PD2781).

♦ PD2743 Poor quality of domestic livestock

Nature Livestock may not reach an acceptable level of excellence for meat, milk, hides, wool, eggs and other purposes for which the animals are reared. The problem may arise through lack of agricultural education, lack of capital funds, lack of availability of methods of cross-breeding (or through the virulence of disease such as trypanosomiasis which make this impossible), overuse of breeding animals, inadequate control of disease, inadequate nutrition, or general agricultural mismanagement.

Broader Underdevelopment of agricultural and livestock production (#PD0629).
Narrower Inefficient use of resources (#PE5001)
Inefficient animal commodity production (#PG5739).
Related Dependence of industrialized countries on import of resources (#PE0537).
Aggravates Poverty in developing countries (#PC0149).
Aggravated by Inadequate feeding of animals (#PC2765)
Limited agricultural education (#PF8835)
Inadequate agricultural capital (#PJ1368)
Decreasing genetic diversity of animals (#PC1408)
Vitamin E deficiency in domestic animals (#PE4760)
Agricultural mismanagement of housed farm animals (#PD2771)
Lack of availability of methods of cross-breeding (#PE8256).

♦ PD2746 Domestic animals as carriers of animal diseases

Domestic animals as vectors of animal disease — Domestic animals as vectors of disease
Refs Hickin, N *Pest Animals in Buildings* (1986); Okerman, Lieve *Diseases of Domestic Rabbits* (1988).
Broader Vectors of animal diseases (#PD2751) Animals as vectors of disease (#PD8360).
Narrower Stray dog populations (#PE0359)
Importation of infected carcass meats as factor in animal diseases (#PE2777).
Related Zoonoses (#PD1770).
Aggravates Fowlpest (#PE1400).
Aggravated by Proliferation of pets (#PD2689)
Difficulty in identifying carriers of animal diseases (#PF2775)
International movement of animals as factor in animal diseases (#PD2755).
Reduced by Excessive animal sanitary regulations in international travel (#PF1555).

♦ PD2748 Insect vectors of animal diseases

Nature Insects, particularly mosquitoes and flies, are the largest category of intermediate hosts for a wide variety of important animal diseases. Controlling the insect population is a major factor in controlling the diseases. Some of the major animal diseases which are transmitted by insects include: encephalitis, tularaemia, bubonic plague, anthrax, trypanosomiasis, distomatosis, babesiosis, theileriosis and African horse sickness.

Broader Insect vectors of disease (#PC3597) Infectious diseases in animals (#PD2732)
Wild animals as carriers of animal diseases (#PD2729).
Narrower Sheep ked (#PG9666) Face flies (#PG9658)
Head flies (#PG7795) Horn flies (#PG9333)
Horse flies (#PE3878) Stable flies (#PG9289)
Buffalo flies (#PG9466) Tick-borne diseases (#PE3897)
Black flies as pests (#PE3646) Siphonaptera as insect pests (#PE3643).
Aggravates Anthrax (#PE2736) Bluetongue (#PE6297) Rift valley fever (#PE7552)
Heartwater disease (#PE2680) Epizootic diseases (#PD2734)
Wesselsbron disease (#PG3959) Nairobi sheep disease (#PE6268)
African horse sickness (#PE1805) Equine infectious anaemia (#PG7882)
Akabane disease of animals (#PG7858) Parasitic diseases in animals (#PD2735)
Crimean-Congo haemorrhagic fever (#PG6985).
Aggravated by Water flea (#PG6595) Disastrous insect invasions (#PD4751)
Difficulty in identifying carriers of animal diseases (#PF2775).
Reduced by Insecticides as pollutants (#PD0983)
Defoliation of insect breeding areas (#PJ4038).

♦ PD2750 Worms as vectors of animal diseases

Nature Earthworms may spread soil-borne diseases through their activity in the soil, as in the case of anthrax, or act as intermediate hosts for virus diseases of pigs and poultry, thus complicating the control of the disease and its eradication.

Incidence Any soil-borne diseases may be spread by the activity of worms. Since worms can live as long as 10 years, they may harbour and spread viral animal diseases over a long period.

Broader Wild animals as carriers of animal diseases (#PD2729).
Aggravates Anthrax (#PE2736) Foot-and-mouth disease (#PE1589).
Aggravated by Soil-borne diseases in animals (#PE2739)
Difficulty in identifying carriers of animal diseases (#PF2775).
Reduced by Insecticides as pollutants (#PD0983).

♦ PD2751 Vectors of animal diseases

Broader Animal diseases (#PC0952) Disease vectors (#PC3595).
Narrower Infected animals (#PG4778) Water-borne animal diseases (#PE2787)
Soil-borne diseases in animals (#PE2739) Human vectors of animal diseases (#PD2784)
Wild animals as carriers of animal diseases (#PD2729)
Domestic animals as carriers of animal diseases (#PD2746).
Aggravates Zoonoses (#PD1770) Fungal diseases (#PD2728)
Viral diseases in animals (#PD2730) Bacterial diseases in animals (#PD2731)
Parasitic diseases in animals (#PD2735).
Aggravated by Difficulty in identifying carriers of animal diseases (#PF2775).
Reduces Insecticides as pollutants (#PD0983).
Reduced by Defoliation (#PD1135) Wildlife extinction (#PC1445)
Excessive use of chemicals to control pests (#PD1207).

♦ PD2752 Spread of animal diseases through factory farming

Nature The environment of factory farming provides ideal conditions for both the existence and expansion of several animal diseases. Because of the large numbers of animals reared in one place, disease can completely destroy a farm's stock or the stock of an area. Diseases may be spread through their virulence in intensive farming units, or through inadequate hygiene measures taken in the disposal of diseased animals or in the sterilizing of implements which have come into contact with them.

Diseases such as Newcastle disease in poultry are particularly well adapted to factory farming conditions. The quantity of animals in a small space means that the disease can spread very rapidly. Animal stress caused by close confining, the need to clip the beaks of poultry to check cannibalism, and general weakness caused by living in unnatural conditions and by being made to produce at an abnormally high rate, leave animals susceptible to disease and with very little resistance. Intensively farmed animals tend to be specially bred for particular qualities and have uniform characteristics, which means that disease can spread more rapidly. Intensive farming units act as a trap for airborne animal diseases and mechanical ventilation ensures that infected particles circulate through the unit many times. Other contributors to the spread of animal disease in factory farming units may be overcrowding and unsanitary conditions.

Incidence Diseases of farm animals are widespread; in addition to diminishing animal yields, diseased animals may also cause illnesses in humans. Furthermore, the extent of certain diseases in developing countries may be so great that the modernization of livestock production is prevented. The destruction of large populations of animals by disease is not a new problem, but only recently have some countries begun to develop extensive programmes to minimize communicable livestock diseases.

Broader Factory farming (#PD1562) Human vectors of animal diseases (#PD2784).
Aggravates Mange (#PE2727) Coccidiosis (#PE2738) Epizootic diseases (#PD2734)
Infectious diseases in animals (#PD2732).
Aggravated by Agricultural mismanagement of housed farm animals (#PD2771).
Reduced by Economic loss through slaughter of diseased animals (#PE8109).

♦ PD2755 International movement of animals as factor in animal diseases

Animals' international movement as factor of animal diseases

Nature International movement of animals may cause the spread of diseases which are enzootic to other countries or continents. Control measures for certain very serious diseases such as trypanosomiasis may be such that the international movement of domestic animals for purposes of cross-breeding and agricultural improvement is severely hindered, giving rise to agricultural backwardness. Many serious epidemics of malaria and bubonic plague have been caused through man's accidental transport of infected animals.

Incidence Diseases caused by the international movement of animals may arise from the migration of wild animals or the import of either wild or domestic animals. The search for improved breeds of domestic stock such as cattle, sheep and pigs leads to cross-breeding and the importation of foreign breeds. The popularity of zoos in developed countries has led to an increase in the importation of wild animals which may set up diseases, as has the increased demand for exotic pets. Diseases from imported livestock in the USA include anthrax, scab, mange, blackleg, tuberculosis, fowl plague, fowl pox, and tick fever.

Broader Human vectors of animal diseases (#PD2784).
Related Inadequate animal quarantine (#PE2756)
Denial to animals of the right to the attention, care and protection of humankind (#PF5121).
Aggravates Zoonoses (#PD1770) Enzootic diseases (#PD2733)
Epizootic diseases (#PD2734) African horse sickness (#PE1805)
Viral diseases in animals (#PD2730) Bacterial diseases in animals (#PD2731)
Contagious bovine pleuropneumonia (#PE1775)
Domestic animals as carriers of animal diseases (#PD2746)
Underdevelopment of agricultural and livestock production (#PD0629).
Aggravated by Proliferation of pets (#PD2689)
Inadequate and insufficient immunization (#PF5969).

♦ PD2760 Animal stress in factory farming

Overcrowding of housed farm animals — Inadequate lighting in factory farming units

Nature Due to the closely confined conditions of factory farming units, some animals die of stress from overcrowding. Productivity under stress tends to be less. Cannibalism increases in overcrowded conditions, particularly with hens.

Incidence According to behaviour studies, the optimum space allowance per hen is 750 sq cm. This is often not reached: average allowance in practice is 450 - 550 sq cm. Hens are often crowded together 4 or 5 to a cage, though 3 hens kept in a 4-hen cage produced 35 extra eggs each per laying season. A pronounced drop in efficiency has been noted among pigs kept under extreme conditions of intensive farming. Debeaking and dim lighting are used to check cannibal-

ism among chickens in overcrowded pens, but not always with success.
Refs Archer, John *Animals under Stress* (1980).
Broader Stress (#PB9165) Factory farming (#PD1562)
Cruelty to animals in factory farming (#PD2768).
Related Denial to food animals of the right to freedom from suffering (#PE3899).
Aggravates Animal deaths (#PE7941)
Inferior meat quality from intensive animal farming units (#PE2770).
Aggravated by Inadequate feeding of animals (#PC2765)
Agricultural mismanagement of housed farm animals (#PD2771)
Inadequate housing and penning of domestic animals (#PE2763).

♦ PD2761 Animal malformation in factory farming
Nature Some animals, especially calves, are kept on wooden slats in stalls where they cannot turn round and where they are tethered by the neck for life; with the result that the joints in the legs become deformed and often when the calf is taken to slaughter it cannot stand or walk. Most pigs after the weaning stage are fattened in houses where they have little more than room to lie down.
Incidence White veal calves are kept in very close confined conditions on a diet specially designed to make them anaemic and therefore produce the 'white meat' which is particularly popular throughout Europe. Although it has been recommended that an animal should at least have sufficient freedom of movement to be able, without difficulty, to turn round, groom itself, get up, lie down and stretch its limbs, in practice the animals are not given this amount of freedom and are kept as immobile as possible in order to put on more weight. The use of slats makes it difficult for animals to stand properly, and as their weight increases so is it badly distributed, causing malformations.
Broader Factory farming (#PD1562) Cruelty to animals in factory farming (#PD2768).
Related Denial to food animals of the right to freedom from suffering (#PE3899).
Aggravates Animal suffering (#PD8812).
Aggravated by Negative effects of gastronomic fashions (#PG5790)
Inadequate legislation for animal welfare (#PE5794)
Agricultural mismanagement of housed farm animals (#PD2771)
Inadequate housing and penning of domestic animals (#PE2763).

♦ PD2762 Disparities in distribution of communication resources and facilities
Nature Communication disparities exist in both developed and developing countries. Some developing countries have TV transmitters which cover only the main cities and immediate surroundings; many villages have no telephones, as the existing networks are for urban populations only. Illiteracy obviously excludes many, as does the limited production and distribution of newspapers, periodicals and books. These basic drawbacks are compounded in many countries by the number of languages used by different population groups, for which it would be economically impossible to provide printed material. In developed countries, it is generally stratification – by sex, age, education, income level, nationality or race, employment, geography – that produces groups which are communication-deprived.
Incidence The United States controls 60 to 70 percent of the market for exporting information by computer and has a near monopoly on scientific-technological information.
Broader Maldistribution of resources (#PB1016).
Narrower Disparity in world telecommunications capabilities (#PD5701).
Related Unequal distribution of goods and services (#PE8603)
Unbalanced application of communications technology (#PE7637).
Aggravates Insufficient communications systems (#PF2350).

♦ PD2768 Cruelty to animals in factory farming
Debeaking of chickens in factory farming
Nature Overcrowding (causing severe stress), inadequate space allowance and immobilization of animals (causing malformations), tethering, use of slats, inadequate lighting, use of wire mesh flooring for poultry and lamb, beak clipping, use of spectacles and dubbing of poultry, inadequate feeding of white veal calves, inhumane slaughter of unwanted male chicks (suffocation in airtight barrels, drowning in boiling water, boxes connected with a car exhaust pipe), are all examples of cruelty to animals in some factory farming.
Broader Maltreatment of animals (#PC0066).
Narrower Animal stress in factory farming (#PD2760)
Animal malformation in factory farming (#PD2761).
Related Denial of rights of animals (#PC5456).
Aggravates Animal suffering (#PD8812).
Aggravated by Commercial exploitation of wild animals (#PD1481)
Inadequate legislation for animal welfare (#PE5794)
Agricultural mismanagement of housed farm animals (#PD2771).

♦ PD2769 Decline in real wages in developing countries
Low level of personal income in developing countries — Decline in earnings in developing countries
Nature The developing world still depends on raw materials for the majority of its export earnings. But in the last ten years, real prices for the developing world's principal commodities, including fuels, minerals, jute, rubber, coffee, cocoa, tea oils, fats, tobacco, and timber have fallen by approximately 30 percent.
Incidence The per capita income of the average Latin American is 9 percent lower today than it was in 1980. In some countries the standard of living has slipped back to what it was 20 years ago. One third of Latin America's population, 130 million people, live in dire poverty.
Broader Low general income (#PD8568) Underpayment for work (#PD8916).
Aggravates Suicide (#PC0417) Underpayment for work in developing countries (#PE9199).
Aggravated by Unemployment in developing countries (#PD0176)
Underemployment in developing countries (#PD8141)
Urban unemployment in developing countries (#PD1551)
Mismatch of national macroeconomic policies among industrialized countries (#PF5000).

♦ PD2771 Agricultural mismanagement of housed farm animals
Agricultural mismanagement animals
Nature Inadequate care of farm animals leads to less than maximum productivity and much animal stress. Inaccurate assessment of demand for animal products leads to surpluses and financial loss. The problem may occur through ignorance or through excessive commercial exploitation of animals, often by industrial concerns which have less knowledge and appreciation of the requirements of animals than does the farming community in general. Surpluses frequently occur in production from intensive farming units, where investors are concerned mainly with a high rate of production and quick turnover, without taking due account of the level of demand. Losses from surplus production are not so detrimental as they can be offset against tax returns; therefore there is less incentive to curb this kind of agricultural mismanagement. Mismanagement of farm animals may include inadequate feeding, inadequate housing (not allowing enough light, space, ventilation), overcrowding, unsanitary conditions, abuse of antibiotics, vaccines, and methods of forced growth. Such mismanagement may lead to a loss of productive capacity but significantly it may be unnecessarily inhumane, leading to severe animal stress and discomfort.
Broader Mismanagement (#PB8406) Agricultural mismanagement (#PD8625).
Narrower Factory farming (#PD1562)
(#PE4760)
Inadequate housing and penning of domestic animals (#PE2763)
Insanitary penning conditions as factor in animal diseases (#PE2764).
Related Inadequate animal welfare (#PC1167) Inadequate feeding of animals (#PC2765)
Decreasing genetic diversity of animals (#PC1408).
Aggravates Animal diseases (#PC0952) Animal infertility (#PC1803)
Animal stress in factory farming (#PD2760) Poor quality of domestic livestock (#PD2743)
Cruelty to animals in factory farming (#PD2768)
Animal malformation in factory farming (#PD2761)
Inefficient use of proteins in factory farming (#PF2758)
Spread of animal diseases through factory farming (#PD2752)
Inferior meat quality from intensive animal farming units (#PE2770).
Aggravated by Commercial exploitation of wild animals (#PD1481)
Excessive commercial exploitation of farm animals by industrial concerns (#PD2772).

♦ PD2772 Excessive commercial exploitation of farm animals by industrial concerns
Nature Use of farm animals for mass production of meat and egg products at the highest turnover rate and with the least expense, leads to inhumane methods of confining and forcing the growth rate of animals, loss of productive capacity owing to insufficient attention given to animal requirements, food surpluses owing to inaccurate assessment of demand, risk of animal and human diseases, increase in drug resistance from abuse of antibiotics for forced growth, and a threat to the farming community from higher relative competitiveness of larger units and surpluses.
Incidence The problem occurs in developed countries, where the demand for meat and egg products is increasingly high. Industrial investors have less knowledge and less concern about the animals which they are using than the farming community; this results in mismanagement and cruelty. Insufficient assessment of demand for meat and egg products and the ensuing financial losses incurred can easily be offset against tax returns. Industrial investors may be less guided by subsidies which serve as an incentive against surpluses. The abuse of antibiotics and vaccines and the use of artificial methods of promoting fast growth encourage drug resistance in animals and humans. Animals which have been successfully treated for disease may still pass on infection in the meat. The risk of epidemics of animal diseases among intensively farmed animals is very high and may cause the spread of disease to other farm animals if adequate measures of disposal of diseased carcasses, disinfection of pens, and general hygiene are not taken.
Broader Economic exploitation (#PC8132) Inadequate animal welfare (#PC1167).
Related Maltreatment of animals (#PC0066) Cruel animal transportation (#PD0390)
Inhumane livestock slaughtering for meat (#PG5816).
Aggravates Surplus domestic animal production (#PJ5775)
Agricultural mismanagement of housed farm animals (#PD2771)
Inferior meat quality from intensive animal farming units (#PE2770).
Aggravated by Inadequate legislation for animal welfare (#PE5794).

♦ PD2774 Inadequate disinfection of pastureland after disease outbreak
Nature Pastureland may remain infected after a disease outbreak and be a source of reinfection for new stock. Adequate disinfection measures are not easy to take since it is difficult to be sure that the agent of disease has been eradicated.
Broader Inadequate control of animal diseases (#PD2781).
Aggravates Blackleg (#PE2737) Zoonoses (#PD1770)
Coccidiosis (#PE2738) Epizootic diseases (#PD2734)
Soil-borne diseases in animals (#PE2739) Infectious diseases in animals (#PD2732)
Harbouring national security offenders (#PE0484).
Aggravated by Airborne viral animal diseases (#PD2741).

♦ PD2776 Diseases of wild animals
Difficulty of controlling disease in wild animals
Nature Difficulty in controlling disease in wild animals may lead to the establishment of enzootic diseases and to outbreaks of epizootic diseases, causing great economic losses and the risk of the spread of zoonoses. Difficulty of controlling disease in wild animals may arise through lack of knowledge of the outbreak in the first place, and through the inaccessibility of wild animal populations. Once the disease is recognized, the only means of control so far available is that of mass extermination, which is undesirable from an ecological and cultural point of view, and total extermination of a disease-carrying species (the only certain way in which to ensure that it will not break out again), is largely impracticable. Immunization measures come up against the problem of tracing infected wild animals and of capturing them, inoculating them successfully, and then being able to retrace them to record results.
Incidence Several hundred thousand of Australia's feral water buffalo have been shot from helicopters in a government programme to wipe out unmustered feral water buffalo in areas where tests show a rate of around 10 percent or higher of bovine tuberculosis, a disease which can affect other cattle and humans. This drastic measure is being taken to protect Australia's cattle industry from the possibility of tuberculosis and its reputation from the suggestion that the disease is common in cattle-raising areas.
Refs Hoff, Gerald L and Davis, John W *Noninfectious Diseases of Wildlife* (1982); Lyubashenko, S Y (Ed) *Diseases of Fur-Bearing Animals* (1983).
Broader Animal diseases (#PC0952) Inadequate health control (#PF9401).
Related Wild birds as vectors of animal diseases (#PE2749).
Aggravates Rabies (#PE1325) Zoonoses (#PD1770) Coccidiosis (#PE2738)
Epizootic diseases (#PD2734) Wild animals as carriers of animal diseases (#PD2729)
Difficulty in identifying carriers of animal diseases (#PF2775).
Aggravated by Wildlife extinction (#PC1445) Lack of funds for veterinary research (#PE8463).

♦ PD2778 Overwork
Excessive overtime — Death from overwork — Karoshi
Incidence Overwork is a characteristic of middle management, especially in industrialized countries where there is a shortage of skilled staff during an economic boom. Workers of all kinds are encouraged, directly or indirectly, to work extensive overtime. In Japan estimates of death from overwork range from the hundreds to the thousands, although the phenomenon has been denied by the government and corporations.
Broader Insufficiency (#PA5473).
Narrower Overworked women (#PD7762).
Aggravates Epilepsy (#PE0661) Headache (#PE1974)
Hallucinations (#PF2249).
Aggravated by Underpayment for work (#PD8916)
Excessive hours of work (#PD0140).

♦ PD2781 Inadequate control of animal diseases
Difficulty in controlling animal diseases
Nature Inadequate measures taken to prevent the spread of animal diseases may result in epidemics and heavy losses, spreading the risk of zoonoses. Such inadequate measures may include lack of quarantine; lack of development of effective vaccine or immunization for certain diseases; lack of research; lack of availability of vaccine or other means of immunization; lack of knowledge of an outbreak; lack of knowledge of the way in which disease can be transmitted; lack of monitoring of wild vectors of disease; inadequate international information on current outbreaks; inadequate international legislation on the traffic in animal products, edible or inedible (such as

hides, wool, bones); lack of disinfection measures for animal housing, equipment, pastureland and human beings; inadequate means of disposal of infected carcasses; confusion of symptoms; ignorance concerning identification of animal diseases; negligence; high cost of adequate means of controlling certain diseases; and difficulty in controlling disease in wild animals because of inaccessiblility.
Broader Animal diseases (#PC0952) Inadequate health control (#PF9401).
Narrower Inadequate animal quarantine (#PE2756)
Unrecognized animal diseases (#PG5768)
Confusion of symptoms in animal diseases (#PF2780)
Difficulty in identifying carriers of animal diseases (#PF2775)
Insanitary penning conditions as factor in animal diseases (#PE2764)
Inadequate disinfection of pastureland after disease outbreak (#PD2774)
Inadequate disinfection measures for humans during animal disease outbreaks (#PE8409).
Aggravates Zoonoses (#PD1770) Infectious diseases in animals (#PD2732)
Airborne viral animal diseases (#PD2741) Contagious bovine pleuropneumonia (#PE1775)
Inadequate disinfection measures for animal housing and equipment (#PE2757).
Aggravated by Lack of funds for veterinary research (#PE8463).

♦ PD2784 Human vectors of animal diseases
Nature Human beings in contact with animals through their occupation may carry infection on their clothing or their hands. In the case of zoonoses, such as anthrax, humans may reinfect animals. Other animal diseases, such as Newcastle disease, may be carried by humans but do not affect them.
Broader Man as vectors of disease (#PD8371) Vectors of animal diseases (#PD2751).
Narrower Spread of animal diseases through factory farming (#PD2752)
Increase in animal disease by increase in aviation (#PJ4036)
Obstacles to the international transfer of corpses (#PG3537)
International movement of animals as factor in animal diseases (#PD2755).
Related Zoonoses (#PD1770)
Denial to animals of the right to the attention, care and protection of humankind (#PF5121).
Aggravates Fowlpest (#PE1400).
Aggravated by Inadequate disinfection measures for humans during animal disease outbreaks (#PE8409).

♦ PD2788 Storm surges
Storm tides — Sea surges
Nature Storm surges are transient, localized disturbances at sea level, resulting from the action of a tropical cyclone, an extratropical cyclone, or a squall over the sea. (They should not to be confused with tsunamis, or tidal waves, which result from seismic disturbances). Such disturbances may be a major cause of the damage associated with cyclonic activity. Protection against storm surges may be achieved by the construction of dyke systems (as in the Netherlands), but since the most damaging surges only recur infrequently (the 1953 surge in the Netherlands which drowned 1,800 people was the first of that height for 300 years), there is an additional economic problem in justifying the construction of dykes of adequate height.
Incidence Regions particularly vulnerable to storm surges include the Gulf of Mexico, the Atlantic Coast of the USA, the Gulf of Bengal, Japan and other islands of the Western Pacific which lie in the typhoon belt, and the coastal regions of the North Sea. Protection is especially difficult where there are numerous and complex estuary systems.
Refs WHO *Present Techniques of Tropical Storm Surge Prediction* (1978).
Broader Meteorological disaster (#PD4065).
Related Tsunamis (#PD0033) Tidal floods (#PE5006).
Aggravates Floods (#PD0452) Tidal water damage (#PJ9589).
Aggravated by Hurricanes (#PD1590).

♦ PD2790 Lack of work commitment
Mistrust of corporate commitment — Minimal vocational commitment — Unwillingness to work — Erosion of work ethic — Discouraged workers — Corporate disloyalty — Lack of worker loyalty to corporations
Nature People are generally suspicious of social groups which require from their members any extensive scale of effort, change in daily activities or personal bonds with other members. They thus deny themselves the power and drive related to bodies based on such commitment, and see powerlessness and inaction as inevitable.
Refs Mowday, Richard et al *Employee–Organization Linkages* (1981).
Broader Lack of commitment (#PF1729).
Related Distrust (#PA8653) Disloyalty (#PJ1895) Rural unemployment (#PF2949)
Non-concerned attitudes (#PF2158).

♦ PD2800 Domestic waste water pollutants
Discharge of dangerous substances into domestic waste water
Nature In addition to the increased production of sewage due to the growth of population, the per capita production of waste water is growing: in many cities it may amount to 600 litres per day per person. At the same time, its content of organic and mineral pollutants is also large and may amount to 10 litres of wet sludge per person daily, or about 50 kg of dry solids per person per year. Domestic and municipal sewage contains decomposable organic matter that exerts a demand on the oxygen resources of the receiving waters. This biochemical oxygen demand (BOD) is a measure of the weight (per unit volume of water or waste water) of dissolved oxygen consumed in the biological processes that degrade organic matter; it is determined by means of a standard test procedure. BOD values range from approximately 1mg/litre (for natural waters) to 300–500mg/litre (for untreated domestic sewage). The organic matter consists primarily of carbohydrates, proteins from animal matter and miscellaneous fats and oils. The specific classes of organic compounds found in sewage include amino–acids, fatty acids, soaps, esters, anionic detergents, amino–sugars, amines, amides, and many others. Much of the impurity in municipal wastes is material capable of settling, which may be deposited at the bottom of receiving waters to form deep layers of organic sludge. Dissolved salts in the form of ions such as sodium, potassium, calcium, manganese, ammonium, chloride, nitrate, nitrite, bicarbonate, sulphate and phosphate are the main inorganic constituents of sewage and other waste–waters. Domestic and municipal sewage invariably contains a variety of micro–organisms, some of which may be pathogenic. Although most human intestinal pathogens do not survive for extended periods outside the body of the host, there is evidence that they may remain sufficiently viable in different types of aquatic environment to be able to infect man.
Broader Biological pollutants (#PC5276) Impurities in waste water (#PD0482)
Discharge of dangerous substances (#PD4542).
Aggravates Lake pollution (#PD8628) Water pollution (#PC0062).
Aggravated by Wasted water (#PD3669) Hotel and restaurant waste (#PE1542).

♦ PD2802 Inadequate structures for achieving global unity
Claim Cooperation among people will be actualized only when the world achieves practical unity. However, current social structures are inadequate to allow the emergence of global unity, and there are no points of cultural similarity through which a global commitment could be consciously realized. Such commitment would have to be based on a practical method for relating diverse peoples to each other at the local level.
Broader Parochial national interests (#PF2600).
Aggravates Lack of international cooperation (#PF0817).

♦ PD2831 Noise in the working environment
Insufficient acoustical treatment in workplaces
Refs Carpentier, James and Cazamian, Pierre *Noise and Vibration in the Working Environment* (1978); European Foundation for the Improvement of Living and Working conditions *Noise, Stress and Work*; ILO *Protection of Workers Against Noise and Vibration in the Working Environment* (1984); Kryter, Karl D *The Effects of Noise on Man* (1985); Yang, S J and Ellison, A J *Machinery Noise Measurement* (1985).
Broader Health hazards of exposure to noise (#PC0268).
Aggravates Occupational deafness (#PD1361).

♦ PD2838 Unhealthy physical posture
Posture
Nature Unhealthy postures are often held for considerable periods of time in the lying, sitting, standing and walking positions. These postures can cramp muscles, strain nerves, and block blood circulation. They may also affect respiration and spiral column alignment. In particular, habitual, injurious postures and movements in the work–place can induce fatigue and consequently affect quality and productivity of work and may increase risk of accidents.
Incidence In the work place, absence of ergonomic designs for equipment, tools, platforms or seats causes undesirable positions for the employee. Absence of procedures to change position frequently through changes in tasks, and absence of exercice breaks, are common.
Refs Bles, W and Brandt, T *Disorders of Posture and Gait* (1986).
Broader Human disease and disability (#PB1044).
Related Deterioration in physical health (#PC0716).
Aggravates Lumbago (#PE1310).

♦ PD2847 Airborne substances harmful to health
Refs ILO *Occupational Exposure to Airborne Substances Harmful to Health* (1985).
Broader Air pollution (#PC0119).
Narrower Airborne diseases (#PD5029) Airborne viral animal diseases (#PD2741).
Aggravates Human disease and disability (#PB1044).

♦ PD2851 Torture of children
Nature Children are systematically tortured in a large number of countries. They are whipped or kicked for hours, given electrical shocks, sexually abused, beaten on the soles of their feet, or their eyes gouged out or mutilated in other ways. In some cases children are made to watch as their parents are tortured. Some times children are tortured in front of their parents to force confessions or as a form of psychological torture. Some children have been imprisoned without charges with their mothers or when their mother was bearing them.
Incidence Torture of children has been reported in the following countries: **Af** Ethiopia, South Africa. **Am** Brazil, El Salvador, Guatemala, Honduras, Peru. **As** Afghanistan, Iran Islamic Rep, Iraq, Israel, Pakistan, Singapore. **Eu** Poland, Turkey.
Broader Human torture (#PC3429) Physical torture (#PD8734)
Victimization of children (#PC5512) Institutionalized torture (#PD6145).
Narrower Disabled victims of torture (#PD0764).
Related Torture of animals (#PC3532) Psychological torture (#PD4559)
Pharmacological torture (#PE4696) State sanctioned torture (#PD0181)
Physical maltreatment of children (#PC2584).
Aggravates Corruption of minors (#PD9481).

♦ PD2860 Politicization of technical debates
Politicization of issues
Broader Ideological conflict (#PF3388) Obstacles to community achievement (#PF7118).
Narrower Politicization of intergovernmental organizational debate (#PD0457).

♦ PD2864 Breakdown of local community cohesion
Limited community cohesion
Nature Communities no longer see themselves as groups of people concerned with others, and no longer have the discipline needed to attain a common vision. Many present–day communities consist of groups of isolated individuals struggling to relate to the demands of urbanization and industrialization, having withdrawn and become isolated from meaningfully taking part in world society. The effect of loss of a sense of destiny, isolationism and a failure to live beyond the moment is also illustrated in the collapse of the family as a unit.
Broader Inhibited capacity to visualize a creative future (#PF2352).
Aggravates Lack of community development (#PF7912)
Deteriorating quality of life (#PF7142).
Aggravated by Unstructured local decision–making (#PF6550).

♦ PD2868 Violation of food taboos
Infringement of dietary taboos — Violation of nutritional taboos
Broader Violating taboos (#PF3976).
Narrower Taboos against eating poultry (#PJ1326).
Related Torture by violation of taboos (#PE1296).
Aggravates Spiritual impurity (#PF6657).
Aggravated by Dietary restrictions (#PJ1933) Unethical catering practices (#PE6615).

♦ PD2872 Discrimination against working women in socialist countries
Denial of right to work for women in socialist countries
Nature In socialist countries, the lot of the working woman is nearly always inferior to that of the man. The level of salaries and wages is lower, the work offered is mainly repetitive and boring and the conditions of work are for the most part unfavourable. Women have little bargaining power, and their prospects of promotion are limited. The level of education and professional qualification of women is inferior to that of men, and they are not often required to do work needing professional qualifications.
Refs Bodrova, Valentina and Anker, Richard *Working Women in Socialist Countries* (1985).
Broader Denial of right to work (#PC5281)
Discrimination against women in employment (#PD0086).
Narrower Discrimination against working mothers (#PD6812).
Related Denial of right to free choice of work (#PE3963).

♦ PD2879 Unethical practices of employers
Irresponsible employers — Negligence by employers — Corruption of employers — Fraud by employers — Irresponsible entrepreneurs
Claim Employers are able to use their privileged position as owners or managers of an organization to place employees at a disadvantage. This may take the form of obliging employees (under threat of losing their jobs): to accept lower wages and unsatisfactory working conditions, to undertake tasks for which they are unqualified (including menial tasks), to contravene industry safety regulations, to carry out unethical or illegal practices, to tolerate abuse and (sexual) harassment, and to deliberately produce products of shoddy workmanship (even when this may

endanger subsequent users). Employers may also engage workers illegally, including children and illegal immigrants. Hiring procedures and employee advancement may be distorted to favour friends and relatives of the employers (or those willing to offer some premium, including sexual favours). They may also harass (to the point of violence or death) those who oppose them in the pursuit of their policies. In carrying out their business they may adopt a variety of techniques to bribe, or otherwise recompense, those who may facilitate their activities. Employers must necessarily seek ways to reduce their tax obligations and the efforts to this end may result in various forms of tax avoidance and tax evasion. In endeavouring to market their services, employers may chose, whether deliberately or by default, to misinform or mislead their customers concerning the nature or quality of the product (notably where customers may be easily swayed because of lack of a basis for comparison, as in developing countries). In dealing with potential competitors, employers may use a variety of techniques, of which some are unethical (including price fixing rings, cartels, underpricing, disinformation or violence) to squeeze them out of profitable markets. Where convenient, employers may acquire goods under exploitative conditions (such as from developing countries unable to protest), market them in an exploitative manner (such as in developing countries with no alternatives), or ensure the diversion of hazardous materials to legally proscribed destinations. Employers may also chose to use their legitimate activities as a front for illegitimate operations such as money laundering, trade in illegal commodities (including drugs), or the manufacture of dangerous products (such as weapon systems components, explosives, or products for chemical and biological warfare).

Refs FAO *Prevention of Post-Harvest Food Losses* (1985).

Narrower Unethical personnel practices (#PD0862).
Related Unethical practices by employees (#PD4334).
Aggravates Cartels (#PC2512) Nepotism (#PD7704) Tax evasion (#PD1466)
Verbal abuse (#PD5238) Labour hoarding (#PE6333) Money laundering (#PE7803)
Fear of reprisals (#PF9078) Abuse of tax havens (#PE2370)
Misleading advertising (#PE3814) Misleading information (#PF3096)
Unreported tax obligations (#PE9061) Exploitation in employment (#PC3297)
Unjust dismissal of workers (#PD5965) Exploitation of child labour (#PD0164)
Exploitation of the unemployed (#PD9347) Predetermined employer evaluation (#PG7714)
Threats against trade union leaders (#PD7471) Abusive traffic in immigrant workers (#PD2722)
Sanctions against trade union workers (#PD0610)
Sexual harassment in the working place (#PE8466)
International trade in chemical weapons (#PD9692)
Circumvention of duties and assessments (#PD4882)
Denial of right to collective bargaining (#PE3970)
Violations of health and safety regulations (#PE4006)
Unethical practices of regulatory inspectors (#PF8046)
Inadequate enforcement of safety regulations (#PD5001)
Government approved employment of war criminals (#PE4697)
Records destruction by transnational enterprises (#PE7061)
Diversion of high technology to hostile countries (#PE7174)
Violation of right of workers to join trade unions (#PE5192)
Suppression of information concerning environmental hazards (#PF4854)
Collusion of trade union leaders with employers and government (#PE8367)
Interference by employers in the affairs of workers' organizations (#PE8384)
Violation of the right of workers organizations to protection against suspension (#PE5793)
Inadequate living and working conditions of immigrant labourers in industrialized countries (#PD3427).

♦ PD2884 Prioritized attachment to security structures

Nature Fear of the unknown and of death blackmails people and enslaves them to security structures by intensifying their attachment to life and the need for security. When mechanisms for maintaining personal security are impinged on, the individual experiences trauma, guilt and a sense of failure. Social mores indicate that the first priority is taking care of one's own personal welfare, making one's world secure. This creates defence systems, idolises success and material possessions, and reinforces the idea "charity begins at home". The energy expended in building structures for personal and family security inhibits capacity to imagine a new future. The result is limited vision, seeing life in a narrow and static "now".

Broader Attachment (#PF6106) Inhibited capacity to visualize a creative future (#PF2352).

♦ PD2886 Unethical documentation practices

Unethical archival practices — Irresponsible librarianship — Irresponsible information system design — Corruption of documentalists

Claim Under pressure from their employers, documentalists may bias the collection of information and the ways through which it may be accessed. These practices may include: the refusal of access, the refusal to hold certain categories of document (if they are considered offensive to certain scientific, moral, religious or aesthetic principles), biased classification of documents (possibly to the point of rendering them unretrievable), tampering with documents (including forgery, theft and destruction of documents), idiosyncratic design of classification systems to render access incompatible with other documentation systems, dispensing inaccurate or misleading information on the contents of the collection and the availability of documents, reservation of documents for users in exchange for financial or other favours.

Narrower Theft of documents (#PD0577) Misleading information (#PF3096)
Inaccessible historical libraries (#PF9046) Tampering with official documents (#PF4699)
Bias in document classification systems (#PF6743)
Destruction of historic documents and public archives (#PD0172).
Aggravates Censorship (#PC0067) Documentary fraud (#PE1110)
Corruption of documents (#PE7900) Misfiled documents and records (#PF4708)
Fragmentation of international documentation (#PF7133).

♦ PD2888 Lack of faith in solutions to ecological issues

Nature Future planning for the use of the earth's resources is hampered by the belief that no solution to current problems exists. When people consider the possible consequences of the ecological imbalance and the inequitable distribution of basic resources, a sense of helplessness often results; this paralyzes the responsible development of resources and continues the ecological imbalance. The possibility that all nations participate in the development and control of resources is thus limited.

Broader Nationalistically determined development of natural resources (#PD3546).

♦ PD2892 Undiversified economies of developing countries

Lack of horizontal commodity diversification in developing countries

Nature One of the characteristics of the less developed countries is the extent to which the growth or extraction of a single commodity tends to dominate the economy. This lack of diversification makes the economy vulnerable to unpredictable factors such as variations in harvests, and price slumps on the world export market. It aggravates short term instability and intensifies the problem of long term growth. The commodities produced may have weak world market prospects, lack a domestic market which would permit actual or potential import substitution, fail to provide additional rural employment and fail to lend themselves to processing by domestic industry. Nearly 50 per cent of the developing countries earn more than 50 per cent of their export receipts from a single commodity. As many as 75 per cent of them earn more than 60 per cent from three primary commodities. In only 6 of these countries do exports of manufactured goods amount to as much as 10 per cent of total exports. Diversification is particularly urgent in the case of commodities subject to over-production or to competition from synthetic substitutes.

Broader Structural rigidity in developing country economies (#PD2970).
Narrower Non-diversification in subsistence fishing economies (#PF2135).
Aggravates Instability in export trade of developing countries producing primary commodities (#PD2968).
Aggravated by Insufficient diversification (#PD0335)
Deterioration of terms of trade for developing countries (#PD2897).

♦ PD2894 Instability of production of food and live animals

Nature Annual fluctuations in production are a major feature of the food situation. Often they tend to offset one another to some extent at the regional or global level, but when there are simultaneous drops in production (or increases below the trend rate) in a large number of major countries and regions, the implications for the world food situation are serious. In assessing the longer-term trends in food production, it is therefore necessary to examine the annual fluctuations above and below this trend, and to attempt to determine whether there has been any cyclical pattern in their magnitude or frequency that has affected the longer-term trend itself.
Variations in seasonal rainfall, temperature, and other climatic factors are generally the principal causes of the fluctuations, although the isolation of their effects from those of the other factors involved is difficult and has been accomplished for only a few crops in some developed countries. Fluctuations in production lead to changes in prices, employment, incomes and demand which in turn influence the course of production, generally accentuating the fluctuations caused by the weather. Changes in price and other agricultural policies can cause production to vary sharply from the trend. Civil disturbance, hostilities, natural disasters unrelated to the weather, and pests and diseases (sometimes linked with weather) are other powerful contributors.

Broader Instability of economic and industrial production activities (#PC1217).
Narrower Instability of food and drink industries (#PE8918).
Related Instability of trade in food and live animals (#PD1434).
Aggravates Overproduction of food (#PD9448)
Fluctuations in food production in developing countries (#PE8188).
Aggravated by Disastrous insect invasions (#PD4751)
Structural rigidity in developing country economies (#PD2970)
Underproduction of primary commodities in developing countries (#PD3042)
Inequality in distribution of natural resources between countries (#PF3043)
Instability in export trade of developing countries producing primary commodities (#PD2968).

♦ PD2897 Deterioration of terms of trade for developing countries

Unequal exchange in trade — Decline in trading position of developing countries

Nature The long-term increase in the price of manufactures which developing countries must import has been accompanied by static prices or price declines for many primary commodities, which are their main export. This deterioration in the terms of trade (a unit value index of exports divided by unit value index of imports) is most visible when the relationship is traced back to the early post-war period; this shows that developing countries are decreasingly able to import goods from funds generated by their exports.

Incidence Nine times the amount of export beef required to import one barrel of oil was needed by developing countries in 1981 than in 1971. Profits from one ton of bananas bought a steel bar in 1970; in 1980 it paid for half a bar of steel.

Broader Underdevelopment (#PB0206) International economic injustice (#PC9112)
Economic and financial instability of the world economy (#PC8073).
Aggravates Undiversified economies of developing countries (#PD2892)
Fragmentation of the international trading system (#PC9584)
Decline in import capacity of developing countries (#PE6861)
Inadequate economic policy-making in developing countries (#PF5964)
Deterioration in external financial position of developing countries (#PE9567)
Instability in export trade of developing countries producing primary commodities (#PD2968)
Vulnerability of national economies to vagaries of external markets for goods and services (#PF9697).

♦ PD2898 Inadequate demand for primary commodities because of rising living standards

Nature Consumer demand for some commodities does not rise as incomes rise (the per capita income elasticity of demand is low in the richer countries). In some cases, most notably with cereals as a source of food calories, rising standards of living tend, in high-income countries, to reduce per capita consumption to the point of offsetting the effects of population growth (which tends to be slow in such countries). This phenomenon leads to an unusually slow increase in demand for the primary commodities produced by the developing countries.

Broader Instability of economic and industrial production activities (#PC1217).
Aggravates Reduction of the share of the developing countries in world exports (#PC2566)
Instability in export trade of developing countries producing primary commodities (#PD2968).

♦ PD2906 Political party manipulation of elections

Abuse of electoral process by political parties — Gerrymandering

Nature The corrupt use of the electoral process may serve various political ends, such as altering the electoral process so that the ruling party is always ensured of victory or, in a single party system, modifying the process so that the candidates most obedient to the party line are chosen. Alternatively, the ruling party may put off the election date until a favourable time for its own purposes, or arrange its policies in such a way that they are popular at election time. Candidates may be chosen and campaigns conducted in accordance with what the party feels will make it most popular or powerful. Manipulation of elections also includes corrupt practices such as manipulation of the ballot and bribery during electoral campaigns. It may equally include tactics such as referenda which give a distorted picture of public opinion. Gerrymandering includes the rearrangement of voting districts, or changing the number of people in such districts, by the party in power to ensure its victory in subsequent elections.

Incidence Election manipulation occurs most frequently at the lowest electoral levels, but it can also occur at national levels.

Broader Lack of political integrity (#PF0796) Unequal parliamentary constituencies (#PD2167).
Narrower Unjust election timing (#PD2907) Unjust electoral campaigns (#PD2919)
Bias in selection of political candidates (#PD2931).
Related Political injustice (#PC2181) Totalitarian democracy (#PD3213)
Limitations on right to vote (#PF2904).
Aggravates Unfair elections (#PC2649) Political stagnation (#PC2494).

♦ PD2907 Unjust election timing

Nature An election may be timed to suit the ambitions of a ruling party, and there may be inconvenient polling hours. In agricultural countries, the season during sowing or harvest is an unsuitable time for elections; in industrialized countries, holiday periods are unsuitable. At other times of the year, insufficient polling time may be allowed after a working day. Ruling parties may choose an unsuitable election time if they think it will be to their advantage, or they may choose a time when national problems are temporarily abated.

Broader Unfair elections (#PC2649) Self interested manipulation of timing (#PF5529)
Political party manipulation of elections (#PD2906).
Related Political injustice (#PC2181).
Aggravates Social conflict (#PC0137).

♦ PD2910 Forced participation in politics

Nature Participation in politics may be enforced by public or social pressure to vote during elections or referenda, to join political clubs or youth movements, or to to work for a party or 'national' effort. Enforced participation is a feature of many dictatorships and single party systems to channel potentially dissident political energy away from subversive activities. Such participation may produce conformism and apathy, and possibly alienation. It serves to strengthen a dictatorship and is an inherently unstable political situation.
Broader Compulsory indoctrination (#PD3097)
Forced participation in social processes (#PC5387).
Narrower Manipulative use of referenda (#PF2909).
Related Forced assimilation (#PC3293).
Aggravates Conformism (#PB3407).
Aggravated by Unethical practices in politics (#PC5517).
Reduces Political stagnation (#PC2494).

♦ PD2912 Political blackmail

Nature Knowledge of private misconduct or misconduct in public office can be used to blackmail public officials into granting concessions either in cash or in other privileges. Political blackmail is used in espionage as a means of extracting information, but it may also be used by individuals or groups with a vested interest. The latter leaves a public official equally vulnerable to espionage tactics.
Broader Coercion (#PC3796) Corruption in politics (#PC0116)
Threatening public servants (#PD0540).
Related Lack of political integrity (#PF0796).
Aggravates Discrimination against homosexuals (#PE1903).

♦ PD2916 Unethical industrial practices

Irresponsible industrial practices — Negligence in industry — Industrial malpractice — Corruption in industry
Broader Corruption (#PA1986) Institutionalized corruption (#PC9173).
Narrower Malpractice in the construction industry (#PD9713).
Aggravates Environmental pollution (#PB1166) Urban-Industrial pollution (#PC8745)
Discharge of dangerous substances (#PD4542)
Environmental hazards from economic and industrial products (#PC0328).
Aggravated by Corruptive crimes (#PD8679) Unethical commercial practices (#PC2563).

♦ PD2918 Police corruption

Corruption of law enforcement officials
Nature Police officers may accept bribes or other favours offered by those who would otherwise face prosecution. They may be intimidated by organized crime and in turn may use intimidation against people who oppose the interests of their benefactors or who threaten to expose their complicity. Corruption can be large-scale where the illegal interest protected by police corruption is considerable and where the benefits of complicity are highly rewarding.
Incidence Early in 1985, Mexico began investigating claims that many of its top police officials were being bribed for their participation in crime cover-ups, particularly in relation to the illegal drug trade. Several were fired and arrested. In the same year 37 police generals were dismissed in Peru. In a five year period in the mid-1980's 170,000 Soviet police officers were fired for offenses from misconduct to murder. Between 1979 and 1987 some 27 police officers in the state of Queensland, Australia received about $2.5 million in bribes. In the Philippines most of the big political murders, including that of opposition leader Mr Aquino in 1983, remain unsolved, because the army is reluctant to use its authority to investigate military abuses in the Marcos era.
Refs Duchaine, Nina *The Literature of Police Corruption* (1979); Sherman, Lawrence W *Scandal and Reform* (1978).
Broader Official corruption (#PC9533) Corruption in prisons (#PD9414)
Unethical practices by police forces (#PD9193).
Related Institutionalized corruption (#PC9173).
Aggravates Meaninglessness (#PA6977) Police brutality (#PD3543)
Mistrust of police (#PF8559) Inadequate firearm regulation (#PD1970)
Official fabrication of evidence (#PD8716).
Aggravated by Bribery (#PC2558) Underpayment of police (#PJ8601)
Ineffective monitoring (#PF2793) Bribery of public servants (#PD4541)
Use of undue influence to obstruct the administration of justice (#PE8829).

♦ PD2919 Unjust electoral campaigns

Nature The cost of publicity used in electoral campaigns may make contesting an election impossible for individuals or parties with only moderate means. Party finance may be provided by wealthy individuals or corporations with a vested interest against other contenders. Access to media may be unequally distributed. Political opportunism and even sabotage may be used against opponents. Electoral campaigns may mislead the public as to the true intentions or degree of integrity of the parties or individuals concerned.
Broader Political injustice (#PC2181) Unequal political representation (#PC0655)
Political party manipulation of elections (#PD2906).
Related Unfair elections (#PC2649).

♦ PD2922 Military espionage

Nature Covert gathering of information on weapons, military strategy and tactics of an enemy or an ally, in peacetime or in war, may be by highly technical methods, including the use of computers, microphotography, wiretapping, bugging and other devices, or it may be personal and psychological in the sense of blackmail or corruption of key officials.
Incidence It is suggested that of the 2000 Soviet officials stationed in the USA, 30 to 40 per cent are engaged in espionage at least part of the time.
Background Espionage systems increased with the growth of nation states in Europe following the Treaty of Westphalia in 1648. Until after the French Revolution they mainly concentrated on military and not political intelligence. With the Napoleonic wars, the whole nation rather than the elite class became involved in war and so the scope of espionage broadened, but the military basis for espionage activities remained the more important until the 20th century, when intelligence bureaux began to have as much power and autonomy in certain cases as the military or the politicians.
Broader Espionage (#PC2140).
Related International political espionage (#PC1868)
Revealing national security information to a foreign power (#PE4343).
Aggravates Counter-espionage (#PD2923).
Aggravated by Militarism (#PC2169).
Reduced by Failure of disarmament and arms control efforts (#PF0013).

♦ PD2923 Counter-espionage

Nature The protection of a country's official and industrial secrets against foreign espionage may involve a high degree of domestic political espionage, providing the opportunity for key individuals to extend their personal power to the detriment of the public well-being. Methods of counter-espionage may include bugging, wiretapping, censorship, indoctrination, torture, the compiling of political dossiers, and the creation of confusion.
Broader Espionage (#PC2140).
Aggravated by Military espionage (#PD2922) Industrial espionage (#PC2921)
Abuse of international cultural, diplomatic and commercial exchanges (#PF3099).

♦ PD2926 Restrictions on the distribution of confidential government information

Inaccessible government information services
Nature The restriction of confidential government information to a closed circle of people can lead to unnecessary, wasteful and expensive duplication of research, or to corruption or espionage which is all the more difficult to detect because so few know the nature of the information being abused.
Background In 1943 the British Admiralty was using a newly discovered technique to increase the effectiveness of explosives, and the Ministry of Home Security knew of increased power. However, the secrecy surrounding the technique was so great that the Air Ministry was unaware of it and was thus using less effective explosives.
Broader Official secrecy (#PC1812) Social inaccessibility (#PC0237).
Aggravates Suppression of information (#PD9146).
Aggravated by Misuse of classified communications information (#PD5183).

♦ PD2929 Maintenance of political dossiers on individuals

Nature The compilation by and for government agencies of political dossiers on people, whether nationals or foreigners, who are considered to be a danger to the state, and particularly to the regime in power, is an essential first step in political repression against the activities of certain individuals or groups, and reinforces government control and the effect of indoctrination.
Broader Political police (#PD3542) Political repression (#PC1919)
Infringement of privacy (#PB0284).
Aggravates Undemocratic political organization (#PC1015).
Aggravated by Excessive government control (#PF0304).

♦ PD2930 Misuse of electronic surveillance by governments

Electronic bugging — Electronic espionage — Interception of telecommunications
Nature The use of electronic devices to obtain information clandestinely. The approach may be relatively unsophisticated as in the use of bugs implanted in office equipment or it may be very sophisticated as with the use of world-wide networks of satellites and monitoring services to track communications of a particular type.
Incidence Widely used by governments of both East and West, whether singly or in combination. Used by intelligence agencies, especially in espionage and counter-espionage activities. Increasingly used by corporations for purposes of industrial espionage. Also used by organized crime. In 1985 the US National Security Agency initiated a five-year programme, estimated to cost up to $40 million, to encode most of the millions of electronic messages exchanged by US government and defence contractors to counteract surveillance by the USSR. Because of the success of this programme, it is reported that the US government is resisting the transfer of more sophisticated telecommunications (especially fibre optic systems) to the USSR because it would make it more difficult to track communications and would jeopardize the extensive investment in satellite monitoring facilities.
Background Following the second world war the USA, UK, Canada, Australia and New Zealand signed a secret treaty known as the UKUSA Agreement whereby each became responsible for signals intelligence in a particular part of the world. This network is currently claimed to have some 250,000 people on its payroll with a combined budget of $16 to $18 billion dollars. The Agreement binds together numerous organizations and agencies in a network of written and unwritten agreements, working practices and personal relationships. Listening posts are based in many countries, especially in embassies.
Refs French, S *The Big Brother Game* (1986); Moran, W B *Covert Surveillance and Electronic Penetration* (1986).
Broader Espionage (#PC2140) Political surveillance (#PD8871)
Misuse of government surveillance of communications (#PD9538).
Related Political police (#PD3542) Infringement of privacy (#PB0284)
Interception of communications (#PD7608)
Inappropriate use of telecommunications services (#PE4450).
Aggravates Crime (#PB0001) Political repression (#PC1919).
Aggravated by Power politics (#PB3202) Excessive government control (#PF0304).

♦ PD2931 Bias in selection of political candidates

Oligarchic political control of elections
Nature Political candidates in countries with sophisticated political systems and in countries with state control are usually nominated by a party. Although this may be done with some reference to public consensus ('primaries'), certain initial requirements may exclude many potential candidates. Candidates with more money may be able to promote themselves better, and where campaigning is very costly, independent candidates may find it impossible to enter the contest. In a single party election, party members who comply best with general party doctrine are more likely to be nominated than those with more independent views.
Incidence In the USA, for example, the acting backgrounds of Richard Nixon (amateur) and Ronald Reagan (professional) gave them a telegenic artfulness before it gave them capacity at governance. So too with John Kennedy's reputed 'charisma'. Thus cosmetics play an initial role in candidate selection in the USA. Kennedy and Rockefeller money has been conspicuously spent in the political campaigns of members of those families, and other millionaires have bought themselves or others political offices.
Counter-claim Choice of political candidates solely by popular referendum is one of the greatest dangers to a democracy as it opens the way for demagogues and liars to sway the populace with empty rhetoric and promises. The use of wealth in order to represent the interests of wealth is as legitimate as it is democratic.
Broader Unfair elections (#PC2649) Unequal political representation (#PC0655)
Political party manipulation of elections (#PD2906).
Narrower Single party democracies (#PD2001).
Related Political injustice (#PC2181) Political bias in official appointments (#PE1728).
Aggravates Political alienation (#PC3227).
Aggravated by Unrepresentative electoral systems (#PD9641)
Prohibitive cost of electoral campaigns (#PG5865).

♦ PD2941 Isolated islands

Nature The extreme geographical isolation of some islands makes it difficult for them to benefit from economic and social developments in which they would otherwise participate.
Incidence In the South Pacific Ocean there are numerous isolated islands and island groups. The UK has several protectorates in the Atlantic Ocean which are even more isolated.
Broader Geographical isolation (#PF9023).
Narrower Vulnerability of island developing countries and territories (#PE5700).
Related Neglect of remote regions and islands (#PE5760)
Disaster hazards to island populations (#PE5784).
Aggravates Non-viability of small states and territories (#PD0441).

♦ PD2942 Micro-state participation in international organizations

Nature The participation of micro-states, small in area, population and human and economic

resources, may weaken an international organization. There are many small territorial entities which have achieved political independence and are thus eligible for membership in major intergovernmental organizations such as the United Nations. These organizations, which subscribe to the principle of universality, have been unable to formulate criteria for a lower limit by which to exclude the smallest states which, despite their willingness, do not have the resources to permit them to participate fully in the work of such organizations.
Incidence Countries with a population of one million or less are considered micro–states, or micro–territories. There are presently some 94 that fit this definition: in the African region there are 16; in the Americas, 24; in Asia, 7; in Oceania 23; and in Europe, 24.
Counter–claim A great deal of caution needs to be exercised in listing together States and territories according to the criterion of their resident population. Included in lists of micro–states, for example, is the Vatican, which claims to represent several hundreds of millions of Roman Catholics world–wide, and such affluent states as Luxembourg, Monaco, and the United Arab Emirates. It is difficult to see the purpose in aggregating a list which includes such diversity as Gabon, Bermuda, Malta, Iceland, and all the island republics of Oceania. Obviously many small states can work very well in international organizations. The argument that they cannot is put forward by the bloc of developed countries that fears the democratic power of one nation–one vote international organization politics.
Reduces Non–viability of small states and territories (#PD0441).

♦ **PD2946 Boundary disputes between states**
Border disputes
Nature Boundaries between states, determined as a result of a variety of political, historical and administrative compromises, may be controversial, ambiguous and unstable. Dissatisfaction over a boundary decision may persist long after the signature of the agreement which gave rise to it. Such agreements may be ambiguous as to the exact boundary line if the area is poorly mapped, or if the natural features on which the boundary is based are displaced, as can happen when a river changes its course for instance. This may give rise to boundary disputes, particularly if the displacement process is not recognized.
Incidence More than 80 territorial and border disputes are currently unresolved including: Alto Adige/South Tyrol between Austria and Italy; four of the Kuril Islands occupied by Russia and claimed by Japan; Taba between Israel and Egypt; the McMahon Line recognized by India but disputed by China; and the borders between Mali and Burkina Faso, Saudi Arabia and Oman, Saudi Arabia and South Yemen, India and Bangladesh, India and Pakistan, and Kuwait and Iraq.
Background Until recent times the exact boundary limits of states and their jurisdictions were not defined. Borders were frequently not marked in any way. The need for fixed and marked boundaries only arose with the concept of the nation state, and it was only in recent times that surveying and cartographic techniques permitted boundaries to be established unambiguously.
Refs Day, Alan J (Eds) *Border and Territorial Disputes* (1987); Lee, Yong Leng *Razor's Edge* (1980); Sharma, S P *International Boundary Disputes and International Law* (1977).
Broader Territorial disputes between states (#PC1888)
Conflicting claims by states to territories (#PC2362).
Aggravates Border incidents and violence (#PD2950)
Domination of the world by territorially organized sovereign states (#PD0055).
Aggravated by Annexation (#PE5210) Enclaves and exclaves (#PD2154)
Unnatural boundaries between states (#PF0090).

♦ **PD2948 Restrictions on passage through straits and interoceanic canals**
Nature Difficulties concerning the right of passage through straits and interoceanic canals occur when the two bordering land areas are under the jurisdiction of different countries, or when the straight lies between portions of the land area of the same country but provides the only access by sea to the land of another country beyond the straight. Passage through such straits may be restricted or the strait may be closed under conditions of war or to further economic sanctions. The jurisdiction over straits waters may be subject to dispute, particularly if there is a conflict over the breadth of the territorial water claimed by the states in question.
Incidence Approximately thirty straits are considered important routes for international traffic, and there are three major interoceanic canals (Panama, Suez and Kiel).
Broader Territorial disputes between states (#PC1888).
Aggravated by Conflicting claims concerning off–shore territorial waters (#PD1628).

♦ **PD2950 Border incidents and violence**
Border disputes
Nature Tension may arise and violence erupt on the border between two countries as a result of any of a number of causes, including: infiltration of armed intruders from one country into the other; exchange of fire across the border; seizure of citizens of one country by agents of the other; movement of individuals from one country to the other in search of asylum or to escape imprisonment; smuggling; build–up of armed forces along a border.
Incidence Border incidents tend to occur in cases where the relations between the two countries are strained, for example: North Korea/South Korea; South Africa/Mozambique; China/Vietnam; India/Pakistan; Israel and neighbouring Arab countries; Tanzania/Uganda; India/China.
Broader Aggressive foreign policy (#PC4667) Abuse of government power (#PC9104).
Related Intergovernmental disputes (#PJ5405).
Aggravates Inadequate international map of the world (#PD0398).
Aggravated by Annexation (#PE5210) Divided countries (#PD1263)
Accidental military incidents (#PE4553) Boundary disputes between states (#PD2946)
Unnatural boundaries between states (#PF0090).

♦ **PD2955 Dependence of developing countries on customs revenue**
Nature Compared with developed countries, developing countries depend on customs duties for a much larger proportion of government revenue; and the condition of their administrative apparatus makes it difficult for them to resort to other forms of taxation. This increases the reluctance to accept substantial trade liberalization commitments, particularly with other developing countries.
Broader Economic and social underdevelopment (#PB0539)
Vulnerability of developing countries (#PC6189).
Aggravates Decline in public sector savings in developing countries (#PE4574)
Trade barriers and protectionism between developing countries (#PD2958)
Absence of coordinated customs services in developing countries (#PJ8325).

♦ **PD2957 Theft of data**
Broader Economic crime (#PC5624).
Aggravated by Computer–based crime (#PE4362)
Polycythemia in animals (#PG1715)
Vulnerability of computer systems (#PE8542).

♦ **PD2958 Trade barriers and protectionism between developing countries**
Nature Special preferential arrangements between some developed and developing countries result in imports from other developing countries being faced with more restrictive tariff or non–tariff barriers than competing imports from these developed countries. Lowering of trade barriers among developing countries is more difficult than among developed countries because of the threat to their sensitive balance of payments situation; loss of customs revenue which is a principal form of taxation; and the loss of employment opportunities in activities which have to be abandoned as a result of trade liberalization. That such losses would presumably be compensated by increased output and employment in other sectors is not likely to carry much weight with governments in the conditions of under–employment and under–utilization of capacity which prevail in most developing countries.
Broader Protectionism in international trade (#PC5842)
International trade barriers for primary commodities (#PD0057).
Aggravates Weakness in trade among developing countries (#PC0933).
Aggravated by Dependence of developing countries on customs revenue (#PD2955).

♦ **PD2961 Developed country limiting of trade between developing countries**
Nature Producers in developing countries, even if they are fully price–competitive, are sometimes placed at a disadvantage because goods from developed countries can be offered on more favourable terms. Developed countries can provide more generous facilities with respect to suppliers' credits; they can supply, on concessional terms, certain primary goods for which other developing countries would otherwise be competitive; and they can offer various products under tied aid arrangements. Moreover, where there are special preferential arrangements between some developed and some developing countries, imports from other developing countries are faced with more restrictive tariff or non–tariff barriers than competing imports from these developed countries.
Broader Underdevelopment (#PB0206).
Aggravates Weakness in trade among developing countries (#PC0933).

♦ **PD2963 Disparities between developing countries**
Nature The differences in levels of development among developing countries are often greater than among developed countries – the difference between two developing countries may even be greater than that between a developing and a developed country. This implies that, when trade barriers are to be reduced between developing countries at different levels of development, special measures must be called for if the benefits arising from a particular scheme are to be shared equitably. The most effective among these have financial implications. Yet in the developing part of the world, the more advanced countries also suffer from capital shortage and often possess, within their own borders, areas of particular poverty or backwardness; for such countries it is politically very difficult to assume the necessary responsibilities, especially with respect to what appears to involve financial transfers towards other countries that are relatively less advanced. Obstacles to the use of barter still remain between some developing countries, although this might facilitate trade, increase development and reduce disparities.
Broader Regional disparities (#PC2049).
Aggravates Weakness in trade among developing countries (#PC0933).

♦ **PD2965 Nitrogen compounds as pollutants**
Nitrogen oxides as pollutants
Nature Exposure to oxides of nitrogen may lead to death from any of three different types of pulmonary lesions, due to reduced partial oxygen pressure in the lungs: sudden death may result from bronchospasm and respiratory failure; otherwise, death can arise from delayed pulmonary oedema; or inflammatory changes termed bronchiolitis fibrosa obliterans, an auto–immune response. Exposure may also cause morbidity in new–born children. Damage to the environment include a brown haze in city air, acute injury to plants, and localized destruction of forests near large industrial sources.
Incidence Oxides of nitrogen are ubiquitous pollutants emitted by internal combustion engines, power stations, furnaces and cars.
Background Any oxidation of atmospheric nitrogen at high temperature results in the production of oxides of nitrogen. Such conditions as are prevalent with internal combustion engines, furnaces, incinerators, industrial processes and forest fires, would be likely to lead to such production. Environmentally, the oxides interact by means of photo–chemical oxidation in the presence of hydrocarbons, to produce irritants such as peroxyacylnitrates.
Claim Nitrous oxide is 250 times carbon dioxide's capacity to trap heat in the atmosphere. Its pre–industrial was 280 part per billion and the 1986 level is 305 parts per billion. The current annual increase in 0.2 percent.
Refs Brogan, J C (Ed) *Nitrogen Losses and Surface Run–Off from Landspreading of Manures* (1981); Schneider, T and Grant, L (Eds) *Air Pollution by Nitrogen Oxides* (1982); Winteringham, F P *Soil and Fertilizer Nitrogen* (1985).
Broader Chemical air pollutants (#PD1271) Photochemical oxidant formation (#PD3663)
Chemical pollutants of the environment (#PD1670).
Narrower Ammonia as pollutant (#PG5273) Nitrates as pollutants (#PE1956)
Nitrites as pollutants (#PE6087)
Increase in nitrous oxide in the atmosphere (#PE9443).
Related Acidic precipitation (#PD4904).
Aggravates Lake pollution (#PD8628) Ozone as a pollutant (#PE1359)
Badly laid out work premises (#PJ2468) Environmental plant diseases (#PD2224)
Nitrogen overdosage of plants (#PE4328).
Aggravated by Motor vehicle emissions (#PD0414)
Environmental hazards of coal energy (#PD7541).

♦ **PD2967 Over–production of primary commodities in developing countries**
Nature Production of certain commodities may be in excess of demand as evidenced by the periodic tendency of world stocks to increase faster than consumption. Over–production has been due in part to the over–optimism of production decisions which, in the case of tree crops, have to be taken several years before the resulting supply becomes available. (The excessive production of coffee in the early 1960s, for example, resulted from the heavy plantings made after the coffee boom in the mid–1950s). Over–optimistic production decisions may also be a consequence of price fluctuations, which have the effect of obscuring the underlying trends of supply and demand and therefore make them liable to be misjudged.
Incidence Over–production has been especially important in the case of sugar, coffee, cocoa and cotton.
Broader Structural rigidity in developing country economies (#PD2970).
Aggravates Reduction of the share of the developing countries in world exports (#PC2566).

♦ **PD2968 Instability in export trade of developing countries producing primary commodities**
Instability of commodity prices and developing country earnings — Decline of developing country commodity prices — Continuing low level of prices of commodities exported by developing countries — Excessive income dependence on primary commodities by developing countries — Dependence of developing countries on export of limited range of raw materials — Economic dependence of developing countries on export of primary commodities — Narrow export base of developing countries — Decline in growth of commodity export earnings of developing countries — Commodity related shortfalls in export earnings of developing countries — Cyclic swings in commodity prices
Nature Instability of commodity prices, and associated fluctuations in earnings, pose a number of problems for developing countries. Price instability may make it difficult for producers to take

rational investment decisions and may therefore lead to losses of real income through misallocation of productive resources. It may also weaken the competitive position of natural products against synthetic or other substitutes supplied at more stable prices. Fluctuations in producers' incomes and foreign exchange earnings may destabilize the economy through their influence on savings and consumer spending and through their disruptive effects on import budgets and development plans. The effects are usually self–cancelling except in the case of countries highly dependent on only one or two unstable commodities.
The greater concentration of developing countries production and exports in primary products makes their economy more vulnerable to external factors than that of industrialized countries. The world economic recession of 1980s has been transmitted from the developed countries to the developing countries to the large extent through commodity prices. Declining prices of commodities have increased the debt burden and impaired the ability to both maintain imports and service their debt.
Primary commodities account for most of the total exports of developing countries. Unlike world demand for processed foodstuffs and manufactured goods, demand for most primary commodities is relatively price–inelastic, with the consequence that prices of these commodities tend to be highly sensitive to changes in supply. The supply of primary commodities, moreover, is generally much less responsive than that of processed commodities to change in prices in the short run. For these reasons, and also because of the influence of random factors such as variations in harvests and strikes, prices of primary commodities often show considerable fluctuations both in the long and short–term. All these factors make the economies of developing countries extremely vulnerable to conditions outside their control. In addition, the primary commodities face a sluggish import demand in the countries that constitute their major markets, particularly where substitutions may be made, such as synthetics for natural commodities.
Incidence Recent rates of change from one year to the next include: over 20 percent changes in food and vegetable oils and oilseeds; over 10 percent changes in tropical beverages, agricultural raw materials, and minerals ores and metals. Taken together, these non–fuel primary commodities price levels declined 16 percent in both 1981 and 1982, with only partial recovery shown in 1983 and 1984.
In 1980 ten developing countries relied on only one primary commodity for 90 percent or more of the value of their exports; twelve for 80 percent or more; seven for 70 percent or more; ten for 60 percent or more and nineteen for 50 percent or more. In contrast, only one developed country depended on one commodity for 50 percent or more of its export earnings. Export earnings for developing countries, because of their greater dependence on primary products, are relatively more unstable than those of developed countries – whether because of wider fluctuations in export prices or in export volumes. Generally it has been presumed that the direct and indirect effects of this instability impose substantial losses in welfare and slow down the role of development. Schemes for trade restructuring import substitution, diversification, buffer stocks, compensation, and international commodity agreements have not been very successful. Instability has been important in the case of bananas, bauxite, cocoa, coffee, copper, cotton yarns, hard fibers, iron ore, jute and jute products, manganese, meat, phosphates, rubber, sugar, tea, tropical timber, tin and vegetable oils. Many of these commodity trades have recently seen a deterioration in trade terms, as well.
World Bank calculation in the mid–1980's shows that the price of all groups of commodities from foods to metals has been declining by about 1 percent a year in real terms over the last 35 years.
Claim Developing countries' long dependence in primary product exports and their lack of human and financial capital weakens their position vis–à–vis developed countries in processing domestically and externally locally–produced primary commodities. The development of commodity–related manufacturing and services sectors by developing countries is needed in view of the declining terms of trade of primary commodities vis–à–vis those of manufactured commodities, in view of the need to increase value added and in view of the state of the international trading and financial system characterized by protectionism and by obstacles to access foreign capital. This is particularly important for the cases of coffee, cocoa, tea, bananas, rubber, iron ore, bauxite, phosphates, sugar, manganese, hard fibres, hides and skins.
Refs Chantagul *Instability of Export Earning of LDC Primary Products and the Role of International Commodity Agreement*; MacBean, Alasdair I *Export Instability and Economic Development* (1966); UNCTAD *Compensatory Financing of Export Earnings Shortfalls* (1985); United Nations Conference on Trade and Development *Commodity Export Earnings Shortfalls, Existing Financial Mechanisms and the Effects of Shortfalls on the Economic Development of Developing Countries* (1987).
Broader Economic dependence (#PF0841) Vulnerability of developing countries (#PC6189)
Economic and financial instability of the world economy (#PC8073).
Related Instability of economic and industrial production activities (#PC1217)
Economic dependence of some developing countries on the drug trade (#PE5296).
Aggravates Reversal of development progress (#PF4718)
Instability of production of food and live animals (#PD2894)
Excessive foreign public debt of developing countries (#PD2133)
Inadequate development of enterprises in developing countries (#PE8572)
Burden of servicing foreign public debt by developing countries (#PD3051)
Excessive concentration of export markets of developing countries (#PE9457)
Reduction of the share of the developing countries in world exports (#PC2566)
Inadequate trade in agricultural commodities between developing countries (#PE4523)
Vulnerability of national economies to vagaries of external markets for goods and services (#PF9697).
Aggravated by Strikes (#PD0694) Bad weather (#PC0293)
Natural disasters (#PB1151) Commodity speculation (#PD9637)
Cyclic business recessions (#PF1277)
Obstacles to commodity futures trading (#PF4870)
Undiversified economies of developing countries (#PD2892)
Declining economic growth in developing countries (#PD5326)
Structural rigidity in developing country economies (#PD2970)
Declining economic growth in industrialized countries (#PF1737)
Deterioration of terms of trade for developing countries (#PD2897)
Reduction in demand for primary commodities due to technological change (#PD1276)
Lack of processing industry for primary commodities in developing countries (#PD1554)
Inadequate demand for primary commodities because of rising living standards (#PD2898).

♦ PD2970 Structural rigidity in developing country economies
Rigidities in production structures in developing countries
Nature Mobility of the factors of production, important in enabling appropriate responses to be made to changes in world market conditions, is absent in the case of many developing countries. Evidence of structural rigidity may be provided in a general way by persistently adverse movements in the terms of trade, but it may also manifest itself specifically in over–production or under–production of particular commodities and in excessive dependence on slow–growing export commodities. The process of development is seen as generally involving a shift in the structure of production towards manufacturing activities, producing intermediate and capital goods, and the provision of different kinds of services.
Broader Underdevelopment (#PB0206)
Weakness of infrastructure in developing countries (#PC1228).
Narrower Undiversified economies of developing countries (#PD2892)
Over–production of primary commodities in developing countries (#PD2967)
Underproduction of primary commodities in developing countries (#PD3042).
Aggravates Instability of production of food and live animals (#PD2894)
Excessive foreign public debt of developing countries (#PD2133)
Instability in export trade of developing countries producing primary commodities (#PD2968).
Aggravated by Low complementarity of developing country economies (#PE8184)
Inadequate level of investment within developing countries (#PD0291).

♦ PD2972 Lack of job opportunities for some sectors of society
Nature Society is so stratified that many groups and individuals find themselves unable to make a significant contribution in the job market. This is more apparent in minority groupings of age (both the youngest and the oldest), sex (most often women), and race.
Aggravates Obsolete vocational skills (#PD3548).

♦ PD2993 Bureaucratization of socialism
Broader Socialism (#PC0115) Bureaucracy as an organizational disease (#PD0460).

♦ PD3003 Monopoly of the economy by corporations
Corporate domination of daily life
Nature Large industrial corporations, often with the support of government, monopolize major sectors of the world economy and exert considerable power over large segments of the mass media. Economic surpluses are reserved for industrial production.
Broader Monopolies (#PC0521).
Narrower Blocked global marketing (#PD0930) Economic bias in development (#PF2997)
Ideological overemphasis in economic administration (#PF1040).
Related Maldistribution of resources (#PB1016) Belittling of grant recipients (#PF2708)
Imperialistic distribution system (#PD7374)
Limited approaches to economic planning (#PF3500)
Unregulated ownership of the means of production (#PF2014).

♦ PD3012 Political confiscation of property
Nature The confiscation of the property of political offenders by government and police authorities usually includes books and other material considered subversive, immoral or degrading, though it may extend to other property especially if the accused becomes a long–term prisoner. Property may be confiscated if it is felt to be connected with espionage practices or if it can be used in a political trial. Political confiscation of property may serve to tighten government control and may cause deprivation.
Broader Insecurity of property (#PC1784) Restrictions on property rights (#PD8937).
Narrower Government expropriation of private property (#PD3055).
Related Political prisoners (#PC0562).
Aggravated by Political trials (#PD3013) Excessive government control (#PF0304).

♦ PD3013 Political trials
Nature The trial of political offenders is generally by one of two forms: either secret trial or show trial (to gain the maximum benefit from propaganda). In either case the trial may be staged, with evidence weighted heavily in favour of the prosecution. The evidence may be forced or may be based on false confessions following torture; and judges and jurists may be biased. Show trials are used as a means of indoctrination, propaganda or intimidation and as a justification for the government, and secret trials are a means of tightening government control. Either may be the vehicle for personal vengeance and power struggles.
Broader Unjust trials (#PD4827) Abuse of authority (#PC8689)
Political repression (#PC1919).
Narrower Show trials (#PG5885) Secret trials (#PG3518).
Related Qualified amnesty (#PF3019) Internment without trial (#PD1576).
Aggravates Political confiscation of property (#PD3012).
Aggravated by Totalitarianism (#PF2190) Political police (#PD3542)
False political evidence (#PD3017) Forced political confessions (#PE3016)
Political corruption of the judiciary (#PE0647).
Reduced by Refusal to grant amnesty (#PF0182).

♦ PD3014 Military political prisoners and detainees
Nature Political prisoners with military status, whether prisoners of war, guerrillas, terrorists, or military personnel who have been purged from political or administrative office or who have been accused of war crimes, may be subjected to torture and brainwashing and be sentenced to imprisonment, forced labour, exile or execution by means of false evidence or forced confessions at secret military tribunals or show trials.
Broader Political repression (#PC1919).
Narrower Injustice of military tribunals (#PE0494).
Related Political prisoners (#PC0562)
Civilian political prisoners and detainees (#PD3015).
Aggravated by Military influence (#PD3385) Excessive government control (#PF0304)
Political corruption of the judiciary (#PE0647).
Reduced by Refusal to grant amnesty (#PF0182).

♦ PD3015 Civilian political prisoners and detainees
Nature Political prisoners of civilian status, including students, interned for opposition to the government, may be members of banned political parties or may be suspected of subversive activities, either in the expression of opinion or in active demonstration or in organizational support for opponent groups. They may be subjected to torture, confiscation of property or brainwashing; and may be sentenced to imprisonment, forced labour, exile or execution on the strength of false evidence or forced confessions used in secret or show trials. They may also be interned without trial.
Broader Political repression (#PC1919)
Political prisoners in communist systems (#PD3171).
Narrower Political opposition (#PJ5891).
Related Military political prisoners and detainees (#PD3014).
Aggravated by Refusal to grant amnesty (#PF0182)
Excessive government control (#PF0304)
Political corruption of the judiciary (#PE0647).

♦ PD3017 False political evidence
Nature Evidence which is contrived to ensure the conviction of political prisoners may be derived from corruption, torture and forced confessions, confiscation of property (which may either be used in evidence or suppressed) or intimidation of witnesses who might testify in defence. The successful use of false evidence for political purposes generally necessitates a biased judge and jury. Sentences resulting from false evidence include exile, forced labour, imprisonment and execution. It also serves to strengthen government control.
Broader Falsity (#PF5900) Evasion of issues (#PF7431).
Related Forced political confessions (#PE3016).
Aggravates Political trials (#PD3013).
Aggravated by Intimidation of witnesses (#PJ5894)
Excessive government control (#PF0304)

Tampering with official documents (#PF4699)
Political corruption of the judiciary (#PE0647)
Inadequate evidence to convict known offenders (#PF8661).

♦ PD3018 Violation of amnesty
Nature The reinternment of political prisoners freed after amnesty, and the detention of others for the same political crimes for which amnesty was given, constitutes violation of basic human rights. Governments having held political prisoners against public protest for a long time may release them or give a general or qualified amnesty to abate the protest, and then shortly afterwards start making secret arrests. Alternatively, they may release prisoners, but keep them constantly under police surveillance and restrict them to a certain area far from their home, so that it is very difficult for them to find work.
Broader Political repression (#PC1919).
Related Qualified amnesty (#PF3019) Refusal to grant amnesty (#PF0182).
Aggravates Internment without trial (#PD1576).
Aggravated by Lack of political integrity (#PF0796).

♦ PD3022 Libel
Risk of libelling
Nature Defamation of character in the written word – libel – whether intentional or unintentional, is an actionable offence. However, in the case of unintentional libel, the publisher and the author may offer a suitable apology and correction which will usually be upheld by the law against prosecution. Libel may be a criminal as well as a civil offence and those convicted of it may have to pay fines and damages or serve a prison sentence. Under the law of certain countries (such as Scotland) no distinction is made between libel and slander, which is not usually a criminal offence.
Refs Bezanson, Randall P, et al *Libel Law and the Press* (1987); Forer, Lois G *A Chilling Effect* (1987); Noam, Eli M and Dennis, Everette *The Cost of Libel* (1989); Sack, Robert D *Libel, Slander, and Related Problems* (1980).
Broader Risk (#PF7580) Malice (#PF5901)
Defamation of character (#PD2569).
Narrower Intentional libel (#PG5898) Unintentional libel (#PG5899).
Aggravates Sedition (#PC2414).
Aggravated by Lack of integrity (#PF7992).
Reduced by Restriction of freedom of expression (#PC2162).

♦ PD3023 Slander
Risk of slandering — Calumny
Nature Defamation of character in speech – slander – is not usually a criminal offence (except in some jurisdictions in Australia and the USA) unless offensive words are blasphemous, seditious or obscene. It is not an actionable offence unless proof of 'special damage' is produced, such as the loss of advantage other than reputation. Slander must impute a punishable crime, a loathsome disease, adultery of a woman, or charges which would prejudice the plaintiff's profession. Under certain jurisdictions no distinction is made between libel and slander (as in Scotland) whereby the latter becomes actionable on grounds otherwise reserved for libel. Under other jurisdictions (non–actionable) slander includes written innuendo (such as in the USA)
Refs Sack, Robert D *Libel, Slander, and Related Problems* (1980).
Broader Risk (#PF7580) Malice (#PF5901)
Defamation of character (#PD2569).
Aggravates False accusations (#PE7348).
Aggravated by Lack of integrity (#PF7992).
Reduced by Restriction of freedom of expression (#PC2162).

♦ PD3026 Book censorship
Book burning — Censorship in library acquisitions
Nature Books which are considered immoral, obscene, subversive, pornographic, heretical, politically embarrassing, threatening to national security or otherwise undesirable may be banned. Book censorship may lead to a lack of information and serve to maintain prejudice, alienation, discrimination, racism and nationalism. It may include library censorship (refusal to buy certain books, the removal and destruction of others) and proscribed reading lists in schools and other educational institutions.
Incidence Over the past 50 years, the following books and authors have either been banned in the USA or the banning of the works has been advocated: Of Mice and Men; The Diary of Anne Frank; The Catcher in the Rye; Huckleberry Finn; 1984; Brave New World; Black Like Me; Lord of the Flies; Henry Miller; Allen Ginsburg; and William Burroughs. Publishers in the United States cut passages from text books, for example, 300 lines were deleted from Romeo and Juliet because of ribald or expressly sexual terminology. In 1989 the book, Satanic Verses, was publically burned at a demonstration in the UK.
Refs Jones, Frances M *Defusing Censorship* (1983); Putnam, George H *Censorship of the Church of Rome and Its Influence upon the Production and Distribution of Literature* (1967).
Broader Censorship (#PC0067).
Related Self censorship (#PF6080).
Aggravates Bias in children's literature (#PD4773).
Reduces Obscenity (#PF2634) Pornography (#PD0132).

♦ PD3027 Newspaper and periodical censorship
Nature Restriction of ideas and information expressed in newspapers and journals may occur through editorial repression of articles; or raids on newspaper offices and confiscation of property; through the arrest and internment of journalists, especially to prevent them discovering information; and through refusal to open political meetings or trials to the press or refusal to issue information about political or governmental affairs. Violation of censorship may lead to imprisonment. Strict censorship may lead to the setting–up of an underground press and other subversive activities, or may induce political ignorance and apathy. Newspaper and journal censorship is particularly severe under political dictatorship and totalitarian regimes. In some countries, leading newspapers are subsidized or entirely supported by the government which carries the evil of information control to the doors of the state, but in several market economies large newspaper and communication channel chains have excessive controls over considerable numbers of outlets, and are able to dictate editorial policies to them.
Incidence Recently China banned issues of The World Economic Herald which criticized the Communist party. Kenya band the magazine, the Financial Review for publishing articles about alleged corruption in the Kenyan government. Military commanders censor the news in the US military newspaper, Stars and Stripes.
Broader News censorship (#PD3030) Harassment of the media (#PD0160).
Related Self censorship (#PF6080)
Foreign controls of newspaper and journal propaganda (#PD3041).
Aggravates Underground press (#PD2366)
Inadequate protection of war correspondents (#PE3034).
Aggravated by Excessive government control (#PF0304).
Reduces Newspaper and journal propaganda (#PD0184).

♦ PD3028 Theatre censorship
Banned plays
Nature Public performance of plays considered to be obscene, sacrilegious, subversive or otherwise unacceptable may be banned by the censor, or he may order scenes to be rewritten. Theatres where such plays have been shown may be closed down or refused renewal of licence. Theatre censorship leads to cultural stagnation and may also lead to general apathy and ignorance.
Broader Censorship (#PC0067).
Narrower Refusal to licence theatre performances (#PG5909).
Related Self censorship (#PF6080).
Aggravates Cultural stagnation (#PC8269).
Aggravated by Traditionalism (#PF2676).
Reduced by Private theatre clubs (#PG5911).

♦ PD3029 Radio and television censorship
Nature Radio and television censorship may be achieved by direct or indirect government control. For example, in the UK the BBC Board of Governors is appointed by the government; elsewhere there may be refusal to renew or grant licences to commercial or semi–commercial companies. Broadcasting may also be censored by public protest, by commercialization (where a commercial firm pays for a particular programme or will not place advertising if programmes are not to its liking), and by editors and producers. Particular areas of censorship are news and political programmes, and also those of sociological interest. Restrictions are made on obscenity and profanity. Broadcasting censorship may lead to cultural stagnation, conformism and apathy. It may also lead to ignorance and indoctrination through lack of information.
Incidence An unusual kind of television censorship follows the apartheid system in South Africa: SA Broadcasting Corporation's channel is predominantly intended for whites; channels such as Bophuthatswana TV are for blacks.
Broader Censorship (#PC0067).
Narrower Jamming of satellite communications (#PD1244).
Related Propaganda (#PF1878) News censorship (#PD3030) Self censorship (#PF6080)
Radio and television propaganda (#PD3085) Excessive regulation of television (#PE6982).
Aggravates Apathy (#PA2360) Ignorance (#PA5568) Conformism (#PB3407)
Cultural stagnation (#PC8269) Compulsory indoctrination (#PD3097).
Aggravated by Moralism (#PF3379) Nationalism (#PB0534)
Totalitarianism (#PF2190) Excessive government control (#PF0304)
Proliferation of commercialism (#PF0815) Refusal to grant licences to media (#PF3079).

♦ PD3030 News censorship
Nature Censorship of current information put out on different media (newspapers, journals, radio, television, film) may imply control of what news items are included as well as the manner in which they are presented. Local news agencies may be directly censored, closed or subject self–censorship. Foreign correspondents' stories may be reviewed before being dispatched, credentials removed or journalists expelled. News can be controlled by providing selective access to news sources, like battle fields during war or guided tours of national projects. News emissions are an important vehicle for propaganda. In a country with heavy censorship, little foreign news may be given, and items tend to concentrate on national success stories rather than bad news. It is especially heavy in wartime. News censorship may lead to ignorance and apathy (particularly concerning political matters) and serve to strengthen government control or dictatorship.
Broader Censorship (#PC0067) Denial of access to news (#PF3081)
Distorted media presentations (#PD6081).
Narrower Newspaper and periodical censorship (#PD3027)
Restrictions on news coverage of legal affairs (#PF3073)
Restrictions on direct news coverage of parliamentary affairs (#PF3072).
Related Self censorship (#PF6080) Film and cinema censorship (#PD3032)
Biased presentation of news (#PD1718) Radio and television censorship (#PD3029).
Aggravates Ignorance (#PA5568) Biased government information (#PF0157).
Aggravated by Excessive government control (#PF0304).

♦ PD3032 Film and cinema censorship
Nature Restrictions on the public showing of films may be on the grounds of obscenity, nudity, violence, or general moral undesirability, including political undesirability. Censorship may involve cutting out certain scenes or dialogue, or may mean that the whole film is banned. The most frequent method is classifying films in various levels of acceptability: universal acceptable, adults only, teenage children, children with an adult etc. It may either be refused a licence or classification, or be confiscated. Newsreels may be censored, as may also publicity photographs for films. The banning of films may lead to the growth of private film clubs and underground cinemas. If a banned film is shown publicly contrary to the law, the exhibitor may be imprisoned. Censorship of films may lead to cultural stagnation.
Refs DeGrazia, Edward and Newman, Roger K *Banned Films* (1982).
Broader Censorship (#PC0067).
Related Self censorship (#PF6080) News censorship (#PD3030).
Aggravates Private theatre clubs (#PG5911).
Aggravated by Nudism (#PF2660) Obscenity (#PF2634)
Immorality (#PA3369) Pornography (#PD0132)
Culture of violence (#PD6279) Proliferation of commercialism (#PF0815).

♦ PD3033 Postal censorship
Nature Censorship of mail may be carried out as a means of political control against political suspects or as a general measure. If such mail is considered to be subversive it may be confiscated and added to a political dossier on the person concerned. Mail may also be censored for possible pornographic or obscene content, where the law prohibits postal communication of such items.
Broader Censorship (#PC0067) Inadequacy of postal services (#PF2717).
Related Self censorship (#PF6080) Inappropriate use of the mail service (#PE6754).
Aggravated by Political repression (#PC1919).

♦ PD3035 Erosion of journalistic immunity
Denial of rights to journalists — Forced breach of journalistic confidence
Nature Pressure to make journalists reveal their sources of information may take the form of arrest, trial and imprisonment or general intimidation. Recent examples of threats to journalistic immunity have occurred in USA court cases, with convictions against journalists for the publication of the Pentagon Papers. In countries where political censorship is heavy, journalists may be imprisoned and tortured for subversive activities. Information may concern crime, corruption or other injustices, or be akin to espionage; and the possibility that sources would be revealed would risk making it more difficult or even impossible to obtain this information. The public would therefore be deprived of knowledge about matters which might concern them directly or indirectly.
Broader Harassment of the media (#PD0160)
Restrictions on freedom of information (#PC0185).
Related Threat to parliamentary immunity (#PF0609).
Aggravates Harassment of journalists (#PD3036)
Journalistic irresponsibility (#PD3071)

Restriction of freedom of expression (#PC2162).
Aggravated by Official secrecy (#PC1812).
Reduces Infringement of privacy (#PB0284).

♦ PD3036 Harassment of journalists

Intimidation of journalists

Nature Harassment of journalists may include police brutality, arrest, internment, trial, confiscation of articles and other information, closure of press offices, or even assassination. Journalists may have security dossiers opened on them and their names may appear on extensively circulated black-lists for prejudicial treatment. The restriction of information which results may lead to ignorance and apathy or further violence and subversive activities. It serves to tighten government control and paves the way for indoctrination. Journalists may also be harassed or prevented from obtaining information by hostile individuals and groups.

Incidence Fifteen journalist were killed in 1986. Another 56 were assaulted, and more than 190 were arrested or kidnapped. Thirty-seven were expelled and eight left countries under threat. In 1987 nearly 600 incidents of press abuse took place in 75 countries. The abuses range from temporary, non-violent acts, such as limiting the circulation of a publication, to the murder of a journalist in the pursuit of his or her profession. Twenty-six journalist were killed in ten countries. Eleven of the 26 were killed in the Philippines and 4 in Sri Lanka. Fifty-nine journalists were physically assaulted, more than 185 arrested or kidnapped, 11 forced to leave their own country under threat, and 32 were expelled from countries where they had been working as foreign correspondents. In the first 11 months of 1989, 63 journalists were murdered around the world.

Broader Harassment (#PC8558) Harassment of the media (#PD0160)
Impediments to internationally mobile professionals and experts (#PF1068).
Related Harassment of public officials (#PD4915).
Aggravates Restriction of freedom of expression (#PC2162).
Aggravated by Excessive government control (#PF0304)
Journalistic irresponsibility (#PD3071)
Erosion of journalistic immunity (#PD3035).

♦ PD3039 Unequal opportunities for media reception

Nature Where education includes the use of mass media, individuals or social groups without the means of obtaining the necessary apparatus are handicapped in securing a higher education or the employment or other opportunities arising from it. Those with limited education are less well equipped to understand and make use of the information included in programmes or broadcasts concerning political, social, economic, legal and other questions, even if they have access to it. This problem applies both on a national and an international level; it serves to maintain existing injustices, discrimination, exploitation and conflict, and may result in alienation, apathy and ignorance.

Broader International inequality (#PC9152) Inequality of opportunity (#PC3435).
Related Risk of unintentional nuclear war generated by the strategy of deterrence (#PF4162).
Aggravates Elitism (#PA1387) Ignorance (#PA5568).
Aggravated by Inadequacy of aid to developing countries (#PF0392)
Technology gap between developed countries (#PD0338).

♦ PD3040 International monopoly of the media

Emergence of international media monopolies

Nature Monopoly of communications media by one or a few countries on the international level is particularly notable in satellite transmission, film and television production. It may be supported by restrictive legislation, and is a result of the concentration of wealth and technological expertise in certain countries. Effective monopoly of certain media on an international level facilitates the spreading of propaganda and indoctrination, may give rise to subversive activities and corruption, and strengthens foreign influence and control.

Incidence Cable services are the only source of international news used by a great many newspapers and other media in developing countries. The reportage of UPI, AP, or AFP, however, cannot assess the relevance of news to the interests of particular regions and their coverages may be biased or inadequate.

Broader Monopolies (#PC0521) Monopoly of the media (#PD3101).
Narrower Film monopoly (#PG5925) Newspaper monopoly (#PE0246)
Television monopoly (#PG5924) Satellite transmission monopoly (#PG5929).
Related Industrial gas monopolies (#PE1813)
Restriction of access to news distribution media (#PF3082).
Aggravates Propaganda (#PF1878) Foreign control (#PC3187)
Cultural invasion (#PC2548) Denial of access to news (#PF3081)
Excessive commercialization of the media (#PE4215).
Aggravated by Restrictive legislation (#PD9012)
Technology gap between developed countries (#PD0338).

♦ PD3041 Foreign controls of newspaper and journal propaganda

Nature Wealth or other means of influence may be used to gain access to the press of another country and use it for propaganda purposes either in support of or to the detriment of the government of that country. If propaganda is used in support it will tend to safeguard the position of the foreign power (investments, effective political control, for example) against a neighbouring country or other powers with a conflicting ideology which may gain influence there. If propaganda is used against the regime, it constitutes infiltration and subversive activity. Reprisals such as imprisonment, expulsion and confiscation may be taken and general censorship and repression instigated afterwards to ensure that the same situation does not happen again.

Broader Disruptive foreign influence (#PC3188) Newspaper and journal propaganda (#PD0184).
Related Newspaper and periodical censorship (#PD3027).
Aggravates Foreign control (#PC3187).
Aggravated by Uncontrolled media (#PD0040) Economic imperialism (#PC3198).

♦ PD3042 Underproduction of primary commodities in developing countries

Nature Under-production of primary commodities in developing countries may exist not only where the share of these countries in world exports has fallen because of supply difficulties (as in the case of wheat, coarse grains, rice, and vegetable oils), but also where the price elasticity of demand for a commodity is greater than unity (namely where an increase in the supply would result in a less than proportionate fall in price), so that total sales proceeds would not be increased.

Incidence Examples of under-production are provided by the chronic food shortages in certain heavily-populated developing countries in recent years. In some countries the problem has manifested itself as a failure to maintain exports to traditional markets, which have instead been supplied by developed countries. In other countries, shortages have necessitated heavy imports of essential foodstuffs (chiefly cereals) from developed countries, with consequent strains on the foreign exchange position.

Counter-claim Some under-production of commodities is desirable as less developed countries divert resources into manufacturing for export. Manufacturers are not as susceptible to price fluctuations and must take a greater place in the foreign exchange earnings of developing countries.

Broader Structural rigidity in developing country economies (#PD2970).
Aggravates Unbankruptability of sovereign states (#PF0478)
Instability of production of food and live animals (#PD2894)
Reduction of the share of the developing countries in world exports (#PC2566).

♦ PD3045 Hoarded monetary gold

Hoarding monetary gold

Nature There is a tendency for people and, in some cases, institutions, to hoard monetary gold, usually as a security against socio-economic crises. The process of hoarding in effect reduces the funds circulating within the economy which would otherwise be used to finance new developments.

Incidence Although it is not possible to be certain of the amount of hoarded gold, it has been estimated, for example, that in some developing countries in Southern Asia and the Middle East the hoarded gold may represent 10 per cent of the national income. Some hoarding is done by governments where the law requires that the local currency should be backed to 100 per cent by foreign exchange.

Broader Hoarding of primary commodities (#PD0651).
Aggravates Uncertain status of monetary gold (#PF2342).
Aggravated by Lack of confidence in the international monetary system (#PF3058).

♦ PD3046 Inequitable tax systems in developing countries

Nature When there are considerable economic inequalities, indirect taxes on income and wealth exert less restraint than direct taxes on socially wasteful consumption, yet the majority of developing countries collect 70 per cent or more of their tax revenue from indirect taxation (customs duties, for example). In addition, exemption levels tend to be higher than they need be. In some developing countries, personal incomes up to a multiple of ten or twenty times the average per capita income remain altogether exempt from income tax (compared with from one to three times in developed market economies). In particular, the tax shelter enjoyed by wealthy landlords is an obstacle to overall economic progress.

Broader Human inequality (#PA0844)
Economic and social underdevelopment (#PB0539).
Aggravates Inadequate domestic savings in developing countries (#PD0465).
Reduces Social insecurity in developing countries (#PE4796).

♦ PD3047 Tax discrimination against investment in a foreign country

Nature Discrimination against foreign income arises in many ways. In some instances special taxes are imposed on foreign investment income, and in others, foreign income is taxed in a way so fundamentally different from that in which domestic income is taxed that there is a strong presumption of intention to discriminate. In some countries, domestic inter-company dividends are exempt from profits tax in the receiving company on the grounds that the profits have already suffered the same tax in the paying company; dividends from a foreign company fail to qualify for exemption in this way. In the case of individual shareholders in foreign companies, an indirect form of discrimination may be occasioned by the fact that a special rebate or credit to the shareholder is afforded in his country of residence only in respect of domestic dividends. In each of these cases, the total tax payable on foreign income may be materially greater than for domestic income. It is believed that this is a significant factor in the relative decline in private as against public sector investment in many countries and that it is largely responsible for the failure to achieve an expansion of more broadly based equity financing.

Broader Discrimination (#PA0833)
Tax obstacles to international investment (#PD0673).
Narrower International double taxation (#PD0858).
Aggravates Discrimination against foreign companies (#PD6417)
Ineffective economic structures in industrial nations (#PE4818).

♦ PD3048 Tax discrimination against non-residents of a country

Nature Countries may introduce special taxes which are theoretically applicable to both resident and non-resident enterprises but which in practice impinge almost exclusively on the profits of the non-resident. Examples are: special taxes on profits when dividends are not paid within the country; special taxes on profits when a certain proportion of shares are not owned by residents; special tax on that part of the profits remitted out of the country; denial to non-resident companies of any rebate of tax on profits.

Broader Tax obstacles to international investment (#PD0673).
Narrower International double taxation (#PD0858)
Vulnerability of frontier workers (#PE6833).
Aggravates Ineffective economic structures in industrial nations (#PE4818).

♦ PD3050 Tax barriers to the dissemination of technical knowledge

International double taxation of royalties

Incidence A double taxation problem arises when the person or corporation making the expertise or inspiration available is resident in one country and the user is resident in a second country. Each of the two countries may then claim the rights to tax the royalties and will not afford relief for tax paid in the other country. The burden of the tax paid for the privilege of using copyright, patents, designs, trade marks, secret processes and formulae, know-how and the like, may then be intolerably heavy. The use may have to be foregone in some countries or the user must accept a serious restriction on the profitability of his business. This creates a special difficulty in the case of developing countries.

Broader Tax obstacles to international investment (#PD0673).

♦ PD3051 Burden of servicing foreign public debt by developing countries

Inadequate public debt relief for developing countries — Inconsiderate insistence by creditors on fulfilment of debt service obligations of developing countries

Nature Developing countries may accumulate an external public debt with servicing obligations which cannot readily be met at the time when they fall due. Under some circumstances, and there is no clear definition, the debt burden may be considered 'oppressive' and some form of debt relief is called for. Frequent rescheduling of debt over the years at commercial interest rates has led to a rapid and insupportable growth in some countries' debt problems. When interest rates rise relative to export prices, real debt burden will increase and import capacity will decrease. Rescheduling of debt service and debt principle payments is widely criticized as aggravating the public debt burden of these countries and drawing attention away from solutions to their economic problems. Few alternatives are forthcoming, however.

Incidence Total outstanding debts for developing countries soared from $68.4 billion in 1970 to $835 billion in 1985, with interest payments alone totalling $60 billion annually.

Claim Developing countries are called upon to fulfil their debt service obligations to the international creditor community without consideration of their ability to do so. The burden of debt repayments makes real growth impossible and the poor even poorer. The only realistic way to deal with third world loans is to write them off.

Broader Uncontrolled growth of debt (#PC8316)
Economic and social underdevelopment (#PB0539)
Outflow of financial resources from developing countries (#PC3134).
Aggravates Reversal of development progress (#PF4718)
Inadequacies of the international monetary system (#PF0048)

Disincentive to invest in heavily indebted countries (#PF9249)
Decline in commercial bank lending to developing countries (#PE4655)
Deterioration in external financial position of developing countries (#PE9567)
Aid lenders as beneficiaries of net outflow of funds from developing countries (#PE5338).
Aggravated by Defaults on international loans (#PD3053)
Foreign exchange shortage in developing countries (#PD3068)
Excessive foreign public debt of developing countries (#PD2133)
Decline in concessional financial resources available to developing countries (#PE3812)
Rescheduling of debts of developing countries at market-related interest rates (#PE8110)
Socially irresponsible programmes of transnational banks in developing countries (#PE4360)
Instability in export trade of developing countries producing primary commodities (#PD2968).

♦ PD3053 Defaults on international loans

Debt repudiation — Arrears in international debt payments

Nature Defaults on international loans may take several forms: repudiation of old loans by new governments swept into power during economic crises; failure to earn or unwillingness to expend foreign exchange to effect the transfer despite the fact the revenues are available in the domestic currency; and failure to collect sufficient revenues to cover debt service even when current expenditures have been drastically reduced (namely bankruptcy of the borrower).

Background The last major wave of defaults occurred in the 1930s but still influences thinking in connection with the international capital market. In 1935, 35 percent of the UK holdings of government and municipal securities were in default. In the same year 38 percent (Europe 51 percent, Canada 4.0, South America 77, Latin America 76, others 3.0) of the outstanding portion of foreign dollar bonds originally taken in the USA were in default. In many cases government loans secured by specific revenues were simply repudiated. Solvent corporations were often forced into default by exchange controls. Many developing countries, and particularly those in Latin America, were faced with an unpleasant choice: to cut back drastically on imports or go into default on their debts. After an initial attempt to cut imports, the second course was often adopted. Today the less developed countries are still paying for the consequence of this choice in reduced access to international capital markets. Thus, from the 1930s to the middle 1980s, the problem of defaults dominated discussions of the international capital market and the question of access by less developed countries to that market. The first prescription of every commentator was to 'solve the defaults problem'.

Incidence It is estimated that by the end of 1990 15 developing countries will owe about $524 billion to banks and governments, with nearly another $50 billion in interest payments falling due.

Refs Aggarwal, Vinod K *International Debt Threat* (1987); Campbell, Barry R and Herzstein, Robert E *The International Debt Problem and Its Impact on Finance and Trade* (1984); Giersch, Herbert (Ed) *International Debt Problem* (1986); Kahler, Miles *The Politics of International Debt* (1986); Res, Zannis and Motamen, Sima *International Debt and Central Banking in the 1980s* (1987); United Nations Centre on Transnational Corporations *Issues in Negociating International Loan Agreements with Transnational Banks* (1983).

Broader Economic and social underdevelopment (#PB0539)
Arrears in payment of government financial commitments (#PF1179).
Related Circumvention of duties and assessments (#PD4882)
Irregular payments of international financial obligations (#PD1157).
Aggravates Global financial crisis (#PF3612)
Inadequacies of the international monetary system (#PF0048)
Insufficient creditworthiness of developing countries (#PD3054)
Burden of servicing foreign public debt by developing countries (#PD3051).
Aggravated by Nationalization of foreign investments (#PC2172).

♦ PD3054 Insufficient creditworthiness of developing countries

Low credit rating of developing countries — Uncreditworthiness of developing countries

Nature The external debt of many developing countries is large and growing rapidly. Without doubting the solvency or good faith of the issuer, bond-holders may legitimately wonder whether the foreign exchange will be available to service any loan. The larger a country's debt servicing requirements relative to its foreign exchange earnings, the more vulnerable it is to default in the face of sharp unexpected declines in earnings or increases in import requirements. Uncertainties concerning the attitudes of present or future governments of a country before the maturity of a debt may also cause potential lenders to hesitate to buy bonds issued even by a highly respected borrower, for future payments of interest and amortization could be put in jeopardy by official action.

The creditworthiness of a country is largely determined by its standing with the IMF, especially in the case of debtor countries where negotiations with other donors are underpinned by IMF programmes. Perceptions of reduced creditworthiness of such countries prompt commercial banks to withhold new financing and in many cases to reduce their net claims on poorer debtor countries.

Broader Restrictions on foreign access to capital bond markets (#PD3135).
Narrower Uncreditworthiness of rural communities (#PJ1268).
Aggravates Inadequacy of aid to developing countries (#PF0392)
Excessive anxiety on lending to developing countries (#PF4345)
Decline in commercial bank lending to developing countries (#PE4655).
Aggravated by Defaults on international loans (#PD3053).

♦ PD3055 Government expropriation of private property

Denial of right to freed from arbitrary deprivation of property — Government expropriation of land

Nature Expropriation is a method frequently used by government, particularly in time of war, to obtain control of specific private property. It antedates and differs from nationalization in both motive and degree. Expropriation is generally used when a government needs a particular piece of private property in connection with a well-defined project (for example, the construction of a road). Usually the owner has the right to judicial redress as a remedy for inadequate compensation, but the owner may believe that no financial compensation settlement can be adequate compensation for property with special intrinsic or sentimental value to himself, his family or, in some cases, his community.

Broader Arbitrariness (#PB5486)
Political confiscation of property (#PD3012).
Related Expropriation of land from indigenous populations (#PC3304).
Aggravates Export credit risks (#PF3065)
Exploitation in housing (#PD3465).

♦ PD3056 Unpaid labour

Broader Underpayment for work (#PD8916).
Narrower Uncompensated tenant repair (#PJ0147)
Lack of payment for housework (#PE4789)
Dependence of developing countries on unpaid female labour (#PE4451).
Related Forced labour (#PC0746)
Non-valuation of housework in national accounts (#PF0023).
Aggravates Cheap labour (#PG5988).

♦ PD3057 War debt

Nature Governments may borrow funds from one another in order to finance their engagement in major wars; (war debt usually excludes the debts between defeated nations which are seldom, if ever, honoured, and also excludes obligations imposed by the victors on the vanquished in the form of reparations for war damages). The cost of servicing and repaying such debts once the war is over may give rise to international friction, particularly when some of the victor nations have difficulty in reconstructing their own economies.

Background Major war debts arose from the two World Wars. The gross total of inter-Allied loans extended during World War I amounted to approximately $20,780 million. As of 1962, the total indebtedness to the USA, for example, was still $19,727 million. As a result of this experience and the difficulties to which it gave rise in the inter-war period, the intergovernmental loans extended during World War II were relatively small. Commonwealth countries accumulated claims on the UK amounting to approximately $3,500 million; the USA extended credits to various countries amounting to $1,100 million. Other assistance was provided in the form of lend-lease aid, totalling $47,865 million.

Broader Excessive public debt (#PC2546).
Aggravated by War (#PB0593).

♦ PD3063 Legal impediments to foreign investment

Nature Widespread legal limitations exist on the holding of foreign securities by institutional investors, together with other forms of discrimination against foreign bonds. The impediments to the sale of foreign bonds, typically dating from the inter-war period of default and bankruptcy, were often designed to protect savers from mismanagement by the trustees of their savings. Some of these safeguards are now outmoded in the light of new priorities, especially in Europe where a restructuring of capital markets is being encouraged. Such impediments may take the form of discriminatory taxes; unnecessary restrictions on portfolio selection by savings banks, insurance companies and other institutional investors; or prohibitory laws regarding countries which defaulted many years ago.

Although some of these impediments are technically non-discriminatory, they may effectively discourage or even exclude foreign bond issues by certain less developed countries, by requiring detailed information on prospectuses for any new public issue. Even without taxes or controls, countries may effectively restrict access to their capital markets simply by maintaining a level of interest rates that discourages new foreign issues. These different regulations are not aimed particularly at the developing countries, but in combination with the limitations of the capital markets themselves and with balance of payments restraints on capital outflow, they impose a formidable barrier to new bond issues by developing countries in many national capital markets.

Broader Restrictions on foreign access to capital bond markets (#PD3135).
Aggravated by Diffuseness of regulatory authority (#PF3064).

♦ PD3067 Export credit competition

Nature The amount of sales in the international market depends, to a very great extent, on the facilities for deferred payments. For the importer, suppliers' credit may often be the chief, if not the only, available source of finance, particularly if he belongs to another developing country and is importing consumer manufactures or capital goods. Consequently, the payment terms become a crucial part of the contract, in addition to price and quality, and competition in the field of export credit becomes as important as pure price or quality competition. The severity of export-credit competition is such that a 'race' tends to develop, resulting in excessive borrowing on the wrong terms. This is particularly difficult for developing countries which may find themselves willing to ignore over-pricing and lower quality in favour of longer periods of credit.

Claim The significance of this element of non-price competition in the export market is hard to exaggerate, particularly for the developing countries trying to increase their exports of capital goods. Owing to an enormous increase in the output of capital goods, importers appear to be in a strong position in the buyers' market for obtaining credit from the exporters of capital goods. In some developing countries facing balance of payments difficulties, the ease with which import licences can be obtained is often a function of how liberal the terms of deferred payment are.

Broader Obstacles to world trade (#PC4890).
Aggravates Decline in export credits to developing countries (#PE3066).
Aggravated by Foreign exchange shortage in developing countries (#PD3068).

♦ PD3068 Foreign exchange shortage in developing countries

Foreign currency shortage in developing countries — Hard currency shortage in developing countries

Nature Availability of foreign exchange is crucial in development planning since all developing countries are forced to purchase imported equipment, spare parts and critical materials without which their own resources cannot be pressed into service. The developing countries face problems both in increasing their earnings of foreign exchange and in the increasing claims on available foreign exchange of rising debt payments and other essential commitments.

Broader Insufficient financial resources (#PB4653)
Chronic shortage of foreign exchange (#PC8182).
Related Dependence of island developing countries on imports (#PD5677).
Aggravates Export credit risks (#PF3065)
Export credit competition (#PD3067)
Foreign exchange restrictions (#PF3070)
Over-rapid timber exploitation (#PD9235)
Inadequate development of enterprises in developing countries (#PE8572)
Burden of servicing foreign public debt by developing countries (#PD3051).
Aggravated by Speculative flight of capital (#PC1453)
Non-convertibility of currencies (#PF3069)
Degradation of developing countries by tourism (#PF4115).

♦ PD3071 Journalistic irresponsibility

Harassment by journalists — Harassment by the media

Nature The abuse of the powers of the press may be dictated by governments, media owners, key advertisers, publishers, managing editors, or department chiefs, but be carried out only, by working reporters, broadcasters and writers. Journalistic disregard of ethical conduct in spreading calumny, incitement to hatred or to war, leakage of official secrets, inaccuracy and the spreading of rumour as if it were fact, may produce a credibility gap with the public, or it may encourage political repression, harassment and a threat to journalistic immunity. Unethical journalists may bait or entrap subjects into actions or statements; may use illegal surveillance techniques; or may invade the right of privacy of individuals. Exposés may even cause the death of sometimes blameless individuals, by suicide or by self-appointed avengers or mobs. Journalists serving international news cable services may report biased accounts or claim to witness events at which, in fact, they were not present.

Incidence 'Cheque-book journalism' is the practice of media executives to pay the actors in sensational events, sometimes criminals, to give an exclusive, inside story to a particular newspaper, magazine or broadcasting station, and to deny access of journalistic competitors. A journalist is assigned to ghost-write the story in the name of the key actor, or in the case of a broadcast, the journalist scripts or coaches the narrator beforehand. Public understanding of AIDS, of race relations and of sexual relations are some of the arenas in which press sensationalism has greviously distorted, creating unnecessary suffering.

Broader Irresponsibility (#PA8658)
Unethical media practices (#PD5251).
Narrower Student press weakness (#PE0628)
Unfair trials due to pre-trial publicity (#PE1692).
Related Inaccuracy (#PF7905).
Aggravates Propaganda (#PF1878)
Misleading information (#PF3096)

Harassment of the media (#PD0160) Infringement of privacy (#PB0284)
Harassment of journalists (#PD3036) Invasion of privacy by media (#PD9603)
Lack of freedom of the press (#PE8951)
Restrictions on freedom of information (#PC0185)
Restrictions on news coverage of legal affairs (#PF3073)
Restrictions on direct news coverage of parliamentary affairs (#PF3072).
Aggravated by Junk food journalism (#PE1750) Unidentified news resources (#PG9995)
Erosion of journalistic immunity (#PD3035).
Reduced by Censorship (#PC0067).

♦ PD3076 Alienation from work

Alienation of human labour

Nature Industrialization produces new wealth, but it also imposes rigid controls over human behaviour. Machines require workers to service them, calling for an adjustment of natural rhythms of the body to their mechanical processes. When labour became a mechanically regulated commodity, man lost part of himself. The worker, having lost control over both the conditions of his labour and the product of his work, became alienated from himself. The disintegrative, negative character of industrial society thus lies in its alienation of human labour and its denial of opportunities for men to fulfil themselves in meaningful work. This leads to a serious fall in morale which, with the boring and degrading conditions of work in modern industry, account for restrictions on output, wildcat strikes, outright sabotage and perhaps most common, in feelings of detachment from the entire work process. Rapid technological change dilutes old skills, makes others obsolete and creates demands for new ones, with the associated displacement of industries and the creation of depressed areas in prosperous economies. Men are dehumanized not only by the work situation but also by the ends for which society uses their work, primarily consumption for its own sake.
Narrower Workers alienation in socialism (#PE3594).
Aggravates Fragmentation of the human personality (#PA0911).
Aggravated by Increasing job monotony (#PD2656).

♦ PD3080 Conflict of laws on international restriction of information

Nature In countries where freedom of information is severely restricted, foreign correspondents researching information which would be available and free for distribution in their own country, may be interned or deported and their articles or photographs confiscated. Lack of information may result from government secrecy and the use of propaganda, and may serve to mask injustices, inequality and exploitation. International ill-feeling may result from a conflict of laws concerning information, or a conflict of information between nationalist and foreign viewpoints.
Broader Conflict of laws (#PF0216)
Restrictions on freedom of information (#PC0185).
Related Conflicting labour laws (#PE5135).
Aggravates Conflict of information (#PF2002)
Restrictions on international freedom of information (#PC0931).
Aggravated by Excessive government control (#PF0304).

♦ PD3085 Radio and television propaganda

Nature Television and radio broadcasts may aim at spreading the influence of a political ideal or government policies. In a single party state or dictatorship only one political or ideological line may be broadcast, but in a constitutional democracy several may be broadcast, creating confusion from conflicting information. Propaganda may be government information, unrelated or misleading. It may induce apathy, ignorance, conformism, prejudice, idealism, alienation, confusion and conflict. It may also strengthen government control, and existing injustices, inequalities, and exploitation.
Incidence Television was a powerful stimulant for change in Eastern European countries because of the picture of a prosperous Western society which it carried across the borders.
Broader Propaganda (#PF1878) Compulsory indoctrination (#PD3097).
Narrower Jamming of satellite communications (#PD1244).
Related Censorship (#PC0067) Government propaganda (#PC3074)
Misleading information (#PF3096) Radio and television censorship (#PD3029)
Excessive regulation of television (#PE6982).
Aggravated by Excessive government control (#PF0304)
Excessive television viewing (#PD1533).

♦ PD3086 Photographic propaganda

Nature The use of photographs to influence public opinion in favour of a political ideal or government policies may be on a national or international level. Photographs may be used as posters, in newspapers and journals, and as stills from propaganda films. Some propaganda using photographs may be in advertising form. They may convey official, government or subversive information, which may be misleading or intentionally doctored to show a scene which did not exist in reality. Photographic propaganda may induce conformism, apathy, idealism or alienation and conflict. It may strengthen government control and existing injustices, inequalities and exploitation. Its particular significance lies in the fact that it can convey propaganda to people who are illiterate or who do not like to read.
Broader Propaganda (#PF1878) Compulsory indoctrination (#PD3097).
Related Censorship (#PC0067) Photographic bias (#PF9707)
Misleading information (#PF3096).

♦ PD3088 Theatre propaganda

Nature The use of plays and other theatre entertainment to influence public opinion in favour of a political ideal or government policies may be on a national scale or internationally through cultural exchange programmes. This can be a very subtle form of propaganda. Plays and other entertainments may be satirical or convey a message of moralism and utopianism. They may contain government or official information misleadingly presented. Presentations may be folkloric promoting the indigenous charms of the country in question. Theatre propaganda may lead to cultural stagnation, conformism and idealism, ignorance and prejudice. If it is the result of subversive activities it may lead to imprisonment or deportation and the closure of the theatre. Alternatively, it may induce subversive activities, conflict and propaganda. It may strengthen government control or dictatorship.
Broader Propaganda (#PF1878) Compulsory indoctrination (#PD3097).

♦ PD3089 Film propaganda

Nature The use of films to influence public opinion, nationally or internationally, in favour of a political ideal or government policies. They may include the use of art, television and photographic propaganda, war, racist or religious propaganda, or official and government information, which may be misleading, inaccurate, utopian or moralistic. Film propaganda may sharpen political and international conflict or strengthen government control and dictatorship or lead to apathy, conformism, idealism, hero worship, ignorance, prejudice and alienation. Government industries or commercial production companies may consciously intend to either strengthen national or group solidarity or change or subvert opinion in a hostile or neutral group. These aims are at their strongest in times of war. In combatant countries every effort is made to denigrate the enemy.
Background Propaganda film came of age during the First World War. Every major power commissioned official films and most ended the war with some kind of government department responsible for co-ordinating film propaganda.
Broader Propaganda (#PF1878) Compulsory indoctrination (#PD3097).
Related Censorship (#PC0067).

♦ PD3090 Book propaganda

Nature The use of books to influence public opinion in favour of a political ideal or government policy, whether nationally or internationally, may include art and photographic, racist and religious propaganda, government and official information, all of which may be misleading, inaccurate, idealistic, moralistic or unrelated to its context. Most significantly book propaganda includes textbook indoctrination which affects the most susceptible members of society and young people, and which may be the only 'factual' information available for studies. Other versions may be censored. Book propaganda sharpens international and political conflict and if it is subversive it may be severely repressed. The effects of official book propaganda may be less obviously visible, but it contributes considerably to feelings of nationalism, prejudice, discrimination, elitism and social conflict, to political apathy, and conformism.
Broader Propaganda (#PF1878) Compulsory indoctrination (#PD3097).
Related Censorship (#PC0067).

♦ PD3092 War and pre-war propaganda

Denial of right to freedom from war propaganda

Nature Incitement to hatred and to war, either in war-time or in peace, may include racist and religious propaganda and may be official or subversive. The information may be false or lacking in content with the aim of guarding secrecy. It may be over-idealistic; occur in all kinds of propaganda media; cause war or prolong it; or mask military atrocities, genocide and the extent of death and destruction. It also tends to obscure the reasons for war and its effects on different sections of international and national society.
Claim States may attempt to condition both behaviour and thinking in anticipation of war. In the case of an aggressive state, bent on secretly developing a war-machine capability, it will foresee large-scale mobilization of the citizenry into fighting forces and attempt to prepare for this by ensuring that the civilian population is already accustomed to acts of brutality and violence. Three fields of activity lend themselves to this: repression of political dissent; persecution of minorities; and encouragement of player and spectator violence at sporting events. For all of these, propaganda will be generated as well. The state propaganda apparatus may stage two events: the first in which national territory, rights or civilian or military personnel are said to have been outrageously violated, possibly with some alleged atrocities; the second, a real and brutal reprisal, the 'justice' of which prepares the public for acceptance of additional and more extensive crimes against humanity.
Counter-claim Without using techniques for motivating the general public such as those developed by the film industry a government would be irresponsible to the nation's decision to be at war. Neutral, anti-war or anti-government propaganda would greatly influence people, weakening the will of the country to win and endangering the nation and its people.
Broader Propaganda (#PF1878) Compulsory indoctrination (#PD3097)
Denial of right to security (#PD7212).
Related Destabilizing international telecommunications (#PD0187).
Aggravates War (#PB0593) Nationalism (#PB0534) Enjoyment of war (#PF4034)
Incitement to hatred (#PE5952) Glorification of war (#PF9312).

♦ PD3093 Racist propaganda

Nature Incitement to racial hatred within the same nation or against other nations on a racial basis may take the form of war propaganda or propaganda aimed at keeping one or more racial groups in suppression. It may mask inequalities, injustices, exploitation and subversive activities.
Broader Propaganda (#PF1878) Compulsory indoctrination (#PD3097).
Related Censorship (#PC0067).
Aggravates Racism (#PB1047) Racial conflict (#PC3684)
Bias in children's literature (#PD4773).
Aggravated by Legalized racial discrimination (#PC3683).

♦ PD3094 Religious propaganda

Nature Incitement to religious hatred on an international or national level may be in the form of war propaganda or propaganda aimed at maintaining political control in the hands of a religious elite against other religions in the same society. It may use all kinds of propaganda media. It may cause war, terrorism, rioting and violence, and mask injustices and inequalities, exploitation and atrocities, inducing apathy and alienation through ignorance.
Broader Propaganda (#PF1878) Compulsory indoctrination (#PD3097).
Related Censorship (#PC0067).
Aggravates Religious conflict (#PC3292) Religious vilification (#PD5534).
Aggravated by Religious deception (#PF3495) Religious intolerance (#PC1808)
Religious indoctrination (#PD4890).

♦ PD3097 Compulsory indoctrination

Dependence on indoctrination

Nature The instilling of doctrine systems or pastiches of ideological, religious or political beliefs particularly in opposition to contrary creeds may result in intolerance and fanaticism; or in conformism and fear. It may sharpen international and political, religious or ideological conflict and encourage subversive activities and espionage. It may serve to strengthen dictatorship, government control, injustice, inequality and exploitation. Indoctrination techniques include the use of propaganda, brainwashing, censorship and other restrictions on freedom of expression and information, advertising, angled phrasing and contents of government and official information, with monopoly of the media. The most effective means of indoctrination are through parental influence and institutional education.
Counter-claim Some degree of indoctrination is inevitable in any education system, if only because learning has to start somewhere, time is too short for everybody to discover everything for themselves, and reason is not sufficient to establish in each person *'de novo'* the complex matrix of beliefs, values, and attitudes that make civilized life possible.
Refs Nagai, Michio *Education and Indoctrination* (1976); Winn, Kenneth *The Manipulated Mind; brainwashing, conditioning and indoctrination* (1983).
Broader Ideological conflict (#PF3388) Incorrect information (#PB3095).
Narrower Film propaganda (#PD3089) Book propaganda (#PD3090)
Art as propaganda (#PF3087) Racist propaganda (#PD3093)
Theatre propaganda (#PD3088) Religious propaganda (#PD3094)
Government propaganda (#PC3074) Photographic propaganda (#PD3086)
Political indoctrination (#PD1624) Religious indoctrination (#PD4890)
War and pre-war propaganda (#PD3092) Radio and television propaganda (#PD3085)
Forced participation in politics (#PD2910)
Promotion of negative images of opponents (#PF4133)
Propaganda by intergovernmental organizations (#PE3077)
Destabilizing international telecommunications (#PD0187)
Attitude manipulation of children through play (#PF2017).
Related Compulsory education (#PJ2615) Intellectual conflict (#PC3390)
Misleading information (#PF3096) Newspaper and journal propaganda (#PD0184)
Religious discrimination in education (#PD8807).

Aggravates Moralism (#PF3379) Prejudice (#PA2173)
Ethnocide (#PC1328) Extremism (#PB3415)
Propaganda (#PF1878) Ideological war (#PC3431)
Uncontrolled media (#PD0040) Ethnic discrimination (#PC3686)
Dependence on mysticism (#PF2590) Educating people to lie (#PE3909)
Denial of freedom of thought (#PF3217) Conflict between minority groups (#PC3428)
Inadequate ideological frameworks (#PD0065) Racial discrimination in politics (#PD3329)
Double standards of sexual morality (#PF3259) Underprivileged ideological minorities (#PC3325)
Disruptive secular impact of holy days (#PE7735)
Burdensome cost of religious ceremonies (#PF3313).
Aggravated by Nationalism (#PB0534) Monasticism (#PF2188)
Hero worship (#PF2650) Subversion of democracy (#PD3180)
Biased government information (#PF0157) Radio and television censorship (#PD3029)
Compulsory organization membership (#PD4098) Politically emotive words and terms (#PF3128)
Restriction of freedom of expression (#PC2162)
Restrictions on freedom of information (#PC0185)
Risk of unintentional nuclear war generated by the strategy of deterrence (#PF4162).
Reduces Social discrimination (#PC1864).
Reduced by Underground press (#PD2366).

♦ PD3101 Monopoly of the media

Nature Monopoly of media outlets such as newspapers, film, television, radio, and satellite broadcasting, may be on a national or international level. Monopoly may be commercial or governmental, in which case it may be used directly or indirectly for propaganda purposes. Commercial monopoly may restrict freedom of expression according to the tolerance of its advertising clientele, and may produce material only for the majority audience. Media monopoly may facilitate repression and political control, and the continuance of existing inequalities, injustices and exploitation.
Incidence States have been playing a growing role in orienting, controlling, organizing and dispensing communication activities dealing either with conditions for communication in a country (which is its main prerogative), or with the delivery of messages and contents (which is necessary in particular circumstances, but may also lead to restrictions). Thus, government responsibilities are, on the one hand, discharged through: legislation regulating rights and responsibilities in various fields of communication; the inclusion of resources for communication in overall planning; regulation of conditions governing media ownership and communication activities; attribution of facilities (such as distribution of frequencies) for telecommunications; control of communication channels and carriers; direct public ownership of media and other means of communication, and, on the other hand, through direct involvement in various communication activities, by creating national, regional and local public bodies; measures aiming to prevent the spread of distorted views and the abuse of communication practices; and limitation of imported contents and messages.
Broader Monopolies (#PC0521)
Restrictions on freedom of information (#PC0185).
Narrower Newspaper monopoly (#PE0246) International monopoly of the media (#PD3040).
Related Industrial gas monopolies (#PE1813)
State control of communications mass media (#PD4597).
Aggravates Denial of access to news (#PF3081)
Unethical media practices (#PD5251)
Restriction of freedom of expression (#PC2162).
Reduced by Industrial control (#PG5965).

♦ PD3102 Computer viruses

Program pollution — Software sabotage — Garbageware — Computer virus warfare
Nature Computer virus programs are programmed to attach themselves to some software and replicate themselves onto other systems and spread. Whether the program is a harmless nuisance or a disastrous command for deletion of data, the problem of eradication is the same: how to locate the extent of damage done and how to stop continuing damage. Various vaccine programs, program and diskette tests are very helpful where they are applied, but the nature of the problem is such that one does not know where they should applied until the damage has already been done.
Incidence Computer viruses can be found in any computer that has exchanged information or software with any other computer by means of telephone or other cable, disk, tape or other media, networks or bulletin boards. They have been discovered in personal, mini and mainframe computers in the United States, Europe and the Middle East. They have affected banks, air traffic control radar, insurance firms, hospitals, multinational corporations, military organizations, utility companies and stock brokers.
Background The notion of an electronic virus was born in the earliest days of the computer era. In fact, it was computer pioneer, John von Neuman who laid out the basic blueprint in a 1949 paper predating the first commercial electronic computers by several years. In 1959 three programmers at AT and T's Bell labs created an after hours recreation programme called Core Wars, the first virus like programme. The first subversive software programmes for personal computers were invented in the 1970s with Worm programs that travelled through a computer system occasionally announcing itself. Trojan Horses, the next type to emerge, had hidden agendas waiting for a specific date (like Friday the 13th) or hour or event when they would instruct the computer to destroy all the data on the disk. True Virus programs are the most advanced form of subversive code. Viruses attach themselves to some software and when ever possible they replicate themselves on other systems. Viruses then execute a command, such as deleting all files, sending a message to the screen, or modifying computer memories. Another type is called a Trap Door which collects users' passwords as the user logs on. The passwords will then give access to the Computer. Retroviruses are designed to reappear in systems after their memories have been wiped clean in an attempt to erase the virus. Other virus infect a computer's hardware, speeding up a disk drive, for example, so that it soon wears itself out. Particularly dangerous are bogus antiviral programs that are actually viruses in disguise and spread infection rather than stop it.
Broader Sabotage (#PD0405) Abuse of computer systems (#PD9544).
Related Electromagnetic pollution (#PD4172).
Aggravates Unreliability of computer software (#PE4428).
Aggravated by Computer-based crime (#PE4362)
Vulnerability of computer systems (#PE8542).

♦ PD3106 Lack of individualism in capitalist systems

Nature Capitalism claims to give the individual the greatest opportunity for developing his or her capabilities, and yet the creation of an artificial class, the bourgeoisie, to whom wealth accumulates, precludes the development of individual capabilities among the majority of people. This majority does not control the means of production, nor the price of exchange, and is at the mercy of employers and speculators. As science and technology develop machinery which can replace the labour force, the majority are pushed out of work and are unable to afford goods and services. This sharpens instability and class consciousness.
Broader Capitalism (#PC0564).
Related Liberalism (#PF0717).
Aggravates Anti-science (#PF2685) Capitalist speculation (#PC2194)
Abuse of science and technology in capitalism (#PE3105)
Counterproductive capitalist investment financing (#PF3104).

Aggravated by Social inequality (#PB0514).
Reduced by Paternalism (#PF2183) Collectivism (#PF2553)
Authoritarianism (#PB1638) Political pluralism (#PF2182).

♦ PD3109 Foreign control of natural resources

Denial of sovereignty over natural resources
Nature There are few means by which non-industrialized nations can participate in the decisions which affect the extraction of their resources. Paralysis or confiscation of the extraction system, while effective in some cases, is costly and dangerous.
Broader Production of non-essentials (#PC3651)
Denial of right to national self-determination (#PC1450).
Narrower Elitist control of production (#PD0154).
Related Erosion of sovereignty (#PE5015) Inflexible management patterns (#PF3091)
Industrial processes geared to reduced social needs (#PE3939)
Violation of sovereignty by trans-border broadcasting (#PE0261).
Aggravates Worldwide misallocation of resources (#PB6719).
Aggravated by Local control of resources (#PF5539)
Unregulated global resources (#PF3183)
Restrictions on the sharing of technical research (#PD3154)
Public non-accountability in control of production processes (#PE3780).

♦ PD3112 Alienation in capitalist systems

Nature Alienation is the fundamental contradiction in capitalism. It occurs primarily in the establishment of ownership and exchange value. These are basic alienations from necessity. The creation of an artificial non-productive elite follows, to whom all profits from artificially created demand accrue. As the capitalist system evolves, the alienation of the artificial class from the means of production, and therefore from the working class, increases and expands onto an international level in the form of economic imperialism or colonialism. In addition, the development of machinery increasingly alienates the worker from the act of producing and finally pushes him out of this role altogether, without reallocating to him any of the benefits of production that he formerly acquired.
Broader Capitalism (#PC0564) Alienation (#PA3545).
Aggravates Elitism (#PA1387) Class conflict (#PC1573).
Aggravated by Technocracy (#PF6330) Irresponsibility (#PA8658).

♦ PD3120 Denial of the right to social security in capitalist systems

Nature Because the basis of capitalism is free enterprise uncontrolled by governmental or community authorities, the right to work is not guaranteed and hence neither is the right to social security (unemployment, sickness and disability benefits, pensions). The capitalist ethic of profit motive as an incentive to production implies that any individual can support himself well if he only works hard enough. Therefore he should not be in need of state aid in times of distress and if he is in need of it, the implied reason is that he has been lazy or simply not smart enough.
Broader Capitalism (#PC0564) Social insecurity (#PC1867)
Denial of right to social security (#PD7251).
Related Inadequate social welfare services (#PC0834)
Social insecurity in developing countries (#PE4796).
Aggravates Elitism (#PA1387) Unemployment (#PB0750) Social isolation (#PC1707)
Irresponsibility (#PA8658) Economic conflict (#PC0840).
Reduced by Excessive government control (#PF0304).

♦ PD3122 Restriction of educational opportunities in capitalist systems

Claim Capitalism, in creating an artificial non-productive class elite, the bourgeoisie, to whom wealth and consequent opportunity accumulate, effectively denies the same opportunities to the majority, the working class. Even where education is provided free of charge, budget allocations are influenced by tax contribution considerations and effective political power, which only the wealthy have. Therefore there are likely to be insufficient and inferior schools in poor communities. This applies equally on the wider scale, among poor countries as opposed to rich countries. The developed countries sell technical know-how to the underdeveloped. Payment is made in resources and in profits from enterprises in the underdeveloped countries which return to the developed, thus depriving the former of the benefits.
Counter-claim Appropriate schooling is the joint responsibility of national and local educational authorities in most free-market economies. There is considerable variation and levels of quality achieved among the school system, but they are free to evolve and make choices concerning subjects taught, design of curricula and methods of teaching. State-control of education is abusive of the individual's right to know, to have access to information, and to think. The separation of school and state is as important as that of church (or ideology) and state. Education in underdeveloped countries, where there are shortages of teachers, equipment, funds and everything else except students, may have a brighter future owing to such technologies as computer-assisted learning which have been pioneered in non-socialist lands.
Broader Capitalism (#PC0564) Denial of right to education (#PD8102)
Limited availability of learning opportunities (#PF3184).
Aggravates Denial of education to minorities (#PC3459).

♦ PD3131 Budget deficits in developing countries

Dependence of developing countries on budget deficit financing through external borrowing
Nature Budget deficits in the developing countries are steadily increasing, due to developmental and energy costs on the one hand, and their foreign trade vulnerability on the other. Many developing countries are running balance-of-payments deficits while carrying the burden of large foreign debt as well. The unwillingness or inability of the industrialized world to help has made more noticeable the inadequate levels of cooperation among the less developed countries and regions.
Broader Imbalance of payments (#PC0998) Vulnerability of developing countries (#PC6189).
Related Deterioration in external financial position of developing countries (#PE9567).
Aggravates Budget deficit (#PD5492)
Excessive foreign public debt of developing countries (#PD2133).

♦ PD3132 Private domestic capital outflow from developing to developed countries

Capital flight — Outflow of indigenous capital from developing countries — Outflow of local investment
Nature Private residents in developing countries may take steps resulting in an effective transfer of domestic capital from the developing country to a developed country. Such flows tend to be associated with political events and uncertainties and, in some cases, inflation.
Incidence It is difficult to arrive at precise conclusions regarding the magnitude and trend of the flow of domestic capital out of the developing countries. In many countries, in which the inducements for outward movements of funds are powerful, residents cannot legally export capital without a licence (which the authorities are unlikely to grant). In such cases, any large outward movements of funds are likely to be concealed, taking the form, for instance, of non-repatriation of export proceeds, over-remittance for imports, or purchases of foreign exchange outside official markets from foreign tourists and others. Depending on the methods used in compiling the balance of payments, capital movements of this sort may not even be reflected in the residual item. One study by a New York bank, after examining the balance of payment accounts of 23 debtor nations, discovered that while those nations added $381.5 billion to their foreign debts,

$103.1 billion flowed back out as capital flight in the period from 1978 to 1983. In one case cited, the Philippines added on $19.1 billion in new debt as $8.9 billion left. In 1989, Mexico owed foreigners about $100 billion and had an estimated $84 billion in assets overseas. Venezuela has $58 billion is assets abroad, nearly twice its foreign debt. Argentina had $46 billion is savings outside its borders, Brazil had $31 billion. The president of Zaire and his clan, alone, are reported to have $5 billion invested in foreign bank accounts and real estate. A 1986 study by the Nigerian government showed that in 1978 corrupt officials were transferring $25 million a day abroad. An investigation by the government of Ghana in the early 1980's revealed that the country was losing at least $60 million a year through over–invoicing of imports.

Broader Speculative flight of capital (#PC1453)
Economic and financial instability of the world economy (#PC8073)
Outflow of financial resources from developing countries (#PC3134).
Narrower Non–repatriation of export proceeds (#PG6005).
Related Foreign private investment income outflow from developing to developed countries (#PE8957).
Aggravates Lack of economic and technical development (#PE8190)
Excessive foreign public debt of developing countries (#PD2133)
Decline in public sector savings in developing countries (#PE4574)
Deterioration in external financial position of developing countries (#PE9567).
Aggravated by Inadequate economic policy–making in developing countries (#PF5964)
Exploitative financial policies of transnational corporations in developing countries (#PE6952).

♦ PD3133 Gerontocracy

Monopoly of power by the elderly
Incidence Several asian socialist countries are run by gerontocracies, China, North Korea and Viet Nam are examples.
Broader Monopoly of power (#PC8410).
Narrower Superannuated religious gerontocracy (#PF3858)
Gerontocracy in developing country agriculture (#PD0947).
Aggravates Corruption of government leaders (#PC7587).
Aggravated by Political elitism (#PE3647).

♦ PD3135 Restrictions on foreign access to capital bond markets

Nature Many obstacles remain to the floatation of foreign bonds by developing countries on national capital markets. These include: ingrained market imperfections, restrictions on institutional portfolios, legal balance of payments difficulties, and low credit ratings for many developing countries.
Incidence Bond issues on international capital markets were once the main channel for foreign private investment. They are now much less important than direct investment, and even than export credits. The annual flow of gross savings in industrialized countries is very large compared to the current external requirements of developing countries, so that it is not a question of capital availability but of policy. Policy may dictate such discriminatory measures as prior permission needed to enter bonds by one or more regulatory agencies; excessive disclosure of information requirements; timing limitations on bond issues, and higher taxes or lower tax exemptions on foreign bond interests.
Broader Competition between states (#PC0114) Restrictive monetary practices (#PF8749)
Economic and social underdevelopment (#PB0539).
Narrower Imperfections of capital markets (#PF3136)
Legal impediments to foreign investment (#PD3063)
Insufficient creditworthiness of developing countries (#PD3054).
Related Domination (#PA0839).
Aggravates Capital shortage in developing countries (#PD3137)
Ineffective economic structures in industrial nations (#PE4818).
Aggravated by Imbalance of payments (#PC0998).

♦ PD3137 Capital shortage in developing countries

Nature Capital transfer remains a central problem of the economic growth of the developing world. The disparity between the demand and supply flow of capital is at variance with humanity's global needs. Foreign aid in the form of grants to developing countries has been declining, as have been multinational enterprise investments. The foreign debt problems of seven or eight countries has caused the tightening up of bank loans as well.
Incidence In the capital–importing developing countries in 1983 as compared to 1981, direct investment was down $5,100 million and long–term credits were down $14,100 million.
Claim The main block to development is lack of capital. Developing countries are poor because they are unable to save much, and because they cannot invest and produce much. They are therefore in a self–perpetuating cycle of poverty which can only be broken with th infusion of new capital.
Counter–claim There is quite enough capital in developing countries to establish the appropriate industries and infrastructure necessary to enable the poor majority to produce for themselves the things that they need for a low but reasonable living standard. That capital is currently being absorbed in part by foreign investors to produce highly inappropriate goods. From 85 to 90 per cent of the funds raised by foreign investors to invest in developing countries is raised within the Third World, usually as loans from Third World banks. Within developing countries the capital is usually held by a small privileged class which is uninterested in investing in ventures conducive to more appropriate forms of development of value to the poor majority. Such capital is rendered unavailable because it is lent to foreign investors, used in speculative ventures or to purchase more land or imported luxuries, or sent out of the country to secure foreign banks.
Refs Nurkse, Ragnar *Problems of Capital Formation in Underdeveloped Countries.*
Broader Obstacles to world trade (#PC4890) Insufficient financial resources (#PB4653)
Deficiencies of developing countries (#PC4094).
Narrower Inadequate channels for direct investments between developing countries (#PE4365).
Related Capitalism (#PC0564) Economic and social underdevelopment (#PB0539).
Aggravates Underdevelopment of industrial and economic activities (#PC0880).
Aggravated by Hoarding in developing countries (#PD1751)
Speculation in developing countries (#PD1614)
Non–productive use of cattle and livestock (#PD1802)
Restrictions on foreign access to capital bond markets (#PD3135)
Inadequacy of the domestic market in developing countries (#PD0928)
Decline in commercial bank lending to developing countries (#PE4655)
Deterioration in external financial position of developing countries (#PE9567)
Discrimination in lending by transnational banks in developing countries (#PE4310)
Reinforcement of inappropriate development by privileged classes in developing countries (#PF6670).

♦ PD3138 Decline in foreign direct investment in developing countries

Disincentives to foreign private investment in developing countries
Nature The investment climate in developing countries tends to discourage foreign direct investment. The poor investment climate is made up of the following elements: cumbersome administrative procedures and inefficient decision–making processes which companies encounter when planning an investment; transportation bottlenecks; unfair competition from local companies; lack of balance in policies toward foreign investment accompanied by politically explosive issues which may lead to disruptive and costly litigation between the government and foreign companies; high intrinsic risks often associated with such investment; restrictions on the repatriation of profits; ignorance of investment opportunities; and the limited nature of the markets for manufactured products. Because of the politically sensitive nature of such investment, and despite all the measures taken to stimulate it, direct investment has long been the least dynamic element in the flow of private capital to developing countries.
Incidence World wide flows of foreign direct investment tripled in the period 1984–87, but are concentrated in developed countries more than ever before. Total outflows from all countries increased 38 percent in 1985, 58 percent in 1986 and 44 percent in 1987. At the same time, developing countries' share of investment inflows declined from 27 percent of the total in 1981–83 to 21 percent in 1984–87, while developed countries, particularly the United States and Western Europe, increased their share of total inflows.
Refs United Nations Centre on Transnational Corporations *Trends and Issues in Foreign Direct Investment and Related Flows* (1985).
Broader Obstacles to world trade (#PC4890).
Aggravates Declining economic growth in developing countries (#PD5326)
Inadequate level of investment within developing countries (#PD0291).
Aggravated by Extraterritorial intrusion of jurisdiction (#PE3140)
Foreign intervention in internal affairs of states (#PC3185)
Disincentive to invest in heavily indebted countries (#PF9249).

♦ PD3139 Urban slums

Urban squatters
Nature Internal migration from countryside to the town creates social problems in the urban areas, particularly in the developing countries. The migrants often arrive at a faster pace than the cities are able to absorb them, and the new arrivals pile up in mushroom settlements made of the flimsiest materials, sometimes without any form of municipal administration or public services. Living conditions in these settlements are often materially worse than in the villages from which the migrants came. This concentration of unassimilated migrants tends to encourage juvenile delinquency, adult crime, vice, alcoholism, gambling, mental disorders, and political instability. Children of the slums are both materially and emotionally disadvantaged and underprivileged.
Incidence The population of the world's cities has doubled in the last thirty years and will double again in the next twenty. In the Third World, the result has been skyscrapers of steel and glass surrounded by slums of mud and wood. In both developed and developing countries the last decade in particular saw the rapid growth of unconventional urban settlements – squatter areas, slums, and, of less importance, mobile home parks. These represent the inability of human settlements to house population growth in terms of permanent accommodation at reasonable standards.
Refs Gaskell, Martin *Slums* (1989); Walker, Mabel L *Urban Blight and Slums* (1971).
Broader Urban poverty (#PC5052) Inadequate housing (#PC0449).
Narrower Urban slums in developing countries (#PD3489)
Urban slums in industrialized countries (#PE1887).
Aggravates Crime (#PB0001) Youth gangs (#PD2682) Deculturation (#PJ1034)
Urban overcrowding (#PC3813) Inadequate health care in urban slums (#PE7877).
Aggravated by Family poverty (#PC0999) Housing shortage (#PD8778)
Socio–economic poverty (#PB0388) Uncontrolled urban development (#PC0442)
Unethical real estate practice (#PD5422) Denial of right to sufficient shelter (#PD5254)
Unethical practices in local government (#PD5948)
Illegal occupation of unoccupied property (#PD0820)
Adjustment difficulties of new urban families (#PF1503)
Unfeasible housing alternatives in urban areas (#PE8061).

♦ PD3142 River ice

Ice runs
Nature Some rivers in the high latitudes of the northern hemisphere do not ice over completely in winter to form a stable ice cover, usually due to their high water velocities. Ice may be formed due to low temperatures, but it breaks away and large quantities move downstream, tending to damage bridge structures, producing ice jams and consequent flooding, particularly when such jams break. Ice moving in this way may causes much damage to hydroelectric installation intakes.
Broader Ice accretion (#PD1393).
Related Coastal erosion (#PE6734) Ice–blocked seaways (#PD2498).
Aggravates Floods (#PD0452).

♦ PD3154 Restrictions on the sharing of technical research

Nature Technical research may be restricted to particular nations, particular corporations, or both. Researchers are often not allowed to share their work with others in their field. These restrictions block common information exchange and hamper those trying to coordinate resource usage on a global basis.
Aggravates Foreign control of natural resources (#PD3109)
Disadvantageous terms for technology transfer (#PE4922).

♦ PD3155 Extermination of wild animal natural prey

Nature Human activities may directly or indirectly result in the elimination or extermination of the natural prey of a carnivorous wild animal, thus depriving it of its normal diet.
Broader Killing of animals (#PD8486).
Related Denial to animals of the right to freedom from mass killing (#PE9650).
Aggravates Endangered species of animals (#PC1713)
Endangered species of carnivores (#PD3482).
Aggravated by River pollution (#PD7636) Vulnerability of ecosystem niches (#PC5773).

♦ PD3156 Endangered species of amphibia

Endangered species of frog — Endangered species of toad — Endangered species of salamander
Incidence Due primarily to the impact of man on the natural environment, whether directly or indirectly, many of the 2,000 species of amphibia are in danger of extinction. Populations of frogs, toads and salamanders appear to be declining in many places around the world (including North, Central and South America, Europe, Asia, Africa and Australia), even from wilderness areas apparently unexposed to levels of pollution characteristic elsewhere. This has been variously attributed to acid rain, to pesticides, viruses and to adverse weather conditions. Amphibia are especially vulnerable because they have two life stages, in and out of water, and thus come into contact with a wide variety of substances. Since their skins are porous, heavy metals and other contaminants in soil or water can easily pass into their bodies, as with the chemical residues in the many creatures they eat.
Claim Frogs play an important role in the control of certain crop pests. Frogs, as with the traditional role of the canary in mines, may prove to be an indicator of the increase in the level of water pollution beyond a threshold at which the life of many species can be sustained over long periods.
Broader Endangered species of animals (#PC1713).
Aggravated by Drought (#PC2430).

♦ PD3158 Endangered species of echinodermata

Marine invertebrates
Nature The 5,700 species of echinodermata are marine invertebrates common in temperate or

tropical zones. They play a very important role in marine ecology by changing the shore-line, removing decaying organic matter, regulating the production of mollusks, and producing microscopic larvae which constitute an important food element for slightly larger plankton organisms. It is not known how many of these species are endangered by different forms of marine pollution.
Broader Endangered species of invertebrates (#PD7513).
Narrower Endangered species of starfish (#PG6028)
Endangered species of sea urchins (#PG6029)
Endangered species of brittle stars (#PG6030)
Endangered species of sea cucumbers (#PG6031)
Endangered species of feather stars (#PG6032).
Aggravated by Marine pollution (#PC1117).

♦ PD3159 Endangered species of aschelminthes
Nature The 12,000 species of small, pseudocoelomate wormlike animals which make up the phylum of aschelminthes include a number of lower classes, some of which are themselves often given phylum rank: gastroticha (140 species), rotifera (1,500 species), echinoderida (100 species), priapulida (5 species), nematomorpha (250 species), and nematode or roundworms (10,000 species). It is not known how many of these species are endangered by different forms of water and marine pollution.
Broader Endangered species of invertebrates (#PD7513).
Narrower Endangered species of Rotifera (#PG6034)
Endangered species of Roundworms (#PG6033).
Aggravated by Marine pollution (#PC1117).

♦ PD3160 Endangered species of annelida
Segmented worms
Nature The 7,000 – 8,000 species of segmented worms constitute a major phylum of the animal kingdom which is divided into three classes. Marine worms (5,000 species) play an important role in turning over much of the marine sea-floor sediment and are important in the region of domestic outfall sewers. Earthworm (3,250 species) turn over and aerate the surface soil. Leeches form the third class. It is not known how many of these species are endangered by different forms of pollution.
Broader Endangered species of invertebrates (#PD7513).
Narrower Endangered species of marine worms (#PG6035).
Aggravated by Leeches as pests (#PE3660).

♦ PD3169 Communist closed society
Nature The frontiers of certain communist countries are completely closed to foreign entry, and exit of nationals to non-communist countries is illegal. There is usually heavy censorship. Movement of people and information between communist countries is also restricted. Despite the doctrine of the withering away of the state, the state is the focus for all effort and ideology. Internationalism in any form may be banned. The policies of individual states may lean towards those of a dominating country within a communist world system, which is in effect an extension of the national closed societies.
Broader Communism (#PC0369) Contradictions in communist systems (#PF3179).
Related Denial of freedom of movement in communist systems (#PC3173).

♦ PD3171 Political prisoners in communist systems
Nature Most communist regimes will not permit ideological or political dissent, even if it is in the guise of another form of communism. Those who do dissent may be imprisoned, tortured, committed to hard labour or executed; or classified as insane, brainwashed and politically reindoctrinated. Their property may be confiscated and their characters defamed. Imprisonment may be without trial or following a show or secret trial, with forced confessions or false evidence. Families of political prisoners may suffer severe deprivation and ostracism; and political arrest may result from a defamation of character to the authorities by a rival for reasons of ambition, revenge or fear. Political imprisonment induces apathy, fear and alienation which may affect social and economic development. Alternatively, it may provoke resistance, often with foreign support and indirect pressure.
Broader Communism (#PC0369).
Narrower Civilian political prisoners and detainees (#PD3015).
Related Denial of rights to prisoners (#PD0520).
Aggravates Internment without trial (#PD1576).
Aggravated by Political corruption of the judiciary (#PE0647).

♦ PD3172 Censorship in communist systems
Nature Despite the claim of communism to be fully democratic and to be working with the wholehearted support of the people, communist governments still feel the need to suppress any political and ideological dissent from the standard party line. All media is subject to censorship; it may be confiscated and banned if it does not conform, and those responsible may be arrested. The gap left by the absence of free expression is filled with government propaganda and official information aimed at total indoctrination. Non-political matters such as scientific theory and art forms may also be censored or trimmed to conform with party policy, which may be unimaginative and anti-innovative. Effective censorship requires domestic intelligence surveillance, which may be aided by informants, and may give rise to underground and subversive measures. This situation encourages foreign influence and foreign propaganda (usually put out by radio broadcast) which may be inflammatory and cause international conflict. The current policy of "glasnost" in the Soviet Union is an attempt to counter the most dangerous tendencies of decades of censorship, but the new policy itself illuminates the fact that the citizenry's responses, even to when and how they may dissent, are controlled by the party.
Incidence In 1983 in China, for example, one magazine was suppressed, and two key persons were purged from the 'People's daily' because that newspaper had suggested that the Chinese might be 'alienated' in their own society.
Refs Choldin, Marianne T *A Fence Around the Empire* (1985); Schopflin, George *Censorship and Political Communication in Eastern Europe* (1983).
Broader Communism (#PC0369) Censorship (#PC0067)
Denial of freedom of expression and thought in communist systems (#PC3174).
Related Self censorship (#PF6080).
Aggravates Conflict of information (#PF2002).

♦ PD3180 Subversion of democracy
Communist subversion in capitalist and neutral countries
Nature Communist systems must use violent and subversive methods of persuasion to gain support for their cause in capitalist and neutral countries, even among the working class. Communist agitators tend to be an elite minority and if their activities are fully successful they will carry this elitism into their communist system. Subversive activities include propaganda and indoctrination methods, terrorism and guerrilla warfare, incitement to strike, use of espionage, and economic and military aid.
Broader Communism (#PC0369) Ideological conflict (#PF3388).
Narrower Underground press (#PD2366).
Aggravates Repression (#PB0871) Propaganda (#PF1878)
International insecurity (#PB0009) Limitations of democracy (#PF6608)
Compulsory indoctrination (#PD3097).
Aggravated by Double standards in morality (#PF5225).
Reduced by Subversion of socialism (#PF9485).

♦ PD3182 Limited ownership of productive systems
Nature Ownership of the means of production for the bulk of small businesses is held by an individual or a family. Nationalized companies in theory are owned by all of the citizens of the country, in fact, all of the responsibility of ownership is in the hands of bureaucrats. Corporations owned by stock holders offer some control by the owners but these owners are for the most part other corporations. Employees do not have enough say in the strategic decisions of the businesses for which they work. The level of motivation and creativity is lower than necessary and the productivity of the business is not as high as it could be.
Broader Unregulated ownership of the means of production (#PF2014).

♦ PD3190 Conflicting social service ideologies
Nature The cost of, and time spent in, arguments on the question of which services society should guarantee its members and which are the province of the individual, have resulted in the provision of social services being sacrificed in many countries. The debate includes questions of society's intent in providing services, which social levels most effectively administer which services, and the differences in cultural preferences. While this debate continues, social services are fragmented and inconsistent.
Broader Breakdown in community security systems (#PD1147).
Narrower Social service inconsistencies (#PF2992)
Non-standardized social services (#PF2974)
Misallocation of resources to protect the aged (#PE0648)
Collapsed tension between care and responsibility (#PF5555).
Related Limited image of employability (#PF2896)
Uncontrolled application of technology (#PC0418)
Unsystematic allocation of market facilities (#PD3507).
Aggravates Neglect of the aged (#PD8945).

♦ PD3197 Inadequate working conditions for women
Refs ILO *Conditions of Work, Vocational Training and Employment of Women* (1979).
Broader International imbalance in the quality of life (#PB4993).
Narrower Work practices requiring women to lift heavy loads (#PE4452).
Aggravates Overworked women (#PD7762).

♦ PD3210 Inequitable access to resources
Nature Inequities in access to resources contribute to ecological imbalance: they may slow social and economic development in resource exporting countries; and the overemphasis on profit by government and business agencies leads to the uneven development of resources.
Incidence In particular, hydrocarbons and ores are extracted without considering long range availability.
Broader Nationalistically determined development of natural resources (#PD3546).
Narrower Limited access to external resources (#PF1653).

♦ PD3213 Totalitarian democracy
Nature Totalitarian democracy claims to reduce social and economic inequalities but requires the exerting of extensive control over these areas in order to achieve its aims. Political democracy in the sense of apparent popular participation in decision-making on societal issues, may be arranged by manipulation of voting and election systems.
Incidence For example, in Turkey in 1983, the military junta manipulated elections and defended its concept of a guided democracy, although it had already behind it several years of rule with mass arrests, detentions and executions of political enemies of the state, and other acts of totalitarian repression.
Broader Totalitarianism (#PF2190).
Narrower Enforced participation in community activity (#PD3386).
Related Political party manipulation of elections (#PD2906).
Aggravates Apathy (#PA2360) Conformism (#PB3407).
Aggravated by Excessive government control (#PF0304).

♦ PD3218 Property and occupational discrimination in politics
Nature Property qualifications for voting are no longer widespread but still exist in certain countries, especially at the local government level. In some countries or territories white people of any class may vote, but there are high property or tax-paying qualifications for non-whites. Persons holding property may be entitled to vote both in the locality of their residence and that of their property. Wealthy individuals may use their economic power to exert pressure for the candidate of their choice. Certain categories of occupations may be denied the right to vote, such as those of 'clerical status', public servants such as soldiers and policemen.
Broader Discrimination in politics (#PC0934) Limitations on right to vote (#PF2904)
Unethical practices in politics (#PC5517).
Aggravates Class conflict (#PC1573) Denial of education to minorities (#PC3459).
Aggravated by Discriminatory professionalism (#PC2178).

♦ PD3223 Language discrimination in politics
Nature Citizens of a country may be denied the right to vote because they cannot pass a literacy test in a language which is not their mother tongue. This requirement gives advantages to members of the dominant language group and serves as a way to exclude non-members from participation in government. Language tests for immigration may be more lenient than those for voting. This kind of discrimination may affect naturalized immigrants or members of a plural society.
Incidence Constitutions of multilingual societies such as Botswana, Gambia, Ghana, Guyana, Kenya, Lesotho, Sierra Leone and Uganda require the command of English for election to the central legislative; while French is mandatory in Cameroun, Congo, Mali, and Niger.
Counter-claim So many languages are spoken in the world (upwards of 2,500) and so many language groups are so small (consisting perhaps of only a few hundred persons) that they can scarcely all have equal status. Sufficient ground for differentiation exists, so differentiation is inevitable.
Broader Discrimination in politics (#PC0934).
Related Discriminatory communication (#PD6804).
Aggravates Tribalism (#PC1910) Cultural fragmentation (#PF0536)
Denial of political rights (#PD8276)
Discrimination against immigrants and aliens (#PD0973)
Denial of right to recognition as a person before the law (#PE4716).
Aggravated by Illiteracy (#PC0210) Language conflict (#PG6098)
Prejudice against other languages (#PD8800)
Political discrimination based on illiteracy (#PC3222)
Discrimination against use of accents of a language (#PE5141).
Reduced by Divisive effects of official cultural pluralism (#PF0152).

♦ **PD3224 Denial of the right of association**
Denial of right of association — Denial of freedom of association
Nature Associations and assemblies founded and organized by groups for purposes that a government considers to be subversive (such as political, trade union, or occult associations) may be suppressed, thus maintaining existing inequalities, exploitation and elitism. Where such associations are formed, official discovery may lead to imprisonment and torture.
Refs ILO *The Right to Organize* (1977).
Broader Denial of political and civil rights (#PC0632).
Narrower Banned associations (#PD3536) Compulsory organization membership (#PD4098) Denial of the right of trade union association (#PD0683) Infringement on the functioning of legitimate organizations (#PE5222).
Related Denial of right to life (#PD4234) Denial of freedom of thought (#PF3217) Legalized racial discrimination (#PC3683) Enforced collectivization of agriculture (#PD7443).
Aggravates Apartheid (#PE3681) Conspiracy (#PC2555) Social inequality (#PB0514) Denial of right of assembly (#PC2383).
Aggravated by Strikes (#PD0694) Social injustice (#PC0797) Political repression (#PC1919) Restrictive practices (#PB9136).

♦ **PD3225 Deprivation of nationality**
Nature The rescinding of a person's nationality deprives him of his political and social rights and makes him stateless. In certain countries a woman's nationality may be rescinded if she marries a foreigner, a man's if he serves in foreign armed forces; or the nationality be rescinded of anyone who has voted in a foreign country, who has become the naturalized citizen of a foreign country, or who has not complied with laws concerning dual nationality.
Related Refusal to grant nationality (#PF2657) Restrictions on recognition of nationality (#PE4912).
Aggravates Refugees (#PB0205) Statelessness (#PE2485) Rejection of refugees (#PF3021) Discrimination in politics (#PC0934) Refusal to issue travel documents, passports, visas (#PE0325) Restrictions on freedom of movement between countries (#PC0935).
Aggravated by Social injustice (#PC0797).

♦ **PD3228 Non-violent political revolution**
Bloodless revolution
Nature Political revolution may be achieved by peaceful means without recourse to arms, and may be achieved constitutionally or unconstitutionally. The fact that political revolution is bloodless may not make it any more popular, as it faces the same problems as any other kind of political revolution; it may indicate that very little has in fact changed, simply the person in command.
Broader Political revolution (#PF3237).
Related Violent revolution (#PC3229) Violent political revolution (#PD3230).
Aggravated by Political instability (#PC2677).

♦ **PD3230 Violent political revolution**
Nature The use of armed forces or terrorist activities to overthrow an existing political regime may lead to repression. This in turn may cause social or national disintegration, counter-revolution and foreign intervention, or civil and guerrilla warfare.
Broader Violent revolution (#PC3229) Political revolution (#PF3237).
Related Overthrow of government (#PD1964) Non-violent political revolution (#PD3228).
Aggravates Militarism (#PC2169) Counter revolution (#PF3232).

♦ **PD3238 Political oligarchy**
Nature Political dictatorship, embodied in a ruling clique or military junta, maintains its rule through intimidation and repression, indoctrination and economic control. Political oligopoly tends to encourage existing social inequalities and is in itself elitist. It halts political and social development and is inherently unstable since members may form new cliques and oust the original one.
Broader Dictatorship (#PC1049).
Related Political dictatorship (#PC0845) Disproportionate influence on national economies of limited number of corporations (#PE1922).
Aggravates Elitism (#PA1387) Lack of political development (#PD8673).
Aggravated by Political instability (#PC2677) Undemocratic political organization (#PC1015) Fraudulent nature of inherited titles (#PE5754).

♦ **PD3239 Dictatorship of the majority**
Dictatorship of the proletariat
Nature Majority control in decision-making without taking account of minority interests or rights may lead to conformism and may involve social or political dictatorship, or take the form of forced assimilation of cultural and other groups. Dictatorship of the majority may be nominal, embodied in the policies of a dictator or ruling clique with indirect approval from the majority. It may take the form of totalitarianism.
Broader Conformism (#PB3407) Dictatorship (#PC1049) Collectivism (#PF2553).
Related Minority control (#PF2375) Anti-intellectualism (#PF1929) Political dictatorship (#PC0845) Manipulative use of referenda (#PF2909) Discrimination against minorities (#PC0582).
Aggravates Social fragmentation (#PF1324).
Aggravated by Populism (#PF3410) Forced assimilation (#PC3293).
Reduces Individualism (#PF8393) Lack of participation in development (#PF3339).
Reduced by Conflict between minority groups (#PC3428) Undemocratic political organization (#PC1015).

♦ **PD3241 Social dictatorship**
Nature Dictatorial leadership of social groupings, involving minority control or dictatorship of the majority, may be ethnic, professional or cross-cultural in character. Social dictatorship may lead ultimately to political dictatorship if the group in question is a predominant national community. It may lead to social conflict and disintegration.
Broader Dictatorship (#PC1049) Social injustice (#PC0797).
Aggravates Social conflict (#PC0137) Political dictatorship (#PC0845).
Aggravated by Social discrimination (#PC1864).
Reduced by Failure of individuals to participate in social processes (#PF0749).

♦ **PD3244 Emotional dependency in marriage**
Incidence Emotional dependency in marriage is conditioned by social attitudes and upbringing in most parts of the world.
Claim Emotional dependency in marriage is characterized by fear, which may be expressed in a number of detrimental ways, such as hostility, domination or submission. It is a result of emotional immaturity and since the couple may mature differently and at different rates, conflict may arise later. One partner may become too assertive and if the other does not submit, he or she may seek satisfaction outside the marriage and contribute to its breakdown.
Counter-claim The idea of marriage itself connotes dependencies of various kinds – financial, physical, intellectual, social and emotional. These "dependencies" may simply be the healthy expression of the need and ability to share.
Refs Thorkelson, Lori *Emotional Dependency* (1984).
Broader Human dependence (#PA2159) Marital instability (#PD2103) Negative effects of the nuclear family (#PF0129).
Related Immaturity (#PF8413) Insecurity (#PA0857) Lack of intimate relationships (#PF4416) Dependency of women in marriage (#PD3694).
Aggravates Separation under marriage law (#PF3251).
Aggravated by Dependency of children (#PD2476).

♦ **PD3245 Economic bias of worker benefits**
Claim Both labour and management stress the necessity of economic benefits for employees to such a degree that needs for other benefits may be overlooked. Employment levels suffer from the high cost of maintaining employees, but at the same time, employee well-being suffers for lack of attention to human needs beyond the pay cheque.
Broader Inflexible management patterns (#PF3091).
Related Insufficient job benefits (#PJ7915).

♦ **PD3246 Excessive extralegal powers**
Nature Highly placed political figures always can draw upon greater powers than they are legally entitled to. Royalty and heads of state are permitted their positions, finally, on the basis of trust by the citizenry (or disinterest, whichever is stronger). The result is that where there exists a leader, there exists the possibility of their grabbing power before the legal systems may take effect to stop them.
Broader Abuse of government power (#PC9104).
Related Confusing official responsibility (#PU0186).

♦ **PD3255 Denial of parental affiliation**
Denial of filiation — Denial of paternity — Confusion of paternity
Nature The inability to establish legal filiation (the right and responsibility of parenthood) over an illegitimate child arises in certain countries where filiation of children born out of wedlock is left to the father's discretion. Maternal affiliation is not automatically established by the sole fact of giving birth to the child; and without the acceptance of the father, the mother may not be legally permitted to assume filiation over the child. In other countries, the establishment of paternal affiliation of children born as a result of adulterous or incestuous associations is not allowed.
Incidence This problem occurs particularly in countries where the law provides only one status for parents, whether married or not, regarding rights and responsibilities. The difference between legitimate and illegitimate children regarding legal status then rests on filiation.
Refs Bryant, N J *Disputed Paternity* (1980).
Broader Unmarried parents (#PD3257) Illegitimate children (#PC1874) Narrow legal definition of the family in developing countries (#PD1501).
Narrower Denial of right to inherit property (#PF0886).
Aggravates Paternal negligence (#PD7297) Invasion of privacy through testing (#PJ6946).
Aggravated by Group marriage (#PF3288) Unmarried mothers (#PD0902) Double standards in morality (#PF5225) Double standards of sexual morality (#PF3259) Discrimination against illegitimate children (#PD0943).

♦ **PD3256 Discrimination against unmarried fathers**
Nature The legal status of unmarried parents in certain countries is inferior to that of married parents, and their children's rights are therefore restricted. Lack of child-care facilities may make it difficult for unmarried fathers to earn an adequate living, and in certain countries the legal status of unmarried parents is weighted in favour of women.
Refs Oakland, Thomas P and Terry, Edwin J *Divorced Fathers* (1983).
Broader Unmarried parents (#PD3257) Single parent families (#PD2681) Discrimination against men (#PC3258).
Related Unmarried mothers (#PD0902).
Aggravates Unequal distribution of social services (#PC3437).
Aggravated by Divorce (#PF2100) Desertion in marriage law (#PF3254) Discrimination in family planning facilities (#PD1036).

♦ **PD3257 Unmarried parents**
Nature The legal status of unmarried parents in certain countries is inferior to that of married parents, and the rights of their children are therefore impaired. Where the unmarried parents are also single there are additional emotional and financial problems. Filiation may be denied to parents under the law of certain countries.
Broader Immorality (#PA3369) Single parent families (#PD2681).
Narrower Unmarried mothers (#PD0902) Denial of parental affiliation (#PD3255) Discrimination against unmarried fathers (#PD3256).
Related Discrimination against illegitimate children (#PD0943).
Aggravated by Promiscuity (#PC0745) Lack of family planning (#PF0148) Inadequate sex education (#PD0759) Desertion in marriage law (#PF3254) Inadequate training in decision-making (#PD2036).

♦ **PD3262 Sexual exploitation of women**
Nature Women are often regarded as sex objects to gratify man's sexual desire, and in this respect women may fall victim to unwanted sexual advances from men. Force, sometimes fatal, may be used against a woman by a man in his attempt to force himself upon her: there are laws to deal with such offences but the intricacies of law enforcement frequently prove inadequate or fail completely, and there is often laxity in dealing with the sexual offences.
The exploitation of women for sexual purposes results in slavery, discrimination against women in marriage and generally before the law and in employment. It limits women's freedom indirectly as well as directly and also includes violations such as rape and other sexual assaults. Blackmail and intimidation of women is used by pimps to obtain and keep ascendancy over prostitutes.
Incidence Discrimination against women encourages global sexual exploitation, which is enhanced by double standards of morality and lack of sex education. Among the many aspects of sexual exploitation of women, sex tourism is a growing form of exploitation of Third World women as more and more agencies advertise "pleasure" vacations – particularly to Thailand. Approximately 1.5 million tourists visit Thailand every year, many for the sex alone.
Refs Barry, Kathleen *Female Sexual Slavery* (1986); Perpiñan, Mary Soledad *International Meeting of Experts on the Social and Cultural Causes of Prostitution and Strategies Against Procuring and Sexual Exploitation of Women, Madrid, 1985* (1985).
Broader Sexual exploitation (#PC3261) Exploitation of women (#PC9733).
Narrower Concubinage (#PF2554) Female prostitution (#PD3380).
Related Bride-price (#PF3290) Sexual exploitation of men (#PD3263) Sexual exploitation of children (#PD3267).
Aggravates Miscegenation (#PC1523) Unmarried mothers (#PD0902) Paternal negligence (#PD7297) Double standards of sexual morality (#PF3259).
Aggravated by Trafficking in women (#PC3298) Inadequate sex education (#PD0759).

♦ **PD3263 Sexual exploitation of men**
Nature Men may be exploited in sexual activity or as a result of it. Sexual feelings are played on by advertizing. Unethical business practices may include the offer of a prostitute as an inducement to a prospective buyer or this may be expected on the part of the client. Exploitation also includes the blackmail of men concerning an embarrassing sexual private life or experience (generally in

compliance with prostitutes, male or female). This may lead to the leaking of state secrets if the man is in a high governmental position. Male prostitutes will often intimidate or assault their clients. Boys may be sexually exploited by homosexuals. Male prostitutes may be exploited by pimps, and also lose social status through their profession.
Incidence Sexual exploitation of men is universal but appears to be particularly widespread in 'Western' countries where homosexuality is taboo and advertizing licence and general permissiveness exists.
Broader Sexual exploitation (#PC3261).
Narrower Male prostitution (#PD3381).
Related Sexual exploitation of women (#PD3262)
Sexual exploitation of children (#PD3267).
Aggravates Social stigma (#PD0884) Double standards of sexual morality (#PF3259).
Aggravated by Discrimination against homosexuals (#PE1903).

♦ PD3266 Rape
Sexual assault — Rape pregnancy
Nature Sexual assault involving the forcing of a person to submit to sexual intercourse may be committed on men, women, boys, or girls of almost any age, but statistics show 16 year-old girls to be the most frequent victims. The violation and physical injury caused by rape may lead to death, particularly with very young victims, but this is rare. It may lead to murder because of the assailant's fear of discovery, or to physical injury and mental disorder.
The legal aspect of rape differ between legal systems. Some judicial systems it is virtually not recognized as a crime. Some systems require penetration and some force. Many countries are defining rape as sexual intercourse against the will of the victim whether by force or not. Intercourse with a sleeping, drunk or drugged woman is considered rape, as is intercourse which is consensual where such consent was induced by fraud; such as impersonating her husband or deceiving the victim as to the nature or quality of the act. Because stereotypes of women, women's role in the sex act and women's behaviour during rape, women have difficulty proving an act of rape in nearly all legal systems.
Generally the law distinguishes between rape and sexual assault. Rape involves penetration, however, slight. Sexual assault is any touching of sexual or other intimate parts of the person for the purpose of sexual arousal or gratification.
From the perspective of the victim, their bodies have been assaulted, their integrity, dignity and self-determination has been violated. Rape victims sometimes become pregnant or infected by veneral disease. Many experience depression, guilt, diminished interest in sex, breakup of relationships, obsessive concern for safety, and loss of trust. Studies show that divorces and suicide attempts are fairly common after rape. Many times the men in a victim's life, partners, relatives and work-mates have problems coping with the attack; problems which affect the woman's own recovery. American studies have shown that a rape tends to end an existing relationship.
A man does not suddenly lose control and commit a rape. His anger, hatred, inadequacy, violence or anti-social behaviour is already manifest; by the time a man is convicted for sexual assault he will already have committed several previous offences.
Another aspect of rape is the social believes surrounding it. The false but widely held myths about rape include: women enjoy rape; you can't rape your wife; rapists are insane; only bad women get raped; rapists have an uncontrollable sex drive; women use the charge of rape to get at innocent men; women ask for rape by the way the dress and act; and it is not rape if the woman does not defend herself.
Incidence Rape remains one of the most misunderstood and underreported crimes. The documentation on the extent to which women have suffered rape historically and in the present is only beginning. Since no more than half of all reported rapes are the work of strangers, a great deal of rape happens in family and familiar community settings. Rape of young children by fathers, male relatives and family friends is being increasingly reported to the authorities. Every investigation of rape undertaken so far is uncovering far more rape experience than expected among women who range from toddlers to old women. When comparable investigative and reporting procedures are established in all countries, it may turn out that rape is an almost universal experience for females. The incidence of rape is particularly high under war conditions.
According to recent surveys in the USA, only 3.5 to 10 percent of rapes are reported. At one time the social stigma on rape victims discouraged all but the most courageous from reporting the crime; as this stigma begins to lessen, victims are beginning to use the legal system, reporting their cases and seeing them through to prosecution. The number of reported rapes has steadily risen, rising 35 percent to 99,146 in 1981. Although most authorities feel there is some increase in the actual number of rapes, including gang rapes and rapes of children and men, most of the rise is attributed to more women reporting the attacks on them. Statistics indicate that approximately 25 percent of women report that they were in some way sexually 'abused' by an adult male before they reached the age of 18. In addition, as adults, women face a one in four risk of being raped – by acquaintances as much as by strangers. If marital rape were to be included in those statistics, the percentage would be far higher.
Refs Carter, John M *Rape in Medieval England* (1985); Clark, Lorenne and Lewis, Debra *Rape*; Drake, Marie R *Rape* (1988); Ellis, Lee *Theories of Rape* (1989); Groth, A N *Men Who Rape* (1979); Hursch, Carolyn J *The Trouble with Rape* (1977); Roberts, Catharine *Women and Rape* (1989); Williams, Joyce E and Holmes, Karen A *The Second Assault* (1981).
Broader Assault (#PD5235) Sexual offences (#PD4082) Sexual violence (#PD3276)
Violence against women (#PD0247).
Narrower Gang rape (#PG5670) Unreported rape (#PE5621)
Undesired sexual obligations (#PF4948).
Related Fornication (#PF5434) Sexual intercourse with minors (#PE6522)
Sexual exploitation of children (#PD3267).
Aggravates Infamy (#PB8172) Inadequate assistance to victims of rape (#PE4449)
Children engendered by occupying soldiers (#PD8825).
Aggravated by Sexual unfulfilment (#PF3260) Inadequate riot control (#PD2207)
Double standards of sexual morality (#PF3259).

♦ PD3267 Sexual exploitation of children
Child sex rings — Organized sexual abuse of children — Ritual sexual abuse of children — Satanic child abuse of children
Nature The types of sexual exploitation of children are not reducible to a common denominator; they differ in severity and in significance. Children may be the victims of sexual violence, rape, seduction or simply fondling by adults or other children. In certain tribes fondling by adults as a form of masturbation is common and socially accepted, but in more sophisticated societies, sexual exploitation of children is unacceptable because of the risk of their adverse sexual development. The identification of assailants (usually male) is often complicated by secrecy on the part of children and parents and the adoption of a very 'respectable' demeanour by the assailant himself.
Incidence Whether in Barcelona, Bombay, Macau, Singapore, Amsterdam, Hamburg, Paris, Marseilles, New York, Mexico City or other cities, the market for pornography, with the help of video techniques, has provided a very extensive scope for expansion of the exploitation of the sexuality of children, which is finding increasing outlets in Europe and North America, and more recently in the oil-producing countries. In 1989 there were increasing reports of national and international networks specializing in the overlapping concerns of child pornography and procurement of children for sex or for ritualistic sexual abuse. Evidence exists for an international child abuse network based in Amsterdam, involving senior politicians, head teachers, lawyers, and child care professionals. There are many reports in the UK of child abuse in assessment centres, secure units and children's homes. There are also many examples of those responsible for such units procuring children in their care for abuse by others. In the UK an allegation by one boy led to the discovery of 643 offences against 45 children.
Counter-claim It is misleading to suggest that child abuse in general is linked to satanic cults. The limited number of much publicized cases of ritual abuse tend to involve molesters who use ritualistic trappings. None of these people has been proven to be a practising occultist, rather than perverted individuals (or imaginative psychiatric patients) using dramatic rituals for sexual excitement, as well as to impress and terrify their victims. The satanic abuse reports were initiated by fundamentalist groups in the USA as part of their well-documented campaign to discredit other spiritual paths.
Refs Cook, Mark and Howells, Kevin *Adult Sexual Interest in Children* (1981); Goldstein, S *Sexual Exploitation of Children* (1986); Kempe, C Henry and Kempe, Ruth *The Common Secret* (1984).
Broader Sexual deviation (#PD2198) Sexual exploitation (#PC3261)
Exploitation of children (#PD0635).
Narrower Incest (#PF2148) Child pornography (#PF1349) Child prostitution (#PE7582)
Sexual abuse of wards (#PE7755) Sexual abuse of children (#PE3265)
Trafficking in children for sexual exploitation (#PE6613).
Related Rape (#PD3266) Prostitution (#PD0693) Corruption of minors (#PD9481)
Sexual exploitation of men (#PD3263) Sexual exploitation of women (#PD3262)
Sexual intercourse with minors (#PE6522).
Aggravates Excessive sexual activity (#PF5868) Distrust of interpersonal relationships (#PF4274).
Aggravated by Satanic rituals (#PF7887) Sexual unfulfilment (#PF3260)
Trafficking in women (#PC3298).

♦ PD3276 Sexual violence
Refs McCuen, Gary E *Pornography and Sexual Violence* (1985).
Broader Human violence (#PA0429) Sexual deviation (#PD2198).
Narrower Rape (#PD3266) Sadism (#PF3270)
Sexual abuse of children (#PE3265).
Related Enforced celibacy (#PD3371) Violence as a resource (#PF3994).
Aggravated by Pornography (#PD0132) Homosexuality (#PF3242)
Sexual unfulfilment (#PF3260) Inadequate sex education (#PD0759)
Anti-social personality disorders (#PF1721).

♦ PD3301 Debt slavery
Debt bondage — Bonded labour
Nature The pledging of a person, either the debtor himself or of someone under his control, as security for a debt is in theory a temporary form of slavery ending after the debt has been repaid. But in fact, the labour given by the person in security is not counted towards the liquidation of the debt, and the debt is often inherited on his death by his children.
Incidence The practice of debt bondage is closely linked to a feudalistic land tenure system and as such it persists in Latin American countries which have not achieved adequate land reform and elsewhere where there are feudalistic systems (for example, in Iran). In 1954, the Anti-Slavery Society reported 'It is a very widely spread form of slavery and it is impossible to estimate the numbers affected by it'. In Latin America, the people affected are the Indians, as a direct result of expropriation of their land (first by the Spanish conquerors, but which continues even now) and the lack of adequate policies to integrate them into sophisticated capitalist economic systems. Estimates of the number of bonded labourers in India range from 2.6 to 5 million, a large number being employed in the urban industrial sector, in the brick kilns and construction sites of Punjab and Haryana. If the definition for bonded labour as stated by the Supreme Court of India is used the number of bonded labourers would be 75–80 million about 10 percent of the population. Bonded labourers also exist in the agrarian sector of 10 States in the Indian Union. It is estimated that there are 25 million bonded children in India, Pakistan, Bangladesh, Nepal and Sri Lanka. Besides, in India alone more than 50 million child-labourers work in conditions indistinguishable from slavery.
Background Debt slavery is the earliest recorded kind of slavery dating from 2600 BC in Egypt, where people sold themselves or their children in order to pay debts. After the Babylonian captivity the Jews were forced to sell their children to pay taxes. During the Greek and Roman Empires debt was one of many sources for obtaining slaves. Under the feudal system in Europe, debt bondage was widespread. Under Spanish colonial rule the South American Indians incurred debt bondage, being dispossessed of the land.
Broader Slavery (#PC0146).
Aggravates Trafficking in women (#PC3298) Socio-economic poverty (#PB0388).
Aggravated by Bondservice (#PE6342) Female prostitution (#PD3380)
Expropriation of land from indigenous populations (#PC3304)
Inconsiderate choice of payment times by creditors (#PE3926).

♦ PD3303 White slave trade
Nature The white slave trade primarily involves women for prostitution or concubinage purposes and is mainly effected by means of kidnapping or false enticement. It is subsequently followed by the use of force, threats, drugs, mistreatment or torture in order to break down resistance. Many victims are eventually murdered and some commit suicide. Few ever return to lead normal lives in society.
Incidence White slaves may be destined for harems or brothels in Asia, the Middle East, Africa, or Europe. Alternatively, they may be exploited as prostitutes and call-girls in their own country by pimps of the same nationality who generally recruit them from young runaway girls who have just arrived in the city. White slaves destined for export may be kidnapped or enticed by false 'show-business' advertisements. According to INTERPOL, information was provided by 17 countries on the recruitment of foreign women under 'seemingly normal contracts' for prostitutional purposes.
Refs Moorehead, Caroline (Ed) *Betrayal* (1989).
Broader Slave trade (#PC0130).
Related Chattel slavery (#PC3300).
Aggravates Prostitution (#PD0693) Human dependence (#PA2159)
Socio-economic poverty (#PB0388).
Aggravated by Juvenile desertion (#PD8340) Unethical personnel practices (#PD0862)
Inadequate national law enforcement (#PE4768)
Unaccompanied foreign travel in remote places (#PE9135).

♦ PD3305 Restriction of indigenous populations to reservations
Nature The confinement of indigenous populations to restricted areas of land, usually infertile, is the ultimate form of legalized segregation and allows discrimination full reign. Conditions on reservations are usually sub-standard. Government allocations for the up-keep of reservations are notoriously low and since the land tends to be infertile the opportunities for the improvement of life on reservations are very limited. Prices given for produce may be unfair. Payment for labour contracted on reservations is usually low. The result is poor housing and a low general standard of living, despondency, and (as reported in the USA) resort to alcohol, drugs, promiscuity, and suicide. Adequate educational facilities are not available. The situation perpetuates discrimination

PD3305

and the inhabitants of reservations may find it difficult to obtain work elsewhere or to become integrated into another society.
Incidence Reservations occur most notably in the USA, Canada, South Africa and Australia. Elsewhere indigenous populations have been similarly dispossessed but not confined to a given area, so that there is theoretically more opportunity for them to become integrated into society.
Broader Ethnic segregation (#PC3315)
Endangered tribes and indigenous peoples (#PC0720).
Related Underprivileged racial minorities (#PC0805)
Neglect of remote regions and islands (#PE5760).
Aggravates Inadequate housing among indigenous peoples (#PC3320).
Aggravated by Racism (#PB1047) Colonialism (#PC0798)
Exploitation (#PB3200) Socio-economic poverty (#PB0388)
Legalized racial discrimination (#PC3683) Inadequate educational facilities (#PD0847)
Unsustainable economic development (#PC0495) Unequal distribution of social services (#PC3437)
Cultivation of marginal agricultural land (#PD4273)
Discrimination against indigenous populations (#PC0352)
Expropriation of land from indigenous populations (#PC3304).

♦ **PD3321 Illiteracy among indigenous peoples**
Nature The inability of indigenous people to read and write in the official language of the country in which they are living, results in a perpetuation of their social and economic subjugation and exploitation; they are unable to find work with high enough wages to provide adequate housing, clothing and nourishment.
Broader Illiteracy (#PC0210) Illiteracy in developing countries (#PD8329)
Inadequate education of indigenous peoples (#PC3322).
Aggravates Malnutrition (#PB1498) Exploitation (#PB3200)
Inadequate housing (#PC0449) Socio-economic poverty (#PB0388)
Discrimination against indigenous populations (#PC0352)
Underdevelopment of agricultural and livestock production (#PD0629).
Aggravated by Unwritten language (#PF3470) Social fragmentation (#PF1324)
Ethnic discrimination (#PC3686) Exploitation in employment (#PC3297)
Exploitation of child labour (#PD0164) Inadequate educational facilities (#PD0847)
Ineffective educational policy decisions (#PF2447).

♦ **PD3323 Bird diseases**
Refs Stroud, Robert *Bird Disease by Stroud.*
Broader Animal diseases (#PC0952).
Narrower Fowlpest (#PE1400) Campylobacteriosis (#PG4274)
Fluke infections in birds (#PG4324) Ulcerative enteritis in poultry (#PG7595)
Marble spleen disease of pheasants (#PG4464) Malabsorption syndrome of chickens (#PG9749).
Aggravated by Marine pollution (#PC1117) Pesticide hazards to wildlife (#PD3680).

♦ **PD3326 Racial discrimination in public services**
Nature Discrimination in the field of health, social security, old age pensions, etc, and in public places, on the grounds of race, colour, national or ethnic origin, creates disadvantages and lack of opportunities for all involved. Health service discrimination may occur in the quality of service given (if it is centralized), or unintentionally but effectively against racial groups in a lower income bracket if the system is mainly private. Some social security and pension schemes require eligibility standards which may exclude certain racial groups, especially if related to income or property. Legal systems frequently discriminate racially. Racial discrimination in public places may occur through a legalized system of segregation, such as apartheid, or through ghetto situations and inherent prejudice or traditionalism.
Incidence In many nations a disproportion of prisoners are of oppressed races, for example, in Australia, Aborigines make up less than 1.5 percent of the population but account for 14.8 percent of the prison population and 21 percent of all deaths in custody.
Broader Racial discrimination (#PC0006) Discrimination in public services (#PD8460).
Narrower De facto racial requirement qualifications for public services (#PE7971).
Related Legalized racial discrimination (#PC3683)
Discrimination against women in social services (#PD3691).
Aggravates Social inequality (#PB0514) Racial segregation (#PC3688)
Social fragmentation (#PF1324) Lack of social mobility (#PF2195)
Inequality of opportunity (#PC3435)
Unequal distribution of social services (#PC3437).
Aggravated by Social injustice (#PC0797) Lack of anti-discrimination legislation (#PF7972).

♦ **PD3328 Racial discrimination in education**
Racism in schools
Nature Discrimination in education based on race, colour, national or ethnic origin, leads to inequality of opportunity and causes discrimination in employment and housing. Discrimination may be a legalized form of segregation and repression (as in South Africa), or segregation may occur through ghetto situations or through the isolation of certain ethnic groups. Insufficient public funds may be allocated towards the education of dominated groups, arising perhaps from the fact that they can pay less in taxes. Certain groups may be held to be intellectually inferior and therefore not worth educating to a high level. Vocational guidance may encourage this bias on the strength of discrimination in employment and consequent lack of openings. Indoctrination to the detriment of a dominated group may be found in textbooks and in the attitudes of teachers if they are from another group. Educational policy may seek to eliminate the cultural heritage of certain groups, which may retaliate by setting up their own schools to the exclusion of other groups.
Refs Willie, Charles V and Beker, Jerome *Race Mixing in Public Schools* (1973).
Broader Racial discrimination (#PC0006) Inequality in education (#PC3434).
Related Segregation in education (#PD3441) Legalized racial discrimination (#PC3683).
Aggravates Social fragmentation (#PF1324) Inadequate career advice (#PJ8018)
Denial of education to minorities (#PC3459) Racial discrimination in politics (#PD3329)
Inadequate education of indigenous peoples (#PC3322).
Aggravated by Social injustice (#PC0797) Social inequality (#PB0514)
Racial inequality (#PF1199) Segregation in housing (#PD3442)
Cultural discrimination (#PC8344) Isolation of ethnic groups (#PC3316)
Lack of anti-discrimination legislation (#PF7972).
Reduces Elitism (#PA1387) Domination (#PA0839).

♦ **PD3329 Racial discrimination in politics**
Racial politics
Nature Discrimination in politics on the grounds of race, colour, national or ethnic origin leads to the domination of one group by another. Discrimination may be legalized and take the form of the denial of the right to vote, the right to organize political parties, stand for election, or enter the civil service.
Incidence A few countries still have legalized discrimination of this kind (for example, South Africa), but in a larger number it is more discrete, taking the form of voting qualification, the violation of electoral procedure, or intimidation by employers and other members of a dominant group. In the civil service top posts may be withheld from naturalized citizens, or a 'national' policy may be followed, as in Kenya's Africanization policy. Political parties on the basis of race may be banned.
Broader Racial discrimination (#PC0006) Discrimination in politics (#PC0934).
Related Moralism (#PF3379) Legalized racial discrimination (#PC3683).
Aggravates Social conflict (#PC0137) Racial conflict (#PC3684)
Racial segregation (#PC3688) Political conflict (#PC0368)
Endangered cultures (#PB8613) Limitations on right to vote (#PF2904)
Denial of the right to organize political parties (#PE9110).
Aggravated by Social injustice (#PC0797) Social inequality (#PB0514)
Segregation in housing (#PD3442) Compulsory indoctrination (#PD3097)
Racial discrimination in education (#PD3328).

♦ **PD3335 Discrimination against women in public life**
Denial of right to access to public service for women
Nature The inferior status of women and discrimination against them in public life is revealed in all countries by the comparatively small number of women in high positions or as representatives of organizations whose members are from both sexes. Male leaders are statistically more representative of male non-leaders than are their few women colleagues. Women politicians sacrifice more for their position. They are less often married, have fewer children and are better trained than other women. In Western democracies they are more often recruited by parties of the left, they tend to be more liberal than male colleagues. Being unmarried increases a woman's chances of success in terms of career or public life. Women leaders tend still to be assigned to posts in the stereotypically soft ministries of family, welfare, culture and except where this is considered an important post, education. These posts tend not to lead to further promotion in that they do not provide the experience of high level management of economic or foreign affairs considered to be important in top leadership selection. The same thing can be said about women's departments in government. In state socialist countries, women, also, tend to have responsibilities in areas traditionally viewed as women's concerns: health, culture, education, and social welfare. While socialist male aspirants are involved in party work in a number of geographical areas with experience in a number of administrative tasks, women hopefuls tend to be considerably less geographically mobile and are confined to few specialities, notably those which follow from training in the humanities or education. If state socialist regimes officially support political equality of women they have failed to give this goal the kind of priority given to economic equality. State socialism has replaced control of ownership by men with control of political power by men. Western liberal democracies lag behind even this.
Refs Bernardoni, Claudia and Werner, Vera (Eds) *Wasted Wealth* (1985).
Broader Discrimination against women (#PC0308).
Related Denial of right to hold public office (#PE5608)
Exclusion of women from decision making (#PE9009).
Aggravates Wasted woman power (#PF3690)
Neglect of the role of women in rural development (#PF4959).
Aggravated by Social injustice (#PC0797) Social inequality (#PB0514)
Complications of childbirth (#PC9042) Dependency of women in marriage (#PD3694)
Gender stereotyping of employment (#PJ6290)
Discrimination against women in education (#PD0190).

♦ **PD3336 Discrimination against men in social services**
Nature Discrimination against men in matters of social security, old age, and widowers' pensions, health benefits, industrial protective legislation, and responsibility considerations arises mainly as a result of the legally accepted dependency status of women in married and family life and the general agreement that women are in greater need of protection with regard to employment in particular. Widowers' pensions are not a matter of course unless the man has been demonstrably dependent on his wife, and in many cases he must in fact be physically disabled in order to qualify. Women qualify for old age pensions earlier than men.
Counter-claim In many cases, women receive lower salaries, lower pensions and lower health benefits. Work in the home is not considered of value and implies no pension or social security rights; a wife may be dependent on her husband for such provisions, and often a husband's pension may die with him.
Broader Discrimination against men (#PC3258) Discrimination in social services (#PC3433).
Related Social injustice (#PC0797)
Discrimination against men before the law (#PD3692)
Discrimination against women in social services (#PD3691).
Aggravated by Social inequality (#PB0514) Dependency of women in marriage (#PD3694)
Gender stereotyping of employment (#PJ6290)
Discrimination against men in employment (#PD3338).

♦ **PD3338 Discrimination against men in employment**
Nature Discrimination against recruiting men for occupations of low status or with low pay or with few promotion prospects arises because such occupations are conventionally considered more suitable for women. Dirty or dangerous jobs, felt to be unsuitable for women, will be allocated to men. In developed countries women and male immigrant workers share a large proportion of the most menial and worst-paid jobs. Recruitment for occupations traditionally regarded as 'women's work': secretarial, nursing, teaching and social work, may exclude male applicants; where formerly skilled trades can be turned over to keyboard-style machines, women will be employed in preference since they are relatively cheap labour.
Claim Men should have flexible hours to help care for new-born children, men should be able to retire at the same age as women, and women should be allowed to have more dirty or dangerous jobs. Male office workers should be allowed to wear what they want, with neckties being abolished during warm weather.
Broader Discrimination against men (#PC3258) Discrimination in employment (#PC0244).
Related Discrimination against men before the law (#PD3692)
Discrimination against women in employment (#PD0086).
Aggravates Social role-playing (#PG6300) Destruction inherent in development (#PF4829)
Discrimination against men in social services (#PD3336)
Lack of time flexibility in the labour market (#PE8283)
International imbalance in the quality of life (#PB4993).
Aggravated by Inferior status employment (#PD8996).

♦ **PD3347 Segregation in marriage**
Nature The effective options for marriage may be based on common religion, race, colour, ethnic background or age group. (It has also been questioned whether the number of people should be limited to two and whether they should be of opposite sex). Segregated marriages perpetuate ethnic, religious, racial, colour, age and sexual divisions and conflict. Segregated marriages arise out of existing patterns of discrimination.
Incidence Segregation in marriage is universal and sometimes enforced by national law. Religious reasons for segregation are diminishing in industrialized countries, although they still exist.
Broader Segregation (#PC0031)
Segregation based on religious affiliation (#PC3365).
Related Miscegenation (#PC1523).
Aggravates Cultural fragmentation (#PF0536).
Reduces Threatened sects (#PC1995)
Endangered tribes and indigenous peoples (#PC0720).

♦ **PD3357 Discriminatory religious influence on the law**
Religious influence on the law

Nature Where a state has had predominantly one religion or where national independence has been gained using religion as a motivating force, the laws of the country may be based on religious doctrine. This may preserve traditionalism and hinder progress. Also, even if other religions are not specifically discriminated against under the law, it may make the position in society of their adherents very difficult or isolated. They may effectively be barred from taking public office.
Incidence Most countries have some vestiges of law based on religious doctrine, but in many this has been modified over the years to meet the demands of social change. Countries where religion has a large influence on the law include Ireland (with many Roman Catholic doctrines built into the legal framework) and the UK (where the reigning monarch and anyone in line for the monarchy must be Protestant).
Background During the middle ages in Europe the Catholic Church served as the law maker and the custodian of the law for most countries. With the Reformation, national independence was asserted against the Catholic Church and although Protestant doctrine had an influence on the law, Protestant Churches were usually more controlled by it than the Catholic Church had been.
Broader Undue religious influence on secular life (#PF3358).
Related Theocracy (#PG5111) Injustice of religious courts (#PE0397)
Religious and political antagonism (#PC0030)
Ineffective regulation of electronic messages (#PE6226)
Religious discrimination in the administration of justice (#PE0168).
Aggravates Conflict of laws (#PF0216)
Disruptive secular impact of holy days (#PE7735).

♦ PD3371 Enforced celibacy
Nature The state of being unmarried is generally a voluntary state, so it only arises as a problem if the person in question wants to marry and cannot or lacks self-confidence. Celibacy in the religious sense is imposed on priests, nuns and monks by certain religions, notably Buddhism, Jainism and Catholicism. If priests, nuns or monks wish to marry under these circumstances, they must leave their vocation.
Claim Since priests may not marry it is arguable that they are less capable of understanding their married parishioners.
Broader Deprivation (#PA0831) Inadequacy of religion (#PF2005)
Irrational religious beliefs (#PF6829).
Related Sexual violence (#PD3276)
Religious opposition to public health practices (#PF3838).
Aggravates Insecurity (#PA0857) Lack of self-confidence (#PF0879).
Aggravated by Impediments to marriage (#PF3343)
Psychological inhibition (#PF6339)
Imbalance in the human sex ratio (#PF1128).
Reduced by Cohabitation (#PF3278).

♦ PD3373 Religious sacrifice
Nature The act of offering objects, animals, or humans to a god or gods includes blood offerings (such as the killing of animals or humans) or bloodless offerings (very often food or flowers). Sacrifices may be made at certain times of the year such planting, harvest, the New Year, or at certain times of life – birth, puberty, marriage, and death. They may also be made in times of danger, sickness, crop failure, or at the time of building, in battle, or for giving thanks. Since religious sacrifice is mostly carried out in traditional or tribal society, people may be required to sacrifice food or other goods they can ill afford to give away; slaves and prisoners have been used for the purpose of human sacrifice.
Incidence An example of animal sacrifice is in Santeria, the Spanish name for the Lucumi religion of Africa, which was brought to Cuba by slaves and where it practice of monthly white rooster sacrifices continues. Santeria has spread to Florida (where it has 15,000 to 20,000 adherents), New York, California and Texas. The Santeria initiation ceremony consists of nine days of rituals in which more than 100 goats, rams, roosters and other domestic animals are slaughtered in the name of the supreme god, Oludumare, and the orishas that are protector gods.
Broader Occultism (#PF3312) Superstition (#PA0430).
Narrower Suttee (#PF4819) Human sacrifice (#PF2641)
Childhood martyrdom (#PF8118) Ritual slaughter of animals (#PF0319).
Aggravates Human death (#PA0072).
Aggravated by Tribalism (#PC1910) Traditionalism (#PF2676)
Inauspicious conditions (#PF6683).

♦ PD3380 Female prostitution
Nature Prostitution exists because of the subordination of women in most societies. This subordination is reflected in the double standard of sexual behaviour for men and women, and is carried out in the discrepancy between women's and men's earning: women earn an average of 60% (and often less) of what men earn. The specific reasons that prostitutes themselves have given for choosing to work as prostitutes, have included money, excitement, independence, and flexibility. Studies have also revealed a high incidence of child sexual abuse in the life histories of prostitutes (sexual abuse in general: 50 percent for adult prostitutes, 75–80 percent for juvenile prostitutes; incest: 25 percent for adult prostitutes, 50–75 percent for juvenile prostitutes). The traditional psychoanalytic explanation for the relationship between the childhood sexual abuse and later involvement in prostitution is that the child has come to view sex as a commodity, and that she is masochistic. The connection prostitutes report, however, is that the involvement in prostitution is a way of taking back control of a situation in which, as children, they had no control.
Female prostitutes constitute the largest proportion of all prostitutes but, unlike male prostitutes, they are not usually independent agents, often being controlled, blackmailed, intimidated, and brutally treated by pimps or organized crime. Female prostitution gives rise to a trade in women and children who may be sold to brothels or as concubines. In certain countries prostitution is a crime and may depend on the corruption or the turning of a blind eye by the authorities. In other countries it may be legalized and prostitutes must be registered and submit to regular medical examinations for venereal disease; this registration may make it more difficult for them to return to normal life if they wish to do so.
Incidence Female prostitution is a worldwide phenomenon but is most notable in cities, where populations are mobile and where there is severe poverty and unemployment. A woman may then earn money for herself and even for her family through prostitution when her husband cannot get work. Female prostitution is also sometimes a feature of indigenous populations which have not adapted to a new way of life after cultural invasion. Female prostitutes may be the products of broken or deprived homes and may be coerced into prostitution if alone. Although the forms vary somewhat from country to country – in part due to cultural differences, in part due to differences in the law – the institution itself is strikingly similar. A few countries, including Cuba, the USSR and China, have undertaken enormous projects to 'rehabilitate' prostitutes, and thereby to eliminate prostitution. However, women in all of those countries continue to work as prostitutes, especially in the large urban centers, and especially since there has been an increase in tourism from other countries.
Background Prostitution has existed in every society for which there are written records. For a long period in history, women had only three options for economic survival: getting married, becoming a nun (earlier a priestess), or becoming a prostitute (related to being a priestess during certain periods). The invention of the spinning wheel, around the 13th century, enabled a woman working alone to produce enough thread to support herself, for the first time, as a spinster.
Female prostitution was a religious duty in the ancient civilizations of Babylon, Cyprus and among the Phoenicians and in parts of Western Asia. It was a means of earning a dowry in certain other ancient cultures. Female prostitution existed in the Greek and Roman civilizations. In the Middle Ages prostitutes were tolerated and efforts made by the Church to rehabilitate them. After the Renaissance, the relative rise in women's status as a result of humanism led to increasing restrictions, which became more organized and better enforced with the creation of the police in the 19th century.
Prostitution has tended to increase at times in which the role of women was changing. Thus, the industrial revolution in the 19th century was accompanied by a marked increase in prostitution. This was due, in part, to the dislocation of large numbers of women who moved from rural, agricultural communities to urban, industrial cities. When they could not obtain jobs in the new factories, significant numbers of them were forced to turn to prostitution for survival. A second factor was that women who left their families to work in factories were considered to be immoral, and they were subject to a good deal of sexual harassment at work. Once they were stigmatized for leaving home, the barriers to their becoming involved in prostitution were reduced. A similar pattern can be seen in the newly industrializing nations today, especially countries in Africa, Latin America and Southeast Asia.
For most of history, prostitution has been a stigmatized profession, although it has rarely been prohibited. The status of prostitutes seems to have been tied directly to the general status of women: the more women, as a class, have been confined and treated as chattels, the freer prostitute women have been to work without official harassment. As non-prostitute women have achieved increasing independence, on the other hand, the prostitutes have been more restricted and condemned, often confined to segregated districts, or required to wear special clothing, for example.
Claim Prostitution is a violation of article 1 of the Universal Declaration of Human Rights, which reads "All human beings are born free and equal in dignity and rights", of article 4: "No one shall be held in slavery or servitude" and of article 5: "No one shall be subject to torture or cruel, inhuman or degrading treatment or punishment". The distinction between enforced prostitution and voluntary prostitution is false as is the view that prostitution is an occupation.
Refs Perpiñan, Mary Soledad *International Meeting of Experts on the Social and Cultural Causes of Prostitution and Strategies Against Procuring and Sexual Exploitation of Women, Madrid, 1985* (1985); United Nations *From Peasant Girls to Bangkok Masseuses* (1982).
Broader Prostitution (#PD0693) Sexual exploitation of women (#PD3262).
Narrower Abuse of brothel legislation (#PE6735) Military prostitution of women (#PE3390)
Exploitation of the prostitution of others (#PE5303).
Aggravates Debt slavery (#PD3301) Sexual unfulfilment (#PF3260)
Maladjusted children (#PD0586) Illegitimate children (#PC1874)
Sexually transmitted diseases (#PD0061) Dissatisfaction with skin colour (#PF1741)
Discrimination against children of prostitutes (#PE0392).
Aggravated by Male prostitution (#PD3381) Trafficking in women (#PC3298)
Double standards of sexual morality (#PF3259)
Degradation of developing countries by tourism (#PF4115).

♦ PD3381 Male prostitution
Nature Because of social barriers to homosexual 'marriages' and other factors which work against lasting homosexual relationships, many male homosexuals become promiscuous and find contact with male prostitutes easier than searching out a less mercenary partner. The need for secrecy in homosexuality encourages prostitution. Male prostitutes are less frequently used by women. Male prostitution is closely connected with crime, since the prostitutes are mainly from impoverished backgrounds and it is easy for them to intimidate and blackmail their clients and also rob them. The difference in wealth of client and prostitute adds to resentment, equally the social stigma.
Incidence Male prostitution flourishes in large cosmopolitan cities all over the world where money is plentiful and the population is fluid. A Danish report found a particularly high incidence of male prostitutions around the Mediterranean coast among unemployed and unsettled youths and tourists, and where girls were highly chaperoned. A large proportion of male prostitutes are heterosexual, and they are mostly young.
Broader Prostitution (#PD0693) Sexual exploitation of men (#PD3263).
Narrower Abuse of brothel legislation (#PE6735) Male homosexual prostitution (#PD4402).
Aggravates Sexual deviation (#PD2198) Male homosexuality (#PF1390)
Female prostitution (#PD3380) Social fragmentation (#PF1324).
Aggravated by Degradation of developing countries by tourism (#PF4115).

♦ PD3385 Military influence
Military influence in politics
Nature Military influence in political, social and economic spheres of national or international life may be direct, in the form of dictatorship, or less direct, in influencing the size of defence budgets or the orientation of foreign trade agreements and alignments. Foreign powers may give military aid to others which they wish to influence in political or economic matters.
Broader Militarism (#PC2169) International insecurity (#PB0009).
Narrower Military obstacles to peaceful development of space (#PE8965).
Related Martial law (#PD2637).
Aggravates Foreign control (#PC3187) Undue political pressure (#PB3209)
Unethical practices in politics (#PC5517)
Military political prisoners and detainees (#PD3014).
Aggravated by Political apathy (#PC1917).

♦ PD3386 Enforced participation in community activity
Nature Coerced public or community activity may include slavery and other forms of forced labour, parental enforcement of a child's participation in certain activities, compulsory education, and enforced citizen participation in a number of activities within a state-controlled society.
Broader Totalitarianism (#PF2190) Totalitarian democracy (#PD3213)
Forced participation in social processes (#PC5387).
Related Forced labour (#PC0746).
Aggravates Slavery (#PC0146) Exploitation (#PB3200).
Aggravated by Intimidation (#PB1992) Secret police (#PE6331).

♦ PD3389 Undemocratic pressures
Nature The influence of ideology in political, economic and legal systems may take the form of pressure from organized religion or nontheistic needs. Ideological influence exerted by pressure groups may lead to international conflict and war, to civil war, guerrilla war and subversive activities. Internally it may lead to repression, militarism, conformity and denial of democracy. Many ideologies are not brought completely into open, public debate and work underground, buying influence and moving into strategic positions.
Incidence The two main vehicles of ideological influence are the spoken word (that is, the rhetoric of politics) and the written word (that is, the power of the press). Of the two, the press is much more influential and for that reason, dangerous, when it is the tool of an ideology that leads it to suppress, distort, or otherwise report in a biased way. The ideological influence of the press is incalculable, and by extension, television journalism and book publishing as well.

PD3389

Claim A free press does not guarantee freedom from ideological influence, as the media may be polarized or fragmented with bias in every direction from pressures of owners, advertisers or the community served. On the other hand, the closing of newspapers by government decree, or the prohibition on publishing or distributing certain books, are the acts of a tyrannic ideology that is embraced by government.
Broader Ideological conflict (#PF3388) Undemocratic social systems (#PB8031).
Related Anti-clericalism (#PF3360).
Aggravated by Undue political pressure (#PB3209)
Double standards in morality (#PF5225).

♦ PD3391 Long-range transboundary air pollution

Nature Measures taken by many industrialized countries in the 1970s to control urban and industrial air pollution (high chimney stacks, for example) quite unintentionally sent increasing amounts of pollution across national boundaries in industrialized countries. During transport in the atmosphere, emissions of sulphur and nitrogen oxides and volatile hydrocarbons are transformed into sulphuric and nitric acids, ammonium salts and ozone. They fall to the ground, sometimes hundreds of thousands of kilometers from their origins, as dry particles or in rain, snow, frost, fog, and dew.
Incidence Britain is the biggest source of lead pollution in the North Sea and of pollution causing acid rain in Norway.
Refs United Nations *Effects and Control of Transboundary Air Pollution* (1987).
Broader Air pollution (#PC0119)
Environmental threats to national security (#PC4341).

♦ PD3423 Disintegration of organized religion

Nature The falling apart or destruction of organized religion may result from cultural invasion or technological change, but it does not negate the need for religion. Religious disintegration may be characterized by the springing up of numerous small and short-lived religious sects and denominations. Religious disintegration may also result from the imposition of another nontheistic ideology, such as communism, in which case the need for a 'religion' or ideology may cause bitterness and frustration in severance from tradition.
Incidence Disintegration may happen in the case of primitive religions following cultural invasion and with the missionary influence of another religion. In industrialized society it is marked by religious apathy and a struggle for relevance to modern living. In certain communist countries religion has been suppressed but still exists, very strongly in some places, but mainly among older people.
Broader Disintegration (#PA6858) Social disintegration (#PC3309).
Aggravates Dependence on religion (#PF0150) Destruction of cultural heritage (#PC2114).
Aggravated by Religious apathy (#PC3414) Inadequacy of religion (#PF2005)
Proliferation of technology (#PD2420)
Fragmentation within organized religions (#PF3364).
Reduces Traditionalism (#PF2676).

♦ PD3427 Inadequate living and working conditions of immigrant labourers in industrialized countries

Menial work status of immigrants in industrialized countries
Nature Because of a general improvement in the national standard of living in industrialized countries, nationals no longer want to do menial or dirty work. They tend educate their children to be able to take on office jobs or skilled trades. This results in a demand for immigrants who are willing to do the menial or dirty work. Because of their lack of education and knowledge of industrialized countries, they accept low payment and consequently find themselves living in slums and shanty towns with inadequate facilities. They may also be undernourished, living in severely overcrowded conditions which may prevent families from joining them, thus destroying their family life. Where families are allowed to enter, educational facilities for the children may be inadequate. Many of such immigrants are from developing countries and are of a different race, thus creating racial, colour and class tensions. Because of the dependence of families at home on their income, they are unwilling to return to their own countries, where their status and incomes would be even less. If a high unemployment occurs in such a situation, the conflict increases and the immigrants are accused of stealing the rightful work of the national population or causing a drain on social security. They may be denied the right to join trade unions which would raise their status. Industrialists prefer to use immigrants for menial work rather than mchinery which cannot be laid off in a slack period. As many immigrants enter illegally, firms can exploit them and pay less than the minimum wage.
Incidence The problem is greatest mainly in Western Europe, but also exists the United States, Canada, and some newly industrializing Gulf countries, where non-European immigration has been increasing with post-war industrialization and general improvements in the national standard of living. ue to the world economic crisis, many countries have recently stopped all work-related immigration and now are interested primarily only in family reunion immigration, except in a few cases of specialist work which are generally highly skilled and highly paid. Industrialists in France, West Germany and the Netherlands find it difficult to think of the future of their firms without a large flow of new immigrant workers to do routine jobs of low status that Europeans do not want to do. France and Germany already have over 3 million immigrant workers, mainly from the Middle East and North Africa; and interracial conflict is marked by killings and riots, particularly in large ports.
Broader Poor living conditions (#PD9156) Exploitation in employment (#PC3297)
Inferior status employment (#PD8996).
Related Social stratification (#PB5577).
Aggravates Crime (#PB0001) Human violence (#PA0429) Class conflict (#PC1573)
Racial conflict (#PC3684) Family breakdown (#PC2102) Resettlement stress (#PD7776)
Inequality of opportunity (#PC3435)
Unequal distribution of social services (#PC3437).
Aggravated by Social injustice (#PC0797) Illegal immigration (#PD1928)
Class consciousness (#PC3458) Racial discrimination (#PC0006)
Lack of minimum wage fixing (#PE6726) Proliferation of immigrants (#PD4605)
Unethical practices of employers (#PD2879) Abusive traffic in immigrant workers (#PD2722)
Overcrowding of housing and accommodation (#PD0758)
Discrimination against immigrants and aliens (#PD0973).
Reduced by Restrictions on freedom of movement between countries (#PC0935).

♦ PD3436 Distortionary tax systems

Inequitable tax systems — Unfair taxes — Excessive taxation — Unequal graduation of income tax — Restrictive tax structure — Paralyzing property tax — Inappropriate tax assessment
Nature When government needs to reduce public deficits, the economic cost of raising more revenue is weighed against the cost of public spending. The temptation in the short-run is to rely on ad hoc increases in taxation because they are administratively and politically convenient. But in many countries this has led to complex and highly distortionary tax systems that not only fail to collect sufficient revenue but also damage long-term growth and increase the burden on the poor.
Incidence Taxation on the earnings and property (via direct and indirect taxes) of lower to middle-wage earners may take away anything from one-half to two-thirds of their pay. Increased taxes reduce savings, reduce bank capital available to stimulate economic growth, and have caused bank failures. Similarly, taxes on business for both international and local trade have mounted so drastically that 80 to 90 percent of gross income after costs and expenses may go to various taxes, and in the case of international trade, to various governments as well. Many governments maintain very low taxes on rural land allowing underuse, on the other hand settlers are allowed to establish title to "virgin" land – forests – by converting it to farmland.
Claim All tax systems are considered inequitable by those taxed; although few dispute the need for taxes, they do dispute the amount that they must pay. However there is a difference among the classes in that the majority of citizenry do not earn enough to be able to save for crises such as major illnesses, or socially desirable improvements in their circumstances (such as their children's education), or for retirement. All tax systems are inequitable when taxing authorities, whether local or national governments, cannot be restrained by budget and public debt limits, and are not accountable for armaments, while the poor suffer from lack of adequate food, shelter and medical care, and while unemployment is left to so-called market forces to rectify.
Refs Hadenius, Axel *Crisis of the Welfare State?* (1986).
Broader Human inequality (#PA0844) Inadequate fiscal policies (#PF4850).
Narrower Taxation of the poor (#PE4601) Tax impediments to marriage (#PJ6283)
Undue taxation of certain goods and services (#PE8444)
Inappropriate taxation of not-for-profit and philanthropic charitable organizations (#PF3049)
Distorting effects of commodity taxes on the transaction of goods and nonfactor services (#PE9220).
Related Unequal distribution of social services (#PC3437)
Limiting effect of individual survivalism (#PF2602)
Undeveloped potential of informal leadership (#PF1196)
Complex regulations paralyzing small communities (#PF2444).
Aggravates Famine (#PB0315) Cohabitation (#PF3278)
Obsolete deliberative systems (#PD0975) Unequal property distribution (#PC3438)
Ineffective economic structures in industrial nations (#PE4818)
Inadequate road and highway transport facilities in developing countries (#PD0543).
Aggravated by Tax evasion (#PD1466) Social injustice (#PC0797)
Inequality before the law (#PC1268).

♦ PD3440 Segregation in social services

Broader Segregation (#PC0031) Discrimination in social services (#PC3433).
Aggravates Social conflict (#PC0137) Social fragmentation (#PF1324)
Unequal distribution of social services (#PC3437).

♦ PD3441 Segregation in education

Nature Segregation in education on the basis of sex, race, religion, age, and class leads to inequality of opportunity and contributes to the persistence of prejudice and discrimination, and in the case of age, to inflexibility in the educational system.
Counter-claim It has been shown that girls achieve better academic results when educated in single-sex schools than in co-educational schools.
Refs Stangvik, Gunnar *Self-Concept and School Segregation* (1979); Stephan, Walter G and Feagin, Joe R *School Desegregation* (1980).
Broader Segregation (#PC0031) Social inequality (#PB0514).
Narrower Sexually segregated schools (#PG3650)
Segregation of handicapped children in education (#PE8424).
Related Inequality in education (#PC3434) Denial of education to minorities (#PC3459)
Racial discrimination in education (#PD3328).
Aggravates Educational elitism (#PC1527) Inequality of opportunity (#PC3435)
Underprivileged religious minorities (#PC2129)
Discrimination against women in education (#PD0190)
Attitude manipulation of children through play (#PF2017).
Aggravated by Inadequate educational facilities (#PD0847).

♦ PD3442 Segregation in housing

Racial discrimination in housing — Racially separated residences
Nature Discrimination in the sale, renting and occupation of housing on the basis of race, colour, national or ethnic origin, leads to segregation in housing, ghettos and slums, and encourages racial conflict and division. Discrimination may be legalized (as in the South African system of apartheid where special areas are designated for Blacks or for Whites) or it may have evolved in practice (for example, at one time only 11 per cent of privately-let property advertised in London did not specifically exclude coloured people). Discrimination may in the latter case take the form of refusal to rent, refusal to sell, or refusal to grant loans and mortgages. This may be based on the assumption that the person involved is not earning well, or the excuse may be along cultural lines, implying noise, gregariousness, dirty habits, etc, and the fact that this will reduce the value of the property in the vicinity.
Broader Segregation (#PC0031) Racial discrimination (#PC0006).
Narrower Refusal to let because of applicant's race (#PE8137).
Related Discrimination in housing (#PD3469) Legalized racial discrimination (#PC3683).
Aggravates Social conflict (#PC0137) Racial conflict (#PC3684)
Housing shortage (#PD8778) Racial intimidation (#PC2936)
Social fragmentation (#PF1324) Exploitation in housing (#PD3465)
Lack of social mobility (#PF2195) Racial discrimination in politics (#PD3329)
Racial discrimination in education (#PD3328) Inadequate care of community space (#PF2346)
Overcrowding of housing and accommodation (#PD0758)
Segregation of poor and minority population in urban ghettos (#PD1260).
Aggravated by Property speculation (#PD8202) Unequal property distribution (#PC3438).

♦ PD3443 Segregation in employment

Broader Segregation (#PC0031).
Narrower Gender stereotyping of employment (#PJ6290).
Related Abusive traffic in immigrant workers (#PD2722).
Aggravates Slavery (#PC0146) Unemployment (#PB0750) Social conflict (#PC0137)
Social fragmentation (#PF1324) Maladjustment to disciplines of employment (#PD7650).
Aggravated by Discrimination in employment (#PC0244)
Denial of the right of trade union association (#PD0683).
Reduced by Lack of anti-discrimination legislation (#PF7972).

♦ PD3444 Age segregation

Nature Old people need other old people, but they also need contact with the young, which increasingly they do not get. A family relationship of great significance for both parties, for example, was once the intimate association between the very young and the very old. Frequently, in past times, they have been left together at home while the able-bodied have gone forth to earn the family living. These old people have protected and instructed the little ones, while the children in turn have acted as the 'eyes, ears, hands and feet' of the old. Care of the young has thus very generally provided the aged with a useful occupation and a vivid interest in life. Such mutual interchange between young and old is disappearing, as old people become more isolated.
Broader Segregation (#PC0031) Age discrimination (#PC2541).
Related Ethnic segregation (#PC3315) Household segregation by age group (#PF6136).
Aggravates Social conflict (#PC0137) Waste of human resources (#PC8914)
Inadequate housing for the aged (#PD0276).
Aggravated by Inequality of opportunity (#PC3435).

♦ PD3446 Maldistribution of electrical energy
Nature The large scale generation of electricity has built Western nations might and dramatically assisted in elevating its stand of living, all at the expense of our global environment. Meanwhile, the lack of inexpensive electrical power is a part of keeping developing countries poor, relatively unproductive, debt ridden and hungry. Maldistribution across time zones exacerbates peak and valley demands. Maldistribution between developed and developing countries hinders world trade and contributes to the debt crisis and low standards of living in the third world.
Broader Maldistribution of resources (#PB1016)
Maldistribution of energy consumption (#PC5038).
Related Ideological war (#PC3431).
Aggravates Hunger (#PB0262) Neutrality (#PF0473) Nuclear waste (#PD4396)
Deforestation (#PC1366) Dependence on oil (#PJ4398) Acidic precipitation (#PD4904)
Excessive public debt (#PC2546) Environmental pollution (#PB1166)
Prohibitive cost of electricity (#PG8518) Unsustainable population levels (#PB0035)
Competitive acquisition of arms (#PC1258).
Aggravated by Inadequate electricity infrastructure (#PD9033).

♦ PD3453 Superstitious persecution of animals
Broader Persecution (#PB7709) Superstition (#PA0430).
Related Animal worship as a barrier to development (#PD2330).
Aggravates Endangered species of animals (#PC1713).
Aggravated by Taboo (#PF3310).

♦ PD3465 Exploitation in housing
Exploitative rental of accommodation — Exploitation of housing tenants
Broader Political exploitation (#PC7356) Exploitative property development (#PD8492).
Narrower Uncompensated tenant repair (#PJ0147).
Related Exploitation in employment (#PC3297) Denial of right to sufficient shelter (#PD5254).
Aggravates Alienating public housing assignments (#PJ9479)
Uncontrolled household rent increases (#PG5510)
Inadequate housing among indigenous peoples (#PC3320).
Aggravated by Housing shortage (#PD8778) Segregation in housing (#PD3442)
Unethical real estate practice (#PD5422)
Government expropriation of private property (#PD3055).

♦ PD3469 Discrimination in housing
Housing prejudice
Nature Discrimination in housing is a serious obstacle to social justice and social progress. Such discrimination may originate either in law or in practice and, in either case, may be direct or indirect in nature. The right to housing occupies a key position among the fundamental human rights enumerated in the Universal Declaration of Human Rights. The realization of this right is, however, fraught with difficulties. Housing problems largely attributable to scarcity of inadequacy of resources arise in all parts of the world, causing deficiencies in quantity and quality, particularly in areas of high population growth. When, due to underemployment or growing unemployment in rural areas, there is a mass exodus from the affected areas to the urban centres, existing shortages in those centres become extreme and, at the same time, available facilities within the reach of the new arrivals fall even further short of minimum safety standards and health requirements.
Incidence Although legally guaranteed equality in housing, many Aboriginal Australians live in inadequate shelters, due to their deprived socio–economic circumstances which prevent them from acquiring lands or homes in the normal way. These circumstances tend to make Aboriginals unwelcome neighbours in many parts of the general community, resulting in the emergence of Aboriginal slums in most large cities. In South Africa, the government discriminates against, Bantus and other non–whites by "legal" means such as "Black spot", squatter, and group area removals, all of which remove peoples from their homes without possible resistance and retaliation, as the government has decreed resistance to be illegal.
Refs Ansley, Robert E *Discrimination in Housing* (1979).
Broader Prejudice (#PA2173) Discrimination (#PA0833)
Unethical real estate practice (#PD5422).
Narrower Segregation of poor and minority population in urban ghettos (#PD1260).
Related Age discrimination (#PC2541) Segregation in housing (#PD3442).
Aggravates Denial of right to sufficient shelter (#PD5254).
Aggravated by Housing shortage (#PD8778) Inadequate standards of living (#PF0344)
Prohibitively expensive housing for the poor (#PE8698).

♦ PD3471 Endangered species of arthropoda
Nature The arthropoda constitute a vast assemblage of invertebrates comprising about two million species; several times the number of all other animal species combined. Their adaptive diversity has enabled representatives to survive in virtually every habitat, and in many respects they are the most successful invaders of the land.
The distinguishing feature of arthropods, and one to which many other evolutionary innovations are related, is the chitinous exoskeleton or cuticle. This covers the entire body and is divided into inter–connecting plates which permit movement. Growth is facilitated by periodic moulting.
The phylum consists of insects, spiders, and crustaceans; their existence is threatened by such factors as: fishing; damage to natural habitats; agricultural, commercial and residential development; exploitation for biomedical use; and the demand for unusual and exotic pets.
Broader Endangered species of insects (#PC2326)
Endangered species of invertebrates (#PD7513).
Narrower Endangered species of crustacea (#PD3472)
Endangered species of arachnida (#PD3473).

♦ PD3472 Endangered species of crustacea
Nature Over 38,000 crustacean species are known, most of which are marine, although some are found in fresh water and some are terrestrial. The majority of planktonic animals throughout the world are crustaceans, but the group is also well represented in bottom fauna from coastal waters to abyssal depths. For marine species, there is currently little evidence that any species are threatened with extinction, but there are ample data indicating overfishing of food species. Freshwater and terrestrial crustaceans often have restricted ranges and are subject to habitat loss and pollution.
Incidence Crustaceans are a vital link in the food chains of aquatic ecosystems. Crayfish, for example, are often major croppers of aquatic vegetation and their disappearance could cause premature aging of lakes. Although many freshwater and terrestrial crustaceans are of interest to biologists on account of their relict distributions and evolutionary history, the main value of crustaceans to man is as a food resource; and they are a major element in both commercial and subsistence fisheries. Prawn and shrimp fisheries are of greatest economic importance, lobster is the most popular crustacean, and crayfish are considered a luxury food in Europe, where they are eaten in great quantities.
Broader Endangered species of arthropoda (#PD3471).
Narrower Endangered species of lobsters (#PS6456).

♦ PD3473 Endangered species of arachnida
Spiders, scorpions
Nature The 65,000 species of arachnida include spiders, scorpions, ticks, mites and many other species. Because of the difficulty of observing their activities, their importance to man may easily be underestimated, particularly in terms of their function in controlling the size of insect populations. It has been estimated that an acre of woodland in the temperate zone may well contain 14,000 spiders. It is not known how many arachnid species are endangered by man's impact on the natural environment.
Broader Endangered species of arthropoda (#PD3471).

♦ PD3474 Endangered species of platyhelminthes
Nature The platyhelminthes are a group of wormlike soft–bodied animals comprising about 25,000 species. They have no body cavity or special respiratory structures and respiration occurs throughout the body surface. The gut usually has only one opening which serves as both mouth and anus. The majority are parasitic, including the flukes (class Trematoda), the monogeneans (class Monogenea) and the tapeworms (class Cestoda), but there is a fourth class, (Turbellaria) which consists of about 4,000 free–living species. Many platyhelminths are dorso–ventrally flattened which accounts for their common name, flatworm.
Very little information is available concerning survival threats to platyhelminthes, as it is very difficult to obtain population estimates. Water pollution seems to pose the largest threat.
Counter–claim Due to the great capability of adaptation and to the resistance of many platyhelminthes species to extreme environment conditions, man's alteration of the environment (such as the creation of dams) has often unintentionally favoured their propagation.
Broader Endangered species of invertebrates (#PDp513).

♦ PD3475 Endangered species of protozoa
Nature Protozoa are a diverse assemblage of over 65,000 minute, mobile, single–called organisms. Moisture is their prime ecological requirement and the majority are found in the sea or freshwater, although others (particularly flagellates, small amoebae and ciliates) are ubiquitous in soil. Protozoa play an extremely important role in many ecological processes, making up a large portion of the biomass of many aquatic systems. Due to their complex taxonomy and the fact that only half of the existing species has been described it is difficult to know the exact threats that endanger their survival. But it is clear nevertheless, that survival is influenced by food supply, light (for certain species), temperature, oxygen, carbon dioxide, acidity of the medium and water (for active life).
Broader Endangered species of invertebrates (#PD7513).

♦ PD3476 Endangered species of sponge
Porifera — Porphyria
Nature 5,000 living sponge species are known, of which 150 occur in freshwater and the rest in the sea. The survival of sponges, the most primitive of multicellular animals, may be threatened due to pollution, disease, exploitation, or hurricanes.
Background For centuries, sponges have been employed for both personal and household purposes; in the first half of the 19th century sponge fishing began on a commercial basis in the Mediterranean, and by the end of that century it was well established in the USA and some Caribbean islands. Currently Tunisia is the world's largest supplier of sponges. In 1986, a fungal disease attacked the sponge fishing areas of the eastern Mediterranean nearly destroying the population.
Broader Endangered species of invertebrates (#PD7513).

♦ PD3477 Endangered species of bryozoa
Polyzoa
Nature The Bryozoa consist of over 4,000 living species which form small sessile colonies displaying a wide range of form, including encrusting calcareous or gelatinous patches, long branching chains, apparently jumbled heaps, or bushy branching growths, some of which superficially resemble corals. As little research has been done on the ecology of this group, it is unknown whether any particular species is under threat of extinction, but it can be inferred that many species could be vulnerable to a variety of adverse factors, such as: pollution; increased turbidity, sedimentation, and siltation; changes in salinity and temperature; and the loss of substrate caused by dredging and kelp harvesting.
Broader Endangered species of invertebrates (#PD7513).
Reduces Fouling of water supply systems (#PJ6457).

♦ PD3478 Endangered species of molluscs
Mollusca
Nature After insects, molluscs are probably the most familiar invertebrate, with more than 100,000 species living to–day and a long fossil history. Of all the invertebrates, molluscs are most valued by man. They are a major food source in many parts of the world, their shells provide a variety of products, and the diverse and beautiful forms that these may take have led to molluscs becoming important elements in the art, culture and traditions of many races. Threats to their survival include exploitation, pollution, habitat alteration or destruction, introduced species, and over–collecting.
Broader Endangered species of invertebrates (#PD7513).
Narrower Endangered species of snails (#PG6458)
Endangered species of squid octopus (#PG6459).
Reduces Trematode diseases (#PE6461) Fouling of water supply systems (#PJ6457).

♦ PD3479 Endangered species of insectivores
Insectivora
Nature The order of insectivora includes moles, tree shrews and hedgehogs. Common throughout the temperate and tropical zones, a number of species – notably the Cuban solenodon and the Haitian solenodon – are endangered, primarily due to land development and deforestation caused by a rapidly expanding human population.
Broader Endangered species of mammals (#PC1326).
Narrower Endangered species of shrews (#PG6468)
Endangered species of tenrecs (#PG6463)
Endangered species of hedgehogs (#PG6466)
Endangered species of solenodons (#PG6462)
Endangered species of otter shrews (#PG6464)
Endangered species of golden moles (#PG6465)
Endangered species of elephant shrews (#PG6467)
Endangered species of moles and desmans (#PE8750).

♦ PD3480 Endangered species of rabbits
Lagomorpha
Incidence The volcano rabbit, found in limited distribution in subalpine areas on volcanoes in central Mexico, is rapidly decreasing due to being hunted for food and sport and also due to habitat destruction.
Broader Endangered species of mammals (#PC1326).
Narrower Endangered species of pikas (#PG6470)
Endangered species of rabbits and hares (#PE8794).
Aggravated by Feral dogs (#PG6473) Myxomatosis (#PG6472).

♦ **PD3481 Endangered species of rodents**
Rodentia
Nature There are currently 34 species of rodentia identified as being already extinct, endangered, or rare; the reasons include habitat destruction or alteration, competition with other species, hunting, indirect interference by man (such as the construction of suburban housing developments), and deliberate elimination campaigns (Utah prairie dog, in the USA).
Broader Endangered species of mammals (#PC1326).
Narrower Endangered species of murids (#PG6483)
Endangered species of gundis (#PG6503)
Endangered species of beavers (#PG6477)
Endangered species of dormice (#PG6484)
Endangered species of jerboas (#PG6488)
Endangered species of capybara (#PG6492)
Endangered species of pacarana (#PG6493)
Endangered species of squirrels (#PG6474)
Endangered species of tucotucas (#PG6498)
Endangered species of springhaas (#PG6479)
Endangered species of spiny rats (#PG6500)
Endangered species of bamboo rats (#PG6482)
Endangered species of chinchillas (#PG6495)
Endangered species of jumping mice (#PG6487)
Endangered species of pocket gophers (#PG6475)
Endangered species of rat chinchillas (#PG6499)
Endangered species of African mole rat (#PE8253)
Endangered species of African cane rat (#PE8684)
Endangered species of mice and hamsters (#PE8967)
Endangered species of old world porcupine (#PE8845)
Endangered species of agoutis and acouchis (#PE8738).
Endangered species of new world porcupines (#PE8847)
Endangered species of palaearctic mole rats (#PE9103)
Endangered species of bush rat and rock rat (#PE9120)
Endangered species of scaly-tailed squirrels (#PG6478)
Endangered species of cavies and dolichotids (#PE8683)
Endangered species of zagoutis and coypu and hutia (#PE8935)
Endangered species of pocket mice and kangaroo mice (#PE8674).
Aggravated by Rodents as pests (#PE2537).

♦ **PD3482 Endangered species of carnivores**
Carnivora
Nature There are currently 21 species of Carnivora identified as being already extinct, endangered, or rare; including wolves, bears, ferrets, otters, ocelots, cougars, mountain cats and jaguars. The reasons for their actual or threatened extinction include: conflict with man (via attacks either on man or his livestock), habitat destruction and alteration, eradication campaigns (in the case of wolves), hunting for sport or fur, decline in prey, and highway mortality (the Florida cougar).
Broader Endangered species of mammals (#PC1326).
Narrower Endangered species of cats (#PD1763)
Endangered species of bear (#PD3483)
Endangered species of hyenas (#PG6508)
Endangered species of racoons (#PG6505)
Endangered species of dog wolves and foxes (#PE8739)
Endangered species of genets, civets and mongooses (#PE9121)
Endangered species of weasels, badgers, skunks, otters (#PE8936).
Aggravated by Extermination of wild animal natural prey (#PD3155).

♦ **PD3483 Endangered species of bear**
Ursidae
Incidence The spectacled bear of the Andes has seen its numbers depleted from hunting, trapping and loss of habitat. Its survival rests on enforcement of game codes and protection in large national parks and reserves. The Baluchistan bear is endangered. It is believed to be confined to a relatively small pocket of south–central Baluchistan, but may also occur in south–east Iran. Its former range and numbers have declined drastically since the 1930's as a result of hunting, because of its depredations on crops and domesticated stock. The bear has no legal protection. The Mexican grizzly bear is possibly extinct; a remnant group, supposedly of Mexican grizzly bears, may still exist on a cattle ranch in the upper Yaqui Basin of Sonora. The animal was apparently exterminated in the Sierra del Nido in the early 1960's by hunting and poisoning. It is legally protected but protection has not been enforced. The barren–ground grizzly bear of the Canadian tundra, a rare subspecies of the grizzly bear, is at risk from illegal killing. In Europe the bears along the Pyrenees are on the verge of extinction. The immediate threat to the giant panda is not man but nature. In their high–altitude retreats in the remote regions of central China, giant pandas survive almost entirely on bamboo, which comprises 99.9 percent of their diet. Unfortunately, a staple of their diet, the arrow bamboo, is now undergoing one of its periodic blossomings. When this happens, once every 45 or 50 years, a whole mountainside of bamboo may erupt in flowers, scatter seeds and then perish. The bamboo will regrow in a few years to sufficient size (perhaps 3 ft high) to provide fresh food for the pandas. Meanwhile, they must find other food. In 1975–76 a similar flower–and–die disaster led to the deaths by starvation of 138 animals in a panda habitat on the border of Sichuan and Gansu provinces. There are only a thousand or so giant pandas still found in the wild.
Broader Endangered species of carnivores (#PD3482).
Narrower Endangered species of ursus (#PG6513)
Endangered polar bear species (#PE1320)
Endangered species of tremarctos (#PG6511)
Endangered species of selenartos (#PG6512).
Aggravated by Crop damage by wildlife (#PC3150)
Degradation of semi-natural and natural habitats of flora and fauna (#PC3152).

♦ **PD3488 Uncontrolled urbanization in developed countries**
Unplanned urbanization in industrialized countries
Nature Europe's total population was 325 million in 1920, 490 million in 1984, and will perhaps be 513 million in the year 2000. Virtually all the increase is reflected in Europe's urban population: 150 million in 1920, but which may number 437 million at the century's end, while the rural population may undergo an appreciable decrease. Other more developed regions have been gaining population at an even faster rate than Europe, and again the tendency is for rapid urbanization to be accompanied by a decline in rural populations.
Broader Uncontrolled urban development (#PC0442)
Poverty in developing countries (#PC0149).
Aggravated by Deficiencies in national statistics (#PF0510).

♦ **PD3489 Urban slums in developing countries**
Bidonvilles — Shanty–towns — Transitional urban settlements — Urban fringe poverty
Nature Fringe poverty has resulted from the imbalance between labour force and employment opportunities in many developing countries. Increasing rural poverty has driven millions of people to the cities in search of work. Many of them are badly undernourished, but not dying of starvation; they are not officially unemployed but only work a part of the year, trying to keep a foot in both rural and urban worlds. They lack basic needs and their income is insufficient to secure enough food for the family, to send their children to school, or to secure medicine when they fall sick. Overcrowding of premises in slums and shanty–type construction are typical. There is enormous pressure on water supplies and the arrangements for waste disposal. Malnutrition and diseases add to the burden on medical services. Schools are over crowded and anti–social behaviour is common. Other social problems linked with urban environmental factors are the unbalanced distribution of population by age group in urban and suburban areas, non–adaptation of rural migrants, dissatisfaction due to instability of employment opportunities, and difficulty in integrating youth. Sickness and disease, along with high mortality rates, are commonplace. Because the people in these areas lack the economic mobility to escape, this micro–environment becomes their life–time experience with the physical environment. What has emerged in the cities of the developing world are thus vast areas of despoiled landscape that provide the physical and life space for one of the worst human environments created by man. Poverty is creating unique micro–environments, which in turn are significantly affecting the total environment of cities. Together, these changes are altering not only the natural environment but the very condition of man. The ultimate consequences are severe biological problems and acute social and political unrest.
Incidence An increasing percentage of the world's population is living in transitional settlements. The fabric of urban life and contemporary society in many developing countries is threatened when these populations occupy 30 to 50 per cent or more of urban areas, as in some cities of Latin America and elsewhere. In most developing countries it has not been possible to provide in advance a rational arrangement of space for living, transportation and recreation, or to provide rapidly enough for housing, water, sewage disposal, education, or the other amenities of urban life.
In developing countries, the urbanization process is accelerating. The annual average growth rate of the urban population between 1960 and 1970 was 4.1 percent in South Asia, 3.3 percent in East Asia, 4.8 percent in Africa and 4.3 percent in Latin America; the projections for 1970–75 indicated, respectively, 4.3 percent, 4.0 percent, 4.9 percent and 3.9 percent. These figures mean that in twenty years the urban population of these regions doubled or even tripled. Back in 1970 there were 11 cities in developing countries with more than 5 million inhabitants; by the year 2000 their number is likely to increase to 335.
Claim In many developing countries, accelerated urban growth is concentrated in a few big cities, where the population is increasing quite inordinately. In one of these countries, for example, the nine largest towns represent today almost one–quarter of the total population. The development of infrastructures cannot keep pace with expansion at such a high rate. Hence the mushrooming of shanty towns where non–integrated population groups crowd into makeshift, insanitary shelters which lack water, drainage, gas and electricity; and where the lack of protection against the hazards of fire and flood breeds a sense of insecurity.
The man–made environment in which individuals have to live is degrading and is increasingly determining the physical environment of the entire urban area. Streams are polluted, land is laid waste and hillsides are eroded through overcrowding and the lack of even the most rudimentary public services. Rats and vermin are ubiquitous throughout the urban slums. Trees and vegetation are all but eliminated by the cutting for use and sale of all available timber.
Counter–claim Increasingly, these unconventional urban settlements also represent practical and effective methods of coping with accelerated urbanization. Uncontrolled settlements on the periphery of cities that formerly were viewed as detrimental because of their haphazard construction and their insanitary conditions, have been recognized to have some positive elements. It is not only that their scale and life style provide a more familiar environment and a sense of community to the rural migrant, but they contain a highly motivated group of people striving towards self–improvement. The trend to transitional settlements, which appears to be an inevitable part of spreading urbanization, can be guided and converted into a positive development factor instead of an impediment to healthy urban growth.
Refs Hardoy, Jorge E and Satterthwaite, David *Squatter Citizen* (1989); Lowder, Stella *Inside Third World Cities* (1986); United Nations *The Aging in Slums and Uncontrolled Settlements* (1977).
Broader Urban slums (#PD3139) Underdevelopment (#PB0206).
Aggravates Pollution in developing countries (#PC2023)
Imbalance in city sizes within a country (#PF2120).
Aggravated by Rural depopulation (#PC0056) Landlessness in developing countries (#PC0990)
Introduction of high-yield crop varieties (#PF3146)
Rural unemployment in developing countries (#PD0295)
Inadequate housing in developing countries (#PE0269)
Inappropriate education in developing countries (#PF1531).

♦ **PD3491 Excessive childhood dependency in developing countries**
Nature In the developing countries the ratio of children to adults is typically very high because of the high birth rate. The burden of childhood dependency may be lightened by putting the children to work at an early age, but the employment of children is not compatible with a high standard of education and therefore is not a satisfactory solution. In so far as the birth rate continues to remain the same, any addition to the adult population implies a proportionate increase in the number of births. If death rates fall while birth rates remain constant, the size of each successive generation will be larger but the average number of dependent children per adult will be practically unchanged. The developing countries will continue to suffer the economic handicap of heavy childhood dependency as long as birth rates remain unchanged.
Broader Unproductive dependents (#PC1420).
Aggravates Ignorance (#PA5568).
Aggravated by High human fertility in developing countries (#PF0906).

♦ **PD3492 Incendiary weapons of massive destructiveness**
Nature Incendiary weapon delivery systems include bombs, rockets, shells, grenades, bullets, flamethrowers, and land–mines. These are used to carry the incendiary agents that affect their targets primarily through the action of heat and flame derived from self–propagating exothermic chemical reactions. They include: petroleum–based incendiaries such as napalm compositions; metal incendiaries such as magnesium, pyrotechnic incendiaries such as white phosphorus; and certain organometallic compounds which ignite spontaneously in air. Such weapons are capable of causing massive destruction to both the rural and urban environment. Such destruction is often unavoidably and even deliberately indiscriminate, and may be particularly detrimental to the civilian rather than the military component of a society. Incendiary warfare is particularly cruel in its effects, notably because of the long period of recovery required for survivors and the high probability of permanent deformity with consequent emotional disorders. The relatively low cost of incendiary weapons adds to the danger, since they are available to even the poorest countries.
Incidence Napalm kills not only by burning but by asphyxiating or poisoning its victims. Its adhesiveness, high burning temperature, and prolonged burning time lead to deep burns. As napalm is scattered over target areas in large clumps, victims are usually struck by a substantial mass of it; by attempting to remove the napalm from the skin, or strip off their burning clothes, they spread it over other parts of their bodies, particularly their hands. Napalm often causes not only third but fourth and fifth–degree burns, which completely char the skin and extend into the deep tissue of the body, damaging muscles and reaching even to the bones and internal organs. For adequate medical treatment of burn casualties, vast facilities would be required to meet the

emergencies caused by large-scale attacks. It has been estimated that to treat 1,000 wartime casualties having 30 per cent burns, the following would be needed: 8,000 litres of plasma, 6,000 litres of blood, 16,000 litres of a balanced salt (lactate) solution, 250 trained surgeons and physicians, approximately 1,500 skilled attendants, and hospital beds for each patient for up to four or five months. Even in the developed countries, such requirements would be difficult to mobilize on any scale; in the developing countries it would be virtually impossible.
Most incendiary bombs used in the Second World War consisted mainly of magnesium or thermite (a mixture of ferric oxide and aluminium). Incendiaries were extensively used in German raids against London, 1940 causing 600 deaths, by the R.A.F. against Hamburg in 1943 causing 50,000 deaths, and Dresden in 1945 causing 80,000 to 135,000 deaths, and by the U.S.A.F. against Tokyo in 1945 causing 83,000 deaths. During the Korean War over 32,000 tons of napalm were used and during the Vietnam War more than 100,000 tons had been dropped by March 1968.
Claim While it may sometimes be convenient to classify incendiary agents as a separate category of weapons, they are capable of acting in ways similar to chemical weapons and therefore should be treated as such by international conventions limited to chemical-biological weapons.
Refs Stockholm International Peace Research Institute *Incendiary Weapons* (1975).
Broader Weapons (#PD0658) Inhumane and indiscriminate weapons (#PD1519).
Narrower Napalm (#PG6516).
Aggravates Burns and scalds (#PE0394).

♦ PD3493 Accidents to nuclear weapons systems

Accidental loss of nuclear weapons — Unsafe nuclear missiles — Broken arrows
Nature Accidents involving nuclear weapons systems could result from some kind of mechanical failure, or from the miscalculation or insubordinate behaviour of members of the military forces who operate the weapons delivery systems. Such accidents might start a nuclear war if one country detonated a bomb by accident on the territory of a nuclear power or a nuclear power's ally; it might also happen if it dropped a bomb on its own territory and another country was suspected. Accidental detonation, even if it did not start a nuclear war, could do great damage if it were detonated over a populated area. There is also the possibility of radioactive contamination from weapons which are damaged or destroyed but not detonated. A nuclear attack could be provoked in the time taken to detect and rectify the effects of the accidents.
The USA definition of nuclear weapon accident is any unexpected event involving nuclear weapons or nuclear components which results in any of the following: (a) accidental or unauthorized launching, firing or use, by USA forces or US-supported allied forces, of a nuclear-capable weapon system which could create the risk of outbreak of war; (b) nuclear detonation; (c) non-nuclear detonation/burning of a nuclear weapon; (d) radioactive contamination; (e) seizure, theft or loss of a nuclear weapon or nuclear component, including jettisoning; and (f) public hazard, actual or implied. A nuclear weapon incident is any unexpected event involving nuclear weapons or nuclear components which does not fall in the nuclear weapon accident category but which: (a) results in evident damage to a nuclear weapon or nuclear component to the extent that major rework, complete replacement or examination or recertification by the Energy Research and Development Administration (ERDA) is required; or (b) requires immediate action in the interest of safety or which may result in adverse public reaction (national or international) or premature release of information.
Incidence There have been an unspecified number of accidents when intercontinental ballistic missiles, presumably fitted with nuclear warheads, have been destroyed by fire or explosion. Anti-aircraft missiles have misfired or been accidentally launched on several occasions. The risk of accident per nuclear weapon system deployed may be diminishing owing to the shift to missiles and improved safety systems, but the number of weapon systems deployed is increasing rapidly. From 1945 through March 1968, there had been at least 32 'major' accidents: those involving the complete destruction of a nuclear weapon delivery system (aircraft, missile, ship and so on) containing a nuclear weapon, and with the destruction, loss or other involvement of the nuclear warhead itself. In addition, a semi-official study completed in 1960, carried out with some access to classified information, indicated that there had further been 'about 50 lesser accidents' involved in the maintenance, transport or modernization of USA nuclear weapons between 1945 and 1960.
A less definitive source reported that President Kennedy had been told subsequent to a 1961 investigation (following the Goldsboro, North Carolina, B-52 accident) that there had been 'more than 60' accidents involving USA nuclear weapons as of that date. As of mid-1976, the official USA figures include 27 major USA nuclear weapon accidents and 70 incidents - with no specific identification of its items in either category except for the 13 major accidents identified in 1968. There is one group of accidents in which nuclear weapons were believed to have been destroyed or seriously damaged. The 13 nuclear weapon accidents specifically identified by the Pentagon are included in this list and they are identified by source. The total number of accidents listed in this group is 32. Another group lists the incidents in which nuclear weapons were present or involved and in which they may have been placed in danger of destruction or serious damage. The total number of USA incidents in this group is 59. Yet another group lists a number of additional accidents or incidents which could fall into either category. In each of these events, nuclear weapons may have been present. Sufficient information is not available to confirm their presence or involvement. The total number of events listed in this group is 17. The final group lists a total of six Soviet nuclear weapon accidents, and 16 Soviet incidents, eight British nuclear weapon incidents, and four French incidents.
In a study of naval accidents from 1945, Greenpeace notes that the world's five nuclear navies have been involved in over 1200 accidents at sea and in ports. As a result, 50 nuclear weapons and 9 nuclear reactors lie abandoned on the seabed, presenting a serious long-term threat to the marine environment. The study documents 2,000 incidents including sinking, grounding, fire and collision but does not include incidents on or over land.
Broader Nuclear accidents (#PD0771) Injurious accidents (#PB0731).
Aggravates Nuclear war (#PC0842) Radioactive contamination (#PC0229)
False nuclear warfare alerts (#PF1236).
Aggravated by Mechanical failure (#PC1904) Insubordinate behaviour (#PJ6517)
Human errors and miscalculations (#PF3702) Proliferation of strategic nuclear arms (#PD0014)
Environmental hazards of nuclear weapons industry (#PE5698).

♦ PD3495 Theft of nuclear materials

Nature The world-wide growth in the use of nuclear materials, whether for weapons systems or in peaceful applications, leads to an increasing risk of theft of such materials by organized criminal groups, terrorists or even governments. The major risk lies with the possible theft of plutonium or uranium 233 or 238. Such materials may be stolen from nuclear power plants, uranium enrichment plants, or from the plants which prepare fuel elements for nuclear reactors. The main danger of theft occurs when such metals are moved from one location to another, but materials may also be stolen from the factories themselves.
Incidence It is estimated that almost 5 million pounds of plutonium and enriched uranium are in storage around the world. The Atomic Energy Commission in the USA records losses of as much as 100 pounds of uranium and 60 pounds of plutonium per year, mostly due to inaccurate inventory taking, but the amounts involved are enough to manufacture more than 10 bombs. An estimated 2 tons of plutonium, enough for 400 bombs, are unaccounted for by British authorities.
Broader Theft of property (#PD4691).
Aggravates Nuclear accidents (#PD0771)
Ease of manufacture of nuclear bombs (#PF3494)
Illegal exports of nuclear materials (#PE3968).
Aggravated by Proliferation of nuclear weapons and technology (#PD0837).

♦ PD3496 Foreign military presence

Extra-territorial military bases and intelligence centres — Erosion of national sovereignty by foreign military presence
Nature The presence of a foreign military power in a country may take the form of access to and use of military facilities (usually in the form of a military, paramilitary or clandestine base), or the actual presence of organized units of military personnel in foreign countries, or the deployment and permanent activity of fleets outside their own territorial waters. The military presence of the great powers in foreign territories is part of the mutual confrontation between the two opposing military and political blocs. In some cases they also serve to prevent political changes which are undesirable to the deploying country. The host country may have no jurisdiction over the military base and little control over the military or clandestine personnel in the country.
Incidence It is estimated that more than 20 countries maintain military forces and bases abroad. Among these, the USA, USSR, UK and France hold dominant roles, both in numbers of military forces and naval presence, and impact on the world's military balance and strategic planning. It is recognized that many bases also exist for use by big powers in wartime, under secret agreements, in which no foreign troops are deployed in peacetime. Covert (legal) and clandestine (illegal) presences of fighters, trainers, technical advisers, equipment maintainers, or military and civilian intelligence agents, evade the communication networks of the host country, whether pro- or anti-government, in order to pursue their own purposes. For example, the USA interventions in Latin America are frequently covert, and its base in Panama for many years served as the springboard for interventions to the south.
Refs Duke, Simon *United States Military Forces and Installations in Europe* (1989).
Broader Occupied nations (#PC1788)
Denial of right to national self-determination (#PC1450).
Related Erosion of sovereignty (#PE5015) Foreign military intervention (#PD9331).
Aggravates Prostitution (#PD0693) Military offences (#PC0742)
European insecurity and vulnerability (#PD1863)
Unaccountable government intelligence agencies (#PF9184)
Environmental degradation through military activity during peace-time (#PE0736).
Aggravated by Non-viability of small states and territories (#PD0441).
Reduces Political instability (#PC2677) National insecurity and vulnerability (#PB1149).

♦ PD3497 Arms trade with developing countries

Nature Savings on arms spending in the North could represent a promising prospect for massive increases in official development assistance to the South. Yet, instead of providing examples of restraint in this area, developing countries have been spending a greater proportion of their GDP's on the armed forces than the developed countries, with the added burden that most of their arms expenditures take the form of payment in hard currencies for imported weapons. In most cases, these countries are not responding to greater defence needs as much as to the intensive campaigns of the arms exporters, along with the importers' illusion that greater quantities of arms may bring greater security and stability. The Third World as a whole spends more on weapons than it spends on education and health combined; this obviously results in lower socio-economic growth rates and in greater chances of armed conflict.
Incidence The war between Iran and Iraq was able to continue for so many years partly due to the provision of arms from Europe and the United States. Between 1981 and 1988, Iran bought $17.5 billion worth of arms, Iraq bought $47.3 billion worth and developing nations as a whole purchased $301.4 billion worth. In 1988, the Soviet Union sold $9.9 billion, the United States sold $9.2 billion, and France and China, each, sold $3.1 billion worth of arms to developing countries. Some of the other nations involved in arms sales include: Czechoslovakia, North Korea, Britain, Italy, West Germany, Austria, Belgium, The Netherlands, Norway, Sweden, and Brazil.
Claim Developing countries are often misled into believing that increased armament stockpiles will result in a parallel increase in national security, rather than realizing that armaments lead to greater chances of conflict; and that the only benefits go to those who export arms, in general for reasons only of their own short-term interests.
Counter-claim It is unrealistic to assume that any single developing country will reduce arms spending unilaterally while its neighbours continue the arms race. In order for this to happen, there must be a mutual conviction of peaceful settlement, which is becoming increasingly unlikely as mutual trust steadily erodes.
Refs Brzoska, Michael and Ohlson, Thomas *Arms Transfers to the Third World, 1971-1985* (1987).
Broader International arms trade (#PC1358).
Aggravated by Militarization in developing countries (#PD9495).

♦ PD3501 Infanticide

Female infanticide
Nature The murder of young children, either individually or en masse, is usually a result of social attitudes. Infanticide has historically been associated with the birth of a deformed or abnormal child, of twins, or of a female child. The major contributing factor is generally poverty, although illegitimacy also plays a large role, especially in more contemporary societies. Infanticide also occurs during movements of people - warrior societies in the past, and refugee movements in recent times - and as a means of evading governmental policies such as China's "one child per family". Instead of outright infanticide, the child may be abandoned (though few are found alive), while other cases take the form of ritual sacrifices.
Background Infanticide has been a common practice since earliest prehistoric times among pagan, primitive tribes throughout the world. An example of the latter is the Jagas tribe of West African which killed all its own children, maintaining the population by adopting the children of its victims. Infanticide was strongly recommended by both Plato and Aristotle as a means of ensuing the stability of the population of Athens. In traditional Japanese culture, the term used was "thinning the family", whereby daughters were not killed at birth but were allowed to grow to an age when their strength or intelligence became apparent (2 to 4 years). The weaker were then drowned or decapitated. Infanticide was extensively practised in ancient China, especially in times of famine. It was a common practice in Rome where the father had the right to dispose of unwanted children by throwing them in the Tiber.
Incidence Examples of infanticide can be found in all societies, both past and present, ranging from isolated individual instances to general policy. Much attention has been focused on the practice in modern China where the number of infanticides (generally females) may exceed 10,000 per year, according to the US National Academy of Sciences.
Claim All human beings derive their essential value not from society, or from their parents but from God who gave them their life an to who they are infinitely precious. Society is judged by the extent to which it cares or fails to care for its weaker members. The parents may feel profoundly sorry for their handicapped child; and not only sorry for themselves, as is sometimes the case; but the decision to kill the child even for what they deem to be his or her 'own good' is one which they are not morally competent to make. The right to life is the infant's, and it is their own subjective feelings by which they judge that it would be better for the child to die. It is never morally permissible deliberately and directly to kill any innocent person. Morally there is no difference

between infanticide and murder.
Counter-claim The painless killing of newborn children who have gross physical or mental handicap is in the interest of their own family and of society as a whole because they would otherwise become an increasing social and economic burden to the community. It is in the infants' own interest and is an act of compassion as they cannot expect to enjoy the pleasures and opportunities available to normal children and adults. It is kinder to spare them the frustrations and hardships they must otherwise inevitably experience. Infanticide should be regarded as a less heinous offence than the murder of a grown child or adult, because an infant cannot experience fear or terror or even pain in a comparable degree, nor does its removal impose any significant hardship or loss on the family circle.
In some societies dealing with harsh environments, infanticide is an ecologically adaptation to reduce population to a size sustainable by the environment; maintain balance sex ratios among adults as young male hunters have high mortality rates; and helps preserve more older persons who are more productive than infants, and their accumulated knowledge increases information storage and retrieval for the group.
Refs Hausfater, Glenn and Hrdy, Sarah B (Eds) *Infanticide* (1984); Pakrasi, Kanthi Bhusan *Female Infanticide in India* (1976); Saxena, R K *Social Reforms* (1975).
Broader Homicide (#PD2341).
Related Genocide (#PC1056) Prenatal wrongful death (#PD6967).
Aggravates Infant mortality (#PC1287) Shortage of marriageable women (#PE0427).
Aggravated by Incest (#PF2148) Psychoses (#PD1722)
Polyandry (#PF3289) Abnormal births (#PG6522)
Primitive cultures (#PG6520) Children's diseases (#PD0622)
Irregular marriages (#PG6521) Religious offerings (#PG6518)
Neglected young children (#PE4245) Poverty in developing countries (#PC0149)
Deliberate imbalancing of population sex ratio (#PF3382).
Reduced by Induced abortion (#PD0158) Gender abortions (#PD3947)
Inadequacy of contraceptive methods (#PD0093).

♦ PD3507 Unsystematic allocation of market facilities
Nature The continuation of past economic attitudes which affirmed that the economy is limitless, self-sustaining and controlled by natural law, has resulted in there being no perceived necessity for systematic planning, organization and coordination of market facilities. This non-political operating stance is being radically called into question by the present collapse in the economy, as demonstrated in the misuse of the environment and the multitudes of poor people throughout the world.
Broader Breakdown in community security systems (#PD1147).
Narrower Self-conception of ethical void (#PF1178)
Non-political market allocation (#PD1211)
Entrenchment of vested interests (#PD1231)
Hierarchical control of market facilities (#PD7461).
Related Limited image of employability (#PF2896)
Conflicting social service ideologies (#PD3190)
Uncontrolled application of technology (#PC0418).

♦ PD3514 Slowness of sensori-motor activities in the aged
Nature Aged persons show simultaneous weakness of the sensory organs (ear, eye, etc) and slowness in motoricity, prolongation of integrating time, weak concentration and bad balance. All these symptoms are the consequence of ageing in general and, more particularly, of the ageing of the central nervous system.
Broader Human ageing (#PB0477).
Related Mental disorders of the aged (#PD0919).
Aggravates Social withdrawal of aged (#PD3518)
Inadequate housing for the aged (#PD0276)
Age discrimination in employment (#PD2318).
Aggravated by Loss of capacity with age (#PC8310)
Ageing of world population (#PC0027)
Susceptibility of the old to physical ill-health (#PD1043).

♦ PD3515 Rigidity and inadaptability in the aged
Broader Human ageing (#PB0477) Mental disorders of the aged (#PD0919)
Susceptibility of the old to physical ill-health (#PD1043).
Related Ageing of world population (#PC0027).
Aggravates Age discrimination (#PC2541) Social withdrawal of aged (#PD3518)
Criminalization of euthanasia (#PF2643) Inadequate housing for the aged (#PD0276)
Age discrimination in employment (#PD2318).
Aggravated by Social disadvantage of the aged (#PD3517)
Inadequate welfare services for the aged (#PD0512)
Decline in cognitive ability with ageing (#PE9620).

♦ PD3517 Social disadvantage of the aged
Nature People chronologically advanced in age suffer substantial disadvantages caused by biological ageing. In favourable societal circumstances, however, the aged would be able to enjoy a considerable period of good health, well-being and self-fulfilment, and to continue to contribute to societal activities; in fact, most cannot do so because of unfavourable conditions, particularly poverty. Their often premature physiological decline severely restricts the extent to which they can contribute to society; and further restrictions arise from societal conditions. In harsh economic situations, respect for age declines substantially, and the commercialization of peasant farming deprives the aged of their traditional role as policy and decision-makers. With the breakdown of the family and of community institutions, the aged suffer isolation and lack of motivation.
Broader Human ageing (#PB0477).
Narrower Social withdrawal of aged (#PD3518).
Aggravates Fear of death (#PF0462) Age discrimination (#PC2541)
Rapidly changing cultures (#PF8521) Ageing of world population (#PC0027)
Inadequate income in old age (#PC1966) Criminalization of euthanasia (#PF2643)
Inadequate housing for the aged (#PD0276) Inadequate recreational facilities (#PF0202)
Rigidity and inadaptability in the aged (#PD3515)
Susceptibility of the old to physical ill-health (#PD1043)
Retirement as a threat to psychological well-being (#PF1269).
Aggravated by Ageing war disabled (#PD0874).

♦ PD3518 Social withdrawal of aged
Turned-in elders' mindset — Elders' communication handicap
Nature Elderly people often suffer from a reduced level of communication brought about by a reduction in their mobility, and a related reduction in their ability to use body language. Because of this diminished mobility, many require help in personal care and thus they are forced to allow others to invade their private zones. The effect can be withdrawal, which may result in a diagnosis of dementia. Subsequently, the persons around the "patient" may then behave in a way which makes it obvious they consider he is demented, and that they are no longer interested in communicating with him. Many with a diagnosis of senile dementia are, in fact, under-stimulated or suffering from sensory deprivation. They react by withdrawal or by aggression which is then taken as a further symptom of their dementia.
Broader Human ageing (#PB0477) Social disadvantage of the aged (#PD3517)
Stifled potential for social interaction between different age groups (#PF6570).
Related Social integration handicap (#PE6779)
Fragmentation of social structures in depressed areas (#PD1566).
Aggravates Fear of death (#PF0462) Criminalization of euthanasia (#PF2643)
Social isolation of the elderly (#PD1564).
Aggravated by Bereavement (#PF3516) Ageing war disabled (#PD0874)
Mental disorders of the aged (#PD0919) Inadequate income in old age (#PC1966)
Inadequate housing for the aged (#PD0276) Inadequate recreational facilities (#PF0202)
Rigidity and inadaptability in the aged (#PD3515) Decline in cognitive ability with ageing (#PE9620)
Susceptibility of the old to physical ill-health (#PD1043)
Slowness of sensori-motor activities in the aged (#PD3514)
Retirement as a threat to psychological well-being (#PF1269).

♦ PD3519 Racial discrimination by security forces
Racism amongst military personnel — Racism within police forces
Claim Military life tends to appeal to people with strong prejudices, especially concerning colour and race. Racism is widely reported amongst military units. The issue has been avoided by limiting recruitment of certain races to particular units. Within mixed race units, information on the extent of racism is limited because of the consequences to the affected individuals of reporting on it. For this reason officers have little knowledge of the extent of day-to-day discriminatory practices although there may be awareness of discrimination in promotion.
Aggravates Military atrocities (#PD1881) Brutalization of military personnel (#PD7602).

♦ PD3520 Legal segregation
Broader Segregation (#PC0031) Racial segregation (#PC3688).
Related Apartheid (#PE3681).
Aggravates Inequality before the law (#PC1268).

♦ PD3521 Inappropriate labelling
Inadequate labelling of packages — Inadequate labelling of goods — Misleading labelling in packages
Nature Inadequate labelling of packages includes labelling which results in difficulty in determining: the weight of the contents (as opposed to that of the packet plus the contents); the unit price of the contents (as opposed to the price of the packaged goods); the date until which the goods may be safely used (if applicable); the exact composition of the packaged substance in chemical terms (if applicable); the conditions under which the package and/or contents should be stored; and the conditions under which it may be safely used, particularly if it is for human consumption. These deficiencies constitute either deliberate or inadvertent misrepresentation of the product. Sometimes, although the labelling of the product is accurate, the message is useless since it has been written in a language unknown to the final consumer or the measures are not those used in the country where the product is sold. All this can lead to a potentially hazardous use of the product or packaging.
Incidence In USA since 1985 food manufacturers have been allowed to claim just about anything for their products. The result is a confusing cacophony and a possibility to use dubious medical studies in selling the products. Oat bran seemed to lower blood cholesterol and so in labels it was almost declared to prevent heart disease.
Broader Consumer vulnerability (#PC0123) Unethical commercial practices (#PC2563).
Narrower Stereotypes (#PF8508) Fraudulent labelling (#PJ8826)
Inadequate labelling of dangerous substances (#PD2468).

♦ PD3522 Household pests
Pest infestations of buildings
Nature Household pests can make floors collapse, spread disease, and destroy furnishings.
Incidence In the UK, household pests cause millions of pounds of damage each year, with pest control a £100 million industry. Infestations of some sort are common in homes. It is estimated that 25 per cent of homes in the UK have dry rot or wet rot, and that half have woodworm; a further 67 pests are common. These figures are expected to increase with the effects of global warming.
Broader Pests (#PC0728).
Narrower Bedbugs (#PE3617) Wood rots (#PE4455) Carpet beetle (#PG1142)
Rodents as pests (#PE2537) Termites as pests (#PE1747) Houseflies as pests (#PE3609)
Wood-boring beetles (#PE0886) Insect pests of wood (#PD3586)
Lice as insect pests (#PE1439) Cockroaches as pests (#PE1633).
Aggravated by Floods (#PD0452) Humidity (#PD2474)
Global warming (#PC0918).

♦ PD3528 Corporate crime
Refs Purvis, R N *Corporate Crime* (1979).
Broader Unethical commercial practices (#PC2563).
Narrower Corporate crime in the pharmaceutical industry (#PE6618).
Aggravates Frauds, forgeries and financial crime (#PE5516).
Aggravated by Limited criminal liability of corporations (#PF7293).

♦ PD3529 Discrimination against foreigners in employment
Restrictions on employment of foreigners
Narrower Refusal of work permits (#PG6605)
Refusal to issue travel documents, passports, visas (#PE0325).
Related Refusal of entry to foreign workers' families (#PE8423).
Aggravates Discrimination (#PA0833) Illegal immigration (#PD1928).
Aggravated by Prejudice (#PA2173)
Restrictions on freedom of movement between countries (#PC0935).

♦ PD3533 Conflicts of labour law
Refs Morgenstern, Felice *International Conflicts of Labour Law. A Survey of the Law Applicable to the International Employment Relation* (1986).
Broader Conflict of laws (#PF0216).

♦ PD3535 Banned trade unions
Nature A situation in which the workers in a country are unable to form and join trade union organizations of their own choosing for the protection of their interests, is contrary to generally recognized principles relating to freedom of association. Sometimes freedom of association is denied on the basis of occupation, sex, colour, race, beliefs, nationality, political opinion, etc, not only to workers in the private sector of the economy but also to civil servants and employees of public services in general. Even if freedom to establish trade unions has long been recognized in a great many countries, authorization to establish associations is not in itself sufficient to guarantee their effective development, due to restrictions such as single trade unions, government favouritism of a trade union, or compulsory union membership.
Background The development of trade unions, largely a product of the Industrial Revolution, began in England in the 1820s and by the late 19th century, had spread to most of Western Europe. Significant efforts to form workers' associations were made in the 1920s and 1930s in Latin America, particularly among rural workers, and after the Second World War in Africa and Asia. In contrast to the situation in Europe, in the developing countries the organization of rural workers, largely employed in the European-imposed plantation system of agricultural production,

preceded or paralleled the organization of urban workers. The formation of trade unions provoked the reaction of employees and governments; it was only after World War II that trade unions were recognized as key partners in modern labour relations.
Broader Banned associations (#PD3536)
Denial of the right of trade union association (#PD0683).
Related Banned political parties (#PJ2274) Persecution of religious sects (#PF3353).
Aggravated by Conspiracy (#PC2555) State-controlled trade unions (#PG6615)
Violation of trade union rights (#PD4695)
Competition between intergovernmental organizations for scarce resources (#PE0063).

♦ PD3536 Banned associations
Broader Denial of the right of association (#PD3224).
Narrower Banned trade unions (#PD3535) Banned political parties (#PJ2274)
Persecution of religious sects (#PF3353).
Related Infringement on the functioning of legitimate organizations (#PE5222).
Aggravated by Conspiracy (#PC2555).

♦ PD3537 Infringement of trade secrets
Infringement of proprietary information
Incidence In the USA a growing number of States impose criminal liabilities on thefts of industrial secrets, in addition to the practice of awarding civil damages to plaintiffs whose proven proprietary trade secrets are lost through breach of employee confidentiality. Incidences include Japanese and Russian industrial espionage, in addition to the domestic variety endemic in some industries such as toys, fashion and other consumer goods, and in the high technology and computer related sectors.
Background Commercial property can include: unpatented inventions, processes and technologies; marketing, pricing, promotion, and distribution data; strategic plans; and internal corporate financial statements. All of these need to be kept secret to avoid offering competitors, domestic or foreign, unintentional help.
Refs Practising Law Institute *Protecting Trade Secrets 1986* (1986).
Broader Industrial espionage (#PC2921).
Related Avoidance of copyright (#PD0188).

♦ PD3542 Political police
Nature Under totalitarian regimes, thought police may be created to monitor publications and other expressions of opinion. Censorship and suspension of publishing, broadcasting, printing and even photocopying becomes common. Political figures, editors, writers and other dissenters may be arrested, sent to internal exile or even murdered. Political police, under whatever name, are the counterparts to political prisoners. They are the agents of brutal regimes whose vague directives against 'enemies of the state' encourage excessive and criminal mistreatment of those whose fault is to desire freedom of thought and expression.
Incidence Political policing has existed within the national security police forces of revolutionary France, Czarist Russia, the Third Reich, Communist East European countries, imperial Iran, militarily-led modern Greece, and under a number of dictatorships in Latin America, Africa, Asia, and Oceania.
Broader Repression (#PB0871) Political repression (#PC1919).
Narrower Police state (#PD7910)
Maintenance of political dossiers on individuals (#PD2929).
Related Vigilantism (#PD0527) Police intimidation (#PD0736) Military police abuse (#PG6627)
Unauthorized police search (#PD3544)
Misuse of electronic surveillance by governments (#PD2930).
Aggravates Political trials (#PD3013) Abuse of police power (#PC1142).
Aggravated by Political corruption of the judiciary (#PE0647).

♦ PD3543 Police brutality
Nature Police brutality is the use of excess force to arrest criminals and to control crowds and of torture to interrogate detainees and prisoners. It is used to intimidate individuals. The methods of police brutality are like those of torture: beating; threats of violence; denial of food, drink, or rest to the person interrogated; prolonged questioning and the like. Because the methods are often used in secret it is impossible to determine their extent. When used against crowds or individuals in public places, witnesses are intimidated or killed, inquiries are suppressed or guilty officers are let off with minor sentences.
Incidence Police brutality is widespread and in some places expected and even encouraged by governments.
Broader Brutality (#PC1987) Abuse of police power (#PC1142)
Unethical practices by police forces (#PD9193).
Narrower Forced confession (#PE8947).
Related Torture (#PB3430) Corruptive crimes (#PD8679) Police intimidation (#PD0736)
Forced confessions with drugs (#PE4888) Inhumane methods of riot control (#PD1156).
Aggravates Fear of police (#PF8378) Civil disorders (#PC2551)
Mistrust of police (#PF8559)
Proliferation of weapons in civilian hands (#PE2449).
Aggravated by Police corruption (#PD2918) Inadequate riot control (#PD2207).

♦ PD3544 Unauthorized police search
Broader Abuse of police power (#PC1142) Infringement of privacy (#PB0284)
Unethical practices by police forces (#PD9193).
Narrower Search without warrant (#PG6631) Arbitrary street search (#PG6632).
Related Political police (#PD3542).
Aggravates Police intimidation (#PD0736) Internment without trial (#PD1576).
Aggravated by State custody of deprived children (#PD0550).

♦ PD3546 Nationalistically determined development of natural resources
Parochial planning for natural resource extraction — Natural resources used for national self-interest
Nature There tends to be a feeling of helplessness when faced with the complexity of an increasingly unbalanced ecological system, and little capability of creating an equitable distribution of basic raw materials. Little research is being done in the area of developing access to resources, which tends to be determined by nationalistic, short-term plans with no regard for local community considerations or global requirements. The result is an overemphasis on profits, leading to uneven resource development and extraction.
The absence of a comprehensive global plan for the extraction, distribution and sharing of natural resources inhibits their equitable utilization. Operating without a system of priorities, and without global standardization and control, results in obsolescence, waste and misuse. A sense of responsibility to overall needs is lacking, and the world's natural resources are misused and exploited with little regard for the claim of future generations upon available reserves.
There is an underlying problem of national self-interest in the development and distribution of natural resources. In their use of natural resources, many developed nations adopt the attitude that they are unrelated to the rest of the world; they see no overall responsibility for sharing and refining such natural resources for the benefit of all nations. This is exemplified by nations which, by using large nets and factory processing methods aboard ship, are depleting the world's oceans of fish, one of the most important food natural resources.
Broader Non-inclusive management decisions (#PF2754).
Narrower Short-term profit maximization (#PF2174)
Inequitable access to resources (#PD3210)
Lack of faith in solutions to ecological issues (#PD2888)
Over-emphasis on immediate solutions in resource development research (#PE4059).
Related Obsolete vocational skills (#PD3548)
Limited access to natural resource use decisions (#PF2882)
Profit motivated utilization of construction technology (#PF2464).
Aggravates Destruction inherent in development (#PF4829).
Aggravated by Parochial national interests (#PF2600).

♦ PD3548 Obsolete vocational skills
Outmoded work-study programmes — Obsolete employment training — Outdated job training
Nature There are increasing numbers of people unemployed or underemployed because, despite having skills which were once relevant, they lack appropriate and up-to-date occupational skills. They are, in fact, penalized by the progress of technology. Skilled workers can seldom utilize their full potential; others perform poorly because they do not have the necessary training. There are 'holes' in the job market and unmet needs among services for which the skills have not yet been determined.
Broader Disparity between workers skills and job requirements (#PC1131).
Narrower Outmoded functional skills in rural communities (#PF2986).
Related Inadequate technical training (#PE8716)
Limited access to natural resource use decisions (#PF2882)
Failure to employ skills of home-bound educated women (#PD8546)
Profit motivated utilization of construction technology (#PF2464)
Nationalistically determined development of natural resources (#PD3546).
Aggravates Maladjustment to disciplines of employment (#PD7650)
Limited means of marketing employable skills (#PE7344)
Lack of opportunities for practical training in communities (#PF2837).
Aggravated by Social isolation (#PC1707) Limited image of employability (#PF2896)
Unimaginative educational vision (#PF3007) Non-inclusive management decisions (#PF2754)
Social bias in planning of training programmes (#PF2885)
Lack of job opportunities for some sectors of society (#PD2972).
Reduced by Lost knowledge (#PF5420).

♦ PD3555 Family dependence on patriarchal role of the man
Dependence of wife on husband — Dependence of children on father
Nature An economic tie traditionally binds both wife and children to the father as the provider. It is this bond which enables the father to act in an authoritarian manner, subjugating the wife and any children. This bond is a major factor hindering the emancipation of the woman.
Aggravates Social subjugation of women (#PD4633)
Economic barriers to women's access to the judicial process (#PE1198).

♦ PD3557 Excessive employment of married women
Adverse social consequences of excessive employment of married women
Incidence A 1975 ILO report on the participation of married women aged 15 and over in the labour force of selected countries showed, for example, that the percentage of married women among active women was 34 for France, 61 for Great Britain, 57 for the USA, 36 for Argentina, 68 for Hungary, 30 for Chile, 57 for Sweden.
Claim Married women who work are taking jobs away from men who really need to work.
Counter-claim Many married women need to earn a substantial portion of their families' income and many more provide the whole support to their families.
Broader Social inequality (#PB0514)
Discrimination against women in employment (#PD0086).
Narrower Gender stereotyping of employment (#PJ6290).
Aggravates Fatigue (#PA0657) Social breakdown (#PB2496) Mental disorders (#PD9131)
Marital instability (#PD2103) Juvenile delinquency (#PC0212)
Inadequate child welfare (#PC0233) Discrimination against women (#PC0308)
Inadequate housing for the aged (#PD0276) Inadequate child day-care facilities (#PD2085).
Aggravated by Leave of absence (#PG6639) Social injustice (#PC0797)
Material expectations (#PG6637) Socio-economic poverty (#PB0388)
Prohibitive cost of living (#PF1238)
Disruptive maternity leave of employees (#PG6638).

♦ PD3568 Volcanic eruptions
Volcanoes
Nature Volcanic eruptions in inhabited regions result in loss of life and destruction of habitat and agricultural land. Destruction is caused by the ejection of molten lava (either as flows or as fountains thrown hundreds of feet into the air), angular blocks, mud, ash (possibly incandescent), dust and gas. The speed of the lava flows (up to 160 km/hour) and the ash may mean a community in the neighbourhood is obliterated in just a few seconds. An eruption may also be prolonged for several years or even centuries. The most terrible of all volcanic manifestations are glowing avalanches or nuées ardentes which give little forewarning, travel at hurricane speeds and are hot enough both to kill instantly by cadaveric spasm and to carbonize the corpse by destructive distillation.
Incidence It is probable that every part of the earth's surface has been the site of volcanic activity at some stage. The active and most recently active volcanoes are however at present concentrated largely in a great belt encircling the Pacific Ocean and in a shorter belt extending from the Solomon Islands through New Guinea and Indonesia; the active volcanoes of the Mediterranean region are regarded as lying on an extension of the latter belt. There are also several active volcanoes in the Atlantic. Of the world's 543 volcanoes with known recorded activity, 90 erupted between 1966 and 1970. Of the 760 catalogued active volcanoes of the world, only 95 are known to have caused casualties during historic times. In the casualty league Japan tops the list with 14 disastrous eruptions; Java comes second with 12; the Philippines and Central America are equal third with 6 each. However, it was Mt Etna in Sicily, which caused the greatest loss of life during an eruption, when in 1669 some 100,000 people died in a particularly violent outburst. A similar event today could result in as many as two million deaths. As many as 20,000 people died in central Columbia when the Nevado del Riuz volcano erupted in November 1986. It was one of the worst volcano eruptions of this century.
Refs Blong, R J *Volcanic Hazards* (1984); Booth, B, et al *Source-Book for Volcanic-Hazards Zonation* (1984).
Broader Bad omens (#PF8577) Geological hazards (#PC6684).
Narrower Lava flows (#PE3937) Volcanic dust (#PE5109).
Aggravates Dust (#PD1245) Tsunamis (#PD0033) Soil pollution (#PC0058)
Rock avalanches (#PG0476) Evolutionary catastrophes (#PF1181)
Stratospheric ozone depletion (#PD6113).
Aggravated by Global cooling (#PF1744).

♦ PD3574 Pesticide destruction of soil fauna and micro-organisms
Nature Modern agriculture implies the application of large amounts of potent chemicals and pesticides onto the land. Some are applied directly to the soil, others to foliage, but most of the pesticide residues reach the top three inches of soil where the majority of soil fauna are found. Many of the soil fauna and micro-organisms that are essential in the processes of soil formation

and maintenance of soil fertility, are killed on contact with these pesticides. The main soil contaminants are arsenic and the chlorinated hydro-carbon insecticides such as DDT, dieldrin and BHC, along with their toxic by-products. Many organic phosphates, when abundant, reduce plant growth.
Incidence DDT has been found in agricultural soils at rates up to 10–100 pounds per acre.
Broader Pesticides as pollutants (#PD0120).
Aggravates Soil degradation (#PD1052) Pesticide damage to crops (#PD2581)
Diseases of beneficial insects (#PD2284)
Health risks to workers in agricultural and livestock production (#PE0524).

♦ PD3577 Fungicide damage to crops
Nature Nearly all fungicides are injurious to plants as well as to fungi. If misused they can damage crop yields.
Refs Dekker, J, et al *Fungicide Resistance in Crop Protection* (1982).
Broader Pesticide damage to crops (#PD2581).
Aggravated by Fungicides as pollutants (#PD1612).

♦ PD3585 Pests and diseases of trees
Forest pests
Nature Many insect pests and fungus diseases attack trees, causing great damage or death. The nature of the problem is different for ornamental and forest trees. For the former, concern is focused on the effect of disease or pests on an individual tree; but for the latter it is the effect on stands of trees which is important. The former are grown for their appearance, the latter for their timber, and so diseases affect their worth differently.
Incidence From seed to maturity forest trees are subject to a succession of diseases. Timber losses due to pests and fire amount to 92 percent of which 45 percent is accounted for by disease. One reason for this high incidence of destruction is the intensiveness and uniformity of most commercial forests. Pests and diseases are no less a problem on ornamental trees, many of which are exotic species, often unsuited to the local environment and with no developed resistance to local pests and pathogens.
Pests and diseases are rarely detected during the incipient stages of damage or infection. Tree destruction or mortality is usually first noticed when very obvious and in some instances the pest may have already disappeared. Several pests may occur at the site, and it is difficult to identify the primary pathogen (s). Another problem is identifying the particular pest since many species are undescribed. A number of insects and caterpillars feed on the foliage of trees; if totally or partially defoliated, a tree will be weakened and more susceptible to attack by disease. Insect infestations of the twig, leaf, and bark may lead to localized necrosis, but in susceptible species the whole tree or even complete stands may be destroyed. The nun moth, for example, has caused extensive damage in European forests, and in North America forests have been virtually wiped out over vast areas by bark beetles and by the spruce budworm. Sucking insects, such as aphids, mealy bugs and scale insects, which are parasitic on trees, severely weaken and reduce the vitality of their hosts. Nearly all tropical forest insect pests belong to one of five orders: Coleoptera, Hemiptera, Isoptera, Lepidoptera and Orthoptera. Forest insect damage may be classified into four broad categories: defoliation (direct or indirect) caused largely by larvae of Lepidoptera and by a few Coleoptera and Orthoptera; boring or mining inside seeds, bark, wood and shoots, mainly due to species of Coleoptera and Isoptera and to a few Lepidoptera; chewing of bark and wood largely by Coleoptera; and necrosis and wilting due to chemicals secreted by some Hemiptera.
Numerous caterpillars defoliate forest trees: in Europe, those of the nun moth on spruce, winter moth and mottled umber moths on broad-leaved trees and pine looper and pine shoot moth; in North America, various tent caterpillars, and those of the gypsy moth, brown-tail moth, spruce budworm and larch case-bearer. The North America fall webworm is a major pest of shade trees in Europe. The Japanese beetle has infested a large area in the United States, defoliating shade and fruit trees. Weevils attack the growing shoots of pines. Wood-boring beetles cause a great deal of damage to park, woodland and forest trees. Indian forests suffer extensive damage from the sal borer.
Other insects, which do little primary damage to trees, are responsible for considerable secondary damage because they are vectors of tree diseases. For example, the Dutch elm disease fungus is introduced into the vascular system of healthy trees by the elm bark beetle. Most of the important tree diseases are fungal, such as chestnut blight, a canker disease which, since its introduction into North America in 1904, has practically annihilated the native chestnut. A similar scourge is white pine blister rust, which spreads rapidly through the tissues of an infected tree, and from one tree to another, invariably killing its hosts. Fungal diseases can take a number of forms of which the most important are wilts, cankers, rusts, heart rots and leaf spots.
Refs Gibson, J A *Diseases of Forest Trees Widely Planted As Exotics in the Tropics and Southern Hemisphere*; Sinclair, Wayne A, et al *Diseases of Trees and Shrubs* (1987).
Broader Pests (#PC0728) Endangered forests (#PC5165).
Narrower Borer insects (#PE6286) Heart rot fungi (#PE6269)
Infestation of seeds (#PE6271) Fungal plant diseases (#PD2225)
Scale insects as pests (#PE3612) Absence of mycorrhizae (#PE6291)
Beetles as insect pests (#PE1679) Pests and diseases of tea (#PE2980)
Pests and diseases of elm (#PE2982) Pests and diseases of oak (#PE2984)
Hemiptera as insect pests (#PE3615) Orthoptera as insect pests (#PE3641)
Pests and diseases of palms (#PE2981) Pests and diseases of cocoa (#PE2979)
Lepidoptera as insect pests (#PE3649) Pests and diseases of coffee (#PE2218)
Pests and diseases of rubber (#PE2977) Pests and diseases of olives (#PE2978)
Pests and diseases of poplars (#PE4413) Pests and diseases of chestnut (#PE2983)
Pests and diseases of citrus fruit (#PE2976)
Pests and diseases of deciduous fruit (#PE3591).
Related Crops pests and diseases (#PE7783) Pests and diseases of groundnut (#PJ1181).
Aggravates Forest decline (#PC7896).
Aggravated by Forest fires (#PD0739) Pests of plants (#PC1627)
Insect vectors of viral diseases of plants (#PD3600).

♦ PD3586 Insect pests of wood
Nature Various insects can cause extensive damage to wood products.
Incidence The particular enemies of structural timbers and furniture are anobiid and lyctid beetles, notably the furniture beetle, the deathwatch beetle and the powder-post beetles. The majority of termites are earth-dwelling species that attack any wood in contact with the ground; but the dry-wood termites, which are independent of the ground, include several destructive and widely transported species. In many parts of the world, the old house borer, wood wasps and wood-boring caterpillars seriously infest timber and wooden structures.
Broader Insect pests (#PC1630) Household pests (#PD3522)
Insect damage to stored and manufactured goods (#PD3657).
Narrower Teredos as pests (#PE3624) Termites as pests (#PE1747)
Beetles as insect pests (#PE1679).
Related Pests of plants (#PC1627) Wood deterioration and decay (#PD2301).
Aggravates Structural failure (#PD1230).

♦ PD3587 Plant diseases in storage and transit
Nature Tubers, fruits, fresh vegetables, seeds and grain are subject to deterioration and spoilage during storage and transit due to a number of pathogenic and non-pathogenic agents.
Incidence The losses due to disease during transit and storage sometimes equal those that occur while the plants are growing. This is especially true of fresh fruit and vegetables, and seeds, such as those of wheat, corn, barley, soybeans and flax, which are stored in bulk for months or even years. Storage diseases are divided into those caused by nonpathogenic factors and those caused by living organisms. But damage beginning from nonpathogenic causes may be increased greatly by subsequent invasion of the tissues by bacteria and fungi able to cause rapid decay. Fruits and vegetables in storage also suffer from a number of physiological diseases caused by an excess of gases which are self generated. Common fungi, such as botrytis, penicillium, rhizopus and Sclerotinia, invade and rot many fruits and vegetables. Losses of up to 25 per cent between harvest and consumption are common in oranges, apples, peaches, pears, plums, potatoes, sweet potatoes, tomatoes and peppers. Bacteria, or a combination of fungi and bacteria, often rot stored potatoes and root vegetables. Grains stored in bulk are subject to invasion by a number of fungi, principally those in the genus aspergillus. These reduce germinability of seeds, which is important in those to be used for planting or malting, and may reduce the quality of grains for processing
Broader Dissemination of plant diseases by man (#PD3593).
Narrower Insect damage to stored and manufactured goods (#PD3657).
Related Fruit rot (#PE5373) Wood deterioration and decay (#PD2301).

♦ PD3593 Dissemination of plant diseases by man
Human vectors of plant disease
Nature Man is an important agent in the dissemination, over both short and long distances, of plant pathogens. Frequently in his commercial, agricultural or recreational activities, man has carried plant pathogens, previously confined to a particular area, to new localities where they have caused tremendous damage, even though they may have been relatively harmless in their native habitat.
Broader Plant disease vectors (#PD3596) Man as vectors of disease (#PD8371).
Narrower Plant diseases in storage and transit (#PD3587).
Related Bird vectors of plant disease (#PD3601) Plant vectors of plant disease (#PD3599).
Aggravates Crop vulnerability (#PD0660).
Aggravated by Inadequate plant quarantine (#PE0714)
Obstacles to the international transfer of corpses (#PG3537).

♦ PD3596 Plant disease vectors
Nature Many of the pathogenic organisms that cause disease in plants are transferred from plant to plant, from field to field and from region to region by another living organism. In fact, some diseases require for their propagation some such intermediary, for their causal pathogens are not adapted for dissemination by wind, water or other inanimate means. Many vectors of plant disease not only transport pathogens, but also introduce these organisms directly into the plant by biting or sucking or in other ways puncturing or rupturing the surface of the plant.
Incidence Insects are by far the most important vectors of plant disease. Other anthropods are also important. Among higher animals, birds and man are important in disseminating plant pathogens. A distinction can be made between man as a vector of plant disease and man as a disseminator of other vectors and diseased plants. Several plant viruses, notably that of American peach mosaic, are transmitted by mites (eriophyidae), and nematodes spread a number of plant viruses of economic importance. The spores of the fungus olpidium brassicae transmit two viruses that affect tobacco and lettuce plants.
Broader Disease vectors (#PC3595).
Narrower Bird vectors of plant disease (#PD3601)
Plant vectors of plant disease (#PD3599)
Insect vectors of plant diseases (#PD7732)
Dissemination of plant diseases by man (#PD3593).
Related Plant pathogens (#PD1866).
Aggravates Plant diseases (#PC0555) Crops pests and diseases (#PE7783)
Pests and diseases of groundnut (#PJ1181).
Aggravated by Pests of plants (#PC1627) Crop vulnerability (#PD0660).

♦ PD3598 Inadequate weed control in developing countries
Nature Weed control is a particularly acute problem in developing countries. Weed plants grow more vigorously and regenerate more quickly in tropical than in temperate zones because of the heat and higher light intensity. Much more is known about weeds in the developed than in the developing countries, and the techniques of control – mechanical and chemical – are often unsuitable to the physical, social and economic conditions of developing countries. Biological controls are a relatively cheap self perpetuating control and are available and proven in some countries for major tropical weeds including water hyacinth. The problem of water weeds has been aggravated by development projects such as reservoirs, irrigation canals and dams. The greater quantities of human effluent and fertilizers in these waters make them richer them natural waters and thus weeds flourish more easily. The effects of an abundance of weeds include slower water flow (by as much as 80 percent) and blockage of shipping and fishing.
Incidence The most destructive weeds are purple nutsedge, Bermuda grass, barnyard grass, junglerice, goosegrass, Johnson grass, Guinea grass, water hyacinth, cogon grass, and lantana. They occur in every major agricultural area in the warmer regions of the world.
Broader Pests (#PC0728) Underdevelopment (#PB0206).
Aggravates Weeds (#PD1574) Aquatic weeds (#PD2232).
Reduces Destruction of weeds (#PE3987).

♦ PD3599 Plant vectors of plant disease
Nature The outbreak and spread of plant diseases and pests creates a serious threat to agricultural products of great importance to international trade. Consignments of plants, plant products or other articles or commodities moving in international traffic may act incidentally as carriers of pests and diseases of plants.
Broader Plant disease vectors (#PD3596).
Narrower Fungi as vectors of plant disease (#PE8138).
Related Bird vectors of plant disease (#PD3601) Insect vectors of plant diseases (#PD7732)
Dissemination of plant diseases by man (#PD3593).
Aggravates Plant diseases (#PC0555).

♦ PD3600 Insect vectors of viral diseases of plants
Broader Insect vectors of plant diseases (#PD7732).
Aggravates Viral plant diseases (#PD2227) Pests and diseases of trees (#PD3585)
Pests and diseases of cocoa (#PE2979) Pests and diseases of sugar-beet (#PE2975).

♦ PD3601 Bird vectors of plant disease
Avian vectors of plant diseases
Nature Several birds are important as agents of dissemination of plant pathogens.
Incidence Though comparatively small in number, bird vectors of plant disease may have a great impact. For example, endothia parasitica, the fungus which causes Chestnut Canker, produces sticky spores which adhere to the feet of birds. These birds carry the spores over considerable distances: in fact, they made it impossible to eradicate the fungus after it had been accidentally introduced into the eastern United States from the Orient.
Broader Plant disease vectors (#PD3596).

Related Plant vectors of plant disease (#PD3599)
Insect vectors of plant diseases (#PD7732)
Dissemination of plant diseases by man (#PD3593).
Aggravates Ornithosis (#PE2578) Plant diseases (#PC0555)
Pests and diseases of chestnut (#PE2983).
Aggravated by Birds as vectors of disease (#PE6659).

♦ PD3603 Endangered species of edentates

Edentata

Nature Threatened species of Edentata include the giant anteater, the Brazilian three–toed sloth, the giant armadillo, the Brazilian three–banded armadillo, the pink fairy armadillo, and Burmeister's armadillo. These inhabit parts of Central and South America and reasons for possible extinction include habitat loss (mainly through deforestation, settlement, and agriculture), being hunted for food, or being preyed upon by local dogs.
Broader Endangered species of mammals (#PC1326).
Narrower Endangered species of sloths (#PG6662)
Endangered species of anteaters (#PG6661)
Endangered species of armadillos (#PG6663).

♦ PD3625 Lack of natural resources in developing countries

Broader Lack of natural resources (#PC7928).
Aggravates Inadequate production capacity in developing countries (#PD4219)
Inadequate development of enterprises in developing countries (#PE8572).
Aggravated by Degradation of the environment in developing countries (#PD3922).

♦ PD3634 Insect pests of plants

Nature Practically all of the crops grown by man have important insect pests. Insects inflict damage on plants by their feeding and reproductive activities.
Incidence No figures exist for the global damage inflicted on crops by insects but it has been estimated that, for the USA, insects cause an annual loss to crops of about $2 billion, nearly 5 percent of the potential production. Insects can damage crops directly, or indirectly as vectors of disease. The direct damage can also provide an avenue for bacterial or fungal infection. Examples are the mealy bug and the green spider mite. About 15 years ago, these two South American insects were accidentally carried into Zaire and Uganda. Since then they have eaten their way across Africa, destroying up to two–thirds of the cassava crop in some places. To the African, cassava is vital; some 200 million people eat the cassava's large leaves and use its tuberous roots for flour, bread, tapioca, and even alcohol. Without it, the people starve.
Refs Annecke, D P and Moran, V C *Insects and Mites of Cultivated Plants in South Africa* (1983); Croft, B A and Hoyt, S C (Eds) *Integrated Management of Insect Pests of Pome and Stone Fruit* (1983); Martineau, R *Insects Harmful to Forest Trees* (1984).
Broader Insect pests (#PC1630) Pests of plants (#PC1627).
Narrower Bee pests (#PG5249) Water flea (#PG6595)
Borer insects (#PE6286) Mites as pests (#PE3639)
Beet leaf miner (#PG6915) Termites as pests (#PE1747)
Crickets as pests (#PE3640) Defoliating insects (#PE4999)
Fruit flies as pests (#PE3607) Beetles as insect pests (#PE1679)
Homoptera as insect pests (#PE3614) Hemiptera as insect pests (#PE3615)
Dermaptera as insect pests (#PE3618) Lepidoptera as insect pests (#PE3649)
Thysanoptera as insect pests (#PE3619) Grasshoppers as insect pests (#PE3642).
Related Weeds (#PD1574) Animal pests (#PD8426) Birds as pests (#PD1689)
Rodents as pests (#PE2537) Flies as insect pests (#PE2254)
Insect damage to stored and manufactured goods (#PD3657).
Aggravates Plant diseases (#PC0555).
Aggravated by Crop vulnerability (#PD0660) Inadequate crop rotation (#PF3698)
Inadequate plant quarantine (#PE0714) Insect resistance to insecticides (#PD2109)
Introduction of new species of insect pests (#PF3592).

♦ PD3647 Alkaline soil

Shallow alkaline topsoil

Nature The term alkaline soil has been rather loosely applied to all soils containing sufficient amounts of soluble salts to cause injury to plant life. When correctly used, the term alkaline refers only to those soils that have a high pH value, usually through the presence of alkaline sodium salts, particularly sodium carbonate. Saline soils are not necessarily very alkaline, this depends on the type of salt. Soil alkalinity tends to be associated with low rainfall: soluble bases released by weathering are not leached from the soil and consequently the soil becomes more alkaline. The fertility of such soils is usually variable and often quite low. Strongly alkaline soils are sticky, impervious to water, and unfavourable to agriculture.
Incidence Alkaline soils are widely distributed in the drier areas of the world. A great deal of foodstuff is produced on soils where salts may be a problem. In regions of low rainfall, salts accumulate where drainage is poor. The tolerance of various plants to soluble salts in soils differs greatly. Among the less tolerant agricultural crops are beans, peas, clover, vetch, oats and peaches. The presence of high concentrations of salts limits the intake of water by plant roots. Yields are reduced approximately in proportion to the osmotic pressure of the nutrient solution. Concentrations of more than 0.2 percent of salts in a soil will harm crop yields. Alkaline salts are toxic at even lower concentrations. Many irrigation projects have failed because the soil has become alkaline through lack of proper drainage facilities or failure to use enough water to move excess salts down and out of the soil. Others have failed because of the high concentration of sodium in the irrigation water.
Broader Soil infertility (#PD0077).
Related Soil salinization (#PE1727).
Aggravates Infertile land (#PD8585) Environmental plant diseases (#PD2224)
Destruction of land fertility (#PC1300) Deficiency diseases in plants (#PD3653)
Underutilization of natural resources (#PF1459).
Aggravated by Inadequate land drainage (#PD2269)
Silting of water systems (#PD3654)
Inadequate irrigation system (#PD8839)
Unsustainable short–term improvements in agricultural productivity (#PE4331).

♦ PD3650 Sand storms

Nature When dry sandy surfaces, unprotected by vegetation, are exposed to winds, the coarser material, the particle size of sand, may be carried along close to the ground, transported by a combination of the rolling and leaping of the individual grains. Where the sand is abundant and the wind is very strong with speeds of more than 15 or 20 mph, a sandstorm results, with the air filled with flying sand up to a height of several feet, having the appearance of 'streaming'.
Incidence The impact of the bouncing, jumping grains severely abrades the surface of the ground. This abrasion breaks down clods, destroys stable crusts, and wears down vegetative residues and living vegetation. Thus the sand in a sand storm is a major agent in erosion, both of the soil and of objects, natural and artificial, in its path. It can damage buildings, constructions, machines, crops and even barriers erected to control sand movement. The harsh sand can cause intense pain and even injury to animals and people in its path. The sandstorm continues until the wind abates (it may often last for days) or until some obstacle is encountered, then the sand accumulates into heaps and mounds, known as dunes. Fields, buildings, roads, and even villages have been known to be overwhelmed and obliterated. Sand storms are obviously most common in areas where there is much loose sand on the ground, in particular in deserts, on sandy coasts and near rivers which vary in volume, leaving sand beds exposed during the dry season. Deserts in which sand storms frequently occur include the Sahara, the Gobi, the Mongolian, the Arabian desert, and those of Iran and Turkistan. Wind–blown sands comprise 11 percent of the Sahara and 30 percent of the Arabian desert. Several decades ago, the streets of Swakopmund in South–West Africa were swept over by a sandstorm that piled up sand dunes 6 metres high.
Broader Inhospitable climate (#PC0387).
Related Dust storms (#PD3655) Migrating sand dunes (#PD0493).
Aggravates Desert advance (#PC2506) Destruction of land fertility (#PC1300)
Deterioration in atmospheric visibility (#PE2593).
Aggravated by Wind (#PE2223) Defoliation (#PD1135)
Deforestation (#PC1366) Soil erosion by wind (#PE3656).

♦ PD3653 Deficiency diseases in plants

Nature Certain elements are essential for plant growth. If these elements are absent from the soil, or present in insufficient quantities, plants will exhibit limited or stunted growth, reduced yields, and sometimes deficiency symptoms. By the time deficiencies appear, plant growth and yield are usually irretrievably retarded. As well as reduced yields, food and fodder crops may also, if the soil is deficient, lack certain elements essential for animal and human health. Thus people and grazing animals may develop deficiency diseases through eating deficient crops.
Incidence Most higher green plants require the following elements: carbon, hydrogen, oxygen, nitrogen, potassium, phosphorus, sulphur, calcium, iron, magnesium, boron, manganese, zinc, copper, chlorine and molybdenum. Those elements must be present in the soil in sufficient quantities (though not in excess, as this may prove lethal to plants). The necessary quantities vary: copper, molybdenum, manganese, zinc, boron, iron, and chlorine are required in trace quantities only. A deficiency of any one of the 16 essential elements results in stunted growth and reduced yield. The elements whose lack is most likely to have a limiting effect on growth are nitrogen, phosphorus and potassium. Some generalizations can be made about the deficiency symptoms of these three main elements. When nitrogen moves out of the older, hence lower, leaves of a plant, the deficiency is generally characterized by yellowing of these leaves. Phosphorus deficiency is characterized by a purpling of the stem, leaf, or veins on the underside of the leaves. Potassium deficiency results in burn or scorch of the margin of the leaves, particularly the older, lower leaves.
Broader Plant pathogens (#PD1866).
Narrower Dehydration (#PE8062).
Related Plant cancer (#PE1899) Animal diseases (#PC0952)
Parasites on plants (#PD4659) Nutritional diseases (#PD0287)
Viral plant diseases (#PD2227) Fungal plant diseases (#PD2225)
Bacterial plant diseases (#PD2226) Environmental plant diseases (#PD2224).
Aggravates Suffering of plants (#PC7825).
Aggravated by Acidic soils (#PD3658) Alkaline soil (#PD3647)
Soil compaction (#PD1416) Soil infertility (#PD0077)
Soil salinization (#PE1727) Inadequate crop rotation (#PF3698)
Destruction of land fertility (#PC1300).

♦ PD3654 Silting of water systems

Siltation of rivers — Siltation of dams — Sedimentation of dams

Nature Water systems may be blocked, or the flow of water seriously reduced, by the deposition of fine material carried by the water. This may curtail the use of water systems for transport, for fishing, as reservoirs, for irrigation, or for drainage.
Incidence The incidence of silting is high when the flow of water is slow and when there is a large amount of material suspended in the water. The salinity of the water and tidal flow affect as well the rate of sedimentation. The material, or silt, may arise from excessive soil erosion or maybe ordinary land run–off, or sewage. Silting occurs in natural water systems, but in man–made water systems results through bad planning and construction. The flow of water may be reduced by weeds, whose growth has been stimulated by the excess of nutrients suspended in the water. The reduced flow allows increased sedimentation, which further stimulates weed growth. Silting may be part of the complete eutrophication of a water system. Reduced flow, causing sedimentation, may also result from mechanical blockage from grills or channels being blocked by debris.
As well as directly reducing the usefulness of a water course, silting may have a number of indirect adverse effects. If drainage and irrigation channels are blocked, land may be flooded, or the water–table may rise, leading to the waterlogging of productive land, to the creation of marsh-land or saline or alkaline soils. Siltation in reservoirs may block the intakes of the turbines used to generate hydroelectric power cause dams to be decommissioned, for example the Sanmenxia Reservoir in China had to be decommissioned four years after it was completed and the Laoying reservoir silted up before its dam was completed. Silting by contaminated sediments and their excavation may also cause problems.
Refs White, W R (Ed) *Sedimentation Problems in River Basins* (1982).
Broader Unsustainable development of fresh waters (#PD6923)
Ecosystem modifications due to creation of dams and lakes (#PD0767).
Related Natural environment degradation (#PB5250)
Environmentally harmful dam construction (#PD9515).
Aggravates Floods (#PD0452) Alkaline soil (#PD3647) Aquatic weeds (#PD2232)
Soil salinization (#PE1727) Desiccation of rivers (#PJ4867)
Inadequate land drainage (#PD2269)
Inadequate water system infrastructure (#PD8517)
Declining productivity of agricultural land (#PD7480)
Reduction of soil fertility downstream due to impoundment (#PE8782).
Aggravated by Soil erosion (#PD0949) Deforestation (#PC1366)
Water pollution (#PC0062) Coastal erosion (#PE6734)
Soil erosion by water (#PD2290) Tropical deforestation (#PD6204)
Impurities in waste water (#PD0482) Disruption of the hydrological cycle (#PD9670)
Failure of governments to fulfil international reporting obligations (#PE2215).

♦ PD3655 Dust storms

Nature During wind erosion, the finer soil particles, when set in motion by the wind, are quickly carried upward, sometimes to great heights, by turbulent air currents, and are transported in suspension as a powdery dust. Under suitable conditions this may result in a dust storm, which may travel long distances. There are two essential requirements for the initiation of a dust storm. Wind speed at the ground must exceed a certain critical speed, ranging between 13 and 30 mph depending on the shape, size, specific gravity, dampness and temperature of the ground particles.
Incidence Dust storms are, at the least, disagreeable, but may also cause considerable damage. People in villages, towns, and cities as well as rural inhabitants undergo inconveniences and sometimes serious illness or suffocation. Fences, ditches, and channels are blocked or buried and farmsteads are rendered uninhabitable. Grass, trees, shrubs, and hedges may be smothered or buried. Insects and weed seeds are often carried to clean fields. Railways and roads are sometimes blocked by dust. Covering of established crops or pasturage by drifting dust may result in crop damage. Dust storms have other serious consequences in obstructing vision and in causing soil erosion, electrostatic discharges and radio static.
Refs Hurt, R Douglas *The Dust Bowl* (1981).

Broader Storms (#PD1150) Inhospitable climate (#PC0387).
Related Sand storms (#PD3650) Migrating sand dunes (#PD0493).
Aggravates Dust (#PD1245) Desert advance (#PC2506)
Destruction of land fertility (#PC1300).
Aggravated by Wind (#PE2223).

♦ PD3657 Insect damage to stored and manufactured goods
Nature Many insects, notably beetles and moths, subsist on stored or manufactured products of vegetable or animal origin. Most have been widely distributed through commerce.
Incidence Insects consume and destroy large quantities of foodstuffs. In so doing, they pollute them with substantial amounts of their excreta, which further damages the product and poses the risk of infection during handling or consumption. At least 30 million tons of bread grains and rice in storage are lost to insects each year. The granary and rice weevils, lesser grain borer and angoumois grain moth prefer whole grains; the saw-toothed grain beetle, flour beetles, meal-worms, Mediterranean flour moth and Indian meal moth prefer flour or coarse ground cereals. The cacao moth and almond moth attack nuts and dried fruit, the bruchid beetle attacks beans, and the khapra beetle eats peanuts and grains. The drugstore beetle and the cigarette beetle infect tobacco, drugs and spices; the larder beetle and cheese skipper are important pests of dairy products, such as bacon and cheese. Goods made of animal hair, skins, wool, or feathers are attacked by clothes moth larvae and carpet beetles.
Broader Insect pests (#PC1630) Plant diseases in storage and transit (#PD3587).
Narrower Insect pests of wood (#PD3586) Wood deterioration and decay (#PD2301).
Related Insect pests of plants (#PD3634).
Aggravates Insect vectors of disease (#PC3597).
Aggravated by Crickets as pests (#PE3640) Cockroaches as pests (#PE1633)
Beetles as insect pests (#PE1679) Thysanura as insect pests (#PE3620)
Dermaptera as insect pests (#PE3618) Lepidoptera as insect pests (#PE3649).

♦ PD3658 Acidic soils
Soil acidification
Nature Certain soils are naturally acidic, and others tend, through use, to become acidic. Soil acidity is common where rainfall exceeds evapo-transpiration. Soils in humid regions gradually become more acidic, under natural or cultivated conditions, as the hydrogen ions, from slightly acidic rain and from the decomposition of soil organic matter, gradually replace calcium and other cations in the soil, which are then leached away. Virgin soils in humid temperate regions may have a pH as low as 4.8 (pH 7 is neutral); such soils have a very low natural fertility.
Incidence Crop plants vary widely in their tolerances of soil acidity. For example oyoya and many clovers require essentially neutral soils, having a minimum pH of 6.5. Soil acidity adversely affects the availability of soil nutrients for plants, it also increases the level of toxic ions in the soil. In particular, acidity leads to the presence of elevated levels of soluble aluminium, which is a common cause of stunted root systems. Very small amounts of aluminium, less than one part in a million in water, can sharply reduce root growth and uptake of water and nutrients. The fish are dying from a toxic aluminium build-up in the lake and river waters in these areas.
Refs Dost, H and VanBreemen, M *Bangkok Symposium on Acid Sulphate Soils* (1982); Schneider, T (Ed) *Acidification and Its Policy Implications* (1986).
Broader Soil infertility (#PD0077).
Aggravates Forest decline (#PC7896) Infertile land (#PD8585)
Metal contamination of soil (#PD3668) Environmental plant diseases (#PD2224)
Deficiency diseases in plants (#PD3653)
Unsustainable agricultural development (#PC8419).
Aggravated by Inadequate land drainage (#PD2269)
Destruction of land fertility (#PC1300).

♦ PD3663 Photochemical oxidant formation
Urban smog — Photochemical smog — Volcanic smog — Vog
Nature Photochemical oxidants are formed through the concentration of a variety of highly reactive gases in the atmosphere and are often implicated in problems of smog, crop damage and the degradation of works of art. Smog is air pollution consisting of smoke and fog. Photochemical 'smog' is composed of a number of toxic compounds, including ozone, nitrogen dioxide, and small particles, which are often referred to as oxidants. They are called secondary pollutants because they are formed in the atmosphere as a result of reactions between certain organic compounds such as hydrocarbons and nitrogen oxides. These two types of substance are the primary pollutants or 'precursors' to the components of photochemical smog. The main types are London smog and Los Angeles smog. The London type consists mainly of sulphur compounds, aerosols and carbon monoxide, leading to bronchial irritation, coughing and a marked rise in mortality. Following radical "smoke-free zone" measures, London now experiences incidences of photochemical pollution-haze, rather than heavy smog. The Los Angeles type is induced by photochemical action of sunlight, consisting mainly of organic compounds, nitrogen oxide and carbon monoxide; it causes temporary eye irritation, but the other effects on humans are not known. Under normal environmental conditions, vegetation can be injured.
As the result of volcanic action resulting in the discharge of gases (especially sulphur dioxide) into the atmosphere, an effect similar to urban smog is created. It can affect people with asthma, allergies and sinus, and can contributes to acid precipitation. Such volcanic smog is also known as vog.
Incidence Photochemical smog is now a rather common, regional-scale phenomenon in many parts of the world. The principal man-made sources of photochemical smog are emission from motor vehicles and photochemical reactions of oxides of nitrogen and reactive hydrocarbons. The severity of smog is most often judged by the ground level concentration of ozone. The results can include possible asthmatic attacks or impaired pulmonary function in diseased people and lachrymation. Damage can occur to materials; visibility is reduced and amenity is affected. Photochemical pollution has been associated with impaired performance of athletes, increased likelihood of car accidents, and absenteeism in work and school. In 1984, the widespread damage to German forests is suspected has having been due to the impact of photochemical oxidants in combination with a wide range of physiological factors.
Refs OECD *Photochemical Smog* (1982).
Broader Degradation of the atmosphere (#PD9413)
Particulate atmospheric pollution (#PD2008).
Narrower Ozone as a pollutant (#PE1359) Smoke as a pollutant (#PD2267)
Aerosols as industrial hazards (#PE1504) Nitrogen compounds as pollutants (#PD2965).
Related Hydrocarbons as pollutants (#PE0754) Urban-Industrial air pollution (#PJ5532)
Chemical pollutants of the environment (#PD1670).
Aggravates Air pollution (#PC0119) Eye irritation (#PE6785)
Throat irritation (#PG6761) Amenity destruction (#PC0374)
Acidic precipitation (#PD4904) Road traffic accidents (#PD0079)
Lung disorders and diseases (#PD0637) Environmental plant diseases (#PD2224)
Plant-pathogenic air pollutants (#PE0155) Deterioration in atmospheric visibility (#PE2593)
Obstruction of astronomical observation by environmental pollution (#PE7244).
Aggravated by Fog (#PE1655) Domestic fires (#PJ6766) Industrial emissions (#PE1869)
Nitrates as pollutants (#PE1956) Motor vehicle emissions (#PD0414).

♦ PD3664 Traffic noise
Motor vehicle noise
Nature Noise from motor vehicles is the most pervasive source of urban noise. Except for locations near to airports where aircraft noise is excessive or near large industrial centers, motor vehicle noise is the controlling factor in setting the background noise levels of the environment. In those few situations where extensive community noise surveys have been made, vehicular traffic noise controlled the noise environment in more than 85 percent of the locations. Automobiles, through their total numbers, are the largest total source of urban noise. Motorcycles are quickly becoming an important part of the total vehicular noise picture. In recent years, the registration of new motorcycles has been growing at an ever increasing rate. Not only are they used increasingly as a means of transportation among certain age groups, but the mini-cycle used off the street on trails and in parks is becoming an important source of high intensity noise. Diesel trucks are clearly the noisiest vehicles on streets and highways; they are considerably larger in size and more complex as noise sources than passenger cars.
Refs Alexandre, A et al *Road Traffic Noise* (1975).
Broader Health hazards of exposure to noise (#PC0268).
Related Traffic congestion (#PD0078) Motor vehicle emissions (#PD0414).
Aggravates Deafness (#PD0659) Sleep disorders (#PE2197)
Stress in human beings (#PC1648).
Aggravated by Inadequate traffic control (#PE8266).

♦ PD3666 Marine dumping of wastes
Ocean disposal of waste — Discharge of dangerous substances at sea — Marine dumping of sewage
Nature Ocean dumping is currently used to dispose of industrial wastes, sewage sludge, garbage, construction debris, derelict vessels and dredged material. Most of these waste materials of our modern society are generated on land. There are three principal modes of transferring these substances from the land into the sea: river discharge; atmospheric transport, followed by washout with rain; and coastal discharges through outfalls. Vessels contribute to a certain amount of pollution through discharge of sewage and garbage. Increasing quantities of material are being transported out to sea for dumping at designated dump sites; however, the largest amount of material will continue to be released into the sea through coastal outfalls.
Incidence Atmospheric transport of pollutants is not a insignificant part of the transfer process from land to sea. First, the movement of fission products from nuclear weapons' tests through the stratosphere, with fallout of these radionuclides on virtually all parts of the globe, demonstrated how widely dispersed substances released into the atmosphere can become. Then followed the DDT problem with virtually every marine animal found anywhere near the sea surface exhibiting measurable quantities of DDT. This widespread distribution of a substance, used mainly in certain agricultural and forested parts of the world, was shown to be a result of rapid dissemination through the atmosphere. The entry of about 20,000 metric tons of lead into the sea from automobile emissions is clearly an atmospheric transport and washout phenomenon. There is some evidence that other metals, such as mercury and cadmium, are transported around the globe by atmospheric currents and may be washed out into the sea with rain. About 5,000 tons of mercury enter the sea each year, mostly as a result of industrial dumping. This about equals the natural contribution from the land. Mercury can be highly toxic, but it depends on the concentrations and the organisms involved. Sometimes inorganic mercury is altered by natural processes to become poisonous methyl mercury. The case of the Minamata deaths in Japan is an example. Sulphur dioxide emissions from coal-burning thermal plants and petroleum refineries in England, West Germany, and other parts of Western Europe have allegedly given rise to acidic rain and a low pH in the soft waters of some Scandinavian lakes and rivers. Although pH shift is not likely to become a problem in highly-buffered sea water, the acidic rain situation does demonstrate what can happen through atmospheric processes.
The amount of solid materials entering the sea annually can be in the millions of tons. While this type of pollutant is largely cosmetic in character and affects mainly the amenities, there are other uses of the sea which are also affected. For example, large polypropylene or nylon ropes, floating just below the sea surface, can become entangled in the propellers of vessels, causing damage and possibly accidents at sea. Plastic sheets can also become entangled in ships' propellers, as well as clogging sea water intake systems for engine cooling and other purposes. Fishing gear may become fouled by netting and ropes, and on large solid objects deposited on the bottom of the ocean. From the ecological point of view, solids deposited on the bottom can adversely affect the benthic habitat. Plastic sheets can smother organisms beneath them, because of lack of oxygen replacement due to elimination or reduction of water exchange, and adversely affect the substrate for settlement of larvae. Plastic bags have been found on the heads of fur seals; plastic and rubber spherules have been found in the intestines of fish.
Of wastes dumped deliberately into the sea, as much as 80 percent consist of 'dredge spoils', the materials scraped from river and harbour bottoms to open channels and facilitate navigation. These materials are natural, for the most part, made up of sand, silt, clay and rock. They are usually barged to offshore areas and dumped, often grossly disrupting these undersea locations for a period of time. These 'spoils' often include a high proportion of deposited sewage sludge and accumulated industrial waste, with its potential toxicity.
Although about half of the oil transported in the world each year is carried by tankers, and tanker oil is dumped as part of routine operations, (such as illegal deballasting and tank washing, which flush oil into the sea), most of the oil in the sea comes from a combination of land-based industrial and municipal sources. As much as 350 million gallons (1.3 billion litres) of used automobile crankcase oil is dumped each year into water drainage systems that reach the sea. About 600,000 metric tons of oil run into the sea each year a result of petroleum carried into the atmosphere from poorly tuned automobile engines. Of the twenty largest metropolitan areas in the world, sixteen are coastal cities, or cities on rivers which empty relatively quickly into the ocean. About 30 billion gallons (113 billion litres) of industrial and municipal wastes are discharged into the coastal areas of the United States alone every year.
The chief danger of ocean sewage dumping lies in the spread of viral and bacterial diseases – directly to bathers, indirectly through fishery products. On the basis of the toxicity and the volume of many wastes entering the sea it seem likely that sea life could be seriously disrupted in some places more than others, but the threat is global.
Claim The sea was the first sewage treatment plant. It has processed the organic waste and the dead husks of living things since life evolved there 3.5 billion years ago. The system thrives on this natural excrement. The animal life of the great depths would not survive without the nourishment of waste matter raining down from more populated surface waters. Gray whales contribute more faecal matter to the sea worldwide than the Los Angeles sewer system does; so do anchovies. There are also natural oil leaks in the sea. Seepage from sub-sea deposits wells up in many places around the world. Off the California coast, one natural seepage contributes about 660,000 tons of petroleum to the sea every year.
Because the sea and its organisms can handle this load of natural wastes quite well, the misconception has arisen that there is no limit to what the sea can dispose of in its self-purification systems. The human waste from a large city like Los Angeles is greatly concentrated in one location, not dispersed widely, and it is usually spiked with toxic substances that the sea may be unable to recycle naturally. There is natural seepage of petroleum into the sea from ocean floor deposits, but human-caused oil pollution is now about ten times that rising naturally from the ocean bottom. Perhaps the most pernicious misconception about the sanitation system of the ocean, however, is that coastal dumping is a negligible contribution to an environment so vast and deep, covering nearly three quarters of the planet's surface. The fact is that most of the sea's life

is concentrated along the world's coastlines, where nutrient-rich currents pass across convenient hiding and attachment surfaces shallow enough to be touched by sunlight. Of the 140 million square miles (363 million square kilometers) of ocean surface, it is only the 14 million square miles (36 million square kilometers) near shore that contain the sea's most important living habitats. About 90 percent of the world's food fish spawn, mature, and are caught, in these coastal areas. Unfortunately, the land adjacent to these productive waters is and where human populations gather, where many industries locate, where the greatest mass of destructive waste matter is dumped.

Counter-claim Dumping was at sea is preferable to expanding disposal sites on land. Land dumping is more expensive, more difficult to gain approval of local authorities and more likely to meet with political opposition.

Refs International Maritime Organization *Inter-Governmental Conference on the Convention on the Dumping of Wastes at Sea*; United Nations *The Law of the Sea* (1985); United Nations *An Oceanographic Model for the Dispersion of Wastes Disposed of in the Deep Sea* (1986); Wolfe, Douglas A and O'Connor, Thomas P *Urban Wastes in Coastal Marine Environments* (1988).

Broader Hazardous waste dumping (#PD1398) Discharge of dangerous substances (#PD4542) Dumping of consumer waste products (#PD8942).
Narrower Marine disposal of obsolete weapons (#PD7574).
Aggravates Food pollution (#PD5605) Marine pollution (#PC1117) Radioactive wastes (#PC1242) Oil as a pollutant (#PE2134) Toxic metal pollutants (#PD0948) Wildlife pollution hazard (#PJ3387) Infectious and parasitic diseases (#PD0982) Marine pollution by plastic waste (#PE3741).
Aggravated by Industrial effluent (#PG6771) Sewage as a pollutant (#PD1414) Industrial waste water pollutants (#PD0575) Unethical practices in transportation (#PD1012).

♦ PD3668 Metal contamination of soil

Nature A number of elements at elevated concentrations are commonly regarded as contaminants of soil. They are also referred to as "heavy" metals, although they include metalloids and non-metals. The main elements implicated as contaminants include: arsenic, cadmium, chromium, copper, fluorine, lead, mercury, nickel and zinc. Beryllium, bismuth, selenium and vanadium may occasionally also be geochemically enriched in some soils. Contamination may result from the weathering of geological parent materials (where element concentrations exceed natural abundance values), soils contamination by industrial activities (usually by wet and dry deposition) and soil applications of waste such as sewage sludge or pig slurry, as well as fertilizer applications.

Broader Soil pollution (#PC0058) Toxic metal pollutants (#PD0948).
Narrower Contaminated pastureland (#PJ5737).
Aggravates Food pollution (#PD5605) Infertile land (#PD8585) Soil degradation (#PD1052) Pesticide damage to crops (#PD2581) Environmental plant diseases (#PD2224).
Aggravated by Acidic soils (#PD3658) Sewage as a pollutant (#PD1414) Pesticides as pollutants (#PD0120) Industrial waste water pollutants (#PD0575) Environmental hazards from fertilizers (#PE1514).

♦ PD3669 Wasted water

Heated effluent waters

Nature Due to reasons such as increasing urbanization, industrial expansion and intensified agriculture, water is frequently used heedlessly, beyond an economically and socially sensible level, and is an undue recipient of harmful pollution loads. Also, the volume of waste water is continuously increasing.

Incidence Pollution due to waste will always eventually involve the water system of the earth, especially because the pollutants emitted into the air and those present in the soil are washed out by precipitations. It is possible to subdivide pollutants according to the effects they exert on the water system and to the degree of harm they do to the environment: pollutants which are capable of self-purification and/or inserting themselves into natural cycles, such as biodegradable organic substances, ammonia, nitrates, phosphate, fluorides; toxic or harmful pollutants which do not accumulate in organisms, such as slow biodegradable organic substances, cyanides, phenols, mineral oils, aldehydes, surfactants, boron and zinc; pollutants such as mercury, lead, pesticides and solvents, which not only have a pronounced toxic effect, but may also accumulate in organisms and pass from one trophic level to another; and other pollutants such as pathogens, phenols, viruses, radioactives, paints and dyes.

Refs FAO *Potential Uses of Waste Waters and Heated Effluents* (1973); FAO *Report of the Symposium on New Developments in the Utilization of Heated Effluents and of Recirculation System for Intensive Aquaculture, Stavanger, 1980* (1981).

Broader Unproductive use of resources (#PB8376) Long-term shortage of natural resources (#PC4824).
Narrower Agricultural effluent (#PE8504) Stagnant surface water (#PE2634).
Related Industrial waste water pollutants (#PD0575).
Aggravates Dysentery (#PE2259) Typhoid fever (#PD1753) Soil pollution (#PC0058) Water pollution (#PC0062) Amenity destruction (#PC0374) Water-borne diseases (#PE3401) Sewage as a pollutant (#PD1414) Hazards to human health (#PB4885) Detergents as pollutants (#PE1087) Phosphates as pollutants (#PE1313) Wildlife pollution hazard (#PJ3387) Soil-transmitted diseases (#PD3699) Pollution of inland waters (#PD1223) Long-term shortage of water (#PC1173) Contamination of drinking water (#PD0235) Domestic waste water pollutants (#PD2800) Eutrophication of lakes and rivers (#PD2257) Damage by degradable organic matter (#PJ6128) Water pollution in developing countries (#PD3675).
Aggravated by Impurities in waste water (#PD0482) Lack of water conservation (#PJ3480).

♦ PD3672 Irritant fumes

Nature Irritant gases and vapours are chemicals characterized by the action they exert on the respiratory system and conjunctivae. The intensity and severity of the irritant action depends on the substance's chemical structure, the concentration in the respired air, and the length of exposure. The substance's solubility may be a factor since it determines the region of the respiratory tract that is exposed to the irritant.

Broader Air pollution (#PC0119) Urban-Industrial pollution (#PC8745).
Aggravates Ulcers (#PE2308) Eczema (#PE2465) Trachoma (#PE1946) Keratitis (#PE6789) Eye irritation (#PE6785) Nose irritation (#PG6784) Throat irritation (#PG6761) Chronic bronchitis (#PE2248) Blood circulation disorders (#PE3830) Diseases of the skin and subcutaneous tissue (#PC8534).

♦ PD3673 Endangered species of marine mammals

Refs FAO *Mammals of the Seas* ; FAO/UNEP *Marine Mammals* (1985).

Broader Endangered species of mammals (#PC1326).
Narrower Endangered species of whale (#PD1593) Endangered species of porpoises (#PE3806) Endangered species of river dolphins (#PG3801).
Related Depletion of fish reserves by marine mammals (#PE4913).
Aggravated by Hunting of marine animals (#PE0439).

♦ PD3675 Water pollution in developing countries

Nature The developing countries are the scene of rapid urbanization and industrial development; and there is a growing demand for water for domestic and industrial purposes, as well as an increase in water pollution, which tends to reduce the available water resources. Water pollution is a particularly acute problem in countries that have scanty water resources, and many of the developing countries are in this situation. Some include arid regions, and in others the rain falls only during a short season and the water cannot be economically stored; much of it runs away to the sea, and much is lost by evaporation. If, in addition to these losses, the natural water resources become increasingly unusable because of pollution, the net reserves of the country are continually reduced. The control of pollution is thus linked closely with the management of water resources. In some areas, ground water has been contaminated by domestic sewage and industrial effluents to such an extent that it has had to be abandoned as a source of supply. If it is not too greatly polluted, water undergoes self-purification, but when pollution is excessive this process is slow and uncertain.

Incidence Barely 70 percent of the urban population in developing countries has access to a water supply: that is, to running water, tube wells, etc. The figure is significantly lower in rural areas, where only 12 percent of the population has reasonable access to a water supply. Some of the serious communicable and parasitic diseases in the developing countries, affecting millions of people, can be directly related to the water conditions prevailing in their immediate environment; according to a WHO report, about 80 percent of all known diseases are related to water misuse. Malaria, a parasitic disease reported to be increasing, is transmitted through insect vectors breeding in water. In developing areas there are 514 million people still without specific protection against this disease, and about 60 percent of them live in the African region. Schistosomiasis is a water-based disease, common in regions lacking water supply and sanitation facilities. According to WHO estimates, more than 600 million people are exposed to the disease, and some 200 million people in the developing countries are suffering from it. Onchocerciasis, transmitted through the larvae of a black fly vector breeding in watercourses, causes visual impairment and blindness. Trachoma, considered to be the single most important cause of preventable blindness in the world, is common under conditions of poor hygiene, sanitation and nutrition. Filariasis, cholera and bacterial enteric infections, the latter a major cause of death among children in the tropics, are also related to water supply and sewage facilities. The lack of these fundamental services provokes great centres of infection, which are characteristic of the developing countries. Chemical pollutants, including pesticides and herbicides, are becoming a danger in many areas. Some chemicals are highly persistent in water and, unlike 'conventional' contaminants, are not amenable to natural purification. This situation is worse in arid zones, where there is little chance of dilution.

Broader Water pollution (#PC0062).
Aggravates Food pollution (#PD5605) Animal diseases (#PC0952) Intestinal diseases (#PD9045) Water-borne diseases (#PE3401) Insect vectors of disease (#PC3597) Rodent vectors of disease (#PE3629) Pests and diseases of fish (#PD8567) Human disease and disability (#PB1044) Inadequate irrigation system (#PD8839) Contamination of drinking water (#PD0235).
Aggravated by Wasted water (#PD3669) Inadequate waste treatment (#PD6795) Lack of water conservation (#PJ3480) Inappropriate sanitation systems (#PD0876) Industrial waste water pollutants (#PD0575).

♦ PD3680 Pesticide hazards to wildlife

Nature Pesticides present many hazards to wildlife – birds, fish and small mammals have been poisoned by pesticides used to control insect outbreaks. These pesticides are passed up the food chains from herbivores to carnivores, and in the process their concentration is often increased. The effect of insecticides on wildlife involves a number of factors, such as toxicity and persistence of a finer compound; stability as it is transferred up the food chain; type of vegetation in which it is applied; and the species of wildlife itself.

Broader Destruction of wildlife habitats (#PC0480).
Related Wildlife pollution hazard (#PJ3387).
Aggravates Bird diseases (#PD3323) Animal diseases (#PC0952) Eggshell thinning (#PE6290) Extinction of species (#PB9171) Pests and diseases of fish (#PD8567) Underutilization of biocontrol (#PF6229) Loss of beneficial plants and animals (#PE8717).
Aggravated by Pesticides as pollutants (#PD0120) Fungicides as pollutants (#PD1612) Insecticides as pollutants (#PD0983) Rodenticides as pollutants (#PE3677) Excessive use of chemicals to control pests (#PD1207) Concentration of noxious substances in food chains (#PE8154).

♦ PD3691 Discrimination against women in social services

Nature Discrimination against women in social services arises in matters of social security, old age and widows' pensions, health benefits, and taxation, as well as in general dependency considerations, such as the legal requirement for the husband's permission (in the case of married women) or the father's permission (in the case of minors) to take advantage of certain public services: for example, the opening of a bank account. Laws are based on functions; and because a woman's functions change, she may lose out on benefits, such as insurance. Allocations in case of illness, accident or occupational disease are lower to women than to men, as they are usually established on the basis of marital status; pensions are usually less; and there is a shortage of institutions to care for the children of working mothers.

Incidence In many developed countries there is widespread inequality of treatment between the sexes both in the state and in many privately-run pension schemes. Examples of this discrimination are, for example, the difference in retirement age – 60 for women and 65 for men, which means that employers can insist that female employees retire five years early, possibly against their will. There is the problem of survivors benefit: although most pension schemes automatically provide a pension for a widow of a male employee, fewer than half provide the same for the widower of a female employee. Just as important, is that the whole structure of pension schemes discriminates against the pattern of most women's working lives, where there is often a break, mostly a long one, to bring up a family. Certain rights to benefits of unemployment, old age and invalidity are denied to women because they obtain protection as a dependent person. In reality some 35 percent of households in the world are now headed by women. But the man is considered the head of the family while the woman brings up the children and her professional activity is considered incidental.

Broader Discrimination against women (#PC0308) Discrimination in social services (#PC3433).
Narrower Unequal health benefits for women (#PE6835) Unequal distribution of old age pensions between men and women (#PE7942).
Related Racial discrimination in public services (#PD3326) Discrimination against men in social services (#PD3336).
Aggravates Dependency of women (#PC3426).
Aggravated by Social injustice (#PC0797) Dependency of women in marriage (#PD3694) Discrimination against women in religion (#PD0127) Discrimination against women in education (#PD0190) Discrimination against women in employment (#PD0086).

♦ **PD3692 Discrimination against men before the law**
Incidence A study published in the USA revealed that men are more severely punished than women for identical crimes. Given the disproportionate level of males in the European prison system, an important issue to study is the possible inequitable treatment of men by European courts. Male prisoners are often given inferior treatment in comparison to female prisoners, especially in terms of personal safety, physical comfort, and access to their children.
    Broader Discrimination against men (#PC3258)   Discrimination before the law (#PC8726).
    Narrower Unequal pension rights (#PJ2030).
    Related Unequal property rights (#PJ2031)
    Discrimination against men in employment (#PD3338)
    Discrimination against women before the law (#PD0162)
    Discrimination against men in social services (#PD3336)
    Ineffective regulation of electronic messages (#PE6226).
    Aggravated by Dependency of women (#PC3426).

♦ **PD3694 Dependency of women in marriage**
Nature Because the man has been traditionally regarded as the head of the family, exercising marital and parental authority over the person and over the property of his wife and of his children, the married woman, in many countries, is deprived of a number of personal and property rights. Discrimination against women in private law lies in the subordinate status of married women. The fact that the role and work carried out for the family by the wife is unrecognized and undervalued by society reinforces the economic basis for her submission: women are thus treated as unproductive, second-rate citizens.
Even within countries that provide a legal basis for women to have personal and property rights, they frequently are deprived of sufficient income. In a recent study of poverty within marriage in the UK suggests that married women and their children are frequently far poorer than realized even if their husbands earn a good wage.
Incidence The status of women is most seriously affected in a great number of countries in respect to her nationality, her right to choose or maintain a residence or domicile, her rights and duties with regard to her children, and her property rights and civil capacity. Under many legal systems there is an automatic change of nationality if the woman marries a man of a different nationality: a woman national of such a country who marries an alien automatically loses her own nationality and acquires the nationality of her husband. In most countries the selection of marital residence is, in practice, one of mutual agreement; but in countries where the man is the head of the family, if agreement is not reached he makes the final decision. Parental authority under most legal systems belongs primarily to the father when both parents live together in the family home, although usually both parents are obliged to support their children according to their means. When the father dies, in most countries provision is made for the mother's right to succeed to the father's authority, although her rights as the surviving parent are not always equal to those of a father; and in some countries, the mother does not succeed to parental power. Under a community property regime, all or part of the property of the spouses is jointly owned, administered by the husband and divided equally by the spouses at the end of marriage; but in countries applying Hindu or Moslem law, distinction is made between the right of a surviving wife and the right of the surviving husband to take share of the deceased spouse's estate.
Marriage restricts the independent exercise of a women's right to prosecute and defend in court in Chile, Brazil, Ecuador, Haiti, some US States, Quebec (Canada), Belgium and the Philippines. In other countries, a woman may be specially restricted with respect to certain types of contracts, particularly those involving trusts, guarantees and pledges of future incomes. Where the right of the wife to engage in independent work is unconditionally subject to her husband's authority, the wife requires his full consent before undertaking outside work or starting in business; if consent is refused, the wife has no recourse in court and the husband need not state his reasons. Consent may be withdrawn at any time. When the husband has limited authority, the wife has recourse to the courts.
    Broader Dependency of women (#PC3426)
    Discrimination against women before the law (#PD0162).
    Related Dependency of children (#PD2476)   Emotional dependency in marriage (#PD3244)
    Denial of right of family planning (#PE5226).
    Aggravates Suicide (#PC0417)   Wasted woman power (#PF3690)
    Marital instability (#PD2103)   Double standards of sexual morality (#PF3259)
    Discrimination against women in education (#PD0190)
    Discrimination against women in public life (#PD3335)
    Discrimination against men in social services (#PD3336)
    Discrimination against women in social services (#PD3691).
    Aggravated by Marriage markets (#PD7282)
    Discrimination against women in religion (#PD0127).

♦ **PD3695 Insecticide damage to crops**
Nature The misapplication of insecticides may result in damage to crops.
Incidence Evidence is accumulating to show that the excessive or continuous use of insecticides may damage crop yields. Continuous use of chlorinated hydrocarbons, such as DDT, which decompose very slowly in the soil, may increase chemical levels to a point where crop growth, especially of sensitive plants such as cucumbers and potatoes, is seriously retarded. The organophosphorus aphicides can also adversely affect plant growth and development. These insecticides, particularly Thionazin, when mixed with sand on field soil, delayed the germinations of wheat seeds by two weeks, produced abnormalities among those that did grow, and appreciably lowered their growth. In beans, effects can be seen at a concentration of 27 ppm of chemical – for example, necrotic leaf lesions where the insecticide is accumulating. Growth of the seed leaves of beet is inversely related to the concentration of insecticide applied. Germination is also delayed, and with 125 ppm of phosphate or Thionazim, germination is reduced to 60 to 70 percent.
    Broader Pesticide damage to crops (#PD2581).
    Aggravated by Nematicides as pollutants (#PJ1961)
    Insecticides as pollutants (#PD0983).

♦ **PD3696 Pest resistance to pesticides**
Nature Ever since chemicals have been used to keep pests such as insects, weeds, fungi and worms from ruining crops, the pests have fought back by developing resistance to the chemicals. A similar problem has long plagued efforts by public health officials to eradicate disease-carrying pests such as mosquitoes and rodents. Pests are the major cause of post-harvest losses and current storage systems and technology dictate that control of these pests is based largely on their resistance to pesticides and thus is a matter of considerable concern: there is no chemical that can immediately replace malathion, currently the most widely used material. Those that may be cleared for international use in the near future are unlikely to have similar spectra of activity, economy and other attributes. Inevitably resistance will impose a considerable economic burden on the industries concerned - a burden which will increase as resistances spread and intensify and infestation levels tend towards those that existed before malathion was introduced.
Incidence The number of pesticide-resistant insect pest species worldwide has increased, with many such insects being resistant to even the newest chemicals. The variety and severity of pest infestations increase, thus increasing the threat to agriculture in the areas concerned. There is increasing concern throughout the world over the problems of pesticide resistance in pests which affect stored grain. Resistant strains of these pests have been shown to be moving actively in world trade and there is a serious threat to the effective use of chemicals in maintaining stored foodstuffs in a sound and insect-free condition. Insecticide resistance is now so widespread in certain species that it has become essential that resistant strains are used in the evaluation of candidate alternative materials. Currently, more than 200 species are known to be resistant to insecticides. The number of pesticide-resistant insects and mites has doubled in 12 years. The extent to which resistance can be met by increased dosages is severely limited by the level of residues internationally acceptable in stored foodstuffs. Countermeasures to resistance will thus necessitate the use of alternative insecticides and possibly fumigants; but because of stringent requirements with respect to chemical residues, there are comparatively few materials that can be used for the control of grain pests. These few materials, however, after repetitive use have led in many instances to the development of resistance.
Stored grain insects are unique in that most of the major species are cosmopolitan and readily move about in domestic and international trade. Hence, resistant strains are also moving throughout the world, reaching countries where resistance had not been suspected. Fumigation has long been regarded as a basic method of controlling stored product insects and one which would be of material assistance in delaying the development of resistance to the unrelated residual pesticides. The method itself, by virtue of the low variability in response of individual insects to the commonly-used materials such as methyl bromide, is also usually considered to be less prone to resistance development than the normal methods with residual pesticides. It is disturbing then to note the increasing prevalence of resistance to fumigants and the resultant weakening of one of the most powerful tools available in stored product pest control for delaying or preventing development of resistance. The emergence of resistance to fumigants under practical conditions is a matter for particular concern. With major world dependence on fumigation both as a routine disinfestation treatment and as a means of combating insecticide-resistant strains, the occurrences reported, although as yet limited in number and often at marginal resistance levels, are of considerable significance and pose a real threat to a continued ability to store grain safely.
Only about 1 percent of the applied pesticide actually reaches the target pest. The other 99 percent enters the ecosystem. The ecological impact of pesticides has been well documented and the list includes the near extinction of the peregrine falcon, osprey, and bald eagle, as well as many other creatures. A less well-known but far-reaching effect of pesticides is on bee populations. As early as 1944, there was evidence that pesticides were seriously affecting honeybee and wild bee populations.
Background Since the World Health Organization began to use DDT against Anopheles mosquitoes, the carriers of malaria, the lives of millions of people in malaria-infested parts of the world have been saved by eliminating mosquitoes, and similar successes were reported for other pests. The future seemed bright for both the farmers and the pesticide industry until the first warning came in 1947, when houseflies began to develop resistance to DDT. Insects with genes giving them resistance to DDT survived the massive sprayings in fields, gardens and homes, and these insects reproduced offspring that were also resistant. As more species of insects evolved into DDT-resistant populations, other chemicals were tried, such as organophosphates and carbamates, but many insects developed resistance to these as well.
Claim Pesticides are not applied to pests: they are applied to ecosystems that happen to include the pests. Humans are also part of the ecosystem. Indiscriminate spraying does more to eliminate the natural enemies of pests than the pests themselves. A continuing spiral of increasing doses of greater varieties of insecticides foster the development of more and more robust pests requiring more insecticides.
Counter-claim There are comparatively few materials that are safe for use in the control of pests of foodstuffs in storage. They comprise essentially a limited range of persistent insecticides and non-persistent fumigants. These few materials have been used widely and intensively and it is probable that the present requirements of freedom from insects would have been unattainable if these materials, and particularly malathion, had not been available. International trade in many of the world's basic foodstuffs, such as cereals, does in fact have almost complete dependence on pesticides to meet the insect tolerance limits. Overall losses would increase about 9 percent (to approximately 42 percent of production) if pesticides were banned.
Refs Champ, B R and Dyte, C E *Report of the FAO Global Survey of Pesticide Susceptibility of Stored Grain Pests* (1976); FAO *Recommended Methods for Measurement of Pest Resistance to Pesticides* (1980); FAO *Pest Resistance to Pesticides and Crop Loss Assessment* (1977).
    Broader Mutation (#PF2276)   Drug resistance (#PF9659).
    Narrower Fungal resistance to fungicides (#PE4456)
    Insect resistance to insecticides (#PD2109)
    Rodent resistance to rodenticides (#PE3573).
    Related Resistant bacteria (#PE6007)   Drug resistant viruses (#PG6399)
    Mutagenic effects of drugs (#PE4896)
    Encouragement of drug resistant diseases (#PJ3767).
    Aggravates Pests (#PC0728)   Insect pests (#PC1630)   Plant diseases (#PC0555)
    Pests of plants (#PC1627)   Crop vulnerability (#PD0660)
    Vector-borne diseases (#PD8385)   Human disease and disability (#PB1044).
    Aggravated by Pesticides as pollutants (#PD0120)
    Underutilization of biocontrol (#PF6229).

♦ **PD3699 Soil-transmitted diseases**
Nature Human disease may result from contaminated soil, due to unsanitary practices for disposal of excreta, improper or inadequate sewage treatment, or unfavourable climatic conditions. Basically, the types of diseases emanating from soil contamination can be divided into three main categories. Man-soil-man diseases result from contamination of soil by human excreta from which disease is contracted by either direct skin contact or consumption of food grown in such soil; examples include enteric bacteria and protozoa, and parasitic worms (helminths). Animal-soil-man diseases result from contact with soil previously contaminated with excreta of animal carriers, cadavers, and any part of infected animal bodies. Among this latter category can be found anthrax, leptospirosis, and Q fever. The third category of disease is a result of fertile climatic conditions inducing the proliferation of a pathogen from microorganisms growing in the soil. The mycoses, tetanus and botulism join this subdivision. Many of the diseases result from inadequate pretreatment of soil reused as fertilizer, or waste water reclaimed for irrigation purposes. Conventional sewage treatment processes cannot remove all the pathogenic organisms, although success of removal generally parallels removal rates for coliform organisms. For practical purposes, it cannot be assumed that even a well-run biological sewage treatment plant can consistently remove more than 90 percent of the pathogens from the sewage.
Incidence As an indication of the widespread nature of the problem, it has been estimated that about one third of the world's population is infected by hookworm, while one out of every four people in the world may be infected with Ascaris lunbricoides.
    Broader Zoonoses (#PD1770)   Soil pollution (#PC0058).
    Related Water-borne diseases (#PE3401).
    Aggravates Malnutrition (#PB1498)   Food pollution (#PD5605).
    Aggravated by Wasted water (#PD3669)   Contradictions (#PF3667)
    Agricultural wastes (#PC2205)   Sewage as a pollutant (#PD1414).

♦ **PD3712 Inappropriate management of development projects**
Inadequate implementation of development programmes — Ineffective execution of development projects
    Broader Mismanagement (#PB8406)
    Inadequate implementation of plans and programmes against problems (#PF1010).

Aggravates Destruction inherent in development (#PF4829)
Inadequate control of development projects (#PF9244)
Excessive foreign public debt of developing countries (#PD2133).
Aggravated by Inadequate sense of time (#PF9980)
Inappropriate design of development projects (#PF4944)
Inappropriate use of innovative management techniques (#PJ0447).

♦ PD3714 Protectionism in developing countries
Incidence As in developed-market economy countries, protection in developing countries has involved efficiency losses in those cases in which it has been excessive. **Counter-claim** The case against protection, in developing countries should be tempered by the fact that the motivations for protection in developing countries are quite different from those in industrialized countries. Broadly speaking, the adoption by developing countries of measures with the potential to restrict trade would appear to be designed to serve one or more of the following purposes: revenue collection, balance-of-payments protection, and infant-industry protection. To the extent that such infant-industries eventually become internationally competitive, such protection serves to effect structural change and not to arrest it, as is frequently the case with protection in industrialized countries from "market disruption". The structure of protection in developing countries tends not to be discriminatory, in contrast with those adopted by developed-market countries. Developing countries tend to be foreign exchange constrained and their import levels are typically determined by their export earnings rather than by their protective policies. Finally, unlike the developed market economies, there would appear not to have been any trend towards increased protection in recent years.
Refs Indian Institute of Foreign Trade *Growing Protectionism in Developed Countries* (1980).
Broader Trade protectionism (#PC4275).

♦ PD3722 Children neglected by teachers
Nature A large number of children are for the most part neglected by their teachers. These students dread being asked questions, hate praise, and do everything they can to evade attention. They camouflage their true character at school with evasion and retreat. Teachers, on the other hand, give the bulk of their time and attention to a handful of charismatic or demanding children, while large numbers of diffident and undemanding pupils pas through their hands unnoticed.
Incidence A study by a British teacher found in a fortnight of 70 lessons, the most active pupil had 123 contacts of some sort with the teacher and the least active pupil has just seven. The most likely to be neglected by teachers are girls. In the same study, the 12 boys in the class achieved 626 contacts and the 16 girls achieved only 489, an average of 52 contacts and 31 contacts respectively.
Broader Social invisibility (#PD8204).

♦ PD3736 Corruption in the entertainment industry
Payola
Incidence Disc jockeys and radio station executives may receive cash or drugs payments for adding some records to playlists or getting involved in chart-rigging.
Broader Corruption (#PA1986).

♦ PD3763 Torture by deprivation
Nature Deprivation is used to torture prisoners for the purpose of breaking down resistance, demonstrate absolute control by the torturers and humiliate the prisoner. The range of deprivation is from total sensory deprivation to being denied access to professional service like medical and legal. Sleep, food, and water deprivation are common. Confinement to cells without the opportunity of exercise or even social contact is used in some cases. Relatives are denied access to prisoners or even the knowledge of their imprisonment.
Incidence Torture by deprivation has been reported in countries including: South Africa, Chile, Colombia.
Broader Psychological torture (#PD4559).
Narrower Torture through waiting (#PE3927) Torture through confinement (#PD4590)
Social isolation as torture (#PD6810) Torture through sensory deprivation (#PE6797)
Depriving prisoners of medical treatment (#PD1480).
Aggravates Soul murder (#PF4213).

♦ PD3765 Economic civil war
Intra-national trade war
Broader Civil war (#PC1869) Economic conflict (#PC0840).

♦ PD3771 Endangered species of elephant
Nature The destruction of the African elephant does far more harm than the reduction of a species to a few thousand protected in parks and zoos. It means the end of a major force that shapes the ecology of forests and the savanna woodlands. In an elephant's constant search for 300 pounds of vegetation every day, it kills small trees and underbush and pulls branches off big trees as high as its trunk will reach. This creates open spaces in both deep forest and in the woodlands. This patchwork of vegetation in various stages of regeneration, in turn creates a greater variety of forage that attracts a greater variety of other vegetation-eaters than would otherwise be the case.
Incidence As recently as the 1930's there were 10 million African elephants. In 1979, there were 1.4 million elephants in Africa. Poaching has reduced this to 650,000 in 1989. At the present rate of killing, some 70,000 a year, there will be no African elephants at the turn of the century. Ivory prices are $200 per kilo and the average tusk is 5 kilograms. There are only 30,000-40,000 Asian elephants left in the wild, and those survivors have been hemmed in by the teeming populations and suffer harassment, death and injury in the conflicts with people. Ivory poaching, which has gone on for centuries, has devastated the large tuskers in Asian countries.
Refs Jackson, P (Ed) *Elephants and Rhinos in Africa* (1982).
Broader Endangered species of mammals (#PC1326).
Aggravated by Illegal ivory trade (#PE4991).

♦ PD3773 Developmental disabilities
Physical developmental defects
Refs Rubin, I Leslie and Crocker, Allen C (Eds) *Developmental Disabilities* (1989).
Broader Physically handicapped persons (#PD6020).
Narrower Autism (#PE1222) Epilepsy (#PE0661) Cerebral paralysis (#PE0763).
Aggravates Hypersensitivity (#PE6898)
Economic and social losses due to disability (#PE4856).

♦ PD3774 Ignorance of history
Limited historical method
Nature The prevailing method of viewing history is inadequate to appropriate the cultures of the past. A new posture needs to be adopted which not only looks into the past but also anticipates the future and enables full social engagement. Until this occurs, people will remain trapped in traditional practices and customs.
Broader Ignorance (#PA5568) Parochial national interests (#PF2600).
Aggravates Risk-aversion strategy (#PF4612) Ignorance of cultural heritage (#PF1985)
Failure to profit from patterns of history (#PF1746).
Aggravated by Historical misrepresentation (#PF4932)
Silence about historical situations (#PF0608).

♦ PD3775 Subsiding coastal areas
Broader Land subsidence (#PD5156) Decreasing land mass (#PF7435).
Aggravates Coastal erosion (#PE6734)
Unsustainable development of coast zones (#PD4671).

♦ PD3784 Mental depression in children
Refs Chiles, John *Teenage Depression and Suicide* (1986); Trad, Paul V *Infant and Childhood Depression* (1987).
Broader Mental depression (#PC0799).
Aggravated by Chronic illness (#PD8239).

♦ PD3785 Violation of rights of vulnerable groups during states of emergency
Lack of protection of vulnerable groups during disasters
Broader Lack of protection for the vulnerable (#PB4353)
Suspension of rights during states of emergency (#PD6380).
Narrower Vulnerability of women and children in emergencies (#PD1078)
Vulnerability of the elderly under states of emergency (#PD0096)
Vulnerability of the disabled during states of emergency (#PD0098).
Related Inadequate protection of civilians in armed conflict (#PE8361).
Aggravated by Lack of commitment to the protection of vulnerable groups (#PF4662).

♦ PD3789 Absence of management training
Broader Inadequate education (#PF4984)
Lack of opportunities for practical training in communities (#PF2837).
Narrower Neglected organization training (#PG5231).
Aggravates Mismanagement (#PB8406) Incompetent management (#PC4867)
Poor managerial communications (#PF1528).

♦ PD3794 Criminal trespass
Trespassers — Trespass
Nature Criminal trespass is illegally entering into or remaining in a building or structure. This is distinguished from burglary by the fact that no other crime is intended.
Broader Criminals (#PC7373).
Narrower Chemical trespass (#PE9363) Trespassing livestock (#PE4898)
Illegal occupation of unoccupied property (#PD0820).
Related Criminal intrusion (#PE6771).
Aggravates Excessive punishment for trespass (#PU1406).

♦ PD3823 Deception by management
Broader Deception (#PB4731).
Aggravated by Mismanagement (#PB8406).

♦ PD3825 Drug dependence
Drug addiction — Substance dependence
Nature The state of drug dependence or addiction arises from repeated administration of a drug on a periodic or continual basis. Drugs are substances which have a debilitating effect on the mind, central nervous system and general health. Addiction undermines the physical, mental and spiritual competence of individuals. The addictive process usually begins with what is considered to be normal or accepted within society. To take alcohol or chemicals to alleviate strain and stress has a long and almost universal pedigree. It is in the nature of addictive substances that an increasing dosage is required which activates the desire to take the next dose. This is the accumulative effect of drug taking.
Incidence Drug dependence has become a world-wide problem, but it seems what was a problem among the lost generation of drop-out youth in the West has turned East, presenting Asia with a vast home-grown problem. In Malaysia, for example, the government has declared the heroin problem the 'No 1 enemy'. Sixty percent of the world's heroin output is now consumed in Asia, and despite increasingly punitive laws against its use, heroin addiction is still increasing at an alarming rate among Asia's teenagers and young adults. In 1984, in Malaysia there were known to be 9,000 addicts, though the actual figure is reckoned to be nearer 500,000; in Pakistan the number has grown to 100,000 in only three years; in Thailand, of 33,000 addicts in treatment centres, nearly 30,000 are heroin cases. In 1990 it was estimated that abut 5.5 million people in the USA required treatment for drug dependency, namely over 2 percent of the population. It was further estimated that the costs to the USA of drug-related crimes was $5 billion in tangible losses to victims of 9 million drug-related crimes, and a further $30 billion in other annual costs relating to such crimes.
Claim Alcohol use and abuse is seen as the most serious element in the whole spectrum of drug or substance abuse. Alcoholism, drug abuse and smoking are increasing at an alarming rate in every nation. Young people and ethnic minorities are amongst those most heavily effected and they continue to be a target of the promotional efforts which seek to increase the sales and profitability of these substances. The root cause of this appalling global problem may be found in the social and economic conditions which lead to fear, unemployment, isolation, loneliness, dissatisfaction and the need to be socially accepted. The economic and political forces behind the traffic of alcohol and drugs are powerful and efforts to restrict, control or legislate the traffic have met strong resistance, not in the least because of the financial implications of decreased production and consumption. Few commodities are advertised as ruthlessly as alcohol and tobacco. A recent WHO Study reveals massive advertising and exploitation in the third world: 'Indeed, in the third world countries, the alcohol and drug problems constitute a serious obstacle to socio-economic development and threaten to overwhelm the health services'. (WHO 1980)
Counter-claim The social use of alcohol, tobacco and tranquillizing drugs is old and widespread. The majority of moderate users show no signs of any ill-effects. The economy of many nations would be seriously disrupted by the ending of trade in alcohol and tobacco. The profitability of large pharmaceutical industries in some developing countries depends on the widespread use of dangerous drugs.
Refs Adkins, Virgil R *The Static Position of Classifying Alcoholism and Drug Addiction As Identical Illnesses* (1986); National Academy Press *Drug and Alcohol Problems* (1988).
Broader Drug abuse (#PD0094) Mental illness (#PC0300).
Narrower Abuse of plant drugs (#PD0022) Abuse of medical drugs (#PD0028)
Abuse of hallucinogens (#PD0556) Polysubstance dependence (#PG4678)
Abuse of sedatives and tranquillizers (#PE0139)
Solvent and methylated spirits drinking (#PE1349)
Inhaling of solvents and anaesthetic drugs (#PE1427)
Increasing drug addiction in drug producing countries (#PJ0680).
Aggravates Overdose (#PJ1995).
Aggravated by Inadequate drug control (#PC0231).

♦ PD3835 National insecurity in developing countries
Political insecurity in developing countries

Broader National insecurity and vulnerability (#PB1149).
Aggravates Militarization in developing countries (#PD9495).

♦ **PD3837 Criminal gangs**
Gangsterism
Broader Organized crime (#PC2343).
Related Youth gangs (#PD2682).
Aggravates Vigilantism (#PD0527).

♦ **PD3841 Insider dealing**
Insider trading — Illicit trading — Stock scandal — Securities fraud — Insider trade
Nature Any individual or corporation who has price-sensitive information about a public corporation which is inaccessible or not available to the public and who uses that information to trade in stocks or bonds to his own benefit is guilty of insider trading. Inside trading flourishes best at a time, when there is a high proportion of takeovers and mergers, for those are the events which most effect share prices.
Incidence Insider trading scandals have recently involved Pechiney SA and Triangle Industries in France; County NatWest WoodMac Securities and Guinness in the UK; Drexel Burnham, the Chicago Mercantile Exchange and The Chicago Board of Trade, and the Butcher brothers banks in the US; Operadora de Bolsa in Mexico; and the Recruit Cosmos Co., Nippon Steel and Sankyo Seiki, and Kyodo Shiryo in Japan.
Background For until perhaps the 1960s in UK and 1930s in USA insider dealing was not just legal; within limits it was perfectly acceptable. Gradually insider dealing came to be recognized for what it is: a form of theft. In Britain insider dealing was made a crime in 1980.
Refs Practising Law Institute *Trading on Inside Information* (1984).
Broader Financial scandal (#PD2458) Unethical commercial practices (#PC2563)
Securities and commodities exchange violations (#PD4500).
Related Documentary fraud (#PE1110) Commodities trading fraud (#PD3917)
Unethical financial practices (#PE0682).

♦ **PD3865 Learning disorders**
Learning disabilities
Nature A learning disability is one or more disfunctions in a persons capacity to learn. A learning disability is not a lack of intelligence. It is 1) a breakdown in dynamic functioning, in the sense that everything that one does is for an audience of self and others. The capacity to reflect metacognitively is lacking. The individual is not actively involved in learning or in ensuring that outcomes are as favourable as possible for self and audience. They do not recognize their own impact on outcomes (low power of potency, low sense of efficacy), so they do not request clarification, or alter productions in response to feedback. 2) People with learning disorders are poor at prioritizing, focusing and editing, which is a breakdown in the perception that outcomes are mutually construed, and that priorities are audience and context specific. 3) They do not notice schema, form, patterns or scripts in area in which they are deficit. They therefore learn material as distinct entities rather in terms of their relationships which overloads the memory span and precludes the effecting of connections, generalizations and taking of short-cuts. They have a poor sense of rhythm of schema so consequently, poor pace and distribution. 4) They have poor use of inner language to: edit and repair language and behaviour; reflect on attribution, meanings and explanations, assigned, and ask self whether there are alternative possibilities, particularly other-centred ones; rehearse situation before they occur and while they are in process so that productions will be concise, focused, pleasant, protectful of self and others, other-centred and have outcomes that are as comfortable for self and audience as possible; reflect on adequacy of one's productions in terms of the view one wants self and others to have of self and compare own productions with those of others; and plan in advance and preventatively. 5) People with learning disorders are impervious to internal and external nuance, detail, subtle differences, and some aspects of implicit or intentional information in deficit areas. 6) They have poor ability to differentiate most important issues from those that are less important, be it in determining which values can be compromised, what to attend to, what to remember, amount of time to spend, energy to expend, and so on, so they tend to be overwhelmed by complexity and disorganized. 7) They have a limited repertoire of possibilities and do not realize that there might be alternatives. They tend to be rigid, concrete and literal on some level. They are, therefore, poor risk-takers, resist change and try to control rather than negotiate change.
Refs Bakker, D J and Vlught, H van der (Eds) *Neuropsychological Correlates and Treatment* (1989); Cruickshank, William M and Hallahan, Daniel P *Perceptual and Learning Disabilities in Children*; Dumont, J J and Nakken, H (Eds) *Cognitive, Social and Remedial Aspects* (1989); Galloway, David M and Goodwin, Carole *Educating Slow-Learning and Maladjusted Children* (1979); Kohen-Raz, R *Learning Disabilities and Postural Control* (1986); Kronick, Doreen *Social Development of Learning Disabled Persons* (1981); Lynn, Robert E *Chicorel Abstracts to Reading and Learning Disabilities* (1984); McIlroy, Ken *Casebook of Learning Problems* (1986); Silverman, H, et al *Early Identification and Intervention* (1979); Thompson, Robert J Jr *Behavior Problems in Children with Development and Learning Disabilities* (1986); Wallach, Gerldine P and Butler, Katherine G *Language Learning Disabilities in School-Age Children* (1983); World Rehabilitation Fund *Transitions and Adults with Learning Disabilities* (1985).
Broader Stigmatized diseases (#PD7279) Human disease and disability (#PB1044).
Narrower Dyslexia (#PE3866) Agraphia (#PE0280) Language disorders (#PE3886)
Korsakoff syndrome (#PE0333) Reading disabilities (#PD1950)
Developmental expressive writing disorder (#PE0330)
Ignorance of nonverbal communication skills (#PE0533).
Related Innumeracy (#PC0143).

♦ **PD3879 Disobedience of judicial order**
Nature The refusal to obey a legal judicial preliminary or final injunction or restraining order. Normally this is a misdemeanour and the sentence is usually a fine.
Broader Disobedience (#PA7250) Contempt of judicial process (#PD9035).
Aggravated by Passive resistance (#PF2788) Collapse of judicial system (#PJ0761).

♦ **PD3888 Brothel slavery**
Involuntary prostitution in brothels
Nature Women are kidnapped or bought from slave dealers and are denied food until they service customers.
Broader Slavery (#PC0146) Prostitution (#PD0693).

♦ **PD3907 Childhood aggression**
Violent children
Nature Aggression is childhood is the emotional trait that is the strongest predictor of later maladjustment. If not shown other ways than anger and violence to get what they are seeking, they are prone to have difficulties later in life, ranging from trouble with the law to depression and neuroses.
Shared traits of aggressive children are the inability to imagine ways to react in the heat of anger other than to strike out. Such children also tend to perceive slights where none are intended. Boys are twice as likely as girls to suffer from this problem.
Refs Crowell, David H, et al *Childhood Aggression and Violence* (1987).
Broader Maladjusted children (#PD0586).
Aggravates Loneliness of children (#PC0239).

♦ **PD3917 Commodities trading fraud**
Insider trading in commodities
Broader Unethical commercial practices (#PC2563).
Related Insider dealing (#PD3841) Documentary fraud (#PE1110).
Aggravated by Unethical financial practices (#PE0682).

♦ **PD3922 Degradation of the environment in developing countries**
Depletion of natural resources in developing countries — Environmental degradation in developing countries
Nature In the process of development, the natural resource of developing countries are being depleted at an alarming rate. Human pressure on woods, pastures, streams, arable land, flora and fauna is leading to deforestation, soil erosion, lowering of the water table and loss of thousands of species.
Claim Economic activity depends heavily on the natural resource base in most developing countries. Unless such countries can increase the productivity of that resource base, their growing populations will be unable to attain substantially higher living standards.
Refs Biswas, Asit K and El-Hinnawi, Essam (Eds) *Third World and the Environment* (1987).
Broader Environmental degradation (#PB6384) Natural environment degradation (#PB5250)
Unconstrained exploitation of natural resources (#PF2855).
Aggravates Long-term shortage of natural resources (#PC4824)
Lack of natural resources in developing countries (#PD3625).
Aggravated by Inadequate economic policy-making in developing countries (#PF5964).

♦ **PD3934 Accumulation of pollutants in marine wildlife**
Accumulation of contaminant residues in marine animals and fish
Nature Environmental pollution is characterized by the accumulation of toxic metals, organochlorine residues and radionuclides in marine wildlife. Radionuclides are accumulated as a result of fallout from nuclear weapons testing and from nuclear reactor accidents.
Broader Accumulation of pollutants in plants and animals (#PD5021).

♦ **PD3947 Gender abortions**
Misuse of results of amniocentesis — Amniocentesis leading to abortion — Female foeticide
Nature Amniocentesis is a procedure which analyses a sample of amniotic fluid to determine whether genetic anomalies exist in a foetus. Because the test also indicates the sex of the foetus, pre-natal diagnostic techniques the tests are widely used in countries and societies where males are preferred to females as a prelude to abortions of females and in a few countries where females are preferred to males.
Incidence In Bombay alone, 258 private centres for amniocentesis have sprung up over the past few years, and 16 government-supported clinics provide the service as well. It is estimated that between 1978 and 1982, there were over 78,000 cases of amniocentesis followed by abortion of females in India as a whole.
Claim Abortion is a far safer and less traumatic way of ridding the community of unwanted female babies than most means traditionally used. Amniocentesis ensures that no wanted males are aborted by mistake. For a woman it is better to choose a male child than to be cruelly punished or even killed for not producing sons.
Amniocentesis is partly a solution to overpopulation. Women may stop having children once they have had their son. And with less girls, less children will be born.
Counter-claim Dealing with the cultural devaluation of girls by preventing their birth is a gross capitulation to sexism. Instead, China, India and other countries must fight the attitudes and traditional values that lead to the undervaluing of daughters.
Broader Abuse of science (#PC9188) Induced abortion (#PD0158)
Sexual discrimination (#PC2022).
Aggravates Shortage of marriageable women (#PE0427).
Aggravated by Deliberate imbalancing of population sex ratio (#PF3382).
Reduces Infanticide (#PD3501).

♦ **PD3954 Consumer debt**
Private debt — Personal debt — Over-commitment to credit financing
Nature Many individuals through out the world are burdened with debt beyond their means to repay. Many live fear being disconnected from service like gas and electricity, being evicted from their homes for being behind on rent or mortgages or having household goods repossessed. Many debtors are neither feckless nor wilful but find themselves with low incomes and a sudden problem like unemployment, crop failure, divorce, illness or death.
Incidence One in ten consumers in the UK faces difficulties paying debts and more two million have borrowed more than they can afford to repay. In the US, consumer instalment-credit outstanding (which excludes borrowing for house purchase) more than doubled in the 1980's, from $300 billion to $660 billion. Credit card debt was $50 billion in 1981. By January 1989, it had risen to $180 billion.
Broader Uncontrolled growth of debt (#PC8316).
Narrower Leveraged buy-outs (#PE4963).
Aggravated by Unethical consumption practices (#PD2625).

♦ **PD3978 Diseases of the digestive system in animals**
Broader Animal diseases (#PC0952).
Narrower Scours (#PE6253) Amoebiasis (#PE6782) Giardiasis (#PE4811)
Duck plague (#PG7496) Coccidiosis (#PE2738) Hexamitiasis (#PG9766)
Salmonellosis (#PE7562) Colic in horses (#PG5184) Vagus indigestion (#PG7816)
Bloat in ruminants (#PG9822) Thrush in chickens (#PE4497) Campylobacteriosis (#PG4274)
Colitis-X in horses (#PG6906) Potomac horse fever (#PG9543)
Pharyngeal paralysis (#PG9645) Gastritis in animals (#PG4408)
Ruminal parakeratosis (#PG3729) Rattle belly in lambs (#PG9361)
Peritonitis in animals (#PG3923) Bovine liver abscesses (#PG7444)
Bovine viral diarrhoea (#PG9719) Bovine winter dysentery (#PG6998)
Peritoneal fat necrosis (#PG9636) Mouth diseases of animals (#PE9702)
Enteric diseases in foals (#PG9600) Enteric diseases in sheep (#PG9262)
Enteric diseases in swine (#PE3936) Trichomoniasis in poultry (#PG4423)
Liver diseases in animals (#PG6225) Rectal prolapse in animals (#PG4406)
Perineal hernia in animals (#PG3772) Enteric diseases in horses (#PE6708)
Chronic diarrhoea in horses (#PG7656) Abomasal disorders of cattle (#PE9364)
Oesophageal spasm in animals (#PG7247) Traumatic reticuloperitonitis (#PG6986)
Salivary disorders in animals (#PE7570) Necrotic enteritis of poultry (#PG4550)
Simple indigestion in animals (#PG9250) Neonatal diarrhoea in ruminants (#PG4499)
Ulcerative enteritis in poultry (#PG7595) Oesophagogastric ulcers in swine (#PG9726)
Coronaviral enteritis of turkeys (#PG9619) Rotaviral infections in chickens (#PG9408)
Digestive disorders of the rumen (#PE3004) Gastroenteric diseases in animals (#PE7423)
Haemorrhagic enteritis of turkeys (#PG5040) Paralysis of the tongue in animals (#PG9273)
Malabsorption syndromes in animals (#PG7588) Acute pancreatic necrosis in animals (#PG9797)
Gastrointestinal parasites of animals (#PE2436)
Diseases of the oesophagus in animals (#PE4630)
Acute intestinal obstructions in animals (#PG9808)
Congenital anomalies of the digestive system in animals (#PE3711).

♦ **PD3982 Obstruction of elections**
Nature Obstructing the electoral process through false registration of voters, bribing voters, or taking bribes as a voter is a crime. In most instances it is a misdemeanour and punishable with a fine.
Broader Unfair elections (#PC2649) Violation of political processes (#PD5457).
Narrower Electoral fraud (#PD5214).

♦ **PD3984 Geographical illiteracy**
Ignorance of world geography
Nature Significant numbers of students entering university are unable to locate major land masses on a map, do not know the major countries of the world, and cannot name the capitals of most nations.
Incidence In the United States, a survey of the University of Miami, 30 percent of the students could not locate the Pacific Ocean on a world map. A recent survey of 5,000 high school students in eight major U.S. cities revealed 25 percent of the students in Dallas could not name the country that borders the United States on the south, 50 percent of the students of Hartford, Connecticut were unable to name three countries in Africa, and 45 percent of those in Baltimore could not shade in the area representing the United States on a map.
Broader Illiteracy (#PC0210).
Related Cultural illiteracy (#PD2041) Computer illiteracy (#PG2575) Mathematical ignorance (#PD6728).
Aggravates Low self image due to illiteracy (#PF9098).

♦ **PD3996 Plagiarism**
Plagiarizing
Nature The intentional use and taking credit for ideas that are someone else's ranges from bad taste to infringement of copyright and theft.
Refs Mallon, Thomas *Stolen Words*; St Onge, K R *The Melancholy Anatomy of Plagiarism* (1988).
Broader Avoidance of copyright (#PD0188) Vulnerability of intellectual property (#PF8854).
Narrower Conceptual plagiarism (#PD1284).

♦ **PD3998 Defective product manufacture**
Production rejects — Unreliable products — Defective manufactured goods
Nature Production of articles, parts and other equipment that does not correspond in quality to standards, technical specifications and other technical norms. Irreparable rejects are articles whose defects are technically impossible or economically disadvantageous to eliminate.
Faulty manufactured goods cause accidents in the home and workplace. Manufacturers who wish to improve product quality cannot do so without risking competitiveness in the marketplace. This is especially difficult because manufacturing costs are already rising to meet the price of insuring against customers' lawsuits.
Incidence According to the OECD, Australia has the world's most exhaustive mechanisms for product ban and recall. Nevertheless, some 25 children per 100,000 population die annually from injuries or poisonings and about 2500 per 100,000 are admitted to hospital from these causes. Home leisure and consumer product-related injuries alone account for at least 30 percent of all accidents and 20 percent of handicaps.
Broader Risk (#PF7580) Unethical commercial practices (#PC2563).
Narrower Product tampering (#PD8804) Defective medical devices (#PG8737) Defects in machinery design (#PE2462).
Related Childhood accidents (#PD6851) Electric current accidents (#PG2862).
Aggravates Mechanical failure (#PC1904) Electronic equipment failure (#PD1475) Occupational domestic accidents (#PE4961) Prohibitive cost of product liability protection (#PE4404) Dumping of defective products by industrialized countries (#PJ1634).

♦ **PD4002 Natural resource depletion due to high-level consumption**
Excessive environmental demand per capita
Nature As world population increases and incomes rise, the growing demand for goods and services results in greater derived demands for non-renewable natural resources. Not only is agricultural land used more intensively and extensively, but mineral resources (including liquid resources, such as water and petroleum, and gaseous resources, such as natural gas) are used more intensively. The extraction of land resources and their transformation into products reduce the concentrations of these resources in the earth's crust, making it more and more difficult to obtain them.
Claim People would like to consume more than they do; and rising consumption levels are among the economic aims of most governments, even if, in the short run, priority is given to building up an industrial base by means of capital formation. This aim is understandable in countries where a large part of the population is badly housed and undernourished. But the same aim is pursued in countries where a large part of the population has reached at least a modest degree of affluence. It is clear that per capita consumption cannot rise indefinitely, indeed that every form of growth is essentially sigmoid rather than exponential. In view of the many pressures on resources and the problems (pollution, congestion and so on) to which high-level consumption gives rise, it is necessary to examine the effects of different kinds of consumption on the community at large. Conserving the earth's resources means an 'energy transition' from non-renewable to renewable resources.
Broader Long-term shortage of natural resources (#PC4824) Unconstrained exploitation of natural resources (#PF2855).
Narrower Environmental poverty (#PD5261) Environmental prodigality (#PF7318).
Aggravates Ecologically unsustainable development (#PC0111).
Aggravated by Unethical consumption practices (#PD2625) Excessive consumption of goods and services (#PC2518).

♦ **PD4004 Inadequate medical facilities**
Insufficient physical infrastructure for health care delivery — Inadequate health equipment
Nature In both developed and developing countries the physical facilities for adequate health care delivery is lacking. In developed countries the shortage is in the inner cities and in rural areas. In the developing countries the shortage is general. The lack of hospitals, clinics, emergency equipment, and factories for manufacturing medical supplies and drugs results in the failure to provide adequate medical care for millions of people.
Broader Inadequate health services (#PD4790) Limited availability of health resources (#PD7669) Weakness of infrastructure in developing countries (#PC1228).
Narrower Overcrowded public clinics (#PG5393) Inadequate hospital facilities (#PE5058) Inadequate rehabilitation facilities (#PD1089) Prohibitive cost of hospital facilities (#PE4154) Lack of facilities for the physically disabled (#PD8314).
Aggravated by Inadequate medical resources (#PD7254) Prohibitive cost of necessities in rural communities (#PF2385).

♦ **PD4007 Depletion of natural resources due to population growth**
Environmental pollution due to over-population — Negative ecological impact of overpopulation — Imbalance between population growth and resource development
Nature Growth in human numbers and in material living standards lead to increased production which, given the technologies that are nowadays employed, result in a rapid depletion of many natural resources and to the production of numerous pollutants which are not only disagreeable and dangerous but are also, in some cases, employed on a scale which cannot be absorbed and dissipated by the natural environment.
Incidence The demands made by the increasing population were previously assumed to be well within the capacity of the Earth, as far as concerned its ability to supply the physical and chemical requirements for continued life and to absorb waste products. However, the late 1970s brought into focus the finite nature of non-renewable resources and the Earth's limited carrying capacity. The more people there are on Earth, the more will be the demand for the limited natural resources to support life and development, and the more will be the attendant environmental pressures. This will certainly continue to be the case for decades to come, especially in developing countries, where population growth is especially great in the cities, 17 of which top the 10 million mark and are expanding in a chaos of unplanned, under-serviced housing. If present trends continue, the populations in urban areas will double in the next decade, and many of these new citizens will live in squatter settlements.
Population pressure is forcing traditional farmers to work harder, often on shrinking farms on marginal land, simply to maintain a subsistence income. In Africa and Asia the rural population nearly doubled between 1950 and 1985, with a corresponding decline in land availability. This trend can only continue.
Claim The situation is sometimes expressed by saying that the societies in which we live can no longer be regarded, as they have been in the past, as 'frontier' societies. By this is meant that 'our' society, whoever 'we' may be, has a vast world lying outside its frontier which can supply at approximately current prices whatever we demand and cannot supply for ourselves; and furthermore, that there is another frontier over which we can dump anything that is disagreeable to us in the confident expectation that we shall never be troubled by it again. Nowadays, however, the implicit assumptions of this view are challenged. The growth of world population, consumption levels and the depletion of natural resources contradict the first; and the inability of the natural world to absorb every kind of pollutant poured out on the contemporary scale, contradicts the second. We have to recognize that the earth, its resources and its capacities are limited, and that we live in a closed world which is daily becoming more cramped.
Refs Barrat, John and Louw, M H H *International Aspects of Over-Population* (1972).
Broader Long-term shortage of resources (#PB6112) Natural environment degradation (#PB5250) Unconstrained exploitation of natural resources (#PF2855).
Related Malthusianism (#PF4606) Long-term shortage of natural resources (#PC4824).
Aggravates Deforestation (#PC1366).
Aggravated by Environmental poverty (#PD5261) Unsustainable population levels (#PB0035).
Reduced by Declining birth rate (#PD2118).

♦ **PD4011 Structural rigidities in labour markets**
Labour immobility — Lack of labour mobility — Low labour mobility — Labour market inflexibility — Lack of labour force mobility — Restrictions on labour mobility
Nature Mobility can take three forms: geographical (the movement of workers from one part of a country to another); occupational (movement from one job to another); and social (movement from one class of job to another). But labour is not perfectly mobile; wage differentials and relativities can be maintained as a result of a lack of any of these types of mobility. Thus wages may be higher in one part of a country than another, but equalization is prevented by the reluctance of workers to move (because, perhaps, of different customs or languages elsewhere). A factory worker may earn more than a farm labourer, but farm labourers may be prevented from moving to the better-paid job because of difficulties in obtaining the necessary skills; and in many countries, there is considerable social immobility - so that it is much easier, for example, for the son of a university professor to become a doctor than it is for the son of a farm labourer or docker. Two forms of mobility are important from the individual's point of view: mobility between occupational and status groups and mobility between industries. A somewhat different aspect of labour mobility is mobility within organizations arising from recruitment, promotion and retirement rules, and from wastage rates.
Both European and American labour inflexibility grow out of workers' efforts to create a stable social existence in insecure and uncertain economic environments. In Europe, this has been done by directly imposing employment security through legal restrictions in businesses' ability to lay off and discharge workers. In the United States, unions sought and won the control of processes through which scarce jobs are distributed among the labour force by restricting both how jobs are allocated among workers and what a manager can ask any given worker to do.
Broader Lack of social mobility (#PF2195).
Narrower Wage rigidity in labour markets (#PF4028).
Aggravates Unemployment (#PB0750) Social inequality (#PB0514) Age discrimination (#PC2541) Lack of economic and technical development (#PE8190).
Aggravated by Ineffective economic structures in industrial nations (#PE4818).
Reduced by Privatization of public services (#PE3391).

♦ **PD4012 Economic unrest**
Industrial unrest — Industrial conflict — Industrial disputes — Labour disputes
Nature Industrial relations vary greatly at different times and places. After a period of comparative tranquillity, unrest may develop and flare up into stoppages of work occasioned by strikes or, less usually, lock-outs. Stoppages can be analysed in terms of their principal cause; by far the most frequent are those described as wage disputes and, in particular, claims for increases in wages. However, since the increases in wage rates now regularly demanded are not realistic, in the sense that they could not be achieved in real terms even under conditions ideal from the point of view of those who demand them, it would appear that although, under present conditions, wage claims provide a socially acceptable basis for conflict, the true basis lies elsewhere. Many types of institution are involved in industrial conflict: trade unions, employers' organizations and a variety of boards, tribunals and courts.
Refs Kornhauser, Arthur, et al *Industrial Conflict* (1977).
Broader Economic conflict (#PC0840).
Narrower Strikes (#PD0694).
Aggravates Panic consumer buying (#PJ0542).

♦ **PD4027 Substance intoxication**
Broader Substance abuse (#PC5536).
Narrower Caffeine abuse (#PE0618) Opioid intoxication (#PG9743) Cocaine intoxication (#PG6889) Inhalant intoxication (#PG7346) Cannabis intoxication (#PG7872) Sedative intoxication (#PG9279) Amphetamine intoxication (#PG3942) Uncomplicated alcohol withdrawal (#PE0375).
Related Public drunkenness (#PE2429).

♦ **PD4029 Still-birth**
Still-birth mortality — Still-born babies — Antenatal foetal death
Nature If a new born baby shows no signs of life it is termed a still-birth and as such does not enter into neonatal or infant mortality statistics. The pregnancy must have been at least 28 weeks

PD4029

of gestation, otherwise it is termed an abortion. Still-birth can be either fresh or macerated. Macerated still-births are foetuses which have died in the womb and have remained their long enough to start decomposing. A macerated still-birth indicates that death occurred a considerable time before onset of labour, due to antenatal causes and maternal disorders (diabetes mellitus, anaemia, hypertension, placental insufficiency, syphilis, herpes). The great majority of still-births occur during labour, due to complications of labour and prematurity. Other factors include congenital malformations, ante-partum haemorrhage, eclampsia, and maternal illness (malaria, tuberculosis, chronic renal disease).
Incidence Statistics are not readily available, especially in developing countries. One study showed 30 per 1000 births in Africa.
Refs DeFrain, John, et al *Stillborn* (1986); Porter, Ian H and Hook, Ernest B *Human Embryonic and Fetal Death* (1980).
Broader Perinatal morbidity and mortality (#PD2387).
Aggravated by Herpes (#PE8615) Anaemia (#PD7758)
Malaria (#PE0616) Diabetes (#PE0102)
Syphilis (#PE2300) Stowing away (#PE0595)
Tuberculosis (#PE0566) Toxoplasmosis (#PE3659)
Heart diseases (#PD0448) Premature birth (#PD1947)
Kidney disorders (#PE2053) Placental anomalies (#PG2046)
Complications of childbirth (#PC9042) Human physical genetic abnormalities (#PD1618)
Haemorrhage of pregnancy and childbirth (#PE4894).

♦ PD4031 Animal abnormalities
Deformed animals — Animal defects — Increase in animal genetic defects
Refs Foley, C W, et al *Abnormalities of Companion Animals* (1979); Kalter, Harold (Ed) *Issues and Reviews in Teratology* (1984); Szabo, Kalman T *Congenital Malformations in Laboratory and Farm Animals* (1988).
Broader Animal diseases (#PC0952).
Narrower Congenital anomalies of the skin (#PG7626)
Congenital porphyria erythropoietica (#PG9832)
Conformational abnormalities of eyelids in animals (#PG4212)
Congenital diseases of the nervous system in animals (#PE0965)
Congenital anomalies of the urinary system in animals (#PG9411)
Congenital anomalies of the digestive system in animals (#PE3711)
Congenital anomalies of the animal musculoskeletal system (#PE7808)
Congenital anomalies of the reproductive system in animals (#PG7447)
Congenital anomalies of the cardiovascular system in animals (#PE5569).
Aggravates Animal infertility (#PC1803).
Aggravated by Radioactive contamination (#PC0229)
Genetic defects and diseases (#PD2389)
Decreasing genetic diversity of animals (#PC1408).

♦ PD4037 Decadent standard of living
Extravagant use of wealth — Extravagant life style — Conspicuous consumption
Nature The extravagent use of expensive goods or services in order to demonstrate status and wealth leads to a large proportion of a country's economic resources being allocated to the production of luxury goods and consumer durables which need to be periodically replaced.
Incidence This term was heavily used by Thorstein Veblen in his Theory of the Leisure Class, published in 1899.
Claim Men are dehumanized not only by the work situation but also by the ends for which society uses work, chiefly consumption for its own sake. The unproductive acquisition of goods has become the primary means of achieving social status in the community. The yearning for achievement and the instinct for workmanship tends more and more to shape itself into a straining to excel others in pecuniary achievement. If consumption then becomes "conspicuous", it is because in an increasingly heterogeneous and differentiated society, there is no ready means of acquiring status except by spending money and acquiring goods. The industrial revolution has thus replaced all workmanship with labour and the result has been that products have become objects for consumption rather than things which are there to be used.
Broader Decadence (#PB2542).
Narrower Conspicuous consumption by international civil servants (#PE3457)
Reinforcement of inappropriate development by privileged classes in developing countries (#PF6670).
Related Deteriorating quality of life (#PF7142)
Excessive consumption of goods and services (#PC2518).
Aggravates Grave robbing (#PF0491) Meaningless recreation (#PF0386)
Loss of cultural identity (#PF9005) Shortage of domestic servants (#PJ9711)
Competitive acquisition of arms (#PC1258).
Aggravated by Environmental prodigality (#PF7318)
Unethical consumption practices (#PD2625)
Unnecessary personal consumption (#PF5931)
Unethical practices in the apparel industry (#PD8001)
Lack of relationship between wealth generation and the public good (#PF4730).

♦ PD4049 Vulnerability of nuclear defence control systems
Nuclear decapitation
Nature The command and control systems for nuclear defence are much more vulnerable than has been officially proclaimed. Ground-based systems could be relatively easily overrun, airborne systems are vulnerable to radioactive dust and to destruction of launch sites, submarines can be isolated, and satellite systems can be destroyed by nuclear explosions in space.
Related Risk of unintentional nuclear war generated by the strategy of deterrence (#PF4162).
Aggravates Slowing growth in food production (#PC1960).

♦ PD4060 Manipulation of debates
Curtailment of discussion of issues in meetings — Avoidance of controversial issues
Broader Inadequate meeting methods (#PF8939).
Related Political displacement activity (#PF5360).
Aggravates Inappropriate arguments (#PF2152)
Lack of parliamentary time to approve needed legislation (#PF8876).
Aggravated by Secrecy (#PA0005) Unreported scandals (#PF5340).

♦ PD4063 Economic dislocations in developing countries
Nature Economic dislocations, whether due to natural disasters, civil disruption, influx of refugees or regional warfare, distort economic projections and necessitate the diversion of resources from planned development programmes.
Aggravates Inaccurate forecasting (#PF4774)
Excessive foreign public debt of developing countries (#PD2133).
Aggravated by Natural disasters (#PB1151) Unforeseen environmental crises (#PF9769).

♦ PD4064 Racial discrimination in sexual preferences
Incidence Twenty-five thousand, mostly male, Chinese spent five years building the Tan-Zam railroad linking Zambia's copper belt with the Tanzanian port of Dar es Salaam. Not one Chinese man was involved with an African women.
Broader Racial discrimination (#PC0006).

♦ PD4065 Meteorological disaster
Broader Natural disasters (#PB1151).
Narrower Storms (#PD1150) Lightning (#PD1292) Storm surges (#PD2788)
Ice accretion (#PD1393) Climatic cold (#PD1404) Thunderstorms (#PE3881).
Related Drought (#PC2430) Climatic heat (#PC2460).
Aggravates Global warming (#PC0918) Global cooling (#PF1744).

♦ PD4067 Unconvicted war criminals
Nature War criminals and people who commit crimes against humanity frequently find refuge in countries where legal, political or social conditions offer them safety. War criminals are known to reside in the UK and Canada where local courts do not have the jurisdiction to prosecute war criminals from World War II. They reside in countries in Latin America where frequently they are effectively immune from detection or have political influence.
Broader Criminals (#PC7373) Inadequate evidence to convict known offenders (#PF8661).
Related Post-revolutionary re-employment by government of security services of the ousted repressive regime (#PF1015).
Aggravated by Legal havens (#PE0621) Ineffective war crime prosecution (#PD1464)
Government approved employment of war criminals (#PE4697).

♦ PD4068 Immune system diseases in animals
Broader Animal diseases (#PC0952).
Narrower Atopic diseases (#PE9509).

♦ PD4069 Escapist family life styles
Nature In order to avoid taking responsibility for the larger society and to minimize fear of social ambiguity and personal insecurity families retreat into escapist life styles. In depth care of the individuals is thought the responsibility of outside agents like teachers, clergy and psychologist. Time is filled with shallow activities designed to avoid questions of meaning. The physical living space reflects a desire to be uninvolved. Relationships between the family and the larger society are the responsibility of individual family members. Meal times are used to meet physical needs.
Broader Escapism (#PF7523) Lost family role in society (#PF7456).
Related Breakdown in covenants for life (#PD1026)
Refusal of family possibilities (#PF0846)
Family adaptation of community status quo (#PE5408)
Refusal of families to participate globally (#PF1006).
Aggravates Meaningless recreation (#PF0386).

♦ PD4082 Sexual offences
Sex crimes
Incidence In UK the number of sex offenders within the prison system has been rising rapidly to almost fifth of the total. More sexual offences are being reported (though probably only a fraction of those committed) and the courts are sending convicted offenders to prison for longer periods than earlier.
Refs Bailey, F Lee and Rothblatt, Henry B *Crimes of Violence* (1973); Cameron, D and Frazer, E *The Lust to Kill* (1987); Drummond, Harold P *Sex Offenses* (1987).
Broader Violent crime (#PD4752) Moral offences (#PD9179).
Narrower Rape (#PD3266) Sodomy (#PE3273) Adultery (#PF2314)
Seduction (#PG5097) Sexual abuse of wards (#PE7755)
Sexual offences by juveniles (#PD9394) Sexual intercourse with minors (#PE6522).
Related Incest (#PF2148) Offences involving danger to the person (#PD5300).

♦ PD4089 Denial of rights to soldiers
Refs Janowitz, Morris *Professional Soldier* (1964).
Broader Denial of rights to vulnerable groups (#PC4405).
Narrower Ill treatment of prisoners of war (#PD2617)
Denial of rights to wounded military personnel (#PE4758).

♦ PD4091 Engaging in riot
Nature Participating with a minimum of 3 to 10 person (depending on the jurisdiction) in a public disturbance which by violent conduct creates a grave danger or injury to persons or property. Most definitions of riot use the minimum of three persons but increasingly this is seen as an unhelpful minimum and more governments are using the higher figure of 5 and some 10 persons are required to constitute a riot.
Broader Civil disorders (#PC2551).

♦ PD4098 Compulsory organization membership
Incidence In Cuba, membership of people's organizations from Pioneers in the primary schools to student, trade-union and other organizations, constitute the first level of politico-ideological integration, which is a prerequisite for functioning in society. Such organizations are instruments for controlling the life of citizens, particularly their leisure time.
Broader Denial of the right of association (#PD3224).
Aggravates Compulsory indoctrination (#PD3097).

♦ PD4100 Narrow range of food crops
Limited food grain species — Dependence on plant species — Undiversified feed crops — Limited farm crops
Nature Although there are at least 20,000 species of edible plants worldwide, fewer than 150 are cultivated widely. This creates a great deal of vulnerability to plant diseases which could ravage the commonly used species. A major epidemic of plant disease striking any of the major cereal crops (wheat, rice, and maize) would result in famine on a massive scale. Research grants, however, are not available for crops which may offer alternatives, but in which few are currently interested.
Incidence It is reported that of the 150 cultivated species, only 22 species provide the vast majority of protein for the world's population. An alternative estimate is that 95 per cent of human food come from 30 plants, and 75 per cent from only 8 of these thus restricting the species to an extremely narrow dietary base.
Broader Inappropriate application of traditional values (#PF2256)
Underdeveloped approaches to local food production (#PF6493)
Underdevelopment of food and live animal production (#PF2821).
Aggravates Limited food variety (#PF0479) Monoculture of crops (#PC3606)
Neglected food resources (#PF7808).
Aggravated by Extinction of species (#PB9171).

♦ PD4101 Inadequate staple food supply in developing countries
Nature The reliance of many developing countries on imported staple foods has markedly increased over the last 25 years. Import dependence has been encouraged by the concessionary price programmes of exporting countries; however, availability under such subsidies is declining, and developing countries must absorb increasing amounts of commercial supplies.
The reasons for increased staple imports are complex. Crop failures, disasters, droughts and wars increase short-term needs in various countries, but do not themselves create dependence on foreign supplies. Rapid population growth and urbanization is a main cause; domestic production

and marketing has not expanded in step with this growth and many governments choose imports to fill the gap. The enterprises (both national and transnational) which participate in this process build plants and adopt technologies designed to handle the imported product, often with financial assistance from the surplus countries. Technical problems arise when countries try to reduce dependence, once the factories that use imports are in place. Retraining workers and re-equipping the industries is a prerequisite to overcoming this institutionalized dependence.
Broader Food insecurity (#PB2846).
Aggravates Dependence of developing countries on food imports (#PE8086)
Underdevelopment of industrial and economic activities (#PC0880)
Import-dependency in food staples in developing countries due to transnational corporations (#PE1806).
Aggravated by Dumping of food in developing countries (#PE0607).

♦ **PD4116 Illegal exports**
Illicit exports
Nature For the purposes of national and international security several treaties limit the export of a number of products. High technology equipment, nuclear technology, and certain fire arms are restricted. A number of companies have violated these restriction, sometimes with the approval of their governments and thus jeopardizing the agreements and many lives.
Incidence A number of Japanese companies have sold high technology equipment to China and the Soviet Union. German companies have exported nuclear and chemical warfare technology to the Middle East. A Norwegian company has sold computer equipment to the Soviet Union. French companies have exported sophisticated machine tools to the Soviet Union. All of these are violations of international treaties.
Broader Unlawful business transactions (#PC4645)
Crimes against national security (#PC0554).
Narrower Illicit export of works of art (#PE9004)   Illegal exports of nuclear materials (#PE3968).

♦ **PD4125 Rural poverty in developing countries**
Economic stagnation due to rural poverty in developing countries — Subsistence poverty economy — Continued subsistence living — Difficulties in establishing changes in rural economic patterns
Nature One of the major goals of every nation's agricultural policy is the provision of adequate food supply to the population. Currently, however, the world has seen food deprivations, chronic malnutrition and famine afflict an increasing number of victims, especially in the poorest of developing countries. There is a growing consensus among experts that world malnutrition, hunger, and starvation are only the most visible aspects of the basic problems of poverty. The landless and the near landless in the rural areas and elsewhere simply lack adequate purchasing power to sustain equitable terms of trade between the agricultural sector and the industrial sector of the nation. Stated somewhat differently, poverty among rural peasants who also make up the majority of the population in most developing countries not only retards the development of a viable industrial sector but also accounts for the stagnation or decline of production in the agricultural sector itself – thus opening the population to hunger and malnutrition. From then on it is a vicious circle – a circle of poverty.
Despite general recognition that fate does not require people to live at a bare subsistence level, long-established patterns of existence and traditional style of life of many Third World rural communities are such that providing the means of day-to-day living overrides making plans for the future. The result is that ancient modes of agriculture and agricultural barter mechanisms are being questioned by the trend toward a cash economy. However, rising expectations on living standards reveal the absence of economic infra-structures such as available capital, saving mechanisms and usable credit. The lack of a well operating marketing system, of an indigenous agricultural research system and of a physical rural infrastructure help maintain subsistence economies. And the frustration of negotiating a change in economic patterns while maintaining a vital cultural heritage serves to discourage the necessary changes from taking place.
Refs ILO *Employment and Poverty in a Troubled World* (1985); Srinivasan, T N and Bardhan, Pranab K *Rural Poverty in South Asia* (1988).
Broader Socio-economic poverty (#PB0388)   Inadequate social reform (#PF0677)
Underdevelopment of industrial and economic activities (#PC0880).
Narrower Limited accumulation of capital (#PF3630)
Unnecessary education expenditure (#PJ0626)
Overemphasis on immediate superficial needs (#PF3243).
Related Time consuming procedures (#PJ8206).
Aggravates Lack of savings structures (#PF1348)
Diminishing capital investment in small communities (#PF6477).
Aggravated by Apathy (#PA2360)   Economic stagnation (#PC0002)
Unproductive subsistence agriculture (#PC0492)
Dumping of food in developing countries (#PE0607)
Subsistence agricultural income level in rural communities (#PE8171).

♦ **PD4131 Segregation through language**
Linguistic dis-integration — Linguistic discrimination
Nature Language has great power to create and to draw asunder social groups. In many situations, language is used as a social divider.
Incidence The English-speaking world has long used the pronunciation of English as a very sharp divider of social classes. Mrs. Thatcher is reputed to have had diction training in the hope that her middle-class pronunciation might reach the heights to which her ambition aspired. Japanese is an extreme example of language that is different not only in content but in syntax and grammar when spoken by a male or a female. Japanese language alters in its terms of self-reference and address, its sentence particles, its verb forms and its interjections, for males, for females and for children. Many societies align their cultural, political and/or economic divisions with language divisions.
Broader Discriminatory communication (#PD6804)
Irrelevance of educational curricula (#PF0443)
Sexually discriminating job terminology (#PF6014).
Aggravated by Lack of English language to describe female experience (#PF7383).

♦ **PD4134 Religious terrorism**
Nature Religious terrorism is based on a fundamentalistic and sometimes literalistic belief in sacred texts and their interpretation by religious leaders. The goal of such terrorism is not freedom or democracy but conversion of the unbelievers and if that is not possible, their death. Religious terrorists are not concerned with their own death because that is release from this world.
Broader Terrorism (#PD5574).
Aggravated by Religious extremism (#PF4954).

♦ **PD4143 Collapsing physical structures**
Broader Structural failure (#PD1230)   Disastrous accidents (#PC6034).
Narrower Mine disasters (#PD2278)   Subsidence from mining (#PE4393)
Collapsing public works (#PG2869).
Related Impoldering risks (#PE6347).

♦ **PD4148 Cruelty to plants**
Torture of plants

Nature Increasing evidence indicates that plants respond to maltreatment in addition to physical abuse. Plants are trimmed, torn up, cut down, pruned and treated in brutal ways without consciousness on the part of the people doing the destruction. Bonsai plants have their growth intentionally stunted and the bodies shaped in outlandish ways. Flowering plants have their reproductive organs plucked before their function is fulfilled. Vegetables are torn from the ground even before they have a chance to flower. Grain producing plants are cut down just as their seeds ripen, ready to fall to the ground. Perhaps more brutal is the indifference and neglect of householders. House and garden plants are left unwatered and unfed during vacations and periods of forgetfulness. Some are brought from special ecological niches and forced to live and die in some living room at the wrong temperature and humidity and fed the wrong nutrients. Each new year millions of plants are dug up and thrown out because their natural annual cycles do not suit the gardener.
Broader Torture (#PB3430).
Related Human torture (#PC3429)   Torture of animals (#PC3532)
State sanctioned torture (#PD0181).
Aggravates Suffering of plants (#PC7825)   Destruction of weeds (#PE3987).

♦ **PD4150 Denial of economic rights**
Lack of economic liberty
Broader Denial of human rights (#PB3121).
Narrower Denial of the right of health (#PJ2269)   Inadequate standards of living (#PF0344)
Violation of trade union rights (#PD4695)   Restrictions on property rights (#PD8937)
Denial of right of equal pay for equal work (#PD1977)
Discrimination against traditional economies (#PD4252).
Related Economic repression (#PC8471).

♦ **PD4158 Offences of general applicability**
Generalized types of crimes
Broader Statutory crime (#PC0277).
Narrower Felonies (#PE1153)   Misdemeanours (#PE5594)   Criminal attempt (#PD5321)
Criminal conspiracy (#PD1767)   Criminal concealment (#PF8000)
Criminal facilitation (#PD6845)   Criminal solicitation (#PD7676)
Violation of regulatory codes (#PD4539).

♦ **PD4171 Endangered species of medicinal plants**
Nature Some of the world's rarest and most interesting medical plants are in danger of extinction. Pharmaceutical firms are cutting them for research purposes without assuring their continued survival. Population growth and the consequent increase in land under cultivation is threatening many plants. The destruction of forests and jungles is also endangering many species.
Broader Endangered species of plants and animals (#PB1395).
Aggravated by Extinction of species (#PB9171).

♦ **PD4172 Electromagnetic pollution**
Electronic smog — Electronic noise — Intermittent electrical signals
Nature Electronic smog occurs when electromagnetic waves from equipment like computers and electronic game machines or discharges of static electricity send signals to equipment causing it to malfunction.
Incidence Such pollution has caused deaths and injuries. Airport radar screens have been jammed. Production line robots have functioned faultily. Roller coasters have crashed. Railway switches have malfunctioned and train doors have opened inadvertently. Computers can malfunction or lose their memories and their components damaged. The magnetic fields surrounding high voltage transmission lines may increase a child's risk of leukaemia according to one study by the New York State Health Department.
Broader Pollutants (#PC5690).
Narrower Radio noise of natural origin (#PF1676)
Radio noise of industrial origin (#PE2473)
Hazards of environmental electromagnetism (#PE1304)
Environmental hazards from electromagnetic pulses (#PE6360)
Environmental hazards of extremely low frequency electromagnetic radiation (#PE7560).
Related Computer viruses (#PD3102)   Unreliability of computer software (#PE4428).
Aggravates Leukaemia (#PE0639)
Obstruction of astronomical observation by environmental pollution (#PE7244)
Health hazards of electromagnetic fields generated by electrical appliances (#PE7879).

♦ **PD4178 Competitiveness in education**
School pressure — Stress due to examinations — Competitive burden on students
Nature Students whose parents wish them to succeed in academy need to begin their quest early enough to compete for limited places in name universities. This includes entrance examinations for children as early as kindergarten level. Many young children have difficulty in academic learning at an early age, while they are still thriving on the concrete learning of play. Nevertheless, some succeed very well up to age 10, and then begin a period of "burnout". Some parents keep their children out of school for a year or so to postpone this traumatic time, but this only permits the tension in the classroom to grow. While the academic tension is surely unhealthy for the child, parents maintain, it is not as terrible as the possibility of academic inadequacy and its accompanying lack of career and social success. This inhumane perspective bodes ill for the generation to come.
Refs Albas, Daniel C and Albas, Cheryl M *Student Life and Exams* (1984).
Broader Stress in human beings (#PC1648).
Aggravates School phobia (#PE4554)   Juvenile stress (#PC0877)
Stress among children (#PE4421).
Aggravated by Inappropriate selection and examination procedures in education (#PF1266).

♦ **PD4182 Unethical practice of meteorology**
Irresponsible meteorologists — Negligence by meteorologists — Malpractice in meteorology — Corruption of meteorologists — Underreporting of weather hazards
Claim Meteorologists, under pressure from their employers, have adopted practices which lead to the underreporting of air pollution and weather hazards to avoid public panic. This has resulted in public unpreparedness, with loss of life and property, in the event of hazardous weather patterns. Meteorologists are intimately involved in experiments in the use of weather modification to the advantage of particular areas and to the disadvantage of others (especially for military or economic warfare purposes) through the creation of patterns of drought, cold or flooding.
Related Corruption of sports and athletic competitions (#PE3754).
Aggravates Hostile environmental modification (#PD7941).

♦ **PD4189 War casualties**
Nature More wars were fought in 1987 than in any year previously on record. Apart from the conflict between Iran and Iraq, these are not conflicts between nations, but between peoples within the same nation. Three million people have died in these wars, four-fifths of them civilians. Since the end of 1945, about 17 million people have died in wars, rebellions and uprisings. This number is about half the world death toll for the 2nd World War.
Incidence In one study through 1987, the following war deaths were catalogued: Latin America: Columbia, about 100 deaths a year since 1958; El Salvador, 65,000 deaths since 1979;

Guatemala, 138,000 since 1966; Nicaragua, 30,000 since 1981; Peru, 10,000 since 1981. Middle East: Iran–Iraq, 377,000 live in eight years; Lebanon, 52,000 since 1982. (The study does not include countries in which uprisings have taken fewer than 1,000 life a year, as in Israel.) Africa: Angola and Namibia, 213,000 since 1975; Chad, 7,000 since 1980; Ethiopia, 500,000 by war and related famine since 1980; Western Sahara, 10,000 since 1975; Mozambique, 400,000 since 1981; South Africa 4,000 since 1985; Sudan, 10,000 since 1984; Uganda, 102,000 since 1981. Asia: Afghanistan, 14,000 Soviet troops and 85,000 Afghans since 1979; Burma, 2,000 since 1985; Separatist violence in India, 5,000 since 1984; Indonesia (no reliable estimates from East Timor); Cambodia, 24,000 since 1979; the Philippines, 60,000 since 1970; Sri Lanka, 6,000 since 1984.
Broader Human death (#PA0072).   Disastrous consequences of war (#PC4257).
Related Housing destruction in war (#PE2592)   War damage in civilian areas (#PD8719)
Industrial destruction by war (#PD8359).

♦ **PD4191 Fraudulent commodities**
Sale of non-existent commodities
Nature Each year $1 billion worth of false commodities, securities and merchandise is sold over the telephone. Tens of millions more are sold by newspaper ads, direct mail and door-to-door solicitation. The products include nonexistent gold, bogus oil and gas wells and counterfeit "signed" prints by famous artists.
Broader Fraud (#PD0486).

♦ **PD4194 Corruption of the judiciary**
Jury nobbling — Harassment of juries
Nature The corruption of the judiciary may involve corruption of judges or other court officials and improperly influencing juries. The corruption of a jury so that a desired verdict is reached. The number of jurors corrupted varies from country to country depending on the number needed for a guilty verdict. Jurors may be influenced through bribery, threats to the timid, pleas of innocence to the impressionable (usually by an attractive member of the opposite sex) or pleas of being falsely accused to the anti-establishment.
Incidence Police estimate that in London alone as many as 12 cases a year are halted and re-tried because jury nobbling has been discovered. There is no estimate of the number of trials in which nobbling remains undetected.
Broader Official corruption (#PC9533)   Hindrance of law enforcement (#PD5515)
Institutionalized corruption (#PC9173).
Related Abuse of authority (#PC8689)   Political injustice (#PC2181)
Harassment of the judiciary (#PE5487)   Political corruption of the judiciary (#PE0647)
Lack of impartiality of the judiciary (#PE7665).
Aggravates Compounding a crime (#PE1485).
Aggravated by Bribery (#PC2558)   Intimidation (#PB1992)
Illicit drug trafficking (#PD0991)   Bribery of public servants (#PD4541)
Use of undue influence to obstruct the administration of justice (#PE8829).

♦ **PD4208 Evasion of the law**
Broader Irresponsibility (#PA8658)   Loss of institutional credibility (#PF1963).
Narrower Avoidance of legal obligations by politicians (#PD4556)
Exploitation of regulatory loopholes in countries with underdeveloped legislation (#PE4339).
Aggravated by Legal havens (#PE0621).

♦ **PD4217 Killing of plants**
Instruction of vegetation — Floracide — Vegecide
Nature Tens of millions of plants are killed by and for human beings every year. They are dug up, cut down, burned, poisoned, mutilated, starved and deprived of water and sunlight for the sole purpose of benefiting humankind. While this may be necessary for the continuance of the species, it is done without regard for the plants, without conscience and without consciousness.
Background To anyone who believes that life itself has some purpose, or is even its own reason for being, the wanton destruction of plants is inappropriate. The destruction of any life is not to be taken lightly, or presumed to be isolated in the scheme of things. It requires careful consideration of the responsibilities and alternatives. Some vegetarians therefore prefer to eat fruits and vegetables only at the peak of ripeness, when a cycle of the plant's life has been completed.
Counter-claim It is a fact of life that humanity's present state of evolution calls for the eating of plants, whether directly or indirectly, in order to ensure survival. Until nutrition can be converted directly from sunlight, it will have to be taken from the plant kingdom in order to do the least amount of harm to sentient beings.
Broader Killing non-human life (#PF6359).
Aggravates Suffering of plants (#PC7825).
Aggravated by Defoliation (#PD1135)   Deforestation (#PC1366)
Lack of social conscience (#PF9144)   Repression of self-consciousness (#PC1777).

♦ **PD4218 Children of alcoholics**
Nature Children of alcoholic parents often display abnormal social behaviour. One in four become alcoholics. They tend to become unconsciously addicted to another person's dysfunctional behaviour. They deny that there is a problem of alcohol in the family. Women children of alcoholics have above average gynaecological problems, and men children of alcoholics are prone to frequent surgery. Adult children of alcoholics often are unclear what normal behaviour is, have difficulty completing jobs, lie when the truth would be just as easy, judge themselves harshly, find having a good time hard, are serious about themselves, find intimate relationships hard, react strongly to changes over which they have little or no control, seek approval constantly, feel that they are different from others, are either overly responsible or overly irresponsible, are very loyal even when continued loyalty is uncalled for, and tend to adhere to a course of action without consideration of consequences.
Refs Ackerman, Robert J *Children of Alcoholics* (1987); Kritsberg, Wayne *The Adult Children of Alcoholics Syndrome* (1988); Smith, Ann *Grandchildren of Alcoholics* (1988); Wood, Barbara L *Children of Alcoholism* (1987).
Broader Victimization of children (#PC5512).
Narrower Foetal alcohol syndrome (#PE3853).
Aggravates Corruption of minors (#PD9481).
Aggravated by Alcohol abuse (#PD0153)   Juvenile alcoholism (#PD1611)
Excessive parental drunkenness (#PE7700).

♦ **PD4219 Inadequate production capacity in developing countries**
Inadequate manufacturing capacity in developing countries — Inadequate industrial capacity in developing countries
Nature Though there is scope for further expansion of demand in the major importing countries for the manufactured exports of developing countries, there are difficult supply capacity problems that remain to be resolved.
Incidence Although industry now assumes a crucial position in the development strategies of all but the very poor and small developing countries, lack of resources, shortage of skilled manpower, insufficient integration and balance within the industrial structure, unsatisfactory technological progress and infrastructural bottlenecks continue to be major constraints on industrial supply capabilities in many developing countries. Given population growth rates, a five to tenfold increase in manufacturing output will be needed just to raise developing-world consumption of manufactured goods to industrialized world levels by the time population growth rates level off in the next century.
Aggravated by Excess production capacity (#PD0779)
Lack of natural resources in developing countries (#PD3625)
Weakness of infrastructure in developing countries (#PC1228)
Use of inappropriate technologies in developing countries (#PF0878)
Restricted growth in export markets of developing countries (#PF1471)
Inadequate research and development capacity in developing countries (#PE4880)
Processing in developed countries of commodities exported by developing countries (#PD0425)
Development by industrialized countries of products substituting for commodities exported by developing countries (#PD7682).

♦ **PD4221 Unlawful trafficking in taxable objects**
Nature Trafficking in taxable objects knowing that the objects have been manufactured, transported, used or sold in violation of government revenue laws or regulations. Alcohol, tobacco, precious metals, drugs, electronic goods are the most common items.
Broader Tax evasion (#PD1466).
Narrower Possession of unlawful distilled spirits (#PE5283).

♦ **PD4233 Romantic separation**
Breakup of affective relationship — Cessation of romance — Ending of love affair
Refs Dilman, Ilham *Love and Human Separateness*; Ottens, Allen J *Coping with Romantic Breakup* (1987).
Broader Personal life crises (#PD4840).
Aggravates Lovesickness (#PF3385)   Attempted suicide (#PE4878).
Aggravated by Dependence on romantic love (#PF7418).

♦ **PD4234 Denial of right to life**
Claim The right to life is the fundamental right of human beings and all other rights make sense only in so far as the right to life is respected. The fundamental reason for retention of the death penalty by some countries is it is a deterrent for those who might attack life and the property of the State. The fact, however, is that in countries where the death penalty has existed for centuries, crimes against life and property has not diminished. As regards the State security argument, the State might well confuse the security of society with the security of the group currently in power, namely, the Government of the day.
Broader Denial of political and civil rights (#PC0632).
Denial to people of control over their own lives (#PC2381).
Narrower Homicide (#PD2341)   Extra-legal executions (#PE6366)
Inhumanity of capital punishment (#PF0399)   Denial of the right of unborn children (#PF6616).
Related Human torture (#PC3429)   Abuse of police power (#PC1142)
Violation of civil rights (#PC5285)   Denial of right to liberty (#PF0705)
Denial of religious liberty (#PD8445)   Denial of freedom of thought (#PF3217)
Denial of the right to procreate (#PC6870)   Denial of the right of association (#PD3224)
Restriction of freedom of expression (#PC2162)   Inequitable administration of justice (#PD0986)
Denial of right to national self-determination (#PC1450)
Denial of economic and social rights to refugees (#PE6375)
Restrictions on freedom of movement within countries (#PE8408).
Aggravated by Forced disappearance of persons (#PD4259).
Reduced by Criminalization of abortion (#PF6169).

♦ **PD4238 Harmful natural foodstuffs**
Refs Fernando, R *Traditional and Non-Traditional Foods* (1981).
Broader Environmental hazards from food and live animals (#PC1411).
Related Natural food poisons (#PD1472).

♦ **PD4239 Falsification of public records**
Broader Corruptive crimes (#PD8679)
Crimes against the integrity and effectiveness of government operations (#PD1163).
Narrower Perjury (#PD2630)   False statements (#PF4583)
Tampering with official documents (#PF4699)
Undocumented violations of human rights (#PF4062).
Related Denial of evidence (#PD7385).

♦ **PD4248 Psychochemical agents**
Psychotomimetica
Nature Psychochemical agents are substances that produce symptoms, or subjective experiences, similar to psychoses.
Broader Chemical warfare (#PC0872).

♦ **PD4252 Discrimination against traditional economies**
Devalorization of traditional economies — Denial of right to traditional economies to indigenous populations — Marginalization of traditional economies — Destabilization of indigenous economies — Disruption of native trade patterns
Broader Denial of economic rights (#PD4150)
Discrimination against indigenous populations (#PC0352).

♦ **PD4259 Forced disappearance of persons**
Involuntary disappearances of persons — Enforced disappearances of persons — Unacknowledged detention — Secret imprisonment
Nature Victims are taken into custody and then "disappear": their friends and relatives cannot find out where they are held or what has happened to them. Sometimes the victims are later discovered in prison, or released; sometimes it is learned that they have been killed. Bodies are recovered from secret graveyards in a state of decomposition that makes it impossible to ascertain their identities. Corpses have been found at roadsides far from where abductions took place, so badly mutilated that identification is difficult or impossible. The practice of leaving bodies in public places is intended to terrorize potential opposition, as does the open reporting in the press of assassinations and the finding of bodies. The secrecy surrounding "disappearances" serves to hide the scale of extrajudicial executions. Another purpose of disappearances is to intimidate the local population.
Incidence In Columbia, human rights organizations have reported 672 cases of missing persons. In 1988, non-government human rights organizations in Peru received complaints concerning 423 cases of disappeared persons. On 14 February 1989 the Philippine Commission on Human Rights had created a task force to study the 413 outstanding cases of disappearances, most of which had been inherited from the Marcos era. After the 1976 coup in Argentina, abductions followed by "disappearances" virtually replaced formal arrest and imprisonment in political cases. General Roberto Viola, Commander-in-chief of the army from 1976 to 1979 and later President, admitted in March 1981 that there were between 7,000 and 10,000 dead and "disappeared". In some cases, victims were flown out to sea to the point where the Gulf Stream would ensure the disappearance of the bodies. Then the prisoners were thrown alive out of the planes. The United Nations Working Group on Enforced or Involuntary Disappearances of the UN reported nearly 18,000 cases of disappearances. A Latin American non-government organization reported 120,000 disappear-

DETAILED PROBLEMS

PD4348

ances have occurred there. The situation is all the more alarming because the true number of disappearances far exceed the cases reported.

While difference criteria and definitions exist between countries and groups reporting disappearances of persons, the following countries have outstanding cases of disappearances: **Af** Angola, Chad, Ethiopia, Guinea, Morocco, Mozambique, Seychelles, Uganda, Zaire, Zimbabwe. **Am** Argentina, Bolivia, Brazil, Chile, Colombia, Cuba, Dominican Rep, Ecuador, El Salvador, Guatemala, Haiti, Honduras, Mexico, Nicaragua, Paraguay, Peru, Uruguay. **As** Afghanistan, China, India, Indonesia, Iran Islamic Rep, Iraq, Lebanon, Nepal, Philippines, Sri Lanka, Syrian AR, Viet Nam. **Eu** Cyprus.

Refs Simpson, John and Bennett, Jana *The Disappeared and the Mothers of the Plaza* (1985).

Broader Missing persons (#PD1380) Internment without trial (#PD1576).
Narrower Forced disappearances of children (#PD5129)
Forced disappearances of trade union leaders (#PE5882).
Related Political prisoners (#PC0562) Extra-legal executions (#PE6366)
State sanctioned torture (#PD0181) Denial of right to security (#PD7212).
Aggravates Denial of right to life (#PD4234).
Aggravated by Political repression (#PC1919) Incommunicado detention (#PE8004)
Inadequate legal counsel for political dissidents (#PF0732)
Denial of human rights in the administration of justice (#PD6927).

♦ PD4265 Unethical practice of chemistry
Irresponsible chemists — Negligence by chemists — Malpractice in chemistry — Corruption of chemists — Underreporting of chemical pollution

Claim Chemists are used to research and manufacture chemical and biological weapons. Chemists, under pressure from their employers, have adopted practices which lead to the underreporting of hazards of chemical pollution, failure to recommend adjustments toxicity thresholds in the light of potential hazards, and failure to investigate adequately the nature of such hazards. Examples include hazards of long-term exposure to low-level chemical pollutants in factories, and environmental consequences of uses of certain chemicals. In each case chemists have denied or suppressed evidence indicative of the levels of danger.

Aggravates Chemical pollutants of the environment (#PD1670).

♦ PD4268 Abandonment of the dying
Nature Among many primitive tribes it is customary to abandon the dying to their fate.

Broader Abandonment (#PA7685).

♦ PD4270 Unstable supply of raw materials
Raw materials — Lack of raw materials

Nature The developed countries depend on a stable supply of raw materials for their industries. Over the last decade this has been threatened by inadequate investment in the production of raw materials. Unstable supplies result in inflationary pressures in the developed market economies, setting off a spiral from which no one escapes. In such a situation it is in the interests of developed countries to ensure stability of prices and supplies.

Incidence Total resource requirements are increasing rapidly over the entire world. In developed countries, although population is increasing slowly, per capita use is increasing rapidly, while the opposite is happening in developing countries. Traditionally, raw materials have been classified as non-renewable resources, but a distinction may be important between 'losable' resources, such as oil and coal, and 'non-losable' resources, such as metals, which can be used several times over by recycling processes.

Counter-claim Any statement regarding the availability of raw materials for future generations should take into consideration that exploitation of raw materials is dependent on the state of technical knowledge, which is not a constant. Thus, with the improvement of technological capability, man has been able to utilize lower concentrations and mine an increasing number of regions of the world.

Refs Bennett, James T and Williams, Walter E *Strategic Minerals* (1981).

Aggravates Neo-colonialism (#PC1876).
Aggravated by Non-destructible packaging and containers (#PD1754).

♦ PD4271 Trafficking in children for medical exploitation
Traffic in children as a source of organ transplants — Traffic in children for medical experiments

Nature Children are purchased or kidnapped from parents and sold to medical facilities for the purposes of medical experimentation and as sources of organ transplants. Frequently these are infants who have handicaps.

Broader Trafficking in children (#PD8405).
Aggravated by Unethical medical practice (#PD5770)
Abusive experimentation on humans (#PC6912)
Lack of human organs for transplantation (#PE7530).

♦ PD4272 Acculturation as a dilution of cultural heritage
Nature This process of cultural change involves the assimilation by one group or member of a group of the cultural patterns of another group, thus diluting their own cultural heritage.

Broader Ignorance of cultural heritage (#PF1985).
Aggravates Cultural illiteracy (#PD2041).

♦ PD4273 Cultivation of marginal agricultural land
Forced relocation of peasants onto marginal lands

Incidence Prime agricultural land in developing countries has been taken over for the production of cash-crops which tends to push peasants onto marginal lands which, more often than not, is totally unsuitable for farming. The results have frequently been disastrous. Peasant farmers have been forced to cultivate mountain sides so steep that it can not be cultivated except by hand and the resulting erosion has been devastating. In Africa nomadic farmers have been pushed into grazing the arid and inhospitable margins of the Sahara desert. Marginal lands are often especially vulnerable to disasters, such as drought, landslides or flooding (whether river or tidal).

Aggravates Infertile land (#PD8585) Flood plain settlement (#PE0743)
Settlement of unprotected coastlines (#PE3813) Deforestation of mountainous regions (#PD6282)
Restriction of indigenous populations to reservations (#PD3305)
Disruption of ecosystems in marginal agricultural lands (#PD6960).
Aggravated by Deprivation of peasantry (#PC8862)
Shortage of cultivable land (#PC0219)
Involuntary mass resettlement (#PC6203)
Inappropriate cash crop policy (#PF9187)
Expropriation of land from indigenous populations (#PC3304)
Limited availability of land for low-income and disadvantaged groups (#PF5008).

♦ PD4277 Unethical practice of marine sciences
Irresponsible oceanographers — Negligence by oceanographers — Malpractice in oceanography — Corruption of oceanographers — Underreporting of marine pollution

Claim Oceanographers, under pressure from their employers, have adopted practices which lead to the underreporting of marine pollution and of hazards associated with semi-stable ocean current systems (such as El Nino) in the light of the risk of global warming, and the failure to investigate adequately the nature of such hazards. Such practices have contributed to the deterioration of inland seas (such as the Caspian) and enclosed seas (such as the Mediterranean).

Aggravates Marine pollution (#PC1117).

♦ PD4278 Dangerous countries
Unsafe countries

Nature A number of countries are dangerous to travel to for a variety of reasons. Some pose immediate threats to life and liberty. Others are dangerous because of high crime rates in cities, risky legal systems, potential terrorist activity, few health facilities, and extremely different customs and practices that may endanger the unwary traveller.

Incidence The US State Department in 1989 listed 9 countries as extremely dangerous and 25 as potentially dangerous.

Broader Travel risks (#PD7716) Personal physical insecurity (#PD8657).

♦ PD4282 Denial of academic freedom
Denial of right to investigate

Nature Academic freedom is being increasingly challenged both by government and by students. Since universities have allowed themselves to become dependent upon public money for their survival, those who work within them can no longer resist pressures from governments which apportion the money or from students who represent those supplying the money.

Incidence During the two decades preceding 1988 there have risen an alarming tendency to undermine, restrict or suppress academic freedom and autonomy of institutions of higher education. This has a direct relation to a contracting system of higher education justified most often in terms of economic austerity and/or political expediency.

Broader Restrictions on freedom (#PC5075) Denial of freedom of thought (#PF3217).
Narrower Erosion of university autonomy (#PG6036).
Aggravated by Politicization of scholarship (#PF7220).

♦ PD4287 Forced social intimacy
Obligatory social kissing — Imposition of intimate forms of address — Unauthentic intimacy

Nature In some social or working contexts a degree of forced intimacy is imposed upon people. This may take the form of use of familiar forms of address, use of given names, or the use of familiar forms of speech (in languages which make such distinctions). It may also take the form of excessive proximity, physical contact or social kissing. When such intimacy is not readily accepted, it constitutes an invasion of personal space and privacy and for people of some cultures, or of particular social situations, may be considered deeply offensive. Situations in which such heightened intimacy is tolerated may be abused to a degree which amounts to sexual harassment.

Incidence Certain cultures, notably the Anglo-American, impose use of first names in many situations. In cultures where the language has a familar form of speech, this is usually required in both work situations and their wider institutional context, whether this be a government administration, a corporate environment, an armed force or a student body. Social kissing may be required as an appropriate form of greeting in certain work situations.

Aggravates Infringement of privacy (#PB0284).
Reduces Lack of intimate relationships (#PF4416).

♦ PD4316 Unequal distribution of agricultural production
Incidence The index numbers of agricultural production for 1978 indicate a per capita total agricultural production of 105 for the world, 100 for developing market economies, 111 for developed market economies, and 111 for the centrally planned economies.

Aggravated by Stagnated development of agricultural production (#PD1285).

♦ PD4322 Unequal distribution of meat production
Broader Underproductivity (#PF1107)
Unequal distribution of production between countries (#PF4336).

♦ PD4334 Unethical practices by employees
Irresponsible workers — Negligent employees — Corruption of workers

Nature Employees, whether of government or of the private sector, are able to exploit such a position to their personal advantage or to that of others they may favour. Such practices may range from borrowing equipment, pilferage of office materials, sloppy workmanship, to systematic diversion of materials or funds. Such a position may be used to facilitate acquisition of a job by a relative, friend or stranger, possibly against suitable compensation for the favour. Privileged access to information may be exploited by making it available to other interested parties, either commercial or governmental, possibly to a degree amounting to espionage, whether industrial or otherwise. Access to files may allow the employee to modify their content, whether by adding or removing information, either to further the interests of other bodies against appropriate compensation or purely to create mischief. Where the employee is in direct contact with outsiders requesting services (processing documents, supplying licenses, approving applications, etc), such services may only be provided (rapidly) following "under-the-table" payments.

Narrower Dishonest employees (#PD9397)
Unethical practices of regulatory inspectors (#PF8046)
Unethical practices by public service employees (#PE6702).
Related Unethical practices of employers (#PD2879).
Aggravates Nepotism (#PD7704) Espionage (#PC2140)
Absenteeism (#PD1634) Embezzlement (#PD2688)
Theft of property (#PD4691) Abuse of government employment (#PE4658)
Deterioration in product quality (#PD1435) Vulnerability of computer systems (#PE8542)
Unethical use of social welfare benefits (#PD8859)
Proliferation of nuclear weapons and technology (#PD0837)
Diversion of high technology to hostile countries (#PE7174)
Unauthorized access to computer information systems (#PE9831).
Aggravated by Organized crime (#PC2343) Unethical personnel practices (#PD0862)
Unethical trade union practices (#PD4341).

♦ PD4341 Unethical trade union practices
Irresponsible trade unions — Trade union fraud — Corruption in trade unions

Broader Corruption (#PA1986).
Narrower Collusion of trade union leaders with employers and government (#PE8367).
Aggravates Unethical practices by employees (#PD4334)
Interference of trade unions with contract performance (#PE8273).
Aggravated by Organized crime (#PC2343).

♦ PD4348 Desecration of monuments
Broader Desecration (#PF9176)
Endangered monuments and historic sites (#PD0253).
Aggravates Desecration of holy spaces (#PF6385).
Aggravated by Defacement of urban structures (#PD5305).

♦ **PD4361 Military-industrial malpractice**
Defence scandal — Military inefficiency
Nature For every dollar diverted by fraud in the military-industrial relationship, there are hundreds wasted by political patronage and micro-management, military frequent changes in design specifications, intransigence and inefficiency, contractor mismanagement, product substitution, kickbacks, falsified records, bid rigging and cost padding, and pure bureaucratic bungling by military commands. When weapons are purchased, not enough attention is paid to how they measure up to operational test results and actual operating experience. Testing procedure are generally designed to reflect favourably on the equipment being tested. Results are often written up so that faults are covered up. Within the military, promotions of programme directors are often based on successful procurement of weapons even if they are faulty.
Incidence There is substantial evidence that the Contras have helped finance their war by large-scale drug-trafficking to the US with the co-operation of the CIA. In an attempt to build an new generation of attack submarines the US Navy approved a flawed designed which could neither meet the speed requirement expected of it nor dive to required depths. The contractor bid $1.2 billion to build 18 of the submarines but, after a series of production and management foul-ups, ended up submitting bills to the military for double that amount. The company claimed that the cost overruns were the fault of 35,000 design changes made by the Navy. The Navy argued that the changes were no larger than in previous programs. When the company threaten to stop production an agreement was reached to save the company.
Claim Faulty weapons are but a symptom, not a disease. The whole military system is sick. It provides weapons that are designed for ideal, simple conditions when war is chaotic, stressful, confusing and full of surprises and ambiguities. The people who operate these systems are undertrained and what training they get is unrealistic. The people in the armed forces are given ineffective weapons, the taxpayers who pay for defence systems are cheated and the nations are left without a real defence.
Broader Fraud (#PD0486) Scandal (#PC8391)
Unethical professional practices (#PC8019).
Related Undue political pressure (#PB3209).
Aggravates Military incompetence (#PJ1069).
Aggravated by National defence procurement procedures (#PE4097)
Malpractice in the construction industry (#PD9713).

♦ **PD4365 Religious prejudice**
Religious bigotry — Racial bigotry — Prejudice against non-worshippers — Discrimination against religious unbelievers — Bigotry against atheists
Broader Bigotry (#PC7652) Prejudice (#PA2173).
Aggravates Consanguineous marriage (#PC2379)
Religious discrimination (#PC1455)
Underprivileged religious minorities (#PC2129).
Aggravated by Atheism (#PF2409) Unbelievers (#PF8068)
Religious extremism (#PF4954)
Decreasing participation in collective religious worship (#PF8905).

♦ **PD4368 Impermanent living conditions**
Dislocated life style — Makeshift settlements — Unstructured home life
Nature Those whom war, adverse climate and/or lack of food force to relocate continually have great difficulty making the permanent relationships and arrangements needed for economic and social life to prosper. Physical and psychological health suffer, families are pulled apart and a cycle of continuing rootlessness ensues.
Broader Poor living conditions (#PD9156) Obstacles to family life (#PF7094)
Fragmented conduct of community operations (#PF1205).
Related Uncomfortable living conditions (#PU0251).

♦ **PD4374 Chronic fatigue**
Broader Fatigue (#PA0657) Chronic illness (#PD8239).

♦ **PD4379 Economic refugees**
Claim Persons who are unwilling to return to their countries of origin because they prefer idleness to facing the hardships of helping in the economic development of their countries, or who intend to settle in other countries for purely economic reasons. Economic refugees may be refused to allow to settle in the foreign country and may be denied assistance by international organizations which assist political refugees who have left their countries because of war or persecution.
Counter-claim There is no such thing as an economic refuge; the economic misery the vast majority of the people face is due to political oppression. Authorities antagonistic to political refugees may deliberately attempt to define them as economic refugees to justify lack of sympathy for the persons concerned.
Broader Refugees (#PB0205).
Aggravated by Idleness (#PA7710).

♦ **PD4383 Media cover-up**
Incidence For example: whilst a major strike of Siberian miners was extensively reported in the North American press in 1989, a strike of US coal miners, of equivalent proportions, during the same period received little mention in the same media.
Broader Secrecy (#PA0005) Unethical media practices (#PD5251)
Distorted media presentations (#PD6081).

♦ **PD4387 Increasing atmospheric carbon dioxide**
Carbon dioxide impact on agriculture in developing countries
Nature Climatic modification induced by the enrichment of the atmosphere with carbon dioxide has direct effects on agriculture. Photosynthesis, the mechanism through which plants convert inorganic substances such as carbon dioxide sulphates, nitrate and water into organic compounds, is the basis not only of all present life and current biomass production by also of fossil fuels, the accumulated products of past photosynthesis. Therefore, any factor which would influence the rate of photosynthesis will have a strong effect of foods, fibres and fuels. The following are likely effects of any doubling of carbon dioxide concentration: average earth temperature would increase by 2 to 3 degrees C; precipitation would increase in some regions, though the change in rainfall patterns is uncertain (some scientists suggest that regions between middle and high latitudes would be affected, which is where most developing countries are located); and evaporative tendencies would increase in parts of the world, perhaps more than rainfall.
Background Carbon dioxide had a pre-industrial level of 275 parts per million (ppm) and the 1986 level was 346 parts per million. The current annual increase is 1.5 ppm.
Claim The annual increase will go higher, because the population will double and perhaps nearly triple again in the next century and the world economy will grow five to ten times larger. Carbon dioxide levels will double from their level prior to the Industrial Revolution by about 2040, latest by 2070.
Counter-claim Any attempt to cut carbon dioxide emissions from cars and power plants would place intolerable restrictions and demand unwanted sacrifices on society.
Refs Fantechi, R and Ghazi, A (Eds) *Carbon Dioxide and Other Greenhouse Gases* (1989).
Broader Air pollution (#PC0119)
Increase in trace gases in the atmosphere (#PD1354).
Aggravates Global warming (#PC0918).
Aggravated by Exploitation of fossil fuels (#PE4891)
Environmental hazards of coal energy (#PD7541).

♦ **PD4392 Loneliness in single people**
Ineffective means for searching for a partner — Lonely hearts
Nature Millions of single and divorced people have no acceptable way of finding marriage partners or even friends. Meeting someone at a bar or night club implies willingness to engage in casual sex. Dating agencies seem unnatural and are often expensive. Advertisements in news papers have the danger of meeting someone quite unacceptable. In modern society after leaving school there are no neutral places for people interested in meeting members of the opposite sex.
Broader Loneliness (#PF2386) Social alienation (#PC2130).
Related Living alone (#PF3089).
Aggravates Loneliness in old age (#PD0633).

♦ **PD4396 Nuclear waste**
Nature Radioactive residue from power stations is a result of their production process, but that production technology has got very much ahead of the waste disposal technology. The result is a growing difficulty in finding ways to deal with the waste. Some of this, like strontium 90, caesium 137, carbon 14 and trans-uranic elements like plutonium, have very long half-lives. They will continue to be capable of radiating the environment for millions of years. No repository currently exists which can securely store such elements, although there are many potentially useful schemes in progress. Further, there is no known way at all to prevent the escape of radioactive gases, tritium or uranium byproducts.
Incidence The US Department of Energy estimates that $1.8 billion is needed every year for 20 years to assure safe storage of nuclear waste from production of weapons grade plutonium, another $14 billion to comply with modern pollution laws and a further $78 billion to clean up abandoned nuclear facilities in the US alone. In the UK waste radioactive nuclear fuel totalled 6,219 tons in 1970: by 1985 it had grown to over 59,000 tons. Accumulated radioactivity dumped into the Atlantic by Britain and other countries totals over 1 million curies and has raised radiation levels near Windscale four thousand time. By the year 2000, there will be 100,000 lorry loads of radioactive waste in store in the UK alone.
Refs Burkholder, H C *High Level Nuclear Waste Disposal* (1986); Byers, R B (Ed) *De-Nuclearization of the Oceans* (1986); Murdock, Steve H, et al *Nuclear Waste* (1983); United Nations *Management of Gaseous Wastes from Nuclear Facilities* (1981).
Broader Radioactive wastes (#PC1242) Unproductive use of resources (#PB8376).
Narrower Waste of nuclear warheads (#PE4199).
Aggravated by Hazardous waste dumping (#PD1398)
Maldistribution of electrical energy (#PD3446)
Environmental hazards of nuclear weapons industry (#PE5698).

♦ **PD4401 Defrauding of secured creditors**
Nature Defrauding creditors is a wide range of criminal activity. The intentional refusal to pay for loans, mortgages or other forms of credit. The difficulty with determining whether a crime is being committed is whether or not the accused is attempting to extend the period of payments or is actually attempting to defraud the creditor.
Broader Economic crime (#PC5624) Theft of property (#PD4691).
Related Frauds, forgeries and financial crime (#PE5516).

♦ **PD4402 Male homosexual prostitution**
Nature Male homosexual prostitution is one of the possible sequels to child prostitution. Boy prostitutes pick up their partners in the street or in public parks in particular districts, in station lavatories, or in hotels or bars. "Call boys", who earn more, offer their services by placing personal advertisements in the same publications which, covertly or overtly, advertise the wants of clients. Magazines or radio stations controlled by homosexual groups inevitably create an additional potential clientele for this type of prostitution. However, unless they have rich customers or are kept by homosexuals on a long-term basis, male prostitutes' earnings are fairly small since the gay world abounds in possibilities for free homosexual relations. One of several alternatives is to turn to transvestite prostitution.
Broader Male prostitution (#PD3381).
Narrower Transvestite prostitution (#PD4525).
Aggravated by Juvenile prostitution (#PD6213)
Degradation of developing countries by tourism (#PF4115).

♦ **PD4403 Overexploitation of underground water resources**
Misuse of artesian water supplies — Misuse of nonrenewable fossil water reserves — Overexploitation of aquifers — Lowering of water table — Lowering of water level — Shortage of groundwater resources
Nature Huge underground reservoirs are being drained at alarming rates in many parts of the world. Arid climates aggravated by lower than average rain fall, complex groundwater systems that receive limited recharges, population growth, uncontrolled drilling, increasing per head use of water, mismanagement of water supplies and little knowledge about underground water resources exacerbate the problem. Properly irrigating the main arid and semi-arid regions of the world would require the use of the total continental run-off (a physical impossibility) and water projects so large as to risk substantial changes in regional global climate. Present rapid depletion of ground-water resources throughout the world will soon lead to widespread local shortages. It is now also believed that irrigation will in most areas contribute too little and too late.
Incidence Only about 2.5 million square kilometres (2 per cent of all ice-free land) is now irrigated, at great cost and with many unwelcome side-effects. As an example, Libya is constructing a 500 km pipeline to take water, pumped out of hundreds of wells dug beneath the Sahara. There is concern that this $25 billion project will drain wells supplying Egypt and the Sudan. The prairie region of the USA, responsible for 23 per cent of agricultural production, is irrigated by an aquifer which is drying up. By 2010 it may be too low to pump.
Broader Shortage of fresh-water sources (#PC4815)
Disruption of the hydrological cycle (#PD9670)
Unimaginative vision of resource utilization (#PF1316).
Aggravates Hunger (#PB0262) Desiccation of rivers (#PJ4867)
Long-term shortage of water (#PC1173) Human disease and disability (#PB1044)
Declining productivity of agricultural land (#PD7480).
Aggravated by Deforestation (#PC1366)
Inadequate water system infrastructure (#PD8517)
Environmental impacts of coal conversion plants (#PE8453).
Reduces Rising water level (#PD8888).

♦ **PD4418 Self-defeating behaviour**
Self-defeating personality disorder
Nature Individuals who sabotage their own success are in severe emotional trouble. Self-esteem, image and social harmony are all enhanced through self-defeat. The intricate gamesmanship

involves accepting blame or a loss of one sort in order to avoid the risk of a setback that seems even more threatening. For instance, someone who says he missed an important interview because he lost track of time, may be more able to accept the appearance of temporary incompetence than the risk of failing in the interview. A person may give himself a handicap in order to maintain the illusion of success without having to risk losing it. Self-defeating people rely so often on excuses and self-imposed handicaps that they become entrapped by them. Children who are excessively praised before they do something may grow up with an inflated image of themselves that they feel they must protect against realistic tests. They take on a handicap to protect the image. When a handicap becomes a permanent reason for failure it changes from a useful handicap to a pathological one. Another form of self-defeating behaviour is pathological excuse making. Using the same excuse over and over or inventing excuses too often are clues to this behaviour. When the excuse-maker sees himself as tragically flawed because of the condition that provides the excuse, the excuse becomes a self-fulfilling prophecy. Self-defeating excuses tend to be too involved and grand for the transgression they are meant to smooth over. Some of the most severe self-defeating behaviour is a result of the person's deep feeling that he is a victim and if he stops being a victim he loses his identity.
Broader Emotional disorders (#PD9159) Personality disorders (#PD9219).
Narrower Inhibited self-promotion (#PJ1544).
Aggravated by Low self esteem (#PF5354) Lack of self-confidence (#PF0879).

♦ **PD4424 Regional underdevelopment**
Regional deprivation
Nature The economic, social and cultural underdevelopment of a region of a nation or extra-national political unit. This might result from geographic isolation, sparse population, cultural political or social differences. The underdevelopment of a region results in further declines in jobs and frequently declines in population due to migration to more prosperous regions.
Broader Regional disparities (#PC2049).
Aggravated by Urban bias (#PF9686).

♦ **PD4425 Political violence**
Nature The use of violence to achieve political ends, the overthrow of an oppressive regime, the abolition of unjust laws, the modification of an discriminatory society, and the elimination of institutionalized violence is likely to end in an equally oppressive, unjust, discriminatory, and violent society. The means, in these situations, lead, as often as not, to civil war. The chief victims of violence are not those who advocate it but, thousands if not millions of innocent civilians caught in the cross fire between the establishment and those in revolt. Those leaders who are attempting to bring about change without resorting to violence are undermined. Any lasting attempt at violent confrontation with a government requires massive support from outside forcing the rebellion to choose between major political powers. **Counter-claim** To suggest that political, religious or other forms of extremism is not justified by history, contemporary reality, logic or law, as did a US commission on violence in 1969, is plain silly. Many people and governments may believe in liberty, equality and justice but their behaviour is something else. And many people and governments simply do not hold these beliefs. Have not all peaceful demonstrations, sit-ins, freedom rides and marches, defiance campaigns, petitions, letters to editors and legislative representatives been singularly unsuccessful for some people. Have not decades spent, knocking, patiently, moderately and modestly at a closed and barred door been spent by some minorities in vain. The resort to violence to achieve political ends is a tactic that embodies a cry for help; that seeks reforms; that wants attention to grievances and demands; and that demands a response to a deep and abiding sense of iniquity and inequity. There is a point where violence is a necessary and creative response to institutionalized violence; it is saying no, no longer, no more, not again loud enough and clearly enough to be heard. If people are barred from using the sophisticated instruments of the established order for their ends, they will find another way. To the people involved in the riot, the civil war, these are far less lawless and far more representative than the system of arbitrary rules and prescribed channels which they confront every day. These are far less violent than the ongoing brutality of living in a slum, shanty town or backward village maintained by a social and political system geared to maintain the status quo.
Refs Fanon, Frantz *The Wretched of the Earth* (1966); Han, Henry H *Terrorism, Political Violence and World Order* (1984); Lentz, Harris M *Assassinations and Executions* (1988); Merkl, Peter H *Political Violence and Terror* (1986); Zimmerman, Ekkart *Political Violence, Crisis, and Revolution* (1983).
Broader Human violence (#PA0429).
Related Terrorism (#PD5574).
Aggravates Political prisoners in mental institutes (#PE4430).
Aggravated by Violence as a resource (#PF3994).

♦ **PD4426 Disastrous technological failures**
Broader Disastrous accidents (#PC6034).
Narrower Nuclear accidents (#PD0771) Sanitation system accidents (#PE4570)
Large-scale industrial accidents (#PD2570).

♦ **PD4445 Super-power monopoly of advanced nuclear warfare technology**
Claim Although the number of nations possessing nuclear weapons is growing, not all countries have the sophisticated delivery systems to deploy these weapons defensively or offensively. ICBMs and long-range bombers, themselves protected by on-board tactical nuclear missiles and massive squadrons of fighter planes, and by radar-jamming and other electronic counter-measures, are only in the aggressive repertoire of the superpowers. The recent expectation that nuclear armed satellites will soon appear indicates the intent of world domination by each of these two giants.
Counter-claim Approximately 130 States have signed the Non-Proliferation Treaty, acknowledging the the possession of nuclear weapons is not required for their own security. The security of many third world countries is grave and the task of enhancing regional security in many areas of the world is of evident importance but it is difficult to see how the spread of nuclear weapons would advance such security.
A number of states not a party of the Non-Proliferation Treaty justify their rejection the Treaty by the argument that it is discriminatory in nature. Given the existence of nuclear weapons, the only non-discriminatory solution would be that of allowing any nation that desires nuclear weapons to have them - a highly destabilizing solution.
It is of significance that those countries taking a particularly negative stand on deterrence and the possession of nuclear weapons wish to maintain the nuclear option for themselves.
Broader Proliferation of strategic nuclear arms (#PD0014).
Related Fortified frontiers (#PD5972) Discriminatory nuclear trade (#PD8124)
Monopoly of nuclear power techniques (#PD1741).

♦ **PD4448 Impairing military effectiveness**
Broader Crimes against national security (#PC0554).
Narrower Mutiny (#PD2589) Sabotage (#PD0405)
Military disobedience (#PD7225).
Aggravates Military incompetence (#PJ1069).

♦ **PD4458 Denial of right to retirement**
Broader Exploitation in employment (#PC3297) Exploitation of the elderly (#PD9343).
Reduced by Employment discrimination against the elderly (#PD4916).

♦ **PD4461 Discontinuity of employment**
Refs ILO *Continuity of Employment of Seafarers* (1976).
Aggravates Inadequate sense of time (#PF9980)
Insecurity of employment (#PD8211)
Maladjustment to disciplines of employment (#PD7650).
Aggravated by Seasonal unemployment (#PC1108).

♦ **PD4469 Criminal coercion**
Criminal blackmail — Dependence on extortion — Extortion of protection money — Criminal extortion
Nature Criminal coercion is intending to compel another person to action or inaction by threatening to commit a crime, accusing someone of a crime, exposing a secret whether true or false in order to subject another to hatred, contempt, or ridicule or in order to impair their credit or business. Public servant who take or withhold official action for the same purposes are also guilty of blackmail. Modern criminal codes have tended to expand the crime of extortion to include what is generally regarded as blackmail, however the crime is named.
Background Historically, blackmail is tribute in corn, cattle, other kind or money levied from the farmers and small owners in the border counties of England and Scotland, and along the Highland border, by freebooting chiefs in return for immunity from pillage. The term is now applied to the extortion of money, or other valuable consideration, by intimidation, by the unscrupulous use of official or social position, or of political influence or vote, by persons upon those whom they have it in their power to help or injure. Originally, the Camorra of the mezzogiorno, the Ndranghata of Calabria, the Mafia of Sicily and similar organizations in the United States and China were blackmail bodies.
Refs Council of Europe *Extortions under Terrorist Threats* (1986).
Broader Coercion (#PC3796) Cheapness (#PA7193) Appropriation (#PA5688).
Narrower Ransom (#PJ1977) Blackmail by government officials (#PD9842).
Aggravates Defamation of character (#PD2569) Criminal harm to property (#PD5511).
Aggravated by Intimidation (#PB1992) Criminal threat (#PE4661)
Illegal immigration (#PD1928).

♦ **PD4475 Physical unfitness**
Reduced physical exertion — Inadequate exercise
Broader Deterioration in physical health (#PC0716).
Narrower Unhealthy lack of daily physical activity in urban environments (#PF6182).
Aggravates Acne (#PE3662) Diabetes (#PE0102) Tonsillitis (#PE2292)
Haemorrhoids (#PE3504) Constipation (#PE3505)
Inadequate maintenance of physical health (#PF1773).
Aggravated by Sedentary habits (#PJ6551).

♦ **PD4478 Participation in torture**
Nature Victims of torture are frequently forced to actively or passively participate in the torture of other victims. Men are forced to have sexual relations with women prisoners. Prisoners hold others while they are tortured. They may be told if they confess another prisoner will not be tortured implying guilt for any harm done.
Broader Induction of incongruent actions (#PF3790)
Forced participation in social processes (#PC5387).
Narrower Inhumane participation of the medical profession in torture (#PE4015).
Related Military and police personnel participation in torture (#PE4119).

♦ **PD4488 Felonious restraint**
Nature Felonious restraint is knowingly abducting another, restraining another under terrorizing circumstances or restaining another with the intent of holding him as a servant. This is generally regarded as a lesser crime than kidnapping and a greater crime than unlawful imprisonment.
Broader Kidnapping (#PD8744).
Related Unlawful imprisonment (#PD4489).

♦ **PD4489 Unlawful imprisonment**
False imprisonment
Nature Unlawful imprisonment is knowingly imprisoning or restraining a person where no further harm is done or threatened. Confinement may, of course, be accomplished by actual physical restrain. How ever, the application of force is not essential; the confinement may also be accomplished by threats. Confinement is not simply preventing a person from going in one direction if they are free to go in another.
Refs Bowker *Prison Victimization* (1980).
Broader Kidnapping (#PD8744) Imprisonment (#PD5142)
Internment without trial (#PD1576).
Narrower Denial of right to freedom from imprisonment for failure to fulfil a contractual agreement (#PE7235).
Related Felonious restraint (#PD4488).

♦ **PD4500 Securities and commodities exchange violations**
Stock-exchange and bourse related crimes — Commodity exchanges related crimes
Nature Security and commodity exchange violations vary greatly from country to country but generally are considered forms of fraud. Signatures may be forged to registration statements of stocks, unregistered stocks may be sold. Stocks may be publicized without disclosing receipt of payment for the advertisement. Stocks may be traded on the basis of illegally obtained or privileged information or the information may be sold. Ownership of stocks may be hidden through the use of middle men or dummy corporations. Stock market crimes are committed by highly sophisticated professionals who know the laws and to a large extent are self-regulating. They are tempted by hugh gains, relative easy access, difficult detection and lenient sentences.
Broader Economic crime (#PC5624)
Violations against economic regulations (#PD7438).
Narrower Insider dealing (#PD3841).
Aggravated by Vulnerability of stock markets (#PD5676)
Lack of international coordination among supervisors of financial stock markets (#PE4508).

♦ **PD4501 Covert intelligence agency operations**
Clandestine intelligence operations
Refs Blackstock, Paul W and Schaf, Frank *Intelligence, Espionage, Counterespionage and Covert Operations* (1978); Johnson, Loch K *America's Secret Power*, Treverton, Gregory F *Covert Action* (1989).
Broader Espionage (#PC2140) Unaccountable government intelligence agencies (#PF9184)
Foreign intervention in internal affairs of states (#PC3185).
Narrower Covert smear campaigns by government (#PD7171).
Related Secret military operations (#PF7669).
Aggravates Front organizations (#PE4358) Government sanctioned killing (#PD7221)
Government complicity in illegal activities (#PF7730)

Government seizure of foreign nationals in foreign countries (#PE6564).
Aggravated by Government action against regimes with alternative policies (#PF2199).

♦ PD4502 Destruction of archaeological sites
Salvage archaeology
Nature Archaeological sites are an irreplaceable, nonrenewable resource. Once destroyed, they are lost forever. Many cultural activities modify the earth's surface, destroying sites. City, road, and reservoir construction and deep plowing in the past century have drastically accelerated destruction. Air pollution decays materials which have survived millennia and commercial art markets inspire pothunters and grave robbers to loot countless sites. Since prehistoric people tended to live in the same favoured locations as modern settlements, the need to preserve archaeological sites too often is lost to progress.
Broader Endangered monuments and historic sites (#PD0253)
Inadequate protection and preservation of cultural property (#PF7542).
Aggravated by Environmentally harmful dam construction (#PD9515).

♦ PD4506 Ignorance of workers
Broader Ignorance (#PA5568).
Aggravates Incompetent workers (#PD4535).

♦ PD4511 Disruption of financial markets
Stock market crash — International collapse of stock exchanges and bourses
Nature Stock market declines and even crashes are the result of more investors offering stocks for sale than offering to buy stocks. Investor sell stocks because they believe that their price will decline or their returns will decline, i.e. dividends will be lower because of decreasing profits, increasing interest rates or depreciating corporate assets.
Background Since 1948, there have been 10 major market declines in the US stock markets. The largest drop before October 1987, from December 1972 to September 1974, was by 46 percent. Declines have averaged 23 percent and have been gradual; it has taken stock prices an average of 14 months to hit bottom from their peaks. About one half of these declines were followed by recessions in the economy; generally economist believe that stock market declines are not the cause of recessions. After reaching a high of 2722 in August 1987 the market crashed by over a third from that peak.
Claim A financial crash on the stock markets temporarily dislocates economic activity, disastrously effects those whose incomes and wealth are based on financial assets, reverse government policies aimed at selling assets to finance current spending, frighten away small or nervous investors in markets, and dramatically demonstrate the risk in market investments including private pension schemes.
In the months and years before the great crashes or 1929 and 1987 a large number of people and institutions were in the market only because it was going up, and they aimed to get out before it went down. The overwhelming factor was greed, which turned suddenly into fear.
Broader Economic inefficiency (#PF7556).
Aggravates Economic loss (#PE9013) Global financial crisis (#PF3612)
Cyclic business recessions (#PF1277).
Aggravated by Financial destabilization of world trade (#PC7873)
Lack of international coordination among supervisors of financial stock markets (#PE4508).

♦ PD4514 Drug abuse at work
Nature Drug abuse is not only a recreational activity, but also pervades factories and offices. Like worker alcoholism, drug addiction is a serious threat to the safety and productivity of industry.
Incidence Drug users spend 200 million pounds or more each year in London alone. According to the Confederation of British Industry, a good portion of this amount goes for drugs used in the workplace.
Refs Newcomb, Michael D *Drug Use in the Workplace* (1988).
Broader Drug abuse (#PD0094) Substance abuse at work (#PD9805).
Related Alcoholic intoxication at work (#PE2033).

♦ PD4517 Biased allegations against governments
Partisan reporting of governmental action
Broader Misleading information (#PF3096).
Narrower Misinformation concerning infringement of human rights (#PF9794).

♦ PD4518 Excessive consumption of animal flesh
Excessive consumption of meat
Claim Ethical considerations aside, meat provides an excessive amount of fat to the body and is therefore dangerous to the health.
Counter–claim It is not meat that is dangerous but meat fat. If the cost of lean meat is excessive, a meat diet can be complemented with cereals, vegetables and pulses.
Broader Excessive consumption of specific foodstuffs (#PC3908).
Related Human consumption of animals (#PC7644).
Aggravates Killing of animals (#PD8486)
Denial to animals of the right to a natural death (#PE8339)
Denial to food animals of the right to freedom from suffering (#PE3899).
Aggravated by Use of agricultural resources for production of animal feed (#PD1283).

♦ PD4519 Politicization of health standards
Commercial distortion of toxicity thresholds — Inappropriate legal definition of health risks
Nature Increasingly safety limits, especially for new products where detailed knowledge of their long–term effects is unavailable, are set as the result of a compromise between a reasonable estimate of the probable health risks and the commercial or political interests of those concerned with their manufacture. The compromise may be deliberately based on inadequate testing, or tests known to be obsolete or insensitive. Excellent tests may however be used which only test for short–term effects. Such results are then used as the basis for the legal definition of acceptable levels of toxicity. The legal definition is then used to provide general assurance in the absence of ability to communicate hard facts.
Incidence When a comparison is made between the toxicity levels considered acceptable in different countries, it is clear that some countries base their legal definitions on old methods of testing particular chemicals. The methods for testing may in some countries be treated as classified information.
Broader Inadequacy of international standards (#PF5072).
Related Conflicting standards for protection against chemical occupational hazards (#PE5651).
Aggravates Withholding of information (#PF8536)
Human disease and disability (#PB1044)
Uncertain toxicity thresholds (#PF5188)
Negative effects of rejection (#PF4351)
Placement of the burden of proof on the disempowered (#PF3918).
Aggravated by Inconclusiveness of scientific and medical tests (#PD7415).

♦ PD4522 Neglected children
Nature Children who are not adequately cared for by their families, relatives, friends nor government or social agencies. They may be physically neglected: left alone, ignored, not properly fed or clothed, never washed, and force to live in filthy conditions. They may be psychologically neglected: looking well cared for but never receiving any love or attention from their parents or guardians. As they grow older, these children roam around during the day, sleep in make shift accommodations like doorways. They may steal, sell illegal goods, resort to prostitution for income, or do temporary labour for income. They may live at home or be orphans, run–aways, abandoned or thrown out of their homes.
Broader Cruelty to children (#PC0838) Victimization of children (#PC5512).
Narrower Orphan children (#PD7046) Neglected young children (#PE4245).
Aggravates Corruption of minors (#PD9481) State custody of deprived children (#PD0550).
Aggravated by Paternal negligence (#PD7297) Inadequate child welfare (#PC0233).

♦ PD4524 Domestic polluters
Broader Pollution (#PB6336).

♦ PD4525 Transvestite prostitution
Nature Young male prostitutes adopt female hair–styles, make–up and dress to offer themselves to men who want sexual relations with a man dressed as a woman.
Incidence Transvestite prostitution is spreading with great rapidity to the point where it is offering serious competition to female prostitution. In one medium–sized European capital, transvestite prostitution, which is practised mainly in outlying woods and forest areas, is engaged in by hundreds of young people, some whom are nationals of the country in question, but most of whom come from abroad with tourist visas. There is reason to believe that they are controlled by a powerful organization which arranges for their passports and visas, as well as for their hormone treatment and possibly surgical operations.
Broader Prostitution (#PD0693) Male homosexual prostitution (#PD4402).
Aggravated by Transvestism (#PE6348) Transsexualism (#PF3277)
Organized crime (#PC2343).

♦ PD4532 Concealed government subsidies
Bribery by government
Incidence As an example, the UK government deliberately concealed from the European Commission, Parliament and the general public, details of a 38 million pound "sweetener" given secretly to British Aerospace to encourage them to purchase the Rover Group as part of the government's privatization programme. This effectively allowed Rover to be purchased at 100 million pound less than the market value, which was arranged by discouraging other potential purchasers.
Broader Bribery (#PC2558) Deception by government (#PD1893)
Unreported government spending (#PF2990).
Related Business bribery (#PD8449).

♦ PD4535 Incompetent workers
Broader Incompetent management (#PC4867).
Aggravated by Ignorance of workers (#PD4506) Shortage of skilled labour (#PD0044).

♦ PD4537 Unsustainable rural development
Nature Unsustainable rural development is the failure to sustain the development of the renewable and nonrenewable resources of natural, modified and cultivated environments, namely excluding built environments. Due to their renewability, renewable resources offer the greatest opportunity for sustainable development. This opportunity is missed or destroyed by inadequate conservation or the separation of of development and conservation efforts.
Broader Unsustainable development (#PB9419) Destruction inherent in development (#PF4829).
Aggravates Rural poverty (#PC4992) Decay of rural communities (#PD9504)
Ecologically unsustainable development (#PC0111)
Destruction of rural subsistence economy (#PC2237)
Degradation of agricultural land by cash crops (#PE8324)
Disruption of ecosystems in marginal agricultural lands (#PD6960).

♦ PD4538 Morbidity
Morbidness
Broader Fear (#PA6030) Disease (#PA6799).
Aggravated by Smoke as a pollutant (#PD2267) Pesticide intoxication (#PE2349).

♦ PD4539 Violation of regulatory codes
Nature The violation of regulatory codes including any statute, regulation, rule or order of a municipal, provincial, state, region, nation or other legally constituted governing body is a punishable offence. The punishments may include criminal sanctions, forfeitures or civil penalties.
Broader Offences of general applicability (#PD4158).
Narrower Flouting regulatory authority (#PD7792)
Wilful violation of regulatory codes (#PE3809)
Violations of health and safety regulations (#PE4006)
Violations of regulatory codes without culpability (#PE0671).

♦ PD4541 Bribery of public servants
Bribery of government officials — Unlawful rewarding of public servants — Corruption of public servants — Graft
Nature Bribery includes the offering, demanding, giving and receiving of any payments or gifts to a public official or other public person in connection with his official functions, with the intention of improperly influencing a governmental decision. Unlawful rewarding of a public servant is making or receiving payment for an official action or violating a legal duty in the past. This is similar to a bribe but follows the act rather than preceding it. Payment for past favours implies the possibility of reward for future favours and as such corrupts officials. Bribery is sometimes disguised as (illegal) political contributions, payment of royalties, favours, or irregular loans. Investigation of bribery charges are exceptionally difficult, as illicit payments can be committed under the guises of various legitimate transactions; for example giving an extra percentage of profit or interest, or other soft terms. It is a crime against legal justice, according to which an official is bound to promote the common good of the community. It is also against distributive justice, by which rulers are bound to act toward persons and groups in accord with their merits, needs and capacities. Elected public office holders are the usual offenders, though the injustice may be committed by those who are appointed to public office rather than elected.
Refs Lethbridge, H J *Hard Graft in Hong Kong* (1985); Mansukhani, H L *Corruption and Public Servants* (1979).
Broader Giving bribes (#PC4631) Receiving bribes (#PC4701)
Corruptive crimes (#PD8679)
Crimes against the integrity and effectiveness of government operations (#PD1163).
Narrower Trading in public office (#PE6948)
Unlawful compensation for assistance in government matters (#PE7176).
Related Vice (#PA5644) Frauds, forgeries and financial crime (#PE5516).
Aggravates Police corruption (#PD2918) Corruption of the judiciary (#PD4194)
Abuse of bureaucratic procedures (#PF2661).

Aggravated by Resignation towards bribery (#PF8611)
Proliferation of public sector institutions (#PF4739)
Unethical practices of regulatory inspectors (#PF8046).

♦ PD4542 Discharge of dangerous substances
Illicit discharges of hazardous substances
Broader Hazardous waste dumping (#PD1398).
Narrower River pollution (#PD7636) Sewage as a pollutant (#PD1414)
Marine dumping of wastes (#PD3666) Domestic waste water pollutants (#PD2800)
Laboratory waste water pollutants (#PE9813) Industrial waste water pollutants (#PD0575).
Aggravated by Unethical industrial practices (#PD2916).

♦ PD4543 Grant frauds
Fraudulent procurement or use of government or private grants — Subsidies fraud
Incidence One two year operation worth £11 million in illegal subsidies involved a trader who imported "prime beef" from South America described as "offal" to escape EEC farm import levies. The offal was then exported as prime beef which attracted export subsidies. In another case, chicken scraps were exported as beef and the exporter applied for beef export subsidies.
Broader Economic crime (#PC5624).
Related Financial frauds (#PE2414).

♦ PD4552 Proliferation of printed matter
Paper proliferation due to computerization
Nature Information technology was supposed to let us taper off paper. But we have not. The idea of the paperless office, the bookless library, the printless newspaper, the cashless, checkless society have been proven illusionary. Computers have created the need for more paper, lots more rather than less. Computers are capturing much more information than was ever saved before and storing it incredibly compactly. Because reading things on a computer screen is relative inefficient, about 20 to 30 percent slower than print, people want information on paper. Paper will last for decades and a simple power surge can erase a computer's memory. The number of people whose work generates documents have increased. More business, government and professional people are requiring access to information which means the physical distribution of paper. The distinction between originals and copies has been blurred because of photocopying and lazer printers.
Incidence From 1959 to 1986 U.S. consumption of writing and printing paper increased from 6.38 million tons to 21.99 million. or 320 percent, while the real gross national production rose 280 percent. It is estimated between 1981 and 1984 American business use of paper grew from 850 billion pages to 1.4 trillion. From 1936 to 1986, the volume of U.S. mail increase from 80 billion pieces a year to 146 billion pieces and the Postal Service estimates volume of 170 billion pieces by 1990. In 1985 U.S. banks processed 40 billion to 45 billion checks more than 66 times the number of electronic transfers.
Broader Proliferation of information (#PC1298).
Narrower Proliferation of documents from international organizations (#PF5992).
Related Excessive paperwork (#PF5856).

♦ PD4556 Avoidance of legal obligations by politicians
Circumvention of the law by politicians
Nature Elected office holders, particularly those leading nations, are corrupted by the believe that they are above the law and have the right or obligation to interpret it in their own way. While those who are self seeking far out number those who have the nation's interests as their central concern, both step beyond the traditions, spirit and letter of the law. Presidents and prime ministers attempts to escape accountability for their deeds threaten the very institutions of rule by law. Reducing the interpretation of law to their own perceptions without standing before the judgements of others is the way of the despot.
Broader Political crime (#PC0350) Evasion of the law (#PD4208).
Related Egoism (#PA6318) Self-interest (#PA8760).

♦ PD4559 Psychological torture
Incidence Psychological torture has been reported in the following countries: **Af** Madagascar. **Am** Colombia, Honduras, Peru. **As** China. **Eu** Bulgaria, USSR, Yugoslavia.
Broader Human torture (#PC3429) State sanctioned torture (#PD0181)
Institutionalized torture (#PD6145).
Narrower Verbal abuse (#PD5238) Death threats (#PD0337)
Torture by deprivation (#PD3763) Induction of incongruent actions (#PF3790)
Torture through sensory overload (#PE5259) Threats against family or friends (#PE3308)
Torture through behavioural regulation (#PE6914)
Torture through destroying cherished things (#PE5279).
Related Physical torture (#PD8734) Torture of children (#PD2851)
Pharmacological torture (#PE4696).

♦ PD4560 Contraceptive availability in developed countries
Nature The provision of birth control information and supplies to young teenagers takes place in most developing countries without the permission of the children's parents. Coupled with sex education offered from childhood onwards, young people are positively encouraged to view the major ethical issue in sexuality as the extent to which contraception is effective.
Aggravates Birth prevention (#PE3286).

♦ PD4563 Cultivation of illegal drugs
Government sanctioned cultivation of illegal drugs
Claim Government subsidies for sugar, vegetables, fruits, grains, wine and tobacco to farmers in industrial countries effectively leave many farmers in developing countries with the need to grow poppies, coca leaf and marijuana to survive. The markets of legal crops are unstable, blocked with protectionist measures and often offering prices below production costs. Illegal crops have guaranteed markets, high profits, and little risk to farmers.
Aggravates Chemicalized farming (#PD7993) Tropical deforestation (#PD6204)
Hazardous waste dumping (#PD1398)
Increasing drug addiction in drug producing countries (#PJ0680)
Economic dependence of some developing countries on the drug trade (#PE5296).
Aggravated by Protectionism in developed countries against agricultural products from developing countries (#PE8321).

♦ PD4571 Trauma
Traumatic stress — Traumatic neurosis
Nature The disruption or breakdown occurs when when sudden stimuli, reminding of the traumatic event, are too powerful to be dealt with or assimilated in the usual way. The recurrent dreams and flashbacks result in anxiety and helplessness, ranging from apathy and withdrawal to panic. Traumas frequently follow criminal assault, accidents and catastrophic events, especially those outside the usual range of human experience.
Refs Altura, Burton M et al *Handbook of Shock and Trauma* (1983); Figley, Charles R *Trauma and Its Wake*.
Broader Stress in human beings (#PC1648).
Narrower Combat trauma (#PE7912) Puberty trauma (#PJ2206)
Childhood trauma (#PD5597) Post abortion syndrome (#PE5846)
Post-traumatic stress disorder (#PE0351)
Residual traumatic pains due to torture (#PE3798).
Aggravates Diseases of metabolism (#PC2270).

♦ PD4572 Socially ineffective family units
Nature The family unit at one time cared for both the larger society and for its members. Children, elders, and incapacitated members were the responsibility of the extended family. The nuclear family and single parent families while perhaps economically viable are not capable of caring for the larger needs of the larger family.
Broader Reinforced parochialism of internal values and images (#PF1728).

♦ PD4575 Slaughter of animals for pelts
Slaughter of animals for fur industry — Breeding of animals for fur
Incidence In 1989 it was estimated that, world-wide, 70 million animals are killed annually for their fur. Some animals are slaughtered whilst still young as in the case of seal pups and Karakul lambs. The latter are slaughtered within 48 hours of birth to make caps and lapels.
Claim Many fur-bearing animals, raised in captivity, are accustomed to movement and extensive territories. Forced confinement is thus a severe source of stress. To protect the fur, such animals are slaughtered using painful methods (poisoning, electrocution, gassing). Such animals rarely receive the protection legally afforded to livestock killed for meat.
Broader Hunting of animals (#PC2024) Killing of animals (#PD8486).
Aggravated by Trapping of animals (#PE5735) Human use of animal by-products (#PF1964)
Trade in animal products of endangered species (#PD0389).

♦ PD4590 Torture through confinement
Nature Other than beating, various types of confinement are the most frequently reported forms of torture. Victims are held in solitary confinement, sometimes without light, adequate air, sanitation facilities, or contact with other people. They may be held in cells so small they are unable to lie, sit or stand. Not infrequently, prisoners are crowded into cells or transportation vehicles so that they are crushed or asphyxiated. In rural areas, holes are dug and covered with bamboo or other materials where prisoners are kept for days or weeks. They may be bound with rope, chains, wire, handcuffs or fetters, sometimes so tight that circulation is impaired. Hooding involves placing a covering over the head so that the victim cannot see. It may be wet so that breathing is difficult. In other cases, gas masks are used.
Incidence Confinement has been reported in the following countries: **Af** Benin, Burundi, Cameroon, Chad, Côte d'Ivoire, Guinea-Bissau, Lesotho, Morocco, Niger, Zaire. **Am** Argentina, Bolivia, Colombia, Honduras, Paraguay, Peru. **As** China, India, Indonesia, Iran Islamic Rep, Israel, Lao PDR, Malaysia, Saudi Arabia, Taiwan (Rep of China), Viet Nam. **Eu** Albania, USSR.
Broader Torture by deprivation (#PD3763).
Narrower Solitary confinement (#PE4056).
Related Incommunicado detention (#PE8004).
Aggravates Disabled victims of torture (#PD0764)
Disease and injury from physical confinement (#PE5763)
Death and disability from inhumane confinement (#PE5648).

♦ PD4597 State control of communications mass media
Nature Governments step into the mass media for three reasons: first, it is ideologically and politically helpful for government to use the mass media to communicate with and influence citizens; second, in many cases only the public sector can provide the massive financial outlay required especially as start-up costs where new media technology is initiated; third, government has a moral responsibility to monitor the quality of information, entertainment, culture and education offered. This necessary involvement too often, however, becomes a stranglehold when government sponsorship prohibits the free competition of alternative opinions and programming.
Narrower Conflict between government and the news media (#PE1643).
Related Monopoly of the media (#PD3101) Political media events (#PD5207)
Biased government information (#PF0157) Newspaper and journal propaganda (#PD0184)
Restriction of access to news distribution media (#PF3082).
Aggravates Unethical media practices (#PD5251)
Insufficient communications systems (#PF2350)
Excessive politicization of the media (#PD5475).
Reduces Excessive portrayal of negative information by the media (#PE1478).

♦ PD4605 Proliferation of immigrants
Immigration overload — Immigrants
Nature Immigrants deprive nationals of jobs and increase unemployment and social security costs.
Counter-claim Immigrants do not cause unemployment of nationals, even among low-paid and minority groups. They take jobs but they also create as many jobs through the enterprises that they initiate. Immigrants do not benefit abusively from welfare services at the expense of nationals since they are typically young and healthy when they arrive and tend to pay more taxes than nationals. Immigrants are typically as well-educated and occupationally skilled as nationals and bring valuable technical knowledge with them, often at a post-graduate level. Immigrants demonstrate desirable economic traits, with a tendency to save more than nationals. Immigrants increase the flexibility of the economy since they are unusually mobile both geographically and occupationally. Immigration alleviates the problem of financing the social security costs of the elderly because they tend to be entering the prime of their work lives and their tax-paying years.
Broader Lack of racial identity (#PF0684).
Narrower Nomadism (#PF3700) Abusive traffic in immigrant workers (#PD2722).
Related Desert nomadism (#PD2520).
Aggravates Rejection of refugees (#PF3021) Plural society tensions (#PF2448)
Discrimination against immigrants and aliens (#PD0973)
Inadequate cultural integration of immigrants (#PC1532)
Inadequate living and working conditions of immigrant labourers in industrialized countries (#PD3427).

♦ PD4621 Military desertion
Deserters
Refs Chase, Salmon P *Reclamation of Fugitives from Service* (1847).
Broader Military disobedience (#PD7225).
Aggravated by Aiding deserters (#PE4909).

♦ PD4629 Misuse of pesticides
Use of inappropriate pesticides — Use of illegal pesticides
Nature The developing countries use some dangerous pesticides that are barred or restricted in the countries that export them. Heavy use of pesticides eliminates the natural enemies of the pests than the pests themselves. These "monster bugs" can no longer be chemically killed.
Broader Misuse of chemicals (#PD5904).
Narrower Misuse of rodenticides (#PG6647).
Aggravates Diseases of beneficial insects (#PD2284).

♦ **PD4632 Dependency on middlemen**
Dependency on intermediaries — Stifling middleman profits — Expensive middlemen cycle — Unnecessary market middlemen — Prohibitive cost of intermediaries
Narrower Middleman control of rural marketing (#PE3528).
Aggravates Disintermediation (#PF3961) Narrow profit margin (#PJ9737)
Prohibitive start-up costs for businesses (#PG9244)
Lack of economic and technical development (#PE8190).
Aggravated by Profiteering (#PC2618) Archaic marketing methods (#PF6465)
Inadequate management skills in rural communities (#PF1442)
Diminishing capital investment in small communities (#PF6477)
Inadequate local expertise in business practices in developing countries (#PE7313).

♦ **PD4633 Social subjugation of women**
Biological subjugation of women — Subjugation of women to domestic service — Subjugation of women to child-rearing
Nature The traditionally inferior position of women in relation to political, social and artistic dimensions of society derives from the "biological tragedy" of women, namely the requirement that they bear and rear children. As such the woman has little opportunity for other than domestic chores (cleaning, cooking, laundry, etc). This tragedy is a function not only of the woman's sex but of the social structure.
Broader Exploitation of women (#PC9733).
Narrower Early marriage (#PE7628).
Aggravated by Family dependence on patriarchal role of the man (#PD3555).

♦ **PD4643 Exhibitionism**
Nature Person has sexually arousing fantasies involving the exposure of his genitals to a stranger. Acting on these urges does not involve any attempt at further sexual activity with the unexpected observer.
Broader Perversion (#PB0869) Sexual deviation (#PD2198).
Narrower Sexual harassment of men (#PE1293) Sexual harassment of women (#PF3271).
Related Nudism (#PF2660).
Aggravates Sexual harassment (#PD1116).
Aggravated by Transvestism (#PE6348).

♦ **PD4651 Political scandal**
Implication of government leaders in scandal
Nature Political scandal is an act or opinion that transgresses the behaviour standards expected of a politician or political system. Political scandals may involve a crime but to be a scandal it only has to be a serious affront of public morality. A politician exposed as having a mistress would be seen as scandalous in one nation and ignored in another.
Broader Scandal (#PC8391).
Related Financial scandal (#PD2458).
Aggravated by Abusive distribution of political patronage (#PF8535).

♦ **PD4652 Accidents caused by fires**
Broader Injurious accidents (#PB0731).

♦ **PD4659 Parasites on plants**
Broader Plant pathogens (#PD1866).
Narrower Dodders (#PE1323) Witchweed (#PE6604) Broomrapes (#PE3795)
Nematoid plant diseases (#PD2228).
Related Plant cancer (#PE1899) Viral plant diseases (#PD2227)
Fungal plant diseases (#PD2225) Bacterial plant diseases (#PD2226)
Environmental plant diseases (#PD2224) Deficiency diseases in plants (#PD3653)
Parasitic diseases in animals (#PD2735).

♦ **PD4666 Violent deaths**
Refs Kaplan, David W *Violent Deaths in Childhood and Adolescence* (1985).
Broader Human violence (#PA0429).

♦ **PD4671 Unsustainable development of coast zones**
Nature Coastal zones are rapidly deteriorating due to intense and increasing human demand. Coastal and nearshore waters and continental shelf areas are a vital source of food, yielding all but 10 per cent of the world's fishing catch. These resources are threatened by poorly planned and regulated urban, industrial, and agricultural development.
Incidence 70 per cent of the population live within 80 kilometres of the coastal waters, and almost 50 per cent of the world's cities with populations in excess of one million are near tidal estuaries. Coastal engineering and development projects are modifying coastal ecosystems on a very large scale. More than 90 per cent of all chemicals, refuse and other material entering continental shelf and coastal waters remain there in sediments, marshes, mangroves, fringing reefs, mudflats, and other coastal ecosystems.
Refs Bird, Eric C *Coastline Changes* (1985).
Broader Unsustainable development (#PB9419) Destruction inherent in development (#PF4829).
Aggravated by Coastal erosion (#PE6734) Coastal water pollution (#PD1356)
Subsiding coastal areas (#PD3775) Uncontrolled urban development (#PC0442)
Inadequate empolderment of wetlands (#PD5110)
Coastal erosion resulting from dams (#PJ7396)
Unsustainable agricultural development (#PC8419)
Lack of finance in coastal communities (#PE2425)
Lack of coastal development in island countries (#PE5689)
Negative consequences of shifting ecology on coastal communities (#PE2305)
Obstacles to the utilization of coastal and deep sea water resources (#PF4767).

♦ **PD4673 Disabled workers**
Incidence In the European Community in 1988 it was estimated that over 30 million nationals suffer from some form of lasting or serious mental or physical disability, requiring a budget of ECU 19 million for the period 1988–1991.
Broader Human disability (#PC0699).
Narrower Disabled migrant workers (#PE0769).
Aggravated by Industrial accidents (#PC0646).

♦ **PD4680 Arrogation of rights**
Narrower Unjust customary rights (#PE5152).
Aggravated by Elitism (#PA1387) Arrogance (#PA7646).

♦ **PD4684 Malpractice in education**
Corruption in training
Broader Corruption (#PA1986) Unethical professional practices (#PC8019).

♦ **PD4691 Theft of property**
Larceny — Stealing — Pilferage
Nature Knowingly taking, transferring, using, retaining or disposing property with the intention of depriving the owner through deception or threat is theft of property. Larceny is the wrongful taking of property from the possession of another person who has a superior right to its possession with the intention of permanently depriving the owner of possession of that property. Larceny is usually limited to tangible goods, but is increasingly extended to include such services as electricity.
Refs Walsh, Marilyn E *The Fence* (1976).
Broader Theft (#PD5552) Financial scandal (#PD2458).
Narrower Robbery (#PD5575) Poaching (#PD2664) Burglary (#PD2561)
Shoplifting (#PE1113) Petty theft (#PG5037) Bag snatching (#PG3445)
Employee theft (#PE3684) Cargo insecurity (#PE5103) Theft of vehicles (#PE4826)
Theft of documents (#PD0577) Theft of works of art (#PE0323)
Misuse of trust funds (#PG1741) Theft of public property (#PG1104)
Theft of nuclear materials (#PD3495) Office and work-place thefts (#PE1106)
Defrauding of secured creditors (#PD4401)
Obtaining property by false pretences (#PE5076).
Related Embezzlement (#PD2688) Theft of services (#PD4711)
Industrial espionage (#PC2921)
Archaeological and anthropological looting (#PD1823).
Aggravated by Drunkenness (#PE8311) Unethical practices by employees (#PD4334)
Police crimes during narcotic investigations (#PE5037).

♦ **PD4695 Violation of trade union rights**
Violation of right of workers to freedom of association — Denial of right to trade union activity — Abuse of trade union rights — Unfair labour practices
Nature Unfair labour practices include: infringing upon a worker's right to organize, bargain collectively, and engage in concerted actions; dismissing, transferring, demoting or treating in other adverse ways a worker in retaliation for the organizations of a union, membership in a labour organization, collective bargaining, strikes, or other legitimate union activity; requiring a worker sign, as a condition of employment a promise to withdraw from or not to join a union; refusing, as an employer, without sufficient cause, to bargain collectively; controlling of or interfering in the formation or administration of a union, including the provision of financial assistance to the union; and dismissing or treating in other adverse ways a worker for requesting relief from unfair labour practices through a labour relations commission or for testifying before such a commission.
Refs International Confederation of Free Trade Unions *Annual Survey of Violations of Trade Union Rights*; International Confederation of Free Trade Unions *Trade Union Rights* (1986).
Broader Abuse of law (#PC5280) Trade unionism (#PF8493)
Denial of economic rights (#PD4150).
Narrower Shopfloor militancy (#PG4982) Prohibition of trade union meetings (#PD7210)
Denial of right to collective bargaining (#PE3970)
Denial of the right of trade union association (#PD0683)
Government discrimination against trade unions (#PE4860)
Denial of right to union activity for special groups (#PE1355)
Violation of the right of trade unions to function freely (#PE1758)
Violation of the right of trade unions to engage in political action (#PE4745)
Violation of the right of workers organizations to protection against suspension (#PE5793).
Related Discrimination against trade unions (#PC4613).
Aggravates Banned trade unions (#PD3535).

♦ **PD4711 Theft of services**
Nature Theft of service is knowingly obtaining or disposing of services which are only available for compensation by deception, threat or other means of avoiding payment is theft of services. When payment of services is normally paid upon receipt, such as hotels or restaurants, absconding without payment or making provision to pay is evidence of theft of the service.
Counter-claim Refusing to pay for services which are honestly believed to be poor in quality is not a crime.
Broader Theft (#PD5552).
Narrower Computer-based crime (#PE4362).
Related Theft of property (#PD4691) Frauds, forgeries and financial crime (#PE5516).

♦ **PD4714 Incitement to war**
Excitement to war — Warmongers
Broader Offences against the peace and security of mankind (#PC6239).
Aggravates Enemies (#PF8404).
Aggravated by War psychosis (#PE7867) Enjoyment of war (#PF4034)
Incitement to violence (#PJ2068).

♦ **PD4721 Unethical practice of the zoosciences**
Irresponsible zoologists — Negligence by zoologists — Malpractice in zoology — Corruption of zoologists — Underreporting of hazards to animal populations
Claim Zoologists, under pressure from their employers, have adopted practices which lead to the underreporting of hazards to animal populations, especially as a consequence of excessive use of fertilizers and pesticides, and of environmental pollution and radiation. Zoologists have failed to investigate adequately the nature of such hazards in the process of further developing products such as pesticides. There is little peer control of irresponsible genetic manipulation and introduction of exotic animal species. Zoologists participate in monopolistic practices designed to create user dependency on high-yield hybrids. Zoologists have adopted questionable practices involving experiments of debatable value on large numbers of live animals. In many cases the pain caused has been considered as irrelevant.
Broader Unethical practice of the biosciences (#PD7731).
Narrower Unethical veterinary practice (#PD7726).
Aggravates Cruel treatment of animals for research (#PD0260)
Monopolistic control of new animal forms (#PD7501)
Irresponsible introduction of new species of animals (#PD1290).

♦ **PD4728 Denial of rights of businesses**
Restrictions on commercial transactions — violation of entrepreneurial activities
Broader Denial of rights (#PB5405).
Narrower Denial of right to business growth (#PE2700)
Suppression of private enterprise in socialist countries (#PD2048).

♦ **PD4734 Intimidation of public officials**
Broader Intimidation (#PB1992)
Crimes against the integrity and effectiveness of government operations (#PD1163).
Narrower Threatening public servants (#PD0540)
Retaliation against public servants (#PD5399).
Related Harassment of public officials (#PD4915).

♦ **PD4738 Conscientious objection**
Conscientious objectors
Refs Postman, Neil *Conscientious Objections* (1988); Seeley, Robert A *A handbook for conscientious objectors* (1981); United Nations *Conscientious objection to military service* (1985).
Aggravated by Pacifism (#PF0010).
Reduced by Denial of right of conscientious objection to military service (#PD1800).

♦ **PD4747 Immuno-deficiency virus**
Bovine immuno-deficiency virus (BIV) — Simian immuno-deficiency virus (SIV) — Feline immuno-deficiency virus (FIV)
Incidence In addition to the HIV virus in humans, 3 others are currently known for cattle, monkeys and cats, in each case with a long latency period. The existence of others is suspected.
Narrower Feline immuno-deficiency virus (#PE7848).
Related AIDS (#PD5111).

♦ **PD4749 Self absorption of political leaders**
Broader Narcissism (#PF7248).

♦ **PD4751 Disastrous insect invasions**
Insect swarms — Difficulty in controlling insect populations
Broader Insect pests (#PC1630) Biological disasters (#PC5489).
Narrower Locust plagues (#PE0725)
Introduction of new species of insect pests (#PF3592).
Related Animal pests (#PD8426).
Aggravates Pests of plants (#PC1627) Insect vectors of animal diseases (#PD2748)
Instability of production of food and live animals (#PD2894)
Fluctuations in food production in developing countries (#PE8188).
Aggravated by Insufficient pest control (#PJ1086)
Excessive use of chemicals to control pests (#PD1207).
Reduced by Defoliation of insect breeding areas (#PJ4038).

♦ **PD4752 Violent crime**
Crimes of violence
Refs Bailey, F Lee and Rothblatt, Henry B *Crimes of Violence* (1973).
Broader Crime (#PB0001) Human violence (#PA0429).
Narrower Sexual offences (#PD4082) Malicious physical disablement (#PE3733).

♦ **PD4755 Neurological effects of torture**
Mental anguish of torture victims
Nature Since the intent of most torture is to break down the self-confidence and humiliate the victims, they are left with deep psychological problems that may never be resolved. In addition to the physical aspect of torture, the most difficult to deal with are those of psychological torture. Victims may feel they have been stripped of their dignity, distrust the world, feel guilty about being tortured, have lost their self-confidence. They retain an incredibly detailed and clear memory of their experiences, and this forms the basis of the anxiety with which they are left. Everyday situations may provoke attacks of fear or panic, for example, the sound of a car door slamming, the sight of a policeman, transport in a lift or visiting the doctor.
Broader Mental illness (#PC0300) Disabled victims of torture (#PD0764).
Narrower Fatigue due to torture (#PE4229) Depression due to torture (#PE0885)
Irritability due to torture (#PE1520) Restlessness due to torture (#PE4074)
Loss of memory due to torture (#PE6593) Anxiety resulting from torture (#PE0969)
Sleep difficulties due to torture (#PE0451) Emotional instability due to torture (#PE4687)
Sexual dysfunction of torture victims (#PE3932) Inability to concentrate due to torture (#PE3716)
Self-imposed social isolation due to torture (#PE4703).

♦ **PD4769 Shortage of firewood**
Shortage of wood fuel — Inaccessibility of fuelwood — Shortage of charcoal for fuel — Shortage of wood in the rough
Nature There is an acute scarcity of wood, which is the main source of cooking and heating fuel for nine-tenths of the people in most of Asia. Africa and parts of Latin America. Some 2,000 million people - who make up roughly the poorer half of mankind - depend on fuelwood as their sole or principal source of energy for cooking food and other household needs. Fuelwood accounts for at least half of all the wood used in the world each year, and for more than 85 percent of wood used in Third World countries. By any measure fuelwood is thus the single largest demand upon the forest, in particular in the rural areas of the developing countries where its use is concentrated. The prospects for fuelwood are alarming. Its supply must be essentially local, as its bulk and relatively low value do not permit long-distance transportation.
Incidence In 1979 fuelwood consumption in developing countries (excluding China) was 1,300 million m3, already 100 million m3 short of requirements; an estimated 250 million people live in areas of fuelwood shortage. By the year 2000 potential demand, at present per capita levels, rises to 2,400 million m3, but actual needs to satisfy minimum requirements would reach 2,600 million m3. There is some parallel with the gap between demand for food and nutritional requirements. Because of the shrinking of the resource base, fuelwood production may be only 1,500 million m3 - 1,100 million m3 short of requirements. As a result, some 3,000 million people would face acute fuelwood shortages by the end of the century, with the result that many poor people will not be able to cook their food adequately. This can have serous nutritional and health consequences. The digestibility of food will decrease and the incidence of parasites ingested with insufficiently-cooked meat will rise; there are already reports of this happening.
Present levels of planting programmes do not offer much hope of alleviating the fuelwood situation. A recent survey of fifteen developing countries estimated that they would have to plant 669,000 ha a year to meet domestic fuel requirements in the year 2000, whereas current programmes cover only 63,500 ha, less than a tenth of what is needed.
Refs Eckholm, Erik P *The Other Energy Crisis* (1975); Energy Probe Research Foundation *Fuelwood* (1983); FAO *Fuelwood Supplies in the Developing Countries* (1983); United Nations *Proceedings of the ESCAP-FAO-UNEP Expert Group on Fuelwood and Charcoal.*
Broader Energy crisis (#PC6329).
Related Long-term shortage of wood, lumber and cork (#PE1372).
Aggravates Endangered forests (#PC5165) Parasites of the human body (#PE0596)
Human disease and disability (#PB1044).
Aggravated by Deforestation (#PC1366) Prohibitive cost of fuel (#PJ0346)
Inadequate cooking stoves (#PE7904).
Reduced by Underutilization of fuelwood energy (#PJ0031).

♦ **PD4773 Bias in children's literature**
Racism in child and youth literature
Nature Children's and young people's literature reflects the social and cultural values of a society. Thus, it also reflects the ideologies, prejudices and clichés, which characterize a certain historical period. The discrimination against children, which has been typical for Western civilization, often leads to children's and young people's books which are hastily composed. Thus, the content of such a book frequently discloses a state of mind and prejudices of which the European, British Commonwealth or American author is often not fully aware. Moreover in the Western world, children's and young people's literature is embedded in the capitalist system. As a consequence, publishers are likely to accept for publication only those manuscripts which reflect or strengthen generally accepted values.
Two stereotypes can be found today in the majority of children's books. Either the representative of a coloured race is seen as a happy natural human being in an exotic environment, living in a primitive jungle village far away from the modern civilization with which he is unable to compete, or he is looked at as the white man's servant. In young people's books, the coloured person appears as an underdog and an object on whom the white man can expend humanity from time to time when he is pricked by his conscience. Also, obedience, faithfulness and submissiveness rank at the top of the depicted qualities of the coloured, while the active role is usually reserved for the white people who take the initiatives and perform feats of leadership throughout such stories.
Incidence Racism against coloured people is still very common in Western Europe, South Africa, the USA and Brazil. For more than a century children's and young people's literature with its stereotyped discrimination against coloured people has perpetuated white racism; either euro-centrism or gringoism projects inferiority feelings into coloured people. Extremely racist groups such as those that exist in the USA, the UK, the Union of South Africa, Germany and France can readily take advantage of children's literature for the purpose of openly preaching racial hatred.
Background The image of coloured people in Western European literature was delineated by a few basic types before the children's books emerged during the 18th century. Child and youth literature adopted existing stereotypes. The stereotype of the coloured man as the noble savage appears in a letter about a journey to Brazil in 1500. Later, the noble savage emerges in occidental philosophical novels from Voltaire's Ingenu up to A. Huxley's Brave New World. This false idea of a primitive yet noble manner of being also provided the basis for Rousseau's philosophy and lay the seed for the turning away from Europe and the enthusiasm for America among some of the European romanticists. The image of the noble savage can be found in youth literature as well as in literature originally written for adults but afterwards mainly read by young people. Famous examples are the Indians in the novels by James Fennimore Cooper and the Indian Winnetou in the novels by Karl May.
Refs Klein, Gillian *Reading into Racism* (1986).
Broader Racism (#PB1047).
Related Deliberate misrepresentation in educational materials (#PF1183).
Aggravates Prejudice (#PA2173).
Aggravated by Book censorship (#PD3026) Racist propaganda (#PD3093)
Biased literature (#PJ1679).

♦ **PD4775 Uncontrolled tropical diseases**
Incidence In 1990 it was estimated that nearly 10 per cent of the world population suffers from tropical diseases, and the number is expected to increase steadily since remedial action is inhibited by civil unrest in many of the countries where such diseases are most prevalent. Most of the infected live in countries were per capita incomes are less than $400 per year and governments are so poor that they spend no more than $4 per person on their entire health systems.
Control of malaria and other parasitic diseases is beyond the reach of many tropical countries, not only because they lack resources but also because of gaps in knowledge and the absence of a proper health technology to make effective use of what is already known. The more important tropical diseases where control is needed include: malaria, schistosomiasis (snail fever), trypanosomiasis (African sleeping sickness and Chaga's disease in South America), leishmaniasis, leprosy and filariasis, including onchocerciasis (river blindness).
The process of development itself tends to spread disease because the people who arrive in undeveloped areas, where many diseases are carried by animals, become new hosts for the parasites or other disease carriers. Development of water storage and irrigation systems, promotes the dissemination of parasites living in water.
Broader Inadequate health control (#PF9401) Human disease and disability (#PB1044).
Narrower Leprosy (#PE0721) Malaria (#PE0616) Elephantiasis (#PE1601)
Onchocerciasis (#PE2388) Schistosomiasis (#PE0921) Chagas' disease (#PE0653).
Related Detrimental effect of jungle environment in tropical villages (#PE2235).

♦ **PD4779 Unjust punishments for crimes**
Broader Injustice (#PA6486) Punishment (#PA5583)
Denial of right to freedom from cruel, inhumane or degrading punishment (#PC3768).
Narrower Punitive amputations (#PE8724) Collective punishment (#PD6970).
Related Denial of rights to prisoners (#PD0520).

♦ **PD4782 Violence along internal borders**
Internal language, religious or ethnic divisions
Nature Countries may be crossed by internal borders or may contain recognized enclaves that respect differing languages, differing ethnicities or differing religious beliefs. This can occur in respect to all three conditions, for which examples exist over all the world. Such divisions have arisen when nations have been occupied and the indigenous peoples segregated, or following political settlements after wars, near wars, civil strife or unrest. The ill-feelings that create such divisions persist and often focus on the borders, where incidents of personal violence or violence against property occur. Unofficial borders occur frequently as well, and are easily seen in large cities where barrios or ghettos exist. Social tension may be even stronger on or near unofficial borders.
Incidence A few examples of groups involved in ethnic violence include: Hindus and Muslims, and Hindus and Sikhs in India; Jews and Muslims in Israel; Armenians and Azerbaijanis in the Soviet Union; Tamils and Ceylonese in Sri Lanka; Tibetans and Chinese in Tibet; Moros and Filipinos in the Philippines; Somalis and the majorities of Kenya and Ethiopia; the Eritreans and Ethiopians; Catholic and Protestants in Northern Ireland; and Basques and other Spaniards in Spain.
Claim Such internal borders in a country are invitations to continued conflict. Obstacles to their climination, however, lie in the legal, economic and social insecurity of citizens in societies that have become increasingly pluralistic, but which are dominated by one language, one ethnicity or one religious creed.
Related Violence as a resource (#PF3994).

♦ **PD4790 Inadequate health services**
Inadequate medical care infrastructure — Undeveloped health services — Uninitiated health services — Distant health services — Remote health care services — Unutilized health services — Uncoordinated health services — Unused health services — Deficient health services — Uncoordinated health care planning — Undeveloped health care systems
Nature The degree of general health improvement achieved by public and private health services is not as high as might be desired. Although technical knowledge for achieving better health is available, in most countries this knowledge is not being put to the best advantage of the greatest number. Health resources are allocated mainly to sophisticated medical institutions in urban areas. Rather than better health for the average person, "improvement of health" tends to be equated with the provision of medical care dispensed by growing numbers of specialists, using narrow medical technologies for the benefit of the privileged few. At the same time, access of large segments of the world's population to health services is limited or non-existent; disadvantaged groups throughout the world have no access to any permanent form of health care. These groups probably total four-fifths of the world's population, living mainly in rural areas and urban slums. In some countries, even though health facilities are located within easy reach, inability to pay or cultural taboos put them out of bounds. To complicate matters, health systems are often devised outside the mainstream of social and economic development, frequently restricting themselves to medical care, although industrialization and deliberate alteration of the environment are creating health problems whose proper control lies far beyond the scope of medical care. Such services

operate in an isolated manner, neglecting other factors contributing to human wellbeing such as education, communications, agriculture, social organization, community motivation and involvement. This ignores the fact that health cannot be attained by the health sector alone. In developing countries in particular, economic development, anti–poverty measures, food production, water, sanitation, housing, environmental protection and education all contribute to health and have the same goal of human development. The pace of technological and economic development requires an intensified release of human energy, placing heightened importance on physical stamina as a precondition. However, although the current diet upon which people exist may appear to be ample, it lacks the nutritional balance to sustain regular participation in a modernized society. In addition, a whole complex of issues such as safe water, refrigeration and basic hygiene remain relatively undeveloped and therefore continues to perpetuate illness that drains vitality. The sheer number of people in the care of one doctor, the remoteness of proper medical facilities and the high cost of treatment prevent early detection of disease; continuation of energy–draining low–grade infections results in either long–lasting or permanently chronic defects. The care of the physical well–being of rural people when called upon to make such efforts at development is a crucial factor that cannot be neglected.
Incidence Huge inequalities characterize the current picture of global health. In the Third World, health problems are related to malnutrition, poverty and lack of access to basic needs. In the First World, the health problems of the ageing populations present the greatest challenge. Health services in the developing countries have often been based on European or North American models, centering on highly technological, cost–intensive urban hospitals focused on curative rather than preventive health care. Conservative estimates of the annual cost of running a primary health care system give a figure of $12.50 per person per year. However, the current level of expenditure on health care is less than $2 per person in many of the poorest countries. About $50,000 million per year would need to be invested to develop primary health care systems in the developing countries In the developed countries, the difference between the free–on–demand system typified by the British National Health Service and the fee–for–service system typified by the United States medical care is perhaps best brought out by a comparison of children's consultation rates; children's consultations tend to be more in the nature of preventive medicine than adults. In the United States, children in families in the highest income group are approximately twice as likely to consult a doctor as children in families in the lowest income group. In the United Kingdom, the consultation rates for children in all social classes are about the same.
Claim Most conventional health care systems are becoming increasingly complex and costly and have doubtful social relevance. They have been distorted by the dictates of medical technology and by the misguided efforts of a medical industry providing medical consumer goods to society. People have become cases without personalities, and contact has been lost between those providing medical care and those receiving it. Even some of the most affluent countries have come to realize the disparity between the high care costs and low health benefits of these systems. Obviously it is out of the question for the developing countries to continue importing them.
Broader Inadequate infrastructure (#PC7693) Inadequate social welfare services (#PC0834)
Uncoordinated social services in urban areas (#PF1853)
Weakness of infrastructure in developing countries (#PC1228).
Narrower Complex health delivery (#PJ8159) Unconsensed health needs (#PG9032)
Inadequacy of psychiatry (#PE9172) Inadequate health control (#PF9401)
Prohibitive medical expenses (#PE8261) Inadequate medical facilities (#PD4004)
Insufficient health personnel (#PD0366) Dehumanization of health care (#PD7821)
Inadequacy of medical science (#PF8326) Inadequate primary health care (#PE8553)
Inadequate radiological services (#PE7544) Limited access to health services (#PF6577)
Neglect of adolescent health care (#PF6061) Irregular hospital transportation (#PJ8984)
Outdated forms of community health (#PF1608)
Inadequate health care in urban slums (#PE7877)
Limited availability of health resources (#PD7669)
Inadequate health care in family planning (#PD1038)
Inadequate health services following nuclear war (#PD6265)
Inadequate health care in least developed countries (#PE9242)
Inadequate nutrition education in least developed countries (#PE0265).
Related Lack of care (#PF4646)
Inadequate supply of pharmaceutical products in developing countries (#PE4120).
Aggravates Human suffering (#PB5955) Health inequalities (#PC4844)
Nutritional deficiencies (#PC0382) Neglected health practices (#PD8607)
Protein–energy malnutrition (#PD0339) Inequality in mortality rates (#PC9586)
Ignorance of health and hygiene (#PD8023) Infectious and parasitic diseases (#PD0982)
Inadequate mental health services (#PJ4379) Blindness in developing countries (#PD5139)
Limited access to society's resources (#PF6573) Fragility of maintaining basic health (#PJ2524)
Non–use of available health facilities (#PF9683)
Delay in administration of medical care (#PD5119)
Disrupted mechanisms for community health (#PF2971)
Underdeveloped potential of basic resources (#PF3448)
Unexplored alternatives for commercial development (#PF6548)
Incomplete understanding of new societal service systems (#PF2212)
Inadequate working conditions in health and medical services (#PE7718)
Cumulative depletion of corporate initiative in rural communities (#PE3296).
Aggravated by Citizen apathy (#PF2421) Underdevelopment (#PB0206)
Medical materialism (#PJ7913) Limited food variety (#PF0479)
Insufficient doctors (#PE8303) Refusal of medical care (#PF4244)
Inadequate medical care (#PF4832) Infrequent doctor contact (#PJ0362)
Fragmentation of health service (#PE5721) Shortage of fresh–water sources (#PC4815)
Lack of refrigeration facilities (#PG0524) Prohibitive cost of disease control (#PF2779)
Unnecessary health system referrals (#PE0952) Increasing public health expenditures (#PF6234)
Prohibitive cost of private medical care (#PF8016)
Institutionalized callousness of public services (#PF2006)
Ineffective self–regulation in the health care sector (#PE9048)
Crimes committed in hospitals and health care facilities (#PE8420).

♦ PD4798 Political infiltration
Political party entryism
Nature Political conflict may involve the unethical placement, on one party's orders, of one or more of its adherents into another party's membership. Infiltration of political parties for the purposes of espionage and political sabotage is relatively easy in countries whose political parties rely heavily on volunteers, or whose treasuries are low. The political spy may work himself, or, by contributions, buy himself in. This is easier at local party levels, but once in at the local level the spy may be placed in a position of trust, or elected as an officer and thus have access to the higher echelons of party planning and personnel. Political infiltration may be associated with criminal acts of illegal entry, theft and illegal wiretapping.
Incidence Political infiltration is a tactic regularly employed by national communist parties in non–communist ruled nations, as has been witnessed in the UK, France, Italy and a number of other countries, where over many decades, communists have infiltrated socialist, social–democratic and liberal parties to attempt to move them to the Marxist extreme–left. In other countries, like the USA, excluding the activities of the communists, political infiltration is sporadic and short–lived at the election campaign times for the key offices of President, Senator, or Governor. In countries with a history of political instability, infiltration of the ruling party may be preparatory to coup or revolution.
Aggravates Undemocratic political organization (#PC1015).
Aggravated by Political surveillance (#PD8871).

♦ PD4800 Excessive government intervention in the private sector
Government interference in the national economy — Proliferation of government regulations
Nature According to the interventionist outlook, the state is not to be excluded from the strategic sectors of the economy; and political and trade union problems are thought to be best avoided, whatever the consequences to profitability, by keeping companies on the brink of bankruptcy from going out of business. In many countries, the major result has been a massive and debt–ridden state share–holding sector, often easily susceptible to political manipulation.
Claim The market price system never works perfectly, least of all in developing countries; the question is whether to rely on imperfect markets or imperfect governments. On the whole, the countries that have grown fastest kept inflation under control by pursuing prudent, unambitious monetary and fiscal policies; they promoted exports mainly by refraining from discriminating against exporters; they left their economies open to foreign competition, which spurred internal efficiency; they left their domestic price systems largely intact, instead of supplanting them with marketing boards and other state monopolies; they allowed their financial system to provide adequate returns to savers; and they gave the private sector a big role in deciding where those savings should be used. Intervention breeds intervention; a quota here creates a shortage there; shortages push prices up, so price ceilings are necessary; price ceilings put firms in difficulty, so some are given subsidized credit; others are not, and want to contract, so sackings are forbidden; and so on.
Counter–claim Government intervention has contributed to growth by reducing rigidities and correcting for market failures often resulting from the absence in developing countries of the array of conditions favourable to development. The case against intervention, in particular protection, in developing countries should be tempered by the fact that the motivations for protection in developing countries are quite different from those in industrialized countries. Broadly speaking, the adoption by developing countries of measures with the potential to restrict trade would appear to be designed to serve one or more of the following purposes: revenue collection, balance–of–payments protection, and infant–industry protection. Certainly, South Korea and Japan, let alone, the US and most European countries can claim to be non–interventionists.
Broader Excessive government control (#PF0304).
Narrower Complex trade regulations (#PF4722) Restrictions against small enterprise (#PD5584)
Complex regulations paralyzing small communities (#PF2444)
Excessive government participation in the economies of developing countries (#PD1902).
Aggravates Cooptation of private sector initiatives by government (#PE0780).
Aggravated by Proliferation of public sector institutions (#PF4739).

♦ PD4811 Violent repression of demonstrations
Violent police intervention in meetings
Broader Repression (#PB0871) Denial of right of assembly (#PC2383).
Narrower Violent repression of trade union demonstrations (#PE2007).
Aggravates Imprisonment (#PD5142) Arrest of trade union leaders (#PD7630).
Aggravated by Demonstrations (#PD8522).

♦ PD4812 Overgrazing in developing countries
Broader Degradation of semi–natural and natural habitats of flora and fauna (#PC3152).

♦ PD4814 Denial of state's rights
Denial of rights to sovereign nations
Broader Denial of rights (#PB5405).
Narrower Denial of right to state succession (#PE0241)
Denial of right to equality between states (#PE4712)
Denial of the right to national sovereignty (#PE7906).
Related Disruption of territorial integrity (#PC2945).

♦ PD4823 Neglect of victims of crime
Inadequate assistance to victims of crime
Nature In order to protect society, the main emphasis of legal systems traditionally has been upon the detection of crime and the punishment of the offender. More enlightened programs aim to support the offender in an effort to prevent recidivism. Very little attention, however, has been paid, either by voluntary or by statutory bodies, to assess and supply the needs of the victim of crime. The main role of the victim at law has been as a source of evidence to secure a conviction against the offender, yet he or she may deserve restitution or recompense, and may also require legal, medical, psychiatric or social welfare assistance as a result of the crime. While the problems of victims, or likely victims, may be referred to incidentally, there has as yet been no concerted attempt to bring together and elaborate the measures needed on their behalf and to develop further approaches and techniques designed to improve their plight. International conventions do not presently articulate explicit rights to protection, reparation or justice for victims of crime.
Background The concern for victims of crime is not new. The Hammarabi Code provided for reparation centuries ago. Further, many of the customary practices that were not changed by colonization promote reconciliation and reparation. However, the rights of victims were curtailed in the 19th century in many industrialized countries for such reasons as controlling unofficial retaliation and guaranteeing fine revenues for the state. Few countries afford victims comprehensive participation in the judicial process.
Refs Austern, David *The Crime Victim's Handbook* (1987); Karmen, Andrew *Crime Victims* (1984); Louis Harris and Associates *Victims of Crime* (1983); Stark, James and Goldstein, Howard *The Rights of Crime Victims* (1985).
Broader Social neglect (#PB0883)
Legal discrimination in favour of offenders (#PD9316)
Inadequate assistance to victims of human rights violations (#PD5122).
Narrower Disabled victims of crimes (#PD0762)
Employment of criminals in policy–making contexts (#PE4439).
Related Grief (#PF5654) Inadequate assistance to victims of abuse of power (#PE7390).
Aggravated by Criminals (#PC7373) Intimidation of victims of crimes (#PJ1543).

♦ PD4827 Unjust trials
Biased judges and juries — Denial of right to fair and public trial
Incidence Allegations of unfair trial procedures have been made with respect to many countries. In most cases, such reports concern countries where a state of siege or emergency is in force, entailing the use of military courts or special courts. In several countries, the usual practice in respect of trials of political opponents appears to be that the accused is notified of the date and time of the trial only two or three hours in advance, thus considerably reducing the possibilities for his defence. The accused and his defending counsel are often informed of the charges against him only during the hearing or, again, the accused may not even be present at the hearing.
Criminal justice systems in democratic societies with long traditions of trial by jury are far from immune to punishing innocent people and failing to free these people even after overwhelming evidence has been uncovered proving their innocence. What sometimes happens is built into the criminal justice system. Unable to find the true perpetrator of a crime, the hard–pressed police

DETAILED PROBLEMS

PD4900

create in their minds a suspect in whose guild they come to believe. Lacking probative evidence, they manufacture it by way of false confessions, suborning of witnesses, loss of documents helpful to the defence, and enlisting informants within the prisons in which the accused is staying, in the deluded belief they are seeing justice done. The trial judge and jury having to choose between the word of the police and the defendant, invariable favour the police. The accused is convicted and sentenced. The case goes to appeal, but the Appeal Court, having no means of knowing that the jury reached their verdict on false evidence, and where deliberations in any event are cursory, invariably dismiss the appeal.
Broader Violation of civil rights (#PC5285)
Deficiencies in national and local legal systems (#PF4851).
Narrower Political trials (#PD3013) Injustice of trials in absentia (#PE0424)
Lack of impartiality of the judiciary (#PE7665) Unfair trials due to pre-trial publicity (#PE1692)
Bias in jury trials in small jurisdictions (#PE4733).
Aggravates Internment without trial (#PD1576).
Aggravated by Complex trials (#PE3916) Denial of right to fair and public trial (#PE3964).

♦ PD4840 Personal life crises
Negative life events — Stressful life experiences — Upheavals in private life
Refs Deits, Bob *Life after Loss* (1988); Ford, C E and Snyder, C R *Coping With Negative Life Events* (1987); Moos, Rudolf H (Ed) *Coping with Life Crises* (1986).
Broader Deteriorating quality of life (#PF7142).
Narrower Divorce (#PF2100) Birth trauma (#PE8911) Puberty trauma (#PJ2206)
Emotional crises (#PE3407) Romantic separation (#PD4233) Male mid-life crisis (#PD5783)
Female mid-life crisis (#PD5675)
Psychological impediments to marriage (#PF3344).
Aggravates Loneliness (#PF2386) Alienation (#PA3545)
Scepticism (#PF3417) Frustration (#PA2252)
Value erosion (#PA1782) Mental illness (#PC0300)
Emotional insecurity (#PD8262) Stress in human beings (#PC1648).
Aggravated by Human death (#PA0072) Emotional immaturity (#PJ5907)
Complications of childbirth (#PC9042).

♦ PD4843 Gang warfare
Gang war
Nature A gang is, by definition, a conflict group at war with other groups or with the forces of organized society. It may originally have formed spontaneously, but it is integrated through conflict. Boy-groups or gangs in many cities are an important factor in juvenile delinquency, in the beginnings of criminal careers, in organized crime, and in political corruption.
Incidence In the UK in 1988, 3 per cent of male murders (and 1 per cent of female) were attributed to gang warfare.
Counter-claim The result of collective gang behaviour is the development of tradition, unreflective internal structure, esprit de corps, solidarity, morale, group awareness, and attachment to a local territory.
Broader War (#PB0593).
Related Racial war (#PD8718).
Aggravates Theft (#PD5552) Robbery (#PD5575) Enemies (#PF8404)
Homicide (#PD2341) Property damage (#PD5859) Criminal intrusion (#PE6771)
Illicit drug trafficking (#PD0991) Firearms and explosives crimes (#PE1108)
Criminally life endangering behaviour (#PD0437).
Aggravated by Youth gangs (#PD2682) Organized crime (#PC2343).

♦ PD4846 Political feuding
Nature Political parties who have not been elected for some time tend to lose touch with the possibility of actually ruling their group or district, and focus their efforts on promoting the right ideas among their own members. Thus groups and splinter groups proliferate, while the group which is in power is forced to make decisions on their own without the benefit of alternative perspectives.
Broader Uncommunicativeness (#PA7411) Behavioural deterioration (#PB6321).
Narrower Unrepresentative international organizations (#PD4873).
Related Feuds (#PE8210) Political conflict (#PC0368)
Biased presentation of news (#PD1718).
Aggravates Propaganda (#PF1878).

♦ PD4858 Illegal international arms shipments
Arms smuggling — Gun running
Nature Shipments of weapons are made illegally by governments and private arms dealers. Government shipments are sometimes covert military assistance and may break international treaties. Illegal private arms deals are done for profit; dealers may supply governments, gangs or revolutionary groups. Terrorist organizations and drug trafficking gangs have established their own illegal arms distribution networks. Arms illegally shipped have ranged from jet fighters, radar, computer-controlled weaponry and intelligence gathering systems to bombs, bullets and explosives of all kinds. No instance of illegal nuclear or biological weapons shipments is known, but the control of uranium and its by-products used for nuclear weaponry is lax and it is suspected that there has been illegal appropriation of nuclear materials from energy-generating and research projects to weaponry research, if not to weapon manufacture.
Incidence In September 1990, the Germany FR was investigating 60 companies suspected of making arms-related sales to Iraq, including the precursors and technology for chemical weaponry.
Refs Karp, Aaron (Ed) *Shades of Grey* (1989).
Broader International arms trade (#PC1358) Unfulfilled treaty obligations (#PF2497)
Private international arms dealers (#PD2107).
Related Illicit drug trafficking (#PD0991)
Lack of business opposition to the arms race (#PF7088).
Aggravated by War (#PB0593) Unethical practices in transportation (#PD1012)
Imbalance of conventional armed forces (#PC5230).

♦ PD4862 Socialist colonialism
Russification
Nature Under the guise of international socialist cooperation some states dominate others militarily. Through their military capabilities they are enabled to dictate favourable economic treaties or agreements and eventually culturally dominate what is then a neo-colonized nation or region. One very evident form of this is the imposition of the dominating country's language.
Incidence Both the National Socialist Government of Germany and the government of the Union of Soviet Socialist Republics colonized adjoining nations and regions. The former Baltic States of Estonia, Latvia and Lithuania and the former sovereign nations of Georgia and Armenia are losing their ethnic and linguistic identity. Afghanistan is currently being colonized. In the Caribbean, socialist colonization is extending from Cuba, and in North Africa from Libya.
Broader Socialism (#PC0115) Communism (#PC0369) Imperialism (#PB0113).

♦ PD4873 Unrepresentative international organizations
Nature Organizations claiming to represent a constituency but which have not arisen democratically, have no legitimacy. Such organizations may be motivated either by good intentions or personal drives for power of fraudulent intent, or may even have as their purpose opposition to the goals of the proposed constituents.
Broader Deception (#PB4731) Political feuding (#PD4846)
Undemocratic organizations (#PC8676).
Narrower Unrepresentative international nongovernmental organizations (#PE7021).
Aggravates Manipulation of students (#PE5777) Elitist control of global economy (#PC3778)
Fragmentation of organized students (#PF5753).

♦ PD4879 Deception in business
Commercial deception — Business lies — Misleading commercial information
Refs Comer, Michael J, et al *Bad Lies in Business* (1988).
Broader Lying (#PB7600) Misleading information (#PF3096).
Narrower Misleading advertising (#PE3814)
Misleading accounting information due to inflation (#PE4285).
Aggravates Unethical financial practices (#PE0682).

♦ PD4882 Circumvention of duties and assessments
Underpayment of duties and taxes — Tax avoidance
Nature The methods employed in underpaying or avoiding local, national and international assessments of various kinds correspond to the method of assessment. Most illegal acts in the private sector deal with concealment or under-valuation of assets, whether these be a farmer's crops, a corporation's income, or a private individual's inherited wealth. Other acts of avoidance apply to manufacturing, sales and usage taxes, and similarly are of the nature of non-declaration or understatement of quantities manufactured, sold, or used, and their worth. An industry for detecting avoidance exists, but it is almost entirely a government monopoly, administered by inland revenue or treasury offices. In some countries a reward is paid to private persons giving information on avoidance. Corruption in the form of bribe-taking by collectors or inspectors is a recurrent phenomenon.
The payment of duties is also circumvented in many ways, including illegal shipment or smuggling. The losses to local or national governments by tax subterfuges, illegal business, or recordkeeping, and by the various other methods of avoiding payment is significant, and as a result penalties are severe. Avoidance also hurts the private sector of the economy when taxes and duties relating to imports are circumvented, to the extent that such taxes and duties are protectionist and are designed to discourage a high level of importation of a commodity or product. Finally, governments which themselves may be taxed or proportionally assessed, for example, for their share of international costs of projects, or international organizational budgets, may also seek avoidance in the form of truant payments or defaults.
Refs Bracewell-Milnes, B, et al *International Tax Avoidance* (1979).
Broader Immorality (#PA3369).
Narrower Tax evasion (#PD1466) Abuse of tax havens (#PE2370)
Unreported tax obligations (#PE9061)
Abuse of status of religious institutions (#PJ1156)
Evasion of shipping regulations and taxes by flags of convenience (#PE5873).
Related Corruptive crimes (#PD8679) Defaults on international loans (#PD3053)
Inequitable tax treaties between developed and developing countries (#PD1477).
Aggravated by Ineffective tax systems (#PF1462)
Unethical practices of employers (#PD2879).

♦ PD4890 Religious indoctrination
Broader Compulsory indoctrination (#PD3097).
Aggravates Religious war (#PC2371) Religious conflict (#PC3292)
Childhood martyrdom (#PF8118) Religious propaganda (#PD3094)
Forced religious conversion (#PD6637) Lack of religious discipline (#PF8010)
Religious and political antagonism (#PC0030).
Aggravated by Dogmatism (#PF6988).

♦ PD4900 Unsustainable development of forest lands
Conflicting demands on forests
Nature The total demand for forest products is expected to double every ten years. Clearing forests is often considered a pre-requisite for economic development in countries where large forest tracts still exist. Pressure increases from the demands for more agricultural land (particularly in the tropics where shifting cultivation is increasingly practised); the establishment of new human settlements; and the development of water impoundments, transportation systems, etc. The world's forestry resources are shrinking at an alarming rate; in Latin America, between 5 and 10 million hectares are felled annually for agriculture. In some areas conflicting needs can be satisfied without reducing the long-term productive capacity of forests and without deteriorating other natural resources or the environment in general; recreation, grazing, and aesthetic considerations usually constrain timber production. However, in other areas, such as tropical regions, arid regions, regions of dense population or those adjacent to major industrial concentrations, forest depletion and degradation are taking place at an accelerated rate; adverse changes in micro-climates, soils, and water cycles sometimes result, and both the quality of the environment and the productive capacity of other natural resources are then affected by local and possibly regional changes in climate, the increased frequency of floods, accelerating soil erosion by wind and runoff (and subsequent silting of water bodies), and the destruction of the natural habitat of wildlife.
The forest products industry can place a particularly heavy burden upon the environment. The chemicals and organic matter in the waste liquid from pulp mills are normally disposed of in adjacent water bodies (inorganic salts, mercury, and heat are all released); unpleasant odours and gases, some of which are toxic in concentrated forms, are produced; and sulphur dioxide, sulphides, and particulate matter are also released. Pollution – particularly noxious smoke and particulate contamination of the air – is also created by the mechanical woodworking industries such as sawmilling, and plywood, particle-board, and fibreboard manufacture.
Background At the world level, forests have a direct and beneficial influence on all parts of the biosphere as a result of photosynthesis, heat capacity, conductivity and reflectivity, aero-dynamic roughness, influence on the water cycle, and emissivity in the infra-red band. They act as buffer zones between man-made ecosystems; and they represent half of the world's photosynthetic fixation of carbon from the atmosphere, with its concurrent release of oxygen. They serve as the source of wood and wood products, and they harbour valuable wild plant and animal species. At local levels, forests contribute to regulation of water catchment and release; protection of soil against erosion by wind and water and against other forms of soil degradation; protection of wildlife; recreational resources; and the improvement of living conditions both in and around human settlements through the control of nuisances such as noise and air pollution, improved aesthetics, psychological relief, provision of shade (particularly in the tropics), agricultural protection and improvement when introduced as shelter belts and windbreaks. Trees serve to moderate wind velocities and improve the micrometeorological and soil moisture conditions in adjacent fields.
Refs Repetto, Robert and Gillis, Malcolm (Eds) *Public Policy and the Misuse of Forest Resources*

Broader Conflict (#PA0298) Unsustainable development (#PB9419).
Narrower Forest fragmentation (#PD9490) Unethical practices in forestry (#PD6701)
Misuse of tropical rain forests for agricultural development (#PE5274).
Aggravates Endangered forests (#PC5165).
Aggravated by Destruction inherent in development (#PF4829).

♦ PD4904 Acidic precipitation
Acid rain — Toxic rain — Acid deposition — Acidic snow

Nature Sulphur and nitrogen oxides and other acid precursors emitted by natural and man-made sources can travel long distances in the atmosphere, undergoing chemical transformations leading to the formation of sulphuric and nitric acids and returning to the earth as precipitation. In sensitive areas, this increases the acidity of water bodies and the soil, and can damage aquatic ecosystems, crops and forests. Wet and dry depositions are absorbed into the soil where they can break down other naturally present minerals and leach away nutrient sources necessary for the health growth of trees, plants and crops. Through the groundwater it eventually enters nearby bodies of water, often carrying toxic metals such as aluminium that can deform or kill aquatic life. The phosphates, which nourish phytoplankton and other aquatic plants, attach themselves to the aluminium and become less available as a nutrient. As the water gets more acid still, other toxic metals like cadmium, zinc, lead and mercury also become increasingly soluble and may be taken up by water life through food chains. Acidic water, by dissolving lead water pipes, can introduce unhealthy levels of lead in drinking water. The acid precursors are mainly produced by fossil fuel combustion in power plants, by smelting industries and from motor vehicle exhausts. Their effects are felt not only in the neighbourhood of the sources, but also at distances of hundreds of kilometres. Corrosion is accelerated in most materials used in construction of buildings, bridges, dams and industrial equipment. It can also severely damage monuments and historic buildings. Remedial action is considerably handicapped by the fact that the population in the source area may have different priorities than that where the acid is deposited, and may place quite different values both on the costs of any damage and on the costs of various control strategies.

There is increasing evidence that toxic substances like DDT, lindane, toxaphene, eldrin, dieldrin, benzene, toluene, napthalene and polychlorinated biphenols (PCBs) are also being transmitted through the atmosphere, i.e. toxic rain.

Incidence Fossil fuels contain chemical elements including carbon, hydrocarbon, sulphur, and nitrogen among others. These chemicals are released into the atmosphere as waste products when the fuels are burned. Oxygen combines with the chemicals to produce oxides, such as sulphur dioxide and nitrogen oxides, the main pollutants which cause acid rain. Sulphur dioxide is emitted principally by power stations and industrial and commercial installations when burning coal and oil, and by metal smelters when burning iron and other metallic ores. Nitrogen oxides: nitric oxide, nitrous oxide and nitrogen oxide, produced during the burning of coal, oil and petroleum. They are emitted both by stationary sources, for instance, power stations and by vehicles. Once emitted, some of the oxides fall directly onto surfaces of plants, trees, soils, lakes and buildings. This is dry deposition which turns to acid when it becomes wet by the action of dew, rain or falling into bodies of water. Oxygen in the atmosphere transforms the remaining oxides into sulphuric and nitric acids which are deposited as rain, snow, hail or dew. This is wet deposition. Dry deposition generally occurs close to the point of emission. Wet deposition often occurs thousands of kilometres downwind of emission sources.

Other air pollutants, not strictly contributing to acid rain but often included under the heading are hydrocarbons, low-level ozone and ammonia. A significant proportion of hydrocarbons is emitted by cars and oil refineries, and during production and use of solvents. They combine with nitrogen oxides in the presence of sunlight, a photochemical reaction, to form photo-oxidants, of which ozone is the most harmful at heights below the stratosphere. Ammonia is a combination of nitrogen and hydrogen and is produced by both industry and agriculture. Sources include nitrogen fertilizer factories and intensive livestock farming.

Acid rain has been observed throughout the world, particularly in Scandinavia and in parts of North America. Much of it results from pollutants travelling over 500 km. The effects occur in countries which are major emitters of the gases and in distant countries receiving the acid deposition as a result of prevailing wind patterns. An estimated 5 to 10 million sq.km in Europe and North America are being turned acid. In 1982 one third of West Germany's forests are estimated to have suffered damage, with three quarters of fir trees damaged in 1983. European rain water should have a pH value between 5 and 6, but over large areas it is now between 4 and 5. Most fish cannot reproduce in water with a pH of 4.5. In Sweden, damage to fisheries attributed to acidification has been observed in 2,500 lakes, and is assumed to have taken place in another 6,500 where signs of the process have been found. The average precipitation today is 100 times more acidic than 180-year old ice cores from Greenland.

Counter-claim Claims of acid rain damage are based on inconclusive scientific evidence. Acid generated by nature is far greater than that contributed by industrially generated acid rain. Natural processes such as volcanic eruptions, forest fires, and bacterial decomposition of organic matter also produce acidic sulphur and nitrogen compounds.

Refs Adams, Donald D and Page, Walter P *Acid Deposition* (1985); Beilke, S and Elshout, A J (Eds) *Acid Deposition* (1983); DOE Technical Information Center Staff *Acid Precipitation* (1983); International Institute for Environment and Development *Acid Earth* (1985); Lier, Irene H van *Acid Rain and International Law* (1981); McCormick, John *Acid Earth* (1990); Pawlick, Thomas *Killing Rain*; Postel, Sandra *Air Pollution, Acid Rain, and the Future of Forests* (1984); Toribara, T Y et al *Polluted Rain* (1980).

Broader Endangered forests (#PC5165) Degradation of the atmosphere (#PD9413).
Narrower Deterioration of stained glass due to acid rain (#PE0082).
Related Ammonia as pollutant (#PG5273) Nitrates as pollutants (#PE1956)
Nitrites as pollutants (#PE6087) Sulphur dioxide as a pollutant (#PE1210)
Nitrogen compounds as pollutants (#PD2965) Particulate atmospheric pollution (#PD2008).
Aggravates Deforestation (#PC1366) Forest decline (#PC7896)
Water pollution (#PC0062).
Aggravated by Photochemical oxidant formation (#PD3663)
Maldistribution of electrical energy (#PD3446).

♦ PD4907 Over-diversification of manufactured goods
Over-diversification of services
Aggravates Inadequate standardization of procedures and equipment (#PC0666).
Reduces Insufficient diversification (#PD0335).

♦ PD4915 Harassment of public officials
Incidence A report issued in the US criticized the Reagan administration for a nationwide pattern of harassment of black politicians and government officials.
Broader Harassment (#PC8558) Intimidation (#PB1992).
Related Police intimidation (#PD0736) Harassment of journalists (#PD3036)
Intimidation of public officials (#PD4734) Fear of retaliation by authorities (#PF3707)
Threats against trade union leaders (#PD7471)
Official cover-up of government harassment of political activists (#PF3819).
Aggravated by Racial intimidation (#PC2936) Political intimidation (#PC2938).

♦ PD4916 Employment discrimination against the elderly
Denial of rights to equal opportunities of employment for elderly
Nature In most areas of the world, efforts by older people to participate in work and economic activities, both to satisfy their need to contribute to the life of the community and to benefit society as a whole, meet with difficulties. Age discrimination is prevalent: many older workers are unable to remain in the labour force or to re-enter it because of age prejudice. In some countries this situation tends to affect women more severely.

Where decreasing numbers of young people are coming into the job market, Britain, West Germany, France, Italy, the United States and Japan companies are starting to recruit retired people and people near retirement age. Many of these are part-time or job-sharing and are for the most part second class jobs, so the discrimination is becoming more sophisticated.

Refs ILO *Problems and Opportunities of Employment and Re-employment of Older Workers in Commerce and Offices* (1974).
Broader Age discrimination (#PC2541) Discrimination in employment (#PC0244).
Narrower Discrimination against women at retirement age (#PE6069).
Related Denial of equal benefits to elderly workers (#PE1625).
Aggravates Unemployment of older people (#PE5951).
Aggravated by Age discrimination in employment (#PD2318).
Reduces Denial of right to retirement (#PD4458).

♦ PD4918 Inhibited grief process
Frozen mourning — Inhibited bereavement
Nature The process of mourning runs through different stages. Typically they begin even before the death. Most deaths come slowly through illness, allowing time for emotional preparation. At this point some of those close to the dying person may be unable to acknowledge that death is near. These people can feel confused or inexplicable angry. In extreme cases, such people avoid the dying person, which can lead to intense remorse after the death. If the approach of death is acknowledged, it can present an opportunity to go over the events of life and reconcile any grievances. The emotional turmoil just after the death revolves around wanting to do something to protect or please the deceased. When this stage goes awry, the reaction to the death may involve panic, with the bereaved person overwhelmed to the point of incoherence or becoming dissociated in which they protect themselves through loss of recent memories. Next, mourners commonly enter a phase in which they turn away from their feelings by avoiding reminders of the death. Yet, the dead person may seem alive in dreams. This normal denial comes at an emotional cost: the mourners may feel numb to all emotions. But this is a necessary prelude in which they regain a sense of equilibrium that will allow them to confront the loss. At this stage some people go to extreme effort to put the death out of mind; sometimes they abuse drugs or alcohol or throw themselves into a frenzy of work, athletics or sexual activity. It is not until the next phase, when the mourners go through a mental review of their life with the deceased, that they actually begin to the loss. Extreme reactions at this point include recurring nightmares or even night terrors when they awake screaming. The person may also be flooded by overwhelming rage, despair, shame, guilt or fear. Once this stage is complete, an intense yearning for the company of the dead person ordinarily develops signifying a last ditch effort to deny the death. This yearning gradually yields to an acceptance of the death. For those who do not reach this point there may be a marked inability to work, to be caring or creative, or to even experience pleasant feelings. They may be plagued by anxiety, depression or rage, followed by shame or guilt.

Incidence The amount of time allocated to bereavement is increasingly determined by the personnel policies of enterprises. In the USA most enterprises set rigid standards (usually 3 days, often including any weekend) governing how much time an employee can officially grieve before returning to work, irrespective of the nature or extent of that grief. This is to be contrasted with the recommended formal mourning period for a widow in the USA in 1927 (3 years), in 1950 (6 months), and in 1972 ("within a week or so").

Counter-claim People are prone to being too judgemental of someone who does not seem to be distressed enough immediately after a loss or who stays upset for too long. Many people, while sad, do not exhibit intense distress. There is not compelling evidence that these people are denying the loss or did not truly love the person. They do not go out of their way to avoid thoughts of the loved one, nor are they unloving in recalling the relationship. The absence of extreme distress can be a sign of resilience. Many of these people have a set of beliefs that give them a broader, often spiritual, perspective that lets them see the loss in a way they can accept. Five different studies of widows and widowers in the US have found that between a quarter and two-thirds of those who are grieving are not greatly distressed. On the other hand, a study showed that people who lost a child or spouse in an auto accident were likely to be depressed and anxious years later.

Broader Psychological inertia (#PF0421) Psychological inhibition (#PF6339).
Related Emotional crises (#PE3407) Psychological conflict (#PE5087)
Psychological withdrawal (#PJ2329) Post-traumatic stress disorder (#PE0351).
Aggravates Anger (#PA7797) Shame (#PF9991) Guilt (#PA6793)
Anxiety (#PA1635) Nightmares (#PE6958) Bereavement (#PF3516)
Mental depression (#PC0799) Separation anxiety (#PE2401).
Aggravated by Selective perception of facts (#PF2453)
Excessive prolongation of the dying process (#PF4936).

♦ PD4925 Accumulation of pollutants in freshwater wildlife
Accumulation of contaminant residues in freshwater animals, fish and birds
Nature Environmental pollution is characterized by the accumulation of toxic metals, organochlorine residues and radionuclides in freshwater wildlife. Radionuclides are accumulated as a result of fallout from nuclear weapons testing and from nuclear reactor accidents.
Broader Accumulation of pollutants in plants and animals (#PD5021).

♦ PD4945 Military brutality
Incidence On a regular basis, Palestinians are being shot in cold blood, randomly killed and maimed, detained without trial and beaten and humiliated by soldiers acting on orders from the Israeli government.
Broader Brutality (#PC1987).
Related Military police abuse (#PG6627).
Aggravates Military atrocities (#PD1881).
Aggravated by Military reprisals (#PJ4986) Brutalization of military personnel (#PD7602).

♦ PD4952 War between socialist states
Nature Whatever disagreements among socialists and Marxists may have existed until recently, there was agreement that war between socialist states is inappropriate. This premise is no longer the case. This changes the conception of the international socialist movement from that of a single force moving in general harmony—although with different means—to that of a number of inimical forces competing for power. This transforms the perception of the conflict between socialist and capitalist states.
Broader Ideological war (#PC3431).

♦ PD4957 Xenophobia
Dependence on xenophobia
Nature This condition of disliking or fearing individuals or groups considered to be foreign may refer to 'groups' consisting of: an entire continent (as with anti-American or anti-European feelings); a neighbouring family of immigrants; or even migrants from another part of the country if regarded as intrusive. Xenophobia commonly takes an ethnic form and in its most extreme and widespread forms of expression may reflect the paranoid state of those in power, as it did with Hitler and Stalin.
Broader Fear (#PA6030) Exclusion (#PA5869).
Narrower Xenophobia with regard to migrant workers (#PC5017).
Related Hate (#PA7338) Narrowmindedness (#PA7306).

Aggravates Discriminatory communication (#PD6804)
Discrimination against foreign companies (#PD6417)
Biased media-image of foreign groups and peoples (#PE8802)
Discrimination against foreign nationals in the military service (#PE6422).
Aggravated by Undue attachment to a social group (#PF1073).

♦ PD4962 Unequal income distribution

Income inequality — Maldistribution of revenues — Inequitable distribution of revenues

Nature Distinct differences in income result from several factors, including sex, age and nationality. Inherited wealth and social position also play a part. In industrialized market economies, wage decisions determine the extent of inequalities between various categories of workers. In both developed and developing countries, income distinctions are affected by the policy decisions of employers, trade unions and governments.

Background The latter 18th and most of the 19th centuries witnessed the flourishing of ideas that advocated no significant interference in the process of production and income distribution. In their extreme, these ideas even favoured the existence of poverty on the grounds that it was a necessary condition for ensuring incentives to work.

20th century thought and policy have tended to reject extremes of equality and inequality of incomes. Most 'egalitarian' thinkers, for example, do not advocate complete equality of incomes as an immediate aim of policy. Marx and the Marxist philosophers deferred the idea, 'from each according to his ability, to each according to his need', to a later and higher phase of socialism. In several socialist countries efforts have been made to modify systems of basic wage tariffs with a view to achieving greater inter-branch uniformity and providing appropriate incentives for workers to acquire higher skills. Another innovation is team contract work whereby a team of workers enters into a contract to complete a specified project for a fixed sum of money. Modern 'non-egalitarian' thinkers, on the other hand, generally reject the idea that poverty is necessary and desirable and accept the proposition that a certain redistribution of income through public policy is desirable and feasible.

In developing countries the attack on unequal income distribution has been focused on raising often desperately low wages. The main instrument in this is statutory minimum wages. At the same time there has been a move away from highly selective and variegated approaches to minimum wage fixing and towards uniform structures of general minimum wages of broad coverage.

Counter-claim Ideas favouring complete equality of income have generally come to be regarded as unfeasible at least as an immediate aim of public policy. Such a distribution would imply a much less than optimum allocation of productive resources and have a serious negative effect on incentives to the acquisition of skills by workers and to the efficiency, drive and productive innovation of managers and owners; it would consequently result in a serious handicap to economic growth.

Refs Lecaillon Jacques, et al *Income Distribution and Economic Development* (1986); Nygård, Fredrik and Sandström, Arne *Measuring Income Inequality* (1982); Phelps Brown, E H *The Inequality of Pay* (1978).

Broader Human inequality (#PA0844) Social injustice (#PC0797).
Narrower Rural-urban income differential (#PE5022)
Unequal income distribution within developing countries (#PD7615)
Unequal income distribution in industrialized countries (#PE6891)
Increasing income disparity in developing countries due to transnational corporations (#PE1660).
Aggravates Urban poverty (#PC5052) Uncontrolled markets (#PF7880).
Aggravated by Fluctuations in real value of money (#PD9356)
Inappropriate education in developing countries (#PF1531).

♦ PD4966 Children in poverty

Child poverty

Nature Relationships have been established between poverty and high infant mortality rates, high child morbidity, wide-spread prevalence of protein-calorie malnutrition, and high rates of school abandonment. For those children faced with the reality of poverty, the consequences of their situation may become the factors that maintain and perpetuate it; they may even result in counteracting a child's mental development. Poor children grow up in conditions of poor health and malnutrition which affect their overall behaviour: reaction to stimuli is weakened; so is their mental activity, ability to concentrate, natural curiosity, and learning motivation.

The worst effects of economic tightening in the industrialized world are being passed on, often multiplied many times over, to the poorest nations, and within these nations, it is the poor, and especially their children, who are the hardest hit. In other words, the weakest members of the human race - young children in poor countries - are the ones left to bear the heaviest burden.

Incidence Latin America had a total of 65 million children under six years of age in 1980. Of these, 35 million lived in poverty. By the end of the century the number will be 97 million, of whom 51 million will be poverty-stricken. In the US one in six children lived below the official poverty line in 1989. More than 7 million American children have no health insurance. Some 100,000 are homeless on any given night of the year. Selective evidence suggests that in most countries significant sections of the children are suffering as one of the consequences of recent economic setbacks. For example, data from Zambia's poorer northern regions indicate that there has been a decline in height-for-age in all age categories up to 15 years old; in Latin America, the number of children treated for severe malnutrition in Costa Rica doubled in three years, while in the state of Sao Paulo in Brazil, there is a pattern of increasing low birth weight babies as well as a significant increase in the number of children given up by their parents because of poverty; even in the United States, infant mortality rates have increased in some areas. The number of children living in poverty in the UK has tripled in the past 20 years. Rises in infant mortality rates are documented for parts of India, Sri Lanka, Bangladesh, Brazil and Costa Rica; in Brazil the incidence of low birth weight and child abandonment has increased.

Broader Socio-economic poverty (#PB0388) Victimization of children (#PC5512).
Aggravates Child beggary (#PG1103) Child prostitution (#PE7582).
Aggravated by Inadequate care for children of prisoners (#PF0131).

♦ PD4974 Individualistic disposition of productive property

Private ownership

Nature The ownership, use and sale of productive property has become a right of the individual rather than a privilege. The context out of which decisions are made about ownership has been reduced to a tension between individual or corporate self-interest and benevolent good will. Both of which are often two sides of self satisfaction. Concern for the larger economy, global ecology or other transnational issues is a function of other pressures, like political awareness of the environment.

Broader Unregulated ownership of the means of production (#PF2014).
Aggravates Maldistribution of agricultural land (#PD9189).
Reduces Denial of the right to ownership (#PE8411).
Reduced by Insecure land tenure (#PD9162).

♦ PD4976 Riverine floods

Downstream flooding

Nature Riverine floods are caused by precipitation over large areas, or by the melting of the winter's accumulation of snow, or both. Riverine floods take place in river systems whose tributaries may drain large geographic areas and encompass many independent river basins. Floods on large river systems may continue for periods ranging from a few hours to many days. Flood flows in large river systems are influenced primarily by variations in the intensity, amount and distribution of precipitation. The condition of the ground - amount of soil moisture, seasonal variations in vegetation, depth of snow cover and imperviousness due to urbanization - directly affects flood runoff. Three characteristics of river channels - channel storage, changing channel capacity, and timing - control the movement of riverine flood waves. As a flood moves down the river system, temporary storage in the channel reduces the flood peak. As tributaries enter the main stream, the river gets larger and larger downstream. Tributaries are not of the same size nor are they uniformly spaced; therefore, their flood peaks reach the main stream at different times. The difference of timing tends to modify peaks as a flood wave moves downstream.

Broader Floods (#PD0452).
Aggravated by Deforestation (#PC1366)
Deforestation of mountainous regions (#PD6282).

♦ PD4977 Environmental hazards of nuclear power production

Proliferation of nuclear power — Unsafe nuclear reactors

Nature The nuclear fuel cycle consists of the process of mining and milling of uranium, conversion to fuel material, usually including enrichment in the isotope U-235, fabrication of fuel elements, utilization of the fuel in nuclear reactors, reprocessing of spent fuel and recycled utilization of recovered fissile materials, transportation of material between fuel-cycle installations, and disposal or radioactive wastes. Almost all the radioactivity associated with the fuel cycle is present in stored, spent fuel elements and in well contained fractions separated from the fuel during the reprocessing operations. However, at each step of the fuel cycle, releases of small quantities of radioactive material into the environment may occur, although these are only of local or regional concern because their half-lives are short compared to the time required for dispersion to greater distances. Some radionuclides, on the other hand, having longer half-lives or being more rapidly dispersed, can become globally distributed. In addition to the small releases of radioactive material during normal operation the possibility exists that additional amounts of radioactive materials may be accidentally discharged.

The problems of the proliferation of nuclear power include the disposition of long-term radioactive waste ('long-term' being without precedent in recorded human history - the half-life of plutonium-239, a significant and inevitable by-product of nuclear reactors, is 24,000 years); the location of power reactors in regards to surrounding human populations and the preparation for emergencies associated with accidental radioactive releases; and the ecological impacts from the use of salt, clay, and hard rock geological formations.

Incidence The Soviet nuclear disaster at Chernobyl is thought to have occurred through a blunder by the reactor's operators. Many have died as a direct result of the accident, many thousands more will die prematurely of cancers and birth defects induced by the radioactivity released there. A very large area of the Soviet Union has been severely contaminated, economic damage to the Soviet Union and surrounding countries is immense. Data from Chernobyl indicates that at least 30 per cent and probably more than half of the reactor's inventory of radioactive isotopes of caesium and iodine were released. The total installed nuclear electric generating capacity in the world in 1976 was 79.8 GW from 188 reactors operating in 19 countries. The predicted capacity for the year 2000 is 2000 GW.

As of 1981, there either existed, were in a state of construction, or were in advanced planning, 762 nuclear reactors spread throughout 42 countries. The most significant number of reactors include the USA with 174, USSR 73, France 70, Germany FR and German DR 53, UK 44, Japan 43, Canada 28, Spain 18 and Sweden 12.

Counter-claim 1. The net impact of Chernobyl is to make nuclear power very much safer throughout the world. Fission power should not be permanently removed from the array of energy-mobilizing options for the future. It is quite possible that reactors that are sufficiently resistant to catastrophic accident and sabotage can be designed, built and operated, and that satisfactory methods for disposing of nuclear wastes can be developed.

2. The extensive use of nuclear power may prove to be the most effective way of diminishing the risk of global warming by reducing the use combustion of fossil fuels that release carbon dioxide.

Refs Flavin, Christopher *Reassessing Nuclear Power* (1987); Shrader-Frechette, Kristin *Nuclear Power and Public Policy* (1980); United Nations *Environmental Effects of Cooling Systems at Nuclear Power Plants* (1975).

Broader Environmental pollution by nuclear reactors (#PD1584)
Environmental hazards from energy production (#PD6693).
Narrower Deterioration of nuclear power plants (#PE5260)
Hazardous locations for nuclear power plants (#PD2718)
Environmental hazards of decommissioned nuclear power plants (#PE7539).
Aggravates Nuclear accidents (#PD0771) Environmental pollution (#PB1166)
Hazards to human health (#PB4885).
Aggravated by Risks in power production (#PE4835)
Vulnerability of nuclear power sources (#PD0365)
Aggression against nuclear power sources (#PE0403).
Reduced by Insufficient nuclear power stations (#PD7663).

♦ PD4982 Deliberate lying by corporation officials

Misrepresentation of facts by corporate leaders — Deliberate distortion of corporate news and information — Corporate over-reporting and under-reporting — Perjury by corporation representatives — Fabrication of reports on corporate competitors — Corporate slander

Nature A company makes false and defamatory statements about it's competitor which are harmful to the competitor's reputation.

Incidence The global pharmaceutical companies and dozens of generic firms, that make and sell the less-expensive, generic versions of big companies' drugs, are engaged in a fierce competition over the markets. The big companies have underwritten scientific studies damning the safety of generic medicines, spread unsubstantiated tales of ill effects of these drugs.

Broader Domination (#PA0839) Institutional lying (#PD2686).
Aggravates Terminological deception (#PF5383).
Aggravated by Avoidance of negative feedback (#PF5311).

♦ PD5001 Inadequate enforcement of safety regulations

Unenforced safety ordinances — Non-reinforced safety factors — Disregard for safety principles and techniques — Inadequate safety precautions

Broader Fragmented planning of community life (#PF2813)
Discrepancies between principles and practice (#PF4705).
Narrower Unenforced sanitation codes (#PJ7979).
Related Inadequate national law enforcement (#PE4768)
Fragmented conduct of community operations (#PF1205)
Insufficient provision of public services for communication (#PF2694).
Aggravates Travel risks (#PD7716) Tourist hazards (#PE8966)
Inadequate prevention of disabilities (#PF0709)
Violations of health and safety regulations (#PE4006)
Export of hazardous industries to developing countries (#PE6687)
Inadequate legislation against environmental pollution in developing countries (#PE7141).

PD5001

Aggravated by Unethical practices of employers (#PD2879)
Unethical practices in transportation (#PD1012)
Unethical practices of regulatory inspectors (#PF8046).

♦ PD5021 Accumulation of pollutants in plants and animals

Accumulation of contaminant residues in plants and animals
Broader Environmental pollution (#PB1166).
Narrower Accumulation of pollutants in marine wildlife (#PD3934)
Accumulation of pollutants in terrestrial plants (#PD0381)
Accumulation of pollutants in freshwater wildlife (#PD4925)
Accumulation of pollutants in terrestrial wildlife (#PD5278).

♦ PD5025 Exploitation of women refugees

Nature Women refugees are often the least well equipped to cope with their difficult circumstances and, as about 30 percent of the female population in the refugee camps is of childbearing age, pregnancy can add intolerably to their other anxieties. These women have little or no access to family planning services or contraception and there is a high percentage of unwanted pregnancies and births, thus increasing the number of refugee children. Violations of their security, one of the most serious problems affecting refugee women, have included rape, physical violence, sexual abuse or harassment and prostitution. Single women and women heads of families are more liable to such violence. Lack of sanitation and privacy is another aspect of the specific hardships refugee women have to endure. In cities, most refugee women need to supplement the family income, particularly women who are heads of households (that is, who earn 50 percent or over of the total earnings). In many cities, opportunities for these women are extremely limited.
Incidence So far, Hong Kong has resettled (in the United States) more than half of all the 'boat people' who have arrived since 1978, but the problem is made more difficult by the large number of Vietnamese babies born in the refugee camps. In the last two years a total of 2,575 births has been recorded – an average of 21 per week. Of the 22,109 refugees remaining in Hong Kong in 1984, 9,212 (or 38 percent) were children under the age of 15.
Broader Refugees (#PB0205) Exploitation of women (#PC9733)
Violence against women (#PD0247).
Aggravated by Social injustice (#PC0797)
Denial of the right to work to refugees (#PE3751).

♦ PD5029 Airborne diseases

Nature Airborne diseases (tuberculosis, pneumonia, diphtheria, bronchitis, whooping cough, meningitis, influenza, measles, chicken pox) spread by breathing in the airborne respiratory secretions of infected persons, represent the second major cause of morbidity in developing countries. Several million persons per year die of acute respiratory diseases. Together with diarrhoea, this is the most common cause of death of children under five in developing countries.
Broader Airborne substances harmful to health (#PD2847).

♦ PD5033 Unpredictable barriers to trade

Unpredictable introduction of protectionist measures
Nature Strategies among developing countries for the diversification of their economies and for increases in manufactured exports instead of primary commodities have to be planned and implemented under adverse conditions. Sharp fluctuations in developed countries' activity together with uncertainty resulting from the frequency of the introduction of new, often discriminatory, protectionist measures complicates policy making as regards both the short and the longer term. Unpredictable barriers to trade discriminates mainly against those countries which had successfully embarked upon export-oriented industrial development and those which aimed at an increase in exports of primary commodities or of manufactures which would otherwise have been absorbed domestically.
Broader Trade protectionism (#PC4275).
Aggravates Economic uncertainty (#PF5817).

♦ PD5034 Proliferation of advertising

Advertising explosion — Advertising clutter
Incidence In the United States recent studies show that the typical customer is bombarded by 5,000 advertising messages a day, a total of nearly 2 million a year. Consumers remember only 1 to 3 percent without prompting. The number of ads is expected to increase steadily for the foreseeable future.
Broader Proliferation of information (#PC1298).
Narrower Proliferation of direct mail advertising (#PE1810).
Related Misuse of advertising (#PE4225).

♦ PD5045 Labour shortage in developing countries

Loss of labour force in developing countries
Nature A lack of persons who have skills, education and experience is typical in developing countries and critical for their economic and political development. The manpower shortages of developing countries fall into several categories:
There is likely to be a shortage of highly educated professional manpower such as scientists, agronomists, veterinarians, engineers and doctors. Such persons, moreover, usually prefer to live in the major cities rather than in the rural areas where in many cases their services are most urgently needed. Thus their shortage is magnified by their relative immobility and their skills are seldom used effectively. For example, graduate engineers may be found managing the routine operation of an electric power sub-station or doing the work of draftsmen, and doctors may spend long hours making the most routine medical tests. The obvious reason is that the shortage of technicians, nurses, agricultural assistants, technical supervisors and other sub-professional personnel is generally even more critical than that of fully qualified professionals. This is because the modernizing countries usually fail to recognize that the requirement for this category of manpower exceed by many times those for senior professional personnel. Also, the few persons who are qualified to enter a technical institute may also be qualified to enter a university, and they prefer the latter because of the higher status and pay which is accorded the holder of a university degree; and finally, there are often fewer places available in institutions providing intermediate training than in universities.
The shortage of top-level managerial and administrative personnel in both the private and public sectors, is almost universal, as is the dearth of persons with entrepreneurial talents. Teachers are almost always in short supply, and their turnover is high because they tend to leave the teaching profession if and when more attractive jobs become available in government, politics, or private enterprise. This shortage is generally most serious in secondary education, and particularly acute in the fields of science and mathematics. It is a 'master bottleneck' which retards the entire process of human resource development. In most modernizing countries there are also shortages of craftsmen of all kinds, senior clerical personnel such as book-keepers, secretaries, stenographers and business machine operators, and of other personnel such as radio and television specialists, airplane pilots, accountants, economists and statisticians.
Broader Labour shortage (#PC0592).
Aggravated by Abusive traffic in immigrant workers (#PD2722).

♦ PD5046 Imbalance in world food economy

World food crisis
Nature The major world food issue, previously posed as one of scarcity, has emerged in recent years as a major development issue, with widespread poverty as the real cause and not just a physical shortage of food. The problem is twofold. There is first the recurrent threat of famine and food shortages that arise from periodical fluctuations in production, leading to sudden changes in food prices and supplies. Secondly, and this is in a sense the real food problem, there is the chronic hunger and malnutrition of large segments of the world population. Even when the world as a whole has plenty of food, and grain prices are stable, at least 500 million are perpetually hungry and malnourished. Large numbers of countries that were once net exporters of foodgrain are becoming increasingly dependent on food imports. This imbalance in the world food economy has serious ramifications for both the food-deficit developing countries and the food-surplus countries alike. How these surpluses will be used or sold in the food-deficit countries is at the heart of the world hunger/food surplus paradox.
Incidence Many countries of the developing world are trying desperately to meet their food needs, while other countries of Europe, North America, Oceania and South America are producing bumper harvests of cereal grains and other food crops. Developing countries, with almost 75 percent of the world population, have only 55 percent of its cultivable land. The poorest 45 countries, with one-third of world population and the bulk of malnourished people, have only one-fourth of the cultivable land.
Counter-claim The world is producing more food that was ever before believed possible. Between 1971 and 1982 world agricultural output rose 25 percent, the output in the less-developed countries was up 33 percent and in developed countries it was up 18 percent. Per capita food production went up 16 percent in South America and 10 percent in Asia during that time. The rate of productivity growth continues to rise.
Refs Lawrence, Peter (Ed) *World Recession and the Food Crisis in Africa* (1986).
Broader Unequal global distribution of economic growth (#PC5601).
Aggravates Food insecurity (#PB2846)
Inadequate mechanisms for securing sufficient food supplies (#PF2857).
Aggravated by Unethical food practices (#PD1045).

♦ PD5076 Nuclear arms race

Nuclear rivalry between countries
Nature Relationships between the world's major military alliances are in danger of becoming more confrontational. There is a lack of political contact and communication among all the nuclear weapons powers. This blind nuclear arms race prevents securing any real progress on disarmament. In this context of heightened tensions and a continuing build-up of nuclear arsenals, the future of civilization as we know it could be threatened. No countries nor peoples would be insulated from that fate.
Claim No arms control or disarmament treaty currently envisaged will in any way restrain the nuclear arms race. Currently foreseen developments in nuclear weapon technologies will lead to perceptions that a considerable advantage, even a "nuclear victory" can be had from a pre-emptive nuclear attack. The acquisition of a first-strike nuclear capability will therefore considerably increase the risk of nuclear war even if there is an East-West détente.
Counter-claim If all nuclear weapons evaporated today, peace would not be the result. Political opponents would find alternative methods for fighting each other with chemical or biological weapons. The problem is neither the hardware not the race to produce more but the systems of political rivalry, the political, economic and social systems that support it and the psychological needs that it meets.
Broader Competition (#PB0848) Competitive acquisition of arms (#PC1258).
Aggravates Exterminism (#PF7401)
Obstacles to unilateral nuclear disarmament (#PF7052)
Government secrecy concerning nuclear weapons testing (#PF4450).
Aggravated by Surface to surface missiles (#PE4515).

♦ PD5077 Inadequacy of training for human settlements in developing countries

Nature Most developing countries lack institutions and personnel with adequate skills in human settlements and there is a shortage of trained personnel in almost every profession and skill involved in settlements development and management. While there are a number of training activities focused on human settlements, many of those involved in the settlement building process do not take part in such programmes. The gap is most pronounced at the grass-roots and intermediate levels, where training is concentrated almost exclusively in metropolitan areas. Even for high-level professionals, training opportunities are grossly inadequate: present training programmes are often divorced from reality, particularly in relation to the needs of low-income populations, and professionals are not equipped to deal with related social and environmental concerns. Agencies engaged in the various types of training focused on human settlements tend to perpetuate an uncritical transfer of basic concepts and training schemes from industrialized countries to situations that are inherently different and which require new approaches.
Broader Inappropriate education in developing countries (#PF1531).

♦ PD5078 Discrimination against minority languages

Nature The disuse of a minority language in teaching when this is the children's mother tongue is discriminating against the minority language population within a nation in that it denies the cultural identity of this minority population.
Incidence There are some 5000 minority languages worldwide all of which are in danger of extinction.
Broader Discrimination against minorities (#PC0582).
Related Dialect discrimination (#PF6016).
Aggravates Endangered cultures (#PB8613).
Aggravated by Prejudice against other languages (#PD8800)
Underprivileged linguistic minorities (#PC3324)
Excessive use of foreign programmes for media (#PE9643)
Discrimination against use of accents of a language (#PE5141).

♦ PD5081 Violence as entertainment

Nature Some persons are emotionally stimulated by viewing scenes of violence in entertainment, such as television films based on crime, horror movies or violent sports such as boxing or bull-fighting. The fear thus engendered can produce an intense emotional satisfaction.
Claim The viewing of violence as entertainment is dangerous for adolescents as violence can become their main emotional gratification and thus encourage delinquency. There seems to be growing evidence that TV violence induces a sense of paranoia and fear in the general public. There is 10 times more violence on US television, for example, than there is in life and this is leading large numbers of people – particularly women, the elderly, non-whites and disadvantaged citizens – to think of themselves as victims. It can be argued that when people think of themselves as victims, they act as victims, and thereby tend to be victimized.
Broader Human violence (#PA0429).
Related Culture of violence (#PD6279) Violence as a resource (#PF3994).
Aggravates Youth violence (#PF7498).

♦ **PD5086 Morbid preoccupation with death**
Thanatomania
Narrower Death instinct (#PF3849) Mind-induced death (#PF7918).
Related Fear of death (#PF0462).
Aggravated by Human death (#PA0072).
Reduces Avoidance of a confrontation with death (#PF1586).

♦ **PD5092 Deterioration of domestic food production in developing countries**
Nature In many developing countries policies are biased against domestic production and in favour of imports of staple foods, which are also often themselves subsidized, or the expansion of cash crops, which means increased area devoted to their cultivation. Some food aid policies and subsidized exports by developed countries compound the biases against sound agricultural development.
Broader Constraints to increased agricultural output in developing countries (#PD5114).
Aggravates Imbalance between agricultural exports and imports in developing countries (#PE4956).
Aggravated by Lack of agricultural machinery (#PF4108)
Inappropriate cash crop policy (#PF9187)
Dumping of food in developing countries (#PE0607)
Contempt for agricultural labour in developing countries (#PD1965)
Inadequacy of agricultural education in developing countries (#PE9096)
Neglect of agricultural and rural life in developing countries (#PF7047)
Domestic agricultural price policy difficulties in developing countries (#PE2890)
Unavailability of land for agricultural purposes in developing countries (#PE5024)
Insufficient fertilizers for agricultural development in developing countries (#PE4140).

♦ **PD5105 Restrictions on freedom of worship**
Broader Denial of right to manifest religion (#PF2850).
Related Denial of freedom of thought (#PF3217) Restriction of freedom of expression (#PC2162)
Denial of freedom of movement in communist systems (#PC3173)
Restrictions on freedom of movement between countries (#PC0935)
Denial of freedom of expression and thought in communist systems (#PC3174)
Restrictions on international freedom of movement for national advantage (#PD0351).
Aggravates Inaccessible places of worship (#PE6795).
Aggravated by Excessive government control (#PF0304).

♦ **PD5107 Pre-marital sexual intercourse**
Claim In many parts of the world abstinence from sexual activity before marriage is valued. In other societies adults (and many adolescents) are unwilling to sacrifice the principle of abstinence even if general practice is different.
Refs Kirkendall, Lester A *Premarital Intercourse and Interpersonal Relationships* (1984).
Broader Fornication (#PF5434).
Narrower Adolescent sexual intercourse (#PD7439).
Aggravates Cohabitation (#PF3278) Paternal negligence (#PD7297)
Adolescent pregnancy (#PD0614).
Aggravated by Excessive portrayal of sex in the media (#PE7930).
Reduced by Early marriage (#PE7628).

♦ **PD5110 Inadequate empolderment of wetlands**
Inadequate empolderment of coastal lowlands — Inadequate empolderment of peat land
Nature Initially, wetlands are empoldered to make the land productive for agriculture, or to establish new human settlements, or both. Wetlands are empoldered in marine flood plains below mean sea level or in river flood plains at higher elevations. In their natural state, higher-lying wetlands serve as flood regulators. They absorb water during wet periods and release it slowly in times of drought. Empoldering destroys this function, and results in a much sharper fluctuation in river levels. Sudden spates become more frequent, the volume of river discharge increases, and the farms and townships in the lower reaches of the valleys are faced with new threats of flooding. To correct this, the river is often trained by embankments, its channel is straightened, and its bottom dredged to help the water get away. Riverside communities thus have forced upon them a stereotyped landscape with fewer natural amenities than before and an impoverishment of plant and animal life.
The same effects are felt within the empoldered land. The elimination of higher-lying wetlands results in a marked lowering of the river level in dry seasons. This can jeopardize the supply of fresh river water to those needing it. In some places, this problem is met by building reservoirs, which is a costly solution and may involve the loss of valuable land. In other places, reliance is being placed on boreholes and wells. At some of these sites, however, the groundwater resources are quickly being depleted. The replenishment of groundwater depends partly on the presence of wetlands, where the water can infiltrate into the soil. If these areas are eliminated by empoldering, the water will no longer permeate, hence groundwater reservoirs will not be recharged.
A river flowing between empoldered lands is unable to expand laterally in times of flood. As a consequence, its flow will accelerate, leading to a possible scouring and deepening of the river channel. After the flood has receded, the river, in its deepened channel, will be at a lower level than before. This can cause the empoldered lands to drain excessively into the river, or can lead to the intrusion of a salt-water tongue far upstream into the river. A river in flood carries a heavy load of silt. The low-lying lands, deprived of their natural enrichment, may then need costly applications of artificial fertilizer. In the meantime, the flood waters carry the silt to the sea, where it is thrown down as banks and bars at the estuary, encumbering the channel and comprising a hazard to navigation. Lower river levels can also affect navigation. They make it necessary to build weirs and locks to raise river levels and so maintain river traffic. Lower river discharges sometimes lead to higher concentrations of pollutants from industrial or domestic waste, high enough in many cases to constitute a threat to fisheries and public health.
The empolderment of coastal lowlands with peat or potentially acid sulphate soils is a hazardous undertaking. Peat soils are extremely unstable when not kept wet. They may shrink or dry irreversibly, offering agricultural land of poor quality. They may even disappear altogether through oxidation upon exposure to air and sunshine. Potentially acid sulphate soils can become extremely acidic upon exposure to oxygen, rendering agriculture virtually impossible. Such soils, under natural conditions, provide gathering grounds for natural products, and breeding grounds for fish and sea animals. The empolderment of such lands may damage these assets with little economic return.
Claim The occasional empolderment of a wetland need not necessarily have a severe effect on the environment. The increasing world population, however, and the growing demand for agricultural produce have led to massive empolderings that have upset hydrological and ecological equilibrium. In many cases, the ill-effects are still accumulating, since the cure to one problem is often the cause of several more.
Aggravates Vulnerability of wetlands (#PC3486)
Unsustainable development of coast zones (#PD4671).

♦ **PD5111 AIDS**
Acquired human immunodeficiency syndrome — HIV - human immunodeficiency virus — ARC - AIDS-related complex
Nature The acquired immune deficiency syndrome (AIDS) is caused by HIV-virus manifested by opportunistic infections and/or malignancies, and the mortality rate is very high. The syndrome results from a breakdown of the body's disease-fighting mechanism that leaves it defenceless against infections, for instance Pneumocystis pneumonia and Kaposi's sarcoma. No effective treatment is available. A striking feature of AIDS is the wide spectrum and frequency of infections with life-threatening pathogens seldom seen in normal hosts. The illness may begin with insidious signs and symptoms, and the process may be more diffuse than when the same conditions are seen in other immune-compromised patients. These findings are consistent with a limited immune response and an inability to contain infection.
Four patterns of disease occur in AIDS patients. The pulmonary pattern, the central nervous system pattern, the gastrointestinal pattern, and the pattern of fever of unknown origin. Most patients who recover from a given opportunistic infection subsequently either have a relapse or develop a new type of infection. Many patients continue to have a wasting syndrome and experience such infections as oral thrush. Feelings of depression and isolation are common among AIDS patients and can be intensified if health care workers display fear of the syndrome.
Incidence Cases of acquired immune deficiency syndrome, first identified in the United States in 1981, have now been reported all over the world. By the end of 1989 a cumulative total of 203,599 cases of AIDS has been reported from 152 countries to World Health Organization. Taking into account the effects of less than complete case detection and reporting, as well as reporting delay, the world-wide cumulative total of AIDS cases was thought to be closer to 600,000. World Health Organization has estimated that by the end of 1991 this cumulative global total of AIDS cases could reach one million or more; at least six to eight million persons are infected with HIV-virus worldwide and the cumulative total will be 15 to 20 million HIV-infected persons by the year 2000. It is spreading rapidly and is switching from the disease of the industrialized world to the disease of the developing nations.
In industrialized countries, AIDS has mainly stroke the homosexual male who takes many partners, as well as the intravenous drug user. However, in cases from the Caribbean and Equatorial Africa, the ratio of female patients to males is much higher than that in the United States and Western Europe; the mode of transmission appears to be different. It is possible that sexually transmitted infections that cause genital ulcers, like syphilis or herpes, facilitate the transmission of the AIDS virus.
Counter-claim AIDS has become the disease of the decade, while other diseases, many of which cause more deaths and suffering, have been pushed to the background.
Refs Alexander, Nancy J, et al (Eds) *Heterosexual Transmission of AIDS* (1990); Bailey, M R, et al (Eds) *The Global Impact of AIDS* (1989); Bint *AIDS and AIDS-related Infections*; Pawlowski, A; Seminara, D and Watson, R R (Eds) *Alcohol, Immunomodulation and AIDS* (1990); Smith, Roberts A *HIV and Other Highly Pathogenic Viruses* (1988).
Broader Epidemics (#PC2514) Stigmatized diseases (#PD7279)
Sexually transmitted diseases (#PD0061) Infectious and parasitic diseases (#PD0982).
Narrower Vulnerability of children to AIDS (#PE4276)
Risk of contracting AIDS through kissing (#PJ5441).
Related Immuno-deficiency virus (#PD4747).
Aggravates Suicide (#PC0417) Distrust (#PA8653) Infectious revenge (#PD5168)
Defective human immunity system (#PE3355) Invasion of privacy through testing (#PJ6946)
Discrimination against HIV-infected persons (#PE4299).
Aggravated by Crack (#PE2123) Herpes (#PE8615)
Syphilis (#PE2300) Sexual mutilation (#PD5718)
Female sexual mutilations (#PE6055) Ignorance concerning disease (#PD8821)
Vice and sex traffic offences (#PD8910) Abuse of tourism for sexual purposes (#PE4437).

♦ **PD5114 Constraints to increased agricultural output in developing countries**
Lack of incentives for the increase of agricultural output in developing countries — Vulnerability of agriculture in developing countries to future declines in production
Nature Major constraints to increased agricultural output in the developing world are the unavailability of adequate land and a lack of incentives for farmers to produce more. In many developing countries, small farmers produce the basic food crops; and it is not possible to increase food production significantly without giving them priority in the distribution of land, water, credit, fertilizers, energy and other production inputs. Although some countries could increase output by concentrating on larger modern farms, this would not benefit the income or employment prospects of small farmers.
In many developing countries governments have manipulated the level of food imports and food prices in favour of urban dwellers. This has often occurred at the expense of fair returns for the rural population. Cheap food policies and large imports of wheat have resulted in increased imbalance between rural and urban incomes and discourage farmers from growing more food. As a whole, the developing countries' agricultural sectors need improvements in transport, housing, water and education. In many cases national policies discriminate against agriculture and discourage production.
Incidence On the basis of movement of surplus potential food production within countries, 55 countries (out of a total of 117 studied) have insufficient land resources to meet the food needs of their 1975 population with low level of inputs. The number of critical countries would rise to 65 by the year 2000. Of these critical countries, for the year 2000, 29 would need to raise their inputs to at least intermediate level, on all their potentially cultivable land, if they are to meet food requirements from their own land resources. A further 17 would need to raise inputs to the high level to attain food self-sufficiency. 19 countries will be unable to meet their food needs, from national land resources, even with high level of inputs.
Preliminary land degradation assessments indicate that, unless conservation measures are introduced on all cultivable land, 544 million hectares of potentially productive rainfed cropland - more than one-sixth of the total - could be lost.
Claim Assuming no movement of surplus potential food production and labour among individual-country length of growing period zones, 2,480 million hectares - 38 percent of the total land area - are unable to produce sufficient food for their 1975 populations with low inputs. In these zones, no less than 1,156 million people - 58 percent of the total - were living on land resources able to sustain only 597 million people.
In evaluating output increases or decreases accompanying land reform, the influence of farm price levels on investments and output cannot be disputed: the level of product prices influences the amount produced. Farmers may shift from one crop to another, or may decrease the use of inputs given lower farm product prices (or anticipated lower prices). Land tenure arrangements influence farmer response to changing prices. An FAO (1963) study concludes that price response was usually greater among owner-operators than among tenants. Tenants paying a fixed rent were likely to benefit more from price incentives and therefore to show a greater response to price changes than sharecroppers. Producer price policies were therefore generally more successful where they had been preceded by land reform measures. Therefore perhaps more important than prices per se, under most circumstances of agricultural development in the world today, is the incentive structure provided by the tenure system.
Counter-claim The lands of the developing world as a whole (excluding East Asia) are capable of producing sufficient food to sustain twice their year 1975 population and one and a half times their year 2000 population, even with low level of inputs. With application of intermediate level of inputs to all cultivable areas, these lands would be able to meet the food needs of more than four times their projected population of 2000. These aggregated developing world findings presuppose massive and unrestricted movement of surplus potential food production and labour

within and between all five regions.
Narrower Deterioration of domestic food production in developing countries (#PD5092).
Aggravates Stagnated development of agricultural production (#PD1285).
Aggravated by Desert advance (#PC2506)
Neglect of agricultural and rural life in developing countries (#PF7047).

♦ PD5115 Unequal employment opportunities for women
Insufficient female employment
Nature Most countries throughout the world do not offer equal employment opportunities for women; this limitation seems to stem from traditional ideas about women's capabilities.
Incidence While women in developed as well as developing countries find few points of access to the traditionally male sectors of the economy, the increasing number of American women in managerial and executive positions is a hopeful precedent.
Refs ILO *Equality of Opportunity and Treatment for Women Workers* (1975); Mies, Maria *Patriarchy and Accumulation on a World Scale* (1986); OECD *Equal Opportunities for Women* (1979); Rosen, Ellen I *Bitter Choices* (1987).
Broader Limited employment options (#PF1658)
Discrimination against women in employment (#PD0086).
Narrower Discrimination against black working women (#PE6245).
Related Unequal pay for women (#PD0309).

♦ PD5119 Delay in administration of medical care
Hospital waiting lists — Delayed surgery
Incidence In the United Kingdom, in 1986, over 661,000 people were on waiting list for hospital treatment. Almost 40,000 people had been waiting for over a year for non-urgent orthopaedic operations and 10,000 had been waiting for over a month for urgent orthopaedic operations including hip replacements. In some districts 95 percent of non-urgent patients must wait over a year for treatment. Some official believe that some people on waiting lists no longer need treatment because they have moved or have died.
Broader Inadequate medical care (#PF4832).
Related Delay in delivery of requested services (#PE8157).
Aggravates Limited psychiatric out-patient care (#PE0540).
Aggravated by Inadequate health services (#PD4790)
Maldistribution of health personnel (#PF4126)
Unethical practices of health services (#PE3328)
Excessive waiting times in government facilities (#PF5120).

♦ PD5122 Inadequate assistance to victims of human rights violations
Nature There are many international mechanisms to judge the perpetrators of human rights violations, but few instruments to guarantee support for the physical and mental rehabilitation of the victims of atrocities. They are totally forgotten, but the violations have had an lasting effect on the mental and physical health of the victims and on the the structure of the society itself.
Broader Legal discrimination in favour of offenders (#PD9316).
Narrower Neglect of victims of crime (#PD4823)
Inadequate assistance to victims of rape (#PE4449)
Inadequate assistance to victims of torture (#PE6936)
Lack of legal protection of extortion victims (#PE8240)
Inadequate assistance to victims of accidents (#PE4086)
Inadequate assistance to victims of abuse of power (#PE7390)
Non-payment of compensation to victims of motor accidents (#PE8824).
Related Neglect of dependents of war victims (#PD2092).

♦ PD5129 Forced disappearances of children
Political abductions of children
Incidence In Argentina between 1976 and 1983 under the military dictatorship hundreds of children disappeared. Most of these children had been born or were presumed to have been born of mothers who were themselves missing and reportedly held in secret detention centres at the time of their delivery. Military did not want the children of "subversive elements" to be returned to their families lest they grew up in the same moral and political climate as their parents. In some cases children were executed and their remains hidden. On the other hand, appropriations had been carried out systematically by some members of military. Many of the children found after their parents' disappearance were abandoned, or left in hospitals or orphanages, in the hands of neighbours, given to adoptive parents who either knew the circumstances or was one of the captors or handed over to their grandparents.
Refs Moorehead, Caroline (Ed) *Betrayal* (1989).
Broader Victimization of children (#PC5512) Forced disappearance of persons (#PD4259).

♦ PD5133 Misuse of grassland and rangeland
Deterioration of rangelands and grasslands
Nature Traditional grazing methods entailing the minimum of grazing are questionable, as too are traditional methods of grass conservation, including hay making. Uncontrolled grazing must essentially be wasteful of feed, for at peak periods of growth the pasture becomes over-mature, and this results in poor quality fodder of low value in terms of animal production. The loss of nutrients in hay as normally cured on the ground is again high and the whole process is therefore wasteful. Hay of a kind can be made even in bad weather, but the labour required for making it is excessive and the loss of nutrients serious. Overexploitation of grazing lands triggers soil erosion, desertification, and other processes of degradation. Productivity is quickly diminished, with far-reaching consequences on the local and national economy and on the well-being of the peoples concerned. Traditionally, livestock production is the main use of these areas, but other considerations, such as tourist value of these spacious lands, are also important.
Incidence Grazing lands cover about one third of the world's land surface. They include many areas in arid and semi-arid regions, as well as mountainous and high altitude zones which are too steep or too hot, too cold or too dry for intensive cultivation. These lands have low productivity per unit area, and are inherently fragile.
Background Grass is the foundation of any sound agricultural system. Existing knowledge with regard to botanical composition, fertilizer and management treatments of grassland is not used to its full extent, thus inhibiting increases in animal production and the improvement of grass production as related to its very seasonal growth pattern. Present-day techniques of grassland evaluation through the animal are not wholly adequate.
Broader Agricultural wastes (#PC2205).
Related Overstocking (#PC3153).
Aggravates Desert advance (#PC2506).

♦ PD5136 Displaced children
Refugee children
Nature In the crisis of refugee dislocation or unwanted migration, it is children who suffer most severely. The sudden displacement of a family and its transfer to a refugee camp or temporary settlement can combine to disrupt the child's security, interrupt his schooling, expose him to serious health hazards and malnutrition, and mar his sense of confidence in his fellow man – all at a critical stage of his intellectual, moral and physical development.
Refs Moorehead, Caroline (Ed) *Betrayal* (1989).
Broader Refugees (#PB0205) Displaced persons (#PD7822)
Victimization of children (#PC5512).
Aggravates Family breakdown (#PC2102) Maladjusted children (#PD0586).
Aggravated by Inadequate child welfare (#PC0233)
Inadequate rehabilitation facilities (#PD1089)
Children engendered by occupying soldiers (#PD8825)
Vulnerability of women and children in emergencies (#PD1078).

♦ PD5139 Blindness in developing countries
Nature The causes of blindness in developing countries ae still mainly infectious and nutritional disorders, such as trachoma, river blindness and vitamin A deficiency.
Incidence A comparison of blindness rates between developing countries and highly industrialized, developed countries shows that the rates are consistently higher, often ten to twenty times in the former. An average representative blindness rate for a developed country with good medical facilities, including eye-health care, is approximately 0.1 – 0.2 percent, using the internationally accepted definition of blindness. The corresponding figure in many developing countries in Africa or Asia is 1 – 2 percent or even higher. The explanation for such pronounced differences is complex, depending both on the causes of blindness prevailing, and the preventive and curative measures that are being undertaken to combat blindness.
Broader Physical blindness (#PD0568) Tolerated atrocities (#PC4710).
Aggravated by Inadequate health services (#PD4790).

♦ PD5142 Imprisonment
Confinement — Detention — Ineffectiveness of correctional institutions
Nature Prisons, penitentiaries, reformatories or correctional institutions, are virtually the only form of legal punishment today. Those who overstep the laws of society are no longer deported, publicly humiliated, or inflicted with brutal corporal punishments, and rarely are they executed. They are shut away from society in prisons which, with few exceptions, provide a punitive, negative environment, in which offenders serve out their terms in a state of demoralizing idleness.
The traditional, reforming role of imprisonment is being increasingly questioned, and the morale of prison staff has suffered as a result. The initial ideal of rehabilitation has fallen prey to all the complexities of the modern crime problem. With the enormous increase in criminal activity, prisons are over-crowded. They are often little more than warehouses of despair where unhealthy and inhumane conditions erupt in violence, rioting and insurrection. With two or three people crowded into the space intended for one, there is little incentive for introspection and reflection. There are other conditions that militate against rehabilitation. For example, the deprivation of personal security, of mobility and of privacy. In some institutions there is never quiet. In others the lights are never turned off. The link with the outside world is often tenuous, with visits limited and correspondence censored, delayed, or sometimes thrown away altogether, according to the inclination of the current administering officials. These seemingly petty matters can cause extreme psychological damage.
Background Imprisonment on a large scale is a relatively recent innovation. The first true penitentiaries (as opposed to penal institutions used solely for punishment) were constructed in America in the 1700s and were the product of two religious streams of thought – the Puritans in New England, who, in line with their prevailing ethic, established workhouses; and Quakers in Pennsylvania who constructed for the first time large conglomerates of solitary cells, seeking to offer 'wrong doers' the time and opportunity for introspection and reflection upon their deeds. Both groups, particularly the Quakers, viewed this new concept of punishment as a means to the larger end of rehabilitation and reformation. The ideas of imposed labour and isolated confinement were proposed with the best of intentions, and were visioned as constructive tools to lift a person from the environment of crime and propel him or her towards a change of heart and direction.
Claim Prison is the great leveller; the mass murderer and the petty thief are often confined together. Prisons teach wariness, distrust and cynicism and put the prisoner on constant guard against the unexpected. Prison life is grim, violent and endlessly boring.
Counter-claim Rule 1 of the prison service reflects a specific ideal: 'The purpose of the training and treatment of convicted prisoners shall be to encourage and assist them to lead a good and useful life'.
Refs Bottoms, Anthony E and Light, Roy *Problems of Long-Term Imprisonment* (1987); Kesse-Adu, K *Politics of Political Detention*; Malik, S Surendra *Supreme Court on Preventive Detention from 1950 to Present* (1985); National Center for State Courts *Alternatives to Incarceration* (1981); Tomasic, Roman *Failure of Imprisonment* (1979).
Broader Political repression (#PC1919).
Narrower Wrongful detention (#PD6062) Detention of mothers (#PG4924)
Unlawful imprisonment (#PD4489) Needless incarceration (#PE5112)
Denial of rights to prisoners (#PD0520) Inadequate correctional systems (#PF5172).
Related Inadequate prevention of crime (#PF4924).
Aggravates Religious extremism (#PF4954) Repression of intellectual dissidents (#PD0434).
Aggravated by Unemployment (#PB0750) Inadequate education (#PF4984)
Violent repression of demonstrations (#PD4811).

♦ PD5156 Land subsidence
Nature Subsidence is the lowering or collapse of the land surface either locally or over broad regional areas. Subsidence is usually not spectacular or catastrophic in itself but can cause great economic losses. It is caused by a large number of natural and man-made activities. Natural processes causing subsidence include: the dissolving of limestone and other soluble materials; earthquakes; and volcanic activity. Man-induced subsidence occurs mainly with the withdrawal of oil, gas or water; and has increased dramatically since 1940. Because underground fluids fill intergranular spaces and support sediment grains, removal of such fluids results in a loss of grain support, reduction of intergranular void spaces, and compaction of clays. The land surface commonly subsides wherever widespread sub-surface compaction has taken place, causing damage to canals, aqueducts, sewer systems and pipelines, and increasing the probability of flooding in some areas.
Incidence Land subsidence causes several tens of millions of dollars in damages annually in the United States. The Houston TX and Beacon Hill area of Boston MA, are undergoing vastly detrimental and expensive changes due to subsidence.
Broader Decreasing land mass (#PF7435).
Narrower Subsiding coastal areas (#PD3775).
Aggravates Hazardous locations for nuclear power plants (#PD2718).
Aggravated by Earth surface faulting (#PE5096).

♦ PD5157 Endangered animal and plant life due to radioactive contamination
Nature The harmful effects of ionizing radiation apply to all living organisms, but the sensitivity to radiation varies from species to species over a wide range. Generally, the higher the species on the evolutionary scale the greater the sensitivity. In plants the sensitivity appears to be related to the volume occupied by the chromosomes in the nucleus of the cell; the larger the chromosome volume, the smaller the dose required to produce a given degree of damage. As a consequence of this, after exposure to a high dose of radiation some animal or plant species will suffer much more than others and this may seriously upset the ecological balance. For example, the killing of birds may result in a large increase in insect populations, which are much less sensitive to radiation, and this in turn would cause enormous damage to plants. A rapid increase in

DETAILED PROBLEMS PD5212

insect population, particularly disease vectors, would also have serious effects on the health of both man and animals.
Incidence The main effect of fall-out on animals is to cause external exposure by the gamma rays emitted from radioactive substances, but the consumption of contaminated grass would in addition produce an internal exposure from the beta-rays. Beta-burns may also occur if fall-out particles remain on the skin of the animal. They could cause mucosal burns in the mouths of ruminants, leading to starvation. As in man, whole body exposure of animals to doses in the range of 1-10 Gy may result in death within a few weeks. The LD-50, that is, the dose which will cause 50 percent mortality in the irradiated animals, varies between the species of domestic animals, being lowest in sheep and highest in poultry.
The main difference between animals and plants is the larger range of sensitivities to radiation observed in plants. Different species may differ in their sensitivities by a factor of 500; if algae are included the variation may be 5,000 fold. Moreover, a given species may itself have a wide range of sensitivities, up to a factor of 50, depending on the different stages of growth. Apart from the dose itself, the effect of radiation on plants may be influenced by many environmental factors. In the case of fall-out, and additional factor is the season of its occurrence; for example, food crops irradiated in the seedling stage will be exposed for a longer time and will therefore receive a larger dose of radiation than if the fall-out occurred near harvesting time. On the other hand it would be impossible to gather in the harvest if the fall-out came down at that time. Exposure to large doses of radiation will kill plants. The lethal doses are much higher than for animals: for food crops, even for the most sensitive plants, the LD-50 is about 10 Gy and it goes up to about 200 Gy. At smaller doses the effects of exposure are reduced yield and height; both are dose-rate dependent, the effect being smaller the lower the rate at which the dose was delivered. Yield is more severely affected in the early reproductive state. Plants with a growing season limited by climatic conditions may produce no yield at all, even if they survive. Flowering and ripening of fruit is delayed by exposure to radiation. Exposure of seeds produces mutations, most of which are deleterious. Among trees, conifers are very sensitive to radiation, whereas deciduous trees are less sensitive. The LD-50 values for exposed trees range from 20 to 100 Gy. Grasses are more radiation-resistant and a dose of 200 Gy is needed to destroy grassland.
Broader Radioactive contamination (#PC0229)
Endangered species of plants and animals (#PB1395).
Narrower Radioactive contamination of plants (#PD0710)
Radioactive contamination of animals and animal products (#PD1119).
Related Denial to animals of the right to a natural death (#PE8339)
Denial to animals of the right to the attention, care and protection of humankind (#PF5121).

♦ PD5168 Infectious revenge
Intentional spread of sexually transmitted diseases
Nature Intentionally infecting others with diseases such as AIDS and herpes. Revenge against society or a individual has lead people to bite and spit, to sell blood, to share needles and have sex with as many people as possible to infect them. Some male and female prostitutes have spread the disease because they do not care what happens to the client. Continuing with an active sex life for some is a way of denying the fact they have the disease.
Broader Juvenile delinquency (#PC0212) Intentional infecting with disease (#PD2651).
Related Attempted murder (#PG0411).
Aggravates Distrust of interpersonal relationships (#PF4274).
Aggravated by AIDS (#PD5111) Herpes (#PE8615)
Unsafe sex (#PE9776) Sexually transmitted diseases (#PD0061).

♦ PD5173 Inadequately heated shelters
Lack of physical warmth — Poorly heated homes
Broader Inadequate housing (#PC0449).
Aggravated by Homelessness (#PB2150).

♦ PD5174 Crop shortfalls
Incidence Over the world as a whole, cereal production rose from 1,315 million tonnes in 1971 to 1,596 million tonnes in 1978; the production of pulses, fruits, nuts, total meat and milk also increased. In 1979, however, world cereals production fell, due largely to shortfalls caused by droughts and adverse climate in the USSR, South Asia, and many African countries.
Aggravated by Monoculture of crops (#PC3606) Vulnerability of food chains (#PB2253).

♦ PD5176 Inadequate trade between developing countries
Broader Weakness in trade among developing countries (#PC0933).
Narrower Inadequate trade in agricultural commodities between developing countries (#PE4523).
Aggravated by Bipolarization of trade between developed and developing countries (#PE4190).

♦ PD5177 Inhibited human physical growth
Short people — Inadequate human physical height
Claim Short children are more likely to be teased and bullied and start failing at school. Later they find it harder to get jobs and to adjust socially as adults.
Broader Human physical genetic abnormalities (#PD1618).
Narrower Inhibited growth of malnourished children (#PE4921).
Related Dwarfism (#PE2715) Infant growth failure (#PE6909).
Aggravated by Turner's syndrome (#PG3943)
Discrimination against dwarfs and midgets (#PE2635).

♦ PD5178 Vulnerability of marine environment to catastrophic warfare damage
Nature Various marine warfare activities constitute a series of threats to the marine environment, the most serious one being the SNS, the strategic nuclear submarine. Nuclear-powered submarines and surface ships routinely release a certain amount of radioactivity into the sea; the quantity is relatively insignificant, when compared to the potential contamination that could result from an accidental or intentional sinking. The long-lived isotopes produced in a reactor build up while a submarine is in continuous operation. If a nuclear submarine were to sink after about fifty days of continuous operation, it could contaminate the sea with long-lived radioactivity equivalent to what a twenty-kiloton atomic bomb would release (the Hiroshima bomb was thirteen kilotons). This calculation does not account for the possibility that nuclear warheads aboard the vessel could be detonated or that, in the case of an intentional sinking, nuclear-armed torpedoes or missiles might have been the enemy's offensive weapon.
Other effects of warfare on the undersea world are: (1) Undersea explosions. These take a serious toll on marine animals, especially ray-finned bony fish, which comprise about 95 percent of the fish in the sea. The air bladders of these fish are easily ruptured by underwater explosions. In addition, some of the materials used in explosives are poisonous. (2) Oil. A threat is posed by the enormous increases in tanker size since the 1940s, and the emplacement of vulnerable offshore oil platforms around the world: oil contamination of the sea could be significant in a major war. (3) Herbicides. As a result of massive herbicide spraying of mangrove forests along the coast of South Vietnam during the Vietnam War, coastal habitats were devastated. Scientists surveying offshore areas found that the loss of these marine nurseries and breeding grounds had caused a severe decline in the populations of fish, planktonic organisms, and shellfish. Recovery, they estimated, could take more than a hundred years. (4) Testing of nuclear weapons. Since the 1940s these have been associated directly with the sea. More than 1,000 nuclear bombs have been detonated; of these, 373 were exploded in the atmosphere (about 80 percent of the long lived components of these blasts probably reached the sea), 35 were exploded along the sea surface, and 6 were detonated undersea.
Incidence The importance of the ocean to military strategists is actually on the increase. Since the 1960s there has been a significant acceleration in the growth rate of the fleets operated by the world's 51 navies. Naval stocks - a measurement combining the number of vessels, their tonnage, and their military capabilities - have doubled since World War II. There are presently about 2,300 large and small naval vessels afloat and about 260 large nuclear submarines in operation (about 240 of them US or Soviet). As this fleet expands, the possibilities for accidents multiply. At least six nuclear-powered submarines have been lost. Two US nuclear subs have sunk in the Atlantic (the Thresher in 1963 and the Scorpion in 1968), and it is believed the Soviets lost as many as four between 1968 and 1971, two in the Atlantic, one in the Mediterranean, and one in the Pacific.
If a major nuclear war erupts, the instruments of greatest violence will probably emerge from the sea. The effects of wartime nuclear contamination in the sea could be devastating. Long-lived radioactive products would be distributed by ocean currents throughout the sea. Several of such products are taken up by marine organisms and distributed through the food web, especially strontium-90 (which follows the same course in organisms as calcium) and caesium-137 (which acts like potassium). Plutonium is also taken up by many marine organisms, some of which concentrate it to levels from a thousand to as much as ten thousand times higher than its initial concentration in seawater.
Claim Nuclear ships sinking, or nuclear weapons being used as torpedoes and as depth charges, could diminish life in the sea, disrupt delicate balances, contaminate ocean fisheries, perhaps even warp the nature of undersea life indefinitely.
Counter-claim Despite the potential for marine destruction that is associated with the world's military forces, there are at least two benefits: (1) a substantial proportion of the ocean research under way around the world is supported by navies and carried out by naval scientists; and (2) during times of war, some detrimental ocean activities cease. After World War II, for example, fish catches along the Atlantic coast of Europe were three times greater than before the war: the lack of access to fisheries had allowed stocks in these intensely exploited areas to build up.
Broader Marine pollution (#PC1117) Environmental degradation (#PB6384).

♦ PD5183 Misuse of classified communications information
Leaking of official secrets — Leak of military data — Premature disclosure of government reports
Nature Leakage of restricted government information can impair state security, hinder development and weaken government policies, whether it is leaked to a foreign power, to the public via the press or other media, or to private business interests. Leakage may arise out of inefficient security measures or from complicity of those with a legitimate access. A leakage of official secrets to a foreign power may go undiscovered for a comparatively long time if complicity is involved and if knowledge of the information is not sufficiently distributed for supervision to be effective. Leakage of industrial secrets can affect whole populations in several countries for generations.
Claim In order to protect the state's economic power, foreign security and defence capabilities, it is the duty of every citizen who has access to classified information to keep such information secret, and this for the public good.
Counter-claim Only a small proportion of leaks, usually involving genuine national defence secrets, are truly deplorable. The rest are harmless or positively desirable. Leaks can be used by low-level employees to make facts public that are being suppressed by their superiors. Frequently, leakage of official secrets serves to expose abuses and unconstitutional activities. As for political effects, only incumbents are worried about leakages, which often actually promote public understanding and democracy.
Refs American Bar Association *National Security Leaks* (1986); Flaherty, David, et al (Eds) *Privacy and Access to Government Data for Research* (1979).
Broader Official secrecy (#PC1812) Uncontrolled media (#PD0040).
Related Corruption in politics (#PC0116)
Crimes related to national security information (#PE3997).
Aggravates Invasion of privacy through testing (#PJ6946)
Restrictions on the distribution of confidential government information (#PD2926).
Aggravated by Journalistic muckraking (#PG5862).
Reduces Suppression of information (#PD9146) Abuse in government policy (#PF8389)
Restrictions on freedom of information (#PC0185).

♦ PD5193 Financial and industrial oligarchy
Misuse of financial and industrial power
Nature The control of economic and financial resources by multinational corporations and banks enables them to realize their financial and political aims. Their main means of control have changed from direct use of power to structural power, of which the main aspects are: exploitation; fragmentation; and penetration. Exploitation involves benefiting from interaction. The technique of fragmentation has three facets: the prevention of horizontal economic interaction between dominated countries; the prevention of multilateral interaction among dominated and dominating countries; and the permitting of only a minimum amount of interaction between the dominated country and its environment. Penetration has two aspects: (a) the elites of the dominated countries form a bridgehead through which the elites of dominating countries can spread their influence; and (b) the degree of inequality is considerably higher in the peripheral countries than in the central nations, which facilitates penetration. Multinational banks and corporations have at their disposal several mechanisms for exploiting, fragmenting and penetrating, and at the same time are driving forces in advocating the use of structural power.
Broader Economic imperialism (#PC3198) Abuse of economic power (#PC6873).
Narrower Bribery by transnational enterprises in developing countries (#PE0322)
Control of national economic sectors by transnational enterprises (#PE0042)
Social service quality negated by oligarchic control of decision-making (#PE1488).

♦ PD5204 Chemical torture
Nature Chemical compounds are used in torture in a variety of ways. Acids are use to burn. Salt or pepper is poured into open wounds. Tear gas is fired into cells containing prisoners or directly into faces. Hoods are impregnated with noxious chemicals like insecticides and then placed on prisoners until they nearly suffocate. Carbonated water is forced into nasal passages.
Incidence Chemical torture has been reported in the following countries: **Af** Madagascar, Zambia. **Am** Canada, El Salvador, Guatemala, Mexico, USA. **As** India, Sri Lanka.
Broader Physical torture (#PD8734) Misuse of chemicals (#PD5904).
Narrower Anti-personnel use of toxic substances in peacetime (#PE9294).

♦ PD5207 Political media events
Broader Delayed development of regional plans (#PF2018).
Related State control of communications mass media (#PD4597).

♦ PD5212 Seasonal fluctuations in agriculture
Short growing season — Seasonal variability in food supplies — Weather-induced fluctuations in food — Seasonally determined diets — Fluctuation in availability of food — Seasonal malnutri-

tion
Nature The availability of food varies from season to season and may be severely affected by the weather, especially when this gives rise to floods, heat waves (droughts), and cold spells. Environmental degradation can increase vulnerability to these effects. The growing dependence on a limited number of crop varieties over large areas may amplify the effects of weather and pest damage.
Broader Inhospitable climate (#PC0387) Seasonal fluctuations (#PF8163).
Aggravates Food insecurity (#PB2846) Seasonal unemployment (#PC1108)
Unbalanced family diets (#PJ0953) Unstable fishing season (#PJ9570)
Nutritional deficiencies (#PC0382) Fluctuating agricultural markets (#PG9369)
Lack of economic and technical development (#PE8190)
Fluctuations in food production in developing countries (#PE8188).
Aggravated by Drought (#PC2430) Vulnerability of crops to weather (#PE5682)
Underutilization of natural resources (#PF1459)
Individualistic retaining of local tradition (#PF1705)
Transfer of business from small communities to larger towns (#PF6540).

♦ PD5214 Electoral fraud
Ballot rigging — Vote rigging — Vote buying
Broader Obstruction of elections (#PD3982).

♦ PD5218 Violation of land rights of a people
Nature The concept that land can be owned, bought and sold is relatively new. Many a people had never even heard of land ownership until white people came to their areas and started laying claims on land.
Broader Expropriation of land from indigenous populations (#PC3304).
Related Denial of right to a people to freely dispose of natural wealth (#PE6955).

♦ PD5227 Inorganic salts as pollutants
Broader Pollutants (#PC5690).

♦ PD5228 Poisoning in animals
Broader Animal diseases (#PC0952).
Narrower Snakebite (#PG4663) Mycotoxicoses (#PE9458) Eggshell thinning (#PE6290)
Enzootic calcinosis (#PG5320) Poisoning by oak buds (#PG9484)
Sweet clover poisoning (#PG6227) Annual ryegrass staggers (#PG9532)
Ethylene glycol poisoning (#PG9322) Lead poisoning in animals (#PE9228)
Salt poisoning in animals (#PG7164) Algal poisoning in animals (#PG9198)
Copper poisoning in animals (#PG9550) Dioxin poisoning in animals (#PG5237)
Perennial ryegrass staggers (#PG9497) Arsenic poisoning in animals (#PG5416)
Cyanide poisoning in animals (#PG7564) Nitrate poisoning in animals (#PG9488)
Senecio poisoning in animals (#PG7883) Sorghum poisoning in animals (#PG7173)
Coal-tar poisoning in animals (#PG9425) Fluoride poisoning in animals (#PG4144)
Photosensitization in animals (#PG5561) Selenium poisoning in animals (#PG0112)
Herbicide poisoning in animals (#PG9301) Molybdenum poisoning in animals (#PG5221)
Strychnine poisoning in animals (#PG6981) Cantharidin poisoning in animals (#PG9264)
Insecticide poisoning in animals (#PG9601) Rodenticide poisoning in animals (#PG0829)
Bracken fern poisoning in animals (#PG3945) Metalaldehyde poisoning in animals (#PG7190)
Pesticide poisoning in small animals (#PG4457) Iron dextran toxicity in newborn pigs (#PG0232).
Related Mercury as a pollutant (#PE1155).

♦ PD5229 Discriminatory imposition of standards
Broader Discrimination (#PA0833) Organization of human thought (#PF5301).
Narrower Discriminatory unwritten codes of behaviour (#PE7017).
Related Double standards of sexual morality (#PF3259)
Inequitable labour standards in developing countries (#PD0142)
Inadequate standardization of procedures and equipment (#PC0666).
Aggravates Inadequacy of international standards (#PF5072)
Contempt for traditional modes of behaviour (#PC4321)
Resistance to internationally agreed standards (#PC4591).
Aggravated by Violating taboos (#PF3976) Excessive standardization (#PF2271)
Uncritical acceptance of dogmas and standards (#PF2901).
Reduced by Unenforced behaviour standards (#PS9251).

♦ PD5235 Assault
Battery
Nature The terms "assault and battery" are often used together, although traditionally they represent different and distinct crimes. Battery is the unlawful application of force to the person of another. Criminal assault may be either an attempt to commit battery or intentionally placing another under fear of battery. Simple assault is considered a misdemeanour, but aggravated assault (with intent to kill, rob, rape), possibly with a deadly weapon, may be considered a felony.
Broader Criminally life endangering behaviour (#PD0437).
Narrower Rape (#PD3266) Simple assault (#PE1144) Aggravated assault (#PD0583)
Public assaults on police (#PE7659).
Aggravated by Drunkenness (#PE8311).

♦ PD5238 Verbal abuse
Abusive language — Invective — Verbal attack — Insult — Vituperation
Refs Flynn, Charles P *Insult and Society* (1976).
Broader Cruelty (#PB2642) Disrespect (#PA6822)
Psychological torture (#PD4559).
Narrower Satire (#PJ5950) Ridicule (#PJ7386) Contumely (#PJ9783)
Detraction (#PE4394) Imprecation (#PF3746).
Related Slang (#PF5213).
Aggravates Teasing (#PE4187) Intimidation (#PB1992)
Discriminatory language (#PF7299)
Verbal sexual harassment of women in public (#PE0756).
Aggravated by Cynicism (#PF3418) Blasphemy (#PF5630)
Profanity (#PF7427) Unparliamentary behaviour (#PF4550)
Unethical practices of employers (#PD2879).

♦ PD5244 Employee disobedience
Worker disobedience — Employee insubordination
Broader Disobedience (#PA7250).
Narrower Conscientious objection at the factory (#PE7007).
Related Insubordinate behaviour (#PJ6517).
Aggravates Strikes (#PD0694).

♦ PD5251 Unethical media practices
Irresponsible media practices — Negligence by the media — Corruption in the media
Narrower Media cover-up (#PD4383) Misuse of advertising (#PE4225)
Invasion of privacy by media (#PD9603) Distorted media presentations (#PD6081)
Journalistic irresponsibility (#PD3071)
Manipulation of the individual by mass media (#PE7448).
Related Misrepresentation of information to consumers (#PE6877).
Aggravates Deterioration of media standards (#PD5377)
Excessive portrayal of negative information by the media (#PE1478).
Aggravated by Uncontrolled media (#PD0040) Monopoly of the media (#PD3101)
State control of communications mass media (#PD4597).

♦ PD5253 Denial of right to a people to live in peace
Broader Denial of right of a people to be self-determining (#PC6727).

♦ PD5254 Denial of right to sufficient shelter
Denial of the right to adequate housing
Incidence In 1988 it was estimated that more than 1 billion people live in adequate houses or have no shelter at all.
Broader Inadequate standards of living (#PF0344)
Unethical real estate practice (#PD5422).
Related Inadequate housing (#PC0449) Exploitation in housing (#PD3465)
Denial of right to sufficient food (#PE0324) Denial of right to social security (#PD7251)
Inadequate social welfare services (#PC0834) Denial of right to economic security (#PD0808)
Denial of right to sufficient clothing (#PE7616)
Denial of right to adequate medical care (#PD2028).
Aggravates Urban slums (#PD3139) Homelessness (#PB2150).
Aggravated by Discrimination in housing (#PD3469).

♦ PD5255 Military blocs
Nature This includes NATO, the Warsaw Pact, etc.
Related Alliance system (#PU4214).

♦ PD5256 Hazardous combinations of substances
Toxic mixtures
Refs FAO *Water Quality Criteria for European Freshwater Fish* (1980).
Broader Toxic substances (#PD1115).
Narrower Long-term hazards of exposure to chemicals (#PE4717)
Carcinogenic consequences of food preparation (#PE6619).

♦ PD5258 Maldistribution of wealth within developing countries
Disparity in distribution of wealth within developing countries
Broader Inequitable distribution of wealth (#PB7666).
Narrower Unequal income distribution within developing countries (#PD7615).
Related Maldistribution of land in developing countries (#PD0050).

♦ PD5261 Environmental poverty
Overuse of environmental resources by the poor
Nature A condition in which people are forced to deplete resources and degrade environments, because they have such limited opportunities to change their economic behaviour. It becomes economically "rational" for them to destroy their resources, even though they themselves will bear much of the costs of doing so. Thus they are unable to respond adaptively to external change (such as drought). It is both a symptom of lack of development and a consequence of unsustainable development. It is both cause and consequence of unsustainable rates of population growth, as well as being the main agent of land degradation in developing countries.
Incidence In many parts of the world poor people are forced to overuse environmental resources to survive from day to day. Their impoverishment of their environment further impoverishes them, making their survival even more difficult and uncertain. In addition, the prosperity attained in some parts of the world is often precarious, as it has been secured through farming, forestry and industrial practices that bring profit and progress over the short term only.
Broader Underdevelopment (#PB0206)
Natural resource depletion due to high-level consumption (#PD4002).
Aggravates Waste of non-renewable resources (#PC8642)
Depletion of natural resources due to population growth (#PD4007).
Aggravated by Socio-economic poverty (#PB0388)
Ecologically unsustainable development (#PC0111).

♦ PD5265 Accidents to agricultural workers
Refs ILO *Guide to Health and Hygiene in Agricultural Work* (1979); ILO *Guide to Safety in Agriculture* (1969).
Broader Injurious accidents (#PB0731).

♦ PD5267 Unethical practices in psychotherapy
Unethical practices in psychiatry — Corruption in psychotherapy — Irresponsible psychotherapists — Psychiatric malpractice
Incidence One study indicates that 70 per cent of therapists had seen for treatment at least one patient who had had sex with a previous therapist. Of the latter, 96 per cent were male.
Refs Robertson, Jeffrey D *Psychiatric Malpractice* (1988).
Broader Unethical professional practices (#PC8019).
Narrower Abusive psychosurgery (#PE1951)
Bogus psychiatrists and personal counsellors (#PJ1180).
Related Abusive treatment of patients in psychiatric hospitals (#PD0584).
Aggravates Abusive detention in psychiatric institutions (#PE2932).

♦ PD5272 Cross border military operations
Cross border anti-terrorist raids
Nature Governments send military or paramilitary units into the territory of other nations to kill or kidnap individuals believed to be involved in terrorist activities.
Incidence South Africa and Israel admit the use of this method to control terrorists.
Broader Foreign military intervention (#PD9331)
Offences against the peace and security of mankind (#PC6239).
Related Defence information uncertainty (#PE7679)
State-supported international terrorism (#PD6008)
Military manoeuvres in sensitive border areas (#PE3704)
Military expeditions against distant objectives (#PE8207)
Physical insecurity of refugees and asylum-seekers (#PD6364).

♦ PD5277 Agricultural poisons
Broader Pesticides as pollutants (#PD0120)
Chemical pollutants of the environment (#PD1670).
Narrower Herbicides as pollutants (#PD1143) Rodenticides as pollutants (#PE3677).
Related Fungicides as pollutants (#PD1612).

♦ PD5278 Accumulation of pollutants in terrestrial wildlife
Accumulation of contaminant residues in terrestrial animals
Nature Environmental pollution is characterized by the accumulation of toxic metals, organochlorine residues and radionuclides in terrestrial wildlife. Radionuclides are accumulated as a result of fallout from nuclear weapons testing and from nuclear reactor accidents.
Broader Accumulation of pollutants in plants and animals (#PD5021).

♦ **PD5300 Offences involving danger to the person**
Crimes which endanger people
Broader Risk (#PF7580) Moral offences (#PD9179) Statutory crime (#PC0277) Criminally life endangering behaviour (#PD0437).
Narrower Kidnapping (#PD8744) Aerial piracy (#PD0124) Criminal threat (#PE4661) Criminal menacing (#PE4467) Malicious physical disablement (#PE3733).
Related Torture (#PB3430) Homicide (#PD2341) Hooliganism (#PD1109) Sexual offences (#PD4082).
Aggravated by Crimes of passion (#PG6526) Violence against prostitutes (#PE6209).

♦ **PD5305 Defacement of urban structures**
Graffiti
Claim Cleaning the graffiti covered surfaces in public places, like in trains, gives the public a feeling of security and that the authorities are in control.
Refs Cooper, Martha and Chalfant, Henry *Subway Art* (1984); Prada, Manuel Ganzalez *Grafitos*

Broader Vandalism (#PD1350).
Related Uncontrolled local vandalism (#PJ0154).
Aggravates Desecration of monuments (#PD4348) Desecration of religious buildings (#PD7278).
Aggravated by Anti-social behaviour of university students (#PE7370) Insalubrity of animal excrement in urban environments (#PE4685).

♦ **PD5308 Disobedience of children**
Disobedient children — Insubordinate children
Broader Disobedience (#PA7250).
Aggravated by Disrespect (#PA6822).

♦ **PD5314 International inaccessibility of justice**
Broader Inaccessibility of justice (#PD8334).

♦ **PD5315 Proliferation of legislation**
Related Proliferation of commercialism (#PF0815).
Aggravates Conflict of laws (#PF0216) Obsolete legislation (#PF5435) Ignorance of the law (#PG3516) Restrictive legislation (#PD9012) Non-compliance with the law (#PG2334) Proliferation of litigation (#PF0361).
Aggravated by Secret laws (#PC6757) Inadequate laws (#PC6848) Meaninglessness (#PA6977) Ineffective legislation (#PC9513) Political barriers to effective legislation (#PC3201).

♦ **PD5317 Denial of right to appeal**
Violation of right to review of conviction by a higher tribunal
Broader Denial of human rights in the administration of justice (#PD6927).
Aggravated by Abuse of right to appeal (#PG4693).

♦ **PD5321 Criminal attempt**
Nature When an individual intentionally takes a significant step toward committing a crime, acting in ways required for the commission of the crime then the person is guilty of criminal attempt whether or not the attempt succeeds.
Broader Offences of general applicability (#PD4158).
Related Criminal conspiracy (#PD1767) Criminal facilitation (#PD6845) Criminal solicitation (#PD7676).

♦ **PD5326 Declining economic growth in developing countries**
Collapse of growth in developing countries — Economic stagnation in developing countries
Nature Economic stagnation exists in developing countries when the total output (or output per capita) of a country remains constant or rises only sluggishly. It also exists when unemployment is chronic or increasing. It may persist because the economy is dominated by unchanging traditional patterns in which there is no incentive for change. Alternatively, the economy may be held in a state of static equilibrium at low levels of income.
Incidence Growth has slowed substantially in developing countries. By early 1985, the economic and financial positions of most major debtor developing countries had failed to improve to the extent originally expected, despite a relatively strong cyclical upturn in the United States economy in 1984. In many of these countries the view began to prevail that the strategy in force required them to sacrifice growth indefinitely. Some African and highly indebted countries have suffered significant declines in incomes per capita. Their investments have fallen to levels at which even minimal replacement may no longer occur in important sectors of their economies. Their debts are growing, but they still face negative net resource transfers because debt service obligations exceed the limited amounts of new financing available. In some developing countries the severity of this prolonged economic slump already surpasses that of the Great Depression in the industrial countries., and in many countries poverty is on the rise.
Real output of the non-oil developing countries grew at only about a 2 percent annual average in the early 1980's, and almost all of any output gains realized were absorbed in reducing current account deficits due to trade losses and high interest payments. Per capita incomes have fallen in those countries, at great social cost.
Broader Economic stagnation (#PC0002).
Narrower Financial paralysis of developing countries (#PD9449) Disparities in economic growth in developing countries (#PD7257).
Aggravates Stagflation (#PC2536) Economic hardship (#PD9180) Poverty in developing countries (#PC0149) International economic recession (#PF1172) Political instability of developing countries (#PD8323) Decline in import capacity of developing countries (#PE6861) Instability in export trade of developing countries producing primary commodities (#PD2968).
Aggravated by Iron dextran toxicity in newborn pigs (#PG0232) Decline in foreign direct investment in developing countries (#PD3138) Inadequate research and development capacity in developing countries (#PE4880).

♦ **PD5335 Unpaid wages**
Failure to make payroll payments
Nature The failure to make payroll payments may be due to poor business practices, accidents or crimes. In any case both the employer and the employees are hurt.
Aggravates Interruption risk (#PF9106) Incomplete access to development capital (#PF6517).

♦ **PD5344 Parental permissiveness**
Parental toleration of drug addiction in children — Parental toleration of alcoholism in children — Parental toleration of sexual activity of children — Lax parental discipline — Inadequate childhood discipline — Lack of constructive family discipline for young people — Permissive child-rearing
Nature Present-day families do not create a systematic discipline for their younger members which would allow the free participation of every individual within a framework obedience to the total family. Without this discipline, no corporate wisdom out of which the policy and actions of the family can be developed, is possible. When this is combined with a lack of willingness of parents to allow their children to risk themselves in significant engagement in society, the young people are left with excessive amounts of time and energy to expend on activities of their own improvisation.
Broader Permissiveness (#PF1252) Rigidly entrenched social traditions in rural areas (#PF1765).
Narrower Juvenile alcoholism (#PD1611) Drug abuse by adolescents (#PD5987).
Related Exclusion of pre-adults from family decisions (#PE2268).
Aggravates Permissive education (#PF0593) Lack of responsible involvement in community affairs (#PF6536).
Aggravated by Frustration (#PA2252) Ineffective parenting techniques (#PJ0551).

♦ **PD5345 Inadequate secondary education**
Nature Secondary schools are not preparing people to live in the twentieth and twenty-first centuries. Those students who do not go on to post-secondary education are not able to function effectively on the job, in the social sphere, or within new families. Many graduates of secondary schools are functionally illiterate. They do not have the skill require by the increasingly service oriented work place: interpersonal skills, teamwork skills, logical skills, problem-solving skills, critical thinking skills or the ability to learn. Many have no concept of social responsibility or rights.
Broader Inadequacy of formal education (#PF4765).
Narrower Inadequate sex education (#PD0759) Inadequate political education (#PJ7906) Inadequate vocational education (#PF0422) Inadequate education concerning the nature of problems (#PE8216).
Aggravates Uncritical thinking (#PF5039) Ignorance of health and hygiene (#PD8023) Inadequate family planning education (#PD1039) Inadequate application of available knowledge to solve problems (#PF8191).
Aggravated by Inadequate education (#PF4984) Inadequate educational facilities (#PD0847) Ineffective educational policy decisions (#PF2447).

♦ **PD5347 Spitting in public places**
Expectoration
Nature Spittle was historically believed to have magical properties. Early people experienced themselves at the centre of the universe, their own bodies being connected with cosmic bodies, gods, demons, the colours, plants, elements and directions. Spittle, blood, sperm, sweat, nails and hair became magical substances not only as a result of this unity but also because, after leaving the person's body, they still retained something of that person's essence. Connections are still made between our bodily fluids and our feelings: anger makes the blood "boil", and one "spits" out angry words. The AIDS virus has of course intensified this mythology in the contemporary mythology. Spitting has taken on a very negative connotation, pointing to a harsh rejection of an individual or situation. Spitting as a way of clearing one's throat is also unsanitary, especially in areas in which tuberculosis is endemic.
Broader Unsociable human physiological processes (#PF4417).

♦ **PD5349 Recklessness**
Criminally reckless endangerment — Endangerment with awareness of high risk
Nature Reckless endangerment is creating the circumstances that another person is at risk of serious bodily harm or death. Tampering with motor carriers, dams, nuclear facilities, etc carry the danger of death or disability and are crimes. Reckless driving is also a dimension of this crime. Recklessness, involving the awareness of high risk, is contrasted with negligence in which there should have been awareness of the existence of such risk.
Broader Criminally life endangering behaviour (#PD0437).
Narrower Reckless driving (#PE9334) Reckless children snowmobile drivers (#PG9220).
Related Negligence (#PA2658).

♦ **PD5350 Industries in difficulty**
Declining industrial sectors — Older industries — Ageing industrial sectors
Refs ILO *Industries in Trouble* (1981).
Narrower Declining local businesses (#PG9007).
Aggravates Industrial accidents (#PC0646).

♦ **PD5351 Export of inflation**
Aggravates Economic inflation (#PC0254).

♦ **PD5361 Conflicts over fishing rights**
Fish wars
Nature A variety of treaties and conventions determine what nation's vessels have the right to fish in what waters. When the status of these treaties changes conflicts over fishing right can result. France and Canada, Denmark and the United Kingdom during the past few years have had conflicts.
Broader Conflicting use of resources (#PF6654) Obstacles to the utilization of coastal and deep sea water resources (#PF4767).

♦ **PD5370 Governmental polluters**
Broader Pollution (#PB6336).

♦ **PD5372 Civil disturbances**
Broader Civil violence (#PC4864) Civil disorders (#PC2551) Man-made disasters (#PB2075).
Related Inhumane methods of riot control (#PD1156).
Aggravates Crimes committed during civil unrest (#PG1179).
Aggravated by Demonstrations (#PD8522).

♦ **PD5373 Military aircraft accidents**
Broader Air accidents (#PD1582) Transport accidents (#PC8478).

♦ **PD5377 Deterioration of media standards**
Deterioration of press standards — Deterioration of television programming standards — Deterioration of film standards — Deterioration of video standards
Claim In UK the press world is becoming far more competitive in the race to provide ever more avant-garde titillating, salacious, intrusive and offensive material which presses against the bounds of acceptability, truth, good taste, common decency and traditional journalistic ethics.
Counter-claim The general public is only too eager to read stories they claim to find distasteful.
Aggravates Excessive television viewing (#PD1533) Restrictions on freedom of information (#PC0185).
Aggravated by Uncontrolled media (#PD0040) Exploitative films (#PE6328) Culture of violence (#PD6279) Unethical media practices (#PD5251) Parochial escapist media entertainment (#PD0917) Excessive portrayal of sex in the media (#PE7930)

PD5377

Excessive commercialization of the media (#PE4215)
Excessive portrayal of crime in the media (#PE7354)
Excessive portrayal of substance abuse in the media (#PE3980).

♦ PD5380 Unethical practices in the legal profession
Irresponsible lawyers — Negligence by lawyers — Corruption in the legal profession — Legal malpractice — Legal errors
Incidence In the UK in 1989 it was estimated that solicitors made mistakes in more than 50 per cent of the divorce case petitions filed in some courts. Error rates of up to 32 per cent occurred in issuing summonses or applications to start legal proceedings, 23 per cent in warrants of execution, and 39 per cent in written evidence supporting divorce petitions. Of 55,000 complaints from the public, 25 per cent were concerned with delays, 15 per cent were of negligence, 10 per cent about overcharging, and 109 per cent about alleged shoddy work.
Refs Anderson, David C *Crimes of Justice* (1988); Mallen, Ronald E and Smith, Jeffrey M *Legal Malpractice* (1988).
Broader Unethical professional practices (#PC8019).
Aggravates Legal prevarication (#PF9756)   Prohibitive legal fees (#PF0995)
Miscarriage of justice (#PF8479)   Restrictive legal practices (#PD8614)
Inequitable administration of justice (#PD0986).

♦ PD5384 Undeclared strikes
Illegal strikes — Wildcat strikes
Broader Strikes (#PD0694).
Narrower Undeclared strikes in socialist countries (#PD1882).

♦ PD5393 Environmental destruction by communist regimes
Communist environmental insensitivity
Nature Communist governments have shown themselves to be singularly slow in making efforts to balance the ecological consequences of rapid industrialization. They have been slow to admit that the problem even exists. A government that violates the natural environment to suit some target figures in their long-range plan is a menace.
Incidence In Western Poland entire areas have been rendered unfit for breathing. Czechoslovakia has created moonscapes where ancient forests stood. The Hungarian authorities have begun building a hydroelectric power system that diverts the river Danube so drastically that drinking water for the Budapest area is endangered. Romania's Ceausescu has built a canal between the Danube and the Black Sea and the Danube delta was being bulldozed for further construction. In Soviet Union the Aral Sea has shrunk by two-fifths since 1960, leaving behind a salty, man-made desert.
Broader Inadequate social discipline in socialist countries (#PD6893).

♦ PD5399 Retaliation against public servants
Nature Retaliation against a public servant is causing harm to a someone, their family or their property in retaliation for the service of a public servant, witness or informant.
Broader Retaliation (#PF9181)   Intimidation of public officials (#PD4734).

♦ PD5422 Unethical real estate practice
Real estate fraud — Housing fraud — Mortgage fraud — Land fraud — Real estate corruption — Negligence by real estate agents
Nature There are four principle ways in which agents cheat customers: failing to tell sellers of higher bids when lower bids provide the agents with more commission through mortgage and insurance needs; switching second bidders to other properties when buyers are in short supply; selling unnecessary insurance or the wrong types of mortgages because they offer better commissions; and leaking to potential buyers the lower price to which sellers are ready to agree.
Broader Fraud (#PD0486)   Unethical commercial practices (#PC2563)
Unethical professional practices (#PC8019).
Narrower Discrimination in housing (#PD3469)   Obtaining property by false pretences (#PE5076)
Denial of right to sufficient shelter (#PD5254).
Related Unethical practices by housing tenants (#PE7169).
Aggravates Urban slums (#PD3139)   Bad property loans (#PE4545)
Exploitation in housing (#PD3465)   Risky rental agreements (#PG1425)
Neglect of property maintenance (#PD8894).
Aggravated by Absentee ownership (#PD2338).

♦ PD5426 Product extortion
Food poisoning as sabotage — Threats to poison food — Terrorist food poisoning
Incidence In the UK in 1989 some 2,000 cases of product extortion were reported, an increase from 683 (1987), 725 (1988).
Broader Sabotage (#PD0405)   Human poisoning (#PD0105)   Product tampering (#PD8804).
Related Food poisoning through negligence (#PE0561)
Contamination of public water supplies by sabotage (#PE1458).

♦ PD5429 Mismanagement of environmental demand
Ineffectiveness of conservation programmes
Nature Sectoralism and separation of conservation and development undermine attempts to manage demands on the environment. Most efforts to develop resources, maintain the resource base, and protect the environment, are pursued separately in uncoordinated and narrowly sectoral ways.
Incidence Pollution control and environmental protection rely heavily on regulation of outputs, rather than on incentives to manage inputs better. As populations and economies grow, governmental micro-management of pollution and other harmful impacts of resource development is becoming increasingly expensive, cumbersome and impractical.
Broader Mismanagement (#PB8406).
Aggravates Environmental warfare (#PC2696)   Environmental pollution (#PB1166)
Destruction of wildlife habitats (#PC0480)   Eutrophication of lakes and rivers (#PD2257)
Destruction of environmental oxygen (#PE5196)
Ecologically unsustainable development (#PC0111)
Environmental degradation of desert oases (#PD2285)
Inevitable destruction of natural environment by mankind (#PE2443)
Degradation of mountain environment by leisure activities (#PE6256)
Degradation of the environment through the destruction of species (#PE5064).

♦ PD5432 Underpopulation
Nature A number of countries and communities do not have large enough a population to survive. In some cases this is due to birthrates being below 2.1 children per woman of childbearing age, the acceptable figure for sustained reproduction of a population. In other cases, a net out migration has caused the under population.
Aggravates Ageing of world population (#PC0027)
Declining economic growth in industrialized countries (#PF1737).
Aggravated by Childlessness (#PC3280)   Induced abortion (#PD0158)
Birth prevention (#PE3286)   Human infertility (#PC6037)

Declining birth rate (#PD2118)   Uncontrolled migration (#PD2229)
Inadequate child day-care facilities (#PD2085).
Reduced by Government-enforced maternity (#PE3601)
Ineffective population control (#PF1020)
Unsustainable population levels (#PB0035).

♦ PD5433 Illegally-obtained funds
Dirty money — Funny money
Incidence There are no precise indications as to the amount of illegally-obtained funds in the world's banking system. The flow of funds from offshore centres to western banks provides some indication. This rose from $350 billion in 1986 to $500 billion in 1989. Specialists in narcotics trade believe that over 50 per cent of this constitutes illegally-obtained funds, the rest being unpaid taxes and flight capital.
Aggravates Money laundering (#PE7803).
Aggravated by Crime (#PB0001).

♦ PD5453 Circulatory system diseases in animals
Cardiovascular diseases in animals — Lymphatic system diseases in animals
Broader Animal diseases (#PC0952).
Narrower Anaplasmosis (#PE6275)   Bovine leukosis (#PE1243)
Anaemia in animals (#PE9554)   Feline lymphosarcoma (#PG9320)
Vasculitis in animals (#PG9826)   Thrombosis in animals (#PG7446)
Polycythemia in animals (#PG1715)   Heart disease in animals (#PE0471)
Lymphadenitis in animals (#PE9844)   Polyarteritis in animals (#PG9812)
Canine malignant lymphoma (#PG5474)   Dissecting aneurysm in turkeys (#PG9514)
Round heart disease in poultry (#PG3727)   Haemostatic disorders in animals (#PG4410)
Sudden death syndrome in chickens (#PG4526)   Cyclic neuropenia in grey Collie dogs (#PG4509)
Haemorrhagic anaemia syndrome in chickens (#PG5506)
Parasites of the cardiovascular system in animals (#PE9427)
Congenital anomalies of the cardiovascular system in animals (#PE5569).

♦ PD5457 Violation of political processes
Broader Criminal violation of civil rights (#PD8709).
Narrower Obstruction of elections (#PD3982)
Misuse of personal authority for political purposes (#PE4635)
Deprivation of government benefits for political purposes (#PE5667).

♦ PD5475 Excessive politicization of the media
Nationalistic media
Broader Distorted media presentations (#PD6081)
Thwarted technological communications (#PF0953).
Aggravated by State control of communications mass media (#PD4597).

♦ PD5492 Budget deficit
Unbalanced budgets — Deficit spending
Nature A country's current account deficit is the gap between its domestic savings and investment. A budget deficit is a form of dis-saving. Large budget deficits generate inflationary pressures.
Counter-claim Some countries have adopted laws that require the national budget to be balanced. Electoral promises to balance the budget may also be made. The economic rationale for this is questionable. A budget designed to achieve targets consistent with an appropriate response to inflation, public debt, and private sector growth does not necessarily require a balanced budget, even if this can be achieved in practice. Balanced budget laws are relatively easy to circumvent in practice by excluding certain items, such as state-controlled enterprises. Preoccupation with balanced budgets can also complicate fiscal planning. They also provide the finance ministry with a ready-made excuse for resisting calls for public spending.
Refs Cebula, Richard *The Deficit Problem in Perspective* (1987).
Related Excessive external trade deficits (#PC1100).
Aggravated by Budget deficits in developing countries (#PD3131).

♦ PD5493 Hunting of humans
Bounty hunting — Murder by contract
Incidence Aside from war and the normal pattern of crime, humans continue to be hunted down and killed under various forms of "contract". Governments, corporations, crime rings and individuals may contract to have an individual killed. Indigenous people in various countries have, in the past, been victimized by legislation encouraging (or ignoring) the organization of hunts to kill them. This remains an issue in the Amazon basin and in the USA.
Refs Leyton, Elliott *Hunting Humans* (1986).
Broader Homicide (#PD2341).
Related Mercenary troops (#PD2592).

♦ PD5494 Inappropriate personal habits
Bad habits
Narrower Smoking (#PD0713)   Recidivists (#PE5581)   Sedentary habits (#PJ6551)
Lack of regular habit (#PG6564)   Disoriented habitual modes (#PG9595)
Habitual lifestyle routines (#PG8709)   Inadequate personal hygiene (#PD2459)
Ingrained segregation habits (#PG8789)   Overpowering traditional habits (#PJ0453)
Dietary deficiencies in developed countries (#PD0800).
Related Excessive television viewing (#PD1533).
Aggravates Religious backsliding (#PF6826).

♦ PD5501 Abusive treatment of the aged
Negligent old age care — Mistreatment of the senile
Nature A growing number of elderly people are being abused physically, psychologically or through neglect. In the United States some 86 percent of the cases of abuse are by the victim's own family.
Reports concerning ill and elderly patients have frequently revealed situations in geriatric wards where the standard of care has been unacceptably low. Examples of poor standards included lack of consideration of patients' feelings; failure to maintain dignity, privacy and personal identity; and lack of social, remedial and recreational stimulation. Emphasis is consistently placed on the failure to provide for the psychological and social needs of patients. Cultural conditioning, whereby the elderly are often disregarded, discourages the flow of resources and produces poor working conditions, which in turn influence the day-to-day experience of nurses and students, and consolidate attitudes of indifference and even despair.
Claim The root cause of the geriatric problem is not shortage of money, equipment or personnel, but a defective attitude to old age, in which the medical and nursing professions share.
Refs Anetzberger, Georgia J *The Etiology of Elder Abuse by Adult Offspring* (1987); Quinn, Mary J and Tomita, Susan K *Elder Abuse and Neglect* (1986); Schlesinger, Ben and Schlesinger, Rachel *Abuse of the Elderly* (1988).
Broader Cruelty (#PB2642).
Related Inadequate income in old age (#PC1966)
Inadequate welfare services for the aged (#PD0512).

DETAILED PROBLEMS PD5575

Aggravates Senilicide (#PJ7124).
Aggravated by Juvenile stress (#PC0877) Unproductive dependents (#PC1420)
Dehumanization of health care (#PD7821).

♦ PD5510 Accumulated junk
Visible accumulated junk
Broader Environmental pollution (#PB1166) Neglected of public space (#PF6578).
Narrower Pollution of orbital space (#PD0089)
Disused machinery accumulation in the seas (#PE5309).

♦ PD5511 Criminal harm to property
Crimes against property
Nature Criminal harm to property may be described as the act of someone, with the purpose of obtaining a material benefit, knowingly misleading another person to disadvantageously dispose of his own or someone else's property, or taking advantage of a mistake or an inability to adequately understand the action taken.
Broader Statutory crime (#PC0277).
Narrower Fraud (#PD0486) Theft (#PD5552) Robbery (#PD5575)
Property damage (#PD5859) Criminal intrusion (#PE6771)
Receiving stolen property (#PE8364) Frauds, forgeries and financial crime (#PE5516)
Violations against economic regulations (#PD7438).
Aggravates Prohibitive cost of business security (#PJ7767).
Aggravated by Criminal coercion (#PD4469) Insecurity of property (#PC1784).

♦ PD5515 Hindrance of law enforcement
Forcible obstruction of law enforcement — Obstruction of justice — Hindering apprehension of a felon — Hindering prosecution of a felon — Misprison of felon — Offences against the administration of justice
Nature A person is guilty of hindering law enforcement if he intentionally interferes with, delays or prevents the discovery, apprehension, prosecution, conviction or punishment of another. This may be by concealing the other; providing means of avoiding discovery or arrest including money, weapons, transport or disguise; or concealing, altering or destroying material with this intention. It may be by warning the other of their impending discovery in order that they by avoid arrest.
Broader Corruptive crimes (#PD8679) Obstruction of government function (#PD6710)
Crimes against the integrity and effectiveness of government operations (#PD1163).
Narrower Perjury (#PD2630) Arming rioters (#PE5327) Criminal contempt (#PD5705)
Compounding a crime (#PE1485) Public assaults on police (#PE7659)
Corruption of the judiciary (#PD4194) Illegally obtained evidence (#PE9309)
Tampering with physical evidence (#PF7291)
Tampering with witnesses and informants in proceedings (#PE6781)
Limited local respect for regional and global legislation (#PF2499)
Use of undue influence to obstruct the administration of justice (#PE8829).
Aggravated by Legal prevarication (#PF9756) Inadequate national law enforcement (#PE4768).

♦ PD5534 Religious vilification
Religious smear campaigns
Incidence The interaction between rival religious groups tends in part to be characterized by vilification of one by the other, whether or not this is communicated to the rival group. Attributing negative qualities to a rival group is a natural means of ensuring the coherence and motivation of a group. In the case of religious groups this can typically extend to labelling the activities of the other as "works of the devil".
Aggravates Discriminatory language (#PF7299) Religious discrimination (#PC1455).
Aggravated by Religious deception (#PF3495) Religious propaganda (#PD3094).

♦ PD5535 Diseases of senses in animals
Eye diseases in animals — Ear diseases in animals
Broader Animal diseases (#PC0952).
Narrower Eyeworms in animals (#PG7525) Periodic ophthalmia (#PG2526)
Deafness in animals (#PG9212) Glaucoma in animals (#PG5818)
Cataracts in animals (#PG5615) Blepharitis in animals (#PG1830)
Dermatitis of the pinna (#PG0870) Otitis interna in animals (#PG6794)
Otitis externa in animals (#PE5444) Chorioretinitis in animals (#PG8917)
Necrotic ear syndrome of swine (#PG7839) Haematoma of the ear in animals (#PG7421)
Diseases of the cornea in animals (#PG8009) Inherited retinopathies in animals (#PG2323)
Bovine ocular squamous cell carcinoma (#PG9494)
Tumours of the external ear in animals (#PG8705)
Diseases of the conjunctiva in animals (#PG6599)
Diseases of the anterior uvea in animals (#PG3952)
Infectious keratoconjunctivitis in animals (#PG5251)
Diseases of the lacrimal apparatus in animals (#PG4093)
Conformational abnormalities of eyelids in animals (#PG4212).

♦ PD5540 Counterfeit drugs
Fake drugs — Counterfeit medicines — Phoney non prescription drugs
Nature Fake medicines produced mostly by European drug firms are responsible for the death and suffering of untold numbers of patients throughout the Third World. A wide range of counterfeit medicines, including antibiotics, steroids, analgesics and heart drugs, allegedly made by reputable pharmaceutical companies are sold. In some cases, high quality generic drug are packaged and sold as originals but more frequently little or none of the active ingredients are present. They cost up to 60 percent less than their authentic counterparts. The escalating trade in fake medicines is estimated to cost the British pharmaceutical industry more than 10 million pounds a year. By far the largest supplier of these type of medicines is Italy and a large number of the manufactures and middle men are involved with organized crime.
Refs Ginneken, Wouter Van and Garzuel, Michel *Unemployment in France, the Federal Republic of Germany and the Netherlands. A Survey of Trends, Causes and Policy Options* (1985).
Broader Counterfeiting (#PD7981).

♦ PD5541 Terrorist havens
Havens for war criminals
Incidence Paraguay has granted asylum, at a price, to former dictators, Nazi war criminals, right-wing extremists, drug traffickers and common criminals. The terrorists against Western industrialized countries have been given safe havens in Syria, Libya, Iraq, Iran, Germany DR and Hungary.
Broader Ineffective prevention of terrorism (#PF4240)
State-supported international terrorism (#PD6008).
Related Legal havens (#PE0621).

♦ PD5543 Declining productivity in industrialized countries
Decline in the rate of productivity increase
Broader Declining economic productivity (#PC8908)
Deterioration of industrialized countries (#PD9202)
General unproductivity of capitalist systems (#PF3103).

♦ PD5552 Theft
Risk of theft — Property acquisition crimes
Nature Theft is the unjust taking of what belongs to another. Stealing, embezzling, cheating in buying and selling, the falsifying of weights and measures, and larceny, come under the same condemnation as theft. Especially aggravating circumstances are the commission of theft by a particularly dangerous recidivist or on a large scale (in theft of state or social property). There is a pervasive trend in modern criminal codes to consolidate most property acquisition offences (including larceny, embezzlement, false pretences, and receipt of stolen property) under theft as a single offence, previously unknown as such under common law (robbery is often treated as an exception).
Counter-claim There are circumstances in which what belongs to a man can be taken from him without injustice, and such taking must not be called theft. For example, an individual's objection to having food belonging to him taken by a starving man to save himself from death would be unreasonable.
Refs International Council of Shopping Centers *Store Security* (1984); Shukla, K S *Adolescent Thieves* (1979).
Broader Risk (#PF7580) Corruptive crimes (#PD8679)
Criminal harm to property (#PD5511).
Narrower Piracy (#PD1877) Petty theft (#PG5037) Pickpocketing (#PE0559)
Theft of property (#PD4691) Theft of services (#PD4711)
Receiving stolen property (#PE8364) Crimes against intangible property (#PE6486).
Related Robbery (#PD5575) Burglary (#PD2561) Expropriation (#PJ6566).
Aggravates Unretrievable documents (#PF4690).
Aggravated by Gang warfare (#PD4843) Insecurity of property (#PC1784)
Police crimes during narcotic investigations (#PE5037).

♦ PD5554 Diplomatic fraud
Fraud by international civil servants
Broader Abuse of privileges and immunities by diplomats (#PF5649).
Related Fraud by government agents (#PD8392).
Aggravates Inefficient public administration (#PF2335).

♦ PD5560 Maltreatment of civilians
State-supported violence against citizens — Beatings of civilians
Refs Lopez, George A, et al *Testing Theories of State Violence, State Terror, and Repression* (1988).
Narrower Excessive imposition of states of emergency (#PE4363).
Related Inadequate protection of civilians in armed conflict (#PE8361).
Aggravated by Unrestrained use of force in administration of justice (#PE8881).

♦ PD5574 Terrorism
Nature Terrorism can be defined as (a) indiscriminate military violence or as (b) indiscriminate military violence by non-state organizations such as revolutionary movements or resistance or, as it is usually understood, as (c) use of terror as a political or military instrument. Because of the disagreements over who is a terrorist, there has been only feeble international cooperation on dealing with terrorism.
Claim The arbitrary nature of dishonesty, the abuse of power, authoritarianism, the subjection of minorities and direct action in the form of coups, components of a general climate of implicit or explicit violence, form the destabilizing cultural context which serve as an internal inducement for the growth of terrorism. Combating terrorism without attacking this cultural context or, even worse, combating terrorism on the basis of this context is a fruitless task. It may end terrorism momentarily, but it will leave untouched the conditions for its re-emergence. The struggle against terrorism can bear fruit only if it is undertaken as a struggle by the whole of society against the roots of its own cultural degradation and lawlessness. The great mistake is to call for the survival of authoritarian structure as a form of prevention of terrorism. To do so would mean giving terrorism the conditions for its own reproduction.
Counter-claim Terrorists are murderers, cold-blooded killers and if people start looking for root causes, devising explanations or making excuses, they are only wasting their time.
Refs Burton, John *Deviance, Terrorism and War* (1979); Chand, Attar *Terrorism, Political Violence and Security of Nations* (1986); Chomsky, Noam *Pirates and Emperors* (1987); Dobson, Christopher and Payne, Ronald *Counterattack* (1984); Evron, Y (Ed) *International Violence*; Han, Henry H *Terrorism, Political Violence and World Order* (1984); Herman, Edward S *Real Terror Network* (1984); Janke, Peter and Sim, Richard *Guerilla and Terrorist Organizations* (1983); Janke, Peter and Sim, Richard *Guerrilla and Terrorist Organizations* (1983); Lakos, Amos *International Terrorism* (1986); Lodge, Juliet *The Threat of Terrorism* (1987); Micklous, Edward F and Flemming, Peter A *Terrorism Nineteen Eighty to Nineteen Eighty-Seven* (1988); O'Neill, Michael J *Terrorist Spectaculars* (1986); O'Sullivan, Noel *Terrorism, Ideology and Revolution* (1986); Ochberg, Frank *Victims of Terrorism* (1981); Ontiveros, Suzanne R *Global Terrorism* (1986); Szumski, Bonnie *Terrorism* (1986); Trager, Oliver C (Ed) *Fighting Terrorism* (1986); Wardlaw, Grant *Political Terrorism*; Wilcox, Laird *Terrorism, Assassination, Espionage and Propaganda* (1988); Wolf, John B *Fear of Fear* (1981).
Broader Civil violence (#PC4864) Political crime (#PC0350)
Guerrilla warfare (#PC1738).
Narrower Air terrorism (#PE4089) Urban terrorism (#PD9997)
Narco-terrorism (#PE0210) Terrorist bombing (#PE2368)
Suicide terrorists (#PG4023) Religious terrorism (#PD4134)
Political assassination (#PE5614) Ineffective prevention of terrorism (#PF4240)
Terrorists armed with nuclear weapons (#PE3769)
State-supported international terrorism (#PD6008)
Terrorists armed with biochemical weapons (#PE9207)
Terrorism targeted against business corporations (#PE5944).
Related Assassination (#PD1971) Hostage taking (#PE4108)
Political violence (#PD4425).
Aggravates Vulnerability of telephone system (#PE8254)
Socially unsustainable development (#PC0381).
Aggravated by Outlaws (#PE4409) Errant nationals (#PE0812)
Security risk people (#PD6818) Passive public support of terrorism (#PF6846)
Excessive portrayal of terrorist activity in the media (#PE6844)
Inadequate international cooperation in reducing terrorism (#PF4366).
Reduced by Informers (#PD8926).

♦ PD5575 Robbery
Mugging — Aggravated larceny
Nature Robbery is the overt theft of another's personal property or of state or social property, committed without violence or with violence that does not endanger another person's life or health. Robbery differs from theft because the removal of property is overt.
Refs Fielding, Henry *An Enquiry into the Causes of the Late Increase of Robbers* ; Hart, David M *Banditry in Islam* (1987); Slatta, Richard W (Ed) *Bandidos* (1987); Walsh, Dermot *Heavy Business* (1986).
Broader Theft of property (#PD4691) Criminal harm to property (#PD5511).
Narrower Looting (#PE4152) Banditry (#PD2609) Grave robbing (#PF0491)

Armed robbery (#PE4739)   Train and armoured car robberies (#PG1580).
Related Theft (#PD5552).
Aggravated by Gang warfare (#PD4843).

♦ **PD5582 Violence by fanatical environmentalists**
Environmental terrorism — Ecotage — Direct action environmentalists — Sabotage by environmental activists — Environmental vigilantism — Ecological sabotage
  Broader Human violence (#PA0429).
  Related Violence as a resource (#PF3994)   Intrusive animal-rights campaigners (#PE4438).
  Aggravates Assassination of environmental activists (#PE5005).

♦ **PD5584 Restrictions against small enterprise**
Punitive regulations — Commercial favouritism — Large farm favouritism
  Broader Excessive government intervention in the private sector (#PD4800)
  Paralyzing patterns between villages and administrative structures (#PF1389).
  Narrower Discouraging conditions for small business (#PD5603).
  Aggravates Inadequate development of enterprises in developing countries (#PE6572).

♦ **PD5586 Grievances of employees**
Complaints of workers — Protest by workers
Refs ILO *Grievance Arbitration. A Practical Guide* (1985).
  Broader Grievance (#PF8029).

♦ **PD5590 Political mass murder**
Nature Liquidation of political opposition by governments may sometimes mean the killing of tens and hundreds of thousands of people sharing a political or religious belief or belonging to one ethnic group. These purges are more often carried out by security forces, although in some cases the participation of civilians that have been set against the designated victims may be decisive, particularly to establish a mock legitimacy for such an ordeal. Staged incidents, such as encounter killings, fictitious escapes from prison, or alleged resistance to police forces, are other forms of concealing the true nature of these killings.
Incidence In Kampuchea, under Pol Pot's rule, 1 million people were killed, often by fellow countrymen. Political mass murders regularly occur in Central America, often under the guise of disappearances.
  Broader Government sanctioned killing (#PD7221)
  Offences against the peace and security of mankind (#PC6239).
  Related Genocide (#PC1056).
  Aggravates Torture (#PB3430).

♦ **PD5597 Childhood trauma**
Nature Unsatisfying or painful experiences of early childhood decisively affect the adult personality. Critical events include weaning, toilet training and resolution of the Oedipus complex.
Refs Johnson, Kendall *Trauma in the Lives of Children* (1989).
  Broader Trauma (#PD4571).

♦ **PD5603 Discouraging conditions for small business**
Nature A large number of government and social restriction to the creation of small business discourage the development of new businesses and the slow down and stagnation of the national economy. The tax structure of the country may deter investment in small businesses. High corporate taxes deter the creation of new companies. High personal taxes for executive discourage individual from becoming managers. Future profits may be taxed. Penalties for underestimating tax bills may be high. Government repayment of over paid taxes may be slow, further disrupting cash flow. Inheritance taxes may be high. The employment structure may discourage the growth of companies. An extremely large number of professions needing to be certified slows the growth of the professions and burdens employers. Small businesses may be required to hire only union members. Social costs and employee taxes may prohibit hiring additional employees, stopping growth. Commercial skills may be lacking. Technically qualified entrepreneurs may lack business skills. The education system may no offer sufficient commercial or business education. The financial climate may discourage new businesses by limiting access to loans.
  Broader Restrictions against small enterprise (#PD5584).
  Related Denial of right to business growth (#PE2700)
  Economic inefficiencies in developing countries due to restrictive business practices (#PE2999).
  Aggravates Business bankruptcy (#PD2591)   Small business failures (#PE9405)
  Continuing commercial decline (#PJ8726)
  Confined scope of business operations (#PF2439).
  Aggravated by Incompetent management (#PC4867)
  Complex business regulations (#PG7894)
  Scarcity of business collaterals (#PG8008)
  Prohibitive cost of business security (#PJ7767)
  Distrust of business by the community (#PE8963)
  Diminishing capital investment in small communities (#PF6477)
  Ineffectiveness of traditional small business methods (#PF3008)
  Inadequate local expertise in business practices in developing countries (#PE7313).

♦ **PD5605 Food pollution**
Food contamination
Refs Morton, Andrew B *Food Contamination* (1987).
  Broader Environmental pollution (#PB1166).
  Narrower Food-borne diseases (#PE2515)   Toxic food additives (#PD0487)
  Chemical contaminants of food (#PD1694)   Biological contamination of food (#PD2594)
  Food poisoning through negligence (#PE0561)
  Infected animal, meat and animal product shipments (#PE7064).
  Aggravates Fear of food contamination (#PF3904)
  Contamination of human body (#PF9150).
  Aggravated by Soil pollution (#PC0058)   Motor vehicle emissions (#PD0414)
  Pesticides as pollutants (#PD0120)   Fungicides as pollutants (#PD1612)
  Marine dumping of wastes (#PD3666)   Soil-transmitted diseases (#PD3699)
  Metal contamination of soil (#PD3668)
  Water pollution in developing countries (#PD3675)
  Polychlorinated biphenyls as a health hazard (#PE2432)
  Environmental hazards from food and live animals (#PC1411)
  Concentration of noxious substances in food chains (#PE8154)
  Environmental hazards from food processing industries (#PE1280).

♦ **PD5616 Privileged families**
  Narrower Reinforcement of inappropriate development by privileged classes in developing countries (#PF6670).

♦ **PD5628 Forced exercise**
Nature Prisoners are tortured by being forced to do physical exercises or labour until they drop from exhaustion. Some are forced to do military drill. Some are forced to run or crawl with heavy loads on their backs and beaten if they stop or fall. Disabled people are required to do the work of an able bodied person.
Incidence Exercise has been reported in the following countries: **Af** Benin, Ghana, Mali, Zambia, Zimbabwe. **Eu** USSR.
  Broader Physical torture (#PD8734).
  Aggravates Human disease and disability (#PB1044).

♦ **PD5636 Inconclusive convictions for fraud**
Nature People convicted of fraud are less likely to be imprisoned – or of imprisoned have been given shorter sentences – than other non-violent criminals who have stolen substantial sums. They have also more often been sent to open prisons and have been more frequently paroled. Recent research has also shown that senior executives rate a suspended prison sentence for a businessman as a less significant punishment than even publicity on its own or a large fine.
  Broader Inadequate evidence to convict known offenders (#PF8661).

♦ **PD5641 Water surface pollution**
Nature Microscopic plants and animals congregate in the top millionth of a metre in any body of water. Toxic pollutants also accumulate in this microlayer, in concentrations many times those of the water below. Scientists theorize that pollution from the air settles on the surface, some components of sewage and industrial waste float to the surface, and pollutants which had previously fallen to the seabed rise up to the surface when the seabed is disturbed by currents or the movements of animals.
Refs Ellis, K V *Surface Water Pollution and Its Control* (1989).
  Broader Water pollution (#PC0062).

♦ **PD5642 Inefficiency of state-controlled enterprises**
Inefficiency of government monopolies — Inefficiency of public enterprises — Monopolization of resources by public sector — Lack of enterprise in parastatals
Nature Government owned and run enterprises are notoriously known for bureaucratic constraints, lack of investment incentives, pricing controls, centralized decision making, and restriction on hiring and firing workers. Control of businesses by government bureaucracies or by legislation is at best done in the context of the needs of the whole nation and at the worst in the context of personal ambition leading to wide spread corruption. In either case decisions are slow and unresponsive to the needs of the company. Profits often revert back to the government creating disincentives for management and labour to work more profitably. Investments are frequently insufficient. Price controls coupled with rising costs of labour and raw materials make unit costs high than prices, discourage efficient running of the company. Employment practices are dictated by political considerations rather than what benefits the corporation. Management jobs are often dependent upon approval of government overseers further eroding concern about effective business practices.
Incidence In almost every developing country the public sector undertakes a significant share of its production and investment through state-controlled enterprises. In the past their budgetary deficits have often been hidden by a lack of consolidated financial data, opaque budgetary procedures, extrabudgetary financing, implicit subsidies and protection from competition. More recently tight budget constraints, limits on domestic and external financing, and the effects of devaluation and trade liberalization have exposed the weakness of their finances and their negative effects on the fiscal stability of developing countries. Their contributions to rising public sector debt and growing foreign indebtedness are increasingly recognized as key issues in public finance.
Counter-claim The government is responsible for the welfare to its citizens and state owned and ran industries provide them with the dignity of work. By not participating in the productivity of society the unemployed are made second class citizens. The costs of subsidizing industries, even if they are inefficient is cheaper than making welfare payments.
  Broader Government inefficiency (#PF8491).
  Narrower Inefficient labour use in socialist countries (#PE7908)
  Abuse of monopoly power of state-owned or state-controlled enterprises (#PE0988).
  Aggravates Undervaluation of public assets (#PF1001)
  Frustrated personal effectiveness (#PJ7892)
  Arrears in financial payments between government agencies (#PE9700).
  Aggravated by Incompetent management (#PC4867)
  Governmental incompetence (#PF3953)
  Complex government regulations (#PF8053)
  Abuse of government employment (#PE4658)
  Abuse of bureaucratic procedures (#PF2661)
  Inefficiency of financial markets (#PF6980)
  Proliferation of public sector institutions (#PF4739)
  Opaque budgetary procedures in the public sector (#PF5374).
  Reduced by Privatization of public services (#PE3391).

♦ **PD5669 Environmental human diseases**
Nature Biological pollution from community wastes reaches drinking water sources in both developed and developing countries. In the former, mainly rural areas are affected; in the latter, cities, towns and villages are exposed, in addition to the rural areas. As well as human wastes, livestock wastes add to the production of contaminants. Convection by water is augmented by soil-borne, air-borne and directly contacted bacteria, fungi, parasites and viruses. Food may be contaminated as well. Inadequate sanitary and hydraulic engineering remain the greatest threat to preserving large numbers of the world's population from communicable diseases.
Incidence At least 500 million people each year are affected; many of these are infants and children.
Refs Rees, A R and Purcell, H J (Eds) *Society for the Environmental Therapy, Inaugural Conference*, 1981.
  Broader Human disease and disability (#PB1044)
  Infectious and parasitic diseases (#PD0982).
  Narrower Cholera (#PE0560)   Schistosomiasis (#PE0921).
  Related Zoonoses (#PD1770)   Environmental plant diseases (#PD2224).
  Aggravated by Sewage as a pollutant (#PD1414).

♦ **PD5674 High minimum wages**
Prohibitive minimum wages
  Broader Prohibitive labour costs (#PF8763).
  Related High severance pay for top managers (#PE3872).
  Aggravated by Unrestrained wage increases (#PE5305).

♦ **PD5675 Female mid-life crisis**
  Broader Mental suffering (#PB5680)   Personal life crises (#PD4840)
  Ignorance of lifelong human development (#PF5759)
  Related Mental illness (#PC0300)   Male mid-life crisis (#PD5783).
  Aggravated by Menopause (#PF5918).

♦ **PD5676 Vulnerability of stock markets**
Excessive interdependence of stock markets
Incidence Between January 1981 and September 1987, the monthly average correlation between the 23 biggest national stockmarkets was just 0.222 (a correlation of zero meaning that the markets move independently of one other). In October 1987 the average correlation was 0.755, meaning that the markets were close to move in line.

Broader Capitalist speculation (#PC2194).
Aggravates Economic inefficiency (#PF7556)
Securities and commodities exchange violations (#PD4500).

♦ PD5677 Dependence of island developing countries on imports

Nature The economies of island developing countries cannot be balanced, nor can complex internal linkages be expected. There is, and will inevitably continue to be, a heavy dependence on imports and a corresponding requirement for the export of goods and services. Economic events overseas beyond the control of the island countries have immediate and widespread effects on the local economy, in particular with respect to sudden fluctuations in foreign exchange receipts.
Broader Dependence on external resources (#PC0065).
Related Foreign exchange shortage in developing countries (#PD3068)
Vulnerability of island developing countries and territories (#PE5700)
Import-dependency in food staples in developing countries due to transnational corporations (#PE1806).

♦ PD5693 Destabilizing governments

Destabilization by government
Nature Conflict among nations can take the form of efforts to create chaos within the enemy's domestic government. This can take the form of violent acts against popular figures, and encouragement and financing of any of an array of opposition forces, from the domestic political scene to neighbouring enemies. Additionally, public image campaigns can be mounted to destroy the reputation of the forces in power.
Incidence Since the inception of the National Security Act in the United States in 1947, there has been a proliferation of Central Intelligence Agency activity in weapons, drugs, shadow government and destabilization of and within a number of countries.
Broader Destabilization of social systems (#PB5417).

♦ PD5701 Disparity in world telecommunications capabilities

Broader Disparities in distribution of communication resources and facilities (#PD2762).
Narrower Inadequate telecommunications in island developing countries (#PE5655).
Aggravates Insufficient communications systems (#PF2350).

♦ PD5705 Criminal contempt

Nature Criminal contempt is obstructing or attempting to obstruct the lawful administration of justice by a court of law whether by a plaintiff, the accused, a witness, a member of the court or a onlooker. It is also criminal contempt to disobey or resist a court's lawful writ or order.
Broader Contempt (#PF7697) Hindrance of law enforcement (#PD5515)
Contempt of judicial process (#PD9035).
Aggravates Jihad (#PF5681).
Aggravated by Instability of trade in petroleum and petroleum products (#PD0909).

♦ PD5706 Hazards to plants

Nature Plants are subject to a number of natural dangers through disease, adverse weather and insect depredations; man-made disasters such as pollution and excessively rapid development take their toll as well. The preservation of plant species and plant genetic diversity is a matter of scientific principle or aesthetics to some, others recognize plants as nature's pharmacy and plant-derived medicines continue to be discovered. Some plants belong in very important ecological chains, and without being crops themselves, their absence would affect such important natural activities as pollination, insect reduction by birds, and soil conditioning.
Refs World Conservation Union *Plants in Danger* (1986).
Broader Vulnerability of plants and crops (#PD5730).
Narrower Crops pests and diseases (#PEp783) Vulnerability of crops to weather (#PE5682).
Related Plant diseases (#PC0555) Pests of plants (#PC1627)
Crop vulnerability (#PD0660).
Aggravates Suffering of plants (#PC7825).
Aggravated by Bad weather (#PC0293) Environmental pollution (#PB1166)
Landscape disfigurement (#PC2122)
Degradation of semi-natural and natural habitats of flora and fauna (#PC3152).

♦ PD5707 Citizen disobedience

Citizen resistance under the Nuremberg obligation
Nature The belief that an individual may have international legal obligations which override his national obligations, and which require him to passively, actively, or violently resist his country's behaviour – whether or not that behaviour has been embodied in national law – arises from the example of the findings of the Nuremberg Tribunal, which sentenced 22 Nazi defendants in 1945–1946. Thus, citizens who hold that their country is waging, or about to wage, an unjust war, may refuse to support that war, may refuse conscription, and may harm the war effort psychologically or materially in acts ranging from anti-war protest and polemic, to destruction of property, and physical attacks on individuals deemed to have some responsibility in making or carrying out policy.
Incidence Resistance can be by civilians or military; the form of verbal dissent is the most common. The American protestors against the Vietnam war are best known, but subsequent to this, there have been Russian protestors against the Afghanistan mission, and British protestors against the Maldive-Falklands exercise and Northern Ireland pacification. Protest has recently moved from the verbal mode, and from the passive occupation of public property, to fasting and fasting to death, and there seems to be a proliferation of willingness for self-violence in advocating causes, with some movement towards acts of sabotage and, among the unstable, towards violence against their societies in general.
Broader Disobedience (#PA7250).
Narrower Military disobedience (#PD7225).
Related Pacifism (#PF0010) Civil disobedience (#PC0690)
Ineffective war crime prosecution (#PD1464).
Aggravated by Passive resistance (#PF2788).

♦ PD5718 Sexual mutilation

Nature Mutilation by modification of the genitalia may cause permanent full or partial sexual impairment to its victims (male or female) depending on the kind of disfigurement performed. In some cases, death may follow. During war and civil strife, sexual mutilation may be inflicted as an act of torture, reprisal or as a victory ritual on prisoners and their families. In peacetime and in certain societies or groups, specific mutilations (circumcision, castration, clitoridectomy, etc) constitute institutionalised practices addressed to certain or all of its members with prophylactic, therapeutic, repressive, discriminatory, religious or statutory aims. When inflicted unlawfully, maliciously, and without consent, deprivation of a member of the body by disablement disfigurement, the rendering of it useless, is considered a criminal act (mayhem).
Incidence The best known and most widespread forms of genital mutilation are circumcision and castration. In certain western sub-cultures, individuals pierce the clitoris, penis or nipples in order to insert padlocks, rivets or wire devices for decorative or symbolic purposes.
Broader Sadism (#PF3270).
Narrower Male sexual mutilation (#PE6054) Female sexual mutilations (#PE6055).
Aggravates AIDS (#PD5111).

Aggravated by Ignorance (#PA5568).

♦ PD5719 Loss of micro-organic proteins

Nature Pollutants are doubly offensive, firstly as contaminants and secondly as lost sources of protein. With the use of bacteria, fungi and yeasts, sewage and waste effluents from such industries as paper mills can be converted into animal feed, as can waste newspapers.
Broader Food insecurity (#PB2846) Inadequate protein supply (#PC1916).

♦ PD5725 Official privilege

Corporate privileges — Perks
Broader Accumulation of privileges (#PF8025).

♦ PD5727 Violations of private law

Civil law transgressions
Nature Any law concerning a society's internal affairs that is not concerned with crime, that is, with grievous offences by an individual against the moral standards of society – and therefore an offence against the state – will probably come into the area of law governing private controversies between persons, or organizations construed as legal entities. Trials in these cases are held in civil as opposed to criminal courts. Offences include those applying to contracts between parties, as for example, breach of contract; acts of negligence which result in personal injuries or property damages; and intentional performances of acts resulting in injuries or damage. Intentional and negligent acts come under the heading of tort law. Examples of offences or torts are: battery or physical contact with another without his consent; slander; libel; false arrest; and malicious prosecution. Another kind of tort is the unreasonable use of one individual's property to the detriment of others, such as creating a nuisance. The violations of contract, tort and other private law, particularly those applying to marriage, clog the civil judicial system in many countries. Present trends in offences seem to indicate either a lessening of moral standards, or a change in standards which makes some civil laws obsolete.
Broader Crime (#PB0001).
Narrower Violations of contract law (#PE5786)
Substandard housing and accommodation (#PD1251).
Aggravated by Inadequate national law enforcement (#PE4768).

♦ PD5730 Vulnerability of plants and crops

Unknown crop dangers
Nature Vegetation can include the edible, the medicinal, the ornamental and also, from the anthropocentric viewpoint, the economic: that which provides wood, fibre and chemicals. Vegetation that is non-edible is also essential for crop production to prevent erosion, provide wind-breaks or shade, or to mulch or condition the soil. Vegetation provides oxygen for the planet. Without vegetation most of organic life would die, and from nature's point of view (perhaps less biased than man's) the vegetable kingdom is therefore not deemed inferior to the animal. In scientific classification, vegetation is one of the three kingdoms, the others being the animal and the mineral. All three share the tendency to deteriorate with age and to be subject to environmental influence. The hazards which animals and plants share include: adverse weather, disease, old age, genetic accidents and even plant war. Like the smallest organisms, the largest and most intelligent plants are involved in the struggle for life among themselves and are the natural prey of the animal kingdom. On the other hand, plants have their defences and there are even plant predators.
Broader Vulnerability of organisms (#PB5658) Non-recognition of problems (#PF8112).
Narrower Damage to crops (#PJ3949) Hazards to plants (#PD5706)
Crop vulnerability (#PD0660).
Aggravates Food insecurity (#PB2846) Destruction of weeds (#PE3987)
Endangered species of plants (#PC0238)
Long-term shortage of food and live animals (#PE0976)
Instability of trade in food and live animals (#PD1434).
Aggravated by Pests (#PC0728) Bad weather (#PC0293)
Soil degradation (#PD1052) Man-made disasters (#PB2075)
Landscape disfigurement (#PC2122) Genetic defects and diseases (#PD2389)
Over-intensive soil exploitation (#PC0052)
Ecologically unsustainable development (#PC0111)
Environmental hazards from live animals (#PD0788)
Decreasing genetic diversity in cultivated plants (#PC2223)
Inadequate dissemination and use of available information (#PF1267)
Degradation of semi-natural and natural habitats of flora and fauna (#PC3152).

♦ PD5734 Abandoned children

Child destitution
Nature Abandoned children, a rapidly increasing and serious social problem, is an extreme form of child neglect. In addition to the physical health problems it perpetuates, abandonment presents the threat of improper personality development and creates conditions in which can be bred the negative social patterns of behaviour which are exhibited in delinquents and criminals. Abandonment stems from many sources – family breakdown (which itself can be the result of inadequate housing and social services, poverty and unemployment); irresponsible fatherhood (such as exhibited by the thousands of American servicemen and civilians who sired children during the Vietnam War); premature motherhood (where the mothers feel themselves unequipped to assume the responsibilities of motherhood). It may be due to illegitimate birth, death of both parents (as a result of war, civil violence or natural disasters. Family breakdown may be due to the death or desertion of one parent and be exacerbated by poverty, migration, chronic unemployment.
Incidence Reliable information on the causes, magnitude and prevalence of child abandonment is scarce. An estimated 8,000 to 15,000 children were born of American/Vietnamese liaisons during the Vietnam War, most of whom were left behind; 7,000 American children are abandoned annually in the USA; Venezuela and Colombia have thousands of abandoned children per year; Brazil acknowledges child abandonment as one of its main social problems; and Zimbabwe recently had up to 20 babies per day left at paediatric hospitals; in Romania tens of thousands of children were abandoned, because mothers were forced to have children against their will.
Background The problem is not new in human history. In 19th century London, ragamuffins were a familiar element of the urban scene, as were 'street arabs' in New York. 'Children of the Sun' by Morris West records their survival in Naples in the 1950s. What is new is the scale of the problem. The present-day numbers of street children in single cities like Calcutta may be equal to the total population of those cities in the last century.
Refs Boswell, John *The Abandonment of Children in Western Europe from Late Antiquity to the Renaissance*; Moorehead, Caroline (Ed) *Betrayal* (1989); Zenzinov, Vladimir M *Deserted* (1975).
Broader Homelessness (#PB2150) Family poverty (#PC0999)
Victimization of children (#PC5512).
Narrower Children engendered by occupying soldiers (#PD8825).
Related Abandoned wives (#PD1030).
Aggravates Street children (#PD5980) Repressive detention of juveniles (#PD0634)
State custody of deprived children (#PD0550).
Aggravated by Family breakdown (#PC2102) Disowned children (#PJ0827)

Dependence within extended families (#PD0850) Impediments to adoption of children (#PF7353)
Narrow legal definition of the family in developing countries (#PD1501).

♦ PD5750 Criminal investment in youth market

Delinquency as a side–effect of the drug trade and drug addiction

Nature The selling of illicit drugs, alcohol and pornography to minors are large sources of income for criminal elements in the societies of developed countries. Criminal operators seek to entrap youth in a state of dependence in their wares, and cross–connect their illicit service so that alcohol abuse and pornography may lead to drug addiction, prostitution and equally serious crimes.

Broader Organized crime (#PC2343).
Aggravates Pornography (#PD0132) Prostitution (#PD0693)
Alcohol abuse (#PD0153) Juvenile delinquency (#PC0212)
Corruption of minors (#PD9481) Drug abuse by students (#PE5507)
Illicit drug trafficking (#PD0991) Drug abuse by adolescents (#PD5987).
Aggravated by Pandering to the youth market (#PF5726).

♦ PD5769 Destruction of coral reefs

Nature Coral reefs are major ecosystems in that they host an enormous variety of molluscs, sea anemones, sponges, sea–squirts and similar primitive sea life that serves as food for another wider group of fish and sea creatures in the ecological chain. Corals themselves are tiny polyps colonizing the hard skeletons they jointly build, which is commonly known as coral. This build–up of coral, which chemically is a mass of carbonate of lime, takes several forms, such as barrier reefs and atoll rings, among others. All forms may attain great size and serve to protect shores from the ocean. Coral reefs are under attack by enormous numbers of starfish of the variety called 'crown–of–thorns', technically Acanthaster planci. These fish eat the soft tissues of the reefs by inverting their stomachs, liquifying, and absorbing the coral. Another threat is mats of green bubble algae produced by the action of pollutants.

Incidence The Great Barrier Reef off of Australia; the reefs of Hawaii in Kaneoke Bay and Waikiki Bay; reefs off Guam, the Virgin Islands, Jamaica, and Bermuda; and reefs in the Indian Ocean: all are in various stages of destruction. African states with endangered reefs in the Western Indian Ocean include Kenya, Tanzania, Mozambique, Mauritius and the Seychelles.

Refs Sheppard, C R C and Wells, S M *Coral Reef Directory* (1984).

Broader Environmental degradation (#PB6384) Vulnerability of ecosystem niches (#PC5773).
Aggravates Inviability of tropical island developing countries (#PE5808).

♦ PD5770 Unethical medical practice

Medical fraud — Immoral medical practice — Lack of medical ethics — Medical negligence — Medical malpractice — Abuse of medical and psychological techniques — Degradation of the medical profession — Abusive medical practices — Medical irresponsibility — Unethical practice by physicians — Incompetent doctors

Nature A small percentage of physicians and others rendering health care services may be doing so unethically, with a wide variety of abuses such as: practising without the proper educational qualifications; practising without required licences and registrations; over–charging; negligence; erroneous or uncertain prescriptions, treatments or surgical procedures; supervising, monitoring or conduct torture. There may be unethical disclosure of a patient's medical history to employers, credit investigators, banks, attorneys and others; and sexual contact may be initiated by the practitioner with his or her patients. Practitioners may accept bribes or excessive fees for expert testimony, and they may make narcotics and other substances that can be misused available to those in their care. They may perform illegal abortions or treat unreported gunshot wounds.

Incidence In the UK in 1986 it was reported that the amount cost of medical negligence or incompetence was 19 million pounds. Abuses have been reported in hospitals and health care facilities, and in the private offices or treatment facilities of practitioners. Specialists, as opposed to general practitioners, are most cited and frequently those treating women. The problems of ageing and incurable terminal conditions and diseases are also susceptible to perpetuations of medical fraud. Ineffectual cures may be offered for everything from cancer to baldness. In the USA in 1989 it was reported that there was far more medical malpractice than indicated by malpractice suits, with over 1 per cent of hospital patients surveyed having been treated negligently. Of those so treated only 3 per cent filed lawsuits because of the expense, the need for expensive expert witnesses and the years required to complete the trial.

Refs Bernstein, Arthur H *Avoiding Medical Malpractice* (1987); Bernzweig, Eli *The Nurse's Liability for Malpractice*; Brushwood, David B *Medical Malpractice* (1986); Charles, Sara C and Kennedy, Eugene *Defendant* (1985); Harney, David M *Medical Malpractice* (1987); Inlander, Charles, et al *Medicine on Trial* (1988); Lewis *Anesthesiology Malpractice* (1989); Lewis, Scott M and McCutchen, Jeffrey R *Emergency Medical Malpractice* (Date not set); Tobias, Andrew *Treating Malpractice* (1986); Zimmerman, Roy R *Malpractice I* (1984).

Broader Unethical practices (#PC8247) Abuse of scientific power (#PF2692)
Unethical professional practices (#PC8019).
Narrower Psychiatrism (#PF6351) Medical quackery (#PD1725)
Surgical malpractice (#PE4736) Dehumanization of health care (#PD7821)
Unauthorized medical practice (#PE5637) Prescription of inappropriate drugs (#PE3799)
Unnecessary health system referrals (#PE0952)
Excessive medical intervention in childbirth (#PE7705)
Bogus psychiatrists and personal counsellors (#PJ1180)
Medical practitioners refusing to treat patients (#PE5027)
Unethical experiments with drugs and medical devices (#PD2697)
Administering of medical drugs for non–medical purposes (#PE3828).
Related Irresponsibility (#PA8658) Induced abortion (#PD0158)
Inhumane scientific activity (#PC1449) Unethical ophthalmic practice (#PE1369)
Inhumane participation of the medical profession in torture (#PE4015).
Aggravates Torture (#PB3430) Iatrogenic disease (#PD6334)
Inadequate medical care (#PF4832) Unnecessary health tests (#PF5679)
Human disease and disability (#PB1044) Malpractice in plastic surgery (#PE4429)
Secrecy of medical facts and records (#PF5983) Unethical practices of health services (#PE3328)
Unethical medical experimentation on prisoners (#PE4889)
Deliberate imbalancing of population sex ratio (#PF3382)
Unethical experimentation using aborted foetuses (#PE4805)
Trafficking in children for medical exploitation (#PD4271)
Increase in insurance claims for medical negligence (#PE4329).

♦ PD5774 Consumerism

Self–indulgent consumerism — Overemphasis on the consumer role — Diverted consumer spending

Nature In wealthier countries, replacement purchases for clothes, entertainment products, automobiles and home–furnishings are made with a frequency that is not justified by the wear of these products. It is the profligacy of a self–indulgent consumerist ethic that is responsible for the excessively high retail turnover. Although it may be argued that this creates jobs, it nevertheless depletes resources, and may also indicate an imbalance of wealth in a society where there are people starving or being denied essential social services, while at the same time department stores are registering record sales. This ethic leads to self–indulgence and a cheapening of human values.

The average individual is most often addressed not as a citizen, a worker, a thinker, or any of his other roles, but as a consumer. This eventually becomes the role in which the individual sees himself most clearly. This image of being a consumer limits the ability to relate to others in appreciation, friendship or even, finally, mutual respect.

Broader Materialism (#PF2655) Abuse of economic power (#PC6873)
Behavioural deterioration (#PB6321).
Related Repression of self–consciousness (#PC1777)
Disjointed patterns of community identity (#PF2845).
Aggravates Proliferation of commercialism (#PF0815)
Static and unrelated social roles (#PF1651).
Aggravated by Self–indulgent societies (#PF5466)
Ineffective mechanisms for functional training (#PF1352).
Reduced by Anti–consumerism (#PF3511).

♦ PD5783 Male mid–life crisis

Male climacteric — Male menopause

Nature The male climacteric is an infrequent and pathologic accompaniment to the process of ageing. True organic or gonadal, endocrine aetiology, with testicular atrophy and degeneration is possible, accounting for physiological causes. Psychoneurotic factors include breakdown revolving around general anxiety or psychogenic impotence, and over–reaction to life and mid–life changes and events, including destabilizing changes in family make–up or relationships, vocational changes including job loss, and deaths of friends or close relatives. The major symptoms include loss or marked decrease of energy, libido and purpose. These may be incident to depression and, in aggravated cases, can lead to suicide, psychosis or violence. The sufferer may destroy what remained initially unaffected; it is frequently his marriage, either legally by divorce or separation, or qualitatively. Thus the combination of female menopause and male mid–life in the same marriage is a time requiring special, mutual sensitivity and understanding, and possibly professional counselling.

Counter–claim The mid–life crisis is nothing more than a myth. In middle age people are not trying to strike out new directions in their lives, but instead they become closer to their spouses, children and friend and find tranquillity and compassion.

Refs Anderson, Kenneth *Symptoms after Forty* (1988); Ruebsaat, Helmut J and Hull, Raymond *The Male Climacteric* (1975).

Broader Mental suffering (#PB5680) Personal life crises (#PD4840)
Ignorance of lifelong human development (#PF5759).
Related Mental illness (#PC0300) Female mid–life crisis (#PD5675).
Aggravates Involutional depression (#PE0655).

♦ PD5788 Vulnerability of land–locked developing countries

Dependency of land–locked developing countries — Insecurity of land–locked developing countries — Land–locked countries

Nature The land–locked developing countries are among the very poorest of the developing countries. Of the 21, 15 are also classified by the United Nations as least developed. Additional obstacles resulting from their geographical situation render their economies particularly vulnerable and reduce their overall ability to cope with the mounting challenge of extreme under–development which all least developed countries currently face. Land–locked countries have to incur high export and import expenses for transport and communications, and other costs that result from the maintenance of higher levels of inventories and thus of greater storage facilities owing to the unpredictability of transit traffic flows, as well as from the development and maintenance of alternative transit routes. The lack of territorial access to the sea also means that the seaborne trade of a land–locked country depends unavoidably on transit through another country.

There are several, less–visible, adverse consequences inherent in such dependence, such as that transit costs incurred by land–locked countries have to be made in foreign exchange, and moreover, land–locked countries are dependent on the transport policy of coastal countries and often on their transport enterprises and transport facilities. Although the implications of this cannot be generalized, the interests of land–locked countries may not always be similar to those of coastal states and thus the former may be disadvantaged as a result of policy measures such as those related to the pricing of transport facilities, the protection of particular modes of transport and the limitation of access to routes and facilities. Typically, the land–locked countries have no influence on the development of the transport infrastructure in the coastal countries and since some of these countries are themselves among the least developed and have limited resources for investment they may not necessarily give transit, communications and port facilities within their borders the kind of priority which the land–locked countries – from their point of view – would accord.

The land–locked position of a country may seriously inhibit the expansion of its trade and economic development. Land–locked developing countries are generally among the very poorest of the developing countries. Their position hampers them in their efforts to take advantage of the international measures in favour of developing countries, because of their isolation from markets caused by high transport costs, and absence of transit rights and facilities in neighbouring coastal states. The land–locked state may be obliged to compete with its neighbouring transit state, often itself a developing country, for internal and external resources and for export markets, with the former always at the mercy of unilateral measures imposed by the latter to maintain its own trading advantage. This situation raises, in principle, the possibility of some degree of monopolistic exploitation, irrespective of whether deliberately practised (in the pricing of transport facilities and their use or in the limitation of access to routes, facilities, or ports) or whether it is done inadvertently by giving less attention to services to the land–locked country than to national traders or authorities. Rather different difficulties occur when the immediate neighbour of a land–locked country is a developed country (as in the case of Lesotho, Botswana, and Swaziland).

Incidence Some 20 per cent of the nations of the world, most of them developing, are land–locked.

The land–locked country closest to the sea is Swaziland with a transit route of 220 kilometres to the port of Maputo in Mozambique, followed by Bolivia with a route of 450 kilometres to the port of Arica in Chile, and the Lao People's Democratic Republic with a 670–kilometre route to the port of Bangkok in Thailand. The most remote are Afghanistan, Chad and Rwanda, all of which have a minimum transit distance of nearly 2000 kilometres.

The following countries are land–locked: **Af** Botswana, Burundi, Central African Rep, Chad, Lesotho, Malawi, Mali, Niger, Rwanda, Swaziland, Uganda, Upper Volta, Zambia. **Am** Bolivia, Paraguay. **As** Afghanistan, Bhutan, Laos, Nepal, Sikkim. **Eu** Andorra, Austria, Czechoslovakia, Liechtenstein, Luxembourg, Switzerland, Vatican.

Refs Glassner, M J *Bibliography on Land–locked States* (1986); United Nations *Land–Locked States* (1987).

Broader Vulnerability of developing countries (#PC6189)
Geographically disadvantaged countries (#PF9247)
Non–viability of small states and territories (#PD0441).
Narrower Military threats to land–locked countries (#PE5837)
Inefficient shipping procedures and documentation (#PF5872)
Vulnerability of lakes and rivers in land–locked countries (#PE5813)
Weakness of infrastructure in land–locked developing countries (#PE7000)
Unreliable transit services for land–locked developing countries (#PE5836)
Excessive costs of sea access for land–locked developing countries (#PE5812)
Inadequate roads and transport in land–locked developing countries (#PE5824)
Inadequate air transport service for land–locked developing countries (#PE5800)

Imbalances in exports and imports of land-locked developing countries (#PE5920)
Inadequate rail transport in developing land-locked and transit countries (#PE5848)
Unreliable telecommunication services for land-locked developing countries (#PE5860)
Inadequate port and storage facilities for land-locked developing countries (#PE5908)
Lack of adherence to international transit conventions for land-locked countries (#PE5789)
Excessive costs and unsuitability of insurance for land-locked developing countries (#PE5896).
Related Underdevelopment (#PB0206) Competition between states (#PC0114)
Abuse of land-locked and island countries as havens (#PE5861).
Aggravated by Uncoordinated international river basin development (#PD0516)
Conflicting claims concerning off-shore territorial waters (#PD1628)
Lack of skilled workers in the transport sectors of land-locked developing countries (#PE5884).

♦ PD5803 Corrosion of ships
Fouling and rust of marine equipment
Nature The action of salt on metal, oxidation, incrustation and fouling by objects, cause high costs to ocean transport due to ship damage and preventative maintenance. Sea-water, spray, and atmospheric corrosion, as well as rust, can reach all parts of ships interiors and exteriors, while incrustation of ship hulls, notably the underside, by algae, bacteria and other sea-borne organisms destroys paint, baring the metal to corrosion, and by heavy layering, slows ship movement and causes higher fuel costs. Object pollution of harbour beds causes anchor losses, hull scrapings, and effluent attacks on metal. Weakened or encrusted equipment and structures cause accidents, impair navigational ability, and may result in the total loss of a vessel.
Refs Acker, Robert F et al (Eds) *Proceedings of the third international congress on marine corrosion and fouling* (1974); Deere, Derek H *Corrosion in Marine Environment International Sourcebook I* (1977).
Broader Obstacles for international ocean shipping (#PD5885).
Aggravated by Wrecks and derelicts as hazards (#PE5340).

♦ PD5807 Excessive power and independence of transnational corporations
Broader Monopoly of power (#PC8410) Transnational corporation imperialism (#PD5891).
Narrower Non-accountability of transnational enterprises (#PF1072)
Political intervention by transnational corporations (#PE0032)
Concentration of power by transnational corporations (#PE0766)
Erosion of national sovereignty by transnational enterprises (#PE1539)
Domination of developing countries by transnational corporations (#PE0163)
Domination of economic integration by transnational corporations (#PE1322)
Ineffective international regulation of transnational corporations (#PF0691)
Inadequate relationship between transnational corporations and specialized agencies of the United Nations (#PE0106).
Aggravates Vulnerability of socio-economic systems from globalization (#PF1245).
Reduced by Burden of conflicting national regulations on transnational corporations (#PE2200).

♦ PD5830 Protectionism in agriculture and the food production industries
Agricultural trade restrictions
Nature A broad indicator of the levels of protection granted by governments to different food products is the divergence between domestic and world prices. These price gaps are assumed to reflect cumulative and interactive tariff and non-tariff barriers, price supports, and various special programmes and stabilization measures.
Incidence A study by the FAO of figures covering five years (1977–1982) considered the ratio of domestic producer to international prices. Nine commodities were covered in the EEC, the USA and Japan. A ratio higher than 1.0 indicated that world export prices were higher than domestic farm support prices, as a result of 'ad valorem' tariff equivalents. In other words, commodities with high ratios were selling at a premium in the export market, or there were very low domestic support levels. From the EEC, all commodities studied except pork and rice (wheat, barley, maize, soybeans, butter, beef and sugar) had ranges from 1.1 to 1.6. From the USA all grains were less than 1.0; butter was 1.6 (beef, pork and sugar did not figure in the export market). From Japan, however, wheat, barley and soybeans ranged from 3.9 to 4.6, sugar was 3.0, and butter 3.2. Beef was 2.3 and only pork showed higher domestic prices.
There are wide differences in the prices paid to farmers in the EEC, Japan and the United States, partly reflecting the entirely different agricultural environment and cost structures in these countries. Prices for cereals in the USA in 1980 and 1981 were on average 50 percent to 75 percent lower than those paid in the EEC and about one-seventh of those paid in Japan. Unlike the EEC and Japan, the US farm support prices for cereals and soybeans were also well below world export prices. In addition there are fundamental differences in the techniques of supporting farm incomes. In the United States, direct intervention in cereal market prices is comparatively minimal and, in periods of tight supply, its producer support has been restricted to sporadic deficiency and disaster payments. Recently, however, considerable government resources have been allocated in the form of farm lending through the Commodity Credit Corporation (CCC). Between 1981/82 and 1982/83, the levels of price support for wheat rose by 25 percent, for maize by 15 percent, and for rice by 14 percent. While such price increases were insufficient to offset the declining trend in farm incomes they contributed to the building up of stocks of cereals and dairy products. As regards sugar, the US Government has periodically operated price support measures through loan or purchase programmes. The differential between the world price and the domestic price for raw sugar averaged 15 cents per pound in the third quarter of 1982. With the fall in world sugar prices the United States government increased import fees in April 1982. When this proved insufficient to raise internal prices to the market stabilization level, it resorted to import quotas.
Farm support operations in the EEC are carried out through a system of guaranteed prices covering a large proportion (about three-quarters in recent years) of the Community's total agricultural production, and a system of variable import levies. By incorporating a variable charge into the delivered price of imports from third countries, the levies maintain foreign prices at or above those received by domestic producers. It has been estimated that for nine main agricultural products of the Community, the additional nominal protection from levies is 45 percent, over three times the average tariff rate of 14 percent. By January 1983, import levies for wheat represented about 55 percent of the support (intervention) price for bread wheat and levies for coarse grain about 55 to 60 percent of the intervention price. Levies for cereals, in particular wheat and barley, rose sharply in 1982, reflecting the decline in international prices. The effects of these mechanisms, insulating farmers from international competition, are reflected by the high levels of support prices granted by the Community in relation to international prices. High prices and relatively stable domestic demand have resulted in increasing overproduction of a number of commodities including wheat, sugar, dairy products, beef and veal. While the Community remains a net agricultural importer, it has also become the second largest agricultural exporter after the United States.
In Japan, farm support is provided through payments from tax and government bond revenues, through public corporations and through income transfers from consumers who pay prices often several times higher than world market prices. About half of the subsidy expenditure is related to the rice programme, under which about half of the country's rice crop is purchased at supported prices and then resold to wholesalers at a loss. However, other agricultural products, including soybeans and wheat, benefit from the government's efforts to move away from overdependence on rice. For livestock products, the profits from the sale of imported beef for which a quota system is applied provide subsidies and low interest loans to livestock producers. They also subsidize the storage of surplus production. A similar system is operated on the sales of imported wheat, barley and rice, the benefits of which are being used to help finance cereal subsidy programmes. When added together the effects of the different programmes on domestic prices are considerable. Japan supports its food production at higher levels than any other major importing country, while still leaving scope for imports. However, some levelling off in support prices has recently taken place, reflecting budgetary strains, supply/demand adjustments, and the decline in international prices.
Broader Protectionism in international trade (#PC5842).
Narrower Over-subsidized agriculture in industrialized countries (#PD9802).
Related Plantation agriculture (#PD7598)
Restrictive practices in the food and live animals trade (#PE0342).
Aggravates Disincentives against farming (#PD7536)
Instability of trade in tobacco and tobacco manufactures (#PE0572)
Inadequate mechanisms for securing sufficient food supplies (#PF2857).

♦ PD5833 Inequality inducing effects of television
Aggravates Human inequality (#PA0844) Declining sense of community (#PF2575).

♦ PD5843 Gender stereotyping
Sexual stereotyping
Refs Andree, Michel *Down with Stereotypes* (1987); Gilman, Sander L *Difference and Pathology* (1985); Hägglund, Solveig *Sex-Typing and Development in an Ecological Perspective* (1986); The Documentation of the Council of Europe (Eds) *Sex Stereotyping in Schools* (1982); Williams, Christine L *Gender Differences at Work*.
Broader Stereotypes (#PF8508) Sexual discrimination (#PC2022).
Narrower Gender-differentiated toys (#PE4664).
Aggravates Discrimination against women in education (#PD0190).

♦ PD5845 Regional environmental degradation
Regional ecosystems degraded by exploitative development
Nature During the development process, local disbenefits, such as in the form of negative ecological consequences, tend to accumulate and to become apparent eventually as regional degradation.
Broader Regional disparities (#PC2049) Natural environment degradation (#PB5250).
Aggravated by Destruction inherent in development (#PF4829).

♦ PD5859 Property damage
Risk of damage to premises — Destruction of property — Criminal property damage — Uncontrolled property damage
Nature The damaging or destruction of property through the intention of a person or persons or through an act of nature. Intentional destruction of property is distinguished from theft by the fact that the person is none the richer for his injustice: he has taken nothing that can be returned.
Broader Human destructiveness (#PA0832) Criminal harm to property (#PD5511)
Deteriorating community identity (#PF2241).
Narrower Arson (#PE5505) Damage to goods (#PE4447) Tidal water damage (#PJ9589)
Unrestrained animal damage (#PJ9217) Uncontrolled waterfront damage (#PG8619)
Deliberate destruction of equipment (#PG8408).
Aggravates Interruption risk (#PF9106) Lack of community participation (#PF3307).
Aggravated by Sabotage (#PD0405) Vandalism (#PD1350)
Drunkenness (#PE8311) Gang warfare (#PD4843)
Injurious accidents (#PB0731) Inadequate safeguards against fire (#PD1631).

♦ PD5865 Economic imbalances among industrialized countries
Incidence These are characterized by the USA federal budget deficit (associated with the current account deficit and high interest rates), the disproportionately low level of domestic demand in Japan and Germany FR, and the high level of current account surpluses maintained by the newly industrialized countries of East Asia.
Broader Faltering structural adjustment in the world economy (#PF9664).
Aggravated by Mismatch of national macroeconomic policies among industrialized countries (#PF5000).

♦ PD5876 Subversion of international agreements
Nature Decisions or principles that are adopted at the international level are apparently being ignored, countermanded or eroded in many instances through unilateral regulation by states. One reason is disenchantment over the leisurely pace at which international conventions or other agreements are implemented or indeed are left to rest idle or seem to be subverted by one means or another. Another reason is the insidious world-wide effect of the economic recession of recent years, which, in the absence of effective international countervailing measures, has encouraged unilateral initiatives. Commercial tension between states is serious in many trades and has raised a spectre of uncooperativeness on a global scale. Unnecessary decisions already taken at the international level cause other areas of outstanding dispute to be of more unmanageable proportions. Threats and counter-threats against unilateral activities have been announced by many countries; if pursued, these would further aggravate international relations and retard the creation and implementation of instruments of international cooperation.
Broader Lack of international cooperation (#PF0817).
Narrower Disagreement within alliances (#PD2620).
Related Limited acceptance of international treaties (#PF0977).
Aggravates International insecurity (#PB0009).
Aggravated by Competition between states (#PC0114).

♦ PD5885 Obstacles for international ocean shipping
Nature Instability in the shipping industry in terms of cargo values and tons shipped cannot be separated from the major economic variables of production of manufactured goods and primary commodities. Shipping and shipbuilding industries also have a mutual dependence. Increased trade causes demand for tonnage; decreased trade means unutilized capacity. On the other hand, under-tonnaging and overtonnaging in relation to demand seems to be an industry characteristic, causing instability in freight rates. Tensions and uncooperative attitudes arise from competition between traditional maritime countries and organizations of major shipowners, and countries, many of them developing, and their shipowners and shippers seeking either to enter the industry or obtain fairer rates. Protectionism and politics play their parts in obstructing development of a better organization of international shipping. Within the shipping industry problems are generated by the instabilities in the commodities trades affecting cargoes. There are specific problems for tanker shipments of liquids, and for dry cargoes of which over 40 percent are bulk commodities.
Incidence The 1980s have brought crisis to the industry. During 1982 the total tonnage laid-up reached a level which was three times that in 1981, and comprised about 7 percent of the world fleet.
Broader Inadequate laws of the sea (#PF5923).
Narrower Corrosion of ships (#PD5803) Ocean shipment delays (#PE5886)
Hazards to navigation (#PE3868) Restrictive shipping practices (#PD0312)
Inequities in marine insurance (#PE5802) Flag discrimination in shipping (#PD0700)

PD5885

Wrecks and derelicts as hazards (#PE5340)
Instability of ocean freight rates (#PE5850)
Imbalances in types of ships built (#PE5874)
Unfair surcharges in ocean freight (#PF5922)
Inequities in ship owner registration (#PE5875)
Protectionism in the shipping industry (#PE5888)
Unfair shipping practices in bulk trades (#PE5849)
Excess capacity in the shipping industry (#PE5897)
Imbalances in the dry bulk shipping industry (#PE5839)
Instability of the maritime shipping industry (#PE5791)
Imbalance in shipbuilding industry distribution (#PE5863)
Obstacles for developing countries ocean shipping (#PE5909)
Transnational corporation control of bulk shipping (#PE5804)
Over-concentration of ownership of maritime fleets (#PE5825)
Imbalance of developing countries share in shipping (#PF5912)
Ineffective self-regulation in the shipping industry (#PF5840)
Evasion of shipping regulations and taxes by flags of convenience (#PE5873)
Violation of the right to the international freedom of movement of shipping (#PE8899)
Lack of technical infrastructure for maritime commerce in developing countries (#PE5814).
Related Lack of world maritime integration (#PE5801)
Lack of capital investment in developing countries (#PE5790)
Domination of the shipping industry by transnational corporations (#PE1620).
Aggravated by Oil pollution (#PE1839)
Maritime fraud (#PE4475)
Cargo insecurity (#PE5103)
Marine accidents (#PD8982)
Transfer pricing (#PE1193)
Seizure of cargoes (#PE0125)
Restrictive transport insurance practices (#PD0881)
Political interference with port operations (#PE5827)
Obstacles to efficient port utilization and operation (#PE5921)
Instability of economic and industrial production activities (#PC1217).

♦ PD5891 Transnational corporation imperialism

International businesses — Transnational enterprises — Multinational enterprises
Refs Baumer, Jean-Max and Gleich, Albrecht von *Transnational Corporations in Latin America* (1982); Réiffers, Jean-Louis and Cartapanis, André *Transnational Corporations and Endogenous Development* (1982); United Nations *Transnational Corporations - A Selective Bibliography 1983-1987* (1988); United Nations Centre on Transnational Corporations *Bibliography on Transnational Corporations* (1978); United Nations Centre on Transnational Corporations *Environmental Aspects of the Activities of Transnational Corporations* (1985); Zurawicki, L *Multinational Enterprises in the West and East* (1979).
Broader Dictatorship (#PC1049)
Economic imperialism (#PC3198).
Narrower Transfer pricing (#PE1193)
Corporation financial secrecy (#PE1571)
Inappropriate transfer of technology (#PE5820)
Lack of consumer influence on industry (#PE1940)
Labour tensions involving transnationals (#PE5927)
Unfair pricing by transnational corporations (#PE5855)
Non-accountability of transnational enterprises (#PF1072)
Monopolization of information within organizations (#PF2856)
Political intervention by transnational corporations (#PE0032)
Social irresponsibility of transnational corporations (#PE5796)
Retarding of development by transnational corporations (#PE0234)
Domination of advertising by transnational corporations (#PE2193)
Monopolization of technology by transnational corporations (#PE1918)
Restrictive pricing policies of transnational corporations (#PE2396)
Domination of labour relations by transnational corporations (#PE1187)
Erosion of national sovereignty by transnational enterprises (#PE1539)
Restrictive business practices of transnational corporations (#PE5915)
Manipulation of transfer prices by transnational corporations (#PE0245)
Excessive power and independence of transnational corporations (#PD5807)
Domination of the copper industry by transnational corporations (#PE2084)
Control of industries and sectors by transnational corporations (#PE5831)
Domination of developing countries by transnational corporations (#PE0163)
Domination of economic integration by transnational corporations (#PE1322)
Domination of the shipping industry by transnational corporations (#PE1620)
Ineffective international regulation of transnational corporations (#PF0691)
Domination of the automobile industry by transnational corporations (#PE1469)
Participation of transnational corporations in the apartheid system (#PE1996)
Restriction of free market competition by transnational corporations (#PE0051)
Disruption of domestic social policies by transnational corporations (#PE1957)
Inadequate supply of pharmaceutical products in developing countries (#PE4120)
Burden of conflicting national regulations on transnational corporations (#PE2200)
Discrimination in lending by transnational banks in developing countries (#PE4310)
Coercive use of economic power by transnational enterprises against labour (#PE0207)
Aggravation of instability in exchange rates by transnational corporations (#PE0980)
Domination of agricultural equipment industry by transnational corporations (#PE1448)
Consequences of restrictive business practices of transnational enterprises (#PE1799)
Disproportionate control of global economy by limited number of corporations (#PE0135)
Control of marketing and distribution channels by transnational corporations (#PE2397)
Minimal export promotion by transnational corporations in developing countries (#PE1598)
Disproportionate influence on national economies of limited number of corporations (#PE1922)
Foreign currency manipulations in accounting records of transnational corporations (#PE2145)
Destabilization of monetary systems and exchange notes by transnational corporations (#PE5903)
Insensitivity of transnational corporations to consumer needs in developing countries (#PE1011)
Limited market access due to the product differentiation of transnational corporations (#PE2328)
Interference of transnational banks' off-shore borrowing with domestic monetary policies (#PE4315)
Inadequate negotiation of entrance terms for transnational corporations in developing countries (#PE0853)
Inadequate relationship between transnational corporations and local industry in developing countries (#PE1511)
Domination by transnational corporations of the domestic name-brand food sector in developing countries (#PE1796).
Related Imperialistic distribution system (#PD7374)
Transnational corporation control of bulk shipping (#PE5804)
Abuse of monopoly power of state-owned or state-controlled enterprises (#PE0988).
Aggravates Socially unintegrated expatriates (#PD2675)
Ineffective industry self-regulation (#PF5841).

♦ PD5904 Misuse of chemicals

Refs Banghawe, A F; Mngola, E N and Maina, G (Eds) *Use and Abuse of Drugs and Chemicals in Tropical Africa* (1975); Kruus, P and Valeriote, I M *Controversial Chemicals* (1984).
Narrower Chemical torture (#PD5204)
Misuse of pesticides (#PD4629)
Herbicide damage to crops (#PD1224)
Environmental hazards from fertilizers (#PE1514).

♦ PD5905 Currency black market

Incidence In the face of the chronic shortage of foreign exchange in some countries, the currency black market becomes a thriving business sector. Companies are forced to resort to illegal methods to obtain the dollars needed to maintain their operations. Those businesses which manage to earn some foreign exchange 'salt' it away abroad for fear of a freeze on foreign currency accounts. This has become extensive in certain countries such as the Philippines. With the flow of fresh foreign loans and investments reduced to a trickle and the ensuing strict trade and exchange controls, only companies involved in priority industries, such as importers of oil and food and manufacturers for export, have access to legal dollars. For non-priority companies, about the only source of dollars for imports is the black market. Another example is China, where there is a flourishing foreign currency black market and speculative trading in scarce products. There is a small but lucrative black market operating in actual foreign currencies, especially US and Hong Kong dollars. Some official organizations may also have been involved in currency trading and speculation, using money obtained through Bank of China loans.
Broader Underground economy (#PC6641).
Narrower Currency black market in socialist countries (#PD2413).
Aggravated by Mismanagement of exchange rate system (#PF1874).

♦ PD5907 Denial of cultural rights

Nature People are denied the right to participate in the cultural life of the community, to enjoy the arts and to share in scientific advancement and its benefits.
Refs Szabo, L *Cultural Rights* (1974); United Nations *Cultural Rights as Human Rights* (1970).
Broader Denial of human rights (#PB3121).
Narrower Loss of cultural identity (#PF9005)
Denial of right to education (#PD8102)
Denial of right to a cultural life (#PE6561)
Denial of right to benefits of science (#PF6077)
Denial of right of peoples to use their own language (#PE2142)
Exclusion of disabled persons from social and cultural life (#PD0784).
Related Cultural repression (#PC8425).
Aggravates Ethnocide (#PC1328)
Endangered cultures (#PB8613).
Aggravated by Forced assimilation (#PC3293).

♦ PD5936 Declining breeds of cultivated plants

Related Endangered species of plants (#PC0238).
Aggravated by Human contingency (#PF7054).

♦ PD5941 Crimes related to military service obligations

Nature It is a crime for a civilian to obstruct the recruitment into, cause insubordination within or evade meeting obligations to the military services. Generally this is applied only in war time.
Broader Military disobedience (#PD7225)
Crimes against national security (#PC0554).
Narrower Draft evasion (#PD0356)
Causing insubordination in the armed forces (#PE5782)
Obstruction of recruiting or induction into armed forces (#PE3912).

♦ PD5948 Unethical practices in local government

Irresponsible local government — Negligence in local government — Corruption in local government — Municipal corruption — Urban corruption
Broader Official corruption (#PC9533)
Institutionalized corruption (#PC9173)
Unethical practices of government (#PD0814).
Aggravates Urban slums (#PD3139).
Aggravated by Conflict of interest among parliamentarians (#PE3735).

♦ PD5962 Endangered family farms

Family use farming
Nature Although the vital economic and social importance of family farming is now realized to a greater extent than in the past, a 1983 European study emphasized that governments are not pursuing an agricultural policy designed to secure the economic survival of family farms, although most European governments recognize these farms as the model for their agricultural policies. Land area and soil quality are the main problems of family farms in many regions, namely because of a lack of programmes for the protection and maintenance of agricultural land, insufficient, measures to preserve the unity of farms, and a lack of directives to facilitate land purchase by farmers - granting of lands, tax exemptions, right of pre-emption. Modern farm leasing legislation also constitutes a hindrance. The maintenance and development of family farms is closely bound up with cooperation and producers' associations, and there is a lack farmers' collaboration with these organizations.
Broader Vulnerability of farming (#PC4906)
Precarious basis for family economics (#PF1382).
Related Uneconomic size of farms (#PJ2079).

♦ PD5965 Unjust dismissal of workers

Narrower Redundancy of workers (#PD8007)
Debasing retirement possibilities (#PG8268)
Haphazard provision of consumer services (#PF2411)
Discrimination against women at retirement age (#PE6069)
Unemployment caused by environmental conservation (#PD0467)
Dismissal of workers to prevent legal strike action (#PE7620).
Aggravates Unemployment (#PB0750).
Aggravated by Unethical practices of employers (#PD2879).

♦ PD5971 Diarrhoea

Infectious diarrhoeas — Diarrhoeal disease acute — Acute diarrhoeal disease
Nature Although diarrhoea is such a common condition that everyone has had it at some time or another, it is also a killer disease causing the death of between four and five million children each year. The frequent watery stools drain the body of its essential water and salts, and if this condition is not corrected, the diarrhoea sufferer may die. Germs which infect the bowel are the main cause, and this is the consequence of poor sanitation and bad hygiene - both environmental and personal hygiene. In the many communities which have inadequate housing, and poor water supplies, cooking facilities and sanitation systems, it is difficult to prevent these infections. The condition is usually contracted by swallowing minute amounts of excreta from people who have infectious diarrhoea. Although an average adult takes in about 2 litres of fluid per day by mouth, and excretes about one tenth of that (0.2 litres) in the stools, large amounts of digestive juices are poured into the tubular tract, so that about 9 litres of fluid enter the system and are absorbed again every 24 hours. The germs causing diarrhoea usually stimulate excessive secretion into the bowel, but the ability of the gut to absorb is often not increased. The serious consequences of diarrhoea are thus the loss of essential water and salts in the liquid stools, in other words dehydration and electrolyte loss.
Incidence Diarrhoeal diseases are most widespread in the developing countries; they are transmitted by human faecal contamination of soil, food and water. Only about a third of the people in the world's least developed countries have dependable access to a safe water supply and adequate sanitary facilities. One out of every 20 children born into the developing world dies from acute diarrhoea before reaching the age of five.
Refs Chen, Lincoln C and Scrimshaw, Nevin S *Diarrhea and Malnutrition* (1983); Gorbach, Sherwood L (Ed) *Infectious Diarrhea* (1986).
Broader Tolerated atrocities (#PC4710)
Infectious and parasitic diseases (#PD0982).
Narrower Colitis (#PE7970)
Diarrhoea in children (#PE9751).
Related Intestinal infectious diseases (#PE9526).
Aggravated by Allergy (#PE1017)
Common cold (#PE2412)
Teeth disorders (#PD1185)
Enteric infections (#PD0640)
Hazards of bottle-feeding (#PE4935)
Parasites of the human body (#PE0596)
Protein-energy malnutrition in infants and early childhood (#PD0331).

♦ PD5972 Fortified frontiers

Frontier barriers — Fenced frontiers

Nature Ditches, barbed wire fences, brick and stone walls are all means of building obstacles between two peoples, sometimes to keep one group in and sometimes to keep the other group out. The Great Wall of China, the Maginot Line, and the Berlin Wall are examples.
Incidence The United States is currently digging a ditch 1.5 metres deep and 4 metres wide near the main border crossing with Mexico at Tijuana to "address a security and safety concern".
Related Illegal movement across frontiers (#PC2367)
Super-power monopoly of advanced nuclear warfare technology (#PD4445).

♦ PD5980 Street children
Unprotected street youth
Nature Children for whom the street or the city has become a habitual abode and/or source of livelihood are often inadequately protected, supervised or directed by adults. They are obliged to scavenge, steal, drug and prostitute themselves in an attempt to face a life that offers little hope. As a result they suffer mental and physical deprivation for which society, in the long run, must pay.
Incidence A 1982 UNESCO report estimated the number of street youths as 200,000 in Istanbul; 10,000 in Bogota; and 2 million in Rio de Janeiro. In all there are an estimated 70 million such children world-wide, of whom 40 million are in Latin America. An alternative estimate is for 30 million street children living by their wits in the world's cities.
Refs Angelli, Susanna *Street Children* (1986); Moorehead, Caroline (Ed) *Betrayal* (1989).
Aggravates Lack of education (#PB8645) Juvenile delinquency (#PC0212)
Juvenile prostitution (#PD6213).
Aggravated by Family poverty (#PC0999) Family violence (#PD6881)
Family breakdown (#PC2102) Abandoned children (#PD5734)
Physical maltreatment of children (#PC2584).

♦ PD5984 Government deficits
Public sector deficits — Fiscal deficits — Unsustainable fiscal deficits — Imprudent budget policies
Nature An excessive one year deficit in governmental accounts or a pattern of continuing deficits over a number of years creates public or national debt. The remedy of reducing expenditures is rarely effective, some elements being inviolable in some countries, as, for example, social welfare spending, civil service payrolls, or defence and foreign aid budgets. The government may also be constrained not to raise taxes, so that financing of the deficits must be done through the creation of fiat money or by borrowing on the non-governmental money markets. Deficits indicate excessive governmental spending. Governmental borrowings to finance them reduce the financial resources available for private investment. Governmental increases in taxation reduce the incentives to invest. If there are no disincentives, deficit borrowings will cause interest rates to go up as there will be a demand for capital exceeding supply. Unsustainable fiscal deficits provoke capital flight because domestic savers anticipate a coming crisis that is likely to involve a major devaluation and new taxes on income and consumption.
Incidence Deficits can be more easily absorbed by countries with high rates of domestic private savings and well-developed capital markets. Thus a relatively high deficit need not cause problems in an efficient high-saving economy, whereas in a low-saving, highly distorted one, even a small deficit might be destabilizing.
Claim Fiscal deficits are a principal cause of the international debt crisis, both directly, because they mean greater public borrowing, and indirectly, because they encourage the private sector to send its capital to elsewhere.
Counter-claim Deficits in themselves do not automatically imply macroeconomic problems. If the use of public resources is sufficiently productive, future income can be generated to cover the servicing costs of any debts incurred. If expenditures rise owing to temporary factors, such as wars or natural disasters, then deficits may be justified as a way to spread the cost over several years.
Aggravates Speculative flight of capital (#PC1453)
Over-reliance of government on money creation (#PF9560)
Dependence of developing countries on external financing for development programmes (#PE7195).

♦ PD5987 Drug abuse by adolescents
Youth drug abuse — Drug addicted children
Nature Adolescence is a time when many problems occur (social, family and economic) as young people seek to become established and achieve independence from their parents. Such problems are likely to be even greater among non-students than among students, since these youngsters often move away from home and have the support neither of their parents nor of schools and teachers. Feelings of alienation, low self-worth and resentment are higher among the illiterate and unemployed than among those who have been successful at school. Young people therefore, in particular the disadvantaged, are susceptible to a variety of social and psychological problems which make them open to the possibility of drug abuse as a means of relief. Factors associated with high risk for drug abuse are: unemployment; living away from home; migration to cities; relaxed parental controls; broken homes and one-parent families; alienation from families; early exposure to drugs; leaving school early; poor use of drugs; family use of drugs. Risky environments include: high urbanization; high rates of crime or vice; areas where drugs are sold, traded or produced; areas where there are drug-using gangs; occupations connected with, for example, tourism, drug production, drug sale; and areas where delinquency is common.
Measures involving the message "don't take drugs" have proved ineffective and unrealistic, particularly in societies where there is virtually unrestricted usage of alcohol and tobacco; emphasis is now being placed on the individual's capacity to manage his use of drugs and make reasonable choices based on factual knowledge about the effects produced by them. Changing attitudes seem to have produced an apparent increased tolerance of drug abuse despite knowledge of the unfortunate, often disastrous, effects this may have. The political will to provide treatment and advice and the money to back such measures is less urgent than in the past, and there are no signs that the drug-abuse explosion is being controlled.
Claim Adolescents are denied the pleasures that adults come to take for granted. They are either given or expected to earn, very small sums of pocket money, and exist under authoritarian pressures at school, at work, at home, and in the community. Sensual pleasures in particular are denied them, specifically sexual and gustatory, the latter implying free, wide-ranging choice of food and drink, and use of tobacco. They have restricted access to the pain-killers or analgesics and stress-relieving tranquillizers that adults may take routinely. They are obliged to sleep under supervised conditions as to places, times and accompaniment, and may also be obliged to attend all family social functions, do chores at home, and attend worship services. The adolescent is thus deprived of freedom of choice when the maturation of his or her personality is demanding a degree of latitude and independence. Secretiveness is the result; secret sexuality and secret 'foods' which include all substances taken into the body, the ones giving pleasant sensations being preferred. Hence the teenage vulnerability to drugs, pushed at them by organized crime, degenerates, and agents of some developing countries whose main crops include the raw materials for drugs and whose hatred of the industrialized nations receives satisfaction in leading hundreds of thousands of youth into addiction.
Refs Beschner, and Friedman, *Youth Drug Abuse* (1979); Wötzel, Horst *Trip into Illusion* (1975).
Broader Drug abuse (#PD0094) Parental permissiveness (#PD5344)
Victimization of children (#PC5512).
Related Children of drug addicts (#PE4609).
Aggravates Youth violence (#PF7498).
Aggravated by Juvenile stress (#PC0877) Corruption of minors (#PD9481)
Neglect of adolescent health care (#PF6061) Criminal investment in youth market (#PD5750).

♦ PD5995 Economic backlash from the richer nations
Incidence The economic threat to industrialized countries from the newly industrializing countries has drawn a sharp backlash from the richer nations. In the USA, steel producers have filed unfair trade-practices complaints against such advanced developing countries as South Korea, Mexico, Argentina and Romania. Other domestic industries, from textiles to petrochemicals, are clamouring for protection. Similar pressures are emerging in Japan and Europe. These pressures are likely to intensify as the developing countries, burdened by large foreign-debt loads, push to increase their exports. This could lead to a sharp increase in trade restrictions by the richer nations, and consequently to a cutback of trade opportunities worldwide.
Aggravates Protectionism in international trade against exports from developing countries (#PD9679).

♦ PD5998 Religious censorship
Nature Censorship of media, school materials (including textbooks, displays, audio-visual products), and public communications of any kind such as speeches, billboard messages, posters, signs, hand-out pamphlets, circulars or circulated private letters, may be effected by religious bodies in cooperation with public officials, at national or at local levels. Religious censorship is aimed at defending dogmas by obliterating dissent; at shielding the purity and the souls of both believers and non-believers in the community even by withholding scientific information or world or local news; and at protecting the institutions of religion, whether the hierarchy personally or its extensions, from loss of political, economic or social advantages.
Incidence Religious censorship is endemic in all state religions and in areas where strong religious sentiment of a homogeneous nature characterizes the electorate. In the southern USA for example, the 'Bible-belt Mid-West', and Puritan-influenced New England, strong feelings against science has prompted excessive reactions to propagation of ideas such as the equality of women with men, legal abortion for medical reasons, and the evolution of man in nature, to name a few 'dangerous' concepts. Since the same behaviour is exhibited in many Islamic countries, notably Iran, it can be concluded that religious censorship is facilitated wherever there is a holy scripture of venerable age. The writings of Marx and Lenin may also be viewed in this light. Pre-literate religious censorship is exhibited in the system of taboos, whose vestiges may remain sub-consciously in modern society.
Refs Putnam, George H *Censorship of the Church of Rome and Its Influence upon the Production and Distribution of Literature* (1967); Puttman, George *The Censorship of the Church of Rome*
Broader Censorship (#PC0067).
Related Self censorship (#PF6080).

♦ PD6002 Abuse of dominant market position in international trade
Nature Anti-competitive and restrictive practices of market-dominant manufactures may include exclusive purchasing and dealing arrangements. Dealers who would sell one manufacturers' products on an exclusive basis may be offered higher margins, free training or other services, and thereby distribution outlets for competitive products may be restricted. Exclusive purchasing arrangements may restrict access to supplies by other manufacturers. Tying arrangements may also be employed, requiring dealers to purchase supplies, replacement parts or services exclusively from manufacturers entering into these agreements.
Incidence In high technology industries, an introduction of a new product offers a stimulus to manufacture of supplies for it or replacement or add-on parts. A leading manufacturer may wish to dominate this secondary market as well, and may withhold technical information and product introduction schedules, as in the information processing, communications and automobile manufacturing industries.
Broader Domination (#PA0839) Abuse of economic power (#PC6873)
Irresponsible international trade (#PC8930).
Aggravated by Capitalism (#PC0564).

♦ PD6005 Maltreatment of prisoners
Nature Corporal punishment, punishment by solitary confinement in a dark cell, and other cruel, inhuman or degrading treatment may be used as punishments for disciplinary offences. Punishment by close confinement or reduction of diet, by stripping in cold weather, and the use of instruments of restraint, such as handcuffs, chains, irons and straitjackets, are often a part of the penitentiary environment for both common criminals and political prisoners. Prison mistreatment may also include insults, mockery and non-delivery of letters.
Incidence Systematic torture and other cruel treatment is inflicted in many countries. Detention incommunicado and detention without trial for long periods create a maximum risk of ill-treatment and torture, as basic procedural rights and avenues of redress appear to be denied to persons thus detained. Disregard of the human rights of detained persons occurs most frequently under states of siege, emergency or exception. Furthermore, even where no state of siege or emergency is officially proclaimed, situations of social and political unrest and civil strife often lead to gross violations of the rights of detainees. They are alleged to occur particularly as regards government repression of national liberation movements, in countries under foreign occupation and in certain situations involving minority groups.
Broader Denial of rights to prisoners (#PD0520) Denial of rights to vulnerable groups (#PC4405).
Narrower Beating of prisoners (#PD2484) Abuse of prison labour (#PD0165)
Ill treatment of prisoners of war (#PD2617)
Unsanctioned maltreatment of prisoners (#PE0998)
Discrimination against prisoners' families (#PE5043)
Maltreatment of prisoners by fellow inmates (#PE0428)
Discriminatory treatment of foreign prisoners (#PE6883)
Unethical medical experimentation on prisoners (#PE4889)
Denial to animals of the right to conditions of life and liberty proper to their species (#PE6270).
Aggravates Prison suicides (#PD8680).
Aggravated by Cruelty (#PB2642) Human torture (#PC3429).

♦ PD6008 State-supported international terrorism
State sponsored terrorism — State terrorism — Government incitement to terrorism
Nature Terrorists receive refuge and both indirect and direct support from their own or sponsoring government. Assistance may include international transportation, false identity papers, training, weapons, money and other support within a host country, or abroad through its embassies or agents.
Incidence Terrorists who attempted to assassinate the Pope had links to Bulgaria. In Poland governmental employees murdered a priest. In the Middle East and North Africa, Iran, Iraq, Syria and Libya export terrorists. Cuba who has been providing mercenaries to fight overseas has also been linked to the terrorist network. Until Israel's invasion of Lebanon, that country was serving as a major training and organizational centre for a spectrum of international terrorist groups. The intelligence agencies of the superpowers have also been linked to terrorist, covert actions.

Refs Cline, Ray S; Alexander, Yonah and Denton, Jermiah *Terrorism As State Sponsored Covert Warfare* (1986); Hippchen, Leonard J and Yim, Yong S *Terrorism, International Crime, and Arms Control* (1982); Wolpin, Miles D *State Terrorism and Repression in the Third World*.
Broader Terrorism (#PD5574).
Narrower Terrorist havens (#PD5541).
Related Cross border military operations (#PD5272)
Military expeditions against distant objectives (#PE8207).
Aggravates Narco-terrorism (#PE9210).
Aggravated by Inadequate international cooperation in reducing terrorism (#PF4366).

♦ PD6009 Missing children
Nature Children who disappear from home or from boarding schools or institutions or during travel may have been abducted or may be runaways. Abduction may be kidnapping for ransom or for revenge or other motives. Abducted children may be sexually violated, tortured and murdered. Runaways chance eventual abduction and victimization. In war zones and occupied countries children frequently become missing or separated from their parents.
Incidence The number of cases world-wide of missing children exceeds two million annually.
Broader Missing persons (#PD1380) Victimization of children (#PC5512).
Aggravated by Satanic rituals (#PF7887) Disowned children (#PJ0827)
Juvenile desertion (#PD8340).

♦ PD6017 Social outcasts
Underclass — Outcasts
Nature The underclass are those poor people whose status as citizens has become undermined and who are excluded from mainstream society. The members of the underclass do not share a common destiny; they are a mass of individuals, each with personal problems and a personal history of failure. It is not really a social class. There is no solidarity among its members because there is no shared reason for being in it – it is a matter of individual fate. Nevertheless, developments in the workplace are largely responsible for the increasing numbers of its members. The introduction of new technologies has led to a demand for more highly qualified workers, which means not only fewer jobs, but also that those with fewer skills are those who lose out. Unemployed youth fall through the social net that supports only those who subscribe to the work ethic. Foreigners and/or natives constitute another class of social outcasts. Many countries import large numbers of foreign workers but do not integrate them or give them a clearly defined status. They occupy an uncertain social position. They have claims and needs that their host countries are unable to fulfil; they are mostly unwanted, a social burden and often scapegoats for what goes wrong.
Incidence Social outcasts have always made up the lower depths of society, but today they assume new proportions, affecting millions in many countries.
Broader Deviance (#PB1125) Social deviation (#PC3452).
Narrower Socially handicapped refugees (#PD1507)
Imbalanced distribution of knowledge (#PF0204).
Aggravates Outlaws (#PE4409) Ostracism (#PF1009)
Social invisibility (#PD8204).
Aggravated by Leprosy (#PE0721) Boycott (#PE8313)
Caste system (#PC1968).

♦ PD6019 Discrimination against non-union workers
Closed union shops
Nature Employers may make contracts with organized labour to only hire members of specified trade unions; in other words, to form a closed shop. Once a closed shop has been agreed on, all trade employees are obliged to join the union. Non-union workers may not be hired or may be hired conditional on their obtaining a union card within a stipulated period of time. Dismissal may result upon refusal. Another disadvantage of the closed shop is that it lends itself to abuse by the union by giving it too much power over individuals in that they may be denied union membership initially or dropped from the union rolls subsequently, and therefore become ineligible for employment.
Incidence In the UK the agency shop, and in the USA the union shop, are the kind of closed shop that allows non-union workers employment conditional on their becoming union members later. In the UK, those electing not to become members have to contribute union dues. Nonetheless the incidence of closed shops is higher in some developing countries of the South than in the Northern market economies. However in some Communist industrialized nations, a Party card is equivalent to a union card and non-Party members may be discriminated against.
Broader Discrimination in employment (#PC0244)
Restrictive trade union practices (#PD8146).
Related Closed professions (#PD8629).

♦ PD6020 Physically handicapped persons
Physical disability
Refs WHO *Sexuality and People with Physical Disabilities* (1987).
Broader Human disability (#PC0699).
Narrower Developmental disabilities (#PD3773) Visually handicapped persons (#PD2542)
Physically handicapped children (#PD0196).
Aggravates Personal care disabilities (#PE6770).
Aggravated by War crimes (#PC0747).

♦ PD6029 Language domination by developed countries
Nature The conceptual terms employed by the international community, whether in diplomacy, commerce, intergovernmental organizations, or the media, are introduced and reinforced by the developed countries using relatively few languages. These terms are culturally biased against the wider range of concepts current in the larger variety of languages used in developing countries.
Incidence The term "development" for example, has a perjorative connotation when expressing the objectives of the South. It masks the root problems that have to be overcome. Thus while this term was articulated and disseminated by the industrialized nations, some in the Third World wanting to themselves name the critical target, have expressed it as liberation. There are not in this view developing or under-developed nations and regions, but ones in which suppression of freedoms and rights calls for a term that encompasses the need for radical change.
Claim More and more leaders from underdeveloped areas are coming to regard 'development' as the lexicon of palliatives. Their recourse to the vocabulary of liberation is a vigorous measure of self-defence, aimed at overcoming the structural vulnerability which denies them control over the economic and political forces which impinge upon their societies.
Broader Domination (#PA0839).
Aggravates Lack of international cooperation (#PF0817).

♦ PD6033 Coitus as a cancer risk
Nature Some cancer specialists blame permissiveness, promiscuity, and the 'copulation explosion' for the higher incidence of cervical cancer.
Incidence A study of cases at a British hospital in 1983 found that the number of women under the age of 35 with the disease had increased from 6 in 1972 to 83 in 1983.
Aggravates Malignant neoplasms (#PC0092)
Malignant neoplasms of female genital organs (#PE1905).
Aggravated by Sex (#PF9109).

♦ PD6062 Wrongful detention
False arrest
Broader Imprisonment (#PD5142) Violation of civil rights (#PC5285)
Politically motivated arrests (#PD9349).
Narrower Concentration camps (#PD0702) Detention of children (#PE6636)
Incommunicado detention (#PE8004) Internment without trial (#PD1576)
Arrest of trade union leaders (#PD7630) Repressive detention of juveniles (#PD0634)
Detention of refugees and asylum-seekers (#PE6376)
Abusive detention in psychiatric institutions (#PE2932).
Aggravated by Miscarriage of justice (#PF8479) Delay in administration of justice (#PF1487)
Denial of human rights in the administration of justice (#PD6927).

♦ PD6065 Endangered children
Nature The negative impact of society in developed countries on its children is mainly because of two important changes which have occurred in the last twenty years, one being the spiralling divorce rate, and another the Women's Movement. Presently, in the USA for example, one out of two marriages ends in divorce. This fact places millions of children each year in the traumatic situation of family breakdown. The Women's Movement resulted in the introduction of millions of women into the workforce, including 60 percent of mothers, and 80 percent of stepmothers. Women now receive greater rewards for career than for effective parenting. This phenomenon results in the fact that children have fewer and fewer resource people available. This is further exacerbated by the exclusionary and often unstable dynamics of stepfamilies and remarriage. Loveless homes also constitute an endangering environment for children, a type of surroundings that breeds delinquency. In loveless homes, children are constantly exposed to traumatic experiences which have a definite influence in the development of their personalities. Furthermore, it is an established fact that such loveless homes have an even more deleterious effect upon the children's mental development than the so-called 'broken homes'.
Claim Twenty years ago, most children were endowed with at least two or more people who focused their main attention upon them. Today's child often does not have one full-time person concerned with his well-being. Not only is our society facing economic downward mobility, but also our children are presented with a caring downward mobility.
Broader Victimization of children (#PC5512).
Aggravates Inadequate child welfare (#PC0233).

♦ PD6075 Decrease in consumer choice
Narrow range of goods
Nature Largely as a consequence of the economic consequences of mass production techniques and economies of scale, the variety in certain classes of goods effectively available to the consumer has decreased.
Broader Consumer vulnerability (#PC0123) Grievances of consumers (#PD7567)
Ineffective means for goods supply and distribution (#PF6495).
Narrower Lack of consumer choice in centrally-planned economies (#PD0515).
Reduced by Proliferation of consumer products (#PF5057).

♦ PD6081 Distorted media presentations
Bias in the media — Biased media coverage of news — Confiscation of news media
Nature The replacement of facts by inaccuracies or intentional untruths in news coverage is an example of how vulnerable the public is to the media. Other distortions occur by the use of stereotypes and perjorative adjectives in slanted interpretations; by the dilution of news coverage of significant events with irrelevant matters; by the invention of falsely described comprehensive reporting on a subject where there is incomplete knowledge; and by a number of other practices ranging from silence to useless information. Distortion affects the contents of all media and all messages, not just the news. It affects more than contents as well, for example: frequency, timing and continuity. In unscrupulous private or public control, the media is a tool for domination, as there are few regulations preventing distortion.
Background Media presentations involve messages of all kinds: hard news; soft news; commentaries; reports; articles; analytical surveys; political cartoons and other political humour; documentaries 'eye-witness' specials; institutional and ideological advertising; consumer product commercials; entertainment; and practical information. The government may also be presented in the media during elections with paid advertising, and interviews and news coverage arranged by public relations staffs.
Broader Distortion (#PA6790) Misleading information (#PF3096)
Unethical media practices (#PD5251).
Narrower Media cover-up (#PD4383) News censorship (#PD3030)
Culture of violence (#PD6279) Biased presentation of news (#PD1718)
Excessive politicization of the media (#PD5475)
Parochial escapist media entertainment (#PD0917)
Biased portrayal of women in mass media (#PE7638)
Excessive portrayal of sex in the media (#PE7930)
Excessive portrayal of crime in the media (#PE7354)
Excessive portrayal of substance abuse in the media (#PE3980)
Excessive portrayal of terrorist activity in the media (#PE6844)
Excessive portrayal of negative information by the media (#PE1478)
Excessive portrayal of perspectives of industrialized cultures in media (#PE3831).
Aggravates Narrow media options (#PJ0094) Insufficient cultural media (#PJ8476)
Psychological pollution by mass media (#PD1983)
Decline in public interest broadcasting (#PF5622)
Manipulation of the individual by mass media (#PE7448).
Aggravated by Uncontrolled media (#PD0040)
Excessive commercialization of the media (#PE4215)
Media theatricalization of public life and politics (#PF9631).

♦ PD6103 Use of agricultural land for fuel production
Energy cropping
Nature Energy cropping – agricultural production for the purpose of manufacturing ethanol for fuel – by traditional food exporters is likely to reduce exportable food surpluses and increase international food prices. By reducing domestic food supplies (thus resulting in upward pressures on domestic food prices), wages would be forced up, real incomes be reduced, particularly among the poor, and social unrest would result.
Claim In a world where millions of people do not have enough food, the use of limited agricultural land for energy cropping and conversion of food commodities into liquid fuel is irrational.
Counter-claim High oil prices enhance the attractiveness of energy cropping, and with this new technology comes the chance for non-oil rich countries to become more independent.
Broader Land misuse (#PD8142) Energy crisis (#PC6329).
Aggravates Unavailability of agricultural land (#PC7597).

♦ PD6113 Stratospheric ozone depletion
Nature The temperature, structure and dynamic processes in the stratosphere are to a large extent determined by the absorption of solar ultraviolet radiation by ozone. Atmospheric ozone

plays an important ecological role since it filers out most biologically harmful ultraviolet radiation. This layer, however, is threatened by the emissions of fluorocarbons used as aerosol propellants, blowing agents in foam production, solvents and refrigerants; and by nitrous oxide emissions from both organic and inorganic nitrogen fertilizers. The above mentioned increase of ultraviolet radiation could cause an increase in the incidence of skin cancers, but potential damage to food crops and fish may well prove to be more significant a problem.

Incidence It has been estimated that a reduction of 1 per cent in total ozone produces a 2–4 per cent increase in biological effects. Current predictions of ozone depletion at the 1978 contaminant input rates are 4 per cent. Such depletions may become much larger if the halocarbon release rates increase so much that chlorine mixing ratios in the stratosphere become of the order of 20 p.p.b.v., namely if the present manufacturing rate were to be doubled. Is is believe that volcanic eruptions may also result in ozone depletion.

Claim The effects of ozone depletion will not be felt for some 25–45 years, when it will be too late to remedy the situation.

Refs International Environment Liason Centre *International Environment–Development Facts March 1989* (1989); International Institute of Refrigeration *Status of CFCs* (1988); Jones, Russell R and Wigley, T (Eds) *Ozone Depletion* (1990); Shea, Cynthia P *A Vanishing Shield* (1988); Siebert, Horst (Ed) *Global Environmental Resources* (1981); United Nations Environment Programme *The Ozone Layer* (1987).

Broader Degradation of the atmosphere (#PD9413)
Increase in trace gases in the atmosphere (#PD1354).
Aggravates Global warming (#PC0918).
Aggravated by Volcanic eruptions (#PD3568)
Chlorofluorocarbons as an environmental hazard (#PE4378).
Reduced by Ozone as a pollutant (#PE1359).

♦ PD6120 Excessive size of metropolitan regions

Nature As metropolitan regions grow ever larger it is hard for their governing bodies to be responsive to the needs of all of the people. Nation–states with a megapolitan capital city tend to override neighbourhoods and neighbourhood culture. This pattern is repeated in other metropolises. Thus the different languages, customs and cultures that exist at present may disappear in their areas over time. Efficiency is the only value that emerges in these vast conglomerations; and mechanization and routine demand the subordination of the individual to the masters of the city–state.

Broader Excess (#PB8952).
Narrower Spatial imbalance of human settlements (#PD6130).

♦ PD6124 Environmental degradation from high-speed roads

Nature High–speed roads are capable of causing considerable damage, especially when they are badly placed. They divide neighbourhoods; damage or obliterate homes and farms; cut off access to the countryside; and, above all, create excessive noise for bordering communities.

Broader Urban road traffic congestion (#PD0426).
Related Sterile working environment (#PD6133)
Intimidation of pedestrians by vehicles (#PE6139)
Environmental degradation by automobiles (#PE6142)
Inadequate integration of transport systems (#PF6157)
Inaccessibility of countryside to city dwellers (#PF6140)
Inadequate arrangement of housing with respect to common land (#PF6146).
Aggravates Deforestation (#PC1366) Inaccessibility of water for recreation (#PF6138).

♦ PD6125 Vulnerability of small towns

Nature It is increasingly hard for small towns to continue to exist in the face of massive migration to cities.

Incidence During the last 30 years, 30 million rural Americans have been forced to migrate to large cities, and such migration continues at the rate of 800,000 people a year; and in numbers of other countries – in Ireland and India for example – many people also move to towns in search of more interesting work, and for a better life – looking for information, for a better connection with their own culture. This is invariably having bad effect on small towns. For example, 50 percent of those people remaining in rural America live on less than $3,000 a year, and there is little money to revitalize the towns they live in.

Broader Spatial imbalance of human settlements (#PD6130).
Narrower Ineffectiveness of individual participation in large communities (#PF6127)
Inhibition of individual psychological development through life cycle (#PF6148).
Related Artificiality of parkland (#PF6135).

♦ PD6128 Vulnerability of sacred sites

Violation of sacred sites — Desecration of ancient burial sites

Nature Religious sites embody the relationship of a people to the land and to the past. Mountains are marked as places of special pilgrimage; rivers and bridges become holy; a building or a tree, or rock or stone, may take on the power through which people can connect themselves to their own past. Such sites tend to be bulldozed, developed, or changed, for political and economic reasons, without regard for the emotional continuity of traditional societies. Destruction of sites which have become part of the communal consciousness creates a pathological condition in the communal body.

Incidence Burial sites are also desecrated by anthropologists and archaeologists in search of information. Indigenous populations whose beliefs attach special importance to ancestral remains are subject to the indignity of attempting to recover such remains from the museums to which they have been transferred.

Counter–claim Such concerns lead to the absurdity of the mass reburial of American Indian remains of a museum collection of skeletons. It implies that Tasmanian skeletons should be repatriated for reburial by their descendants, although they have been exterminated. Although such skeletons have been thoroughly studied, new techniques tend to emerge which permit them to be restudied for more information, especially about genetic evolution.

Broader Destruction of cultural heritage (#PC2114)
Endangered monuments and historic sites (#PD0253).
Narrower Misappropriation of sacred objects (#PD8041).
Related Inaccessibility of quiet zones in an urban environment (#PF6160).
Aggravates Sacrilege (#PF0662) Abuse of relics (#PF5107)
Desecration of holy spaces (#PF6385) Endangered indigenous cultures (#PC7203).
Aggravated by Vandalism (#PD1350) Fetishism (#PF8363)
Ancestor worship (#PD2315).

♦ PD6130 Spatial imbalance of human settlements

Nature If the population of a region is weighted too far toward small villages, urban civilization can never emerge; but contemporary trends for people to leave their farms, small towns and villages and crowd into the cities leaves vast areas depopulated and undermaintained. The population is weighted too far towards big cities, letting the resources of the land go to ruin.

Refs United Nations Statistical Office *A Global Review of Human Settlements and Statistical Annex* (1976).

Broader Excessive size of metropolitan regions (#PD6120)
Obstacles to availability of community space (#PF7130).
Narrower Artificiality of parkland (#PF6135) Vulnerability of small towns (#PD6125)
Inaccessibility of countryside to city dwellers (#PF6140).

♦ PD6131 Unbalanced urban population density gradients

Nature People want to be close to shops and services, for excitement and convenience, and they want to be away from shops and services for quietness and enjoyment of the countryside. The exact balance of these two desires varies from person to person, but in the aggregate it is the balance of the two desires which determines the ideal gradient of housing densities in a neighbourhood. However, under present–day conditions when density gradients are usually not stable, most people are forced to live where the balance of peace and quiet with activity does not correspond to their wishes or their needs; this is because the total number of available houses and apartments at different distances is inappropriate.

Broader Uncontrolled urban development (#PC0442)
Maldistribution of population within countries (#PC8192).
Narrower Impersonality of high density accommodation (#PF6156)
Inadequate arrangement of housing with respect to common land (#PF6146).
Related Insufficient separation between urban subcultures (#PF6137)
Unattractive pedestrian environments in urban areas (#PE6151)
Ineffectiveness of individual participation in large communities (#PF6127).
Aggravated by Urban overcrowding (#PC3813).

♦ PD6133 Sterile working environment

Claim If people spend eight hours of their day at work, and eight waking hours at home, there is no reason why their workplace should be any less of a community than their home. Yet this is often not the case. Most workplaces have an atmosphere that reflects only their function as places where money is made. For workplaces to function as communities, certain factors are critical. For instance, they must not be too scattered, nor too agglomerated, but clustered in manageable groups. Workplaces should be decentralized, but not so separated that a single workplace is isolated from others. Work communities need to be small enough so that people know each other, at least by sight; they should not be too specialized either, but should contain a mixture of manual jobs, desk jobs, craft jobs etc, so as to create a variety. Lack of common land within the work community, to unite the individual workshops and offices and where people can sit, eat lunches and make contact with one another, produces a sterile environment. Work communities should also be interlaced with the larger community in which they are located, possibly sharing services like restaurants, cafes and libraries.

Broader Obstacles to community achievement (#PF7118)
Artificial separation of home and workplace (#PF6122).
Narrower Inaccessibility of quiet zones in an urban environment (#PF6160)
Inhibition of communication between non–proximate offices (#PF6197).
Related Household segregation by age group (#PF6136)
Unidentifiable urban neighbourhoods (#PF6147)
Excessive use of land by automobiles (#PF6152)
Impersonality of public squares in cities (#PF6165)
Unconvivial hotel environments for travellers (#PF6196)
Insufficient common land in urban environments (#PE6171)
Environmental degradation from high–speed roads (#PD6124)
Inhibition of exploration by children of urban environment (#PF6159)
Inhibition of individual psychological development through life cycle (#PF6148).
Aggravates Unhealthy lack of daily physical activity in urban environments (#PF6182).
Aggravated by Inaccessibility of water for recreation (#PF6138).

♦ PD6145 Institutionalized torture

Broader Torture (#PB3430)
Denial of right to freedom from cruel, inhumane or degrading punishment (#PC3768).
Narrower Torture of children (#PD2851) Psychological torture (#PD4559)
Pharmacological torture (#PE4696) Disabled victims of torture (#PD0764).
Related Physical torture (#PD8734) Religious torture (#PC7101).

♦ PD6150 Obsolescence of suburban mode of human settlement

Claim Suburbs are very wasteful of land and tend to isolate families in their own houses and gardens. They are the prototype dwelling places of the bourgeois nuclear family, and the symbol of maintenance of the emotional equilibrium of the owners' parents.

Counter–claim Suburbs relieve the congestion of cities and afford clean and healthy living conditions for growing children. Family life outside the cities, closer to natural recreation areas, provides a low–stress, positively reinforced environment in which family and inter–personal community relationships can grow and contribute to human development and happiness.

Broader Preservation of obsolete systems (#PC8390)
Inaccessibility of countryside to city dwellers (#PF6140).
Narrower Artificiality of parkland (#PF6135).

♦ PD6204 Tropical deforestation

Destruction of rain forests

Nature As people seek land to cultivate, wood to burn, and raw materials for their industries, they turn to tropical forests – which are being destroyed at an unprecedented rate. Numerous economic, social, and ecological problems are being created by this loss and degradation and it is the world's poorest people who are the most severely affected. Major repercussions of deforestation include: intensified seasonal flooding with resultant loss of lives and property; water shortages in dry seasons; accelerated erosion of agricultural lands; silting of rivers and coastal waters; the disappearance of plant and animal species; and local and regional climate modifications. In many tropical forests, the soils, terrain, temperature, patterns of rainfall and distribution of nutrients are in precarious balance, and neither trees nor grasses will grow again once they are disturbed by extensive cutting. Even in those place where regrowth is possible, extensive clearance destroys the ecological diversity the original forest offered.

Background Since pre–agricultural times, the world has lost 20 per cent of its forest resources, with a reduction from 12 billion to 10 billion acres. In the past, most of forest losses were in the temperate forests of Europe, Asia and North America. In recent years, it is the tropical forests of Latin America, Asia and Africa that have been disappearing most rapidly.

Incidence In 1990 it was estimated that each year 16 to 20 million hectares (40 to 50 million acres) of tropic forest have been vanishing as trees are cut for timber and land is cleared for agricultural development. The rate of loss in 1987 was nearly 50 per cent greater than in 1980. FAO has estimated that tropical forests are being removed at the rate of 7.3 million hectares per year. Brazil, with the largest remaining forest area, is experiencing the most rapid losses (12.5 to 22.5 million acres per year). The economic, ecological, social, and other costs to be paid for this loss are as yet unassessed. On the basis of the current rates of deforestation, it is plausible that natural tropical forests will largely disappear over the next 100 years. However their conversion into vast expanses of wasteland with little vegetative cover of economic value is not entirely improbable. These changes would imply extensive regional and global changes in climate.

Claim Tropical forests are the world's richest biological zones and are estimated to contain as much as 40 percent of all the terrestrial species on the planet. In addition, tropical forests produce a significant proportion of the world's oxygen and provide a wide range of useful products (fuelwood, building materials, pulpwood, food, pharmaceuticals, resins, gums, dyes) of economic

significance for both developing and developed countries. Undisturbed tropical forests are also home to millions of the world's tribal peoples.
Refs Guppy, N *Tropical Deforestation* (1984); United Nations *Conservation and Development of Tropical Forest Resources* (1982).
Broader Deforestation (#PC1366).
Aggravates Floods (#PD0452) Drought (#PC2430) Ethnocide (#PC1328) Soil erosion (#PD0949) Desert advance (#PC2506) Global warming (#PC0918) Extinction of species (#PB9171) Silting of water systems (#PD3654) Unsustainable agricultural development (#PC8419) Vulnerability of world genetic resources (#PB4788) Low productivity of agricultural workers in developing countries (#PE5883).
Aggravated by Corruption in politics (#PC0116) Cultivation of illegal drugs (#PD4563) Malnutrition in developing countries (#PD8668).

♦ PD6213 Juvenile prostitution
Adolescent prostitution — Teenage prostitution
Nature There has always been a demand for young adolescents for sexual purposes because of their freshness and simplicity. This is particularly true of the present permissive and, at the same time, surfeited and sexually vulgarized, age which seeks all kinds of erotic refinements, with its attempt to renew jaded sensuality through the sexuality of a child. The effects of tourism on child prostitution has led to dangerous developments. The presence of children of both sexes ready to satisfy the sexual appetites of organized bodies of tourists is very often an additional attraction, with paedophile tourists being able to obtain many profusely illustrated guides containing extensive information, including addresses, hotels, rates, local agents, local practices and traditions, and also the "legal limits" in each country.
Incidence A recent study estimated that 5000 boys and 3000 girls below the age of 18 are involved in prostitution in Paris. The stereotypical account of how a child gets into prostitution in the USA is that he or she is usually a runaway newly arrived in a big city bus station and picked up by a sweet–talking pimp who treats him/her well for a while and then expects favours in return. Brazilian girls make more money as prostitutes than their fathers do as factory workers, thus they often support their entire family on their wages and find it almost impossible to leave such a lucrative profession.
Broader Prostitution (#PD0693).
Aggravates Male homosexual prostitution (#PD4402) Adolescent sexual intercourse (#PD7439).
Aggravated by Juvenile stress (#PC0877) Street children (#PD5980) Child prostitution (#PE7582) Trafficking in children for sexual exploitation (#PE6613).

♦ PD6216 Degenerative diseases
Nature Patients, their immediate families, and close friends feel a demoralizing hopelessness when told that everything possible has been done for them and that there is no other possible avenue to explore in hopes of curing their degenerative disease.
Claim As morale is one of the most necessary factors for combating any illness, a patient should never be told that there is no more hope left, for with that dictum comes the decline in inspiration and will, without which no cure can be effected.
Counter–claim Degenerative diseases are none other than a retaliatory effect of man's destructive manipulation of his environment, and until man learns to respect and peacefully coexist with that environment, he will always be prey to attracting inflictions to himself.
Broader Human disease and disability (#PB1044).

♦ PD6223 Obstacles to international trade in services
Nature International trade in services, an essential adjunct to commerce in raw materials and manufactured goods, has become an increasingly important factor in world trade. But although services have been included in postwar trade liberalization from the outset, many restrictions remain. These obstacles include: refusal or delay in responding to an application for a licence or operating permit; terms under which foreign companies must operate (often there is differential treatment of foreign and national firms); special taxes; recruitment requirements; and outright prohibitions, such as in travel employment.
Broader Protectionism in international trade (#PC5842).

♦ PD6265 Inadequate health services following nuclear war
Nature A nuclear war would result in human death, injury and disease on a scale that has no precedent in history, dwarfing all previous plagues and wars. There is no possible effective medical response after a nuclear attack – in one major city alone, in addition to the hundreds of thousands of sudden deaths, there would be hundreds of thousands of people with severe burns, trauma, and radiation sickness – all demanding intensive care. Even if all medical resources were intact, the care of these immediate survivors would be next to impossible. In fact, most hospitals would be destroyed, medical personnel among the dead and injured, most transportation, communication and energy systems inoperable, and most medical supplies unavailable.
Refs Chivian, Eric, et al *Last Aid* (1982).
Broader Nuclear war (#PC0842) Inadequate health services (#PD4790).

♦ PD6273 Conflicting roles of women
Dual roles of women
Nature Girl students and women who have a job outside the home are overwhelmed with the twofold burden of work. Girls who are able to attend school often find the burden of family duties consumes much of their time and energy, so that it is difficult if not impossible to complete even their primary schooling; and a women who must be, or chooses to be, economically active outside the home will in effect perform two jobs: her paid employment and keeping the home. Although in the industrialized countries the amount of housework has decreased, it still constitutes a heavy burden, especially for women who go out to work.
Incidence In the Soviet Union, where female labour represents one–half of the total labour force, seven women out of ten spend more than two hours a day on housework, whereas only three men in ten do this. In Japan the working day of a couple engaged in agricultural occupations is made up as follows: work, 10 hours 46 minutes in the case of the man and 9 hours in the case of the woman; housework, 11 minutes in the case of the man and 4 hours 11 minutes in the case of the woman. In France where the average working week is 45 hours for men and women, a woman engaged in an occupational activity actually works 73 hours a week if she has no children, and 83 hours if she has one or two children.
Counter–claim To mix the roles of man and woman is hostile to the laws of nature. Driving woman to man's work at the expense of her feminity with which she is naturally provided for a natural purpose in life.
Broader Static and unrelated social roles (#PF1651).
Related Elimination of the socio–cultural role of western women (#PE1046).
Aggravated by Neglect of the role of women in rural development (#PF4959).

♦ PD6279 Culture of violence
Excessive portrayal of violence in the mass media — Media violence — Glorification of violence in the media
Nature Although it is true that the media communicate news of real events, it is also true that news is changed by very fact of its passage through the media, whatever the efforts to present it objectively. In particular, the media are often accused of giving undue importance to events involving violence. There is also concern for the growing tendency to mix fact and fiction in presentations designed to give heightened interest to the news. The media can not only reinforce impressions gained from other sources, they can also create the impressions. The commonest charge made is that by portraying violence the media lead people to imitate what they read, hear or see. It is even more often claimed that the violence gives rise to imitation and thus leads to juvenile delinquency or crime. The long–term effects of violence in media far outweigh its immediate effects, but are more difficult to assess.
Claim The portraying of violence by the media contributes to violent behaviour by conferring approval, by spreading information, by suggesting violence as a form of problem solving, by shifting levels of tolerance, by associating certain groups with violence so that they felt they had to live up to their image, by exaggerating the problem, and by over–simplifying alternatives. The media may lead to violence in other ways that by depicting violence. Feelings of frustration can be aroused or reinforced by an emphasis on materialistic aims which are not attainable by the audience.
For media war is an activity in which the mutilation and death of hundreds of thousands or millions of people is a necessary by–product. War sells television news, films, literature and newspapers.
Counter–claim Violence existed before the mass media; it is therefore unfair and illogical to blame the media for violent behaviour. Using the mass media as the scapegoat is merely a way of avoiding recognizing the real roots of the problem. The portrayal of violence could be seen as a form of communication in itself, perhaps as a danger signal, perhaps leading to increased awareness of a problem, perhaps enabling people to come to terms with the violence in themselves and in society.
Broader Human violence (#PA0429) Social insecurity (#PC1867) Distorted media presentations (#PD6081).
Narrower Video violence (#PE2224) Enjoyment of war (#PF4034) Television violence (#PE4260) Glorification of war (#PF9312) Violence in comic books (#PG4262).
Related Violence as entertainment (#PD5081).
Aggravates Apathy (#PA2360) Youth violence (#PF7498) Exploitative films (#PE6328) Film and cinema censorship (#PD3032) Deterioration of media standards (#PD5377).
Aggravated by Excessive portrayal of crime in the media (#PE7354).

♦ PD6282 Deforestation of mountainous regions
Deforestation of upland watersheds — Inappropriate cultivation of steep slopes — Destruction of watersheds — Watershed deterioration — Vulnerability of mountainous environments
Nature Forests in mountainous regions are particularly vulnerable to serious losses at the hands of local farmers, logging operations, and charcoal producers. Deforestation most severely disrupts such areas and the ecosystems that depend upon them. The uplands influence precipitation. The state of their soil and vegetation systems influence how this precipitation is released into the streams and rivers and onto the croplands on the plains below. The need for more agricultural land forces villagers to cultivate marginal, unproductive land on steep slopes and hillsides and to denude vast areas of forest cover. Such agricultural techniques and deforestation lead to soil erosion and ultimately to the loss of cultivated land. The problem is aggravated by increasing population pressures and the subsequent developments such as road building. The growing numbers and severity of both floods and droughts have been linked to such deforestation.
Incidence Mountainous regions account for a quarter of the earth's land surface and provide a home for a tenth of the world's people, while a further forty percent live in adjacent lowland areas (including some of the great fertile plains of the world) whose future is intimately bound to developments on the slopes above. Environmental deterioration in many mountain regions of the world is proceeding irreversibly, and the widening circle of destruction originating in the mountains is spreading to the plains, river systems and harbours. Damage to basic life support systems is accelerating in practically every mountain area of Asia, Africa and Latin America. There has been a marked increase in the destructive clearance of forests, in flood damage, silting, soil erosion and the spread of pests.
Population pressure in Asia is causing deforestation of the Himalayan foothills in Nepal, India and Pakistan, and a consequent loss of the soil. Demand for agricultural land for farming and wood for firewood and furniture and construction materials, as well as over–grazing by animals, causes the soil loss. The same process is occurring in the Andes in Peru, Colombia and all the other Andean countries. In Africa, the Ethiopian Amhara plateau, and the Kenyan, Tanzanian and Ugandan highlands, are all being deforested. In the Alps, the hill agricultural populations have drastically declined in France, Switzerland, Germany, Austria, Italy and Yugoslavia, while the tourist population increases. Developments have already destroyed much of the land.
Soil erosion is increasing in many hilly sections of both developed and developing countries through the extension of cultivation onto sleep slopes in an inappropriate manner, especially when accompanied by extensive deforestation.
Refs Bagh, Hazari *Challenge of Watersheds*; Cépède, M and Abensour, E S *Rural Problems in the Alpine Region* (1961); FAO *Guidelines for Watershed Management* (1983); FAO *Watershed Development* (1985); FAO *Management of Upland Watersheds* (1983); Golubev, G N (Ed) *Environmental Management of Agricultural Watersheds* (1983).
Broader Deforestation (#PC1366) Environmental degradation (#PB6384) Vulnerability of ecosystem niches (#PC5773).
Narrower Degradation of mountain environment by leisure activities (#PE6256).
Related Destruction of the countryside (#PE3914).
Aggravates Landslides (#PD1233) Soil erosion (#PD0949) Riverine floods (#PD4976) Degradation of semi–natural and natural habitats of flora and fauna (#PC3152).
Aggravated by Forest fires (#PD0739) Shortage of cultivable land (#PC0219) Unequal property distribution (#PC3438) Depopulation of mountainous regions (#PD1908) Cultivation of marginal agricultural land (#PD4273) Disruption of ecosystems in marginal agricultural lands (#PD6960).

♦ PD6283 Arctic air pollution
Nature A heavy reddish haze is a regular feature of the Arctic in winter and early spring, often reducing visibility to less than 6 miles (about 10 kilometers). A haze so far from the industries and vehicles that pollute the air of major cities seems to require a wholly natural explanation, but evidence has begun to accumulate that a part of the arctic haze may have its origins 6,000 miles (10,000 kilometers) away in the same polluted air that produces acid rain over the United States and Europe. According to a study undertaken in 1976, it would seem that, due to air flow patterns and other factors, the major source of the pollutants in the arctic haze is the Soviet Union. Europe and England are the next largest sources of the pollution. North American countries contribute little because of air flow patterns.
Incidence Air pollution threatens an area of the Arctic equal in size to the North American continent, jeopardizing the fragile Arctic ecosystem. Pollutants are carries north by dry air masses from industrial areas, and in winter there is no precipitation to remove them from the atmosphere. Increasing pollution may affect not only Arctic life forms, but also global climate patterns. The pollution extends from Alaska eastward to Norway – about half of the circumference of the

polar ice cap – and is suspended from ground level to as high as 25,000 feet from November through April.
Broader Air pollution (#PC0119).
Related Polar pollution (#PE5993).

♦ PD6284 Parasitic plants
Nature Parasitic plants are a major problem in some areas, especially the family of mistletoes (Loranthaceae). Such plants can attack a wide range of hosts; heavy attacks can kill the host plant.
Broader Parasites (#PD0868).
Narrower Witchweed (#PE6604).
Related Nematoid plant diseases (#PD2228).

♦ PD6298 Politicization of education
Ideological domination of the humanities — Politicization of the curriculum
Claim The notion that any course in Western culture can be enriched by adding to its core list of readings, books by women and persons of colour, not because they are the best or most appropriate for the subject at hand, but because of the race or sex or class of the author (a kind of affirmative action programme in scholarship) is both patronizing and ludicrous from the standpoint of honest scholarship. The folly is compounded by the obligation to recruit women and people of colour to teach the ideas and aspects of culture that involve them. This is nothing more than a sophisticated revival of the old folk fantasy that only like can understand like.
Counter–claim Existing educational curricula are already politicized in that basic texts tend to have been composed by Western white males and that, in consequence, the required courses are intrinsically infected with racism, sexism and imperialism. This can be corrected by replacing, or supplementing, the relevant traditional texts by books composed by women and people of colour and by representatives of the oppressed classes of the past and of the present, as well as by encouraging the study of foreign cultures in greater depth. To ensure that courses are appropriately taught, personnel should be recruited from minorities and women, wherever possible.
Refs Englund, Tomas *Curriculum as a Political Problem* (1986).
Aggravated by Anti–intellectualism (#PF1929).
Reduces Inadequate political education (#PJ7906).

♦ PD6305 Declining breeds of domesticated animals
Nature As farming has become more industrialized and energy–intensive, a small number of breeds have become dominant - specialized animals that respond well to high–energy diets, confined living conditions and intense medical treatment. In Canada Holstein cow stands in 95 per cent of the dairy stalls and there are only six commercial poultry breeds. This specialization develops narrow gene pools which may be susceptible to diseases.
Claim Special breeding for rate–of–gain production rates and narrow gene pool may hurt the people if consumers' taste, climatic patterns and diseases change in the future.
Refs FAO *Declining Breeds of Mediterranean Sheep* (1978).
Broader Neglected food resources (#PF7808).
Related Endangered species of animals (#PC1713)
Destruction of rural subsistence economy (#PC2237).
Aggravated by Geographical isolation (#PF9023).

♦ PD6306 Hearing defects
Refs Martin, Frederick N *Hearing Disorders in Children* (1987); Northern, Jerry L and Downs, Marion P *Hearing in Children* (1984); Taylor, I G and Markides, Andreas *Disorders of Auditory Function, III* (1981); Thomas, Alan J *Acquired Hearing Loss* (1985); Tweedie, Joyce *Children's Hearing Problems* (1987).
Broader Human disease and disability (#PB1044).
Narrower Deafness (#PD0659) Head noise (#PG5106).
Aggravates Inadequate welfare services for the deaf (#PD0601).
Aggravated by Diseases of the ear (#PD2567) Health hazards of exposure to noise (#PC0268).

♦ PD6324 Addiction
Nature Addiction is the physiological and/or emotional dependence upon a substance, an activity, or a modus operandi that is so strong as to have a harmful physical and/or emotional effect, and which keeps the individual from dealing effectively with his own life and with interactions with society. The addict often loses his power of self–control and his behaviour becomes determined by the source of his addiction.
Refs Carone, Pasquale, et al *Addictive Disorders Update* (1982); Milkman, Harvey and Shaffer, Howard J (Eds) *The Addictions* (1984).
Narrower Stress addiction (#PE4951) Exercise addiction (#PF0958)
Pathological gambling (#PG4237) Excessive television viewing (#PD1533).
Related Sexual craving (#PE7031).
Aggravated by Substance abuse (#PC5536).

♦ PD6334 Iatrogenic disease
Diseases of medical practice — Physician–induced diseases — Drug induced diseases — Infections acquired in hospital — Inadequate hospital hygiene — Hospital–acquired infections
Nature Iatrogenic diseases may be induced by or attributed to doctors themselves, or the medicines, drugs and treatments they prescribe. Illness and suffering are often inflicted by wrong diagnosis, ignorance or negligence, and this professional incompetence is compounded by lack of understanding and sympathy. The presence of or need to see a doctor instils certain fear in people, and doctors are increasingly representing an unresponsive, alien and sometimes hostile factor in the daily lives of those people already beset by fear.
Along with specialized care for illness, hospitals have the less salubrious trait of infecting patients with diseases they did not have when they arrived. This happens when serious infections are spread from one patient to the other. While this is sometimes caused by poor hygiene in a hospital, it can also occur when infections exist unknown to hospital personnel, who do not know that radical hygiene is required in a particular case. Associated with the incidence of infections are: the use of invasive techniques such as catheterization, impaired defences of the body arising from the use of drugs or radiation and increased use of antibiotics.
Incidence In the USA, 40 million people enter the hospital each year, 5 percent acquire infections there, and 120,000 of them die. Hospital–acquired infections are responsible for the expenditure of around $1000 million yearly. In UK one in 10 patients catch an illness in hospital. The most infamous iatrogenic disease is the congenital absence of limbs in the children of mothers who received thalidomide during pregnancy.
Improper or excessive medication, spreading of an infection by patients, redundant surgery, and neuroses engendered by a psychiatrist are all examples of iatrogenic diseases. Medical institutionalization is also a significant contributory ingredient, with hospitals sometimes doing more harm than good. The depersonalization of modern medicine, X–rays, gadgetry, mechanical contrivances and computer diagnosis also facilitate the occurrence of iatrogenic diseases.
Haemophiliacs are a large proportion of those who have contracted AIDS through hospital transfusions of infected blood.
Refs Ayliffe, G A J and Collins, B *Hospital Acquired Infection* (1989); Bristow, M R *Drug Induced Heart Disease* (1981); D'Arcy, P F and Green, J P *Iatrogenic Disease* (1986); David, M *The Complications of Modern Medicine* (1963); Meyler, L and Peck, H M *Drug Induces Disease* (1962); Moser, Robert H *Diseases of Medical Progress* (1969); Siedek, Shelby V *Jurisprudence and Iatrogenic Problems* (1988).
Broader Human disease and disability (#PB1044)
Inadequacy of medical science (#PF8326).
Narrower Laboratory testing errors (#PF5304).
Related Allopathy (#PF7703).
Aggravated by Unethical medical practice (#PD5770)
Inappropriate basic hygiene (#PD8294)
Unethical practices of health services (#PE3328)
Inconclusiveness of scientific and medical tests (#PD7415).

♦ PD6346 Denial of rights to students
Abuse of students — Violation of student's rights
Nature Students can be denied the right to accessible education, the right to organize themselves at all levels of the educational structures, the right to free and independent expression of opinions on political, economical and educational issues and the right to a satisfactory work according to one's education.
Broader Denial of rights to vulnerable groups (#PC4405).
Narrower Discrimination against adult students (#PE6258).
Aggravated by Inadequate research on problems (#PF1077)
Anti–social behaviour of university students (#PE7370).

♦ PD6361 Discrimination against foreigners
Nature Domestic laws may discriminate against individuals who are not citizens of the country in which they live, especially regarding the rights to freely leave and re–enter the country, to be joined by family members, and to benefit from government social welfare schemes.
Broader Discrimination (#PA0833).
Narrower Discrimination against immigrants and aliens (#PD0973)
Discriminatory treatment of foreign prisoners (#PE6883)
Biased media–image of foreign groups and peoples (#PE8802)
Discrimination against foreign nationals in the military service (#PE6422).
Aggravated by Undue attachment to territory (#PF3390).

♦ PD6363 Zoonotic bacterial diseases
Broader Zoonoses (#PD1770) Infectious and parasitic diseases (#PD0982).
Narrower Plague (#PE0987) Anthrax (#PE2736) Tetanus (#PE2530)
Glanders (#PE2461) Vibriosis (#PG5278) Tularaemia (#PE6872)
Erysipelas (#PG5619) Ornithosis (#PE2578) Brucellosis (#PE0924)
Melioidosis (#PG0774) Listeriosis (#PE3779) Yersiniosis (#PG7862)
Tuberculosis (#PE0566) Lyme disease (#PG6489) Leptospirosis (#PE2357)
Salmonellosis (#PE7562) Rat–bite fever (#PG6702) Cat scratch fever (#PE5105)
Campylobacteriosis (#PG4274).

♦ PD6364 Physical insecurity of refugees and asylum–seekers
Military attacks on refugee camps and settlements
Nature Military attacks on camps or settlement areas for refugees and displaced persons take the form of raids by regular or paramilitary armed forces; random bombings; militarization of camps; forcible recruitment into regular and irregular armed forces; and savage killings or abductions, including women children and old persons. These attacks are aimed at people who have already once risked their lives to flee their original countries, and who are now completely unable to defend themselves. In addition, the attacks also endanger the lives of innocent citizens living near refugee camps.
Incidence Serious attacks have occurred in Guatemala, El Salvador, and Honduras; Angola, South Africa, Mozambique, Ethiopia, and Uganda; Jordan, Israel, Lebanon, and Syria; Indonesia, Malaysia, Kampuchea, Laos, and Viet Nam.
Background The 4 May 1978, military attack on the Kassinga area in Southern Africa began what has been a series of vicious attacks on refugee camps. It has been recommended by the UNHCR that refugee camps be moved farther from the frontier so that further attacks can be avoided, a UNHCR presence should be in all refugee camps.
Counter–claim The fault is not always so much with the attacking forces as with terrorist groups who deliberately use refugee camps as their strongholds, thus hoping to render themselves immune from reprisals against their own savage behaviour.
Broader Refugees (#PB0205) Military atrocities (#PD1881).
Related Cross border military operations (#PD5272).
Aggravates Rejection of refugees (#PF3021).

♦ PD6380 Suspension of rights during states of emergency
Restrictions of rights under martial law — Restriction of rights during war — Violation of rights under state of emergency
Nature In cases of invasion, grave disturbance of the public peace, national disasters or any other exceptional circumstance which may place society in grave danger, a government may suspend throughout the nation or in particular places, those rights that may constitute obstacles to re–establishing the anterior situation. Although states of emergency are legally intended to be as short as possible and not directed against any particular individual, they are often misused by government authorities who use them as legal pretexts to violate human rights. Special powers tend to facilitate abuses such as arbitrary arrest, torture, 'disappearances' and extrajudicial executions. Suspects can be held on vague suspicions; crimes against the state are given broad, elastic definitions. The suspension of habeas corpus and other legal remedies, trials of political detainees in military courts, as well as the suspension of the rights to strike or of assembly may be alienated under states of emergency. In some countries no formal state of emergency is declared, prevailing conditions are nevertheless tantamount to an emergency; that is, the constitution is suspended, parliament dissolved, all political activity banned and various special measures taken affecting the rights of prisoners and detainees.
Incidence In 1988, there were states of emergency or comparable situations in more than 30 countries, in Africa, the Americas, Asia and Europe. In Paraguay the state of emergency has already lasted for 30 years.
Broader Disaster unpreparedness (#PF3567).
Narrower Violation of rights of vulnerable groups during states of emergency (#PD3785).
Related Unpreparedness for food emergencies (#PC5016).
Aggravated by Excessive imposition of states of emergency (#PE4363).

♦ PD6392 Inciting riot
Nature Encouraging, urging, commanding, instructing or directing 5 or more persons to riot is inciting riot.
Broader Complicity (#PF4983) Civil disorders (#PC2551).
Related Incitement to violence (#PJ2068).

♦ PD6417 Discrimination against foreign companies
Nature Since the juridical personality of a company determines the degree of legal personality the company will enjoy within a territory, including the right to engage in various types of

commercial activity and the ability to appear before the local courts as plaintiffs or defendants, the non–recognition of foreign companies severely restricts their ability to engage in commercial activity and compete with the local companies.
Broader Discrimination (#PA0833).
Narrower Discrimination against transnational banks in developing countries (#PE4320).
Related Discriminatory business practices (#PD8913).
Aggravated by Xenophobia (#PD4957) Foreign ownership (#PE4738)
Nationalization of foreign investments (#PC2172)
Tax discrimination against investment in a foreign country (#PD3047).

♦ PD6421 Disease transmission by international travel
Nature The transmission of communicable diseases by international travelling constitutes a menace to countries by the spread or potential spread of diseases across frontiers. When travelling, persons may come in contact with diseases for which no measures of control have been taken and thus propagate them directly or indirectly to one or more countries. The more serious case, because less obvious, is when carriers become new reservoirs for animal vectors of endemic diseases, enabling their extension to countries where they are unknown or had already been eradicated. Since no public health measures exist in such cases, the diseases rapidly turn into epidemics causing loss of human life and a strain on health resources and infrastructure that must rapidly adapt to the new situation.
Background When Europeans travelled to the New World they brought with them many diseases unknown to the indigenous population which resulted in the wiping–out of entire villages. Wars fought in a different geographical and cultural backgrounds, pilgrimages (the massive annual migration to the Mecca in Islamic countries), migration, tourism and educational exchanges all, inevitably, contribute to a 'trade' of diseases.
Aggravated by Restrictions on freedom of movement between countries (#PC0935).

♦ PD6423 Acanthosis in animals
Broader Skin diseases in animals (#PD9667).

♦ PD6439 Space warfare
Star wars
Refs Stockholm International Peace Research Institute *Outer Space* (1978); United Nations *Satellite Warfare*.
Broader War (#PB0593) Unconventional war (#PC8836).
Narrower Space weapons arms race (#PD0087).

♦ PD6482 Soil mismanagement in developing countries
Inadequate soil conservation developing countries
Refs FAO *Soil Conservation for Developing Countries* (1981); FAO *Soil Conservation and Management in Developing Countries* (1985).
Broader Mismanagement (#PB8406) Over–intensive soil exploitation (#PC0052)
Mismanagement in developing countries (#PD8549).
Aggravates Infertile land (#PD8585).

♦ PD6520 Inadequate maintenance equipment
Lack of maintenance tools — Inappropriate maintenance equipment
Nature There is a general lack of basic modern equipment to meet requirements in rural communities for maintaining electrical generators, water and sewer systems, and public roads. These inadequacies mean the community feels it has to rely on outside sources to maintain its public works and that it cannot provide needed physical maintenance by using its own resources.
Broader Lack of essential local infrastructure (#PF2115)
Inadequate maintenance of infrastructure (#PD0645).
Narrower Insufficient heavy equipment (#PJ0259)
Lack of agricultural machinery (#PF4108)
Lack of fire–fighting facilities (#PJ2437)
Inadequate electricity infrastructure (#PD9033).
Related Limited access to social benefits (#PF1303)
Prohibitive cost of farm machinery (#PF2457)
Underprovision of basic services to rural areas (#PF2875).
Aggravates Inadequate maintenance (#PD8984) Inadequate cooking stoves (#PE7904)
Instability of water supply (#PD0722) Inadequate irrigation system (#PD8839)
Inadequate equipment maintenance (#PD1565)
Acquisition of inappropriate equipment by developing countries (#PE6640)
Insufficient and inappropriate energy equipment in developing countries (#PE8592).
Aggravated by Lack of trained firefighters (#PJ0277)
Unethical maintenance practices (#PD7964)
Growth of anti–systemic movements (#PJ0051)
Prohibitive cost of transportation (#PE8063).

♦ PD6571 Dependence on sophisticated technology for development
Nature There is a gap between the significant minority with access to technological benefits and the vast majority without. Many rural communities in the technically advanced nations are without certain fundamental benefits: some have open ditch sewage systems which serve as a breeding ground for mosquitoes and give off unpleasant odours; others may be located by rivers which threaten homes and businesses with seasonal flooding; in others, houses and farms still depend on trucked–in water or cisterns for supply. Such conditions, often falling short of minimum government standards, make these areas unattractive for new business and architectural development. Similarly, farmers need to up–date their farm machinery to remain competitively productive; those without sufficient land or equipment are forced out of farming. Costs for improved systems and keeping up with advanced technology are unrealistically high for small rural communities. Residents aim for impossibly complex solutions; intermediate options are vague and treated as unserious. There seems no real hope of getting ahead, and a sense of local creative initiative is lost.
Broader Irresponsible scientific and technological activity (#PC1153).
Narrower Lack of community development (#PF7912).
Related Technological monoculture (#PF4741).
Aggravates Unnecessary gadgets (#PE3745) Inappropriate sanitation systems (#PD0876).
Aggravated by Stagnant surface water (#PE2634)
Unresponsive river experts (#PG9002)
Prohibitive cost of living (#PF1238)
Proliferation of computers (#PE3959)
Limited specialized farming (#PG9059)
Proliferation of technology (#PD2420)
Limited merchandising interest (#PG9145)
Prohibitive cost of farm labour (#PG9069)
Restricted farming alternatives (#PG9079)
Unresearched adaptable products (#PG9154)
Deteriorated building conditions (#PG7716)
Insufficient plant life information (#PG9158)
Crippling dependence on mechanization (#PG9088)
Uncoordinated international river basin development (#PD0516).

♦ PD6576 Professional secrecy
Professional secretism
Broader Secrecy (#PA0005).
Aggravates Secrecy in scientific research (#PF1430)
Prevention of the exchange of ideas (#PD8731).
Aggravated by Professional jealousy (#PD8488).

♦ PD6579 Marginal level of family income
Limited family income
Nature The rise in cost of municipal services in some communities means that, despite high taxes, local funds are insufficient to cover basics. People on fixed income in these communities fear rising taxes and inflation and find themselves unable to pay for essentials like home repairs. Their houses fall into disrepair, and they themselves sink into a hidden subsistence lifestyle of marginal diet, restricted cashflow, and increasing dependency.
Broader Low general income (#PD8568).
Narrower Prohibitive medical expenses (#PE8261).
Aggravates Underground economy (#PC6641) Undiversified tax base (#PJ7897)
Increased fear of taxation (#PG8725) Inadequate income in old age (#PC1966)
Prohibitive cost of home maintenance (#PJ9022).
Aggravated by Prohibitive cost of accommodation (#PD1842)
Mass unemployment of human resources (#PD2046)
Underdeveloped sources of income expansion (#PF1345)
Excessive proportion of income spent on food (#PE8659)
Dependency on unpredictable sources of income (#PF3084).

♦ PD6585 Inappropriate size of school classes
Overcrowded school classes
Broader Obstacles to education (#PF4852)
Unrecognized socio–economic interdependencies (#PF2969).
Narrower Overcrowding in schools (#PE3757).

♦ PD6589 Inadequate security system
Broader Insecurity (#PA0857).
Narrower Vulnerability of computer systems (#PE8542).
Related Social insecurity (#PC1867).
Aggravates Insecurity of property (#PC1784).

♦ PD6612 Denial of right to confidentiality
Broader Infringement of privacy (#PB0284).
Related Interception of communications (#PD7608)
Denial of right to private home life (#PE6168).
Aggravated by Invasion of privacy through testing (#PJ6946).
Reduces Lack of international cooperation (#PF0817).

♦ PD6620 Denial of rights to territories
Broader Denial of right of a people to be self–determining (#PC6727).

♦ PD6626 Unethical practices in the social sciences
Irresponsible social scientists — Negligence by social scientists — Malpractice in the social sciences — Corruption of social scientists — Underreporting of social hazards
Claim Social scientists, under pressure from their employers and peers, have adopted practices which lead to the underreporting of dangerous social trends. They have failed to investigate adequately the nature of such hazards in the process of further developing knowledge about societies and cultures. There is little peer control of irresponsible intervention in minority cultures with the associated introduction of exotic norms, artefacts and modes of behaviour. Social scientists have lent their skills to covert operations designed to subvert existing governments, especially in developing countries, or to maintain repressive regimes. Their skills have also been extensively employed by commercial and political interests to manipulate public opinion, especially through advertising campaigns and appropriate disinformation.
Broader Unethical professional practices (#PC8019).
Narrower Unethical practice of anthropology (#PD2623).
Related Reductionism (#PF7967).
Aggravates Defective reasoning (#PF5711).

♦ PD6627 Indoor air pollution
Sick building syndrome — Unhealthy buildings
Nature New problems of indoor air pollution are arising from new materials and structural techniques, and from forms of treatment whose constituents are potentially toxic or irritant and hence likely to produce allergic responses. The causes of such pollution are not precise. The term may also be extended to cover tobacco smoke and body odour. It may include the combined effects of office machines and carpeting. Recent reports suggest that there is both direct and circumstantial evidence that exposure of the public to indoor air pollutants are large enough and widespread enough to account for a substantial amount of sickness and premature death.
Sick building syndrome is a general physical dissatisfaction related to office working often resulting in the symptoms of lethargy, headaches, sore throats, watery eyes and others. Complaints may include rashes, dizziness, nausea or even short–term memory loss. The symptoms usually disappear within a few hours of leaving the environment. The malaise is associated with but not necessary caused by the physical condition of the building. The classic sick building has large open spaces, tinted glazed windows and air conditioning. It is not known what are the causes of sick buildings but factors that contribute are air and surface temperatures, the level of humidity, air movement especially in the local work area, air purity and amount of fresh air, the organization of the space, the colour schemes and lighting. Buildings with monotonous environments are more likely to be sick. Public sector buildings are more effected than private sector buildings. The attitude of people is a contributing factor. If the general feel of the building is good it is more likely to not be sick.
Incidence Schools in Britain have been forced to close because of vapour from cavity–wall insulation; a mysterious form of rot has caused the disintegration of carpets; people in several communities in the USA have been forced to move out of their homes because they get ill inside them. In the USA it has been estimated that passive smoking aggravates indoor pollution to the extent of tens of billions of dollars in medical costs and absenteeism. The risks of of indoor pollution from tobacco smoke may be twice as great as the danger from radon gas and more than 100 times as great as from cancer–causing outdoor pollutants.
Background The need for energy conservation has governed the rate of air change in a building, with the result that higher levels of pollutants now occur.
Refs Godish, T *Indoor Air Pollution Control* (1989).
Broader Air pollution (#PC0119).
Narrower Badly laid out work premises (#PJ2468)
Unappealing business buildings (#PG5309)
Health hazards of passive smoking (#PE5146)
Social inadequacy of large buildings (#PF6194).
Related Poor social environment (#PJ8742)
Monolithic architecture of high–rise buildings (#PE1925).

Aggravates Stress in human beings (#PC1648) Human disease and disability (#PB1044)
Psychological stress of urban environment (#PE6299)
Disorientation stress in large building complexes (#PF6174)
Environmental stress on inhabitants of tall buildings (#PE4953).
Aggravated by Contamination by natural radiation (#PC1299)
Health hazards of modern insulating materials (#PE1499).
Reduced by Uninsulated buildings (#PE0242).

♦ PD6628 Intergovernmental organization mismanagement

Nature Mismanagement of inter-governmental organizations may include: inattention to administrative detail; careless formulations of redundant programs; programme ineffectiveness; overlapping, dispersion, and fragmentation of activities; and lack of clarity in the definition and achievement of goals.
Incidence The USA and the UK cited some of the above reasons for their decisions to leave UNESCO.
Broader Mismanagement (#PB8406) Human errors and miscalculations (#PF3702).
Aggravates Ineffectiveness of international organizations and programmes (#PF1074).
Aggravated by Mismanagement by intergovernmental organization leadership (#PE6947).

♦ PD6632 Ideological offences

Broader Deviance (#PB1125).

♦ PD6637 Forced religious conversion

Incidence Pressure from missionary groups can force indigenous peoples to convert from a traditional religion. This may be associated with taking of children from families and putting them into missionary schools. The problem can also take more subtle forms when individuals are under obligation to convert in order to be more effectively assimilated into social groups of their choice, whether to be able to marry or to avoid impediments to the advancement of a career in business or in politics.
Broader Denial of religious liberty (#PD8445).
Aggravated by Religious indoctrination (#PD4890)
Denial of right to freedom of religion of indigenous peoples (#PE4332).

♦ PD6645 Illiteracy in the fourth world

Illiteracy amongst the destitute
Nature The predicament of illiterates in industrialized countries is undoubtedly more difficult than it is in the Third World, as an adult in an economically and technologically advanced country will suffer more from the isolation resultant from illiteracy than will someone similarly handicapped in a country where the adult illiteracy rate is high. In addition, many Third World countries have a rich community life and oral tradition which serve as a source of support for the illiterate person. The illiterate in the Fourth World finds himself in a position of inferiority in a number of everyday situations and is also confronted with the complexity of life in industrially developed countries. Being unable to read street signs, the names of underground stations, bus and tram destinations, it is very difficult to get around in a city. An illiterate is unable to compare labels in a store; cannot calculate the price of foodstuffs; cannot fill in application forms for jobs, for assistance, or for schooling for his children.
Incidence UNESCO statistics for 1980 revealed that 2.5 percent of the industrialized countries' populations (22.5 million people) admitted to being illiterate. In most countries, illiteracy rates for women were 2-3 times those for men, and illiterate persons are over-represented among the unemployed and among prisoners, the latter suffering extreme consequences of their illiteracy, as writing is an important means of communication with the outside world.
Background Many industrialized countries which initiated literacy programs years ago assume the problem has been solved, thus the question does not appear on census questionnaires and is not brought to public attention. For newly arrived immigrant workers, few industrialized countries have a comprehensive literacy education policy.
Claim While it may be embarrassing for First World countries to admit a common problem with Third World nations, there must be political determination to solve the problem of Fourth World illiteracy. More effective identification of illiterates and their environment is necessary as well as a fuller understanding of the causes of different forms of illiteracy. Given the capital and technology available in all industrialized nations, it is a disgrace that millions of their inhabitants live in fear, shame and isolation caused by a problem for which there is a solution.
Broader Illiteracy (#PC0210) Illiteracy in developed countries (#PC1383).
Narrower Illiteracy among women (#PE4380).
Aggravated by Obstacles to education (#PF4852).

♦ PD6651 Human disease vectors

Vectors of human disease — Human vectors
Broader Disease vectors (#PC3595).
Narrower Rodent vectors of disease (#PE3629) Animals as vectors of disease (#PD8360)
Insect vectors of human disease (#PE3632) Arthropods as intermediate hosts (#PG6593).
Aggravates Endemic typhus (#PG5389).

♦ PD6655 Aiding consummation of crime

Nature A person is guilty of aiding in the consummation of a crime if he intentionally aids another to secrete, disguise or convert the proceeds of a crime for his own benefit. This includes possession and fencing of stolen goods, and receiving or transmitting ransom money or goods.
Broader Complicity (#PF4983) Obstruction of government function (#PD6710).
Narrower Receiving stolen property (#PE8364).
Aggravates Lack of information (#PF6337).

♦ PD6656 Intellectual terrorism

Terrorism against an idea
Broader Denial of freedom of opinion (#PD7219).
Aggravates Socio-economic poverty (#PB0388) Lack of job satisfaction (#PF0171).

♦ PD6663 Man-made diseases

Dependence on man-made diseases
Nature Man-made diseases may result from biological weapons in several ways: a disease which is normally initiated by, say, the bite of an infected mosquito, may manifest itself quite differently when caused by inhalation of the pathogen; the natural defences of man towards respiratory infection can be impaired by exposure to lung irritant chemicals, so that the dissemination of a mixed chemical-biological aerosol might lead to signs and symptoms not normally associated with the pathogen used; and confusion may result from simultaneous infection by two or more different pathogens, which may in addition have a synergetic effect upon one another. In addition, new fears arise from the increasing arsenal of genetic engineering techniques which breed strains of well-known pathogens that have increased antibiotic resistance or altered biochemical and immunological characteristics that could ultimately lead to a strain of pathogen so different from its parent that it would be classified as a new disease agent.
Broader Environmental degradation (#PB6384) Human disease and disability (#PB1044).
Aggravated by Human errors and miscalculations (#PF3702).

♦ PD6667 Extra-legal conscription

Impressment into military service
Refs Cohen, Eliot A *Citizens and Soldiers* (1985).
Broader Conscription (#PF6051).
Aggravates Kidnapping (#PD8744) Militarization of children (#PE5986).

♦ PD6693 Environmental hazards from energy production

Refs United Nations *Risks and Benefits of Energy Systems* (1985); United Nations Environment Programme *Comparative Data on the Emissions, Residuals and Health Hazards of Energy Sources* (1985).
Broader Environmental degradation (#PB6384).
Narrower Environmental hazards of coal energy (#PD7541)
Environmental hazards from hydropower (#PE6292)
Environmental hazards from electricity (#PE1412)
Adverse effects of power production on weather (#PE9134)
Environmental hazards of nuclear power production (#PD4977)
Environmental hazards from mineral fuels, lubricants and related materials (#PE1346).

♦ PD6701 Unethical practices in forestry

Mismanagement of forests — Forestry malpractice
Broader Mismanagement (#PB8406)
Unsustainable development of forest lands (#PD4900).
Aggravates Endangered forests (#PC5165).

♦ PD6710 Obstruction of government function

Nature Obstruction of government function is when a person, intending to obstruct in any way the administration of a law or other government function, interferes physically. It is a crime.
Broader Corruptive crimes (#PD8679)
Crimes against the integrity and effectiveness of government operations (#PD1163).
Narrower Hindrance of law enforcement (#PD5515)
Aiding consummation of crime (#PD6655)
Escape from official detention (#PE6260)
Failure to appear for trial after release (#PE4577)
Flight to avoid prosecution or giving testimony (#PE5267)
Introducing or possessing contraband useful for escape (#PE5390).

♦ PD6728 Mathematical ignorance

Nature Students do not learn mathematics adequately because rote learning is emphasized at the expense of fundamental concepts and problem solving strategies.
Broader Ignorance (#PA5568) Scientific ignorance (#PD8003).
Related Ignorance of grammar (#PD7566) Geographical illiteracy (#PD3984).
Aggravates Innumeracy (#PC0143).

♦ PD6733 Inadequate maternity protection in employment

Nature Maternity protection is an increasingly important matter as more women, particularly married women and single mothers, are employed. In a strict sense, maternity protection includes the right to maternity leave, the right to benefits, and the right to job security during prescribed periods. In a broader sense, it also includes the right to on-premises nursing facilities and the subsequent necessary workbreaks. Denial of these rights may jeopardize the health of both the mother and her child and add additional financial burdens to individual women and their families.
Refs Cook, Alice H *The Working Mother* (1978).
Broader Discrimination against women in employment (#PD0086).
Narrower Denial of right to maternity leave (#PE3951)
Denial of right to job protection during maternity leave (#PE7119)
Denial of right to periods to nurse infants during working hours (#PE3044).
Related Disruptive maternity leave of employees (#PG6638).
Aggravated by Maternity (#PJ1893).

♦ PD6744 Segmented labour markets

Nature Migration is a prominent feature on the debate surrounding segmented labour markets. In most neo-classical models, migration is seen as an equilibrating factor as individuals respond rationally to the comparative incentives of different sectors. On the other hand, migrants are often seen as an identifiable group in the labour market who are either discriminated against or favoured. This discrimination is one factor leading to the emergence of segmented labour markets. Migration pattern itself may raise barriers to competitive forces within urban labour markets – by limiting entry into particular occupations, restricting access to training and influencing wage determination – and hence result in segmentation.
Aggravated by Ineffective economic structures in industrial nations (#PE4818).

♦ PD6758 Wife abuse

Battered women — Wife beating
Nature Wife beating is prevalent in all societies and crosscuts all racial, cultural and socio-economic lines. Despite its prevalence wife abuse is largely a secret crime: the legal system and the public see it as a private matter. Societies keep marital violence invisible because its existence contradicts the idealized image of the family as a haven for love, security and loyalty.
Incidence In industrialized countries reports of wife beating are on the increase. In the UK in 1989 a survey concluded that 28 per cent of married women had been hit by their husbands and 33 per cent hit or threatened with physical violence. The proportion rises to 63 per cent among women who were divorced or separated. In the USA, a survey indicated that 40 per cent of men considered it normal to hit their partners. In developing countries, especially in rural areas and in those cultures in which the male has a strongly dominant role, beating the wife has been considered the right of husbands.
Counter-claim In some cultures the wife assumes that being beaten is an indication that the husband cares for her.
Refs Beaudry, Micheline *Battered Women* (1985); Brainard, Willard T *Women and Spouse Abuse* (1988); Gondolf, Edward W and Roy, Maria *Men Who Batter* (1985).
Broader Family violence (#PD6881) Violence against women (#PD0247).

♦ PD6760 Medical experimentation on socially vulnerable groups

Nature The use of human beings for purposes of experimentation includes individuals or groups of individuals drawn from sectors of society which are particularly vulnerable to gross abuses of power: prison inmates, psychiatric patients, old people, terminally ill patients, and the poor. Acts of abuse include the subjection of psychiatric patients to electric-shock experimentation; the administration of hallucinogens and poisons, such as LSD and curare; bombardment with tape-recorded messages; and the injection of prison inmates with live cancer cells for the purpose of studying the effects of the disease; injections of plutonium, thorium or radium to measure human retention.
Claim If issues such as: the limits to be placed on scientific research; who should hold the authority to establish such limits; and for what aims these limits should be established, are not resolved, then medical and pharmacological experimentation could well be licence for collective and

PD6760

individual victimization.
Broader Abusive experimentation on humans (#PC6912)
Lack of protection for the vulnerable (#PB4353)
Unethical experiments with drugs and medical devices (#PD2697).

♦ PD6795 Inadequate waste treatment

Substandard waste treatment — Inadequate waste disposal
Refs Association of Bay Area Governments *The Disposal of Hazardous Waste by Small Quality Generators – Magnitude of the Problem* (1985); Greer-Wootten, Bryn, et al *Waste Management* (1986).
Broader Inappropriate sanitation systems (#PD0876).
Narrower Hazardous waste dumping (#PD1398)
Inadequate carcass disposal of diseased animals (#PE2778).
Aggravates Radioactive wastes (#PC1242)
Water pollution in developing countries (#PD3675)
Unsanitary environment for basic health in small rural villages (#PD2011).
Aggravated by Limited accountability of public services (#PF6574).

♦ PD6800 Manipulation of nongovernmental organizations

Nature Non-governmental organizations are increasingly vulnerable to manipulations from established institutions on whom they are made to depend. These manipulative schemes result in an immense expenditure of time and resources on the part of NGOs who are reduced to becoming the 'water carriers' for development's elite corps.
Incidence Many retired university professors and government officials 'give' seminars, which actually amount to little more than demanding extensive reports from NGO's which are then compiled and published under the consultants' authorship. This disrupts the priority work of the NGO's as local meetings must be quickly organized and reports be hastily published.
Broader Obstacles to effective international nongovernmental organizations (#PF7082).

♦ PD6804 Discriminatory communication

Nature Communication, both verbal and non-verbal, can enrich or dehumanize life. Discriminatory communication can be injurious to racial, ethnic, female, older persons, children, and disabled groups, and exists in the mass media, at work, and in face-to-face meetings.
Incidence With older age groups, mass media tends to patronize and stereotype. Stories dwell on the problems and foibles of old age rather than its rewards and joys. It is invariable portrayed as a time of waning physical and mental powers, typified by canes, wheelchairs and persons who are child-like, self-indulgent, helpless and a bother to have around.
Claim Non-discriminatory titles such as flight attendants, firefighters, letter carriers, sales clerks, spouses and chairperson go a long way to contributing to non-sexist communication.
Broader Discrimination (#PA0833).
Narrower We / they language patterns (#PG9219)
Segregation through language (#PD4131)
Prejudice against other languages (#PD8800)
Exclusive nationally-oriented language systems (#PD2579).
Related Pursuit of personal prestige (#PF8145) Language discrimination in politics (#PD3223)
Prejudice against communication by visual imagery (#PF0076).
Aggravates Insufficient communications systems (#PF2350).
Aggravated by Xenophobia (#PD4957) Multiplicity of official languages (#PF6027)
Ethnic and social discrimination in foreign language teaching (#PF5929)
Antisocial attitudes in the planning of second language policy (#PF5958).
Reduced by Lack of an international language (#PG1739).

♦ PD6808 Lock-out

Lock-outs
Nature Lockout is a means used by industrial management to force workers to accept lower wages and worse conditions than they demand, that is, the lowering of the price of the only commodity which the workers have to offer, their labour. The factory is either closed down or operated by 'blackleg' labour and the workers are thus 'locked out'.
Incidence Growth in the strength of the trade union and labour movements has made it very difficult if not impossible to use the lockout, at least in Britain, but it is occasionally still used in continental Europe.
Related Factory closures (#PE3537).
Aggravated by Industrial intimidation (#PC2939).

♦ PD6810 Social isolation as torture

Nature Isolation from other human beings is used in a number of ways as torture, including solitary confinement. Generally, in its extreme form the prisoner never sees another human being for months at a time and when he does it is his torturers. In less extreme forms prisoners are denied the right to receive visitors or access to medical or legal assistance as punishments. In some cases access to the prisoner is only made available after a confession or other incriminating evidence has been obtained and formal legal charges have been made.
Incidence Social isolation has been reported in the following countries: **Af** South Africa, Zambia. **Am** Guyana, Haiti, Uruguay. **As** Afghanistan, Bangladesh, United Arab Emirates.
Broader Social isolation (#PC1707) Torture by deprivation (#PD3763).
Narrower Incommunicado detention (#PE8004).
Related Solitary confinement (#PE4056).

♦ PD6812 Discrimination against working mothers

Incidence More than 50 percent of US children have mothers who work outside the home.
Claim The influence of mothers' work outside the home does not pose any negative threat to the nature of children's daily experiences and their social, emotional, and cognitive development. Maternal employment, by itself, is neither good nor bad for all children in all circumstances. Other factors such as income, family structure, individual characteristics of the child, mother's education, and the availability of support services are much more important in shaping children's growth aid development.
Broader Discrimination against women in employment (#PD0086)
Discrimination against working women in socialist countries (#PD2872).
Aggravated by Discrimination against part-time work (#PE6241).

♦ PD6818 Security risk people

Aggravates Crime (#PB0001) Sabotage (#PD0405) Terrorism (#PD5574)
Aerial piracy (#PD0124).

♦ PD6823 Venomous animals

Nature The animals which can inflict injury on man due to their venom include spiders, scorpions, and sun spiders; ticks and mites; centipedes; snakes and lizards; and certain types of fish.
Refs Bucherl, Wolfgang et al *Venomous Animals and Their Venoms* (1971); Habermehl, G G *Venomous Animals and Their Toxins* (1981).
Broader Harmful wildlife (#PC3151).

♦ PD6845 Criminal facilitation

Nature Knowingly providing substantial assistance to a person intending to commit a felony who then commits the crime using the assistance provided is criminal facilitation.
Broader Offences of general applicability (#PD4158).
Narrower Criminal omission to act (#PG5228).
Related Criminal attempt (#PD5321) Criminal conspiracy (#PD1767)
Criminal solicitation (#PD7676).

♦ PD6850 Destruction of alluvial forests

Nature Destruction of alluvial forests can be attributed to the following factors: the working of gravel pits; establishment of industrial areas and ports; creation of leisure facilities; modification of watercourses; land reorganization; hydraulic development; enlargement of watercourses and canalisation; the building of dams and reservoirs; and pollution.
Background Alluvial forest used to extend throughout the flood plains of large rivers and their main tributaries, particularly those whose water supply during growth periods was provided by rapid floods not causing root asphyxiation. Their destruction started several centuries ago and the large valleys have been marked by man since at least the Neolithic period. It has gathered pace abruptly since the development, for the production of hydro-electric power, and canalisation on a European scale of the major water-courses: Rhine, Rhône, Danube, etc. (50 percent of surfaces destroyed for the Rhine, 25 percent for the Danube).
Broader Deforestation (#PC1366).
Aggravates Soil degradation (#PD1052).
Aggravated by Endangered forests (#PC5165) Forest damage by wildlife (#PD0500).

♦ PD6851 Childhood accidents

Playground accidents
Nature Progress in techniques, mechanization in the home and the impact of machines on our daily life, increasingly provide new dangers to be faced by the 'unprotected' in society, in particular children who have little experience and few means of self-defence against such perils.
Incidence In most highly developed countries, accidents have become the number one killer of children from a year old and upwards and there are reasons to suggest that, with the increasing demand for automation and sophisticated techniques, the number of accidents in childhood will continue to increase.
Refs Garling, Tommy and Valsiner, Jaan (Eds) *Children within Environments* (1985).
Broader Human disease and disability (#PB1044).
Related School accidents (#PE1990) Defective product manufacture (#PD3998)
Occupational domestic accidents (#PE4961).
Aggravated by Lead as a pollutant (#PE1161).

♦ PD6880 Occupational psychopathology

Nature The physical, psychological and general social atmosphere at the workplace and in an occupation as a whole, as well as the external conditions, play a considerable part in occupational psychopathology. While mental illness is rare, neurosis, especially depression, is fairly common.
Incidence If individual needs and differences are not taken into account, division of work may complicate relations among workers. Monotonous, repetitive and excessively standardized jobs are disliked, while work with automatic equipment provokes a sense of isolation. Wage rates and the basis on which wages are paid are also a source of conflict and frustration, as are inadequate communications.
Claim Work is an essential factor in the preservation of both physical and mental health, but a worker must feel that he is doing something worthwhile and he must enjoy confident relations with his fellows.
Broader Psychoses (#PD1722) Occupational diseases (#PD0215).
Aggravated by Unhealthy environment (#PJ1680).

♦ PD6881 Family violence

Domestic violence — Conjugal violence — Mate abuse — Spouse and cohabitant abuse — Problem families
Nature This general term includes all forms of destructive behaviour, physical and emotional, committed between persons who are married, living together or who otherwise enjoy a close personal relationship. Such physically abusive behaviour may include slapping, pushing, pounding with fists, or attacking with a lethal weapon. Emotional abuse does not have physical battering concomitants and is usually demonstrated by teasing, coercion or ultimatums.
Family violence is cruelty exercised on those nearest, most vulnerable, least able or inclined to defend themselves; in short it is usually practised on women and children. Within the confines of the home – which is supposed to be a refuge of warmth and security – private violence takes the form of child abuse, spouse beatings (usually of the wife), rape and sometimes even murder. And once violence becomes routine, there is no way of stopping it. The husband beats the wife, the wife may then learn to beat the children, the bigger siblings learn that it is permitted to hit the little ones, and the family pet may be the ultimate recipient.
Mate abuse is a crime and is legally referred to as assault and battery. Assault is the attempt to commit injury while battery is the actual use of force. Assault and battery are usually punishable as a misdemeanour; however, they can be charged as a felony, depending upon the amount of injury involved or the instrument used. Mayhem is charged when the attack results in permanent damage. Both mayhem and assault with a deadly weapon are usually punishable as felonies.
Incidence Officially reported cases of battering do not reflect the entire extent of the abuse problem since victims are usually unwilling to report for fear of reprisal from their mate, or because social support systems are unable to provide adequate services to protect victims. Nonetheless, it is conservatively estimated that at least two million women are beaten each year in the United States. These figures would indicate that as many as 16 percent of all married women residing in the USA are beaten annually. 25 percent of all reported murders are family related and over half of these murders involved spouse killings. Although many batterings are perpetrated by men against women there is an alarming increase in the number of reported incidents where women have murdered their mates as a retaliatory self-defence measure for beatings they have suffered over an extended period of time. The above figures reflect conditions from all parts of the USA. In California alone it is estimated that 100,000 women are seriously beaten every year. In the UK one in ten women who do report being battered refuse to prosecute or change their story. This usually happens after they have contact with their spouse or other family members.
Background Family violence is an age-old phenomenon which traditionally has not been questioned. The right of the family head to abandon disabled children, to expose and kill newborn children as a means of birth control, and severe and repeated physical punishment in education are well-known historical facts. A husband's right to physically punish his wife was considered to be a natural and legitimate means of maintaining order, which had a solid foundation in society and the legal system. Today's violence may go unchecked and unreported because battered women have been told by their ministers and their families that a good woman can change a man. Women may be nurturing, may want to help, may want successful marriages, may see the good in their mate. In addition, professionals themselves can be an essential part of the violence syndrome by failing to note signs of abuse (seeing a woman's or child's injury as caused by a fall or accident), or by labelling battered women as neurotic, prescribing tranquillizers for them, and telling them to go home.
Already during the reign of Queen Elizabeth I of England the problem of mate abuse was partially

dealt with by forbidding husbands to beat their wives after 10 pm because of the noise.
Refs Blackman, Julie *Intimate Violence* (1989); Bolton, Frank G and Bolton, Susan R *Working with Violent Families* (1987); Brainard, Willard T *Women and Spouse Abuse* (1988); Edelman, marian W *Families in Peril* (1987); Finkelhor, David, et al *The Dark Side of Families* (1983); Gelles, Richard J and Cornell, Claire P *International Perspectives on Family Violence* (1983); Hotaling, Gerald T, et al *Family Abuse and Its Consequences* (1988); Hotaling, Gerald T, et al (Eds) *Coping with Family Violence* (1988); Nordquist, Joan (Ed) *Domestic Violence* (1986); Okun, Lewis *Woman Abuse* (1985).
Broader Cruelty (#PB2642) Human violence (#PA0429) Aggravated assault (#PD0583).
Narrower Wife abuse (#PD6758) Battered husbands (#PG1898).
Related Violence as a resource (#PF3994).
Aggravates Youth violence (#PF7498) Street children (#PD5980) Violence against women (#PD0247) Physical maltreatment of children (#PC2584).
Aggravated by Drunkenness (#PE8311) Male domination (#PC3024) Domestic quarrels (#PE4021) Failure to recognize uniqueness of family members (#PF1750).

♦ PD6885 Occupational dangers in developing countries
Nature The industrialization of developing countries, necessary in order to raise the standard of living of their people, nevertheless results in occupational dangers which are more prevalent for their workers than for workers in developed countries. This is due to the pressures to increase production, which disregard the norms for hours of work, training of workers, guarding of machines, and provision of personal protective equipment. Non-observance of safety measures, resulting in accidents, is either taken as the normal outcome in industries or results in minor warnings or punishments which have no serious effect. Higher level management as well as the workers themselves do not realize that safety and production cannot be separated.
Broader Occupational hazards (#PC6716).

♦ PD6893 Inadequate social discipline in socialist countries
Nature Open disregard for law, coupled with an apparent decline of the fear of authority, seem to have become a pervasive phenomenon in socialist countries. There is an overall erosion of respect for Party discipline, with little fear of sanctions; breaches of labour discipline are not only rampant, but officially tolerated; and there is open disregard of printed accusations and court sentences.
Incidence A survey conducted in the Soviet Union revealed that 42 percent of respondents admitted that they sometimes violated labour discipline. 49 percent thought that labour discipline was something to be regarded everywhere, while 48 percent thought it depended on the circumstances.
Background De-terrorization throughout the population, combined with better education and a certain expansion of contacts with the outside world, have led to a somewhat freer and franker climate in some socialist countries. This climate has generated a gradual erosion of fear of authorities which has contributed towards the growth of corruption and the weakening of labour discipline, both of which are contributing factors to the damage of the economy and the fabric of society.
Narrower Environmental destruction by communist regimes (#PD5393).

♦ PD6896 Elitist intergovernmental groupings
Nature Global policy is strongly determined by groupings of a small number of self-selected governments or of their leaders and advisers, whether within formal frameworks such as the United Nations Security Council or within more informal groupings. Such groupings tend to maintain their distance from more widely representative bodies, informing them of decisions rather than consulting with them. Members may prefer to divert resources to such bodies rather than increase their participation in larger groupings, thus increasing the effectiveness of the former at the expense of the latter.
Incidence In addition to the 5 permanent members of the Security Council, such groupings include: Group of 5 (Ministers of Finance of most industrialized countries), occasionally extending its membership to form the Group of 7, or to form the Group of 10. The Organization for Economic Cooperation and Development is an institutionalized grouping of the most industrialized countries. More informally there exist such groups as the Group of 33, the Bilderberg Group, and the Trilateral Commission, all of which are especially concerned with the contained development of industrialized countries.
Counter-claim A primary role of such groupings is to provide constructive leadership for the world socio-political system, avoiding the confusion associated with inconclusive debates within arenas in which more governments are represented.
Broader Elitism (#PA1387).
Aggravates Elitist control of global economy (#PC3778).

♦ PD6907 Accumulation of cultural property
Broader Accumulation of property (#PC8346).
Aggravates Misappropriation of cultural property (#PE6074).

♦ PD6923 Unsustainable development of fresh waters
Environmental degradation of fresh water sources
Incidence Global water use doubled between 1940 and 1980, and is expected to double again by 2000. Yet 80 countries, with 40 per cent of the world's population, already suffer serious water shortages. There will be growing competition for water for irrigation, industry and domestic use. River water disputes have already occurred in North America (the Rio Grande), South America (the Rio del la Plata and Parana), South and Southeast Asia (the Mekong and Ganges), Africa (the Nile), and the Middle East (the Jordan, Litani, Orontes, and the Euphrates).
In most countries, agriculture is the main consumer of water, accounting for about 70 per cent of world water use. The irrigated land area has almost tripled since 1950, and supplies one third of the world's food. But only 37 per cent of the water supplied for irrigation contributes to the growth of crops; the rest is wasted. Farmers have little incentive to use the water efficiently because, in general, they pay only 10 to 20 per cent of the cost of supplying it. In most countries, water policies and laws are inadequate to manage an increasingly scarce resource. Consequently they lose millions of dollars in wasted freshwater and crop production. Water is in many cases allocated inefficiently and inequitably and used inefficiently.
Broader Unsustainable development (#PB9419).
Narrower Lake pollution (#PD8628) River pollution (#PD7636) Silting of water systems (#PD3654) Uncontrolled river erosion (#PJ9518) Eutrophication of lakes and rivers (#PD2257).
Aggravated by Water pollution (#PC0062) Vulnerability of wetlands (#PC3486) Shortage of fresh-water sources (#PC4815) Uncoordinated international river basin development (#PD0516).

♦ PD6927 Denial of human rights in the administration of justice
Denial of right to due process of law
Nature Human rights are denied in the administration of justice when a person is detained arbitrarily without an explicit reason; when he is subjected to torture or to cruel, inhuman and degrading treatment or punishment; and when free legal counsel and a prompt and impartial trial are refused.
Broader Denial of right to justice (#PC6162) Inequitable administration of justice (#PD0986).
Narrower Denial of right to appeal (#PD5317) Suspension of habeas corpus (#PG3794) Government inducement to crime (#PD6943) Lack of access for prisoners' defence (#PE8637) Denial of right to fair and public trial (#PE3964) Denial of right to freedom from double jeopardy (#PF6158) Denial of right to freedom from retroactive laws and punishments (#PE4743) Denial of right to freedom from imprisonment for failure to fulfil a contractual agreement (#PE7235).
Related Denial of right to freedom from cruel, inhumane or degrading punishment (#PC3768).
Aggravates Wrongful detention (#PD6062) Forced disappearance of persons (#PD4259).
Aggravated by Repression (#PB0871) Undemocratic political organization (#PC1015).

♦ PD6930 Exploitation of casual workers
Discrimination against casual labour — Exploitation of contract labour
Nature Casual workers, of whom there is a disproportionately high number of women, are paid less than their regularly employed counterparts; cannot participate in employees benefits; are largely excluded from the protection of labour legislation and trade union affiliation; and are shown, due to the declining purchasing power of their low wages, to have a high incidence of malnutrition. Entrepreneurs using casual workers tend to have difficulties in forecasting their budgets and to exploit such workers, who may well be a new kind of marginality that could easily lead to social unrest.
Refs ILO *Social Problems of Contract, Sub-contract and Casual Labour in the Petroleum Industry* (1974).
Broader Exploitation in employment (#PC3297).
Aggravates Discrimination against women in employment (#PD0086).

♦ PD6943 Government inducement to crime
Incrimination through entrapment — Government conspiracy to incriminate — False incrimination
Nature Civil or military police, or government agents, may entice or lure individuals or groups into breaking the law. This may be done to apprehend known criminals by having an undercover officer party to the incriminating act and therefore able to testify concerning it, if in fact, the individual concerned is not caught 'red-handed' in the act itself. Trapping individuals in this manner may also be used to incriminate and convict enemies of the state and political dissenters, by leading them to break laws concerning the national interest or censorship. Entrapment is illegal in many countries, in particular because it may be abused; it may cause an individual to commit a crime in order to punish him, when in fact the crime would not otherwise have been committed at all.
Where individuals cannot be tricked or entrapped into breaking the law, and are innocent of wrongdoing, law enforcement agents may introduce or plant evidence of law-breaking in the individuals residence, office, automobile, or other property, or in clothing; or they may falsely testify that they found incriminating evidence in the individual's possession. Where the nature of the act does not require physical evidence, agents may conspire to incriminate individuals by giving false witness and perjuring themselves, or by using paid or black-mailed individuals in their place to testify against the innocent. These acts are criminal under all legal systems but occur nevertheless.
Incidence Entrapment includes such devices as 'sting operations' in the USA where police officers pose as receivers of stolen goods or 'fences' and encourage burglars and other criminals to bring them stolen property which they purchase and hold as evidence. It is also related to the actions of government agitators in political parties, labour unions, or social action groups, where they encourage violence or other law breaking in order to discredit or incriminate the entire organization. A government or regime may also be entrapped (to bring it down or to blackmail it), by agents of foreign powers.
Broader False evidence (#PF5127) Tampering with physical evidence (#PF7291) Denial of human rights in the administration of justice (#PD6927).

♦ PD6945 Atrocities
Dismemberment of bodies
Broader Evil (#PF7042).
Narrower Military atrocities (#PD1881).
Aggravates Dying a bad death (#PF1421).
Aggravated by War (#PB0593) Torture (#PB3430) Brutality (#PC1987).

♦ PD6960 Disruption of ecosystems in marginal agricultural lands
Degradation of fragile ecosystems
Nature In developing countries population pressure expands against a limited land resource base and pushes cultivation out onto more fragile resources, thereby rapidly destroying forests, grasslands and rainfed uplands. Increased production on fragile lands increases soil erosion, causes flooding and deforestation and reduces agricultural production.
Refs Farnworth, E G and Golley, F B (Eds) *Fragile Ecosystems* (1974).
Broader Natural environment degradation (#PB5250) Vulnerability of ecosystem niches (#PC5773).
Narrower Disruption of arid zone ecosystems (#PD7096).
Related Vulnerability of wetlands (#PC3486) Destruction of the countryside (#PE3914).
Aggravates Deforestation of mountainous regions (#PD6282).
Aggravated by Inappropriate cash crop policy (#PF9187) Unsustainable rural development (#PD4537) Cultivation of marginal agricultural land (#PD4273).

♦ PD6963 Political myopia
Nature Political myopia stems from the inability of policymakers to give serious attention to problems whose effects may not be fully visible for several years. If effect are unlikely to become apparent prior to a politician's's departure, then the problem gets put aside in favour of those whose effects can be used to bolster prestige. The long-term result of political myopia is that problems are passed on to succeeding generations.
Incidence An example of political myopia is the failure to instigate a ban on the manufacture and stockpiling of nuclear weapons, following the explosions at Hiroshima and Nagasaki.
Aggravates Political over-reaction (#PF4110).

♦ PD6967 Prenatal wrongful death
Nature The death of a foetus, through malpractice in amniocentesis for example, is not recognized as a cause for legal action in most common law jurisdictions. This is primarily because there is a requirement that one have had live birth in order to be a legal person. Because laws concerning the unborn are still in major transition in many countries, wrongful death for a foetus is permitted.
Broader Illegal induced abortion (#PD0159).
Related Homicide (#PD2341) Infanticide (#PD3501).

♦ PD6970 Collective punishment
Nature Collective punishment includes policies and practices which punish groups of people indiscriminately as a method of collective repression.

PD6970

Incidence The Israelis have placed the Palestinian villages under a curfew for 19 to 24 days, during which they were deprived of foodstuffs and medicines. The electricity supply and phones were cut off, houses were demolished and the inhabitants were prevented from engaging in agricultural work. Markets, shops and businesses were forced to close in reprisal for the national strikes or protest that had been ordered by the Palestinian leadership. The supply of foodstuffs between towns and villages was interrupted.
Broader Punishment (#PA5583) Unjust punishments for crimes (#PD4779)
Narrower Curfew (#PG5582) Politically motivated mass arrests (#PJ1301)
Demolition of homes by government authorities (#PE9337).
Aggravates Closure of social institutions (#PF3831).

♦ **PD7001 Protectionism in the computer services industry**
Protectionism in the computer and office machine industries
Nature Measures in many developed and a number of developing countries which affect computer services impede the movement of software and data across national borders. A frequently used measure is the requirement that data-processing functions have to be performed as far as possible within the country. Thus, transnational corporations may be required to maintain local data-processing facilities which could involve higher operating costs. There is a growing tendency in many countries to require that data files remain within the country rather than be transmitted to computers and computer banks located abroad. Other measures may affect access to government-controlled communication lines required for the transmission of data and the establishment of international computer networks. In some countries, foreign firms face a surcharge that is added to the normal leasing rates for communication lines. With regard to tariffs to be levied on imports of software, several developed countries have made efforts to establish workable valuation methods, involving a formula which would allow import duties to be levied on the value of the data. To date, tariffs have been confined to valuations of data communication hardware.
Broader Protectionism in the services industries (#PD7135).
Aggravates Narrow range of practical skills (#PF2477).

♦ **PD7012 Protectionism in the insurance industry**
Nature In many countries, legislation prohibits the insuring abroad of certain risks. Frequently, only locally incorporated companies may do business in the domestic market, thus excluding the provision of insurance directly by the parent company overseas. Placing insurance abroad may, however, be permitted when the local market does not have the necessary insurance capacity. Insurance with companies located abroad may be discouraged through fiscal measures. A number of countries tax premiums paid locally for imported insurance, for instance marine insurance, whereas premiums paid for insurance effected domestically are tax deductible. In the reinsurance sector, the placement of reinsurance directly with an institution located abroad is prohibited or limited in some developing countries. All reinsurance, or a fixed percentage thereof, has to be placed with a local reinsurance entity, normally a public or semi-public institution. As regards freight insurance, another measure is the requirement that imports have to be insured in the domestic market of the importing country. In some countries, a similar regulation exists for exports, requiring insurance in the exporting. Many governments limit direct foreign investment in their domestic insurance sectors. Some require that locally-established companies must be owned and managed entirely by nationals. Frequently, prohibitions are directed against a further expansion of established foreign insurance companies.
Limitations on foreign equity in insurance firms are also a common feature of many national insurance markets. The establishment of foreign reinsurers is excluded in some countries, where all reinsurance must be place with a designated indigenous institution. Foreign insurance companies may, like banking institutions, be required to meet higher capital and reserve requirements than national insurance firms. A number of developing and developed countries do not allow foreign insurance firms to offer certain types of insurance, such as life insurance. Foreign companies may also be subject to higher taxation of premium income. In some countries, requirements and regulations affect the operations of foreign insurance companies by influencing consumer choice in favour of national companies. Thus, it may be obligatory for a person who enjoys any government incentive or subsidy to insure with a national company.
Broader Protectionism in the services industries (#PD7135).

♦ **PD7046 Orphan children**
Orphans
Broader Neglected children (#PD4522) Victimization of children (#PC5512)
Inadequate laws of adoption (#PD0590) Neglect of dependents of war victims (#PD2092).
Narrower Decreasing number of adoptive parents (#PF4205)
Decreasing number of adoptable children (#PF4200)
Children engendered by occupying soldiers (#PD8825)
Unlimited practice of human embryo storage (#PE5623).
Aggravates Corruption of minors (#PD9481) Untenable orphan care (#PJ9718)
State custody of deprived children (#PD0550).
Aggravated by Family breakdown (#PC2102)
Narrow legal definition of the family in developing countries (#PD1501).

♦ **PD7049 Protectionism in the construction and engineering services industries**
Nature Governments, particularly in developing countries, increasingly require some form of cooperation with local enterprises by foreign firms providing construction and engineering services. Thus, foreign firms may have to engage in joint ventures with local enterprises or conduct their transactions through local subsidiaries or affiliates, which act as their representatives in the host countries.
Broader Protectionism in the services industries (#PD7135).

♦ **PD7051 Societal over-commitment to learning**
Claim The ultimate limit to societal learning may lie in the consequences of unrestricted societal over-commitment to learning. As enthusiastically described by some organizations, learning is not limited by its relationship to other social pressures but this leads via the 'eternal student' to a society dedicated to the consumption of information and totally unable to focus that learning for action (on the world problematique for example). This raises the question as to what extent learning systems do, or should, empower users to act.
Broader Limits to societal learning (#PF7074) Obstacles to community achievement (#PF7118).
Related Inadequate education (#PF4984).
Aggravated by Lack of commitment (#PF1729).

♦ **PD7060 Protectionism in the entertainment products and film industries**
Nature Measures affecting the importing of films, such as licensing requirements or import quotas, apply in many developed and developing countries. Screen quotas require cinemas and television studios to allocate a certain proportion of their screen time to the showing of domestic films. A further measure which impedes the import of films can be the requirement that prohibits foreign films from being dubbed into the local language. Governments may also impose higher admission taxes and other levies on foreign than on domestic films, and they may require all films to be imported and distributed through a public distribution organization which has a monopoly position.

Such centralization of purchase and distribution is found particularly in developing countries.
Broader Protectionism in the services industries (#PD7135).

♦ **PD7065 Divided cities**
Incidence There are a number of cities divided legally or de facto by claims from two or more nations of sovereignty over the separate parts: such as Berlin, Jerusalem and Nicosia. Cities like Belfast and Beirut have been divided by hostile religious factions engaged in mutual murders and activities; some bilingual or multilingual cities may have formal or informal language zones or show commercial preferences for one official language above another, giving rise to considerable tensions, as exampled in Montreal, Brussels and Miami. Some governments may have a federal district whose residents have one set of laws and advantages but they are situated within a larger metropolitan region with other local ordinances, taxes and amenities, such as happens in Washington and Canberra. Considerable conflict arises because of disparities between these areas. On a smaller scale, university towns pose the adversity of 'town versus gown', not unlike towns in which there is one major supporting industry where interests of the town corporation and the business corporation do not always coincide.
Claim The origins of cities, as recorded in the histories concerning Lycurgus, Solon and Numa for example, show that they were inhabited from the beginning, if not created, by confederations of tribes, clans or classes. Thus cities were, from the start, frequently ethnically and economically pluralistic. Jerusalem too had its Hebrews, Jebusites, Hittites and others, as well as classes of priests, scribes, nobles, merchants, farmers, artisans and slaves. Cities represent the opportunity to form the nucleus of nations by bringing diverse elements into cooperation; to be divided is unnatural.
Broader Divergence (#PA5573).
Related Divided countries (#PD1263).

♦ **PD7073 Protectionism in public accounting and auditing services**
Nature In both developing and developed countries, regulations curtail the right of foreign professional accountants to practice locally, or even completely prohibit them from doing so. Audits may have to be supervised by locally registered, qualified accountants. Foreigners may be required to possess the requisite professional degree from a local university, or pass a special examination, as well as be a member of a local professional body.
Broader Protectionism in the services industries (#PD7135).
Aggravates Creative accounting (#PE6093).

♦ **PD7086 Inadequate use of visual imagery for societal learning**
Nature Comprehension of complex or voluminous data is handicapped by a reliance on the catalogue or 'laundry-list' approach of linear, successive itemization. Mnemonic systems and other memory or machine retrieval techniques in effect are merely tricks that allow for subject and data apprehension, but not necessarily for comprehension, as this involves a representation of relationships among data. Successive items on a vertical list may indicate mutually exclusive classes, or with indentation or numbering, classes within classes within classes, but they cannot contain much more than a name and a class relationship. The considerable intellectual and financial investment in the hardware and software of non-image oriented computerized information systems makes it unlikely that any useful link to image manipulating systems (including map-generating devices) can be established. Parallel systems may well be developed which fragment what should be an integrated approach.
Broader Limits to societal learning (#PF7074)
Ineffective methods of practical education (#PF2721).
Aggravates Data-oriented education (#PF1217).

♦ **PD7087 Conflicting claims over shared inland water resources**
Nature The quantity and quality of available water resources for human, animal and plant consumption is limited and fixed. Population increase, technological advances and higher standards of living make the competition for securing water supplies a vital issue. National sovereignty over these waters allows a state to divert waters while in its territory, often causing injury to other states sharing the same resource. Problems are, therefore, increasing over the use of such shared inland water resources, and disputes are present in many parts of the world.
Incidence The Indus, Ganges, underground water tables between Mexico and the USA, the Nile and the Jordan river basins are examples of problem areas in the more than 250 surface international drainage basins in the world and numbers of international aquifers; that is, those underground waters shared between two or more states. International agreements for sharing the benefits deriving from these waters are few. As a consequence, the potential for conflicts cannot but increase.
Claim In order to cope with this problem, states should accept the principle of limited territorial sovereignty over the shared water resources located on or under their territories, and thus refrain from acting unilaterally without consulting and/or cooperating with other basin states.
Counter-claim It is unrealistic to expect states to accept the principle of limited territorial sovereignty over the shared water resources located in their territories.
Aggravates Inadequate water system infrastructure (#PD8517).

♦ **PD7089 Excessive consumption of protein**
Broader Excessive consumption of specific foodstuffs (#PC3908).
Reduces Protein-energy malnutrition (#PD0339).

♦ **PD7096 Disruption of arid zone ecosystems**
Endangered drylands — Destruction of arid ecosystems by irrigation
Nature Waterlogging and increased salinity and/or alkalinity can defeat the very purpose of the irrigation scheme. In almost every irrigation zone, significant species shifts in soil fauna and plants can be found. Enlarged insect populations spread diseases such as malaria.
Broader Disruption of ecosystems in marginal agricultural lands (#PD6960).
Aggravates Desert advance (#PC2506) Excessive use of land for agriculture (#PD9534).
Restriction of wild animal range size (#PC0475)
Endangered lifestyles of nomads and pastoralists (#PE8077).

♦ **PD7108 Protectionism in the advertising industry**
Nature In the advertising sector, a number of developed and developing countries do not permit the import of television and radio commercials. Apart from total prohibitions, there are various controls affecting the import of advertising material, such as import licensing and other clearance requirements, import duties and taxes and regulations prohibiting commercials using foreign models and languages.
Broader Protectionism in the services industries (#PD7135).
Aggravates Domination of advertising by transnational corporations (#PE2193).

♦ **PD7120 Protectionism in the banking industry**
Nature At the broadest level, control of international banking transactions involves limitations on the acceptance of foreign-owned deposits, on lending by resident banks to non-resident customers and on borrowing by residents from banks which operate outside the country. Measures affecting foreign-owned deposits and foreign lending and borrowing are part of the broader framework applied by countries to international capital movements through which their

markets communicate with each other and with the Euro-markets. The measures may not, therefore, be specifically aimed at influencing bank lending, but more generally at controlling inward and outward movements of capital by residents and controlling the scope for speculation in the local currency. Frequently, a more liberal attitude is taken towards capital inflows, as long as they do not threaten domestic monetary stability. Government involvement can take the form of a variety of measures, such as fixing maximum - even negative - interest rates on foreign-owned deposits and placing ceilings on the amounts of funds which domestic banks may lend to foreign clients, or residents may borrow from abroad. Export credits, that is, financing related to manufactured exports, are normally exempt from such measures. A further measure, frequently encountered, is the prohibition of branch banking under the direct control of the parent bank overseas. Such prohibitions also involve, to some extent, constraints on 'across-frontier' operations.

Some countries with private banking sectors, among them small developing economies, do not permit the entry of foreign banks or only allow them to establish representative offices which cannot extend loans or accept deposits. A number of developed and developing countries have prohibited the further expansion of foreign banks in their domestic banking sectors; new foreign banks may not enter and established foreign banks are not permitted to set up new branch offices. A less prohibitive approach chosen by some other countries stipulates that the opening of any additional branch by a foreign bank, already maintaining one or more branches in the host country, requires a new licence. Many countries apply limitations on the extent of foreign equity in a banking institution, often limiting foreign participation to a minority holding. Branch banking under the direct control of the overseas parent bank is frequently prohibited, making the establishment of capitalized - fully or partially owned - subsidiaries obligatory.

On account of limitations on the scope of operations, foreign banks may not be in a position to offer customers the same services that 'full service' commercial banks can offer. They may not be permitted to accept certain types of deposits, such as demand and savings deposits, or to enter more widely into retail banking and instalment credit business. Moreover, access to Central Bank rediscount facilities can be excluded or more limited than for domestic banks. Operations may also be limited to transactions related to international trade and to foreign currency project lending. The acceptance of local currency deposits and the extension of local currency loans are, then, prohibited. Such limited access is, for instance, encountered in some socialist countries of Eastern Europe, which do not normally allow the entry of Western banks. Foreign financial institutions are, in these cases, considered as vehicles to facilitate international trade. Foreign banks may be subject to stricter capital and reserve requirements and liquidity ratios than locally-owned banks. Branches of non-national banks must often be capitalized as if they were independent banks, thus excluding from consideration the capital of the parent bank. They may also be required to invest a larger proportion of their portfolios in government bonds, long-term securities of development banks or other government financial institutions, or to invest specified portions of earnings in long-term loans to local industry. Furthermore, foreign banks may be required to maintain fixed amounts in foreign currency deposits with the Central Bank. Sometimes a certain amount has to be deposited for each branch office established. Ceilings on the volume of credits which the bank may extend may also be made dependent upon the bank's foreign currency deposit with the Central Bank.

Broader Protectionism in the services industries (#PD7135).
Narrower Discrimination against transnational banks in developing countries (#PE4320).

♦ PD7121 Protectionism in the franchising services industry

Nature Many countries, particularly developing countries, do not allow the establishment of franchise companies which are fully foreign-owned. Furthermore, a number of developing countries require foreign firms to disclose know-how and technology before franchising operations are permitted. Firms may, for instance, be required to lay open the process technology used in fast-food production. In other cases, the allocation of foreign exchange for remittances of royalties and other earnings is made dependent on how much know-how is transferred to the franchise holder under the franchise agreement.
Broader Protectionism in the services industries (#PD7135).

♦ PD7132 Protectionism in the air transportation industry

Nature On most routes, the civil aviation market is apportioned by bilateral agreements between countries designating specific carriers on particular routes and the frequency of the service. Some developing countries also use bilateral agreements to control the number of flights and passengers that a foreign airline may fly to or from their territories to protect their own, much smaller, national airlines from competition from airlines of developed countries. At the non-governmental level, air fares are agreed upon in tariff conferences of the International Air Transport Association (IATA), subject to the approval of the governments concerned. Despite bilateral agreements and other controls, it would seem that the influence of IATA on the setting of tariffs has been declining in recent years, notably due to a policy of reduced intervention by the United Nations. Almost every country reserves for its national airline all flights wholly within its territory. Some regulations and practices in both developed and developing countries frequently favour local airline companies. Foreign airlines may be subject to differential taxation, such as higher business and sales taxes. National carriers are also frequently granted preferential airport user rates on such items as landing fees and hangar and parking charges. Furthermore, national carriers are accorded preference in airport facilities and services. Ground handling services may be the monopoly of the national airline or some other designated national company. The access of foreign airline companies to automated reservation and ticketing systems may be limited.
Broader Protectionism in the services industries (#PD7135).

♦ PD7135 Protectionism in the services industries

Nature A full understanding of the role of services in domestic economies and international trade and investment activities has yet to be developed. Consequently, present economic policies often do not take adequate account of the potential effect that service industries may have on growth, development and export expansion. While government intervention in the production of goods has been primarily through commercial policies, this has not been the case with services. Most services are intangible in nature and hence do not lend themselves to traditional border measures such as tariffs or quotas. Consequently, regulation of foreign investment becomes a particularly important element in government policy regarding the service sector. Foreign investment policies often take the form of measures relating both to initial establishment and to operations after establishment. In addition, government measures affecting services are often for reasons unrelated to trade and are not directed to the maintenance of internationally uncompetitive industries. This is due primarily to the nature of many service activities. For example, though national banking laws may directly affect a foreign bank's operation within a host country, the laws exist primarily for domestic reasons (such as protection of the consumer, protection against monopolies, and control over monetary policy). Regulations concerning the operations of insurance companies and investment management companies, for example, specifying the areas where they can invest their funds, are also for the purpose of protecting the consumer from undue risk. These regulations become trade/investment policies only when they treat foreign operations differently from domestic ones.

Governmental and non-governmental measures bearing upon international service operations are likely to affect a broad range of key economic variables: domestic prices, production and consumption of services; the volume of trade in services; allocation of resources; distribution of income; and social welfare. Governmental measures can, in particular, have a potential influence on patterns and structural change in the national and international services sector and on the choice made by transnational service firms between export, direct foreign investment and contractual arrangements with companies abroad. Restrictive business practices of transnational service firms can influence the formation of oligopolistic market structures, with adverse implications for global efficiency and welfare. Preferential trading arrangements among countries such as free-trade areas, customs unions and common markets, also affect transnational service activities and may lead to trade creation or trade diversion in the services sector which will either improve or worsen resource allocation and welfare. Government policies which affect the import of services or the inflow of foreign capital are other primary determinants of how enterprises provide their services to foreign markets. In certain instances a host country's policies may require local establishment for the provision of the service or local participation in the establishment. In other cases, host-governments may try to induce direct foreign investment through various subsidization schemes. This is illustrated by the recent trend to move film production abroad in order to benefit from financial incentives provided by host countries hoping to develop this sector in their domestic economy. Similarly, foreign investment decisions are influenced by tax considerations, including the practice of transfer pricing for the services provided. The licensing-investment choice will often depend on the transaction costs involved in transferring the proprietary advantage relative to the net profit which can be obtained from direct investment. It also depends on host country policies regarding foreign equity holdings. The trend in the international hotel sector towards management contracts to operate local properties is due, among other things, to the increasingly high construction and financing costs of hotel properties and to host government policies directed at a continued of foreign shareholdings.

Government policies affecting investment and trade in services therefore can have a significant impact on the manner in which a firm markets its services abroad. To the extent that investment policies deter the establishment of a foreign subsidiary, a firm will seek to pursue its export option. Where trade policies prevent the export of a service, the firm is likely to seek sales through direct investment in a foreign subsidiary. In cases where both policies are pursued, or where the firm has only one viable alternative, such policies may prevent the service from entering into the international market altogether.

Broader Trade protectionism (#PC4275).
Narrower Protectionism in legal services (#PD7097)
Protectionism in the banking industry (#PD7120)
Protectionism in the insurance industry (#PD7012)
Protectionism in the advertising industry (#PD7108)
Protectionism in the computer services industry (#PD7001)
Protectionism in the air transportation industry (#PD7132)
Protectionism in the franchising services industry (#PD7121)
Protectionism in public accounting and auditing services (#PD7073)
Protectionism in free-zone international financial services (#PE7987)
Protectionism in the entertainment products and film industries (#PD7060)
Protectionism in the construction and engineering services industries (#PD7049).
Related Restrictive legal practices (#PD8614).

♦ PD7171 Covert smear campaigns by government

Government dirty tricks policies — Black propaganda — Vilification by government
Nature Governments may disinform public, the press and politicians in order to smear and damage the opponents' reputation and to bolster their own supporters. These campaigns involve forged documents or photographs, fake confessions, false bank statements, bogus letters and leaflets.
Broader Disinformation (#PB7606)   Unethical practices in politics (#PC5517)
Covert intelligence agency operations (#PD4501).
Related Psychological warfare (#PC2175).
Aggravates Rumour (#PF5596)   Discriminatory language (#PF7299).
Aggravated by Political smear campaigns (#PD9384).

♦ PD7179 Uncompensated damages

Non-payment of compensation for damages — Delay in payment of compensation for damages — Lack of compensation for damages
Refs Riley, Tom *Proving Punitive Damages* (1981).
Narrower Non-payment of reparations by government (#PE4446)
Non-payment of compensation to victims of crime (#PE3913)
Non-payment of compensation for damages to consumers (#PE0290)
Non-payment of compensation to victims of malpractice (#PE0811)
Non-payment of compensation to victims of catastrophes (#PE5229)
Non-payment of compensation to victims of motor accidents (#PE8824).
Aggravated by Non-restitution of property (#PC7859).

♦ PD7183 Release of genetically engineered micro-organisms

Nature Such micro-organisms are being designed for agricultural purposes such as extending the range of plants which harbour nitrogen-fixing bacteria in their roots. Release of such organisms creates the risk that once released they may exhibit some previously unknown pathogenicity, might take over from some naturally occurring bacteria (possibly having other positive functions which thus are lost) or pass on some unwanted trait to such indigenous bacteria. There is also concern that an uncontrolled genetic mutation could produce, from such an engineered micro-organism, a form with hazardous consequences for the environment.
Incidence Genetically engineered micro-organisms were under trial in 1989. None are as yet known to have been deliberately released.
Broader Irresponsible genetic manipulation (#PC0776).

♦ PD7187 Parental punishment

Parental beating of children — Smacking children
Broader Physical maltreatment of children (#PC2584).
Related Paternalistic punishment (#PJ3700).

♦ PD7205 Bioaccumulation of toxic substances

Refs FAO *Report of the FAO/UNEP Meeting in the Toxicity and Bioaccumulation of Selected Substances in Marine Organism, Rovinj (Yugoslavia), 1984* (1985).
Aggravated by Toxic substances (#PD1115).

♦ PD7210 Prohibition of trade union meetings

Broader Violation of trade union rights (#PD4695).
Related Violation of the right of trade unions to function freely (#PE1758).
Aggravated by Unethical financial practices (#PE0682).

♦ PD7212 Denial of right to security

Broader Denial of political and civil rights (#PC0632).
Narrower Death threats (#PD0337)   Incitement to hatred (#PE5952)
War and pre-war propaganda (#PD3092).
Related Forced disappearance of persons (#PD4259).

♦ **PD7215 Metal failure**
Metal fracture — Metal fatigue
Refs Frost, N E, et al *Metal Fatigue* ; Gifkins, R C (Ed) *Strength of Metals and Alloys (ICSMA 6) Proceedings of the 6th International Conference, Melbourne, Australia, August 16–20, 1982* (1982); Gurney, T R *Fatigue of Welded Structures* (1980); Kocanda, S *Fatigue Failure of Metals* (1978); Tanaka, T M and Komai, K (Eds) *Current Research on Fatigue Cracks* (1987).
Broader Failure of materials (#PD2638).

♦ **PD7216 Stress in materials**
Mechanical stress
Broader Stress (#PB9165).
Narrower Stress in industry (#PE6996).
Aggravates Mechanical failure (#PC1904).

♦ **PD7219 Denial of freedom of opinion**
Broader Denial of freedom of thought (#PF3217).
Narrower Intellectual terrorism (#PD6656).
Related Denial of religious liberty (#PD8445) Denial of freedom of conscience (#PD7612) Restriction of freedom of expression (#PC2162) Restrictions on freedom of information (#PC0185) Denial of right to correct misinformation (#PE7349).

♦ **PD7221 Government sanctioned killing**
Government sanctioned murder — Staged encounter killings — Government sanctioned assassination — Death squads — Political murders authorized by governments — Extrajudicial executions — Political killings
Nature Police may stage false encounters with criminals or political opponents in order to eliminate them because they cannot obtain sufficient evidence to bring them to justice before the country's courts of law. Staged incidences may include 'suicide whilst under cross-examination', 'battle encounters', 'escape from prison' and 'resistance to arrest'. In fact these are terms intended to mask official murder. Governments may also arrange an assassination of one of the leaders of an enemy country or international movements or organizations by secret police or some illegal underground organization.
Political murders authorized by governments may take the form of disappearances, staged incidents, "suicides" or death caused by gang warfare. Governments often try to dismiss such events, whether by denying that they have taken place at all, by attempting to attribute them to opposition forces, or by alleging they resulted from armed confrontation with government forces, or that the victim was murdered while attempting to escape from custody. The killings are often accompanied by intimidation of witnesses and relatives of victims, and suppression of evidence. Victims may be chosen for their political beliefs or activities, religion, ethnic origin, sex, colour or language. There is occasionally a show trial to prove lack of government involvement and, given the nature of the complicity, it is virtually impossible to appeal against the verdict or otherwise hope for justice.
Incidence In India, the killing of suspected dacoits (robbers) was common during the 1960s. In 1983, the President of the Human Rights Commission of El Salvador, Marianela Garcia, was killed by security forces in what was called an 'encounter with guerrilla forces'. This was in fact the elimination of a redoubtable enemy, while at the same time attempting to defame her political character and to reduce world disapproval for her murder.
In examples of political murder Amnesty International has cited the following countries: **Af** Ghana, Guinea, Morocco, Mozambique, Namibia, South Africa, Togo, Uganda, Zambia, Zimbabwe. **Am** Argentina, Brazil, Chile, Colombia, El Salvador, Grenada, Guatemala, Haiti, Honduras, Mexico, Paraguay, Peru, Suriname, Uruguay. **As** Afghanistan, India, Indonesia, Iran Islamic Rep, Iraq, Israel, Lebanon, Philippines, Sri Lanka, Syrian AR.
Claim These killings flout the absolute principle that governments must protect their citizens against arbitrary deprivation of life, which cannot be abandoned under any circumstances, however grave. Such killings are crimes for which governments and their agents are responsible under national and international law. Their accountability is not diminished by opposition groups committing similar abhorrent acts. Nor does the difficulty of proving who is ultimately answerable for a killing lessen the government's responsibility to investigate unlawful killings and take steps to prevent them. It is the duty of governments not to commit or condone political killings, but to take all legislative, executive and judicial measures to ensure that those responsible are brought to justice.
Refs Amnesty International *Political Killings by Governments* (1983).
Broader Homicide (#PD2341) Assassination (#PD1971) Unethical practices of government (#PD0814).
Narrower Vigilantism (#PD0527) Political mass murder (#PD5590).
Related Military atrocities (#PD1881) Crimes against humanity (#PC1073) Political assassination (#PE5614) Corporation-sanctioned assassination (#PE6356) Official cover-up of government harassment of political activists (#PF3819).
Aggravated by Covert intelligence agency operations (#PD4501).

♦ **PD7225 Military disobedience**
Military insubordination
Broader Citizen disobedience (#PD5707) Impairing military effectiveness (#PD4448).
Narrower Military desertion (#PD4621) Inadequate army discipline (#PD2543) Crimes related to military service obligations (#PD5941).
Related Insubordinate behaviour (#PJ6517) Inadequate government control of military (#PF9542).
Aggravates Military atrocities (#PD1881) Military demoralization (#PE6639) Drunkenness of military personnel and troops (#PE8495).
Aggravated by Mutiny (#PD2589) Passive resistance (#PF2788) Injustice of military tribunals (#PE0494) Drug abuse by military personnel (#PE5579) Deficiencies in military codes of justice (#PE8300) Conflicting military obligations for persons of multiple nationality (#PE6678).

♦ **PD7231 Deceptive misuse of research**
Misrepresentation of socially unacceptable activity as research — Misuse of research as a cover for illegal activity — Research as disinformation — Monitoring as a cover for inaction
Incidence The most striking example is the insistence by the governments of Japan and Iceland that continuation of whaling is essential for research purposes. More generally, space research is frequently used as a cover for missile development. A number of countries engage in research on nuclear physics which, despite denials over an extended period, subsequently becomes apparent as being the means through which they were able to develop nuclear weapons. It is reported that experimental uses of drugs on convicted criminals has been used as means for developing biochemical weapons.
Broader Abuse of science (#PC9188).
Aggravates Scientific fraud (#PF1602) Unreported research (#PF9141) Ineffective monitoring (#PF2793).
Aggravated by Unlawful government action (#PF5332).

♦ **PD7238 Physically dependent people**
Broader Dependence (#PA4565).
Aggravates Family rejection of physically handicapped (#PE2087) Stress on families of the physically or mentally handicapped (#PD1405).
Aggravated by Senility (#PE6402) Human disability (#PC0699) Susceptibility of the old to physical ill-health (#PD1043) Inadequate community care for handicapped persons (#PE8924).

♦ **PD7251 Denial of right to social security**
Refs ILO *Maintenance of Rights in Social Security.*
Broader Denial of social rights (#PC0663) Inadequate standards of living (#PF0344).
Narrower Denial of the right to social security in capitalist systems (#PD3120).
Related Denial of right to sufficient food (#PE0324) Inadequate social welfare services (#PC0834) Denial of right to economic security (#PD0808) Denial of right to sufficient shelter (#PD5254) Denial of right to sufficient clothing (#PE7616) Denial of right to adequate medical care (#PD2028).

♦ **PD7254 Inadequate medical resources**
Rationing of medical treatment — Triage
Nature The demand for transplants has outstripped the supply of hearts, livers and kidneys by so much that doctors have to single out candidates they think will benefit most and who have the best chance of long-term survival. Patient selection can be based on utilitarian criteria, especially estimations of social worth or then on ethical criteria: lottery, queuing or "first-come, first-served" among medically suitable candidates.
Broader Rationing (#PF9026) Limited availability of health resources (#PD7669).
Narrower Insufficient medical supplies (#PE1634) Maldistribution of medical resources (#PD2705).
Aggravates Overcrowded public clinics (#PG5393) Inadequate medical facilities (#PD4004) Inadequate hospital facilities (#PE5058) Prohibitive cost of hospital facilities (#PE4154).
Aggravated by Alcohol abuse (#PD0153).

♦ **PD7257 Disparities in economic growth in developing countries**
Differential economic performance among developing countries
Nature The incidence of external shocks and of the evolution of the external environment has depended on the type and degree of exposure to external trade and finance, in particular a country's initial position as regards the size and structure of its trade and debt. However, developing countries have not been merely passive victims of exogenous shocks. Their own policies have had a significant influence on the extent to which external factors have affected their economic performance. Countries have also differed in their underlying potential for adjustment owing to differences in the size and diversity of their tradeable goods sectors and the level of income, consumption and investment, as well as the overall size, geographic location, climate, natural resource base and population.
Broader Declining economic growth in developing countries (#PD5326).

♦ **PD7258 Desecration of cemeteries**
Burial site desecration
Broader Desecration (#PF9176).
Related Desecration of holy spaces (#PF6385).
Aggravates Infamy (#PB8172) Undignified treatment of corpses (#PF5857).

♦ **PD7261 Military expeditions against friendly powers**
Nature A person is guilty of military expeditions against friendly powers if he launches an air attack against a friendly power, organizes a military expedition to engage in armed hostilities against a friendly power, or joins or knowingly provides substantial resources or transportation to a military expedition to engage in hostilities against a friendly power.
Broader War (#PB0593).
Related Crimes related to foreign relations and trade (#PE5331) Military expeditions against distant objectives (#PE8207).

♦ **PD7266 Trafficking in children for economic exploitation**
Broader Trafficking in children (#PD8405).
Aggravates Exploitation of child labour (#PD0164).

♦ **PD7276 Political tourism**
Foreign intervention through tourism — Subversive tourism
Broader Social environmental degradation from recreation and tourism (#PD0826).

♦ **PD7278 Desecration of religious buildings**
Broader Desecration (#PF9176).
Related Desecration of holy spaces (#PF6385).
Aggravated by Defacement of urban structures (#PD5305).

♦ **PD7279 Stigmatized diseases**
Nature The community holds the victim responsible for his illness and sees the condition as a punishment for his earlier actions.
Broader Social stigma (#PD0884).
Narrower AIDS (#PD5111) Mumps (#PE2356) Herpes (#PE8615) Plague (#PE0987) Obesity (#PE1177) Cholera (#PE0560) Anthrax (#PE2736) Measles (#PE1603) Leprosy (#PE0721) Dwarfism (#PE2715) Deafness (#PD0659) Smallpox (#PE0097) Club foot (#PE3836) Cleft palate (#PE5117) Yellow fever (#PE0985) Tuberculosis (#PE0566) Poliomyelitis (#PE0504) Typhoid fever (#PD1753) Speech disorders (#PE2265) Physical blindness (#PD0568) Learning disorders (#PD3865) Human sexual disorders (#PD8016) Sexually transmitted diseases (#PD0061) Congenital syndromes affecting multiple systems (#PE9324).
Related Mental illness (#PC0300) Deterioration in physical health (#PC0716).

♦ **PD7282 Marriage markets**
Nature Marriage markets transfer girls and young women from the third world to industrialized societies which often deliberately lead them to prostitution.
Broader Traffic in persons (#PC4442) Involuntary mass resettlement (#PC6203).
Aggravates Prostitution (#PD0693) Early marriage (#PE7628) Marriage by proxy (#PG6160) Dependency of women in marriage (#PD3694).
Aggravated by Forced marriage (#PD1915) Arranged marriage (#PF3284).

♦ **PD7286 Grievances of employers**
Broader Grievance (#PF8029).

♦ **PD7297 Paternal negligence**
Absent fathers — Irresponsible fatherhood — Male sexual irresponsibility — Paternal deprivation
Claim Men, who have sexual relationships with emotionally immature women without marriage

and without commitment, conceive children and move on to new conquests. For them lechery is glamorous and the successful philander is envied by younger men. Those in positions of influence, politician, sportsmen, media stars, teachers, doctors and bosses exploit these roles, much like the dominate male in a baboon troop. By their actions, these men teach that it is acceptable to use women like prostitutes; sexual irresponsibility is a rite of manhood. The mothers who bear their children often struggle in squalor to find food and clothing for their babies these men cavalierly ignore.
Broader Childlessness (#PC3280).
Aggravates Crime (#PB0001) Sexual harassment (#PD1116) Unmarried mothers (#PD0902)
Neglected children (#PD4522) Adolescent pregnancy (#PD0614)
Infant growth failure (#PE6909) Starving for attention (#PF1113).
Aggravated by Lust (#PA4673) Seduction (#PG5097) Sexual immorality (#PF2687)
Sexual exploitation of women (#PD3262) Denial of parental affiliation (#PD3255)
Pre-marital sexual intercourse (#PD5107) Ignorance of reproductive processes (#PD7994)
Excessive portrayal of sex in the media (#PE7930)
Lost covenantal understanding of sexuality (#PF1764)
Individualistic perception of sexual activity (#PF1682).

♦ PD7298 Destruction of cultural property during warfare
Broader Inadequate protection and preservation of cultural property (#PF7542).
Aggravated by War (#PB0593).

♦ PD7302 Decreasing diversity of biological species
Incidence Key areas of concern are: wild relatives of domesticated species which are essential to maintain the genetic variety of the domesticated breeds; harvested species and the threat to them through unsustainable rates of harvesting, leading to their biological extinction; totemic species, namely those held to be of special socio-cultural value to particular peoples; species of special importance to science; animal species that are sufficiently important and endangered to warrant captive breeding (especially in the light of the limited facilities for such breeding); wide ranging species that move between noncontiguous ecosystems, because of their ecological value and because an exclusively ecosystem-oriented approach will not conserve them; and indicator species because of their role in indicating the effectiveness of ecosystem maintenance.
Claim Habitats are being destroyed and species presumed lost faster than at any other period of human existence and probably in the life of the planet. There is not enough time or resources to undertake any species-by-species approach to maintain species diversity.
Broader Erosion of biological diversity (#PB9748)
Decreasing diversity of biological habitats (#PD5386).
Aggravates Vulnerability of world genetic resources (#PB4788).
Aggravated by Extinction of species (#PB9171) Unsustainable harvesting rates (#PD9578)
Vulnerability of ecosystem niches (#PC5773).

♦ PD7307 Diseases of respiratory system in animals
Broader Animal diseases (#PC0952).
Narrower Calf diphtheria (#PG7366) Laryngeal oedema (#PG7893)
Mycotic pneumonia (#PG4254) Lungworm infection (#PG9652)
Aspiration pneumonia (#PG4084) Hypostatic pneumonia (#PG1124)
Influenza of poultry (#PG6994) Laryngitis in animals (#PG4096)
Pharyngitis in animals (#PG9403) Respiratory diseases of sheep (#PE9438)
Respiratory diseases of swine (#PE7553) Pulmonary emphysema in animals (#PG5032)
Respiratory diseases of cattle (#PE5524) Respiratory diseases of horses (#PE5182)
Respiratory diseases of poultry (#PE5567)
Respiratory diseases of small animals (#PE9854).

♦ PD7320 Insensitive urban renewal
Narrower Widespread urban destruction (#PU3028).
Aggravated by Environmental degradation of inner city areas (#PC2616).

♦ PD7330 Emotional abuse of children
Refs Covitz, Joel D *Emotional Child Abuse* (1986).
Broader Cruelty to children (#PC0838) Victimization of children (#PC5512).

♦ PD7352 Bourgeois deviationism
Bourgeois liberalization — Frivolity of bourgeois democracy
Nature Bourgeois liberalization denies old virtues of thrift, hard work and Marxist classics and advocates pernicious Western ideas that aim at a Western style democracy using weapons such as pornography and rock music to achieve it.
Aggravated by Counter revolution (#PF3232).

♦ PD7359 Hindering proceedings by disorderly conduct
Nature A person who intentionally disrupts or attempts to disrupt an official government proceeding by making noise or behaving violently or creating a disturbance is guilty of hindering proceeding by disorderly conduct.
Broader Disorderly conduct (#PD9178) Contempt of judicial process (#PD9035).

♦ PD7360 Unethical military practices
Corruption in the armed forces — Military corruption
Broader Corruption (#PA1986) Official corruption (#PC9533)
Institutionalized corruption (#PC9173).
Aggravated by Illicit drug trafficking (#PD0991).

♦ PD7374 Imperialistic distribution system
Nature The distribution system at the global level is imperialistic in nature based on corporate empires rather than national. Production and distribution of products is based on limited operating images of society's and company needs. International, national and local economic and business relations are structurally blocked by tax, financial, and legal policy.
Broader Capitalism (#PC0564) Capitalist economic imperialism (#PC3166).
Narrower Entrenchment of vested interests (#PD1231)
Undercapitalized waste use schemes (#PD1410)
Gap between material and technological needs and demands (#PF1321).
Related Transnational corporation imperialism (#PD5891)
Limited approaches to economic planning (#PF3500)
Monopoly of the economy by corporations (#PD3003)
Short-term planning of product life cycles (#PF1740)
Unregulated ownership of the means of production (#PF2014).
Aggravates Risk of capital investment (#PF6572).
Aggravated by Concentration of power by transnational corporations (#PE0766).

♦ PD7385 Denial of evidence
Suppression of evidence
Incidence Conflicting claims of experts in such recent cases as the tobacco/smoking controversy, the environmental depredations of acid rain, the health effects of long-term exposure to low-level radiation, the levels of safety of nuclear reactors, the health hazards of food additives and of habitual levels of consumption of specific foodstuffs, have all created a situation in which political, scientific and corporate groups consider it legitimate to deny the existence of any significant evidence in each case.
Broader Deception (#PB4731) Evasion of issues (#PF7431).
Narrower Destruction of scientific records (#PF4633)
Unproven relationships between problems (#PF7706).
Related Lying (#PB7600) False evidence (#PF5127)
Falsification of public records (#PD4239).
Aggravated by Intellectual arrogance (#PF7847) Suppression of information (#PD9146)
Illegally obtained evidence (#PE9309) Tampering with official documents (#PF4699).

♦ PD7399 Urban crime
Refs Gomez Buendia, Hernando (Ed) *Urban Crime* (1989).
Broader Crime (#PB0001).
Related Locales of high crime rates (#PE7311).

♦ PD7407 Aiding national security criminals
Broader Complicity (#PF4983) Crimes against national security (#PC0554).
Narrower Aiding deserters (#PE4909) Harbouring national security offenders (#PE0484)
Aiding escape of prisoners of war or enemy aliens (#PE1200).

♦ PD7415 Inconclusiveness of scientific and medical tests
Misuse of scientific and medical test results — Inappropriate scientific and medical testing
Nature Certain phenomena may not lend themselves to conclusive results when tests are designed for them, whether because of the inadequacies of the test or because of the difficulty of detecting the phenomena. This uncertainty can then be misused to assert that there is no detectable evidence for the phenomena.
Incidence This is clearly evident in the determination of thresholds of toxicity of chemicals or radiation on humans, especially in the case of long-term effects of low exposure levels. Inappropriate tests can be used to deny the existence of hazards, at least in the short-term, without revealing the inadequacies of such tests in the case of long-term, low level exposure. Such tests are then used to justify legal thresholds which ultimately prove hazardous to health. The situation is complicated by the different degree to which people are affected by such tests. Some people have a higher tolerance to exposure than others.
Broader Uncertainty (#PA6438) Environmental hazards constraining scientific research (#PF1789).
Narrower Inadequate testing of drugs (#PD1190).
Aggravates Iatrogenic disease (#PD6334) Inappropriate arguments (#PF2152)
Medication side effects (#PD9807) Uncertain toxicity thresholds (#PF5188)
Politicization of health standards (#PD4519)
Long-term hazards of exposure to chemicals (#PE4717)
Long-term hazards of exposure to radiation (#PE4057)
Procrastination of science in the face of the unexplained (#PF3682)
Conflicting standards for protection against chemical occupational hazards (#PE5651).
Aggravated by Scientific fraud (#PF1602) Suppression of information (#PD9146)
Withholding of information (#PF8536).

♦ PD7420 Metabolic diseases in animals
Metabolic disturbances in animals
Broader Animal diseases (#PC0952).
Narrower Fat cow syndrome (#PG4372) Puerperal tetany (#PG4087)
Ketosis in cattle (#PG5091) Obesity in animals (#PG4092)
Myopathies in animals (#PE9774) Porcine stress syndrome (#PG5197)
Hypoglycaemia of piglets (#PG7683) Lactation tetany in mares (#PG4797)
Parturient paresis in cows (#PG1381) Parturient paresis in ewes (#PG9634)
Pregnancy toxaemia in ewes (#PG7782) Pregnancy toxaemia in cattle (#PG7505)
Photosensitization in animals (#PG5561) Transport tetany in ruminants (#PG9285)
Postparturient haemoglobinuria (#PG9682) Hypomagnesemic tetany in cattle (#PG9231)
Congenital porphyria erythropoietica (#PG9832)
Sudden death syndrome of feeder cattle (#PG7866).

♦ PD7424 Diseases of musculoskeletal system in animals
Animal bone diseases
Broader Animal diseases (#PC0952).
Narrower Sarcocystosis (#PG9569) Elbow dysplasia (#PG9477)
Lameness in sheep (#PE5617) Lameness in swine (#PE5050)
Patellar luxation (#PG6295) Rickets in animals (#PG9546)
Lameness in cattle (#PE4161) Lameness in horses (#PE7826)
Bursitis in animals (#PE6555) Enzootic calcinosis (#PG5320)
Osteitis in animals (#PG5220) Myopathies in animals (#PE9774)
Hypertrophic osteopathy (#PG9352) Osteochondritis dissecans (#PG9462)
Hypoparathyroidism in animals (#PG7818) Arthritis disorders in animals (#PE5913)
Rupture of the Achilles tendon (#PG6409) Diaphragmatic hernia in animals (#PG9192)
Rupture of the cruciate ligaments (#PG9648)
Primary hyperparathyroidism in animals (#PG4390)
Disorders of skeletal system of poultry (#PG9644)
Nutritional hyperparathyroidism in animals (#PG3730)
Renal secondary hyperparathyroidism in animals (#PG9681)
Congenital anomalies of the animal musculoskeletal system (#PE7808)
Neuromuscular paresis in cows following parturient paresis (#PG7851).

♦ PD7438 Violations against economic regulations
Broader Criminal harm to property (#PD5511).
Narrower Abuse of credit (#PF2166) Collusive tendering (#PE4301)
Crimes against trade (#PG6404) Banking law violations (#PE1208)
Hoarding of primary commodities (#PD0651)
Securities and commodities exchange violations (#PD4500)
Fraud concerning economic situation and corporate capital of companies (#PE5021).
Related Cartels (#PC2512).

♦ PD7439 Adolescent sexual intercourse
Teenage sex
Broader Pre-marital sexual intercourse (#PD5107).
Aggravates Adolescent pregnancy (#PD0614) Adolescent induced abortions (#PD1302).
Aggravated by Juvenile stress (#PC0877) Juvenile prostitution (#PD6213)
Excessive portrayal of sex in the media (#PE7930).

♦ PD7441 Trade harassment
Nature This includes harassment of trading partners under laws and regulations which are technically consistent with multilateral obligations.
Broader Protectionism in international trade (#PC5842).

♦ PD7443 Enforced collectivization of agriculture
Broader Inappropriate modernization of agriculture (#PF4799).
Related Denial of the right to ownership (#PE8411)
Denial of the right of association (#PD3224)
Denial of right to free choice of work (#PE3963)

Restrictions on freedom of movement within countries (#PE8408).
Aggravates Involuntary mass resettlement (#PC6203).
Aggravated by Communism (#PC0369).

♦ **PD7450 Discriminatory design of information systems**
Biased computer software design
Broader Abuse of computer systems (#PD9544)
Restrictions on freedom of information (#PC0185).
Aggravates Lack of specifically designed software in developing countries (#PF6067).

♦ **PD7461 Hierarchical control of market facilities**
Nature Market facilities both for the buyer and for the seller are dominated through a hierarchy by a wealth elite who control capital, corporations and technology. Any attempt of equal distribution of marketing facilities is blocked by the structure of this hierarchy and elite.
Broader Unsystematic allocation of market facilities (#PD3507).

♦ **PD7463 Abusive technological development under capitalism**
Paradoxes in technology transfer programmes — Perverse impact of technological development on capitalist society
Nature The historical development of capitalism has gone far beyond the initial separation of workers from the means of production. Technological development has in fact so altered the production process that it is now enough to have control of the vital points of a technology system to control production. The transfer of any technology from one system to another raises the question of social appropriateness of the technology transferred. Extensive studies have shown that large-scale capital-intensive technologies are often unsuitable to the employment needs and resource requirements of developing economies. Meanwhile, as the driving force behind technology transfer is increasingly the large, multinational corporations, the technology so transferred is more likely to make a receiving economy a market of the multinational who controls the key technological factors than to become a competitor in the technology transferred.
Broader Capitalism (#PC0564) Irresponsible scientific and technological activity (#PC1153).
Related Technological revolution (#PC3234)
Technology gap between developed countries (#PD0338)
Abuse of science and technology in capitalism (#PE3105).
Aggravates Destruction inherent in development (#PF4829)
Minimal access to appropriate technology (#PF3554)
Uncontrolled environmental impact of technology (#PC1174)
Increasing development lag against technological growth (#PE3078)
Use of inappropriate technologies in developing countries (#PF0878)
Insufficient access to technology for agricultural upgrading (#PF3467).
Aggravated by Haphazard transmission of practical technology (#PF3409)
Protectionism in the high-technology industries (#PE8458)
Restrictive business practices in technology transactions (#PE1978)
Monopolization of technology by transnational corporations (#PE1918).

♦ **PD7465 Genetic inbreeding**
Related Decreasing genetic diversity in cultivated plants (#PC2223).
Aggravates Friedreichs ataxia (#PE8605) Genetic defects and diseases (#PD2389)
Decline in human genetic endowment (#PF7815)
Decreasing genetic diversity of animals (#PC1408)
Vulnerability of world genetic resources (#PB4788).
Aggravated by Consanguineous marriage (#PC2379).

♦ **PD7466 Export of unemployment**
Aggravates Unemployment (#PB0750).

♦ **PD7470 Structurally blocked scientific co-operation**
Nature Co-operation between scientists and scientific bodies is blocked structurally. The educational processes of training scientists over emphasizes individual achievement promoting competitiveness, suspicion and fear. The specialized vocabulary of the sciences mediate against mutual understanding or comprehension by lay people.
Broader Undirected technological expansion (#PC1730).

♦ **PD7471 Threats against trade union leaders**
Intimidation of workers' representatives — Harassment of workers' organization leadership
Broader Sanctions against trade union workers (#PD0610).
Related Harassment of public officials (#PD4915).
Aggravated by Unethical practices of employers (#PD2879).

♦ **PD7473 Urban malnutrition**
Malnutrition in slums
Nature The differences in diet between urban and rural populations lead to urban malnutrition. Traditional foods are not available to the urban dweller. Thus, the dietary wisdom developed over centuries in rural areas is not applicable in the city. The urban dweller resorts to felt need and price as the basis of purchasing food and eating. This often results in high carbohydrate intake and few vitamins, mineral and protein.
Incidence Studies show that urban malnutrition may be more widespread than in villages; that its severity is often sharper. Urban slums frequently present the largest concentrations of malnourished people in a country. Child malnutrition is seen to occur at an earlier stage in cities than than it does in rural areas.
Broader Malnutrition (#PB1498).
Aggravates Inadequate health care in urban slums (#PE7877).

♦ **PD7480 Declining productivity of agricultural land**
Loss of agricultural crops — Low crop yields
Nature Soil productivity is reduced by loss of soil nutrients and organic matter and the water-holding capacity of soil. To increase the yields developing countries need to invest in irrigation, mechanization, pesticides, fertilizers and storage that require expensive energy and resources.
Incidence Among other things erosion have lowered the 1980 yields of four African countries to less than in 1950.
Refs FAO *Assessment and Collection of Data on Pre-Harvest Foodgrain Losses* (1983).
Broader Declining economic productivity (#PC8908)
Unproductive subsistence agriculture (#PC0492).
Related Loss of animal productivity (#PD8469) Unproductive labour resources (#PC6031)
Low productivity of agricultural workers in developing countries (#PE5883).
Aggravates Crop failures (#PG3363) Uneconomic size of farms (#PJ2079).
Aggravated by Floods (#PD0452) Soil erosion (#PD0949)
Air pollution (#PC0119) Desert advance (#PC2506)
Soil compaction (#PD1416) Water pollution (#PC0062)
Soil infertility (#PD0077) Unreliable rainfall (#PD0489)
Silting of water systems (#PD3654) Shortage of fresh-water sources (#PC4815)
Destruction of crop pollinating species (#PE3880)
Subsistence approach to capital resources (#PF6530)
Degradation of agricultural land by cash crops (#PE8324)
Overexploitation of underground water resources (#PD4403)
Increasing proportion of land surface devoted to urbanization (#PE5931)
Unsustainable short-term improvements in agricultural productivity (#PE4331).

♦ **PD7486 Vulnerability of economies to import penetration**
Nature Increased import penetration and loss of competitiveness of domestic industries can lead to mounting pressures for protectionism.
Aggravates Trade protectionism (#PC4275).

♦ **PD7500 Benefit overpayments**
Nature Societies which pay their members cash benefits find it hard to recover overpayments. It is difficult to discern which overpayments result from human error and which from intentional claimant fraud. Further, those required to repay the excess benefits tend to claim lack of resources to do so.
Incidence In 1987, the total debt owed to local social security offices in the UK was 87 million pounds. This is more than twice the debt which existed in 1984, and it continues to grow.
Related Unethical use of social welfare benefits (#PD8859).
Aggravated by Dependence on social welfare (#PD1229).

♦ **PD7501 Monopolistic control of new animal forms**
Irresponsible patenting of genetically transformed animals
Broader Monopolistic control of new life forms (#PD7840).
Aggravated by Unethical practice of the zoosciences (#PD4721).

♦ **PD7506 Endangered species of farm animals**
Incidence In UK this century at least 20 breeds of British farm animals have died out.
Broader Endangered species of animals (#PC1713).

♦ **PD7513 Endangered species of invertebrates**
Nature Invertebrates which outnumber man up to 10 million to one have hundreds of species threatened with extinction by means of both natural events and human activities (habitat alteration, pollution, introduction of exotic animals and plants, and exploitation). Invertebrate organisms play the key role in supporting the complex food chains of the seas and oceans, only on land do large grazing animals dominate. Any disturbance of ecological balance can lead to a situation in which some species of invertebrates form uncontrolled over-expanding population damaging other species.
Refs Collins, N M; Pyle, R M and Wells, S M *IUCN Invertebrate Red Data Book* (1983).
Broader Endangered species of animals (#PC1713).
Narrower Endangered species of sponge (#PD3476)
Endangered species of bryozoa (#PD3477)
Endangered species of annelida (#PD3160)
Endangered species of protozoa (#PD3475)
Endangered species of molluscs (#PD3478)
Endangered species of arthropoda (#PD3471)
Endangered species of echinodermata (#PD3158)
Endangered species of aschelminthes (#PD3159)
Endangered species of platyhelminthes (#PD3474).

♦ **PD7516 Unstable shifting agriculture**
Shifting cultivation
Nature Shifting cultivation is a traditional method of cultivating tropical upland soils, used mainly for subsistence purposes. During the fallow periods of rest intervening between crops, the natural fertility of the soil is restored for renewed utilization in a subsequent period of crop growth. This traditional system of cultivation is in ecological balance with the environment and does not irreversibly degrade the soil resource, provided a sufficient length of fallow is allowed for soil restoration. However, increasing population pressures have necessitated more intensive use of land, particularly in the humid tropics of Asia and in the savanna and forest zones in Africa. The consequence is extended cropping periods and shortened fallows, which are inadequate to restore the soil's productive capacity. The subsistence farmers in the tropics are thus caught in a cycle of increasingly falling yields, more poverty and even less opportunity to subsist, let alone to improve their standard of living. In addition to this perpetuation of human misery, shifting cultivation, as currently practised in many areas, is wasteful of scarce land resources and frequently leads to intolerable erosion, particularly of hillsides and sloping lands.
Counter-claim Stable shifting agricultural systems should be allowed to maintain themselves since they maintain ecological processes and biological diversity. Unstable systems should be helped to stabilize or (if the population is growing too quickly) helped to change to a sustainable and more productive system.
Refs FAO *Shifting Cultivation and Soil Conservation in Africa* (1978); FAO *Institutional Aspects of Shifting Cultivation in Africa* (1984); FAO *The Future of Shifting Cultivation in Africa and the Task of Universities* (1985); FAO *Changes in Shifting Cultivation in Africa* (1984).
Aggravates Unsustainable agricultural development (#PC8419)
Lack of economic and technical development (#PE8190).
Aggravated by Nomadism (#PF3700).

♦ **PD7520 Offences against public order**
Broader Statutory crime (#PC0277).
Narrower Beggars (#PD2500) Arming rioters (#PE5327) Moral offences (#PD9179)
Illegal marriage (#PE7935) Unlawful assembly (#PG5067) Juvenile desertion (#PD8340)
Disorderly conduct (#PD9178) Gambling and wagering (#PF2137)
Incitement to violence (#PJ2068) Illicit drug trafficking (#PD0991)
Desertion in marriage law (#PF3254) Criminal killing of animals (#PJ1158)
Vice and sex traffic offences (#PD8910) Firearms and explosives crimes (#PE1108).
Related Civil disorders (#PC2551).

♦ **PD7522 Instability in relations between allies of superpowers**
Disputes between allies
Nature Given the widespread conviction that regardless of first strike, both superpowers would be destroyed, it is probable that the superpowers (the USSR and the USA) will not fight over disagreements between themselves, but will be dragged into war by disagreements between their allies.
Background Some historians maintain that World War I was not brought about by deep hostility between Russia and Germany, but by the instability in relations among the lesser nations to which these powers were allied.
Related Intergovernmental disputes (#PJ5405).
Aggravated by Competition between states (#PC0114).

♦ **PD7526 Underutilized animal genetic resources**
Nature Governmental infrastructure for improving services for recording and progeny-performance testing is weak or non-existent in developing countries, and some potentially viable animal breeds and strains in the developing countries are much underutilized, even though the animals could play an important role in economic and social development.
Incidence The FAO includes the following animals as underutilized: Boran and Sahiwal cattle,

Awassi sheep, Shami goats, water buffalo, and camelidae (llama and alpaca).
Broader Underdeveloped use of agricultural resources (#PF2164).

♦ PD7536 Disincentives against farming

Economic difficulties of farmers
Refs FAO *Incentives and Disincentives for Farmers in Developing Countries* (1976).
Narrower Subsistence life style (#PF1078) Demeaning farmer image (#PJ9781)
Underdeveloped farming skills (#PJ0729)
Inadequate conditions of work of agricultural workers (#PE4243).
Related Protectionism in developed countries against agricultural products from developing countries (#PE8321).
Aggravated by Inappropriate policies (#PF5645)
Lack of credit facilities for agricultural producers (#PE8516)
Over-subsidized agriculture in industrialized countries (#PD9802)
Protectionism in agriculture and the food production industries (#PD5830)
Distortion of international trade by selective domestic subsidies (#PD0678)
Distortion of international trade by export subsidies and countervailing duties (#PE1961).

♦ PD7541 Environmental hazards of coal energy

Nature Environmental problems caused by coal energy may be regarded in terms of various components: mining, transportation, and combustion and waste disposal. The environmental considerations include the emission of dust to the atmosphere and chemicals to groundwater; land subsistence in underground mining and land reclamation in surface mining; disposal of mine waste; miners' safety; and water consumption, especially in dry areas. Transportation problems include dust and water pollution (mainly from storage piles); traffic noise and congestion; and accident risk. Combustion and waste disposal elicit the most concern due to the oxides, monoxides, dioxides, trace minerals and volatile elements of which the flue gases are composed.
Counter-claim Advances in mining technology over the past two decades have greatly reduced the overall impact of coal mining on the environment. Measures such as: mine drainage control; compacting, landscaping and replanting of waste piles, and restoration of surface mines with the development of recreation areas or new communities on the site; the improvement of mine safety through better underground technology and more attention to precautionary procedures and equipment, have all contributed to improving the situation.
Refs OECD Staff *Coal* (1983).
Broader Environmental hazards from energy production (#PD6693).
Related Coal fly-ash (#PG5363).
Aggravates Sulphur compounds as pollutants (#PG6442)
Nitrogen compounds as pollutants (#PD2965)
Increasing atmospheric carbon dioxide (#PD4387).
Aggravated by Risks in power production (#PE4835).

♦ PD7550 Self-inflicted physical suffering

Self-torture — Flagellation — Fasting — Martyrdom — Asceticism — Immolation
Incidence Self-inflicted suffering continues to form a part of some traditional religious ceremonies and private practices (as recommended by Opus Dei, for example). It is also used widely used, in the form of fasting (and occasionally immolation), to highlight political issues (with or without religious overtones).
Counter-claim The mythology of martyrdom transcends divisions between secualr and fundamentalist Muslims. The belief is fostered that those who die for Islamic causes (such as the liberation of Palestine) will ascend directly to heaven. The burial of such martyrs is an occasion both for mourning and celebration during which the the family of the deceased is congratulated.
Broader Non-violent weapons (#PF9327) Human physical suffering (#PB5646).
Related Jihad (#PF5681) Corporal punishment (#PD8575)
Childhood martyrdom (#PF8118).
Aggravated by Religious extremism (#PF4954).

♦ PD7565 Preponderance of non-food crops in tropical economies

Nature Non-food crops (such as tea, coffee, cocoa, flowers, sisal and sugarcane) take up vast tracts of land; and while they may sustain many tropical nations' economies, they also make those nations largely dependent on developed, temperate zone countries - usually the USA and the USSR - to provide their daily staples.
Incidence More than 50 percent of the Caribbean's arable land is planted with crops for export; 66 percent in Guadeloupe; and 30 percent of land in East Java grows sugarcane.
Claim Even though there is sufficient cropland to feed the world, people in developing countries go hungry because much of their land belongs to the rich, and the rich grow luxury cash crops for export to the middle and upper classes in the developed countries. (This practice was established by colonial rule in the 17th and 18th centuries). Even if developing countries were to achieve zero population growth, their people would still go hungry unless the land was distributed to local farmers and cash export crops were replaced by local food crops.
Broader Inappropriate cash crop policy (#PF9187).

♦ PD7566 Ignorance of grammar

Broader Ignorance (#PA5568).
Related Mathematical ignorance (#PD6728).
Aggravates Discrimination against use of accents of a language (#PF5141)

♦ PD7567 Grievances of consumers

Consumer complaints
Broader Grievance (#PF8029).
Narrower Decrease in consumer choice (#PD6075)
Unsafe design of consumer products (#PF1379)
Haphazard provision of consumer services (#PF2411)
Environmentally unfriendly consumer products (#PD9310)
Incorporation of carcinogens into consumer goods (#PE8934)
Non-payment of compensation for damages to consumers (#PE0290).
Related Misrepresentation of information to consumers (#PE6877).

♦ PD7571 Delay in civil litigation

Broader Delay in administration of justice (#PF1487).

♦ PD7574 Marine disposal of obsolete weapons

Incidence Because it is such a convenient hiding place for wastes, the sea has long been a receptacle for obsolete weaponry. After World War II more than thirty thousand bombs and canisters containing poison gases - along with other unwanted munitions, mostly German - were dumped into the southern Baltic Sea. In 1969 these rusting canisters came back to haunt Baltic shorelines; leaking mustard gas injured fishermen and panicked bathers. Danish fishermen caught at least sixteen mustard-gas bombs in their nets and suspected contamination from these materials caused the boycotting of thousands of tons of fish. In 1976, mustard-gas bombs washed ashore along the Welsh coast, the debris from the dumping of British chemical munitions off the coasts of Scotland and Ireland between 1945 and 1956. The United States has disposed of obsolete chemical warfare agents in the sea on at least three known occasions, in 1967, 1968 and 1970. The materials were embedded in concrete within steel vaults and carried several hundred miles out to sea aboard obsolete ships, which were scuttled. The effects these agents might have on the sea is not known.
Broader Marine dumping of wastes (#PD3666).
Aggravated by Naval arms race (#PD8412).

♦ PD7576 Torture through mutilation

Nature Mutilations are used as punishment, intimidation or a way of inflicting pain. Finger and toe nails are pulled out or things inserted under them including matches which are then lit. Hair is pulled out. Victims are forced to walk or crawl on glass. Limbs and hands are amputated. Victims are castrated. Eyes are removed or stabbed.
Incidence Torture by mutilation has been reported in the following countries: **Af** Ghana. **Am** Bolivia, Guatemala, Honduras, Suriname.
Broader Physical torture (#PD8734)
Mutilation and deformation of the human body (#PD2559).
Narrower Punitive amputations (#PE8724).
Related Child mutilations (#PG3726) Torture by wounding (#PE4078)
Male sexual mutilation (#PE6054) Female sexual mutilations (#PE6055)
Punishment of criminals by mutilation (#PE3488).
Aggravates Death threats (#PD0337).

♦ PD7579 Nuclear reactor accidents

Reactors accidents
Nature The ultimate nuclear reactor accidents would result from an uncontrollable chain reaction, although the possibility of such a reaction arising is virtually nil, due to control and protection systems. Ultimate nuclear reactor nightmare is a meltdown, that occurs when overheated reactors boil away the massive water reservoirs, and the molton core burns its way deep into the earth's crust with repercussions on a scale never experienced before. However, localized brief periods of criticality may arise which in themselves can cause serious threats to health.
Incidence The following nuclear reactor accidents have occurred: 1957, Windscale, UK; 1965, Idaho Falls ID, USA; 1966, Detroit MI, USA; 1969, Saint-Laurent, France; 1975, Brown's Ferry AL, USA; 1979, Three Mile Island PA, USA; 1982, Kozluduj, Bulgaria; 1983, Constituyentes, Argentine; 1984; Greifswald, Germany FR; 1986, Gore, OK, USA; 1986, Chernobyl, Soviet Union. According to the International Atomic Energy Agency (IAEA) more than 250 nuclear reactor accidents have been kept secret. The costs of such accidents are very high - the Three Mile Island accident was estimated to cost $500 million to clean-up, plus the costs to those inhabitants who lived nearby and left the area either permanently or for a short time, which amounted to $19 million. Nuclear explosion at Chernobyl caused evacuation of 135,000 people, death of 31 workers and immediate fall-out and the poisoning of agricultural land in the Soviet Union and Europe from relatively long-lived radioactive isotopes. The number of fatal cancers that might develop worldwide as a result of the accident could range from zero to 17,400, among local population between 5,000 and 25,000.
Claim Nuclear reactor accidents may cause severe diseases or death, as well as profound psychological stress.
Refs Milne, Teddy *The Unseen Holocaust* (1987); OECD Staff *International Comparison Study on Reactor Accident Consequence Modeling* (1984); OECD Staff *Nuclear Reactor Accident Source Terms* (1986); OECD Staff and NEA Staff *Severe Accidents in Nuclear Power Plants* (1986).
Broader Nuclear accidents (#PD0771).
Aggravates Radioactive contamination (#PC0229)
Vulnerability of nuclear power sources (#PD0365).
Aggravated by Defects in machinery design (#PE2462)
Radiation damage to materials (#PD1206)
Inadequate nuclear reactor safeguards (#PF6084)
Aggression against nuclear power sources (#PE0403)
Hazardous locations for nuclear power plants (#PD2718)
Human fatigue during control of complex equipment (#PE5572)
Non-verificability of compliance with nuclear power safeguards (#PF4455).

♦ PD7585 Over-use of designated wilderness areas

Excessive use of natural parks — Degradation of countryside by careless walkers — Degradation of wilderness areas by campers
Nature Wilderness areas are large natural environments without roads or permanent buildings. With increasing population and transport facilities, such areas are now subject to a level of use which endangers the local ecosystems.
Incidence In 1983, 244 million people visited US national parks. In the backcountry of Wyoming's Yellowstone National Park, the increasing traffic has forced the grizzly bear into a shrinking area and is threatening its existence there.
Claim Excessive numbers of people seeking solitude and nature are threatening to destroy the very qualities they drive hours to find. Alarming numbers of nature lovers flock to wilderness areas, only to discover more people than deer, a mini-village of orange tents, public campsites equipped with portable television hookups, and a wilderness damaged by the ignorance of those who create their own paths, leave their litter behind, or take part of the wilderness (plants, rocks, flowers) home with them.
Broader Vulnerability of ecosystem niches (#PC5773)
Decreasing diversity of biological habitats (#PD5386).
Related Environmental degradation by off-road and all-terrain vehicles (#PE1720).
Aggravates Destruction of wildlife habitats (#PC0480)
Vulnerability of protected natural areas (#PC4764)
Degradation of semi-natural and natural habitats of flora and fauna (#PC3152).
Reduced by Inaccessible wilderness areas (#PF9360)
Obliteration of footpaths by development (#PE3874)
Inaccessibility of countryside to city dwellers (#PF6140).

♦ PD7592 Exploitation of dependence on food aid

Claim Superpower nations may initially sell underpriced food to developing countries for a definite length of time, and when the countries are dependent on that food, when the demand is high, the food may then be sold at regular market prices. Other hidden subtleties in food shipments usually centre around the largest amount of food aid going to political allies rather than to the neediest nations. Therefore, if a country wants to feed its people, it may be coerced into accepting some other country's political philosophy along with their economic aid.
Broader Misuse of food as a political weapon (#PF6202)
Lack of protection for the vulnerable (#PB4353).
Aggravates Malnutrition (#PB1498) Compassion fatigue (#PF2819).
Aggravated by Toughness (#PA6976) Ignorance (#PA5568)
Misinformation (#PD8523) Economic exploitation (#PC8132)
Unethical food practices (#PD1045) Vulnerability of developing countries (#PC6189).

♦ PD7598 Plantation agriculture

Agribusiness — Capitalist plantation system
Nature The capitalist plantation system is a system of large-scale agriculture involving the cultivation of industrial and food crops, primarily tropical and subtropical ones such as sugarcane,

coffer, cacao, tea, rice, bananas, pineapples, tobacco, cotton, rubber trees and indigo. The plantations are often owned by large food monopolies which may provide inadequate accommodation for the workers, require gruelling hours of back-breaking work, and offer little pay or security. Workers are often 'forced' to stay on the plantations by the fear of not being able to find work elsewhere, where conditions could be better.
Background The first plantations were established by the Spanish in the early 16th century, in the West Indies; from there they spread to Brazil, Mexico, North America and Indonesia in the 16th, 17th and 18th centuries.
Refs Sajhau, J P and von Muralt, J *Plantations and Plantation Workers* (1987).
Broader Inappropriate modernization of agriculture (#PF4799).
Related Abuse of agricultural techniques (#PG5746)
Protectionism in agriculture and the food production industries (#PD5830)
Environmental hazards from agricultural and livestock production (#PD0376)
Domination of agricultural equipment industry by transnational corporations (#PE1448).
Aggravates Slavery (#PC0146) Unused land (#PF4670)
Inadequate conditions of work of agricultural workers (#PE4243).

♦ PD7602 Brutalization of military personnel
Dehumanization of soldiers — Bullying amongst military personnel
Incidence Soldiers are forced to carry on training in full battle dress, with a hat, rifle and pack. In the USA 11 servicemen died from heat exhaustion in the past decade and 400 were injured. Brutality and negligence by officers of the USSR military have claimed the lives of an estimated 15,000 soldiers in the period 1985-90. Bullying by officers and sergeants, together with appalling living conditions, have been responsible for a very high level of suicides among recruits, with 3,900 dying in 1989 alone. It is claimed that thousands have died or been maimed because of negligence or dangerous exercises with live ammunition. Cruelty and bullying may be exacerbated by racial tensions between soldiers of different ethnic origin.
Most military units have a tradition of imposing some sort of informal initiation test on new recruits. This may range from efforts to get them drunk or to undertake some degrading task. A vicious element may be added if the process involves sexual humiliation, whether coating the genitals with some substance, performing with a specially hired prostitute, or the rare, but well-publicized, instances of buggery with a broomstick.
Counter-claim Senior military officers argue that if war is accepted as a natural state of man, then bullying must also be accepted in an aggressively male group.
Aggravates Military brutality (#PD4945) Military atrocities (#PD1881).
Aggravated by Hazings (#PF5392)
Racial discrimination by security forces (#PD3519).

♦ PD7608 Interception of communications
Denial of right to private correspondence
Nature The intentional interception of communications whether by mail, phone, computer, facsimile, telex, or telegraph or listening in on private conversation is a violation of the right to privacy and a crime.
Broader Infringement of privacy (#PB0284) Criminal violation of civil rights (#PD8709).
Narrower Interception of correspondence (#PE5093)
Misuse of telephone surveillance by governments (#PD1632).
Related Denial of right to confidentiality (#PD6612)
Denial of right to private home life (#PE6168)
Misuse of electronic surveillance by governments (#PD2930).
Aggravated by Socialism (#PC0115) Capitalism (#PC0564).

♦ PD7609 Denial of right of complaint
Denial of right to petition authority — Lack of legal recourse
Nature The refusal of government, business or professional organizations to receive, judge or act on complaints against its or its representatives or members activities. The denial of the right of complaint takes many forms. A bureaucratic procedure requiring years and massive expenditure may cause the applicant delays, or bankruptcy or may intimidate him. People with complaints may be threatened or harassed directly or indirectly. complaints may be heard and automatically disregarded.
Broader Denial of political rights (#PD8276).
Related Denial of right to hold public office (#PE5608)
Denial of right to participate in government (#PE6086).
Aggravates Apartheid (#PE3681) Human torture (#PC3429)
Underproductivity (#PF1107) Infringement of privacy (#PB0284)
State sanctioned torture (#PD0181) Exploitation in employment (#PC3297)
Official evasion of complaints (#PF9157) Legalized racial discrimination (#PC3683)
Underutilization of legal rights (#PF3464)
Expropriation of land from indigenous populations (#PC3304).

♦ PD7612 Denial of freedom of conscience
Broader Denial of freedom of thought (#PF3217).
Narrower Denial of right of conscientious objection to military service (#PD1800).
Related Denial of freedom of opinion (#PD7219) Harassment of human rights monitors (#PF1585)
Denial of right to correct misinformation (#PE7349).
Aggravates Inadequate social welfare services (#PC0834).

♦ PD7615 Unequal income distribution within developing countries
Income inequality within developing countries
Incidence In Brazil the poorest fifth of the population have just 2 per cent of all income, the richest fifth have 67 per cent.
Broader Human inequality (#PA0844) Unequal income distribution (#PD4962)
Maldistribution of wealth within developing countries (#PD5258).
Related Unequal income distribution between countries (#PC2815).
Aggravated by Imbalance between rural and urban incomes in developing countries (#PE8584)
Excessive social costs of structural adjustment in debtor developing countries (#PD8114).

♦ PD7617 Interpersonal rivalry
Broader Competition (#PB0848).
Narrower Sibling rivalry (#PG2399).
Aggravates Professional burnout (#PF4833).
Aggravated by Capitalism (#PC0564).

♦ PD7625 Excessively large families
Large families
Nature Despite the apparent advantages of many children, it is not clear that, from a strictly economic point of view, parents gain. Children may end up costing more than parents expected; for households close to subsistence levels, food clothing and housing many children is a chief concern, and the need in some countries to provide girls with a dowry is an additional burden. Support in the parents' old age is not certain – some children will die and, of those who survive, daughters may move to another village with their husbands, while sons who move away to work may not prove as supportive as was hoped often because they face difficulties in obtaining reasonably paying jobs. Even when parents seem to gain from large families, children may lose. This is obviously true when births are closely spaced; the resulting harm to the health and nutrition of mothers can cause low birth weight, early weaning, and poor health of children in the critical early years. Older children may also be handicapped. Even in developed countries, studies show that children in large families and those born close together tend to be physically and intellectually inferior to other children.
Incidence Studies in France, the USA and the Netherlands have shown that a large number of children in a family has a negative effect on classroom performance and test scores.
Broader Unsustainable population levels (#PB0035).
Related Excessive family size (#PJ5277).
Aggravates Consanguineous marriage (#PC2379).
Aggravated by Inability of governments to regulate family size (#PF0401).
Reduced by Family rejection of children (#PC8127)
Denial of the right to procreate (#PC6870).

♦ PD7630 Arrest of trade union leaders
Detention of workers representatives — Ill-treatment of trade union leaders
Broader Wrongful detention (#PD6062) Politically motivated arrests (#PD9349)
Sanctions against trade union workers (#PD0610).
Aggravated by Violent repression of demonstrations (#PD4811).

♦ PD7636 River pollution
Stream pollution — Discharge of dangerous substances into rivers
Nature River pollution, which results from: industrial, mineral, agricultural and watercraft wastes; sediment and erosion; oil and hazardous substances spillage; and human littering, is a major source of pollution. Illnesses, both mild and severe, result from this pollution and unless something is done to clean up the world's rivers, these sources of usable water and reserves for human recreation may well soon be unusable.
Incidence The River Ganges, faced in 1985 with a $250 million antipollution campaign, is befouled by chemical wastes, human excrement, cremated bodies, and cattle carcasses. Faithful Hindus go daily to bathe in it and drink its water, thus making themselves susceptible to various diseases, the two most common of which are gastrointestinal disorders and infectious hepatitis.
Broader Pollution of inland waters (#PD1223) Discharge of dangerous substances (#PD4542)
Unsustainable development of fresh waters (#PD6923).
Aggravates Marine pollution (#PC1117) Pollution of sediments (#PE5539)
Eutrophication of lakes and rivers (#PD2257)
Extermination of wild animal natural prey (#PD3155).

♦ PD7650 Maladjustment to disciplines of employment
Lack of discipline in work
Broader Lack of social discipline (#PF8078).
Related Occupational difficulties in adolescence (#PU3051).
Aggravates Exploitation in employment (#PC3297)
Limited availability of permanent employment in inner-cities (#PE1134).
Aggravated by Youth unemployment (#PC2035) Segregation in employment (#PD3443)
Limited employment options (#PF1658) Obsolete vocational skills (#PD3548)
Discontinuity of employment (#PD4461) Discrimination in employment (#PC0244)
Inappropriate employment incentives (#PD0024).

♦ PD7661 Internal armed conflicts
Nature Internal armed conflict has resulted in the killing of many people in a number of countries. Killings may be carried out by government as well as opposition forces, and there are frequent reports of the indiscriminate killing of non-combatant civilians by government forces in areas where guerrillas are active. Intensive counter-guerrilla operations may use the strategy of emptying areas of the entire civilian population in order to eliminate any possible support for the guerrillas, and often indiscriminate killings occur, involving entire village populations including women and children. Villagers are also abducted and killed by 'death squads' under military control; torture and mutilation are routinely practised. In some cases, people trying to flee the areas of armed conflict or those who reached refugee camps in neighbouring countries are indiscriminately attacked by government forces; many are killed. Often it is claimed that those villages and refugee camps were infiltrated by guerrilla forces and that death occurred in armed clashes between government troops and guerrilla forces. In a number of countries a state of siege or a state of emergency is imposed and constitutional guarantees for human rights are suspended or severely curtailed. In some countries heavy security measures are enforced and arbitrary arrest and detention of those suspected of their involvement with guerrilla movements frequently result in the execution of detainees.
Broader War (#PB0593) Civil war (#PC1869).

♦ PD7663 Insufficient nuclear power stations
Inadequate infrastructure for nuclear power generation — Poor viability of nuclear power plants
Nature Nuclear power, although a potential major energy source, is losing some of its support; and the exploitation of the future potential of this power is associated with a great deal of uncertainty. Delays and revisions of projections reflect a reduction in the estimates of electricity demand as well as increasing public and private opposition which has, in some case, been strong enough to lead to the postponement or cancellation of nuclear projects. If this downward spiral continues, the confidence of utility planners in nuclear power may be further eroded and the viability of the nuclear power industry will be endangered.
Broader Energy crisis (#PC6329) Inadequate electricity infrastructure (#PD9033).
Aggravates Deterioration of nuclear power plants (#PE5260).
Aggravated by Monopoly of nuclear power techniques (#PD1741)
Irrational rejection of nuclear power (#PF8531)
Vulnerability of nuclear power sources (#PD0365)
Restrictive regulation of nuclear power (#PF8654)
Prohibitive cost of nuclear power plants (#PF7543)
Aggression against nuclear power sources (#PE0403)
Hazardous locations for nuclear power plants (#PD2718)
Environmental hazards of decommissioned nuclear power plants (#PE7539)
Non-verificability of compliance with nuclear power safeguards (#PF4455).
Reduces Environmental hazards of nuclear power production (#PD4977).

♦ PD7669 Limited availability of health resources
Inaccessible health resources
Broader Inadequate health services (#PD4790)
Underprovision of basic urban services (#PF2583).
Narrower Inadequate medical resources (#PD7254)
Inadequate medical facilities (#PD4004).
Aggravated by Decline in government health expenditure (#PF4586).

♦ PD7676 Criminal solicitation
Nature To command, induce, entreat or in any way attempts to persuade another person to commit a particular felony and an individual makes an overt act to accomplish the intent of the solicitation is to be guilty of criminal solicitation.
Broader Complicity (#PF4983) Offences of general applicability (#PD4158).

Related Criminal attempt (#PD5321) Criminal conspiracy (#PD1767)
Criminal facilitation (#PD6845).
Reduced by Fragmentation (#PA6233).

♦ PD7682 Development by industrialized countries of products substituting for commodities exported by developing countries
Import substitution by developed countries
Broader International economic injustice (#PC9112).
Narrower Protectionism in developed countries against agricultural products from developing countries (#PE8321).
Aggravates Inadequate production capacity in developing countries (#PD4219)
Vulnerability of national economies to vagaries of external markets for goods and services (#PF9697).

♦ PD7692 Career interruption due to pregnancy
Career disruption due to childbearing
Nature Regardless of the advances in the cause of equal opportunity between the sexes, the fact remains that they have really been quite marginal in changing the division of responsibility. Women are the ones who bear children, thus most of the problems emanating from career interruption affect them.
Incidence In addition to the sexism of the premise that men have the right and even the duty to work while women have to fit home duties in between their careers, it could also be financially unwise for women to interrupt their careers. A recent UK researcher has estimated that mothers with average earnings potential lose $70,000 in lifetime earnings by raising two children, and with three children, that figure rises to $85,000.
Claim The societal and governmental barriers which in effect make it so difficult to combine motherhood with career, are a means of inflicting cash penalties on becoming a mother and result in a wasteful under-utilization of female resources in the labour market.
Refs Fogarty, Michael P, et al *Sex, Career and Family.*
Broader Imposed career interruptions (#PF4128).
Aggravated by Complications of childbirth (#PC9042)
Discrimination against women in employment (#PD0086).

♦ PD7699 Human consumption of animal products
Human consumption of dairy products
Refs Delmont, J (Ed) *Milk Intolerance and Rejection* (1983).
Broader Human use of animal by-products (#PF1964).
Narrower Denial to food animals of the right to freedom from suffering (#PE3899)
Denial to animals of the right to the attention, care and protection of humankind (#PF5121).
Related Human consumption of animals (#PC7644).
Aggravated by Unethical consumption practices (#PD2625).

♦ PD7704 Nepotism
Nepotism in public office — Dynastic politics
Broader Corruption in politics (#PC0116)
Unequal distribution of fame and honours (#PF3439).
Narrower Nepotism in socialist countries (#PF6013)
Nepotism in developing countries (#PD1672)
Corruption amongst relatives of government leaders (#PE9140).
Aggravates Use of undue influence to obstruct the administration of justice (#PE8829).
Aggravated by Resignation towards bribery (#PF8611)
Unethical practices by employees (#PD4334)
Unethical practices of employers (#PD2879).

♦ PD7707 Environmental sexual harassment
Nature Creating a working environment in any way that has the purpose of substantially interfering with an individual's working performance or that is intimidating, hostile or offensive and is directed toward a single sex is sexual harassment. A employee of one sex continually asking for a date a fellow employee of another sex creates pressure to have sex for the sake of stopping the requests. A company which allows lewd or obscene behaviour or remarks on a day to day basis is encouraging an atmosphere in which members of the opposite sex are continually annoyed because of their sex.
Broader Sexual harassment (#PD1116).

♦ PD7714 Military destabilization of developing countries
Broader Destabilization of social systems (#PB5417).
Aggravates Excessive foreign public debt of developing countries (#PD2133).
Reduced by Militarization in developing countries (#PD9495).

♦ PD7716 Travel risks
Vulnerability of traveller — Dangers to travellers
Broader Risk (#PF7580).
Narrower Hitchhiking (#PE5448) Air accidents (#PD1582)
Travel delays (#PE1977) Marine accidents (#PD8982)
Railway accidents (#PD0126) Illegal roadblocks (#PE9605)
Dangerous countries (#PD4278) Pedestrian accidents (#PD0994)
Road traffic accidents (#PD0079)
Harassment of travellers by immigration officials (#PE7780).
Related Tourist hazards (#PE8966).
Aggravated by Inadequate emergency medical services (#PD1428)
Inadequate enforcement of safety regulations (#PD5001).

♦ PD7719 Disintegration of technological capacity
Loss of technical skills by society — Loss of technical know-how
Broader Disintegration (#PA6858).
Aggravates Human suffering (#PB5955) Human contingency (#PF7054)
Fragmentation of technological development (#PC1227).

♦ PD7723 Low-quality construction work
Shoddy construction work — Minimal construction standards
Broader Unfocused design of community space (#PF1546)
Inadequacy of international standards (#PF5072)
Debilitating deterioration of physical environment (#PD2672).
Related Substandard housing and accommodation (#PD1251).
Aggravates Structural failure (#PD1230) Unsafe construction techniques (#PG9831)
Malpractice in the construction industry (#PD9713).
Aggravated by Counterfeit machine parts (#PE5319)
Inadequate building standards (#PF8829).

♦ PD7726 Unethical veterinary practice
Irresponsible veterinarians — Negligence by veterinarians — Veterinary malpractice — Corruption of veterinarians — Underreporting of animal diseases
Claim Veterinarians, under pressure from their employers, have adopted practices which lead to the underreporting of animal diseases, especially as a consequence of factory farming methods and pressures to protect the financial investments therein. They have failed to investigate adequately the extent or nature of such hazards in the process of further increasing the yields of animal husbandry. Veterinarians have condoned the restrictive conditions under which animals are kept for such purposes or for entertainment (as in zoos, aquaria or circuses). They have also condoned the development of breeds which cannot survive without surgically assisted births (as in the case of certain breeds of dogs).
Broader Unethical practice of the zoosciences (#PD4721).
Related Unethical ophthalmic practice (#PE1369).
Aggravates Maltreatment of zoo animals (#PE4834)
Maltreatment of performing animals (#PE4810).

♦ PD7727 Deflation
Deflation of the world economy — Self-reinforcing contractionary economic spiral
Nature Deflationary tendencies impede the efforts of developing countries to improve resource allocation and efficiency, thus jeopardizing the success of such efforts. Deflation thus prevents many developing countries from reconciling debt-servicing with growth, despite the vigorous adjustment efforts and painful sacrifices they must make. It also obstructs the adjustment of the massive and unsustainable current account imbalances of certain developed-market countries. In the absence of more buoyant demand kin the world economy, national trade and exchange rate objectives remain in conflict, damaging the prospects for exchange rate stability. The adverse consequences are not confined to the short-term. High real interest rates add significantly to structural budget deficits in both developed and developing countries. Low levels of investment harm the development prospects of developing countries by limiting capacity expansion and structural adjustment. The persistence of slack may also turn the cyclical rise of unemployment and fall in commodity prices into structural and secular problems.
Incidence The world economy now displays a wide variety of symptoms indicative of strong deflationary pressures: (a) In real terms international interest rates remain high, both by past standards and compared to rates of growth; (b) Commodity markets remain glutted, with a tendency for prices to weaken even more. To the extent that inflation is negligible, further drops in commodity prices tend to widen profit margins and lessen aggregate spending, while increasing interest rates for debtors; (c) In many developed-market economies, unemployment rates remain extraordinarily high. Poor sales prospects are deterring firms from enlarging capacity despite higher profit margins. consequently the use of labour and raw materials is especially sluggish. (d) Markets for numerous products, including some of the most advanced technologically, are slack; this is reflected both in price movements and in a variety of practices that further close up the trading system; (e) Export markets in developing countries remain depressed, denying sales top businesses in both developed and developing countries.
Broader Economic uncertainty (#PF5817).
Aggravates Trade protectionism (#PC4275).
Aggravated by International economic injustice (#PC9112)
Decline in import capacity of developing countries (#PE6861).
Reduces Economic inflation (#PC0254).

♦ PD7731 Unethical practice of the biosciences
Irresponsible biologists — Negligence by biologists — Irresponsible geneticists — Irresponsible microbiologists — Irresponsible bacteriologists — Irresponsible biochemists — Malpractice in biology — Corruption of biologists — Underreporting of hazards to biological systems
Claim Biologists, under pressure from their employers, have adopted practices which lead to the underreporting of hazards to biological systems, especially as a consequence of excessive use of fertilizers and pesticides, and of environmental pollution and radiation. Bioscientists have failed to investigate adequately the nature of such hazards in the process of further developing products such as herbicides and pesticides. There is little peer control of irresponsible genetic manipulation and introduction of exotic species by bioscientists. Biologists participate in monopolistic practices designed to create user dependency on non-reproducing high-yield hybrids. They are also intimately involved in the design of biochemical weapons systems, notably for defoliation, and in the design of addictive drugs.
Refs Ribes, Bruno *Biology and Ethics* (1978).
Broader Irresponsible scientific and technological activity (#PC1153).
Narrower Unethical practice of the zoosciences (#PD4721).
Aggravates Irresponsible genetic manipulation (#PC0776)
Monopolization of agricultural genetic resources (#PE6788)
Environmental hazards of new species introduction (#PC1617).

♦ PD7732 Insect vectors of plant diseases
Refs Maramorosch, K and Harris, K F *Leafhopper Vectors and Plant Disease Agents* (1979); Weiser, J *Atlas of Insect Diseases* (1977).
Broader Plant disease vectors (#PD3596) Insect vectors of disease (#PC3597).
Narrower Mites as vectors of plant disease (#PE9017)
Insect vectors of viral diseases of plants (#PD3600).
Related Mites as pests (#PE3639) Aphids as pests (#PE3613)
Homoptera as insect pests (#PE3614) Bird vectors of plant disease (#PD3601)
Plant vectors of plant disease (#PD3599).
Aggravates Domination (#PA0839).

♦ PD7758 Anaemia
Nature Anaemia, or blood deficiency, includes a groups of diseases whose primary characteristic is a reduction in the content of haemoglobin in the erythrocytes, in the number of erythrocytes per unit volume of blood for a person of a given age and sex, and in the total amount of blood in the organism. Anaemia causes pathological changes which result from disruption of the organism's oxygen supply; the degree of manifestation dependent upon the degree of anaemia and its developmental speed. The most important general symptoms of anaemia are weakness, pale skin colour, shortness of breath, dizziness, and a tendency to faint; the primary causes are loss of blood, impairment of blood formation, and increased destruction of blood.
Incidence Anaemia is a frequent result of pregnancy and childbirth and as such is a prominent problem in developing countries where high-frequency pregnancies are standard.
Refs Bank, Arthur, et al (Eds) *Fifth Cooley's Anemia Symposium* (1985).
Broader Diseases of blood and blood-forming organs (#PF8026).
Narrower Aplastic anaemia (#PE7894) Nutritional anaemia (#PD0321)
Sickle cell disease (#PE3724) Anaemia of pregnancy (#PG9577)
Iron deficiency anaemias (#PE7854) Alimentary anaemia in stock (#PG5189)
Acquired haemolytic anaemias (#PG3810) Hereditary haemolytic anaemias (#PG5035).
Related Anaemia in animals (#PE9554).
Aggravates Thrombosis (#PE5783) Still-birth (#PD4029).
Aggravated by Alcohol abuse (#PD0153).

♦ PD7762 Overworked women
Incidence The World Bank estimates that Kenyan women work 13-14 hours a day. They have little time or energy to imagine a better life, let alone to work for one.
Broader Overwork (#PD2778).
Aggravates Lack of payment for housework (#PE4789)
Work practices requiring women to lift heavy loads (#PE4452)

Failure to employ skills of home-bound educated women (#PD8546).
Aggravated by Separation of family members (#PE4959)
Inadequate working conditions for women (#PD3197).

♦ PD7773 Dependence on the media
Communication isolation
Nature Increasingly life will be lived within four walls, with people relying on the media for their images of the outside world, and with little incentive to venture as a tourist through a transportation network to experience the natural environment.
Aggravates Social isolation (#PC1707)
Manipulation of the individual by mass media (#PE7448).
Aggravated by Loneliness (#PF2386) Social isolation of the elderly (#PD1564)
Parochial escapist media entertainment (#PD0917)
Personal isolation in communities of industrialized countries (#PD2495).

♦ PD7776 Resettlement stress
Stress in immigrants
Broader Stress in human beings (#PC1648) Disruptions due to migration (#PC0018).
Aggravates Deculturation (#PJ1034) Juvenile stress (#PC0877).
Aggravated by Illegal immigration (#PD1928) Involuntary mass resettlement (#PC6203)
Socially inappropriate housing (#PD8638) Expulsion of immigrants and aliens (#PC3207)
Discrimination against immigrants and aliens (#PD0973)
Inadequate cultural integration of immigrants (#PC1532)
Insensitivity to diversity of cultural traditions (#PF8156)
Inadequate living and working conditions of immigrant labourers in industrialized countries (#PD3427).

♦ PD7792 Flouting regulatory authority
Nature Wilful and persistent disobedience of any body of related regulations, rules, statutes, or orders is a criminal offence.
Broader Violation of regulatory codes (#PD4539).
Aggravated by Contempt for authority (#PF5012).

♦ PD7799 Reproductive system diseases in animals
Broader Animal diseases (#PC0952).
Narrower Brucellosis (#PE0924) Mammary tumours (#PG9597)
Egg drop syndrome (#PG5400) Agalactia syndrome (#PG9703)
Cystic hyperplasia (#PG7888) Animal infertility (#PC1803)
Phimosis in animals (#PG9570) Orchitis in animals (#PG2704)
Pyometra in animals (#PG9400) Dystocia in animals (#PG9596)
Mastitis in animals (#PG4592) Contagious agalactia (#PE9795)
Vaginitis in animals (#PG7508) Bovine trichomoniasis (#PG5353)
Prostatitis in animals (#PG9401) Paraphimosis in animals (#PG9614)
False pregnancy in dogs (#PG7671) Equine coital exanthema (#PG7290)
Endemic abortion of ewes (#PG1537) Epidemic bovine abortion (#PG9574)
Udder acne of dairy cows (#PG7831) Low fertility of turkeys (#PG9833)
Abortion in large animals (#PG7740) Acute metritis in animals (#PG6830)
Prostatic cysts in animals (#PG7223) Balanoposthitis in animals (#PG9255)
Contagious equine metritis (#PG9552) Luteal cystic ovary disease (#PG9491)
Uterine prolapse in animals (#PG5125) Vaginal prolapse in animals (#PG7200)
Retained placenta in animals (#PG3466) Bovine ulcerative mammillitis (#PG4595)
Prolonged gestation in cattle (#PG9772) Prostatic neoplasms in animals (#PG5056)
Vaginal hyperplasia in animals (#PG7464) Follicular cystic ovary disease (#PG0809)
Ulcerative posthitis in animals (#PG9801) Ovine genital campylobacteriosis (#PG5775)
Bovine genital campylobacteriosis (#PG5215)
Transmissible canine venereal tumour (#PG7622)
Congenital anomalies of the reproductive system in animals (#PG7447).

♦ PD7821 Dehumanization of health care
Degrading medical treatment — Abusive treatment of women in labour — Undignified medical treatment
Nature Patients are being perceived and treated as things or nonpersons or lesser persons without any warmth in human interaction. The same social forces that contribute to dehumanization in economic and political milieus lead to dehumanization in health care: aggregation of services, bureaucratization of services, secularization of values, professionalization of skills, and proliferation of technologies.
Broader Inadequate health services (#PD4790) Unethical medical practice (#PD5770).
Aggravates Dehumanization of death (#PF2442) Neglected young children (#PE4245)
Abusive treatment of the aged (#PD5501) Non-use of available health facilities (#PF9683).
Aggravated by Inadequate health care in least developed countries (#PE9242).

♦ PD7822 Displaced persons
Deportees — Deportation
Nature Displaced persons are those who have been deported from, or have been obliged to leave their country of nationality or former habitual residence, such as persons who are compelled to undertake forced labour or who are deported for racial, religious or political reasons.
Narrower Displaced children (#PD5136) Forced repatriation (#PD8099)
Forced mass expulsion (#PD0531).
Related Refugees (#PB0205) Internal exile (#PJ9588)
Expulsion of immigrants and aliens (#PC3207).
Aggravated by War (#PB0593) Involuntary mass resettlement (#PC6203).

♦ PD7835 Environmental degradation in industrialized countries
Environmental degradation in developed countries
Incidence The emissions of sulphur and nitrogen oxides and concentrations of atmospheric ozone have increased, agricultural and industrial wastes have been accumulating, lake acidification has taken place, forests and water quality have been declining.
Broader Environmental degradation (#PB6384)
Deterioration of industrialized countries (#PD9202).
Narrower Obliteration of footpaths by development (#PE3874).

♦ PD7840 Monopolistic control of new life forms
Irresponsible patenting of life forms
Narrower Monopolistic control of new animal forms (#PD7501)
Monopolization of agricultural genetic resources (#PE6788).
Aggravated by Irresponsible genetic manipulation (#PC0776).

♦ PD7841 Diseases of nervous system in animals
Broader Animal diseases (#PC0952).
Narrower Rabies (#PE1325) Scrapie (#PE5330) Louping ill (#PG6452)
Pseudorabies (#PE9799) Tick paralysis (#PG4545) Puerperal tetany (#PG4087)
Feline dysautonomia (#PG1394) Botulism in animals (#PG2994)
Equine grass sickness (#PG1883) Meningitis in animals (#PG7785)
Polioencephalomalacia (#PG7460) Oedema disease in swine (#PG9698)
Equine encephalomyelitis (#PG9701) Paralysis of the forelimb (#PG6092)
Facial paralysis in animals (#PG9740) Encephalomyelitis in poultry (#PG9615)
Transport tetany in ruminants (#PG9285) Neonatal maladjustment syndrome (#PG9351)
Hypomagnesemic tetany in cattle (#PG9231) Haemophilus septicaemia of cattle (#PG0354)
Eastern encephalitis in pheasants (#PG9396) Porcine congenital tremor syndrome (#PG9391)
Porcine enteroviral encephalomyelitis (#PG5030)
Coronaviral encephalomyelitis of swine (#PG3788)
Diseases of the spinal column in animals (#PE4604)
Nervous system diseases caused by parasites (#PE4683)
Congenital diseases of the nervous system in animals (#PE0965)
Bovine progressive degenerative myeloencephalopathy of brown Swiss cattle (#PG6591).
Related Tetanus (#PE2530).

♦ PD7875 Restrictive scientific practices
Broader Restrictive practices (#PB9136) Restrictive professional practices (#PD8027).

♦ PD7876 Contract fraud
Broader Fraud (#PD0486) Unethical commercial practices (#PC2563).
Aggravates Documentary fraud (#PE1110).
Aggravated by Unethical financial practices (#PE0682).

♦ PD7885 Investigatory malpractice
Witch hunt
Aggravates Scapegoats (#PF3332).

♦ PD7910 Police state
Broader Repression (#PB0871) Political police (#PD3542).
Narrower Secret police (#PE6331).
Aggravates Injustice (#PA6486) Fear of police (#PF8378)
Abuse of police power (#PC1142) Demeaning community self-image (#PF2093).

♦ PD7918 Long-term effects of war
Broader Disastrous consequences of war (#PC4257).
Related Housing destruction in war (#PE2592).

♦ PD7924 Diseases of the respiratory system
Diseases of the nose — Diseases of the throat — Respiratory failure — Respiratory infections — Respiratory virus diseases — Respiratory illness
Refs Ballenger, John J (Ed) *Diseases of the Nose, Throat, Ear, Head and Neck* (1985); Becker, W, et al *Ear, Nose and Throat Diseases* (1988); Kendig, Edwin L and Chernick, Victor *Disorders of the Respiratory Tract in Children* (1983); Lee, Douglas H *Environmental Factors in Respiratory Disease* (1972); Meyer, *Secondary and Functional Rhinoplasty* (1988).
Broader Human disease and disability (#PB1044).
Narrower Asthma (#PD2408) Pleurisy (#PG6121) Influenza (#PE0447)
Pneumonia (#PE2293) Epistaxis (#PG6688) Laryngitis (#PE2653)
Pneumoconiosis (#PD2034) Chronic bronchitis (#PE2248)
Acute respiratory infections (#PE7591) Diseases of upper respiratory tract (#PE7733)
Benign neoplasm of respiratory system (#PE3637)
Malignant neoplasm of respiratory system (#PE7572)
Symptoms referable to respiratory system (#PE7864).
Aggravates Diseases of the ear (#PD2567).
Aggravated by Cement dust (#PE2854) Virus diseases (#PD0594)
Adenovirus infections (#PE2355)
Protein-energy malnutrition in infants and early childhood (#PD0331).

♦ PD7929 Unstable family life
Broader Obstacles to family life (#PF7094)
Fragmented patterns of extended family relationships (#PF1509).
Aggravates Domestic quarrels (#PE4021).
Aggravated by Failure to recognize uniqueness of family members (#PF1750).

♦ PD7940 Hyperinflation
Incidence Latin American countries have demonstrated that hyperinflation goes far beyond the economic field and affects all aspects of society. In recent years, countries such as Brazil, Argentina, Bolivia and Peru have been devastated psychologically and socially by a currency in which their users have little confidence. Over and above the economic consequences of daily devaluation (financial speculation, a chronic disease in productive investments and a systematic deterioration of real wages), constant inflation, with annual rates of three or even four digits, erodes a people's faith in their country, and gives rise to deep uncertainty about the future. This acute deterioration in confidence, along with a sense of uncertainty and scepticism create a phenomenon which is difficult to reverse and an environment where innovative alternatives capable of overcoming an inflationary crisis are almost impossible to generate.
Claim Inflation might be easier, not harder, to stop once it has become hyperinflation. It can collapse all at once under its own weight. When economy hits rock-bottom, it is also more likely that the popular will to do something will appear.
Broader Economic uncertainty (#PF5817)
Overlooked potential for industrial development in rural communities (#PF2471).
Aggravates Scepticism (#PF3417)
Loss of confidence in government leaders (#PF1097).

♦ PD7941 Hostile environmental modification
Hostile climate modification — Hostile weather modification
Nature Use of techniques to produce substantial environmental modification. This may be designed to inhibit agricultural production and to render the affected country more dependent on other countries, possibility even leading to its destabilization.
Broader Anthropogenic climate change (#PC9717).
Related Inadvertent modifications to climate (#PC1288).
Aggravated by Volcanic dust (#PE5109) Unethical practice of meteorology (#PD4182)
Aerial explosions of unknown origin (#PF9167).

♦ PD7952 Manslaughter
Nature Manslaughter can be divided into two categories: reckless and voluntary. Reckless manslaughter is causing the death of another person by consciously and unjustifiably disregarding the risk while engaging in dangerous behaviour. Voluntary manslaughter is causing the death of another person under circumstances which would be considered murder, except the death is caused under extreme emotional disturbance for which there is a reasonable excuse. Someone so disturbed by his sister being verbally attacked that he strikes and causes the death the person doing the taunting could be considered guilty of voluntary manslaughter. A person so upset by the politics of a city mayor that he assassinates him is not guilty of voluntary manslaughter but of murder.
Broader Homicide (#PD2341).
Narrower Negligence in manslaughter (#PE0437).

♦ PD7963 Domestic cartel
Nature The cartel is a form of monopolistic agreement among companies, usually belonging to

one economic sector, for the purpose of extracting monopolistic profits through quotas regulating the volume of production and of products marketed for all its basic participants.
Broader Cartels (#PC2512) Restrictive trade practices (#PC0073).
Narrower Industrial gas monopolies (#PE1813).
Related Monopolies (#PC0521).

♦ PD7964 Unethical maintenance practices

Irresponsible maintenance practices — Negligence of maintenance — Corruption in maintenance practices
Broader Inadequate maintenance (#PD8984).
Aggravates Inadequate maintenance equipment (#PD6520)
Inadequate maintenance personnel (#PJ0088)
Prohibitive cost of equipment maintenance (#PE1722).
Aggravated by Prohibitive cost of maintenance (#PF0296)
Unethical practices in transportation (#PD1012)
Wastage of highly skilled personnel in the routine maintenance of complex systems (#PE1396).

♦ PD7966 Bureaucratic opposition

Bureaucracy blocks appeals
Broader Unrecognized socio-economic interdependencies (#PF2969)
Paralyzing patterns between villages and administrative structures (#PF1389).
Aggravated by Bureaucratic factionalism (#PF7979)
Bureaucracy as an organizational disease (#PD0460)
Unchecked power of government bureaucracy (#PD8890).

♦ PD7977 Excessive environmental heat

Narrower Heat as an occupational hazard (#PE5720)
Diseases of the lacrimal apparatus in animals (#PG4093).
Related Climatic heat (#PC2460).
Aggravates Heat disorders (#PE2398) Hazards of strong toxic substances (#PD0122).
Aggravated by Global warming (#PC0918) Thermal pollutants (#PC1609).

♦ PD7978 Development of informal sector in developing countries

Refs Turnham, David; Salomé, Bernard and Schwarz, Antoine (Eds) *Development Centre Seminars* (1990).
Broader Underground economy (#PC6641).

♦ PD7981 Counterfeiting

Counterfeit products and services — Product piracy — Fake products — Counterfeit goods — Forgery of brand-name merchandise — Trade in counterfeit goods
Nature Counterfeiting occurs where there is deliberate copying of a product or the packaging, label or its trade-mark. It is the unauthorized use of a legitimate product's commercial presentation or any protected indication.
Incidence The crime of counterfeiting has expanded beyond the realm of printing false paper money. Present-day counterfeiters take everything from designer fashions to birth control pills to oil filters to tractor parts. Unlike printing money, the counterfeiting of consumer products may have personal detrimental effects and even cause fatalities.
Counterfeit goods from software to watches and novels to tennis shoes undermine trade and incentives to produce, reduce profits and leave customers with shoddy products. Copied auto and aircraft parts have been implicated in several fatal accidents, pseudo face creams have damaged skin, false birth control pills have resulted in unwanted pregnancies, counterfeit fertilizers and herbicides have lost farmers' crops, fake drugs can be relatively harmless or potentially lethal. Counterfeited products cost US companies an estimated $20 billion in lost sales in 1984, compared with $3 billion lost in 1978. In 1986 it was estimated that the worldwide cost of counterfeit and theft of intellectual property was $23.8 billion, divided as follows: scientific and photographic goods (21 per cent), computers and software (17), electronics (10), motor and vehicle parts (9), entertainment (8), pharmaceuticals (8), chemicals (6), and petroleum products (6). An alternative estimate puts the value of such fraudulent goods at between 3 and 5 per cent of world trade, or between $90 and $150 billion per year costing 130,000 jobs in US and more than 100,000 in European Community countries.
Refs American Institut Psychology *The psychology of the master counterfeiters* (1986); Mills, John F and Mansfield, John M *The Genuine Article* (1982); Practising Law Institute *Product Counterfeiting* (1984).
Broader Forgery (#PD2557) Vulnerability of intellectual property (#PF8854).
Narrower Computer piracy (#PE6625) Counterfeit drugs (#PD5540)
Avoidance of copyright (#PD0188) Counterfeit machine parts (#PE5319)
Forgery of shares and bonds (#PG1146)
Counterfeit money and government securities (#PE5503).

♦ PD7986 Psychotronic warfare

Psionic warfare
Nature Extremely low frequency magnetic and electric fields of 6.67 Hz, 6.26 Hz and lower can be used to modify behaviour in humans and animals producing symptoms of confusion, anxiety, depression, tension, fear, mild nausea, headaches, and hemispheric EEG desynchronization. The established physics of radio propagation suggest that vast geographical areas can be readily mood-manipulated by transmission of such energy within the earth-ionosphere waveguide.
Incidence There are documented reports of efforts to test transmissions with these effects.
Broader War (#PB0593).
Related Psychic warfare (#PF4866) Psychological warfare (#PC2175)
Psychic interference in decision-making (#PF0508)
Environmental hazards of extremely low frequency electromagnetic radiation (#PE7560).

♦ PD7989 Animal extinction

Broader Extinction of species (#PB9171).
Related Animal deaths (#PE7941) Wildlife extinction (#PC1445)
Periodic mass extinctions of species (#PF4149)
Denial to animals of the right to freedom from mass killing (#PE9650).
Aggravated by Inadequate animal welfare (#PC1167).

♦ PD7993 Chemicalized farming

Chemical fertilizers — Intensive farming techniques — Hazards of agricultural chemicals
Nature The use of intensive farming techniques are creating not only localized ecological disasters but in many places having impacts across way sections of the countryside. Intensive farming designs limit the range of habitats, inputs to the system particularly those used for soil fertility and pest control determine the impact on species within the habitats. Intensive farming substitutes rather than enhances the natural biological production processes. Livestock and arable enterprises are separated breaking the cyclical processes characteristic of natural ecosystems. The lack of diversity of crops and animals narrows the range of natural plants and animals on the land. Autumn and winter-sown cereals decrease the availability of food for wild birds at the start of the crucial breeding season and thus limiting the range of wild birds, hares and other small mammals. The large farm size further limits the size of animal populations because field boundaries are smaller, fewer and farther apart. Habitats within boundaries, especially hedgerows, are important to many wild flowers, insects, birds and mammals. Oversowing land decreases the populations of insects. Soil fertility is decreased because of intensive farming techniques. Soil invertebrates are fewer where inorganic fertilizers are used and crop residues are burned. Weed control is more efficient further limiting the range of species of plants and animals. Intensive farming causes more nitrate and phosphate pollution of surface waters. Soil erosion is greater. Leaching rates are greater. Animal wastes are in surplus where animals are raised intensively. Soils are effected negatively also. Soil organic matter levels are lowered decreasing the availability of water and some nutrients to plants. Trace elements are lost or used up. The buffering capacity of the soil is disrupted increasing pH fluctuations. Substrate is lost further decreasing fertility. Continuous cropping of autumn and winter-sown cereals promote erosion because the soil is exposed during the wetter part of the year. Soil flora and fauna are lost with their ability to cycle nutrients within the soil and their contribution to the suppression of plant diseases. The continuous use of chemicals has enabled the development of pesticide resistant insects, weeds and microbes. Plants that are chemically fertilized may look lush, but lush growth produces watery tissues, which become more susceptible to disease; and the nutritive quality suffers.
Claim Agriculture as it is currently practised is a global disease, because it wastes the soil through erosion, burns up more energy than it produces in food and requires the extensive use of agricultural chemicals, whose side-effects are largely unknown. Studies show that farmers who apply few or no chemicals to crops are usually as productive as those who use synthetic fertilizers and pesticides.
Counter-claim 1. The use of agricultural chemicals is not in itself harmful. The level of use remains quite low in many regions where the response rates are high and the environmental consequences of residues are not yet a problem. The issue is that the growth in the use of these chemicals tends to be concentrated where they tend to do more overall harm than good.
2. Chemicals are used because they ensure an economic return. Organic farming is generally accepted as a niche market unable to satisfy the agricultural needs of a country. Organic farming is labour intensive at a time when agricultural is faced with a shortage of labour in industrialized countries, because of the low level of farm incomes and despite the subsidies in those countries.
Broader Over-intensive soil exploitation (#PC0052)
Inappropriate modernization of agriculture (#PF4799).
Aggravates Soil infertility (#PD0077) Decline in nutritional quality of food (#PE8938)
Unsustainable agricultural development (#PC8419).
Aggravated by Cultivation of illegal drugs (#PD4563)
Over-subsidized agriculture in industrialized countries (#PD9802).
Reduces Shortening of fallow periods on agricultural land (#PE9407).

♦ PD7994 Ignorance of reproductive processes

Ignorance of the cause of conception — Ignorance concerning sex
Incidence It has been argued that such ignorance was once widespread and was possibly general in the history of man. In certain tribes, especially those practising cohabitation before puberty, conception may be believed to result by more or less magical means and apart from sexual intercourse. This is reinforced by the disproportion of pregnancies to the frequency of such intercourse. In the absence of sexual education in modern schools, such ignorance may persist into early adulthood, especially for those living in isolated areas.
Broader Ignorance (#PA5568) Ignorance of health and hygiene (#PD8023).
Aggravates Paternal negligence (#PD7297) Adolescent pregnancy (#PD0614)
Health risks of teenage sex (#PE6969) Ignorance concerning disease (#PD8821)
Deception between sexual partners (#PE4890).

♦ PD8001 Unethical practices in the apparel industry

Unethical practices in the clothing industry — Corruption in the clothing industry
Claim The nature of the industry is such as to favour the development of "sweat shop" working conditions in which unprotected workers (illegal immigrants, moonlighters, minors) are employed at very low wages, usually for very long hours, and under the constant threat of losing a job which may be their only possible source of income. The extent of such practices has been limited in industrialized countries although they continue, especially as a characteristic of any underground economy. Such practices are widespread in developing countries. The highly competitive nature of the apparel industry also favours illicit copying of styles, counterfeiting of trade marks and labels, and exceptionally high mark-ups. The industry is strongly dependent on encouraging the consumer to adopt particular new styles and to reject old ones as unfashionable, thus leading to excessive allocation of resources to apparel which is only worn for a limited period before being discarded. Style changes are determined by a small elite whose motivation is to undermine the appropriateness of old styles, whether or not the new styles are appropriate, and to encourage conspicuous consumption on apparel.
Broader Unethical commercial practices (#PC2563).
Aggravates Decadent clothing (#PE5607) Decadent standard of living (#PD4037)
Denial of right to sufficient clothing (#PE7616)
Denial of freedom of expression in clothing (#PE5409)
Instability of textile and clothing industries (#PE1008)
Protectionism in the textile and apparel industries (#PE5819)
Environmental hazards from textile and clothing industries (#PE1103).

♦ PD8003 Scientific ignorance

Scientific illiteracy — Lack of scientific knowledge — Inadequate public understanding of science
Nature People feel victimized by not being able to participate in the decisions affecting their lives. Appropriation of new knowledge into an understandable framework is impossible; the individual finds even the task of keeping expertise up-to-date faltering. Although communications technology is highly developed, and storage and retrieval systems are achieving a promising stage of performance, and the store of scientific knowledge is accumulating at an alarming rate, people are still woefully uninformed, to the point where their ability to affect their own social survival is paralyzed.
Incidence In the UK, surveys indicate that 30 per cent of the public believe that the sun goes around the Earth and more than 50 per cent believe' that antibiotics kills viruses as well as bacteria.
Counter-claim Scientific knowledge, as defined by scientists, is often of little practical use to people. By the time it is translated into useable knowledge, it is no longer recognizable to scientists as science.
Broader Ignorance (#PA5568) Undirected technological expansion (#PC1730).
Narrower Mathematical ignorance (#PD6728) Ignorance of health and hygiene (#PD8023).
Aggravates Inadequate warning of disasters (#PF3565)
Inadequate models of socio-economic development (#PF9576).

♦ PD8005 Disorders of consciousness

Nature Disorders of consciousness include those in which the perception of external objects and spatial and temporal orientation are disrupted, thinking is disordered, events are not fixed in the memory, and alienation from the real world sets in. Stupor is frequently encountered, manifested in retardation, somnolence, impoverished psychic life, and elevated threshold for external irritants. Cases range from mild to extremely severe, characterized by sopor and coma. Delirium

may also result, characterized by illusions, hallucinations, affective disorders, acute delirium, and motor excitation.
Broader Mental disorders (#PD9131).

♦ PD8007 Redundancy of workers
Dismissal of workers to improve profitability
Nature Workers may be dismissed for reasons of redundancy because of a reduction in the amount of work of an enterprise, or as the result of the introduction of computers and automated equipment which eliminate certain jobs. In deciding which workers are to be retained, preference is usually given to workers of higher qualification and labour productivity, though redundancy usually strikes at those with less time in the job, who are often immigrants and young people.
Refs Grais, Bernard *Lay-Offs and Short-Time Working in Selected OECD Countries* (1983).
Broader Unjust dismissal of workers (#PD5965).
Narrower Reduction of workforce in enterprises (#PE7515).
Aggravated by Asset stripping (#PE9224) Narrow profit margin (#PJ9737)
Short-term profit maximization (#PF2174)
Unemployment caused by environmental conservation (#PD0467).

♦ PD8016 Human sexual disorders
Sexual dysfunction
Refs Huhner, M *The Diagnosis and Treatment of Sexual Disorders in the Male and Female, Including Sterility and Impotence* (1946).
Broader Stigmatized diseases (#PD7279) Human disease and disability (#PB1044).
Narrower Sex guilt (#PJ2396) Vaginismus (#PG1752) Dyspareunia (#PG9647)
Transsexualism (#PF3277) Sexual frigidity (#PE6408) Orgasm disorders (#PG5418)
Sexual unfulfilment (#PF3260) Sexual impotence of men (#PF6415)
Sexual desire disorders (#PE9607) Sexual arousal disorders (#PG9473)
Human pseudo hermaphroditism (#PE2246) Sexual dysfunction of torture victims (#PE3932).
Related Sexual craving (#PE7031).
Aggravates Sexual deviation (#PD2198) Total sexual abstention (#PE3298).
Aggravated by Sex (#PF9109).

♦ PD8017 Governmental disregard for people as human beings
Official disregard for individuals — Impersonality of bureaucracy — Unresponsive social service bureaucracy
Claim The inflexible traditional bureaucracy of social care services limits both the accessibility of services and the coverage of people's needs. There is a lack of creative exchange and a narrow range of alternatives in the political process, and the individual is no longer able to control his access to services, nor is there consensus on the maintaining of systems of access.
Broader Bureaucratic inaction (#PC0267) Inadequate political networks (#PD2213)
Bureaucracy as an organizational disease (#PD0460).
Aggravates Inefficient public administration (#PF2335)
Limited access to social benefits (#PF1303)
Denial of right to develop as human beings (#PF2364)
Excessive waiting times in government facilities (#PF5120).
Aggravated by Marxism (#PF2189)
Proliferation of public sector institutions (#PF4739).

♦ PD8021 Occupied territories
Military occupation — Belligerent occupation
Refs Harkavy, Robert E *Bases Abroad* (1989).
Broader Alien domination of peoples (#PC7384)
Denial of right to national self-determination (#PC1450)
Denial of right of a people to be self-determining (#PC6727).
Related Colonialism (#PC0798) Occupied nations (#PC1788).
Aggravates Informers (#PD8926)
Children engendered by occupying soldiers (#PD8825).
Aggravated by Neutrality (#PF0473) Territorial expansionism (#PC9547).

♦ PD8023 Ignorance of health and hygiene
Limited health knowledge — Inadequate understanding of hygiene — Ineffective health education — Inadequate health education — Misunderstanding of preventive health care — Unawareness of symptoms of illness — Lack of sanitation knowledge — Lack of sanitation expertise
Broader Scientific ignorance (#PD8003)
Inadequate technical advice and consultation on problems (#PF1981).
Narrower Nutritional ignorance (#PE5773) Limited medical knowledge (#PD9160)
Unawareness of health problems (#PG8949) Ignorance of reproductive processes (#PD7994)
Misunderstanding of veterinary approach (#PG5402)
Ignorance of traditional herbal remedies (#PE3946)
Ignorance of women concerning primary health care (#PD9021).
Aggravates Tuberculosis (#PE0566) Neglected health practices (#PD8607)
Ignorance concerning disease (#PD8821) Outdated forms of community health (#PF1608)
Inadequate maintenance of physical health (#PF1773)
Undeveloped channels for commercial initiative (#PF6471)
Inadequate health care in least developed countries (#PE9242)
Health risks to workers in agricultural and livestock production (#PE0524).
Aggravated by Inadequate health services (#PD4790)
Unprepared adult leadership (#PF6462)
Inadequate secondary education (#PD5345)
Narrow range of practical skills (#PF2477)
Educational gap between generations (#PF6497)
Limited availability of functional information (#PF3539)
Outmoded functional skills in rural communities (#PF2986)
Incomplete understanding of new societal service systems (#PF2212)
Inadequate dissemination and use of available information (#PF1267)
Insufficient government spending on cost-effective activities (#PE5302).

♦ PD8027 Restrictive professional practices
Professional restrictiveness
Broader Restrictive practices (#PB9136).
Narrower Restrictive legal practices (#PD8614) Restrictive medical practices (#PD8831)
Restrictive scientific practices (#PD7875).
Aggravates Non-equivalence of national educational qualifications (#PC1524).

♦ PD8030 Inappropriate foreign investment
Disruptive foreign investment — National insecurity due to excessive foreign investment — Vulnerability of countries to destabilization from foreign investment
Claim Foreign investment has the net effect of draining considerable sums of capital out of developing countries and in the long run tends to slow economic growth down rather than to boost it. The reasons include: profits flow back to the rich countries, local firms and craftsmen lose their livelihood while the investment creates only few jobs, the transferred technology has little use outside the foreign subsidiary and it imports required inputs. Often corporations are allowed, for a certain period of time, to operate without paying any taxes and afterwards they can use the transfer pricing mechanism to avoid taxes.
Narrower Dislocation of productive units by foreign investment (#PE8787)
Excessive foreign investment in traditional industries of developing countries (#PD0765)
Direct foreign investment by transnational enterprises as a restrictive business practice (#PE0161).
Aggravates Maldevelopment (#PB6207).
Reduced by Foreign intervention in internal affairs of states (#PC3185)
Foreign private investment income outflow from developing to developed countries (#PE8957)
Distortion of international trade by restrictive controls over foreign investment (#PE8525).

♦ PD8034 Refugees by boat
Boat people
Broader Refugees (#PB0205).
Related Environmental refugees (#PE3728).
Aggravated by Piracy at sea (#PD8438) Environmental stress (#PC1282).

♦ PD8041 Misappropriation of sacred objects
Broader Vulnerability of sacred sites (#PD6128).
Related Misappropriation of cultural property (#PE6074).

♦ PD8047 Apathy in developing countries
Undeveloped work ethic in developing countries — Lack of work commitment in developing countries
Broader Apathy (#PA2360) Deficiencies of developing countries (#PC4094).
Related Inadequate supply of appropriate trained manpower in developing countries (#PE6243).
Aggravates Rivalry and disunity within developing regions (#PD0110).

♦ PD8050 Health hazards of radiation
Radiation damage to the human body — Radiation carcinogenesis in humans
Nature Biological radiation damage may be acute or long-term. Acute damage occurs soon (within days or weeks) after exposure to high doses of radiation delivered over a short period of time, and may range from slight and temporary reddening of the skin to dramatic and often lethal syndromes which involve the major body systems. Pathological acute effects arise after exposure to doses hundreds of times higher than those likely to be received from environmental contamination, except in major accidents or nuclear warfare.
Long-term damage may be genetic or somatic. Genetic damage affects the germ cells of the irradiated individual, is transmitted to his or her descendants and may not appear for generations, eventually to result in hereditary diseases of various degrees of seriousness. Gene mutations may either be changes in single genes or gross chromosome anomalies which are due to loss, duplication, or rearrangement of chromosomes. Most of the known defects associated with chromosome anomalies are so severe as to preclude reproduction of the individuals affected. Somatic damage appears clinically in the irradiated individuals only years after exposure and consists largely of an increased frequency of malignancies (mostly leukaemias and tumours of the thyroid and bone) indistinguishable from those that occur spontaneously in the general population.
While experiments have made it possible to describe the mutational effects of irradiation, they do not provide adequate evidence (that could be applied to man) regarding the manner or rate with which induced gene mutations would be eliminated from the population, or the proportion of mutations that would have serious consequences. It is not, therefore, possible to assess how many crippled or mentally defective individuals descended from irradiated persons would appear in any generation, and the total number summed over all generations is also highly uncertain. It is reasonably certain that a population which had been irradiated at an intensity sufficient to kill even a few per cent of its members would suffer important long-term consequences.
Claim Radiation prematurely ages everyone, which means they get the diseases they might have had in old age at a much younger age, including cancers. They get arthritis, diabetes and serious allergies in their fifties now instead of late old age. Up to twenty years or more after exposure to radiation people become susceptible to leukaemia and other cancers. Men and women become sterile. Exposure during pregnancy can cause abortion, deformities or mental damage. Children are now being born weakened by radioactivity, prone to enzyme disorders, allergies and asthma directly caused by cell mutations. This new weakened generation is less able to cope with an ever increasing dose of radiation in the environment. Within the next 5 generations, children born into the post nuclear age the damage to the entire gene pool will be very clear.
Radiation levels are justified in terms of background levels of radiation but these natural levels have been increasing. In 1940 levels were 40 millirems a year, in the 1950's and 1960's they were 100 and in the 1980's they are 200 millirems, mainly due to weapons testing.
Refs Kriegel, H, et al *Developmental Effects of Prenatal Irradiation* (1982); United Nations Environment Programme *Radiation* (1986).
Broader Hazards to human health (#PB4885) Radioactive contamination (#PC0229)
Contamination by natural radiation (#PC1299)
Health hazards of environmental pollution (#PC0936).
Narrower Health hazards of irradiated food (#PD0361)
Radio frequencies as a health hazards (#PE5099)
Microwave radiation as a health hazard (#PE6056)
Excessive exposure of medical patients to radiation (#PE1704).
Aggravates Mutation (#PF2276) Leukaemia (#PE0639)
Neoplastic diseases (#PC3853) Genetic defects and diseases (#PD2389).

♦ PD8052 Job fatigue
Sleepiness at work — Sleeping during work — Dozing on the job
Incidence One-third of the one million truck drivers on the US highways may be too tired to stay awake while driving or to respond quickly enough to avert an accident.
Broader Fatigue (#PA0657).
Narrower Human fatigue during control of complex equipment (#PE5572).
Aggravates Oil pollution (#PE1839) Nuclear accidents (#PD0771)
Industrial accidents (#PC0646).
Aggravated by Boredom (#PA7365) Hypersomnia (#PG4415)
Sleep deprivation (#PE2741).
Reduced by Unnatural urban environments inhibiting sleeping in public (#PF6193).

♦ PD8054 Fires
Fire risk — Fire
Nature All fires involve the chemical reaction of the burning substance with oxygen in the air, the release of great amounts of heat, and the rapid conversion of the combustion products into gases. Uncontrolled burning may destroy property and human life. Fires are most frequently caused by carelessness, misuse of industrial equipment, spontaneous combustion of raw materials and finished products, static electricity, lightning, and arson.
Incidence Fires extinguished in their early stages and involving no losses constitute about 25 percent of the total number of fires. Only a small percentage of fires are major fires occurring in depots, warehouses, stores, and other places where goods are concentrated; nevertheless, the material damage resulting from these fires is considerable.
Refs Bonadio, George R *Fires* (1988); Rushbrook, Frank *Fire Aboard* (1979); Wolffsohn, A *Fire Control in Tropical Pine Forests* (1981).
Broader Disastrous accidents (#PC6034).

Narrower Urban fires (#PD2211)
Forest fires (#PD0739)
Domestic fires (#PJ6766)
Toxic modern furniture materials (#PE4675).
Aggravates Air pollution (#PC0119)
Damage to goods (#PE4447)
Burns and scalds (#PE0394)
Interruption risk (#PF9106)
Vulnerability of telephone system (#PE8254)
Endangered monuments and historic sites (#PD0253)
Deterioration of the physical condition of art objects (#PD1955)
Degradation of semi-natural and natural habitats of flora and fauna (#PC3152).
Aggravated by Smoking (#PD0713)
Drought (#PC2430)
Lightning (#PD1292)
Earthquakes (#PD0201)
Waste paper (#PD1152)
Rodents as pests (#PE2537)
Domestic refuse disposal (#PD0807)
Solid wastes as pollutants (#PD0177)
Transport of dangerous goods (#PD0971)
Beryllium as a health hazard (#PE2209)
Occupational hazards of benzene (#PE1849).

♦ PD8056 Maldistribution of water

Uneven distribution of water — Abusive restrictions on use of water — Inequitable right of access to water — Privately owned water
Broader Maldistribution of resources (#PB1016)
Underprovision of basic services to rural areas (#PF2875).
Aggravates Long-term shortage of water (#PC1173)
Boundary disputes between neighbours (#PE7903)
Inadequate water supply in the rural communities of developing countries (#PD1204).
Aggravated by Lack of accord on water use (#PF4839)
Unethical practice of hydrology (#PD2586)
Privatization of public services (#PE3391)
Fragmented planning of community life (#PF2813)
Inadequate water system infrastructure (#PD8517).

♦ PD8075 Scarcity of residential land

Broader Scarcity (#PA5984)
Limited available land (#PC8160)
Subsistence approach to capital resources (#PF6530)
Personal isolation in communities of industrialized countries (#PD2495).
Aggravates Inadequate housing (#PC0449)
Illegal occupation of unoccupied property (#PD0820)
Prohibitively expensive housing for the poor (#PE8698).
Aggravated by Excessive land usage (#PE5059).

♦ PD8076 Personal physical disfigurement

Facial disfigurement
Nature Deviations from facial stereotypes have become synonymous with uncertainty and discomfort. Certain facial features (such as clear skin, rosy complexion, well-defined eyes) are sufficient in themselves to enhance social standing. Disfigurement distorts and disorganizes the signalling mechanism in interpersonal relations. Facial movements (such as smiles or frowns) are misinterpreted or unacknowledged. The absence of good looks is frequently associated with suspicious character.
Claim People's responses to disfigurement makes victims vulnerable. It robs them of privacy and dignity. Apprehension and dread, in addition to the normal social anxieties and nervousness, accompany disfigured people wherever they go as they wait for the inevitable questions: jokey, curious, pitying. Disfigurement is associated with the darker side of culture. It is the damaged, the deformed, the bad and the ugly. Attitudes are struck in responses of ridicule, rejection, contempt, humiliation and defeat or a grossly maudlin and misdirected pity.
Counter-claim Rejection of the disfigured is at its basis a biological drive to reproduce only the best of the species. It is a survival instinct. While any one individual does not have to succumb to this instinct most are quite unaware of it and simply follow it.
Broader Ugliness (#PA7240)
Distortion (#PA6790)
Imperfection (#PA6997).
Narrower Forced depilation (#PE5697)
Disfiguring impairments (#PE6776).
Aggravated by Smallpox (#PE0097)
Malicious physical disablement (#PE3733).

♦ PD8099 Forced repatriation

Claim Asylum-seekers must be treated as refugees – even the "economic migrants" – if they have a well-founded fear of persecution upon return.
Counter-claim International treaties provide for the repatriation of illegal immigrants to their country of origin.
Broader Displaced persons (#PD7822)
Restrictions on freedom of movement between countries (#PC0935).
Narrower Unlawful repatriation of minors (#PE6854)
Forced repatriation of prisoners of war (#PD0218).
Aggravated by Rejection of refugees (#PF3021)
Ill treatment of prisoners of war (#PD2617).

♦ PD8102 Denial of right to education

Denial of literacy and education
Broader Denial of cultural rights (#PD5907).
Narrower Dual school systems (#PG8759)
Restricted higher education (#PJ0572)
Denial of right to educational choice (#PE4700)
Discrimination against women in education (#PD0190)
Ethnic and social discrimination in foreign language teaching (#PF5929)
Restriction of educational opportunities in capitalist systems (#PD3122).
Related Inequality in education (#PC3434)
Insufficient primary education (#PC6381)
Denial of education to minorities (#PC3459).
Aggravates Inequality of employment opportunity in developing countries (#PD1847).
Aggravated by Unequal access to education (#PC2163)
Discrimination against women (#PC0308)
Exploitation of child labour (#PD0164)
Limitations on school admission (#PJ1364)
Denial of rights of children and youth (#PD0513).

♦ PD8108 Shortage of trained teachers

Shortage of qualified teachers — Insufficient skilled teachers
Broader Obstacles to education (#PF4852).
Narrower Lack of vocational teachers (#PJ9603)
Unavailability of trained teachers in the rural areas of developing countries (#PE8429).
Aggravates Inadequate education (#PF4984)
Inadequate vocational education (#PF0422)
Narrow range of practical skills (#PF2477)
Inappropriate education in developing countries (#PF1531).
Aggravated by Unmotivated teachers (#PF5978)
Underpayment of teachers (#PE8645)
Inadequate teacher training (#PJ1327)
Non-resident school personnel (#PG8764)
Inadequate working conditions of teachers (#PE7165)
Inaccessibility of decision-makers in multinational enterprises (#PE0573).

♦ PD8113 Unbalanced social life

Broader Deteriorating quality of life (#PF7142)
Modern disruption of traditional symbol systems (#PF6461).

♦ PD8114 Excessive social costs of structural adjustment in debtor developing countries

Neglect of human resource development in debtor developing countries — Development with an inhuman face — Regressive social development — Reduction in public expenditure on human resources in developing countries — Negligent implementation of austerity measures
Nature Debtor developing countries have to bear the burden of arduous structural adjustment with insufficient external support in order to benefit from the facilities offered by the International Monetary Fund in rescheduling their debts. The restructuring required by the IMF is designed to restimulate the economy. However the austerity measures associated with this process include reduction of government spending on: welfare, subsidized food and housing, wages and credit facilities. All these measures impose increased hardships on the poor. They redistribute wealth and opportunities to the wealthy who have capital to invest, in order to encourage them to generate more economic activity, notably exports. These measures tend to have a serious impact on social development and well-being, as education and health needs remain unfulfilled, especially amongst the most vulnerable groups. Human resource development is neglected because although economic growth is a necessary condition for attacking poverty, complementary policies focusing resources and programmes towards the more deprived tend to be given a low priority through force of circumstances.
Background It is generally recognized that the development of human resources both benefits from and contributes to the development process. Effective programmes to promote education and training, science and technology, and popular participation in economic activity, including the participation of women, are therefore key elements in development strategies.
Incidence Austerity measures and general recessionary conditions have brought sharp declines in per capita incomes. Unemployment, especially of the young, continues to increase in debtor countries. Hunger and poverty in absolute numbers continue to rise. This forces more people back into subsistence agriculture, where they draw heavily on the natural resource base and thus degrade it. Austerity programmes inevitably include government cutbacks in both the staff and expenditure of new and unproven environmental agencies, undermining even the minimal efforts being made to bring ecological considerations into development planning. The possibility of addressing serious environmental concerns is thwarted, thus affecting long-term development prospects. The persistence of acute poverty is another dimension of the problem. The population living below a minimally acceptable level has increased in Latin America and Sub-Saharan Africa. Vast poverty persists in Asia.
Claim Economic restructuring measures, although quite successful in getting an economy going (in narrow terms at least), is primarily of advantage to transnational corporations and to consumers of exported products in industrialized countries. The cost is borne by the impoverished who can least afford to pay it. IMF prescriptions are designed by and for the developed capitalist countries and are inappropriate for developing countries of any kind.
Counter-claim Social and human costs of adjustment are transitional.
Refs Feinberg, Richard E and Kallab, Valerina *Adjustment Crisis in the Third World* (1984).
Broader Economic hardship (#PD9180).
Narrower Underutilization of human resources (#PF3523)
Lack of long-term development assistance (#PF5181)
Enforced curtailment of living standards in borrower nations (#PE7136).
Aggravates Hunger (#PB0262)
Natural environment degradation (#PB5250)
Poverty in developing countries (#PC0149)
Inappropriate education in developing countries (#PF1531)
Inadequate domestic savings in developing countries (#PD0465)
Unequal income distribution within developing countries (#PD7615)
Decline in public sector savings in developing countries (#PE4574)
Inadequate research and development capacity in developing countries (#PE4880).
Aggravated by Deprivation (#PA0831)
Economic bias in development (#PF2997)
Short range planning for long-term development (#PF5660)
Excessive foreign public debt of developing countries (#PD2133)
Lack of commitment to the protection of vulnerable groups (#PF4662)
Deterioration in external financial position of developing countries (#PE9567)
Policy cross-conditionality restrictions in multilateral development aid (#PF9216).

♦ PD8122 Water system contamination

Unsafe water — Dirty water — Lack of clean water — Unsanitary water supply — Unsanitary water system — Untreated water supply
Incidence Over 80 per cent of all illness in the developing world is directly or indirectly associated with poor water supply and sanitation. It is estimated that the majority of people in developing countries use domestic water from sources that are subject to contamination. For example, in 5 villages from widely separated parts of India it was found that from 23 per cent to 75 per cent of the people were infected with roundworm, hook worm, pin worm, dwarf tapeworm and intestinal amoebas. Many villagers are consequently sick a great deal of the time. They are unable to absorb all the food the eat because of the damage done by parasites to the intestinal membranes and because part of their food goes to feed the worms and protozoa that infest their intestines. In some Latin American cities 60 per cent of the children die before the age of 5 because of water-borne diseases.
Broader Water pollution (#PC0062)
Pollution of inland waters (#PD1223)
Inadequate water system infrastructure (#PD8517).
Narrower Pollution of groundwater (#PD2503)
Fouling of water supply systems (#PJ6457)
Pollution of water by infected faeces (#PE8545)
Infection of industrial water supplies (#PJ6198)
Disease-causing microbes in drinking water (#PG3400)
Contamination of public water supplies by sabotage (#PE1458).
Aggravates Infant mortality (#PC1287)
Human disease and disability (#PB1044)
Contamination of drinking water (#PD0235).
Aggravated by Impurities in waste water (#PD0482)
Physically inaccessible services (#PC7674)
Lack of essential local infrastructure (#PF2115)
Disrupted mechanisms for community health (#PF2971).

♦ PD8123 Limited verbal skills

Limited language fluency
Claim Children born into large families have poorer verbal skills than children from small families.
Broader Restrictions on the acquisition of knowledge (#PF1319).
Narrower Inadequate national language proficiency (#PG8362).
Related Semilinguism (#PF2789).
Aggravates Corruption of meaning (#PB2619)
Irresponsible expression of emotions equated with free speech (#PF7798).
Aggravated by Profanity (#PF7427)
Speech disorders (#PE2265)
Limited development of functional abilities (#PF1332)
Discrimination against use of accents of a language (#PE5141)
Denial of right of peoples to use their own language (#PE2142).

♦ PD8124 Discriminatory nuclear trade

Broader Obstacles to world trade (#PC4890)
Discriminatory business practices (#PD8913).
Related Super-power monopoly of advanced nuclear warfare technology (#PD4445).

♦ **PD8130 Family stress**
Refs Figley, Charles R and McCubbin, Hamilton I *Stress and the Family* (1983).
Broader Stress in human beings (#PC1648).
Narrower Marital stress (#PD0518)
Stress on families of the physically or mentally handicapped (#PD1405).
Aggravates Juvenile stress (#PC0877) Domestic quarrels (#PE4021).
Aggravated by Inadequate facilities for children's play (#PD0549).

♦ **PD8141 Underemployment in developing countries**
Nature The absence of unemployment benefits forces people to find some way of earning money, for example as a shoeshine boy. Labour in the informal sector is typically characterized by low levels of productivity and income, and must face a high degree of instability of employment. Remuneration tends to be too low to provide workers with an adequate standard of living for themselves and their families.
Related Rural unemployment in developing countries (#PD0295).
Aggravates Wasted woman power (#PF3690)
Decline in real wages in developing countries (#PD2769).

♦ **PD8142 Land misuse**
Broader Restrictive use of available land (#PF6528).
Narrower Unused land (#PF4670) Excessive land usage (#PE5059)
Excessive use of land by automobiles (#PF6152)
Use of agricultural land for fuel production (#PD6103)
Defective land use planning in developing countries (#PD1141)
Alienation of land through acquisition by foreigners (#PE0896)
Increasing proportion of land surface devoted to urbanization (#PE5931).
Aggravates Soil compaction (#PD1416) Soil degradation (#PD1052).
Aggravated by Soil erosion (#PD0949).

♦ **PD8146 Restrictive trade union practices**
Restrictive union practices
Broader Trade unionism (#PF8493) Restrictive practices (#PB9136).
Narrower Union protectionism (#PG6833) Restricted union entry (#PG8925)
Union monopoly and violence (#PG1735)
Trade union opposition to profit-sharing (#PE6722)
Discrimination against non-union workers (#PD6019)
Lack of unionization among working women (#PE8345)
Interference of trade unions with contract performance (#PE8273)
Restrictive trade union policies concerning employment (#PF6046).
Related Discriminatory professionalism (#PC2178).
Aggravates Exploitation in employment (#PC3297).

♦ **PD8149 Hereditary regression**
Atavism — Degeneration
Nature Hereditary reappearance of features, usually physical, absent from parents and recent ancestors. Such features may be characteristic of more distant ancestors, or they may constitute a return to characteristics not typical of the species or race at all, but found in allied species or races having a common evolutionary ancestry (as in the case of strongly pronounced canine teeth in humans).
Counter-claim Atavism does not necessarily represent a deterioration. It may be the re-emergence of valuable genetic characteristics of less-specialized progenitors. Many phenomena have been labelled atavisms which admit of other interpretations such as the consequences of arrested development due to malnutrition, a qualitative variation in some vestigial organ, or the predictable consequences of hybridization.
Broader Genetic defects and diseases (#PD2389).
Related Human racial regression (#PF0411).
Aggravates Decline in human genetic endowment (#PF7815).

♦ **PD8153 Armed crimes against national security**
Broader Crimes against national security (#PC0554).
Narrower Vigilantism (#PD0527) Armed insurrection (#PD8284).

♦ **PD8168 Political unrest**
Incidence In the major oil-producing countries, for example, the probability of major political unrest in the 1990s could lead to significant shortages in world oil supply. The unrest in 1990 in 5 of the oil-producing countries suggest that in the context of a more fragile oil supply-demand equilibrium, the underlying socioeconomic tensions could quickly assume major importance for world oil supplies.
Mounting unrest in black African nations in 1990 signals an era of turmoil and worsening deprivation as the post-colonial political order collapses. There exists a widespread resentment against the corruption and tribalism of authoritarian rulers.
Aggravates Assassination (#PD1971) Terrorist bombing (#PE2368)
Excessive government control (#PF0304).
Aggravated by Newspaper and journal propaganda (#PD0184)
Forced repatriation of prisoners of war (#PD0218)
Religion as a reinforcement of nationalism (#PF3351).

♦ **PD8179 Visual deficiencies**
Vision impairment — Eye disorders — Refractive disorders of the eye
Broader Eye diseases and disorders (#PD8786).
Narrower Myopia (#PE7771) Presbyopia (#PG8968) Astigmatism (#PG9442)
Night blindness (#PG4565) Colour deficiencies (#PE6343)
Impairment of visual-spatial ability (#PG3467).
Aggravates Visually handicapped persons (#PD2542).
Aggravated by Diabetes (#PE0102) Eye irritation (#PE6785)
Prohibitive cost of eye-glasses in developing countries (#PE7548).

♦ **PD8183 Depressed regions in developed countries**
Underdevelopment in developed countries
Broader Underdevelopment (#PB0206)
Deterioration of industrialized countries (#PD9202).

♦ **PD8200 Environmentally induced diseases**
Incidence In three regions of the Asia-Pacific zone there is remarkably high incidence of amyotrophic lateral sclerosis, parkinsonism, and progressive pre-senile dementia. Studies have singled out heavy use of the neurotoxic seed of cycad plants for food and/or medicine as a likely key etiological factor.
Refs Rees, A R and Purcell, H J (Eds) *Society for the Environmental Therapy, Inaugural Conference, 1981.*
Broader Human disease and disability (#PB1044).
Narrower Wind-borne animal diseases (#PG5736)
Poisonous, allergenic and biologically active wood (#PE6867).
Related Hazards to human health in the natural environment (#PC4777).

♦ **PD8201 Least developed countries**
Nature Among the developing countries, the least developed countries (LDCs) are particularly handicapped and ill-equipped to develop their domestic economies and to ensure adequate living standards for their population. Their average GDP per capita is slightly higher than $200, which is less than one quarter of that of the developing countries as a whole and only about 2 per cent of that of the developed market-economy countries. Average agricultural productivity is less than one half that of other developing countries. Only a very small proportion of cultivated land has assured irrigation facilities. Moreover, most of these countries suffer from one or more important geographical or climatological handicaps, such as drought and desertification or natural disasters.
Extremely low per capita GDP levels make LDCs unable to generate domestically the savings needed to finance investment for developing purposes. Inadequate physical, technical and social infrastructures create major bottlenecks which undermine their capacity to broaden and modernize their production structure. Their small domestic markets justify the local production of only limited number of goods, thus adding to their import burden. With a narrow export base, LDCs are at the mercy of the vagaries of world markets for their export earnings. Their share of world exports declined from 1.6 per cent in 1950 to 0.4 per cent in 1985, principally on account of three major factors: structural handicaps (shortage of skilled manpower and of administrative and managerial capacity), an adverse international economic environment (exports are few primary products for which demand increases slowly) and inadequate domestic policies (tendency towards overvaluation of currency). Their marginal place in world economy limits the leverage they have to negotiate technology contracts profitably and to secure access to international capital markets. Their capacity to absorb foreign assistance has in some cases been questioned.
Refs UNCTAD *The Least Developed Countries* (1986).
Broader Underdevelopment (#PB0206).
Narrower Unemployment in least developed countries (#PE9476)
Minimal exports in least developed countries (#PE8306)
Natural disasters in least developed countries (#PE0299)
Geographic barriers for least developed countries (#PE8049)
Minimal manufacturing in least developed countries (#PE0282)
Insufficient forestry in least developed countries (#PE8244)
Inadequate health care in least developed countries (#PE9242)
Inadequate agriculture in least developed countries (#PE8082)
Dependence of least developed countries on foreign aid (#PE8116)
Environmental destruction in least developed countries (#PE8401)
Underutilization of livestock in least developed countries (#PE8595)
Inadequate nutrition education in least developed countries (#PE0265)
Deterioration of the quality of life in least developed countries (#PE7734)
Insufficient use of natural resources in least developed countries (#PE0273)
Inadequate infrastructure and services in least developed countries (#PE0289)
Inadequate human resources development in least developed countries (#PE9764)
Inadequate investment of transnational corporations in least developed countries (#PE7892).

♦ **PD8202 Property speculation**
Refs Sandercock, Leonie *Land Racket* (1979).
Broader Capitalist speculation (#PC2194) Unfavourable capital risk (#PF6045).
Aggravates Segregation in housing (#PD3442) Accumulation of property (#PC8346).

♦ **PD8204 Social invisibility**
Unrecognized people — Invisible people
Incidence So called guest workers are often treated as if they do not exist or have any feelings.
Narrower Children neglected by teachers (#PD3722).
Aggravates Ostracism (#PF1009).
Aggravated by Social outcasts (#PD6017) Officially nonexistent people (#PF7239).

♦ **PD8211 Insecurity of employment**
Job insecurity
Broader Economic insecurity (#PC2020) Lack of job satisfaction (#PF0171).
Narrower Job insecurity for pregnant women (#PE5316).
Aggravates Problems of migrant labour (#PC0180).
Aggravated by Seasonal unemployment (#PC1108)
Discontinuity of employment (#PD4461)
Fluctuations in real value of money (#PD9356)
Coercive use of economic power by transnational enterprises against labour (#PE0207).

♦ **PD8217 Mistrust of system of justice**
Lack of confidence in administration of justice
Aggravated by Inequitable administration of justice (#PD0986)
Use of undue influence to obstruct the administration of justice (#PE8829).

♦ **PD8227 Misconduct in public office**
Unethical behaviour by leaders — Ethical misconduct of leaders — Moral imperfections of leadership
Nature Politicians are known to have taken bribes, payoffs, letting companies or other outside interests to pay speaking fees, trips, accommodation, "gifts" and "income" from joint business ventures.
Claim Ethics cannot be enforced by laws only, because in a legal-minded society too often bad laws legitimize any activity not made illegal and because the norms must be absorbed.
Refs Noorani, A G *Ministerial Misconduct* (1972); O'Brien, David W *Misconduct Cases Book* (1985).
Broader Ethical decay (#PB2480) Statutory crime (#PC0277)
Moral imperfection (#PB7712).
Narrower Unparliamentary behaviour (#PF4550)
Government leaders associated with sex scandals (#PE7937).
Related Corruption of government leaders (#PC7587).
Aggravates Deception by government (#PD1893)
Inability to define moral standards (#PF7178).

♦ **PD8239 Chronic illness**
Chronic diseases
Nature Chronic diseases span many months, even years intruding upon the lives of patients and their families. Many chronic diseases are uncertain in prognosis, phasing the disease, and response to treatment. Cure is problematic or impossible, so treatment concentrates in making the patient comfortable or relieving pain. Often patients suffer from multiple diseases or side effects of medical treatments lead to additional chronicity. Long-time use of drugs, routine monitoring and crisis requiring hospitalization make chronic diseases expensive.
Incidence In USA it is estimated that approximately half of the population suffers from one or more chronic illnesses, such as heart conditions, arthritis, rheumatism, and impairments of back and spine.
Refs Burish, Thomas G and Bradley, Laurence A *Coping with Chronic Disease* (1983); Chilman, Catherine S, et al (Eds) *Chronic Illness and Disability* (1988); Locker, David *Disability and Disadvantage* (1984); Mehl, Lewis E *Mind and Matter* (1986); Stein, Ruth *Care of Children*

with Chronic Illness (1988).
Broader Intractable diseases (#PC8801).
Narrower Chronic cough (#PG4472) Chronic fatigue (#PD4374)
Chronic bronchitis (#PE2248) Chronic terminal illness (#PE4906)
Chronic pelvic inflammatory disease (#PG2024).
Aggravates Chronic pain (#PE2694) Mental depression in children (#PD3784).
Aggravated by Alcohol abuse (#PD0153).

♦ PD8262 Emotional insecurity
Lack of emotional security
Refs English, O Spurgeon and Pearson, Gerald H J *Emotional Problems of Living*.
Broader Insecurity (#PA0857) Psychological impediments to marriage (#PF3344).
Aggravates Emotional crises (#PE3407).
Aggravated by Personal life crises (#PD4840) Emotional immaturity (#PJn907)
Insecurity of property (#PC1784).

♦ PD8276 Denial of political rights
Lack of political liberty
Broader Violation of human rights (#PB3860).
Narrower Undemocratic policy-making (#PF8703)
Limitations on right to vote (#PF2904)
Denial of right of complaint (#PD7609)
Denial of right to hold public office (#PE5608)
Denial of right to participate in government (#PE6086).
Related Political discrimination in politics (#PC3221).
Aggravates Insecurity (#PA0857).
Aggravated by Excessive government control (#PF0304)
Inaccessible decision makers (#PF2452)
Language discrimination in politics (#PD3223)
Religious discrimination in politics (#PC3220)
Ideological discrimination in politics (#PC3219)
Political discrimination based on illiteracy (#PC3222).

♦ PD8284 Armed insurrection
Insurrection — Insurgency
Nature It is a crime to engage in armed insurrection with intent to overthrow, supplant or change the form of government of a legally constituted government. It is equally a crime to participate in the leadership of such an insurrection or to organize, provide substantial support or advocate it.
Refs Fauriol, Georges (Ed) *Latin American Insurgencies* (1985); Khairallah, David *Insurrection Under International Law* (1973); O'Neill, Brad E, et al (Eds) *Insurgency in the Modern World* (1980).
Broader Civil violence (#PC4864) Seizure of power (#PC8270)
Armed crimes against national security (#PD8153).
Aggravates Revolution (#PA5901) Civil disorders (#PC2551).

♦ PD8285 Homelessness in industrialized countries
Homeless in developed countries
Incidence In the USA it is estimated that 500,000 young Americans live in the streets throughout the year. Almost all have been sexually or physically abused. Most cannot write well enough to fill out a job application and would not know how to go to work if they obtained a job. Many will die in the streets or in prison, in some cases after killing others. In the UK in 1990, it was estimated that the level of homelessness had doubled to over 300,000 (126,000 households) in 11 years. Between 1982 and 1989, the number of households in temporary accommodation quadrupled. Such figures only cover those who apply to local authorities and do not include those not having a priority need.
Claim Practices like real estate speculation treat property as a market commodity without social value and push up the prices so that the poorest cannot afford a home.
Refs Torrey, E Fuller *Nowhere to Go* (1988).
Broader Homelessness (#PB2150)
Deterioration of industrialized countries (#PD9202).

♦ PD8286 Consumption of alcoholic beverages
Manufacture of alcoholic beverages — Trade in alcoholic beverages
Incidence Total production of alcoholic rose by almost 50 per cent between 1965 and 1980. Two-thirds of the world's output of alcohol during this time was produced in Europe, the USA, and Canada. Over the period alcohol consumption increased very rapidly in Western countries, but also in Japan, Mexico and Korea Rep. International trade in alcoholic beverages accounts for less than 1 per cent of the total. Nevertheless, in monetary terms, this amounted to almost $8,730 million in 1980. The main trade flows are between relatively wealthy developed countries.
Broader Unnecessary personal consumption (#PF5931).
Narrower Alcohol abuse (#PD0153) Medicines containing alcohol (#PE2047)
Susceptibility of women to alcohol (#PE7161).
Aggravates Haemorrhoids (#PE3504)
Lack of responsible involvement in community affairs (#PF6536).
Aggravated by Immoral public policy (#PF4753) Lack of self-confidence (#PF0879)
Unethical consumption practices (#PD2625)
Illicit production of alcoholic beverages (#PE7188)
Excessive portrayal of substance abuse in the media (#PE3980)
Limited availability of public services in the small towns of developed countries (#PF6539)
Dependence of government revenues on exploitation of environmentally inappropriate products (#PD1018).

♦ PD8290 Unethical practice of radiology
Irresponsible radiologists — Negligence by radiologists — Malpractice in radiology — Corruption of radiologists — Underreporting of radiation risks
Claim Radiologists, under pressure from their employers, have adopted practices which lead to the underreporting of radiation hazards, failure to adjust radiation thresholds in the light of potential hazards, and failure to investigate adequately the nature of such hazards. Examples include hazards of long-term exposure to low-level radiation in nuclear installations, improper recommended X-ray dosages for medical purposes, and long-term consequences of exposure to nuclear weapons tests. In each case radiologists have denied or suppressed evidence indicative of the levels of danger.
Broader Unethical professional practices (#PC8019).
Aggravates Uncertain toxicity thresholds (#PF5188)
Abusive experimentation on humans (#PC6912)
Inadequate radiation monitoring systems (#PF6635)
Long-term hazards of exposure to radiation (#PE4057)
Excessive occupational exposure to radiation (#PD1500)
Excessive exposure of medical patients to radiation (#PE1704)
Excessive exposure to radiation from consumer goods and electronic devices (#PE1909)
Environmental hazards of extremely low frequency electromagnetic radiation (#PE7560).

♦ PD8294 Inappropriate basic hygiene
Broader Unhygienic conditions (#PF8515).
Narrower Inadequate personal hygiene (#PD2459).
Aggravates Iatrogenic disease (#PD6334).
Aggravated by Ineffective systems of practical education (#PF3498).

♦ PD8297 Amnesia
Loss of memory
Nature Memory disorders usually result following brain surgery, encephalitis (brain inflammation), electroconvulsive therapy, senility, or physical injury, or it can be of psychogenic origin. Those of the first case may well resemble those of alcoholic psychosis known as Korsakoff's syndrome. However the individual in both cases recognizes the memory difficulty. Frequent electroconvulsion sometime leads to exaggerated forgetfulness for day-to-day events. Psychogenic amnesia is a defensive process that operates continuously to cope with emotionally disturbing experiences.
Analogous phenomena may occur in the case of collective memory as a result of physical destruction of portions of a manual or electronic information system, uncontrolled hyper-development of such a system, or the subjection of such a system to frequent overload beyond the breakdown threshold.
Refs Butters, Nelson and Cermak, Laird S *Alcoholic Korsakoff's Syndrome* (1980); Mayes, Andrew R *Human Organic Memory Disorders* (Date not set); Parkin, Alan J *Memory and Amnesia* (1987).
Broader Memory defects (#PD8484) Mental illness (#PC0300)
Diseases and injuries of the brain (#PD0992).
Narrower Global amnesia (#PF0306) Hypnotic amnesia (#PE0314)
Infantile amnesia (#PG0325) Traumatic amnesia (#PE0357)
Retrograde amnesia (#PE0354) Hysterical amnesia (#PE0321)
Korsakoff syndrome (#PE0333) Alternating amnesia (#PE0298)
Transient global amnesia (#PE0355) Post-traumatic stress disorder (#PE0351).
Related Paramnesia and confabulation (#PE0349)
Symptoms referable to nervous system (#PE9468)
Inadequate sense of personal identity (#PF1934).
Aggravated by Senility (#PE6402) Alcohol abuse (#PD0153).

♦ PD8301 Gland disorders
Glandular disorder — Nutritional and metabolic diseases endocrine — Endocrine deficiency — Diseases of the endocrine glands — Endocrine gland disturbances
Refs Mendelsohn, Geoffrey, et al *Diagnosis and Pathology of Endocrine Diseases* (1988); Shah, Nandkumar S and Donald, Alexander G *Psychoneuroendocrine Dysfunction* (1984).
Broader Human disease and disability (#PB1044).
Narrower Diabetes (#PE0102) Ovarian dysfunction (#PG6775)
Testicular dysfunction (#PG2456) Thyroid gland disorders (#PE0652)
Diseases of thymus gland (#PE7708) Pituitary gland disorders (#PE2286)
Parathyroid gland disorders (#PG2721) Dysfunction of the adrenal glands (#PG5181)
Benign neoplasm of endocrine glands (#PG7527).
Related Diseases of metabolism (#PC2270).
Aggravates Obesity (#PE1177) Hypertension (#PE0585) Nutritional anaemia (#PD0321).
Aggravated by Heart diseases (#PD0448).

♦ PD8314 Lack of facilities for the physically disabled
Broader Inadequate medical facilities (#PD4004).
Narrower Lack of facilities for severely deformed people (#PD0211).
Related Inadequate public facilities (#PG9297).

♦ PD8323 Political instability of developing countries
Nature Frequent changes in governments, or attempts thereof, internal dissensions and labour disputes and involvement in border clashes with neighbouring countries contribute to the political instability of many developing countries.
Broader Political instability (#PC2677).
Aggravates Violation of civil rights (#PC5285) Abusive traffic in immigrant workers (#PD2722).
Aggravated by Dissidents (#PC9695) Deprivation (#PA0831)
Socio-economic burden of militarization (#PF1447)
Declining economic growth in developing countries (#PD5326).

♦ PD8328 Aircraft environmental hazards
Aviation risks — Flying hazards — Aircraft hazards — Natural hazards to air traffic
Nature Natural hazards to airborne traffic include adverse wind conditions, lightning, poor visibility, freezing temperatures, hail, birds, volcanic ash clouds and snow weight. Runway hazards of a natural kind include flooding and icing, large animals, storm debris, snow drifts and, in arid regions, sand drifts.
Broader Risk (#PF7580).
Narrower Aerial piracy (#PD0124) Aircraft noise (#PE5799)
Unsafe aircraft (#PE1575) Icing on aircraft (#PE8059)
Aircraft harassment (#PE6006) Pollution by aircraft (#PE4802)
Aircraft pilot fatigue (#PE3870) Unidentified flying objects (#PF1392)
Threat of birds to aircraft safety (#PD1111).
Related Air accidents (#PD1582)
Environmental hazards from transport equipment (#PE0738).
Aggravated by Fog (#PE1666) Deterioration in atmospheric visibility (#PE2593).

♦ PD8329 Illiteracy in developing countries
Claim In developing countries, illiteracy is a major obstacle to sustainable development. At current rates of population and education growth, there will be more than 900 million illiterate people by the year 2000.
Broader Illiteracy (#PC0210).
Narrower Illiteracy among indigenous peoples (#PD3321)
Illiteracy in least developed countries (#PE8978)
Illiteracy among women in developing countries (#PE8660).

♦ PD8334 Inaccessibility of justice
Lack of access to justice
Broader Injustice (#PA6486).
Narrower International inaccessibility of justice (#PD5314).
Aggravated by Prohibitive legal fees (#PF0995) Lack of legal aid facilities (#PF8869).

♦ PD8340 Juvenile desertion
Runaway adolescents — Runaway youth — Runaway children
Incidence Driven by drug abuse, sexual or physical abuse, financial pressures and a growing number of single-parent families, approximately half a million adolescents are running away from or being forced out of their homes every year in the US.
Refs Janus, Mark D, et al *Adolescent Runaways Causes and Consequences* (1987).
Broader Abandonment (#PA7685) Missing persons (#PD1380)
Offences against public order (#PD7520).

PD8340

Aggravates Missing children (#PD6009)
Disowned children (#PJ0827)
Aggravated by Juvenile stress (#PC0877).
White slave trade (#PD3303)
Student absenteeism (#PE4200).

♦ PD8347 Benign tumours
Benign neoplasms
Broader Neoplastic diseases (#PC3853).
Narrower Haemangioma (#PG7810)
Benign neoplasm of ovary (#PE0368)
Benign neoplasm of uterus (#PE7922)
Benign neoplasm of nervous system (#PG5284)
Benign neoplasm of throat and mouth (#PG4748)
Benign neoplasm of digestive system (#PG4117)
Benign neoplasm of endocrine glands (#PG7527)
Benign neoplasm of respiratory system (#PE3637)
Benign neoplasm of bone and cartilage (#PG0819)
Benign neoplasm of kidney and other urinary organs (#PE9086).
Benign neoplasm of skin (#PG9649)
Benign neoplasm of breast (#PG5371)
Benign neoplasm of genital organs (#PE9705)
Aggravates Malignant neoplasms (#PC0092).

♦ PD8359 Industrial destruction by war
Broader Disastrous consequences of war (#PC4257).
Related War casualties (#PD4189)
Housing destruction in war (#PE2592)
Poverty as a consequence of war (#PE5252)
Prisoners of war (#PC8848)
War damage in civilian areas (#PD8719)
Environmental consequences of war (#PC6675).

♦ PD8360 Animals as vectors of disease
Animals as transmitters of disease — Animal vectors of disease
Broader Disease vectors (#PC3595).
Narrower Birds as vectors of disease (#PE6659)
Wild animals as carriers of animal diseases (#PD2729)
Domestic animals as carriers of animal diseases (#PD2746).
Aggravates Mycosis (#PE2455)
Virus diseases (#PD0594).
Aggravated by Zoonoses (#PD1770)
Human disease vectors (#PD6651).
Faecal transmission of disease (#PG4438)
Toxoplasmosis (#PE3659)
Vampire bats (#PE1890)
Health risks to workers in agricultural and livestock production (#PE0524).

♦ PD8362 Limited purchasing power
Insufficient buying power — Limited buying power
Broader Insufficient financial resources (#PB4653).
Confined scope of business operations (#PF2439)
Strained capital resources in small communities (#PF3665).
Narrower Lack of purchasing power in developing countries (#PD8707).
Aggravates Malnutrition (#PB1498).
Aggravated by Rural unemployment (#PF2949)
Underdeveloped capacity for income farming (#PF1240)
Ineffective means for goods supply and distribution (#PF6495).

♦ PD8371 Man as vectors of disease
Broader Disease vectors (#PC3595).
Narrower Human vectors of animal diseases (#PD2784)
Dissemination of plant diseases by man (#PD3593)
Vulnerability of intravenous drug users to AIDS (#PE5798)
Infected animal, meat and animal product shipments (#PE7064).
Aggravates Virus diseases (#PD0594).

♦ PD8375 Psychological disorders
Refs Ross, Alan O *Psychological Disorders of Children* (1979).
Broader Human disease and disability (#PB1044).
Narrower Mental disorders (#PD9131)
Occupation handicap (#PE6778)
Psychological conflict (#PE5087)
Physical independence handicap (#PE6773)
Tumours of the external ear in animals (#PG8705).
Related Frustration (#PA2252).
Aggravated by Human sexual inadequacy (#PC1892)
Chemical imbalances in the brain (#PE4715).
Enjoyment of war (#PF4034)
Orientation handicaps (#PE6772)
Social integration handicap (#PE6779)
Economic self-sufficiency handicap (#PE6780)

♦ PD8380 Corruption of ruling classes
Broader Institutionalized corruption (#PC9173).
Narrower Unethical practices by public service employees (#PE6702).
Related Corruption of government leaders (#PC7587).
Aggravates Congenital anomalies of musculoskeletal system (#PE8589)
Denial of right to periods to nurse infants during working hours (#PE3044)
Use of undue influence to obstruct the administration of justice (#PE8829).

♦ PD8385 Vector–borne diseases
Nature A vector is a transmitter of disease from one animal to another and to man. For many diseases, no effective immunizing agent is at present available and for others, specific chemo-therapeutic agents do not exist or have serious disadvantages. A number of these vectors are becoming resistant to pesticides and insecticides.
Incidence Most of the vector–born disease are found in tropical areas of the world; however, some are distributed to the Arctic. The most important of vector–born disease is malaria of which there are an estimated 400 million cases. Other important disease are sleeping sickness, the vector is the tsetse fly; Chagas disease, the vector is a triatomid bug; onchocerciasis which is transmitted by a black fly, the Simulium; Bancroftian and Brugian Filariases which is transmitted by various mosquitoes; dengue and dengue haemorrhagic fever transmitted by a specific type of mosquito; yellow fever transmitted by a genus of mosquitoes; and schistosomiasis or bilharzia is transmitted by a small snail living in water.
Broader Disease (#PA6799).
Aggravates Yellow fever (#PE0985)
Elephantiasis (#PE1601)
Chagas' disease (#PE0653)
Aggravated by Disease vectors (#PC3595)
Pest resistance to pesticides (#PD3696)
Irresponsible introduction of new species of fish (#PF3602).
Dengue fever (#PE2260)
Onchocerciasis (#PE2388)
Schistosomiasis (#PE0921).
Domestic refuse disposal (#PD0807)

♦ PD8388 Lack of training
Broader Lack of education (#PB8645)
Lack of opportunities for practical training in communities (#PF2837).
Narrower Undomesticated men (#PF0551)
Limited medical knowledge (#PD9160)
Lack of apprentice training (#PJ8498)
Untrained clerical personnel (#PG8787)
Underdeveloped farming skills (#PJ0729)
Inadequate training in decision–making (#PD2036).
Aggravates Insufficient skills (#PC6445)
Inadequate education (#PF4984)
Restricted job training (#PJ0617)
Lack of agro–urban training (#PG7992)
Untrained security personnel (#PG8069)
Insufficient special training (#PG9609)
Limited agricultural education (#PF8835)
Incompetent management (#PC4867)
Shortage of skilled labour (#PD0044)
Inadequate practical training in rural areas (#PF6472)
Health risks to workers in agricultural and livestock production (#PE0524).
Aggravated by Unimaginative educational vision (#PF3007)
Nonavailability of technical training (#PJ0121)
Lack of local services for community leadership training (#PF2451).

♦ PD8392 Fraud by government agents
Fraud by government bureaucrats — Embezzlement by civil servants — Corruption practised by government agents — Official profiteering
Broader Fraud (#PD0486)
Corruption of government leaders (#PC7587).
Related Diplomatic fraud (#PD5554).
Aggravated by Governmental incompetence (#PF3953).
Profiteering (#PC2618)

♦ PD8399 Dependency of the elderly
Dependence of the aged
Refs Munnichs, J M A and Heuwel, J A van den (Eds) *Dependency or Interdependency in Old Age* (1976); Phillipson, Chris, et al (Eds) *Dependency and Interdependency in Old Age* (1986).
Broader Lack of protection for the vulnerable (#PB4353).
Aggravates Being a burden (#PF9608)
Aggravated by Human ageing (#PB0477)
Ageing populations (#PB8561).
Neglect of the aged (#PD8945).
Incontinence (#PE4619)

♦ PD8402 Misuse of medicines
Broader Inadequate medical care (#PF4832)
Narrower Analgesia (#PG2701)
Overprescription of drugs (#PE9087)
Excessive proliferation of medical drugs (#PD0644).
Aggravates Human poisoning (#PD0105).
Aggravated by Misdiagnosis (#PF8490).
Inhumane scientific activity (#PC1449).
Irritating drugs (#PJ4491)
Violent purgatives (#PG6554)
Prescription of inappropriate drugs (#PE3799)

♦ PD8405 Trafficking in children
Sale of children — Enslavement and exploitation of minors
Nature Children are bought from mothers, especially in countries where there is great poverty, who, perhaps, hope their children will have a better future. Children are also kidnapped to be sold for the purposes of economic exploitation, criminal activities, begging, child pornography or child prostitution, medical experimentation and organ transplants and adoption.
Incidence In Romania babies were sold for export to earn foreign currency. Foreigners who wanted to adopt children had to pay thousands of dollars different expenses and bribes.
Broader Slave trade (#PC0130) Traffic in persons (#PC4442).
Narrower Trafficking in children for adoption (#PF3302)
Trafficking in children for sexual exploitation (#PE6613)
Trafficking in children for medical exploitation (#PD4271)
Trafficking in children for economic exploitation (#PD7266).

♦ PD8412 Naval arms race
Maritime nuclear weapons systems
Refs Anthony, Ian *The Naval Arms Trade* (1990); Fieldhouse, Richard (Ed) *Security at Sea* (1990); Fieldhouse, Richard and Taoka, Shunji *Superpowers at Sea* (1989); United Nations *The Naval Arms Race* (1986).
Broader Competitive acquisition of arms (#PC1258).
Aggravates Naval warfare (#PJ7824)
Marine disposal of obsolete weapons (#PD7574)
Obstacles to unilateral nuclear disarmament (#PF7052).
Marine accidents (#PD8982)

♦ PD8418 Arid developing countries
Refs Burley, Jeffrey *Obstacles to Tree Planting in Arid and Semi–Arid Lands* (1983).
Broader Infertile land (#PD8585)
Geographically disadvantaged countries (#PF9247).

♦ PD8422 Underpayment of government officials
Underpaid bureaucrats — Underpayment of the judiciary — Limited public salaries
Broader Underpayment for work (#PD8916).
Narrower Underpayment of police (#PJ8601).
Aggravates Bureaucratic corruption (#PC0279)
Ineffective use of external relations relating to sportsmen (#PE6515).
Aggravated by Excessive salaries of corporate executives (#PF7578)
Disparity in remuneration between public and private sector employees (#PE7760).
Abuse of government employment (#PE4658)

♦ PD8426 Animal pests
Animal pest control
Refs Drummond, R O and Kunz, S E (Eds) *Arthropod Pests of Livestock* (1988).
Broader Pests (#PC0728).
Narrower Birds as pests (#PD1689)
Protozoa as pests (#PE6741).
Related Insect pests (#PC1630)
Disastrous insect invasions (#PD4751).
Aggravates Pests and diseases of rice (#PE2221)
Endangered species of animals (#PC1713).
Aggravated by Irresponsible introduction of new species of animals (#PD1290).
Rodents as pests (#PE2537)
Insect pests of plants (#PD3634)

♦ PD8438 Piracy at sea
Incidence The International Maritime Bureau has estimated that at least one pirate attack per week is likely in the Strait of Malacca, the key strategic passage between the Pacific and Indian Oceans.
Refs Akindele, R A and Vogt, M A (Eds) *Smuggling and Coastal Piracy in Nigeria* (1983); International Chamber of Commerce *Piracy at Sea*.
Broader Piracy (#PD1877).
Aggravates Refugees by boat (#PD8034).

♦ PD8439 Restrictive religious practices
Broader Restrictive practices (#PB9136).
Related Peoples perceiving themselves as specially chosen (#PF4548).

♦ PD8445 Denial of religious liberty
Denial of right to freedom of religion — Denial to right of freedom of belief
Nature Denial of religious liberty takes many forms. An individual may be denied the right to manifest his religion in public or in private. He may be forbidden to have a specific religion. He may be denied the right to change religious belief. He may be denied the right worship, teach or practice as his religious belief dictates. Parents may be denied the right to provide their children with religious or moral education consistent with their own. It is in intruding on a person's relationship with the ultimate realities of his existence that the greatest danger to his self understanding and to his culture is present.

Claim Monotheistic religions such as Judaism, Christianity and Islam are guilty of violating the conscience of people of other faiths. Lasting peace in a number of regional conflicts will only be attained when world religions openly declare that religious liberty is a basic human right. No true democracy can be established when the manifestation of any religion in teaching, practice, worship and observance is denied. The denial leads in fact to inhuman superiority, ugly oppression, blatant racism, and homicide.
Counter–claim Any religious practice that endangers pubic safety, order, health, or morals or infringes on the fundamental rights and freedoms of others must be limited.
Refs Swidler, Leonard *Religious Liberty and Human Rights in Nations and in Religions* (1986).
Broader Religious intolerance (#PC1808) Violation of civil rights (#PC5285)
Denial of right to liberty (#PF0705).
Narrower Forced religious conversion (#PD6637)
Denial of right to change religion (#PE6397)
Denial of right to manifest religion (#PF2850)
Discrimination against women in religion (#PD0127)
Denial of right to freedom of religion of indigenous peoples (#PE4332).
Related Religious repression (#PC0578) Denial of right to life (#PD4234)
Denial of freedom of opinion (#PD7219)
Denial of right to correct misinformation (#PE7349).
Aggravates Apostasy (#PE9018).

♦ **PD8449 Business bribery**
Nature Although bribes and influence peddling have always been part of the international and local business scenes for thousands of years, sophisticated business cheating is on the increase. It is not only the criminal elements involved in bribery but well–placed individuals, high government officials and reputable corporations. It is a global problem caused by greed. Relegation of religious and ethical values to the background, the desire for material wealth and glorification of unethical practices through the media have all made their contributions. Profit has become an end in itself. Public relation campaigns with their business junkets, gifts to potential buyers or legislators, donations to political parties have turned grey and debatable ethical considerations. Ethical judgements are nearly impossible. A history of bribery and influence peddling during the colonial period or the purposes of maintaining power and profiteering eroded the ethics of both the colonial power and the colony. Object poverty set in the daily context on the media of wealth and power further erode ethical standards.
Broader Bribery (#PC2558) Unfair competition (#PC0099)
Unethical commercial practices (#PC2563).
Narrower Bribery by transnational enterprises in developing countries (#PE0322).
Related Concealed government subsidies (#PD4532).

♦ **PD8460 Discrimination in public services**
Broader Discrimination (#PA0833).
Narrower Racial discrimination in public services (#PD3326)
Discrimination against men in public services (#PE8507).

♦ **PD8461 Intracerebral and other intracranial haemorrhage**
Broader Diseases and injuries of the brain (#PD0992).

♦ **PD8465 Covert violation of international treaties**
Collusion between governments
Broader Unfulfilled treaty obligations (#PF2497).
Aggravates Limited acceptance of international treaties (#PF0977)
Non–verifiability of compliance with nuclear arms treaties (#PF4460).
Aggravated by Government treachery (#PF4153)
Secret international agreements (#PF0419)
Ineffective international agreements (#PF6992).

♦ **PD8468 Politicization of decision–making**
Unscientifically based policy formulation — Selective avoidance of facts in decision making
Aggravates Inappropriate policies (#PF5645).

♦ **PD8469 Loss of animal productivity**
Lack of productivity in farm animals
Broader Declining economic productivity (#PC8908).
Narrower Economic loss through slaughter of diseased animals (#PE8109)
Economic loss through reduced productivity of diseased animals (#PE8098).
Related Unproductive labour resources (#PC6031)
Declining productivity of agricultural land (#PD7480).
Aggravated by Inadequate housing and penning of domestic animals (#PE2763).

♦ **PD8484 Memory defects**
Memory abnormalities
Nature Defects of the memory can be divided into two broad types, organic and psychogenic amnesia. Memory defects of an organic nature may be transitory or enduring and include: Transitory global amnesia is an abrupt loss of memory lasting a few seconds to several hours without the loss of consciousness and is usually the result of a transient blockage of the blood supply to the brain. Traumatic amnesia lasts from a few minutes to weeks and is usually the result of a blow to the head. Post traumatic automatism results from a light blow to the head and is the loss of memory of the period following the blow the without loss of consciousness and without a change of behaviour. Memory loss do to electroconvulsive therapy. Korsakoff syndrome may be do to chronic alcoholism, a variety of toxic and infectious brain illnesses, nutritional disorders as deficiency of the B vitamins and cerebral tumours. The main psychological feature is the gross defect in recent memory, sometimes producing moment to moment consciousness where people story new information for a few seconds and report no continuity between one experience and the next. In addition the person experiences retrograde amnesia, remember experiences they never had which is called confabulation, and deny their memory is in any way affected. Encephalitis may cause memory defects closely resembling Korsakoff's syndrome except the person has good insights into his memory loss and generally does not confabulation. Brain surgery to the temporal lobes causes a degree of memory loss. Diffuse diseases of the brain such as senility and cerebral arteriosclerosis may cause impairment of the memory. Psychogenic amnesia can be hypnotic amnesia or hysterical amnesia. Hysterical amnesia is of two types: the failure to recall specific past events or a period of one's life and the failure to register and therefore the inability to recall events in one's ongoing life.
Another form of memory abnormality is hyperamnesia, the enhancement of the memory function.
Refs Gadow, Kenneth D and Swanson, H Lee *Memory and Learning Disabilities* (1987); Kihlstrom, J F and Evans, F J (Eds) *Functional Disorders of Memory* (1979).
Broader Diseases and injuries of the brain (#PD0992).
Narrower Amnesia (#PD8297) Absent mindedness (#PF0424)
Paramnesia and confabulation (#PE0349).
Related Chemical imbalances in the brain (#PE4715).
Aggravated by Health hazards of aluminium (#PE4969).

♦ **PD8486 Killing of animals**
Killing of animals by humans — Slaughter of animals by humans
Broader Killing by humans (#PC8096) Hunting of animals (#PC2024)
Killing non–human life (#PF6359).
Narrower Hunting tourism (#PE3008) Ritual slaughter of animals (#PF0319)
Inhumane killing of animals (#PE0358) Criminal killing of animals (#PJ1158)
Slaughter of animals for pelts (#PD4575) Pollution–induced fish diseases (#PE7584)
Inhumane killing of stray animals (#PE2759)
Extermination of wild animal natural prey (#PD3155).
Aggravates Lack of slaughter facilities (#PE9767)
Economic loss through slaughter of diseased animals (#PE8109).
Aggravated by Excessive consumption of animal flesh (#PD4518)
Abuse in control of wild animal populations (#PE7995).

♦ **PD8488 Professional jealousy**
Broader Jealousy (#PF5013).
Aggravates Professional secrecy (#PD6576) Prevention of the exchange of ideas (#PD8731).
Aggravated by Conceptual plagiarism (#PD1284).

♦ **PD8492 Exploitative property development**
Broader Unethical commercial practices (#PC2563)
Destruction inherent in development (#PF4829).
Narrower Exploitation in housing (#PD3465).
Aggravated by Eugenics (#PC2153).

♦ **PD8517 Inadequate water system infrastructure**
Insufficient water delivery — Inadequate water services — Inadequate water control — Inadequate watermains locations — Lack of water systems — Unmarked public water outlets
Broader Inadequate infrastructure (#PC7693)
Inadequate dissemination and use of available information (#PF1267)
Underdeveloped provision of basic services in developing countries (#PF6473).
Narrower Water system contamination (#PD8122)
Fluoridation of drinking water (#PE2871)
Mismanagement of irrigation schemes (#PE8233)
Deterioration in water supply systems (#PD9196)
Inadequate facilities for the transport of water supplies (#PD1294).
Aggravates Maldistribution of water (#PD8056) Inadequate care of community space (#PF2346)
Reluctant claims on external resources (#PF1226)
Inaccessibility of water for recreation (#PF6138)
Overexploitation of underground water resources (#PD4403).
Aggravated by Water softness (#PE0199) Water pollution (#PC0062)
Water salinization (#PE7837) Rising water level (#PD8888)
Silting of water systems (#PD3654) Impurities in waste water (#PD0482)
Instability of water supply (#PD0722) Lack of accord on water use (#PF4839)
Long–term shortage of water (#PC1173) Prohibitive irrigation costs (#PG8730)
Shortage of fresh–water sources (#PC4815) Eutrophication of lakes and rivers (#PD2257)
Fragmented planning of community life (#PF2813)
Conflicting claims over shared inland water resources (#PD7087)
Inadequate water supply in the rural communities of developing countries (#PD1204).
Reduces Health hazards from water development schemes (#PE8692).

♦ **PD8522 Demonstrations**
Protest meetings
Refs Mason, Henry L *Mass Demonstrations Against Foreign Regimes* (1966).
Broader Civil disobedience (#PC0690).
Narrower Violent demonstrations (#PG5872).
Aggravates Civil violence (#PC4864) Civil disturbances (#PD5372)
Prevailing community insecurity (#PD9044) Violent repression of demonstrations (#PD4811)
National insecurity and vulnerability (#PB1149).
Aggravated by Grievance (#PF8029) Passive resistance (#PF2788)
Inadequate riot control (#PD2207) Inhumane methods of riot control (#PD1156).
Reduces Lack of community participation (#PF3307)
Failure of individuals to participate in social processes (#PF0749).

♦ **PD8523 Misinformation**
Broader Miseducation (#PA6393) Uncommunicativeness (#PA7411).
Aggravates Ignorance (#PA5568) Mental pollution (#PB6248)
Exploitation of dependence on food aid (#PD7592).

♦ **PD8529 Inappropriate education**
Broader Inadequate education (#PF4984).
Narrower Impractical education (#PF3519) Educating people to lie (#PE3909)
Inappropriate role learning (#PG5343) Inappropriate education of graduates (#PF1905)
Inappropriate education in developing countries (#PF1531)
Lack of appropriate education for children of immigrants (#PE5020).
Aggravates Underemployment (#PB1860) Uncritical thinking (#PF5039)
Uncritical acceptance of authority (#PF8596)
Uncritical preservation of the status quo (#PF1688)
Uncritical acceptance of dogmas and standards (#PF2901)
Inappropriate selection and examination procedures in education (#PF1266).
Aggravated by Rote learning (#PJ5437)
Inability of educational systems to keep pace with technological advancement (#PF7806).

♦ **PD8544 Disruptive behaviour**
Unchecked destructive behaviour
Broader Human destructiveness (#PA0832) Behavioural deterioration (#PB6321).
Narrower Disruptive behaviour in schools (#PE9092).
Related Loss of civility (#PC7013).
Aggravates Human violence (#PA0429).

♦ **PD8546 Failure to employ skills of home–bound educated women**
Domestic deskilling of women
Broader Educated unemployed (#PD8550)
Underutilization of locally available skills (#PF6538).
Related Obsolete vocational skills (#PD3548).
Aggravated by Overworked women (#PD7762) Social isolation of women at home (#PE8681)
Lack of opportunities for practical training in communities (#PF2837).

♦ **PD8549 Mismanagement in developing countries**
Broader Mismanagement (#PB8406) Incompetent management (#PC4867).
Narrower Soil mismanagement in developing countries (#PD6482)
Inadequate management skills in rural communities (#PF1442).
Aggravates Undirected expansion of economic base (#PF0905)
Excessive foreign public debt of developing countries (#PD2133).
Aggravated by Lack of management skills in developing countries (#PE0046).

♦ **PD8550 Educated unemployed**
Refs Morio, Simone and Zoctizum, Yarrise *Two Studies on Unemployment Among Educated*

Young People (1980).
Broader Unemployment (#PB0750)
Mass unemployment of human resources (#PD2046).
Narrower Unemployed educated youth (#PE1379)
Unemployment of educated older people (#PE9071)
Graduate and post-graduate unemployment (#PD1162)
Failure to employ skills of home-bound educated women (#PD8546).
Aggravated by Inappropriate employment incentives (#PD0024).

♦ PD8557 Inadequate road maintenance
Broken surfaces of streets and roads — Unrepaired roads — Roadway pot-holes — Street dust — Postponement of street repair
Nature Sub-surface water may, by erosion or freezing and melting, cause cracks and breaks in roadways. Pits and grooves in road surfaces are the first signs of highway degradation. These may form as holes deep enough to trap a tyre and cause an axle to break. If left unrepaired, small fissures can allow water to erode the foundation of a road, and before long the entire road will require reconstruction. Roads in developing countries, and in heavily trafficked major urban areas in developed countries, tend to be subject to insufficient road maintenance which permit rutted and pitted surfaces and dangerously deep pot-holes to develop. A road surface can last thirty years with proper maintenance.
Incidence Developing countries need to spend as much as $45 billion to rebuild roads fallen into disrepair through lack of maintenance.
Refs International Bank for Reconstruction and Development *Road Deterioration in Developing Countries*; OECD *Heavy Trucks, Climate Change and Pavement Damage* (1988); OECD Staff *Impacts of Heavy Freight Vehicles* (1983).
Broader Inadequate maintenance (#PD8984)
Inadequate maintenance of infrastructure (#PD0645)
Underprovision of basic services to rural areas (#PF2875).
Narrower Unmaintained bridges (#PE2471).
Related Underdeveloped provision of basic services in developing countries (#PF6473)
Limited availability of public services in the small towns of developed countries (#PF6539).
Aggravates Road hazards (#PD0791) Road traffic accidents (#PD0079)
Inadequate road and highway transport facilities (#PD0490)
Restricted delivery of essential services to developing country rural communities (#PF1667).
Aggravated by World anarchy (#PF2071) Prohibitive cost of maintenance (#PF0296)
Excessive speed of motor vehicles (#PE2147) Prohibitive cost of basic services (#PF6527)
Prohibitive cost of road construction (#PJ1070)
Unorganized development of work forces (#PF2128)
Self-defeating style of community planning (#PF6456).

♦ PD8561 Ageing populations
Increasing requirements for aged persons
Nature Decreasing birth rates and mortality rates are combining to increase dramatically the average age of populations and the number of elders in a population. In developed countries one person in ten is over 65 and in some countries it is twice that ratio. The social implications are many. Post-retirement income needs to be ensured. Basic health care services needs to be accessible. Health and nutritional education must be provided. Retirement institutions, personal care services for those at home, social supports, counselling and leisure-time services and activities must be made available. The care of the elderly has largely shifted from a family responsibility to one of the government. In many countries governmental structures have lagged behind these social realities.
Refs Brzezínski, Z J; Heikkinen, E and Waters, W E (Eds) *The Elderly in Eleven Countries* (1983); ILO *From Pyramid to Pillar*, United Nations *The World Aging Situation* (1985).
Broader Human ageing (#PB0477) Inadequate social welfare services (#PC0834).
Related Maldistribution of world population (#PF0167).
Aggravates Dependency of the elderly (#PD8399).

♦ PD8566 Environmental hazards from plastic materials
Nature Plastic is virtually indestructible: it resists dissolution by water, air or sunlight, is unpalatable to the microorganisms that break down natural materials and burning it gives off noxious gases.
Broader Environmental hazards from chemicals (#PC1192).
Narrower Non-biodegradable plastic waste (#PD1180)
Marine pollution by plastic waste (#PE3741)
Environmental hazards from manufacture of plastic products (#PE8651).

♦ PD8567 Pests and diseases of fish
Refs Ahne, W (Ed) *Cooperative Programme of Research on Aquaculture (Fish Diseases) - 3rd Session, Munich (Germany FR), 23 Oct 1979, fish diseases* (1980); FAO *Fish Diseases* (1965); FAO *Report of the Symposium on the Major Communicable Fish Diseases in Europe and Their Control* (1974); FAO *Parasites, Infections and Diseases of Fish in Africa* (1980); Kabata, Z *Parasites and Diseases of Fish Cultures in the Tropics* (1985); Mawdesley-Thomas, Lionel E, et al *Diseases of Fish* (1974).
Broader Animal diseases (#PC0952).
Narrower Pollution-induced fish diseases (#PE7584).
Aggravated by Water pollution (#PC0062) Marine pollution (#PC1117)
Leeches as pests (#PE3660) Pollution of inland waters (#PD1223)
Pesticide hazards to wildlife (#PD3680) Industrial waste water pollutants (#PD0575)
Water pollution in developing countries (#PD3675).

♦ PD8568 Low general income
Insufficient personal income — Low income potential — Minimal excess income — Denial of right to just income — Increasing low-income population
Broader Strained capital resources in small communities (#PF3665).
Narrower Underpayment for work (#PD8916) Lack of minimum wage fixing (#PE6726)
Marginal level of family income (#PD6579)
Inadequate income of rural communities (#PJ9649)
Denial of right of equal pay for equal work (#PD1977)
Decline in real wages in developing countries (#PD2769).
Related Inadequate income in old age (#PC1966)
Deteriorating community identity (#PF2241)
Non-diversification in subsistence fishing economies (#PF2135)
Constricting level of capital development in rural areas (#PE1139).
Aggravates Slow rate of income expansion (#PF6478)
Unattractive fishing business (#PG5399).
Aggravated by Underdeveloped capacity for income farming (#PF1240)
Excessive proportion of income spent on food (#PE8659)
Dependency on unpredictable sources of income (#PF3084).

♦ PD8575 Corporal punishment
Nature Corporal punishment used judicially usually takes the form of flogging as a penalty for crimes such as robbery with violence or sexual assault. Does it deter ? There is a widespread tendency to exaggerate its deterrent effect and evidence that it does not deter. Does it reform ? Only those who need reform the least. Is is suitable retribution ? Only if it is suitable to be physically harsh with those who have been physically harmful. Corporal punishment may be effective when administered by someone in a sustained, predictable and loving relationship with the person punished. But the question remains, is it really necessary ?
Broader Physical torture (#PD8734).
Narrower Beating of prisoners (#PD2484) Corporal punishment in schools (#PE0192).
Related Self-inflicted physical suffering (#PD7550).
Aggravates Repressive detention of juveniles (#PD0634).
Aggravated by Paternalistic punishment (#PJ3700).

♦ PD8585 Infertile land
Land infertility — Inferior land — Infertile deserts
Broader Soil degradation (#PD1052).
Narrower Arid developing countries (#PD8418).
Related Animal infertility (#PC1803).
Aggravates Desert nomadism (#PD2520) Unproductive subsistence agriculture (#PC0492)
Decreasing agricultural growth per capita (#PF4326).
Aggravated by Soil erosion (#PD0949) Acidic soils (#PD3658)
Alkaline soil (#PD3647) Soil pollution (#PC0058)
Desert advance (#PC2506) Soil compaction (#PD1416)
Soil infertility (#PD0077) Soil salinization (#PE1727)
Hazardous waste dumping (#PD1398) Metal contamination of soil (#PD3668)
Demineralization of the soil (#PD9227) Destruction of land fertility (#PC1300)
Over-intensive soil exploitation (#PC0052) Industrial waste water pollutants (#PD0575)
Cultivation of marginal agricultural land (#PD4273)
Soil mismanagement in developing countries (#PD6482)
Reduction of soil fertility downstream due to impoundment (#PE8782).

♦ PD8588 Diseases of unknown aetiology
Diseases of unknown cause — Diseases of unknown origin
Broader Intractable diseases (#PC8801).

♦ PD8604 Lack of capital development
Broader Dependency on unpredictable sources of income (#PF3084)
Lack of awareness of potential for investment in small, inner-city enterprises (#PF2042).
Aggravates Insufficient financial resources (#PB4653).
Aggravated by Lack of venture capital (#PG7833).

♦ PD8607 Neglected health practices
Neglect of personal health
Broader Neglect (#PA5438) Inadequate health control (#PF9401).
Narrower Neglect of sexual health of women (#PF5147)
Inadequate maintenance of physical health (#PF1773).
Related Age discrimination (#PC2541).
Aggravates Human illness (#PA0294).
Aggravated by Inadequate health services (#PD4790)
Insufficient health payments (#PG8867)
Ignorance of health and hygiene (#PD8023)
Decline in government health expenditure (#PF4586)
Inadequate dissemination and use of available information (#PF1267).

♦ PD8614 Restrictive legal practices
Restrictive practices in the courts — Protectionism in legal services
Nature Licensing regulations in both developed and developing countries severely limit the areas in which non-nationals may practice law. By and large, foreign law firms are, in these cases, only permitted to provide advice on their respective national law or on multinational aspects of the law. Appearances in court proceedings are generally still excluded. Furthermore, an elaborate screening and reviewing process can make it difficult to obtain work permits. In a number of countries, citizenship of the country is required for inscription on the professional register, which is necessary in order to practice.
Broader Restrictive practices (#PB9136) Outmoded legal systems (#PF2580)
Restrictive professional practices (#PD8027).
Narrower Unnecessary verbosity of legal documents (#PF7137)
Economic barriers to access to the legal profession, the judiciary and jury membership (#PE0803).
Related Biased legal systems (#PF8065)
Protectionism in the services industries (#PD7135).
Aggravates Obstruction of international criminal investigations (#PF7277).
Aggravated by Unethical practices in the legal profession (#PD5380).

♦ PD8621 International trade in biological weapons
Trade in biological warfare technology
Nature The 1972 Biological and Toxin Weapons Convention bars development and stockpiling of such agents. But that prohibition applies only to states, not private individuals or companies who can legally develop, produce and export biological weapons.
Broader International arms trade (#PC1358).
Related Trade in products for chemical warfare (#PE3808).
Aggravates Biological warfare (#PC0195).

♦ PD8622 Discrimination against unmarried women
Unmarried women — Spinsterhood
Refs Jansen, Willy *Women without Men* (1986).
Broader Discrimination against women (#PC0308)
Narrow legal definition of the family in developing countries (#PD1501).
Aggravates Discrimination in family planning facilities (#PD1036).
Aggravated by Ageing women (#PE6784)
Discrimination against women without children (#PE8788).

♦ PD8625 Agricultural mismanagement
Agricultural mismanagement of pasture and arable land
Broader Mismanagement (#PB8406) Incompetent management (#PC4867).
Narrower Inappropriate cash crop policy (#PF9187)
Mismanagement of irrigation schemes (#PE8233)
Agricultural mismanagement of housed farm animals (#PD2771).
Related Mismanagement of food resources (#PE6115).
Aggravates Over-intensive soil exploitation (#PC0052)
Unavailability of agricultural land (#PC7597).

♦ PD8627 Discrimination against physically disabled
Broader Denial of rights to disabled (#PC3461) Discrimination against the disabled (#PD9757).
Narrower Family rejection of physically handicapped (#PE2087).
Aggravated by Normalism (#PF3758).

♦ PD8628 Lake pollution
Broader Pollution of inland waters (#PD1223)
Unsustainable development of fresh waters (#PD6923).
Aggravates Pollution of sediments (#PE5539) Eutrophication of lakes and rivers (#PD2257).

DETAILED PROBLEMS

Aggravated by Agricultural effluent (#PE8504) Sewage as a pollutant (#PD1414)
Nitrates as pollutants (#PE1956) Nitrites as pollutants (#PE6087)
Pesticides as pollutants (#PD0120) Domestic waste water pollutants (#PD2800)
Nitrogen compounds as pollutants (#PD2965)
Environmental hazards from fertilizers (#PE1514).

♦ **PD8629 Closed professions**
Discrimination against non-members of professional bodies
Broader Discrimination in employment (#PC0244).
Related Discrimination against non-union workers (#PD6019).
Aggravated by Non-equivalence of national educational qualifications (#PC1524).
Reduces False qualifications (#PF0704) Lack of qualifications (#PG6377).

♦ **PD8638 Socially inappropriate housing**
Culturally insensitive house design — Architecture insensitive to needs of women
Nature Houses designed according to a standard, low-cost model, tend to be insensitive to the special needs of women with children, the elderly, the disabled, or those with many dependents. In the case of home, housing projects often use a gridiron layout that does not allow women to work in their house (whether domestic or making articles for sale) and at the same time keep an eye on their children. Such housing design also tends to be insensitive to the needs of those with a particular cultural emphasis (respect for elders, need of a secluded space for women, religious constraints, etc).
Aggravates Resettlement stress (#PD7776) Loss of cultural identity (#PF9005)
Inadequate housing for the aged (#PD0276).
Aggravated by Involuntary mass resettlement (#PC6203)
Obsolete basis of cultural identity (#PF0836).

♦ **PD8647 Manipulation of commodity markets**
Broader Economic manipulation (#PC6875).

♦ **PD8655 Extortionate bureaucracy**
Broader Abuse of government power (#PC9104)
Bureaucracy as an organizational disease (#PD0460).
Aggravates Denial of right to periods to nurse infants during working hours (#PE3044).

♦ **PD8657 Personal physical insecurity**
Claim Insecurity feeds on itself. Residents remain indoors in fear, emptying the streets at sundown, and thereby adding to the already prevalent insecurity. Such an environment can be expected where community development programmes do not directly focus upon the contradictions due to deep social insecurity.
Broader Insecurity (#PA0857).
Narrower Dangerous countries (#PD4278)
Personal physical insecurity of women (#PE7750)
Personal physical insecurity in developing countries (#PE5391)
Personal physical insecurity in industrialized countries (#PE4639).
Aggravated by Crime (#PB0001) Social insecurity (#PC1867)
Insecurity of property (#PC1784) Breakdown of police protection (#PF8652).

♦ **PD8668 Malnutrition in developing countries**
Nature One of the major reasons for malnutrition in developing countries is the process of urbanization. Individuals and families leave the rural areas and move to a city. Traditional foods are frequently not available and so foods that are available are substituted. While these foods may have enough calories they frequently do not provide a balanced diet resulting in protein and vitamin deficiencies.
Broader Malnutrition (#PB1498).
Aggravates Tropical deforestation (#PD6204).
Aggravated by Factory farming (#PD1562) Underdevelopment (#PB0206)
Soil infertility (#PD0077) Nutritional diseases (#PD0287)
Maternal malnutrition (#PE1085) Infectious and parasitic diseases (#PD0982)
Lack of purchasing power in developing countries (#PE8707)
Uncontrolled urbanization in developing countries (#PD0134)
Inadequate nutrition education in least developed countries (#PE0265)
Low productivity of agricultural workers in developing countries (#PE5883).

♦ **PD8673 Lack of political development**
Underdevelopment of political systems — Limited political developments
Broader Underdevelopment (#PB0206)
Ineffective structures of local consensus (#PF6506).
Narrower Underdevelopment of legal infrastructure (#PF4836).
Related Political alienation (#PC3227).
Aggravates Political apathy (#PC1917).
Aggravated by Tribalism (#PC1910) Social conflict (#PC0137)
Occupied nations (#PC1788) Political oligarchy (#PD3238)
Political opportunism (#PC1897) Political dictatorship (#PC0845).
Disruption of development by tribal warfare (#PD2191).

♦ **PD8679 Corruptive crimes**
Broader Corruption (#PA1986) Statutory crime (#PC0277).
Narrower Theft (#PD5552) Drug smuggling (#PE1880)
Gambling and wagering (#PF2137) Illicit drug trafficking (#PD0991)
Bribery of public servants (#PD4541) Hindrance of law enforcement (#PD5515)
Falsification of public records (#PD4239) Obstruction of government function (#PD6710)
Criminal violation of civil rights (#PD8709).
Related Bribery (#PC2558) Tax evasion (#PD1466) Police brutality (#PD3543)
Abuse of authority (#PC8689) Vice and sex traffic offences (#PD8910)
Crimes against national security (#PC0554)
Circumvention of duties and assessments (#PD4882).
Aggravates Corruption of minors (#PD9481) Corruption in politics (#PC0116)
Unethical industrial practices (#PD2916) Corruption of government leaders (#PC7587)
Corruption in developing countries (#PD0348).
Aggravated by Vigilantism (#PD0527) Organized crime (#PC2343)
Underground economy (#PC6641) Abuse of banking secrecy (#PF5991).

♦ **PD8680 Prison suicides**
Jail suicides — Gaol suicides
Incidence In UK the number of suicides averaged less than 20 a year until 1986, when the prison population drastically increased. The figures rose to almost 50 in 1987 and fell to about 33 in 1988.
Counter-claim Prison officers cannot hope to eliminate suicide totally, any more than it can be prevented in the outside world. Prison inmates are much more likely to exhibit characteristics which are known to give rise to suicide risk. However it is impossible to keep every prisoner under surveillance, even if this were desirable.
Refs Burtch, Brian E and Ericson, Richard V *Silent System* (1979).

PD8744

Broader Suicide (#PC0417).
Aggravated by Maltreatment of prisoners (#PD6005)
Denial of rights to prisoners (#PD0520).

♦ **PD8696 Drug abuse by government officials**
Drug abuse by decision-makers — Drug abuse by policy-makers
Broader Drug abuse (#PD0094).
Aggravates Inappropriate policies (#PF5645).

♦ **PD8697 Arbitrary enforcement of regulations**
Ambiguous enforcement procedures — Code enforcement difficulties — Inadequate code enforcement — Random code enforcement — Subjective law enforcement
Broader Arbitrariness (#PB5486)
Unfocused style of community operations (#PF6559)
Ineffective organization of community action (#PF6501).
Aggravates Havens for environmental pollution (#PE6172)
Inadequate national law enforcement (#PE4768)
Limited local respect for regional and global legislation (#PF2499).
Aggravated by Vagueness of laws (#PF9849) Inadequately worded agreements (#PF5421)
Unethical practices of regulatory inspectors (#PF8046)
Use of undue influence to obstruct the administration of justice (#PE8829).

♦ **PD8709 Criminal violation of civil rights**
Nature Conspiracy against the rights of citizens is to injure, oppress, threaten or intimidate any citizen, resident or visitor in the free exercise of their rights as citizens of a country or because they have exercised these rights.
Broader Statutory crime (#PC0277) Corruptive crimes (#PD8679)
Violation of civil rights (#PC5285).
Narrower Strikebreaking (#PF5805) Interception of communications (#PD7608)
Violation of political processes (#PD5457)
Political contributions by agents of foreign principals (#PE6715).
Related Abuse of authority (#PC8689).

♦ **PD8716 Official fabrication of evidence**
Planting of evidence — Police falsification of evidence — Official misinterpretation of evidence
Broader False evidence (#PF5127).
Aggravated by Police corruption (#PD2918) Deception by government (#PD1893)
Inadequate evidence to convict known offenders (#PF8661)
Use of undue influence to obstruct the administration of justice (#PE8829).

♦ **PD8718 Racial war**
Broader War (#PB0593).
Related Gang warfare (#PD4843) Religious war (#PC2371)
Ideological war (#PC3431).
Aggravates Enemies (#PF8404).
Aggravated by Racism (#PB1047)
Race as a reinforcement of nationalism (#PF3352).

♦ **PD8719 War damage in civilian areas**
Broader Disastrous consequences of war (#PC4257).
Narrower Destruction of economy due to war (#PE8915)
Destruction of civilian populations and institutions (#PE8564).
Related War casualties (#PD4189) Prisoners of war (#PC8848)
Housing destruction in war (#PE2592) Industrial destruction by war (#PD8359)
Poverty as a consequence of war (#PE5252) Environmental consequences of war (#PC6675).
Aggravated by War (#PB0593) Nuclear war (#PC0842).

♦ **PD8723 Functional illiteracy**
Adult illiteracy — Learning-disabled college students — Mal-educated school leavers — Uneducated school leavers
Nature Inability of a person to engage in those activities in which literacy is required for effective functioning of his group or community and also to enable him to continue to use reading, writing and calculation for his own and the community's development.
Incidence An estimated 13 per cent of America's 17-year-olds are functionally illiterate. In Denmark, with one of the most advanced educational systems in the world, there are 500,000 functional illiterates, representing almost 10 per cent of intelligent adults. In India almost 70 per cent of adults are illiterate, impeding national development by causing stagnation in the economy.
Refs Edgerton, Robert B and Meyers, C Edward *Lives in Process* (1984); Fuchs-Bruninghoff, E, et al *Functional Illiteracy and Literacy Provision in Developed Countries* (1987).
Broader Illiteracy (#PC0210) Illiteracy in developed countries (#PC1383).
Narrower Incorrect spelling (#PE4302).
Aggravates Shortage of skilled labour (#PD0044).
Aggravated by Narrow range of cultural exposure (#PF3628)
Educational gap between generations (#PF6497).

♦ **PD8725 Abdominal pain**
Abdominal cramp
Broader Pain (#PA0643).
Aggravated by Allergy (#PE1017).

♦ **PD8731 Prevention of the exchange of ideas**
Restrictions on the exchange of ideas
Broader Restrictions on freedom of information (#PC0185).
Aggravated by Professional secrecy (#PD6576) Conceptual plagiarism (#PD1284)
Professional jealousy (#PD8488).

♦ **PD8734 Physical torture**
Broader Human torture (#PC3429).
Narrower Cold torture (#PE3734) Water torture (#PE4792)
Sexual torture (#PE5108) Dental torture (#PE7201)
Dietary torture (#PE4371) Forced exercise (#PD5628)
Chemical torture (#PD5204) Electrical torture (#PE9000)
Torture by burning (#PE2678) Corporal punishment (#PD8575)
Torture by wounding (#PE4078) Torture of children (#PD2851)
Torture through mutilation (#PD7576) Stationary position torture (#PE4471)
Torture by exposure to weather (#PE3915) Torture by exposure to animals (#PE7327)
Intentional infecting with disease (#PD2651).
Related Religious torture (#PC7101) Torture of animals (#PC3532)
Psychological torture (#PD4559) Pharmacological torture (#PE4696)
State sanctioned torture (#PD0181) Institutionalized torture (#PD6145).

♦ **PD8744 Kidnapping**
Risk of being kidnapped — Abduction

PD8744

Nature Abducting another person, or having abducted another person continues to hold them is kidnapping. The victim may be held for ransom or reward. They may be used as a shield or hostage. They might be held as slaves or involuntary servants. The victim may held to be terrorized or to terrorize another. They may be held so a crime can be committed or a government action prevented.
Incidence Up to one million children have been kidnapped for the sex market. The sex tour industry and the presence of military bases in developing countries has lead to an enormous increase of child prostitution which aggravates kidnapping. Children from Latin American, Asian and African slums are kidnapped and sold to rich clients in the north to be used privately or in brothels.
Background Abduction of women, now considered a serious crime throughout the world is probably a survival of one of the most primitive forms of marriage, marriage by capture. At the beginning of the nineteenth century it was still in full force among some Slavs. Teutonic and Scandinavian tribes regularly resorted to forcible abduction of women for wives. As did the early Greeks and Romans. Wars were frequently carried out to capturing wives. Traces of the same custom are found in the Old Testament. Many traces of this practice are still found in the marriage rituals. In some cases the bride groom is expected to go through the motions of capturing his bride. The bride in some places hides until found by the groom. In some ceremonies the bride is expected to resist and struggle. The bride is carried over the threshold of the new household.
Refs Cassidy, E L *Political Kidnapping* (1986).
Broader Risk (#PF7580) Offences involving danger to the person (#PD5300).
Narrower Hostage taking (#PE4108) Child abduction (#PE6154)
Felonious restraint (#PD4488) Parental kidnapping (#PE6075)
Unlawful imprisonment (#PD4489) Political hostage–taking (#PD1886)
Kidnapping of pet animals (#PE0805) Extraterritorial abduction (#PE2113)
Abduction by extraterrestrials (#PF3881).
Related Missing persons (#PD1380).
Aggravates Changelings (#PE9453).
Aggravated by Aerial piracy (#PD0124) Chattel slavery (#PC3300)
Child prostitution (#PE7582) Sexual abuse of children (#PE3265)
Extra–legal conscription (#PD6667) Militarization of children (#PE5986)
Trafficking in children for sexual exploitation (#PE6613).

♦ **PD8762 Racial repression**
Broader Repression (#PB0871) Racial discrimination (#PC0006).
Related Religious repression (#PC0578).
Aggravated by Genocide (#PC1056).

♦ **PD8775 Diseases of female genital organs**
Diseases of uterus
Refs Hurley, Patricia S *Female Genital Diseases* (1985); Wilkinson, Edward J *Pathology of the Vulva and Vagina* (1986).
Broader Diseases of the genito–urinary system (#PC4575).
Narrower Leukorrhoea (#PG5233) Female sterility (#PE9302)
Diseases of cervix (#PG9823) Menopausal symptoms (#PG0795)
Uterovaginal prolapse (#PG9787) Malposition of uterus (#PG9763)
Disorders of menstruation (#PG6762) Infective diseases of uterus (#PE9286)
Benign neoplasm of genital organs (#PE9705).

♦ **PD8778 Housing shortage**
Broader Inadequate housing (#PC0449).
Narrower Insufficient rural housing (#PF6511)
Unfeasible housing alternatives in urban areas (#PE8061).
Aggravates Urban slums (#PD3139) Homelessness (#PB2150)
Exploitation in housing (#PD3465) Discrimination in housing (#PD3469)
Illegal occupation of unoccupied property (#PD0820).
Aggravated by Lack of war relief (#PF0727) Segregation in housing (#PD3442)
Housing destruction in war (#PE2592) Insufficient housing funds (#PG8768).

♦ **PD8779 Invasion**
Broader Defence (#PA5445) Intrusion (#PA6862)
Foreign intervention in internal affairs of states (#PC3185).
Narrower Extraterrestrial invasion (#PF4444).
Related Cultural invasion (#PC2548).
Aggravates Foreign control (#PC3187) Foreign dictatorship (#PC3186).

♦ **PD8785 Communist repression**
Claim Communist party machine crushes revolts anytime it sees a threat to its monopoly: tanks in Budapest and Prague, martial law in Warsaw and gunfire in Beijing.
Broader Communism (#PC0369).

♦ **PD8786 Eye diseases and disorders**
Infection of the eye — Eye damage — Ophthalmic diseases — Inflammatory diseases of the eye
Refs Bohigian, George M *Handbook of External Diseases of the Eye* (1987); Caird, F I and Williamson, J *Eye and Its Disorders in the Elderly* (1986); Char, Devron H *Thyroid Eye Disease* (1985); Clayton, R M, et al *Problems of Normal and Genetically Abnormal Retinas* (1983); Cullinan, Tim *Visual Disability in the Elderly* (1986); Sandford–Smith, J *Eye Disease in Hot Climates* (1986); Sundmacher, R *Herpetic Eye Disease* (1981).
Broader Animal diseases (#PC0952) Diseases of the sense organs (#PC9623).
Narrower Scotoma (#PJ9214) Trachoma (#PE1946) Glaucoma (#PE2264)
Cataract (#PE6817) Pterygium (#PG9496) Hordeolum (#PG8350)
Keratitis (#PE6789) Strabismus (#PG3837) Blepharitis (#PG7531)
Photophobia (#PG5450) Xerophthalmia (#PE2538) Conjunctivitis (#PE7974)
Dacryocystitis (#PG7100) Corneal opacity (#PG9354) Diseases of retina (#PE4584)
Physical blindness (#PD0568) Visual deficiencies (#PD8179)
Inflammation of retina (#PG8672) Oculomotor disturbance (#PG6827)
Hereditary opticatrophy (#PG4816) Inflammation of the uveal tract (#PG1678)
Inflammation of lacrimal glands and ducts (#PG4547).
Aggravated by Measles (#PE1603) Houseflies as pests (#PE3609).

♦ **PD8800 Prejudice against other languages**
Broader Prejudice (#PA2173) Discriminatory communication (#PD6804).
Related Discriminatory language (#PF7299).
Aggravates Language discrimination in politics (#PD3223)
Discrimination against minority languages (#PD5078).

♦ **PD8804 Product tampering**
Nature Both random mass poisoning and the deliberate poisoning of food and drugs as an instrument in commercial blackmail is modern phenomena born in 1980s. The criminal adulteration of food by retailers, mixing chalk with flour or dilute milk with water, is not new.
Incidence The first case of mass food tampering was in 1982 in USA, when four people died after swallowing the pain–relief drug Tylenol, laced with cyanide. There was no blackmail, no claim of responsibility and no arrest.
Claim To poison food is to break all the taboos about what is natural and good. Our dependency on pre–packaged and hygienic food has made mass poisoning possible. From the criminal's point of view, it has the merits of anonymity and remoteness: he need never come into personal contact with his victim.
Broader Adulteration (#PD9433) Defective product manufacture (#PD3998).
Narrower Product extortion (#PD5426).

♦ **PD8807 Religious discrimination in education**
Discriminatory effects of religious and denominational education
Nature Religious education may be the means whereby religion gains influence in political matters or over society at large. It usually involves segregation according to denomination, serves to reinforce existing prejudices, and may ultimately lead to discrimination, conflict and even war.
Incidence Denominational education exists world wide and is particularly important for Islam, Christianity and Judaism. It may be the main source of education and therefore be equated with prestige, making many converts for this reason and thus being a foundation for nationalism and other political policies. It may have an influence on the law. Religious teaching in schools is a matter of controversy in Western industrialized countries, especially where immigrant populations prefer religious teaching in their own traditions, rather than those of the host country.
In the case of Christianity, for example, most new catechisms and religious instruction books have no section on the basic beliefs of other religions. A vast majority of Christians do not gain insight in the spirituality and ethics of others.
Broader Inequality in education (#PC3434) Religious discrimination (#PC1455)
Undue religious influence on secular life (#PF3358).
Related Compulsory indoctrination (#PD3097) Religious discrimination in politics (#PC3220)
Denial of right to educational choice (#PE4700)
Inadequate integration of religions into society (#PF3403)
Religious discrimination in the administration of justice (#PE0168).
Aggravates Anti–clericalism (#PF3360) Religious rivalry (#PC3355)
Religious conflict (#PC3292) Denial of education to minorities (#PC3459)
Religious and political antagonism (#PC0030)
Segregation based on religious affiliation (#PC3365).
Aggravated by Religious intolerance (#PC1808).

♦ **PD8812 Animal suffering**
Broader Suffering (#PA7690).
Narrower Boredom of captive and domesticated animals (#PF7681)
Denial to food animals of the right to freedom from suffering (#PE3899).
Related Denial to working animals of limitation of working hours (#PE6427)
Denial to working animals of restorative nourishment and rest (#PE4793)
Denial to experimental animals of the right to freedom from suffering (#PE8024).
Aggravated by Cruelty to animals in factory farming (#PD2768)
Animal malformation in factory farming (#PD2761).

♦ **PD8821 Ignorance concerning disease**
Medical ignorance — Ignorance concerning disease transmission — Ignorance concerning sexually transmitted diseases
Broader Ignorance (#PA5568).
Aggravates AIDS (#PD5111).
Aggravated by Nutritional ignorance (#PE5773) Ignorance of drug users (#PJ2389)
Ignorance of health and hygiene (#PD8023) Ignorance of reproductive processes (#PD7994)
Ignorance of women concerning primary health care (#PD9021).

♦ **PD8825 Children engendered by occupying soldiers**
Abandoned children of foreign soldiers
Nature During periods of war or occupation of territories by foreign armed forces, or in association with military bases during peacetime, soldiers establish temporary or semi–permanent liaisons with local women, usually during their brief recreation periods from military duties. When children are born of such relationships, the fathers may not accept paternity or may already have moved to other locations or have returned to their home country. Such children grow up without a father (although a proportion also lose their mothers as casualties of war), often with the additional stigma of being physically distinct from their peers, if the father was of a different racial type. Their emotional growth may be further distorted by the vain hope that they may be successfully reunited by the father, and the rejection they experience from their peers.
Incidence Such children are engendered as a result of most extended military conflicts. In past soldiers have been tacitly encouraged to force the women of the occupied territories to bear children bearing the genes of the occupying force as a means of extending and maintaining the domination of that culture. The most striking recent case is that of the children of members of the USA armed forces based in Viet–Nam or vacationing in Bangkok. Such children born in the 1970s, are now adolescents, and recognizably non–Asiatic although speaking little of their fathers language. Many live in the hope of being recognized by fathers they have never seen who will care for them in the USA.
Claim One of the tragic consequences of the epidemic of illicit sex in the West has been the legacy of illegitimate children abandoned by American soldiers in Japan, Korea, Taiwan, Philippines, Thailand and Viet–Nam. It is no exaggeration to claim that American soldiers fighting in Korea and Viet–Nam have fathered sufficient numbers of illegitimate children to equal the population of several good sized towns.
Broader Orphan children (#PD7046) Abandoned children (#PD5734)
Discrimination against illegitimate children (#PD0943).
Related Discrimination against children of prostitutes (#PE0392)
Vulnerability of children during armed conflict (#PE8174).
Aggravates Miscegenation (#PC1523) Social stigma (#PD0884)
Displaced children (#PD5136) Neglect of dependents of war victims (#PD2092)
Discrimination against mixed race children (#PE5183).
Aggravated by War (#PB0593) Rape (#PD3266) Occupied territories (#PD8021)
Prejudice in children (#PD8973).

♦ **PD8831 Restrictive medical practices**
Restrictive health practices
Broader Restrictive practices (#PB9136) Inadequate medical care (#PF4832)
Restrictive professional practices (#PD8027).
Aggravates Human disease and disability (#PB1044).

♦ **PD8832 Minimal community services**
Minimal community organization
Broader Obstacles to community achievement (#PF7118)
Incomplete implementation of community decisions (#PF2863).
Narrower Lack of local services for community leadership training (#PF2451).
Related Declining sense of community (#PF2575).
Aggravates Lack of community development (#PF7912).

♦ **PD8834 Economic gap**
Broader Disparity between industrialized and developing countries (#PC8694).
Aggravated by Economic intimidation (#PC3011)

General unproductivity of capitalist systems (#PF3103)
Conflicting roles of commodities in capitalism (#PF3115)
Conflicting roles of money in capitalist systems (#PF3114)
Counterproductive capitalist investment financing (#PF3104)
Contradictions of capitalism in developing countries (#PF3126)
Denial of effective national self-determination by capitalist exploitation (#PE3123).

♦ **PD8837 Lack of care for animals**
Lack of concern for animal welfare — Unknowledgeable animal care — Uninformed animal sanitation
Refs Hood, D E and Tarrant, P V (Eds) *Problem of Dark Cutting in Beef* (1981).
Broader Lack of care (#PF4646) Inadequate animal welfare (#PC1167).
Related Denial to animals of the right to the attention, care and protection of humankind (#PF5121).
Aggravated by Inadequate legislation for animal welfare (#PE5794)
Stagnated development of agricultural production (#PD1285)
Restrictive effects of external capital on development (#PF3318)
Lack of skilled manpower in rural areas of developing countries (#PE5170).

♦ **PD8839 Inadequate irrigation system**
Undeveloped irrigation system — Fragmented irrigation distribution — Lack of irrigation — Incomplete irrigation system — Unavailability of irrigation water
Broader Lack of essential local infrastructure (#PF2115).
Related Physically inaccessible services (#PC7674).
Aggravates Alkaline soil (#PD3647) Aquatic weeds (#PD2232)
Desert nomadism (#PD2520) Schistosomiasis (#PE0921)
Soil infertility (#PD0077) Disunity in urban villages (#PF1257)
Slow rate of income expansion (#PF6478) Unproductive subsistence agriculture (#PC0492)
Underdeveloped approaches to local food production (#PF6493)
Underproductive methods of agricultural management (#PF6524).
Aggravated by Inadequate maintenance equipment (#PD6520)
Water pollution in developing countries (#PD3675)
Minimal access to appropriate technology (#PF3554)
Obsolete methods of agricultural production (#PF1822).

♦ **PD8841 Occupational illness in developing countries**
Nature The working environment itself causes occupational disease and injuries, fatigue, unsatisfactory man/machine relationships and physiological stresses. Risks are greater among the migrant workers who have no vocational training or industrial experience and whose illiteracy and ignorance make it difficult to understand written or oral instructions. Exposure to typical high temperatures or to high humidity causes heat stress. Special problems may also arise in connection with work at high altitudes or in thick forests.
Broader Occupational diseases (#PD0215).

♦ **PD8844 Food wastage**
Food waste — Dumping food products as waste
Broader Unproductive use of resources (#PB8376)
Distortion of international trade by dumping (#PD2144).
Related Dumping of food in developing countries (#PE0607).
Aggravates Food insecurity (#PB2846) Shortage of cultivable land (#PC0219).
Aggravated by Hunting of animals (#PC2024) Food grain spoilage (#PD0811)
Overproduction of food (#PD9448) Unethical food practices (#PD1045)
Hotel and restaurant waste (#PE1542) Inhumane killing of animals (#PE0358)
Unethical catering practices (#PE6615)
Inefficient use of proteins in factory farming (#PF2758).

♦ **PD8851 War-time disruption of economies and production facilities**
Nature Under war-time conditions the production facilities and economies of territories involved in the hostilities are severely disrupted as a result of conversion of such facilities to war-time needs, destruction of such facilities, loss of foreign investment, decrease in international trade, disequilibrium in balance of payments, and disruption of long-range balanced growth. On cessation of hostilities, it may take an extended period of time to return to economically viable peace-time production.
Broader War (#PB0593).
Related Destruction of economy due to war (#PE8915).
Aggravates Economic stagnation (#PC0002) Imbalance of payments (#PC0998)
Unequal global distribution of economic growth (#PC5601).

♦ **PD8856 Homelessness in developing countries**
Broader Homelessness (#PB2150).
Aggravates Improvisational housing in developing countries (#PE4386).

♦ **PD8859 Unethical use of social welfare benefits**
Social security fraud — Welfare benefits fraud — Unemployment benefits fraud — Social welfare corruption
Incidence Some companies knock social security contributions off their employees' salaries but tell to the officials that they have smaller staffs than is the case. Multinational companies pay enhanced salaries for working overseas and declare only a "phantom" wage.
Refs Hutton, Gary W *Welfare Fraud Investigation* (1985).
Broader Fraud (#PD0486).
Related Benefit overpayments (#PD7500).
Aggravated by Unethical practices by employees (#PD4334).

♦ **PD8871 Political surveillance**
Internal surveillance — Secret investigation
Nature Surveillance is used to get information about groups and individuals who are considered a potential threat to public order or security, such as ethnic and religious minorities, political activists, dissidents and foreigners.
Broader Infringement of privacy (#PB0284)
Misuse of government surveillance of communications (#PD9538).
Narrower Misuse of postal surveillance by governments (#PD2683)
Misuse of electronic surveillance by governments (#PD2930)
Secret government security vetting of job applicants (#PE9441).
Related Misuse of telephone surveillance by governments (#PD1632).
Aggravates Informers (#PD8926) Political infiltration (#PD4798)
Surreptitious entry during surveillance (#PE3973).
Aggravated by Abdication of government ministerial control (#PF7342)
Unaccountable government intelligence agencies (#PF9184).

♦ **PD8872 Elimination of traditional skills**
Vanishing artistic skills
Broader Lost knowledge (#PF5420).
Aggravates Lack of job satisfaction (#PF0171) Lagging training in social skills (#PF8085)
Underdeveloped technological skill (#PF8552)
Limited development of functional abilities (#PF1332)
Lack of artisans and craftsmen in developed countries (#PE4804).
Aggravated by Decreased skill transference (#PG5492)
Untransposed significance of cultural tradition (#PF1373)
Incompatibility of traditional and new technologies (#PE3337)
Unclarified procedures for transposing ancient traditions (#PF6494).
Reduces Outmoded functional skills in rural communities (#PF2986).

♦ **PD8877 Political aggression**
Broader International aggression (#PC7559) Abuse of government power (#PC9104).
Aggravates Civil war (#PC1869)
Undirected expansion of economic base (#PF0905).

♦ **PD8888 Rising water level**
Rising level of underground water — High groundwater levels — Waterlogging of land
Nature The amount of water leaving the soil must be at least equal to the amount entering it. The water should not be allowed to accumulate, but perennial irrigation invariably raises the water table. In some areas the groundwater tables are rising at a rate of 3 to 5 metres a year. This is caused primarily by the water lost through seepage from irrigation channels or simply by overuse of irrigation water. As waterlogging sets in, so the inevitable process of salinization begins.
Broader Disruption of the hydrological cycle (#PD9670).
Aggravates Soil salinization (#PE1727) Water-borne diseases (#PE3401)
Inadequate water system infrastructure (#PD8517)
Endangered monuments and historic sites (#PD0253).
Aggravated by Inadequate land drainage (#PD2269)
Mismanagement of irrigation schemes (#PE8233)
Unimaginative vision of resource utilization (#PF1316)
Seepage water losses from irrigation systems (#PE9088).
Reduced by Overexploitation of underground water resources (#PD4403).

♦ **PD8890 Unchecked power of government bureaucracy**
Broader Abuse of government power (#PC9104).
Narrower Over-development of bureaucracy in ex-colonial countries (#PD2511)
Excessive bureaucratic requirements for welfare benefits (#PE8893).
Aggravates Fear of officialdom (#PD9498) Administrative delays (#PC2550)
Bureaucratic opposition (#PD7966) Bureaucratic superiority (#PC1259)
Suspicion of bureaucracy (#PF8335) Abuse of bureaucratic procedures (#PF2661)
Inefficient public administration (#PF2335)
Abusive distribution of political patronage (#PF8535).
Aggravated by Over-centralization (#PF2711) Bureaucratic ignorance (#PF8582)
Government insensitivity (#PF2808) Governmental incompetence (#PF3953)
Institutionalized corruption (#PC9173)
Proliferation of public sector institutions (#PF4739).
Reduced by Government limitations (#PF4668).

♦ **PD8894 Neglect of property maintenance**
Inadequate building maintenance — Unsafe home maintenance — Unkept rental property — Neglected personal property — Unplanned maintenance margin
Broader Inadequate maintenance (#PD8984).
Narrower Neglect of school buildings (#PG9829)
Depressing effect of poor housing construction (#PF1213).
Aggravates Structural failure (#PD1230) Demeaning community self-image (#PF2093)
Stagnated images of community identity (#PF6537)
Insufficient care of community property (#PF1600)
Individualistic practices of local business (#PF1176)
Poor condition of open spaces in urban communities (#PF1815).
Aggravated by Uncontrolled urban development (#PC0442)
Unethical real estate practice (#PD5422)
Unethical practices by housing tenants (#PE7169)
Inadequate maintenance of infrastructure (#PD0645).

♦ **PD8897 Economic rivalry**
Industrial rivalry — Economic competition — Commercial rivalry — Corporate rivalry — Rivalry in international markets
Refs Brenner, Reuven *Rivalry* (1990).
Broader Competition (#PB0848).
Narrower Restricted growth in export markets of developing countries (#PF1471).
Aggravates War (#PB0593) Industrial espionage (#PC2921)
International aggression (#PB0968) Pursuit of corporate prestige (#PF7983)
Competitive acquisition of arms (#PC1258) Competition for scarce resources (#PC4412)
Inadequacy of international legislation (#PF0228).

♦ **PD8903 Inequality in employment**
Broader Lack of job satisfaction (#PF0171) Inequality of opportunity (#PC3435).
Narrower Inequality of employment opportunity in developing countries (#PD1847).

♦ **PD8904 Misuse of wild animals**
Broader Maltreatment of animals (#PC0066).
Narrower Bird netting (#PG6093) Trapping of animals (#PE5735)
Cruel culling of seals (#PE3484)
Capture and use of wild animals as pets (#PD1179)
Commercial exploitation of wild animals (#PD1481)
Abuse in control of wild animal populations (#PE7995).

♦ **PD8909 Discrimination against men in education**
Broader Discrimination against men (#PC3258) Sexual discrimination in education (#PD1468).
Related Discrimination against women in education (#PD0190).

♦ **PD8910 Vice and sex traffic offences**
Broader Moral offences (#PD9179) Traffic in persons (#PC4442)
Offences against public order (#PD7520).
Narrower Obscenity (#PF2634).
Related Prostitution (#PD0693) Corruptive crimes (#PD8679).
Aggravates AIDS (#PD5111).

♦ **PD8913 Discriminatory business practices**
Broader Discrimination (#PA0833).
Narrower Credit discrimination (#PE7902) Discriminatory nuclear trade (#PD8124)
Refusal of sale because of buyer's race (#PE8823)
Distortion of international trade by selective domestic subsidies (#PD0678)
Discrimination against transnational banks in developing countries (#PE4320)
Discrimination in lending by transnational banks in developing countries (#PE4310).
Related Discrimination against foreign companies (#PD6417).
Aggravates Middleman control of rural marketing (#PE3528).

♦ **PD8916 Underpayment for work**
Decline in real wages — Unattractive industrial wages — Unmotivating low wages — Low tenant wages — Excessively low wages — Inadequate wages — Underpaid employees — Unfair remuneration for labour — Low pay — Cheap labour — Non-competitive business wages

PD8916

Refs Bluestone, Barry, et al *Low Wages and the Working Poor* (1973).
Broader Low general income (#PD8568) Inequitable range of salaries (#PD9430)
Undirected expansion of economic base (#PF0905).
Narrower Unpaid labour (#PD3056) Underpayment of police (#PJ8601)
Underpayment of teachers (#PE8645) Lack of payment for housework (#PE4789)
Underpayment of government officials (#PD8422)
Underpayment for work in developing countries (#PE9199)
Decline in real wages in developing countries (#PD2769).
Aggravates Strikes (#PD0694) Overwork (#PD2778)
Socio-economic poverty (#PB0388) Lack of job satisfaction (#PF0171)
National economic recession (#PD9436) Slow rate of income expansion (#PF6478)
Limited accumulation of capital (#PF3630) Prohibitive cost of accommodation (#PD1842)
Obstacles to community achievement (#PF7118)
Disorganized attempts of upgraded employment (#PF6458)
Transfer of business from small communities to larger towns (#PF6540).
Aggravated by Rural unemployment (#PF2949) Lack of minimum wage fixing (#PE6726)
Inadequate education of indigenous peoples (#PC3322)
Constricting level of capital development in rural areas (#PE1139)
Limited availability of permanent employment in inner-cities (#PE1134).
Reduces Prohibitive labour costs (#PF8763).

♦ PD8926 Informers

Tattletales — Squealer — Stool pigeon — Denunciation — Whistle-blowers
Nature It is accepted that a person with knowledge that a crime, or some other form of abuse, has been committed should convey that information to appropriate authorities in order that remedial measures may be taken and justice done. Where the authorities are themselves perceived as unjust or the abuse is defined by laws or regulations which are perceived as unjust, the informer is viewed as acting against the interests of the social group within which the so-called abuse took place. Informers are especially suspect when they act not for the promotion of the welfare of the society, however it is defined, but for payment or in a spirit of personal revenge. The negative consequences of their activities are increased when the information conveyed is false or deliberately misleading and when authorities actively encourage people to inform against each other.
Incidence Informers have existed in most cultures. The difficulties they create were recognized in the Code of Manu and especially from the early Roman Empire. Extensive use of networks of informers has traditionally been made by despots, dictators and repressive regimes. Their use has been encouraged in occupied countries and territories as a means of controlling the population and establishing a reign of fear. Such informers are subsequently judged as traitors to their country. Networks of informers have acquired a more organized form on the initiative of modern intelligence services.
Counter-claim Withholding information which could remedy an abuse or bring a criminal to justice makes a person party to that abuse. In many cases informers advance the cause of law and order when otherwise crime would flourish. Often the use of informers is the only means to break criminal networks and to prevent subversive or terrorist activities.
Aggravates Treason (#PD2615)
Harassment of human rights monitors (#PF1585).
Aggravated by Unjust laws (#PC7112) Dictatorship (#PC1049)
Secret police (#PE6331) Industrial espionage (#PC2921)
Political repression (#PC1919) Occupied territories (#PD8021)
Political surveillance (#PD8871)
Unaccountable government intelligence agencies (#PF9184).
Reduces Crime (#PB0001) Secrecy (#PA0005) Terrorism (#PD5574)
Subversive activities (#PD0557).

♦ PD8937 Restrictions on property rights

Violation of premises — Violation of property rights
Broader Denial of economic rights (#PD4150).
Narrower Denial of the right to ownership (#PE8411)
Unequal property rights for women (#PE4018)
Political confiscation of property (#PD3012).
Aggravates Discrimination against women in politics (#PC1001).
Aggravated by Inadequate laws of adoption (#PD0590).

♦ PD8941 Child malnutrition

Refs Walker-Smith, J A and McNeish, A S *Diarrhea and Malnutrition in Childhood* (1986).
Broader Malnutrition (#PB1498) Victimization of children (#PC5512).
Narrower Unbalanced infant diets (#PE0691)
Protein-energy malnutrition in infants and early childhood (#PD0331).
Aggravates Child mortality in developing countries (#PE5166)
Inhibited growth of malnourished children (#PE4921)
Damage to infant brains from malnutrition and insufficient stimuli (#PE4874).
Aggravated by Hazards of bottle-feeding (#PE4935)
Ignorance of women concerning primary health care (#PD9021).
Reduced by Dependence on breast feeding (#PE7627).

♦ PD8942 Dumping of consumer waste products

Throwaway society — Proliferation of consumer waste — Unrecycled consumer waste
Refs Ferranti, M P and Ferrero, G L (Eds) *Sorting of Household Waste and Thermal Treatment of Waste* (1985).
Broader Unproductive use of resources (#PB8376).
Narrower Hazardous waste dumping (#PD1398) Marine dumping of wastes (#PD3666)
Environmentally unfriendly consumer products (#PD9310).
Aggravated by Proliferation of consumer products (#PF5057).

♦ PD8945 Neglect of the aged

Reduction in motivation to care for the aged
Nature Individual and social motivation to care for the elderly are reduced to guilt or political expediency. What care is given encourages the elderly to become dependent on their need for care.
Broader Apathy (#PA2360) Social neglect (#PB0883).
Narrower Fear of growing old (#PJ9144) Erosion of elders' wisdom (#PF1664).
Aggravates Inadequate social welfare services (#PC0834)
Inadequate welfare services for the aged (#PD0512).
Aggravated by Dependency of the elderly (#PC8399)
Generation communication gap (#PF0756)
Conflicting social service ideologies (#PD3190).

♦ PD8973 Prejudice in children

Broader Prejudice (#PA2173).
Aggravates Social stigma (#PD0884) Discrimination (#PA0833)
Physical intimidation by children (#PE2876)
Children engendered by occupying soldiers (#PD8825).

♦ PD8982 Marine accidents

Risk of maritime accidents — Sea disasters — Capsizing ferries — Collisions at sea — Shipwrecks — Ship collisions — Naval accidents
Nature The collision of seagoing vessels and ships running aground or capsizing may result in a grave loss of life, property, and in some cases in ecological disasters. Some of these disasters may be the result of unfavourable weather conditions; a defect which affects the safety of the ship; failure to follow steering and sailing rules of navigation and right-of-way; or negligence with regard to the observance of good seamanship, including the duty to carry lights or signals and safety equipment, to keep a proper look-out, or to take any other precautions required by the ordinary practice of seamen or the special circumstances of the voyage. Some collisions may have their origin in other accidents, particularly if a damaged ship does not sink but becomes a drifting object constituting a menace to other vessels that might strike it.
Incidence In the decade ending 1989, 3,302 ships were lost worldwide (on average, one every day). In addition, many more vessels were victims of other marine casualties in which thousands of people were killed and many more injured. In 1988, of the fewest incidents since the 1960s, 230 ships sank and 760 people died in 52 of them.
Every year the equivalent of about a million tons of ocean-going vessels are lost at sea; this represents a greater fleet of ships than the individual fleets of nearly one hundred nations of the world. Although more precise navigational equipment and techniques are available today, larger vessels, increasingly crowded waters, and dangerous cargoes, aggravate the risk factors in shipping.
Refs Mankabady, Samir *The International Maritime Organisation* (1986).
Broader Travel risks (#PD7716) Injurious accidents (#PB0731)
Transport accidents (#PC8478).
Related Risk (#PF7580) Oil pollution (#PE1839).
Aggravates Wrecks and derelicts as hazards (#PE5340)
Obstacles for international ocean shipping (#PD5885).
Aggravated by Fog (#PE1655) Port congestion (#PE4766) Naval arms race (#PD8412)
Hazards to navigation (#PE3868) Sea traffic congestion (#PD1486)
Substandard shipping vessels (#PE6630)
Military manoeuvres in sensitive border areas (#PE3704).

♦ PD8984 Inadequate maintenance

Absence of maintenance system
Broader Inadequate infrastructure (#PC7693).
Narrower Insufficient land upkeep (#PJ8634) Inadequate road maintenance (#PD8557)
Neglect of property maintenance (#PD8894) Unethical maintenance practices (#PD7964)
Inadequate equipment maintenance (#PD1565) Inadequate electrical maintenance (#PG0584)
Inadequate maintenance of infrastructure (#PD0645)
Irregular repair services in the countryside (#PE8997).
Related Substandard housing and accommodation (#PD1251).
Aggravates Inadequate care of community space (#PF2346).
Aggravated by Inadequate maintenance personnel (#PJ0088)
Inadequate maintenance equipment (#PD6520)
Limited accountability of public services (#PF6574)
Undeveloped channels for public and private resources (#PF3526)
Ineffective operation of community networks in urban ghettos (#PF1959).

♦ PD8992 Political rivalry

Political competition
Refs Bogdanov, A *Triangle of Rivalry*.
Broader Competition (#PB0848) Political fragmentation (#PF3216).
Aggravates Political opportunism (#PC1897).

♦ PD8994 Declining international competitiveness

Declining industrial competitiveness — Fluctuations in international competitiveness — Decline in competitiveness of domestic industries — Declining international market share
Broader Competition between states (#PC0114).
Related Reduction of the share of the developing countries in world exports (#PC2566).
Aggravates Trade protectionism (#PC4275)
Loss of international market leadership (#PF8081)
Imposition of trade quotas for political reasons (#PE9762)
Vulnerability of national economies to vagaries of external markets for goods and services (#PF9697).
Aggravated by Declining economic productivity (#PC8908)
Socio-economic burden of militarization (#PF1447).
Reduces Lack of international cooperation (#PF0817).

♦ PD8996 Inferior status employment

Broader Lack of job satisfaction (#PF0171).
Narrower Inadequate living and working conditions of immigrant labourers in industrialized countries (#PD3427).
Aggravates Discrimination against men in employment (#PD3338).
Aggravated by Social stratification (#PB5577) Pursuit of personal prestige (#PF8145).

♦ PD9012 Restrictive legislation

Broader Inadequate laws (#PC6848).
Narrower Vagrancy (#PE5460) Criminalization of drug use (#PF4735)
Criminalization of abortion (#PF6169) Criminalization of euthanasia (#PF2643)
Criminalization of prostitution (#PF6231)
Criminalization of sexual relations out of wedlock (#PE9122).
Aggravates International monopoly of the media (#PD3040).
Aggravated by Proliferation of legislation (#PD5315)
Limited local respect for regional and global legislation (#PF2499).
Reduces Social inequality (#PB0514) Social discrimination (#PC1864).

♦ PD9021 Ignorance of women concerning primary health care

Ignorance of women concerning appropriate child care — Ignorance of women concerning child bearing — Ignorance of women concerning weaning of infants — Ignorance of women concerning infant nutrition — Inadequate maternal education
Nature Infants and children under conditions of deprivation die a slow silent death as a consequence of: malnutrition of mothers, inattention during pregnancy and birth, and neglect of natural practices like breastfeeding; ignorance about when to wean and the foods suitable for that stage; helplessness in responding to common illnesses such as diarrhoea in infants and young children; lack of knowledge of, and lack of access to, immunization against common childhood illnesses; absence of awareness about deficiencies in iron, iodine and vitamin A which lead to debility, cretinism and blindness; indifference to, or ignorance about, the need to periodically watch the growth, in weight and height, of children, as an expression of their status of nutrition and health.
Broader Ignorance of health and hygiene (#PD8023).
Aggravates Infant mortality (#PC1287) Child malnutrition (#PD8941)
Maternal malnutrition (#PE1085) Diarrhoea in children (#PE9751)
Nutritional deficiencies (#PC0382) Ignorance concerning disease (#PD8821)
Inappropriate infant feeding strategies (#PD9661)
Inadequate and insufficient immunization (#PF5969)
Smoking during pregnancy and breast-feeding (#PE5026)

Protein–energy malnutrition in infants and early childhood (#PD0331)
Substitution of inappropriate foodstuffs for breast feeding (#PE8255)
Damage to infant brains from malnutrition and insufficient stimuli (#PE4874).
Aggravated by Gender discrimination in developing countries (#PD9563)
Illiteracy among women in developing countries (#PE8660)
Inadequate nutrition education in least developed countries (#PE0265).

♦ **PD9022 Torts**
Wrongful acts — Wrongful omissions
Nature A wrongful act or omission (under English law), not arising exclusively out of breach of contract or trust, but which gives rise to an action for damages at the suit of the injured party. A breach of contract is however not of itself a tort, but many situations which give rise to an action in contract may also be torts. An act, not otherwise unlawful, will become a tort only when it has been done with a mischievous motive.
Broader Crime (#PB0001).
Narrower Nuisance (#PJ7837)   Defamation of character (#PD2569).
Related Fraud (#PD0486)   Negligence (#PA2658)   Conspiracy (#PC2555).

♦ **PD9033 Inadequate electricity infrastructure**
Insufficient electrical power — Limited electrical capacity
Refs Flavin, Christopher *Electricity for a Developing World* (1986).
Broader Inadequate infrastructure (#PC7693)   Inadequate maintenance equipment (#PD6520).
Narrower Inadequate electricity supply (#PJ0641)
Inadequate electrical maintenance (#PG0584)
Insufficient nuclear power stations (#PD7663)
Inadequate electrical power supply in developing countries (#PE1900)
Inadequate facilities for the transport of electrical energy (#PE1048).
Aggravates Electrical power failure (#PE1341)   Prohibitive cost of electricity (#PG8518)
Maldistribution of electrical energy (#PD3446)
Long–term shortage of electric energy (#PE1216)
Environmental hazards from electricity (#PE1412)
Restricted delivery of essential services to developing country rural communities (#PF1667).
Aggravated by Unresearched electrical source (#PU0625)
Insufficient skilled electricians (#PU0338).

♦ **PD9035 Contempt of judicial process**
Contempt of court
Refs Miller, C J *Contempt of court* (1988).
Broader Contempt (#PF7697)   Incorrect information (#PB3095).
Crimes against the integrity and effectiveness of government operations (#PD1163).
Narrower Criminal contempt (#PD5705)   Refusal to testify (#PE4349)
Withholding of information (#PF8536)   Disobedience of judicial order (#PD3879)
Soliciting obstruction of proceedings (#PD0790)
Hindering proceedings by disorderly conduct (#PD7359)
Failure to appear as witness to produce information or to be sworn (#PE1756).
Aggravated by Collapse of judicial system (#PJ0761).

♦ **PD9044 Prevailing community insecurity**
Broader Insecurity (#PA0857)   Obstacles to community achievement (#PFp118).
Aggravates Scarce options for involvement in culture (#PF6535).
Aggravated by Demonstrations (#PD8522)   Breakdown of police protection (#PF8652).

♦ **PD9045 Intestinal diseases**
Refs Alexander-Williams, J and Irving, M *Intestinal Fistulas* (1982); Walker-Smith, John A *Diseases of the Small Intestine in Childhood* (1988).
Broader Human disease and disability (#PB1044)
Diseases and deformities of the digestive system (#PC8866).
Narrower Colitis (#PE7970)   Peritonitis (#PE2663)   Constipation (#PE3505)
Chronic enteritis (#PG4039)   Peritoneal adhesions (#PG2025)
Intestinal obstruction (#PG4858)   Intestinal perforation (#PG4859)
Diverticula of intestine (#PE8117)   Gastrointestinal diseases (#PE3861)
Acute intestinal diseases (#PG6532)   Intestinal schistosomiasis (#PG3101).
Aggravated by Water pollution in developing countries (#PD3675).

♦ **PD9051 Inaccessible educational facilities**
Remote training resources
Broader Unequal access to education (#PC2163)
Physically inaccessible services (#PC7674)
Restrictions on the acquisition of knowledge (#PF1319).
Narrower Unequal school distribution (#PJ9458)   Distant schooling facilities (#PJ7693).
Related Inadequate welfare services for the aged (#PD0512).
Aggravates Lack of vocational teachers (#PJ9603).
Aggravated by Closure of schools (#PJ8556)
Underutilization of locally available skills (#PF6538).

♦ **PD9057 Cerebral infarction**
Broader Diseases and injuries of the brain (#PD0992)
Disrupted mechanisms for community health (#PF2971).

♦ **PD9082 Unsustainable exploitation of fish resources**
Overfishing — Overexploitation of fish resources — Indiscriminate fishing methods — Ocean stripping — Fisheries depletion — Deployment of excessively efficient fishing nets — Depleted fish resources — Reduction of fish yields — Excessive use of drift nets — Destructive fishing — Cyanide fishing
Nature Overfishing is taking out of the sea more than natural population growth can sustain. Overfishing has a number of causes. Fishermen, nations and international bodies are reluctant to limit catches. In some circumstances, a fishery will be more profitable by catching every fish possible and investing the proceeds. When catches are limited, the form of limitation may encourage overfishing some species. When whale quotas were set in Blue Whale Units - one BWU is equivalent to two fin whales, two and a half humpbacks and six sei whales - it was more profitable to take a blue whale when ever it showed up even when the whaler was seeking smaller species.
Some fishing techniques, like the drift nets yield not only tons of fish but kill millions of birds, whales and seals and catch millions of fish not intended. Small net holes often capture juvenile fish who never have a chance to reproduce. Millions of fish may be killed because of being caught by lost or abandoned fishing equipment. Some forms of equipment destroy natural habitats, for example bottom trawling may destroy natural reefs. The difficulty of negotiating international conservation legislation makes the process of limiting overfishing drawn out, sometimes to the point of endangering species. Other destructive techniques are illegal dynamite and cyanide fishing, the latter killing coral reefs and few moths later the captured goldfish.
Background There are different types of fishermen. Some are explorers, forever seeking out new fishing ground but failing to exploit them optimally. Others play it safe and over-fish a stock they know rather than taking the risk of looking for a new one. Some gauge success by the number of fish that land on the deck of their ship, even if most must be thrown back and if a larger mesh net would yield higher profits. Where the number of fishermen is not managed, a high yield one year may dramatically increase the size of the fleet causing overfishing the next year. In some parts of the world, especially South East Asia, fishing is the occupation of last resort. The landless farmer goes to sea to become locked into the fishing industry.
Incidence Most familiar varieties of fish which have been a traditional food source, and which provide 95 per cent of the world's fish catch, are now threatened by overfishing. Fisheries, whether coastal or oceanic, are fundamental to the diets of many countries. For some countries, fishing is a key economic sector, and overfishing poses immediate danger to several national economies. Traditional fishermen, such as in South Asia, are being forced out of their occupations by mechanized trawlers providing for markets mainly in industrialized countries. The indiscriminate fishing methods are leading to declining fish stocks and thus to a decline in protein consumption among those who have virtually no other source of protein. The rapid growth in demand for fish puts pressure on the nations to, if not encourage, turn a blind eye to overfishing. One of the fastest growing demands for fish is for animal feeds. One third of the global catch is used for oil and fishmeal for this purpose.
Refs FAO *Catches and Landings 1983–1984*; FAO *Approaches to the Regulation of Fishing Effort* (1984); FAO *Fishery Commodities 1983–1984*.
Broader Unsustainable harvesting rates (#PD9578)
Environmental hazards from fishing industry (#PD0743).
Related Trapping of animals (#PE5735)
Degradation of the environment through the destruction of species (#PE5064).
Aggravates Environmental stress (#PC1282)   Hunting of marine animals (#PE0439)
Shortage of animal protein (#PC4998)   Vulnerability of marine ecosystems (#PC1647)
Ecological disruption of animal breeding grounds (#PJ3994)
Conflicting claims concerning off-shore territorial waters (#PD1628).
Reduces Sharks (#PJ3738).

♦ **PD9090 War-time conditions and pressure**
Broader Disastrous consequences of war (#PC4257).
Related Poor living conditions (#PD9156).
Aggravates Mutiny (#PD2589).

♦ **PD9094 Bacterial disease**
Water–borne pathogenic bacteria — Bacterial infections — Bacteria — Bacteria causing disease — Microbes
Refs Baldry, P E *The Battle Against Bacteria* (1976); Evans, Alfred S and Feldman, Harry A *Bacterial Infections of Humans* (1982); Shipley, Elizabeth H *Bacterial Infections* (1987).
Broader Microbial diseases (#PC7492)   Infectious and parasitic diseases (#PD0982).
Narrower Angina (#PE5204)   Leprosy (#PE0721)   Tetanus (#PE2530)
Diphtheria (#PE8601)   Erysipelas (#PG5619)   Septicaemia (#PE9422)
Scarlet fever (#PG2757)   Whooping cough (#PE2481)   Actinobacillosis (#PG9855)
Spirochaetal diseases (#PE3254)   Bacterial plant diseases (#PD2226)
Meningococcol meningitis (#PG4602)   Bacterial diseases in animals (#PD2731)
Haemolytic streptococcal infection of the throat (#PG3063).
Aggravates Asthma (#PD2408)   Liver diseases (#PE1028)
Diseases of metabolism (#PC2270)   Food poisoning through negligence (#PE0561).
Aggravated by Resistant bacteria (#PE6007)   Biological contamination of water (#PD1175).
Reduced by Air pollution (#PC0119).

♦ **PD9108 Dangerous cargo handling**
Risk during loading and unloading
Nature The periods when a cargo is most likely to be damaged is when it is being loaded onto and off of its means of transportation.
Broader Transport of dangerous goods (#PD0971).
Related Inadequate cargo transportation (#PE0430).

♦ **PD9113 Insufficient trained labour**
Unprepared local labour — Undertrained labour force — Inexperienced labour force — Unskilled work force — Little employable skills
Broader Depleted expertise of the rural labour force (#PF2973)
Disparity between workers skills and job requirements (#PC1131).
Narrower Apprentice experience vacuum (#PU8623).
Aggravates Limited means of marketing employable skills (#PE7344)
Inadequate practical training in rural areas (#PF6472).
Aggravated by Lagging training in social skills (#PF8085)
Unattractive locale for economic development (#PF3499)
Lack of opportunities for practical training in communities (#PF2837).

♦ **PD9117 Instability of manufacturing industries in developing countries**
Unstable growth in manufacturing in developing countries
Broader Instability of manufacturing industries (#PC0580).

♦ **PD9118 Destruction of agricultural land**
Damage to agricultural land
Broader Destruction of land fertility (#PC1300).
Narrower Degradation of agricultural land by cash crops (#PE8324).
Related Destruction of land resources (#PJ8397).
Aggravates Unavailability of agricultural land (#PC7597).
Aggravated by Natural disasters (#PB1151)   Man–made disasters (#PB2075)
Over-spacing of suburban housing (#PE1708)   Excessive use of land for agriculture (#PD9534).

♦ **PD9131 Mental disorders**
Mental disorder
Refs Barrett, James and Rose, Robert M *Mental Disorders in the Community* (1986); Fanon, Frantz *The Wretched of the Earth* (1966); Linton, Ralph *Culture and Mental Disorders* (1956); Oshimata, Tana *Therapy of Mental Disorders* (1987); Rogow, Arnold *The Psychiatrists* (1971); Snyder, Solomon H *Biological Aspects of Mental Disorder* (1980); Sullivan, H S *Schizophrenia as a Human Process* (1962); Szasz, Thomas *The Myth of Mental Illness* (1967).
Broader Psychological disorders (#PD8375).
Narrower Psychoses (#PD1722)   Sexual deviation (#PD2198)
Psychological inertia (#PF0421)   Disorders of consciousness (#PD8005)
Anti-social personality disorders (#PF1721)
Physiological malnutrition arising from mental factors (#PE8925).
Aggravates Social stigma (#PD0884)   Abuse of sedatives and tranquillizers (#PE0139).
Aggravated by Fear (#PA6030)   Magic (#PF3311)
Sadism (#PF3270)   Occultism (#PF3312)
War crimes (#PC0747)   Alienation (#PA3545)
Social alienation (#PC2130)   Wasted woman power (#PF3690)
Political indoctrination (#PD1624)   Inhumane interrogation techniques (#PD1362)
Excessive employment of married women (#PD3557)
Abusive detention in psychiatric institutions (#PE2932)
Abusive treatment of patients in psychiatric hospitals (#PD0584).

♦ **PD9145 Parental lying**
Lying of parents to children

Broader Lying (#PB7600).
Aggravates Ineffective parenting techniques (#PJ0551).
Aggravated by Deception between sexual partners (#PE4890).

♦ PD9146 Suppression of information
Suppression of information by corporations — Suppression of information by government
Broader Inappropriate arguments (#PF2152)
Restrictions on freedom of information (#PC0185)
Restrictions on international freedom of information (#PC0931).
Narrower Unreported tax obligations (#PE9061) Unreported government spending (#PF2990)
Suppression of scientific information (#PF1615)
Suppression of information by security classification (#PF4050)
Suppression of information concerning social problems (#PF9828)
Suppression of information concerning environmental hazards (#PF4854).
Related Incorrect information (#PB3095) Misleading information (#PF3096).
Aggravates Denial of evidence (#PD7385) Loss of information (#PF9298)
Underreported issues (#PF9148)
Inconclusiveness of scientific and medical tests (#PD7415).
Aggravated by Censorship (#PC0067) Withholding of information (#PF8536)
Proliferation of information (#PC1298) Restriction of freedom of expression (#PC2162)
Incomplete access to information resources (#PF2401)
Restrictions on the distribution of confidential government information (#PD2926).
Reduced by Misuse of classified communications information (#PD5183).

♦ PD9154 Diseases of male genital organs
Broader Diseases of the genito-urinary system (#PC4575).
Narrower Orchitis (#PG6129) Phimosis (#PG6095) Hydrocele (#PG9722)
Male sterility (#PG7878) Prostate diseases (#PE8372)
Sexual impotence of men (#PF6415) Benign neoplasm of genital organs (#PE9705).

♦ PD9155 Insufficient social security in the agricultural sector
Inadequate social protection of farmers — Lack of social security for farmers — Lack of social security for peasant farmers
Nature Farmers are often excluded from social security protection for sickness, maternity, invalidity, old age, death, occupational injuries and diseases, and family commitments. They experience serious difficulties through the loss of employment and the absence of new employment opportunities in farming areas.
Broader Social insecurity (#PC1867) Economic insecurity (#PC2020)
Inadequate social welfare services (#PC0834).
Aggravates Inadequate conditions of work of agricultural workers (#PE4243).

♦ PD9156 Poor living conditions
Bad living conditions
Narrower Impermanent living conditions (#PD4368)
Substandard housing and accommodation (#PD1251)
Inadequate living and working conditions of immigrant labourers in industrialized countries (#PD3427).
Related War-time conditions and pressure (#PD9090)
International imbalance in the quality of life (#PB4993).
Aggravates Strikes (#PD0694) Trachoma (#PE1946) Dystrophy (#PE3506)
Rheumatic fever (#PE0920) Children's diseases (#PD0622)
Mental disorders of the aged (#PD0919)
Inadequate welfare services for the blind (#PD0542).
Aggravated by Subsistence life style (#PF1078) Prohibitive cost of living (#PF1238)
Ill treatment of prisoners of war (#PD2617)
Deterioration of living standards of workers (#PE9075).

♦ PD9159 Emotional disorders
Affective disorders — Emotional instability — Emotional disturbance — Mood disorders
Refs Buckholdt, David R and Gubrium, Jaber F *Caretakers* (1985); Cantwell, D and Carlson, G (Eds) *Affective Disorders in Childhood and Adolescence* (1983); Carone, Pasquale et al *The Emotionally Troubled Employee* (1976); Davis, John M and Maas, James W (Eds) *The Affective Disorders* (1985); Epanchin, Betty C and Paul, James L *Emotional Problems of Childhood and Adolescence* (1987); Gardos, George and Casey, Daniel E *Tardive Dyskinesia and Affective Disorders* (1984); Knight, Edwin W *Emotional Illness* (1979); McRoy, Ruth G, et al *Emotional Disturbance in Adopted Adolescents* (1988); Whitlock, F A *Symptomatic Affective Disorders* (1983); Whybrow, Peter C, et al *Mood Disorders* (1984).
Broader Mental illness (#PC0300).
Narrower Alexithymia (#PG5045) Anaclitic depression (#PE1883)
Self-defeating behaviour (#PD4418) Lack of intimate relationships (#PF4416).
Aggravates Asthma (#PD2408) Speech disorders (#PE2265) Emotional crises (#PE3407)
Multiple sclerosis (#PE1041) Emotional immaturity (#PJ5907)
Inadequate community care for transient urban populations (#PF1844).
Aggravated by Fear (#PA6030) Rheumatic fever (#PE0920)
Cerebral paralysis (#PE0763) Loneliness in old age (#PD0633)
Natural human abortion (#PD0173) Dependency of children (#PD2476)
State custody of deprived children (#PD0550).

♦ PD9160 Limited medical knowledge
Inadequate medical training — Limited medical skills
Broader Lack of training (#PD8388) Inadequate education (#PF4984)
Ignorance of health and hygiene (#PD8023).
Narrower Nutritional ignorance (#PE5773) Limited paramedic training (#PG8035)
Unavailability of first aid (#PJ1261) Malpractice in plastic surgery (#PE4429)
Inadequate family planning education (#PD1039)
Ignorance of traditional herbal remedies (#PE3946)
Inadequate knowledge of incubation periods for animal diseases (#PE8133).
Aggravates Defective medical devices (#PG8737)
Inadequate nutrition education in least developed countries (#PE0265).
Aggravated by Limited access to practical education (#PF2840).

♦ PD9162 Insecure land tenure
Unclear property rights — Traditional property rights — Short term land tenure — Encroachment on communal land — Ill-defined property rights
Nature Nomads, pastoralists and certain subsistence farmers have property rights defined by tradition. This does not necessarily give them a clearly defined right to the land that they use. This lack of any clear legal tenure may be abused by those with appropriate influence and expertise. As a result such groups may be driven off their lands onto more marginal lands.
In slum areas around major cities in the developing countries ill-defined property rights increase the risk of buying and selling a site and discourages squatters from improving their buildings.
Refs Powelson, John P *The Story of Land* (1987).
Broader Restrictive use of available land (#PF6528).
Related Underdeveloped capacity for income farming (#PF1240)
Stagnated development of agricultural production (#PD1285).
Aggravates Gypsy persecution (#PE1281) Unequal property rights (#PJ2031).
Aggravated by Maldistribution of land in developing countries (#PD0050).
Reduces Individualistic disposition of productive property (#PD4974).

♦ PD9163 Unfair air transport practices
Air transport monopolies — Airline monopolies — Air transport cartels
Incidence Under existing agreements, the number of airline seats between European capitals or regional airports is generally divided precisely between the national carriers. This cartel is enshrined by law, competition is restricted by governments, and one airline would never permit another to gain a bigger share of the custom.
Broader Monopolies (#PC0521) Unfair transport practices (#PD1367).
Aggravates Inadequate air transport service (#PJ0260).

♦ PD9168 Excessive foreign public debt of industrialized countries
Broader Disproportionate external public debt (#PC3056).

♦ PD9170 International imbalance of quality of working life
Poor work environment — Poor working conditions — Poor working environment — Denial of right to favourable work conditions — Adverse working conditions
Refs Delamotte, Yves and Takezawa, S *Quality of Working Life in International Perspective* (1984); ILO *Bibliography of Periodicals on the Quality of Working Life* (1983).
Broader International imbalance in the quality of life (#PB4993).
Narrower Inadequate working conditions of employees of commerce and offices (#PE8251).

♦ PD9178 Disorderly conduct
Nature Disorderly conduct is intentionally harassing, annoying or alarming another person. Fighting in public, making loud or unreasonable noise, using abusive or obscene language in a public place or making an obscene gesture are examples. Other types of behaviour considered disorderly are obstructing vehicular or pedestrian traffic, preventing the use of a public facility, persistently following a person in a public place, loitering in a public place in order to solicit sexual contact, or acting in any way which creates a hazardous, physically offensive or seriously alarming conditions and has no legitimate purpose.
Counter-claim Disorderly conduct is a catch-all law used by the police to harass anyone they please.
Broader Offences against public order (#PD7520).
Narrower Vagrancy (#PE5460) Hindering proceedings by disorderly conduct (#PD7359).

♦ PD9179 Moral offences
Broader Deviance (#PB1125) Offences against public order (#PD7520).
Narrower Sexual offences (#PD4082) Vice and sex traffic offences (#PD8910)
Moral offences in heterosexual pairing (#PG6445)
Offences involving danger to the person (#PD5300).
Related Victimless crime (#PC5005).
Aggravated by Inability to define moral standards (#PF7178).

♦ PD9182 Deception by natural scientists
Lying by scientists
Broader Lying (#PB7600) Exploitation of trust (#PC4422).
Narrower Medical deception (#PD9836).
Aggravates Scientific fraud (#PF1602) Abuse of science (#PC9188).

♦ PD9183 Discrimination against mentally disabled
Broader Denial of rights to disabled (#PC3461) Discrimination against the disabled (#PD9757).
Narrower Denial of rights of mental patients (#PD1148)
Denial of the right to procreate to the severely mentally handicapped (#PE4544).
Aggravated by Normalism (#PF3758) Mental deficiency (#PC1587).

♦ PD9189 Maldistribution of agricultural land
Lack of agrarian reform — Inadequate land reform
Incidence It is estimated that 80 percent of Latin America's land is owned by less than 10 percent of its people, or that 50 percent of the farm land in many parts of Asia is owned by less than 10 percent of farmers.
Claim Land reform would lead to production increases ranging from 10 percent in Pakistan, 20 percent in Colombia to 80 percent in north-eastern Brazil.
Refs Christodoulou, Demetrios *The Unpromised Land* (1989); Koo, Anthony Y *Land Market Distortion and Tenure Reform* (1982); Montgomery, John D *International Dimensions of Land Reform* (1984).
Broader Inadequate social reform (#PF0677) Unequal property distribution (#PC3438).
Aggravates Hunger (#PB0262) Maldistribution of land in developing countries (#PD0050).
Aggravated by Lack of appreciation of cultural differences (#PF2679)
Individualistic disposition of productive property (#PD4974).
Reduces Racism (#PB1047).

♦ PD9193 Unethical practices by police forces
Police malpractice — Police negligence
Broader Abuse of police power (#PC1142) Unethical professional practices (#PC8019).
Narrower Police brutality (#PD3543) Police corruption (#PD2918)
Police intimidation (#PD0736) Military police abuse (#PG6627)
Unauthorized police search (#PD3544) Abusive police interrogations (#PG3615)
Abuses by private police forces (#PE4847) Unethical interrogation methods (#PE3553)
Police indifference to community (#PF8125)
Police crimes during narcotic investigations (#PE5037)
Military and police personnel participation in torture (#PE4119)
Unrestrained use of force in administration of justice (#PE8881).

♦ PD9196 Deterioration in water supply systems
Inadequate maintenance of water systems — Failure to maintain sewage systems — Deficient water lines — Leakage from water supply systems — Leaks
Incidence As an indication, the market for sealants (to seal buildings and fittings against leakage) in the UK in 1986 was £18 million.
Broader Inadequate water system infrastructure (#PD8517).
Aggravates Water losses from irrigation systems (#PE8796).
Aggravated by Fragmented planning of community life (#PF2813).

♦ PD9197 Environmental pollution in socialist countries
Nature Communist development has favoured heavy industry like steel and chemicals, which produce copious toxic waste. Lacking price signals, the planned economies became highly inefficient users of energy, other source of pollution. The main fuel is also the dirtiest: coal, especially brown coal or lignite. Besides, there were or are no pressure from the public on the planners to heed the consequences of pollution.
Incidence The industrial fallout levels in Eastern Europe are 10 to 20 times greater than in Western Europe.
Broader Environmental pollution (#PB1166).

DETAILED PROBLEMS

♦ PD9202 Deterioration of industrialized countries
Reversal of development process in developed countries
Narrower Homelessness in industrialized countries (#PD8285)
Depressed regions in developed countries (#PD8183)
Dietary deficiencies in developed countries (#PD0800)
Declining productivity in industrialized countries (#PD5543)
Environmental degradation in industrialized countries (#PD7835)
Unequal income distribution in industrialized countries (#PE6891)
Personal physical insecurity in industrialized countries (#PE4639)
Personal isolation in communities of industrialized countries (#PD2495)
Institutional obsolescence in modern industrialized societies (#PE2862).
Aggravates Decadence (#PB2542)   Poverty in developed countries (#PC0444)
Illiteracy in developed countries (#PC1383).
Aggravated by Maldevelopment (#PB6207).

♦ PD9219 Personality disorders
Nature Inflexible and maladaptive personality traits can cause significant impairment in social or occupational functioning or subjective distress.
Broader Mental illness (#PC0300).
Narrower Paranoia (#PE0435)   Narcissism (#PF7248)
Self-defeating behaviour (#PD4418)   Schizoid personality disorder (#PG6805)
Avoidant personality disorder (#PE3901)   Sadistic personality disorder (#PG4642)
Dependent personality disorder (#PE6696)   Hysterical personality disorder (#PE4561)
Schizotypal personality disorder (#PG5464)   Borderline personality disorders (#PE4396)
Anti-social personality disorders (#PF1721)
Passive aggressive personality disorder (#PG9329)
Obsessive compulsive personality disorder (#PE7632).

♦ PD9227 Demineralization of the soil
Depletion of soil nutrients
Nature Minerals are removed from the soil in harvested crops or forests or by leaching aggravated by erosion. With the loss of minerals, soils become impoverished. Soil microorganisms are then unable to obtain the nutrients they need to reproduce. As a result the amount of stable humus produced is reduced thus rendering the soil increasingly unable to sustain plants (and especially trees) and animals.
Claim Demineralization leads to the ultimate impoverishment of the soil on which the fate of the biosphere depends.
Broader Soil infertility (#PD0077).
Aggravates Forest decline (#PC7896)   Infertile land (#PD8585).
Aggravated by Soil erosion (#PD0949).

♦ PD9235 Over-rapid timber exploitation
Forest overcutting — Excessive logging
Nature The need for foreign exchange encourages many developing countries to cut timber faster than forests can be regenerated. This overcutting not only depletes the resource that underpins the world timber trade, it causes loss of forest-based livelihoods, increases soil erosion and downstream flooding, and accelerates the loss of species and genetic resources.
Broader Endangered forests (#PC5165).
Aggravates Deforestation (#PC1366)   Unavailability of timber (#PG5299)
Environmental colonialism (#PE3447)   Underutilization of natural resources (#PF1459).
Aggravated by Foreign exchange shortage in developing countries (#PD3068).

♦ PD9252 Military rivalry
Rivalry between armed forces
Refs Bogdanov, A *Triangle of Rivalry*.
Broader Competition (#PB0848).
Aggravates Inadequate government control of military (#PF9542).

♦ PD9293 Urinary system diseases in animals
Broader Animal diseases (#PC0952).
Narrower Urolithiasis (#PE4656)   Bovine cystitis (#PG9549)
Porcine cystitis (#PG4587)   Uroperitoneum in foals (#PG7751)
Disorders of micturition (#PG3893)   Disease of the glomerulus (#PG7594)
Swine kidney worm infection (#PG9211)   Acute renal failure in animals (#PG6737)
Obstructive uropathy in animals (#PG5357)   Chronic renal failure in animals (#PG9680)
Tumours of the kidney in animals (#PG6582)
Giant kidney worm infection in the mink (#PG5485)
Metabolic disease of the kidney in animals (#PG3925)
Tumours of the lower urinary tract in animals (#PG9304)
Congenital anomalies of the urinary system in animals (#PG9411).

♦ PD9297 Health fraud
Claim With the extensive increase in the range of health related products, whether pharmaceutical or advocated by the alternative health community, preoccupation with personal health can lead to debilitating psychological consequences. Some practitioners, especially of alternative cures, actively feed people's anxieties, without necessarily being able to focus on specific illnesses. Whereas practitioners of conventional medicine focusing on specific conditions, overlook the emotional, dietary and environmental causes which may subsequently engender others symptoms. Both approaches are distorted by a need to maximize the financial exploitation of the patient.
Refs Sullivan-Fowler, Micaela *Alternative Therapies, Unproven Methods, and Health Fraud* (1988).
Broader Unethical professional practices (#PC8019).
Narrower Food fads (#PD1189)   Medical quackery (#PD1725)
Unnecessary surgery (#PE9271)   Malpractice in plastic surgery (#PE4429).
Related Scientific fraud (#PF1602).
Aggravated by Escaping reality through popular psychological screens (#PF1112).

♦ PD9306 Understaffing of basic facilities
Undermanning of critical equipment
Broader Labour shortage (#PC0592).
Narrower Understaffed health clinics (#PJ7980)   Insufficient health personnel (#PD0366)
Shortage of military manpower (#PE4920)   Insufficient supervisory personnel (#PG9449).

♦ PD9310 Environmentally unfriendly consumer products
Unrecyclable disposable products
Broader Grievances of consumers (#PD7567)   Dumping of consumer waste products (#PD8942)
Environmental hazards from economic and industrial products (#PC0328).
Reduced by Consumer boycotts (#PE1213).

♦ PD9316 Legal discrimination in favour of offenders
Undue consideration for criminals — Legal bias against victims of crime — Legal indifference to victims of crime
Broader Discrimination before the law (#PC8726).
Narrower Neglect of victims of crime (#PD4823)
Inadequate assistance to victims of rape (#PE4449)
Inadequate assistance to victims of torture (#PE6936)
Inadequate assistance to victims of accidents (#PE4086)
Lack of legal protection of extortion victims (#PE8240)
Employment of criminals in policy-making contexts (#PE4439)
Inadequate assistance to victims of abuse of power (#PE7390)
Inadequate assistance to victims of human rights violations (#PD5122).
Aggravates Inadequate evidence to convict known offenders (#PF8661).

♦ PD9330 Bias in information systems
Biased computerized information displays
Incidence Two American airlines were rigging their computerized reservation systems, on which travel agency business is dependent, so that their flight information received more display-screen prominence than competitors' flights.
Related Bias in document classification systems (#PF6743).

♦ PD9331 Foreign military intervention
Refs Byers, R B and Leyton-Brown, D (Eds) *Superpower Intervention in the Pesian Gulf* (1982).
Broader Foreign intervention in internal affairs of states (#PC3185).
Narrower Military blockade (#PJ5234)   Secret military operations (#PF7669)
Cross border military operations (#PD5272)
Military expeditions against distant objectives (#PE8207)
Related Foreign military presence (#PD3496).
Aggravated by Government support for repressive regimes (#PF4821)
Government action against regimes with alternative policies (#PF2199).

♦ PD9338 Inappropriate use of financial resources
Excessive costs
Narrower Prohibitive labour costs (#PF8763)   Unnecessary education expenditure (#PJ0626)
Low return on investment in developing countries (#PE9811).
Related Socio-economic burden of militarization (#PF1447).
Aggravates Insufficient financial resources (#PB4653)
Excessive growth of social expenditure (#PC6215)
Shortage of financial resources for action against problems (#PF0404).
Aggravated by Worldwide misallocation of resources (#PB6719).

♦ PD9343 Exploitation of the elderly
Fraud against the elderly — Vulnerability of elders' property
Refs Costa, Joseph J *Abuse of the Elderly* (1984).
Broader Fraud (#PD0486)   Lack of protection for the vulnerable (#PB4353).
Narrower Denial of right to retirement (#PD4458)
Denial of equal benefits to elderly workers (#PE1625).
Aggravated by Juvenile stress (#PC0877)   Disrespect for elders (#PF3979).

♦ PD9347 Exploitation of the unemployed
Incidence Unemployed women and men are being persuaded by relative large sums of money to act as subjects in multiple drug studies.
Broader Lack of protection for the vulnerable (#PB4353).
Aggravated by Unethical practices of employers (#PD2879).

♦ PD9349 Politically motivated arrests
Broader Political injustice (#PC2181).
Narrower Wrongful detention (#PD6062)   Arrest of trade union leaders (#PD7630)
Politically motivated mass arrests (#PJ1301).
Related Internment without trial (#PD1576).

♦ PD9356 Fluctuations in real value of money
Broader Lack of assimilation (#PF2132).
Aggravates Strikes (#PD0694)   Economic inefficiency (#PF7556)
Insecurity of resources (#PB8678)   Insecurity of employment (#PD8211)
Unequal income distribution (#PD4962).
Aggravated by Instability of prices (#PF8635)   Exchange rate volatility (#PE5930)
Uncontrolled growth of debt (#PC8316)   Inefficiency of financial markets (#PF6980).

♦ PD9357 Exhaustion of mineral resources
Depletion of mineral resources — Long-term shortage of mineral resources
Nature Industrial development has placed increasing demand on the available mineral resources. These resources are non-renewable and methods of recycling tend to be inadequate.
Broader Long-term shortage of natural resources (#PC4824).

♦ PD9362 Insecurity of leadership
Aggravates Avoidance of negative feedback (#PF5311)
Leadership impaired by illness (#PF8387)
Extreme detachment from represented constituency (#PF0889).
Aggravated by Loss of international leadership (#PF8353).

♦ PD9366 Ill-defined health conditions in animals
Broader Animal diseases (#PC0952).
Narrower Dehydration (#PE8062)   Gout in animals (#PG9735)
Burns in animals (#PG7260)   Lightning stroke (#PG9435)
Shock in animals (#PG3776)   Accidents of animals (#PE4247)
Frostbite in animals (#PG9460)   Hypothermia in animals (#PG4061)
Hyperthermia in animals (#PG6523)   Air sac mite of poultry (#PG9407)
Ammonia burn of poultry (#PG9463)   Pendulous crop of poultry (#PG1421)
Motion sickness of animals (#PG0898)   Breast blisters of poultry (#PG5578)
High-mountain disease of cattle (#PG3920)   Stray voltage in animal housing (#PJ9234).

♦ PD9370 Commercial exploitation of education
Commercially sponsored education — Commercially biased educational materials — Intrusion of advertising into education — Commercial exploitation of students
Nature The cost of financing education has encouraged the development of commercial sponsorship schemes which may involve sales promotion (including leaflets, posters, and direct mail) directed at parents or students, whether for immediate profit or as a longer-term strategy of developing a loyal clientele. Sponsored stationery, textbooks, teaching packs and videos tend to contain a promotional bias and may well be inaccurate.
Broader Proliferation of commercialism (#PF0815).

♦ PD9376 Personal insolvency
Personal bankruptcy
Broader Insolvency (#PC6154).
Related Business bankruptcy (#PD2591).

♦ **PD9384 Political smear campaigns**
Negative election campaigning — Political vilification
Nature Because voters do not study the detail of policies, but go for the general impressions, it is important for political parties to build a picture of unity and affinity with popular prejudices. There is no place for civilized debate when weaknesses of political rivals are exploited.
Broader Unfair elections (#PC2649) Unethical practices in politics (#PC5517).
Aggravates Discriminatory language (#PF7299) Impoverishment of political debate (#PF4600) Covert smear campaigns by government (#PD7171) Avoidance of issues in political campaigns (#PE6311).
Aggravated by Promotion of negative images of opponents (#PF4133).

♦ **PD9389 Economic retaliation**
Trade reprisals — Threat of commercial reprisals — Fear of economic reprisals
Broader Retaliation (#PF9181).
Aggravates Economic conflict (#PC0840).

♦ **PD9394 Sexual offences by juveniles**
Incidence In USA from 1976 to 1986 the arrest rate for 13 and 14–year–olds accused of rape doubled to 40 arrests per 100,000 children. For sex offences like exhibitionism, grabbing and fondling in the same age group arrests increased by 80 per cent.
Broader Sexual offences (#PD4082).
Aggravated by Sexual abuse of children (#PE3265).

♦ **PD9397 Dishonest employees**
Risk of employee infidelity — Criminal workers
Broader Criminals (#PC7373) Unethical practices by employees (#PD4334).
Related Risk (#PF7580) Disloyalty (#PJ1895) Interruption risk (#PF9106).
Aggravates Frauds, forgeries and financial crime (#PE5516).

♦ **PD9398 Sex scandal**
Homosexual scandals
Broader Scandal (#PC8391).
Narrower Government leaders associated with sex scandals (#PE7937).

♦ **PD9412 Inadequate economic integration between regional groupings of developing countries**
Refs Mashbits, Ya G and Utkin, G N (Comp) *Problems of Economic Regionalization in the Developing Countries* (1984).
Broader Fragmented regional cooperation (#PF9129) International economic fragmentation (#PC0025).

♦ **PD9413 Degradation of the atmosphere**
Narrower Acidic precipitation (#PD4904) Stratospheric ozone depletion (#PD6113) Photochemical oxidant formation (#PD3663) Atmospheric corrosion of materials (#PE9525) Deterioration in atmospheric visibility (#PE2593) Increased reflection of solar radiation (#PF5069) Increase in trace gases in the atmosphere (#PD1354) Increase in concentration of atmospheric water vapour (#PE9446).

♦ **PD9414 Corruption in prisons**
Incidence In Mexico two convicted drug traffickers had bribed prison officials to allow them to turn the cell–blocks into virtual villas where they continued their drug–trafficking operations and lived in luxury together with bodyguards and relatives.
Broader Corruption (#PA1986) Official corruption (#PC9533) Institutionalized corruption (#PC9173).
Narrower Police corruption (#PD2918).
Related Bureaucratic corruption (#PC0279).

♦ **PD9430 Inequitable range of salaries**
Disproportionate salary scales within countries — Unfair salary scales
Incidence In the UK, for example, the person earning the highest corporate salary earns as much in a day as the person earning the lowest salary earns in a year. From 1980 to 1989, salaries of company directors rose by up to 856 per cent in the UK at a time when the less privileged were exposed to reductions of: state pensions for a retired couple from 43.3 per cent of average earnings to 32.7 per cent (for a single person, from 27 to 20.5 per cent); unemployment benefits from 21.5 to 16.3 per cent for a single person; industrial benefit from 27.7 to 20.5 for death; maternity (34.7 to 25.2 per cent for a couple; and invalid care allowance from 26 to 19.6 per cent for a couple.
Narrower Underpayment for work (#PD8916) Prohibitive labour costs (#PF8763) Ineffective use of external relations relating to sportsmen (#PE6515).
Aggravated by Lack of minimum wage fixing (#PE6726) Government bias in wage bargaining (#PF6745).

♦ **PD9433 Adulteration**
Nature It is the use of cheaper materials in the production of an article so as to transform it into an inferior article. The consumer or the purchaser can not distinguish this as inferior.
Incidence In 1981 in Spain more than 600 people died of adulterated rape seed oil.
Narrower Product tampering (#PD8804) Adulteration of illicit drugs (#PE0456).

♦ **PD9436 National economic recession**
Domestic recession
Nature Generally defined by economists as two quarters of negative growth in the gross national product, although because of delays in collecting data it is usually difficult to determine when a recession actually begins. An alternative definition is a decline in growth that is so serious that it begins to feed on itself. A recession is characterized by rising unemployment, sharply contracting imports, and falling real wages and living standards. Lower living standards are then unavoidable when the previous level has been artificially raised by unsustainable policies. Such a recession can be damaging to future growth if it is too deep or to prolonged. The blow to the confidence of domestic investors may inhibit necessary new investment. The decline in the economy can also strain financial system and impair its ability to finance new growth. Excessive cuts in spending risk a downward spiral of continually falling output.
Incidence During the recession of the beginning of 1980s the rate of unemployment in the United States and Western Europe escalated from under 6 per cent in 1979 to over 10 per cent in the last quarter of 1982. Output fell for two consecutive years in Italy, UK and Germany FR. Capacity utilization in manufacturing, where most of the rise in unemployment took place, fell drastically. Interest rates rose sharply with the tightening of monetary policy.
Refs Lewin, Keith *Educational Finance in Recession*; United Nations *Economic Recession and Specific Population Groups* (1986).
Broader International economic recession (#PF1172).
Aggravates Crime (#PB0001) Environmental degradation of inner city areas (#PC2616).
Aggravated by Unemployment (#PB0750) Underpayment for work (#PD8916) Deteriorating quality of life (#PF7142).

♦ **PD9437 Government loan defaults**
Unpaid government loans
Broader Restrictive effects of external capital on development (#PF3318) Arrears in payment of government financial commitments (#PF1179).
Aggravates Global financial crisis (#PF3612) Arrears in financial payments between government agencies (#PE9700).

♦ **PD9448 Overproduction of food**
Agricultural overproduction
Incidence In some parts of the world (or in some parts of the same country), too much food is grown. Such overproduction is expensive economically and ecologically. Subsidized disposal of the surpluses (often as food aid) depresses markets for commodities such as rice and sugar, and undermines the economies of the developing countries that depend on them. Overproduction and its attendant environmental and economic ills are greatest in North America and Europe.
Broader Over–production of commodities (#PD1465).
Narrower Surplus domestic animal production (#PJ5775).
Aggravates Food wastage (#PD8844) Unsustainable agricultural development (#PC8419) Limiting effect of individual survivalism (#PF2602) Economic disadvantages of excessive food production in developing countries (#PF4130).
Aggravated by Instability of production of food and live animals (#PD2894).

♦ **PD9449 Financial paralysis of developing countries**
Broader International economic injustice (#PC9112) Declining economic growth in developing countries (#PD5326).
Aggravated by Excessive foreign public debt of developing countries (#PD2133) Deterioration in external financial position of developing countries (#PE9567).

♦ **PD9480 Deformation of plant life**
Plant abnormalities — Deformed plants — Increase in plant genetic defects
Aggravated by Radioactive contamination of plants (#PD0710).

♦ **PD9481 Corruption of minors**
Broader Corruption (#PA1986).
Narrower Sexual abuse of children (#PE3265) Sexual intercourse with minors (#PE6522).
Related Exploitation of children (#PD0635) Victimization of children (#PC5512) Sexual exploitation of children (#PD3267).
Aggravates Youth gangs (#PD2682) Child pornography (#PF1349) Juvenile alcoholism (#PD1611) Juvenile delinquency (#PC0212) Drug abuse by adolescents (#PD5987).
Aggravated by Orphan children (#PD7046) Corruptive crimes (#PD8679) Neglected children (#PD4522) Torture of children (#PD2851) Children of alcoholics (#PD4218) Family rejection of children (#PC8127) Physical maltreatment of children (#PC2584) Criminal investment in youth market (#PD5750).

♦ **PD9489 Speculation on money markets**
Exchange rate speculation — Foreign currency speculation
Incidence Foreign exchange markets daily turnover was almost $180 billion in 1989. Only about 20 per cent of that daily business is concerned with paying for imports and exports, the rest is speculative.
Broader Capitalist speculation (#PC2194) Financial destabilization of world trade (#PC7873).
Aggravates Exchange rate volatility (#PE5930) Misalignment of currencies (#PF6102) Speculative flight of capital (#PC1453).
Aggravated by Mismanagement of exchange rate system (#PF1874).

♦ **PD9490 Forest fragmentation**
Nature With the spread of human activity, and the need to clear land for agricultural activity, roads and dwellings, large forests are gradually cut up into smaller and smaller parts. This has considerable impact on the wildlife, reducing their territories, increasing their competition for food, and increasing their exposure to other predators and parasites (as in the case of forest–dwelling birds).
Broader Endangered forests (#PC5165) Unsustainable development of forest lands (#PD4900).

♦ **PD9495 Militarization in developing countries**
Nature Militarization in developing countries often has an important influence on capital intensity and import dependence of their economies. One purchase of military equipment and weapons systems often enhances propensities for subsequent additional expenditures. Both indigenous military production and production of military equipment and components under arrangements for the transfer of technology require substantial imports and other expenditures of foreign exchange.
Incidence It has been estimated that the arms imports of a group of 20 countries with relatively large external debts were equivalent to 20 per cent of the rise in their external debt between 1976 and 1980.
Refs Akinyemi, A B et al *Disarmament and Development*; Wolpin, Miles D *Militarization, Internal Repression and Social Welfare in the Third World.*
Broader Militarization (#PD1897).
Narrower Proliferation of nuclear weapons in developing countries (#PE9052).
Aggravates Arms trade with developing countries (#PD3497) Excessive foreign public debt of developing countries (#PD2133).
Aggravated by National insecurity in developing countries (#PD3835).
Reduces Military destabilization of developing countries (#PD7714).

♦ **PD9498 Fear of officialdom**
Fear of government agents — Fear of bureaucracy
Broader Fear (#PA6030).
Narrower Fear of police (#PF8378).
Aggravates Inefficient public administration (#PF2335) Lack of cooperation with officialdom (#PF8500).
Aggravated by Fear of retaliation by authorities (#PF3707) Unchecked power of government bureaucracy (#PD8890) Proliferation of public sector institutions (#PF4739).

♦ **PD9504 Decay of rural communities**
Village decay
Nature Typical small villages are being engulfed by new suburban development or ribbon development stretching out along country roads or haphazard scattered development in the countryside. Villages are in a danger to become commuter or retirement dormitories accessible only to those on higher incomes.
Aggravated by Unsustainable rural development (#PD4537) Demoralizing images of rural community identity (#PF2358).

♦ **PD9515 Environmentally harmful dam construction**

Narrower Dam failures (#PE9517) Aquatic weeds (#PD2232)
Inundation of forests through dams (#PE7855) Coastal erosion resulting from dams (#PJ7396)
Destruction of fisheries through dams (#PE8148)
Reservoir induced increases in seismicity (#PG7404)
Inundation of agricultural land through dams (#PE3786)
Inundation of wildlife habitats through dams (#PE7794)
Submergence of historical sites through dams (#PJ9372)
Ecosystem modifications due to creation of dams and lakes (#PD0767)
Reduction of soil fertility downstream due to impoundment (#PE8782)
Increase in pests and diseases through perennial irrigation (#PE7832).
Related Silting of water systems (#PD3654).
Aggravates Involuntary mass resettlement (#PC6203)
Inappropriate cash crop policy (#PF9187)
Loss of water to industrial uses (#PE7433)
Destruction of archaeological sites (#PD4502)
Export of hazardous industries to developing countries (#PE6687)
Increasing proportion of land surface devoted to urbanization (#PE5931).
Aggravated by Neglect of environmental consequences of government policies (#PE9295).

♦ PD9534 Excessive use of land for agriculture
Uncontrolled growth of arable land
Incidence Until quite recently the major form of investment in rural productivity took the form of extensive land clearance and land reclamation. In the 120 years preceding 1978, the total world expansion of agricultural land has been 852 million hectares. In the period 1972–80 in the USA, cropland was increased by 24 million hectares, and it is estimated that a further 24 to 28 million hectares of pasture, rangeland and forest will be converted to crops in the following 30 years. Such massive conversions have had profound effects on soil conditions and biological diversity.
Refs Brown, Lester R *Food or Fuels* (1980).
Aggravates Soil erosion (#PD0949) Deforestation (#PC1366)
Destruction of agricultural land (#PD9118)
Vulnerability of world genetic resources (#PB4788).
Aggravated by Vulnerability of wetlands (#PC3486)
Disruption of arid zone ecosystems (#PD7096)
Unavailability of agricultural land (#PC7597).

♦ PD9538 Misuse of government surveillance of communications
Abusive monitoring of communication by governments
Refs French, S *The Big Brother Game* (1986).
Narrower Political surveillance (#PD8871)
Misuse of postal surveillance by governments (#PD2683)
Misuse of satellite surveillance by governments (#PF3701)
Misuse of telephone surveillance by governments (#PD1632)
Misuse of electronic surveillance by governments (#PD2930)
Misuse of radio transmission surveillance by governments (#PJ3739).

♦ PD9544 Abuse of computer systems
Abuse of information systems — Misuse of computer databases
Nature Abuse ranges from game–playing on office computers to unauthorized access into computer systems (hacking), possibly with the intent to commit a crime, to engage in some form of espionage, or to sabotage data. Sabotage, with the deliberate destruction of computer files, tends to result from disgruntled employees. Misuse may also take the form of improper disclosure of information to other parties.
Narrower Computer viruses (#PD3102) Computer–based crime (#PE4362)
Uncontrolled use of computer data (#PF4176)
Discriminatory design of information systems (#PD7450)
Unauthorized access to computer information systems (#PE9831).
Aggravated by Vulnerability of computer systems (#PE8542).

♦ PD9557 Governmental disregard for legitimate protests
Governmental rejection of citizen grievances
Aggravates Grievance (#PF8029) Grievances of citizens (#PJ5035).
Aggravated by Government arrogance (#PF8820).

♦ PD9563 Gender discrimination in developing countries
Sexual discrimination in developing countries
Nature In male–dominated societies typical of many developing countries, discrimination against females is characterized by: a mortality rate for female infants which is generally higher than that for females; a literacy level that is lower for girls and women; lower female life expectancy; a health and nutritional status which is worse than that for males; a death rate which is higher partly due to very high maternal mortality rates; an adverse population sex ratio; lower food intake by females; lower access to helath care; lower employment opportunities and income levels.
Refs Anker, Richard and Hein, Catherine *Sex Inequalities in Urban Employment in the Third World* (1986).
Broader Sexual discrimination (#PC2022).
Aggravates Ignorance of women concerning primary health care (#PD9021).
Aggravated by Male domination (#PC3024)
Deliberate imbalancing of population sex ratio (#PF3382).

♦ PD9578 Unsustainable harvesting rates
Overharvesting — Excessive economic exploitation of animal species — Over–exploitation of biological resources — Unregulated harvesting
Nature Biological resources include wild organisms harvested for subsistence, commerce, or recreation (such as fish, game, timber or furbearers); domesticated organisms raised by agriculture, aquaculture and silviculture; and ecosystems (such as rangeland) cropped by livestock. Irrespective of the condition of their habitats, excessive rates of harvesting, especially of animal species, can lead to their extinction for economic purposes, and in some cases their biological extinction. Intensive harvesting of species can also endanger their genetic diversity.
Narrower Unjustified game cropping (#PE1327)
Unsustainable exploitation of fish resources (#PD9082).
Related Excessive land usage (#PE5059).
Aggravates Shortage of animal protein (#PC4998)
Endangered species of plants and animals (#PB1395)
Decreasing diversity of biological species (#PD7302).

♦ PD9603 Invasion of privacy by media
Press abuses of privacy
Broader Infringement of privacy (#PB0284) Unethical media practices (#PD5251).
Aggravated by Journalistic irresponsibility (#PD3071).

♦ PD9628 Discrimination against women executives
Discrimination against women in business — Discrimination against women non–manual workers — Unequal promotion of women — Discriminatory requirement for over–qualification of women
Nature The demand for higher qualifications for the same job from a woman than from a man, because of prejudice and general practice, means that fewer top opportunities are open to women. Women with high qualifications often have to accept lower positions with less chance of responsibility and promotion than a man of the same calibre. Promotion opportunities through retraining programmes which are open to men, are not to women on an equal basis. As competition among women increases for lower positions, regarded as 'women's work', the qualification requirements increase out of proportion with the requirements in the equivalent 'men's work' sector.
Incidence Women account for only 5 per cent of American expatriate managers. Reasons for not sending women managers abroad include: women lack technical qualifications, they do not apply or have conflict with spouse's career and local prejudices against female managers.
Refs Bloomfield, Horace R *Female Executives and the Degeneration of Management* (1983); ILO *Problems of Women Non–Manual Workers* (1981).
Broader Discrimination against women in employment (#PD0086).
Related Unequal pay for women (#PD0309).
Aggravates Unequal promotion (#PG6294).

♦ PD9637 Commodity speculation
Speculation on commodities futures markets
Nature Commodity speculation may be undertaken through the buying, holding and reselling of physical commodities or through the buying and selling of commodity futures contracts – not the commodity itself. Speculators hope to make profits by anticipating price movements. If they expect the price of a commodity to go higher they buy futures contracts, which is known as taking the "long" position, whereas if they think the price will fall they will sell futures, or sell "short". Commodity markets are widely used also as a means of foreign exchange speculation for the reason that, in the case of commodities traded internationally, the effects of the exchange rate movement will be combined with the underlying shifts in the value of the commodity.
Refs Belveal, L Dee *Speculation in Commodity Contracts and Options* (1985).
Broader Capitalist speculation (#PC2194) Inaccurate forecasting (#PF4774)
Obstacles to commodity futures trading (#PF4870).
Aggravates Instability in export trade of developing countries producing primary commodities (#PD2968).

♦ PD9641 Unrepresentative electoral systems
Biased voting systems — Unjust voter registration procedure — Undemocratic voting systems
Incidence In the 1980 New York Senate Election, a conservative receiving only 45 percent the vote won, not because the state wanted a conservative, but because of the 2 other candidates, both liberals, one received 44 percent of the vote and the other 11 percent. If there had been only 2 candidates, one conservative facing one liberal, it is highly likely (and supported by an ABC News survey of voters) that the liberal would have been sent to Washington.
Claim The approval voting system often works against the majority–wanted candidate when more than 2 candidates are in a race. The voter aware of this contradiction comes to realize that casting a ballot for his first choice might mean throwing his vote away, so he comes to improvise a voting strategy chosen to minimize losses rather than to express his sincere choice.
Broader Unfair elections (#PC2649).
Aggravates Political stagnation (#PC2494)
Bias in selection of political candidates (#PD2931).

♦ PD9651 Inter–cultural trade barriers
Broader Non–tariff barriers to international trade (#PC2725).

♦ PD9653 Excessive accumulation of wealth by government leaders
Hypocritical accumulation of personal wealth by rulers of countries — Corrupt acquisition of offshore assets by heads of state — Personal greed of government leaders — Excessive salaries of government leaders
Incidence In the case of a number of developing countries, life presidents, absolute rulers, senior civil servants, military leaders and other members of ruling elites have diverted funds intended for the acceleration of the development process (received in the form of aid grants and bilateral or multilateral loans). In many of these countries capital flight remains the greatest constraint on achievement of agricultural and commercial self–sufficiency. In 1990 the President of Zaire, who came to power in 1965, was estimated to have a amassed a personal fortune of $6 billion (including much real estate in Europe), although he only claimed $50 million. Studies by the World Bank have identified hundreds of millions of dollars unreported in the national budget of Zaire, and more gold and coffee smuggled out of the country than legally exported.
Claim Unless all governments cooperate in making their leaders fully accountable to their people and prevail on offshore banking centres to trace, and report to competent national authorities, corruptly gained flight capital movements, new financial aid to many Third World countries, coupled with debt reduction, will not achieve the goals set.
Broader Personal wealth (#PC8222) Accumulation of capital (#PC5225)
Corruption of government leaders (#PC7587).
Related Government hypocrisy (#PF9050).
Aggravates Insufficient role models (#PF8451) Lack of models of equality (#PF8639)
Leadership as symbolic of wealth (#PF2870) Mediocrity of government leaders (#PF3962)
Inequitable distribution of wealth (#PB7666).
Aggravated by Abusive national leadership (#PD2710).

♦ PD9654 Endocrine diseases in animals
Broader Animal diseases (#PC0952).
Narrower Addison's disease (#PG2161) Hirsutism in animals (#PE5478)
Hypothyroidism in animals (#PG5976) Hyperthyroidism in animals (#PG9622)
Diabetes mellitus in animals (#PG4040) Pituitary tumours in animals (#PG9780)
Panhypopituitarism in animals (#PG5065) Diabetes insipidus in animals (#PG9521)
Hyperadrenocorticism in animals (#PG9537)
Humoral hypercalcaemia of malignancy (#PG0891)
Functional islet cell tumours in animals (#PG3791).

♦ PD9655 Superpower rivalry
Refs Hoxha, Enver *Superpowers*; MacFarlane, S N *Superpower Rivalry and Third World Radicalism* (1985).
Broader Competition between states (#PC0114).
Aggravated by Super–power chauvinism (#PE7778).

♦ PD9661 Inappropriate infant feeding strategies
Improper infant weaning
Narrower Substitution of inappropriate foodstuffs for breast feeding (#PE8255).
Aggravated by Ignorance of women concerning primary health care (#PD9021).

♦ PD9667 Skin diseases in animals
Broader Animal diseases (#PC0952).
Narrower Fowlpox (#PG7028) Saddle sores (#PG9581) Wool maggots (#PG9392)
Dermatophilosis (#PG9258) Interdigital cysts (#PG0230) Alopecia in animals (#PG9568)

Eczema nasi of dogs (#PG0899)
Urticaria in animals (#PG1377)
Acanthosis in animals (#PD6423)
Exudative epidermitis (#PG9439)
Parafilaria infection (#PG9257)
Hypodermosis of cattle (#PG9238)
Flea allergy dermatitis (#PG9287)
Pityriasis rosea in pigs (#PG7180)
Dermatophytosis in animals (#PG4581)
Contagious ecthyma in animals (#PG7491)
Ulcerative dermatosis of sheep (#PG4644)
Congenital anomalies of the skin (#PG7626)
Cuterebra infestation in small animals (#PG5894).
Pyoderma in animals (#PG5996)
Sweet itch in horses (#PG9200)
Dermatitis in animals (#PG5935)
Helminths of the skin (#PE5380)
Pediculosis in animals (#PG9723)
Pox diseases in animals (#PE7304)
Parakeratosis in animals (#PG4123)
Papillomatosis in animals (#PG4795)
Lumpy skin disease of cattle (#PG7207)
Photosensitization in animals (#PG5561)
Necrotic dermatitis of poultry (#PG9606)
Aggravated by Bedbugs (#PE3617).

♦ PD9669 Microbial contamination of food
Claim People who wonder about the safety of the food are usually worried about pesticides and food additives while there has been an explosion of diseases, such as gastrointestinal illnesses and salmonella, caused by food-borne microorganisms in poultry, meat, seafood and eggs.
Broader Microbial diseases (#PC7492) Biological contamination of food (#PD2594).
Aggravated by Environmental hazards from food processing industries (#PE1280).

♦ PD9670 Disruption of the hydrological cycle
Disruption of natural water systems
Nature Human activities may change various aspects of the hydrological cycle. Among other consequences this may contribute to climate modification over large areas of the Earth. Direct changes in surface hydrology over large areas may occur as a result of changes in the vegetative cover or in the area covered by irrigation systems. Tropical deforestation results in significant modifications to surface hydrology, since any disruption in vegetative cover effects the water-storage capacity of the land and modifies the water supply. To the extent that runoff is increased, evapotranspiration is decreased, modifying the time and spatial characteristics of rainfall.
Broader Natural environment degradation (#PB5250).
Narrower Rising water level (#PD8888)
Overexploitation of underground water resources (#PD4403).
Aggravates Silting of water systems (#PD3654).
Aggravated by Deforestation (#PC1366) Mismanagement of irrigation schemes (#PE8233)
Water losses from irrigation systems (#PE8796)
Sunlight inhibition by nuclear warfare soot (#PE6350).

♦ PD9679 Protectionism in international trade against exports from developing countries
Inaccessibility of markets of developed countries to developing countries — Decline in exports of developing countries to developed-market economies
Nature Increased unemployment in developed countries has encouraged protection against labour-intensive imports from developing countries, especially traditional products. Increased protectionism has had a direct influence on the growth performance of developing countries by lowering the demand for their exports and exerting a downward pressure on export prices and earnings. It has also hampered efforts in developing countries to diversify away from traditional products.
Incidence The total exports of developing to developed market-economy countries declined by about one fifth between 1981 and 1985.
Claim If developing countries are to reconcile a need for rapid export growth with a need to conserve the resource base, it is imperative that they achieve access to the markets of industrialized countries in the case of those non-traditional exports where they have a comparative advantage.
Broader International economic injustice (#PC9112)
Protectionism in international trade (#PC5842).
Narrower Obstacles to trade in fertilizer products (#PE4525)
Protectionism in developed countries against agricultural products from developing countries (#PE8321).
Aggravates Vulnerability of national economies to vagaries of external markets for goods and services (#PF9697).
Aggravated by Economic backlash from the richer nations (#PD5995).

♦ PD9685 Exploitation of the mentally handicapped
Exploitation of the mentally ill
Incidence Mentally ill who are left to fend for themselves are tricked in to doing work and then dumped without wages, sexually and physically abused or they allowances are stolen.
Broader Denial of rights of mental patients (#PD1148)
Lack of protection for the vulnerable (#PB4353).

♦ PD9692 International trade in chemical weapons
Trade in chemical warfare technology
Broader International arms trade (#PC1358).
Narrower Trade in products for chemical warfare (#PE3808).
Aggravated by Unethical practices of employers (#PD2879).

♦ PD9699 Criminal insanity
Nature Criminal insanity means that a defendant was impaired by mental illness or retardation lacking the basic capacity to engage in morally reprehensible behaviour, at the time of the crime.
Refs Monahan, John and Steadman, Henry J *Mentally Disordered Offenders* (1983); Templin, Linda A *Mental Health and Criminal Justice* (1984).
Broader Criminality (#PA9226).
Narrower Psychotic violence (#PE7645).
Related Diminished personal capacity (#PF9429).
Aggravated by Alcohol abuse (#PD0153) Mental illness (#PC0300)
Mental deficiency (#PC1587)
Inadequate facilities for mentally disabled criminals (#PE8900).

♦ PD9713 Malpractice in the construction industry
Broader Unethical industrial practices (#PD2916).
Aggravates Structural failure (#PD1230) Military-industrial malpractice (#PD4361).
Aggravated by Low-quality construction work (#PD7723)
Absence of accountability in construction planning (#PF2804)
Ineffective self-regulation in the housing construction industry (#PE8265).

♦ PD9736 Inadequately cooled shelters
Poorly cooled homes
Broader Inadequate housing (#PC0449).
Aggravated by Homelessness (#PB2150).

♦ PD9742 Diseases of breast
Diseases of fallopian tube — Diseases of ovary — Diseases of parametrium
Refs Devereux, D and Greco, G *Color Atlas on the Treatment of Breast Diseases* (1989); Grundfest-Broniatowski and Esselstyn *Controversies in Breast Disease* (1988); Haagensen, C D *Diseases of the Breast* (1986).
Broader Diseases of the genito-urinary system (#PC4575).
Narrower Mastitis (#PJ1369) Salpingitis (#PG7551)
Benign neoplasm of breast (#PG5371).

♦ PD9750 Neonatal mortality
Nature Infant mortality is a composite index which includes deaths of infants 0–1 weeks of age, namely neonates, and those of infants 7 days to 12 months of age. The causes of death in these two periods are characteristic and call for different modes of intervention. A new born baby is termed a neonate only if it shows life immediately after birth, otherwise it is termed a still-birth. In the neonatal period, deaths may be due to antenatal and birth causes. The single most important cause of neonatal deaths is prematurity and its related problems, followed closely by birth trauma, maternal diseases during pregnancy, congenital abnormalities (particularly neural tube and cardiac defects) and infections.
Incidence At levels of infant mortality in excess of 100 per 1000, about one third of the deaths occur in the neonatal period. At lower levels of infant mortality, neonatal deaths become very important and may account for 70 percent of infant deaths. The former is the situation in developing countries like Kenya, and the latter is the case in industrialized countries.
Broader Infant mortality (#PC1287) Perinatal morbidity and mortality (#PD2387).
Aggravated by Infection (#PC9025) Birth trauma (#PE8911)
Premature birth (#PD1947) Pregnancy disorders (#PD2289)
Health risks of teenage sex (#PE6969)
Human physical genetic abnormalities (#PD1618).

♦ PD9757 Discrimination against the disabled
Discrimination against the handicapped — Stigmatization of the handicapped
Broader Denial of rights to disabled (#PC3461).
Narrower Discrimination against mentally disabled (#PD9183)
Discrimination against physically disabled (#PD8627)
Segregation of handicapped children in education (#PE8424)
Immigration barriers for handicapped family members (#PE4868)
Unequal employment opportunities for disabled persons (#PE0783).
Aggravates Teasing (#PE4187)
Stress on families of the physically or mentally handicapped (#PD1405).
Aggravated by Human disability (#PC0699).

♦ PD9765 Toxic waste smuggling
Illicit movement of toxic products
Refs OECD *Transfrontier Movements of Hazardous Wastes* (1985).
Broader Transport of dangerous goods (#PD0971)
Evasion of customs and excise duties (#PD2620).
Aggravates Hazardous waste dumping (#PD1398).
Aggravated by Hazardous wastes (#PC9053) Unethical practices in transportation (#PD1012).

♦ PD9792 Political destabilization of developing countries
Broader Destabilization of social systems (#PB5417).
Aggravates Excessive foreign public debt of developing countries (#PD2133).

♦ PD9802 Over-subsidized agriculture in industrialized countries
Excessive subsidies to farmers in developed countries — Over-dependence on farm subsidies in developed countries — Counter-productive price supports and subsidies — Domestic agricultural price difficulties in developed countries
Nature In industrialized countries the agricultural sector is marked by a competitive structure, low price and income elasticity of supply and demand, rapid technical progress, and relatively low mobility of resources employed. One consequence is that farmers' returns on investments cannot be maintained by price policies alone. Despite this, most developed importing countries maintain farm prices well above international levels, often with the dual objective of supporting incomes and reducing reliance on imported foodstuffs. Once established, support prices can only be lowered with considerable loss to those who have invested in the industry in the expectation of higher returns, and have purchased land and equipment at prices which have been inflated by the support levels. Productivity of land is generally high and productivity of labour, though rising, is low. Established farmers expect their governments to control imports, a pressure which is often difficult to resist. The relative decline in the cost of producing certain commodities may enable governments to resist raising levels of price support, although this approach is incapable of solving basic structural problems.
Background At the beginning of the 1970s, farmers' incomes were approaching the parity objective of the 'green laws' and the fair standard of living provided for in the Treaty of Rome. These objectives are receding in the 1980s and farm policies have been powerless in most European countries to prevent farm incomes from getting out of step with those in other sectors.
Incidence The cost of direct or indirect farm subsidies in Europe and North America has risen considerably and now covers almost the entire food cycle. In the EEC such costs have grown from $6.2 billion in 1976 to $21.5 billion in 1986, and in the USA they have grown from $2.7 billion in 1980 to $25.8 billion in 1986. In 1990 it was estimated that farm subsidies and import barriers cost the 17 major industrialized countries $72 billion per year in lost income. Each agricultural job saved by farm supports costs the equivalent of $13,000 in lost household income in Japan, $20,000 in the European Community and the USA, and almost $100,000 in Canada.
Claim The policies reduce economic efficiency, destabilize world markets, exacerbate tensions between industrialized countries, and threaten progress towards further multilateral trade liberalization.
Refs Franklin, Michael *Rich Man's Farming.*
Broader Protectionism in agriculture and the food production industries (#PD5830)
Distortion of international trade by selective domestic subsidies (#PD0678)
Detrimental international repercussions of domestic agricultural policies (#PF2889).
Related Unknown availability of subsidies (#PG9905)
Limited availability of financial credit (#PF2489)
Inappropriate modernization of agriculture (#PF4799).
Aggravates Economic conflict (#PC0840) Chemicalized farming (#PD7993)
Agricultural surpluses (#PC2062) Disincentives against farming (#PD7536)
Destruction of the countryside (#PE3914) Environmental hazards from fertilizers (#PE1514)
Lack of economic and technical development (#PE8190)
Inappropriate agricultural subsidies for chemicalized farming (#PE1785)
Protectionism in developed countries against agricultural products from developing countries (#PE8321).
Aggravated by Distortion of national economies from food subsidies (#PE7413)
Inappropriate government intervention in agriculture (#PE1170)
Domestic agricultural price policy difficulties in developing countries (#PE2890).

♦ PD9805 Substance abuse at work
Nature Substance abuse affects job safety, productivity, absenteeism, morale, and the quality of the products and services companies provide.
Incidence Alcohol and illegal drug use by employees is estimated to cost US industry over $100 billion per year.
  Broader Substance abuse (#PC5536).
  Narrower Drug abuse at work (#PD4514)   Alcoholic intoxication at work (#PE2033)
  Substance abuse during control of complex equipment (#PE0680).
  Aggravates Injurious accidents (#PB0731).
  Aggravated by Increasing job monotony (#PD2656).

♦ PD9807 Medication side effects
Side effects of pharmaceutical products
Claim Even odds of a fatal reaction of a million to one are unacceptable.
Counter-claim Hundreds of thousands of patients have benefitted immeasurably from drugs which in rare cases trigger a potentially fatal idiosyncratic reaction.
Refs Dukes, M *Side Effects of Drugs*.
  Aggravates Use of animals in toxicological experiments (#PE9611).
  Aggravated by Inadequate testing of drugs (#PD1190)
  Inconclusiveness of scientific and medical tests (#PD7415).

♦ PD9836 Medical deception
Withholding of medical information — Lying by medical professionals
Nature The patient may be told nothing at all or only part of the story, or relevant information may be couched in technical or euphemistic language, calculated to deceive.
Claim The medical ethics require that the doctors are concerned for the well-being of their patients, and this is not constrained by the demand for truthfulness. It is better that the patients are able to enjoy what time they have left free from anxiety and fear.
Counter-claim Patients are generally in favour of being told the truth about their condition. Uncertainty and confusion can often be more unsettling than the truth which is crucial to the restoration of some control over one's affairs.
Refs Fletcher, Joseph and Menninger, Karl *Morals and Medicine* (1979).
  Broader Lying (#PB7600)   Deception by natural scientists (#PD9182).
  Aggravated by Medical paternalism (#PF5397)   Inadequate testing of drugs (#PD1190).

♦ PD9842 Blackmail by government officials
Official blackmail — Extortion by public officials
Nature Common law extortion is the corrupt collection of an unlawful fee by a public officer through the power of his office. It is a misdemeanour.
  Broader Criminal coercion (#PD4469).
  Aggravated by Official secrecy (#PC1812)   Deception by government (#PD1893)
  Governmental incompetence (#PF3953).

♦ PD9848 Complacency in science
  Broader Complacency (#PA1742).

♦ PD9869 Attractive city jobs
Nature Through out the world and especially in developing countries jobs in urban areas are more attractive for many reasons. They tend to be more respectable. Urban areas are seen to have more jobs available. Farming is often seen as backward, dirty and for the uneducated. Urban areas offer many cultural advantages: entertainment, education, anonymity and breaks from the past.
  Broader Rural underdevelopment (#PC0306).
  Aggravates Uncontrolled urban development (#PC0442)
  Unperceived relevance of formal education in rural communities (#PF1944)
  Emigration of trained personnel from developing to developed areas (#PD1291).
  Aggravated by Attraction of city life (#PJ7861)   Mobility of village populations (#PE1848).

♦ PD9997 Urban terrorism
  Broader Terrorism (#PD5574).

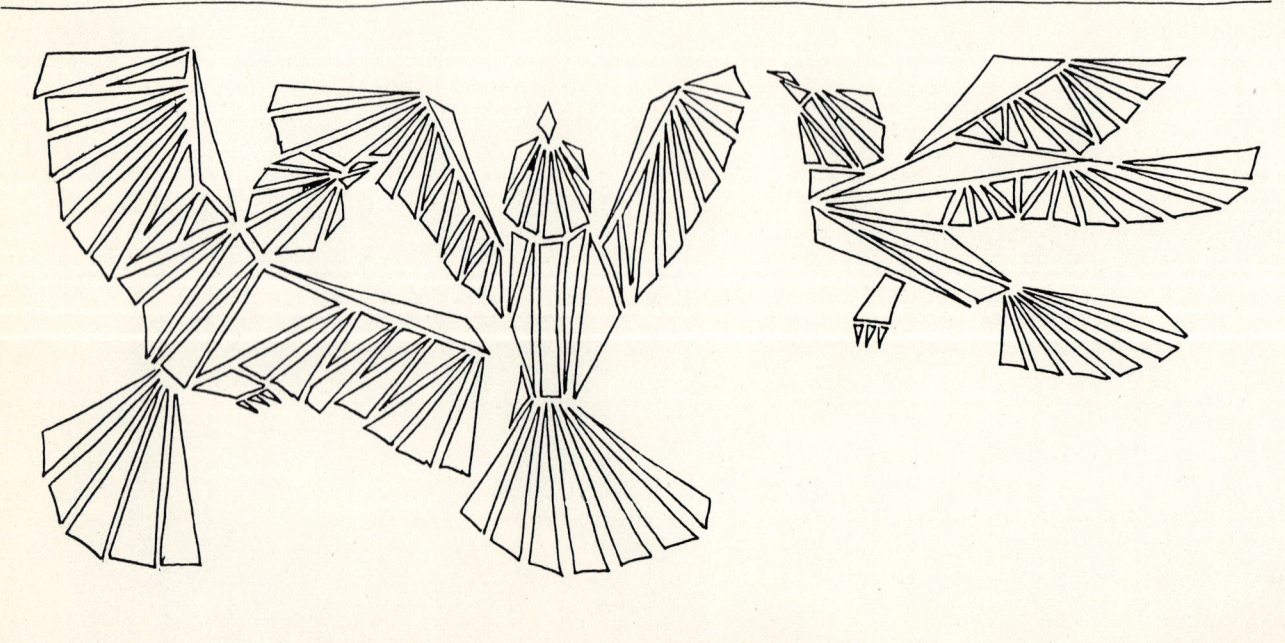

Emanations of other problems

PE

Content

This section groups together detailed and sectorally-specialized problems. The problems in this section tend to be permutations and combinations of the broader problems described in the previous sections (Section PB, PC or PD).

Many of the problems in this section are parts of sets or series resulting from such combinations. Whereas the earlier sections aim to be comprehensive in covrage, this section does not necessarily include all potential problems forming part of such series.

Also included are the very detailed problems which would otherwise be allocated to Section G (index only). These are included when the information available, or the pattern of cross-references, suggests that it would be of value to see them as part of the problem network.

Note that further information relevant to an understanding of the problem may be present in other problems cross-referenced in the entry consulted, especially any broader problems.

This section groups 3,106 problems for which there are 14,835 cross-references.

Rationale

Whereas the problems allocated to earlier sections tend to be the subject of distinct studies, conventions or organizational programmes, those in this section tend to emerge from the paragraphs and sub-paragraphs of documents which may only incidentally be problem-focused. Problems at this level of detail frequently escape information collection procedures and are easily ignored as side effects of broader problems. It is also the case that these problems may be more readily detected in practical situations

Method

The entries are based on information obtained from international organizations, from a wide variety of reference books, or as reported in the international media. The procedures for identifying world problems are described in Section PZ.

Index

A keyword index to entries is provided in Section PX.

Comment

Detailed comments are given in Section PZ at the end of this volume.

Reservations

The emphasis throughout this volume has been placed on providing descriptions of less well-known problems, particularly when the extensive material available on the better known problems contained neither succinct descriptions of them nor descriptive material which could easily be reduced to succinct descriptions. The problem descriptions here represent a compilation of views from published documents (usually from international organizations). The text provided does not necessarily constitute the best possible description of the problem, since a compromise has had to be struck between availability of information, the resources to process it, and the space available in this volume.

In a number of cases a problem could have been allocated to another section. Inclusion of a problem in this section, rather than in a preceding or following section, has been based on a number of factors. The position of the problem in one or more hierarchies of cross-references was a major factor in determining its allocation to this section.

Possible future improvements

There is much scope for improving the quality of problem entries through feedback from interested bodies. More bibliographic references could be included where appropriate, as well as references to major resolutions concerning those problems recognized by the United Nations. There is also much scope for improving the pattern of cross-references, both between problems, to other sections of this volume (eg values) and to the 20,000 internationally-active bodies in the companion series (*Yearbook of International Organizations*)

ENTRY CONTENT AND ORGANIZATION

Ordering of entries
Entries are in **numeric order**. Entry numbers have been **allocated randomly**; they have no significance other than as a permanent point of reference to facilitate indexing, cross-referencing, and updating between editions.

Index access to entries
The location of an entry in this sub-section may be determined from the **Volume Index** (Section PX) on the basis of keywords in the name of the entry or its alternate titles.

Structure of entries
Entries may be composed of the following descriptive elements:

(a) **Entry number** This number has **no significance**, except as a convenient method of identifying the entry (particularly for indexing purposes), of filing information on it, and as an identifier to which cross-references from other entries (possibly in other Sections) may refer in this and future editions. The first letter of the entry number refers to the section of this volume in which the sub-section, denoted by the second letter, is located.

(b) **Problem name** This is printed in bold characters. It is the name selected as best indicating the nature of the problem. It may be followed by alternative problem names.

(c) **Nature** Description of the problem which attempts to identify the nature of the disruptive processes involved. The information included here, and in the following paragraphs, is compiled directly, to the extent possible, from available published documents. Where appropriate the text included may be reproduced, in a minimally edited form, from the publications of international organizations, such as those of the United Nations or its Specialized Agencies.

(d) **Incidence** Summary description of the extent of the problem which makes it of more than national significance.

(e) **Background** Describes briefly when and how the problem's importance was recognized initially, and how this recognition has evolved over time.

(f) **Claim** Stresses the special importance of this problem and why action is particularly urgent. This paragraph offers means of including statements which may deliberately exaggerate claims for the unique importance of the problem.

(g) **Counter-claim** Stresses, where appropriate, the relative insignificance or erroneous conception of the problem as described. This paragraph offers a means of including statements which may deliberately exaggerate the arguments refuting the evidence for the existence of the problem. Absence of such arguments from the text does not mean that they do not exist.

Cross-referencing of entries
At the end of any entry, there may be cross-references to other entries. These indicate the number and name of the cross-referenced entry, whether within this Section or in other Sections. There are 3 types of **hierarchical** cross-references between problems:

    Broader = Broader problem: more general problems of which the problem described may be considered a part. The described problem may be considered an aspect of several broader problems
    Narrower = Narrower problem: more specific problems which may be considered a part of the described problem
    Related = Related problem: problems that may be considered as associated in a hierarchically undefined way with the described problem.

There are 4 types of **functional** cross-references between problems:

    Aggravates = Problems aggravated by the described problem: a forward or subsequent negative causal link
    Aggravated by = Problems aggravating the described problem: a backward or prior negative causal link
    Reduces = Problems relieved, alleviated or reduced by the described problem: a forward or subsequent positive causal link
    Reduced by = Problems relieving or alleviating the described problem: a backward or prior positive causal link

♦ **PE0004 Inadequacy of telecommunication facilities in developing countries**
Nature Economic growth in developing countries is hampered by 'gross inadequacy of telecommunication facilities'. The accelerating development in the technology of semiconductors, optical fibres and satellites means that developing countries, while making large investments in communications, are getting ever farther behind. Only a few years ago the most advanced integrated circuit consisted of about 10,000 components on a single chip. Today that complexity is in the range of 100,000 components per chip, and by 1990, will go up to a million. Telecommunication networks may be considered as 'the central nervous systems of complex societies'. In particular it is impossible, without adequate telephone service, to meet the growing communication needs of business, government and the professions in developing countries. Present demand so far exceeds supply that business and government entities are often forced to build uneconomic private systems for their own use.
Incidence In 1983, 1,200 million television sets and telephones concentrated largely in nine countries. The top nine telephone user list read: USA (80 telephones per 100 population); Sweden (78); UK (48); Japan (48); West Germany (44); Italy (31); Hong Kong (30); Spain (29); Barbados (21). Those densities, are in sharp contrast to a country such as Ethiopia which has about 0.2 per hundred. In 1971, the developing countries of Africa, Asia and Latin America had only 7% of all telephones, serving 61% of the world's population. This represented an average of one telephone per 100 population, compared with 55 in North America, 18 in Western Europe and 22 in Japan. In developing countries most telephones are concentrated in a few cities, with little or no service in many areas. Typically, there is a large gap between demand and supply; the waiting list for new connections is often greater than the number of telephones in service, and the waiting time is often several years. In most countries, traffic exceeds capacity, resulting in overloaded facilities and deteriorating service.
Counter-claim Telecommunications is a means of fulfilling the basic need to communicate. Although it is a challenge still largely to be faced, past and present cooperation, and future possibilities arising from further developments, should lead to continuing improvements in telecommunications for the Third World. These will not only be in quantifiable economic gain, but also in social benefits related to education, medicine and the sense of being part of a wider community; and they will be out of all proportion to the initial investment. The very lack of modern technology in many Third World countries may be an advantage, as it presents a possibility of choice not open to many developed nations.
Broader Weakness of infrastructure in developing countries (#PC1228).
Narrower Maldistribution of radios (#PF4142)
Inadequate telecommunications in island developing countries (#PE5655)
Unreliable telecommunication services for land-locked developing countries (#PE5860).

♦ **PE0015 Unemployment of premature school leavers in developing countries**
Nature Most developing countries are faced with arduous problems over placement, in the various sectors of organized activity, of young people between 14 and 18 years of age who do not have the requisite qualifications for any specific employment. The competition for jobs drives wages down to below subsistence levels for these youths and leads to their exploitation. Such large pools of unskilled, cheap labour have been unincorporated in national and local developmental programmes, although their accumulated efforts could bring substantial social improvements. Governmental programmes are lacking, which is not surprising since cheap labour is in the interests of profiteers: only a large number of unemployed youth guarantees low wage levels.
Broader Unemployment (#PB0750)
Inappropriate education in developing countries (#PF1531).
Aggravates Prostitution (#PD0693) Juvenile delinquency (#PC0212).
Aggravated by Educational wastage (#PC1716).

♦ **PE0032 Political intervention by transnational corporations**
Nature Transnational corporations may occasionally actively promote subversive political intervention in the domestic affairs of host, particularly developing, countries. Such intervention is incompatible with the long-term existence of transnational corporations in host countries and infringes upon national sovereignty. Regional political influence, if not intervention, may also be attained by foreign investors, and this may be facilitated by the assistance and incentives given to transnational corporations by colonial governments, either by increasing the number of transnationals or by strengthening them directly and indirectly.
Background The classic cases of foreign corporate control of governments were the so-called 'banana republics' of Latin America.
Claim Covert action may be taken by transnational corporations to influence foreign and domestic policy by utilizing their broad financial power and their often close relationship with government cadres. They may lobby for or against governments of host countries, depending on whether or not they receive specially favourable terms or treatment. In host countries, the affiliates can seek to support particular political parties of their choice and they can rally against groups advocating social reforms.
Counter-claim The national willingness to subordinate local welfare to imported claims by transnational corporations no longer exists as it did earlier in this century. Those transnational corporations which stay in business do so because they provide benefits for their host countries commensurate with the profits they receive. At the forefront of the much-heralded "global village" which is so enriching to society today are transnationals. Where others approach global harmony in abstraction, these are families and individuals who practically engage in the face-to-face effort to build bridges between cultures and nations. Perhaps the most important force that exists today to counter the wanton destructiveness of the chauvinistic nation state is the transnational corporation.
Broader Economic dictatorship (#PC3240) Transnational corporation imperialism (#PD5891)
Excessive power and independence of transnational corporations (#PD5807).
Related Control of national economic sectors by transnational enterprises (#PE0042).
Aggravates Undue political pressure (#PB3209) Destabilization of social systems (#PB5417)
Erosion of national sovereignty by transnational enterprises (#PE1539).

♦ **PE0042 Control of national economic sectors by transnational enterprises**
Foreign ownership
Nature Direct foreign investment by transnational enterprises can reduce the control that a country has over its own economy, particularly in the case of developing countries where this is the most usual form in which direct foreign investment takes place. Movement of capital, technology transfer, commodity movements and changes in the pattern of industrial organization may and do all take place within an enterprise but across national borders. This may lead to a three-cornered conflict among the enterprise, the government of its base country and the government of the country in which it is investing. A more balanced international legal framework, acceptable both to the corporations and to the countries in which they are investing, is essential.
Incidence Japanese investors increasingly hold economic power in countries outside of Japan, and are finding themselves having to confront labour, cultural and political issues unknown at home.
Broader Foreign ownership (#PE4738) Economic dictatorship (#PC3240)
Financial and industrial oligarchy (#PD5193).
Narrower Control by transnationals of the global communications industry (#PE8640).
Related Political intervention by transnational corporations (#PE0032).
Aggravated by Inadequate participation in the control of joint venture (#PE3869).

♦ **PE0043 Abuse of animal drugs**
Nature Hormones which are intended for occasional therapeutic use are injected into young animals such as calves to fatten them faster. Like athletes taking steroids, this practice results in fast flesh, but the side effects are unclear. Some single out specific hormone levels and combinations as dangerous while others insist that any hormone additive is potentially dangerous to human consumers. Practically it is an open question whether beef raised with no hormones at all can compete with the quality and market price of beef raised with some hormones.
Incidence Hormones banned in the EEC are held by some farmers to be necessary to the production of high-quality beef at competitive prices. At a time of beef surplus, the EEC banned all hormones, reputedly to discourage farmers from producing beef. There is not a fundamental conviction among farmers in the EEC that the total ban is necessary or economically viable, and this makes the ban difficult to enforce, however needed it may be for public health reasons.
Broader Drug abuse (#PD0094).
Narrower Abuse of antibiotics and vaccines in factory farming (#PE8383).
Related Abuse of medical drugs (#PD0028)
Denial to animals of the right to the attention, care and protection of humankind (#PF5121)
Denial to animals of the right to conditions of life and liberty proper to their species (#PE6270).
Aggravates Zoonoses (#PD1770) Drug smuggling (#PE1880)
Drug resistance (#PF9659).
Aggravated by Illicit drug trafficking (#PD0991).

♦ **PE0046 Lack of management skills in developing countries**
Inadequate administrative skills in developing countries
Nature The capacity for industrialization and economic growth depends, to a large extent, on a nation's ability to develop the productivity of its manpower, in particular that of individuals in top managerial positions. In most developing countries there is a shortage of managerial skills. Typically such countries start with great handicaps: experienced administrators are scarce, and those who are available at the time when the development effort is launched have been trained with a narrow and restricted view of their functions; few have the background or experience needed to master the complex responsibilities which the development task imposes. In dealing with the international transfer of management skills, trainers have to face several peripheral problems, such as the mutual understanding between managers with different cultural backgrounds, education, behaviour, belief, or mother-tongue. Disregarding such problems often results in misunderstandings or in the ineffective application of management techniques and methods. A related problem is that until very recently, industry in developing countries was as a general rule represented by family enterprises. To develop a family enterprise into a manageable industrial set-up can be a major task, which it may take generations to complete.
Incidence Although India has the third largest pool of skilled scientific personnel in the world and the capability of space and nuclear power high technology, it is proving immensely difficult to organize critical national projects. These include potable drinking water for the majority of people, a telephone and telecommunications system, immunization of children from basic diseases, oil plant production and literacy. The country's rich capability in scientific research is in no way matched by the ability to organize and utilize its resources.
Background In most former colonial nations, managerial posts prior to independence were held by expatriates, so that the process of decision-making on critical issues was the prerogative of the foreigners who controlled business. For many cultures, this action-oriented style of the Western industrial revolution conflicted with images of good breeding and responsible living.
Refs Agarwal, A N *Problem of Management Grasp in Underdeveloped Countries* (1970).
Broader Incompetent management (#PC4867) Deficiencies of developing countries (#PC4094)
Weakness of infrastructure in developing countries (#PC1228).
Narrower Inadequate cartographic skills in developing countries (#PJ8291)
Inadequate local expertise in business practices in developing countries (#PE7313).
Aggravates Mismanagement (#PB8406)
Mismanagement in developing countries (#PD8549)
Inadequate development of enterprises in developing countries (#PE8572)
Inadequate institutional structures for local government in developing countries (#PE0365).

♦ **PE0051 Restriction of free market competition by transnational corporations**
Nature The nature of transnational corporations dictates certain patterns of behaviour which may restrain competition. While the allocation of markets may be rational from the viewpoint of an enterprise when it is engaged in activities across national boundaries, it is almost certain to clash with the interests of some countries. Mergers involving foreign firms may be beneficial to the enterprises involved, but the resulting changes in industrial structure may be contrary to the domestic or international public interest. In establishing affiliates in host countries, transnational corporations may find themselves competing with local firms. This increased competition may be beneficial, but it may also result in the take-over or elimination of local firms, which for various economic, political and social reasons may be an undesirable development. Some transnational firms retain acquired companies' names to give the appearance that the market is not dominated by one giant brand. The low profile approach also de-fuses resentment against foreign business. Indeed, many transnational corporations' methods may be termed covert, in the light of increasing sensitivity to their presence.
Domination of a market, or restrictive business practices within it, by transnationals, may therefore go undetected until a rather late stage when considerable harm may have already been done.
Refs United Nations Centre on Transnational Corporations *Transnational Corporations and International Trade* (1985).
Broader Restrictive trade practices (#PC0073) Transnational corporation imperialism (#PD5891)
Restrictive business practices of transnational corporations (#PE5915).
Narrower Domination of transnational corporations in the coffee industry (#PE3781)
Excessive control of raw materials markets by transnational corporations (#PE0194)
Control of marketing and distribution channels by transnational corporations (#PE2397)
Limited market access due to the product differentiation of transnational corporations (#PE2328).
Aggravates Monopolistic activity by transnational enterprises (#PE0109).
Aggravated by Disproportionate control of global economy by limited number of corporations (#PE0135).

♦ **PE0060 Excessive exploitation of raw material reserves by transnational enterprises**
Broader Resource wastage in capitalist systems (#PC3108)
Control of industries and sectors by transnational corporations (#PE5831).

♦ **PE0063 Competition between intergovernmental organizations for scarce resources**
Rivalry between intergovernmental organizations — Inter-union competition
Nature Decisions by specialists as to which international agency to join, and decisions by governments as to which to fund, reflect the scarcity of two vital resources for intergovernmental organizations: talented staff and money. Considerable wastage of both is due to overlapping and duplicating structures and programmes. In another sense intergovernmental organizations and agencies are competing for the world's attention and hence for governments' and peoples' time. The excessive number of publications put out during a United Nations Year by the UN and other intergovernmental organizations is an example of this continuing competition.

Broader Competition between international organizations for scarce resources (#PC1463).
Aggravates Banned trade unions (#PD3535)
Inefficient location of facilities of international organizations (#PE3538).
Aggravated by Bureaucratic factionalism (#PF7979)
Inadequate coordination of the intergovernmental system of organizations (#PE0730)
Proliferation and duplication of intergovernmental organizations and coordination bodies (#PE2417).

♦ PE0064 Jurisdictional conflict and antagonism between international nongovernmental organizations

Nature A certain amount of 'territorial imperative' is seen in the rivalry between INGOs. In the main inspired by differing geographical regional bases, conflict may also arise on ideological grounds, as between socialist and non-socialist camps. Aggravating these tendencies are the confused policies of 'recognition' accorded these organizations by intergovernmental bodies.
Broader Obstacles to effective international nongovernmental organizations (#PF7082)
Jurisdictional conflict and antagonism between international organizations (#PD0138).
Aggravates Inadequate coordination of international nongovernmental organizations and programmes (#PE1209).

♦ PE0069 Lack of legal provision for international nongovernmental organizations

Broader Inadequate legislation relating to action against problems (#PF1645)
Obstacles to effective international nongovernmental organizations (#PF7082).
Narrower Lack of international legal provision for nongovernmental organizations (#PF7058)
Lack of national legal provision for international nongovernmental organizations (#PF7034).
Aggravates Inadequate relationship between international governmental and nongovernmental organizations and programmes (#PE1973)
Inadequate relationship between international nongovernmental organizations and the specialized agencies of the United Nations (#PE0777).

♦ PE0075 Proliferation and duplication of United Nations information systems

Lack of coordination of United Nations information systems
Nature Despite their broad similarities, numerous differences prevent UN bodies from having a unified information system. The problem starts with the absence of a unified policy or systematic organizational and procedural approach to cooperation. There does not yet exist a framework within which UN inter-agency programmes can make cooperative decisions, nor a system for managing cooperation activities. Decisions affecting development cooperation resources and activities are made in numerous places without sufficient knowledge of, or reference to, one another. As a result, most information systems or system design efforts have suffered from the same fragmentation. This fragmented approach, in turn, leads each Agency to classify differently data relating to the same or similar objectives and to the same or similar types of activities. As a result, government bodies and senior officials throughout the UN system do not have the information necessary to make rational system-wide policy and programme decisions. In short, there are simply too many separate, inconsistent, incomplete information systems relating to some facet of development cooperation activities, and these systems are undirected or uncoordinated by any central authority.
Counter-claim The Advisory Committee for the Co-ordination of Information Systems (ACCIS) was created in 1983 to address such problems. ACCIS works in many ways, chief among them the selection of standards and the recommendation of guidelines throughout the system. One example is the registration of all serial publications of the UN into the ISDS, the International Serials Data System.
Refs Advisory Committee for the Coordination of Information Systems *Directory of United Nations Databases and Information Systems* (1985).
Broader Proliferation and duplication of international information systems (#PE0458).
Aggravated by Proliferation and duplication of organizational units and coordinating bodies of the United Nations system (#PE1579).

♦ PE0078 Cystitis

Nature The main symptom of cystitis, or inflammation of the bladder, is the frequent and desperate need to urinate, accompanied by an agonizing burning sensation and a feeling of incomplete emptying of the bladder. The most frequent cause is transfer of the common colon bacillus E. Coli. from the large intestine to the bladder. Bruising and the presence of irritants may also trigger the disease. Cystitis often beings during a period of intense sexual activity, and may then develop following any sexual intercourse.
Incidence For anatomical reasons, women are 10 times more likely to suffer from this infection.
More than half the adult women experience urinary tract infections at some time.
Refs Shreeve, Caroline *Cystitis* (1987).
Broader Urinary bladder disorders (#PE2307).

♦ PE0081 Monopoly power due to advertising

Nature Advertising leads to economic concentration by serving as a barrier to new product entry by creating brand loyalties. In the early stages of the concentration process the disappearance of the small firms proceeds more or less automatically, while in the later stages competition takes on the character of a war, with each firm being prepared to incur losses in order to keep its territory from being intruded upon by others. If such concentration is not justified by the existence of economies of large-scale production, concentration brought about by advertising results in higher prices to the consumer because of the increase in the degree of monopoly power enjoyed by those in the market.
Broader Monopoly of power (#PC8410).
Aggravates Domination of advertising by transnational corporations (#PE2193).
Aggravated by Unfair competition (#PC0099).

♦ PE0082 Deterioration of stained glass due to acid rain

Nature Acid rain occurs when emissions from coalburning factories and power plants are transformed into sulphuric and nitric acids and fall to earth. In Europe, more than 100,000 stained glass objects, some of them more than 1,000 years old, are threatened by acid rain. Documentary records show that the stained glass in Europe was in a relatively good condition up to the turn of the century and even the Second World War caused practically no damage since all European stained glass was sheltered during that time, but in the last 30 years the deterioration process has accelerated to the extent that a total loss is expected within a few decades. Glass dating from the 8th to the 17th centuries is particularly endangered. Sulphuric acid has an etching effect on the surface and the resulting salts form a chalky crust that accelerate the decomposition process, allowing the paint to peel off. The glass substance finally splits and disintegrates into minute particles. Sulphur compounds also seriously crack old organically treated leather and paper produced after 1750.
Broader Acidic precipitation (#PD4904)
Damage to cultural artefacts by environmental pollution (#PD2478).

♦ PE0083 Distortion of international trade by discriminatory requirements with respect to product standards and measures

Nature Barriers in the field of standards arise in cases where there is a significant divergence in the mandatory standards adopted in various countries as well as in cases where methods used for enforcement of such standards are unduly rigorous. Some developing countries are exposed to standards for electrical machinery and equipment involving difficulties to their exporters, particularly as the procedures adopted for prior testing and inspection are time-consuming and costly in some countries. They may also experience difficulties because of certain provisions in the technical regulations regarding product content, or processes used in the manufacture of enamel products, pencils, fruit juices, etc.
Broader Distortion of international trade by discriminatory customs and administrative entry procedures (#PE2603).

♦ PE0088 Injustice of special courts

Nature It is in relation to trials for political offences that departures are most often made from the principle that everyone has a right to be tried by his natural judge; special courts have from time to time been set up on an ad hoc basis in various countries to try such offences, especially in conditions of political turmoil. Such establishment of a special court to try one person or a group of persons is to be regarded with the gravest suspicion. Even the establishment of special courts having continuous jurisdiction over crimes against the security of the State may carry with it the possibility of discrimination on political grounds.
Broader Injustice (#PA6486) Political repression (#PC1919)
Denial of right to fair and public trial (#PE3964).
Narrower Injustice of mass trials (#PE0597) Injustice of military tribunals (#PE0494).

♦ PE0097 Smallpox

Nature Smallpox was an acute infectious disease, now eradicated globally. It was characterized by sudden onset with fever, malaise, followed by characteristic skin eruptions which, after passing through stages of macules, papules, vesicules and pustules, dried up leaving scars (pockmarks). Two principle varieties of smallpox were recognized: variola minor (alastrim) and variola major (classical smallpox). In variola major, the fatality rate among unvaccinated was 15 to 40 percent, while outbreaks of variola minor were associated with a fatality rate of 1 percent. Although the disease has been eradicated, the virus is still in existence in research laboratories and could conceivably cause an outbreak of the disease in a population no longer protected against it.
Incidence Formerly a world-wide disease decimating the human population. The WHO global programme culminated in a world-wide eradication of smallpox. The last known case of naturally acquired variola major occurred in Bangladesh in October 1975. Variola minor persisted in the Horn of Africa until October, 1977. Since that date, two cases occurred in 1978 in Birmingham (UK), related to a research laboratory. Global eradication of smallpox was certified by the WHO in 1979 and confirmed by the World Health Assembly in May 1980. Since smallpox has now been eradicated routine vaccination against smallpox is no longer justified. International certificates of vaccination against smallpox are no longer required by any travellers.
Stocks of the virus are maintained in 1990 by the USA and the USSR but may be destroyed by agreement between both countries.
Refs Fenner, F, et al *Smallpox and Its Eradication.*
Broader Epidemics (#PC2514) Virus diseases (#PD0594)
Stigmatized diseases (#PD7279) Infectious and parasitic diseases (#PD0982).
Narrower Variola major (#PG1911) Variola minor (#PG1912).
Related Laryngitis (#PE2653).
Aggravates Pleurisy (#PG6121) Pneumonia (#PE2293)
Physical blindness (#PD0568) Chronic bronchitis (#PE2248)
Enforced vaccination (#PF1916) Natural human abortion (#PD0173)
Personal physical disfigurement (#PD8076)
Swelling of the mucous membrane larynx and windpipe (#PG8962).
Aggravated by Natural disasters (#PB1151) Unhygienic conditions (#PF8515)
Overcrowding of housing and accommodation (#PD0758)
Health risks to workers in agricultural and livestock production (#PE0524).

♦ PE0101 Military expropriation of satellites

Nature Unarmed communication satellites and research space stations may be diverted to the purposes of general intelligence gathering and military surveillance or incorporated in early warning and other space defence systems. Expropriation may include arming with a number of devices including nuclear space mines. The expropriation of privately-owned satellites, particularly when ownership may be international, poses unique problems of compensation.
Broader Misuse of satellite surveillance by governments (#PF3701).

♦ PE0102 Diabetes

Diabetes mellitus
Nature Diabetes is a condition in which the body is unable to metabolize sugar and other foods efficiently, due to an insulin deficiency. Persons with diabetes suffer from extreme thirst, hunger, weakness and loss of weight. They excrete abnormally large quantities of urine of high specific gravity, containing sugar and other substances not normally present. Very susceptible to infection, the diabetic's most serious complication is retroacidosis with or without coma, formerly the major cause of death, but absent in well-treated patients.
Incidence Diabetes is a common disease from which no age-group is exempt. Type I (WHO 1980) with rapid onset and more severe course, requiring insulin from diagnosis on, occurs predominantly in young persons; type II with more gradual onset, frequently associated with obesity, occurs preponderantly in older persons, in women more than in men. Hereditary predisposition for diabetes is well-established, more so in type II than in type I. In type I, susceptibility may be conveyed by genes close to the histocompatibility genes DR-3 and DR-4 on chromosome 6. Manifestation of this type seems to require additional exogenous influences. The proportion of persons with diabetes in the population is estimated to be two or three percent, about half of whom are unaware of their disease, but only 5 in 100 of these are primarily insulin dependent (type I).
Background The relationship of the pancreas to diabetes was first suggested by Cawley, an English physician, in 1788. In 1921, insulin was found by Banting and Best to revert clinical symptoms. Before insulin was available, fewer than 25% of patients with the severe form of the disease lived for more than 10 years, children rarely for more than one year. Today, life expectancy and health prognosis are limited by vascular and neural complications with progressive dysfunction of kidneys, heart and coronary vessels, retina, autonomic and peripheral nervous system. Knowledge of the disease is still rudimentary. Insulin is an essential drug for millions of persons. Yet, in many parts of the world insulin is scarce, of uneven quality, or priced beyond the reach of the insulin-dependent person. A shortage of nursing and allied health professional who play a major role in the delivery of diabetes health care further exacerbates the problem.
Refs Ahmed, Paul I and Ahmed, N (Eds) *Coping with Juvenile Diabetes* (1985); Ahuja, M M S (Ed) *Epidemiology of Diabetes in Developing Countries* (1979); Covelli, Pasquale and Wiedman, Melvin *Diabetes* (1988); Crepaldi, Gaetano, et al *Diabetes, Obesity and Hyperlipidemias* (1983); Jovanovic, L *Controversies in Diabetes and Pregnancy* (1988); Kaplan, D W *Diabetes in Childhood and Adolescence* (1986); Kritzinger, Erna E and Taylor, Kenneth G *Diabetic Eye Disease* (1984); Kushi, Michio and Mann, John D *Diabetes and Hypoglycemia* (1985); Mngola, E N (Ed) *Diabetes Nineteen Eighty-Two* (1983); Pierce, Grant N, et al *Heart Dysfunction in Diabetes* (1988); Smith, Wrynn *Diabetes, Liver and Digestive Diseases* (1988).
Broader Gland disorders (#PD8301) Diseases of metabolism (#PC2270)
Genetic defects and diseases (#PD2389).

Related Human ageing (#PB0477)
Foetal malformation in diabetic pregnancies (#PE4808).
Aggravates Ulcers (#PE2308) Mycosis (#PE2455) Pneumonia (#PE2293)
Still–birth (#PD4029) Tuberculosis (#PE0566) Encephalitis (#PE2348)
Hypoglycaemia (#PE1926) Heart diseases (#PD0448) Arteriosclerosis (#PE2210)
Kidney disorders (#PE2053) Visual deficiencies (#PD8179) Periodontal diseases (#PE3503)
Majority rule mindset (#PF0851) Psychological conflict (#PE5087)
Diseases of the arteries (#PE2684) Urinary bladder disorders (#PE2307)
Gangrene, carbuncles and boils (#PG1923) Perinatal morbidity and mortality (#PD2387)
Diseases of blood and blood–forming organs (#PF8026)
Diseases of the skin and subcutaneous tissue (#PC8534)
Diseases and deformities of the digestive system (#PC8866).
Aggravated by Obesity (#PE1177) Alcohol abuse (#PD0153)
Physical unfitness (#PD4475) Autoimmune disease (#PE5214)
Consanguineous marriage (#PC2379) Diseases of the pancreas (#PE1132)
Increased food consumption (#PJ1931).
Reduced by Malnutrition (#PB1498) Dietary restrictions (#PJ1933).

♦ PE0106 Inadequate relationship between transnational corporations and specialized agencies of the United Nations

Broader Excessive power and independence of transnational corporations (#PD5807)
Inadequate relationship between international governmental and nongovernmental organizations and programmes (#PE1973).

♦ PE0109 Monopolistic activity by transnational enterprises

Nature The predominance of transnational enterprises, including transnational corporations, in world economic affairs can often create monopolistic structures which reduce world efficiency and may displace or prevent alternative activities. The concentration of transnational enterprises on the production and consumption of certain types of products and services not only influences consumption patterns but, in developing countries, often responds mainly to the demand of small segments of the population.
Incidence One very specific example occurs in the seed industry, in basic plant breeding research and patenting of new breeds of agricultural crops. These activities are dominated by a small number of corporations, many of which are transnational. According to a recent OECD survey, there were over 600 enterprises engaged in the international trade of agricultural seed in 37 leading countries. In 1979 the FAO identified nearly 1800 public and private entities engaged in plant breeding or seed trading. Of these, some 550 have been acquired by transnational corporations, and, in addition, some 300 seed houses, are linked to TNCs by contracts and other arrangements. The dominant transnationals are the major petroleum companies and pharmaceuticals houses.
Counter–claim The size and spread of transnational enterprises imply increased productive efficiency and reduction of risks, both of which have positive effects from the point of view of the allocation of resources.
Broader Monopolies (#PC0521)
Restrictive business practices of transnational corporations (#PE5915).
Narrower Transnational monopolies in developing countries (#PE5751).
Aggravates Inefficiency (#PB0843).
Aggravated by Restriction of free market competition by transnational corporations (#PE0051).

♦ PE0117 Inadequate barter system in international trade

Underutilization of non–monetary foreign exchange
Nature Barter, or countertrade as it has come to be known, is the oldest form of human commerce. It is growing worldwide. Shortages of hard currency, complex and sever exchange restrictions, the debt crisis, weak commodities prices, the collapse of the oil market — all of these factors have contributed. In 1973 only 15 countries bartered regularly. Now 88 do so. Since barter is subject to much less regulation than monetary trade, no one knows for sure what proportion of global trade takes place in kind. Some experts estimate that the proportion could be as high as 30 per cent. The revival of bartering is seen as a disaster by those who seek to promote free monetary trade. It is a declaration that the system is unviable. The World Bank opposes countertrade because it can deprive Third World nations of the cash and capital resources they need for development. The General Agreement on Tariffs and Trade (GATT) sees it as a major factor in the universal trend toward protectionism, in part because a countertrade deal can force one of its participants to accept goods it would not otherwise buy.
Incidence Since 1983, anyone selling more than $500,000 in goods to Indonesia is legally obligated to accept Indonesian products in payment. Ecuador, Malaysia, Thailand, Mexico, Brazil and South Africa strongly encourage such deals. The Soviet Union and Eastern Bloc nations will not buy foreign products unless the buyer takes bloc goods as at least partial payment. Nigeria's bartering of its excess oil in recent years of declining oil prices has kept the economy from collapse. At the same time, however, it has made a sham of OPEC's ability to negotiate world oil prices.
Background Barter as a vehicle for contemporary world trade got its start in postwar Europe, where such trades as Austrian matches for Italian black–market coffee promoted reconstruction. Eastern Europe, long strapped for convertible currency, has long been unable to pay cash for Western–made goods but has offered home–made goods in exchange. The quality of these goods have been often too poor to sell in West European markets. Traders in Western Europe have developed an ingenious system of trading and remanufacturing for using Eastern bloc merchandise. Many of these goods now find their way to Black Africa in return for raw materials.
Claim Countries with liquidity or hard currency foreign exchange shortages are not able to increase trade quickly by bartering commodities, goods, services, and physical and economic resource development rights owing to the lack of development of common bartering standards and mechanisms. The process of bartering is susceptible to uncertainties and delays as a result.
Broader Obstacles to world trade (#PC4890)
Fragmentation of the international trading system (#PC9584).
Reduced by Countertrade (#PE6621).

♦ PE0125 Seizure of cargoes

Nature Financial losses to shippers, carriers and consignees from cargo seizures are a risk that adds to the costs of shipping. Consignees in developing countries who pay shipping costs may be uninsured against certain kinds or all kinds of seizure. Port authorities, national customs agents, public health agents and police are all frequently empowered for cargo seizure. In addition, banks and other creditors represented by marshals or other public enforcement officials, may also effect seizure. Cargo seizure reason may range from governmental political actions to impounding of illegal or hazardous shipments, or to illegal acts of cargo detention or appropriation by governmental or non–governmental organizations, or piracy.
Broader Cargo insecurity (#PE5103).
Aggravates Inequities in marine insurance (#PE5802)
Obstacles for international ocean shipping (#PD5885).

♦ PE0133 Environmental hazards from meat and meat preparations

Nature The inner flesh of meats of poultry and fish from healthy animals contain few or no micro–organisms, although they may be present in other parts of the carcase. Contamination can occur, however, during slaughtering, handling and processing.
Broader Environmental hazards from food and live animals (#PC1411).

♦ PE0135 Disproportionate control of global economy by limited number of corporations

Escalating control of international trade by transnational corporations — Undisclosed influence of transnational corporations on global policy–making — Emerging oligopolistic world trading system
Nature Even more impressive than the growth in trade itself has been the growth in the role of a limited number of giant transnational corporations in world trade. This is due not only to the large percentage of trade which takes place within and among such corporations, but also to the fact that recognition of the role of transnational corporations in the development of new technologies in production innovation, and in the penetration of export markets for goods and services, has led to their being seen as "national champions", whose interests are increasingly seen to coincide with national goals. Increased participation by transnationals has had implications for the structure of international trade, including the growing oligopolistic nature of world trade, transfer pricing and the greater impact of restrictive business practices in the international trading system. Despite efforts to regulate such excesses, none of them have effectively imposed obligations upon the transnationals to ensure that their activities conform to the overall goals of the international community, especially that of economic development.
Incidence Some 200 global corporations now control nearly 80 per cent of the productive assets of the non–socialist world. It is predicted that within 25 years they will own production assets in excess of $4 trillion, or 54 per cent of the economic wealth of the planet. In 1971 the total value of production controlled internationally by these corporations was US $330 billion, compared to a total of US $310 billion of world exports. Also about 25 per cent of the exports of major industrial countries consisted of transactions within companies. This implies a reduction of the power of governments to control vital areas of economic policy. It has also meant that transnational corporations have been able to restrict the growth of competition by a variety of monopolistic practices. In 1982 the total revenue of the top 200 transnationals in the services sector was $1,192 billion, compared to $1,853 billion for the manufacturing transnationals.
Claim The future of everyone is becoming increasingly dependent on decisions made by, and in the interests of, a limited number of major corporations, many of which now control resources in excess of those of many individual countries.
Broader Oligopolies (#PC3825) Elitist control of global economy (#PC3778)
Transnational corporation imperialism (#PD5891).
Narrower Disproportionate influence on national economies of limited number of corporations (#PE1922).
Related Economic dictatorship (#PC3240)
Incomplete access to information resources (#PF2401).
Aggravates Transfer pricing (#PE1193) International economic injustice (#PC9112)
Vulnerability of socio–economic systems from globalization (#PF1245)
Over–reliance on economic interest groups by policy agencies (#PF1070)
Inaccessibility of decision–makers in multinational enterprises (#PE0573)
Restriction of free market competition by transnational corporations (#PE0051).
Aggravated by Interlocking corporate directorates (#PF5522)
Collusive international trade arrangements (#PE0396).

♦ PE0139 Abuse of sedatives and tranquillizers

Abuse of barbiturates and non–barbiturate hypnotics — Abuse of prescription drugs
Nature Barbiturates, derivatives of barbituric acid, are the most important of the group of hypnotics, sedatives and tranquillizers. Although they may be considered 'safe' if used as sleeping pills or sedatives in small doses or in large doses under medical supervision, as in anaesthesia, after regular and prolonged use there is great danger of dependence. Although the behavioural effects closely resemble alcoholic intoxication, abuse is far more dangerous with a high possibility of unintentional overdose. With chronic abuse a rapid tolerance develops and both physical and psychic dependence occur. Abrupt withdrawal is dangerous and the convulsions which follow can be fatal. A period of mental confusion, delirium, hallucination and temporary psychotic, often paranoid reactions, may follow. Pharmacological dangers are increased by the character of those who use the drug. Social, emotional and personality deterioration are associated with chronic abuse. If the drugs are injected large ulcers develop at the site. Barbiturates are commonly used in conjunction with other substances, the most widespread and dangerous use being with heroin, alcohol and stimulants. 'Street drugs' (for example, marijuana, mescaline) are sometimes doctored with animal tranquillizers (for example, PCP).
Incidence Since large quantities are used for therapy it is much more difficult to evaluate the extent of abuse of barbiturates and other hypnotics and sedatives than that of narcotic drugs or other psychotropic substances. In a number of countries they account for about 10 per cent of prescriptions. The general picture is complicated by the fact that the majority of people who are addicted to barbiturates are also dependent on other substances. 'Chasing the dragon' with a mixture of heroin barbiturate is fairly widespread in Asia. Barbiturates are the most commonly used group of addictive drugs.
Background Rauwolfia root was used as a tranquillizer in India as early as 1000 BC. Bromide, paraldehyde barbiturate sedatives were introduced for medical purposes between 1857 and 1903 AD.
Broader Drug dependence (#PD3825) Abuse of medical drugs (#PD0028).
Narrower Sedative withdrawal (#PG4462) Sedative intoxication (#PG9279).
Related Abuse of amphetamines (#PE1558).
Aggravates Suicide (#PC0417) Encephalitis (#PE2348) Drug smuggling (#PE1880)
Mental depression (#PC0799).
Aggravated by Neurosis (#PD0270) Mental disorders (#PD9131)
Inadequate drug control (#PC0231) Illicit drug trafficking (#PD0991).

♦ PE0141 Restrictive practices in mineral fuels trade

Nature The world's demand for ever larger quantities of more kinds of minerals is growing. Although most of the demand still comes from the industrially advanced nations, the growing economies of the developing nations require minerals in greater quantity and variety. At the same time, older, long–known mines are slowly being exhausted, while new ones are being developed in areas further and further removed from former centres of consumption. As a consequence, not only does the problem of transport over long distances at economic rates grow more acute, requiring for its solution many technological developments; but the situation is further complicated by the necessity of having to penetrate into regions hitherto remote and inaccessible.
There are two major factors which retard the general adoption of modern transportation technology and management and the development of an integrated transportation system. One is the existence of national policies, laws and regulations concerning transportation. These laws govern procedures and the economics of individual modes of transportation as well as the competition between modes. Regulation of transportation inevitably affects the characteristics of national and international transportation systems. Regulations governing road transportation, for example, may be so severe and limiting as to discriminate against roads in favour of railways in situations where this would not otherwise happen. Similarly, the development of long–distance slurry pipelines for coal has been blocked by regulations favouring railways. The other potentially retarding factor is of an institutional nature. The handling and transport of minerals has been so fragmented among agents and carriers that in most cases no single operator has full control over the system. From

the time the product leaves the mine or mill to its arrival at destination it may well have come under the control of several dozen agents and organizations. This multiplicity creates obstacles to achieving integration within the system.
Broader Restrictive trade practices (#PC0073).

♦ PE0155 Plant-pathogenic air pollutants

Nature Air pollutants that are pathogenic to plants arise from a variety of natural and man-related phenomena. They can be divided into primary and secondary types. Primary pollutants originate at the source in a form toxic to plants; secondary pollutants develop as a result of reactions among pollutants.
Incidence The types of pollutants injurious to plants may be classified as necrotic, chlorotic, or atrophic. Specific effects include changes in cell-wall permeability, plasmolysis, changes in tissue pH, interference with cell-wall synthesis, acceleration of respiration, inhibition of enzymes, and a reduction in the rate of photosynthesis. A particular cause of concern is the damage to forests by acid rain, due to sulphur emission. Transboundary agreements on reduction of annual sulphur emission by at least 30 percent of the 1980 level by 1993 had been agreed by 18 countries in Europe as of June 1984.
Background The deleterious effects of our pollution on vegetation has been recognized for more than 100 years. Ozone and hydrogen chloride (from the LeBlanc soda process) where two of the first air pollutants to be recognized as phybotoxic. Concern over the effects of air pollutants on plants has grown with the increasing damage caused.
Claim Since vegetation, through the process of photosynthesis, represents the basis for life on this planet, it is essential that effects of various air pollutants on vegetation be recognized, understood and, where possible, curbed.
Refs World Meteorological Organization *Air Pollutants, Meteorology and Plant Injury* (1969).
Broader Air pollution (#PC0119).
Related Endangered urban trees (#PD2025) Pesticide damage to crops (#PD2581).
Aggravates Crop vulnerability (#PD0660) Environmental plant diseases (#PD2224).
Aggravated by Photochemical oxidant formation (#PD3663).

♦ PE0161 Direct foreign investment by transnational enterprises as a restrictive business practice

Nature The preference of foreign business when investing in developing countries is undoubtedly for equity-based control. The aversion of international enterprises for shared ownership has lessened, particularly in the face of threats of nationalization, or legal requirements for significant local equity participation. However, where the transnational enterprise holds less than 50 percent, it can still possess control if the other equity is distributed in smaller shares among several interests. Export oriented industries having a large foreign exchange component, and industries requiring a steep investment in research and development, and operations needing support from high-technology and other specialists, are more readily dominated by the transnationals. A particular device favoured to extend control is the construction of a vertically integrated chain from raw material supply, through distribution to final sale point, with the transnational enterprise being able to manipulate every step and exert a compelling leverage on operations where it is technically a minority presence.
Counter-claim Direct foreign investment increase employment. A conservative estimate of direct employment by multinationals is 65 million people or 3 percent of the world's work force. Add indirect employment, such as jobs created by suppliers and the general lift to an economy that multinationals provide and such companies may generate 6 percent of the world's labour force. They develop skills, deliver export income and provide import substitutions for the host country. A substantial number of the Transnational Corporations are not giant multinational but are small and medium sized firms. In 1984 23 percent of Japanese multinationals employed less than 300 people; and in 1981 78 percent of British firms with direct investment overseas employed fewer than 500 people.
Broader Restrictive trade practices (#PC0073) Inappropriate foreign investment (#PD8030)
Restrictive business practices of transnational corporations (#PE5915).

♦ PE0163 Domination of developing countries by transnational corporations

Intimidation of developing countries by transnational corporations
Nature The sheer size, the seeming power, and the evident alternatives of transnational enterprises fill the leaders of developing countries with a sense of diluted control over the economic life of their own countries. As seen through the eyes of national leaders, transnational enterprises seem to have more options than the countries that the leaders represent. Mutual suspicion continues to exist because of the asymmetry in bargaining power and disagreements concerning the process of technology transfer, the development of natural resources, and the use of the environment.
Counter-claim 1. The existence of these options in the hands of the multinational enterprise is not necessarily harmful to the developing countries. As long as the options are not exercised, the operations of the foreign-owned enterprise may well be benign. Even when they are exercised, the economic consequences may be helpful to the developing countries. It may be that developing countries are bound to feel a sense of vulnerability to world markets, irrespective of their policies towards multinational enterprises. It is not at all evident, for instance, that developing countries which exclude multinational enterprises from their territory do very much more than change the quality and form of their dependence. All still face the need to mobilize resources, internal and external.
2. Transnational corporations play an important role as owners, as partners in joint enterprises, and as suppliers of technology in the mining and manufacturing sectors of many developing countries. Their role has been viewed more positively in recent years by developing countries, in response to their need for foreign exchange and their awareness of the value of foreign investment. Effective cooperation has proved possible by strict observance of the principle of sovereignty and a recognition that profit-seeking objectives must be pursued within a framework of long-term sustainable development.
Refs Bornschier, V and Chase-Dunn, C *Transnational Corporations and Under Development* (1985); Dev, Som *Multinational Corporations and the Third World* (1986).
Broader Economic intimidation (#PC3011) Transnational corporation imperialism (#PD5891)
Excessive power and independence of transnational corporations (#PD5807).
Narrower Inadequate supply of pharmaceutical products in developing countries (#PE4120)
Minimal export promotion by transnational corporations in developing countries (#PE1598)
Inadequate investment of transnational corporations in least developed countries (#PE7892)
Harmful effects of advertising by transnational corporations in developing countries (#PE2004)
Insensitivity of transnational corporations to consumer needs in developing countries (#PE1011)
Exploitative financial policies of transnational corporations in developing countries (#PE6952)
Increasing income disparity in developing countries due to transnational corporations (#PE1660)
Interference in labour relations in developing countries by transnational corporations (#PE7355)
Import-dependency in food staples in developing countries due to transnational corporations (#PE1806)
Disruption of cultural and social identities in developing countries by transnational corporations (#PE1082)
Inadequate relationship between transnational corporations and local industry in developing countries (#PE1511).
Related Domination by transnational corporations of the domestic name-brand food sector in developing countries (#PE1796).
Aggravates Retarding of development by transnational corporations (#PE0234)
Inadequate negotiating capacity of developing countries (#PE9646).
Aggravated by Inadequate participation in the control of joint venture (#PE3869).

♦ PE0168 Religious discrimination in the administration of justice

Nature In some countries admission to the judiciary or some parts thereof is limited, either by law or in practice, to persons belonging to a particular religion. Where the law requires the taking of an oath by judges (upon their taking office), or by jurors, assessors or lawyers (before acting) this, as such, discriminates against those who object to taking oaths, on religious grounds, and those who profess no religion, to the extent that the formula for the oath is incompatible with their beliefs.
Similarly, where the law requires the taking of an oath before giving evidence in court, a person who objects to taking oaths in a manner which he deems to be incompatible with his religious or non-religious beliefs may be impeded in his defence in criminal proceedings or in presenting his case in civil or administrative matters. An accused or a litigant may similarly suffer if one of his witnesses or an interpreter who is to assist him is under the same disadvantage.
Broader Religious discrimination (#PC1455) Discrimination before the law (#PC8726)
Undue religious influence on secular life (#PF3358).
Narrower Injustice of religious courts (#PE0397).
Related Religious conflict between sects (#PC3363)
Religious discrimination in politics (#PC3220)
Inequitable administration of justice (#PD0986)
Religious discrimination in education (#PD8807)
Discriminatory religious influence on the law (#PD3357).
Aggravates Injustice (#PA6486) Anti-clericalism (#PF3360)
Religious repression (#PC0578) Religious intimidation (#PC2937)
Persecution of religious sects (#PF3353) Underprivileged religious minorities (#PC2129).
Aggravated by Religious intolerance (#PC1808) Multidenominational society (#PF3368)
Corruption in organized religion (#PC3359)
Negative effects of claims of religious infallibility (#PF3376).

♦ PE0174 Inflexible intermediary political implementation networks

Nature The mid-level structures for implementing political decisions are to a large degree inflexible and traditional. They are closed to alternative approaches to carrying out decisions. The tensions between local concerns and general policy directions are not held by intermediary networks.
Broader Inadequate political networks (#PD2213).

♦ PE0175 Poisonous animals

Refs Freiberg, Marcos A and Walls, Jerry G *The World of Venomous Animals* (1984); Halstead, Bruce W and Saunders, Paul *Poisonous and Venomous Marine Animals of the World* (1988).
Related Prejudice against animals (#PC0507).
Aggravates Human poisoning (#PD0105).

♦ PE0176 Humanism

Secular humanism
Claim Humanism fails to acknowledge the value of animals, plants and the inanimate creation by which humanity is surrounded and supported.
Refs Ehrenfeld, David *The Arrogance of Humanism* (1978).
Broader Insufficient provision of public services for communication (#PF2694).

♦ PE0179 Proliferation and duplication of international nongovernmental organization and coordination bodies

Broader Obstacles to effective international nongovernmental organizations (#PF7082)
Proliferation and duplication of international organizations and coordinating bodies (#PE1029).
Aggravates Competition between international nongovernmental organizations for scarce resources (#PE0259)
Proliferation and duplication of international nongovernmental organization information systems (#PE0362).

♦ PE0185 Feral mammals

Feral livestock
Nature Feral animal are descendants from once domesticated animals that have escaped or were set free and may have interbred with native species. They are interlopers in their environment. They often have no natural predators to control populations and can destroy whole ecosystems.
Incidence Feral pigs have contributed to the destruction of one third of the bird species on the Hawaiian Islands and to the disappearance of the land iguanas and the poor survival rate of Galapagos tortoise on the Galapagos Islands. In Australia, more than 20 kinds of feral animals, including camels, dogs (dingoes), and water buffaloes are destroying habitat and threatening domestic animals.
Refs McKnight, Thomas L *Feral Livestock in Anglo-America*; World Conservation Union *Feral Mammals* (1984).
Broader Lack of community planning (#PF2605).
Narrower Feral dogs (#PG6473) Feral cats (#PE5396)
Feral children (#PJ7453).
Aggravated by Excessive stray animal populations (#PE5776).

♦ PE0189 Blacklisting

Censured people — Blacklists
Nature The maintenance of lists, written or unwritten, of persons whose position, behaviour or beliefs are disapproved of; who are to be boycotted or economically deprived; or who are to be the subject of any number of diverse punitive measures, have been perennial activities in individual and organizational history. When this is engaged in by governmental agencies, civil rights are violated and such blacklisting can lead to more serious governmental sanctions and crimes against individuals.
Incidence Blacklisting in Britain is a reputable industrial service, reputed to protect industries against spies and industrial saboteurs. The services help companies to vet potential employees.
Refs Foley K, Sue *The Political Blacklist in the Broadcasting Industry* (1979).
Broader Economic discrimination (#PC2157).
Aggravated by Secret government security vetting of job applicants (#PE9441).

♦ PE0191 Distortion of international trade by discriminatory application of antidumping regulations

Nature Dumping is the process by which products of one country are introduced into the commerce of another country at less than the normal value of the products, thus causing or threatening material injury to established industries in the territory of the second country, or materially retarding the establishment of a domestic industry. Anti-dumping procedures have been elaborated to counteract such activities which can however be applied in a discriminatory manner: a lengthy period may be required to process antidumping cases; and appraisal for customs purposes may be withheld until a definitive determination has been made as to whether there were

sales at less than the domestic market price and, if so, whether material injury resulted. Such practices create considerable uncertainty in trading relations and thereby bring about an appreciable decline in the trade of certain types of goods.
Broader Distortion of international trade by discriminatory customs and administrative entry procedures (#PE2603).

♦ PE0192 Corporal punishment in schools
Flogging in schools — Beating school students
Refs Ontario, Ministry of Education *Corporal Punishment in the Schools* (1981).
Broader Corporal punishment (#PD8575) Obstacles to education (#PF4852)
Physical maltreatment of children (#PC2584).
Related Beating of prisoners (#PD2484).

♦ PE0194 Excessive control of raw materials markets by transnational corporations
Nature For many raw materials the world market (excluding the centrally planned economies) is controlled by a few giant companies based in developed countries. The developing countries which face this concentration of buying power are unorganized and poor. Not surprisingly, their share of the profits from their exports is small. Most of the money stays with the corporations in developed countries.
Incidence For example, five companies account for 75 per cent of the world tea market; six companies control 50 per cent of manganese ore capacity; three companies control 60 per cent of banana imports (90 per cent in the United States); and six companies control 76 per cent of the world's alumina production capacity. In the major importing countries the market control is even higher. Five companies have 78 per cent of the tyre manufacturing capacity in the United States; four of them have 70 per cent of the coffee market, another four share 70 per cent of the refined copper market, and yet another four control 77 per cent of the chocolate manufacturing capacity.
Broader Control of industries and sectors by transnational corporations (#PE5831)
Restriction of free market competition by transnational corporations (#PE0051).

♦ PE0197 Discrimination against women in sports
Broader Discrimination against women (#PC0308).
Narrower Exclusion of women from athletic competition (#PE4242)
Discrimination against women in athletic training (#PE4246)
Discrimination against women in payment and prizes for athletic events (#PG4236).
Related Discrimination against men in sports (#PE4232).
Aggravated by Athletic competition (#PE4266).

♦ PE0199 Water softness
Nature Areas supplied with soft drinking-water almost consistently experience a significantly higher prevalence of arteriosclerotic heart disease or degenerative heart disease, hypertension, sudden deaths of cardiovascular origin, or a combination of these.
Aggravates Hypertension (#PE0585) Heart diseases (#PD0448)
Degenerative heart disease (#PG2089) Arteriosclerotic heart disease (#PG2088)
Inadequate water system infrastructure (#PD8517).

♦ PE0207 Coercive use of economic power by transnational enterprises against labour
Nature Multinational enterprises tend to use the existence of their international network of manufacturing facilities as a means of coercing trade union negotiators in a given country to accept inferior agreements on the occasion of disputes and collective bargaining. Such corporations can transfer existing production or new investment to other countries, thus depriving workers of jobs to the advantage of more amenable workers in other countries.
Counter-claim Irresponsible behaviour by labour, provoked by political ideologists, has led to extortionist demands impossible for enterprises to meet while at the same time maintaining full employment under profitable working conditions. Most transnational enterprises have heavy capital investment in non-moveable plants and equipment in their host countries and are, therefore, seriously motivated partners in national development. Virtual conspiracy of some militant international labour unions with socialist international civil servants in intergovernmental organizations has created a climate of hostility to foreign investment by transnationals with all its attendant consequences. As a direct result labour itself suffers.
Broader Unethical commercial practices (#PC2563)
Transnational corporation imperialism (#PD5891)
Labour tensions involving transnationals (#PE5927).
Aggravates Insecurity of employment (#PD8211)
Transnational strike action by trade unions (#PE1541).

♦ PE0208 Excessive frontier formalities in international travel
Nature Travellers from or to some countries experience considerable difficulty and delay in obtaining travel authorization (usually visas). Some countries charge exit visa that hinders the travelling of e.g students. Clearance through customs can be a major ordeal with lines causing an hour or more of travel delay. Delay may also be experienced at the frontier when further questions may be raised which have to be answered, in many cases in detail on special forms. These difficulties may be considerably increased in the case of travellers from countries which do not have diplomatic relations with the country visited. In the extreme case, travellers from such countries may be obliged to initiate requests for visas many months in advance, ensuring that the visa for the last country visited prior to the return is obtained before that for the previous country, and so on for each country back to that first visited after departure from the traveller's home country. While these problems may be, in the main, of an administrative nature, there may be political discrimination against citizens of various countries, involving a number of further additions to the formalities.
Claim Visa and frontier formalities frequently give rise to occasions for bribery. Visa possession and ability to cross a frontier has been a matter of life or death in numerous countries for those seeking refuge from injustice. The advent of computer-controlled crossing points may impose additional problems including the threat to freedom of movement.
Counter-claim Frontier controls are essential to curb drug traffic, terrorists, illegal immigration and other serious crime which often come from outside the country. In many cases, the management of these problems requires controls, restrictions and special police powers.
Broader Restrictions on emigration (#PC3208).
Related Excessive customs formalities (#PG2102)
Refusal to issue travel documents, passports, visas (#PE0325)
Excessive animal sanitary regulations in international travel (#PF1555).
Aggravates Restrictions on freedom of movement between countries (#PC0935).
Aggravated by Non-recognition of foreign governments (#PF8040).

♦ PE0223 Invasion of privacy by compulsory telecommunications
Claim Modern life requires telecommunications. In some countries, telephones outnumber people, while lower ratios (as in Asia where there is one telephone to about 33 people) indicate, when seen as the equivalent of one phone to ten families, that the rest of the world is catching up. The integration of home telecommunications in developed countries to include interactive telephone-television-computer capability, which offers shopping and banking services, points to the inevitability of consumer home interaction with local and national government. At first this will be through the electronic banking capability of paying taxes, but since tax information must also be supplied, the possibility exists of answering government questions beyond those related to taxes: examples might be referendums, censuses and surveys. Inevitably, the home telecommunication capability will need a legal requirement, at which point privacy will have been invaded to the extent of state control over individual life. The black box and big screen in everyone's living room may become a watcher, time-clock and mind-checker, monitoring both personal behaviour and private thoughts.
Broader Infringement of privacy (#PB0284)
Inappropriate use of telecommunications services (#PE4450).

♦ PE0225 Inadequate regulation of the restrictive business practices of state enterprises
Broader Lack of control (#PF7138) Restrictive trade practices (#PC0073).
Narrower Collusive tendering in international trade (#PE7072).
Aggravates Inadequate development of enterprises in developing countries (#PE8572).

♦ PE0226 Destabilizing financial action of transnational enterprises
Nature The degree of stability of exchange rates and the adjustment mechanism provided by the international monetary system affect the policies of transnational enterprises. Such enterprises are able to move massive amounts of funds across borders in a short space of time, aggravating monetary crises.
In addition to the problem of the high levels of profit repatriation when there is a fear of devaluation in a country, another tactic which can be used by the transnational enterprise is to defer payment from its affiliates in the country concerned to the subsidiary or perhaps even to the parent company for goods that had been purchased from it. In addition, the subsidiary could be required to make immediate payment for all purchases made from affiliates, instead of permitting a normal 30 or 90 day payment period.
Furthermore, transnational corporations tend to move funds into and out of currencies to maximize their profits by taking advantage of interest rate differentials and tax advantages which are available in one country compared with another. This potential has been greatly enhanced by dramatic changes in international banking and consortia arrangements. Such action also contributes to the fundamental disequilibria in the balance of payments of some major industrial countries
Incidence For example, it was found in a survey of 115 foreign-owned subsidiaries in the United Kingdom that close to 30 per cent of such firms in 1964 and 1965, when devaluation of the sterling seemed imminent, remitted over 100 per cent of their earnings, whereas in the previous three or four years no dividends had been paid. In so doing, the subsidiaries in question paid dividends out of accumulated profits and, in a few cases, virtually all of the retained earnings were remitted.
Broader Destabilization of monetary systems and exchange notes by transnational corporations (#PE5903).

♦ PE0234 Retarding of development by transnational corporations
Nature When transnational corporations introduce into a host developing country a package of resources and capabilities which they own or control, the impact may be such as to retard the development process. Foreign capital may divert profits into unjustifiably large outflows of dividends and service payments to the corporate headquarters. New technology and machine-intensive processes may not always be appropriate for local needs such as employment creation. Managerial and marketing decisions may divert resources from where they are most needed to where they are most profitably sold. Transnational corporations may serve as carriers of modernization, or they may place the host countries in a situation of even greater dependency.
Claim The extraction of natural resources may generate few processing industries or do little to raise the level of local skills. Branch plants which operate purely as off-shoots of their parent companies, such as component manufacturers, are unlikely to integrate fully into the local economy. The attempts of host countries to raise taxes or to place limitations on foreign exchange remittances can be negated by vertically or horizontally integrated multinational corporations through transfer pricing and the use of tax havens. The negative non-economic impact may also be important. The very cultural identity and the entire social fabric may be at stake, especially if transnational corporations attempt to transplant their own models of social development to the host country.
Counter-claim Foreign capital augments the resources of the host country and relieves bottlenecks in foreign exchange. New technology improves the utilization of resources. Managerial and marketing skills enhance productivity and the availability of goods. Access is provided to the capital, technology, skills and markets of the global network of the corporation and employment is created.
Broader Transnational corporation imperialism (#PD5891)
Social irresponsibility of transnational corporations (#PE5796).
Aggravated by Domination of developing countries by transnational corporations (#PE0163).

♦ PE0236 Cruelty to animals in food preparation
Incidence Animals of many kinds are killed in the actual process of meal preparation: lobster, shrimp and crab are boiled alive, frog's legs are removed whilst the frog is alive, snakes are skinned alive, live monkey's skulls are opened at the dining table, and drunken shrimp and sea-urchin are eaten alive. Some insects may be fried alive. Certain forms of paté are obtained by force feeding animals to provoke an organic disorder from which the paté is derived. These process cause pain and suffering to the animals and are unnecessary to the food preparation.
Broader Maltreatment of animals (#PC0066).
Narrower Inhumane killing of animals (#PE0358).
Related Denial of rights of animals (#PC5456)
Exploitation of animals for amusement (#PD2078)
Cruel treatment of animals for research (#PD0260).
Aggravated by Unethical food practices (#PD1045)
Human consumption of animals (#PC7644)
Unethical catering practices (#PE6615).

♦ PE0241 Denial of right to state succession
Denial of right of nations to select their own leaders
Broader Denial of state's rights (#PD4814).
Aggravated by Abusive national leadership (#PD2710).

♦ PE0242 Uninsulated buildings
Inadequate home insulation — Underheated school buildings
Broader Substandard housing and accommodation (#PD1251)
Undirected expansion of economic base (#PF0905).
Aggravates Draught (#PJ0600) Unsustainable development of energy use (#PC7517).
Aggravated by Inadequate governmental energy conservation policies (#PF0037).
Reduces Indoor air pollution (#PD6627)
Health hazards of modern insulating materials (#PE1499).

♦ **PE0245 Manipulation of transfer prices by transnational corporations**
Nature Transnational corporations tend to fix prices of goods and services traded between the corporation and its affiliates located in different countries. Intracorporate transfer pricing by a multiregional company within a country may matter little to a national government, unless it is a restrictive business practice, since all the benefits of the transaction are retained domestically. When engaged in by transnational corporations, however, it affects the distribution of the benefits of their activities between countries, and may stifle local competition.
Incidence Research has shown that, although intracorporate trade in goods within transnational corporations is concentrated within certain industries, such as motor vehicles and chemicals, more than one quarter of the value of all international trade in goods appears to be of an intragroup character. In addition, although much less well-documented, there is the provision of intracorporate services, for example: research and development, rentals of equipment, administration and loans. The scope for price manipulation is therefore quite extensive.
Background Transfer prices may be distorted for internal motives: the varying degree of ownership in its subsidiaries may induce the parent company to make profits appear where its ownership is relatively large; there may be an incentive to reduce the apparent profits in a particular affiliate for purposes of wage bargaining; transfer pricing may be an indirect way of allocating markets if, for instance, the prices charged to an affiliate are such as to make its exports non-competitive. The manipulation of prices may also respond to external factors: the diversity among countries in the rates of taxation or in the rules of assessment; the difference in taxation even in the same country on the various forms of remuneration of capital, dividends, interests and royalties, and the ensuing tendency to transform taxable income into non-taxable costs; the varying rules of exchange control by some host countries regarding the remittance of those various types of remuneration; the risk of changes in exchange rates; and finally, the risk of nationalization or expropriation.
Broader Economic manipulation (#PC6875) Transnational corporation imperialism (#PD5891)
Unfair pricing by transnational corporations (#PE5855).

♦ **PE0246 Newspaper monopoly**
Concentration of press ownership
Nature The concentration of press ownership puts freedom of expression in jeopardy for a number of reasons, particularly the need to please advertisers and employers. The latter may be foreign, governmental, industrial, financial, for example; but they all usually have a central motive of profit or propaganda. Articles and other content may be censored by the owners and the tone of a number of different newspapers within the same group may be conformist. Such newspapers and hence the conglomerates that own them may exert undue political influence, leading to a lack of democracy and political instability. Where the monopoly is exerted by a national or foreign company, governmental reprisals may be taken, especially if undue political pressure has been exerted against the regime. Concentration of press ownership and the consequent increased conformism may constitute minority discrimination particularly against strong ethnic groups using diverse languages.
Broader Uncontrolled media (#PD0040) Monopoly of the media (#PD3101)
International monopoly of the media (#PD3040).
Related Propaganda (#PF1878) Industrial gas monopolies (#PE1813).
Aggravates Underground press (#PD2366) Denial of access to news (#PF3081)
Restriction of freedom of expression (#PC2162).
Aggravated by Misuse of advertising (#PE4225) Proliferation of commercialism (#PF0815).

♦ **PE0250 Damage caused by space objects**
Nature Damage caused by space objects may include loss of life or injury to persons or loss of or damage to property, either on the earth (including aircraft in flight) or in outer space (including other space objects or installations). Difficulties may arise in establishing liability for damage caused by space objects or in obtaining prompt payment of a full and equitable measure of compensation to the victims of such damage.
Aggravates Death of living creatures (#PF7043).
Aggravated by Pollution of orbital space (#PD0089)
Restriction of outer space benefits to a limited number of developed countries (#PD0530).

♦ **PE0252 Assassination of trade union leaders**
Assassination of workers representatives
Incidence Nearly 650 labour union members were murdered in 1988, nearly double the number reported in 1987.
Broader Assassination (#PD1971)
Sanctions against trade union workers (#PD0610).
Related Political assassination (#PE5614)
Forced disappearances of trade union leaders (#PE5882).
Aggravated by Corporation-sanctioned assassination (#PE6356).

♦ **PE0258 Seasonal affective disorder**
Winter depression
Nature A recurrent winter depression is brought on in some people by the deprivation of light which exists in northern latitudes around the time of the winter solstice. Seasonal affective disorder (SAD) is triggered by lack of bright light penetrating the retina to the pineal gland in the brain. Darkness produces the sleep-inducing hormone melatonin, which makes people lethargic, oversleep, crave carbohydrates and feel depressed. Women sufferers outnumber men four to one. Exposure to strong, continuous artificial light sources can immediately alleviate the symptoms, but the disorder has no known cure.
Broader Mental depression (#PC0799) Seasonal fluctuations (#PF8163).

♦ **PE0259 Competition between international nongovernmental organizations for scarce resources**
Rivalry between international nongovernmental organizations
Broader Obstacles to effective international nongovernmental organizations (#PF7082)
Competition between international organizations for scarce resources (#PC1463).
Aggravates Inadequate funding of international nongovernmental organizations and programmes (#PE0741).
Aggravated by Inadequate coordination of international nongovernmental organizations and programmes (#PE1209)
Proliferation and duplication of international nongovernmental organization information systems (#PE0362)
Proliferation and duplication of international nongovernmental organization and coordination bodies (#PE0179).

♦ **PE0261 Violation of sovereignty by trans-border broadcasting**
Foreign-controlled direct satellite broadcasting to individual receivers — Uncontrolled satellite broadcasting
Nature Although the transmission of television (and to a lesser extent, radio) broadcasts by means of satellites directly to domestic receivers is one of the most promising trends in the use of outer space for the needs of man, it may have unfortunate political, economic and human consequences. Direct broadcasting penetrates the national frontier and in so doing contributes to the weakening of those features which each country has retained in order to preserve its individuality, security and national sovereignty. The modern economy is based on an interplay of commercial and industrial forces. Through television advertising (which is not permitted in a number of countries), the use of satellites may bring about disequilibrium in trade balances, or the modification in actual practice of commercial trade treaties. Transmissions to a country, made without the consent of that country, may contain propaganda of violence and horror or may be hostile to the internal or external policies of the nation, may exhibit habits or customs at variance with the population's standards of morality. Such transmissions may be viewed as undermining the foundations of the local civilization and culture and prejudicing the cause of safeguarding international peace and security.
Claim Satellite Television broadcasts beamed by foreign companies and governments will promote cultural subversion. Direct broadcast satellite TV may contain radical political ideas, unacceptable cultural values, fanatical religious influences or unobtainable economic standards. For many countries, radio and TV broadcasts are means of nation building and not as a means to capture consumer advertisers.
Counter-claim Radio messages which are divorced from the local supporting evidence and background are believed to have relatively little effect on primitive societies. The local government of an underdeveloped country is likely to be far more effective in its propaganda than remote outsiders. In more advanced societies, where propaganda might be more effective, the scope for counter measures is likely to be greater, too.
With satellites the ability of a country to defend itself against intruding broadcasts is greater than with conventional broadcasting. In addition to the possibility of jamming, there is the possibility (for the major powers at least) of silencing or even destroying the offending satellite without violating the territory of the state from which the original transmission is coming.
Broader Denial of right to national self-determination (#PC1450)
Risk of unintentional nuclear war generated by the strategy of deterrence (#PF4162).
Narrower Denial of the right to national sovereignty (#PE7906).
Related Colonialism (#PC0798) Mercenary troops (#PD2592)
Colonization of information (#PF4894) Foreign control of natural resources (#PD3109)
Misuse of satellite surveillance by governments (#PF3701)
Denial of right to national self-determination in communist systems (#PC3177).
Aggravates Propaganda (#PF1878) Erosion of sovereignty (#PE5015)
National insecurity and vulnerability (#PB1149)
Copyright barriers to transfer of knowledge (#PE8403).
Aggravated by Unilateral declarations of independence by extra-territorial bases (#PF1066).
Reduces Ignorance (#PA5568).
Reduced by Jamming of satellite communications (#PD1244)
Limited number of geostationary satellite orbits (#PF0545)
Allocation of television frequency bands for satellite transmission (#PF3703).

♦ **PE0264 Sarcoidosis**
Refs James, D Geraint and Williams, W Jones *Sarcoidosis and Other Granulomatous Disorders* (1985).
Broader Infectious and parasitic diseases (#PD0982).

♦ **PE0265 Inadequate nutrition education in least developed countries**
Nature A fragmentary approach to nutrition education has not brought about lasting developments; an effective nutrition education strategy is essential. Low literacy rate makes it more difficult to promote health through an understanding of the importance of nutrition and hygiene.
Refs Somogyi, J C (Ed) *World-Wide Problems of Nutrition Research and Nutrition Education* (1982).
Broader Inadequate education (#PF4984) Least developed countries (#PD8201)
Inadequate health services (#PD4790).
Related Inappropriate education in developing countries (#PF1531).
Aggravates Nutritional ignorance (#PE5773) Malnutrition in developing countries (#PD8668)
Ignorance of women concerning primary health care (#PD9021).
Aggravated by Limited medical knowledge (#PD9160).

♦ **PE0269 Inadequate housing in developing countries**
Housing infrastructural weakness in developing countries
Nature Housing conditions still leave much to be desired in most countries, especially for the rural and the low-income urban groups. This is particularly true in the less developed countries, especially in regard to the necessary supporting infrastructure, especially sanitation, clean water, toilet facilities and sewerage. Population growth, and particularly the increase in urban areas caused by natural population growth and the exodus from the rural areas, has made the housing situation of the less developed countries much worse than it was 10 years ago.
Incidence In the world's poor countries, given the scarcity of government resources for the housing sector in relation to need, attempts to provide finished public housing for the urban and rural poor have almost without exception produced units too few in number to meet the demand and too expensive to be within the grasp of the poor.
In most developing countries, less than two houses per thousand inhabitants are being built annually; some countries are building at less than 10 per cent of their requirements. In Latin America and Asia the housing deficit is over 100 million units. The problem of housing and increasing population is especially grave in the urban areas, where 33 per cent of the world's population lives. By the year 2000 51 per cent will be in urban areas. In Africa alone, population growth as a whole is estimated at 2.4 per cent per year, but the urban population is increasing at 5 per cent per year. By the year 2000, it is estimated that 30 per cent of Africa's population will be urbanized. Investment in new housing construction is less than half of what is required, which may run to 5 percent of the national income in some countries. The total funds required has tended to discourage realistic approaches to the shelter problem.
Background Only a few of the less developed countries have announced housing policies and only a few have any housing programme integrated in the national development plan. Housing schemes are carried out on an ad hoc basis as and when funds can be made available. Housing has a low priority and is still regarded in the majority of countries as a social service. Only in countries where the exploitation of natural resources has increased the national income is housing receiving what might be described as a 'fair share' in the allocation of national income to different sectors.
Broader Inadequate housing (#PC0449)
Weakness of infrastructure in developing countries (#PC1228).
Narrower Insufficient rural housing (#PF6511)
Improvisational housing in developing countries (#PE4386).
Related Inadequate maintenance of infrastructure (#PD0645).
Aggravates Inappropriate sanitation systems (#PD0876)
Urban slums in developing countries (#PD3489)
Overcrowding of housing and accommodation (#PD0758).
Aggravated by Inadequate electricity supply (#PJ0641).

♦ **PE0273 Insufficient use of natural resources in least developed countries**
Nature Insufficient use is made of the local resources within developing countries and inadequate encouragement is given to local firms to develop their own indigenous skills. Foreign currency is expended on equipment and services for which support, in terms of maintenance facilities, training and spares, is insufficient.

Background The constraints which impede sufficient utilization include: technological; markets and marketing (including commercial power and the response of developed economies); and external (such as infrastructure and long-term effects).
Broader Least developed countries (#PD8201) Underutilization of natural resources (#PF1459).
Aggravates Inadequate development of enterprises in developing countries (#PE8572).

♦ PE0274 Cold disorders

Frostbite — Hypothermia
Nature Prolonged exposure to climatic cold may result in various cold injuries. Frostbite is the most common and results from the freezing of tissues. Frostbite is damage to body tissues as a result of cold. It usually affects the lower extremities and can set in at non-freezing temperatures if the body's resistance is lowered from starvation, intoxication, or loss of blood from a wound. Wind and increased humidity also promote the development of frostbite, which varies in intensity from first degree (remedied at home with all ill-effects gone after several days) to fourth degree which requires antibiotics and hospitalization, sometimes surgery. Hypothermia results from general body cooling and may occur rapidly in water or more slowly in air (death occurs due to heart failure). Elderly people are particularly susceptible to the latter condition, partly because those most susceptible are least sensitive to the cold, and partly because there is a drop in deep body temperature with age. Open-air work during the winter season causes inadequately protected hands to lose dexterity. Drivers with numb hands, and others working with machinery, electricity and tools, are susceptible to accidents as a result of lost manipulative ability.
Incidence Although no definite information is available, it is thought that the incidence of hypothermia as a major contributory factor to death is much higher than indicated by available statistics, because hypothermia is less evident than the final cause, such as heart failure.
Refs Lloyd, Evan L *Hypothermia - Cold Stress* (1986); Mountfort, Paul *Exposure on Hypothermia.*
Broader Disease and injury from exposure to weather (#PF5739).
Narrower Cold as an occupational hazard (#PF5744).
Related Burns and scalds (#PE0394).
Aggravated by Climatic cold (#PD1404).

♦ PE0280 Agraphia

Dysgraphia
Nature Agraphia is the loss of the ability to express ideas through writing.
Refs Benson, D Frank *Aphasia, Alexia and Agraphia* (1979).
Broader Learning disorders (#PD3865) Diseases and injuries of the brain (#PD0992) Symptoms referable to nervous system (#PE9468).

♦ PE0282 Minimal manufacturing in least developed countries

Nature The world economic crisis in the beginning of 1980s has halted the previously steady progress made by developing countries in increasing their manufacturing output. It has brought many developing countries to the brink of disaster. Outside of food-processing and textiles the modern manufacturing sector is highly import intensive. A consequence of this dependence on imported inputs is that there is a lack of linkages between the industrial sector and the rest of economy. Once the least developed countries started to experience balance of payment deficits, as they have done throughout the 1980s with commodity prices being low, then there simply is not the foreign exchange to obtain the inputs, and hence there is a reduction in capacity utilization if not outright closure. The manufacturing sector depends for the supply of raw materials in part on local agriculture and for the sale of local manufactured goods in domestic markets, the income of farmers is vital, because the markets are very small. But the farming sector has failed to grow to levels beneficial to the economy. Besides weak infrastructure in least developed countries there is a shortage of entrepreneurs and qualified labour in science and technology, management, finance, accountancy and marketing. Manufacturing is dependent on external financial resources, but the huge losses in the wake of the international debt crisis have made export credit agencies more cautious.
Incidence In the end of 1980s MVA (Manufacturing Value Added) accounted for no greater percentage of total GDP (Gross Domestic Product) in the least developed countries than it did in the 1970s. If the rapid growth of population is taken into account, than the picture becomes even more dismal: sixteen least developed countries experienced negative growth rates in MVA per capita during the first half of the 1980s.
Broader Least developed countries (#PD8201) Underdevelopment of manufacturing industries (#PF0854).
Aggravates Weakness in trade in manufactured goods among developing countries (#PE2966).

♦ PE0289 Inadequate infrastructure and services in least developed countries

Nature The underdeveloped transport network has hampered the effective exploitation of the greater part of the natural resources in the least developed countries and thus set serious limitations to their use in the over-all development effort; The operational capability of the existing infrastructure is poor for a number of reasons: the acute shortage of skilled manpower, poor skills of workers, scarcity of operating funds, at the operational and management level poor planning leads to weaknesses in resolving the problems of complementarity and competitiveness between the various modes of transport. The efficiency is also constrained by poor transport planning and operational arrangements at the regional and subregional levels between LDCs and neighbouring countries.
Broader Inadequate infrastructure (#PC7693) Least developed countries (#PD8201) Weakness of infrastructure in developing countries (#PC1228).
Related Inadequate maintenance of infrastructure (#PD0645) Weakness of infrastructure in island developing countries (#PE5772).
Aggravates Inadequate development of communication services in the least developed countries (#PF4297).

♦ PE0290 Non-payment of compensation for damages to consumers

Inadequate consumer compensation — Inadequate redress for consumers' loss
Broader Uncompensated damages (#PD7179) Consumer vulnerability (#PC0123) Grievances of consumers (#PD7567).
Aggravated by Restrictive trade practices (#PC0073).
Reduced by Consumer boycotts (#PE1213).

♦ PE0296 Fragmentation and complexity of the United Nations system

Inadequate coordination of the United Nations system
Nature Within the United Nations system of organizational sub-entities which are concerned with promoting the economic and social goals set out in the United Nations Charter, and between them and the specialized agencies, which are working towards the same ends, continuing differences of view regarding their respective competences exist. There continue to be cases of duplication and overlapping, of lack of cooperation among organizations and their staffs, of failures to consult, and divergencies of objectives at headquarters, regional and field levels. The extreme decentralization of the system, deliberate at the outset and then aggravated by the establishment of new organs, has not been counterbalanced by a coordination imposed on agents that had no desire to be coordinated.
Incidence The structural complexity is indicated by the number of legally independent entities attached to the main bodies within the system: 20 for WHO, 18 for FAO, 10 for UNESCO, 10 for ILO, 13 for UNDP and 15 for the United Nations. Furthermore, within any of these bodies, the degree of independence of a division (which may have its own committee of experts or intergovernmental organ) or of a field office, is often at least as great as that of the legal entity. The degree of coordination and hierarchical structure varies within each organization but is in most cases very weak. The same situation prevails with respect to regional offices, programmes and field projects. Because of the very ambitious scope of the programme coverage, there is extreme fragmentation of the resources available for any particular project. Because of the system of disbursement of funds, in a single country which is a recipient of aid, some 15 different organizations intervene simultaneously to organize their projects there, which may be extended to 30 when bodies attached to the United Nations are included.
The complexity of intergovernmental machinery and experts reflects the number of bodies and programmes. Moreover, because of the number of member States, the main committees, with a representative on each committee, cannot examine all problems in detail. This has led to the creation of smaller committees, specialized subsidiary organs, and a system of relationships has been established between them. A whole network of coordination machinery has been imposed on this structure. The vagueness of the terms of reference; the similarity of jurisdiction of organs as important as the Economic and Social Council, UNCTAD, the Second and Third Committees of the General Assembly; and the number and repetition of general debates repeated in committee after committee whose relative status is not clearly defined, have created in the United Nations particularly a state of confusion which in spite of countless efforts it has been found impossible to remedy.
Claim The very complexity and extraordinary diversity of the UN system, and often merely apparent lack of coherence in its activities, are sources of member frustration, as is the sense among the major contributors that the regular UN budgets, and the programmes financed under those budgets by mandatory assessments, escape their control.
The structural complexity is aggravated by a number of factors, including: the proliferation of external intergovernmental organs, many with overlapping mandates and almost all of unmanageable size; the proliferation of highly independent voluntary trust funds for purposes not necessarily corresponding to established high priorities; the soaring costs for UN and agency tasks which may not always be justified from the standpoint of benefit; the numerous obstacles to comparing and therefore to coordinating the future plans of different agencies; the involvement of so many agencies, including the organs of the United Nations, in almost every undertaking; the independent public information and public relations offices for most of the agencies and United Nations programmes; the 'tangle' of United Nations and agency regional and subregional structures which makes system-wide action at those levels so difficult; the over-frequent and uncoordinated visits by officials of different organizations to the capitals of developing countries and the excessive time and effort which coordinating processes, where they exist, seem to require.
Underlying such conditions, but partly independent of them, is the seriously fragmented character of the UN system and the possibility of increasing fragmentation in very important fields such as population, food, trade, environment and technology, if current trends are not arrested.
Counter-claim The United Nations has already been able to assist economic development and social progress throughout the world without being seriously inhibited by organizational incoherence. Some duplication and overlapping of activities and arrangements, some failures to cooperate, some conceptual differences in regard to objectives (all of which are common phenomena in national administrations) are unavoidable in a dynamic, growing and pioneering international system. Furthermore, they are part of the price that realistically will have to be paid by the United Nations for the advantage of being able, through the international functional agencies, to mobilize the active participation and support of the relevant technical ministries and professional groups in each country.
It would be easy to show how all parts of the decentralized system have learned to work in concert on an ever-widening series of broad programmes in a way never envisaged in 1945, and, still more striking, how in major emergencies such as the Congo operation of 1960-1964, the Biafra situation in 1969, the aftermath of the Bangladesh conflict in 1971-1972, and the Sudano-Sahelian drought from 1973, as well as in numerous operations for relief and reconstruction after sudden natural disasters, the specialized agencies and United Nations programmes have worked in concert under United Nations leadership.
Refs United Nations Institute for Training and Research *Towards Greater Order, Coherence and Coordination in the United Nations System* (1975).
Broader Inadequate coordination of the intergovernmental system of organizations (#PE0730).
Aggravates Ineffectiveness of the United Nations system of organizations (#PF1451).
Aggravated by Jurisdictional conflict and antagonism between the specialized agencies of the United Nations (#PE2486).

♦ PE0298 Alternating amnesia

Nature In this condition two separate states of consciousness alternate with one another, during each of which there is no memory for events that occurred during the other. Each state is a complex set of memories, attitudes and behaviours with distinctive characteristics, and is sequentially and disjointedly manifested. (This is a form of hysterical amnesia). Analogous phenomena may occur with collective memory, segments of which may be activated and expressed sequentially but without reference to each other. One possible indication of this is when a body of individuals convene in one mode and subsequently reconvene in another mode wearing different hats in which it is inappropriate to make any reference to the previous occasion.
Broader Amnesia (#PD8297).

♦ PE0299 Natural disasters in least developed countries

Nature A natural event may be an Act of God but the disaster which follows is caused or magnified by human and environmental mismanagement. In the rural Third World, growing populations are forced by poverty to overcultivate, overgraze and deforest their land. This makes drought and floods more destructive. In mushrooming cities of the least developed countries, the poor live on the most dangerous ground: in shantytowns on river flood plains or coastal mudflats, an in heavy mud-brick shacks on steep hillsides and ravines. This multiplies the casualties from floods and earthquakes. In cases like prolonged drought, major floods or cyclones, the affected countries' ability to renew their developmental efforts regresses by several years.
Broader Natural disasters (#PB1151) Least developed countries (#PD8201).
Narrower Disaster hazards to island populations (#PE5784).

♦ PE0302 Inappropriate food aid

Disruptive food imports
Nature Food aid reduces local incentives to produce food, causes damaging shifts in food habits (away from local staples to imported wheat; away from breastfeeding to powdered milk), and undermines the cultural cohesiveness of the recipient country by introducing new cultural concepts which may be readily adopted but intrinsically disruptive to the traditional way of life.
Counter-claim Properly managed, food aid can lead to food self-sufficiency. For example, food can be used as wages in food-for-work projects, or in land development and land improvement projects. It can also be used as interim supplies for land settlement and land reform schemes and in supplementary feeding for vulnerable groups. In addition, it can be purchased in other developing countries, thus contributing to regional self-sufficiency in food.

Broader Imbalance between agricultural exports and imports in developing countries (#PE4956).
Related Inadequacy of food aid (#PF3949).

♦ **PE0303 Long-term shortage of clothing**
Nature Production of natural fibres for cloth used in garments may not be able to keep pace with demand from a growing world population. This is true of plant fibres, from cotton for instance, and animal hairs like sheep's wool and many furs. Production of artificial fibres such as the polyesters, and fabrics such as vinyl, may be subordinated to needs for chemicals in construction and in industrial applications. In addition, world weather is showing continued instability and seasonal cycles are interrupted by cold spells, hot spells, storms and droughts. This increases the demand for clothing that normally would not often be worn in certain latitudes, for example: tropical, light weight fabrics in the northern temperate zone; and heavier garments in the southern hemisphere.
Broader Long-term shortage of miscellaneous manufactured articles (#PE0613).
Aggravates Inappropriate clothing (#PJ6604).

♦ **PE0304 Apraxia**
Refs Roy, E A, et al (Ed) *Neuropsychological Studies of Apraxia and Related Disorders* (1985).

♦ **PE0310 Trade restrictions due to voluntary export restraints**
Trade restrictions due to intergovernmental arrangements — Discriminatory orderly marketing arrangements
Nature There is an apparent trend towards greater use of trade restrictions imposed as a result of negotiated 'voluntary export restraints' both between governments and between enterprises. Such restraints on trade frequently involve enterprises in importing and exporting countries agreeing among themselves on future levels of imports and exports from the country whose exports are to be restrained; and the reaching of agreement by enterprises in the latter country on how the quantity agreed upon is to be shared out among them. In certain cases, agreement is also reached on the prices of such exports. Such restraints have continued in 'sensitive' areas such as steel, electronics, machine tools, and canned food, and have increased and strengthened in respect of motor vehicles. The actual extent of such restrictions is unknown because of the secrecy which surrounds them.
Background Despite agreement in principle, as envisaged in the GATT Tokyo Round declaration of 1979, to provide safeguards for the multilateral trading system, the nature of appropriate escape clauses continues to be debated, thus preventing implementation of an improved system. The issue concerns whether protective measures can be applied on a discriminatory or selective basis. Existing legislation of certain major trading countries permits such action, particularly through the application of voluntary export restraints or orderly marketing arrangements. There is concern that when a safeguard system is negotiated it will serve to legitimize discrimination through such export restraints, now a preferred form of safeguard action, rather than eliminate them.
Broader Ineffective industry self-regulation (#PF5841)
Imposition of trade quotas for political reasons (#PE9762).
Related Grant-back provisions (#PE5306) Challenges to validity (#PF1200)
Restrictions on research (#PF0725) Restrictions on publicity (#PF1575)
Restrictions on adaptations (#PF5248) Restrictions on use of personnel (#PF3945)
Patent pool or cross-licensing agreements (#PE4039)
Exclusive sales and representation agreements (#PE4581)
Payments after expiration of industrial property rights (#PF5292).

♦ **PE0313 Instability of olive oil trade**
Nature The essential feature of the olive oil market lies in the irregularity of harvests and in that of supplying the market; these irregularities result in fluctuations in the value of production and in the instability of prices and of receipts from exportation, as well as in considerable differences in the incomes of producers.
The proportion of oil produced which enters into world trade is small, most of it being consumed in the producing country. However, consumption in a number of countries has been decreasing, and there have been temporary surpluses. There are a number of difficulties in preventing the deterioration of olive oil in storage.
Incidence Olive oil production is the main source of income of millions of families who are wholly dependent on the measures taken for maintaining and developing the consumption of its products. Control of production is made more difficult by the long time span involved. The tree begins to produce at ages varying from 6 to 15 years and reaches maturity between the age of 80 and 120 years. About one–and–a–half million metric tons of olives are produced annually. The largest growers, in order, are: Italy, Spain, Greece, Turkey, Tunisia, Portugal and Morocco, some of whose export ratios are very high in relation to domestic consumption (for example, Tunisia, Morocco, Spain). All of the above are exporters of olive oil, to which may be added Argentina, Algeria and France.
Broader Instability of trade in fixed vegetable oils and fats (#PE0861).
Aggravated by Synthetic food products (#PG2222).

♦ **PE0314 Hypnotic amnesia**
Nature Amnesia may be induced by the use of suggestion usually under hypnosis (in a trance state). Memory of the trance state is vague and fragmentary, especially if the suggestion is that it should be forgotten. Analogous phenomena may be found with collective memory. Public opinion may be hypnotized by suitable processes (perhaps fascist propaganda is an extreme example) and these may well cause awareness of the hypnotized state to be collectively repressed.
Broader Amnesia (#PD8297).

♦ **PE0321 Hysterical amnesia**
Nature One form of this involves failure to recall particular past events, possible in a particular period. In another form there is failure to register current events and subsequently to recollect them. In both cases the memories may influence behaviour although they resist efforts at recollection. Such memories are usually painful and are repressed as a psychological defence. (It is characteristic that they may be recovered under hypnosis). Analogous phenomena may occur in the case of collective memory. In bureaucratic environments they are associated with the process of burying some unpleasant file of information, if only by severely restricting its distribution. Or alternatively it will be carefully arranged that no lies is created in the first place. The media and politicians often act on the assumption that the public will forget some unpleasant item of information.
Broader Amnesia (#PD8297).

♦ **PE0322 Bribery by transnational enterprises in developing countries**
Nature The business world is particularly active in promoting corrupt practices among politicians and higher officials in developing countries. Developed country business interests competing for markets in developing countries or embarking on direct investments in industrial enterprises there (either independently or in joint ventures with indigenous firms or governments), make use of bribery of higher officials and politicians in order to facilitate a business deal. It is often necessary to bribe both high and low officials in order to run the enterprises without any obstacles. Some transnational firms are abetted by their own headquarters' country governments in covertly or corruptly influencing local authorities. Multinationals may also bribe their own consulate officials to engage in unethical behaviour to further their business interests.
Broader Business bribery (#PD8449) Financial and industrial oligarchy (#PD5193)
Social irresponsibility of transnational corporations (#PE5796).
Aggravates Corruption in developing countries (#PD0348).

♦ **PE0323 Theft of works of art**
Stolen artworks — Looting of works of art
Nature Art thefts are increasing. Apart from representing a menace to the cultural heritage, it makes added demands on local and international law enforcement.
Incidence In 1989 some 5,000 works of art were reported stolen, representing only a small fraction of the total. Many smaller museums do not report thefts, whether because the theft is not discovered or because they have inadequate information to prove that it has occurred. It is estimated that the annual value of stolen art is more than $5 billion. In Italy, for example, an average of 16 works were stolen every day in 1971 from museums, churches and galleries, for a total of 7,560. In 1987 9,417 art works were stolen in Italy. In France 440 pictures were reported stolen in 1969, 1,500 in 1970 and 3,000 in 1971. In 1983, 400 works were listed stolen in one month.
Broader Theft of property (#PD4691) Destruction of cultural heritage (#PC2114).
Related Art vandalism (#PE5171) Financial scandal (#PD2458)
Avoidance of copyright (#PD0188) Illicit export of works of art (#PE9004)
Frauds, forgeries and financial crime (#PE5516)
Archaeological and anthropological looting (#PD1823).
Aggravated by Looting (#PE4152) Vandalism (#PD1350)
Social insecurity (#PC1867)
Inadequate documentation of works of art (#PE8088).

♦ **PE0324 Denial of right to sufficient food**
Denial of right to freedom of hunger — Denial of entitlement to food
Nature Every human being has the right to freedom from hunger, that is the right to a nutritionally adequate and safe diet and is based on the human right to dignity as a person. This right to food implies a right to certain means of producing/acquiring food.
The lack of sufficient food is related to food production, distribution and income levels. Advances in food production methods have made it possible for vast regions to produce sufficient food for their populations yet they still go hungry. Over 50 million children in South Asia are undernourished, despite the regions food surplus. Some 25 million children in Latin America are inadequately feed even though their region has become, after the United States the world's major food exporter. In these situations millions of people do not have the income to buy food, the means to grow it or the goods to exchange for it.
Refs Byron, William J *On the Protection and Promotion of the Right to Food* (1988).
Broader Denial of right to dignity (#PE6623) Inadequate standards of living (#PF0344).
Narrower Maternal malnutrition (#PE1085).
Related Inadequate social welfare services (#PC0834)
Denial of right to social security (#PD7251)
Denial of right to economic security (#PD0808)
Denial of right to sufficient shelter (#PD5254)
Malnutrition among indigenous peoples (#PC3319)
Denial of right to sufficient clothing (#PE7616)
Denial of right to adequate medical care (#PD2028).

♦ **PE0325 Refusal to issue travel documents, passports, visas**
Refusal of visas
Nature Refusal to issue essential travel documents without which a person can lose his right of sojourn in a foreign country, his right to return to his own country, and – in practice – the right to leave his own country, may be on the basis of discrimination (racial, sexual, religious, political) within the country issuing the passport or on the part of the authorities of a foreign country issuing a visa. Refusal may also be based on the criteria of national security, national interest, public order abroad, misconduct of the applicant either at home or abroad, legal or mental incapacity, lack of supporting health documents.
Incidence The inability of citizens to receive passports is associated with countries charged with human rights violations and where political instability exists. Other conditions include national poverty and economic crisis, where both citizens and their capital are controlled as to movement. The inability of non–political civilians to receive visas is often based on fear that temporary visitors will seek to reside illegally in the country, seek welfare benefits, and take jobs away from the nationals. In Socialist Europe, travel and emigration visas are usually denied East Germans, Jews in the USSR, and other nationals who may take their skills outside the region. In Africa, South African blacks may not be permitted to travel for political reasons. In West Europe, North Africans and Asians encounter high obstacles to their entry. In the USA, Mexican nationals are discriminated against due to labour protectionism, Philippine and Korean nationals are denied visas because of a high number who illegally immigrate, and others may be denied visas to the USA based on past or current political beliefs.
Broader Restrictions on emigration (#PC3208)
Discrimination against foreigners in employment (#PD3529)
Restrictions on freedom of movement between countries (#PC0935).
Narrower Financial security requirements (#PU2249)
Prohibitive cost of travel documents (#PU2248).
Related Harassment (#PC8558) Unreported births (#PF5381)
Foreign exchange restrictions (#PF3070)
Excessive frontier formalities in international travel (#PE0208).
Aggravates Illegal immigration (#PD1928) Rejection of refugees (#PF3021)
Illegal movement across frontiers (#PC2367) Delay in issue of travel documents (#PE9123).
Aggravated by Discrimination (#PA0833) Deprivation of nationality (#PD3225)
Conflicting multiple nationalities (#PE6677) National insecurity and vulnerability (#PB1149)
Loss of civil capacity for married women (#PE8720).

♦ **PE0330 Developmental expressive writing disorder**
Nature This impairment interferes with academic achievement and daily living as a inability to express oneself in written texts without spelling mistakes, grammatical and punctuation errors and poorly organized paragraphs.
Broader Mental illness (#PC0300) Learning disorders (#PD3865).
Related Innumeracy (#PC0143) Reading disabilities (#PD1950)
Developmental receptive language disorder (#PE9300).

♦ **PE0333 Korsakoff syndrome**
Nature This is a complex syndrome defined by four possible conditions. – gross defect in recent memory (associated with retrograde amnesia) although memory for remote events and didactically learned facts remains intact; – a spatial or temporal disorientation; – some degree of confabulation; – false recognition. It occurs in a wide variety of toxic and infectious brain illnesses as well as in association with some nutritional disorders. The syndrome may be so severe as to produce moment–to–moment consciousness, with information only being retained for a few seconds and providing no continuity between one experience and the next. Learning may thus be severely

limited or impossible. The condition can be transitory or chronic. Analogous phenomena may occur in the case of collective memory in social conditions of extreme deprivation or disorganization.
Refs Butters, Nelson and Cermak, Laird S *Alcoholic Korsakoff's Syndrome* (1980).
Broader Amnesia (#PD8297) Learning disorders (#PD3865).
Aggravated by Alcohol abuse (#PD0153).

♦ PE0341 Negative effects of family planning education on children
Excessive youth sex education
Claim Permissive sex education has placed before young people through books, youth clubs, clinics and classrooms all the facts, in four letter words, including the morally squalid kind, about every conceivable kind of sex. It takes sex education out of the context of traditional values, which in the ideal at least, have related sex to marriage and loyalty. By providing sex education without a moral context it implicitly approves of any form of sexual activity that a young person wants or thinks they want. It, not infrequently, introduces sexual behaviour unknown about by youth.
Broader Inadequate family planning education (#PD1039).

♦ PE0342 Restrictive practices in the food and live animals trade
Claim Among the more notable food industries where international restrictive practices exist, with governments and the private sectors acting in virtual collusion, are the grains and fishing sectors. In the former, government intervention determines harvest levels and therefore prices, some of which may be subsidized. In the latter, government gun-boats may defend unilaterally proclaimed territorial fishing limits, and catch and fish-products may themselves be government exports, at subsidized prices.
Counter-claim A number of restrictions on importations have non-commercial purposes. For the protection of health there may be restrictions on certain foods. To protect endangered species there may be importation proscriptions on some kinds of animals and animal products.
Refs United Nations *Banned Products*.
Broader Restrictive trade practices (#PC0073).
Related Protectionism in agriculture and the food production industries (#PD5830).

♦ PE0345 Motor aphasia
Ataxic aphasia
Nature Motor aphasia is the loss of the ability to speak without necessarily the loss of intelligence, or the ability to read or write.
Refs Luria, A R *Traumatic Aphasia* (1970).
Broader Speech disorders (#PE2265) Language disorders (#PE3886).
Related Sensory aphasia (#PE4234).
Aggravated by Diseases and injuries of the brain (#PD0992).

♦ PE0346 Restrictive business practices in relation to patents and trademarks
Nature The licensor can, through the bargaining process, place restrictions on the licensee which limit his competitive impact. In addition, the licensor is assured of specific territorial protection from the competition of the licensee by the system of national patent and trademark laws.
A significant number of foreign collaboration agreements, most of which presumably involve some sort of a licensing arrangement, have been entered into with firms in developing countries. These have involved independent firms, or firms in which the foreign licensor has a minority participation, or firms which are subsidiaries of a foreign firm. Regardless of these explicit contractual restrictions and implicit restrictions on export through equity control, certain restrictions on export are inherent in the present system of protection of industrial property (mainly patents and trade marks).
Patent law generally confers, as a means of encouraging inventions and their exploitation, the right to preclude third parties from making and selling the patented production, and in the case of a process patent, from applying the patented process without the agreement of the patent owner. This right may give the patent owner a monopoly position to the extent that there are no competing products (substitutes) but it is rare that a single patent gives its owner a monopoly position with regard to a specific product. Where, however, a large number of related patents are owned by one firm or where the patents of the leading firms in an industry are pooled, patents are likely to result in monopoly positions in a market. Unpatented know-how has been defined as technical knowledge of industrial significance which has been built up in one organization and is not in the public domain. By contrast to the case of patents, there is at present no generally accepted view as to the scope of legal protection of unpatented know-how. But it is clear that the protection of know-how does not include a right to exclude others from utilizing the know-how concerned and from selling the resultant products. Rather, it is limited to safeguarding against unlawful communication to others or disclosure to the public, and against 'stealing'. The protection of unpatented know-how does not therefore involve inherent restrictions on exports. Such restrictions may, however, be stipulated in licensing agreements under which such know-how is made available. Licences frequently cover both patents and unpatented know-how since, in many cases, the licensee would be unable to make use of the patent without having access to the related unpatented know-how. In such cases it is often the possession of the unpatented know-how rather than the patent itself which confers upon the licensor his strong bargaining position vis-à-vis the licensee. This situation may force the licensee to accept wider restrictions on his activities, in particular with regard to exports, than might be the case if only a patent were involved.
A trademark is any visible sign serving to distinguish the goods of one enterprise from those of other enterprises. Trademark laws generally give the owner of a trademark the right to preclude others from using that trade mark in commerce for goods or services in respect of which the mark is registered. Like a patent, a trademark thus involves inherent restrictions on the sale of products, but the exclusive right applies only to the use of the particular mark in association with the goods or services concerned. The reputation of a trademark may, however, in particular cases, be so strong that other enterprises may have difficulties in selling the same or similar products without that mark or with a different mark. In consequence, they may be forced to enter into a licence agreement with the trademark owner. In such cases the trademark owner is often in a similar strong bargaining position as the licensor of important unpatented know-how and may be able to impose substantial restrictions, such as export restrictions, on the licensee.
The fact that in the case of agreements involving patents and trademarks, not only do contractual restrictions exist, but also certain inherent restrictions resulting from these two forms of industrial property as to the production and or marketing of goods, is of vital importance in considering possible ways and means of removing export restrictions.
Broader Restrictive trade practices (#PC0073).
Aggravated by Suppression of information by security classification (#PF4050).
Reduced by Vulnerability of intellectual property (#PF8854).

♦ PE0347 Distortion of international trade by discriminatory government and private procurement policies
Nature Governments are major purchasers of both commodities and services. Procurement policies may be established or applied in such a way as to discriminate against some bidders, particularly those from other countries. Bids may be solicited amongst a group of suppliers selected by the purchasing authority, or under single tender from one supplier only. As opposed to public tender, these methods are the most widely used, and lend themselves most easily to discriminatory practices. Discrimination may also be achieved against foreign firms by failure to provide adequate information concerning bidding opportunities; announcements of government purchasing intentions may not be widely publicized or may be placed in publications that are not well known to most foreign firms; furthermore, the time limit for submission of bids may be so short as to constitute a major barrier to foreign concerns.
Bids may be evaluated and contracts awarded in a discriminatory manner on the basis of non-economic criteria, particularly in order to favour domestic producers. Little information on the reasons for accepting or rejecting bids may be made available after the contract is awarded (partly in order to avoid the possibility of collusion between bidders). Foreign participation in public procurement negotiations is also inhibited by factors such as special residency requirements, and special technical requirements may also be used, with government encouragement, by major private firms. Developing countries may also be obliged to restrict the manner in which they spend aid funds in foreign markets. Bilateral aid may be fully tied to procurement in the donor country, even when the lowest prices are available elsewhere.
Broader Inappropriate policies (#PF5645)
Distortion of international trade as a result of government participation (#PD2029)
Distortion of international trade by discriminatory customs and administrative entry procedures (#PE2603).

♦ PE0349 Paramnesia and confabulation
Nature These are errors and illusions in memory and their reproduction. They may consist of. treatment of fantasies as genuine events, belief that events similar to a unique event have previously occurred (reduplication of memory), or belief that an event identical to a previous event has previously occurred (déjà vu). Whilst all remembering depends heavily on reconstruction rather than on mere reproduction alone, confabulation is a highly error prone form of production of spurious memories and fabrications. Analogous phenomena may be found with collective memory particularly in some of the abuses of speech-makers and the media whereby facts are invented which appear to fit the context of a presentation. These processes also be associated with the important phenomena of rumour in establishing public opinion.
Broader Memory defects (#PD8484).
Related Amnesia (#PD8297).

♦ PE0351 Post-traumatic stress disorder
Post-traumatic amnesia
Nature The stressor producing this disorder is usually experienced with intense fear, terror and helplessness and it can be natural disasters, disastrous accidents, bombing, torture or death camps. The traumatic event can be reexperienced in dreams, flashbacks or in events that resemble the cause of trauma. Besides, the person tries to avoid thoughts or feelings about the trauma and may even lose the ability to recall some important aspects of it. Soon after the traumatic event the person starts suffering from "emotional anaesthesia", decreased ability to have affectionate feelings towards others. People may exhibit various physical symptoms, such as difficulty falling or staying asleep, irritability or outbursts of anger or hypervigilance.
Refs Eth, Spencer and Pynoos, Robert S *Post-Traumatic Stress Disorder in Children* (1985); Picquet, D Cheryn and Best, Reba A *Post-Traumatic Stress Disorder, Rape Trauma, Delayed Stress and Related Conditions* (1986).
Broader Trauma (#PD4571) Amnesia (#PD8297).
Narrower Combat trauma (#PE7912) Loss of memory due to torture (#PE6593).
Related Inhibited grief process (#PD4918).
Aggravates Nightmares (#PE6958).

♦ PE0353 Long-term shortage of metalliferous ores and metal scrap
Nature The ratios of ore and scrap reserves to annual production and consumption have been dropping irreversibly.
Incidence For example, recent calculations in the USA indicated that country would exhaust its own iron ores in 100 years. Exhaustion of world iron reserves was estimated at 350 years. Metals for which reserves will be exhausted in the twenty-first century include many essential for industrial production.
Broader Long-term shortage of manufactured goods (#PE0802)
Long-term shortage of inedible crude non-fuel materials (#PE0461).
Narrower Silver ores shortage (#PG2279) Platinum ores shortage (#PG2280)
Iron ore and concentrates shortage (#PG2277) Shortage of non-ferrous metal scrap (#PG2278)
Uranium ores and concentrates shortage (#PG2281)
Thorium ores and concentrates shortage (#PG2282)
Long-term shortage of non-ferrous metal ores (#PE0824).

♦ PE0354 Retrograde amnesia
Nature This consists of loss of memory for events that occurred at a time when brain function was unimpaired. It is therefore generally due to failure of retrieval although this is usually very selective – islands of memory in a sea of amnesia often emerge. An analogous phenomenon may be encountered in collective memory. For example particular incidents may be recalled, and reflected in various repositories, although there is loss of memory concerning the processes which connected them together and with the present.
Broader Amnesia (#PD8297).

♦ PE0355 Transient global amnesia
Nature This consists of an abrupt loss of memory, lasting from a few seconds to several hours, without any loss of consciousness. No information is stored for that period and thus there is complete loss of memory. Such attacks may be recurrent and are thought to result from temporary reductions in blood supply to specific areas of the brain (possibly presaging a stroke). Actions may continue to be performed automatically during the attack (traumatic automatism). Analogous phenomena may occur in the case of collective memory.
Broader Amnesia (#PD8297).

♦ PE0357 Traumatic amnesia
Nature Following recovery if consciousness after cerebral trauma caused by a head injury a person is typically dazed, confused, and imperfectly aware if his whereabouts and circumstances. During this state it is not possible to store new memories. On recovery the person may be unable to recall this period (post-traumatic amnesia) and may exhibit memory failure concerning brief or long periods into the past (retrograde amnesia). Subsequently memories may gradually return and be interrelated in an appropriate time sequence. Analogous phenomena may occur in the case of collective memory as a result of natural disaster or major social upheaval (war, revolution, etc.), or as a result of damage to some particular repository of collective memory.
Broader Amnesia (#PD8297).

♦ PE0358 Inhumane killing of animals
Cruel and inefficient slaughterhouse practices
Nature Animals continue to be subjected to avoidable stress during the stages of slaughter for two reasons: either unsuccessful attempts are made to institute modern scientific slaughter, or

no attempt at modernization is made and the traditional unscientific methods are allowed to continue.
Abuses in modern processes include: overcrowded transport facilities with lack of adequately trained and supervised attendants; infection from unrestricted contact with diseased animals; inadequate feeding arrangements and gastric upsets from overfeeding by owners prior to sale; rough handling by attendants; disturbance from noise or careless handling; inadequate protection from extremes of climate; inadequate ante-mortem examination facilities; inefficient stunning from lack of training of abattoir personnel; inaccurate cutting or stabbing because an animal is recovering consciousness (following electric stunning or carbon dioxide anaesthesia) due to delay; poor hygiene; unskilled meat inspection and inefficient utilization of by-products leading to waste.
Abuses in traditional slaughter processes include; throwing fully conscious animals to the ground with ropes or chains prior to killing; hoisting fully conscious cattle into the air by means of a chain fastened to a hind leg; stabbing into the heart and major blood vessels of fully conscious pigs; somersaulting conscious cattle; cutting into the throat of conscious sheep and lambs; bleeding to death of conscious poultry, which is a common system in many of the world's poultry packing stations; puncturing the spinal cord or driving a blunt instrument into the skull of meat animals by unskilled persons.
Background Of the approximately 1,000 million animals slaughtered each year in the world, at least 500 million (and possibly 600 million) continue to be killed without effective steps being taken to render them insensitive prior to being cut and bled. Of the 597 million killed on which data are available, the situation is as follows: Europe (humanely 140 million, inhumanely 60 million); Asia (humanely 3 million, inhumanely 77 million); Latin America (humanely 4 million, inhumanely 70 million); North America (humanely 138 million, inhumanely 15 million); Australasia (humanely 14 million, inhumanely 76 million). Of the approximately 317 million slaughtered in other regions on which less adequate data is available, the situation is estimated to be as follows: USSR (humanely 60 million, inhumanely 40 million); China (humanely 50 million, inhumanely 50 million); Eastern Europe (humanely 44 million, inhumanely 40 million); Africa (humanely 10 million, inhumanely 23 million).
Claim Animals slaughtered with spiked hammer blows may be struck in the eye or have their heads opened. Several blows may be required. They may fall or remain standing in crowded pens smothered in blood. Animals may be hung from chains fully conscious until they have bled to death or are flung into boiling water. Sometimes the eyes of horses burst before they are killed. The human effect of these practices is that slaughterhouse and abattoir workers may be traumatized by such violence and that all people with knowledge of these atrocities may be brutalized.
Counter-claim In some countries it is believed that the more painful the death the more potent the taste of meat.
Refs FAO *Slaughterhouse Cleaning and Sanitation* (1985).
Broader Killing of animals (#PD8486) Cruelty to animals in food preparation (#PE0236)
Denial to food animals of the right to freedom from suffering (#PE3899).
Narrower Cruel culling of seals (#PE3484) Inhumane killing of stray animals (#PE2759).
Related Ritual slaughter of animals (#PF0319).
Aggravates Food wastage (#PD8844).

♦ PE0359 Stray dog populations
Nature Unrestrained dogs living alone, in small groups or in large packs cause the spread of disease, inflict wounds on people, damage property, create a hazard on roads, make excessive noise.
Incidence Problem is worldwide, particularly in urban areas. Diseases spread by stray dog populations include rabies, hydatid disease and, to a lesser extent, anthrax, pasteurellosis and leptospiroses. Human injuries inflicted by stray dogs include bites, (which may lead to infection), fractures and sprains in old people and children, and road accidents. Stray dogs may damage property, particularly gardens, with their faeces; and litter the streets by overturning dustbins, which also increases the risk of disease and encourages rats. Domestic animals may be injured or killed by packs of stray dogs. Barking and howling causes a nuisance to the community. Origins of stray dog populations include lost dogs, abandoned dogs, dogs which are sheltered by people but not owned by them, dogs bred from stray dogs. The size of the stray dog population varies in accordance with the social and cultural characteristics of the human population, nature of the area, and season of the year. In the UK, some 90,000 strays dogs are destroyed each year.
Claim Because of the current breakdown of urban law-enforcement, more and more people feel forced to keep guard dogs; many of these dogs escape and help to swell the already startlingly large feral dog population. If city authorities are to be ready to combat this threat, it is imperative to find out much more about the ecology and behaviour of the canine community.
Broader Excessive stray animal populations (#PE5776)
Domestic animals as carriers of animal diseases (#PD2746).
Related Cruel methods of destruction of stray dogs (#PE0360).
Aggravates Rabies (#PE1325) Zoonoses (#PD1770) Feral dogs (#PG6473)
Animal injuries (#PC2753) Viral diseases in animals (#PD2730)
Inadequate animal welfare (#PC1167) Bacterial diseases in animals (#PD2731).
Aggravated by Difficulty in identifying carriers of animal diseases (#PF2775).
Reduced by Inhumane killing of stray animals (#PE2759).

♦ PE0360 Cruel methods of destruction of stray dogs
Nature The surplus dog populations in many cities, particularly in Asia, is such that effective and economic methods are required to destroy them. Methods most frequently employed, such as the use of strychnine poisoning or mass electrocution, fulfil these criteria but involve considerable suffering to the animals.
Broader Inhumane killing of stray animals (#PE2759).
Related Stray dog populations (#PE0359).

♦ PE0362 Proliferation and duplication of international nongovernmental organization information systems
Nature In the face of the impressive information systems now being planned and implemented for industry-wide data banks and governmental data integration, the network of non-governmental organizations is in a position of weakness. Each non-governmental organization, because it is required to have a competency in a specific area of interest to serve its purpose, builds data-banks, sometimes fully computerized, of facts and figures and other information resources that it considers necessary to have on hand.
Inevitably, organizations duplicate each other in these efforts and this contributes to the proliferation of reports generated from each data-bank. In addition, the effort going into the building of proprietary information systems detracts from their achieving comprehensiveness or breadth, as well as from achieving depth of necessary detail. The isolated climate in which these organizational efforts are conducted is conducive to information hoarding; an unwillingness to share any output that the information system can produce. Thus the net result of these attitudes is incomplete and inefficient multiple information systems that stand alone unlinked by computer telecommunications or cooperative research efforts.
Incidence NGO information duplication and inefficiency occurs in almost every vital area: peace research and armaments statistics, health, population growth, labour and employment, economic data, social justice, and industry and trade sectors, for example.
Broader Proliferation and duplication of international information systems (#PE0458)
Obstacles to effective international nongovernmental organizations (#PF7082).
Aggravates Competition between international nongovernmental organizations for scarce resources (#PE0259).
Aggravated by Proliferation and duplication of international nongovernmental organization and coordination bodies (#PE0179).

♦ PE0364 Basidiomycetes
Broader Fungal plant diseases (#PD2225).
Narrower Dodders (#PE1323) Blights (#PE3919) Cankers (#PE0640)
Wood rots (#PE4455) Rust fungi (#PE6255) Fairy rings (#PF4187)
Rots in plants (#PE3363) Galls in plants (#PE3715) Cereal root rots (#PE4453)
Snowmold of lawns (#PG0875) Mistletoe in trees (#PE4139)
Smut diseases of plants (#PE0857).

♦ PE0365 Inadequate institutional structures for local government in developing countries
Inadequate local authority legal structures in developing countries — Inadequate financing of local government in developing countries
Nature The local authority structures and legal frameworks in most developing countries tend to be based on those designed for rural and agricultural communities of the colonial period and are completely inappropriate to both the local culture and to the challenges of rapid urbanization. Since these structures are based on models from industrialized countries, they have tended to favour the development of energy- and material-intensive cities that are dependent on imports.
Claim Local authorities in developing countries have not been given the political power, decision-making capacity and access to revenues that are essential for them to fulfil their role. This has led to frustration, to continuing criticism of local authorities for insufficient and inefficient services, and to a downward spiral of weakness feeding on weakness.
Aggravated by Lack of management skills in developing countries (#PE0046)
Uncontrolled urbanization in developing countries (#PD0134).

♦ PE0366 Inadequate emergency blood supply
Limited blood transfusion systems
Nature Although formerly limited to whole blood only and to emergency blood transfusion, the need for human blood in medicine has grown greatly in recent decades due to specific usage of blood components. In industrialized countries where this therapy is common practice, the notion of its cost arises in a health system in which self-sufficiency is being sought in the face of commercial practices. An entire national organization must be set up to safeguard the donors and the correct preparation of blood products for patients, often acting through specific legislation. Most developing countries simply do not have the basic infrastructure for producing and delivering blood and blood products. Motivated professionals are scarce. Political and financial backing is insufficient. International assistance is difficult to implement. Because of this mortality and morbidity are higher in these countries.
In developing countries, where the structures for collecting, testing and distributing blood and blood products exist the immediate main problem of national blood programmes is to inform their populations with a view to recruiting blood donors. The blood needed for transfusion can only be supplied by the portion of the population which is healthy, neither too old nor too young, not pregnant nor nursing a baby. Paid donors, whose ranks essentially include poor people of low socio-economical standards, greatly increase transfusion risks. Blood and plasma are liable to transmit infectious diseases: viral hepatitis, AIDS, syphilis and malaria. The clinical and biological control of the blood donor must be compulsory and there are far fewer risks in the voluntary contingent.
In developed countries the lack of recognition of transfusion medicine as a medical specialty is the most important problems facing the field.
Although shortages and crises of supply (especially of the rarer blood types) were common in the past, during holidays and on week-ends, plastic containers and new storing solutions now make it possible to prolong the life span of red blood cells and platelets. Still, there may be acute shortage during the period leading up to Christmas or during flu epidemics.
Background Blood transfusion is one of the cornerstones of modern public health systems. Since World War II is has greatly contributed to the progress in human therapeutics. Cancer chemotherapy, supportive care of blood coagulation or immune disorders, organ-tissue transplants and innumerable surgical procedures would not exist to matured without blood transfusion.
An efficient blood programme requires: appropriate operational structures for collecting blood; systems for testing, preparing components and distributing blood products; adequate staff with appropriate training; and sufficient numbers of voluntary donors who are healthy.
Incidence In United States, regions with surplus blood supplies have long sent blood to areas with shortages, but it is also increasingly turning to import from Europe. While demand there has increased, donors have remained steady at about 6 per cent of the population.
Broader Social neglect (#PB0883)
Inadequate emergency medical services (#PD1428)
Limited availability of therapeutic substances of human origin (#PF6751).
Related Lack of human organs for transplantation (#PE7530).
Aggravates Malaria (#PE0616) Syphilis (#PE2300) Hepatitis (#PE0517)
Diseases of blood and blood-forming organs (#PF8026).
Aggravated by War (#PB0593) Injurious accidents (#PB0731)
Unsafe blood-related products (#PJ4536).

♦ PE0368 Benign neoplasm of ovary
Broader Benign tumours (#PD8347).

♦ PE0375 Uncomplicated alcohol withdrawal
Nature Cessation of prolonged drinking of alcohol or reduction in alcohol ingestion is followed by tremor of hands and nausea, sweating, depressed mood, illusions or headache.
Broader Alcohol abuse (#PD0153) Substance intoxication (#PD4027).
Related Delirium alcoholicum (#PG4121).
Reduces Draining of resources due to alcohol (#PE8865).

♦ PE0379 Energy deficient developing countries
Nature The rate of progress of energy resource development in many developing countries has been quite inadequate both in relation to their potential endowment as well as their requirements for various kinds and types of energy. This is particularly the case for the least developed, low-income, energy deficient developing countries where even the basic knowledge of resource occurrence is extremely sparse and fragmentary.
Background Developing countries have, almost without exceptions, faced considerable economic difficulties during the last few years, which have overshadowed all other developmental concerns, including energy. Also, many have lacked the financial resources necessary for a certain degree of autonomy in planning the development of the energy sector, and have also lacked the manpower resources to staff the various functions needed to support integrated energy policy-making.
Incidence Per capita commercial energy consumption is only 2 per cent of that of the developed world.

Refs Smil, V and Knowland, W E *Energy in the Developing World* (1980).
   Broader Compliance∗complex (#PA5710).
   Maldistribution of energy consumption (#PC5038).
   Aggravated by Inadequate cooking stoves (#PE7904).

♦ **PE0383 Instability of sugar trade**
Instability of trade in sugar, sugar preparations and honey
Nature There are two world sugar markets: one protected by regional commercial agreements at relatively high prices; the other, unprotected, and characterized by sales at relatively lower prices. One of the major reasons for the chronic instability of the latter market is that it is open to commercial dumping. A certain number of countries consider that their profits are insufficient and therefore accumulate surpluses until forced to dump them on the free market. Other countries limit their purchases, particularly when the price rises, because of lack of sufficient foreign exchange.
The world sugar market is such that of some 100 million tons produced, 70 percent are consumed in the producing country, 30 percent are exported. About 50 per cent of the export trade is governed by the Commonwealth Sugar Agreement, the US Sugar Act, agreements governing Cuban exports to the centrally planned economies and the EEC's imports from the Lomé countries. The remaining tonnage constitutes the free market, which is governed by the International Sugar Agreement. (There are also higher priced sales to the US under its current legislation, which could be called 'special arrangements', though they are considered under the ISA as part of the free market).
The free market is limited in size as a result both of special arrangements and of protection for domestic production in most developed countries; and is partly taken up by developed country exports, where protective programs have resulted in more than self–sufficiency, these exports being sold at prices far below cost and often heavily subsidized. Efforts to improve the situation through International Sugar Agreements have not been successful in the past decade. Although admittedly flawed in some of its particulars, the essential reason for the lack of success of the ISA of 1977 was the absence of the largest exporter to the market it attempted to regulate; and efforts in 1983–84 to negotiate a new agreement failed because of the inability of major exporters to agree on their own roles in the proposed agreement. The United Nations Conference in which the negotiations were held ultimately settled for an administrative agreement which is intended to prepare the way for a new economic agreement when attitudes are more propitious. The result of this whole situation has been chronically low foreign exchange earnings for most developing country exporters, relieved only by short bursts of high prices at wide intervals and, more importantly, higher prices under the particular arrangements referred to above.
Claim Annual fluctuations in the free market have a considerable effect. The situation is especially difficult because the developed producing countries treat the free market as a means of dumping their surpluses and are not especially concerned about the price impact on developing producer countries which do not have large internal markets to absorb most of their production. Efforts towards national self–sufficiency and increases in the production of beet sugar, particularly in Europe, aggravate the difficulties. As recently as 1976 the EEC has moved from a net importer to by far the largest exporter to the free market in each of the past five years. It has been estimated that removal of the trade barriers instituted by the protectionist countries in Western Europe would result in a very substantial increase in their import requirements and increase the annual export earnings of the developing countries.
Refs Coote, Belinda *The Hunger Crop* (1987).
   Broader Instability of trade in food and live animals (#PD1434).

♦ **PE0385 Instability of wheat trade**
Nature Wheat is now the staple food of nearly half the world's population, even in those countries where it is not an indigenous crop. In developing countries demand is expanding rapidly, often outpacing domestic production and leading to growing import requirements. These in turn burden the countries' balances of payments, forcing many of them even further into debt. At the same time, the rising volume of imports has overstressed the capacity of their ports, storage facilities and transportation networks, leading to expensive delays and considerable waste. In the poorest countries low production and the inability to make adequate imports often affect consumption levels.
Wheat is mostly exported by developed countries, whose producers face rising costs of inputs and may plant alternative crops if market price relationships are favourable. Producers, though efficient, therefore often need expensive government support if output is to be maintained. Those countries have the facilities to store vast amounts of grain, but find it expensive to do so. To reduce their surpluses, exporting countries may provide export credit or subsidies, leading to deteriorating trade relationships with their competitors. Periods of very low prices tend to be followed by production cutbacks and the risk of shortage if import demand unexpectedly increases. Centrally–planned countries, although themselves major wheat producers, also import on a large scale. But the variability and unpredictability of their demand can destabilize international markets.
Incidence World wheat production in the mid–1980s rose to about 500m tons, having doubled in only 20 years. Most of the increase was due to higher yields: the result of improved varieties, increased irrigation, and greater use of pesticides and fertilizers. World consumption has been growing at a comparable rate – about 3.5% a year. Developing countries, whose consumption is growing at over 4% a year, now account for more than half of world wheat imports, which averaged 100 m tons in the early 1980s. The CIF cost of their commercial imports of wheat and flour is now over US$ 8 billion a year, compared with under $ 2 billion in the early 1960s. Recently, the fastest growth in imports has been in Africa and Near East Asia.
The United States is the largest exporting country, with about 45% of world trade. Argentina, Australia, Canada and the European Economic Community together usually account for around 50%. World end–of–season carryover stocks stood at about 120m tons in 1984. Developed countries accounted for about some 75m of the total, of which half was in the United States. Developing countries held less than 20m tons and centrally–planned countries the rest. The last period of short wheat supplies was in the mid–1970s (at the time of the so–called 'world food crisis'): prices of some grades increased threefold in 18 months. Production subsequently rose rapidly and prices fell, although there was a brief return to higher prices in 1980–81, the market subsequently eased again, and by mid–1984 prices stood at levels which in real terms were probably the lowest for 50 years.
   Broader Instability of trade in cereals and cereal preparations (#PE1769).
   Narrower Excessive wheat surpluses (#PE2902)   Long-term shortage of wheat (#PE2903).
   Aggravated by Bad weather (#PC0293).

♦ **PE0386 Instability of trade in oil-seeds, oil nuts and oil kernels**
Nature Synthetics are substituted for major tropical oils in non–food uses. In addition, protectionist measures by the USA and the EEC countries tend to destabilize the oils trade with developing nation producers.
   Broader Long-term shortage of inedible crude non-fuel materials (#PE0461)
   Instability of trade in inedible crude non-fuel materials (#PD0280).

♦ **PE0392 Discrimination against children of prostitutes**
   Broader Discrimination against illegitimate children (#PD0943)
   Discrimination and harassment of children in public life (#PE6922).
   Related Children engendered by occupying soldiers (#PD8825).
   Aggravated by Female prostitution (#PD3380).

♦ **PE0394 Burns and scalds**
Burns
Nature Burns and scalds are thermal injuries in which a portion of the body surface is exposed to either dry or moist heat of sufficiently high temperature to cause local and systemic reactions. Burns may also be caused by chemical substances, electricity and ionising radiations.
The severity of a cutaneous burn wound depends not only upon the depth of the burn but also upon the extent of the body surface area affected, which determines the loss of fluids and heat and therefore the impact on the major physiological systems of the body.
Burns which only injure the epidermis, such as mild sunburn, are known as first degree burns. They result in temporary erythema (redness), due to dilation of the capillaries, and oedema (swelling). Burns which extend into the living layer of the skin, the dermis, are much more serious. It is common to distinguish between second degree, or partial thickness, burns, and third degree, or full thickness, burns. Second degree burns are those in which necrosis extends into the dermis, but with the survival of a sufficient foundation of such skin appendages as sweat glands and hair follicles to ensure that the skin regenerates without having to heal from the edges of the wounds. Third degree, or full thickness, burns are those in which all the dermis is destroyed. In addition there may be destruction of the underlying fat, muscle, bone and other tissues. The terms fourth and fifth degree burns are sometimes used to describe such injuries.
Incidence Every month hundreds of persons in the world die from burns and thousands are crippled and disfigured. The cost of care for burn patients is very high. As an indication, the market for bite and burn remedies in the UK in 1986 was £2 million.
Refs Feller, Irving *International Bibliography on Burns*; Muir, I F, et al *Burns and their Treatment* (1987); National Institute for Burn Medicine Staff *International Bibliography on Burns*.
   Broader Human disease and disability (#PB1044).
   Related Cold disorders (#PE0274).
   Aggravates Protein–energy malnutrition (#PD0339).
   Aggravated by Fires (#PD8054)   Urban fires (#PD2211)
   Injurious accidents (#PB0731)   Fungicides as pollutants (#PD1612)
   Rodenticides as pollutants (#PE3677)
   Incendiary weapons of massive destructiveness (#PD3492).

♦ **PE0395 Unsanitary and inhumane urban food animal conditions**
Nature Before the advent of commercial pasteurization of milk and its transport under protected conditions to consuming centers, cows were kept inside the cities to provide this perishable commodity within a few hours of its production. In almost all developing countries, this pattern persists. Due to the high cost of land and services and to the lack of facilities, the city cattle are kept in congested and insanitary stables. They may be in badly ventilated underground basements of residences, into which sunlight never enters. In both developed and developing countries poultry markets feature live chickens and sometimes pigeons and other birds. Most poultry is kept in confined, unclean quarters and suffer from hunger and thirst. Markets may also house live rabbits, turtles, lambs, or kids, and inhumane slaughter as well as maltreatment may occur.
Incidence In India alone it is probable that over 2 million cows and buffaloes are kept inside the cities and large towns, and the number is steadily increasing.
   Broader Maltreatment of animals (#PC0066).
   Related Insanitary penning conditions as factor in animal diseases (#PE2764)
   Denial to food animals of the right to freedom from suffering (#PE3899).
   Aggravates Biological contamination of food (#PD2594)
   Environmental hazards from food processing industries (#PE1280).

♦ **PE0396 Collusive international trade arrangements**
Secret trade cartels — International cartels
Nature Collusive arrangements at national and international levels with regard to exports and imports clearly affect international trade. Collusion at the international level is often aimed at retaining exports by combating newcomers; and at maximizing gains, in particular with respect to governments' procurement of imports through collusive tendering. Intrafirm transactions by transnational corporations can also have adverse effects on international trade, in particular when imports by subsidiaries from parent companies are at prices above those prevailing in the world market. Moreover, refusal to supply, or selling at discriminatory prices, singly or collectively, can also have adverse effects on trade. Exclusive dealing arrangements in international trade transactions also tend to rigidify trade patterns by making it difficult for newcomers to enter markets, especially where the purchasers or distributors who are parties to such arrangements hold a dominant position of market power.
Incidence In the case of international cartels and national export cartels, which are permitted in most countries, there continues to be a general lack of transparency even about their number, let alone the nature and extent of their operations. This lack of transparency is a major hindrance to ensuring effective control, including that by countries which suffer from the effects of such practices.
Refs Edwards, Corwin, et al *Economic and Political Aspects of International Cartels* (1976); Mason, Edward S *Controlling World Trade*.
   Broader Cartels (#PC2512)   Restrictive trade practices (#PC0073).
   Aggravates Excessive external trade deficits (#PC1100)
   Vulnerability of socio–economic systems from globalization (#PF1245)
   Disproportionate control of global economy by limited number of corporations (#PE0135).
   Aggravated by Secret international agreements (#PF0419).

♦ **PE0397 Injustice of religious courts**
Nature In some countries there exist religious courts which deal with matters pertaining to the personal status of members of the religion in question. These courts may have exclusive jurisdiction over matters such as succession, inheritance, wills, legacies, gifts, marriage, divorce and family relations. Individuals appearing before a religious court may be denied the rights and procedural guarantees which exist in other courts. Additional hardships may occur if a person does not belong to any of the religions whose courts have exclusive jurisdiction over the matter in question.
   Broader Injustice (#PA6486)   Undue religious influence on secular life (#PF3358)
   Religious discrimination in the administration of justice (#PE0168).
   Narrower Religious or civil refusal of divorce (#PF3248).
   Related Religious and political antagonism (#PC0030)
   Discriminatory religious influence on the law (#PD3357).
   Aggravates Anti-clericalism (#PF3360)   Religious conflict between sects (#PC3363).
   Aggravated by Religious intolerance (#PC1808)   Multidenominational society (#PF3368)
   Corruption in organized religion (#PC3359)
   Negative effects of claims of religious infallibility (#PF3376).

♦ **PE0402 Unemployment in developed countries resulting from participation of developing countries in manufacturing**
Nature Liberalization by the developed market economy countries of their tariff and non–tariff barriers to imports from developing countries would cause injury to many of the factors of

production which are employed in affected import–competing industries in the developed world. It is likely that the scope of the generalized system of preferences and other forms of trade liberalization have been, or will be, adjusted by governments of developed market economy countries because of anticipated injury to domestic factors of production. Furthermore, it is possible that the actual intent of such liberalization measures may be at least partially offset by 'escape clauses' and similar provisions designed to protect vulnerable sectors of production in import–competing industries. Since the share of the developing countries in the world trade in manufactures is very small, the over–all effect of the increase in imports from the developing countries is not likely to constitute a serious problem. However, an increase in such imports may have a significant impact on employment in certain industries and certain regions of a developed country.

Studies to indicate the likely impact on employment of increasing imports in the Federal Republic of Germany, the UK and the USA, indicate that the overall magnitude of labour displacement would not be large. For many imports, the required increase in demand to offset the assumed increase in imports from the developing countries would be insignificant. For others, however, it is high and not likely to be attained, especially in view of the low–income elasticity of the demand for these products, which include clothing and footwear, leather and leather products, and wood and cork products. The labour displacement effect of increased imports is also modest and would seem far more manageable than the displacement effect resulting from increases in labour productivity in these three developed countries.

Counter–claim There is considerable scope for the developed countries to expand their imports of industrial products, capital–intensive as well as labour–intensive, from developing countries without seriously reducing employment in the industries directly affected. For some industries, in which labour productivity is substantially below the overall manufacturing average, an increase in competing imports would provide a salutary impetus to shifting their workers, especially workers in the least efficient producing units, to new or expanding industries with labour productivity higher than average, or to industries producing goods for export to developing countries. This process of labour transfer would be facilitated by the vigorous application of measures of structural adjustment. Such structural employment problems as may arise from increased import competition from developing countries seem far more manageable than the employment problems presented by rising labour productivity resulting from technological progress.

A far more significant loss of jobs has been due to the debt crisis. Some 8 million jobs have been lost in the North by a reduction of imports by Latin America and Africa. In part, this has increased protectionism further reducing the capacity for third world countries to import from the North.

Broader Unemployment in developed countries (#PC9718).

♦ PE0403 Aggression against nuclear power sources

Broader Aggression (#PA0587) Inadequate nuclear reactor safeguards (#PF6084)
Aggressive uses of natural energy resources (#PD0408).
Aggravates Nuclear reactor accidents (#PD7579)
Insufficient nuclear power stations (#PD7663)
Environmental hazards of nuclear power production (#PD4977).

♦ PE0405 Legal profession's monopoly of court proceedings

Nature Only barristers and solicitors may speak in court on behalf of another person.
Broader Monopolies (#PC0521).

♦ PE0412 Experimental surgery on animals

Nature Animals are subjected to experimental burns, parabiosis (the surgical connection of two individuals), grafting of limbs or organs (possibly from other species), displacement of limbs to other locations on the body, and mutilation of genitalia to observe changes in mating habits. Demonstration surgery on animals may also be performed for educational purposes.
Broader Cruel treatment of animals for research (#PD0260).

♦ PE0413 Exclusive dealing arrangements

Nature Exclusive dealing arrangements in international trade are aimed at restricting the channels through which particular goods may be imported or exported, thereby ensuring control over the distribution, sale and resale of the products. Where trade marks are used in respect of the products in question and the parties to the exclusive dealing arrangement can invoke trade mark rights to prevent parallel imports by third parties of such products, the importer/distributor has an effective monopoly over the distribution of the product in the importing country.

Exclusive dealing arrangements enable an enterprise, when in a dominant position of market power in respect of the supply of a product, to engage in differential pricing policies as between each importing market, the price being set on the basis of the highest that each market can bear. Such policies can have detrimental effects on the balance of payments of the importing country. Exclusive dealing arrangements, when engaged in by enterprises in a dominant position of market power, are also likely to adversely affect the ability of new entrants to penetrate the market, either as domestic manufacturers or as competing importers/distributors.

Broader Restrictive trade practices (#PC0073).
Related Grant–back provisions (#PE5306) Challenges to validity (#PF1200)
Restrictions on research (#PF0725) Restrictions on publicity (#PF1575)
Restrictions on adaptations (#PF5248) Restrictions on use of personnel (#PF3945)
Patent pool or cross-licensing agreements (#PE4039)
Exclusive sales and representation agreements (#PE4581)
Payments after expiration of industrial property rights (#PF5292).

♦ PE0424 Injustice of trials in absentia

Nature There can be no doubt that rules permitting trial in absentia have at times operated to the disadvantage of persons accused of political offences, or of persons regarded as politically undesirable by the regimes of their countries, who may be tried in their absence for an offence without the proper observance of guarantees for their defence.
Broader Injustice (#PA6486) Unjust trials (#PD4827) Political repression (#PC1919).
Related Imbalance (#PA0224) Political discrimination in the administration of justice (#PE1828).

♦ PE0427 Shortage of marriageable women

Shortage of girls
Incidence In China, bachelors were found to outnumber single women by 10 million within the 29 to 49 year–old age group (based on the 1982 census and a 1987 survey). In Korea Rep there will be a 120 men per 100 women in the year 2000, because of the traditional preference for boys.
Related Shortage of marriageable men (#PE9379).
Aggravated by Infanticide (#PD3501) Gender abortions (#PD3947)
Deliberate imbalancing of population sex ratio (#PF3382).

♦ PE0428 Maltreatment of prisoners by fellow inmates

Bullying amongst prisoners — Harassment amongst prisoners — Intimidation of prisoners by other inmates — Crime in prisons — Inmate killing
Nature Under certain conditions prisoners act violently against each other. In addition to many forms of intimidation, forced homosexuality, and general harassment, this may include knifing and murder. Sex offenders are especially vulnerable.
Incidence In the USA the prisons themselves are so rife with crime, feuds and violence that about 8 per cent require protective custody for fear of retribution for sex offences (some committed against other prisoners), for fear of homosexual attacks, because of unpaid debts amongst prisoners. In 1980, rioting prisoners in the Santa Fe penitentiary used blow torches and knives to slaughter (and in some cases decapitate) 36 other inmates.
Broader Maltreatment of prisoners (#PD6005).
Aggravated by Homosexuality in prisons (#PE1363)
Denial of rights to prisoners (#PD0520).

♦ PE0429 Neurological rage

Explosive anger — Violent rage
Nature Neurological impairment can result in a form of explosive rage which is distinct from ordinary anger. It is a sudden and unpredictable storm of overwhelming fury which may be triggered by a trivial event. It is often out of character, with the person subsequently embarrassed at the realization at having been out of control.
Incidence The condition is fairly common, being a symptom of any disease that causes damage to brain cells. It has been estimated that a third of those with Alzheimer's disease exhibit uncontrollable rage, which provides one of the main reasons for which families feel they are unable to care for relatives so afflicted.
Aggravated by Alzheimer's disease (#PE7623) Diseases and injuries of the brain (#PD0992).

♦ PE0430 Inadequate cargo transportation

Shipping risk
Nature The risk of cargo not being delivered, not being delivered on time or being delivered damaged is increasing costs. While a portion of the reason is due to accidents and delays that can not be prevented, much of the problems is due to theft.
Broader Risk (#PF7580) Insufficient transportation infrastructure (#PF1495).
Narrower Unsafe transport of perishable foodstuffs (#PG4679).
Related Cargo insecurity (#PE5103) Dangerous cargo handling (#PD9108).
Aggravates Poor communications networks in rural areas (#PF6470).
Aggravated by Mechanical failure (#PC1904) Injurious accidents (#PB0731)
Inadequate air transport service (#PJ0260).

♦ PE0435 Paranoia

Delusional insanity — Paranoid traits — Paranoid personality disorder — Paranoid states — Persecutory paranoid state — Paranoid schizophrenia
Nature Paranoia refers to gradually developing, systematized delusional states, without hallucinations or general personality deterioration but with preservation of intelligence, and with emotional responses and behaviour that remain congruous with and appropriate to the persecutory or grandiose delusions.
Incidence Paranoid patients are relatively rare in the mental hospital population. They constitute about 0.5 percent of first admissions and less than 1.5 percent of resident patients in mental hospitals in the United States. However, these figures probably underestimate the incidence of the illness in the population at large; many patients are able to control the socially disruptive manifestations of their delusions and are never hospitalized; others, especially where the disorder is less severe, are tolerated at home and at work as eccentrics.
Background The term paranoia is one of the oldest in the history of psychiatry. It was used in pre–Hippocratic times and, in ancient Greek literature, is the term for mental derangement. It was reintroduced into medicine in 1764, and was applied to various conditions. In 1883, a New York psychiatrist, E C Spitzka, gave paranoia its present definition.
Broader Psychoses (#PD1722) Schizophrenia (#PD0438)
Personality disorders (#PD9219) Psychological conflict (#PE5087)
Mental disorders of the aged (#PD0919).
Narrower Megalomania (#PF2108) Jealous paranoia (#PG4309)
Somatic paranoia (#PG2714) Erotomanic paranoia (#PG6789)
Persecutory paranoia (#PG6942).
Aggravated by Amphetamine withdrawal (#PG9778).

♦ PE0436 Misuse of nonprofit associations as front organizations by government

Incidence In the superpowers' cold war, 'Trojan horse' organizations have been created to mask espionage, subversion and propaganda. Organizations involved have been trade and professional associations: for example, those of journalists, trading companies, research institutions, universities and cultural organizations.
Broader Front organizations (#PE4358) Abuse of government power (#PC9104)
Obstacles to effective international nongovernmental organizations (#PF7082).
Related Bogus firms (#PF0326)
Ineffective governmental use of nongovernmental resources (#PF4095).
Aggravates Bogus public interest groups (#PE7575)
Unethical practices of philanthropic organizations (#PE8742)
Jurisdictional conflict and antagonism between international organizations (#PD0138).
Aggravated by Repression (#PB0871).

♦ PE0437 Negligence in manslaughter

Negligent murder — Homicide by neglect
Nature Causing the death of another person through neglect of their care is a form of homicide.
Broader Homicide (#PD2341) Negligence (#PA2658) Manslaughter (#PD7952).

♦ PE0439 Hunting of marine animals

Whaling — Scientific whaling
Incidence An estimated 500,000 small whales, porpoises and dolphins are killed annually by hunting or through being caught in nets. Most evident is the destruction of whales and dolphins, although the latter are often killed inadvertently in the process of netting tuna or through the use of drift nets. Also widely publicized is the clubbing to death of seal pups for the fur trade. Despite the 1986 moratorium against whaling, a process of "scientific whaling" has been continued by Norway, Japan and Iceland. This is stated to signify the killing of whales to assess their age and sex so that stocks can be calculated. The whales are nevertheless processed in the usual way so that there is strong suspicion that the term is merely a cover–up for normal commercial whaling practices. To compensate for the reduction in supplies of whale meat, the hunting of porpoises has increased since 1986. Thus in three years, Japanese fishermen have killed 70 per cent of the estimated population of 105,000 Dall's porpoises in Japanese waters.
Counter–claim Fish–eating whales are vermin which should be kept to a minimum because they endanger those whose welfare depends on the fishing industry. The worst offenders are the minke whale and the harp seal. In the case of Norway, if the stock of sea mammals was reduced to extinction, that country's estimated annual fisheries income would increase from £100 million to £250 million per year. Employment would increase from 25,200 jobs to 65,500. However, it is recognized that it would not be economic to hunt the whales to extinction because of the cost of locating the last few specimens.
Broader Hunting of animals (#PC2024).
Aggravates Endangered species of porpoises (#PE3806)
Endangered species of marine mammals (#PD3673)

Endangered species of rough-toothed and white dolphins (#PE8302).
Aggravated by Unsustainable exploitation of fish resources (#PD9082).
Reduces Depletion of fish reserves by marine mammals (#PE4913).

♦ PE0447 Influenza

Flu — Grippe

Nature An acute infectious disease, characterized by a sudden onset, fever and generalized aches and pains, influenza usually occurs in epidemics and pandemics. There are at least 3 types of influenza virus known as A,B and C, with numerous sub-types or strains within each. Infection with one type gives no protection against another; and the virus is continually changing its character, making it difficult to prepare an effective vaccine. In most cases the disease lasts 2–4 days and is characterized by headache, generalized aching and prostration symptoms. Even after an attack of only average severity there tends to be a period of weakness and depression. The A-type disease also effects animals and birds. Some influenza-A viruses may be classed as zoonoses, as it appears that new strains affecting man may have originated in animals. On the other hand the swine influenza virus, such as occurred in the mid-western USA, is considered to have its origin in the much earlier devastating human flu epidemic following Worldwar I.

Incidence The flu virus changes its genes constantly and each change produces a new epidemic. Pandemics and epidemics may be as frequent as at ten year intervals; for example 1947, 1957, 1968. Earlier series include 1889, 1918, and the mid 1930s. In the 1918–1919 pandemic, 20 million people perished, and 1,000 million people suffered from the disease but recovered.

Refs Kilbourne, Edwin D *Influenza* (1987); Patterson, K David *Pandemic Influenza Seventeen Hundred to Nineteen Hundred* (1987).

Broader Epidemics (#PC2514) Virus diseases (#PD0594)
Infectious and parasitic diseases (#PD0982) Diseases of the respiratory system (#PD7924).
Narrower Cyanosis (#PG2361).
Aggravates Absenteeism (#PD1634) Human ageing (#PB0477)
Encephalitis (#PE2348) Lung disorders and diseases (#PD0637).
Aggravated by Natural disasters (#PB1151).

♦ PE0451 Sleep difficulties due to torture

Nature Difficulty in sleeping is the second most frequent difficulty reported by victims of torture. Many find it difficult to go to sleep and have nightmares. Some victims scream in their sleep.
Broader Neurological effects of torture (#PD4755).
Aggravates Nightmares (#PE6958) Sleep disorders (#PE2197).

♦ PE0453 Underdevelopment of textile and clothing industries

Broader Underdevelopment of manufacturing industries (#PF0854).
Narrower Footwear manufacture underdevelopment (#PU2388)
Wearing apparel manufacture underdevelopment (#PU2387).
Aggravates Inappropriate clothing (#PJ6604).

♦ PE0456 Adulteration of illicit drugs

Nature Drugs sold on the street or under casual circumstances for illicit use are often adulterated with less expensive drug or non-drug substances. Marijuana and mescaline sold as 100 per cent pure have shown additives of a powerful animal tranquillizer (PCP) in analysis. Drug testing services in USA estimate that half of the drugs tested are either something else altogether or doctored with other substances. Alum (toxic) is sometimes added to heroin when cheap non-toxic substances such as quinine and lactose are not available. Under casual circumstances it is relatively easy for buyers to be deceived. They may be inexperienced and ignorant about drugs. Results of adulterated drugs may be relatively harmless in the instance of a 'bad trip', although even this may have long-term psychological repercussions. However, the user runs the risk of poisoning and in very severe cases perhaps even death.
Broader Adulteration (#PD9433) Illicit drug trafficking (#PD0991).
Related Drug smuggling (#PE1880).
Aggravated by Ignorance of drug users (#PJ2389)
Limited access to pharmaceutical drugs (#PE1278).

♦ PE0458 Proliferation and duplication of international information systems

Refs Mowlana, Hamid *International Flow of News* (1985).
Broader Lack of control (#PF7138) Human errors and miscalculations (#PF3702).
Narrower Proliferation and duplication of United Nations information systems (#PE0075)
Proliferation and duplication of international nongovernmental organization information systems (#PE0362).
Aggravates Competition between international organizations for scarce resources (#PC1463).
Aggravated by Underemployment (#PB1860) Unreported crimes (#PF1456)
Proliferation of information (#PC1298)
Inadequate integration of international information systems (#PE8066)
Proliferation and duplication of international organizations and coordinating bodies (#PE1029).

♦ PE0459 Instability of forestry and logging

Nature The growing and felling of trees generates products that are chiefly desired as roundwood, measured in cubic metres of solid volume without bark; wood pulp; newsprint; and paper and paper board. The instability of supplies and prices for these products is a function of forest vulnerability to over-cutting, fire and pests, and under-planting to compensate for these.

Incidence Some 32 percent as the world land area is classified as forest land. Asia and Europe have 1.1 hectares per capita of population, South America has 4.9, USSR 3.8, and North America, 3.4 hectares. In volume the world's standing timber is estimated at over 235 milliard (billion) cubic metres. The annual harvest is about 2 milliard; just under 7 percent. Fuel wood consumption accounts for nearly half; and industry uses over 1 milliard cubic metres annually.
Broader Instability of economic and industrial production activities (#PC1217).
Aggravates Deforestation (#PC1366).

♦ PE0461 Long-term shortage of inedible crude non-fuel materials

Broader Long-term shortage of commodities (#PC1195)
Insufficient availability of goods (#PB8891).
Narrower Paper shortage (#PE1616) Skins and fur skins shortage hides (#PE8327)
Long-term shortage of wood, lumber and cork (#PE1372)
Long-term shortage of metalliferous ores and metal scrap (#PE0353)
Instability of trade in oil-seeds, oil nuts and oil kernels (#PE0386)
Long-term shortage of crude fertilizers and crude minerals, excluding coal, petroleum and precious stones (#PE1353).

♦ PE0471 Heart disease in animals

Broader Circulatory system diseases in animals (#PD5453).
Narrower Arrhythmia in animals (#PG9277) Myocardium diseases in animals (#PG9800)
Pericardium diseases in animals (#PG9731) Endocardium diseases in animals (#PG3914)
Cardiac insufficiency in animals (#PG9656).

♦ PE0472 Abnormal blood pressure

High blood pressure — Low blood pressure

Refs Kesteloot, H and Joossens, J V (Ed) *Epidemiology of Arterial Blood Pressure* (1980).
Broader Diseases of blood and blood-forming organs (#PF8026).
Aggravated by Hypertension (#PE0585).

♦ PE0481 Environmental hazards from coffee, tea, cocoa, spices and their manufacture

Nature Grinding mills, roasting and other processing of spices, cocoa, coffee and tea present a number of hazards. They include flying particles (such as hulls, stems, leaves and other impurities) and dusts and powders. Strong aromas produced in storage or manufacture can be considered air pollutants as they cause some people to become nauseated.

Incidence Green coffee beans arriving bagged from the plantations may have been fumigated with bromomethane or other halogenated or organophosphorous pesticides. Solvent extraction of caffeine employs trichloroethylene. Pesticide and solvent poisoning both occur to coffee handlers. Disposal of solid wastes from instant coffee, tea, cocoa, and similar preparations may cause pollution of water courses. Environmental hazards arising from cultivation of these crops include destruction of forests and wildlife. Additional damage is done in the human environment by excessive consumption of caffeine-bearing liquids: coffee, tea, cocoa and some soft drinks.
Broader Environmental hazards from economic and industrial products (#PC0328).

♦ PE0483 Environmental hazards from tobacco and tobacco manufactures

Incidence Tobacco cultivation causes the loss of forests; constant handling of fresh tobacco leaves can cause nicotine poisoning; processing of dry leaf causes tobacco dust, giving rise to respiratory diseases among workers; consumption of tobacco products is linked to cancer; and discarded remnants of cigarettes and cigars are a major source of litter and environmental disfigurement. In addition, burning discards cause home and forest fires; and discards spread disease germs and add to the costs of all public services, since they must be removed and the receptacles provided for them cleaned. Service workers are exposed to health risks in performing such janitorial tasks.
Broader Environmental hazards from economic and industrial products (#PC0328).
Related Environmental hazards from food processing industries (#PE1280).

♦ PE0484 Harbouring national security offenders

Nature It is a crime against the national interest to harbour or conceal anyone who has committed or is about to commit treason, sabotage, espionage or the murder of national leaders.
Broader Aiding national security criminals (#PD7407).
Aggravated by Coccidiosis (#PE2738) Heartwater disease (#PE2680)
Wild birds as vectors of animal diseases (#PE2749)
Vitamin E deficiency in domestic animals (#PE4760)
Inadequate carcass disposal of diseased animals (#PE2778)
Inadequate disinfection of pastureland after disease outbreak (#PD2774).
Reduces Excessive cost of animal protein (#PE4784).

♦ PE0494 Injustice of military tribunals

Military tribunals

Nature Members of the armed forces are, in many if not most legal systems, accorded an inferior type of procedural justice. Charges are brought under the authority of the accused's commanding general or equivalent superior officer, and the same officer often appoints the members of the court and counsel for both prosecution and defence, all from among members of his own command, and also has the power to review the findings of the court. It is questionable whether the personnel mentioned can always exercise complete freedom of judgement and action if they are dependent on their commanding officer for their efficiency ratings, promotions, allocation of duties, and leave rights.

Military personnel accused before military tribunals often have no freedom of choice of counsel, even from among military personnel. They may face particular difficulties in securing evidence in their favour and in having access to all the evidence brought before the tribunal by the prosecution. Hearings by military tribunals are held in what may in some cases be an unnecessary degree of secrecy. Procedures followed by military tribunals tend to stress the importance of dealing with cases speedily, and this factor may also operate to the detriment of accused military personnel.

Military tribunals also have jurisdiction over civilians in a number of countries; and the offences which they are empowered to try are often of a political character.
Broader Injustice of special courts (#PE0088)
Military political prisoners and detainees (#PD3014).
Related Political repression (#PC1919)
Political discrimination in the administration of justice (#PE1828).
Aggravates Military disobedience (#PD7225).
Aggravated by Deficiencies in military codes of justice (#PE8300).

♦ PE0497 Distortion of international trade by restrictive customs valuation practices

Nature In some countries, special valuation procedures exist under which value for customs purposes is determined, taking into account domestic prices prevailing in the importing country or current domestic value for the same product in the exporting countries. In particular, the system of levying customs duty on the basis of FOB value or current domestic value, whichever is higher, prevalent in some countries, makes it difficult for exporters to know in advance the amount of duty payable; resulting uncertainty has adverse effects on exports. These systems act particularly to the disadvantage of developing countries as in a large number of cases, because of structural imbalances, supply scarcities and other factors, the domestic prices in these countries are maintained at artificially high levels. In addition, in some cases, goods which are produced by specially established export-oriented industries are not sold in the domestic market, which creates special difficulties in ascertaining comparable current domestic value.
Broader Distortion of international trade by discriminatory customs and administrative entry procedures (#PE2603).

♦ PE0500 Instability of trade in chemical elements and compounds

Broader Instability of chemicals trade (#PD0619).
Narrower Radioactive and associated materials trade instability (#PE8843)
Trade instability in inorganic chemicals, elements, oxides and halogen salts (#PE9076).

♦ PE0501 Downy mildews in plants

False mildews

Nature Downy mildew makes white, grey or violet patches on the leaves.
Broader Oomycetes (#PE5465).

♦ PE0502 Occupational rheumatism

Nature Rheumatic diseases occur more frequently in certain occupations than in others.

Incidence Surveys in coal mines have indicated a greater loss of work from rheumatic complaints than in the general population; from the age of 30 miners lose more work than non-miners and prolonged incapacity is not unusual. Miners suffer mainly in the low-back, thighs and knees, with associated degenerative changes in the lower dorsal and lumbar intervertebral discs and in the knee joints. Radiological studies have shown moderate or severe changes of dicc degeneration more than twice as often in miners as in other manual workers and six times as often as clerks and other office staff. Building operatives have similar symptomatology to miners but lose more work despite fewer radiological changes. On the other hand, foundry workers, especially those who are exposed to radiation from molten metal have fewer complaints and lose less work than miners despite a rather greater frequency of radiological change, particularly in the lumbar spine. Osteoarthrosis of the elbows is increased in miners, building operatives and workers using pneumatic drills. Cervical dicc degeneration is common in dentists but is also a feature of rheumatic disease in workers in certain countries (for example Jamaica) where it is customary to carry heavy loads on the head; despite severe radiological changes, there are few symptoms. In farmers in temperate climates sciatic pain appears to be a more frequent complaint. In cotton spinners and weavers, the finger joints are mainly affected, the changes being those of osteoarthrosis, often with Heberden's nodes; symptoms, however, are infrequent and the incapacity slight.

Broader Rheumatic diseases (#PE0873) Occupational diseases (#PD0215).
Related Lumbago (#PE1310).
Aggravates Absenteeism (#PD1634) Age discrimination in employment (#PD2318).
Aggravated by Sciatica (#PE2428) Dystrophy (#PE3506)
International imbalance in the quality of life (#PB4993).

♦ PE0503 Lack of accountability in the disposal of wealth

Claim Those who have accumulated more wealth than is necessary for the needs of their lifetime are not held accountable for the use of that resource in relation to global needs. This creates needless suffering among both wealthy and poor.

Broader Social unaccountability (#PC1522)
Unregulated ownership of the means of production (#PF2014).
Aggravates Environmental prodigality (#PF7318).

♦ PE0504 Poliomyelitis

Polio — Infantile paralysis

Nature Poliomyelitis is an acute infectious disease due to a virus which attacks the nervous system, primarily the anterior horn cells of the spinal cord and brain-stem. The usual route of infection is the alimentary tract via personal contact with healthy carriers and abortive cases, and via the faecal contamination of food. The disease is characterized by symptoms ranging from a mild nonparalytic infection to an extensive flaccid paralysis of voluntary muscles. It occurs most often in summer and fall and generally attacks children under 5 years of age.

Incidence The first epidemic of poliomyelitis to be recorded occurred in Sweden in 1881. Since then severe epidemics have been reported in many countries and islands including North, Central and South America, Europe and Africa. The disease reached epidemic proportions in some region of the US almost every year after the beginning of the 20th century until vaccine became available. The greatest recorded sustained incidence in the United States was in 1942–43, and in 1950 there were 33,344 cases in the US. In 1952 severe epidemics occurred in Denmark, Germany and Belgium. In Asia, outbreaks have been reported in Bombay, Singapore, Japan, Korea and the Philippines.

Since the widespread introduction of vaccination against poliomyelitis, around 1957, the incidence of this killing and crippling disease has fallen to insignificant proportions in North America, parts of Europe, and several countries in other parts of the world; in 1968 there were only 24 cases in England and Wales, compared with 3,200 in 1956. In Sweden and Finland, paralytic poliomyelitis has disappeared following immunization programs with multiple doses of inactivated vaccine that covered close to 100 percent of the population. However, in large areas of Africa, Asia and Latin America the incidence appears to have been rising disconcertingly during recent years, and large outbreaks are being reported there with increasing frequency. In many tropical countries paralytic poliomyelitis is a far more common disease than had been appreciated. The World Health Organization estimates that 220,000 children are paralysed and 23,000 killed by polio every year in developing countries.

Background There is evidence that the disease has existed for many centuries. However, the first clear description of poliomyelitis came from Jakob von Heine in 1840. It was not recognized in its epidemic form until Oskar Medin in 1887 reported observations on an outbreak in Sweden. In addition to the usual spinal form, he described the bulbar and certain less common types of the disease. C S Caverly, in 1896, recognized that the disease may occur in an abortive or nonparalytic form, a fact confirmed by others later. While earlier outbreaks centred largely in Sweden, epidemics have occurred in many parts of the world and became particularly prominent in the United States after the beginning of the 20th century.

Counter-claim The global elimination of polio is technically feasible and is targeted as one of the six vaccine-preventable diseases by the World Health Organization's Expanded Programme on Immunizations.

Refs Laurie, G and Raymond, J *Proceedings of Rehabilitation Gazette's Second International Post-Polio Conference and Symposium on Living Independently with Severe Disability* (1984); Laurie, G and Raymond, J *Proceedings of GINI's Third International Polio and Independent Living Conference, May 10–12, 1985, St Louis, Missouri* (1986); Munsat, Theodore L *Post-polio Syndrome* (1989).

Broader Paralysis (#PD2632) Virus diseases (#PD0594)
Stigmatized diseases (#PD7279) Infectious and parasitic diseases (#PD0982).
Aggravates Child mortality in developing countries (#PE5166).
Aggravated by Houseflies as pests (#PE3609) Cockroaches as pests (#PE1633)
Flies as insect pests (#PE2254) Contamination of drinking water (#PD0235)
Carelessness in dealing with infectious patients (#PE9105).

♦ PE0505 Environmental hazards from dairy products and eggs

Nature Milk and dairy products and eggs can be contaminated in various ways, in which case they constitute a hazard to public health. Milk contains few bacteria when it leaves the udder of the healthy cow, but is liable to contamination from the exterior of the animal, especially the exterior of the udder and adjacent parts. Micro-organisms of manure, soil and water may enter from this source. Automatic milking apparatus and various milk utensils may add contaminants after the milk leaves the udder. Contamination can also occur from the tanker truck and various utensils and equipment at the market milk plant, cheese factory, condensery or other processing plant. Other possible sources of contamination are the hands of the milker or other dairy workers, an flies which may add spoilage organisms or pathogens.

Although the majority of freshly laid eggs are sterile inside, the shells become contaminated by faecal matter from the hen, by washing water if the eggs are washed, and by handling. Micro-organisms, including pathogenic salmonella, can penetrate a cracked shell and enter the egg.

Broader Environmental hazards from food and live animals (#PC1411).

♦ PE0509 Instability of construction industry

Nature Activity in the construction industry, whether measured by new home starts, floor areas of office space built in a period, or employment of construction workers, exhibits both seasonal cycles and sporadic industry-wide recessions. The recessions are usually part of general economic down-turns. They may be due to excessively high mortgage interest rates or unavailability of lending money or both. Construction is affected by prices and by shortages of supplies, notably steel and wood. The instability of the industry exposes millions of construction workers to part-time employment and frequent lay-offs.

Broader Instability of economic and industrial production activities (#PC1217).

♦ PE0511 Long-term shortage of furniture

Nature Population increases and resulting demands on diminishing resources give priority to industrial and community needs. Private property such as home furniture may be only obtainable at exhorbitant prices.

Claim Wood will vanish as a resource. Metals will be diverted to non-furniture needs. Factory produced synthetic materials will be produced at premium prices with synthetic furniture becoming a luxury for very rich individuals and corporations. Inadequate numbers of local craftsmen will not be able to provide quantities of furniture made from junk, garbage, stone, clay and earth.

Broader Long-term shortage of miscellaneous manufactured articles (#PE0613).

♦ PE0517 Hepatitis

Infectious hepatitis — Jaundice — Infectious jaundice — Viral hepatitis A — Viral hepatitis B — Serum hepatitis — Hepatitis C

Nature Hepatitis, also known as viral or virus hepatitis, is defined as an acute inflammation of the liver caused by one of the hepatitis viruses: Viral hepatitis A (formerly known as infectious hepatitis); viral hepatitis B (formerly known as serum hepatitis); viral hepatitis non-A non-B (other viruses). Although they differ etiologically and epidemiologically they are similar clinically. The characteristic symptoms and signs include anorexia, nausea, vomiting, abdominal distress, liver enlargement and jaundice. In typical cases of viral hepatitis there is a pre-icteric and an icteric stage, although unicteric cases (cases without jaundice) are common. The duration of illness is usually between two to eight weeks but may be longer. Attacks are more severe in the old than the young, in whom the disease may be mild and brief.

Hepatitis A is usually spread by the intestinal-oral route; infection occurs readily in conditions of poor sanitation and overcrowding. A common source of epidemic outbreak is contaminated food and/or water. The incubation period varies from between 15 to 50 days. In countries with poor standards of hygiene and sanitation the majority of infections occur in children under the age of ten, often without any clinically apparent disease. In countries with high standards of hygiene and sanitation, only 10–15 percent of infections occur during childhood. Viral hepatitis B is spread parentally with an incubation period of between 50 to 160 years. In areas of low endemicity the majority of cases occur in adults, whereas in hyperendemic areas the infection predominates in infancy and childhood. Persistent or chronic active hepatitis may follow hepatitis B or hepatitis non-A non-B, but not hepatitis A. This occurs especially when infection is acquired in infancy and childhood. Viral hepatitis non-A non-B can be diagnosed only when infection by viruses of hepatitis A and hepatitis B have been ruled out. The faecal-orally transmitted or water-borne form has a short incubation period and resembles viral hepatitis A, whereas the blood-borne form, which has a long incubation period, resembles viral hepatitis B. Delta hepatitis is an additional form of hepatitis caused by defective delta virus. This a unique agent capable of replication only in the presence of hepatitis B virus. Although safe and effective vaccines have been developed, they are expensive.

Incidence The case fatality rate of hepatitis is between 1 in 500 to 1 in 1000; exceptions are hepatitis B following blood transfusion and, in some countries, hepatitis during pregnancy. The direct cost of hepatitis in countries surveyed by the World Health Organization is estimated at US $20,000 to $75,000 per 100,000 population. Two thirds of these costs can be attributed to hepatitis A.

Hepatitis is an occupational hazard among health care personnel and the staff of closed institutions. High rates of infection have been reported in drug abusers, prostitutes, and male homosexuals.

Hepatitis A has worldwide distribution. In Europe, about 50% of the cases are children under 15 years of age. As for hepatitis B, there are at present at least 200 million persistent carriers worldwide. The prevalence of hepatitis B carriers in northern Europe is 0.1% or less, in central and eastern Europe up to 5%, and in southern Europe yet higher still. The highest prevalence is in the 20–40 year age group. Transmission from hepatitis B carrier mothers to their babies appears to be the single most important factor for the high prevalence in some areas. The degree of risk depends on the proportion of mothers who are carriers, which may be as high as 40% in some countries.

Claim While public and media concern has been heavily preoccupied with AIDS and the search for a cure, hepatitis – 100 times more infectious – have been almost forgotten.

Refs Gerety, R J *Non-A, Non-B Hepatitis* (1981); Gerety, Robert J *Hepatitis B* (1985); Gerety, Robert J *Hepatitis A* (1984); Melnick, J L and Maupas, P *Hepatitis B Virus and Primary Liver Cancer* (1981); Millman, Irving, et al *Hepatitis B* (1984); Regamey, R H, et al *International Symposium on Viral Hepatitis* (1975); Tarizzo, M L *Field Methods for the Control of Trachoma* (1973); WHO Meeting Geneva, 1974 *Viral Hepatitis* (1975); WHO Scientific Group Geneva, 1972 *Viral Hepatitis* (1973); Zuckerman, Arie J *Viral Hepatitis and Liver Disease* (1988).

Broader Epidemics (#PC2514) Virus diseases (#PD0594)
Infectious and parasitic diseases (#PD0982).
Narrower Amoebic hepatitis (#PG4512) Canine viral hepatitis (#PG2452).
Related Viral diseases in animals (#PD2730).
Aggravates Encephalitis (#PE2348) Post hepatitis disorders (#PG2454).
Aggravated by Overcrowding (#PB0469) Alcohol abuse (#PD0153)
Liver diseases (#PE1028) Natural disasters (#PB1151)
Cuts and abrasions (#PG7877) Blowflies as pests (#PE3627)
Mycoplasmal arthritis (#PG2302) Cirrhosis of the liver (#PE2446)
Fungicides as pollutants (#PD1612) Inadequate personal hygiene (#PD2459)
Inappropriate sanitation systems (#PD0876) Recreational contact with sewage (#PE6685)
Inadequate emergency blood supply (#PE0366) Haemolytic disease of the new born (#PE2399)
Damage by degradable organic matter (#PJ6128).

♦ PE0522 Distortion of international trade by embargoes and similar restrictions

Nature Governmental prohibitions of all or certain kinds of shipments to foreign countries, to serve national political, military or economic interests, frequently are ineffective, unjustified and destabilizing to international free trade. Intergovernmental organization embargoes against aggressor nations have had little impact.

Claim Among useless embargoes were those imposed by the League of Nations on oil shipments to Italy, and by the United Nations on armaments to China and North Korea during the Korean War. The only consequences of a grain embargo by the United States were losses for its own economy.

Refs Lundborg, Per *The Economics of Export Embargoes* (1987); Shihata, Ibrahim *Case for the Arab Oil Embargo* (1975).

Broader Economic conflict (#PC0840)
Non-tariff barriers to international trade (#PC2725)
Political interference with port operations (#PE5827).
Related Distortion of international trade by discriminatory customs and administrative entry procedures (#PE2603).
Aggravates Profiteering (#PC2618).

♦ PE0523 Inadequate marketing of products of developing countries

Inadequate export promotion of developing country small industry products

Nature Effective exploration and exploitation of export opportunities requires certain supporting national organizations to carry out export promotion services on a continuing basis. Without these, a concerted and sustained national export promotion effort is not possible. In addition to the government level, the task of export promotion in most developing countries is requisite authority and responsibility for coordinating the formulation and supervision during the implementation phase of a cohesive national programme for export promotion. In most countries, there exist, long before import substitution is completed, numerous smallscale industries and handicrafts producing goods many of which, if certain conditions were fulfilled, could be directed towards the export market. Yet, with a few exceptions, even the countries having reached relatively advanced stages of industrialization are paying little attention to the export promotion of these goods.
In fact, most small-scale industries and handicrafts in the developing countries turn out products of a type, quality and price which do not meet satisfactorily the requirements of the domestic market, let alone those of consumers in the advanced countries. In the less developed countries, the frequently noted preference for the imported product as against the corresponding domestic one, is more often than not due to appreciable shortcomings of the latter. These shortcomings are due to the well-known weaknesses of small-scale industries and handicrafts - lack of technological and managerial knowledge, inadequate skill of labour, primitive or antiquated equipment, unsatisfactory premises and working conditions, use of poor raw materials, lack of information on markets - which tend to perpetuate themselves for lack of financial resources and inadequacy or non-existence of assistance, servicing, training and financing facilities.
The weaknesses and handicaps of small-scale industries and handicrafts create even greater obstacles to the export of their products than to their sale on the home market. The difficulties of producing, packaging, shipping and selling goods of a type, style, grade, quality and price fulfilling demand requirements and trade conditions in a variety of foreign countries, especially in the advanced ones, are indeed formidable for individual small entrepreneurs frequently unable to meet satisfactorily less exacting conditions in their own markets. In the absence of arrangements between small producers and specialized servicing institutions, these problems are usually solved - as far as the small entrepreneur is concerned - in a wasteful and uneconomic way by wholesalers, middlemen and export traders who provide them with materials and finance, give them some guidance on products and processes, and secure an outlet for their output.
Background By conveying information to consumers, distributors and processors, and by speeding up adjustment to technological or economic change, market promotion reduces imperfections in market operations. It has a important role in the competitive process, increases consumer satisfaction and - by expanding demand- results in higher prices and/or a larger volume of sales for producers. Unlike "brand" promotion, which is aimed at expanding a given producer's share of the existing market, "generic" promotion is aimed at expanding the size of the market and so must be financed internationally. Existing international financial institutions, however, have not filled this role: very little has been done in terms of horizontal diversification to address the problem of commodities in global oversupply, no generic market promotion programmes have been financed, and generic research and development efforts have deliberately been concentrated on food crops. International financial institutions and individual developing countries are geared to looking at projects in a country rather than global commodity context. Besides, the financial institutes have been unable to lend to commodity organizations. Although in a few cases producers' associations have been able to carry out generic research and development and market promotion, in most cases financial resources and the necessary organizational structure has been lacking.
Claim If the developing countries are to secure the foreign exchange they need, and achieve fuller employment, they must substantially increase their manufactured exports. It is here that world demand grows most rapidly. And there is a variety of products which these countries can produce at competitive cost; products that generally have either a high labour content, or that utilize domestic materials. Growth in exports of manufactured goods from the developing countries has already been rapid. They started, however, from such a small base that the share of the manufactured exports of the developing countries remains less than 20 per cent of the manufactured imports of the developed nations, and one-third of one per cent of their GNP.
Broader Economic and social underdevelopment (#PB0539).
Narrower Inadequate international marketing of jute products of developing countries (#PE4522).
Aggravated by Minimal export promotion by transnational corporations in developing countries (#PE1598).

♦ PE0524 Health risks to workers in agricultural and livestock production

Nature The main health hazards relating to work in agriculture include: zoonotic diseases, pesticide poisoning, accidents, and respiratory diseases due to organic and vegetable dusts. Health conditions in agriculture depend on the nature and level of development of agricultural production.
On large state and cooperative farms making wide use of agricultural machinery, the main health concern lies with machine operators. Although the problem of accidents in agriculture is not a new one, in a number of countries the number of accidents is growing in direct proportion to the increase in the use of agricultural machinery. On the other hand, in a small-farm economy, with its low level of mechanization, each individual performs a variety of different jobs, so that the work has no clearly distinguishable occupational character and is heavy and exhausting.
Regardless of the size of the farm, a distinctive feature of agriculture is its seasonal nature and the urgency of different forms of field work depending on the character of the crop under cultivation. As a result, one of the main tenets of occupational physiology - a regulated working day - is often violated. Long or irregular hours of work lead to over-exhaustion, to a reduction in productivity and, in a number of cases, to fatigue followed by a reduction in the immune reaction to infection.
Domestic animals are a common source of infection and infestation. At the present time, the transmission of brucellosis, tuberculosis and other diseases is recorded in many countries, and in a number of places also tularaemia. Throughout the world, tetanus and other diseases connected with agriculture are encountered. The widespread use of pesticides has led in many countries to cases of intoxication, both acute and chronic, though the number of acute poisonings has not been very great.
Because of the large number of workers employed in agriculture and in the processing of agricultural products, exposure to vegetable and other organic dusts is widespread. Several occupational diseases due to such exposure have been described, and some are included in the statutory lists of notifiable diseases in certain countries, for example: byssinosis, farmers' lung, bagassosis, and occupational asthma. Many dusts and their health effects have not been systematically investigated. Exposure to dusts of grain, rice, cocoa, coconut fibres, tea, kapok, tobacco and wood is common in the countries where these products are grown; and there is evidence that obstructive respiratory disease and asthma may result. Respiratory disability has been observed among villagers exposed since childhood to flax dusts at home. Investigations in almost all countries where textile materials made of cotton, flax, or soft hemp are produced or processed have revealed a significantly high rate of byssinosis.
Incidence Although farm workers comprise only about 4.4% of the work force in the USA, about 16% of total occupational deaths and 9% of all occupational injures occur in this group. With the increasing use of agricultural machinery occupational injuries are becoming more frequent in developing countries; and in some industrialized countries the occupational accident rate for agriculture now ranks third, after those for mining and building. In some countries field surveys among spraymen exposed to agricultural chemicals revealed an average prevalence of symptoms of poisoning in up to 40% of them during a spraying period. Up to 90% of textile workers exposed to cotton, flax or hemp dust have been affected by byssinosis; elsewhere, depending on the dust concentration and duration of exposure, a prevalence of from 20% to 40% has frequently been observed; permanent pulmonary disability has been found in up to 20% of the workers affected.
Broader Occupational risk to health (#PC0865).
Related Hunting of animals (#PC2024)
Environmental hazards from forestry and logging (#PE1264).
Aggravates Fatigue (#PA0657) Tetanus (#PE2530) Smallpox (#PE0097)
Brucellosis (#PE0924) Tuberculosis (#PE0566) Underproductivity (#PF1107)
Industrial accidents (#PC0646) Occupational diseases (#PD0215)
Animals as vectors of disease (#PD8360).
Aggravated by Zoonoses (#PD1770) Bad weather (#PC0293)
Organic dusts (#PG2461) Lack of training (#PD8388)
Aggressive animals (#PG2463) Unmaintained bridges (#PE2471)
Gaseous air pollutants (#PG2460) Inadequate medical care (#PF4832)
Pesticides as pollutants (#PD0120) Parasites of the human body (#PE0596)
Defects in machinery design (#PE2462) Badly laid out work premises (#PJ2468)
Heat as an occupational hazard (#PE5720) Ignorance of health and hygiene (#PD8023)
Inadequate equipment maintenance (#PD1565) Health hazards of exposure to noise (#PC0268)
Environmental hazards from fertilizers (#PE1514)
Improper lighting as an occupational hazard (#PE5780)
Pesticide destruction of soil fauna and micro-organisms (#PD3574)
Instability of trade in mineral fuels, lubricants and related materials (#PD0877).

♦ PE0525 Stress and trauma in a context of civil violence

Nature In many places in the world today there are civil violence and war-like conditions such that the people there live under the constant threat that each day may be their last. The strain of living in constant fear exacts a heavy emotional toll. People who might be healthy under peaceful conditions often develop severe emotional and psychosomatic illnesses. Sleep disorders, depression, suicide and stress-related medical problems are widespread. The most vulnerable and tragic victims are children born into an environment where death and destruction are the norm. Some children develop severe learning disabilities and become pathologically frightened and withdrawn.
Incidence Lebanon, El Salvador and Northern Ireland are examples of such zones of random violence.
Broader Stress in human beings (#PC1648).
Aggravated by Civil violence (#PC4864).

♦ PE0526 Health risks to workers in construction industry

Nature Risks to construction workers include occupational hazards and accidents. There is particular vulnerability to accidents in this industry because turnover at very high levels does not allow for proper safety training or thorough implementation of safety procedures. Work with dangerous machines, with heavy loads in constant movement, and work in inclement outdoor weather, provide numerous opportunities for accidents. Occupational hazards include: noise; vibration; heat exhaustion; exposure to ultraviolet radiation from welding processes; lung diseases arising from cement and concrete dusts; and skin irritations, allergies and toxic reactions due to a variety of chemical hazards on the construction sites.
Broader Occupational risk to health (#PC0865).
Aggravates Industrial accidents (#PC0646) Occupational diseases (#PD0215).

♦ PE0527 Agoraphobia

Nature It is a fear of being in places or situations from where escape might be difficult or embarrassing or help not available in the case of attack of dizziness, vomiting, loss of bladder or bowel control or cardiac distress. As a result a person may avoid being outside of the house alone, being in a crowd, waiting in a line or travelling.
Refs Gelder, Michael G; Johnston, Derek W and Matthews, Andrew M *Agoraphobia* (1981); Mathews, Andrew M, et al *Agoraphobia and Treatment* (1986).
Broader Phobia (#PE6354).
Aggravates Panic (#PF2633) Loneliness (#PF2386).

♦ PE0529 Powdery mildews in plants

Nature Scattered spots of white appear on aerial parts of plants enlarging until they may cover most of the plant surface.
Broader Ascomycetes (#PE4586).

♦ PE0533 Ignorance of nonverbal communication skills

Insensitivity to nonverbal messages — Nonverbal communication deficiency
Nature Nonverbal communications skills are required in many social situations. These involve reading the emotions revealed in a tone of voice, sensing how close to stand when talking to someone, facial expressions, assessing the mood of others. Where such skills are lacking, possibly because of disruption in the socialization process, this may sufficiently important to impair their social or academic functioning. When trying to make friends, any approach made without such skills, may be rejected. Unpopular people may not even realize that they are initiating many of the negative reactions they receive from their peers. They may even inadvertently communicate overeagerness that their peers interpret as aggression. Since most emotional messages between people are communicated nonverbally, by gesture or tone of voice, the inability to read such messages adeptly is a major social handicap.
Incidence Recent studies of children in the USA found that up to 10 percent may have nonverbal communication problems severe enough to impair social or academic functioning. Studies of more than 1,000 children aged from 9 to 11 showed that those scoring lowest in such skills tended to be among the least popular in the class.
Broader Learning disorders (#PD3865).
Aggravates Apathy (#PA2360) Frustration (#PA2252) Mental depression (#PC0799)
Personal unpopularity (#PF4641)
Insensitivity to diversity of cultural traditions (#PF8156).

♦ PE0535 Chemical contamination of water

Chemical water pollutants
Nature Water courses and ground water are contaminated chemically primarily due to industrial and agricultural activities. Industrial effluents and toxic chemical residues that are buried and which seep into water tables are the chief sources of contamination. Agricultural activities, mainly involving pesticides, herbicides and chemical fertilizers, are the cause of contaminated water run-offs brought about by rain or irrigation. Rivers and estuaries from which drinking water is drawn and purified have also to contend with contamination from boat and ship emissions of oil and petrol. Open areas of water may also, on occasion, be contaminated by precipitated particles in rainfall.
Broader Water pollution (#PC0062) Pesticides as pollutants (#PD0120)
Impurities in waste water (#PD0482) Dysfunctional public utilities (#PE7647).
Narrower Oil as a pollutant (#PE2134) Zinc as a pollutant (#PE7229)

Copper as a pollutant (#PG6587) Nickel as a pollutant (#PE1315)
Nitrites as pollutants (#PE6087) Chlorine as a pollutant (#PG7454)
Chromium as a pollutant (#PE4072) Suspended matter in water (#PE0579)
Radioactive contamination of water (#PE2441) Environmental hazards from fertilizers (#PE1514)
Toxic organic compounds as water pollutants (#PE0617).
Aggravated by Cadmium as a pollutant (#PE1160)
Toxic metal pollutants (#PD0948)
Environmental impacts of coal conversion plants (#PE8453).

♦ PE0537 Dependence of industrialized countries on import of resources
Resource import dependence of developed countries — Dependence of industrialized countries on imports of primary commodities
Incidence Industrialized countries depend increasingly on the import of of resources from developing countries. Fuel imports represented 16 percent of consumption in 1959–60 but rose to 43 percent in 1980–81. Dependence on other mineral imports rose from 19 percent of consumption to 30 percent over the same period. Non-renewable resources, such as fuel and minerals, as well as manufactured goods, are now far more important in the flow of primary products from developing to industrialized countries.
Broader Dependence on external resources (#PC0065).
Related Poor quality of domestic livestock (#PD2743).
Aggravates Economic imperialism (#PC3198)
Lack of economic and technical development (#PE8190).

♦ PE0538 Instability of chemical and petrochemical industry
Broader Instability of manufacturing industries (#PC0580).
Narrower Instability of paint industry (#PG6356) Instability of plastics industry (#PE5143).

♦ PE0540 Limited psychiatric out-patient care
Nature In the early 1980's new drug therapies allowed psychiatric hospitals to release patients earlier, and former patients began returning to local communities where they were to be assisted by welfare institutions. But these institutions are inadequately organized, staffed or funded to provide the care needed. The patients are usually poor, single and without family support. They find it difficult to apply for help because of the indifference of staff, complexity of application and rigidity of procedures. Some are so heavily medicated that they find it hard to attend to ordinary things like cooking, paying rent, or washing. They need help finding places to live and jobs. They need someone to smooth over a dispute with a shopkeeper, or to calm them after a psychotic episode brought on by simply being on a crowded tram. They need friendship and support. They may withdraw, fearing condemnation of others, even before they experience it, or antagonize people with their odd behaviour. Many end up living on the street or returning to hospital simply for the food, shelter and friendship.
Broader Inadequate mental hospitals (#PF4925).
Aggravates Human disease and disability (#PB1044).
Aggravated by Inadequate mental health services (#PJ4379)
Delay in administration of medical care (#PD5119).

♦ PE0546 Environmental hazards from inedible crude non-fuel materials
Broader Environmental hazards from economic and industrial products (#PC0328).
Narrower Environmental hazards of textile fibres and waste (#PE9074)
Environmental hazards of metalliferous ores and metal scrap (#PE8906)
Environmental hazards of oil nuts, oil kernels and oil-seeds (#PE8250)
Environmental hazards from undressed hides, skins and fur skins (#PE0828)
Environmental hazards from non-metallic mineral products industries (#PE0890).

♦ PE0553 Instability of trade in metalliferous ores and metal scrap
Incidence World production, consumption and trade in iron ore in 1982 and 1983 were the lowest for a decade as a result of the sharp contraction in the steel industry in the developed market-economy countries. Though economic recovery is now well under way in the United States, and gross output is also expanding in Japan and the European Economic Community, the expansion has not so far had a significant impact on the world market for iron ore, particularly as regards demand and prices. As a consequence of the fall in demand, world iron ore shipments were reduced by some 12 per cent, and though prices had risen by a similar proportion the increase in unit values was relatively small, so that total export earnings from iron ore shipments, in terms of US dollars, fell by almost 10 per cent. A recent feature of the market has been a tendency for buyers to shift away from long-term contracts to greater purchases on a short-term basis including some spot purchases, generally at lower prices than those paid for deliveries under long-term contracts.
The many uncertainties surrounding the market outlook, and the fact that overall capacity is expected to remain substantially in excess of production, are likely to restrain investment. Additional uncertainty is caused by the shift to short-term contracts, since this could affect access to funds for new mine development, for the financing of beneficiation plants and for the purchased of bulk ore carriers.
Broader Instability of trade in inedible crude non-fuel materials (#PD0280).
Narrower Iron and steel scrap trade instability (#PE8312)
Silver and platinum ores trade instability (#PE8556)
Iron ore and concentrates trade instability (#PE8772)
Uranium ores and concentrates and thorium trade instability (#PE8212)
Instability of trade in ores and concentrates of non-ferrous base metals (#PE8131).

♦ PE0558 Debasement of works of art
Nature Fine art is being increasingly used to advertise products. Paintings are used in advertisements. Products may be inserted into the painting or a painting may be altered to make an advertising point. Classical music is used in television commercials. Great literature may be quoted in ads. Pictures of sculpture may be used.
Claim The use of works of art to advertise is demeaning to the art and the artist and creates a false impression of quality or luxury for the products advertised.
Counter-claim Advertisements using art reflect the increasing importance of art for the public. The ads help educate the general public by exposing more people to fine art. Tampering with some painting for comic effect is harmless and many artists would have been amused, even flattered.
Broader Art vandalism (#PE5171).
Narrower Dismembered works of art (#PE6252).
Related Deterioration of the physical condition of art objects (#PD1955).
Aggravates Representative arts (#PF0981).
Aggravated by Indecent art (#PE5042).

♦ PE0559 Pickpocketing
Refs Gupta, Krishna K *Pickpockets* (1987).
Broader Theft (#PD5552).
Related Bag snatching (#PG3445).

♦ PE0560 Cholera
Nature Cholera is an acute infection of the intestine caused by a two biotypes of bacteria: Vibrio Cholerae and El Tor vibrio (now called V cholerae, biotype eltor). The two biotypes sometimes co-exist. Cholera has no host other than man and is quickly transmitted, mainly through water and food, to persons with poor personal hygiene in areas where sanitation is deficient. The incubation period is short: from less than one day to five days. The disease is characterized by profuse painless diarrhoea and vomiting causing dehydration and acidosis. In all newly infected areas, adults of both sexes are usually affected, while in endemic situations cholera is mainly a paediatric problem. In spite of the possibility of efficient treatment of cholera (which consists in rehydration by replacement of water, salts and alkali, supported by antibiotic therapy) there is a problem of logistics, as cholera often occurs in areas where few treatment facilities are available. Excessive quarantine measures and restrictions on traffic and trade imposed in panic by different countries cause harassment, economic losses, and encourage the suppression of information which, in turn, favours the spread of the disease.
Background Cholera is one of the ancient diseases of mankind which earned the notoriety of being a great killer during the nineteenth century when it reached Europe for the first time and caused six large pandemics. After the sixth pandemic, cholera retreated to its homeland in Asia, particularly in the deltas of the Ganges and Brahmaputra rivers, making occasional sorties such as the epidemic in Egypt in 1947. In 1961, El Tor vibrio moved out of its endemic sites, probably because of population movements in Asia and across the Pacific Ocean, to areas which had been free from cholera for many years and had few public health workers with experience in the diagnosis, treatment and control of the disease. The seventh pandemic thus began to spread from one country to another adjacent one in a predictable manner. It reached India in 1964 and almost replaced the classical V cholerae. The westward march continued until 1966, and after a temporary lull in 1967–68 the disease became widespread in some Mediterranean and neighbouring countries and in 1970 invaded West Africa, a territory which had always remained practically free from the disease. In the wake of the westward extension of the seventh pandemic a few European countries also experienced small outbreaks; others like Japan and Australia detected the importation of infection in time to prevent indigenous cases. This is one of the diseases which could be exploited for biological warfare.
Refs Ouchterlony, O and Holmgren, J *Cholera and Related Diarrheas* (1980); Pollitzer, R *Cholera* (1959).
Broader Stigmatized diseases (#PD7279) Environmental human diseases (#PD5669)
Infectious and parasitic diseases (#PD0982).
Narrower Cholera El Tor (#PG9575) Vibrio cholerae (#PG2512).
Related Enteric infections (#PD0640).
Aggravates Encephalitis (#PE2348).
Aggravated by Famine (#PB0315) Overcrowding (#PB0469)
Natural disasters (#PB1151) Houseflies as pests (#PE3609)
Sewage as a pollutant (#PD1414) Flies as insect pests (#PE2254)
Contamination of drinking water (#PD0235) Inappropriate sanitation systems (#PD0876)
Damage by degradable organic matter (#PJ6128)
Calculated delays in releasing controversial news items (#PE0598).

♦ PE0561 Food poisoning through negligence
Nature The term food poisoning covers gastrointestinal disorders caused by the consumption of foods that are toxic in nature or that contain bacterial poisons or toxic impurities. Disorders caused by food poisoning are noncontagious, have a sudden onset, and are short-lived. Bacterial food poisoning is called intoxication or alimentary toxinfection. Nonbacterial food poisoning is often caused by poisonous plants. The organic and inorganic chemicals most commonly causing food poisoning are arsenic, copper, and sodium nitride. The most common organisms causing bacterial food poisoning are salmonellae (responsible for typhoid, paratyphoid, and other gastro-intestinal diseases); staphylococcal food poisoning, tape worm, trichosis and infectious hepatitis are also important causes of food-borne diseases. Conditions leading to food poisoning can be found throughout the world, but most frequently in developing countries. They include: lack of hot water; rodent and insect infestation; food handling by infected persons; improperly cleaned equipment; contamination of food by overhead sewage pipes; and incorrect temperature control in steam tables and refrigerators.
Incidence The risk of poisoning increases with the growth of the human and animal population. Mass production and distribution of food and the development of international trade contribute to the danger. Mass catering has frequently resulted in food poisoning. In England and Wales, for example, of more than 1,000 outbreaks of food poisoning intensively investigated between 1970 and 1979, at least 60% were caused by holding food for more than 6 hours without adequate temperature control. The number of cases reported increased from 12,000 in 1981 to 41,000 cases in 1988. In the spring of 1981, the consumption of adulterated edible oils poisoned thousands of people in the central area of Spain.
Refs Monahan, John *Food Poisoning* (1987).
Broader Food pollution (#PD5605) Human poisoning (#PD0105)
Bacterial diseases in animals (#PD2731).
Narrower Botulism (#PE7766) Listeriosis (#PE3779)
Bacterial food poisoning (#PE9374).
Related Product extortion (#PD5426) Natural food poisons (#PD1472).
Aggravates Gastritis (#PG2250) Salmonellosis (#PE7562).
Aggravated by Bacterial disease (#PD9094) Unethical food practices (#PD1045)
Unethical catering practices (#PE6615) Spoilage of agricultural products (#PC2027)
Damage by degradable organic matter (#PJ6128).

♦ PE0564 Unauthorized pharmaceutical manufacture and distribution
Broader Unethical commercial practices (#PC2563).
Narrower Manufacture of illicit drugs (#PE2512)
Marketing of banned pharmaceutical drugs in developing countries (#PE6036).

♦ PE0566 Tuberculosis
Consumption — Phthisis — Hectic fever — Lupus — Scrofula — Wasting sickness
Nature Tuberculosis is a disease caused by the infectious Mycobacterium tuberculosis or Koch bacilli which occurs in human, bovine and avian varieties. It is transmitted mainly by airborne droplets of sputum and saliva containing mycobacteria; and gains access to the body by three possible channels: inoculation, inhalation and ingestion. In humans, tuberculosis mainly affects the lungs (pleurisy or pulmonary tuberculosis) but may invade almost any organ, particularly the lymph nodes, kidneys and bladder, and skin. The severity of the disease varies considerably according to the organ attacked; tuberculosis affecting the membranes of the brain and causing meningitis is a particularly dangerous disease, particularly in infants and young children. The bovine variety of tuberculosis bacteria is pathogenetic for all mammals and to a lesser extent for birds. The human variety causes tuberculosis in horses, dogs, swine, cats, sheep, birds, and cattle. The avian variety affects birds, swine, horses, dogs, and sometimes cattle. Tuberculosis in animals is chronic but may be acute in young animals after extensive infection.
During the last 30 years increasingly active drugs for the treatment of tuberculosis have changed both the conditions of treatment and also the impact on employment. However, the part played by silicotic dust in the mining environment encourages tubercular complications as well as the development of atypical mycobacteria. Studies have shown that the risk of contracting a mycobacteriosis is 30 times greater in a silicotic than in a subject whose lungs are not affected by dust.
Incidence According to World Health Organization (WHO) estimates, at least three million people

die each year of tuberculosis. Every year four to five million new infectious cases are added to the statistics, and the same number of noninfectious cases develop, many of which will later become infectious. The tuberculosis mortality per 100,000 population in 1970 was 5.4 in (Germany) 8.2 in France, 15.in Japan, 36 in Hong Kong, and 82 in the Philippines. In the Philippines it is the second leading cause of death. In France, the mortality from tuberculosis is three to five times higher among miners, sailors and fishermen than among persons engaged in the professions. In the USA, the incidence and mortality rates of tuberculosis among negroes, Indians, Puerto Ricans, and other nonwhite groups are three to four times higher than among whites. Incidence and mortality are high among New Zealand aborigines, and among Australian aborigines relocated to regions where living conditions are unfavourable. In India approximately 10 million adults have active cases. Tuberculosis holds third or fourth place among the main causes of death in many developing countries, and eight or ninth place in developed countries. Tuberculosis of animals is widespread, especially in Western Europe where it causes substantial economic losses. More than 55 species of domestic and wild mammals and about 25 species of birds are susceptible to tuberculosis.

Background Ancient Egyptian and Indian manuscripts mention tuberculosis symptoms, and traces of tuberculosis can be found in Egyptian mummies. It did not become endemic, however, until the rise of urban centres where the disease could spread easily. In London in the 17th and 18th centuries the annual mortality rate from tuberculosis was over 800 per 100,000 population. Elsewhere in Europe it accounted for 20% of all deaths. The causative bacillus was first isolated in 1882 by R Koch.

Refs Blau, Sheldon P *The Body Against Itself* (1984); Schlossberg, D *Tuberculosis* (1988).

Broader Children's diseases (#PD0622) Stigmatized diseases (#PD7279)
Lung disorders and diseases (#PD0637) Zoonotic bacterial diseases (#PD6363)
Infectious and parasitic diseases (#PD0982).

Narrower Lupus vulgaris (#PG2525) Silicotuberculosis (#PG7869)
Fibroid tuberculosis (#PG2530) Pulmonary tuberculosis (#PE2526)
Tuberculous meningitis (#PG4600) Fibro–caseus tuberculosis (#PG2529)
Disseminated tuberculosis (#PG9737) Tuberculous lymphadenitis (#PG5624)
Tuberculosis of intestines (#PG9233) Tuberculosis of bones and joints (#PE9846)
Tuberculosis of genito-urinary system (#PG4280).

Related Bacterial diseases in animals (#PD2731).

Aggravates Fever (#PD2255) Pleurisy (#PG6121) Pneumonia (#PE2293)
Meningitis (#PE2280) Laryngitis (#PE2653) Still–birth (#PD4029)
Heart diseases (#PD0448) Diseases of the spleen (#PE6155)
Skeletal system disorders (#PE2298) Unethical consumption practices (#PD2625)
Child mortality in developing countries (#PE5166)
Individually defined operating structure of marriage (#PD2294).

Aggravated by Dust (#PD1245) Measles (#PE1603)
Diabetes (#PE0102) Silicosis (#PE1314)
Malnutrition (#PB1498) Unclean food (#PJ2532)
Pneumoconiosis (#PD2034) Natural disasters (#PB1151)
Chronic bronchitis (#PE2248) Pregnancy disorders (#PD2289)
Inhospitable climate (#PC0387) Increasing pace of life (#PF2304)
Uncontrolled urban development (#PC0442) Dust as an occupational hazard (#PE5767)
Uncontrolled industrialization (#PB1845) Ignorance of health and hygiene (#PD8023)
Overcrowding of housing and accommodation (#PD0758)
Health risks to workers in agricultural and livestock production (#PE0524).

♦ PE0567 Sexual bigotry in organized religion

Gender bias in theology

Refs Ranke–Heinemann, Uta *Eunuchs for Heaven* (1990).

Broader Bigotry (#PC7652).
Aggravates Misuse of spiritual authority for sexual purposes (#PE1348).
Aggravated by Dogmatism (#PF6988).

♦ PE0572 Instability of trade in tobacco and tobacco manufactures

Tobacco manufactures instability

Broader Instability of trade in beverages and tobacco (#PE1641).
Related Underdevelopment of food processing industries (#PD0908).
Aggravated by Smoking (#PD0713)
Protectionism in agriculture and the food production industries (#PD5830).

♦ PE0573 Inaccessibility of decision–makers in multinational enterprises

Refs ILO *Access to decision–makers in multinational and multi–plant enterprises* (1985).

Broader Social inaccessibility (#PC0237) Inaccessible administrative agencies (#PF2261).
Narrower Inaccessible decision makers (#PF2452).
Aggravates Shortage of trained teachers (#PD8108)
Lack of worker participation in business decision–making (#PF0574).
Aggravated by Domination of labour relations by transnational corporations (#PE1187)
Disproportionate control of global economy by limited number of corporations (#PE0135).

♦ PE0576 Instability of trade in dairy products and eggs

Broader Instability of trade in food and live animals (#PD1434).
Narrower Butter trade instability (#PU2556) Cheese and curd trade instability (#PU2557).

♦ PE0579 Suspended matter in water

Silt

Nature Large amounts of suspended matter in water reduce light penetration and thereby interfere with photosynthesis in plant life in bodies of water. The solids can settle into sludge blankets which can smother the normal bottom dwelling organisms and destroy fish spawning beds. Suspended solids are often associated with oxygen demanding matter such as sewage, further polluting the water.

Broader Impurities in waste water (#PD0482) Chemical contamination of water (#PE0535).
Related Oxygen demanding matter (#PS7763).

♦ PE0580 Intranational competition

Incidence Global corporations are increasingly decentralized and willing to manufacture, design or assign managerial authority wherever they can best serve the customer, irrespective of government interests. There is now as much intranational competition as international competition. Cities and regions within the same county compete against each other to attract investment, both from domestic and from foreign sources.

Counter–claim Only a global regionalism that ignores national boundaries makes sense. Local communities develop natural partners, whether geographically close or remote, and prosper to the extent that there are alert.

Broader Competition (#PB0848).
Related Rivalry and disunity within developing regions (#PD0110).
Aggravated by Competition for scarce resources (#PC4412).

♦ PE0585 Hypertension

Chronic hypertension

Nature Hypertension refers to abnormally high blood pressure without known cause. In adults it is arbitrarily defined as a systolic pressure equal to or greater than 160 mmHg (21.3 kPa) and/or a diastolic pressure (fifth phase) equal to or greater than 95 mmHg (12.7 kPa). Factors which seem to influence hypertension include: age (more frequent in those over 40); heredity; obesity; high salt intake; high alcohol intake; psychological and social conditions (such as stress or low income); and renal disease.

The evidence that psychological factors play a primary role in human essential hypertension is inconclusive, but there is overwhelming clinical and experimental evidence that renal disease in man and a variety of experimental procedures in animals that affect renal function are associated with or induce hypertension. The mechanism by which the blood pressure–obesity correlation arises is unknown. Possible mechanisms include increased sodium intake and retention, increased tubular reabsorption of sodium due to increased insulin levels, increased oestrogen levels, disproportion between body mass and renal size, disproportion between increased blood volume and vascular capacity, and increased sympathetic nerve activity due to increased energy consumption.

There are many dietary and other factors that distinguish populations without hypertension from those in which hypertension exists. The isolated groups in whom blood pressure does not rise with age also tend to have a low energy intake. The relationship between energy consumption and blood pressure is obscure, but a reducing diet that leads to weight loss is associated with a substantial fall in blood pressure, measured as intra–arterial pressure. A reduction in total energy consumption can therefore reduce blood pressure and this may partially explain the fall in cardiovascular mortality in populations with impaired food supplies, such as in the Netherlands in the Second World War.

Refs Brunner, and Gravas, *Clinical Hypertension and Hypotension* (1982); Fregly, Melvin and Kare, Morley *The Role of Salt in Cardiovascular Hypertension* (1982); Gavras, H and Gavras, I *Hypertension in the Elderly* (1983); Genest, Jacques, et al *Hypertension* (1983); Giovannelli, G et al *Hypertension in Children and Adolescents* (1981); Hardine, Rosetta R *Psychology of Hypertension* (1985); Hart, Julian T *Hypertension* (1987); Hatano S, et al (Eds) *Hypertension and Stroke Control in the Community* (1976); Iwai, Junichi *Salt and Hypertension* (1982); Kaplan, Norman M et al *The Kidney in Hypertension* (1987); Messerli, Franz H *The Heart and Hypertension* (1987); Messerli, Franz H (Ed) *The Heart of Hypertension* (1987); Onesti, Gaddo and Klimt, Christian R *Hypertension* (1978); Orloff, M J and Stipa, S *Medical and Surgical Problems of Portal Hypertension* (1981); Paquet, K J, et al *Portale Hypertension* (1982); Rosenthal, J (Ed) *Arterial Hypertension* (1982); Saito, H et al *Neurotransmitters as Modulators of Blood* (1987); Sharp, F and Symonds, M *Hypertension in Pregnancy* (1987).

Broader Diseases of metabolism (#PC2270) Diseases of the circulation system (#PC8482).
Narrower Essential hypertension (#PG2569).
Related Trace element imbalance in the human body (#PE5328).
Aggravates Stroke (#PE1684) Glaucoma (#PE2264) Heart diseases (#PD0448)
Abnormal blood pressure (#PE0472) Diseases of the arteries (#PE2684)
Diseases of the lymphatic system (#PD2654).
Aggravated by Obesity (#PE1177) Water softness (#PE0199)
Gland disorders (#PD8301) Kidney disorders (#PE2053)
Bright's disease (#PE2272) Pregnancy disorders (#PD2289)
Coarctation of the aorta (#PG2571) Health hazards of exposure to noise (#PC0268)
Human physical genetic abnormalities (#PD1618) Mental, physical and emotional strain (#PG2572).

♦ PE0595 Stowing away

Nature Intentionally remaining on board a vehicle like a bus, ship or plane for the purposes of obtaining transportation without paying is stowing away.

Broader Criminal intrusion (#PE6771).
Aggravates Still–birth (#PD4029).

♦ PE0596 Parasites of the human body

Incidence Among diseases caused by parasites of the human body, intestinal parasitic infections (especially ascariasis, ancylostomiasis, necatoriasis, amoebiasis and giardiasis) are widespread and remain among the commonest infections in the world. In 5 villages from widely separated parts of India it was found that from 23 per cent to 75 per cent of the people were infected with roundworm, hook worm, pin worm, dwarf tapeworm and intestinal amoebas. Many villagers are consequently sick a great deal of the time. They are unable to absorb all the food the eat because of the damage done by parasites to the intestinal membranes and because part of their food goes to feed the worms and protozoa that infest their intestines.

Refs Maunder, J W *Human Ectoparasites* (1988).

Broader Parasites (#PD0868) Infectious and parasitic diseases (#PD0982).
Narrower Hookworm (#PE3508) Giardiasis (#PE4811)
Ascariasis (#PE2395) Toxoplasmosis (#PE3659)
Helminthiasis (#PE6278) Ticks as pests (#PE1766)
Dracunculiasis (#PE3510) Sarcosporidiosis (#PG9770)
Trematode diseases (#PE6461) Tapeworm as a parasite (#PE3511).
Related African trypanosomiasis (#PE1778).
Aggravates Plague (#PE0987) Eczema (#PE2465) Malaria (#PE0616)
Dysentery (#PE2259) Dystrophy (#PE3506) Liver rot (#PG2606)
Diarrhoea (#PD5971) Filariasis (#PE2391) Trichinosis (#PE2311)
Convulsions (#PG2611) Elephantiasis (#PE1601) Liver diseases (#PE1028)
Onchocerciasis (#PE2388) Paragonimiasis (#PG2609) Schistosomiasis (#PE0921)
Hydatid disease (#PE2354) Chagas' disease (#PE0653) Physical blindness (#PD0568)
Nutritional anaemia (#PD0321) Children's diseases (#PD0622)
Cardiovascular diseases (#PE6816) Urinary bladder disorders (#PE2307)
Diseases of the lymphatic system (#PD2654) Diseases and injuries of the brain (#PD0992)
Health risks to workers in agricultural and livestock production (#PE0524).
Aggravated by Shortage of firewood (#PD4769) Inadequately cooked foods (#PJ2614)
Solid wastes as pollutants (#PD0177).

♦ PE0597 Injustice of mass trials

Nature Under no system of law designed to protect the rights of every individual accused has it proved possible to conduct a mass trial fairly. There are two significant points about mass trials which run counter to the principle of the fair administration of justice. First, they tend to shift the weight of repressive law from the political ring–leaders to the rank and file of any political resistance movement. Thus they may be used to deter people from joining political organizations with which the government may clash. Further, these trials, because of their inevitable length, constitute a great burden on the accused and their families. However independent the judiciary is, and however fairly the trials are conducted, there is no effective way of preventing or redressing the injury suffered by the prolonged nature of the trial. The award of costs against the prosecution is rarely experienced and there is often little if any form of legal aid for the accused.

Broader Political repression (#PC1919) Injustice of special courts (#PE0088).

♦ PE0598 Calculated delays in releasing controversial news items

Postponement of bad news — Delay in disseminating negative information

Broader Self interested manipulation of timing (#PF5529).
Aggravates Cholera (#PE0560) Non–recognition of problems (#PF8112).
Aggravated by Issue avoidance (#PF1623) Avoidance of negative feedback (#PF5311).

♦ PE0607 Dumping of food in developing countries

Dumping of agricultural products in developing countries

Claim The third world's local food production efforts are being undermined by the reckless and

indiscriminate dumping of subsidized European and North American farm products on developing countries. Food shortages in Africa, underline experts at the World Food Conference in April 1988, cannot be solved by shipping subsidized food surpluses to Africa. While these practices may alleviate shortages in the short term generally they are disastrous in the long term. Dumping also forces down world food prices making it more difficult for farmers in developing countries.
Broader Distortion of international trade by dumping (#PD2144).
Related Food wastage (#PD8844) Inadequacy of food aid (#PF3949).
Aggravates Rural poverty in developing countries (#PD4125)
Instability of food prices in developing countries (#PE4986)
Inadequate agriculture in least developed countries (#PE8082)
Inadequate staple food supply in developing countries (#PD4101)
Deterioration of domestic food production in developing countries (#PD5092).
Aggravated by Unethical food practices (#PD1045).

♦ PE0612 Lack of funds for medical research
Broader Prohibitive medical expenses (#PE8261)
Shortage of funds for research (#PF5419)
Shortage of financial resources for action against problems (#PF0404).
Related Lack of funds for veterinary research (#PE8463).
Aggravates Inadequate medical care (#PF4832).

♦ PE0613 Long-term shortage of miscellaneous manufactured articles
Broader Long-term shortage of commodities (#PC1195)
Insufficient availability of goods (#PB8891).
Narrower Long-term shortage of clothing (#PE0303)
Long-term shortage of furniture (#PE0511)
Shortage of professional, scientific and controlling instruments (#PE7961)
Shortage of sanitary plumbing, heating and lighting fixtures and fittings (#PE7940).

♦ PE0614 Zygomycetes
Broader Fungal plant diseases (#PD2225).
Narrower Blights (#PE3919) Rots in plants (#PE3363)
Damping-off disease of plants (#PE4787).

♦ PE0616 Malaria
Ague — Paludism — Jungle fever — Marsh fever — Periodic fever
Nature Malaria is a disease caused by infection with parasites of the genus Plasmodium, transmitted by the bite of infected anopheline mosquitoes, and characterized clinically by recurrent paroxysms of chills, fever and sweating. In man, malaria is produced by four specific parasites: Plasmodium vivax, P malariae, P ovale and P falciparum. The last mentioned has a fatality rate in untreated cases of 25 per cent. Malaria, a major threat to health and development, is primarily an environmental and socio-economic problem and as such demands appropriate solutions; past reliance on narrower strategies is increasingly seen as the reason for the resurgence of this debilitating disease.
Background Malaria has existed in most countries of the world, from northwest Russia to Argentina. Though it was common in southern Europe up to the 1950s, in the last half-century malaria has been more or less confined to tropical and sub tropical regions.
Incidence Malaria continues to be the most important single disease in sub-Saharan Africa, and one of the most significant elsewhere in the tropics. It is endemic almost everywhere in the tropics, where approximately 2,582 million people live in affected areas. In 1990 it was estimated that 270 million people were infected. Anti-malarial activities have protected only about 11 percent of the population, mainly in urban and peri-urban localities. The disease has claimed millions of lives in several countries over recent decades, and continues to account for a very high morbidity and mortality in the non-immune inhabitants of endemic areas. It is the most important single cause of febrile convulsions in children in tropical Africa, and a major cause of death in early life. Approximately 50 percent of children up to the age of three are infected. It is estimated that one million people die from malaria every year and at any given moment, 160 million people suffer from the disease.
General trends in the total number of reported malaria cases indicate a stabilization after 1979, following the control of the major resurgence in Southeast Asia in the mid-1970s; the Southeast Asia region and China show a slowly declining trend in the prevalence of the disease (6.5 million cases outside Africa in 1982 compared to 10.7 million in 1977) while the rest of the Western Pacific Region, the Eastern Mediterranean and European Regions and the Region of the Americas show an increasing trend (700,000 in the Americas in 1982 compared to 280.000 in 1973). Malaria has saturated the middle of the African continent, but other areas of the world are now experiencing major malarial epidemics: Brazil (100,000 in 1977; 560,000 in 1988), Afghanistan (47,000 in 1980; 422,000 in 1986, aided by the war).
Refs Bruce-Chatt, L J and DeZulueta, Julian *The Rise and Fall of Malaria in Europe* (1980); Killick-Kendrick, R and Peters, W *Rodent Malaria* (1978); Kreier *Malaria* (1980); Kreier, Julius P *Malaria* (1980); Kreier, Julius P *Malaria* (1980); Shuler, Alexanderina V *Malaria* (1985); WHO *WHO Expert Committee on Malaria* (1986).
Broader Tolerated atrocities (#PC4710) Arthropod-borne diseases (#PE7796)
Uncontrolled tropical diseases (#PD4775) Infectious and parasitic diseases (#PD0982).
Narrower Vivax malaria (#PG2625) Ovale malaria (#PG2627)
Simian malaria (#PG2630) Malariae malaria (#PG2626)
Drug-resistant malaria (#PG2638) Recurrent induced malaria (#PG2629).
Related Yellow fever (#PE0985) Encephalitis (#PE2348).
Aggravates Fever (#PD2255) Still-birth (#PD4029) Hyperpyrexia (#PG2631)
Liver diseases (#PE1028) Blackwater fever (#PG2628)
Child mortality in developing countries (#PE5166).
Aggravated by Swamps (#PG2633) Natural disasters (#PB1151)
Environmental warfare (#PC2696) Stagnant surface water (#PE2634)
Parasites of the human body (#PE0596) Common use of hypodermic needle (#PG2637)
Mosquitoes as vectors of disease (#PE1923) Inadequate emergency blood supply (#PE0366)
Anopheline mosquitoes as vectors of disease (#PE3622)
Increase in pests and diseases through perennial irrigation (#PE7832).
Reduced by DDT as a pollutant (#PE5028).

♦ PE0617 Toxic organic compounds as water pollutants
Refs Branson, D R and Dickson, K L *Aquatic Toxicology and Hazard Assessment (Fourth Conference)- STP 737* (1981).
Broader Impurities in waste water (#PD0482) Chemical contamination of water (#PE0535).
Narrower Polychlorinated biphenyls as a health hazard (#PE2432).

♦ PE0618 Caffeine abuse
Caffeinism — Caffeine intoxication
Nature Excess drinking of coffee may lead to addiction to caffeine as a stimulant, resulting in tension, headache and in severe cases, malnutrition through loss of appetite. Caffeine is used medically in tablet form as a stimulant, against headache and in certain cases for the treatment of asthma, but its abuse induces tolerance and mental dependence. Caffeine abuse, which is often coupled with nicotine abuse, may go unrecognized because coffee drinking is commonplace and socially acceptable; it is fully recognized as a drug problem.
Refs Dews, P B (Ed) *Caffeine* (1984).
Broader Drug abuse (#PD0094) Substance abuse (#PC5536)
Substance intoxication (#PD4027).
Related Smoking (#PD0713).
Aggravates Tension (#PB6370).
Aggravated by Ulcers (#PE2308) Mental depression (#PC0799).

♦ PE0621 Legal havens
Havens from prosecution
Broader Impediments to extradition (#PF5947).
Narrower Abuse of tax havens (#PE2370)
Abuse of land-locked and island countries as havens (#PE5861).
Related Terrorist havens (#PD5541).
Aggravates Draft evasion (#PD0356) Evasion of the law (#PD4208)
Unconvicted war criminals (#PD4067)
Obstruction of international criminal investigations (#PF7277).
Aggravated by Ultraviolet radiation as a hazard (#PE5672)
Improper lighting as an occupational hazard (#PE5780).

♦ PE0628 Student press weakness
Nature Student newspapers may suffer from fragmentation, inter-student faction fighting, lack of finance, or control by authorities who censor them or use them as propaganda tools. Alternatively, student newspapers may have undue political influence in relation to the relative isolation of students from the rest of society. Student opinion may be idealist or cynical, elitist or extremist and it may incite political unrest or violence and hence cause repression or revolution. Statements may be inaccurate or exaggerated to the point of propaganda.
Broader Journalistic irresponsibility (#PD3071) Fragmentation of organized students (#PF5753).
Related Underground press (#PD2366).
Aggravates Student revolt (#PC2052) Subversive activities (#PD0557)
Political impotence of students (#PE5729).
Aggravated by Extremism (#PB3415) Censorship (#PC0067)
Repression (#PB0871).

♦ PE0630 Failure of methods to appropriate data
Absence of method to appropriate available data
Nature To function effectively in the modern world people need technologies to appropriate data. All of the past processes by which people learned data have become ineffective. The ever increasing amount and diversity of information available to any individual simply overwhelms simplistic methods of data retention. Past wisdom is seemingly irrelevant, decision are confusing, learning is difficult and capacity to operate is reduced.
Broader Human wisdom unrelated to daily life (#PD1703)
Lack of central planning structures in small communities (#PF2540).
Aggravates Inadequate statistical information and data on problems (#PF0625).

♦ PE0635 Resource-intensive packaging
Energy-intensive packaging
Incidence The packaging demanded for products requires large amounts of energy, in addition to the chemicals, wood and plastics needed to produce boxes, padding, bags and decorative materials which increase the environmental cost. It has been estimated that over 360 billion BTUs of energy could have been saved in the USA if the 1986 per capita packaging consumption had remained at the per capita levels of 1958.
Broader Unsustainable development of energy use (#PC7517).
Aggravated by Non-destructible packaging and containers (#PD1754).

♦ PE0639 Leukaemia
Refs Baker, Lawrence, et al *Biology and Therapy of Acute Leukemia* (1985); Büchner, T, et al (Ed) *Acute Leukemias* (1987); De Vries, Jan *Cancer and Leukaemia* (1988); Gale, R P and Hoelzer, D (Eds) *Acute Lymphoblastic Leukemia* (1990); Gale, Robert and Golde, David *Leukemia* (1985); Neth, R and Gallo, R C (Eds) *Modern Trends in Human Leukemia VII* (1987); Schmalzl, F, et al (Eds) *Preleukemia International Workshop* (1979); Symposium on Comparative Leukemia Research, 7th International Copenhagen, October 1975 *Comparative Leukemia Research 1975* (1976).
Broader Malignant neoplasms (#PC0092) Diseases of the lymphatic system (#PD2654)
Neoplasms of lymphatic and haematopoietic tissue (#PE4637).
Narrower Lymphatic leukaemia (#PE2686) Monocytic leukaemia (#PG9501)
Myelogenous leukaemia (#PG2685).
Aggravates Mycosis (#PE2455) Diseases of the spleen (#PE6155).
Aggravated by LSD abuse (#PG2510) Radiation accidents (#PD1949)
Electromagnetic pollution (#PD4172) Health hazards of radiation (#PD8050)
Occupational hazards of benzene (#PE1849)
Long-term hazards of exposure to radiation (#PE4057).

♦ PE0640 Cankers
Diebacks
Nature A canker is a localized lesion or diseased area resulting in an open wound and usually on a woody structure killing the water-conducting tissues so that the symptom becomes a dieback.
Broader Oomycetes (#PE5465) Ascomycetes (#PE4586) Basidiomycetes (#PE0364)
Deuteromycetes (#PE4346).
Narrower Hypoxylon canker of trees (#PG4453) Botrytis diseases in plants (#PG1668)
Fusarium diseases in plants (#PG5461).

♦ PE0647 Political corruption of the judiciary
Nature Since most judges are appointed either by or on the advice of elected persons, or are themselves elected, they rarely profess political opinions generally disfavoured in the country in question, although the degree of tolerance shown in this connection varies from country to country. Provided that the country's electoral processes are free and democratic and that the law does not specifically exclude from membership of the judiciary persons of a specific political persuasion, or make other provisions unfavourable to them, it is difficult to say how far discrimination exists in these instances. However, the independence of the judiciary is clearly undermined when candidates for judicial offices are appointed or elected to such office as a reward for political services, or when political considerations enter into the promotion, transfer or dismissal of judges or into decisions concerning any other aspect of their status.
Broader Official corruption (#PC9533) Institutionalized corruption (#PC9173).
Related Political appointees (#PF2031) Political opportunism (#PC1897)
Corruption in politics (#PC0116) Corruption of the judiciary (#PD4194)
Frauds, forgeries and financial crime (#PE5516).
Aggravates Political crime (#PC0350) Political trials (#PD3013)
Political police (#PD3542) Political prisoners (#PC0562)
False political evidence (#PD3017) Forced political confessions (#PE3016)
Lack of internal political independence (#PF3194)
Political prisoners in communist systems (#PD3171)
Mis-classification of political prisoners (#PF3020)
Military political prisoners and detainees (#PD3014)

Civilian political prisoners and detainees (#PD3015).
Aggravated by Political ignorance (#PC1982) Political inequality (#PC3425) Political intimidation (#PC2938) Undue political pressure (#PB3209).

♦ PE0648 Misallocation of resources to protect the aged

Nature The aged are not getting the care, protection and basic necessities that society can afford. Resources are assumed to be scarce yet both public and private spending patterns indicate otherwise. Band-aids solution to elder's needs then clog delivery systems and mediate against finding real solutions and allocating funds effectively.
Broader Conflicting social service ideologies (#PD3190).
Related Inadequate income in old age (#PC1966).
Aggravates Inadequate welfare services for the aged (#PD0512).

♦ PE0651 Parenticide

Killing of parents — Parricide — Patricide — Matricide
Broader Homicide (#PD2341).
Related Senilicide (#PJ7124) Criminalization of euthanasia (#PF2643).
Aggravated by Disrespect for elders (#PF3979).

♦ PE0652 Thyroid gland disorders

Thyroid glands tumorous overgrowth — Thyroid glands inflammatory lesions
Nature Disorders of the thyroid gland include tumorous growths and inflammatory lesions. Causes may be dietary or hereditary.
Refs Andreoli, M; Monaco, F and Robbins, J *Advances in Thyroid Neoplasia*; Clark, O and Roeher, H D *Thyroid Tumors* (1988); Udupa, K N, et al *Disorders of the Thyroid Gland in Tropics* (1983).
Broader Gland disorders (#PD8301) Nutritional diseases (#PD0287) Human disease and disability (#PB1044).
Narrower Cretinism (#PE7905) Myxoedema (#PG3408) Thyroiditis (#PG2722) Endemic goitre (#PE1924) Thyrotoxicosis (#PG9248).
Aggravates Malignant neoplasms (#PC0092).
Aggravated by Inherited diseases (#PG2727) Iodine deficiency disorders (#PD2726).

♦ PE0653 Chagas' disease

South American trypanosomiasis — American trypanosomiasis
Nature Chagas disease can be either acute or chronic. It is caused by the pathogenic hemoflagellate Trypanosoma cruzi, and is transmitted to man by blood-sucking reduviid bugs who live mostly in or near human dwellings. Domestic animals and humans in rural areas and in slums become infected by contamination of the bite wound with the insects' faeces, which are infected with the invasive trypanosomes. Transmission by blood transfusion may also occur where no prevention measures exist. The acute form of Chagas disease occurs predominantly in young children and is characterized by fever, adenopathy, spleen and liver enlargements, and in some forms facial oedema. The brain is often affected and convulsions may result in permanent mental damage or even in death. The chronic form may be mild, but cardiopathy due to parasite development in the heart muscles may be extremely serious, provoking in some cases sudden death.
The economic harm caused by the disease is considerable. In the first place, the incapacitating symptoms of the chronic forms of the disease generally develop in the second half of life when the individual is making his greatest contribution to society. Secondly, the disease is found principally in rural areas where those affected are often rendered incapable of the heavy physical work demanded of them. Equally costly is the hospitalization and subsequent rehabilitation of patients with Chagas disease.
Incidence South American trypanosomiasis is restricted to Central and South America, but it occurs in every country there, the highest prevalence being in Argentina, Brazil, Chile, and Venezuela. Present estimates suggest that some 90 million people live under conditions favouring transmission, the number of infected people being probably 16–18 million.
Broader Arthropod-borne diseases (#PE7796) Uncontrolled tropical diseases (#PD4775).
Related Trypanosomiasis (#PE5725) African trypanosomiasis (#PE1778) Parasitic diseases in animals (#PD2735).
Aggravates Debilitated working capacity (#PG2728).
Aggravated by Protozoan parasites (#PE3676) Unhygienic conditions (#PF8515) Vector-borne diseases (#PD8385) Parasites of the human body (#PE0596) Inadequate housing construction (#PG0561) Kissing bugs as vectors of disease (#PE3616).

♦ PE0655 Involutional depression

Nature Involutional depression occurs when women are forty to fifty-five and men fifty to sixty-five years, when the activity of endocrine glands decreases and the reproductive capacity is lost and there occur changes in relationships and work.
Broader Mental depression (#PC0799).
Aggravated by Menopause (#PF5918) Male mid-life crisis (#PD5783).

♦ PE0661 Epilepsy

Falling sickness — Epileptics — Post epileptic automatism
Nature Epilepsy is, in fact, not a disease at all. It is a symptom of some other problem. All people with epilepsy have one feature in common, a disturbance in the normal pattern of the electrical activity of the brain. Such disturbance is sudden and episodic, and frequently recurrent. It can involve impairment of thought and awareness or responsiveness, or both, and may involve convulsions or automatic movements.
Incidence Epilepsy is of unexpectedly high frequency and severity in many parts of the world. In developing countries, where medical care has been insufficient, the total prevalence may reach 4 per 100 of the population, and frequent severe seizures have been reported in 1 per 100. Epilepsy presents an important mental health and public health problem, not only because of its serious economic implications but also through its social impact on the family and the community. For example, the economic cost of epilepsy in the United States in 1975 is estimated to be over 3 billion dollars, including unemployment, underemployment, excessive mortality, treatment costs, care for the severely disabled, drug costs, vocational rehabilitation, special education and research in epilepsy. Furthermore, severe epilepsy causes serious disability and has a high mortality. For various reasons, including superstition, many patients are not brought for treatment. In one study nearly three quarters of the people studied had intellectual, behavioural or neurological handicaps. Children with epilepsy only have a few problems at school and those with additional problems have many difficulties. A higher percentage of people with epilepsy are unemployed, in part because of the additional risk of injury on the job and transportation to and from work. Approximately 50 percent of persons with epilepsy die directly or indirectly because of the condition. With modern methods the condition can probably be controlled in 75 percent of patients. An unnecessary burden of disability therefore exists.
Refs Aicardi, Jean *Epilepsy in Children* (1986); Edwards, Felicity, et al (Eds) *Epilepsy and Employment - A Medical Symposium on the Current Law* (1986); Hopkins, Anthony (Ed) *Epilepsy* (1987); Janz, Dieter, et al (Eds) *Epilepsy, Pregnancy and the Child* (1982); Nistico, Giuseppe et al *Neurotransmitters, Seizures, and Epilepsy III* (1986); Wolf, Peter, et al (Eds) *Advances in Epileptology* (1987).
Broader Developmental disabilities (#PD3773) Diseases of the central nervous system (#PE9037).
Aggravates Fear of ostracism (#PF2776).
Aggravated by Anxiety (#PA1635) Overwork (#PD2778) Drunkenness (#PE8311) Alcohol abuse (#PD0153) Consanguineous marriage (#PC2379) Diseases and injuries of the brain (#PD0992).

♦ PE0667 Interference in labour relations of industrialized countries by transnational corporations

Refs ILO *Employment Effects of Multinational Enterprises in Industrialised Countries* (1985).
Broader Domination of labour relations by transnational corporations (#PE1187).

♦ PE0669 Tying of supplies to subsidiaries by transnational enterprises

Restrictions on raw materials and components utilization
Nature Transnational enterprises tend to require that their subsidiaries import from within the corporation's structure the raw materials and intermediate goods required rather than utilize similar domestically produced goods where these are available. Such restrictive practices distort international trade patterns, often to the disadvantage of the developing countries which are host to the transnational corporation subsidiaries. Restrictions have been particularly marked in the sectors of electrical goods and machinery and machine tools.
Broader Restrictive trade practices (#PC0073).
Related Grant-back provisions (#PE5306) Challenges to validity (#PF1200) Restrictions on research (#PF0725) Restrictions on publicity (#PF1575) Restrictions on adaptations (#PF5248) Restrictions on use of personnel (#PF3945) Patent pool or cross-licensing agreements (#PE4039) Exclusive sales and representation agreements (#PE4581) Payments after expiration of industrial property rights (#PF5292).

♦ PE0670 Cabotage

Nature The provision of an internal service in one country by an enterprise registered in another country (as when a UK registered lorry takes goods to Germany and then takes on further goods there for delivery to another part of Germany).
Incidence Cabotage is illegal in many countries.
Broader Unfair transport practices (#PD1367).

♦ PE0671 Violations of regulatory codes without culpability

Nature A person who unknowingly violates a penal regulation is guilty of an infraction, a much lesser crime.
Broader Violation of regulatory codes (#PD4539).

♦ PE0680 Substance abuse during control of complex equipment

Flying under the influence of alcohol — Drug use by pilots
Incidence Although alcohol has been traced to very few commercial aviation accidents, there is evidence that drinking among pilots is more widespread than the accident statistics indicate.
Counter-claim Since 1983 only 16 accidents in the USA involved drug or alcohol abuse on the part of the pilot. None of the accidents involved a major airline.
Broader Substance abuse at work (#PD9805).
Narrower Drunken driving (#PE2149).
Aggravates Draining of resources due to alcohol (#PE8865) Human fatigue during control of complex equipment (#PE5572) Information overload during control of complex equipment (#PF6411).
Aggravated by Boredom (#PA7365) Mental impairment (#PF4945).

♦ PE0681 Instability of woodworking industries

Broader Instability of manufacturing industries (#PC0580) Instability of trade in wood, lumber and cork (#PE2521).

♦ PE0682 Unethical financial practices

Irresponsible financial practices — Banking malpractice — Misleading borrowers of funds — Misinformation concerning loans
Refs Barksdale, Byron L *Investment Broker Malpractice* (1987).
Narrower Bank fraud (#PE1398).
Related Insider dealing (#PD3841) Financial scandal (#PD2458) Unethical insurance practices (#PE1826).
Aggravates Bank failure (#PE0964) Contract fraud (#PD7876) High interest rates (#PF9014) Payment of interest (#PF5514) Banking law violations (#PE1208) Abuse of banking secrecy (#PF5991) Commodities trading fraud (#PD3917) Prohibition of trade union meetings (#PD7210).
Aggravated by Deception in business (#PD4879).

♦ PE0688 Health risks to workers in commerce

Broader Occupational risk to health (#PC0865).
Aggravates Industrial accidents (#PC0646) Occupational diseases (#PD0215).

♦ PE0689 Experimental exposure of animals to radiation

Nature In order to determine the acceptable levels of radiation exposure for humans, animals are exposed to both mild and severe doses of radiation. Radiation is also used on rodents, dogs and monkeys to discover the physical effects of induced cancers.
Incidence In the UK in 1986, 87,686 experiments were performed on animals using exposure to ionizing radiation.
Broader Cruel treatment of animals for research (#PD0260).

♦ PE0691 Unbalanced infant diets

Inadequate weaning diet
Broader Child malnutrition (#PD8941) Unbalanced family diets (#PJ0953) Nutritional deficiencies (#PC0382). .
Related Substitution of inappropriate foodstuffs for breast feeding (#PE8255).
Aggravates Infant mortality (#PC1287) Kwashiorkor disease (#PE2282) Infant growth failure (#PE6909) Protein-energy malnutrition in infants and early childhood (#PD0331) Damage to infant brains from malnutrition and insufficient stimuli (#PE4874).
Aggravated by Neglected young children (#PE4245) Outdated forms of community health (#PF1608).

♦ PE0696 Instability of rice trade

Nature International trade plays only a marginal role (4 per cent) in meeting the world's rice requirements, but a considerable group of countries remains heavily dependent on rice exports as a source of foreign exchange to finance development programmes. The stability of international trade since 1955 has been largely due to the high degree of government control exercised on both sides of the market. Yet the longer range outlook is very uncertain. The upward trend in imports in past years has been largely a reflection of failure of importing countries to meet their own production objectives. The slow increase of imports into developed countries is unlikely to offset the loss of outlets in the traditional deficit areas (developing regions of the Far East, West Africa, Near East and Latin America) which could occur if these countries succeed in their production objectives, thus leading to accumulation of surplus stocks.

Claim Rice is not only the principal source of foreign exchange for several developing countries, especially in the Far East, it is also a major import item in trade balances of a number of others. Moreover, in many developing countries rice is the key crop, on which plans of agricultural and economic development hinge. Instability in price, uncertainty of export outlets and insecurity in obtaining import requirements reduce the exercise of forward planning to a gamble for many developing countries.
Broader Instability of trade in cereals and cereal preparations (#PE1769).
Aggravated by Lack of purchasing power in developing countries (#PE8707).

♦ PE0697 Teratogens
Nature Teratogens are physical or chemical agents which interfere with normal embryonic development leading to congenital malformations. Such malformations are not hereditary (in contrast to those resulting from changes in genetic material). The period of maximum teratogenic sensitivity is the embryonic. The foetus becomes less susceptible the more it grows; however teratogens can still disturb its development.
In addition to general malformations, some teratogenic agents affect particular organ systems: thalidomide affects the skeleton and limbs; X-rays cause malformation of nerve and eye tissues; rubella causes heart deformities, cataracts and deafness.
Incidence It has been shown that a very broad range of chemical agents can produce, under the proper conditions, some serious type of developmental deviation. Substances already found to be embryopathic in animals range from highly toxic substances, such as anti-tumour agents, to commonplace consumer items such as aspirin, and completely inert materials. There are 33 known and proven teratogens. Known agents include: chemical factors (methylmercury, aminopterin, thalidomide, iodine deficiency, steroid hormones with androgenic activity, carbon monoxide); infective micro-organisms (rubella virus, herpes virus, cytomegalovirus, toxoplasma, syphilis); physical factors (ionizing radiation, trauma and perhaps high temperatures).
Claim With the increasing number of women employed in industry, environmental pollution due to industrial wastes, the use of insecticides, herbicides and defoliants constitute a definite potential hazard for the human embryo.
Broader Human disease and disability (#PB1044).
Related Mutagens (#PD1368) Carcinogenic chemical and physical agents (#PD1239).
Aggravates Human physical genetic abnormalities (#PD1618).
Aggravated by Biochemical warfare (#PC1164) Environmental warfare (#PC2696).

♦ PE0701 Instability of trade in crude synthetic and reclaimed rubber
Broader Instability of trade in manufactured goods (#PE0882)
Instability of trade in inedible crude non-fuel materials (#PD0280).
Aggravated by Reduction in demand for primary commodities due to technological change (#PD1276).

♦ PE0706 Unequal opportunities for disabled persons
Nature To achieve the goals of 'full participation and equality', rehabilitation measures aimed at the disabled individual are not sufficient. Experience shows that it is largely the environment which determines the effect of an impairment or a disability on a person's daily life: a person is handicapped when he or she is denied the opportunities generally available in the community that are necessary for the fundamental elements of living, including family life, education, employment, housing, financial and personal security, participation in social and political groups, religious activity, intimate and sexual relationships, access to public facilities, freedom of movement and the general style of daily living.
Refs United Nations *Disability* (1986).
Broader Human disability (#PC0699) Inequality of opportunity (#PC3435).
Narrower Unequal employment opportunities for disabled persons (#PE0783).

♦ PE0707 Structural barriers for disabled persons
Nature Many are excluded from active participation in society because of doorways that are too narrow for wheelchairs; steps that cannot be mounted leading to buildings, buses, trains and aircraft; telephones and light switches that cannot be reached; and sanitary facilities that cannot be used. Similarly they can be excluded by communication barriers, for example oral communication which ignores the needs of the hearing impaired and written information which ignores the needs of the visually impaired. Such barriers are the result of ignorance and lack of concern; they exist despite the fact that most of them could be avoided by inexpensive yet careful planning. Although some countries have enacted legislation and launched campaigns of public education to eliminate such obstacles, the problem remains a crucial one.
Claim Structural barriers turn the man-made environment into a frustrating obstacle course for an appreciable number of the world's citizens. The aged, mothers with prams, wheel-chair users, pregnant women, people with chronic heart or bronchial conditions, the person with a crutch, all are daily impeded, inconvenienced and endangered by architectural barriers.
Counter-claim The problem is exaggerated. The handicapped do after all constitute only a minority of the population, and the most severely handicapped – those forced to sit in a wheel-chair – account for only a minute part of the population. Even if we were to mobilize all allies such as infants, the aged, the poor-sighted, pregnant women etc, it would still be difficult to get up to as much as 10% of the population.
Broader Human disability (#PC0699).

♦ PE0712 Unidentified submarine objects
Unidentified submarine patrols
Incidence Numerous incidents of unidentified submarine patrols are reported in coastal areas. To fishing vessels, these constitute a navigational hazard, especially since the government responsible is reluctant to acknowledge responsibility for any damage to nets or to fishing vessels. To governments, the presence of unidentified submarines in coastal waters is a sever threat to security, as is the case for Sweden.
Broader Hazards to navigation (#PE3868).
Related Unidentified flying objects (#PF1392)
Invasion of airspace by foreign aircraft (#PE3972).
Aggravates National insecurity and vulnerability (#PB1149).

♦ PE0714 Inadequate plant quarantine
Nature Through the international exchange of plants and plant products, many harmful insects and destructive plant diseases have travelled with their hosts to distant lands. Inadequate plant quarantine services have failed to exclude such pests and diseases, many of which have become established with even greater vigour in their new environments.
Incidence Plant pests and diseases destroy over 20 percent of the potential world harvest annually. Among the hundreds of noxious insects and plant diseases that countries seek to exclude through quarantine regulations and controls are the golden nematode of potatoes and tomatoes, pink bollworm of cotton and cotton boll weevil, harmful species of fruit flies, such as the Mediterranean oriental and Mexican fruit flies and the melon fly, Colorado potato beetle, citrus canker and various virus diseases. Because there is greater impetus behind species introduction, commercial and otherwise, and behind international transport, than there is behind quarantine controls, the execution of controls remains inadequate, and plant pests and diseases are carried to new regions on plant products, seed and nursery stock. The worldwide distribution of many major crop and forest pests – the Hessian fly, Japanese beetle, Colorado potato beetle, grape Phylloxera, spruce sawfly and gypsy moth; tree diseases – Dutch elm disease and chestnut blight; and crop diseases – golden nematode of potatoes, potato blight and vine powdery mildew – testify to the ability of man to spread plant pests and disease.
Background Prior to 1870 there was little realization by national governments of the dangers inherent in a free exchange of the plants and plant products that are hosts to injurious plant pests. The Colorado potato beetle was responsible for the first legislation affecting the international movement of a plant product, after its introduction into Germany. Following eradication of the infestation, a decree was issued by the German government in 1875, forbidding the further importation of potatoes and potato sacks. In the same year France also imposed exclusion measures against the beetle.
About 1859 the grape Phylloxera, a native of the eastern USA, was introduced into France through the medium of imported grapevine cuttings; during the following 20 years it spread throughout Europe. Within 25 years of its discovery in France it had destroyed nearly one third of the French vineyards. Similar shipments of vine cuttings from France about 1872 spread the Phylloxera to Australia, where legislacion to suppress the infestation was passed in 1877. This act gave power to quarantine and even to eradicate vines and destroy vineyards. In France regulation relating to Phylloxera was proclaimed in 1878. The first international action in the field of plant quarantine was in 1881, when a conference held in Bern, Switzerland, drafted an agreement known as the Phylloxera convention, preventing the introduction of the Phylloxera from the US and restricting the movement of grapevines and grape products to prevent its further spread between European countries.
Other major landmarks in the development of plant quarantine were the International Convention for the Protection of Plants of 1929 and the International Plant Protection Convention drafted in 1951. These both sought closer liaison between national plant quarantine services.
Claim Quarantines are effective, preventing the introduction of many pests and diseases and retarding the movement of others, giving scientists time to combat such pests and diseases before they become well established; and annually save agriculture large amounts of money, more than sufficient to offset the business and trade losses due to, and the costs of, quarantine control measures.
Counter-claim As man is unable to effectively prevent the movement of microscopic pathogens, many quarantines are scientifically unsound and ineffectual. Occasionally, quarantines have been used as economic sanctions in restraint of free trade and have caused unnecessary economic losses.
Broader Crop vulnerability (#PD0660) Inadequate quarantine (#PE2850).
Aggravates Plant diseases (#PC0555) Pests of plants (#PC1627)
Viral plant diseases (#PD2227) Insect pests of plants (#PD3634)
Dissemination of plant diseases by man (#PD3593)
Introduction of new species of insect pests (#PF3592)
Environmental hazards of new species introduction (#PC1617).

♦ PE0715 Disastrous failure of natural dams
Nature Dams can be formed naturally, typically as a result of landslides, avalanches, mud flows, or glacier activity. Most such dams fail rapidly and the sudden release of water temporarily formed into a lake can prove disastrous.
Broader Natural disasters (#PB1151).
Aggravates Floods (#PD0452).
Aggravated by Landslides (#PD1233) Avalanches (#PD1146)
Earthquakes (#PD0201) Hazardous glaciers (#PE6824).

♦ PE0718 Inaccessible commercial and financial services
Broader Physically inaccessible services (#PC7674).
Narrower Inadequate financial services (#PJ8366)
Inaccessible market and supply centres (#PF8299)
Remoteness of legal services in developing countries (#PE9001)
Inaccessibility of insurance for island developing countries (#PE5665).

♦ PE0721 Leprosy
Hansen's disease
Nature Leprosy is a chronic communicable disease which affects particularly the nerves, the skin and mucosae. Damage to and ultimate destruction of the nerves leads to the deformities which used to be regarded as typical of the disease. These involved the upper limb (claw-hand, wrist-drop), the lower limb (claw-toes, foot-drop), the eye (anaesthesia of the cornea, paralysis of the eyelids) and ulceration of the larynx; and resulted in neglected cases of extensive ulceration of hands and feet, with loss of digits, hoarseness of voice and severe impairment of vision or total blindness.
Leprosy attacks at any age and either sex, and the effects show the widest possible variation – from complete refractoriness to high susceptibility. Hence leprosy may be a small localized patch in the skin, or a progressive generalized disease affecting the skin of the entire body, all the superficial nerve trunks of limbs and face, and the lining of the nose. In any population, a small proportion succumb to the latter form; most seem to be refractory to infection. At present, only one patient in five is able to get treatment.
Because of misunderstanding, ignorance and prejudice, leprosy is often regarded as medically unique, and because of its long incubation period, inauspicious onset, slow and painless development, and slight tendency to shorten life, it is surrounded by an aura of mystery. The deformities and disfigurements, blindness and mutilations habitually associated with leprosy are mainly the result of damage sustained by tissues (muscles and skin) whose nerves have been destroyed as the result of leprosy. It is a major crippling disease, and hence is of economic importance.
Incidence Leprosy occurs in most countries (hot or cold, wet or dry, low-lying or mountainous) but the bulk of the up to 10-12 million sufferers live in the developing countries.
Background The earliest written references to true leprosy are from India (600 BC), whence it was probably introduced into the West (with Alexander the Great's troops, 327-326 BC). Enormous confusion existed in the literature before leprosy was delimited (in 1847) as a specific disease and as caused by a specific micro-organism, Mycobacterium leprae (1873). References to 'leprosy' in the Old Testament of the Bible (Authorized Version) are a mistranslation. Leprosy has virtually disappeared from north-west Europe, but still exists in countries of Southern Europe. Imported cases comprise a manageable problem in the industrialized West.
Claim Because leprosy does not kill (unlike smallpox or tuberculosis), or assume epidemic proportions, and is not easily preventable (by vaccination or general health measures) and because, until recently, it could not be treated rapidly or cheaply, it has tended to be neglected and to be given a low priority in official thinking. There still persists a certain shame about leprosy, and about admitting its existence. The deformed beggars, often a major social problem, represent a small and innocuous proportion of sufferers. They are not contagious for the most part and their hurt is psychological as well as physical. Hence, since leprosy does not constitute a problem of obvious urgency or outstanding importance in any country, it has tended to be neglected and ignored. However, recent recommendations concerning the use of a combination of drugs, given for only a limited period of time (in many cases as short as six months) are changing this attitude and giving new impetus to the efforts of both governments and voluntary agencies to control leprosy in the countries where it is at present a major public health problem. To succeed, these

efforts involve both political will and a highly motivated health staff as well as a stable infrastructure and financial resources. The participation of the community and, where appropriate, the primary health care system are equally important. The priority is to break the chain of transmission of the infection by the treatment of the infectious cases and so in the long term to reduce the human suffering and physical disability caused by the disease. The need for custodial care for the permanently crippled and for reconstructive surgery and rehabilitation for those likely to benefit therefrom must be taken into account but at a lower priority rating.
Refs Chaffee, Judy K *Leprosy* (1987); Hastings, Robert C (Ed) *Leprosy* (1986); ILO *Vocational Rehabilitation of Leprosy Patients Report on the ILO–DANIDA Asian Seminar Bombay, India (26 October–6 November, 1981)* (1982); Pontificia Accademia delle Scienze *Immunology, Epidemiology and Social Aspects of Leprosy* (1984).
   Broader Bacterial disease (#PD9094)   Stigmatized diseases (#PD7279)
   Uncontrolled tropical diseases (#PD4775).
   Aggravates Ulcers (#PE2308)   Superstition (#PA0430)   Social outcasts (#PD6017)
   Physical blindness (#PD0568)   Skeletal system disorders (#PE2298)
   Genetic susceptibility to disease (#PG2834)
   Mutilation and deformation of the human body (#PD2559).
   Aggravated by Cockroaches as pests (#PE1633)
   Inadequate personal hygiene (#PD2459)
   Overcrowding of housing and accommodation (#PD0758).

◆ **PE0725 Locust plagues**
Grasshopper plagues
Nature Locusts are any of certain insects of the family Acrididae, which at times multiply greatly and migrate long distances in destructive swarms. In Europe the term 'locust' connotes large size; smaller acridids are called grasshoppers. In North America 'locust' and 'grasshopper' are used for any acridid.
Certain kinds of locusts such as the Red (Nomadacris) and the African Migratory species, a variant of Locusta migratoria, have reasonably well-defined outbreak areas. They can often be detected and suppressed before they reach plague proportions. But Schistocerca is the exception. It is nomadic. It can emerge in countless millions from the breeding and concentration of insects widely scattered, in altitudes from sea level to 10,000 feet, as well as from swarms that survive a recession period. A major swarm of these metallic black-and-yellow locusts can shadow an area of 500 square kilometres or more, in layers up to a kilometre deep. There will be ten thousand million insects in such a swarm, a flying mass of 50,000 or 100,000 tons. A swarm can fly 3,000 miles overland, normally in 10-hour daily flights at air-speeds of only a few miles an hour. The locusts hold reserves of fat adequate even for overseas flights (wind-tunnel tests indicate flight times of up to 17 hours).
Incidence The outbreak areas of the migratory locust of the old world Locust migratoria, are of four ecological types: (1) Deltas of rivers entering the Caspian and Aral seas and Lake Balkhash (and similar situations in China and Africa), surrounded by arid sand tracts; there the extent of the grassland habitat of the locust changes greatly as a result of irregularities of floods. (2) Grassland areas adjoining deserts, subject to extreme fluctuations in precipitation with corresponding changes in extent of habitable area. (3) Islands of dry warm soil in the central USSR, a region generally too cold and wet for the species; there overcrowding occurs after several exceptionally warm dry years. (4) Grasslands produced by periodic burning in the unfavourable, humid forestlands of the Philippines; their extent varies greatly, leading to overcrowding and production of the gregarious phase.
Claim The outbreak of locust swarms presents a serious threat to agriculture and therefore to the well-being of millions persons dependent on the crops for their food and livelihood.
   Broader Pests (#PC0728)   Disastrous insect invasions (#PD4751)
   Grasshoppers as insect pests (#PE3642).
   Aggravates Destruction of land fertility (#PC1300)
   Deterioration in atmospheric visibility (#PE2593).
   Aggravated by Introduction of new species of insect pests (#PF3592).

◆ **PE0730 Inadequate coordination of the intergovernmental system of organizations**
Claim The problems within the United Nations system represent only part of the problem of bringing greater order, coherence and coordination into international economic and social activities as a whole, among United Nations and non-United Nations, world-wide and regional organizations and between United Nations bodies and the bilateral aid activities of individual 'donor' Governments. To separate the part from the whole has great disadvantages, since duplication and even conflict between United Nations and non-United Nations bodies are themselves among the most serious and conspicuous causes of criticism, and solutions may well affect the structure, functioning and programmes of organs within, no less that those outside, the United Nations system.
   Broader Institutional fragmentation (#PC3915)
   Inadequate coordination of international organizations and programmes (#PD0285).
   Narrower Fragmentation and complexity of the United Nations system (#PE0296)
   Inadequate coordination between regional intergovernmental organizations with common membership (#PE1184).
   Aggravates Ineffectiveness of intergovernmental organization and programmes (#PF0074)
   Competition between intergovernmental organizations for scarce resources (#PE0063).
   Aggravated by Failure to adapt general initiatives to specific needs (#PF1578)
   Inefficient location of facilities of international organizations (#PE3538)
   Jurisdictional conflict and antagonism between intergovernmental organizations (#PE7901).

◆ **PE0735 Instability of trade in animal and vegetable oils and fats**
   Broader Instability of the primary commodities trade (#PC0463).
   Narrower Instability of trade in animal oils and fats (#PE8896)
   Instability of trade in fixed vegetable oils and fats (#PE0861).

◆ **PE0736 Environmental degradation through military activity during peace-time**
Environmental destruction on military bases
Incidence The environmental impacts of preparations for war include: indirect impacts made through the diversion of resources from environmental development and through the impact of the armaments industry; and direct impact through weapons testing and military operations and through the profiferation of nuclear technology. Military bases, especially airfields, may require large areas of ecologically valuable land and require the destruction of any fauna perceived as hindering such activity (such as birds congregating on runways). The construction of such bases tends to destroy such environments irretrievably, especially in the case of island bases which are turned into concrete wastelands. Large sectors of most countries are reserved for military exercises, especially tank manoeuvres, tactical missile exercises, bombing exercises and artillery practice, in addition to those reserved in some countries for chemical and biological warfare exercises. In those countries manufacturing weapons, such areas may also be used for testing missiles, chemical and biological warfare products, as well as nuclear weapons testing. All of these activities severely degrade the natural environment and tend to be treated as exceptions to any regulatory measures to protect the environment.
   Related Obsolete military bases (#PG8043)   Nuclear weapons testing (#PC2201)
   Environmental consequences of war (#PC6675)
   Military manoeuvres in sensitive border areas (#PE3704).

   Aggravated by Foreign military presence (#PD3496)
   Excessive military usage of land (#PE3402)
   Socio-economic burden of militarization (#PF1447).

◆ **PE0738 Environmental hazards from transport equipment**
Nature Transport systems consume scarce natural resources of land and energy, kill many people in accidents, and cause severe local pollution of air and water. New advances in the field of transport have exposed users to a whole series of hazards that, although not new, are becoming more and more striking because of the constantly increasing number of persons affected by them.
The rapidity of air transport, for example, results in disturbances of the circadian rhythm in aircrews and passengers who have to travel long distances. Variations in pressure during the flight may result in injuries to the ear, the sinuses, and sometimes the teeth. Airsickness, and fainting following hyperventilation provoked by anxiety, are not uncommon, but the most frequent hazard is constituted by the accidents, usually slight but sometimes serious, caused by turbulence. On a more general health level, the contact between distant populations resulting from the growth of air travel facilitates the spread of communicable diseases and raises numerous health problems both in flight and at airports. In addition there are the difficulties of adapting to sudden changes of climate, way of life, and food, which especially affect tourists.
Risks due to air travel are relatively insignificant compared with the dramatic toll of road accidents. Indeed, among the accidents resulting from the development of air, land, sea and river transport, road traffic accidents predominate both in respect of their frequency and seriousness and in terms of human and economic cost: more than 150,000 dead and 6,000,000 injured each year. In 1969 in the USA the total cost of motor vehicle accidents was estimated at $12,500,000,000.
Incidence Transport is a great consumer of energy accounting in the United States for about 55 percent of liquid fuels (petroleum products); in Europe, about 31 percent; and in Kenya, more than 65 percent (reflecting the fact that in the developing countries the bulk of liquid fuels is used in the transport sector). The efficiency of energy utilization is variable. Railroads and waterways, for example, are more efficient than aircraft or automobiles. The latter are the least efficient, and they account for the bulk of energy consumption in the transport sector.
The most familiar environmental impacts of road transport are those from air pollution. Petrol-burning vehicles emit carbon monoxide, hydrocarbons, and oxides of nitrogen. Diesel engines emit relatively little of these, but produce more particulates and smoke. Alkyl lead is still added to vehicle fuel in many countries, and is emitted in the exhaust as very small particles. In confined spaces (like tunnels or very narrow streets) carbon monoxide concentrations can rise to levels hazardous to health, especially to people with heart or lung weakness. Nitrogen oxides and hydrocarbons, on the other hand, are not directly toxic, but interact in the presence of sunlight to produce an oxidant smog which irritates the eyes and lungs and damages sensitive plants. This kind of smog made Los Angeles notorious before emissions were controlled, and remains a problem in many large cities such as Tokyo.
Aircraft and railway locomotives together emit a far smaller volume of air pollutants than road vehicles do. However, high-flying aircraft release oxides of nitrogen directly into the lower stratosphere where they may become involved in chemical reactions which could reduce the concentration of ozone, important as a screen against ultraviolet radiation from the sun. Aircraft also contribute carbon monoxide, hydrocarbons, particulates, and nitrogen oxides at lower levels, locally, around airports. The best available evidence suggests, however, that none of these impacts is significant.
Noise is one of the most widely recognized and resented environmental consequences of the increase in road and air transport. It interferes with work, prevents sleep, has psychological effects, and can even damage hearing.
The expansion of rail and marine transport in the past, and more recently of road and air transport, has obviously consumed much land, both for new tracks and highways and around ports and airports. New transport corridors also have a secondary effect on land use because they attract industrial development. New highways can sever wildlife habitats and stimulate uncontrolled and inappropriate rural development, while new seaports are often placed in coastal areas also valued for wildlife and recreation. Because the density of development is lower in cities built to accommodate road traffic, public services may be more costly to operate there. But there are also problems when poorly designed roads are overburdened with vehicles. Drainage systems can be interfered with, and costs of maintenance of both roads and vehicles can become unacceptably high.
   Narrower Environmental hazards from road motor vehicles (#PE8336).
   Related Aircraft environmental hazards (#PD8328).

◆ **PE0740 International indebtedness arising from insurance transactions in developing countries**
Nature A lack of national insurance markets and institutions causes countries to lose foreign exchange and investment capital. Insurance costs involve a net external debit in normal years to practically all developing countries, whether measured in terms of net indebtedness or of physical flows of funds. The incidence of such flows is clearly and inseparably bound up with the lack of a significant local insurance sector, the absence of a substantial statutory framework for insurance, and with shortages of adequate local forms of investment for an insurance portfolio. These deficiencies are bound up one with another, and their rectification tends often to be almost simultaneous.
Reinsurance is essentially an extension of the principle of spreading risks as widely as possible, and reducing them to known costs. By retaining only a part of risks directly underwritten, and ceding the remainder to other direct insurers acting in the capacity of reinsurers or to professional reinsurers, the insurers themselves are widening the spread of risks and reducing part of the risk element to a cost. As its nature would suggest, reinsurance is a major international aspect of insurance and one that involves very large sums. Apart from reinsurance, the other main overseas flow to which non-life insurance operations may give rise (in those developing countries exercising any significant degree of regulation over insurance) consists of surpluses or profits earned by foreign agencies, branches and subsidiaries.
The concept of profit in insurance is extremely contentious. It is difficult to suppress suspicions that some of the profit and loss accounts do as much to obscure still further the true underwriting profit achieved as to clarifying the issue. Provisions for reserves over and above those required legally or even by the canons of prudent insurance practice are one source of obscurity. Life insurance business can also involve a substantial drain if the funds generated by it are invested out of the country. The loss of the physical use of life insurance funds (or mathematical reserve, since it is calculated on actuarial principles) can be of potentially great importance to countries in the basic stages of economic development when investment capital is scarce.
   Broader Economic and social underdevelopment (#PB0539).

◆ **PE0741 Inadequate funding of international nongovernmental organizations and programmes**
   Broader Inadequate funding of international organizations and programmes (#PF0498)
   Obstacles to effective international nongovernmental organizations (#PF7082).
   Narrower Lack of financial information systems for international nongovernmental organizations (#PE7009).

Aggravates Inadequate facilities for international nongovernmental organization action (#PF2016)
Ineffectiveness of international nongovernmental organizations and programmes (#PF1595).
Aggravated by Compassion fatigue (#PF2819)
Competition between international nongovernmental organizations for scarce resources (#PE0259).

♦ PE0742 Substance abuse by role models
Drug abuse by community leaders — Alcohol abuse by people of authority
Broader Substance abuse (#PC5536).
Narrower Substance abuse by physicians (#PE7568).

♦ PE0743 Flood plain settlement
Building on flood plains — Construction on flood plains
Incidence In many river valleys, areas chronically liable to floods are now farmed through lack of more suitable cultivable land.
Aggravates Floods (#PD0452).
Aggravated by Shortage of cultivable land (#PC0219)
Unequal property distribution (#PC3438)
Cultivation of marginal agricultural land (#PD4273).

♦ PE0751 Insufficient preventive medicine
Broader Inadequate medical care (#PF4832)
Inadequate maintenance of physical health (#PF1773)
Underprovision of basic services to rural areas (#PF2875).
Aggravates Inadequate prevention of disabilities (#PF0709)
Inadequate community care for transient urban populations (#PF1844).
Aggravated by Curative health mindset (#PU5393).

♦ PE0752 Unjust financing of political parties
Nature Political parties may be financed by individuals or corporations with a vested interest of keeping a certain party in power. Bribes may be accepted by politicians on a local or national basis. The party in power may misappropriate public funds for its own use. Unjust financing of political parties may lead to elite control, political dictatorship, apathy, alienation and stagnation, or to political conflict and revolution.
Broader Political bribery (#PC2030) Corruption in politics (#PC0116).
Aggravates Political conflict (#PC0368).
Aggravated by Power politics (#PB3202).

♦ PE0753 Spouses of torture victims
Nature The spouses of torture victims are victims of the situation in which their mates were tortured. They are mostly women and are often held responsible for their husbands arrest by their neighbours. When the victim returns home with dramatic changes in behaviour they do not understand why. These wives have had to solve problems while alone because their husbands were in prison. They have had to find work, take care of the children, in many cases, carry on family obligations of large extended families, and deal with neighbours. Often the children of torture victims suffer psychological problems and place additional burdens on the family.
Broader Family of torture victims (#PE2119).
Related Children of torture victims (#PE4579).
Aggravated by Susceptibility of the old to physical ill–health (#PD1043).

♦ PE0754 Hydrocarbons as pollutants
Nature Sources of hydrocarbons which pollute the air are common to industrialized areas: motor vehicles; petrol stations; industrial processes; domestic cleaning agents. There are also natural sources.
Refs Müller, Helga *Hydrocarbons in the Freshwater Environment* (1987).
Broader Chemical air pollutants (#PD1271)
Chemical pollutants of the environment (#PD1670).
Narrower DDT as a pollutant (#PE5028).
Related Oil as a pollutant (#PE2134) Gaseous organic compounds (#PJ5013)
Photochemical oxidant formation (#PD3663).
Aggravates Ozone as a pollutant (#PE1359) Environmental plant diseases (#PD2224).
Aggravated by Motor vehicle emissions (#PD0414).

♦ PE0755 Instability of trade in meat and meat preparations
Broader Instability of trade in food and live animals (#PD1434).
Narrower Instability of trade in dried salted or smoked meat (#PE8257)
Instability of trade in fresh, frozen or chilled meat (#PE8591).

♦ PE0756 Verbal sexual harassment of women in public
Broader Sexual harassment of women (#PF3271).
Aggravated by Profanity (#PF7427) Verbal abuse (#PD5238).

♦ PE0757 Requisitioning of workers to prevent strike action
Broader Violation of the right to strike (#PE5070).

♦ PE0760 Instability of trade in crude fertilizers and crude minerals, excluding coal, petroleum and precious stones
Broader Instability of trade in inedible crude non–fuel materials (#PD0280).
Narrower Instability of trade in sulphur (#PG6078)
Sand and gravel trade instability stone (#PE8612)
Instability of trade in crude fertilizers (#PE9049).
Related Shortage of pesticides (#PE4853).

♦ PE0763 Cerebral paralysis
Spastics — Cerebral palsy — Cerebral spastic infantile paralysis
Nature Cerebral palsy is the medical term covering a whole group pf neurological conditions. These conditions vary one from another as the clinical features of spasticity, athetosis, ataxia, tremor, rigidity or atony predominate, but having one feature in common – a disorder of motor control. All forms of cerebral palsy result from injury to or developmental anomaly or disease of the brain, which usually arises before or during, but sometimes after, birth. These disorders of motor control may be further complicated by associated mental, visual, auditory and speech defects, by emotional instability and by epilepsy. The physical handicap in any particular case may range from slight lack of control in one limb to complete physical helplessness. The physical disabilities, though usually the most obvious, are not necessarily the only, or even the most serious, problem. Any degree of mental ability or disability may be associated with any degree of physical handicap; conversely, severe physical disability does not necessarily indicate severe mental disability.
Refs Bishop, Beverly *Spasticity* (1977).
Broader Paralysis (#PD2632) Developmental disabilities (#PD3773)
Diseases of the central nervous system (#PE9037).
Narrower Paresis (#PG2947) Diplegia (#PG2942) Hemiplegia (#PG2943)
Tetraplegia (#PE2946) Little's disease (#PG2938) Infantile monoplegia (#PG2944)
Intracranial paralysis (#PG2940) Spastic infantile paraplegia (#PG2939).
Related Diseases of the spine (#PD2626).
Aggravates Paraplegia (#PE2945) Speech disorders (#PE2265)
Emotional disorders (#PD9159).
Aggravated by Meningitis (#PE2280) Encephalitis (#PE2348)
Diseases and injuries of the brain (#PD0992) Haemolytic disease of the new born (#PE2399)
Defective oxygen supply to the foetus (#PG9126)
Precipitate or ill–judged forceps delivery (#PG2953).

♦ PE0766 Concentration of power by transnational corporations
Economic imperialism of transnational enterprises
Nature The power concentrated in the hands of the transnational enterprises and their actual or potential use of it, their ability to shape demand patterns and values and to influence the lives of people and policies of governments, as well as their impact on the international division of labour, is dangerous. This is especially so, since there is no systematic process of monitoring their activities and discussing them in an appropriate forum.
While the international division of labour is influenced by the existing international trade and monetary regimes, it may be strongly affected, intentionally or unintentionally, by transnational corporations. Their large capabilities for moving products and inputs across borders are important instruments in affecting the actual division of labour. At the same time, the apprehension that host countries may be turned into 'branch–plant' economies may not be limited to developing countries. The organizational, productive and distributive complexes created by transnational corporations often assign a peripheral and dependent role to affiliates in many host countries, while the centres of top decision–making and scientific research remain in a few highly industrialized countries.
Although the locational pattern of transnational corporations reflects the uneven distribution of the factor endowments, it is in many cases also moulded by artificial administrative devices employed by home and host governments (tariffs, subsidies, etc), as well as by the corporations themselves. In today's complex economy, the 'invisible hand' of the market is far from the only force guiding economic decisions. To a considerable extent, conscious planning, both public and private, has played an increasing role in decision–making. Increasingly, basic decisions on the allocation of resources with respect to what, how and for whom to produce are being concentrated in corporate planning mechanisms. The growth of transnational corporations gives them increasing control over resources and thus augments their capacity to re–allocate them. Such decisions, when taken exclusively from the point of view of the interests of the enterprise pose serious problems.
Claim There is no indication, as is sometimes suggested, that transnational corporations are evolving into real internationalized entities whose ownership, management and objectives are truly global, and within which all nations and their citizens are treated equitably and world welfare is truly maximized.
Counter–claim The transnational corporations have developed distinct advantages which can be put to the service of world development. Their ability to tap financial, physical and human resources around the world and to combine them in economically feasible and commercially profitable activities, their capacity to develop new technology and skills and their productive and managerial ability to translate resources into specific outputs have proven to be outstanding. The importance of the foreign private sector to the development of developing countries has been recognized in UN International Development Strategy.
Broader Economic imperialism (#PC3198)
Excessive power and independence of transnational corporations (#PD5807).
Related Concentration of investment power (#PC5323).
Aggravates Imperialistic distribution system (#PD7374)
Non–accountability of transnational enterprises (#PF1072).

♦ PE0768 Punch drunk syndrome
Nature Boxing can cause chronic brain damage. In early stages it includes unsteadiness in gait and slight mental confusion, later, distinct leg dragging, hand tremors, general slowing of muscular movements, hesitant speech, nodding movements of the head, and finally, facial characteristics of Parkinson's syndrome, tremors, staggering gait and severe mental deterioration.
Incidence A boxer's fist travels at some 20 mph at the moment of impact, delivering a force of some 0.4 tons, or as a maximum roughly half a ton. Single blow or cumulative blows to head have caused at least 10 to 15 percent, according to other studies, 87 percent of professional boxers to display some or all these symptoms of cerebral dysfunction.
Broader Prizefighting (#PE8766).

♦ PE0769 Disabled migrant workers
Nature Workers disabled while employed abroad often find themselves in difficult situations resulting from differences in environment, inadequate knowledge of the local language, prejudice and discrimination, deficient vocational training, or inadequate living conditions. The special position of migrant workers in the country of employment exposes them and their families to health hazards and increased risk of occupational accidents. The situation of disabled migrant workers may be further aggravated when they need to return to the country of origin, where, in most cases, special services and facilities for the disabled are very limited.
Broader Disabled workers (#PD4673).

♦ PE0777 Inadequate relationship between international nongovernmental organizations and the specialized agencies of the United Nations
Broader Obstacles to effective international nongovernmental organizations (#PF7082)
Inadequate relationship between international governmental and nongovernmental organizations and programmes (#PE1973).
Aggravated by Lack of legal provision for international nongovernmental organizations (#PE0069).

♦ PE0778 Distortion of international trade by discriminatory requirements with respect to marks of product origin
Nature Compulsory requirements for labelling products with the country of origin create an unwarranted restriction on trade, partly because of the added costs of fulfilling the marking requirements of each importing country, and partly because the legibility requirement often significantly reduces the attractiveness of the product.
Broader Distortion of international trade by discriminatory customs and administrative entry procedures (#PE2603).

♦ PE0779 Class discrimination in education
Educational snobbery
Refs Barton, Len and Walker, Stephen *Race, Class and Education* (1983).
Broader Inequality in education (#PC3434).

♦ PE0780 Cooptation of private sector initiatives by government
Nature Small private enterprises wishing to attain greater levels of self–reliance and autonomy are permanently threatened by the cooptive strategies of the state, political parties and other institutions which operate according to a logic of power. Cooptation is critical in shaping the articulations between local organizations and global processes. It is achieved through the identification and political manipulation of social actors. This invariably leads not only to loss of their identity but also to actions that ultimately defeat their endogenous objectives. The organizations lose control over their own resources and their own destiny.

Incidence Especially in developing countries, economic micro-organizations and social movements are frequently neutralized by a political landscape dominated by pyramidal structures in which struggles for hegemony are constantly taking place.
Aggravates Front organizations (#PE4358).
Aggravated by Excessive government intervention in the private sector (#PD4800).

♦ PE0783 Unequal employment opportunities for disabled persons
Discrimination against the handicapped at work — Discrimination against the disabled in employment
Nature Many persons with disabilities are denied employment or given only menial and poorly remunerated jobs. This is true even though it can be demonstrated that with proper assessment, training and placement, the great majority of disabled persons can perform a large range of tasks in accordance with prevailing work norms. In time of unemployment and economic distress, disabled persons are usually the first to be discharged and the last to be hired. In some industrialized countries experiencing the effects of economic recession, the rate of unemployment among disabled job-seekers is double that of able-bodied applicants for jobs.
Programmes have been developed and measures taken in many countries to create jobs for disabled persons, including sheltered and production workshops, sheltered enclaves, designated positions, quota schemes, subsidies for employers who train and subsequently engage disabled workers, cooperatives of and for the disabled, etc; but the actual number of disabled workers employed in either regular or special establishments is far below the number of employable disabled workers. Wider application of ergonomic principles would lead to adaptation of the work place, tools, machinery and equipment at relatively little cost and thus help to widen employment opportunities for the disabled.
Many disabled persons, particularly in the developing countries, live in rural areas. When the family economy is based on agriculture or other rural occupations and when the traditional extended family exists, it may be possible for most disabled persons to be given some useful tasks to perform. As more families move from rural areas to urban centres, as agriculture becomes more mechanized and commercialized, as money transactions replace barter systems and as the institution of the extended family disintegrates, the vocational plight of disabled persons becomes more severe. For those living in urban slums, competition for employment is heavy, and other economically productive activity is scarce. Many disabled persons in such areas suffer from enforced inactivity and become dependent; others must resort to begging.
Refs ILO *Adaptation of New Jobs and the Employment of the Disabled* (1984); ILO *Employment of Disabled Persons. Manual on Selective Placement* (1984).
Broader Discrimination against the disabled (#PD9757)
Unequal opportunities for disabled persons (#PE0706).
Related Denial of right to free choice of work (#PE3963).
Aggravated by Human disability (#PC0699).

♦ PE0785 Rubella
German measles
Nature The rubella virus usually infects children and is of limited danger to them. However, when it infects a pregnant woman, there is a very high probability that the development of the foetus will be retarded or defective, or that the mother will abort.
Refs Van Dijk, J *Rubella Handicapped Children* (1982).
Broader Virus diseases (#PD0594) Infectious and parasitic diseases (#PD0982).
Related Diseases of the lymphatic system (#PD2654).
Aggravates Deafness (#PD0659) Glaucoma (#PE2264)
Heart diseases (#PD0448) Natural human abortion (#PD0173)
Diseases of metabolism (#PC2270) Psychomotor retardation (#PG2961)
Foetal infection and death (#PE2041) Perinatal morbidity and mortality (#PD2387)
Diseases of blood and blood-forming organs (#PF8026).

♦ PE0789 Domestic bias in the regulation of restrictive business practices
Nature Double standards in national policies, namely, on the one hand, that restrictive business practices affecting the domestic market should be strictly controlled and in many cases prohibited, since they impede economic growth and efficiency, while, on the other hand, ignoring or fostering the use of such practices when they have effects outside the domestic economy, notably in respect of exports, have increasingly been used as a government commercial policy device to regulate world trade. These double standards have enabled governments of a number of developed market-economy countries to resort to growing protectionism via the back door. Instead of raising tariffs or introducing non-tariff barriers to regulate import competition, industries and governments in certain importing developed market-economy countries have brought pressure on industries and governments in exporting countries to limit (and fix) prices for exports, leading to the establishment or strengthening of export cartel arrangements, in order to implement so-called "voluntary export restraints" or orderly marketing arrangements agreed between the parties concerned.
Broader Restrictive trade practices (#PC0073).

♦ PE0793 Aggressive honey bees
Killer bees
Nature The African honey bee is very aggressive. It has the proven ability to subvert domesticated honey-bee colonies and frequently launches vicious and unprovoked attacks on people and livestock.
Incidence Despite incomplete reporting, deaths from bee stings in Brazil, for example, are between 300 to 400 per year, as compared to 100 in the USA, towards which the bee population is spreading. The problem originated in Brazil, where queens were imported from South Africa in 1956 on the theory that the bee families they generated would be more productive in tropical climates that those of the European species. In 1957, 26 African swarms escaped. The Africanized bee reproduces faster than European varieties, and the worker bees fly longer distances in search of nectar. After only a few years they were flourishing, with up to 100 swarms per square kilometre. From the Brazilian interior, the bees spread southward into Paraguay and Argentina and thousands of miles across the Amazon basin. By 1979, they were in Venezuela and Colombia. They are well established in Costa Rica, where they arrived from Panama in 1982. Swarms are advancing up the Atlantic and Pacific sides of the Central American isthmus at up to 400 miles (650 kilometers) a year and are heading toward Mexico and the United States.
This spreading of the African bee has had catastrophic effects on commercial beekeeping. In Venezuela, production has dropped 80 percent. If the Africanized honey bee spreads to all 11 US States that have 240 frost-free days a year, direct damage to the beekeeping industry could be more than $50 million a year. The new bee threatens annual sales of $150 million in honey, wax, pollen and other products.
Broader Insect pests (#PC1630)
Environmental hazards of new species introduction (#PC1617).
Related Bee pests (#PG5249).

♦ PE0802 Long-term shortage of manufactured goods
Shortage of industrial goods
Broader Long-term shortage of commodities (#PC1195)
Insufficient availability of goods (#PB8891).
Narrower Paper shortage (#PE1616)
Long-term shortage of wood, lumber and cork (#PE1372)
Long-term shortage of non-ferrous metal ores (#PE0824)
Shortage of textile, yarn fabrics, made-up articles (#PE8436)
Long-term shortage of metalliferous ores and metal scrap (#PE0353)
Shortage of leather, miscellaneous leather manufactures and dressed fur skins (#PE9130)
Long-term shortage of crude fertilizers and crude minerals, excluding coal, petroleum and precious stones (#PE1353).
Aggravated by Excessive consumption of goods and services (#PC2518).

♦ PE0803 Economic barriers to access to the legal profession, the judiciary and jury membership
Nature Access to the legal profession is limited by the cost of the necessary training, and also by the fact that in some countries the payment of a fee, either annually or on admission to the profession, to the appropriate professional association is required. To the degree that the judiciary is recruited from the legal profession, economic obstacles to an individual's becoming a lawyer also affect his hopes of becoming a judge.
Property or tax-paying qualifications are sometimes a prerequisite for membership of the judiciary. De facto economic barriers to access to the judiciary also exist; in particular, lack of finance limits access to the requisite educational processes. Property, tax-paying or salary qualifications are also often a prerequisite for jury service. As a result of these qualifications for being a juror, jury service is largely confined to the middle and upper classes while the poorer classes are often not represented.
Broader Restrictive legal practices (#PD8614)
Economic discrimination in the administration of justice (#PE1399).
Aggravates Lack of impartiality of the judiciary (#PE7665).

♦ PE0805 Kidnapping of pet animals
Ransoming of pet animals — Theft of pets
Incidence Estimated two million dogs and cats are stolen each year in the USA and supplied to animal research institutes for experimentation.
Broader Kidnapping (#PD8744) Unethical practices with domesticated animals (#PE4771).
Aggravates Maltreatment of pet animals (#PD1265).
Aggravated by Disproportionate allocation of resources to pet animals (#PF1085).

♦ PE0806 Instability of trade in manufactured fertilizers
Broader Instability of chemicals trade (#PD0619).
Narrower Potassic fertilizers trade instability (#PU2981)
Phosphatic fertilizers trade instability (#PU2980)
Nitrogenous fertilizers trade instability (#PU2979).
Related Shortage of pesticides (#PE4853).

♦ PE0810 Professional discrimination between educators
Nature Hierarchies, whether official or unofficial (depending on the society in question), are apparent in the structures of education. The terms primary, secondary, higher, professional, technical, or scientific are loaded with discriminatory overtones. The teaching profession is structured in terms of prestige; lowest status being accorded to the primary school teacher and the highest status to the university lecturer.
Broader Discriminatory professionalism (#PC2178).
Aggravated by Social injustice (#PC0797).

♦ PE0811 Non-payment of compensation to victims of malpractice
Delay in payment of compensation to victims of malpractice
Broader Uncompensated damages (#PD7179).

♦ PE0812 Errant nationals
Aggravates Terrorism (#PD5574) Espionage (#PC2140)
Tax evasion (#PD1466) Mercenary troops (#PD2592)
International crime rings (#PG5512).
Aggravated by Restrictions on nationals returning to their country (#PE8817).
Reduced by Restrictions on nationals leaving their own country (#PE8414).

♦ PE0814 Instability of trade in miscellaneous manufactured articles
Broader Instability of manufacturing industries (#PC0580)
Instability of the primary commodities trade (#PC0463).
Narrower Instability in trade of travel goods (#PE8920)
Instability of textile and clothing industries (#PE1008)
Photographical and optical goods trade instability (#PE8470)
Scientific and controlling instruments trade instability professional (#PE8169)
Instability of trade in sanitary plumbing, heating, lighting fixtures and fittings (#PE8272).

♦ PE0818 Heterosexism
Nature People who prefer to have sexual experiences with the sex other than the one to which they belong tend to project this as the norm for all people. There is a propensity to exploit those whose sexual experiences are preferably with members of their own sex.
Broader Cultural fragmentation (#PF0536) Plural society tensions (#PF2448)
Inter-cultural misunderstanding (#PF3340).
Related Traditionalism (#PF2676) Lack of racial identity (#PF0684).
Aggravates Threatened sects (#PC1995) Cultural barriers (#PB2331)
Ethnic disintegration (#PC3291) Inflexible social structure (#PB1997)
Inadequate ideological frameworks (#PD0065).

♦ PE0824 Long-term shortage of non-ferrous metal ores
Broader Long-term shortage of manufactured goods (#PE0802)
Long-term shortage of metalliferous ores and metal scrap (#PE0353).
Narrower Instability of tin trade (#PG1422) Instability of lead trade (#PG1402)
Instability of zinc trade (#PG2010) Instability of copper trade (#PG2472)
Instability of nickel trade (#PG2532).

♦ PE0828 Environmental hazards from undressed hides, skins and fur skins
Broader Environmental hazards from manufactured goods (#PE1344)
Environmental hazards from inedible crude non-fuel materials (#PE0546).

♦ PE0835 Decline in communal spirit and village solidarity in developing countries
Nature Local industrialization gives an added impetus to the commercialization of services which were formerly carried out within the subsistence economy as social duties. Customs of mutual aid and communal cooperation, even among kinfolk, tend to be undermined by the penetration of the exchange system upon which development of secondary industry necessarily depends. With increased competition and individualism, a decline in communal spirit and village solidarity seems on occasion to have been part of the social price paid for expansion of industry.
Counter-claim In the urban environment, manufacturing industry may often be a stabilizing influence, creating: regular employment in place of the irregularities of tribal agriculture; new loyalties to new communities; new skills; and a new pride in work. It may make possible the building up of social amenities such as schools, libraries and cinemas, which are an integrating force.

PE0835

Broader Apathy (#PA2360) Obstacles to community achievement (#PF7118).
Related Unorganized development of work forces (#PF2128)
Decline in rural customs and traditions in developing countries (#PD1095).
Aggravated by Uncontrolled industrialization (#PB1845).

♦ PE0849 Environmental hazards from beverages
Broader Environmental hazards from economic and industrial products (#PC0328).
Narrower Instability of alcoholic beverages trade (#PE2585)
Environmental hazards of non-alcoholic beverages (#PE3035).

♦ PE0853 Inadequate negotiation of entrance terms for transnational corporations in developing countries
Nature The terms on which transnational corporations gain entry into a host country are obviously a matter of considerable importance. Influenced by the once widely view held that developing countries should open their doors to foreign capital to enhance their development, many of these terms were not sufficiently carefully negotiated.
Many developing countries have felt that their bargaining position in dealing with multinational corporations is weak. There has been the assumption that transnational corporations, with the exception of certain resource-based industries, can choose their location for production according to the country offering them the most attractive environment and most favourable terms.
The initial agreement concluded with transnational corporations thus tends to include a large number of special concessions. Later, as circumstances change, the concessions appear to be too onerous and the host country may deem it necessary to redress the situation. In such cases foreign affiliates could be treated in a discriminatory fashion or could even be expropriated. Such treatment, though it may be directed towards particular transnational corporations, inevitably creates an atmosphere of mistrust which operates against the long-term interests of both host countries and corporations. Moreover, concern about future unfavourable treatment may lead transnational corporations to attempt to extract the most out of their investment in the least possible time. These, and other uncertainties, make transnational corporations reluctant to invest in some developing countries unless their prospects are distinctly more attractive than those expected in developed countries.
Broader Transnational corporation imperialism (#PD5891)
Restrictive business practices of transnational corporations (#PE5915).
Aggravates Underdevelopment (#PB0206)
Inadequate investment of transnational corporations in least developed countries (#PE7892).
Aggravated by Inadequate negotiating capacity of developing countries (#PE9646).

♦ PE0857 Smut diseases of plants
Nature Fungi causing smuts are plant parasites and the disease is generally characterized by black, dusty masses of spores. The fungi are very specific in their selection of hosts and host organs: they attack only the stems, flowers, anthers or ovules of their host and no other part or other host.
Broader Basidiomycetes (#PE0364).
Narrower Smuts of grasses and cereals (#PE2234).

♦ PE0861 Instability of trade in fixed vegetable oils and fats
Broader Instability of trade in animal and vegetable oils and fats (#PE0735).
Narrower Instability of olive oil trade (#PE0313)
Rape and mustard oils trade instability (#PE8159).

♦ PE0864 Environmental hazards from woodworking industries
Nature There are numerous hazards inherent to the woodworking industries. Among them are: exposure (either directly by contact or indirectly by splashes or mists) to the highly toxic chemicals used to treat wood against insect and mould attack; contraction of parasitic diseases borne by wood; accidents with the slicing, peeling, chipping, or disintegrating machines employed in the industry; exposure to adhesives (especially of the formaldehyde family) which are used in the bonding of man-made panels and which could cause skin disease or systemic intoxication; and the ever-present hazard of fire and explosion due to the highly flammable nature of wood (especially in the form of dust or shavings) and of the various other products used (solvents, adhesives, paints, varnishes, and lacquers).
Broader Environmental hazards from manufacturing industries (#PD0454).
Narrower Environmental hazards of furniture and fixtures manufacture (#PE8165).

♦ PE0873 Rheumatic diseases
Arthritis
Nature Rheumatic diseases are disorders of the musculoskeletal system not due to trauma or infection. Such disorders as rheumatic fever, rheumatoid arthritis, spondylitis, degenerative joint disease and gout are, by general consent, included in this category, as well as a group of frequent muscular and synovial painful conditions classified as fibrositis or muscular rheumatism.
Rheumatic diseases have widespread social and economic consequences in all societies. This varied group of diseases includes at least 100 different conditions, and constitutes a significant public health problem in countries of North America and Europe, in Japan, and in some urban centers of South America and Australia. As far as two-thirds of the world population is concerned, concern of public health authorities for rheumatic diseases is rarely more than marginal and is frequently nonexistent. There is increasing evidence that rheumatoid arthritis and related diseases are a major cause of disability and incapacity for work. Genetic predisposition and a variety of environmental factors may be involved in their aetiology.
The incidence of rheumatoid arthritis increases with age. Women are affected more frequently than men. It is a remittent disease, but each exacerbation adds to the degree of irreversible damage in the joints, so that the prevalence of the arthritis and the resulting crippling disability rises rather steeply with age. Perhaps the most fundamental difficulty with regard to rheumatic diseases today is that the problem is insufficiently appreciated and understood.
Refs Ansell, Barbara M *Rheumatic Disorders in Children* (1980); Dick, W C *Immunological Aspects of Rheumotology* (1981); Gerber, Lynn (Ed) *Psoriatic Arthritis* (1985); Gibson, Terry *Rheumatic Diseases* (1986); Kovarsky, Joel *Arthritis* (1987).
Broader Diseases of the spine (#PD2626) Human disease and disability (#PB1044)
Diseases of connective tissue (#PD2565).
Narrower Lumbago (#PE1310) Polymyositis (#PE5509) Dermatomyositis (#PE5086)
Tumours of joints (#PG4619) Traumatic arthritis (#PG3057) Neurogenic arthritis (#PG3058)
Rheumatoid arthritis (#PE5081) Occupational rheumatism (#PE0502)
Nonarticular rheumatism (#PG3060) Arthritis accompanying psoriasis (#PG3054)
Arthritis accompanying ulcerative colitis (#PG3055)
Arthritis accompanying foreign protein reactions (#PG3056).
Related Diseases of metabolism (#PC2270) Disorders of joints and ligaments (#PD2283).
Aggravates Headache (#PE1974) Absenteeism (#PD1634)
Heart diseases (#PD0448) Kidney disorders (#PE2053)
Diseases of the spleen (#PE6155) Unethical consumption practices (#PD2625)
Age discrimination in employment (#PD2318) Diseases and injuries of the brain (#PD0992).
Aggravated by Allergy (#PE1017) Gonorrhoea (#PE1717)
Rheumatic fever (#PE0920) Injurious accidents (#PB0731)
Skeletal system disorders (#PE2298)
Haemolytic streptococcal infection of the throat (#PG3063).

♦ PE0875 Health risks to workers in service industries
Broader Occupational risk to health (#PC0865).
Aggravates Industrial accidents (#PC0646) Occupational diseases (#PD0215).

♦ PE0882 Instability of trade in manufactured goods
Broader Instability of the primary commodities trade (#PC0463).
Narrower Instability of trade in dressed fur skins (#PE8740)
Instability of trade in non-ferrous metals (#PE1406)
Instability of paper and printing industries (#PE1927)
Instability of trade in wood, lumber and cork (#PE2521)
Iron and steel manufactures trade instability (#PE8969)
Instability of machinery and equipment industries (#PE1852)
Leather and leather manufactures trade instability (#PE8320)
Instability of non-metallic mineral products industry (#PE2599)
Instability of trade in crude synthetic and reclaimed rubber (#PE0701)
Instability of trade in textile yarn fabrics, made-up articles (#PE8970).
Related Instability of manufacturing industries (#PC0580).

♦ PE0885 Depression due to torture
Depression in torture victims
Nature Torture victims suffer from a variety of psychological symptoms. In a recent study about 20 percent suffered from depression. They maybe lethargic, or listless, or may have a low self-estimation and have fits of crying and low moods.
Refs Ochberg, Frank M and Hamburg, David A *Post-Traumatic Therapy and Victims of Violence* (1988).
Broader Mental depression (#PC0799) Neurological effects of torture (#PD4755).

♦ PE0886 Wood-boring beetles
Woodworm
Incidence In the UK, up to 50 per cent of the homes are' believed to be infested with woodworm. In the case of the common furniture beetle (Anobium punctatum), which causes up to 80 per cent of the damage, it lays eggs on unplaned joists and floorboards. The grub bores for three years and then pupates and flies away. The many other kinds of wood-boring beetle cause the remaining 20 per cent of the damage.
Broader Household pests (#PD3522) Beetles as insect pests (#PE1679).
Aggravates Structural failure (#PD1230).

♦ PE0890 Environmental hazards from non-metallic mineral products industries
Nature Non-metallic mineral products industries provide natural and prepared rock products for construction use. Rock and many other mineral products are quarried, while some non-metallic minerals may be sub-surface mined. The impact of these operations affects the natural environment (land, atmosphere, water) and also the human environment (physical, social and economic).
Non-metallic and metallic mineral mining together probably disturb some 500,000 hectares of land per year world-wide. Land impact includes excavation, waste dumping and water table interference or pollution, with consequent damage to plants, and the ecological niches for animals. Air pollution is caused by dust, sulphur oxides, nitrogen oxides and hydrogen sulphide. Sulphur oxides cause acid rain. The human environment may be contaminated by proximity to some mines, mills and smelters; settlements may be moved as mines expand; and the natural landscape may be destroyed.
Broader Environmental hazards from manufactured goods (#PE1344)
Environmental hazards from manufacturing industries (#PD0454)
Environmental hazards from inedible crude non-fuel materials (#PE0546).
Narrower Environmental hazards of glass and glass products manufacture (#PE8473)
Environmental hazards from china, pottery and earthenware industries (#PE9019).

♦ PE0891 Exploitation of animals in spectator sports
Gambling on the performance of animals
Incidence Animals play a central role in three major spectator sports which involve gambling: horse-racing, dog-racing and, to a lesser extent, cock-fighting. Because of the large amounts of money staked on the outcome of such events, the animals are subject to many abuses including drugging, biological manipulation and torture. For example, greyhounds, although gentle by nature, are often trained to become aggressive through systematic starvation. It has been estimated that 90 per cent of greyhound trainers use live lures (such as rabbits) to stimulate the dogs to higher performance, resulting in some 100,000 small animals being torn apart annually. When they are no longer of value for racing they may be sold to research laboratories (an estimated 1,500 to 2,000 per year in Massachusetts alone). In the process of racing or fighting, many animals suffer severe injuries and have to be destroyed.
Broader Exploitation of animals for amusement (#PD2078)
Gambling on sports and athletic competitions (#PE4576).
Narrower Bullfighting (#PG4834) Animal fighting sports (#PE4893).

♦ PE0893 Disrelationship between production and work force needs
Nature The inability of production staff supervisors to build an effective work force in the face of the development of instruments of production with capacities beyond the wildest of dreams and the frustratingly high national and international rates of unemployment in spite of unbelievable human need across the world points to the disrelationship between production capabilities and work force needs. The shear number of types of specialization and its increasingly narrowness generates ambiguity about one's work role and one's role in the future of the planet. At the time of the early labour movements and the communist revolutions workers saw that they were in the vanguard of social creativity. Being a worker was caring for the planet. This is largely gone leaving a gap between working and living.
Broader Over-specialized supervisory personnel (#PF3588).

♦ PE0895 Restrictions on export activity due to licensing arrangements
Nature A considerable number and variety of restrictive provisions can be and are included in licensing agreements and contracts. Such restrictions can be classified into two categories: restrictive provisions on exports; and restrictive provisions affecting the export potential and activity of a firm in the developing country concerned. The restrictions that may be placed on exports are: global bans; prohibition of exports to specified countries; permission to export only to specified countries; the requirement of prior approval from the licensor for exports; export quotas; price controls on exports; the restriction of exports to specified products; permission to export only to, or through, specified firms; and prohibition of exports of substitute products.
The use of export restrictions enables the licensor to regulate the competitive impact of the licensee's activities upon his own interests in other markets. The different types of export restrictions vary in intensity and can be used individually or in combination with one another. The most restrictive is the global export prohibition. In such cases, the economic activity of the licensee is limited to his domestic market; and frequently the licensee is also prohibited from selling the products covered by the licence to any third party who would export them. The requirement that the licensee obtain the licensor's prior approval to export is less restrictive, in that the possibility of exporting is not ruled out; but the licensor retains control over exports, and reserves the option

of a global ban. As an alternative, approval for exports to specified countries may be provided for in the licence agreement. The extent of the restrictive nature of this provision will depend largely on the relative bargaining powers of the licensor, the licensee, and the government of the licensee's country.
Not all export restrictions are territorial constraints. A licensor can place a ceiling on the licensee's exports by means of an export quota, expressed in physical or monetary terms. This can either be coupled with a territorial limitation or used by itself. In the latter case, the competitive impact of the licensee's export activity is limited rather than directed. In addition, a licensor can use, singly or in connection with some form of territorial constraint, a provision restricting exports to specified products or specified product forms, and a provision prohibiting the exportation of products that are similar to, or substitutes for, the licensed goods. The latter restriction is frequently related to prohibition of the use of a trademark in export sales. Finally, the licensor can attempt to use the licence contract for the purpose of retaining control over the export (as well as of the domestic) price of the licensed product.
Other potential restrictive provisions deal with the tying of purchases of essential inputs to the licensor; restrictions on production patterns; payments of a minimum royalty; restrictions on the use of the acquired technology after the termination of the contract; restrictions on disclosure of the technology used; agreements not to challenge the validity of patents and trademarks transferred; and the mutual exchange, or unilateral grant-back, of technical improvements.
Incidence Restrictions contained in licensing agreements are the most common form of restrictive business practice to which developing countries are subjected. The extent to which provisions are restrictive varies from case to case.
Broader Restrictive trade practices (#PC0073).

♦ PE0896 Alienation of land through acquisition by foreigners
Nature In countries with areas of great natural beauty, beaches, a pleasant climate, or a pleasant local culture, land or buildings may be purchased by foreigners. These acquisitions may then be exploited in a manner which prevents residents of the country from deriving benefit from them.
Incidence Examples include: purchase by US nationals of land on the shores of Lake Muskoka in Canada, denying access of locals to the lake; purchase of high country land in New Zealand by US nationals, together with fishing and shooting rights. Some 3 per cent of Swiss real estate is owned by foreigners.
Broader Land misuse (#PD8142) Shortage of cultivable land (#PC0219)
Lack of international cooperation (#PF0817).
Related Social environmental degradation from recreation and tourism (#PD0826)
Natural environmental degradation from recreation and tourism (#PE6920).
Aggravated by Unethical commercial practices (#PC2563).

♦ PE0897 Unjustified restrictions on the free movement of commercial vehicles
Road traffic restrictions
Nature There are general national or regional traffic prohibitions on Sundays and on public holidays, during summer holiday period in some countries, additional restrictions of limited duration are also enforced at other times of the year on certain high density road, local traffic prohibitions restrict night-time driving or transit through certain towns, and special prohibitions are applied to the transportation of dangerous goods and outside loads.
Incidence Annual cost of these restrictions has been estimated in the UK at over £200 million.
Aggravates Road and highway traffic congestion (#PD2106).
Reduces Intrusive truck traffic on residential streets (#PU9137).

♦ PE0904 Endangered species of llamas
Endangered species of vicuña
Refs FAO *Vicuña Conservation Legislation* (1971).
Broader Endangered species of mammals (#PC1326).

♦ PE0906 Maltreatment of marine show animals
Maltreatment of performing dolphins
Incidence Marine shows and dolphinaria where dolphins, whales, seals and other marine animals perform for paying customers are becoming an increasingly important industry.
Claim In addition to the constraints of captivity, the initial training of these animals often involves deprivation and suffering. The idea that dolphins and whales benefit from being in the presence of humans is to be challenged as species-ist and chauvinistic.
Broader Maltreatment of animals in aquaria (#PE5461)
Maltreatment of performing animals (#PE4810).

♦ PE0912 Abuse of khat
Catha edulis — Abuse of qat
Nature The euphoric or pleasure-giving use of khat (qat) steadily increasing. Some authorities consider that its continuous consumption does much harm from physiological, psychological and social points of view - particularly the latter, since the product is expensive and the habitual consumer spends a large part of his income on it, with consequent neglect of his family and job. Effects of khat are similar to those produced by the amphetamine group but milder. Physical dependence and tolerance do not occur. Recognition of this problem has arisen only in recent years. Khat is not covered by the schedules annexed to international conventions on narcotic drugs or by the Convention on Psychotropic substances.
Background Khat is a tree which grows in East Africa and Arabia whose leaves and twigs contain active chemical compounds which provoke excitement and general euphoria. It is used for magical and euphoric purposes. Depending on the degree of freshness the leaves and twigs are either chewed or drunk in an infusion, sometimes known as 'Abyssinian tea'. A study of khat and its properties has been completed by the World Health Organization at the request of the UN Economic and Social Council, which has drawn attention of the governments concerned to the report for any action they may consider necessary.
Incidence Some urban Yemenis spend up to one third of their income on khat.
Broader Substance abuse (#PC5536) Abuse of plant drugs (#PD0022).
Aggravates Magic (#PF3311).
Aggravated by Tribalism (#PC1910) Superstition (#PA0430).

♦ PE0915 Instability of trade in coffee, tea, cocoa, spices and manufactures thereof
Broader Instability of trade in food and live animals (#PD1434).
Narrower Instability of tea trade (#PE2054) Instability of cocoa trade (#PE1549)
Instability of coffee trade (#PE0950) Instability of spices trade (#PE1619)
Instability of trade in tea and mate (#PG1706).

♦ PE0916 Sooty mould in plants
Nature Sooty mould is a black coating on surface of leaves or fruit caused by fungi living on insect exudate.
Broader Ascomycetes (#PE4586) Deuteromycetes (#PE4346).

♦ PE0920 Rheumatic fever
Rheumatic heart disease
Incidence The peak incidence of rheumatic fever and rheumatic heart disease is in children between 5 and 15 years of age, and is most prevalent in developing countries. In India alone, it is estimated that over six million children are afflicted. General medical treatment is only symptomatic or supportive and the dramatic advances in cardiac surgery have not greatly reduced morbidity and mortality. At best, only a small proportion of patients can have these expensive operations and few developing countries can afford to provide them.
Refs Markowitz, Milton and Gordis, Leon *Rheumatic Fever.*
Narrower Rheumatic pericarditis (#PG4659) Valvular diseases heart (#PG2363).
Aggravates Fever (#PD2255) Chorea (#PG3096) Heart diseases (#PD0448)
Rheumatic diseases (#PE0873) Emotional disorders (#PD9159)
Diseases of the arteries (#PE2684) Disorders of joints and ligaments (#PD2283).
Aggravated by Tonsillitis (#PE2292) Poor living conditions (#PD9156)
Streptococcus infection (#PE3098) Diseases of connective tissue (#PD2565).

♦ PE0921 Schistosomiasis
Bilharziasis — Blood flukes
Nature Schistosomiasis is a bloodworm infection maintained by poor agricultural practices and unsanitary habits. Adult worms of Schistosoma japonicum live in the veins of human intestines, where they lay vast numbers of eggs that work their way into the intestinal tract and are eventually passed out in the faeces. In many areas of the world, people defaecate in rivers and irrigation canals. Once in the water, each egg hatches into a small worm called miracidium which penetrates the body of a snail, maturing and multiplying within it over several weeks, and then producing new worm forms called cerceriae, which emerge from the snail into the water and penetrate human skin, eventually passing through the heart, lungs, and liver before ending up in the veins of the intestines.
People with the worm infections develop ballooning stomachs caused by water retention, along with large livers and spleens. The infection causes a variety of intestinal and liver complications which weaken the sufferer and eventually may cause severe disability and death. Although the direct mortality from infection is low, the importance of the disease lies in the sheer size of the epidemiological phenomenon. As human population grows, the problem intensifies on two counts - increased excretal pollution of water, and the extended use of irrigation land for crop-growing - while the creation of man-made lakes has encouraged the spread of the disease.
Incidence Schistosomiasis is endemic in some 71 countries, where some 200 million people are infected. It is estimated that over 600 million people are exposed to the risk of infection. Intestinal helminthic and protozoal infections affect many millions of people in Africa, Asia, and Latin America. Development projects have increased the incidence of schistosomiasis owing to drainage and irrigation canals providing a habitat for the snails. Large-scale water impoundments, artificial man-made lakes, and population habits, are all factors favouring the transmission or recrudescence of this disease.
Refs Ansari, N (Ed) *Epidemiology and Control of Schistosomiasis* (1973); Hoffman, Donald B and Warren, Kenneth S *Schistosomiasis IV* (1978).
Broader Helminthiasis (#PE6278) Environmental human diseases (#PD5669)
Uncontrolled tropical diseases (#PD4775)
Parasites of the cardiovascular system in animals (#PE9427).
Narrower Haemic schistosomiasis (#PG3104) Vesical schistosomiasis (#PG3100)
Oriental schistosomiasis (#PG3102) Intestinal schistosomiasis (#PG3101).
Aggravates Absenteeism (#PD1634) Reduced resistance to disease (#PG3105)
Child mortality in developing countries (#PE5166).
Aggravated by Natural disasters (#PB1151) Vector-borne diseases (#PD8385)
Problems of migrant labour (#PC0180) Parasites of the human body (#PE0596)
Inadequate irrigation system (#PD8839) Snail vectors of animal diseases (#PE2747)
Disrupted mechanisms for community health (#PF2971)
Increase in pests and diseases through perennial irrigation (#PE7832).

♦ PE0924 Brucellosis
Brucellosis in cattle — Contagious abortion — Bang's disease — Infectious abortion in ruminants — Epididymitis of rams
Nature Brucellosis is a highly contagious infection of many animals which can be transmitted to man. Infection often begins without clinical signs and subsequently localizes in lymph nodes, spleen, reproductive organs, tendon sheaths, joints and other organs, where it persists for long periods. It affects mainly goats, cattle and swine, but is also found in sheep, horses, dogs, camels and other domestic animals, and in a large variety of wild animals, notably bison, elk, caribou and reindeer. Other groups of animals have been found to have had brucellosis but not to epidemic proportions. These include domestic poultry and wild birds, hares and rabbits, and other wildlife such as the American desert wood rat. These groups play a role as carriers of the disease, and, as in the case of hares in Denmark (where the disease had been eradicated) provide a threat of reinfection.
Brucellosis affects many different organs in animals and signs of the disease are influenced by the nature and extent of the infection and the species involved. These may be abscesses (horses, cattle and reindeer), lameness and swelling of joints, chronic infections of the bones and joints in livestock and reindeer. In ruminants, brucellosis commonly induces abortions in the latter half of gestation and may result in infertility. Important economic losses arise from aborted calves, lambs, etc and through subsequent slaughter of livestock affected by the disease. In man it may cause a variety of symptoms from mild headache to crippling joint pains, incapacity or general apathy. In particular it affects people whose occupations bring them in contact with infected animals. The disease can also be contracted through unpasteurized milk and milk products, and improperly cooked animal products.
Incidence Reported throughout the world, this is one of the diseases being explored because of its value in biological warfare.
Refs Madkour, Monir M (Ed) *Brucellosis* (1989); Nicoletti, Paul *Diagnosis and Vaccination for the Control of Brucellosis in the Near East* (1982); Rementsova, M M *Brucellosis in Wild Animals* (1987); WHO *Joint FAO/WHO Expert Committee on Brucellosis* (1986).
Broader Zoonotic bacterial diseases (#PD6363) Infectious and parasitic diseases (#PD0982)
Reproductive system diseases in animals (#PD7799).
Related Cattle diseases (#PD0752) Occupational diseases (#PD0215)
Bacterial diseases in animals (#PD2731).
Aggravates Encephalitis (#PE2348)
Economic loss through slaughter of diseased animals (#PE8109)
Economic loss through reduced productivity of diseased animals (#PE8098).
Aggravated by Epidemics (#PC2514) Rodent vectors of disease (#PE3629)
Lack of slaughter facilities (#PE9767)
Wild birds as vectors of animal diseases (#PE2749)
Wild animals as carriers of animal diseases (#PD2729)
Infected animal, meat and animal product shipments (#PE7064)
Health risks to workers in agricultural and livestock production (#PE0524).
Reduced by Inadequate carcass disposal of diseased animals (#PE2778).

♦ PE0930 Road traffic violations
Road traffic offences — Driving law violations
Refs Viski, L *Road Traffic Offenders and Crime Policy* (1982).

Broader Criminally life endangering behaviour (#PD0437).
Narrower Drunken driving (#PE2149) Driving delinquency (#PE6119)
Excessive speed of motor vehicles (#PE2147).
Aggravated by Inadequate traffic control (#PE8266).

♦ PE0938 Excessive dependence on export credits by developing countries
Nature While commercial credits play an important and useful role in world trade, excessive reliance on these credits may lead to debt-service crises and has, in some cases, already done so. There are many instances in which developing countries have accepted suppliers' credits which were characterized not only by very short maturities and high nominal interest charges but also by substantial over-pricing of the goods supplied, so that extremely high effective interest charges were in fact incurred. It has been suggested, on the one hand, that developing countries ought not to accept suppliers' credits on such terms (notwithstanding the heavy pressures to do so where official aid is not forthcoming in sufficient quantity); and on the other hand, that the developed countries cannot legitimately escape responsibility for preventing the worst excesses involved in the system of commercial credits, especially since the primary object of such credits is not to aid developing countries but to encourage the marketing of exports.
Broader Economic and social underdevelopment (#PB0539)
Vulnerability of developing countries (#PC6189).
Aggravates Excessive foreign public debt of developing countries (#PD2133).

♦ PE0948 Non-repatriation of prisoners of war
Nature Following a cease-fire, and especially in the absence of a peace agreement, prisoners of war from the opposing forces may be held in camps with no prospect of release (in explicit contravention of the Geneva Convention). Soldiers are kept apart from their families, whom they may not have seen since the war began. The prisoners effectively become hostages held against the day when peace negotiations commence. Furthermore, unless the prisoners are registered with the Red Cross (which may be obstructed), there is no way in which their families can no that they are still alive.
Incidence As an example, a year after the Gulf War ended in 1989, from 69,000 to 109,000 Iranian and Iraqi prisoners of war remained in camps with no prospect of release.
Broader Prisoners of war (#PC8848).
Related Political hostage-taking (#PD1886).
Aggravated by Persistence of a technical state of war following cease-fire agreements (#PE2324).

♦ PE0950 Instability of coffee trade
Nature Coffee has experienced wide price fluctuations. There is excess production capacity which tends to reduce prices progressively to the special disadvantage of the many countries highly dependent on coffee trade for their export earnings. Such countries have attempted to group together at various times to introduce a country-by-country export quotas and to restrict the total amount of coffee traded and maintain higher prices. Such agreements are often by-passed by black market trading as a result of pressure from importers. The situation is complicated by the influence of the related market for soluble coffee which may or may not be manufactured in the producing country prior to export.
Incidence Over 20 million people in 41 producing countries derive their income from the growing, processing, transportation and export of coffee. 15 developing countries earn from 30 to 80 per cent of their foreign exchange from the export of coffee.
Broader Instability of trade in coffee, tea, cocoa, spices and manufactures thereof (#PE0915).
Reduced by Bad weather (#PC0293).

♦ PE0952 Unnecessary health system referrals
Fraudulent medical referrals
Broader Unethical medical practice (#PD5770).
Aggravates Inadequate health services (#PD4790).

♦ PE0958 Gastrointestinal infections in animals due to parasites
Broader Gastrointestinal parasites of animals (#PE2436).
Narrower Hookworm (#PE3508) Strongyloidosis (#PG6574)
Canine trichuriasis (#PG2794) Stomach worm infections of swine (#PG4281)
Small-strongyle infection of horses (#PG6315) Large-strongyle infections of horses (#PG6104)
Intestinal trichostrongylosis in sheep (#PG5977).
Related Ascariasis (#PE2395) Trematode diseases (#PE6461)
Tapeworm as a parasite (#PE3511).

♦ PE0961 Instability of trade in fruit and vegetables and preparations thereof
Broader Instability of trade in food and live animals (#PD1434).
Narrower Instability of soybean trade (#PJ1686) Instability of vegetable trade (#PE1711)
Instability of trade in preserved fruit (#PE8741)
Instability of fresh fruit and edible nut trade (#PE2587)
Instability of trade in vegetables, roots and tubers (#PE8208).

♦ PE0962 Health hazards of radiation in aircraft
Incidence Pilots and flight attendants on many airline routes are exposed to more radiation than most workers in nuclear power plants, although the effects of such low doses are uncertain and nearly impossible to measure in the population. The risk is greater at high altitudes and closer to the poles. It is considered to be especially dangerous to pregnant women, especially if they fly frequently or during peaks of the 11-year sunspot cycle. For 100,000 crew members spending 20 years flying 960 hours a year on routes of more than 3 hours, 59 to 61 premature cancer deaths could be expected to occur. For 100,000 passengers flying 480 hours a year on the same route, namely about 9 hours a week, 29 to 39 additional cancer deaths could be expected (over the 22,000 which would otherwise be expected for that group).
Broader Contamination by natural radiation (#PC1299).
Aggravated by Medium-term cyclic variations in solar radiant energy (#PE9528).

♦ PE0964 Bank failure
Failure of savings institutions
Nature Bank failures are common during severe depressions. Bank liquidity disappears, depositors who wish to withdraw are kept waiting for days, weeks, or months, and runs on banks ultimately cause their closure. Hundreds of banks failed in the Great Depression following the stock market crash of 1929 in the USA, and thousands of depositors lost their savings. Recent strains in the international monetary system have caused some failures, such as Continental Illinois in the USA.
Incidence From 1982 to 1988 some 620 American banks have failed or have been forced to merge and 100 savings and loan institutions have been closed and another 350 have been declared insolvent. These thousand-odd banking failure have involved 7 percent of the total domestic deposits in US savings institutions. It has been estimated that the cost to the government of protecting investors may total over $500 billion over the next 30 years.
Claim In 1990, ever respectably-sized bank in the USA has a balance sheet replete with bad or doubtful property loans, Third World debt, junk bonds, and doubtful loans to the corporate sector.
Broader Insolvency (#PC6154)
Instability of economic and industrial production activities (#PC1217).
Narrower Small bank failure (#PE1815) Failure of state banks (#PE9465).
Aggravates Excessive corporate debt (#PE1879)
Financial destabilization of world trade (#PC7873).
Aggravated by Bank fraud (#PE1398) Bad property loans (#PE4545)
Risk of capital investment (#PF6572) Unethical financial practices (#PE0682).

♦ PE0965 Congenital diseases of the nervous system in animals
Broader Animal abnormalities (#PD4031)
Diseases of nervous system in animals (#PD7841).
Narrower Cerebral defects in animals (#PG4554)
Spastic diseases of animals (#PG9511)
Defects of cerebellum in animals (#PG9254)
Disorders of the nervous system of dogs (#PG4308).
Aggravated by Inadequate coordination of action on intergovernmental programmes at national level (#PE1375).

♦ PE0969 Anxiety resulting from torture
Torture victim's fear
Incidence In a recent study of victims of torture nearly 25 percent of those studied suffered from anxiety or fear even after several years.
Broader Anxiety (#PA1635) Neurological effects of torture (#PD4755).
Aggravated by Fear of police (#PF8378)
Rivalry and disunity within developing regions (#PD0110).

♦ PE0972 Instability of trade in fish, crustacea and molluscs and preparations thereof
Refs OECD *Problems of Trade in Fishery Products* (1985).
Broader Instability of trade in food and live animals (#PD1434).

♦ PE0976 Long-term shortage of food and live animals
Agricultural shortages
Nature The remarkable increases in food production in industrial and developing countries in 1970s and 1980s have come in part at the expense of soil and water resources. There are no major new technologies waiting in the wings to improve the food output. Food scarcity and higher prices may dominate the 1990s.
Broader Food insecurity (#PB2846) Long-term shortage of commodities (#PC1195)
Insufficient availability of goods (#PB8891).
Narrower Long-term shortage of livestock (#PG5779)
Long-term shortage of dairy products and eggs (#PE8225)
Long-term shortage of meat and meat preparations (#PE1490)
Long-term shortage of cereals and cereal preparations (#PE1218)
Long-term shortage of sugar and honey and preparations thereof (#PE1120)
Long-term shortage of animal feedstuffs excluding unmilled cereals (#PE8514)
Long-term shortage of fruit and vegetables and preparations thereof (#PE1013)
Long-term shortage of coffee, tea, cocoa, spices and manufactures thereof (#PE1197)
Long-term shortage of salt-water fish, crustacea and molluscs and preparations thereof (#PE1783).
Related Long-term shortage of natural resources (#PC4824).
Aggravates Famine (#PB0315) Unaesthetic foodstuffs (#PD1126)
Nutritional deficiencies (#PC0382).
Aggravated by Overstocking (#PC3153) Vulnerability of plants and crops (#PD5730)
Stagnated development of agricultural production (#PD1285).

♦ PE0978 Drug abuse in prisons
Drug abuse by prisoners
Broader Drug abuse (#PD0094).
Aggravates Impaired vigilance (#PF6863).

♦ PE0980 Aggravation of instability in exchange rates by transnational corporations
Nature The system of trade and payments and the policies of national governments may have a major influence on the behaviour of transnational corporations as well as on policies of individual host governments. Tariff and trade policies of developed countries affect the level of exports from developing countries. Instability in exchange rates may lead multinational corporations to move funds across national borders in a way which tends to accentuate this instability.
The manipulation of intra-firm trade and transfer pricing by transnational corporations can have an adverse effect on exchange rates. Transnational corporations also effect interest rates and therefore exchange rates by borrowing on local markets.
The international monetary system also has an important bearing on the operations of transnational corporations, in areas such as choice of location and financial flows. Apart from its influence on national and regional policies on production and trade, the degree of stability of exchange rates and the adjustment mechanism provided by the system affects the policies of transnational corporations. Events in recent monetary crises have directed attention to the possible role of transnational corporations in the volatile, short-term movements that have occurred, in addition to the fundamental disequilibria in the balance of payments of several major industrial countries.
Claim Although convulsions in the international monetary system may not be caused by speculative activities of transnational corporations, the ability of these enterprises to move massive amounts of funds across borders is unquestionable and such movements can undoubtedly aggravate the situation. This potential has been greatly enhanced by dramatic changes in international banking and consortia arrangements. A vigilant monitoring of surveillance by central banks of movements of funds of transnational corporations across borders has not been effected. Moreover, in discussion of a new monetary system, the role of transnational corporations is benignly neglected.
Broader Transnational corporation imperialism (#PD5891)
Destabilization of monetary systems and exchange notes by transnational corporations (#PE5903).
Aggravates Mismanagement of exchange rate system (#PF1874).

♦ PE0984 Waste of electricity
Claim If every citizen of the USA turned off a single 100-watt bulb for 1.5 hours per day, it would have the same impact in reducing the carbon dioxide responsible for global warming as planting 1 billion new trees.
Broader Unsustainable development of energy use (#PC7517).

♦ PE0985 Yellow fever
Yellow Jack — Vomito amarilli
Nature Yellow fever is an acute disease of certain tropical regions, characterized by fever and jaundice. It occurs in tropical Africa and South and Central America, primarily as a sylvatically-cycled infection of primates. Transmission from host to host is accomplished via species of Aedes mosquitoes in Africa and Haemagogus and Aedes in South America.
Broader Virus diseases (#PD0594) Stigmatized diseases (#PD7279)
Infectious and parasitic diseases (#PD0982).
Related Malaria (#PE0616) Dengue fever (#PE2260)
Viral diseases in animals (#PD2730).
Aggravates Fever (#PD2255) Liver diseases (#PE1028).

Aggravated by Epidemics (#PC2514) Cockroaches as pests (#PE1633)
Vector-borne diseases (#PD8385) Mosquitoes as vectors of disease (#PE1923)
Mosquito resistance to insecticides (#PE3582)
Aedes mosquitoes as vectors of disease (#PE3621).

♦ PE0987 Plague

Bubonic plague — Black death

Nature Primarily a disease of rodents, plague in human beings originates by contact with infected rodents, most commonly rats, or their fleas. The disease in man has three clinical forms: bubonic, characterized by swelling of the lymph nodes; pneumonic, in which the lungs are extensively involved; and septicaemic, in which the blood stream is so strongly invaded by Pasteurella pestis that death ensues before the bubonic or pneumonic forms have had time to appear.

The natural foci scattered over almost the whole world provide plague with ample opportunities to re-emerge, and the appearance of resistance phenomena gives rise to the fear that present methods of combating plague may eventually lose their effectiveness. Although the third plague pandemic, which began at the end of the nineteenth century and, aided by steamships, invaded the entire world, has now come to an end and although continuous world-wide surveillance of human and rodent plague is imperative if great human and economic distress is to be avoided, modern communications bring a growing danger that outbreaks of plague will occur in areas hitherto free from the disease. The new technique of freighting in containers presents a definite threat. Under this system, cargo packed in infected areas cannot be inspected or treated en route, and there may be no facilities for inspection and treatment at its destination. If plague-infected rodents and fleas are present in the containers, they may well survive and cause human or rodent plague. The transport of such containers by air may present a particular hazard. With the growing use of high-speed passenger aircraft, moreover, there is an ever increasing possibility that travellers in the incubation phase of plague will disembark in a hitherto plague-free area.

Incidence Sporadic cases and epidemics are reported in countries with known natural foci of the disease; the total number of cases recorded in 1982 was much higher than in 1981. In 1982, 713 cases of human plague, including 36 deaths, were announced to the World Health Organization. A total of 250 cases and 31 deaths were recorded in Africa (Uganda, Madagascar, Tanzania, Zimbabwe). After being dormant for 10 years, human plague was noted in South Africa. In Asia, human plague was notified only from Burma and Viet Nam. However, local outbreaks could appear in any country of South-East Asia, especially in those areas which are enzootic for wild rodent plague and where disinsectization of populated localities is not carried out regularly enough. Human plague was also noted in four countries in the Americas (Brazil, Bolivia, Peru, USA). This is one of the diseases being explored because of its value in biological warfare.

Claim All that separates the virulent and non-virulent forms of the disease is a mutation in the bacteria. The bacteria mutates to its virulent form, spreads through high density populations along trade routes, migration routes and in times of war. It then subsides and mutates back to its non-virulent form to await the next pestilent opportunity.

Refs Butler, Thomas C *Plague and Other Yersinia Infections* (1983); Godfried, Robert S *The Black Death* (1983); Tenney, Louise *Modern Day Plagues* (1987); WHO Expert Committee Geneva, 1969, 4th *WHO Expert Committee on Plague* (1970).

Broader Epidemics (#PC2514) Stigmatized diseases (#PD7279)
Zoonotic bacterial diseases (#PD6363) Infectious and parasitic diseases (#PD0982).
Narrower Pneumonic plague (#PG3175) Septicaemic plague (#PG3176).
Related Bacterial diseases in animals (#PD2731).
Aggravates Encephalitis (#PE2348).
Aggravated by Rats as pests (#PE3177) Natural disasters (#PB1151)
Cockroaches as pests (#PE1633) Rodent vectors of disease (#PE3629)
Parasites of the human body (#PE0596) Siphonaptera as insect pests (#PE3643)
Rodent resistance to rodenticides (#PE3573).

♦ PE0988 Abuse of monopoly power of state-owned or state-controlled enterprises

Unaccountability of decentralized government agencies

Broader Monopoly of power (#PC8410) Abuse of economic power (#PC6873)
Inefficiency of state-controlled enterprises (#PD5642).
Related State monopoly (#PJ4242) Economic exploitation (#PC8132)
Transnational corporation imperialism (#PD5891).
Aggravates Inefficient mobilization of government revenue (#PF4197)
Inadequate development of enterprises in developing countries (#PE8572).
Aggravated by Proliferation of public sector institutions (#PF4739)
Social environmental degradation from recreation and tourism (#PD0826).

♦ PE0989 Mental illness in adolescents

Mentally ill children — Psychotic children

Nature In discussing behavioural disturbances in adolescents and young people, two aspects of failure in social adaptation are clearly distinguishable. The first includes problems arising from the mutual relationship between youth and society, manifested by widescale rebellion against existing social customs, educational systems and so on, sometimes dramatized as 'the crisis of modern youth'. These phenomena are of social origin and cannot be adequately explained from psychological or biological points of view. The second aspect of maladaptation embraces the psychological disturbance of the individual, and here socio-psychological and biological causes are of considerable importance.

It is by no means clear what special combination of endogenous and exogenous factors determines the occurrence of psychotic illness rather than of those milder and much more common disorders that are located on the borderline between mental health and mental illness. To the latter group belong psychogenic maladjustment and the pathological development of personality, although a diagnosis of psychopathy should be made with caution at a time when the personality is still in process of formation. Psychogenic maladjustment is facilitated by the psychological changes during puberty, although, because of the emotional instability, intensified self-awareness, and suggestibility that accompany it, it may be argued that the pubertal phase of psychological maturation is inevitably marked by transient maladjustments and behavioural difficulties. Since the group they constitute is clinically so indistinct, the bulk of mental disorders associated with puberty occupy a position intermediate between what is psychologically 'normal' and what is pathological, and may include both masked psychopathic and neurotic conditions, as well as examples of the initial manifestations of a psychosis.

Essential differences have been established between the clinical aspects of schizophrenia in adolescents in whom physical maturation is early and those in whom it is delayed. Unsynchronized development or disturbances of sexual maturation appear to be significantly associated with neurotic states or with pathological personality formation and psychopathic behaviour.

The importance of 'minimal brain damage' as a factor contributing to maladjustment has tended to be overlooked, though it may give rise to various psychopathological syndromes, especially under adverse environmental conditions. The primary symptoms of cerebral disorder such as hyperkinesia, clumsiness, defects of visuomotor performance, impulsive behaviour, or distractibility, may serve to frustrate or irritate the parents, causing them to react with anxiety or rejection towards the child. The child's own feelings of security are therefore impaired, with further delay and disturbance of psychological maturation.

The resulting clinical picture may therefore be complicated by the continuous interaction of neurological abnormalities and environmental factors. Disabilities which may have their basis in the minimal organic lesion, such as impulsivity, short attention span, or dyslexia, may lead to problems at school as well as at home. There are cases in which minimal brain damage in itself seems to be the main factor underlying delinquent behaviour, though more often there are associated factors such as hereditary loading, early childhood deprivation, or disturbed family relations. The manifestations of mental disorders in adolescence very often follow a common pattern of antisocial, psychopathic-like behaviour, which makes for difficulty both in classification and in a systematic approach to prevention and treatment.

Refs Looney, John G (Ed) *Mental Illness in Children and Adolescents.*

Broader Mental illness (#PC0300).
Related Mental deficiency in children (#PD0914).
Aggravates Juvenile delinquency (#PC0212).
Aggravated by Juvenile stress (#PC0877) Excessive television viewing (#PD1533)
Neglect of adolescent health care (#PF6061).

♦ PE0993 Instability of mining and quarrying industry

Broader Instability of economic and industrial production activities (#PC1217).
Narrower Petroleum and natural gas production instability (#PE8955)
Instability of miscellaneous mining and quarrying (#PE8809).
Aggravates Excessive external trade deficits (#PC1100).

♦ PE0997 Long-term shortage of manufactured fertilizers

Nature In the present state of agricultural technology, chemical fertilizers constitute one of the most important single means to increase food production. Increasing application of fertilizers is a key element of the package that has made possible most of the increase in agricultural productivity achieved over recent years. The successes of the high-yielding varieties of wheat and rice derive from the fact that they are highly fertilizer-responsive. Use of fertilizers in developing countries, although still relatively low, has been doubling every five years. A serious deficit in the physical supplies of fertilizers available for the continuing expansion of usage in developing countries, therefore, must be considered as nothing less than a major world problem. That such a problem is developing is indicated by the current shortage and high prices of fertilizers, as well as by the underlying longer-term trends of supply and demand.

In theory, supply gaps could be filled either by additional production in the developing world (but not necessarily in each individual country), or by imports from the developed countries. However, it now seems unlikely that the developed countries will be in a position to fill the entire fertilizer gap in developing countries. The recent spurt in prices has generated large new investment in fertilizer capacity in developed countries; but there is still a large uncovered gap between expected requirements and likely supply, particularly in developing countries. The actual magnitude of this gap would of course depend on the rate at which the expansion of fertilizer capacity actually takes place in the next 4-5 years. The situation is changing very rapidly, and it is very hazardous to forecast the actual gaps. There are, however, a variety of factors which will determine the future supply/demand balance, including the persistence of high fertilizer prices, the ability of traditional producing countries to assure themselves of feedstock supplies on a long-term contract basis, and the initiatives taken to build new capacity in both developing and developed countries.

Broader Long-term shortage of chemicals (#PE1261).

♦ PE0998 Unsanctioned maltreatment of prisoners

Maltreatment of prisoners by prison officers

Claim Prison officers are able to use their freedom of action to harass particular prisoners. This may range from surreptitious beatings, arbitrary deprivation of privileges to acts of humiliation such as urinating in the food or beds of prisoners in solitary confinement. Any protest, if possible, is treated as suspect and used against the prisoner.

Broader Maltreatment of prisoners (#PD6005).
Related Abuse of prison labour (#PD0165).
Aggravated by Denial of rights to prisoners (#PD0520).

♦ PE1008 Instability of textile and clothing industries

Instability of wearing apparel industries

Nature The clothing industry is affected by seasonal cycles and fashion changes which are peculiar to it. It is labour intensive, and because of this and the growing internationalization of styles and manufacture, there is considerable regional and inter-regional competition to which the costs of labour and manufacture are the key. In the textile industry, fluctuations in availability and prices of natural materials occur, particularly as regards cotton, but also with respect to wood, leather and fur. Both the clothing and textile industries are objects of protectionist action by governments.

Broader Instability of manufacturing industries (#PC0580)
Instability of trade in miscellaneous manufactured articles (#PE0814)
Instability of trade in unprocessed textile fibres and their waste (#PE1550).
Narrower Instability of footwear industries (#PU3204).
Aggravated by Unethical practices in the apparel industry (#PD8001).

♦ PE1011 Insensitivity of transnational corporations to consumer needs in developing countries

Nature Producing the goods which respond to the real needs of individuals in the light of their social and economic conditions is a general problem. It is particularly important however, in developing countries. Since products of transnational corporations are often geared to the consumption patterns of advanced countries, the needs of the majority of the population in poor countries may not be fulfilled. Consumers may be induced through intensive advertising to buy goods which otherwise they would not have felt they needed. Given the limited financial means of the great majority of the population of developing countries, such practices may lead to the diversion of scarce resources from basic needs to less basic ones.

Broader Consumer vulnerability (#PC0123) Transnational corporation imperialism (#PD5891)
Domination of developing countries by transnational corporations (#PE0163).
Related Underdevelopment (#PB0206).
Aggravates Lack of production for domestic consumers in developing countries (#PJ8860).

♦ PE1013 Long-term shortage of fruit and vegetables and preparations thereof

Broader Long-term shortage of food and live animals (#PE0976).
Narrower Instability of fresh fruit and edible nut trade (#PE2587).
Related Slowing growth in food production (#PC1960).

♦ PE1017 Allergy

Allergic reactions

Nature Allergy is a 'supersensitivity' or a form of exaggerated immunological reactivity to some organic substance in the air, in food or on an inert surface, or to certain industrial chemicals. Allergens are usually harmless in themselves, but the reactions they sometimes excite may cause considerable physical distress ranging from a runny nose, a skin rash or wheezing to, at worst, coma. The allergies include asthma, hayfever, urticaria, eczemas, local and systemic anaphylaxis, sensitization to industrial chemicals, and strong reactivity towards tuberculin. Allergic reactivity can be transferred from one individual to another by means of transfusions of blood serum containing the appropriate antibody.

Incidence All around the world, even in developing countries, allergies are on the increase.

PE1017

Claim Some allergies are the result of increased pollution of the environment, including food, air and water.
Refs Asthma and Allergy Foundation of America Staff and Norback, Craig (Eds) *The Allergy Encyclopedia* (1981); Bock, S Allan *Food Allergies* (1988); Kay, A B *Allergy and Inflammation* (1987); King, H C and Marchan (Eds) *Otolaryngolic Allergy* (1981); Lessof, M H *Allergy* (1988); Maragos, G D *Seminar on Pediatric Allergy* (1977); Roth, Alexander (Ed) *Allergy in the World* (1978); Swineford, Oscar *Asthma and Hay Fever and Other Allergic Diseases for Victims and Their Families* (1973); WHO *Health Aspects of Chemical Safety* (1983).
Broader Human disease and disability (#PB1044).
Narrower Eczema (#PE2465) Hay fever (#PE6197) Urticaria (#PG4480)
Anaphylaxis (#PG4894) Allergy inducing food (#PE3225)
Allergy inducing pollens (#PG3219) Allergy inducing bacteria (#PG3224).
Related Asthma (#PD2408) Hypersensitives (#PE5169).
Aggravates Ulcers (#PE2308) Rashes (#PG3215) Headache (#PE1974)
Diarrhoea (#PD5971) Abdominal pain (#PD8725) Bright's disease (#PE2272)
Rheumatic diseases (#PE0873).
Aggravated by Pollutants (#PC5690) Hazards of cosmetic use (#PE4895)
Environmental pollution (#PB1166)
Poisonous, allergenic and biologically active wood (#PE6867).

♦ PE1024 Denial of rights of indigenous people to be self–governing
Broader Bogus public interest groups (#PE7575)
Denial of right of a people to be self–determining (#PC6727).
Aggravates Secession (#PD2490).
Aggravated by Discrimination against indigenous populations (#PC0352)
Expropriation of land from indigenous populations (#PC3304).

♦ PE1028 Liver diseases
Liver fatty diseases — Liver damage — Liver disease
Nature Infectious hepatitis, liver cancer and alcoholic liver diseases are major causes of morbidity and mortality.
Refs Chopra, Sanjiv *Disorders of the Liver* (1988); Mowat, Alex P *Liver Disorders in Childhood* (1987); Santos, Adolfo L *Liver Diseases* (1987); Williams, Roger *Liver Failure* (1986).
Broader Human disease and disability (#PB1044)
Diseases and deformities of the digestive system (#PC8866).
Narrower Liver rot (#PG2606) Liver cancer (#PE3233) Liver abscess (#PG3229)
Liver congestion (#PG3230) Kwashiorkor disease (#PE2282)
Acute yellow atrophy (#PG3232) Cirrhosis of the liver (#PE2446)
Enlargement of liver and spleen (#PG4671).
Aggravates Dropsy (#PG3235) Dysentery (#PE2259) Hepatitis (#PE0517)
Haemorrhoids (#PE3504).
Aggravated by Malaria (#PE0616) Syphilis (#PE2300)
Ascariasis (#PE2395) Malnutrition (#PB1498)
Yellow fever (#PE0985) Alcohol abuse (#PD0153)
Viral infections (#PG3236) Bacterial disease (#PD9094)
Diseases of metabolism (#PC2270) Abuse of coca and cocaine (#PD2363)
Parasites of the human body (#PE0596)
Environmental hazards from chemicals (#PC1192)
Polychlorinated biphenyls as a health hazard (#PE2432).

♦ PE1029 Proliferation and duplication of international organizations and coordinating bodies
Broader Politicization of intergovernmental organizational debate (#PD0457).
Narrower Proliferation and duplication of intergovernmental organizations and coordination bodies (#PE2417)
Proliferation and duplication of international nongovernmental organization and coordination bodies (#PE0179)
Proliferation and duplication of organizational units and coordinating bodies of the United Nations system (#PE1579).
Aggravates Proliferation and duplication of international information systems (#PE0458)
Competition between international organizations for scarce resources (#PC1463).
Aggravated by Endemic abortion of ewes (#PG1537)
Organizational empire–building (#PF1232)
Loss of credibility in international institutions (#PE8064).

♦ PE1032 Pecuniary guarantees as a condition of provisional release in criminal cases
Nature Inability to furnish pecuniary guarantees or to find someone willing to do so on their behalf sometimes prevents accused persons from being released from custody pending or during criminal proceedings against them. Imprisonment pending or during trial causes a number of hardships, some of which hit with particular severity precisely those so imprisoned because they lack financial means. Imprisonment may cause loss of income; it carries with it a social stigma; and the person imprisoned may lose his job as a result of this stigma or due to economic factors operating on his employer. If the imprisoned person is a breadwinner, the family also suffers from his loss of income or occupation.
Other hardships result from imprisonment during or pending trial: the expense and difficulty of defence is increased, since the accused's counsel has to visit him, instead of the reverse; facilities in prison for consultation between counsel and accused may not be such as to assure that such consultations are adequate and secret; the accused is unable to seek evidence personally; his anxiety is increased by unfamiliar surroundings, by the absence of his relatives and friends and by his own absence from his family; he may be subject to squalor, enforced idleness and criminalizing influences while under detention.
Broader Economic discrimination in the administration of justice (#PE1399).

♦ PE1040 Scab diseases in plants
Nature Scab is an overgrowth of tissue in a limited area, especially in fruit trees and vegetables. It is caused mainly by microscopic pathogenic fungi and, sometimes by actinomycetes and bacteria.
Broader Ascomycetes (#PE4586) Deuteromycetes (#PE4346).
Narrower Apple scab (#PG1100) Fusarium diseases in plants (#PG5461).

♦ PE1041 Multiple sclerosis
Demyelinating diseases of the central nervous system
Nature Multiple sclerosis is one of the most common organic diseases affecting the nervous system. It is important not only because it is frequently encountered but also because it almost invariably affects young people. While rarely fatal in its early stages, it may be progressive and cause long incapacity. Moreover, no treatment is available.
The disease is termed 'multiple' because it usually affects many parts of the nervous system, and it is characterized by frequent relapses followed by periods of partial and sometimes complete recovery. It is thus multiple both anatomically and chronologically. The spinal cord is perhaps the most frequently affected; its involvement may cause periods of partial to complete paralysis of the legs, and, at times, of the trunk and arms. With or without this weakness or paralysis, there may be lack of coordination, staggering and tremor, or there may be poor coordination plus tremor of the extremities, sometimes of the body and head. Numbness, tingling, and various sensory changes may also occur. Eye symptoms are also common, with nystagmus, periods of double or blurred vision, and even temporary to permanent blindness in one or both eyes. In advanced cases there may be slurred speech and sometimes difficulty with bladder and bowel control.
Incidence The disease is most common in cold, damp climates. In Europe, it is found most frequently in the Scandinavian and low countries, the Baltic region, Northern Germany, and Great Britain. It is rare in the Mediterranean countries. In the United States, incidence is considerably greater in the North than in the South. The disease usually begins between the ages of 20 and 40; the average age at onset is 30 years for females and 34 years for males. The age at onset tends to be lowest in those areas with the highest incidence.
Refs Caroscio, James T *Amyotrophic Lateral Sclerosis* (1986); Confavreux, C, et al *Trends in European Multiple Sclerosis Research* (1988); Kuroiwa, Yoshigoro (Ed) *Multiple Sclerosis in Asia* (1976); Mims, Cedric, et al *Viruses and Demyelinating Diseases* (1984); Scarlato, G and Matthews, W B *Multiple Sclerosis* (1984); Tsubaki, T and Yase, Y (Eds) *Amyotrophic Lateral Sclerosis* (1988).
Broader Critical illnesses (#PE9038)
Diseases of the central nervous system (#PE9037).
Aggravates Paralysis (#PD2632).
Aggravated by Fatigue (#PA0657) Human illness (#PA0294)
Climatic cold (#PD1404) Autoimmune disease (#PE5214)
Emotional disorders (#PD9159).

♦ PE1046 Elimination of the socio–cultural role of western women
Long–term historical effect of witch hunting
Nature The social and cultural roles of western women have been destroyed by a number of factors. The process of urbanization isolated her from the extended family and removed her from the farm. Many of the roles such as midwife have been turned to professions. Her role as medium and healer has been eliminated by science. The long and bloody period of witch–burning resulted in the torture and murder of six to nine million wise women, left a deep mark of terror and eradicated untold knowledge among western women.
Broader Static and unrelated social roles (#PF1651).
Related Conflicting roles of women (#PD6273).
Aggravates Wasted woman power (#PF3690) Unused female talents (#PJ0833).
Reduces Neglect of the role of women in rural development (#PF4959).

♦ PE1048 Inadequate facilities for the transport of electrical energy
Broader Inadequate electricity infrastructure (#PD9033).
Narrower Inadequate electrical power supply in developing countries (#PE1900).
Related Insufficient transportation infrastructure (#PF1495).

♦ PE1054 Long–term shortage of coal
Nature Although the world's total coal reserves considered together are sufficient for many centuries beyond the expected date of depletion of petroleum, some major producing countries are experiencing a shortage of certain coals, including those used for high–grade coke which is needed for the iron and steel industry and for fuel.
Broader Long–term shortage of mineral fuels and lubricants (#PE1712).
Aggravated by Inadequate cooking stoves (#PE7904).

♦ PE1055 Underdeveloped road network
Road network unconnected with major roads — Insufficient good roads — Poor access roads — Inadequate road access to village — Inadequate access roads
Broader Ineffective rural transport (#PF2996) Dysfunctional public utilities (#PE7647)
Unprofitable scope of industrial operations (#PF1933)
Inadequate transportation facilities for rural communities in developing countries (#PE6526).
Related Unemployment in developing countries (#PD0176)
Unrecognized socio–economic interdependencies (#PF2969)
Inadequate road and highway transport facilities (#PD0490)
Negative consequences of shifting ecology on coastal communities (#PE2305).
Aggravates Prohibitive cost of basic services (#PF6527).

♦ PE1056 Wilt diseases of plants
Nature Wilting is due to either too rapid transpiration or continued loss of water beyond the recovery point, or due to disease organisms which reduce or inhibit water conduction.
Refs Mace, M E and Bell, A A *Fungal Wilt Diseases of Plants* (1981).
Broader Oomycetes (#PE5465) Ascomycetes (#PE4586) Deuteromycetes (#PE4346).
Narrower Tomato wilt (#PG4456) Verticillium wilt in plants (#PG1352)
Fusarium diseases in plants (#PG5461).

♦ PE1066 Anosmia
Lack of sense of smell
Incidence About 1 percent of the American population either smell nothing at all or cannot detect some of the 100,000 distinguishable odours.
Broader Human disability (#PC0699).
Reduces Unpleasant odours (#PJ5522).

♦ PE1082 Disruption of cultural and social identities in developing countries by transnational corporations
Nature The effect of transnational corporations on the social institutions and cultural values of host countries may be especially striking if the tenor, tradition and stage of development of these countries differ considerably from those of the home countries. For example, the "business culture", with its emphasis on efficiency, may be considered too impersonal in traditional societies. The very cultural identity and the entire social fabric may be at stake, especially if transnational corporations attempt to transplant their own models of social development to the host country.
Broader Social irresponsibility of transnational corporations (#PE5796)
Domination of developing countries by transnational corporations (#PE0163).
Related International paternalism (#PF1871).

♦ PE1085 Maternal malnutrition
Protein–energy malnutrition in women during pregnancy and nursing — Denial of sufficient nutrition for women
Nature Pregnant and lactating women form a vulnerable group in whom the effects of low protein intakes may manifest themselves in various ways: the normal increase in body–weight may be considerably reduced; complications of pregnancy such as miscarriage, stillbirth or premature birth are more frequent.
Although mothers on relatively poor diets seem able to provide sufficient milk for their children, this may have debilitating long–term consequences for the woman, as she is forced to consume her own body reserves. The total protein content and amino–acid composition of milk from malnourished mothers are not significantly different from well–nourished mothers, with the result that the extra protein requirement must be at the expense of the mothers' tissues. For example, in India lactating women were found to have an average deficit of two grams of protein per day. Women with very inadequate diets or frank malnutrition, have a reduced milk volume, and the energy content of the milk may be lowered by a decline in the lipid constituents. The protein and carbohydrate components of the milk appear to be relatively unaffected, even in extreme cases,

though the concentrations of some vitamins may be significantly decreased. The poor milk supply associated with maternal malnutrition can in turn lead to unsatisfactory infant nutrition, especially among the low–birth–weight babies who need to compensate for inadequate intrauterine growth. There is, therefore, a link between maternal malnutrition and subsequent malnutrition in the offspring.

In affluent societies most women have an adequate diet and sufficient nutritional stores to compensate for the increased demands of breast feeding, but in poorer societies this is often not the case. In developing countries, women are subjected to a variety of nutritional stresses caused by inadequate diets, the demands of recurrent pregnancy and lactation associated with high fertility, the energy requirements of hard physical work, the adverse nutritional effects of infection and parasitic diseases, and traditional customs which proscribe certain foods or lead to an unequal distribution of food within the family. All these factors interact and produce a 'continuous, cumulative nutritional drain', which has been described as the 'maternal depletion syndrome'. Lactation plays an important role in the genesis of this chronic maternal under–nutrition, and in extreme cases lactation may even precipitate frank malnutrition. Furthermore, these poorly nourished mothers tend to have low birth weight infants who suffer from an increased risk of morbidity and mortality.

Refs Xanthou, M *New Aspects of Nutrition in Pregnancy, Infancy and Prematurity* (1987).
Broader Malnutrition (#PB1498) Denial of right to sufficient food (#PE0324)
Protein-energy malnutrition in vulnerable groups (#PD0363).
Aggravates Malnutrition in developing countries (#PD8668).
Aggravated by Dependence on breast feeding (#PE7627)
Discrimination against women (#PC0308)
Unequal health benefits for women (#PE6835)
Ignorance of women concerning primary health care (#PD9021).

♦ PE1087 Detergents as pollutants

Incidence About 25 per cent of the total phosphorus discharged into British waters originates from detergents.
Broader Chemical pollutants of the environment (#PD1670).
Aggravates Eczema (#PE2465) Hay fever (#PE6197)
Air pollution (#PC0119) Aquatic weeds (#PD2232)
Water pollution (#PC0062) Domestic hazards (#PG2144)
Eutrophication of lakes and rivers (#PD2257) Symptoms referable to sense organs (#PE2665).
Aggravated by Wasted water (#PD3669) Industrial waste water pollutants (#PD0575).

♦ PE1093 Excessive burden on the poor due to legal delays

Nature The poor are less able than others to afford the results of judicial delay. Although speed in the judicial process should not be such as to deprive the parties, especially accused persons, of adequate time to prepare their cases, nor be such as to result in a perfunctory hearing, nevertheless much can and should be done to remedy the causes of delay, of which the following are those most commonly reported:
(i) insufficiency of courts or of judges in relation to case loads; (ii) lack of industry, vigilance or competence on the part of the bench, the granting of excessive time for parties to take procedural steps and undue length of time taken to render decisions; (iii) insufficiency or inefficiency of auxiliary court personnel, such as stenographers, clerks, registrars and bailiffs, which delay the administrative machinery of justice; (iv) insufficiency or inefficiency of police needed for the making of investigations; (v) insufficiency of lawyers; (vi) delaying tactics of lawyers or parties, including excessive pleading and seeking of adjournments; (vii) delay in locating witnesses and securing their attendance at court.
Broader Socio–economic poverty (#PB0388) Delay in administration of justice (#PF1487)
Economic discrimination in the administration of justice (#PE1399).
Related Excessive length of pre–trial internment (#PE4887).

♦ PE1101 Unsystematic use of powerful relationships by rural communities

Nature Many rural communities concerned with development have extended family through out many government and private institutions. A community seriously undertaking comprehensive development uses these extended relationships to obtain various forms of support. Most of these communities do not take advantage of these in any regular or systematic way. Voting patterns often leave a community low on political priority lists. Many communities wait for bureaucracies to initiate programs rather than begin the dialogue themselves. Unless communities put forth a strong corporate voice to take advantage of its multitude of extended relations, they will remain dependent on sporadic development efforts from the outside.
Broader Unrecognized socio–economic interdependencies (#PF2969).
Narrower Spiritual impurity (#PF6657) Low commune priority (#PG9873)
Limited political power (#PG9865) Limited local initiative (#PJ9554)
Ineffective political ties (#PG9892) Unprofitable local politics (#PG9848)
Distrusted local politicians (#PG9908) Insufficient councilman support (#PG9765)
Politically determined services (#PG9837) Noninclusive council representation (#PG9227).
Related Blocked minority opinion (#PD1140)
Paralyzing patterns between villages and administrative structures (#PF1389).
Aggravates Overdependence on government (#PF9530)
Limited availability of financial credit (#PF2489).

♦ PE1103 Environmental hazards from textile and clothing industries

Environmental hazards of wearing apparel manufacture
Nature The textile industry is one of the largest and oldest employers of labour. Textile mills usually last for up to 100 years, and some old textile mills are using machines which are over 40 years old. As a result, there are many hazards indigenous to the textile and clothing industries.
Incidence The most common causes of injury and fatality are: (1) Fire. Fires in textile mills usually spread with great rapidity due to highly inflammable loose textile fibre and accumulated fibre dust. Most textile mills are not of modern fire–resistant construction and floors are usually made of wood, often impregnated with oil dripping from the machines and thereby increasing fire risk. (2) Lifting and carrying. The heavy loads which must be handled and carried can cause excessive strain and damage to health. (3) Bleaching kiers. Cotton cloth is sometimes bleached in kiers – vertical tanks about 3m in diameter and 4m in height. One or more workers, usually youths, may have to go into the kiers (drained of the boiling water, bleaching solution and alkaline liquor used in the process) to stack or remove pieces of cloth. Deaths have occurred when one worker has been in the kier and another worker unknowingly opened valves to admit the deadly liquids. (4) Temperature, humidity and ventilation. Temperatures and humidity levels vary depending upon whether the textile factory is in a cold or temperate climate or in a tropical region. In hot and humid climates, textile factories' humidity sometimes reaches 95% (Bangkok and Shanghai) with temperatures reaching 35 C and inadequate ventilation. These three factors lead to overheating and exhaustion. (5) Dust. Weaving and yarn preparation workrooms emit the toxic chemical asbestos into the air, and byssinosis, a respiratory disease, is prevalent among workers in cotton, flax, and hemp mills. (6) Lighting. Good lighting is rare in textile factories even though workers must often thread thousands of intricate threads per day, taxing their eyesight. Windows are rarely kept clean and rooms are not painted in a colour with a high reflection factor, thus leading to eyestrain and possible adverse effects on production and quality control. (7) Noise. It is impossible for workers to hear one another without shouting into each other's ears, due to the high noise level of shuttle looms, and the noise is often in excess of 85dB. (8) Hours of work. In the past, working hours in textile mills were excessive: 13 hours a day, 7 days a week. That has been alleviated by the introduction of shift labour (4 six–hour shifts per day) but shift labour has initiated its own problems, with male workers being locked into permanent night labour with no hope of changing over to day work, due to labour laws which prohibit women from being employed after 11 p.m.
Broader Environmental hazards from manufactured goods (#PE1344)
Environmental hazards from manufacturing industries (#PD0454).
Narrower Environmental hazards of footwear manufacture (#PU3315).
Aggravates Byssinosis (#PE2319).
Aggravated by Unethical practices in the apparel industry (#PD8001).

♦ PE1106 Office and work–place thefts

Refs Mars, Gerald *Cheats at Work* (1983).
Broader Theft of property (#PD4691).

♦ PE1108 Firearms and explosives crimes

Violations against firearms and explosives laws
Broader Offences against public order (#PD7520).
Narrower Illegal firearms (#PE2470) Trafficking in illegal firearms (#PE7711)
Proliferation of weapons in civilian hands (#PE2449).
Aggravated by Gang warfare (#PD4843).

♦ PE1110 Documentary fraud

Incidence Traders suffer losses running into millions of dollars each year from documentary fraud during the course of a business transaction.
Broader Fraud (#PD0486).
Related Insider dealing (#PD3841) Commodities trading fraud (#PD3917)
Counterfeit money and government securities (#PE5503).
Aggravates Maritime fraud (#PE4475)
Evasion of customs and excise duties (#PD2620).
Aggravated by Contract fraud (#PD7876) Commercial fraud (#PD2057)
Unethical documentation practices (#PD2886).

♦ PE1113 Shoplifting

Nature Shoplifters are those people who take articles from retail stores. Items stolen may be as small as buttons, as bulky as a stuffed toy animal; as inexpensive as a pencil or as costly as a diamond; and the shoplifter may be a minor, an adult, or a person of advanced age. The shoplifter may combine this illegal activity with legitimate shopping (typically in food stores) and the crime can be a one–time affair or a habitual pattern of behaviour. Some shoplifters are professional criminals who later offer the stolen goods for sale.

A major problem is that much shoplifting appears to be involuntary. Current methods of display in retail stores may create too much temptation which the individual finds unable to resist even though he or she may wish to do so. Or the individual may have normal respect for property in all other situations but somehow does not see shoplifting as a crime. It becomes a habit that is difficult to break
Incidence It has been estimated that in the USA, one in 52 supermarket customers steals something and that the annual value of goods stolen is in excess of US $2,000 million. In the UK, the total loss from all stores and markets is estimated at over 1 percent of retail turnover and increasing at a rate of 15 percent a year. Shoplifting costs Australian retailers more than A$1000 million a year and is increasing at a rate of 20 percent a year.

Up to one–half of 'Inventory shrinkage' attributed to shop–lifting may be due to employee theft and poor accounting.
Refs Francis, Dorothy B *Shoplifting* (1980); MATCOM *Shoplifting* (1980).
Broader Theft of property (#PD4691).
Aggravated by Boredom (#PA7365) Economic inflation (#PC0254).

♦ PE1117 Slow economical growth in the socialist countries

Incidence The growth of net material product fell from over 6 per cent in the early 1970s to 2 per cent in the beginning of 1980s. The earlier high rates were made possible by heavy investment in basic industries, notably engineering, metallurgy, electric power generation and chemicals, as well as by the abundance of labour and raw materials. How ever, as labour reserves became exhausted, investment became increasingly capital–intensive. Moreover, investment requirements in fuel and raw material exploitation rose as new deposits had to be worked. Although the external sector of the socialist countries of Eastern Europe is small, there is no doubt that the slower growth of their trade owing to recession in the developed market–economy countries and the growth of protectionism has affected the overall economic performance. In addition, increased indebtedness as a result of export credits accorded by the developed market–economy countries, together with high interest rates, has exerted pressure on the external accounts of some socialist countries, which have faced larger annual interest payments and heavy refinancing requirements, as loans come to maturity.
Aggravated by International economic recession (#PF1172).

♦ PE1120 Long–term shortage of sugar and honey and preparations thereof

Broader Long–term shortage of food and live animals (#PE0976).
Narrower Shortage of sugar confectionery (#PU3325).
Related Slowing growth in food production (#PC1960).

♦ PE1121 Parole violation

Parole failure
Refs Westhuizen, J van der and Oosthuizen, H *Prediction of Parole Failure and Maladjustment.*
Broader Arbitrariness (#PB5486) Abuse of government power (#PC9104)
Deficiencies in the criminal justice system (#PF4875).
Related Breach of promise (#PF7150).

♦ PE1122 Prohibitive cost of maintaining comprehensive document collections

Excessive cost of comprehensive information systems
Related Reduced scope of intergovernmental development assistance (#PF4794)
Dependence of island developing countries on official assistance (#PE5724).
Aggravates Global amnesia (#PF0306)
Incompatibility of document classification systems (#PE7753)
Weakness in trade between different economic systems (#PC2724)
Destruction of historic documents and public archives (#PD0172).
Aggravated by Prohibitive cost of knowledge and information (#PF0703).

♦ PE1127 Asbestos as a pollutant

Asbestos dust
Nature Asbestos is a mixture of magnesium and iron silicates in fibrous form. It appears as dust in the form of fine fibres in the air. Asbestos enters the body by inhalation, and fine dust containing fibres with a diameter of less than 5 microns and length greater than 5 microns may be deposited in the alveoli. These fibres are insoluble. The dust deposited in the lungs causes fibrosis or

asbestosis. Asbestosis results in impaired lung function after five to ten years. The symptoms are: shortness of breath, chest pain; and later bronchitis with increased sputum, and clubbing of the fingers. Radiography of the lungs shows certain characteristic changes. Other specific diseases associated with asbestos are; cancers of the bronchi, pleura and peritoneum and probably other organs; and asbestos corns of the skin. All these, with the exception of corns, are due to the inhalation of asbestos fibres and consequently any process which gives rise to large amounts of asbestos dust may constitute a health hazard.
Incidence Occupational exposure occurs in asbestos mines and wherever asbestos or asbestos products are used, for instance, in handling asbestos-cement products used in the building industry (roofing sheets, wallboard and pipes). Exposure may also occur in the textile industry in the manufacture of fireproof materials, such as asbestos clothes or brake linings for motor vehicles. Asbestos is also used for insulation and fire protection purposes in ship-building, house-building, and in the undersealing of cars.
From 1940 to 1980, some 21 million American were exposed to microscopic fibers of asbestos from insulation and other building materials. The fibers entered their lungs and, in a high proportion of cases, caused cancer. From that group, the best estimate is that in each year until the turn of the century, 8,000 to 10,000 people will die of asbestos disease. American manufactures were aware since the 1930's of the dangers to employees and customers but largely suppressed the knowledge.
Broader Mineral dusts (#PE4679) Chemical air pollutants (#PD1271)
Chemical pollutants of the environment (#PD1670).
Narrower Health hazards of asbestos (#PE3001).
Aggravates Occupational risk to health (#PC0865).
Aggravated by Dust (#PD1245) Motor vehicle emissions (#PD0414)
Dust as an occupational hazard (#PE5767).

♦ PE1129 Discrimination against the poor in judicial sanctions
Nature A poor person, when convicted, is often imprisoned because of inability to pay the alternative of a fine. Imprisonment carries with it a greater social stigma than the payment of a fine. A person imprisoned because of inability to pay a fine also suffers some of the hardship undergone by persons imprisoned during or pending a trial because of inability to furnish pecuniary guarantees. Within the area of payment of fines there is great de facto inequality, since a fine of little significance to a rich man may be ruinous to a poor one.
Broader Economic discrimination in the administration of justice (#PE1399).
Aggravated by Collapse of judicial system (#PJ0761).

♦ PE1132 Diseases of the pancreas
Refs Burns, Gerald P and Bank, Simmy *Disorders of the Pancreas* (1989); Hollender, L F, et al *Acute Pancreatitis* (1983); Volk, Bruno W and Arquilla, Edward R *The Diabetic Pancreas* (1985).
Broader Human disease and disability (#PB1044).
Narrower Pancreatitis (#PG3330) Cystic fibrosis (#PE3331).
Aggravates Diabetes (#PE0102) Malnutrition (#PB1498)
Lung disorders and diseases (#PD0637) Vitamin deficiencies in diet (#PD0715)
Diseases and deformities of the digestive system (#PC8866).
Aggravated by Mumps (#PE2356).

♦ PE1134 Limited availability of permanent employment in inner-cities
Nature Urban inner-city unemployment has reached a much higher than average level as the work skills needed in modern industry have moved beyond the reach of inner-city residents. The technological revolution has radically altered the job market around the world. Labour saving, sophisticated business practices have diminished the market for unskilled personnel with limited education. It is a common practice for training programs in low socio-economic areas to use out-dated equipment, such as letter presses and manual typewriters, which do not prepare the student for competing in the labour market where offset presses and electric or correcting typewriters are used. Also, industry has found it more profitable to move out of the inner city, which adds a complex, expensive and time-consuming burden to the inner city resident.
Those who continue to yearn for the personal relationships of the rural style experience shifting to the urban style an overwhelming risk. The style and idiom of the streets which has equipped ghetto residents for survival is not easily translated into the sophistication of the business world. Low pay, long hours, hard work and the inequitable nature of manual labour result in available jobs going unfilled while the unemployed stand by, hoping for something better. The inner city resident is further discouraged by hiring practices which, in effect, promote discrimination.
Broader Limited employment options (#PF1658).
Narrower Restricted union entry (#PG8925) Inaccessible job market (#PE8916)
Undesirable manual labour (#PG8875).
Related Rural unemployment (#PF2949) Insufficient job options (#PJ0070).
Aggravates Underpayment for work (#PD8916)
Limited means of marketing employable skills (#PE7344).
Aggravated by Second employment (#PF6908) Clandestine employment (#PC7607)
Blocked skills advancement (#PG8864) Discrimination in employment (#PC0244)
Inappropriate employment incentives (#PD0024)
Maladjustment to disciplines of employment (#PD7650).

♦ PE1136 Underdevelopment of paper and printing industries
Broader Underdevelopment of manufacturing industries (#PF0854).

♦ PE1137 Traumatic shift in life-styles of mining communities
Nature While some local communities today have shifted from secluded rural villages to factory working suburbs, from remote hamlets to bustling urban centres, former mining towns have lost their economic role. They are experiencing a new financial autonomy calling for additional economic acumen.
Claim In the past, mining towns were thriving places with good housing, good schools, good businesses and employment. Now that many mines have closed, most residents have moved away and the remaining miners commute to nearby jobs. Social life has undergone a basic change. Festivals and holiday celebrations once expressed a vital community spirit arising from a practical relationship to day-to-day struggles. Although this spirit is still alive, its impoverished expression only reflects the greatness of the past. The once lively and hospitable inns are either closed or partially occupied as residences. There may be a nucleus in the community which is promoting new businesses and inviting basic industries, but the existing businesses covering a broad range of dry goods and foodstuffs are deteriorating. Some business people carefully maintain what they have rather than risking new directions.
Broader Obstacles to community achievement (#PF7118).
Narrower Declining public image (#PG8266) Divided community spirit (#PJ0399)
Inhibiting business outlook (#PG8263) Insufficient cultural heroes (#PF8623)
Uncertainty of miners' future (#PG8265) Inexperienced financial autonomy (#PG8267)
Debasing retirement possibilities (#PG8268)
Detrimental story of community future (#PF6575).
Related Unchallenging world vision (#PF9478).
Aggravates Demeaning community self-image (#PF2093).
Aggravated by Declining school enrolment (#PJ7844).

♦ PE1139 Constricting level of capital development in rural areas
Nature The growth of national and international corporations in the last 40 years has created enormous concentrations of capital throughout the world. Developing villages have begun to experience the impact of the interchange of these funds, but have little or no way to participate in it. In fact, they often experience more goods and services going out than coming in, thus crippling a solid financial base. Despite the proximity of credit sources, the community experiences difficulty in attracting capital for business development due to high interest rates, lack of understanding of rules and regulations and inconsistent means of repayment. The generation of capital within the community is restricted by low wages of workers and the almost immediate out-flow of resources to the city for purchase of goods and services. It is increasingly evident that unless the flow of money is reversed towards the village, there can be little significant economic development as a self-sustaining community.
Broader Limited local availability of capital reserve (#PF2378).
Narrower Restrictive farm capital (#PJ0788) Excessive capital outflow (#PG9661)
Unviable commercial ventures (#PG9728)
Inadequate income of rural communities (#PJ9649)
Uncreditworthiness of rural communities (#PJ1268)
Limited availability of investment capital for urban renewal (#PF3550).
Related Low general income (#PD8568) Small local markets (#PJ8200)
Limited availability of financial credit (#PF2489).
Aggravates Underpayment for work (#PD8916).
Aggravated by High interest rates (#PF9014).

♦ PE1144 Simple assault
Nature Simple assault is willfully causing bodily injury (but not serious) to another person or by negligently causing bodily injury by means of a weapon which would likely cause death or serious bodily injury.
Broader Assault (#PD5235).
Related Aggravated assault (#PD0583).

♦ PE1145 Vibrations as a health hazard
Vibration in the working environment
Nature Vibrations are motions, usually unintentional, which occur frequently in rotating as well as stationary machines and structures. Prolonged exposure to vibration at a level well in excess of the threshold of perception (20-1000 Hz) leads to physiological disorders, and in some cases to occupational diseases (for example, as a result of using pneumatic drilling equipment or power saws). The very low-frequency vibrations (under 2 Hz) produced by various forms of transport can cause distress in some people. The disturbances originate in the central nervous system and are due to stimulation of the labyrinth. The degree of disturbance may vary with circumstances and particularly the weather. Low-frequency vibration (1.5-16 Hz) produced by vehicles is transmitted through the body to all the internal organs. The effects of such vibration are cumulative and include postural reactions, neurological symptoms, behavioural changes and problems of concentration. In addition to their effect on the human body, vibrations may cause damage to buildings, and in particular to buildings of historic interest which were not constructed in anticipation of such effects.
Incidence In all sectors of modern industry, machinery, equipment and power-driven tools can be encountered which generate intense vibration that may be transmitted to the workers who operate them.
Refs Bianchi, G; Prolow, K and Oledzki, A (Eds) *Man under Vibration* (1981); Carpentier, James and Cazamian, Pierre *Noise and Vibration in the Working Environment* (1978); Dupuis, H and Zerlett, G *Effects on Whole-Body Vibration* (1986); ILO *Protection of Workers Against Noise and Vibration in the Working Environment* (1984).
Broader Occupational hazards (#PC6716) Human destructiveness (#PA0832)
Hazards to human health (#PB4885).
Narrower Vibration sickness (#PE7788).
Related Environmental hazards of vibration (#PJ2171)
Health hazards of exposure to noise (#PC0268).
Aggravates Injuries (#PB0855) Mechanical failure (#PC1904)
Endangered monuments and historic sites (#PD0253).
Aggravated by Uncontrolled urban development (#PC0442).

♦ PE1153 Felonies
Nature Crimes have been traditionally classified into felonies and misdemeanours (with treason categorized separately). Common law felonies are those offences punishable by total forfeiture of land, goods or both. Such felonies include murder, manslaughter, sodomy, rape, robbery, arson, burglary and larceny. Under modern statutes, felonies may include those felonies for which a defendant is punishable by death or imprisonment.
Broader Offences of general applicability (#PD4158).
Related Misdemeanours (#PE5594).

♦ PE1154 Dutch elm disease
Ceratcystis ulmi
Nature Dutch elm disease is a fungal plant disease. The fungus is most frequently carried from tree to tree by female bark beetles. The leaves on one or more branches of a stricken tree suddenly wilt, turn dull green to yellow or brown then curl and finally drop off. On young trees the progress of the disease is rapid as the fungus spreads quickly through vascular tissue and the tree generally dies within two months; older, less vigorous trees take several years to die.
Background The disease probably originated in Asia. It was found in Romania in 1910, in France in 1919, in the Netherlands in 1920, the UK in 1927 and the US in 1930. By the late 1930s the disease had peaked after destroying thousands of trees. In 1970 an new outbreak was discovered in the UK.
Broader Ascomycetes (#PE4586) Fungal plant diseases (#PD2225)
Pests and diseases of elm (#PE2982).
Related Wood rots (#PE4455) Heart rot fungi (#PE6269) Cereal root rots (#PE4453)
Rusts of grasses and cereals (#PE2233) Smuts of grasses and cereals (#PE2234).
Aggravated by Beetles as insect pests (#PE1679)
Insect vectors of disease (#PC3597).

♦ PE1155 Mercury as a pollutant
Mercury poisoning in animals
Nature Mercury, the only common liquid metal, is released directly into the air during mining and smelting, from weathering of paint, and from losses during industrial processing. It leaches into water from old batteries and other discarded products and into soils directly from industrial uses, pesticides, and waste disposal. Proportionally, soils receive the most mercury, and erosion carries a significant amount to waterways. In water, bacterial action can change mercury into methylmercury, which is easily ingested by fish of all types and is accumulated in their fatty tissue. Mercury evaporates and also leaches into water from natural ore bodies at rates that are sometimes higher than those from human activity. Industrial discharges, however, are generally more concentrated and are released in populated areas The clinical picture of poisoning by alkylmercury compounds is well known. The patient may complain of headache; paraesthesia of the tongue, lips, fingers, and toes; and other nonspecific dysfunction. Early signs of more severe poisoning include

fine tremors of the extended hands, loss of side vision, and slight loss of coordination. Uncoordination may progress to the point of inability to stand or to carry out other voluntary movements. Occasionally there is muscle atrophy and flexure contractures. In other cases, there are generalized myoclonic movements. There may be difficulty in understanding ordinary speech. Irritability and bad temper are frequently present and may progress to mania. Occasionally the mental picture deteriorates to stupor or coma. Especially in children, mental retardation may be added to the symptoms of poisoning already mentioned. Many patients gradually become much worse after their illness has been recognized and exposure stopped. Even in those cases in which recovery occurs in the course of months or years, there may be little or no real neurological improvement, only an adaptation and re-education. The duration of illness in fatal cases has ranged from about one month to 15 years. Intercurrent infection, aspiration pneumonia, or inanition are the immediate causes of death in protracted cases.

Incidence An average of 4 million pounds of mercury is brought into commercial use each year. Since 1954, 100 million pounds of mercury have been mined, used, and reused or disposed of. This figure does not include the large amount released from processes other than those involved in production of the metal, for example, from coal combustion.

Acute poisoning by organomercurial compounds has been only infrequently reported in man, although methyl and other alkyl compounds have caused such poisoning. There have been many cases of chronic poisoning involving organomercurial. Most chronic cases caused by known organic chemicals have been associated with repeated exposure in connection with the manufacture of alkyl compounds, their use for treating seed, or the eating of treated seed. Such use of alkylmercury-treated seed as food has produced epidemics of poisoning in man, although this is less widespread than the poisoning of domestic animals due to this cause. The exact amount of any alkylmercury compound necessary to produce poisoning in man is not known, but is obviously small. The average intake of persons poisoned by dressed seed has been estimated at 2.7 mg per person per day and the highest intake at 8.2 mg per person per day. Since the late 1950s, nearly 1,080 Japanese have died and 2,871 have been recognized as seriously affected by Minamata disease, caused by eating fish polluted with methylmercury. The concentration of alkylmercury in fish and shellfish that led to this poisoning was in the range of 5 to 20 mg/kg calculated as mercury. Those who became sick gave a history of eating fish between 0.5 and 3 times per day. It has been estimated that the intake of mercury in fatal cases was at the rate of about 1.64 mg per person per day. Illness began to appear on the shore of the Minamata Bay about two years after mercury-containing waste from a plastics factory was diverted into the bay. In the United States, parts of two rivers, the North Fork of the Holston and the Shenandoah in Virginia, remain closed to fishing because of mercury contamination Silver dental fillings are only 30 percent silver and 50 percent mercury. The mercury is released in vapour form when people chew. Mercury is also released by the natural process of corrosion. The limited number of published scientific research experiments on amalgam biocompatibility have all strongly suggested that mercury and even the amalgam itself pose potential health problems of sufficient risk to warrant further extensive research.

Refs D'itri, P A and D'itri, F M *Mercury Contamination* (1977); Daniel, J W, et al *Mercury Poisoning, No 2* (1972); Nriagu, J O (Ed) *Biogeochemistry of Mercury in the Environment* (1979); Tsubaki, Tadao and Irukayama, Katsuro (Eds) *Minamata Disease* (1977); United Nations *Mercury Contamination in Man and His Environment* (1973); WHO *Health Effects of Methylmercury in the Mediterranean Sea* (1987); Watkin, J E (Ed) *Mercury in Man's Environment* (1971).
Broader Toxic metal pollutants (#PD0948) Chemical contaminants of food (#PD1694)
Environmental hazards from metals (#PE1401).
Related Poisoning in animals (#PD5228).
Aggravates Encephalitis (#PE2348) Water pollution (#PC0062)
Hazards to human health (#PB4885) Occupational risk to health (#PC0865)
Human physical genetic abnormalities (#PD1618).
Aggravated by Industrial emissions (#PE1869) Pesticides as pollutants (#PD0120)
Industrial waste water pollutants (#PD0575).

♦ PE1158 Dangerous toys
Undesirable influence of toys
Nature Many toys are capable of causing physical injury to children. Some may also cause psychological damage. (Examples are: do-it-yourself guillotine kit, simulated human organs dripping with gore, make believe hypodermic needles, and games involving killing and wounding.) Toys simulating weapons of modern warfare may orient the children to the acceptability of war as a method of resolving conflict, thus influencing their actions as adults and their attitude to foreigners. Realistic toy guns have been mistaken for real guns resulting in death or bodily harm.
Broader Risk (#PF7580) Consumer vulnerability (#PC0123)
Unethical entertainment (#PF0374).
Narrower Violent interactive toys (#PE4297).
Related Recreational war games (#PF1406) Gender-differentiated toys (#PE4664).
Aggravates War (#PB0593) Human violence (#PA0429)
Attitude manipulation of children through play (#PF2017).

♦ PE1159 Health risks to workers in electricity, gas, water and sanitary services
Broader Occupational risk to health (#PC0865).
Aggravates Industrial accidents (#PC0646) Occupational diseases (#PD0215)
Inadequate infrastructure (#PC7693).

♦ PE1160 Cadmium as a pollutant
Nature Cadmium occurs in zinc-lead-copper ores. Zinc ores constitute the main industrial source of cadmium and the metal is fractionated during the smelting or electrolytic processes employed for the refining of zinc. There is, therefore, the possibility of a considerable release of cadmium into the environment during zinc-refining operations and, to a lesser extent, during lead and copper smelting. Man has, therefore, been releasing cadmium into the environment from the time he was first able to smelt and refine these metals thousands of years ago. Cadmium is now used industrially as an antifriction agent, as a rust proofer, in plastics manufacture, in alloys, as an orange colouring agent in enamels and paints, in alkaline storage batteries, and for many other purposes.

World production of cadmium has increased heavily during this century. Only marginal amounts of cadmium were refined before 1920; in the late 1970s, world production was of the order of 15,000 tons per annum. The marked increase in cadmium use during the last three decades has caused a corresponding increase in environmental contamination and in problems caused by exposure at different states. The main pathways of cadmium to man are via inhalation or food intake. Cadmium is released into the air as a result of incineration or disposal of cadmium-containing products (for example, rubber tyres and plastic containers) and as a byproduct in the refining of other metals, primarily zinc.

Incidence Near smelters, atmospheric cadmium concentration can be as high as 0.5 micro g/m3 (microgram per cubic metre). High concentrations have also been encountered in some working environments, although more typical levels are now around 0.05–0.02 micro g/m3. One study showed that of 58 cities in the United States of America, cadmium was found in the air of 36, in concentrations ranging between 0.002 and 0.370 micro g/m3 (38). Of 29 nonurban areas, 17 showed cadmium levels of 0.004 to 0.026 micro g/m3. In nearly all cases cadmium was associated with zinc. Cadmium concentrations are also higher in soil and fresh water around smelters and industries processing materials that contain the metal. The amount of waterborne cadmium is affected by acidity, but concentrations up to 10 micro g/l are found in some mining areas, while in neutral or alkaline waters suspended particulate matter can contain as much as 700 micro g/l of cadmium. Concentrations in sediments can exceed 100 micro g/g. Plants vary in their ability to take up cadmium from soil, but some grasses, wheat, and lettuce do so fairly readily, and plant/soil ratios for most crops range between 0.5:1 and 2:1. When rice is grown in an environment highly contaminated by cadmium, concentrations can reach 0.5–1 micro g/g, which is 10 to 15 times higher than in noncontaminated areas. The sensitivity of plants to cadmium also varies, spinach, lettuce, and soya bean being affected when levels reach 3 to 4 micro g/g, while other species may tolerate concentrations ten or a hundred times greater. Aquatic organisms vary more widely in their sensitivity The main impact of prolonged human exposure to cadmium is on the kidney, although obstructive lung disorders can also result from respiratory exposure. The effect on the kidney is due to accumulation of cadmium in the renal cortex, leading to tubular protein urea. Cadmium absorbed by the body is only slowly excreted; as a consequence, cadmium toxicity is markedly cumulative so that there is the possibility of chronic cadmium poisoning among industrial workers regularly exposed to this metal or its compounds. An epidemiological survey of workers exposed to cadmium dust found excessive protein urea due to kidney damage in 68 percent of a group of male workers with over 20 years' exposure. Occupational exposure to cadmium oxide dust has been suggested to increase the risk of prostate cancer in man.
Refs Dekker, Lies, et al *Management of Toxic Materials in an International Setting* (1987); FAO *Water Quality Criteria for European Freshwater Fish* (1977); Joint Group of Exports on the Scientific Aspects of Marine Pollution *Cadmium, Lead and Tin in the Marine Environment* (1985); Mislin, Hans and Ravera, Oscar (Eds) *Cadmium in the Environment* (1986); Nath, R *Environmental Pollution of Cadmium* (1986).
Broader Toxic metal pollutants (#PD0948) Chemical air pollutants (#PD1271)
Chemical contaminants of food (#PD1694) Environmental hazards from metals (#PE1401).
Aggravates Water pollution (#PC0062) Chemical contamination of water (#PE0535).
Aggravated by Pesticides as pollutants (#PD0120)
Industrial waste water pollutants (#PD0575).

♦ PE1161 Lead as a pollutant
Lead in petrol — Lead in gasoline
Nature Lead is one of the most serious of environmental pollutants; no other toxic chemical pollutant has accumulated in man to average levels so close to the threshold for overt clinical poisoning. The possibility of a causal relationship between lead absorption and mental retardation has been suggested, but has also been disputed. However, acute encephalopathy, a very serious manifestation of lead poisoning in young children, is followed by permanent neurological consequences in at least 25 per cent of cases. Organic lead compounds are primarily neurotoxic. Most lead accumulates in bone and kidney with potential long-term effects.
Incidence Man-made sources of lead include lead smelting and refining, the combustion of leaded fuel, the production of storage batteries, the manufacture of alkyl lead and lead points and the application of lead-based pesticides. Lead pipes, lead-glazed earthenware and flaking lead points are possible sources of lead in the domestic environment. The predominant source of atmosphere lead appears to be from the use of 'antiknock' agents in petrol. Lead pollutes the air (above quiet roads the concentration may be in the range 0.25 – 1.2 g/m3 ; busy roads 2.5 – 4.5 g/m3 ; and congested roads up to 50 g/m3); it also pollutes fresh water (1–10 g/litre; this figure may be much higher in areas with lead pipes and soft, slightly acidic water), sea water (0,01 – 0,3 g/litre) and food (an average of 0,2 g/kg) Food is the major source of lead intake in adults who are not occupationally exposed. The contribution of airborne lead to the total daily absorption as compared to average dietary intake is more difficult to estimate, as it depends upon the concentration, particle size and solubility of the lead. Some scientists suggest that airborne lead is much more dangerous and that about 50 per cent of it may be absorbed on inhalation. Although symptoms of clinical lead poisoning in adults do not appear at levels in the whole blood below about 80 g/100g, the inhibition of certain enzymes involved in the synthesis of haem can be shown to occur at levels now present among urban populations. Russian workers have reported changes in conditioned reflex at low levels of exposure. Young children in urban and industrial areas are much more prone to lead poisoning than other sectors of the population. Such children may ingest lead from paint, roadside dust and industrial pollution. In New York and Chicago 1–2 per cent of children in low income areas were shown to have blood levels indicative of lead poisoning, whereas a further 25 per cent had levels above 40 g/100g. A study in London showed that 40.9 per cent of children living within 100–400m of a lead factory had blood lead levels in excess of 40 g/100g.
Claim Lead contributes to the high rate of osteoporosis, the brittle-bone disease that bends the backs, shortens the stature and breaks the hips of many older women. At high levels, lead in the blood of a pregnant woman can lead to a miscarriage or premature birth and at lower levels it can hold back the development of an unborn baby's nervous system and brain. Childhood is a risky time because exposure to lead during the first four years can damage nerve cells and retardation. Students with high levels of lead are more likely to be distracted and easily bored. Hyperactivity in a young child often turns out to be a symptom of lead poisoning, and hyperactive children are more likely to become delinquent.
Refs Farmer, P *Lead Pollution from Motor Vehicles, 1974–86* (1987); Joint Group of Exports on the Scientific Aspects of Marine Pollution *Cadmium, Lead and Tin in the Marine Environment* (1985); Nriagu, J O (Ed) *Biochemistry of Lead in the Environment* (1978); Nriagu, J O (Ed) *Biogeochemistry of Lead in the Environment* (1978).
Broader Toxic metal pollutants (#PD0948) Chemical air pollutants (#PD1271)
Chemical contaminants of food (#PD1694).
Related Environmental hazards from metals (#PE1401).
Aggravates Encephalitis (#PE2348) Water pollution (#PC0062)
Childhood accidents (#PD6851) Juvenile delinquency (#PC0212)
Lead as a health hazard (#PE5650) Occupational risk to health (#PC0865)
Diseases and injuries of bone (#PE3822).
Aggravated by Arsenic as a pollutant (#PE1732)
Motor vehicle emissions (#PD0414)
Pesticides as pollutants (#PD0120).

♦ PE1162 Misguided missiles
Refs Rosenblum, Simon *Misguided Missiles* (1985).

♦ PE1169 Jurisdictional conflict and antagonism within international nongovernmental organizations
Broader Obstacles to effective international nongovernmental organizations (#PF7082)
Jurisdictional conflict and antagonism within international organizations (#PD0047).
Aggravates Ineffectiveness of international nongovernmental organizations and programmes (#PF1595).

♦ PE1170 Inappropriate government intervention in agriculture
Nature The criteria underlying the planning of interventions lack an ecological orientation and are often dominated by short-term considerations, thus encouraging environmentally unsustainable farm practices. Agricultural policy tends to operate within a national framework (uniform criteria; prices and subsidies) which is insensitive to regional variations and needs, especially with

regard to environmental vulnerability. Overprotection of farmers (subsidies, tax relief, price controls), and the associated overproduction, encourage the degradation of the agricultural resource base.
Incidence Government intervention in agriculture is widely practised in both industrialized and developing countries. In industrialized countries it tends to be overprotective of agriculture whereas in developing countries such intervention tends to be weak and provide inadequate incentives to farmers, exposing them to a high degree of uncertainty.
Broader Excessive government control (#PF0304).
Aggravates Unsustainable agricultural development (#PC8419)
Distortion of national economies from food subsidies (#PE7413)
Over-subsidized agriculture in industrialized countries (#PD9802)
Inappropriate agricultural subsidies for chemicalized farming (#PE1785).
Aggravated by Lack of economic and technical development (#PE8190).

♦ PE1171 Experimental battering of animals
Nature Experiments are conducted on animals, especially to test safety features for humans, which involve beating or crushing limbs, and induction of a state of shock by battering an unanaesthetized animal in a revolving drum.
Incidence A common piece of equipment is the Noble–Collip drum which, like a tumble–drier, tumbles the unanaesthetized animal in a revolving drum in which there are triangular projections such that within 10 minutes the animal has endured 800 falls, usually resulting in broken teeth, concussion, bruising of internal organs, haemorrhaging from mouth and anus, engorgement of bowels, kidneys and lungs.
Broader Cruel treatment of animals for research (#PD0260).
Aggravates Experimental exposure of animals to pain (#PE1670).

♦ PE1174 Maltreatment of animals in aggression experiments
Nature In order to study the nature of aggressive behaviour, especially the implications for human behaviour, predators are used in prey–killing experiments. The predators may be partially mutilated prior to the experiment. This can involve castration, cutting the olfactory nerve (to prevent smell), or blinding. To stimulate aggression, electric shocks may be administered through metal grids or electrodes planted in the brain. The animals most frequently used are monkeys, cats and rodents.
Broader Cruel treatment of animals for research (#PD0260).

♦ PE1175 Breast cancer
Malignant neoplasm of breast
Refs Ames, Frederick C, et al (Eds) *Current Controversies in Breast Cancer* (1984); Ariel, Irving M and Cleary, Joseph *Breast Cancer* (1987); Baum, Michael *Breast Cancer* (1988); Brunner, S and Langfeldt, B (Eds) *Breast Cancer* (1987); Hayward, J L *Hormones and Human Breast Cancer* (1970); Lippman, Marc E, et al *Diagnosis and Management of Breast Cancer* (1988); Scarff, R W and Torloni, H *Historical Typing of Breast Tumours* (1977).
Broader Malignant neoplasms (#PC0092).

♦ PE1177 Obesity
Corpulence — Overnutrition — Overweight obesity
Nature Obesity becomes a health problem when it leads to disability, illness or increased risk of death, and these effects are difficult to quantify. Although serious overweight decreases life expectancy, it is by no means clear why. Obesity predisposes to maturity–onset diabetes, but the evidence linking it to hypertension is controversial and per se it appears to be a relatively minor risk factor for coronary heart disease. It does, however, cause disabilities, including osteo–arthritis of weight–bearing joints, varicose veins, increased surgical and obstetric risks and numerous psycho–social problems.
Incidence Obesity was recognized as a serious health problem during the 1970s in industrialized countries and among affluent groups elsewhere, and its prevalence appears to be increasing. In the United Kingdom, body fat has increased by about 10 percent in all age groups over the last four decades; in Czechoslovakia there was a striking increase in the incidence of obesity in men and women between 1956 and 1972; and in a national survey in Canada more than 60 percent of people of both sexes aged over forty were said to be overweight. From 1960 to 1980 the percentage of American white women aged 25 to 34 who were obese increased from 13.3 to 17.1 percent of the population and the percentage of obese American black women increased from 28.8 to 31 percent. Obesity also increased in some countries in the South Pacific region following dietary changes (especially increased consumption of refined carbohydrates), and the incidence of hypertension, diabetes and gout rose in parallel.
Claim People in the West on average take in around 40 percent more calories than they need for their physical wellbeing.
Refs Crepaldi, Gaetano, et al *Diabetes, Obesity and Hyperlipidemias* (1983); Enzi, G et al *Obesity* (1981); Kowalski, Charlotte J *Obesity* (1987).
Broader Stigmatized diseases (#PD7279) — Human disease and disability (#PB1044)
Negative emotions and attitudes (#PA7090).
Narrower Childhood obesity (#PE1814) — Obesity due to low metabolism (#PE4463).
Related Substance abuse (#PC5536).
Aggravates Eczema (#PE2465) — Diabetes (#PE0102)
Hypertension (#PE0585) — Heart diseases (#PD0448)
Mental deficiency (#PC1587) — Chronic bronchitis (#PE2248)
Lung disorders and diseases (#PD0637) — Perinatal morbidity and mortality (#PD2387)
Diseases of the skin and subcutaneous tissue (#PC8534).
Aggravated by Decadence (#PB2542) — Myxoedema (#PG3408)
Inactivity (#PB7991) — Over–eating (#PE5722)
Human ageing (#PB0477) — Alcohol abuse (#PD0153)
Gland disorders (#PD8301) — Emotional crises (#PE3407)
Diseases of metabolism (#PC2270) — Pituitary gland disorders (#PE2286)
Excessive consumption of sugar (#PE1894)
Dietary deficiencies in developed countries (#PD0800)
Environmental hazards of nuclear weapons industry (#PE5698).
Reduced by Dietary restrictions (#PJ1933).

♦ PE1178 Criminal association
Broader Criminal conspiracy (#PD1767) — Unethical personal relationships (#PF8759).
Aggravates Vigilantism (#PD0527) — Recidivists (#PE5581)
Organized crime (#PC2343) — Juvenile delinquency (#PC0212).

♦ PE1181 Criminal usury
Loan–sharking — Enticement into debt
Nature Engaging in criminal usury is to engage in or to finance directly or indirectly extension of credit at a rate of interest more than what is enforceable through civil courts. Enforceable rates vary from 0 percent in some countries whose laws are based on the Koran to over 50 percent per year. In a more severe form, loan sharks entice people into debt at interest rates up to 100 per cent, then threaten or inflict violence upon them, their families, or their property if payments are not made. Most victims fail to report such violence for fear of retribution.
Incidence Loan–sharking is a characteristic of poverty in inner cities where many have to borrow money in order to meet day to day or urgent needs. Although people may be receiving support from welfare services, significant portions of such payments may be deducted before they receive it, at times when they have other urgent needs.
Refs Mark, Jeffrey *An Analysis of Usury* (1980).
Aggravated by Urban poverty (#PC5052).

♦ PE1182 Distortion of international trade by minimum pricing regulations and other measures to regulate domestic prices
Nature Protection and commodity price support by governments have resulted in widely divergent producer prices in different countries. Although this intervention has brought stability to national markets, it tends to hinder the process of adjustment to supply and demand. The policy decisions of official agencies will often be widely different from the responses of producers, traders, and consumers in a free economy. Import demand is rendered less sensitive to offers. Exporting countries are strongly motivated to protect their producers from serious price falls in the international market.
Broader Non–tariff barriers to international trade (#PC2725)
Distortion of international trade by discriminatory customs and administrative entry procedures (#PE2603).
Related Unfair pricing by transnational corporations (#PE5855).

♦ PE1184 Inadequate coordination between regional intergovernmental organizations with common membership
Broader Inadequate coordination of the intergovernmental system of organizations (#PE0730).
Aggravates Fragmented regional cooperation (#PF9129).
Aggravated by Jurisdictional conflict and antagonism between regional intergovernmental organizations with common membership (#PE1583).

♦ PE1186 Cannabis abuse
Abuse of marijuana and hashish
Nature Abuse of cannabis resin in the form of marijuana, hashish or other preparations results in toxicity which varies widely depending on the degree of THC (tetrahydrocannabinal) present. This in turn depends on ecological conditions, the part of the plant used, the preparation process and transport and storage conditions. Long term use may give rise to conjunctivitis, bronchitis and chronic catarrhal laryngitis. It may heighten the possibility of cerebral atrophy. It is associated with a high incidence of psychosis in the Middle East. However, no conclusive evidence connects these conditions with cannabis abuse, nor with an increase in crime and violence rates. Reliable evidence points to variation in perception of time and space, disinhibition, fragmentation of thought and an altered sense of identity, although clinical studies are often complicated by the intake of alcohol and other drugs by the cannabis users under observation. In small doses cannabis acts as a mild euphoriant and sedative; in large doses as a hallucinogen. Pharmacological tolerance develops in time. There are no withdrawal symptoms, but users often show a psychological dependence. Its use is associated with multi–drug abuse, but no conclusive evidence exists to show that it leads on to abuse of stronger drugs such as LSD, heroin, cocaine. Marijuana has been adulterated with other drugs and sold on the street to unsuspecting users.
Incidence It has been estimated that one third of US university students have tried marijuana; that one seventh are habitual users; and that there are 200 million users in the world.
Background Cannabis resin has been used as an intoxicant for the last 4000–5000 years. It has been linked with the traditional way of life in Islamic countries where alcohol is prohibited. In eastern countries it has been traditionally used in medicine and continues to be used for medical purposes in India and Pakistan.
Counter–claim There is no record in the extensive medical literature describing a proven, documented cannabis–induced fatality. Marijuana has been demonstrated to overcome the side effects of chemotherapy and for relieving the symptoms of multiple sclerosis.
Refs Auld, John *Marijuana Use* (1981); DuToit, Brian M *Cannabis in Africa* (1980); Fehr, Kevin and Kalant, Harold (Eds) *Cannabis and Health Hazards* (1983); Gold, M S *Marijuana* (Date not set); Kalant, O J, et al (Eds) *Cannabis* (1983); Mechoulam, Raphael *Marijuana* (1973); Miller, L L *Marijuana* (1974).
Broader Substance abuse (#PC5536) — Abuse of plant drugs (#PD0022).
Narrower Cannabis intoxication (#PG7872).
Aggravates Drug smuggling (#PE1880)
Inadequate sense of personal identity (#PF1934).
Aggravated by Illicit drug trafficking (#PD0991).

♦ PE1187 Domination of labour relations by transnational corporations
Nature The immobility and fragmentation of labour organization in respect to ability to cross national boundaries, and the greater geographical flexibility and centralized decision making of many transnational corporations, allows a favourable balance of bargaining power sharply in favour of the corporations. Decisions having repercussions on working conditions and the social rights of the employees are often made outside the country in which they are implemented, and the employees usually have no access to the decision makers.
Broader Transnational corporation imperialism (#PD5891)
Labour tensions involving transnationals (#PE5927).
Narrower Interference in labour relations in developing countries by transnational corporations (#PE7355)
Interference in labour relations of industrialized countries by transnational corporations (#PE0667).
Aggravates Inaccessibility of decision–makers in multinational enterprises (#PE0573).

♦ PE1188 Long–term shortage of animal and vegetable oils and fats
Broader Long–term shortage of commodities (#PC1195)
Insufficient availability of goods (#PB8891).
Narrower Shortage of fixed vegetable oils and fats (#PE8277)
Shortage of animal and vegetable oils and fats and waxes (#PE6498).

♦ PE1193 Transfer pricing
Profit repatriation concealed by transfer pricing
Nature Intra–firm trade is becoming an increasingly important component of world trade; and the dominant market position which is held by transnational corporations in certain clusters of industries provides substantial scope for differential pricing policies which may result in an abuse of such a position.
Such an abuse occurs through manipulation of prices for goods and services sold within the corporation involving transactions both between parents and subsidiaries and among individual subsidiaries themselves, and is possible because of the oligopolistic and monopolistic power exercised by transnational corporations in respect of the production and distribution of particular goods. Transnational corporations frequently hold such market power not only in the national markets of both developed and developing countries, but also in the world market. Moreover, such abuses of a dominant position in transfer prices for intra–firm trade may also be reflected in the prices charged subsequently to third parties by the selling unit of the corporation, and occur in respect both of the prices of goods for resale without further transformation and of goods which are further manufactured using inputs supplied on an intra–firm basis. Abuses of a dominant position deriving from manipulations of transfer prices are in consequence likely to have an adverse effect on market and industrial structures and on the balance of payments of either the home or the host countries in which transnational corporations operate.

Broader Restrictive trade practices (#PC0073)   Transnational corporation imperialism (#PD5891)
Unfair pricing by transnational corporations (#PE5855).
Aggravates Restrictive shipping practices (#PD0312)
Obstacles for international ocean shipping (#PD5885)
Outflow of financial resources from developing countries (#PC3134).
Aggravated by Evasion of shipping regulations and taxes by flags of convenience (#PE5873)
Disproportionate control of global economy by limited number of corporations (#PE0135).

♦ **PE1197 Long-term shortage of coffee, tea, cocoa, spices and manufactures thereof**
Broader Long-term shortage of food and live animals (#PE0976).
Narrower Spices shortage (#PU3457)   Chocolate shortage (#PU3455)
Tea and mate shortage (#PU3456).
Related Slowing growth in food production (#PC1960).

♦ **PE1198 Economic barriers to women's access to the judicial process**
Nature Women are less likely than men to possess the assets needed to finance litigation to protect their individual rights. Fewer women than men are employed and the women who are employed often receive less pay for the same work. Furthermore, many husbands, either by law or de facto, exercise control over household assets and expenditures.
Broader Economic discrimination in the administration of justice (#PE1399).
Related Discrimination against women in employment (#PD0086).
Aggravated by Collapse of judicial system (#PJ0761)
Family dependence on patriarchal role of the man (#PD3555).

♦ **PE1200 Aiding escape of prisoners of war or enemy aliens**
Nature It is a crime to aid in the escape of a prisoner of war or an enemy alien in the time of war. This includes interfering with hindering delaying or preventing the discovery or capture of such a person. This not only returns a person to the enemy but demoralizes those fighting against the enemy.
Broader Aiding national security criminals (#PD7407).
Aggravates Animal injuries (#PC2753)   Endangered species of animals (#PC1713).

♦ **PE1208 Banking law violations**
Nature An employee or owner of a bank who knowingly violates banking regulations, such as, not declaring deposits in excess of stipulated amounts is guilty of banking law violations.
Broader Economic crime (#PC5624)
Violations against economic regulations (#PD7438).
Narrower Money laundering (#PE7803).
Aggravated by Unethical financial practices (#PE0682).

♦ **PE1209 Inadequate coordination of international nongovernmental organizations and programmes**
Nature Irrespective of whether INGOs duplicate each others' activities, INGOs with complementarity programmes, preoccupations, common positions, or common operational problems have considerable difficulty in linking together in some coordinated activity of other than a token nature. The absence of powerful inter-INGO federations with a common position considerably weakens their ability to act under certain circumstances and makes it easy to out-manoeuvre their separate actions and difficult to support their common position.
Counter-claim It is questionable whether the organizational models for such confederations are adequate for the complexity of the pressures which they are expected to bring into focus and reconcile. In addition, emphasis is now being placed on functional and regional decentralization. INGO disarray is paralleled by that of the agencies of the UN system; weak and inadequate systems of coordination are in fact a general problem of the times.
Broader Obstacles to effective international nongovernmental organizations (#PF7082)
Inadequate coordination of international organizations and programmes (#PD0285).
Aggravates Ineffectiveness of international nongovernmental organizations and programmes (#PF1595)
Competition between international nongovernmental organizations for scarce resources (#PE0259).
Aggravated by Jurisdictional conflict and antagonism between international nongovernmental organizations (#PE0064)
Jurisdictional conflict and antagonism between the specialized agencies of the United Nations (#PE2486).

♦ **PE1210 Sulphur dioxide as a pollutant**
Sulphur dioxide as an occupational hazard
Nature Concentrations of sulphur dioxide (SO2) in the atmosphere are increasing, particularly in certain urban areas but also in regions remote from industrial emissions. It often occurs together with fine particulate matter, emission of which is also increasing. In the presence of water vapour and certain particulate matter in the atmosphere it is transformed into a fine mist of sulphuric acid.
Sulphur dioxide and its transformation products in the atmosphere may cause: haze conditions which reduce visibility; increased rate of corrosion of metals; damage to certain species of plants; and aggravate emphysema and chronic bronchitis in man. At the concentrations sometimes reached near oil refineries, the level (3 ppm) can be high enough to produce an irritating odour. At lower concentrations (0.3 ppm), and with 8-hour exposure, injury to certain species of plants can be caused. At even lower concentrations (0.12ppm) in humid conditions, corrosion rates have been shown to be 50 per cent higher than in dry climates. Even at relatively low concentrations (0.10 ppm) there may be reduced visibility (8 kms) if the sulphur dioxide is produced over a large area causing a haze which is lightly scattered along the whole path between the object and viewer.
The main source of sulphur dioxide production is the burning of fossil fuel, particularly coal and oil. Both coal and oil vary in their sulphur content depending on the region of origin. Commonly used coal can vary from 0.5 to 6 percent sulphur content by weight, while crude oil varies from virtually zero to 4.5 percent sulphur content by weight. In the process of oil refining, most of the sulphur remains in the heavy fraction which is used in large scale heating installations. The other major sources are industrial and include smelting, sulphuric acid manufacture and petroleum refining.
Sulphur dioxide has a limited life in the atmosphere. It is oxidized within a few days, and the major process for removal is by precipitation in rain or snow. Nevertheless, the general expectation that requirements for energy will double in the next decade, means that despite the increasing use of low sulphur natural gas and nuclear energy, the use of oil and coal for combustion will increase and, hence, there will be increased pollution unless control measures are taken. Furthermore, depending on meteorological conditions, large masses of air with high concentrations may be carried over long distances (several hundred kilometers) to cause pollution conditions far from the source of the initial pollution. In this context, a particular concern is the phenomenon of acid rain which is causing widespread damage to forests, particularly in Sweden and Germany, as a result of sulphur dioxide emission in industrial areas of Western Europe. This transfer of pollution across national boundaries is beyond the control of the individual countries affected.
There is, at the present time, no certain proof of effects on man's health at the levels generally occurring in cities, nevertheless, there is an appreciation that when occurring in the presence of high concentrations of particulate combustion products, it may give rise to a synergistic effect on man which may be more injurious than the sulphur dioxide itself.

Occupational exposure may occur in oil refineries, certain mines for sulphur or sulphur-containing ore, in smelters where sulphur-containing ore is roasted, in the paper and pulp industry, in factories manufacturing sulphuric acid, in some chemical plants where sulphur dioxide is used for organic synthesis, and in any work near chimneys or furnaces where coal, oil or other fossil fuel is burned. Sulphur dioxide acts as a powerful irritant to the mucous membranes of the eyes and the upper respiratory tract. It causes rapid, acute irritation of the eyes with tears and redness; its action on the upper respiratory tract causes cough, shortness of breath and spasm of the larynx.
Claim Some 600 million people live in urban areas where the average level of sulphur dioxide pollution endangers their lives.
Counter-claim In many areas of the world, atmospheric sulphur dioxide has been a major source of sulphur for vegetation, including agricultural crops. As pollution regulations restrict the amount of sulphur dioxide emitted to the atmosphere, it is becoming increasingly necessary in many countries to provide essential plant nutrient sulphur to crops by other means, usually by incorporating sulphur into fertilizers.
Broader Occupational hazards (#PC6716)   Chemical air pollutants (#PD1271)
Chemical pollutants of the environment (#PD1670).
Related Acidic precipitation (#PD4904)   Sulphur compounds as pollutants (#PG6442).
Aggravates Occupational diseases (#PD0215)   Occupational risk to health (#PC0865).
Aggravated by Industrial emissions (#PE1869)   Motor vehicle emissions (#PD0414).

♦ **PE1213 Consumer boycotts**
Incidence American trade unions have boycotted consumer products of big employers. More and more mass protests are organized by middle-class zealots supporting complicated and divisive issues like abortion, experiments on animals, the environment or political repression in foreign countries. Campaigns have been organized, for example, to stop canners selling tuna fish caught in nets that also trap and kill dolphins, against hamburger chain importing beef from ranches that had destroyed tropical rainforests, against companies experimenting on animals and against wearing fur coats.
Broader Boycott (#PE8313).
Reduces Unsafe design of consumer products (#PF1379)
Lack of consumer influence on industry (#PE1940)
Environmentally unfriendly consumer products (#PD9310)
Non-payment of compensation for damages to consumers (#PE0290).

♦ **PE1216 Long-term shortage of electric energy**
Broader Long-term shortage of natural resources (#PC4824)
Long-term shortage of mineral fuels and lubricants (#PE1712).
Aggravated by Inadequate electricity infrastructure (#PD9033)
Underutilization of oil shale as an energy source (#PF0445).

♦ **PE1218 Long-term shortage of cereals and cereal preparations**
Broader Long-term shortage of food and live animals (#PE0976).
Narrower Long-term shortage of rice (#PG5759)   Long-term shortage of wheat (#PE2903)
Meal and flour of wheat or meslin shortage (#PE8980)
Meal and flour other than wheat or meslin shortage (#PE8606).
Related Slowing growth in food production (#PC1960).

♦ **PE1221 Spina bifida**
Refs McLaurin, Robert L, et al *Spina Bifida* (1986).
Broader Diseases of the spine (#PD2626)   Human disease and disability (#PB1044)
Human physical genetic abnormalities (#PD1618).
Narrower Congenital hydrocephalus (#PG6205).
Related Hydrocephalus (#PJ3190).
Aggravates Paralysis (#PD2632)   Kidney disorders (#PE2053)
Mental deficiency (#PC1587)   Urinary bladder disorders (#PE2307).

♦ **PE1222 Autism**
Infantile autism — Autistic children
Nature Infantile autism is characterized by extreme self-isolation and, more generally speaking, describes any non-communicating behaviour in children. Autistic children find it very difficult to use or comprehend language or any kind of communication. The autistic child always takes longer to acquire language, and some never develop any language skills at all. About half are not speaking any words by the age of five. When language does come it is usually halting and stilted. The child often uses words in unusual ways or even makes up his own words. The apparent withdrawal of the autistic child has traditionally been seen as resulting from an unresolved conflict between fear and curiosity in social encounters. Recent evidence, however, leads researchers to believe there are genetic preconditions, such as a "fragile X" chromosome that increase the likelihood of the condition.
Incidence Studies in the United States and in England have established an incidence of about four psychotic children per 10,000 births. Since true autism seems to represent about ten percent of children loosely called autistic, about one child in 25,000 could be a realistic estimate of classical autism.
Refs Schopler, Eric and Mesibov, Gary *Autism in Adolescents and Adults* (1983); Schopler, Eric and Mesibov, Gary B *Communication Problems in Autism* (1985).
Broader Schizophrenia (#PD0438)   Developmental disabilities (#PD3773).
Related Mental deficiency in children (#PD0914).

♦ **PE1234 Ignorance of administration**
Broader Ignorance (#PA5568).
Narrower Ignorance of procedures (#PJ1219)
Underdeveloped administrative abilities (#PG7799).
Aggravates Paralyzing patterns between villages and administrative structures (#PF1389).

♦ **PE1235 Instability of trade in undressed hides, skins and fur skins**
Broader Instability of trade in inedible crude non-fuel materials (#PD0280).
Narrower Instability of the fur trade (#PE1474).
Instability of trade in undressed hides and non-fur skins (#PE8106).
Aggravated by Reduction in demand for primary commodities due to technological change (#PD1276).

♦ **PE1243 Bovine leukosis**
Bovine lymphosarcoma — Bovine malignant lymphoma — Bovine leukaemia
Refs European Communities *International Symposium on Bovine Leukosis* (1985).
Broader Viral diseases in animals (#PD2730)
Circulatory system diseases in animals (#PD5453).

♦ **PE1253 Long-term shortage of beverages and tobacco**
Broader Long-term shortage of commodities (#PC1195)
Insufficient availability of goods (#PB8891).
Narrower Shortage of tobacco and tobacco manufactures (#PE7926).

♦ PE1255 Forced to witness torture

Nature Forcing prisoners to view the torture of other prisoners or members of their own family is intended to inflect pain, a sense of guilt and of powerlessness. Victims are often told that it is their fault that the person is being tortured. Children are sometimes required to watch the torture of their parents to further torment the parents. Couple can be tortured together, for example the wife being raped while the husband watches.
Incidence Witnessing torture has been reported in a number of Islamic countries.
Broader Induction of incongruent actions (#PF3790).

♦ PE1256 Methane gas emissions from landfill sites

Nature Putrescible matter producing methane gas in landfills has caused explosions resulting in damaged property, caused injury and death. The danger may present from 10 to 20 years for any given site.
Broader Hazardous wastes (#PC9053)
Increase in atmospheric concentration of methane (#PE8815).
Narrower Damage by degradable organic matter (#PJ6128)
Inadequate facilities for storing, transporting and processing solid wastes (#PE2829).
Aggravates Environmental pollution (#PB1166) Contamination of drinking water (#PD0235).
Aggravated by Inappropriate sanitation systems (#PD0876).

♦ PE1261 Long-term shortage of chemicals

Broader Long-term shortage of commodities (#PC1195)
Insufficient availability of goods (#PB8891).
Narrower Medicinal products shortage (#PE3502)
Shortage in tanning and dyeing materials (#PE8501)
Long-term shortage of manufactured fertilizers (#PE0997)
Shortage of coal-derived tar and crude chemicals (#PE8242).

♦ PE1264 Environmental hazards from forestry and logging

Nature The mismanagement of forest lands and forest resources over centuries, mainly due to the need to provide land on which to grow food, has led to a situation where the forest is now in rapid retreat. The main aspects of the situation are: serious shortages in the supply of industrial wood; the catastrophic erosion and floods accompanying the stripping of forests from mountainous land - especially in the Andes and the Himalayas; the acute shortages of fuel wood in much of the developing world; the spread of desert conditions at an alarming rate in the arid and semi-arid regions of the world; and the many environmental effects of the destruction of tropical rain forests.
Background Much of the history of mankind has consisted in driving back the boundaries of the forest - first for grazing land and shifting cultivation, then for permanent agriculture. Most of our present cultivated and inhabited lands have been derived from forest. In favourable sites - such as the deltas of the Mekong and Nile, or on the fertile volcanic soils of Java, the agricultural experiment has been successful. In unsuitable sites, on the other hand, it has led to land degradation; and in some cases, over-exploitation has been shown to cause catastrophic erosion and the collapse of civilizations, as that of the Mayans in Guatemala. For centuries, forests have been despoiled of desirable species, and this has sometimes led to the near-extermination of some varieties; the cedars of Lebanon and of Cyprus were an early and notorious example. Forest soils were thought of as only there to be cleared and tilled; timber, fuelwood, animals for meat and skins, and other forest produce were there to be harvested. The numerous environmental benefits - the moderation of local climates, clear and regular water, freedom from soil erosion, protection from avalanches, and abundant virginal surroundings - were part of the natural order of things and taken for granted. The values of forests only began to be appreciated when they had nearly disappeared. Greek foresters are only now working to undo the damage done in the 3rd century BC. Sustained management began in English woodlands in the Middle Ages when only sufficient woodlands were left to meet the needs of local markets. The Swiss forest law was passed in the late 19th century in response to the alarming effects of over-exploitation of the forests in the alpine valleys; at the much the same time, the incoming British administration began forest restoration in Cyprus as a reaction to the devastated state of the forest there. Each of these was a local reaction and confined to the forest values considered important at that place and time - timber supply, soil erosion, water, danger of avalanches.
Claim A rapidly expanding world population is making increasing inroads into forest lands. Yet every country in the world also depends on the goods and benefits provided by forests, which in the future will have to be met from an ever decreasing area. Forest lands are urgently needed for three purposes: for new agriculture; for the production of timber and fuelwood; and for a whole range of environmental benefits (catchment protection, the preservation of genetic resources, wildlife conservation, amenity and pleasure). Unplanned clearing and exploitation of forests continues to lead to land degradation much faster than well planned development brings land into sustainable productive use - either for agriculture or forestry. There is also the spread of desert conditions and the degradation of mountain catchments. In addition, the speed of change is bringing new preoccupations about the loss of wild species and genetic diversity.
It is no coincidence that the forests of all the countries with major crop failures in recent years due to droughts or floods - Bangladesh, Ethiopia, India, Pakistan, the Sahel countries - had been razed to the ground and had not been replaced by woody vegetational cover.
Broader Ecologically unsustainable development (#PC0111).
Related Health risks to workers in agricultural and livestock production (#PE0524).
Aggravates Deforestation (#PC1366).

♦ PE1265 Destructive action of mould in tropical climates

Nature A combination of high temperature (25-30 degrees C) and abundant moisture (70 per cent relative humidity) creates the optimal conditions for cryptogamic growth (fungi, algae, mosses and lichens). These develop on any kind of manufactured organic material, complicating cleaning operations without which the object is permanently discoloured and eventually disintegrates. Mould action is particularly serious in connection with the preservation of paper, books and other records.
Related Corrosion in tropical climates (#PE1811)
Detrimental effect of jungle environment in tropical villages (#PE2235).
Aggravates Inadequate protection and preservation of cultural property (#PF7542).
Aggravated by Humidity (#PD2474) Inhospitable climate (#PC0387).

♦ PE1274 Distortion of international trade by discriminatory formulation of health and sanitary regulations for agricultural and pharmaceutical products

Nature Almost all countries have legislation for the protection of the health and safety of their human, animal and plant populations. The basic objective of these laws and regulations is to prevent introduction and spread of human, animal and plant diseases, to ensure adequate sanitary conditions and minimum health standards in units producing food and other articles for human and animal consumption, and to protect consumer interests by prohibiting the sale of unwholesome and contaminated food. However, such regulations may be structured so that deliberately or inadvertently they act as a barrier to trade, by being unreasonable or unnecessarily restrictive, inconsistent with regulations in other countries, or requiring excessive amounts of documentary evidence.
Broader Distortion of international trade by discriminatory customs and administrative entry procedures (#PE2603).

♦ PE1275 Fraudulent impersonation

False personation
Broader Unethical personal relationships (#PF8759).
Related Impersonating officials (#PE7687).
Aggravates False statements (#PF4583) False qualifications (#PF0704).

♦ PE1277 Social isolation of computer users

Terminal isolation — Atomization of computer-based work — Social isolation of computer-based recreation
Aggravates Computer stress (#PE5053)
Downgrading of jobs due to computerization (#PE5014)
Health hazards of computer visual display units (#PE5083).
Aggravated by Computer obsession (#PF1288) Proliferation of computers (#PE3959).

♦ PE1278 Limited access to pharmaceutical drugs

Unavailability of medicines
Refs Medawar, Charles *Drugs and World Health* (1984).
Broader Insufficient medical supplies (#PE1634).
Aggravates Adulteration of illicit drugs (#PE0456).
Aggravated by Outdated forms of community health (#PF1608).

♦ PE1280 Environmental hazards from food processing industries

Environmental hazards of food products
Refs United Nations *First World-wide Study on the Food Processing Industrie.*
Broader Environmental hazards from food and live animals (#PC1411)
Environmental hazards from manufacturing industries (#PD0454).
Related Environmental hazards from tobacco and tobacco manufactures (#PE0483).
Aggravates Unclean food (#PJ2532) Food pollution (#PD5605)
Fear of food contamination (#PF3904) Microbial contamination of food (#PD9669)
Health hazards of irradiated food (#PD0361) Food manufacturing industry wastes (#PE8702)
Carcinogenic consequences of food preparation (#PE6619).
Aggravated by Malodorous fumes (#PD1413) Toxic food additives (#PD0487)
Food spoilage in storage (#PD2243) Unethical food practices (#PD1045)
Predominance of fast food (#PF5940)
Underdevelopment of food processing industries (#PD0908)
Infected animal, meat and animal product shipments (#PE7064)
Unsanitary and inhumane urban food animal conditions (#PE0395).

♦ PE1281 Gypsy persecution

Romanies — Discrimination against gypsies
Nature Gypsies constitute a tribe or race of uncertain origin; they are usually nomadic and therefore underprivileged and often suffer general discrimination. In general, society associates gypsies with petty crime, theft, confidence tricks and begging. They are also felt to be insanitary and for this reason gypsies are frequently ostracized by the local community. Regarding theft, gypsies have a different code of morality, though this is very strict. This may be that it is a crime to steal from someone poorer than yourself (the tradition in Serbia) or that it is a crime to steal something you do not need (the tradition in England). Because there is very little common land left in Western European countries, gypsies in these states have difficulty in finding camping sites and when they do, they are usually driven off by the local community within a short time. Because of constantly moving, gypsy children tend to have an inadequate education, which confines them later to traditional gypsy occupations, such as being musicians, horse dealers, hawkers, blacksmiths, coppersmiths, fairground entertainers, and beggars. Those who want to leave the nomadic life may meet discrimination in housing and employment. For those who remain nomadic, their isolated and close-knit society makes it difficult for public authorities to administer a reasonable and acceptable standard of social welfare.
Incidence Of the estimated world population of 1,000,000 gypsies, more than half live in Romania and Hungary and at least 150,000 in Bulgaria, Macedonia and Yugoslavia. Between 250,000 - 500,000 gypsies died in the concentration camps or massacres in the Second World War. According to the Nazis the gypsies were a genetically criminal people who had polluted their Aryan blood.
Broader Nomadism (#PF3700) Underprivileged racial minorities (#PC0805)
Restrictions on freedom of movement between countries (#PC0935).
Narrower Land enclosure (#PJ3523).
Aggravates Superstition (#PA0430) Inadequate social welfare services (#PC0834).
Aggravated by Social injustice (#PC0797) Social fragmentation (#PF1324)
Insecure land tenure (#PD9162) Homogenization of cultures (#PB1071)
Isolation of ethnic groups (#PC3316) Discrimination against minorities (#PC0582)
Lack of variety of social life forms (#PE8806).

♦ PE1282 Carbamate insecticides as pollutants

Nature The first carbamic acid derivatives having insecticide properties were synthesized in 1947. The most commonly used carbamate insecticide in agriculture is carbaryl. It is a systemic poison which produces moderately severe acute effects when ingested, inhaled or absorbed through the skin. It may cause local skin irritation. Being a cholinesterase inhibitor, it is much more active in insects than in mammals. This compound is only partly broken down in plants to non-toxic compounds, and metabolites with anticholinesterase properties can become translocated to a certain extent into plant tissues.
Incidence More than 1,000 carbamic acid derivatives are how known. More than 50 of them are used as pesticides, herbicides, fungicides and nematocides. Total diet studies indicate a daily intake of 0.02 mg of carbaryl from meat, fish and poultry.
Broader Insecticides as pollutants (#PD0983) Chemical contaminants of food (#PD1694).

♦ PE1289 Icebergs

Sea ice
Nature Icebergs should be distinguished from polar pack ice, which is frozen sea water, and from ice islands, which are large areas of consolidated multi-year pack ice. An iceberg is a large mass of ice which has broken off and drifted from parent glaciers or ice shelves along polar seas. Icebergs drift with ocean currents into shipping lanes where they constitute a hazard. In polar seas their presence, together with pack ice, may prevent normal shipping movements.
Incidence It is estimated that approximately 16,000 icebergs are 'calved' annually in the Northern Hemisphere. Of these, approximately 90 per cent are produced by glaciers along the coast of Greenland. These may weigh over two million tons. About 400 icebergs drift past Newfoundland each year, the remainder having been stranded or eroded in the north. The Northern icebergs are rarely more than 2,000 feet in width or over 400 feet above water level. Antarctic icebergs are much larger and may reach over 100 miles in length. The heaviest concentration of icebergs in the region of the Grand Banks or Trans-Atlantic shipping lanes occurs from April to June each year.
Broader Hazards to navigation (#PE3868).
Related Ice accretion (#PD1393).
Reduced by Reduction of glacier size (#PE4256).

♦ PE1293 Sexual harassment of men
Sexual provocation by women — Sexual provocation of men — Female sexual exhibitionism
Broader Exhibitionism (#PD4643)   Sexual harassment (#PD1116).
Aggravated by Disobedient wives (#PF4764)   Decadent clothing (#PE5607).

♦ PE1295 Discrimination against juveniles in judicial proceedings due to protective legislation
Nature In some jurisdictions special procedures are applicable to minors which, although originally intended to protect them and spare them from certain painful situations (in particular in the case of criminal trials), in effect permit decisions and denials of liberty which would not be tolerated, and in fact would be inconceivable, in relation to adults. Discrimination against juveniles may therefore arise out of the very legislation that was originally intended to protect them by shielding them against the harshness of criminal justice.
Broader Injustice (#PA6486)   Discrimination before the law (#PC8726).
Aggravates Repressive detention of juveniles (#PD0634).
Aggravated by Collapse of judicial system (#PJ0761).

♦ PE1296 Torture by violation of taboos
Nature People are torture by violating religious and cultural taboos. Devout Muslims and Jews are forced to eat or are covered with pork. Devout Muslims are denied the opportunity to pray and forced to live in cells with the opposite sex. Victims are forced to eat excrement or eat from plates on which guards have urinated. Women prisoners are force to have sex with other prisoners and guards or are forced to stand naked in public while menstruating.
Broader Violating taboos (#PF3976)   Induction of incongruent actions (#PF3790).
Related Violation of food taboos (#PD2868).

♦ PE1304 Hazards of environmental electromagnetism
Broader Electromagnetic pollution (#PD4172)
Environmental hazards of non-ionizing radiation (#PE7651).
Related Environmental hazards of extremely low frequency electromagnetic radiation (#PE7560).

♦ PE1309 Knee pain
Problematic knees
Refs Cailliet, Rene *Knee Pain and Disability* (1983); Macnicol, M F *The Problem Knee* (1986).
Broader Symptoms referable to musculoskeletal system (#PE4566).

♦ PE1310 Lumbago
Back-ache — Back pain
Nature Lumbago affects the muscles of the lower part of the back and can cause disabling pain. It is generally regarded to be of rheumatic origin and does not respond readily to treatment, so that it is likely to affect future as well as present earning capacity and health. Once a patient has had one attack of lumbago or sciatica, he is four times as likely to have another; in addition, a number of occupations may lead to earlier onset of spinal degeneration.
Incidence Disorders of the lumbar region of the spine are the commonest of all lesions affecting the joints of the human body. Millions of work-days are lost to industry each year by back disorders and injuries. For example, a recent industrial survey mostly involving coal mining, found an average sickness absence of 12 weeks arising from all lumbar disorders including injuries.
Refs Davis, Peter R (Ed) *Industrial Back Pain in Europe* (1985); Deyo, R A (Ed) *Back Pain in Workers* (1988); Golding, D N and Barrett, J *The Practical Treatment of Backache and Sciatica* (1984); Hukins, D W and Mulholland, R C (Eds) *Back Pain* (1986); Kraus, Hans *Backache Stress and Tension* (1984); Rowe, M Laurens *Backache at Work* (1983).
Broader Rheumatic diseases (#PE0873)   Diseases of the spine (#PD2626).
Narrower Fibrositis (#PG3542).
Related Occupational rheumatism (#PE0502).
Aggravates Disc disorders (#PG5560)   Kidney disorders (#PE2053).
Aggravated by Strain (#PJ3545)   Paraesthesia (#PG3547)
Climatic cold (#PD1404)   Muscular wasting (#PG3548)
Injurious accidents (#PB0731)   Loss of tendon reflex (#PG3549)
Unhealthy physical posture (#PD2838).

♦ PE1311 Fluorides as pollutants
Nature Atmospheric fluorides reduce or interfere with photosynthesis as a result of fluoride accumulation or alteration in plant function, even before the leaves show damage. Exposure to fluoride also causes plants to undergo changes in growth, metabolism, tissue degeneration and inhibition of certain enzyme activity.
Fluoride emissions from industrial sources must be carefully controlled because prolonged exposure to ambient air concentration of less than one million parts of air, by volume, may create a hazard to sensitive agricultural crops and forest species or to farm animals feeding on exposed forage. In this respect, fluorides are more than 100 times as toxic as sulphur dioxide.
The major sources of atmospheric fluorides from industrial operations are the manufacture of aluminium metal products and phosphate fertilizers. However, waste products from smelting, production of pig iron, clay and ceramic industries, brick and glass manufacture also contain fluorides. In fact, any process that involves new materials taken from the earth's crust and subjected to heating at high temperatures will liberate fluorides. Both gaseous and particulate fluorides may be discharged as waste products. Of these, the most injurious to vegetation is the gaseous hydrogen fluoride. Particulate fluorides are much less toxic, and calcium fluoride, because of its insolubility, is the least toxic.
Incidence Plants vary greatly in their sensitivity to injury by fluorides. Pine is among the most sensitive species and cotton among the most resistant.
Refs WHO *Fluorine and Fluorides* (1984).
Broader Chemical air pollutants (#PD1271).
Aggravates Soil pollution (#PC0058)   Water pollution (#PC0062)
Occupational risk to health (#PC0865).
Aggravated by Industrial emissions (#PE1869)   Industrial waste water pollutants (#PD0575).

♦ PE1313 Phosphates as pollutants
Nature Phosphates are used as chelating agents for detergents. The most effective and the most widely used compound today is sodium tripolyphosphate (STP), which produces a marked improvement in detergent properties but provides a source of environmental damage, including agricultural run-off and sewage. Ordinary sewage treatment processes do not remove phosphates, and the presence of this chemical has created a major source of pollution in marine and other aquatic environments, in that phosphates are conducive to excessive growth of aquatic plants, depletion of oxygen, loss of fish and general degradation of water quality. Phosphates are also said to contribute largely to the eutrophication process.
Broader Chemical pollutants of the environment (#PD1670)
Environmental hazards from fertilizers (#PE1514).
Aggravates Eutrophication of lakes and rivers (#PD2257).
Aggravated by Wasted water (#PD3669)   Contradictions (#PF3667).

♦ PE1314 Silicosis
Broader Pneumoconiosis (#PD2034)   Occupational diseases (#PD0215).
Narrower Grinders' rot (#PG3553)   Potters' asthma (#PG3552).
Aggravates Tuberculosis (#PE0566)   Chronic bronchitis (#PE2248).
Aggravated by Dust (#PD1245)   Smoking (#PD0713)
Dust as an occupational hazard (#PE5767).

♦ PE1315 Nickel as a pollutant
Refs FAO *Water Quality Criteria for European Freshwater Fish* (1984).
Broader Toxic metal pollutants (#PD0948)   Chemical contamination of water (#PE0535)
Environmental hazards from metals (#PE1401).
Aggravates Water pollution (#PC0062)   Occupational risk to health (#PC0865).
Aggravated by Industrial emissions (#PE1869)   Motor vehicle emissions (#PD0414)
Pesticides as pollutants (#PD0120)   Industrial waste water pollutants (#PD0575).

♦ PE1320 Endangered polar bear species
Nature Hunting of the polar bear, particularly for the fur trade, has increased in recent years and rendered the species vulnerable. Part of the problem is that such hunting is an intrinsic part of the Eskimo way of life, and the number of Eskimos is now increasing more rapidly than are the polar bears. In addition, the increasing exploitation of the mineral and fossil fuels of the arctic regions poses a continuing threat of encroachment and environmental degradation. The level of PCBs, polychlorinated biphenyls, in marine mammals' fat is rising worldwide and can cause infertility in polar bears, because they are meat-eaters in the end of the food chain.
Incidence It has been estimated that no more than 30,000 – 40,000 polar bears remain. Research indicates that the annual sustainable yield may be as low as 3–4% of the population. There are several areas where polar bears are being hunted even though the population is inadequate.
Broader Endangered species of bear (#PD3483).
Aggravated by Polychlorinated biphenyls as a health hazard (#PE2432).

♦ PE1322 Domination of economic integration by transnational corporations
Nature The transnational corporation process of world integration, which concentrates strategic "overhead" in the home countries, creates international dependence through hierarchy as well as through regional inequality. Furthermore, it raises questions about how the world community should share in financing the world overhead, since all but a few core countries participate at only a "lower echelon " of the productive activities integrated through transnational corporations operations. Transfer pricing (including profit and royalty remissions), and international tax avoidance through transfer pricing, redistribute world income and finance the concentrated development of transnational corporations' overhead. In the longer run, perhaps more important than the unequal distribution of transnational corporations' surplus is the unequal accumulation of productive and technological capabilities in the home countries of the transnational corporations' parents. That concentration and captivity of key productive activities in the home countries could make host economies more dependent on those enterprises as they move through production, deepening into areas which could potentially create higher domestic value added (for example, from the last touches of import substitution to intermediate products and capital goods, or from commodity and mineral production to resource-based industries). The higher value-added activities, particularly those which involve more than returns to large-scale financial capital, might demand a high component of decision-making as well as highly skilled personnel. Through the historical performance of the transnational corporations, such capabilities might have been kept captive by the parent firms in their country of origin. This more complex form of dependency is particularly relevant to countries which attempt production-deepening through regional cooperation.
Background The enormous spread and growth in the activities of transnational corporations during the past 20 years has been a major factor in establishing the present structure of world industrialization and trading patterns. Their activities cover all economic sectors – agriculture, mining, manufacturing, services including banking, insurance and shipping, as well as wholesale and retail trading. Their power in these sectors has become dominant, enabling them to influence world supply and demand patterns, especially through their ownership of or control over capital, technology and its development, their generation of particular management skills, and the close tie-up in their activities between production, marketing and distribution.
Transnational corporations play an important role in existing world trade and industrialization structures. While their activities are heavily concentrated in the developed countries, those in developing countries are by no means insignificant in terms of either their industrialization or their trade. Such corporations are major channels for the transfer of technology and for the provision of investment capital to developing countries. They provide a substantial part of their imports and have made an important contribution to their exports of manufactures, although these are still small. It is on account of the dominant market power which they hold in both developed and developing country markets that they are able to resort to a variety of restrictive business practices which limit the actual and potential benefits accruing to developing countries from activities undertaken in their territories.
Refs UNCTAD *Current Problems of Economic Integration* (1986).
Broader Transnational corporation imperialism (#PD5891)
Excessive power and independence of transnational corporations (#PD5807).

♦ PE1323 Dodders
Love vine — Strangle weed — Gold thread — Hairweed — Devil's hair — Devil's ringlet — Pull down — Clover silk — Hell-bind
Nature Dodders are leafless parasites on stems and other parts of cultivated or wild plants. Parasites twine and spread orange or yellowish tendrils over the host and moving from one plant to another.
Broader Basidiomycetes (#PE0364)   Parasites on plants (#PD4659).

♦ PE1325 Rabies
Nature Rabies is an acute viral infection of the central nervous system which affects virtually all mammals, including man, and occasionally occurs in birds. It is characterized by nervous derangement with paralysis in the final, and sometimes in the intermediate stages, resulting in death. The virus is present in the saliva of affected animals, and is transmitted by biting. The average incubation period is 2–12 weeks, rarely less than 10 days, rarely more than 6 months; the incubation period is shorter for young animals. There are two distinct forms; dumb and furious rabies. Dumb rabies symptoms are: apathy; nervousness; lack of coordination, finally paralysis, especially of the lower jaw and hind quarters in dogs; ultimate death. In furious rabies, the excitation period (which also occurs in dumb rabies) is more pronounced, manifesting restlessness, viciousness, difficulty in chewing and swallowing; the apathy period at the start is less pronounced, equally the final stage of paralysis may be shorter. In humans, the disease is virtually always fatal. Although in terms of the traditional measures of incidence and mortality in man, rabies cannot be considered an important disease in comparison with many other conditions, the economic costs of rabies cannot be neglected. First, even where incidence is zero, significant and continuing costs are incurred for the surveillance and prevention of outbreaks, such as quarantine regulations, publicity campaigns and the investigation of persons exposed outside the country. Second, sizeable resources are committed in many countries to control the incidence and spread of infection among animals, both domestic and wild. Third, in infected areas there may be significant indirect costs, whereby the presence of rabies lowers the national income from tourism

and various forms of agriculture, because of fear of infection.
Incidence This disease is of great importance to livestock in the South and Central American zones where paralytic rabies is transmitted by blood–lapping bats. Thousands of cattle die annually of the disease and expenditure on vaccination and other control measures is high. In central Europe, wildlife rabies is the predominant form of the disease: foxes are the main victims and at the same time also the main vectors of the infection in this region. But canine rabies is the prevailing source of the disease. Dogs, particularly stray dogs, are an important reservoir of infection in Africa, Central and South America, Asia and also in Greece. Bites by rabid dogs are responsible for the vast majority of all human cases of rabies – 90 percent or more – and a similar proportion of post–exposure treatment occurs in countries affected with canine rabies. Increased urbanization, the more widespread practice of camping, and increased international travel by man and transfer of animals, increase the rabies risk. Organ transplants, particularly corneal transplants, have also been a concealed means of transmission of rabies to man.
Background One of the oldest recorded infectious diseases, rabies was first described by Democritus (500 BC). It was first noted in Asia and Europe, the first case being recorded in the USA in 1812, in Brazil in 1911.
Refs Baer, George M *The Natural History of Rabies* (1975); Regamey, R H, et al (Eds) *International Symposium on Rabies, 2* (1974).
Broader Zoonoses (#PD1770) Virus diseases (#PD0594)
Viral diseases in animals (#PD2730) Infectious and parasitic diseases (#PD0982)
Diseases of nervous system in animals (#PD7841).
Aggravates Encephalitis (#PE2348) Ecological imbalance (#PG3026)
Economic loss through slaughter of diseased animals (#PE8109).
Aggravated by Stray dog populations (#PE0359)
Diseases of wild animals (#PD2776)
Inadequate animal quarantine (#PE2756).
Reduced by African trypanosomiasis (#PE1778)
Inadequate carcass disposal of diseased animals (#PE2778).

♦ PE1327 Unjustified game cropping
Broader Inadequate animal welfare (#PC1167) Unsustainable harvesting rates (#PD9578)
Long-term shortage of natural resources (#PC4824).
Related Cruel culling of seals (#PE3484).
Aggravates Degradation of semi–natural and natural habitats of flora and fauna (#PC3152).

♦ PE1329 Abuse of opiates
Refs Harding, Geoffrey *Opiate Addiction, Morality and Medicine* (1987); Kuschinsky, K *Opiate Dependence* (1977); Lenz, George, et al *Opiates* (1986); United Nations *Report of the International Narcotics Control Board for 1985* (1986).
Broader Substance abuse (#PC5536) Abuse of plant drugs (#PD0022).
Narrower Opium abuse (#PG3576) Codeine abuse (#PG3579)
Morphine abuse (#PG3577) Abuse of heroin (#PE1776)
Opioid withdrawal (#PG5201) Opioid intoxication (#PG9743).
Aggravates Drug smuggling (#PE1880).
Aggravated by Illicit drug trafficking (#PD0991) Manufacture of illicit drugs (#PE2512).

♦ PE1331 Environmental hazards from animal feedstuffs, excluding unmilled cereals
Broader Environmental hazards from food and live animals (#PC1411).
Narrower Contaminated animal feed (#PE3896).

♦ PE1334 Wind damage to structures
Nature The possibility of wind damage is a continuing and important factor in the design of structures such as bridges, cooling towers, high buildings, and roof attachments. In addition, even adequately designed tall buildings may cause extremely annoying wind movements at ground level at the base of the building. Structural failure due to winds varies geographically. Hurricanes, cyclones and typhoons are frequent in some areas and structures in equatorial regions may have greater exposure to atmospheric turbulence than elsewhere.
Broader Wind (#PE2223) Bad weather (#PC0293).
Aggravated by Tornadoes (#PD1739) Hurricanes (#PD1590)
Thunderstorms (#PE3881).

♦ PE1335 Tsetse flies as pests
Incidence The tsetse fly which transmits nagana to cattle, is found in approximately 11 million sq km in Africa. Of this area, 7 million sq km could support a population of 140 million cattle, if cleared of the tsetse. Only 20 million cattle are currently in this area. The potential meat production would be equal to an additional 1.5 million tons per year.
Background Tropical regions have a greater diversity of insect life than any other climatic region. Some of these insects transmit diseases to man and livestock and are major pests of agricultural crops. Trypanosomiasis is a major endemic tropical disease, and is one of the most important livestock diseases in Africa. It causes a fatal disease in humans (sleeping sickness) and enormous losses in livestock (nagana).
Refs Cavalloro, R (Ed) *Integrated Tse-tse Fly Control, Methods and strategies* (1987); FAO *Insecticides and Application Equipment for Tsetse Control* (1977); FAO *The Environmental Impact of Tsetse Control Operations* (1980).
Broader Parasites (#PD0868) Flies as insect pests (#PE2254)
Insect vectors of disease (#PC3597) Insect vectors of human disease (#PE3632).
Aggravates African trypanosomiasis (#PE1778) Insect bites and stings (#PE3636).

♦ PE1337 Inappropriate transplantation of industrialized country methods to developing countries
Nature The political, economic and educational systems of many developing countries are strongly influenced by the legacy of the colonial powers. Such systems, or transplanted models, are therefore geared to the conditions of countries already highly industrialized, whereas the transition from a traditional type of society (pastoral or agricultural) towards an industrial type is still in its infancy in a good many of the newly independent countries. The systems are therefore ineffective in responding to the real needs of the countries.
Broader Human errors and miscalculations (#PF3702)
Economic and social underdevelopment (#PB0539).
Aggravated by Domination (#PA0839) Colonialism (#PC0798).

♦ PE1339 Inequality of life expectancy by gender
Shorter life expectancy of men — Social inequalities in life expectancy — Inequality of mortality rates by gender
Nature Progress in the prevention of avoidable deaths has been distributed unequally among the population of the two sexes, and very unevenly among different age and social groups.
Incidence World statistics for 1982 indicate an average male/female life expectancy of 57.5 / 60.3 years in the world, 48.2 / 51.3 years in Africa, 53.5 / 53.8 years in South Asia, 61.8 / 66.5 years in Latin America, 65.9 / 70.1 years in East Asia, 65.5 / 69.9 years in Oceania, 66.5 / 75.4 years in the USSR, 70.4 / 78.1 years in North America, and 70.0 / 76.6 years in Europe. Thus in most parts of the world, men have a lower life expectancy than women – and the gap is increasing. The average world life expectancy in the period 1980–1985 was 60.3 years for women and 57.5 years for men. In the developed regions, the 5.7 years difference of 1950–1955 jumped to 7.3 years for 1970–1980, and in the period 1980–1985, women could expect to live 76.9 years compared with 69.4 years for men. In Western Europe, current life expectancy is 69.8 years for men and 76.0 years for women, and in the USSR there is a difference of more than 9 years (65.0 for men and 73.3 for women). In these industrialized countries, mortality rates for males exceed those for females at all age groups. The least difference generally occurs in the youngest and oldest groups while the greatest difference occurs during adulthood. The sex difference in life expectancy is less pronounced in the developing world, women living an average of 1.8 years longer than men in the years 1975–1980; while for the period 1980–1985, life expectancies were 57.7 years for women and 55.5 years for men. For these areas, the regional variations are greater (more than 6 years in temperate South America, less than 3 years in North Africa and China), but the gap is also generally increasing. There are, however, some developing countries in which the male mortality rate is less than the female for some age groups. According to the World Health Organization, this is true in North Africa for the ages 1 to 4 years. There are at least four countries in Asia for which the mortality rate for women in certain age groups is sufficiently higher than for men to generate a life expectancy at birth lower for women than for men. There are also areas, for example in Latin America, where until recently, in the childbearing ages and sometimes in adolescence, the life expectancy of women was less than that of men. The reasons for differences in the death risks of men and women are many. Greater gains in life expectancy have accrued to women than to men. Various social, environmental and sanitary factors also cause a different incidence of deaths at different ages.
After the Second World War, social security systems were more or less completed in all industrialized countries, life expectancy had increased dramatically on average and it was widely believed that inequality with respect to death for all social groups was disappearing. The unprecedented and widely distributed progress in levels of living in the 1950s, 1960s and 1970s was a powerful reason for optimism. Yet the most recent studies in the developed countries suggest that the differences in mortality rates between social groups have not been narrowing in recent years. They rather suggest a great stability of social differences in life expectancy in a context of general and rapid improvement for all social groups.
Broader Human death (#PA0072) Sexual discrimination (#PC2022)
Inequality in mortality rates (#PC9586).
Related Unequal regional distribution of deaths (#PC4312)
Unequal morbidity and mortality between countries (#PC6869).
Aggravated by Malnutrition (#PB1498) Complications of childbirth (#PC9042)
Human disease and disability (#PB1044).

♦ PE1341 Electrical power failure
Power supply system overload — Power blackouts
Nature Breakdowns can be caused by incorrect operation of automatic devices in the system and by errors on the part of the operating personnel. Dangers include overloading, overvoltages, short circuits, damage to insulating and supporting structures, mechanical damages and ground fault.
Broader Injurious accidents (#PB0731) Human errors and miscalculations (#PF3702)
Long–term shortage of natural resources (#PC4824).
Aggravated by Natural disasters (#PB1151) Inadequate electricity infrastructure (#PD9033).

♦ PE1344 Environmental hazards from manufactured goods
Broader Environmental hazards from economic and industrial products (#PC0328).
Narrower Environmental hazards from metals (#PE1401)
Environmental hazards of wood and cork manufactures (#PE8091)
Leather and leather manufactures environmental hazards (#PE8447)
Environmental hazards from paper and printing industries (#PE1425)
Environmental hazards from textile and clothing industries (#PE1103)
Environmental hazards from undressed hides, skins and fur skins (#PE0828)
Environmental hazards from non–metallic mineral products industries (#PE0890).

♦ PE1346 Environmental hazards from mineral fuels, lubricants and related materials
Broader Environmental hazards from energy production (#PD6693)
Environmental hazards from economic and industrial products (#PC0328).
Narrower Environmental hazards from petroleum (#PE1409).
Related Environmental hazards from electricity (#PE1412).

♦ PE1347 Geographic discrimination in the administration of justice
Nature In some countries, criminal and civil cases may, under certain circumstances, be decided in the absence of defendants; and a person who is prevented or impeded by financial factors from attending court is normally at a disadvantage in not being present to defend his innocence or his interests. This may happen when the location of the court necessitates travel or when attendance at court would cause the defendant to lose earnings for the time spent there.
People in rural areas may experience difficulties due to the expense of reaching the seat of the nearest appropriate court. Having arrived there they may incur further expense by way of accommodation costs and loss of earnings while waiting for their case to be reached by the court. Any person involved as a party in judicial proceedings may suffer expense due to the necessity of waiting, together with counsel and witnesses, for a court to reach his case.
The financial problems that arise out of the geographical inaccessibility of courts may be aggravated by the fact that the seats of certain local courts, which once coincided with the main centres of population, no longer do so. Thus, in England and Wales, the Queen's Bench Division of the High Court of Justice, which has original criminal and civil jurisdiction over more serious cases and a supervisory and appellate jurisdiction over certain lower courts, sits not only in London, but also in certain assize towns, which must be visited periodically; the choice of assize towns was made some centuries ago, however, and no longer adequately reflects the distribution of population, especially as affected by the industrial revolution.
Broader Discrimination before the law (#PC8726)
Inequitable administration of justice (#PD0986).
Related Isolation of ethnic groups (#PC3316).

♦ PE1348 Misuse of spiritual authority for sexual purposes
Incidence There are numerous reports, notably in the UK, of Christian clergymen and priests who have misused their authority, and perverted their victims sense of spirituality, through various forms of sexual abuse, notably with children.
Broader Unethical practices of priesthood (#PF8889).
Related Homosexuality within the priesthood (#PG4151).
Aggravated by Sexual bigotry in organized religion (#PE0567).

♦ PE1349 Solvent and methylated spirits drinking
Adulteration of commercial alcohol
Nature The drinking of solvents, silver polish and other cleaning fluids, methylated spirits and other raw alcohols may cause severe damage to internal organs and ultimately death. As the alcoholism of the poor, who cannot afford more refined and relatively less harmful forms of alcohol, it is associated with problems of abject poverty, namely homelessness, malnutrition, disease and total alienation from society.
Broader Alcohol abuse (#PD0153) Substance abuse (#PC5536)
Drug dependence (#PD3825).
Aggravates Mental depression (#PC0799).

♦ **PE1353 Long-term shortage of crude fertilizers and crude minerals, excluding coal, petroleum and precious stones**
Nature Crude fertilizers and crude minerals are in danger of long-term shortage, but little is being done to counteract that probability.
Incidence Examples of minerals most in danger of being depleted, with their most common industrial uses are: antimony (batteries, semiconductors); bismuth (fire safety devices); cobalt (jet aircraft engines, mining tools); copper (electric wiring and equipment); lead (storage batteries for transportation, communications); mica (insulation, stereo speakers); silver (photography, dental alloys); tin (anticorrosive in many alloys); and zinc (brass, bronze, galvanized iron).
Claim The greatest single physical obstacle to increased food production, particularly in developing countries, is probably the shortage of fertilizer: in the short-term of nitrogen but, in the longer-term, of non-renewable phosphate. At the same time, recent studies have shown that many soils used for vegetable growing in the UK, for example, contain up to four times as much phosphate and potassium as is needed. This is an obvious example of how fertilizer saving could have broad implications.
Broader Long-term shortage of manufactured goods (#PE0802)
Long-term shortage of inedible crude non-fuel materials (#PE0461).
Narrower Sulphur shortage (#PU3599) Abrasives shortage (#PU3600)
Shortage of sand and gravel (#PU3598).

♦ **PE1355 Denial of right to union activity for special groups**
Broader Violation of trade union rights (#PD4695).
Narrower Denial of trade union rights to public employees (#PE6888)
Denial right to union activity for rural workers (#PE3910).
Aggravates Deafness (#PD0659) Inefficiency (#PB0843)
Mental illness (#PC0300) Stress in human beings (#PC1648).

♦ **PE1359 Ozone as a pollutant**
Nature At ground level ozone is a pollutant. Ozone scars lung tissue, which worsens respiratory diseases and can cause breathing problems especially for young children. It also makes eyes sting and throats itch. Ozone is one of the causes of acid rain. Ozone is a summer gas, baked by strong sunlight from a mixture of unburnt fuel, oxygen and nitrogen oxides, much of which comes from car exhausts. High concentration of ozone have tended to be found not so much in the cities but in the surrounding countryside where the conversion process is completed. Ozone breaks down the outer cuticle of the needles of trees and opens them up for further attack from acid precipitation.
Depletion of stratospheric ozone may be compensated by increases at lower altitudes, enabling the lower stratosphere to receive more solar heating and to efficiently trap outgoing heat radiation.
Incidence High ozone levels seriously affected the USA in 1988 and Western Europe in 1989.
Refs National Research Council, Division of Medical *Ozone and Other Photochemical Oxidants* (1977).
Broader Photochemical oxidant formation (#PD3663)
Chemical pollutants of the environment (#PD1670)
Increase in trace gases in the atmosphere (#PD1354).
Aggravates Ulcers (#PE2308) Mutagens (#PD1368) Corrosion (#PD0508)
Air pollution (#PC0119) Forest decline (#PC7896)
Lung disorders and diseases (#PD0637) Environmental plant diseases (#PD2224).
Aggravated by Motor vehicle emissions (#PD0414)
Hydrocarbons as pollutants (#PE0754)
Nitrogen compounds as pollutants (#PD2965).
Reduces Stratospheric ozone depletion (#PD6113).

♦ **PE1363 Homosexuality in prisons**
Rape in prisons — Unreported rape in prisons
Nature Homosexuality is often rampant in jails and prisons where, for example, men work out their frustrations on new prisoners in gang-rape. Often the "gang rapers" do not see themselves as homosexuals but commit their crimes as a need to exert their virility which has no other opportunity for outlet. The men who are thus attacked (most often those of small build), having had their bodies defiled, their manhood degraded, and their will broken, are released into society ashamed, confused, and filled with hatred. They often leave prison with more aggressive psychological disorders than they had initially, thus becoming potential criminals of a more severe nature.
Claim Many of the penal systems today where homosexuality and gang rape are prevalent, impose a more cruel and harsh punishment on a criminal than could ever be imposed by a court of justice.
Refs Wooden, Wayne S and Parker, Jay *Men Behind Bars* (1982).
Broader Homosexuality (#PF3242).
Related Unreported rape (#PE5621).
Aggravates Maltreatment of prisoners by fellow inmates (#PE0428).
Aggravated by Denial of rights to prisoners (#PD0520).

♦ **PE1364 Manganese as a health hazard**
Nature In occupational exposure, manganese is absorbed mainly through inhalation, but may also enter the gastrointestinal tract with contaminated food and water. If entering the body in sufficient quantities, this can lead to manganese poisoning, with its nervous disorders and possible disfiguration of the joints.
Incidence Intoxication by manganese is reported among workers in the mining and processing of manganese ores, and in the production of manganese alloys, dry cell batteries, welding electrodes, varnishes, and ceramic tiles. Mining of ore is still the major occupational hazard, with the ferromanganese industry as the next most important source of risk. The operations that produce the highest concentrations of manganese dioxide dust are those of drilling and shotfiring. Consequently, the most dangerous job is high-speed drilling.
Refs WHO *Manganese* (1981).
Broader Occupational hazards (#PC6716) Toxic metal pollutants (#PD0948)
Chemical air pollutants (#PD1271) Chemical contaminants of food (#PD1694)
Environmental hazards from metals (#PE1401).
Aggravates Occupational risk to health (#PC0865).
Aggravated by Industrial emissions (#PE1869) Motor vehicle emissions (#PD0414).

♦ **PE1369 Unethical ophthalmic practice**
Irresponsible opticians — Negligence by opticians — Malpractice by opticians — Corruption of opticians
Claim Opticians, under pressure from their employers, have adopted practices which lead to overprescription of corrective spectacles.
Refs Rosenwasser, Harvey M *Malpractice and Contact Lenses* (1988).
Broader Unethical professional practices (#PC8019).
Related Unethical medical practice (#PD5770) Unethical veterinary practice (#PD7726).

♦ **PE1371 Slime moulds in plants**
Nature Slime moulds are an intermediate between bacteria and fungi and they are found in rotting logs or in lawns.
Broader Fungal plant diseases (#PD2225).

♦ **PE1372 Long-term shortage of wood, lumber and cork**
Broader Long-term shortage of manufactured goods (#PE0802)
Long-term shortage of inedible crude non-fuel materials (#PE0461).
Narrower Wood shaped or simply worked shortage (#PE7920).
Related Shortage of firewood (#PD4769).
Aggravated by Endangered forests (#PC5165).

♦ **PE1375 Inadequate coordination of action on intergovernmental programmes at national level**
Nature There is an absence of coordination at the national level in regard to the policies and programmes of United Nations organizations. Donor can complicate policy making because they deal directly with individual spending ministries. This can lead to duplication of effort and impede central control of the host country's budget. Donor representatives may also be under pressure from their own organizations to lend and disburse which may lead them to seek special treatment, for example through the separation of counterpart funds from the budget or exempting projects from the normal procedural checks. The weakness and rareness of such policy coordination in many governments has led not only to divergent positions being taken by the representatives of the same country in different organizations but also not infrequently to divergent decisions actually being reached by those organizations themselves.
Incidence Poor coordination in respect of international activities in certain fields - one of which is the application of science and technology - usually reflects poor coordination on those subjects at the national level. Again, proliferation and overlapping of programme activities (which is perhaps more obvious at the national and regional levels than at the global level) arise because, in so many broad fields, different ministries and divisions are involved in each country; and initiatives varying slightly in approach but having very similar objectives can well be launched and carried through simultaneously by two or more international agencies. For example, the United Nations regional and community development programme is closely related to UNESCO's work on functional literacy, FAO's agricultural extension work, aspects of the public health work of WHO and aspects of the ILO programme on handicrafts and cottage industries. UNESCO's early work on fundamental education and that of the United Nations on community development, each with its separate antecedents, were found to be so similar in aim that they were eventually merged under arrangements proposed by the secretariats concerned.
Claim Most countries still lack systematic arrangements even for keeping the activities of international organizations under central review, let alone for developing coordinated positions on issues coming before them. Nothing approaching full coherence and coordination in the United Nations system will occur so long as this situation prevails.
Broader Inadequate coordination of international organizations and programmes (#PD0285).
Aggravates Congenital diseases of the nervous system in animals (#PE0965)
Ineffectiveness of international organization and programme action at the country level (#PE9124).
Aggravated by Lack of a world government (#PF4937).

♦ **PE1379 Unemployed educated youth**
Refs Das, A K *Unemployment of Educated Youth in Asia* (1981).
Broader Youth unemployment (#PC2035) Educated unemployed (#PD8550).
Aggravates Underdeveloped sources of income expansion (#PF1345).

♦ **PE1396 Wastage of highly skilled personnel in the routine maintenance of complex systems**
Nature Complex computer systems are designed and implemented by a team of relative experts. Because they know ever part if the system, they end up becoming the maintenance people for the system, a boring, dead end job. They will spend years in this position and when they finish they will be untrained in new generations of hardware and software. They will have lost their chances for promotion within in their own company and will have difficulty in finding jobs elsewhere.
Broader Underutilization of intellectual ability (#PF0100).
Aggravates Eugenics (#PC2153) Unethical maintenance practices (#PD7964).
Aggravated by Inadequate maintenance personnel (#PJ0088).

♦ **PE1397 Persons missing in military action**
MIA's — Missing in action
Nature Service personnel who are missing during a military action and have not been confirmed as dead nor as prisoners of war are listed as missing in action.
Incidence Some 78,000 American in World War II, 8,000 during the Korean War and nearly 2,400 in the Vietnam War are considered as missing in action although officially listed as dead. The Soviet Union lists 313 people as missing in action in Afghanistan.
Broader Missing persons (#PD1380).
Related Prisoners of war (#PC8848) Suspicious deaths during detention (#PE6367).
Aggravates Non-recognition of foreign governments (#PF8040)
Uncertainty of death of missing persons (#PF0431).
Aggravated by Persistence of a technical state of war following cease-fire agreements (#PE2324).

♦ **PE1398 Bank fraud**
Fraud in savings banks — Loan fraud — Fraudulent loans
Incidence Widespread fraud in the USA savings and loan institutions during the 1980s is complicating government attempts to rescue thousands of institutions declared insolvent. In 1990 it was estimated that 60 per cent of the savings and loans institutions seized in order to protect investors were the victims of fraud.
Broader Financial scandal (#PD2458) Unethical financial practices (#PE0682).
Aggravates Bank failure (#PE0964).

♦ **PE1399 Economic discrimination in the administration of justice**
Inequitable justice for the poor — Educational discrimination in the administration of justice
Nature Low social justice is often accompanied by poor education, which tends to put the individual concerned in a position of psychological inferiority in relation to court systems, which represent the state apparatus. In criminal trials, the court and the prosecution tend not to be seen as separate entities. This state of mind is aggravated by: the presence of uniformed police in the courts, in excess of what is reasonably needed for the physical restraint of the accused; the practice of placing an accused in the dock during trial; and the housing of minor criminal courts in the same building as a police station. Lack of education usually includes ignorance of the law in general and of procedural law in particular, and those who suffer from such lack need adequate legal advice even more than the rest of society. It is also beyond dispute that the less sophisticated are more easily induced to sign "confessions" of guilt.
Lack of finance may affect the capacity of an accused or of a party in a civil action to secure the evidence, particularly scientific evidence, that he needs to locate and secure the attendance of witnesses (especially if they are not easily accessible geographically), to pay their travelling expenses and compensation for loss of earnings during court hearings, and to pay the fees of expert witnesses. A poor person may also face difficulties in preparing his case because certified

copies of documents may be available to him only on payment of a fee.
Refs Petersilia, Joan *Racial Disparities in the Criminal Justice System* (1983).
Broader Economic discrimination (#PC2157) Discrimination before the law (#PC8726)
Inequitable administration of justice (#PD0986).
Narrower Prohibitive legal fees (#PF0995)
Excessive burden on the poor due to legal delays (#PE1093)
Discrimination against the poor in judicial sanctions (#PE1129)
Economic barriers to women's access to the judicial process (#PE1198)
Ineffective protection of individual rights due to excessive court costs (#PF0922)
Pecuniary guarantees as a condition of provisional release in criminal cases (#PE1032)
Economic barriers to access to the legal profession, the judiciary and jury membership (#PE0803).
Related Injustice (#PA6486) Cultural discrimination in the administration of justice (#PE6529).
Aggravates Inadequate legal counsel for minorities (#PF1219).

♦ PE1400 Fowlpest

Newcastle disease — Avian pneumoencephalitis — Newcastle disease of poultry
Nature An infectious disease of poultry (mainly of chickens and turkeys) and pigeons. Other domestic poultry and various species of wild birds may be affected, also man, though in man it usually appears as simply a very mild form of conjunctivitis. Symptoms in poultry include: marked drop in egg yield, somnolence, later paralysis, gasping inhalation, mucous discharge, diarrhoea, nervous twitching. High mortality rate, may be as much as 100 per cent among young birds.
Incidence Has caused, and continues to cause, very heavy losses to the poultry industries of many countries. Generally high worldwide incidence, persisting wherever poultry are kept and particularly virulent where broiler techniques and close housing are used. Virus does not exist easily outside the living bird, but it is particularly hardy at low temperatures and can exist in frozen birds for more than 2 years. People, dogs and cats as well as certain wild birds may be unrecognized carriers of the disease, since they show little or no evidence of it. Symptoms can be confused with fowl plague, fowl cholera, fowl pox.
Background Fowlpest was first recorded near Newcastle (UK) in 1926, by Doyle who distinguished it from fowl plague, which it resembles. First recorded in 1945 in USA in the states of New Jersey and New York also occurred in California in the 1970s.
Refs Allan, W H; Lancaster, J E and Toth, B *Newcastle Disease Vaccines* (1978); Shortridge, K F (Ed) *Newcastle Disease and its Control in Southeast Asia* (1982).
Broader Bird diseases (#PD3323) Virus diseases (#PD0594)
Viral diseases in animals (#PD2730).
Related Infectious laryngotracheitis (#PG5712).
Aggravated by Factory farming (#PD1562) Human vectors of animal diseases (#PD2784)
International trade in endangered species (#PC0380)
Domestic animals as carriers of animal diseases (#PD2746).
Reduced by Inadequate carcass disposal of diseased animals (#PE2778).

♦ PE1401 Environmental hazards from metals

Nature Metals have been used for centuries and are fundamental to major industries, yet some have the potential to damage human health and disturb the balance of environmental systems if they are allowed to reach excessive concentrations in air, water, soil, or food.
Only a few metals are important environmentally: those most likely to cause concern include copper, cadmium, mercury, tin, lead, vanadium, chromium, molybdenum, manganese, cobalt and nickel. In addition, the metalloids (including antimony, arsenic and selenium), which have some metallic properties, may cause environmental problems; uranium, plutonium and other actinides also have metallic properties, and are a cause for concern.
A metal is regarded as toxic if it injures the growth or metabolism of cells when it is present above a given concentration. Almost all metals are toxic at high concentrations, and some are severe poisons even at very low concentrations. Copper, for example, is a micronutrient, a necessary constituent of all organisms, but if the copper intake is increased above the proper level, it becomes highly toxic. Like copper, each metal has an optimum range of concentration, in excess of which the element is toxic. The toxicity of a metal depends on its route of administration and the chemical compound with which it is bound. The combining of a metal with an organic compound may either increase or decrease its toxic effects on cells. On the other hand, the combination of a metal with sulphur to form a sulphide results in a less toxic compound than the corresponding hydroxide or oxide, because the sulphide is less soluble in body fluids than the oxide. Toxicity generally results: when an excessive concentration is presented to an organism over a prolonged period of time; when the metal is presented in an unusual biochemical form; or when the metal is presented to an organism by way of an unusual route of intake. Less well understood, but perhaps of equal significance, are the carcinogenic and teratogenic properties of some metals.
Incidence Man has been exposed more and more widely to metallic contaminants in his environment, resulting from the products of industry. There are three main sources of metals in the environment. The most obvious is the process of extraction and purification: mining, smelting, and refining. The second is the release of metals from fossil fuels (such as coal or oil), when these are burned. Cadmium, lead, mercury, nickel, vanadium, chromium, and copper are all present in these fuels, and considerable amounts enter the air, or are deposited in ash, from them. The third and most diverse source is production and use of industrial products containing metals, which is increasing as new applications are found. The modern chemical industry, for example, uses many metals or metal compounds as catalysts; metal compounds are used as stabilizers in the production of many plastics; and metals are added to lubricants, and so find their way into the environment.
Metals follow many pathways and cycles in the environment, and some of them undergo transformations in the process – like the conversion of inorganic mercury to the more toxic methyl form, and the subsequent accumulation of the latter by fish. Where metal–rich mine drainage enters fresh waters there are often obvious ecological effects, including a great reduction in the invertebrate fauna and the absence of fish. Some plants and invertebrate animals also accumulate metals of potentially toxic levels. Once toxic concentrations have been reached, it may take a long time to reduce them to nontoxic levels. Pathways within man and other targets are also crucially important. The rates and mechanisms of absorption and excretion, and the extent to which metals are deposited in such tissues as bone or the kidney cortex and then only slowly removed, need to be known if risks are to be assessed. The biological half–life of methyl mercury in man, for example, is about 70 days, that of cadmium around 20 years, and that of lead only a few weeks in blood and soft tissue, but at least ten years in bone. The risk from cadmium appears limited largely to groups of people consuming food produced in areas where the soil or irrigation water is contaminated, although there is concern about rising cadmium levels in the environment. Mercury is a problem where populations eat large amounts of fish taken from contaminated waters. Lead is the most widespread potential hazard, since quite small increases in lead consumption can raise blood lead levels to the point where biochemical changes are detectable. Children appear to be more sensitive to exposure to heavy metals than adults, and are consequently the focus of concern. Today, increasing emphasis is being placed on the carcinogenic effects of metals. Chromium, nickel, lead, and cadmium are all proven or suspected causes of certain cancers associated with industrial processes. Large doses of cadmium and nickel are teratogenic in animals, but this effect of metals is not well established in man.
Background Smelting of ores and refining of metals has been going on for a long time, introducing metals into air and water, but human exposures were usually local; during the past 50 years they have become fairly general. Many metals have been known for centuries to be toxic. Attention to the effects of metals on man focused first on acute poisoning following industrial exposure, or through diet. Inhalation of mercury vapour in both the mining and felt hat industries used to cause many cases of damage to the central nervous system. Lead poisoning was for decades a well–known hazard to smelters, and later to those engaged in storage battery production. Inhalation of manganese has been known for many years to cause irreversible damage to the central nervous system. Cadmium, mercury, tin, lead, vanadium, chromium, molybdenum, manganese, cobalt, and nickel are all known to pose hazards to those working with them. Since the 1960's, attention has focused especially on lead, cadmium, and mercury, because they have also been shown to cause more general environmental hazards, in each case mainly through the ingestion of excessive quantities of the metal.
Refs Salánki, J (Ed) *Heavy Metals in Water Organisms* (1985).
Broader Environmental hazards from manufactured goods (#PE1344).
Narrower Tin as a pollutant (#PE1438) Nickel as a pollutant (#PE1315)
Cobalt as a pollutant (#PE2339) Mercury as a pollutant (#PE1155)
Cadmium as a pollutant (#PE1160) Selenium as a pollutant (#PE1726)
Vanadium as a pollutant (#PE2668) Antimony as a health hazard (#PE1989)
Manganese as a health hazard (#PE1364) Beryllium as a health hazard (#PE2209).
Related Lead as a pollutant (#PE1161).

♦ PE1404 Mail fraud

Nature The intentional use of postal, telegraph, facsimile, computer or other public or private communications service to defraud is considered mail fraud.
Broader Fraud (#PD0486) Inappropriate use of the mail service (#PE6754).

♦ PE1406 Instability of trade in non–ferrous metals

Broader Instability of trade in manufactured goods (#PE0882).
Narrower Instability of tin trade (#PG1422) Instability of lead trade (#PG1402)
Instability of zinc trade (#PG2010) Instability of copper trade (#PG2472)
Instability of nickel trade (#PG2532) Instability of trade in tungsten (#PE7621)
Instability of trade in aluminium (#PG6019)
Silver and platinum group trade instability (#PE7956)
Instability of trade in U235–depleted uranium (#PE8560).

♦ PE1407 Government inaction on alleged human rights violations

Government non–assistance to endangered peoples
Nature Reports of extreme violations of human rights in a country, especially those which take the form of the massacre of civilians, do not necessarily lead to action on the part of other governments. Whilst governments may express disapproval, whether individually or collectively (such as through the United Nations), and urge the restoration of peace, concrete measures to assist the endangered peoples are seldom taken unless the incident is perceived to serve the national interests of the assisting government. The principal reason for such non–involvement results from the principle of non–interference of one government in the internal affairs of another. A second reason is that any such interference can be viewed and labelled by other countries as a form of aggression calling for counter–measures which would escalate the incident.
Incidence Pleas to western countries by the peoples of Hungary and Czechoslovakia on the occasion of the USSR suppression of their revolt were not acted upon. No action was taken by outside governments to counteract the massacres of Cambodian civilians by the Khmer Rouge, or the massacres of Ugandan civilians by the regime of Idi Amin. Governments were careful not to act in response to the massacres of civilians on the occasion of the revolution in Romania. In response to the Tianamen Square massacre, US human rights groups consider that the US President enacted the minimum sanctions that an outraged US public would permit whilst actively working against further sanctions that Congress sought to impose.
Broader Government inaction (#PC3950)
Internationally non–cooperative governments (#PF9474).
Aggravates Impunity of violators of human rights (#PF3474)
Inadequate enforcement of human rights (#PC4608).
Aggravated by Limited acceptance of human rights treaties (#PE7300)
Connivance of authorities in human rights abuses (#PF9288).

♦ PE1409 Environmental hazards from petroleum

Broader Environmental hazards from chemicals (#PC1192)
Environmental hazards from mineral fuels, lubricants and related materials (#PE1346).
Narrower Oil as a pollutant (#PE2134).
Aggravates Air pollution (#PC0119) Marine pollution (#PC1117).

♦ PE1412 Environmental hazards from electricity

Environmental hazards of electric energy — Health risks of electricity — Degradation of the environment from electrical power generation
Nature Electricity generation in large centralized installations, whether using fossil fuel or nuclear power, has several environmental impacts. It is thermodynamically inevitable that about 60 percent of the heat energy generated is 'waste' and, unless it can be used as low–grade heat to warm buildings nearby, it has to be released into the environment. Other impacts arise from electricity transmission. It is very costly to place transmission lines underground (where access for maintenance is also more difficult) and since the 1970s this has virtually only been done in areas of particularly outstanding landscape quality. Generally above–ground transmission lines have been built, their length naturally being related to the type of power station and its location. Hydropower schemes, usually being remote from the consumer, require especially long lines. The main impact of such lines comes from their visual intrusion, but they also restrict agriculture on bands of land 30–120 m wide and may cause some interference with nearby radio and television reception. It has been postulated that the lines may affect bird behaviour (especially water fowl), perhaps because they sway or hum in the wind: other hypotheses involve the effects of electrical fields.
Several studies indicate exposure to week electrical and magnetic fields may result in cancer, lethargy and loss of sex drive. Due to the difficulty of actually measuring the amount of exposure to these fields exact relationships have not been proven.
Counter–claim Where lines traverse forests they produce useful fire–breaks and also provide permanently maintained clearings bearing herbaceous or scrub vegetation, which increases ecological diversity and provides grazing for herbivores.
Refs OECD *Environmental Effects of Electricity Generation* (1985).
Broader Hazards to human health (#PB4885)
Ecologically unsustainable development (#PC0111)
Environmental hazards from energy production (#PD6693).
Related Environmental hazards from electromagnetic pulses (#PE6360)
Environmental hazards from mineral fuels, lubricants and related materials (#PE1346).
Aggravated by Risks in power production (#PE4835)
Inadequate electricity infrastructure (#PD9033).

♦ PE1417 Religious riots

Broader Civil disorders (#PC2551).
Related Prison riots (#PE1675) Student revolt (#PC2052).

♦ PE1418 Human psychological regression
Libidinal regression — Ego regression
Nature Psychological regression means a return to a more developmentally immature level of mental functioning.
Aggravates Ego defect (#PG0728).
Aggravated by Psychic conflict (#PF4968) Massive psychic traumatization (#PE6968).

♦ PE1424 Instability of fishing industry
Broader Instability of economic and industrial production activities (#PC1217).

♦ PE1425 Environmental hazards from paper and printing industries
Nature Hazards from paper and printing industries include accidents, and environmental, noise, dust and chemical hazards.
Accidents are responsible for the greatest amount of lost time. In addition to the usual tripping, falling, striking and being-struck accidents, must be added those from moving and revolving machinery. In the pulp industry, bark removing machinery, log cutting and shredding of pulp present serious hazards. Serious accidents can also be caused by the falling of badly stacked paper rolls.
Environmental problems are aggravated by the hot humid atmosphere of the drying rooms; and the high temperatures in the boiling, washing, and sulphate recovery areas expose operators to high levels of relative humidity and considerable temperature variations. Bronchitis and other respiratory ailments may result. Noise hazards result from the high noise levels at every stage of pulp and paper production (80dB to over 100dB); dust accumulations lead to increased fire and explosion possibilities; and the most serious chemical hazard results from the handling of lime and the lime kiln.
Broader Environmental hazards from manufactured goods (#PE1344)
Environmental hazards from manufacturing industries (#PD0454).
Narrower Environmental hazards of publishing and printing industries (#PE8095).

♦ PE1427 Inhaling of solvents and anaesthetic drugs
Glue sniffing — Inhalant abuse — Volatile substance abuse
Nature The solvents concerned include alcohols, aliphatic and aromatic hydrocarbons, aldehydes, ketones, chlorinated hydrocarbons and carbon disulphide. Inhalation may be: an occupational hazard (degreasing of metals in machine industry, extraction of fats and oils in chemical and food industry, dry cleaning, painting, plastics industry); use for medical purposes (for example, the inhalation of benzedrine to clear nasal congestion, overdose of ether or other anaesthetic, or inhalation by anaesthetists); or drug abuse for 'kicks'. Abuse of solvents as a potentially dangerous form of social challenge apparently begins when one or more members of a group of children or adolescents discover that prolonged inhalation makes them giddy. The symptoms of most forms of solvent abuse are fairly easy to detect and the practice should easily be discovered. Its undetected persistence over a long time points to a lack of supervision by parents or teachers. The extreme youth of those who abuse solvents (the average age is less than 19) brings the added danger that adolescents may be induced to begin experimenting with narcotic drugs or other psychotropic substances.
The chief dangers of inhaling solvents are death by suffocation, the development of psychotic behaviour, and the state of intoxication these substances produce. It is known that many solvents can damage the kidneys, liver, heart, blood and nervous system. Results of inhalation include fatigue, headache, vertigo, vomiting, skin irritation, unconsciousness, even death. Toxicity to the liver and kidneys may cause jaundice and anaemia. Prolonged exposure to benzene may cause leukaemia and anaemia. Carbon disulphide may cause atherosclerosis, ischaemic heart disease, and psychoses.
Incidence As yet the sniffing or inhaling of paints, glues and other volatile solvents is reported by relatively few countries. Statistical estimates or survey data were found for only 5 countries and verbal estimates for "some" abuse in only 11 others. The total for all countries is only 7,000, giving a global rate of 0.002 per 1 000. Only one country has moderate rates of abuse, Bolivia (0.66 per 1 000) and no country has a rate over 1 per 1,000.
Refs Watson, Joyce M *Solvent Abuse* (1986).
Broader Drug abuse (#PD0094) Substance abuse (#PC5536)
Drug dependence (#PD3825).
Narrower Inhalant intoxication (#PG7346).
Related Occupational hazards (#PC6716) Solvents as an occupational hazard (#PE5708).
Aggravates Mental depression (#PC0799).
Aggravated by Inadequate drug control (#PC0231)
Inadequate information on drugs (#PF0603).

♦ PE1429 Quelea
Nature Swarms of quelea birds regularly cause considerable damage to crops of sorghum, millet, rice, and other small grains. In recent years this damage has extended to newly introduced crops of wheat and barley.
Incidence Quelea attacks occur in East and West African countries, particularly in the drought-prone region of the Sahel. Nearly 1,000 million quelea birds are destroyed each year in Africa by the bombing and poison-spraying of their nightly roosting sites. This costly and hazardous operation often brings about only temporary, local reductions in the numbers of this pest. It seems that there is one very large and highly mobile population of these birds able to concentrate in different areas depending on rainfall. Control efforts have had little detectable effect since the population of birds is rapidly replenished.
Broader Birds as pests (#PD1689) Crop damage by wildlife (#PC3150).

♦ PE1431 Radioactive contamination of the marine environment and of fisheries products
Nature Radioactive fallout leads to the collection of radioactive materials in the sea and in freshwater environments, whether by direct deposition or by the outflow of such materials from land to sea or into freshwater sources. Such contamination may also derive from radioactive wastes resulting from nuclear energy operations.
Incidence The total sea-water burden of artificial radio-activity is less than 1 percent of the natural level in activity units. According to figures published by Unesco in 1977, the contributions from nuclear explosions in 1970 dominated those from reactor wastes and from the reprocessing of fuels; by the year 2000 they will still be of higher magnitude, assuming atmospheric nuclear testing will continue to take place at the 1968–70 rate. Artificial radio-activity will not be distributed as homogeneously within the ocean system as the two most important naturally occurring radionuclides, potassium-40 and rubidium-87. Higher values of all fall-out species are found in the northern than in the southern, hemisphere, as the greater number of nuclear-bomb detonations took place there. Similarly, the higher values of radio-active wastes associated with the production and use of nuclear fuels will be found in marine waters near the waste discharge sites. In such cases the levels of a given isotope are far in excess of the open sea values where entry has been primarily from atmospheric fallout.
Background Radioactivity was perhaps the first form of global marine pollution. It initially rose from the early post-World War II nuclear weapons tests by the United States of America in the South Pacific. Transported by stratospheric air currents, the fission products from a sub-aerial nuclear explosion were distributed through the earth's atmosphere and deposited on the earth's surface, mainly in temperate zones.
Refs Baranov, V I and Khitrov, L M (Eds) *Radioactive Contamination of the Sea* (1966); National Research Council *Contaminated Marine Sediments* (1990).
Broader Radioactive contamination (#PC0229).
Related Radioactive contamination of water (#PE2441).

♦ PE1436 Long-term shortage of machinery and transport equipment
Nature A decline in capital equipment production is associated with the post-industrial society in the developed countries; manufacturers are hesitant to design and produce efficient machinery in the absence either of home demand or of an export market with the ability to pay. It can therefore be expected that, for the forseeable future, developing countries will continue to be in short-supply of the latest industrial technology and be forced to employ antiquated methods and equipment. Developing countries suffer in particular from long-term shortages of equipment in the distribution sector; while all nations, producers included, will suffer from a long-term shortage of equipment in the manufacturing sector.
Broader Long-term shortage of commodities (#PC1195)
Insufficient availability of goods (#PB8891).
Narrower Electric machinery apparatus and appliances shortage (#PE8878).

♦ PE1437 Defeating size requirements
Claim Large size and the drive to increase size are basic causes of many of the ills of our times. The notion that big is good and bigger better has held sway too long among politicians, economists, business people, journalists and TV commentators. Industry is considered more viable if big. It creates massive demands through colossal expenditure on advertising so that it can produce more with larger machines so that it can have even larger markets. Small and medium companies are destroyed in the process taking control of local economies and jobs away from these communities. The Gross National Product has become a state God, putting strains on people, using up resources and creating pollution. Persuaded by politicians and others that all is needed is faster economic growth, people do not face up to the moral issue of just distribution of the present excess of goods and services. Science and technology is trapped in the bigger is better syndrome. The urban environment including, schools, building, cars, roads and houses has become to large to be compatible with human dignity, well being and spiritual growth.
Aggravates Unexplored opportunities for community education (#PF6512).

♦ PE1438 Tin as a pollutant
Stannosis
Nature Tin may be encountered as a contaminant in tinned food products; the presence of nitrate in such products may contribute to increased amounts of tin being dissolved from the container. A number of acute poisoning incidents have resulted from ingestion of fruit juices containing tin.
Refs FAO *Arsenic and Tin in Food* (1979); Joint Group of Exports on the Scientific Aspects of Marine Pollution *Cadmium, Lead and Tin in the Marine Environment* (1985).
Broader Toxic metal pollutants (#PD0948) Chemical contaminants of food (#PD1694)
Environmental hazards from metals (#PE1401).

♦ PE1439 Lice as insect pests
Anoplura
Nature Anoplura are the sucking lice that live in the hair of mammals and feed upon their blood. They are parasitic upon man, his domestic animals, dogs and wild animals.
Broader Parasites (#PD0868) Insect pests (#PC1630) Household pests (#PD3522)
Insect vectors of disease (#PC3597).
Aggravates Trench fever (#PG3689) Typhoid fever (#PD1753)
Relapsing fever (#PE7787) Insect bites and stings (#PE3636).
Aggravated by Louse resistance to insecticides (#PE3576).

♦ PE1444 Irresponsible introduction of new plant species
Broader Environmental hazards of new species introduction (#PC1617).
Aggravates Substitution of fast growing plant species (#PE6396).

♦ PE1448 Domination of agricultural equipment industry by transnational corporations
Nature The oligopolistic character of the farm machinery industry and the prevailing production technology that has enabled the major manufacturers to rationalize geographically their production operations and to achieve a measure of global standardization of parts and components, have also contributed to considerable inertia as regards changes in standard product designs not instigated by the manufacturers' perception of what is necessary for developed country markets. Thus, new products specifically designed to be convenient and economically available to small-scale farmers unaided by government subsidies have not originated from transnational corporations but from small, independent local manufacturers in developed and certain developing countries.
Transnational corporations have the capacity to meet most, if not all, the farming mechanization needs of developing countries. The above analysis of the situation seems to make it clear, however, that the primary orientation of transnational corporations towards developed country markets will continue to dominate strategic decisions, especially with regard to product design, global rationalization of the production process, and the standardization of parts and components to facilitate external sourcing from ancillary industry. Largely due to the inertia born of the established (and until recently profitable) order of operations, therefore, transnational corporations are (with minor exceptions) unlikely to make any special efforts to design and manufacture equipment within the budget of small-scale peasant farmers in developing countries without large subsidizing by the host government.
Broader Ineffective industry self-regulation (#PF5841)
Transnational corporation imperialism (#PD5891)
Control of industries and sectors by transnational corporations (#PE5831).
Related Plantation agriculture (#PD7598).

♦ PE1458 Contamination of public water supplies by sabotage
Nature Water supplies may be deliberately contaminated with chemical or biological agents. This may be done by aerial attack or by sabotage on the ground by introducing the agent into the water source at the intake or treatment works, a raw-water or treated water reservoir, or a transmission main by injection. Such introduction could be effected by suborning the works staff and sabotaging the control and detection procedures. The rapid dissemination of such agents to the unsuspecting population could have devastating effects.
Incidence Estimates suggest that in a large industrial city in a temperate climate, one kilogram of Botulin A toxin could lead to the death of 28,000 people, while 10 kilograms of LSD could incapacitate the same number.
Broader Sabotage (#PD0405) Occupied nations (#PC1788)
Water system contamination (#PD8122).
Related Product extortion (#PD5426).
Aggravates Impurities in waste water (#PD0482).

♦ **PE1469 Domination of the automobile industry by transnational corporations**
Nature The international auto industry is one of the most important in the world in terms of its size, economic significance and employment generation. Transnational corporations will largely affect what will be produced where and how. The activities of these firms have important consequences not only for developed countries but also for developing countries where transnational corporations operate or where there is aspiration for participation in the industry.
Incidence There are few industries in the world where a small number of corporations account for so much of the world production and trade as in the international auto industry. At the beginning of the 1980s, 22 transnational corporations operating in the market economies accounted for about 90 percent of total world–wide production in the industry. These transnational corporations produced some 27 million vehicles in their respective countries of origin and about 7 million in foreign locations. Accordingly, international production (not including assembly activities) came to 20 percent of their total production and 25 percent of their home production. Within the group of the 22 transnational corporations, 13 had foreign production facilities; all had foreign assembly in their own affiliates or licensees.
Broader Transnational corporation imperialism (#PD5891)
Control of industries and sectors by transnational corporations (#PE5831).

♦ **PE1474 Instability of the fur trade**
Incidence The Soviet Union and in the USA are the leading fur exporters. At least 20 million animals are killed annually the USA for their pelts. The fur trade depends on the trapper: of some 825,000 in the USA few are professionals, licensed or controlled. Overkilling of species is rampant and some species may disappear completely.
Broader International trade in endangered species (#PC0380)
Instability of trade in undressed hides, skins and fur skins (#PE1235).
Aggravated by Trade in animal products of endangered species (#PD0389).

♦ **PE1478 Excessive portrayal of negative information by the media**
Media emphasis on bad news — Excessive unsupportive criticism by the media — Excessive portrayal of destabilizing information by the media
Broader Criticism (#PF4530) Distorted media presentations (#PD6081).
Aggravates Insufficient cultural media (#PJ8476)
Psychological pollution by mass media (#PD1983)
Excessive commercialization of the media (#PE4215)
Manipulation of the individual by mass media (#PE7448)
Conflict between government and the news media (#PE1643)
Reduced impact of impersonal and repetitive news items (#PJ4911)
Excessive portrayal of perspectives of industrialized cultures in media (#PE3831).
Aggravated by Unethical media practices (#PD5251).
Reduced by State control of communications mass media (#PD4597).

♦ **PE1479 Psychological barriers to the judicial protection of individual rights**
Nature Protection of individual rights by the courts operates only if the individual appreciates the possibility of legal remedy. Cultural habits and fears deeply rooted in tradition may preclude uneducated individuals from recourse to the courts.
Broader Underutilization of legal rights (#PF3464).
Aggravates Denial of human rights (#PB3121) Inequality before the law (#PC1268).
Aggravated by Lack of self–confidence (#PF0879).

♦ **PE1483 Underdevelopment of chemical and petrochemicals industry**
Broader Underdevelopment of manufacturing industries (#PF0854).
Narrower Petroleum and coal products manufacture underdevelopment (#PE7960).
Aggravates Underdevelopment of industrial and economic activities (#PC0880).

♦ **PE1485 Compounding a crime**
Nature Failure to prosecute a person for a felony or misdemeanour in exchange for some consideration.
Broader Hindrance of law enforcement (#PD5515).
Aggravates Secularization (#PB1540).
Aggravated by Corruption of the judiciary (#PD4194).

♦ **PE1488 Social service quality negated by oligarchic control of decision–making**
Nature Social services, often perceived as charitable gestures by the rich toward the poor, may be maintained for their own sake, not for their contribution to the overall quality of life. Decisions are correspondingly made by those who primarily provide the funding.
Broader Financial and industrial oligarchy (#PD5193).

♦ **PE1490 Long–term shortage of meat and meat preparations**
Broader Long–term shortage of food and live animals (#PE0976).
Narrower Shortage of meat and meat preparations (#PE8693)
Shortage of fresh, chilled or frozen meat (#PE8224).
Related Slowing growth in food production (#PC1960).
Reduced by Vitamin E deficiency in domestic animals (#PE4760).

♦ **PE1496 Excessive external trade deficits of developing countries**
Nature Developing countries' efforts to reduce their trade deficits took place in an environment in which the prices of the majority of primary commodities had moved against them since the late 1970s.
Broader Excessive external trade deficits (#PC1100).
Aggravates Organizational empire–building (#PF1232)
Disincentives for financial investment within developing countries (#PF3845).

♦ **PE1499 Health hazards of modern insulating materials**
Nature Fibre glass and other modern loft insulation materials are as dangerous as white asbestos, according to a UK government report.
Broader Hazards to human health (#PB4885).
Aggravates Indoor air pollution (#PD6627).
Aggravated by Jurisdictional conflict and antagonism between government agencies within each country (#PE8308).
Reduced by Uninsulated buildings (#PE0242).

♦ **PE1504 Aerosols as industrial hazards**
Nature Aerosols are suspensions of solid or liquid particles in gas. They may be or contain toxic materials released by work operations and processes into the working environment.
In industrial hygiene surveys of air contaminants in the working environment a distinction is made between dusts, fumes, smokes, and mists and fogs. Dusts are generally formed by disintegration processes, such as in mining and ore–reduction operations. Examples are silica and asbestos dusts. Fumes usually result from chemical reactions, such as oxidation, or from sublimation or distillation processes followed by condensation. Examples are oxides of iron and copper. Smokes result from the combustion of fossil fuels, asphaltic materials, and wood. Smokes consist of soot, liquid droplets and, as in the case of wood and coal, a significant material–ash fraction. Mists and fogs consist of liquid droplets produced by atomization or condensation processes. Examples are oil mists from cutting and grinding operations, mists from spraying operations.
Broader Photochemical oxidant formation (#PD3663).
Aggravates Ulcers (#PE2308) Asthma (#PD2408) Heart diseases (#PD0448).

♦ **PE1506 Denial of right to social welfare services for indigenous peoples**
Denial of the right to social security for indigenous peoples — Social insecurity of indigenous peoples
Broader Social insecurity (#PC1867) Inadequate social welfare services (#PC0834)
Discrimination against indigenous populations (#PC0352).
Aggravates Ineffectiveness of intergovernmental organization and programmes (#PF0074)
Vulnerability of indigenous populations to introduction of diseases (#PE3721).

♦ **PE1508 Non–settled refugees living outside camps**
Nature The non–settled refugee population living outside camps comprises a group of refugees who have already achieved some degree of economic integration, particularly in respect of employment. Some of these refugees suffer from physical or social handicaps, but they are still capable of taking care of their own households and earning a sufficient income to pay a certain rent. For a variety of reasons, however, such as the continued overall shortage of housing in certain areas and the marginal character of refugee employment, most of these refugees, even among the non–handicapped, still live in dwellings which, according to local criteria, are in sub–standard categories. Quite apart from legal reasons (such as the fact that refugees do not possess the citizenship of their country of residence) which may in themselves be an obstacle to getting adequate housing, these refugees are unlikely, for many years, to benefit from the normal improvement of housing conditions on account of what may be called their 'marginality' as a social and economic category.
The refugees in this group may be classified into three categories: households wishing to emigrate, households where a (potential) breadwinner is unemployed, underemployed or requires (re) training; and households which need only adequate accommodation, but do not have the means to rent it.
The problem is further complicated by the fact that new refugees continue to arrive. This factor is particularly important in those areas where comprehensive country clearance programmes are planned to clear the backlog of earlier refugees.
Broader Non–settled refugees (#PC0519).

♦ **PE1510 Instability of cotton trade**
Nature The proportion of cotton in world fibre consumption (in cotton equivalent weight) is declining. Synthetic substitutes are taking over particularly in developed market economies, but inroads by synthetics have also occurred in cotton–producing developing countries. This has not only resulted in a stagnation in demand for cotton but has also in some cases caused cotton prices to rise higher than the price of synthetics. The high cotton prices have forced many mills out of business or caused them to change to production of synthetics, and this move is difficult to reverse. Despite high world prices, producer developing countries have not benefitted proportionately because they have failed to switch to strains of cotton which blend more easily with synthetic fibres. These trends in consumption and prices are threatening the well–being and jeopardizing the economic and social development large areas of the developing world.
Incidence About 55 per cent of cotton fibre consumption is accounted for by the developing countries.
Counter–claim Classifying the cotton trade as unstable is inaccurate. Since the 1960's, imports and exports have generally been in balance. While there are short–term fluctuations in both production and consumption, the price mechanism, coupled with the fact that cotton is grown in both hemispheres, has meant that imbalances have ben quickly corrected. Cotton consumption is volume terms has risen uninterrupted since the Second World War, by an average of 3.5 percent per annum.
Broader Instability of trade in unprocessed textile fibres and their waste (#PE1550).
Aggravated by Reduction in demand for primary commodities due to technological change (#PD1276).

♦ **PE1511 Inadequate relationship between transnational corporations and local industry in developing countries**
Broader Transnational corporation imperialism (#PD5891)
Social irresponsibility of transnational corporations (#PE5796)
Domination of developing countries by transnational corporations (#PE0163).
Related Underdevelopment (#PB0206).

♦ **PE1513 Instability of vegetable fibre trade, excluding cotton and jute**
Broader Instability of trade in unprocessed textile fibres and their waste (#PE1550).
Narrower Instability of trade in hard fibres (#PJ8431)
Instability of trade in manila fibre (#PE8682).
Aggravated by Reduction in demand for primary commodities due to technological change (#PD1276).

♦ **PE1514 Environmental hazards from fertilizers**
Excessive reliance on mineral fertilizers — Excessive use fertilizers — Overprescription of fertilizers — Misuse of fertilizers — Overuse of nitrate fertilizers
Nature In developed countries, the consumption of meat and meat products has risen markedly. Animals convert plant protein to animal protein at low efficiency, and hence there has been a parallel increase in the demand for grain and livestock feed concentrates. The change in human dietary habits has therefore led to a substantial rise in fertilizer use. The increased application of chemical fertilizers supplying the plant nutrients (nitrogen, phosphorus, and potassium) is an essential component of modern agriculture. Plants rarely use more than 50–60 percent of the nitrogen in fertilizers or 30 percent of that in animal manure. The residual nitrogen (nitrate) is liable to pollute ground and surface waters, causing over–enrichment (eutrophication), while some may be converted to nitrogen oxides.
The use of fertilizers changes both the quantity and the quality of crops. For instance, nitrogen fertilizers increase the protein content in maize and wheat grains, just as the phosphorus content of crops increases with the application of phosphate fertilizers. However, added nutrients may be detrimental in the case of legumes. If the plant takes up an excessive dose of nitrogen and cannot convert it into protein, the nitrogen remains in ionic form in the plant and might cause harm to the human body, particularly if it is transformed into a nitrate compound, and consumed as such. Furthermore, the interaction of nutrients can considerably change the composition of food. In certain vine experiments in 1982, the excessive use of potassic fertilizers resulted in a magnesium deficiency. This proves that food quality is greatly influenced by the primary and additional nutrient content of fertilizers. The general practice is to increase the nutrient content of fertilizers so as to reduce costs and to improve the efficiency of transport.
The micro–elements contained in fertilizers are no less important. Depending on the origin of raw materials and on how far they have been processed, different quantities of micro–elements can be found both in phosphorus and potassic fertilizers. Because of overdoses of phosphate fertilizers, plants take up zinc in insufficient quantities. The lack of zinc results in deficiency symptoms, and can cause illness in animals fed with such fodder.
Incidence Excessive reliance on mineral fertilizers has resulted in pollution of surface and

subsurface waters and tends to lead to loss of soil structure. FAO estimates that world consumption will increase to about 84 million tonnes in the 1985/1986 period. The developing countries (70 percent of the human population) use only about 15 percent of the world's fertilizers today, but this proportion is certain to change in the near future.
Counter-claim Mineral fertilizers are essential to achieve the required increases in food production.
Refs FAO *Effects of Intensive Fertilizer Use on the Human Environment* (1978).
    Broader Misuse of chemicals (#PD5904)   Chemical contamination of water (#PE0535)
    Environmental hazards from chemicals (#PC1192).
    Narrower Nitrates as pollutants (#PE1956)   Silicates as pollutants (#PG6267)
    Phosphates as pollutants (#PE1313)   Water pollution by fertilizers (#PE8729).
    Related Misuse of rodenticides (#PG6647).
    Aggravates Soil pollution (#PC0058)   Lake pollution (#PD8628)
    Metal contamination of soil (#PD3668)   Environmental plant diseases (#PD2224)
    Eutrophication of lakes and rivers (#PD2257)
    Health risks to workers in agricultural and livestock production (#PE0524).
    Aggravated by Unethical practice of soil sciences (#PD1110)
    Over-subsidized agriculture in industrialized countries (#PD9802).

◆ **PE1520 Irritability due to torture**
Incidence In a recent study of victims of torture some 30 percent of those studied suffer from some form of irritability as a result of the torture. These attacks often place great strain on family, friends and fellow workers.
    Broader Neurological effects of torture (#PD4755).
    Related Irritability (#PJ2736).

◆ **PE1535 Limited meeting facilities**
Small gathering places — Insufficient meeting space
Refs International Environment Liason Centre *International Environment-Development Facts March 1989* (1989).
    Broader Lack of essential local infrastructure (#PF2115).
    Related Inadequate meeting methods (#PF8939)
    Undeveloped potential of informal leadership (#PF1196).
    Unformed structures of community organization (#PF2810)
    Unexercised responsibility for external relations (#PF6505)
    Restrictive effects of traditional community decision-making (#PF3454).
    Aggravates Unexplored alternatives for commercial development (#PF6548).
    Aggravated by Fragmented planning of community life (#PF2813).

◆ **PE1536 Denial of right to a people to pursue development**
    Broader Denial of right of a people to be self-determining (#PC6727).
    Narrower Denial of rights to development for indigenous peoples (#PE4972).
    Reduces Destruction inherent in development (#PF4829).

◆ **PE1538 Exhausting commuter travel**
Refs European Foundation for the Improvement of Living and Working conditions *The Journey Between Home and Work*; European Foundation for the Improvement of Living and Working conditions *A European Study of Commuting and Its Consequences*.
    Broader Fragmented patterns of community activity (#PF6504).
    Related Aircraft pilot fatigue (#PE3870).

◆ **PE1539 Erosion of national sovereignty by transnational enterprises**
Nature Most of the problems connected with transnational enterprises stem from their distinctive transnational features in a world that is divided into separate sovereign states. Transnational corporations have developed important capacities which can be put to the service of world development, yet these same capacities can also be used in ways which may conflict with the interests of individual states. While governments may pursue a variety of economic and non-economic objectives to advance the welfare of their citizens, the chief goals of transnational corporations, like those of all business enterprises, are profit and growth. The differing objectives of nation-states and transnational corporations suggest that their respective decisions will not always be in harmony with each other.
The exercise of direct control over the allocation of one country's resources by residents of another makes the task of harmonizing varying interests and the promotion of the public good by governments especially complex. Advances in communications technology allow many transnational corporations to pursue global strategies which, rather than maximizing the profits or growth of individual affiliates, seek to advance the interest of the enterprise as a whole. Lack of harmonization of policies among countries, in monetary or tax fields for example, allows transnational corporations, on occasion, to utilize their transnational mobility to circumvent national policies or render them ineffective. It is in this context that countries may find their national sovereignty infringed upon and their policy instruments blunted by the operations of transnational corporations.
Since the objectives of nation-states and transnational corporations are frequently different, their respective power to attain them assumes particular importance. Under any form of social organization, the power exerted by individuals, corporations, pressure groups or nation-states is basically determined by the extent to which their opinions or decisions affect others. Because of their size and the transnational nature of their activities, transnational corporations, particularly the very large ones, possess considerable power and influence. In the process of conducting their normal business activities, transnational corporations make decisions which may have far-reaching consequences for the societies in which they operate. They affect patterns of consumption and the direction of innovation; they orient technological change and investment; and they own or produce most of the basic commodities used in industry and commerce. Intentionally or unintentionally, they can affect political processes of both home and host countries.
Decisions on the allocation of resources, with respect to what, how, and for whom to produce, are usually made by corporate planning mechanisms situated in a few industrial countries. The size and scope of the larger transnational corporations make it possible for a few large firms to control substantial shares of local and sometimes world markets. Because of this, and their transnational flexibility, they can engage in export market allocation, price discrimination, and transfer pricing; place stringent conditions on the transfer of technology and patents; and enter into cartel agreements that reduce competition.
Some of the problems posed by the activities of transnational corporations are similar to those which may arise in a modern or emerging industrial society from the activities of large and dominant corporations that are wholly national. In view of their transnational character, however, a policy framework which may be adequate for dealing with national corporations needs to be modified when dealing with transnational ones. In developing countries particularly, where transnational corporations may often be the only large enterprises, legislation and other institutional checks and balances such as public control and trade unionism, may not have developed sufficiently to cope with the power of those corporations.
Counter-claim However sacred and inviolable national sovereignty may be from the political point of view, few national boundaries correspond to economic demarcation lines and few states are self-contained economic entities.

    Broader Transnational corporation imperialism (#PD5891)
    Excessive power and independence of transnational corporations (#PD5807).
    Narrower Denial of the right to national sovereignty (#PE7906).
    Related Erosion of sovereignty (#PE5015).
    Aggravated by Political intervention by transnational corporations (#PE0032).

◆ **PE1541 Transnational strike action by trade unions**
Nature Due to the coercive use of economic power by transnational enterprises, successful strike action must be carried out in several countries to have its desired effect, although the international orientation of such strike action on the part of labour may lead to conflict with states. Problems may arise in that: efforts to influence a bargaining process by citing better conditions elsewhere may not always be welcome in the host country; sympathetic strikes in support of labour disputes in other countries are illegal in many places. In addition, labour has economic interests beyond working conditions; for example, barriers erected against certain imports may be considered beneficial to labour in a particular nation but detrimental to labour in the exporting country. Restrictions on the outflow of capital and technology may be favoured by labour in a home country, but undesirable from the point of view of labour in other countries.
    Broader Strikes (#PD0694)   Labour tensions involving transnationals (#PE5927).
    Aggravated by Coercive use of economic power by transnational enterprises against labour (#PE0207).

◆ **PE1542 Hotel and restaurant waste**
    Broader Unproductive use of resources (#PB8376).
    Aggravates Food wastage (#PD8844)   Domestic waste water pollutants (#PD2800).

◆ **PE1549 Instability of cocoa trade**
Nature Production is subject to large seasonal variations owing to changes in weather conditions and liability to pests and diseases. Because demand is relatively unresponsive to price, these variations in output tend to induce even greater price changes in the opposite direction. In addition, the need to export the entire crop each year (owing to cocoa's unsuitability for storage in tropical climates), and market speculation based on crop expectations before the final crop is known, tend to accentuate the amplitude of price fluctuations. Longer-term difficulties arise from the delay between planting and maturity which make it difficult to adjust supply. In the 1960s a steep fall in prices had severely damaging effects on the economies of some producing countries (in the case of Ghana, leading to a 50 per cent devaluation) and a general loss of confidence in the crop. Production capacity was severely reduced with the consequence that, in contrast to other commodities, insufficient cocoa is produced and countries are reluctant to increase production. The world's leading producer, the Ivory Coast, and the leading consumer nation, the USA have not joined the International Cocoa Agreements of 1972, 1975 and 1980. Insufficient provision for buffer stocks, other deficiencies in the Agreements' price-defence mechanisms, and lack of full international support perpetuate the instability of the trade.
    Broader Instability of trade in coffee, tea, cocoa, spices and manufactures thereof (#PE0915).
    Aggravated by Pests (#PC0728)   Bad weather (#PC0293).

◆ **PE1550 Instability of trade in unprocessed textile fibres and their waste**
    Broader Instability of trade in inedible crude non-fuel materials (#PD0280).
    Narrower Instability of jute trade (#PE1794)   Instability of wool trade (#PE2056)
    Instability of silk trade (#PG5966)   Instability of cotton trade (#PE1510)
    Instability of textile and clothing industries (#PE1008)
    Instability of trade in synthetic and regenerated fibres (#PE8950)
    Instability of vegetable fibre trade, excluding cotton and jute (#PE1513).

◆ **PE1552 Instability of date trade**
Nature World trade in dates is declining, mainly due to competition from other fruits. This decline primarily affects high-quality dates imported by developed countries. Low quality dates are exported principally by Iraq and imported by China and India. Two thirds of the international trade is between developing countries.
Incidence At least one million people live solely from the proceeds of date production which is estimated to be about 2 million tons annually.
    Broader Instability of fresh fruit and edible nut trade (#PE2587).

◆ **PE1553 Alienation of skilled and committed personnel from international organizations and programmes**
    Broader Ideological conflict (#PF3388)   Negative emotions and attitudes (#PA7090).
    Related Lack of continuity amongst personnel of international organizations (#PF3434).
    Aggravates Ineffectiveness of international organizations and programmes (#PF1074).
    Aggravated by Endemic abortion of ewes (#PG1537)
    Double standards in morality (#PF5225)
    Loss of credibility in international institutions (#PE8064).

◆ **PE1556 Cardio-pulmonary disease due to torture**
Incidence In a recent study of victims of torture 22 percent suffered from cardio-pulmonary diseases due to torture.
    Broader Heart diseases (#PD0448)   Lung disorders and diseases (#PD0637)
    Somatic and psychosomatic effects of torture (#PE5294).

◆ **PE1557 Excessive institutional debts**
Highly indebted corporations
    Aggravates Insolvent institutions (#PE6431).
    Aggravated by Leveraged buy-outs (#PE4963).

◆ **PE1558 Abuse of amphetamines**
Nature Amphetamines constitute a group of synthetic chemical stimulant drugs. They have been used medically to relieve catarrhal congestion (nasal), as general stimulants, and as slimming agents in cases of obesity. Initially amphetamines produce wakefulness, lessening of fatigue, increased energy and self-confidence and euphoria. This effect is only temporary and as tolerance builds up rapidly, increased dosages are necessary for the same effect. Ultimately these drugs produce restlessness, irritability, anxiety, slurred speech, unsteadiness and dryness of the mouth. They may also result in hallucinations, psychosis, paranoia or schizophrenia. Side-effects include malnutrition. Dependence is psychological rather than physical, but overactivity under the influence of the drug induces exhaustion afterwards which in extreme cases may lead to death. The combination of use of amphetamines with that of other drugs such as alcohol or barbiturates can be catastrophic. Amphetamines may be inhaled, taken orally or injected.
Incidence Recent total numbers of abusers are 2,284,209 for a global rate of 0.5 per 1000. Except for Japan (1.6 per 1000) all countries with extensive abuse are in the Americas: Canada (1.7), Cayman Islands (4.5), Mexico (2.0), Saint Lucia (9.1) and the United States (8.9 per 1000). Moderate abuse countries are in the Americas (Brazil and Bolivia) and Europe (Finland and Sweden). 68 countries and territories report at least some abuse of amphetamines. 86% of the world's abusers are from the United States, and another 9% from Japan. Most of the remaining 5% are from 8 other countries with high and moderate abuse rates.
    Broader Abuse of medical drugs (#PD0028).

Narrower Amphetamine withdrawal (#PG9778) Amphetamine intoxication (#PG3942).
Related Abuse of sedatives and tranquillizers (#PE0139).
Aggravates Psychoses (#PD1722) Hallucinations (#PF2249).
Aggravated by Mental depression (#PC0799).

♦ PE1570 Endangered species of non-human primates

Nature Rhesus monkeys, chimpanzees and other primates used in laboratory research are being seriously depleted in the wild by irresponsible collecting for medical and space research. Capture of predominantly young animals (in some cases 4–9 individuals are killed for each one caught) and females of any primate population reduces the breeding potential of the population to the point where the future availability of the species becomes seriously endangered.
Incidence During the 6 peak years of polio vaccine production in the 1950's, the USA imported more than 1.5 million rhesus monkeys. As most of these were juveniles, the projected population loss is five times higher, or about 7 million animals. In 1960, a survey in India showed that 63 per cent of the villages and temples surveyed had lost their populations of rhesus monkeys during the preceding 5 years. Nine Asian and 8 African species of primates, plus 7 entire genera from the Americas are now potentially threatened by laboratory use. It is estimated that only half of those collected are used for meaningful research.
Broader Endangered species of mammals (#PC1326).
Narrower Endangered species of lemurs (#PG3774)
Endangered species of aye-aye (#PG3778)
Endangered species of gibbons (#PG3784)
Endangered species of tarsiers (#PG3780)
Endangered species of great apes (#PG3785)
Endangered species of tree shrews (#PG3775)
Endangered species of potto and loris (#PE7953)
Endangered species of new-world monkeys (#PG3781)
Endangered species of old world monkeys (#PE8991)
Endangered species of tamarins and marmosets (#PE7933)
Endangered species of sifakas, avahis and indri (#PE8189).
Related Inhumane use of non-human primates in research (#PE1621).
Aggravated by Shortage of experimental non-human primates (#PF5073)
Degradation of semi-natural and natural habitats of flora and fauna (#PC3152).
Reduces Crop damage by wildlife (#PC3150).

♦ PE1571 Corporation financial secrecy

Nature Business enterprises disclose as little information as possible concerning: the identity of stockholders (frequently hidden by the use of nominees) and the amount of their holdings; annual financial accounts; the location and identity of subsidiary and associated corporations; and the relations with other corporations in the same business (particularly when mergers are envisaged). The need for secrecy is invoked by corporations because of the competitive disadvantages resulting from untimely disclosure. However, in many cases the owners of a corporation are able to remain totally anonymous, to the point that it is impossible or very difficult to determine who controls its operations and what their relationship is to other, supposedly competing, enterprises or official bodies. Since such secrecy is not complete, networks of privileged individuals with differing degrees of access to such information are built up. The information they obtain is often used, in some cases illegally, to the advantage of privileged clients and to the disadvantage of the small investor.
Incidence Although a great deal of information is made available on corporations in the USA, even there it is only with great difficulty and after much study that it is possible to determine, for example, how centralized is the control of major corporations. Less information is made available in other English-speaking countries. In continental Europe it is not common practice to make the same amount of information available. The owners of many corporations are able to remain totally anonymous and few financial details are made available to the public. In many other countries, governments have no reliable information on the bank accounts and incomes of their own taxpayers since banking secrecy could be invoked against most tax administrations.
Counter-claim If one analyses the financial statements of corporation, there is a wealth of data available. Governments throughout the world have enacted laws which require ample information to be published regularly. Publicly quoted entreprises (on stock exchanges) are submitted to even more detailed publication requirements. The law offers exceptions to companies where information is too sensitive and could be detrimental to their competitive position. These companies have in return also accepted an unlimited and personal liability of the shareholders.
Apart from the need for secrecy in a competitive environment, the system does permit investors who wish to remain anonymous to have their affairs handled by third parties. Disclosure of identity can be considered as an infringement of privacy. In addition, the system avoids considerable legal and administrative complexities, to the advantage of general flexibility.
Refs Walter, Ingo *Secret Money* (1985).
Broader Secrecy (#PA0005) Transnational corporation imperialism (#PD5891).
Related Inadequate international standards of financial accounting and reporting (#PF0203).
Aggravates Immorality (#PA3369) Tax evasion (#PD1466)
Ineffective tax systems (#PF1462) Ineffective industry self-regulation (#PF5841)
Undisclosed control of national economies by limited number of individuals (#PF2344)
Disproportionate influence on national economies of limited number of corporations (#PF1922).
Aggravated by Distortion of corporation financial statements (#PE2032).

♦ PE1575 Unsafe aircraft

Faulty aircraft
Nature Lack of aircraft safety usually stems from economic considerations. Inadequate escape precautions (such as jammed doors, buckled seats, torn-up floors, badly placed fuel pipes, and poorly arranged escape chutes) are asserted to be a main cause of potentially lethal incidents.
Counter-claim Aircraft are one of the safest type of vehicles in which to travel.
Broader Consumer vulnerability (#PC0123) Aircraft environmental hazards (#PD8328).
Aggravates Air accidents (#PD1582).
Aggravated by Mechanical failure (#PC1904).

♦ PE1578 Treachery by double agents

Betrayal by double agents
Incidence One renowned double agent, George Blake, claimed in 1990 that he had betrayed 600 agents working for the UK intelligence services.
Broader Espionage (#PC2140).

♦ PE1579 Proliferation and duplication of organizational units and coordinating bodies of the United Nations system

Nature There is a progressively increasing number of intergovernmental and expert organs within the United Nations system, many with overlapping competences. This proliferation complicates the task and consumes the time of governments and delegates. With it is associated a dispersal of authority among the key units of the United Nations and a consequent difficulty of harmonizing their activities. In some cases geographical separation of the units compounds the coordination problems resulting from the political and administrative fragmentation.
Incidence The dispersal of authority is not an entirely new phenomenon (in the area of narcotic drug control, for example, divided authority was a cause of constant friction for many years), but it has become much more common with the development of trust funds, the natural desire to meet the views of generous donors, and the equally natural desire to enlist the services of prestigious persons of the right nationalities to help ensure the success of the enterprise concerned. Year by year, units and posts are being set up for special tasks, with little regard for their relationship to units already in existence. An example is provided by the arrangements to deal with population questions. For administrative reasons, the United Nations Fund for Population Activities was set up in 1967 entirely independent of the existing Population Division of Economic and Social Affairs (ESA). It was officially placed under the supervision of the Governing Council of UNDP in 1969 but it has its own funds and has engaged in an active and independent programme throughout the world with its own coordinators in different countries in which projects are being undertaken. It was given main responsibility for preparations for World Population Year, 1974, of which the World Population Conference was the highlight. An independent Secretary-General and staff for the Conference were appointed, whose relationships with the Fund (or indeed with the Population Division) were quite tenuous, although the Director of that Division was named one of the Assistant Secretaries-General of the Conference.
One early case of duplication is particularly interesting in relation to the respective roles of governmental bodies and secretariats: the case of UNCTAD and GATT, the latter regarded in most of the Third World as a bastion of the great trading countries. UNCTAD represented a direct challenge to GATT, which, apart from its traditional work on tariffs, had already embarked on a number of the activities assigned to UNCTAD by the General Assembly after the first Conference on Trade and Development, in December 1964. In response to the challenge, in February 1965, the Contracting Parties to GATT added a Fourth Part, on Trade and Development, to the General Agreement, giving GATT a mandate (as well as subsidiary bodies) on subjects falling directly within UNCTAD's domain. Two great intergovernmental institutions with largely overlapping membership were thus, by deliberate decision of their governing organs, engaged in open duplication.
Background The change in the political balance in the General Assembly was mainly responsible for a great fragmentation of the United Nations economic machinery that occurred in the mid-1960s: namely, the establishment by the General Assembly, under strong pressure from the developing countries, of important new organizations within the United Nations in the fields of trade and development (UNCTAD) and industrial development (UNIDO). The form they were given (that of 'an organ of the General Assembly' (UNCTAD) or an 'autonomous' organization within the United Nations (UNIDO), and not of specialized agencies) was influenced by other considerations: first, the urgent search for economy on the part of the major contributors, and, secondly, doubts, largely in the same quarters, as to the efficacy of the arrangements between the United Nations and the specialized agencies in enabling the Council and the General Assembly to exercise their responsibilities under the Charter for co-ordination. It was hoped that, under the direct authority of the General Assembly, their expenditures could be better held in check and their activities better coordinated. Given the circumstances in which they were established, considerations of good administrative order were perhaps bound to be secondary. But the then Secretary-General felt it necessary to observe that the creation of autonomous units within the Secretariat, and therefore under his jurisdiction as Chief Administrative Officer, raised serious questions of organizational authority and responsibility. Moreover, such a trend was perceived as inconsistent with the concept of a unified secretariat working as a team towards the accomplishment of the main goals of the Organization and tending to have the adverse effect of pitting one segment of the Secretariat against another in competition for the necessary financial and political support for its own work programmes.
Broader Proliferation and duplication of international organizations and coordinating bodies (#PE1029).
Aggravates Proliferation and duplication of United Nations information systems (#PE0075).

♦ PE1581 Health risks to workers in transport, storage and communication industries

Broader Occupational risk to health (#PC0865).
Aggravates Industrial accidents (#PC0646) Occupational diseases (#PD0215).
Aggravated by Overloaded vehicles (#PE4127).

♦ PE1583 Jurisdictional conflict and antagonism between regional intergovernmental organizations with common membership

Nature Some of the difficulties arising from jurisdictional disputes between intergovernmental agencies are compounded at the regional level, particularly in the case of the United Nations Regional Economic Commissions. These are specially prone to jurisdictional clashes, since they cover the whole array of economic and social matters and so impinge on the areas of competence of most Specialized Agencies.
Claim The irrelevant organization of the United Nations regional economic commissions, which is largely responsible for the fact that so many governments have preferred to solve multilateral economic problems outside the United Nations, arises from the failure to lay down at the outset any consistent doctrine and consequent structure with regard to the relationship of United Nations regional activities with those of the Organization as a whole. Particularly unfortunate has been the United Nations doctrine, cherished since the foundation of the Organization, which regards political blocs as illegitimate bodies which should have no part in United Nations institutions, regional or otherwise. Not only are such blocs a part of reality which cannot be ignored except at the cost of failure, but they are in fact a positive necessity to the democratic structuring of any international body composed of sovereign states with varying interests.
In Europe and in Asia, at all events, governments look for political responsiveness and for effective regional action to the non-UN regional organizations. The regional economic commissions are left with those scraps of the whole which are not already being dealt with elsewhere. In Africa, for example, the political preference of many governments for the Organization of African Unity as compared with the United Nations Economic Commission for Africa is only partly outweighed by the superior equipment and experience of the latter in its own field.
Broader Jurisdictional conflict and antagonism between international organizations (#PD0138).
Aggravates Fragmented regional cooperation (#PF9129)
Inefficient location of facilities of international organizations (#PE3538)
Inadequate coordination between regional intergovernmental organizations with common membership (#PE1184).
Aggravated by Bureaucratic factionalism (#PF7979)
Jurisdictional conflict and antagonism between government agencies within each country (#PE8308).

♦ PE1589 Foot-and-mouth disease

Hoof-and-mouth disease
Nature Foot-and-mouth disease is a contagious viral infection of cloven-footed domestic and wild mammals and man, characterized by the formation of vesicles or blisters on the mucous membrane covering the various parts of the mouth, including the tongue, lips, gums, dental pad, and the palate.
Vesicles are also found above the claws of the feet and on dew claws, muzzle and nostrils, also teats and udder of nursing cows. The blisters rupture leaving a raw surface which is prone to other infections. Symptoms include an increase in body temperature, loss of appetite, lassitude and profuse slobbering. Chewing is painful and the animal loses condition and weight, and often becomes lame.
Milk flow drops or stops, abortion, mastitis and sterility may occur. Mortality in adult animals is usually less than 5 per cent but is higher in young animals. In severe outbreaks mortality of 50

per cent has been recorded, following severe inflammation and degeneration of the heart muscle. Cattle, hogs, sheep and goats are most frequently affected, in that order, though deer, camels, giraffes, hedgehogs, tapirs and elephants have also been found with the disease. Other animals such as dogs, rats, rabbits, guinea pigs, and mice have contracted the disease artificially under laboratory conditions but not naturally. Common carriers of the disease are earthworms and birds. The disease can be transmitted to man but this is rare. It can be transmitted by man, by contagion, and by wind. The most common means of infection is the ingestion of contaminated foodstuffs. The virus is shed in all excretions of sick animals including urine, faeces, saliva, milk and semen. This occurs before clinical evidence of the disease is present.
Incidence Incidence is higher during wet weather, as in the UK in 1967. It is generally considered enzootic, or frequently epizootic, in most of the major livestock producing countries of the world, except in North America, Central America, Australia and New Zealand. It is prevalent in many African game areas, also among affected wildlife species in any part of the world.
The 1981–1982 biennium was one of the periods of the least recurrence of foot-and-mouth disease since the commencement of the national programmes for control of the disease. In 1982 recorded occurrences of the disease reached the lowest historical levels in the chronological series, during that time marking a decrease of 38 percent compared to 1981, 61 percent compared to the 1978–1980 triennium, and 75 percent compared to the 1976–1977 period. During 1983 the situation was also notably favourable in Peru where, after the introduction of virus A from the country's northern border at the end of 1982, the occurrence of foot-and-mouth disease was only sporadic. In Venezuela recorded occurrences in 1983 were approximately one-half of the average recorded in recent years. Lastly, the situation in Uruguay may be mentioned, where for the third successive year foot-and-mouth disease occurred in the form of isolated and sporadic cases.
Background First recorded outbreak in the USA: 1870, last: 1929. Last Australian outbreak: 1872. Confusion of symptoms with bovine vesicular disease resulted in unnecessary slaughter in the UK in Dec 1972.
Refs Machado, Manuel A *An Industry in Crisis*; Regamey, R H, et al (Eds) *International Symposium on Foot-and-Mouth Disease* (1968).
    Broader Virus diseases (#PD0594)   Viral diseases in animals (#PD2730).
    Aggravates Economic loss through slaughter of diseased animals (#PE8109)
    Economic loss through reduced productivity of diseased animals (#PE8098).
    Aggravated by Worms as vectors of animal diseases (#PD2750)
    Wild birds as vectors of animal diseases (#PE2749)
    Importation of infected carcass meats as factor in animal diseases (#PE2777).
    Reduced by Inadequate carcass disposal of diseased animals (#PE2778).

♦ **PE1597 Inequitable allocation of rights to exploit sea-bed and marine resources**
Discriminatory exploitation of sea-bed resources
Claim Developing countries that are land-locked and otherwise geographically disadvantaged in relation to access to the sea are discriminated against in proposals, including those by maritime developing countries, for the sharing of the world's deep sea mineral and other resources. Exclusive economic zones (EEZ) in the sea, extending 200 miles, and in some cases 550 miles, depending on the length of the continental shelf, give an unfair allocation of rights.
    Broader Territorial disputes between states (#PC1888)
    Denial of right to equality between states (#PE4712).

♦ **PE1598 Minimal export promotion by transnational corporations in developing countries**
    Broader Transnational corporation imperialism (#PD5891)
    Social irresponsibility of transnational corporations (#PE5796)
    Domination of developing countries by transnational corporations (#PE0163).
    Related Economic and social underdevelopment (#PB0539).
    Aggravates Inadequate marketing of products of developing countries (#PE0523).

♦ **PE1599 Gastric disorders**
Stomach upsets
Nature Stomach upsets of various kinds are a frequent cause of absence of workers.
Incidence A survey in the UK suggests that 28 million man-hours are lost each month due to stomach upsets. As an indication, the market for stomach remedies in the UK in 1986 was £10 million, with a further £28.4 million for indigestion remedies.
    Broader Human disease and disability (#PB1044)
    Diseases of the oesophagus, stomach and duodenum (#PE8624).
    Narrower Dyspepsia (#PE4724)   Gastrointestinal diseases (#PE3861).
    Aggravates Absenteeism (#PD1634).
    Aggravated by Adenovirus infections (#PE2355).

♦ **PE1601 Elephantiasis**
Bancroftian filariasis — Elephantiasis arabum — Barbados leg — Boucremia — Bancroft's filariasis
Nature One of the most widespread and most disfiguring diseases - filariasis or elephantiasis - is caused by a thread-like nematode - Wuchereria Bancrofti. The disease is usually spread by the mosquito Culex fatigans which, when it bites an infected individual, takes up the filariae worms with the blood meal and injects these into the blood of another victim. Elephantiasis is characterized by infection of the lymph glands which causes swellings throughout the body, mostly of the legs, scrotum and the breasts which sometimes attain dimensions of several feet.
Incidence Elephantiasis occurs in almost every tropical and sub-tropical country from Charleston, South Carolina in the United States, Southern Spain in Europe, to Brisbane in Australia. It is common in India and China, where in some areas 50 percent of the inhabitants are infected, and in the West Indies, South America and Africa. In some of the Pacific Islands, the infection rate is 70 percent. In 1990 it was estimated that some 90 million people were infected.
    Broader Paralysis (#PD2632)   Uncontrolled tropical diseases (#PD4775)
    Infectious and parasitic diseases (#PD0982).
    Related Chyluria (#PG3813)   Filariasis (#PE2391).
    Aggravated by Vector-borne diseases (#PD8385)
    Parasites of the human body (#PE0596)
    Mosquitoes as vectors of disease (#PE1923)
    Aedes mosquitoes as vectors of disease (#PE3621)
    Anopheline mosquitoes as vectors of disease (#PE3622).

♦ **PE1602 Excreting in public places**
Urinating in public — Defecating in public
    Broader Unsociable human physiological processes (#PF4417).
    Aggravated by Incontinence (#PE4619)   Inappropriate sanitation systems (#PD0876).

♦ **PE1603 Measles**
Rubeola
Nature Measles is an acute infectious disease occurring mostly in children and characterized by an eruptive fever.
Incidence Measles is a major source of unnecessary suffering, premature mortality, and expense. Except in isolated populations, measles is nearly universal, most persons being infected before reaching the age of 15. Measles, under any circumstances, can cause serious complications. Among these are diarrhoea, encephalitis, otitis media, pneumonia, and exacerbation of protein-energy malnutrition. Therapy for measles and its complications is a major drain on medical care resources in most parts of Africa, Asia, and Latin America. It has been estimated that approximately 900,000 deaths from measles occur each year in the developing world. In the Inter-American Investigation of Mortality in Childhood it was found that measles is either the main cause of death or the second most frequent cause in children aged 1–4 years in several cities in Latin America. Measles outbreaks in Africa and Asia have case-fatality rates of 5–20% among children, especially malnourished ones. Measles complications may also result in developmental retardation, lifelong handicaps, and both direct and indirect economic loss. Furthermore, in children in the developing world, measles interacts with diarrhoeal disease and malnutrition to increase the morbidity and mortality from these conditions. In the developed nations, where the disease is less severe and there are facilities for saving lives, it is still important to eliminate measles.
    Broader Virus diseases (#PD0594)   Stigmatized diseases (#PD7279)
    Infectious and parasitic diseases (#PD0982).
    Related Laryngitis (#PE2653)   Rinderpest (#PE2786).
    Aggravates Thirst (#PE3818)   Deafness (#PD0659)   Headache (#PE1974)
    Gangrene (#PG5675)   Pneumonia (#PE2293)   Meningitis (#PE2280)
    Sore throat (#PE4651)   Tuberculosis (#PE0566)   Encephalitis (#PE2348)
    Chronic bronchitis (#PE2248)   Diseases of the ear (#PD2567)
    Eye diseases and disorders (#PD8786)   Child mortality in developing countries (#PE5166)
    Inflammatory infections of the respiratory organs (#PE9151).

♦ **PE1604 Discrimination against developing countries by the formation of regional groupings of developed countries**
Nature The difficulties faced by developing countries in exporting have been heightened with the formation of regional groupings among developed countries and the consequent removal of barriers to their intra-trade. Among the countries outside these groupings, the developing countries tend to be most vulnerable to the resultant differential tariff and non-tariff treatment, given their initial competitive disadvantages. As a result of the formation of such groupings and other preferential arrangements, almost two-fifths of the intra-trade in manufactured and semi-manufactured products among the developed market economy countries are already on a preferential basis. With the enlargement of the EEC, the share of preferential intra-trade among developed market economy countries is even greater. In addition, trade in industrial products between EEC member states and other European developed market economy countries is increasingly on a preferential basis.
    Broader Imperialism (#PB0113)   Military and economic hegemony (#PB0318)
    Domination of the world by territorially organized sovereign states (#PD0055).
    Related Domination (#PA0839)   Discrimination (#PA0833).
    Aggravates Regional disparities (#PC2049)   Fragmented regional cooperation (#PF9129).

♦ **PE1605 Health risks to workers in manufacturing industries**
Claim Mass production takes for granted the human suffering and boredom of assembly lines and the risks of physical injury.
    Broader Occupational risk to health (#PC0865).
    Related Underdevelopment of manufacturing industries (#PF0854).
    Aggravates Industrial accidents (#PC0646)   Occupational diseases (#PD0215).

♦ **PE1606 Dry rot of wood**
Nature Spread by airborne spores which settle on damp wood in badly ventilated rooms. The mycelia of Serpula lacrymans can grow through bricks, mortar and wood. Infected wood turns dull brown and breaks up in cube shapes. The rot may travel over 2 metres a year, causing major structural damage to buildings. It tends to occur where there has been flooding, rising damp or leaking plumbing.
Incidence In the UK, it has been estimated that 25 per cent of houses have either dry rot or wet rot.
    Broader Wood rots (#PE4455).
    Aggravates Structural failure (#PD1230).

♦ **PE1607 Black mildew in plants**
Nature Black mildew is a parasitic fungi that have a superficial dark mycelium.
    Broader Ascomycetes (#PE4586)   Fungal plant diseases (#PD2225).

♦ **PE1616 Paper shortage**
Long-term shortage of pulp and waste paper
Nature Paper is a material consumed in vast quantities, whose price in recent years has spiralled out of proportion to the general world-wide inflation. Shortage in supply of paper is due to considerable increase in consumption and, in the case of the developing countries, to inadequate local production and to a shortage of foreign exchange necessary for purchasing paper abroad.
The problem is most often publicized in terms of newsprint, but it is having a more lasting effect in limiting the production of school textbooks, and indeed books of all kinds, in developing countries, and in preventing or delaying the modernization of textbooks in desired directions. The recurrent difficulties arising from paper shortage have many serious consequences, particularly as regards educational, scientific and cultural progress in developing countries.
    Broader Long-term shortage of manufactured goods (#PE0802)
    Long-term shortage of inedible crude non-fuel materials (#PE0461).
    Related Waste paper (#PD1152).
    Aggravates Imbalanced distribution of knowledge (#PF0204)
    Shortage of books and textbooks in developing countries (#PF0118).
    Aggravated by Proliferation of information (#PC1298).

♦ **PE1619 Instability of spices trade**
    Broader Instability of trade in coffee, tea, cocoa, spices and manufactures thereof (#PE0915).
    Narrower Instability of pepper trade (#PE3000).

♦ **PE1620 Domination of the shipping industry by transnational corporations**
Nature International seaborne trade in the major bulk commodities is characterized by a high degree of concentration. The 10 major exporters of dry bulk cargoes and the 10 major importers account for over three-quarters of the tonnage.
The importing countries own a substantial percentage of the world fleet of bulk and combined carriers, and the exporting countries only a minor share. The major bulk commodities all present a natural field for vertically integrated operations by transnational corporations. Thus there is a natural tendency for key firms to control each stage in the chain, from mining (in the case of the metals and minerals) or collection (in the case of grain) to the consumer, either by owning or by holding a controlling interest in the companies performing these operations. Where a transnational corporation controls such a chain, the transportation of the raw material from the exporting to the importing country becomes an "intra-firm" transaction. In broad terms, it has been estimated that over one third of world trade flows are intra-firm transactions of transnational corporations. Even though developing countries own their mines, transnational corporations may still control the chain of marketing and distribution operations. This has been illustrated in the oil industry, where the nationalization of oil wells still left the transnational oil companies in the position of buying FOB

under production participation agreements or long-term contracts, and hence controlling ocean transport.
Ocean transport is an important link in the chain of coordination, but it is also a profit-making activity; in fact, where mineral transport involves the "dedication" of specific bulk carriers to specific routes over long periods, the profitability of the shipping operation is virtually guaranteed. The transnational corporations thus have a double reason for becoming involved in ocean transport. An added, and equally important, reason is that in vertically integrated operations ocean transport can be used for "transfer pricing" with respect to freight costs, by which a transnational corporation can minimize tax payments.
Broader Ineffective industry self-regulation (#PF5841)
Transnational corporation imperialism (#PD5891)
Control of industries and sectors by transnational corporations (#PE5831).
Narrower Transnational corporation control of bulk shipping (#PE5804).
Related Obstacles for international ocean shipping (#PD5885).

♦ PE1621 Inhumane use of non-human primates in research
Nature Monkeys and apes have been used for biomedical purposes on an increasing scale during the last 40 years, but the greatest increase in recent years occurred as a result of the introduction of poliomyelitis vaccine. Monkeys have traditionally been used in large numbers for both production and testing of polio vaccine and continue to be used in many countries, even though diploid cells can replace monkey kidneys at the production stage, and the inactivated (Salk) vaccine does not require the monkey neurovirulence test.
The main source of monkeys and apes used in the laboratory has been wild populations, but although action has been taken to conserve stocks, some species are already in short supply and this source of supply cannot be relied upon in the future. Many primate species are caught by the shooting of mother primates so that their infants, who cling to their mother's dead body, can be easily caught. Primates caught by this method include gorillas, orang utans, chimpanzees, gibbons, and many arboreal species. Many mothers and babies die for each baby ape successfully brought into captivity, and sometimes protective group members are also killed. Between trapping and receipt in the laboratory, many animals die and the tissues of many others have to be discarded owing to illness or infection. The animals may experience considerable pain and suffering during the experiments performed upon them and before they are finally discarded.
Incidence The greatest exploitation of any species of primates occurred during the 1950s when laboratories were racing to produce poliomyelitis vaccine using monkey kidneys. It was found convenient to use Rhesus monkeys from India. During the years of this boom about 2.5 million Rhesus monkeys were exported from India, mostly for vaccine production. For a time they left at the rate of 250,000 per year.
In 1968 it was estimated that total world trade amounted to 200,000 primates. The USA alone took 124,000 monkeys, 60 per cent coming from Latin America and the remainder from Africa and Asia. In 1978, India entirely stopped export of Rhesus monkeys, after learning that monkeys exported from India were being used in warfare experiments by the US military. In January 1979, Bangladesh stopped monkey exports for the same reason, as did Malaysia in 1984. Many countries feel that many of the uses to which primates are put overseas violate their religious traditions or national culture and principles.
Claim Although chimpanzees are now identified as threatened with extinction, demand for chimpanzees continues, especially for hepatitis research and vaccine testing. Although the research and testing is not in itself painful, large numbers of surplus chimpanzees infected with or carrying hepatitis result, and many become unsocialized due to having lived in isolation during experiments. Considerable controversy exists over whether these animals should be destroyed, used in fatal experimentation such as AIDS and cancer studies, or maintained, preferably at the expense of the user institutions or companies. Recently, there has been an increasing world-wide protest over use of primates in allegedly inhumane experiments. Pro-primate demonstrations are now frequent and many animal protection organizations include pro-primate work in their programs.
Broader Cruel treatment of animals for research (#PD0260).
Related Endangered species of non-human primates (#PE1570)
Shortage of experimental non-human primates (#PF5073).

♦ PE1622 Government intimidation of governments
Threats by governments against governments — Government bullying of foreign governments
Nature The aim is to accumulate military power for translation into territorial (Nazis in the 1930s), economic and political gain through extortion and, if necessary, war.
Broader Intimidation (#PB1992).
Narrower Military threats to land-locked countries (#PE5837)
Military threats to island countries and territories (#PE5785).
Related Foreign intervention in internal affairs of states (#PC3185).

♦ PE1625 Denial of equal benefits to elderly workers
Violation of right to equal work benefits for the aged
Broader Age discrimination (#PC2541) Exploitation of the elderly (#PD9343).
Related Employment discrimination against the elderly (#PD4916).

♦ PE1626 Concrete fatigue
Nature Concrete is one of the principal construction materials. Depending on the use to which it is put, concrete eventually fails as a material due to excessive loading, stresses, weathering and chemical action, or a combination of these. In compression it becomes stiffer and stronger when subjected to repeated loading; but it can be weakened if critical tensile stresses arise under loading conditions, and it may soften or crack. Concrete volume contains about 15 percent of minute empty spaces. It is therefore liable to absorb destructive substances such as acid rain, and in the case of concrete roadways and bridge decks, de-icing salts. When used over steel, expansion of the metal due to rusting creates pressure on the concrete. Another source of stress on concrete is during the sway of tall buildings.
Broader Fatigue in materials (#PD1391).
Aggravates Structural failure (#PD1230).

♦ PE1633 Cockroaches as pests
Nature Cockroach or roach, is the name applied to members of the Blattidae, a family of orthopterous insects. Some species live in human dwellings and are among the most abundant and persistent of household pests. In addition to eating human food supplies, cockroaches will eat soap, paper, books, shoes and clothing, even attacking the eyelashes and fingernails of sleeping humans. Cockroaches have an amazing ability to survive. A roach's heart will beat for 30 hours after the insect has been decapitated, and a decapitated female will find a place to lay her eggs before dying. They can sense poison via tiny hairs without ingesting it and will then avoid such poison for the rest of their lives. Their wings fold to allow them to hide in very small spaces and their tight shell retains moisture so they can survive for long periods without water. Conventional insecticides make little or no impact on the cockroach population.
The cockroach is fast and clever. They are nocturnally active, and a sudden room illumination may find the walls swarming with hundreds. The cockroach has had homage paid to it by Kafka in a story of that name and in the popular song 'La Cucaracha'.
Incidence Four species: the German, found world-wide; the American; the Oriental; and the brown-banded; are those which plague mankind, and have been in existence for 350 million years.
Claim Roaches, along with ants and rodents, cause vast amounts of anti-pest poisons with persistent toxicity to be used in dwellings. From kitchens and pantries, poison traces are ingested by humans due to contaminated cooking utensils, dishes, silverware and foodstuffs.
Broader Household pests (#PD3522) Orthoptera as insect pests (#PE3641)
Insect vectors of human disease (#PE3632).
Aggravates Plague (#PE0987) Leprosy (#PE0721) Yellow fever (#PE0985)
Poliomyelitis (#PE0504) Typhoid fever (#PD1753) Coxsackie virus (#PE3852)
Tapeworm as a parasite (#PE3511)
Insect damage to stored and manufactured goods (#PD3657).
Aggravated by Cockroach resistance to insecticides (#PE3579).

♦ PE1634 Insufficient medical supplies
Shortage of medical drugs
Broader Inadequate medical resources (#PD7254).
Narrower Lack of vaccines (#PE4657)
Limited access to pharmaceutical drugs (#PE1278)
Shortage of biological specimens for medical study (#PF5097)
Limited availability of therapeutic substances of human origin (#PF6751).
Aggravates Inadequate medical care (#PF4832).
Aggravated by Disorganized liaison with formal support (#PF2947)
Lack of means for local technological development (#PF6454)
Marketing of banned pharmaceutical drugs in developing countries (#PE6036).

♦ PE1635 Executive stress
Managerial stress — Stress in entrepreneurs
Incidence Executive stress is estimated to cost corporations $50 billion annually for counselling, disability claims and absenteeism.
Claim Many executives struggle with an almost neurotic inability to communicate their desires and goals at work. These habits can hinder the efficiency, productivity and profitability of their corporations.
Broader Occupational stress (#PE6937).

♦ PE1641 Instability of trade in beverages and tobacco
Broader Instability of the primary commodities trade (#PC0463).
Narrower Instability of trade in beverages (#PE1680)
Instability of trade in tobacco and tobacco manufactures (#PE0572).

♦ PE1643 Conflict between government and the news media
Broader Conflict (#PA0298) State control of communications mass media (#PD4597).
Narrower Denial of access to news (#PF3081)
Restrictions on news coverage of legal affairs (#PF3073)
Restrictions on direct news coverage of parliamentary affairs (#PF3072).
Aggravates Religious and political antagonism (#PC0030).
Aggravated by Proliferation of public sector institutions (#PF4739)
Excessive portrayal of negative information by the media (#PE1478).

♦ PE1646 Deliberate deformation of childrens' bodies
Broader Mutilation and deformation of the human body (#PD2559).
Aggravated by Child beggary (#PG1103).

♦ PE1651 Misuse of business enterprises as front organizations by government
Broader Front organizations (#PE4358).
Aggravates Bogus firms (#PF0326).

♦ PE1655 Fog
Foggy atmosphere
Broader Bad weather (#PC0293).
Narrower Haze (#PG3864) Mist (#PG3865).
Aggravates Common cold (#PE2412) Road hazards (#PD0791)
Marine accidents (#PD8982) Chronic bronchitis (#PE2248)
Hazards to navigation (#PE3868) Aircraft environmental hazards (#PD8328)
Photochemical oxidant formation (#PD3663) Deterioration in atmospheric visibility (#PE2593).

♦ PE1656 Endangered species of seals
Pinnipedia
Nature Many seal species depend on the same species of fish as are utilized by man. The rapid development of intensive fisheries in many parts of the world, leading to high catches of such species, threatens the food supply of certain seal species, preventing the maintenance of reasonable population levels, whether or not these seals are exploited by man. Because such seals have habits inimical to fishing interests (consumption of fish, damage to nets, hosts to fish diseases), there is considerable commercial pressure to maintain seal populations at levels below their natural size, leading to the risk of local extirpation. Seals are also threatened by marine (and particularly oil) pollution and the modification of estuaries for water storage. Human activity on seal hauling-out grounds which will increase with tourism, and particularly disturbance of nursing mothers and their young, can cause significant mortality among seal populations.
Background Indiscriminate killing of seals, whether for pelts or to protect fishing grounds, would soon reduce their numbers sharply and perhaps jeopardize their existence. This was demonstrated in the late 19th century when the herds of fur seals in the North Pacific, whose numbers in the 1860s had reached almost 2,000,000 fell to about 200.000 in 1911 after commercial hunters had relentlessly pursued them. As a result of protection their numbers rose again to 1,600,000 in 1941.
Perhaps the greatest danger to seals at the present time is pollution of marine habitats which is causing a decrease in their ability to fight off infections. About one half of the European seal population died of a form of distemper because of their weakened condition from living in the highly polluted North Sea.
Refs Busch, Briton Cooper *War Against the Seals* (1984).
Broader Endangered species of mammals (#PC1326).
Narrower Endangered species of phoca (#PG3873)
Endangered species of zalophus (#PG3870)
Endangered species of monachus (#PG3874)
Endangered species of odobenus (#PE8288)
Endangered species of arctocephalus (#PG3869).
Related Cruel culling of seals (#PE3484).
Aggravated by Marine pollution (#PC1117)
Polychlorinated biphenyls as a health hazard (#PE2432).
Reduces Over-population of seals (#PG3872).

♦ PE1657 Carbon monoxide as a health hazard
Carbon monoxide as an occupational hazard
Nature The best understood biological effect of CO is its combination with haemoglobin to form carboxyhaemoglobin (COHb). Carbon monoxide competes with oxygen for the binding sites of the haemoglobin molecules. The affinity of human haemoglobin for CO is about 240 times that of its

affinity for oxygen. The formation of COHb has two undesirable effects: it blocks oxygen carriage by inactivating haemoglobin, and its presence in the blood shifts the dissociation curve of oxyhaemoglobin to the left so that the release of remaining oxygen to tissues is impaired. Because of this latter effect, the presence of any percentage level of COHb in the blood interferes with tissue oxygenation considerably more than an equivalent reduction of haemoglobin concentration, e.g. through bleeding. The toxicity of carbon monoxide can be increased considerably by the presence of other pollutants. Among the substances that exert a synergistic action with carbon monoxide are nitrogen oxides, hydrocarbons, and hydrogen sulphide – a highly significant fact because all of these pollutants are present in the urban atmosphere.

Carbon monoxide is produced when organic material, such as coal, wood, paper, oil, gasoline, gas, explosives or any other carbonaceous material, is burned in a limited supply of air or oxygen. Small quantities of CO are produced within the human body from the catabolism of haemoglobin and other haemo-containing pigments leading to an endogenous COHb saturation of about 0.3–0.8% in the blood. Endogenous COHb concentration is increased in haemolytic anaemias and after bruises or haematomas which result in increased haemoglobin catabolism. CO is easily absorbed through the lungs into the blood.

Carbon monoxide is thought to be the most common single cause of poisoning in industry. Occupational exposure occurs in mines after explosions, in the iron and steel industry, where carbon monoxide is used to reduce the iron oxide to iron, and in gas plants. Automobile emissions expose garage workers, traffic police, and street vendors to appreciable amounts of carbon monoxide. Poisoning can be fatal, or if not, it may cause permanent damage to the central nervous system. Immediate symptoms are headache, dizziness and nausea, followed by unconsciousness in the case of exposure of longer than 10 to 45 minutes to carbon monoxide concentrations between 1,000 and 10,000 ppm; higher concentrations lead to death within a few minutes.

Incidence Carbon monoxide is thought to be by far the most common single cause of poisoning both in industry and in homes. Thousands of persons succumb annually as a result of CO intoxication. The number of victims of non-fatal poisoning that suffer from permanent central nervous system damage is estimated to be even larger. The magnitude of the health hazard due to carbon monoxide, both fatal and non-fatal, is huge and poisonings are probably more prevalent than is generally recognized. A sizeable proportion of the workforce in any country has a significant occupational CO exposure. CO is an ever-present hazard in the automobile industry, garages and service stations. Road transport drivers may be endangered if there is a leak of engine exhaust gas into the driving cab. Occupations with potential exposure to CO are numerous, e.g. garage mechanics, charcoal burners, coke oven workers, cupola workers, blast furnace workers, blacksmiths, miners, tunnel workers, Mond process workers, gas workers, boiler workers, pottery kiln workers, wood distillers, cooks, bakers, firemen, formaldehyde workers, and many others. Welding in vats, tanks or other enclosures may result in production of dangerous amounts of CO if ventilation is not efficient. The explosions of methane and coal dust in coal mines produce 'afterdamp' which contains considerable amounts of CO and carbon dioxide. If ventilation is decreased or CO emission increases owing to leaks or disturbances in process, unexpected CO poisonings may occur in industrial operations that usually do not create CO problems.

Automobiles produce the largest amount of carbon monoxide. Reduction of CO emissions from recent-model automobiles has been offset by a 34% increase since 1970 in the number of vehicles on the road and by an increase in the number of miles driven. Without pollution controls, the total carbon monoxide emitted would have increased significantly. Overall, CO emissions have changed little since 1970.

All appliances for heating and for production of hot water can produce carbon monoxide in some circumstances. In Brussels, Belgium, these account for the largest number of poisoning admissions in some hospitals.
Broader Occupational hazards (#PC6716)   Chemical air pollutants (#PD1271).
Aggravates Occupational diseases (#PD0215)   Occupational risk to health (#PC0865)
Badly laid out work premises (#PJ2468).
Aggravated by Motor vehicle emissions (#PD0414).

♦ **PE1659 Restrictions on socialist citizens working abroad**
Nature One of the most deeply-rooted characteristics of socialist states is the closing of state frontiers and the prevention of citizens from maintaining contact with foreign countries. They are prevented especially from travelling to, or working in, capitalist countries. There have, however, been experiments in allowing citizens of socialist countries to work in other socialist countries.
Incidence Polish people living in areas bordering on East Germany and Czechoslovakia were at one time permitted to work in those countries. The largest project of this kind was an agreement between Czechoslovakia and Vietnam for some 26,000 Vietnamese to work in Czechoslovakia.
Broader Denial of freedom of movement in communist systems (#PC3173).
Related Travel restrictions (#PC8452).

♦ **PE1660 Increasing income disparity in developing countries due to transnational corporations**
Nature The local procurement practices of transnational corporations have in some cases contributed to increasing income disparities and land concentration in rural areas by favouring large farmers and capital intensive technologies. Transnational corporations active in supplying technology and agricultural inputs may also contribute to increasing such disparities.
Broader Unequal income distribution (#PD4962)
Social irresponsibility of transnational corporations (#PE5796)
Domination of developing countries by transnational corporations (#PE0163).
Aggravated by Urban bias (#PF9686).

♦ **PE1666 Military use of animals**
Exploitation of marine mammals by the military — Military experiments on animals
Incidence Animals have traditionally been used in warfare, especially horses, camels, elephants, mules, donkeys and dogs. Pigeons have been used to carry messages (with some 20,000 having been killed in the First and Second World Wars) and have been proposed as guidance devices for missiles. Bats have been tested as delivery systems for incendiary devices. Dogs, cats and pigs have been tested for the delivery of explosives and for torpedo guidance. In the light of research in public aquaria, marine mammals are increasingly exploited by the military whether for torpedo recovery, espionage or suicide missions.
Animals are extensively used in military research: animals are shot to determine the damage capacity of bullets and explosive devices; after being subjected to lethal doses of radiation, cats, dogs and other animals are run to death in rotary drums to determine their survival capacity; other animals are tested with biological and chemical weapons and then exposed to similar endurance tests.
Broader Cruel treatment of animals for research (#PD0260).
Aggravated by Maltreatment of animals in aquaria (#PE5461).

♦ **PE1669 Deterioration of stored documents and archives**
Paper deterioration — Inadequate preservation of books
Nature Books, periodicals, documents and reports are the main means of recording and storing knowledge. Contemporary mass-produced paper, although possessing many qualities, frequently lacks that of endurance. One of the main reasons of paper deterioration is the cross linking of molecular and supermolecular fibre structures (so called 'hornification'). This, together with the acidic degradation of cellulose chains, leads to the brittling of fibers and in consequence of paper. The main reasons are low purity of cellulose, use of alum to fix the sizing (which gives rise to acidity in the paper) and the increasing amount of acid fumes in the environment.
Incidence A survey of books published between 1900 and 1939 suggested that nearly 40 per cent could not survive 25 years, even in moderate use, and most would be unusable in the 21st century. A later survey showed books published between 1940 and 1949 to be even shorter lived than newsprint. In the case of books of the 19th century, the deterioration can be indicated by the percentage categorized as needing restoration. Only 5 per cent from the first fifty years are in this class, rising to 10 for the period 1850–1869, and to 48 for the last decade, due to use of new techniques.
Broader Fatigue in materials (#PD1391)
General obstacles to problem alleviation (#PF0631).
Aggravated by Humidity (#PD2474)   Air pollution (#PC0119)
Destruction of historic documents and public archives (#PD0172).

♦ **PE1670 Experimental exposure of animals to pain**
Trauma-inducing experiments on animals — Experimental induction of psychological stress in animals
Nature As a means of obtaining insights in human psychology, animals are subjected to prolonged isolation as well as being exposed to pain and stress. Such experiments may include withholding postoperative pain-killers. Electric shocks may be administered, whether to observe psychological reactions or to condition the animals to respond to various stimuli.
Incidence In the UK in 1986, 25,650 experiments were conducted involving the deliberate induction of psychological stress in animals. Restraining chairs and racks are used for monkeys, cats and dogs to clamp them into a fixed position for periods extending up to months. A wide range of mechanical devices are used to batter animals with hammers or crush them with clamps. Other devices in standard experiments are used to inflict electric shock, extreme heat and extreme cold.
Broader Cruel treatment of animals for research (#PD0260).
Aggravated by Experimental battering of animals (#PE1171).

♦ **PE1675 Prison riots**
Disturbances in correctional institutions
Refs Morris, Roger *Riots and Disturbances in Correctional Institutions* (1982).
Broader Civil disorders (#PC2551).
Related Religious riots (#PE1417).
Aggravated by Denial of rights to prisoners (#PD0520).

♦ **PE1679 Beetles as insect pests**
Coleoptera
Nature In the animal kingdom beetles form by far the largest major group or order. At least 250,000 kinds are known. They constitute more than a quarter of all kinds of animals. Most of them feed either upon other animals or upon plants, but some eat decaying matter of various kinds as well as a variety of other organic substances.
Many beetles are injurious, either as larvae or adults by being vectors of disease and as pests. Among those which attack farm crops, the wireworm is important. Wireworms are most prevalent in newly ploughed fields and attack the supervening crops especially cereals and roots. Adult and larvae flea beetles cause great damage to turnip crops. The asparagus beetle is a pest to that plant in North America and Europe. The Colorado potato beetle is destructive to potatoes in the eastern half of North America and in the Bordeaux district of France. The Japanese beetle is injurious to the foliage of fruit and other trees and its larvae damage lawns. The larvae of the May or June beetles attack the roots of grasses, potatoes, strawberries, etc. and the adults feed on foliage of trees. The boll weevil is a serious problem to cotton crops. The granary weevil and the meal worms attack stored grain, meal and other dried products. The allied apple blossom weevil is very destructive to unopened blossom buds. The palm weevil injures toddy and coconut palms and the pine weevil attacks young conifers. The alfalfa weevil is a serious enemy of clovers and alfalfa.
Counter-claim Many species of beetles are used to control pest of commercial crops. Lady beetles or ladybirds and ground beetles are of particular benefit. The Australian ladybird has been imported to most citrus growing areas of the world to control fluted scale. Another species is used to control mealy bugs. The European ground beetle preys upon the caterpillars of the gypsy moth and the brown-tail moth. Beetles save millions of dollars in crop damage around the world.
Broader Insect pests of wood (#PD3586)   Insect pests of plants (#PD3634)
Pests and diseases of trees (#PD3585).
Narrower Carpet beetle (#PG1142)   Wireworms as pests (#PE3645)
Wood-boring beetles (#PE0886).
Aggravates Dutch elm disease (#PE1154)   Maize pests and diseases (#PE3589)
Pests and diseases of elm (#PE2982)   Pests and diseases of oak (#PE2984)
Pests and diseases of rice (#PE2221)   Pests and diseases of wheat (#PE2222)
Pests and diseases of palms (#PE2981)   Pests and diseases of potato (#PE2219)
Pests and diseases of chestnut (#PE2983)   Pests and diseases of sugar cane (#PE2217)
Pests and diseases of sugar-beet (#PE2975)   Pests and diseases of grain sorghum (#PE3590)
Insect damage to stored and manufactured goods (#PD3657).

♦ **PE1680 Instability of trade in beverages**
Broader Instability of trade in beverages and tobacco (#PE1641).
Narrower Instability of alcoholic beverages trade (#PE2585)
Instability of trade in non-alcoholic beverages (#PE8748).

♦ **PE1683 Instability of trade in miscellaneous food preparations**
Broader Instability of trade in food and live animals (#PD1434).
Narrower Instability of margarine trade (#PG6004).

♦ **PE1684 Stroke**
Apoplexy
Nature Apoplexy, or a stroke, is the sudden insensibility or bodily disablement caused by a diseased condition of the brain, when a blockage or a leak in a blood vessel interrupts the blood supply, causing the death of brain cells.
Incidence In the USA, for example, 400,000 people suffer a stroke each year, killing about 160,000 of them. It is the third most common cause of death in the USA, after heart disease and cancer.
Refs Banks, Moira A *Stroke* (1986); Dunkle, Ruth E and Schmidley, James *Stroke in the Elderly* (1987); Milikan, Clark H, et al *Stroke* (1987).
Broader Human disease and disability (#PB1044)
Diseases and injuries of the brain (#PD0992).
Aggravates Paralysis (#PD2632)   Human death (#PA0072)
Speech disorders (#PE2265)   Mental depression (#PC0799)
Limited individual attention span (#PF2384).
Aggravated by Gout (#PG3061)   Over-eating (#PE5722)
Hypertension (#PE0585)   Whooping cough (#PE2481)

PE1684

Kidney disorders (#PE2053) Bright's disease (#PE2272)
Overheated rooms (#PG3890) Emotional crises (#PE3407)
Diseases of the arteries (#PE2684) Degenerative changes in the heart (#PG3887)
Environmental hazards of solar radiation (#PE3883)
Disruption of internal balance of the human body (#PE6603).

♦ PE1690 Animal road deaths

Incidence It has been estimated that in the USA about one million animals are killed every day by vehicles.
Broader Animal deaths (#PE7941).
Related Road traffic accidents (#PD0079)
Denial to animals of the right to conditions of life and liberty proper to their species (#PE6270).
Aggravated by Inadequate animal welfare (#PC1167).

♦ PE1692 Unfair trials due to pre-trial publicity

Nature Premature publicity concerning criminal trials (especially trials by jury) when critical of the accused, undermines his enjoyment of the right to a fair hearing. This is particularly true of accused persons belonging to racial or other groups against whom prejudice exists in the community in whose midst the trial is being held or will be held. In some countries, committal proceedings may be reported in the press, yet the accused may have reserved his defence and only the prosecution's case may, therefore, have been heard; such one-sided publicity may prejudice the attitude of persons later serving on a jury in the trial of the accused. In any type of judicial proceeding, and at any stage of the hearing, unfavourable and unfair publicity may work to the particular detriment of members of groups against whom prejudice already operates.
Broader Unjust trials (#PD4827) Uncontrolled media (#PD0040)
Journalistic irresponsibility (#PD3071).
Related Infringement of privacy (#PB0284) Bias in jury trials in small jurisdictions (#PE4733).
Reduces Restrictions on news coverage of legal affairs (#PF3073).

♦ PE1693 Denial of right to minimum work age

Lack of minimum age law — Failure to enforce minimum age laws
Broader Denial of rights of children and youth (#PD0513).
Aggravates Exploitation of child labour (#PD0164).

♦ PE1702 Monopolization by interest groups in development of community priorities

Nature Global participation is often lost when interest groups determine the priorities for the future, rather than allowing grass roots expression of concern to point to the priorities. When such decisions are monopolized by a few leaders, trust is undermined and so, therefore, is the participation of the community in acting out the decisions.
Broader Obstacles to community achievement (#PF7118)
Inadequate means for upholding global concern (#PF1817).
Aggravates Conflicting community priorities (#PJ7975).

♦ PE1704 Excessive exposure of medical patients to radiation

Incidence At a British hospital over 200 cancer patients received radiation overdoses.
Claim In general, X-ray examinations are over-used throughout the world. Far too frequently, such examination, whether of the chest, the skeleton, or the abdomen, yields little information that is clinically useful. Many are given routinely, or without medical justification, and thus are 'not worth-while', according to a recent report by a group of leading radiologists convened by the World Health Organization (WHO). So over-used and mis-used are X-rays that they constitute a major source of population exposure to man-made ionizing radiation. They are also costly, accounting for an estimated 6 to 10 percent of a country's health expenditure.
Broader Inadequate medical care (#PF4832) Health hazards of radiation (#PD8050).
Related Radioactive contamination (#PC0229) Denial of rights of medical patients (#PD1662).
Aggravated by Unethical practice of radiology (#PD8290).

♦ PE1708 Over-spacing of suburban housing

Nature Housing in suburban settlements may be too widely spaced for genuine community development, too demanding of extended and thus expensive services, too wasteful of land and too sprawling to permit any real safeguarding of open space or providing of sufficient areas of recreation. These problems are at present a phenomenon of industrial society; but public policy confronts similar difficulties of over-extended services, access to work and potential sprawl, in the planning and upgrading of fringe communities around new cities in the developing countries.
Broader Ineffective space usage (#PE5458).
Narrower Obstacles to availability of community space (#PF7130).
Related Scattered housing locations (#PJ0760)
Prohibitively expensive housing for the poor (#PE8698)
Inaccessibility of city centres to suburban residents (#PF6132).
Aggravates Limited public space (#PJ9066) Destruction of agricultural land (#PD9118).

♦ PE1711 Instability of vegetable trade

Broader Instability of trade in fruit and vegetables and preparations thereof (#PE0961).
Narrower Instability of trade in fresh potatoes (#PE8258)
Instability of trade in fresh tomatoes (#PE8688)
Roots and tuber trade instability vegetable products (#PE8901)
Instability of trade in beans, dried, peas, lentils and other leguminous vegetables (#PE8259).

♦ PE1712 Long-term shortage of mineral fuels and lubricants

Broader Long-term shortage of commodities (#PC1195)
Insufficient availability of goods (#PB8891).
Narrower Long-term shortage of coal (#PE1054)
Long-term shortage of electric energy (#PE1216)
Long-term shortage of natural and manufactured gas (#PE8045)
Long-term shortage of petroleum and petroleum products (#PE8626).

♦ PE1717 Gonorrhoea

Nature Gonorrhoea is an inflammatory disease mainly affecting the mucous membrane of the urethra in the male and that of the vagina in the female, but spreading also to other parts. The disease is directly contagious usually by sexual intercourse from another person already suffering in this manner. The infecting agent is the gonococcus or Neisseria gonorrhoea.
As yet it has proved practically impossible to produce a vaccine against gonorrhoea because gonococci provoke only a very limited immunological response in humans. At one time, treatment with penicillin was very successful. Although mildly penicillin-resistant gonococci had been encountered, until recently none had been found which was able to break down the antibiotic and thus become almost completely insensitive to the drug. However, during the last few years, gonococci which can destroy penicillin have been encountered. As penicillin was the drug of choice for gonorrhoea, this development has exacerbated what the World Health Organization admits to be a global epidemic wildly out of control in many countries.
Refs Hethcote, H W and Yorke, J A *Gonorrhea Transmission Dynamics and Control* (1984).
Broader Sexually transmitted diseases (#PD0061).
Narrower Gonococcal urethritis (#PG1872).
Aggravates Peritonitis (#PE2663) Heart diseases (#PD0448)
Human infertility (#PC6037) Physical blindness (#PD0568)
Rheumatic diseases (#PE0873) Natural human abortion (#PD0173).

♦ PE1720 Environmental degradation by off-road and all-terrain vehicles

Nature Off-road vehicles are predominantly recreational and include motorcycles, minibikes, trail bikes, snowmobiles, dune buggies, all-terrain vehicles, hovercraft and possibly also motor boats. They are motorized and designed for travel on or over all forms of natural terrain. Their increasing popularity is leading to damage to soil and vegetation. Wildlife is harassed and its habitat disrupted. The distances one must travel to find terrain in which to use such vehicles encourages use of them in some places which are unsafe, where ecological damage can be caused, or which are otherwise unsuitable.
Incidence Experts in the Glacier National Park in the USA argue that use of a 4-wheel drive vehicle there would damage the ground for ten years, because the ecology is in such a delicate state of balance.
Refs Leitch, D A *Environmental Impact and Regulation of Recreational All-Terrain Vehicles in Manitoba* (1975); Webb, R H and Wilshire, H G (Eds) *Environmental Effects of Off-Road Vehicles* (1983).
Broader Environmental degradation (#PB6384).
Narrower Snowmobiles on roads (#PG9341).
Related Over-use of designated wilderness areas (#PD7585).
Aggravates Deforestation (#PC1366) Permafrost instability (#PD1165)
Health hazards of exposure to noise (#PC0268)
Environmental degradation from recreational use of unsurfaced country roads and tracks (#PE7403).
Aggravated by Unpreparedness for surplus leisure time (#PF5044).

♦ PE1722 Prohibitive cost of equipment maintenance

Prohibitive cost of product repair
Claim Many of the products currently manufactured, especially those based on sophisticated technologies, are built so that, in the vent of breakdown, they cost more to repair than to replace. This is also true of larger products which are increasingly manufactured as complete units (such as doors, windows, staircases).
Broader Prohibitive cost of maintenance (#PF0296).
Aggravates Inadequate equipment maintenance (#PD1565)
Prohibitive cost of home maintenance (#PJ9022)
Acquisition of inappropriate equipment by developing countries (#PE6640).
Aggravated by Planned obsolescence (#PC2008)
Unethical maintenance practices (#PD7964).

♦ PE1726 Selenium as a pollutant

Nature Selenium is a chemical element closely allied in physical and chemical properties with sulphur. It occurs in rocks and soils all over the world. There are no true deposits of selenium anywhere, and it cannot be economically recovered from the earth directly. The highest concentrations known are in native sulphur from volcanoes, which contains up to 8.350 ppm. Selenium does, however, occur together with tellurium in the sediments and sludges left from electrolytic copper refining and the chief world supplies are from the copper refining industries of Canada, the United States, and Zimbabwe where the slimes contain up to 15 percent selenium The manufacture of selenium rectifiers, which convert alternating current to direct current, accounts for over half the world's production of selenium. It is also used for decolourizing green glass and making ruby glass, for allying with stainless steel and copper, as an additive in the natural and synthetic rubber industries, as an insecticide, as 75-Se for the radioactive scanning of the pancreas and for photostat and X-ray xerography The elemental forms of selenium are probably harmless; its compounds, however, are dangerous and their action resembles that of sulphur compounds. There is some evidence that they may be teratogenic for certain species. Selenium compounds may be absorbed in toxic quantities through the lungs, intestinal tract or damaged skin. Many selenium compounds will cause intense burns of skin and mucous membranes, and chronic skin exposure to light concentrations of dust from certain compounds may produce dermatitis and paronychia. The sudden inhalation of large quantities of selenium fumes, selenium oxide or hydrogen selenide may produce pulmonary oedema due to local irritant effects on the alveoli; this oedema may not set in for 1–4 hours after exposure. Exposure to atmospheric hydrogen selenide concentrations of 5mg is intolerable; however, this substance occurs in only small amounts in industry (for example, due to bacterial contamination of selenium-contaminated gloves), although there have been reports of exposure to high concentrations following laboratory accidents. Skin contact with selenium oxide or selenium oxychloride may cause burns or sensitization to selenium and its compounds, especially selenium oxide. Selenium oxychloride readily destroys skin on contact, causing third degree burns, unless immediately removed with water Certain plants are known to accumulate selenium when grown in soils containing a high natural content of this element or where selenium has entered the soil from industrial waste. Cereals such as wheat can accumulate up to 30mg/kg without impairment of growth. It is thus possible for food crops to be contaminated by selenium.
Incidence In certain areas (such as parts of the mid-Western USA) some plants (for example, loco-weed) accumulate selenium to such an extent that grazing animals display toxic symptoms.
Broader Toxic metal pollutants (#PD0948) Chemical contaminants of food (#PD1694)
Environmental hazards from metals (#PE1401).
Aggravates Water pollution (#PC0062).

♦ PE1727 Soil salinization

Salty land — Saline soils
Nature Saline soils are those that are high in non-alkaline salts, such as sodium chloride and sodium sulphate. Soils with a high concentration of alkaline salts are referred to as alkaline soils (see separate entry). Restricted drainage, caused either by slow permeability or by a high water table, is the principal factor in the formation of saline soils. Because all irrigation waters contain dissolved salts to some extent, nonsaline soils may become saline unless water is applied in addition to that required to replenish water losses by plant transpiration and evaporation, in order to leach out the salt that has accumulated during previous irrigations and due to the addition of fertilizer.
Salinization occurs because, in general, irrigation waters have a high salt content, and because, in irrigated areas, rainfall is generally too low to leach accumulated salts. Salts that most frequently predominate are the carbonates, chlorides and sulphates of sodium, magnesium and calcium, and mixtures of two or more of these salts. Reductions in growth and yield of crops due to saline conditions may go unnoticed. But with higher salt concentrations, plant growth is even more restricted, and the characteristic symptoms (burning and firing of leaves) of saline poisoning are shown. This is effected by the high osmotic pressure of the soil solution and the reduced solubility of some nutrients at higher pH values. When sodium salts accumulate in the soil, they usually impart undesirable physical characteristics to the soil, such as reduced permeability to water and reduced aeration. This increases the susceptibility of the soil to further salinization.
Salty ground is unsuitable for plant growth. In semi-arid and arid regions, where irrigation is widely practised, there is always a danger that high concentrations of salts may accumulate in the soil, reducing its productivity or even rendering it completely unproductive.

Incidence Salinity is regarded as a problem when the concentration of salts reaches 0.2 per cent of the dry weight of a loam soil; most species of plants die when the concentration is around 2 per cent. Throughout the world, there are more than 10 million hectares of now saline land which was once fertile, and every year over 100,000 hectares of irrigated land fall victim to salinization. In 1985 it was estimated that up to 20 million hectares of irrigated land around the world may be affected by salinization.
In Pakistan, 15 million acres of the 37 million acres under irrigation are estimated to be salinized. Of the area earmarked to receive water via China's giant Yangtse Diversion scheme, 2.7 million hectares already suffer from salinization. An FAO study reported that 35 percent of Egypt's cultivated surface is afflicted by salinity. 500,000 acres in Syria, half of the country's irrigated land, are waterlogged or salinized. In Iran, of 16.8 million hectares of arable land, 7.3 million are estimated to be saline. In India, the amount of land devastated by water and salt has been estimated at between 6 million and 10 million hectares, almost a quarter of the 43 million hectares under irrigation. In the US, 25 to 35 percent of the country's irrigated land suffers from salinity and the problem is getting worse.
Refs FAO *Man's Influence on the Hydrological Cycle* (1976); FAO *Water and the Environment* (1975); Szabolcs, I *Review of Research on Salt-Affected Soils* (1979).
Broader Soil infertility (#PD0077).
Related Alkaline soil (#PD3647).
Aggravates Infertile land (#PD8585) Soil degradation (#PD1052) Environmental plant diseases (#PD2224) Destruction of land fertility (#PC1300) Deficiency diseases in plants (#PD3653) Endangered monuments and historic sites (#PD0253).
Aggravated by Rising water level (#PD8888) Inadequate land drainage (#PD2269) Silting of water systems (#PD3654) Water pollution by fertilizers (#PE8729) Mismanagement of irrigation schemes (#PE8233) Unsustainable short-term improvements in agricultural productivity (#PE4331).

♦ PE1728 Political bias in official appointments
Inadequate appointment systems
Incidence Protest has been registered against the ideological or political bias exhibited by presidents, prime ministers and even the Pope, when making appointments to major posts supposedly outside the political arena. In the case of the USA, recent presidents have steadily continued to fill the federal courts with judges expected to construe laws narrowly, side more frequently with the police and defer to Congress.
Related Bias in selection of political candidates (#PD2931).

♦ PE1729 Understated debts
Incidence The Soviet Union and many key borrowers in the developing countries owe more to major industrialized countries than was previously known, because they have received large-scale government-backed credits besides the normal bank loans.
Broader Uncontrolled growth of debt (#PC8316).

♦ PE1732 Arsenic as a pollutant
Nature Arsenic as a chemical element occurs in both grey and yellow crystalline forms. The most common commercial product is white arsenic (arsenious oxide), which is usually prepared as a by-product of the roasting of various ores. The highly toxic nature of many arsenic compounds has led to their extensive use in insecticides and weed killers.
Incidence Arsenic occurs in almost all soils in small amounts. It is also found in variable quantities in natural waters, depending on geographical location. Air also contains arsenic, but usually in very small concentrations. Environmental pollution by arsenic may arise from agricultural practices (weed-killers, fungicides, sheep dips, rodenticides, insecticides) and from industry. Atmospheric emissions of arsenic are associated with metal smelters refining arsenical ores and with the use of arsenic-containing pesticides; reports from Czechoslovakia and India indicate that coal burning is another possible source. Ambient air concentrations of arsenic are of the order of 0,02 micro g/m3; but these may be much higher in the vicinity of arsenic-emitting sources. Arsenic in topsoils, since it is exposed to atmospheric oxygen, is usually present in the less reactive, pentavalent form. Industrially produced arsenic is in the more toxic trivalent form. Arsenic may be present in small amounts in drinking-water, and a tentative safety limit of 0,05 micro g/l is recommended by WHO. Arsenic occurs naturally in foods and beverages and is normally present in relatively high concentrations in crustacea and other shellfish. It was detected in shrimp as long ago as 1935 in amounts of 42-174mg/kg. The naturally occurring form found in marine animals is probably not very toxic. Arsenic can occur as a contaminant if employed as a pesticide.
Refs FAO *Arsenic and Tin in Food* (1979); Fowler, B A (Ed) *Biological and Environmental Effects of Arsenic* (1984); WHO *Arsenic* (1981).
Broader Toxic metal pollutants (#PD0948).
Aggravates Encephalitis (#PE2348) Water pollution (#PC0062) Lead as a pollutant (#PE1161) Occupational risk to health (#PC0865) Ill-considered pressure to eliminate nakedness in developing countries (#PF3350).
Aggravated by Pesticides as pollutants (#PD0120).

♦ PE1736 Denial of right to change nationality
Broader Resistance to change (#PF0557) Restrictions on recognition of nationality (#PE4912).

♦ PE1743 Prohibitive cost of linguistic interpretation legal proceedings
Denial of right to access to interpreters in judicial hearings
Nature Where a country has linguistic minorities, they suffer difficulties in court unless proper provision is made on their behalf. This problem is particularly acute in countries where numerous local languages or dialects exist. Aliens also may find themselves faced with linguistic difficulties in court. The problem of language may affect not only accused persons or other parties to judicial proceedings, but also witnesses.
In any proceedings, a person financially able to make his own provisions is, on the whole, likely to command assistance of a higher calibre. Expenses may also arise due to linguistic problems not only in court in connection with oral interpretation but also out of court, for instance in connection with the translation of relevant documents. Furthermore, a person who is preparing to go to court may need to address his counsel through an interpreter while preparing his case.
While most legal systems attempt to provide for interpretation, free or otherwise, for persons without a sufficient knowledge of the language of the court, this is not an invariable rule. Official provision of aid is less automatic in civil and administrative, than in criminal, cases; and although, in criminal proceedings, the great majority of legal systems make some provisions for interpretation for the benefit of those without a sufficient command of the language or languages used in court and for assistance to those who are handicapped in speech or hearing, and provide such aid free (even if no legal requirement of such aid exists the courts nearly always make ad hoc arrangements in criminal proceedings), in several countries the accused may be condemned to pay the cost of interpretation if he is found guilty. In Madagascar, if the condemned person does not have the means to pay such cost, he is imprisoned for a debt after his sentence runs out.
Broader Multiplicity of languages (#PC0178) Denial of right to a legal defence (#PE4628).
Aggravates Obstruction of international criminal investigations (#PF7277).

♦ PE1747 Termites as pests
Nature Some termite species attack living trees, doing considerable damage to forests exploited for commercial purposes and constituting a serious brake on forest development, particularly in tropical areas of developing countries. The termites causing maximum damage to plantations belong to the genera Odontotermes and Microtermes, which are fungus-growing species. They commence their attack in the upper 15-25cm of the soil. The tap-root is ring barked, often extending for some distance above the soil, which usually results in the death of the plant. In addition, other species attack wood used for fencing or housing purposes.
Other species of termites, particularly the large mound-building termite, impact grasslands in two ways, contributing to the desertification of North Africa. Termites can consume as much plant material as livestock and game animals combined thus increasing competition for forage, especially during drought. Termites also modify large quantities of soil through mound building and recycling of organic matter.
Incidence A survey in Australia, for example, showed that losses due to termites far exceeded those due to decay or fire and accounted for up to 80 per cent of total losses. Calculated on a lost-royalty basis, rejected logs cost forest management from $12 to $120 per acre.
Refs Roonwal, M L *Termite Life and Termite Control in Tropical South Asia* (1979).
Broader Household pests (#PD3522) Insect pests of wood (#PD3586) Insect pests of plants (#PD3634).
Aggravates Desert advance (#PC2506) Endangered forests (#PC5165) Structural failure (#PD1230) Wood deterioration and decay (#PD2301).

♦ PE1750 Junk food journalism
News inflation — Junk news journalism
Nature The journalists are not giving the public straight, well-documented, understandable information about those things that make societies ill economically, politically or physically. Instead, the news consist of sensationalized, personalized, and homogenized trivia.
Broader Biased presentation of news (#PD1718).
Aggravates Journalistic irresponsibility (#PD3071).

♦ PE1753 Unemployed skilled labour
Underutilized labour skills — Lack of skilled jobs — Lack of skills opportunities
Broader Unemployment (#PB0750) Mass unemployment of human resources (#PD2046) Arrested development of labour potential (#PF6532).
Narrower Unemployment of educated older people (#PE9071).
Related Unprofitable scope of industrial operations (#PF1933).
Aggravated by Limited employment options (#PF1658).

♦ PE1756 Failure to appear as witness to produce information or to be sworn
Nature A person who is in possession of evidence to give an official judicial, administrative or legislative proceeding, is lawfully summoned to give that evidence and fails to appear, to produce the evidence to be sworn as a witness is guilty of a crime. It is not possible to judge the merits of a case without all the evidence available and to hinder the gathering of such evidence endangers the decision making processes of government.
Broader Perjury (#PD2630) Contempt of judicial process (#PD9035).

♦ PE1758 Violation of the right of trade unions to function freely
Interference by public authorities in the functioning of workers organizations — Violation of the right of trade unions to freely elect their representatives
Broader Violation of trade union rights (#PD4695).
Narrower Violation of right of trade unions to publish (#PE5312) Deprivation of trade union funds and property (#PE8170) Violent repression of trade union demonstrations (#PE2007) Seizure of trade union property by public authorities (#PE7581) Occupation of trade union premises by public authorities (#PE5462) Interference by employers in the affairs of workers' organizations (#PE8384).
Related Prohibition of trade union meetings (#PD7210) Sanctions against trade union workers (#PD0610).

♦ PE1764 Human flatulence in public
Belching — Burping — Farting — Disagreeable bodily conditions
Nature Accumulation of gases in the stomach or bowels as a result of indigestion. That in the stomach may be expelled from time to time through the mouth as a belch, whereas that in the bowels may be expelled through the anus.
Incidence Normal people fart between 300ml and 2 litres of gas a day, the average being half a litre. Normal average is 13.4 farts a day.
Counter-claim In certain cultures belching after a meal may be considered a polite indication of satisfaction with the meal.
Broader Unsociable human physiological processes (#PF4417).
Related Bad breath (#PE6558) Disagreeable human body odour (#PE4481).
Aggravated by Dyspepsia (#PE4724).

♦ PE1766 Ticks as pests
Nature Ticks are among the most significant parasites of large wild and domestic animals, as regards both economic loss and transmission of disease. While no species of tick is primarily a human parasite, some species do attack man when the opportunity presents itself. The tick-borne diseases of livestock constitute a complex of several diseases whose etiological agents may be protozoal, rickettsia, bacterial or viral.
Incidence Over 60 tick-borne agents may be pathogenic to livestock throughout the world. Tick-borne diseases are widely distributed throughout tropical countries. In general, the vecto ixodid ticks favour humid and subhumid zones. These diseases cause some of the most serious economic losses of ruminants in Africa (assuming that rinderpest is routinely controlled). In many countries, they are the major health impediments to efficient livestock production. On a global basis, the economic toll caused by tick-borne diseases is staggering. Exotic breeds of livestock are being imported into tropical and sub-tropical regions in increasing numbers, often in spite of their poor adaptation to local climatic conditions. Such cattle are especially susceptible to tick-borne diseases and the effects of tick infestation.
Broader Arachnida pests (#PE3986) Disease vectors (#PC3595) Insect vectors of disease (#PC3597) Parasites of the human body (#PE0596) Insect vectors of human disease (#PE3632).
Narrower Fowl ticks (#PG6107).
Aggravates Q fever (#PE2534) Babesiosis (#PE6288) Tularaemia (#PE6872) Encephalitis (#PE2348) Theileriasis (#PE3996) Anaplasmosis (#PE6275) Typhoid fever (#PD1753) Relapsing fever (#PE7787) East Coast fever (#PE7946) Eperythrozoonosis (#PG3997) Texas cattle fever (#PG3990) Haemorrhagic fevers (#PE5272) Tick-borne diseases (#PE3897) Spirochaetal diseases (#PE3254) Rocky Mountain spotted fever (#PG3899).

♦ PE1769 Instability of trade in cereals and cereal preparations
Instability of trade in unmilled cereals
Nature Developed countries increasingly have the capacity to produce more grain than can be

absorbed in commercial markets. This is in part due to the more general problem of immobility of agricultural resources, in part to inconsistencies in agricultural policy objectives pursued by exporters and in the policy instruments used. Overcapacity is aggravated by technological advances which enable countries which previously imported grain to become new and competitive exporters. The adverse effects of weather on crops is a variable factor beyond control in all regions, although some suffer from a more permanent vulnerability which results in periodic crop failures. The basic overcapacity therefore continues to be accompanied by recurring short-term scarcities of an emergency nature.
Incidence About 50 per cent of the 1,052 million hectares of cultivated land in the world is used in the production of cereals, which constitute 80 per cent of man's intake of food calories.
Broader Instability of trade in food and live animals (#PD1434).
Narrower Instability of rice trade (#PE0696) Instability of wheat trade (#PE0385)
Meal and flour of wheat or meslin trade instability (#PE8489).
Aggravates Agricultural surpluses (#PC2062).

♦ PE1771 Adverse effect of transnational corporations on balance of payments
Broader Economic uncertainty (#PF5817)
Destabilization of monetary systems and exchange notes by transnational corporations (#PE5903).

♦ PE1774 Instability of banana trade
Nature Producers and exporters of bananas face three principal problems. First: the outlook for bananas on the basis of the expansion programmes already initiated is that export availability may outstrip the growth in world import demand. Estimates of surpluses vary from 10 to 70 per cent of world import requirements. The resulting competition for markets may lead to drastic falls in prices. This is particularly serious for those developing countries for which the fruit is the mainstay of the economy. Second: a problem arises from preferential trading arrangements which give effective protection to certain highly dependent producing countries having traditional links with major consuming countries in Europe. Third: constraints on trade may have the incidental result of inhibiting the expansion of consumption. For example, some importing nations have up to 100 percent ad valorem duties.
Incidence Next to citrus fruits, bananas are the most important item in international fruit trade. More than 90 percent of world banana imports are absorbed by only 42 countries. In addition to this role in world trade, the banana is a staple food for people in tropical zones, accounting for 80 per cent of world production.
Broader Instability of fresh fruit and edible nut trade (#PE2587).

♦ PE1775 Contagious bovine pleuropneumonia
Nature Contagious bovine pleuropneumonia is probably directly responsible for the death of more cattle than any other single disease. Incubation period: 3 weeks to 6 months; symptoms: rise in temperature, short painful cough, disturbed breathing, indigestion, tympany, constipation, diarrhoea, abortion, emaciation, weakened heart. Death usually follows in 2 or 3 weeks after symptoms have become pronounced and acute. Recovery is often more apparent than real. Mortality rate is usually 50–70 per cent. Cattle, buffaloes and related species such as reindeer, yak and bison are susceptible. Housed cattle are more susceptible than those in the open, and animal housing may remain infected for long periods.
Additional economic losses are due to the need to slaughter affected animals because of the difficulty of controlling contagious pleuropneumonia, and possible diagnosis of the disease may necessitate slaughter before it can be confirmed.
Incidence It has been present in Asia and Africa. Outbreaks have occurred in Australia and South America. Most countries in Europe are free of it with the exception of the Iberian peninsula; and the USA has been free of it since 1892. It is now restricted to the semi-arid tropical zone of Africa.
Refs Hudson, J R *Contagious Bovine Pleuropneumonia* (1971).
Broader Respiratory diseases of cattle (#PE5524).
Aggravates Inadequate carcass disposal of diseased animals (#PE2778)
Economic loss through reduced productivity of diseased animals (#PE8098).
Aggravated by Inadequate control of animal diseases (#PD2781)
Confusion of symptoms in animal diseases (#PF2780)
International movement of animals as factor in animal diseases (#PD2755).

♦ PE1776 Abuse of heroin
Refs Agar, Michael *Ripping and Running* (1973); Stimson, Gerry V and Oppenheimer, Edna *Heroin Addiction* (1982).
Broader Abuse of opiates (#PE1329).

♦ PE1778 African trypanosomiasis
Sleeping sickness
Nature Trypanosomiasis is a protozoan parasitic disease of vertebrate animals and man, with a high mortality rate among domestic animals. Trypanosomiases are single cell parasites which are transmitted from the faeces of infected animals to flies which reinfect animals and man. Flies may also become reinfected from the blood of infected animals. The disease in wild animals is usually benign, as they develop a certain immunity, although this is rarely complete. Domestic animals, especially those raised for meat and milk-production are highly susceptible. Those severely affected are cattle, camels and pigs.
Incidence There are many species of trypanosomiasis which exist in all continents. The highest incidence of trypanosomiasis (gambiense and rhodesiense in man; vivax, congolense and simiae in animals) is in Africa, where it is transmitted particularly by the tsetse fly. There are four main varieties of infection: 'nagana' and 'surra' which particularly affect domestic animals; 'dourine' which affects equidae; 'sleeping sickness' which affect man. 'Nagana', 'sleeping sickness' and 'dourine' are the most important African varieties. There are 50 million people at risk and it is estimated that there are about 25,000 new infections in man per year.
Background Trypanosomes were first identified as belonging to the genus Trypanosoma by Gruby in 1843 and to the family Trypanosomatidae by Doflein in 1901. International importance was given to the disease by the League of Nations, which organized the first international conferences on the subject in London 1925 and Paris 1928.
Refs FAO *The African Trypanosomiasis* (1979); FAO *African Animal Trypanosomiasis* (1983); Lee, C W and Maurice, J M *African Trypanosomiasis* (1983).
Broader Protozoan parasites (#PE3676) Arthropod-borne diseases (#PE7796)
Parasitic diseases in animals (#PD2735) Infectious and parasitic diseases (#PD0982).
Related Chagas' disease (#PE0653) Trypanosomiasis (#PE5725)
Parasites of the human body (#PE0596).
Aggravates Encephalitis (#PE2348).
Aggravated by Tsetse flies as pests (#PE1335) Insect vectors of human disease (#PE3632).
Reduces Rabies (#PE1325).

♦ PE1783 Long-term shortage of salt-water fish, crustacea and molluscs and preparations thereof
Nature Fish and shellfish provide an important source of protein for much of the world's population and yet the increasing human population and the increasing pollution of productive coastal waters are, in combination, intensifying the pressures on the world's remaining fishery resources, thereby raising the likelihood of overexploitation and resource failure. Inefficient management of marine fisheries results in a lower fisheries food supply and a failure to protect the fisheries against depletion.
Incidence World marine fisheries landings increased during the 1970s, but in 1980 they were 15 to 20 million tons less than they would have been had management been more competent. There were noticeable shifts in the distribution of the world fish catch. Between 1970 and 1976, the total catch of the developed countries grew at 2.5 percent a year, but that of developing countries actually declined by an average of 1.4 percent a year. Much of the drop was due to the collapse of the anchovy fishery in the southeast Pacific. But between 1977 and 1981 the roles were reversed. The catch of developing countries grew at 3.1 percent a year, while that of the developed countries expanded by only 0.6 percent a year. Western Europe's catch fell by 1.6 percent a year. FAO estimated the 1988 world fish catch at 94 million tonnes, based on preliminary data, 2 million tonnes more than in 1986 and 1987.
Broader Long-term shortage of food and live animals (#PE0976).
Related Slowing growth in food production (#PC1960).

♦ PE1785 Inappropriate agricultural subsidies for chemicalized farming
Inappropriate agricultural subsidies for the sales of pesticides — Inappropriate irrigation subsidies in developing countries — Inappropriate agricultural subsidies for livestock production — Inadequate subsidies for organic farming
Nature The economic support structure for agriculture is based on producing more food without taking into account its quality or the environmental damage caused. Subsidizing the use of pesticides harms the health of human and other species and builds up the resistance of the pests. Heavy irrigation aggravates floods, salinization and waterborne diseases. In Latin America millions of hectares of tropical forest, despite its rapid pasture deterioration and low carrying capacity, have been lost to subsidized ranching investments. Conventional farming does not currently bear the cost of cleaning nitrates out of the water resulting from use of fertilizers, or of loss of soil through soil erosion, or of destruction of habitats. Such costs are however absorbed, or avoided, in unsubsidized organic farming but which is nevertheless handicapped in competing with subsidized conventional farming. An organic farm avoids the use of agrochemicals by the use of rotation of crops in a fertility building cycle. It is the cost of this cycle that undermines the economics of organic production because between 40 and 60 per cent of the land at any one time has to be devoted to "non-productive" fertility building (or fertility maintenance). But it is precisely this rotational factor that prevents environmental damage.
Incidence Despite $6.1 billion in subsidies for fertilizers and irrigation, the prairie region of the USA (responsible for 23 per cent of USA agricultural production) is becoming progressively more difficult to farm.
Aggravates Prohibitive cost of nutritious food (#PF1212).
Aggravated by Endangered species of tamarins and marmosets (#PE7933)
Inappropriate government intervention in agriculture (#PE1170)
Over-subsidized agriculture in industrialized countries (#PD9802).

♦ PE1793 Administrative difficulties in new states
Broader Underdevelopment (#PB0206).
Narrower Metropolitan-type administration (#PS4020).
Related Socially unintegrated expatriates (#PD2675)
Non-viability of small states and territories (#PD0441)
Over-development of bureaucracy in ex-colonial countries (#PD2511).
Aggravates Disruptive foreign influence (#PC3188)
Loss of institutional credibility (#PF1963).
Aggravated by Colonialism (#PC0798) Abuse of authority (#PC8689)
Cultural imperialism (#PC3195) National political dependence (#PF1452).
Reduces Lack of political independence (#PF0297).

♦ PE1794 Instability of jute trade
Broader Instability of trade in unprocessed textile fibres and their waste (#PE1550).
Aggravated by Reduction in demand for primary commodities due to technological change (#PD1276).

♦ PE1796 Domination by transnational corporations of the domestic name-brand food sector in developing countries
Nature The world-level quality image established by the transnational corporation is important in differentiating its products and constitutes a continuing advantage in building market shares against local rivals. The firms with the greatest transnational activity in specific name-brand food industries or product lines are, with few exceptions, also the leading firms in their home country markets, in the industry world-wide, and in numbers of developing country affiliates. The products of these affiliates have secured and held their market positions largely through advertising and brand-name promotion. Local industries for name-brand foods in most developing countries tend to be both highly concentrated and dominated by transnational corporation affiliates. The effectiveness of the acquisition and promotion strategies of transnational corporations is at least one important reason for the observed concentration. It is a practice in several industries for the leading firms to cross-license their brand names and products and to have joint ventures with rivals. Some of the marketing techniques utilized by transnational corporation affiliates to expand the market for their branded products suggest that considerations of nutritional appropriateness are secondary.
Broader Food monopolies (#PE8018) Transnational corporation imperialism (#PD5891)
Control of industries and sectors by transnational corporations (#PE5831).
Related Domination of developing countries by transnational corporations (#PE0163).
Aggravated by Unethical food practices (#PD1045).

♦ PE1799 Consequences of restrictive business practices of transnational enterprises
Nature Although restrictive business practices attributed to transnational enterprises may not differ in form from those operated by purely national firms, their international character means that their impact on trade and competition is more significant. The reasons for this difference are to be found in their superior size, superior economic power, easier access to international financial markets and raw materials, as well as in their superior technological and management standards. They also play a greater role in the process of concentration and in concentrated markets.
Broader Restrictive trade practices (#PC0073) Transnational corporation imperialism (#PD5891).

♦ PE1805 African horse sickness
Nature African horse sickness is an acute or subacute febrile disease of equines found in humid or subhumid zones of tropical Africa caused by a virus with a number of antigenially different strains. It is transmitted by night-flying Culicoides spp. and probably has a reservoir host. It results in the impaired function of the circulatory and respiratory systems, and gives rise to serious effusions accompanied by haemorrhages in various organs and tissues of the body. Recovered animals develop immunity to the particular strain of virus. Because of the association with the insect vector, the disease prevails in certain areas (warm, moist, low-lying or coastal regions, valleys, swamps or river basins) and during certain seasons (rainy season – intermittent rain followed by summer temperatures).
Incidence Horse sickness originated in the African continent and is widely distributed over the central and southern parts. In the Near East a major epidemic occurred in 1959 and spread to

Pakistan and India; mortality was high.

Background The first historical reference to Horse sickness was in Yemen in 1327. It was mistaken for anthrax (1881 and 1895), then for a malaria strain (1888, 1895, 1900, 1911), then heartwater (1900, 1904), because of the high degree of mortality. M'Fadyean (1900), Theiler (1901) and Nocard (1901) isolated the causal agent as belonging to the group of ultravisible viruses. First attempts to immunize horses: 1900, 1918, 1925.

Claim High fatality rate: 50–95 per cent. In enzootic areas disruption to agriculture and transport has been met by the replacement of horse–drawn traffic by mechanization or other means, but when the disease appears for the first time, agriculture is disrupted with resulting widespread hardship (Iran, India, Middle East, Pakistan). Development of irrigation systems and conservation of surface water, creating favourable conditions for the perpetuation of insect populations, provides a serious threat to the continued use of horses and mules in farming systems.

Broader Viral diseases in animals (#PD2730) Infectious diseases in animals (#PD2732).
Aggravates Underdeveloped use of agricultural resources (#PF2164)
Underproductivity of draught animal power in developing countries (#PF0377).
Aggravated by Insect vectors of animal diseases (#PD2748)
Increase in animal disease by increase in aviation (#PJ4036)
International movement of animals as factor in animal diseases (#PD2755).
Reduced by Defoliation of insect breeding areas (#PJ4038).

♦ PE1806 Import–dependency in food staples in developing countries due to transnational corporations

Nature The principal issues which have arisen in relation to transnational corporation operations in the staple–food sector grow out of the role they have played in fostering import–dependent processing, and in impeding improvements in the production and marketing of indigenous staple foods. Problems may also arise when transnational corporations in these industries seek to develop local supply sources. Increased local production of the new product can usually be achieved only at the expense of the output of the new staple foods. Even where local farmers are encouraged to produce the imported staple (for instance, wheat) significant amounts of land may be diverted from crops better suited to the local climate or of wide popular consumption. A related issue arises when the replacement crop is intended as animal feed, and displaces production of staples for direct human consumption. Typically, the final consumer of meat is at a higher income level than the individual for whom the staple product was intended. Products fostered by transnational corporations may thus tend to heighten issues of distribution (of nutrition as well as national income) raised by this kind of agricultural activity.

Broader Dependence of developing countries on food imports (#PE8086)
Control of industries and sectors by transnational corporations (#PE5831)
Domination of developing countries by transnational corporations (#PE0163).
Related Dependence of island developing countries on imports (#PD5677).
Aggravates Shortage of cultivable land (#PC0219).
Aggravated by Mismanagement of food resources (#PE6115)
Inadequate staple food supply in developing countries (#PD4101).

♦ PE1810 Proliferation of direct mail advertising

Junk mail — Junk fax mail

Incidence In one medical institute in Sweden, doctors on the staff received in one day 5,775 advertising items in the mail.

Broader Misuse of advertising (#PE4225) Proliferation of advertising (#PD5034)
Inappropriate use of the mail service (#PE6754).
Aggravates Inadequacy of postal services (#PF2717).
Aggravated by Unethical commercial practices (#PC2563)
Inappropriate use of telecommunications services (#PE4450).

♦ PE1811 Corrosion in tropical climates

Nature Tropical conditions of damp heat do not, in themselves, offer a serious threat to the stability of metal objects. They do however operate to intensify corrosion when it has already begun to take place. Galvanic action becomes intensified under these conditions, particularly when salts are present in incrustations or, in general, when there is a form of moist contamination on metallic surfaces bearing stains of rancid fatty matter, residues of foodstuffs, etc. To this extent, problems of the preservation and conservation of metals in the humid tropics tend to be more acute than in temperate regions.

Broader Corrosion (#PD0508).
Related Destructive action of mould in tropical climates (#PE1265)
Detrimental effect of jungle environment in tropical villages (#PE2235).
Aggravates Inadequate protection and preservation of cultural property (#PF7542).
Aggravated by Inhospitable climate (#PC0387).

♦ PE1813 Industrial gas monopolies

Nature The business of supplying nitrogen, oxygen and other gases for industrial use has been from its start limited to five big companies. Contracts can extend over 15 years with stability of earnings, prices and products. There are high barriers to entry and a lack of technological change. There is very strong client stability because the product is piped to the client's sites. A change of supplier would involve construction of a new piping system.

Broader Cartels (#PC2512) Monopolies (#PC0521) Domestic cartel (#PD7963).
Related Food monopolies (#PE8018) Newspaper monopoly (#PE0246)
Aerospace monopolies (#PE7747) Aluminium monopolies (#PG5191)
Monopoly of the media (#PD3101) International monopoly of the media (#PD3040)
Insurance monopolies in capitalist countries (#PE8006)
Excessive concentration of business enterprises (#PD0071).

♦ PE1814 Childhood obesity

Fat children

Nature Early mortality in adulthood has been shown to be nearly double in those who had been obese in childhood, and is most frequently associated with heart disease. Birthweight tends to correlate with the mother's degree of obesity, but consumption of greater quantities of food than the recommended norm is not a major factor.

Broader Obesity (#PE1177).

♦ PE1815 Small bank failure

High risk policy of smaller banks

Incidence In the USA in 1990, the most serious problem in the banking system occur with the smaller regional banks which are heavily exposed to decline in property markets. It has been estimated that 720 USA banks, with combined assets of $72 billion, are technically insolvent although still operating.

Broader Bank failure (#PE0964).
Aggravated by Risk of capital investment (#PF6572).

♦ PE1817 Disparagement of indigenous cultures

Broader Discrimination against indigenous populations (#PC0352).
Aggravates Degradation of indigenous cultures (#PJ4963).

♦ PE1821 Dependence on minor tranquillizers

Abuse of minor drugs — Abuse of anti–anxiety agents — Abuse of sedatives — Abuse of anxiolytics — Abuse of hypnotics — Abuse of sleeping pills

Nature Drugs of this kind "damp down" the activity of the brain so that the effect of naturally calming substances is increased and naturally arousing substances is stopped. In the long run the brain's own mechanisms for controlling anxiety and tension can no longer function. They are addictive to the extent that after using them for three months, there is a 50 per cent probability of becoming dependent on them. It is claimed that withdrawal is then as difficult as from hard drugs such as heroin.

Incidence It is thought that one million people are addicted to tranquillisers in UK.

Claim More women than men are using tranquillizers or sleeping pills. When men and women report similar psychological or psychosomatic symptoms, men are more likely to be given physical and laboratory tests, and women are more likely to be given drugs. Women are also more likely to be given a repeat prescription once they have been prescribed a minor tranquillizer.

Refs Hartmann, Ernest *The Sleeping Pill* (1978); Mendelson, Wallace B *The Use and Misuse of Sleeping Pills* (1980); Petursson, Hannes and Lader, Malcolm *Dependence on Tranquilizers* (1984); Smith, Mickey C *Small Comfort* (1985).

Broader Abuse of medical drugs (#PD0028).

♦ PE1826 Unethical insurance practices

Unethical life assurance practices — Insurance fraud

Incidence American insurance companies estimate that private and government insurers paid $60 billion in 1989 for health insurance claims that were fraudulent or abusive.

Related Unethical financial practices (#PE0682).

♦ PE1828 Political discrimination in the administration of justice

Incidence Governments may decide not to prosecute those guilty of offences because the offenders are tied to the present administration. Alternately, prosecution may be delayed, held 'in camera, or acquittal assured, or penalties reduced to wrist–slapping. Others may become targets of judicial action because they are opponents of incumbent regimes. Another instance of the maladministration of justice is in lower court discrimination against proponents of politically unpopular ideologies such as the peace movement, women's liberation, and environmentalism.

Refs Okolicsanyi, L, et al (Ed) *Assessment and Management of Hepatobiliary Disease* (1987).

Broader Discrimination in politics (#PC0934) Discrimination before the law (#PC8726)
Inequitable administration of justice (#PD0986).
Related Injustice of trials in absentia (#PE0424) Injustice of military tribunals (#PE0494).
Aggravates Political repression (#PC1919).
Aggravated by Political prejudice (#PC8641).

♦ PE1832 Irresponsibility of young people towards the family

Nature Pre–adults can be very creative and productive, but the current position of young people in the family structure tends to be seen as preparation for life rather actually than living life. Under these circumstances they put off assuming real responsibility, and become very self–centered. They fail to provide the family with insights and the creative approach necessary to experience life in new ways. When it thus loses the vision of the new, the family stagnates; and the young people are driven to other surroundings where they can live life on their own terms.

Broader Irresponsibility (#PA8658)
Exclusion of pre–adults from family decisions (#PE2268).

♦ PE1837 Female criminals

Women offenders

Refs Bardsley, Barney *Flowers in Hell* (1988); DuPerron, William *Annotated Research Bibliography on the Female Offender* (1978); Rothman, Sheila M and Rothman, David J (Eds) *Women in Prison* (1986); Sturgeon, Susan and Rans, Laurel *The Woman Offender* (1975).

Broader Crime (#PB0001).
Related Detention of children (#PE6636).
Aggravates Detention of mothers (#PG4924)
Inadequate care for children of prisoners (#PF0131).

♦ PE1839 Oil pollution

Oil spillage — Oil tanker disasters — Offshore oil disasters — Oil spills

Nature Every year 1.2 billion tonnes of oil cross the seas in tankers. And since navigation is a human endeavour, there will always be human error. Oil spills are an utterly predictable cost of doing business. The main threat posed to living creatures by the persistent residues of spilled oils and water–in–oil emulsions is one of physical smothering leading to death through the prevention of feeding, respiration and movement. As damage is caused by physical contact, the animals and plants at most risk are marine mammals and reptiles, birds that feed by diving or flock at on the sea, marine life on shorelines and plants and animals in aquaculture facilities and storage pens in tidal area. Lethal concentrations of toxic components are relatively rare, localized and short–lived and are associated with spills of light refined products and fresh crude; animals and plants at risk are those living in areas of poor water exchange and in other situations where high concentrations of toxic components persist for longer periods. Sedentary animals in shallow waters such as oysters, mussels and clams that routinely filter large volumes of seawater to extract food are especially susceptible to accumulation of components of oil spills.

Accidents have increased because of corner–cutting driven by slumping revenues in 1980s. As crude oil prices crashed from $40 a barrel in 1981 to as low as $10 a barrel five years later, the industry hacked spending, reduced tanker crews, hired cheaper and fewer sailors, and begun to stretch equipment, delayed maintenance and repairs. When the market expanded again in the mid–1980s, it cost owners millions to restore the badly maintained tankers. Still majority of large tankers are out of date, overage and often poorly maintained. New tankers have been built with 10 per cent less steel and when they hit the ground they do not have the strength to take the impact.

Incidence The average amount of oil spilled annually from tankers is now approaching 1.5 million tons. The wreck of the Amoco Cadiz supertanker in 1981 spilled over 1.5 million barrels of crude oil and fouled more than 160 kilometres of shoreline. This ruined the annual harvest of oysters, lobsters, fish and seaweed, and caused some one million or more tourists to avoid the Brittany coast. In 1988 the explosion and subsequent burning of the Piper Alpha drilling platform in the North Sea killed 170 people. More than 170,000 gallons of petroleum fuels escaped the Bahia Paraiso, an Argentine Navy supply ship, after it ran aground in January 1989 threatening wildlife in the Bismarck Strait of the Antarctic Peninsula. The Exxon Valdez ran aground in March of 1989 spilling 10.1 million gallons of oil into Prince William Sound spreading over 1,200 square miles of water and fouling 1000 miles of shoreline.

Refs Bilal, J and Kuehnhold, W W *Marine Oil Pollution in Southeast Asia, Regional* (1980); Clark, R B *The Long–Term Effects of Oil Pollution on Marine Populations, Communities and Ecosystems* (1982); Gargas, Eivind, et al *Lectures in Coastal Pollution* (1976); Inkley, F A *Oil Loss Control in the Petroleum Industry* (1986); International Tanker Owners Pollution Federation (Ed) *Use of Oil Spill Dispersants* (1982); International Tanker Owners Pollution Federation (Ed) *Aerial Application of Oil Spill Dispersants* (1982); International Tanker Owners Pollution Federation (Ed)

PE1839

Action (1986); International Tanker Owners Pollution Federation (Ed) *Effects of Marine Oil Spills* (1985); International Tanker Owners Pollution Federation (Ed) *Contingency Planning for Oil Spills* (1985); International Tanker Owners Pollution Federation (Ed) *Recognition of Oil Shorelines* (1983); International Tanker Owners Pollution Federation (Ed) *Use of Booms in Combating Oil Pollution* (1981); International Tanker Owners Pollution Federation (Ed) *Aerial Observation of Oil at Sea* (1981); ; OECD *The Cost of Oil Spills* (1982); Waters, W G; Heaver, Trevor D and Verrier, T *Oil Pollution from Tanker Operations*.
Broader Injurious accidents (#PB0731) Environmental degradation (#PB6384).
Narrower Environmental hazards of sea transportation (#PJ3036).
Related Marine accidents (#PD8982).
Aggravates Water pollution (#PC0062) Marine pollution (#PC1117)
Oil as a pollutant (#PE2134) Wrecks and derelicts as hazards (#PE5340)
Obstacles for international ocean shipping (#PD5885).
Aggravated by Job fatigue (#PD8052) Sea traffic congestion (#PD1486)
Unethical practices in transportation (#PD1012).

♦ **PE1840 Parochial telecommunications standards**
Nature Setting electronic, procedural, software and hardware standards at by a nation to protect telecommunications interests from competition. This denies telephone equipment manufactures the economic advantages of producing for an large homogeneous international market. It discourages the development of new international services carried by telephone lines. It makes communications more time-consuming and less efficient for organizations exchanging a lot of data.
Broader Inadequacy of international standards (#PF5072)
Inadequate standardization of procedures and equipment (#PC0666).
Aggravates Excessive standardization (#PF2271)
Insufficient communications systems (#PF2350).
Aggravated by Limited enforceability of international standards (#PF8927)
Ineffective self-regulation in the telecommunications sectors (#PF5877).

♦ **PE1843 Market indicators exclusion of human requirements**
Nature Production systems are designed to take account of market indicators that determine whether goods produced will sell, rather than the indicators of human requirements.
Broader Production of non-essentials (#PC3651).

♦ **PE1848 Mobility of village populations**
Nature Despite vigorous efforts by some rural communities to attract and hold permanent residents, the pattern of residency in other villages may be remarkably mobile as every weekend those who work or study in urban centres return to their village, where they may take part in its social life but play little role in its economic vitality. The daily exodus of students of middle and upper schools and adults working in nearby industry leaves the village virtually empty during the day. Because of the anticipated temporary return of relatives, many houses have several guest rooms well-equipped but unused. Full-time year-round village residents are outnumbered several times over by those who still consider the village as their home, but they are seldom present to lend their support to effective development. All of these factors lend a sense of impermanence and incompleteness to the social fabric, resulting in a sense of inertia, a feeling that the permanent residents are simply acting as caretakers for those who come only occasionally.
Broader Excessive population mobility (#PF3806).
Narrower Second employment (#PF6908) Massive urban emigration (#PG9773)
Limited daytime encounters (#PG9805) Holiday residency patterns (#PG9820)
Insufficient young families (#PG9794) Distant schooling facilities (#PJ7693)
Inadequate economic incentives (#PF8554) Individualistic construction methods (#PG9842).
Aggravates Attractive city jobs (#PD9869) Discrimination against part-time work (#PE6241).

♦ **PE1849 Occupational hazards of benzene**
Nature Benzene gives off very toxic, inflammable vapours and grave risks are thus associated with its industrial use. It exerts the acute narcotic action common to many hydrocarbons and has a local irritant effect on the skin and mucous membranes. The outstanding feature of benzene is its ability to damage blood-forming tissues of chronically exposed persons, with resulting hyporegenerative anaemia of various degree. The onset of chronic benzene poisoning is extremely insidious and its ultimate injury potentially incurable.
Incidence Even though hyporegenerative anaemia resulting from chronic exposure to benzene has been known for over a century, it was not until 1928 that the action of benzene was shown to be a cause of leukaemia. Since then, approximately 200 cases of benzene leukaemia have been reported, either as single cases or as outbreaks. An outbreak of acute leukaemias from benzene exposure has been recently reported among Turkish workers chronically exposed to high concentrations of benzene. Recent epidemiological studies carried out in the United States and in Japan confirm an increased risk of leukaemias in workers with chronic exposure to benzene. Some French and American reports have also attributed cases of chronic leukaemias to benzene.
Background Benzene is a constituent of coal tar from which benzol is obtained by distillation. The designation benzol 90/100 indicates a substance containing 90% of hydrocarbons distilling below 100 deg C. Benzol is extracted from coal gas and coke oven gas by a stripping or scrubbing operation. In the petroleum industry large quantities of benzene are produced by catalytic reforming, dealkylation and dehydrogenation processes or by cyclization and aromatization of paraffin hydrocarbons. In industry, benzene is used as a fuel, as a chemical reagent and as a solvent. In certain parts of the world a major use of benzene is as an additive of motor fuel and large quantities are used for this purpose. Benzene is chemically reactive and serves as a raw material for a great number of chemical syntheses.
Broader Occupational hazards (#PC6716).
Aggravates Fires (#PD8054) Leukaemia (#PE0639) Leukopenia (#PG5543)
Myelotoxicity (#PG4058) Thrombocytopenia (#PG4057) Hepatic disorders (#PG4059)
Nutritional anaemia (#PD0321) Occupational risk to health (#PC0865).

♦ **PE1851 Underdevelopment of non-metallic mineral products industries**
Broader Underdevelopment of manufacturing industries (#PF0854).
Narrower Glass and glass products manufacture underdevelopment (#PE8715)
China, pottery and earthenware manufacture underdevelopment (#PE8427).

♦ **PE1852 Instability of machinery and equipment industries**
Broader Instability of manufacturing industries (#PC0580)
Instability of trade in manufactured goods (#PE0882).
Narrower Metal products industry instability (#PJ4066)
Machinery and equipment industries underdevelopment (#PJ3130).

♦ **PE1858 Underdevelopment of mining and quarrying industries**
Nature Mining and quarrying are vital sectors in many developing countries, but because the industry is often dominated by foreign transnationals, and may also be located in isolated rural areas, it remains largely underdeveloped. The bulk of production is exported in a raw or semi-processed state and as a result, the sector has not developed into a dynamic factor for industrial development.
Broader Underdevelopment of industrial and economic activities (#PC0880).
Narrower Natural gas production underdevelopment (#PU4072)
Crude petroleum production underdevelopment (#PU4071).

♦ **PE1859 Environmental hazards from machinery and equipment industries**
Broader Environmental hazards from manufacturing industries (#PD0454).
Narrower Machinery and equipment industries underdevelopment (#PJ3130)
Environmental hazards of electrical equipment industries (#PE8097).

♦ **PE1862 Cultural diversity ignored in social service agencies**
Nature When attempting to provide help for a growing diversity of complex needs, social service institutions tend to pass over the possible contribution which could be made by the diverse cultures which they serve. Past traditions, strange anomalies, uniquenesses and eccentricities, particular interpretations of the meaning of life, and even factual information about the world, are all untapped. This frequently results in unnecessary subversion of otherwise helpful social programmes.
Incidence Much of Britain's continual disagreement with the European Community results from bitterness over the loss of British anomalies, such as their non-metric monetary and measurement systems. This suppression of senseless but precious eccentricity has created unnecessary resentment.
Broader Repression of self-consciousness (#PC1777).

♦ **PE1869 Industrial emissions**
Narrower Industrial and domestic heating emissions as air pollutants (#PE2824).
Aggravates Air pollution (#PC0119) Soil pollution (#PC0058)
Malodorous fumes (#PD1413) Thermal pollutants (#PC1609)
Smoke as a pollutant (#PD2267) Nickel as a pollutant (#PE1315)
Mercury as a pollutant (#PE1155) Fluorides as pollutants (#PE1311)
Vanadium as a pollutant (#PE2668) Manganese as a health hazard (#PE1364)
Sulphur dioxide as a pollutant (#PE1210) Photochemical oxidant formation (#PD3663).
Aggravated by Waste paper (#PD1152).

♦ **PE1879 Excessive corporate debt**
Claim In 1990, economic models prepared in the USA indicate that in the event of a recession, the effects on corporations would be catastrophic because of their current level of debt. Corporations would be unable to generate sufficient cashflow to cover the interest on the billions of dollars they have borrowed during the 1980s. In the USA, debt service costs as a percentage of cash flow for non-financial corporations reached an all-time high in 1990 at 31 per cent, having risen from 18.5 per cent in 1985.
Aggravated by Bank failure (#PE0964) International economic recession (#PF1172).

♦ **PE1880 Drug smuggling**
International drug trade
Nature The illegal transportation of prohibited narcotics from one country to another in violation of customs legislation. The major contemporary drugs being smuggled are marijuana, heroin and cocaine. This is possible because of the huge profits being made.
Incidence It is likely that the greatest value of drug smuggling is between South America and the United States. Other major routes are from the Middle East and South East Asia to Europe and the United States. Most heroin is produced in Southeast Asia, with West Asia and Central America producing the bulk of the rest. It is shipped to Europe, Australia and North America with the largest part going to the United States. Cocaine is produced in Colombia, Peru and Bolivia and shipped to Europe and North America. The largest producer of marijuana is the United States where most of it is consumed. It has been estimated that the annual drugs business is worth $300 billion to the world banking system, namely as much as oil. These illicit substances are now only exceeded by armaments in total value as an internationally traded commodity.
Refs MacDonald, Scott B *Dancing on a Volcano* (1988); Sujan, M A and Trivadi, V D *Smuggling* (1976).
Broader Corruptive crimes (#PD8679) Illicit drug trafficking (#PD0991)
Evasion of customs and excise duties (#PD2620).
Narrower Narco-terrorism (#PE9210).
Related Adulteration of illicit drugs (#PE0456) Illicit trade in prescribed drugs (#PE4946).
Aggravates Basuco (#PE5245).
Aggravated by Drug abuse (#PD0094) Cannabis abuse (#PE1186)
Abuse of opiates (#PE1329) Abuse of animal drugs (#PE0043)
Abuse of hallucinogens (#PD0556) Inadequate drug control (#PC0231)
Abuse of coca and cocaine (#PD2363) Abuse of sedatives and tranquillizers (#PE0139)
Unethical practices in transportation (#PD1012)
Government complicity in illegal activities (#PF7730).

♦ **PE1883 Anaclitic depression**
Nature Some infants around nine months of age suffer from sadness, weeping, withdrawal, and refusal to eat. It is caused by the absence of the mother figure for at least three months.
Broader Mental depression (#PC0799) Emotional disorders (#PD9159).
Related Maternal deprivation (#PC0981).

♦ **PE1887 Urban slums in industrialized countries**
Refs United Nations *The Aging in Slums and Uncontrolled Settlements* (1977).
Broader Urban slums (#PD3139).
Aggravated by Family poverty in industrialized countries (#PD1998)
Environmental degradation of inner city areas (#PC2616).

♦ **PE1890 Vampire bats**
Nature Vampire bats (desmodontidae) constitute a serious economic, veterinary and public health problem in Latin America, not only because they can transmit rabies and other diseases to domestic animals and to man, but also because their sole food is the blood of birds and mammals, including man: the vampire drinks about 20 cc of blood from its prey.
Incidence Because vampire bats are often rabid or spread diseases such as encephalomyelitis, they can cause the destruction of as many as one million cattle each year. It is estimated that these bats cause the destruction of US $1,250 million worth of livestock annually in Latin America.
Broader Pests (#PC0728).
Aggravates Animals as vectors of disease (#PD8360).
Reduced by Endangered species of chiroptera (#PE3604).

♦ **PE1894 Excessive consumption of sugar**
Sugar craving — Environmental hazards from sugar and honey and preparations thereof
Nature Sugar consumption may be a public health danger, particularly for infants and children who consume large quantities and can form an addiction to it. Sugar can be condemned for its proven ill effects on teeth, but there is also evidence to show that it is a causative factor in heart disease, either on its own or as a result of interaction with the effects of coffee consumption and smoking. In some people, sugar consumption leads to a complex hormonal response which may eventually result in chronically high insulin levels. The latter correlates with coronary disease. Sugar has also been accused of contributing to obesity, diabetes, indigestion, poor eyesight, dermatitis, early

puberty, ulcers, gall stones, gout and premature ageing.
Incidence Annual sugar consumption has increased (in developed countries) from 4 to 120 pounds per head over the past 200 years, now providing up to one sixth of daily caloric intake. Two thirds of this consumption is via a wide range of factory prepared foods. As an indication, the market for artificial sweeteners and sugar substitutes in the UK in 1986 was £24.5 million.
Refs Poleszynski, Dag *Food, Social Cosmology and Mental Health* (1982).
Broader Excessive consumption of specific foodstuffs (#PC3908)
Environmental hazards from food and live animals (#PC1411).
Related Toxic food additives (#PD0487).
Aggravates Obesity (#PE1177) Heart diseases (#PD0448) Teeth disorders (#PD1185)
Mucopolysaccharide diseases (#PE4414).

♦ PE1899 Plant cancer
Broader Plant pathogens (#PD1866).
Related Animal cancer (#PG1900) Parasites on plants (#PD4659)
Viral plant diseases (#PD2227) Fungal plant diseases (#PD2225)
Bacterial plant diseases (#PD2226) Environmental plant diseases (#PD2224)
Deficiency diseases in plants (#PD3653).

♦ PE1900 Inadequate electrical power supply in developing countries
Nature With increasing mechanization, manufacturing industry comes to depend more and more upon local power facilities. Although large concerns may be capable of generating their own power, this procedure is usually inappropriate to less developed countries, partly because it involves new establishments and larger capital charges; partly because the resulting energy is more costly; and partly because it is the lighter industries and smaller units which are usually pioneers in the industrialization process.
Broader Inadequate electricity infrastructure (#PD9033)
Inadequate facilities for the transport of electrical energy (#PE1048)
Underdevelopment of the power industry in developing countries (#PF4135).
Aggravates Underdevelopment of industrial and economic activities (#PC0880).

♦ PE1901 Metal deficient diets
Nature Behavioural abnormalities, in some cases, are caused by deficiencies of metals in the diet. For example, anorexia nervosa is recognized as stemming from zinc deficiency and possibly abnormal zinc metabolism.
Broader Nutritional deficiencies (#PC0382).
Aggravates Social conflict (#PC0137).
Aggravated by Aggression (#PA0587).
Reduced by Moralism (#PF3379).

♦ PE1903 Discrimination against homosexuals
Criminalization of homosexuality — Violation of the rights of homosexuals — Denial of rights of homosexuals
Nature Homosexuals may be discriminated against in employment or cut off from social benefits. Discrimination is particularly prevalent in Anglo-American culture, though there is a current movement to counter this. Male homosexuals are generally more despised and discriminated against than lesbians, as they are thought to constitute a greater threat to society. Although homosexuality is fairly widely accepted in Arab and Asian countries, American influence has had the effect of reducing this tradition. As a result most homosexual affairs are conducted clandestinely and may be accompanied by a guilt complex, while surface appearances are kept up.
Counter-claim Homosexuality cannot lead to procreation and as such is both antisocial and hedonistic. When a homosexual couple is formed is rarely lasts. In one study, homosexual couples were found to last an average of three years and, for men, the average number of partners is 16 per year. Homosexuals are more vulnerable to blackmail and therefore should not be given positions of responsibility. Homosexuals in positions to influence and directly contact children set an abnormal example. Homosexuals are potential paedophiles, for, having gone beyond one moral and social barrier, they are ready to go beyond another. Full equality for homosexuals and heterosexuals is incompatible with the protection with the family. Sexual relations between men is an important factor in spreading sexually transmitted diseases, homosexuals account for 70 percent of recent cases of syphilis in England because of the large number of indiscriminate partners. And because they have heterosexual relations they contaminate their female partners. By far the largest numbers of AIDS carriers in the developed world are homosexual men. Leaving aside moral and religious reasons, homosexuals should be discriminated against for the sake of the larger society.
Refs Brongersma, Edward *Loving Boys* (1988); Katz, Jonathan *Government Versus Homosexuals* (1975).
Broader Violation of the rights of sexual minorities (#PD1914).
Narrower Violation of the rights of male homosexuals (#PE3882)
Violation of the rights of female homosexuals (#PE5741).
Related Homosexuality (#PF3242).
Aggravates Male homosexuality (#PF1390) Social fragmentation (#PF1324)
Sexual exploitation of men (#PD3263).
Aggravated by Conformism (#PB3407) Social stigma (#PD0884)
Political blackmail (#PD2912) Religious intolerance (#PC1808)
Discrimination against women (#PC0308).

♦ PE1904 Desynchronization of bodily rhythm by international travel
Jet-lag — Circadian dysrhythmia — Disruption of biological rhythms
Nature Travel by aircraft through several time zones leads to desynchronization of internal bodily rhythms and a phase shift between physiological and psychological functions. Neurotics seem to be particularly sensitive to this temporary dislocation. The disorientation creates a special difficulty for international diplomatic and business missions since the travelling party is at a disadvantage in any meeting following arrival. There is also some evidence that suggests the existence of cumulative long-term effects which shorten the life span in the case of frequent travel of this kind.
Jet lag, which is partly sleep deprivation and partly bodily confusion, causes efficiency problems for transnational corporations and military organizations which need to be assured that their representatives (businessmen, soldiers) are fully alert after a long journey across time zones. Airline pilots spending a week or longer on international flight schedules with long hours and multiple crossings of time zones have difficulty staying awake while flying and sleeping during layovers. For holiday travellers, the reorientation period inconveniently cuts into their allotted vacation time.
Incidence Jet lag has become a real and persistent problem in industrialized societies. A frequently cited example is the failure of USA negotiations with Egypt in 1950, because of the jet lag of the Secretary of State, resulting in the contract for the Aswan Dam being given to the USSR, thus providing the USSR with its first base of operations in Africa.
Broader Separation from nature (#PF0379) Human disease and disability (#PB1044).
Aggravates Sleep-wake schedule disorder (#PG9626).
Aggravated by Multiplicity of time standards (#PF2621).

♦ PE1905 Malignant neoplasms of female genital organs
Cervical cancer — Uterine cancer — Ovarian cancer
Refs Bender, H G and Beck, L *Cancer of the Uterine Cervix* (1985); Lopez de la Osa, L (Ed) *Aspects and Treatment of Vulvar Cancer*, McBrien, David C and Slater, Trevor F *Biomedical and Clinical Aspects of Cancer of the Uterine Cervix* (1984); Mould, R F and Bose, Aranbinda *International Strategies for the Eradication of Carcinoma of the Cervix in Developing Areas* (1987).
Broader Diseases of cervix (#PG9823)
Malignant neoplasm of genito-urinary organs (#PE5100).
Aggravated by Smoking (#PD0713) Coitus as a cancer risk (#PD6033)
Health hazards of passive smoking (#PE5146)
Inadequate health care in family planning (#PD1038).

♦ PE1907 Unwanted children
Broader Victimization of children (#PC5512).
Aggravated by Unwanted pregnancies (#PF2859)
Inadequacy of contraceptive methods (#PD0093)
Discrimination in family planning facilities (#PD1036).

♦ PE1909 Excessive exposure to radiation from consumer goods and electronic devices
Nature Many millions of units of various types of consumer products containing deliberately incorporated radionuclides are in everyday use around the world. These include: radioluminous products; electronic and electrical devices; antistatic devices; gas and aerosol (smoke) detectors; ceramic, glassware, alloys, etc, containing uranium or thorium; and other devices including scientific instruments. Estimates of doses in individuals resulting from the use of such products show that in all cases these doses are small. The highest calculated whole-body doses result from the wearing of radioluminous watches, which are the most widespread radioactive consumer product. Other sources of radiation are more or less widespread throughout the population and include alarm clocks, television sets, and the continually increasing number of applications of radioactive substances (as found in fluorescent lamps and electronic components of various devices). Singly, they produce only small gonadal doses, but their total contribution may eventually constitute a danger.
Broader Radioactive contamination (#PC0229) Human disease and disability (#PB1044)
Environmental hazards of non-ionizing radiation (#PE7651).
Narrower Microwave radiation as a health hazard (#PE6056)
Harmful biological effects of ionizing radiation (#PE6294).
Aggravated by Unethical practice of radiology (#PD8290).

♦ PE1910 Anthracnose diseases of plants
Nature Anthracnose diseases are characterized by distinctive lesions on stem, leaf, or fruit, often accompanied by dieback. The diseases is worldwide and of significance in humid, temperate zones.
Broader Ascomycetes (#PE4586) Deuteromycetes (#PE4346).

♦ PE1911 Financial and economic disputes between states and nationals of other states
Transfrontier private investment disputes
Nature In the absence of an agreement to the contrary between the foreign investor and the host government, an investment is subject to the laws of that government (local law); and the redress of grievances which the investor may seek by direct access to that government is equally determined by local law. If the investor feels aggrieved by actions of the host government he may invoke the diplomatic protection of his national State or he may request his national State to espouse his case and bring a claim before an international tribunal. In some countries, the foreign investor may, as a condition of entry, be required to waive diplomatic protection; and even if the national State is willing to espouse the investor's case, the host government may be unwilling to submit to the jurisdiction of an international tribunal.
However, even in the absence of these obstacles, the present situation may be regarded as unsatisfactory because of the investor's inability to proceed with an international claim directly against the host government. The necessity of espousal of his case by his national government before an international claim can be lodged introduces a political element. An investor may well find that his national government refuses to espouse a meritorious case because it fears that to do so would be regarded as an unfriendly act by the host government. And this consideration is even more likely to cause the national government to refrain from acting if the merits of the investor's case are not wholly clear in its view, thus withholding from the investor an opportunity to have his case judged by an impartial tribunal. In an attempt to overcome these difficulties, some investors, mostly large corporations (especially in the field of extractive industry), have been able to negotiate arbitration agreements with host governments, providing for detailed rules regarding the selection of arbitrators, the arbitration procedure and, in some cases, the law to be applied by the arbitration tribunal. It is quite clear that only a few investors can be in a position to negotiate such agreements. Moreover, the validity of such agreements is sometimes questioned. If the government refused to proceed with the arbitration, the investor's remedy would once again be either a request to his national State for diplomatic intervention or for an espousal of his case before an international tribunal.
The absence of adequate machinery for international conciliation and arbitration often frustrates attempts to agree on an appropriate mode of settlement of disputes. Tribunals set up by private organizations, such as the International Chamber of Commerce, are frequently unacceptable to governments and the only public international arbitral tribunal (the Permanent Court of Arbitration) is not open to private claimants.
Broader Lack of international cooperation (#PF0817).
Aggravates Domination of the world by territorially organized sovereign states (#PD0055).

♦ PE1914 Chronic fatigue syndrome
Chronic fatigue immune dysfunction syndrome — Low natural killer cell syndrome — Myalgic encephalomyelitis
Nature Controversial myalgic encephalomyelitis (ME), sometimes called Yuppie flu, is characterized by a range of symptoms including fatigue, fevers, lymph-node swelling, persistent diarrhoea, joint and muscular pain, inability to think and loss of short-term memory.The illness lasts sometimes for few months or it can linger for years. It is believed to be caused by water-borne virus. Some doctors recognize the condition, others are sceptical about its existence.
Broader Virus diseases (#PD0594) Defective human immunity system (#PE3355).
Related Encephalitis (#PE2348).

♦ PE1918 Monopolization of technology by transnational corporations
Nature The global strategy of transnational corporations undeniably generates technological dependence in the host country. For one thing, national firms cannot match the global research and development of a transnational corporation. While transnational corporation research and development activities are highly centralized in the parent company, any local activities in affiliates are mostly limited to adapting the final product to consumer preferences and do not lead to any substantial innovation. In addition, when transnational corporations expand abroad through acquisitions or take-overs, the research and development activities of the acquired local firms are either integrated into the global system or totally discontinued.

While transnational corporations play an essential role in disseminating and diffusing technological knowledge world-wide, their general unwillingness to licence technology to third parties results in a large percentage of their technology sales consisting of internal trade between parent companies and foreign subsidiaries. In addition, the available evidence strongly suggests that transnational corporation parent companies get more royalty payments for low-technology industries, and that relatively little new technology is transferred from parent companies to their foreign affiliates. The actual host country cost of internal transnational corporation technology transfers is considerably higher than that recorded in payments of royalties for knowledge and trademarks and financial charges for technical assistance. Recorded payments do not include the considerable indirect cost to affiliates (and thus to their host country) which may not be immediately apparent under the elaborate internal accounting procedures practised by most transnational corporations. Also, there may be additional costs due to restrictions upon the use of this imported technology, dictated by the global interests of the transnational corporation parent.

In their strategy of product differentiation, most transnational corporations appear to devote more resources to lesser innovations used in the manufacture of new products than to search for new production processes. Furthermore, transnational corporations concentrate on consumption-oriented, rather than production-oriented, technology in developing countries.

Claim Technology transfers within transnational corporations that tend to standardize production, management and marketing techniques generally reflect the patterns of the highly advanced countries, endanger the research and development activities of local enterprises, and run counter to developing countries' attempts to develop technologies more 'appropriate' to their own needs and their factor endowment. Not only have research and development expenditures in transnational corporations been unevenly distributed between parent and affiliates, but worse, foreign affiliates often have to finance new products and processes without deriving any benefit from them, since transnational corporations prefer to exploit their product advantage by exporting rather than by transferring their latest technologies to affiliates.
  Broader Inappropriate transfer of technology (#PE5820)
  Transnational corporation imperialism (#PD5891).
  Aggravates Abusive technological development under capitalism (#PD7463).

♦ **PE1920 Haemophilia**
Haemophilia
Nature Haemophilia is a hereditary disease manifested by increased bleeding. Even slight bruises can cause extensive haemorrhages, both subcutaneous and intramuscular. Repeated haemorrhages in the joints result in serious changes in them, which are characteristic of haemophilia. Normally non-severe injuries (cuts, for example) and surgery (tooth extraction) are accompanied by life-threatening bleeding. Haemophiliacs must stringently curtail their activities, which means that they, especially when young, may be isolated from their peer groups. Many haemophiliacs were infected with AIDS through the blood transfusions upon which they depend before 1985, when new methods of screening rendered the treatment safe from this infection.
Refs Egli, H and Inwood, M J *The Haemophiliac in the Eighties* (1981).
  Broader Coagulation disorders (#PE2373)
  Diseases of blood and blood-forming organs (#PF8026).
  Aggravates Nutritional anaemia (#PD0321)   Disorders of joints and ligaments (#PD2283).
  Aggravated by Genetic defects and diseases (#PD2389).

♦ **PE1922 Disproportionate influence on national economies of limited number of corporations**
Undisclosed control by major corporations on national policy-making — National oligopolistic trading systems
Nature Through the use of networks of holding companies and cross-linking directorships, a relatively small number of major corporations or financial institutions can in effect control a large percentage of a national economy. Because of corporation secrecy on these matters, it is very difficult to determine the extent of such control and to hold the controllers responsible for policies which they attempt to implement to the disadvantage of the smaller enterprise.
  Broader Oligopolies (#PC3825)   Transnational corporation imperialism (#PD5891)
  Disproportionate control of global economy by limited number of corporations (#PE0135).
  Related Political oligarchy (#PD3238)   Ineffective industry self-regulation (#PF5841).
  Aggravates Elitist control of global economy (#PC3778).
  Aggravated by Corporation financial secrecy (#PE1571)
  Interlocking corporate directorates (#PF5522)
  Undisclosed control of national economies by limited number of individuals (#PF2344).

♦ **PE1923 Mosquitoes as vectors of disease**
Mosquitoes as vectors — Mosquito vectors
Nature Life and death are determined for millions of human beings by their relationship to the mosquito. Blood-sucking mosquitoes act as vectors in the transmission of serious diseases of both man (yellow fever, malaria, filariasis, dengre) and animals (virus encephalitis, yellow fever). Often painful allergic responses follow the salivary injection made by the insect. Methods of control, whereby breeding grounds are destroyed, have often proved difficult and ineffective. The use of chemicals has given rise to new insecticide-resistant populations, leading to the additional cost of research and production of replacement insecticides, with the knowledge that in a few years time these will also have become ineffective. Biological control through natural agents has not been as successful as hoped; but the worst aspect of the problem is that many drugs to treat or to prevent mosquito-borne diseases are no longer having any effect.
Incidence Approximately 3,400 species are distinguished, many of which are vectors of a same disease or may become so under particular conditions.
  Broader Parasites (#PD0868)   Disease vectors (#PC3595)
  Insect vectors of human disease (#PE3632).
  Narrower Aedes mosquitoes as vectors of disease (#PE3621)
  Culicine mosquitoes as vectors of disease (#PE3623)
  Anopheline mosquitoes as vectors of disease (#PE3622).
  Aggravates Malaria (#PE0616)   Filariasis (#PE2391)   Yellow fever (#PE0985)
  Dengue fever (#PE2260)   Encephalitis (#PE2348)   Avian malaria (#PG4158)
  Elephantiasis (#PE1601)   Botflies as pests (#PE3635)   Mammalian malaria (#PG4156)
  Flies as insect pests (#PE2254)   Insect bites and stings (#PE3636).
  Aggravated by Mosquito breeding ponds (#PG9038)
  Mosquito resistance to insecticides (#PE3582).

♦ **PE1924 Endemic goitre**
Graves' disease
Nature Rarely acknowledged because of its benign appearance, (enlargement of the thyroid gland), endemic goitre can in fact be deleterious to the wellbeing of large populations, notably due to its consequences: cretinism and endemic deaf-mutism. Politicians have often failed to recognize the gravity of the physical and mental consequences of goitre, and the adverse effects which subclinical hypothyroidism has on general health and productivity.

There are many different causes of goitre: disease; developmental defects; and environmental conditions. Iodine deficiency has dramatic effects on goitre prevalence rates and on endemic cretinism. Other factors involved in the causation of the disease are found in a number of commonly consumed foods: cabbage and other members of the cruciferae family, turnips, various staple foods (maize, cassava, millet, sweet potatoes, lima beans), onions and garlic, olive oil and other vegetable oils, and milk. Water, too, plays a part in the aetiology of endemic goitre. Significant factors include: industrial pollution or artificial contamination of water by the potent antithyroid disulfides of saturated and unsaturated aliphatic hydrocarbons and phthalates; the antithyroid organochlorines detected in chlorinated effluent from plants treating domestic sewage; and the relationship between the geological composition of watersheds and goitre prevalence – a high prevalence occurring in towns located down-stream from sedimentary deposits (shales and coals) rich in organic matter, and a low prevalence where drinking-water is taken from streams flowing across igneous rocks. The goitrogenic effect of a high calcium intake may be masked in the presence of iodine deficiency. Drinking-water heavily polluted by certain bacteria also makes for a high goitre prevalence.
Incidence An estimated one billion people are exposed to the risk of goitre because of a deficiency of iodine, and most of them dwell in the tropical regions of the Third World. There are 400 million in Asia alone (excluding China), of whom 80 million are actually suffering from subclinical hypothyroidism.
Background Goitre has existed since ancient times in some regions, for example in China and in parts of Europe, and it was endemic in the Himalayan foothills 350 years ago. It was prevalent in the Andes in the 18th and early 19th centuries. Fifty years ago, it was observed that in some Himalayan villages almost the entire population was affected, including over half of all breast-fed infants. In such circumstances cretinism, deaf-mutism, and mental retardation were common.
  Broader Nutritional diseases (#PD0287)   Thyroid gland disorders (#PE0652)
  Malnutrition among indigenous peoples (#PC3319).

♦ **PE1925 Monolithic architecture of high-rise buildings**
Socially inadequate housing estates
Nature In both industrialized and developing countries a certain tendency to put up monumental buildings has produced undesirable results, in terms both of human need and of natural environment. In some areas, this tendency has been intensified by unregulated urban land markets and inflated land values. Massive high-rise buildings can be designated as offices in such a way as to sterilize city life. As apartments, they can break up local 'corner-store' relationships, imprison mothers and young families, and impoverish the imaginative life of a whole generation of children. Such high-rise apartments are particularly unsuitable for first generation migrants from rural areas; they are also unduly costly compared with alternative public policies of land settlement, site preparation, and self help in housing.
  Broader Inadequate housing (#PC0449)
  Stultifying homogeneity of modern cities (#PF6155)
  Monotonous and unaesthetic architecture and design (#PF0867).
  Related Ugliness (#PA7240)   Indoor air pollution (#PD6627)
  Socially sterile rental accommodation (#PF6195)
  Inaccessibility of city centres to suburban residents (#PF6132).
  Aggravates Social neglect (#PB0883)   Alienating child-birth environments (#PF6161)
  Impersonality of high density accommodation (#PF6156)
  Unconvivial hotel environments for travellers (#PF6196)
  Impersonality of mass market shopping facilities (#PE6153)
  Insufficient separation between urban subcultures (#PF6137)
  Inhibition of communication between non-proximate offices (#PF6197)
  Excessive intensification of the parent-child relationship (#PF6186)
  Unhealthy lack of daily physical activity in urban environments (#PF6182)
  Lack of places in urban environments encouraging unstructured public access (#PF6190)
  Inadequate recognition by institutions of the transition through adolescence (#PF6173).
  Aggravated by Prohibitive cost of land (#PE4162)
  Uncontrolled land markets (#PG4161)
  Excessive use of land by automobiles (#PF6152)
  Social inadequacy of large buildings (#PF6194)
  Unrelated buildings in urban environments (#PF6199)
  Wastage of open space in urban environments (#PE6163)
  Disorientation stress in large building complexes (#PF6174)
  Unattractive pedestrian environments in urban areas (#PE6151).

♦ **PE1926 Hypoglycaemia**
Low blood sugar
Refs Andreani, D et al *Hypoglycemia* (1987); Kushi, Michio and Mann, John D *Diabetes and Hypoglycemia* (1985); Ruggiero, Roberta *The Do's and Dont's of Low Blood Sugar* (1988).
  Aggravated by Diabetes (#PE0102).

♦ **PE1927 Instability of paper and printing industries**
  Broader Instability of manufacturing industries (#PC0580)
  Instability of trade in manufactured goods (#PE0882).

♦ **PE1928 Chlamydid infections in animals**
Refs Aitken, I D *Chlamydial Diseases of Ruminants* (1986).

♦ **PE1935 Accidental poisonings**
Incidence Accidental poisoning deaths, the fifth leading cause of unintentional fatal injuries in the United States, jumped from 4,331 in 1980 to 5,740 in 1986. Drug poisoning, often by overdoses of illegal drugs, accounted for 4,187, some 73 percent, of the deaths in 1986.
  Broader Human poisoning (#PD0105).

♦ **PE1936 Trace element deficiencies in soils**
Refs FAO *Trace Elements in Soils and Agriculture* (1979).
  Broader Soil infertility (#PD0077).

♦ **PE1940 Lack of consumer influence on industry**
Disregard of consumer interests by transnational corporations
Claim The consumer has no significant means of influencing industrial development, thus leaving development of industry to its own resources which are poorly geared to serve society as a whole.
  Broader Transnational corporation imperialism (#PD5891)
  Underprovision of basic urban services (#PF2583).
  Related Economic influence (#PG4965).
  Reduced by Consumer boycotts (#PE1213).

♦ **PE1944 Underground construction obstacles**
Nature Existing pipes and other underground metal objects are a big problem for companies in the building community.
  Broader Obstacles to building construction (#PF7106).

♦ **PE1945 Corrosion of iron and steel**
Rust
Nature Iron and steel exposed to moist air or oxygenated water are corroded, leaving a red encrustation of iron oxide on the surface. Progressive rusting is a major source of failure of

unprotected structural materials. It is aggravated by bad design which leaves moisture traps in the structure. Rust may pit small holes in a surface, or uniformly progress over its area. It may attack joints and crevices. Despite their universal vulnerability to penetration, iron and steel materials are insufficiently protected, both in manufacture and in maintenance.

Incidence The importance of rust is particularly evident in the automobile industry. In the UK it has been estimated that automotive rust costs some £260 million per year, decreasing the value of each automobile by £1 every week. In countries exposed to ice and snow on the roads, the use of salt to clear it accelerates rusting. In the UK, where the use of salt in this way is estimated to cause 50 per cent of rusting, it therefore costs £130 million per year. In the USA, repairs and replacements due to corrosion and rust damage may be worth nearly 5 percent of the gross national product. High humidity countries or locations experience the worst rusting: Suriname, Abu Dhabi and Indonesia top one list of rust-prone climates. In 1990 it was reported that the 1,300 kilometre trans-Alaska pipeline (designed to be rustproof for 30 to 40 years) was seriously corroding because of failure to corrode it adequately. Repairs were expected to cost from $600 to $1,500 million.

Refs American Society for Testing and Materials *Atmospheric Corrosion Investigation of Aluminum-Coated, Zinc-Coated and Copper-Bearing Steel Wire and Wire Products* (1975); Cihal, V *Intergranular Corrosion of Steels and Alloys* (1984); Karpenko, G V *Stress-Corrosion Cracking of Steels* (1979); Tonini, J *Corrosion of Reinforcing Steel in Concrete* (1980).

Broader Atmospheric corrosion of materials (#PE9525).
Related Wear (#PB1701).
Aggravates Structural failure (#PD1230) Mechanical failure (#PC1904)
Failure of materials (#PD2638) Fatigue in materials (#PD1391).
Aggravated by Snow (#PG4193) Humidity (#PD2474)
Planned obsolescence (#PC2008).

♦ PE1946 Trachoma

Granular conjunctivitis

Nature Trachoma, with its complications, is still today the single most important cause of preventable blindness and loss of vision in the world. The burden of trachoma for a community is heavy: human suffering requiring treatment and assistance, and a severe handicap for education and work. Trachoma can be transmitted by direct or indirect contact. Overcrowding, lack of clean water, and unsanitary habits all contribute to its spread. Repeated exposure to infection, reinfections and relapses play a role in increasing the severity of the disease. This chronic infection is caused by micro-organisms (chlamydiae) which are very similar to bacteria but which, like viruses, are intracellular parasites. The clinical manifestations of trachoma range from a severe disease causing blindness, to a relatively mild condition which evolves towards spontaneous cure.

Incidence This is one of the most common diseases in the world, with an estimated 450 million people suffering from trachoma and at least 2 to 5 million blinded because of the disease. Trachoma is found mainly in northern Africa, the countries south of the Sahara, the Middle East and in most Asian and Latin American countries.

Broader Virus diseases (#PD0594) Eye diseases and disorders (#PD8786)
Infectious and parasitic diseases (#PD0982).
Related Conjunctivitis (#PE7974).
Aggravates Physical blindness (#PD0568) Partial loss of vision (#PE4196).
Aggravated by Onchocerciasis (#PE2388) Irritant fumes (#PD3672)
Flies as insect pests (#PE2254) Adenovirus infections (#PE2355)
Poor living conditions (#PD9156)
Overcrowding of housing and accommodation (#PD0758)
Loss of humility in relation to the environment (#PF2527)
Underdevelopment of industrial and economic activities (#PC0880).

♦ PE1951 Abusive psychosurgery

Surgical manipulation of the brain

Nature Psychosurgery can be considered an extreme and destructive attempt at controlling human behaviour, involving at least partial death of the personality, irreversible mutilation, or a surgical manipulation of the brain for purely behavioural aberrations of unknown aetiology.

Psychosurgery may be used to deliberately destroy certain areas of the human brain to produce more amenable individuals. The technique is occasionally used to treat violent or criminal behaviour. Post-operative blunting of the personality, apathy and irresponsibility are frequent, and some researchers mention deterioration of the intellect in those of high intelligence. Psychosurgery is subject to strong criticism because of the impossibility of obtaining the informed consent of the psychotic or mentally handicapped patient to such treatment; because of its evolving nature; and because of the lack of a clear indication that alternative treatments have been exhausted.

Claim Although after the operation, the patients may be less subject, or no longer subject at all, to anxieties, fears, or symptoms of violence; but they are reduced to a state (variously describe as 'buffoons', 'clowns', or 'human vegetables') considered by some to be incompatible with human dignity.

Counter-claim While psychosurgery cannot guarantee results (good results are only obtained in about 40 percent of cases), in carefully screened patients it offers the only hope for ameliorating extreme behavioural disturbances such as uncontrollable violence. It enables some otherwise intractable patients to adapt to society or to become more manageable in institutions.

Broader Inhumane scientific activity (#PC1449) Unethical practices in psychotherapy (#PD5267).

♦ PE1953 Soot

Nature Soot is produced by inefficient combustion. Its most obvious effect is on the cleanliness of buildings. As an efficient absorbant of solar radiation it contributes significantly to the deterioration in atmospheric radiation balance.

Incidence Soot has contributed significantly to pollution of the Arctic atmosphere in the spring season, which is expected to deteriorate further with the industrialization of the USSR.

Broader Air pollution (#PC0119).
Aggravates Global warming (#PC0918).
Aggravated by Sunlight inhibition by nuclear warfare soot (#PE6350).

♦ PE1954 Leaf spots

Nature Leaf spots are delimited necrotic lesions. When numerous spots grow together the disease becomes a blight, a blotch, or scorch.

Broader Ascomycetes (#PE4586) Deuteromycetes (#PE4346).
Narrower Tar spot of maple (#PG1373) Alternaria diseases in plants (#PJ1685).

♦ PE1956 Nitrates as pollutants

Peroxyacetyl nitrates

Nature Nitrates are used in chemical fertilizers. When applied to fields much of nitrates often filter through the soil to the water table where they contaminate the drinking supply. Nitrates turn to nitrites in water and in human stomach and is suspected of causing stomach cancer and other diseases. In rivers, streams, estuaries and seas nitrates encourage the proliferation of algae, which stifle other, more various plants and fish. The concentration of 50mg of nitrates per litre of water is considered by the European Commission to be the maximum permissible level for water for human consumption.

Refs WHO *Health Hazards from Nitrates in Drinking Water* (1985).

Broader Nitrogen compounds as pollutants (#PD2965)
Chemical pollutants of the environment (#PD1670)
Environmental hazards from fertilizers (#PE1514).
Related Acidic precipitation (#PD4904).
Aggravates Lake pollution (#PD8628) Water pollution (#PC0062)
Photochemical oxidant formation (#PD3663).

♦ PE1957 Disruption of domestic social policies by transnational corporations

Nature The disruptive impact of the activities of transnational corporations on the social structure and patterns of society has to be viewed in the context of a country's over-all socio-economic objectives. In developing countries, such objectives relate particularly to the amelioration of poverty through increased employment and more equitable income distribution, greater regional development, and rapid growth of local skills and capability to meet essential socio-economic needs. The interaction between such developmental objectives and various social aspects such as the structure of social institutions - family, religion, tribe, social class - and social norms and values differ from country to country. The role of transnational corporations in affecting social patterns and conditions can, however, be very significant through their influence on employment and industrial growth and on the pattern of production and consumption.

The introduction of certain consumer goods may require or lead to changes in national consumption patterns, which in turn have implications for the way incomes are spent, levels of nutrition and health, and cultural values. The social effects related to the supply of consumer goods in developing countries are of obvious concern.

The production process, that is, the mix of production factors (capital, land and labour) and the characteristics of the technology (labour-intensive or capital-intensive) used by the affiliates of transnational corporations may have repercussions on employment levels, on the distribution of incomes (both between capital and labour and between different types of labour) and on the local absorption of technology. Important factors conditioning the effects of a transnational corporation's production process are the choice of location within a country, the type of labour employed and the training opportunities made available to local employees. Transnational corporations may also affect working conditions - wages, hours, benefits, health and safety measures.

Broader Transnational corporation imperialism (#PD5891)
Social irresponsibility of transnational corporations (#PE5796).

♦ PE1961 Distortion of international trade by export subsidies and countervailing duties

Nature Export subsidies to domestic producers may be either generally or selectively applied, covering all commodity exports, or specific products only. General export subsidies are used by government to increase employment or to improve the balance of payments. Subsidized producers are able to undercut foreign competitors by offering their output at less than its real social cost and are thereby able to increase their share of world markets. Although general export subsidies may be effective in correcting a payments imbalance, they cause a misallocation of world resources. The subsidy encourages producers to expand export production to levels at which the marginal social costs in the subsidizing country exceed the free international prices of export goods. This leads to a decline in potential real income in the world economy. Selective subsidies are also sometimes employed to improve the balance of payments, but most often they are expected to provide special economic assistance to particular industries. They also bring about an inefficient allocation of world resources. Countervailing duties are imposed by some countries in the case of imported products for which domestic producers receive an export subsidy.

Claim Incentive schemes which result in indiscriminate use of export subsidies may lead to the establishment of uneconomic and inefficient production in developing countries; and encourage damaging competition among these countries in export markets, which could be particularly detrimental to the interest of those countries not in a position to resort to export subsidies to the same extent.

Counter-claim As far as developing countries are concerned, there is some recognition and acceptance that properly formulated and implemented export incentive schemes have a role to play in promotion and development of their exports of manufactures.

Broader Distortion of international trade by selective domestic subsidies (#PD0678)
Distortion of international trade as a result of government participation (#PD2029).
Related Unknown availability of subsidies (#PG9905)
Limited availability of financial credit (#PF2489)
Distortion of international trade by discriminatory customs and administrative entry procedures (#PE2603).
Aggravates Disincentives against farming (#PD7536)
Excessive foreign public debt of developing countries (#PD2133).

♦ PE1973 Inadequate relationship between international governmental and nongovernmental organizations and programmes

Broader Inefficiency (#PB0843) Ideological conflict (#PF3388)
Obstacles to effective international nongovernmental organizations (#PF7082).
Narrower Inadequate relationship between transnational corporations and specialized agencies of the United Nations (#PE0106)
Inadequate relationship between international nongovernmental organizations and the specialized agencies of the United Nations (#PE0777).
Aggravates Inadequate coordination of international organizations and programmes (#PD0285).
Aggravated by Double standards in morality (#PF5225)
Human errors and miscalculations (#PF3702)
Lack of legal provision for international nongovernmental organizations (#PE0069)
Jurisdictional conflict and antagonism between international organizations (#PD0138).

♦ PE1974 Headache

Nature Headache is one of the most common ills that plague mankind. It is often a source of concern because neither its intensity nor its location on the head denotes whether it is due to a serious intracranial disease or to a minor and superficial disorder. In approximately 9 out of 10 patients afflicted with headache, the condition stems from readily reversible changes in cranial arteries or in skeletal muscles of the scalp and neck; these changes are usually physiological responses to adverse attitudes and stressful living situations.

The major types of chronic headaches are: 1) migraine is usually a throbbing, severe pain that occurs on one side of the head. The classic migraine is preceded by a pinwheel or shimmering disturbance in the field of vision. Sufferers often become nauseated and may vomit. Many become sensitive to light or noise. Women victims outnumber men by four to one, and their migraine attacks frequently happen around the time of menstruation. The onset of chronic migraines often occurs before the age of 12. 2) Tension headaches are a dull and more generalized pain, often affecting the entire head and sometimes the neck. Many sufferers feel a vice-like sensation of squeezing around the hatband area. They tend to be less severe than migraines and may last for days. They are the most common. 3) Cluster headaches are the most excruciating, occur most often on one side of the head with an intense, stabbing pain near the eye, which frequently drips tears. The sufferer may also have a blocked or runny nose. They can occur several times a day, for weeks or months and then disappear, sometimes forever. Men, especially smokers, are most apt to get them. Other types of headaches are: 1) Organic headaches can be caused by a tumour

or stroke. 2) Traumatic headaches follow a head or neck injury. 3) Sex headaches are a dull ache or sudden, intense pain during or after sexual activity. 4) Sinus headaches are caused by severe sinus infections. 5) Hangovers occur after bouts of drinking. And 6) hunger headaches, a generalized ache, are due to going without food for long periods.
Incidence An estimated 45 million Americans suffer from chronic, recurrent headaches. They make more than 50 million office visits a year to doctors and spend more than $400 million on over-the-counter pain relievers. Industry loses at least $55 million a year due to absenteeism and medical expenses caused by headaches. American children lost 1.3 million days of school in 1986 because of headaches.
Refs Aggerholm, Paula N *Headache* (1987); Barlow, Charles F *Headaches and Migraine in Childhood* (1985); Sicuteri, F L, et al *Trends in Cluster Headache* (1987); Wolff, Harold G *Wolff's Headache and Other Head Pain* (1987).
    Broader Human disease and disability (#PB1044)
    Ill defined health conditions (#PC9067).
    Narrower Migraine (#PE6357).
    Aggravated by Allergy (#PE1017)   Measles (#PE1603)
    Anxiety (#PA1635)   Uraemia (#PG2241)
    Overwork (#PD2778)   Gastritis (#PG2250)
    Pneumonia (#PE2293)   Toothache (#PG4232)
    Meningitis (#PE2280)   Common cold (#PE2412)
    Constipation (#PE3505)   Menstruation (#PE4838)
    Teeth disorders (#PD1185)   Malodorous fumes (#PD1413)
    Rheumatic diseases (#PE0873)   High altitude stress (#PD2322)
    Intracranial disease (#PG4228)   Diseases of metabolism (#PC2270)
    Infection of the sinuses (#PG4577)   Diseases and injuries of the brain (#PD0992).

♦ **PE1976 Acid mists**
    Broader Dust (#PD1245).
    Narrower Sulphuric acid mist (#PE4825).
    Aggravates Hazards of strong toxic substances (#PD0122).

♦ **PE1977 Travel delays**
Incidence American airlines waste an average of 2,000 hours a day in delays for passengers, at a loss of $1 billion a year.
    Broader Delay (#PA1999)   Travel risks (#PD7716).
    Aggravated by Aerial piracy (#PD0124).

♦ **PE1978 Restrictive business practices in technology transactions**
Nature Restrictive business practices applied in technological transactions include: territorial market restraints; tying transferred technology to the purchase of goods or services within the transnational corporation system; restricting an affiliate from entering into agreements involving competing or complementary technology; restrictions on research and development; and restrictions on adaptation or innovation of technology.
Refs UNCTAD *Control of Restrictive Practices in Transfer of Technology Transactions* (1982).
    Broader Restrictive trade practices (#PC0073).
    Aggravates Abusive technological development under capitalism (#PD7463).

♦ **PE1980 Planning disrelationship between political and industrial sectors**
Nature The absence of a responsible, effective liaison between the industrial sector and political centers of power creates disrelationships between them. This has lead to inadequate, even capricious treatment of human need in underdeveloped countries. Economic planning at the national and global levels is uncoordinated and arenas of global concern, such as: the environment, resource management, population growth, and aging are not adequately considered.
    Broader Industrial processes geared to reduced social needs (#PE3939).

♦ **PE1989 Antimony as a health hazard**
Nature The principal hazard of antimony is that of intoxication by ingestion, inhalation or skin absorption. The respiratory tract is the most important route of entry since antimony is so frequently encountered as a fine airborne dust, although ingestion may occur through swallowing dust or through contamination of beverages, food or tobacco. Skin absorption is less common but may occur when antimony is in prolonged contact with skin.
Incidence The dust encountered in antimony mining may contain free silica and cases of penumoconiosis termed 'silico–antimoniosis' have been reported among antimony miners. During processing, the antimony ore, which is extremely brittle, is converted into fine dust more rapidly than the accompanying rock, leading to high atmospheric concentrations of fine dust during such operations as reduction and screening. Furnacemen refining metallic antimony and producing antimony alloy, and workers setting type in the printing industry are all exposed to antimony metal dust and fume **Background** Antimony is stable at room temperature but, when heated, burns brilliantly giving off dense white fumes of antimony oxide. It is closely related, chemically, to arsenic. It readily forms alloys with arsenic, lead, tin, zinc, iron and bismuth. High-purity antimony is employed in manufacture of semi-conductors. Normal purity antimony is used widely in the production of alloys to which it imparts increased hardness, mechanical strength, corrosion resistance and a low coefficient of friction.
    Broader Toxic metal pollutants (#PD0948)   Environmental hazards from metals (#PE1401).

♦ **PE1990 School accidents**
Refs Kigin, Denis J *Teacher Liability in School–Shop Accidents* (1983); Leibee, Howard C *Tort Liability for Injuries to Pupils* (1965).
    Broader Injurious accidents (#PB0731).
    Related Childhood accidents (#PD6851).

♦ **PE1996 Participation of transnational corporations in the apartheid system**
Nature The role of transnational corporations in South Africa is closely related to South Africa's international position and domestic economic structure. The international position of South Africa has been defined largely by its reliance on major developed market economies and its dominance in the southern African region. The pattern of dependency and regional strength can be seen in the structure of South Africa's foreign trade. The developed market economies are the main trading partners from which manufactured goods are imported and to which raw materials are exported. South Africa's exports to developing countries consist mainly of manufactured goods and processed raw materials. This pattern has started to change with the industrialization of the country, but is still prevalent. Industrialization in South Africa has advanced to the stage where many capital goods are produced locally by South African firms and by subsidiaries of transnational corporations. An important feature of the process of economic development in South Africa has been the import of foreign capital and technology, in many cases through the subsidiaries of transnational corporations. Because South Africa is in political conflict with the majority of members of the international community, its government has attempted to strengthen the economy and reduce dependency on foreign imports to enhance independence and minimize the potential for external influence. There is little sign that transnational corporations are decreasing their investment in South Africa. A few firms have sold or reduced their holdings, but others have expanded or newly entered the market.
Incidence The transnational corporations in, or trading heavily with South Africa are from the NATO, EEC and OECD countries, with the UK, the USA, the German Federal Republic, and Japan's corporations leading. About 60 percent of South African trade is with the developed market economies.
Refs Lipton, Merle *Sanctions and South Africa* (1988); United Nations *Transnational Corporations in South Africa and Namibia*.
    Broader Transnational corporation imperialism (#PD5891)
    Social irresponsibility of transnational corporations (#PE5796).
    Aggravates Discrimination (#PA0833).

♦ **PE2000 Increasing development lag against information growth**
Increasing inability to integrate available knowledge
    Broader Increasing development lag against socio-economic growth (#PC5879).
    Narrower Increasing lag in education against growth in knowledge (#PE7139).
    Aggravated by Information gap (#PF3397)   Language barriers (#PF6035)
    Proliferation of information (#PC1298)   Unorganized transfer of skills (#PJ8603)
    Ignorance of cultural heritage (#PF1985)   Unavailability of legal information (#PJ8698)
    Copyright barriers to transfer of knowledge (#PE8403)
    Limited availability of functional information (#PF3539)
    Barriers to transfer between educational facilities (#PD0084).

♦ **PE2004 Harmful effects of advertising by transnational corporations in developing countries**
Nature Advertising does more than merely sell products and form consumption patterns: it informs, educates, changes attitudes, and builds images. Advertising may facilitate the transfer of consumption patterns of developed countries to developing ones, by introducing needs which may not be appropriate, given the income and demand structure in these countries. Advertisements may even create needs by suggesting unnecessary uses for products, or through needless product differentiation and packaging. Sometimes the elaborate packaging characteristics of products of transnational corporations, associated with attempts to achieve brand distinctiveness, accounts for a significant part of the cost of the product in developing countries.
Other unfavourable implications can arise for the developing countries as a result of misleading advertising which does not reveal the harmful effects of some products of transnational corporations which, although banned in the developed market economies, are available in the developing countries because of insufficient regulation. In addition, with respect to the advertising of drugs and food products, numerous cases of exaggerated claims about the benefits of particular products have been reported in many developing countries, including false and misleading statements and inadequate directions for use.
Furthermore, aggressive advertising campaigns by transnational corporations in developing countries may have adverse effects on local competition. The affiliates of transnational corporations may be able, through advertising, to displace domestic enterprises in the consumer goods industry, where transnational corporations do not always have major advantages to offer to the domestic economy.
    Broader Misuse of advertising (#PE4225)
    Domination of developing countries by transnational corporations (#PE0163).
    Aggravated by Unethical commercial practices (#PC2563).

♦ **PE2007 Violent repression of trade union demonstrations**
Violent police intervention in workers meetings — Military intervention in workers meetings
    Broader Violent repression of demonstrations (#PD4811).
    Violation of the right of trade unions to function freely (#PE1758).

♦ **PE2032 Distortion of corporation financial statements**
Nature It is a commonly accepted practice for management or directors to manipulate the information presented in corporate financial accounts so that advantages or even tax relief may be obtained from the tax administration. The information may then be presented in a modified manner for the stockholders and general public. A separate presentation is frequently developed for the management of the corporation. The process is sometimes known as 'window-dressing'.
Counter–claim Widely divergent taxation laws make it impossible today to make a simple and uniform tax declaration across boundaries. The complexity of taxation law and the multitude of incentives and exception force companies to hire tax specialists.
A minority of companies seek the most taxed way. It is a commonly accepted practice to seek the least taxed way.
The tax forms prescribe a certain form of presentation of the accounts, which are far different from the information required by shareholders, banks, labour unions and the general public. It is therefore necessary to present the financial information for each target group in the format and with the details that they need. It is not window dressing, but rather responding to the needs of each target group.
    Broader Immorality (#PA3369)   Tax evasion (#PD1466)   Economic crime (#PC5624).
    Related Frauds, forgeries and financial crime (#PE5516).
    Aggravates Corporation financial secrecy (#PE1571).

♦ **PE2033 Alcoholic intoxication at work**
Alcohol abuse in business — Alcohol-related absenteeism
Nature Consumption of alcohol decreases accuracy, and where attention and skill are required a blood alcohol concentration as low as 0.01 g/l will increase the number of errors. Relatively small doses of alcohol also increase reflex time and impair judgement, thus increasing the possibility of accidents.
Background Only over the last 30 years has industry recognized the disastrous effects of alcohol on its workers and found that excessive drinkers have a higher rate of sickness absenteeism and a higher accident proneness (the risk increases 50% for a blood alcohol over 0.25 g/l), that their professional ability is impaired, and they age and die prematurely. The absence of a parallel between mental and physical disorders is frequent: there are cases where, although the physical damage has reached an advanced stage, mental and especially artistic faculties are apparently unaffected; conversely there are heavy drinkers whose physical strength seems undiminished and who continue to do very heavy muscular work. Nevertheless, modern industry demands such high psychosensory and mental capacities, and places such heavy responsibility on the individual, that the alcoholic's reduced physical and mental ability makes him less and less employable in industry. Sooner or later he is left behind and becomes a burden upon society. While the cost of alcohol-related disabilities is very difficult to estimate, it is nevertheless clear that the costs which arise from its consumption by a fairly large proportion of the active population greatly exceed the returns which may result from alcohol in the form of employment, commercial profits and state taxes.
Incidence In the USA it has been estimated that more than 2.5 million workers, namely 3 per cent of the labour force, are alcoholics. The cost to industry in absenteeism, fringe benefits, loss of trained manpower, inefficiency, loss of production and accidents, is estimated to exceed US $2,000 million annually. In the UK, the cost to industry of alcohol-related sickness, absenteeism and deaths in 1987-8 was estimated at 800 million pounds.
    Broader Alcohol abuse (#PD0153)   Substance abuse at work (#PD9805).
    Related Drug abuse at work (#PD4514).

Aggravates Absenteeism (#PD1634) Inefficiency (#PB0843)
Underproductivity (#PF1107) Industrial accidents (#PC0646)
Human errors and miscalculations (#PF3702) Draining of resources due to alcohol (#PE8865).
Aggravated by Drunkenness (#PE8311) Loss of the significance of work (#PF3676)
Inappropriate employment incentives (#PD0024).

♦ PE2038 Limestone dust

Nature During production and use of lime and limestone dust is created that is hazardous to health. Limestone dust mixed with free silica as found in limestone quarries can cause pharyngitis, bronchitis and emphysema. Pure lime dust is known to cause bronchitis and pneumonia.
Broader Mineral dusts (#PE4679).
Aggravates Pneumonia (#PE2293) Chronic bronchitis (#PE2248)
Diseases of upper respiratory tract (#PE7733).

♦ PE2041 Foetal infection and death

Damage and death to the foetus
Refs Kurjak, A (Ed) *The Fetus As a Patient* (1985); WHO Expert Committee, Geneva *Prevention of Perinatal Mortality and Morbidity* (1970); WHO Seminar, Tours *Prevention of Perinatal Mortality and Morbidity* (1972).
Broader Natural human abortion (#PD0173) Perinatal morbidity and mortality (#PD2387)
Denial of the right of unborn children (#PF6616).
Narrower Foetal erythroblastosis (#PG1921) Respiratory distress syndrome (#PE7159).
Related Violation of the rights of foetuses (#PE6369).
Aggravated by Rubella (#PE0785) Infection (#PC9025).

♦ PE2042 Physical malformation of foetus

Deformation of foetus
Refs Bergsma, Daniel (Ed) *Cytogenetics, Environmental Malformation Syndromes* (1976); Brock, D J *Early Diagnosis of Fetal Defects* (1983); Rockford, Doris E *Drug Effects on the Fetus* (1987).
Broader Natural human abortion (#PD0173) Complications of childbirth (#PC9042).
Narrower Foetal malformation in diabetic pregnancies (#PE4808).
Related Violation of the rights of foetuses (#PE6369).
Aggravated by Deliberate imbalancing of population sex ratio (#PF3382).

♦ PE2047 Medicines containing alcohol

Nature Many medicines for children include levels of alcohol which parents might find disturbing. Many medicines shipped to Arab countries seem to have a particularly high alcohol content in spite of the fact that Muslim law forbids drinking alcohol in any form.
Broader Consumption of alcoholic beverages (#PD8286)
Religious opposition to public health practices (#PF3838).
Aggravated by Unknown alcohol information (#PS0298).

♦ PE2053 Kidney disorders

Kidney diseases — Kidney tumours — Renal glycosuria — Renal failure — Renal asthma — Renal diseases — Enterotoxaemia
Refs Amerio, Alberto, et al *Acute Renal Failure* (1987); Brenner, Barry M and Lazarus, J Michael *Acute Renal Failure* (1983); Cummings, Nancy B and Klahr, Saulo *Chronic Renal Disease* (1985); Olsen, Steen *Tumors of the Kidney and Urinary Tract* (1985); Schmicker, R, et al *Metabolic Disturbances in the Predialytic Phase of Chronic Renal Failure* (1988); Schrier, Robert W and Gottschalk, Carl W *Diseases of the Kidney* (1988); Strauss, José (Ed) *Acute Renal Disorders and Renal Emergencies* (1984).
Broader Asthma (#PD2408) Diseases of metabolism (#PC2270)
Human disease and disability (#PB1044).
Narrower Pyelitis (#PG4286) Nephrosis (#PG4287) Nephroptosis (#PG4288)
Kidney injuries (#PG4284) Renal sclerosis (#PG7861) Bright's disease (#PE2272)
Suppuration within the kidney (#PG4282) Tuberculosis of genito–urinary system (#PG4280)
Benign neoplasm of kidney and other urinary organs (#PE9086).
Related Rickets (#PG2295) Skeletal system disorders (#PE2298).
Aggravates Stroke (#PE1684) Dropsy (#PG3235) Uraemia (#PG2241)
Still–birth (#PD4029) Hypertension (#PE0585)
Urinary bladder disorders (#PE2307).
Aggravated by Lumbago (#PE1310) Diabetes (#PE0102)
Spina bifida (#PE1221) Rheumatic diseases (#PE0873)
Pregnancy disorders (#PD2289) Stress in human beings (#PC1648)
Diseases of the arteries (#PE2684) Diseases of the lymphatic system (#PD2654).

♦ PE2054 Instability of tea trade

Nature There is a tendency for the growth of exportable supplies to outstrip the slow growth in demand, bringing a downward trend in world prices and pressure on export earnings of producer countries. There is an insufficient identity of interest in these countries to counteract such trends.
Incidence Nearly 2 million metric tons of tea were produced in 1983. India produced over one–third; mainland China one–quarter; Sri Lanka, Kenya and Indonesia produced 7 to 10 percent each; Bangladesh and Argentina produced under 3.0 percent each. Smaller producers included Malawi and Brazil. Nearly 50 percent of production entered world trade. India, Sri Lanka, China and Kenya were the dominant exporters but Malawi, Mauritius, Mozambique and Rwanda exported nearly 100 percent of their smaller. Other nations that exported more than 50 percent of their production included Indonesia, Bangladesh, Tanzania, Zaire and Brazil. The big consumer–importers are the UK, and the Islamic Middle–East and North Africa, these together taking nearly half of tea exports. The USA takes about 10 percent and Pakistan 8 to 9 percent.
Broader Instability of economic and industrial production activities (#PC1217)
Instability of trade in coffee, tea, cocoa, spices and manufactures thereof (#PE0915).
Related Instability of trade in tea and mate (#PG1706).

♦ PE2056 Instability of wool trade

Nature Wool has been subject to intensive competition, due to price and technical characteristics, from synthetic fibres, which resulted in slowly falling wool prices as manufacturers switched to synthetics. As a result, producers were forced to reduce the size of sheep flocks. Such reductions, followed in recent years by droughts in Australia which killed many sheep (thus effectively cutting back production even further), have resulted in dramatic price increases. In addition, wools from developing countries have gradually lost ground to those from developed countries because of inadequate supplies and quality.
Broader Instability of trade in unprocessed textile fibres and their waste (#PE1550).
Aggravated by Reduction in demand for primary commodities due to technological change (#PD1276).

♦ PE2061 Trading in products containing toxic substances with developing countries

Dumping toxic wastes in developing countries — Dumping of dangerous and illegal products on developing countries
Nature The industrialized countries are producing enormous quantities of toxic wastes and they do not have enough room to bury them. Disposal have been made even more difficult by pressures from ecologist movements and by draconian regulations. Furthermore, the cost of "clean" disposal of wastes, by high–temperature incineration is very high. The industrialized countries have therefore sought to get rid of their wastes on other continents and found out that sending them to developing countries was 10 times cheaper than other method. In the countries of destination, the local trading partners fell into the trap and it had thus been possible to offload substantial shipments.
Incidence European companies are allowed to manufacture soap containing mercuric iodine for export to Africa, where it is used to lighten skin colour. The soap can cause foetal damage, anaemia and kidney failure.
Refs Norris, Ruth, et al *Pills, Pesticides and Profits* (1982).
Broader Hazardous waste dumping (#PD1398) Unethical commercial practices (#PC2563)
Irresponsible international trade (#PC8930).
Related Marketing of banned pharmaceutical drugs in developing countries (#PE6036).
Aggravated by Hazardous wastes (#PC9053) Dissatisfaction with skin colour (#PF1741)
Havens for environmental pollution (#PE6172)
Inadequate industrial trade in developing countries (#PF4160).

♦ PE2062 Torture schools

Broader State sanctioned torture (#PD0181).
Aggravates Human torture (#PC3429) Dehumanization (#PA1757)
Military atrocities (#PD1881).
Aggravated by Abuse of government power (#PC9104)
Abuse in government policy (#PF8389).

♦ PE2069 Aeroallergens

Broader Air pollution (#PC0119).
Narrower Moulds (#PG4295) Allergy inducing pollens (#PG3219).
Related Dust (#PD1245) Biological air pollutants (#PD0450).
Aggravates Occupational risk to health (#PC0865).

♦ PE2076 Disordered behaviour

Behaviour disturbances
Refs Abel, Ernest L *Behavioral Teratology* (1985); Gunn, John C (Ed) *The Mentally Disordered Offender* (1990); Rutherford, Robert B, et al *Severe Behavior Disorders of Children and Youth* (1987); Vandenberg, Steven G, et al *The Heredity of Behavior Disorders in Adults and Children* (1986).
Aggravated by Encephalitis (#PE2348).

♦ PE2082 Blind children

Refs Scholl, Geraldine (Ed) *Foundations of Education for Blind and Visually Handicapped Children and Youth* (1986).
Broader Physical blindness (#PD0568) Physically handicapped children (#PD0196).

♦ PE2083 Deaf children

Refs Fraser, George R *The Causes of Profound Deafness in Childhood*; Liben, Lynn S *Deaf Children* (1979); McInnes, John M and Treffry, Jacqueline *Deaf-Blind Infants and Children* (1982).
Broader Deafness (#PD0659) Physically handicapped children (#PD0196).

♦ PE2084 Domination of the copper industry by transnational corporations

Nature Despite the considerable diffusion of ownership during the past three decades, transnational corporations still play a major role in the international copper industry and certain barriers to entry remain significant. Transnational corporations specializing in copper production and processing still account for a considerable share of all activities in the industry, and many of the new entrants are, themselves, large transnational corporations.
Barriers to entry into the industry affect several developing countries with copper resources. Exploration, which is a risk venture, especially for smaller corporations, acts as one such barrier. Lack of managerial expertise in bringing a project online seems to be more of a barrier to entry than does lack of the technical know–how needed. The trend towards massive mining operations requires enormous amounts of capital. However, the financing pattern in the international copper industry has shifted to an increasing element of debt financing.
Broader Transnational corporation imperialism (#PD5891)
Control of industries and sectors by transnational corporations (#PE5831).

♦ PE2087 Family rejection of physically handicapped

Broader Family rejection of children (#PC8127)
Discrimination against physically disabled (#PD8627)
Inadequate community care for handicapped persons (#PE8924).
Aggravates Infant growth failure (#PE6909).
Aggravated by Physically dependent people (#PD7238)
Physically handicapped children (#PD0196).

♦ PE2095 Nuclear explosions underground

Refs Bolt, Bruce A *Nuclear Explosions and Earthquakes* (1976).
Broader Nuclear weapons testing (#PC2201).
Aggravates Earthquakes (#PD0201).

♦ PE2113 Extraterritorial abduction

Abduction by government agents acting in foreign countries
Incidence Governments, notably the USA, tolerate the abduction by their agents of people in foreign countries that are required for interrogation or trial. In the case of the USA, the courts hold that a fugitive can stand trail even if the arrest overseas was improper.
Broader Kidnapping (#PD8744).

♦ PE2119 Family of torture victims

Nature The family of victims of torture are victims also. Because torture is frequently arbitrary and always dehumanizing the victim and the family experience themselves as guilty of some terrible but unknown wrong. In many cases at the time of arrest the family is victimized. The husband may be beaten and forced to watch the mistreatment of the rest of the family. The wife may be beaten and raped. The children are terrorized by being forced to watch their parents being brutalized, pets killed, toys like dolls destroyed. Children may be kidnapped and sold or "adopted" or left to fend for themselves. The family may be shunned by neighbours and family who are afraid they too might be victims.
Broader Disabled victims of torture (#PD0764).
Narrower Spouses of torture victims (#PE0753) Children of torture victims (#PE4579).

♦ PE2123 Crack

Crack cocaine
Nature The crack high reinforces feelings of power and aggression rather than the blissful lassitude of heroin. It can cause paranoid psychosis, hallucinations and chronic anxiety and in rare cases can provoke heart attacks or strokes.
Crack is popular with women. It is cheaper than other forms of cocaine, does not require syringes

or other drug paraphernalia and and the euphoria is intense and quickly reached. The "high" lasts only 30 minutes followed by a deep and frightening depression which leads to addiction. This high–low cycle of an hour or less leaves no time to attend to small children.
A mother's use of crack cocaine during pregnancy cuts off the blood supply to the baby, it causes serious damage: strokes, seizures, paralysis, prematurity, deformed hearts and lungs, abnormal genital and intestinal organs and permanent brain damage. The crack babies have 10–15 times the risk of Sudden Infant Death Syndrome.
Crack is distributed by young, wild, heavily armed gangs. They arrogantly intimidate whole communities and make war on each other to control the lucrative business. Crack dealing involves more adolescents than the heroin trade ever did, offering them money enough to have clothes, jewellery, cars, guns and power.
Crack has forced criminal justice to spend billions for police, prosecutors, courts, judges and penal institutions. California has 81,000 people locked up; from 1983 to 1989 it has built 21,000 new prison beds and plans 16,000 more. The total cost is $3.2 billion. In the same period, New York state has spent $900,000 to build 17,780 cells. President Bush pledged $1 billion to build 24,000 federal prison cells, largely for drug violators. More than half the males arrested in nine major American cities in 1988 tested positive for cocaine.
Hospitals and clinics are being overwhelmed by addicts, their children and casualties of crack wars. Each crack baby costs about $90,000 causing the US some $2.5 billion. The annual cost of treating gunshot wounds most of which are drug related is $1 billion, 85 percent of which is at US tax-payer expense. Health officials blame crack for a new outbreak of syphilis in cities due to prostitution and casual sex in "crack houses". Syphilis causes skin lesions and so facilitates the spread of AIDS.
At the same time vigilantism has begun to increase as citizens demand control of crack related crime: some crack houses have been burned by angered residents.
Refs American Health Research Institute Staff *Medical Subject Research Index of International Bibliography Concerning Cocaine* (1982); Wallace, Joseph A *Crack busters workbook* (1988); Weiss, Roger D and Mirin, Steven M *Cocaine* (1986).
Broader Abuse of coca and cocaine (#PD2363).
Related Basuco (#PE5245).
Aggravates AIDS (#PD5111) Family breakdown (#PC2102)
Excessive sexual activity (#PF5868) Sudden unexpected infant death (#PD1885).

♦ PE2125 Down's syndrome

Congenital acromicria — Mongolism
Nature It is an inherited disorder characterized by a redundant third copy of the 21st chromosome, distinctive appearance, cluster of medical symptoms, small brain and some degree of mental retardation. Almost all Down syndrome children have problems with receptive and expressive language.
Incidence The chances of a woman having a baby with Down syndrome are one in 365 at 35 years one in 109 at 40 years and one in 40 at 44 years of age.
Refs Byrne, Elizabeth; Cunningham, Cliff and Sloper, Patricia *Families and Their Children with Down's Syndrome*; Dmitriev, Val and Oelwein, Pat *Advances in Down Syndrome* (1988); Gath, Ann *Down's Syndrome and the Family* (1978); Gibson, D *Down's Syndrome* (1979); Pueschel, and Steinberg, *Down Syndrome* (1980).
Broader Congenital syndromes affecting multiple systems (#PE9324).
Aggravated by Neurosis (#PD0270).

♦ PE2131 Antisemitism

Discrimination against Jews
Nature Antisemitism consists of hostile expressions or actions directed against the interests, legal rights, religious practices, or lives of Jews. It has been a more or less constant feature of Jewish life since the Diaspora. The term was coined by Ernest Renan or Wilhelm Marr in the 1970s. It connoted the new form of Jew–baiting generated during the era of Jewish emancipation, which stressed racial and socio–economic antagonisms above the religious issues which earlier had dominated Jewish–gentile controversies. Nevertheless, the heritage of ancient and mediaeval hostilities remained **Background** Anti–Jewish hostility existed in ancient Alexandria where local tensions led to the propagation of many anti–Jewish slander and was concomitant of ancient ethnocentrism with its conviction of Greek and later Roman superiority Antisemitism in the church and as a religious reality has its roots in the earliest teachings of the Church fathers, Epiphanius and Origen. To them the Jews were guilty of deicide, or treachery, and of failure to recognize the Messiah. Christians appeared to Jews as heretics, and Jews appeared to Christians as recalcitrant unbelievers. As Christianity grew in numbers and power, Christians subjected Jews to many forms of discrimination, sometimes to outright persecution with fratricidal fury. In the Middle Ages the Jews in Europe were generally consigned to ghettos and regarded as aliens within the feudal order.
Incidence Since the attempt by the Nazis to eliminate European Jewry, the burden of guilt and the fear of being associated with the views which led to that catastrophe have been a potent force in European politics, leading in several countries to legislation against the incitement to racial hatred and internationally inclining the countries of Western Europe and the USA, towards the side of Israel. In the United States incidents of antisemitism rose in 1988 to the highest level in 5 years. Antisemitism is still a potent force in East European political life and, under the guise of hostility to Zionism, in Arab states **Counter–claim** Any intentional or unintended criticism of Israel, Judaism or Jews is called antisemitic. Jewish national emotions permeate the cultural life of whole countries. They are a negative influence on disarmament negotiations, on trade agreements and on international scientific and cultural relations. They provoke demonstrations and strikes. The Jewish issue acquires an incomprehensible power over people's minds, overshadowing the problems of other minorities.
Refs Bergmann, Werner (Ed) *Without Trial* (1987); Curtis, Michael (Ed) *Antisemitism in the Contemporary World* (1985); Epstein, Simon *Cry of Cassandra* (1986); Fein, Helen (Ed) *Current Research on Antisemitism, Vol 1* (1987); Furet, François (Ed) *Unanswered Questions* (1989); Gerber, David A (Ed) *Anti–Semitism in American History* (1986); Pulzer, Peter *The Rise of Political Anti–Semitism in Germany and Austria* (1988); Strauss, Herbert A and Bergmann, Werner (Ed) *Current Research on Antisemitism* (1987); United Nations *The United Nations and Human Rights* (1984).
Broader Neo–fascism (#PF2636) Ethnic discrimination (#PC3686)
Religious discrimination (#PC1455).
Narrower Pogroms (#PJ2093) Christian antisemitism (#PE4369).
Aggravates Refugees (#PB0205) Genocide (#PC1056)
Scapegoats (#PF3332) Segregation (#PC0031)
Forced labour (#PC0746) Statelessness (#PE2485)
Concentration camps (#PD0702)
Forced repatriation of prisoners of war (#PD0218).
Aggravated by Zionism (#PF0200) Fascism (#PF0248)
Prejudice (#PA2173) Nationalism (#PB0534)
Lack of communication (#PF0816) Racial discrimination (#PC0006).

♦ PE2134 Oil as a pollutant

Nature Sudden, massive oil pollution can occur when jumbo tankers break up, on and offshore wells blowout, or pipelines fail. As increasing demand and dwindling supplies necessitate exploiting reserves in increasingly inhospitable arenas, the likelihood of serious accidents increases. But most oil pollution (80 to 90 percent) occurs during everyday shipping, refining, processing, and burning of hydrocarbons. The most important form of oil pollution is failout of airborne hydrocarbons. When hydrocarbons reach the oceans, they are diluted and dispersed. Eventually, they disappear through microbial degradation, evaporation, oxidation, and deposition. Along the way, however, great damage may occur to ocean life. Sometimes this damage is obvious, sometimes it is subtle and indirect. Thus, hydrocarbons can destroy vital parts of some food chains directly (for example, aquatic insects); can accumulate and magnify in food chains (for example, in large fish and birds near tops of food chains); and can interfere with the communication systems of organisms (for example, disruption of chemical 'messages' from rivers to fish returning to spawn).
Incidence A 1975 report estimated that about 6 million metric tons of petroleum enter the ocean each year. Recent research shows that tanker contributions alone could account for as much as 6 million metric tons. But although about half of the oil transported in the world each year is carried by tankers, most of the oil accumulating in the sea comes from a combination of land–based industrial and municipal sources: 350 million gallons (1.3 billion litres) of used automobile crankcase oil is dumped each year into water drainage systems that reach the sea; and about 600,000 metric tons of oil run into the sea each year as a result of petroleum carried into the atmosphere from poorly tuned automobile engines.
Refs FAO *Impact of Oil and the Marine Environment* (1977); IMCO/UNEP *The Status of Oil Pollution and Oil Pollution Control in the West and Central African Region* (1982); Kuiper, J and Brink, W J van den (Eds) *Fate and Effects of Oil in Marine Ecosystems* (1987); UNEP/ASEAMS *Oil Pollution and Its Control in the East Asian Seas Region* (1988).
Broader Chemical contamination of water (#PE0535)
Environmental hazards from petroleum (#PE1409).
Related Hydrocarbons as pollutants (#PE0754).
Aggravates Energy crisis (#PC6329) Marine pollution (#PC1117)
Pollution of inland waters (#PD1223)
Carcinogenic chemical and physical agents (#PD1239).
Aggravated by Oil pollution (#PE1839) Marine dumping of wastes (#PD3666).

♦ PE2138 Underdevelopment of fishing industry

Broader Lack of coastal development in island countries (#PE5689)
Underdevelopment of industrial and economic activities (#PC0880).

♦ PE2142 Denial of right of peoples to use their own language

Denial of right of indigenous peoples to use their own language
Broader Denial of cultural rights (#PD5907).
Aggravates Endangered cultures (#PB8613) Limited verbal skills (#PD8123).

♦ PE2143 Stomatitis in animals

Broader Mouth diseases of animals (#PE9702).
Narrower Papular stomatitis of cattle (#PG5602) Mycotic stomatitis in animals (#PG5368)
Gangrenous stomatitis in animals (#PG4610) Ulcerative stomatitis in animals (#PG4567)
Ulcenomembranous stomatitis in animals (#PG0686).

♦ PE2145 Foreign currency manipulations in accounting records of transnational corporations

Nature For the purpose of preparing consolidated statements, the financial statements of foreign subsidiaries expressed in terms of the various host countries' currencies need to be translated into the currency of the parent company. At a time when exchange rates are subject to substantial fluctuations, the method of translation can significantly affect the results. The simplest method, which is still widely used by companies in Western Europe, is the application of the closing rate to all items in the subsidiaries' financial statements. More sophisticated methods involve the use of different rates for the translation of different items.
Broader Creative accounting (#PE6093) Transnational corporation imperialism (#PD5891)
Destabilization of monetary systems and exchange notes by transnational corporations (#PE5903).
Related Inadequate international standards of financial accounting and reporting (#PF0203).

♦ PE2147 Excessive speed of motor vehicles

Nature The speed of a vehicle is, or should be, entirely under the control of the driver. Proceeding at a speed that is excessive in view of the traffic conditions is a common contributory cause of accidents.
Broader Road traffic violations (#PE0930) Human errors and miscalculations (#PF3702).
Aggravates Injuries (#PB0855) Reckless driving (#PE9334)
Road traffic accidents (#PD0079) Inadequate road maintenance (#PD8557).

♦ PE2149 Drunken driving

Driving under the influence of alcohol — Driving under the influence of drugs
Nature The presence of excessive amounts of alcohol in the blood of the driver plays a substantial role in the production of accidents that cause fatal or serious injuries to vehicle occupants and pedestrians, since, in addition to the affected driver, innocent people who have not been drinking (passengers, pedestrians, and people in other cars) are often victims. A person who is severely affected by alcohol may be physically incapable of driving a motor vehicle; one who is less affected but who is still obviously under the influence of alcohol constitutes a danger as a driver. Further, in subclinical intoxication, in which a person has consumed alcohol but shows no clinical signs whatever, his judgement may be significantly impaired.
Incidence At a blood alcohol concentration of about 50mg/100ml, a statistically significant impairment of performance is observed in more than 50 percent of cases. Such impairment may, however, be present from 0–50mg/100ml; and between 50–100mg/100ml, the deterioration in performance increases rapidly. In Canada, for example, studies indicate than 50 per cent of all drivers and pedestrians fatally injured were drinking shortly before the crash; 78 percent of those drivers had blood alcohol concentrations of 0.10 per cent by weight or greater. Compared to the nondrinking driver, a driver with a concentration of 0.10 per cent is 8 times more likely to be involved in a fatal crash. At the 0.15 per cent level, the risk of crash involvement increases by a factor of 25 or more. In the USA, from one–third to two–thirds of drivers involved in fatal crashes have a record of previous arrests for public drunkenness, for drunken driving, or have registered as having a drinking problem.
Refs De Gier, J J and O'Hanlon, J F *Drugs and Driving* (1986); Goldberg, Leonard (Ed) *Alcohol, Drugs and Traffic Safety* (1981); West, L H T and Hore, T (Eds) *Analysis of Drunk Driving Research* (1980).
Broader Road traffic violations (#PE0930)
Substance abuse during control of complex equipment (#PE0680).
Related Reckless driving (#PE9334) Motor vehicle driver fatigue (#PG7482).
Aggravates Road traffic accidents (#PD0079) Draining of resources due to alcohol (#PE8865).
Aggravated by Drunkenness (#PE8311) Alcohol abuse (#PD0153)
Public drunkenness (#PE2429).

♦ PE2156 Unaesthetic location of advertising hoardings and billboards

Broader Ugliness (#PA7240) Misuse of advertising (#PE4225)
Landscape disfigurement (#PC2122).

♦ **PE2176 Decline in competition due to entrance barriers**
Nature The analysis of entry barriers generally revolves around four factors: economy of scale; minimum capital requirements; product differentiation; and access to special resources or privileges. It is widely accepted that economies of scale are the single most important determinant of concentration levels in modern industry. For given technological and marketing conditions, and for a given size of the final market, the higher the minimum scales for achieving such economies, the more concentrated an industry is likely to be.
Aggravates Restrictive trade practices (#PC0073).

♦ **PE2185 Beriberi**
Thiamine deficiency
Nature Beriberi is a nutritional disorder caused by a deficiency in vitamin B1, presenting cardiac and neurological symptoms. Infantile beriberi, an important problem in breast-fed infants whose mothers milk is deficient in thiamine, can lead to heart failure. The disease has two forms: in dry beriberi, there is a gradual degeneration of the long nerves, of the legs and the arms, with associated atrophy of muscle and loss of reflexes; in wet beriberi, a more acute form, there is oedema resulting in large part from cardiac failure and poor circulation.
Incidence Thiamine deficiency subsists in situations where imbalanced diets are prevalent (rice-eating countries, chronic alcoholism, etc).
Background Beriberi was first shown conclusively to be a nutritional disease by the Japanese naval surgeon Admiral Takaki. In 1878 the diet in the Japanese navy considered almost entirely of milled rice, and about 30 percent of the sailors suffered from beriberi each year.
Refs Williams, Robert R *Toward the Conquest of Beriberi* (1961).
Broader Nutritional diseases (#PD0287) Vitamin deficiencies in diet (#PD0715).
Aggravated by Diseases of metabolism (#PC2270).

♦ **PE2192 Gossip**
Gossip-oriented communication
Nature Gossip is talk often critical or malicious about other people's affairs and functions as a means of maintaining social control and preserving group boundaries or as a weapon in factional disputes.
Refs Rosnow, R L and Fine, G A *Rumor and Gossip* (1976).
Broader Biased presentation of news (#PD1718).
Aggravates Rumour (#PF5596) Infringement of privacy (#PB0284).
Aggravated by Declining sense of community (#PF2575)
Insufficient communications systems (#PF2350).

♦ **PE2193 Domination of advertising by transnational corporations**
Nature Transnational advertising agencies, like the transnational corporations they serve, derive their competitive power from their capacity to use finance, technology and marketing skills in an integrated way on a world scale. The worldwide brand or name-trade is now a common phenomenon, and is an example of the mastery of the techniques of marketing that is characteristic of the transnational corporations. Transnational corporations place heavy emphasis on brand name and image when selecting the message to advertise. The advantage of the brand system is that a brand name may be effective in cases where a patent is not – for example, for products which cannot be patented, or which are freely licensed, or for which the patent has expired. Transnational corporations are also more inclined to use advertising aggressively. Studies have shown, for instance, that transnational corporations on the average outspend local firms in terms of advertising expenditures at the ratio of about six to one.
Incidence According to one study, some products of transnational corporations sold in Kenya are known only by their brand names rather than by their generic names.
Refs United Nations Centre on Transnational Corporations *Transnational Corporations in Advertising* (1979).
Broader Transnational corporation imperialism (#PD5891)
Control of industries and sectors by transnational corporations (#PE5831).
Aggravated by Monopoly power due to advertising (#PE0081)
Protectionism in the advertising industry (#PD7108).

♦ **PE2197 Sleep disorders**
Insomnia — Hyposomnia — Sleeplessness — Sleep disturbance
Nature The many abnormalities of sleep include excessive sleep, inability to sleep, restless sleep, nightmares, bed-wetting, sleep paralysis, and other problems. Neurologists classify sleep disorders under three headings: hypersomnia, insomnia, and nocturnal behavioural symptoms.
Incidence Surveys in many cultures show that sleep disorders of some sort bother about one-third of the population, often being associated with workers whose sleep schedules are disturbed by rotating shifts.
Refs Dooley, Tricia C *Insomnia* (1988); Ehrenberg, B *Sleep and Sleep Disorders* (1989); Guilleminault, Christian *Sleeping and Waking Disorders* (1982); Langen, Dietrich *Speaking of Sleeping Problems* (1983); Mendelson, Wallace B, et al *Human Sleep and Its Disorders* (1977).
Broader Mental illness (#PC0300) Human disease and disability (#PB1044).
Narrower Insomnia (#PE3924) Nightmares (#PE6958) Hypersomnia (#PG4415)
Sleepwalking disorder (#PG9221) Sleep terror disorder (#PG4073)
Sleep-wake schedule disorder (#PG9626).
Aggravates Hallucinations (#PF2249) Sleep deprivation (#PE2741).
Aggravated by Fatigue (#PA0657) Anxiety (#PA1635)
Ascariasis (#PE2395) Human ageing (#PB0477)
Schizophrenia (#PD0438) Traffic noise (#PD3664)
Mental depression (#PC0799) Monotonous activity (#PG4419)
High altitude stress (#PD2322) Diseases of metabolism (#PC2270)
Diseases of the nervous system (#PC8756) Sleep difficulties due to torture (#PE0451)
Diseases and injuries of the brain (#PD0992) Diseases of the circulation system (#PC8482)
Health hazards of exposure to noise (#PC0268).

♦ **PE2200 Burden of conflicting national regulations on transnational corporations**
Nature Transnational corporations' foreign investments and technology transfers, particularly needed in developing countries, are obstructed by the number of local regulations they have to abide by. These regulations may differ so widely from country to country that transnational corporations compliance in each instance becomes very costly.
Broader Transnational corporation imperialism (#PD5891).
Reduces Excessive power and independence of transnational corporations (#PD5807).

♦ **PE2203 Contradictions within the growth and partnership model of developed countries**
Claim It is beyond question that the capitalist world system is in the grip of a global crisis: its symptoms and its consequences are clearly visible. Although the crisis has its familiar economic, political, social, and ecological aspects in the western industrialized countries, it is in fact a crisis of the world-capitalist system as a whole – and consequently of the countries of the Third World too. The socialist countries have also not been spared – depending on how much they have become integrated into the capitalist world-system. At the origin of the crisis is the gradual erosion of US global hegemony, one of its outcomes is the development of internal contradictions or difficulties within the growth and partnership model in the industrialized countries.
Broader Contradictions (#PF3667).
Aggravated by Inadequate models of socio-economic development (#PF9576).

♦ **PE2206 Paralysis agitans**
Parkinsonism — Shaking palsy — Parkinson's disease
Nature Parkinson's disease is a chronic, progressive, incurable disorder of the central nervous system, marked by slow movement and muscular rigidity, weakness and tremor at rest. It involves the destruction of cells in a tiny region of the brain called the substantia nigra. The loss of cells leads to a lack of a brain chemical, or neurotransmitter, called dopamine, which is important in the control of movement.
Incidence In the USA there are approximately 500,000 suffering from the disease.
Refs Jankovic, Joseph and Tolosa, Eduardo *Parkinson's Disease and Movement Disorders* (1988); Nakanishi, T (Ed) *Long Term Clinical Care of Parkinson's Disease, Tokyo, April 1987* (1988).
Broader Paralysis (#PD2632) Diseases and injuries of the brain (#PD0992)
Diseases of the central nervous system (#PE9037).
Aggravates Deterioration of the mind with age (#PE4649).

♦ **PE2209 Beryllium as a health hazard**
Nature Beryllium and its compounds are highly toxic substances. Beryllium enters the body almost entirely by inhalation, but the toxic reaction is body-wide rather than in the lung alone, in much the same way that lead may enter the body by inhalation and cause systemic disease. Experimental evidence suggests that little beryllium is absorbed through the intestinal wall. Traumatic introduction of beryllium and its compounds subcutaneously can give rise to local damage, but beryllium does not enter the body through the unbroken skin. Beryllium intoxication can cause bronchitis, pnemonitis, dermatitis, acute pneumonitis, chronic pulmonary granulomatosis etc.
Incidence This is an intoxication arising from the inhalation of any beryllium compound (beryl excepted). The precise quality and quantity of the disease-producing dose is at present unknown, although there is some knowledge of both certainly harmful and probably safe dose levels gained from a registry of case records established in 1952 in the United States.
Broader Toxic metal pollutants (#PD0948) Chemical air pollutants (#PD1271)
Environmental hazards from metals (#PE1401).
Aggravates Fires (#PD8054).

♦ **PE2210 Arteriosclerosis**
Atheroma
Nature Arteriosclerosis produces death or disability by narrowing and occluding the arteries supplying such vital organs as the heart, brain and kidneys.
Incidence The disease tends to occur in industrialized societies where people consume much animal protein.
Refs Suckling, K E and Groot, P H *Hyperlipidaemia and Atherosclerosis* (1988); Wolman, M *Healing and Scarring of Atheroma* (1984).
Broader Diseases of the arteries (#PE2684).
Related Bright's disease (#PE2272).
Aggravates Glaucoma (#PE2264) Disorders of joints and ligaments (#PD2283).
Aggravated by Diabetes (#PE0102).

♦ **PE2214 Over-dependency on international financial institutions by developing countries**
Nature There is an excessive indebtedness of the developing countries to the public and private international financial institutions, and linked with this, a foreign dependency amounting to virtual total subordination to the international financial system and the forces behind it.
Incidence Indebtedness has now reached such dimensions (US $810 billion at the end of 1983), that the mechanism of debt service now drains the bulk of the economic surplus (that is, the portion of national income otherwise available for net investment) of many countries.
Broader Deterioration in external financial position of developing countries (#PE9567)
Dependence of developing countries on external financing for development programmes (#PE7195).
Aggravates Unaccountability of international financial institutions (#PF1136).
Aggravated by Autocracy of intergovernmental financial institutions (#PE2805).

♦ **PE2215 Failure of governments to fulfil international reporting obligations**
Non-response of governments to international surveys — Governmental delays in fulfilling international reporting obligations
Broader Unfulfilled treaty obligations (#PF2497).
Aggravates Unrecorded knowledge (#PF5728) Silting of water systems (#PD3654).
Aggravated by Avoidance of negative feedback (#PF5311).

♦ **PE2217 Pests and diseases of sugar cane**
Nature Sugar cane is attacked by a number of diseases and insects which are transferred from country to country with seed canes. Of the more than 60 diseases, 9 cause serious losses in susceptible varieties though fortunately none of them are distributed world-wide. Apart from this, the incidence of certain weeds is so severe that crop values are seriously reduced.
Incidence Viruses cause mosaic, Fiji disease and ratton stunting disease; bacteria cause gummosis and leaf scald; fungi cause red rot, smut, downy mildew and root rot. The highly destructive mosaic disease at one time threatened to destroy entirely the sugar cane industry in such places as Java, Puerto Rico, Argentina and Louisiana. Gummosis is widely distributed, occurring in Brazil, Australia, Mauritius and several other countries. Fiji disease and leaf scald are both prevalent in the Pacific area. Insect pests include various moth borers and beetles, froghoppers, the woolly aphid and sapsucking leafhoppers. In the Western hemisphere and to some extent in India, Java and the Philippines, moth borers have been especially destructive. In the Pacific area, particularly Australia, the grubs of several species of beetles have also caused severe damage. Of the weed pests, the most severe and widespread are Cyperus rotundus, Cynodon dactylon, Digitaria sanguialis, Portufaca oleracea, Eleusine indica and Echinochloa colonum. Of these the most important is Cyperus rotundus which is a principal weed of sugar cane wherever it is grown.
Broader Crops pests and diseases (#PE7783).
Related Infestation of seeds (#PE6271) Maize pests and diseases (#PE3589)
Pests and diseases of rice (#PE2221) Pests and diseases of wheat (#PE2222)
Pests and diseases of vines (#PE2985) Pests and diseases of cotton (#PE2220)
Pests and diseases of potato (#PE2219) Pests and diseases of sugar-beet (#PE2975)
Pests and diseases of grain sorghum (#PE3590).
Aggravated by Weeds (#PD1574) Aphids as pests (#PE3613)
Viral plant diseases (#PD2227) Beetles as insect pests (#PE1679)
Bacterial plant diseases (#PD2226) Lepidoptera as insect pests (#PE3649).

♦ **PE2218 Pests and diseases of coffee**
Nature Coffee is attacked by over 300 diseases and a number of insects. In certain areas the coffee industry is no longer viable because of pests, and in all other areas pests represent a serious threat to yields.
Incidence The coffee-bean borer (referred to in Brazil as the 'coffee plague'), coffee 'leaf-miner',

and Mediterranean fruit fly are the most common insect pests. Of these, the 'leaf-miner' is the most serious pest in many Latin American countries, causing up to 50 per cent loss in plant efficiency. Of the many diseases, by far the most destructive is Hemileia vastatrix, a fungal leaf rust which causes spectacular defoliation and crop failure. **Background** The leaf rust fungus is a native of Northeast Africa but during the 19th Century it spread to many other parts of the eastern tropics. In 1869 it appeared in Ceylon and in just 20 years most of the millions of plantation trees were destroyed. By 1876 the rust had spread to Java, where it devastated the coffee industry. The nascent coffee industry of India was also destroyed.
Refs Firman, I D and Waller, J M *Coffee Berry Disease and Other Colletotrichum Diseases of Coffee* (1977).
Broader Pests and diseases of trees (#PD3585).
Related Pests and diseases of tea (#PE2980); Pests and diseases of elm (#PE2982); Pests and diseases of oak (#PE2984); Pests and diseases of cocoa (#PE2979); Pests and diseases of palms (#PE2981); Pests and diseases of rubber (#PE2977); Pests and diseases of olives (#PE2978); Pests and diseases of poplars (#PE4413); Pests and diseases of chestnut (#PE2983); Pests and diseases of citrus fruit (#PE2976); Pests and diseases of deciduous fruit (#PE3591).
Aggravated by Borer insects (#PE6286); Fruit flies as pests (#PE3607); Fungal plant diseases (#PD2225); Lepidoptera as insect pests (#PE3649).

♦ PE2219 Pests and diseases of potato

Nature The potato, the fourth most important food plant of mankind, is subject to a recorded 70 pathogens and 198 pests. Fortunately only a few are of widespread economic importance, the seriousness of which varies in different regions and in different years according to the prevailing weather. The potato is especially subject to disease because it is propagated vegetatively. This permits transmission of pathogens (especially viruses) from generation to generation, and produces genetically homogenous populations which favour epidemic diseases.
Incidence In the temperate zones of North America and Europe the Colorado potato beetle is the principal insect pest. In warmer climates the potato tuber moth causes major losses in the field and stores. Similarly, cyst nematodes, which originated in the Andes of southern Peru, are important pests throughout the warm regions of the world. Late blight is the most serious disease of potatoes during cool, wet weather in most of the 129 countries where the potato is grown. Soft rot and wilt diseases caused by bacteria, and wilt diseases of fungal origin may also cause significant losses. Early blight may be destructive under warm, dry conditions. The international trade in certified seed tubers owes its importance to the need to replace virus degenerated seed stocks. The potato spindle tuber viroid is of special quarantine significance since there is no known resistant varieties. In some developed countries rigid control practices minimize the effects of a particular pest or pathogen.
Late blight is the most destructive disease of potato; it is a fungal disease which is practically universal, and it causes destruction both of the foliage and the tuber. A devastating epidemic of late blight fungus on potatoes beginning in Europe in 1845 brought about the famine, particularly in Ireland, that caused starvation, death and mass migration. Of the other fungal diseases the most important is early blight, which may be as destructive as late blight under hot, dry conditions. Viruses cause the most economically important diseases. Over two centuries ago, English farmers realized that when locally grown seed was used year after year, yields of potato deteriorated. This was due to the spread of virus diseases, of which 20 have been recorded as occurring in potatoes. Of these, it is generally accepted that leaf roll is responsible for the largest reductions in yields. It is found practically everywhere potatoes are grown commercially. Other virus diseases include various mosaics and spindle tuber. A few of the bacterial diseases of potato are black–leg and various rots and wilts.
Refs Rich, Avery E *Potato Diseases* (1983); Théberge, R L (Ed) *Common African Pests and Diseases of Cassava, Yam, Sweet, Potato and Cocoyam* (1985).
Broader Crops pests and diseases (#PE7783).
Related Infestation of seeds (#PE6271); Maize pests and diseases (#PE3589); Pests and diseases of rice (#PE2221); Pests and diseases of wheat (#PE2222); Pests and diseases of vines (#PE2985); Pests and diseases of cotton (#PE2220); Pests and diseases of sugar cane (#PE2217); Pests and diseases of sugar–beet (#PE2975); Pests and diseases of grain sorghum (#PE3590).
Aggravated by Pests of plants (#PC1627); Aphids as pests (#PE3613); Viral plant diseases (#PD2227); Fungal plant diseases (#PD2225); Nematoid plant diseases (#PD2228); Beetles as insect pests (#PE1679); Bacterial plant diseases (#PD2226); Lepidoptera as insect pests (#PE3649).

♦ PE2220 Pests and diseases of cotton

Nature Cotton is subject to attack by numerous insects, pathogenic fungi, bacteria, virus and nematodes. Its succulent foliage, large flowers and extended fruiting period make cotton an attractive host for many insects and the diseases they carry. In all, 500 species of insects attack cotton and among them are some of the most destructive known to agriculture, including the boll weevil, jassid, whitefly, pink bollworm, bollworm, cotton aphid, cotton stainer, cotton fleahopper, cotton leafworm, spider mites, grasshoppers and tarnished plant bugs. The most widespread disease which attacks growing seedlings is a complex of disease organisms, the predominant organism differing with environmental conditions. A number of weeds also compete with cotton, damage the crop, and reduce its value. The average crop losses worldwide due to pests and diseases are 60 percent of potential production.
Incidence All cotton–producing areas suffer high losses from insects and diseases. Due to insects, the annual losses in the United States alone has averaged more than US $243 million since 1929 with an all time high, in 1950, of over $900 million. To this must be added the cost of insect control measures which in the USA vary between $58 million and $95 million annually.
Losses due to disease have been estimated to be as high as 50 per cent in Brazil and some African countries. Losses in India have fluctuated between 7 percent and 20 percent and in the USA have been reported as high as 15 percent. One widespread disease is bacterial, known as angular leaf spot, and in African countries this is the most serious disease. Fungus diseases which attack the growing cotton plant include fusarium and verticillium wilt and Texas root rot. Fusarium wilt is an important disease in the USA and the major disease in Egypt. Texas root rot is limited in distribution to areas of Mexico and the USA. Of the insect pests, the pink bollworm is the most widely distributed. It has become established in the 8 the countries that produce 90 percent of the world's cotton. In the USSR, it has caused severe losses. In India and Egypt the average annual loss caused by pink bollworm is between 15 and 25 percent and in Brazil, between 20 and 25 percent. It is the most serious cotton pest in China. The most serious cotton pest in the US is the boll weevil. It is restricted to North and Central America, but another closely related species occurs in South America. Insects which feed on the leaves and buds, including the cotton leafworm and the sucking insects (fleahoppers, leafhoppers and aphids), do considerable damage to crops throughout the world. Of the weed pests, the most severe and widespread are cyperus rotundus, cynodon dactylon, and portulaca oleracea.
Counter–claim Cotton is a hardy plant capable of resisting many, if not all, of the destructive agents listed. Resistant strains to specific pests are currently bred in both developed and developing countries. Cotton is naturally resistant to insects and small animals because of the poison gossypol contained in the leaves. The plant has other morphological and physiological protection.
Refs FAO *Guidelines for Integrated Control of Cotton Pests* (1984).
Broader Crops pests and diseases (#PE7783).
Related Infestation of seeds (#PE6271); Maize pests and diseases (#PE3589); Pests and diseases of rice (#PE2221); Pests and diseases of wheat (#PE2222); Pests and diseases of vines (#PE2985); Pests and diseases of potato (#PE2219); Pests and diseases of sugar cane (#PE2217); Pests and diseases of sugar–beet (#PE2975); Pests and diseases of grain sorghum (#PE3590).
Aggravated by Mites as pests (#PE3639); Pests of plants (#PC1627); Aphids as pests (#PE3613); Crickets as pests (#PE3640); Leaf curl of cotton (#PG4455); Viral plant diseases (#PD2227); Fungal plant diseases (#PD2225); Nematoid plant diseases (#PD2228); Bacterial plant diseases (#PD2226); Hemiptera as insect pests (#PE3615); Grasshoppers as insect pests (#PE3642).

♦ PE2221 Pests and diseases of rice

Nature Rice, the staple food of most of the world's population, is attacked by a array of insect, disease, weed and vertebrate pests. Since rice is still a subsistence crop to many people, and there is often a close balance between production and consumption, even small losses can lead to conditions approaching famine.
Incidence In temperate regions, rice is grown only during the summer months as a single crop and insect pest populations are suppressed by winter temperatures. In the tropics, where the expansion of irrigation systems prompted by the development of photoperiod insensitive varieties has not yet occurred and rice is grown as a single rainy season crop, the principal insect pests are striped, pink, and dark headed stem borers which have alternative hosts on other crops such as sugarcane, maize and sorghum, or the white stem borer which remains dormant in the rice stubble during the off season. The first monsoon rains signalling the beginning of the rainy season break the dormancy and the moths emerge during the start of the rice planting season. With modern varieties and dry season irrigation, rice is double cropped, and there is a shift in insect pest species in favour of those that feed only on rice and do not undergo a dormant period, but instead have dispersal powers to move to nearby fields being planted. The yellow stem borer is the most common species in tropical Asia, however its damage is tolerated because of the high tillering habit of the modern varieties. The rice brown planthopper and rice green leafhopper, both feeding only on rice and having robust dispersal powers, are prevalent throughout Asia and have produced many wide scale epidemics as vectors of virus diseases. Blast, a fungal disease, continues to pose serious threats to rice production, especially in areas where it is upland, highlands in the tropics, or temperate regions. Sheath blight, sheath rot, leaf scald and some other fungal diseases have threatened rice production in the humid tropics where agronomic practices have been improved to induce dense crop stands and succulent plants. Bacterial diseases, especially rice bacterial blight, caused epidemics in South and Southeast Asia in the 60's and are still threatening today's rice crop. In China, bacterial blight is a major threat to rice production. The major weeds of rice in South and Southeast Asia are members of the Echinochloa crus–galli complex, Echinochloa colona, Cyperus differmis, Cyperus iria, Fimbristylis miliacea, and Monochoria vaginalis. Many of these are important in other areas of the world where rice is grown under flooded conditions. Wet–seeded rice is more susceptible to weed competition than transplanted rice and weed control with herbicides is more difficult because the rice and the weeds are at the same stage of development and as a result selectivity is often marginal. Wetseeding has increased in importance in some areas in recent years because of such factors as increasing in importance due to increasing production costs and the decreasing availability of labour.
Refs FAO *Guidelines for Integrated Control of Rice Insect Pests* (1979); Mather, T H *Environmental Management for Vector Control in Rice Fields* (1985); Ou, S H *Rice Diseases* (1985).
Broader Crops pests and diseases (#PE7783).
Related Infestation of seeds (#PE6271); Maize pests and diseases (#PE3589); Pests and diseases of wheat (#PE2222); Pests and diseases of vines (#PE2985); Pests and diseases of potato (#PE2219); Pests and diseases of cotton (#PE2220); Pests and diseases of sugar cane (#PE2217); Pests and diseases of sugar–beet (#PE2975); Pests and diseases of grain sorghum (#PE3590).
Aggravated by Weeds (#PD1574); Animal pests (#PD8426); Borer insects (#PE6286); Plant pathogens (#PD1866); Beetles as insect pests (#PE1679); Hemiptera as insect pests (#PE3615); Insect vectors of disease (#PC3597); Lepidoptera as insect pests (#PE3649).

♦ PE2222 Pests and diseases of wheat

Nature Wheat, the most important food crop in the Western world, is subject to attack by a number of pests, which can seriously diminish yields.
Incidence Of the diseases, the rusts can cause very serious losses. The three that attack wheat are stem or black stem rust, leaf or brown rust and yellow or stripe rust. Together they produce average annual losses of 10 percent, though losses may be as high as 50 percent in epidemic years. Black stem rust, which is distributed world wide, is the most destructive of all the wheat rusts, doing more harm than all the other diseases put together. Yellow rust is common in the more temperate areas, particularly Europe. Brown rust is very widespread but it only causes losses of economic importance in areas which have high humidity and high temperatures during the summer, such as parts of the USA and India. Another group of diseases of wheat are the smuts. The important species are bunt or stinking smut and loose smut. Bunt is the best known, having been responsible for serious crop losses and for contamination of flour, but now the damage done by this and other smut diseases is relatively small. The principal insect pest is the wivaworm which is most important in Northern Europe, USSR, and Canada, though less so in southern regions such as Turkey, Spain, Argentina, South Africa and Australia. Local pests are to be found (such as the hessian fly which causes serious damage) in the eastern half of the USA. Of the weeds, those most difficult to control include the species of Nabricaria, Polyganum and Chrysanthemum segatum, and arable weeds such as Agropyron repens (couch), Avena sativa (wild oat) and Alopecuris pratense (black–grass).
Background Rusts of wheat have always plagued man. Similarly, bunt was responsible for considerable losses of wheat until the end of the 18th century, but since then methods of controlling it have become so efficient that it is now rarely seen. The insect pests and weeds have generally occurred originally on grassland. For example, wireworms become pests of crops when grassland is converted to arable land, as happened during the two world wars.
Broader Crops pests and diseases (#PE7783).
Related Infestation of seeds (#PE6271); Maize pests and diseases (#PE3589); Pests and diseases of rice (#PE2221); Pests and diseases of vines (#PE2985); Pests and diseases of potato (#PE2219); Pests and diseases of cotton (#PE2220); Pests and diseases of sugar cane (#PE2217); Pests and diseases of sugar–beet (#PE2975); Pests and diseases of grain sorghum (#PE3590).
Aggravated by Weeds (#PD1574); Plant diseases (#PC0555); Wireworms as pests (#PE3645); Beetles as insect pests (#PE1679); Hemiptera as insect pests (#PE3615); Rusts of grasses and cereals (#PE2233); Smuts of grasses and cereals (#PE2234).

♦ PE2223 Wind

Wind storms
Broader Storms (#PD1150).
Narrower Wind shear as a hazard (#PE5085); Wind damage to structures (#PE1334); Wind–borne animal diseases (#PG5736).

Aggravates Dust (#PD1245) Anemophobia (#PG1013) Sand storms (#PD3650)
Dust storms (#PD3655) Radioactive fallout (#PC0314) Biochemical warfare (#PC1164)
Soil erosion by wind (#PE3656).

♦ PE2224 Video violence
Trivialization of death and suffering on videos — Snuff movies
Refs Barlow, Geoffrey and Hill, Alison *Video Violence and Children* (1986); Hardy, Phil *The Encyclopedia of Horror Movies* (1987).
Broader Human violence (#PA0429) Culture of violence (#PD6279)
Unethical entertainment (#PF0374).

♦ PE2229 Fire blight
Nature Fire blight is a bacterial plant disease which attacks apple, pear and quince trees. The bacteria, carried by bees, enter blossoms, killing the tissue, and on susceptible species may move down the stem producing an extensive canker, where they overwinter. Certain apple and pear varieties have been driven out of intensive cultivation because of their susceptibility to this disease.
Background Fire blight first developed as a devastating disease a century ago in the apple and pear orchards of the USA.
Broader Bacterial plant diseases (#PD2226).
Related Crown gall (#PE2230) Bacterial soft rot in plants (#PE4706).
Aggravates Pests and diseases of deciduous fruit (#PE3591).

♦ PE2230 Crown gall
Nature Crown gall is a bacterial plant disease with a very wide host range. The bacteria enter through wounds and live between cells, which they stimulate to grow, producing the characteristic knobbly growth. Eventually the gall may hinder the transmission of water and food supplies, and the plants may wilt and die.
Incidence The disease occurs widely throughout the world on such economically important plants as apple, peach, pear, raspberry, rose and sugarbeet.
Broader Bacterial plant diseases (#PD2226).
Related Fire blight (#PE2229) Bacterial soft rot in plants (#PE4706).
Aggravates Pests and diseases of sugar-beet (#PE2975)
Pests and diseases of deciduous fruit (#PE3591).

♦ PE2233 Rusts of grasses and cereals
Nature Rusts are microscopic fungi and are strictly parasitic, producing disease in grasses and cereal crops. In the USA alone, about 125 different species of rusts attack grasses and nearly 400 species of grasses are among the hosts of rusts. Some of the rusts attack only one of a few grasses but others can attack a great many. Similarly, some of the rusts of grasses are also destructive to cereals and most of the so-called cereal rusts have numerous grass hosts.
Because the rust fungi are parasites, their development on a host plant is at the expense of that plant and the nutrients that they take would otherwise go into seed, forage, or both. A light infestation of rust is not likely to cause noticeable effects on yield of seed or forage, but a heavy infestation definitely will. In grasses grown for seed, a heavy infestation will result in low test weight of the seed because of the direct effect of parasitism in sapping nutrients from the host and because of water loss through the numerous open rust pustules on the leaves and stems. Rust likewise affects the production of grasses for forage, chiefly in lower yields as a result of reduced vigour.
Secondary or indirect adverse effects may also occur. Heavy attacks of rusts on grasses will make them more likely to succumb to other factors that are always more severe on the already weakened plant - drought, winter injury, root rot, snow mould, and perhaps other diseases.
Incidence Although rusts extensively attack grasses in general, it is on cereal crops that they inflict the greatest economic damage. Stem rust of wheat causes the most spectacular and perhaps the greatest losses, which range up to 85 and 90 per cent. Leaf rust of wheat and crown rust of oats occur more frequently and affect large acreages, and so may cause greater average losses year in and year out. Furthermore, leaf rust of wheat and crown rust of oats occur wherever wheat and oast, are grown. Crown rust is the most destructive disease of oats, cutting yields by 20 to 50 per cent. Stem and leaf rusts are the most destructive diseases of wheat in North America, whereas in Europe, stripe rust is the most common and most destructive. Besides wheat, it also attacks barley and rye, and it occurs in Africa, South America, Japan, China and India.
Wheat, oats, barley, and rye may be attacked by eight distinct species or subspecies of rust fungi. Wheat is subject to stem rust Paccinia graminis tritici), leaf rust (P rubigovera), and stripe rust (P glumarum). Oats are attacked by a stem rust (P graminis avenae) and by crown rust (P coronata avenae). Barley may fall prey to the same stem rust (P graminis tritici) that attacks wheat and a leaf rust (P hordei). A stem rust (P graminis secalis) and a leaf rust (P rubigovera) attack rye. Each of the eight rusts is made up of several or many different races, which may attack certain varieties of a particular cereal crop but not others.
Background Cereal rusts are among the oldest known diseases of food crops. The Romans attributed it to the god Robigus, who imposed this damage upon the crops of wicked people.
Narrower Stem rust of wheat (#PG1868).
Related Wood rots (#PE4455) Heart rot fungi (#PE6269) Cereal root rots (#PE4453)
Dutch elm disease (#PE1154) Smuts of grasses and cereals (#PE2234).
Aggravates Maize pests and diseases (#PE3589)
Pests and diseases of wheat (#PE2222)
Pests and diseases of grain sorghum (#PE3590).
Aggravated by Rust fungi (#PE6255).

♦ PE2234 Smuts of grasses and cereals
Nature Smuts are fungi that are parasitic on grasses and cereals. In the USA alone nearly 140 different species of smuts attack approximately 300 species of grasses.
Many smut fungi exhibit a remarkable degree of specialization not only to certain species of plants but also to certain varieties or strains within those host species. Furthermore there are often strains or races of the smut fungi to contend with. The common stinking smut, for example, has nearly 30 known strains or races, each capable of attacking different varieties of wheat and different strains or varieties of wheat grasses and related grasses.
The smut fungi have a more adverse effect directly (and perhaps indirectly) on their hosts than do the rust fungi. The smuts that attack all or parts of the flowering structures generally destroy the seeds entirely. The leaf smuts and the stem smuts, while only occasionally involving the flowering structures, do nevertheless generally suppress these structures and likewise result in a more or less complete loss of seed on affected plants. The smuts that attack the vegetative structures (that is, the leaf smuts and stem smuts) have a decidedly weakening effect on their host plants and make them more susceptible to other sinister factors in their environment.
Background Smuts have plagued man ever since crops were first cultivated. They were among the first cereal diseases to come under the scrutiny of early writers on plant diseases and to be studied by plant scientists.
Broader Smut diseases of plants (#PE0857).
Related Wood rots (#PE4455) Heart rot fungi (#PE6269) Cereal root rots (#PE4453)
Dutch elm disease (#PE1154) Rusts of grasses and cereals (#PE2233).
Aggravates Pests and diseases of wheat (#PE2222)
Pests and diseases of grain sorghum (#PE3590).

♦ PE2235 Detrimental effect of jungle environment in tropical villages
Nature In contrast to the generally increased control man has over his immediate physical environment, the life-style of people in tropical villages tends to be dictated by the rapidly growing jungle. Very little equipment is available in the village for managing the environment and means of repairing equipment are limited. The resource of outside consultants or government aid is still often unknown and plans rarely emphasize how to secure information or financial help. Unkept coconut groves are a constant reminder of the losing battle to care for resources so close at hand. The continued presence of pests like monkeys, wild pigs and crabs cause crop damage, water pollution and erosion. This discourages local people from risking large farm plots and they neither use, nor seek information on, pest eradication methods. In addition, drainage systems, which are a particular hazard to health in tropical areas, remain disorganized and left to individual planning and application.
Narrower Unknown outside successes (#PS9503)
Unkempt coconut growing land (#PS9531)
Undetermined land crab control (#PS9573)
Unchecked crab damage to crops (#PS9545)
Undeveloped jungle pest control (#PU9621)
Unresearched crab control methods (#PS9517)
Small animal numbers in jungle areas (#PS9598).
Related Corrosion in tropical climates (#PE1811) Uncontrolled tropical diseases (#PD4775)
Underutilized government resources (#PE9325)
Destructive action of mould in tropical climates (#PE1265).
Aggravates Inadequate land drainage (#PD2269).
Aggravated by Unrecognized benefits from cooperatives (#PF9729).

♦ PE2238 Inadequate increase in employment in the manufacturing industries in developing countries
Nature An important objective of an industrialization strategy should be to create additional employment opportunities in non-traditional activities for the growing labour force. However, the experience of developing countries indicates that demand for labour in the manufacturing sector has not expanded at a satisfactory rate, given the rate of growth of output. Among the reasons for this lack of employment expansion are: the labour-saving bias embedded in modern industrial technology; changes in the composition of manufacturing output; the evolution of relative factor prices; and the growth in production.
Broader Underdevelopment of manufacturing industries (#PF0854).

♦ PE2239 Haemorrhage
Haemorrhage
Narrower Haemorrhage of pregnancy and childbirth (#PE4894).
Related Loss of blood (#PJ2230).
Aggravates Nutritional anaemia (#PD0321).
Aggravated by Ulcers (#PE2308) Typhoid fever (#PD1753)
Pregnancy disorders (#PD2289).

♦ PE2240 Continued operation of unsafe motor vehicles
Nature Preventive maintenance and corrective repairs for motor vehicles are frequently inadequate, and a substantial number of vehicles are operated on public thoroughfares in an unsafe condition. International road traffic is impeded by a lack of universal safety standards.
Incidence Commonly existing discrepancies between national safety standards allow for: unsafe, worn tyres; inadequate headlamps, passing and signal lights; and the absence of other safety features such as belts, roll bars, and fire extinguishers.
Broader Consumer vulnerability (#PC0123) Human errors and miscalculations (#PF3702).
Aggravates Injuries (#PB0855) Road traffic accidents (#PD0079).
Aggravated by Motion sickness of animals (#PG0898)
Defects in machinery design (#PE2462).

♦ PE2246 Human pseudo hermaphroditism
Nature An individual may suffer the malformation of having both male and female sexual characteristics. This may occur in any of four ways: bilateral, an ovary and testis on each side; lateral, an ovary on one side and a testis on the other; ovatesticular, an ovatestis on one or both sides; unilateral, combination of an ovatestis on one side with an ovary or testis on the other. Malformation may occur in the embryo or in puberty.
Refs Jirasek, Jan E and Cohen, M Michael *Development of the Genital System and the Male Pseudohermaphroditism.*
Broader Malformation (#PE4460) Human sexual disorders (#PD8016)
Human physical genetic abnormalities (#PD1618).
Narrower Puberty trauma (#PJ2206) Embryogenesis malfunction (#PG4461).
Related Transsexualism (#PF3277) Sexual deviation (#PD2198)
Congenital anomalies of genital organs (#PE4249).
Aggravates Social stigma (#PD0884).

♦ PE2248 Chronic bronchitis
Persistent cough — Emphysema — Pulmonary emphysema
Nature Chronic bronchitis is a chronic or recurrent cough together with expectoration occurring on most days for at least three months in the year during at least two years. Chronic productive cough followed after some years by the onset of breathlessness is initially episodic and may follow an infection such as influenza or pneumonia; but it later becomes persistent and, after 20 or 30 years with recurrent acute illnesses or exacerbations, respiratory failure ensues and progresses until death with or without congestive heart failure (cor pulmonale).
Incidence The many studies of mortality and of morbidity carried out in different countries indicate causative factors operative in both developed and developing countries. Comparing emphysema in different countries has not progressed beyond the confirmation of male preponderance and a relationship to smoking. In relation to chronic obstructive bronchitis, five factors have been identified; three which can be partly quantified and two which do not lend themselves to measurement. The three most significant factors are smoking, air pollution and occupation.
Refs Bignon, J and Scarpa, G L (Eds) *Biochemistry, Pathology and Genetics of Pulmonary Emphysema* (1981); Fraser, et al *Diagnosis of Diseases of the Chest, Vol II* (1988); Orie, N G and Lende, R van der (Eds) *Bronchitis III* (1970); Taylor, Joseph C and Mittman, Charles *Pulmonary Emphysema and Proteolysis, 1986* (1987).
Broader Chronic illness (#PD8239) Human disease and disability (#PB1044)
Diseases of the respiratory system (#PD7924).
Aggravates Asthma (#PD2408) Pneumonia (#PE2293) Tuberculosis (#PE0566)
Typhoid fever (#PD1753) Heart diseases (#PD0448)
Lung disorders and diseases (#PD0637).
Aggravated by Fog (#PE1655) Dust (#PD1245) Damp (#PG4479)
Ulcers (#PE2308) Smoking (#PD0713) Obesity (#PE1177)
Measles (#PE1603) Rickets (#PG2295) Smallpox (#PE0097)
Silicosis (#PE1314) Urticaria (#PG4480) Hay fever (#PE6197)
Byssinosis (#PE2319) Human ageing (#PB0477) Climatic cold (#PD1404)
Air pollution (#PC0119) Virus diseases (#PD0594) Irritant fumes (#PD3672)
Limestone dust (#PE2038) Pneumoconiosis (#PD2034) Inadequate housing (#PC0449)
Smoke as a pollutant (#PD2267) Psychological conflict (#PE5087)
Symptoms referable to sense organs (#PE2665).

♦ PE2251 Cholecystitis

Inflammation of the gallbladder

Nature Cholecystitis is an inflammation of the gallbladder that frequently occurs after viral hepatitis and other infectious diseases. Stoppages or changes in the composition of the bile (which may be related to a person's diet), are contributing factors.

Broader Diseases of gallbladder (#PE9829) Human disease and disability (#PB1044).
Aggravates Peritonitis (#PE2663).
Aggravated by Typhoid fever (#PD1753) Bile–duct obstruction (#PG4857).

♦ PE2254 Flies as insect pests

Diptera

Nature The fly (any two–winged insect belonging to the order Diptera) affects human welfare more extensively than any other insect. Flies may transmit disease directly, as when they contaminate food; the larvae of certain flies invade the bodies of men and animals; and other larvae feed on, and thus damage, plants.

Incidence It is while feeding that flies inflict their damage. Larvae which live inside growing plant material and causing damage and loss of crops include those of gall midges, frit flies and fruit flies. Other fly larvae live in carrion and in the flesh of animals, including that of man. For example, the larvae of various bot flies and screwworms are parasitic within domestic animals. The adult flies feed upon fluids, which they draw up through a tubular proboscis. The basic food of most flies is carbohydrates and water. To get these substances, flies frequent flowers, pools of water, fresh dung, and most damp, rotting or fermenting matter. It is during these activities that flies may mechanically transmit disease organisms through the contamination of food. All the adults in some species and only the females in some other species supplement the basic carbohydrate diet with protein. These flies have mouth parts that can pierce the skin of another insect or of a vertebrate animal in order to suck blood. Bloodsucking flies include mosquitoes, black flies, buffalo gnats, sand flies, biting midges, horseflies, tsetse flies, and stable flies. Such flies are a nuisance and often cause physical pain and severe irritation. But their greatest concern to man is their ability to transmit disease–producing organisms found in the blood of man and other animals. Some flies live as external parasites in the hair of mammals or in the feathers of birds. They feed upon the blood of their hosts and in birds are known to transmit diseases.

Broader Parasites (#PD0868) Insect pests (#PC1630).
Narrower Sheep ked (#PG9666) Face flies (#PG9658)
Head flies (#PG7795) Horn flies (#PG9333)
Horse flies (#PE3878) Cattle grubs (#PG5534)
Stable flies (#PG6720) Stable flies (#PG9289)
Buffalo flies (#PG9466) Beet leaf miner (#PG6915)
Botflies as pests (#PE3635) Blowflies as pests (#PE3627)
Houseflies as pests (#PE3609) Fruit flies as pests (#PE3607)
Black flies as pests (#PE3646) Tsetse flies as pests (#PE1335)
Rodent–or–rabbit bot fly (#PG3839) Screw–worm flies as pests (#PE8150).
Related Insect pests of plants (#PD3634) Insect vectors of disease (#PC3597).
Aggravates Cholera (#PE0560) Mycosis (#PE2455) Trachoma (#PE1946)
Poliomyelitis (#PE0504) Typhoid fever (#PD1753) Pappataci fever (#PG4504)
Sweet itch in horses (#PG9200) Insect bites and stings (#PE3636).
Aggravated by Mosquitoes as vectors of disease (#PE1923).

♦ PE2259 Dysentery

Bacillary dysentery — Bloody flux — Amoebic dysentery

Refs WHO *Dysentery* (1969).

Broader Intestinal infectious diseases (#PE9526).
Aggravates Encephalitis (#PE2348) Amoebic hepatitis (#PG4512)
Dysenteric arthritis (#PG4510).
Aggravated by Shigella (#PG4520) Human ageing (#PB0477)
Malnutrition (#PB1498) Wasted water (#PD3669)
Liver diseases (#PE1028) Blowflies as pests (#PE3627)
Houseflies as pests (#PE3609) Unhygienic conditions (#PF8515)
Parasites of the human body (#PE0596) Flies as vectors of diseases (#PE4514)
Contamination of drinking water (#PD0235)
Damage by degradable organic matter (#PJ6128)
Pollution of water by infected faeces (#PE8545)
Overcrowding of housing and accommodation (#PD0758).

♦ PE2260 Dengue fever

Break–bone fever — Dandy fever — Three–day fever

Nature An acute viral disease with fever, rash, lymphadenopathy and extreme pain and stiffness in the joints that temporarily is completely incapacitating. The fatality–rate of this mosquito–transmitted disease, although not very high, is on the increase.

Incidence Dengue may occur in any country where the mosquito carriers (mainly Aedes aegypti) breed. Outbreaks occur chiefly in Africa, India, the Far East, and also in Hawaii, the Philippines and Caribbean Islands, reaching a peak during the rainy season. Cases are found mainly in cities and towns, although there is a rise in the number of reported cases in rural areas. 90 percent of the victims are under 5 years of age.

Broader Virus diseases (#PD0594) Infectious and parasitic diseases (#PD0982).
Related Yellow fever (#PE0985).
Aggravates Fever (#PD2255).
Aggravated by Vector–borne diseases (#PD8385)
Mosquitoes as vectors of disease (#PE1923)
Aedes mosquitoes as vectors of disease (#PE3621).

♦ PE2264 Glaucoma

Nature The principal symptom of this eye disease is an increase in intra–ocular pressure, with a reduction of vision. Glaucoma usually develops in the elderly and is the most common cause of blindness.

Incidence Glaucoma is responsible for 20 percent of blindness in Pakistan and 14 percent of blindness in both eyes in the United Kingdom and 11.6 percent in the U.S.

Refs Greve, E L (Ed) *Glaucoma and Cataract* (1986); Luntz, Maurice H (Ed) *Glaucoma* (1989); McAllister, James and WIlson, R *Glaucoma* (1986); Merté, H–J (Ed) *Genesis of Glaucoma* (1978); Ritch, et al *The Glaucomas* (1988); Winograd, Jesse S *Glaucoma* (1987).

Broader Eye diseases and disorders (#PD8786).
Narrower Primary glaucoma (#PG4523) Secondary glaucoma (#PG4524)
Congenital glaucoma (#PG4522).
Aggravates Physical blindness (#PD0568).
Aggravated by Rubella (#PE0785) Hypertension (#PE0585)
Arteriosclerosis (#PE2210).

♦ PE2265 Speech disorders

Language disorders — Speech defects — Dysphasia — Stuttering — Stammering — Twangs — Aphonia — Cluttering

Nature Mutism, autism and dysphasia are among the severe impairments of communication abilities. Specific impairment of voice production may be due to laryngeal deficiency, artificial larynx, laryngeal palsy or from deficiencies in other speech organs. The indistinct speech that results may be characterized by drawling, mumbling, slurring, etc. Impairments of voice function include absent or irregular voice modulation, pitch, intonation or volume; and harsh or other deficient qualities. Impaired speech forms includes stuttering, stammering, and similar disorders.

Claim As is generally known, speech disorders may have psychological causes. Thus, just as there are normal behavioural and speech ranges, atypical speech may be viewed as a disorder. In this class fall excessively copious and rapid speech, unusual speech patterning, phonation and resonation. This class of speech disorders may also include distorted grammar, lack of logical connection, sudden irrelevancies, interminable digressions, and answering off the point. Non–social speech, in this category, is talking, muttering or whispering out loud; and impaired conversation is exhibited by simultaneous talking or talking out of turn. Impaired speech content includes idiosyncratic uses of phrases and terms, neologisms, excessively recurrent punning, rhyming, joking or impaired humour, singing and irrelevant location.

Refs Agranowitz, Aleen and McKeown, Milfred R *Alphasia Handbook* (1975); Brain, Walter R *Speech Disorders*; Cantwell, Dennis and Baker, Lorian *Developmental Speech and Language Disorders* (1987); Code, Chris and Muller, Dave J (Eds) *Second International Aphasia Rehabilitation Congress* (1987); Grunwell, P *The Nature of Phonological Disability in Children* (1981); Keith, R W (Ed) *Central Auditory and Language Disorders in Children* (1985); Ludlow, Christy L and Cooper, Judith A *Genetic Aspects of Speech and Language Disorders* (1983); McDowell, Elizabeth V *Educational and Emotional Adjustments of Stuttering Children*; Peterson, H and Marquardt, J *Appraisal and Diagnosis of Speech and Language Disorders* (1981); Preus, Alf Edward *Identifying Subgroups of Stutterers* (1981); Sparks, S N *Birth Defects and Speech–Language Disorders* (1984); St Louis, Kenneth O *The Atypical Stutterer* (1985); Stromsta, Courtney *Elements of Stuttering* (1986); Wyke, Maria A *Developmental Dysphasia* (1979).

Broader Mental illness (#PC0300) Stigmatized diseases (#PD7279)
Human disease and disability (#PB1044).
Narrower Mutism (#PE4526) Motor aphasia (#PE0345) Language disorders (#PE3886)
Occupational diseases of the voice (#PE6866) Developmental articulation disorder (#PE9712)
Developmental receptive language disorder (#PE9300)
Developmental expressive language disorder (#PE5545).
Aggravates Limited verbal skills (#PD8123) Personal and social maladjustment (#PC8337).
Aggravated by Stroke (#PE1684) Deafness (#PD0659)
Hysteria (#PE6412) Laryngitis (#PE2653)
Cleft palate (#PE5117) Mental deficiency (#PC1587)
Cerebral paralysis (#PE0763) Emotional disorders (#PD9159)
Diseases and injuries of the brain (#PD0992) Facial or oral injury or deficiency (#PE8505).

♦ PE2268 Exclusion of pre–adults from family decisions

Nature Pre–adults are excluded from the family decision making process. They are regarded as possessions of their parents. The pre–adult years are regarded as non–productive and therefore children are exempted from responsibility for the family.

Broader Undemocratic policy–making (#PF8703) Lost family role in society (#PF7456)
Possessive attitude of parents (#PD1317).
Narrower Irresponsibility of young people towards the family (#PE1832)
Parental control of children's thoughts and reflections (#PF2060).
Related Parental permissiveness (#PD5344) Reduced interior structure of families (#PF3783)
Individualistic perception of sexual activity (#PF1682)
Individually defined operating structure of marriage (#PD2294).
Aggravates Insecurity (#PA0857) Loneliness (#PF2386).
Aggravated by Anxiety (#PA1635) Frustration (#PA2252)
Bereavement (#PF3516)
Failure to recognize uniqueness of family members (#PF1750).

♦ PE2272 Bright's disease

Nephritis — Glomerulonephritis

Refs Kincaid–Smith, Priscilla, et al *Progress in Glomerulonephritis* (1979).

Broader Kidney disorders (#PE2053)
Diseases of the genito–urinary system (#PC4575).
Related Arteriosclerosis (#PE2210) Infection of the throat (#PG4579)
Infection of the tonsils (#PG4580) Infection of the sinuses (#PG4577)
Systemic lupus erythematosus (#PG5084)
Haemolytic streptococcal infection of the throat (#PG3063).
Aggravates Stroke (#PE1684) Asthma (#PD2408) Uraemia (#PG2241)
Pneumonia (#PE2293) Hypertension (#PE0585) Heart failure (#PE3958)
Diseases of retina (#PE4584) Nutritional anaemia (#PD0321)
Urinary bladder disorders (#PE2307).
Aggravated by Allergy (#PE1017) Tonsillitis (#PE2292)
Climatic cold (#PD1404) Scarlet fever (#PG2757)
Diseases of the arteries (#PE2684).

♦ PE2274 Fragmented forms of care at the neighbourhood level

Nature Urbanization across the world with its resultant growth of anonymity and destruction of the extended family has created an urgent need for new interdependent structures to care for individuals and families at the local level. Neighbourhood based employment created both income opportunities but played a vital role in providing daily social contact among members of the community. These provided an invisible and unstructure but nonetheless effective form of community care. Public services when seen as beneficial to small segments of a community further divide. Housing projects tend to create division between those in the projects and those outside them. Isolation from other neighbourhoods further creates tensions.

Broader Lack of care (#PF4646) Social fragmentation (#PF1324).
Narrower Inadequate dental care (#PJ5478) Unused gathering places (#PG9389)
Mothers' self–imposed isolation (#PJ9425) Insufficient supervisory personnel (#PG9449)
Unavailability of community centres (#PG9399) Preoccupation with isolated problems (#PF6580)
Insufficient care of community property (#PF1600)
Inadequate medical care for pregnant women (#PE4820)
Inadequate community care for transient urban populations (#PF1844).
Related Inadequate care of community space (#PF2346)
Inadequate child day–care facilities (#PD2085)
Inflexible social care structures in developing countries (#PF2493).
Aggravates Limited youth activities (#PJ0106).
Aggravated by Inadequate welfare services for the aged (#PD0512).

♦ PE2280 Meningitis

Nature An inflammation of the meninges and brain, meningitis is caused by a variety of agents, including viral, bacterial and protozoon, and in non–infectious forms, by irritation of different sources (head injuries, spinal injections, overuse of alcoholic beverages). The clinical features are fever, petechial rash (only in the mericoccal form), vomiting, stiff neck and headache. Death may occur within 24 hours. Complications are common, especially when the diagnosis is delayed or treatment is inadequate. Collection of fluid in the subdural area, abscess formation, hydrocephalus, and various types of neurological problems, along with mental retardation, are some of the common sequelae.

Incidence The main form of meningitis, during epidemics, meningococcal meningitis has been causing increasing concern in recent years owing to its changing patterns and the rise in prevalence in several parts of the world where it was previously not considered to be a public health problem. Meningococcus infection may occur at any season, although most cases occur in late winter and spring. It is primarily a disease of youth and especially of children under 10 years, though all ages may be affected. From 1939 to 1972, nearly a million people were affected by meningitis and over 150,000 died in the African countries that are wholly or partly located in that area of the Sahel and the savannah known as the 'meningitis belt' (where the rainfall is more than 300 mm and less than 1,100 mm per year). During 1982–1983 there were again serious outbreaks of cerebrospinal meningitis affecting a number of countries in both the tropical and the temperate zones. In 1989, an estimated 10,000 people died of a particularly virulent strain of meningitis in Ethiopia.

Refs Vedros, Neylan A (Ed) *Evolution of Meningococcal Disease* (1987); WHO Study Group *Cerebrospinal Meningitis control* (1976); Williams, J D and Burnie, J *Bacterial Meningitis* (1987).

Broader Infectious and parasitic diseases (#PD0982)
Diseases of the central nervous system (#PE9037).
Narrower Leptomeningitis (#PG4597) Pyogenic infections (#PG5610)
Syphilitic meningitis (#PG4601) Tuberculous meningitis (#PG4600)
Pneumococcol meningitis (#PG4598) Meningococcol meningitis (#PG4602).
Aggravates Rickets (#PG2295) Headache (#PE1974) Cerebral paralysis (#PE0763).
Aggravated by Mumps (#PE2356) Measles (#PE1603)
Syphilis (#PE2300) Ringworm (#PD2545)
Mastoiditis (#PG3510) Tuberculosis (#PE0566)
Malnutrition (#PB1498) Typhoid fever (#PD1753)
Pneumoconiosis (#PD2034) Whooping cough (#PE2481)
Bad ventilation (#PG4606) Adenovirus infections (#PE2355)
Unhygienic conditions (#PF8515) Naso-pharyngeal catarrh (#PG4605)
Diseases and injuries of the brain (#PD0992).

♦ PE2281 Leishmaniasis

Nature Leishmaniasis are mutilating and difficult to treat, and the disfigurement caused by cutaneous forms has a lifelong psychological impact. If left untreated, death may follow in some forms. These diseases represent a great danger for the health of children since this age group is more vulnerable and is also the group in which the risk of failure of diagnosis is the highest. Control efforts are insufficient and hampered by the wide diversity of transmission situations, each of which may require different control approaches. Moreover, leishmaniases control is usually hampered by ignorance of the true prevalence of the diseases and underestimation of the human suffering and invalidity they cause. In cutaneous leishmaniasis (Oriental sore, Baghdad boil) caused by leishmania tropica, the parasites invade the skin and subcutaneous tissue provoking deep ulcers. The mucocutaneous form caused by leishmania brazihensis, affects the mucous membranes of the nose and mouth causing severe ulceration and disfigurement. The more severe, generalized and often fatal form is visceral leishmanias; and organs rich in reticuloendethial cells are destroyed by leishmania donovani.

Incidence Leishmaniasis is found in tropical and subtropical areas all over the world. The disease is widely disseminated in South America and spreading elsewhere, notably in Africa; serious outbreaks took place in India and Kenya in 1977. The disease is spreading to areas never affected before, sometimes causing epidemics of alarming proportions. Because of their zoonotic aspects, both the cutaneous and visceral forms are liable to increase with urbanization and with the reclamation of vast expanses of forest or desert, as in the Amazon basin or in the USSR. Movements of migrants, temporary labourers, and any large-scale population displacement may result in a high incidence of infection among such groups. Increasingly, tourists visiting endemic areas are reported to return infected. Among the population in endemic areas, the permanent risk of an epidemic creates constant and genuine fear. The World health Organization estimates about 12 million cases of leishmaniasis in most parts of the world. More than 400,000 cases are reported each year. Some 350 million people in 80 countries are at risk of this disease.

Refs WHO *The Leishmaniases* (1984).

Broader Arthropod-borne diseases (#PE7796) Parasitic diseases in animals (#PD2735)
Infectious and parasitic diseases (#PD0982).
Narrower Visceral leishmaniasis (#PG4609) American leishmaniasis (#PG4611)
Leishmaniasis of the skin (#PG4608).
Aggravated by Sandfly (#PG4612) Uncontrolled urban development (#PC0442).

♦ PE2282 Kwashiorkor disease

Nature Kwashiorkor is one of the most important causes of ill-health and death among children in the tropics. Caused by protein deficiency, it typically affects the small child weaned to a diet consisting chiefly of starchy foods. Its onset is characterized by anaemia. The child is apathetic, the hair reddish-orange, the skin dry. Oedema develops and the liver is often enlarged. Protein malnutrition increases vulnerability to infectious diseases; it may lead to an adult predisposition to certain illnesses and to lasting impairment of brain development. Other causes of Kwashiorkor include intestinal malabsorption, chronic alcoholism, kidney disease, and other trauma resulting in the abnormal loss of body protein.

Incidence It is estimated that between 100 and 270 million children, commonly in the 6 month to 3 year age group, are afflicted; many die. It occurs throughout the world, mostly in developing areas.

Broader Liver diseases (#PE1028) Nutritional diseases (#PD0287)
Protein-energy malnutrition (#PD0339).
Aggravates Blood circulation disorders (#PE3830).
Aggravated by Hookworm (#PE3508) Ascariasis (#PE2395)
Unbalanced infant diets (#PE0691) Protein deficiency in cereals (#PE3147).

♦ PE2286 Pituitary gland disorders

Injury to pituitary tissue — Tumours of pituitary tissue

Broader Gland disorders (#PD8301) Human disease and disability (#PB1044).
Narrower Dwarfism (#PE2715) Gigantism (#PE3837).
Aggravates Obesity (#PE1177).
Aggravated by Over-activity of a benign or malignant tumour (#PS4634).

♦ PE2287 Pellagra

Niacin deficiency

Nature Pellagra is a nutritional disorder showing a number of nervous, digestive and skin symptoms. It is due to deficiency of the nicotinic acid component of the vitamin B in association with deficiency of protein. It occurs where people live on a diet of maize without adequate first class protein. The cause of the disease lasts many years with digestive disturbances including loss of appetite, diarrhoea or constipation, headache and irritability. The skin symptoms last about 2 weeks and consist of redness, the skin then remains rough, thickened and permanently brownish. For several years the disease may recur in the spring, gradually becoming more severe, with the patient slowly growing emaciated and in some cases completely paralytic or demented.

Incidence Pellagra is still prevalent in areas of Africa south of the Sahara where maize is the staple cereal. Once a serious problem in Egypt, it has now largely or completely disappeared. This decline has been mainly due to varied diet containing more wheat and other cereals and less maize. Pellagra has been observed in people eating another cereal, jowar (Sorghum vulgar), especially in the Indian city of Hyderabad. Pellagra continues to be endemic in Mediterranean countries, the Far East, Africa, Mexico and the southern United States; and it can be seen as a complication of chronic alcoholism in any part of the world.

Broader Nutritional diseases (#PD0287) Vitamin deficiencies in diet (#PD0715).
Aggravated by Vitamin B deficiency (#PG4635) Diseases of metabolism (#PC2270)
Protein-energy malnutrition (#PD0339).

♦ PE2292 Tonsillitis

Nature Tonsillitis, an inflammation of the tonsils, may be acute or chronic, chronic tonsillitis frequently following episodes of catarrhal tonsillitis or of other infectious diseases such as scarlet fever, measles, and diphtheria. Acute tonsillitis is self-limited and usually lasts five days; chronic tonsillitis may require the surgical removal of the tonsils, called tonsillectomy.

Broader Children's diseases (#PD0622) Acute respiratory infections (#PE7591)
Diseases of upper respiratory tract (#PE7733)
Respiratory diseases of small animals (#PE9854).
Related Adenovirus infections (#PE2355).
Aggravates Asthma (#PD2408) Quinsy (#PG4646) Sore throat (#PE4651)
Rheumatic fever (#PE0920) Bright's disease (#PE2272) Diseases of the ear (#PD2567)
Peritonsillar abscess (#PG4647) Infection of the sinuses (#PG4577).
Aggravated by Diphtheria (#PE8601) Malnutrition (#PB1498)
Scarlet fever (#PG2757) Physical unfitness (#PD4475)
Malignant neoplasms (#PC0092) Inadequate fresh air (#PJ4652)
Execution of inappropriate orders (#PF2418).

♦ PE2293 Pneumonia

Bronchopneumonia

Refs Lambert, H P and Caldwell, A D *Pneumonia and Pneumococcal Infections* (1980); Stannic, Dominic L *Pneumonia* (1987).

Broader Lung disorders and diseases (#PD0637)
Bacterial diseases in animals (#PD2731)
Diseases of the respiratory system (#PD7924).
Narrower Acute congestion of lungs (#PG2675).
Aggravates Headache (#PE1974) Pleurisy (#PG6121)
Heart diseases (#PD0448) Woolsorters' disease (#PG4653)
Occupational diseases (#PD0215) Sudden unexpected infant death (#PD1885)
Child mortality in developing countries (#PE5166)
Individually defined operating structure of marriage (#PD2294).
Aggravated by Measles (#PE1603) Mycosis (#PE2455)
Rickets (#PG2295) Smallpox (#PE0097)
Diabetes (#PE0102) Ringworm (#PD2545)
Dystrophy (#PE3506) Ornithosis (#PE2578)
Tularaemia (#PE6872) Common cold (#PE2412)
Human ageing (#PB0477) Tuberculosis (#PE0566)
Typhoid fever (#PD1753) Alcohol abuse (#PD0153)
Pneumoconiosis (#PD2034) Whooping cough (#PE2481)
Limestone dust (#PE2038) Bright's disease (#PE2272)
Chronic bronchitis (#PE2248) Injurious accidents (#PB0731)
Systemic lupus erythematosus (#PG5084).

♦ PE2298 Skeletal system disorders

Refs Abramson, D I and Miller, D S *Vascular Problems in Musculoskeletal Disorders of the Limbs* (1981).

Broader Diseases of connective tissue (#PD2565).
Related Kidney disorders (#PE2053).
Aggravates Rickets (#PG2295) Rheumatic diseases (#PE0873)
Malignant neoplasms (#PC0092) Generalized osteitis fibrosa (#PG4683).
Aggravated by Leprosy (#PE0721) Syphilis (#PE2300)
Tuberculosis (#PE0566) Malnutrition (#PB1498)
Fungal diseases (#PD2728).

♦ PE2300 Syphilis

Neapolitan disease — French pox — Great pox

Nature Syphilis, if untreated, is a serious chronic disease caused by the spirochete Treponema pallidum; it is usually transmitted by direct sexual contact. The final and most serious stage of the disease (more or less 10 years after infection) is characterized by: ulceration of skin, palate, and bones, with scarring and disfigurement; involvement of the main blood vessel (aorta), leading to angina pectoris and serious or fatal heart disease; involvement of the nervous system, with resulting epileptic fits, insanity, blindness, deafness, unsteadiness in walking, crippling joint involvement, bladder disturbance, ulcers on the soles, and palsies. Congenital syphilis may result in miscarriage, stillbirth, or an infant born with syphilis who may develop deafness and blindness as well as other debilitating illnesses. In poor communities, syphilis can be a non-venereal disease, transmitted by overcrowding and by the use of common eating and drinking vessels (nonvenereal treponematoses include endemic syphilis, yaws and pinta). As social conditions improve and the chances of non-venereal contact diminish, syphilis evolves into a venereal disease, although the persistence of unhygienic practices or deteriorating social conditions may result in mixed venereal and non-venereal syphilis.

Incidence High prevalence rates for early syphilis occur wherever there is social disruption and mass population movements, as in some parts of South-East Asia and Africa. In developed countries, there has been a rise in the incidence of infectious syphilis. Congenital syphilis is still a serious disease resulting in foetal wastage, neonatal mortality and infant morbidity in countries where the services dealing with sexually transmitted diseases and with maternal and child health are poorly developed. A disconcertingly high prevalence of venereal syphilis has been observed as a consequence of the eradication of yaws in countries where syphilis control has not been carried out simultaneously.

Homosexual transmission is an epidemiological factor of increasing importance. In the USA the proportion of men with infectious syphilis who named other men as sexual partners has recently increased by almost 200 percent in cases of primary and secondary syphilis infections in males, and in Australia in 1973 homosexuals accounted for 73.2 per cent of the total male cases. In the United Kingdom, the proportion of early syphilis cases which had been acquired as a result of homosexual activity increased from 42.4% in 1971 to 54% in 1977. This association between homosexuality and syphilis transmission appears to be more pronounced in developed countries, although some reports from developing countries - for example Sri Lanka and India - also make reference to the importance of homosexual transmission. The high infection and reinfection rates among homosexuals make this relatively small group an important reservoir of infection which may contribute significantly to the transmission of syphilis in the community at large. Syphilis and other sexually transmitted infections that cause genital ulcers facilitate the spread of HIV-virus through homosexual and heterosexual contact.

Refs Brown, William J et al *Syphilis and Other Venereal Diseases.*

Broader Sexually transmitted diseases (#PD0061).
Narrower Early syphilis (#PG4573) Congenital syphilis (#PG7275)
Cardiovascular syphilis (#PG0121) Syphilis of central nervous system (#PG1801).
Aggravates AIDS (#PD5111) Ulcers (#PE2308) Deafness (#PD0659)
Paralysis (#PD2632) Meningitis (#PE2280) Laryngitis (#PE2653)
Still-birth (#PD4029) Encephalitis (#PE2348) Mental illness (#PC0300)
Heart diseases (#PD0448) Liver diseases (#PE1028) Physical blindness (#PD0568)
Diseases of the arteries (#PE2684) Skeletal system disorders (#PE2298)

Disorders of joints and ligaments (#PD2283) Perinatal morbidity and mortality (#PD2387)
Diseases and injuries of the brain (#PD0992).
Aggravated by Pregnancy disorders (#PD2289) Inadequate emergency blood supply (#PE0366)
Discharge from sores of infected persons (#PE8510).

♦ PE2305 Negative consequences of shifting ecology on coastal communities

Nature Increasingly shifting ecology of the coastal terrains is a challenge to local communities in developing practical ways of adapting to the constant changes in their environment. Rechannelling upstream of a shallow river allows sea water to to flow further inland. Tidal flooding causes major erosion around houses and salinates much agricultural land, leaving it useless for many crops. The problem of drainage becomes progressively more acute. Shallow wells have a higher salt content, and fresh water may be found in only a few wells further inland. Road construction is impracticable, as it would require many major bridge spans; and tides determine water transport by anything larger than small fishing boats, implying adjustments to schedules and making transportation less adaptable to the community's needs. People are forced to locate farmland away from their villages and thus must travel back and forth. Finally, people begin to leave their villages altogether and relocate near their newly acquired land.
Broader Obstacles to community achievement (#PF7118).
Narrower Salty padi fields (#PG9573) Tidal water damage (#PJ9589)
Shallow river restrictions (#PG9599) Decreasing coconut production (#PG9547)
Tide determination of transportation (#PG9533).
Related Water salinization (#PE7837) Underdeveloped road network (#PE1055).
Aggravates Unsustainable development of coast zones (#PD4671).

♦ PE2306 Bombardment

Bombing
Nature Bombing is to attack with a destructive device containing explosive or incendiary material or gas by throwing, firing from heavy guns, dropping from an aircraft or placing in position.
Broader Conventional warfare (#PC4311).
Narrower Terrorist bombing (#PE2368).

♦ PE2307 Urinary bladder disorders

Cancer of the bladder — Chronic urinary infection
Nature The urinary bladder is subject to anomalies, obstructions, inflammations, calculi, fistulae, and tumours. Although bladder cancer is a relatively rare disease, there have been many reports of occupational bladder cancer in the chemical industry in many different countries. Aniline, benzidine, and 1– and 2– naphthylamine are likely causative agents. The manufacture of auramine and magenta presents excess risk as a bladder cancer hazard.
Refs Freeman, R M and Malvern, John *The Unstable Bladder* (1988); Hueper, Wilhelm C *Occupational and Environmental Cancers of the Urinary System*; Smith, Phillip H and Prout, George R *Bladder Cancer (BIMR Urology)* (1983).
Broader Infection (#PC9025) Occupational diseases (#PD0215)
Diseases of the genito–urinary system (#PC4575)
Malignant neoplasm of genito–urinary organs (#PE5100).
Narrower Cystitis (#PE0078) Urethritis (#PG9767).
Aggravates Peritonitis (#PE2663) Susceptibility to infection (#PG4694).
Aggravated by Smoking (#PD0713) Diabetes (#PE0102)
Spina bifida (#PE1221) Kidney disorders (#PE2053)
Bright's disease (#PE2272) Parasites of the human body (#PE0596).

♦ PE2308 Ulcers

Gastric ulcers — Duodenal ulcers — Peptic ulcers — Gastrojejunal ulcer
Nature An ulcer, which can be benign or malignant, is a lesion of the skin, any mucous membrane (throat, stomach, intestine, bladder, etc), or of the aorta and valves of the heart.
Gastric, duodenal or peptic ulcers is a chronic disease characterized by the formation of ulcerations in the stomach or duodenal wall.
Incidence They may occur at any age but develop most frequently in men between the ages of 25 and 50.
Broader Human disease and disability (#PB1044)
Diseases of the oesophagus, stomach and duodenum (#PE8624).
Aggravates Eczema (#PE2465) Peritonitis (#PE2663)
Perforation (#PG4700) Haemorrhage (#PE2239)
Loss of blood (#PJ2230) Caffeine abuse (#PE0618)
Chronic bronchitis (#PE2248) Malignant neoplasms (#PC0092)
Incapacity from work (#PG4696) Disorders of the bowel (#PE6553)
Lung disorders and diseases (#PD0637).
Aggravated by Gout (#PG3061) Dropsy (#PG3235)
Leprosy (#PE0721) Allergy (#PE1017)
Diabetes (#PE0102) Syphilis (#PE2300)
Human ageing (#PB0477) Alcohol abuse (#PD0153)
Irritant fumes (#PD3672) Whooping cough (#PE2481)
Nutritional anaemia (#PD0321) Ozone as a pollutant (#PE1359)
Diseases of metabolism (#PC2270) Aerosols as industrial hazards (#PE1504).

♦ PE2309 Government imposition of rural cooperative projects

Government organized cooperatives — Failure of rural cooperatives
Aggravated by Unrecognized benefits from cooperatives (#PF9729).

♦ PE2310 Trichomoniasis

Nature Trichomoniasis is an inflammatory disease of the urogenital organs, caused by Trichomonas vaginalis, a flagellate protozoon. It is usually transmitted by sexual intercourse, and causes vaginitis, a superficial ulceration, inflammation and discharge. It may also occur in animals, causing miscarriages and sterility in cows (and resultant economic loss); poultry are also affected.
Incidence It has been estimated that about 20 percent of the female population of the United States is affected. The infection is more prevalent in institutionalized patients and lower socio–economic classes than in other groups.
Broader Sexually transmitted diseases (#PD0061)
Infectious and parasitic diseases (#PD0982).
Aggravated by Inferior classes (#PF7428) Institutionalized patients (#PG4705).

♦ PE2311 Trichinosis

Trichiniasis
Nature Trichinosis infects both man and animals, and can be fatal. Man usually becomes infected by eating uncooked or inadequately cooked infected meat (generally pork). The disease has been observed in more than 100 species of domestic and wild animals, including herbivores, carnivores and rodents, but is most common in swine.
Incidence The number of cases of trichinosis worldwide is about 30 million; almost all are in the USA, with minor outbreaks in Europe, Canada, Chile and Australia.
Refs Campbell, William C *Trichinella and Trichinosis* (1983).
Broader Ascariasis (#PE2395) Helminthiasis (#PE6278) Nutritional diseases (#PD0287).
Related Parasitic diseases in animals (#PD2735).
Aggravates Pleurisy (#PG6121).
Aggravated by Rats as pests (#PE3177) Unhygienic conditions (#PF8515)
Inadequately cooked foods (#PJ2614) Rodent vectors of disease (#PE3629)
Parasites of the human body (#PE0596).

♦ PE2317 Denial of the right to legal services for indigenous populations

Broader Biased legal systems (#PF8065) Denial of right to justice (#PC6162)
Discrimination against indigenous populations (#PC0352).

♦ PE2319 Byssinosis

Cannabosis
Nature Byssinosis is a chronic respiratory disease of cotton, flax and soft hemp workers. Recently it has been found also to occur in sisal workers preparing fibres prior to processing them. Byssinosis is characterized by chest tightness and breathlessness at work after the weekend break or other absence. In its late stages, which usually occur after many years of exposure to dust, the worker is severely disabled with symptoms of chronic bronchitis and emphysema. It occurs principally among those who clean and prepare fibres for spinning.
Incidence In the cotton industry byssinosis may affect workers in the ginneries where the seeds are removed, the bale pressing plants and the mixing and card rooms where the fibres are cleaned and combed. As a result of the introduction of mechanical picking, which has increased the contamination of cotton with plant debris, and the speeding up of all processes, dust concentrations in workrooms have risen and the disease has recently been found in spinners, winders and weavers. Among flax workers in factories making linen, ropes and twines, byssinosis occurs only in the preparatory processes.
Byssinosis has been described among hemp workers in Spain, where it is called cannabosis. The risk of the disease among hemp workers appears to be confined to the processing of soft hemp, which is a fibre from the stem of the plant used for making ropes and twines.
In developing countries the risk of byssinosis is likely to increase. The building of new textile factories to process the natural fibres which they grow in abundance is an important part of their economic expansion.
Broader Pneumoconiosis (#PD2034) Lung disorders and diseases (#PD0637).
Aggravates Chronic bronchitis (#PE2248).
Aggravated by Smoking (#PD0713) Air pollution (#PC0119)
Inhospitable climate (#PC0387)
Environmental hazards from textile and clothing industries (#PE1103).

♦ PE2320 Bursitis

Nature Bursitis, an inflammation of a bursa (especially in the elbow and knee joints) is frequently caused by activities performed at work. The pain of bursitis can disable the worker, and at the very least, its cure demands cessation of that activity.
Broader Disorders of joints and ligaments (#PD2283).
Narrower Parson's knee (#PG4716) Miner's elbow (#PG4806)
Housemaid's knee (#PG4717).
Aggravated by Injuries (#PB0855).

♦ PE2324 Persistence of a technical state of war following cease–fire agreements

Cease–fire violations — Failure to sign a peace treaty following war — Persistence of hostilities following cease–fire agreements
Nature An armistice does not necessarily put an end to the state of war, especially if the parties fail to make arrangements for the time after the cease–fire. Belligerents may continue to blockade, seize goods, etc. until such time as final peace agreements are made. A serious violation of a cease–fire agreement gives the other party the right to resume hostilities.
Incidence A number of countries are still in a technical state of war, even though the wars in question were terminated many years ago.
Broader War crimes (#PC0747) Abuse of government power (#PC9104)
Unfulfilled treaty obligations (#PF2497).
Aggravates War (#PB0593) Persons missing in military action (#PE1397)
Non–repatriation of prisoners of war (#PE0948)
Non–recognition of foreign governments (#PF8040).
Aggravated by Enjoyment of war (#PF4034).

♦ PE2328 Limited market access due to the product differentiation of transnational corporations

Nature Product differentiation is an obstacle to market entry in that potential suppliers must either engage in advertising, which would require large and long–term capital expenditure, or fix their selling prices below those of established suppliers, which would then not be sufficient to cover average costs. Transnationals are in a better position to overcome market entry barriers caused by product differentiation since the exploitation of advantages arising from specific know–how in this field is one of the main reasons for the internationalization of production.
Broader Transnational corporation imperialism (#PD5891)
Restrictive business practices of transnational corporations (#PE5915)
Restriction of free market competition by transnational corporations (#PE0051).

♦ PE2329 Hydrogen sulphide as a pollutant

Hydrogen sulphide as a health hazard
Nature Although hydrogen sulphide is detectable at concentrations far below those considered toxic, due to its disagreeable odour, even at low concentrations it has an irritant action on the eyes and respiratory tract. Hydrogen sulphide is a toxic and strongly odorous substance formed in a number of industrial processes, notably petroleum refining, manufacture of pulp and paper, tanning, and sulphur dye manufacture. Intoxication may be hyperacute, acute, subacute or chronic. Inhalation of massive quantities of hydrogen sulphide will rapidly produce anoxia resulting in death by asphyxia; epileptiform convulsions may occur and the individual falls apparently unconscious, and may die without moving again. This is a syndrome characteristic of hydrogen sulphide poisoning in sewermen; however, in such cases, exposure is often due to a mixture of gases including methane, nitrogen, carbon dioxide and ammonia.
Broader Malodorous fumes (#PD1413)
Chemical pollutants of the environment (#PD1670).
Related Thiols and mercaptans as pollutants (#PE5436).
Aggravates Human disease and disability (#PB1044).
Aggravated by Badly laid out work premises (#PJ2468).

♦ PE2332 Albinism

Nature Albinism, which is the complete absence of pigmentation in the skin and the retina of the eye provoked by a congenital defect of mutational origin in the pigmentary system, leaves the sufferer without the light screen, radiation protection, and, for animals, protective colouration which normal pigmentation provides. Defective vision, photophobia and nystagmus are additional handicaps. As a result, albino animals are rarely able to survive in the wild. In man, at least three different forms of albinism occur, depending on the degree of lack of melanin: universal complete albinism, involving the skin, hair and eyes; ocular albinism, in which only the eyes are affected; and localized albinism, the commonest form, in which small areas of the body lack pigment (white locks, spotlings). Albinism is a serious disablement, especially among non–white groups, because an albino's physical appearance is completely in contrast with that of the other

members of community. exposing him to mockery, discrimination and social exclusion.
Broader Bad omens (#PF8577) Congenital disorders of amino–acid metabolism (#PE9291).
Aggravates Discrimination (#PA0833) Mental deficiency (#PC1587)
Physical feebleness (#PG4732).
Aggravated by Genetic defects and diseases (#PD2389).

♦ PE2339 Cobalt as a pollutant

Nature Inhalation of cobalt fumes and absorption of cobalt salts produces systemic poisoning with myocardial disorders and irritant effects on the eyes, and on the respiratory and digestive tracts; inhalation of cobalt dust produces an asthma–like disease and fibrotic pulmonary lesions; and allergic dermatitis has been reported in workers exposed to cobalt. In the process of concentration of the cobalt ore, workers are exposed to dust and fumes containing both cobalt and other metals and metalloids such as arsenic and nickel. Carbon monoxide is formed during melting, and hydrogen sulphide is used for the precipitation of copper. Melting and pouring cobalt before pelletizing also produces cobalt fumes. Dust containing cobalt together with tungsten, titanium and tantalum is a potential hazard in the production of cemented tungsten carbides and the grinding and sharpening of cemented carbide tools. Radioactive cobalt does not exist in nature but is prepared in nuclear reactors and is used as a gamma–emitter in industry and medicine.
Incidence Cobalt is a relatively rare metal. The most important mineral sources are the arsenides, the sulphides, and various oxidized forms. The main producers are Zaire, Canada, Morocco, Finland, the USSR and Zambia. The principal consumer of cobalt is the United States which uses about half of world production.
Broader Toxic metal pollutants (#PD0948) Chemical contaminants of food (#PD1694)
Environmental hazards from metals (#PE1401).
Aggravates Occupational risk to health (#PC0865)
Human disease and disability (#PB1044)
Carcinogenic chemical and physical agents (#PD1239).

♦ PE2348 Encephalitis

Viral encephalitis — Equine encephalitis — St Louis encephalitis — Japanese B encephalitis — Murray Valley encephalitis — California encephalitis — Russian spring–summer encephalitis — Tick–borne encephalitis
Nature Encephalitis refers to any inflammatory process involving the brain.
Epidemic encephalitis, popularly known as sleeping sickness, was first described in Vienna in 1917; by 1918 the epidemic had reached Germany and Great Britain, and by 1920 the whole world. There was another peak incidence in 1924, but if the virus that is presumed to have been the etiologic agent now exists at all, it is not known to have produced an epidemic since 1927.
Japanese encephalitis, is virus infection of the brain that occurs most often in summer epidemics. The brain stem, basal ganglia, and white matter of the cerebral hemispheres are mainly involved. Mortality is high (50–60%) but recovery, when it does occur, is rapid (10–14 days) and ordinarily complete. Mosquitoes are the vector for the virus.
Incidence Japanese encephalitis is a public health problem of increasing concern to countries of South–East Asia and the Western Pacific. Thousands of cases are occurring annually, with high case–fatality rates. The disease is expanding into new areas.
Refs Alvord, Elssworth C Jr, et al (Eds) *Experimental Allergic Encephalomyelitis* (1984); WHO *Tick–Born Encephalitis and Haemorrhagic Fever with Renal Syndrome in Europe* (1986); WHO *Tick–Borne Encephalitis and Haemorrhagic* (1986).
Broader Virus diseases (#PD0594) Diseases and injuries of the brain (#PD0992)
Diseases of the central nervous system (#PE9037).
Related Malaria (#PE0616) Chronic fatigue syndrome (#PE1914)
Viral diseases in animals (#PD2730).
Aggravates Paralysis (#PD2632) Peculiar gait (#PG4752)
Mental deficiency (#PC1587) Cerebral paralysis (#PE0763)
Disordered behaviour (#PE2076) Alcohol idiosyncratic intoxication (#PG9365).
Aggravated by Plague (#PE0987) Rabies (#PE1325)
Measles (#PE1603) Cholera (#PE0560)
Diabetes (#PE0102) Morphine (#PG4760)
Syphilis (#PE2300) Influenza (#PE0447)
Dysentery (#PE2259) Hepatitis (#PE0517)
Brucellosis (#PE0924) Louping ill (#PG6452)
Typhoid fever (#PD1753) Leptospirosis (#PE2357)
Toxoplasmosis (#PE3659) Ticks as pests (#PE1766)
Natural disasters (#PB1151) DDT as a pollutant (#PE5028)
Lead as a pollutant (#PE1161) Von Economo's disease (#PG4747)
Mercury as a pollutant (#PE1155) Arsenic as a pollutant (#PE1732)
African trypanosomiasis (#PE1778) Inoculation for viral diseases (#PJ4755)
Contamination of drinking water (#PD0235) Mosquitoes as vectors of disease (#PE1923)
Infectious and parasitic diseases (#PD0982) Abuse of sedatives and tranquillizers (#PE0139)
Aedes mosquitoes as vectors of disease (#PE3621)
Culicine mosquitoes as vectors of disease (#PE3623).

♦ PE2349 Pesticide intoxication

Pesticide poisoning
Nature All the available data indicate that acute pesticide poisoning will continue to be a cause of ill–health and death, especially in developing countries which are predominantly agricultural. Environmentalists and health officials, weighing the advantages of biodegradable organophosphorous pesticides against the less toxic but more persistent organochlorines, have officially rejected many organochlorines for ecological reasons and substituted other compounds with higher mammalian toxicity. As a result, more cases of human pesticide poisoning are to be expected.
Severity and frequency of intoxication depend upon the mode of pesticide application and its handling. Industrial hazards in the manufacture, packing and transport, as well as public health hazards, result from the chemical nature of the pesticide and its method of application. Industrial hazards also involve contact with other toxic chemicals that contribute to the ultimate synthesis of the end product. Among factory workers, systemic uptake is likely to occur by inhalation and through the skin, or by mouth. In the countryside the farm workers are often working in the fields after the use of pesticides and their families living near fields receive a dose of chemicals sprayed by air. The possible long term effects of pesticide poisoning are carcinogenic, teratogenic, mutagenic and environmental.
Incidence According to the World Health Organization, about one million people throughout the world are unintentionally poisoned by pesticides every year. Most nonsmokers in developed countries probably receive their highest exposure to harmful chemicals from pesticide residues in the food they eat. In developing countries an estimated 80,000 people die each year from pesticide poisoning, and about 400,000 suffer acutely.
Recently there have been an alarming number of ecological disasters caused by pesticides. One, for example, was a major fish kill in Thailand involving many millions of fish due to contamination of rivers by the herbicide paraquat. This kill wiped out the major source of protein for many of Thailand's people. Another disaster in the central Indian city of Bhopal in December 1984 may have been the deadliest industrial accident world–wide. An American built insecticide plant leaked poisonous gas that within hours killed or fatally injured at least 1,200 local residents, and blinded, sterilized or otherwise sickened thousands of others. Many victims were people living adjacent to the plant. More than half of the patients admitted to UK hospitals between 1970 and 1982 for poisonings were children under 5 years of age and the bulk of them had consumed garden and home pesticides left around the house.
Refs Sim, Foo Gaik *The Pesticide Poisoning Report* (1985).
Broader Hazards to human health (#PB4885) Pesticides as pollutants (#PD0120).
Narrower Insecticide poisoning (#PG6682).
Related Pesticide damage to crops (#PD2581).
Aggravates Morbidity (#PD4538) Human poisoning (#PD0105)
Occupational diseases (#PD0215)
Destruction of crop pollinating species (#PE3880).

♦ PE2353 Actinomycosis

Nature Actinomycosis probably occurs in most parts of the world, although complete data are not available. The mode of transfer of the organism from one animal to another, from man to man, or from other environment to animal or man is not known. No consistent source has been found outside of animals or man, and because the organisms are more frequently found in the mouth of man and animals, the source of infection is believed to be endogenous. The transfer in this case must be by contact with carriers of the organisms and with fomites. Infected individuals no doubt also act as passive carriers at least.
Broader Mycosis (#PE2455) Fungal diseases (#PD2728).
Aggravated by Ringworm (#PE2545) Contaminated saliva (#PG4765).

♦ PE2354 Hydatid disease

Echinococcis
Nature Hydatid disease attacks mainly the liver and lungs. It can seriously affect sheep. Humans generally become infected by handling infected sheep or dogs.
Refs Mufti, El M *Hydatid Disease* (1989).
Broader Zoonoses (#PD1770) Infectious and parasitic diseases (#PD0982).
Aggravated by Parasites of the human body (#PE0596).

♦ PE2355 Adenovirus infections

Adenoids
Nature Adenovirus infections can affect the respiratory system, gastrointestinal system, glands, eyes or other organs, and can be fatal.
Broader Human disease and disability (#PB1044)
Diseases of upper respiratory tract (#PE7733).
Related Tonsillitis (#PE2292).
Aggravates Trachoma (#PE1946) Deafness (#PD0659)
Meningitis (#PE2280) Convulsions (#PG2611)
Gastric disorders (#PE1599) Diseases of the nervous system (#PC8756)
Diseases of the respiratory system (#PD7924).

♦ PE2356 Mumps

Epidemic parotitis — Cynanche parotidea — The branks
Nature Although mumps, an acute infectious disease, is not usually serious, rare and dangerous complications can occur. These include encephalitis, meningoencephalitis, and damage to the inner ear which leads to deafness. Children from age 5 to 15 are most susceptible.
Broader Virus diseases (#PD0594) Stigmatized diseases (#PD7279)
Infectious and parasitic diseases (#PD0982).
Aggravates Deafness (#PD0659) Meningitis (#PE2280)
Meningoencephalitis (#PG4775) Diseases of the pancreas (#PE1132).

♦ PE2357 Leptospirosis

Weil's disease — Canicola fever — Mud fever — Pea picker's disease — Swineherds disease
Nature Leptospirosis, an infectious disease produced by numerous antigenically distinct and morphologically identical bacteria called leptospires, infects humans and animals through contaminated water, food, or contact with a carrier. It is an occupational hazard for people involved in animal husbandry. It can be fatal.
Incidence A moist, warm environment, combined with alkaline to neutral pH of soil and water, favour the growth and maintenance of leptospires outside the animal host. Those working in such conditions are at risk.
Sex differences are not apparent where equal risk is present. However, men are more frequently affected than women because their occupations may bring them more often into contact with infected animals and a contaminated environment. The age group most commonly affected is that between 20 and 30 years. Individuals and young children may be affected if they swim or play in contaminated ponds or streams.
Refs Ellis, W A and Little, T W (Eds) *The Present State of Leptospirosis Diagnosis and Control* (1986).
Broader Occupational diseases (#PD0215) Spirochaetal diseases (#PE3254)
Zoonotic bacterial diseases (#PD6363).
Narrower Leptospirosis in dogs (#PG6563) Leptospirosis in cattle (#PG9246).
Aggravates Fever (#PD2255) Encephalitis (#PE2348).
Aggravated by Damp (#PG4479) Rats as pests (#PE3177)
Infected animals (#PG4778) Natural disasters (#PB1151)
Rodent vectors of disease (#PE3629) Contamination of drinking water (#PD0235).

♦ PE2365 Congenital heart disease

Refs Anderson, and Becker, *Pathology of Congenital Heart Disease* (1982); Bloor, Colin M and Liebow, Averill A *The Pulmonary and Bronchial Circulations In Congenital Heart Disease* (1980); Taussig, Helen B *Congenital Malformations of the Heart* (1960).
Broader Human physical genetic abnormalities (#PD1618).

♦ PE2368 Terrorist bombing

Incidence Terrorist bombings occur all over the world, and a multitude of groups are responsible. The IRA in Northern Ireland has one of the longest records of bombings. They have been responsible for the death of one of British royal family, Lord Mountbatten, and for an attempt on the life of the Prime Minister in 1984. There have been numerous civilian casualties as a result of indiscriminate bombing of public places, primarily in London.
Broader Terrorism (#PD5574) Bombardment (#PE2306).
Narrower Home–made bombs (#PG4790).
Related Mail bombs (#PG1130).
Aggravates Fear (#PA6030) Human death (#PA0072) Retaliation (#PF9181).
Aggravated by Political unrest (#PD8168) Suicide terrorists (#PG4023).

♦ PE2370 Abuse of tax havens

Nature Tax havens are countries (usually small) which introduce special tax legislation to permit individuals and transnational business enterprises to establish a base for financial operations involving the transfer of funds to and from other countries. It is to the advantage of corporations to establish their nominal headquarters in such countries so that, in the consolidation of their international accounts, the maximum income is declared in relation to the tax haven operations where the tax is minimal. In addition to the feature of no, or nominal taxes, tax haven countries usually offer political stability, good international communications, and financial, legal and other professional services. Thus they are centres of financial activity free of exchange controls. Abuse

of tax havens permits some enterprises to bypass tax treaty safeguards between two countries. Tax haven countries may also have more flexible laws regarding corporate organization and financial controls. When abused, these give rise to promotional and investment schemes that may be inadequately managed, or on occasion, to schemes of a fraudulent nature.
Counter–claim Tax havens play an important role in the development and operation of transnational corporations. They have provided a favourable basis for the development of an entirely new concept of investment in the form of offshore funds.
Broader Legal havens (#PE0621) Capitalist speculation (#PC2194)
Circumvention of duties and assessments (#PD4882).
Aggravates Tax evasion (#PD1466).
Aggravated by Unethical practices of employers (#PD2879).

♦ PE2371 Elevated blood cholesterol
Nature Cholesterol is a fat like substance produced mostly in the liver and important in the structure of cells throughout the body as well as the manufacture of various hormones. Cholesterol also constitutes a major building block of the waxy atherosclerotic deposits that grow inside arteries. When these deposit develop in crucial arteries, such as, the coronary vessels of the heart they can lead to heart attacks and strokes. There are two ways cholesterol gets to the blood stream. Consumption of foods with high amounts of cholesterol like eggs, and rich cuts of beef and pork add to the cholesterol level. When large amounts of saturated fats, in foods such as butter, beef and bacon, people's livers produce increased levels of cholesterol.
Counter–claim Men who attempt to lower cholesterol levels are more likely to be murdered, have fatal automobile accidents or commit suicide than those who do not.
Broader Excessive consumption of fats (#PE4261).
Aggravates Heart diseases (#PD0448).
Aggravated by Over–eating (#PE5722).

♦ PE2373 Coagulation disorders
Broader Diseases of blood and blood–forming organs (#PF8026).
Narrower Haemophilia (#PE1920) Haemorrhagic fibrinolysis (#PJ4186).
Aggravates Heart diseases (#PD0448).

♦ PE2380 Scurvy
Vitamin C deficiency
Incidence Scurvy, formerly associated primarily with long sea voyages and polar expeditions, is still common in areas where drought has destroyed fruit and vegetable crops. A decline in the habit of breast–feeding infants has also led to an increased susceptibility to scurvy.
Refs Carpenter, Kenneth J *The History of Scurvy and Vitamin C* (1988).
Broader Nutritional diseases (#PD0287) Vitamin deficiencies in diet (#PD0715).
Aggravates Natural human abortion (#PD0173).
Aggravated by Ascariasis (#PE2395) Diseases of metabolism (#PC2270).

♦ PE2382 Art forgery
Nature An art forgery causes a purchaser to be defrauded and the public deceived. The forgery may have been created for the purposes of fraud, or it may have been incorrectly attributed through ignorance. In some cases duplicates are made as duplicates and later used to deceive purchasers. The increased interest in art collecting and record breaking prices for art and antiques is serving the interests of con artists and fraudsters.
Incidence The biggest obstacle to collecting African art is the number of fakes available through tourist shops. African masks may be carved and aged by placing them in termite hills, smoked over cooking fires, stained with soot and water, rubbed with shoe polish, and sanded smooth or rubbed with sweat marks to show wear.
Refs Koobatian, James (Comp) *Faking It* (1987).
Broader Forgery (#PD2557).
Narrower Literary forgery (#PE6188).
Related Historical forgery (#PE5051).

♦ PE2388 Onchocerciasis
River blindness
Nature Apart from its generally debilitating effects, onchocerciasis frequently causes eye lesions leading to impairment of vision. The disease is caused by a nematode worm parasite, Onchocerca volvulus, transmitted mainly by members of the Simulium damnosium complex of blackfly species found near rapid rivers. Repeated superinfections result in blindness and also in a skin condition known as 'craw–craw', which causes intense itching. Resistance to certain larvacides renders the control of the vector flies a very difficult task. Moreover, water impoundment schemes have created new breeding places thereby intensifying endemicity.
Incidence Human onchocerciasis occurs in Africa and in the Americas. In Africa the disease is present in a wide belt south of the Sahara from Angola in the west to Tanzania in the east. In the Americas, important foci exist in Mexico, Guatemala, Colombia and Venezuela, and also among small human groups of the Amazonian forest. Blindness rates may attain dramatic proportions in heavily infected villages, particularly in the older age groups. In some areas up to 20 percent of villagers aged between 30 and 60 years are blind. The greatest economic losses occur when horses and cattle are affected. In 1990 it was estimated that 40 million people were infected by leprosy, Chagas' disease and river blindness.
Refs WHO *WHO Expert Committee on Onchocerciasis Third Report* (1987).
Broader Filariasis (#PE2391) Helminths of the skin (#PE5380)
Uncontrolled tropical diseases (#PD4775).
Aggravates Trachoma (#PE1946) Keratitis (#PE6789)
Shoroiditis (#PG4837) Physical blindness (#PD0568)
Inflammation of the uveal tract (#PG1678)
Diseases of the skin and subcutaneous tissue (#PC8534).
Aggravated by Black flies as pests (#PE3646) Vector–borne diseases (#PD8385)
Parasites of the human body (#PE0596) Flies as vectors of diseases (#PE4514)
Fly–infested rivers and streams (#PG4841)
Ecosystem modifications due to creation of dams and lakes (#PD0767)
Increase in pests and diseases through perennial irrigation (#PE7832).

♦ PE2390 Irresponsible pharmaceutical advertising
Aggressive promotion of pharmaceutical products
Nature In their campaigns to promote their products, pharmaceutical companies may make inflated promises, offer "rewards" to purchasers or promoters (such as "educational seminars" or even cash bonuses), and provide incomplete and thus misleading information or instructions. Advertising aimed directly at the physician often lacks information on the contra–indications to, and toxicity and hazards of, the preparations concerned. Non–technical advertising aimed at the public tends to take advantage of the gullibility of many people (for example, in the case of advertising of medicines for slimming, for constipation, for sexual impotence or for loss of hair).
Incidence The problem occurs everywhere, but is particularly severe in the developing world. A well–known example is the decline in breast–feeding due to the promotion of powdered infant formulae which has increases the incidences of malnutrition and disease.
Broader Falsity (#PF5900) Irresponsibility (#PA8658)
Misuse of advertising (#PE4225).
Related Inadequate drug quality control (#PD2392).
Aggravates Violence against women (#PD0247) Inadequate information on drugs (#PF0603)
Excessive cost of medical drugs (#PE5755)
Excessive proliferation of medical drugs (#PD0644)
Marketing of banned pharmaceutical drugs in developing countries (#PE6036).
Aggravated by Inadequate testing of drugs (#PD1190).

♦ PE2391 Filariasis
Filariae — Filarial infection
Nature Filariasis comprises several diseases including blindness and disfigurement causing economic, social and personal impairment. The economic loss to a country from filariasis has never been calculated, but it would seem reasonable to assume that it must be considerably greater than the cost of an effective control programme. The disease is caused by several species of nematodes of the Filariidae family and is transmitted by haemophagous arthropods, such as mosquitoes, blackflies, and deerflies.
Incidence Filariasis affects at least 250 million people in the developing countries. Major foci continue to exist in Africa. Southeast Asia and Oceania. In the endemic foci in tropical Africa, in Yemen, and in Middle and South America, filariasis is estimated to number more than 30 million cases. The disease is relatively severe in Guatemala, Mexico, Venezuela, Brazil, Colombia and Northern Yemen.
Refs WHO *Control of Lymphatic Filariasis* (1987); WHO *Lymphatic Filariasis* (1984); WHO Expert Committee, Athens *WHO Expert Committee on Filariasis* (1973).
Broader Helminthiasis (#PE6278) Parasitic diseases in animals (#PD2735)
Infectious and parasitic diseases (#PD0982).
Narrower Onchocerciasis (#PE2388) Dracunculiasis (#PE3510).
Related Elephantiasis (#PE1601).
Aggravates Physical blindness (#PD0568) Diseases of the lymphatic system (#PD2654).
Aggravated by Parasites of the human body (#PE0596)
Mosquitoes as vectors of disease (#PE1923)
Culicine mosquitoes as vectors of disease (#PE3623)
Anopheline mosquitoes as vectors of disease (#PE3622)
Increase in pests and diseases through perennial irrigation (#PE7832).

♦ PE2395 Ascariasis
Roundworm infestation — Strongyloidiasis
Nature Because ascariasis is a very common disease, it often goes unrecognized or ignored not only by the infected person and his family, but also by medical workers; yet the complications of an ascariasis infection can be serious and even fatal. It is an infectious disease caused by a roundworm, the ascaris, sometimes nearly 30 cm in length, which is parasitic in the human intestine and in horses and swine. Among the grave consequences of ascariasis are surgical complications, liver abscess and bile–duct obstruction. In several regions the commonest cause of intestinal obstructions in children in Ascaris infection; these obstructions can be fatal.
Incidence Ascariasis is found throughout the world, but the degree of its seriousness varies. Ascaris is prominent parasite in both temperate and tropical countries; it is most common in warm countries where sanitation is poor or lax. Approximately 900 million of the world's population are infected. Although ascaris occurs at all ages, it is mostly found among children – who are more frequently exposed to contaminated soil than adults – and slightly more often among males. Prevalence rates vary greatly, even within the same country. In the early 1980s some of the highest prevalences reported included: Philippines 85–90 percent, Malaysia 82 percent, Thailand 70 percent, Indonesia 83 percent, Taiwan 50 percent, Brazil 58 percent, Colombia 59 percent, Costa Rica 40 percent, Nigeria 30 percent, India 20 percent, Republic of Korea 58 percent, Vietnam 45 percent, Iran 98 percent, Ethiopia 58 percent, South Pacific countries 35 percent.
Refs Crompton, D W, et al *Ascariasis and Its Prevention and Control* (1988); Grove, D I (Ed) *Strongyloidiasis* (1989).
Broader Helminthiasis (#PE6278) Parasites of the human body (#PE0596)
Nervous system diseases caused by parasites (#PE4683).
Narrower Trichinosis (#PE2311) Visceral larva migrans (#PG9206)
Cutaneous larva migrans (#PG3885)
Manson's eyeworm infection in poultry (#PG7126).
Related Parasitic diseases in animals (#PD2735)
Gastrointestinal infections in animals due to parasites (#PE0958).
Aggravates Scurvy (#PE2380) Peritonitis (#PE2663) Appendicitis (#PG2327)
Xerophthalmia (#PE2538) Liver abscess (#PG3229) Liver diseases (#PE1028)
Sleep disorders (#PE2197) Kwashiorkor disease (#PE2282)
Bile–duct obstruction (#PG4857) Medical complications (#PE2863)
Intestinal obstruction (#PG4858) Intestinal perforation (#PG4859)
Haemorrhagic pancreatitis (#PG4860) Diseases and injuries of the brain (#PD0992)
Diseases and deformities of the digestive system (#PC8866).
Aggravated by Hookworm (#PE3508) Inadequate personal hygiene (#PD2459)
Inadequate standards of living (#PF0344).

♦ PE2396 Restrictive pricing policies of transnational corporations
Nature Restrictive business practices arise in the context of the pricing policies of transnational corporations in two ways: the prices charged for the products sold on domestic markets; and the prices charged for the products supplied by way of intra–company transactions within the corporation (including affiliates).
Regarding the prices it charges on domestic markets, an important element is the corporation's ability to isolate one market from another by way of its territorial market allocation arrangements and associated restrictive practices. To the extent that it holds a monopoly or near monopoly position in a market, the prices charged will essentially be what the corporation considers each market will bear and, as such, can involve abuse of market power. Where a transnational corporation finds itself in an oligopoly market situation, it can use its market power in either of two ways. First it can set prices at such a level that its competitors will follow suit (namely price leadership). Alternatively, it can deliberately adopt lower prices than its competitors, and, if necessary, sustain losses for the sale of the product in question, with the object of increasing its market share and perhaps eliminating certain of its competitors. The latter is frequently called predatory pricing.
Broader Transnational corporation imperialism (#PD5891)
Unfair pricing by transnational corporations (#PE5855)
Restrictive business practices of transnational corporations (#PE5915).

♦ PE2397 Control of marketing and distribution channels by transnational corporations
Nature The very fact that transnational corporations operate in more than one market and generally hold dominant market power as producers and traders has enabled them to acquire extensive control of marketing and distribution channels. In a number of cases their power may have limited the formation or expansion of alternative channels, given their ability to maintain or raise barriers to the entry of outsiders. The marketing and distribution of commodities and manufactures in developing countries is being increasingly controlled by specialized transnational marketing or trading corporations. In fact, it has been said that marketing is the key contribution of transnational corporations to developing countries and that marketing barriers constitute the principal obstacle to their exports.

Broader Transnational corporation imperialism (#PD5891)
Control of industries and sectors by transnational corporations (#PE5831)
Restriction of free market competition by transnational corporations (#PE0051).
Related Restrictive business practices of transnational corporations (#PE5915).

♦ **PE2398 Heat disorders**
Sunstroke — Prickly heat — Heatstroke — Heat fatigue — Heat stress
Nature A high level of climatic or environmental heat, possibly accompanied by high workload, may result in a variety of bodily heat disorders. These include heat stroke, heat exhaustion (circulatory, water, salt or sweat deficiency), skin disorders (prickly heat), and psychoneurotic disorders (heat fatigue). A tendency to heat disorders may prevent movement and employment of unacclimatized people in regions of great climatic heat.
Permissible heat exposure threshold limit values have been recommended in terms of the Wet Bulb–Globe Temperature Index which most nearly correlates with the deep body temperature. As an example, the maximum temperatures for light work load are 30 deg C (continuous working), 32.2 deg C (25 per cent work and 75 per cent rest); for a heavy work load they are 25 and 30 deg C respectively.
Refs Hales, J R S and Richards, D A B (Eds) *Heat Stress* (1987); Khogali, Mustafa M, et al *Heat Stroke and Temperature Regulation* (1984).
Broader Stress in human beings (#PC1648) Human disease and disability (#PB1044).
Narrower Heat as an occupational hazard (#PE5720).
Aggravated by Climatic heat (#PC2460) Inhospitable climate (#PC0387)
Excessive environmental heat (#PD7977)
Environmental hazards of solar radiation (#PE3883).
Reduced by Absence of direct sunlight (#PG1918).

♦ **PE2399 Haemolytic disease of the new born**
Nature Haemolytic disease can be fatal to a newborn child. It is caused by the incompatibility of the mothers' blood with that of the foetus with respect to the rhesus factor. The most severe form of haemolytic disease, the edematous form, generally causes stillbirth, or death a few hours after birth (which is often premature). The icteric form, if not adequately treated, may retard the child's development. The mildest form, congenital anaemia, is curable if the necessary medicine and expertise is available. The risk of haemolytic disease increases with succeeding pregnancies, and also in pregnancies following abortion.
Aggravates Hepatitis (#PE0517) Cerebral paralysis (#PE0763)
Nutritional anaemia (#PD0321) Perinatal morbidity and mortality (#PD2387).
Aggravated by Rh factor (#PS4863).

♦ **PE2401 Separation anxiety**
Separation anxiety disorder
Nature The children of this disorder manifest persistent reluctance or refusal to travel independently away from the house or from other familiar areas. Even at home they may avoid being alone and stay all the time close to the parent, following him or her around the house and insisting someone to stay with them until they fall asleep. When separated from parents these children have unrealistic and persistent worry of never being reunited again with the parents or that some calamitous event will make the reunification impossible. Complaints of physical symptoms, e.g headache, nausea, or excessive distress are common when separation is anticipated or occurs.
Refs Gardner, Richard A *Separation Anxiety Disorder* (1985).
Broader Mental illness (#PC0300).
Related Fear of death (#PF0462).
Aggravated by Inhibited grief process (#PD4918).

♦ **PE2412 Common cold**
Dripping nose
Nature The various viral infections of the respiratory tract referred to as colds, and their complications, are responsible for much discomfort and disability, including work absenteeism and vast expenditures for drugs and medical services. Colds are caused by many different kinds of viruses. About 30 to 50 per cent are caused by viruses of the rhinovirus family, which includes more than 100 known kinds.
Incidence Colds are responsible for more lost work days than any other viral infection. On average adults have two colds a year and children five or six. And it occurs in wet and cold weather. People with impaired or weak immunity systems, such as people with AIDS, the very young and the very old are at particular risk. Smokers tend to get more frequent and severe colds as a result of damaged lungs. Babies may be at risk of death because their narrow airways are easily plugged by secretions pouring out in response to infection. Introverted people have more colds and spread more viruses when they are infectious. As an indication, the market for cold remedies (including inhalants and rubs) in the UK in 1986 was £34 million.
Refs Andrewes, Christopher *In Pursuit of the Common Cold* (1973); Cheslock, Charles J *Understanding the Common Cold* (1987).
Broader Virus diseases (#PD0594) Acute respiratory infections (#PE7591)
Unsociable human physiological processes (#PF4417).
Narrower Naso–pharyngeal catarrh (#PG4605).
Related Cough (#PE6825).
Aggravates Deafness (#PD0659) Headache (#PE1974)
Gastritis (#PG2250) Pneumonia (#PE2293)
Diarrhoea (#PD5971) Sinusitis (#PG4873)
Laryngitis (#PE2653) Absenteeism (#PD1634)
Underproductivity (#PF1107) Diseases of the ear (#PD2567).
Aggravated by Fog (#PE1655) Draughts (#PG4883) Overcrowding (#PB0469)
Malnutrition (#PB1498) Inhospitable climate (#PC0387).

♦ **PE2414 Financial frauds**
Incidence Advance–fee fraud is one of the most common financial frauds. A conman, posing as a agent for a bank, oil money etc, offers a business loan on favourable terms for an up–front fee. These "loans" are costing small businesses, farmers and even large corporations hundreds of millions of dollars a year in the USA.
Broader Fraud (#PD0486) Economic crime (#PC5624)
Frauds, forgeries and financial crime (#PE5516).
Narrower Unethical practices of philanthropic organizations (#PE8742).
Related Grant frauds (#PD4543).

♦ **PE2417 Proliferation and duplication of intergovernmental organizations and coordination bodies**
Nature There is widespread concern in both intergovernmental organizations and in governments at the haphazard proliferation of the former and of their activities. This concern arises from the increasing burden placed upon governments, particularly the smaller ones, by the multiplicity of intergovernmental organizations and their expanding activities; and from indications that amidst the present confusion the results achieved are neither commensurate with the effort involved, nor sufficiently productive.
In addition to the commitment of governments in financing these organizations, there exists the further burden of financing their participation, and the complex business of providing permanent or temporary missions, delegations to conferences, teams of specialists and experts to attend committees often of a highly technical nature, and of maintaining political control over all of this governmental supporting activity. In this process is involved the need to provide within the governmental machine, the machinery for briefing for all of these activities and the complex business of coordinating the activities of different departments of government. The providing of these services is an expense which, unlike direct contributions to budgets, cannot be scaled down and which thus weighs disproportionately upon the smaller governments.
However, more important for all governments than the financial burden is the difficulty of finding the necessary skilled and expert manpower (diplomatic, administrative or specialist) to perform such manifold tasks. In this respect, a majority of governments have already reached the limits of their capacity in an area where they themselves are in need of such people for their own internal purposes, and only the very largest are able to make the necessary provisions without difficulty.
Background An understanding of the problems which lie concealed behind the word 'cooperation', as applied to intergovernmental organizations is fundamental to any understanding of the problems which beset governments in an attempt to use a number of such bodies in an efficient manner. It is perhaps more accurate to assume that one intergovernmental organization as such cannot by its nature have any meaningful relationship with another, and to refer instead to relations between their secretariats within the limited range of the secretariats' functions and competence. Such cooperation between secretariats is certainly of value, though it is not an end in itself, but only an assistance to governments in the decision–making process. Though such cooperation is thus useful, experience shows that the results to be obtained from it are very limited.
Claim Both in finance and in qualified manpower, the rapid and uncontrolled growth of intergovernmental organizations tends to widen, rather than to narrow, the gap between the performance of the larger and the smaller governments, and the latter tend to be less and less able to participate effectively in intergovernmental work. This is precisely the opposite of what ought to be the case; multilateral organization ought to be one of the means by which smaller governments can increase their participation in the life of the international community; the present situation makes it increasingly difficult for them to do this, while it leaves the largest governments little affected.
The charge so often levelled against intergovernmental organizations by governments and others, that they 'duplicate' one another's work is in fact only a negative way of approaching the problem of cooperation. If two secretariats are frank with one another, then information can be put before the relevant intergovernmental organs of the bodies concerned which, theoretically, should enable to governments constituting them to take decisions which would avoid 'duplication' of effort. All experience shows, however, that under present conditions governments are unlikely to act in this way.
Counter–claim True 'duplication' of the kind for which no excuse could be found, that is to say two organizations being engaged in carrying out precisely similar projects in identical groups with identical objectives, is so rare as to be almost unheard–of. The same subject matter may be dealt with in one organization from a legal standpoint, in another from a financial one and in a third from a technical one, and the boundaries of these activities will inevitably overlap and will be difficult to define. Alternatively it may be alleged that whereas organization 'A' is dealing with many member states with widely varying interests and therefore in rather general terms, organization 'B,' with a much smaller and more homogeneous group of Member States, is studying it in depth. All of these differentiating factors may be perfectly valid.
Broader Proliferation and duplication of international organizations and coordinating bodies (#PE1029).
Aggravates Competition between intergovernmental organizations for scarce resources (#PE0063).

♦ **PE2425 Lack of finance in coastal communities**
Nature In some fishing communities, although there is full awareness of the need to expand existing but depleted fishing grounds, the small boats are incapable of covering the distance to the sea or of using large enough nets. Residents of coastal communities are frustrated by lack of capital which prevents them increasing individual potential or starting a new trade to replace the declining fishing industry. There is similar difficulty in acquiring improved equipment or new investments in more diverse economic arenas. Agriculture, considered the most practical alternative to fishing, is stifled by local people's limited knowledge on obtaining the necessary loans for equipment. Building of roads and bridges, to remove barriers of physical isolation and increase inter–village commerce, cannot be funded by the community's meagre income. The low restrictive capital sources discourage storekeepers from venturing in the area of slow turn–over merchandise or buying in large enough quantities to decrease their long–term investments. For those villagers with any assets, the local economy is seen as a bad business risk, forcing investments elsewhere.
Broader Insufficient financial resources (#PB4653)
Limited availability of investment capital for urban renewal (#PF3550).
Narrower Restricted store capital (#PG9542) Inadequate improvement funds (#PG9506).
Aggravates Unsustainable development of coast zones (#PD4671).
Aggravated by Absentee ownership (#PD2338).

♦ **PE2427 Economic exploitation of developing countries by industrialized countries**
Claim Although industrialized countries may not deliberately set out to exploit poorer countries, they tend to practice an approach to development which is almost entirely in their own national interest. In so doing they involve the privileged classes of Third World countries as willing accomplices. The productive capacity of the developing countries is then organized so as to produce for these groups and only incidentally, if at all for the impoverished classes of those countries. **Counter–claim** The richer nations do not exploit, rather they facilitate the development of the poorer countries by encouraging a free enterprise system based on the profit motive to stimulate the economic involvement of the impoverished classes in the development process.
Broader Economic exploitation (#PC8132).
Aggravates Disparity between industrialized and developing countries (#PC8694)
Reinforcement of inappropriate development by privileged classes in developing countries (#PF6670).

♦ **PE2428 Sciatica**
Refs Golding, D N and Barrett, J *The Practical Treatment of Backache and Sciatica* (1984).
Broader Diseases of nerves and peripheral ganglia (#PE8932).
Aggravates Occupational rheumatism (#PE0502).

♦ **PE2429 Public drunkenness**
Acute alcoholic intoxication
Incidence Many individuals arrested for being drunk in public are alcoholics, homeless, or both.
Broader Drunkenness (#PE8311) Alcohol abuse (#PD0153)
Victimless crime (#PC5005).
Narrower Drunken pedestrians (#PG8377).
Related Substance intoxication (#PD4027).
Aggravates Injuries (#PB0855) Hangover (#PJ5020) Drunken driving (#PE2149).
Aggravated by Mental illness (#PC0300) Discrimination (#PA0833)
Social conflict (#PC0137) Socio–economic poverty (#PB0388).

♦ PE2432 Polychlorinated biphenyls as a health hazard

PCBs

Nature In people occupationally exposed to PCBs a broad spectrum of adverse health effects have been reported. These effects may be generally explained by the induction or the inhibition of the activity of a large number of enzymes which upset quantitatively normal biological processes. The prevalence of the adverse health effects increased with the concentration of PCBs in the working environment and thus in the workers' tissues. Reported effects are: changes in the skin and mucous membranes; swelling of the eyelids, burning of the eye, and excessive eye discharge; burning sensation an oedema of face and hands; simple erythematous eruptions with pruritus; acute eczematous contact dermatitis (vesiculo-erythematous eruptions); chloracne (an extremely refractory form of acne); hyperpigmentation of skin and mucous membranes (palpebral conjunctiva, gingiva); discoloration of finger nails; and thickening of the skin. Irritation of the upper respiratory tract is frequently seen. A decrease in forced vital capacity, without radiological changes, was reported in a relatively high percentage of the workers exposed in a capacitor factory. Digestive symptoms such as abdominal pain, anorexia, nausea, vomiting, jaundice, with rare cases of coma and death, may occur. At autopsy, acute yellow atrophy of the liver has been sporadically reported in lethal cases. Neurological symptoms such as headache, dizziness, depression, nervousness, etc, and other symptoms such as fatigue, loss of weight, loss of libido and muscle and joint pains has been found in various percentages of exposed people. Carcinogenicity of PCBs has been shown in animals, experimentally exposed. Marine mammals have a rising quantity of PCBs in their blubber.

PCBs get into the environment by incomplete burning, leaching from the soil, vapourizing from paints, coatings and plastics, by illegal dumping or by accident.

Refs United Nations *The Determination of Polychlorinated Biphenyls in Open Ocean Waters* (1985).

Broader Occupational hazards (#PC6716) Chemical contaminants of food (#PD1694)
Toxic organic compounds as water pollutants (#PE0617).
Aggravates Liver diseases (#PE1028) Food pollution (#PD5605)
Water pollution (#PC0062) Marine pollution (#PC1117)
Pesticides as pollutants (#PD0120) Wildlife pollution hazard (#PJ3387)
Occupational risk to health (#PC0865) Endangered species of seals (#PE1656)
Endangered species of whale (#PD1593) Endangered polar bear species (#PE1320)
Concentration of noxious substances in food chains (#PE8154).
Aggravated by Sewage as a pollutant (#PD1414)
Industrial waste water pollutants (#PD0575).

♦ PE2435 Sonic boom generated by supersonic aircraft

Risk of sonic bangs

Nature When an aircraft flies at supersonic speed it produces a boom corridor some 60 to 100 kms wide in which the pressure experienced on the ground rises very sharply to a higher value through a shock wave. The pressure then drops until it is approximately as much below ambient pressure as it was above it. A second shock wave brings the pressure back to its normal atmospheric value. While absolute pressure change on the ground caused by the shock wave is relatively small, its onset is very rapid, creating a characteristic sharp report that is known as the sonic boom, and which can be heard (depending on atmospheric conditions, location of observer, flight altitude, or ground configuration), as anything between a sharp crack of a rifle shot and a distant rumble of thunder.

Claim Available evidence points to sonic booms as the source of considerable annoyance and complaints in the communities exposed to them at frequent and regular intervals. Sonic booms startle, interfere with sleep and concentration, cause irritation and affect people's aesthetic enjoyment of life. Sonic booms cause damage to property, particularly windows and plaster. The effects of sonic booms on the natural environment (particularly on triggering snow avalanches and injecting water vapour into the atmosphere) also pose problems.

Counter-claim Although supersonic aircraft have flown for over thirty years, until now they have not affected a significant portion of man's environment. This is because the use of supersonic aircraft has been largely limited to military aviation and their unintended effects on the environment in the form of sonic booms have been sporadic and confined, in most countries, to relatively small and isolated areas of land. The introduction of the supersonic aircraft into commercial service adds a new dimension to the problem.

Broader Aircraft noise (#PE5799).
Aggravates Human disease and disability (#PB1044).

♦ PE2436 Gastrointestinal parasites of animals

Broader Diseases of the digestive system in animals (#PD3978).
Narrower Digestive tract helminthiasis of poultry (#PG9355)
Gastrointestinal infections in animals due to parasites (#PE0958).

♦ PE2441 Radioactive contamination of water

Nature Except where water is drawn from deposits of highly radioactive minerals, its radioactivity from natural causes is usually low and of no immediate health significance. Pollution by radioactive wastes, however, may be highly dangerous. Radioactive material may be ingested directly through water supplies, but may also be present in more concentrated forms in fish, shellfish, or in plants irrigated with contaminated water

Artificial radioactive substances in water are derived from the fallout from nuclear testing, discharges from nuclear power reactors and reprocessing plants, and the disposal of radioactive wastes. The radionuclides of importance are strontium-90, caesium-137 and to some extent, iodine-131; but the concentrations of these radionuclides in drinking-water are normally very low. Radioactive contamination of reservoirs can result from underground explosions and nuclear fallout.

Broader Impurities in waste water (#PD0482) Radioactive contamination (#PC0229)
Chemical contamination of water (#PE0535).
Related Radioactive contamination of the marine environment and of fisheries products (#PE1431).
Aggravates Water pollution (#PC0062).

♦ PE2443 Inevitable destruction of natural environment by mankind

Broader Fatalism (#PF6430) Human destructiveness (#PA0832).
Aggravated by Mismanagement of environmental demand (#PD5429)
Environmental destruction in least developed countries (#PE8401)
Adverse consequences of scientific and technological progress (#PF3931).

♦ PE2446 Cirrhosis of the liver

Nature Cirrhosis of the liver is a chronic progressive disease involving destruction of the liver substance and injury to all its structural elements. Depending on the form and stage of the disease, individuals may suffer from weakness, emaciation, jaundice, fever and bleeding. Clubbed fingers, anaemia and leukaemia may also occur, and ultimately, death, which usually results from hepatic coma or haemorrhages from dilated veins of the oesophagus and stomach.

Background The term cirrhosis was proposed by R Laennec in 1819, to designate diseases in which the liver becomes tawny-coloured, wrinkled, and compressed. Classification is based on several features of the disease, including the cause, functional condition of the liver, and morphological picture of the lesions. It may be caused by infectious diseases (most notably hepatitis), poisoning (including alcoholic intoxication), protein deficient diets, and constitutional genetic characteristics.

Refs Boyer, J L, et al (Eds) *Liver Cirrhosis* (1987).

Broader Liver diseases (#PE1028)
Diseases and deformities of the digestive system (#PC8866).
Aggravates Dropsy (#PG3235) Hepatitis (#PE0517)
Diseases of the spleen (#PE6155) Unethical consumption practices (#PD2625).
Aggravated by Human ageing (#PB0477) Malnutrition (#PB1498)
Alcohol abuse (#PD0153).

♦ PE2449 Proliferation of weapons in civilian hands

Sale of weapons to civilians — Supplying firearms for criminal activity

Nature Knowingly supplying firearms or explosives to an individual or criminal organization who intend to commit a crime with the aid of the weapons is guilty of a crime. Supplying weapons for criminal activity greatly increases the possibility of injury and death in the progress of the crime.

Incidence Handguns are used to kill 21,000 American annually; an additional 12,000 die from handguns through suicide or accident. More Americans died from handguns in the two years 1987-88 than during the during 16 years of the war in Vietnam. A gun in an American home is 18 times more likely to kill a household member than an intruder.

Counter-claim Many people live where government cannot, or will not, enforce its proper monopoly on the use of force. Given this failure, or abdication, by government, it is discrimination that the safe majority deny the endangered minority the handguns needed for self-help. When the police enforce the law in the ghettos and other crime infested areas, then guns should be outlawed.

Refs Wright, James D; Rossi, Peter H and Daly, Katleen *Under the Gun* (1983).

Broader Firearms and explosives crimes (#PE1108).
Aggravates Armed robbery (#PE4739) Gunshot wounds (#PE9111)
Firearm accidents (#PE2857).
Aggravated by Police brutality (#PD3543) Abuse of police power (#PC1142)
Inadequate firearm regulation (#PD1970)
Imbalance of conventional armed forces (#PC5230).

♦ PE2451 Demoralizing constraints on housing rehabilitation

Broader Apathy (#PA2360).
Narrower Insecure lease tenure (#PJ8344) Deteriorated vacant houses (#PJ8678)
Restrictive building codes in urban areas (#PE8443)
Depressing effect of poor housing construction (#PF1213).
Related Alienating public housing assignments (#PJ9479)
Demoralizing image of urban community identity (#PF1681)
Demoralizing images of rural community identity (#PF2358).
Aggravates Inadequate housing (#PC0449).
Aggravated by Temporary residence (#PJ3760) Prohibitive cost of home maintenance (#PJ9022).

♦ PE2455 Mycosis

Mycoses — Mycotic disease

Nature Mycoses are caused by a variety of fungi which attack diverse structures of the body, and are divided accordingly. Superficial mycoses attack the stratum corneum of the epidermis only. Their natural habitat is either man, animal or soil. Infection is due to direct contact of the skin with the infected surroundings. In most instances, intermediate mycoses attack the epidermis, but they may also attack deeper structures. One of the most common offending organisms is a Candida species which is present in the digestive tract of man as a saprophyte but may be found elsewhere. Infection takes place by direct contact or by inhalation and spreads via the blood-stream. Deep mycoses attack primarily deeper structures but may metastatically also invade the epidermis. Most of the fungi causing these diseases have been found in soil.

Refs CAB International Mycological Institute *Mycoses of the Eye and Orbit*; CAB International Mycological Institute *Mycoses of the Heart*; WHO *Mycotic Diseases in Europe* (1987).

Broader Fungal diseases (#PD2728) Human disease and disability (#PB1044)
Infectious and parasitic diseases (#PD0982).
Narrower Mycetoma (#PG5054) Candidosis (#PE4923)
Moniliasis (#PG5292) Nacardiosis (#PG4931)
Madura foot (#PG5044) Coccidiosis (#PE2738)
Phycomycosis (#PG5053) Mucormycosis (#PG9715)
Geotrichosis (#PG0787) Balstamycosis (#PG4925)
Blastomycosis (#PE4928) Actinomycosis (#PE2353)
Aspergillosis (#PE5212) Chromomycosis (#PG9572)
Cryptococcosis (#PE4932) Sporotrichosis (#PG4935)
Histoplasmosis (#PG3696) Dermatomycosis (#PG9771)
Bhynosporidiosis (#PG4936) Rhinosporidiosis (#PG9835)
Chromoblastomycosis (#PG4922) Paracoccidioidomycosis (#PG4934)
Nocardiosis in animals (#PG6309) North American blastomycosis (#PG5620).
Aggravates Ringworm (#PD2545) Pneumonia (#PE2293)
Diseases of the ear (#PD2567).
Aggravated by Sores (#PG4940) Diabetes (#PE0102)
Leukaemia (#PE0639) Flies as insect pests (#PE2254)
Animals as vectors of disease (#PD8360)
Insalubrity of animal excrement in urban environments (#PE4685).

♦ PE2461 Glanders

Farcy

Nature An infectious disease found in animals and man with an acute or chronic course caused by the glanders bacillus. The sources of the infection are domestic animals. Glanders begins with chills, fever, and pain in the muscles and joints. Numerous pustules, which subsequently develop into ulcers, appear on the skin, especially on the face. They then develop in the internal organs. Hospitalization is imperative. The outcome is often fatal in its acute form.

Broader Zoonotic bacterial diseases (#PD6363) Human disease and disability (#PB1044)
Bacterial diseases in animals (#PD2731) Infectious diseases in animals (#PD2732).

♦ PE2462 Defects in machinery design

Unsafe automobile design and construction — Faults in reactor design — Unexplained technical faults

Broader Defective product manufacture (#PD3998)
Human errors and miscalculations (#PF3702).
Narrower Non-ergonomic design (#PG3697) Unsafe design of consumer products (#PF1379).
Aggravates Structural failure (#PD1230) Mechanical failure (#PC1904)
Road traffic accidents (#PD0079) Nuclear reactor accidents (#PD7579)
Electronic equipment failure (#PD1475)
Unreliability of equipment and machinery (#PC2297)
Continued operation of unsafe motor vehicles (#PE2240)
Health risks to workers in agricultural and livestock production (#PE0524).

♦ PE2465 Eczema

Dermatitis — Radiation dermatitis — Contact dermatitis — Dermatitis venenanta

Nature Eczema is a superficial disease of the skin. It is of an inflammatory nature, associated with

itching or even pain; and embraces about one-half of all cases of skin disease. It is often more generally called 'dermatitis'.
Occupational contact dermatitis, or occupational eczema, is a skin disease caused or favoured by exposure to chemical, physical or biological agents present in the work environment. From the legal standpoint, occupational eczema is any eczema produced or contributed to by noxious agents listed in the regulations concerning occupational diseases.
Incidence Occupational dermatoses account for over 50 percent of occupational diseases receiving compensation, while contact forms of dermatitis account for 80–90 percent of occupational dermatoses; of the latter, over 50 percent are attributed to allergy.
Refs Fregert, Sigfrid *Manual of contact dermatitis* (1981); MacKie, Rona M *Eczema and Dermatitis* (1983).
Broader Allergy (#PE1017) Diseases of the skin and subcutaneous tissue (#PC8534).
Related Asthma (#PD2408) Allergy inducing food (#PE3225).
Aggravates Malignant neoplasms (#PC0092).
Aggravated by Ulcers (#PE2308) Obesity (#PE1177)
Irritant fumes (#PD3672) Detergents as pollutants (#PE1087)
Fungicides as pollutants (#PD1612) Radioactive contamination (#PC0229)
Parasites of the human body (#PE0596) Health hazards of irradiated food (#PD0361)
Environmental hazards from chemicals (#PC1192).

♦ **PE2467 Anoxemia**
Oxygen deficiency in the blood
Refs Heath, Donald (Ed) *Aspects of Hypoxia* (1986).
Narrower Defective oxygen supply to the foetus (#PG9126).

♦ **PE2470 Illegal firearms**
Nature There are many people who are forbidden to possess firearms because of previous criminal activity or mental incapacitation. Knowingly supplying these people with firearms, explosives or ammunition is criminal offence.
Broader Firearms and explosives crimes (#PE1108).

♦ **PE2471 Unmaintained bridges**
Disrepair of bridges — Non-durability of bridges — Unrailed public bridges
Incidence It is estimated that at least 60 percent of New York's 800 bridges are in a dangerous condition.
Broader Inadequate road maintenance (#PD8557)
Inadequate maintenance of infrastructure (#PD0645).
Aggravates Structural failure (#PD1230)
Limited accountability of public services (#PF6574)
Health risks to workers in agricultural and livestock production (#PE0524).

♦ **PE2473 Radio noise of industrial origin**
Radio interference of industrial origin
Nature Clear reception of radio signals depends upon the ability to exclude unwanted noise. The chief cause of man-made noise is electrical circuit transients which cause radiation from portions of circuits where sparks occur. Typical causes are: switches, motors, ignition systems, X-ray apparatus, diathermy machines, industrial precipitators, high-frequency heating appliances, domestic appliances, etc. The problem may be aggravated by the use of a large number of such devices in the same area.
Broader Electromagnetic pollution (#PD4172) Environmental degradation (#PB6384).
Related Jamming of satellite communications (#PD1244).
Aggravates Lack of communication (#PF0816) Radio frequency interference (#PD2045).
Aggravated by Mechanization (#PU3486).

♦ **PE2481 Whooping cough**
Hooping cough — Pertussis — Chin-cough
Nature Whooping cough is an acute, highly communicable respiratory disease, worldwide in distribution and among the most acute infections of children. It can severely damage their health, growth, and resistance, and is the cause of more school absenteeism than any other infectious disease.
Refs Hemert, P A van; Ramhorst, J D van and Regamey, H (Eds) *International Symposium on Pertussis* (1970).
Broader Bacterial disease (#PD9094).
Aggravates Stroke (#PE1684) Ulcers (#PE2308) Pneumonia (#PE2293)
Meningitis (#PE2280) Convulsions (#PG2611)
Prolapse of the rectum (#PG4955) Child mortality in developing countries (#PE5166)
Inflammatory affections of the bronchial tubes and lungs (#PE8822).

♦ **PE2485 Statelessness**
Involuntary loss of nationality — Stateless persons — Involuntary loss of citizenship
Nature The condition of being without a nationality or without a legal right to domicile may arise from the refusal to grant nationality; deprivation of nationality; or expulsion (usually for political reasons, though also for misconduct). Statelessness involves homelessness, loss of property, unemployment, separation of family through nationality complications, general disorientation and conflict in countries which give asylum to stateless persons in large numbers.
Claim Stateless persons constitute a largely unprotected, vulnerable group of people, to whose problems the international community generally gives insufficient attention.
Refs Lloyd, Peter J *International Trade Problems of Small Nations*; Mutharika, A Peter *The Regulation of Statelessness Under International and National Law* (1977); Rapaport, Jacques, et al *Small States and Territories*.
Broader Restrictions on recognition of nationality (#PE4912).
Narrower Statelessness of women (#PE4016) Separation of family members (#PE4959)
Involuntary loss of nationality of children (#PE6676).
Related Homelessness (#PB2150) Unreported births (#PF5381)
Officially nonexistent people (#PF7239).
Aggravates Social conflict (#PC0137).
Aggravated by Refugees (#PB0205) Antisemitism (#PE2131)
Conflict of laws (#PF0216) Rejection of refugees (#PF3021)
Deprivation of nationality (#PD3225) Refusal to grant nationality (#PF2657)
Declining sense of community (#PF2575) Refusal to grant citizenship (#PE5453)
Conflict of laws over nationality (#PF8953) Expulsion of immigrants and aliens (#PC3207)
Forced repatriation of prisoners of war (#PD0218)
Immigration barriers for handicapped family members (#PE4868)
Restrictions on freedom of movement between countries (#PC0935).

♦ **PE2486 Jurisdictional conflict and antagonism between the specialized agencies of the United Nations**
Nature Although the proliferation of intergovernmental organizations rarely leads to pure duplication of activities, there is a considerable problem of overlapping. Two or more intergovernmental organizations may carry out somewhat similar projects which cover much but not all of the same ground, or which are executed in a group of states most but not all of which participate in both organizations.
This kind of 'duplication', more properly called overlapping, is to be found on every hand and is no doubt to some extent inevitable. The mandates of international organizations themselves overlap and, in any case, most organizations tend, with the passage of time, to enlarge their mandates beyond the intentions of their founders. This is most likely to occur in areas which have become 'fashionable'. There is a scramble by international secretariats to establish their organizations in what seems to be a promising area of work, and for this purpose political differences between governments are exploited in favour of this or that organization, as are differences in membership of competing bodies. The unnecessary element in the problem of overlapping thus arises from defects in the structure, functions and control by governments of intergovernmental organizations. The complementary vice of 'overlapping' is 'lacunae', or areas of problem solution which are neglected because no intergovernmental body has paid attention to them. Both sets of problems are inadequately understood, and both merit study by governments.
Incidence The ILO provides a host of illustrations. Its main fields of activity are employment promotion, vocational guidance, social security, safety and health, labour laws and labour relations, labour administration, workers' education, cooperatives, rural and related institutions. Each of these fields contains 'grey areas' where other agencies are also much concerned. Furthermore, the ILO is empowered by the Philadelphia Declaration of 1944, which is incorporated in its revised Constitution, to undertake if it wishes, a considerably broader range of international economic and social responsibilities. A number of cases of overlap and proliferation of machinery have therefore occurred. For example, the improvement of living conditions has involved the ILO in aspects of rural development generally, including land reform (FAO) and the conditions of indigenous populations; training is closely linked to education (UNESCO) ; manpower to questions of small industries (FAO and UNIDO); workers' health standards with health standards generally (WHO). Such situations are of course met so far as possible by various coordinating devices. Some programmes, such as the World Employment Programme, are carried out in cooperation with all or most members of the United Nations family.
UNESCO provides illustrations no less striking. UNESCO has a general responsibility in matters of science, education and culture which entitles it to contribute, should it wish, to almost every field of international endeavour. Under 'science', the organization has not only played a part in the technical work of the ILO, FAO, IAEA, WMO and ITU, as well as the United Nations itself, but, since the term has been understood to include social as well as natural sciences, it has been concerned from time to time with such subjects of major concern to the United Nations as development theory and the peaceful settlement of disputes. On the basis of its general scientific mandate, it has established an International Oceanographic Commission, while FAO (fishery), IMO (marine pollution), IAEA (pollution through atomic waste), and of course the United Nations itself, have all developed their own networks in related aspects. A 1984 report by th US General Accounting Office showed that 30% of UNESCO's programmes were duplicating other work, and various UN emergency aid agencies have been proven duplicative to the point of developing antagonisms which have impeded the delivery of aid to African famine victims.
Claim Most of the United Nations Specialized Agencies have now become the equivalent of principalities, free from any centralized control. Over the years, like all such institutions, they have learned to safeguard and increase their powers, to preserve their independence, and to resist change. Difficulties arise from jurisdictional disputes between them. Lacking any central control, they have naturally advanced independent sectoral policies, often without due regard to the interests of either the developing countries or the United Nations system. Some of these difficulties are compounded at the regional level, particularly in the case of the Regional Economic Commissions – which are also more prone to jurisdictional clashes, since they cover the whole array of economic and social matters and so impinge on the areas of competence of most Specialized Agencies.
Broader Jurisdictional conflict and antagonism between international organizations (#PD0138).
Aggravates Fragmentation and complexity of the United Nations system (#PE0296);
Inadequate coordination of international nongovernmental organizations and programmes (#PE1209).

♦ **PE2488 Bilingualism in national settings and regional linguistic controversies**
Nature Use of two or more official languages within a country or regional federation, creates considerable difficulty because of the rivalry and the friction between the two linguistic communities and their concern to ensure that no minor advantage to one be used as a basis for progressively eliminating the other. As a result, every official communication must be made in two or more languages, and excessive attention is given to the linguistic balance within the governmental or intergovernmental bureaucracy. The situation is further complicated when countries are divided into language zones, because of the implications for speakers of the minority language in any such zone, and particularly for the language in which their children will be educated in schools funded by the government.
Claim Official bilingualism is the limit of accommodation for three or more languages in a country or regional federation. It is unrealistic to attempt to perpetuate equal usage for more than two languages, but political sensitivities prevent an early agreement on language limitation (at least for diplomatic and other external needs). For example, an impediment to closer cultural and political integration in Western Europe is the number of languages spoken. Only a common language can bridge the gap, which would need to be official in each country and taught to school children as a second language. Unofficially English, French or both, have fulfilled the 'second language' requirement.
Broader Multiplicity of languages (#PC0178).

♦ **PE2501 Poisonous algae**
Noxious algae — Toxic algae — Red tide — Sea slime — Algae blooms
Nature A wide variety of species of algae grow in the oceans and some species annually bloom into hugh floating masses of slime. These patches of algae appear every summer and have up to an estimated 100 million organism per litre of water. Increasingly these blooms are occurring in unprecedented places. The dead sink to the bottom. As they putrefy, they absorb oxygen from the water, strangling many other forms of life. Some blooms are so thick they block sunlight and force out other small organism that are the start of the marine food chain. Some species give off toxic wastes killing fish and other marine life. Some varieties, when consumed by shellfish can kill people. They contribute to acid rain by producing various sulphur compounds. They effect the weather by absorbing carbon dioxide from the atmosphere and reflect sunlight. They destroy tourist beaches by making the sea dangerous to swim in. The increase in algae blooms is, in some places, caused by pollution coming from cities, industries and farms.
Incidence Red tides have been reported off the east coast of America, Tasmania, Taiwan, Guatemala, Korea, Hong Kong, Venezuela, Iceland, the UK, Papua New Guinea, Sabah, Brunei, the Philippines, Scandinavia, Germany and Italy.
Broader Pests (#PC0728) Biological pollutants (#PC5276)
Environmental degradation (#PB6384).
Aggravates Hunger (#PB0262) Enteritis (#PE4973) Human poisoning (#PD0105)
Coastal water pollution (#PD1356).
Aggravated by Marine pollution (#PC1117) Sewage as a pollutant (#PD1414).

♦ **PE2512 Manufacture of illicit drugs**
Nature Manufacturing drugs chemicals that are available from almost any chemical company and rudimentary laboratory equipment that is widely available all around the world. Processing and dilution are relatively cheap compared to the street prices of illicit drugs.

Broader Unethical commercial practices (#PC2563)
Unauthorized pharmaceutical manufacture and distribution (#PE0564).
Aggravates Abuse of opiates (#PE1329) Abuse of hallucinogens (#PD0556)
Abuse of coca and cocaine (#PD2363).

♦ PE2515 Food-borne diseases

Food-borne infections and intoxications
Refs Riemann, Hans and Bryan, Frank L *Food-Borne Infections and Intoxication* (1979).
Broader Food pollution (#PD5605).

♦ PE2521 Instability of trade in wood, lumber and cork

Broader Instability of trade in manufactured goods (#PE0882)
Instability of trade in inedible crude non-fuel materials (#PD0280).
Narrower Instability of woodworking industries (#PE0681)
Fuel wood and charcoal trade instability (#PE8119).

♦ PE2522 Instability of wine trade

Nature In most countries wine is produced mainly for the domestic market. World trade in wine only represents 10 per cent of production. The North African countries constitute an exception since their production is mainly for export. However, with the independence of these countries and the establishment of the European Economic Community, their access to European markets has been restricted. The international barriers to trade in wine hinder expansion in trade by the North African developing countries which are confronted with the possibility of surpluses.
Broader Instability of alcoholic beverages trade (#PE2585).

♦ PE2524 Usury in developing countries

Exorbitant debt interest — Deepening debt cycle in rural communities
Nature In the agricultural regions which form the major part of developing countries, agricultural credit to farmers is supplied mainly by local money-lenders and traders. Exorbitant rates of interest, amounting to barely disguised usury, are charged. In addition, the lender is usually in a position to apply, or to threaten, a wide variety of sanctions which, although usually intangible, may maintain the borrower in a situation bordering on servitude, since it may be very difficult for him to pay off his accumulated debts. The lender is often also the local commodities broker-speculator.
Counter-claim The organized money-market, even if developed in agricultural regions, would tend to find that loans by verbal agreement, without security, and dependent on ability to repay based on income constantly subject to climatic hazards and price fluctuations, involve an unacceptably high risk. The unorganized money-market has the advantage of making cash readily available under conditions which official institutions would not accept.
Refs Mark, Jeffrey *An Analysis of Usury* (1980).
Broader High interest rates (#PF9014).
Aggravates Monopolies (#PC0521) Business bankruptcy (#PD2591)
Diminishing capital investment in small communities (#PF6477).
Aggravated by Capitalism (#PC0564) Unethical commercial practices (#PC2563).

♦ PE2525 Excessive land usage by transportation systems

Nature The provision of routes for transportation whether by road, rail, wire or pipeline is consuming ever greater areas of land yearly, with resultant costs of land acquisition and losses through land alienation. The tendency has been to provide diverse routes for road, rail, wire and pipeline. This has resulted in duplicate aquisition of strips of land between end points and unnecessary alienation of productive land from alternative uses. The situation is particularly critical in areas of rough terrain and in densely settled areas where land is at a premium.
Suburban-inner city links, particularly where they are based upon private transportation, can disrupt settled local communities and tear out usable housing as a result of the overriding priority given to expressways and intersections. Modernizing lands have the opportunity of devising transport systems of a less ruthless type.
Broader Excessive land usage (#PE5059) Shortage of cultivable land (#PC0219).
Narrower Airport location (#PG5014).
Related Insufficient transportation infrastructure (#PF1495).
Aggravated by Underutilization of facilities due to daily or seasonal peaks (#PF0827).

♦ PE2526 Pulmonary tuberculosis

Respiratory tuberculosis
Broader Tuberculosis (#PE0566) Lung disorders and diseases (#PD0637).
Narrower Tuberculous laryngitis (#PG5617).

♦ PE2530 Tetanus

Lockjaw
Nature This acute disease of man is induced by the toxin of the tetanus bacillus growing anaerobically at the site of an injury, and is characterized by prolonged, violent and agonisingly painful contractions of the voluntary muscles of the jaw, neck, abdomen and extremities. Despite intensive pharmacological and therapeutic research, the prognosis for each case of tetanus is doubtful. The infectious agent, Clostridium tetani, is excreted by infected animals, especially horses. The immediate source of infection may be soil, dust, or animal and human faeces.
Incidence The organism has a world-wide distribution, though the occurrence and severity of tetanus are determined by the amount of toxin produced and the resistance of the host. The tetanus case-fatality rate is rather high (30-80 percent): it is calculated that tetanus causes at least 50,000 deaths per year throughout the world. It is an occasional disease among farmers, especially following the contamination of wounds with manured soil.
Refs Veronesi, *Tetanus* (1981).
Broader Bacterial disease (#PD9094) Clostridial infections (#PE7769)
Zoonotic bacterial diseases (#PD6363).
Related Bacterial diseases in animals (#PD2731)
Diseases of nervous system in animals (#PD7841).
Aggravates Perinatal morbidity and mortality (#PD2387)
Child mortality in developing countries (#PE5166).
Aggravated by Injuries (#PB0855)
Health risks to workers in agricultural and livestock production (#PE0524).

♦ PE2534 Q fever

Q-fever
Nature Q fever is caused by the rickettsia Coxiella burnetti. Most human infection arises from direct or indirect association with cattle, sheep or goats. Rickettsiae may be present on soil and dust, where they can survive for long periods, since they are highly resistant to drying. This is particularly important in those countries where ewes are brought into yards to lamb. Very high concentrations of rickettsiae may then be present in the dust of such yards, which is therefore highly infective when sheltered from direct sunlight.
Incidence The disease occurs in nearly every country in the world. In some rural areas in Africa and the Middle East, and probably elsewhere, infection is almost universal in young children; adults there are thus immune and the acute illness is seen only in visitors or migrants.
Refs WHO *Mosquito-Borne Haemorrhagic Fevers of South-East Asia and the Western Pacific* (1966).
Broader Rickettsiosis (#PE5530) Human disease and disability (#PB1044)
Infectious diseases in animals (#PD2732).
Aggravates Fever (#PD2255).
Aggravated by Rickettsiae (#PE2572) Ticks as pests (#PE1766)
Rodent vectors of disease (#PE3629).

♦ PE2537 Rodents as pests

Rodents
Nature Rodents are of public health importance as reservoirs of numerous diseases of man, and of economic importance because of their destruction of growing crops and stored food. Rats are difficult to catch or poison. Although they are not particularly physiologically resistant to poisons, they cannily avoid eating unfamiliar food, or anything which they suspect as being harmful. In addition, single dose rodenticides have tended to be too persistent for widespread use; the body of the dead rat is itself toxic for other animals. The alternative cumulative poisons are anticoagulants, but anticoagulant resistances have been increasing.
Incidence Almost all countries within the temperate and tropical zones are engaged in active rodent control. The damage done by rodents to man's food, his animals and other property (including structural damage by gnawing through wood pipes and cabling) is extensive, and the list of human diseases transmittable by these animals or their parasites is not yet complete. The best known, but not necessarily the most important, include plague, salmonellosis, tularaemia and brucellosis, the relapsing fevers, leptospirosis and rat-bite fever, all caused by bacteria; scrub typhus, Q fever and rickettsial pox, due to rickettsiae; and diseases such as certain haemorrhagic fevers, caused by viruses about which little is known at present. In the USA it has been estimated that up to 20 per cent of unexplained fires may be started by rats or mice biting through electricity cables.
Refs FAO *Rodent Control in Agriculture* (1982); FAO *Rodent Pest* (1977).
Broader Animal pests (#PD8426) Pests of plants (#PC1627)
Household pests (#PD3522).
Narrower Rats as pests (#PE3177).
Related Insect pests of plants (#PD3634) Rodent vectors of disease (#PE3629).
Aggravates Fires (#PD8054) Structural failure (#PD1230)
Food spoilage in storage (#PD2243) Rodenticides as pollutants (#PE3677)
Endangered species of rodents (#PD3481) Insect vectors of human disease (#PE3632).
Aggravated by Rodent resistance to rodenticides (#PE3573).

♦ PE2538 Xerophthalmia

Vitamin A deficiency
Nature Xerophthalmia is a term used to cover all the ocular manifestations of vitamin A deficiency, including not only the structural changes affecting the conjunctiva, cornea and occasionally the retina, but also the biophysical disorders of retinal rod and cone function that are attributable to vitamin A deficiency. Xerophthalmia, while clearly not synonymous with blindness resulting from vitamin A deficiency, does denote an advanced degree of vitamin A depletion which constitutes a potential threat to sight. Vitamin A deficiency, of necessity, includes xerophthalmai but has much wider implications. It relates to any state in which the vitamin A status is subnormal. Although this is not capable of precise definition, it can be presumed to occur when the habitual intake of total vitamin A is markedly below the recommended dietary intake (RDI).
Incidence Vitamin A deficiency and xerophthalmia are among the most widespread and serious nutritional disorders that affect mankind. For many years the problem remained unchecked and continued to exact a devastating toll in blindness and death among young children. More than half a million children become blind every year, two-thirds of them die within weeks of becoming blind, through vitamin A deficiency.
Serious deficiencies in vitamin A have been reported in the following countries: **Af** Angola, Benin, Burundi, Burkina Faso, Chad, Ethiopia, Ghana, Kenya, Malawi, Mali, Mauritania, Mozambique, Niger, Nigeria, Sudan, Tanzania UR, Uganda, Zambia. **Am** Bolivia, Brazil, El Salvador, Haiti, Mexico. **As** Afghanistan, Bangladesh, Burma, India, Indonesia, Kampuchea Dem, Laos, Nepal, Oman, Philippines, Sri Lanka, Vietnam.
Refs Sommer, Alfred *Nutritional Blindness* (1982).
Broader Nutritional diseases (#PD0287) Eye diseases and disorders (#PD8786)
Vitamin deficiencies in diet (#PD0715).
Narrower Night blindness (#PG4565).
Aggravates Physical blindness (#PD0568).
Aggravated by Ascariasis (#PE2395) Protein-energy malnutrition (#PD0339).

♦ PE2572 Rickettsiae

Nature Rickettsiae are microorganisms which resemble bacteria but which grow only within susceptible cells - not in cell-free culture media, as bacteria do. The rickettsiae are mainly parasites or symbiotes of arthropods - especially lice, mites, and ticks; from these hosts the rickettsiae may spread to man and domestic animals, causing serious disease.
Incidence Epidemics of louse-borne typhus are still a threat. Among the other major groups of rickettsioses, murine typhus, scrub typhus, and Rocky Mountain spotted fever commonly occur in endemic areas.
Broader Human disease and disability (#PB1044)
Parasites of the cardiovascular system in animals (#PE9427).
Aggravates Q fever (#PE2534) Anaplasmosis (#PE6275) Typhoid fever (#PD1753)
Rickettsiosis (#PE5530) Eperythrozoonosis (#PG3997) Equine ehrlichiosis (#PG9732)
Rocky Mountain spotted fever (#PG3899).
Aggravated by Insect vectors of disease (#PC3597)
Vertebrate reservoirs of disease (#PG5136).

♦ PE2578 Ornithosis

Psittacosis
Nature Ornithosis is an infection communicable by animals to man (zoonosis), caused by an organism belonging to the Chlamydia group. As carriers, parrots were first recognized; later, procellarians, pigeons, song birds and poultry (ducks, chickens, turkeys) were also incriminated. However, neither from the microbiological nor from the clinical standpoint is it possible to differentiate between the carriers. Consequently today it is no longer justifiable to distinguish between psittacosis (the disease communicated by psittacine birds) and ornithoses (infections communicated by other birds). As a particular form of ornithosis, psittacosis is now largely of historical interest only.
Incidence The infection is communicated to man either by direct contact with infected birds or indirectly by inhaling dry dust from the plumage or excrement of such birds. Infection can in principle be passed from man to man, but this results in a considerable decline in virulence. Most infections occur in persons having close contact with birds - breeders, handlers, fanciers, pet owners and so on. Ducks, turkeys and pigeons are also major sources of the disease in man. Infection from these sources has been reported among farmers, poultry processors and rendering-plant workers, some of whom have been severely infected more than once. Several outbreaks of an endemic nature have been reported in workers at poultry slaughterhouses, particularly in persons working in air containing plumage dust. Agents prevalent in birds are highly contagious and virulent for man. The fact that there are, in animals and man, serologically related strains of

Chlamydia which show relatively low virulence may explain why, in spite of a very high incidence of Chlamydia infection in the population (30–40%), only relatively few cases of actual disease occur.
Broader Zoonotic bacterial diseases (#PD6363) Human disease and disability (#PB1044).
Aggravates Pneumonia (#PE2293).
Aggravated by Virus diseases (#PD0594) Bird vectors of plant disease (#PD3601)
International trade in endangered species (#PC0380).

♦ PE2585 Instability of alcoholic beverages trade
Broader Instability of trade in beverages (#PE1680)
Environmental hazards from beverages (#PE0849).
Narrower Instability of wine trade (#PE2522) Instability of beer trade (#PG6257)
Instability of spirits trade (#PG5143).

♦ PE2587 Instability of fresh fruit and edible nut trade
Broader Long–term shortage of fruit and vegetables and preparations thereof (#PE1013)
Instability of trade in fruit and vegetables and preparations thereof (#PE0961).
Narrower Complicity with evil (#PF0926) Instability of date trade (#PE1552)
Instability of banana trade (#PE1774)
Instability of ground nut and peanut trade (#PJ8727).

♦ PE2592 Housing destruction in war
Broader Disastrous consequences of war (#PC4257).
Related War casualties (#PD4189) Prisoners of war (#PC8848)
Long–term effects of war (#PD7918) Hazardous remnants of war (#PF2613)
War damage in civilian areas (#PD8719) Industrial destruction by war (#PD8359)
Poverty as a consequence of war (#PE5252) Environmental consequences of war (#PC6675).
Aggravates Housing shortage (#PD8778) Inadequate housing (#PC0449)
Illegal occupation of unoccupied property (#PD0820).

♦ PE2593 Deterioration in atmospheric visibility
Lack of visibility — Visibility degradation
Nature Visibility in the atmosphere decreases when solar radiation in the visible band is scattered by particles or gases. The most efficient scattering particles have sizes similar to those of visible light. These include sulphate aerosol, graphitic soot and hydrocarbons. Combustion of fossil fuels, wood burning and agricultural biomass burning produce light–scattering particles.
Broader Inhospitable climate (#PC0387) Degradation of the atmosphere (#PD9413).
Narrower Obstruction of astronomical observation by environmental pollution (#PE7244).
Aggravates Injuries (#PB0855) Road hazards (#PD0791)
Hazards to navigation (#PE3868) Aircraft environmental hazards (#PD8328).
Aggravated by Fog (#PE1655) Dust (#PD1245) Sand storms (#PD3650)
Air pollution (#PC0119) Locust plagues (#PE0725)
Smoke as a pollutant (#PD2267) Motor vehicle emissions (#PD0414)
Photochemical oxidant formation (#PD3663) Particulate atmospheric pollution (#PD2008).

♦ PE2598 Excessive injury to export interests developing countries due to export cartels
Nature Export cartels of firms in developed market–economy countries may in theory affect the export interests of developing countries in three ways. (1) They can discriminate against developing countries, in terms of price or otherwise, in the sale of such products, or they can refuse to sell production equipment, vital raw materials or intermediate goods which the developing countries need for their export industries. (2) Adverse effects may occur when exporters from developing countries are confronted in their export markets with powerful export cartels of firms in developed market–economy countries. These firms may apply monopolistic practices, such as predatory prices, to exclude developing countries' exporters. (3) Export cartels of firms in developed market–economy countries may be detrimental to the export interests of developing countries where such cartels allocate export markets and where this allocation includes subsidiaries of the parties located in developing countries.
Broader Restrictive trade practices (#PC0073).
Aggravated by Cartels (#PC2512).

♦ PE2599 Instability of non–metallic mineral products industry
Broader Instability of manufacturing industries (#PC0580)
Instability of trade in manufactured goods (#PE0882).
Narrower Glass and glass products industries instability (#PE9084)
China, pottery and earthenware industries instability (#PE8928).
Aggravated by Inadequacy of civil defence (#PF0506).

♦ PE2601 Instability of basic metal industries
Broader Instability of manufacturing industries (#PC0580).
Narrower Iron and steel basic industries instability (#PE8070)
Non–ferrous metal basic industries instability (#PG5166).

♦ PE2603 Distortion of international trade by discriminatory customs and administrative entry procedures
Excessive customs and trade formalities
Nature Objects, other than personal effects, which are moved across frontiers are subject to customs regulations. The complexity of such regulations, particularly when objects have to be moved through several countries in succession, creates considerable obstacles to legitimate movement and may cause considerable delay. This aggravates problems of trade but also seriously hinders the movement of educational, scientific and cultural materials into the country in question.
Incidence All modes of transport are now going through a period of technological change, notably in the areas of air freight and containerization. Consequently goods are being moved much more quickly than hitherto, so much so that in an increasing number of cases they are arriving at critical points in the movement cycle, such as airports or customs clearance centres, before the necessary formalities required to release them can, under present circumstances, be completed. Frustrating and costly delays then result. In addition, quite apart from the regular procedures and documentation connected with the levying of duties on goods across frontiers, there are still a number of onerous formalities and requirements such as special customs invoices or declarations demanded by some countries which may also be discriminatory.
Background Use of the term 'formalities' indicates that the problem is not simply one of documents but also of the procedures which give rise to them. These procedures originate from various sources. From the commercial aspect there is the obvious need for a supplier of goods to specify his sale and render a demand for payment. Conversely a receiver of goods needs to verify what he is receiving. There may also be a need for a document of title to enable the goods to be resold during transit whilst in many cases proof of shipment is required to give effect to documentary credits or to secure bank finance for the exporter. During the transport phase there is the corresponding need for carriers and any others, such as port authorities and insurers, involved in the transaction to know what is being carried, where to, and on what terms, in order properly to fulfil their contract for services being rendered. Lastly governments, in regulating their external trade by financial and fiscal measures, unavoidably have to create documentary procedures. The creation of burdensome formalities tends primarily to stem from the actions of governments which, when taken independently, lead to proliferation of paperwork.
Claim International trade is overwhelmed by a mass of formalities. Some of these are unavoidable but often they are merely traditional and remain a hindrance to the expansion of world trade. Furthermore compliance involves the wasteful expenditure of much time and money.
Refs Mansukhani, H L *Jungle of Customs Law and Procedures* (1974).
Broader Discrimination (#PA0833) Restrictive trade practices (#PC0073)
Non–tariff barriers to international trade (#PC2725)
Distortion of international trade as a result of government participation (#PD2029).
Narrower Distortion of international trade by quantitative restrictions (#PE9027)
Absence of coordinated customs services in developing countries (#PJ8325)
Distortion of international trade by restrictive customs valuation practices (#PE0497)
Distortion of international trade by restrictive controls on movement of labour (#PE8882)
Distortion of international trade by restrictive controls over foreign investment (#PE8525)
Distortion of international trade by state–trading and government monopoly practices (#PE8267)
Distortion of international trade by discriminatory application of antidumping regulations (#PE0191)
Distortion of international trade by discriminatory government and private procurement policies (#PE0347)
Distortion of international trade by discriminatory formulation of equipment safety regulations (#PE9073)
Distortion of international trade by discriminatory requirements with respect to marks of product origin (#PE0778)
Distortion of international trade by discriminatory requirements with respect to product standards and measures (#PE0083)
Distortion of international trade by minimum pricing regulations and other measures to regulate domestic prices (#PE1182)
Distortion of international trade by discriminatory formulation of health and sanitary regulations for agricultural and pharmaceutical products (#PE1274).
Related Discriminatory exchange rate policies (#PE8583)
Distortion of international trade by dumping (#PD2144)
Distortion of international trade by selective domestic subsidies (#PD0678)
Distortion of international trade by embargoes and similar restrictions (#PE0522)
Distortion of international trade through obstacles to patent protection (#PD0455)
Distortion of international trade by discriminatory preference agreements (#PD0340)
Distortion of international trade by export subsidies and countervailing duties (#PE1961).

♦ PE2611 Motion sickness
Travel sickness — Kinetosis — Seasickness — Car sickness — Air sickness
Nature Motion sickness is not normally a pathological condition but a normal response to certain motion stimuli with which the individual is unfamiliar and to which he is therefore unadapted; only those without a functioning vestibular apparatus of the inner ear are truly immune.
The essential characteristics of all motion stimuli that induce motion sickness is that they generate discordant information from the sensory systems which provide the brain with information about the spatial orientation and motion of the body. The principal feature of this discord is a mismatch between the signals provided by the eyes and inner ear and those that the central nervous system 'expects' to receive and be correlated.
Incidence The incidence of motion sickness may be illustrated by the following examples: 98% of occupants of life rafts in rough seas vomit; 60% of student aircrew suffer from air sickness (and in 15% it is of sufficient severity to interfere with flying training), but less than 0.5% of passengers in civil aircraft are affected. Differences between subjects in their tolerance to provocative motion can, in part, be related to their physical and mental constitution. Infants below the age of 2 years are rarely affected, but with maturation, susceptibility increases rapidly to reach a peak between 4 to 10 years. Thereafter, susceptibility falls progressively so that the elderly (60 years) are relatively immune. In any age group, females are more sensitive than males. Certain dimensions of personality, such as neuroticism, introversion and perceptual style have also been shown to be correlated, albeit weakly, with susceptibility. Motion sickness can also be a conditioned response and a manifestation of phobic anxiety. As an indication, the market for travel sickness remedies in the UK in 1986 was £3.3 million.
Refs Daunton, N G, et al *Mechanisms of Motion–Induced Vomiting* (1983); Reason, J T and Brand, J J *Motion Sickness* (1976).
Broader Human disease and disability (#PB1044).
Narrower Simulator sickness (#PE4929).
Aggravated by Unpleasant sights (#PG5520) Unpleasant odours (#PJ5522).

♦ PE2634 Stagnant surface water
Stagnant waste water — Unmanageable stagnant pools
Broader Wasted water (#PD3669) Pollution of inland waters (#PD1223)
Lack of essential local infrastructure (#PF2115).
Related Pond pollution (#PE5385).
Aggravates Malaria (#PE0616)
Dependence on sophisticated technology for development (#PD6571).
Aggravated by Physically inaccessible services (#PC7674).

♦ PE2635 Discrimination against dwarfs and midgets
Unequal facilities for undersized persons
Nature Discrimination against undersized persons exists, tacitly, in the size ranges of clothing, furniture, vehicles, equipment, stairs, and public conveniences. Employment discrimination is overt, and formal height requirements exist for a number of jobs, although tasks could be developed in some occupations for those of lesser height. Undersized persons may be midgets or dwarfs by individual genetic anomaly, or comparatively or relatively small (pygmies, or average East Asians among average Scandinavians for example), or they may be children.
Incidence Serious inconvenience to the undersized may exist in hospitals, schools or other facilities of institutional life.
Broader Discrimination against minorities (#PC0582)
Discrimination against people of abnormal height (#PE9402).
Narrower Domestic inconvenience for dwarfs, midgets and giants (#PE8036)
Lack of adequate clothing supply for dwarfs, midgets and giants (#PE8496).
Related Denial of rights to disabled (#PC3461) Discrimination against giants (#PE5578).
Aggravates Domination (#PA0839) Lack of self–confidence (#PF0879)
Inequality of opportunity (#PC3435) Inhibited human physical growth (#PD5177).
Aggravated by Dwarfism (#PE2715) Excessive standardization (#PF2271)
Denial of rights of minorities (#PC8999).

♦ PE2653 Laryngitis
Broader Human disease and disability (#PB1044)
Diseases of the respiratory system (#PD7924).
Narrower Acute laryngitis (#PG5614) Tuberculous laryngitis (#PG5617).
Related Measles (#PE1603) Smallpox (#PE0097) Erysipelas (#PG5619)
Scarlet fever (#PG2757).
Aggravates Speech disorders (#PE2265).
Aggravated by Smoking (#PD0713) Syphilis (#PE2300)
Common cold (#PE2412) Tuberculosis (#PE0566)
Alcohol abuse (#PD0153).

♦ PE2663 Peritonitis
Inflammation of the peritoneum
Broader Intestinal diseases (#PD9045) Human disease and disability (#PB1044)
Diseases and deformities of the digestive system (#PC8866).
Aggravates Constipation (#PE3505).
Aggravated by Ulcers (#PE2308) Injuries (#PB0855)
Gonorrhoea (#PE1717) Ascariasis (#PE2395)
Appendicitis (#PG2327) Typhoid fever (#PD1753)
Cholecystitis (#PE2251) Abdominal surgery (#PG5640)
Abscess of the ovary (#PG5639) Disorders of the bowel (#PE6553)
Urinary bladder disorders (#PE2307).

♦ PE2665 Symptoms referable to sense organs
Broader Diseases of the sense organs (#PC9623)
Ill defined health conditions (#PC9067).
Narrower Scotoma (#PJ9214) Dizziness (#PE5101) Head noise (#PG5106)
Photophobia (#PG5450) Oculomotor disturbance (#PG6827).
Aggravates Asthma (#PD2408) Chronic bronchitis (#PE2248).
Aggravated by Detergents as pollutants (#PE1087).

♦ PE2668 Vanadium as a pollutant
Nature Vanadium oxides, particularly vanadium pentoxide and its derivative, ammonium metavanadate, can produce severe toxic effects. Initial symptoms are profuse lacrimentation, nose or throat soreness, bronchitis, and chest pain. Severe reactions include fatal pneumonia, emphysema or chronic bronchitis. The tongue may turn green and cigarette ends of vanadium workers may show a greenish tinge.
Incidence Ferrovanadium is used in tool and high speed steel making. Vanadium pentoxide is an important catalyst. Their common usage assures some pollution and health hazards from this source.
Broader Toxic metal pollutants (#PD0948) Environmental hazards from metals (#PE1401).
Aggravates Occupational risk to health (#PC0865).
Aggravated by Industrial emissions (#PE1869) Motor vehicle emissions (#PD0414).

♦ PE2678 Torture by burning
Nature Victims of torture are frequently burned using a variety of methods including pouring hot oil or water over them and burning soles of feet, the head, sexual organs, palms of hands, rectums and other sensitive part of the body with cigarettes, blowtorches, hot coals, electric cattle prods, acids and hot irons.
Incidence Burning has been reported in the following countries: **Af** Egypt, Ethiopia, Madagascar, Mali, Mauritania, Morocco, Tunisia, Uganda. **Am** Argentina, Bolivia, Chile, El Salvador, Guatemala, Mexico, Suriname, Uruguay. **As** Bahrain, Bangladesh, India, Iran Islamic Rep, Iraq, Pakistan, Philippines, Syrian AR. **Eu** Italy, Poland.
Broader Physical torture (#PD8734).

♦ PE2680 Heartwater disease
Cowdriosis
Nature Heartwater, a febrile septicaemia of cattle, sheep, goats and wild ungulates, is caused by Rickettsia ruminantium. Transmission is by ixodid ticks of the genus Amblyomma, which are three-host ticks.
Incidence The disease is found in semi-arid rangelands and the tick is difficult to eradicate or control. Some species are known to survive off the host up to 18 months. Nonapparent infections in certain wild antelope, in addition to range stock, are thought to provide a reservoir in bushveld areas for new generations of young ticks since infection does not pass through tick eggs. Strains of varying virulence cause subacute or acute disease with high mortality.
Broader Rickettsiosis (#PE5530) Parasitic diseases in animals (#PD2735).
Aggravates Harbouring national security offenders (#PE0484).
Aggravated by Insect vectors of animal diseases (#PD2748).

♦ PE2684 Diseases of the arteries
Diseases of veins
Refs Stipa, S and Cavallaro, A *Peripheral Arterial Diseases* (1982); Tulenko, Thomas and Cox, Robert H *Recent Advances in Arterial Diseases* (1986).
Broader Human disease and disability (#PB1044)
Diseases of the circulation system (#PC8482).
Narrower Aneurysm (#PE3597) Endarteritis (#PG5673) Haemorrhoids (#PE3504)
Varicose veins (#PE6864) Arteriosclerosis (#PE2210)
Generalized necrotizing arteritis (#PG5087).
Aggravates Stroke (#PE1684) Gangrene (#PG5675) Heart diseases (#PD0448)
Kidney disorders (#PE2053) Bright's disease (#PE2272)
Lung disorders and diseases (#PD0637).
Aggravated by Gout (#PG3061) Diabetes (#PE0102)
Syphilis (#PE2300) Hypertension (#PE0585)
Rheumatic fever (#PE0920) Excessive consumption of fats (#PE4261).

♦ PE2686 Lymphatic leukaemia
Lymphocytic leukaemia — Lymphoid leukaemia
Refs Catovsky, D and Foa, R *The Lymphoid Leukaemias* (1989).
Broader Leukaemia (#PE0639).
Aggravates Diseases of the spleen (#PE6155)
Short range planning for long-term development (#PF5660).

♦ PE2690 Abusive behaviour modification
Misapplied behaviourism — Behaviourism
Nature Coercive behaviour modification techniques may be used on aggressive, manipulative or 'independent' individuals, usually in prisons. Such techniques aim to weaken, undermine or remove the supports of old patterns of behaviour and old attitudes, either by removing the individual physically and preventing any communication with those he cares about or who can give him any reinforcement, or by 'proving' to him that those ideals and persons whom he respects are not worthy of it and should in fact be actively mistrusted. Methods used include social disorganization and the creation of mutual mistrust (achieved by spying on individuals and reporting back private material), tricking people into written statements which are then shown to others to convince individuals to trust no one, segregation of natural leaders, and drug assaults. The latter reduce prisoners to a vulnerable state of mind in which they are unable to respond with emotion (side-effects and after-effects may include: induction of a catatonic-like state, nausea, loss of appetite, impotence, liver damage, and hypertension severe enough to cause cardiac arrest). Other techniques used in adjunct to behavioural modification include torture, electric shocks, emetics, and surgical destruction of parts of the brain.
Claim The infiltration of behaviour modification techniques into schools poses as an innocuous pedagogical innovation when, in fact, it is an attempted medical treatment for behavioural disorders applied without the parents' permission and without medical supervision. The tacit sanctioning of behaviour modification techniques in schools by school boards and uninformed parents opens the way for manipulation of students' minds. Coercive pressures from superiors or peers in a variety of organizations, such as business, religious and military, induces a spectrum of involuntary behavioural changes. Psychiatric patients may be made guinea pigs in behavioural experiments.
Refs Bellack, Alan S, et al (Eds) *International Handbook of Behavior Modification and Therapy* (1982); Benson, Hazel B *Behavior Modification and the Child* (1979); Martin, Gray and Pear, Joseph *Behavior Modification and What it is and How to do it* (1988).
Broader Inhumane scientific activity (#PC1449)
Irresponsible scientific and technological activity (#PC1153).
Related Hedonism (#PF2277)
Lack of opportunities for practical training in communities (#PF2837).

♦ PE2691 Harmful effects of comic strips and picture-story books
Nature Modern comic strips are gaining in popularity throughout the world as they offer a concise and easily understood story, and have an attractive, seemingly innocuous appearance, often syndicated throughout the world to millions of adherents. However, due to their world-wide publication, they may conflict, in some countries, with cherished moral or cultural values; may present a falsified account of history; have racist or militarist overtones; or encourage a fascination with violence which has anti-social effects.

♦ PE2693 Bird shooting
Broader Cruel sports (#PD1323).
Aggravates Lead poisoning in animals (#PE9228)
Endangered species of birds (#PD0332).

♦ PE2694 Chronic pain
Intractable pain
Refs Doyle, D (Ed) *International Symposium on pain Control, 1986* (1988).
Broader Pain (#PA0643).
Aggravated by Chronic illness (#PD8239).

♦ PE2700 Denial of right to business growth
Restrictions on private economic expansion
Nature Business, like any other organism, can survive only if it has the capacity to develop to its full potential. Government regulations, import and export bottlenecks, and excessively complex legal machinery can strangle business, especially new ones.
Broader Denial of rights of businesses (#PD4728).
Related Discouraging conditions for small business (#PD5603).
Aggravates Economic stagnation (#PC0002)
Distortion of international trade by selective domestic subsidies (#PD0678).
Reduces Expansionism (#PB5858).

♦ PE2715 Dwarfism
Microsomia
Nature Dwarfism, or restricted growth, is due to genetic changes arising in the children of normal parents. It may also be developed as a result of organic disease or determined by environmental factors in the case of a genetically susceptible individual. Microsomia has three forms: in the first one there is only a deficiency in stature; in the second one additional serious hypogenesis of the skeleton except for a large skull is present; and in the third one includes a combination of infantilism and premature senility. Racial dwarfism occurs in some groups whose physiological bodily functions are otherwise quite normal. Sub-normal height creates a multitude of special problems for such individuals in relating to others and in using facilities designed for those of average height.
Incidence The normal height of adult men has been arbitrarily set at 150 cm, and on this basis dwarf populations exist in Central Africa, the Andaman Islands, Philippines, and New Guinea. The pygmies of Central Africa have heights down to 120 cm.
Refs Ablon, Joan *Little People in America* (1984); Wood, Edward J *Giants and Dwarfs* (1976).
Broader Stigmatized diseases (#PD7279) Pituitary gland disorders (#PE2286)
Human physical genetic abnormalities (#PD1618).
Related Inhibited human physical growth (#PD5177).
Aggravates Discrimination against dwarfs and midgets (#PE2635).

♦ PE2719 Antagonism between government agencies and officials
Nature Mutual opposition or ill will on the part of agencies and officials is found more or less everywhere, and it aggravates the difficulties any central organ has in taking action, even where it is backed up by a coherent civil service policy, government support, and a satisfactory legislative framework. The reasons for this opposition are at times somewhat specious. Ministries and agencies often cite their autonomy and their responsibility for personnel matters in order to cover up wrongdoings. Civil servants frequently have misgivings, with some justification, about rationalization of personnel management and the institution of controls. Measures taken at the centre – when the centre is not responsible for direct management – sometimes meet with a manifest lack of good will. Rules are either not applied or are misapplied; requests for information from the central organ for, say, the establishment of central files and dossiers, post classification or evaluation of duties, etc, are left unanswered. Some central organs are confronted with a wall of antipathy or even hostility, to the point where centralization of personnel administration, in spite of its drawbacks, seems in the long run preferable to any other solution. On the whole this opposition must perhaps not be exaggerated; but it does exist everywhere in varying degrees: it is an obstacle to collaboration with ministries and agencies; it puts a brake on action by the central organs.
Broader Conflict (#PA0298).
Aggravates Bureaucracy as an organizational disease (#PD0460).
Aggravated by Governmental incompetence (#PF3953).

♦ PE2727 Mange
Nature Mange is a parasitic disease caused by mites which live in or under the skin and affect all kinds of animals and man, causing scabby eczema and depilation. Emaciation may occur in chronic cases and under cold conditions or with poor nutrition; heavily infested cattle may die from psoroptic mange. Sarcoptic mange may cause mastitis in milking cows.
Incidence Mange is a world-wide problem, and there are three kinds of mange mites: psoroptic, chorioptic and sarcoptic which attack a variety of animals. Cattle are also affected by demodectic mange mites. Within these groupings there are many different species. Transmission is generally from one animal to another and for this reason among domestic animals there is a far higher incidence during the winter when they are confined in close quarters. The mites eggs may remain on barn walls, fence posts, cattle trucks or any other equipment which comes into contact with diseased animals and they may be transmitted in this way.
Broader Parasitic diseases in animals (#PD2735).
Narrower Mange in sheep (#PG4137) Mange in swine (#PG7868)
Mange in horses (#PG9788) Mange in cattle (#PG6179)
Parasitic otitis externa (#PG7387) Nasal acariasis in animals (#PG5521)
Cutaneous acariasis in animals (#PG9510) Cheyletiella infestations in animals (#PG4207).
Aggravates Mastitis in milking cows (#PG5698)

hair (#PE7111)
Economic loss through reduced productivity of diseased animals (#PE8098).
Aggravated by Mites as pests (#PE3639) Inadequate feeding of animals (#PC2765)
Spread of animal diseases through factory farming (#PD2752)
Inadequate housing and penning of domestic animals (#PE2763)
Inadequate disinfection measures for animal housing and equipment (#PE2757).
Reduced by Forced depilation (#PE5697).

♦ PE2736 Anthrax

Nature An infectious bacterial disease, characterized by a high fever, enlarged spleen, swelling of the throat in some animals, exudation of tarry blood, and resulting in death, anthrax may attack all domestic animals and man; it also affects numerous wild animals, and is most common among herbivora. It most commonly affects livestock, although it is occasionally transmitted to humans. It has three forms, affecting the skin (most common), the intestines or the respiratory tract. Anthrax bacteria produce spores that are highly resistant to disinfectant and therefore remain viable long after the death of the animal, even adhering to the handicraft goods made from the skins.
Incidence Anthrax is a widespread disease, occurring in all parts of the world, but particularly in tropical and sub-tropical areas. In temperate climates, it occurs only spasmodically and affects fewer animals in each outbreak than are affected in the tropics. The disease is most commonly contracted by way of the mouth or alimentary system, and is most often soil-borne (and water-borne). It may be transmitted by birds or flies which have fed on infected animals or on carcasses and which subsequently infect the trees and shrubs which are eaten by herbivores, or (in the case of insects) directly infect by biting; the latter especially affect man. Animals that wallow in mud or dry soil may carry the infected soil from one place to another. Anthrax spores may also be carried by surface drainage or in wind-borne fragments of disintegrated carcasses. Occurrence of anthrax is influenced by climatic and ecological conditions. Rainy weather followed by hot days appears to be most favourable for the disease but it also frequently occurs during hot dry summers, or drought when animals are forced to graze close to the soil. Anthrax spores are very hard to destroy and may remain alive in the soil for 10 years after an outbreak. Worms are thought to contribute in the spread of the disease. It can be contracted sometimes as a result of breathing dust from affected animals (woolsorters' disease).
Broader Epidemics (#PC2514) Stigmatized diseases (#PD7279)
Zoonotic bacterial diseases (#PD6363) Bacterial diseases in animals (#PD2731).
Related Occupational diseases (#PD0215).
Aggravates Economic loss through reduced productivity of diseased animals (#PE8098).
Aggravated by Blowflies as pests (#PE3627) Houseflies as pests (#PE3609)
Soil-borne diseases in animals (#PE2739) Insect vectors of animal diseases (#PD2748)
Worms as vectors of animal diseases (#PD2750)
Wild birds as vectors of animal diseases (#PE2749).

♦ PE2737 Blackleg

Nature Blackleg is an acute febrile noncontagious infection of deer, cattle, sheep and pigs, characterized by sudden onset, fever and emphysematous serohemorrhagic swellings in the subcutaneous tissues and heavy musculature. Death usually occurs within 24 hours.
Incidence Animals of all ages in all parts of the world may be affected by the disease but young animals between the ages of 6 months and 2 years are particularly susceptible. The blackleg bacillus is one of the anaerobic bacteria which only grow in the absence of oxygen. It forms spores which are highly resistant to soil and which lodge in the intestinal tracts of animals. Infection may occur through contaminated wounds, particularly small punctures made by thorns or barbed wire.
Broader Clostridial infections (#PE7769) Bacterial diseases in animals (#PD2731).
Aggravates Economic loss through slaughter of diseased animals (#PE8109).
Aggravated by Soil-borne diseases in animals (#PE2739)
Inadequate carcass disposal of diseased animals (#PE2778)
Inadequate disinfection of pastureland after disease outbreak (#PD2774).

♦ PE2738 Coccidiosis

Coccidioidomycosis
Broader Mycosis (#PE2455) Epidemics (#PC2514) Fungal diseases (#PD2728)
Parasitic diseases in animals (#PD2735)
Diseases of the digestive system in animals (#PD3978).
Narrower Coccidiosis in poultry (#PG9814) Cryptosporidiosis of animals (#PG9839).
Aggravates Harbouring national security offenders (#PE0484).
Aggravated by Ringworm (#PD2545) Diseases of wild animals (#PD2776)
Spread of animal diseases through factory farming (#PD2752)
Inadequate disinfection of pastureland after disease outbreak (#PD2774).

♦ PE2739 Soil-borne diseases in animals

Nature Animal diseases can develop from sources of infection that are transmitted via the soil. Transmission may occur through the disintegration of infected carcasses. The disease is contracted by the animals grazing close to the ground or during wet seasons when infection may occur through absorption of surface moisture. Soil-borne diseases include anthrax and blackleg.
Broader Vectors of animal diseases (#PD2751).
Related Denial to animals of the right to a natural death (#PE8339).
Aggravates Anthrax (#PE2736) Blackleg (#PE2737)
Nocardiosis in animals (#PG6309) Malignant oedema in animals (#PG9283)
Bacterial diseases in animals (#PD2731) Worms as vectors of animal diseases (#PD2750)
Bacillary haemoglobinuria in animals (#PG2359).
Aggravated by Weather as a factor of animal disease (#PD2740)
Inadequate carcass disposal of diseased animals (#PE2778)
Inadequate disinfection of pastureland after disease outbreak (#PD2774).

♦ PE2741 Sleep deprivation

Sleep loss
Nature A major cause of sleep deprivation is the complexity of daily life, the increase in pace and the shortage of time. Sleep is the most expendable activity in comparison with other priorities. Although mild sleep deprivation is of little concern, people are less productive, ill-humoured and dissatisfied with life if they fail to get a full complement of sleep each night. When sleep deprivation becomes chronic and extensive, it can lead to impaired judgement and accident proneness.
Incidence Surveys indicate that the majority of people in industrialized countries (who normally sleep 6.5 hours) are sleeping at least 60 to 90 minutes less than required to overcome daytime sleepiness. It is reported that sleepiness is second only to drunkenness as a cause of traffic accidents. In the USA 40,000 road accidents a year may be sleep-related, especially since more than 20 per cent of drivers report having fallen asleep more than once while driving. Many major industrial accidents occurred at night when it is probable that workers were not optimally alert.
Aggravates Fatigue (#PA0657) Job fatigue (#PD8052) Injurious accidents (#PB0731).
Aggravated by Lack of time (#PC4498) Sleep disorders (#PE2197)
Increasing pace of life (#PF2304).

♦ PE2746 Otosclerosis

Refs Beales, Philip *Otosclerosis* (1981); Gapany-Gapanavicius, B *Otosclerosis* (1975).
Broader Diseases of the ear (#PD2567).
Aggravates Deafness (#PD0659).

♦ PE2747 Snail vectors of animal diseases

Nature Snails act as intermediate hosts for certain animal diseases, notably liver flukes, blood flukes and tapeworm diseases. The part played by snails as intermediate hosts complicates the eradication of the disease and its control.
Broader Wild animals as carriers of animal diseases (#PD2729).
Aggravates Liver fluke (#PE2785) Schistosomiasis (#PE0921).
Aggravated by Difficulty in identifying carriers of animal diseases (#PF2775).
Reduced by Excessive use of chemicals to control pests (#PD1207).

♦ PE2749 Wild birds as vectors of animal diseases

Nature Animal diseases may be transmitted by wild birds which have eaten from infected carcasses, and which provide a complicating factor in the control of the disease, as in the case of anthrax. Anthrax is transmitted from birds having eaten from infected carcasses to trees and other plant life which provide fodder for susceptible animals. Parasitic and fungal animal diseases can be transmitted by birds to domestic poultry or other animals. Most birds support sizeable groups of parasites, and while wild birds may develop a certain immunity to the diseases caused by them, domestic poultry reared in artificial conditions are more susceptible to epidemics. Wild birds also usually support parasitic fungi, but to a lesser extent than they do parasitic insects. Parasitic insects include ticks, mites, lice, flies, intestinal worms, roundworms (filaria) and protozoa, which may be present in vast quantities.
Broader Birds as vectors of disease (#PE6659)
Wild animals as carriers of animal diseases (#PD2729).
Related Diseases of wild animals (#PD2776).
Aggravates Anthrax (#PE2736) Brucellosis (#PE0924)
Foot-and-mouth disease (#PE1589)
Harbouring national security offenders (#PE0484).
Aggravated by Difficulty in identifying carriers of animal diseases (#PF2775).

♦ PE2756 Inadequate animal quarantine

Nature Inadequate measures taken for isolation of infected animals entering a country or an area may cause disease to spread from one country to another or from an infected area to uninfected parts of the same country. Inadequate isolation measures may arise from a faulty knowledge of the incubation period for a disease, or insufficient knowledge concerning new and more virulent strains of a well-known disease, such as rabies. Inadequate quarantine measures may occur where there is no quarantine policy for certain diseases.
Incidence In the UK, in order to avoid outbreaks of rabies, a six-month quarantine period was normal for imported dogs and cats, until a dog died of rabies 3 months after being released from quarantine. The period was then extended to 9 months and subsequently to 12; but finally reduced again to 6 months, with vaccination on entering quarantine and again one month afterwards.
Broader Inadequate quarantine (#PE2850)
Inadequate control of animal diseases (#PD2781).
Narrower Inadequate carcass disposal of diseased animals (#PE2778).
Related International movement of animals as factor in animal diseases (#PD2755).
Aggravates Rabies (#PE1325) Irresponsible introduction of new species of animals (#PD1290)
Difficulty in identifying carriers of animal diseases (#PF2775).
Aggravated by Unrecognized animal diseases (#PG5768)
Inadequate knowledge of incubation periods for animal diseases (#PE8133).

♦ PE2757 Inadequate disinfection measures for animal housing and equipment

Nature Animal diseases may be transmitted from infected cages and pens, animal transport trucks, and other equipment that might have been used for diseased animals. If disinfection measures are not carried out sufficiently thoroughly after a disease outbreak, reinfection of new stock may take place.
Broader Inadequate housing and penning of domestic animals (#PE2763).
Related Lack of slaughter facilities (#PE9767)
Denial to animals of the right to the attention, care and protection of humankind (#PF5121).
Aggravates Mange (#PE2727) Zoonoses (#PD1770) Rinderpest (#PE2786)
Epizootic diseases (#PD2734) In-home animal shelter (#PJ9800)
Viral diseases in animals (#PD2730).
Aggravated by Inadequate control of animal diseases (#PD2781)
Inadequate disinfection measures for humans during animal disease outbreaks (#PE8409).

♦ PE2759 Inhumane killing of stray animals

Nature Unwanted animals such as dogs and cats are inefficiently and inhumanely killed by means of strychnine poisoning, beating, inadequate gas chambers, inadequate electrocution apparatus, inadequate decompression machines, drowning, strangling, burning, stoning, traps, and caustic or acid.
Broader Killing of animals (#PD8486) Inadequate animal welfare (#PC1167)
Inhumane killing of animals (#PE0358).
Narrower Cruel methods of destruction of stray dogs (#PE0360).
Related Torture of animals (#PC3532).
Aggravated by Surplus domestic animal production (#PJ5775)
Excessive stray animal populations (#PE5776).
Reduces Stray dog populations (#PE0359).

♦ PE2763 Inadequate housing and penning of domestic animals

Nature Animals may be given insufficient space, light, ventilation and comfort, causing stress, malformation, loss of productivity and infertility. Intensive farming units rarely give animals sufficient space, light or comfort, and in certain cases, such as with 'sweat houses' for pig fattening, ventilation is also restricted. Under such conditions stress and malformation occur, as well as loss of productivity. Inadequate housing may also occur through negligence or through ignorance of the optimum conditions required and of animal psychology and behaviour. With breeding animals it may lead to infertility, rather than the desirable peak condition. Zoos may not provide enough space for wild animals, nor the environment which is conducive to their well-being.
Broader Inadequate animal welfare (#PC1167)
Agricultural mismanagement of housed farm animals (#PD2771).
Narrower In-home animal shelter (#PJ9800)
Inadequate disinfection measures for animal housing and equipment (#PE2757).
Related Maltreatment of zoo animals (#PE4834).
Aggravates Mange (#PE2727) Animal infertility (#PC1803)
Loss of animal productivity (#PD8469) Lack of slaughter facilities (#PE9767)
Animal stress in factory farming (#PD2760)
Animal malformation in factory farming (#PD2761).
Aggravated by Prohibitive cost of land (#PE4162)
Commercial exploitation of wild animals (#PD1481)
Inadequate legislation for animal welfare (#PE5794).

♦ PE2764 Insanitary penning conditions as factor in animal diseases

Nature Insanitary penning conditions, particularly for domestic animals, may play an important part

PE2764

in the spread of animal disease, particularly as many infections are transmitted through faeces and other excreta. Coccidiosis and foot-and-mouth disease are encouraged by the existence of insanitary conditions.
Broader Inadequate animal welfare (#PC1167)
Inadequate control of animal diseases (#PD2781)
Agricultural mismanagement of housed farm animals (#PD2771).
Related Unsanitary and inhumane urban food animal conditions (#PE0395)
Denial to food animals of the right to freedom from suffering (#PE3899).
Aggravates Epizootic diseases (#PD2734) Infectious diseases in animals (#PD2732).

♦ PE2770 Inferior meat quality from intensive animal farming units
Nature Animals for meat production are required to produce less fat and more lean meat to meet consumers' wishes and should be produced at the lowest cost which requires a fast rate of growth. This has been achieved by selective breeding, intensive feeding systems and supplementing diets with growth promotion agents. Such animals, particularly pigs, tend to be more susceptible to stress producing poor quality meat after slaughter. Stress increases with overcrowding and inadequate space Residues of hormone preparations and antibiotics are suspected of causing harm to humans and may develop resistance to drugs. resistance. Infections from diseases to which intensively farmed animals are susceptible may remain in the meat after slaughter.
Narrower Use of artificial methods of promoting fast livestock growth (#PE9125).
Related Denial to food animals of the right to freedom from suffering (#PE3899).
Aggravates Food tastelessness (#PJ5809).
Aggravated by Inadequate feeding of animals (#PC2765)
Animal stress in factory farming (#PD2760)
Vitamin E deficiency in domestic animals (#PE4760)
Agricultural mismanagement of housed farm animals (#PD2771)
Excessive commercial exploitation of farm animals by industrial concerns (#PD2772).

♦ PE2777 Importation of infected carcass meats as factor in animal diseases
Nature Animal diseases may be spread internationally by the importation of carcass meats from areas where a disease is enzootic.
Incidence The problem occurs particularly with diseases where the causative agent is resistant to freezing, such as in the case of the foot-and-mouth disease virus, and that of fowl pest. The foot-and-mouth disease virus has been known to remain active in the bone marrow of pork and beef for 76 days, even at freezing temperatures. Such carcasses can infect human beings who may act as carriers of the disease to animals. The fowl pest virus can exist for 2 years in frozen birds.
Broader Domestic animals as carriers of animal diseases (#PD2746).
Aggravates Zoonoses (#PD1770) Foot-and-mouth disease (#PE1589)
Viral diseases in animals (#PD2730).
Aggravated by Inadequate hygiene restrictions on carcass meat exports (#PE8398).

♦ PE2778 Inadequate carcass disposal of diseased animals
Undisposed dead animals
Nature Certain very virulent diseases, including anthrax, Newcastle disease, foot-and-mouth disease, and rinderpest, may be transmitted through inadequate disposal of infected carcasses. Carcass disposal methods include incineration and burial in quick-lime, and if this is adequately done, there should be no risk of infection. However, when the disease has affected wild animals, it may be difficult to find carcasses before they are eaten by carrion birds or other animals which may then themselves become carriers of the disease; this happens most frequently in the case of anthrax and rinderpest.
Broader Inadequate waste treatment (#PD6795) Inadequate animal quarantine (#PE2756)
Disjointed patterns of community identity (#PF2845).
Related Disposal of corpses (#PG4007).
Aggravates Blackleg (#PE2737) Zoonoses (#PD1770)
Rinderpest (#PE2786) Epizootic diseases (#PD2734)
Viral diseases in animals (#PD2730) Bacterial diseases in animals (#PD2731)
Soil-borne diseases in animals (#PE2739) Infectious diseases in animals (#PD2732)
Harbouring national security offenders (#PE0484).
Aggravated by Contagious bovine pleuropneumonia (#PE1775).
Reduces Rabies (#PE1325) Fowlpest (#PE1400) Brucellosis (#PE0924)
Foot-and-mouth disease (#PE1589).

♦ PE2785 Liver fluke
Distomatosis
Nature Liver fluke is a parasitic disease affecting the liver and bile ducts of a large variety of warm-blooded animals, domestic and wild, and of man, causing enlargement and thickening of the walls of the bile ducts and fibrosis of the liver tissue, and resulting in loss of condition, digestive disorders, anaemia, and other symptoms of parasitism. In sheep, extensive liver damage caused by flukes contributes directly to 'black disease', (infectious necrotic hepatitis), which does not occur in healthy animals. Economic losses arise from emaciation of domestic animals, and from production of liver unfit for human consumption.
Incidence Liver fluke is a widespread disease present on all continents. Liver flukes are hermaphroditic and release their eggs into the bile ducts of the host, which are then passed out through the faeces. In the three most common species, the eggs embryonate when they reach water, where they hatch in 2 to 6 weeks and find suitable snails in which to mature. Transmission comes via the snails, from which the fluke larvae are deposited on grass or other vegetation or under the surface of the water, and are then ingested by the host animals. Carnivores or man may become infected from eating the livers of diseased animals, or from water intake. There are many species of liver flukes and since all need intermediate hosts, it is hard to develop control measures which apply to all. Four common species are: Fasciola hepatica (common liver fluke), fascioloides magna (large American liver fluke), fasciola gigantica (giant liver fluke), and dicrocoelium dendriticum (lancet fluke). Lancet fluke differs from the other three in so far as the eggs are eaten by land snails in which they mature into cercariae which escape from the snails and are eaten by ants. The ants in turn are eaten by cattle and other host animals while grazing. Adult flukes have a long life cycle inside the infected animal.
Broader Zoonoses (#PD1770) Parasitic diseases in animals (#PD2735).
Narrower Infectious necrotic hepatitis (#PG0371).
Aggravated by Snail vectors of animal diseases (#PE2747).
Reduced by Wildlife extinction (#PC1445) Pesticides as pollutants (#PD0120).

♦ PE2786 Rinderpest
Cattle plague
Nature A viral contagion of cattle, buffaloes, other ruminants and pigs, characterized by fever, discharge from the mucous membranes, constipation followed by diarrhoea in the final stages, and mouth ulcers, rinderpest results in a mortality rate of up to 90 per cent. The disease is acute, lasting from 4 - 10 days.
Incidence In India, rinderpest still occurs and is accentuated by the vast numbers of cattle and domestic buffalo and by the difficulties in maintaining strict quarantine measures. The cow in India is a sacred animal; this increases the difficulties of stringent control. In some instances, particularly in vaccinated herds where the disease is no longer self-evident by its high morbidity rate, rinderpest may be confused with other diseases, such as mucosal disease, so that adequate laboratory services must be available for differential diagnosis. In South-East Asia, the elimination of rinderpest has progressed, but the condition is still present in Vietnam, Laos and Democratic Kampuchea. Rinderpest, like food-and-mouth disease, is a major obstacle to international meat trade.
Background An ancient scourge, it originated in Asia or Eastern Europe and was brought to Western Europe by invading barbarian tribes. It ravaged Europe for fifteen centuries until the 19th century when it was eradicated. Importation of animals (cattle in particular) helped to spread the disease in Africa and Asia. Importation of wild animals from Africa and Asia has given rise to outbreaks in zoos.
Broader Viral diseases in animals (#PD2730).
Related Measles (#PE1603).
Aggravates Canine distemper (#PG5710).
Aggravated by Leeches as pests (#PE3660)
Inadequate vaccination of domestic animals (#PG5840)
Wild animals as carriers of animal diseases (#PD2729)
Inadequate carcass disposal of diseased animals (#PE2778)
Inadequate disinfection measures for animal housing and equipment (#PE2757).
Reduced by Economic loss through slaughter of diseased animals (#PE8109).

♦ PE2787 Water-borne animal diseases
Nature Animal diseases may be transmitted by infected water either when used for drinking or when deposited on pastureland during the wet season. Water-borne diseases may be parasitic, viral or bacterial. Parasites such as liver flukes depend on water for the maturation of their larvae. The larvae may be ingested through water intake. Incidence of coccidiosis is higher in wet and marshland areas in the natural state when pastureland may become contaminated from the integration of infected faeces into the soil. Drinking water may become infected by diseased animals, particularly domestic animals kept in close and insanitary conditions. Foot-and-mouth disease as an airborne infection has a higher incidence during wet weather, since this is the means by which the virus descends to ground level. Anthrax as a soil-borne disease may also be water-borne, infected from the soil, and carried by surface drainage.
Broader Water-borne diseases (#PE3401) Vectors of animal diseases (#PD2751)
Weather as a factor of animal disease (#PE2740).
Related Zoonoses (#PD1770).
Aggravates Viral diseases in animals (#PD2730) Bacterial diseases in animals (#PD2731)
Parasitic diseases in animals (#PD2735).
Aggravated by Ecosystem modifications due to creation of dams and lakes (#PD0767).
Reduced by Insecticides as pollutants (#PD0983)
Economic loss through slaughter of diseased animals (#PE8109).

♦ PE2805 Autocracy of intergovernmental financial institutions
Claim Measures intended to intensify integration into the capitalist world system, confirm the primacy of capital and guarantee its global valorization, result in an increasing number of countries, particularly in the Third World, being forced by intergovernmental financial institutions such as the World Bank and the International Monetary Fund to adopt measures which are in direct opposition to the interests of the majority of their populations.
Broader Restrictive conditions on loans to developing countries through intergovernmental facilities (#PE9116).
Aggravates Unaccountability of international financial institutions (#PF1136)
Over-dependency on international financial institutions by developing countries (#PE2214).

♦ PE2820 Inadequate budgetary coordination within the United Nations systems
Lack of centralized financial control within the United Nations systems
Nature The lack of a centrally controlled and consolidated budget in the United Nations family of organizations, resulting in part from the fact that some agencies came into existence before the United Nations itself, leads as a natural consequence to an absence of authority in relation to programming. Each agency scrambles for funds using the vested interests which support it in national capitals. The result is a situation which can be correctly described as uncontrollable. This further exacerbated by incomplete payment of contributions by some of the States members of the organizations and from delays in payment of contributions.
Broader Inefficiency (#PB0843)
Ineffectiveness of international organizations and programmes (#PF1074).
Aggravated by Bureaucracy as an organizational disease (#PD0460).

♦ PE2824 Industrial and domestic heating emissions as air pollutants
Nature The combustion of fuel for heating and energy production represents the most common and most widespread source of atmospheric pollution. In contrast, the emissions of specific pollutants by industry are much more restricted both in area and numbers of sources. However, industrial activities are a major source of pollutants in many places. The effects of point sources of pollutants, such as industrial plants, depend on many local factors, among which are topography, weather conditions, stack height, location, control equipment, raw materials used, and the type of process. In addition to sulphur dioxide, suspended particulate matter, and oxides of nitrogen, industrial air pollutants include lead, cadmium, mercury, beryllium, the mercaptans and hydrogen sulphide, fluorides, chlorine, asbestos, and many other wastes and by-products of technological processes.
Broader Thermal pollutants (#PC1609) Industrial emissions (#PE1869)
Exploitation of fossil fuels (#PE4891).
Narrower Domestic emissions (#PG3368).

♦ PE2829 Inadequate facilities for storing, transporting and processing solid wastes
Nature The great increase in the production of wastes is causing storage, collection, and transport difficulties as well as problems of treatment and final disposal. Existing houses and apartments do not usually provide satisfactory storage for the volume and type of wastes now being produced. Furthermore, in many countries it is difficult to recruit adequate staff for the collection of refuse, and increasing traffic congestion creates difficulties and delays in transport. Inadequate storage and collection arrangements can create health and safety hazards and neighbourhood blight. On-site incinerators, which can substantially reduce refuse volume, can also cause an unacceptable degree of air pollution unless they are properly designed and operated. Kitchen grinders help to reduce the volume of putrescible refuse to be collected, but still leave up to 85 per cent or more of the total volume to be collected, treated, and disposed of.
Present methods of refuse disposal are dumping, controlled tipping or sanitary land-fill, incineration, composting, and discharge into the sea. In some cases refuse is mechanically reduced in volume before being deposited in a tip or land-fill. Dumping, which is widely practised, is an unsanitary method that creates public health hazards, nuisance, and severe pollution of the environment. Sanitary land-fill or controlled tipping is an acceptable method in suitable locations, if the possibility of pollution of surface water and ground water is eliminated. However, in most cases it is difficult to find suitable land in or near urban areas. In congested areas, the absence of the substantial areas of land required for this operation has forced the adoption of other systems.
Large world-wide variations in conditions relating to storage and collection of solid waste can render a method that is optimum for one country totally unsuitable for another. Examples of such variations are the nature and density of the wastes, access to properties, economic conditions,

and climate. In some developing countries, collection methods are unsuitable because they involve excessively laborious and unsanitary handling of the wastes. The best methods of collection in less developed countries may involve handcarts or animal–drawn carts, and agricultural tractors pulling trailers. Incineration is used successfully in several of the industrialized countries, particularly in larger cities. This method reduces the volume of domestic refuse eventually to be disposed of, usually by tipping, to about 15–20 per cent of the original. In many older plants, however, devices for removing pollutants from the exhaust gases are inadequate to comply with present air pollution standards or those likely to be established. The increasing quantities of bulky waste such as old furniture and other household equipment, also create problems in most existing incinerators; a few installations have incorporated special equipment for breaking up such large items before incineration.
Composting of domestic refuse, sometimes together with sewage sludge and other organic wastes such as market refuse and agricultural waste, has been adopted in several countries. In many cases the fraction of domestic refuse that can be suitably incorporated in compost is decreasing, resulting in increasing quantities of material (up to 50 per cent in some cases) that must be dealt with in other ways. In addition, there is some anxiety that municipal compost might have certain toxic properties, and many large composting schemes involving heavy machinery have failed because of the high cost of producing the compost, or an insufficient market. Disposal of refuse at sea, some distance from shore, is practised by some large seaside communities.
There are many ways in which components of solid waste that have some value can be recycled. Food wastes can be fed to animals; metals can be recovered and remelted, and paper can be repulped. Glass and plastics may also be removed. In some countries solid wastes have a fuelvalue that can be utilized. Alternately, methane gas can be withdrawn from well–engineered landfill sites and used as fuel. But such separation of the wastes must be controlled if health hazards, degradation of the environment and impairment of disposal operations, are to be avoided.
Incidence The collection and transportation of solid wastes are very costly and may account for up to 80 per cent of the total cost of solid wastes disposal, which in turn may absorb as much as 0.5 per cent of the gross national product. Efficient methods are therefore vital and several unconventional systems are being developed. Solid wastes are not being made full use of as potential resource materials. In the developing countries, fermentable organic matter forms a high proportion of domestic wastes and could readily be transformed into a compost, which in many countries would be a valuable soil conditioner and a minor source of plant nutrients. In the industrialized countries the improper management of solid wastes can lead to loss of valuable mineral resources such as copper, lead, tin, zinc, aluminium, iron, and crude oil. When a nation's reserve of one or more of these mineral resources has declined to the extent that practically all of its needs are imported from other parts of the world, it does not make economic sense for that nation to use these materials to produce items that are used briefly and then discarded and lost in a buried mass of heterogeneous materials.
Refs Bonomo, Luca and Higginson, A E (Eds) *International Overview on Solid Waste Management* (1988); Holmes, John R (Ed) *Managing Solid Wastes in Developing Countries* (1984); Noll, Kenneth E, et al *Recovery, Recycle and Reuse of Industrial Waste* (1985); Suess, M J, et al (Eds) *Ambient Air Pollutants from Industrial Sources* (1985).
Broader Methane gas emissions from landfill sites (#PE1256).

♦ PE2832 Denial of right to grievance procedures
Broader Denial of right to collective bargaining (#PE3970).
Aggravates Ineffectiveness of international organizations and programmes (#PF1074).

♦ PE2844 Inadequate adjustment assistance to industries and labour affected by developing country exports
Nature In the developed countries, adjustment assistance leading to trade liberalization would involve the re–allocation of production factors from less efficient into more efficient domestic industries. Temporary deterioration in the developed country's balance of payments following trade liberalization might be expected, since, even with very efficient adjustment assistance programmes, the immediate increase in imports following trade liberalization would normally occur before personnel had been fully retrained and re–allocated into more efficient export–oriented industries. Re–allocation might be expected to cause psychological and social hardships for the affected workers, as well as possible loss of income and other employment rights. The scrapping of not fully depreciated capital equipment would involve costs for the affected firms.
Claim Adjustment assistance programmes for the adaptation of domestic industries could eliminate inefficiency and competition of the industrial sector and would facilitate trade liberalization, particularly with respect to relaxation and eventual elimination of tariff and non–tariff barriers applied to imports from the developing countries. Protection of existing production and structure of employment is applied in developed market economy countries without paying due consideration to its burden on the domestic economy. If more effective adjustment assistance programmes are not soon introduced, the outlook for significantly reducing nontariff distortions of trade as well as for making further major tariff cuts is not favourable. Not only are the major remaining high–duty product lines in industries that would feel strong repercussions from tariff cuts, but many important nontariff trade–distorting measures now protect these industries.
Counter–claim The adjustment problems that result from increased imports of manufactures from developing countries vary considerably from industry to industry. However, the overall magnitudes, measured in terms of labour displacement, are not very large. Even in the case of the very large increases in imports of manufactures from developing countries assumed by this study, total labour displacement is no more than 0.7 per cent of total employment in manufacturing industry. Bearing in mind that industries have in the past adapted without too much difficulty, partly through their own effort and partly through government measures, to the much larger labour displacement created by rising labour productivity and increased imports from other developed countries, and taking into account the fact that some increases in domestic demand are likely to take place in the future, the overall impact of the increased imports is not likely to present a serious problem. Certain industries appear to be more sensitive to imports from the developing countries. The sensitive industries seem to differ slightly from country to country, but will usually include the footwear and clothing industry, the leather goods industry and, to a lesser extent, the basic metal manufacturing and electrical machinery industries.
Broader Economic and social underdevelopment (#PB0539).

♦ PE2850 Inadequate quarantine
Broader Inadequate health control (#PF9401).
Narrower Inadequate plant quarantine (#PE0714)
Inadequate animal quarantine (#PE2756).
Aggravates Pests (#PC0728) Disease vectors (#PC3595).
Aggravated by Prohibitive cost of disease control (#PF2779)
Excessive animal sanitary regulations in international travel (#PF1555).

♦ PE2853 Falls of elderly
Nature After the age of 65, one person out of three falls at least once a year. This proportion rises with age. Seventeen percent of these falls result in significant physical trauma. Another 50 percent are responsible for psychological sequelae, resulting far too often in institutionalization of these patients. Most of the time, a multiplicity of intricate causes are to blame: intrinsic causes (hypertension, neurological problems, gait disorder, visual impairment, etc.) or extrinsic factors (side effects of drugs, inadequate domestic environment and so on).
Broader Accidental falls (#PE7113).

♦ PE2854 Cement dust
Nature Cement is used as a binding agent in morter and concrete, a mixture of cement, gravel and sand. In mining for raw materials and in manufacturing and using cement, concentrations of dust can reach as 384 mg per cubic metre of air. Individuals exposed to much lower levels of concentration can contract respiratory tract diseases, digestive disorders, and skin diseases.
Broader Mineral dusts (#PE4679).
Aggravates Occupational dermatoses (#PE5684)
Diseases of the respiratory system (#PD7924)
Diseases of the skin and subcutaneous tissue (#PC8534)
Diseases and deformities of the digestive system (#PC8866).

♦ PE2857 Firearm accidents
Unsafe firearm storage — Careless firearm use
Broader Injurious accidents (#PB0731).
Aggravates Gunshot wounds (#PE9111).
Aggravated by Proliferation of weapons in civilian hands (#PE2449).

♦ PE2858 Misuse of free production zones and export enclaves
Nature Factories for world market oriented (semi–)manufacture in free production zones, export enclaves and other sites, with a structure of production which is competitive on the world market (not merely the local protected market), is very fragmented, highly susceptible to trade fluctuations and basically parasitic on the local economy and society. Nevertheless a world market for production sites is developing, on which the traditional industrial countries and the developing countries are forced to compete with and against each other to retain or attract world market oriented manufacturing industry. Although capital uses and needs the state to fulfil a variety of functions, this does not necessarily mean it has to be reliant on one particular state.
Incidence The international competitiveness of manufacture at traditional sites is threatened by lower–cost manufacture at new sites increasingly located in the developing countries and centrally planned economies. Examples can be found in synthetic fibres, textiles and garments, leather and footwear, steel–making, ship–building, watchmaking, optical industry, and sections of the mechanical and electrical engineering industries.
Claim In free trade zones cheap goods are produced which flood Western markets, displace jobs, and unleash protectionist outcries. It is also in these zones that multinational firms exploit cheap, unskilled labour, practise transfer pricing and other techniques of disguised profiteering. Technology and skill–transfer are not a consequence of setting up free trade zones because they are used primarily in the assembly stages of the production cycle.
Refs ILO *Economic and Social Effects of Multinational Enterprises in Export Processing Zones.*
Broader Abuse of economic power (#PC6873).

♦ PE2862 Institutional obsolescence in modern industrialized societies
Nature Failures in recent generations to modernize the major institutional systems of advanced societies has brought those institutions into conflict with cultural aspirations and impaired their capacities to respond to changes in material and social conditions.
Broader Preservation of obsolete systems (#PC8390)
General obstacles to problem alleviation (#PF0631)
Deterioration of industrialized countries (#PD9202).

♦ PE2863 Medical complications
Surgical complications
Broader Human disease and disability (#PB1044).
Related Pregnancy disorders (#PD2289) Complications of childbirth (#PC9042).
Aggravates Injurious accidents (#PB0731).
Aggravated by Ascariasis (#PE2395).

♦ PE2866 Ageing industrial plants and processes
Nature Current trends show a drop or stagnation in investment in industrial plants in the industrial countries ('investment gap'). There are rising or comparatively high shares of replacement investment and investment for rationalization and phase–out, coupled with falling or comparatively small shares or investment for extending capacity in the industrial countries.
Broader Inadequate maintenance of infrastructure (#PD0645).

♦ PE2871 Fluoridation of drinking water
Nature Public drinking–water may be enriched with fluoride in an effort to utilize its powerful caries–preventing effect. Doubts persist, however, concerning the application of this trace element on account of its toxic effect. Fluoride doses only twice those which effectively prevent caries may cause slight disturbances of the calcification of enamel, and doses about 20 times greater, when taken over a long period, are known to cause damage to the skeletal system. Recent research indicates that the use of chlorine (of the same chemical group as fluorine) in drinking water (to kill bacteria) may increase the risk of leukaemia in children.
Counter–claim Water fluoridation is a practical, safe and efficient public health measure. Fluoride is a normal component of human and animal tissues, found mostly in teeth and bones. It occurs naturally in varying amounts of all water supplies and in most foods. Studies made in different parts of the world show that dental caries is much less common where the fluoride content is around one part fluoride per million parts of water. A recent survey in the U.S. showed that 49.9 percent of all children has no decay in their permanent teeth, as against 36.6 percent in a similar 1979–1980 study and an estimated 28 percent in the early 1970s.
Refs WHO *Experience on Water Fluoridation in Europe* (1987).
Broader Contamination of drinking water (#PD0235)
Inadequate water system infrastructure (#PD8517).

♦ PE2876 Physical intimidation by children
Bullying by children — Bullying in schools
Incidence In the first six months of 1985, over 250 cases of bullying involving violence or intimidation were reported in Japan. In that same period two 13 year old girls and two 14 year old boys committed suicide because they could not stand the torment. In the UK in 1984 a survey of 4,000 children indicated that more than two thirds had been bullied at some time and 38 per cent were being regularly bullied. Another survey in the UK indicated that 18 per cent of secondary children and 17 per cent of middle–school children claimed to have been bullied; 8 per cent said it happened once a week, and 5.5 per cent said it happened several times per week. More than half make no attempt to report it to anybody, making it easy for teachers to assume that very little occurs. Another survey suggests that 70 per cent of children are bullied at some time; one in 7 are chronically and severely bullied.
Claim Bullying damages self–esteem, shatters lives and detrimentally affects the quality of a child's education.
Counter–claim The extent of bullying is exaggerated. Bullying is a natural part of the socialization

process whereby the child discovers how to find its place in society.
Refs Olweus, Dan *Aggression in the Schools* (1978).
  Broader Physical intimidation (#PC2934)   Victimization of children (#PC5512).
  Narrower Harassment in playgrounds (#PE7768).
  Aggravates School phobia (#PE4554)   Juvenile suicide (#PE5771)
  Cruelty to children (#PC0838).
  Aggravated by Social stigma (#PD0884)   Prejudice in children (#PD8973).

♦ **PE2890 Domestic agricultural price policy difficulties in developing countries**
Adverse effects of high-yield grain
Nature In the past, the cost of growing grain (wheat, rice, maize, barley), compared with alternative forms of production has generally been low, whereas the value of production (in relation to alternative use) has been high. What is now beginning to happen is that the real value of the grain output in developing countries is being transmitted to the farmer who, with the aid of high-yield varieties, irrigation and fertilizers, is responding to the challenge. These techniques have met with greater response amongst the large and medium scale farmers, with the danger that small farmers will be largely confined to subsistence production. This would intensify social and employment problems, and widen disparities between incomes. There still remains the difficulty of ensuring that incentive prices to farmers do not cause hardship to consumers. As developing countries approach self-sufficiency and begin to produce grain surpluses in favourable seasons, their policy problem of adapting to international market requirements at competitive prices, becomes the same as that of the developed countries.
  Broader Detrimental international repercussions of domestic agricultural policies (#PF2889).
  Aggravates Over-subsidized agriculture in industrialized countries (#PD9802)
  Neglect of agricultural and rural life in developing countries (#PF7047)
  Deterioration of domestic food production in developing countries (#PD5092).

♦ **PE2902 Excessive wheat surpluses**
Nature Periodically a combination of good weather and other factors gives rise to surpluses of wheat stocks which seriously depress the price of wheat on the international market and disrupt the economies of exporting countries. The accumulation of such surpluses has been a recurring feature of world markets for 50 years.
Claim Wheat surpluses are used by North American and European countries as a political weapon. Past occasions have seen threats to withhold surplus wheat exports from countries whose own stocks are inadequate. Due to the international political benefits of having excess grain, domestic governmental policies may be to support wheat over-production.
  Broader Agricultural surpluses (#PC2062)   Instability of wheat trade (#PE0385)
  Limiting effect of individual survivalism (#PF2602).
  Aggravated by Bad weather (#PC0293).
  Reduces Long-term shortage of wheat (#PE2903).

♦ **PE2903 Long-term shortage of wheat**
Nature Because wheat is a basic food for many parts of the world, wheat shortages and the high prices resulting from a shortage, can seriously affect the economies of importing countries. In a developing country whose local crops have failed, such a shortage may lead to famine.
  Broader Instability of wheat trade (#PE0385)
  Long-term shortage of cereals and cereal preparations (#PE1218).
  Aggravates Famine (#PB0315).
  Aggravated by Bad weather (#PC0293).
  Reduced by Excessive wheat surpluses (#PE2902).

♦ **PE2932 Abusive detention in psychiatric institutions**
Misuse of psychiatric diagnosis — Unjust commitment to psychiatric hospitals
Nature Involuntary detention of a person in a psychiatric institution may not always be motivated by their mental disorder. Its duration may also be extended unnecessarily. Abusive internment may occur on basis of mental retardation, family disputes or in order to escape severe punishment for a criminal act. Political prisoners may be committed to a psychiatric hospital rather than a normal prison, for brainwashing and indoctrination to combat ideological deviation. In the case of a well-known personality, commitment to a psychiatric hospital may more easily discredit his or her views, and avoid political martyrdom. It may also make the confiscation of property seem more justifiable, particularly if the person has held a responsible position.
Incidence Countries whose legislation does not allow for a precise definition of the terms 'mental illness' or 'mental disorder', nor for the protection of 'mental patients' against family and state, and where psychiatric hospitals are repressive institutions, offer greater possibilities for abuse. The commitment of dissidents in USSR to psychiatric hospitals, and the physical treatment they receive in these institutions, is clearly an advanced weapon of political warfare. In the early 1950s and again since 1965, it has been clear that psychiatric diagnoses of political dissenters are not based on clinical impressions or objective tests but on official instructions.
  Broader Wrongful detention (#PD6062)   Political prisoners (#PC0562)
  Political repression (#PC1919)
  Excessive institutionalization of vulnerable groups (#PF8209).
  Aggravates Mental disorders (#PD9131)   Inadequate mental hospitals (#PF4925).
  Aggravated by Social isolation (#PC1707)   Inhumane scientific activity (#PC1449)
  Denial of rights to prisoners (#PD0520)   False diagnosis of mental disorder (#PG5868)
  Denial of rights of mental patients (#PD1148)   Unethical practices in psychotherapy (#PD5267)
  Unethical practices of health services (#PE3328).
  Reduced by Political martyrdom (#PS5869).

♦ **PE2945 Paraplegia**
Locomotor ataxia
Refs Bromley, Ida *Tetraplegia and Paraplegia* (1985); Mooney, Thomas O, et al *Sexual Options for Paraplegics and Quadriplegics* (1975); Rogers, Michael A *Living with Paraplegia* (1986).
  Broader Paralysis (#PD2632)   Critical illnesses (#PE9038).
  Narrower Friedreichs ataxia (#PE8605)   Spastic infantile paraplegia (#PG2939).
  Aggravated by Cerebral paralysis (#PE0763).

♦ **PE2946 Tetraplegia**
Paralysis of anatomical extremities — Quadriplegia
Refs Bromley, Ida *Tetraplegia and Paraplegia* (1985); Mooney, Thomas O, et al *Sexual Options for Paraplegics and Quadriplegics* (1975).
  Broader Paralysis (#PD2632)   Cerebral paralysis (#PE0763).

♦ **PE2951 Weakness in primary commodity trade amongst developing countries**
Nature The volume of trade among developing countries represents only a small proportion of their total foreign trade. This pattern of trade is in sharp contrast to that of the developed countries: the bulk of exports of primary commodities (other than fuels) from developed market countries, and of those from socialist countries, consists of their respective intra-trade. The pattern of primary commodity imports of developing countries is also in striking contrast to that of developed countries. Whereas the greater part of all imports of primary commodities (other than petroleum), into the developed market economies consists of their intra-trade, the intra-trade of developing countries furnishes only about one third of their total primary commodity imports (other than petroleum).

Incidence Although the intra-trade between developing countries covers a wide variety of primary commodities, it is fairly heavily concentrated, in terms of value, in a limited number of items. In 1979, for example, the value of such intra-trade accounted for only one quarter of their total merchandise exports: for non-fuel primary commodities it was about 24 per cent; but for primary commodities, the corresponding proportion was less than one seventh. Petroleum is by far the most important single item, which in 1979 accounted for over 50 per cent of the total intra-trade in primary commodities. Food commodities (rice, fats and oils, tea, sugar, meat, etc) account for a further 12 per cent of the total. The fact that certain of these, such as tea and sugar, are among the more important commodity exports to developed countries would seem to indicate that some part of the present network of intra-trade among developing countries is a by-product of the main stream of trade between developing and developed countries. This is seen more clearly in the case of petroleum, where production capacity and related capital investment are stimulated by developed country demand.
  Broader Weakness in trade among developing countries (#PC0933).
  Aggravates Excessive external trade deficits (#PC1100).

♦ **PE2953 Weakness in trade between socialist and developing economies**
Nature The desired expansion of trade between the socialist and the developing countries continues to be limited by the inadequate complementarity of their respective economic structures.
Incidence In 1972, despite improvement in recent years, the foreign trade of the socialist countries with developing countries only amounted to 14.1 per cent of their total export trade, and 10.0 per cent of their import trade. In 1982, total foreign trade of socialist with developing countries amounted to 18.0 percent; while total foreign trade of developing with socialist countries only amounted to 3.7 percent.
  Broader Weakness in trade between different economic systems (#PC2724).

♦ **PE2954 Weakness in trade between socialist and developed market economies**
Nature The differences in the trading practices between these two economic systems has acted and continues to act as an impediment to trade expansion. Consistent trade practices are especially hindered by political intervention into the market, such as the US grain embargo of 1980. Rapid changes cannot be expected because of such differences.
Incidence In 1972, despite considerable expansion in East-West trade over the past decade, 63.5 per cent of the export trade of the Eastern European socialist countries was to socialist countries, and 22.4 per cent was to developed market economy countries. In 1982 Western, industrialized states counted 4.8 percent of their trades as being with socialist economies. Socialist economies counted 31.7 percent of their trade with the Western market economies.
  Broader Weakness in trade between different economic systems (#PC2724).

♦ **PE2964 Sulphur trioxide as a pollutant**
Nature Sulphur trioxide in solid, liquid or as a mist can cause severe damage to living tissue. Used in manufacturing of dyes, anhydrous nitric acid, and explosives and for sulphonation of organic acids. Exposure to sulphur trioxide, while normally confined to those working with it, can result in chronic respiratory tract damage, metabolic damage and acid burns.
  Broader Pollutants (#PC5690).
  Related Sulphur compounds as pollutants (#PG6442).

♦ **PE2966 Weakness in trade in manufactured goods among developing countries**
Nature The volume of trade in manufactured goods among developing countries represents only a small proportion of their total foreign trade. This is part of the vicious circle of dependency on markets in developed countries and the need for export earnings in stable and readily convertible currencies. The bartering of some manufactured goods among developing countries has not yet been fully exploited as a way out of old trading patterns.
  Broader Weakness in trade among developing countries (#PC0933)
  Underdevelopment of manufacturing industries (#PF0854).
  Aggravated by Minimal manufacturing in least developed countries (#PE0282).

♦ **PE2975 Pests and diseases of sugar-beet**
Nature Sugar-beet is attacked by many pests which may cause direct injury or may introduce virus diseases to the plant, causing severe economic loss. In most parts of the world virus diseases and nematodes appear to be the principal problem; and, generally speaking, the same pests and diseases occur wherever sugar-beet is grown, although the predominant ones of Europe and neighbouring countries differ from those in America.
Incidence The insect pests which cause direct injury to sugar-beet generally do so at the seedling stage. Soil insects like wireworms and pygmy beetles are destructive on crops in America and northern Europe. In the more southern beet-growing regions such as Spain, Italy and Turkey, the beet flea beetle is more important. The beet leaf miner or beet fly is a very common pest of sugar beet in northern Europe, with injuries consisting of blisters in the leaves caused by the feeding of the larvae. The beet cyst nematode is an important pest, present in all established beet-growing countries.
The most important insect pests of sugar beet are aphids, which attack both the root and seed crops. They occur throughout Europe south to Turkey, and can lower the yield of beet and ruin the seed crop entirely.
Apart from this they are also important vectors of disease, notably mosaic and the yellow viruses. Virus yellows is the predominant disease of northern Europe, though it is also present in southern Europe and the USA. The incidence of this disease is extremely variable from year to year, depending on the winter conditions of green peach aphid and virus.
The curly top disease is the predominant disease in North America, where it is a controlling factor in beet cultivation. It is also prevalent in South America, but absent in Europe. The disease is caused by a virus which is transmitted by leaf hoppers. In the USA a bad attack can reduce the normal average yield of 15 tons of beet per acre to 5 tons or less. Severe outbreaks in the west of the USA have forced farmers to give up sugar-beet growing.
Rhizomania is a virus disease transmitted by the fungus Polymyxa betae which can stay alive without a host crop for more than 15 years. The root system of the sugar beet is ruined by this disease, which can lead to complete crop failure. Initially rhizomania occurred mainly in the south of Europe and Japan, but is spreading in northern Europe and America. Until now no sugar beet growing was possible in severely infected fields, but progress in breeding for tolerance and resistance solved this problem to some extent.
  Broader Crops pests and diseases (#PE7783).
  Related Infestation of seeds (#PE6271)   Maize pests and diseases (#PE3589)
  Pests and diseases of rice (#PE2221)   Pests and diseases of wheat (#PE2222)
  Pests and diseases of vines (#PE2985)   Pests and diseases of cotton (#PE2220)
  Pests and diseases of potato (#PE2219)   Pests and diseases of sugar cane (#PE2217)
  Pests and diseases of grain sorghum (#PE3590).
  Aggravated by Crown gall (#PE2230)   Pests of plants (#PC1627)
  Aphids as pests (#PE3613)   Beet leaf miner (#PG6915)
  Wireworms as pests (#PE3645)   Viral plant diseases (#PD2227)
  Nematoid plant diseases (#PD2228)   Beetles as insect pests (#PE1679)
  Bacterial plant diseases (#PD2226)
  Insect vectors of viral diseases of plants (#PD3600).

♦ PE2976 Pests and diseases of citrus fruit

Nature Citrus is attacked by innumerable pests and pathogens. Although many of the pathogens are widespread, their relative importance may vary from region to region. Thus, citrus is subject to attack by a number of diseases, each of very great regional importance, but often minor importance elsewhere. Of the insect pests, citrus being a non-rotational crop, provides an ideal host for those which are sedentary and immobile.

Incidence The most important insect pests of citrus are the scale insects. The red scale which originated in Japan is the most important pest in California, South Africa and the Mediterranean region. The black scale occurs in all the main citrus-growing regions including South America. Scale insects invade the entire tree, attacking the wood, the foliage and the fruit. With such an infestation, heavily attacked trees shed their leaves and the branches die back. The second most important pests are mealybugs, which are widely distributed throughout the world. Other major insect pests include the white fly, which occurs in the USA, the black-fly which is important in India and in the Far East, and the Mediterranean Fruit Fly which is a serious pest in the Mediterranean region and South Africa. Red spider is also serious both in the USA and South Africa. In the Mediterranean region its place is taken by the rust mites.

Of all citrus diseases, a fungal disease, brown rot gummosis, or root rot, is the most widely distributed; the two alternative names describing the two most obvious symptoms. If the attack is severe, trees may die within a year. It caused very serious losses in California and Florida and in Sicily, where the citrus trees were wiped out between 1863-70. Melanose is another fungal disease which can cause very heavy losses. It disfigures the fruit of all varieties of citrus, particularly grapefruit. Melanose is very widely distributed in regions with an early summer rainfall such as Florida. Citrus Tristeza is a serious viral disease in South America, South Africa and Australia. Trees which have been attacked may die within 3 months. In Brazil and Argentina during the 1930s and 1940s citrus growers suffered the appalling loss of some 20 million trees due to this disease.

Some 30 different species of fungi are recorded as causing rotting of citrus fruits. The chief offenders are two species of Penicillium, the common green mould and the blue contact mould. The two moulds have a world-wide distribution and are found wherever there are citrus fruits; in the plantation, in the pack-house and in the markets.

Broader Pests and diseases of trees (#PD3585).
Related Pests and diseases of tea (#PE2980)
Pests and diseases of elm (#PE2982)
Pests and diseases of palms (#PE2981)
Pests and diseases of rubber (#PE2977)
Pests and diseases of poplars (#PE4413)
Pests and diseases of deciduous fruit (#PE3591).
Pests and diseases of oak (#PE2984)
Pests and diseases of cocoa (#PE2979)
Pests and diseases of coffee (#PE2218)
Pests and diseases of olives (#PE2978)
Pests and diseases of chestnut (#PE2983)
Aggravated by Mites as pests (#PE3639)
Fruit flies as pests (#PE3607)
Fungal plant diseases (#PD2225)
Nematoid plant diseases (#PD2228)
Lepidoptera as insect pests (#PE3649).
Viral plant diseases (#PD2227)
Black flies as pests (#PE3646)
Scale insects as pests (#PE3612)
Bacterial plant diseases (#PD2226)

♦ PE2977 Pests and diseases of rubber

Nature Rubber trees are subject to attack by a number of fungal diseases. One, the South American leaf disease, effectively bars the commercial production of rubber in its native continent, South America. On the whole insect pests do not pose a serious problem for rubber planters, though they can be destructive in combination with certain diseases. The rubber tree is deciduous, and the new foliage which grows after the 'wintering' period is susceptible to attack by fungi, insects and mites, in combination. In a severe attack, the young leaves are killed and drop off, a process known as secondary leaf fall. The defoliation can be as severe as 50 to 100 per cent, with a consequent reduction in rubber yield of a half.

Incidence The most serious diseases of rubber are root diseases which are caused by fungi that attack and eventually kill the roots. The diseases persist in the soil on infected dead roots. Freshly planted rubber trees remain healthy until their roots come into contact with pieces of diseased root. From a few foci of infection, a whole plantation can be destroyed.

Several fungi can attack and damage the tapping panels. The most important is mouldy rot, which first appeared in Malaya in 1916 and in Java in 1920. A potential threat to the world's rubber industry is the South American leaf blight. This attacks wild rubber trees which occur in the tropical forests of Central and South America. Various attempts, such as those by the Ford and Goodyear companies, have been made to establish rubber plantations in these areas. All have been abandoned because of the devastation caused by leaf blight. The disease is a serious potential threat to the Asian rubber plantations.

Broader Pests and diseases of trees (#PD3585).
Related Pests and diseases of tea (#PE2980)
Pests and diseases of elm (#PE2982)
Pests and diseases of palms (#PE2981)
Pests and diseases of olives (#PE2978)
Pests and diseases of chestnut (#PE2983)
Pests and diseases of deciduous fruit (#PE3591).
Pests and diseases of oak (#PE2984)
Pests and diseases of cocoa (#PE2979)
Pests and diseases of coffee (#PE2218)
Pests and diseases of poplars (#PE4413)
Pests and diseases of citrus fruit (#PE2976)
Aggravated by Mites as pests (#PE3639)
Fungal plant diseases (#PD2225).

♦ PE2978 Pests and diseases of olives

Nature The olive tree is subject to attack by a number of pests and diseases. Because the olive tree can grow to a great age and because there are many wild olive trees scattered in olive growing districts, the systematic control of certain pests is very difficult. The main animal pests are: dacus eleae, or olive fruit fly; prays oleae, or olive kernel borer, or olive moth; saissetia oleae, or black scale. The olive tree is also attacked by cryptogamic and bacterial diseases such as olive lenot and olive leaf spot.

Incidence The most important pest of olives, especially in Italy, Spain and Turkey, is the olive moth. It occurs in all the Mediterranean olive-growing countries. The caterpillar of the moth feeds on the flower organs, especially the pistil and ovary, and later attacks the young fruit. In some years 50 per cent of the flowers are destroyed, followed by a heavy fall of fruit. The other principal pest is the olive fly which is similar in activity to the notorious Mediterranean fruit fly. The eggs of the olive fly are deposited inside the fruit; the larvae feed on the pulp of the fruit, causing maggoty olives. The olive fly occurs throughout the Mediterranean region, especially in Italy. About 20 per cent of olives are attacked and, where attacks are heavy, they are unsuitable as table olives.

Olive leaf spot is a fungal disease which is found around the Mediterranean and in Eritrea, South Africa, California, Argentine and Chile. The fungus infects the leaves, severe attacks resulting in partial to complete defoliation of the tree, which seriously reduces the crop in the following season. Olive knot is a bacterial disease which occurs in the Mediterranean region, in California and Argentina. The disease causes knots to develop on the young shoots which eventually kill the shoots. The bacteria can also infect the leaves, causing premature leaf drop.

Broader Pests and diseases of trees (#PD3585).
Related Pests and diseases of tea (#PE2980)
Pests and diseases of elm (#PE2982)
Pests and diseases of palms (#PE2981)
Pests and diseases of rubber (#PE2977)
Pests and diseases of chestnut (#PE2983)
Pests and diseases of deciduous fruit (#PE3591).
Pests and diseases of oak (#PE2984)
Pests and diseases of cocoa (#PE2979)
Pests and diseases of coffee (#PE2218)
Pests and diseases of poplars (#PE4413)
Pests and diseases of citrus fruit (#PE2976)
Aggravated by Borer insects (#PE6286)
Fungal plant diseases (#PD2225)
Bacterial plant diseases (#PD2226)
Fruit flies as pests (#PE3607)
Scale insects as pests (#PE3612)
Lepidoptera as insect pests (#PE3649).

♦ PE2979 Pests and diseases of cocoa

Nature Pests and diseases cause heavy losses in cocoa production, making cocoa estates uneconomic in both the original home of the plant, South America, and the new centre of world cocoa production, West Africa.

Incidence The most important pests of cocoa in West Africa are capsid bugs; they feed on the young new growth, holding up the natural growth of the tree and preventing the development of pods. The extensive use of insecticides to control capsids and mealy bugs, the vector of virus diseases, has led to a disturbing increase in other insect pests, notably a pod miner and a shoot borer.

The most serious problem facing the West African cocoa industry is the menace of a complex of virus diseases referred to as swollen shoot. So far, some fifty distinct viruses have been distinguished as part of the swollen shoot virus complex. Many of these isolates are littoral, the time taken to kill the trees varying from three to six years. No method is known of curing an infected tree, so that there is no alternative to the destruction of infected trees in the control of the disease. The incidence of the problem has increased rapidly over the last 40 years. In 1936, when work on swollen shoot began, the outbreaks were limited but scattered over a large area. The outbreaks grew larger and fresh sites became infected in succeeding years. In the Eastern region of Ghana the output of dry cocoa fell from 116,000 tons in 1936-7 to 38,000 ton in 1955-6, and now cocoa production in much of the region has been wiped out. Cutting out first started in 1941; and in 1948, following the advice of a Commission of Enquiry, a general policy of cutting out infected trees was adopted in West Africa. This policy led to 63 million trees, 10 percent of the total, being cut out in the period 1946-57 in Ghana alone. In Nigeria from 1946-50 a total of 1,500,000 trees were cut out; but thereafter large areas were abandoned as being too heavily infected. Although the campaign has been reasonably effective it has proved politically unpopular especially with farmers and is liable to interruption.

Another important cocoa disease is black pod, which, it has been estimated, reduces the world production by 7 percent, but in some areas, such as Nigeria and the Cameroons, losses may be as high as 90 percent.

The most damaging disease of cocoa in South America is Witches' broom disease, a fungal disease, which has completely destroyed whole crops. The original home of the disease was the Upper Amazon Valley. From there it has spread outwards to the neighbouring South American countries – Peru, Ecuador, Colombia, Venezuela and the Guianas. In the 1920s it practically wiped out cocoa growing in Surinam, Trinidad and Ecuador. It also infiltrated the West Indies, reaching Trinidad in 1928, Tobago in 1939, and Grenada in 1948.

Refs Baker, R E D and Holliday, P *Witches' Broom Disease of Cacao* (1957).

Broader Pests and diseases of trees (#PD3585).
Related Pests and diseases of tea (#PE2980)
Pests and diseases of elm (#PE2982)
Pests and diseases of coffee (#PE2218)
Pests and diseases of olives (#PE2978)
Pests and diseases of chestnut (#PE2983)
Pests and diseases of deciduous fruit (#PE3591).
Pests and diseases of oak (#PE2984)
Pests and diseases of palms (#PE2981)
Pests and diseases of rubber (#PE2977)
Pests and diseases of poplars (#PE4413)
Pests and diseases of citrus fruit (#PE2976)
Aggravated by Borer insects (#PE6286)
Fungal plant diseases (#PD2225)
Lepidoptera as insect pests (#PE3649)
Insect vectors of viral diseases of plants (#PD3600).
Viral plant diseases (#PD2227)
Hemiptera as insect pests (#PE3615)

♦ PE2980 Pests and diseases of tea

Nature The principal pests and diseases of tea all occur in the Far East, the most important being blister blight disease. Pests in other parts of the world are mainly local ones, and the plantations in East Africa are relatively disease free.

Incidence The two most damaging pests are tea mosquito bug and the red spider mite. In severe attacks, the feeding of the bugs may entirely suppress development of new growth and whole areas of planted tea may go out of production. The most dangerous tea disease is blister blight, a fungal disease which quickly spreads until practically all the young foliage is diseased. Although it was first recorded in 1868, it was relatively unimportant until 1946 when it suddenly spread to Southern India and Sri Lanka, where it would have wiped out the tea industry had fungicides not been available. From it spread to Sumatra in 1949, Malaya in 1950 and Java in 1951. Root diseases of tea are also a serious problem in most of the tea growing areas of the world. At least nine different causal fungi have been recognized as being lethal to the plant.

Broader Pests and diseases of trees (#PD3585).
Related Pests and diseases of oak (#PE2984)
Pests and diseases of cocoa (#PE2979)
Pests and diseases of rubber (#PE2977)
Pests and diseases of coffee (#PE2218)
Pests and diseases of chestnut (#PE2983)
Pests and diseases of deciduous fruit (#PE3591).
Pests and diseases of elm (#PE2982)
Pests and diseases of palms (#PE2981)
Pests and diseases of olives (#PE2978)
Pests and diseases of poplars (#PE4413)
Pests and diseases of citrus fruit (#PE2976)
Aggravated by Mites as pests (#PE3639)
Hemiptera as insect pests (#PE3615).
Fungal plant diseases (#PD2225)

♦ PE2981 Pests and diseases of palms

Nature Of the two economically important palms, oil and coconut, the coconut is the most susceptible to pests and diseases. Four major epidemic diseases have killed millions of coconut palms; in the absence of a visible pathogen these diseases are attributed to viruses. Animal pests and fungal diseases attack both types of palm.

Incidence The most important animal pest of palms is the rhinoceros beetle, occurring in the Far East, in Africa, in Central America and the Pacific Islands. It attacks and severely damages palms of all ages but is most important in young plantations. In the South Pacific Islands up to 50 per cent of young palms have been destroyed. Several species of leaf-feeding caterpillars, principally on oil palms, can cause extensive defoliation.

Four major diseases (of uncertain origin) of coconut all kill affected trees. Bronze leaf wilt, which is found in Trinidad and the Guianas, kills the palm within 4-6 months of the onset of the disease. Lethal yellowing, which has been recognized in Jamaica and the Cayman Islands for over a hundred years, kills affected trees in 3 months. Palm seedlings replanted in the same area usually succumb to the disease after 2 or 3 years. Cadang-cadang kills palms in 5-6 years. It occurs in the Philippines where it has been present for at least 50 years, but its incidence has increased recently. It has been responsible for the death of 14 million coconut palms. Kerala wilt has killed 10 million coconut palms in Kerala State (India) alone. It kills the tree over a period of 3-15 years.

The diseases of oil palms are very dependent on cultural practices. When the palms are grown in small scattered units, the main disease is a trunk rot, which is caused by various species of ganoderma, but when they are grown in plantations fusarium wilt is liable to cause serious losses. The fungal disease, but rot, kills infected trees.

Refs CAB International Mycological Institute *Diseases of Oil Palm*; Turner, P D *Oil Palm Diseases and Disorders*.

Broader Pests and diseases of trees (#PD3585).
Narrower Pests and diseases of the coconut palm (#PG5999).

Related Pests and diseases of tea (#PE2980)
Pests and diseases of elm (#PE2982)
Pests and diseases of rubber (#PE2977)
Pests and diseases of coffee (#PE2218)
Pests and diseases of chestnut (#PE2983)
Pests and diseases of deciduous fruit (#PE3591).
Pests and diseases of oak (#PE2984)
Pests and diseases of cocoa (#PE2979)
Pests and diseases of olives (#PE2978)
Pests and diseases of poplars (#PE4413)
Pests and diseases of citrus fruit (#PE2976)
Aggravated by Viral plant diseases (#PD2227)
Beetles as insect pests (#PE1679)
Fungal plant diseases (#PD2225)
Lepidoptera as insect pests (#PE3649).

♦ **PE2982 Pests and diseases of elm**
Nature The elm tree is subject to attack by numerous pests and diseases: aphids, bark beetles, blight, leaf spot, moths, scale insects and sooty mould.
Broader Pests and diseases of trees (#PD3585).
Narrower Dutch elm disease (#PE1154).
Related Pests and diseases of tea (#PE2980)
Pests and diseases of cocoa (#PE2979)
Pests and diseases of rubber (#PE2977)
Pests and diseases of coffee (#PE2218)
Pests and diseases of chestnut (#PE2983)
Pests and diseases of deciduous fruit (#PE3591).
Pests and diseases of oak (#PE2984)
Pests and diseases of palms (#PE2981)
Pests and diseases of olives (#PE2978)
Pests and diseases of poplars (#PE4413)
Pests and diseases of citrus fruit (#PE2976)
Aggravated by Aphids as pests (#PE3613)
Fungal plant diseases (#PD2225)
Beetles as insect pests (#PE1679)
Viral plant diseases (#PD2227)
Scale insects as pests (#PE3612)
Lepidoptera as insect pests (#PE3649).

♦ **PE2983 Pests and diseases of chestnut**
Nature The chestnut tree is attacked by numerous pests and diseases: anthracnose, beetles, blight and leaf spot. The most important of these is chestnut blight, a canker disease, which since its introduction to North America in 1904 has eliminated the native chestnut as a commercial species.
Broader Pests and diseases of trees (#PD3585).
Related Pests and diseases of tea (#PE2980)
Pests and diseases of oak (#PE2984)
Pests and diseases of palms (#PE2981)
Pests and diseases of rubber (#PE2977)
Pests and diseases of poplars (#PE4413)
Pests and diseases of deciduous fruit (#PE3591).
Pests and diseases of elm (#PE2982)
Pests and diseases of cocoa (#PE2979)
Pests and diseases of coffee (#PE2218)
Pests and diseases of olives (#PE2978)
Pests and diseases of citrus fruit (#PE2976)
Aggravated by Fungal plant diseases (#PD2225)
Beetles as insect pests (#PE1679)
Bacterial plant diseases (#PD2226)
Bird vectors of plant disease (#PD3601).

♦ **PE2984 Pests and diseases of oak**
Nature The oak tree is attacked by a large number of pests and diseases including: anthracnose, beetles, blight, borers, brown rot, cankers, chlorosis, galls, leaf miners, leaf rollers, leaf scorch, mealy bugs, mildew, red spider, root rot, rust, scale, thrips and wood rot. Perhaps the most notorious disease is oak wilt, which kills within a matter of weeks by plugging the tree's vascular system. The fungus is disseminated through root grafts and by insects.
Broader Pests and diseases of trees (#PD3585).
Related Pests and diseases of tea (#PE2980)
Pests and diseases of cocoa (#PE2979)
Pests and diseases of coffee (#PE2218)
Pests and diseases of olives (#PE2978)
Pests and diseases of chestnut (#PE2983)
Pests and diseases of deciduous fruit (#PE3591).
Pests and diseases of elm (#PE2982)
Pests and diseases of palms (#PE2981)
Pests and diseases of rubber (#PE2977)
Pests and diseases of poplars (#PE4413)
Pests and diseases of citrus fruit (#PE2976)
Aggravated by Borer insects (#PE6286)
Fungal plant diseases (#PD2225)
Insect vectors of disease (#PC3597)
Thysanoptera as insect pests (#PE3619).
Viral plant diseases (#PD2227)
Beetles as insect pests (#PE1679)
Hemiptera as insect pests (#PE3615)

♦ **PE2985 Pests and diseases of vines**
Nature The grapevine and its fruit are seriously attacked by a number of insects and diseases. Diseases cause more damage than pests, particularly in the more humid regions.
Incidence Two of the more important insect pests are the grape berry moth and la cochylis. Both are well known in Europe, and also occur in Egypt, North Africa and Japan. Damage is caused by the caterpillars feeding on the developing fruit. In the USA the grape leaf-hopper is the most important pest. The Vine phylloxera, a type of aphid, occurs in most vine-growing parts of the world, in Europe, North Africa, South Africa, South America, Australia, Mexico and the Middle East. The insect appears to have been introduced into Europe from America about 1860. In California it caused devastation in the vineyards, and in France it destroyed at least 1,500,000 hectares and threatened the total extinction of the industry.
Of the numerous diseases that attack vines, two fungal diseases are of outstanding importance: vine downy mildew and powdery mildew. Both are natives of the USA. Vine powdery mildew, which lives on both the leaves and fruit, crossed the Atlantic in 1845 (the same year as potato blight) when it was first recognized in England. It reached France in 1848 and then spread steadily through Europe. It is now found wherever grapes are grown. Vine downy mildew reached France in 1878 on rootstock imported from America because of its aphid resistant qualities (the destructive aphid, Phylloxera, had also been introduced from North America) and soon reached epidemic proportions. The disease is of particular importance in the more humid regions such as France, Italy and parts of Spain, but negligible in the drier climates of California. The most important vine disease in the USA is black rot; it is estimated to cause losses of about 20 per cent in the south, and up to 10 per cent in the north.
Broader Crops pests and diseases (#PE7783).
Related Infestation of seeds (#PE6271)
Pests and diseases of rice (#PE2221)
Pests and diseases of cotton (#PE2220)
Pests and diseases of sugar cane (#PE2217)
Pests and diseases of grain sorghum (#PE3590).
Maize pests and diseases (#PE3589)
Pests and diseases of wheat (#PE2222)
Pests and diseases of potato (#PE2219)
Pests and diseases of sugar-beet (#PE2975)
Aggravated by Phylloxera as pests (#PE3611)
Lepidoptera as insect pests (#PE3649).
Fungal plant diseases (#PD2225)

♦ **PE2999 Economic inefficiencies in developing countries due to restrictive business practices**
Nature Restrictive business practices tend to interfere with the efficient use of economic resources and thus reduce economic growth and welfare, with particularly adverse effects on developing countries. Such practices include: cartel activities; export prohibitions; agreements on market distribution and allocation; tying of the supply of inputs including raw materials and components; restrictions specified in contracts for the transfer of technology; arbitrary transfer pricing between the parent company and its affiliates; monopoly practices.
Claim Both the International Monetary Fund and the World Bank impose upon developing countries draconian measures designed to promote economic growth and stability, most often at the price of extreme human suffering and environmental damage. Such measures fly in the face of the best learnings of the developed world about the need to manage economic development carefully if the populace and the environment are to benefit in the long run.
Broader Economic inefficiency (#PF7556).
Related Discouraging conditions for small business (#PD5603)
Ineffective economic structures in industrial nations (#PE4818).

♦ **PE3000 Instability of pepper trade**
Nature Since the mid-1950s, pepper has shown the highest degree of annual fluctuation in price of any primary commodity. This instability has reflected considerable short-term fluctuation in production owing to changes in weather conditions and incidence of plant diseases, as well as short-term speculative transactions. An additional factor for price instability is that, as a result of a relatively long gestation period, increases in production resulting from earlier periods of higher prices result in price declines which, in turn, eventually lead to reduced output. These cyclical movements in production often have drastic repercussions on prices.
Incidence The relative importance of pepper in world trade is small, the total value amounting to about $60 million each year. However, in the three principal exporting countries, pepper constitutes a major source of cash income for a large part of the population in particular localities.
Broader Instability of spices trade (#PE1619).

♦ **PE3001 Health hazards of asbestos**
Asbestos as an occupational hazard — Asbestosis
Nature Asbestos is a mixture of magnesium and iron silicates in fibrous form. It appears as dust in the form of fine fibres in the air. Occupational exposure occurs in asbestos mines and wherever asbestos or asbestos products are in use; for instance, in handling asbestos-cement products used in the building industry (roofing sheets, wallboard and pipes). Exposure may also occur in the textile industry in the manufacture of fireproof materials, such as asbestos clothes or brake linings for motor vehicles. There is increasing concern about water-borne asbestos in communities which have used asbestos-and-cement pipe to transmit water from reservoirs ad wells. Asbestos is also used for insulation and fire protection purposes in ship-building, house-building, and in the undersealing of cars. Asbestos enters the body by inhalation, and fine dust containing fibres of diameters less than 5 microns and lengths greater than 5 microns may be deposited in the alveoli. The fibres are insoluble. The dust deposited in the lungs causes fibrosis, pleural plaques, mesothelioma and lung cancer. Asbestosis results in impaired lung function after five to ten years. The symptoms are shortness of breath, chest pain, and later bronchitis with increased sputum and clubbing of the fingers. Radiography of the lungs shows certain characteristic changes.
Incidence In the USA, for example, projections indicate that in the second half of this century, about 2 million workers exposed to asbestos will have died of from cancers it has caused. In the United Kingdom, it is estimated that as many as 500,000 workers will die from asbestos related disease over the next 30 years. In 1985, Martin Marietta, a major US corporation, agreed to pay restitution charges to the employees and their families who were injured by asbestos inhalation resulting from their work. While countries like the US are considering a total ban on asbestos, Third World nations such as Malaysia, Thailand and China are actually buying more because it is a cheap building material easily mixed with cement. Canada is the largest exporter and is actively promoting its use.
Background Asbestos was first used in Finland about 2500 BC to strengthen clay pots. In classical times, the indestructible shrouds in which in the eminent were preserved were woven from asbestos (the word comes from the Greek meaning inextinguishable or indestructible), and it is still use for lamp wicks.
Refs ILO *Safety in the Use of Asbestos* (1984); McCulloch, Jock *Asbestos*; Peters, George A and Peters, Barbara J *Sourcebook on Asbest Diseases, Vol II* (1986); Selikoff, Irving and Lee, Douglas H *Asbestos and Disease* (1978).
Broader Occupational hazards (#PC6716)
Asbestos as a pollutant (#PE1127).
Hazards to human health (#PB4885)
Aggravates Mesothelioma (#PG5882)
Occupational cancer (#PE3509)
Occupational diseases (#PD0215)
Lung disorders and diseases (#PD0637)
Pleura and peritoneum cancers of the bronchi (#PE8228).
Pneumoconiosis (#PD2034)
Environmental hazards (#PC5883)
Occupational risk to health (#PC0865)
Aggravated by Smoking (#PD0713)
Dust as an occupational hazard (#PE5767).

♦ **PE3004 Digestive disorders of the rumen**
Broader Diseases of the digestive system in animals (#PD3978).
Narrower Grain overload (#PG7288)
Ruminal parakeratosis (#PG3729)
Vagus indigestion (#PG7816)
Simple indigestion in animals (#PG9250).

♦ **PE3008 Hunting tourism**
Broader Killing of animals (#PD8486)
Social environmental degradation from recreation and tourism (#PD0826).
Hunting of animals (#PC2024)
Narrower Stag hunting (#PJ3556)
Disruptive airplane hunting (#PG0382).
Animal trophy hunting (#PE5644)
Related Denial to animals of the right to freedom from mass killing (#PE9650).
Aggravated by Human use of animal by-products (#PF1964).

♦ **PE3016 Forced political confessions**
Nature Confessions of guilt, extorted by means of torture, brainwashing and inhumane treatment for the purposes of conviction in civil and political trials (secret or show), may be sufficient to condemn an individual to a long prison sentence, forced labour, exile or execution. The defendant may in fact not be guilty and may be condemned unjustly. His confession, true or false, may endanger the lives of others. Confessions may serve as propaganda for indoctrination purposes in show trials, or as an administrative convenience and salving of conscience in secret trials.
Broader Forced confession (#PE8947)
Inhumane interrogation techniques (#PD1362).
Narrower Show trials (#PG5885).
Related Political repression (#PC1919)
False political evidence (#PD3017).
Political indoctrination (#PD1624)
Aggravates Political trials (#PD3013).
Aggravated by Police intimidation (#PD0736)
Political corruption of the judiciary (#PE0647).
Excessive government control (#PF0304)

♦ **PE3034 Inadequate protection of war correspondents**
Nature Inadequate provisions may be made for safeguarding the security of war correspondents and for their access to information. War correspondents risk death and injury, but may also be interned as prisoners of war and political prisoners and subjected to torture, brainwashing and other injustices. This may serve to restrict available information and to mask war crimes and other atrocities: propaganda may be substituted for faithful reporting.
Broader Inadequate protection of civilians in armed conflict (#PE8361).
Related Political prisoners (#PC0562).
Aggravates Restriction of freedom of expression (#PC2162).
Aggravated by Espionage (#PC2140)
Newspaper and periodical censorship (#PD3027).

♦ **PE3035 Environmental hazards of non-alcoholic beverages**
Carcinogenic soft drinks — Cancer-causing additives in mineral waters — Carcinogenic carbon-

ated drinks
Incidence In California legislation has been introduced clearly stating the cancer-related dangers of the consumption of soft drinks. In India in 1990, soft drinks containing brominated vegetable oil (a known carcinogenic emulsifier, long identified by WHO as unfit for human consumption) continued to be sold in the bazaars throughout the country. As an indication, the market for carbonated drinks in the UK in 1986 was £1760 million.
Broader Environmental hazards from beverages (#PE0849).

♦ **PE3044 Denial of right to periods to nurse infants during working hours**
Broader Inadequate maternity protection in employment (#PD6733).
Aggravated by Extortionate bureaucracy (#PD8655)
Corruption of ruling classes (#PD8380).

♦ **PE3066 Decline in export credits to developing countries**
Burden of export credit financing upon developing countries — Lack of credit guarantee facilities for developing country exports
Nature The growth of developing country exports has necessitated the extension of credit to their foreign customers, including developed countries. The amount of credit thus provided by developing countries to developed countries in connection with exports on deferred-payments terms may be such as to impose a heavy strain on the balance of payments position of the developing country.
Incidence Export credit agencies have encouraged short-term credits, particularly to the countries which did not experience debt-servicing difficulties or which did have payments problems but were implementing adjustment programmes. As a result, total outstanding officially-supported export credits rose by $7.8 billion in 1985 compared with $3 billion in 1983. Export credits with longer maturities continued on a downward trend throughout the first half of the 1980s, and collapsed in 1985 to a net $1.9 billion, or less than half the previous year's level.
Broader Export credit risks (#PF3065).
Aggravates Imbalance of payments (#PC0998)
Deterioration in external financial position of developing countries (#PE9567).
Aggravated by Export credit competition (#PD3067).

♦ **PE3077 Propaganda by intergovernmental organizations**
Nature Information issued by intergovernmental organizations tends to promote or support the group identity. The main media used for propaganda is the press and also international cultural exchange exhibitions, delegations and conferences. The group may seek to promote a political bias or an economic community. Propaganda may serve to increase international tension and the possibility of conflict, which may include subversive activities and terrorism.
Broader Propaganda (#PF1878)   Compulsory indoctrination (#PD3097).
Related Official secrecy (#PC1812).
Aggravates Conflict of information (#PF2002).
Reduced by Uncontrolled media (#PD0040).

♦ **PE3078 Increasing development lag against technological growth**
Broader Increasing development lag against socio-economic growth (#PC5879).
Narrower Lack of technical development and excess of manpower in developing countries (#PE4933).
Aggravated by Inappropriate transfer of technology (#PE5820)
Technology gap between developed countries (#PD0338)
Disadvantageous terms for technology transfer (#PE4922)
Abusive technological development under capitalism (#PD7463).

♦ **PE3083 Senile dementia**
Presenile dementia
Nature Short- and long-term memory impairment is the most prominent symptom of dementia. It ranges from inability to learn new information to inability to remember past personal history. Besides, there could be impairment in abstract thinking or in judgement. Dementia also involves disturbances of higher cortical function affecting language and motor activities. Personality change happens often involving either an alteration or an accentuation of premorbid traits.
Broader Psychoses (#PD1722)   Mental disorders of the aged (#PD0919).
Narrower Alzheimer's disease (#PE7623)   Dementia associated with alcoholism (#PG9268).
Aggravated by Health hazards of aluminium (#PE4969).

♦ **PE3098 Streptococcus infection**
Refs Read, Stanley E and Zabriskie, John B *Streptococcal Diseases and the Immune Response* (1980); Skinner, F A and Quesnel, L B *Streptococci* (1978).
Broader Infection (#PC9025)   Bacterial diseases in animals (#PD2731).
Narrower Haemolytic streptococcal infection of the throat (#PG3063).
Aggravates Rheumatic fever (#PE0920).

♦ **PE3105 Abuse of science and technology in capitalism**
Claim The selling of scientific and technological knowledge to developing countries forms the basis for economic imperialism and foreign control through debt, and the development of science and technology applied to the production and evolution of machinery gradually eliminates the need for human labour, thus producing unemployment. Since those who are displaced by machines cannot easily buy the goods that are produced, resources, including human resources, are wasted. Alienation of the majority from the production process sharpens class consciousness.
Counter-claim Freedom and free-market economies produced the health, human and physical sciences we know today. Some socialist countries, such as the USSR, cannot feed themselves and they desperately try to acquire Western agricultural technology. In the developing countries, millions of persons are saved from disease and death by capitalist medicine and public health measures. The countries of the South are industrializing and trying to increase their exports of manufactured goods, as well as trying to build infrastructures and extend human rights. The science, technology and machinery of developed countries are being exported to the South to help meet these needs.
Broader Capitalism (#PC0564)   Irresponsible scientific and technological activity (#PC1153).
Narrower Surplus labour (#PG5971).
Related Elimination of jobs by automation (#PD0528)
Abusive technological development under capitalism (#PD7463).
Aggravates Capitalist speculation (#PC2194)
Uncontrolled application of technology (#PC0418)
General unproductivity of capitalist systems (#PF3103)
Disadvantageous terms for technology transfer (#PE4922)
Counterproductive capitalist investment financing (#PF3104).
Aggravated by Social inequality (#PB0514)
Lack of individualism in capitalist systems (#PD3106)
Excessive demand for goods in capitalist systems (#PC3116).

♦ **PE3123 Denial of effective national self-determination by capitalist exploitation**
Claim Capitalism is by nature acquisitive, with excess profits accruing to a small, non-productive minority or elite. Because the motivation for production is that of profit, property-owners seek and find ways of making more profit from overseas investment where labour costs are low and raw materials easily available. To ensure that the profit flows to them and is not diverted to the majority in the countries concerned, political control may have been or may be established, directly as colonialism or indirectly as economic imperialism and the manipulation of foreign debts and capital control. The people of such countries are thus effectively denied the right to national self-determination. This may cause alienation and retard development, and is likely to lead to revolution or guerrilla warfare which may escalate if larger powers take part on either side.
Broader Capitalism (#PC0564)   Exploitation in capitalist systems (#PC3117)
Denial of right to national self-determination (#PC1450).
Related Domination (#PA0839)   Foreign control (#PC3187)
Capitalist economic imperialism (#PC3166)
Denial of right to national self-determination in communist systems (#PC3177).
Aggravates Economic gap (#PD8834)
Benign neoplasm of bone and cartilage (#PG0819).

♦ **PE3140 Extraterritorial intrusion of jurisdiction**
Nature A country may introduce legislation requiring its corporations' subsidiaries in foreign countries to conform to regulations at home concerning, for example, anti-trust matters, trading with proscribed countries, and the repatriation of profits. This amounts to an intrusion of the base country jurisdiction into that of the country in which the subsidiary is located, effectively infringing upon its sovereignty. While this illustrates unilateral intended intrusion of jurisdiction there are a number of multilateral intended intrusions that have originated within intergovernmental organizations, such as the UN and the EEC for example.
Incidence In the past, this issue has arisen primarily in connection with US investments in other industrialized countries and with attempts to enforce US anti-trust law or policy with respect to communist countries. Recently the problem has come to be of more general concern with the introduction of regulations designed to assist the balance of payments position of the capital exporting country. Intergovernmental organizations may be the vehicle for proposed intrusions of sovereignty and legal jurisdiction under the most high-sounding reasons. For example, one of the nation-state members of the EEC forcefully advanced an idea for a European Judiciary Area to assure that there would be no frontiers for terrorists: common laws and legal processes would be enacted, and possibly a European court would be created to try terrorists under this proposal. It was defeated by the defence of member nations of their jurisdictional sovereignty despite some 700 terrorist acts per year in NATO Europe.
Broader Obstacles to world trade (#PC4890)   Lack of international cooperation (#PF0817).
Aggravates Decline in foreign direct investment in developing countries (#PD3138).

♦ **PE3147 Protein deficiency in cereals**
Nature Cereals provide the largest single source of protein in the diet of most people in the world. However, cereals in general are not a good balanced food. They often lack sufficient proteins, and the proteins have an inherently unsatisfactory nutritional balance of amino acids such as lysine and tryptophan, which are essential for proper growth and health. A large proportion of people living in tropical and subtropical countries in Central and South America, Africa and Asia, exist on diets consisting largely of cereals and consequently suffer varying degrees of protein deficiency. Although some attempts have been made to alleviate this deficiency by the use of animal protein, leaf protein, fish meal and legume protein, the impact on the problem has been insignificant.
Broader Malnutrition (#PB1498).
Aggravates Kwashiorkor disease (#PE2282).

♦ **PE3149 Evasion of social costs by companies**
Nature Corporations and business avoid paying health, retirement, and other benefits by falsifying employment records, hiring people part-time, hiring people and then firing them before social benefits have to be paid, and hiring illegal residents.
Broader Tax evasion (#PD1466)   Economic crime (#PC5624).
Aggravated by Increasing cost of social security (#PF7911).

♦ **PE3153 Accidental explosions**
Risk of accidental explosion
Nature The sudden and violent expansion of gases or liquids due to an internal chemical reaction or other source of pressure can cause damage to life and property.
Refs Baker *Explosion Hazards and Evaluation* (1983); Bartknecht, W *Explosions* (1980); International Center for Mechanical Sciences Staff *Introduction to Gasdynamics of Explosions* (1972).
Broader Risk (#PF7580)   Large-scale industrial accidents (#PD2570).
Narrower Aerial explosions of unknown origin (#PF9167).
Aggravates Damage to goods (#PE4447)   Interruption risk (#PF9106).
Aggravated by Mechanical failure (#PC1904)   Transport of dangerous goods (#PD0971).

♦ **PE3157 Animal influenza**
Broader Infectious and parasitic diseases (#PD0982).
Narrower Swine influenza (#PG5342)   Equine influenza (#PG6609)
Influenza of poultry (#PG6994).

♦ **PE3170 Inadequate working conditions for professionals**
Refs International Labour Organization *Conditions of Work and Employment of Professional Workers. Tripartite Meeting on Conditions of Work and Employment of Professional Workers* (1977).
Broader International imbalance in the quality of life (#PB4993).
Related Inadequate working conditions in health and medical services (#PE7718).

♦ **PE3177 Rats as pests**
Rats — Rat-infested areas
Broader Rodents as pests (#PE2537).
Aggravates Plague (#PE0987)   Trichinosis (#PE2311)   Typhoid fever (#PD1753)
Leptospirosis (#PE2357).
Aggravated by Environmental degradation of inner city areas (#PC2616).

♦ **PE3196 Restrictive market divisions by transnational corporations**
Nature The most significant restrictive practice of transnational corporations is their territorial production and market allocation arrangements. In many cases such arrangements form the central feature of the global plan for the corporation's operations as a whole, whereas in other cases, such arrangements have simply evolved as a result of a series of not necessarily connected decisions by the parent company concerning manufacturing and sales operations in various countries. In the latter case, the principal export activities of the corporations are often retained by the parent company. Additional restrictive business practices are frequently used to maintain and reinforce such territorial production and market allocation arrangements, and include controls by transnational corporations on the use made of patents, trade marks and copyrights.
Common features of cartel arrangements are market sharing, namely, the assignment of particular markets to certain members; quotas for each member for particular markets; and agreed prices for sales. Multinational enterprises, pursuing long-run global strategies of profit and growth maximisation as well as risk minimisation, may individually restrict the range of products to be produced or the markets to be served by their component units in particular countries. The means

used include decisions respecting prices, investments and international commodity trade and other forms of action, and patent and trademark licensing restrictions are often employed. Such restrictions can have deleterious effects on the economy of the country where the unit is located. Its industry may be prevented from producing goods for home consumption or export, which it is capable of producing competitively. Where production for home consumption is involved, the country may have to import products which could be made domestically at lower cost. It may furthermore be prevented from importing products from the lowest cost sources, with consequent adverse effects upon domestic price levels and productive efficiency. On the other hand, since multinational enterprises may wish to create subsidiaries solely to exploit foreign markets and may be unwilling to create intra-enterprise competition within their existing markets, a ban on the control of subsidiaries might deter their creation and thus decrease foreign investment.
Multinational firms sometimes consolidate their economic power and draw advantages from it not only individually but also by means of agreements or concerted actions with other enterprises, particularly in oligopolies where most multinationals are to be found. Such agreements are facilitated by the possibility in many countries of legalising certain types of cartels: for example, rebate, rationalisation, import and specialisation cartels in which the subsidiaries of multinational firms may participate and which may and in fact sometimes do, serve as the nucleus for an international system of restrictive agreements. This may in particular be the case for national and international export cartels.
Broader Restrictive trade practices (#PC0073)
Restrictive business practices of transnational corporations (#PE5915).

♦ PE3225 Allergy inducing food
Food allergy
Refs Orenstein, Neil S and Bingham, Sarah L *Food Allergies* (1988).
Broader Allergy (#PE1017).
Related Eczema (#PE2465) Food intolerance (#PE9541).
Aggravates Hypersensitivity (#PE6898).

♦ PE3233 Liver cancer
Refs Bottino, Joseph C, et al *Liver Cancer* (1985); Melnick, J L and Maupas, P *Hepatitis B Virus and Primary Liver Cancer* (1981).
Broader Liver diseases (#PE1028) Malignant neoplasms (#PC0092)
Malignant neoplasm of digestive organs (#PE4303).

♦ PE3247 Inadequate provision of alimony
Inadequate provision of palimony — Inadequate child maintenance
Nature Inadequate provision of maintenance for deserted, separated or divorced wives and for the children of the marriage or union may result from poverty or the inability to trace the man responsible, especially in the case of desertion or where the two parties were not married. Inadequate alimony from divorce or separation proceedings may derive from poverty, or alimony may be legally restricted on the grounds of adultery, cruelty or other misconduct by the woman. Inadequate alimony also includes the unjust division of marital property. The result is usually poverty and may aggravate the adjustment problems of both wife and children. Under Muslim law or African customary law, the wife's dowry after divorce may not be returned; or even if returned, it is often inadequate to maintain her or her children.
Incidence Inadequate alimony provision can occur wherever there is divorce, separation or desertion.
Broader Divorce (#PF2100) Separation under marriage law (#PF3251)
Conflict concerning legal custody of children (#PF3252).
Related Inadequate system of child support enforcement (#PF6076).
Aggravates Single parent families (#PD2681) Religious or civil refusal of divorce (#PF3248).
Reduces Excessive alimony costs (#PJ8379).

♦ PE3254 Spirochaetal diseases
Spirochaetosis
Broader Bacterial disease (#PD9094) Infectious and parasitic diseases (#PD0982).
Narrower Yaws (#PE6857) Pinta (#PJ9332) Leptospirosis (#PE2357)
Vincent's angina (#PG4474).
Aggravated by Ticks as pests (#PE1766).

♦ PE3265 Sexual abuse of children
Pedophilia — Sexual love for children — Seduction of children — Pederasty — Molestation of children
Nature Sexual abuse is the involvement of dependent, developmentally immature children and adolescents in sexual activities they do not fully comprehend, to which they are unable to give informed consent, or that violate the social taboos of family roles. It includes paedophilia, an adult's preference for or addiction to sexual relations with children; rape and incest. The term pedophilia implies the non-violent love of children by an adult for sexual purposes. Sexual intercourse with children results from an abnormal erotic attraction to children. Pedophilia is often connected with homosexuality, usually male which is pederasty, but most victims of pedophilia are girls.
Secrecy and conscious or unconscious condoning of such behaviour by parents and children may complicate the indictment of the crime. The abusers are predominantly members of the victims family or household.
Incidence In 'Western' countries pedophilia occurs individually or within families as incest (father-daughter relationship is most common). Statistics show that the incidence of pedophilia with young children is rising but this may be because the secrecy surrounding such acts is becoming less effective. In Britain it is estimated that one child in ten in sexually abused. In the Federal Republic of Germany, expert estimate that 150,000 children suffer constant sexual abuse from family members and relatives.
Claim Sexual abuse robs children and adolescents of their developmentally determined control over their own bodies; and of their own preference, with increasing maturity, for sexual partners on an equal basis. This is so whether the child has to deal with a single, over, and perhaps violent act, usually committed by a stranger; or with incestuous acts, forceful or otherwise, often continued over many years.
Child abuse is more likely to occur in domestic situations where traditional norms are broken. Dramatic increase in the number of divorces, cohabitation, illegitimacy and one-parent families will increase the cases of sexual abuse of children.
Refs Barnard, George W *The Child Molester* (1989); Bernard, F *Paedophilia* (1985); Kempe, C Henry and Kempe, Ruth *The Common Secret* (1984); Schlesinger, Benjamin *Sexual Abuse of Children* (1982); Willich, Ray *Troubled Ones* (1979); Wilson, Glenn D and Cox, David N *Child-Lovers* (1983); Young, Mary de *Child Molestation* (1987).
Broader Sexual violence (#PD3276) Cruelty to children (#PC0838)
Corruption of minors (#PD9481) Victimization of children (#PC5512)
Sexual exploitation of children (#PD3267).
Related Incest (#PF2148) Male homosexuality (#PF1390)
Sexual abuse of wards (#PE7755) Genetic sexual attraction (#PE5925)
Sexual intercourse with minors (#PE6522)
Trafficking in children for sexual exploitation (#PE6613).
Aggravates Infamy (#PB8172) Kidnapping (#PD8744) Soul murder (#PF4213)
Child pornography (#PF1349) Child prostitution (#PE7582)
Sexual offences by juveniles (#PD9394).
Aggravated by Homosexuality (#PF3242) Satanic rituals (#PF7887)
Sexual deviation (#PD2198) Sexual abuse by women (#PJ1767)
Human sexual inadequacy (#PC1892) Unethical personal relationships (#PF8759)
Degradation of developing countries by tourism (#PF4115).

♦ PE3272 Voyeurism
Scopophilia — Peeping toms
Nature Voyeurism refers to sexual pleasure obtained by looking at the genitals of another or by witnessing others' sexual activity. Voyeurs may obtain pleasure by watching people undress or by watching sexual intercourse. Those who are not satisfied with shows such as strip tease, may go to immense trouble to spy on other people and cause themselves injury in the process. Voyeurism stems from immaturity and fear, and indicates repressed sexual tendencies. It appears to be derived from the infantile desire to look – children are intensely curious about sexual matters, and even young monkeys are absorbed by the sight of their mother's genitals.
Incidence In adults, the incidence is indicated to some extent by the number of strip or similar clubs where copulation or simulated copulation is on show. Pornography and blue movies are associated with voyeurism but not the ultimate object of a voyeur's desires. Voyeurs may be prosecuted under common law.
Background The mediaeval legend of Godiva and peeping Tom, gave rise to the epithet of Peeping Tom for voyeurs. The Justices of the Peace Act (England) of 1361 provides for the prosecution of voyeurs.
Broader Sexual deviation (#PD2198) Human sexual inadequacy (#PC1892).
Related Pornography (#PD0132) Sexual harassment of women (#PF3271).
Aggravated by Strip tease (#PG6142) Sexual immaturity (#PJ6143)
Sexual repression (#PF2922) Sexual impotence of men (#PF6415).

♦ PE3273 Sodomy
Anal intercourse
Nature Sodomy encompasses male homosexuality, and various forms of genital contact (except for genital to genital contact); though it is generally taken to mean anal intercourse. Although it may be accepted as normal in certain primitive tribes, in more sophisticated societies it is considered unacceptable when it concerns two males. If sodomy is committed by force, the passive partner may suffer painful injuries. Sodomy committed by force, threat of force, or impairment of victim's ability to decide through drugs or alcohol or with an under-age child is considered to be a crime parallel to rape. Venereal disease contracted through anal intercourse may go unrecognized until it is too late for treatment (for example, syphilis).
Incidence Sodomy occurs between men in tribal society, often as part of an initiation rite and among male homosexuals in sophisticated societies; It may occur because of lack of opportunity for vaginal intercourse, the fixation of libido at the stage where sexual interest is centred on the anus, or as a dominance phenomenon.
Broader Sexual offences (#PD4082) Sexual deviation (#PD2198)
Victimless crime (#PC5005).
Related Bestiality (#PE3274).
Aggravates Infamy (#PB8172) Male homosexuality (#PF1390)
Sexually transmitted diseases (#PD0061).
Aggravated by Sexual immaturity (#PJ6143) Sexual unfulfilment (#PF3260).

♦ PE3274 Bestiality
Human sexual intercourse with animals
Nature Bestiality is sexual relations with animals, often goats, horses or dogs.
Incidence Bestiality appears in pornographic literature, live sex shows, and in some primitive societies, where it may constitute acceptable behaviour.
Broader Perversion (#PB0869) Sexual deviation (#PD2198)
Human sexual inadequacy (#PC1892).
Related Sodomy (#PE3273) Denial to animals of the right to dignity (#PE9573).
Aggravates Social stigma (#PD0884).
Aggravated by Sexual unfulfilment (#PF3260)
Unethical practices with domesticated animals (#PE4771).

♦ PE3286 Birth prevention
Birth control — Contraception
Nature The use of physical or chemical means to prevent sexual intercourse from resulting in the conception of a child.
Claim The only certain and acceptable means of birth control is abstinence. The use of artificial birth control means that people arrogate to themselves the divine right to determine which actions will have the power to generate life.
Refs Kellhammer, U and Überla, K (Eds) *Long-Term Studies on Side-Effects of Contraception* (1978); Spicker, S F, et al (Eds) *Contraceptive Ethos* (1987).
Broader Victimless crime (#PC5005).
Narrower Compulsory sterilization (#PF3240).
Related Induced abortion (#PD0158).
Aggravates Adultery (#PF2314) Underpopulation (#PD5432).
Aggravated by High human fertility in developing countries (#PF0906)
Contraceptive availability in developed countries (#PD4560).
Reduces Complications of childbirth (#PC9042).
Reduced by Lack of family planning (#PF0148) Inadequacy of contraceptive methods (#PD0093)
Inadequate family planning education (#PD1039)
Opposition to population control and family planning (#PF1021).

♦ PE3296 Cumulative depletion of corporate initiative in rural communities
Nature The shifting of social services to the urban centres leaves rural villages without access to the type of expertise they once enjoyed. New services may be nearby but are nevertheless out of reach; water, health and sanitation services are available only in the larger regional cities. Village residents may be aware of modern ways but are without the means for being self-sufficient. To the villager, what was once a contented, quiet existence has become remote, poor and dissatisfying. Community attempts at implementing local initiatives are often inhibited or frustrated. The people of rural communities – the educated, the farmers and the others – live together, forming part of the development of two million rural villages and hamlets. However, social and economic development have not yet touched these communities in any significant way; some changes appear to hinder development rather than help it, and plans for others have simply not materialized. For example, land reform may have distributed the land on an equitable basis in some countries, but in many cases it has also limited the use to which land is put; river basins which have been designated for reclamation remain without irrigation; resources and trained personnel leave villages without a corresponding return in services or training. As a result, communities are increasingly dependent on outside programmes and agencies which seem remote and unresponsive.
Broader Insufficient communications systems (#PF2350)
Reluctance to join in community action (#PF1735).
Narrower Dependence on cities (#PG5298) Delimited desert use (#PG5362)

Unavailability of pumps (#PG5361) Restraints on legal land (#PG5359)
Isolated village mentality (#PG9576) Dominated women's behaviour (#PG5306)
Longterm external dependency (#PG5263) Unresponsive social services (#PG5344)
Lack of community self-worth (#PF3512) Uncorporate economic structure (#PG5341)
Unsuccessful economic ventures (#PG5349) Inaccessible government agencies (#PF5351).
Aggravates Declining community confidence in its ability to change (#PF9066).
Aggravated by Frustrated past goals (#PF5272) Inadequate medical care (#PF4832)
Inadequate health services (#PD4790).

♦ PE3298 Total sexual abstention
Celibacy
Nature The conscious decision not to marry, or, since the turn of the century, not to engage in sexual relationships denies future generations of an important dimension of the genetic pool.
Refs Ford, J Massingberd *Trilogy on Wisdom and Celibacy*.
Aggravated by Human sexual disorders (#PD8016).
Reduces Inadequacy of male contraceptive methods (#PF1069).

♦ PE3308 Threats against family or friends
Torture of relatives or friends
Nature The family and friends of prisoners have been tortured or threatened with torture. Children are tortured as are parents. Family members are tortured to force wanted persons to give up. Torture of friends is used to extract confessions. Wives and daughters are raped in front of relatives. In some cases tortured prisoners are shown to their families so that the families would persuade the prisoners to confess to avoid further torture.
Incidence Torture or threats of torture of the family or friends of prisoners has been reported in the following countries: **Af** Mali, South Africa. **Am** Colombia, Peru, Uruguay. **As** Iran Islamic Rep, Pakistan, Syrian AR. **Eu** Yugoslavia.
Broader Racism (#PB1047) Psychological torture (#PD4559)
Arbitrary external interference in family life (#PE4058).
Related Intimidation (#PB1992) Racial inequality (#PF1199).
Aggravates Racial conflict (#PC3684) Racial exploitation (#PC3334).
Aggravated by Social injustice (#PC0797) Racial discrimination (#PC0006)
Inappropriate selection and examination procedures in education (#PF1266).

♦ PE3328 Unethical practices of health services
Irresponsible health care services
Narrower Crimes committed in hospitals and health care facilities (#PE8420).
Related Abusive treatment of patients in psychiatric hospitals (#PD0584).
Aggravates Iatrogenic disease (#PD6334) Compulsory health care (#PF4820)
Inadequate mental hospitals (#PF4925) Inadequate hospital facilities (#PE5058)
Delay in administration of medical care (#PD5119)
Abusive detention in psychiatric institutions (#PE2932).
Aggravated by Unethical medical practice (#PD5770).

♦ PE3331 Cystic fibrosis
Mucoviscidosis
Refs CIBA Foundation Symposium *Fibrosis* (1986).
Broader Diseases of the pancreas (#PE1132).

♦ PE3337 Incompatibility of traditional and new technologies
Refs ILO *Blending of New Technologies with Traditional Activities* (1985).
Broader Wasted woman power (#PF3690).
Related Inequality of opportunity (#PC3435).
Aggravates Frustration (#PA2252) Unprofitable traditional skills (#PJ1031)
Elimination of traditional skills (#PD8872) Disoriented traditional occupations (#PG5431)
Discrimination against women in education (#PD0190)
Ineffectiveness of traditional small business methods (#PF3008).
Aggravated by Prejudice (#PA2173) Social injustice (#PC0797)
Social inequality (#PB0514) Dependency of women (#PC3426)
Customary working patterns (#PG9278) Limited traditional markets (#PG9299)
Traditional hiring practices (#PG8822) Traditional market practices (#PJ9742)
Traditional purchasing habits (#PG8540).

♦ PE3354 Accumulation and misuse of religious property
Nature The possession of land, investments, and other assets by the Church may involve its administration along similar lines to that of a commercial enterprise. Much Church property is provided by donation and it therefore seems exploitative and corrupt that it should be used on any other basis than for social need and welfare. Donation to Church funds may contribute to persistent poverty, since some of the most ardent benefactors are the poor and uneducated.
Incidence Tenants have been evicted from Church owned housing for not paying rents, despite poverty and with no provision of alternative accommodation.
Background The question of possession of property has caused religious schism since the Reformation.
Broader Accumulation of property (#PC8346) Religious and political antagonism (#PC0030).
Related Corruption in organized religion (#PC3359).
Aggravates Religious schism (#PF1939) Anti-clericalism (#PF3360).

♦ PE3355 Defective human immunity system
Refs Perk, Kalman, et al *Immunodeficiency disorders and retroviruses* (1988).
Broader Human disease and disability (#PB1044).
Narrower Hypersensitives (#PE5169) Autoimmune disease (#PE5214)
Chronic fatigue syndrome (#PE1914).
Aggravated by AIDS (#PD5111).

♦ PE3359 Scabies
Refs Mellanby, K *Scabies* (1972).
Aggravated by Natural disasters (#PB1151).

♦ PE3363 Rots in plants
Nature Rotting is either a dry decay or soft and squashy decomposition or disintegration of plant tissue.
Broader Oomycetes (#PE5465) Ascomycetes (#PE4586) Zygomycetes (#PE0614)
Deuteromycetes (#PE4346) Basidiomycetes (#PE0364).
Narrower Wood rots (#PE4455) Cereal root rots (#PE4453)
Botrytis diseases in plants (#PG1668) Fusarium diseases in plants (#PG5461)
Alternaria diseases in plants (#PJ1685) Sclerotinia diseases of plants (#PG4655).

♦ PE3377 Inflected loss of vision
Blindings
Broader Aggravated assault (#PD0583) Torture by wounding (#PE4078).
Related Physical blindness (#PD0568).
Aggravated by Inhumane methods of riot control (#PD1156).

♦ PE3383 Radioactive contamination of soil
Broader Soil pollution (#PC0058) Radioactive contamination (#PC0229)
Radioactive contamination of plants (#PD0710).
Aggravates Soil degradation (#PD1052).

♦ PE3390 Military prostitution of women
Military brothels
Broader Female prostitution (#PD3380).

♦ PE3391 Privatization of public services
Privatization of state monopolies — Privatization of satellite systems
Nature Where economies of scale result in a natural monopoly, privatization results in the creation of a private monopoly with few gains in efficiency and competitiveness and with a deterioration in public accountability.
Claim In response to problems of prison administration throughout the world, some governments have considered implementing programmes of private incarceration. This creates four problems of particular concern: (a) a state may argue that there has been a shift of liability: that prisoners' claims of mistreatment are claims against private entity, not against the government, even if under human rights law the government is responsible for maintaining prisoners' rights; (b) private incarceration will be structured toward financial gain, which will come true when the greatest possible number of prisoners were incarcerated over the longest possible period of time consuming the smallest possible quantity of resources, producing the greatest possible quantity of labour under conditions requiring the cheapest possible maintenance and supervision costs; (c) the sequestration of such a vital political phenomenon as imprisonment from public or international scrutiny threatens to distance citizens from their right and duty to participate in the penal process; (d) private entities may not held to the same standards regarding prisoners' rights as are state prison administrators.
Counter-claim Privatization can be used effectively to reduce the size of state monopolies, particularly where such enterprises have been judged to have entered into or continued activities which private firms could carry out more efficiently. It can be used to counteract the tendency for publicly owned firms to be managed on excessively bureaucratic principles, inattentive to costs and market opportunities, overmanned, lacking in entrepreneurship and unwilling to innovate and take risks. It can be used to reduce labour market rigidities.
Refs Vickers, John and Yarrow, George *Privatization* (1988).
Broader Capitalism (#PC0564).
Aggravates Maldistribution of water (#PD8056) Undervaluation of public assets (#PF1001).
Aggravated by Denial of rights to prisoners (#PD0520).
Reduces Nationalized industries (#PG6114) Structural rigidities in labour markets (#PD4011)
Proliferation of public sector institutions (#PF4739)
Inefficiency of state-controlled enterprises (#PD5642).

♦ PE3394 Maltreatment of livestock
Maltreatment of transport animals — Cruelty to herd animals — Beating of draught animals — Maltreatment of horses
Incidence Animals such as mules, donkeys, oxen, camels and horses are widely used in developing countries, both on farms and in urban areas. They are frequently maltreated and beaten mercilessly. In industrialized countries, horses, once they have ceased to be useful for their original purpose (racing, draught, amusement) because of age or any other reason, are frequently maltreated in the process of transporting them to a slaughterhouse.
Broader Maltreatment of animals (#PC0066).
Related Denial of rights of animals (#PC5456).

♦ PE3401 Water-borne diseases
Water-borne disease — Waterborne diseases
Nature Fluorosis, skin infection, trachoma, guinea worm and schistosomiasis are some of the diseases transmitted by water. Large water development schemes, dams and irrigation, invariably favour the spread of diseases, if the projects are not implemented with proper drainage and water management systems.
Broader Biological contamination of water (#PD1175).
Narrower Water-borne viral disease (#PG3399) Water-borne animal diseases (#PE2787).
Related Soil-transmitted diseases (#PD3699).
Aggravated by Wasted water (#PD3669) Contradictions (#PF3667)
Water pollution (#PC0062) Rising water level (#PD8888)
Water pollution in developing countries (#PD3675).

♦ PE3402 Excessive military usage of land
Incidence Military authorities are in continuing competition with agricultural interests in an effort to extend the land available for military bases, weapons testing grounds and training areas. In the USA in 1990 military training areas occupied 25 million acres and efforts were being made to extend this by a further 6 million acres in 16 states. In the UK facilities for military war games exclude the public unfairly from vast areas of national parkland, although the exact amount is unknown.
Counter-claim 1. Modern aircraft need approximately 50 miles in which to turn and require longer bomb runs and more room to practice dogfights than in earlier combat situations which governed the choice of existing training areas. Modern firepower enables a battalion to defend ten times as much ground as those of the 1940s, thus requiring more room to train and engage in realistic manoeuvres with other battalions.
2. Exclusion policies protect rare species, habitats and archaeological monuments from possible destruction in other hands. Communities near such military-controlled land benefit from supplying the military with essentials.
Broader Compulsory acquisition of land by government (#PC1005).
Aggravates Environmental degradation through military activity during peace-time (#PE0736).
Aggravated by Obsolete military bases (#PG8043).

♦ PE3407 Emotional crises
Emotional flare-ups — Violent emotions
Broader Personal life crises (#PD4840) Psychological conflict (#PE5087).
Related Inhibited grief process (#PD4918).
Aggravates Stroke (#PE1684) Obesity (#PE1177).
Aggravated by Emotional disorders (#PD9159) Emotional insecurity (#PD8262).

♦ PE3445 Failed migration
Early return migration
Nature Due to loneliness, unfavourable working conditions, or economic and familiar problems, migrants may return to their own country prematurely.
Incidence About 20 percent of Sri Lankan housemaids working in the Middle East return within 18 months, thus ending prematurely their 2-year contract.
Broader Disruptions due to migration (#PC0018).
Aggravated by Uncontrolled migration (#PD2229).

♦ PE3447 Environmental colonialism
Eco-colonialism

PE3447

Broader Neo-colonialism (#PC1876).
Aggravated by Over-rapid timber exploitation (#PD9235).

♦ PE3450 Denial of right to cultural asylum
Related Rejection of refugees (#PF3021) Denial of right to economic asylum (#PE6212)
Lack of individual rights to political asylum (#PF1075).

♦ PE3457 Conspicuous consumption by international civil servants
Conspicuous consumption by international delegations
Broader Decadent standard of living (#PD4037).

♦ PE3463 Denial of right to leave any country
Nature Sometimes states penalize or harass persons seeking to exercise the right to leave any country, including their own. In retaliation they can be deprived of their nationality and their right to return to their own country.
Broader Over-qualification (#PF3462)
Restrictions on freedom of movement between countries (#PC0935).
Related Denial of the right to return to country of residence (#PJ7425).
Aggravates Inflexible social structure (#PB1997).

♦ PE3468 Cruelty to insects
Incidence Insects may have their legs or wings removed whilst alive, whether for amusement or in the process of food preparation. Children devise painful ways to kill insects, such as burning them alive. In the preparation of silk, 15 silk moths are either boiled or steamed alive to produce one gram of woven silk.
Broader Maltreatment of animals (#PC0066).

♦ PE3484 Cruel culling of seals
Nature The population of certain species of seal using well-defined breeding grounds increases to the point that culling is considered necessary to prevent over-crowding and to maintain a stable population. Such culls may be performed regularly or irregularly and may involve the slaughter of adults or of seal pups. The usual practice, in the case of young seals, is to kill them with one or more blows to the head using a club. Death may not be instantaneous and some of the animals may be, possible inadvertently, skinned before death takes place (which is difficult to establish) or before the animal is unconscious. When adults are culled, mothers may be shot leaving the pups abandoned.
Refs Busch, Briton Cooper *War Against the Seals* (1984).
Broader Misuse of wild animals (#PD8904) Inhumane killing of animals (#PE0358).
Related Unjustified game cropping (#PE1327) Endangered species of seals (#PE1656).
Aggravated by Over-population of seals (#PG3872).

♦ PE3488 Punishment of criminals by mutilation
Incidence This has been a traditional practice in many countries. It continues to be used unofficially during the course of torture. Islamic law prescribes certain punishments involving mutilation of the body, especially cutting off limbs.
Broader Mutilation and deformation of the human body (#PD2559).
Related Torture through mutilation (#PD7576) Inhumanity of capital punishment (#PF0399).
Aggravated by Unjust laws (#PC7112).

♦ PE3502 Medicinal products shortage
Broader Inadequate medical care (#PF4832) Long-term shortage of chemicals (#PE1261).
Narrower Unavailability of quality medicine (#PG5321).

♦ PE3503 Periodontal diseases
Gingivitis — Peridontitis — Periodontitis — Periodontosis
Nature Pathological processes involving the supporting structures of the teeth (periodentium) destroy a large part of the natural dentition and deprive many people of all their teeth long before old age. Exogenous factors such as plague, oral debris, mechanical irritation, traumatic occlusion and irritation from crown and filling materials, and some systemic factors such as vitamin and protein deficiencies, hormonal disturbances, and some blood dyscrasias, produce inflammatory (gingivitis, periodontitis), degenerative and neoplastic processes.
Incidence Periodontal disease is one of the most widespread diseases of mankind. No nation and no area of the world is free from it and in most it has a high prevalence, affecting in some degree approximately half the child population and almost the entire adult population.
Refs Frandsen, *Public and Health Aspects of Periodontal Disease* (1984).
Broader Teeth disorders (#PD1185) Human disease and disability (#PB1044)
Diseases and deformities of the digestive system (#PC8866).
Narrower Gingivosis (#PG6536) Traumatism (#PG6539)
Periodontal atrophy (#PG6540) Periodontal disease (#PG3418).
Aggravated by Bruxism (#PE5685) Diabetes (#PE0102)
Metal poisoning (#PJ6543) Hormonal disorder (#PE4431)
Inadequate dental care (#PJ5478) Nutritional deficiencies (#PC0382)
Diseases of blood and blood-forming organs (#PF8026).

♦ PE3504 Haemorrhoids
Piles — Internal piles
Nature Piles are distended veins inside or outside the anus causing bleeding and itching. With a thrombosed pile, sitting and walking becomes a nightmare.
Refs Wood, Clive *Haemorrhoids* (1979).
Broader Diseases of the arteries (#PE2684).
Narrower Mixed piles (#PG6549) External piles (#PG6547).
Aggravated by Over-eating (#PE5722) Constipation (#PE3505)
Liver diseases (#PE1028) Sedentary habits (#PJ6551)
Violent purgatives (#PG6554) Physical unfitness (#PD4475)
Pregnancy disorders (#PD2289) Disorders of the bowel (#PE6553)
Consumption of alcoholic beverages (#PD8286).
Reduced by Use of agricultural resources for production of animal feed (#PD1283).

♦ PE3505 Constipation
Nature The failure to have regular and satisfactory bowel motions causes a mild poisoning of the system, with certain side-effects (foul tongue, bad breath, headache, lassitude and loss of appetite). The common causes of constipation are: habit; a colon which absorbs water too quickly; a spastic colon (the muscle of which remains in a state of spasm); lack of tone of the colon muscle (sometimes because of too little vitamin B in the diet); a diet which has not enough 'roughage' in it to stimulate the intestine to activity; or constant neglect to respond to the sensation in the rectum which indicates that it is full and needs emptying (which leads to retention of faeces which then become dry and hard). The condition may be aggravated by the use of purgatives.
Refs Moyle, Alen *Conquering Constipation.*
Broader Intestinal diseases (#PD9045).
Aggravates Acne (#PE3662) Headache (#PE1974) Lassitude (#PG6560)
Haemorrhoids (#PE3504) Loss of appetite (#PG6559)
Diseases and deformities of the digestive system (#PC8866).
Aggravated by Peritonitis (#PE2663) Appendicitis (#PG2327)
Physical unfitness (#PD4475) Malignant neoplasms (#PC0092)
Vitamin B deficiency (#PG4635) Lack of regular habit (#PG6564)
Disorders of the bowel (#PE6553) Nutritional deficiencies (#PC0382).
Reduced by Use of agricultural resources for production of animal feed (#PD1283).

♦ PE3506 Dystrophy
Nature Dystrophy is the result of starvation of the organism, and most often affects children. It is accompanied by a loss of body weight, retarded growth, suppressed neuropsychic and physical development, reduced resistance to infection, and changes in the general reactivity of the organism.
The most frequent cause of dystrophy is improper diet. Unfavourable living conditions and failure to observe the rules of the child's personal hygiene and of the hygiene of his dwelling may also contribute to the development of dystrophy. Frequent infectious and non-infectious diseases are also an important cause of development of dystrophy in children. A particularly significant part is played by acute intestinal diseases (dyspepsias) and various infectious and parasitic diseases. Development of dystrophy is favoured by tuberculosis, rickets, sepsis, malaria and other chronic diseases which extremely emaciate the child. A dystrophic child becomes more susceptible to various diseases. The child particularly frequently develops pyodermas, pneumonia and gastro-intestinal disorders. Since such children become very weak, the signs of concurrent diseases are not very clearly pronounced and are revealed with difficulty.
In children dystrophy may develop very rapidly, but it lasts very long and recovery takes place extremely slowly.
Refs Doury, P et al *Algodystrophy* (1981).
Broader Children's diseases (#PD0622).
Narrower Toxicosis (#PG6525).
Aggravates Pyoderma (#PG6528) Gastritis (#PG2250)
Pneumonia (#PE2293) Occupational rheumatism (#PE0502).
Aggravated by Malnutrition (#PB1498) Improper feeding (#PG6531)
Poor living conditions (#PD9156) Acute intestinal diseases (#PG6532)
Parasites of the human body (#PE0596) Human disease and disability (#PB1044).

♦ PE3508 Hookworm
Ancylostomises – Nectoriasis — Hookworm disease — Ancylostomiasis
Nature Hookworm diseases are helminthic, caused by parasites in the human intestine, and can lead to iron-deficiency anaemia, vomiting, and various other debilitating effects. The parasitic roundworm Ancyclostoma duodenal causes ancyclostomiasis; Necator americanus causes necatoriasis. Infection occurs mainly through vegetables, fruits, etc, which are contaminated by soil containing the larvae, deposited there from the faeces of an infected person. Nectoriasis infection can also be caused by going barefoot, because nector larvae enter the host through the skin.
Incidence Hookworm diseases are particularly widespread in tropical and subtropical zones, especially Latin America, Africa and Asia.
Refs Schad, G A and Warren, K S (Ed) *Hookworm Disease* (1989).
Broader Helminthiasis (#PE6278) Parasites of the human body (#PE0596)
Gastrointestinal infections in animals due to parasites (#PE0958).
Narrower Cutaneous larva migrans (#PG3885).
Aggravates Ascariasis (#PE2395) Strongyloidosis (#PG6574)
Trichocephaliasis (#PG6573) Nutritional anaemia (#PD0321)
Kwashiorkor disease (#PE2282) Cardiovascular diseases (#PE6816).
Aggravated by Humidity (#PD2474) Unhygienic conditions (#PF8515)
Excrement as fertilizer (#PG6579).

♦ PE3509 Occupational cancer
Nature Occupational cancers are those that are due to exposure to chemical or physical carcinogens in the workplace. Clinically, cancer due to occupation is at present undistinguishable from cancer due to other causes. The occupational origin is essentially considered as proven when there is an unequivocally identified exposure causally associated with an increased occurrence of a cancer.
Incidence Adequately documented cases of occupational cancer represent only a fraction of 1 percent of cancer in Western countries although occasionally it can represent a very large percentage of cancer in a single manufacturing plant. For the latter reason, occupational cancer has frequently been detected through the clustering of otherwise frequently occurring cancers (mainly of the skin, lung or bladder) or through the occurrence of otherwise extremely rare cancers (angiosarcoma of the liver after exposure to vinylchloride monomer, to arsenic or to thorotrast). The real incidence of occupational cancer must inevitably be higher than the adequately documented cases. Initiation and interpretation of epidemiologic investigations may be difficult because frequently only small numbers of workers are involved, because of multiple exposures in the working environment, because of the movement of workers from one job to another or from one country to another, and, above all, because of the lack of detailed cancer registries in many industrialized countries.
Refs ILO *Control and Prevention of Occupational Hazards Caused by Cancerogenic Substances and Agents* (1974); ILO *Occupational Cancer* (1988); ILO *Prevention of Occupational Cancer – International Symposium* (1982).
Broader Malignant neoplasms (#PC0092) Occupational diseases (#PD0215).
Aggravates Medicinal use of occupational carcinogens (#PG6568)
Carcinogenic chemical and physical agents (#PD1239)
Incorporation of carcinogens into consumer goods (#PE8934).
Aggravated by Lack of knowledge (#PF8381) Health hazards of asbestos (#PE3001)
Dust as an occupational hazard (#PE5767)
Unwillingness to divulge information by industrial concerns (#PE8230)
Inadequate application of available knowledge to solve problems (#PF8191).

♦ PE3510 Dracunculiasis
Guinea-worm — Dracunculus infections
Nature Guinea worm disease is a painful, incapacitating affliction caused by the parasite Dracunculus Medinensis. About one year after a person has drunk contaminated water, one or more adult female worms – each up to 1 metre long – emerge through the skin, usually of the lower limbs. When the affected part is immersed in water, the worm expels hundreds of thousands of tiny larvae, which are then ingested by cyclops to continue the cycle. The adult worm emerges slowly over several weeks. The irritation from the emerging worm and from secondary infection of the site is so painful that victims are often unable to walk or work, depending on which part of their body is affected. Sometimes the ulcer which surrounds the emerging worm is secondarily infected by tetanus, and the victim dies. Victims do not become immune. The overwhelming majority, who do not contract tetanus, live to suffer the infection year after year.
Incidence The disease probably affects an estimated 10–50 million people in poor, rural areas of Africa and Asia. In some villages of West Africa, the disease may affect up to 40 percent of farm workers and, because the transmission cycle is often confined to the rainy season, the victims are put out of action during the very time of year when they must plant or harvest their crops. Therein lies much of the disease's economic and social significance. Among schoolchildren, guineaworm is one of the major reasons for absenteeism. Dracunculiasis still afflicts millions

of rural villagers in India and Pakistan and in a broad band of African countries from Mauritania and Senegal in the west, through Mali, Niger, Burkina Faso (formerly Upper Volta), Ivory Coast, Ghana, Togo, Benin, Nigeria, Cameroon, and Chad, to Sudan, Uganda and Ethiopia. Fewer than five percent of the cases are routinely reported.
Claim A group of Nigerian villagers questioned about their priorities for outside assistance, listed electrification first, and elimination of guinea worm disease second; ahead of roads, hospitals, jobs, water, and several other critical needs.
Broader Filariasis (#PE2391) Helminths of the skin (#PE5380)
Parasites of the human body (#PE0596).
Aggravated by Water flea (#PG6595) Contamination of drinking water (#PD0235).

♦ PE3511 Tapeworm as a parasite
Taeniasis — Cestoda
Nature There are more than 3,000 known specied of tapeworm, all of which can infect man. The pork tapeworm, Taenia solium, ranging in length from three to eight metres, causes taeniasis. Unless this disease is detected and treated quickly, human cysticercosis may set in. T solium embryos penetrate the capillaries and spread throughout the body, particularly into the brain, and within a few months develop into cysticerci of about 5 to 20 mm in diameter containing a tapeworm head in a bladder filled with fluid. Cysticercosis can provoke convulsions, cerebral hypertension and psychiatric disorders. If not treated, it may be fatal.
The eating of infected pork causes enormous suffering to millions of people in the Third World. The spread of cysticercosis is facilitated by two main factors, both of which could be controlled by human effort - inadequate sanitation, which contaminates the environment with faecal material and taenia eggs; and the breeding of pigs in unsanitary conditions, including feeding them on human waste.
Incidence In the 1970s, cysticercosis was responsible for 28 percent of all neurological cases in Mexico and about one percent of all deaths. In many Latin American countries human brain cysticercosis occurs in more than 0.1 percent of the population; the fatality rate of untreated cerebral cysticercosis in Latin America exceeds 50 percent. In Africa, cysticercosis is common in Zimbabwe, Gambia, Guinea, Togo, Rwanda, Burundi, Malawi, Swaziland, Madagascar and Zaire. In Asia, it is common in northern India, where over 10 percent of the human population of labour colonies and slums are affected with T. solium. Taeniasis and human cysticercosis are not infrequent in Indonesia and the Republic of Korea. In general, cysticercosis is prevalent in countries where people live in unsanitary conditions and in close proximity to pigs. In Moslem countries where no pork is eaten it is almost non-existent.
Background Cysticercosis is a complication of Taenia solium taeniasis. T solium is the pork tapeworm; people contract taeniasis by ingesting T. solium cysticerci in raw pork. The adult T solium is a segmented worm, generally between two and three metres long but sometimes as long as eight metres. Taenia may persist in humans for many years, with affected individuals spreading hundreds of thousands of tapeworm eggs every day both to other humans and to pigs. T solium eggs from human faeces survive in the environment (the household, pastures, water) for up to several months. Pigs become infected after ingestion of a tapeworm segment or eggs present in an environment contaminated by man. The biological cycle of T solium therefore consists of two stages in three hosts - human, pig, human. Expelled from humans, the eggs of T solium develop in pigs into cysticerci, and then humans are again infected by eating raw or undercooked meat from the pig. But humans can also infect themselves, members of their family or other people with T solium eggs, which cause human cysticercosis.
Broader Helminthiasis (#PE6278) Parasites of the human body (#PE0596)
Nervous system diseases caused by parasites (#PE4683).
Narrower Sparganosis (#PG7777) Gid parasite (#PG6589)
Beef tapeworm (#PE6586) Pork tapeworm (#PG9319)
Dwarf tapeworm (#PG6585) Echinococcosis (#PE7518)
Digestive tract helminthiasis of poultry (#PG9355).
Related Parasitic diseases in animals (#PD2735)
Gastrointestinal infections in animals due to parasites (#PE0958).
Aggravated by Infected dogs (#PG6590) Sheep diseases (#PE6594)
Houseflies as pests (#PE3609) Cockroaches as pests (#PE1633)
Inadequately cooked foods (#PJ2614) Inappropriate sanitation systems (#PD0876)
Arthropods as intermediate hosts (#PG6593) Vertebrates as intermediate hosts (#PG6594).

♦ PE3520 Neurasthenia
Nature Neurasthenia is a condition of nervous exhaustion in which the patient becomes incapable of sustained exertion although he suffers from no definite disease.
Refs Kleinman, Arthur *Social Origins of Distress and Disease* (1986).
Broader Neurosis (#PD0270).
Aggravated by Retirement as a threat to psychological well-being (#PF1269).

♦ PE3528 Middleman control of rural marketing
No alternatives to the dominance of middle men in marketing village products — Farm tenant's dependence on middlemen
Nature Many Third World farmers do not market their products but rather depend on "middlemen" to be the link between the farm and the market. By creating a false market in which local prices are manipulated below the level of the actual market, the middleman effectively reduces the profit margin for the farmer. This manipulation of the local market creates an annual crisis of indebtedness when, in order to sustain family and farm, the farmer is forced to borrow capital at high rates of interest, which intensifies a debilitating cycle. Probably more damaging than the financial crisis is the position of subservience in which the farmer is placed: he is not it control of his own destiny or the destiny of his land.
Refs Chaturvedi, M K *Rural Middlemen* (1986).
Broader Dependency on middlemen (#PD4632)
Surrendered control of marketing systems (#PF6533)
Lack of economic and technical development (#PE8190).
Narrower Lack of dairy markets (#PG8471) Lack of contract power (#PG8510)
Middle-man price control (#PG8493) Exploitation in rural pricing (#PG8423)
Marketing skills which reinforce a self image of being a victim of circumstances (#PE7752).
Related Paralyzing patterns between villages and administrative structures (#PF1389).
Aggravates Unreliable freight transport (#PG1026).
Aggravated by Inadequate marketing knowledge (#PJ9659)
Discriminatory business practices (#PD8913).

♦ PE3537 Factory closures
Plant shutdowns — Industrial failures — Insolvency in industry
Refs Burchell, Robert W and Sternleib, George *Plant Closings in the New Industrial Revolution* (1988); Perrucci, Robert, et al *Plant Closings* (1988).
Related Lock-out (#PD6808) Closure of social institutions (#PF3831).
Aggravates Unemployment (#PB0750).
Aggravated by Strikes (#PD0694) Business bankruptcy (#PD2591)
Insolvent institutions (#PE6431) Excess production capacity (#PD0779).

♦ PE3538 Inefficient location of facilities of international organizations
Inter-state rivalry for secretariats of intergovernmental organizations — Costly rotation of activities between intergovernmental facilities
Aggravates Excessive complexity of intergovernmental organizations (#PF2806)
Inadequacy of intergovernmental decision-making process (#PF2876)
Inadequate coordination of the intergovernmental system of organizations (#PE0730)
Inadequate coordination of governmental representation in intergovernmental organizations (#PE4344).
Aggravated by Competition between states (#PC0114)
Intergovernmental suspicion (#PC2089)
Competition between intergovernmental organizations for scarce resources (#PE0063)
Jurisdictional conflict and antagonism within intergovernmental organizations (#PE9011)
Jurisdictional conflict and antagonism between intergovernmental organizations (#PE7901)
Jurisdictional conflict and antagonism between regional intergovernmental organizations with common membership (#PE1583).

♦ PE3540 Unethical pharmaceutical practices
Irresponsible pharmaceutical practices — Negligence by pharmacists
Incidence In some countries pharmacists adhere strictly to regulations, whereas in others drugs which require prescriptions are available freely without such prescriptions, if only to regular customers.
Aggravates Inconsistent medical practices (#PF1624).

♦ PE3547 Isolating effects of seasonal variations on undeveloped transportation
Seasonal roadway adequacy — Confined seasonal transportation
Nature Many rural communities in developing nations are isolated both from outlying hamlets and from the remainder of the country for several months each year. Hills and ridges, impassable mud tracks, water courses, snow and ice can prevent travel completely at some times of the year. Such conditions can last for up to four months, with consequent falls in school attendance, sharp rises in the cost of consumer goods, and dangerous shortages of vital food supplies.
Broader Seasonal fluctuations (#PF8163).
Narrower Lack of modern production (#PG8448) Irregular transport services (#PE5345).
Related Lack of school transportation (#PJ7849).
Aggravates Distant schooling facilities (#PJ7693).
Aggravated by Bad weather (#PC0293) Road hazards (#PD0791)
Unmotivated teachers (#PF5978).

♦ PE3553 Unethical interrogation methods
Improper police questioning — Oppressive police questioning
Nature Interrogation of suspects may be undertaken using oppressive and underhand tactics to obtain information and convictions. The situation is aggravated by arrogance in assuming pssession of interviewing skills and by pressures to ensure convictions, irrespective of the quality of the investigation.
Incidence Courts in some countries, notably the UK and the USA, rule confessionable evidence as inadmissable, even when interrogations have been taped. There are many examples of officers losing their tempers during interrogation, shouting at suspects, and using verbal abuse (including graphic advice concerning what should be done with them).
Broader Unethical practices by police forces (#PD9193).
Related Inhumane interrogation techniques (#PD1362).

♦ PE3569 Unavailability of scholarship funds for students
Narrower Unequal distribution of fellowships (#PG7984)
Unavailability of scholarship funds for developing country students (#PE8883).
Related Unbudgeted child education (#PG1363) Unavailability of training costs (#PJ8395).
Aggravates Illiteracy (#PC0210) Lack of formal education (#PF6534)
Unequal access to education (#PC2163) Prohibitive cost of education (#PF4375).
Aggravated by School supply costs (#PJ0686) Lack of school transportation (#PJ7849)
Unnecessary education expenditure (#PJ0626).

♦ PE3571 Infantile neurosis
Nature Infantile neurosis is characterized by internalized conflict that arouses anxiety and it forms the basis for a later adult neurosis. The concept of infantile neurosis have been utilized to describe oedipal conflict.
Broader Neurosis (#PD0270).
Aggravates Childhood neurosis (#PE3717).
Aggravated by Psychic conflict (#PF4968).

♦ PE3572 Flea resistance to insecticides
Nature Among the most important chemical weapons against plague are insecticides to kill the flea vectors, but their effectiveness has been threatened by the increasing resistance of fleas.
Incidence Human fleas first showed resistance to pesticides in 1949, cat and dog fleas in 1952, and oriental rat fleas in 1959.
Broader Pests (#PC0728).
Related Louse resistance to insecticides (#PE3576)
Blackfly resistance to insecticides (#PE3578)
Mosquito resistance to insecticides (#PE3582)
Housefly resistance to insecticides (#PE3583)
Cockroach resistance to insecticides (#PE3579).
Aggravates Siphonaptera as insect pests (#PE3643).

♦ PE3573 Rodent resistance to rodenticides
Broader Drug resistance (#PF9659) Pest resistance to pesticides (#PD3696).
Related Resistant bacteria (#PE6007) Drug resistant viruses (#PG6399).
Aggravates Plague (#PE0987) Marine typhus (#PG6648) Rat-bite fever (#PG6702)
Rodents as pests (#PE2537) Rodent vectors of disease (#PE3629).
Aggravated by Misuse of rodenticides (#PG6647)
Rodenticides as pollutants (#PE3677).

♦ PE3575 Panic disorder
Nature A mental illness through which people experience dread at the prospect of being exposed to the daily routine of community life (driving, shopping, etc). It causes heart palpitations, smothering sensations, faintness, sweating and gasping. The sufferers can be so anguished that they think they are dying or going mad.
Incidence In the USA it is estimated that some 2 per cent of the population are susceptible to this disorder.
Refs Ballenger, J C *Neurobiology of Panic Disorder* (1990).
Broader Mental illness (#PC0300).
Aggravates Panic (#PF2633) Suicide (#PC0417) Substance abuse (#PC5536).

♦ PE3576 Louse resistance to insecticides
Nature Typhus and relapsing fever are spread by the body louse, Pediculus humanus, which has shown increasing resistance to insecticides, thus threatening the effectiveness of typhus control campaigns. Some other species of lice which threaten the health of domestic animals have also shown signs of resistance.
Incidence The body louse first showed resistance to insecticides in 1950 in South Korea and Japan. By 1955 resistant strains had appeared in South America and West Africa, and by 1958

in India.
Broader Insect resistance to insecticides (#PD2109).
Related Flea resistance to insecticides (#PE3572)
Blackfly resistance to insecticides (#PE3578)
Mosquito resistance to insecticides (#PE3582)
Housefly resistance to insecticides (#PE3583)
Cockroach resistance to insecticides (#PE3579).
Aggravates Lice as insect pests (#PE1439).

♦ PE3578 Blackfly resistance to insecticides
Nature Blackflies are vectors of onchocerciasis and other diseases, the control of which are threatened by the increasing resistance of blackflies to insecticides.
Incidence Resistant strains of blackfly were first reported in 1963, and have been observed in many areas including Japan, Quebec and Ghana.
Broader Insect resistance to insecticides (#PD2109).
Related Flea resistance to insecticides (#PE3572)
Louse resistance to insecticides (#PE3576)
Mosquito resistance to insecticides (#PE3582)
Housefly resistance to insecticides (#PE3583)
Cockroach resistance to insecticides (#PE3579).
Aggravates Black flies as pests (#PE3646).

♦ PE3579 Cockroach resistance to insecticides
Nature Cockroaches, important disease vectors, have developed resistance to the various insecticides used in their control.
Incidence The German roach, Blattella germanica, first showed signs of resistance in 1951 in Texas (USA). By 1958 resistant strains were evident in Europe and the Caribbean. The oriental roach, Blatta orientalis, first developed resistance in Germany in 1958.
Broader Insect resistance to insecticides (#PD2109).
Related Flea resistance to insecticides (#PE3572)
Louse resistance to insecticides (#PE3576)
Blackfly resistance to insecticides (#PE3578)
Mosquito resistance to insecticides (#PE3582)
Housefly resistance to insecticides (#PE3583).
Aggravates Cockroaches as pests (#PE1633).

♦ PE3580 Abortion-related deaths
Inadequate abortion procedures — Unsanitary abortion practices — Self-abortion — Unsafe abortion
Nature As a growing number of non-communist countries outlaw abortion, the services available for women who need them are outside the purview of public health services. The inspection, advice and cooperation processes which ensure that health services maintain a suitable level of quality are not applied to abortion clinics.
Incidence At least 200,000 woman die annually as a result of abortions carried out by unauthorized practitioners, more than half of which occur in developing countries. It is estimated that illegal abortions worldwide kill a woman every three minutes, whilst for every one that dies 30–40 others suffer severe health problems.
Broader Induced abortion (#PD0158)
Illegal induced abortion (#PD0159).
Aggravated by Naval warfare (#PJ7824)
Lack of abortion facilities (#PF8481)
Criminalization of abortion (#PF6169).

♦ PE3582 Mosquito resistance to insecticides
Incidence At least 38 species of anopheline mosquito have developed resistance to one or more insecticides.
Broader Insect resistance to insecticides (#PD2109).
Related Mosquito breeding ponds (#PG9038)
Flea resistance to insecticides (#PE3572)
Louse resistance to insecticides (#PE3576)
Blackfly resistance to insecticides (#PE3578)
Housefly resistance to insecticides (#PE3583)
Cockroach resistance to insecticides (#PE3579).
Aggravates Yellow fever (#PE0985)
Mosquitoes as vectors of disease (#PE1923).

♦ PE3583 Housefly resistance to insecticides
Nature Houseflies, vectors of enteric and ophthalmic diseases, are showing increasing resistance to insecticides.
Incidence Resistant strains of flies were first reported in 1946 in Sweden.
Broader Insect resistance to insecticides (#PD2109).
Related Flea resistance to insecticides (#PE3572)
Louse resistance to insecticides (#PE3576)
Blackfly resistance to insecticides (#PE3578)
Mosquito resistance to insecticides (#PE3582)
Cockroach resistance to insecticides (#PE3579).
Aggravates Houseflies as pests (#PE3609).

♦ PE3584 Fumigant damage to crops
Nature Fumigants are toxic to most plants, and if time is not allowed for these chemicals to decompose or volatilize, they will injure or even kill crop plants.
Broader Pesticide damage to crops (#PD2581).
Aggravated by Fumigants (#PG6650).

♦ PE3589 Maize pests and diseases
Nature Maize is subject to attack by a number of pests and diseases, mainly in insect pests (stalk borers and armyworms) and soil pests (wireworms and rootworms).
Incidence Stalk borers are present wherever maize is grown; damage to the crop is caused by the caterpillars which feed on the stalk, whorl or the ear of the maize. Some idea of the amount of damage caused can be gauged from an experiment in East Africa, in which yields of maize were increased 2 fold as a result of insecticidal treatment. The European corn-borer Ostrinia nubilalis has a world wide distribution and causes great economic damage. In the Americas, main borers are in genera Zeadiatraea, Diatraea and Elasmopalpus; in Africa, species of Chilo, Sesamia, and Busseola, and in Southeast Asia Chilo, Sesamia and Ostrinia furnicalis. In the armyworms, the fall armyworm Spodoptera frugiperda is the most important budworm in the tropical Americas, and species of Pseudaletia and Mithimna are important in the Americas and Southeast Asia respectively. The corn earworm in the genus Heliothis is another very important pest recorded in North and South America, Europe, India, the Far East, Africa and Australia. In many regions, especially highlands of South America and the Far East, it is considered to be the most important insect pest. Wireworms, which are also serious pests of maize, have a more limited distribution than ear and stalk borers. Wireworms and rootworms in the genus Diabrotica occur mainly in the more northern latitudes of Europe and in the USA. Injury is caused by the feeding of the larval stage on parts of the plant below the soil surface, especially the seed and the stem.
Stored-grain insects can also produce great damage. Species in the genera Ephestia, Rhyzoperta, Sitophilus, Tribolium, Sitotroga, Cathartus, Dinoderus and Trogoderma have the widest distribution and cause major damage in stored grains. Maize rusts (especially Puccinia sorghi and P polysora) and leaf blights (especially Helminthosporium turcicum and H maydis) are the most damaging diseases. Rust due to Puccinia sorghi is found in the USA and is common throughout highlands of subtropical areas of Latin America and Africa. Previously regarded as of minor importance, P sorghi rust became increasingly prevalent in the USA during the 1950s when open pollinated varieties of maize were replaced by P sorghi susceptible hybrids. The rust caused by P polysora is of some importance in southern USA, but is more common in the tropical areas of the Americas. Since 1948, it has spread with remarkable speed. In 1949 it was found in Sierra Leone, and in 1950 in the Ivory Coast, Ghana and Southern Nigeria. Then it moved on to the Cameroons, to East Africa and far as Zululand. A few months later it w; spreading through the scattered islands of the Indian Ocean, reaching Mauritius, Reunion, Madagascar and the Seychelles. Finally from these islands it invaded North Borneo, Thailand, Malaya, and the Philippines. The effect of the maize varieties grown in Africa and Asia was devastating, with up to 70 per cent crop losses. Thus it was and still is a fearful threat to all those Africans and Asians dependent on maize for much of their food. Helminthosporium turcicum leaf blight is common in cool, moist areas, and may decrease yields mainly when infection occurs at the silking stage of the plant. H maydis race T caused great losses on maize production of USA during the early 70s due to susceptible germplasm included in commercial hybrids through the incorporation of Texas male-sterile cytoplasm source. Probably this has been one of the most devastating maize diseases ever known.
Other economically important maize diseases are stalk rots and ear rots. Stalk rots cause wilting of the plants before or after pollination thus decreasing their yield. Several ear-rotting fungi are known to affect the yield and quality of grain produced. Several mycotoxins have been identified produced by fungi, mainly in the genus Aspergillus and Fusarium which affect birds and mammals fed with infected kernels. Of major concern are the downy mildews of maize caused by several species in the genera Sclerospora, Sclerophthora and Peronosclerospora. These diseases were originally reported as limiting factors in maize production in Southeast asian countries, and later the disease spread to African countries where maize and sorghum are cultivated. Since 1964 the disease was reported in southern Texas and northeast Mexico and has expanded rapidly in humid tropics of the Americas, probably through contaminated seed. Diseases caused by viruses (such as Maize Streak transmitted by leafhoppers Cicadulina spp in countries of tropical subecuatorial Africa, Rayado Fino transmitted by the leafhopper Dalbulus maidis in Latin America and southern USA, Maize Dwarf Mosaic Virus transmitted either through infectious sap or the aphid Rhopalosiphum maidis) are creating concern in maize growers around the world. Spiroplasma-caused diseases like Corn Stunt transmitted mainly by the leafhopper Dalbulus maidis is important in tropical Latin America, the Caribbean and Southern USA.
Refs FAO *Guidelines for Integrated Control of Maize Pests* (1979).
Broader Crops pests and diseases (#PE7783).
Related Infestation of seeds (#PE6271)
Pests and diseases of rice (#PE2221)
Pests and diseases of wheat (#PE2222)
Pests and diseases of vines (#PE2985)
Pests and diseases of cotton (#PE2220)
Pests and diseases of potato (#PE2219)
Pests and diseases of sugar cane (#PE2217)
Pests and diseases of sugar-beet (#PE2975)
Pests and diseases of grain sorghum (#PE3590).
Aggravated by Borer insects (#PE6286)
Wireworms as pests (#PE3645)
Army worms as pests (#PE3644)
Viral plant diseases (#PD2227)
Corn borers as pests (#PE3648)
Fungal plant diseases (#PD2225)
Beetles as insect pests (#PE1679)
Insect vectors of disease (#PC3597)
Lepidoptera as insect pests (#PE3649)
Rusts of grasses and cereals (#PE2233).

♦ PE3590 Pests and diseases of grain sorghum
Nature Sorghum is attacked by a number of pests and diseases which frequently may be limiting factors in its commercial production. Because it is predominantly a peasant crop, its pests and diseases and the losses they cause have not been as well documented as for other cereal crops.
Incidence The insect pests which attack sorghum are, in most cases, those which also attack maize. The two main pests are stalk borers and soil pests, principally wireworm. Stalk borers are present wherever sorghum is grown but wireworms occur mainly in the more northern latitudes, especially in the USA.
All the important diseases of sorghum are fungal and they can be classified into (1) those that rot the seed and kill seedlings; (2) those that attack the leaves making the plants less valuable for forage; (3) those that attack and destroy the grain; and (4) those that attack the roots and stalks. The most important of these are smuts which attack the grain. Smut losses vary greatly and although damage as high as 100 percent has been recorded, the overall figure would appear to be between 5 and 10 percent. The peasant culture of sorghum makes it difficult to assess the relative importance of the different diseases attacking the crop. A survey of the grain losses in former British Commonwealth countries in Africa gave an overall loss due to disease of at least 9 percent, of which an estimated 6.6 percent was due to smut. The overall figure may well be on the low side as diseases such as leaf blight, anthracnoze and bacterial blights are very widely distributed and take a steady annual toll.
Nutritional disorders are a major limitation to high productivity because grain sorghum is grown in areas with impoverished soils and because farmers generally use little fertiliser.
Background Grain sorghum is the sixth most important source of dietary calories for the world's population, after rice, wheat, sugar (beet and cane), maize, and Solanum potatoes. It is the fifth most important cereal grain on a world production basis, after wheat, maize, rice and barley. It is the major food grain of many low-income people living in the semi-arid tropical regions of Africa and Asia, and is also used extensively throughout the world as a livestock feed, either as green forage, dry straw, or grain concentrate.
Refs CAB International Mycological Institute *Diseases of Sorghum*; FAO *Elements of Integrated Control of Sorghum Pests* (1979); Frederiksen, R A; Girard, J C and Williams, R J *Sorghum and Pearl Millet Disease Identification Handbook* (1978); ICRISAT *Sorghum Diseases, a World review* (1980).
Broader Crops pests and diseases (#PE7783).
Related Infestation of seeds (#PE6271)
Maize pests and diseases (#PE3589)
Pests and diseases of rice (#PE2221)
Pests and diseases of vines (#PE2985)
Pests and diseases of wheat (#PE2222)
Pests and diseases of cotton (#PE2220)
Pests and diseases of potato (#PE2219)
Pests and diseases of sugar-beet (#PE2975)
Pests and diseases of sugar cane (#PE2217).
Aggravated by Borer insects (#PE6286)
Wireworms as pests (#PE3645)
Beetles as insect pests (#PE1679)
Lepidoptera as insect pests (#PE3649)
Rusts of grasses and cereals (#PE2233)
Smuts of grasses and cereals (#PE2234).

♦ PE3591 Pests and diseases of deciduous fruit
Nature The number of recorded insects and allied pests of deciduous fruit (mainly apples, pears, plums, cherries and peaches) in different parts of the world is legion. Many of the most important pests are common to all the main fruit-growing countries of the world, as are some of the important diseases. For the predominant deciduous fruit, apple, average annual world wide crop losses due to pests and diseases are 30 percent of potential production.
Incidence Apples suffer more from pests than other kinds of fruit. The main apple pest is the codling moth which occurs wherever apples are grown. Certain species of mites, including the fruit tree red spider mite, have become serious apple pests over the last 30 years. In the warmer parts of Europe, the San Jose scale ranks as a pest of first importance. Pears, peaches, plums and cherries have pest problems similar to, but usually milder than apples. The mediterranean fruit fly is the most important pest of pears and peaches. Many fungi can also attack deciduous fruit. On apple, two diseases are of outstanding importance, apple scab and powdery mildew, of which the

latter is the most serious. On pears, pear scab is a major disease, while powdery mildew is rare. A bacterial disease, fire blight, is a serious problem for apple and pear growers particularly in the USA. The most important disease of peaches is leaf curl, a fungal disease which occurs wherever the crop is grown. The fruit is also susceptible to attack by the brown rot fungi which invade the fruit as it begins to ripen and are sometimes responsible for serious losses.
Refs Alford, David V *A Colour Atlas of Fruit Pests* (1984).
Broader Pests and diseases of trees (#PD3585).
Related Pests and diseases of tea (#PE2980)   Pests and diseases of oak (#PE2984)
Pests and diseases of elm (#PE2982)   Pests and diseases of cocoa (#PE2979)
Pests and diseases of palms (#PE2981)   Pests and diseases of coffee (#PE2218)
Pests and diseases of rubber (#PE2977)   Pests and diseases of olives (#PE2978)
Pests and diseases of poplars (#PE4413)   Pests and diseases of chestnut (#PE2983)
Pests and diseases of citrus fruit (#PE2976).
Aggravated by Crown gall (#PE2230)   Fire blight (#PE2229)
Mites as pests (#PE3639)   Viral plant diseases (#PD2227)
Fruit flies as pests (#PE3607)   Fungal plant diseases (#PD2225)
Scale insects as pests (#PE3612)   Bacterial plant diseases (#PD2226)
Hemiptera as insect pests (#PE3615)   Lepidoptera as insect pests (#PE3649).

♦ **PE3592 Credit card fraud**
Cheque card fraud
Incidence In the UK, the annual cost of credit card fraud was estimated to be in excess of 1.5 billion pounds in 1989.
Broader Fraud (#PD0486).

♦ **PE3594 Workers alienation in socialism**
Nature Alienation is a significant issue in any assessment of workers in a socialist state. Working conditions, and the worker's attitudes towards his conditions, are the focal point of this widespread feeling of alienation. The work itself is unsatisfying, being both monotonous and an inadequate source of livelihood, and workers are isolated from one another because of a sense of competition fostered by the managers, who live in a world equally isolated from all workers.
Broader Alienation from work (#PD3076).

♦ **PE3597 Aneurysm**
Refs Bergan, John J and Yao, James (Eds) *Aneurysms* (1981); Fox, J L *Intracranial Aneurysms* (1983).
Broader Diseases of the arteries (#PE2684).

♦ **PE3601 Government-enforced maternity**
State-imposed functional breeding of humans
Nature In order to remedy perceived deficiencies in national population, governments may institute measures obliging women to have children against their will.
Incidence The classic example is that of Nazi Germany where women conforming to the preferred Aryan stereotype were encouraged or obliged to have as many children as possible, whether with their husbands or with partners chosen for them. A childless woman was labelled a disgrace, an outcast or even a traitor. Soldiers on leave were exhorted to find women and breed children. In the USSR under communism, it was unofficially estimated that in 1964 41 per cent of the women over 21 years were unmarried, and that as a consequence the country was underpopulated. In response it was agreed to encourage women, especially unmarried women, to have children and to hand them over to state-run institutions where they would remain until university age. In Romania under Ceausescu, married women were obliged to have children. Use of contraceptives and abortions was forbidden. Official pressure was placed on women who failed to produce children. Some of these children were handed over to state-run orphanages, following the policy originally developed in the USSR.
Broader Eugenics (#PC2153).
Reduces Underpopulation (#PD5432)   Lack of eugenic measures (#PD1091).

♦ **PE3604 Endangered species of chiroptera**
Bats
Incidence Endangered species of chiroptera in the USA include the Indiana bat, Hawaiian hoary bat, ozark big-eared bat, Virginia big-eared bat, and the spotted bat. The causes of their possible extinction include vandalism, commercialization of caves, insecticide poisoning, loss of habitat, and capture for laboratory use.
Broader Endangered species of mammals (#PC1326).
Narrower Endangered species of fruit bats (#PG6664)
Endangered species of smoky bats (#PG6675)
Endangered species of golden bat (#PG6677)
Endangered species of common bats (#PG6678)
Endangered species of bulldog bats (#PG6667)
Endangered species of true vampires (#PG6673)
Endangered species of false vampires (#PG6660)
Endangered species of horseshoe bats (#PG6670)
Endangered species of sac-winged bats (#PG6666)
Endangered species of slit-faced bats (#PG6668)
Endangered species of leaf-nosed bats (#PG6671)
Endangered species of spear-nosed bats (#PG6672)
Endangered species of disc-winged bats (#PG6676)
Endangered species of free-tailed bats (#PG6680)
Endangered species of mouse-tailed bats (#PG6665)
Endangered species of funnel-eared bats (#PG6674)
Endangered species of New Zealand short-tailed bat (#PG9028).
Aggravated by Insecticide poisoning (#PG6682).
Reduces Vampire bats (#PE1890).

♦ **PE3607 Fruit flies as pests**
Nature Many of the species of fruit flies are of considerable economic significance, their larvae attacking various cultivated fruits.
Incidence The Mediterranean fruit fly lays its eggs in citrus and other fruits, the larvae tunnelling into the flesh of the fruit, making it unfit for human consumption. The apple maggot, the larva of Rhagoletis promonella, burrows in apples causing the fruit to become spongy and discoloured. This species and the closely related cherry fruit fly (R cingulata) cause extensive losses in the northeastern United States. Other widespread pests include the Mexican fruit fly (Anastrepha ludens), which attacks citrus crops; the Oriental fruit fly (Dacus dorsalis), which infests many kinds of subtropical fruit; and the olive fruit fly, which destroys olives in the Mediterranean region.
Refs Economopoulos, A P (Ed) *Fruit Flies* (1987).
Broader Flies as insect pests (#PE2254)   Insect pests of plants (#PD3634).
Aggravates Pests and diseases of olives (#PE2978)
Pests and diseases of coffee (#PE2218)
Pests and diseases of citrus fruit (#PE2976)
Pests and diseases of deciduous fruit (#PE3591).

♦ **PE3609 Houseflies as pests**
Nature The housefly (Musca Domestica) is a common dipterous insect of the family muscidae, which constitutes a major nuisance and hazard to public health. Female houseflies often deposit more than 100 eggs at a time on decomposing organic wastes such as horse manure or fermenting garbage. The young adult houseflies, when they fly from their filthy breeding sites into human habitations, may carry on their feet many million bacteria. If they contaminate human food, they can cause bacillary dysentery. In this way, flies act as vectors of cholera, typhoid fever, amoebic dysentery, poliomyelitis, anthrax, eggs of parasitic worms, and cysts of Entomoeba histolytica and other organisms.
Incidence The housefly is virtually cosmopolitan and comprises about 90 percent of all flies occurring in human habitats. It is a health hazard in virtually all human communities, ranging in size between small farms and large cities, wherever food substances suitable for the development of its maggots are allowed to accumulate.
Broader Household pests (#PD3522)   Flies as insect pests (#PE2254).
Aggravates Cholera (#PE0560)   Anthrax (#PE2736)   Parasites (#PD0868)
Dysentery (#PE2259)   Poliomyelitis (#PE0504)   Typhoid fever (#PD1753)
Enteric infections (#PD0640)   Tapeworm as a parasite (#PE3511)
Eye diseases and disorders (#PD8786).
Aggravated by Myiasis (#PE3633)   Housefly resistance to insecticides (#PE3583).

♦ **PE3611 Phylloxera as pests**
Nature Phylloxera are aphidlike plant pests of the order homoptera. The most notorious, because of its devastation of vineyards, is the vine phylloxera, or grape louse.
Broader Homoptera as insect pests (#PE3614).
Aggravates Pests and diseases of vines (#PE2985).

♦ **PE3612 Scale insects as pests**
Coccids
Nature Scale insects are severe tree pests, attacking many fruit trees, shrubs, forest trees and woody ornamentals.
Broader Homoptera as insect pests (#PE3614)   Pests and diseases of trees (#PD3585).
Aggravates Pests and diseases of elm (#PE2982)
Pests and diseases of olives (#PE2978)
Pests and diseases of citrus fruit (#PE2976)
Pests and diseases of deciduous fruit (#PE3591).

♦ **PE3613 Aphids as pests**
Nature Aphids, or plant lice, are destructive, tiny insects of the order homoptera. They are among the most common insect pests of plants throughout the growing season, and there is hardly a species of plant, cultivated or wild, that escapes aphid infestation.
Incidence Aphids are very abundant and diverse in the temperate zone of the northern hemisphere: North America, Europe and Eastern Asia are rich in aphid fauna. Aphids are phytophagous (plant eating) insects, which extract vital sap from the leaves and stems of plants. They usually do not kill their host plants, but they frequently reduce plant vigour, distort leaves and cause other malformations.
Broader Homoptera as insect pests (#PE3614).
Related Insect vectors of plant diseases (#PD7732).
Aggravates Viral plant diseases (#PD2227)   Fungal plant diseases (#PD2225)
Bacterial plant diseases (#PD2226)   Pests and diseases of elm (#PE2982)
Pests and diseases of potato (#PE2219)   Pests and diseases of cotton (#PE2220)
Pests and diseases of sugar cane (#PE2217)   Pests and diseases of sugar-beet (#PE2975).

♦ **PE3614 Homoptera as insect pests**
Nature Homoptera are a major group of sucking insects: many of them are serious pests of plants, trees and crops. They inflict damage through their feeding activities which may directly injure and enervate plants or may be the mode of transmission of important plant diseases.
Broader Insect pests of plants (#PD3634).
Narrower Aphids as pests (#PE3613)   Phylloxera as pests (#PE3611)
Scale insects as pests (#PE3612).
Related Insect vectors of plant diseases (#PD7732).
Aggravates Viral plant diseases (#PD2227)   Fungal plant diseases (#PD2225)
Bacterial plant diseases (#PD2226).

♦ **PE3615 Hemiptera as insect pests**
Bugs
Nature Hemiptera, the true bugs, are a major order of insects, many of which are important pests. Almost all are phytophagous, except for a small number which feed on the blood of other animals. Hence many species are of importance as plant pests and vectors of disease, both of plants and animals.
Broader Parasites (#PD0868)   Insect pests (#PC1630)   Disease vectors (#PC3595)
Insect pests of plants (#PD3634)   Pests and diseases of trees (#PD3585).
Narrower Bedbugs (#PE3617)   Chinch bug (#PG6686)
Kissing bugs as vectors of disease (#PE3616).
Aggravates Insect bites and stings (#PE3636)   Pests and diseases of tea (#PE2980)
Pests and diseases of oak (#PE2984)   Pests and diseases of rice (#PE2221)
Pests and diseases of wheat (#PE2222)   Pests and diseases of cocoa (#PE2979)
Pests and diseases of cotton (#PE2220)
Pests and diseases of deciduous fruit (#PE3591).

♦ **PE3616 Kissing bugs as vectors of disease**
Triatominae
Nature The kissing bugs are exclusively blood suckers and hence important vectors of disease, notably American Trypanosomiasis, or Chaga's disease.
Incidence Several species of kissing bugs regularly feed on man in the American tropics: panstronglyus megistus, rhodmus prolix, triatoma infestans, and others. The bugs are commonly infected with tryponosoma cruzi which causes Chaga's disease: up to 30 percent are infected in places where the disease is endemic, in most rural areas of Central and South America, especially in Brazil, Argentina and Chile. The trypanosome which causes the disease is picked up by the bug while sucking blood but is transmitted through the faeces. Faecal contamination may be at the place of the bite or in the conjuctiva of the eye – the bites are often about the face and the faeces are rubbed into the eye. The bugs infest huts and rural dwellings made of mud or adobe, hiding in cracks.
Broader Parasites (#PD0868)   Hemiptera as insect pests (#PE3615)
Insect vectors of human disease (#PE3632).
Aggravates Chagas' disease (#PE0653)   Insect bites and stings (#PE3636).

♦ **PE3617 Bedbugs**
Nature Bedbugs are small blood-sucking insects, ectoparasites, which feed upon warm-blooded animals, including man, bats and birds.
Incidence The exact number of species is uncertain; approximately 36 are recognized. Only two species, Cimex lecturarius and Cimex rotundatas, attack man. Cimex lectularius is the common species of the temperate and sub-tropical regions, while the second species is restricted to

tropical regions of Asia and Africa. Cimex columbaruis infests pigeons in Europe; Cimex pilosellus and Cimex pipistrelli occur on bats in North America and Europe respectively; Oeciacus vicaruis and Oeciacus hirundinis occur on birds and will attack man; and Haematosiphon inodorus, the poultry bug, is a serious pest of domestic birds and may invade human dwellings. Bedbugs are among the most cosmopolitan of human parasites. During the day they conceal themselves within mattresses, the joints of bedspreads, or cracks in the wall. At night they suck the blood of man, and after feeding retreat to their hiding places. They can exist up to a year without food. Their bites may cause allergic reactions and itching, and are possible sites for secondary infection.
Broader Parasites (#PD0868) Household pests (#PD3522)
Hemiptera as insect pests (#PE3615).
Aggravates Insect bites and stings (#PE3636) Skin diseases in animals (#PD9667).

♦ PE3618 Dermaptera as insect pests
Earwigs
Nature Dermaptera, or earwigs, are omnivorous feeders; a few are plant pests, destructive to tender foliage and flowers.
Incidence The most widespread species and the most troublesome pest is the European earwig, which occurs in temperate and sub–tropical regions. Commerce–borne from its native Europe, it has become established in North and South America and Australasia, and seems destined to become cosmopolitan. It is a garden and household pest. The earwig feeds mainly on green plants, but prefers anything more nourishing that it can find, hence its presence in the kitchen, where its activities resemble those of the cockroach.
Broader Insect pests (#PC1630) Insect pests of plants (#PD3634).
Aggravates Insect damage to stored and manufactured goods (#PD3657).

♦ PE3619 Thysanoptera as insect pests
Thrips
Nature Thrips are small phytophagous insects of the order thysanoptera. Many thrips are economically important as serious pests of cultivated plants, since they injure or destroy the plants in feeding and breeding activities.
Incidence Thrips occur in every part of the world where plants grow. In addition to inflicting injury directly by extracting plant sap or juice, some thrips transmit virus diseases of plants. Some may bite man and cause dermatitis. Their main importance, though, is as plant pests: vegetables and fruit trees are especially subject to damage by these insects.
Broader Insect pests (#PC1630) Insect pests of plants (#PD3634).
Related Viral plant diseases (#PD2227).
Aggravates Pests and diseases of oak (#PE2984).

♦ PE3620 Thysanura as insect pests
Firebrats — Silverfish
Nature Thysanura is an order of wingless insects. A major family of this order – lepismatidae – are common household pests, called silverfish, which feed on starch and cellulose in paper and paper products.
Broader Insect pests (#PC1630).
Aggravates Insect damage to stored and manufactured goods (#PD3657).

♦ PE3621 Aedes mosquitoes as vectors of disease
Yellow–fever mosquitoes
Nature Aedes is one of the most important genera of disease–carrying mosquito, transmitting viruses which cause yellow fever, dengue, equine encephalitis and Bancroft's filariasis.
Incidence The genus Aedes contains one hundred species. It is abundant and has worldwide distribution; certain species being serious pests in Arctic and subarctic latitudes. Most of the medically important species of Aedes belong to the subgenus Stegomyia. They include Aedes scutellaris, the main vectors of human filariasis over much of the Pacific area; Aegypti, the cosmopolitan carrier of yellow fever and other human diseases; Africanus which maintains an important reservoir of yellow fever among monkeys; Aedes circumluteolus, carrier of several diseases in South Africa; and Aedes scapularis which carries several diseases in the Western hemisphere.
Broader Mosquitoes as vectors of disease (#PE1923).
Aggravates Yellow fever (#PE0985) Dengue fever (#PE2260)
Encephalitis (#PE2348) Elephantiasis (#PE1601).

♦ PE3622 Anopheline mosquitoes as vectors of disease
Malaria mosquitoes
Nature Anopheles is a genus of medically very important mosquitoes. About 50 species are important as disease vectors in transmitting malaria and the nematodes that cause elephantiasis. During the process of taking a blood meal, the disease entity, present in either the proboscis or salivary glands, is injected with the saliva.
Broader Mosquitoes as vectors of disease (#PE1923).
Aggravates Malaria (#PE0616) Filariasis (#PE2391) Elephantiasis (#PE1601).

♦ PE3623 Culicine mosquitoes as vectors of disease
Common house mosquitoes
Nature Culicine mosquitoes are members of one of the largest genera of medically important mosquitoes. They are biting pests and vectors of disease, with worldwide distribution. In particular they transmit various forms of virus encephalitis, mainly diseases of domestic and wild animals, but they also occur in epidemic form in man.
Broader Mosquitoes as vectors of disease (#PE1923).
Aggravates Filariasis (#PE2391) Encephalitis (#PE2348).

♦ PE3624 Teredos as pests
Shipworms — Pipeworms
Nature Teredos are a genus of highly specialized wormlike molluscs that are extremely destructive to wooden vessels and waterfront installations.
Incidence Teredos are worldwide in distribution, generally living in marine or brackish water, though occasionally in fresh water. Since earliest times they have been of concern to maritime people. The damage that teredos do is not readily seen; the outer surface of the wood may look undamaged and yet the interior may be completely riddled; hence the description 'termites of the sea'. Teredos are able to survive within wood for long periods of time during adverse conditions, including reduced salinity, pollution and silting. Furthermore, despite many advances made during this century in teredo control, there is still no permanent protection against them.
Background Before the advent of man, teredos lived in fallen trees in the sea. When man began building wooden ships and wharves, teredos became pests. Archimedes of Syracuse protected his ships by a sheathing of lead. Copper sheathing was first used by the British in 1758 to protect their ships (from teredos, borers and fouling organisms). In the days of the clipper ships, nearly all vessels sailing in tropical seas were sheathed with copper, a very expensive practice, but with the advent of metal ships the problem of teredos ended. Teredos are still a problem for wooden ships, small fishing and pleasure boats, and specialized ships such as minesweepers. For waterfront installations, teredos are a significant though decreasing problem as more and more wharves and piers are constructed of steel and concrete. The Dutch were the first to be greatly concerned at such destruction. In the 18th century, severe infestations of teredos, attacking the wooden parts of locks and dykes, threatened the existence of much of their countryside. Of the most recent attacks by teredos, the most famous was that in San Francisco bay during 1917–21, when $25 million worth of destruction was caused to wharves and jetties.
Broader Insect pests of wood (#PD3586) Wood deterioration and decay (#PD2301).
Related Mollusc pests (#PG2849).

♦ PE3627 Blowflies as pests
Nature Blowflies are large flies of the family calliphoridae. Although the larvae of blowflies usually feed on decaying meat, they sometimes infest open wounds, eating healthy tissue. They may live within the flesh of man and his domestic animals, forming tumours.
Incidence The most dangerous species are the screw worms, which attack both man and livestock, especially sheep and goats. They occur in tropical regions and the southern USA. Several species of blowfly attack, and will kill, livestock in Australia, where these is a $150 million–a–year blowfly strike problem in sheep. The Old World screw worm is widely distributed in the Pacific regions, Australia and Asia, occurring in enormous numbers, breeding in decaying vegetation, manure, human excrement and carcasses. It is a most important vector of dysentery, jaundice and anthrax.
Counter–claim Blowfly maggots were introduced into battlefield wounds by military surgeons from the U.S. Civil War until the 1930's as a means of cleaning wounds. Many thousands of soldiers's lives and maybe untold thousands of non–military lives were saved using this procedure. Blowfly maggots have in their guts large populations of benign bacteria which exude two powerful bactericides, phenylacetic acid and phenylacetaldehyde, to which they are resistant. In the environment of the open wound these substances are not stable enough to be protective. The maggots eating debris and infectious microbes allows the benign bacteria to act as a filter, resulting in bacteriological cleanliness, rapid healing and removal of unnecessary scar tissue.
Broader Parasites (#PD0868) Flies as insect pests (#PE2254).
Related Insect vectors of disease (#PC3597).
Aggravates Anthrax (#PE2736) Dysentery (#PE2259) Hepatitis (#PE0517)
Insect bites and stings (#PE3636).
Aggravated by Myiasis (#PE3633).

♦ PE3629 Rodent vectors of disease
Nature Rodents are important vectors and reservoirs of many serious human diseases. Rodents can be so numerous that, once transmission of a disease has been established in a particular area (especially in the tropics and where sanitation is poor), many or most of the human population in that area will be affected.
Incidence Rodents cause much greater harm as vectors of disease than as annoying pests; the cost to man is impossible to estimate. The list of human diseases for which rodents are either vectors or carriers of insect vectors is long and as yet incomplete. The best known include plague, salmonellosis, tularaemia and brucellosis, all caused by bacteria; the relapsing fevers, leptospirosis and rat–bite fever, caused by spirochaetes; scrub typhus, Q fever and rickettsial pox, due to rickettsia; and diseases such as certain haemorrhagic fevers, caused by viruses about which little is known at present. Most of the diseases are widespread; for example leptosirosis and rat bite fevers have a worldwide distribution. Plague and murine typhus are endemic in many cosmopolitan areas.
Such rodent–borne diseases as leptospirosis and rat–bite fever are usually transmitted directly to man. In the case of many others, the disease organism reaches man via an intermediate host, usually an arthropod. Thus the intermediate host of plague is a flea; of scrub typhus, a trombiculid mite; and of the spotted fevers, a tick.
Some diseases, such as plague, are spread from one territory to another by commensal rodents and may then be passed on to field rodents of another species and lie undetected for years. Later they may find their way back to man directly or, more often, via a commensal rodent. Two factors that often make it easier for such diseases to survive for long periods are the very favourable temperature and humidity conditions of deep burrows of rodents and the very considerable anthropod fauna that these support.
Broader Disease vectors (#PC3595) Human disease vectors (#PD6651).
Related Rodents as pests (#PE2537) Insect vectors of disease (#PC3597).
Aggravates Plague (#PE0987) Q fever (#PE2534) Tularaemia (#PE6872)
Brucellosis (#PE0924) Trichinosis (#PE2311) Scrub typhus (#PG1989)
Leptospirosis (#PE2357) Murine typhus (#PG6700) Salmonellosis (#PE7562)
Rickettsiosis (#PE5530) Rat–bite fever (#PG6702) Histoplasmosis (#PG3696)
Relapsing fever (#PE7787) Haemorrhagic fevers (#PE5272).
Aggravated by Water pollution (#PC0062) Rodent resistance to rodenticides (#PE3573)
Water pollution in developing countries (#PD3675).

♦ PE3632 Insect vectors of human disease
Nature Insects that are alternate hosts for a disease are essential in its transmission.
Incidence Insects can be so numerous that, once transmission of a disease has been established in a particular area, especially in the tropics and where sanitation is poor, many or most of the members of the human communities in that area will be affected. Perhaps the best example is provided by malaria in areas where eradication of the disease has not as yet been achieved or is in its early stages. In many parts of West Africa, there is probably no child who has not been infected once maternal immunity has been lost. Consequently, there are areas where as much as 10 percent of the mortality and a major part of the morbidity occurring among young children may be due to malaria. Such a high incidence of disease can be accounted for when it is realized that many anopheline mosquitoes may feed on a person every night, thus making it almost certain that at least one of them will carry the malaria infection.
Broader Human disease vectors (#PD6651) Insect vectors of disease (#PC3597).
Narrower Ticks as pests (#PE1766) Cockroaches as pests (#PE1633)
Black flies as pests (#PE3646) Tsetse flies as pests (#PE1335)
Siphonaptera as insect pests (#PE3643) Mosquitoes as vectors of disease (#PE1923)
Kissing bugs as vectors of disease (#PE3616).
Related Mites as pests (#PE3639).
Aggravates Virus diseases (#PD0594) African trypanosomiasis (#PE1778)
Human disease and disability (#PB1044).
Aggravated by Rodents as pests (#PE2537) Inappropriate sanitation systems (#PD0876).

♦ PE3633 Myiasis
Nature Myiasis is the infestation of vertebrates by the larvae, or maggots, of numerous species of flies. These larvae may invade different parts of an animal's body or may appear externally. Some invertebrates, such as spiders, may be also attacked by species of Sarcophagidae, the flesh flies. Myiasis of such domestic animals as horses, sheep or cattle, is of considerable economic importance. Cattle afflicted with cutaneous myiasis produce poor–grade hides full of small perforations; while infestation by screw worms, if untreated, results in death. Horses afflicted with the stomach bot become emaciated and may die.
Incidence In cutaneous myiasis, the larvae are found in or under the skin. There may be a migration of some species of these larvae through host tissues, resulting in swelling and intense itching. Intestinal myiasis in man is usually the result of accidentally swallowing the eggs or larvae of these flies. It commonly occurs in many herbivores who ingest the eggs while feeding on contaminated herbage. The larvae settle in the stomach or intestinal tract of the animal host.

Broader Parasites (#PD0868) Infectious and parasitic diseases (#PD0982).
Aggravates Botflies as pests (#PE3635) Blowflies as pests (#PE3627)
Houseflies as pests (#PE3609).

♦ PE3635 Botflies as pests

Nature Botfly is the name applied to dipterous insects of the families gasterophilidae, cuterebridae and oestridae, the larvae of which are parasitic on mammals. In tropical America, the human botfly (which attacks livestock and deer but rarely man), causes great loss of beef and hides.
Broader Parasites (#PD0868) Flies as insect pests (#PE2254).
Narrower Chin fly (#PG6715) Nose fly (#PG6716) Warble fly (#PG6719)
Deer nose fly (#PG6717).
Aggravates Insect bites and stings (#PE3636).
Aggravated by Myiasis (#PE3633) Stable flies (#PG6720)
Insect vectors of disease (#PC3597) Mosquitoes as vectors of disease (#PE1923).

♦ PE3636 Insect bites and stings

Deadly insect bites and stings

Nature As well as transmitting many important diseases, insect bites produce in their host certain adverse reactions resulting from toxins peculiar to the insect itself. Most insect bites, such as those of the fly, mosquito and flea, cause an almost immediate reaction in humans, consisting of local irritation of the skin which usually subsides without trace within a day. Stinging insects such as the bee, hornet or wasp give rise to painful, elevated skin lesions, that are also transitory for most of the population. For the small percentage which is allergic to bee stings, however, one sting can well be fatal. It is not hard to determine beforehand whether one is allergic or not, but this is not a common procedure. Multiple stings may also give rise to alarming systemic illness, including fainting, difficult breathing, and collapse of the circulatory system, and may result in death. Blood sucking lice, bedbugs, and fleas are more intimately associated with man and live on his body as parasites. Bedbugs live in mattresses, walls, and furniture, and attack their host only in the dark. Lice fix their eggs on hair or clothing and periodically feed on their host. Their bites result in severely itchy skin lesions, which may become infected and result in widespread complications.

Incidence As an indication, the market for bite and burn remedies in the UK in 1986 was £2 million, with a further £2 million for insect repellants.
Broader Insect pests (#PC1630).
Aggravates Itch (#PE3940) Insect vectors of disease (#PC3597).
Aggravated by Bedbugs (#PE3617) Botflies as pests (#PE3635)
Blowflies as pests (#PE3627) Lice as insect pests (#PE1439)
Black flies as pests (#PE3646) Tsetse flies as pests (#PE1335)
Flies as insect pests (#PE2254) Hemiptera as insect pests (#PE3615)
Siphonaptera as insect pests (#PE3643) Mosquitoes as vectors of disease (#PE1923)
Kissing bugs as vectors of disease (#PE3616).

♦ PE3637 Benign neoplasm of respiratory system

Broader Benign tumours (#PD8347) Diseases of the respiratory system (#PD7924).
Narrower Benign neoplasm of throat and mouth (#PG4748).

♦ PE3638 Myasthenia gravis

Refs De Baets, M, et al (Eds) *Myasthenia Gravis* (1988).
Broader Diseases of connective tissue (#PD2565).

♦ PE3639 Mites as pests

Acarina

Nature Acarina, an order comprised of mites and ticks, are of the class arachnida. Mites live in varied habitats, and while some are highly destructive of plants, others are parasitic on and in animals, including man. Different species of ticks may be important vectors of disease of man, animals and plants.

Incidence The parasitic mites occur on a wide range of animals and in diverse habitats: they may be external or internal; and may infest the lungs, nasal passages, stomach, or body tissues. Some important parasitic mites include the chicken mite and the rat mite – both will attack man – the lung mites of monkeys, and the nasal mites of dogs. Itch mites burrow into the skin of animals including man, dogs, pigs and sheep, and cause serious injury. The chigger tick is parasitic on man and animals only during its nymphal stages. Its bites produce a severe dermatitis and may also transmit scrub typhus. Mites of the family demodicidae live in the pores of animals, including man, and may result in serious injury and death. Plant feeding mites destroy billions of dollars worth of crops each year. They cause damage by feeding on leaves and by transmitting virus diseases, such as those which distort leaves and fruits. Spider mites are perhaps the most notorious of the acarina plant pests, especially to fruit growers, as they can completely defoliate plants. Other mites infest stored agricultural products, such as grain and copra. The grain mites also cause dermatitis on those who handle stored products which the mites have damaged.
Broader Parasites (#PD0868) Arachnida pests (#PE3986)
Insect pests of plants (#PD3634).
Narrower Feather mite (#PG6584) Chicken mites (#PG3725)
Common chigger (#PG6871) Depluming mite (#PG6692)
Scaly leg mite (#PG9741) Turkey chigger (#PG7696)
Subcutaneous mite (#PG4638) Northern fowl mite (#PG9386)
Tropical fowl mite (#PG9335) Air sac mite of poultry (#PG9407).
Related Disease vectors (#PC3595) Insect vectors of human disease (#PE3632)
Insect vectors of plant diseases (#PD7732).
Aggravates Mange (#PE2727) Scrub typhus (#PG1989) Typhoid fever (#PD1753)
Mange in sheep (#PG4137) Mange in swine (#PG7868) Mange in horses (#PG9788)
Mange in cattle (#PG6179) Mite-borne typhus (#PG3896) Itch mite infestation (#PG1719)
Parasitic otitis externa (#PG7387) Pests and diseases of tea (#PE2980)
Nasal acariasis in animals (#PG5521) Pests and diseases of cotton (#PE2220)
Pests and diseases of rubber (#PE2977) Cutaneous acariasis in animals (#PG9510)
Pests and diseases of citrus fruit (#PE2976) Cheyletiella infestations in animals (#PG4207)
Pests and diseases of deciduous fruit (#PE3591).

♦ PE3640 Crickets as pests

Nature Crickets are orthopterous insects of the family gryllidae; many are plant and household pests.

Incidence Crickets vary in their feeding habits; many are omnivorous. The field cricket, acheta assimilis, attacks most crops and is a serious pest of cotton in the USA. The Mormon cricket (anabrus simplex) is a pest on crops and rangelands in the northwestern USA. A number of subterranean crickets subsist largely upon roots and are quite injurious when abundant in crops, gardens and young forest plantations; for example, the mole cricket (gryllotalpa hexadactylo) eats the roots of seedlings and is particularly destructive of tobacco and vegetable crops in North America and the West Indies. The slashes of the snowy tree cricket (oceanthus nneus), weaken the twigs and canes of various fruits.
Broader Insect pests of plants (#PD3634) Orthoptera as insect pests (#PE3641).
Aggravates Pests and diseases of cotton (#PE2220)
Insect damage to stored and manufactured goods (#PD3657).

♦ PE3641 Orthoptera as insect pests

Nature Orthoptera is an order of insects, many of which are at times serious pests of crops and nuisance household pests.

Incidence Orthoptera are found in various habitats throughout the world. Blattida, or cockroaches, are serious cosmopolitan pests of buildings. Phasmidae, or stick insects live in trees and certain species (diapheromera femorata) may be so abundant as to completely defoliate trees. The gryllidae or crickets are mainly agricultural pests. To the family acrididae belong the majority of crop destroying orthoptera: they are the grasshoppers and locusts, and are the most universally distributed and most abundant members of this order.
Broader Insect pests (#PC1630) Pests and diseases of trees (#PD3585).
Narrower Crickets as pests (#PE3640) Cockroaches as pests (#PE1633)
Grasshoppers as insect pests (#PE3642).

♦ PE3642 Grasshoppers as insect pests

Acrididae

Nature Grasshoppers are any of the leaping insects belonging to the orthoptera family acrididae. They are almost entirely herbivorous, and among them are some serious crop pests. They constitute the most abundant and widely distributed pests in the order orthoptera.

Incidence Grasshoppers are virtually cosmopolitan, being found in a variety of habitats – mountains, deserts, temperate forests and grasslands – but they occur in greatest numbers in lowland tropical forests, semiarid regions, and grasslands. Some species are restricted in their feeding habits to certain plants, but most feed on any suitable vegetation. Natural controls (mainly predators, birds, frogs, snakes and the larvae of certain flies) check the populations of most species, and they usually occur in insufficient numbers to cause serious damage. A small number of species known as locusts, experience massive population explosions and because of their strong migratory habits, they cause widespread and terrible destruction. A number of other species, without reaching such devastating proportions, are still highly destructive of crops and rangelands especially in North and South America, Africa and Asia.
Broader Insect pests of plants (#PD3634) Orthoptera as insect pests (#PE3641).
Narrower Locust plagues (#PE0725).
Aggravates Pests and diseases of cotton (#PE2220).

♦ PE3643 Siphonaptera as insect pests

Fleas

Nature Fleas are wingless insects of the order siphonaptera. They are bloodfeeders and live as ectoparasites in the hair of mammals and the feathers of birds, and are thus nuisance pests. Furthermore, because they readily transfer from one host to another, they are very important vectors of disease.

Incidence Some 1,600 species and subspecies of fleas are known. All the adults feed avidly and repeatedly on blood. Some species of fleas, such as shrew fleas, are highly host specific, but others will parasitize a variety of animals, for example, the cat flea which infests domestic cats, leopards, mongooses, foxes, civets, dogs and opossums, and in their absence, will readily feed on human beings. Severe dermatitis and intense pruritis may follow infestation by fleas. Often the urticarial eruption resulting from flea bites in allergic persons is mistakenly diagnosed as hives and attributed to some kind of food. Species attacking man and his domestic animals include the human flea (pulex irritans), the cat flea, and the stickbight flea (echidnophaga gallinacea). In heavy flea infestations, animals may be severely damaged or killed by the effects of flea bites and by the considerable loss of blood. Poultry may be parasitized by the western chicken flea (ceratophyllus niger) and the European chicken flea.
Two species of fleas, the stickbight flea and the sand flea (tunga penetrans) penetrate the skin of the host and embed prior to egg deposition. The sand flea burrows into the soft skin of the feet of man. Intense itching accompanies the growth of the pregnant flea to pea-size. Itching and irritation may lead to secondary infections, and a number of deaths have occurred from gas gangrene and tetanus. The stickbight flea is similar to the sand flea, but attacks the heads of poultry, cats, and dogs. Certain fleas that primarily feed on rodents and birds will, if need be, attack man, as when hungry oriental rat fleas (xenopsylla cheopis) abandon their hosts dying of bubonic plague, thereby transmitting the plague bacillus to man. Other species of fleas, such as xenopsylla brasiliensis and nosopsyllus fasciatus, also transmit plague, and are widespread among rodents and small mammals. Natural infection among animals with the plague cacillus has been demonstrated in more than 90 species of fleas, representing 45 genera and 9 families. Fleas are believed to be the principal vectors of murine typhus to man from rats and mice; of various enzootic infections among animals, including tularaemia and Russian spring-summer encephalitis; of myxomatosis, a virus disease of rabbits; of a filorial worm of dogs; and of a common tapeworm (diplydium caninum) of dog, cats and, occasionally, children.

Refs Traub, R and Starcke, H (Eds) *Fleas* (1980).
Broader Parasites (#PD0868) Insect pests (#PC1630)
Insect vectors of human disease (#PE3632) Insect vectors of animal diseases (#PD2748).
Aggravates Plague (#PE0987) Typhoid fever (#PD1753) Infected dogs (#PG6590)
Murine typhus (#PG6700) Sylvatic plague (#PG6726)
Insect bites and stings (#PE3636) Flea allergy dermatitis (#PG9287).
Aggravated by Flea resistance to insecticides (#PE3572).

♦ PE3644 Army worms as pests

Nature The name army worm is generally applied to the larvae of certain types of lepidoptera that sometimes migrate to new feeding grounds in large armies, destroying corn, small grains, sugar cane, cotton and other crops as they move. The moths deposit their eggs on the underside of grass leaves and when the larvae hatch they grow slowly, eating small amounts of leaf tissue and causing little, if any, damage. But when they enter their last larval stage, they exhibit tremendous appetites and it is during this phase, which usually lasts 10 days, that farm crops in an area of several hundred square miles can be destroyed.
Broader Lepidoptera as insect pests (#PE3649).
Related Corn borers as pests (#PE3648).
Aggravates Maize pests and diseases (#PE3589).

♦ PE3645 Wireworms as pests

Nature The larvae of various species of the beetle family elateridae live in the soil and are very serious pests of a range of crops.

Incidence Wireworms are prevalent in recently ploughed grassland and attack subsequent crops, especially cereals and roots. Seeds, germinating seedlings, stems below the ground, and roots, are all attacked by wireworms. Wireworms tend to be most important in temperate agriculture, and are the most important pests of temperate cereals especially in Northern Europe, the USSR and Canada. In the USA and Australia, wireworms are pests of sugarcane.
Broader Beetles as insect pests (#PE1679).
Aggravates Maize pests and diseases (#PE3589)
Pests and diseases of wheat (#PE2222)
Pests and diseases of sugar-beet (#PE2975)
Pests and diseases of grain sorghum (#PE3590).

♦ PE3646 Black flies as pests

Buffalo gnats

Nature Black fly, or buffalo gnat, is the name applied to any member of the dipterous family simuliidae, which suck blood from birds and mammals, including man.
Incidence About 300 kinds of black fly are known, being widely distributed throughout the world. Only the females bite, causing extreme, sometimes fatal, loss of blood. In tropical America and Africa, black flies may be parasitized by the nematode worm, onchocerca volvulus, which can be transferred to other victims causing the disease onchocerciasis, resulting in blindness if the worms settle in the eye.
Broader Parasites (#PD0868) Flies as insect pests (#PE2254)
Insect vectors of human disease (#PE3632) Insect vectors of animal diseases (#PD2748).
Aggravates Onchocerciasis (#PE2388) Insect bites and stings (#PE3636)
Pests and diseases of citrus fruit (#PE2976).
Aggravated by Blackfly resistance to insecticides (#PE3578).

♦ PE3647 Political elitism
Refs Putman, Robert D *Comparative Study of Political Elites* (1976).
Broader Elitism (#PA1387).
Aggravates Gerontocracy (#PD3133).

♦ PE3648 Corn borers as pests
Nature The corn borer or European stalk borer (pyrausta mibilalis) is the larval stage of a pyralid moth which is highly destructive of a range of crops. The caterpillar is a stalk borer.
Incidence The corn borer has a very wide geographic distribution. It occurs, and attacks maize and sorghum, throughout the southern latitudes of Europe, including the USSR, and in many parts of Asia and the South Pacific, and has been introduced into North America. In Europe it is known also as an enemy of hops and hemp. Considerable losses are also recorded on sweet corn, on other vegetables and on various ornamental plants. There are more than 200 species of grains, weeds, vegetables and flowers that can act as alternate hosts for the borers.
Broader Borer insects (#PE6286) Lepidoptera as insect pests (#PE3649).
Related Army worms as pests (#PE3644).
Aggravates Maize pests and diseases (#PE3589).

♦ PE3649 Lepidoptera as insect pests
Caterpillars
Nature Lepidoptera are butterflies and moths, the larvae of which have extremely destructive feeding habits. In fact, many of the most destructive agricultural, forest and household pests are caterpillars.
Incidence The great majority of Lepidoptera feed on plants, chiefly on the foliage. Most larvae chew up whole leaves, though some, the leaf miners, tunnel into the leaves. Other groups mine under the stems of plants and the barks of trees; species of many families bore into buds or soft, succulent stems.
Broader Insect pests of plants (#PD3634) Pests and diseases of trees (#PD3585).
Narrower Army worms as pests (#PE3644) Corn borers as pests (#PE3648).
Related Defoliating insects (#PE4999).
Aggravates Maize pests and diseases (#PE3589)
Pests and diseases of elm (#PE2982)
Pests and diseases of rice (#PE2221)
Pests and diseases of cocoa (#PE2979)
Pests and diseases of palms (#PE2981)
Pests and diseases of vines (#PE2985)
Pests and diseases of coffee (#PE2218)
Pests and diseases of olives (#PE2978)
Pests and diseases of potato (#PE2219)
Pests and diseases of sugar cane (#PE2217)
Pests and diseases of citrus fruit (#PE2976)
Pests and diseases of grain sorghum (#PE3590)
Pests and diseases of deciduous fruit (#PE3591)
Insect damage to stored and manufactured goods (#PD3657).

♦ PE3656 Soil erosion by wind
Nature Soil erosion by wind may occur wherever dry, sandy or dusty surfaces, inadequately protected by vegetation, are exposed to strong winds. Erosion involves the picking up and blowing away of loose fine grained material within the soil. Damage from wind erosion is of numerous types. The dust storms resulting therefrom are very disagreeable and the land is robbed of its long-term productivity. Crop damage, particularly in the seedling stage, by blowing soil is often a major concern. Serious stand and subsequent yield and quality losses are incurred and, in the extreme, tender seedlings may be completely killed. Often, sufficient soil is removed to expose the plant roots or ungerminated seed, and this results in complete crop failure. Covering of established crops or pasturage by drifting soil is another common result. These are but a few of the more evident results of wind erosion. The most serious and significant by far, however, is the change in soil texture caused by wind erosion. Finer soil fractions (silt, clay, and organic matter) are removed and carried away by the wind, leaving the coarser fractions behind. This sorting action not only removes the most important material from the standpoint of productivity and water retention, but leaves a more sandy, and thus a more erodible, soil than the original.
Incidence Few attempts have been made in the past to assess the extent of damage caused by wind erosion and the annual losses of land and production which have taken place. All countries admit that the problem does exist to varying degrees, but few can provide exact figures on its magnitude. With the expansion of agricultural production in lesser developed countries in recent years has come fuller appreciation of past damage, and recent surveys indicate that wind erosion is far more widespread than was commonly believed. Countries of South America, North Africa, the Near East, and Asia all report wind erosion in varying degrees. Damage results both from the erosion and the consequent dust storms. Saskatchewan farms have 1000 tons of topsoil per acre, of which 15 tons are lost annually, due to wind erosion.
Background Successive removals eventually create a soil condition wherein plant growth is minimized and erodibility greatly increased. Control becomes more and more difficult. In the extreme, the sands begin to drift and form unstable dunes which encroach on better surrounding lands. Throughout recorded history, huge agricultural areas have been ruined for further agricultural use in this manner. It is this degradation of soil resources that constitutes the most serious aspect of wind erosion. Areas so affected become totally unsuited for cultivation. Through difficult and expensive reclamation they may provide, at best, limited grazing. Uncontrolled, they become a menace to adjacent productive lands.
Refs FAO *Soil Erosion by Wind and Measures for its Control on Agricultural Lands* (1978).
Broader Soil erosion (#PD0949).
Related Land erosion brought about by site development (#PE7099).
Aggravates Dust (#PD1245) Sand storms (#PD3650) Desert advance (#PC2506)
Rock avalanches (#PG0476) Soil infertility (#PD0077)
Migrating sand dunes (#PD0493) Destruction of land fertility (#PC1300)
Slowing growth in food production (#PC1960).
Aggravated by Wind (#PE2223) Deforestation (#PC1366)
Diplomatic errors (#PF1440).

♦ PE3657 Delay in connection of telecommunications facilities
Delay in connection of telephones
Aggravates Maldistribution of telecommunications facilities (#PF4132).
Aggravated by Prohibitive cost of connection to public utilities (#PJ9426).

♦ PE3659 Toxoplasmosis
Nature Toxoplasmosis is a common parasitic disease affecting man and animals in all parts of the world. The disease can produce lymphadenopathy and in some parts of the world is reported to be responsible for up to 15 percent of otherwise unexplained cases of this condition. It can also cause more severe symptoms – fever, rash, malaise, muscular pains, pneumonia, myocarditis or meningoencephalitis. By far the most serious form of the infection, however, is congenital toxoplasmosis, which is acquired by the foetus during a mild infection of the mother. The baby may be stillborn, or it may suffer from hepatosplenomegaly, purpura, jaundice, lesions of the central nervous system, or destroyed areas of the retina. These conditions may be present at birth or may appear weeks or months afterwards. Once tissue has been destroyed – for example, in the brain or the eye – the effects do not regress.
Incidence No region is exempt from the toxoplasma parasite except possibly the Antarctic. It has been estimated that between 25 and 50 percent of the adult human population in most places in the world have been infected, as evidenced by serological studies. Several epidemiologic and serologic studies of the disease in animals and man, and studies identifying environmental reservoirs of the disease, have suggested that agricultural workers are at increased risk of acquiring this disease as compared to the general population. Many species of wildlife and domestic cat and certain other members of the family Felidae are the only hosts in which the parasite will complete the sexual stage of the life cycle. The cat is an important source for the widespread dissemination of the organism in the environment. Infection in cats and other animals is transmitted either by ingestion of oocysts in faeces, in soil, or by consumption of infected animals which have toxoplasma cysts in their tissues. Infection is quite common in cattle, sheep, goats, swine and chickens. Serological surveys in several countries have shown that the average positive reaction rate recorded for the food animal species ranged from 22 to 39 percent, with sheep being the highest.
Broader Protozoan parasites (#PE3676) Parasites of the human body (#PE0596)
Infectious and parasitic diseases (#PD0982).
Narrower Acquired toxoplasmosis (#PG6735) Congenital toxoplasmosis (#PG6734).
Related Parasitic diseases in animals (#PD2735).
Aggravates Still-birth (#PD4029) Encephalitis (#PE2348)
Hydrocephalus (#PJ3190) Physical blindness (#PD0568)
Pre-natal infections (#PG6733).
Aggravated by Pica (#PG4505) Protozoa as pests (#PE6741)
Animals as vectors of disease (#PD8360)
Government action against regimes with alternative policies (#PF2199).

♦ PE3660 Leeches as pests
Bloodsuckers
Broader Parasites (#PD0868).
Aggravates Rinderpest (#PE2786) Human death (#PA0072)
Animal diseases (#PC0952) Trypanosomiasis (#PE5725)
Pests and diseases of fish (#PD8567) Endangered species of annelida (#PD3160)
Infectious and parasitic diseases (#PD0982).
Aggravated by Contamination of drinking water (#PD0235).

♦ PE3662 Acne
Nature A chronic skin disease, acne affects the sebaceous glands of the forehead, nose, chin, chest, back of shoulders and outer side of the thighs. Occurring in persons with sebarrhoea, acne is found in both sexes, usually in the age group 14–20. There is an individual predisposition to the disease, dependent upon the development of the sebaceous glands which takes place at puberty. The condition of acne is associated with dyspepsia, constipation, lack of fresh air, lack of exercise, and in women it tends to worsen during menstruation. Black spots indicate that the mouth of the small sebaceous ducts are clogged with dirt; little pustules, slightly inflamed, grow, burst and heal. Hard lumps, up to half an inch across, last for weeks or months, slowly suppurate, and leave a permanent hardness. Although not a serious health hazard, acne causes widespread psychological discomfort, especially among teen-agers.
Incidence There are millions of acne sufferers all over the world. In France alone, it was recently estimated that 5 million people suffer from acne.
Refs Cullen, S I (Ed) *Focus on Acne Vulgaris* (1985); Plewig, G and Kligman, A M *Acne* (1975).
Broader Unsociable human physiological processes (#PF4417)
Diseases of the skin and subcutaneous tissue (#PC8534).
Narrower Scarring (#PG6749).
Aggravates Embarrassment (#PF7950) Discrimination (#PA0833)
Lack of self-confidence (#PF0879).
Aggravated by Constipation (#PE3505) Menstruation (#PE4838)
Puberty trauma (#PJ2206) Hormonal disorder (#PE4431)
Physical unfitness (#PD4475) Inadequate fresh air (#PJ4652)
Nutritional deficiencies (#PC0382) Inadequate personal hygiene (#PD2459).

♦ PE3676 Protozoan parasites
Protozoan diseases
Refs Kreier *Parasitic Protozoa* (1977).
Broader Parasites (#PD0868).
Narrower Giardiasis (#PE4811) Babesiosis (#PE6288) Hexamitiasis (#PG9766)
Besnoitiosis (#PG5724) Toxoplasmosis (#PE3659) Sarcosporidiosis (#PG9770)
Coccidiosis in poultry (#PG9814) African trypanosomiasis (#PE1778)
Blood sporozoa of birds (#PE4284) Histomoniasis in poultry (#PG9380)
Trichomoniasis in poultry (#PG4423) Visceral leishmaniasis in animals (#PG7809).
Related Protozoa as pests (#PE6741).
Aggravates Chagas' disease (#PE0653).
Aggravated by Sewage as a pollutant (#PD1414).

♦ PE3677 Rodenticides as pollutants
Nature Rodenticides are toxic chemicals used for the control of rats, mice and other pest species of rodents. Poisoned baits are the most generally effective and widely used means of formulating rodenticides, but some are used as 'contact' poisons (such as dusts, foams and gels), where the toxicant adheres to the fur of the animal and is ingested during subsequent grooming, while a few are applied as fumigants to burrows or infested premises.
Incidence Although toxicity levels of rodenticides may vary between target and non-target species, all poisons must be presumed to be lethal to man. Acute poisons are potentially more dangerous than chronic ones because they are rapid in action, non-specific, and generally lack effective antidotes. Anticoagulants, on the other hand, are slow and cumulative, allowing adequate time for the administration of the reliable antidote, vitamin K. The concentrations of active ingredients in contact formulations of a given poison are higher than those in bait preparations, thus making operator hazard considerably greater. Fumigants present a special danger when used to treat infested premises, holds of ships, etc, and should only be used by trained technicians. The gassing of rodent burrows, although less hazardous, must also be carried out with extreme

caution.
Refs FAO *Rodenticides* (1979).
Broader Agricultural poisons (#PD5277)
Pesticides as pollutants (#PD0120).
Aggravates Alopecia (#PG4264)
Polyneuritis (#PG6802)
Hepatonephritis (#PG6799)
Burns and scalds (#PE0394)
Hazards to human health (#PB4885)
Blood circulation disorders (#PE3830)
Pesticide hazards to wildlife (#PD3680)
Rodent resistance to rodenticides (#PE3573).
Aggravated by Rodents as pests (#PE2537).

♦ PE3680 Viruses
Refs Berns, Kenneth I *The Parvoviruses* (1984); Bishop, David H and Compans, Richard W *Nonsegmented Negative Strand Viruses* (1984); Compans, Richard W and Bishop, David H *Segmented Negative Strand Viruses* (1984); Epstein, M A *The Epstein-Barr Virus* (1986); Fenner, Frank, et al *The Orthopoxviruses* (1988); Ginsberg, Harold S *The Adenoviruses* (1984); Hanshaw, J B; Plowright, W and Weiss, K E *Cytomegloviruses* (1968); Horzinck, Marian C *Non-Arthropod Borne Togaviruses* (1981); Hoyle, Fred, et al *Viruses from Space* (1986); Martyn, E B (Ed) *Additions and corrections to phytopathological paper no 9, 1968, and newly recorded plant virus names* (1971); Pandey, R (Ed) *Nononcogenic Avian Viruses* (1989); Pullman, Maynard E, et al *Retroviruses and Disease* (1989); Schlesinger, R Walter *The Togaviruses* (1980); Smith, Roberts A *HIV and Other Highly Pathogenic Viruses* (1988); Vesenjak-Hirjan, J; Portersfield, J S and Arslanagic, E (Eds) *Arboviruses in the Mediterranean Countries* (1980).
Aggravates Virus diseases (#PD0594).

♦ PE3681 Apartheid
Slavery-like practices of apartheid
Nature Apartheid, the South African government's policy of 'separate development' for all groups 'within their own communities', is a series of systematic acts of oppression and discrimination against the overwhelming majority of the population of South Africa. It renders non-whites political and social outcasts in their own fatherland; violates human rights, especially the right to self-determination; and permeates all aspects of life. Blacks, as part of the policy of total social segregation, are forced to live in 'homelands', a move which not only causes great suffering, but also breeds crime, violence and oppression caused by rivalries between 'haves' and 'have-nots'; and hunger, disease, and starvation have become the marks of many communities. As apartheid translates into extreme poverty for the black community, it is the women and children who suffer most. Able-bodied men often work and live in urban areas, while their wives and families are unable to secure passes for these areas and must remain on the homelands, thus negating prospects for a normal family life, and also engendering high numbers of illegitimate children. 'Pass laws' are strictly enforced in the white urban areas and the fines imposed on those who illegally employ blacks has had a marked effect on the scope for employment of women. In order to remain in the area, married women have to be included in the residential rights of their husbands, who must be present after the 'lawful entry' of the wives. If a black woman becomes widowed or divorced, she may lose her eligibility to live and work in the urban area.
Forced removal to the homelands has resulted in uneven population growth; huge tracts of the country could well become perpetual wastelands if development is neglected in the areas made barren by the removals. On the homelands, large numbers of people have no land to cultivate and must spend most of their money on imported basic foodstuffs. Children are often forced to work in an effort to counter the shortage of male labour, and are severely exploited; those who are not offspring of farm labourers live in squalid communal huts and have to cook for themselves. They have an inadequate diet and receive no education.
Apartheid views blacks as cheap labour without rights of their own. Black agricultural workers are excluded from unemployment and sickness benefits; black miners must work in mines where lax or non-existant safety measures result in many accidents, some fatal, and those who refuse to work in unsafe mines may be fired without notice; violations of trade union freedoms are rampant; and all blacks, regardless of their trade, have little or no hope for job improvement. Dissenters of apartheid, even women and children, are often subjected to brutal and sophisticated torture; political trials may last 3-4 years, with the accused spending all that time in goal; suspicious deaths of detainees are known; the press is forbidden to publish reports or photographs of people under 'banning orders', and those banned cannot communicate with more than one other person at a time.
Incidence Examples of the effects of apartheid include: (1) A 1982 ILO report showing that out of every 1000 white South Africans, 16 had not progressed beyond primary school; of every 1000 Asian South Africans, 257 had not progressed beyond primary schools; for coloureds and blacks the numbers were 590 and 840 respectively. Again, 16.7 percent of whites had received university diplomas or degrees, 1.3 percent of coloureds, and 2 percent of blacks. (2) A 1983 WHO study showed that the white community had a mortality pattern identical to that of developed countries (12 per 1000), whereas that of the black South African, particularly the rural black South African, is less than that found in a typical Third World African country, with widespread malnutrition leading to a heavy toll of infant deaths from infectious disease. (3) A WHO 1983 study in the Ciskei revealed that in both urban and rural groups, 60% of malnourished children were illegitimate, whereas 80% of well-nourished children were legitimate. Less than half of the malnourished children were looked after by their mothers; such widespread malnutrition leads to the heavy toll of infant deaths from infectious diseases. (4) As an example of the imprisonment of dissidents, Nelson Mandela was imprisoned since 1962 for inciting violence against the apartheid regime; and his wife was briefly arrested for meeting, while she was under a banning order, with more than one other person; and for refusing to leave her home to live in another area of the country.
Background From the days of the first encounters between blacks and whites in South Africa, the white man has dominated because of his technical superiority and his possessions of firearms. In 1910, in the Act of the Union, whites in South Africa and Britain agreed that 'the native question' was the crucial issue, and ultimately white political parties were judged and elevated to government or thrown out of government because of their particular view of that question. At that time, the rationale for the exclusion of blacks was that they were uncivilized and unfit to participate in government. In 1948 the National Party came to power and began its policy of apartheid; in 1956 black voters were removed from the rolls; the Bantu Self-Government Act of 1959 set up the machinery to create a number of separate 'homelands'; and Mr P W Botha's election to the office of Prime Minister, in 1978, assured continuation of a strict racist regime.
Claim Apartheid, as a system of government which has the colour of a person's skin as a determining factor enshrined in the law and constitution of the State, is not only an attack upon basic rights, it strikes at human dignity and the right to be and be recognized as an individual. Not only is it an intolerable affront to the coloured races of the world, but also to any concept of humanity. With all its facets taken into account, the criminal effects of apartheid are without comparison regarding denial of human rights and actually amount to a policy bordering on genocide.
Counter-claim Blacks in South Africa are actually living under conditions better than many of their fellows in neighbouring countries, many of whom are struggling for immigration into South Africa. Blacks there at least have the right to protest and demonstrate, and famine and starvation has not yet affected South Africa, as it has so many other African countries which are under inept (often black) rule. Racial policy in South Africa is no more destructive in itself than that of Australia toward its Aboriginals; it is only more visible because South African Blacks are a majority and Aboriginals in Australia a tiny minority.
Refs Mandela, Nelson *L'Apartheid* (1985); Pyatt, Sherman E *Apartheid* (1989); Sawyer, Roger *Slavery in the Twentieth Century* (1986); Uhlig, Mark (Ed) *Apartheid in Crisis* (1986); United Nations *Apartheid, Poverty and Malnutrition* (1982).
Broader Racial conflict (#PC3684)
Racial segregation (#PC3688)
Legalized racial discrimination (#PC3683).
Narrower Job reservation under the apartheid system (#PE6246).
Related Legal segregation (#PD3520).
Aggravates Social injustice (#PC0797)
Detention of children (#PE6636)
Internment without trial (#PD1576).
Aggravated by Destiny (#PF3111)
Lack of social conscience (#PF9144)
Denial of right of complaint (#PD7609)
Violation of the right to strike (#PE5070)
Denial of the right of association (#PD3224)
Undemocratic political organization (#PC1015)
Restrictions on freedom of movement between countries (#PC0935)
Disregard for internationally imposed economic sanctions (#PF1976).
Reduces Extremism (#PB3415).
Reduced by Guerrilla warfare (#PC1738).

♦ PE3684 Employee theft
Incidence In the UK in 1989 it was estimated that some 830 million pounds of goods was stolen by employees.
Refs Barefoot, J K *Employee Theft Investigation* (1990).
Broader Theft of property (#PD4691).

♦ PE3700 Devaluation of money
Currency devaluation — Depreciation of currency
Nature Currency devaluation's primary effects are to raise import prices in terms of domestic currency, thus causing a reduction in the demand for imports, and to reduce export prices in terms of foreign currency, thus causing an increase in external demand for exports.
Incidence For countries with a large trade sector and substantial industrial capacity it is easier to generate a given rise in export earnings relative to domestic production by real currency depreciation and ensure an effect on a large proportion of the output. It therefore needs a smaller real devaluation to generate a given swing in the trade balance than for an economy with a small trade sector. For these reasons, and because it does not entail a substantial decline in real wages, devaluation can provide a growth stimulus without setting off an inflationary spiral. However, where the trade sector is small, the depreciation needed to generate the same amount of net foreign exchange revenues is much sharper and is associated with large changes in income distribution, with disruptive effects on out put and price instability. The major exporters of manufactures in Latin America have corresponded more to the latter type and those of East Asia to the former. In a number of countries where the external debt of private firms was large, devaluations substantially weakened the financial position of firms.
Broader Faltering structural adjustment in the world economy (#PF9664).
Aggravates Speculative flight of capital (#PC1453)
Decline in public sector savings in developing countries (#PE4574).

♦ PE3704 Military manoeuvres in sensitive border areas
Provocative military exercises — War games — Naval manoeuvres in sensitive areas
Nature From a military point of view it is important to exercise forces in the terrain where they will operate in the event of armed conflict, namely, border areas. Political reasons for training military forces in a border area include: demonstrating military strength with a view to deterring other countries from aggression; exerting pressure on neighbouring countries; strengthening the confidence of the local population. Such manoeuvres increase the tensions between countries, particularly when they are carried out without advance notice as to the size and composition of the forces involved, the purpose of their movement, and the areas in which the manoeuvre will take place. They may also provoke similar displays of force on the other side of the border. Such military alerts and the confrontations to which they give rise increase both the difficulty of peaceful negotiations and the risk of armed conflict.
Broader Threat of war (#PF8874).
Related Cross border military operations (#PD5272)
Environmental degradation through military activity during peace-time (#PE0736).
Aggravates Conflict (#PA0298)
Aggression (#PA0587)
Surprise attack (#PE3705)
Marine accidents (#PD6982)
European insecurity and vulnerability (#PD1863).

♦ PE3705 Surprise attack
Nature The possibility of a surprise military attack by one country or bloc of countries on another increases international tension and hinders discussion of disarmament and arms control measures. Such attacks may take the form of long-range ballistic missiles or the use of tactical weapons by forces in close proximity (such as in the European region).
Counter-claim Satellite reconnaissance has provided some means to guard against massive conventional attack by surprise. The second-strike capability which both the USA and the USSR possess has largely decreased the probability of a surprise nuclear attack. These arguments are only valid in connection with the military alliances centred on Europe. Surprise attack remains an important problem for non-aligned countries, in both military and political terms.
Aggravates European insecurity and vulnerability (#PD1863).
Aggravated by Military manoeuvres in sensitive border areas (#PE3704).
Reduced by Misuse of satellite surveillance by governments (#PF3701).

♦ PE3711 Congenital anomalies of the digestive system in animals
Broader Animal abnormalities (#PD4031)
Diseases of the digestive system in animals (#PD3978).
Narrower Clefts (#PG7294)
Atresia in animals (#PG5440)
Abnormalities of teeth (#PG9665)
Dilatation of oesophagus (#PG1248)
Abdominal hernias in animals (#PG9730)
Rectovaginal fistula in animals (#PG9317).

♦ PE3715 Galls in plants
Nature Galls are local swellings of plant tissue caused by insects, bacteria, fungi, viruses or physiological factors.
Broader Oomycetes (#PE5465)
Ascomycetes (#PE4586)
Basidiomycetes (#PE0364)
Deuteromycetes (#PE4346).

♦ PE3716 Inability to concentrate due to torture
Nature Victims of torture frequently experience an inability to concentrate even years after the events. The ability to learn new skills and to adjust to new situations is impaired. Physical damage as a direct result of blows to the head, structural change in the brain due to extreme stress or non-physical psychological problems may be the sources of this incapacity.
Incidence In a recent study of torture victims 45 percent complained of losses of memory or concentration. Most of the victims were young adults who were interviewed some time after being tortured.
Broader Neurological effects of torture (#PD4755)
Limited individual attention span (#PF2384).

♦ PE3717 Childhood neurosis
Nature Childhood internal conflict manifests itself in eating or excretory problems, antisocial

behaviour, learning disabilities, and so on. Symptoms very from anxiety and phobias to tics and rituals.
Broader Neurosis (#PD0270).
Aggravated by Infantile neurosis (#PE3571).

♦ **PE3719 Torture by crushing**
Nature Hands, legs and arms are crushed as a form of torture is several countries. Heavy rollers, sometimes on specially designed equipment, are applied to victims legs. Bullets, pencils and other instruments are placed between fingers which are then crushed together. Genitals are squeezed.
Incidence Crushing has been reported in the following countries: **Af** Ethiopia, Zaire, Zimbabwe. **As** India, Pakistan, Philippines. **Eu** Italy.
Broader Torture by wounding (#PE4078).

♦ **PE3720 Rural violence**
Refs Prasad, N *Rural Violence in India* (1985).
Broader Human violence (#PA0429).
Related Urban violence (#PJ9519).

♦ **PE3721 Vulnerability of indigenous populations to introduction of diseases**
Massacre of indigenous populations through imported diseases
Incidence The pattern of European colonization resulted in the introduction of diseases into the colonized territories to which there was no natural resistance among the indigenous populations. The Europeans brought with them smallpox, influenza, cholera, tuberculosis, plague, malaria, and typhus, among other illnesses. Even such minor ailments as measles and chickenpox proved fatal and killed millions of American Indians. During the first century of European contact with the Americas, the indigenous populations were assaulted by a succession of major epidemics which killed 50 to 90 per cent of their populations. Several of these epidemics were intentionally spread by Europeans (for example, in the case of the British, by the deliberate donation to Indians of infected blankets from a smallpox hospital). In less than a century, all the tribal peoples of the West Indies had been exterminated.
The process has continued through the 1980s in the case of isolated tribes in the Amazon basin, especially where the land is coveted for agricultural or mining purposes. Those of European origin are still intentionally introducing killer microbes to breakdown local resistance to seizure of indigenous lands.
Related Biological warfare (#PC0195).
Aggravates Genocide (#PC1056)
Endangered tribes and indigenous peoples (#PC0720).
Aggravated by Denial of the right to health for indigenous populations (#PE4459)
Denial of right to social welfare services for indigenous peoples (#PE1506).

♦ **PE3724 Sickle cell disease**
Sickle cell anaemia
Refs Bohrer, Stanley P and Alavi, Abass *Bone Ischaemia and Infarction in Sickle Cell Disease* (1981); Serjeant, Graham R *Sickle Cell Disease* (1988).
Broader Anaemia (#PD7758).
Aggravates Natural human abortion (#PD0173).

♦ **PE3728 Environmental refugees**
Nature Although the immediate cause of of any mass movement of people may appear to be political upheaval and civil or military violence, the underlying causes often include the deterioration of the natural resource base and a significant reduction in its capacity to support the population.
Incidence In the case of Ethiopia, for example, it has been reported that the primary cause of recent famines was not drought but rather a combination of long-continued bad land use and steadily increasing human and livestock populations.
Refs El-Hinnawi, Essam *Environmental Refugees* (1985); Jacobson, Jodi L *Environmental Refugees* (1988); United Nations Environment Programme *Environmental Refugees* (1985).
Broader Refugees (#PB0205).
Related Refugees by boat (#PD8034).
Aggravated by Environmental stress (#PC1282)
Inequality in distribution of natural resources between countries (#PF3043).

♦ **PE3732 Joint pains due to torture**
Nature Victims of torture frequently suffer from pains in their joints. Connecting tissue may have been destroyed, torn or damaged in some way, limbs or other bones may have been pulled out of their sockets by being hung or through mechanical means, bones may have been broken, and joints may have been damaged due to stress from prolonged standing, squatting or other uncomfortable positions.
Broader Pain (#PA0643) Disorders of joints and ligaments (#PD2283)
Somatic and psychosomatic effects of torture (#PE5294).

♦ **PE3733 Malicious physical disablement**
Malicious personal disfigurement — Mayhem
Nature Under traditional common law, mayhem consisted of malicious deprivation of the use of another's members such as to render him less able to fight. This was later extended to cover injuries that disfigured but did not disable the victim. Disablement thus involves loss of use of a major part of the body (leg, foot, arm, hand, eye, tooth or testicle), whether or not it is completely removed. Disfigurement involves alteration of the victim's face or body such as to change the normal appearance (severing or slitting the nose, lip, ear or tongue). Originally the crime was punishable by comparable mutilation.
Broader Violent crime (#PD4752)
Offences involving danger to the person (#PD5300).
Aggravates Human disability (#PC0699) Personal physical disfigurement (#PD8076).

♦ **PE3734 Cold torture**
Torture by exposing to freezing temperatures
Nature Prisoner are tortured by exposure to extremely cold temperatures. They left outside without adequate clothing in cold weather. In some cases they are plunged into ice cold baths, sprayed with ice water, or exposed to cold drafts for long periods of time.
Incidence Torture by cold has been reported in the following countries: **Af** South Africa. **Am** Argentina. **As** Israel, Syrian AR. **Eu** Italy.
Broader Physical torture (#PD8734).

♦ **PE3735 Conflict of interest among parliamentarians**
Conflict of interest among peoples representatives
Claim Parliament members spend a lot of time harnessing a business or trade union interest to their political work. There is no guarantee that a parliamentarian will not put the directorship or consultancy or other private interests before the public good.
Counter-claim Parliamentarians outside interests should be increasing, because otherwise they do not know the real world outside the parliament.
Broader Conflicts of interest (#PF9610).
Aggravates Corruption of government leaders (#PC7587)
Unethical practices in local government (#PD5948).

♦ **PE3741 Marine pollution by plastic waste**
Incidence Sea life is drowning under waves of plastic refuse which will never die and which cannot be destroyed. As many as 100,000 marine mammals die each year as a result of ingesting plastic, and members of over 50 species of seabird, many already endangered, have died after ingesting plastic pellets. 100 million pounds of plastic trash enter the world's oceans each year, not including lost fishing nets and other gear. The 70,000 ships in the world fleet dump up to 640,000 plastic containers daily, about 10 percent of the total, as part of the six million tons of solid waste attributed to ships each year.
Broader Marine pollution (#PC1117) Unproductive use of resources (#PB8376)
Environmental hazards from plastic materials (#PD8566).
Aggravated by Marine dumping of wastes (#PD3666).

♦ **PE3745 Unnecessary gadgets**
Technological household gimmicks
Nature Millions are spent each year to develop new products. Many of these are not new products at all, but rather more interesting, alluring and desirable forms of the same old products, which make people buy new things. Kitchens and bathrooms are where gadgets are most commonly found.
Broader Uncontrolled application of technology (#PC0418).
Aggravated by Misuse of advertising (#PE4225) Blind faith in technology (#PF4989)
Dependence on sophisticated technology for development (#PD6571).
Reduced by Insufficient modern technology (#PJ0996).

♦ **PE3749 Mishandling national security information**
Nature The reckless disregard to the potential injury to national security by knowingly revealing national security information to persons not authorized to receive it, or not protecting such information when in the position of a responsible public servant, or not reporting the unlawful loss, theft, destruction or compromise of it, or refusing to deliver it on demand to a public servant entitled to receive it.
Broader Espionage (#PC2140) Crimes related to national security information (#PE3997).

♦ **PE3751 Denial of the right to work to refugees**
Inability of refugees to obtain employment
Nature The rights of refugees do not stop with determination of status and admission for asylum purposes. The Universal Declaration of Human Rights asserts: "Everyone has the right to work, to free choice of employment, to just and favourable conditions of work and to protection against unemployment." Refugees frequently do not enjoy this right. Many countries place serious legal and practical obstacles before refugees who wish to work. They are treated as aliens, or even illegal immigrants, often confined to closed camps under harsh conditions and thereby denied legal and practical access to the job market. The drawn-out nature of procedures for determining their refugee status, sometimes a deliberate unofficial policy to deter applications, prevents access to work opportunities up to two years. Even where employment rights are enjoyed, working can depend on prevailing official attitudes. In the majority of countries where refugees may work, work permits are required. This requirement is often used to curtail work options through limiting work permits both in duration and in scope, i.e. to a specific job only. The annual cost of the permit is, in some instances, a further inhibiting factor. Some refugees experience discrimination in employment or exploitation as a source of cheap manual labour. The widespread requirement that diplomas or degrees be recognized, be validated or be subject to reciprocal acceptance frequently discriminates against professionals.
Broader Denial of economic and social rights to refugees (#PE6375).
Aggravates Exploitation of women refugees (#PD5025).

♦ **PE3754 Corruption of sports and athletic competitions**
Rigging sports contests — Bribery in connection with sports and athletic competitions
Nature Rigging a sporting contest is to prevent a publicly-exhibited sporting contest from being conducted in accordance with the rule of the contest. Generally this is done through the corruption of officials and players through bribery or threat or through interference with animals, players or equipment.
Refs Sabljak, Mark and Greenberg, Martin H *Sports Babylon* (1988).
Broader Fraud (#PD0486) Bribery (#PC2558) Corruption (#PA1986).
Related Unethical practice of meteorology (#PD4182)
Politicization of international sports events (#PF4761)
Conformational abnormalities of eyelids in animals (#PG4212).
Aggravates Athletic competition (#PE4266).
Aggravated by Violent sports fans (#PE6281) Unethical entertainment (#PF0374)
Gambling on sports and athletic competitions (#PE4576)
Commercialization of athletic activities and sports events (#PE4222).

♦ **PE3757 Overcrowding in schools**
Broader Inappropriate size of school classes (#PD6585).
Narrower Overcrowding of schools by transients (#PG8014).
Aggravates Inappropriate study patterns (#PJ9371)
Inappropriate education in developing countries (#PF1531).
Reduced by Limitations on school admission (#PJ1364).

♦ **PE3760 Occupational hazards in farming**
Nature Tractor rollovers are the biggest killer on farms. Farmers have the highest rate of skin disorders, including skin cancer resulting from too much sun and eczema from handling detergents and chemicals. They inhale clouds of dust, animal dender, even deadly fumes. Many suffer hearing loss from the clatter of machinery. Farmer's lung is an allergic reaction to moldy hay; Silo-filler's disease kills farmers unaware of the buildup of nitrogen oxides produced by fermenting forage; Milker's knee is arthritis caused by years of squatting.
Health risks to farmers
Broader Occupational hazards (#PC6716).

♦ **PE3761 Premenstrual tension**
Premenstrual syndrome
Nature The hormonal upset prior to menstruation results in not only in aches and pains but also in extreme depression for a large proportion of women. Since these symptoms are perceived as psychosomatic or ignored by the medical profession, the problem is largely unmentioned and untreated.
Claim A large proportion of women suffer from PMT, and society too is suffering as a result.
Refs Dalton, Katharina *The Premenstrual Syndrome and Progesterone Theory* (1983); Fisher, H W *The Pre-Menstrual Syndrome* (1988); Ginsburg, Benson E and Carter, Bonnie F *Premenstrual Syndrome* (1987); Gise *Premenstrual Syndromes* (1987); Liehaus, Jacob L *Premenstrual Syn-*

drome (1988); O'Brien, P M S *The Premenstrual Syndrome* (1987).
Related Menstruation (#PE4838) Mental tension (#PB6302).

♦ PE3762 Endangered species of butterfly
Incidence In Britain, three–quarters of 59 surviving native species are declining rapidly. Butterflies are easy to count and are good indicators of ecological health, and their reduction probably signifies heavy losses in other species.
Broader Endangered species of insects (#PC2326).

♦ PE3769 Terrorists armed with nuclear weapons
Nature Nuclear weapons are becoming smaller, deadlier and more precise. Information on their fabrication and use is more readily available. Many weapons can be relatively easily constructed by hobbyists or those with access to the appropriate equipment or materials. The essential components of a neutron bomb can be carried in a back pack, and the rest are obtainable from local electronics and hardware stores.
Incidence In 1970 the city of Orlando, Florida, made ready to pay a $11 million ransom to a terrorist who claimed to have stolen fissionable material from the Atomic Energy Commission and made a hydrogen bomb. The culprit turned out to be a 14 year–old boy. With his ransom note he sent a workable diagram for a hydrogen bomb, though this does not mean he had the capability to build one. There were 43 less credible threats in the US between 1970 and 1979.
Refs Beres, Louis R *Terrorism and Global Security* (1987).
Broader Terrorism (#PD5574).
Related Terrorists armed with biochemical weapons (#PE9207).
Aggravated by Illegal exports of nuclear materials (#PE3968).

♦ PE3770 Conduct disorder
Nature Children and adolescents of the disorder violate the basic rights and have no concern for the feelings and well–being of the other people. Physical aggression is common: they initiate fights, use weapons, are cruel to animals and people, steal with confrontation of the victim and in later ages they may rape or kill. Other common features are lying, cheating, destroying others' property and stealing.
Broader Mental illness (#PC0300).
Aggravates Anti–social personality disorders (#PF1721).

♦ PE3779 Listeriosis
Listerellosis — Circling disease
Refs Ralovich, B *Listeriosis Research* (1984); Seeliger, Heinz P *Listeriosis* (1961).
Broader Zoonotic bacterial diseases (#PD6363) Bacterial diseases in animals (#PD2731)
Food poisoning through negligence (#PE0561).

♦ PE3780 Public non–accountability in control of production processes
Nature Individuals and groups responsible for quality and other controls of production processs are not accountable to the public. Where regulations exist they are frequently violated and not enforced effectively. The consumer is left unprotected.
Broader Social unaccountability (#PC1522).
Aggravates Foreign control of natural resources (#PD3109).

♦ PE3781 Domination of transnational corporations in the coffee industry
Nature Growing concentration in the ownership of the processing and distribution industry is a result of major coffee firms' strategy to merge with and/or acquire their rivals in order to increase their market shares. This strategy aims primarily to either facilitate entry to a national market or to reduce the number of rivals in an established national market. Besides, they use product differentiation, price discrimination and technology control. These strategies of leading transnational corporations in the coffee industry are constraints on the independent entry of other coffee producers (particularly the small ones) into major consumer country markets and on the increase of their share of coffee value added.
Broader Restriction of free market competition by transnational corporations (#PE0051).

♦ PE3786 Inundation of agricultural land through dams
Loss of agricultural land due to artificial flooding
Broader Environmentally harmful dam construction (#PD9515).
Related Floods (#PD0452).
Aggravates Unavailability of agricultural land (#PC7597).

♦ PE3795 Broomrapes
Nature Broomrapes are parasitic seed plants, leafless herbs, living on roots of other plants.
Broader Parasites on plants (#PD4659).

♦ PE3797 Speaking in opposition out of personal responses
Nature The processes of free speech are endangered by those who speak in opposition to established political structures and politicians out of unreflected upon, personal and immediate concerns.
Broader Irresponsible expression of emotions equated with free speech (#PF7798).

♦ PE3798 Residual traumatic pains due to torture
Nature Victims of torture, if they live, may be in physical pain for years following the torment. The specific pains depend on the type of torture they had to endure. Muscles may have been torn, crushed, damaged through exposure to electricity, stretched beyond their capacity, or even destroyed. Bones may have been broken and left to mend on their own or repeatedly damaged. Tendons may be damaged. Nerves may be destroyed through exposure to electricity, heat, repeated blows, or chemicals. Ears may ring or hurt due to blows to both ears at the same time. Headaches may be the result of repeated blows to the head.
Broader Pain (#PA0643) Trauma (#PD4571)
Somatic and psychosomatic effects of torture (#PE5294).

♦ PE3799 Prescription of inappropriate drugs
Claim Drug utilization is to a large extent irrational and related to teachings by opinion leaders and to influences of marketing efforts by the pharmaceutical industry. As a result of the so called pharmacological revolution an immense body of knowledge about modern drugs has accumulated. It is not used optimally, however, because training in pharmacology is insufficient and because information about drugs is neither well organized nor readily available. This problem can be improved only by developing an appropriate infrastructure with teaching, service and research as important functions. This infrastructure will have to overcome resistances by physicians, pharmacologists and pharmacists.
Broader Misuse of medicines (#PD8402) Unethical medical practice (#PD5770).
Aggravates Drug abuse (#PD0094).
Aggravated by Misdiagnosis (#PF8490) Overprescription of drugs (#PE9087).

♦ PE3802 Denial of right to choose moral and religious education
Broader Denial of right to educational choice (#PE4700).

♦ PE3804 Carbon black
Nature Carbon black is manufactured for use in rubber products to reduce abrasion and wear, printing inks, phonograph records, protective coatings, carbon paper and batteries. While there is no substantial proof that carbon black is carcinogenic it does cause some diseases of the skin.
Broader Dust (#PD1245).

♦ PE3806 Endangered species of porpoises
Incidence 7 species; none endangered.
Refs Klinowska, Margaret *Dolphins, Porpoises, and Whales of the World* (1988).
Broader Endangered species of whale (#PD1593)
Endangered species of marine mammals (#PD3673).
Aggravated by Hunting of marine animals (#PE0439).

♦ PE3808 Trade in products for chemical warfare
Claim Strict export controls are necessary to block deliberate export of equipment and chemicals needed to manufacture gases for chemical warfare.
Counter–claim Many of the substances sold to countries with chemical weapons programs are of "dual use", meaning they can have legitimate applications in pesticides, dyes or even ink for ballpoint pens. Developing countries have sometimes portrayed the calls for export controls as "colonialism" by Western countries and advocate instead a worldwide ban on chemical weapons.
Broader International trade in chemical weapons (#PD9692).
Related International trade in biological weapons (#PD8621).
Aggravates Chemical warfare (#PC0872).
Reduced by Restrictive practices in trade in chemicals (#PE8600).

♦ PE3809 Wilful violation of regulatory codes
Nature The wilful violation of a regulatory statute, rule, regulation or order is a criminal offence.
Broader Violation of regulatory codes (#PD4539).

♦ PE3812 Decline in concessional financial resources available to developing countries
Inappropriate loans available to developing countries — Erratic flows of concessional and non–concessional lending to developing countries
Nature Results in dramatic shifts in the debt structure from concessional loans to non–concessional loans with harder lending terms.
Aggravates Reallocation of aid funds to alternative priorities (#PF0648)
Excessive foreign public debt of developing countries (#PD2133)
Burden of servicing foreign public debt by developing countries (#PD3051)
Deterioration in external financial position of developing countries (#PE9567).
Aggravated by Inadequate diversification of loans to developing countries (#PE4305)
Deteriorating terms of financial loans to developing countries (#PE4603).

♦ PE3813 Settlement of unprotected coastlines
Exposure of settlements to tidal flooding
Incidence Subsistence farmers lacking cultivable land are increasingly forced to settle marginal lands exposed to tidal flooding.
Aggravated by Cultivation of marginal agricultural land (#PD4273).

♦ PE3814 Misleading advertising
Exaggerated advertising claims for products — Fraudulent advertising practices
Nature Advertisements may be framed so as to abuse the confidence of the consumer or exploit his lack of experience or knowledge. They may appeal to superstition or may play unjustifiably on fear. They may contain statements or visual presentations which directly or by implication, omission or ambiguity are likely to mislead the consumer, including claims purporting to be statements of fact. Advertisements may mislead as to: the qualities of the product advertised (for example, its composition, construction, utility or suitability, or its commercial or geographical origin), the price or value of the product or the terms of purchase; the services accompanying purchase (including delivery, exchange, return, repair and maintenance); the contents and the value of the guarantee attached to the product; the existence of any patent protection or other industrial property right, or medals, prizes, etc; the qualities, the price, the value or the terms of purchase of other products on the market and the services accompanying the purchase of such products; the trustworthiness of statements made by other advertisers. Scientific terms, statistics and quotations from technical literature and the like may be used without a proper sense of responsibility to the consumer (for example, statistics with a limited validity may be presented in such a way as to make it appear that they are universally true). Advertisements which support some charitable cause may mislead as to the share of the proceeds which will in fact go to the charity. Advertisements may contain references to persons or organizations without due permission, or to competitors and their products in such a way as to bring them into contempt or ridicule, or to exploit the goodwill attached to them, or to imitate their style in such a way as to create confusion among consumers. Advertisements may also be designed so as not to be clearly distinguishable as such (possibly to give the impression of being an editorial opinion in the case of newspaper advertising).
Broader Unfair competition (#PC0099) Deception in business (#PD4879)
Misleading information (#PF3096).
Narrower Misleading endorsement advertising (#PE7502).
Related Promotion of negative images of opponents (#PF4133).
Aggravates Medical quackery (#PD1725)
Non–destructible packaging and containers (#PD1754)
Misrepresentation of information to consumers (#PE6877)
Substitution of inappropriate foodstuffs for breast feeding (#PE8255).
Aggravated by Unethical commercial practices (#PC2563)
Unethical practices of employers (#PD2879)
Lack of legislative control on advertising (#PE8467).

♦ PE3815 Eichornia crassipes
Water hyacinth
Nature Eichornia crassipes is the world's worst aquatic weed. It constitutes a serious hindrance to river traffic and affects fishing and irrigation. Large masses of it floating or immersed, accumulating in hydro–electric installations, behind dams and in reservoirs, may create various problems by clogging grids and speeding up silting processes.
Incidence Eichhornia crassipes occurs in many streams, lakes, reservoirs and swamps of the tropics. It is in the Nile and Zaire rivers, in the delta of the Mississippi, and is found all along the coast of Southeastern USA, and is distributed all across South Asia. In parts of the world this weed has always been a major problem. But Eichornia crassipes is still spreading to new areas; when they enter a stream they are carried up the river by boats. The wind, floods and currents push them into back–waters and pools and swamps. The river people use water hyacinth plants as pads in canoes, and to plug holes in their charcoal sacks as they are transported from the bush. The seeds may be carried in mud on the legs and fur of wildlife. Finally, man is the worst offender. He has carried it around the world because he liked the flowers. The first infestation was seen in the Zaire

river in 1952 and by 1955 it covered 1,500 kilometers. Using herbicides, several thousand kilometers of the river were cleared by 1957. But after all this effort, it was estimated that 150 tons per hour were still passing Kinshasa down near the sea. The first infestation was seen in the Nile in 1958. The government of Sudan created a Water Hyacinth Section of the Plant Protection Department which took charge of control measures. Previously, the weed was contained above the Jebel Aulia dam near Khartoum. But seedlings have now been found far downstream. It can be commonly seen floating in the Nile in Cairo. As it was found that to eradicate the weed in Sudan is impossible, current activities concentrate on its conversion into animal feed, fertilizer, and for energy production. Chemical and mechanical means for controlling water hyacinth have proved either ineffective, too expensive, or environmentally hazardous. In the USA, Australia, Sudan and some other countries, substantial reduction in water hyacinth production has resulted from introducing biological control using the weevil Neochetina eichhorniae. A programme for the mass culture and release of these insects in the Sudan was commenced in 1979, and they are now successfully established and dispersing in the White Nile. More biocontrol agents are being tested.
Claim Eichornia crassipes stops ships, and villagers on rivers who need the protein from fish to supplement their grain diets cannot reach their fishing grounds. Hydro–electric schemes and irrigation pumps are affected. Bridges are pushed over, and in the Far East the floods cause great islands of the weed to go crashing through the fences placed in rivers for fish culture. Insects which are vectors of human and animal diseases are harboured in the weeds, and the dangers from snakes and crocodiles are increased. Fishing is reduced because there is little light and oxygen under the thick weed mats.
Counter–claim Recently, benefits of water hyacinth have been recognized. It can be put to use in a freshwater environment to remove chemical pollutants from waste water, and can be used for feed, fertilizer or biogas production. In specially built treatment ponds, organically rich waste water is efficiently purified by this plant and the water exits clear, odourless and essentially unpolluted. One hectare of water hyacinth has the potential to extract about 4 tonnes of nitrogen and 1 ton of phosphorous each year. Water hyacinth can also be used as mulch to prevent water losses from evaporation. Other uses, such as manure, compost, and addition to animal feeds have been applied with varying success.
Refs United Nations, Geneva *Water Hyacinth. Proceedings of the International Conference on Water Hyacinth* (1984).
Broader Aquatic weeds (#PD2232).
Aggravated by Deforestation (#PC1366) Forest decline (#PC7896).

♦ PE3816 Non–productive capitalist elites
Nature The creation of a non–productive elite follows from private ownership of the means of production. Therefore most of the nation's wealth accrues to its least productive members, who may render the economy unstable through speculative ventures and investment abroad. As science and technology replace manpower as the means of production, so the social and economic gap widens between the elite and the proletariat who face rising unemployment without redistribution of wealth.
Broader Capitalism (#PC0564).
Narrower Reinforcement of inappropriate development by privileged classes in developing countries (#PF6670).

♦ PE3818 Thirst
Refs Rolls, B J and Rolls, E T *Thirst* (1982).
Broader Starvation (#PB1875).
Aggravated by Measles (#PE1603).

♦ PE3822 Diseases and injuries of bone
Osteoporosis — Porous bones
Nature The five main aspects of diseases and injuries of bone are congenital defects, metabolic disturbances, infections, tumours and fractures. Osteoporotic fracture is caused by little or even no impact cracking a brittle bone. Osteoporosis is a major health issue for women as levels of oestrogen fall after the menopause, the bone becomes porous and fragile, and the body's bone mass diminishes. This causes bones to fracture more easily, with fractures of the hip, wrist and spine being the most common. Osteoporosis can cause pain, loss of activity and restricted movement.
Refs Beck, William A and Avioli, Louis V *Osteoporosis* (1988); Cohn, D V and Martin, T J *Osteoporosis* (1987); Singer, Frederick *Paget's Disease of Bone* (1977).
Broader Human disease and disability (#PB1044)
Diseases of connective tissue (#PD2565).
Narrower Flat foot (#PG5603) Metatarsalgia (#PG5605) Osteochondrosis (#PG6134)
Curvature of spine (#PE4113)
Osteomyelitis, periostitis and other infections involving bones and acquired deformities of bone (#PE8912).
Aggravated by Lead as a pollutant (#PE1161).

♦ PE3827 Disparity between share prices and underlying asset values
Aggravates Capitalist speculation (#PC2194).

♦ PE3828 Administering of medical drugs for non–medical purposes
Denial of the right to medical consent
Nature 'Behaviour modification' techniques involving drugs are increasingly used on groups of people who are not always in a position, or given the opportunity, to give their free and informed consent to such procedures. The three major groups involved are hyperkinetic children, the mentally ill and prison inmates. The techniques may be used mainly or exclusively to control dissidents, including political dissenters.
Incidence Amphetamines and methyl phenidrate have been administered to children diagnosed as having minimal brain dysfunction, a state characterized by hyperactivity, inattentiveness, diminished perception and physical and social awkwardness. It has been argued that such treatment has produced dramatic effects in alleviating these problems and enabling the hyperkinetic child to achieve greater success both academically and socially. Tranquillizers have been used in hospitals for the mentally ill and the aged, and various drugs have been used to modify the behaviour of prison inmates. Such uses, for non–medical purposes, raise questions regarding the human rights of the persons involved.
Broader Denial of right to liberty (#PF0705) Unethical medical practice (#PD5770).

♦ PE3829 Liner shipping cartels
Multimodal transport monopoly
Nature Many problems connected with liner shipping relate equally to multimodal transport. Unchecked spread of containerization can lead to consortia power, reinforcing the cartellist grip of liner conferences on their ocean trades and inland transport infrastructures. However, restatement in the Multimodal Transport Convention of the public rights of states to regulate and control multimodal transport operations at the national level, and to take all other steps in the national, economic and commercial interest, mitigates the potential of the consortia, although these developed mainly as a result of the introduction of container systems, which necessitate very high capital investments in vessels and equipment and therefore will not readily dissolve. International multimodal transport operators (MTOs) of developed countries have come to dominate as containerization has spread. There is an inadequate participation of indigenous MTOs in the trade and insufficient protection of shippers' interests in developing countries. Also, a number of questions as to the liability of MTOs for loss/damage/delay to cargo, and correct documentation remain open. The Convention on International Multimodal Transport Goods, adopted by consensus in 1980, was addressed to these concerns which impact the developing countries. However, these countries have insufficient financial and technical resources to create or improve their own national transport capabilities especially in order to facilitate the implementation of the Multimodal Convention.
Broader Cartels (#PC2512) Monopolies (#PC0521).

♦ PE3830 Blood circulation disorders
Pulmonary oedema — Cerebral oedema
Refs Fernandez, Clara M *Edema Research* (1987); Inaba, Y, et al *Brain Edema* (1986); Luisada, Aldo A *Pulmonary Edema in Man and Animals* (1970); WHO Scientific Group – Geneva 1971 *Inherited Blood Clotting Disorders* (1972).
Broader Lung disorders and diseases (#PD0637)
Diseases and injuries of the brain (#PD0992).
Aggravated by Irritant fumes (#PD3672) Kwashiorkor disease (#PE2282)
Fungicides as pollutants (#PD1612) Rodenticides as pollutants (#PE3677)
Halogen compounds as pollutants (#PE4483).

♦ PE3831 Excessive portrayal of perspectives of industrialized cultures in media
Inadequate coverage of developing country perspectives in the media — Excessive emphasis on developing country weaknesses in the media
Broader Cultural imperialism (#PC3195) Distorted media presentations (#PD6081).
Aggravated by Excessive use of foreign programmes for media (#PE9643)
Excessive portrayal of negative information by the media (#PE1478).

♦ PE3832 Conditional membership in international organizations
Nature Membership in some international organizations or agencies is conditioned upon being a member in good standing of the United Nations. A state banned or expelled from membership in the United Nations generally, for political reasons, is not allowed to benefit from its participation in some international bodies denying thereby development resources to the people of that state.

♦ PE3836 Club foot
Refs Imhauser, G *Club Foot* (1986).
Broader Stigmatized diseases (#PD7279) Foot diseases and disabilities (#PD2647)
Disorders of joints and ligaments (#PD2283)
Human physical genetic abnormalities (#PD1618).

♦ PE3837 Gigantism
Giantism
Refs Wood, Edward J *Giants and Dwarfs* (1976).
Broader Pituitary gland disorders (#PE2286).
Aggravates Discrimination against giants (#PE5578).

♦ PE3847 Disabled elderly persons
Nature In most countries the number of elderly people is increasing, and already in some as many as two thirds of disabled people are also elderly. Most of the conditions which cause their disability (for example, arthritis, strokes, heart disease and deterioration in hearing and vision) are not common among younger disabled people and may require different forms of prevention, treatment, rehabilitation and support services.
Broader Human disability (#PC0699).

♦ PE3851 Sexual masochism
Nature Some people with this disorder are bothered by their fantasies while others act on the masochistic sexual urges by themselves (through binding themselves or self–mutilation) or with a partner (restraint, whipping, cutting, infibulation, humiliation).
Broader Masochism (#PF3264) Sexual deviation (#PD2198).
Related Sexual sadism (#PE6748).

♦ PE3852 Coxsackie virus
Refs Bendinelli, M and Friedman, H *Coxsackie Viruses* (1988).
Broader Virus diseases (#PD0594) Viral diseases in animals (#PD2730).
Aggravated by Cockroaches as pests (#PE1633).

♦ PE3853 Foetal alcohol syndrome
Refs Abel, Ernest L *Fetal Alcohol Syndrome and Fetal Alcohol Effects* (1984); Abel, Ernest L *Fetal Alcohol Syndrome* (1981).
Broader Children of alcoholics (#PD4218) Human physical genetic abnormalities (#PD1618)
Alcoholism amongst indigenous peoples (#PE7242).
Aggravated by Drunkenness (#PE8311) Alcohol abuse (#PD0153)
Substance abuse during pregnancy and breast–feeding (#PE3858).

♦ PE3858 Substance abuse during pregnancy and breast–feeding
Refs Canadian Institute of Child Health *Effects of Alcohol, Tobacco and Caffeine on the Fetus* (1979); Do It Now Foundation *Drugs, Alcohol and Pregnancy* (1988); Gartner, Leslie P *Alcohol and Pregnancy* (1984); Pinkert, Theodore *Current Research on the Consequences of Maternal Drug Abuse* (1985).
Related Smoking during pregnancy and breast–feeding (#PE5026).
Aggravates Foetal alcohol syndrome (#PE3853).

♦ PE3860 Bronchial asthma
Nature Bronchial asthma, named after the Greek word for breathlessness and panting, is an inflammatory disease of the lungs. The wheezing, coughing and gasping to get air through the narrowed airways is the result of inflammation an damage which makes the airways hyper–responsive. Cells in the airways are triggered to release chemical factors that ultimately cause the airways to become inflamed and infiltrated with special types of white blood cells. In chronic severe asthma, the epithelial lining of the airways is progressively destroyed, leaving them hyper–responsive and susceptible to severe spasm at the slightest provocation. This may be from an allergen, such as house dust or animal fur, or simply through exercise, breathing cold air or the night–time occurrence of broncho–constriction.
Broader Asthma (#PD2408).
Aggravated by Stress in human beings (#PC1648).

♦ PE3861 Gastrointestinal diseases
Gastroenteritic disturbances — Gastro–intestinal infections — Bleeding from the gastro–intestinal tract — Dysfunction of the gastrointestinal tract
Refs Bennett, John; Cuschieri, A and Hennessy, T P J *Reflux Oesophagitis* (1989); Coffman,

Derek A, et al *Gastrointestinal Disorders* (1987); Dykes, P W and Keighley, M R *Gastrointestinal Haemorrhage* (1981); Gryboski, Joyce *Gastrointestinal Problems in the Infant*; Levin, Bernard (Ed) *Gastrointestinal Cancer* (1987); Sodeman, W A (Ed) *Acute Gastrointestinal Problems* (1986); Whitehead, William E and Schuster, Marvin M *Psychophysiological Gastrointestinal Disorders* (1985).
Broader Gastric disorders (#PE1599) Intestinal diseases (#PD9045).
Narrower Gastro-intestinal disorders due to torture (#PE6724).
Aggravates Nutritional anaemia (#PD0321).
Aggravated by Rickets (#PG2295) Stress in human beings (#PC1648)
Recreational contact with sewage (#PE6685)
Malnutrition among indigenous peoples (#PC3319).

♦ PE3865 Premature drying of plants
Nature Extreme dryness can be caused by a disruption of the water balance, so that the evaporation exceeds the absorption from the soil, or by overheating. In trees the leaves wither while retaining their green colour, in cereals the grains are undersized and wrinkled.

♦ PE3866 Dyslexia
Alexia
Nature Dyslexia is a learning disorder arising from abnormality in the part of the brain that organizes written information. Dyslexics have difficulties in reading and writing. The disability may include words and not letters, figures and not letters, and so on.
Refs Colheart, M; Marshall, J C and Patterson, K (Eds) *Deep Dyslexia* (1987); Evans, Martha M (Ed) *Dyslexia* (1982); Gilroy, E and Miles, T R *Dyslexia at College* (1986); Malatesha R N and Whitaker, Harry A (Eds) *Dyslexia* (1984); Powills, Leo I *Dyslexia* (1987); Young, Peter and Tyre, Colin *Dyslexia or Illiteracy?* (1983).
Broader Mental illness (#PC0300) Learning disorders (#PD3865)
Reading disabilities (#PD1950).
Related Language disorders (#PE3886).
Aggravated by Limited individual attention span (#PF2384).

♦ PE3868 Hazards to navigation
Shipping hazards
Broader Obstacles for international ocean shipping (#PD5885).
Narrower Icebergs (#PE1289) Mined international waterways (#PG2262)
Unidentified submarine objects (#PE0712) Wrecks and derelicts as hazards (#PE5340).
Aggravates Marine accidents (#PD8982).
Aggravated by Fog (#PE1655) Deterioration in atmospheric visibility (#PE2593).

♦ PE3869 Inadequate participation in the control of joint venture
Nature Many developing countries have come to learn through experience that the degree of participation in equity by itself tells little about how the effective control of the enterprise is shared by the partners. Inadequate participation may be attributable to the minority control of the company, to the passivity of the local shareholders, to the reservation of the right to appoint a smaller number of directors than would be justified by the share in the equity. It may also happen that the foreign interest deliberately takes advantage of the lack of experience of the local partner and introduces into the basic agreement legal devices which limit the decision-making power of the latter. Another common practise is that of drawing a distinction between the board of directors and management the foreign partner having effective control over the day-to-day management. In the sophisticated industries effective control and management are likely to pass into the hands of foreign partner if the local directors and executives do not possess the necessary technical knowledge, which is rare in most developing countries.
Aggravates Domination of developing countries by transnational corporations (#PE0163)
Control of national economic sectors by transnational enterprises (#PE0042).

♦ PE3870 Aircraft pilot fatigue
Nature Airline pilots are frequently falling asleep in the cockpits in flight, particularly on long ocean crossings. Some pilots are spending as much as 10 days away from home, often making flights of 10 hours or more. Their biological clocks are disrupted, making it difficult for them to get proper sleep during layovers. New airplanes are extending flight time because of their longer ranges. Other equipment have made it possible to have two rather than three member crews. The smaller crews are more liable to fall asleep. Computerized navigation equipment and autopilots have lightened pilots workloads, giving them little to do when cruising except monitor instruments. As a result fatigued pilots are more vulnerable to falling asleep. While there has been no evidence that fatigued crews have been responsible for airline crashes there is a great deal of concern.
Refs International Civil Aviation Organization *Flight Crew Fatique and Flight Time Limitations* (1984).
Broader Increasing job monotony (#PD2656) Aircraft environmental hazards (#PD8328)
Human fatigue during control of complex equipment (#PE5572).
Related Exhausting commuter travel (#PE1538).
Aggravates Human errors and miscalculations (#PF3702).

♦ PE3872 High severance pay for top managers
Nature Managing directors and other top managers are often guaranteed from one half to one full year's salary at the conclusion of their contracts. This could be more than $1 million in addition to salary after only 2 or 3 years work. There is little evidence that a single manager adds this level of profitability to any company.
Claim High severance pay for managers is just a way for fellow directors to pave the way for their own high severance pay. It buys the loyalty for a year of the manager while they are at another company thus assuring secrets from one company don't go to another along with the executive.
Counter-claim High severance pay of a departing manager reassures managers staying with the company that they too will be treated well when they leave or retire.
Broader Non-comprehensive wage scales (#PD1133).
Related High minimum wages (#PD5674)
Excessive salaries of international civil servants (#PE6388).

♦ PE3874 Obliteration of footpaths by development
Disappearance of walking trails
Incidence Since the 1930's, urban development has obliterated 30 percent of America's footpaths. Some 200,000 miles of trails led through national forests in 1940: only half as many remain.
Broader Destruction inherent in development (#PF4829)
Environmental degradation in industrialized countries (#PD7835).
Aggravates Inaccessibility of countryside to city dwellers (#PF6140).
Reduces Over-use of designated wilderness areas (#PD7585).

♦ PE3878 Horse flies
Deer flies
Refs Chvála, Milan, et al *Horse Flies of Europe (Diptera, Tabanidae)* (1972).
Broader Flies as insect pests (#PE2254) Insect vectors of animal diseases (#PD2748).

♦ PE3880 Destruction of crop pollinating species
Destruction of crop pollinating insects
Aggravates Declining productivity of agricultural land (#PD7480).
Aggravated by Pesticide intoxication (#PE2349).

♦ PE3881 Thunderstorms
Nature Thunderstorms begin as little fair-weather clouds that usually form and disappear without producing rain or electrical effects. When the air is thermally unstable, a few of these clouds undergo a rapid change: they increase in height and rain and electric fields appear.
Broader Meteorological disaster (#PD4065).
Related Lightning (#PD1292).
Aggravates Tornadoes (#PD1739) Electrical storms (#PJ4133)
Wind damage to structures (#PE1334).

♦ PE3882 Violation of the rights of male homosexuals
Discrimination against male homosexuals — Denial of the rights of male homosexuals
Broader Discrimination (#PA0833) Discrimination against homosexuals (#PE1903).
Aggravates Male homosexuality (#PF1390).

♦ PE3883 Environmental hazards of solar radiation
Sunburn — Sunstroke — Overexposure to the sun
Incidence As an indication, the market for protective suntan creams in the UK in 1986 was £37 million.
Broader Radioactive contamination (#PC0229).
Aggravates Stroke (#PE1684) Heat disorders (#PE2398)
Malignant neoplasm of skin (#PE5016).
Reduced by Absence of direct sunlight (#PG1918).

♦ PE3886 Language disorders
Aphasia
Nature Aphasia is a general term for all disturbances of language due to brain lesions but not a result of faulty innervation of the speech muscles, involvement of the organs of articulation, or general mental or intellectual deficiency. Language, here, refers to the expression or communication of thought by word, writing, gesture, to the reception and interpretation of such acts when carried out by others, and to the retention, recall and visualization of the symbols involved.
Refs Benson, D Frank *Aphasia, Alexia and Agraphia* (1979); Cantwell, Dennis and Baker, Lorian *Developmental Speech and Language Disorders* (1987); Code, Chris (Ed) *Characteristics of Aphasia* (1989); Luria, A R *Traumatic Aphasia* (1970); Peterson, H and Marquardt, J *Appraisal and Diagnosis of Speech and Language Disorders* (1981); Sarno, Martha Taylor and Höök, Olle (Eds) *Aphasia* (1980); Tibbits, Donald F *Language Disorders in Adolescents* (1982).
Broader Speech disorders (#PE2265) Learning disorders (#PD3865)
Diseases and injuries of the brain (#PD0992)
Symptoms referable to nervous system (#PE9468).
Narrower Agnosia (#PE5150) Motor aphasia (#PE0345) Sensory aphasia (#PE4234).
Related Dyslexia (#PE3866).

♦ PE3889 Crimes related to immigration, naturalization and passports
Broader Statutory crime (#PC0277).
Narrower Fraudulent acquisition or use of passports (#PE4496).
Related Illegal immigration (#PD1928) Illegal movement across frontiers (#PC2367)
Abusive traffic in immigrant workers (#PD2722)
Commercialization of nationality acquisition (#PF0699).

♦ PE3895 Non-epidemic typhus
Broader Typhoid fever (#PD1753) Arthropod-borne diseases (#PE7796).
Narrower Endemic typhus (#PG5389) Mite-borne typhus (#PG3896).

♦ PE3896 Contaminated animal feed
Broader Environmental hazards from animal feedstuffs, excluding unmilled cereals (#PE1331).
Aggravates Contaminated pastureland (#PJ5737).

♦ PE3897 Tick-borne diseases
Broader Insect vectors of animal diseases (#PD2748).
Aggravates Lyme disease (#PG6489) Sweating sickness (#PG9662)
Lyme disease in dogs (#PG6391).
Aggravated by Ticks as pests (#PE1766).

♦ PE3898 Acute endocarditis
Refs Finch, R G, et al (Eds) *Infective Endocarditis* (1988); Finch, R G, et al *Infective Endocarditis* (1988).
Broader Heart diseases (#PD0448).

♦ PE3899 Denial to food animals of the right to freedom from suffering
Nature Animals being reared for food are fed, housed, transported and slaughtered without consideration of their right to feeding, housing, transportation and slaughter befitting their importance to humankind.
Broader Animal suffering (#PD8812)
Human consumption of animal products (#PD7699)
Denial to animals of legal protection of their rights (#PE8643).
Narrower Inhumane killing of animals (#PE0358).
Related Animal stress in factory farming (#PD2760)
Animal malformation in factory farming (#PD2761)
Denial to animals of the right to life (#PF8243)
Denial to animals of the right to dignity (#PE9573)
Undesirable effects of animal feed additives (#PD1714)
Denial to animals of the right to a natural death (#PE8339)
Unsanitary and inhumane urban food animal conditions (#PE0395)
Denial to working animals of limitation of working hours (#PE6427)
Inferior meat quality from intensive animal farming units (#PE2770)
Insanitary penning conditions as factor in animal diseases (#PE2764)
Denial to animals of the right to freedom from mass killing (#PE9650)
Denial to working animals of restorative nourishment and rest (#PE4793)
Denial to experimental animals of the right to freedom from suffering (#PE8024)
Denial to animals of the right to the attention, care and protection of humankind (#PF5121)
Denial to animals of the right to conditions of life and liberty proper to their species (#PE6270).
Aggravated by Unethical food practices (#PD1045)
Excessive consumption of animal flesh (#PD4518).

♦ PE3901 Avoidant personality disorder
Nature People with this order often don't have any close friends and aren't willing to get involved with others unless given an uncritical acceptance. Timidity causes them to avoid social or occupational activities that involve interpersonal contact or deviation from the normal routine. As a result of failing to develop social relations they experience depression and anger at them-

PE3901

selves.
Broader Personality disorders (#PD9219).

♦ **PE3903 Low bank interest rates**
Nature When interest rates paid by banks to savers is low, investors will withdrawn their savings and may spend it rather than reinvesting it. This causes a strain on the economy.
Broader Inadequate economic incentives (#PF8554).
Aggravates Inadequate savings (#PC0327)   Insufficient capital investment (#PF2852).

♦ **PE3909 Educating people to lie**
Teaching children to lie — Subverting the minds of the young
Claim In totalitarian countries, a principal lesson for personal survival is that for those who are fortunate enough to know the truth it is usually better to deny such knowledge or claim the opposite.
Broader Inappropriate education (#PD8529).
Aggravates Lying (#PB7600).
Aggravated by Dogmatism (#PF6988)   Propaganda (#PF1878)
Compulsory indoctrination (#PD3097).

♦ **PE3910 Denial right to union activity for rural workers**
Broader Denial of right to union activity for special groups (#PE1355).

♦ **PE3911 Sandy soils**
Refs FAO *Sandy Soils* (1979).
Broader Soil infertility (#PD0077).

♦ **PE3912 Obstruction of recruiting or induction into armed forces**
Nature To prevent or attempt to prevent another person from being recruited or inducted into the armed forces during war time is a crime.
Broader Crimes related to military service obligations (#PD5941).
Related Draft evasion (#PD0356).

♦ **PE3913 Non-payment of compensation to victims of crime**
Delay in payment of compensation to victims of crime
Incidence In the UK claims for compensation take on average 18 months to process. This can be extended for a further year if the victim disputes the award and calls for a full hearing.
Broader Uncompensated damages (#PD7179).
Aggravates Non-payment of compensation to victims of motor accidents (#PE8824).

♦ **PE3914 Destruction of the countryside**
Broader Landscape disfigurement (#PC2122)   Natural environment degradation (#PB5250).
Related Deforestation of mountainous regions (#PD6282)
Destruction of hedges and hedgerow trees (#PD1642)
Disruption of ecosystems in marginal agricultural lands (#PD6960).
Aggravates Degradation of semi-natural and natural habitats of flora and fauna (#PC3152).
Aggravated by Inappropriate modernization of agriculture (#PF4799)
Over-subsidized agriculture in industrialized countries (#PD9802).

♦ **PE3915 Torture by exposure to weather**
Nature Exposure to various weather conditions is used as a form of torture. Some victims have their arms and legs bound and are left exposed to the hot sun for prolonged periods or forced to stand in the rain for days. Some are left out in the cold at night.
Incidence Torture by exposure to weather has been reported in the following countries: **Af** Mali, Morocco, Zaire. **Am** Colombia, Uruguay. **As** Israel.
Broader Physical torture (#PD8734).

♦ **PE3916 Complex trials**
Trials exceeding competence of juries — Lengthy trials
Claim Jurors have many difficulties if a crime has been committed, for example, in the sophisticated world of high finance or international trading that is unfamiliar to the jury. Especially fraud cases are complex and usually last for weeks if not months and they exceed the limits of comprehension and memory of jurors. The result may be conviction of an innocent or acquittal of a guilty person.
Counter-claim The jury is well equipped to decide the dishonesty or honesty of the defendant. The responsibility to make the issues clear to the jury is for the prosecution, defence lawyers and the judge.
Aggravates Unjust trials (#PD4827)   Miscarriage of justice (#PF8479).

♦ **PE3919 Blights**
Nature Blight is characterized by sudden and conspicuous disease or injury resulting in withering, cessation of growth and death of parts. It does not include rotting nor disturbances in vascular system and the dead areas are not definitely delimited.
Broader Oomycetes (#PE5465)   Ascomycetes (#PE4586)   Zygomycetes (#PE0614)
Deuteromycetes (#PE4346)   Basidiomycetes (#PE0364).
Narrower Late blight of potato (#PG3346)   Botrytis diseases in plants (#PG1668)
Fusarium diseases in plants (#PG5461)   Alternaria diseases in plants (#PJ1685).

♦ **PE3924 Insomnia**
Nature The people suffering from insomnia disorders complain of difficulty in initiating or maintaining sleep or of not feeling rested after sleeping adequate amount of time.
Broader Sleep disorders (#PE2197).

♦ **PE3926 Inconsiderate choice of payment times by creditors**
Excessive and unnecessary burden of interest repayments on developing countries
Broader Self interested manipulation of timing (#PF5529).
Related Profit-oriented interest payments (#PD2552).
Aggravates Debt slavery (#PD3301).
Deterioration in external financial position of developing countries (#PE9567).

♦ **PE3927 Torture through waiting**
Nature Prisoner are tortured by simply left to wait for long periods of time, often with no human contact. This creates a sense of isolation, foreboding and humiliation.
Broader Torture by deprivation (#PD3763).

♦ **PE3928 Delays in delivery of goods and services**
Risks due to delays in delivery of goods and services
Nature Delays in the delivery of goods or services can result in the loss of a perishable shipment, failure meet contract deadlines and increased expenses for labour and delivery guarantees.
Broader Risk (#PF7580)   Delay (#PA1999).
Narrower Delays in delivery of goods (#PF8268)
Delay in delivery of requested services (#PE8157).
Related Communication delays (#PF4453).

♦ **PE3932 Sexual dysfunction of torture victims**
Nature Victims of torture have a high rate of sexual dysfunction. Victims have reduced desire or requirement to engage in sexual activity, decreased ability to obtain and maintain a penile erection during sex or delayed or absent intervaginal ejaculation.
Incidence In a recent study of Greek men who had been tortured 29 percent experienced some form of sexual dysfunction.
Broader Human sexual disorders (#PD8016)   Neurological effects of torture (#PD4755).
Related Sexual unfulfilment of the disabled (#PE5197).

♦ **PE3933 Symptoms referable to cardiovascular system**
Symptoms referable to lymphatic system
Broader Ill defined health conditions (#PC9067)   Diseases of the circulation system (#PC8482).
Narrower Dropsy (#PG3235)   Cyanosis (#PG2361)   Fainting (#PG7452)
Palpitation (#PG9738).

♦ **PE3934 Water pollution from animal production**
Agricultural effluents from animal husbandry — Contaminated farm slurry — Pollution of water by fish farms
Nature The fish farms pollute the water with nutrients, methane and hydrogen sulphide which threaten both farmed fish and other marine life. Dangerous pesticides have been used to treat infestations of sea lice.
Broader Agricultural effluent (#PE8504)   Agricultural pollution (#PD0563).
Related Water pollution by fertilizers (#PE8729).

♦ **PE3936 Enteric diseases in swine**
Broader Diseases of the digestive system in animals (#PD3978).
Narrower Swine dysentery (#PG7824)   Parasitosis in swine (#PG6447)
Oedema disease in swine (#PG9698)   Rotaviral enteritis in swine (#PG1343)
Enteric salmonellosis in swine (#PG9524)   Enteric colibacillosis in swine (#PG9445)
Porcine proliferative enteritis (#PG7499)   Transmissible gastroenteritis in swine (#PG4138)
Clostridium perfringens type C enteritis in swine (#PG1168)
Mesenteric torsion of the small intestine in swine (#PG0675).

♦ **PE3937 Lava flows**
Nature Lava flows are known to destroy life and property. When crop lands are covered hundreds of years are required before the land is usable.
Broader Volcanic eruptions (#PD3568).

♦ **PE3938 Exploitative use of consultants**
Nature Consultants may be requested to submit proposals (with compensation), in competition with other consultants, in order that the requesting agency can profit from their thinking, even though allocation of the project had already been decided.

♦ **PE3939 Industrial processes geared to reduced social needs**
Nature Industrial processes, the extraction of resources, production of goods and services and their distribution are narrowly focused on the economic processes that they fail to recognize responsibility for other global needs. These processes have a large impact on crime, the environment, social roles, education, law, politics, communications, transportation and the quality of life and yet most claim no responsibility or at best reduce their responsibility to paying taxes and making donations. Liaison between economic and political structures is ineffective at the global level and largely adversarial at every other level. The consumer and national governments are largely impotent in demanding accountability of industry. It is left to the dangers of short term planning and an excess profit-oriented use of resources.
Broader Misuse of resources (#PB5151)   Production of non-essentials (#PC3651).
Narrower Self-interested industrial vision (#PF3679)
Imbalance in distribution of industrial processes (#PE7479)
Inadequate systems for monitoring industrial growth (#PF2905)
Planning disrelationship between political and industrial sectors (#PE1980).
Related Inflexible management patterns (#PF3091)
Foreign control of natural resources (#PD3109)
Over-specialized supervisory personnel (#PF3588).

♦ **PE3940 Itch**
Incidence Itching is of considerable concern to both humans and animals in environments infested with fleas, lice or other skin pests, whether or not they bite. Itching can also be a major preoccupation during the healing process when recovering from a wound. Itching is one of the symptoms of certain skin diseases. Genital itch, for example, is one problem associated with sexually-transmitted diseases.
Broader Diseases of the skin and subcutaneous tissue (#PC8534).
Aggravated by Ringworm (#PD2545)   Itchgrass (#PE6668)
Insect pests (#PC1630)   Insect bites and stings (#PE3636)
Sexually transmitted diseases (#PD0061).

♦ **PE3946 Ignorance of traditional herbal remedies**
Ignorance of traditional medical practices — Ignorance of traditional plant remedies
Nature Between 25,000 and 75,000 flowering plant species have been used in traditional medicine throughout the world. Only about one percent of these have been acknowledged through scientific studies to have real therapeutic value. Today, however, when buying a medicinal preparation, there is a one chance in four that the product owes its origin to one of the plants from this small, already explored group.
Incidence Plants have yielded strong analgesics, cardiovascular drugs, powerful antibiotics, anti-parasitic and anti-cancer compounds, laxatives and diuretics. Plants have also provided precursors for the production of oral contraceptives, and anti-inflammatory drugs. Products derived from plants are also used as suspending agents in the pharmaceutical industry, in the production of spermicidal jellies, toothpaste and skin lotions. Traditional medicine is still used by 70 percent of the African population.
Broader Lost knowledge (#PF5420)   Limited medical knowledge (#PD9160)
Ignorance of health and hygiene (#PD8023).
Related Ignorance (#PA5568)   Poisonous plants (#PD2291)
Neglected food resources (#PF7808).
Aggravates Lack of integration of traditional and Western medicine (#PF4871).

♦ **PE3948 Failure to understand the necessity of creative establishment and disestablishment tensions**
Nature Society fails to understand the need for the pro-establishment, those forces in society which tend to maintain the status quo, to be in tension with the dis-establishment, those forces which tend to force change on society. If the pro-establishment gains over-all control society becomes rigid and crystallized and a stifling social conformity results. If the dis-establishment becoming the dominate force in society a social anarchy in which the strongest and most ruthless dominate and the weak are enslaved or die would result. When tension is maintained so that neither dominates society can grow and stability is present.
Broader Irresponsible expression of emotions equated with free speech (#PF7798)

♦ **PE3951 Denial of right to maternity leave**
Broader Inadequate maternity protection in employment (#PD6733).

♦ **PE3957 Occupational hazards in commerce and offices**
Refs ILO *Occupational Hazards and Diseases in Commerce and Offices* (1985).
Broader Occupational hazards (#PC6716) Occupational diseases (#PD0215).

♦ **PE3958 Heart failure**
Congestive heart failure
Refs Bussman, W D and Beisel, A *Acute and Chronic Heart Failure* (1986); Eisenberg, Mickey, et al *Sudden Cardiac Death in the Community* (1984).
Aggravated by Typhoid fever (#PD1753) Bright's disease (#PE2272).

♦ **PE3959 Proliferation of computers**
Overdependence on computers as universal panacea — Overutilization of personal computers
Nature For developing countries the computer makes it possible to leapfrog certain stages of development by slashing paper bureaucracy, streamlining production and improving health care. But the computer explosion in developing countries are in many cases not being effectively used. Computers don't clothe, cure, or feed the poor. Their power begins and ends with information.
Incidence Many countries are rushing into computer without examining their needs. China, in 1984, imported $300 million worth of computer components to make 120,000 computers. By 1985 at least half of them were unused because of shortages of skilled users and software programmes. In some African francophone countries it is not unusual for government data-processing offices to close down due to a lack of paper.
Claim The proliferation of computers reduces productivity. They enable people to create, use and store considerably more information. However much of this is unnecessary at best and counter-productive at worst. Much time is wasted creating it and then reacting to it once created. In the USA, for example, most computers are used in the service sector, where productivity has grown neglibly, not in the manufacturing sector where productivity has grown dramatically.
Broader Uncoordinated use of computers and automation (#PE8721).
Narrower Socially disruptive effects of video games (#PF6345).
Related Excessive consumption of goods and services (#PC2518).
Aggravates Computer obsession (#PF1288) Declining economic productivity (#PC8908)
Social isolation of computer users (#PE1277)
Dependence on sophisticated technology for development (#PD6571).
Aggravated by Proliferation of information (#PC1298)
Dependence of developing countries on imported technology (#PF1489).

♦ **PE3963 Denial of right to free choice of work**
Broader Denial of right to work (#PC5281)
Denial to people of control over their own lives (#PC2381).
Narrower Discrimination against part-time work (#PE6241).
Related Enforced collectivization of agriculture (#PD7443)
International imbalance in the quality of life (#PB4993)
Denial of the right to work in capitalist systems (#PC3119)
Unequal employment opportunities for disabled persons (#PE0783)
Discrimination against working women in socialist countries (#PD2872).

♦ **PE3964 Denial of right to fair and public trial**
Broader Violation of civil rights (#PC5285)
Denial of human rights in the administration of justice (#PD6927).
Narrower Injustice of special courts (#PE0088) Denial of right to a legal defence (#PE4628)
Extrajudicial courts and tribunals (#PE7664) Denial of right to trial by a court (#PE4737)
Excessive length of pre-trial internment (#PE4887)
Denial of right to be informed of criminal charges (#PE7336)
Denial of right to presumed innocent until proven guilty (#PE7393)
Denial of right to freedom from testifying against oneself (#PE6633).
Aggravates Unjust trials (#PD4827).
Aggravated by Bias in jury trials in small jurisdictions (#PE4733).

♦ **PE3968 Illegal exports of nuclear materials**
Covert trade in nuclear bomb-making materials — Clandestine trade in nuclear weapons — Nuclear smuggling
Nature In contravention of the Non-proliferation Treaty weapons grade nuclear materials have been sold to countries which are secretly developing nuclear weapons.
Background In 1942, some three years before the USA had tested its first atomic bomb, Manhattan Project officials were negotiating covert purchase agreements with the Congo, India, Brazil, and other nations in a surreptitious attempt to obtain a post-war US monopoly over the world's known uranium supplies. In 1988, West German and Belgian officials are thought to accept bribes to conceal and falsify documentation of shipments of radioactive material. The British government admitted that shipments of civilian plutonium to the US for military purposes. India is thought to receive heavy water from China for use in building a nuclear arsenal.
Incidence Several countries are known to be circumventing supplier-country restrictions on importation of nuclear weapons material manufacturing facilities by means of purchasing such components piece by piece through a series of dummy trading companies, anonymous purchasing agents, and falsified intent of means statements. Clandestine trade has played a central role in advancing the nuclear weapons capability of Argentina, Brazil, India, Iraq and Pakistan.
Broader Illegal exports (#PD4116).
Aggravates Terrorists armed with nuclear weapons (#PE3769)
Proliferation of nuclear weapons and technology (#PD0837).
Aggravated by Theft of nuclear materials (#PD3495)
Secret international agreements (#PF0419).

♦ **PE3969 Inadequate working conditions in construction industry**
Refs ILO *The Improvement of Working Conditions and of the Working Environment in the Construction Industry* (1983).

♦ **PE3970 Denial of right to collective bargaining**
Violation of the right of workers organizations to bargain collectively — Restrictions on collective bargaining by workers organizations — Violation of the right of trade unions to negotiate freely with employers
Nature The right to bargain collectively can be restricted by an obligation to obtain a negotiation licence or by exclusion for certain categories of workers or by the state's interference in free collective bargaining during the time of serious economic and social problems.
Refs ILO *Collective Bargaining Problems and Practices on Plantations and the Exercise of Trade Union Rights*; ILO *Collective Bargaining* (1984); United Nations *Collective Bargaining in Industrialized Countries* (1978).
Broader Violation of trade union rights (#PD4695).
Narrower Denial of right to grievance procedures (#PE2832).
Related Denial of the right to picket (#PE8712) Violation of the right to strike (#PE5070)
Sanctions against trade union workers (#PD0610)
Denial of right to organize trade unions (#PE5398).
Aggravated by Unethical practices of employers (#PD2879).

♦ **PE3972 Invasion of airspace by foreign aircraft**
Aerial surveillance by foreign powers — Over-flying by foreign aircraft
Related Unidentified submarine objects (#PE0712)
Misuse of satellite surveillance by governments (#PF3701).

♦ **PE3973 Surreptitious entry during surveillance**
Refs French, S *The Big Brother Game* (1986).
Related Burglary (#PD2561).
Aggravated by Political surveillance (#PD8871).

♦ **PE3974 Psychogenic physical disorders**
Broader Mental illness (#PC0300).
Narrower Vaginismus (#PG1752) Dyspareunia (#PG9647)
Hyperventilation (#PJ4225) Sexual impotence of men (#PF6415).

♦ **PE3977 Fragmented social structures for environmental protection**
Nature The environment is endangered by the very structures created for its protection. The structures which are based on geography are exclusive and uncoordinated, for example the agencies dealing with north sea pollution number in the dozens but none in fact deal with the whole geographic area effecting it. Structures which are formed along functional lines, like oil pollution or endangered species are based on obsolete categories and are often to closely related to those causing the damage. In both cases these agencies do not have the power to effect change. The illusion that something is being done about the environment creates and even greater danger.
Broader Social fragmentation (#PF1324) Institutional fragmentation (#PC3915)
Uncontrolled application of technology (#PC0418).

♦ **PE3980 Excessive portrayal of substance abuse in the media**
Glorification of substance abuse in the media — Excessive portrayal of drugs, alcohol and cigarettes in the media — Irresponsible advertising for alcoholic beverages — Irresponsible liquor advertising
Nature Liquor advertising seeks to promote a wider acceptance and use of alcoholic beverages. Even though the consumption of such beverages leads to alcoholism, motor vehicle accidents, a physiological and/or psychological dependence and the loosening of sex and other inhibitions, liquor advertising does not mention these hazards.
Liquor advertising occurs in all forms of mass media: printed in magazines, newspapers, and journals; heard live on the radio; viewed on television; it may be covertly slipped in as a name brand mentioned in an article or on a talk show.
Claim Liquor advertising, in the light of good advertising practice, is untruthful in that it leads the public to believe that alcohol consumption is beneficial; none of the dangers are presented; liquor advertising is indiscriminate in that, while minors may be forbidden by law to purchase alcohol, that are constantly inundated by advertising promoting its usage, and minors are more susceptible to propaganda than are adults; liquor advertising also filters into homes located in 'dry' areas, thus attempting to break down the expressed will of communities which are against such consumption.
Refs Chapman, Simon *Pushing Smoke* (1988).
Broader Misuse of advertising (#PE4225) Distorted media presentations (#PD6081).
Aggravates Youth violence (#PF7498) Substance abuse (#PC5536)
Deterioration of media standards (#PD5377) Consumption of alcoholic beverages (#PD8286)
Draining of resources due to alcohol (#PE8865).

♦ **PE3981 Endangered species of marsupial moles**
Incidence 1 species; not known to be endangered.
Broader Endangered species of marsupials (#PD1762).

♦ **PE3986 Arachnida pests**
Broader Pests (#PC0728).
Narrower Ticks as pests (#PE1766) Mites as pests (#PE3639).

♦ **PE3987 Destruction of weeds**
Nature Gardeners and farmers indiscriminately destroy weeds in spite of the present urgent and worldwide need for conservation and diversification rather than elimination and extinction of plant and animal life. Weeds are a part of the complex web of plant life, and clearly should be respected as such, even if they are unwanted guest in a garden plot or farmer's field. It is estimated that over a thousand different kinds of plants are lost each year from our gardens, either because they go out of fashion, are destroyed by disease or are dropped by the nursery trade because of propagation problems. At the same time it is estimated that fifty acres of tropical rain forest are lost every minute. Herbalists and homeopaths find value in weeds. The Chinese use three and a half thousand plant species medicinally and are always looking for new ones. What are called weeds today are in part what fed the human species from its beginning through the middle ages.
Broader Endangered species of plants (#PC0238).
Aggravates Inadequate plant genetic resources conservation (#PF3581).
Aggravated by Plant diseases (#PC0555) Pests of plants (#PC1627)
Cruelty to plants (#PD4148) Vulnerability of plants and crops (#PD5730).
Reduces Weeds (#PD1574).
Reduced by Air spora (#PE4732) Inadequate weed control (#PG9029)
Inadequate weed control in developing countries (#PD3598).

♦ **PE3988 Low-intensity conflict**
Small war — Low-level conflict — Police operation — LIC — Limited war
Nature The military doctrine of low-intensity conflict draws on counterinsurgency strategy of the early Vietnam War years, lessons from experiences in Lebanon, Central America, the Falklands war, and Afghanistan. In addition to counterinsurgency activity "pro-insurgency" (support of anti-government insurgents), counterterrorism, narcotics interdiction and police operations of the type used by the United States in Grenada. Specially trained units are developed and prepared for rapid deployment.
Claim If engaging in a limited war results in a series of battle field defeats strong pressure would come to bear on the government to escalate the level of conflict. This could not only lead to a major long term commitment to a war like the U.S. in Vietnam or the U.S.S.R in Afghanistan. In pro-insurgency campaigns may provoke major attacks on countries providing sanctuary for insurgents also escalating the war.
Refs Charters, D and Tugwell, M *Armies in Low Intensity Conflict* (1988); O'Brien, William V *The Conduct of a Just and Limited War* (1981).
Broader Conventional warfare (#PC4311).
Related Pervasive fear of nuclear war (#PC3541)
Risk of unintentional war generated by the arms race (#PF4152).

PE3988

Aggravates Risk of unintentional nuclear war due to international crises (#PF4302)
Risk of unintentional nuclear war generated by developments of strategic doctrine (#PF4156).
Aggravated by Civil war (#PC1869) Guerrilla warfare (#PC1738).

♦ **PE3992 Inadequate guardianship for mentally retarded adults**
Nature In addition to physical care, a mentally retarded adult may need to be directed where and how he or she is to live, how to avoid being exploited socially, and may need assistance with social and sexual relationships.
Broader Human disease and disability (#PB1044).
Aggravates Mental deficiency (#PC1587) Mental deficiency in children (#PD0914).

♦ **PE3993 Ghost employees**
Employment fraud
Nature Individuals are hired, placed on the payroll but they never actually work. This is most common in public sector companies and government offices.
Broader Fraud (#PD0486).
Related Unemployment (#PB0750).

♦ **PE3995 Disparity in facilities for military mobilization and reinforcement**
Nature Many military establishments, including NATO and Warsaw Pact forces, rely heavily on the mobilization of reservists to bring active duty up to strength and to man mobilized formations. Countries with highly controlled social structures and long and intensive training periods of military conscripts permit them to maintain a more significant pool of trained reserve manpower than is maintained by other countries. In cases where long distances or sea crossings have to be travelled make mobilization difficult. The rapid reinforcement of land force is a very complex operation that demands the timely availability of numerous resources, particularly transport aircraft and shipping as well as reception and prepositioned equipment storage facilities. Reinforcement of air forces involves infrastructure and logistic problems of a different but also complex nature, particularly in the areas of survivability and combat support. While a number of reinforcement air squadrons may be available to deploy to a combat zone, they would have to wait for the arrival of their ground crew and support equipment before they can become operational.
Broader Military insecurity and vulnerability (#PC0541).
Aggravates Failure of disarmament and arms control efforts (#PF0013).

♦ **PE3996 Theileriasis**
Theileriases
Broader Parasites of the cardiovascular system in animals (#PE9427).
Narrower East Coast fever (#PE7946).
Aggravated by Ticks as pests (#PE1766) Protozoa as pests (#PE6741).

♦ **PE3997 Crimes related to national security information**
Broader Crimes against national security (#PC0554).
Narrower War-time communications with the enemy (#PE6974)
Mishandling national security information (#PE3749)
Revealing national security information to a foreign power (#PE4343).
Related Espionage (#PC2140) Misuse of classified communications information (#PD5183).

♦ **PE4001 Institutionalized members of society**
Broader Social neglect (#PB0883) Human disease and disability (#PB1044).

♦ **PE4003 Non-implementation of workers wage increases provided for in legislation and collective agreements**
Broader Breach of contract (#PE5762).

♦ **PE4006 Violations of health and safety regulations**
Nature The wilful violation of a penal regulation on health and safety creates a danger to life or health.
Broader Economic crime (#PC5624) Violation of regulatory codes (#PD4539)
Resistance to internationally agreed standards (#PC4591).
Aggravates Human disease and disability (#PB1044).
Aggravated by Suppression of safety records (#PF2714)
Unethical practices of employers (#PD2879)
Inadequate enforcement of safety regulations (#PD5001).

♦ **PE4010 Discrimination against men in parental rights**
Denial of rights as a parent to men
Nature Sole custody of children, as practised in most divorce courts, constitutes legal discrimination against men. Few rights are enforced less stringently than the right of a non-custodial parent (usually the father) to visit his children. Unmarried fathers, a growing proportion of the male population, are generally denied the right to contest custody. Fathers of unborn and 'illegitimate' children have even less control over the fate of their offspring than do other fathers. In cases of legal abortion, usually only the rights of prospective mothers, and sometimes the rights of the foetus, are considered.
Counter-claim Neither parent should have a superior claim to children but, owing to the fact that mothers have always cared for children and will no doubt continue to do so, they should be given custody.
Broader Discrimination against men (#PC3258)
Conflict concerning legal custody of children (#PF3252).
Related Discrimination against women in parental rights (#PE4019).

♦ **PE4015 Inhumane participation of the medical profession in torture**
Nature Health personnel have been known to assist in the interrogation or punishment of prisoners in a way that adversely affects their health. Involvement of doctors in torture, floggings, amputations and executions is not rare in countries where torture is practised. Doctors have been involved in declaring tortured prisoners fit for further interrogation or for other cruel, inhuman or degrading treatment or punishment. They have also amputated limbs by way of punishment; been present at executions by lethal injection; and signed false death certificates. Psychiatrists and psychologists have participated in the torture process by injecting drugs or certifying dissidents as insane. Medical researchers and pharmacologist have developed drugs and sensory deprivation and saturation technologies to encourage victims to give information, experience pain or become disoriented. Such participation, complicity, incitement or attempts to commit torture are a gross contravention of medical ethics.
Incidence It has been estimated that in 60 per cent of cases, medical doctors assist in the torturing of victims, not only pronouncing judgement on the fitness of victims but also signing bogus death certificates obscuring the use of torture. A British Medical Association report describes doctors as not only assisting in torture under duress, but also voluntarily. Doctors in the armed forces and prison services have created techniques to interrogate suspected terrorists in Northern Ireland including hooding, making detainees stand with arms against a wall for prolonged periods, continuous noise during interrogation and a bread and water diet. In Iran, a doctor bled to death 1,000 criminals in a Baghdad prison, having dosed them previously with hypnotic drugs. His explanation was that it permitted the use of their blood before they were executed. Doctors and torture has been reported in the following countries: **Af** Mauritania. **Am** Chile, Colombia, Argentina. **As** Iran Islamic Rep., Pakistan, Singapore. **Eu** Spain, UK.
Broader Participation in torture (#PD4478).
Related Inadequate medical care (#PF4832) Unethical medical practice (#PD5770)
Depriving prisoners of medical treatment (#PD1480)
Inadequate assistance to victims of torture (#PE6936).
Aggravates Torture (#PB3430).

♦ **PE4016 Statelessness of women**
Discrimination against women in the right to a nationality — Denial of right to nationality for women
Nature Each country has its own nationality laws, and in many cases in the past such laws have permitted or provided for discrimination against women, or have deprived women of their nationality upon marriage to, or divorce from, a foreigner. Under the law of some countries a woman still loses her nationality on marriage with a foreigner or upon a change in the nationality of her husband occurring during the marriage. Thus the woman loses the protection of the state and of the law of her original nationality. In the event of a dissolution of the marriage or the death of the husband, the woman does not automatically recover her original nationality. In these countries, women are still not assured of equality with men in the exercise of the right to a nationality and there is no law to prevent them from becoming stateless upon marriage or at its dissolution.
Broader Statelessness (#PE2485).
Related Involuntary loss of nationality of children (#PE6676).

♦ **PE4018 Unequal property rights for women**
Denial of right to property to women
Nature Women in many countries do not have equal rights in the field of civil law, and in particular the right to acquire, administer, enjoy, dispose of and inherit property, including property acquired during marriage. Daughters may not have the same rights as sons to the property of their parents, and wives may not inherit their husband's property – in these cases, when a man dies, his property does not devolve on the widow and their children, but on relatives.
Claim Women and girls are half of the world's population, do two-thirds of the world's work hours, receive a tenth of the world's income and own less than a hundredth of the world's property.
Broader Restrictions on property rights (#PD8937)
Discrimination against women before the law (#PD0162).
Narrower Intestacy (#PE5063).
Aggravated by Unequal property rights (#PJ2031).

♦ **PE4019 Discrimination against women in parental rights**
Denial of rights as parent to women
Nature In some legal systems parental authority belongs exclusively to the father; in some of these countries, in the event of loss of such authority, it does not pass automatically to the mother; and in some instances, on dissolution of marriage, custody of the children is awarded to the father regardless of the merits of the case.
Incidence In Ethiopia, for example, parental rights are customarily exercised by the father in any marriage. The natural guardian is the father, if he himself is not under guardianship, the next person would be one appointed guardian by the father, next the eldest brother, then the grandfather, the parental uncle, and then the parental nephew.
Broader Discrimination against women before the law (#PD0162).
Related Discrimination against men in parental rights (#PE4010).

♦ **PE4021 Domestic quarrels**
Family disagreements
Incidence In 1988 in the UK, 56 per cent of killings took place during quarrels or bouts of temper.
Aggravates Family violence (#PD6881) Family breakdown (#PC2102)
Complications of childbirth (#PC9042).
Aggravated by Family stress (#PD8130) Unstable family life (#PD7929)
Family disorganization (#PC2151) Inadequate family structures (#PF1000).

♦ **PE4026 Violent sports**
Sports violence
Refs Merkl, Peter H *Political Violence and Terror* (1986).

♦ **PE4027 Fuel-air explosive**
Fuel bomb
Nature The warhead contains fuel that is dispersed into the air by an initial explosion. A second explosion ignites the mixture of air and fuel, creating a fireball and shock wave. It can be 10 times more powerful than conventional explosives of the same size.
Broader Weapons (#PD0658).

♦ **PE4031 Lack of political neutrality of civil servants**
Nature Civil servants must be politically neutral and wholly impartial in advice and diligently implement government policies in both the spirit and the letter of the administration in order for the body politic to be served justly. The tendency by elected officials to mislead the public for political reasons can be checked by civil servants.
Aggravates Inefficient public administration (#PF2335)
Violation of privileges and immunities of international civil servants (#PE5488).

♦ **PE4032 Public non-accountability of organizations developing technology**
Nature Many institutions including corporations, military commands, research institutes and universities developing technology are not accountable to the public in any way for the use or consequences of the products they are developing. In large part, the highly competitive atmosphere in which research is done results in great efforts to maintain secrecy during the development process and in some cases after the product is in use; for example, the Stealth Bomber was virtually operational before it came to the public's attention. Also, in many cases the impact of a product is not clear until after the product has been in use for some time. But the main reasons for the failure in accountability to the public are neither the public wants the responsibility nor the organization wants to be accountable.
Broader Social unaccountability (#PC1522) Non-inclusive management decisions (#PF2754).

♦ **PE4033 Corruption of customs and excise officials**
Bribery of customs officials
Broader Official corruption (#PC9533).
Aggravates Evasion of customs and excise duties (#PD2620)
International trade in endangered species (#PC0380).
Aggravated by Absence of coordinated customs services in developing countries (#PJ8325).

♦ **PE4036 Paratyphoid fever**
Refs WHO *Typhoid and Paratyphoid Fevers* (1969).
Broader Intestinal infectious diseases (#PE9526).

♦ PE4039 Patent pool or cross-licensing agreements
Nature Restrictions on territories, quantities, prices, customers or markets arising out of patent pool or cross-licensing agreements or other international transfer of technology interchange arrangements among technology suppliers which unduly limit access to new technological developments or which would result in an abusive domination of an industry or market with adverse effects on the transfer of technology, except for those restrictions appropriate and ancillary to co-operative arrangements such as co-operative research arrangements.
Broader Restrictive trade practices (#PC0073).
Related Collusive tendering (#PE4301) Grant-back provisions (#PE5306)
Challenges to validity (#PF1200) Restrictions on research (#PF0725)
Restrictions on publicity (#PF1575) Restrictions on adaptations (#PF5248)
Exclusive dealing arrangements (#PE0413) Restrictions on use of personnel (#PF3945)
Exclusive sales and representation agreements (#PE4581)
Trade restrictions due to voluntary export restraints (#PE0310)
Payments after expiration of industrial property rights (#PF5292)
Tying of supplies to subsidiaries by transnational enterprises (#PE0669).

♦ PE4045 Agricide
Nature The combination of population growth and agricultural practices which destroy the land are terminating the ability of the planet to support life. Eventual agricide could result.
Broader Ecologically unsustainable development (#PC0111).
Aggravated by Unsustainable population levels (#PB0035)
Unsustainable agricultural development (#PC8419).

♦ PE4048 Iron dust
Nature Iron dust is created in the process of mining iron ore, smelting iron and steel and processing iron and steel products, such as grinding or polishing. Iron dust if inhaled in sufficient quantities can cause pneumoconiosis of a special kind called siderosis. Finely divided freshly reduced iron powder is pyrophoric and ignites on exposure to the air at normal temperatures.
Broader Uncleanness (#PA5459).
Aggravates Siderosis (#PE4274).

♦ PE4051 Unilateral structural disarmament of nuclear weapons
Nature The failure to provide up-to-date infrastructure for development, production, maintenance and disposal of nuclear weapons in effect is unilateral disarmament of nuclear weapons in the west. Development and maintenance of sophisticated, technologically complex weapons require major expenditures and often long periods of time. Low budgetary priorities, little understanding on the part of legislatures, political expediency, deferred maintenance, forgone investment and aging designs have left nuclear weapons systems infrastructure nearly unworkable.
Claim A policy of calculated inaction under current circumstances will have the devastating effect on global security of unilateral structural de-nuclearization.
Broader Insecurity through unilateral structural disarmament (#PE7670).
Reduces Nuclear accidents (#PD0771).

♦ PE4052 Inadequate standard of living in developing countries
Nature The precipitous decline in living standards in many developing countries, especially in Africa and Latin America results in social unrest and threatens democratic political institutions. This decline is the result of the drop in commodity prices, deteriorating terms of trade, barriers in industrialized countries to the import of manufactured goods from developing countries and the onerous external debts of capital which acts as severe constraints on economic and social development.
Incidence The rate of growth in gross national product per capita in Latin America and the Caribbean in 1988 was 1.5 percentage points lower than in 1987. During the seven years ending in 1988, there had been a net transfer of $US 20 billion to $US 30 billion from countries in the region.
Broader Deteriorating quality of life (#PF7142) Inadequate standards of living (#PF0344).
Narrower Deterioration of the quality of life in least developed countries (#PE7734).

♦ PE4054 Abuse of traditional cultural expressions of peoples
Exploitation of artistic expressions of peoples
Nature Exploitation of cultural expressions of peoples, whether in the form of handicrafts, dances, ceremonies, or music, with complete disregard for authenticity and preservation, contributes to what has been termed the prostitution and degeneration of cultures.
Incidence This is especially severe in the case of indigenous peoples whose cultures are vulnerable for other reasons.
Broader Social environmental degradation from recreation and tourism (#PD0826)
Abuse of international cultural, diplomatic and commercial exchanges (#PF3099).
Aggravates Endangered indigenous cultures (#PC7203)
Socio-cultural environment degradation (#PC4588).
Aggravated by Undue attachment to a social group (#PF1073).

♦ PE4056 Solitary confinement
Nature Prisoners are often submitted to solitary confinement as a means of isolation, protection from other prisoners, preventing them from influencing other prisoners or torture. Some of these cells have no light or sanitary facilities and are cold and to small to stand or lie down.
Incidence Solitary confinement has been reported in the following countries: **Af** Lesotho, Liberia, Morocco, Namibia, Rwanda, South Africa. **Am** Bolivia. **As** China, Saudi Arabia, Viet Nam. **Eu** Poland.
Refs Jackson, Michael (Ed) *Prisoners of Isolation* (1983).
Broader Torture through confinement (#PD4590) Denial of rights to prisoners (#PD0520).
Related Social isolation as torture (#PD6810).

♦ PE4057 Long-term hazards of exposure to radiation
Hazards of low-level exposure to radiation
Nature Safety thresholds for radiation have not fully taken into account long-term effects, especially those resulting from low-level exposure for extended periods of time. Whereas the effects for higher levels of exposure, such as those from nuclear weapons have been monitored, far less attention has been accorded to effects which may, for example, only appear to result in a higher vulnerability to leukaemia.
Claim There is a direct, proportional correlation between exposure to penetrating, or ionizing, radiation and cancer risk. In short, no radiation exposure can be considered free of risk.
Counter-claim Circumstantial evidence indicates that some forms of low level radiation may be beneficial. Some tests on laboratory animals exposed to low levels of radiation indicate that they tend to live longer. In regions where the natural radiation is high (India, China, Brazil) there is some indication that such levels of radiation offer protection against cancer.
Refs Sutcliffe, Charles *The Dangers of Low Level Radiation* (1987); United Nations *Biological and Environmental Effects of Low-Level Radiation* (1976).
Broader Radioactive contamination (#PC0229).
Aggravates Leukaemia (#PE0639).
Aggravated by Uncertain toxicity thresholds (#PF5188)
Unethical practice of radiology (#PD8290)
Inconclusiveness of scientific and medical tests (#PD7415).

♦ PE4058 Arbitrary external interference in family life
Unlawful interference in family life — Unlawful interference with the family home
Nature Often people are not protected by law against arbitrary or unlawful interference with their privacy, family, home or correspondence or against unlawful attacks on their honour and reputation.
Broader Arbitrariness (#PB5486).
Narrower Threats against family or friends (#PE3308).
Aggravated by Denial of right to a family (#PE7267)
Denial of right to private home life (#PE6168).

♦ PE4059 Over-emphasis on immediate solutions in resource development research
Emphasis on economic priorities in research
Nature Research in resource development and cooperation is carried out as a response to the immediate situation, and fails to anticipate sufficiently the needs and problems of the future in relation to accessory improvements. Such research tends to be limited by national and corporate interests, to the exclusion of those in greatest need.
Claim Economic and commercial considerations override deep seated global social needs in the organization of research, distorting emphasis and causing science to be dependent upon economic priorities for its existence.
Broader Undirected technological expansion (#PC1730)
Nationalistically determined development of natural resources (#PD3546).
Aggravates Inadequate research and development on problems of developing countries (#PF1120).
Aggravated by Excessive reliance on fashionable solutions to problems (#PF4473)
Domination of government policy-making by short-term considerations (#PF0317).

♦ PE4071 Violation of the right of workers' organizations to establish confederations
Violation of the right of workers' organizations to affiliate internationally
Broader Denial of the right of trade union association (#PD0683).

♦ PE4072 Chromium as a pollutant
Refs FAO *Water Quality Criteria for European Freshwater Fish* (1983); WHO *Chromium* (1988); Williams, J H *Chromium in Sewage Sludge Applied to Agricultural Land* (1988).
Broader Chemical contamination of water (#PE0535).
Aggravates Water pollution (#PC0062).

♦ PE4074 Restlessness due to torture
Nature The victims of torture experience restlessness for years after the events.
Broader Neurological effects of torture (#PD4755).

♦ PE4078 Torture by wounding
Torture by bruising
Incidence Torture by wounding has been reported in the following countries: **Af** Burundi, Mali, Rwanda, Uganda, Zimbabwe. **Am** Peru, Suriname. **As** Afghanistan, Indonesia, Sri Lanka, Syrian AR. **Eu** Italy.
Broader Physical torture (#PD8734).
Narrower The rack (#PE7430) Surgical torture (#PE7547) Torture by crushing (#PE3719)
Torture by shooting (#PE6339) Inflected loss of vision (#PE3377)
Torture through breaking bones (#PE7015).
Related Torture through mutilation (#PD7576).

♦ PE4086 Inadequate assistance to victims of accidents
Accident victims
Nature The victims of accidents usually do not receive the maximum amount of assistance available to them at the time of the accident, such as medical care or other emergency services, in compensation for losses or in long term care they require.
Broader Legal discrimination in favour of offenders (#PD9316)
Inadequate assistance to victims of human rights violations (#PD5122).
Narrower Non-payment of compensation to victims of motor accidents (#PE8824).
Aggravates Imposed career interruptions (#PF4128).
Aggravated by Injurious accidents (#PB0731).

♦ PE4089 Air terrorism
Incidence In 1990 it was estimated that in the previous 5 years, terrorist attacks against planes of 6 countries had killed 1,030 people.
Broader Terrorism (#PD5574).
Related Aerial piracy (#PD0124).

♦ PE4097 National defence procurement procedures
National defence industries
Nature The procurement procedures of the national defence institutions are structured so that waste of resources, over-spending, corruption and spiralling costs and bloated military budgets are the natural consequences. Weapons systems generally cost too much, take too long to develop and perform too poorly. For a large part of business between the military and private enterprise, negotiated cost-plus-fixed-fee contract are used rather than sealed-bid fixed-price contracts. Regardless of the type of contract, they are not binding; fixed price contracts are only tentative and provisional prices, potential losses resulting from errors, poor judgement, and performance failures on the part of contractors are averted by modifications and amendment of contracts. Close relationships between government and defence industries are maintained by hiring practices by which retired military officers readily join contractors' staffs and top industry executives move to political positions within defence departments. Classification of documents and equipment mean that only approved industries are allowed access to contracts. Defence industry lobbyists keep contact with both politically influential people and bureaucrats. Reforms in national defence procurement procedures are often called for but little is done. Elected and appointed officials face incentives to opt for quick fixes; political authorities invest little in follow-up to the changes that are made; elections are always around the corner. Adversely affected parties keep quiet and wait for the storm to pass over. Elected officials, concerned with re-election, are more interested in how a defence contract will benefit their constituency than military effectiveness.
Broader Military-industrial complex (#PC1952).
Aggravates Military-industrial malpractice (#PD4361).
Aggravated by Inflexible military thinking in industry (#PE7040).

♦ PE4102 Discrimination against women workers in multinational enterprises in developing countries
Refs ILO and United Centre on Transnational Corporations *Women Workers in Multinational Enterprises in Developing Countries* (1985); Kim, Linda *Women Workers in Multinational Enterprises in Developing Countries* (1985).

Broader Discrimination against women in employment (#PD0086)
Discrimination against women in developing countries (#PC4898).

♦ **PE4103 Excessive rainfall**
Rain-storms — Monsoon
Nature Rain-storms can cause flooding, land and mud slides, structural damage to building and public works and damage to crops.
Incidence In August 1989, over 750 people died and 2,000 people were missing in fierce rain-storms that lashed several parts of India. In the state of Maharashtra, over 500 people were reported killed in flash floods, collapsing houses, land-slides and electrical accidents caused by incessant rain. Over 1,000 fisherman were caught in storms on the Arabian Sea were among those reported missing.
Refs Das, P K *Monsoons* (1986).
Broader Storms (#PD1150).
Aggravated by Large-scale weather anomalies (#PC4987)
Long-term changes in precipitation patterns (#PE4263).

♦ **PE4104 Asphyxia**
Nature It refers to series of symptoms which follow stoppage of breathing and of the heart's action. It is caused by drowning, poisonous gases, some diseases, such as croup, diphtheria, asthma, tumours in the chest, or swelling of the throat due to wounds or inflammation, suffocation or strangling.
Refs Lacoius-Petruccelli, Alberto *Perinatal Asphyxia* (1987).
Broader Human physical suffering (#PB5646) Human disease and disability (#PB1044)
Ill defined causes of morbidity (#PE5463).
Related Drowning (#PG2857).
Aggravates Human death (#PA0072) Sudden unexpected infant death (#PD1885).

♦ **PE4107 Multiple births**
Risk of multiple birth
Nature Multiple births tend to be associated with a high incidence of prematurity, caesarian deliveries and low birth weight. More specialized hospital care tends to be required than for singletons, meaning that they are sometimes split up and distributed to separate facilities. Handicaps and language development problems are more common in triplets. Bringing up the children requires resources not available to many families.
Incidence Fertility treatment has led to a large increase in multiple births. In the UK the number of triplets has doubled in the decade to 1990.
Broader Risk (#PF7580).
Aggravates Sudden unexpected infant death (#PD1885).

♦ **PE4108 Hostage taking**
Refs Mirron, Murray S and Goldstein, Arnold P *Hostage* (1979).
Broader Kidnapping (#PD8744).
Narrower Child hostages (#PG4028).
Related Terrorism (#PD5574) Political hostage-taking (#PD1886).

♦ **PE4109 Acoustic trauma**
Noise damage to the ear
Nature Acoustic trauma is entirely preventable and totally untreatable. Noise damage is most frequent in industrial societies and is caused by a variety of sources. Radios with earphones and loud music at rock concerts are thought to cause damage. Manufacturing processes are know to damage hearing. Aircraft and trains are knowing to cause hearing loss. In rural settings, simple sugar-cane crushers are very noisy. Men who work with such machines, tinplate workers, and those who hammer out old oil drums for roofs and fences are subject to a level of noise which causes progressive high-frequency hearing loss and eventually deafness for the speech frequencies.
Broader Deafness (#PD0659) Diseases of the ear (#PD2567).

♦ **PE4111 The split between eastern and western Europe**
The east-west division in Europe
Broader Social fragmentation (#PF1324).

♦ **PE4113 Curvature of spine**
Scoliosis — Kyphosis
Refs Bradford, David S, et al *Moe's Textbook of Scoliosis and Other Spinal Deformities* (1987); Zorab, P A and Siegler, David *Scoliosis, 1979* (1980).
Broader Diseases and injuries of bone (#PE3822).

♦ **PE4119 Military and police personnel participation in torture**
Broader Abuse of police power (#PC1142) State sanctioned torture (#PD0181)
Unethical practices by police forces (#PD9193).
Related Participation in torture (#PD4478).
Aggravates Human torture (#PC3429).

♦ **PE4120 Inadequate supply of pharmaceutical products in developing countries**
Inappropriate pharmaceutical products in developing countries — Underdevelopment of the pharmaceutical industry in developing countries — Non-responsiveness of transnational corporations to pharmaceutical needs of developing countries
Nature Developing countries are at different stages in establishing a pharmaceutical industry and the constraints on the growth of the industry are different at each stage. Three principal groups of countries can be identified: (a) developing countries with little or no pharmaceutical manufacturing activity; (b) developing countries with facilities to formulate a range of drugs; and (c) developing countries with facilities to manufacture some of the active ingredients as well as formulate drugs.
At present, the availability of drugs, active ingredients and intermediates for drug manufacture often depends on imports. The growing cost of such imports makes it imperative for most developing countries to consider the local formulation and/or basic manufacture of as wide a range of bulk drugs as possible.
The main constraints on the growth of the pharmaceutical industry in developing countries are: inadequate technological capability; lack of qualified and trained personnel; high cost and limited availability of imported bulk drugs and intermediates; scarcity of financing available on terms and conditions suitable for the industry; and absence of well-defined national policies to promote the growth of the industry. Furthermore, because of the scarce resources available in most developing countries, the decision to develop a domestic pharmaceutical capability has to compete with other national development projects. Hence it is unrealistic for most developing countries to hope that domestic production will be able to make them self-sufficient in drugs, especially since the drug industry is divided into several quite separate therapeutic markets.
Pharmaceutical transnational corporations are not responsive to the real drug needs of developing countries. Importation of pharmaceuticals is one of the fastest growing drains on hard foreign currency for developing countries. Some health ministries are spending over 50 per cent of their budgets on drugs alone. In addition to cost, there are dangers of uncontrolled availability of potentially dangerous drugs. In many countries the great majority of drugs are available over the counter, without prescription.
Incidence Developing countries obtain pharmaceutical supplies from two sources: imports from abroad and domestic output. Both these sources are heavily influenced by technological factors, which are responsible for keeping prices of imported pharmaceuticals well above the capacity of most developing countries to obtain in sufficient volume to satisfy their health care needs. Indicators for several countries suggest that in 1980 pharmaceutical expenditures constituted about one sixth of health expenditures in developed countries but from one third to one half of such expenditures in developing countries. It has been the general experience of the last 10 years that expenditures on pharmaceuticals have been rising by some 13-14 percent annually. Assuming that the past pattern of pharmaceutical expenditures were to prove relevant for the future, it will be seen that total pharmaceutical costs in developing countries would have risen from $15-20 billion in 1980 to $55 billion by 1990 and to the very high amount of some $200 billion by the year 2000.
In the case of inappropriate use of drugs, recent studies have disclosed, for example, that a fifth or North Yemen's drug imports and a quarter of all drugs sold in India are tonics, vitamins or indigestion tablets.
Background Part of this flood of inessential drugs is due to the marketing practices of the international pharmaceutical industry. The WHO says that promotion activities of the drug manufacturers have created a demand greater than the actual needs. In Colombia, for example, the money spent on drug advertising is equivalent to more than half of the country's health budget. In Nepal, Brazil and some Central American countries there is one sales representative for every three doctors; this leads to overprescribing on a massive scale.
Claim The best way to cope with these problems is to take control of drug production and distribution. This is why, in addition to urging the adoption of an essential list of basic drugs, WHO is encouraging an increase in generic prescribing (that is according to what drugs contain, rather than according to trade name) and local manufacture. This would enable countries to cut costs dramatically either by manufacturing the drugs themselves or by bulk-buying cheaper generic drugs on the world market.
Counter-claim International pharmaceutical companies employ tens of thousands of local people in developing countries. These operations generate billions of $US in sales every year, including exports contributing to local trade balances. Pharmaceuticals are the most practical and cost-effective elements of medical technology. Drug companies tailor product lines to meet the needs of the specific populations they serve. The pharmaceutical industry is active in research and production of drugs for disease endemic in developing parts of the world. It is an active partner with government agencies and NGO's in making products available where they are needed. They are making significant progress in supplying medical professionals with literature about their products that is scientifically accurate and includes essential information on safety contraindications and side-effects where applicable.
If the markets of the international pharmaceutical industry are cut back drug companies will not be able to keep up the present volume of research into new drugs.
Refs Gereffi, Gary *The Pharmaceutical Industry and Dependency in the Third World* (1983); United Nations *Guidelines on Technology Issues in the Pharmaceutical Sector in the Developing Countries* (1983); United Nations Industrial Development Organization *The Growth of the Pharmaceutical Industry in Developing Countries*.
Broader Transnational corporation imperialism (#PD5891)
Underdevelopment of manufacturing industries (#PF0854)
Domination of developing countries by transnational corporations (#PE0163).
Related Inadequate health services (#PD4790).
Aggravates Racial discrimination (#PC0006) Human disease and disability (#PB1044)
Marketing of banned pharmaceutical drugs in developing countries (#PE6036).

♦ **PE4127 Overloaded vehicles**
Nature Lorry overloading is fairly widespread, and enforcement programs have shown themselves to be largely ineffective. Road surfaces suffer, such vehicles are involved in a greater number of accidents than those loaded according to their specifications, and the accidents which occur are more likely to result in death. Less than 10 per cent of developing countries have any effective weight control programmes for lorries. A major reason for the poverty of these programmes is that there is a complete absence of simple, reliable and inexpensive weighing equipment.
Related Ineffective rural transport (#PF2996)
Inadequate road and highway transport facilities (#PD0490)
Inadequate roads and transport in land-locked developing countries (#PE5824)
Lack of skilled workers in the transport sectors of land-locked developing countries (#PE5884).
Aggravates Transport accidents (#PC8478)
Health risks to workers in transport, storage and communication industries (#PE1581).
Aggravated by Lack of transport vehicles (#PF0713)
Unprofitable transport business (#PJ8253)
Scarcity of appropriate transport (#PE8551)
Prohibitive cost of transportation (#PE8063).

♦ **PE4129 Inaccurate medical diagnosis due to inadequate equipment**
Nature All medical tests have some margins of error, some quite little and some significant. Equipment can be miscalibrated or damaged in unnoticed ways. Human errors by those administering the test, those doing the laboratory work and those interpreting the result compound the potential mistakes. Patients and doctors tend to view such test as infallible or with little error. Treatments resulting from these diagnosis can be quite disastrous.
Broader Misdiagnosis (#PF8490) Inadequate medical care (#PF4832)
Inadequacy of medical science (#PF8326).
Related False diagnosis of mental disorder (#PG5868).
Aggravates Surgical malpractice (#PE4736).

♦ **PE4131 Alcohol-related crime**
Refs Council of Europe *Alcohol and Crime* (1984).
Broader Crime (#PB0001) Alcohol abuse (#PD0153).
Related Alcohol-related violence (#PE7084).

♦ **PE4139 Mistletoe in trees**
Nature Mistletoes are semiparasites depending on a host tree for water and mineral salts. Dwarf mistletoes are injurious to conifers reducing the quality and quantity of timber and paving the way for bark beetle infestations.
Broader Basidiomycetes (#PE0364).

♦ **PE4140 Insufficient fertilizers for agricultural development in developing countries**
Nature With greatly increased demand for basic food agricultural products to sustain the growing population, particularly in countries where population density is high and continues to rise at a rapid pace, the demand for fertilizers has also increased significantly. Such demand is likely to continue to rise considerably in the next few years, particularly in a number of developing countries where agricultural production will have to be substantially increased to meet domestic demand and to provide export opportunities for particular agricultural products. The availability of chemical fertilizers at reasonable prices, therefore, constitutes an essential feature of agricultural develop-

ment in most countries and particularly in several developing countries which continue to be basically agricultural economies. Domestic production has constantly been below domestic demand and consumption in the developing countries, necessitating large imports of chemical fertilizers by these countries.

Fertilizer production facilities in many developing countries are operating substantially below capacity; in some cases the operating rate is less than half the design rate. There are a number of reasons for this, but they are usually due to a combination of: processing problems, inadequate maintenance, high fixed operating costs, lack of market for the product – or inability to market it, inadequate supplies of feed materials, inadequate transport and storage facilities, lack of operating capital, technical incompetence, poor product quality, and lack of dependable power and water supplies. Also, shipping, storage and handling costs in the developing countries often represent more than half the total cost of getting the fertilizer to the farmer. In the United States, although relatively efficient methods are used in transport and handling, getting the fertilizer from the production point to the farmer accounts for as much as two thirds of the cost.

Incidence World fertilizer consumption has risen above 100 million metric tons (mmts) of nutrients a year. A quarter of this is in developing countries. Though these countries have increased their fertilizer consumption by more than 50 percent since 1973–74, the amount is still insufficient to produce adequate food. Despite the energy crisis, there should be no disagreement about the urgency of sustaining rapid growth in fertilizer consumption in the developing world. Population pressure on land, growing food deficits, depleted soil fertility and the complementarity between proven yield-increasing technologies and high levels of fertilizer application all point up its importance.

Claim The real question concerning the developing world is not whether but how to maintain rapid growth in fertilizer consumption.
   Aggravates Stagnated development of agricultural production (#PD1285)
   Deterioration of domestic food production in developing countries (#PD5092).

♦ **PE4145 Underparticipation of developing countries in the airline industry**
Nature The airline sector is dominated by national and international regulations, the question of which airlines are to serve which routes being largely settled around a conference table. In this connection, the problem of developing countries is one both of lack of bargaining power and of inability (because of insufficient economies of scale) to compete effectively for the main international air routes. This inevitably leads to a degree of dependence by some tourism-intensive countries (for instance, the Caribbean islands) on foreign airlines for the transportation of tourists there. Should the routes involving these countries become less profitable than alternative routes, foreign airlines might cease to supply them, thus causing the host countries severe economic difficulties. Some airlines in the more advanced developing countries, particularly of Southeast Asia have managed to break into the market and grow to a size which allows them to compete effectively with developed country airlines; but in the case of other smaller countries, the economies of scale in international airline operations have been an obstacle in this respect.
   Broader Underdevelopment of industrial and economic activities (#PC0880).
   Reduces Air traffic congestion (#PD0689).

♦ **PE4146 Ineffective official inspection of regulated activities**
Ineffective inspection of toxic waste sites
Incidence In UK there is a multiplicity of understaffed authorities trying to control a highly competitive private-sector industry of toxic waste and other pollution. Local authorities cannot afford to provide plans for dealing with wastes, few sites are able to test the waste they receive, inspections are rare and operators cannot be prosecuted unless they are caught in the act of dumping waste illegally.
   Broader Ineffective monitoring (#PF2793)   Inadequate environmental monitoring (#PF4801).
   Aggravates Hazardous wastes (#PC9053)   Counterfeit machine parts (#PE5319)
   Unfocused style of community operations (#PF6559).

♦ **PE4152 Looting**
Pillage
Incidence Enraged by soaring prices and low wages, crowds, many led by women and children, looted stores and supermarkets in Argentina in 1989. Altogether more than 1,500 looters were arrested nationwide.
   Broader Robbery (#PD5575)   Criminal intrusion (#PE6771).
   Related Lack of war relief (#PF0727).
   Aggravates Theft of works of art (#PE0323).
   Aggravated by War crimes (#PC0747)   Food hoarding (#PJ2225)
   Inadequate riot control (#PD2207).

♦ **PE4154 Prohibitive cost of hospital facilities**
Nature With in a hospital, major cost elements associated with health care are: people, energy, equipment and maintenance. Expenditure associated with people dominates hospital operating costs. Planners have underestimated the capacity of relatively untrained personnel to participate in the delivery of health care. Increasing fuel costs, unnecessary use of energy and inefficient building designs have made energy the second most costly element in health care in hospitals. Medical equipment is becoming obsolete more and more quickly requiring replacement more frequently. Increasingly sophisticated equipment is demanding higher outlays of capital for equipment. Poor designs and increasing complexity increases hospital maintenance costs as health delivery services are disrupted by repairs and routine maintenance.
   Broader Inadequate medical facilities (#PD4004).
   Aggravates Inadequate mental hospitals (#PF4925)
   Human disease and disability (#PB1044)
   Inadequate hospital facilities (#PE5058).
   Aggravated by Inadequate medical resources (#PD7254).

♦ **PE4161 Lameness in cattle**
Refs Greenough, P R; MacCallum, F J and Weaver, A D *Lameness in Cattle* (1981).
   Broader Diseases of musculoskeletal system in animals (#PD7424).
   Narrower Corns (#PG1201)   Foot rot (#PG9779)   Ergotism (#PG7849)
   Fractures (#PG9824)   Laminitis (#PG5410)   Fescue lameness (#PG9336)
   Stable foot rot (#PG7689)   Spastic paresis (#PG7351)   Spastic syndrome (#PG9677)
   Calving paralysis (#PG3895)   Frostbite in calves (#PG9809)   Fractures of cattle (#PG6694)
   Ulceration of the sole (#PG4941)   Degenerative arthropathy (#PG5288)
   Peroneal nerve paralysis (#PG5668)   Vertical hoof wall fissures (#PG4686)
   Horizontal hoof wall fissures (#PG7723)   Dermatitis verrucosa in animals (#PG1884)
   Rupture of the gastrocnemius muscle (#PG6303).

♦ **PE4162 Prohibitive cost of land**
Inflated land values — High cost of land — Overpriced valuable land — Rising land price — Land levelling costs — Cost of clearing land
Nature Rising land prices make it impossible for hard-working families to buy homes. Land prices make it difficult for new businesses to be housed. Authorities cannot afford to build parks or public transportation systems.
Incidence Extreme examples include the indication that the cost of all real estate in Tokyo exceeds the cost of all real estate in the USA. That of Kanagawa prefecture (bordering Tokyo) is equivalent to the cost of all real estate in France, Germany, Italy and England.
Counter-claim A sudden decline in land prices would reduce the availability of investment funds.
   Broader Unattractive locale for economic development (#PF3499)
   Restrictive effects of external capital on development (#PF3318).
   Related Prohibitive ownership costs (#PJ0033).
   Aggravates Deprivation of peasantry (#PC8862)   Prohibitive cost of accommodation (#PD1842)
   Monolithic architecture of high-rise buildings (#PE1925)
   Inadequate housing and penning of domestic animals (#PE2763)
   Diminishing capital investment in small communities (#PF6477).

♦ **PE4167 Capital investments supporting racial discrimination**
Incidence Transnational corporations, banks, insurance companies, firms and other enterprises did not break all the business ties with South-Africa even if they publicly disinvested. They have shut down sales and representative offices and subsidiaries, but through new acquisitions have maintained ongoing relations. They have reduced direct investment, but partial sales amount to nothing more than change of ownership, parent companies retaining hold on the financial and commercial markets. Sales to third parties, local management or to a trust have not decreased the ex-subsidiaries dependence on the parent company.
   Aggravates Racial discrimination (#PC0006).

♦ **PE4170 Trade barriers to manufactured goods from developing countries**
Nature The growth of trade among developed countries has been encouraged by steadily dropping the tariffs and other barriers which they impose on each other's goods. However, the barriers they impose on the manufactured goods of developing countries have not been dropped to the same extent. In fact new barriers have been created in recent years to limit these imports, as developed countries move to protect their economies from foreign competition.
   Broader International trade barriers for primary commodities (#PD0057).

♦ **PE4171 Nervousness**
Refs Beard, George M and Rosenberg, Charles E *American Nervousness, Its Causes and Consequences*.
   Broader Fear (#PA6030)   Agitation (#PA5838)
   Ill defined health conditions (#PC9067).

♦ **PE4173 Denial of right to redress for rights violations**
   Broader Inequality before the law (#PC1268).

♦ **PE4179 Inadequate sense of community and solidarity amongst workers**
   Broader Lack of a sense of community and solidarity at the world level (#PF8704).
   Aggravates Authoritarian division of labour (#PC6089).
   Aggravated by Ineffective worker organizations (#PF1262).

♦ **PE4183 Abuse of rights of criminal suspects**
Withholding of information by police from criminal suspects
   Related Denial of rights to prisoners (#PD0520).

♦ **PE4184 Endangered totemic species**
Nature Certain species, especially animals, have special symbolic value to people as contributing to their socio-cultural, spiritual and emotional lives. Such species may be endangered as a result of the increased impact of man on the environment and through their exploitation, particularly when their value may derive in part from their natural rarity.
   Broader Endangered species of plants and animals (#PB1395).
   Aggravated by Totemism (#PF3421)
   Natural environmental degradation from recreation and tourism (#PE6920).

♦ **PE4187 Teasing**
Taunting — Name-calling
   Broader Harassment in playgrounds (#PE7768).
   Aggravates Juvenile suicide (#PE5771)   Anti-social behaviour (#PC4726).
   Aggravated by Verbal abuse (#PD5238)   Social stigma (#PD0884)
   Sexual harassment (#PD1116)   Physical intimidation (#PC2934)
   Discrimination against the disabled (#PD9757).

♦ **PE4188 Developing country dependence on a single source of finance**
Undiversified forms of credit used by developing countries
Incidence The pattern of borrowing by developing countries on international capital markets in the 1970s revealed the danger of heavy reliance on a single source of credit, especially in the context of fast-changing capital markets. Bank lending has so far transferred most risks to the borrower, including the exchange rate risk, the interest rate risk, and especially the risk of mismatch between foreign exchange earnings and debt service obligations.
   Broader Inadequate economic policy-making in developing countries (#PF5964).
   Aggravated by Policy cross-conditionality restrictions in multilateral development aid (#PF9216).

♦ **PE4190 Bipolarization of trade between developed and developing countries**
Nature Over a long period of history the countries of the third world have had their economies linked to metropolitan countries in a bipolar, two-way relationship, which virtually excluded their trading among themselves. While many of their demands in the realm of international economic relations are directed at removing the constraints that impede their efforts at development, within the bipolar relationship itself there is a growing awareness that the mere intensification of this two-way flow as a means of solving their problems is both undesirable and impracticable. It is undesirable as it may serve to reinforce the dependency relationship; and it is impracticable because it is scarcely conceivable that the existing industrialized countries will provide an unlimited outlet for the vastly increased volume of tradable goods that the developing countries will be able to provide in the future. There will inevitably need to be 'horizontal' links as the developing countries acquire a greater capacity to meet each other's needs. Moreover, the slowing down of the rate of economic growth in the developed countries makes it even more necessary for the developing countries to strengthen their mutual economic links in order to accelerate their own progress.
Incidence In recent years the most dynamic expression of Third World or South-South cooperation has probably been that of 'economic cooperation among developing countries' (ECDC). During the 1970s, trade among Third World countries grew more rapidly than that between them and the industrialized world (at an average rate of 22 percent per annum). Today, however, such trade still constitutes only a bare 4 percent of world trade as a whole, which shows how much still remains to be done. The lack of a real South-South economic front, or of an economic community of the South, is bound to have an impact on the North-South negotiations, through which the South is seeking to negotiate with a group of countries which by contrast is indeed organized in an economic community. Unlike the South, nearly 70 percent of the North's trade is with the North, let alone the multiplicity of existing technical and communications links among the industrialized countries.

Broader Imbalance in international trade patterns (#PC8415).
Aggravates Weakness in trade among developing countries (#PC0933)
Inadequate trade between developing countries (#PD5176).

♦ **PE4196 Partial loss of vision**
Refs Ainlay, Stephen *Day Brought Back My Night* (1989).
Broader Physical blindness (#PD0568).
Aggravated by Trachoma (#PE1946).

♦ **PE4199 Waste of nuclear warheads**
Nature Agreements to reduce nuclear arms means that ways of safely neutralizing some 12,000 high yield atomic warheads have to be found. There are four things governments cans do. 1) Keep the warheads intact, violating the treaty or at least violating the intent of the treaty. 2) Use the fissionable material for a new generation of weapons, which would achieve exactly the opposite of a nuclear arms treaty. 3) Store the plutonium, which would become the target of terrorist and tempt weapons makers for tens of thousands of years. And 4) burning the material in nuclear power generating reactors, producing nuclear waste. All but the last option would be a waste of resources and the later would produce nuclear waste threatening the environment.
Broader Nuclear waste (#PD4396) Unproductive use of resources (#PB8376).

♦ **PE4200 Student absenteeism**
Absence from school — Irregular school attendance — Erratic student attendance — Truancy — Inadequate enforcement of school attendance — Low school attendance — Unaccounted for truancy
Nature Educational systems are increasingly experienced as unsatisfactory by a significant proportion of students. Whether or not they play truant, they leave with pleasure. Teachers are unable to make their classes interesting enough to hold the attention of pupils and maintain their motivation, in part because lack of resources prevents teachers from presenting exciting material.
Incidence In the UK in 1989 it was estimated that 20 per cent of school experience serious truancy problems. In the inner cities 25 per cent of secondary schools are losing 10 per cent of their senior pupils; 12 per cent of schools had absentee rates of more than 20 per cent. Only 48 per cent of pupils claimed that they had never played truant. 5 per cent of teenagers was defined as a serious truant, missing school for several days, or even weeks, with the proportion doubled in inner city schools. Attendance rates of 70 to 80 per cent were common among 15 and 16–year olds, with some pupils losing as much as 50 per cent of their senior pupils. 70 per cent of the inner city students admitting to truancy finally left school without any examination passes and were more likely to be unemployed.
Broader Absenteeism (#PD1634) Obstacles to education (#PF4852).
Related Limiting effect of individual survivalism (#PF2602)
Inappropriate application of traditional values (#PF2256).
Aggravates Educational wastage (#PC1716) Reading disabilities (#PD1950)
Declining school enrolment (#PJ7844)
Unrealized use of education structures (#PF2568)
Maldistribution of students enrolled in school (#PE8733)
Inadequate management skills in rural communities (#PF1442).
Aggravated by Boredom (#PA7365) School phobia (#PE4554)
Juvenile desertion (#PD8340)
Decline in government expenditure on education (#PF0674).

♦ **PE4203 Collapsed relationships between work and society**
Nature The social function of work has been reduced to that of providing income to the individual, his family and to the state. Any sense of labour contributing to the betterment of the community, region or nation is forgotten by both the worker and by society. The meaning of work is defined by income. A hierarchy of material significance in which the wealthy are the most important and the poor are the least important, regardless of what either contributes to society.
Broader Limited image of employability (#PF2896).

♦ **PE4206 Political boycott of international sports events**
Broader Politicization of international sports events (#PF4761)
Abuse of international cultural, diplomatic and commercial exchanges (#PF3099).

♦ **PE4208 Symptomatic heart disease**
Broader Heart diseases (#PD0448).
Narrower Cardiac asthma (#PG4866).

♦ **PE4214 Denial of the right to receive information**
Broader Restrictions on freedom of information (#PC0185).

♦ **PE4215 Excessive commercialization of the media**
Nature A worldwide trend towards commercialization of the mass media is apparent. In the printed media this is reflected in a substantial increase in the commercial content of newspapers and magazines. Although on television the time devoted to advertising is not so significant, this is partially due to the fact that the amount of advertising time allowed on television is regulated in most countries. The increasing commercialization and dependence of media upon advertising revenues implies the production and diffusion of items which have mass appeal. In addition, although advertising revenues can lead to lower cost of newspapers, magazines and broadcasting, commercialization of the media has the potential to exercise a numbing effect on the content of news reporting and commentaries, as the media may be wary of offending sponsors by being too controversial. The possibility of withdrawing or threatening to withdraw advertising gives advertisers the potential to influence the policies of specific media which depend heavily on advertising revenues. Although in practice advertisers are interested in reaching as great an audience as possible, and in achieving this aim they often ignore the overall political orientation of the media they use, such potential for influence does exist, although particular cases are hard to document. Moreover, in countries where professional standards or regulations are weak, the dividing line between editorial commentary and advertisement can be blurred.
Broader Proliferation of commercialism (#PF0815).
Aggravates Distorted media presentations (#PD6081)
Deterioration of media standards (#PD5377)
Decline in public interest broadcasting (#PF5622).
Aggravated by International monopoly of the media (#PD3040)
Excessive portrayal of negative information by the media (#PE1478).

♦ **PE4222 Commercialization of athletic activities and sports events**
Broader Proliferation of commercialism (#PF0815).
Aggravates Athletic competition (#PE4266)
Corruption of sports and athletic competitions (#PE3754).

♦ **PE4224 Immune–deficiency diseases in animals**
Immunodeficiency diseases in animals
Refs Perk, Kalman, et al *Immunodeficiency disorders and retroviruses* (1988).
Broader Atopic diseases (#PE9509).
Narrower Simian acquired immunodeficiency syndrome (#PG9720)
Combined immunodeficiency disease in animals (#PG9784).

♦ **PE4225 Misuse of advertising**
Irresponsible advertising — Negative economic and social effects of advertising — Harmful advertising effects — Insufficient awareness by advertisers of social responsibility
Nature The arguments against advertising refer to its negative economic and social impact and centre around: the inappropriate size of advertising expenditures, given the scarce economic resources of developing countries; the stimulating of demand for goods that are not appropriate, given the income and demand structure; and their contribution to a cultural standardization among countries.
Advertising has both a social and economic impact. Its socio–cultural effects relate to whether advertising creates or reflects the values and life–styles of members of society. Advertising can have adverse effects when it persuades consumers to buy things they do not absolutely need, when the differences among competing producers are slight or non–existent, and when it induces false beliefs in the consumer about the capabilities of particular products. Criticism of the economic effects of advertising is generally based on the resultant misallocation of economic resources, since consumers spend more than they would if they had a genuine choice. This is particularly the case with products characterized by high advertising–to–sales ratios and a low level of informational content.
Institutional and commercial advertising may be powerfully suggestive in their use of motivational psychology. Advertisements can affect voting behaviour, purchasing habits, styles of dress, speech and mannerisms, family and social relations, use or non–use of stimulants, and many other aspects of life. Advertising can thus create, destroy or change values. The power of advertising is abused when it is intentionally deceptive and when it withholds information needed for informed judgement. In the health area, advertising may have harmful effects on nutritional intake; on recovery from illness; on occurrence of accidents and some crimes; and on emotional and mental health.
Incidence One of the most lucrative sectors of the communication industry is advertising, with national and transnational ramifications and channels. Although the colossal size and ever–growing extent of advertising firms in the United States creates the impression that it is primarily an American phenomenon, it has become an enormous world–wide activity. Annual expenditure on advertising is now approximately $64 billion a year. More than half of this is spent in the USA, but several other countries – Britain, France, Federal Republic of Germany, Japan, and Canada – account for over one billion dollars each. The dependence of the mass media on advertising is also growing. Few newspapers in the world of private enterprise could survive without it. As for radio and television, advertising provides virtually the sole revenue for the privately–owned broadcasting companies which are dominant in the USA and in Latin America and is an important source of financing in various other countries.
Claim As a means of supplying information, advertising is biased, in the sense that it concentrates on particular features to the exclusion of others. Indeed what distinguishes advertising from the editorial content of newspapers and from radio or television programmes is that its avowed purpose is that of persuasion; a balanced debate in advertising is a contradiction in terms. Because advertising is overwhelmingly directed toward the selling of goods and services which can be valued in monetary terms, it tends to promote attitudes and life–styles which extol acquisition and consumption at the expense of other values. A particular material possession is elevated to a social norm, so that people without it are made to feel deprived or eccentric.
Advertisers in all branches of industry are failing to make responsible contributions, by sufficiently legitimate and complete descriptions and claims, to social needs. Proof of this is provided by the growth of consumer movements and the increase in international governmental investigations and regulatory enactments. The social needs neglected by advertisers include population planning, use of natural resources, levels of safety, pollution control, and individual freedom of choice. The consumer movement also calls attention to the irresponsibility of some governmental advertising, press conferences and programme publicity. Special cases are the advertising and promotion engaged on by the military–industrial complexes for armament expenditures, the political education and propaganda advertising used by centrally planned countries, and censorship. Commercially, the equivalent may be the highly selective data that appears on package labelling.
Advertisements addressed to children or young people may contain statements or visual presentations which might result in physical, mental or moral harm to them. Such advertisements may take advantage of the natural credulity of children or the lack of experience of young people and may strain their sense of loyalty. Advertisements may also take advantage of the hopes of persons suffering from illness, or of an impaired ability on the part of such persons to judge critically an advertisement holding out the promise of a cure or recovery from illness
Counter–claim Advertising has positive features. It is used to promote desirable social aims, like savings and investment, family planning, and purchases of fertilizer to improve agricultural output. It provides the consumer with information about possible patterns of expenditure (in clothing and other personal needs, in house purchase or rental, in travel and holidays, to take obvious examples) and equips him to make choices; this could not be done, or would be done in a much more limited way, without advertising. Small–scale 'classified' advertising – which, in the aggregate, fills almost as much space in some newspapers as 'display' advertising by major companies – is a useful form of communication about the employment market, between local small businesses and their customers, and between individuals with various needs. Finally, since the advertising revenue of a newspaper or a broadcaster comes from multiple sources, it fosters economic health and independence, enabling the enterprise to defy pressure from any single economic interest or from political authorities.
Broader Propaganda (#PF1878) Unethical media practices (#PD5251).
Narrower Irresponsible pharmaceutical advertising (#PE2390)
Proliferation of direct mail advertising (#PE1810)
Irresponsible tobacco and cigarette advertising (#PE9093)
Excessive portrayal of substance abuse in the media (#PE3980)
Unaesthetic location of advertising hoardings and billboards (#PE2156)
Harmful effects of advertising by transnational corporations in developing countries (#PE2004).
Related Indecent advertising (#PD2547) Proliferation of advertising (#PD5034)
Newspaper and journal propaganda (#PD0184)
Destabilizing international telecommunications (#PD0187).
Aggravates Newspaper monopoly (#PE0246) Unnecessary gadgets (#PE3745)
Violence against women (#PD0247) Reduced civic awareness (#PJ1835)
Dissatisfaction with skin colour (#PF1741)
Abusive exploitation of cultural heritage (#PC7605)
Production serving false consumption needs (#PF2639)
Incomplete understanding of new societal service systems (#PF2212)
Marketing skills which reinforce a self image of being a victim of circumstances (#PE7752).
Aggravated by Unethical commercial practices (#PC2563).

♦ **PE4226 Politicization of public service in developing countries**
Nature Few governments readily accept political neutrality on the part of their civil servants. They consider that no government employees should be neutral with regard to government aims and objectives and that they should be committed to the basic goals as expressed through the political organizations and processes. Some governments expect the civil/public servants, especially those occupying senior positions, not only to toe the party line but to be active party

members. There is a growing concern for bureaucracy's awareness of and support for the substance of the government's policies, and this requirement is tending to extend an imperative for commitment to the governing party's ideology.
The other factor which has eroded civil service neutrality is the new role imposed on the service by the requirements of development. Development imposes a multiple role on the public servant because what is intended is to emphasize the spirit of cooperation that should characterize the public service. Development requires that civil servants come out openly in support of government policies and even try to legitimize these policies to the public so that they will have a real chance of success. Political neutrality of public servants denies society the benefit of making full use of the few educated and enlightened people who are concentrated in the public services. Anonymity ceases to be a virtue and public servants, especially those in the upper echelons, are encouraged, indeed required, to come out of their protective shells and acquire conviction through active participation in politics. In fact, in some cases they are even required to assume a proselytizing role in the interest of recruiting the widest political support for the party and its ideology.
Incidence Politicization of the public-service system has general implications. First, party and bureaucratic positions tend to be interchangeable. Second, public servants are required to be highly politicized and may even engage in proselytizing activities, with the result that their role is no longer limited to the execution of policy. Third, as noted above, promotions tend to be granted not only on merit and seniority principles but also according to the individual's contribution to the activities of the party. Fourth, low-ranking public servants may accordingly enhance their prospects for promotion by becoming active agents of the party. Fifth, senior public servants get a chance to improve the quality of policy-making since they are better educated and better trained than many of the politicians. And finally, party positions may in the end be filled, to some degree at least, on the basis of education and long service.
Broader Obstacles to community achievement (#PF7118).

♦ PE4229 Fatigue due to torture
Nature For years after the events victims of torture experience fatigue following the least exertion.
Incidence In a resent survey of torture victims, 14 percent of them suffered from fatigue or asthenia.
Broader Fatigue (#PA0657) Neurological effects of torture (#PD4755).

♦ PE4230 Underparticipation of socialist countries in international data systems
Nature The socialist countries participate only marginally in the international on-line data-base market. As of 1981, only 4 or 5 data bases were internationally accessible in the socialist countries.
Broader Inadequate statistical information and data on problems (#PF0625).
Aggravated by Ineffective data systems (#PF3671).

♦ PE4231 Excessive consumption of salt
Incidence As an indication, the market for salt in the UK in 1986 was £19 million.
Refs Iwai, Junichi *Salt and Hypertension* (1982).
Broader Excessive consumption of specific foodstuffs (#PC3908).

♦ PE4232 Discrimination against men in sports
Broader Discrimination against men (#PC3258).
Related Discrimination against women in sports (#PE0197).
Aggravated by Athletic competition (#PE4266).

♦ PE4234 Sensory aphasia
Nature A person is unable to perceive words in a conversation, although he or she can hear the sound of the speech.
Broader Language disorders (#PE3886).
Related Agnosia (#PE5150) Motor aphasia (#PE0345).

♦ PE4235 Blotch diseases in plants
Nature Blotches are irregular or indefinite large or small necrotic areas on leaves or fruit. The symptoms are intermediate between blights and leaf spots.
Broader Ascomycetes (#PE4586) Deuteromycetes (#PE4346).
Narrower Alternaria diseases in plants (#PJ1685).

♦ PE4241 Imbalance of population growth between developed and developing countries
Unequal global distribution of family planning education and facilities
Nature In 1988 the world's population reached five billion. The figure suggests a veritable population boom, but the explosion is not being heard everywhere. Throughout the 80's, the population growth rate will be negative or near-zero in most European countries, the US and Japan. In both the US and France, the birth rate plunged from 3.6 to 1.8 births per woman (the replacement rate is 2.1). In Germany the figure has evolved from 2.6 to less than 1.3. Italy is lower than Germany; Spain, Greece and Portugal are following suit. In communist countries the story is the same. Hungarians, Czechs, and Bulgarians are reproducing at the same level as people in the free-market countries. The USSR is keeping up a birth rate equal to renewal of generations, but only because of the Asian and Muslim communities. The only exception is Poland, where the rate is a bit higher than minimum for renewal. The results will be an elderly population and diminished resistance to Third World migration. Constant medical advances are prolonging life. There will be more women among the elderly population and the number of very old people (80 and up) will create medical and social problems. Migrants are settling in developed countries and their population is becoming younger. In developing countries, the rates differ dramatically. In ten years, the birth rate in China has plunged by 50 percent. Hong Kong, Singapore, Taiwan and South Korea are nearing the level of renewal. Thailand, the Philippines, Malaysia and even Indonesia seem to be moving in the same direction. In Latin America, countries like Mexico and Brazil are experiencing brutal falls in birth rates. At the same time, in other developing areas: these include sub-Saharan Africa, Pakistan, Bangladesh, Burma, Vietnam, Cambodia, Afghanistan, Haiti and others the projected growth rate is so high as to be scarcely believable. Africa has an annual increase of 2.9 percent; Latin America has an increase of 2.7 percent; and Asia and Oceania has an increase of 1.9 percent. These are among the poorest and least stable countries.
Refs Caldwell, John C (Ed) *Persistence of High Fertility*.
Broader Imbalance (#PA0224).
Aggravates Increasing development lag against population growth (#PF3743).
Aggravated by Population decrease (#PF6441) Declining birth rate (#PD2118)
Ineffective population control (#PF1020) Unsustainable population levels (#PB0035)
High human fertility in developing countries (#PF0906).
Reduced by Unequal mortality of the elderly between countries (#PE4354).

♦ PE4242 Exclusion of women from athletic competition
Broader Discrimination against women in sports (#PE0197).

♦ PE4243 Inadequate conditions of work of agricultural workers
Inadequate working conditions of peasant farmers — Inadequate working conditions of plantation workers — Inadequate working conditions of agricultural workers
Nature Plantation workers are cut off from all civilization: there is no medical assistance, no trade unions, 16 hours of work and a small salary that pays the food from the store of the plantation owner.
Refs ILO *Improvements of the Conditions of Life and Work of Peasants, Agricultural Workers and Other Comparable Groups.*
Broader Disincentives against farming (#PD7536)
International imbalance in the quality of life (#PB4993).
Narrower Inadequate working conditions for land tenants (#PE8633).
Aggravated by Plantation agriculture (#PD7598)
Insufficient social security in the agricultural sector (#PD9155).

♦ PE4245 Neglected young children
Inadequate infant welfare — Deficient infant care — Infant mortality due to parental abuse and chronic neglect
Nature Newborns and infants, particularly within the least advantaged socio-economic strata, are at particular risk of developmental deficit or disability, from injury, disease, malnutrition and lack of access to health care during their earliest life.
Incidence A 1982 study of violent and suspicious childhood deaths suggests that two to three children under 6 years of age are dying each week in New York City as a result of parental abuse or chronic neglect and the failure of private and city agencies to intervene. City officials acknowledged that confirmed cases of child abuse and neglect in the city had soared 140 per cent in the last decade to more than 20,000 children, of whom 112 died of parental abuse or neglect. The report said most of the deaths occurred in impoverished families with high rates of criminal activity and domestic violence and with one or both parents affected by drugs, alcohol, mental illness or retardation. Nearly all the families had extensive contacts with government and private agencies responsible for detecting, reporting and helping to prevent the abuse and neglect of children.
Refs Greenland, Cyril *Preventing CAN Deaths* (1987).
Broader Neglected children (#PD4522) Inadequate child welfare (#PC0233)
Victimization of children (#PC5512).
Aggravates Infanticide (#PD3501) Infant mortality (#PC1287)
Unbalanced infant diets (#PE0691) Sudden unexpected infant death (#PD1885)
Inadequate results of formal schooling (#PF6467).
Aggravated by Drunkenness (#PE8311) Family rejection of children (#PC8127)
Dehumanization of health care (#PD7821)
Inadequate health care in family planning (#PD1038)
Inadequate medical care for pregnant women (#PE4820)
Inappropriate application of traditional values (#PF2256).
Reduces Preoccupying child care (#PU0956).
Reduced by Non-parental custody of children (#PF3253).

♦ PE4246 Discrimination against women in athletic training
Nature The very traits that set women apart as successful athletes: an aggressively competitive nature and natural strength and size, are the same that are generally regarded as masculine and therefore not to be cultivated by women. This attitude in the West along with the belief that women are physically inferior result in fewer training facilities, less money and poor quality coaching for women athletics.
Broader Discrimination against women in sports (#PE0197)
Discrimination against women in education (#PD0190).
Aggravated by Athletic competition (#PE4266).

♦ PE4247 Accidents of animals
Broader Ill-defined health conditions in animals (#PD9366).
Narrower Drowning (#PG2857) Wounds of animals (#PG9237)
Electric shock in animals (#PG7234) Thoracic trauma of animals (#PG9516).

♦ PE4249 Congenital anomalies of genital organs
Refs Money, John *Sex Errors of the Body.*
Broader Human physical genetic abnormalities (#PD1618).
Related Human pseudo hermaphroditism (#PE2246).

♦ PE4250 Drug abuse among athletes
Doping
Nature A number of athletes dope themselves by using drugs to artificially and temporarily increase their natural capacity. Doping is a fraudulent method of obtaining an advantage over competitors and it is the cause of accidents that are often severe and sometimes fatal. A wide variety of drugs have been used, either singly or in combination, including: digitalin, because a champion normally has a congenitally slow pulse rate; strychnine, ephedrin and caffeine, to delay the onset of fatigue; and anorexic drugs, to aid an athlete to lose weight so that he can get into a lower weight group. In most common use, however, are amphetamines and related drugs, which are taken as stimulants; or tranquillisers, which include barbiturates and meprobromate, to combat anxiety, particularly in shooting competitions. Even strong analgesic drugs such as morphine or dextromonoramide (palfium) may be used to reduce pain due to excessive effort, as in cycling, or due to injury, as in boxing. The ease with which chemical tests for doping can be made has tended considerably to restrict the use of these drugs, but in recent times they have been replaced by the corticoids (cortisone affects the cerebral nerve cells and also plays a role in resisting stress, thereby enabling competitors to maintain long periods of effort); and especially by the anabolic steroids, which by synthesizing the tissue substances, particularly proteins, promote the growth of muscle volume when taken during training. The latter substances are sometimes known as 'muscle fertiliser'.
Incidence Drugs used may be pain-killers which give rise to euphoria, and suppress the sensation of fatigue which is, physiologically, a danger sign. The athlete surpasses himself - which is the object he is aiming for - but he has made too great an effort and thereby suffers from acute overstrain. In other cases, in a cycle race for example, euphoria makes the athlete oblivious to danger, and reflex changes (or more serious mental disorders) may lead to serious accidents. Finally, amphetamines contribute to the development of hyperthermia which can result in heat-stroke, sudden collapse during a race, psychological disorders, heart failure and coma. The use of corticoids can lead to duodenal ulcers, oedema, intermittent or premature heartbeats, osteoporosis, and mental troubles. Continued heavy dosing with anabolic steroids can lead to aggressive character modifications, hypertension, hypertrophy of the prostate (with a risk of cancer), and alterations in the liver - in some cases even to adenomas or to cancer of the liver. Testosterone, which is a natural anabolic hormone, may lead to sterility. There is also a risk of virilism in the sportsman or arrest of growth in adolescents.
Claim Since antiquity, man has always sought to improve his natural performance; however, it is the public's craze for sporting records together with the progress of pharmacology that have made doping a social evil during recent years. Secretly and widely used during sporting competitions, reputedly miraculous products are often sold at exorbitant prices by the trainers. It has been called the 'cancer of sport', and what is more important, young people are often tempted to follow the example set by the professionals.
Refs Meer, Jeff *Drugs and Sports* (1987); Weiner, Betty *Abuse in Sports* (1985).

Broader Drug abuse (#PD0094).
Narrower Drug abuse in sports (#PJ9593).
Aggravated by Athletic competition (#PE4266) Abuse of coca and cocaine (#PD2363).

♦ PE4256 Reduction of glacier size

Nature With the rise in the global mean temperature, glaciers tend to retreat. Although no global trends are evident, there is some indication that glaciers in wet maritime environments have tended to increase in mass in recent decades whereas some glaciers in dry, continental climatic areas have shown a tendency to lose mass.
Incidence The alpine ice masses have been reduced by about 50 percent since 1850.
Related Degradation of mountain environment by leisure activities (#PE6256).
Aggravated by Global warming (#PC0918).
Reduces Icebergs (#PE1289) Hazardous glaciers (#PE6824).

♦ PE4258 Social maladjustment of children of migrants

Nature Children of migrants in Europe and the US face a variety of social, political and economic barriers before being integrated into these societies. The language programmes provided by host countries, with the exception of Sweden, are fairly useless. Indeed there is a lack of any kind of orientation, let alone language lessons, before or after arrival for most countries. This results in the migrants remaining at the bottom of a stratified economic and social system rather than gradually narrowing the gap between them and their hosts. Many migrant families and their host countries assume that the migrants will return to their home countries. For the most part this is illusionary. Many young people from immigrant families are less active in applying for apprenticeships, additional schooling or jobs than their host country counterparts. Numerous children attend less schooling in the host countries, achieve lower capacity in the host country's language and are culturally and socially unsure of themselves. Parents may force values and practices from the home country that are at odds with the host country's culture, placing the students in the position of having to reconcile the differences. This results in youth who are not socially and culturally adept in either country. The process of adaptation is often counter-acted by spatial and social segregation.
Refs Moorehead, Caroline (Ed) *Betrayal* (1989).
Broader Personal and social maladjustment (#PC8337).
Narrower Lack of appropriate education for children of immigrants (#PE5020)
Inhibition of personality development in exiled children (#PE7931).
Aggravated by Inadequate education for nomadic children (#PE6206)
Inadequate cultural integration of immigrants (#PC1532).

♦ PE4260 Television violence

Effect of violence on television and films — Effect of television and film violence
Refs Gunter, Barrie *Dimensions of Television Violence* (1985); Milavsky, J Ronald et al *Television and Aggression* (1982); Signorielli, Nancy and Gerbner, George *Violence and Terror in the Mass Media* (1988); Voort, T H A van der *Televison Violence* (1986).
Broader Human violence (#PA0429) Culture of violence (#PD6279)
Excessive television viewing (#PD1533).
Aggravates Torture (#PB3430) Human torture (#PC3429)
Psychological conflict (#PE5087)
Attitude manipulation of children through play (#PF2017).

♦ PE4261 Excessive consumption of fats

High fat diets — Fatty food — Consumption of excessive saturated fats
Nature Diets high in saturated fat and dietary cholesterol are linked to heart disease, strokes, diabetes, some cancers and exacerbate diseases related to overconsumption of alcohol. The choice of diet influences an individual's long-term health more than any other single factor.
Incidence Of 2.1 million Americans who died in 1987, a poor diet was associated with the cause of death in two-thirds of the cases. Dietary fat accounts for an average of 37 percent of the daily diet in the USA.
Counter-claim Eskimos living on a traditional diet have the highest fat intake in the world and have none of the diseases associated with high fat diets. The substance conjugated linoleic acid, found in fatty meats and cheeses, is a potent anti-carcinogen. It has a molecular structure that tends to attract and immobilize free oxygen radicals, rare forms of oxygen molecules found in blood and tissue that have been strongly implicated in the initiation of cancers and degenerative conditions like arthritis, heart disease and aging. In 1990 evidence was presented suggesting that reduction of fat intake increased susceptibility to violent death (accidents, suicides or murders) by altering brain chemistry.
Refs FAO *Dietary Fats and Oils in Human Nutrition* (1980).
Broader Dietary deficiencies in developed countries (#PD0800)
Excessive consumption of specific foodstuffs (#PC3908).
Narrower Elevated blood cholesterol (#PE2371).
Aggravates Thrombosis (#PE5783) Malignant neoplasms (#PC0092)
Diseases of the arteries (#PE2684).

♦ PE4262 Sports accidents

Sports injuries — Athletic injuries
Incidence Accidents resulting from sports most often affect limbs (80% of all cases), particularly the lower limbs (50%); while the small proportion of injuries to the skull, thorax or abdomen results in the low mortality rates. But, depending on the country, damage to the spinal cord in sporting accidents accounts for between 3 and 12 percent of spinal cord damage. The most frequent causes are shallow water diving, rugby football, horse-riding, mountaineering, and gymnastics.
Refs Bronsen, Hugo H *Sports and Athletic Injuries* (1985); Riffer, Jeffey K *Sports and Recreational Injuries* (1986).
Broader Tourist hazards (#PE8966).
Related Long-term injuries from sports (#PE5686).

♦ PE4263 Long-term changes in precipitation patterns

Reduction in rainfall — Increase in rainfall
Incidence During the last 30-40 years precipitation has increased in the higher latitude zones (35-70 degrees N), but has decreased in the lower latitudes.
Aggravates Excessive rainfall (#PE4103) Unreliable rainfall (#PD0489).
Aggravated by Global warming (#PC0918) Anthropogenic climate change (#PC9717).

♦ PE4266 Athletic competition

Over competitive sports
Broader Competition (#PB0848) Untransposed community structures (#PF6450).
Related Violent sports fans (#PE6281) Animal fighting sports (#PE4893)
Drug use in animal sports (#PJ1135)
Collapsed tension between care and responsibility (#PF5555).
Aggravates Cruel sports (#PD1323) Drug abuse among athletes (#PE4250)
Long-term injuries from sports (#PE5686) Non-productive athletic activities (#PF4202)
Discrimination against men in sports (#PE4232)
Discrimination against women in sports (#PE0197)
Discrimination against women in athletic training (#PE4246)
Excessive claims for human development through sports (#PG4881)
Interference of school athletic activities with education (#PG4256)
Discrimination against women in payment and prizes for athletic events (#PG4236).
Aggravated by Short duration of athletic careers (#PG4186)
Gambling on sports and athletic competitions (#PE4576)
Politicization of international sports events (#PF4761)
Corruption of sports and athletic competitions (#PE3754)
Excessive expense of athletic training programmes (#PF4196)
Excessive expense of international athletic competitions (#PF4192)
Commercialization of athletic activities and sports events (#PE4222)
National bias among judges of international athletic competitions (#PF4216).

♦ PE4274 Siderosis

Nature Siderosis is a special form of pneumoconiosis caused by inhaling iron dust over long periods of time. The disease is seen in metal polishers, ochre workers and arc-welders and above all among iron miners. Siderosis is observed only after long exposure to heavy quantities of dust. Sixty percent of miners with siderosis have been exposed for 25 years or more. Probably not more than 5 percent of all miners are so affected.
Broader Pneumoconiosis (#PD2034).
Aggravated by Iron dust (#PE4048).

♦ PE4276 Vulnerability of children to AIDS

Children born with AIDS — Pediatric AIDS
Nature Babies born with AIDS face a short life, during which they will be acutely ill most of the time. Nearly all will die before reaching two years of age. Most child victims are infected before they were born. Recent studies show that approximately half of all HIV-infected women will pass the virus to their babies while still in the womb or during childbirth. Usually, these women are unaware of their infection when they become pregnant. Small numbers of older children are also infected, mainly through contaminated blood received in transfusions or through skin piercing procedures. Adolescents are affected in small but growing numbers, usually through sexual intercourse.
Incidence In some African countries, as many as one third of all cases occur in the very young. In some parts of Africa, 25 percent of urban women of reproductive age are infected with HIV, mening that approximately one in every ten urban children is being born with the AIDS virus.
Broader AIDS (#PD5111) Human physical genetic abnormalities (#PD1618).

♦ PE4280 Dependence of developing countries on foreign insurance

Nature Insurance companies cede part of their risks to reinsurers; thus, a reinsurance company is an insurance company's insurance company. This situation enables insurers to underwrite a larger amount of business than would otherwise be possible, by protecting them both from a series of small losses and from a single big loss arising out of a major catastrophe.
Dependence on foreign reinsurance is much more persistent than dependence on foreign insurance. With an increasing number of countries limiting or excluding direct foreign participation in national insurance markets and restricting the foreign ownership of insurance companies, international reinsurance is gaining in importance in the international arena. Even in countries where the insurance sector has been totally nationalized, dependence on foreign reinsurance cannot be eliminated altogether. This is particularly the case in developing countries, where the premium income is relatively small and where large risks cannot be covered by the premium receipts generated there. Developing countries are concerned about the cost of foreign reinsurance for two main reasons. The first is the cost in foreign currency and the second is the negative impact that excessive recourse to reinsurance has on the growth and development of the domestic insurance industry.
Incidence Reinsurance is increasingly becoming the main sector of international insurance cooperation, and it is in this sector that most of the new problems may arise. The world reinsurance market is at present characterized by the existence of 'over-capacity', that is, by the eagerness with which a very large number of international reinsurers and brokers offer cover for all types of risks on very competitive terms and conditions. So far, this temporary 'over-capacity' - which is often of a cyclical nature - may have assisted the emerging insurance companies of the Third World, since these companies are generally highly dependent upon reinsurance protection, and since they have been able to buy such protection easily, on favourable terms. In the long run, however, such a situation tends to be harmful, for the following reasons: (a) reinsurance terms that are too profitable encourage local insurance companies to reinsure abroad much more than they would normally need to, thus depriving themselves and their national market of the bulk of their premiums; (b) easy placing of reinsurance prevents direct insurance companies from selecting their risks with adequate care, and from rating them correctly; (c) serious doubts must be expressed concerning the solvency of those who make these excessively favourable reinsurance offers, which might result in catastrophic bankruptcies, particularly detrimental to the weaker insurance markets of developing countries.
Broader Vulnerability of developing countries (#PC6189).

♦ PE4284 Blood sporozoa of birds

Broader Protozoan parasites (#PE3676) Parasitic diseases in animals (#PD2735)
Parasites of the cardiovascular system in animals (#PE9427).
Narrower Leucocytozoonosis in birds (#PG6614).

♦ PE4285 Misleading accounting information due to inflation

Nature In an inflationary period, the basic accounting principle of recording all assets, liabilities and transactions at historical cost distorts the relationship between various items in the financial statements to a varying degree, depending upon the prevailing rates of inflation. In the case of transnational corporations, such distortions are often very significant because of differing rates of inflation in the host countries of foreign subsidiaries. Most companies retain historical cost data in their primary financial statements, but a few supplement the basic statements with statements showing date adjusted for changes in the general price level or giving information on the replacement cost of assets. There is no generally accepted practice with respect to accounting for the effects of inflation.
Broader Deception in business (#PD4879) Self interested manipulation of timing (#PF5529).
Aggravates Inadequate international standards of financial accounting and reporting (#PF0203).
Aggravated by Creative accounting (#PE6093).

♦ PE4289 Undermining of multilateral forums by industrialized countries

Incidence A number of developed countries engage in processes to undermine multilateral forums, thereby restricting participation of developing countries in decisions on major international themes.
Broader Misuse of international forums (#PF2216)
International economic injustice (#PC9112).
Aggravates Lack of international cooperation (#PF0817)
Over-centralization of global decision-making (#PF5472).
Aggravated by Inadequate funding of international organizations and programmes (#PF0498).

♦ PE4290 Tax holidays

Nature Many investors insist upon a tax holiday at the beginning of a project. This may, in certain

cases, have relatively small financial implications for governments if the period is limited and if start-up expenses are high and profits low in the first few years. More crucial are the questions of whether or not depreciation and amortization of production expenses are required to be taken from the beginning of the tax holiday period, and whether or not initial losses incurred during the tax holiday can be carried forward after the holiday has ended. The foreign investor will probably want to use that period to build up costs (and losses) that can be used later to lower the tax liability after the holiday period is over. Thus, although the host country authorities may not lose much revenue by granting a short initial holiday, there could be considerable loss of revenue if costs and losses from the holiday period are carried forward to reduce taxes after the holiday has ended. Even if the host country does grant only a short initial holiday, however, it may want to protect itself by establishing a ceiling on the amount of the investment that can be recouped during the holiday period. The value of a tax holiday to the investor will also depend on whether the host government has a tax-sparing treaty with the home country of the investor. If it does not, the investor may simply have to pay to the home authorities what he saves under the terms of a tax holiday. This represents a loss of revenue to the host country and provides no advantage to the foreign investor.
Broader Ineffective tax systems (#PF1462).

♦ **PE4293 Congenital disorders of carbohydrate metabolism**
Broader Diseases of metabolism (#PC2270).
Narrower Galactosemia (#PG4551).   Glycogen storage diseases (#PG5513).

♦ **PE4294 Periods of high crime rate**
Broader Crime (#PB0001).
Narrower Civil crimes committed during war (#PG1112)
Crimes committed during civil unrest (#PG1179)
Crimes committed during high unemployment (#PJ1139).

♦ **PE4296 International adoptions**
Nature The effect of the institution of full adoption in international legal relationships is to completely break the links of the child with his country of origin and remove all traces of such links. Such adoption gives rise to a series of specific problems concerning the circumstances of the abandonment, the consent to the adoption, the capacity for adoption of the potential adopters, the international competence of the judge making the adoption order, and the reliability of the guarantees offered. At the moment, there are no internationally recognized principles on these various points. The result is that the demand for adoption of foreign children often gives rise to trafficking and international legal abductions. It is also noteworthy that there is an absence of legal cooperation between states to settle these various questions, with the result that the adoption of foreign children is currently settled solely at the domestic level.
Broader Inadequate laws of adoption (#PD0590).
Aggravates Trafficking in children for adoption (#PF3302).

♦ **PE4297 Violent interactive toys**
Nature Playthings which can communicate electronically with specially created television programs, enabling youngsters with hand-held weapons to zap onscreen villains and even coordinate battles at home with those depicted on the television. In addition to pre-empting children's imagination in play, these toys actively promote aggressive violence.
Incidence Out of 13 studies conducted in the US on war toys, 12 have found harmful increases in both verbal and physical violence to other children. An additional 34 studies have been completed on violent cartoons on normal children in the US and six other countries. Thirty-one of these report at least some harmful effects.
Broader Dangerous toys (#PE1158).
Related Gender-differentiated toys (#PE4664).
Aggravates Human violence (#PA0429)
Socially disruptive effects of video games (#PF6345).

♦ **PE4299 Discrimination against HIV-infected persons**
Social stigmatization of AIDS-infected persons — Psychological effects of AIDS — Social segregation of AIDS virus victims — Officially endorsed homophobia — Travel restrictions on AIDS-infected persons
Nature In an effort to combat AIDS and avoid being infected governments, institutions and individuals are violating the civil rights of people carrying the Human Immunodeficiency Virus in ways that have more to do with racist, xenophobic and homophobic paranoia than any concern with public health. Public insensitivity to those with AIDS or who have tested HIV positive has aggravated the problem. Infected persons have been expelled from jobs; shunned by work or school mates, employers or neighbours; denied mortgages or life assurance at reasonable premiums; run out of towns or hunted down by the police; restricted entering some foreign countries. Children with the disease have been denied access to schools. Homosexuals have been attacked and discriminated against because of their association with the disease. Many victims are embarrassed by stigma of having a "gay disease" or by having their sexual behaviour or drug usage know by family, friends and acquaintances.
In some countries health officials must report the names of anyone who has tested positive. In others attempts are being made to trace sexual relations back over 10 to 15 years for each victim of the disease. Most of the people who are at risk of testing or who have tested positive conceal the possibility of their having the disease.
Some institutions are being discriminated against. Sperm banks are suffering from a lack of donors. Imported blood has been banned in some countries.
Incidence The US military personnel who test HIV positive are denied access to classified information, denied foreign postings and have their security clearances revoked. Some 1,500 Third World students who receive a grant from the Belgian are required to have blood tests for AIDS and those who decline will have their grants withdrawn. The Bavarian state government requires compulsory testing of AIDS of all Turks, Yugoslavs and Eastern Europeans applying for a residence permit. Africans, Asians, North and South Americans and Australians planning to stay for more than 3 months also have to be tested.
Aggravates Separation of family members (#PE4959).
Aggravated by AIDS (#PD5111)   Social stigma (#PD0884)
Invasion of privacy through testing (#PJ6946).

♦ **PE4301 Collusive tendering**
Bid rigging — Price fixing — Fare-fixing
Refs UNCTAD *Collusive Tendering* (1985).
Broader Economic crime (#PC5624)   Frauds, forgeries and financial crime (#PE5516)
Violations against economic regulations (#PD7438).
Narrower Price fixing in the commodity markets (#PJ5528).
Related Grant-back provisions (#PE5306)   Challenges to validity (#PF1200)
Restrictions on research (#PF0725)   Restrictions on publicity (#PF1575)
Restrictions on adaptations (#PF5248)   Restrictions on use of personnel (#PF3945)
Patent pool or cross-licensing agreements (#PE4039)
Exclusive sales and representation agreements (#PE4581)

Payments after expiration of industrial property rights (#PF5292).
Aggravates Monopolies (#PC0521)   Price warfare (#PJ4045).
Aggravated by Cartels (#PC2512)   Conspiracy against the public (#PF4198).

♦ **PE4302 Incorrect spelling**
Inability to spell
Broader Functional illiteracy (#PD8723).
Aggravated by Complicated spelling (#PF6218)   Instability of orthographic standards (#PF0552).

♦ **PE4303 Malignant neoplasm of digestive organs**
Refs Correa, P and Haenzel, W (Eds) *Epidemiology of Cancer of the Digestive Tract* (1982).
Broader Malignant neoplasms (#PC0092)
Diseases and deformities of the digestive system (#PC8866).
Narrower Liver cancer (#PE3233)   Gastric cancer (#PE7509)
Colon-rectal cancer (#PE9399).

♦ **PE4304 Hazardous industrial installations**
Refs ILO *Major Hazard Control*.
Aggravates Occupational hazards (#PC6716).

♦ **PE4305 Inadequate diversification of loans to developing countries**
Nature An ability to diversify among transnational bank lenders is a key element in improving the bargaining position of a developing country. This ensures more reliable access to private capital markets and more competitive terms for loans. Little systematic conceptual and empirical work exists on the way in which developing countries can maximize their negotiating capability when borrowing from transnational banks. A developing country's ability to diversify among lenders and to obtain better costs and other terms of credit depends on three overall economic conditions at the time of borrowing: (a) the general supply and demand for loanable funds in international capital markets; (b) the country's creditworthiness; (c) the competitive structure of the global banking industry and the motivations behind the external lending of individual transnational banks. All three conditions change over time. The most essential ingredient for developing countries is detailed current information on each condition. Such information is, on the whole, lacking at present.
Broader Compliance*complex (#PA5710)
Excessive foreign public debt of developing countries (#PD2133).
Aggravates Decline in concessional financial resources available to developing countries (#PE3812).

♦ **PE4310 Discrimination in lending by transnational banks in developing countries**
Nature In their local operations in developing countries transnational banks give preference, for instance allocate more credit, to subsidiaries of transnational corporations and joint ventures.
Broader Discriminatory business practices (#PD8913)
Transnational corporation imperialism (#PD5891)
Restrictive practices of transnational banks (#PE9657).
Related Discrimination against transnational banks in developing countries (#PE4320).
Aggravates Capital shortage in developing countries (#PD3137).
Aggravated by Control of industries and sectors by transnational corporations (#PE5831).

♦ **PE4314 Endangered species of flowering plants**
Incidence In Europe conservationists estimate that some 1,000 flowering plants are in danger, of which only 244 are officially recognized as endangered.
Broader Endangered species of plants (#PC0238).

♦ **PE4315 Interference of transnational banks' off-shore borrowing with domestic monetary policies**
Nature A matter of concern to governments, both of developed and developing countries, is the effect on domestic monetary policies of off-shore borrowings of foreign branches for their parent banks. For example, where central banks emphasize the use of reserve requirements to control aggregate increases in the money supply and credit, and when such inflows are not included in the liability base for the requirement, these can give rise to unexpected credit expansion.
Broader Restrictive monetary practices (#PF8749)
Transnational corporation imperialism (#PD5891)
Restrictive practices of transnational banks (#PE9657).

♦ **PE4319 Denial of defendant's right of silence**
Nature Innocent people held in police custody can be frightened, disorientated and at a low physical and mental ebb, which can lead them into false confessions.
Incidence Among western democracies, the UK has abolished the right to silence in Northern Ireland.
Broader Denial of right to freedom from testifying against oneself (#PE6633).

♦ **PE4320 Discrimination against transnational banks in developing countries**
Nature Banking regulations and other banking policies of developing countries reflect the infant industry protection approach. The objective of these policies is twofold: to minimize the access of transnational banks to local savings; and to promote national financial institutions. Access to local savings by transnational banks is curbed by imposing restrictions on the number, location and services of their branch offices. A variety of supplementary policies are designed to promote national institutions. The restrictions on transnational banks range from outright prohibition of any further branches or only the rarest of exceptions, to outright prohibition, which restricts foreign branches to one per transnational bank and limits their location to the nation's capital.
Broader Discriminatory business practices (#PD8913)
Protectionism in the banking industry (#PD7120)
Discrimination against foreign companies (#PD6417).
Related Antiquated regulations in the banking industry (#PF4370)
Control of industries and sectors by transnational corporations (#PE5831)
Discrimination in lending by transnational banks in developing countries (#PE4310).
Aggravates Limited availability of loans in developing countries (#PE4704).
Reduced by Restrictive practices of transnational banks (#PE9657).

♦ **PE4325 Interference of transnational banks in domestic economic policies**
Nature Transnational banks generally seem to prefer to lend to a country that is in agreement with the International Monetary Fund about the policies to be pursued under stand-by agreements. There are two reasons for this. First, lending to a nation that is in compliance with International Monetary Fund arrangements helps the transnational banks with their home country banking authorities. Second, without directly setting economic policy, the banks can be reassured that it meets the rather well-known requirements of the International Monetary Fund for balance-of-payments solvency. When a nation is unwilling to submit to International Monetary Fund stand-by agreements and commitments, however, the banks very often make their exposure conditional on commitments as to the particular economic policy that the borrowing country will put into effect in order to secure further loans.
Broader Antiquated regulations in the banking industry (#PF4370).

♦ **PE4328 Nitrogen overdosage of plants**
Aggravated by Nitrogen compounds as pollutants (#PD2965).

♦ **PE4329 Increase in insurance claims for medical negligence**
Nature Medical negligence claims in the developed world have become so numerous that doctors often refuse to take difficult cases or perform unnecessary procedures rather than risk the danger of a lawsuit. Other doctors are paying exorbitant rates to insure themselves against this risk. These factors contribute to already skyrocketing health costs, without providing any appreciable increase in the quality of healthcare.
Largely because of the increase in negligence claims one in four births in the US are performed by Caesarean section. Also in the US insurance companies are talking about premiums of $50,000 a year for midwives who have an average income of $25,000 to $30,000 a year. Many general practitioners are refusing to have anything to do with delivering babies.
Broader Negligence (#PA2658) Prohibitive medical expenses (#PE8261).
Related Prohibitive cost of product liability protection (#PE4404).
Aggravates Excessive medical intervention in childbirth (#PE7705)
Medical practitioners refusing to treat patients (#PE5027).
Aggravated by Inadequate medical care (#PF4832)
Unethical medical practice (#PD5770).

♦ **PE4330 Domination of restrictive project loans by transnational banks in developing countries**
Incidence Banks grouped by size of assets, show that those with the smallest assets have an inclination to provide the riskier types of loan: free disposition and refinancing.
Background Project loans are for a bank the relatively safest form of finance. Such loans are tied to a precise activity in which there is an economic return to cover repayment which can be evaluated on an ex-ante basis. Moreover, project loans carry a substantial element of built-in discipline because the use of resources is tied to project proposals, and project implementation may also benefit from the expertise of an established private foreign or local supplier or contractor. Characteristics similar to project lending can also be found in the financing of capital goods imports. Offering greater risk are freely disposable loans and refinancing credits. The riskiest loan is one of free disposition, since a bank has no ex-ante insurance that the resources will be employed wisely and generate returns sufficient to permit repayment.
Broader Restrictive monetary practices (#PF8749)
Restrictive practices of transnational banks (#PE9657)
Control of industries and sectors by transnational corporations (#PE5831).
Related Antiquated regulations in the banking industry (#PF4370).

♦ **PE4331 Unsustainable short-term improvements in agricultural productivity**
Nature Efforts to improve agricultural productivity in the short-term may be designed to succeed at the expense of long-term sustainability, whether this takes the form of ecological stress, loss of genetic diversity in standing crops, salinization and alkalization of irrigated lands, nitrate pollution of ground-water, or pesticide residues in food.
Aggravates Alkaline soil (#PD3647) Soil salinization (#PE1727)
Pollution of groundwater (#PD2503) Pesticide residues in food (#PE6480)
Natural environment degradation (#PB5250)
Declining productivity of agricultural land (#PD7480)
Decreasing genetic diversity in cultivated plants (#PC2223).
Aggravated by Short-term gain (#PF8675) Short-term profit maximization (#PF2174)
Short-term planning of product life cycles (#PF1740)
Domination of government policy-making by short-term considerations (#PF0317).

♦ **PE4332 Denial of right to freedom of religion of indigenous peoples**
Prohibition of the practice of indigenous religion — Repression of indigenous spirituality — Inappropriate transfer of forms of worship
Nature Indigenous are prohibited from practising their religion, subjected to forced conversions, and may have their children taken away and placed in missionary schools.
Incidence During the missionary period, foreign liturgies, hymns and rites were imported into Africa, for example, thus stifling indigenous spirituality. In countries dominated by Buddhism, Hinduism, Islam and tribal religions, relating Christian worship to such religious traditions constitutes a major difficulty.
Claim Prefabricated and imported liturgies reflect particular cultural traditions, which are not necessarily suited to indigenous cultures. In such cultures there may be emotional depths to which such liturgies cannot reach. Any such liturgy is bound to be a frustration to the worshipper.
Broader Denial of religious liberty (#PD8445)
Discrimination against indigenous populations (#PC0352).
Related Imbalance in shipbuilding industry distribution (#PE5863).
Aggravates Forced religious conversion (#PD6637)
Endangered indigenous cultures (#PC7203).

♦ **PE4335 Restrictions of home countries on transnational banking activities in developing countries**
Nature Since 1974, banking authorities in the home countries of transnational banks have voiced increasing concern about the growth of bank assets in developing countries. Granted that there is some basis for this concern, home countries could facilitate flows to developing countries by widening the pool of potential lenders instead of restricting it. Home governments could also help to increase the information about developing countries available to small banks, thus encouraging the move away from the dominance of the largest transnational banks. If home banking authorities doubt the capacity of the smaller banks to evaluate risk, they might consider permitting them to rely on co-financing schemes with multilateral development banks such as the World Bank or the regional development banks. This means that co-financing may require renewed support as a major method of reaching the second tier of banks with international operations, many of which are smaller transnational banks.
At times, home country authorities have little incentive to facilitate the flow of funds to developing countries. Their primary concern is to protect the home financial system. Even those among them that are becoming more familiar with country analysis may at times have little understanding of the needs of developing countries for finance or of the development process. Thus, greater direct contact between home country banking authorities and borrowers from developing countries could have a salutary effect. A precedent for this exists in the cooperative efforts of central banks from countries in the pacific Basin. In some cases, and to the extent that the policies of the transnational banks reflect those of their home country governments, it may be useful to draw the attention of those governments to the possible drawbacks of tied lending of any sort for borrowers in developing countries. The transnational banks that restrict themselves to financing the exports of their home countries are tying financial flows in a manner reminiscent of tied development assistance. As transnational banks with special relationships to their home governments supplant transnational banks which do not show such ties, the posture of their home governments toward lending to developing countries will become more important.
Home country governments can also affect the context in which transnational banks operate in numerous indirect ways: examples of this are their management of aggregate real demand and their foreign exchange controls.
Related Control of industries and sectors by transnational corporations (#PE5831).

♦ **PE4339 Exploitation of regulatory loopholes in countries with underdeveloped legislation**
Evasion of national law regulations
Broader Irresponsibility (#PA8658) Evasion of the law (#PD4208)
Ineffective legislation (#PC9513).
Aggravated by Inadequate national law enforcement (#PE4768)
Deliberate governmental avoidance of legislative reform (#PF5736)
Limited local respect for regional and global legislation (#PF2499).

♦ **PE4343 Revealing national security information to a foreign power**
Criminal espionage
Nature Criminal espionage is the act of revealing information to a foreign power or their agents with the intention of the information be used to cause danger to one's own country. In the time of war it is gathering, recording or communicating information of national importance to the enemy.
Broader Espionage (#PC2140) Crimes related to national security information (#PE3997).
Related Military espionage (#PD2922) International political espionage (#PC1868).

♦ **PE4344 Inadequate coordination of governmental representation in intergovernmental organizations**
Nature Member States of intergovernmental bodies are obliged to organize a network of representation at the headquarter offices of the bodies and to coordinate their activities in relation to numerous committees. The complexity and often the relative unimportance of the problems dealt with makes it impossible for a central administration to give instructions on all subjects. Furthermore the foreign ministries responsible for this coordination do not have all the powers they need to supervise the representatives of technical ministries. The result is that in a very large number of cases, in spite of considerable paperwork, the personality of the representatives has greater impact on the official position taken up by governments than the directives sent from the capitals themselves. Representatives of the same government may then take different positions according to the organization to which they are accredited. In order to overcome the difficulties inherent in the complexity of the whole, delegations are forced not only to learn how the machinery conditions the interplay of influence, but also to follow closely the most trivial administrative matters.
Aggravated by Inefficient location of facilities of international organizations (#PE3538).

♦ **PE4346 Deuteromycetes**
Fungi imperfecti
Broader Fungal plant diseases (#PD2225).
Narrower Blights (#PE3919) Cankers (#PE0640) Leaf spots (#PE1954)
Fruit spots (#PG1121) Leaf scorch (#PG3786) Rots in plants (#PE3363)
Galls in plants (#PE3715) Moulds in plants (#PE8051) Snowmold of lawns (#PG0875)
Sweet potato scurf (#PJ1915) Blackleg of crucifers (#PG4759)
Sooty mould in plants (#PE0916) Scab diseases in plants (#PE1040)
Wilt diseases of plants (#PE1056) Needle casts of conifers (#PG1795)
Blotch diseases in plants (#PE4235) Spot anthracnose in plants (#PG4914)
Botrytis diseases in plants (#PG1668) Fusarium diseases in plants (#PG5461)
Verticillium wilt in plants (#PG1352) Alternaria diseases in plants (#PJ1685)
Anthracnose diseases of plants (#PE1910).

♦ **PE4349 Refusal to testify**
Nature During an official government proceeding including judicial, legislative or administrative a person who refuses to answer questions after being directed to do so by the legal authority is guilty of the crime refusal to testify.
Broader Contempt of judicial process (#PD9035).

♦ **PE4354 Unequal mortality of the elderly between countries**
Incidence The life expectancy of men aged 65 is over 16 years in Japan; 15 to 16 years in Cuba, Greece, Switzerland and Canada; 14 to 15 years in France, Spain, Sweden, USA, Argentina, Australia, Sri Lanka, the Netherlands, Portugal, and Austria; 13 to 14 years in Denmark, Federal Republic of Germany, Italy, Finland, Uruguay, Belgium, England and Wales; 12 to 13 years in Eire, Northern Ireland, Scotland, German Democratic Republic and Poland.
Broader Inequality in mortality rates (#PC9586)
Unequal morbidity and mortality between countries (#PC6869).
Reduces Imbalance of population growth between developed and developing countries (#PE4241).

♦ **PE4355 Domination of loan negotiations by transnational banks**
Nature Many developing countries seek to diversify their borrowing sources as far as possible. This is thought to ensure more reliable access to credit markets and the most competitive cost and maturity structure. Little systematic conceptual and empirical work has been done, however, on the way in which developing countries can maximize their negotiating capability when borrowing from transnational banks. A developing country's ability to diversify external loans and obtain better costs and terms depends on three overall economic conditions at the time of borrowing: (a) the nation's debt level and debt-servicing capacity; (b) the general supply of or demand for loanable funds in international capital markets; and (c) the characteristics of the transnational banks and the competitive structure of the global banking industry. All three conditions change over time. The most essential ingredient from the point of view of the developing countries is detailed current information on each condition.
Broader Restrictive practices of transnational banks (#PE9657)
Control of industries and sectors by transnational corporations (#PE5831).

♦ **PE4358 Front organizations**
Nature Front organizations are credible cover operations for other activities which are illegal, unfashionable or philosophically eccentric. Some front organizations use issues of genuine public concern or needs for social service as a springboard to advance their particular causes. Others use a cover of providing needed services in order to perform or to fund illegal activities.
Incidence The Lyndon LaRouche political network, based in the United States, uses companies such as Campaigner Publications, Fusion Energy Foundation, New Benjamin Franklin House Publishing Company, Club of Life, and the Executive Intelligence Review, to fund and promote fanatical conspiracy theories.
Broader Fraud (#PD0486) Deception (#PB4731).
Narrower Misuse of business enterprises as front organizations by government (#PE1651)
Misuse of nonprofit associations as front organizations by government (#PE0436).
Related Bogus public interest groups (#PE7575).
Aggravated by Covert intelligence agency operations (#PD4501)
Cooptation of private sector initiatives by government (#PE0780).

♦ **PE4360 Socially irresponsible programmes of transnational banks in developing countries**
Nature Lending across borders to borrowers in developing countries is now big business for many transnational banks. To the extent that the international market offers fewer opportunities for growth than in the early 1970s, many transnational banks recognize the importance of developing countries as a market. However, although transnational banks are increasingly engaged in

socially responsible programmes in their home countries, they have not yet adequately extended similar programmes to the new market, although there are some examples (very few in number) in which transnational banks have extended this line of activity abroad. The growing importance of developing countries as a market for transnational banks, especially for those with constituent or integral operations in developing countries, requires the extension of socially responsible programmes to these countries, such as, for instance, financing of training courses, small businesses and the like.

Broader Social irresponsibility of transnational corporations (#PE5796).
Related Irresponsibility (#PA8658)
Antiquated regulations in the banking industry (#PF4370).
Aggravates Burden of servicing foreign public debt by developing countries (#PD3051).
Aggravated by Restrictive practices of transnational banks (#PE9657).

♦ PE4362 Computer-based crime

Risk of computer fraud — Computer-aided fraud

Nature Computers facilitate the theft of money and property and the destruction of data when there are inadequate controls against their misuse. Crimes perpetrated by unauthorized access to keyboards, terminals and communications devices generally can be described as thefts, misapplications of assets, or destruction of information. These terms may apply to the misappropriation of money and real property, or of proprietary information and intangible assets. The misuse of the computer may involve the forgery of computer signatures such as authorizing codes; the creation of false accounts payable to disburse cheques; improper use of personal information; the creation of "virus" or "rogue" programs which interfere in software operations and destroy data. All of these crimes include programming the erasure of any evidence of the computer crime perpetrated. Probably the fastest growing category of computer related crime is that involving electronic fund transfer systems. The most significant types of computer crime were: arson, sabotage and malicious damage of computer installations; system penetration, or "hacking"; unauthorized use of computer time; thefts of assets, including software; embezzlement of funds; defrauding of consumers and investors; and destruction or alteration of data (including college transcripts and diplomas) and software. The motive is usually personal financial gain, anger or revenge but another significant impetus is 'the intellectual challenge' associated with computer crime.

The absence of, or inadequate provision for, documentation and access controls for computer installations, facilitates computer crime. Unauthorized access to software and hardware is almost exclusively the means of crime perpetration. With authorized access, but with criminal collusion, two or more persons may commit crimes unnoticed, until financial audits, inventories, and computer operation system checks uncover the fraud or misuse. In the case of theft of intangible properties such as computer-stored patents of engineering, chemical or other designs, processes, or marketing and strategic data, the crime is exposed, if at all, by inferences drawn from the activities, products or knowledge shown by competitors.

Incidence Studies in the United States indicate that about one-third of such crimes were committed by staff or consultant data-processing personnel, almost exclusively below management level; but the bulk were committed by non-data-processing personnel with normal, job-related access to computers. A recent United States study has concluded that computer-related crime now rivals white collar crime in cost and seriousness. The report was based on a survey of 283 corporations and government agencies and among the conclusions were the following: about 48 percent of those surveyed reported some form of computer crime during 1983 with total annual losses estimated to be $145 million to $730 million. More recent estimates based on actual reported crime place losses in the US at $3 billion per annum. An Australian computer expert has estimated that there were about 4000 computer-related frauds over the period 1975-1983. Pranksters in Canada in the past have re-routed the entire delivery system for Pepsi-cola; in 1971 the New York-Penn Central Railroad Company discovered 200 of its box-cars had been re-routed and ended up near Chicago and another 200 cars were found to be missing. In New Jersey seven young people, all under 18 years, were charge with conspiring to use their home computers for exchanging stolen credit card numbers, information on how to make free phone calls, and to call coded phone numbers in the Pentagon. They were found with codes capable of changing the position of communication satellites. One youngster had run up a large phone bill at home and when berated by his parents, he proceeded to break into the phone billing system and cancelled the charge.

Few computer crimes are actually reported. In the US is required by law to report computer crimes; many commentators agree that only about 15 percent of computer-related crimes are actually reported. The Australian Computer Abuse Research Bureau argues that only one in 20 cases of the $2 million or so worth of computer-related offences in Australia during 1980 was in fact reported. The reasons given are that there is not much faith in the legal system and its ability to prosecute a case of computer crime successfully; that companies fear that to declare publicly that their system has been breached and that their assets are not as secure as once thought will cause a flight of capital in shareholder's funds and deposits; and that there is a reluctance to expose the company's records and systems to public scrutiny and competitors.

In the UK in 1989, it was estimated that the cost to industry of computer-based crime was over 400 million pounds per year. The average annual incidence was 9 incidents per 100 companies (rising in some cases to 1 in 2), costing on average 46,000 pounds per incident. Other estimates put the cost of computer crime as high as 2 billion pounds per year. The most vulnerable sector is the communications industry in which there were 192 incidents per 100 companies. Fraudulent input of information, notably on payroll systems, accounted for 4.1 million pounds of losses.

Refs Asimov, Isaac, et al (Eds) *Computer Crimes and Capers* (1983); BloomBecker, Jay (Ed) *Computer Crime, Computer Security, Computer Ethics* (1986); Chesterman, John and Lipman, Andy *The Electronic Pirates*; Cornwall, Hugo *Datatheft* (1987); OECD *Microelectronic, Robotics and Jobs* (1983); Sieber, Ulrich *The International Handbook on Computer-Related Crime and the Infringements of Privacy* (1986).

Broader Economic crime (#PC5624) Theft of services (#PD4711)
Abuse of computer systems (#PD9544).
Narrower Illegal long-distance telephone calls (#PE4367).
Related Risk (#PF7580).
Aggravates Theft of data (#PD2957) Computer viruses (#PD3102).
Aggravated by Vulnerability of computer systems (#PE8542)
Unreliability of computer software (#PE4428)
Unauthorized access to computer information systems (#PE9831).

♦ PE4363 Excessive imposition of states of emergency

Undeclared states of emergency

Nature The declaration of states of emergency by governments allow them to suspend, limit or abolish essential judicial guarantees which can degenerate into serious, systematic violations of human rights. Excessive numbers of such declarations can become commonplace and are liable to become perennial.

Incidence While in some cases states of emergency are justified, the following countries were reported in the period between August 1987 to November 1988 to have proclaimed, extended or continued states of emergency: **Af** Cameroon, Chad, Egypt, Lesotho, Senegal, Sierra Leone, South Africa, Sudan, Zimbabwe. **Am** Argentina, Bolivia, Chile, Colombia, Ecuador, El Salvador, Haiti, Honduras, Nicaragua, Panama, Paraguay, Peru, Suriname. **As** Bangladesh, Brunei Darussalam, Burma, Israel, Jordan, Malaysia, Pakistan, Singapore, Sri Lanka, Syrian AR, Taiwan (Rep of China). **Au** Fiji, Papua New Guinea. **Eu** France, Turkey, UK.

Broader Maltreatment of civilians (#PD5560).
Aggravates Suspension of rights during states of emergency (#PD6380).

♦ PE4365 Inadequate channels for direct investments between developing countries

Nature An international framework to facilitate the flow of financial resources among the countries of the Third World could significantly strengthen their resource base. At present, many oil-producing countries have a surplus of financial resources that require investment outlets; in the future, there may be other developing countries in a similar position. Nevertheless, the organized investment outlets today are almost exclusively in the industrialized countries, a situation that results in a net transfer of capital from the third world to these countries. The oil-producing countries have responded to the needs of other developing countries through an encouraging expansion of official aid flows, but they have still too few opportunities for productive investments that afford them both security and an adequate rate of return. To some extent, multilateral financial institutions like the International Monetary Fund and the World Bank have served as a channel for the recycling of the surpluses, and the petro-dollar has played a similar role. Although there is scope for the continuance and even strengthening of these links, there is also a need for some direct mechanisms for providing the investing countries with access to investment outlets in the developing countries. Such a development will hardly occur spontaneously; it will need to be facilitated by new mechanisms and institutions within the Third World that would guarantee adequate rates of return and security of investment. The establishment of appropriate instruments could constitute an important element in a system of cooperation among developing countries.

Incidence To cover growing deficits, non-oil exporting countries in the Third World and some socialist countries have increasingly turned to the private international capital markets, which have re-cycled some of the OPEC surplus funds to them and have additionally lent them other funds at high rates of interest that found no borrowers in the industrial countries where investment has been low. The extension of these loans and particularly their roll-over rescheduling to finance the growing debt service when the borrowers are unable to pay have become the basis of stringent economic and political conditions that the private banks and/or the International Monetary Fund (IMF) acting as their intermediary have imposed on Third World (and some socialist and developed) countries. The standard 'conditionality' to the IMF package that governments are obliged to accept in their 'letter of intent' before being certified to receive further loans always includes devaluation of the currency, reduction of government expenditures especially on consumer subsidies and popular welfare, the reduction of the wage rate through various devices, and more favourable treatment for private and especially foreign capital. These conditions have sometimes led to 'IMF riots' as the people sought to resist the enforced curtailment of their standards of living. It has been said that the IMF has overthrown more governments than Marx and Lenin put together. An important political economic consequence, if not rationale, of these IMF promoted government policies in the Third World is to promote 'export led growth' by cheapening Third World labour and its fruits for international capital and foreign importers (by lowering the price of Third World wages and currencies) and to lend support to the domestic forces in these Third World countries that have an economic interest in export promotion. Thus, the international financial system and the financing of the Third World debt serves to perpetuate the mechanism of the emerging international division of labour based on Third World export promotion. The political consequences of all these economic policies are that it is necessary to repress the labour force in order to keep wages low or to reduce wages.

Broader Capital shortage in developing countries (#PD3137).

♦ PE4367 Illegal long-distance telephone calls

Nature A fast-growing multimillion-dollar crime is selling and making illicit long-distance telephone calls. The fraud usually begins with a computer hacker who penetrates a telephone company's electronic files and extracts secret billing codes, the numbers used to determine who makes a call and who to bill for the call. The hackers then sell the code to middle men who distribute them to hustlers. The hustlers at bus, rail and air terminals seek out prospective customers at banks of pay telephones with offers of calls of unlimited duration at black-market rates. The victims are the long distance phone companies and their millions of customers who pay for the losses.

Incidence In the US some $500 million is lost to illegal calls.

Broader Computer-based crime (#PE4362)
Inappropriate use of telecommunications services (#PE4450).

♦ PE4369 Christian antisemitism

Nature Christians, since their own breaking away from Judaism at the time of the apostles, have been opposed to Judaism that its proponents did not do likewise. To them, the Jews are guilty of deicide, treachery and failure to recognize the Messiah. This prejudice has stood at the root of the social practices of antisemitism for centuries.

Broader Antisemitism (#PE2131).

♦ PE4371 Dietary torture

Nature The use of food and drink to torture people is quite common. Prisoners are given inadequate or no food or drink. Soapy or salt water or urine may be the only liquid available to the victim. In other cases, victims are forced to drink large quantities of water. They may be forced to eat excrement. Religiously or culturally forbidden foods, like pork for Jews or Muslims or meat for Hindus may be served after long periods of starvation.

Incidence Dietary torture has been reported in the following countries: **Af** Cameroon, Djibouti, Mali, Mauritania, Morocco, Namibia, Rwanda, South Africa, Tanzania UR, Uganda, Zaire. **Am** Argentina, Bolivia, Canada, Colombia, El Salvador, Guyana, Paraguay, Peru, Suriname. **As** Afghanistan, Bangladesh, Indonesia, Israel, Malaysia. **Eu** Italy, USSR, UK, Yugoslavia.

Broader Physical torture (#PD6734).
Aggravates Malnutrition (#PB1498).

♦ PE4372 Endometriosis

Refs Schenken, Robert S *Endometriosis* (1988); Teoh, Eng-Soon, et al (Eds) *Endometriosis* (1987).

♦ PE4378 Chlorofluorocarbons as an environmental hazard (CFC's)

Nature Chlorofluorocarbons are used in industry for cleaning electronic components, filling refrigerators and air-conditioning systems, producing plastic foams, and propelling aerosol sprays. CFC's slowly diffuse upward into the stratosphere, where intense ultraviolet radiation causes the molecules to release free chlorine atoms. These free chlorine atoms break down ozone. Since the chlorine atoms are regenerated in the reaction, they are free to repeat the cycle significantly depleting the ozone layer, which protects the Earth from the harmful effects of excess ultraviolet rays from the sun which can cause skin cancer and damage crops.

Incidence CFC's currently are critical chemical in 100 million refrigerators, 90 million cars and trucks, 40,000 supermarket display cases, and 100,000 commercial building air conditioners. The total ban of CFC's will render useless or require alterations to capital equipment valued at $135 billion.

Background Fluorocarbon 12 has 20,000 carbon dioxide's capacity to trap heat in the atmos-

phere. It is widely used in automobile air-conditioning units. Its pre-industrial level was 0 and the 1986 level was 400 parts per trillion. In 1986 the annual increase was 5 percent.
Fluorocarbon 11 has 17,500 carbon dioxide's capacity to trap heat in the atmosphere. It is widely used in commercial refrigerators, and in insulating foam in which it is trapped. Its pre-industrial level was 0 and the 1986 level was 230 parts per trillion. In 1986 annual increase was 5 percent.
Claim Only some of the 17m metric tons of CFCs produced by the end of 1988 is released in stratosphere, lots more is still locked up in foam and refrigeration units etc.
Counter-claim The scientific community simply does not have the capacity of predicting accurately what will happen if CFC's continue to be released into the atmosphere. Also there is no evidence at all that any person has yet been harmed by letting more ultra-violet through from the sun, because this has not been measured in populated regions.
Broader Chemical air pollutants (#PD1271)
Increase in trace gases in the atmosphere (#PD1354).
Aggravates Stratospheric ozone depletion (#PD6113).

♦ PE4380 Illiteracy among women
Illiteracy of women
Nature Girls and women are still in the minority in education in many countries and the proportion of girls and women tends to decrease progressively as the level of education rises. Several economic, social and cultural factors impede, in different contexts, the access of girls and women to, and their participation in, education. The number of illiterate women is actually increasing, due not only to the rise in population but also to the fact that schoolchildren who have not been able to consolidate their knowledge sufficiently, relapse into illiteracy. Efforts to improve community sanitation, economic production, child welfare, family planning, and housing, to name a few, are badly impeded by the cost and logistical difficulties of educating women who are illiterate about development possibilities. Further, female illiteracy confirms society's worst notions about the inutility of women in the community. Because of their lack of education, women can be the most insidious opponents of precisely those programs intended to improve their lives and those of their children. They are often not permitted to attend community training sessions and meetings, where their ignorance can embarrass community members in front of government officials or other agents of change. Their questions, left unanswered, turn into passive resistance to new practices.
Incidence The proportion of women illiterates is steadily growing. In 1960, 58 percent of illiterates were women; by 1970 this percentage had risen to 60 percent. In those ten years the number of illiterate men rose by 8 million and that of illiterate women by 40 million. World statistics for 1980 indicate that the male/female adult illiteracy rate was 23/34 in the world, 50/71 in Africa, 30/45 in Asia, 18/23 in South America, 9/11 in Oceania, 33/38 in Europe and 1/1 in North America. It is projected that the number of illiterate women will reach an estimated 539 million by 1990.
Claim Two out of three of the world's illiterates are women. In many parts of the world women are shown to have no business, indeed no right, to associate with books.
Broader Illiteracy (#PC0210) Wasted woman power (#PF3690)
Illiteracy in the fourth world (#PD6645).
Narrower Illiteracy among women in developing countries (#PE8660).
Aggravates Denial of right of family planning (#PE5226).

♦ PE4386 Improvisational housing in developing countries
Makeshift dwellings in developing countries
Nature Human settlements research shows that the true architects, builders and planners of Third World settlements are the people themselves. In rapidly growing cities, most additions to the housing stock are made outside any plan or officially approved housing project. They are largely built by the people themselves, since there is no other way they can get a house. Many houses or shacks are technically 'illegal', that is, they do not meet official standards and are usually built on illegally occupied or subdivided land because most city dwellers cannot afford to rent or buy a house, room or housing plot. In small towns and rural areas, acquiring a plot on which to build is usually less of a problem and people manage the construction of their own house, still often building most or all of it themselves.
The task confronting policymakers is not to envisage large publicly funded house construction programmes – in the past, such programmes have failed to solve the poor's housing problems. The cost of each unit so produced is too high for the low income majority. If subsidies are given so the poor can afford them, few houses can be built and those that are constructed rarely go to those most in need. If the houses are not subdivided, then they are too expensive for all but the rich. While people can manage their own housing construction more cheaply and efficiently than any government agency, they have difficulty obtaining certain key resources. In cities, perhaps the most difficult resource to get is a legal housing plot on which to build. In all settlements, people need water, sanitation, garbage removal and roads, footpaths, electricity and basic social services. They may also need technical advice and small loans. These necessities are too short supply and their cost is far too high: it is here that policymakers need to focus their attention.
Incidence The UN estimated that in 1970, four out of every five households in the developing countries lived in low-quality accommodation in the rural areas and urban slums. The period 1970–80 witnessed a further deterioration, with their numbers increasing 14 percent. Nor were ancillary services more adequate in terms of needs. It was estimated that around 1980, less than 50 percent of rural population of the developing world had access to safe drinking water, while nine out of ten rural households lacked any form of waste disposal facilities, while only 4 percent of all rural households in Africa and 15 percent in Asia had access to electric supply. International recognition of the gravity of the problem is evidenced by the declaration of 1987 as the "International Year of Shelter for the Homeless" and the inauguration of the "International Drinking Water Supply and Sanitation Decade".
Claim The strategies, levels of investment and types of technology hitherto adopted are incapable of meeting the basic needs of shelter and ancillary services in the rural areas, given the high growth rates of rural population.
Broader Inadequate housing in developing countries (#PE0269).
Aggravated by Homelessness in developing countries (#PD8856).

♦ PE4391 Elimination of jobs due to automation in developing countries
Elimination of jobs in developing countries due to introduction of new technologies
Nature The assembly operations will be increasingly automated in electronics industry. This, in turn, implies a drastic reduction of low-skill job opportunities, combined with a drastic increase in the investment outlays required per new job created.
Refs ILO *Automation in Developing Countries* (1974).
Broader Elimination of jobs by automation (#PD0528).

♦ PE4393 Subsidence from mining
Nature Building and in some cases whole villages or towns have been abandoned because of the sinking and shifting of ground surface due to sub-surface mining. Mines under towns can cause extensive damage to buildings, roads, and the gas, water and sewerage systems.
Broader Collapsing physical structures (#PD4143).

♦ PE4394 Detraction
Sullying reputations
Nature Detraction does not aim to rob a man of his honour, but to blacken his reputation. It springs from envy rather than from anger.
Broader Verbal abuse (#PD5238).

♦ PE4395 Television viewer fatigue
Reduces Exploitative commercial television (#PD0433).

♦ PE4396 Borderline personality disorders
Refs Gunderson, John G *Borderline Personality Disorder* (1987).
Broader Self disorders (#PF4843) Personality disorders (#PD9219).
Related Neurosis (#PD0270) Psychoses (#PD1722) Schizophrenia (#PD0438).

♦ PE4398 Frustrated talent in government posts
Lack of scope for intellectual ability in the public sector
Broader Lack of job satisfaction (#PF0171).
Aggravates Abuse of government employment (#PE4658).
Aggravated by Governmental incompetence (#PF3953)
Mediocrity of government leaders (#PF3962)
Limited ways of matching talent and jobs (#PF2792)
Underutilization of intellectual ability (#PF0100)
Proliferation of public sector institutions (#PF4739).

♦ PE4399 Denial of right to a people to their own means of subsistence
Broader Denial of right of a people to be self-determining (#PC6727).

♦ PE4404 Prohibitive cost of product liability protection
Nature The producer, importer or supplier of a product passes on to the consumer the high cost of protecting himself against litigation for damage cause by a defective product. Legal costs include insurance and fees for lawyers. In the United States, awards by courts for injury has risen at an exorbitant rate. Defensive measures increase the cost of a product. Additional and unnecessary medical tests might be conducted to protect against medical malpractice suits. Industries may be made internationally uncompetitive because of these increased costs.
Refs Bass, Lewis J D *Product Liability* (1986); Cartwright, Robert E and Phillips, Jerry J *Products Liability* (1986); Cartwright, Robert E, et al *Products Liability* (1986); Freedman, Warren *International Products Liability* (1986); Herzog, Peter E *Harmonization of Laws in the European Communities* (1983); Phelan, Richard J and Ross, Kenneth *Product Liability* (1984).
Broader Prohibitive cost of insurance (#PE9632)
Prohibitive administrative overhead costs (#PF8158).
Related Increase in insurance claims for medical negligence (#PE4329).
Aggravated by Defective product manufacture (#PD3998).

♦ PE4407 Sham executions
Nature A frequent form of psychological torture is mock or sham executions of prisoners. Prisoners have a gun pointed at them, the trigger is pulled and there is no bullet in the chamber. In other cases, prisoners are partially asphyxiated by wrapping their heads in a wet towels or plastic sheets or having their heads held under water which sometimes contains excrement, urine or other filth added.
Incidence Sham executions has been reported in the following countries: **Af** Lesotho, Namibia, Somalia, South Africa. **Am** Argentina, Bolivia, Canada, Colombia, El Salvador, Paraguay, Peru, Uruguay. **As** Iran Islamic Rep, Iraq.
Broader Death threats (#PD0337).

♦ PE4409 Outlaws
Background The welfare and the existence of a community against outside world and its enemies depended on the maintenance of the peace that is safeguarded by custom and reason. The one who broke the peace had no more place within the community which he had imperilled and polluted. He had to be slain or expelled. Outlaw was totally expelled from human society and treated like an animal. Thus, in its earlier forms outlawry included all punishments in one.
Claim Outlaws are fugitives from the law. Turning them into romantic anti-heroes standing up to the social and psychological oppression of the establishment through the myths of Bonnie and Clyde, Robin Hood and Jessie James or into evil non-humans through the over-responses of Western governments and news-media has, to a large degree, given certain types of crime legitimacy. One man's terrorist is another's freedom fighter. Indiscriminate killing, while not acceptable, is understandable. When outlaws are seen as people who are criminals, outside the law, anti-social and brutal misfits, society will find a way of progressing toward effective responses to their dangerous activities.
Aggravates Banditry (#PD2609) Terrorism (#PD5574)
Ineffective prevention of terrorism (#PF4240).
Aggravated by Boycott (#PE8313) Ostracism (#PF1009)
Social outcasts (#PD6017).

♦ PE4413 Pests and diseases of poplars
Refs FAO *Breeding Poplars for Disease Resistance* (1985).
Broader Pests and diseases of trees (#PD3585).
Related Pests and diseases of tea (#PE2980) Pests and diseases of oak (#PE2984)
Pests and diseases of elm (#PE2982) Pests and diseases of cocoa (#PE2979)
Pests and diseases of palms (#PE2981) Pests and diseases of rubber (#PE2977)
Pests and diseases of olives (#PE2978) Pests and diseases of coffee (#PE2218)
Pests and diseases of chestnut (#PE2983) Pests and diseases of citrus fruit (#PE2976)
Pests and diseases of deciduous fruit (#PE3591).

♦ PE4414 Mucopolysaccharide diseases
Nature Mucopolysaccharide, MPS, diseases are inherited from healthy parents who have no idea they carry the genetic defect. The diseases are rare and life-threatening which cause severe progressive handicap, often both physical and mental. Victims of theses diseases are unable to produce certain enzymes necessary for the correct chemical changes to take place in their bodies, consequently complex sugars become stored in all types of connective tissue causing progressive damage. They appear normal at birth and by the time the disease manifests itself sufficiently to be recognized they are toddlers most of whom will die before reaching adulthood. There are six main types: Hurler's Syndrome, Hunter's Syndrome, Sanfilippo Syndrome, Morquio Syndrome and Maroteaux-Lamy Syndrome and three associated diseases: Mucolipidosis, Fucosidosis and Sialic Acid Disease.
Broader Genetic defects and diseases (#PD2389).
Aggravated by Excessive consumption of sugar (#PE1894).

♦ PE4415 Snoring
Noisy sleeping habits
Nature The snorting or groaning sounds made in the breath while sleeping is an important symptom of many health disorders, causes stress in marriage and in its extreme form may cause

death. It is the sound generated when the airway is obstructed. It can be an indicator of severe allergies, respiratory diseases, insomnia or even depression. It grows worse as people age, and afflicts men twice as often as women. It can cause irreparable damage to the tissues of the throat, making the snoring even noisier and affecting breathing during the day.
The worst snoring disorder is sleep apnea, an extreme form of snoring in which the victim chokes hundreds of times a night. More common in overweight people, sleep apnea can put enough strain on the heart and lungs to cause or aggravate heart disease, and sometimes it kills people. It also causes daytime drowsiness and, in some case, narcolepsy – the tendency to fall asleep suddenly during the daytime.
Snoring is caused or aggravated by the shape of the face, tongue and uvula, lack of sleep, consumption of alcohol and caffeine and position of sleeping. It may be alleviated by loss of weight, muscle-toning drugs or a uvulopalatopharyngoplasty, a form of surgery that scoops out extra tissue that blocks the trachea. Smoking tends to moderate the problem by inducing muscle tone in the throat.
Broader Symptoms referable to respiratory system (#PE7864)
Unsociable human physiological processes (#PF4417).

♦ PE4421 Stress among children
Childhood burnout — Stressed-out kids — Overworked children — Academic stress in children
Nature Stress-related problems in children are increasing. Children's hospitals report more injuries associated with repetitive training exercises. Second and third grade children are being referred to paediatricians for attention problems and hyperactivity as a result of being stressed and being expected to do things they are not ready to do. Examinations and preparation for examinations begin before school age, in preschool or kindergarten. Parents expecting high results from children as young as two or three, schools developing curriculum to meet these expectations, textbook publishers that see preschoolers as a growth market and increasing competition within schools and internationally contribute to levels of stress for children from preschool age through university.
Claim There is no evidence that the academic instruction found in many kindergartens will produce long-term gains in achievement. In fact, the competitive activities in nursery schools can do lasting damage: the children may become chronic underachievers and engage in delinquent behaviour, they may suffer from mental disturbance or emotional disorder.
Refs Humphrey, James H (Ed) *Stress in Childhood* (1984); Youngs, Bettie, B *Stress in Children* (1986).
Broader Juvenile stress (#PC0877) Stress in human beings (#PC1648).
Related Marital stress (#PD0518).
Aggravates Juvenile suicide (#PE5771).
Aggravated by Competitiveness in education (#PD4178).

♦ PE4426 Masturbation
Autoeroticism
Claim Whether practised by females or males, masturbation – the stimulation of one's own sexual organs for pleasure with or without reaching orgasm – is condemned by Roman Catholic church as a grave moral evil, because it is not directed toward procreation. Earlier male masturbation was viewed as comparable to abortion, because it was thought that the male seed contained the whole offspring.
Broader Sexual deviation (#PD2198).
Aggravates Masturbation guilt (#PF5609).

♦ PE4428 Unreliability of computer software
Computer bugs — Programming defects — Hazardous computer software errors — Unreliability of computer processing — Risk of computer error
Nature A computer programme may consist of millions of lines of computer code, written by hundreds of people who each work on small segments of the programme and an error as tiny as a misplaced semicolon can cause a system to malfunction. Corporate computer programmers spend 80 percent of their time repairing software and updating it to keep it running. Software projects are typically 100 percent over budget and a year behind schedule. The best programmers can be 25 times as competent as the worst, with many software design supervisors unable to evaluate or even understand their programmer's' work. If management changes during the development of a programme, the programmers may find that their product is unwanted when it is completed. Confusion also develops in data-processing departments because programmers have written computer code without documenting how they approached the problem; the individual maintaining the programme 10 or 20 years later, after the initial programmer has left, is out of luck. Such problems are likely to grow as industry and the military increasingly rely on software to run systems of phenomenal complexity.
Incidence Examples include: patients given fatal doses by malfunctioning hospital computers, 22 fatal crashes by the fly-by-wire UH-60 helicopter, 104 failures in a day at a single air traffic control location in 1989, and the failure during the 1970s and the 1980s of observation satellites to detect atmospheric ozone depletion (due to a programming error). A specific example of the problems with software is in its use in the Strategic Defence Initiative. An SDI programs must make certain assumptions about target and decoy characteristics; but those characteristics are controlled by the attacker. It must also make assumptions about the structure of the attack. Those assumptions make it easier to overload the system by using an attack strategy that violates those assumptions. Realistic testing of the integrated hardware and software after deployment is impossible, and there will be no opportunity to modify the software during or after its first battle. Bugs have been the cause of death or maiming to individuals, serious financial harm to corporations, and nearly caused the collapse of the U.S. government-securities market.
Claim Computer systems are inherently flawed and too unreliable for critical or vital tasks. They cannot be designed without the ever-present threat of life-endangering malfunctions because their very complexity makes thorough testing for errors impossible. The way they are built means that they are prone to total catastrophic failure, rather than partial failure. As they become more complex, so the level of the catastrophe increases. For these reasons they are dangerous when used in sensitive areas such as intensive care wards, the nuclear industry, air trafic control and early warning and strike command systems. Checking a typical power station's computer programme to ensure it is error free under all conditions would take software testers literally trillions of years.
Refs Ward, Robert *Debugging* (1986).
Broader Human errors and miscalculations (#PF3702).
Narrower Unreliability of weapons systems (#PF7801).
Related Electromagnetic pollution (#PD4172).
Aggravates Computer-based crime (#PE4362) Pervasive fear of nuclear war (#PC3541)
Vulnerability of computer systems (#PE8542).
Aggravated by Computer viruses (#PD3102)
Excessive reliance on infallibility of equipment (#PE4742).

♦ PE4429 Malpractice in plastic surgery
Untrained cosmetic surgeons — Incompetent plastic surgery — Cosmetic surgery quackery
Nature Cosmetic operations – breast implants, nose, chin and eye restructuring, facelifts and the like – are being performed by surgeons who are are not trained in plastic surgery, resulting in painful disfigurement. These services are often promoted by salespeople who allow the customer inadequate time to check the credentials of the surgeons or companies involved.
Incidence In the 1989 over 600,000 Americans had plastic surgery according to the American Society of Plastic and Reconstructive Surgeons (ASPRS), probably more than twice as many operations actually took place. ASPRS requires its members to have a medical degree, to have completed three years' surgical training and a three year residency in plastic surgery, and to have passed written and oral tests. But membership in ASPRS or any other recognized body is required to practice plastic surgery. Some companies offer to train doctors how to do chemical peels, a legitimate procedure to remove wrinkles or facial scars, and to market their services in one week. In the UK anyone who is qualified as a doctor is eligible to carry out cosmetic operations. Plastic surgeons having a certificate of accreditation from the Royal College of Surgeons spend a minimum of eight years in training.
Claim Quite apart from essential cosmetic surgery for accident or birth defect victims, the body of people who seek such radical means of altering their appearance are already a pitiable lot indeed. The last thing such folks need is to be further victimized by criminals of this type.
Broader Health fraud (#PD9297) Limited medical knowledge (#PD9160).
Aggravates Mutilation and deformation of the human body (#PD2559).
Aggravated by Cult of youth (#PF6766) Unethical medical practice (#PD5770).

♦ PE4430 Political prisoners in mental institutes
Nature People are forcibly confined for months or years to psychiatric hospitals for political reasons. Sometimes they are punished by the used of powerful drugs such as triftazin, sulfazin and aminazin. In special psychiatric hospitals, where regimes are harsher, inmates are severely beaten by convicted criminals employed as orderlies.
Incidence Repression of political dissent though abusive internment has been used in many countries, prominent amongst which is the Soviet Union, where people have been detained in psychiatric institutions for expressing criticism of the authorities or for wanting to leave the country.
Broader Political prisoners (#PC0562) Political injustice (#PC2181).
Aggravates Mental illness (#PC0300).
Aggravated by Political violence (#PD4425).

♦ PE4431 Hormonal disorder
Hormone disturbance
Refs Keller, Paul J *Hormonal Disorders in Gynecology* (1981).
Aggravates Acne (#PE3662) Sexual deviation (#PD2198)
Periodontal diseases (#PE3503).

♦ PE4432 Black lies
Nature Telling an untruth and attributing it to a false source.
Broader Lying (#PB7600).
Related Grey lies (#PF3098) White lies (#PF7631).
Aggravates Breach of promise (#PF7150) Exploitation of trust (#PC4422).

♦ PE4437 Abuse of tourism for sexual purposes
Travel for sexual purposes
Nature Tourism for the purpose of casual homosexual or heterosexual encounters, particularly in the third world involves billions of dollars and hundreds of thousands of people.
Broader Prostitution (#PD0693) Tourist hazards (#PE8966)
Social environmental degradation from recreation and tourism (#PD0826).
Aggravates AIDS (#PD5111).
Aggravated by Sexual immorality (#PF2687).

♦ PE4438 Intrusive animal-rights campaigners
Intolerant antivivisectionists
Nature As regards human beings, the animals rights issues are a matter of consciousness-raising, attempting to bring to human consciousness new respect and concern and to change behaviour toward animals. In order to make the required leap of attitudes, animals rights activists find it necessary to over-correct their views, making animal consciousness not just one issue among the many which a person is concerned about, but rather the major focus of existence. Very few other issues—most of them religious in some respect—consistently drive adherents to the public abuse of offenders and to violent confrontation and terrorist activities.
Incidence The animal rights movement is a loose coalition of groups ranging from the Society for the Prevention of Cruelty to Animals, which uses advertising and telephone campaigns to encourage better treatment of animals to extremist groups like the Animal Liberation Front, whose members believe that animals are entitled to the same rights as people and that violent and illegal tactics, like fire-bomb attacks on stores selling fur products, are justified to disrupt cruelty against animals. These extremist groups insist on strict vegetarianism, denounce any use of animals in laboratories and even foreswear wool clothing because sheep can be nicked in shearing. The US FBI has declared the Animal Liberation Front a terrorist group.
Claim To eat flesh and wear the fur and skin of other beings is to proclaim one's own barbarity, leaving oneself open to treatment befitting barbarians.
Broader Extremism (#PB3415).
Narrower Hunt sabotage (#PE4454).
Related Denial of rights of animals (#PC5456)
Violence by fanatical environmentalists (#PD5582).
Aggravated by Maltreatment of animals (#PC0066).

♦ PE4439 Employment of criminals in policy-making contexts
Employment of war criminals in policy-making contexts
Nature Individuals (often of considerable recognized competence) convicted as criminals may subsequently be employed in policy or decision-making contexts calling for a sense of social responsibility. Although such individuals may have fulfilled any obligations through carrying out the sentence, they constitute an unnecessary risk in sensitive situations and may take advantage of the situation to further their own ends in ways dangerous to society.
Background Such employment raises issues which are fundamental to the relation between crime, punishment and what constitutes adequate retribution, and whether a criminal can be considered to have been rehabilitated on completion of a sentence. It also raises the question as to what activities judged as terrorism are to be considered (or will come to be considered) justifiable in terms of the ends to which such terrorism is purportedly directed.
Incidence As an example, the executive director of a steering and policy committee of the USA Congress in 1989 ("arguably the most powerful staff member on Capitol Hill") had been convicted in 1973 for assaulting and wounding a woman and served 27 months of a 15 year sentence, 7 of which were suspended.
Counter-claim When an individual has paid his debt to society for past failures, he should be considered trustworthy until the contrary is proven. It is questionable whether a conviction for assault in the distant past affects a person's judgement in decision-making situations, especially since many in decision-making situations may themselves have participated in military actions calling for even greater violence, or allowing indulgence in such violence. The government leadership of a number of countries, including Israel, acquired their positions through violent

actions judged (before their success), as terrorism.
Broader Neglect of victims of crime (#PD4823)
Legal discrimination in favour of offenders (#PD9316).
Narrower Government approved employment of war criminals (#PE4697).
Related Post-revolutionary re-employment by government of security services of the ousted repressive regime (#PF1015).
Aggravates Inappropriate policies (#PF5645).
Aggravated by Recidivists (#PE5581).

♦ **PE4446 Non-payment of reparations by government**
Delay in payment of government reparations — Delay in governmental payment of compensation
Incidence In 1988, the US government agreed to pay $120 million a year for 10 years in reparations to Americans of Japanese descent who were interned during World War II. President Reagan then requested from Congress $20 million for the payments, one sixth of the amount promised. The House of Representatives doubled the figure, still only one third of the amount promised. In 1989, the Senate appropriated nothing for that year but created an entitlement beginning in fiscal 1991, so the money would be paid automatically. In 1990 Austria agreed to pay $25 million in reparations to Jewish survivors of the Holocaust.
Broader Government inaction (#PC3950) Uncompensated damages (#PD7179)
Arrears in payment of government financial commitments (#PF1179).
Narrower Non-payment of compensation for forced relocation (#PE8898).
Related Delay (#PA1999).
Aggravated by Delay in administration of justice (#PF1487).

♦ **PE4447 Damage to goods**
Damaged merchandise — Risk of damage to stock
Nature The retailer, wholesaler, manufacturer and customer lose from having merchandise damaged. Goods can be damaged by vandals or disgruntled employees, by accidents or by acts of nature.
Broader Property damage (#PD5859).
Aggravated by Fires (#PD8054) Lightning (#PD1292)
Mechanical failure (#PC1904) Injurious accidents (#PB0731)
Radiation accidents (#PD1949) Accidental explosions (#PE3153).

♦ **PE4449 Inadequate assistance to victims of rape**
Broader Legal discrimination in favour of offenders (#PD9316)
Inadequate assistance to victims of human rights violations (#PD5122).
Aggravated by Rape (#PD3266).

♦ **PE4450 Inappropriate use of telecommunications services**
Unethical use of telephone service — Irresponsible use of telecommunications service
Narrower Electronic pornography (#PE5402) Obscene telephone calls (#PE5757)
Inconsiderate telephone use (#PG9378) Illegal long-distance telephone calls (#PE4367)
Invasion of privacy by compulsory telecommunications (#PE0223).
Related Destabilizing international telecommunications (#PD0187)
Misuse of telephone surveillance by governments (#PD1632)
Misuse of electronic surveillance by governments (#PD2930).
Aggravates Proliferation of direct mail advertising (#PE1810).

♦ **PE4451 Dependence of developing countries on unpaid female labour**
Nature A large part of female employment takes place at home in the agricultural and artisanal sectors in the Third World where the female, beside domestic duties, produces marketable goods which are sold by the males of the family. Domestic duties, largely unrecognized as economic activity, includes cooking, driving, cleaning, baby sitting, laundering and ironing. In the agricultural sectors, women plant, water, weed, harvest, process grains, preserve fruit and vegetables, and prepare animal products. In the artisanal sector, most of the output is actually provided by females at home. These include: pickles, home made sandwiches, pasties, pastries, sweets, desserts, home brewed beer, small loom textiles, rugs, towels, and tailored goods. The women who perform these jobs are, for the most part, unpaid, adding billions of dollars to the Gross National Product of every nation of the world.
Broader Unpaid labour (#PD3056) Exploitation of women (#PC9733).

♦ **PE4452 Work practices requiring women to lift heavy loads**
Claim Women who have to lift heavy machinery in their work can miscarry or have problems in labour.
Broader Inadequate working conditions for women (#PD3197).
Aggravated by Overworked women (#PD7762).

♦ **PE4453 Cereal root rots**
Broader Basidiomycetes (#PE0364) Rots in plants (#PE3363).
Related Wood rots (#PE4455) Heart rot fungi (#PE6269) Dutch elm disease (#PE1154)
Rusts of grasses and cereals (#PE2233) Smuts of grasses and cereals (#PE2234).

♦ **PE4454 Hunt sabotage**
Sabotage of hunting expeditions by direct action environmentalists
Nature Dressed in camouflage gear, hunt saboteurs follow hunters as they stalk wild game. When the hunter attempts to shoot an animal, saboteurs blow air horns that frighten away the game throughout the vicinity. Land managers stalk the saboteurs, and the press delightedly stalks them all.
Broader Intrusive animal-rights campaigners (#PE4438).

♦ **PE4455 Wood rots**
Broader Basidiomycetes (#PE0364) Rots in plants (#PE3363)
Household pests (#PD3522).
Narrower Dry rot of wood (#PE1606) Wet rot of wood (#PE4767).
Related Cereal root rots (#PE4453) Dutch elm disease (#PE1154)
Rusts of grasses and cereals (#PE2233) Smuts of grasses and cereals (#PE2234).
Aggravates Heart rot fungi (#PE6269).

♦ **PE4456 Fungal resistance to fungicides**
Refs Groupement International des Associations Nationales de Fabricants de Produits Agrochimiques *Coping with Resistance to Fungicides* (1981).
Broader Drug resistance (#PF9659) Pest resistance to pesticides (#PD3696).
Related Resistant bacteria (#PE6007) Drug resistant viruses (#PG6399).
Aggravates Fungal plant diseases (#PD2225).

♦ **PE4459 Denial of the right to health for indigenous populations**
Broader Discrimination against indigenous populations (#PC0352).
Aggravates Human disease and disability (#PB1044)
Vulnerability of indigenous populations to introduction of diseases (#PE3721).
Aggravated by Denial of the right of health (#PJ2269).

♦ **PE4460 Malformation**
Refs Bergsma, Daniel (Ed) *Limb Malformations* .
Broader Ugliness (#PA7240) Distortion (#PA6790) Nonconformity (#PA5878).
Narrower Human pseudo hermaphroditism (#PE2246).

♦ **PE4463 Obesity due to low metabolism**
Nature Obesity is a body weight that is over 20 percent above what is considered the ideal body weight for a particular height. Many people are obese because they have low metabolism which is inherited.
Broader Obesity (#PE1177).
Related Over-eating (#PE5722).

♦ **PE4465 Reduced activity of malnourished children**
Nature A child's first and instinctive reaction to lack of food – that is, to lack of energy intake – is to reduce energy output. And by conserving health and growth at the expense of activity, the child can even maintain a normal appearance. A European child may spend up to two-and-a-half times as much energy on walking and running as a child in a malnourished community in Africa. The reduced activity of the malnourished child comes at an age when play and exploration of the environment is particularly important to the development of physical and mental skills.
Incidence In 1984 the average estimated energy output for an African child was 79 units compared with 98 units for a European child (Unit = Kcal/Kg bodyweight/day). One study in Mexico has shown that by the age of three years, a group of malnourished children were already one year behind their well-nourished contemporaries in language development.
Broader Protein-energy malnutrition in infants and early childhood (#PD0331).

♦ **PE4466 Terrorizing**
Nature Terrorizing is threatening to commit any crime of violence or dangerous act or falsely informing another that a dangerous situation exists with the intent of keeping another person in sustained fear, causing evacuation of a building, hall, facility or transportation means, or causing serious other public inconvenience. This includes phone and mail threats as well as bomb scares.
Broader Criminal threat (#PE4661).
Aggravates Intimidation (#PB1992).

♦ **PE4467 Criminal menacing**
Nature Menacing is knowingly placing or attempting to place another person in fear (but not sustained fear) by threatening imminent serious bodily harm.
Broader Harassment (#PC8558)
Offences involving danger to the person (#PD5300).
Related Criminal threat (#PE4661).

♦ **PE4471 Stationary position torture**
Nature A variety of tortures involve placing people in a fixed position and forcing them to remain there. Some victims are hung, straining joints, muscles, and tendons. Victims are hung by their hands, wrists and ankles for minutes or hours at a time. They are tied to a chair and hung upside down. A method call the Parrot's Perch, Pau de Arara, the Roast Chicken, the Swing or the Jack in which the victim is tied into a crouching position, with the arms hugging the knees, a pole being then passed through the narrow gap behind the bent knees and in front of the elbows. The ends of the pole is then rested on two desks or trestles. The victims hangs head down and the tortures can beat him, use electrical prods or other instruments. Some victims are tied in positions which are uncomfortable, cuts off circulation and strains muscles, tendons and joints. The use of arm or leg irons for prolonged periods can lead to disability. They are tied in contorted positions. Victims are made to stand, occasionally in awkward positions, or squat in one position for hours or even days. In Feto, the victim is forced to remain for hours in a foetal position. In the Guardia, the victim is placed upright in a box with holes to enable him to breath. In Secadera, the victim is wrapped in a plastic sheet and placed in a metal cylinder. A technique called the Chancho is where prisoners lie parallel with the floor supported only by the head and the tips of their toes – if they fall or move they are beaten. Some are forced to hold heavy weights above their heads. Prisoners are forced to sit straddling iron or wooden bars which cut cruelly into the groin. Detainees are sometimes locked into cages with several others so that only a few can lay down and the rest must remain standing, this can last up to months at a time.
Incidence Torture stationary position has been reported in the following countries: **Af** Benin, Congo, Djibouti, Egypt, Ethiopia, Mauritania, Morocco, Somalia, South Africa, Tunisia. **Am** Argentina, Bolivia, Brazil, Canada, Chile, Colombia, Haiti, Paraguay, Peru, Suriname, Uruguay. **As** Bangladesh, India, Iraq, Israel, Korea Rep, Pakistan, Sri Lanka, Syrian AR, Taiwan (Rep of China). **Eu** Turkey, UK.
Broader Physical torture (#PD8734).

♦ **PE4475 Maritime fraud**
Barratry
Nature Maritime fraud exists whenever there is intentional deception as to some fact or circumstance in connection with maritime activities which enables the unjust obtaining of money or goods. It frequently involves the misuse of commercial contracts and documents, such as bills of lading, charter parties and marine insurance policies.
Falsification of bills of lading causes the largest proportion of losses due to fraud. The importance of close controls on the preparation and transference of bills of lading arises from their service, not only as receipts, but as proofs of the transport contract and of ownership of the goods. A bogus bill may be sold along with forged accompanying documents to a buyer who may later find that the cargo and perhaps even the ship does not exist. Other major types of fraud include shipowner theft of cargo and/or fraud against insurance companies sometimes involving the scuttling of overinsured vessels in which case additional crimes are perpetrated, such as against personal property and life. Fraud is also committed by shipping agents billing customers for services never performed. There is also falsification of mortgage documents to enable intermediaries to pocket part of the mortgage repayments. Understated invoicing to avoid customs taxation or overstated invoicing to take foreign currency out of the buyer's country in violation of foreign exchange controls make governments victim of these maritime related crimes.
Incidence There has been an alarming increase of reported instances of maritime fraud and related acts occurring in the field of international shipping and trade in the last few years. Losses have been estimated by the shipping community at the equivalent of $1 billion per year. Developing countries are the most frequent victims of maritime fraud because they have not yet acquired the same level of experience in shipping and trade as exists elsewhere. However, the means by which they become victims derive from the same loopholes and inadequacies in the existing structure in which international shipping and trade are carried on.
Broader Fraud (#PD0486) Unethical practices in transportation (#PD1012).
Aggravates Inequities in marine insurance (#PE5802)
Obstacles for international ocean shipping (#PD5885).
Aggravated by Documentary fraud (#PE1110).

♦ **PE4476 Dual exchange rate systems**
Incidence Such systems are especially common in low-income African countries. They effectively tax exports.
Counter-claim Unifying the exchange rate system may help the export sector, but the temporary loss of revenue can lead to larger fiscal deficits and higher inflation.
Broader Mismanagement of exchange rate system (#PF1874).
Reduces Economic inflation (#PC0254).

♦ **PE4477 Short sighted decisions about intersocial interaction**
Nature Organizations concerned with interaction between societies operate out of short range visions of the future. The rich cultural heritages available in most of the world's cities, for example, are used in limited and unimaginative ways. Visions of the future and long-range plans often regard these potential sources of social creativity as dangerous or disruptive.
Broader Fragmented decision-making (#PF8448)
Misuse of international forums (#PF2216).

♦ **PE4479 Static grassroots involvement in planning the economy**
Nature Local people do not actively participate in planning the economy. Capitalism encourages this by limiting forms of participation to consumption, thereby encouraging materialism. Socialism, disfranchises local people encouraging cynicism.
Broader Dominance of economic motives (#PF1913).
Aggravated by Technological monoculture (#PF4741)
Fixation on partial solutions to problems (#PF9409)
Unconvincing alternatives to existing societies (#PF3826).

♦ **PE4481 Disagreeable human body odour**
Perspiration
Incidence As an indication, the market for deodorants and anti-perspirants (excluding perfumes) in the UK in 1986 was £125 million. That for male deodorants was £27.3 million.
Broader Unsociable human physiological processes (#PF4417).
Related Bad breath (#PE6558) Human flatulence in public (#PE1764).
Aggravated by Incontinence (#PE4619) Menstruation (#PE4838)
Inadequate personal hygiene (#PD2459).

♦ **PE4483 Halogen compounds as pollutants**
Nature Fluorine, chlorine, bromine, iodine and astatine make up the family of elements know as the halogens. The gases, fluorine and chlorine and the vapours of bromine and iodine are irritants of the respiratory system. They can damage lung tissue resulting in pulmonary oedema which can prove fatal.
Broader Environmental hazards from chemicals (#PC1192).
Aggravates Blood circulation disorders (#PE3830).
Aggravated by Badly laid out work premises (#PJ2468).

♦ **PE4484 Unlawful recruiting for and enlistment in foreign armed forces**
Nature A person is quality of unlawful recruiting for and enlistment in foreign armed forces if he enters or agrees to enter the armed forces of a foreign nation, or recruits or attempts to recruit another for the armed forces of another nation.
Broader Crimes related to foreign relations and trade (#PE5331).
Aggravates Mercenary troops (#PD2592).

♦ **PE4486 Copper dust**
Nature The inhalation of dusts, fumes and mists of copper salts can cause congestion of the nasal and mucous membranes and ulceration of perforation of the nasal septum.
Broader Metal dust and fumes (#PE5439).
Aggravated by Copper as a pollutant (#PG6587).

♦ **PE4493 Inadequate conditions of work in the hotel and catering industries**
Refs ILO *Conditions of Work and Life of Migrant and Seasonal Workers Employed in Hotels, Restaurants and Similar Establishments. Report II. 2nd Tripartite Technical Meeting fro Hotels, Restaurants and Similar Establishments* (1974).
Broader International imbalance in the quality of life (#PB4993).
Aggravated by Unethical catering practices (#PE6615).

♦ **PE4496 Fraudulent acquisition or use of passports**
Forged travel documents
Nature To obtain a passport through deception or to use a passport issued to another is a crime. The most common practices are substituting a page with a legitimate visa into a passport without a visa, substituting a photograph on a stolen passport and stamping a forged visa on a passport.
Incidence In one instance, a Portuguese consul stole 2000 passports which he sold to wealthy Chinese who used them to enter the EEC as Portuguese citizens from Macau. Officials believe that a high proportion of the 50,000 Dutch passports that go missing each year end up with forgers.
Broader Misleading information (#PF3096)
Crimes related to immigration, naturalization and passports (#PE3889).
Related Commercialization of nationality acquisition (#PF0699).
Aggravated by Unethical practices in transportation (#PD1012).

♦ **PE4497 Thrush in chickens**
Thrush in turkeys — Candidiasis — Moniliasis — Sour crop
Broader Fungal diseases (#PD2728)
Diseases of the digestive system in animals (#PD3978).

♦ **PE4507 Denial of right to enjoyment of arts**
Related Denial of right to benefits of science (#PF6077).

♦ **PE4508 Lack of international coordination among supervisors of financial stock markets**
Need for increased cooperation of central banks
Aggravates Global financial crisis (#PF3612) Disruption of financial markets (#PD4511)
Securities and commodities exchange violations (#PD4500).

♦ **PE4513 Nausea**
Vomiting
Refs Davis, C J, et al *Nausea and Vomiting* (1986).
Broader Symptoms referable to digestive system (#PE9604).

♦ **PE4514 Flies as vectors of diseases**
Flies as vectors of disease
Refs Greenberg, Bernard *Flies and Diseases* (1970); Zumpt, F *Stomoxyine Biting Flies of the World* (1973).
Broader Disease vectors (#PC3595) Insect vectors of disease (#PC3597).
Narrower Screw-worm flies as pests (#PE8150).
Aggravates Dysentery (#PE2259) Onchocerciasis (#PE2388).

♦ **PE4515 Surface to surface missiles**
Ballistic missiles
Nature The development, sale and use of ballistic missiles with ranges from less than a hundred kilometers to several thousand kilometers. These weapons can equipped with nuclear, chemical, biological or conventional war heads.
Incidence The major producing nations of missile systems are The United States, The Soviet Union, China, Brazil and France. In September 1989, at least 22 Third World nations were trying to buy or build ballistic missiles. Many had succeeded. Among the nations building, buying, attempting to acquire missiles or have advanced aerospace industries are: **Af** Liberia, South Africa. **Am** Argentina, Brazil, Canada, USA. **As** China, India, Iran Islamic Rep, Iraq, Israel, Japan, Korea DPR, Korea Rep, Saudi Arabia, Syrian AR, Taiwan (Rep of China). **Eu** Belgium, France, Germany FR, Netherlands, Sweden, USSR, UK.
Broader Competitive acquisition of arms (#PC1258).
Aggravates Chemical warfare (#PC0872) Nuclear arms race (#PD5076)
Biochemical warfare (#PC1164) Proliferation of strategic nuclear arms (#PD0014)
Inadequate governmental energy conservation policies (#PF0037).

♦ **PE4522 Inadequate international marketing of jute products of developing countries**
Nature The jute industries in Asian countries have no organized system for market information, analysis, planning and actions. Several factors inhibit proper marketing techniques and strategy: (a) exports are handled by a very large number of mills and merchant exporters; (b) there is no close contact between the the users in the importing countries and the manufacturers in the exporting countries, nor is there adequate consumer service; (c) vigorous and intense marketing efforts require large funds which the industries in the exporting countries cannot afford; (d) up-to-date knowledge of the market and existing market surveys are insufficient; (e) there is not enough cooperation between consumers and manufacturers on development and introduction of new and improved jute products; (f) there can be wide variations in quality and so consumer satisfaction is difficult to obtain.
Broader Inadequate marketing of products of developing countries (#PE0523).

♦ **PE4523 Inadequate trade in agricultural commodities between developing countries**
Refs FAO *Promoting Agricultural Trade Among Developing Countries* (1984).
Broader Inadequate trade between developing countries (#PD5176).
Aggravated by Instability of the primary commodities trade (#PC0463)
Processing in developed countries of commodities exported by developing countries (#PD0425)
Instability in export trade of developing countries producing primary commodities (#PD2968).

♦ **PE4524 Corrosion of glass**
Refs Clark, David, et al *Corrosion of glass* (1979).
Broader Atmospheric corrosion of materials (#PE9525).

♦ **PE4525 Obstacles to trade in fertilizer products**
Nature The expansion of fertilizer production capacity by the developing countries requires large-scale integrated production facilities which depend on exports and liberal access to the markets of developed countries. In this connection the level and the structure of tariffs at all stages of production are crucial. The positive effects of the generalized system of preferences (GSP) can be nullified through the application of various limitations, such as country specific tariff quotas, ceilings, maximum country amounts, competitive need criteria, denial of beneficiary status, etc. Besides, different non-tariff measures can be imposed on fertilizer products exported by the developing countries, such as licensing measures, technical standard quotas, anti-dumping duties and marking and packaging requirements. Market-economy countries producers exert high degree of market power in international markets, they also own a large number of process patents as well as non-patented know-how that makes it difficult for the developing country producers. Industry associations in developed countries have a decisive influence on regulating trade in fertilizers. Other factors which can have important trade effects include, transportation, marketing, finance, choice and availability of appropriate technology and discriminatory industrial collaboration arrangements.
Broader Protectionism in international trade against exports from developing countries (#PD9679).

♦ **PE4526 Mutism**
Muteness — Elective mutism
Nature The children refuse to speak in one or more major social situations, but may communicate with gestures or monotone utterances. These children understand spoken language and have normal language skills, although some speech disorder or abnormality of articulation may be present.
Refs Kratochwill, Thomas R *Selective Mutism* (1981).
Broader Mental illness (#PC0300) Speech disorders (#PE2265).
Narrower Deaf mutism (#PE5261).
Related Deafness (#PD0659).

♦ **PE4531 Denial of right to benefits to survivors**
Broader Denial of right to economic security (#PD0808).
Narrower Inefficient support for widowers (#PJ7997).
Related Unequal pension rights (#PJ2030) Neglect of dependents of war victims (#PD2092)
Denial of right to adequate medical care (#PD2028)
Denial of right to benefits for invalids (#PE5211)
Denial of right to economic security during periods of unemployment (#PE5406).
Aggravated by Burden on society of widows (#PF6149).

♦ **PE4538 Bulimia nervosa**
Nature People with bulimia nervosa have recurrent episodes of binge eating and once eating has begun they can't control their behaviour anymore. They regularly engage in self-induced vomiting that allows continued eating or termination of the binge. Eating binges are followed by self-criticism and depressed mood, strict dieting or fasting, vigorous exercise or use of diuretics in order to prevent weight gain.
Refs Duker, Marilyn and Slade, Roger *Anorexia and Bulimia Nervosa* (1988); Fichter, Manfred M (Ed) *Bulimia Nervosa* (1990).
Broader Eating disorders (#PE5187).
Related Over-eating (#PE5722) Anorexia nervosa (#PE5758).
Aggravated by Mental depression (#PC0799).

♦ **PE4544 Denial of the right to procreate to the severely mentally handicapped**
Sterilization of the mentally disabled
Nature The use of sterilization of mentally handicapped for what ever reason is a denial of a foundational human right, the right to procreate. Without this right the human race is unable to continue. The power to sterilize in the hands of evil or even misguided people, can cause racial, religious, or other undesirables inimical to the dominant group in society to disappear.

Incidence Sterilization has been used in Nazi Germany against Jews and other racial minorities to purify the race. Criminals, particularly sex offenders, in the United States and other countries have been sterilized as part of their punishment. In India the poor and poorly educated have been given hysterectomies or vasectomies, without knowing what they were, in order to lower birth rates. In none of these cases did the individual concerned have an informed right to say no.
Counter-claim While there are many instances of the misuse of sterilization, there are a few cases where this must be an option available to authorities for the sake of the good of society and for the sake of the health and happiness of the individual. Due to society's and the medical profession capacity for the mentally handicapped more and more are able to function effectively within the community in the care of relatives. Severely mentally handicapped individuals cannot care for themselves are certainly unable to care for an infant or a child. Where such a person cannot use other forms of contraception because they are unsuitable, inappropriate, dangerous or not completely effective sterilization is the only answer in the best interests of the individual.
Where a severely mentally handicapped person is member of a family that for generations has been severely mentally subnormal, in all likelihood these traits are transmittable and generations to come would be s subnormal sterilization must be a possibility. Society must have the power to sterilize for the biological benefit of humankind.
For biological reasons or for reason of the happiness of the individual involved all alternatives must be taken into account because an individual sterilized against his will is deprived of a basic human right. That right must be weighed against the other rights of the individual and of the society as a whole.
Refs Macklin, R and Gaylin, W *Mental Retardation and Sterilization* (1981).
Broader Denial of the right to procreate (#PC6870)
Discrimination against mentally disabled (#PD9183).
Aggravated by Inadequate rehabilitation facilities for the mentally handicapped (#PE8151).
Reduces Mental deficiency (#PC1587).

♦ PE4545 Bad property loans
Inadequately secured commercial property loans
Incidence Competitive pressure on banks during the 1980s, especially in the USA, forced them to increase their investment in higher risk loans on commercial property resulting in a building glut and a corresponding increase in bad loans.
Claim Loans on commercial property are emerging in the 1990s as the equivalent to the Third World debt problem of the 1980s.
Aggravates Bank failure (#PE0964).
Aggravated by Unethical real estate practice (#PD5422).

♦ PE4553 Accidental military incidents
Nature Accidental entry into national territory of adversary because of adverse weather or instrument malfunctions, troop or ship manoeuvres in regions of tension together with mistaken identities increase the risk of conflict or war.
Narrower Risk of unintentional nuclear war due to accidents (#PF4346).
Aggravates Border incidents and violence (#PD2950).

♦ PE4554 School phobia
Fear of school — School refusers
Nature For some children, schools is primarily associated with loneliness, bullying, stress concerning school work, and problems with teachers. These may cause the pupil to want to quit, to indulge in absenteeism, or to become suicidal.
Refs Blagg, Nigel *School Phobia and Its Treatment* (1987); Reid, Ken *Disaffection from School* (1986).
Aggravates Educational wastage (#PC1716) Student absenteeism (#PE4200).
Aggravated by Inhibiting shyness (#PF1278) Competitiveness in education (#PD4178)
Physical intimidation by children (#PE2876).

♦ PE4555 Lung-damaging agents
Phosgene — Chlorine
Nature In high concentrations, chlorine gas, phosgene, ammonia, sulphur dioxide and hydrogen chloride are lethal, because they damage the lung tissues. Chlorine and phosgene both irritate the eyes, the gullet, the respiratory tract and the lungs. After heavy exposure, the cause of death is often a combination of suffocation, due to reduced lung functions, and of excessive strain on the heart.
Broader Chemical warfare (#PC0872).

♦ PE4559 Seed-borne diseases
Refs Nene, Yeshwant Laxman and Aggarwal, V K *Some Seed Borne Diseases and their Control* (1978); Richardson, M J *An Annotated List of Seedborne Diseases* (1979).

♦ PE4561 Hysterical personality disorder
Histrionic personality — Labile personality
Nature These people are constantly seeking attention, praise or reassurance and are always drawing attention to themselves. They are self-centered and tend to have little or no tolerance for delayed satisfaction or normal routine. Their emotions are rapidly shifting and often expressed with inappropriate exaggeration while the style of speech remains impressionistic and lacking in detail.
Broader Personality disorders (#PD9219).

♦ PE4566 Symptoms referable to musculoskeletal system
Symptoms referable to limbs — Symptoms referable to joints
Refs Blakeslee, Berton *The Limb-Deficient Child* (1963).
Broader Ill defined health conditions (#PC9067) Diseases of connective tissue (#PD2565).
Narrower Cramps (#PG3717) Knee pain (#PE1309) Peculiar gait (#PG4752)
Transient paralysis of limb (#PG9217).

♦ PE4570 Sanitation system accidents
Broader Disastrous technological failures (#PD4426).
Aggravated by Inappropriate sanitation systems (#PD0876).

♦ PE4571 Arrears in payments for housing
Arrears in rental payments — Arrears in home loan payments — Arrears in mortgage payments
Incidence In the UK in 1989 it was reported that the number of households between 6 and 12 months in arrears on mortgage payments rose by 29 per cent to 58,380, although the number of people failing to keep up monthly mortgage payments could be 10 times that amount. No official figures were available on arreas of less than 6 months, but according to some estimates between 6 and 7 per cent of customers, approximately 450,000 to 600,000 families could be 2 months in arrears.
Aggravates Risky rental agreements (#PG1425).
Aggravated by Unethical practices by housing tenants (#PE7169).

♦ PE4574 Decline in public sector savings in developing countries
Inadequate government savings capacity in developing countries
Incidence Measures taken to redress external imbalances in payments have also affected the savings capacity of governments. Currency devaluation has been an important factor since it raises the domestic currency cost of servicing the external public debt, and hence aggravates the budget problem created by high interest rates and debt-service obligations. Where domestic interest rates have been raised substantially in order to prevent a collapse of the currency and forestall capital flight, internal debt-service obligations have also risen. In debtor developing countries, many of which have been experiencing falling real incomes, it has not been possible to offset these higher burdens to any significant extent through cuts in expenditure or higher taxation. As a result, the government budget balance deteriorated and public sector savings fell. The situation was further aggravated in those countries, a majority, for which receipts from tariffs represent an important source of revenue. Where imports were reduced as part of an adjustment programme, government revenues fell correspondingly, again reducing public sector savings.
Aggravated by High interest rates (#PF9014) Devaluation of money (#PE3700)
Dependence of developing countries on customs revenue (#PD2955)
Inadequate economic policy-making in developing countries (#PF5964)
Private domestic capital outflow from developing to developed countries (#PD3132)
Excessive social costs of structural adjustment in debtor developing countries (#PD8114).

♦ PE4576 Gambling on sports and athletic competitions
Clandestine sports betting
Broader Gambling and wagering (#PF2137).
Narrower Exploitation of animals in spectator sports (#PE0891).
Aggravates Athletic competition (#PE4266)
Corruption of sports and athletic competitions (#PE3754).
Aggravated by Unethical entertainment (#PF0374).

♦ PE4577 Failure to appear for trial after release
Bail jumping
Nature A person who has been released on bail, has agreed to appear before the court or a juridicial officer and fails to appear is guilty of bail jumping. This is not an offence if the person does not appear for reasons beyond his own control.
Broader Obstruction of government function (#PD6710).

♦ PE4579 Children of torture victims
Nature Sometimes children are arrested together with their parent or both parents, or they are born in the prison or they are left on their own, at home or in the streets after the arrestation of their parents. The children of torture victims live under traumatic circumstances: the arrest of the parents, the absence of one parent, separation from both parents, witnessing the arrest, torture or even execution of a parent. Some children are tortured in front of their parents.
Incidence In a recent study of chilean refugee children whose parent or parents had been tortured showed 78 percent of them suffered emotional or physical effects of the torture. From the study it is evident that such children suffer serious and perhaps life long psychological and social problems.
Broader Family of torture victims (#PE2119).
Related Spouses of torture victims (#PE0753).

♦ PE4581 Exclusive sales and representation agreements
Nature Requiring the acquiring party to grant exclusive sales or representation rights to the supplying party or any person designated by the supplying party, except as to subcontracting or manufacturing agreements wherein the parties have agreed that all or part of the production under the technology transfer arrangements will be distributed by the supplying party or any person designated by him.
Broader Restrictive trade practices (#PC0073).
Related Collusive tendering (#PE4301) Grant-back provisions (#PE5306)
Challenges to validity (#PF1200) Restrictions on research (#PF0725)
Restrictions on publicity (#PF1575) Restrictions on adaptations (#PF5248)
Exclusive dealing arrangements (#PE0413) Restrictions on use of personnel (#PF3945)
Patent pool or cross-licensing agreements (#PE4039)
Trade restrictions due to voluntary export restraints (#PE0310)
Payments after expiration of industrial property rights (#PF5292)
Tying of supplies to subsidiaries by transnational enterprises (#PE0669).

♦ PE4584 Diseases of retina
Detachment of retina — Diseases of optic nerve
Refs Ryan, Stephen J, et al *Retinal Disease* (1985).
Broader Eye diseases and disorders (#PD8786).
Aggravates Physical blindness (#PD0568).
Aggravated by Bright's disease (#PE2272).

♦ PE4585 Declining area of irrigated land
Nature In recent years the world's two leading food producers have experienced a decline in irrigation. China's irrigated area has shrunk by some 2 percent since 1978. The United States' use of irrigation water dropped 9 percent between 1980 and 1985. The Soviet Union in 1986 stopped plans to divert rivers now flowing into the Arctic Ocean into central Asia and while investment in irrigation continues the expected gain in the Soviet Union are modest.
Aggravates Slowing growth in food production (#PC1960).

♦ PE4586 Ascomycetes
Broader Fungal plant diseases (#PD2225).
Narrower Blights (#PE3919) Cankers (#PE0640) Apple scab (#PG1100)
Leaf spots (#PE1954) Fruit spots (#PG1121) Leaf scorch (#PG3786)
Leaf blisters (#PG4292) Ergot of grain (#PG4913) Rose blackspot (#PG5469)
Rots in plants (#PE3363) Peach leaf curl (#PG3936) Galls in plants (#PE3715)
Lichen in plants (#PG0608) Moulds in plants (#PE8051) Tar spot of maple (#PG1373)
Dutch elm disease (#PE1154) Black rot of grape (#PG5524) Black knot of plum (#PG3820)
Sycamore anthracnose (#PG5536) Sooty mould in plants (#PE0916)
Black leaf spot of elm (#PG1852) Black mildew in plants (#PE1607)
Scab diseases in plants (#PE1040) Wilt diseases of plants (#PE1056)
Needle casts of conifers (#PG1795) Powdery mildews in plants (#PE0529)
Hypoxylon canker of trees (#PG4453) Brown rot of stone fruits (#PG1821)
Blotch diseases in plants (#PE4235) Beech bark disease complex (#PG0894)
Spot anthracnose in plants (#PG4914) Sclerotinia diseases of plants (#PG4655)
Anthracnose diseases of plants (#PE1910).

♦ PE4588 Mutagenic chemicals
Toxic genetic chemicals — Radiomimetic chemicals
Nature Certain chemicals, especially those used in agriculture (insecticides, fungicides), are radiomimetic in that they mimic the genetic effects of nuclear radiation.
Claim Annual use of toxic genetic chemicals in the USA in the 1970s averaged 453,000 tons, which has been estimated to cause genetic damage equivalent to 72,000 bombs of the Hiroshima type. Such damage is indicated by the number of mentally retarded at 15 percent of live births,

by reduction in the sperm count of adult Americans by 30 percent from 30 years ago (attributable to chlorinated hydrocarbon pesticides), and by 25 percent of American male college students being sterile.
Aggravates Mutation (#PF2276).

♦ PE4589 Social reinforcement of shallow personal meaning
Socially–reinforced shallow perception of personal meaning
Nature Society is reenforcing a personal sense of meaningless by continuing to use former images and styles which have little or no connection to personal experience. Images point to an absolute, unambiguous universe. Powerful Twentieth Century symbols, rites and mythologies point to shallow spirit journeys. A person experiences his perceptions and experiences as suspect. Few individuals have the courage to explore the depth of their lives so most live with little meaning or greatness.
Claim The social forms that once created meaningful, reinforcing images and life styles no longer do so. Personal images and symbols are shallow and inadequate to express individual consciousness and the meaning of life. In an absence of situations where the meaning of symbols is acted out, people lack consciously appropriate perceptions, both of awe ana of encounter with their own inner depths. They fail to recognize the greatness and significance of their own existence.
Broader Defensive life stance (#PF0979).

♦ PE4593 Schizmogenesis
Nature Cleavage within a group resulting from behaviour associated with different social roles. The process of separation is cumulative as the role behaviour tends to continue to produce reactions that intensify the corresponding patterns of behaviour. Complementary schismogenesis is when one behaviour—for example, male aggressive, boasting and competitive behaviour—is associated with the opposite type of behaviour, such as female submissive, quiet and cooperative behaviour. Symmetrical schismogenesis is the process of role behaviour resulting in similar, rather than contrary, behaviour, as when aggressiveness stimulates aggressiveness.
Broader Social fragmentation (#PF1324) Cultural fragmentation (#PF0536).
Related Class consciousness (#PC3458).

♦ PE4596 Training inappropriate to structural and technological changes
Refs ILO *The Adaptation of the Training of Managerial Staff and Employees to Structural and Technological Changes in Hotels, Restaurants and Similar Establishments* (1983).
Broader Inadequate education (#PF4984).
Aggravates Worker maladjustment to technology (#PC7041).
Aggravated by Inability of educational systems to keep pace with technological advancement (#PF7806).

♦ PE4601 Taxation of the poor
Broader Distortionary tax systems (#PD3436).

♦ PE4602 Disproportionately long prison sentences
Excessive length of prison sentences — Delay in processing parole applications
Nature Some punishments of criminals are so inherently degrading and therefore disproportionate to the criminal's culpability, as to preclude them from ever being used. In some cases these are internationally recognized, such as, torture, death by painful and lingering methods and expatriation, forfeiture of citizenship. Some sentences that are not inherently cruel and unusual, in particular cases, are grossly and outrageously disproportionate to the defendant's conduct and to the crime committed. Pre-trail detention can be so long as to constitute cruel and inhumane punishment.
Broader Denial of rights to prisoners (#PD0520)
Denial of right to freedom from cruel, inhumane or degrading punishment (#PC3768).
Aggravated by Denial of right to freedom from double jeopardy (#PF6158).
Reduced by Excessive leniency in sentencing of offenders (#PF4723).

♦ PE4603 Deteriorating terms of financial loans to developing countries
Deterioration in the cost of external finance for developing countries — Increases in interest rates on long–term debts to developing countries
Nature Developing countries are exposed to deteriorating terms of borrowing, including sudden increases in interest rates paid on long–term debts, particularly commercial loans. In addition grace and repayment periods may be reduced.
Background Besides trade, the cost and availability of international finance are the main external determinants of the economic performance of developing countries. The debt crisis has had a profound impact. Developing countries have traditionally been net importers of capital; their domestic savings are generally insufficient to meet their investment needs. The availability and cost of such external finance depend mainly on the overall size of the pool of exportable savings in capital–surplus countries and on the competing claims on that pool. During the 1980s both moved against the developing countries.
Claim Every time the US raises its interest rates thousands die in developing countries because money that would be used for health care and food is sent outside these countries to pay the debt.
Refs United Nations Economic Commission for Latin *External Debt in Latin America* (1985).
Broader Deterioration in external financial position of developing countries (#PE9567).
Narrower Uncertainty of development programmes due to short–term loans (#PF4300)
Uncertainty of development expenditures due to floating–rate loans (#PF4295)
Rescheduling of debts of developing countries at market–related interest rates (#PE8110).
Aggravates Excessive foreign public debt of developing countries (#PD2133)
Decline in concessional financial resources available to developing countries (#PE3812).

♦ PE4604 Diseases of the spinal column in animals
Diseases of the spinal cord in animals
Broader Diseases of nervous system in animals (#PD7841).
Narrower Canine distemper (#PG5710) Polyradiculoneuritis (#PG9591)
Granulomatous meningoencephalomyelitis (#PG6398).

♦ PE4605 Dust explosions
Refs Cross, Jean and Farrer, Donald *Dust Explosions* (1982); Eckhoff, Rolf Kristian *Dust Explosions in the Process Industry* (1990); Nagy, and Verakis, *Development and Control of Dust Explosions* (1983).

♦ PE4609 Children of drug addicts
Foetal damage from drug use — Drug threat to the young
Nature Babies born to crack addicts tend to suffer low birth weight, brain damage and malformation. As they grow, they do not respond to usual human signals such as eye contact and smiling, have difficulty learning and fail to understand simple problems. They ignore toys and do not play, have no concept of right and wrong, do not make friends, are easily frustrated and often violent. Children of crack addicts are at extreme risk of neglect and abuse. In one report 73 percent of all children who died of battering in New York City in 1988 had parents who used cocaine or crack.
Damage to the foetus from cocaine exposure could include prenatal strokes and lasting brain damage, premature birth, retarded foetal growth, breathing lapses, absence of part of the gut, structural abnormalities in genital and urinary organs and seizures after birth. Babies that had been exposed to marijuana are likely to be smaller than normal and to show such neurological difficulties as an abnormal startle reflex, an increase in tremors and an inability to shut out disturbing stimuli. At high levels of use, alcohol can cause serious malformation and at moderate levels of use it is associated with an increased risk of mental or physical damage to the foetus.
Incidence One survey in the US showed at least 11 percent of women in the sample hospitals had used illegal drugs during pregnancy. The data suggests that 375,000 newborns a year face the possibility of health damage from their mother's drug abuse.
Broader Drug abuse (#PD0094).
Related Drug abuse by adolescents (#PD5987).

♦ PE4619 Incontinence
Infant incontinence
Incidence As an indication, the market for disposable infant nappies and towelling nappies in the UK in 1986 was £262.5 million.
Broader Unsociable human physiological processes (#PF4417).
Narrower Enuresis (#PE5431) Random defecation (#PG6578).
Aggravates Dependency of the elderly (#PD8399)
Excreting in public places (#PE1602)
Disagreeable human body odour (#PE4481).
Aggravated by Susceptibility of the old to physical ill–health (#PD1043).

♦ PE4620 Irreversible coma
Brain death
Refs Plum, Fred and Posner, Jerome *The Diagnosis of Stupor and Coma* (1980).
Broader Symptoms referable to nervous system (#PE9468).

♦ PE4624 Denial of right to a name
Imposition of husband's name on married women
Broader Denial of rights of children and youth (#PD0513).
Aggravates Inadequate sense of personal identity (#PF1934).

♦ PE4627 Obstacles to political union among island developing countries
Nature Although small island countries could benefit considerably from the strength and influence which a political union would afford, they find it difficult to carry out. In any group of islands, there are one or a few which are somewhat better off or larger. They fear the possibility of taking on problems of smaller territories which they might not have on their own. For all parties, national interest must be very strong to survive as an entity at all. And yet this same national interest blocks the idea of union with others, even be it a limited one. Finally, small island politicians and populaces are accustomed to a level of personalism in their dialogue that union with outside territories cannot permit.
Broader Vulnerability of island developing countries and territories (#PE5700).

♦ PE4628 Denial of right to a legal defence
Violation of right to a defence against criminal charges
Broader Denial of right to fair and public trial (#PE3964).
Narrower Denial of right to examine witnesses (#PE7489)
Denial of right to confront accusers (#PE7217)
Denial of right to time to prepare a trial defence (#PE7624)
Prohibitive cost of linguistic interpretation legal proceedings (#PE1743).

♦ PE4630 Diseases of the oesophagus in animals
Broader Diseases of the digestive system in animals (#PD3978).
Narrower Choke (#PG3833) Grain overload (#PG7288)
Oesophagitis in animals (#PG9684) Oesophageal diverticulum (#PG8770)
Cricopharyngeal achalasia (#PG9359) Oesophageal stenosis in animals (#PG2927)
Foreign bodies in the oesophagus of animals (#PG7761).

♦ PE4634 Oppositional defiant disorder
Nature The disorder usually appears before puberty and features frequent negativistic, hostile and defiant behavior. The children lose very often their temper, swear, annoy other people deliberately and refuse adult requests. They also have a tendency to blame others for their mistakes and justify their behavior as a response to unreasonable circumstances.
Broader Mental illness (#PC0300).

♦ PE4635 Misuse of personal authority for political purposes
Nature Misuse of personal authority for political purposes is for a person in government service to use his authority to discharge, promote or degrade or to threaten to these things to secure political contributions, activities or votes from public servants.
Broader Abuse of authority (#PC8689) Violation of political processes (#PD5457).
Aggravated by Abusive distribution of political patronage (#PF8535).

♦ PE4637 Neoplasms of lymphatic and haematopoietic tissue
Cancer of lymphatic tissue
Broader Malignant neoplasms (#PC0092).
Narrower Leukaemia (#PE0639) Polycythemia (#PG5546)
Lymphosarcoma (#PE5548) Multiple myeloma (#PE5289).

♦ PE4639 Personal physical insecurity in industrialized countries
Broader Personal physical insecurity (#PD8657)
Deterioration of industrialized countries (#PD9202).

♦ PE4640 Indiscriminate use of tear gas
Nature Tear gases are gaseous substances or aerosols that cause strong irritations and pain in the eyes, producing a flow of tears. At the same time, it becomes difficult to keep the eyes open. A further effect is a burning sensation in the mouth, throat and respiratory tract. High concentrations of tear gas produce breathing difficulties and pressure on the chest. Vomiting may also result. A burning, prickling sensation may be experienced on the skin. Warm moist skin reacts more strongly, and inflammations may occur. In high concentrations, tear gases can be dangerous for people with allergies, for the elderly, for those with heart or lung defects, for children and for pregnant women. If exposure occurs in enclosed places, and at high concentrations, the consequences can be serious, even fatal.
Claim Tear gases are chemical weapons and should be banned as such.
Broader Chemical warfare (#PC0872).

♦ PE4645 Abuse of expense accounts
Broader Tax evasion (#PD1466).
Aggravated by Unethical commercial practices (#PC2563).

♦ **PE4647 Inefficiency due to mismatch of religious or national holidays**
Nature The large variety of national, religious, and local holidays creates hugh losses of money and time for any individual or organization operating across international or interreligious barriers. Ether holidays are ignored and thus the occasion or person they are ment to honour is dishonoured or all are observed and time and money is wasted.
Broader Inefficiency (#PB0843).

♦ **PE4649 Deterioration of the mind with age**
Diminishing mental capacity with age
Broader Loss of capacity with age (#PC8310).
Aggravates Senility (#PE6402).
Aggravated by Paralysis agitans (#PE2206) Alzheimer's disease (#PE7623)
Mental disorders of the aged (#PD0919).

♦ **PE4650 Mania**
Nature An interest in something so great that it causes an individual to lose perspective. The maniac tends to ignore other possibilities.
Broader Manic-depressive psychosis (#PD1318).

♦ **PE4651 Sore throat**
Soreness of throat — Throat inflammation — Inflammation of pharynx
Refs Gerrick, David J *Sore Throat-The Danger Within* (1978).
Broader Acute respiratory infections (#PE7591).
Aggravates Deafness (#PD0659).
Aggravated by Measles (#PE1603) Tonsillitis (#PE2292).

♦ **PE4655 Decline in commercial bank lending to developing countries**
Incidence The decline throughout the 1980s, at a time when the financing needs of these countries were growing because of weakness in their export markets, demonstrated the pro-cyclical character of such lending. The simultaneous action by banks, a form of herd instinct, owes much to the practice of syndication. Thus rather than help to insulate developing countries receiving such flows from external pressures, bank lending became an added source of disturbance. Total annual bank lending (including short-term) to developing countries dropped precipitously in 1982–83 and by 1985 was approximately 25 per cent of the 1981 level.
Aggravates Capital shortage in developing countries (#PD3137)
Lack of capital investment in developing countries (#PE5790)
Limited availability of loans in developing countries (#PE4704).
Aggravated by Inappropriate loans (#PF4580)
Excessive foreign public debt of developing countries (#PD2133)
Insufficient creditworthiness of developing countries (#PD3054)
Burden of servicing foreign public debt by developing countries (#PD3051)
Inadequate external debt management capacity within developing countries (#PE7184).

♦ **PE4656 Urolithiasis**
Refs Sutton, R A L (Ed) *Urolithiasis* (1987).
Broader Urinary system diseases in animals (#PD9293).
Narrower Feline urolithiasis (#PG4975) Canine urolithiasis (#PG9798)
Urolithiasis in horses (#PG9674) Urolithiasis in ruminants (#PG7863).

♦ **PE4657 Lack of vaccines**
Incidence More than 3.5 million children died in 1987 because they were not vaccinated against common children's diseases. Measles kill two million a year, tetanus one million and whooping cough 600,000. Polio kills some 50,000 people a year and cripples hundreds of thousands more.
Broader Insufficient medical supplies (#PE1634).
Aggravates Inadequate and insufficient immunization (#PF5969).
Reduces Inoculation for viral diseases (#PJ4755).

♦ **PE4658 Abuse of government employment**
Ghost workers — Phantom workers
Nature The most blatant abuses of government employment policy are those cases where a person receives government wages but either does not exist or is not employed in the position for the employment is made. In some situations it is possible for those in a bureaucracy to create positions, whether to inflate the importance of their departments or to provide employment for a relative or for some other consideration. This leads to a high level of underemployment, to the point that many people may share the same desk at which they are assumed to be working. Of a totally different nature is the practice whereby one person, usually at an executive level, cumulates the salaries associated with different positions, if that person can arrange to be appointed to those positions simultaneously.
Incidence Ghost or phantom workers were identified at one stage as constituting approximately 7 per cent of civil service employment in the Central African Republic and Guinea. A common practice is for a person to be on the government payroll but to spend most working hours engaged in some private enterprise in order to provide an additional source of income. Cumulation of functions is not uncommon in intergovernmental organizations and contributes to the drain on their resources.
Counter-claim In countries with high unemployment or where government salaries are comparatively low, such practices offer a means of employment and ensure levels of income which would otherwise make it impossible for skilled personnel to work for the government at all.
Aggravates Inefficiency of state-controlled enterprises (#PD5642).
Aggravated by Underemployment (#PB1860) Accumulation of functions (#PF4174)
Unethical practices by employees (#PD4334) Underemployment of skilled workers (#PJ5489)
Underpayment of government officials (#PD8422)
Frustrated talent in government posts (#PE4398).

♦ **PE4661 Criminal threat**
Broader Harassment (#PC8558)
Offences involving danger to the person (#PD5300).
Narrower Terrorizing (#PE4466).
Related Criminal menacing (#PE4467) Criminal harassment (#PD2067).
Aggravates Criminal coercion (#PD4469).
Aggravated by Organized crime (#PC2343).

♦ **PE4664 Gender-differentiated toys**
Broader Gender stereotyping (#PD5843) Unethical entertainment (#PF0374).
Related Dangerous toys (#PE1158) Violent interactive toys (#PE4297).

♦ **PE4665 Excessive consumption of vitamins**
Refs Wahlqvist, Mark, et al *Use and Abuse of Vitamins.*
Broader Excessive consumption of specific foodstuffs (#PC3908).

♦ **PE4669 Lack of freedom of movement of children of separated parents**
Broader Travel restrictions (#PC8452) Victimization of children (#PC5512).

♦ **PE4675 Toxic modern furniture materials**
Lethal fumes in modern house fires
Nature Modern homes are potentially lethal because of the toxic gases given off by burning furniture and other household materials. The most deadly fumes are from burning foam used to stuff furniture. Burning foam may produce gases reaching 1000 degrees centigrade.
Incidence In 1985 there were 62,000 house fires in Britain which claimed 700 lives. Almost 500 of the deaths were caused by smoke.
Broader Fires (#PD8054).
Aggravated by Flammable construction materials (#PG4448).

♦ **PE4679 Mineral dusts**
Broader Dust (#PD1245).
Narrower Coal dust (#PG5093) Cement dust (#PE2854) Limestone dust (#PE2038)
Asbestos as a pollutant (#PE1127).

♦ **PE4683 Nervous system diseases caused by parasites**
Broader Diseases of nervous system in animals (#PD7841).
Narrower Ascariasis (#PE2395) Trematode diseases (#PE6461)
Tapeworm as a parasite (#PE3511).

♦ **PE4685 Insalubrity of animal excrement in urban environments**
Dog faeces on pavements — Excrement of birds
Nature Bird excrement, especially in urban environments, is a vehicle for the transmission of disease as well as contributing to the defacement of buildings and monuments.
Broader Birds as vectors of disease (#PE6659).
Aggravates Mycosis (#PE2455) Defacement of urban structures (#PD5305).

♦ **PE4687 Emotional instability due to torture**
Introversion due to torture
Broader Neurological effects of torture (#PD4755).

♦ **PE4693 Fare evasion**
Aggravated by Prohibitive cost of transportation (#PE8063).

♦ **PE4694 Vulnerability of animals during states of emergency**
Lack of protection for domestic animals during disasters — Lack of protection for wild animals during disasters
Broader Disaster unpreparedness (#PF3567).

♦ **PE4696 Pharmacological torture**
Neuropharmacological torture
Nature Pharmacological torture is the enforced application of psychotropic or other drugs for the purposes of punishing, yielding information or causing profound mental destruction, anxiety and psychological disturbances, pain, immobilization or disorientation. Morphine is administered to prisoners until they are addicted. Once they are addicted the drugs are stopped and interrogation begins at the onset of withdrawal symptoms. If they give enough information they are given another fix. Ether or other pain causing chemical are injected. Victims are given a slow acting poison like thallium and then released so they die sometime after they leave the prison. Psychopharmacological drugs are used to blunt senses or as psychotropics, including hallucinogens. Neuromuscular blocking agents (curare compounds) are abused to paralyse fully-awake subjects, causing total panic because they are unable to breath. The victims are not offered oxygen until the point of suffocation. Victims have been given insulin shock therapy as a form of punishment.
Incidence Neuropharmacological torture has been reported in the following countries: **Af** South Africa, Zaire. **Am** Chile, Colombia, El Salvador, Uruguay. **As** Iraq, Israel. **Eu** USSR.
Broader Human torture (#PC3429) State sanctioned torture (#PD0181)
Institutionalized torture (#PD6145).
Related Physical torture (#PD8734) Religious torture (#PC7101)
Torture of children (#PD2851) Psychological torture (#PD4559)
Forced confessions with drugs (#PE4888).

♦ **PE4697 Government approved employment of war criminals**
Broader Employment of criminals in policy-making contexts (#PE4439).
Related Post-revolutionary re-employment by government of security services of the ousted repressive regime (#PF1015).
Aggravates Unconvicted war criminals (#PD4067)
Ineffective war crime prosecution (#PD1464).
Aggravated by Unethical practices of employers (#PD2879).

♦ **PE4700 Denial of right to educational choice**
Violation of freedom of parents to choose their children's education
Broader Denial of right to education (#PD8102)
Denial to people of control over their own lives (#PC2381).
Narrower Inequitable education selection (#PG8023)
Denial of right to choose moral and religious education (#PE3802).
Related Restricted higher education (#PJ0572) Insufficient primary education (#PC6381)
Religious discrimination in education (#PD8807).
Aggravates Undersupported school system (#PU8326).

♦ **PE4703 Self-imposed social isolation due to torture**
Nature For the torture victims it is an extreme stress to suffer from mental symptoms due to torture, more than from physical injury, because family and friends tend to react negatively when confronted with difficult behaviour, while obvious physical injuries elicit sympathy and understanding. Victim feels mentally changed and often "shut in", lonely and dominated by anxiety. Often the victim's mental state makes him or her impossible to live with.
Broader Loneliness (#PF2386) Neurological effects of torture (#PD4755).

♦ **PE4704 Limited availability of loans in developing countries**
Restrictive access to loans — Inaccessibility of bank loans — Limited access to loans — Limited availability of loan capital — Difficult development loans
Broader Restrictive monetary practices (#PF8749)
Limited accumulation of capital (#PF3630)
Restricted flow of local economy (#PF6451).
Narrower Restrictive farm capital (#PJ0788) Inaccessible housing loans (#PG8648).
Related Lack of venture capital (#PG7833).
Aggravates Subsistence approach to capital resources (#PF6530)
Lack of economic and technical development (#PE8190)
Diminishing capital investment in small communities (#PF6477).
Aggravated by Decline in commercial bank lending to developing countries (#PE4655)
Discrimination against transnational banks in developing countries (#PE4320)
Deterioration in external financial position of developing countries (#PE9567).

PE4706 Bacterial soft rot in plants
Nature Soft rots attack a large number of plants throughout the world causing more damage to vegetables than any other group of related diseases. Once bacteria has gained entrance through wounds and temperatures are warm and moisture is high, it will multiply in the intercellular spaces.
Broader Bacterial plant diseases (#PD2226).
Related Crown gall (#PE2230)   Fire blight (#PE2229).

PE4712 Denial of right to equality between states
Denial of equality between nations
Broader Denial of state's rights (#PD4814).
Narrower Inequitable allocation of rights to exploit sea-bed and marine resources (#PE1597).
Aggravates Second class states (#PD0579)
Domination of the world by territorially organized sovereign states (#PD0055).

PE4715 Chemical imbalances in the brain
Nature Psychological disorders can be linked to chemical imbalances in the brain. Excessive dopamine production in the brain, for example, is thought to be one cause of schizophrenia, especially when the disease is accompanied by paranoia.
Broader Diseases and injuries of the brain (#PD0992).
Related Memory defects (#PD8484).
Aggravates Psychological disorders (#PD8375).

PE4716 Denial of right to recognition as a person before the law
Disenfranchisement — Restricted franchise
Broader Inequality before the law (#PC1268).
Aggravated by Language discrimination in politics (#PD3223).

PE4717 Long-term hazards of exposure to chemicals
Hazards of low-level exposure to chemicals — Health hazards of long-term, low-level exposure to toxic mixtures of non-toxic chemicals
Nature Most tests of chemical products, especially drugs and foodstuffs, are based on short-term effects, especially the LD-50 test which determines the amount of a substance fed to a test group of animals which will cause 50 percent of them to die within a given length of time (usually measured in days or weeks). Such short-term data is then used to set legal standards covering long-term exposure to lower levels of the chemical.
Refs Block, J Bradford *The Signs and Symptoms of Chemical Exposure* (1980); Patosaari, P *Chemicals in Forestry* (1987); WHO *Health Aspects of Chemical Safety* (1983); WHO *Health Aspects of Chemical Safety* (1984); WHO *Health Aspects of Chemical Safety* (1983).
Broader Environmental hazards from chemicals (#PC1192)
Hazardous combinations of substances (#PD5256)
Health hazards of environmental pollution (#PC0936).
Aggravated by Uncertain toxicity thresholds (#PF5188)
Inconclusiveness of scientific and medical tests (#PD7415).

PE4718 Bride burning
Dowry deaths
Nature Ever-escalating dowry demands represent a real financial burden to the parents of unwed daughters. Young brides may take severe punishment from their get-rich-quickly husbands, if promised money or goods do not materialize. Sometimes dowry harassment ends in the wife's suicide or murder, that is usually committed by setting her afire with kerosene and then claiming she died in a kitchen accident.
Refs Krishnamurty, S *Dowry Problem* (1981).
Broader Violence against women (#PD0247).
Related Suttee (#PF4819).
Aggravated by Bride-price (#PF3290).

PE4724 Dyspepsia
Indigestion — Nervous dyspepsia — Catarrh of the stomach — Mucous gastritis — Fermentative dyspepsia
Incidence As an indication, the market for indigestion remedies in the UK in 1986 was £28.4 million.
Refs Roth, June *Living Better with a Special Diet* (1983); Smith, M J *Ulcer and Non-Ulcer Dyspepsias* (1987).
Broader Gastric disorders (#PE1599)
Diseases and deformities of the digestive system (#PC8866).
Aggravates Human flatulence in public (#PE1764).
Aggravated by Appendicitis (#PG2327).

PE4725 Unsanitary disposal of human remains
Medical waste — Hospital waste
Nature Waste from hospitals, including human residue, syringes, and other infectious material, is generally disposed of separately from other waste. When the disposal systems are ineffective or the medical waste dumped illegally, the public is exposed to highly infectious debris.
Incidence New York City produces 800 lbs of potentially infectious medical waste each week and the rest of New York State produces an additional 1000 lbs some of which is being improperly disposed.
Broader Hazardous waste dumping (#PD1398).
Aggravates Coastal water pollution (#PD1356)   Human disease and disability (#PB1044).

PE4732 Air spora
Pollens of grasses, weeds and trees
Nature Proliferation of airborne fungus spores, pollen grains and microorganisms.
Reduces Destruction of weeds (#PE3987).

PE4733 Bias in jury trials in small jurisdictions
Nature Jurors may be swayed by sympathy or prejudice and, in extreme cases, public hysteria is true of all jury trial but in small jurisdictions, such as small island nations, it is more liable to occur and is difficult to prevent. Impartiality of juries may be against the accused or be too ready to acquit. Jurors may know each other or be related and be embarrassed to oppose each other. Jurors may be biased against a foreigner, whether as the accused or as the victim of a crime. Changes of venue are not always possible and are frequently unreasonably expensive. Delays in trials long enough to diminish prejudice may lead to violations of human rights.
Broader Unjust trials (#PD4827).
Related Unfair trials due to pre-trial publicity (#PE1692).
Aggravates Denial of right to fair and public trial (#PE3964).
Aggravated by Lack of impartiality of the judiciary (#PE7665).

PE4736 Surgical malpractice
Surgical errors — Operating theatre mistakes
Incidence At least 1000 patients a year in Britain die as a result of operating theatre errors by surgeons and anaesthetists. Inappropriate operations and mistakes in preparing patients for operations are the most serious mistakes, for example, junior doctors performing operations without supervision from the consultants or surgeons undertake surgery outside their speciality.
Broader Inadequate medical care (#PF4832)   Unethical medical practice (#PD5770)
Human errors and miscalculations (#PF3702).
Narrower Unnecessary surgery (#PE9271).
Related Inadequacy of medical science (#PF8326).
Aggravated by Misdiagnosis (#PF8490)
Inaccurate medical diagnosis due to inadequate equipment (#PE4129).

PE4737 Denial of right to trial by a court
Broader Violation of civil rights (#PC5285)   Denial of right to fair and public trial (#PE3964).
Narrower Extrajudicial courts and tribunals (#PE7664).

PE4738 Foreign ownership
Claim Foreign ownership troubles those who want a comfortable, assured life, it increases prejudice and fears, it threatens local pride and smooth operation of the economy, disturbs the public order, adversely affects the country's competitors, and may impair the safety of the general public.
Narrower Control of national economic sectors by transnational enterprises (#PE0042).
Aggravates Absentee ownership (#PD2338)
Discrimination against foreign companies (#PD6417).
Aggravated by Social environmental degradation from recreation and tourism (#PD0826).

PE4739 Armed robbery
Refs Gabor, Thomas, et al *Armed Robbery* (1987).
Broader Robbery (#PD5575).
Aggravates Homicide (#PD2341).
Aggravated by Proliferation of weapons in civilian hands (#PE2449).

PE4740 Corrosive groundwater
Corrosion and encrustation in water wells
Refs FAO *Corrosion and Encrustation in Water Wells* (1980).
Broader Corrosive substances (#PE6887).
Aggravates Pollution of groundwater (#PD2503)   Fouling of water supply systems (#PJ6457).

PE4742 Excessive reliance on infallibility of equipment
Broader Human errors and miscalculations (#PF3702).
Aggravates Unreliability of computer software (#PE4428).

PE4743 Denial of right to freedom from retroactive laws and punishments
Broader Denial of human rights in the administration of justice (#PD6927).

PE4744 Health hazards of air pollution for plants
Harmful effects of air pollution on agricultural crops
Nature Air pollution, can cause harmful effects to agricultural crops and other species of vegetation. Apart from sulphur dioxide, the principal pollutants of serious concern are fluorides, ozone, photochemical smog products, ethylene and various aerosols occurring as suspended particular matter. Gases such as chlorine, hydrogen sulphide, oxides of nitrogen, and ammonia may constitute a hazard or cause damage to vegetation on occasion in localized areas. Suspended particular matter, consisting of small particles of smoke, sulphuric acid mist, soot and various metallic dusts, may be transported by wind for considerable distances from strong sources such as urban and industrial areas. These pollutants accumulate on leaf surfaces and clog the stomata to produce direct damage or retardation in growth.
Refs Anderson, F K and Treshow, M *Plant Stress from Air Pollution* (1989); Dässler, H G and Böritz, S (Eds) *Air Pollution and Its Influence on Vegetation* (1987); WHO *Review of the Present Knowledge of Plant Injury by Air Pollution* (1976).
Broader Air pollution (#PC0119)   Inhibition of crop growth by pollution (#PE8476).
Aggravates Plant diseases (#PC0555).

PE4745 Violation of the right of trade unions to engage in political action
Broader Violation of trade union rights (#PD4695).

PE4758 Denial of rights to wounded military personnel
Refs Rivkin, Robert S and Stichman, Barton F *The Rights of Military Personnel* (1981).
Broader Denial of rights to soldiers (#PD4089).
Related Ill treatment of prisoners of war (#PD2617).

PE4759 Degradation of the environment through contamination
Refs International Environment Liason Centre *International Environment-Development Facts March 1989* (1989).
Broader Environmental pollution (#PB1166)   Environmental degradation (#PB6384).
Aggravated by Pollutants (#PC5690).

PE4760 Vitamin E deficiency in domestic animals
Dietary deficiencies in domestic animals
Incidence Vitamin E deficiency occurs in pigs, cattle, sheep, rabbits, chickens and other animals raised for food. Muscular dystrophy is the most frequent, and, from a diagnostic point of view, the most important manifestation of the vitamin E-deficiency syndrome in domestic animals. Liver necrosis seems to be restricted to the pig. Other symptoms and lesions such as anaemia, yellow fat, gastric ulcers, reproduction disturbances, and neonatal mortality occur sporadically. Species-differences in the absorption vitamin E supplied in feeds are highly probable.
Broader Animal diseases (#PC0952)   Inadequate feeding of animals (#PC2765)
Agricultural mismanagement of housed farm animals (#PD2771).
Aggravates Animal infertility (#PC1803)   Poor quality of domestic livestock (#PD2743)
Harbouring national security offenders (#PE0484)
Inferior meat quality from intensive animal farming units (#PE2770).
Aggravated by Floods (#PD0452)   Drought (#PC2430)
Bad weather (#PC0293)   Insect pests (#PC1630)
Criminal conspiracy (#PD1767).
Reduces Excessive cost of animal protein (#PE4784)
Long-term shortage of meat and meat preparations (#PE1490).
Reduced by Undesirable effects of animal feed additives (#PD1714).

PE4766 Port congestion
Inadequate seaport facilities — Port traffic congestion
Nature Port service can be said to be inadequate when traffic has to be restricted for lack of port capacity or when traffic uses a port only at excessive cost. The consequences of a failure to provide proper port capacity before the increased traffic arrives are clearly illustrated by the recent congestion in many ports of the world, in particular in developing countries. The enormous sums of money lost through congestion would often have been sufficient to build a lavish system of

modern ports.
Broader Traffic congestion (#PD0078).
Aggravates Marine accidents (#PD8982).
Aggravated by Unethical practices in transportation (#PD1012).

♦ PE4767 Wet rot of wood
Nature Commonly found in external joinery with relatively high moisture content. As paintwork cracks, spores of Coniophora puteana enter the wood and the fungus can reduce it to powder. The affected wood must be destroyed and the causes, such as faulty plumbing, eliminated.
Incidence In the UK it has been estimated that 25 per cent of the houses have dry or wet rot.
Broader Wood rots (#PE4455).
Aggravates Structural failure (#PD1230).

♦ PE4768 Inadequate national law enforcement
Failure to enforce judgements and orders — Failure in the service of legal process — Unstructured law enforcement — Unenforced laws — Non-application of existing laws
Nature A relatively large number of national and local laws are not adequately enforced. Changes in moral value have resulted in the non-application of a number of existing laws, so the public is no longer quite sure what is permitted and what is not. Some legal systems maintain laws on the books so that they may be selectively enforced. Some enforcement agencies are so overwhelmed by the amount of crime taking place in their jurisdiction that it must select what laws to enforce. Some legal systems are so complex that neither the law enforcement agencies nor the public knows what is legal and what is not. Some police are not so much hampered by a lack of laws as by a lack of political backing for action. In some cases, the rights of the accused may be so restrictive to law enforcement that attempts to enforce the law are hampered by restrictions of investigations, producing evidence, finding witnesses that for practical purposes the laws are enforced in such narrow ways as to be unenforced.
Broader Lack of control (#PF7138)
Deficiencies in national and local legal systems (#PF4851).
Narrower Unenforced animal control (#PG8642) Unenforced littering laws (#PJ0380)
Unenforced civic ordinances (#PG5217) Inadequate local enforcement (#PF0336)
Inadequate system of child support enforcement (#PF6076)
Government failure to prosecute offenders effectively (#PE9545).
Related Deficiencies in international law (#PF4816)
Inadequate international law enforcement (#PF8421)
Inadequate enforcement of safety regulations (#PD5001).
Aggravates Poaching (#PD2664) Injustice (#PA6486)
Concubinage (#PF2554) Forced labour (#PC0746)
Chattel slavery (#PC3300) Secret societies (#PF2508)
White slave trade (#PD3303) Trafficking in women (#PC3298)
Inadequate drug control (#PC0231) Violations of private law (#PD5727)
Violations of contract law (#PE5786) Exploitation in employment (#PC3297)
Hindrance of law enforcement (#PD5515) Inadequate prevention of crime (#PF4924)
Havens for environmental pollution (#PE6172) Trafficking in children for adoption (#PF3302)
Inequitable administration of justice (#PD0986)
Inadequate enforcement of human rights (#PC4608)
Limited local respect for regional and global legislation (#PF2499)
Exploitation of regulatory loopholes in countries with underdeveloped legislation (#PE4339).
Aggravated by Unsolved crimes (#PF6911) Government inaction (#PC3950)
Lack of enforcement power (#PF0223) Law enforcement complexity (#PF2454)
Diminished personal capacity (#PF9429) Inadequately worded agreements (#PF5421)
Arbitrary enforcement of regulations (#PD8697) Ineffective monitoring of illegal activity (#PF7264)
Unethical practices of regulatory inspectors (#PF8046)
Inadequate evidence to convict known offenders (#PF8661)
Excessive cost of effective prosecution of offenders (#PE6059)
Deliberate governmental avoidance of legislative reform (#PF5736)
Use of undue influence to obstruct the administration of justice (#PE8829).
Reduces Psychotic violence (#PE7645).

♦ PE4771 Unethical practices with domesticated animals
Unethical practices with pet animals — Unethical practices with farm animals — Unethical practices with circus animals — Unethical practices with zoo animals
Claim Animals may be kept in environments, or under conditions, which are inappropriate to their well-being. This is especially the case with animals kept for purposes of personal prestige, entertainment or companionship (pets, circus and zoo animals), where the animals normally require large surface area, or a particular vegetation and climate.
Narrower Kidnapping of pet animals (#PE0805)
Disproportionate allocation of resources to pet animals (#PF1085).
Aggravates Bestiality (#PE3274) Unwanted pet animals (#PD2094)
Maltreatment of animals (#PC0066) Ritual slaughter of animals (#PF0319)
Maltreatment of animals in aquaria (#PE5461)
Capture and use of wild animals as pets (#PD1179)
International trade in endangered species (#PC0380)
Environmental hazards of new species introduction (#PC1617).
Aggravated by Proliferation of pets (#PD2689).

♦ PE4780 Excessive dependence of local communities on outside services
Lack of neighbourhood technology
Nature Communities or neighbourhoods in large cities depend on centralized services for information, energy, mass transportation, food distribution, water, sewage systems and a number of other services that are controlled from a distance. Inefficiencies due to the large-scale of such services are manifested in inadequate administration and excessive cost, while the alternative of generating some local services through the development of appropriate technologies is rarely considered, despite the potential efficiencies that this would offer. In fact, decentralization opportunities exist wherever cities have fallen behind in utilizing contemporary technology. This might apply, for example, to centralized large main-frame computers, where small, powerful local networks might be a possible alternative. Another example is the huge private power plants required to support the needs of a whole city: local areas could utilize solar energy, high-energy producing windmills, photovoltaic batteries, or other practical means, to generate most if not all of a neighbourhood's energy requirements. Other local technology available includes: in-house toilet and waste-disposal systems; locally-produced and operated electric cars; hydroponic and other food-growing technologies for fresh fruits and vegetables; and high-production protein ponds using specially bred fish. Such emphasis on local technology could lead to the development of carpentry and bricklaying trades, and small scale manufacturing and light assembly.
Broader Obstacles to community achievement (#PF7118).
Related Lack of artisans and craftsmen in developed countries (#PE4804).
Aggravates Low productivity of agricultural workers in developing countries (#PE5883).

♦ PE4781 Inhumane medical experimentation during war-time
Nature Suspension of many civil rights and violations of international accords are characteristic of declared and undeclared wartime conditions, during which captured enemy forces, citizens of occupied countries or political prisoners may be used for scientific and unscientific experimentation.
Incidence The physician Celsus of the first century A.D. approved the vivisection of condemned criminals by his Egyptian predecessors, Heoophilus and Erasistratus. His defence was that the suffering of the few is justified by the benefit of the multitude. By the late 1930s, the professional clinical investigator was established on the medical scene and research had become an integral part of hospital practice in the UK and the US. During World War II, Nazi Germany conducted infamous experiments in the concentration camps. During the Korean War and recently in Northern Ireland, medical-psychological experimentation has been reported as an aspect of torture.
Broader Abusive experimentation on humans (#PC6912)
Obstacles to medical experimentation (#PF4865)
Unethical experiments with drugs and medical devices (#PD2697).
Aggravates Torture (#PB3430).
Aggravated by Concentration camps (#PD0702).

♦ PE4784 Excessive cost of animal protein
Nature Animal protein foods are very popular but their production is very inefficient in its utilization of land (as compared to grain or vegetable production). Consequently, the price of animal protein foods is high and beyond the reach of many low-income populations. As the amount of available agricultural land per capita decreases, the inefficiency of livestock raising becomes an increasing burden.
Broader Food insecurity (#PB2846) Economic inflation (#PC0254)
Malnutrition among indigenous peoples (#PC3319).
Aggravated by Prohibitive livestock husbandry costs (#PJ4075)
Underdevelopment of food and live animal production (#PF2821).
Reduced by Harbouring national security offenders (#PE0484)
Vitamin E deficiency in domestic animals (#PE4760).

♦ PE4785 Obstructions to international personnel exchanges and cultural cooperation
Narrower Obstacles to international cultural exchange (#PF4857)
Obstructions to international athletic exchange (#PE4809).
Related Lack of international cooperation (#PF0817)
Politicization of international sports events (#PF4761)
Abuse of international cultural, diplomatic and commercial exchanges (#PF3099).
Aggravated by War (#PB0593) Ideological conflict (#PF3388).

♦ PE4786 Maltreatment of animals for the media
Maltreatment of animals for the cinema — Maltreatment of animals used in theatre acts — Maltreatment of animals for advertising
Claim Animals are widely used in advertising, television and the cinema. Their participation often involves prolonged captivity, stress, heat exhaustion and even brutal treatment (as in battle scenes) or death before the camera (if the plot so requires it). Live birds crushed and killed during magicians performances and the confinement and handling of animals by ill-trained personnel in theatres and television studios exemplify the dangers to animals used as entertainment, which range from simple discomfort to terror and death.
Broader Maltreatment of performing animals (#PE4810).

♦ PE4787 Damping-off disease of plants
Nature Soil organisms destruct young seedlings either before the sprouting seeds break through the soil or after they emerge from the soil. It is due to soil fungi functioning in cold, wet soils or due to warm, humid weather and overcrowdedness of seedlings.
Broader Zygomycetes (#PE0614).

♦ PE4789 Lack of payment for housework
Uncompensated housewives' labour — Dependence on unpaid household work — Unpaid household work
Nature The amount of time taken for various household tasks varies, partly with the number in the family and the ages of the children, partly with the size and location of the residence and partly with the technology available. It is generally agreed that household work is time-consuming and usually takes up as many hours as full-time work outside the home, sometimes more. Yet the women who put in this time are generally held in low economic esteem and the importance of their function is given little recognition. Because their work is unpaid and not evaluated in monetary terms, its economic and social value is seldom appreciated at its true worth. Because they tend to be regarded simply as dependents, they have low social visibility and seem scarcely to exist in their own right. They are frequently badly disadvantaged as regards such things as social security, property and taxation.
Incidence If the value of housework were to be calculated in monetary terms, it would make a considerable difference to the Gross National Product of every country. It would also affect family decisions about how the work is done and the feeling of personal worth and dignity of the housewife herself.
In one study in the US, in 1979 concluded that the average value of housework in a family with two teenage children was US$ 14,500 if the wife was not otherwise employed and US$ 10,500 if the wife was otherwise employed.
Refs Goldschmidt-Clermont, Luisella *Economic Evaluations of Unpaid Household Work. Africa, Asia, Latin America and Oceania* (1987); Goldschmidt-Clermont, Luisella *Unpaid Work in the Household* (1989); Kurian, George and Ghosh, Ratna *Women in the Family and the Economy* (1981).
Broader Unpaid labour (#PD3056) Exploitation of women (#PC9733)
Underpayment for work (#PD8916).
Aggravated by Overworked women (#PD7762).

♦ PE4791 Borderline mental retardation
Backwardness — Deficientia intelligentiae
Refs Robson, Kenneth S *The Borderline Child* (1982).
Broader Mental deficiency (#PC1587).

♦ PE4792 Water torture
Tortures using water
Nature A variety of methods for torturing victims have been developed. In the Bathtub or Bañera, which is contaminated with hair, vomit, urine or excrement, the victim's head is held under water until the point of suffocation, then the victim is removed from the water and further interrogated. Or the subjects are immersed in water with a hood over their heads or while tied in a cloth sack that prevents breathing when the victims are taken out of the water. Wet towels can also be wrapped around the victim's face to prevent breathing. They are bound and squirted with water under high pressure in the mouth and nose. Another form is to expose the victim's body to dripping water for extended periods of time. During interrogation the naked prisoners can be regularly drenched with cold water. Victims are held naked for long periods of time in cells partially filled with water. They are forced to sleep on wet floors.
Incidence Water torture has been reported in the following countries: **Af** Djibouti, Gabon, Kenya. **Am** Argentina, Bolivia, Chile, Colombia, Mexico, Paraguay, Peru, Uruguay. **As** China, Indonesia, Iran Islamic Rep, Korea Rep, Syrian AR.
Broader Physical torture (#PD8734).

♦ PE4793 Denial to working animals of restorative nourishment and rest
Nature Horses, buffaloes and other working animals are used until their strength fails, with little consideration for their needs as sentient beings.
Broader Denial of rights of animals (#PC5456)
Denial to animals of legal protection of their rights (#PE8643)
Denial to experimental animals of the right to freedom from suffering (#PE8024).
Related Animal suffering (#PD8812) Scant animal feed (#PG5264)
Inadequate feeding of animals (#PC2765) Prohibitive livestock husbandry costs (#PJ4075)
Denial to animals of the right to life (#PF8243)
Denial to animals of the right to dignity (#PE9573)
Undesirable effects of animal feed additives (#PD1714)
Denial to animals of the right to a natural death (#PE8339)
Denial to working animals of limitation of working hours (#PE6427)
Denial to animals of the right to freedom from mass killing (#PE9650)
Denial to food animals of the right to freedom from suffering (#PE3899)
Underproductivity of draught animal power in developing countries (#PF0377)
Long-term shortage of animal feedstuffs excluding unmilled cereals (#PE8514)
Instability of trade in animal feedstuffs, excluding unmilled cereals (#PE8816)
Denial to animals of the right to the attention, care and protection of humankind (#PF5121)
Denial to animals of the right to conditions of life and liberty proper to their species (#PE6270).

♦ PE4796 Social insecurity in developing countries
Inadequate social insurance in developing countries — Inadequate public assistance in developing countries — Inadequate social security and welfare services in developing countries
Nature Social security in developing countries is not always based on basic social rights guaranteed by law to all human beings, whether living from the results of their work or during their temporary or permanent inability to work; nor does the law always guarantee the same application of social rights to everyone without discrimination of race, nationality, religion, sex and age, and without distinction between contract and independent workers belonging to any part of the economy. Social security frequently does not guarantee against all social difficulties and risks, such as sickness, old age, invalidity, work accidents, occupational diseases, unemployment, family difficulties, the death of the family provider, poor crops, or natural calamities and epizootic diseases which destroy produce and livestock. Free medical, psychological, pharmaceutical and family planning services through social institutions such as hospitals, dispensaries, mothers' and infants' welfare centres, and social centres are not always ensured. There may be a lack of special attention to the prevention of work accidents and occupational diseases; the system may not grant sufficient allowances to ensure a means of existence to all insured persons and their families; may not be financed by employers and the state without heavy contributions from those insured; and may not be managed by representatives of those insured, under state supervision.
In addition, the lack of realistic statistical data hampers the preparation of general, rational local and national social service plans. Such data is needed by governmental and non-governmental agencies and organizations, such as trade unions, cooperatives, employers programmes, charities, private hospitals and schools, banks and developmental consortia.
Industrialization is accompanied by an influx of workers from areas with traditional ways of life into urban centres, and this change gives rise to feelings of insecurity. The workers may fear that sickness or accident, invalidity or old age will deprive them of their means of existence, as they can no longer easily turn to a family or communal group for support and aid.
The decline of traditional social security arrangements is more obvious in cities but it also affects the countryside. The depopulation of villages by emigration of men results in a lessening of communal and family solidarity and in an increasing need to assist the traditional structure of society to care for the helpless, the aged, the blind, the infirm and orphaned children. Moreover, the governments involved face a fundamental problem regarding the extent to which their limited facilities and financial resources should be applied to the destitute, who represent only the fringe of a vast population that is, in any case, always on the edge of destitution; and the extent to which those facilities and resources should be applied to preventative health measures and measures for improvement of general standards of living.
Incidence Many third world suburban communities are pervaded by a feeling of social insecurity, and this atmosphere is immediately sensed by an outsider who walks into the community. An environment of uncertainty has become a part of what it means to live in such areas. The local people are insecure; their day-to-day operations are carried out in (usually unconscious) fear. This social anxiety is manifest in many ways: inadequate safeguards against devastating fires or theft; inadequate services for accidents; individuals disabled by undiagnosed or untreated disease; threat of personal assault; and simple fear of unlighted roads and homes.
Claim The low state of social security in the developing countries is due mainly to their poverty. This in turn arises from the imbalance in the wealth of the population. Lack of educational democratization, and a class system of a privileged few and underprivileged many, creates serious obstacles to the institution and maintenance of social security insurance and welfare service programs as an inadequate infrastructure is maintained.
Refs ILO *Improvement and Harmonisation of Social Security Systems in Africa* (1977); Mouton, Pierre *Social Security in Africa* (1976).
Broader Social insecurity (#PC1867) Economic insecurity (#PC2020).
Related Unequal coverage by social security (#PF0852)
Inadequate welfare services for the deaf (#PD0601)
Inadequate welfare services for the aged (#PD0512)
Inadequate welfare services for the blind (#PD0542)
Personal physical insecurity in developing countries (#PE5391)
Denial of the right to social security in capitalist systems (#PD3120)
Narrow legal definition of the family in developing countries (#PD1501).
Aggravates Family poverty (#PC0999) Poverty in developing countries (#PC0149)
Limited access to social benefits (#PF1303) Negative effects of family allowances (#PF0107).
Aggravated by Inadequate insurance (#PF8827) Increasing cost of social security (#PF7911)
Economic and social underdevelopment (#PB0539)
Disruption of family system in developing countries (#PD1482)
Loss of traditional forms of social security in developing countries (#PD1543).
Reduced by Weakness in trade among developing countries (#PC0933)
Inequitable tax systems in developing countries (#PD3046)
Weakness of infrastructure in developing countries (#PC1228).

♦ PE4802 Pollution by aircraft
Nature The increasing size of aircraft, the emission of black smoke during take-off, and the density of air traffic at major airports, have directed attention to pollution by aircraft. Airports are usually situated in relatively open country, so that aircraft fumes are unlikely to constitute a significant health hazard. The problem is mainly one of amenity; there may be complaints of smell and there is inevitably concern over noise. Further pollution by aircraft arises from the jettisoning of spare fuel after take-off. Under such circumstances, it must be released at a height sufficient to allow it to vaporize so that it does not reach the ground in liquid form. Some countries have already taken action to prevent the jettisoning of fuel except in emergency. Concern has also been expressed that the increasing use of supersonic aircraft flying at high altitudes may lead to increasing pollution of the upper air, where pollutants may accumulate since natural dispersion at such heights is not very effective.
Broader Aircraft environmental hazards (#PD8328).
Narrower Aircraft noise (#PE5799).

♦ PE4804 Lack of artisans and craftsmen in developed countries
Insufficient manually skilled workers
Nature A lack of economic incentives has caused a dearth of manpower to enter the crafts in many high-technology oriented countries. There are shortages of skilled carpenters, plumbers, electricians, bricklayers and others, particularly in many major urban labour markets.
Broader Shortage of skilled labour (#PD0044).
Related Excessive dependence of local communities on outside services (#PE4780).
Aggravated by Elimination of traditional skills (#PD8872).

♦ PE4805 Unethical experimentation using aborted foetuses
Research on foetal cadavers and foetal tissue — Research on human embryos
Nature For certain branches of medical research the use of foetal tissues is claimed to be indispensable, and the lost or discarded foetus serves as an obvious source of such material. However, the use for medical research of foetal tissues, or even of whole but non-viable foetuses, may be conducted unethically. Research with foetal tissue denies the special respect due to the human foetus. Especially problematic is non-therapeutic research on the foetus in utero in anticipation of abortion. The aborted foetus can be regarded as a dead person, and may be treated with all the respect that is normally accorded to the dead, or it can be regarded as a sort of tumour or parasite of the womb of the woman who conceived it. In aborted foetal experimentation, the medical profession follows the legal loophole that the woman who conceived it has no rights concerning its disposition, and that the state does not regard it as a human being or even an animal. However, there is also the medical eventuality that some aborted foetuses may be made viable ex-utero, and in the meantime the foetus' right to life should be protected.
Some distinctions are important. Are the foetuses alive or dead ? Is the research being done *in utero* or *ex utero*; on foetuses to be aborted or to be brought to term; or on pre-viable or viable foetuses ? Is the foetus spontaneously aborted or is the abortion induced ? Is fertilization *in utero* or *ex utero*? Is the research to benefit the subject as well as others, therapeutic research; or to only benefit others, non-therapeutic research ? Is the risk to the subject minimal, moderate or serious ?
Claim Many sectors of the public and some physicians find the idea of research on whole living foetuses, or even on living foetal tissue, as indefensible on moral, religious or emotional grounds. The only relevant guideline that could be proposed in this case is that when foetal research is undertaken it should be only on subjects that would, by common consent, have no hope whatsoever of continued extra-uterine existence.
Counter-claim The use of human foetal tissue, or of the whole foetus, is indispensable for some medical research, including the culture of certain pathogenic viruses, immunological and chromosomal studies, the study of foetal development, and the preparation of certain vaccines. For example, in 1965, John Enders and Thomas Weller were awarded the Nobel Prize for growing poliomyelitis virus in cells cultured from human foetal tissues. Foetal tissue is more versatile than older people's tissue. Implants of it might one day remedy a wide range of disorders, from Alzheimer's disease, through stroke and epilepsy, to injuries of the spinal cord and even blindness caused by destruction of the optic nerve. A discovery in any of these problem areas would be a real breakthrough. When the aborted foetus, perhaps temporarily still living, suffers no harm or no pain from being the subject of an experiment, there is no logical reason to object to its use, which may be the means of arriving at a better understanding of the processes of life and disease.
Broader Inhumane scientific activity (#PC1449)
Obstacles to medical experimentation (#PF4865).
Related Undignified treatment of corpses (#PF5857).
Aggravates Unlimited practice of human embryo storage (#PE5623).
Aggravated by Unethical medical practice (#PD5770)
Denial of the right of unborn children (#PF6616).

♦ PE4807 Mine dust
Refs ILO *Fifth International Report on the Prevention and Suppression of Dust in Mining, Tunelling and Quarrying, 1968-1972* (1980); ILO *Guide to the Prevention and Suppression of Dust in Mining, Tunelling and Quarrying* (1965).
Broader Dust as an occupational hazard (#PE5767).

♦ PE4808 Foetal malformation in diabetic pregnancies
Incidence The high incidence of foetal malformation is one of the key problems in diabetic pregnancies. More babies die from lethal malformation in these cases than from all other causes during the perinatal period in a standard population, and the proportion of perinatal deaths due to malformation rises from 9 to 20 percent as the total perinatal mortality is reduced. Cerebral malformation, including acrania, hydrocephalus, meningocele and monster may occur in up to 25 percent of such malformed infants, all leading to the death of the baby.
Refs Hare, J W (Ed) *Diabetes Complicating Pregnancy* (1989).
Broader Pregnancy disorders (#PD2289) Physical malformation of foetus (#PE2042)
Human physical genetic abnormalities (#PD1618).
Related Diabetes (#PE0102) Premature birth (#PD1947).
Aggravates Sudden unexpected infant death (#PD1885).
Aggravated by Lack of eugenic measures (#PD1091)
Inadequate medical care for pregnant women (#PE4820).
Reduces Perinatal morbidity and mortality (#PD2387).

♦ PE4809 Obstructions to international athletic exchange
Broader Obstructions to international personnel exchanges and cultural cooperation (#PE4785).
Narrower Politicization of international sports events (#PF4761).
Related Obstacles to international cultural exchange (#PF4857).

♦ PE4810 Maltreatment of performing animals
Maltreatment of circus animals — Maltreatment of rodeo animals
Nature Cruelty to performing and circus animals includes over-discipline involving excessive beatings and withholding of food, excessive hours of training, cramped quarters (often small cages on trailers), and extermination when biologically fit but ageing. Transportation between locations subjects the animals to extreme changes in temperature and disruption of natural cycles, causing considerable stress.
Claim Learning of tricks goes against an animal's natural instincts, especially when taught under demeaning conditions. During any performance animals may be subject to ridicule and contempt in order to conform to the public's stereotype of them.
Counter-claim By experience and training circus animals become accustomed with many things other animals, inexperienced in circus, would find distressing.
Broader Exploitation of animals for amusement (#PD2078).
Narrower Maltreatment of marine show animals (#PE0906)
Maltreatment of animals for the media (#PE4786).
Related Maltreatment of zoo animals (#PE4834) Maltreatment of animals in aquaria (#PE5461).
Aggravated by Unethical veterinary practice (#PD7726).

♦ PE4811 Giardiasis
Lambliasis
Nature Giardiasis is an infectious disease caused by one celled protozoan parasite called Giardia.

Symptoms range through diarrhoea, weakness, weight loss, abdominal cramps, nausea, greasy stools, abdominal distension, flatulence, vomiting, belching and fever. As well as intestinal disorders, there might be severe malnutrition and problems with the bile ducts and gall–bladder or even the pancreas. As a faecal infection, giardiasis spreads from person to person, particularly in families and amongst those living in close–knit environments. It is also transmitted from animals to humans. Giardia are not effected by chlorination and remain viable in water for a long time.
Background Giardia was first described in 1681 by a Dutch scientist, Antoni van Leeuwenhoek, and as such was the first pathogenic protozoan seen. It went unrecognized until 1859, when Dr. Vilem Dusan Fedorovic Lambl accidentally discovered the organism. Giardiasis was not effectively treated until 1937.
Incidence Giardiasis may affect two percent of the population in Europe and up to 20 percent in some tropical countries.
Refs Erlandsen, Stanley L and Meyer, Ernest A *Giardia and Giardiasis* (1984).
Broader Protozoan parasites (#PE3676) Parasites of the human body (#PE0596)
Intestinal infectious diseases (#PE9526)
Diseases of the digestive system in animals (#PD3978).
Related Amoebiasis (#PE6782).

♦ PE4814 Neurofibromatosis
Elephant Man's disease
Incidence The disease afflicts about 1.5 million people worldwide, of which some 100,000 are in the USA.
Refs Riccardi, Vincent M and Eichner, June E *Neurofibromatosis* (1986).
Broader Congenital anomalies of nervous system (#PE9296).
Aggravated by Consanguineous marriage (#PC2379).

♦ PE4818 Ineffective economic structures in industrial nations
Nature The economic structures of industrial nations are failing to close the gap between their economic potential and their actual performance. Protectionism, including agricultural, industrial and other subsidies reduce competition in national and international markets for goods and services. High marginal taxes, currency restrictions, financial market regulations, discriminatory tax rates applied to similar transactions, and excessively steep progressive income tax rates restrict capital flows. Centralized collective bargaining, employment regulation, social security restrictions, and government control of training and education have created rigid labour markets.
Broader International economic injustice (#PC9112)
Underdevelopment of industrial and economic activities (#PC0880).
Related Economic inefficiencies in developing countries due to restrictive business practices (#PE2999).
Aggravates Labour hoarding (#PE6333) Segmented labour markets (#PD6744)
Prohibitive labour costs (#PF8763) Insufficient capital investment (#PF2852)
Declining economic productivity (#PC8908) Underutilization of labour force (#PF6293)
Structural rigidities in labour markets (#PD4011)
Arrested development of labour potential (#PF6532)
Lack of time flexibility in the labour market (#PE8283)
Diminishing capital investment in small communities (#PF6477).
Aggravated by Rural unemployment (#PF2949) Trade protectionism (#PC4275)
Economic isolationism (#PC2791) Distortionary tax systems (#PD3436)
Absence of a long–range, world–wide capital flow plan (#PF2865)
Tax discrimination against non–residents of a country (#PD3048)
Restrictions on foreign access to capital bond markets (#PD3135)
Tax discrimination against investment in a foreign country (#PD3047)
Unimaginative vision of existing international economic structures (#PF2699)
Inequitable tax treaties between developed and developing countries (#PD1477)
Distortion of international trade by restrictive controls on movement of labour (#PE8882)
Distortion of international trade by selective indirect taxes and import charges (#PE8867).
Reduced by Disorganized labour force (#PU0929).

♦ PE4820 Inadequate medical care for pregnant women
Lack of prenatal care
Broader Fragmented forms of care at the neighbourhood level (#PE2274).
Related Inadequate maternal and child health care (#PE8857)
Excessive medical intervention in childbirth (#PE7705).
Aggravates Premature birth (#PD1947) Pregnancy disorders (#PD2289)
Neglected young children (#PE4245)
Foetal malformation in diabetic pregnancies (#PE4808).
Aggravated by Unmarried mothers (#PD0902) Denial of right of family planning (#PE5226)
Resistance to incorporating midwives in medical care systems (#PE4901).
Reduces Perinatal morbidity and mortality (#PD2387).

♦ PE4821 Feeblemindedness
Mild mental retardation
Nature Children with this level of mental retardation are often not distinguishable from normal children until their teens, when they are able to acquire academic skills up to sixth–grade level. Later they can achieve the necessary skills for minimum self–support, but need assistance to cope with unusual stress situations.
Refs Goldstein, Herbert and Goldstein, Marjorie T *The Reasoning Ability of Mildly Retarded Learners* (1980).
Broader Mental deficiency (#PC1587).
Aggravated by Consanguineous marriage (#PC2379).

♦ PE4825 Sulphuric acid mist
Nature Sulphuric acid in vapour form causes intense irritation and chemical burn of the mucous membranes of the respiratory and digestive tracts, the teeth, eyes and skin. Inhalation of vapours produces the following symptoms: nasal secretion, sneezing, a burning feeling in the throat; these are followed by cough, respiratory distress, sometimes accompanied by spasm of the vocal cords, a burning sensation in the eyes with lacrimation and conjunctival congestion. High concentration may cause a bloody nasal secretion and sputum, haematemesis, gastritis, etc. Dental lesions are common.
Broader Acid mists (#PE1976).
Aggravates Acidosis (#PG4564).

♦ PE4826 Theft of vehicles
Risk of car theft — Automobile theft
Nature Virtually every type of vehicle, mechanical, motorized, or otherwise powered, has been and still is subject to theft. Even babycarriages, biplanes and boats have been absconded with, but the thefts of bicycles, motorcycles, automobiles, utility vehicles and trucks is an international phenomenon. Some of this activity may involve organized crime and thus represent an illegal industry of large proportions. Other thefts are perpetrated by adolescents seeking the excitement of ownership and operation. Insufficient safeguards to prevent vehicular theft is a contributory factor.
Broader Theft of property (#PD4691).

♦ PE4828 Birth injuries
Refs Kobayashi, Michele Y *Birth Injuries* (1987).
Related Human physical genetic abnormalities (#PD1618).
Aggravates Perinatal morbidity and mortality (#PD2387).

♦ PE4833 Exploitation of athletic competition for commercial or political ends
Broader Political exploitation (#PC7356) Exploitative entertainment (#PD0606).
Narrower Politicization of international sports events (#PF4761).

♦ PE4834 Maltreatment of zoo animals
Zoo mismanagement — Inadequate zoo facilities
Nature Animals are frequently placed in zoos in unnatural habitats. Confined areas, little natural light, inadequate temperature control, improper diet and stress–inducing proximity of spectators, are some of the conditions experienced. Inadequate companionship of their kind and abuse by keepers are additional factors contributing to animal illness and mortality in zoological parks or research centres open to the public.
Claim 1. Of the tens of thousands of zoos worldwide, there are less than 50 with sufficient funds and expertise to provide adequate conditions for their animals. Large zoos are often willing to buy critically endangered species to enhance the prestige of their collection, even though they are aware that the species can only be obtained illegally. Private zoos have frequently been proved to be fronts for illegal animal–trafficking operations.
2. The life of most animals in captivity is an unusual one. Most animals are forced to make considerable adjustments to captive life. Furthermore, many of these creatures are captured as infants, which precludes satisfactory parental and hierarchical relationships during periods crucial for normal psychological development. This environmental upheaval is likely to have an important effect on behaviour, morbidity and mortality. For example, hypertension, gastric ulcers, eclampsia and remarkable cerebral arteriosclerosis are not unusual in captive primates.
Counter–claim Captive breeding programmes in zoos are often the only means whereby an endangered species can be saved, especially when their habitat is imminent danger of being destroyed.
Refs Batten, Peter *Living Trophies* (1976).
Broader Mismanagement (#PB8406) Inadequate animal welfare (#PC1167)
Exploitation of animals for amusement (#PD2078).
Narrower Zoosadism (#PG2204).
Related Maltreatment of performing animals (#PE4810)
Inadequate housing and penning of domestic animals (#PE2763).
Aggravated by Unethical veterinary practice (#PD7726).

♦ PE4835 Risks in power production
Nature All forms of power production carry measurable risks which must be weighed against the benefits resulting from the power thus made available. Conversely, the complete elimination of these risks, possible only if the power sources were not used, needs to be weighed against the consequent disadvantage of having no power. Compared to the hazards involved in nuclear power generation, there appear to be considerably greater hazards in the case of coal, somewhat greater for oil and possibly somewhat less for natural gas. Hydro–electric, solar and other forms of power generation cannot be considered on the same scale.
Aggravates Environmental hazards of coal energy (#PD7541)
Environmental hazards from electricity (#PE1412)
Environmental hazards of nuclear power production (#PD4977).

♦ PE4838 Menstruation
Menstrual bleeding — Negative attitudes towards menstruation
Nature Periodic discharge, mainly of blood, from the womb, beginning between the age of 13 to 15 and lasting from 2 to 8 days. It recurs approximately every 28 days, ceasing during pregnancy and lactation, and continues until the age of 44 to 50 years. It causes considerable inconvenience, whether physical or through increase in tension. It may be associated with menstrual pains and headaches. It may provoke various forms of discrimination against the woman at that time, or be used to justify such discrimination at any time.
Background A number of cultures have prescribed strict adherence to ritual behaviour coordinated with the menstrual cycle. For instance, during the first few days of her menstrual period a Hindu woman may not mount a horse, an ox, or an elephant, or drive a vehicle; and many peasants of Central and Eastern Europe persist in the belief that a menstruating woman should not bake bread, churn butter, or spin thread. The Orthodox Jewish and Muslim cultures specify a number of restrictions on menstruating women; including sexual abstinence and separation from society.
In early Western cultures, menstruation was believed to render a woman periodically dangerous, and numerous and varied social restrictions were created to limit her contact with her husband and with members of her community. During the 19th century, it was the opinion of some physicians that menstruation had no purpose whatsoever and that, indeed, it was a pathological condition, which had not existed in pre–Biblical times. This concern about the dangers of menstrual blood prevailed until as late as 1945.
Incidence As an indication, the market for female sanitary protection ((including female hygiene tampons and sanitary towels) in the UK in 1986 was £123 million.
Claim Throughout history, menstrual bleeding has been seen as a supernatural event, and generally one with evil consequences. The belief that blood carried some basic life principle led to the prevalent fear of menstrual flow and consequently to numerous restrictions and taboos on the activities of women.
Refs Blumenthal S J and Osofsky, H J *Premenstrual Syndrome* (1986); Golub, Sharon (Ed) *Lifting the Curse of Menstruation* (1983).
Related Premenstrual tension (#PE3761).
Aggravates Acne (#PE3662) Headache (#PE1974) Nutritional anaemia (#PD0321)
Toxic shock syndrome (#PE5106) Disagreeable human body odour (#PE4481).

♦ PE4847 Abuses by private police forces
Nature Private police, whether detective, security, protection, detention or vigilante, maintain their effectiveness by not obtaining public police power hence not being restricted by constitutional obligations. In many instances these groups carry out their activities with implicit cooperation from public agencies; they are given the right to bear arms, the right to detain or arrest; and their use of violence lead to abuses.
Broader Human violence (#PA0429) Abuse of police power (#PC1142)
Unethical practices by police forces (#PD9193).

♦ PE4853 Shortage of pesticides
Nature Some pesticides, herbicides and fungicides are going out of production because of declining markets in the richer countries.
Related Instability of trade in manufactured fertilizers (#PE0806)
Instability of trade in crude fertilizers and crude minerals, excluding coal, petroleum and precious stones (#PE0760).

♦ PE4856 Economic and social losses due to disability

Excessive costs and cultural burdens of human illnesses and disabilities

Nature Disability often leads to poverty or disruptive change in economic status for the disabled individual and for his or her family. The effect on the distribution of income between households may be doubly adverse: the disabled person loses his or her income; the need to care for the disabled person may cause some other family member to stop working. Disability can have serious financial implication for industry in terms of labour turnover and the retention of new workers. Disability may force people to remain idle and dependent. In countries with high unemployment, the disabled may be relegated to reserve labour force status, to be employed only when demand for labour is very high and to be laid off as soon as demand falls.

Disability may reduce the active work force capability of a nation with a resultant effect on the support of the social benefit system. The costs of disability are greatest in those nations which are in need of an increased active work force. In some industrialized nations with ageing populations and increased numbers of disabled persons, there is a trend towards reduced numbers of active workers supporting each recipient of social benefits. In other nations, the population in younger age groups is increasing. These trends have long-term effects on the financial bases of social benefit schemes. In developing countries with normally low rates of employment, planners may erroneously conclude that it is unnecessary to include the disabled in their labour policy.

Narrower Melioidosis (#PG0774) Being a burden (#PF9608).
Aggravates Absenteeism (#PD1634) Prohibitive cost of living (#PF1238).
Aggravated by Disease (#PA6799) Social neglect (#PB0883)
Mental deficiency (#PC1587) Health inequalities (#PC4844)
Environmental pollution (#PB1166) Personal care disabilities (#PE6770)
Developmental disabilities (#PD3773) Human disease and disability (#PB1044)
Inadequate social welfare services (#PC0834).
Reduces Labour shortage (#PC0592).

♦ PE4859 Unsafe industrial, laboratory and medical equipment

Nature Equipment may be electrically unsafe, explosion hazardous, radiation unsafe or environmentally unprotected. Safety cannot usually be built into equipment alone, but calls for a combination of measures extending also to installation rules and maintenance and application of the equipment. A number of isolated first faults may eventually cause a hazard. Faults are due to designs having an insufficiently large safety factor and lack of redundancy techniques or protective devices. Manufacturers and installers do not take into account a safety factor that not only allows for initial mechanical and electrical strength but also the effect of use and wear, production methods and transport and storage conditions.

Incidence Considering scientific laboratories in particular, their variety, size, type and complexity preclude simple generalizations on the health and safety of laboratory work. Laboratory workers are selected and employed primarily because of their specialized education, knowledge and skills, and not because of any qualifications related to health or safety interest. Unless the laboratory is intimately integrated with a manufacturing facility, the degree of regulation and control actually enforced is usually lower than that for production operations. In addition, there is a feeling which has been fostered by the academic community itself that little should be done to interfere with 'academic freedom', no matter how serious the consequence of that freedom. In addition, research laboratories in academic institutions and in government and industrial establishments are frequently at the frontiers of knowledge both of science and of hazards. For this reason, laboratory workers are often the first persons to be exposed to new chemical and physical dangers, and they may suffer unexpected injury unless effective control, monitoring and medical supervision are integrated into the planning of the laboratory operations.

Narrower Counterfeit machine parts (#PE5319).
Aggravates Large-scale industrial accidents (#PD2570).

♦ PE4860 Government discrimination against trade unions

Denial of trade union rights by governments

Broader Violation of trade union rights (#PD4695)
Discrimination against trade unions (#PC4613).
Narrower Anti-labour laws (#PG8050).

♦ PE4863 Unfit legal defendants

Nature Mentally ill or mentally handicapped persons who are apprehended for alleged commission of crimes may be unfit to deal with the criminal justice system. They may be unable to participate in reasonable and comprehending dialogue with police officers and investigators and with private or publicly appointed lawyers for the purposes of defence. Inability may extend to standing trial and pleading innocent or guilty. The treatment of such persons varies significantly from country to country. The mentally ill or handicapped person may be forced to stand trial without any special assistance, or may be interned without trial in a psychiatric hospital from which release is a difficult and prolonged process. In some countries such as the UK, there is provision for medical and mental treatment before trial.

Incidence A recent case in the UK involved a mentally deficient youth being committed to prison for stealing a milk bottle.

Broader Social injustice (#PC0797).

♦ PE4866 Imbalance between capital and technical assistance

Nature Much of the aid administered by the United Nations Development Programme is concentrated in capital investment. Only about 20 percent goes toward technical assistance for developing the human and institutional infrastructures required to support such lending on a practical basis. Of the World Bank loan commitments in 1982, about 11 percent were applied to technical assistance. The insufficiency of technical assistance is in the range of one-half; that is, technical assistance funding needs to be doubled if strictly capital investments are to be better secured and social development forwarded.

Broader Imbalance (#PA0224).
Related Inadequate technical advice and consultation on problems (#PF1981).
Aggravates Inadequate social welfare services (#PC0834)
Weakness of infrastructure in developing countries (#PC1228)
Disparity in social development within developing countries (#PD0266).

♦ PE4868 Immigration barriers for handicapped family members

Handicapped refugees

Nature A handicapped refugee, in addition to suffering the loss of ties with his family, country, and national community that all refugees face, has, in addition, the stigma while living in a refugee camp awaiting emigration, of being labelled 'unemigrable'. He must then often decide whether to break up his family, by sending away those members who can leave, or have them all suffer the physical and moral degeneration inherent in refugee camp life.

Incidence Only the Scandinavian countries, Belgium, Switzerland and France have emphasized the successful integration of badly handicapped refugees into their communities.

Counter-claim Handicapped refugees have proven adaptable in new countries and modern medical methods and modern methods of rehabilitation permit the fairly rapid re-adaptation of handicapped individuals to the requirements of modern societies in developed countries.

Broader Restrictions on immigration (#PC0970) Discrimination against the disabled (#PD9757).
Related Socially handicapped refugees (#PD1507).
Aggravates Statelessness (#PE2485) Rejection of refugees (#PF3021)
Refusal to grant nationality (#PF2657).
Aggravated by Social neglect (#PB0883) Physically handicapped children (#PD0196)
Vulnerability of the disabled during states of emergency (#PD0098)
Inadequate vocational rehabilitation facilities for disabled persons (#PE7317).

♦ PE4869 Death threats against trade union leaders

Death threats against workers representatives

Broader Sanctions against trade union workers (#PD0610).

♦ PE4874 Damage to infant brains from malnutrition and insufficient stimuli

Nature Malnutrition during the critical brain growth period, which lasts from the seventh foetal month until the age of 18 to 24 months, can have serious effects. Recent research has shown that children under two years of age are particularly vulnerable to brain damage as a result of poor nutrition and that such damage is often irreversible. At the same time, a child's development can be deeply affected by the lack of mental stimulation which is often associated with large families and a poverty-stricken environment.

Incidence In developing countries the young child suffers more than the adult from an unbalanced diet. Children between 1 and 3 years of age frequently suffer from protein deficiency, and they often suffer from the double insult of undernutrition and environmental deprivation – factors which almost certainly act synergistically in impairing brain growth.

Refs Winick, Myron *Malnutrition and Brain Development* (1976).

Broader Neglect (#PA5438) Inhibited growth of malnourished children (#PE4921)
Protein-energy malnutrition in infants and early childhood (#PD0331).
Aggravated by Boredom (#PA7365) Child malnutrition (#PD8941)
Unbalanced infant diets (#PE0691)
Ignorance of women concerning primary health care (#PD9021).

♦ PE4875 Religious dissent

Broader Atheism (#PF2409).

♦ PE4877 Inadequate food storage facilities

Nature Foods may become spoilt as a result of lack of adequate storage, preservation or packaging. This leads to abnormal flavour, colour, odour, or consistency, although these may not necessarily be harmful. Inadequate food storage facilities in particular cause a great deal – in many areas upwards of 20 percent – of food produced to be lost due to rodents, other pests and general deterioration before it reaches the consumer. Better storage offers one of the quickest ways to increase food supply. At the national level, field experience indicates that much could be done to improve home and village level capacity for food storage and protection.

Broader Underdevelopment of food and live animal production (#PF2821).
Aggravates Food grain spoilage (#PD0811) Food spoilage in storage (#PD2243)
Unethical catering practices (#PE6615).

♦ PE4878 Attempted suicide

Parasuicide — Non-fatal deliberate self-harm

Nature Attempted suicide is usually an impulsive response to an intolerable social situation. For instance, attempted suicide is often provoked by a family quarrel, and the overdose serves as a dramatic expression of feelings. Perhaps it is the attitudes in society which condone this fashionable way of expressing distress that need to be better understood before today's rising suicide rates can be curtailed.

Incidence None of the many theories to explain the spectacular rise in attempted suicide rates has found general acceptance, nor led to successful experiments in primary prevention. Variations in the rates of attempted suicide across major demographic and social groups show remarkably consistent patterns, changing little over the years or from one country to another. Females outnumber males in all age groups. The overall ratio of females to males is about 1.4 to 1. It is among teenagers and young adults that the rise in rates has been most marked, reaching levels that are now many times higher than among the middle-aged and elderly. There is a steep gradient of increasing rates with lower social class. The unskilled lowest social class male group has more than eight times the rate of attempted suicide as the professional highest social class male group. Divorcees have substantially higher rates than the single, married or widowed. A large proportion of patients in most series have alcohol problems, criminal, unemployment and debt records, and a history of family violence. Half of those who attempt suicide have done so before, and one fifth will attempt it again within 12 months. Again, these figures for repetition are remarkably consistent from year to year and in different centres, as is the frequency with which suicide follows attempted suicide – 1 percent per annum according to many follow-up studies. Underestimation of suicide attempts is likely to be much greater than for actual suicides. Surveys in general practice in Scotland and the Netherlands showed that hospital admissions underestimated the number of cases known to general practitioners by 30 percent. Field studies, in which everyone in a population sample was screened, suggested that 50–60 percent of those who admitted to suicide attempts during the previous year had not reported the fact to a doctor. In countries where primary medical care and hospital provision is less available for these cases, the errors will be even greater.

Refs Arngrim, Torben *Attempted Suicide* (1975).

Broader Criminally life endangering behaviour (#PD0437).
Aggravated by Pain (#PA0643) Drunkenness (#PE8311)
Family breakdown (#PC2102) Mental depression (#PC0799)
Romantic separation (#PD4233) Marital instability (#PD2103).

♦ PE4880 Inadequate research and development capacity in developing countries

Inadequate capacity in developing countries for science and technology

Nature In many developing countries there are scarcities of scientists, few and fragmented research projects, inadequate science education, and social structures discouraging to science. There is a lack of indigenous scientific groups to apply science to national development by converting information into suitable practical advice in the required form; and where there are groups, they may not be able to keep in touch with up-to-date developments nor are they equipped to modify findings from abroad to provide answers that are applicable to local requirements. National and local governmental budgeting for scientific education, applied research projects and laboratories, is low priority. Besides, the scale of research and development is likely to have been reduced in many countries as part of the response to the external financial stringency widely experienced during the 1980s. The result is that scientific consultants from aboard are employed to study problems immediately affecting the economy, such as key export crops, exported natural resources and critical intermediate and finished export products, while other scientific applications to the economic and social development of these countries are deferred or neglected.

Broader Shortage of adequately trained personnel to act against problems (#PF0559).
Related Unemployed intellectuals in developing countries (#PD1273)
Barriers to the international flow of knowledge and educational materials (#PF0166).
Aggravates Underdevelopment (#PB0206) Social underdevelopment (#PC0242)
Declining economic growth in developing countries (#PD5326)

Inadequate production capacity in developing countries (#PD4219)
Inadequate research and development on problems of developing countries (#PF1120).
Aggravated by Educational wastage (#PC1716)
Maldistribution of science and technology (#PC8885)
Inappropriate education in developing countries (#PF1531)
Shortage of books and textbooks in developing countries (#PF0118)
Emigration of trained personnel from developing to developed areas (#PD1291)
Excessive social costs of structural adjustment in debtor developing countries (#PD8114).
Reduced by Use of inappropriate technologies in developing countries (#PF0878).

♦ PE4882 Public displays of sexuality
Claim Public displays of nudity, promiscuity, eroticism, rape, homosexuality and sodomy are disgusting and degrading and the unrepentant transgressors deserve the harshest penalties society can inflict.
Related Group sex (#PG6147).
Aggravated by Debauchery (#PE8923) Sexual deviation (#PD2198)
Excessive portrayal of sex in the media (#PE7930).

♦ PE4883 Inadequate immune responses in malnourished persons
Nature Malnutrition dilutes the body's natural immunity and lowers the resistance to disease. Many clinical and biological studies have been devoted to the problem of the relationships between infection and immune responses among malnourished children living in those regions where there is a high incidence of protein–calorie malnutrition. The extensive involvement of the cellular and humoral defence mechanisms explains both the frequency and gravity of certain infections among malnourished children (as, for example, in measles patients), and contributes to the understanding of why immune responses to immunization are often inadequate.
Incidence A talk with the mother of a typically malnourished three–year old in a community of the developing world, says UNICEF, would reveal that in its short life the child had suffered perhaps 6 to 16 bouts of diarrhoea, 7 or 8 infections of the upper respiratory tract, 2 or 3 attacks of bronchitis, as well as measles and conjunctivitis and maybe – depending on the exact location of the village – an attack of malaria or meningitis. The detailed health record of 45 such three–year–olds, studied in the villages of Central America, showed an average, for each child, of one illness every three weeks. It is this mutually reinforcing relationship between invisible malnutrition and infection which is responsible for the majority of the 40,000 deaths every day among the developing world's infants and children.
Claim In probably half of all cases, child malnutrition is precipitated not by the lack of food itself but by infection – especially diarrhoeal infection – which depresses the appetite and causes food to pass too quickly through the gastro–intestinal tract to be efficiently absorbed. That is why a child can sometimes be malnourished when there is food in the household and adequately nourished brothers and sisters are playing outside. More usually it is the interaction between malnutrition and infection which does the damage. Studies in Guatemala, for example, have shown that even a moderately malnourished child is three times more likely to contract a diarrhoeal infection than a child who is well fed.
Broader Malnutrition (#PB1498).
Aggravated by Infection (#PC9025)
Trace element imbalance in the human body (#PE5328).

♦ PE4887 Excessive length of pre–trial internment
Imprisonment with delayed trial — Denial of right to speedy trial — Excessive imprisonment on remand
Nature The proportion of defendants in criminal cases imprisoned pre–trial is indicative of the efficiency and equity of a nation's penal and judicial system. Where pre–trial prisoners are high in ratio to convicted prison inmates, it is a danger signal that human rights are being violated. While fluctuating crime rates and average lengths of sentences affect this ratio, it obviously bears social and legal investigation when, for example, numbers of prisoners have not been brought to trial even after months of incarceration.
Incidence In the UK for example, there is an ever increasing number of prisoners kept in less than humane conditions in police cells, and delays for prisoners in custody from arrest to trial have increased from 23 days in 1979 to 41 days in 1982. Around 1,400 prisoners on any one day have been in custody and awaiting trial for over 110 days. Other examples are Belgium, with recent figures giving 1,456 pre–trial prisoners out of its 5,844 inmates, representing 25 percent; and Venezuela, where the proportion of cases held pre–trial is excessive: 11,412 out of 14,425 inmates – almost 80 percent.
Broader Violation of civil rights (#PC5285) Denial of rights to prisoners (#PD0520)
Denial of right to fair and public trial (#PE3964).
Related Excessive burden on the poor due to legal delays (#PE1093).

♦ PE4888 Forced confessions with drugs
Involuntary narco–analysis
Nature Certain chemicals affect consciousness and volition in ways that might be used to lead a person to reveal information he would not otherwise disclose. Truth drugs such as scopolamines, sodium pentathol and sodium amytal, are relaxant agents that release inhibiting controls. Their use to force criminal and political confessions has been attested.
Broader Forced confession (#PE8947) Inhumane scientific activity (#PC1449)
Inhumane interrogation techniques (#PD1362).
Related Police brutality (#PD3543) Pharmacological torture (#PE4696).
Aggravates Substance abuse (#PC5536).

♦ PE4889 Unethical medical experimentation on prisoners
Nature When drug or other therapeutic research reaches the stage when it is ready for human trial on a controlled, scientifically–observed experimental basis, particularly under conditions of segregation, the prison population is often turned to. Experimentation in jails and penitentiaries may be done without the prisoners knowledge and consent, or with consent but with little or no understanding of the risks involved. In some cases the experiments are dangerous, the risks being surgical mutilation, pharmaceutically or chemically induced organ impairment, or personality disorders.
Incidence Although the use of prisoners for biomedical research is not explicitly debarred by any of the international declarations when all appropriate safeguards are followed, arguments on both sides are persuasive and such contradictory ethical evaluations provide no basis for an international recommendation. The Nuremberg Code is a set of principles especially applicable to controlling but also to authorizing ethical research on prisoners. Nevertheless, the tradition in Europe since World War II has been to avoid all medical research on prisoners. This strong position has been a reaction to the Nazi atrocities. Also, it is generally believed that prisoners cannot be expected to give a free consent to experimental projects. This position is contrary to the whole purpose of the Nuremberg Code. Up to recent years, the USA was in clear disagreement with the European position. Prisoners were used in widespread fashion all over the USA in therapeutic drug trials and other medical and behaviour experiments. Now, however, there has been a rapid reversal, due to law suits and adverse publicity engendered by civil rights activist groups. Approximately two–thirds of the American states now ban research on prisoners and the ban has been extended to federal penitentiaries. The recommendations of the National Commission on Protection of Human Subjects would virtually end all research in the United States using prison inmates.
Background The idea of medical experiments on volunteer prisoners is by no means new, for as long ago as 1721 King George I of England offered a free pardon to such inmates of a London prison as would offer to submit themselves to smallpox inoculation (vaccination came over 70 years later). Six prisoners volunteered. These prisoners, by volunteering, saved their necks and won their freedom, and also acquired a new immunity to a highly dangerous disease that was then a universal threat. Unfortunately in the next two centuries there were few such happy endings and ethical problems still remain.
Claim The consent of members of a captive population cannot be valid in that it is influenced by the hope of advantageous benefits such as earlier parole, and that it is purchased by this and other expectations rather than given freely.
Counter–claim Prisoners are particularly suitable subjects for medical experimentation in that they are living in a standard physical – and, indeed, psychological – environment; that they have the time to participate in long–term experiments that is not available to socially active populations; and that the prisoners themselves regard such participation as a means of escaping from the tedium of prison life, or demonstrating their social worth, and of earning a small income.
Broader Maltreatment of prisoners (#PD6005) Inhumane scientific activity (#PC1449)
Abusive experimentation on humans (#PC6912).
Related Obstacles to medical experimentation (#PF4865)
Unethical experiments with drugs and medical devices (#PD2697).
Aggravated by Unethical medical practice (#PD5770).

♦ PE4890 Deception between sexual partners
Lying concerning sexual relations
Incidence Aside from the many classic reasons for deception, a 1989 study in the USA has shown that potential sex partners will lie as to whether they have AIDS or are at risk. Over 30 per cent of the men and 10 per cent of the women surveyed indicated that they had told a lie in order to obtain sex. When asked hypothetically whether they would lie about testing postively for the AIDS virus, 20 per cent of the men and 4 per cent of the women said they would.
Broader Lying (#PB7600).
Aggravates Unsafe sex (#PE9776) Parental lying (#PD9145)
Marital stress (#PD0518) Marital instability (#PD2103).
Aggravated by Adultery (#PF2314) Sexual immorality (#PF2687)
Sexual unfulfilment (#PF3260) Human sexual inadequacy (#PC1892)
Undesired sexual obligations (#PF4948) Ignorance of reproductive processes (#PD7994)
Effects of AIDS on sexual behaviour (#PE5376)
Psychological inconsistency of marriage partners (#PF9818).

♦ PE4891 Exploitation of fossil fuels
Dependence on fossil fuels — Fossil fuel emissions
Nature Use of fossil fuels as a source of energy contributes significantly to air pollution in the form of: sulphur dioxide, nitrogen oxides, carbon monoxide, suspended particles (fly ash), and various volatile organic compounds. All of these may be inuriocus to health and to the environment. Combustion of such fuels contributes to the increase of carbon dioxide in the atmosphere and thus increases the risks of global warming.
Narrower Industrial and domestic heating emissions as air pollutants (#PE2824).
Related Motor vehicle emissions (#PD0414).
Aggravates Air pollution (#PC0119) Urban–Industrial air pollution (#PJ5532)
Increasing atmospheric carbon dioxide (#PD4387).
Aggravated by Energy crisis (#PC6329).

♦ PE4893 Animal fighting sports
Nature Animal fighting sports are a human entertainment usually associated with gambling on the outcome of a fight to the death between two or more animals. Fights may consist of a snake versus a mongoose (in Asia), a dog versus a dozen large rats (in Europe), any single captured wild animal against a pack of dogs (in the Americas), or a man versus a bull (in Hispanic countries). However, because in such matches there is a natural disadvantage on one side, more sport is considered to exist in pitting a pair or more of specimens from the same species against each other. This is reflected in animal racing where camels, horses, elephants, dogs and many other species are placed in contest among their own kind. Its cruel version is in the fight to the death or near–death, with cockfighting and dogfighting among the best known examples. Frequently a pit is employed from which the animals cannot escape and they may be repeatedly struck or prodded to madden them. Cockfighting is known on at least three continents and the birds are usually specially bred. In the fight the cocks wear metal leg spikes long enough to fatally puncture the bodies of their opponents. Breeders may also service the frequently illegal dogfighting market and different species are preferred in different countries. In England and the USA, for example, it may be the small but powerful pit bull terrier.
Incidence In the USA and particularly in the UK, dog fighting is an illegal – though much practised – "country" sport sometimes involving large amounts of money (up to $50,000) and secrecy. Fights take place in obscure settings (anywhere from lone country barns or woods to boats and the empty swimming pools of country estates) and in the UK, a video film of a dog fight was recently sold to those who wanted to participate, but from the safety of their own homes.
Background Dog fighting can be traced to the Middle Ages, when blood lust and large amounts of money were its main attractions. Today's dogs – usually American pit bull terriers – are specially trained, from the start, to recognize the flesh of other dogs as food and their blood as drink. This is encouraged by days of starvation, after which they are placed in a cage with a weakened animal (the other animal is bleeding from razor slashes); during their training, the fighting dogs' endurance is built up by placing them on specially designed treadmills with live kittens or freshly slaughtered meat hanging in front of them.
Claim As in other forms of cruelty to animals, animal fighting sports serve as schools in cruelty, educating human beings towards a desensitivity to suffering, both animal and human. Psychologists in America have observed that children brought to animal fights can experience serious emotional effects from the experience. In addition, due to the large financial stakes and the illegal status of most animal fighting sports (dogfighting is illegal in the UK and the USA; four US states still recognize cockfighting as legal), the Mafia has become associated with it, thus giving organized crime another milieu in which it can profit.
Broader Cruel sports (#PD1323)
Exploitation of animals in spectator sports (#PE0891).
Narrower Fox hunting (#PG3555) Stag hunting (#PJ3556)
Bullfighting (#PG4834) Dog fighting (#PE7786)
Cock fighting (#PG3558).
Related Athletic competition (#PE4266) Criminal killing of animals (#PJ1158)
Cruel treatment of animals for research (#PD0260)
Denial to animals of the right to dignity (#PE9573).
Aggravates Dangerous pet animals (#PE7175).
Aggravated by Organized crime (#PC2343) Gambling and wagering (#PF2137)
Human use of animal by–products (#PF1964).

♦ PE4894 Haemorrhage of pregnancy and childbirth
Broader Haemorrhage (#PE2239) Pregnancy disorders (#PD2289)
Complications of childbirth (#PC9042).
Aggravates Still–birth (#PD4029) Maternal mortality (#PD2422).

♦ **PE4895 Hazards of cosmetic use**
Allergy inducing cosmetics
Nature A large number of chemical substances may be used in a single cosmetic product. Most of them comprise one of the following: emulsions, colours, perfumes, preservatives, or special ingredients. Inflammable solvents consisting principally of alcohols and esters are used in several emulsifying agents. Many of them are volatile and evolve inflammable concentrations of vapour at ordinary temperatures. All colouring agents are capable of producing sensitizing reactions. In particular 'para-type' anthraquinone (used in hair dyes) and azo dyes (face and nail preparations) produce not only sensitization reactions but also cross sensitivity to each other. Perfumes often produce contact dermatitis and melanosis and the reaction may be acute and severe. Contact dermatitis may be caused by ionene, balsam of Peru, cloves, oil of bergamot, benzyl alcohol and pine terpenes. Pigmentation may be caused by any perfumed cosmetic and may be localized or diffuse. An ideal preservative should be colourless, odourless, stable and non-toxic, but this is rarely the case. For example, benzoic acid and propionic acid are mild irritants, salicylic acid is a strong irritant and monochloroacetic acid produces severe local reaction of the skin, eye or respiratory tract. Benzaldehyde has an irritant effect on the skin and formaldehyde causes dermatitis, cough, lacrimation and injury to the bronchi.
Claim It is the believe that people who are not beautiful are physically handicapped that the billion-dollar cosmetics industry draws its profits. Each year vast amounts of money are spent on a lie in which most women and increasing numbers of men willingly co-operate. While everyone agrees that nothing can erase wrinkles, a great deal of effort is spent trying. At best, cosmetics can temporarily plump out the skin. At worst, they can actually do damage. Millions of dollars are spent each year to reinforce the common knowledge that the sun's rays, pollution, bad diet, lack of exercise and anxiety are all bad for the skin. Yet women continue to refuse to take responsibility for their own health and keep turning to over-priced potions and surgeons for relief. In the end it is laziness that keeps the profits of cosmetics companies coming in.
Aggravates Allergy (#PE1017).
Reduced by Use of animals in toxicological experiments (#PE9611).

♦ **PE4896 Mutagenic effects of drugs**
Hazardous genetic effects of medications
Nature Although it is not known how much mutagenesis in man is due to drugs, concern that a serious problem may exist stems from a number of observations. Many substances have been shown to cause gene mutations in micro-organisms and in insects. The induction of mutations by some drugs has been demonstrated in mammalian systems and in human somatic cells. In man, significant frequencies of chromosome aberrations have been observed in spontaneous abortions and in newborn infants, although it is not known what part drugs might have played. The use of drugs that affect nucleic acids is increasing, both in children and in adults with non-malignant disorders, such as psoriasis, virus infections, and conditions associated with altered immunological reactivity.
Incidence Because of the genetic complexity of man, there are immense possibilities for mutation. There is no valid information regarding recent changes in human mutation rates because no systematic population monitoring has been performed. Genetic diseases are, however, becoming relatively more important owing to the reduction in the incidence and severity of parasitic and bacterial diseases. Two main types of genetic damage are recognized: chromosome aberrations and gene mutations. These may affect either somatic cells or germinal cells. Mutations occurring in somatic cells may lead to cancer or other degenerative diseases in exposed individuals, and a foetus may have congenital malformations. Although damage to either cell population may have serious consequences, from the public health stand point mutations in germinal cells are of paramount importance, as they present a hazard to future generations. Indeed, if mutations occur in a germ cell, they may be transmitted to the next generation and give rise to an individual all of whose somatic cells carry the mutant gene. This individual may then transmit the mutant to his offspring and later generations. Thus an increase in mutation rate will lead to an accumulation of harmful mutant genes in the population, and this in turn to increased incidence of hereditary diseases and a decrease in general health and well-being, with severe economic consequences to the society as a whole. While the relation between the ability of a chemical to produce mutations in experimental test systems and its ability to affect humans is not firmly established, the potential hazard to the population is of great magnitude.
Broader Mutagens (#PD1368).
Related Irritating drugs (#PJ4491) Pest resistance to pesticides (#PD3696).
Aggravates Genetic defects and diseases (#PD2389)
Decline in human genetic endowment (#PF7815).

♦ **PE4897 Unsafe port facilities**
Dangerous harbours and docks
Broader Risk (#PF7580) Obstacles to efficient port utilization and operation (#PE5921).
Aggravates Industrial accidents (#PC0646).
Aggravated by Unethical practices in transportation (#PD1012).

♦ **PE4898 Trespassing livestock**
Trespassing cattle
Broader Criminal trespass (#PD3794).
Aggravated by Non-productive use of cattle and livestock (#PD1802).

♦ **PE4901 Resistance to incorporating midwives in medical care systems**
Nature In many nations, two different medical/health systems exist: the modern system, consisting of scientific medicine, doctors, nurses, clinics, and hospitals; and the traditional system, consisting of herbalists, mystics, massage, herbal cures and midwives. The latter is less visible and is often overlooked by planning officials. But in many developing nations, the traditional system, especially midwives, has higher credibility and is more fully utilized for many health purposes, than the modern system.
Incidence Traditional midwives are found in almost every village and in many urban neighbourhoods in Asia, Africa, and Latin America. In these areas, they deliver the majority of births. An estimated two-thirds of all babies born in the world today are delivered by traditional midwives.
Claim Acceptable to the population, accessible in sufficient numbers where they are needed, capable of absorbing training, cost-effective, midwives or traditional birth attendants afford the surest means by which the health of mothers and babies can be improved in many areas of developing countries. Yet all too often their position with respect to the law remains anomalous.
Broader Inadequate medical care (#PF4832)
Lack of integration of traditional and Western medicine (#PF4871).
Aggravates Inadequate maternal and child health care (#PE8857)
Inadequate medical care for pregnant women (#PE4820).
Aggravated by Excessive medical intervention in childbirth (#PE7705).

♦ **PE4903 Residues of veterinary drugs in foods**
Refs FAO *Residues of Veterinary Drugs in Foods* (1985).
Broader Chemical contaminants of food (#PD1694).
Aggravated by Undesirable effects of animal feed additives (#PD1714).

♦ **PE4905 Restrictions on industrial and economic development due to environmental policies**
Nature Economic development and preservation of the environment are often considered to pose a dilemma; environmental issues inevitably have their economic components. Environmental legislation and standards may cause inflation, increased costs and prices, generate unemployment, bring about plant closures, or discourage new investment. The growing pressure of public opinion in many industrialized countries has resulted in newly enacted environmental laws which have delayed a number of industrial projects.
Counter-claim Environmental regulations are seen to confer substantial societal benefits including a healthier environment; employment is stimulated through the creation of jobs in the pollution control equipment industry and jobs for those who operate and maintain this equipment. The measured costs of pollution control are probably upwardly biased, because the input-output computation overstates the cost by not considering input coefficient changes in response to environmental control standards and costs. Moreover, studies include cost increases in major polluting industries, but exclude any cost savings conferred through increased industrial productivity. The negative effects of environmental regulations on rate of growth are likely to be small, and overall costs of pollution control measures in developed market economy countries are of the order of 0.75 - 2.0 percent of GNP.
Refs Kapp, K William *Social Costs, Economic Development and Environmental Disruption* (1983).
Broader Inappropriate policies (#PF5645)
Economic and social underdevelopment (#PB0539)
Inappropriate design of development projects (#PF4944).
Aggravates Inappropriate development policy (#PF8757)
Lack of economic and technical development (#PE8190).

♦ **PE4906 Chronic terminal illness**
Broader Chronic illness (#PD8239) Critical illnesses (#PE9038).
Aggravates Dehumanization of death (#PF2442).

♦ **PE4907 Racial bias in sentencing offenders**
Racial discrimination by the judiciary
Incidence In the USA in 1989, it was estimated that 23 per cent of the population of black men in the 20 to 29 age group was in prison, on parole or on probation. This exceeded the number of black men enrolled in higher education in 1986. This figure compared with 6.2 per cent in the case of white men and 10 per cent in the case of Hispanics.
Broader Racial discrimination (#PC0006) Lack of impartiality of the judiciary (#PE7665).
Related Legalized racial discrimination (#PC3683).
Aggravated by Legal inconsistency (#PF5356).

♦ **PE4909 Aiding deserters**
Nature Knowingly aiding a member of the armed forces to desert, to attempt to desert or to hinder the enforcement of laws against desertion is a criminal offence.
Broader Aiding national security criminals (#PD7407).
Aggravates Military desertion (#PD4621).

♦ **PE4912 Restrictions on recognition of nationality**
Denial of right to nationality
Nature A person may be without a nationality due to not acquiring one at birth or by losing it due to marriage, legitimation, recognition by their natural parents or adoption. Such persons are generally in less favourable positions than those provided for refugees with regard to wage-earning employment and right of association.
Broader Denial of political and civil rights (#PC0632).
Narrower Statelessness (#PE2485) Refusal to grant nationality (#PF2657)
Denial of right to change nationality (#PE1736).
Related Deprivation of nationality (#PD3225)
Conflicting military obligations for persons of multiple nationality (#PE6678).
Aggravates Inadequate sense of personal identity (#PF1934).
Aggravated by Conflict of laws over nationality (#PF8953)
Conflicting multiple nationalities (#PE6677).

♦ **PE4913 Depletion of fish reserves by marine mammals**
Refs FAO *World Review of Interactions Between Marine Mammals and Fisheries* (1984).
Related Endangered species of marine mammals (#PD3673).
Reduced by Hunting of marine animals (#PE0439).

♦ **PE4914 Racketeering**
Broader Crime (#PB0001).
Aggravated by Organized crime (#PC2343)
Ineffective legislation against organized crime (#PE6699).

♦ **PE4915 Export earnings instability**
Export earnings shortfalls — Insufficient export benefits
Nature Export earnings instability in developing countries is the result of a number of factors. First, many developing countries have specialized on the export of primary commodities, which are peculiarly susceptible to shifts in supply and demand, as well as being more price inelastic than are, for example, manufactured goods. The transmission of instability of demand for developing country exports through the business cycles in the industrialized countries or fluctuations in the quantities supplied for export may thus be one potential source of instability in export revenues. Second, the exports of many developing countries are not only concentrated by sector (commodities) but also geographically, with obvious implications when linked to factors affecting demand in the importing countries. Third, the markets for products in which developing countries have specialized are often characterized by speculation on the one hand and oligopsony on the other. Instability affects development through such variables as imports, savings, investment, employment, government revenues and private income.
Incidence Instability in developing country commodity export prices, volume and earnings has been in most cases greater in the 1970s than earlier. Out of 36 commodities examined in the 1971-80 period, 19 had an instability index of over 15 per cent for export earnings, 20 for export unit values (prices) and 10 for export volumes.
Aggravates Diminishing capital investment in small communities (#PF6477).

♦ **PE4920 Shortage of military manpower**
Shortage of soldiers — Failure of military recruitment — Under-strength military forces — Insufficient military personnel
Incidence The British army 1989 was under-strength by 4,100 trained officers, mostly due to low retention rates of people in their early thirties. In the US the Army and the Marine Corps missed their overall recruiting goals in 1989 for the first time in 10 years. An increasing number of least qualified people are being accepted into the services. The West German army is experimenting with shrinking 3 of its 36 brigades by nearly one-third because of few people of recruiting age and an extreme shortage of non-commissioned officers.
Broader Labour shortage (#PC0592) Understaffing of basic facilities (#PD9306).

Aggravates Military unreadiness (#PF5933).
Aggravated by Draft evasion (#PD0356) Shortage of skilled labour (#PD0044)
Imbalance in the human sex ratio (#PF1128).

♦ PE4921 Inhibited growth of malnourished children

Wasting in growth of children — Stunting in growth of children

Nature Wasting is a symptom of acute under-nutrition and is usually a combined consequence of insufficient food intake and a high incidence of infectious diseases, especially diarrhoea. In many developing countries, wasting most severely affects children during the second year of life, when they are no longer breast-fed. The missing benefits of breast milk are often not compensated by nutritious weaning foods. And in the absence of breast milk, children lose their protection against infection at just the time when they are becoming increasingly mobile and coming into direct and frequent contact with an often unhealthy environment.

Stunting, on the other hand, is a consequence of chronic under-nutrition arising from insufficient food intake and exposure to infections over a long period. Given the cumulative effect of these conditions, the longer they prevail the more likely a child is to be stunted. Stunting is a more widespread condition than wasting, and as many children seem to adapt to it and to function at or near normal levels, there is some debate about how severe stunting really is. In the long term, however, this condition has a number of serious consequences. For example, women who themselves were stunted during childhood are more likely to develop complications when they have their own babies an to give birth to low-birth-weight children.

Broader Inhibited human physical growth (#PD5177).
Narrower Muscular wasting (#PG3548)
Damage to infant brains from malnutrition and insufficient stimuli (#PE4874).
Related Infant growth failure (#PE6909).
Aggravates Low birth-weights (#PF5970).
Aggravated by Malnutrition (#PB1496) Child malnutrition (#PD8941)
Infectious and parasitic diseases (#PD0982).

♦ PE4922 Disadvantageous terms for technology transfer

Decline in technology transfer to developing countries — Impasse in technology transfer to developing countries

Nature Innovation in industry in the developing countries is severely constricted by low investment rates, high interest rates and low profit rates. Relocation of production processes to low-cost areas abroad is only viable as long as profitability holds up, but quite the reverse is taking place.

Incidence Commercial technological flows have fallen or stagnated for all groups of developing countries from the peak reached in 1981, in contrast to dynamic growth during much of the 1970s. Most indicative of this slow-down is the behaviour of imports of capital goods which actually declined by 10 percent for developing countries as a group between 1981 and 1986 as compared with the period 1970-1981 during which an average growth of over 20 percent per annum was experienced. Geographically, the decline was concentrated in developing America and Africa. Asian countries as a group continued to experience an increase in capital goods inflows, but at a drastically reduced rate compared with the 1970s.

Refs World Intellectual Property Organization *Licensing Guide for Developing Countries* (1977).

Broader Reversal of development progress (#PF4718).
Related Technology gap between developed countries (#PD0338).
Aggravates Economic underdevelopment (#PC0281)
Natural environment degradation (#PB5250)
Increasing development lag against technological growth (#PE3078).
Aggravated by Secrecy in scientific research (#PF1430)
Abuse of science and technology in capitalism (#PE3105)
Restrictions on the sharing of technical research (#PD3154).

♦ PE4923 Candidosis

Refs MacFarlane, T W and Samaranayake, L P *Oral Candidosis* (1989).
Broader Mycosis (#PE2455) Fungal diseases (#PD2728).

♦ PE4927 Missing public signs

Nature In hundreds of thousands of villages there are no public signs, hindering communications. When villages were small, close knit and isolated there was no need for signs. With increased interaction between villages and the larger society the need is two fold. Residents are increasingly aware public signs are marks of progress, among other and even more important things. Visitors are put off by having to ask residents constantly for directions. Public signs include, street signs and number, traffic signs, commercial signs, public notice boards and service signs.

Broader Social underdevelopment (#PC0242) Lack of local information systems (#PF6541).

♦ PE4928 Blastomycosis

Refs CAB International Mycological Institute *Blastomycosis*.
Broader Mycosis (#PE2455).
Aggravated by Ringworm (#PD2545).

♦ PE4929 Simulator sickness

Nature Simulator sickness is suffered by pilots, co-pilots, and other crew members after their simulator training sessions. Symptoms range from nausea, dizziness, and chills, to visual flashbacks, loss of motor skills, and spinning sensations. It is caused by sensory conflict.

Broader Motion sickness (#PE2611).

♦ PE4930 State immunity

Nature There are two different legal approaches to the state immunity: restrictive, which grants immunity only to the public acts of states, and absolutist rule, which says that immunity must also be afforded to their private acts, such as trading contracts and ownership of property.

Refs United Nations *Materials on Jurisdictional Immunities of States and Their Property*.
Narrower Police immunity (#PE5832).

♦ PE4931 Domestic animal bites

Bites of animal pets — Dog bites

Nature In the developed world of urbanized societies, man has maintained a close association with domesticated animals. One of the most prevalent health risks with regard to this association is the wounds produced by animal bites. Apart from the trauma they cause, they also prove a hazard to health because of the danger of wound infection.

Incidence 76 percent of animal bites occur in the face, head and neck and may result in a serious injury or disfigurement. In France there are an average of 500,000 reported attacks by dogs, with 60,000 of those bitten requiring hospitalization and plastic surgery. In the UK, dogs cost an estimated 70 million pounds damage a year, including 250,000 animal bites.

Aggravated by Proliferation of pets (#PD2689) Maltreatment of pet animals (#PD1265).

♦ PE4932 Cryptococcosis

Refs CAB International Mycological Institute *Cryptococcosis*.
Broader Mycosis (#PE2455) Fungal diseases (#PD2728).
Aggravated by Ringworm (#PD2545).

♦ PE4933 Lack of technical development and excess of manpower in developing countries

Nature Technical development, so vital for developing countries to improve their economic status as well as keep pace with developed countries, is severely lacking. This lack stems from a variety of causes, including shortages of funds and adequately trained manpower; inability to maintain technological advancements; unwillingness to change older, established modus operandi; and a reluctance of industrialized countries to freely give technological equipment and advice, as that would lessen their hold over the developing countries.

Counter-claim New technologies replace older technologies already installed; they create hardship, particularly for unskilled labour and illiterate women and for others who depend on traditional handicrafts for their livelihood. Traditional technologies have evolved over centuries as part of society's social patterns. They are a source of cultural pride and are the backbone of socio-economic activities in most developing nations.

Broader Labour surpluses in developing countries (#PD0156)
Increasing development lag against technological growth (#PE3078).
Related Antagonism between employment policy and technical advance (#PE5104)
Inadaptation of technology to man in the industrialized societies (#PE5023).
Aggravates Labour displacement (#PE6843) Underutilization of labour force (#PF6293)
Emigration of trained personnel from developing to developed areas (#PD1291).
Aggravated by Technology gap between developed countries (#PD0338)
Lack of means for local technological development (#PF6454)
Deterioration of the quality of life in least developed countries (#PE7734).

♦ PE4934 Discrimination in employment against immigrant workers

Nature One of the most frequent forms of discrimination existing in employment is that practised against foreign migrant workers, who constitute a sort of ethnic minority within a country, with inadequately protected rights and a differing economic and social situation from that of indigenous workers. Aside from the fact that in most countries some occupations are the subject of regulations and are closed to foreigners, access to employment is, for migrant workers, subject to the limitations of the employment situation, which determines and curtails their chances and frequently confines them to the most arduous or disagreeable occupations (building, mining, etc) which have been abandoned by the national workers. Occupational advancement depends on learning the language, and the lack of opportunity in this respect may amount to discrimination. Where the migrant workers are from former colonies or dependent territories, stereotypes associated with colonialism may continue or be reactivated. Restrictions as regards access to employment also include the age limits for the entry of migrant workers laid down by some countries. These limits often hinder the arrival of a family group.

Background The extensive mixing of populations which occurred after the Second World War, followed by an intensification of the movement of migrant workers within Europe, helped to create this situation of discrimination.

Broader Discrimination in employment (#PC0244)
Discrimination against immigrants and aliens (#PD0973).

♦ PE4935 Hazards of bottle-feeding

Nature In the poor communities of the world, lack of sanitation, clean water, refrigeration, sterilizing equipment, fuel and time for boiling water, of money for adequate quantities of milkpowder, or education to read instructions, can contribute to make bottle-feeding a baby killer. Especially in warm countries, milk-powders can be over-diluted with unclean water in an unsterilized feeding-bottle which is then often left to stand in the heat. The result is that babies who are bottle-fed are many times more likely than breastfed babies both to be malnourished and to contract infections, thus setting up the cycle of malnutrition and infection at an even earlier and more vulnerable age.

Incidence Separate studies in India and Canada have found that artificially fed infants were three times more likely to contract diarrhoeal infections and twice as likely to suffer from respiratory infections (the two main causes of child death) as infants who were breastfed. In Chile, artificially fed infants have been found to be two or three times more likely to die in the first year of life. In the United States, a study in New York State has found that bottle-fed infants were hospitalized three times more frequently than breastfed infants during their first year of life. In a hospital in the Philippines, the decision to encourage breastfeeding instead of bottle-feeding of new-born babies reduced clinical infections by 88 percent diarrhoeal infections by 93 percent and infant mortality by 95 percent In rural Jamaica, it was found that diarrhoea incidence increased directly with the degree of bottle-feeding, and was three times as high among exclusively bottle-fed compared with exclusively breast-fed infants. In Congo, a UNICEF-sponsored study echoed the evidence accumulating from all over the world that infant malnutrition is frequently co-related with a reduced period of breast-feeding, and warned that the social changes that bring this about are far more complex than is sometimes implied by the 'breast versus bottle' protagonists. During the 1970s, large numbers of women in the developing world abandoned breastfeeding for the use of powdered formulas. Studies published in 1983 confirmed that breast-feeding still predominates in the rural areas of almost all developing countries, but that in the growing metropolitan areas bottle-feeding is on the rise. Estimated global sales of infant formulas rose from US $1.5 billion in 1978 to US $4 billion in 1983. Most of the increase in formula sales occurred in the more rapidly industrializing developing countries.

Broader Hazards to human health (#PB4885).
Related Dependence on breast feeding (#PE7627)
Substitution of inappropriate foodstuffs for breast feeding (#PE8255).
Aggravates Diarrhoea (#PD5971) Child malnutrition (#PD8941)
Lung disorders and diseases (#PD0637).

♦ PE4938 Endangered migratory bird species

Nature The danger of extermination threatens certain species of birds, particularly migratory birds, due to: the killing of birds during their breeding seasons and (for migratory birds) during their return flight to their nesting ground; the capture or killing of birds or the collecting of their eggs for commercial exploitation; the use of methods of mass killing or capture of birds (which also often causes them unnecessary suffering); the rapid disappearance of suitable breeding grounds for birds as a result of human intervention; and the increasing destruction of birds by hydrocarbons, water pollution, insecticides and other poisons.

Claim It is in the interests of science, the protection of nature, and the economy of each nation to protect these species of birds.

Broader Endangered species of birds (#PD0332).

♦ PE4946 Illicit trade in prescribed drugs

Nature There is a rapid expansion of the underworld trade and abuse of so-called mood-shaping drugs. These chemical compounds are manufactured on a massive scale, largely in the great drug laboratories of the industrial world, and increasingly consumed in the Third World.

There are hundreds of different kinds designated psychotropic, which range from tranquillisers like valium and librium to hallucinogens including LSD, barbiturates (downers) and stimulants (uppers) such as amphetamine, known universally as 'speed'. Although many are legally prescribed, several are banned or in heavily restricted official use in most countries, notably the hallucinogen

family of which LSD is the father, and hypnotics (used as sleep aids) such as Mandrex. All of them figure in widespread abuse in the Third World, often in tandem with traditional home–grown drugs such as cannabis or opiates. Although a convention on psychotropic substances was adopted in Vienna in 1971, aimed at controlling the manufacture and trade of some of the most dangerous substances, many important manufacturing countries including Britain, Switzerland, Italy and Japan, have not yet signed it. The pharmaceutical industry has often lobbied against curbs and drug control agencies have been reluctant to identify and blacklist companies that do not adequately police the distribution of their products.
Incidence Massive seizures of illicit supplies are reported regularly all over Africa, Asia and the Middle East. Illicit trade in these drugs in the USA is put at a retail value of up to $21 billion annually, considerably more than twice the retail value of the heroin trade. Up to 250 million dosage units of controlled drugs production in the USA are diverted to the illicit trade.
Claim The diversion of prescribed drugs has been described as a much greater health hazard than the use of any illegal drug, including heroin.
Broader Illicit drug trafficking (#PD0991).
Related Drug smuggling (#PE1880).

♦ PE4947 Discrimination against rural women

Nature Legal obstacles to the maximum utilization of women's economic potential are interrelated and tend to be mutually reinforcing. Several types of customary, religious and civil laws limit women's access to land and other forms of income producing property. Customary land tenure systems frequently give women fewer rights than men to land and valuable livestock. Customary and religious laws often prohibit women from inheriting real property or allow them a much smaller share of inherited property than male heirs. Customary and religious marriage and family laws exist which deny women the legal capacity to own or administer property or act in commercial matters in their own name. As a result women are not able to acquire land by tenancy or sales contracts. Women's rights to landownership are also limited by the nonrecognition, by customary, religious and civil legal systems, of women's labour within the family. This results in women not having rights to property acquired during marriage through the labours of both parties. In certain cases, civil laws also limit women's property rights by requiring courts to apply customary law. The areas of customary and religious law most frequently incorporated into civil law, either through judicial decisions or by civil codes, are in the areas of marriage and inheritance which are the two major sources of laws which discriminate against women in relation to property ownership.
These factors frequently result in rural women not owning of having the legal capacity to administer the land which they cultivate. Aside from the inequities of the situation there are various negative consequences for development. Lack of landownership also limits women's participation in cooperative organizations, both credit generating and other types of collective organizations which are important elements of development. Cooperative regulations and by–laws frequently limit membership to landowners and or heads of family, thereby eliminating an important source of credit and participation for rural women in development at the grass roots level. In many countries where land has been nationalized and landownership is no longer a factor, membership in cooperatives and collective organizations is often limited to heads of family, thereby eliminating the participation of the majority of rural women. The head of family concept is prevalent in land reform legislation thereby eliminating the majority of women from receiving the full benefits of these reforms. Land reform legislation which merely prohibits discrimination on the basis of sex does not ensure land access to women.
Land is still the most important form of collateral for agricultural credit. If, due to lack of landownership, women are not able to obtain credit for agricultural inputs, and must resort to high interest unsecured loans or to the mortgaging or advanced sale of crops, the land they cultivate will be less economically productive. This may result in women contributing less to rural economies and having less economic incentive to better utilize their productive resources.
Claim Recognition of the need to fully integrate rural women in development is essential if women are to receive their equitable share of development benefits. In addition to equitable and moral considerations, the underutilization of one half of the rural labour force does not make economic sense, especially when increasing human productivity is a major objective of development efforts.
Broader Discrimination against women (#PC0308).
Aggravated by Neglect of the role of women in rural development (#PF4959).

♦ PE4949 Lactose intolerance

Refs Martens, Richard A and Martens, Sherlyn *Milk Sugar Dilemma* (1987).

♦ PE4950 Inadequate dietary fibre

Low–fibre diets — Inadequate roughage in diet
Nature People with diets low in amounts of insoluble fibre have a greater risk of colon–rectal cancer. Nearly all cases of this type of cancer are believed to begin with the formation of benign growths call polyps in the colon or rectum. Over a period of years, the polyps can enlarge and some may gradually become malignant. Diets rich in cereals, such as bran, reverse the usual progression by inhibiting a pre–malignant lesion.
Refs Trowell, H C, et al *Dietary Fibre, Fibre–Depleted Foods and Disease* (1985).
Broader Nutritional deficiencies (#PC0382).
Aggravates Teeth disorders (#PD1185) Colon–rectal cancer (#PE9399).

♦ PE4951 Stress addiction

Nature A number of activities and situations increase the flow of the brain's opiates, creating a sense of euphoria or excitement. One of these is stress. When a person is in a stressful situation endorphins are produced reducing pain and creating a sense of joy. Endorphins are much like morphine in their effect. Sports people, religious practitioners and military personnel use this effect to enhance their activities.
Broader Addiction (#PD6324).
Aggravates Occupational stress (#PE6937).
Reduces Stress in human beings (#PC1648).

♦ PE4953 Environmental stress on inhabitants of tall buildings

Nature Where more people are crowded into cities, there are more buildings. As more land is built up, structures start getting taller in order to use vertical space, thus intensifying crowding, economic viability, transportation and service problems. This situation leads to environmental stress or discomfort in human beings occupying tall buildings, sometimes even to neuroses, delinquency, and violence.
Background The changing vertical dimension in relation to human habitations has evolved in importance as a consequence of the worldwide phenomenon of enormous migrations from rural farms to the large industrialized centers, the cities, in a process known as urbanization; which in turn results in the social pressure for more working and living space than is available from the ordinary two–dimensional aspects of these city areas. The tall building solution tends to contribute to the already overwhelming problems of the city and its environment.
Claim Tall buildings have no genuine advantages, except in speculative gains for banks and land owners. They are not cheaper, do not help create open space, destroy the townscape, destroy social life, promote crime, make life difficult for children, are expensive to maintain, wreck the open spaces near them, and damage light, air and view.
Counter–claim Tall buildings are a logical solution to encompass more people, save space, and create a more harmonious environment for humans and nature.
Refs Conway, Donald J *Human Response to Tall Buildings* (1982).
Broader Uncontrolled urban development (#PC0442)
Psychological stress of urban environment (#PE6299).
Related Long–term shortage of water (#PC1173).
Aggravates Overcrowding (#PB0469) Juvenile stress (#PC0877).
Aggravated by Indoor air pollution (#PD6627).

♦ PE4956 Imbalance between agricultural exports and imports in developing countries

Nature Countries that are not self–sufficient in food have to rely on imports to feed their people. Developed countries in this position can generate foreign exchange to buy food by exporting the products of their industrial sectors. In contrast, non–oil–exporting developing countries have a limited capacity for earning foreign exchange, as the terms of trade of their main, primary commodity exports have generally declined. The balance of payments position of most developing countries has become increasingly serious. The value and volume of their own agricultural exports have not kept pace with the amount and the cost of the food they need to import. Recent trends in world commodity prices have generally not favoured developing countries.
Incidence The increasing imbalance between agricultural exports and imports is particularly serious in Africa, where agricultural exports declined in constant value by an average of 3 percent per year during the 1970s while the value of imports increased by more than 7 percent per year. The medium and long–term debts of the non–oil–exporting developing countries increased nearly fourfold between 1973 and 1980.
Broader Fiscal and trade imbalances (#PC4879).
Narrower Inappropriate food aid (#PE0302) Export of nutritious food (#PJ1365).
Aggravated by Deterioration of domestic food production in developing countries (#PD5092).

♦ PE4959 Separation of family members

Family separation
Incidence Especially in developing countries, members of a family (particularly the husband) leave the rest of the family in the rural areas to subsist on peasant agriculture in order to be able to work in the towns or on mines. In some parts of Africa as many as 2 rural families in 5 are headed by women. Husbands may visit occasionally if at all, leaving the women trapped in a work pattern of up to 13 to 14 hours a day.
Broader Statelessness (#PE2485).
Aggravates Overworked women (#PD7762).
Aggravated by Discrimination against HIV–infected persons (#PE4299).

♦ PE4960 Mental dullness

Nature Mentally dull children are those children who, at about the middle of their school career and without being mentally deficient, are behind the normal average of their age to the extent of two years or two classes. This corresponds to a retardation of between 15 and 20 per cent of their age or to about twice the 'standard deviation', and roughly to an IQ of between 80 and 85.
Broader Mental deficiency (#PC1587).

♦ PE4961 Occupational domestic accidents

Accidents at home — Safety in the home
Nature Domestic accidents result from an interaction of three main factors: the person (host), the agent, and the (home) environment. The nature of occupational domestic accidents most frequently encountered have been cuts (33 percent), dislocations and fractures (12 percent), contusions (12 percent), sprains and strains (9 percent), and burns (4 percent). Non–fatal injuries occur most frequently in the kitchen.
Incidence Accidents in the home account for a large proportion of accidental deaths and accidents in general, and are one of the three leading causes of death in many countries. The most common causes of home accidents are falls (40 percent), burns (24 percent), poisoning (6 percent), firearms (5 percent), gassing (4 percent), asphyxia (8 percent), others (electrocution, falling objects, etc) (13 percent). Domestic accidents strike women more often than men. Furthermore, children under 4 years of age and people over 65 have more accidents than all other age groups taken together. Recent figures show that in one year 20 million people in the United States alone were injured in domestic accidents – five times more than the number injured on the roads. Of these 20 million injured, 28,500 died within the year and 110,000 suffered permanent disabilities. Each year in the UK 6,000 people die as a result of accidents in the home, 100,000 are admitted to hospital and some 1 million receive treatment from their doctor. In France, 5,000 deaths are reported annually in this accident class. In the European Community as a whole some 45 million people are injured a year and 80,000 killed in home and leisure accidents. Canada reports 2,000 deaths a year due to home accidents.
Broader Injurious accidents (#PB0731) Disastrous accidents (#PC6034).
Related Childhood accidents (#PD6851).
Aggravated by Drunkenness (#PE8311) Defective product manufacture (#PD3998)
Inadequate safeguards against fire (#PD1631).

♦ PE4963 Leveraged buy–outs

Junk bonds
Nature Junk bonds are securities that pay high interest because they are issued by companies with little collateral, as tends to be the case with new businesses. A bond is an IOU which promises to pay a fixed amount on a fixed date, with (usually) interest payments in between. Junk bonds are bonds that yield high returns on the loan with higher risks. Government and blue chip companies issue bonds but the risk is lower because the chance of default is less. Junk bonds are issued by companies which are not monitored for their credit ratings. The term junk bonds originally applied to sound companies too small to merit a credit rating. The risk was not much greater than normal bonds because the companies assets were generally greater than the value of the bonds issued. Junk bonds are now being issued for more than the company is worth. The risk is worth taking if cash flow in the company allows interest to be paid and the assets of the company are rising. When this is not the case the bonds become worthless. Such bonds may also be used to acquire other businesses under hostile and controversial circumstances. Otherwise health companies may then face ruin because they are unable to manage the debt resulting from such takeover operations. **Counter–claim** Junk bond is a bad term for a good investment opportunity. When investors recognize the risk and are willing to take it, the investment opportunity is like any other with greater risks for greater rewards. Such bonds provide an important source of finance where banks are reluctant to provide it. Only 5 per cent of American companies have high credit ratings. Cutting off junk bonds, removes a vital credit opportunity for the remaining 95 per cent, usually including those run by minority groups and women.
Broader Consumer debt (#PD3954).
Aggravates Uncontrolled growth of debt (#PC8316)
Excessive institutional debts (#PE1557).
Aggravated by Capitalist speculation (#PC2194).

♦ PE4964 Discrimination against domestic servants

Denial of right to freedom from servitude

Nature In many countries, domestic service is the main source of employment for young girls, who, often on their own, must contend with various difficulties relating to lack of respect, excessive hours, poor lodging, and lack of free time and holidays. This occupation is not governed by regulations and the girls are rarely covered by social security. They lack status and social advantages equivalent to those of young people in other occupations.
Incidence Domestic employment in private households is almost non-existent in socialist countries, it is moderate and decreasing in Western Europe and North America, and is on a fairly wide but indeterminate scale in many developing countries. In countries of North Africa, for example, young girls under the age of 15 come from the countryside to work as servants in cities for very low wages, or sometimes no wages at all.
Broader Denial of right to liberty (#PF0705) Discrimination in employment (#PC0244).
Aggravates Shortage of domestic servants (#PJ9711).

♦ **PE4965 Environment policy as restriction on trade in developing countries**
Nature Environmental laws in developed countries have a direct or indirect influence on trade in developing countries, mainly because of certain restrictions on food imports. For instance, several West European countries will no longer import fruit and vegetables showing traces of DDT or other pesticides.
Broader Inappropriate policies (#PF5645).

♦ **PE4966 Desiccation of human skin**
Dry skin
Incidence As an indication, the market for skin moisturizing creams in the UK in 1986 was £101 million.
Broader Unsociable human physiological processes (#PF4417).

♦ **PE4969 Health hazards of aluminium**
Aluminium poisoning
Incidence Aluminium poisoning, with loss of memory, tremor and jerkiness, was first reported in 1921, and there is evidence that aluminium can be a neuro-toxin. It has been implicated as a factor in patients with senile and pre-senile dementia of the Alzheimer type and in parkinsonism-dementia. It is widely accepted as the major toxic factor in renal dialysis encepathology, leading to speech disorders, dementia and convulsions. Manufacturers continue routinely and unnecessarily to add aluminium to their products and some water authorities continue to treat water with aluminium sulphate.
Counter-claim Outside the industrial environment, aluminium normally presents no health hazard.
Aggravates Memory defects (#PD8484) Senile dementia (#PE3083)
Alzheimer's disease (#PE7623).

♦ **PE4972 Denial of rights to development for indigenous peoples**
Broader Discrimination against indigenous populations (#PC0352)
Denial of right to a people to pursue development (#PE1536).

♦ **PE4973 Enteritis**
Inflammation of the intestines
Refs Macdermott, R P (Ed) *Inflammatory Bowel Disease – Current Status and Future Approach* (1988).
Broader Bacterial diseases in animals (#PD2731)
Intestinal infectious diseases (#PE9526).
Aggravated by Poisonous algae (#PE2501).

♦ **PE4985 Insecurity of Western marriages**
Nature Insecurity in Western marriage occurs both on the part of women and of men. Liberation of both sexes has caused the demise of long-held social roles, and while those roles may have been degrading they offered a socially accepted identity which often went unquestioned. Insecurity has been one result of the challenging and changing of those norms. Another is that while married people are still more likely than single people to report they are "very happy" (a sociological fact since at least the early '70s), the gap has narrowed considerably.
Broader Insecurity (#PA0857) Marital instability (#PD2103).

♦ **PE4986 Instability of food prices in developing countries**
Nature In general, the poorer the country, the greater the impact of rising domestic food prices; this is reflected in the heavy burden of food purchases on the budgets of low-income families. Political violence, natural disasters and curfews can all exacerbate this problem with little or no notice at all. In many developing countries, food accounts for 50 to 70 percent of consumer spending.
Broader Instability of prices (#PF8635).
Aggravates Food insecurity (#PB2846).
Aggravated by Dumping of food in developing countries (#PE0607).

♦ **PE4990 Marginalization of second-generation immigrants**
Nature Immigrants may find that they are cut off from their roots but without succeeding in acquiring the cultural identity of the host country. Consequently they live in a cultural "no man's land". Whereas adult immigrants, whose personalities have already been formed by a set of values, moral codes, customs, myths and symbols, can to a certain degree handle the difficulties of their new situation, second-generation immigrants cannot escape exposure to the effects of a double sub-culture. The result is that immigrants constitute an anxious, disturbed population. Initial insecurity and inequalities lead to a loss of interest in acquiring literacy, with a consequent structural form of illiteracy and illiteracy due to revulsion, both amongst adult immigrants and their children. This places them at a disadvantage as regards access to employment or job advancement.
Incidence Some host countries give immigrants no security of long-term residence and maintain them in a permanently uncertain, temporary legal situation. This situation has repercussions on the second generation. In five European countries (Belgium, France, the Federal Republic of Germany, Sweden and Switzerland), 3 million persons under 25 have not formally acquired the citizenship of the host country.
Claim Second generation 'immigrants' have never emigrated from anywhere.
Aggravated by Inadequate cultural integration of immigrants (#PC1532).

♦ **PE4991 Illegal ivory trade**
Banning of ivory trade
Nature Poaching elephants for their ivory is profitable because the price of ivory has risen from about $50 per kilo to nearly $300 per kilo. Around 94 percent of all the ivory being traded internationally if from poached elephants. Around 20 percent of ivory in "legal" when it leaves Africa because governments legalize poached ivory when it is confiscated. The remaining 80 percent in laundered into the "legal" ivory system so that by the time carved ivory reaches the streets of Hong Kong legal and illegal ivory are indistinguishable.
Incidence From 1979 to 1987, 6,828 tonnes of ivory were exported from Africa representing from 680,000 to 760,000 elephants, which may be a low estimate. The biggest importers are Japan, with nearly 40 percent of the trade, Singapore, Hong Kong and Belgium. From 1973 to 1982, 300 tons of ivory were exported from Burundi to Belgium, representing the death of 25,000 elephants in neighbouring Zaire and Tanzania, the countries of origin of the tusks.
Claim It is legal ivory trade which has consumed the ivory from one million African elephants in the past 10 years up to 1989.
Counter-claim The probable effect of a ban will be to drive the price of ivory up faster. As the legal supply from managed herds of elephants is stopped the poaching will continue and the increased demand will drive the price up as buyers turn to smugglers for supplies. The real solution to the decline of the African elephant is to manage herds and control trade like Botswana, where herds have grown; and South Africa, Zimbabwe, Malawi and Namibia where herds are stable. The revenue from their exports of ivory and elephant leather helps pay for conservation work. Zimbabwe has long found big-game hunting an even more lucrative use for its elephants. 100 - 200 elephants are allowed to be killed by American or West German hunters who spend $15,000 all told, some of which goes to local guides and bearers.
Broader Unlawful business transactions (#PC4645)
Trade in animal products of endangered species (#PD0389).
Aggravates Endangered species of elephant (#PD3771).
Aggravated by Government complicity in illegal activities (#PF7730).

♦ **PE4994 Destabilization of national insurance markets by offshore insurers**
Nature Offshore insurers are insurers incorporated in a country on the strict understanding that they will refrain from underwriting the local risks of that country, all their activities being oriented towards other countries. They can destabilize the markets of neighbouring countries and the market of the country where they are incorporated by resorting to "fronting" or concealing of unauthorized direct insurance operations.
Broader Aggressive economic destabilization of countries by external forces (#PE9420).

♦ **PE4995 Health hazards of smoking for women**
Nature The combination of taking oral contraceptives and smoking substantially increases the risk of heart attack and stroke, and there is evidence that smoking can increase the likelihood of menstrual disorders. A study of more than 66,000 nurses between the ages of 30 and 55 indicated that smoking lowers the age of the natural menopause. Furthermore, there is strong evidence of an association between smoking and cancer of the cervix. Smoking also has more immediately bad effects on pregnant women, their foetuses, and nursing mothers. It increases the likelihood of miscarriage and stillbirth, and may result in low birth-weight and backward babies. Both the quality and quantity of breast milk may be reduced if the mother smokes.
Incidence The results of a Canadian survey showed that the relative risk of disability was 1.25 times higher in women who were current or former smokers compared with non-smokers. Almost 12 percent of the days taken off work because of sickness in Canada are attributable to smoking: smoking accounts for 19.4 million 'disability' days every year, or 2.65 days per year for each woman aged between 15 and 64. Similarly, in the United States, it was estimated that cigarette smoking accounted for 18 percent of all newly diagnosed cancers in women and for a quarter of all cancer deaths in 1980. Statistics about women and smoking in the developing world are strikingly similar to those seen in surveys of Western women 20 years ago. In poor countries it is literate women, possibly of the middle, professional classes, who smoke and who see the habit as a symbol of being modern and moving with the times. Industrialization and more money to spend also affect the smoking habits of women. In Nigeria, for example, there have been few women smokers up to now because there are many socio-cultural influences that inhibit the habit among women, but as women's literacy and their access to cash increases, the influence of Western culture grows, and the latest surveys in Nigeria show a considerably higher incidence of smoking among women than before.
Refs Chavkin, Wendy *Double Exposure* (1984).
Broader Smoking (#PD0713).
Narrower Smoking during pregnancy and breast-feeding (#PE5026)
Aggravated by Unequal health benefits for women (#PE6835).

♦ **PE4996 Smoking in developing countries**
Nature As smoking is becoming increasingly a minority habit in the Western world, the multinational tobacco companies have begun to look for new markets and have found them in the developing countries where the appeal of the cigarette is growing, largely thanks to advertising and the association in people's minds of smoking with modernity and sophistication. Many Third World nations have welcomed the new 'prosperity' that tobacco agriculture has brought and have provided economic initiatives to attract these companies. The industry has thus enjoyed special privileges and freedom in many parts of Africa and the Far East. By contrast with North America and Europe, in many developing nations the tobacco manufacturers are free to advertise cigarettes with no legal restrictions, 'voluntary' promotion codes, or other government measures. So while there has been a steady decline in 'tar' and nicotine levels in the Western world, the Third World has become a dumping ground for cigarettes with high tar and nicotine contents not permissible elsewhere. The 20-year lag between the first use of the cigarette and the definite evidence of disease means that the people of the developing world will soon display symptoms of smoking-related diseases.
Incidence In 1980 one-third more Africans smoked than in 1970. In Latin America 20 percent more people were smoking and in Asia the figure was 23 percent.
Refs Consumers' Association of Pinang *Fighting Tobacco in the Third World*; Uma Ram Nath *Smoking* (1986).
Broader Smoking (#PD0713).

♦ **PE4997 Inadequate system of political checks and balances**
Uncontrolled executive branch action — Unaccountability in government — Non-accountability in government action — Unlawful government — Extra-legal government initiatives
Incidence Between 1798 and 1970, U.S. presidents sent troops into imminent hostilities or transferred arms or other war material abroad without congressional authorization, contrary to the constitutional requirement that Congress alone has the right to declare war.
Counter-claim In many countries the general public does not have the civic education to responsibly deal with a free press or multi-party elections.
Broader Social unaccountability (#PC1522).
Aggravates Official secrecy (#PC1812) Governmental incompetence (#PF3953)
Failure of government intelligence services (#PF8819).
Aggravated by Inadequate public finance statistics (#PF7842)
Limited accountability of public services (#PF6574)
Proliferation of public sector institutions (#PF4739)
Opaque budgetary procedures in the public sector (#PF5374).

♦ **PE4999 Defoliating insects**
Nature Overgrazing by defoliating insects can disrupt natural forest ecosystems. Defoliation seriously interferes with energy flows: the ingestion of leaves that otherwise would have fallen and decomposed, interferes with normal nutrient cycles. Periodic severe outbreaks by insects such as the gypsy moth, spruce budworm, tent caterpillar, and cankerworm can significantly alter species composition and influence succession.

Broader Insect pests of plants (#PD3634).
Related Lepidoptera as insect pests (#PE3649).

♦ **PE5001 Inefficient use of resources**
Broader Poor quality of domestic livestock (#PD2743)
Inefficient use of proteins in factory farming (#PF2758).
Narrower Under-utilized raw materials (#PF6590)
Underutilization of human resources (#PF3523).
Aggravates Worldwide misallocation of resources (#PB6719).

♦ **PE5005 Assassination of environmental activists**
Incidence Game wardens in many countries have been killed by poachers (10 in in Zambia, 23 in Rwanda). Examples of assassinations include: Guy Bradley for efforts to protect egrets in Florida (1905), the chief warden of the Naorongoro reserve in Tanzania (1979) Roger Edward by poachers in Zimbabwe (1980), Joy Adamson for efforts to to prevent poaching in Kenya (1980), Valery Rinchinov in Siberia (1982), Dian Fossey for efforts to protect gorillas in Rwanda (1985), Fernando Pereira, Greenpeace photographer in the explosion of the "Rainbow Warrior", Chico Mendes for protecting Brazil's rainforests from ranchers, mining and timber companies (1988).
Broader Assassination (#PD1971).
Aggravated by Violence by fanatical environmentalists (#PD5582).

♦ **PE5006 Tidal floods**
Nature Tidal floods are overflows of coastal lands bordering an ocean, an estuary, or a lake. These coastal lands, such as bars, spits and deltas affected by the coastal current, occupy the same protective position relative to the sea that flood plains do to rivers. Coastal flooding is primarily due to landward flows caused by high tides, waves from high winds, surges from distant storms, tsunamis, or a combination of these events. Along shores, damage also can be caused by ice driven ashore by wind or wave action. Tidal floods can also be caused by the combination of waves generated by hurricane winds and flood runoff resulting from the heavy rains that accompany hurricanes. Tidal floods may extend over large distances along a coastline. The duration of tidal floods is usually short, being dependent upon the elevation of the tide which rises and falls twice daily in most places. However, maximum tide elevations can be identical on consecutive days. In the case of tidal floods associated with hurricanes, the high velocities of hurricanes winds often produce wave heights about 3 feet higher than the maximum level of the prevailing high tide.
Incidence Most of the severe tidal floods are caused by tidal waves generated by high winds superimposed on the regular cyclic tides. Tropical hurricanes are the primary sources of the extreme winds. Each year, several hurricanes enter the United States mainland, striking along the coasts of the Gulf of Mexico and the Atlantic Ocean.
Broader Water salinization (#PE7837).
Related Storm surges (#PD2788).

♦ **PE5009 Inefficient tax administration in developing countries**
Nature In developing countries the tax administration does not always carry out the intent of tax legislation, but makes its own policy.
Broader Inefficient public administration in developing countries (#PF0903).
Related Ineffective tax systems in developing countries (#PF2124).

♦ **PE5014 Downgrading of jobs due to computerization**
Nature Large numbers of office workers have been affected by computerization. A few move up to prestigious jobs as programmers or operators, but most are forced to move to totally new jobs, often at new locations, where they are given 'de-skilled' jobs; many people accept this frustration rather than being without employment.
Incidence In the UK, a 1983 study revealed a trend whereby women are entering jobs formerly held by men, such as printing, garment-cutting, mail sorting and to a certain extent computer programming. However, in every case, this has been due to the deskilling and downgrading of the job and the substitution of electronic equipment for personal judgement.
Broader Social ill-effects of automation (#PE5134).
Aggravated by Social isolation of computer users (#PE1277).

♦ **PE5015 Erosion of sovereignty**
Infringement of national sovereignty
Nature The principle of national sovereignty is eroding because of increased interdependence between nations, regions and continents and because of increased vigour of sub-national regions.
Incidence The development of the European Community is creating a super-national entity with increasingly important smaller regions, such as Scotland, Wales, Catalonia, Euzkadi, Flanders, Wallonia, Sicily, Piedmont, Venetia, North Rhine-Westphalia, and Bavaria, which will outlast the 12 nations that make up the Community.
Counter-claim National sovereignty is an out of date concept encouraging governments to lose power to narrow internal interests. For example, farmer's pressure groups representing a small number of people prevent rational farm policies in much of the developed world. Transnational problems like drug trafficking, international fraud and environmental pollution are not dealt with transnationally because of the concept of national sovereignty.
Broader Erosion (#PC8193).
Related Foreign military presence (#PD3496) Foreign control of natural resources (#PD3109)
Erosion of national sovereignty by transnational enterprises (#PE1539).
Aggravates Domination of the world by territorially organized sovereign states (#PD0055).
Aggravated by Denial of the right to national sovereignty (#PE7906)
Violation of sovereignty by trans-border broadcasting (#PE0261).
Reduces Inadequate power of intergovernmental organizations (#PF9175).

♦ **PE5016 Malignant neoplasm of skin**
Skin cancers — Melanoma
Incidence Skin cancer is the most common form of cancer in Caucasian populations. Some 10,000 die each year worldwide. Melanoma, one of the most serious forms with respect to fatal outcome, has been increasing over recent decades by 3 to 7 per cent per year.
Refs Ackerman, A Bernard and Maize, John C (Eds) *Malignant Melanoma and Other Melanocytic Neoplasms* (1985); Callagher, R P (Ed) *Epidemiology of Malignant Melanoma* (1986); Veronesi, Umberto, et al *Cutaneous Melanoma* (1987).
Broader Malignant neoplasms (#PC0092)
Diseases of the skin and subcutaneous tissue (#PC8534).
Aggravated by Dissatisfaction with skin colour (#PF1741)
Ultraviolet radiation as a hazard (#PE5672)
Environmental hazards of solar radiation (#PE3883).

♦ **PE5020 Lack of appropriate education for children of immigrants**
Nature Belonging to under privileged socio-economic milieux, deprived of their traditional cultural support, and in many cases having an inadequate command of the host country's language, the children of immigrants have a disturbing school failure rate.
Incidence One report for 1982 cites the degree of failure or backwardness of 60 percent of the children of immigrants on leaving school to be so serious that they could not be rescued by the educational system.
Refs The Documentation of the Council of Europe (Eds) *The Education of Migrant Workers' Children* (1981).
Broader Inappropriate education (#PD8529)
Social maladjustment of children of migrants (#PE4258).

♦ **PE5021 Fraud concerning economic situation and corporate capital of companies**
Broader Economic crime (#PC5624)
Violations against economic regulations (#PD7438)
Limited criminal liability of corporations (#PF7293).
Aggravates Disaster hazards to island populations (#PE5784).

♦ **PE5022 Rural-urban income differential**
Imbalance between rural and urban incomes
Nature The growth of minimum wages in some developing countries during the past 2 decades has exceeded what might have been expected from observation of the labour market. This has contributed to an increasingly inequitable income differential between urban and rural unskilled workers, aggravated by unemployment. The model of dual economy assumes that earnings for unskilled labour in the traditional sector set a floor to wages in the capitalistic sector, and that a gap of 30 percent or more usually exists between the two sectors. The gap, in part 'illusory' because of the higher cost of living and other disadvantages associated with living in urban areas, may nevertheless represent a real difference between urban unskilled and rural earnings, in other words some of the rural-urban differential may be the result of institutional pressures not directly based on labour market conditions.
Incidence Available data for comparisons show average wage incomes to be much higher than average incomes in the non-wage sector (including the money value of subsistence output). A recent ILO investigation into the size of the average rural-urban income gap in 14 countries of sub-Saharan Africa showed in the first stage an average rural-urban gap with values of from 1 to 2 to 1 to 3.3 for the poorest of three groups of countries (including a value of 1 to 3 in Burkina Faso). For the middle group values were much higher, ranging from 1 to 4 (Benin) to 1 to 8 (Lesotho). For the final and wealthiest group, values fell again, mostly to around 1 to 2 (including a value of 1 to 2.4 in the Ivory Coast). This pattern of change follows both the increasing degree of urbanization and the growth of rural nonfarm activities. In the second stage, estimates were made of the difference in rural-urban prices. These showed a price gap ranging from 10 to 40 percent between town and country. The result was naturally to reduce the average rural-urban income gap but not, by and large, to change the observed cross-national pattern. Finally, a range of national data was reviewed in order to place the income and expenditure level of the average agricultural worker (whether employed or self-employed) within the hierarchy of urban occupational earnings. For French-speaking countries the results can be summarized as showing that average agricultural incomes were similar to those of urban informal sector employees or unskilled modern sector workers paid the minimum wage. In Kenya the agricultural-urban wage gap was found to be still significant at 1 to 2.9; in Nigeria and Tanzania, however, the ratio approached unity.
Broader Social inequality (#PB0514) Unequal income distribution (#PD4962)
Inequitable distribution of wealth (#PB7666).
Narrower Imbalance between rural and urban incomes in developing countries (#PE8584).
Aggravated by Urban bias (#PF9686).

♦ **PE5023 Inadaptation of technology to man in the industrialized societies**
Broader Technological revolution (#PC3234) Social ill-effects of automation (#PE5134).
Related Technology gap between developed countries (#PD0338)
Antagonism between employment policy and technical advance (#PE5104)
Lack of technical development and excess of manpower in developing countries (#PE4933).
Aggravates Mental stress due to automation (#PE5164).
Aggravated by Uncontrolled application of technology (#PC0418).

♦ **PE5024 Unavailability of land for agricultural purposes in developing countries**
Nature In developing countries, the expense involved in bringing new land into production generally results in preference for intensified use of existing farmland. Reform of the land tenure system would normally be necessary to make adequate land available for small farmers and landless labourers.
Incidence In 1975 only 21 percent of the potential arable land in the developing countries of Asia was unused, in the Near East the proportion was 37 percent; but in Africa, 70 percent of the potential arable land was unused, and in Latin America the figure was 75 percent.
Broader Compliance*complex (#PA5710) Shortage of cultivable land (#PC0219)
Unavailability of agricultural land (#PC7597).
Aggravates Stagnated development of agricultural production (#PD1285)
Lack of purchasing power in developing countries (#PE8707)
Deterioration of domestic food production in developing countries (#PD5092).
Aggravated by Excessive land usage (#PE5059).

♦ **PE5026 Smoking during pregnancy and breast-feeding**
Nature The foetus and the breast-fed baby can become the victim of a mother who smokes. Indeed, smoking has immediately bad effects on pregnant women, their foetuses, and on nursing mothers. It increases the risk of miscarriage and stillbirth and may result in low birth-weight and backward babies. If the mother smokes during pregnancy, the new-born child weighs an average 200-250g less than it would if she did not smoke. In terms of statistical risks, it seems that the probability of a woman who smokes five to ten cigarettes a day will give birth to a child weighing less than 2.5kg is 50 percent more than that for a woman who does not smoke. The corresponding increase in probability for a woman who smokes twenty or more cigarettes a day is 130 percent. Both the quality and quantity of breast milk may be reduced if the mother smokes, the risk of sudden infant death syndrome is increased, and various aspects of the child's physical growth, intellectual and emotional development and behaviour may be affected.
Broader Health hazards of smoking for women (#PE4995).
Related Abuse of coca and cocaine (#PD2363)
Substance abuse during pregnancy and breast-feeding (#PE3858).
Aggravates Sudden unexpected infant death (#PD1885).
Aggravated by Ignorance of women concerning primary health care (#PD9021).

♦ **PE5027 Medical practitioners refusing to treat patients**
Nature An increasing number of doctors and dentists are refusing to treat certain kinds of patients because of increased financial and physical risk. Particularly the risk of contracting AIDS and the risk of law suits for accidental death or disability are causing medical practitioners to select patients more carefully.
Claim No one is forced to enter training for the life of a doctor, and patients have the right to expect treatment from their doctors when they are sick. Law suits are covered by medical insurance and its cost is passed on to patients. There is no reason for a doctor, dentist or nurse to refuse to treat a patient because of the disease unless it is outside the practitioners field of expertise.

Broader Unethical medical practice (#PD5770) Denial of rights of medical patients (#PD1662).
Aggravated by Increase in insurance claims for medical negligence (#PE4329)
Occupational risks and hazards of the medical profession (#PE5355).

♦ PE5028 DDT as a pollutant

DDT

Nature DDT is a synthetic compound called Dichlorodiohepytrichloroethane and is a chlorinated hydrocarbon. While this broad spectrum insecticide can kill a wide range of insects and other arthropods, it is also persistent in the environment, breaking down very slowly and retaining its effectiveness for long periods after application. Because chlorinated hydrocarbons are not water soluble, they tend to accumulate in fatty tissues when taken by living organisms and may remain in the body indefinitely. DDT interferes with bird's ability to metabolize calcium, resulting in the production of eggs with shells so thin that they break when nesting parents sit on them. Some species of birds in certain environments accumulate enough DDT to cause death. DDT also has potential for causing cancer, mutations and birth defects. DDT and its sister chemicals of lindane and chlordane have tumour–promoting and not tumour–initiating properties.
Incidence The two primary uses of DDT that have resulted in its environmental dispersal are in public health and agriculture. In the former it has largely been employed as a biocide against adult mosquitoes in malarial control. In agriculture, one of its principal applications is as a biocide in the protection of cotton crops. Restrictions upon its agricultural uses have already been enacted in the United States, Canada, Japan, Sweden and the Soviet Union following observations of unwanted mortalities and morbidities among non–target organisms.
Broader Insecticides as pollutants (#PD0983) Hydrocarbons as pollutants (#PE0754).
Aggravates Encephalitis (#PE2348).
Reduces Malaria (#PE0616).

♦ PE5036 Expansive soils

Nature Soils and soft rocks which tend to swell or shrink due to changes in moisture content are commonly known as expansive soils. Two major groups of rocks serve as parent materials of expansive soils. The first group consists of ash, glass, and rocks from volcanic eruptions. The aluminium silicate minerals in these volcanic materials often decompose to form expansive clay minerals of the smectite group, the best known of which is montmorillonite. The second group consists of sedimentary rocks containing clay minerals. Expansive soil causes many millions of dollars worth of damage a year in damage to homes, commercial buildings, roads and highways, buried utilities, and other structures.
Related Soil compaction (#PD1416).
Aggravated by Over–intensive soil exploitation (#PC0052).

♦ PE5037 Police crimes during narcotic investigations

Nature Crime in association with enforcement of drug laws is not rare. There are three major categories: (1) Those crimes in which enforcement authorities engage during the investigation apprehension and trial of suspected drug offenders. Here occur violations of the rights of citizens by illegal electronic surveillance, invasions of privacy, interrogations by beating or torture, false arrest, planting of evidence, perjured testimony and the like. These can be, from the standpoint of the police, morally proper crimes which are seen as being in the service of the law to ensure application of criminal sanctions against offenders. Prosecutions of authorities for such violations appear to be rare, although observations made by outsiders indicate that the violations are unduly prevalent. (2) Crime which occurs when authorities take advantage of their position vis–a–vis a suspect to achieve personal gain. Typically the drug dealer may have his money or possessions stolen; an officer may steal funds intended for an informant but require the informant nevertheless to sign a receipt for full payment; the drugs carried by a suspect may be seized and either resold or given away to others; or female offenders may be exploited sexually. In these and other like instances the offender is in no position to complain and may in fact consider it a fair trade if the police in turn reduce the charges or fail to pursue the investigation further (3) The rarely occurring crime which involves a shift to chronic predatory offences, perhaps in partnership with drug dealers. It is not unknown for enforcement personnel to hijack drug shipments, rob or kill dealers, be paid to betray by reporting on enforcement plans, and otherwise "go rogue". In such instances the officer is at greatest risk of apprehension for he no longer has a sanctuary as regards the offender.
Broader Abuse of police power (#PC1142) Unethical practices by police forces (#PD9193).
Aggravates Fraud (#PD0486) Theft (#PD5552) Homicide (#PD2341)
Receiving bribes (#PC4701) Theft of property (#PD4691)
Violation of civil rights (#PC5285).

♦ PE5042 Indecent art

Obscene art — Decadent art — Erotic art — Blasphemous art — Nudity in art

Nature Indecent art is any depiction of obscene or indecent material, including by not limited to depictions of sado–masochism, homo–eroticism, the sexual exploitation of children, individuals engaged in sex acts; any depiction which denigrates the objects or beliefs of the adherents of a particular religion; or any depiction which denigrates, debases or reviles a person, group or class of people on the basis of race, creed, sex, handicap, age or national origin.
Background A classic example is the case of the painting of the Last Judgement by Michelangelo in the Sistine Chapel. The original was attacked at the time because of the exposure of "genitals in such relief that even in the whore–houses they could not fail to make one close one's eyes". Following criteria for church art defined by the Council of Trent in 1563, a decision was taken to paint over the parts that the priests found offensive. In 1990, concern remained during the restoration of the painting as to whether the concealment should be removed.
Broader Obscenity (#PF2634).
Related Nudism (#PF2660) Pornography (#PD0132) Immoral literature (#PF1384).
Aggravates Adultery (#PF2314) Profanity (#PF7427)
Debauchery (#PE8923) Immorality (#PA3369)
Promiscuity (#PC0745) Ethical decay (#PB2480)
Representative arts (#PF0981) Obscene telephone calls (#PE5757)
Debasement of works of art (#PE0558).
Aggravated by Prudery (#PF5892).

♦ PE5043 Discrimination against prisoners' families

Deprivation of prisoners' families — Discrimination against families of offenders — Discrimination against families of the accused

Nature Welfare authorities being primarily concerned with the well–being of the imprisoned husbands and only secondarily with that of their wives and families, a conflict of interest arises to the disadvantage of the latter. In societies where there is no social security system, communal sympathy and protection may be shattered. The chance of marriage disintegration is exacerbated by the separation and the artificial arrangement for prisoners' wives at visits, where free movement is impossible and noise makes conversation difficult. The wives become integrated into the penal system and are to some extent punished for an offence in which they took no part. They are also alone to bear the family worries and responsibilities with little or no financial aid, often they are socially rejected. The family may become the subject of a kind of social ostracism that destroys friendships, breaks fraternal blood relationships and makes good–neighbourliness a rare exception rather than an accepted mode of behaviour, particularly towards those related to political prisoners. It is almost as though these families suffer from a contagious disease and no one who can possibly avoid it wants to become contaminated.
In many cases, the families of political detainees are actively harassed by local military units who force them out of their homes and deprive them of their belongings. Often wives loose trace of their husbands, either immediately after arrest or as a result of transfer from one place of detention to another. Their fear of their children's security renders wives particularly vulnerable to pressures exerted upon them by military officers.
Refs Gattis, Louie S *Living in The Shadows* (1988); Schneller, Donald P *The Prisoner's Family* (1976).
Broader Injustice (#PA6486) Discrimination (#PA0833)
Maltreatment of prisoners (#PD6005).
Related Inadequate care for children of prisoners (#PF0131).
Aggravated by Police intimidation (#PD0736) Refusal to grant amnesty (#PF0182)
Discriminatory treatment of foreign prisoners (#PE6883)
Discrimination against ex–prisoners and ex–detainees (#PE6929).

♦ PE5048 Multiple personality disorder

Nature Within a person there exists two or more distinct personalities or personality states each having its own unique memories, behaviour patterns and social relationships. At least two of the personalities take full control of person's behaviour and transition may occur suddenly or gradually, over several hours or days.
Incidence The implications of the problem are seen in the extreme case of a 1990 trial in the USA in which a single witness was sworn in 3 times to allow different personalities to testify. Psychiatrists believed that a further 43 personalities inhabited her body.
Refs Bliss, Eugene L *Multiple Personality, Allied Disorders and Hypnosis* (1986).
Broader Neurosis (#PD0270) Mental illness (#PC0300).
Related Fragmentation of the human personality (#PA0911).

♦ PE5050 Lameness in swine

Broader Diseases of musculoskeletal system in animals (#PD7424).
Narrower Laminitis (#PG5410) Foot lesions (#PG9404) Overgrown hooves (#PG9423)
Glässer's disease (#PG6669) Mycoplasmal arthritis (#PG2302)
Mycoplasmal polyserositis (#PG7281) Degenerative joint disease (#PG5219)
Neonatal septic polyarthritis (#PG4104) Proximal femoral epiphysiolysis (#PG3747)
Suppurative arthritis in older animals (#PG9725)
Nutritionally induced skeletal abnormalities in swine (#PG7819).

♦ PE5051 Historical forgery

False historical documents and objects

Nature Forgery, or the creation of a false document or object with the pretence of its recent discovery, is most often done for financial gain, in which case it constitutes criminal fraud. It may also be done to establish or enhance the reputation of an individual, or an institution, or to establish legal claims.
Incidence Forgery of historical documents and objects in recent times has included such diverse items as: the Piltdown Man; the Vinland Map which purported to show North America as discovered by Leif Ericson; the 1903 forgery "The Protocols of the Learned Elders of Zion" used by the Nazis to justify genocide of the Jews; and diaries claimed to have been written by Adolph Hitler.
Broader Forgery (#PD2557) Historical misrepresentation (#PF4932).
Narrower Religious historical forgery (#PG6355).
Related Falsity (#PF5900) Art forgery (#PE2382) Literary forgery (#PE6188)
Misleading information (#PF3096)
Misrepresentation of historical persons (#PF5662)
Production serving false consumption needs (#PF2639).
Aggravates Inaccessible historical libraries (#PF9046).

♦ PE5053 Computer stress

Technostress

Nature The computer is a highly technological tool and its introduction into the lives of people who are unprepared for it causes stress. High levels of stress are especially observed in office environment where the lack of control over job activities introduced by computerization and the increased workload lead to less worker interaction, greater performance monitoring and less worker participation. As control activities decrease, the perception of monotony increases. The computer system technology is designed without regard to the human factor in the system. In essence, the design reflects the computer capabilities and performance functions which are then imposed on the operator. This leads to a dehumanization of the work activity that is comparable to that produced by the introduction of assembly lines in manufacturing industries. In fact, many offices become paper factories with clerical assembly lines in which the work content is simplified to increase production and capitalize on computer capabilities. These jobs produce boredom and dissatisfaction. Such a work environment also brings about worker fears that further automation and computerization of their job may lead to job loss or downgrading to a lower level job.
Working conditions in computer rooms frequently leave much to be desired: they may be too noisy, too warm, or unpleasant (temperature, humidity). Outside the computer room, the display/keyboard workplace, in particular, can cause medical complaints: reflections on the screen and bad contrast can give rise to eye fatigue and headaches. Wrongly designed furniture can cause pain in the neck and back. Aspects of a more psychological nature are: monotonous work in an isolated position, frustration caused by waiting times and the requirement of intense concentration (air traffic control).
Another disturbing aspect is the potential of computers for storing personal information about workers for use (and abuse) by management, thus creating a climate of control and mistrust.
Incidence Computer–related stress leads to absenteeism and a high turnover rate (25 percent to 50 percent in data processing).
Claim Techno–stress manifests itself in two distinct but related ways: in the struggle to accept computer technology, and in over identification of computer technology. The primary symptom of those who are ambivalent about or fearful of computers is anxiety. This anxiety is expressed in many ways: irritability, headaches, nightmares, resistance to learning about computers or outright rejection of the technology. The primary symptom among those who have too successfully identified with computer technology is a loss of the capacity to feel and to relate to others. Signs of this techno–centered state include a high degree of factual thinking, poor access to feelings, an insistence on efficiency and speed, a lack of empathy for others, and a low tolerance for the ambiguities of human behaviour and communication. At its most serious, this form of techno–stress can cause aberrant and antisocial behaviour an the inability to think intuitively and creatively.
Refs Brod, Craig and St John, Wes *Technostress*; Dy, Fe Josefina F *Visual Display Units* (1985); Sethi, Amarjit S et al (Ed) *Strategic Management of Technostress in an Information Society* (1987).
Broader Occupational stress (#PE6937) Mental stress due to automation (#PE5164)
Social ill–effects of automation (#PE5134).
Narrower Health hazards of computer visual display units (#PE5083).
Related Human fatigue during control of complex equipment (#PE5572).
Aggravated by Computer obsession (#PF1288) Social isolation of computer users (#PE1277).

♦ PE5058 Inadequate hospital facilities
Substandard hospital buildings — Inoperable hospital facilities — Unsafe hospitals — Lack of hospital facilities
Broader Inadequate medical facilities (#PD4004).
Related Substandard housing and accommodation (#PD1251).
Aggravates Human disease and disability (#PB1044).
Aggravated by Civil war (#PC1869) Lack of war relief (#PF0727)
Inadequate medical resources (#PD7254) Ill treatment of prisoners of war (#PD2617)
Unethical practices of health services (#PE3328) Prohibitive cost of hospital facilities (#PE4154)
Denial of human rights in armed conflicts (#PC1454).

♦ PE5059 Excessive land usage
Broader Land misuse (#PD8142)
Unconstrained exploitation of natural resources (#PF2855).
Narrower Excessive land usage by transportation systems (#PE2525).
Related Unused land (#PF4670) Unsustainable harvesting rates (#PD9578)
Excessive consumption of resources in developed countries (#PE5551).
Aggravates Limited available land (#PC8160) Inadequate grazing land (#PJ0404)
Scarcity of residential land (#PD8075) Unavailability of building sites (#PJ8549)
Long-term shortage of natural resources (#PC4824)
Insufficient common land in urban environments (#PE6171)
Limited availability of land for low-income and disadvantaged groups (#PF5008)
Unavailability of land for agricultural purposes in developing countries (#PE5024).
Aggravated by Idle private land (#PJ8020) Decreasing land mass (#PF7435)
Shortage of urban land (#PD0384) Shortage of cultivable land (#PC0219)
Destruction of land resources (#PJ8397)
Exploitation of land for the burial of the dead (#PE9095).

♦ PE5063 Intestacy
Nature Dying without leaving a will can be a disaster for women whose husband dies without making a will to protect her interests. Many people assume that all they own will automatically go to their spouses. It is not so. When a person dies intestate brothers, sisters and even cousins may have a claim on what he owned.
Broader Unequal property rights for women (#PE4018).
Aggravates Burden on society of widows (#PF6149)
Denial of right to inherit property (#PF0886).
Aggravated by Human death (#PA0072).

♦ PE5064 Degradation of the environment through the destruction of species
Broader Extinction of species (#PB9171) Natural environment degradation (#PB5250).
Related Destruction of wildlife habitats (#PC0480)
Endangered species of plants and animals (#PB1395)
Unsustainable exploitation of fish resources (#PD9082)
Environmental hazards of new species introduction (#PC1617).
Aggravated by Mismanagement of environmental demand (#PD5429).

♦ PE5066 Ground failures
Nature Ground failures involving landslides, expansive soils, and subsidence are a major threat each year to man and his works.
Broader Topological disaster (#PC5010).
Narrower Ground failures due to liquefaction (#PE5126).
Related Avalanches (#PD1146) Landslides (#PD1233).

♦ PE5067 Endangered species of water fowl
Incidence Water fowl have suffered greatly from habitat encroachment throughout the world; populations in the United States may be as little as one tenth of what they were before the coming of European civilization. For instance, the Redhead, a once popular and fairly common duck species, decreased in numbers from about 100,000 to fewer than 3,000 in the period from 1955 to 1974, a reduction in the population of 97.6 percent.
Broader Endangered species of birds (#PD0332).

♦ PE5070 Violation of the right to strike
Suppression of strikes by public authorities — Denial of right to strike
Broader Denial of the right of trade union association (#PD0683).
Narrower Union busters (#PG1767)
Requisitioning of workers to prevent strike action (#PE0757)
Dismissal of workers to prevent legal strike action (#PE7620).
Related Sanctions against trade union workers (#PD0610)
Denial of right to collective bargaining (#PE3970).
Aggravates Apartheid (#PE3681).
Reduces Strikes (#PD0694).

♦ PE5074 Disruption of work schedule due to computerization
Worker participation excluded from company polity
Nature A computer is an expensive piece of hardware. To make it pay, it must be used intensively, forcing night and weekend work on people who previously worked normal hours.
Broader Inflexible management patterns (#PF3091).
Aggravates Excessive hours of work (#PD0140).

♦ PE5076 Obtaining property by false pretences
Broader Fraud (#PD0486) Theft of property (#PD4691)
Unethical real estate practice (#PD5422).
Aggravated by Forgery (#PD2557).

♦ PE5080 Nuclear accidents in space
Radioactive fallout from accidents to orbiting nuclear-powered reactors
Nature Increasing use is made of nuclear-powered spacecraft which threaten contamination of the environment when their orbits decay and the vehicles are destroyed on re-entry to the atmosphere.
Incidence The nuclear accidents in space that have occurred to date have had widely varying actual or assumed effects. Experts still differ over the radiological implications of several of the most salient events. The environmental effects of accidents involving nuclear-powered spacecraft depend on whether significant radioactivity is returned to the Earth, to what extent it is contained, and how released radioactivity is distributed in the atmosphere or over the surface of the Earth; whether it is dispersed widely or becomes concentrated in specific geographic areas, especially in populated areas or locations in which it might be taken up into biological systems. The geographic coordinates of the re-entry or subsequent Earth impact, the altitude at which radioactivity is released as well as corresponding meteorological conditions, and the choice of reactor design and associated safety systems, are all important in determining the extent of the radiological consequences.
Broader Nuclear accidents (#PD0771).
Aggravates Gamma ray disturbance in space (#PF1441).
Aggravated by Government secrecy concerning nuclear weapons testing (#PF4450).

♦ PE5081 Rheumatoid arthritis
Rheumatic gout
Incidence As an indication, the market for topical analgesics (for muscular aches and pains) in the UK in 1986 was £10.2 million.
Refs Brewer, Earl J *Juvenile Rheumatoid Arthritis* (1982); Brown, Thomas M and Scammell, Henry *Rheumatoid Arthritis* (1988).
Broader Rheumatic diseases (#PE0873).
Narrower Ankylosing spondylitis (#PE8990).
Aggravated by Autoimmune disease (#PE5214) Diseases of connective tissue (#PD2565).

♦ PE5083 Health hazards of computer visual display units
Job stress from video display terminal work
Nature The potential health hazards of working on video display terminals (VDTs) include eye strain, repeated strain injury particularly of the wrist and reproductive hazards.
Counter-claim There is no proof that VDTs cause reproductive problems, eyestrain or backache or any other health problems. There is some evidence that these problems associated with VDTs are the result of working condition, job stress and miss use of the equipment.
Refs Chénier, Nancy Miller *Reproductive Hazards at Work* (1983); Dy, Fe Josefina F *Visual Display Units* (1985); European Foundation for the Improvement of Living and Working conditions *Effect of Introduction of a Visual Display Unit in a Computerised Office on the Health of Operators*; European Foundation for the Improvement of Living and Working conditions *The Working Environment at Visual Display Units*.
Broader Computer stress (#PE5053) Occupational hazards (#PC6716)
Visual strain in modern technology (#PE6865).
Related Health hazards of electromagnetic fields generated by electrical appliances (#PE7879).
Aggravates Occupational stress (#PE6937).
Aggravated by Computer obsession (#PF1288) Social isolation of computer users (#PE1277)
Socially disruptive effects of video games (#PF6345).

♦ PE5085 Wind shear as a hazard
Nature Wind shear is an abrupt and often invisible movement of air, the most dangerous of which is the microburst. A microburst is a shaft of cold air that moves at high speed (up to 100 kph) towards the earth's surface, often from thunderstorms, spreading in all directions as it reaches the ground. Metrologists have been able to predict certain types of wind shears (mainly those associated with warm and cold fronts, low-level jet streams, and mountains) for a number of years, but there has emerged a newly identified and treacherous type of wind shear is a downward surge of air that, striking upon the earth, spreads out in all directions. Wind shears do damage to the environment, to man-made structures, to migratory birds, and are a hazard to aviation. In aviation, wind shears play havoc with an aircraft's aerodynamics by dispersing air in all directions. An aircraft moving into this formation at low level (take off or landing) will encounter winds that will tend to lift it at first, and an unsuspecting pilot will slow the throttle. But once the plane reaches the other side of the shear, where the winds are moving with equal force in the opposite direction, it will be slowed suddenly, often to the point where the pilot cannot throttle quickly enough to avert a crash.
Incidence US scientists blame 28 accidents, accounting for 491 deaths and 206 injuries to windshears since 1964, not including an August 1985 crash which killed 132 people and in which windshear was suspected.
Refs International Civil Aviation Organization *Wind Shear* (1987).
Broader Wind (#PE2223).
Aggravates Air turbulence (#PD2127).

♦ PE5086 Dermatomyositis
Refs Dalakas, Marinos *Polymyositis and Dermatomyositis* (1988).
Broader Rheumatic diseases (#PE0873).
Aggravated by Diseases of connective tissue (#PD2565).

♦ PE5087 Psychological conflict
Mental conflict — Mental disturbance — Psychological disturbance — Psychological distress
Nature Personal or psychological conflict refers to a situation in which a person is motivated to engage in two or more mutually exclusive or incompatible activities. It occurs when the overt, verbal, symbolic, or emotional responses required to fulfil one motive are incompatible with those required to fulfil another. Social existence involves a great number of conflicts. The individual in society, subject to the pressures of the groups to which he belongs and the demands of the roles he must play, often experiences personal conflict. The entire process of the socialization of the child can be viewed as a conflict between the individual and society. Clinical studies show that the concept of conflict is particularly significant in the areas of personal adjustment and mental disorder. Psychological conflicts are of central importance not only in neuroses but also in psychosomatic disease, sexual deviation and functional psychosis. Furthermore, they seem to contribute to various forms of social pathology, such as marital, educational and vocational failure; delinquency, crime and prostitution; and alcoholism and drug addiction.
Claim Freud said that civilization itself is a product of the clash between the incompatible demands of biological urges and social conformity.
Broader Conflict (#PA0298) Psychological disorders (#PD8375).
Narrower Paranoia (#PE0435) Sado-masochism (#PE6137) Emotional crises (#PE3407).
Related Social conflict (#PC0137) Political conflict (#PC0368)
International conflict (#PB5057) Inhibited grief process (#PD4918).
Aggravates Mental illness (#PC0300) Mental impairment (#PF4945)
Chronic bronchitis (#PE2248).
Aggravated by Fatigue (#PA0657) Diabetes (#PE0102)
Impure thoughts (#PF5205) Television violence (#PE4260).

♦ PE5089 Ineffective monitoring of hazardous substances
Claim The task of identifying and regulating toxic, mutagenic, teratogenic, carcinogenic, or otherwise hazardous substances well in advance of their distribution grows steadily more urgent as man manufactures each year more than a thousand substances that did not exist before. Experience with these new substances must be based on certain conclusions: no system is truly closed; toxicity may be hard to determine even after years of tests; the primary effects of man-made substances are virtually as unknown as the rates, routes and reservoirs of their natural distribution, the ways in which they may be concentrated or altered in natural systems, and the ways in which they may interact with each other; it will never be possible to predict the effects of all random combinations of man-made substances in vivo; the ethical questions about the risks and benefits of new substances, the rate at which future risks should be discounted, and the burden of proving risk must be taken into account; and monitoring of the entire biosphere is an important, though belated, line of defence in detecting risks which have not been detected through earlier screening.
Broader Ineffective monitoring (#PF2793) Inadequate environmental monitoring (#PF4801).

♦ PE5090 Inadequacy of staff qualifications in intergovernmental organizations
Nature On average the qualifications of personnel are inadequate to the responsibilities implied by the tasks they are called upon to perform. This has severe negative consequences for the

results of programmes and projects and aggravates other difficulties of such bodies.
Incidence In the United Nations, for example, 25 per cent of those in professional grades (staff engaged in programme design, management, research and drafting) have had no university training and 10 per cent have had less than three years of university studies. In the higher grades, where the implications are much more serious, the percentage of staff members without university education is approximately the same.

♦ PE5093 Interception of correspondence
Nature Interception of correspondence is knowingly destroying, modifying, opening, reading, or divulging the contents of private correspondence undelivered to the addressee and without the consent of the sender or receiver.
Broader Interception of communications (#PD7608).
Related Misuse of postal surveillance by governments (#PD2683).
Aggravates Insufficient communications systems (#PF2350).

♦ PE5096 Earth surface faulting
Nature Surface faulting – the differential movement of the two sides of a fracture at the Earth's surface – is of three general types: strike–slip, normal, and reverse. Combinations of the strike–slip type and the other two types of faulting can be found. Although displacements of these kinds can result from landslides and other shallow processes, surface faulting, as the term is used here, applies to differential movements caused by deep–seated forces in the Earth, the slow movement of sedimentary deposits toward the Gulf of Mexico, and faulting associated with salt domes. Death and injuries from surface faulting are very unlikely, but casualties can occur indirectly through fault damage to structures. Surface faulting, in the case of a strike–slip fault, generally affects a long narrow zone whose total area is small compared with the total area affected by ground shaking. Nevertheless, the damage to structures located in the fault zone can be very high, especially where the land use is intensive. A variety of structures have been damaged by surface faulting, including houses, apartments, commercial buildings, nursing homes, railroads, highways, tunnels, bridges, canals, storm drains, water wells; and water, gas, and sewer lines. Damage to these types of structure has ranged from minor to very severe.
Broader Geological hazards (#PC6684).
Aggravates Earthquakes (#PD0201) Land subsidence (#PD5156).

♦ PE5099 Radio frequencies as a health hazards
Refs ILO *Protection of Workers Against Radio–Frequency and Microwave Radiation. A technical review* (1986).
Broader Health hazards of radiation (#PD8050).
Aggravates Human disease and disability (#PB1044).

♦ PE5100 Malignant neoplasm of genito–urinary organs
Refs Garnick, Marc and Richie, Jerome *Urologic Cancer* (1983); Pavone–Macaluso, M, et al *Testicular Cancer and Other Tumors of the Genitourinary Tract* (1985).
Broader Malignant neoplasms (#PC0092)
Diseases of the genito–urinary system (#PC4575).
Narrower Prostate diseases (#PE8372) Urinary bladder disorders (#PE2307)
Malignant neoplasms of female genital organs (#PE1905).

♦ PE5101 Dizziness
Vertigo — Giddiness
Refs Honrubia, Vicente and Brazier, Mary (Eds) *Nystagmus and Vertigo* (1982); Wright, Anthony *Dizziness* (1987).
Broader Diseases of the ear (#PD2567) Symptoms referable to sense organs (#PE2665).

♦ PE5102 Low confidence in investment and stock markets
Aggravated by Economic inefficiency (#PF7556).

♦ PE5103 Cargo insecurity
Freight pilferage and theft
Nature The handling of cargo occasions theft and pilferage. The latter is considered to be the theft of one or two items out of a package; it is usually committed on impulse, and occurs in terminal areas where goods are handled or stored. On the other hand, a major theft is considered to be the theft of one or more packages, and is rarely committed on impulse. It is usually planned on receipt of information identifying certain shipments, and this often involves collusion between several employees in the transport system. Most major thefts today occur while the goods are in transit.
Claim The enormous sums of money involved in cargo theft are a severe drain on the economy of the countries concerned, and they indicate the enormity of the problem when considered on a world–wide basis.
Broader Theft of property (#PD4691) Insecurity of property (#PC1784).
Narrower Seizure of cargoes (#PE0125).
Related Insecurity of resources (#PB8678) Inadequate cargo transportation (#PE0430).
Aggravates Unreliable freight transport (#PG1026)
Inequities in marine insurance (#PE5802)
Obstacles for international ocean shipping (#PD5885).
Aggravated by Unethical practices in transportation (#PD1012).

♦ PE5104 Antagonism between employment policy and technical advance
Nature The primary aim of an employment policy is the creation of as many jobs as possible, whereas the primary aim of industrial production is the improvement of productivity by advanced working techniques. Quantitative employment is thus in antagonism with qualitative employment.
Counter–claim By taking only a partial view of the impact of new technologies on the level of employment, it is easy to come to overly pessimistic conclusions. However, in addition to reduced cost and improved quality of goods and services, there are long–term positive effects on employment that can offset short–term job losses – the creation of new products and industries, stimulation of demand for cheaper, better products, and spurring economic growth through increased investment in the new technologies.
Broader Unemployment (#PB0750) Inappropriate policies (#PF5645)
Technological revolution (#PC3234).
Related Social ill–effects of automation (#PE5134)
Inadaptation of technology to man in the industrialized societies (#PE5023)
Lack of technical development and excess of manpower in developing countries (#PE4933).

♦ PE5105 Cat scratch fever
Broader Zoonotic bacterial diseases (#PD6363).

♦ PE5106 Toxic shock syndrome
Nature Toxic shock syndrome is a rare condition associated with tampon use. It is caused by bacteria, Staphylococcus aureus. Initial symptoms are similar to those of flu, but as the illness progresses, severe hypotension and kidney failure can occur, followed by respiratory arrest.
Refs National Research Council Staff *Toxic Shock Syndrome* (1982).
Aggravated by Menstruation (#PE4838).

♦ PE5108 Sexual torture
Nature Victims of torture are sexually abused in many ways. Women and men are raped, sometimes in front of spouses. Weight are hung from testicles. Electrical contacts are inserted into the penis or vagina. Hot iron rods, sticks, bottles and irritating organic compounds are inserted into rectums. Cattle prods are used on genitals and breasts. Sexual organs are burned with acids or cigarettes.
Incidence Sexual torture has been reported in the following countries: **Af** Ethiopia, Namibia, Somalia, South Africa, Tanzania UR, Uganda, Zambia. **Am** Bolivia, Chile, Colombia, El Salvador, Mexico, Uruguay. **As** Bangladesh, Indonesia, Iraq, Philippines, Syrian AR. **Eu** Yugoslavia.
Broader Physical torture (#PD8734).

♦ PE5109 Volcanic dust
Nature Volcanic dust poses a number of potentially dangerous situations. Whole cities have been buried under volcanic dust. Large amounts of dust in the air can disrupt the weather. Floating clouds of volcanic ash can pose a serious hazard to international air traffic. So far it has been impossible to alert air traffic promptly enough on dangerous approaching ash clouds.
Broader Volcanic eruptions (#PD3568).
Aggravates Hostile environmental modification (#PD7941).

♦ PE5112 Needless incarceration
Inappropriate imprisonment
Claim Sending many offenders to prison represents an incredible waste in terms of money, manpower, facilities, and the infliction of human suffering. Incarceration may do harm to the individual involved by immersing him into a dependence–prone, criminally oriented counter–culture behind institution walls. Many offenders could perhaps be rehabilitated outside prison walls and could be employed in necessary social services.
Broader Imprisonment (#PD5142) Violation of civil rights (#PC5285).
Related Inadequate prevention of crime (#PF4924)
Inadequate correctional systems (#PF5172).

♦ PE5117 Cleft palate
Cleft lip — Hare lip — Hare–lip
Nature A cleft is a separation of parts of the mouth which are usually joined together during the early weeks of the development of an unborn child. Cleft lip is a separation on one or both sides of the upper lip and, quite often, of the dental ridge as well. Clefts of the palate, or roof of the mouth, may occur in the bony hard palate or in the soft palate at the back of the mouth, or in both. Cleft lip may involve only one side of the upper lip or both. The split may be only in the upper lip, or may extend up into the nostril. Usually the split goes through the outer skin, muscles, and inside of the lip. Feeding is the first problem to be overcome. Since a split in the roof of the mouth makes it difficult for a baby to suck, food backs up through its nose and may cause choking. Defective speech is one of the most serious results of cleft palate and, to a lesser extent, cleft lip. Psychological problems may result from speech difficulties and the child's unusual appearance before clefts are repaired, may require a professional counsellor's help. Clefts have no relation to mental ability or retardation. Ear infections are common: difficulty in swallowing affects air pressure around the inner ear, spreading infection directly through the nose to ear; frequent or severe ear infections may lead to hearing loss.
Incidence In the USA, some 5,000 children (one out of every 700 births) are born with cleft lip or palate each year. About 40 percent have both cleft lip and palate. The defect appears more often among orientals and certain tribes of American Indians than among white Americans. It occurs less frequently among black Americans.
Background There is no one cause for all clefts. Scientists believe that any number of factors such as drugs, diseases, heredity, malnutrition, and adverse environment may act on each other to disturb normal growth. Many infants with cleft palate are premature and have other defects. Heredity appears to play a role in about 25 percent of cases. In the other 75 percent there is no family history of the defect, even among distant relatives. It is known that if both parents are normal and have a child with a cleft, the chances that subsequent babies will have a cleft increases progressively. If either parent has a cleft, there is a four percent chance that their first baby will have a cleft, and the chances increase with each time they have an affected child.
Refs Powers, Gene R *Cleft Palate* (1986); Spriestersbach, D C *Psychosocial Aspects of the "Cleft Palate Problem"* (1973).
Broader Teeth disorders (#PD1185) Stigmatized diseases (#PD7279)
Human physical genetic abnormalities (#PD1618).
Aggravates Speech disorders (#PE2265).

♦ PE5122 Diseases of the nose
Refs Maran, A G *Logan Turner's Diseases of the Nose, Throat and Ear* (1988).
Aggravates Diseases of the ear (#PD2567).

♦ PE5126 Ground failures due to liquefaction
Nature Liquefaction is not a type of ground failure; it is a physical process that takes place during some earthquakes that may lead to ground failure. As a consequence of liquefaction, clay–free soil deposits, primarily sands and silts, temporarily lose strength and behave as viscous fluids rather than as solids. Liquefaction takes place when seismic shear waves pass through a saturated granular soil layer, distort its granular structure, and cause some of the void spaces to collapse. Disruptions to the soil generated by these collapses cause transfer of the ground–shaking load from grain–to–grain contacts in the soil layer to the pore water. This transfer of load increases pressure in the pore water, either causing drainage to occur or, if drainage is restricted, a sudden buildup of pore–water pressure. When the pore–water pressure rises to about the pressure caused by the weight of the column of soil, the granular soil layer behaves like a fluid rather than like a solid for a short period. In this condition, deformations can occur easily.
Incidence Liquefaction is restricted to certain geologic and hydrologic environments, mainly areas where sands and silts were deposited in the last 10,000 years and where ground water is within 30 feet of the surface. Generally, the younger and looser the sediment and the higher the water table, the more susceptible a soil is to liquefaction.
Broader Ground failures (#PE5066).
Aggravates Inadequate earthquake resistant construction (#PE6257).
Aggravated by Earthquakes (#PD0201).

♦ PE5132 Concentration of national governments activity on national affairs
Insufficient work on and awareness of international affairs
Broader National isolationism (#PF2141) Non–recognition of problems (#PF8112).
Aggravates Nationalism (#PB0534) Poor international relations (#PJ5600).

♦ PE5134 Social ill–effects of automation
Negative social effects of introduction of new technologies
Nature The human impact of automation is of a twofold nature. It affects the physiological and psychological functioning of the individual, thus influencing social structures; it also induces a

number of social and cultural changes which have repercussions on the individual. Wherever automation is introduced it leads to an important transformation of human existence in the biological, psychological and social spheres. The computer-type of automation in particular considerably affects human beings, who are more or less psychologically unprepared for changes of this kind. It may, in fact, be less harmful mentally for manual workers to have to accept new forms of mechanization which are only an extension of what they already know, than for office personnel to encounter a development which to some extent seems to menace their status by bringing it nearer to that of the machine operator. For both management and workers, serious problems arise. There is no assurance that the gains from new technology will be equitably distributed. Hardships accrue to individuals – and sometimes to large groups of workers in certain plants, industries, occupations or communities – who are displaced by technological change. The nature of the new work may be such as to create increased tension and loneliness. Management must also face problems of reorganization and adjustment of managerial personnel.

Background The term automation is used to designate certain new forms of mechanization of work which are progressively giving to automatic devices functions previously carried out by human beings. Three types can be characterized as: (1) the expansion of the scope of mechanization by transfer devices that link machine tools in automatic production lines and by advanced techniques of material and product handling and of assembly; (2) the rapid development of techniques of automatic control over manufacturing processes and the application of these techniques to an ever-widening range of industries; and (3) the rapid and automatic processing of an increasing range of technical and business information by the electronic digital computer, with a consequent extension of automatic control to complex manufacturing operations and commercial offices. In other words, a differentiation is made between transfer automation (also called Detroit automation), control automation, and computer automation.

The first type would seem to follow the general principles of the assembly line. The second type which is essentially based on the automatic feed-back of information into a mechanical system, can be conceived as a development of engineering devices of the sort represented by the so-called fly-ball governor in the steam engine, but it has become eminently more applicable since the introduction of electronics. The third type goes even further in replacing certain functions of the human being, in so far as it acts increasingly according to the functioning of the human brain. It can therefore be used in fields of activity which until now were only to a very small extent mechanized. A typical use is for the recording, co-ordination and analysis of information in the administrative field.

Claim 1. Automation is not leading towards a homogeneous society, but towards two incompatible societies. For every job that can be automated will be automated. The only non-automated jobs left will be leadership jobs. This dilemma will be one of the key problems facing humanity at the start of the next century.
2. Firms introducing automation spend 90 percent of the effort on technical and financial issues and less than 10 percent on the human and organizational ones. Usually, they do not pay enough attention to job redesign, machinery layout, training and skill bonuses.

Counter-claim Technological change often confers substantial benefits such as rising standards of living, greater leisure, reduction in the arduousness of work and more pleasant working conditions.

Broader Technological revolution (#PC3234).
Narrower Computer stress (#PE5053)   Mental stress due to automation (#PE5164)
Unpreparedness for surplus leisure time (#PF5044)
Downgrading of jobs due to computerization (#PE5014)
Inadaptation of technology to man in the industrialized societies (#PE5023).
Related Increasing job monotony (#PD2656)
Antagonism between employment policy and technical advance (#PE5104).

♦ **PE5135 Conflicting labour laws**
Nature Issues of conflict of laws in labour matters arise where the relations between an employer and a worker are liable to be affected by the legislation of two or more countries, whether this is the result of the employer and the worker having different nationalities or domiciles, of the worker being sent abroad by the employer for a more or less prolonged period, of the work being carried out in several countries (such as in international transport), of the employer having branches in several countries, or of a host of other eventualities. The variety of such situations and the number of legislations which may be involved have increased substantially in recent years. The question of conflicting labour laws is of particular interest to migrant workers and of daily relevance to undertakings that send workers abroad, have foreign subsidiaries or engage foreign labour; to workers' representatives in such undertakings and to unions bargaining with their management; to labour inspection services and other bodies concerned with the enforcement of labour law; and, of course, to the workers concerned.

Many of the conflicts and uncertainties inherent in the subject would seem to be due to the difficulty of establishing a clear dividing line between the legitimate scope of the law of the place of work and that of the home base (a difficulty compounded by such problems as that of identifying the true employer). The controversial issue of the choice of law by the parties may well be largely a reaction to that difficulty. Since there are already conflicts and uncertainty where these two basic systems confront each other, it is not surprising that the problems should be multiplied in cases in which a larger number of legal systems are involved, usually because there is no clear home base.

Incidence Approaches differ from country to country, and reflect different types of foreign connections in labour matters. In many countries private international law is undergoing a wider process of evolution and reconsideration. There are two main reasons for this. On the one hand, on the national plane the law has to take account of growing state intervention in economic and social affairs; labour relations are one of the fields in which this factor is of considerable relevance. On the other hand, internationally there has been a development of practices and institutions, including the growth of multinational groups of companies, which did not fit in with traditional concepts. As a reflection of these developments, though not as a necessary result, the question is now also being asked in some countries, whether private international law, which had traditionally been seen as providing a neutral method of choosing the system of law to govern a particular relation (a method treating all systems of law as equivalent), should not be so applied as to further social concepts, in particular, by taking account in each case of the results of a choice of law.

Background Under the rules of private international law there are two legal systems which have an important role to play in the labour field. One is that of the place of work: it is widely applied to the typical individual employment relation; its applicability is reinforced by the fact that much of it is mandatory and it is prime importance in collective relations. The second of these legal systems is that of a home base, often (though not necessarily) of both employer and employee: that system is applied not only to travelling personnel such as transport works and to certain categories of mobile managerial staff, but also to a considerable number of workers sent abroad for specific projects or assignments; certain connecting factors used to determine the law applicable to the typical employment relation (such as the common nationality or domicile of the parties, the place of business of the employer and even the place of conclusion of the contract) may lead to the adoption of that system; within its limits, it can apply to collective as well as individual labour relations.

Broader Conflict of laws (#PF0216).
Related Conflict of laws on international restriction of information (#PD3080).

♦ **PE5140 Flash floods**
Nature Flash floods, which have taken many lives and caused great property damage, are local floods of great volume and short duration. A flash flood generally results from a torrential rain or 'cloudburst' on a relatively small drainage area. Cloudbursts, associated with severe thunderstorms, take place mostly in the summer. Violent thunderstorms or cloudbursts usually develop in a short time and produce floods on relatively small and widely dispersed streams. Runoff from intense rainfalls result in high flood waves that can destroy roads, bridges, homes, buildings, and other community developments. Discharges quickly reach a maximum and diminish almost as rapidly. Flood flows frequently contain large concentrations of sediment and debris collected as they sweep channels clean. Flash floods can also result from the failure of a dam or from the sudden breakup of an ice jam. Each can cause the release of a large volume of flow in a short time.

Incidence Flash floods can take place in almost any area but they are particularly common in mountainous areas and desert regions. They are a potential source of destruction and a threat to public safety in areas where the terrain is steep, surface runoff rates are high, streams flow in narrow canyons, and severe thunderstorms prevail.
Refs Hall, A J *Flash Flood Forecasting* (1981).
Broader Floods (#PD0452).

♦ **PE5141 Discrimination against use of accents of a language**
Discrimination against regional accents
Claim In UK people associate the Standard British English or "received pronunciation" with the social class and ideas of intelligence, ambition and occupational status. The accent determines chances in life, success or failure.
Refs Honey *Does Accent Matter?* (1989).
Broader Discrimination (#PA0833).
Aggravates Limited verbal skills (#PD8123)   Language discrimination in politics (#PD3223)
Discrimination against minority languages (#PD5078).
Aggravated by Ignorance of grammar (#PD7566).

♦ **PE5143 Instability of plastics industry**
Incidence The plastics industry has been a rapidly growing one consistently outstripping growth in other industrial sectors. In 1969 world output of plastic materials was about 24 million tons. This increased to 58 million in 1980 and was forecast to grow to over 100 million tons by 1990. The recession of the late 1970s and early 1980s, however, has caused the pace of growth to slacken. The rate of growth of plastics production is also subject to regional differences, and additionally, different patterns of per capita consumption have emerged. For example, whilst the Western European industry grew by 15.4 percent per annum in 1965–70 and then at 6.7 percent per annum in 1970–76, and the US industry by 11.3 percent per annum from 1965–70 and 6.4 percent from 1970–76, the Japanese industry registered annual increases of 23.6 percent in 1965–70 and 0.6 percent in 1970–76. In 1976 the Federal Republic of Germany consumed 90 kg of plastics per capita, the United States 52.3 and Japan 41.7. Plastics production and per capita consumption in any country are closely related to both gross national product and the rate at which traditional materials can be displaced. Highest production and consumption figures are found in the industrialized countries and the lowest rates in the developing countries.
Broader Instability of chemical and petrochemical industry (#PE0538).

♦ **PE5144 Revolt**
Refs Bell, J Bowyer *On Revolt* (1976); Davies, J C *When Men Revolt and Why* (1971).
Broader Seizure of power (#PC8270).
Related Mutiny (#PD2589).

♦ **PE5146 Health hazards of passive smoking**
Health hazards to children of smokers
Nature Tobacco has its victims other than smokers: those living among smokers can be considered as passive smokers because they are exposed to smoke concentration in the atmosphere they live in. The effects of passive smoking are not limited to benign ailments. Recent studies have shown an increased incidence of lung cancer in women married to heavy smokers. The risk of wives developing lung cancer doubled when the husbands smoked over twenty cigarettes a day. There was also an increased incidence of emphysema and asthma, although to a lesser degree. Many studies have also shown that when parents smoke, their children cough. Babies are most at risk, with the highest percentages for bronchitis and lung ailments in infants under a year old.
Incidence In the period 1985–88, it has been estimated that passive smoking accounts each year for up to 5,000 deaths in the USA, 1,000 in the UK, and 500 in Canada. A recent US study has found that leukaemia appears seven times more often among people who have spent their lives with smokers. Among people who had lived with three smokers or more, the risk of breast cancer rose 3.3 times and the risk of cervical cancer increased 3.4 times. A 1990 study concluded that passive smoking accounted for more than 3,000 cases of lung cancer among non-smokers in the USA.
Broader Smoking (#PD0713)   Smoke as a pollutant (#PD2267)
Indoor air pollution (#PD6627).
Related Passivity (#PF6177).
Aggravates Malignant neoplasms of female genital organs (#PE1905).

♦ **PE5150 Agnosia**
Nature In certain diseases of the brain, the patient loses the ability to recognize the character of objects through touch, taste, sight, hearing.
Refs Peach, Richard K *Readings in Agnosia* (1986).
Broader Language disorders (#PE3886).
Related Sensory aphasia (#PE4234).

♦ **PE5152 Unjust customary rights**
Droit du seigneur
Nature The droit du seigneur is a legal or customary right of a feudal lord to have a sexual relations with a vassals bride on her wedding night, whatever the view of the bride or the groom.
Incidence Such practices tend to emerge in groups and bands, such as gangs or groups of bandits, lead by a charismatic figure who arrogates to himself rights such as those of a feudal lord.
Broader Arrogation of rights (#PD4680).

♦ **PE5160 Coal mining environmental hazards**
Refs ILO *Prevention of Accidents Due to Fire Underground in Coal Mines* (1959); ILO *Prevention of Accidents Due to Explosions Underground in Coal Mines* (1974); ILO *Prevention of Accidents Due to Electricity Underground in Coal Mines* (1959).
Broader Environmental hazards from mining (#PD2596).

♦ **PE5161 Calcium deficiency**
Refs FAO *Calcium Requirements* (1968).

♦ PE5164 Mental stress due to automation
Nature The introduction of automation has certain psychological repercussions on those involved, and in some cases these may set off reactions affecting mental health. Two types of reaction can be distinguished: first, emotional reaction to the introduction of an essentially new technological method and in particular to the anticipation of possible consequences of the innovation; second, the reaction of the person who, confronted by new working and living conditions, is exposed to physiological and psychological strain.
Refs Fitter, M *The Impact of New Technology on Workers and Patients in the Health Services* (1987); ILO *Automation, Work Organisation and Occupational Stress* (1984).
 Broader Occupational stress (#PE6937)   Occupational hazards (#PC6716)
  Social ill-effects of automation (#PE5134).
 Narrower Computer stress (#PE5053).
 Aggravated by Technological revolution (#PC3234)
  Inadaptation of technology to man in the industrialized societies (#PE5023).

♦ PE5166 Child mortality in developing countries
Nature Millions of children throughout the world die unnecessarily each year as a result of preventable diseases and other health hazards.
Incidence In 1982 alone, a total of 15 million children – the equivalent of the entire under-5 population in the United States – died as a result of malnutrition, dehydration and disease. According to a 1983 WHO report, 122 million infants are born each year, and of those roughly 10 percent – over 12 million – will die before their first birthday. A further 4 percent will die before their fifth birthday. Simple, curable diarrhoea will take 6 million young lives. Another 5 million will be claimed by measles, whooping cough, polio, tetanus, diphtheria and tuberculosis. Thousands more will die of pneumonia, malaria or schistosomiasis (bilharzia). The survival chances of children decrease with poorly spaced births to teenage mothers and to those with relatively fast childbearing for their age. 'The continuing challenge' says the report 'is to turn this tide of childhood mortality'.
Refs Mackenzie, F *Child Health and Mortality in Sub-Saharan Africa–Annotated Bibl. of 1975-86 Lit. – Sante et Mortalite Infantiles* (1987); United Nations *Socio-Economic Differentials in Child Mortality in Developing Countries* (1985).
 Broader Human death (#PA0072)   Infant mortality (#PC1287).
 Aggravated by Malaria (#PE0616)   Measles (#PE1603)
  Tetanus (#PE2530)   Pneumonia (#PE2293)
  Diphtheria (#PE8601)   Dehydration (#PE8062)
  Tuberculosis (#PE0566)   Poliomyelitis (#PE0504)
  Whooping cough (#PE2481)   Schistosomiasis (#PE0921)
  Child malnutrition (#PD8941)   Human disease and disability (#PB1044).

♦ PE5169 Hypersensitives
Nature Pollutants destroy a number of nutrients vital to our physical and nervous function, the immune system weakens progressively until the tolerance breaks down. The victim can end up hypersensitive to food, drink, fabrics, pollen and dust, petrol and tobacco fumes, industrial emission and natural gas, etc.
 Broader Defective human immunity system (#PE3355).
 Related Allergy (#PE1017).
 Aggravated by Pollutants (#PC5690).

♦ PE5170 Lack of skilled manpower in rural areas of developing countries
Wastage of local skills in developing countries — Absence of skilled workers in the countryside — Limited availability of practical skills in rural communities — Restricted availability of urban skills in local communities
Nature As agriculture is upgraded in technology and new industrial activities are introduced in rural areas the need for systematic training increases. One of the major constraints on rural industrialization is lack of skills, since the complexities of modern farming have made literacy essential. Even programmes to improve crafts have had difficulties, since understanding of improved methods and processes requires some degree of education. Civil work contractors working in rural areas often prefer to bring in semi-skilled workers from urban areas, because rural workers on such projects are unable to understand simple instructions. Not only are educational facilities less developed in rural than in urban areas, there is also a general preference for informal methods of training and on-the-job training, partly because this overcomes the problem of lack of formal education as an entry qualification and partly because such training is less costly. However, it is not comprehensive and standards of achievement vary. Therefore the ex-trainees have little mobility. Over and above this, there are hardly any systematic facilities for the training of operatives such as tractor drivers, truck drivers, cement mixer operators and processing machinery operators. If enterprises are to organize such training themselves they need subsidies to cover the costs incurred. The shortage of facilities in supervisory and management development programmes is still more acute in rural areas except in countries having well-developed cooperatives.
In an age when vocational training is capable of making a wide range of modern technological skills available to any community, the lack of local provision for such training means that the reservoir of practical skills in some rural communities is extremely limited. As a result, residents find themselves ill-equipped for carrying out basic development. Continuance of traditional construction practices often means that houses and pathways are in need of constant repair; few medical skills are available to deal with emergency situations or hygiene problems; inadequate techniques of animal care result in problems of health as well as in offensive odours and dirty paths. Those who leave the village for further academic or technical training rarely return, and if they do they can seldom find employment.
Because rural communities do not have at their disposal the possibility for training in the wide range of practical skills necessary for present-day living, there are few trained people locally available. Those wishing to be trained have the financial burden of seeking such training in urban areas. Many villages are still faced with insufficient business management skills to fulfil their vision of the future.
 Broader Insufficient skills (#PC6445)   Deficiencies of developing countries (#PC4094)
  Limited reservoir of technical skills in rural communities (#PF2848).
 Narrower Lack of trained firefighters (#PJ0277)   Unsafe construction techniques (#PG9831)
  Inadequate vocational education (#PF0422).
 Related Labour shortage (#PC0592)   Limited basic skills (#PJ8138)
  Limited technical skills (#PF8594)   Insufficient health personnel (#PD0366)
  Inadequate maintenance personnel (#PJ0088)   Unorganized community recreation (#PF5409).
 Aggravates Incompetent management (#PC4867)
  Lack of care for animals (#PD8837)
  Youth migration to cities (#PJ7859)
  Narrow range of practical skills (#PF2477).
 Aggravated by Shortage of skilled labour (#PD0044)
  Risk of capital investment (#PF6572)
  Unequal access to education (#PC2163)
  Underdeveloped technological skill (#PF8552)
  Lack of local services for community leadership training (#PF2451)
  Lack of opportunities for practical training in communities (#PF2837).

♦ PE5171 Art vandalism
Nature Brought about by aggression, the violation of works of art can be the individual gesture of a vandal or a collective result of war. Profit-seeking forms of vandalism also cause extensive loss or damage to art objects, such as the destruction of buildings and settings of an artistic nature and their replacement by inaesthetic ones within urban development. The pillaging of historical sites such as ancient tombs and the hacking to pieces of stone carvings for easier illicit transport show the extent of profit-seeking vandalism.
Incidence A recent example occurred China. In 1966, the Great Proletarian Cultural Revolution commenced its effort to root out old ideas and values, and to achieve a spiritual regeneration spearheaded by youth – the Red Guard. In their often uncontrolled hunt for vestiges of 'bourgeois' culture, the Red Guard destroyed or removed works of art and libraries, chiefly those held in private hands.
Background Art vandalism is a universal and perennial problem, practised long before the Vandals sacked Rome in 455 AD. Historically, the greatest vandalism have been perpetrated by armies and partisans in the service of gods and princes.
 Broader Vandalism (#PD1350)   Aggression (#PA0587).
 Narrower Iconoclasm (#PF4923)   Debasement of works of art (#PE0558).
 Related Theft of works of art (#PE0323).
 Aggravated by War (#PB0593)   Uncontrolled urban development (#PC0442).

♦ PE5173 Legg-Perthes disease
Refs Katz, Jacob F *Legg-Calve-Perthes Disease* (1984).

♦ PE5179 Distortion of democratic procedures within international organizations
 Broader Undemocratic social systems (#PB8031).

♦ PE5182 Respiratory diseases of horses
 Broader Diseases of respiratory system in animals (#PD7307).
 Narrower Pleurisy (#PG6121)   Strangles (#PJ9367)   Equine influenza (#PG6609)
  Laryngeal hemiplegia (#PG3742)   Equine viral rhinopneumonitis (#PG7856)
  Diseases of the guttural pouches (#PG2098)
  Chronic obstructive pulmonary disease (#PG4314)
  Exercise-induced pulmonary haemorrhage (#PG9529).

♦ PE5183 Discrimination against mixed race children
Discrimination against coloured children
Refs Wilson, Anne *Mixed Race Children* (1987).
 Broader Discrimination and harassment of children in public life (#PE6922).
 Aggravates Social stigma (#PD0884).
 Aggravated by Miscegenation (#PC1523)
  Children engendered by occupying soldiers (#PD8825).

♦ PE5187 Eating disorders
Incidence Eating disorders are on the increase. A 1984 international conference on the subject disclosed that the number of adolescents with anorexia nervosa and bulimia, eating disorders characterized by starvation or binge eating and purging, have increased dramatically in the past 15 years. In the USA, for example, a third of female high school and college students show tendencies toward anorexia or bulimia, or both.
Refs Boskind-White, Marlene and White, William C Jr *Bulimarexia* (1987); French, Barbara *Coping with Bulimia* (1988); Gibson, Diane *The Evaluation and Treatment of Eating Disorders* (1986); McFarland, Barbara and Baumann, Tyeis B *Feeding the Empty Heart* (1988).
 Broader Mental illness (#PC0300)   Human disease and disability (#PB1044).
 Narrower Pica (#PG4505)   Over-eating (#PE5722)   Bulimia nervosa (#PE4538)
  Anorexia nervosa (#PE5758)   Rumination disorder of infancy (#PG9708).
 Aggravated by Nutritional deficiencies (#PC0382).

♦ PE5188 Progressive muscular atrophy
Refs Gamstorp, Ingrid and Sarnat, Harvey B (Eds) *Progressive Spinal Muscular Atrophies* (1984).
 Broader Diseases of the spinal cord (#PE8813).

♦ PE5190 Vulgar combination of sacred and erotic in advertising
Nature The use of sacred symbols in connection with erotic material to advertise merchandise is vulgar, tasteless and often blasphemous. Most often the use of religious artifacts and low eroticism is intended to offend and shock and so increase impact on the market.
Claim Combining religious icons with pornography to advertise is one more example of the commercialization of society and materialism. The values of the economy, in this case marketing and sales, occupy a higher plane than the profound, the sacred or the authentically sexual.
 Broader Indecent advertising (#PD2547).
 Aggravates Abusive exploitation of cultural heritage (#PC7605).

♦ PE5191 Bovine spongiform encephalopathy (BSE)
Mad cow disease
Nature BSE progressively destroys the brains of infected cattle. It is caused by rearing cattle on processed feed which includes sheep brains infected with scrapie. There is no treatment and they have a very long incubational period of up to eight years, during which time the infected animals appear healthy. There is no proof yet that mad cow disease can be passed on to humans, but other animals eating infected cattle brains develop the disease.
 Related Creutzfeldt-Jakob disease (#PJ1045).
 Aggravated by Scrapie (#PE5330)
  Inefficient use of proteins in factory farming (#PF2758).

♦ PE5192 Violation of right of workers to join trade unions
Union busting
 Broader Denial of the right of trade union association (#PD0683).
 Aggravated by Unethical practices of employers (#PD2879).

♦ PE5194 Imbalance between training and existence of openings in various professions
 Broader Disparity between workers skills and job requirements (#PC1131).
 Aggravates Unemployment (#PB0750)   Obstacles to education (#PF4852).
 Aggravated by Inadequate education (#PF4984).

♦ PE5195 Titanium dioxide as a pollutant
Nature Titanium dioxide is used as a pigment in paints. The production procedure is highly polluting because of the large quantities of wastes which are often thrown into the sea where, by chemical reaction, they can be transformed into 'red mud'. These wastes destroy living organisms such as plankton and they can also impregnate the flesh of fish used for human consumption.
Refs WHO *Selected Radionuclides* (1983).
 Broader Chemical air pollutants (#PD1271)
  Chemical pollutants of the environment (#PD1670).

♦ **PE5196 Destruction of environmental oxygen**
Ecological destruction
Nature All activity requires energy, which normally involves the combustion of oxygen, the indispensable element of life and nature. The present amount of human activity is such that it involves a greater combustion of oxygen than the chlorophyllian function is capable of producing; while water pollution and destruction of green spaces is causing the destruction of oxygen–creating agents, phytoplankton in water (which at present produces 70 percent of the oxygen in air) and vegetation on land (which produces 30 percent of air oxygen). The risk now exists of more oxygen being used than is actually produced.
Incidence The accelerating process of the destruction of oxygen is such that some scientists predict the combustion of all oxygen in the atmosphere in less than 2,000 years.
Claim While man has learned to master the laws of nature, it is now clear that he is also in the process of destroying it.
Broader Human destructiveness (#PA0832) Environmental degradation (#PB6384)
Natural environment degradation (#PB5250).
Narrower Marine oxygen deficiency (#PE6289).
Related Shortage of fresh–water sources (#PC4815)
Long–term change in atmospheric chemistry (#PF1234).
Aggravated by War (#PB0593) Mismanagement of environmental demand (#PD5429).

♦ **PE5197 Sexual unfulfilment of the disabled**
Sexual reproduction by the disabled
Refs Dechesne, B H, et al (Eds) *Sexuality and Handicap* (1986); Shaked, Ami *Human Sexuality in Physical and Mental Illnesses and Disabilities*.
Related Human sexual inadequacy (#PC1892) Sexual dysfunction of torture victims (#PE3932).
Aggravates Sexual unfulfilment (#PF3260) Fear of sexual intercourse (#PF6910).

♦ **PE5199 Underprivileged home environment**
Nature Children from an underprivileged home environment are at a disadvantage socially and educationally. An under–privileged home environment gives rise to an inability to benefit from education and to poor behaviour, possibly eventually leading to delinquency.
Aggravated by Homelessness (#PB2150).

♦ **PE5204 Angina**
Nature Angina is swelling of the throat and is of a variety of types including Tonsillar angina, laryngeal angina, membranous angina and anginal scarlatina.
Refs Julian, Desmond G *Angina Pectoris* (1985).
Broader Bacterial disease (#PD9094) Diseases of upper respiratory tract (#PE7733).

♦ **PE5207 African swine fever**
Refs Becker, Yechiel (Ed) *African Swine Fever* (1986); FAO *Eradication of Hog Cholera and African Swine Fever* (1976).
Broader Viral diseases in animals (#PD2730).
Aggravates Fever (#PD2255).

♦ **PE5210 Annexation**
Foreign annexation — Border encroachments
Broader Political domination (#PC8512)
Offences against the peace and security of mankind (#PC6239)
Denial of right of a people to be self–determining (#PC6727).
Narrower Political annexation (#PG4044).
Aggravates Border incidents and violence (#PD2950)
Boundary disputes between states (#PD2946).
Aggravated by Vulnerability of small nations to foreign intervention (#PD2374).

♦ **PE5211 Denial of right to benefits for invalids**
Broader Denial of right to economic security (#PD0808).
Related Unequal pension rights (#PJ2030)
Denial of right to adequate medical care (#PD2028)
Denial of right to benefits to survivors (#PE4531)
Denial of right to economic security during periods of unemployment (#PE5406).

♦ **PE5212 Aspergillosis**
Refs Bossche, H Vanden et al *Aspergillus and Aspergillosis* (1988); CAB International Mycological Institute *Aspergillosis*.
Broader Mycosis (#PE2455).

♦ **PE5214 Autoimmune disease**
Nature Autoimmune diseases result when the immune system turns against the body. There are over 40 such diseases including juvenile diabetes, multiple sclerosis, rheumatoid arthritis, myasthenia gravis and lupus.
Refs Rose, Noel and Mackay, Ian R (Eds) *The Autoimmune Disease* (1985).
Broader Defective human immunity system (#PE3355).
Aggravates Diabetes (#PE0102) Multiple sclerosis (#PE1041)
Rheumatoid arthritis (#PE5081) Systemic lupus erythematosus (#PG5084).

♦ **PE5222 Infringement on the functioning of legitimate organizations**
Broader Denial of the right of association (#PD3224).
Related Banned associations (#PD3536)
Denial of the right of trade union association (#PD0683)
Denial of the right to organize political parties (#PE9110).

♦ **PE5226 Denial of right of family planning**
Nature The right of couples and individuals to freely decide the number and spacing of their children and the right to information and education that allow them to do so is a basic guarantee of individual liberty. This right is denied in a number of ways. The societal mores stress early marriage and large families. In India the average age of marriage for women is 14.3 and the earlier they become mothers the higher their status in the community. In many cultures, family size and spacing is determined by the husband. In many cultures a cult of "machismo" dictates that a man's masculinity is measured by the number of children he has. Denial of education or of opportunities to work indirectly denies this right. Women who are educated or who work outside the home or farm are more likely to space children, and to care for them more effectively.
Incidence At the present time there are an estimated 500 million people who want to limit their family size, but who lack either the information, access or means to do so.
Claim The right of a woman to control her own body and therefore her own fertility is foundation to the right to freedom. A woman should have all the knowledge and means available to make her choice and exercise her freedom to decide whether and when to have children. She also needs to know the consequences in terms of health, nutrition, shelter, education, and education and social opportunities for herself, her family and her society. The denial of this right endangers the woman. An estimated 500,000 women die each year due to complications associated with childbirth and millions survive with long–term damage to their health. It endangers the infant. Each day more than 42,000 infants die for lack of proper health care and nutrition, many because their parents cannot care for the children they have. The two most important determinants to a child's chance of survival are the length of interval between births of siblings and the educational level of the mother. It endangers the whole family. In cultures where 5 to 10 children is the norm providing physical care is difficult and providing adequate psychological care almost impossible for all of the children.
Broader Denial of right to liberty (#PF0705)
Denial to people of control over their own lives (#PC2381).
Related Forced motherhood (#PE5919) Criminalization of abortion (#PF6169)
Dependency of women in marriage (#PD3694).
Aggravates Lack of family planning (#PF0148)
Inadequate medical care for pregnant women (#PE4820).
Aggravated by Illiteracy among women (#PE4380)
Discrimination against women in politics (#PC1001)
Discrimination against women in education (#PD0190)
Discrimination against women in employment (#PD0086)
Discrimination against women without children (#PE8788).
Reduced by Feminism (#PF3025).

♦ **PE5229 Non–payment of compensation to victims of catastrophes**
Delay in payment of compensation to victims of disasters — Delay in payment of compensation to victims of major accidents
Incidence More than five years after deadly gas leaked from a pesticide plant in Bhopal, India, relief payments were being distributed. Only about 70,000 of estimated 500,000 victims have received payments by mid–1990s.
Broader Uncompensated damages (#PD7179).

♦ **PE5239 Failure to conform to international health standards**
Failure to respect internationally agreed toxicity thresholds
Nature Authorities in one country tend, on specific issues, to fail to conform to internationally accepted health standards, or to conform to weaker standards when more recent research indicates that stronger standards are called for.
Incidence In 1985 in the UK it was reported that 49 pesticides in current use were probably carcinogenic, and that 38 of the most popular pesticides used were banned or severely restricted in other western countries. On examination of the scientific literature concerning the pesticides approved by the Ministry of Agriculture, 164 of 426 approved pesticides were known or suspected of causing a range of problems including cancer, birth defects, genetic defects, or irritations and allergies. About one third of these pesticides are approved for use on food products.
Broader Resistance to internationally agreed standards (#PC4591).

♦ **PE5241 Denial of right to extended family**
Broader Denial of social rights (#PC0663) Denial of right to a family (#PE7267).
Aggravated by Narrow legal definition of the family in developing countries (#PD1501).

♦ **PE5245 Basuco**
Nature Basuco is the waste from refining, cocaine packaged and sold as cigarettes. It is deadlier than crack and cheaper than marijuana. Criminal gangs are now in the process of test marketing basuco in North America after its use in Colombia.
Broader Abuse of coca and cocaine (#PD2363).
Related Crack (#PE2123).
Aggravates Economic dependence of some developing countries on the drug trade (#PE5296).
Aggravated by Drug smuggling (#PE1880).

♦ **PE5246 Non–adaptive local structure of social care**
Nature The local forms of social care are unable to adapt to rapidly changing needs of individuals, families and communities. The small traditional associations based on family, faith and locality are expected to communicate to individuals the principle moral means and ends; to transmit the implicit rules governing society and bestow psychological gratification for participating in the social fabric. These associations are increasingly disrelated from functional relevance to the larger economic, political and cultural decisions of the world. Individuals are left clinging to parochial and wooden structure frantically trying to create the stability of the old order or simply give up any attempt at moral, social or cultural order.
Broader Delayed development of regional plans (#PF2018)
Reinforced parochialism of internal values and images (#PF1728).
Related Inadequate development of new social structures in developing countries (#PD0822).
Aggravates Inadequate social welfare services (#PC0834).

♦ **PE5247 Monetary bloc**
Nature Monetary bloc is created to ensure the monetary and economic hegemony of the country heading the bloc by means of tying the currency of the participant countries to its own.
Broader Elitist control of global economy (#PC3778).
Reduced by Nationalistic attitudes to currency (#PF6094).

♦ **PE5252 Poverty as a consequence of war**
Economic decline caused by war
Nature War impoverishes the people and nations who fight them. The wounded and the poor are left to beg. Shops and stores run out of goods to sell. The clothes of even the middle class become increasingly shabby. Inflation ruin the rich and the poor alike. Fuel supplies become short and power, water and other service are cut. Infrastructure can not be repaired or constructed and so every day things like telephones lines, roads, bridges and port facilities deteriorate. Garbage is left uncollected. Wars destroy economic growth and may turn a prosperous country into a poverty stricken one.
Broader Disastrous consequences of war (#PC4257)
Destruction of economy due to war (#PE8915).
Related Housing destruction in war (#PE2592) War damage in civilian areas (#PD8719)
Industrial destruction by war (#PD8359).
Aggravated by Disruption of food supply due to military activities (#PE8979).

♦ **PE5259 Torture through sensory overload**
Nature Victims of torture have their sense overloaded by being subjected to intense or continuous light or noise.
Broader Psychological torture (#PD4559).
Narrower Torture by continuous noise (#PE5691).

♦ **PE5260 Deterioration of nuclear power plants**
Nature Nuclear reactors, which started commercial operations in the 1960s are wearing out faster than expected. As the plants age, numerous unexpected problems will be inevitable. The potential for a disaster increase with each year.
Broader Inadequate maintenance of infrastructure (#PD0645)
Environmental hazards of nuclear power production (#PD4977).
Related Building depreciation (#PG5286).

Aggravates Nuclear accidents (#PD0771)
Vulnerability of nuclear power sources (#PD0365)
Environmental hazards of decommissioned nuclear power plants (#PE7539)
Non-verificability of compliance with nuclear power safeguards (#PF4455).
Aggravated by Insufficient nuclear power stations (#PD7663).

♦ **PE5261 Deaf mutism**
Refs Rüedi, L (Ed) *Deaf Mutism* (1959).
Broader Mutism (#PE4526) Diseases of the ear (#PD2567).

♦ **PE5267 Flight to avoid prosecution or giving testimony**
Unlawful flight — Fugitive felons
Nature A person is guilty of a crime if he travels across a jurisdictional boundary with the intention of avoiding being prosecuted or giving testimony in a criminal case.
Refs Wright, James D and Rossi, Peter H *Armed and Considered Dangerous* (1986).
Broader Issue avoidance (#PF1623) Obstruction of government function (#PD6710).

♦ **PE5272 Haemorrhagic fevers**
Haemorrhagic fever
Refs WHO *Mosquito-Borne Haemorrhagic Fevers of South-East Asia and the Western Pacific* (1966).
Broader Virus diseases (#PD0594).
Aggravates Fever (#PD2255).
Aggravated by Ticks as pests (#PE1766) Rodent vectors of disease (#PE3629).

♦ **PE5274 Misuse of tropical rain forests for agricultural development**
Nature The growth of population combined with the decreasing availability of arable land, forces poor farmers to seek new land in forests in order to grow more food. Forests cleared in an uncontrolled manner tend to make available land which is of only short-term value for agriculture but has long-term negative ecological effects in addition to leading to extensive soil erosion.
Incidence Tropical rain forests are extremely delicate eco-systems and can not support arable or pastoral farming over a number of years, yet a number of governments are encouraging the development of tropical forest as agricultural land. First and perhaps second year yields may be high but even with the use of pesticides and fertilizers later yields may be one quarter of these. Cleared land is exposed to high rain falls of these areas and nutrients are washed away very quickly.
Counter-claim There is nothing inherently wrong with clearing forests for farming, provided the land is the best there is for new farming, can support the numbers encouraged to settle upon it, and is not already serving a more useful function, such as watershed protection.
Broader Endangered forests (#PC5165) Destruction inherent in development (#PF4829)
Unsustainable development of forest lands (#PD4900).
Narrower Slashburning (#PE6264).
Related Economic exploitation (#PC8132).
Aggravated by Back to the land (#PF4181).

♦ **PE5278 Demotivated civil servants in developing countries**
Nature Public administration personnel policies rarely compare in quality with those of the private or voluntary sectors. Employees are poorly trained and poorly rewarded. They perform accordingly. Civil servants are subjected to overcrowded offices, lack of appropriate working equipment, lack of proper residential accommodations, very low incomes with irregular pay increases, Draconian codes of conduct and expenditure, poor health and welfare conditions. The results of such neglect are absenteeism, corruption and behaviour towards the public that is often obstructive. Further, many public administrations offer but rudimentary systems of labour relations.
Incidence The majority of developing countries have no procedure for the settlement of labour disputes on pay or working conditions. Strikes and industrial action, often the only negotiating tool available, are very often discouraged or forbidden in the public sector. Brazil, Indonesia, the Republic of Korea, Nepal, Pakistan and Thailand, as well as most Arab countries, forbid strikes. Several countries have suspended this right for several years as part of martial law. Individual resistance such as indiscipline, negligence and corruption become the only options available.
Aggravates Inefficient public administration (#PF2335).

♦ **PE5279 Torture through destroying cherished things**
Nature The destruction of cherished things is used as a form of torture. Homes are burned down. Pets are killed. Toyes of children are destroyed.
Incidence Torture through destruction of cherished things has been reported in some countries.
Broader Psychological torture (#PD4559).

♦ **PE5283 Possession of unlawful distilled spirits**
Nature The possession of distilled spirits, knowing that an imposed tax has not been paid, is a crime.
Broader Unlawful trafficking in taxable objects (#PD4221).

♦ **PE5286 Fishing for sport**
Cruelty to fish
Refs FAO *Summary of the Organized Discussion on the Economic Evaluation of Sport Fishing, Dublin, 1967* (1973); FAO *Second European Consultation on the Economic Evaluation of Sport and Commercial Fisheries* (1977).
Broader Maltreatment of animals (#PC0066).
Aggravates Lead poisoning in animals (#PE9228).

♦ **PE5288 Neglect of elderly in institutional care**
Inhumane geriatric wards
Broader Inadequate housing for the aged (#PD0276).

♦ **PE5289 Multiple myeloma**
Refs Azar, Henry A and Potter, Michael *Multiple Myeloma and Related Disorders*.
Broader Neoplasms of lymphatic and haematopoietic tissue (#PE4637).

♦ **PE5290 Lack of biocoenosis**
Nature The interdependence of plants, animals and bacteria in ecological communities is breaking down as humankind interferes with them.
Aggravates Ecological imbalance (#PG3026) Vulnerability of ecosystem niches (#PC5773).

♦ **PE5294 Somatic and psychosomatic effects of torture**
Nature Victims of torture have a loss of hearing due to the telephone torture, simultaneous blows on both ears. Bones are broken, muscles are torn or bruised. They walk in pain due to falanga, the systematic beating of the soles of the feet. Rectums are injured in the form of scars and abscesses after the insertion of bottles, police batons or other instruments in the anus. Teeth are lost from electrical torture to the gums and teeth or being broken. Women have had periods of menstrual disturbances and miscarriages.
Incidence In a study of 135 victims of torture, examined two to seven years after the events showed the following symptoms: headache 36 percent, reduced hearing 15 percent, vision disturbances 14 percent, intolerance to alcohol 11 percent, vertigo 4 percent, gastro-intestinal symptoms 32 percent, pains in the joints 19 percent, cardio-pulmonary symptoms 22 percent, difficulty walking 17 percent, pains related to trauma 13 percent and other symptoms 18 percent.
Broader Disabled victims of torture (#PD0764) Human disease and disability (#PB1044).
Narrower Joint pains due to torture (#PE3732) Skin lesions due to torture (#PE7154)
Gait disturbances due to torture (#PE6791)
Cardio-pulmonary disease due to torture (#PE1556)
Residual traumatic pains due to torture (#PE3798)
Gastro-intestinal disorders due to torture (#PE6724).
Aggravates Psychosomatic disorders (#PD1967).

♦ **PE5296 Economic dependence of some developing countries on the drug trade**
Broader Economic dependence (#PF0841) Illicit drug trafficking (#PD0991)
Vulnerability of developing countries (#PC6189).
Related Instability in export trade of developing countries producing primary commodities (#PD2968).
Aggravates Drug abuse (#PD0094) Inadequate drug control (#PC0231).
Aggravated by Basuco (#PE5245) Economic imperialism (#PC3198)
Cultivation of illegal drugs (#PD4563).

♦ **PE5302 Insufficient government spending on cost-effective activities**
Incidence In many developing countries, too little of the very limited resources is allocated to cheap and cost-effective services. In education there is a pressing need to expand and improve primary education, the socially most profitable form of investment, particularly in the poorest countries. The problem of resource allocation for health and education is partly the result of large across the board subsidies and the lack of any pricing mechanism, particularly in centralized systems. Because of subsidies the private rate of return to higher education for all developing countries exceeds 20 per cent, namely about twice the social rate of return for higher education.
Broader Inappropriate public spending by government (#PF6377).
Aggravates Ignorance of health and hygiene (#PD8023)
Inappropriate education in developing countries (#PF1531).

♦ **PE5303 Exploitation of the prostitution of others**
Pandering — Promoting prostitution — Facilitating prostitution — Procurers — Pimps
Nature Procurers and pimps, who both victimize prostitutes and live off their earnings, are rarely arrested or charged with a crime even though it is they who cause much of the violence and abuse that prostitutes suffer. Almost all the money derived from prostitution is controlled and used by the pimps, and the "prohibitionist" system, as exists in the USA (with the exception of Nevada), is morally hypocritical, as it acknowledges the need for and inevitability of prostitution but still makes it illegal. Thus the women who practice it are punished while the men who derived profit and pleasure from it, the procurers and customers, are virtually ignored in both the eyes of the law and in society.
Broader Prostitution (#PD0693) Traffic in persons (#PC4442) Female prostitution (#PD3380).
Aggravates Infamy (#PB8172).

♦ **PE5304 Unimagined possibility of expanding the preschool**
Broader Unchallenging world vision (#PF9478)
Unperceived relevance of formal education in rural communities (#PF1944).

♦ **PE5305 Unrestrained wage increases**
Inadequate constraints on salary increases — Unrealistic wage claims
Refs Bilson, Beth (Ed) *Wage Restraint and the Control of Inflation* (1987).
Aggravates High minimum wages (#PD5674).

♦ **PE5306 Grant-back provisions**
Nature Requiring the acquiring party to transfer or grant back to the supplying party, or to any other enterprise designated by the supplying party, improvements arising from the acquired technology, on an exclusive basis without offsetting consideration or reciprocal obligations from the supplying party, or when the practice will constitute an abuse of a dominant market position of the supplying party.
Broader Restrictive trade practices (#PC0073).
Related Collusive tendering (#PE4301) Challenges to validity (#PF1200)
Restrictions on research (#PF0725) Restrictions on publicity (#PF1575)
Restrictions on adaptations (#PF5248) Exclusive dealing arrangements (#PE0413)
Restrictions on use of personnel (#PF3945)
Patent pool or cross-licensing agreements (#PE4039)
Exclusive sales and representation agreements (#PE4581)
Trade restrictions due to voluntary export restraints (#PE0310)
Payments after expiration of industrial property rights (#PF5292)
Tying of supplies to subsidiaries by transnational enterprises (#PE0669).

♦ **PE5308 Carcinoma in sita**
Refs Rutherford, Robert B, et al *Severe Behavior Disorders of Children and Youth* (1987).
Broader Neoplastic diseases (#PC3853).

♦ **PE5309 Disused machinery accumulation in the seas**
Submarine junk
Broader Accumulated junk (#PD5510).

♦ **PE5312 Violation of right of trade unions to publish**
Broader Lack of freedom of the press (#PE8951)
Violation of the right of trade unions to function freely (#PE1758).

♦ **PE5316 Job insecurity for pregnant women**
Claim Many women must choose between motherhood and employment. Employers require that women who are pregnant to quit their jobs. For the majority of women this is financial ruin.
Counter-claim Offering pregnancy leave for pregnant women creates more problems than it solves. Women of childbearing age are discriminated against because of the additional burden it places on employers. Small companies simply cannot afford to keep open jobs for people who are gone for 6 or 8 weeks. Pregnancy leave discriminates against men, who are also parents of children.
Broader Insecurity of employment (#PD8211)
Discrimination in the employment of women with family responsibilities (#PE7206).

♦ **PE5319 Counterfeit machine parts**
Nature The use of substandard counterfeit machine parts is becoming alarmingly commonplace in both business and government. Counterfeit ball bearings, nuts and bolts, valves and circuit

♦ **PE5324 Uncertainty of long-term health effects of radioactive fallout**
Panic as a result of nuclear accidents — Fear of a nuclear winter
Nature In addition to the possible physical danger from radioactive fallout, millions of people have been exposed to a lifelong psychological immersion of death: permanent fear of invisible contamination. Radiation cannot be detected by senses and might strike at any time. Radiation disasters have an added aura of dread associated with limitless danger, hypothesis of nuclear winter and images of Hiroshima and Nagasaki. People exposed to radiation have felt to have been exposed to lethal impairment that, if it did not manifest itself in one generation might well make itself felt in the next ones. Nuclear disasters create terrifying rumours about the after effects.
Incidence Chernobyl disaster raised the level of fear and increased the belief that nuclear power casts an evil spell over the world. Quarter of the British public started to oppose the use of nuclear power to provide electricity. The danger of Chernobyl is unclear and will remain unknown for years: estimations of radiation induced cancer deaths in Soviet Union and Europe varies between 1,000 and 500,000.
Refs Grinspoon, Lester (Ed) *The Long Darkness* (1986); WHO *Nuclear Power* (1982).
  Broader Uncertainty (#PA6438)   Inadequacy of medical science (#PF8326).
  Aggravated by Radioactive fallout (#PC0314).

♦ **PE5325 Lipidosis**
Lipoidosis
Refs Salvayre, R et al *Lipid Storage Disorders* (1988).
  Broader Congenital disorders of lipid metabolism (#PE9816).

♦ **PE5327 Arming rioters**
Nature Knowingly supplying a weapon for use in a riot, teaching others in the use of weapons with the intention that they be used in a riot, or possessing a weapon while engaged in a riot is the crime arming rioters.
  Broader Civil disorders (#PC2551)   Hindrance of law enforcement (#PD5515)
  Offences against public order (#PD7520).

♦ **PE5328 Trace element imbalance in the human body**
Trace element deficiency — Excessive trace elements in human body
Nature Insufficient or an abundance of trace elements are a contributing factor in cardiovascular diseases. People living in different geochemical environments may have an enhanced or depressed intake of certain elements reflecting the chemical composition of the environment. These imbalances may also be the result of diets. Populations that are highly prone to atherosclerosis and myocardial infarction show significantly lower tissue chromium concentrations than less coronary–prone populations. This may be due to higher consumption of refined sugars and other refined products. The unrefined products have chromium the refined products lack. Geographical differences in hypertension have been found to be positively correlated with variations in the concentration of cadmium in the kidney.
Claim It has been found out that biochemical differences, abnormal levels of at least one trace element, exist between violent and non–violent criminals and that these differences may cause violent behaviour.
Refs Fowden, L, et al *Trace Element Deficiency* (1982).
  Broader Malnutrition (#PB1498)   Nutritional diseases (#PD0287)
  Human disease and disability (#PB1044).
  Related Hypertension (#PE0585)   Arteriosclerotic vascular disease (#PG4555)
  Dietary deficiencies in developed countries (#PD0800).
  Aggravates Cardiovascular diseases (#PE6816)
  Inadequate immune responses in malnourished persons (#PE4883).

♦ **PE5330 Scrapie**
Tremblante du mouton — Rida
Nature A degenerative brain disease in sheep.
  Broader Diseases of nervous system in animals (#PD7841).
  Aggravates Bovine spongiform encephalopathy (#PE5191).

♦ **PE5331 Crimes related to foreign relations and trade**
  Broader Statutory crime (#PC0277).
  Narrower Diversion of high technology to hostile countries (#PE7174)
  Unlawful recruiting for and enlistment in foreign armed forces (#PE4484).
  Related Unlawful business transactions (#PC4645)
  Military expeditions against friendly powers (#PD7261).

♦ **PE5336 Decreasing rate of development of major agriculture technologies**
Claim There are no technologies waiting in the wings that will lead to the quantum jump in food output of the sort associated with the hybridization of corn, the eightfold increase in fertilizer use between 1950 and 1980, the near tripling of irrigated area in the same period or the rapid spread of the Green Revolution. Farmers are struggling to feed a record 86 million additional people each year, without any major new technologies to draw on.
  Broader Inadequacy of scientific expertise (#PE5055).
  Aggravates Slowing growth in food production (#PC1960)
  Stagnated development of agricultural production (#PD1285).

♦ **PE5338 Aid lenders as beneficiaries of net outflow of funds from developing countries**
Nature Because of the amount of interest paid by developing countries to service their existing indebtedness to industrialized countries, and with the reduction in the level of commercial lending to the developing countries, lenders of funds may in fact become recipients of larger amounts than are currently being sent to the borrowing countries in various forms of development aid.
Incidence Net inflows to the World Bank from the 17 highly indebted middle–income countries began in the fiscal year to 30 June 1988. They were then $1.27 billion and rose to $1.92 billion in the following fiscal year.
Counter–claim Lending to developing countries cannot rise continually, and it is not appropriate to aggregate the results of such operations beyond the country level.
  Aggravated by Burden of servicing foreign public debt by developing countries (#PD3051).

♦ **PE5339 Scrapped automobiles**
Abandoned cars
Nature Abandoned vehicles cost public services millions of dollars a year for their removal, create safty hazards and disfigure the community.
  Broader Landscape disfigurement (#PC2122)
  Limited availability of functional information (#PF3539).
  Aggravated by Planned obsolescence (#PC2008).

♦ **PE5340 Wrecks and derelicts as hazards**
  Broader Hazards to navigation (#PE3868)
  Obstacles for international ocean shipping (#PD5885).
  Aggravates Corrosion of ships (#PD5803).
  Aggravated by War (#PB0593)   Dereliction (#PE5715)   Oil pollution (#PE1839)
  Marine accidents (#PD8982).

♦ **PE5341 Discredited moneyed hereditary class**
Claim The presence of so called old boy network of distinguished families do not only enrich schools, colleges, and regiments; these ancestral connections also enrich trade unions, businesses, and indeed all human organizations and countries.
  Broader Class consciousness (#PC3458).

♦ **PE5345 Irregular transport services**
Infrequent bus service — Uncommunicated bus schedule
  Broader Insufficient transportation infrastructure (#PF1495)
  Underdeveloped potential of basic resources (#PF3448)
  Poor communications networks in rural areas (#PF6470)
  Lack of central planning structures in small communities (#PF2540)
  Isolating effects of seasonal variations on undeveloped transportation (#PE3547).
  Related Limited access to external resources (#PF1653)
  Paralyzing patterns between villages and administrative structures (#PF1389).
  Aggravated by Insufficient communications systems (#PF2350).

♦ **PE5355 Occupational risks and hazards of the medical profession**
Nature The risks of occupational accidents among medical professionals are very numerous and the circumstances vary according to the speciality. They include road accidents in the case of general practitioners, burns suffered by operating room personnel as a result of explosions in operating theatres, septic cuts and scratches for biologists and anatomo–pathologists, the danger of septic liquid or vaccines entering the eyes of general practitioners, surgeons, paediatricians, etc and their assistants, syphilis and AIDS infections among gynaecologists and obstetricians, accidental internal and external irradiation and contamination among radiologists, dentists, dental technicians and physicians working in research laboratories, blows and wounds inflicted on psychiatrists and psychologists by delirious or violent patients, or bites and scratches from test animals. Occupational diseases can be classed under four main headings: infections and contaminations, disorders due to radiation, occupational dermatitis, and psychological disturbances.
Incidence The possibility of contagion is considerable owing to contacts with patients and the need to handle septic objects. The incidence of virus hepatitis among physicians is between 15 and 40 times greater than in the whole of the population. Doctors working in hospitals and clinics and those in laboratories are the more exposed than those in private practice.
  Broader Occupational hazards (#PC6716).
  Aggravates Medical practitioners refusing to treat patients (#PE5027).

♦ **PE5362 Unemployment in the wood industry**
Refs ILO *Accident Prevention on Board Ship at Sea and in Port* (1986); ILO *The Achievement of Full Employment in the Wood Industries* (1985).
  Broader Unemployment (#PB0750).

♦ **PE5369 Increasing lag in education against population growth**
Incidence The number of children aged between 6 and 11 years in sub–Saharan Africa is multiplying at a rate of 3.5 per cent a year. High population growth against limited education budgets virtually guarantees that the number of illiterates in the region will continue to grow.
  Broader Restrictions on the acquisition of knowledge (#PF1319)
  Increasing development lag against population growth (#PF3743).
  Aggravates Illiteracy (#PC0210)
  Increasing lag in education against growth in knowledge (#PE7139).
  Aggravated by Rapid population growth (#PD9153).

♦ **PE5373 Fruit rot**
Nature Fruit grows old, flavourless and mushy as a result of normal ageing processes. This reduces the effective yield of fruit crops.
Counter–claim The decay of fruit releases and feeds seeds, thus transferring the genes of the plant to the following generation.
  Related Food spoilage in storage (#PD2243)   Plant diseases in storage and transit (#PD3587).

♦ **PE5376 Effects of AIDS on sexual behaviour**
Prostitution and AIDS — Restricted sexual liberation by AIDS
Nature Mandated testing for HIV–virus, quarantine, and legislation of morality, sexual behaviour and self–inflicted suffering are considered ineffective or useless in controlling the spread of AIDS. The control lies at the individual level, in changing promiscuous behaviour.
Incidence The Public Health Service of Thailand estimates that six percent of the country's 650,000 prostitutes are infected with HIV.
  Aggravates Deception between sexual partners (#PE4890).
  Aggravated by Risk of contracting AIDS through kissing (#PJ5441).

♦ **PE5380 Helminths of the skin**
  Broader Skin diseases in animals (#PD9667).
  Narrower Elaeophorosis (#PG9500)   Onchocerciasis (#PE2388)
  Dracunculiasis (#PE3510)   Stephanofilariasis (#PG7895)
  Pelodera dermatitis (#PG7853)   Cutaneous habronemiasis (#PG7469).

♦ **PE5385 Pond pollution**
  Broader Pollution of inland waters (#PD1223)
  Underdeveloped provision of basic services in developing countries (#PF6473).
  Related Stagnant surface water (#PE2634).

♦ **PE5390 Introducing or possessing contraband useful for escape**
Aiding escape
Nature Introducing or possessing any contraband: tool, weapon or other object with the intention of providing a prisoner of a lawful detention centre with a means of escape is a crime.
  Broader Obstruction of government function (#PD6710).

♦ PE5391 Personal physical insecurity in developing countries
Nature Many third world suburban communities are pervaded by a feeling of physical insecurity, and this atmosphere is immediately sensed by an outsider who walks into the community. An environment of uncertainty has become a part of what it means to live in such areas. The local people are insecure; their day-to-day operations are carried out in (usually unconscious) fear. This social anxiety is manifest in many ways: inadequate safeguards against devastating fires or theft; inadequate services for accidents; individuals disabled by undiagnosed or untreated disease; threat of personal assault; and simple fear of unlighted roads and homes.
  Broader Personal physical insecurity (#PD8657).
  Related Social insecurity in developing countries (#PE4796).

♦ PE5396 Feral cats
Incidence In UK there are 1.2 million domestic cats living out a feral life.
  Broader Feral mammals (#PE0185).

♦ PE5397 Legal contract system reduced to individual needs
Individualistically reduced legal contract
Nature The systems of legal contract across the world are reduced to meet the needs of individuals rather that those of the society. Individual rights are over emphasized. Structures which give people a sense of responsibility for the law and the legal system are lacking. The legal system tends to be static, difficult to change and out moded.
  Broader Overemphasis on self-sufficiency with respect to interdependence (#PF3460).
  Narrower Outmoded legal systems (#PE2580)   Non-concerned attitudes (#PF2158)
  Misuse of international forums (#PF2216)   Overemphasis on individual rights (#PF2475)
  Structural failure of citizen participation (#PF2347)
  Limited local respect for regional and global legislation (#PF2499).
  Related Exclusion of opposing views (#PF3720)   Parochial national interests (#PF2600)
  Reduced understanding of globality (#PF7071)
  Breakdown in community security systems (#PD1147)
  Economically controlled political power (#PC2510).
  Aggravated by Declining sense of community (#PF2575).

♦ PE5398 Denial of right to organize trade unions
  Broader Denial of the right of trade union association (#PD0683).
  Narrower Denial of right to recruit union members (#PE7036).
  Related Sanctions against trade union workers (#PD0610)
  Denial of right to collective bargaining (#PE3970)
  Denial of trade union rights to public employees (#PE6888).
  Aggravates Victimization of workers' representatives (#PD1846).

♦ PE5399 Methane gas emissions from animal husbandry
Incidence It has been estimated that cattle generate 60 million tons of methane gas each year, contributing significantly to the greenhouse effect.
  Broader Increase in atmospheric concentration of methane (#PE8815).

♦ PE5402 Electronic pornography
Pink messaging services — Pornography by telephone — Electronic sex
Nature Financially, dial-a-porn is good to telephone companies, which keep about a third or half of the revenue the services generate. The privatization of British Telecom and breakup of AT and T in US have generated premium rate services. Nothing in law restricts what two individuals can say to each other over the telephone. So unless the service provider can be judged somehow guilty of "inciting debauchery", the only way to prevent pornographic telephone services is to close down the entire service category – as France did with chat lines that turned out to be pornographic.
Incidence In France "pink messages" consist of porno messages designed to put certain Minitel (videotex network) users in touch with each other, or of advertisements for sexually explicit products or services.
  Broader Pornography (#PD0132)
  Inappropriate use of telecommunications services (#PE4450).
  Related Obscene telephone calls (#PE5757).
  Aggravated by Ineffective regulation of electronic messages (#PE6226).

♦ PE5404 Accidental weed creation by genetic engineering
Nature With the creation of genetically engineered crops comes the risk that some of these engineered plants may acquire the traits of weeds and escape from cultivation with profound effects on agriculture and the environment. They would be especially dangerous if, as in the case of potatoes for example, they were engineered to be tolerant of frosts, blight and herbicides.
  Broader Weeds (#PD1574).
  Aggravated by Irresponsible genetic manipulation (#PC0776).

♦ PE5406 Denial of right to economic security during periods of unemployment
  Broader Denial of right to economic security (#PD0808).
  Related Unequal pension rights (#PJ2030)
  Denial of right to adequate medical care (#PD2028)
  Denial of right to benefits for invalids (#PE5211)
  Denial of right to benefits to survivors (#PE4531).

♦ PE5408 Family adaptation of community status quo
Nature The most readily accessible and frequently adopted model for family structures is the provincially biased status quo of the local community. Methods of personal interaction, symbols of significance, processes of decision-making, visions of the future, articulations of consciousness and all the other social dimensions of being a family are received from the local community. Most often these are the remnants of a society which no longer exists except in the wish dreams of a few.
  Broader Lost family role in society (#PF7456).
  Related Escapist family life styles (#PD4069).

♦ PE5409 Denial of freedom of expression in clothing
Enforcement of standards of dress
Incidence School councils in UK have the power to decide whether school uniforms should be compulsory. If uniforms are to be worn, the school principle can discipline for failing to wear the uniform. In Iran the fundamentalist dress code considers immodest the showing of any hair, or any contour of the body, such as the line of the neck or the waist. The penalties for refusing to submit have been intimidation, threats and at times imprisonment.
  Broader Restriction of freedom of expression (#PC2162).
  Aggravated by Decadent clothing (#PE5607)
  Unethical practices in the apparel industry (#PD8001).

♦ PE5424 Inadmissability of evidence from other jurisdictions
  Broader Evasion of issues (#PF7431).
  Aggravates Illegally obtained evidence (#PE9309)
  Unproven relationships between problems (#PF7706)
  Inadequate evidence to convict known offenders (#PF8661)
  Placement of the burden of proof on the disempowered (#PF3918).

♦ PE5431 Enuresis
Involuntary passage of urine — Unconscious passage of urine — Bed wetting
Refs Heston, Stanley M *Bed Wetting* (1987).
  Broader Incontinence (#PE4619)   Mental illness (#PC0300)
  Symptoms referable to genito-urinary system (#PE9369).

♦ PE5436 Thiols and mercaptans as pollutants
Mercaptans
Nature Thiols have an intensely disagreeable odour and contact with the liquid or vapour may cause irritation of the skin, eyes and mucous membranes of the upper respiratory tract. Liquid thiols can also cause contact dermatitis.
  Broader Malodorous fumes (#PD1413).
  Related Hydrogen sulphide as a pollutant (#PE2329).
  Aggravates Badly laid out work premises (#PJ2468).

♦ PE5439 Metal dust and fumes
  Broader Dust (#PD1245).
  Narrower Zinc dust (#PG5385)   Copper dust (#PE4486).
  Aggravates Lead as a health hazard (#PE5650).

♦ PE5443 Neighbourhood noise
Noisy neighbours
  Broader Health hazards of exposure to noise (#PC0268).
  Aggravates Neighbourhood disputes (#PE5504).

♦ PE5444 Otitis externa in animals
  Broader Diseases of senses in animals (#PD5535).
  Narrower Acute otitis externa (#PG4026)   Chronic otitis externa (#PG9639)
  Mycotic otitis externa (#PG5557)   Parasitic otitis externa (#PG7387).

♦ PE5448 Hitchhiking
Incidence In some countries there is a rising rate of crime associated with hitchhiking. A growing number of hitch-hikers are being killed and a large number of motorists who have picked up hitchhikers have been robbed, beaten or killed.
  Broader Travel risks (#PD7716).

♦ PE5453 Refusal to grant citizenship
Incidence In 1990 in Japan, as an example, 650,000 Koreans still living in Japan three generations after they were brought there as labourers continue to be refused citizenship.
  Broader Discrimination against immigrants and aliens (#PD0973).
  Aggravates Statelessness (#PE2485).

♦ PE5457 Plant tumours
Refs Misra, A, et al *Plant Tumors* (1983).
  Broader Plant diseases (#PC0555).

♦ PE5458 Ineffective space usage
Inefficient space usage
  Broader Obstacles to availability of community space (#PF7130).
  Narrower Scattered housing locations (#PJ0760)
  Over-spacing of suburban housing (#PE1708).
  Aggravates Unfocused style of community operations (#PF6559)
  Constricting effect of educational structures (#PF6476).

♦ PE5460 Vagrancy
Tramps — Hoboes — Criminalization of homelessness
Refs Cook, T *Vagrancy* (1979); Ribton-Turner, C J *A History of Vagrants and Vagrancy, and Beggars and Begging* (1972).
  Broader Disorderly conduct (#PD9178)   Restrictive legislation (#PD9012).
  Aggravated by Homelessness (#PB2150)   Alcohol abuse (#PD0153).

♦ PE5461 Maltreatment of animals in aquaria
Aquarium imprisonment
Claim Capturing animals, especially large marine mammals, is a basic violation of their right to enjoy their lives in a natural habitat. There is intense suffering among those animals who are forced to spend their lives in aquariums, whether small or large, as well as in being transported from one location to another. Social and behavioural needs are not and cannot be provided for in small, sterile pools. Dolphins and whales are especially sensitive to their surroundings and to the stress of living in an unnatural, cramped environment where the chances of disease are increased. A life of boredom, constriction and early death is guaranteed for such mammals. Sentient beings are regarded as "exhibits" or "conversation pieces" – mere commodities. This callous attitude contributes to society's failure to protect natural environments because of a false sense of security, that if it is in captivity, it will not become extinct. Aquariums isolate humans from nature.
Counter-claim Aquaria provide an important educational service and a good environment for the study and appreciation of marine life. Many injured animals are treated in the larger establishments. For some this contact with humans is considered to be beneficial to the species concerned.
  Broader Exploitation of animals for amusement (#PD2078)
  Denial to animals of the right to conditions of life and liberty proper to their species (#PE6270).
  Narrower Maltreatment of marine show animals (#PE0906).
  Related Maltreatment of performing animals (#PE4810).
  Aggravates Military use of animals (#PE1666).
  Aggravated by Unethical practices with domesticated animals (#PE4771).

♦ PE5462 Occupation of trade union premises by public authorities
Raids on trade union premises by public authorities — Seizure of trade union premises
  Broader Violation of the right of trade unions to function freely (#PE1758).

♦ PE5463 Ill defined causes of morbidity
Ill defined causes of mortality — Unknown causes of mortality
  Broader Ill defined health conditions (#PC9067).
  Narrower Asphyxia (#PE4104)   Toxaemia (#PG3957)   Toxicosis (#PG6525)
  Malingering (#PE7701).

♦ PE5464 Shortage of industrial water
  Broader Long-term shortage of water (#PC1173)

Underdeveloped provision of basic services in developing countries (#PF6473).
Aggravates Loss of water to industrial uses (#PE7433).

♦ PE5465 Oomycetes

Broader Fungal plant diseases (#PD2225).
Narrower Blights (#PE3919) Cankers (#PE0640) Rots in plants (#PE3363) Galls in plants (#PE3715) Late blight of potato (#PG3346) White rusts in plants (#PG5346) Downy mildews in plants (#PE0501) Wilt diseases of plants (#PE1056) White rusts of crucifers (#PG0420).

♦ PE5470 Corrosion of stonework

Nature Stonework is corroding through the world mostly due to pollution of the atmosphere. While the rate of corrosion on structurally used stone may be very slow, a centimetre a century, this rate on decorative stonework is disastrous. Priceless statues are being eaten away.
Background Sulphur dioxide formed by burning high-sulphur fossilized fuel is especially damaging to limestone. The calcium carbonate interacts with the sulpher dioxide and becomes calcium sulphate which is more soluble and washes away in the rain.
Broader Corrosion (#PD0508) Atmospheric corrosion of materials (#PE9525).

♦ PE5476 Chlamydia

Chlamydid infections
Refs Mardh, P A, et al *Chlamydia* (Date not set); Oriel, D et al *Chlamydial Infections* (1986).
Broader Sexually transmitted diseases (#PD0061).

♦ PE5478 Hirsutism in animals

Hypertrichosis
Refs Mauvais-Jarvis, P et al *Hirsutism* (1981).
Broader Endocrine diseases in animals (#PD9654).

♦ PE5480 Torture through denial of sanitary facilities

Nature In the process of humiliating victims of torture they are denied the use of sanitary facilities or fresh clothing. Cells are unsanitary and are without toilets or wash basins.
Incidence Denial sanitary has been reported in the following countries: **Af** Mauritania, South Africa. **Am** Argentina. **As** Afghanistan, Israel.
Broader Induction of incongruent actions (#PF3790).

♦ PE5483 Low back pain

Lumbalgia
Refs Bernstein, J; et al *Low Back Pain* (1989).
Broader Vertebrogenic pain syndrome (#PE9461).

♦ PE5484 Denial of the right to an adequate standard of living for indigenous peoples

Broader Inadequate standards of living (#PF0344)
Discrimination against indigenous populations (#PC0352).

♦ PE5487 Harassment of the judiciary

Harassment of judges — Harassment of lawyers
Incidence In the year ending June 1990, 430 judges and lawyers in 45 countries were named as having been harassed or persecuted. Of these, 67 were killed, 167 were detained, 40 were attacked, and 67 received threats of violence.
Related Corruption of the judiciary (#PD4194)
Harassment of human rights monitors (#PF1585).
Aggravates Lack of impartiality of the judiciary (#PE7665).

♦ PE5488 Violation of privileges and immunities of international civil servants

Detention of international civil servants — Threat to diplomatic immunity
Nature Privileges and immunities of international officials performing functions in connection with activities of their organizations are not always duly respected, thereby seriously affecting the proper functioning of the organizations.
Incidence In the beginning of 1987 some 50 United Nations staff members were detained, imprisoned, reported missing - some having died in detention - or held in a country against their will.
Related Threat to parliamentary immunity (#PF0609).
Aggravates Inefficient public administration (#PF2335).
Aggravated by Lack of political neutrality of civil servants (#PE4031)
Abuse of privileges and immunities by diplomats (#PF5649).

♦ PE5503 Counterfeit money and government securities

Illicit coining — Counterfeit financial instruments
Nature Currency is understood to mean paper money (including banknotes) and metallic money, the circulation of which is legally authorized. A person who commits any of the following acts is engaged in counterfeiting currency: fraudulent making or altering of currency, whatever means are employed; fraudulent uttering of counterfeit currency; introduction into a country of, or the receiving or obtaining of counterfeit currency with a view to uttering the same and with knowledge that it is counterfeit; attempts to commit, and any intentional participation in, the foregoing acts; and the fraudulent making, receiving, or obtaining of instruments or other articles peculiarly adapted for the counterfeiting or altering of currency.
Incidence It has been estimated that for every 1 million American notes in circulation, 500 is counterfeit.
Broader Counterfeiting (#PD7981).
Related Documentary fraud (#PE1110).

♦ PE5504 Neighbourhood disputes

Disputes between neighbours
Incidence In the UK in 1988 it was estimated that 25 per cent of households on an estate were suffering from long-standing disputes with a neighbour. Most indicated that the quality of their lives had declined because of their inability to find a solution to the situation, whether noise, children's misbehaviour, broken fences, dumping of rubbish, boundary encroachment, deprivation of light, odours, badly parked cars or other forms of insensitivity. Although the issues may initially be minor, over time they acquire much greater significance, occasionally leading to violence.
Claim Not a few murders every year are caused by minor disputes between people who are forced to go on living close to one another.
Aggravated by Neighbourhood noise (#PE5443) Declining sense of community (#PF2575)
Boundary disputes between neighbours (#PE7903).

♦ PE5505 Arson

Endangering by fire or explosion — Incendiarism — Threat of arson attacks
Nature Arson is the intentional burning of property, namely the dwelling house or business premises of another. The most usual motives for intentionally burning property are to collect fire insurance; to conceal the existence of some other crime; to obtain revenge upon the owner; to punish the owner, generally a merchant for failure to pay demanded extortion money; and to experience the excitement of watching the blaze.
Intentionally starting or maintaining a fire or causing an explosion which endangers a person or a building or causes substantial property damage is a crime. This is distinguished from arson in that destruction of the property is not the intent and generally is considered a lesser offence.
Incidence In the UK, arson attacks resulted in the loss of 350 million pounds of military stores in 1983 and 1988. In the later case 70 per cent of the equipment destroyed was essential for operational effectiveness.
Refs Bennett, Wayne W and Hess, Karen M *Investigating Arson* (1984).
Broader Property damage (#PD5859).
Aggravated by Pyromania (#PG5853) Insecurity of property (#PC1784).

♦ PE5506 Unfair competition from convict-made goods

Unfair competition from workers employed as forced labour
Nature Where convicts are employed productively, the goods they produce may compete unfairly against products manufactured by workers employed under normal wage conditions. In some case labour camps are being used expressly for the production of export manufactures and raw materials.
Broader Unfair competition (#PC0099).

♦ PE5507 Drug abuse by students

Incidence Drug abuse by students is increasing and there has been a shift in the type of drugs they use. In the USA for example, cocaine is the fastest growing drug among high school students. The number of seniors using it nationally has doubled since 1976, according to university statistics, while marijuana use has dropped and alcohol use has remained steady. Experts believe that cocaine is penetrating more deeply at private boarding schools and affluent suburban public high schools, because these students can afford it. They also come out of a social milieu where more and more parents are abusing drugs too. Cocaine is said to be on 99 percent of the campuses.
Broader Drug abuse (#PD0094).
Aggravated by Criminal investment in youth market (#PD5750).

♦ PE5508 Criminal subculture

Underworld society — Criminal sub-culture
Nature Members of the criminal underworld are alienated from their societies and may characterize themselves as such not only by idiom, but by mannerisms, clothes, and a range of behavioural preferences and adaptations.
Incidence A criminal sub-culture is in evidence in most large cities in the industrialized west, and is a phenomenon of growing concern in most of the rest of the world. A recent study in the United States on convicted murderers showed that 60 percent had been abused children, 50 percent came from broken homes, 65 percent never finished high school, 70 percent lived below the poverty line, and 60 percent were unemployed at the time of the commission of their crime.
Claim The criminal sub-culture in every nation is governed by rules and codes of behaviour that do not apply outside the shadows of this underworld. Children brought up in the penumbra of illicitness are exposed to all its immorality which is presented as the normative standard for values and behaviour. The ageing criminal dies among his fellows, so that crime is a way of life for many, from cradle to grave. The criminal sub-culture is much more than organized crime or the aggregate of all active or imprisoned criminals; it is also comprised of the wives, children, grand-children, friends, and places for business meetings or business entertainment. It is the people in collusion with criminals: corrupt lawyers, politicians, police; and people who turn the other way to overlook criminality in those who may be customers, patrons, or contributors to churches, charities, elections or other appeals. These are the people of the twilight; who stand on the edge of the darkness, who feed it, and eventually who are engulfed. That the underworld is indeed a sub-culture is also shown by its own idioms; and a number of lexicons of the languages of the criminal elements in the French, English and American cities, for example, have been compiled
Broader Social deviation (#PC3452).
Related Organized crime (#PC2343) Underworld milieu (#PG1850).
Aggravates Recidivists (#PE5581) Juvenile delinquency (#PC0212)
Increasing female criminality (#PE5592).
Aggravated by Unethical personal relationships (#PF8759).

♦ PE5509 Polymyositis

Refs Dalakas, Marinos *Polymyositis and Dermatomyositis* (1988).
Broader Rheumatic diseases (#PE0873).

♦ PE5516 Frauds, forgeries and financial crime

White-collar crime
Nature Crimes committed by businessmen, politicians and government employees are common to both developed and developing countries. They include: embezzlement and tax evasion; violations of import and export laws; deliberate deception in labelling and packaging; food adulteration; misappropriation of funds in trust; breaches of monopoly regulations; insurance fraud; arson-for-profit; toxic waste violations; labour racketeering and bribery. Corruption, and more specifically political corruption, involves economic and social costs of serious proportions.
Claim Illicit gains from white collar crime far exceed those from all other crime combined. One corporate price-fixing conspiracy may criminally convert more money each year than hundreds of burglaries, larcenies or thefts. In fact, these are the most pervasive, most pernicious and most costly crimes of all in society. The likelihood of getting caught is relatively small, the likelihood of getting convicted is even smaller and sentences for white collar crimes are lower than for example for bank robberies.
Refs Sutherland, Edwin; Geis, Gilbert and Goff, Colin *White Collar Crime* (1983).
Broader Fraud (#PD0486) Exploitation of trust (#PC4422)
Criminal harm to property (#PD5511).
Narrower Financial frauds (#PE2414) Agricultural fraud (#PE5687) Collusive tendering (#PE4301) Misuse of trust funds (#PG1741) Trafficking in government benefit coupons (#PE7271).
Related Monopolies (#PC0521) Tax evasion (#PD1466) Embezzlement (#PD2688) Theft of services (#PD4711) Theft of documents (#PD0577) Theft of works of art (#PE0323) Bureaucratic corruption (#PC0279) Bribery of public servants (#PD4541) Defrauding of secured creditors (#PD4401) Corruption of government leaders (#PCp587) Evasion of customs and excise duties (#PD2620) Political corruption of the judiciary (#PE0647) Distortion of corporation financial statements (#PE2032).
Aggravated by Corporate crime (#PD3528) Dishonest employees (#PD9397)
Corruption in politics (#PC0116).

♦ PE5523 Over-pricing

Incidence Among the most damaging practices for developing countries is that of overpricing - particularly of essential commodities and products such as drugs. Overpricing rates of between 30 and 700 percent in the price of drugs have been documented in several Latin American

countries.
Narrower Excessive cost of medical drugs (#PE5755).

♦ **PE5524 Respiratory diseases of cattle**
Refs Martin, W B (Ed) *Respiratory Diseases in Cattle* (1978).
  Broader Cattle diseases (#PD0752)
  Diseases of respiratory system in animals (#PD7307).
  Narrower Endemic pneumonia in calves (#PG7759)
  Bovine pneumonic pasteurellosis (#PG9381)
  Farmer's lung disease in cattle (#PG5383)
  Acute bovine pulmonary emphysema (#PG7557)
  Contagious bovine pleuropneumonia (#PE1775)
  Infectious bovine rhinotracheitis (#PG9706)
  Tracheal oedema syndrome of feeder cattle (#PG7167).

♦ **PE5530 Rickettsiosis**
Tick–borne rickettsiosis — Rickettsial diseases
Refs Burgdorfer, Willy and Anacker, Robert *Rickettsiae and Rickettsial Diseases* (1981).
  Broader Zoonoses (#PD1770)   Arthropod–borne diseases (#PE7796)
  Biological contamination of food (#PD2594).
  Narrower Q fever (#PE2534)   Rickettsial pox (#PG9834)   Boutonneuse fever (#PG9313)
  Heartwater disease (#PE2680)   Canine ehrlichiosis (#PG2402)   Elokomin fluke fever (#PG4029)
  Salmon poisoning disease (#PG9553)   Rocky Mountain spotted fever (#PG3899)
  Canine infectious cyclic thrombocytopenia (#PG9616).
  Aggravated by Rickettsiae (#PE2572)   Rodent vectors of disease (#PE3629).

♦ **PE5538 Criminal negligence in performing socialist responsibilities**
  Broader Socialism (#PC0115)   Negligence (#PA2658)
  Crimes against the integrity and effectiveness of government operations (#PD1163).
  Related Criminality (#PA9226)   Behavioural deterioration (#PB6321).

♦ **PE5539 Pollution of sediments**
Contamination of sediments by toxic substances
Nature Many metals and organic substances have an affinity for organic matter or mineral particles. Sediments thus provide a suitable medium for the accumulation of contaminants from aqueous sources and atmospheric deposition. The contaminant level is often higher than in the overlying body of water.
Counter-claim Certain elements, often associated with pollution, tend to occur naturally in sediments and their concentrations reflect the geology of the areas (as in the case of lake basin).
  Broader Soil pollution (#PC0058).
  Aggravated by Lake pollution (#PD8628)   River pollution (#PD7636)
  Marine pollution (#PC1117).

♦ **PE5545 Developmental expressive language disorder**
Nature The disturbance in expressive language can be evidenced in children by the use of limited size of vocabulary, vocabulary errors, shortened or limited sentences, simplified grammatical structures, difficulties in the production of long sentences and by the slow rate of language development.
  Broader Mental illness (#PC0300)   Speech disorders (#PE2265).
  Related Developmental receptive language disorder (#PE9300).

♦ **PE5546 Revisionism and anti-marxist crimes**
Nature Revisionism is the revision of the tenets of Marxism–Leninism. It is viewed as a crime against the communist state as it opposes Marxist–Leninist doctrine on all points. It rejects the necessity of revolution and asserts that capitalism should be simply reformed, claiming that the modern scientific and technological revolution is totally reshaping the structure of society and 'erasing' class antagonisms. Revisionism seeks to disarm the working class ideologically and to instil among workers reformist or anarchist views.
Incidence The Soviet Union attributes the Hungarian Revolution of 1956 to the revisionist group founded by I Nagy and G Losonczy.
Background Revisionism arose in the late 1870's in the German Social Democratic Party, which had accepted Marxism. In 1879 Edouard Bernstein, along with two others, proposed a revision of the basic tenets of Marxism, and in the late 1890's, Bernstein came out with his full–fledged programme for revisionism.
Refs Lenin, V I *Marxism and Revisionism.*
  Broader Communism (#PC0369)   Crimes against national security (#PC0554).

♦ **PE5548 Lymphosarcoma**
Hodgkin's disease — Reticuloses — Reticulum–cell sarcoma
Refs Kaplan, Henry S *Hodgkin's Disease* (1980); McElwain, T J and Selby, P *Hodgkin's Disease* (1987).
  Broader Diseases of the lymphatic system (#PD2654)
  Neoplasms of lymphatic and haematopoietic tissue (#PE4637).
  Aggravates Diseases of the spleen (#PE6155).

♦ **PE5550 Infectious mononucleosis**
Pfeiffer's disease
  Broader Virus diseases (#PD0594).
  Aggravates Diseases of the spleen (#PE6155)   Unethical consumption practices (#PD2625).

♦ **PE5551 Excessive consumption of resources in developed countries**
Unsustainable consumption of resources in industrialized countries
Nature The consumption patterns of countries of the developed world have produced demands on the environment highly disproportionate to the relative size of their populations. The natural resource base for such countries often extends far beyond their natural borders and environment.
Incidence According to one estimate, the riches 5 percent of the world's population may exert as much pressure on environmental resources as the poorest 25 percent.
Claim If everyone in the world consumed at the rate of Europeans and North Americans, the natural resource base would collapse and biosphere pollution would be intolerable.
  Broader Excessive consumption of goods and services (#PC2518).
  Related Excessive land usage (#PE5059).
  Aggravates Worldwide misallocation of resources (#PB6719).
  Aggravated by Unethical consumption practices (#PD2625)
  Ecologically unsustainable development (#PC0111).

♦ **PE5556 Possession of drugs**
Nature Possession of dangerous or abusable drugs is considered a crime in most countries. The severity of the punishments range from none to death for possession of large amounts.
  Broader Illicit drug trafficking (#PD0991).

♦ **PE5567 Respiratory diseases of poultry**
  Broader Diseases of respiratory system in animals (#PD7307).
  Narrower Turkey coryza (#PG4939)   Quail bronchitis (#PG7039)
  Gapeworm infection (#PG4831)   Swollen head syndrome (#PG6394)
  Chlamydiosis in poultry (#PG5071)   Aspergillosis in animals (#PG9721)
  Infectious coryza in poultry (#PG6801)   Infectious bronchitis in poultry (#PG4830)
  Infectious laryngotracheitis in poultry (#PG4503).

♦ **PE5569 Congenital anomalies of the cardiovascular system in animals**
  Broader Animal abnormalities (#PD4031)
  Circulatory system diseases in animals (#PD5453).
  Narrower Septal defects in animals (#PG0711)   Aortic stenosis in animals (#PG4578)
  Pulmonic stenosis in animals (#PG4615)   Tetralogy of fallot in animals (#PG5378)
  Persistent ductus arteriosus in animals (#PG4527)
  Persistent right aortic arch in animals (#PG0712).

♦ **PE5572 Human fatigue during control of complex equipment**
  Broader Fatigue (#PA0657)   Job fatigue (#PD8052).
  Narrower Aircraft pilot fatigue (#PE3870)   Motor vehicle driver fatigue (#PG7482).
  Related Computer stress (#PE5053).
  Aggravates Nuclear reactor accidents (#PD7579)
  Information overload during control of complex equipment (#PF6411).
  Aggravated by Boredom (#PA7365)   Unethical practices in transportation (#PD1012).
  Substance abuse during control of complex equipment (#PE0680).

♦ **PE5578 Discrimination against giants**
  Broader Discrimination against minorities (#PC0582).
  Discrimination against people of abnormal height (#PE9402).
  Narrower Domestic inconvenience for dwarfs, midgets and giants (#PE8036)
  Lack of adequate clothing supply for dwarfs, midgets and giants (#PE8496).
  Related Denial of rights to disabled (#PC3461)
  Discrimination against dwarfs and midgets (#PE2635).
  Aggravated by Gigantism (#PE3837)   Excessive standardization (#PF2271).

♦ **PE5579 Drug abuse by military personnel**
Incidence In 1980, out of 5,324 United States servicemen removed from nuclear weapons work, the largest proportion (1,726) went because of drug abuse. A 1982 Pentagon survey showed that 31.4 percent of American servicemen in Europe admitted using drugs including marijuana, cocaine and LSD in 1981. Among young enlisted men the figure exceeded 40 percent. Some scientists believe the drug–taking increases the risk of a serious incident, maybe nuclear, happening by mistake.
  Broader Drug abuse (#PD0094).
  Aggravates Military disobedience (#PD7225).

♦ **PE5580 Excess human body hair**
Unwanted facial hair
Incidence As an indication, the market for male shaving lotions and soap in the UK in 1986 was £114.3 million. That for female shaving lotions and depilatories was £17.9 million.
  Broader Unsociable human physiological processes (#PF4417)
  Cosmetic and health problems related to human hair (#PE7111).

♦ **PE5581 Recidivists**
Habitual criminals
Nature A recidivist is a person who commits a crime after having been previously convicted of a criminal act. This tendency to commit further crimes means that recidivists and habitual criminals present more of a danger to society than do once–off offenders. A recidivist may receive different treatment from first–time offenders, in that he may be refused bail and conditional early release as well as receiving harsher than usual treatment from gaolers and cellmates.
Counter-claim Given the conditions of most penal institutions, it may well be that those institutions themselves turn criminals into recidivists, by immersing them indiscriminately into a sub–culture environment which, even within prison walls, is operated by crime; and then releasing those prisoners back into society without the means and skills necessary earn a decent, lawful living. In some cases it is also in the state's best interest to define a person as a recidivist, thereby gaining justification for his incarceration in prison, a labour camp or an insane asylum.
Refs Maltz, Michael D *Recidivism* (1984).
  Broader Criminals (#PC7373)   Inappropriate personal habits (#PD5494).
  Related Discrimination against ex-prisoners and ex-detainees (#PE6929).
  Aggravates Employment of criminals in policy–making contexts (#PE4439).
  Aggravated by Criminal subculture (#PE5508)   Criminal motivation (#PJ6406).
  Criminal association (#PE1178)   Denial of rights to prisoners (#PD0520)
  Deficiencies in the criminal justice system (#PF4875).

♦ **PE5592 Increasing female criminality**
Nature Changes taking place in the forms and dimensions of female criminality are related to increasing opportunities for equal participation in the mainstream of development and to other changing socio–economic circumstances. In some developing countries, it is suggested that the influx of foreign customs that are not in harmony with indigenous customs have an adverse effect on female behaviour patterns and are conducive to deviance. It is the experience of some developed countries that where the process of development and equalization of opportunities for women has reached a stage of relative stability, likewise the level of female criminality stabilizes but still does not decrease. Adolescent female delinquency is increasing massively. This also seems related to the change in women's role in society.
Incidence Although the incidence of female crime which came to the attention of the authorities during the period 1970–1982 constituted a relatively small portion of the total overall crime figures, it still gave cause for concern. In some developing countries, where rapid transitions were still taking place, it was considered to be a new or emerging phenomenon; while in some developed countries it was a familiar one that was assuming new forms and more serious dimensions. Offences such as infanticide, child abuse, adultery, abortion, shoplifting and other petty theft, prostitution, moral offences, etc, were cited as conventional female crimes in many countries; in others, however, serious concern was expressed with regard to an increasing female involvement in very dangerous forms of conduct, especially drug trafficking, frequently on an international scale, and acts of violence, including terrorism. In particular, it was reported in some countries that females participated at every level of drug trafficking, from criminal association at the highest level to quasi-domestic situations – for example, involving the provision of food for transporters of the drugs. Such trends were becoming particularly serious in a number of countries. In one country, the involvement of suspected female offenders in the drug trade reached a growth rate of over 200 percent during the 1970–1982 period. Terrorist activities were reported as a serious development in some countries, and, in that connection, the involvement of females in terrorism was becoming a common feature of the phenomenon.
  Broader Criminals (#PC7373).
  Related Lesbianism (#PF2640)   Preponderance of male criminal offenders (#PG6485).
  Aggravated by Feminism (#PF3025)   Criminal subculture (#PE5508)
  Criminal motivation (#PJ6406).

♦ **PE5594 Misdemeanours**
Nature Misdemeanour constitutes a legal offence of less gravity than felony. In some countries, a person is guilty of misdemeanour who utters any word, makes any sound or gesture, or exhibits any objects, intending that such word or sound shall be heard, or that such gesture or objects shall be seen by some woman, or intrudes upon the privacy of some woman. The reason that very few cases are taken to court for the above misdemeanours is that women have learned to acquiesce in the case of such insults.
Broader Offences of general applicability (#PD4158).
Related Felonies (#PE1153).

♦ **PE5606 Tenosynovitis**
Repetitive strain injuries — Cumulative trauma syndromes — Occupational cervicobrachial disorders — Policeman's heel
Refs Stevenson, Michael (Ed) *Readings in RSI.*
Broader Foot diseases and disabilities (#PD2647)
Disorders of joints and ligaments (#PD2283).

♦ **PE5607 Decadent clothing**
Provocative dress — Improperly veiled women
Incidence This is especially significant to religious fundamentalist groups, whether Christian or Moslem, with respect to the wearing of inappropriate and especially western clothing. It includes appearing in public without being full covered. Certain modern styles of clothing, whether worn by men or women, may also be viewed as decadent according to some political ideologies.
Claim 1. Women are symbolic of the nation and of the religious tradition and just as unveiling is the first step on a path that leads to her rape, so too the nation and the religion will find itself penetrated and dominated by evil forces. 2. In the Islamic tradition it is said that: When a woman reaches puberty, nothing should be exposed but her face and hands. Physical beauty should only be revealed to near relatives, to domestic servants, or to children.
Counter-claim The suggestion that males can be provoked by the clothes that females wear fails to take account of the sense of responsibility of women in selecting the clothes appropriate to the environment in which they find themselves.
Broader Decadence (#PB2542).
Aggravates Indecent exposure (#PF4317) Sexual harassment of men (#PE1293)
Denial of freedom of expression in clothing (#PE5409).
Aggravated by Disobedient wives (#PF4764)
Unethical practices in the apparel industry (#PD8001).

♦ **PE5608 Denial of right to hold public office**
Violation of right to be elected to government positions — Denial of public service
Broader Violation of civil rights (#PC5285) Denial of political rights (#PD8276)
Denial of right to participate in government (#PE6086).
Related Limitations on right to vote (#PF2904) Denial of right of complaint (#PD7609)
Discrimination against women in politics (#PC1001)
Discrimination against women in public life (#PD3335)
Denial of the right to organize political parties (#PE9110).

♦ **PE5610 Political interference with television news coverage in capitalist countries**
Broader Excessive regulation of television (#PE6982).

♦ **PE5614 Political assassination**
Political murder — Political killings — Terrorism targeted against politicians
Nature A person may be assassinated for what he represents politically, when a government or any grouping chooses to eliminate its political opponents physically rather than fight them otherwise. Cold-blooded murders may also be committed on members of a group as a solution to delicate political issues. (The same ends may be achieved by character assassination or blackmail).
Refs Ford, Franklin L *Political Murder* (1987).
Broader Terrorism (#PD5574) Assassination (#PD1971).
Related Government sanctioned killing (#PD7221)
Vulnerability of diplomatic agents (#PE5683)
Assassination of trade union leaders (#PE0252)
Abuse of privileges and immunities by diplomats (#PF5649)
Terrorism targeted against business corporations (#PE5944).
Aggravated by Corporation-sanctioned assassination (#PE6356).

♦ **PE5617 Lameness in sheep**
Broader Diseases of musculoskeletal system in animals (#PD7424).
Narrower Foot abscess (#PG7812) Benign foot rot (#PG9810)
Septic laminitis (#PG5131) Virulent foot rot (#PG5019)
Ovine interdigital dermatitis (#PG7800).

♦ **PE5621 Unreported rape**
Unreported rape attempts
Nature Given that in most countries rape victims are viewed as instigators, given that most rapes are committed by family members or friends, given that most societies view female rape victims as unclean; given the mental anguish that results from recounting the incident in a public courtroom; and given the fact that only a small percentage or rapists are apprehended and incarcerated, the majority of rape victims do not report the crime.
Incidence Surveys in the UK indicate that only one in 12 women who are raped report the offence to the police. This is partly due to the fact that of those cases reported, less than half are prosecuted through the courts, and only 23 per cent of these result in convictions. Women are further deterred from reporting rape when the attacker is a policeman or a person in authority who will be able to ensure that the charge is not investigated impartially.
Broader Rape (#PD3266) Unreported crimes (#PF1456) Unreported violence (#PF4967).
Related Underreported issues (#PF9148) Homosexuality in prisons (#PE1363).

♦ **PE5623 Unlimited practice of human embryo storage**
Orphaned embryos
Nature There are presently about 10,000 human embryos in frozen storage in embryo banks around the world, with the number rapidly increasing. There is no time limit to this storage, and embryos are often left orphaned when parents die or change their minds about the procedure. There are no common guidelines and very little legislation of this matter.
The death of the parents of live "in vitro" embryos or of embryos emplanted in another's uterus, produced through artificial insemination or other means, raises the question of whether the embryos have legal status and whether they can be subject to donation, disposal, or custody.
Incidence Two orphaned embryos remained frozen and in legal custody pending a decision by the Australian Parliament as to their status. Both parents had died in a plane crash, leaving no instructions in their wills as to what to do with the embryos. Also in question was the inheritance of the $1 million fortune they left.
Broader Orphan children (#PD7046).
Related Commercialization of human embryos (#PE6038).
Aggravated by Eugenics (#PC2153)
Unethical experimentation using aborted foetuses (#PE4805).
Reduced by Lack of eugenic measures (#PD1091)
Non-acceptance of embryo transfer technology (#PE6039).

♦ **PE5634 Crimes against government and public property**
Nature Crimes against government and public property are numerous and may include: the stealing of state or public property; appropriation; embezzlement; extortion; abuse of public office; or the intentional destruction or damaging of public property. All crimes of this nature, on whatever scale, serve to undermine the good of the government and its public services.
Broader Crimes against the integrity and effectiveness of government operations (#PD1163).

♦ **PE5637 Unauthorized medical practice**
Nature Unauthorized medical practice is the illegal practice of medicine as a profession by a person who has not had the appropriate medical education. It is a criminal offence.
Broader Unethical medical practice (#PD5770).
Aggravated by Impairment of physicians' ability (#PE5746).

♦ **PE5644 Animal trophy hunting**
Animal head hunting
Claim The trophy hunter does not kill as natural predators do by selecting the oldest or the weakest. The trophy hunter seeks the finest animals, thus eroding the long-term ability of the species to maintain itself through elimination of the animals with the most valuable genes.
Broader Hunting tourism (#PE3008).
Related Head hunting in tribal societies (#PF2666)
Denial to animals of the right to freedom from mass killing (#PE9650).

♦ **PE5648 Death and disability from inhumane confinement**
Nature Inhumane confinement can result from neglect, but usually is by intent. In deadly heat or cold, the weather may be used just as effectively as a gun to kill or injure the confined. Where weather cannot do this, some of custodial mankind have assured the extinction or disablement of captive humanity by means hardly indistinguishable from torture. Lack of sanitation, space and adequate sleeping provision are frequent, as well as lack of physical security for the individual. Metabolic disorders arise from starvation; nutritional diseases are induced by insufficient intake of vitamins and minerals; bone and joint diseases stem from cramped postures or bondage; and muscle atrophy and corporeal wasting result from lack of exercise. Heart disease is aggravated if not engendered, and even if release is obtained, mental problems, fatigue, depression and premature senescence may be experienced.
Incidence Concentration camps during World War II, and labour camps in Siberia, are the more publicized cases of confinement conditions that lead to death. Devil's Island, now out of operation, was an infamous example of penal conditions with fatal results. Latin American jails, chain gangs of the southern USA; migrant worker enslavement in the southwestern USA; and prisoner-of-war and rebel confinement in Latin America, Africa and Asia, have evidenced inhumane confinement conditions. Institutional confinement in prisons, psychiatric care facilities and facilities for the senile, juvenile and handicapped are sometimes afflicted by inhumane practices. Home confinement where there is criminal abuse and neglect is also known.
Broader Death of living creatures (#PF7043).
Narrower Suspicious deaths during detention (#PE6367).
Related Torture (#PB3430) Inhumane interrogation techniques (#PD1362).
Aggravated by Pain (#PA0643) Fatigue (#PA0657)
Injuries (#PB0855) Human disability (#PC0699)
Concentration camps (#PD0702) Inadequate animal welfare (#PC1167)
Torture through confinement (#PD4590) Human disease and disability (#PB1044).

♦ **PE5650 Lead as a health hazard**
Lead as an occupational hazard — Lead dust — Saturnism — Lead poisoning — Lead contamination of drinking water
Nature Lead appears as dust or fumes in the air of the work-place. Occupational exposure may occur in mines; but more commonly in lead smelters, where lead is produced from lead ore or scrap, and in occupations where lead or lead compounds are used, such as the production and repair of storage batteries and the polishing and welding of lead-coated or lead-painted materials. This may occur in shipyards, car factories, glass and ceramic factories, and printing and paint shops. Lead also exists in air due to pollution of exhaust gases of automobiles. Manifestations of lead poisoning (mainly in occupationally exposed adults) include gastrointestinal disturbances, anaemia, neuropathy and renal damage. Lead has a toxic effect on the human foetus through the mother's employment in lead-using industries. Children in urban areas are at risk because of the greater sensitivity of the developing nervous system to disturbance of the haem synthesis pathway. Higher lead intake occurs in industrial areas, in areas of high traffic density, and in soft water areas with lead plumbing, the latter being a major source of risk to bottle-fed infants.
Refs Annest, Joseph L; Mahaffey, Kathryn and Cox, Kaludia *Blood Lead Levels for Persons Ages 6 Months to 74 Years* (1984); Chisholm, J Julian and O'Hara, David M (Eds) *Lead Absorption in Children* (1982); Mahaffey, Kathryn (Ed) *Dietary and Environmental Lead* (1985); Nriagu, J O (Ed) *Biochemistry of Lead in the Environment* (1978); Nriagu, J O (Ed) *Biogeochemistry of Lead in the Environment* (1978); Quinn, Michael A, et al *Lead Paint Poisoning in Urban Children* (1976); Wedeen, Richard P *Poison in the Pot* (1984); Winder, Christopher *The Development Neurotoxicity of Lead* (1984).
Broader Occupational hazards (#PC6716)
Health hazards of environmental pollution (#PC0936).
Aggravates Metal poisoning (#PJ6543) Occupational diseases (#PD0215)
Occupational risk to health (#PC0865).
Aggravated by Pica (#PG4505) Lead as a pollutant (#PE1161)
Metal dust and fumes (#PE5439) Contamination of drinking water (#PD0235).

♦ **PE5651 Conflicting standards for protection against chemical occupational hazards**
Nature Recommended permissible levels for occupational exposure to hazardous chemicals in various parts of the world sometimes differ by a factor of ten or more. Threshold limit values, permissible doses, or maximum allowable concentrations, as these levels are variously called, have been established on over 700 airborne contaminants and physical chemical agents; but these standards vary between North American, Western European and eastern European versions. Lack of agreement on research findings, and use by some authorities of time-weighted values as opposed to maximum values, obstructs progress in setting universally regarded safeguards for workers exposed to chemical hazards.
Incidence Although the American Pentagon's research on chemical and germ warfare has increased more than fivefold under the Reagan administration, more than 60 percent of those funds are awarded to universities, biotechnology firms and large corporations which are not subject to federal safeguards.
Broader Inadequacy of international standards (#PF5072).
Related Politicization of health standards (#PD4519).
Aggravates Occupational hazards (#PC6716) Occupational diseases (#PD0215)
Occupational risk to health (#PC0865).
Aggravated by Inconclusiveness of scientific and medical tests (#PD7415).

♦ **PE5655 Inadequate telecommunications in island developing countries**
Nature Communications between island developing countries are very costly, especially when considered in proportion to the populations involved. This applies within regions; it is even more the case between regions, since island developing countries are scattered throughout the oceans of the world. The island developing countries are not able to benefit from each other's experience in seeking solutions to the specific problems they share under these conditions. Insufficient attention has been devoted to the potential offered by technological advances in the field of telecommunications, particularly since it is becoming easier to provide small, remote communities with proper telecommunications. Lack of technical and financial assistance for installing or improving telecommunications is an obstacle to sustainable developmental programs for the island countries.
Broader Insufficient communications systems (#PF2350)
Disparity in world telecommunications capabilities (#PD5701)
Inadequacy of telecommunication facilities in developing countries (#PE0004).
Related Weakness of infrastructure in island developing countries (#PE5772).

♦ **PE5656 Combined stresses as occupational hazards**
Nature Workers are often exposed to stresses of more than one kind, and human tolerance varies. Workers may be easily affected by minor degrees of stress because of the lowered 'vital' status of exposed individuals. The environmental stress in this case does not cause the illness, but rather brings about, in vulnerable groups, a rapid shift from the previously tolerated levels of existing illness, or of subclinical impairment, to a state of disability. Such acute events have often occurred as a consequence of different environmental stress factors acting on individuals with quite different kinds of physiological handicaps. In developing countries the employment of children, women, the elderly, and the partially disabled , s common. The degree of tolerance and susceptibility to psychological and physical stresses at work varies in these groups, and may result in health impairment and increased labour turnover.
Incidence With the quickening pace of information technology come the new managerial stresses associated with performance analyses that are performed in hours rather than the weeks or months it used to take to produce these accounts and analyses. Whereas managers once knew they had sufficient time to react, adjust, or prepare excuses, they now find they have no time or place to hide. Also attributed to increase stress is the strain of increased managerial responsibilities as firms slim down staffs when they instal computers. The UK reports 23 million working days are lost annually due to stress.
Broader Occupational stress (#PE6937) Occupational hazards (#PC6716).
Aggravates Occupational diseases (#PD0215) Occupational risk to health (#PC0865).

♦ **PE5664 Deficient transport in remote island developing countries**
Nature Remoteness is one of the roots of the specific problems of island developing countries, which are not found in the same form in small continental States. Most island developing countries are more than 500 kilometres from the nearest continent. Furthermore, there are over 36 which are archipelagic, made up of a number of islands which may be a long way from one another. Remoteness, however, is more than a question of distance in kilometres. It is more a function of the frequency, reliability, and convenience of transport and communications links. This in turn depends on the size of the market – the number of people or the amount of freight offering at a particular point – and on the transport and communication technologies available. Distances to be travelled within some archipelagic countries may represent an even greater remoteness in terms of travel time than that between the capital and neighbouring countries.
There is insufficient research into the development of appropriate types of vessel and aircraft for inter-island and feeder transport services: the transport development costs may be beyond the means of many island communities, but so may be the fixed infrastructure the transport services require. These internal transportation problems are aggravated by the fact that external transport services available to island countries continue to deteriorate, since technical progress in aviation and shipping is biased against small communities where the unit costs of operations tend to be correspondingly higher. Inadequate internal and external transport services are obstacles to island development, affecting tourism, conversion and production, exports, international political and diplomatic action, and scientific, technical and cultural interchange.
Broader Weakness of infrastructure in island developing countries (#PE5772)
Inadequate transportation facilities in developing countries (#PD1388)
Vulnerability of island developing countries and territories (#PE5700).

♦ **PE5665 Inaccessibility of insurance for island developing countries**
Nature Island developing countries may be subject to earthquakes, tsunamis, tropical storms and other natural disasters that destroy crops, food in storage, housing, industry, the environment, the economy, and human lives. These countries cannot find insurance through normal sources to cover these contingencies by providing funds for disaster relief, reconstruction, and commercial obligations on which they would otherwise have to default, thus impairing their credibility and credit worthiness as trade partners.
Broader Inaccessible commercial and financial services (#PE0718)
Vulnerability of island developing countries and territories (#PE5700).
Aggravates Inadequate insurance (#PF8827).

♦ **PE5667 Deprivation of government benefits for political purposes**
Nature Any attempt to influence the political rights of an individual or group of individuals by intentionally withholding or threatening to withhold from another benefits from any government programme, government supported programme, or government contract is a crime.
Broader Violation of political processes (#PD5457).

♦ **PE5671 Military conflict between communist countries**
Claim Conflicts between Marxist-socialist countries are the results of the Marxist-Leninist philosophy which advocates the resort to force to effect change or resolve issues. Border clashes between the Soviet Union and China (though diminishing) dramatize the haunting spectre of a third World War beginning in Asia, while the spectre haunting Europe already has taken lives in Hungary, Czechoslavakia and Poland. Conflict between communist countries can appear anywhere, for example, between China and Vietnam, and indicates the belligerency of the communist system.
Broader War (#PB0593).

♦ **PE5672 Ultraviolet radiation as a hazard**
Nature Ultraviolet radiation represents that portion of the spectrum between X-rays and visible light. The greatest natural source is the sun. Excessive or prolonged exposure to ultraviolet radiation gives rise to skin-burn (or in severe cases to severe blistering), conjunctivitis and keratitis (when the eyes are unduly exposed) and to carcinogenic effects. The earth's atmosphere screens out much of the ultraviolet radiation due to sunlight; but thermal and chemical pollution of the atmosphere diminishes its screening ability, thus exposing the biosphere to a potential hazard of global extent.
Incidence Skin cancer has frequently been reported in people whose occupation requires them to be exposed for long periods to direct solar radiation. Many people in industrialized countries are exposed to additional ultraviolet radiation because of their use of a wide range of apparatus in homes, industries, places of entertainment, health clubs and research establishments. Typical apparatus emitting ultraviolet radiation includes therapeutic lamps, sterilization and welding equipment. Occupational exposure to ultraviolet radiation occurs constantly in arc welding and in the use or proximity to a large variety of testing and quality control machines using ultraviolet radiation, in industries as varied as food processing and machine assembly. Other occupational exposure occurs among attendants at health spas, beauticians and similar places where sun lamps are used. Such radiation mainly affects the eyes, causing intense conjunctivitis and keratitis (welder's flash). Symptoms are redness of the eyes and pain, which usually disappears in a few days. No permanent disability appears to result from this occupational disease; however, during periods of eyesight impairment, driving and other accidents may be caused.
Refs Passchier, W F and Bosnjakovic, B F M (Eds) *Human Exposures to Ultraviolet Radiation* (1987).
Broader Hazards to human health (#PB4885) Radioactive contamination (#PC0229)
Environmental hazards of non-ionizing radiation (#PE7651).
Related Occupational diseases (#PD0215) Occupational risk to health (#PC0865)
Improper lighting as an occupational hazard (#PE5780).
Aggravates Legal havens (#PE0621) Occupational dermatoses (#PE5684)
Malignant neoplasm of skin (#PE5016).

♦ **PE5678 Imbalance in trade of cultural products between capitalist and socialist countries**
Nature Books, films, television programmes, records and similar cultural products are imported by socialist countries in number and worth far exceeding cultural exports. This contributes to trade deficits, but also to promoting the capitalist ideal of an endless appetite for consumption. Such cultural products falsely portray the world's goods as available almost for the asking, and lure the unwary into states of dissatisfaction and alienation.
Broader Obstacles to world trade (#PC4890).
Related Imbalance in trade of cultural products between developed and developing countries (#PE5702).

♦ **PE5682 Vulnerability of crops to weather**
Weather hazards for crops and plants
Nature Adverse weather is the major threat to agricultural food production, primarily affecting plants during the growth phase, but also during storage. The weather conditions causing damage are: precipitation, which includes heavy rain, hail or snow; extremes of heat or cold; excessive sunshine and radiation; excessive wind; insufficient water due to lack of precipitation and ground water level changes; and insufficient or excessive humidity. Practical methods have not been found to assure 100 percent protection, and in developing countries there is often no protection at all. In such countries, whose economies may depend on a single crop or a single harvest, adverse weather is an economic and social disaster.
Incidence Serious droughts occurred in in 1972 in the USSR, the Sahel, India, South America and Australia. The weather was then considered responsible for the death of over a million people in India and Bangladesh alone.
Broader Hazards to plants (#PD5706) Crop vulnerability (#PD0660).
Related Plant diseases (#PC0555).
Aggravates Suffering of plants (#PC7825) Seasonal fluctuations in agriculture (#PD5212).
Aggravated by Bad weather (#PC0293) Unreliable rainfall (#PD0489).

♦ **PE5683 Vulnerability of diplomatic agents**
Nature A fact of diplomatic life, despite supposed protection by international agreements, is the risk to life and limb. Occupation of embassies, kidnappings of consulate or embassy officers or their families, attacks on couriers, wounding and assassination, are threats with which diplomatic people have to live in unfriendly countries, or in countries with hostile elements. Insufficient physical protection is frequent, particularly for the families of diplomatic staff; and even in friendly countries, third parties may find it easy to mount attacks of a terrorist nature.
Refs Selth, Andrew *Against Every Human Law* (1988).
Related Political assassination (#PE5614).
Aggravated by Cardiovascular syphilis (#PG0121).

♦ **PE5684 Occupational dermatoses**
Skin irritants as occupational hazards — Chapped skin
Nature Occupational dermatoses may be caused by organic substances, such as formaldehyde; and by solvents or inorganic materials, such as acids and alkalis, and chromium and nickel compounds. Skin irritants are usually either liquids or dusts; they may have a primary toxic effect, as with solvents, acids and alkalis, or produce an allergic reaction after 3–4 weeks of exposure or longer (chromium and nickel compounds, formaldehyde). Dermatosis or eczema develops mainly on the skin areas exposed at work, such as the hands and forearms, but also on other parts of the body as a result of contact with contaminated clothes. Exposure to fine arsenical powder in the handling of arsenic compounds causes the development of warts on the skin; these may become malignant.
Broader Occupational diseases (#PD0215) Human disease and disability (#PB1044)
Diseases of the skin and subcutaneous tissue (#PC8534).
Aggravated by Cement dust (#PE2854) Ultraviolet radiation as a hazard (#PE5672).

♦ **PE5685 Bruxism**
Nature Bruxism is a disease thought to be mainly psychological in origin although some pathology of the nervous system cannot be ruled out. It is characterized by the grinding of the teeth, or the making of clicking sounds, or kinds of gnashing. Less obviously, teeth clenching and other voluntary or involuntary movements of the lower jaw muscles may be exhibited. The movements can occur around the clock and unconscious teeth grinding by children and adults while sleeping is well-known. Bruxism wears the teeth and bones supporting them and is a factor in peridontal disease. Bruxism in mild forms may be very prevalent in high-stress societies.
Broader Human disease and disability (#PB1044).
Aggravates Periodontal diseases (#PE3503).

♦ **PE5686 Long-term injuries from sports**
Nature Excessive engagement in a particular sport may give rise to permanent physical damage. Familiar examples are tennis elbow, football knee, and basketball back injuries. Sensitive areas in these cases are joints and the spine. More serious hazards are brain and eye damages from boxing; and damage to the groin, pelvis, uterus, kidneys and other internal organs can occur from riding horses or motorcycles. Professional athletes suffering continuous pain from injuries, or experiencing excessive stress due to the competitive nature of their games, may seek relief in drugs, leading to drug abuse and narcotic addiction. Competitors pressured into use of steroids may develop cancer.
Refs McLatchie, Greg *Combat Sports Injuries* (1982); Reid, Stephen E and Reid, Stephen E *Head and Neck Injuries in Sports* (1984); Riffer, Jeffey K *Sports and Recreational Injuries* (1986).
Broader Injuries (#PB0855).
Related Sports accidents (#PE4262).
Aggravated by Athletic competition (#PE4266).

♦ **PE5687 Agricultural fraud**
Broader Fraud (#PD0486) Frauds, forgeries and financial crime (#PE5516).

♦ **PE5689 Lack of coastal development in island countries**
Neglected marine space of island developing states
Nature The development potential of coastal zones includes: fish and products of mariculture for the local population or for export; energy; water; and contributions to land-based agriculture. The conservation of marine life, coastal conservation and the control of marine pollution assume great importance in this context. Lack of effective exploitation of island developing countries coastal zones, including reefs, increases the need for imports and decreases food security; and lack of integration of marine space utilization and sea resources into national development strategies affects a large number of sectoral roles, such as agriculture, energy and environmental management. In many areas of great potential importance to the self-reliant development of island developing countries, such as mariculture, polyculture, energy and coastal transport, development efforts may well take a decade or more to ripen into substantial value-adding and other economic gains.
Broader Vulnerability of island developing countries and territories (#PE5700).
Narrower Underdevelopment of fishing industry (#PE2138).
Related Obstacles to the utilization of coastal and deep sea water resources (#PF4767).
Aggravates Unsustainable development of coast zones (#PD4671).

♦ **PE5691 Torture by continuous noise**
Broader Torture through sensory overload (#PE5259)
Health hazards of exposure to noise (#PC0268).

♦ **PE5696 Biological agents as occupational hazards**
Nature Biological agents in the work environment include viruses, rickettsiae, bacteria, and parasites of various types. Occupational infectious diseases occur in industry and mining. The common infectious and parasitic occupational diseases include anthrax (wool-sorting and handling of infected hides), brucellosis (contact with infected animals), tetanus (from infected wounds), ancylostomiasis (hookworm disease) and schistosomiasis (bilharzia). Fungi from organic dusts, such as bagasse and cocoa, may cause myotic respiratory diseases or skin infections. Diseases transmitted from animal to man in agricultural work are of common occurrence in view of the fact that agriculture is the major economic activity in developing countries. In such work, infectious and parasitic diseases may result from exposure to contaminated water or to insects.
Incidence The case of schistosomiasis as an occupational disease is an example. In endemic areas in Brazil, workers in plantations are exposed at work to infected water and soil. It was found that the proportion of stool-positive cases was 59 percent among exposed workers, as compared with only 10 percent among other workers in the same area who did not come into contact with infected water or soil.
Broader Occupational hazards (#PC6716) Human disease and disability (#PB1044).
Aggravates Occupational diseases (#PD0215) Occupational risk to health (#PC0865).

♦ **PE5697 Forced depilation**
Mandatory hair shaving
Nature The non-voluntary removal or cutting of hair is known in most cultures. Hair may be removed with good intentions for medical treatment, or surgery, or may result, in some cases, from chemotherapy. It is commonly done for cosmetic purposes. Scalp and facial hair is removed from men entering boarding schools, military training or prisons, primarily to reduce possibilities of contagious scalp infection, although there are elements of initiation rites are also present. There are advantages in some sports: martial arts, swimming and wrestling and in some occupations: mining and milling. Shaving heads has also be used for penal purposes.
Broader Personal physical disfigurement (#PD8076).
Reduces Mange (#PE2727).

♦ **PE5698 Environmental hazards of nuclear weapons industry**
Violation of safety regulations in nuclear weapons industry — Pollution from nuclear weapons manufacture
Nature Environmental problems of long standing plague the nuclear weapons industry. Besides soil and water contamination by radioactive materials, some sites suffer contamination by conventional hazardous chemicals used in the production process.
Incidence In the USA in 1990 it was reported that as many as 42 of the 177 underground tanks that are used to store waste from nuclear bomb production are in danger of exploding. Such an explosion could mean the spread of toxic chemicals and radioactive materials over large areas. The risk is due to unforeseen reactions between the chemicals stored and those introduced in an endeavour to consolidate the waste. The ferrocyanide percolating in the tanks is sufficient to cause an explosion equivalent to 36 tons of TNT. In 1957 in the USSR, the explosion of such a nuclear waste storage tank spread radiation over a large area and forced the evacuation of 10,000 people, with some reports that hundreds of people later died. It was also reported that 28 kg of plutonium (equivalent to 7 nuclear bombs) had escaped into the air ducts at the Rocky Flats weapons plant during its 30 years of operation, although it is so toxic that it is usually accounted for in gram quantities.
Refs United Nations *Current Nuclear Power Plant Safety Issues* (1981).
Broader Environmental consequences of war (#PC6675).
Aggravates Obesity (#PE1177) Nuclear waste (#PD4396)
Accidents to nuclear weapons systems (#PD3493).
Aggravated by Proliferation of nuclear weapons and technology (#PD0837).

♦ **PE5700 Vulnerability of island developing countries and territories**
Nature Island developing societies share common problems. Many of these are similar to those of small states and territories, or developing countries in general, although aggravated by transportation and communication difficulties. Unique problems are: exposure to the elements and natural disasters; ecological vulnerability to imported diseases of all types; and particular hardships in maintaining infrastructures in archipelago and multiple island states. Island countries may depend economically on the presence of foreign military bases, and because of difficulty of maintaining coastal defences, may be threatened by military intervention. Dozens of these countries and territories are remotely situated more than 500 kilometres from continents.
Broader Isolated islands (#PD2941) Vulnerability of developing countries (#PC6189)
Geographically disadvantaged countries (#PF9247)
Non-viability of small states and territories (#PD0441).
Narrower Lack of coastal development in island countries (#PE5689)
Inviability of tropical island developing countries (#PE5808)
Excessive emigration from island developing countries (#PE5713)
Lack of skilled workers in island developing countries (#PF5737)
Increasing foreign debt of island developing countries (#PE5748)
Deficient transport in remote island developing countries (#PE5664)
Weakness of infrastructure in island developing countries (#PE5772)
Inaccessibility of insurance for island developing countries (#PE5665)
Obstacles to political union among island developing countries (#PE4627)
Dependence of island developing countries on official assistance (#PE5724)
Inadequate international participation by island developing countries (#PE5761).
Related Neglect of remote regions and islands (#PE5760)
Abuse of land-locked and island countries as havens (#PE5861)
Dependence of island developing countries on imports (#PD5677).
Aggravated by Disaster hazards to island populations (#PE5784)
Military threats to island countries and territories (#PE5785).

♦ **PE5702 Imbalance in trade of cultural products between developed and developing countries**
Nature Hollywood-style films, and the phonorecord and cassette containing contemporary, youth-oriented music, continue to dominate the youth-market in developing countries. Money spent on these ephemeral products represents a substantial drain on the economies of some small states. In addition, importation of the value system that is exposed in these products and alluringly packaged, may subvert local social standards and cultural maintenance.
Broader Imbalance in international trade patterns (#PC8415).
Related Imbalance in trade of cultural products between capitalist and socialist countries (#PE5678).

♦ **PE5708 Solvents as an occupational hazard**
Nature Solvents include aliphatic and aromatic hydrocarbons, alcohols, aldehydes, ketones, chlorinated hydrocarbons and carbon disulpfide. The vapours of organic solvents may be toxic. Occupational exposure can occur in many different processes, such as the decreasing of metals in the machine industry, the extraction of fats or oil in the chemical or food industry, in dry-cleaning, in painting, in the plastics industry, and in the viscose-rayon industry. Solvent vapours enter the body mainly by inhalation, although some skin absorption may occur. The vapours are absorbed from the lungs into the blood, and are distributed mainly to tissues with a high content of fat and lipids, such as the central nervous system, liver, and bone marrow. Most solvent vapours have an anaesthetic effect on the central nervous system. Some solvents may, in addition, cause damage to the liver and kidneys (carbon tetrachloride) or to the blood-forming organs (benzene), or contribute to early atherosclerosis.
The action on the central nervous system causes nervous symptoms, such as fatigue, headache and vertigo. Unconsciousness and death may result from short-term exposure to high concentrations. Toxicity to the liver and kidneys may cause jaundice and uraemia. Prolonged exposure to benzene may cause leukaemia and anaemia. Carbon disulphide may contribute to a high incidence of atherosclerosis and possibly to ischaemic heart disease; and may also cause severe nervous symptoms, including psychoses.
Broader Occupational hazards (#PC6716).
Related Inhaling of solvents and anaesthetic drugs (#PE1427).
Aggravates Occupational diseases (#PD0215) Occupational risk to health (#PC0865).

♦ **PE5713 Excessive emigration from island developing countries**
Nature Some island developing countries are suffering from acute population pressure, one response to which is emigration. Indeed, in some such countries, emigration can be the dominant demographic feature. Emigration is also influenced by cultural factors, among which is the frequent need to seek higher education abroad: small island countries cannot economically provide at home for their full range of educational needs. While the traditional destinations for unskilled emigrants are becoming more difficult to enter, the technically trained have less difficulty in finding a welcome and may be a lost resource for the country as they seek higher standards of living elsewhere. More generally, the high import content of the economy implies an imported cultural model which itself encourages emigration.
Broader Vulnerability of island developing countries and territories (#PE5700).
Related Lack of skilled workers in island developing countries (#PF5737).
Aggravates Disruptions due to migration (#PC0018).

♦ **PE5715 Dereliction**
Nature As industries and demographics change, large industrial and housing areas are being deserted, with no effort being made either to clear the land or rehabilitate the buildings.
Broader Landscape disfigurement (#PC2122).
Related Derelict industrial wastelands (#PE6005).
Aggravates Wrecks and derelicts as hazards (#PE5340).

♦ **PE5720 Heat as an occupational hazard**
Heat stress at work
Nature Exposure to heat is common in work-places in many branches of industry. Acute disorders may result either from excessive demands on, or failure of, the human temperature control mechanism or from a combination of the two. The complex interactions between temperature at the work-place, physical work and climatic conditions are such that each of these types of stress reduces the ability to tolerate the other two. Some individuals may react to hot work with impairments of cardiovascular and renal functions. It is known that heat may adversely affect alertness, reaction times and psychomotor coordination; this would account for a higher accident rate among workers exposed to heat. Accidents are particularly frequent among workers who are not acclimatized.
Heat is one of the commonest potentially harmful physical environments. In industries such as mining, steel or glass manufacture, agriculture and road building, workers are often exposed to severe environmental heat stress, which may even threaten survival. Hard physical work or exercise in hot surroundings is the main hazard, for the total heat load and the body is in fact the sum of environmental and metabolic heat. 'Heat stress' is the burden or load of heat that must be dissipated if the body is to remain in thermal equilibrium, and is represented by the sum of metabolic rate (minus external work) and gain or loss by convection and radiation. 'Heat strain' is the physiological or pathological change resulting from heat stress, including increase in heart rate and body temperature, sweating, heat syncope, or water and salt imbalance.
Incidence The problem of heat stress is more serious in developing countries, since most of these are located in the tropical and subtropical regions. In the tropics, the workers is expected to cope with the combined effects of industrial and climatic heat – with the probability of having to work hard in hot spaces during an eight-hour shift, and of living for many months at a stretch without complete relief from heat stress at any time of the day or night. To complicate matters, his diet may well be inadequate to support him in this ordeal.
Broader Heat disorders (#PE2398) Occupational hazards (#PC6716)
Excessive environmental heat (#PD7977).
Aggravates Occupational stress (#PE6937) Occupational diseases (#PD0215)
Occupational risk to health (#PC0865)
Health risks to workers in agricultural and livestock production (#PE0524).
Aggravated by Human ageing (#PB0477) Inappropriate clothing (#PJ6604)
Lack of acclimatization (#PG6601) Nutritional deficiencies (#PC0382)
Age discrimination in employment (#PD2318)
Poor physiological adaptability of women (#PG6603).

♦ **PE5721 Fragmentation of health service**
Lack of holistic medicine
Nature Patients are often referred to medical specialists by their family doctor, a general practitioner, or clinic. This procedure is intended, in part, to screen the patient through general examinations to get an overview of his or her health and detect related or unrelated health problems that the specialist may or may not be concerned with or qualified to treat. Therefore the

availability of special medical services which do not require a preliminary general health screening may represent a hazard to public health. One example is the public's recourse to opticians and ophthalmologists for correction of eye problems that may by caused by disease. Another example is dental practice which establishes no relationship to the patient's medical history, sometimes with serious consequences. The public has direct access to psychologists and some psychiatrists for personality disorders, but medical screening might show a physiological cause and offer appropriate treatment. Other therapies for which general practitioner referral is not required are physical therapy, massage, chiropracty, acupuncture, hearing aids, some prosthetic services, and nutritional and diet regimes. Holistic medical services for the public are obstructed by tradition and economic considerations, as most holistic methodologies are not covered by insurance policies.
Broader Institutional fragmentation (#PC3915).
Narrower Over-specialization in medical care (#PF5709).
Aggravates Inadequate health services (#PD4790).

♦ PE5722 Over-eating
Gluttony — Bulimia — Immoderate eating and drinking — Overeating — Overfeeding
Nature An attitude towards eating when food becomes an habitual object for the imagination and desire results in greed to obtain food, and voracity both of appetite and consumption. Habitual eating to excess is a characteristic of diverse cultures. In it may be due to the high standard of living where great varieties of food are available, or to a high standard of culinary art, or a combination of these. In poorer cultures, the diet may be rather restricted as to variety, but excessive consumption of foods produced from a single main crop may produce protein, energy, or mineral deficiencies. Over-eating may lead to obesity and a number of illnesses, including bowel, stomach and heart disorders, and as it includes consumption of beverages, can lead to alcoholism. As a cause of being over-weight it can affect glandular functioning and personality. The glutton frequently has no regard for the wastage of food. Pathological gluttony is an incessant, uncontrollable craving for food resulting in gorging. Its medical term is bulemia and the disorder is sometimes found in connection with anorexia nervosa.
Refs Hudson, Jame I and Pope, Harrison G (Eds) *The Psychology of Bulimia*; McFarland, Barbara and Baumann, Tyeis B *Feeding the Empty Heart* (1988); Pope, Harrison G Jr and Hudson, James I *New Hope for Binge Eaters* (1985).
Broader Eating disorders (#PE5187)
Excessive consumption of goods and services (#PC2518).
Related Anger (#PA7797) Substance abuse (#PC5536) Bulimia nervosa (#PE4538)
Anorexia nervosa (#PE5758) Hyperalimentation (#PG5307)
Obesity due to low metabolism (#PE4463).
Aggravates Stroke (#PE1684) Obesity (#PE1177) Haemorrhoids (#PE3504)
Elevated blood cholesterol (#PE2371) Human disease and disability (#PB1044).
Aggravated by Gluttony (#PA9638) Nutritional deficiencies (#PC0382)
Excessive consumption of specific foodstuffs (#PC3908).

♦ PE5723 Formalism in developed countries
Nature Formalism is based on a segmented society of socio-economic classes, or classes and hierarchies within the apparatus of the state and all its bureaucracies. Desirable social progress may be hindered in societies where there is a high degree of formalism or strict adherence to prescribed complex forms of behaviour and communication. Information is passed very slowly upwards and decisions may be made inappropriately at lower levels that block or distort communications or proposals.
Incidence The classic example of obstructive formalism existed in Imperial China. All developed countries are to some degree or other, formalistic. Organizations and corporations may be administered in a formalistic way, although this can be disguised by formal rules for personnel to dress and act in a supposed informal manner.
Aggravates Bureaucratic inaction (#PC0267).

♦ PE5724 Dependence of island developing countries on official assistance
Nature Island developing countries, because they are small and prone to natural disasters, have economies particularly susceptible to to sudden shocks, especially in their external sectors. This makes many of them incapable of being economically viable in present circumstances and therefore inherently dependent on foreign aid.
Broader Vulnerability of developing countries (#PC6189)
Vulnerability of island developing countries and territories (#PE5700).
Related Restrictions imposed on aid to developing countries (#PF1492)
Prohibitive cost of maintaining comprehensive document collections (#PE1122).

♦ PE5725 Trypanosomiasis
Fly disease
Broader Arthropod-borne diseases (#PE7796) Parasitic diseases in animals (#PD2735)
Parasites of the cardiovascular system in animals (#PE9427).
Related Chagas' disease (#PE0653) African trypanosomiasis (#PE1778).
Aggravated by Leeches as pests (#PE3660) Protozoa as pests (#PE6741).

♦ PE5729 Political impotence of students
Nature Thirty million students are enrolled in the world's higher institutions of learning. Many are acquiring knowledge and skills to enable them to contribute to the solution of the social and technical problems facing mankind. At any time, the students own working and living conditions are determined by the existing social and economic system's characteristics. Therefore their interests are linked at all times with working youth in general, whether students or not, and all classes of people, professionals or labours who are the productive element in society. However, particularly in the free-market economies, the political responsibilities and roles of students are not incorporated in the social dialectic, either at the national or, where it should have entered naturally, at the local level.
Broader Manipulation of students (#PE5777).
Narrower Fragmentation of organized students (#PF5753).
Related Denial of rights of children and youth (#PD0513).
Aggravates Student revolt (#PC2052).
Aggravated by Political impotence (#PJ1283) Student press weakness (#PE0628).

♦ PE5735 Trapping of animals
Cruel animal traps
Incidence It is estimated that there are approximately 2 million trappers in the USA alone, and for 50,000 of these it is a major source of income. Trapping is an important activity in northern countries such as Canada, Scandinavia and the USSR. The number of animals killed for the fur trade is estimated to be in excess of 50 million per year, but of these 50 per cent are raised in captivity, the remainder being trapped. Another estimate gives 20 million animals trapped annually in North America using the leghold trap.
Claim The agony and terror of trapped animals is excruciating. In the case of the leghold trap, the animal awaits the arrival of the trapper who finally kills it by clubbing or through strangulation (to protect the fur). Many animals, especially nursing mothers, have been known to chew off their legs to escape.
Counter-claim Since there are few natural predators left in the wild, trappers fulfil an important function in maintaining a natural balance and preventing escalation of population of certain species.
Broader Cruel sports (#PD1323) Hunting of animals (#PC2024)
Misuse of wild animals (#PD8904).
Narrower Bird netting (#PG6093).
Related Unsustainable exploitation of fish resources (#PD9082).
Aggravates Animal injuries (#PC2753) Slaughter of animals for pelts (#PD4575).

♦ PE5741 Violation of the rights of female homosexuals
Discrimination against lesbians — Denial of the rights of female homosexuals
Nature A woman who is suspected or known to be lesbian will be more or less boycotted by her social environment and forced into an isolation which will make her particularly vulnerable to alcoholism. In professional life, a different reason than the lesbianism of the employee will be given for dismissal or withholding promotion. In the event of divorce, they are generally denied the custody of their child. Most of the countries deny the possibility of artificial insemination for those who wish to have a child. Lesbians are one of the favourite targets of small neo-nazi groups.
Broader Discrimination (#PA0833) Discrimination against homosexuals (#PE1903).
Aggravates Lesbianism (#PF2640).

♦ PE5742 Citizen incompetence
Nature Citizenship in free societies is a political enfranchisement that allows adults to constitute an electorate and so determine the leaders and laws of the nation. However, in every country there are many citizens who do not understand the organization and function of government, the legal system of their nation, their national institutions and goals, nor prevailing current issues. The electorate contains many people with no information, little information or wrong information on the personalities and programmes of the political parties, and the achievements or lack of them of incumbent office-holders. This lack of competence leaves them open, at best, to ineffectiveness and a neglect of their own interest; and, at worst, to susceptibility to emotionalism in the exercise of their electoral power that could lead to national disaster. This incompetence is not a failure of formal education but is a product of multiple factors: on the one hand, the lies and confusion in statements of the government, the political parties and their allies, and the media; and on the other hand, the general absence of compulsory political involvement, or at least, voting, at all levels, local through national.
Broader Incompetence (#PA6416).
Related Political ignorance (#PC1982).
Aggravates Inadequate prevention of crime (#PF4924).
Aggravated by Political stagnation (#PC2494).

♦ PE5746 Impairment of physicians' ability
Nature Physicians are exposed to a variety of diseases and hazards of infection. As an occupational disease, however, heart conditions are most associated with the profession, and usually explained as due to its stresses. Less well-known are physicians' mental disorders. Physicians' ill-health may disqualify them from some kinds of medical practice.
Incidence In the United States among male doctors who died aged 39 or less, a study showed that 28 percent were suicides; English, Canadian and Danish physician statistics show high suicide rates as well. In Australia it has been estimated in the medical journal of that country that two to four percent of all doctors commit suicide. A Harvard University study of New England physicians showed that 17 percent of doctors had been hospitalized for psychiatric reasons, 35 percent used mind and mood altering drugs such as tranquillizers regularly, and over 30 percent had consulted psychiatrists for personal help. Drug addiction studies in France, Holland, Germany and England indicate that physicians comprise 17 percent of drug abusers, with nurses and pharmacists comprising another 15 percent. American Medical Association statistics estimate that 6 percent of all American physicians are significantly impaired by addiction to drugs or alcohol. Regarding moral impairment, a survey in the American Journal of Psychiatry reported that 13 percent of the male physicians studied indicated that they had erotic behaviour with their patients.
Claim Physicians should be in a state of well-being while rendering health services. They are generally scrupulous about risks to patients from their own infectious diseases, such as respiratory or skin ailments, but are less careful when maintaining a stressful practice, such as surgery, while they themselves have serious heart conditions. The statistics show that physicians' serious problems of personal, emotional, psychological, nervous or behavioural natures that should bar active practices in fact are no hindrances to their rendering unrestricted health care. Such problems may in fact be known to their medical colleagues as well, indicating that there is insufficient internal regulation in the profession. The absence of careful periodic physical and psychological screening of physicians may reduce the trust the public places in the MD as a qualification, viewing it more in relationship to the drivers licence. In the case of drivers, the licence can be taken away for drunken driving, and at a certain age re-testing is required; in the abuse of the medical licence, patients may be harmed, but physicians fight control, handicapped by a lack of understanding of their own problems.
Broader Occupational diseases (#PD0215).
Narrower Substance abuse by physicians (#PE7568).
Aggravates Suicide (#PC0417) Inadequate medical care (#PF4832)
Unauthorized medical practice (#PE5637).
Aggravated by Alcohol abuse (#PD0153).

♦ PE5748 Increasing foreign debt of island developing countries
Nature The 20 island developing countries for which information is available increased their total outstanding debt at the same rapid rate as did developing countries in general over the period 1975-1982. However, over the period 1979-1982 the acceleration was significantly greater for all island developing countries than the average for all developing countries, and in particular for smaller island developing countries. Total outstanding debt grew by less than the average for all developing countries during these four years in only five island developing countries for which information is available; three of these have populations of over 2 million and the fourth is Bahrain, for which no external debt is recorded for the whole period 1975-82.
Broader Excessive foreign public debt of developing countries (#PD2133)
Vulnerability of island developing countries and territories (#PE5700).

♦ PE5751 Transnational monopolies in developing countries
Claim The right to instal and operate, in a developing country, mining equipment, processing plants, factories, or services such as extensive telecommunications networks, has sometimes been purchased by multinational and international firms with payments to government officials, or effected through unfairly conducted bidding. The gains to be made by such arrangements are often the exaggerated operating profits of a firm which has secured a monopoly position. A monopoly foreign ownership of plant, telephone exchanges and the like can be a multiple threat to developing countries: because of the drain on local currency as money leaves the country and flows to the multinational; because the raw materials processed or the telecommunications systems may be a strategic part of defence capabilities but are not under government regulation; because foreign-owned telecommunications operations may not secure privacy and may become adjuncts of their home country's intelligence services; and because the multinational monopolies may become activity involved in influencing the host countries political affairs.

Counter-claim The capital investment of a multinational form to establish itself is a host country is enormous. Projects may take several years and tie-up a considerable portion of a firms financial, technical, manufacturing and manpower resources. Pay-back may be protracted and even uncertain if political instability enters the area. It is therefore legitimate for the multinational to defend its vital interests; neglect of these could cause the failure of the firm, with adverse economic impact on hundreds of thousands of people among international supplies companies, employees and their families, individual investors and institutions. Self-defence legitimately covers counter-efforts to sabotage, whether it is physical, economic or political. Many operations such as mineral extraction from one site, or telecommunications, cannot be performed by two companies at a time, thus the appearance of monopolization is a false one.
Refs Shchetinin, V D U S *Monopolies and Developing Countries* (1985).
Broader Monopolistic activity by transnational enterprises (#PE0109).

♦ PE5754 Fraudulent nature of inherited titles
Claim Inherited noble and honorific titles impute to descendents the virtues and at least potential for similar achievements of their forebears. To a lesser extent, honorifics are also passed to the successive occupants of venerable or powerful offices, without blood-line descent. The passing of noble and honorific titles that are not earned by the recipients is a fraud against society and hereditary nobility is a racist concept when strictly analyzed. While estates pass and may continue to pass, subject to due taxation for the whole society to benefit, hereditary titles are not expiring quickly enough. Taxing the nobility to penury is not a remedy, for there is moral justification for the retention and transmission of wealth. There is insufficient utilization of one alternative to hereditary privilege, lifetime recognition after the models of the British lifetime peer, the Roman Catholic lifetime prince, or the Soviet Union hero. One may take into account, as well, that titles such as PhD and MS, for example, are not transmittable to another generation. Fortunately, people seeking medical advice do not have to contend with hereditary MDs. Considering the political dimensions, the classical world long ago found out the shortage of able men to lead society from among the ranks of hereditary nobility and in some cases instituted legislative and other offices for representatives of the commoners. The provision that one may become a senator or Lord by birthright rather than ability is an appalling one, equivalent to physicians attaining their practice because their fathers had studied and trained for twelve years. The legislative houses of commons or commoners are a democratic necessity; the houses of 'lords' are anachronisms that should be done away with, along with the trappings of hereditary nobility, which include racist ideas, unjustified political power, and an entrenched mentality that obstructs individual human and social development.
Counter-claim Hereditary offices, hereditary functions that are traditional but are not legally established, and hereditary holdings can provide the cohering force of a nation and preserve the national identity. Hereditary holdings and traditional functions guard and maintain the cultural heritage and the fabric of social relations. Hereditary responsibilities of representing constituencies in public office are the only counter-balance to demagogic manipulation of the electorate. Those who must suffer the vicissitudes of periodic re-election will bow before public opinion and fit themselves to short-term emotional issues, but those whose office is taken, not only from hereditary eligibility but from preferential appointment due to ability, stand fast for the longer-term interests of the whole society. Those that attack what they call privilege are in fact attacking responsibility. The ship of state, no matter what its destination under the captain-of-the-moment's hand, will always need stability to challenge the wind and waves of national and international turbulence. The mechanism of preserving tradition by inheritance is the ballast that assures safe passage. The responsibility of a proud lineage is inseparable from the cultivation of the virtue of patriotism: 'shamed be he who thinks evil of it' (motto of the Order of the Garter).
Broader Elitism (#PA1387) Deception (#PB4731).
Related Insufficient nobility (#PF5778).
Aggravates War (#PB0593) Feudalism (#PF2136) Imperialism (#PB0113) Exploitation (#PB3200) Political oligarchy (#PD3238).

♦ PE5755 Excessive cost of medical drugs
Overcharging for proprietary pharmaceutical products — Excessive pricing of brand-name drugs
Incidence The total value, at retail prices, of all pharmaceutical products sold in the world during 1980 was US $90 billion. The German Federal Republic spent US $55 per person per year; Japan, US $40; and the USA, $36; in contrast, Nigeria spent US $1.50 per person per year; India, US $1.00; and Sri Lanka, less than US $1.00. Yet in many countries the drug budget still represents a sizable proportion of the total health expenditure. In developed countries it ranges from 10 percent to 30 percent of the total cost of health care; but, paradoxically, in some developing countries the percentages are much larger, reaching up to 50 percent or more, particularly because of the sales promotion activities of manufacturers which create a demand greater than actual needs. In fact, the problem is considerably accentuated in developing countries because of the launching of nation-wide programmes for the eradication or control of communicable diseases and the development of primary health care programmes, from the grass-roots level, covering large areas and populations. The expenditure on drugs as a percentage of total health out-lays in Bangladesh is 60 percent; in Nepal, 44 percent; in Thailand, 31 percent; in Burma, 25 percent; and in India, 19 percent. The figure for Sri Lanka is, however, only 8 percent. The low figure for Sri Lanka is the result of introducing a rational approach to the procurement, production, supply and use of pharmaceutical products. In the late fifties, Sri Lanka had, in common with other countries, a bewildering variety of commercial brands of drugs aggressively promoted by drug companies. The government undertook the challenging task of devising a sound system of procurement in the international market. An outstanding example of this has the procurement of diazepam, which could be obtained at one-fiftieth (2 percent) of its previous cost.
Burroughs Wellcome was allowed to claim a use patent as well as 7-year exclusive rights for its AIDS drug AZT under the Orphan Drug Act. These are usually medicines that are not marketed because the afflictions they treat are too rare for drug companies to deal with. The company charged $10,000 a year for the AZT that can delay the onset of full-blown AIDS in people who carry the HIV-virus. Pressure from AIDS lobby has helped to force Wellcome to cut the price to $3,000 per patient per year. Monopolies able drug companies to charge high prices for medicines that must be taken for a lifetime.
Claim Pharmaceutical manufacturers take advantage of public ignorance with the cooperation of medical practicioners, to promote their brand-name products as having special value over the scientifically or generically named formulations. A combination of caffeine and analgesic may regularly sell for two or three times as much under a manufacturer's brand-name. Other brand-name pharmaceuticals have occasionally sold for four or five times the price of the generic version. Since drug quality is the subject of international standardization, there is no difference between generic and proprietaries. Doctors, however, continue to prescribe drugs under their proprietary names which leads consumers to purchase them without consideration of choice. The drug companies spend more money promoting to a single doctor than the average annual income of most families in developing countries. Labelling of generic drugs always indicates the manufacturer, so there is no loss of identity, contrary to what some pharmaceutical companies have maintained in their fight against making generic drugs available at lower costs.
Refs Cook, James *Remedies and Rackets* (1976).
Broader Over-pricing (#PE5523).
Aggravates Inadequate medical care (#PF4832) Inconsistent medical practices (#PF1624).
Aggravated by Inadequate information on drugs (#PF0603)
Irresponsible pharmaceutical advertising (#PE2390).

♦ PE5757 Obscene telephone calls
Salacious telephone services
Nature The making of telephone calls, either live or by employing audio tapes with obscene or profane content, may cause great to distress to the receiver and is legally interdicted with civil or criminal penalties in many countries. But evasion of prosecution is achieved with anonymity; and in some countries, if the language is clinical and falsely alleged to be a survey, or a similar false device is employed, it is difficult to bring charges. Obscene telephone calls are characterized by considerable inventiveness. They may be expressed in a language foreign to the receiver, or may even be innocuous while attended by auto-erotic acts. Some degree of compliance by a small percentage of recipients encourages offenders. Prostitutes may offer to talk explicitly about sex over the telephone to anyone who gives them their credit card number and accepts a charge to it, as is practised in the United States.
Incidence In Britain alone about 15 million malicious phone calls are made every year.
Broader Obscenity (#PF2634) Criminal harassment (#PD2067)
Inappropriate use of telecommunications services (#PE4450).
Related Electronic pornography (#PE5402).
Aggravated by Indecent art (#PE5042).

♦ PE5758 Anorexia nervosa
Wilful starvation — Weight phobia — Loss of appetite
Nature The anorexic, usually an adolescent female, has a phobic fear of what she deems as excess body weight. This results in voluntary malnutrition, sometimes self-induced vomiting after eating, and, in some instances, a state of starvation leading to organic damage or death. It is associated with an endocrine disorder leading to amenorrhoea (cessation of menstruation), and may also be characterized by bradycardia (slow pulse rate), lanugo (excessively fine pubic and body hair), periods of hyper-activity, and bulimia (food gorging).
Incidence Karen Carpenter, a famous American singer, died of heart strain due to anorexia nervosa; and in 1985 Jane Fonda, American actress, author, and activist, disclosed that she had been bulemic during her late teens and early twenties.
Refs Andersen, Arnold E *Practical Comprehensive Treatment of Anorexia Nervosa and Bulimia* (1985); Beumont, P J V and Touyz, S W *Anorexia Nervosa and Bulimia* (1985); Brumberg, Joan J *Fasting Girls* (1988); Darby, Padraig L, et al *Anorexia Nervosa* (1983); Duker, Marilyn and Slade, Roger *Anorexia and Bulimia Nervosa* (1988); Erlanger, Ellen *Eating Disorders* (1987); Stierlin, Helm and Weber, Gunthard *Unlocking the Family Door* (1989).
Broader Phobia (#PE6354) Eating disorders (#PE5187)
Physiological malnutrition arising from mental factors (#PE8925).
Related Obsession (#PA6448) Starvation (#PB1875) Over-eating (#PE5722)
Bulimia nervosa (#PE4538) Causing insubordination in the armed forces (#PE5782).

♦ PE5760 Neglect of remote regions and islands
Nature Geography works to the disadvantage of parts of countries that are either divided from the mainland by water (from a few to a thousand or more kilometres), or isolated by mountains, desert and other inhospitable terrain. These lands may be heavily or sparsely inhabited with a population who are usually full-citizens of the state, yet they do not have a comparable standard of living. In fact their standard may be considerably lower than either that of urban or rural populations. Lack of natural resources, local industry or strategic importance perpetuates the lack of interest of central planners. In some cases these regions are the homelands of an ethnic minority whose neglect by the state occasions little political dissent.
Related Isolated islands (#PD2941)
Restriction of indigenous populations to reservations (#PD3305)
Vulnerability of island developing countries and territories (#PE5700).
Aggravated by Geographical isolation (#PF9023)
Discrimination against minorities (#PC0582).

♦ PE5761 Inadequate international participation by island developing countries
Nature Small island developing countries cannot afford to maintain a full range of foreign-based diplomatic services. Furthermore, domestic officials are hard pressed to find time for collecting and analysing regional and international political, economic and other vital data. Consequently, either by omitting to have national representatives abroad or by the inability to utilize documentation and communications from governments and organizations, the island developing countries are an insufficient presence at international conferences, and miss bilateral and multilateral assistance opportunities.
Broader Non-participation (#PC0588)
Transnational corporation control of bulk shipping (#PE5804)
Vulnerability of island developing countries and territories (#PE5700).

♦ PE5762 Breach of contract
Risk of breach of contract — Breach of agreement — Violation of agreements
Nature Violations of agreements either to provide goods or services, or to compensate for them, occur in numerous ways. A binding written agreement, that is, a contract with legal force, stipulates the terms of the transaction which may be between private individuals or organizations, or between either and public agencies of government. Violations or breaches of contract provisions are one of the most frequent causes of legal actions. Any business transaction, therefore, involving contractual obligations, can be tainted with suspicion, if not apprehension, that breach is all too likely. This increases the rigidity of contract terms and increases interest rates to cover risk, which adds to the cost of doing business.
Broader Risk (#PF7580) Violations of contract law (#PE5786).
Narrower Non-implementation of workers wage increases provided for in legislation and collective agreements (#PE4003).
Related Abuse of credit (#PF2166).
Aggravates Risk of capital investment (#PF6572).
Aggravated by Breach of promise (#PF7150).

♦ PE5763 Disease and injury from physical confinement
Broader Human physical suffering (#PB5646) Human disease and disability (#PB1044).
Aggravated by Torture through confinement (#PD4590).

♦ PE5767 Dust as an occupational hazard
Nature Dusts are solid particles generated by handling, crushing, grinding and disintegrating organic or inorganic materials, such as rocks, ore, metals, coal, wood and grains. The exposure of man to dusts can lead to a variety of respiratory diseases, including pulmonary fibrosis, obstructive lung disease, allergy and lung cancer. Toxic dusts may produce systemic poisoning after inhalation, or act as skin irritants to produce dermatoses, allergic reactions or cancer.
Broader Dust (#PD1245) Occupational hazards (#PC6716).
Narrower Mine dust (#PE4807).
Aggravates Silicosis (#PE1314) Tuberculosis (#PE0566)
Heart diseases (#PD0448) Occupational cancer (#PE3509)

Asbestos as a pollutant (#PE1127) Health hazards of asbestos (#PE3001)
Lung disorders and diseases (#PD0637).

♦ PE5768 Shift work stress as an occupational hazard

Inability to adapt to shift work

Nature Shift work creates a psychosocial working environment that may adversely influence the health of the worker. Night work, and the change of working hours from one shift to another, may subject the workers to certain stresses which affect the nervous system, thus increasing the frequency of nervous symptoms such as fatigue, nervousness, irritation and insomnia. These nervous symptoms are usually related to lack of sleep. Shift workers may have a higher incidence of chronic gastritis and peptic ulcer than day workers. The stress associated with night work was found, in a study carried out in Egypt, to contribute to a heightened incidence of high blood pressure.

Psychological and social factors, whether or not combined with physio–pathological disorders, may play a decisive role in a person's inability to adapt to shiftwork. Generally speaking, attention is drawn to the feeling of being cut off from the community and of not participating in social life and collective responsibilities. Moreover, when members of the same family have different work schedules it is particularly difficult for them to get together and organize their family life. Because of the need to work during weekends and on holidays, collective leisure activities, family and neighbourly relations and participation in group entertainment also suffer from shiftwork.

Incidence Research in the field of shiftwork has led to the conclusion that the accident rate decreases during the night while the seriousness of accidents increases. On the other hand, where environmental factors are particularly relevant, as in the case of road transport, the element of risk is much higher during the night, particularly between midnight and 4 am, than during the day. It has also been found that the mental alertness of drivers of railway locomotives declines towards 3 am in actual working conditions. Looking, for example, at the whole of the metal processing industry in France, 27.5 percent of the employees work three shifts and 11.7 percent four or more shifts. In iron and steelworks alone, 70 percent of the manual staff is employed on shift work. Similar figures are found in the paper industry, compared with 60 percent in the chemical industry, compared with 60 percent in the chemical industry and 45 percent in the glassware industry. All in all, 900 000 workers in France have to work at night.

In virtually all epidemiological studies, the most frequently observed disorders have to do with the digestive system, frequently combined with sleep anomalies. Altogether, some 25 to 30 percent of shiftworkers complain of lack of sleep when undergoing clinical tests, either because of difficulty in going to sleep or a tendency to wake up early. In almost 50 percent of the cases covered by studies of intolerance symptoms connected with shift work, lack of sleep is combined with other pathological disorders. In a survey conducted in France in 1973, of the people surveyed, 30 percent suffered little inconvenience from shiftwork; 25 percent encountered some difficulties; 35 percent encountered more serious problems that had repercussions either on their health or on their family life, usually because the wife also worked and 10 percent developed serious sleep and health disorders and found it difficult to organise their family and social life with the result that crisis situations arose that seemed to be caused by a physiological disequilibrium rendering adaptation impossible.

Refs European Foundation for the Improvement of Living and Working Conditions *The Effects of Shiftwork on Health, Social and Family Life*; European Foundation for the Improvement of Living and Working conditions *Shiftwork*; European Foundation for the Improvement of Living and Working conditions *Shiftwork and Accidents*.

Broader Occupational stress (#PE6937) Occupational hazards (#PC6716)
Separation from nature (#PF0379).
Related Night work (#PE7589).
Aggravates Occupational diseases (#PD0215) Occupational risk to health (#PC0865).

♦ PE5771 Juvenile suicide

Adolescent suicide — Teenage suicide

Nature Adolescent suicide and attempted suicide are known in most cultures; however, their proposed etiologies are often not distinguished in the literature from those of the adult acts. Specific to the adolescent's experiences of probable suicide–related stress are: puberty and post–puberty socio–sexual adjustments; social and peer pressures; expectations to excel at school; and the sense of parental domination. Other factors which may influence juvenile suicide are: the death of parents, brothers or sisters; parental behaviour suggesting that the adolescent is unloved or unwanted; and physical or psychological isolation. Special circumstances that may aggravate youth suicide include: imprisonment or detention; drug or alcohol addiction; sexual abuse or exploitation; economic exploitation with no or little payment for excessively long, hard or hazardous work; and life under repressive regimes or other circumstances where educational, economic, and social opportunities are severely restricted.

Incidence Statistics on all suicides are published neither regularly nor uniformly; juvenile suicide statistics appear sporadically and are also subject to scepticism. The absence of reliable data aggravates the problem of insufficient study for preventive purposes.

Claim The failure of sociologists, psychologists, psychiatrists, physicians, and police to document this affliction and communicate it to the educational and social systems, has resulted in an inability among teachers, parents, ministers, and youth organization volunteers and workers to recognize and take seriously the personality disturbances and social turbulence in adolescent lives that may relate to self–destruction.

Refs Diekstra, René F W and Hawton, Keith E (Eds) *Suicide in Adolescence* (1986); Husain, Syed A and Vandiver, Trish *Suicide in Children and Adolescents* (1984).

Broader Suicide (#PC0417).
Aggravated by Teasing (#PE4187) Juvenile stress (#PC0877)
Stress among children (#PE4421) Physical intimidation (#PC2934)
Stress in human beings (#PC1648) Physical intimidation by children (#PE2876).

♦ PE5772 Weakness of infrastructure in island developing countries

Nature Island developing countries must provide their people with as great a range of services, particularly government services, as any other country. Yet given their remoteness, compounded in most cases by their archipelagic character, these services must be provided to small dispersed communities. Island developing countries are therefore inevitably faced with high overheads, including costs of such major basic infrastructure as hospitals, ports or airports. This is increasingly so as world trends in technological development favour increasing scale (as in international transport) and call for increasing specialization. Another infrastructural consequence of remoteness is that these countries must hold larger stocks of a wide range of goods, including essential ones such as foodstuffs and fuel, than must countries with easier access to supplies. A failure to do so could result in shortages, but stocks and shortages involve economic costs.

Broader Inadequate infrastructure (#PC7693)
Weakness of infrastructure in developing countries (#PC1228)
Vulnerability of island developing countries and territories (#PE5700).
Narrower Lack of skilled workers in island developing countries (#PF5737)
Deficient transport in remote island developing countries (#PE5664).
Related Inadequate telecommunications in island developing countries (#PE5655)
Weakness of infrastructure in land–locked developing countries (#PE7000)
Inadequate infrastructure and services in least developed countries (#PE0289).

♦ PE5773 Nutritional ignorance

Claim Most food technologists, health officials and nutritionists agree that partial understanding of the composition of food can create a many problems as it solves. The biggest problem is that no one has more than a partial understanding of the food sciences. The subject of what is good for you and what is bad for you is a mine field of contradictions cloaked in jargon few understand. Furthermore, conventional wisdom about food changes frequently. For example, bread, once considered bad because of its high starch content is now considered good because of its dietary fibre. Potato skins were considered highly nutritional may contain nerve toxin solanine. Fast foods were considered unhealthy are considered by some as as good as any other normal meal. Diet fads have done considerable harm. Food poisoning, the risk of which can be minimized, aside there is little mortal danger in the industrialized world from the food eaten provided a balanced diet is kept in mind and risks are not taken during preparation.

Broader Limited medical knowledge (#PD9160) Ignorance of health and hygiene (#PD8023).
Aggravates Nutritional diseases (#PD0287) Nutritional deficiencies (#PC0382)
Ignorance concerning disease (#PD8821)
Malnutrition among indigenous peoples (#PC3319)
Inefficient use of proteins in factory farming (#PF2758).
Aggravated by Inadequate nutrition education in least developed countries (#PE0265).

♦ PE5776 Excessive stray animal populations

Narrower Stray dog populations (#PE0359).
Aggravates Feral mammals (#PE0185) Inhumane killing of stray animals (#PE2759).
Aggravated by Unwanted pet animals (#PD2094).

♦ PE5777 Manipulation of students

Claim The fear of legitimate student organization causes representatives of the status quo to exert themselves in subverting the energy and voice of youth. School trustees, administrators and tenured faculty entrench themselves with quasi–legal powers of summary justice. Students may be cowed by threats of political blackmail, grade distortion, suspension, expulsion, scandal, or malicious prosecution for such things as trespass, nuisance, breaking into school property, or damages. Students may also be forced to report on others' words and actions, and even to bully and actually physically harm their fellows. Counsellors and chaplains fail to be students' advocates and the administration uses parents to control the minds and behaviour of their adult offspring. Where military training is given in conjunction with academic subjects, the weight of the military system of injustice is brought to bear, with physical penalties of compulsory exercise, labour or other harassment.

Students may also be manipulated by natural and international political or religious organizations, who infiltrate the student body and subsidize the presentation of their ideologies by supposed students whose real roles, however, are those of missionaries and recruiters. Students are also manipulated by organized crime, which finds the campus an ideal market for drugs. For the individual higher education is an opportunity, but higher educational institutions are a threat.

Broader Manipulation (#PA6359).
Narrower Political impotence of students (#PE5729).
Aggravated by Unrepresentative international organizations (#PD4873).

♦ PE5780 Improper lighting as an occupational hazard

Poor lighting

Nature The assessment of lighting conditions at work must include not only the light intensity but also other characteristics, such as shadows, contrasts and colour. The desirable quantity of light depends on the fineness of the detail and the accuracy required in performance of the task. With regard to the quality of light, many complex factors are involved such as glare, diffusion of light, direction, uniformity and distribution. Dim light associated with high visual demands may lead to eye strain and fatigue. Exposure to the dim light of inadequate illuminated work–places or to the partial darkness of a mine or a darkroom for eight hours a day over a long time, can cause both acute and chronic effects on health. Dim work–places cause headache, eye pain, lachrymation and congestion around the cornea, particularly associated with eye strain from trying to see small objects. Dark work–places cause miner's nystagmus. Distraction from visual tasks and loss of concentration may result from direct glare. This kind of glare is also associated with discomfort, annoyance, and visual fatigue. Intense direct glare may also result in temporary loss of visual ability, as in the case of drivers exposed to direct intense light from on–coming cars at night. Other kinds of glare include the indirect glare from an intense light spot which may cause blurring of vision. Reflected glare from shiny surfaces or dials can obscure details and prevent perception of visual displays. In the occupational environment, intense colours may result in fatigue of certain retinal cones.

Broader Occupational hazards (#PC6716).
Related Ultraviolet radiation as a hazard (#PE5672).
Aggravates Legal havens (#PE0621) Occupational diseases (#PD0215)
Occupational risk to health (#PC0865)
Health risks to workers in agricultural and livestock production (#PE0524).

♦ PE5782 Causing insubordination in the armed forces

Nature Intentionally causing insubordination, mutiny or refusal to perform duties of any military person is a crime.

Broader Crimes related to military service obligations (#PD5941).
Related Draft evasion (#PD0356) Anorexia nervosa (#PE5758).

♦ PE5783 Thrombosis

Blood clots

Refs Chazov, E I and Smirnov, V N *Thrombosis and Thrombolysis* (1986); Hume, Michael et al *Venous Thrombosis and Pulmonary Embolism* (1970); Verstraete, M and Vermylen, J *Thrombosis* (1984); WHO Scientific Group – Geneva 1971 *Inherited Blood Clotting Disorders* (1972); Webber, Kathleen C *Thrombosis* (1987).

Broader Diseases of blood and blood–forming organs (#PF8026).
Aggravated by Anaemia (#PD7758) Human ageing (#PB0477)
Malignant neoplasms (#PC0092) Excessive consumption of fats (#PE4261).

♦ PE5784 Disaster hazards to island populations

Incidence An UNCTAD study of natural disasters (earthquakes, tropical cyclones and floods) mainly between 1977 and 1981 showed that in 11 cases out of 54, over 5 percent of the population of the island country affected was made homeless. Of these 11 cases, eight were island developing countries. Similarly, damage as a percentage of GNP could be estimated in 38 cases. In nine cases it was over 5 percent and, of these, eight were island developing countries.

Broader Natural disasters in least developed countries (#PE0299).
Related Isolated islands (#PD2941).
Aggravates Vulnerability of island developing countries and territories (#PE5700).
Aggravated by Uncoordinated disaster relief (#PF6022)
Uncontrolled costs of disaster relief (#PF6025)
Fraud concerning economic situation and corporate capital of companies (#PE5021).

♦ PE5785 Military threats to island countries and territories

Nature Islands have traditionally been battlegrounds for rival sea powers, or for maritime invaders

versus natives, because of ease of access by sea.
Incidence A great many islands are pluralistic societies. Some are calm, some have continuing fighting factions, some have been repeatedly invaded. Although their histories are very different, major European islands that can be cited are Great Britain, Ireland, Greenland and Iceland in the Atlantic; and Sicily and Cyprus in the Mediterranean. During World War II, the Pacific theatre came to be a predominantly island battle ground. Disputed sovereignty exists or has existed for many islands such as the Falkland Isles (Malvinas) and the Kuril, and argument extends to the Antarctic, which is an entire continent.
    Broader Geopolitical vulnerability (#PF5749).
    Government intimidation of governments (#PE1622).
    Related Military threats to land-locked countries (#PE5837).
    Aggravates Vulnerability of island developing countries and territories (#PE5700).

♦ **PE5786 Violations of contract law**
Nature Legal actions brought in civil courts arising from commerce and trade include failures to comply with provisions of business contracts. This legal process may refer to failures to deliver according to terms, unpaid debts, fraud, or damages wrought by legal or self-appointed agents of the plaintiff. Other litigation applies to controversies involving loan transactions and insurance claims of all types including maritime. The frequency of such contract litigation is an affront to the moral self-image of society, and raises questions concerning the effectiveness of contract law as it applies both to local and international trade.
    Broader Violations of private law (#PD5727).
    Narrower Breach of contract (#PE5762).
    Aggravated by Inadequate national law enforcement (#PE4768).

♦ **PE5787 Harmful effects of sensory deprivation**
Nature The senses are a vital connection between body and mind; they are the link between the mental and the elemental. Loss of contact with the senses – sensory deprivation – may be the cause of a wide range of personal, social and environmental problems. The absence of focused smelling, tasting, touching, seeing and hearing can lead to the mental distraction that results in irritability, anger, fatigue and boredom.
Incidence In a recent research project at Shenley Hospital, London, work in the "neurosis unit" recorded, over an eight month period, a dramatic fall in the use of anti-depressants, tranquillisers and sleeping pills when such techniques as sensory awareness and spiritually creative activities were employed.
Background The ancient Indian vedic school of thought recognizes that the elements, as well as being linked to the senses (water to taste, earth to smell, fire to sight, air to touch, and space (or ether) to hearing) are also representative of positive and negative mental tendencies, emotional aspects and personality characteristics.
    Broader Human physical suffering (#PB5646)   Human disease and disability (#PB1044).
    Aggravated by Torture through sensory deprivation (#PE6797).

♦ **PE5789 Lack of adherence to international transit conventions for land-locked countries**
Nature There are a number of international conventions which are relevant to transit, but to which most of the land-locked countries and their transit neighbours have not adhered. For example, only one land-locked country and two transit countries are contracting parties to the Customs Conventions of 1950 and 1975 on the International Transport of Goods Under Cover of TIR Carnets. Two land-locked countries and four transit countries have adhered to the Customs Convention on Containers of 1956 and 1972. None of the land-locked developing countries and transit countries are contracting parties to the International Convention to Facilitate the Crossing of Frontiers for Goods by Rail of 1952 (TIF). While 14 land-locked countries are contracting parties to the Convention on Transit Trade of Land-locked States of 1965, only four transit countries have so far adhered to it. This situation is disconcerting since these conventions, if adhered to and complied with, would contribute considerably to removing some of the bottlenecks currently constraining regional transit traffic.
    Broader Unfulfilled treaty obligations (#PF2497).
    Vulnerability of land-locked developing countries (#PD5788).

♦ **PE5790 Lack of capital investment in developing countries**
Lack of investment capital for developing countries' shipping fleets
Claim The lack of finance for ship acquisition was and remains a major difficulty of developing countries in expanding their national merchant marines. The fact that soft financing for ships has recently been amply available does not necessarily affect the need for general improvements in financing arrangements. Soft financing instruments now available are basically subsidy programmes for ailing shipbuilding industries designed as a temporary measure. Consequently, similar arrangements are not available for the purchase of second-hand tonnage. Furthermore, these instruments would no longer be available should governments of shipbuilding countries, for whatever reason, decide to cease or change the basis of subsidization.
Counter-claim Shipbuilding by developing countries adds to the overtonnage crisis.
    Broader Insufficient capital investment (#PF2852).
    Obstacles for developing countries ocean shipping (#PE5909)
    Inadequate level of investment within developing countries (#PD0291).
    Related Obstacles for international ocean shipping (#PD5885).
    Aggravated by Abusive traffic in immigrant workers (#PD2722)
    Decline in commercial bank lending to developing countries (#PE4655).

♦ **PE5791 Instability of the maritime shipping industry**
Nature The changing pattern of international trade in 1980s, combined with important technological developments, has led to structural as well as technological changes in the world shipping industry, which is the main means of transport in international trade. The world merchant fleet is unable to respond quickly to a rapid decline of seaborne trade because existing tonnage cannot easily be redeployed, while contractual obligations regarding ships under construction or ordered cannot be broken without substantial financial penalties. This characteristic of maritime transport may lead to excess tonnange situations – as in mid-1970s when seaborne trade started to diminish, especially in the petroleum trades. The resulting inbalance between supply and demand reached its peak in 1983, when the average annual figure of estimated surplus tonnage reached 28.5 per cent of the world fleet. The physical volume of world international trade stood at the lowest level since 1975, at 3.2 billion tons. This declining trend continued until 1988. The main cause of the prolonged shipping crisis was the very high level and the scale of government financial subsidies in the developed market-economy countries as well as indirect support measures (e.g. through fiscal arrangements) for the acquisition, operation and, in particular, building of new ships during the past decade.
    Broader Obstacles for international ocean shipping (#PD5885)
    Instability of economic and industrial production activities (#PC1217).

♦ **PE5792 Inadequate port infrastructure**
Nature Many ships spend over 50 percent of their operational time in ports and, for liner shipping, port costs represent some two-thirds of the total cost of sea transport. Thus, conditions in port have a profound impact on the efficiency of maritime transport. Whereas shippers can often choose between different shipping services, they are generally obliged to use a particular port and are therefore at the mercy of its efficiency. When such a port lacks adequate facilities, trade can ultimately be stifled.
    Broader Insufficient transportation infrastructure (#PF1495).
    Related Inadequate maintenance of infrastructure (#PD0645).
    Obstacles for developing countries ocean shipping (#PE5909)
    Obstacles to efficient port utilization and operation (#PE5921).

♦ **PE5793 Violation of the right of workers organizations to protection against suspension**
Violation of the right of trade unions to protection against dissolution
    Broader Violation of trade union rights (#PD4695).
    Narrower Economic uncertainty (#PF5817).
    Aggravated by Unethical practices of employers (#PD2879).

♦ **PE5794 Inadequate legislation for animal welfare**
Lack of legislation for animal welfare — Lack of animal welfare legislation
    Broader Inadequate laws (#PC6848)   Deficiencies in international law (#PF4816)
    Deficiencies in national and local legal systems (#PF4851).
    Narrower Denial to animals of legal protection of their rights (#PE8643).
    Aggravates Lack of care for animals (#PD8837)   Inadequate animal welfare (#PC1167)
    Inadequate feeding of animals (#PC2765)   Cruelty to animals in factory farming (#PD2768).
    Animal malformation in factory farming (#PD2761)
    Inadequate housing and penning of domestic animals (#PE2763)
    Excessive commercial exploitation of farm animals by industrial concerns (#PD2772).

♦ **PE5795 Story that self-sufficiency is socially responsible**
Nature The major contradiction in the area of corporate welfare is the myth that to be self-sufficient is the defining factor for being human. This extension of the lie of individualism attacks the interdependence of individuals and of societies. Structures that support relationships of mutual dependence are undermined. Individuals who recognized their dependence on society are thought to be suspect. Wanting to help others is at odds with the prevailing morality.
    Broader Overemphasis on self-sufficiency with respect to interdependence (#PF3460).

♦ **PE5796 Social irresponsibility of transnational corporations**
    Broader Irresponsibility (#PA8658)   Transnational corporation imperialism (#PD5891).
    Narrower Retarding of development by transnational corporations (#PE0234).
    Bribery by transnational enterprises in developing countries (#PE0322)
    Participation of transnational corporations in the apartheid system (#PE1996)
    Disruption of domestic social policies by transnational corporations (#PE1957)
    Minimal export promotion by transnational corporations in developing countries (#PE1598)
    Socially irresponsible programmes of transnational banks in developing countries (#PE4360)
    Increasing income disparity in developing countries due to transnational corporations (#PE1660)
    Disruption of cultural and social identities in developing countries by transnational corporations (#PE1082)
    Inadequate relationship between transnational corporations and local industry in developing countries (#PE1511).
    Aggravated by Unaccountability of institutions degrading the environment (#PF3458).

♦ **PE5798 Vulnerability of intravenous drug users to AIDS**
    Broader Man as vectors of disease (#PD8371).

♦ **PE5799 Aircraft noise**
Airport noise — Jet noise
Nature Aircraft noise is environmental pollution. Advances in the quietening of aircraft are entirely negated by increases in the number of aircraft and longer airport operating hours. The main factors determining the nuisance are the noise peak levels, the frequency of disturbance and the time of day.
Incidence Near Heathrow airport in UK more than 700,000 people live above reasonable level of 35 noise units.
    Broader Pollution by aircraft (#PE4802)   Aircraft environmental hazards (#PD8328)
    Health hazards of exposure to noise (#PC0268).
    Narrower Sonic boom generated by supersonic aircraft (#PE2435).
    Aggravates Deafness (#PD0659)   Inefficiency (#PB0843)
    Mental illness (#PC0300)   Stress in human beings (#PC1648).

♦ **PE5800 Inadequate air transport service for land-locked developing countries**
Nature Air transport is an important supplementary mode for alleviating the transit transport problems of land-locked developing countries. Up to now, outgoing air freight from the land-locked developing countries has consisted primarily of either perishable goods, such as fruit, vegetables, fishery products and flowers, or high value goods such as skins or leather, carpets and precious metals. However, because of the unreliability of surface transit transport services, a number of land-locked developing countries do airlift some of their bulkier commodities on an irregular basis. The development of air freight operations is, however, hampered by various obstacles. Considerable investments are required to extend and strengthen the runways at main international airports and to provide adequate navigational aids and cargo handling facilities. These improvements are necessary to permit the gradual introduction of the larger and newer four-engined and wide-bodied jets on scheduled passenger services: such aircraft have larger freight holds and lower costs per tonne-kilometre of capacity than those now used in many of the land-locked countries. An additional obstacle to air freight operations is the lack of an adequate surface collection and distribution system radiating from the major airport (s) of the land-locked states. Without such a system, the hinterland of the airport is severely limited and even goods travelling by air cannot benefit fully from the advantages offered by air transport. Skilled manpower constraints are also a major obstacle. Financial risks for individual airlines are great, partly because of directional imbalance in traffic, and airlines are reluctant to cooperate in land-locked countries' development plans.
    Broader Vulnerability of land-locked developing countries (#PD5788)
    Weakness of infrastructure in land-locked developing countries (#PE7000).
    Related Inadequate air transport service (#PJ0260).

♦ **PE5801 Lack of world maritime integration**
Obstacles to effective use of oceans and waterways
    Broader Fragmentation (#PA6233)
    Inadequate integration of transport systems (#PF6157).
    Narrower Inadequate laws of the sea (#PF5923)
    Imbalance in shipbuilding industry distribution (#PE5863)
    Obstacles to efficient port utilization and operation (#PE5921).
    Related Obstacles for international ocean shipping (#PD5885)
    Lack of technical infrastructure for maritime commerce in developing countries (#PE5814).

♦ **PE5802 Inequities in marine insurance**
Nature There are few internationally agreed rules for marine insurance. Each market uses either its own insurance policy terms and conditions or adopts or amends those of another leading

market drawn up by insurers without formalized consultation with the assured. Inconsistent application of international insurance conventions is aggravated by delays in compliance. There has been a fairly consistent bias favouring shipowners in regard to loss and damage to goods. Broadly speaking, shipowners are able to claim exemptions from liability to compensate shippers for loss, damage or delay to goods without proving that they took all measures that could reasonably be required to avoid the loss. Also the monetary limitation of liability does not account for inflation. Treatment of cargo claims and settlements is unordered, flowing from a regime in whose formulation shippers and shipowners from all countries did not participate on an equal footing.

Broader Restrictive shipping practices (#PD0312)
Restrictive transport insurance practices (#PD0881)
Obstacles for international ocean shipping (#PD5885).
Related Compliance*complex (#PA5710).
Aggravates Inadequate insurance (#PF8827).
Aggravated by Maritime fraud (#PE4475) Cargo insecurity (#PE5103)
Seizure of cargoes (#PE0125).

♦ PE5804 Transnational corporation control of bulk shipping

Nature Where transnational corporations (TNCs) are involved with the extraction of minerals, food growing, and the use or conversion of other natural commodities, it has been in their interest to own or control the means of transporting these high bulk products in order to reduce costs. In bulk trade shipping most indicators point to a high degree of control by TNCs. They exercise their control partly because they are involved in vertically-integrated operations, partly because they have participation agreements or long-term contracts with supplier countries which give them the right to control transport, and in some trades because they operate in a buyer's market and can insist on buying on FOB terms.

Incidence Shipping company affiliates are used by transnational oil companies for the purpose of diverting profits to tax haven countries by means of transfer prices. The affiliation of shipping companies of developed countries to buyers, an aspect of the control being exercised by transnationals and large trading organizations and their integrated operations, is sufficient to create a situation in which developing countries' fleets are effectively excluded from participation. UNCTAD reports indicate that dry bulk cargo movements are also dominated by transnationals using the instrument of chartering, as 'markets', 'exchanges' or 'parties'.

Claim Bulk cargoes appear to a large extent to be captive cargoes of TNCs to the disadvantage of exporting and importing developing countries.

Broader Obstacles for international ocean shipping (#PD5885)
Control of industries and sectors by transnational corporations (#PE5831)
Domination of the shipping industry by transnational corporations (#PE1620).
Narrower Inadequate international participation by island developing countries (#PE5761).
Related Restrictive shipping practices (#PD0312)
Transnational corporation imperialism (#PD5891).
Aggravates Imbalances in the dry bulk shipping industry (#PE5839).

♦ PE5808 Inviability of tropical island developing countries

Broader Vulnerability of island developing countries and territories (#PE5700).
Related Compliance*complex (#PA5710).
Aggravated by Destruction of coral reefs (#PD5769).

♦ PE5812 Excessive costs of sea access for land-locked developing countries

Nature Coastal states in their function as transit states, that is in providing transport services, facilities, or at least port-access, to land-locked countries, are in the position of a monopolistic enterprise. Tariffs for road or rail use, and to a lesser extent for utilization of inland waterways and air fields, and also for use of cargo and port storage, handling and other facilities, are often arbitrary, variable and at excessive levels. An UNCTAD survey revealed wide variations in the relative transit costs for different commodities, with the comparatively low-value high-bulk commodities showing the largest percentage mark-ups of transit costs. The result is reduced incomes for the producers of commodity exports and raised import costs, thus hitting hardest at the broader, poorer section of the population. On the other hand, many coastal states are themselves only now developing, and inefficient or inadequate infrastructures raise the costs for sea access without there being any possibility of subsidizing their neighbours. Hence the excessive costs of transit are, in fact, regional problems.

Refs Glassner, M I *Access to the Sea for Developing Landlocked States* (1970).
Broader Vulnerability of land-locked developing countries (#PD5788).

♦ PE5813 Vulnerability of lakes and rivers in land-locked countries

Nature The small amount of fresh-water in land-locked countries may lead to depletion of fish stock and over-utilization of water areas for recreation. In industrialized land-locked countries, lake and river pollution is a threat.

Broader Vulnerability of land-locked developing countries (#PD5788).

♦ PE5814 Lack of technical infrastructure for maritime commerce in developing countries

Nature The absence of modern ports, efficient cargo handling systems, and internal rail or road distribution networks are obstacles for shippers and shipping industries. The deficiencies are aggravated by a lack of qualified people to create, maintain and manage such an infrastructure.

Broader Inadequate maintenance of infrastructure (#PD0645)
Obstacles for international ocean shipping (#PD5885)
Insufficient transportation infrastructure (#PF1495).
Related Lack of world maritime integration (#PE5801).

♦ PE5815 Port delays

Nature Slowness and inefficiency of port operations is an obstacle to increased foreign trade, particularly for developing countries. Many ports have failed to keep pace with the rate of expansion of the country's overseas and coastal trade. The initial result is port congestion, with significant increases in trading costs. Disruption of trade follows, and eventually trade is forced to adapt to the limited port capacity with a permanent detrimental effect on the national economy. The appearance of a scarcity of tonnage can result from port congestions and slow turnround of ships at ports. This in turn can lead to excessive shipbuilding and an overtonnage situations as port management improves its traffic handling.

Broader Delays in delivery of goods (#PF8268)
Obstacles to efficient port utilization and operation (#PE5921).

♦ PE5819 Protectionism in the textile and apparel industries

Nature Motivated by a desire to protect jobs in their own countries, governments of many industrialized nations have introduced restraints against developing countries' exports of manufactures, primarily in the textile and clothing industries. Such restraints obviously hinder the progress from developing to developed, as textiles and clothing constitute almost 30 percent of the manufactured exports of developing countries. The restraints also, however, entail high costs to the industrialized countries themselves as their governments have had to pay benefits to those workers made redundant.

Incidence In the USA it has been found that a permanent policy of tariff protection would cost, per job, 14 times more than it would provide in private benefit to the individual worker; it would cost US $1 for every 7 cents gained by workers whose jobs were preserved. In Canada, the ratio was 70 to 1; for every 1.5 cents by which the worker would be better off, one Canadian dollar would be wasted.

Background Textile and apparel exports have been essential in the development of many countries, from the UK in the 19th century to Japan in the early 20th, to Korea in the 1960's and 1970's.

Claim Protectionism leads to distortions in the structure of production, consumption and trade, and these are costly to economic growth. Many developed countries are willing to donate "humanitarian" aid to developing countries in need, but they refuse to allow the freedom for those same countries to become economically viable in their own right.

Counter-claim Given the high numbers of unemployed in most industrialized countries today, why should they allow competition from non-industrialized countries to threaten their work forces, many of which, anyhow, are comprised of immigrant workers from the non-industrialized countries themselves.

Broader Protectionism in international trade (#PC5842).
Aggravated by Unethical practices in the apparel industry (#PD8001).

♦ PE5820 Inappropriate transfer of technology

Introduction of inappropriate technology — Adverse impact on technology transfer by transnational corporations

Claim Most developing countries experience an intellectual enslavement and technological dependence on developed countries, largely through the aggressive policies of the latters' transnationals. The purely materialistic approach they follow erodes economic systems, mutilates indigenous skills and craftsmanship, degenerates value systems, and nurtures an infectious "nationhood" concept in the narrow geographical sense.

Refs Tuomi, Helena and Värynen, Raimo *Transnational Corporation, Armaments and Development* (1980); United Nations Centre on Transnational Corporations *Transnational Corporations and Technology Transfer* (1987); United Nations Centre on Transnational Corporations *Measures Strenghtening the Negociating Capacity of Governments in their Relations with Transnational Corporations* (1979).

Broader Transnational corporation imperialism (#PD5891).
Narrower Monopolization of technology by transnational corporations (#PE1918).
Aggravates Underemployment (#PB1860)
Uncontrolled application of technology (#PC0418)
Increasing development lag against technological growth (#PE3078).

♦ PE5823 Inadequate conditions of work in the textile industry

Refs ILO *Conditions of Work in the Textile Industry, Including Problems Related to Organization of Work.*

Broader International imbalance in the quality of life (#PB4993).

♦ PE5824 Inadequate roads and transport in land-locked developing countries

Nature Despite the important role played by the road haulage industry in the international transport between land-locked developing countries, its efficiency is impaired by a number of factors. Basic shortages exist in all-weather, through-road transport routes linking the main commercial centres in the land-locked countries with the seaports, and in the numbers of transport vehicles. The maintenance of the road network is generally poor, due to a shortage of the requisite technical skills, as well as to inadequate supplies of maintenance equipment, facilities and spare parts. The deterioration of the road situation is compounded by the fact that the technical specifications for existing roads are not properly observed because of poor enforcement and surveillance arrangements. In several cases, there are no road transit transport agreements between the land-locked countries and their transit neighbours. Where such agreements do exist, provisions concerning vehicle dimensions, axle loading specifications, the equitable sharing of inter-state traffic, road permits and visa procurement are often cumbersome and difficult to implement. There are frequently no through transport routes from the transit port to the final destination in land-locked countries, so cargo transfers have to be made.

Broader Vulnerability of land-locked developing countries (#PD5788)
Weakness of infrastructure in land-locked developing countries (#PE7000).
Related Overloaded vehicles (#PE4127).

♦ PE5825 Over-concentration of ownership of maritime fleets

Imbalance of power in the shipping industry

Nature In the international shipping industry there are two dominant types of economic power concentration, one on the level of international division of labour in shipping (that is, the concentration of ownership of fleets in privileged hands) and the other on the level of the industry. The developed countries have retained, from colonial times, the highest concentration in shipping. They effectively control some 80 percent of the supply of world shipping. The high concentration of fleet ownership in the hands of the traditional maritime countries has impaired the ability of developing countries to compete and increase their participation in shipping operations. At the level of the industry, concentration in shipping takes place through both internal and external growth, the latter being achieved through merger and acquisition of other firms and also through cooperative agreements such as pools or consortia. It is also achieved through the formation of shipping cartels, the main purpose of which is to reduce price competition while at the same time guaranteeing a certain level of service. Internally within the shipping industry in general, the concentration of power varies, and is more pronounced in the liner sector than in the bulk market.

Broader Obstacles for international ocean shipping (#PD5885).

♦ PE5827 Political interference with port operations

Nature A port entity may receive instructions from the public authorities to give particular treatment to certain national port-users (owners of either ships or cargo). This treatment may limit the port charges which can be applied and hence the level of the benefit that can be reallocated. The government may also issue an order to the port authorities prohibiting the departure of specified ships. Without declarations of war, ports may sometimes be temporarily blockaded, either close at hand or in the sea lanes, in order to obstruct exit or entry, or in the pursuit of those suspected or identified as engaged in illegal or hostile activity. Quarantine may also close port operations down, and may be a political instrument.

Broader Obstacles to efficient port utilization and operation (#PE5921).
Narrower Distortion of international trade by embargoes and similar restrictions (#PE0522).
Aggravates Obstacles for international ocean shipping (#PD5885).

♦ PE5831 Control of industries and sectors by transnational corporations

Refs United Nations Centre on Transnational Corporations *Transnational Corporations in the International Auto Industry* (1983); United Nations Centre on Transnational Corporations *Transnational Corporations Linkages in Developing Countries The case of Backward Linkages via Subcontracting* (1981); United Nations Centre on Transnational Corporations *Transnational Corporations in the International Semiconductor Industry* (1986); United Nations Centre on Transnational Corporations *Transnational Corporations in International Tourism* (1982); United

Nations Centre on Transnational Corporations *Transnational Corporations in the Copper Industry* (1981); United Nations Centre on Transnational Corporations *Transnational Corporations in Food and Beverage Processing* (1981); United Nations Centre on Transnational Corporations *Transnational Corporations Linkages in the Bauxite/Aluminium Industry* (1981); United Nations Centre on Transnational Corporations *Transnational Corporations in the Fertilizer Industry* (1982); United Nations Centre on Transnational Corporations *Transnational Corporations in the Agricultural Machinery and Equipment Industry* (1983); United Nations Centre on Transnational Corporations *Transnational Corporations in the Pharmaceutical Industry of Developing Countries* (1984); United Nations Centre on Transnational Corporations *Transnational Corporations in the Power Equipment Industry* (1982); United Nations Centre on Transnational Corporations *Transnational Corporations and Contractual Relations in the World Uranium Industry* (1983).
Broader Transnational corporation imperialism (#PD5891).
Narrower Unknown relatives (#PF0782)
Transnational corporation control of bulk shipping (#PE5804)
Domination of loan negotiations by transnational banks (#PE4355)
Domination of advertising by transnational corporations (#PE2193)
Domination of the copper industry by transnational corporations (#PE2094)
Domination of the shipping industry by transnational corporations (#PE1620)
Domination of the automobile industry by transnational corporations (#PE1469)
Excessive control of raw materials markets by transnational corporations (#PE0194)
Domination of agricultural equipment industry by transnational corporations (#PE1448)
Excessive exploitation of raw material reserves by transnational enterprises (#PE0060)
Control of marketing and distribution channels by transnational corporations (#PE2397)
Domination of restrictive project loans by transnational banks in developing countries (#PE4330)
Import-dependency in food staples in developing countries due to transnational corporations (#PE1806)
Domination by transnational corporations of the domestic name-brand food sector in developing countries (#PE1796).
Related Instability of the primary commodities trade (#PC0463)
Evasion of shipping regulations and taxes by flags of convenience (#PE5873)
Discrimination against transnational banks in developing countries (#PE4320)
Restrictions of home countries on transnational banking activities in developing countries (#PE4335).
Aggravates Discrimination in lending by transnational banks in developing countries (#PE4310).

♦ **PE5832 Police immunity**
Nature Criminal activities done by police officers are less likely to be punished than those of the general populace. Police are trained in the use of weapons, have personal friends among the officers and lawyers who may be required to testify against them, are well acquainted with the loopholes in criminal law, and can experience situations in which their lives depend on their personal ability to protect themselves. In some places, investigation of police activities is done by the police themselves.
Broader State immunity (#PE4930).
Related Abuse of privileges and immunities by diplomats (#PF5649).

♦ **PE5836 Unreliable transit services for land-locked developing countries**
Nature With the very poor economic performance of land-locked developing countries and the severe shortages of foreign exchange availabilities and domestic resources, these countries continue to fail to meet their requirements for maintaining the operational capability of the existing transit transport facilities. The limited resources they have are generally devoted to such pressing needs as food and fuel, with the result that the maintenance of transit facilities has often had to be deferred. This has led to the deterioration of facilities, which in many cases then need to be totally rehabilitated and overhauled. The transport of goods through a foreign territory risks their diversion into the domestic market, thereby evading import controls and tariffs. This necessitates in several cases the restriction by transit countries of the movement of transit traffic to specific modes of transport and routes and the requirement of a security bond deposit. More often, there are no clear analyses of present and prospective demands made on each existing or potential transit route by either the land-locked or the transit countries, nor the management and maintenance needs for each actual or proposed transit route. Storage capacities on each corridor may be non-existent or inadequate and there is usually insufficient establishment of the procedures for each transit route, as well as inadequate existing legislation, documentation, conventions and working agreements to govern international transport under all conditions. Detailed direct and indirect costs for both operators and users at each stage of each major transit route are not calculated and overall rate structures for routes are inconsistent. There is, characteristically, little concrete forward planning based on the capacity of each transit corridor, including consideration neither of short-term alternative cargo mixes and cargo types in the present system, nor of significant long-term changes in infrastructure, investment patterns, commodity markets and foreign trade.
Broader Vulnerability of land-locked developing countries (#PD5788)
Weakness of infrastructure in land-locked developing countries (#PE7000).

♦ **PE5837 Military threats to land-locked countries**
Claim All land-locked countries are militarily weak. Their greatest unilateral defence is neutrality or insignificance. This is less safe, however, than peaceful cooperation in regional development built on the principal of interdependence of economies.
Broader Government intimidation of governments (#PE1622)
Vulnerability of land-locked developing countries (#PD5788).
Related Military threats to island countries and territories (#PE5785).

♦ **PE5839 Imbalances in the dry bulk shipping industry**
Nature The major bulk commodities all present a natural field for vertically integrated operations by transnational corporations. In the case of iron ore, coal and bauxite, large capital investments are needed to develop mining sites and associated infrastructures, and it is necessary to ensure close coordination among all stages of operations, from mining right through to consumption. Even in the case of grain, where production is often in the hands of individual farmers or farming companies, capital expenditure is needed for collection, storage and port terminal facilities and a similar degree of coordination is required. Ocean transport is an important link in the chain of coordination, but it is also a profit-making activity; in fact where mineral transport involves the dedication of specific bulk carriers to specific routes over long periods, the profitability of the shipping operation is virtually guaranteed. The transnational corporations thus have a double reason for becoming involved in ocean transport.
An added, and equally important, reason is that, in vertically integrated operations, ocean transport can be used for transfer pricing with respect to freight costs by which a corporation can minimize its tax payments. Control also can be exercised through long-term charters and contracts of affreightment, and arrangements with closely related parties can easily provide (covertly, if not overtly) for participation in the profits and for advantageous transfer pricing. In fact, a transnational corporation may achieve the same results by negotiating privately with independent shipowners, especially if it uses open-registry ships whose owners reside outside the jurisdiction of the flag states and are not subject to substantial reporting requirements. It may even be to the advantage of a corporation to use an independent vessel, or one which appears to be independent, rather than a company-owned vessel, in order to avoid allegations of misuse of transfer pricing. The use of independent shipowners also gives transnational corporations greater flexibility in dealing with irregular shipments.

Broader Obstacles for international ocean shipping (#PD5885).
Aggravated by Unfair shipping practices in bulk trades (#PE5849)
Transnational corporation control of bulk shipping (#PE5804).

♦ **PE5844 Frotteurism**
Frottage — Erotic friction
Nature Sexual stimulation by rubbing against other persons is practised by both men and women, and by adults as well as adolescents. As a regular substitute for sexual relations with a voluntary partner, frotteurism, like voyeurism and masturbation, may be a perversion. Apprehensiveness concerning this undesirable experience is a great cause of social inconvenience for many women wishing to avoid crowds. The presence of frotteurs on public conveyances causes additional use of private automobiles by many to evade this form of indecent assault. Many women workers find superiors leaning over them practicing this form of sexual harrassment.
Claim Rubbing against people for sexual stimulation is usually done by rather sick men. It is aggressive and a symbol of intended or existing dominance.
Counter-claim Erotic friction has been observed wherever conditions force people into close proximity, standing in a crowd or sitting. Many consider it a normal and pleasurable part of social dancing.
Broader Sexual deviation (#PD2198).

♦ **PE5846 Post abortion syndrome**
Nature Women who have had abortions suffer from a specific "post-traumatic stress disorder" in which the destruction of their own child is sufficiently traumatic and beyond the range of customary human experience so as to cause significant symptoms of reexperience, avoidance and impacted grieving.
The abortion trauma is reexperienced as recurrent and intrusive distressing recollections of the experience; recurrent distressing dreams of the abortion or of the unborn child; sudden acting or feeling as if the abortion were recurring, accompanied by illusions, hallucinations and dissociative episodes especially upon awakening or when intoxicated; intense psychological distress at exposure to events that symbolize or resemble the experience (eg, clinics, pregnant mothers, subsequent pregnancies); anniversary reactions of intense grieving or depression. There is persistent avoidance of stimuli associated with the abortion trauma or numbing of general responsiveness (not present before the abortion), such as: efforts to avoid or deny thoughts or feelings associated with the abortion; efforts to avoid activities, situations or information that might arouse recollections of the abortion; inability to recall the abortion or an important aspect of the abortion (psychogenic amnesia); markedly diminished interest in significant activities; feeling of detachment or estrangement from others; withdrawal in relations and/or reduced communication; restricted range of effect, eg., unable to have loving or tender feelings; sense of foreshortened future, eg., does not expect to have a career, marriage, children or a long life. Associated persistent symptoms can include: difficulty falling or staying asleep; irritability or outbursts of anger; difficulty concentrating; hypervigilance; exaggerated startle response to intrusive recollections or reexperiencing of the abortion trauma; physiologic reactivity upon exposure to events or situations that symbolize or resemble an aspect of the abortion; depression or suicidal ideation; guilt about surviving when one's unborn child did not; self devaluation and/or an inability to forgive oneself; secondary substance abuse. The disturbance can exceed a month in duration or the onset may be delayed by six months or more after the abortion.
Broader Trauma (#PD4571).
Aggravated by Induced abortion (#PD0158).

♦ **PE5848 Inadequate rail transport in developing land-locked and transit countries**
Nature Many land-locked developing countries depend on rail transport for the movement of high-volume traffic, notably minerals and agricultural products. The railway services are, however, very inadequate. Internal networks to facilitate the movement of commodities from different regions, and rail systems extended from the transit country into the land-locked country in order to avoid the need for trans-shipments, are lacking. Other factors affecting the efficiency of railway systems in most of these countries are: skilled manpower shortages at technical, supervisory and managerial levels; poor corporate planning; unsatisfactory maintenance of railway infrastructure and equipment; sub-optimal wagon turnround cycle times; inadequacy of locomotive power and rolling stock; and poor marshalling facilities, which lead to unnecessary delays at terminals and the need for remarshalling en route. Furthermore, bilateral rail agreements designed to harmonize the pattern of operations are in many cases inadequate. Land-locked countries and their transit neighbours have uncoordinated technical standards, documentation and procedures and management practices, as well as a lack of arrangements for the shared use of railway wagons and of port terminal facilities and the sharing of capital investment required for additional installations in the transit country to accommodate transit traffic.
Broader Vulnerability of land-locked developing countries (#PD5788)
Weakness of infrastructure in land-locked developing countries (#PE7000).

♦ **PE5849 Unfair shipping practices in bulk trades**
Nature Bulk cargoes make up some 80 percent of the total tonnage carried in seaborne trade, much of it transported through vessel pooling arrangements. Bulk pools have developed as the result of an increasing use of bulk carriers (decrease of tramp shipping), and the growing preference for freight contracts and other long-term shipping arrangements (contraction of the spot market) by the dominant shippers. In bulk shipping, the barriers to entry by developing countries are due more to the nature of the economic organization of chartering 'markets' or 'exchanges', and controls exercised by transnational corporations, than to the institutionalized grouping of shipowners in the liner conferences. Internally within the shipping industry in general, the concentration of power varies and is more pronounced in the cargo liner sector than in the bulk market. However this varies from dry bulk to liquid bulk, and in oil transport, for example, the oil companies play a restrictive role in the development of bulk shipping.
Broader Restrictive shipping practices (#PD0312).
Obstacles for international ocean shipping (#PD5885).
Aggravates Imbalances in the dry bulk shipping industry (#PE5839).

♦ **PE5850 Instability of ocean freight rates**
Nature Very frequent and sharp fluctuations in liner freight rates cause grave concern among shippers, particularly in developing countries, who in the last analysis bear the brunt. With liner freight rates changing at frequent intervals, shippers have increasingly found that one of the supposed advantages of the conference system, namely stability of freight rates, has hardly existed. Bunker costs, combined with general inflationary trends, have contributed to a rise in liner freight rates, and another important element is the demand conditions which have generally characterized cargo markets. Because of the peculiar structure of the liner shipping industry, competition is severely limited both in extent and effect, and it is not possible for those concerned with the use of its services to remain inactive and to trust that competition will reduce the prices charged for those services to a reasonable margin above cost levels. Nor is it possible to rely upon pressure from competitors and consumers to force the managements of shipowning enterprises to increase efficiency so that cost levels are reduced to the minimum. Furthermore, the inefficiencies to which excessive costs are attributable are often beyond the control of individual shipping managements. In the liner sector of shipping there is no price competition between shipowners

who are members of a shipping conference, and competition between shipping conferences and non-conference lines is likely to be diminished, if not eliminated entirely, by the prevailing systems of loyalty arrangements which bind shippers to utilize conference vessels.
The shippers who utilize liner services are usually numerous, often widely dispersed, and on shipping matters they usually lack any form of commercial relation with each other - since, unlike the shipowners, they are not in business primarily to deal with shipping. Accordingly, their bargaining power is weak compared with that of the shipowners. Additional problems arise within the liner sector because many shippers do not themselves produce, or own, the goods which they ship, and so long as the level of freight rates is not actually restricting the flow of trade, they have no interest in resisting freight rate increases. Higher freight rates in fact increase the profits of middlemen merchants whose remuneration is based on a percentage of the CIF value of the goods they handle. In a developing country they can simply pass on the freight increases by increasing the market prices of imported consumer goods, and by decreasing the FOB prices which they pay to the producers of export goods. Similarly, international merchant houses which have affiliations with shipping companies may be more concerned with the vessels in which their goods are shipped than with the costs of transport. For these reasons, shippers cannot be expected to exert the type of user pressure that would be needed for the purpose of resisting freight increases and inducing shipowners to maximize their efficiency.
Refs United Nations Conference on Trade and Development *Ocean Freight Rates and Their Effects on Exports of Developing Countries* (1987).
Broader Obstacles for international ocean shipping (#PD5885).

♦ PE5851 Overcrowded port warehousing
Nature A pricing policy which encourages port users to allow their goods to remain in the transit shed instead of being warehoused outside the port, leads to overcrowding of the sheds and ultimately to inadequate utilization of quays and other equipment, delays to ships, waste of gang time, and so on.
Broader Obstacles to efficient port utilization and operation (#PE5921).

♦ PE5855 Unfair pricing by transnational corporations
Broader Transnational corporation imperialism (#PD5891).
Narrower Transfer pricing (#PE1193)
Restrictive pricing policies of transnational corporations (#PE2396)
Manipulation of transfer prices by transnational corporations (#PE0245).
Related Distortion of international trade by minimum pricing regulations and other measures to regulate domestic prices (#PE1182).

♦ PE5860 Unreliable telecommunication services for land-locked developing countries
Nature The inadequacy of communication links between various ports and commercial centres in land-locked developing countries and between ports and overseas markets continues to be a major handicap inhibiting the speedy movement of transit cargo. This often leads to long delays in getting the cargo in and out of the ports, since there is irregular information on the time schedules for the arrival and departure of cargo. The costs caused by such delays can be considerable. This aggravates the already poor transport and communications along the transit corridors.
Broader Vulnerability of land-locked developing countries (#PD5788)
Weakness of infrastructure in land-locked developing countries (#PE7000)
Inadequacy of telecommunication facilities in developing countries (#PE0004).
Related Inadequate development of communication services in the least developed countries (#PF4297).

♦ PE5861 Abuse of land-locked and island countries as havens
Nature The evolution of European countries such as Switzerland, the Vatican, Italy and Luxembourg (all of which are land-locked) to be centres of banking and valuables collections is attributable, at least in part, to singularities in their legal systems that make them havens, away from excessive governmental regulations, for certain kinds of transactions. Similarly, a number of island countries and territories, such as the US Virgin Islands, the Bahamas, Singapore and several others, have constitutions that favour entrepreneurs, investors and speculators. The service economies of finance, banking and legal representation may attract developing island and land-locked countries, but many, however, lack the professional infra-structure for this. Tourism and entertainment industries require less capital investment than energy, food, and resources development projects, and may be the only other alternative.
Claim Small states, land-locked or island, (particularly those that are tropical), may be trapped into being 'happy natives' for the benefit of cruise ships, airlines, or holiday clubs. This may not satisfy national ambitions and there may be strong temptations to offer themselves either as military bases for larger and more aggressive countries; as corporate pawns of the transnationals who can exploit their resources, natural or legal; or as hosts for elements that are marginal to society and for whom lack of extradition treaties or imprisonment may be a boon. Inadequate assistance to these countries by the developed nations endangers world-wide standards for legal and fair international trade, and uniform civil and criminal codes.
Broader Legal havens (#PE0621).
Related Vulnerability of land-locked developing countries (#PD5788)
Vulnerability of island developing countries and territories (#PE5700).

♦ PE5863 Imbalance in shipbuilding industry distribution
Incidence The total world order book for new ships is inequitably divided, with Japan getting about 40 percent; Sweden, the second largest shipbuilder, has less than 10 percent. All the developing countries building ships, taken together, comprise less than 5 percent; indeed their total share is exceeded by that of each of the top eight countries following Japan and Sweden. Imbalances also exist in the proportions of different types of ships built.
Broader Lack of world maritime integration (#PE5801)
Obstacles for international ocean shipping (#PD5885).
Related Imbalances in types of ships built (#PE5874)
Maldistribution of merchant vessels (#PG6316)
Denial of right to freedom of religion of indigenous peoples (#PE4332).

♦ PE5866 Protectionism in steel and basic metal industries
Nature Due to an excess of supply in the steel industry, protectionist trade laws are enacted, mainly against developing countries, in the hopes of curtailing over-production while maintaining jobs in the steel sector of the industrialized countries. These protectionist laws (some are under the guise of "voluntary limits") are more favourable to the large, wealthy nations, and hinder the economic development of the smaller, poorer countries.
Incidence In 1985 the non-Communist world has a capacity to produce 640 million metric tons of steel, while it was estimated that world demand would be only 438 million tons for that year, 441 million tons for 1986, and 467 million in 1995. Some US steel executives estimate that 15 percent of worldwide capacity needs to be cut in order to balance supply and demand.
Counter-claim In the USA alone in 1980-1984, 40,000 steel jobs were lost and 210,000 more were threatened because of the availability of much less expensive imported steel. In the ten years between 1973 and 1983, steel production in the USA and the UK declined by 44 percent for each, while it declined by 37 percent in Belgium and Luxembourg and 30 percent in France. During that same period, steel production increased by 65 percent in Romania, 105 percent in Brazil, and 930 percent in South Korea. Why should so many workers in the industrialized nations forfeit their jobs in order to keep production down, if the non-industrialized nations continue to over-produce ?
Broader Obstacles to world trade (#PC4890) Protectionism in international trade (#PC5842).

♦ PE5873 Evasion of shipping regulations and taxes by flags of convenience
Convenience flagging of ships
Nature Operation of shipping under flags of convenience constitute a device which enables the traditional maritime countries to maintain ownership and control over world shipping despite the fact that they cannot operate economically under their own flags. In principle all aspects of merchant shipping are governed by the International Maritime Organization, which regulates everything from construction standards to marine pollution restrictions. In practice those countries operating flags of convenience are unlikely to enforce the standards to which they have agreed in fleets operating around the world. Convenience flagging of vessels allow companies to reduce their operating costs dramatically with lucrative tax advantages and non-unionized cheap labour (avoiding social security, pension and other costly benefits).
The anonymity of such ownership facilitates the use of transfer pricing and restrictive practices to the detriment of developing countries. Owners of flags of convenience (FOC) ships have no real difficulty in concealing their true identities if they so wish, and so any affiliation between the registered proprietors and transnational corporation (TNC) owners can be concealed. Moreover, the countries of registration of these ships impose no reporting requirements, and rebates can be paid into banks which have no obligation to report to governments.
Apart from considerations relating to transfer pricing, the use of FOC ships provides TNCs with a means of maintaining ownership and control over shipping operations despite the fact that they cannot operate under their own flags, since most are based in countries which can no longer supply economical shipboard labour. Thus the TNCs reap the profits from shipping operations that would make a very substantial contribution to the economies of the developing countries with which they trade. Furthermore, the use of flags of convenience has contributed to the present situation of overtonnaging, as owners operating under these flags have recycled profits in tax haven countries into new tonnage to avoid the taxation they would incur if repatriating their money. The manner in which the transnational oil companies have built up their excessive fleets out of accumulated tax haven earnings is an example. The absence of a set of basic principles concerning the conditions upon which vessels should be accepted on national shipping registers, leaves open the possibility of continuing abuses.
Incidence Almost one-third of the world fleet, mainly in the bulk trades, has at various times been owned by non-nationals who have little or no connection with the states whose flags their ships fly and whose precise role in world shipping is uncertain. Faced with growing competition from developing countries, industrialized countries have been obliged to make increasing use of convenience flagging. For example, less than a third of the UK-owned shipping tonnage is now registered under the UK flag.
Claim Ships sailing under flags of convenience are often sub-standard, badly-crewed, badly-maintained, badly-managed and they have poor safety records. This means that it is not a question of "if" a major environmental disaster occurs, but "when".
Counter-claim While pressing for control and regulation of free-enterprise in the shipping industry, some socialist countries are operating state-owned merchant fleets, which by virtue of their government dependence, may be undercutting world tariffs, thus attacking the mechanisms of free adjustment between tonnage demand and tonnage supply.
Refs ILO *Social Security Protection for Seafarers Including Those Serving in Ships Flying Flags Other Than Those of Their Own Country.*
Broader Inequities in ship owner registration (#PE5875)
Circumvention of duties and assessments (#PD4882)
Obstacles for international ocean shipping (#PD5885).
Related Restrictive shipping practices (#PD0312)
Control of industries and sectors by transnational corporations (#PE5831).
Aggravates Transfer pricing (#PE1193) Substandard shipping vessels (#PE6630).
Aggravated by Unethical practices in transportation (#PD1012)
Deliberate governmental avoidance of legislative reform (#PF5736).

♦ PE5874 Imbalances in types of ships built
Nature Some liner operators in industrial countries have shifted to containerization of their major liner trades, so that the building of conventional liner tonnage has dropped. This reduces the flexibility of shipping services in the liner trades, and can cause an excess of container capacity and a deficiency of conventional liner tonnage for essential commodities that cannot be transported in containers.
Broader Obstacles for international ocean shipping (#PD5885).
Related Imbalance in shipbuilding industry distribution (#PE5863).

♦ PE5875 Inequities in ship owner registration
Nature Some countries maintain an open registry, allowing for foreign-owned vessels to fly their national flags, with a nominal or shadow national shipping company created to be the official owner of record. This makes it difficult to identify real owners or operations in order to make them accountable for all shipping operations. Ships and groups of ships may be registered in more than one country to evade regulations or secure preferential treatments. Flag states, that is those countries allowing open registration, often show little interest in basic regulatory actions and indeed may discriminate against countries ships they do not wish registered.
Broader Compliance*complex (#PA5710)
Obstacles for international ocean shipping (#PD5885).
Narrower Evasion of shipping regulations and taxes by flags of convenience (#PE5873).
Aggravated by Flag discrimination in shipping (#PD0700).

♦ PE5882 Forced disappearances of trade union leaders
Forced disappearances of workers representatives
Broader Forced disappearance of persons (#PD4259)
Sanctions against trade union workers (#PD0610).
Related Assassination of trade union leaders (#PE0252).

♦ PE5883 Low productivity of agricultural workers in developing countries
Nature Excess labour in relation to productive labour opportunities in agriculture not only signifies a low level of production per unit of labour, but is also a direct cause of stagnation in agricultural productivity. The lower the productivity of labour in agriculture, the less is it capable of producing an excess over the subsistence requirements of the workers involved in production and therefore of creating capital required for the purchase of inputs that are essential for increasing yields.
Broader Unproductive labour resources (#PC6031)
Unproductive subsistence agriculture (#PC0492).
Related Declining productivity of agricultural land (#PD7480).
Aggravates Malnutrition in developing countries (#PD8668)
Rural unemployment in developing countries (#PD0295).
Aggravated by Tropical deforestation (#PD6204)
Excessive dependence of local communities on outside services (#PE4780).

♦ **PE5884 Lack of skilled workers in the transport sectors of land–locked developing countries**
Nature In many land–locked developing countries and their coastal neighbours, freight forwarding services are inefficient, due not only to lack of adequate transport but also to the want of expertise and skills. This contributes to the serious delays in moving transit cargo to and from the ports. Transit transport facilities are ineffectively utilized due to an acute shortage of trained manpower at all levels. Where there are training programmes they lack an integrated approach encompassing all levels from the overall and top management to the middle and low–level operators, and regional specialized workshops and seminars held periodically for overall and top management and conducted at the country level are also marginal or non–existent. Training of instructors and provision for training materials have been neglected, aggravating the lack of skilled workers in the sector.
Broader Labour shortage (#PC0592).
Related Overloaded vehicles (#PE4127).
Aggravates Vulnerability of land–locked developing countries (#PD5788).

♦ **PE5886 Ocean shipment delays**
Nature Delays in shipping owing to unavailability of vessels, strikes, or port congestion, may cause excessive storage costs, food spoilage, exposure to weather, exposure to theft and pilfering, delays in overseas production or construction in the absence of materials; and, in the case of needed food, medical, or disaster relief shipments, even death.
Broader Delays in delivery of goods (#PF8268)
Obstacles for international ocean shipping (#PD5885).

♦ **PE5888 Protectionism in the shipping industry**
Nature The current world economic situation is characterized by structural disequilibria whereby the increasing distortion and structural rigidity of the international division of labour have a cause and effect relationship with intensified protectionism. The shipping industry has not been insulated from this crisis, and itself continues to operate under various types of restrictive and distorting business practices.
Broader Restrictive shipping practices (#PD0312)
Protectionism in international trade (#PC5842)
Obstacles for international ocean shipping (#PD5885).

♦ **PE5889 Carbon disulphide**
Nature Carbon disulphide is a important solvent of alkali cellulose, fats, oils, resins and waxes. Besides, it is being used in the production of viscose rayon, in the manufacture of optical glass, as a pesticide and in oil extraction. It is primarily a neurotoxic poison affecting central and peripheral nervous systems. Continued exposure may give rise to, for example, polyneuritis, atrophic gastritis, atherosclerosis, disorders in the sexual sphere, in women more frequent abortions, coronary heart disease.
Broader Malodorous fumes (#PD1413).
Aggravates Heart diseases (#PD0448) Badly laid out work premises (#PJ2468).

♦ **PE5896 Excessive costs and unsuitability of insurance for land–locked developing countries**
Nature The particular problems faced by the land–locked developing countries in the promotion of their international trade – such as the need for transshipment of transit cargo in some cases, and the extensive delays in transit resulting from cumbersome customs inspection procedures – often lead to the loss of and/or damage to the cargo. This, in the end, gives rise to higher premiums for cargo insurance. Moreover, as the loss experience is not favourable, many importers and exporters encounter difficulties in obtaining the types of cover they desire, particularly warehouse–to–warehouse policies. Such non–availability leads them to accept segmented policies, one covering ocean marine transit, the second covering the inland leg in the neighbouring country and the third covering the final transit in the country itself. Comparatively, such segmented cover costs shippers much higher premiums.
The basic difficulties of land–locked countries in the marine cargo insurance field are obviously of a structural nature and are beyond the scope of insurance. Insurers can do very little to overcome them. Measures that could be taken by insurers to ease these constraints are neglected. These include the local covering of external trade risks whenever this is possible, to enable shippers to obtain protection that is tailormade to their requirements; and the implementation of loss prevention programmes in cooperation with insurers in the coastal countries through sharing the business with them on a co–insurance or reinsurance basis. Specialized loss prevention firms which take care of and survey the loading and discharging of goods in transit are not employed and there is no experience rating system for shippers to induce them to take all the necessary measures to reduce their losses and thus, their premiums. The above problems are aggravated by a lack of cooperation among insurers in the land–locked country and between insurers in the land–locked country and those in the coastal countries, through which transshipments usually take place.
Broader Prohibitive cost of insurance (#PE8632)
Vulnerability of land–locked developing countries (#PD5788).
Aggravates Inadequate insurance (#PF8827).

♦ **PE5897 Excess capacity in the shipping industry**
Overtonnaging
Nature A persisting overtonnaging crisis in the shipping industry cannot be dismissed as a cyclical phenomenon, and it gravely affects the long–term interests of both developing and developed countries. Scrapping proceeds apace, while going freights in traditional tramp trades have declined to the point at which some owners, who attempt to operate rather than to scrap their ships or to let them lie idle, have been reduced to accepting rates at well below operating costs (or lay–up levels). Liner freights have fluctuated in many trades, where they may receive sympathetic shipper response at one moment and despair the next as a result of the chaotic tonnage supply situation. At the same time, in other conference trades, the perennial problems with regard to the adequacy of services, freight levels, surcharges, dispensation, cargo share, technological innovations and changing patterns of trade, have intensified, with prominent 'lines' or powerful 'outsiders' continually either seeking or resigning from membership.
Incidence The world–wide economic recession of the 1980s has retracted demand for shipping space to the extent that 64.5 million dwt tons of shipping, representing 7.2 percent of world tonnage, are laid up (mid–1982).
Broader Obstacles for international ocean shipping (#PD5885).

♦ **PE5898 Reduction of ocean shipping services**
Claim Shipowners tend to reduce services, regardless of consequences to shippers, in order to maintain profits in the face of increasing cost for fuel and labour. Speeds reduced by 1 to 1 1/2 knots cause a 16/17 knot cargo liner ship to cut consumption by about 20 percent, because it is at the higher range of speeds that fuel consumption accelerates significantly. Such measures as reducing speed, curtailing the number of sailings and the number of calls made at ports, as well as not calling at intermediate ports, significantly alters the services available. Reduced speed means longer round voyage times, only partly offset by limiting the number of port calls. Such measures increase space demand and difficulties for shippers.
Broader Restrictive shipping practices (#PD0312).

♦ **PE5899 Excessive costs of inefficient port cargo–handling**
Nature The costs of inefficient cargo handling at ports increases transport expenses. Handling costs represent from 20 to 30 percent of the total expenditure for the shipment, some twice the amount required for the fuel. Between 5 and 10 percent of the value of goods is made up of these costs. Handling charges include equipment and manpower usage, but also idle time. Inefficient overtime and night shifts contribute to costs, as do poor preparation of bulk items for shipments – for example, in small packagings instead of pallets or containers. Port regulations often overlook packing efficiencies that shippers can be obliged to follow. Extra costs are borne by the idle time of ships. The inefficiency of cargo–handling lengthens port stays, fuel consumption, shipboard labour costs and overall length of voyages.
Broader Obstacles to efficient port utilization and operation (#PE5921).
Aggravated by Unethical practices in transportation (#PD1012).

♦ **PE5902 Undeveloped financial markets in developing countries**
Broader Lack of economic and technical development (#PE8190).
Aggravates Inadequate development of enterprises in developing countries (#PE8572).

♦ **PE5903 Destabilization of monetary systems and exchange notes by transnational corporations**
Broader Transnational corporation imperialism (#PD5891)
Aggressive economic destabilization of countries by external forces (#PE9420).
Narrower Destabilizing financial action of transnational enterprises (#PE0226)
Adverse effect of transnational corporations on balance of payments (#PE1771)
Aggravation of instability in exchange rates by transnational corporations (#PE0980)
Foreign currency manipulations in accounting records of transnational corporations (#PE2145).

♦ **PE5908 Inadequate port and storage facilities for land–locked developing countries**
Nature Land–locked developing countries require, like other port users, a number of qualities in a sea port, namely: no port congestion and thus no congestion surcharges; adequate berth, handling and storage facilities; facilities adapted to modern shipping and handling technology; minimal loss to cargo through losses, damage and deterioration; rapid transit through the port; and reasonable charges for the services provided. However, because of the unreliability or unavailability of transport facilities between the port and the final destination, the land–locked countries face a particular problem. There is a need for adequate storage space and facilities for goods in transit. In order to avoid congestion such facilities should, wherever possible, be removed from the port area. Storage arrangements are not adequate in most ports used by land–locked countries, and the problem of poor facilities in such ports is compounded by the fact that the transit countries – most of them developing countries – have themselves serious resource constraints and a grave shortage of trained manpower at all levels.
Broader Vulnerability of land–locked developing countries (#PD5788)
Weakness of infrastructure in land–locked developing countries (#PE7000).

♦ **PE5909 Obstacles for developing countries ocean shipping**
Nature Quite apart from the effects of shipping upon trade and industry, developing countries have grounds for concern about the level of freight rates, because they bear the freight costs of both their imports and their exports. As freight levels rise they have to pay more for their imports, and at the same time their producers receive less for the goods they sold overseas. High freight levels have an adverse effect upon their balance of payments by causing a reduction in receipts and an increase in payments, and these adverse effects are accentuated if a developing country relies heavily upon the use of foreign vessels. Land–locked countries must add the difficulties and increasing costs of shipping to the costs and difficulties they face in regard to overland transport. The fact that developing countries own less than 10 percent of tonnage, yet are the major export shippers by volume (over 50 percent), cannot mean an equitable situation. Yet another obstacle is the ultimate ownership of the exports. Much of it is controlled by transnational corporations who exploit resources but contribute disproportionately small returns to their host countries. Traditional maritime countries with protectionist activities, shipping line conferences and TNCs, make formidable external adversaries; but developing countries also lack capital, infrastructure and a history of mutual cooperation to apply to their ocean transport needs.
Background The developing countries that gained political independence from the late 1940s onwards found that the system which they had inherited in shipping was traditionally structured on the colonial concepts of keeping them as passive suppliers of raw materials without affording their indigenous interests reasonable opportunities to participate in downstream activities such as transport. As some of these countries established infant manufacturing industries for the purpose of processing their own and imported raw materials into manufactured or semi–manufactured exports, and of taking advantage of relatively low labour rates, costs and difficulties of transport were found to outweigh the labour savings.
Broader Obstacles for international ocean shipping (#PD5885).
Narrower Lack of capital investment in developing countries (#PE5790)
Imbalance of developing countries share in shipping (#PF5912).
Related Inadequate port infrastructure (#PE5792).

♦ **PE5910 Restrictive practices in cargo airline services**
Nature Although the institutional framework for civil aviation provides for equitable relations between states, discriminatory and unfair practices exist in the relations between developed and developing countries in the matter of civil aviation. Many developing countries in addition, have structural disadvantages that restrict their own airlines, according to UNCTAD studies. Developing countries face many international obstacles against increasing the air transport of their fleets and the development of their export of goods by air.
Broader Unfair transport practices (#PD1367) Restrictive trade practices (#PC0073).
Aggravated by Unethical practices in transportation (#PD1012).

♦ **PE5911 Underpricing of port services**
Incidence In the absence of marked unemployment, labour can always be used in alternative ways and thus always has an economic cost. Land may, or may not, be capable of an alternative use, although in most ports situated in densely populated areas of great economic activity the value of the alternative use of land may be greater than that of its actual use in the port. Capital, on the other hand, frequently has no alternative use. Once a quay is built, it is useless for anything other than transferring goods between ships and inland; a breakwater can provide a sheltered haven and nothing else. Where there is no alternative use, there is no economic cost. Realistically, however, a port is concerned with its own costs in the sense of the annual cash outflow. In addition, port investments are very costly, and technical progress gives them a shorter useful life than was the case in the past. It is particularly important to use them economically, and port charges have a function to perform in this respect. Thus, while in principle, a pricing system has only to deal with economic costs, in practice, it has to provide a cash flow to meet the payments which the port must make, whether for costs which are recognized as economic or for those which are not.
Claim Serious underpricing of port service leads to unprofitable operations and inadequate capital

formation for port improvements. Pricing is usually based on accountants definitions of disbursed costs, but even the indirect elements or overheads neglect another important area. There are also economic costs, that is, the costs to the economy as a whole in terms of the resources which are used to provide services and which, if not used in the ports, could be used elsewhere in the economy.
Broader Obstacles to efficient port utilization and operation (#PE5921).

♦ PE5913 Arthritis disorders in animals
Degenerative joint disease — Polyarthritis — Traumatic osteoarthritis
Broader Diseases of musculoskeletal system in animals (#PD7424).
Narrower Canine hip dysplasia (#PG9745)
Lyme disease in dogs (#PG6391)
Tendinitis in animals (#PG5391)
Spondylosis deformans (#PG9223)
Chlamydial polyarthritis (#PG7397)
Tenosynovitis in animals (#PG6588)
Osteoarthritis in animals (#PG4546)
Viral arthritis in poultry (#PG9334)
Suppurative arthritis in animals (#PG9675).

♦ PE5915 Restrictive business practices of transnational corporations
Broader Transnational corporation imperialism (#PD5891).
Narrower Restrictive practices of transnational banks (#PE9657)
Monopolistic activity by transnational enterprises (#PE0109)
Restrictive pricing policies of transnational corporations (#PE2396)
Restrictive market divisions by transnational corporations (#PE3196)
Restriction of free market competition by transnational corporations (#PE0051)
Limited market access due to the product differentiation of transnational corporations (#PE2328)
Direct foreign investment by transnational enterprises as a restrictive business practice (#PE0161)
Inadequate negotiation of entrance terms for transnational corporations in developing countries (#PE0853).
Related Control of marketing and distribution channels by transnational corporations (#PE2397).

♦ PE5919 Forced motherhood
Forced pregnancy — Government-imposed pregnancy
Incidence While a larger number of countries consider their national fertility level high and would like to reduce it, a number of other countries consider higher rates of fertility desirable. For instance, in the United Nations Fifth Population Inquiry conducted in 1982, 46.7 percent of the developed countries responding indicated that higher rates of fertility were desirable and the remaining 53.3 percent of the countries responding indicated that their fertility level was considered satisfactory. In some countries experiencing a decline in their birth rate, governments are furthering policies to increase it, sometimes even to the extent of enforcing motherhood.
In Romania, for example, where the party discovered that out of 743,000 registered pregnancies only 40 percent resulted in live births, mainly because of abortions, women were obliged to submit to monthly gynaecological examinations to encourage them to become pregnant and to ensure that pregnancies were not illegally terminated. The target was four children per healthy female. Women who failed to produce the required number of children were subject to economic sanctions. The consequences were a significant in the number of illegal abortions and abandoned children. It is estimated that only 5 per cent of the women of child bearing age have a completely healthy uterus.
Claim The World Population Plan of Action underlines the importance of the fundamental right of individuals and couples to decide freely and responsibly the number and spacing of their children.
Broader Eugenics (#PC2153).
Related Denial of right of family planning (#PE5226).
Aggravates Opposition to population control and family planning (#PF1021).
Aggravated by Undesired sexual obligations (#PF4948).

♦ PE5920 Imbalances in exports and imports of land-locked developing countries
Nature Many land-locked developing countries import and export low value-to-weight products, thus paying a disproportionate amount of their costs for freight. The restructuring of the pattern of production with the deliberate intention of reducing sensitivity to transport costs by promoting import substitution industries producing high-bulk low-value products, and developing high-value low-bulk products for export, is hindered by the unavailability of such domestic high-bulk low-value commodities as fertilizers and construction materials, and by the incapacity to produce at reasonable cost through limiting the import content of local production. The dependency of many land-locked developing countries on external sources of supply for such basics as food and energy makes a rapid rate of development unattainable. Exports with a high value-to-weight ratio, such as high-value minerals, have so far played a very minor role in most of the land-locked developing countries. These countries have inadequate mineral exploitation programmes and have not succeed in the processing of some of the mineral and food currently exported manufacturing of products for export. In some cases the developed countries have made their markets difficult to access for semi-processed, processed and manufactured goods from the land-locked countries, and regional cooperation arrangements in the area of trade are not far advanced.
Broader Vulnerability of land-locked developing countries (#PD5788).

♦ PE5921 Obstacles to efficient port utilization and operation
Broader Lack of world maritime integration (#PE5801).
Narrower Port delays (#PE5815)
Unsafe port facilities (#PE4897)
Inconsistent port charges (#PF5887)
Overcrowded port warehousing (#PE5851)
Underpricing of port services (#PE5911)
Political interference with port operations (#PE5827)
Excessive costs of inefficient port cargo-handling (#PE5899)
Ineffective self-regulation in the telecommunications sectors (#PF5877).
Related Inadequate port infrastructure (#PE5792).
Aggravates Obstacles for international ocean shipping (#PD5885).

♦ PE5925 Genetic sexual attraction
Nature Experienced between mothers and sons, fathers and daughters, and between more distant blood relatives, but most common between siblings of opposite sex who bear a close resemblance. It takes the form of an overpowering, almost electrical grip of emotion, associated with an inability to keep away from the other person and an almost primordial sense of having belonged together all their lives. The attraction gives rise to a sense of underlying shame and guilt, together with a feeling of rejection that may prevent effective communication because the emotions are too threatening to share with anyone. This may be compounded by any sexual relationship resulting from the attraction.
Incidence Particularly noted in the case of adopted children who are subsequently reunited with the biological parent or sibling of the opposite sex, seemingly because the normal bonding mechanism has been disrupted.
Broader Unethical personal relationships (#PF8759).
Related Sexual abuse of children (#PE3265).
Aggravates Incest (#PF2148).

♦ PE5926 Restrictive business practices in the markets of developed countries against exports from developing countries
Broader Restrictive trade practices (#PC0073).
Narrower Protectionism in developed countries against agricultural products from developing countries (#PE8321).
Aggravates Lack of economic and technical development (#PE8190)
Excessive foreign public debt of developing countries (#PD2133)
Excessive concentration of export markets of developing countries (#PE9457).

♦ PE5927 Labour tensions involving transnationals
Broader Transnational corporation imperialism (#PD5891).
Narrower Transnational strike action by trade unions (#PE1541)
Domination of labour relations by transnational corporations (#PE1187)
Coercive use of economic power by transnational enterprises against labour (#PE0207).

♦ PE5928 Disruptive effect of household moving
Nature The tendency of humans to move from one city to another, one country to another, or one continent to another, has rapidly accelerated over the past decades, due mainly to employment changes or demands. While moving is said to be the third most stressful event in a woman's life (childbirth and the death of a loved one being the other two), the event can also be very traumatic for the children involved, and can be a causative factor in marital difficulties. (As most "executive" moves are initiated by a husband's career, it is usually the man who is least negatively affected by a move, as he retains his career identity and is very immersed in his work, leaving his wife to attend to the problems of schools, social activities, and acclimatization to different systems and ways). Moving is also very costly, and if there is little or insufficient aid from the employer, the high price can alter individual and family lifestyles.
Broader Stress in human beings (#PC1648).
Aggravates Marital stress (#PD0518).

♦ PE5930 Exchange rate volatility
Instability of exchange rates — Floating exchange rates
Nature When exchange rate movements are comparatively more active they create or fuel uncertainty in the global and national economies and lead to conservatism in terms of investment. Interest rates may be affected in an upwards direction, and developing countries earnings from exports lowered while import prices are raised by these exchange fluctuations. Increased exchange-rate instability has reduced the effectiveness of tariffs as a protective device and made more direct controls more attractive.
Claim Exchange rate volatility may do invisible damage to world trade where global aggregate volumes are not affected, but where specific commodities and trading countries are negatively impacted and covered by offsetting volumes elsewhere.
Refs Krugman, Paul *Exchange Rate Instability* (1988).
Broader Financial destabilization of world trade (#PC7873)
Economic and financial instability of the world economy (#PC8073).
Aggravates Trade protectionism (#PC4275)
Misalignment of currencies (#PF6102)
Fluctuations in real value of money (#PD9356)
Fragmentation of the international trading system (#PC9584)
Excessive foreign public debt of developing countries (#PD2133)
Uncertainty of development expenditures due to floating-rate loans (#PF4295)
Inability of developing countries to adopt appropriate exchange rate policies (#PE7563).
Aggravated by Speculation on money markets (#PD9489)
Mismanagement of exchange rate system (#PF1874)
Mismatch of national macroeconomic policies among industrialized countries (#PF5000).

♦ PE5931 Increasing proportion of land surface devoted to urbanization
Loss of agricultural land to urbanization — Competition between industry and agriculture for land — Loss of land to industrialization
Nature The migration to cities and the high birth rates there are creating urban sprawls and suburbanization which annually consume tens of millions of hectares of open spaces. Vast conurbations with populations of multi-millions are created and continue to increase in size.
Incidence In the USA, for example, 25 percent of the population lives on 1.8 percent of the country's total land surface, creating, on the Eastern Seaboard, a megalopolis extending from Massachusetts to Maryland. In Europe 60 to 80 percent of the population lives in dense conurbations which increasingly take land from the surrounding area. This is true around Paris, in Southeast England, in the built-up areas of the Ruhr, and in the Netherlands Randstadt, for example. In England and Wales, it is calculated that 80 percent of the 3 million new homes required by the year 2000 will urbanize an additional 11 or 12 percent of the land.
Broader Land misuse (#PD8142)
Uncontrolled urban development (#PC0442).
Aggravates Endangered parklands (#PE9282)
Shortage of urban land (#PD0384)
Shortage of cultivable land (#PC0219)
Derelict industrial wastelands (#PE6005)
Unavailability of agricultural land (#PC7597)
Unsustainable agricultural development (#PC8419)
Declining productivity of agricultural land (#PD7480).
Aggravated by Environmentally harmful dam construction (#PD9515).

♦ PE5938 Violation of the right to organize trade unions in government services
Restrictions on trade union rights in public enterprises
Broader Denial of trade union rights to public employees (#PE6888).

♦ PE5944 Terrorism targeted against business corporations
Terrorism targeted against corporate executives — Killing of corporate leaders — Assassination of corporation leaders
Nature Terrorist attacks against corporations, whether through the destruction of property, by the kidnapping of executives, or the use of extortion, may be used to secure funds to finance continuing activities, thus crating a vicious circle.
Broader Terrorism (#PD5574).
Related Political assassination (#PE5614).

♦ PE5951 Unemployment of older people
Nature Discrimination of employment of older people may be based on applicants' declining physical strength, or on employer expectations that the older worker may be more vulnerable to ill health, that he or she will not be able to work as fast, or that they will be less productive. In addition, employer contributions to social security may be higher for older than for younger workers in some countries, aggravating the prejudices that view employment of older people as not cost-effective.
Incidence One worker in three in developed countries, and one in four in the developing countries, is over 45 years of age. Worldwide, these total nearly 480 million, with about one-third being women. Exact statistics on older worker unemployment are lacking but indications are that the rate is lower than the total worker force, but longer lasting and in many cases permanent. ILO statistics indicate that the over 45 age group have the highest unemployment in the manufacturing sector, in the higher blue collar job ratings, as craftsmen, foremen and kindred workers for males, and as operatives and kindred workers in the lower job ratings for females. Statistics from the UK, for example, show the deviation in weeks of unemployment, to climb upwards from age 45, to double, triple, and quadruple the period for those under 30.
Broader Unemployment (#PB0750).

Narrower Unemployment of educated older people (#PE9071).
Aggravates Inadequate income in old age (#PC1966)
Retirement as a threat to psychological well-being (#PF1269).
Aggravated by Compulsory retirement (#PJ2411)
Employment discrimination against the elderly (#PD4916).

♦ PE5952 Incitement to hatred
Slogans of hate — Denial of right to freedom from advocacy to discrimination
Broader Denial of right to security (#PD7212).
Aggravates Hatred (#PA8487) Discrimination (#PA0833).
Aggravated by War and pre-war propaganda (#PD3092).

♦ PE5953 Minor ailments
Incidence Basically healthy people are not necessarily immune to minor ailments. For example, a 1984 survey of 2,000 adults and 500 children in the USA shows that the average American develops a minor health problem as often as once every three days. In the average two weeks, 28 percent complained of overweight problems, the number one cause of anxiety. Equal second were indigestion and muscle aches, which each afflicted 25 percent of those surveyed. Next came eye problems (22 percent), minor fatigue (20 percent), and minor cuts and the common cold (both 19 percent).
Broader Human disease and disability (#PB1044).

♦ PE5959 Cultural dominance in translation
Nature Some languages are virtually impossible to translate accurately into another language, as the nuances of the original tongue are lost in translation. A good writer does not rely only upon the richness of imagery, rather he often draws heavily upon the connotations and associations that only readers of his own background and history can know. Thus, regardless of how excellent a translation may be, the intent of the original work very often loses its own identify and assume some of the culture of the language into which it has been translated.
Broader Language barriers (#PF6035) Interpretation and translation errors (#PF6916).

♦ PE5966 Video censorship
Nature Video censorship may be employed in order to prevent children from watching programs of questionable merit – those with excessive violence or sex, for example – but the notion of merit is subjective, thus video censorship may be seen as a significant threat to the principle of free speech in the arts and media.
Broader Censorship (#PC0067).
Related Self censorship (#PF6080).

♦ PE5975 Embourgeoisement in socialist countries
Nature Socialist countries are not yet the classless society envisaged by Marx and Lenin. Officially, social class has been abolished in the Soviet Union. But, as some sociologists acknowledge, 67 years after the Bolshevik takeover, class-consciousness is reappearing. There is a large and powerful state bureaucracy founded on privilege rather than private property and the nouveaux riches want all the ornaments of class superiority. This re-emergence of class attitudes is a cause of concern to the Soviet authorities, who are anxious that the better-off should not come to regard the 'lower classes' as inferior or even criminal.
Broader Socialism (#PC0115).

♦ PE5986 Militarization of children
Child soldiers — Extra-legal impressment of children into the armed forces — Child combatants
Nature Children may be made regular combat soldiers, or forced into fighting militia or guerrilla forces. They may be employed as paramilitary support in front-line areas, as messengers, scouts, supply carriers and so forth. In these capacities children may face professional, adult fighting forces who will not discriminate by age in identifying and attacking the enemy. Many of these teenagers and pre-teenagers face death and mutilation, but beyond this they face the terror of warfare that they cannot comprehend. Children are taken as prisoners of war, and are also sent to civilian internment camps. As civilians they may survive the destruction of their home and the killing of family members. There are no horrors that children are not exposed to in total war.
Incidence Boy soldiers, aged fifteen or less, appeared in the Nazi forces at the end of the world war in Europe. Children have been recruited by the Viet Cong, by terrorist groups for action against Israel, by the Khomeini regime in Iran to fight the war with Iraq, by radical groups in Northern Ireland and by the Sandinistas in Nicaragua. Children fighters are also reported in Afghanistan, El Salvador, Colombia, Uganda and Ethiopia.
Refs Moorehead, Caroline (Ed) *Betrayal* (1989).
Broader Militarization (#PD1897) Victimization of children (#PC5512)
Exploitation of child labour (#PD0164).
Aggravates Kidnapping (#PD8744)
Vulnerability of children during armed conflict (#PE8174)
Vulnerability of women and children in emergencies (#PD1078).
Aggravated by Detention of children (#PE6636) Extra-legal conscription (#PD6667).

♦ PE5989 Military censorship
Nature Military censorship may include censorship of national and international media by means of mandatory submission of reports before they can be released, of every page of copy needing to be stamped with a censor's stamp before it can be telexed, and by phone tapping. While practising countries maintain this scrutiny is vital to their national interests, it is often used only to serve the best interests of the people in power, not the interests of those whom they serve.
Incidence The US press was forbidden to enter Granada when the USA invaded that country in 1983, and Israel practices most of the censorship techniques mentioned above.
Broader Censorship (#PC0067).
Related Self censorship (#PF6080).

♦ PE5993 Polar pollution
Arctic pollution — Antarctic pollution
Nature A heavy reddish haze is a regular feature of the Arctic in winter and early spring, often reducing visibility to less than 6 miles (about 10 kilometers). A haze so far from the industries and vehicles that pollute the air of major cities seems to require a wholly natural explanation, but evidence has begun to accumulate that a part of the arctic haze may have its origins 6,000 miles (10,000 kilometers) away in the same polluted air that produces acid rain over the United States and Europe. According to a study undertaken in 1976, it would seem that, due to air flow patterns and other factors, the major source of the pollutants in the arctic haze is the Soviet Union. Europe and England are the next largest sources of the pollution. North American countries contribute little because of air flow patterns.
Possible future mining and oil drilling pose the most serious threat to Antarctica. A major oil spillage could take hundreds of years to disperse. Any upset can seriously disturb weather patterns and ocean food supplies. American bases for scientific research have burned combustible waste in open pits with no emission controls, nonburnable waste and toxic chemicals have been dumped at sea.
Incidence The pollution extends from Alaska eastward to Norway – about half of the circumference of the polar ice cap – and is suspended from ground level to as high as 25,000 feet from November through April.
Broader Environmental pollution (#PB1166).
Related Arctic air pollution (#PD6283).

♦ PE6004 Governmental abuse of extradition
Nature The demand for extradition may be the consequence of the existence of special agreements concluded between the states concerned. Governments may exploit extradition arrangements to obtain an individual by lying about their intentions concerning trial and punishment.
Incidence Some countries may add one or several alleged previous offences, criminal or political, to the charges against the prisoner mentioned in the extradition request. Others may assure that punishment will not exceed that provided by the extraditing nation's legislation for similar offences, but following on condemnation and judgement of the criminal they may exact or unusual punishment or even the death penalty.
Broader Abuse of government power (#PC9104).
Aggravates Impediments to extradition (#PF5947).

♦ PE6005 Derelict industrial wastelands
Unallocated industrial land — Unfinished industrial site — Degraded industrial land
Refs Bradshaw, A D and Chadwick, M J *The Restoration of Land* (1980); Bradshaw, A D and Chadwick, M J *The Restoration of the Land* (1980); Hutnik, Russell and Davis, Grant *Ecology and Reclamation of Devastated Land* (1973).
Broader Unused land (#PF4670) Landscape disfigurement (#PC2122)
Fragmented planning of community life (#PF2813).
Related Dereliction (#PE5715).
Aggravates Uncontrolled industrialization (#PB1845).
Aggravated by Increasing proportion of land surface devoted to urbanization (#PE5931).

♦ PE6006 Aircraft harassment
Interference with transiting aircraft
Nature International flight-paths of non-military aircraft can involve over-flights of sovereign land or sea territories and transit through or near foreign airspace. Aircraft may be harassed by local airforce, by buzzing or other threatening behaviour and communications, or by actually being fired upon. Foreign ground-controls may also threaten or withhold cooperation or assistance. Harassment may lead to an unjustified attack with casualties or fatalities. In some countries like the USA where there is considerable private aviation, small planes may be harassed by irresponsible private aviators, for entertainment, and private planes may be harassed by military aircraft if near bases or sensitive locations.
Broader Harassment (#PC8558) Intimidation (#PB1992)
Aircraft environmental hazards (#PD8328).

♦ PE6007 Resistant bacteria
Superbug
Nature Many diseases are caused by pathogenic bacteria. These diseases include various epidemic illnesses and also tuberculosis, blood infection (sepsis), leprosy, and syphilis. Bacteriologists are faced with a relentless increase in the resistance of bacteria to bactericidal drugs, and humans receive some of their most savage attacks by bacteria in the hospital. Such infections are not common, but when they do occur, they can render doctors therapeutically impotent. Antibiotics having been deployed so intensively in hospitals over the past four decades that their environment has provided a breeding ground for resistant bacteria. When a patient succumbs to one of these hospital strains, perhaps after surgery or perhaps when debilitated by another illness altogether, drugs that would normally abolish the infection have little or no effect.
Incidence One strain called methicillin-resistant staphylococcus aureus thrive because competing bacteria have been killed off by the over-use of antibiotics. In the United Kingdom one in 10 patients contract an infection while they are in hospital, one half of which are caused by this bacteria. Over 2000 patients in 99 hospitals were infected in one 12 month period.
Broader Mutation (#PF2276).
Related Drug resistant viruses (#PG6399) Pest resistance to pesticides (#PD3696)
Fungal resistance to fungicides (#PE4456) Insect resistance to insecticides (#PD2109)
Rodent resistance to rodenticides (#PE3573).
Aggravates Bacterial disease (#PD9094) Bacterial plant diseases (#PD2226)
Bacterial diseases in animals (#PD2731).
Aggravated by Abuse of antibiotics (#PE6629)
Encouragement of drug resistant diseases (#PJ3767).

♦ PE6010 Endangered black rhinoceros
Nature Africa's black rhinos are killed for their horns, which are made into highly prized dagger handles. Rhino horn is still consumed in some places as an aphrodisiac or in traditional medicine.
Incidence The effect of the illegal trade of rhino horn on Africa's black rhino population has been catastrophic: down from 65,000 in 1970 to 4,000 in 1989. This smuggling racket involving mainly North Yemen and the Sudan has become the biggest single threat facing Africa's endangered black rhinos.
Refs Jackson, P (Ed) *Elephants and Rhinos in Africa* (1982); Martin, Bradley E *The International Trade in Rhinoceros Products* (1980); Penny, Malcolm; Quinn, David and Werikhe, Micheal *Rhinos* (1988).
Broader Endangered species of mammals (#PC1326).
Aggravated by Human use of animal by-products (#PF1964).

♦ PE6018 Unrepresentativity of trade unions
Inadequate employee participation in collective-bargaining
Nature Under agreements between employers and business agents of unions, one union may be recognized by the employers as representing the interests of all wage earners, whether union or non-union. The unionized worker already has the problem that his union may only be responsive to its national headquarters and to dealing with the employer directly, having little consultation with the local union members. The non-union member may be considered spoken-for by the recognized union, but has no possibility of voting in union decisions. Large numbers of non-union members may not be allowed to constitute themselves into a collective bargaining body. This lack of voice for non-union workers, clericals, or professionals is a cause of unnecessary strife between management and employees.
Broader Undemocratic policy-making (#PF8703) Ineffective worker organizations (#PF1262).

♦ PE6024 Over-population of wild animals
Refs Jewell, Peter A *Problems in Management of Locally Abundant Wild Mammals* (1982).
Narrower Over-population of seals (#PG3872).
Aggravates Degradation of semi-natural and natural habitats of flora and fauna (#PC3152).

♦ **PE6036 Marketing of banned pharmaceutical drugs in developing countries**
Nature Pharmaceuticals banned by industrialized countries' health authorities as deadly, dangerous or ineffective are routinely marketed in the Third World.
Incidence Dipyrone, which can cause fatal blood damage, is freely available without prescription in many countries in Latin America under 100 different brand names. In the Pacific and Southeast Asia it is available under about 30 brand names. Deaths from dipyrone may be in the range of 5 persons per 1000 users, with national usage levels giving a probability of one to several thousands of mortalities per country. On a world-wide scale the deaths from this one drug marketed perhaps in over 60 countries may be estimated as already having exceeded 10,000.
Refs Silverman, Milton, et al *Prescription for Death* (1982).
Broader Irresponsible international trade (#PC8930)
Unauthorized pharmaceutical manufacture and distribution (#PE0564)
Ineffective self-regulation in the pharmaceutical and medical devices industries (#PE8502).
Related Trading in products containing toxic substances with developing countries (#PE2061).
Aggravates Insufficient medical supplies (#PE1634).
Aggravated by Irresponsible pharmaceutical advertising (#PE2390)
Inadequate supply of pharmaceutical products in developing countries (#PE4120).

♦ **PE6038 Commercialization of human embryos**
Nature Human embryos may be sold to couples or to single persons unable to conceive a baby of their own. The fertilized embryo may be transferred from one woman to another, or it may involve the surgical removal of an unfertilized egg which is fertilized in a laboratory dish before being replaced in the original owner. Objections to such practices centre around the commercial aspects (the cost for such procedures in the USA is between $2,500 and $7,000); such a sale of human life is said to cheapen it by turning it into a consumer product.
Broader Proliferation of commercialism (#PF0815).
Related Unlimited practice of human embryo storage (#PE5623).
Aggravated by Inhuman methods of conception (#PE8634).
Reduced by Non-acceptance of embryo transfer technology (#PE6039).

♦ **PE6039 Non-acceptance of embryo transfer technology**
Nature The early development of embryo transfer gave rise to much controversy. Since the birth of the first embryo transfer baby, many questions have been raised about the ethics, legality and marketability of the new technology.
Claim Surrogate motherhood, or womb-leasing, might be regarded as prostitution of motherhood. Also, the for-profit nature of some of the business involved may be considered as cheapening human life and turning it into a consumer product.
Counter-claim Surrogate motherhood should not be prohibited by law because such law would probably be unenforceable, and the practice might be considered justifiable when a couple are unable to have children by any other means. Experts estimate that as many as three million US couples are unable to conceive after a year of trying. The majority could be helped by in-vitro fertilization or artificial insemination, but embryo transfer may be the only hope for women with damaged ovaries or genetic diseases.
Broader Irresponsible scientific and technological activity (#PC1153).
Reduces Commercialization of human embryos (#PE6038)
Unlimited practice of human embryo storage (#PE5623).

♦ **PE6052 Military aid**
Military aid to repressive regimes
Incidence The United States have secretly funded Pol Pot and his Khmer Rouge from the beginning of 1980s.
Claim A significant proportion of all aid by donor countries is military aid, much of it given to governments with repressive records. In general, repressive governments tend to receive more aid than non-repressive governments. Furthermore, industrialized countries have proved unable or unwilling to enforce export controls on technology of military value, even to states whose military ambitions they claim to oppose.
Broader Foreign intervention in internal affairs of states (#PC3185).
Aggravates Military government (#PC0698).
Aggravated by Government support for repressive regimes (#PF4821).

♦ **PE6053 Circumcision as a health hazard**
Nature Circumcision, although relatively simple, does carry operational risks and can be extremely painful for the patient; many contend that it is totally unnecessary for health reasons and that, because it is most frequently performed on children, the suffering involved is inflicted without the patient's consent. The operation consists of cutting off the prepuce of the penis. (While there are instances of similar operations performed on females, such as cutting of the internal labia, the term circumcision is usually limited to males.) It is often a religious rite, notably among Jews and Moslems; some theologians argue, however, that there is no doctrinal reason for it. In developed countries, it is practised mainly for reasons of hygiene and cleanliness. Depending on the conditions accompanying the operation, complications can include infections, gangrene, wound diphtheria, septicaemia, and loss of the penis. There is also a possibility of haemorrhage, which can be fatal. If the patient is a newborn, there may be a deficiency of Vitamin K.
Incidence Not all males are circumcised at birth and the percentage seems to be decreasing. It is estimated that only one seventh of the males worldwide are circumcised.
Background Circumcision is an ancient practice common to various peoples of primitive agriculture, (although not among those of truly primitive culture), living in such disparate locales as Africa, America and Australia. It seems to have been a rite connected with puberty and the entrance into the adult or married state and probably related to fertility rites.
Broader Hazards to human health (#PB4885).
Narrower Male sexual mutilation (#PE6054).
Related Female sexual mutilations (#PE6055).

♦ **PE6054 Male sexual mutilation**
Circumcision — Castration
Incidence The custom of castration is practised among the native peoples of Algeria, Egypt, Ethiopia, South Africa and Australia; Skoptsy, a Russian religious sect, practice castration as a way of mortifying the flesh in order to gain salvation. Circumcision is a common ritual practice in many traditional societies and often represents the achievement of a certain status. In North America and Israel the ablation of the foreskin is a regular procedure for all male newborns irrespective of any therapeutic indication; 400 babies are circumcised every hour in North America alone.
Background Castrated males were recorded in Assyria 3000 years ago; causes were punishment and to produce eunuchs. The Italian practice of castrating boys to make them soprano singers was nominally ended only in 1878 by a papal order. A trade in castrated boys to be used in Muslim harems is believed to still exist.
Claim Circumcision (along with any form of male sexual mutilation) is one of the insanities of religion based upon the ignorance and superstition of primitive man. There is no absolute medical indication for routine circumcision of the newborn; the total helplessness of the newborn makes such an operation even more frightening and unbearable. When routine practice and cultural attitudes violate human rights, they must be abolished.
Counter-claim Circumcision is advisable for hygienic reasons, to make washing easier. It may also help prevent kidney- and urinary-tract infections, which can be quite dangerous. If done in infancy, it is a trivial, harmless operation.
Refs
Broader Sexual mutilation (#PD5718) Circumcision as a health hazard (#PE6053)
Mutilation and deformation of the human body (#PD2559).
Related Torture through mutilation (#PD7576).

♦ **PE6055 Female sexual mutilations**
Female circumcision — Excision — Infibulation — Clitoridectomy
Nature The custom of sexual mutilation of young girls is a violation of the right of the child to be protected from bodily harm and mutilation. The degree of mutilation varies: circumcision, the removal of part or all of the clitoris; excision, the removal of part or all of the clitoris and the labia minora; infibulation, the almost complete removal of the clitoris and of the labia minora and majora, following which the vagina is sewn up to leave only a small opening. Although the severity of the effects on health vary according to the degree of mutilation, the practice is a major health problem. Immediate effects include shock, retention of urine and of menstrual blood with attendant infections, and infections resulting from the operation itself, which is often performed with unsterilized and inadequate equipment. Later the woman may suffer cyst formation, fistulas, sterility, and pelvic and urinary infections. Almost all sexually mutilated women have difficulties in labour, incision of the vulva generally being necessary.
There is no religious basis for female sexual mutilation, and the practice is found in communities of different religions. The cultural importance can however be so great that a girl who does not undergo the operation at puberty will bring shame to her family, will be considered unclean, impure, and probably a prostitute; she will never be considered marriageable and thus will have no future in her own community. A girl who does undergo the operation, however, will probably have poorer health and will never be able to enjoy sexual relations with her husband when she does marry.
Incidence It is estimated that 80 million women are sexually mutilated. The practice is most common in a continuous belt crossing the middle of Africa from west to east and continuing down the Nile. It is estimated that in Djibouti 95 percent of the women are infibulated, and that in the Sudan and Somalia nine out of ten women are circumcised, although the Sudan, along with Kenya, has laws prohibiting female circumcision. Statistics show little reduction in numbers of mutilated women in the last ten years despite increasing awareness of its role as a health problem.
Background The practice can be traced back to ancient Egypt.
Claim This practice is a form of sexual oppression, designed and perpetuated to deny women normal sexual expression and enjoyment. Mutilation does serve its purpose of decreasing the sexual urges of a woman, but only because it makes any sexual act extremely painful for her. Female circumcision often results in bleeding during intercourse and the use of anal intercourse in the first 3 to 9 months of marriage, until the vagina is open enough to permit penetration. This may be the cause of widespread incidence of AIDS among heterosexuals in cultures that practice female circumcision.
Refs Dareer, Asma El *Woman, Why Do You Weep?* (1983); Koso-Thomas, Olayinka *The Circumcision of Women* (1987).
Broader Sexual mutilation (#PD5718) Violence against women (#PD0247)
Mutilation and deformation of the human body (#PD2559).
Related Torture through mutilation (#PD7576) Circumcision as a health hazard (#PE6053).
Aggravates AIDS (#PD5111).

♦ **PE6056 Microwave radiation as a health hazard**
Incidence Microwave radiation is emitted by military and civilian radar installations, satellite ground stations, relay towers for long-distance telephone links and television transmitters, as well as microwave ovens and citizens band radios. Results emerging from a US study in 1984, investigating the biological and health effects of non-ionizing radiation, show glandular changes and a higher rate of cancer among laboratory rats chronically exposed to low-intensity microwaves. These findings could provide an experimental basis for widely reported complaints of headaches, dizziness, memory loss and fatigue from people chronically exposed to microwave radiation.
Refs ILO *Protection of Workers Against Radio-Frequency and Microwave Radiation. A technical review* (1986).
Broader Radioactive contamination (#PC0229) Health hazards of radiation (#PD8050)
Excessive exposure to radiation from consumer goods and electronic devices (#PE1909).

♦ **PE6059 Excessive cost of effective prosecution of offenders**
Aggravates Inadequate national law enforcement (#PE4768).
Aggravated by Prohibitive legal fees (#PF0995).

♦ **PE6069 Discrimination against women at retirement age**
Compulsive early dismissal
Nature Retirement discrimination means that women are compelled to retire at an earlier age than her male colleagues.
Broader Unjust dismissal of workers (#PD5965)
Discrimination against women in employment (#PD0086)
Employment discrimination against the elderly (#PD4916).
Aggravates Ageing women (#PE6784).

♦ **PE6074 Misappropriation of cultural property**
Illicit appropriation of cultural objects — Non-restitution of cultural property — War-time exportation of cultural property from occupied territories — Lost cultural heritage
Nature Throughout history, wars and colonization have led to the appropriation of cultural objects by the conquerors to the detriment of the vanquished or colonized peoples. The transfer of the cultural object outside its culture of origin irreparably deprives it of one of its dimensions. A loss of this kind can occur in a historical context in respect of a work of art which, initially, is more a testimony to the sacred than a form of aesthetic expression.
Incidence The Third World is pressing for the restitution or return of its cultural heritage, but although general public in the West is now aware for the first time of the massive scale of cultural loss suffered by the Third World, the initial response to Third World demands for restitution has not been good. Fearing a massive onslaught on their possessions, the guardians of Europe's and the United States' treasure houses are preparing to prevent the removal of their treasures.
Claim The West is caught in a painful dilemma. To concede the principle of restitution which in terms of its own moral or ethical code it is bound to do, involves the admission that the goods were illicitly acquired in the first place: a harder look at the dark side of Europe's 'civilizing mission'. Colonialism is scarcely a generation away. It will take longer that that for Europe to re-write the history of Empire in the image of the Third World. Seldom raised as a rational objection to the demands for restitution, this factor constitutes a major barrier in Western thinking on the matter. In securing the return of its cultural heritage, the Third World may well have to show greater tolerance towards the psychological problems of their former rulers than was ever extended to them as subject peoples. The transfer of cultural items is the strongest case yet for historical redress: the Third World is asserting the primacy of its own culture against centuries of cultural

domination by the West.
Broader Non-restitution of property (#PC7859) Restriction of freedom of expression (#PC2162).
Narrower Misappropriation of words (#PF5247)
Restrictive commercialization of definitions (#PF4674).
Related Misappropriation of sacred objects (#PD8041)
Abusive exploitation of cultural heritage (#PC7605).
Aggravated by Illicit export of works of art (#PE9004)
Accumulation of cultural property (#PD6907).

♦ PE6075 Parental kidnapping

Nature A child who is in the legal custody of only one of his or her parents (the parents being legally separated) may be kidnapped by the parent without custody and, in some cases, taken outside the country and thus outside the scope of the law. This violates not only the legal rights of the parent with custody but also the basic rights of the child.
Incidence In the UK in 1990 it was estimated that 500 children were abducted and taken abroad each year by one parent, without the consent of the other.
Broader Kidnapping (#PD8744).
Related Child abduction (#PE6154).

♦ PE6086 Denial of right to participate in government

Violation of right to participate in conduct of public affairs
Broader Denial of political rights (#PD8276) Denial of political and civil rights (#PC0632).
Narrower Lack of representation (#PF3468) Limitations on right to vote (#PF2904)
Denial of right to hold public office (#PE5608)
Denial of the right to organize political parties (#PE9110)
Denial of right of indigenous peoples to participate in political processes (#PE7312).
Related Denial of right of complaint (#PD7609)
Discrimination against women in politics (#PC1001).

♦ PE6087 Nitrites as pollutants

Nature Nitrites are found in the soil and formed in the human gut by the action of bacteria on nitrate. In nature is is part of the nitrogen cycle and is rapidly converted to nitrate which is absorbed by plants. In humans it can cause problems. Nitrite combines with haemoglobin preventing the blood from carrying oxygen around the body efficiently. A significant degree of the body may be suffocated from within, a condition called methaemoglobinaemia. This is rare in adults but occurs in babies and small children. Nitrite may also present a cancer risk.
Refs FAO *Water Quality Criteria for European Freshwater Fish* (1984).
Broader Chemical contamination of water (#PE0535)
Nitrogen compounds as pollutants (#PD2965).
Related Acidic precipitation (#PD4904).
Aggravates Lake pollution (#PD8628).

♦ PE6093 Creative accounting

Narrower Foreign currency manipulations in accounting records of transnational corporations (#PE2145).
Aggravates Abuse of credit (#PF2166) Business bankruptcy (#PD2591)
Risk of capital investment (#PF6572)
Misleading accounting information due to inflation (#PE4285).
Aggravated by Protectionism in public accounting and auditing services (#PD7073)
Inadequate international standards of financial accounting and reporting (#PF0203).

♦ PE6115 Mismanagement of food resources

Nature Much of the starvation and malnutrition prevalent today is not due to problems of food production, but rather international food disorder rooted in the mismanagement of food resources. Many countries in the South produce foods which are insufficient in providing their peoples with a balanced diet, thus they are dependent upon the markets of wealthier nations. Agricultural infrastructure, research and finance are aimed at improving the production of export crops rather than at producing foods to be consumed by the countries' peoples; concurrent with this is the forced reduction of crops in industrialized countries, which elevates food prices.
Claim Food is often used in international affairs as a political weapon. This bargaining, interfering and manipulating is counter to the fundamental and basic human right of all people to have adequate food. Every national also has the right to self-determination and self-reliance, and under no circumstances should food supplies – and food disorder – be used to control or limit that right.
Broader Mismanagement (#PB8406).
Related Agricultural mismanagement (#PD8625).
Aggravates Food insecurity (#PB2846) Unnecessary reserves of material (#PF0687)
Unpreparedness for food emergencies (#PC5016)
Inadequate mechanisms for securing sufficient food supplies (#PF2857)
Import-dependency in food staples in developing countries due to transnational corporations (#PE1806).
Aggravated by Worldwide misallocation of resources (#PB6719)
Secrecy of national basic food stocks (#PF4763).

♦ PE6119 Driving delinquency

Young driver accidents
Broader Reckless driving (#PE9334) Road traffic violations (#PE0930).
Aggravates Road traffic accidents (#PD0079).
Aggravated by Mental deficiency (#PC1587).

♦ PE6126 Inadequate night life entertainment facilities

Nature Many cities close down at night, and do not have adequate facilities for the evening entertainment of the population. Conversely, those cities which provide entertainment often do so on a lavish scale that is actually alienating. Those cities that do provide a few facilities often have their cinemas, cafes, ice cream parlours, etc, scattered throughout the community, so that each one cannot, by itself, generate enough attraction; furthermore, such scattered night spots attract crime. There are very few cities that have suitable centres of mutually enlivening night places, such as for instance a cinema, a restaurant and a bar together with a bookshop open till midnight or, say, a launderette, liquor store and cafe near to a meeting hall and beer hall.
Broader Unstimulating entertainment (#PF8105).
Narrower Unconvivial hotel environments for travellers (#PF6196)
Inhibition of collectively organized fantastic happenings (#PF6170).
Related Unattractive pedestrian environments in urban areas (#PE6151)
Inaccessibility of city centres to suburban residents (#PF6132)
Ineffectiveness of individual participation in large communities (#PF6127)
Lack of places in urban environments encouraging unstructured public access (#PF6190).
Aggravated by Closure of recreation areas (#PE6276).

♦ PE6137 Sado-masochism

Homosexual sadomasochism
Nature In the psyche coexist sadistic and masochistic wishes, fantasies, and drive derivatives. Sadomachochistic gratifications can be involved e.g in teasing, sarcasm, slander, clowning and passive aggression. Some people consciously or unconsciously need to suffer, be punished or punish. In sexual behaviour it ranges from relatively normal to the perverse when physical pain is requisite for the sexual relationship.
Counter-claim With increasing sexual liberation, justifications for sadomasochism have been increasingly advanced by the homosexual community. A therapeutic social role has been claimed by social scientists. It has been described as a new dimension of love and as imitating, and thus respecting, nature.
Broader Sadism (#PF3270) Psychological conflict (#PE5087)
Human sexual inadequacy (#PC1892).
Aggravated by Male homosexuality (#PF1390).

♦ PE6139 Intimidation of pedestrians by vehicles

Nature Vehicles are a threat to pedestrians even when the people walking have the legal right-of-way, particularly where the footpath and the road are at the same level. No amount of painted white lines, zebra-crossings, traffic lights, button-operated signals or even traffic police on duty ever alter the fact that a car may strike a pedestrian unless the driver brakes; and brakes and drivers do fail. The absence of raised cross-walks and raised road-side pavements contributes to accidents and pedestrian insecurity.
Broader Physical intimidation (#PC2934).
Related Urban road traffic congestion (#PD0426)
Maldistribution of urban shopping facilities (#PE6144)
Environmental degradation from high-speed roads (#PD6124)
Impersonality of mass market shopping facilities (#PE6153)
Unattractive pedestrian environments in urban areas (#PE6151).

♦ PE6142 Environmental degradation by automobiles

Nature Some of the obvious societal and health problems created by the car are air pollution, noise, danger, ill-health, congestion, parking problems and that of being an eyesore. Cars are not practical means of transport for short trips inside a town, and it is on these trips that they do their greatest damage. In addition, cars and their parts have considerable economic value. Many are stolen and abandoned later as wrecks. Cars are also representative of affluence and as such are targets for the alienated actions of juveniles. Deliberately vandalized, and in some cases, burned, automobiles are abandoned in the streets by their owners.
Refs Renner, Michael *Rethinking the Role of the Automobile* (1988).
Broader Environmental degradation (#PB6384).
Narrower Excessive use of land by automobiles (#PF6152).
Related Environmental degradation from high-speed roads (#PD6124).
Aggravates Environmental degradation of inner city areas (#PC2616).
Aggravated by Urban road traffic congestion (#PD0426)
Vulnerability of computer systems (#PE8542)
Stultifying homogeneity of modern cities (#PF6155)
Maldistribution of urban shopping facilities (#PE6144)
Inaccessibility of countryside to city dwellers (#PF6140)
Inaccessibility of city centres to suburban residents (#PF6132).

♦ PE6144 Maldistribution of urban shopping facilities

Nature In an ideal town, where the shops are seen as part of society's necessities, shops would be widely and homogeneously distributed. At present, large parts of modern towns have insufficient services. New shops which could provide these services often locate themselves near the other shops and major centres, instead of locating themselves where they are needed. It is also true that many small shops are unstable (in the United States, two-thirds of the small shops that open go out of business within a year). Obviously the community is not well served by unstable businesses; once again, economic instability is largely linked to mistakes of location.
Broader Stultifying homogeneity of modern cities (#PF6155)
Artificial separation of home and workplace (#PF6122)
Insufficient separation between urban subcultures (#PF6137).
Narrower Impersonality of mass market shopping facilities (#PE6153).
Related Intimidation of pedestrians by vehicles (#PE6139).
Aggravates Urban road traffic congestion (#PD0426)
Environmental degradation by automobiles (#PE6142).

♦ PE6151 Unattractive pedestrian environments in urban areas

Nature Streets in modern cities tend to be designed for going through not for staying in. This is reinforced by regulations which make it a crime to loiter, and by streets which are too unattractive or have too heavy traffic to socialize in, so that people are virtually forced into their houses.
Claim Just as for centuries the street provided city dwellers with usable public space right outside their houses, present-day streets should be designed for human contact, and not just for moving through.
Broader Uncontrolled urban development (#PC0442)
Stultifying homogeneity of modern cities (#PF6155).
Narrower Impersonality of public squares in cities (#PF6165)
Inhibition of collectively organized fantastic happenings (#PF6170)
Inhibition of exploration by children of urban environment (#PF6159).
Related Intimidation of pedestrians by vehicles (#PE6139)
Impersonality of high density accommodation (#PF6156)
Unbalanced urban population density gradients (#PD6131)
Unconvivial hotel environments for travellers (#PF6196)
Inadequate night life entertainment facilities (#PE6126)
Disorientation stress in large building complexes (#PF6174)
Inaccessibility of city centres to suburban residents (#PF6132)
Ineffectiveness of individual participation in large communities (#PF6127)
Lack of places in urban environments encouraging unstructured public access (#PF6190).
Aggravates Monolithic architecture of high-rise buildings (#PE1925).
Aggravated by Inaccessibility of water for recreation (#PF6138).

♦ PE6153 Impersonality of mass market shopping facilities

Nature There is a tendency for shopping in modern cities, with their big supermarkets, to become increasingly impersonal. When shops are too large, or controlled by absentee owners, they become rather bland and abstract. The larger they become, the less personal their service is, and the harder it is for smaller shops around them to survive. Merchandise in supermarkets, though varied, has much less 'character' than that in individual shops or market stalls, and there is little or no interaction between the citizen and the shopkeeper. Many economic factors tend to exacerbate this problem, such as the policy of central buying and the profit motive on the part of supermarket chains: the convenience of customers is served, but at the price of sacrificing individual personality and human contact in shopping. Franchising is another modern example of 'impersonal' shopping, since the object of franchising is to create shops that look individual, but are in fact part of a chain operation, wherein each shop is the same, or extremely similar.
Incidence Research into people's attitudes towards personal shopping suggests that there is room to keep or restore the small local shop. In Berkeley, California, a survey showed that 80 percent of the customers of a neighbourhood store still happily walked to that store, and suggested that corner groceries should be within three to four blocks, or 1200 feet, of homes. Further studies in San Francisco indicate that a corner grocery shop can survive under circumstances where there are 1000 people within three or four blocks, a net density of at least 20 persons per net acre, or six houses per net acre. Most city neighbourhoods do have this density and San Francisco, for example, supports 638 neighbourhood grocery shops with a population

of 750,000 people. One of the difficulties of maintaining local shops is the size of the capital investment each requires. A solution could be to make them smaller: in Morocco, India, Peru and many other countries, shops are often no more than 50 square feet in area.
Broader Maldistribution of urban shopping facilities (#PE6144)
Insufficient separation between urban subcultures (#PF6137)
Inaccessibility of city centres to suburban residents (#PF6132).
Related Intimidation of pedestrians by vehicles (#PE6139).
Aggravated by Urban road traffic congestion (#PD0426)
Monolithic architecture of high-rise buildings (#PE1925).

♦ PE6154 Child abduction
Illegal appropriation of children — Baby snatching
Refs Sachs, Chritina *Child Abduction* (1987).
Broader Kidnapping (#PD8744) Victimization of children (#PC5512).
Related Parental kidnapping (#PE6075).
Aggravated by Childlessness (#PC3280).

♦ PE6155 Diseases of the spleen
Broader Human disease and disability (#PB1044)
Diseases of blood and blood-forming organs (#PF8026).
Aggravates Leukopenia (#PG5543) Pancytopenia (#PG5541)
Nutritional anaemia (#PD0321).
Aggravated by Leukaemia (#PE0639) Tuberculosis (#PE0566)
Polycythemia (#PG5546) Lymphosarcoma (#PE5548)
Gaucher's disease (#PG5547) Rheumatic diseases (#PE0873)
Lymphatic leukaemia (#PE2686) Cirrhosis of the liver (#PE2446)
Infectious mononucleosis (#PE5550).

♦ PE6163 Wastage of open space in urban environments
Nature Research indicates that people will only use open space if it is in the sun; thus thousands of acres of open space in every city are unused because they are shielded from the sun by buildings – whether public buildings and of private houses. For example, many modern buildings in the Northern hemisphere have plazas and gardens on the north side. At lunchtime the plazas are empty and people eat their sandwiches in the street, on the south side where the sun is. The same is true for small private houses.
Incidence A survey of a residential block in Berkeley, California confirmed that 18 out of 20 people only used the sunny part of their gardens. People living on the north side of the street did not use their back gardens at all, and not one person interviewed indicated a preference for a shady garden. Other research indicates the importance of open space being 'defined' by some kind of enclosure, such as wings of buildings, trees, hedges, fences, arcades, etc. The popularity of European city squares confirms this; those that are particularly used are always partly enclosed, but they are also open to one another, so that each one leads into the next.
Broader Impersonality of public squares in cities (#PF6165)
Obstacles to availability of community space (#PF7130)
Insufficient common land in urban environments (#PE6171)
Inaccessibility of green parkland in an urban environment (#PE6175).
Narrower Unrelated buildings in urban environments (#PF6199).
Related Displacement of natural light in buildings (#PF6198).
Aggravates Monolithic architecture of high-rise buildings (#PE1925).
Aggravated by Inaccessibility of water for recreation (#PF6138).

♦ PE6168 Denial of right to private home life
Broader Infringement of privacy (#PB0284).
Related Interception of communications (#PD7608)
Denial of right to confidentiality (#PD6612).
Aggravates Arbitrary external interference in family life (#PE4058).
Aggravated by Homelessness (#PB2150).

♦ PE6171 Insufficient common land in urban environments
Nature Common land has two specific social functions. First, the land makes it possible for people to feel comfortable outside their buildings and their private territory, and therefore allows them to feel connected to the larger social system – though not necessarily to any specific neighbour. And second, common land acts as a meeting place for people. In pre-industrial societies, common land between houses and workshops existed automatically; but in a society with cars and trucks, the common land which could play an effective social role in knitting people together no longer occurs in this way. There is an argument that agoraphobia is increased among city dwellers by their lack of common land (people may feel that they have no 'right' to be outside their own territories). In fact, estimates suggest that the amount of common land needed in a neighbourhood is about 25 percent of the land held privately. This is enough to provide a meeting ground for the people who live in the vicinity, to contain children's games and to give a meeting place for mothers where, on a good day, they might come together under a big tree or a pergola, to sew or gossip, while their infants sleep in a pram or their runabout children grub around in a play pit.
Broader Limited available land (#PC8160).
Narrower Wastage of open space in urban environments (#PE6163)
Inadequate interaction between humans and animals in the urban environment (#PF6185).
Related Sterile working environment (#PD6133)
Impersonality of high density accommodation (#PF6156)
Disorientation stress in large building complexes (#PF6174)
Inadequate arrangement of housing with respect to common land (#PF6146).
Aggravates Inadequate facilities for children's play (#PD0549)
Unhealthy lack of daily physical activity in urban environments (#PF6182).
Aggravated by Excessive land usage (#PE5059)
Shortage of urban land (#PD0384)
Excessive use of land by automobiles (#PF6152)
Inaccessibility of water for recreation (#PF6138).

♦ PE6172 Havens for environmental pollution
Havens for hazardous products
Incidence Certain countries have little legislation regulating actions which are considered as hazardous to the environment in others. Many countries have legal loopholes permitting some forms of pollution. Many others, even where there is legislation against such pollution, are lax in the implementation of the law. Others are prepared to create exceptions, especially if this is in their financial interest.
Broader Environmental pollution (#PB1166)
Inaccessibility of green parkland in an urban environment (#PE6175)
Inhibition of individual psychological development through life cycle (#PF6148).
Aggravates Hazardous waste dumping (#PD1398)
Trading in products containing toxic substances with developing countries (#PE2061).
Aggravated by Inadequate national law enforcement (#PE4768)
Arbitrary enforcement of regulations (#PD8697).

♦ PE6175 Inaccessibility of green parkland in an urban environment
Incidence A 1971 citizen survey in Berkeley, California showed that the great majority of people living in apartments in towns want access to a) a pleasant, private balcony and b) a quiet public park within walking distance. Further research showed that walking distance, in effect, means that the park should not be more than 2–3 minute walk away. People who live in close proximity to green parkland use it frequently. Those who live more than a 3 minute walk away do so less frequently in direct proportion to the increase in distance. Since the people who live further away obviously need the relaxation a park brings as much as those who live near, it follows that most urban planning penalizes many city dwellers by not providing enough parks. The same research suggested that an adequate park should be as much as 60,000 square feet in area, and at least 150 feet wide in the narrowest direction, in order to enable people to feel in touch with nature, and away from the hustle and bustle.
Broader Inaccessible recreation areas (#PF6503)
Inaccessibility of quiet zones in an urban environment (#PF6160).
Narrower Havens for environmental pollution (#PE6172)
Wastage of open space in urban environments (#PE6163).
Related Unidentifiable urban neighbourhoods (#PF6147)
Insufficient separation between urban subcultures (#PF6137)
Unnatural urban environments inhibiting sleeping in public (#PF6193)
Inadequate interaction between humans and animals in the urban environment (#PF6185).
Aggravates Unhealthy lack of daily physical activity in urban environments (#PF6182).

♦ PE6183 Maldistribution of teachers
Nature Many of the problems concerning children are made worse by the high ratio of children to teachers in schools.
Incidence In American and Western Europe, for example, a common ratio is 30 children to 1 teacher. It is generally recognized that a satisfactory ratio is no more than 10 children to 1 teacher, but this is affordable only by private or special schools, where parents pay fees.
Broader Inadequacy of formal education (#PF4765).
Aggravates Inhibition of exploration by children of urban environment (#PF6159)
Excessive intensification of the parent-child relationship (#PF6186)
Unavailability of trained teachers in the rural areas of developing countries (#PE8429).
Aggravated by Lack of job satisfaction (#PF0171)
Inadequate working conditions of teachers (#PE7165)
Excessive dispersion of community facilities (#PF6141)
Inhibition of individual psychological development through life cycle (#PF6148).

♦ PE6187 Inappropriate accommodations for single people
Nature In modern cities it is almost impossible for architects to build houses suitable for one person to live in because zoning codes and banking practices prohibit very small building plots. Nor does the existing housing market contain many houses or apartments specifically built for one person. Men and women who choose to live alone tend to live in larger houses and apartments originally built for two people or for families, yet for one person such large dwellings are are unwieldy, hard to live in and hard to look after; and lack the compactness and simplicity of the dwelling specifically designed for one person.
Broader Inadequate housing (#PC0449).
Related Socially sterile rental accommodation (#PF6195).
Aggravated by Household segregation by age group (#PF6136)
Negative effects of the nuclear family (#PF0129)
Stultifying homogeneity of modern cities (#PF6155)
Inhibition of adult life in small households (#PF6167)
Inadequate arrangement of housing with respect to common land (#PF6146).

♦ PE6188 Literary forgery
Refs Boller, Paul F Jr and George, John *They Never Said It* (1989); Koobatian, James (Comp) *Faking It* (1987).
Broader Art forgery (#PE2382).
Related Historical forgery (#PE5051).
Aggravates Historical misrepresentation (#PF4932).

♦ PE6197 Hay fever
Pollinosis — Summer catarrh — Autumn catarrh
Refs Turner, Roger N *The Hay Fever Handbook* (1988).
Broader Allergy (#PE1017) Human disease and disability (#PB1044)
Diseases of upper respiratory tract (#PE7733).
Aggravates Fever (#PD2255) Asthma (#PD2408) Chronic bronchitis (#PE2248).
Aggravated by Detergents as pollutants (#PE1087)
Genetic defects and diseases (#PD2389).

♦ PE6201 Increasing drug experimentation in developing countries
Nature Introduction of experimental pharmaceuticals into less developed countries where public health regulations are not restrictive, is an increasing phenomenon.
Incidence Trials of new contraceptive drugs, for example, are often conducted in Third World countries which may have no preclinical requirements or supervision for therapeutic testing; pharmaceutical companies may wish to evade controls imposed for the safety of women's health in such trials. Drug companies in the USA, for example, spend between 15 and 20 percent of their research budgets abroad, mainly in developing countries, for all kinds of products.
Counter-claim Drugs tested in developing countries are often for diseases, deficiencies, and dysfunctions endemic to the region and require on-site testing with the blood-types and other genetic factors of the native population, in their own environment, present on the trials. Developing countries welcome such attempts to supply their medical needs. Some, perhaps, neglect the opportunity to work more closely with the manufactures to build medical and scientific infrastructures, including technician training, clinical laboratories and local drug manufacturing.
Broader Unethical experiments with drugs and medical devices (#PD2697).
Aggravated by Inadequate testing of drugs (#PD1190).

♦ PE6206 Inadequate education for nomadic children
Nature Schools for nomadic children are not based upon the particular nature of the nomadic mode of life as people constantly on the more, but are instead structured on the pattern of schools organized for sedentary children. They are often located long distances from the children's camps (so that parents are thus unwilling to enroll their children), they offer regular school hours although parents are unwilling to allow their children to attend distant schools for long hours, leaving herding tasks to them; they do not take into consideration the difference between the rainy season (when grazing grounds nearby are abundant) versus the dry season (when children must often search hours to graze their animals), and their curricula often stress knowledge which will never be applicable to a nomad's way of life.
Claim Due to the particular nature of the nomadic way of life as people constantly on the move, school provision for nomadic children should be based on the occupational roles and other cultural values of nomads. The development of such education is necessary to ensure progress in the nomadic way of life, with the aim of enabling nomads to respond to modern ways while retaining what is good in their own culture.
Broader Lack of education (#PB8645) Inadequate education (#PF4984).
Aggravates Social maladjustment of children of migrants (#PE4258).

♦ **PE6209 Violence against prostitutes**
Incidence The danger to prostitutes does not come only from pimps, but from customers and police as well. A 1981 study in New York found that 70 percent of prostitutes had been raped. Murder is also a serious problem, with more than 200 prostitutes being killed a year, giving a risk of about four times that of women in general. The police are largely negligent in helping prostitutes and undercover vice officers often feel free to insult and roughly handle the prostitutes they arrest, with the physical abuse ranging from tightly handcuffed hands being pulled roughly up in the back, to outright beatings and kickings. The verbal abuse ranges from specific insults about the individual prostitute's body to taunting about the potential for the police officer to get free sex with no one the wiser, to suggestions that the prostitute give the sheriff sexual favours in order to be released from gaol. Most prostitutes accept this abuse as part of the job, and so the few accounts that surface must be seen to be symptoms of a much larger problem.
Broader Human violence (#PA0429).
Aggravates Offences involving danger to the person (#PD5300).

♦ **PE6212 Denial of right to economic asylum**
Broader Rejection of refugees (#PF3021).
Related Denial of right to cultural asylum (#PE3450)
Lack of individual rights to political asylum (#PF1075).

♦ **PE6214 Over-use of formaldehyde in building materials and personal products**
Nature Formaldehyde is a potent sensitizing agent and an animal carcinogen. As a reactive and inexpensive chemical it is a building block in many manufacturing processes, during which formaldehyde may be given off to react with other substances in the environment to cause, from short-term exposure, respiratory problems, and from long-term exposure, damage to the immune system. Small concentrations may cause allergic reaction and have particularly bad effects on people suffering from respiratory complaints such as asthma.
Incidence Formaldehyde is a byproduct of many combustion processes, including that of natural gas and petroleum, so it is a component of car exhausts; and is also found in building materials, urea-formaldehyde foam insulation, resin treated textiles, and wood particle products glued together with formaldehyde resins. Such products are used and manufactured around the world. Sweden, Germany, Denmark, Belgium and the USSR have set maximum formaldehyde standards for buildings.
Background US consumers complained to manufacturers about urea-formaldehyde foam insulation, mobile homes and conventional homes constructed with wood particles products – they were getting sick in their homes. The complaints multiplied until the US Consumer Product Safety Commission launched an investigation in 1980.
Broader Dangerous substances (#PC6913).

♦ **PE6219 Smoking by adolescents**
Smoking by children
Incidence In the UK, one in five 15 year olds smoke. Over 300,000 children between 11 and 15 smoke regularly and 180,000 smoke occasionally. By the age of 17, 90 percent of young people will have tried smoking. Of these, 20 percent continue. The earlier a person begins to smoke, the more likely he is to become a heavy smoker. Children from homes where there is stress because of divorce, alcoholism or unemployment are more likely to begin smoking regularly than are children from stable families.
Claim Undeterred by the deaths of at least 110,000 people every year in the UK from smoking-related illnesses, the tobacco industry continues to derive some 70 million pounds per year from the sale of cigarettes to children under 16.
Refs Rattray, Jamie and Howells, Bill and Siegler, Irv et al *Kids and Smoking* (1983).
Broader Smoking (#PD0713).

♦ **PE6226 Ineffective regulation of electronic messages**
Unregulated videotext
Related Law enforcement complexity (#PF2454) Criminalization of prostitution (#PF6231)
Diffuseness of regulatory authority (#PF3064)
Discrimination against men before the law (#PD3692)
Discriminatory religious influence on the law (#PD3357).
Aggravates Electronic pornography (#PE5402).

♦ **PE6237 Export-led development**
Export-led industrialization
Nature Export-led industrialization rarely links the production of internationally marketable goods to the basic needs of a developing country. The export goods are often unusable in the country where they are produced or are only sub-assemblies of a final product completed in an industrial country. The technologies used are only partially transferable to domestic needs and few skills are acquired by the workers. In many cases governments have to invest considerable amounts of scarce foreign exchange in the necessary infrastructure to attract the transnational companies, which usually form the basis of export industries. Liberal tax and profit remittance rules may also be needed, thus the yield of foreign exchange is often quite negligible or even negative.
Incidence Although successfully applied by the newly industrializing countries, such as Taiwan and South Korea, during the 1970s, it no longer constitutes a viable model for the appropriate development of other developing countries. The markets of the industrialized countries are increasingly saturated with such exports and are imposing a variety of protectionist barriers. There is therefore little possibility of basing development on such exports.
Aggravates Maldevelopment (#PB6207) Destruction inherent in development (#PF4829).

♦ **PE6240 Disadvantages for homeworking employees**
Disadvantages for workers at home
Nature Homeworking employees or sub-contractors often experience the conditions of sweated labour including very low payment, long hours (frequently including the night), no paid absences or other benefits, and poor working conditions. In addition, tools, equipment or materials may provide hazards for which there is inadequate protection. Especially for women there is the strain of combining employment and housework in the same place. The circumstances of outwork cannot readily be improved by collective action, as the workers are too dispersed. A great number of homeworkers may find the conditions isolating and depressing. Outwork may be a device to return women to the home, and this will reverse progress made towards equal pay and opportunities for women.
Counter-claim Outwork is intended for both men and women. It offers workers flexible hours; no travel time, travel inconveniences or travel fatigue; healthier work environments than congested offices or factories; and opportunities for more contact between the homeworker and his or her family. When desired, decentralized neighbourhood work centres, or other small groupings of workers in homes can be provided.
Aggravates Social isolation (#PC1707).

♦ **PE6241 Discrimination against part-time work**
Exploitative part-time work
Nature Part-time work is often underpaid, unskilled and insecure and without social security benefits. It is frequently used to extend working hour of a business with a minimum increase of costs. It is frequently at the expense of full-time work and salaries.
Incidence In 1989, part-time workers comprise 25.6 per cent of the total workforce in Norway, 24.7 per cent in UK, 20.2 per cent in Australia, 17.3 per cent in US and 12 per cent in Japan. The typical part-time worker is a woman between 25 and 44 years of age with children. In the UK, individuals working less than 16 hours per week, and, in Belgium, 2 hours or less a day, lose their benefits.
Refs ILO *Part-time Work* (1989).
Broader Underemployment (#PB1860) Discrimination in employment (#PC0244)
Denial of right to free choice of work (#PE3963).
Narrower Part-time farm employment (#PJ1074).
Related Insufficient part-time employment (#PJ0108).
Aggravates Discrimination against working mothers (#PD6812).
Aggravated by Mobility of village populations (#PE1848).

♦ **PE6243 Inadequate supply of appropriate trained manpower in developing countries**
Scarcity of skilled manpower in developing countries
Nature The inadequate supply of appropriate trained manpower – that is the right numbers, at the right time, in the right place, and with the right balance of technical knowledge and practical skills – impedes the pace and direction of industrial growth, industrial innovation and economic and social development.
Broader Scarcity (#PA5984) Labour shortage (#PC0592)
Deficiencies of developing countries (#PC4094).
Narrower Inadequate human resources development in least developed countries (#PE9764).
Related Apathy in developing countries (#PD8047)
Limited exchange of skills among developing countries (#PF5006).
Aggravates Inadequate development of enterprises in developing countries (#PE8572).

♦ **PE6245 Discrimination against black working women**
Nature Black working women experience the double humiliation of discrimination because they are women, and discrimination because they are black. They are often (due to discrimination that begins with poverty, education and lack of opportunities) forced to do menial jobs (due to lack of skills); take employment as domestics (which means long hours and low wages); work in industry (where the pay is low, the benefits often non-existent, and the work conditions deplorable); or take work in the agricultural sector, where they are exploited as casual workers.
Broader Racial discrimination (#PC0006)
Discrimination against women in employment (#PD0086)
Unequal employment opportunities for women (#PD5115).

♦ **PE6246 Job reservation under the apartheid system**
Background Job reservation is one means by which the South African government (under the apartheid system of maintaining white supremacy at the cost of the black majority) limited job opportunities for blacks by prohibiting the replacement of workers of one race by workers of another race; by compelling employers to maintain a fixed percentage of workers of a particular race; by reserving any class of work of specific jobs or work generally for members of a specific race; and by fixing maximum, minimum, or average numbers or percentages of persons of a specified race who may be employed in any factory or industry or other places of employment. These restrictions were applied under Section 77 of the Industrial Conciliation Act of 1956; all but five of them were abolished in 1977, but this had little effect as the general industrial colour bar based on long-standing attitudes and attitudes was underpinned by formal closed shop agreements between white trade unions and major companies. These affect the majority of black workers far more seriously.
Claim Although reforms have been instituted on paper, they are, in practice, far removed from genuine change. The attitude of employers and the white trade unions still effectively means that black African workers are barred from access to certain types of jobs which remain firmly in the hands of white workers.
Broader Apartheid (#PE3681).

♦ **PE6251 Environmental degradation by pipelines**
Nature There are enormous latent environmental implications in the laying of pipelines, both for the communities through which they pass and for those to which they transport their energy resources. A pipeline corridor frequently crosses undeveloped areas, sensitive environments, agricultural lands, or timberlands. The heavy construction activity taking place within this corridor may interrupt the migratory patterns of caribou, elk or other large animals, or the fish spawning grounds on which local hunting and fishing activities depend. Gravel mining sites impact hillsides or stream beds with scars; and haul roads built to transport materials and equipment cause impacts both inside and outside of the pipeline corridor.
Incidence A 1982 study estimated that over 133,000 miles of pipeline were being constructed or planned; if stung together, they would stretch around the globe 5 times.
Counter-claim With proper pre-planning the destructive environmental impacts of pipelines can be minimized and their constructive impacts can be used to improve living conditions for people around the world by making access to much-needed oil more readily available.
Broader Environmental degradation (#PB6384).
Aggravates Permafrost instability (#PD1165).

♦ **PE6252 Dismembered works of art**
Nature The dismembering of works of art is a mutilation of the cultural heritage of mankind. Wars, invasions and conquests, the ostentatiousness of famous people, the passion of amateurs and magnates for collecting, and the profit motive of art dealers – all must take some of the blame for such occurrences. Guilty, too, are curators who succumb to the temptation to enrich their museum's collections by acquiring fragments of dismembered works of art.
Background Since time immemorial, conquerors have always striven to destroy a nation's treasures as an effective means of impairing or annihilating the personality of defeated nations while furthering the profits and prestige of the victor.
Broader Debasement of works of art (#PE0558).

♦ **PE6253 Scours**
Nature Losses from scours are considerable in small herds and flocks where hygiene is lacking. Escherichia coli and certain viruses are usually incriminated as the cause of gastro-enteritis infection. Calves deprived of colostrum are at much greater risk. Treatment by antibiotics is of uncertain value. Moving calf pens to fresh clean ground at frequent intervals has been found valuable in Kenya for clearing up outbreaks. In rangelands where the young suckle, perinatal diseases are usually not serious although salmonellosis has been responsible for early deaths in calves, lambs and kids.
Broader Diseases of the digestive system in animals (#PD3978).

♦ **PE6254 Salvinia auriculata**
Nature Salvinia auriculata (water fern, Kariba weed, African pyle) are notorious for their ability to colonize large areas of water in a short amount of time. When they cover a water surface, they

cause more water to be lost by evapotranspiration than would be lost by evaporation from an open water surface of the same size (this excess water loss is always likely to occur whenever floating or emergent vegetation is present on a waterbody). The problem is more serious in areas where water is scarce or when the contents of a reservoir are not quickly replaced. Permanent weed mats can interfere with water transport and pose a hazard to navigation.
Incidence The explosive growth of salvinia auriculata has caused difficulties in North America, Africa, Asia and Australasia for the management of man-made lakes, canals and irrigation schemes in tropical areas. On Lake Kariba (on the Zambia, Zimbabwe border), large boats are unable to penetrate into certain marginal areas of the lake because of the presence of salvinia auriculata weed mats.
Broader Aquatic weeds (#PD2232).

♦ PE6255 Rust fungi
Rusts of plants
Incidence Rust fungi in southern Mexico, Guatemala and the West Indies present a barrier to natural reforestation of the pine trees. The cones are infected shortly after pollination, they increase up to four times in size, and turn bright orange.
Refs CAB International Mycological Institute *Coffee Rust (Hemileia Vastatrix and H Coffeicola).*
Broader Basidiomycetes (#PE0364)
Fungal plant diseases (#PD2225).
Narrower Needle rusts (#PG1325)
Hollyhock rust (#PG0877)
Cedar apple rust (#PG0858)
White pine blister rust (#PG4078).
Aggravates Rusts of grasses and cereals (#PE2233).

♦ PE6256 Degradation of mountain environment by leisure activities
Destruction of mountain ecosystems by ski resorts — Pollution of mountain environments by mountaineers and trekkers
Nature The ski industry brings, along with economic prosperity, destruction to the mountain environment. Tourism devours the landscape and leave behind the rubbish. Skiers leave behind heavy damage; when ski runs are initially established the growing vegetation is removed, leaving raw soil which is subject to erosion and to colonization by pioneer plants; the establishment of ski lifts leaves a succession of bare scars on mountain slopes; the heavy machinery used in the preparation of ski runs impact the snow cover (and as a result the snow cover remains longer, leaving a shorter vegetation period); and ski edges shave off the vegetation on elevated areas and ridges, leaving the turf subject to damage during periods of little snow.
Broader Destruction of wildlife habitats (#PC0480)
Deforestation of mountainous regions (#PD6282)
Natural environmental degradation from recreation and tourism (#PE6920).
Related Reduction of glacier size (#PE4256).
Aggravated by Mismanagement of environmental demand (#PD5429).

♦ PE6257 Inadequate earthquake resistant construction
Nature Increased urban populations and the use of new construction systems whose resistance to earthquakes has not been sufficiently studied have led to an increase of the earthquake hazard. Also, in spite of the considerable achievements of theoretical and experimental research into the origin and the nature of earthquakes the dynamics of soils, the response of structures and soil-construction interaction, the results are not yet generally reflected in existing building codes and regulations.
Incidence The 19 September 1985 Mexico City earthquake left thousands dead and tens of thousands homeless, and a pattern of devastation in its wake. Many buildings made of structural steel designed to sway flexibly during an earthquake remained intact, while immediately next door brick or concrete buildings came crashing down. The city's tallest buildings (the Pemex Tower with 46 stories and the Latin American Tower with 43 stories) suffered little damage, while less than 2 miles away the 5 story Televisa building collapsed, killing 50–60 employees.
Claim Buildings in earthquake-prone areas that do not take into consideration the effects of seismic forces may be considered death traps.
Refs Berlin, G Lennis *Earthquakes and the Urban Environment* (1980).
Broader Structural failure (#PD1230).
Aggravates Dam failures (#PE9517).
Aggravated by Ground failures due to liquefaction (#PE5126)
Unethical practice of earth sciences (#PD0708).

♦ PE6258 Discrimination against adult students
Nature Adult students may be discriminated against by the unthinking attitude expressed in the wording of letters and forms, the way in which queries and difficulties are handled, as well as in the way the written learning material is arranged; they may be treated as inferior to or less important than teenage students. They may be unaccustomed to formal learning and, not knowing how to study, need help in organizing themselves for this purpose, but may be ridiculed for needing such help.
Broader Discrimination (#PA0833)
Denial of rights to students (#PD6346).

♦ PE6260 Escape from official detention
Nature Escape is a person leaving official detention without authorization whether on his own or with the aid of others. Official detention includes: arrest, the period between arrest and being charged with a crime, detention before and during criminal proceedings, or during imprisonment. It also includes failing to return after a temporary leave granted for a specific purpose or period.
Broader Obstruction of government function (#PD6710).

♦ PE6263 Political impotence of youth
Nature In view of the earlier maturity of young people and their increasing involvement in all fields of social activity, the age of political majority in many countries is high, and is therefore an obstacle to active participation by young people in political or parliamentary affairs and to their standing for election. It is significant that the ages at which young people are placed under heavy obligations – for example, the age of criminal responsibility, the age of admission to employment, the call-up age for military service – are generally lower than the age of political or civil majority. In a number of countries the right to stand for election is not granted until an even later age.
Incidence In many countries the age of political majority is still twenty-one, for example in Belgium, Italy, Malta, Monaco, the Netherlands, Turkey, Colombia, Jamaica, Trinidad and Tobago, Australia, India, Pakistan, the Ivory Coast, Ghana, Mauritius, Liberia, Nigeria and Sierra Leone. It is even higher in Italy for senate elections, for which the right to vote is not granted until twenty-five.
Related Political impotence (#PJ1283).

♦ PE6264 Slashburning
Nature Fire is used extensively in the tropics for the destruction of forests. The objectives of burning are site clearing for shifting cultivation, bush and weed control, hunting, grazing, and often simply for fun. Burning is a simple and effective method of land clearing but it has serious consequences for the ecosystem which is being burnt and for its surroundings.
Burning emits large amounts of matter into the atmosphere. Large aerosol particles from burning vegetation spend a short time in the atmosphere but may effectively increase the infra-red re-radiation from the lower atmosphere. Small particles remain for periods which may be as long as several weeks or even years in the upper troposphere. Extensive savanna fires during the dry season lead to heavy dust concentrations in the atmosphere which may even spread into the region of the perhumid equatorial forest. The net effect of smoke pollution at ground surface may be cooling or warming depending on the direction of changes of the surface albedo and on the absorption coefficient of the particles in the atmosphere. Burning also releases nutrients, especially nitrogen, into the atmosphere and into the soil water. Eventually part of the latter will enter the drainage system and be lost. The fate of the former is more complicated and little understood.
Broader Misuse of tropical rain forests for agricultural development (#PE5274).
Aggravates Air pollution (#PC0119)
Smoke as a pollutant (#PD2267).

♦ PE6266 Swine fever
Classical swine fever — Hog cholera
Nature Swine fever, having a high mortality rate, can cause severe economic losses among livestock. It is infectious by contact or through feeding uncooked garbage. The disease often runs an atypical course with only a small number of animals becoming sick. Because clinical symptoms are not severe, outbreaks caused by low virulent strains may pass unnoticed for weeks or months, thus giving rise to a considerable spread of the virus.
Incidence Swine fever is widely spread throughout the tropical countries of Latin America and South East Asia. It is also evident, although controlled, in North America and Europe.
Broader Viral diseases in animals (#PD2730).
Aggravates Fever (#PD2255).

♦ PE6268 Nairobi sheep disease
Nature Nairobi sheep disease is a febrile virus gastroenteritis in sheep and goats, transmitted by Rhipicephalus appendiculatus, the same tick vector that transmits theileriosis in cattle.
Broader Virus diseases (#PD0594)
Viral diseases in animals (#PD2730).
Aggravated by Insect vectors of animal diseases (#PD2748).

♦ PE6269 Heart rot fungi
Nature Diseases caused by heart rot fungi may be classified into trunk, butt and root rots, all of which cause damage of economic significance. Root rot is especially insidious because it is difficult to determine the extent of the infection without seriously harming or killing the host tree. The incidence of heart rot causes much wastage during logging.
Incidence The fungi are most prevalent in forests in areas with high rainfall, such as northern India.
Broader Pests and diseases of trees (#PD3585).
Related Cereal root rots (#PE4453)
Dutch elm disease (#PE1154)
Rusts of grasses and cereals (#PE2233)
Smuts of grasses and cereals (#PE2234).
Aggravated by Wood rots (#PE4455).

♦ PE6270 Denial to animals of the right to conditions of life and liberty proper to their species
Nature Animals of wild species are restricted from their natural environments, and species traditionally living in association with humankind are hindered in the pursuit of their proper lifestyle. Both wild and domesticated species are restricted in reproducing.
Broader Maltreatment of prisoners (#PD6005)
Denial to animals of legal protection of their rights (#PE8643)
Denial to experimental animals of the right to freedom from suffering (#PE8024).
Narrower Maltreatment of animals in aquaria (#PE5461).
Related Animal road deaths (#PE1690)
Abuse of animal drugs (#PE0043)
In-home animal shelter (#PJ9800)
Cruel animal transportation (#PD0390)
Inadequate feeding of animals (#PC2765)
Infectious diseases in animals (#PD2732)
Inexperienced animal husbandry (#PG9643)
Denial of the right to procreate (#PC6870)
Violation of the rights of foetuses (#PE6369)
Restriction of wild animal range size (#PC0475)
Weather as a factor of animal disease (#PD2740)
Denial to animals of the right to life (#PF8243)
Denial to animals of the right to dignity (#PE9573)
Abuse in control of wild animal populations (#PE7995)
Denial to animals of the right to a natural death (#PE8339)
Disruption of animal migration and movement patterns (#PC2279)
Denial to working animals of limitation of working hours (#PE6427)
Denial to animals of the right to freedom from mass killing (#PE9650)
Denial to food animals of the right to freedom from suffering (#PE3899)
Denial to working animals of restorative nourishment and rest (#PE4793)
Denial to animals of the right to the attention, care and protection of humankind (#PF5121).

♦ PE6271 Infestation of seeds
Nature Seeds can be infested with disease or pests, causing a reduction in yield and quality. Sources of infection are smut saccules, sclerotid of fungi, spores and bacteria on the surface or inside the seeds. The most common pests are insects, ticks and worms, and they can infect the seeds either in storage or in the field. In their hidden form, as opposed to the overt, the infestation is not detectable except by special measures.
Broader Crops pests and diseases (#PE7783)
Pests and diseases of trees (#PD3585).
Related Maize pests and diseases (#PE3589)
Pests and diseases of rice (#PE2221)
Pests and diseases of wheat (#PE2222)
Pests and diseases of vines (#PE2985)
Pests and diseases of potato (#PE2219)
Pests and diseases of cotton (#PE2220)
Pests and diseases of sugar cane (#PE2217)
Pests and diseases of sugar-beet (#PE2975)
Pests and diseases of grain sorghum (#PE3590).
Aggravated by Pests of plants (#PC1627)
Fungal plant diseases (#PD2225)
Bacterial plant diseases (#PD2226).

♦ PE6272 Sheep pox
Goatpox
Nature Sheep pox febrile virus disease can cause heavy mortality and severe wool damage. Immunity is short so that annual vaccination is necessary – a costly and difficult procedure with large numbers of sheep on rangelands. The virus is comparatively resistant to physical and chemical agents and disinfection of premises is difficult.
Incidence Sheep pox is widely distributed in the semi-arid tropics of Africa and Asia. Introduced European breeds are especially susceptible.
Broader Pox diseases in animals (#PE7304).

♦ PE6274 Inadequacies of foreign consultants
Nature Faced with a strange language and a strange culture, a foreign consultant may find it difficult to give advice which will be at once professional and acceptable. To understand the country and its traditions he may need longer than his contract allows; in consequence, he may give advice that is either too narrowly restricted to his particular skill and takes insufficient account of national circumstances, or which is too tentative and guarded to be worth while. A large number of short-term consultants from a variety of cultural traditions – a feature of some international teams – may produce confusion.
Aggravates Superficial research (#PF7889)
Irresponsible international experts (#PF0221).

♦ **PE6275 Anaplasmosis**
Gallsickness
Nature Anaplasmosis, sometimes called gallsickness, is a disease of domestic and wild animals, caused by the rickettsial parasite Anaplasma Marginale, which invades and destroys the erythrocytes, causing anaemia and severe digestive disturbance. There can also be a reduction in milk productivity. It is transmitted by a number of species; mechanical transmission by biting flies and unsterilized syringe needles also occurs. Some wild ungulates carry infection without showing symptoms.
Incidence Anaplasmosis is widely distributed through tropical Africa, Asia, Australia and America.
Broader Circulatory system diseases in animals (#PD5453).
Aggravated by Rickettsiae (#PE2572) Ticks as pests (#PE1766).

♦ **PE6276 Closure of recreation areas**
Aggravates Inaccessible recreation areas (#PF6503)
Inadequate recreational facilities (#PF0202)
Inadequate facilities for children's play (#PD0549)
Inadequate night life entertainment facilities (#PE6126).
Aggravated by Unused recreation spaces (#PJ8596)
Unprofitable entertainment market (#PG9384).

♦ **PE6278 Helminthiasis**
Nature Helminthiases, the vast variety of disease of man and animals caused by an infestation of worms, were formerly considered to be limited to the tropical belt, but increasing travel and trade has caused them to spread around the world. The pathogenic effects may not be apparent for a long time, making accurate diagnosis difficult. Infestation occurs in several ways, depending on the type of worm: consumption of vegetables and fruits contaminated by soil or dirty hands; consumption of inadequately cooked foods, including meat and fish; punctures in the skin caused by the bites of blood–sucking insects; or contact with contaminated soil by going barefoot or lying on the ground (some larvae are able to actively penetrate the skin).
Broader Parasites of the human body (#PE0596)
Infectious and parasitic diseases (#PD0982).
Narrower Hookworm (#PE3508) Filariasis (#PE2391)
Ascariasis (#PE2395) Trichinosis (#PE2311)
Schistosomiasis (#PE0921) Sarcosporidiosis (#PG9770)
Trematode diseases (#PE6461) Tapeworm as a parasite (#PE3511)
Visceral larva migrans (#PG9206).

♦ **PE6280 Malignant catarrhal fever**
Malignant head catarrh — Snotsiekte — Catarrhal fever — Gangrenous coryza
Nature Malignant catarrhal fever, an acute febrile virus disease of cattle, causes gastrointestinal and respiratory problems, affects the eyes and the central nervous system, and is frequently fatal. It is usually transmitted by inoculation of infected blood; transmission from beast to beast by contact is unusual.
Incidence The disease appears sporadically in cattle in Kenya, Zimbabwe, Botswana, Brazil and Mexico; is has been recorded, however, in the majority of countries in the world.
Broader Viral diseases in animals (#PD2730).
Aggravates Fever (#PD2255).

♦ **PE6281 Violent sports fans**
Nature Wherever team sports are played, violence among the spectators is increasing, resulting in extensive damage to property and in injuries and death not only to those involved but often also to innocent persons. The aggressiveness of some fans is often exacerbated by alcohol.
Incidence Violence associated with team sports is not new, and is even reflected in the vocabulary of many games. Football (American: soccer) has earned a reputation as the most violent game. In the 1960s a riot in Lima took 318 lives; in 1969 a football match between El Salvador and Honduras was blamed for sparking a brief but genuine war. In May 1985, 53 fans were killed in a fire which destroyed stands in a stadium in northern England; at least 10 police officers and foreigners were injured in Beijing when Chinese fans rioted after losing an international game; and in Brussels 39 people were killed and at least 400 injured in a riot which broke out even before a match between Liverpool and Turin had begun. The next month 8 people were killed and 50 injured when fans stampeded into a stadium in Mexico City.
Broader Human violence (#PA0429).
Related Athletic competition (#PE4266).
Aggravates Corruption of sports and athletic competitions (#PE3754).

♦ **PE6285 Plutonium pollution**
Nature Plutonium, a radioactive waste product of nuclear production, is carcinogenic; and even minute quantities, if inhaled directly, are fatal.
Incidence A 1,000 megawatt nuclear power plant creates about 200 kilograms of plutonium in a year.
Broader Radioactive contamination (#PC0229).
Related Plutonium overproduction (#PD2539).

♦ **PE6286 Borer insects**
Nature Crop borers cause serious damage to staple food crops (maize, sorghum, millet, rice, and cowpeas) and to timber. Abrupt and serious outbreaks may occur after long periods of absence and farmers are often unprepared.
Incidence Crop borers are responsible for 20 to 75 percent of the losses due to pests of maize in much of Africa.
Broader Insect pests of plants (#PD3634) Pests and diseases of trees (#PD3585).
Narrower Corn borers as pests (#PE3648).
Aggravates Maize pests and diseases (#PE3589)
Pests and diseases of oak (#PE2984)
Pests and diseases of rice (#PE2221)
Pests and diseases of cocoa (#PE2979)
Pests and diseases of olives (#PE2978)
Pests and diseases of coffee (#PE2218)
Pests and diseases of grain sorghum (#PE3590).

♦ **PE6287 Lynching**
Nature Lynching the suspected criminals publicly is a reaction to a rising crime rate and reflects dissatisfaction with the police and delays in the application of justice.
Refs Ames, Jessie D *Changing Character of Lynching*.
Broader Homicide (#PD2341) Scapegoats (#PF3332).

♦ **PE6288 Babesiosis**
Redwater — Babesioses
Nature Babesiosis attacks cattle, causing high temperatures, constipation alternating with diarrhoea, weight loss, a decrease in milk productivity and quality, abortions, and about 40 percent mortality. Other animals are also affected, but with less economic effect. The disease is spread by pasture ticks. Immunization can be broken down in the event of stress, such as an attack of rinderpest.
Incidence Babesiosis is widespread wherever pasture ticks are found: mainly in Europe, Middle Asia, and Africa.
Refs Ristic, Miodrag and Kreier, Julius P *Babesiosis* (1981).
Broader Protozoan parasites (#PE3676)
Parasites of the cardiovascular system in animals (#PE9427).
Aggravated by Ticks as pests (#PE1766) Protozoa as pests (#PE6741).

♦ **PE6289 Marine oxygen deficiency**
Nature Thermohaline stratification and its effect on vertical circulation can cause the development of water bodies with low oxygen content.
Incidence In 1976 large scale oxygen depletion in the New York Bight caused mass mortality of benthic fauna. The German Bight and the Central North Sea are also showing indications of oxygen depletion, with resulting damage to fauna.
Broader Marine pollution (#PC1117) Destruction of environmental oxygen (#PE5196).
Related Eutrophication of lakes and rivers (#PD2257).

♦ **PE6290 Eggshell thinning**
Nature Eggs of wild birds show a negative correlation between shell thickness and residues of DDT and DDE.
Broader Poisoning in animals (#PD5228).
Aggravated by Pesticide hazards to wildlife (#PD3680).

♦ **PE6291 Absence of mycorrhizae**
Nature An absence of the fungi that form mycorrhizae with plant roots gives rise to disease symptoms, especially severe nutrition deficiency, thus affecting the establishment of forests and lowering the productivity of existing forests.
Broader Pests and diseases of trees (#PD3585).

♦ **PE6292 Environmental hazards from hydropower**
Nature The construction and implementation of hydropower plants involves considerable environmental costs: relocation of populations; loss of land; dramatic changes in species composition; unsurmountable obstacles to migrating fish, such as salmon; and a multitude of other changes in the ecological balance. In addition, the functioning of the dam is not hazard– free: marginal cultivators denude the watershed of forest cover which eventually leads to erosion of the cultivable soil and siltation of the dam reservoir; and waterborne diseases and water weeds breed easily in the reservoir.
Counter–claim Hydropower plants not only provide reliable, efficient and clean energy, they also permit flood control and prevention, irrigation, increased fish production and recreational facilities.
Broader Environmental hazards from energy production (#PD6693).

♦ **PE6293 Contagious caprine pleuropneumonia**
Nature Contagious caprine pleuropneumonia occurs in epidemics among goat herds, with sever mortality. Immunization is difficult.
Incidence Contagious caprine pleuropneumonia is found mainly in the arid regions of Africa and in nearly all of the Near East. It is also recorded in India.
Broader Respiratory diseases of sheep (#PE9438)
Infectious diseases in animals (#PD2732).

♦ **PE6294 Harmful biological effects of ionizing radiation**
Nature Ionizing radiation causes reversible physiological effects in its early stages, but, when prolonged, biological damages are produced as a result of metabolic processes. Genetic damage resulting in somatic mutation is possible, and cancers, leukaemia, and generally decreased longevity are additional risks, varying according to exposure.
Background Irradiated cells disrupt tissues, especially the mucous membranes of the stomach and intestines which leads to disturbance of digestion and absorption, emaciation, body poisoning with the products of cell decomposition (toxaemia), and the penetration of bacteria living in the intestines into the blood (bacteraemia). The liematogenous system is damaged, which leads to a decrease in the leukocytes in the peripheral blood and to a decline in its defensive properties. Manufacture of antibodies falls off, which further weakens the body's defensive forces. The number of red blood cells decreases, which causes disruption of the respiratory function of the blood. The biological effect of ionizing radiation causes disturbance of sexual function and of sex–cell production to the point of complete sterility in irradiated organisms. A fatal outcome may be determined by disturbances in the central nervous system, causing cessation of heart activity and paralysis of breathing. After–effects are characteristic of the biological effect of ionizing radiation, since the chain of biochemical and physiological reactions initiated with absorption of radiant energy continues for a long time.
Refs Czerski, P, et al (Eds) *Biological Effects and Health Hazards of Microwave Radiation* (1974); Egami, Nobuo (Ed) *Radiation Effects on Aguatic Organisms* (1980); United Nations *Genetic and Somatic Effects of Ionizing Radiation* (1986); United Nations *Ionizing Radiation* (1982).
Broader Radioactive contamination (#PC0229)
Excessive exposure to radiation from consumer goods and electronic devices (#PE1909).
Aggravates Health hazards of irradiated food (#PD0361).

♦ **PE6297 Bluetongue**
Nature Bluetongue is a febrile disease of sheep causing inflammation of mucosae and of the coronary bands, severe loss of condition, and a break in the fleece. Mortality may be as high as 30 percent. Because of the persistence of the causative virus, the disease seems to be virtually impossible to eradicate.
Broader Viral diseases in animals (#PD2730).
Aggravated by Insect vectors of animal diseases (#PD2748).

♦ **PE6299 Psychological stress of urban environment**
Urban distress — Urban insecurity
Nature Modern urban housing, even if reasonably priced and providing the desired conveniences, is often a source of sociological and psychological problems. Concentration, memory, self–confidence, creative thought, and contact with neighbours and with nature decrease; aggression, anxiety and tension increase.
The dangers which threaten existence in an urban environment are such that man needs a secluded refuge today as much as his primitive ancestors did. He is assaulted by a multitude of sensations he would like to exclude from the space he needs to reserve for his own use. Subjection over a long period to vibrations and sound can be psychologically damaging; vibration, in fact, contributes to neuroses.
Modern urban housing, even if reasonably priced and providing the desired conveniences, is often a source of sociological and psychological problems. Concentration, memory, self–confidence, creative thought, and contact with neighbours and with nature decrease; aggression, anxiety and tension increase. Equally, the fear of being overheard by others can cause nervousness. The large panes of glass necessary to meet minimum requirements of natural light can increase the anxiety of people who fear prying eyes and the invasion of their private lives.

Odours, mostly from industrial sources, can be not only disagreeable but positively harmful. Pollution in general is not only a danger to health but is also psychological menace. Crowding makes the pursuit of hobbies that involve space, dirt or noise almost impossible. The green spaces provided by most cities are inadequate and insufficient to fulfil the population's requirements for contact with nature, leading to such effects on personality as: gradual diminution of initiative and creative thought; loss of sensitivity to primary values; false intellectualism; psychosis and psychic tension; emotional disturbances; negative attitudes. In addition, few people and few families are capable of standing the strain of constant adaptation which the mobility of the modern wage–earner often demands.
Refs Benyon, John and Solomos, John *The Roots of Urban Unrest* (1987).
Broader Stress in human beings (#PC1648).
Narrower Disorientation stress in large building complexes (#PF6174)
Environmental stress on inhabitants of tall buildings (#PE4953).
Related Marital stress (#PD0518).
Aggravates Juvenile stress (#PC0877).
Aggravated by Indoor air pollution (#PD6627) Urban–Industrial air pollution (#PJ5532).

♦ PE6308 Grey human hair
Nature Human hair turns grey as a result of ageing or certain diseases. As a symptom of old age and of loss of youthful vitality this is experienced with considerable anxiety.
Incidence Traditionally a problem of special importance to women. Of increasing importance in those societies which make a cult of youthfulness and have difficulty in respecting older generations. This anxiety is reinforced by advertising for pharmaceutical products designed to dye the hair. In many industrialized countries, it is difficult for those in positions of prominence, especially in business or in politics, to avoid pressures to "bring out the natural colour" in hair which would otherwise appear grey. Grey hair is increasingly interpreted as a sign of incompetence relative to younger colleagues.
Broader Cosmetic and health problems related to human hair (#PE7111).
Aggravates Fear of growing old (#PJ9144) Age discrimination in employment (#PD2318).
Aggravated by Human ageing (#PB0477).

♦ PE6311 Avoidance of issues in political campaigns
Vacuous political campaigns
Nature Politicians increasingly rely upon symbols of their general goodness and trustworthiness and the inferiority of their opponents for election campaign material. The issues with which the territory might be concerned are avoided by the zealous politician, as miscalculations could be made with loss of voter support.
Broader Issue avoidance (#PF1623).
Related Political displacement activity (#PF5360).
Aggravated by Political smear campaigns (#PD9384).

♦ PE6313 Creep of concrete
Nature Creep of materials is the slow, continuous plastic deformation of a solid under the action of a constant load or mechanical stress. All solids are to some extent subject to creep; in the case of concrete this can lead to the deformation of buildings and constructions.
Refs ACI Committee Staff *Shrinkage and creep in concrete* (1972).
Broader Failure of materials (#PD2638).
Related Creep in metals (#PG5589).

♦ PE6314 Imbecility
Severe mental retardation
Refs Meyers, C Edward *Quality of Life in Severely and Profoundly Mentally Retarded People* (1978); Stur, O (Ed) *Mental Retardation, 2nd congress, Vienna, 1961* (1963); Whitman, Thomas L, et al *Behavior Modification with the Severely and Profoundly Retarded* (1983).
Broader Unintelligence (#PA7371) Mental deficiency (#PC1587).
Aggravates Impotence (#PA6876) Legal impediments to marriage (#PF3346).

♦ PE6322 Nervous breakdown
Nature Nervous breakdown, the feeling of an inability to cope with the problems of ordinary daily life, is often accompanied by physical symptoms such as headaches, dizziness, vomiting and insomnia. Nervous breakdown may occur in healthy people who had previously coped with life's situations, or may occur in already neurotic people. The causes are diverse and include physiological stress, certain particularly disturbing experiences, and physical illness.
Broader Neurosis (#PD0270) Mental depression (#PC0799).
Aggravated by Marital stress (#PD0518).

♦ PE6323 Vicarious criminal liability
Liability for crimes committed in disregard of orders
Nature Liability for crimes crimes committed by another simply because of the relationship between the parties. In practice vicarious liability is limited to employer–employee situations in which the employer is made vicariously liable for any crimes committed by an employee. This may or may not involve crimes committed in disregard of instructions or orders.
Broader Crime (#PB0001).
Related Complicity (#PF4983).

♦ PE6325 Psoriasis
Nature Psoriasis is a chronic, recurrent, noncontagious skin disorder of man, perhaps caused in part by neuropsychic traumas and metabolic and endocrine disorders. Psoriasis is both uncomfortable and unsightly.
Refs Farber, E M, et al *Psoriasis* (1987).
Broader Diseases of the skin and subcutaneous tissue (#PC8534).
Narrower Pityriasis rosea (#PG4163).

♦ PE6328 Exploitative films
Exploitative imagery
Claim The genres favoured by exploitation films are generally those permitting the maximum of violence, nudity and horror, thus offering virtually no educational merit while instilling and perhaps encouraging the acceptance of such anti–social behaviour.
Broader Exploitative entertainment (#PD0606).
Aggravates Deterioration of media standards (#PD5377)
Psychological pollution by mass media (#PD1983).
Aggravated by Culture of violence (#PD6279)
Biased portrayal of women in mass media (#PE7638)
Excessive portrayal of sex in the media (#PE7930).

♦ PE6331 Secret police
Undercover police operations
Nature In a considerable number of countries, special units of secret police security forces exist, acting outside ordinary legal procedures under the supervision of the authorities or with their approval or connivance. In a number of situations, personnel, sometimes called 'death squads', carry out arrests and detention and in many cases kill suspects without any of the legal formalities required by law and without reference to the judiciary. Their activities in most cases are kept secret and outside the control of the judiciary. Information on arrests or detention is not communicated even to the families of the arrested or detained.
Undercover operations can be risky for the agents who spend a lot of time with the criminals. They may become unable to draw a line between impersonating and becoming lawbreakers. After successful undercover assignments these agents suffer from depression, anxiety, lethargy and lack of motivation, even personality changes. Another risk is that the respect for and trust in the administration of justice may diminish, especially if government knowingly bypasses normal restraints and uses disreputable methods, such as deception or videotapes, to gather evidence.
Incidence Well–known examples are the German Gestapo, Russian KGB and the secret forces which exist in some Latin American countries.
Counter–claim With the political revolutions of 1989 in Eastern Europe, and the opening of frontiers, there remains a need for security organizations to deal with criminal elements seeking to exploit the situation. Such countries do not have time to dismantle the existing security services and create new ones, they are obliged to adapt the operations of the existing services.
Refs Dion, Robert *Crimes of the Secret Police* (1982).
Broader Secrecy (#PA0005) Police state (#PD7910)
Unaccountable government intelligence agencies (#PF9184).
Aggravates Informers (#PD8926) Political purges (#PC2933)
Internment without trial (#PD1576)
Enforced participation in community activity (#PD3386).
Aggravated by Espionage in domestic politics (#PD1787).

♦ PE6333 Labour hoarding
Labour 'hoarding' by employers — Labour hoarding by employers
Nature Labour hoarding is the practice adopted by some companies, during a decline in demand, of retaining their labour force intact, thus reducing the productivity of labour and raising the average variable cost per unit of output.
Aggravates Labour shortage (#PC0592) Shortage of skilled labour (#PD0044).
Aggravated by Unethical practices of employers (#PD2879)
Ineffective economic structures in industrial nations (#PE4818).

♦ PE6336 Manipulative cults
Exclusive sects — Exploitative gurus
Nature Cults are religious groups which are unorthodox or spurious. They are a form of radical mystical individualism, an entirely inward spiritual religious form, indifferent to moral discipline, public worship, and social concerns.
Cults, as opposed to sects, lack authoritative grounds for discerning heresy from orthodoxy because of their epistemological individualism. This precludes stable doctrine, organization and membership. Cults lack creedal and sacramental authority of the church and the ethical rigour of the sect. The antinomian and subject cult creates no community because it possesses neither the sense of solidarity nor the faith in authority which this requires, nor the no less necessary fanaticism and desire for uniformity.
Background Recruits to religious cults often find themselves in a battlefield between true believers and witch–hunters. Cultists often deceive them and frequently defraud them, sometimes corece them and occasionally brainwash them. Anti–cultists, with equally ferocious dedication, spread hysterical propaganda, terrify distraught relatives and occasionally have their relatives kidnapped for brutal, illegal deprogramming.
Claim Cults are religious groups which demand their members' complete commitment to absolutist beliefs and authoritarian leaders, regimented organizations, and deviant ways of life. They subject their recruits to brainwashing, disrupt their families, and exploit their labour and assets. They have been know to sexually exploit members, murder opponents and defraud people. They are especially dangerous when they endeavour to capture children and impressionable young people, indoctrinating and brainwashing them so that they become unquestioning captives and tools of the inner circle of the cult.
Refs Greenberg, Martin H and Waugh, Charles G (Eds) *Cults* (1983).
Broader Social exclusion (#PC0193) Experimental religion (#PF3367).
Aggravates Religious intolerance (#PC1808) Persecution of religious sects (#PF3353)
Unethical practices of philanthropic organizations (#PE8742).

♦ PE6339 Torture by shooting
Incidence Shooting has been reported in a number of countries.
Broader Torture by wounding (#PE4078).

♦ PE6342 Bondservice
Denial of right to freedom from compulsory labour — Violation of freedom from bond service
Nature Bondservice, being bound to work without wages, often involves children who are either hired on a permanent basis for a wage paid to the parents, placed in work or hired out temporarily by the parents, pledged for a debt, or simply sold.
Incidence In Bolivia, Chile, Brazil, Thailand, and the Maghreb, Children are often 'given' away in payment for a debt entered into by the family, or merely to have one less mouth to feed.
Broader Slavery (#PC0146).
Aggravates Debt slavery (#PD3301).

♦ PE6343 Colour deficiencies
Colour blindness
Nature Colour deficiencies include colour blindness, a condition of incomplete colour vision, of which the most common type is Daltonism, the inability to distinguish between red and green. Colour blindness may be genetically determined (innate) in which case there is a sex linkage, which explains why men have about 16 times more chance to be afflicted or acquired by traumatic or toxic damage to the optic sensory path. Colour blindness can result in accidents if people are unable to identify the conventional identification colours used on electric cables, control buttons on machines, gas cylinders and various types of transport – railways, automobiles, aircraft and shipping. In addition, colour blind persons may have difficulties (but not necessarily so) in paint or colour mixing, geology, chemistry, microscopic work or in the operation of nuclear reactors.
Incidence 8 percent of men and 0.5 percent of women suffer from dichromatism, in which vision is based two primary colours rather than the normal three.
Broader Visual deficiencies (#PD8179) Visually handicapped persons (#PD2542).
Related Physical blindness (#PD0568).

♦ PE6347 Impoldering risks
Nature The most serious risk inherent to impoldered lands is flooding (especially in areas with deep polders), often exacerbated by inadequate systems for warning, emergency–repair, evacuation provisions and public information.
Incidence The problem occurs in Egypt, Bangladesh, Thailand, Venezuela, Japan and the Netherlands.
Background During the last few thousand years, man has substantially altered his environment in order to live in safety and ensure his food supply; poldering is a very sophisticated form of land reclamation which drastically alters the natural environment.

Broader Ecosystem modifications due to creation of dams and lakes (#PD0767).
Related Collapsing physical structures (#PD4143).

♦ PE6348 Transvestism

Nature Transvestism, the adoption of the dress and/or behaviour of the opposite sex, is often representative of repressed homosexuality; in addition to its illegal status in many countries, it may be indicative of severe mental imbalance.
For a man or woman who has no doubts as to his or her sex, transvestism is the sometimes erotic pleasure (a form of fetishism) of disguising oneself as a person of the opposite sex.
Broader Sexual deviation (#PD2198).
Related Transsexualism (#PF3277).
Aggravates Exhibitionism (#PD4643) Transvestite prostitution (#PD4525).

♦ PE6350 Sunlight inhibition by nuclear warfare soot

Nuclear war induced winter chill
Nature Nuclear war between the superpowers in the spring or summer would not only contaminate the land with radioactive fall-out, but the nuclear forces expended would drive hundreds of millions of tons of soot, smoke and dust into the atmosphere and, by blocking sunlight, drop the temperature 10 to 25 deg C throughout North America, Europe and Northern Asia, destroying most crops immediately. A nuclear war in January could blacken the sky until June, kill all livestock and induce famine for over a year. There would be few survivors: civil defence plans for food stores do not take into account such a nuclear winter.
Counter-claim Theories about a nuclear winter are speculative. Much nuclear-made soot would be dispersed by winds, and high soot clouds would disperse because of their absorbed heat. The Gulf Stream and other natural, warm thermal currents would help raise local temperatures. Pollution of the upper atmosphere can also create the greenhouse effect of trapping heat; a nuclear winter in the summer, for example, could be followed, eventually, by a frost-free year-round climate. Some scientists have allowed their anxiety about nuclear war to influence their professional evaluation of its implications.
Refs Harwell, M A *Nuclear Winter* (1984); Harwell, M A *Nuclear Winter* (1984).
Broader Nuclear war (#PC0842) Global warming (#PC0918).
Related Absence of direct sunlight (#PG1918).
Aggravates Soot (#PE1953) Evolutionary catastrophes (#PF1181)
Disruption of the hydrological cycle (#PD9670).
Aggravated by Exterminism (#PF7401).

♦ PE6352 Accident proneness

Nature Certain types of people seem to have a propensity to have accidents more often than others. Bring labelled 'accident prone' could significantly limit employment possibilities, make one socially unacceptable, and result in a degrading self-image.
Incidence A recent major study on road accidents has shown that it is possible to lower the accident rate in a transport company by using psychological tests (which have been subjected to repeated cross validation) to screen out the bad accident risk people on the basis of their intelligence, psychomotor functions, personality, social attitudes and interpersonal relationships.
Counter-claim It is difficult to statistically substantiate accident proneness or to rigidly define it. Accident proneness is not a simple phenomenon; it will manifest itself differently in different people and to different degrees. A person could be careful in one area but reckless in another; therefore it is necessary to consider the multi-dimensional complexity of this problem before labelling someone accident prone.
Broader Destiny (#PF3111).
Aggravated by Unluckiness (#PF9536).

♦ PE6354 Phobia

Simple phobia
Nature An individual's intense, irrational fear of an object or situation may lead to compulsive acts to protect himself. The intensity and compulsive quality of such neurotic fear may severely interfere with normal activities.
Incidence Commonly experienced phobias are those of germs, dirt, incurable disease, animals, heights, open and enclosed spaces.
Refs Emmelkamp, Paul M G *Phobic and Obsessive-Compulsive Disorders* (1982); Hand, I and Wittchen, H U (Eds) *Panic and Phobias Two* (1988); Henly, Arthur *Phobias* (1988); Hightower, J Howard and Baker, Eugene *A New Type of Phobia is Isolated* (1987); Morris, Samuel M *Phobias and Disorders* (1988).
Broader Fear (#PA6030) Neurosis (#PD0270) Mental illness (#PC0300).
Narrower Agoraphobia (#PE0527) Anemophobia (#PG1013)
Social phobia (#PE6374) Anorexia nervosa (#PE5758)
Triskaidekaphobia (#PG2200) Morbid fear of illness (#PE9091)
Fear of sexual intercourse (#PF6910).
Aggravated by Anxiety (#PA1635) Alcohol abuse (#PD0153).

♦ PE6356 Corporation-sanctioned assassination

Corporation-sanctioned killing
Incidence There are many reports of assassination of worker activists, human rights activists and environmental activists where their activities were embarrassing to the interests of corporations, and notably multinational corporations. Many such reports concern plantations in developing countries, notably in Latin America. These include reports of large-scale attempts to poison or hunt down indigenous peoples. In industrialized countries highly publicized incidents have been related to the nuclear power industry. In the USA and the Far East, a wider range of incidents is reportedly linked to the development of the business interests of organized crime.
Broader Assassination (#PD1971).
Related Government sanctioned killing (#PD7221).
Aggravates Political assassination (#PE5614) Assassination of trade union leaders (#PE0252)
Endangered tribes and indigenous peoples (#PC0720).
Aggravated by Organized crime (#PC2343).

♦ PE6357 Migraine

Nature Migraine, also called megrim, sick, nervous, or psychosomatic headache, is a severe periodical pain in the head usually on one side only, hence the name. Victims may suffer pains in various bodily limbs, chilled extremities, nausea, vomiting, blurred vision, and blurred mental faculties. Attacks last from one hour to several days and are most commonly instigated by humiliation, anger, insecurity or nervous stress.
Incidence Women suffer more than men and there is usually another victim in the family. Some doctors speak of a migrainous personality as a person who is usually sensitive imaginative, intelligent, serious, and ambitious. Sufferers have included George Eliot, Alexander Pope, and Virginia Woolf and Joan Didion.
Background Migraine is recorded in the Ebers papyrus (1630 BC) of Egypt as a 'sickness of half the head'.
Refs Hanington, Edda *Migraine* (1974); Rose, F Clifford *Advances in Migraine Research and Therapy* (1982); Shore, Susan A *Migraine* (1988).
Broader Headache (#PE1974) Diseases of the central nervous system (#PE9037).
Aggravated by Health hazards of exposure to noise (#PC0268).

♦ PE6360 Environmental hazards from electromagnetic pulses

Nature Environmental hazards of electromagnetic pulses include the killing of marine life and the upsetting of delicate ecological balances.
Incidence The US Navy is installing an electromagnetic pulse simulator in the Chesapeake Bay, close to an existing one, against local concern about their environmental impacts.
Broader Electromagnetic pollution (#PD4172).
Related Environmental hazards from electricity (#PE1412).

♦ PE6366 Extra-legal executions

Summary executions — Arbitrary executions
Nature Governments may sentence individuals to death either without any trial; or with a trial but without a fair and public hearing, right to legal defence, notification of the charge, right not to be compelled to testify against oneself, right to appeal, or right not to be tried twice for the same offence. Arbitrary executions take place in all parts of the world. They have most frequently taken place in internal armed conflict, during excessive or illegal use of force by law enforcement agents, in custody as a result of torture or after death threats made by members of police, military or paramilitary groups.
Incidence In 1984, instances of summary execution were reported in at least 12 countries: Afghanistan, Angola, Cameroon, Guatemala, Iran, Kuwait, Liberia, Malawi, Nigeria, Pakistan, Sudan, and the United Arab Emirates.
Broader Arbitrariness (#PB5486) Denial of right to life (#PD4234)
Inhumanity of capital punishment (#PF0399).
Related Forced disappearance of persons (#PD4259).

♦ PE6367 Suspicious deaths during detention

Nature Official inquiries into the growing number of prison deaths in suspicious circumstances are not pursued or tend to result in a finding of suicide. In many countries prisoners are executed, without charge or trial, for their political opinions, religious beliefs or ethnic origins.
Incidence Death in prison has been reported in the following countries: **Af** Côte d'Ivoire, Guinea-Bissau, Lesotho, Libyan AJ, Mali, Morocco, Niger, Nigeria, Rwanda, South Africa, Tanzania UR, Uganda, Zaire, Zimbabwe. **Am** Argentina, Bolivia, Colombia, Dominica, El Salvador, Guatemala, Guyana, Haiti, Honduras, Mexico, Paraguay, Peru, Suriname, Uruguay. **As** Australia, Afghanistan, Bahrain, Bangladesh, India, Indonesia, Iran Islamic Rep, Iraq, Korea Rep, Lebanon, Nepal, Pakistan, Philippines, Sri Lanka, Taiwan (Rep of China). **Eu** Poland, Spain, Turkey.
Broader Death and disability from inhumane confinement (#PE5648).
Related Persons missing in military action (#PE1397).
Aggravated by Human torture (#PC3429).

♦ PE6369 Violation of the rights of foetuses

Birth chauvinism — Sale of foetuses
Nature Those who have been born deny legal rights to those who have not. Legal processes require that the claimant have been born, leaving the unborn without legal protection.
Broader Denial of rights of children and youth (#PD0513).
Related Induced abortion (#PD0158) Foetal infection and death (#PE2041)
Physical malformation of foetus (#PE2042)
Defective oxygen supply to the foetus (#PG9126)
Denial to animals of the right to conditions of life and liberty proper to their species (#PE6270).
Aggravated by Lack of human organs for transplantation (#PE7530).

♦ PE6374 Social phobia

Nature Fear of social situations in which the person is exposed to possible scrutiny by others and also a fear of doing something embarrassing or humiliating in the presence of other people. It may be inability to talk in public, to urinate in public lavatory, to write without hand-trembling in the presence of others or to answer questions in social situations.
Broader Phobia (#PE6354).

♦ PE6375 Denial of economic and social rights to refugees

Abuse of refugee status
Nature 1) In many countries refugees encounter difficulties in obtaining employment either because legal access to the labour market is restricted, or because in practice preference is given to nationals. Widespread recessionary trends have exacerbated these difficulties. 2) The proliferation of ill-founded or abusive applications for refugee status causes overloading of procedures, long delays, higher administrative costs, and problems for bonafide asylum-seekers.
Refs Miserez, D (Ed) *Refugees - The Trauma of Exile* (1989).
Broader Denial of rights of minorities (#PC8999) Denial of rights to vulnerable groups (#PC4405).
Narrower Denial of the right to work to refugees (#PE3751).
Related Illegal immigration (#PD1928) Denial of right to life (#PD4234).
Aggravated by Rejection of refugees (#PF3021).

♦ PE6376 Detention of refugees and asylum-seekers

Nature A high proportion of the world's refugees, in the process of seeking asylum, suffer some form of deprivation of liberty. This may involve being required to live in an assigned place with certain restrictions on their freedom of movement, or, at the other end of the scale, actually being imprisoned or confined for indeterminate periods under prison-like conditions.
Broader Wrongful detention (#PD6062) Rejection of refugees (#PF3021).

♦ PE6388 Excessive salaries of international civil servants

High costs of staff in intergovernmental organizations
Nature Secretariats of some intergovernmental organizations and their agencies offer higher salaries than in equivalent private or public sector positions. For expatriates, in addition, there may be excessive cost-of-living differential allowances, paid tuition for children's private schooling, excessive home leave and holidays, and income tax advantages not offset by salary adjustments downwards. These or other perquisites and benefits add to the deficits that have to be made up by membership contributions; but beyond this, it is incongruous that those concerned with the societal development issues raised in the agendas of such international bodies should take such inflated salaries and benefits.
Incidence According to the United Nations Staff Union in New York, and the US International Civil Service Commission, UN Secretariat base salaries alone are higher by 22 percent than US private comparable scales. Also since US civil service pay is 18 percent lower than the private sector, UN base pay is 40 percent higher than US civil service. Taking the 22 percent figure, to this may be added benefits of the type indicated. The UN New York pattern may range from at least 25 percent higher in salary and benefits than the private sector and as much as 50 percent higher, in some instances, than the average national US civil servant scales. The EC secretariat salaries in Brussels, including similar benefits for expatriates as noted above, are also in the range of 50 percent higher than comparable private sector salaries, as for example, on the secretariat level.
Counter-claim Workers overseas are always paid higher whether embassy guards, construction

labourers, diplomats or businessmen. It is partly incentive and partly just compensation for the separation from homeland, friends, family and institutions.
Broader Prohibitive labour costs (#PF8763).
Related High severance pay for top managers (#PE3872).
Aggravates Inefficient public administration (#PF2335).
Reduces Disparity in remuneration between public and private sector employees (#PE7760).

♦ PE6390 Unsafe artificial sweeteners
Nature Saccharin, the most widely used artificial sweetener, has been unequivocally proved to be a weak carcinogen. Cyclamate may be carcinogenic and may also cause chromosome breakage and atrophy of the testicles. Aspartame may cause brain damage or changes in brain chemistry and behaviour.
Broader Toxic food additives (#PD0487).
Related Food fads (#PD1189) Cancer–causing foods (#PD0036) Synthetic food products (#PG2222) Health hazards of irradiated food (#PD0361).

♦ PE6396 Substitution of fast growing plant species
Aggravates Soil erosion (#PD0949) Bacterial plant diseases (#PD2226).
Aggravated by Irresponsible introduction of new plant species (#PE1444).

♦ PE6397 Denial of right to change religion
Broader Resistance to change (#PF0557) Violation of civil rights (#PC5285) Denial of religious liberty (#PD8445).
Related Denial of right to manifest religion (#PF2850).

♦ PE6402 Senility
Nature Senility is one of the most distressing burdens of old age. It is a progressive, irreversible illness that affects not only the victims but their families and society as a whole. Adequate research into causes and cures is lacking; and as the old and very old come to define larger and larger segments of the population, the scope of the problem will inevitably grow.
Broader Human ageing (#PB0477) Ill defined health conditions (#PC9067).
Aggravates Amnesia (#PD8297) Age discrimination (#PC2541) Physically dependent people (#PD7238).
Aggravated by Loneliness in old age (#PD0633) Deterioration of the mind with age (#PE4649).

♦ PE6403 Disease–causing viral combinations
Nature Diseases stemming from viral combinations of two viruses (which, of themselves would not be significant enough to cause infection) are potentially move dangerous than diseases caused from a single virus. This discovery – called piggybacking – is forcing medicine to take a new look at formally understood problems.
Broader Virus diseases (#PD0594).
Aggravates Viral plant diseases (#PD2227) Viral diseases in animals (#PD2730).

♦ PE6408 Sexual frigidity
Frigidity — Anorgasmia — Orgasmic dysfunction
Nature Sexual frigidity is the inability of a woman to achieve orgasm during sex. It may a total or partial and is a symptom of an underlying neurotic conflict which can be attributed to any number of causes. Some women may feel tormented by this and go from one partner to another, hoping that each new experience will bring orgasm.
Incidence Some estimates suggest that 50–90 percent of all women are frigid.
Refs Hastings, D W *Impotence and Frigidity* (1963).
Broader Human sexual disorders (#PD8016) Symptoms referable to nervous system (#PE9468).
Related Orgasm disorders (#PG5418).
Aggravates Sexual unfulfilment (#PF3260).
Aggravated by Lesbianism (#PF2640).

♦ PE6412 Hysteria
Hysterical insanity
Nature Hysteria is generally considered to be a neurosis. It is manifested in the presence of or physical symptoms for the sake of advantage, usually unconsciously motivated. Hysterical symptoms are characteristic of extroverted neurotics; epidemic or mass hysteria is said to explain a variety of irrational or bizarre behaviour affecting persons in close proximity to each other.
Refs Abse, D Wilfred *Hysteria and Related Mental Disorders* (1987); McGrath, William J *Freud's Discovery of Psychoanalysis* (1987); Veith, Ilza *Hysteria* (1970).
Broader Neurosis (#PD0270).
Narrower Theomania (#PF6966).
Aggravates Paralysis (#PD2632) Speech disorders (#PE2265) Disorders of joints and ligaments (#PD2283).
Aggravated by Fear (#PA6030) Anxiety (#PA1635) Abuse of plant drugs (#PD0022).

♦ PE6413 Absolute idiocy
Profound mental retardation
Nature Absolute idiocy is a severe subnormality either innate or acquired through brain damage in very early childhood. It is characterized by an almost complete condition of ineducability (the inability to speak or defend oneself against physical danger).
Broader Mental deficiency (#PC1587).

♦ PE6418 Non–recognition of foreign powers of attorney
Nature When a country does not recognize the granting of powers of attorney necessary for the sale or administration of property or the prosecution or defence of lawsuits amongst others, it hinders legal transactions between countries. This entails longer and more expensive legal procedures.
Broader Legal inconsistency (#PF5356).

♦ PE6419 Congenital anomalies due to infectious diseases in animals
Broader Infectious diseases in animals (#PD2732).
Narrower Border disease (#PG4244) Akabane disease of animals (#PG7858).

♦ PE6422 Discrimination against foreign nationals in the military service
Nature Foreign nationals who are members of the military service may not receive the same rights and benefits as national members, such as the rights and benefits derived in case of death or disability.
Incidence The Jews who fought with the British against the Germans in the Middle East did not receive the same compensations as did their British counterparts.
Broader Discrimination against foreigners (#PD6361).
Aggravated by Xenophobia (#PD4957).

♦ PE6424 Failure to notify the imprisonment or death in prison of foreign nationals
Nature Non–notification of the imprisonment or death in prison of foreign nationals may result in a deprivation of the individual's rights and substantial hardship for the person's friends and relatives.
Broader Official secrecy (#PC1812) Unreported births (#PF5381) Dehumanization of death (#PF2442).
Aggravated by Government seizure of foreign nationals in foreign countries (#PE6564).

♦ PE6427 Denial to working animals of limitation of working hours
Nature Working animals are often driven past their capacity because they have no means of recourse. Their suffering is both inhumane and economically self–defeating.
Broader Denial to animals of legal protection of their rights (#PE8643) Denial to experimental animals of the right to freedom from suffering (#PE8024).
Related Animal suffering (#PD8812) Denial to animals of the right to life (#PF8243) Denial to animals of the right to dignity (#PE9573) Denial to animals of the right to a natural death (#PE8339) Denial to animals of the right to freedom from mass killing (#PE9650) Denial to food animals of the right to freedom from suffering (#PE3899) Denial to working animals of restorative nourishment and rest (#PE4793) Underproductivity of draught animal power in developing countries (#PF0377) Denial to animals of the right to the attention, care and protection of humankind (#PF5121) Denial to animals of the right to conditions of life and liberty proper to their species (#PE6270).

♦ PE6431 Insolvent institutions
Bankrupt institutions
Broader Insolvency (#PC6154).
Aggravates Factory closures (#PE3537).
Aggravated by Excessive institutional debts (#PE1557).

♦ PE6443 Use of animals in warfare
Nature Warfare experiments designed to give actual experience in the treatment of wounds, the effects of chemicals and drugs, and general practice in military manoeuvres, are often carried out with animals as both the subjects and targets, as the effects and their reactions come closest to approximating those which will be found among humans.
Incidence From 1978–1984, India, Bangladesh and Malaysia stopped exporting rhesus monkeys to the West when it was discovered that they were used in military experiments. In 1984, NATO troops participated in a military exercise in northern Norway where 6 pigs were shot in the leg from 10 yards away and then shot in the abdomen at close range. Medical treatment followed, and after the operation was completed, the pigs were burned in order to provide experience in destroying bodies on a battlefield.
Claim Such brutal experiments deny animal rights and denigrate man to depths of insensitivity and brutality. It is a sad reflection upon man that he spends his talents and time upon such pursuits as inflicting suffering on innocent and helpless animals. The use of animals for such purposes is also a direct insult to people who regard such depravity as a violation of their own religions traditions and national cultures and principles.
Counter–claim The use of animals is the only realistic way to train surgeons and medical personnel in the treatment of casualties and to deduce probable effects of chemicals, etc, on human populations.
Broader Denial of rights of animals (#PC5456).

♦ PE6461 Trematode diseases
Trematode infection
Broader Helminthiasis (#PE6278) Parasites of the human body (#PE0596) Nervous system diseases caused by parasites (#PE4683).
Narrower Lancet fluke (#PG8570) Paramphistomes (#PG5082) Pancreatic fluke (#PG3965) Giant liver fluke (#PG0623) Common liver fluke (#PG7426) Large American liver fluke (#PG1831) Digestive tract helminthiasis of poultry (#PG9355).
Related Parasitic diseases in animals (#PD2735) Gastrointestinal infections in animals due to parasites (#PE0958).
Reduced by Endangered species of molluscs (#PD3478).

♦ PE6480 Pesticide residues in food
Food contamination by pesticides
Nature In the developed countries over–production of food has been achieved by using pesticides, agrochemicals, fertilizers, antibiotics and hormones. Governments may have approved the use of some pesticides that are known or suspected of causing a range of problems from cancer to allergies. It was found out that potentially toxic pesticides "bound" with cereals making them impossible to detect, destroy or wash away. Pesticide residues, caused by spray drift and soil contamination, have even been found on samples of organically grown food in UK.
Incidence The chemical industry produces the equivalent of three kilograms of pesticide for every man, woman and child of the world each year. USA and many other countries have banned carcinogenic DDT, but the substance is exported to developing countries which use it on products that are later shipped back to developed countries.
Refs Ecobichon, Donald J and Joy, Robert M *Pesticides and Neurological Diseases* (1982); FAO *Pesticide Residues in Food* (1978); FAO *Pesticide Residues in Food (1984)* (1985); FAO *Pesticide Residues in Food (1980)* (1981); FAO *Pesticide Residues in Food* (1981); FAO *Pesticide Residues in Food* (1984); FAO *Pesticide Residues in Food (1982)* (1983); FAO *Pesticide Residues in Food (1979)* (1980); FAO *Pesticide Residues in Food* (1978); FAO *Residues* (1986); FAO *Pesticide Residues in Food (1985)* (1986); FAO *Pesticide Residues in Food (1979)* (1980); FAO *Pesticide Residues in Food (1977)* (1979); FAO *Pesticide Residues in Food (1984)* (1985); FAO *Pesticide Residues in Food* (1982); FAO *Pesticide Residues in Food (1983)* (1985); FAO *Pesticide Residues in Food* (1982); FAO *Pesticide Residues in Food* (1983); FAO *Pesticide Residues in Food*; FAO–WHO Experts on Pesticide Residues Staff *Pesticide Residues in Food*; WHO *Health Aspects of Chemical Safety* (1984).
Broader Pesticides as pollutants (#PD0120) Chemical contaminants of food (#PD1694).
Aggravated by Unsustainable short–term improvements in agricultural productivity (#PE4331).

♦ PE6486 Crimes against intangible property
Broader Theft (#PD5552).
Related Avoidance of copyright (#PD0188).
Aggravated by Vulnerability of intellectual property (#PF8854).

♦ PE6515 Ineffective use of external relations relating to sportsmen
Nature Local people have more difficulty in having their case heard with central authority than powerful lobbies with the ear of those in authority. An example is evident in areas where hunting or shooting of game is prevalent. The rights of local people often go unheeded, while sportsmen have no difficulty in getting their needs attended to by regional or national authorities to whom local residents have difficulty in gaining access.
Broader Inequitable range of salaries (#PD9430) Unrecognized socio–economic interdependencies (#PF2969).
Aggravated by Poaching (#PD2664) Ethical decay (#PB2480) Lack of community voice (#PG9471) Infrequent doctor contact (#PJ0362) Unenforced littering laws (#PJ0380) Disruptive airplane hunting (#PG0382)

EMANATIONS OF OTHER PROBLEMS   PE6619

Inappropriate school lunches (#PJ0335)
Underpayment of government officials (#PD8422)
Vulnerability of government to lobbying (#PF5365).

♦ **PE6522 Sexual intercourse with minors**
Statutory rape — Corruption of minors — Rape of minor with consent — Rape of children
Nature Statutory rape is having sexual intercourse with a minor (under 16 in the U.S.A.) even with their consent by a person substantially older. Sexual experimentation among teenage peers is not considered a criminal act.
  Broader Sexual offences (#PD4082)   Corruption of minors (#PD9481)
  Victimization of children (#PC5512).
  Related Rape (#PD3266)   Sexual abuse of children (#PE3265)
  Sexual exploitation of children (#PD3267).

♦ **PE6526 Inadequate transportation facilities for rural communities in developing countries**
Nature One of the basic requirements for developing communities is the introduction of surfaced roads and motor vehicles to allow both goods and services to enter the community and increase its development. It is crucial that an exchange of ideas occurs within the village itself, between the village and nearby urban centers, and between the village and the world if development is to be practical. At present, most of the narrow, unsurfaced roads are turned to impassable mud by one heavy rain. Often a community owns no vehicles at all – hired four-wheel-drive jeeps brave the roads occasionally, but that is very expensive and also of high risk to the vehicles. Most travel in many villages, including crop transport, is by foot over long distances and difficult terrain. Although most families make one or more trips to market each week, such trips are a major undertaking, and obviously any travel is judged carefully against a number of inhibiting factors. This is especially evident in the limited amount of travel done even within a village, further isolating neighbour from neighbour.
Refs Maggied, H S *Transportation for the Poor* (1982).
  Broader Ineffective rural transport (#PF2996)
  Inadequate transportation facilities in developing countries (#PD1388).
  Narrower Inadequate road conditions (#PJ0860)   Underdeveloped road network (#PE1055)
  Lack of school transportation (#PJ7849)   Scarcity of appropriate transport (#PE8551).
  Related Dangerous paths (#PJ9888)   Limited transportation availability (#PJ1707).
  Aggravates Unreliable freight transport (#PG1026)
  Ineffective means for goods supply and distribution (#PF6495).
  Aggravated by Distant schooling facilities (#PJ7693)
  Insufficient modern technology (#PJ0996)
  Prohibitive cost of transportation (#PE8063).

♦ **PE6529 Cultural discrimination in the administration of justice**
Culturally biased legal systems
Nature Systems of justice do not apply equally to citizens of different class, racial, religious, linguistic and social persuasions. Where special legislation exists to protect minority groups, the majorities are unjustly penalized. Likewise, minority groups suffer from application of legal expectations which do not apply to them in the same way that they do the majority.
Incidence Native Canadians comprise only five per cent of the population of western Canada, but account for 32 per cent of the prison inmates. Tribal elders, upon whose wisdom and judgement the culture depends, are generally ignored by courts of law.
  Broader Cultural discrimination (#PC8344)   Discrimination before the law (#PC8726)
  Inequitable administration of justice (#PD0986).
  Related Economic discrimination in the administration of justice (#PE1399).
  Aggravated by Injustice (#PA6486)   Narrowmindedness (#PA7306)
  Biased legal systems (#PF8065)   Legalized discrimination (#PC8949).

♦ **PE6553 Disorders of the bowel**
Obstruction of the bowel — Obstruction by stricture of the bowel — Obstruction of the pylorus
Refs DeDombal, F T (Ed) *Inflammatory Bowel Disease* (1986).
  Broader Diseases of the oesophagus, stomach and duodenum (#PE8624).
  Narrower Colon–rectal cancer (#PE9399).
  Aggravates Peritonitis (#PE2663)   Haemorrhoids (#PE3504)
  Constipation (#PE3505).
  Aggravated by Ulcers (#PE2308).

♦ **PE6554 Narrow range of business skills**
Inadequate business know–how — Inexperienced business practices — Insufficient business experience — Latent business skills
Nature In many small communities in the developed nations, the increasing complexity of modern economic life has outpaced the structures and procedures established for managing a business with the small–town environment. Current costs and complexities of doing business limit the extent to which the personalized approach to management can be actualized. The restricting of personal service, in turn, discourages the frequent patronage of many customers, thereby reinforcing a sense of failure on the part of many small businessmen. The skills of doing precise cost–projections, analyzing the potential profit over against the cost of doing business, advertising to wider trade areas, and developing new product and customer markets are not utilized. The basic skill of calculating the number of manufactured items required to cover the cost of the machine producing the item, for example, is virtually unknown to small-town residents. Businessmen do not understand or seek out skills or funds required for surviving the first year of operation, which results in a high number of initial failures. Both the lack of business skills and the unawareness that certain skills are needed result in a failure mentality and a reluctance of businessmen to try to establish a business.
  Broader Narrow range of practical skills (#PF2477).
  Narrower Inadequate local expertise in business practices in developing countries (#PE7313).
  Aggravates Limited basic skills (#PJ8138)   Narrow profit margin (#PJ9737)
  Incompetent management (#PC4867)   Incomplete cost projections (#PJ8109)
  Untrained security personnel (#PG8069)   Unexplored potential markets (#PF0581)
  Minimal financial understanding (#PG8129)
  Underutilization of locally available skills (#PF6538).
  Aggravated by Limited exposure to outside influences in rural villages (#PF2296)
  Lack of opportunities for practical training in communities (#PF2837).

♦ **PE6555 Bursitis in animals**
  Broader Diseases of musculoskeletal system in animals (#PD7424).
  Narrower Capped hock (#PG3938)   Fistulous withers (#PG4607).

♦ **PE6558 Bad breath**
Halitosis
Nature Halitosis is caused by putrefaction. Bacteria grow on dead cells an food particles which collect on the tongue and in pockets in the teeth. They generate hydrogen sulphide, methyl mercaptan and dimethyl sulphide. Volatile compounds of onions, alcohol and garlic are absorbed and excreted through the lungs causing bad breath. Usually those with bad breath are unaware of it and those wrongly convinced they have it need psychotherapy.

Incidence As an indication, the market for mouthwashes (whether antiseptic, cosmetic or medicinal) in the UK in 1986 was £10.5 million.
  Broader Unsociable human physiological processes (#PF4417).
  Related Human flatulence in public (#PE1764)   Disagreeable human body odour (#PE4481).
  Aggravated by Inadequate personal hygiene (#PD2459).

♦ **PE6561 Denial of right to a cultural life**
Exclusion from community's culture
  Broader Denial of cultural rights (#PD5907).
  Narrower Limited artistic expression (#PF9735).
  Aggravates Cultural decline (#PC9083).

♦ **PE6564 Government seizure of foreign nationals in foreign countries**
Disregard of international law in the capture of foreign nationals
  Broader Internment without trial (#PD1576).
  Aggravates Failure to notify the imprisonment or death in prison of foreign nationals (#PE6424).
  Aggravated by Secret military operations (#PF7669)
  Covert intelligence agency operations (#PD4501)
  Foreign intervention in internal affairs of states (#PC3185).

♦ **PE6581 Gender identity disorders**
Nature In the case of mild gender identity disturbance, the person is aware of his or her assigned sex, but experiences discomfort and a sense of inappropriateness about it. When disorder is severe, the person has the sense of belonging to the opposite sex. In the childhood girls or boys have a persistent and intense distress about their sex and their anatomic structures. With boys the grossly feminine behaviour may lessen during the adolescence, but up to two–thirds of the boys with the disorder develop a homosexual orientation. Only a minority of females retain the masculine identification and from those only some develop homosexual orientation.
  Broader Mental illness (#PC0300).
  Narrower Transsexualism (#PF3277).

♦ **PE6586 Beef tapeworm**
Cysticercosis
Refs Flisser, Ana et al *Cysticercosis* (1982).
  Broader Tapeworm as a parasite (#PE3511).

♦ **PE6593 Loss of memory due to torture**
  Broader Post–traumatic stress disorder (#PE0351)
  Neurological effects of torture (#PD4755).

♦ **PE6594 Sheep diseases**
Refs Hoff–Jørgensen, R and Pétursson, G (Eds) *Maedi–Visna and Related Diseases* (1989); Martin, W B *Diseases of Sheep* (1983).
  Aggravates Tapeworm as a parasite (#PE3511).

♦ **PE6603 Disruption of internal balance of the human body**
Inflammation of internal parts of the body — Sudden shocks to the body
Nature There are two critical periods in life when the body's internal balance is in danger (thus making us vulnerable to physical and emotional illness). The first period comes during middle age (45–59) and the second in old age (74–89).
  Broader Human disease and disability (#PB1044).
  Aggravates Pain (#PA0643)   Stroke (#PE1684).

♦ **PE6604 Witchweed**
Striga species
Nature Striga species are parasitic weeds attacking the roots of tropical cereal crops, particularly sorghum and millet but also maize, sugarcane and cowpea. They not only sap the host of water and nutrients but also have a poisoning effect which can result in complete crop failure. Damage is done before the weed emerges and cannot be prevented by traditional handweeding.
Striga affects the poorest of farmers, those least able to change their cropping or use advanced methods of control. Traditionally the problem has been avoided by moving village when the land becomes severely infested. Greater pressure on land now prevents this solution and Striga is contributing increasingly to famine in the West African Sahel, Ethiopia etc. The most promising approach is the introduction of resistant crop varieties but few have yet been developed.
Incidence Striga species occur mainly in the semi–arid tropics and affect many million acres of crops across Northern tropical Africa, East and Southern Africa, Indian Ocean, SW Arabia and the Indian sub–continent. The problem is most acute under conditions of low soil fertility and erratic rainfall.
  Broader Parasitic plants (#PD6284)   Parasites on plants (#PD4659).

♦ **PE6613 Trafficking in children for sexual exploitation**
Claim One million young children are kidnapped, sold or otherwise forced onto the international sex market.
  Broader Traffic in persons (#PC4442)   Trafficking in children (#PD8405)
  Sexual exploitation of children (#PD3267).
  Related Sexual abuse of children (#PE3265)
  Aggravates Kidnapping (#PD8744)   Child pornography (#PF1349)
  Child prostitution (#PE7582)   Juvenile prostitution (#PD6213)
  Destruction of cultural heritage (#PC2114).

♦ **PE6615 Unethical catering practices**
Irresponsible restaurant practices — Corruption in catering industry
  Broader Unethical food practices (#PD1045)   Undemocratic social systems (#PB8031).
  Aggravates Food wastage (#PD8844)   Unclean food (#PJ2532)
  Violation of food taboos (#PD2868)   Food poisoning through negligence (#PE0561)
  Cruelty to animals in food preparation (#PE0236)   Decline in nutritional quality of food (#PE8938)
  Inadequate conditions of work in the hotel and catering industries (#PE4493).
  Aggravated by Inadequate food storage facilities (#PE4877)
  Prohibitive cost of nutritious food (#PF1212).

♦ **PE6618 Corporate crime in the pharmaceutical industry**
Nature The fragmented hierarchy of a large international drug company, with separate management structures in marketing, research, and administration, make the identification of fraud or misconduct extremely difficult.
Refs Braithwaite, John *Corporate Crime in the Pharmaceutical Industry* (1984).
  Broader Corporate crime (#PD3528).

♦ **PE6619 Carcinogenic consequences of food preparation**
Mutagenic consequences of food preparation
Nature Foodstuffs, such as vegetables, treated with fertilizer, pesticides and fungicides, may acquire carcinogenic (and mutagenic) properties when cooked or digested, even though the residues in the uncooked food are below toxic levels. This is due to alteration in the properties

PE6619

of the residues during the preparation process or due to interaction between them.
Broader Hazardous combinations of substances (#PD5256)
Carcinogenic chemical and physical agents (#PD1239)
Environmental hazards from food and live animals (#PC1411).
Aggravates Cultural suicide (#PF5957) Impaired vigilance (#PF6863)
Cancer-causing foods (#PD0036).
Aggravated by Toxic food additives (#PD0487)
Environmental hazards from food processing industries (#PE1280).

♦ PE6621 Countertrade
Incidence Countertrade is heavily practised by developing and Eastern European Countries. It is estimated to count up to 10 per cent of world trade.
Claim Countertrade, of which barter is one form, is inefficient, cumbersome and costly. It is not a panacea for the developing world's balance-of-payments problems. But because it puts off the evil day when a country will have to cut its imports to match its available cash and credit, it looks like an attractive expedient.
Reduces Inadequate barter system in international trade (#PE0117).

♦ PE6622 Lawsuits abuse
Nature Greed both among the public in general and in the legal profession in particular can lead to individuals and corporations demanding ridiculously large sums in damages and lawyers charging equally ridiculously large sums in fees. With such money at stake, more people will become lawyers, and create more and more work for themselves. The consequence is delay, expense, and an inability by deserving plaintiffs to get prompt justice.
Incidence Between 1962 and 1982 lawsuits filed at US Federal courts tripled from 85,802 to 238,875 – and that number increases annually. At present, state and local courts have 8 million suits pending, or one for every 20 adult US citizens. Of these, 60,000 are suits against producers of asbestos.
Broader Abuse of law (#PC5280).

♦ PE6623 Denial of right to dignity
Broader Denial of political and civil rights (#PC0632).
Narrower Birth trauma (#PE8911) Dehumanization of death (#PF2442)
Denial of right to sufficient food (#PE0324).
Aggravated by Excessive prolongation of the dying process (#PF4936).

♦ PE6625 Computer piracy
Illegal software copies — Computer cloning — Software piracy
Nature Large number of countries has provided legal protection for software under copyright laws. However, the differences in the scope, duration and object of protection are so wide from one country to the other and the interpretations given by tribunals within the same country sometimes so much at variance that many software firms have to rely on other protective measures or turn to patent protection and trade secrecy. The copiers deprive the software publishers of adequate revenue, the lost sales dissuade companies from developing new products and prevent them from lowering the high prices of software that is one of the biggest reasons for illegal copies. Copying makes it also impossible for the national software industry to get off the ground in many countries.
Broader Counterfeiting (#PD7981) Avoidance of copyright (#PD0188).
Aggravated by Human errors and miscalculations (#PF3702).

♦ PE6629 Abuse of antibiotics
Bacterial resistance to antibiotics
Nature The overuse of such antibiotics as penicillin and tetracycline has dramatically reduced their effectiveness. Some doctors prescribe antibiotics against nonbacterial organisms, notably viruses that cause the common cold, for which the drugs are not effective. In the Third World, antibiotics are a cure-all, and are mostly available without prescription. Repeated exposure of bacteria to drugs can result in the development of resistant strains. These resistant bacteria can then spread throughout the population. Thus, treatment for some diseases is now more difficult and expensive. For example, one type of penicillin, 100 percent effective in the 1940s against the common Staphylococcus aureus bacterium, is now only 10 percent effective. When an antibiotic is no longer effective, others must be developed to deal with the new resistant forms of the disease. This increases costs and the new drugs may have side effects. For example, the new drug needed to successfully treat Staphylococcus aureus bacterium is 10 times as expensive as the original penicillin. Reduced effectiveness of these drugs costs lives, especially in developing countries.
Claim Antibiotics are a uniquely potent defence against bacteria, but they have one fatal weakness: if used to excess, their power fails. This is now happening because of doctors who over-prescribe antibiotics and farmers who regularly add them to animal feed.
Counter-claim Research on bacteria from the frozen corpses of 19th Century arctic explorers has proven that substances other than antibiotics can make bacterial strains become resistant to antibiotics. It is suggested that such resistance may be induced by the body's reaction to heavy metals and that environmental pollution could contribute significantly to the development of this resistance.
Broader Substance abuse (#PC5536) Abuse of medical drugs (#PD0028)
Overprescription of drugs (#PE9087).
Related Drug resistance (#PF9659).
Aggravates Resistant bacteria (#PE6007) Drug resistant viruses (#PG6399)
Bacterial plant diseases (#PD2226)
Encouragement of drug resistant diseases (#PJ3767).

♦ PE6630 Substandard shipping vessels
Inadequate implementation of maritime safety standards
Claim Recent shipping catastrophes suggest that industry standards are worse than even the most pessimistic critics have claimed. Even ships serving the wealthiest industrialized countries endeavour to cut costs by operating under flags of convenience, employing ill-trained crew, and avoiding desirable maintenance.
Refs ILO *Substandard Vessels, Particularly those Registered under Flags of Convenience* (1976).
Broader Inadequacy of international standards (#PF5072).
Aggravates Marine accidents (#PD8982).
Aggravated by Evasion of shipping regulations and taxes by flags of convenience (#PE5873).

♦ PE6633 Denial of right to freedom from testifying against oneself
Broader Denial of right to fair and public trial (#PE3964).
Narrower Denial of defendant's right of silence (#PE4319).

♦ PE6636 Detention of children
Child prisoners
Nature Children of less than 18 years old are held 'at least overnight' in police stations, prisons and jails, in conditions that permit at least visual or auditory communication between children and adult inmates, regardless of the reason for the child being held there (suspected or convicted offender, destitution, child of adult detainee, war prisoner). Contact with adult inmates may have actual or potentially harmful results for the children. A slow decision-making process and a bureaucratic legal framework may prevent a prompt placement or release.
Refs Moorehead, Caroline (Ed) *Betrayal* (1989).
Broader Wrongful detention (#PD6062) Victimization of children (#PC5512).
Related Female criminals (#PE1837).
Aggravates Militarization of children (#PE5986).
Aggravated by Apartheid (#PE3681).

♦ PE6639 Military demoralization
Low military morale
Nature Low military morale may result in a lack of discipline, a high desertion rate, apathy and indifference to the outcome of a war, which could result in a government's or nation's fall.
Incidence Law military morale compounded the US government's efforts in the Vietnam War, and today is responsible for the high drug and alcohol abuse, military personnel crimes, and low re-enlistment rate that is beleaguering the Soviet army.
Broader Demoralization (#PF8446).
Aggravated by Military disobedience (#PD7225).

♦ PE6640 Acquisition of inappropriate equipment by developing countries
Nature Since investment capital is scarce in developing countries the purchase of new, technologically up-to-date, equipment seems prohibitive if, alternatively, satisfactory 'second-hand' or obsolete equipment is available in working order at substantially lower cost. However, spare parts for the latter become increasingly unavailable, maintenance becomes prohibitively costly, and long-term investment costs may be higher than new equipment when both production run units of output and decreased labour productivity are considered. Additional disadvantages are the sometimes difficult transport, installation or adaptation costs; and a lack of manufacturer support, particularly as regards training. Broader issues involved in acquiring used or unwanted equipment concern the economic life of the machinery as further innovations may make it obsolete even at low wage rates, and the demand for the products made may dry up for a variety of reasons.
Incidence Used or obsolete military equipment, which in relation to better equipped antagonists may be termed junk, has been provided to a number of countries by the super powers, such as by the USA to El Salvador in the case of Huey 66 helicopters.
Claim The use of second-hand manufacturing equipment is not a solution to the search for more appropriate technologies. The induced availability of such equipment as part of development aid programmes, to the extent that they may be forced on some countries, is a dis-service to the recipients, and, in effect, may be a covert restrictive trade practice which ensures poor economic performance in sectors of the recipient countries.
Aggravated by Inadequate maintenance equipment (#PD6520)
Prohibitive cost of equipment maintenance (#PE1722)
Dependence of developing countries on external financing for development programmes (#PE7195).

♦ PE6642 Financial manipulation by sects
Incidence Currently under investigation in the USA, Canada, and West Germany, is L Ron Hubbard, the founder of the 'Church of Scientology', who former devotees accuse of secretly diverting more than $100 million from the Church into his own foreign bank accounts. These former members contend that personal facts conveyed during private counselling sessions were later used to blackmail and intimidate them.
Claim Sects may find it very easy to manipulate their large numbers of usually totally devoted and unquestioning members, and to enforce silence concerning such manipulations. Under the guise of religion, some sects have inveigled large financial 'gifts' from their members.
Broader Economic manipulation (#PC6875).

♦ PE6649 Professional stagnation
Nature Professional stagnation is the inability of people trained in one field of knowledge or activity to make use of their talents in a manner which gives them satisfaction in another field. It is a limitation in people's ability to cope with changing situations.
Claim Either the original training in tertiary education needs to be modified, or refresher courses need to be introduced to those who graduated 5, 10 or 15 years before. The education of the non-graduate population also needs modification or the problem of stagnation becomes a societal problem as developments in micro-electronics leave less and less scope for the time-consuming and labour intensive occupations that provide jobs for the major proportion of the adult population and which have, with more or less satisfaction, kept them occupied during more than half their waking hours.
Broader Limited ways of matching talent and jobs (#PF2792).
Aggravated by Resistance to change (#PF0557)
Fear of vocational change (#PJ1318).

♦ PE6650 Break-down in communications due to difference in training
Pollution of information
Claim Most countries education policies are for specialization in increasingly narrow fields of learning, which ultimately leads to difficulty in talking on any field outside one's own. This tendency is sustained by pride (one cannot enter into conversation on a subject one knows little about) and the desire to overcome deficiencies in one's own training by picking up bits of information of those areas outside one's own speciality, and repeating them with authority. This results in half-truths taking the place of truth, innuendo replacing reason, and communication thus becomes largely restricted to the flow of misinformation. Misinformation also flows through today's mass media, which place emphasis on the spectacular rather than the verifiable.
Broader Information gap (#PF3397) Excessive specialization in education (#PC0432).
Aggravates Ineffective communication (#PF6052)
Insufficient communications systems (#PF2350).

♦ PE6658 Extension of the family cycle
Nature The time of the family life-cycle between the time of its formation and the time the last child leaves home has, in historical terms, been lengthening in all societies, the main reason being the decline of mortality. Combined with an early age at marriage and with a longer period of schooling and training for children, the decline of mortality means an increase in the duration of the central phase of the family life cycle, which is still true in most societies.
Incidence An average woman in Asia, Africa or Latin America will marry or live in consensual union before she reaches her twentieth birthday, have a first child soon, continue her child-bearing period in her thirties and still be child-rearing during her forties. During this 25- to 30-year period, her mortality risks have diminished, as well as those of her husband.

♦ PE6659 Birds as vectors of disease
Broader Animals as vectors of disease (#PD8360).
Narrower Wild birds as vectors of animal diseases (#PE2749)
Insalubrity of animal excrement in urban environments (#PE4685).
Aggravates Bird vectors of plant disease (#PD3601).

♦ **PE6660 Excessive office space in cities**
Nature In spite of the general economic decline, office space has tended to increase. This is due to the fact that the building process itself has become a major source of profit; it is now an ultimate goal and no longer a means. However, proliferation of office space has serious effects on a community. Districts which are of interest to the office market are withdrawn from small owners and from the former use by their inhabitants; inhabitants are confronted with expulsion; houses decay; remaining residents have long-term building and construction yards "on their doorsteps"; and when the offices are finally completed, there are resultant heavy traffic flows and parking problems.
Broader Obstacles to building construction (#PF7106).

♦ **PE6668 Itchgrass**
Nature Itchgrass (Rottboellia exaltata) is a rank, aggressive, annual grass creating problems in many regions of the world, and may be considered among the world's worst weeds. It has the ability to grow taller than most crops, can grow at any time during the crop season (in both sun and shade), and produces stiff, barb-like hairs that are extremely irritating to human skin.
Aggravates Itch (#PE3940).

♦ **PE6671 Social symbols dominated by the economy**
Nature The symbols of modern society which seem to point to a meaningful life are dominated by the economic processes. Pay scales create images of relative worth in society. Possession of goods determine self-esteem. Job perks motivate workers. Many of the rituals of society are dominated by the economy. birthdays and Christmas are wonderful if many and costly presents are exchanged. Entrance into adulthood is acquiring car or a job. Retirement may be the most important rite of passage in modern society.
Broader Dominance of economic motives (#PF1913).
Aggravates Decreasing participation in collective religious worship (#PF8905).
Aggravated by Use of offensive symbols (#PF2826).

♦ **PE6676 Involuntary loss of nationality of children**
Stateless children
Nature A child may lose his or her original nationality in the event of the naturalization of the parents, the adoption of the child by foreign persons, or the legitimation or recognition of an illegitimate child by a father of foreign nationality.
Broader Statelessness (#PE2485).   Victimization of children (#PC5512)
Denial of rights of children and youth (#PD0513).
Related Statelessness of women (#PE4016).

♦ **PE6677 Conflicting multiple nationalities**
Dual nationality
Nature A person having two or more nationalities may be regarded as its national by each of the states whose nationality he possesses. This may result in excessive burdens being placed on the individual by virtue of multiple national obligations such as those relating to taxes or military service. With regard to diplomatic protection, a state may not afford this protection to one of its nationals against a state whose nationality such person also possesses. The individual's ability to select one nationality may be subject to limitations or restrictions in terms of his or her ability to renounce the nationality of one state or to retain the nationality of a state in circumstances which give rise to the involuntary loss of nationality under state law.
Aggravates Restrictions on recognition of nationality (#PE4912).
Refusal to issue travel documents, passports, visas (#PE0325).

♦ **PE6678 Conflicting military obligations for persons of multiple nationality**
Nature A person who possesses two or more nationalities may be subject to military obligations in each of the countries of nationality. Military service in one country may result in the loss of the nationality of other countries or may create conflicting obligations for the individual.
Refs Council of Europe *Convention on the Reduction of Cases of Multiple Nationality and Military Obligations in Cases of Multiple Nationality* (1978).
Related Restrictions on recognition of nationality (#PE4912).
Aggravates Military disobedience (#PD7225).

♦ **PE6679 Conflicting national laws governing cheques**
Nature The validity of a cheque; the requirements relating to the drawing and form, the negotiation, the presentation and payment; and the recourse for non-payment, are all governed by national law. Conflicting national laws may lead to different results as to the rights and obligations of the parties concerned in the event of a dispute relating to a bill or a note which has connections with more than one country. This adds a degree of uncertainty to international commercial transactions and may result in a grave injustice for one or more of the parties concerned.
Broader Conflict of laws (#PF0216).

♦ **PE6680 Conflicting laws concerning bills of exchange and promissory notes**
Nature The validity of bills of exchange and promissory notes, and the requirements concerning the issue and form, the endorsement, and the acceptance or non-payment of these instruments, are all governed by national laws. Conflicting national laws may lead to different results as to the rights and obligations of the parties concerned in the event of a dispute relating to a bill or a note which has connections with more than one country. This adds a degree of uncertainty to international commercial transactions and may result in a grave injustice for one or more of the parties concerned.
Broader Conflict of laws (#PF0216).

♦ **PE6681 Tax impediments to international motor traffic**
Nature International motor traffic is hindered by the imposition of taxes on foreign motor vehicles. When a motor vehicle registered in one country circulates temporarily in the territories of another country, taxes or charges may be levied on the circulation or possession of the motor vehicle or on consumption. These taxes or charges discourage international motor traffic.

♦ **PE6682 Environmental hazards from mineral exploitation of seabed resources**
Nature Such environmental hazards include the restricted fishing activities and local redistribution of fish populations which result from offshore structures such as platforms, wellheads, and pipelines. In addition, the disposal of cuttings and losses of drilling mud could taint fish feeding as well as blanket the fauna at the bottom of the sea.
Broader Ecologically unsustainable development (#PC0111).

♦ **PE6685 Recreational contact with sewage**
Bathing near sewage discharge — Bathing in sewage-contaminated water
Nature Ailments which may result from such contact include gastro-enteritis, skin infections, minor eye, ear and throat infections, as well as more serious diseases transmitted via the faecal-oral infection route, including hepatitis.
Related Pollution of water by infected faeces (#PE8545).
Aggravates Hepatitis (#PE0517)   Gastrointestinal diseases (#PE3861).
Aggravated by Sewage as a pollutant (#PD1414)
Coastal water pollution (#PD1356).

♦ **PE6687 Export of hazardous industries to developing countries**
Relocation of hazardous industries to developing countries
Nature Due to lax safety standards and regulations, cheap labour and regulatory costs, and sometimes government subsidies, more and more manufacturers of hazardous substances are locating in developing countries. Often the host countries lack adequate capability to control byproducts of the manufacturing process.
Incidence The Bhopal disaster of 1984 left 2500 Indians dead when a lethal gas leaked from the Union Carbide plant located there. Inadequate safety control was cited as one cause.
Aggravated by Inadequate safety legislation (#PJ2716)
Environmentally harmful dam construction (#PD9515)
Inadequate enforcement of safety regulations (#PD5001)
Inadequate legislation against environmental pollution in developing countries (#PE7141).

♦ **PE6696 Dependent personality disorder**
Nature Dependent people are unable to make any kind of decisions without the advice or reassurance from others and so has great difficulties doing things on their own or initiating projects. They are easily hurt by disapproval and have fears of being abandoned, rejected or alone.
Broader Personality disorders (#PD9219).

♦ **PE6699 Ineffective legislation against organized crime**
Ineffective laws against racketeering
Incidence Organized crime was cheating the US of more than $18 billion a year and costing it more than 400,000 jobs, it raised consumer prices about 0.3 per cent and reduced per capita income more than $77 in 1985.
Broader Inadequate laws (#PC6848)   Ineffective legislation (#PC9513)
Deficiencies in the criminal justice system (#PF4875).
Aggravates Racketeering (#PE4914)   Organized crime (#PC2343)
Limited local respect for regional and global legislation (#PF2499).
Aggravated by Inadequate narcotics legislation (#PF6787)
Use of undue influence to obstruct the administration of justice (#PE8829).

♦ **PE6700 Inadequate evaluation of environmental costs**
Nature Environmental costs arise either through the damage done as a consequence of resource exploitation or through the effort expended to redress the damage. Damage costs may also be imposed in the process of development, through the destruction of certain types of renewable resources such as: large-scale loss of tropical forests, soil degradation due to salinization, and imperfect cultivation of sub-marginal lands.
Incidence Some 30 million square kilometers (19 percent of the earth's land surface) are threatened with desertification, and consequently with huge economic and human losses. From 1970–1980, the economic cost of pollution damage in developed countries varied between 3 and 5 percent of GNP.
Claim The costing of the damage due to the irrational use of natural resources and/or pollution is not complete. Environmental damage is often selective and unequally distributed in time and space and among societies. Many of the physical, biological, and socio-economic consequences of large development projects are inadequately known, and some cannot be assessed. Examples of this are landscape or historic monuments threatened with irreversible change. Even if all the consequences could be enumerated and their likelihood assessed, placing a price tag on them would pose further difficulties.
Narrower Neglect of environmental consequences of government policies (#PE9295).

♦ **PE6702 Unethical practices by public service employees**
Irresponsible civil servants — Negligent public service employees — Corruption of civil servants — Unethical allocation of public contracts
Nature Employees of government are able to exploit such a position to their personal advantage or to that of others they may favour. Where the employee is in direct contact with outsiders requesting services (processing documents, supplying licenses, approving applications, etc), such services may only be provided (rapidly) following "under-the-table" payments.
Broader Corruption of ruling classes (#PD8380)   Unethical practices by employees (#PD4334)
Corruption of government leaders (#PC7587).

♦ **PE6704 Functional obsolescence of roads**
Claim The network of streets in most cities was built for slow traffic, and lacks sufficient roadspace or pedestrian sidewalks. Such patterns are obsolete: cities should adopt a parallel system of roads as in Berne, Switzerland, where there is little traffic congestion; or like the alternating one-way avenues of New York and one-way major streets of downtown San Francisco.
Broader Preservation of obsolete systems (#PC8390).

♦ **PE6705 Decline of street life**
Nature The dominating role of the automobile in modern society, the increase in crime, the proliferation of television and home entertainment, and the telephone, have all contributed to the decline of street life apparent in most inner cities and suburbs. The street has become little more than a parking area for cars, supplanting the children's games, neighbourhood festivities, flea markets and gossip groups that once used it as their backdrop. People now isolate themselves within their homes and have little personal interaction with neighbours or strangers.
Broader Deteriorating quality of life (#PF7142).

♦ **PE6708 Enteric diseases in horses**
Broader Diseases of the digestive system in animals (#PD3978).
Narrower Colitis-X in horses (#PG6906)   Potomac horse fever (#PG9543)
Enteric diseases in foals (#PG9600)   Chronic diarrhoea in horses (#PG7656).

♦ **PE6711 Medical backlash**
Nature The introduction of measures to improve health and the environment sometimes has unexpected consequences. For example, the successful application of biomedical achievements to public health services has generated a medical backlash in the form of unanticipated hazards to health from the increased expectation of life. To anticipate or mitigate such backlash requires the active collaboration of doctors, economists, sociologists.
Broader Adverse consequences of scientific and technological progress (#PF3931).

♦ **PE6715 Political contributions by agents of foreign principals**
Nature Political contributions by foreign organizations and their agents in order to influence the out come of any election, referendum, caucus or political convention is a breach of national sovereignty and as such is a serious crime. Difficulties arise when dealing with this crime when a citizen acts as an agent for a foreign organization and when a transnational enterprise is involved through a local subsidiary.

PE6715

Broader Criminal violation of civil rights (#PD8709).

♦ **PE6722 Trade union opposition to profit-sharing**

Nature Trade union opposition to profit-sharing includes the threat it poses to the position of trade union; the obstacle it presents to higher wages; and the difficulty of securing comparable participation in all undertakings in the same industry.
Broader Restrictive trade union practices (#PD8146).

♦ **PE6724 Gastro-intestinal disorders due to torture**

Nature Victims of torture frequently experience gastro-intentional disorders for years following the events. Symptoms include periodic or permanent ulcer like dyspeptic symptoms, regurgitation pains, depression after eating and irritable spastic colon. Because the vast majority of victims suffer psychological problems the symptoms may be thought to be psychosomatic but other factor play a role. Frequently prisoners are given insufficient liquids or food at the beginning of the torture period. The lack of protein and roughage in prison fare adds to the problem.
Incidence In a recent study of victims of torture some 32 percent of them suffered from gastro-intestinal disorders.
Broader Gastrointestinal diseases (#PE3861)
Somatic and psychosomatic effects of torture (#PE5294).

♦ **PE6725 Pay system erosion**

Nature The difficulty of determining and maintaining accurate standards of performance is one of the principle causes of the erosion of the payment-by-results system. Once inconsistencies are maintained the erosion becomes more severe. Average bonus earnings drift upwards beyond those originally expected or output falls well below that attainable; adherence to customary norms becomes subject to strict informal work group discipline; inequities in effort/bargain relations give rise to continuing disputes and leap-frogging wage claims; and illicit administrative practices proliferate.
Broader Erosion (#PC8193).
Narrower Crisis in long-term pension funds (#PF5956)
Increasing cost of social security (#PF7911).

♦ **PE6726 Lack of minimum wage fixing**

Violation of minimum wage laws — Denial of right to minimum wage — Low wage scale — Nonexistent salary base
Nature Lack of minimum wage fixing increases industrial conflict and discourages the development of stable collective bargaining relationships.
Counter-claim Copious evidence exists that minimum wage laws reduce the job prospects of the least skilled, tend to create high tensions over wage differentials, and often lead to a wage explosion. A minimum wage policy has no place in a programme that is serious about employment, competitiveness and credibility.
Refs ILO *Minimum Wage-fixing and Economic Development* (1975); Starr, Gerald *Minimum Wage Fixing* (1981); Starr, Gerald *Minimum Wage-fixing. An International Review of Practices and Problems* (1981).
Broader Low general income (#PD8568) Exploitation in employment (#PC3297).
Aggravates Underpayment for work (#PD8916) Inequitable range of salaries (#PD9430)
Incomplete access to development capital (#PF6517)
Inadequate living and working conditions of immigrant labourers in industrialized countries (#PD3427).

♦ **PE6734 Coastal erosion**

Refs UNEP/UNESCO/UN *Coastal Erosion in West and Central Africa* (1985); UNEP/UNESCO/UN-DIESA *Bibliography on Coastal Erosion in West and Central Africa* (1985); United Nations *Technologies for Coastal Erosion Control* (1982).
Broader Topological disaster (#PC5010) Decreasing land mass (#PF7435)
Soil erosion by water (#PD2290).
Narrower Coastal erosion resulting from dams (#PJ7396).
Related Floods (#PD0452) River ice (#PD3142) Avalanches (#PD1146)
Landslides (#PD1233) Migrating sand dunes (#PD0493)
Land erosion brought about by site development (#PE7099).
Aggravates Rock avalanches (#PG0476) Silting of water systems (#PD3654)
Unsustainable development of coast zones (#PD4671).
Aggravated by Subsiding coastal areas (#PD3775).

♦ **PE6735 Abuse of brothel legislation**

Nature The laws governing prostitution from country to country do not cover the cases of women prostitutes working as illegal aliens, procured from poorer countries by international syndicates. Further, all-male establishments catering to homosexuals are not effectively served by legislation created for traditional, heterosexual brothels.
Broader Abuse of law (#PC5280) Male prostitution (#PD3381)
Female prostitution (#PD3380).
Aggravated by Traffic in persons (#PC4442).

♦ **PE6740 Unethical practices of interpreters**

Irresponsible interpreters — Negligent interpretation — Corruption of interpreters
Claim Interpreters may take advantage of their privileged position in order to bias, distort or filter the information being communicated, whether under personal, financial or political pressure. In the case of consecutive interpretation, where notes are taken, these may be communicated to third parties such as commercial or intelligence agencies. Negligence or incompetence on the part of interpreters can severely aggravate tensions between opposing parties.
Broader Unethical professional practices (#PC8019).
Aggravates Interpretation and translation errors (#PF6916).
Aggravated by Questionable facts (#PF9431).

♦ **PE6741 Protozoa as pests**

Refs Kreir, Julius P *Parasitic Protozoa* (1977).
Broader Animal pests (#PD8426).
Narrower Water-borne amoeba (#PS2111).
Related Protozoan parasites (#PE3676).
Aggravates Babesiosis (#PE6288) Theileriasis (#PE3996)
Toxoplasmosis (#PE3659) Trypanosomiasis (#PE5725)
East Coast fever (#PE7946) Feline cytauxzoonoses (#PG6901).

♦ **PE6746 Foreign money liabilities**

Nature Disputes may arise in connection with foreign money liabilities when the currency or the rate of exchange for payment has not been clearly specified. Currency fluctuations may result in a substantial prejudice to one of the parties.
Broader Risk (#PF7580).

♦ **PE6748 Sexual sadism**

Nature Some people are bothered by their fantasies of having control over the victim while others act on on the sexual urges with a consenting partner or with a nonconsenting one. The suffering of the victim is sexually arousing.
Broader Sadism (#PF3270) Sexual deviation (#PD2198)
Human sexual inadequacy (#PC1892).
Related Sexual masochism (#PE3851).

♦ **PE6749 Non-uniformity in marking navigable waters**

Nature The lack of a uniform system for the marking of navigable waters presents a serious threat to safe, international navigation. Some countries use the 'lateral system' which means that vessels proceeding in a channel in a given direction (such as from seaward to a port) are guided by seamarks of a specified type to starboard and of a different type to port, without reference to the points of the compass. Other countries use the 'cardinal system' which means that the appearance of sea marks is determined by their position – in terms of the nearest cardinal point (N, S, E or W) – with respect to the shoal they are marking.
Broader Inadequate standardization of procedures and equipment (#PC0666).

♦ **PE6750 Ineffective testamentary dispositions**

Nature Divergent national laws hamper the effective testamentary disposition of property by persons who have connections with more than one country, such as dual nationality, multiple residences in different countries, or the ownership of property in different countries.

♦ **PE6754 Inappropriate use of the mail service**

Unethical use of postal service — Irresponsible use of postal service
Nature Mailing inappropriate articles may present a serious threat to the safety of the article as well as the safety of members of the postal service. Dangerous or inappropriate articles include: (a) articles which, by their nature or their packing, may expose officials to danger, or soil or damage correspondence; (b) opium, morphine, cocaine and other narcotics; (c) articles of which the importation or the circulation is prohibited in the country of destination; (d) living animals; (e) explosives, inflammable or dangerous substances; (f) obscene or immoral articles.
Narrower Mail fraud (#PE1404) Mail bombs (#PG1130)
Proliferation of direct mail advertising (#PE1810).
Related Postal censorship (#PD3033) Inadequacy of postal services (#PF2717)
Misuse of postal surveillance by governments (#PD2683).

♦ **PE6755 Unlawful interference with rights of innocent passage in territorial waters**

Denial to right to innocent passage in territorial waters
Nature It is a violation of international law, and a threat to friendly international relations, for a coastal state to engage in any of the following acts which interfere with rights of innocent passage: (1) to hamper innocent passage through the territorial sea; (2) to fail to give appropriate publicity to dangers to navigation, of which it has knowledge, within its territorial sea; (3) to discriminate amongst foreign ships in the effective exercise of the rights of innocent passage; (4) to suspend the innocent passage of foreign ships through straits which are used for international navigation between one part of the high seas and another part of the high seas or the territorial sea of a foreign state.
Background Ships of all states, whether coastal or not, enjoy the right of innocent passage through the territorial sea of any state. Passage means navigation through the territorial sea for the purpose either of traversing that sea without entering internal waters, or of proceeding to internal waters, or of making for the high seas from internal waters. Passage includes stopping and anchoring, but only in so far as the same are incidental to ordinary navigation or are rendered necessary by force majeure or by distress. Passage is innocent so long as it is not prejudicial to the peace, good order or security of the coastal state. Such passage must take place in conformity with the rules of international rules of international law. For example, submarines are required to navigate on the surface and to show their flag. Passage of foreign vessels may not be considered innocent if they fail to observe the laws and regulations of the coastal state, in particular, laws which prevent these vessels from fishing in the territorial sea.
Broader International aggression (#PB0968).
Related Denial of the right to ownership (#PE8411).
Aggravated by Territorial expansionism (#PC9547).

♦ **PE6756 Accidents involving astronauts**

Nature With the increase in space travel there is a corresponding increase in the risk of accidents involving astronauts. In the event of an accident, a distress situation, or an emergency or unintended landing, the safety of the astronaut would require immediate search and rescue operations. However, the effective conduct of such operations may be impeded by national boundaries if the astronaut is located beyond the territorial limits of the launching state.
Broader Injurious accidents (#PB0731).

♦ **PE6763 Medical experimentation on institutionalized subjects**

Nature The one area from which subjects for medical experimentation are more copiously drawn than any other are institutions. They are more easily available in large numbers in a compact geographical area; remain available over relatively longer periods of time with a lesser risk of drop-out; live under uniform living conditions which can be well controlled; can be subjected to as intense or prolonged observation as required; and in case of any adverse effects of the research procedures, there is a ready availability of emergency and other remedial and restitutive procedures.
Broader Abusive experimentation on humans (#PC6912)
Unethical experiments with drugs and medical devices (#PD2697).

♦ **PE6764 Medical experimentation on children**

Nature While the advantages of research involving child subjects are usually uncontested on medical grounds, many question such research on moral grounds. The ethics of conducting research using human subjects rest on informed consent, which infants and young children can never be sufficiently mature to provide.
Claim No parent or guardian, even with the best intentions, has the moral status to consent for a child to be made the subject of medical investigation. Even if this severely restricts possible advances in childhood medicine, the moral progress of the race is more important than the scientific.
Counter-claim (1) The ethical questions of child medical experimentation can be seen in terms of the future good of many against the possible risk to the few, or the moral duty of medical research to overcome disease threatening children as a group, versus the moral imperative to protect each individual child patient. In the case of orphaned poverty-stricken children with little hope of a satisfying life, perhaps their 'legacy' may well be to provide science with the chance of helping other mankind rather than spending their lives (which, chances are, are already condemned to being unrealistically short) in a parasitic mode.
(2) Child medical research in one form or another has led to the virtual elimination in industrialized countries of the leading causes of child mortality and morbidity, including tetanus, pertussis, diphtheria, poliomyelitis and measles. Though not yet sufficiently implemented, a simple and effective treatment has been found for diarrhoea (the main cause of child death in developing countries), and progress is being made in treating certain forms of cancer, which is among the

largest child killers in the industrialized world.
Broader Victimization of children (#PC5512) Abusive experimentation on humans (#PC6912) Unethical experiments with drugs and medical devices (#PD2697).

♦ PE6767 Dexterity disabilities
Nature Dexterity disabilities refer to difficulties with adroitness and skills in bodily movements, including manipulative skills and the ability to regulate control mechanisms.
Broader Human disability (#PC0699).

♦ PE6768 Body disposition disabilities
Nature Body disposition disabilities refer to an individual's ability to execute distinctive activities associated with the disposition of the parts of the body, and include derivative activities such as execution of tasks associated with the individual's domicile.
Broader Human disability (#PC0699).

♦ PE6769 Locomotor disabilities
Movement disorders
Nature Locomotor disabilities refer to an individual's ability to execute activities associated with moving, both himself and objects, from place to place.
Refs Shah, Nandkumar S and Donald, Alexander G *Movement Disorders* (1986).
Broader Human disability (#PC0699).

♦ PE6770 Personal care disabilities
Nature Personal care disabilities refer to an individual's ability to look after himself in regard to basic physiological activities, such as excretion and feeding, and to caring for himself, such as with hygiene and dressing.
Broader Human disability (#PC0699).
Related Orientation handicaps (#PE6772).
Aggravates Economic and social losses due to disability (#PE4856).
Aggravated by Poverty and disability (#PD0723) Physically handicapped persons (#PD6020).

♦ PE6771 Criminal intrusion
Broader Criminal harm to property (#PD5511).
Narrower Looting (#PE4152) Stowing away (#PE0595).
Related Burglary (#PD2561) Criminal trespass (#PD3794).
Aggravated by Gang warfare (#PD4843).

♦ PE6772 Orientation handicaps
Nature An orientation handicap is the individual's inability to orient himself in relation to his surroundings, and broadly includes being unable to receive signals from his surroundings (via the senses), assimilate those signals, and respond to the assimilation.
Broader Psychological disorders (#PD8375).
Related Occupation handicap (#PE6778) Personal care disabilities (#PE6770) Social integration handicap (#PE6779) Physical independence handicap (#PE6773) Economic self-sufficiency handicap (#PE6780).

♦ PE6773 Physical independence handicap
Nature Physical independence is the individual's inability to sustain a customarily effective independent existence in regards to aids and the assistance of others, and also includes self-care and other activities of daily living.
Broader Psychological disorders (#PD8375).
Related Orientation handicaps (#PE6772).

♦ PE6776 Disfiguring impairments
Nature Disfiguring impairments include those with a potential to interfere with or otherwise disturb social relationships with other people. This includes conditions that may not be the consequence of specific diseases, such as disfigurement, as well as disorders that may impair control of bodily functions in the manner that is customary and socially acceptable.
Broader Personal physical disfigurement (#PD8076).

♦ PE6777 Visceral impairments
Nature Visceral impairments include impairments of internal organs and of their special functions.
Broader Symptoms referable to digestive system (#PE9604).

♦ PE6778 Occupation handicap
Nature Occupation handicap is an individual's inability to occupy his time in the manner customary to his sex, age and culture.
Broader Psychological disorders (#PD8375).
Related Orientation handicaps (#PE6772).

♦ PE6779 Social integration handicap
Nature A social integration handicap adversely affects an individual's ability to participate in and maintain customary social relationships.
Refs United Nations *Integration of Disabled Persons into Community* (1981).
Broader Social isolation (#PC1707) Psychological disorders (#PD8375).
Related Orientation handicaps (#PE6772) Social withdrawal of aged (#PD3518) Limited spheres of relationship (#PD1941).
Aggravated by Fear of intimacy (#PF8012).

♦ PE6780 Economic self-sufficiency handicap
Nature Economic self-sufficiency handicap impairs an individual's ability to sustain customary socioeconomic activity and independence.
Broader Psychological disorders (#PD8375).
Related Orientation handicaps (#PE6772).

♦ PE6781 Tampering with witnesses and informants in proceedings
Nature The use of threat, force, deception or bribery to influence a witness or an informant to withhold information, give false evidence, or elude giving testimony is a crime. It is also a crime for a witness or a informant to solicit a bribe to testify falsely.
Broader Hindrance of law enforcement (#PD5515).
Narrower Intimidation of witnesses (#PJ5894) Intimidation of victims of crimes (#PJ1543).

♦ PE6782 Amoebiasis
Nature Amoebiasis is the infection produced in human beings by a histolytic, or tissue-destroying, parasite, known scientifically as Entamoeba histolytica. This amoeba usually lives and reproduces in the large intestine, and does not always cause disease. In most cases, peaceful coexistence reigns between the parasite and its host, or carrier. Occasionally, however, the amoeba penetrates deeper into the body, damaging tissues and causing amoebic disease, or invasive amoebiasis.
Incidence Amoebiasis can be found all over the world, including the cold and even polar regions: epidemics of amoebic dysentery have been recorded among Eskimos at the North Pole. However, the frequency of invasive amoebiasis varies greatly from one geographical area to another. In the less developed countries, most of which are located in the tropics and subtropics, the proportion of patients with the invasive form is much higher. This does not mean that a tropical climate favours development of the disease – in Mexico, for example, invasive amoebiasis is more prevalent in the central plateau, at an average height of 2000 feet above sea level and in a temperate climate. It is simply that in the tropics most of the human population live under poorer economic conditions, are less well educated and lack adequate health facilities. It can be said, therefore, that amoebiasis is not a tropical disease, but one of poverty and ignorance. Certain zones within the areas where amoebiasis is prevalent are severely affected, and are referred to as 'homelands of amoebiasis'. They include Southeast Asia, East and West Africa, Mexico and the Northwest part of South America. The reasons for the concentration of the disease in these zones is not known.
Background Amoebiasis has existed since remote antiquity and many of the outbreaks of dysentery recorded since the time of Hippocrates were probably of amoebic origin. Clinical descriptions which make it possible to identify amoebiasis with relative certainty have been published since the 17th century, but Fedor Aleksandrovich Lösch actually made the discovery of E histolytica in 1875 in St Petersburg, now Leningrad, in a patient suffering from dysentery. A few years later the amoeba was also shown to be present in liver abscesses.
Refs Martinez-Palomo, A (Ed) *Amebiasis* (1986); Martinez-Palomo, A (Ed) *Ambebiasis* (1986).
Broader Intestinal infectious diseases (#PE9526) Diseases of the digestive system in animals (#PD3978).
Related Giardiasis (#PE4811).

♦ PE6783 Legionnaire's disease
Legionellosis
Nature Outbreaks of Legionnaire's disease in the USA and Europe have focused attention on this problem with its as yet undefined epidemiology and potential serious impact on countries with a large tourist industry. The early symptoms of Legionella pneumonia include malaise, myalgia, anorexia, headache, profuse sweating and chills. In some series patients have developed confusion or delirium during the first week of illness, and diarrhoea, abdominal pain, and vomiting may also result.
The bacteria has been associated with cooling towers and evaporative condensers that form part of the air conditioning system. The cooling towers circulate large quantities of warm water where the bacteria multiplies easily. The escaping water carries the bacteria in the air. The disease can be contracted if people inhale high concentrations of bacteria.
Background The disease was first identified in 1976 after an outbreak at an American Legion convention, which killed 30 people.
Incidence Almost all reported outbreaks of Legionella in the USA and Europe have occurred during the summer and autumn, with a mortality rate at about 10–12.5 percent.
Broader Bacterial diseases in animals (#PD2731).

♦ PE6784 Ageing women
Loss of sex appeal by women — Negative self-image of ageing women
Incidence Among the early Franks in the Middle Ages, the fine for murdering a fertile women was three times that for murdering a man, but after menopause her value became negligible.
Claim Women are enjoined to disguise their age and then ridiculed for doing so. Middle aged men are considered to be at their peak whereas their female peers are constrained to imitate their younger sisters.
Counter-claim Surely men are not such superficial creatures as to be blind to the qualities of women their own age.
Broader Human ageing (#PB0477) Age discrimination (#PC2541).
Aggravates Discrimination against unmarried women (#PD8622).
Aggravated by Fear of growing old (#PJ9144) Discrimination against women at retirement age (#PE6069).

♦ PE6785 Eye irritation
Eye strain
Incidence As an indication, the market for eye strain and eye care products in the UK in 1986 was £9 million.
Aggravates Visual deficiencies (#PD8179).
Aggravated by Irritant fumes (#PD3672) Fungicides as pollutants (#PD1612) Photochemical oxidant formation (#PD3663).

♦ PE6786 Abuse of commercial confidentiality
Abuse of commercial secrecy — Commercial cover-up — Secret deals — Secret commercial agreements
Nature The need for commercial secrecy is used as a justification for withholding information which may be of vital importance in establishing the existence of environmental hazardous products or processes.
Broader Exploitation of trust (#PC4422).
Related Secret international agreements (#PF0419).
Aggravates Suppression of information concerning environmental hazards (#PF4854).

♦ PE6788 Monopolization of agricultural genetic resources
Restricted access to plant seeds — Reduction in genetic diversity of available seeds — Irresponsible patenting of genetically transformed plants
Nature Transnational corporations increasingly gain proprietary rights to improved seed varieties, often without recognizing the rights of the countries from which the plant materials were obtained. Much of this material originates in developing countries.
Background Nations have historically struggled to monopolize certain plants of commercial importance. This struggle was strongest during the colonial era of the last century, when new scientific tools and a network of botanical gardens brought about the transformation of world agriculture. Today, the struggle to monopolize specific genes in the development of agriculture is leading to the development of other scientific tools and institutions; scientists are largely pawns in a much larger and expensive scheme.
Incidence It is estimated that 55 per cent of the world's scientifically stored plant genetic resources is controlled by institutions in industrialized countries, 31 per cent by institutions in developing countries and 14 per cent by international agricultural research centres.
Refs FAO *Water Law in Selected African Countries* (1980).
Broader Monopolistic control of new life forms (#PD7840).
Aggravates Inflated seed costs (#PJ4397) Decreasing genetic diversity in cultivated plants (#PC2223).
Aggravated by Unethical practice of the biosciences (#PD7731).

♦ PE6789 Keratitis
Ulceration of cornea — Inflammation of cornea — Interstitial keratitis — Superficial keratitis — Ulcerative keratitis
Refs Grayson, Merrill *Diseases of the Cornea* (1983); Leibowitz, Howard M *Corneal disorders* (1984).

Broader Eye diseases and disorders (#PD8786).
Aggravated by Irritant fumes (#PD3672)   Onchocerciasis (#PE2388).

◆ PE6791 Gait disturbances due to torture
Nature Many of the victims of torture suffer from difficulty in walking ranging from paralysis to tiredness and some pain in the feet. In addition to breaking bones, crushing feet, bruising and tearing muscles and tendons one of the most frequent forms of torture is falanga, beating the soles of the feet. Immediate symptoms after falanga torture is pains and swelling. The feet become reddened and hot. The swelling can be so pronounced that the feet and ankles become pyramid shaped. The swelling lasts about 14 days. There is a temporary loss of sensitivity and a reduction in the ability to function in the ankles, feet and toes. Victims have difficulty in walking for from a few days to several months. Nearly one half of the victims of falanga have difficulty with their feet for years following the torture.
Broader Somatic and psychosomatic effects of torture (#PE5294).

◆ PE6795 Inaccessible places of worship
Closure of places of worship — Inaccessible places for meditation
Aggravated by Restrictions on freedom of worship (#PD5105)
Decreasing participation in collective religious worship (#PF8905).

◆ PE6797 Torture through sensory deprivation
Nature Sensory deprivation is used to punish prisoners, create disorientation and break down resistance to questioning. Victims are subjected to nearly total deprivation of the senses. Prisoners are frequently blindfolded or have their heads placed in a hood sometimes for months. They are placed in cells without any light for periods over two years. They may be deprived of sleep for days causing mental breakdown. Some are wrapped in blankets and tied in order to prevent any movement. Often prisoners are fed food bland and tasteless; others are deprived of food and drink altogether for days at a time.
Incidence Torture through sensory deprivation has been reported in the following countries: **Af** Gabon, Kenya, Mauritania, Morocco, Namibia, Rwanda, South Africa, Zaire, Zambia, Zimbabwe. **Am** Argentina, Colombia, Guyana, Paraguay, Peru, Suriname, Uruguay. **As** Afghanistan, Bangladesh, China, Israel, Korea Rep, Pakistan, Saudi Arabia, Taiwan (Rep of China). **Eu** Italy, UK.
Broader Torture by deprivation (#PD3763).
Aggravates Harmful effects of sensory deprivation (#PE5787).

◆ PE6806 Occupational blood diseases
Nature Although benzene, ionizing radiation and lead are the most common causal agents in occupational blood diseases, technological development has led to a significant increase in the number of occupational haemotoxic agents.
Incidence The health hazard is particularly increased during work such as the degreasing of metal components, dissolving and reprocessing rubber, the manufacture of solvents for glues, paints and varnishes, the use of printing inks (heliogravure and photogravure) and dry cleaning. Other activities which may be the source of agents producing occupational blood diseases include the leather industry, textile industry, woodworking industry, battery manufacture, and work in the vicinity of sources of ionizing radiation or carbon monoxide.
Background Since these haemotoxic agents are of varied nature (chemical, physical or biological), their haemotoxic activity differs. The haemotropic property depends on the specific characteristics of a toxic substance, its physicochemical properties and its route of entry is into the body. In addition is the significance of age, sex, race and, in particular, acquired or hereditary personal predisposition. The specific features of the work station are related to: the presence and concentration of haemotoxic substances in the environment; time of exposure; effectiveness of the ventilation system; and use of personal protective equipment.
Broader Occupational diseases (#PD0215).

◆ PE6811 Cardiac conditions in work environment
Nature Cardiac conditions may be disparate, functional, or organic, with varying consequences. The lesions may involve the coronary system (the heart vessels), the endocardium (the internal membrane of the heart), the myocardium (the cardiac muscle), and pericardium (the serous envelope of the cardiac muscle). These lesions, or simple functional disorders, may induce symptoms such as palpitations, cardiac rhythm disorders, cardiac insufficiency, precordial pain, angina pectoris, and fainting fits. Injuries may result in heart valve ruptures or heart wounds.
Incidence Approximately 6 percent of all workers suffer from a heart condition, though the incidence is higher among elderly employees. Living and working conditions (hours of work, speed of work, nutrition, smoking) are factors at least as important in the aetiology of heart disease as physical work itself. Coronary diseases are found more frequently amongst sedentary workers or persons with heavy responsibility, such as manager's or executive's coronary disease.
Broader Occupational diseases (#PD0215).

◆ PE6815 Hazardous aquatic animals
Nature Practically all of the phyla of aquatic animals have some which are dangerous to man. Men may encounter these animals in the course of various activities including surface and subaqua fishing, the installation and handling of equipment in connection with the exploitation of petroleum under the sea, underwater construction and scientific research; and thus be exposed to the risk of accidents. Bathers and subaqua enthusiasts are also exposed to the same risk. Most of the dangerous species inhabit warm or temperate waters.
Broader Harmful wildlife (#PC3151).

◆ PE6816 Cardiovascular diseases
Cardiac weakness
Nature Cardiovascular diseases, the main causes of invalidism and premature death in economically developed countries, have been increasing in incidence as average life expectancy increases. It has been shown that urbanization – which increases stress and makes for irregular eating habits (fast food and prepared foods) – plays a significant role in cardiovascular diseases.
Incidence The prevalence of cardiovascular diseases has increased with the average life expectancy (and, thus, an increasingly aging population) and with advances in diagnosis. They are currently responsible for more deaths than any other reported cause in the Americas and Europe.
Background Cardiovascular diseases include endocarditis, which is the main cause of acquired heart diseases; arteriosclerosis, the most widespread arterial disease; arteritis; and myocardial ischaemia, which is the major cause of death attributed to cardiovascular diseases.
Refs Dunning, A J and Wils, W I M (Eds) *Heart and the Future/Future of the Heart* (1987); Dunning, A J and Wils, W I M (Eds) *Heart of the Future/Future of the Heart* (1986); Reiffel, James et al *Psychosocial Aspects of Cardiovascular Disease* (1980); Steptoe, Andrew *Psychological Factors in Cardiovascular Disorders* (1981).
Broader Human disease and disability (#PB1044).
Narrower Cardiac arrhythmia (#PE8219).

Aggravated by Hookworm (#PE3508)   Stress in human beings (#PC1648)
Parasites of the human body (#PE0596)   Health hazards of exposure to noise (#PC0268)
Trace element imbalance in the human body (#PE5328).

◆ PE6817 Cataract
Occupational cataract
Nature Cataract is a disease of the crystalline lens with the clinical appearance of clouding. It is caused by the action of chemical or physical factors disturbing the internal respiration and the metabolic processes in the lens, thus inhibiting its nutrition and giving rise to the atrophy of epithelial elements of the capsule. Cataract reduces vision capacity and accuracy.
Refs Greve, E L (Ed) *Glaucoma and Cataract* (1986).
Broader Occupational diseases (#PD0215)   Eye diseases and disorders (#PD8786).
Aggravates Physical blindness (#PD0568).
Aggravated by Diseases of metabolism (#PC2270).

◆ PE6820 Irritant gases and vapours
Nature Irritant gases and vapours comprise a wide array of chemicals whose common characteristic is the action exerted on the respiratory system and conjunctivae. The intensity and severity of the irritant depends on the chemical structure of the substance, the concentration in the respired air, and the length of exposure.
Narrower Increase in atmospheric concentration of methane (#PE8815).
Related Unpleasant odours (#PJ5522).

◆ PE6824 Hazardous glaciers
Nature Persons working in high-altitude mountain regions may be exposed to the hazards of glaciers. When a glacier snout ends in a steep slope, glacial movement may result in falls of ice blocks that constitute a considerable danger for persons working in their path.
Broader Ice accretion (#PD1393).
Aggravates Disastrous failure of natural dams (#PE0715).
Reduced by Reduction of glacier size (#PE4256).

◆ PE6825 Cough
Incidence As an indication, the market for cough remedies in the UK in 1986 was £47 million.
Refs Korpas, J and Tomori, Z *Cough and other respiratory reflexes* (1979).
Broader Symptoms referable to respiratory system (#PE7864)
Unsociable human physiological processes (#PF4417).
Related Common cold (#PE2412).

◆ PE6831 Contagious ecthyma
Nature Contagious ecthyma is a highly infectious viral disease of sheep and goats; common names for the same disease are orf and sore mouth. The bovine may be accidentally infected, but no other domestic livestock species is susceptible to the disease.
Incidence The disease occurs world-wide wherever sheep and goats are raised. It occurs primarily in lambs 3–6 months of age, but also may affect older sheep. Mortality is usually low, but may affect as much as 90 percent of the flock. Most human cases have histories of direct contact with infected animals.
Broader Virus diseases (#PD0594)   Viral diseases in animals (#PD2730).

◆ PE6833 Vulnerability of frontier workers
Nature Frontier workers (those who live in one country but who commute daily or at frequent intervals across a frontier to another country) are vulnerable to unfair taxation, that is, being taxed in both their home and host countries.
Incidence In 1984, approximately 200,000 EEC nationals were frontier workers.
Broader Lack of trans-frontier cooperation (#PF6855)
Tax discrimination against non-residents of a country (#PD3048).

◆ PE6835 Unequal health benefits for women
Broader Health inequalities (#PC4844)
Discrimination against women in social services (#PD3691).
Aggravates Maternal malnutrition (#PE1085)   Neglect of sexual health of women (#PF5147)
Health hazards of smoking for women (#PE4995).

◆ PE6836 Excessive job control
Incidence In 1981, an American airline company introduced a computerized system into their operations in the Federal Republic of Germany which measured an employee's performance in each work station including the number of reservations made, telephone calls answered, actual working hours, telephone calls per working hour, and ticket sales.
Broader Denial to people of control over their own lives (#PC2381).
Related Elitist control of production (#PD0154).

◆ PE6840 Toxic hazards of cassava
Nature For millions of people in developing countries, cassava is a staple food. Though hardy and easy to grow, it has little nutritional value and is a toxic hazard for men and animals alike. Recent studies suggest that it can be a key factor in the aetiology of goitre in areas where there is a deficiency of iodine in the people's diet.
Refs Théberge, R L (Ed) *Common African Pests and Diseases of Cassava, Yam, Sweet, Potato and Cocoyam* (1985); United Nations *Cassava Toxicity and Thyroid* (1983).
Broader Toxic substances (#PD1115).

◆ PE6841 Inadequate conditions of work in the construction industry
Broader International imbalance in the quality of life (#PB4993).

◆ PE6843 Labour displacement
Displacement of workers in developed countries
Nature Transfer of capital and technology has an effect on labour similar to that produced by technological change. In the developed countries, this transfer displaces workers with low skills and educational attainments and accelerates the necessity of readapting, retraining and relocating displaced workers. In the developing countries, industrial operations that maintain the labour force at the particular level of skills needed for the production of a component, or for a certain industrial process, create a problem. While the immediate employment effect may be positive, the long-term effect may be negative, since the approach does not offer potential for the upgrading and diversification of skills necessary for job betterment.
Aggravated by Lack of technical development and excess of manpower in developing countries (#PE4933).

◆ PE6844 Excessive portrayal of terrorist activity in the media
Nature Terrorists have found a most invaluable ally in the press, which often acts as though it was held hostage itself, by being manipulated by the terrorists' overall strategy to bring down the rule of the law and sap the core of free societies.
Claim Today's terrorists do not kill or plant bombs to weaken the spirit of those groups most

directly attacked by them, but rather carry out their outrages to secure big headlines in newspapers, and radio and television coverage, thus creating a spiralling multiplication effect for their terror. It may even be ascertained that they murder in order to be featured on a particular high-rated television current affairs programme or in a particular large selling journal. Journalists are puppets in the hands of those lunatics out to destroy press freedom and all other types of freedom. Converted into spokesmen of terror, they sometimes seem to be aiding the terrorists against their will and despite themselves.
Counter-claim Journalists are not manipulated against their own will, they allow themselves to be manipulated. By maintaining definitive, unified stances, they could play an active role in undermining much of a terrorist's strategy. Possible tactics include stony silence on the part of the press; biased coverage not assisting terrorist designs; strictly neutral coverage without sensationalism; and/or an aggressive strike treatment which could include destroying terror's public image and prestige; demolishing the terrorists' ideological alibis; standing up to the terrorists; and diminishing the intensity of their message.
Broader Distorted media presentations (#PD6081).
Aggravates Terrorism (#PD5574) Ineffective prevention of terrorism (#PF4240).

♦ PE6853 Inadequate registration of wills
Nature An increasing number of persons make their wills in a place not their home and even in a foreign country. As most States do not require wills to be deposited with a court of law a notary, or another authority, and also have no central register, the heirs are often unaware of the existence or the whereabouts of a will.

♦ PE6854 Unlawful repatriation of minors
Nature A minor is a person who has not attained his or her majority under the applicable law and does not have the right to determine his or her place of residence. The involuntary repatriation of minors may occur when a minor is in the territory of a state against the will of those responsible for protecting the interests of the minor; when the presence of the minor in the territory of a state is incompatible with the interests of the state; when the presence of the minor in the territory of the requested state is incompatible with a measure of protection or re-education taken in respect of him by the competent authorities of the requesting state; or when the presence of the minor is necessary in the territory of the requesting state because of the institution of proceedings there with a view to taking measures of protection or re-education in respect of him or her.
Broader Forced repatriation (#PD8099).

♦ PE6857 Yaws
Nature Yaws is a chronic and infectious spirochetal skin disease. Infection usually occurs by contact through injured skin and is fostered by unsanitary living conditions. The symptoms include granulomatous swellings and ulcerations of the skin; lesions of the bones; and sometimes severe pain.
Incidence Yaws is common in almost all the tropical countries of North and South America, Africa, Asia, and Oceania, usually affecting the native rural population.
Broader Spirochaetal diseases (#PE3254).

♦ PE6860 Dangerous materials in frontier areas
Nature The storage of dangerous materials in frontier areas creates tension in the international relations of the countries and hazards for nationals and property of both countries. The government of a country which seeks to authorize the construction of stores of explosives or other dangerous materials in the immediate vicinity of a neighbouring country, may encounter serious difficulties and delays in obtaining necessary safety information in the neighbouring territory. In the event of an accident causing loss or damage in the territory of the neighbouring state, it may be difficult or impossible to obtain compensation from the country which was responsible for selecting the storage site.
Broader Dangerous substances (#PC6913).

♦ PE6861 Decline in import capacity of developing countries
Developing country import strangulation — Import compression of developing countries
Nature The external financial stringency experienced by the great majority of developing countries has led to widespread attempts to compress imports. The import reductions have frequently affected categories essential to investment and certain types of production. For example, many countries have reduced imports of machinery, equipment and industrial supplies. This, in turn, has led to lower levels of economic activity in most developing countries.
Incidence Import compression after 1982 in several highly indebted developing countries, many of which, especially in Latin America, are major trading partners of the USA, accounted for a large part of the decline of exports from the USA after 1982.
Aggravates Deflation (#PD7727).
Aggravated by Declining economic growth in developing countries (#PD5326)
Deterioration of terms of trade for developing countries (#PD2897)
Disincentives for financial investment within developing countries (#PF3845).

♦ PE6864 Varicose veins
Nature Varicose veins is a change in the veins expressed in sacculate dilation, increase in length, and the formation of convolutions and modular glomeris. The disease develops gradually. Patients complain of a sensation of heaviness in the affected limb, of rapid fatigue and of puffiness. Sometimes a varicose ulcer develops. In addition to the physical limitations demanded by this complaint (curtailment of activities which involve long-term standing or walking), it also cause psychological discomfort due to its appearance.
Incidence The disease most often affects veins of the lower extremities and rectum, and is three times more frequent among women than men.
Refs Borschberg, E *Prevalence of Varicose Veins in the Lower Extremities* (1967); Tibbs, David J *Venous Disorders of the Lower Limbs* (1989).
Broader Diseases of the arteries (#PE2684).

♦ PE6865 Visual strain in modern technology
Nature Considerable increase in eye strain has been the consequence of the introduction of modern techniques, not only in industry but also in the service sector (banking, insurance, public utilities, hospital administration). This strain is due to both the increase in miniaturization (characteristic of electronics and watchmaking, but also part of the service sector as scientific and technical papers are filed on microfilms or microfiches which must be read with the aid of projection systems) and the enormous expansion of information processing and the use of computers.
Background The use of cathode-ray tube (CRT) screens, video screens, microreaders and mounting tables illuminated from below considerably modify the visual and luminous environment characteristic against a bright background under photopic lighting conditions. On the screen, however, the characters displayed are bright against a dark background. The display by projection of electrons onto a fluorescent screen involves temporal and spectral characteristics which differ considerably from those pertaining to the reading of printed documents. The text is presented on an almost vertical surface placed at a fixed distance and subject to reflective glare.
Narrower Health hazards of computer visual display units (#PE5083).
Aggravates Occupational stress (#PE6937).

♦ PE6866 Occupational diseases of the voice
Nature Occupational diseases of the voice fall into two different categories; the first being diseases resultant from conditions of an organ, and the second being functional conditions which may either be pure dysfunctions or occur together with an organic disease.
Incidence The types of person most at risk are those engaged in occupations in which the voice is preponderant, such as teachers, sales agents, barristers, politicians, lecturers, radio and television reporters, show singers, comedians, actors, and opera singers. The type of voice which is most liable to contract diseases of the vocal function is the "projected" voice, that is, the voice which is used with the deliberate intention of exercising an influence on others.
Broader Speech disorders (#PE2265) Occupational diseases (#PD0215).

♦ PE6867 Poisonous, allergenic and biologically active wood
Nature The biological effects of wood give rise to many different morbid symptoms or processes, the nature of which depend on the quantity and composition of the constituent substances. There are primary irritating effects (dermatitis due to the direct action of the wood on the skin or inflammation of the tissues where splinters have entered skin wounds), and allergic conditions. A variety of non-classified miscellaneous symptoms include nose bleeding, nausea, anorexia, vomiting, headache, weakness and vertigo. Lung diseases are relatively rare, though they have sometimes occurred.
Broader Environmentally induced diseases (#PD8200).
Aggravates Allergy (#PE1017).

♦ PE6868 Hazards to young people at work
Occupational hazards of youth workers
Nature Apart from the danger of exposing young and immature bodies to health hazards, heavy loads, etc, there are also some special factors in the causation of industrial accidents that operate in respect of juveniles alone. These include ignorance of factory conditions and hazards; inexperience and natural curiosity; desire to prove oneself and show oneself to be "tough"; contempt for the safety-first principle; liability to fatigue; and boredom leading to "skylarking or horseplay". Also, the young worker or apprentice straight from school may constitute a source of cheap and readily exploitable labour.
Broader Occupational hazards (#PC6716).

♦ PE6872 Tularaemia
Nature Tularaemia is an acute, moderately severe, infectious disease of animals and, secondarily, of man. Man becomes infected by handling rodents or hares that are infected with the disease or that have died of it. Infection may also occur through contact with water, straw, or food products contaminated by such animals, as well as by tick or insect bites. The symptoms are fever, severe headache, insomnia, night sweats, and swelling and tenderness of the lymph nodes.
Incidence Tularaemia has been reported on all continents except Australia, but does not occur to the same degree in all areas. It is more prominent in the Holarctic animal region; the two main regions in which numerous infections of man have occurred are the USA (except Hawaii) and the southern USSR.
Background The disease is named after Tulare County, California, USA, where it was first isolated, in 1911, by G McCoy and C Chapin, in infected ground squirrels.
Broader Epidemics (#PC2514) Zoonotic bacterial diseases (#PD6363)
Bacterial diseases in animals (#PD2731).
Aggravates Pneumonia (#PE2293).
Aggravated by Ticks as pests (#PE1766) Rodent vectors of disease (#PE3629).

♦ PE6874 Traumatic injuries
Nature Trauma is an externally caused injury to the human or animal organism. Depending on the duration of the traumatic event, a trauma may be either acute or chronic. It may be classified according to the circumstances in which it occurred; for example, nonoccupational injuries, industrial accidents, or injuries resulting from athletic activity or military combat.
Background Traumatic injuries include cuts, punctures, lacerations and contusions, shell-type wounds, amputations, and multiple injuries (resulting mainly from traffic and occupational accidents.)
Refs Andreasen, J O *Traumatic Injuries of the Teeth* (1981); Gruber, R T, et al *The Pathophysiology of Combined Injury and Trauma* (1987).
Broader Injuries (#PB0855).
Aggravated by Industrial accidents (#PC0646).

♦ PE6877 Misrepresentation of information to consumers
Consumer fraud — Exploitation of consumers
Nature Consumer fraud is a false or misleading representation of a material fact, whether by words or conduct, that causes a consumer to be deceived. More than 800 categories of consumer frauds have been identified to date; some of those causing concern to countries in different parts of the world are fraudulent sales, including deception as to weights and measures and unsafe or substandard products, adulterated food and obsolete or hazardous drugs, real estate swindles, including fictitious land registration schemes, hoarding and preorganizing, especially as related to black-market operations, collusive bidding, unnecessary repairs, usury and credit fraud, pilferage, smuggling and various other schemes involving misrepresentation, concealment, manipulation, breach of trust, subterfuge or illegal circumvention.
Incidence Polls show that 18 million people in UK have bought items because of "green" advertising, packaging or changes made to products by companies anxious to show they are environmentally aware, although it is difficult to sift the genuine claims from the cosmetic. During sales some shops are inventing higher prices for goods, using the loopholes in a code of practice, so they can offer bogus reductions.
Broader Fraud (#PD0486) Economic crime (#PC5624).
Related Grievances of consumers (#PD7567) Unethical media practices (#PD5251).
Aggravates Unethical consumption practices (#PD2625).
Aggravated by Consumer vulnerability (#PC0123)
Misleading advertising (#PE3814)
Limited consumer knowledge (#PG7662)
Ineffective self-regulation in the consumer goods manufacturing industries (#PE8574).

♦ PE6883 Discriminatory treatment of foreign prisoners
Nature The numbers of individuals in prison in countries other than their own has grown significantly in recent years as a result of the increased mobility of people and the marked differences in economic well-being and opportunities between one country and another. The major problem faced by foreign prisoners is one of language. In order to receive the best medical treatment, it is necessary for prisoners to be able to communicate with their doctors and be able to understand them in return. This is especially true in instances of psychiatric treatment. In discipline areas, the need for prisoners to understand the prison rules and the authorities which enforce such rules is obvious. The negative effects of cultural and social deprivation are widely

recognized, and the deprivation may be even more severe for the foreign prisoner who has limited opportunity for social intercourse because of a language barrier, and is far from family or friends who might visit. Cultural differences pose a second source of problems for individuals in prisons in countries other than their own. Differences in diet can arise from religious rules, and the radical changes in diet caused by cultural differences in the country in which they are incarcerated can cause health problems.
Broader Prisoners of war (#PC8848) Maltreatment of prisoners (#PD6005)
Discrimination against foreigners (#PD6361).
Aggravates Forced repatriation of prisoners of war (#PD0218)
Discrimination against prisoners' families (#PE5043)
Discrimination against ex-prisoners and ex-detainees (#PE6929).

♦ PE6884 Transport and storage of radioactive material
Nature Radioactive materials, when not properly handled, present risks of irradiation and contamination; their radiations, which are capable in varying degrees of penetrating matter, are not directly perceptible by the human senses. Despite the risks, such materials have to be transported and stored so that precautions are simple and do not require ordinary transport workers to possess any special knowledge or use any special appliances. Accordingly, the regulations make safety depend essentially on the packaging of the materials.
Refs United Nations *Packaging and Transportation of Radioactive Materials–Patram 1986* (1987); United Nations *Recommendations on the Transport of Dangerous Goods* (1987); United Nations *Advisory Material for the IAEA Regulations for the Safe Transport of Radioactive Material* (1987); United Nations *Emergency Response Planning and Preparedness for Transport Accidents Involving Radioactive Material* (1988).
Related Radioactive wastes (#PC1242).

♦ PE6886 Cowpox
Pseudocowpox — Pseudo-cowpox — Milker's nodes — Paravaccinia
Nature Cowpox and pseudocowpox are two antigenically and structurally distinct pox viral agents that cause skin infections of the teats and udders of cattle. Cowpox was once common in most countries of the world where cattle were raised, but as incidence of smallpox (thought to be naturally connected with cowpox) decreased in the human population due to vaccination, the incidence of cowpox in the bovine population also decreased. Most pox-like infections in cattle around the world today are thought to be caused by pseudocowpox, which is a relatively common cattle infection. Transmission to man is primarily by direct contact with lesions on cattle. Pseudocowpox infections in man are termed 'milker's nodules' or 'milker's wart'. The infection begins with a lesion which progresses from a papule to a vesicle surrounded by erythema and oedema. The lesions develop into firm, elastic nodules which usually heal spontaneously in four to six weeks. There is no specific treatment but topical antibiotic treatments are sometimes helpful.
Broader Virus diseases (#PD0594) Pox diseases in animals (#PE7304)
Viral diseases in animals (#PD2730).

♦ PE6887 Corrosive substances
Nature Corrosive substances are those which possess the property of severely damaging living (in particular human) tissue and of attacking other materials such as metals and wood. Although the risks to humans from such substances appear more evident with respect to direct tissue damage, the attacking of non-living material is also significant, particularly to those concerned with the transport of dangerous goods and for whom the safety of the carriers (railway trucks, road vehicles, sea and inland waterway vessels, and aircraft) are of prime concern.
Broader Dangerous substances (#PCo913).
Narrower Corrosive groundwater (#PE4740).
Aggravates Corrosion (#PD0508).

♦ PE6888 Denial of trade union rights to public employees
Nature There are very few countries where employers in the public sector can count on a full set of trade union rights. The question of why public employees' rights lag behind is especially crucial since the same authority may combine the role of employer and of the public authority with the legislative machinery to impose its view.
Broader Denial of right to union activity for special groups (#PE1355).
Narrower Violation of the right to organize trade unions in government services (#PE5938).
Related Denial of right to organize trade unions (#PE5398).

♦ PE6891 Unequal income distribution in industrialized countries
Income inequality in developed countries
Incidence The distribution of income from salaries and wages - who gets how much of the national income - has not changed substantially in any of the major industrialized nations. The unequal ownership of wealth - houses, real estate, stocks, bonds and personal possessions - is even more startling.
In the UK, for example, a study found that five per cent of the population owns over 50 per cent of total personal wealth. Those with more wealth are able to make more out of it, to extend their power through investments which give them control over important economic decisions. In the UK in 1990, top executives were receiving incomes 344 times greater than those of the lowest-paid seventh of the labour force. In USA in 1987 the highest fifth of all families had a income share of 43.7 per cent of all national income and the lowest fifth had 4.6 per cent income share.
Broader Unequal income distribution (#PD4962)
Deterioration of industrialized countries (#PD9202).

♦ PE6898 Hypersensitivity
Nature Certain groups within the population become adversely affected by toxic substances or work stresses while others apparently exposed to the same substances and/or performing the same tasks remain unaffected. Hypersensitivity may stem from developmental/ageing effects; nutritional status; genetic factors; pre-existing diseases; and personal/life-style habits.
Broader Human disease and disability (#PB1044).
Aggravates Occupational stress (#PE6937).
Aggravated by Decadence (#PB2542) Malnutrition (#PB1498)
Toxic substances (#PD1115) Food intolerance (#PE9541)
Allergy inducing food (#PE3225) Developmental disabilities (#PD3773)
Genetic defects and diseases (#PD2389).

♦ PE6899 Farmer's lung
Nature Farmer's lung is an allergic disease due to inhaled organic dusts, affecting mainly the alveoli. The main causes of the disease are the spores of certain thermophilic actionmycetes, particularly micropolyspora faeni and, to a lesser degree, thermoactinomyces vulgaris.
Broader Pneumoconiosis (#PD2034).

♦ PE6902 Occupational hazards of female workers
Nature Female workers are subject to problems unknown to their male counterparts and which could, if preventive measures were taken, be greatly reduced. The problems include menstrual disorders which may reduce a woman's concentration and make her accident-prone, and which can be exacerbated by inadequate sanitary facilities geared to the specific needs of a woman during menstruation; prolapse of the uterus which seriously impairs a woman is working capacity and is exacerbated by the heavy postnatal physical work common to farmers and peasants; inflammatory disorders of the small pelvis which can result from tight fitting clothes or chemical pollution of the working environment; gynaecological tumours which may be caused by occupational carcinogenic substances; deformities of the pelvis and lumbar spinal column caused by extreme physical stress during pregnancy; neuro-endocrine disorders linked with the unique responsibilities of women - outside employment, household duties, and possibility motherhood as well; and the recent discoveries that video display units may cause pregnancy disorders and that work involved in creating microchips can lead to miscarriages.
Claim Prevention of overstrain and damage to health of working women could be realized if adequate emphasis was given to socio-political considerations (labour protection laws, childcare provisions); working conditions (pre-employment medical examinations to determine job suitability, perhaps an adjusted work rhythm); and domestic considerations.
Broader Occupational hazards (#PC6716).

♦ PE6904 Occupational hazards of male workers
Nature A variety of hazardous agents that men encounter in their work may have deleterious effects on the genital system and sexual function. Such effects include sexual insufficiency, gonadic insufficiency, scrotal cancer, and infectious diseases of occupational origin.
Broader Occupational hazards (#PC6716).

♦ PE6909 Infant growth failure
Nonorganic failure to thrive — Deprivation dwarfism
Nature Child's social, emotional, or nutritional environment is disturbed to the point where it interferes with normal growth and development. Some children with emotional deprivation and growth retardation may be secreting abnormally low amounts of pituitary growth hormone causing deprivation dwarfism. When these children are removed from their emotionally disturbing environments, they grow rapidly and release of growth hormone returns to normal.
Related Inhibited human physical growth (#PD5177)
Inhibited growth of malnourished children (#PE4921).
Aggravated by Paternal negligence (#PD7297) Maternal deprivation (#PC0981)
Unbalanced infant diets (#PE0691) Family rejection of children (#PC8127)
Family rejection of physically handicapped (#PE2087).

♦ PE6914 Torture through behavioural regulation
Nature The behavioural regulation is used to torture a subject by forcing them to see their dignity and identity completely destroyed. Victims are submitted to a detailed set of regulations and rules, resulting in close supervision where everything (including completely insignificant details) is controlled. Violation of the rules (either real or supposed) is used as an excuse to punish the victim. They are subject to all kinds of humiliation.
Incidence Behavioural regulation as a form of torture has been reported in a number of countries.
Broader Psychological torture (#PD4559).
Related Political indoctrination (#PD1624).

♦ PE6920 Natural environmental degradation from recreation and tourism
Nature Recreation and tourism are becoming popular to the extent that in many countries they have developed into a national industry; they are often accompanied by extensive damage to the environment. Aquatic ecosystems are particularly vulnerable to the effects of an increased tourist trade and the resultant building of hotel accommodations, sewage disposal works, roads, car parks and landing jetties on banks and coastlines; and the increased angling, swimming, water skiing, shooting or use of motor-boats in the water body. These all produce direct deleterious effects when conducted on a massive scale, including shore damage, chemical changes in the water, and sediments and biological changes in the plant and animal communities. Indirect effects are caused when small towns and villages near a water body are visited periodically by a large number of people; or by changes in the land usage within the drainage area often due to an increase in industrial sewage or agricultural run-off that is associated with the local growth of tourism.
Environmental disadvantages accompanying the over-exploitation of tourism include: despoiling of coastlines by construction of tourist facilities; pollution of the sea; loss of historic buildings to make way for tourist facilities; loss of agricultural land for airport development. Ownership of land and the control of components of the tourist industry are increasingly in the hands of non-residents and of companies based elsewhere, giving rise to serious problems of control. Often outside interests acquire the best sites and beaches and then exploit them in such a way that an overall tourist plan cannot be implemented at a later date.
Incidence The Mediterranean is a prime example of the consequences of the over-development of a natural asset (namely the coast) for tourist purposes. Every one of the 6,000 registered beaches in Italy is dangerously polluted according to standards decreed by the Italian government (some have bacteria counts 5 times higher than the limit). On a particular 40-mile stretch of the French coastline there are 195 open drains discharging untreated sewage straight into a tideless sea. Hundreds of miles of coastline have been irremediably ruined by the virtually uncontrolled building of hotels, restaurants, bars and houses. Beaches have been divided into unsightly allotments, and noise from juke boxes, fumes from traffic and sheer human overpopulation have all become indicative of the consequences of over-exploitation of tourism. In some of the larger African game parks, 20-30 vehicles may park around one pride of lions. The character of such parks and their viability as environments for some species is threatened by the volume and behaviour of tourists and the construction of tourist facilities.
Refs OECD *The Impact of Tourism on the Environment* (1980); World Tourism Organization *WTO/UNEP Workshop on Environmental Aspects of Tourism, 1983.*
Broader Landscape disfigurement (#PC2122) Environmental degradation (#PB6384)
Degradation of semi-natural and natural habitats of flora and fauna (#PC3152).
Narrower Degradation of developing countries by tourism (#PF4115)
Degradation of mountain environment by leisure activities (#PE6256).
Related Alienation of land through acquisition by foreigners (#PE0896)
Abuse of international cultural, diplomatic and commercial exchanges (#PF3099).
Aggravates Wildlife extinction (#PC1445) Endangered totemic species (#PE4184)
Health hazards of exposure to noise (#PC0268)
Endangered monuments and historic sites (#PD0253)
Insufficient transportation infrastructure (#PF1495).
Aggravated by Tourist hazards (#PE8966) Inappropriate policies (#PF5645)
Unethical entertainment (#PF0374) Ineffective industry self-regulation (#PF5841)
Long-term shortage of natural resources (#PC4824).

♦ PE6922 Discrimination and harassment of children in public life
Nature Children in many countries are subjected to interrogation usually with respect to activities (alleged or substantiated) of their parents, and suffer subsequent emotional disturbance. They include children of prisoners, activists, and those deemed 'enemies of the state'.
Incidence The children of refuseniks (Soviet Jews whose applications to emigrate have been turned down by Soviet authorities) are subject to constant discrimination and harassment. It has been reported that the KGB has raided the Moscow Jewish Kindergarten several times, terrorizing

the children. In elementary school, children may be called into the principal's office, sometimes in the presence of their parents, and asked to denounce their parents; they are told that their parents are criminals.
Broader Harassment (#PC8558) Victimization of children (#PC5512)
Denial of rights of children and youth (#PD0513).
Narrower Discrimination against mixed race children (#PE5183)
Discrimination against children of prostitutes (#PE0392).
Related Cruelty to children (#PC0838) Exploitation of children (#PD0635)
Physical maltreatment of children (#PC2584).
Aggravated by Discrimination against illegitimate children (#PD0943).

♦ **PE6924 Congenital anomalies of circulatory system**
Refs Peacock, Thomas B *On Malformations of the Human Heart* (1977).
Broader Human physical genetic abnormalities (#PD1618).
Narrower Coarctation of the aorta (#PG2571).

♦ **PE6929 Discrimination against ex-prisoners and ex-detainees**
Denial of rights to ex-convicts — Discrimination against ex-offenders
Nature Life for the released detainee is a mass of difficulties. After years with little or no contact, a released prisoner may return home to find that his wife has remarried or that all his relatives have left without trace. Released detainees often find a community afraid to welcome them back; alienation from friends, neighbours and even close relatives is the result. There may be no state schemes to facilitate the process of adjustment to society. Few people want to take the risk of employing a man or woman with a criminal past, and many jobs are firmly closed to ex-prisoners because they require political clearance papers. In many cases prisoners find themselves second-class citizens, and to the reality of such dire straits is added the possibility of rearrest. Released prisoners are always likely suspects in crimes. Indeed, there often seems to be no alternative to continuing the life of a criminal.
Refs Rudenstine, David *The Rights of Ex-Offenders* (1981).
Broader Discrimination (#PA0833).
Related Recidivists (#PE5581) Denial of rights to prisoners (#PD0520).
Aggravates Discrimination against prisoners' families (#PE5043).
Aggravated by Discriminatory treatment of foreign prisoners (#PE6883).

♦ **PE6934 Government harassment of human rights activists**
Vulnerability of human rights activists — Denial of rights to human rights activists
Nature People working peacefully for human rights may be persecuted under the pretext that they constitute a threat against the national security of a country. In many cases, human rights activists are jailed because they defend or try to help people who have been harassed or detained because of their beliefs. In some countries, people working for human rights are taken to clandestine detention centres and tortured. They are sentenced to long prison terms by secret trials. Their efforts are officially denounced, their houses are raided, their papers are confiscated, and their families intimidated.
Incidence According to the United Nations during 1986–1988 several of human rights leagues or associations were threatened and banned and their activists persecuted, imprisoned and murdered in Algeria, Chile, Colombia, El Salvador, Guatemala, Honduras, Poland, Singapore, Tunisia and Vietnam.
Broader Persecution (#PB7709).
Aggravates Denial of human rights (#PB3121).

♦ **PE6936 Inadequate assistance to victims of torture**
Neglect of victims of torture
Nature Rarely do countries have special funding for care structures for torture victims who often are in need of financial, social, medical and psychological help after release. Practically all victims suffer from multiple mental and physical sequelae to torture that prevent them from leading normal lives at work and at home. The immediate and long-term effects of physical and psychological abuse are oppressive: suicide is a not uncommon result of torture, either in prison to avoid further pain or after release due to the suffering that persists. The after-effects are felt by the families of the victims as well; they are often unqualified to give proper support and care. Because of their fear, some torture victims simply do not request medical help.
Broader Abuse of government power (#PC9104)
Legal discrimination in favour of offenders (#PD9316)
Inadequate assistance to victims of human rights violations (#PD5122).
Related Torture (#PB3430)
Inhumane participation of the medical profession in torture (#PE4015).
Aggravates Disabled victims of torture (#PD0764).

♦ **PE6937 Occupational stress**
Hypertension on the job — Pressured employees — Stress at work
Nature Occupational stress is an often unrecognized health problem that is costly to both workers and businesses. Faulty adaptation to stress, or failure of the body's resistive ability, may lead to various emotional disturbances, headaches, insomnia, chronic fatigue, peptic ulcers, allergic and renal diseases, and even heart attacks.
Workers today must handle numerous arenas of decision-making at once. It is not just the frequency of stress that is increasing, but also the duration. Whereas past generations of workers faced occasional stress periods, people today do not have time in between to recover. The structure of companies, in which there is no opportunity to let off steam, in which people have responsibility but no control, in which a neurotic supervisor can terrorize subordinates, in which computers monitor the hours and productivity of workers, in which workers rarely have a primary group with whom to talk out their troubles—all of these matters contribute to an unprecedented level of stress in the working environment. Some of the most stressful jobs include inner-city high-school teachers, police officers, waitresses, complaint department workers, secretaries, miners, air traffic controllers, medical interns and stockbrokers. The price in absenteeism, medical costs, and legal costs is growing, as is the human price in damaged lives and families.
Refs Cooper, C L and Payne, R *Causes, Coping and Consequences of Stress at Work* (1988); Dalton, Thomas F *The Effects of Heat and Stress on Cleanup Personnel Working with Hazardous Materials* (1984); European Foundation for the Improvement of Living and Working conditions *Physical and Psychological Stress at Work*; Fraser, T Morris *Human Stress Work and Job Satisfaction* (1987); Matthews, Catherine J (Ed) *Police Stress* (1979); Phillips, E Lakin *Stress, Health and Psychological Problems in the Major Professions* (1983); Wills, Geoff and Cooper, Cary L *Pressure Sensitive* (1988).
Broader Socioeconomic stress (#PC6759) Occupational diseases (#PD0215)
Stress in human beings (#PC1648).
Narrower Computer stress (#PE5053) Executive stress (#PE1635)
Stress in industry (#PE6996) Mental stress due to automation (#PE5164)
Combined stresses as occupational hazards (#PE5656)
Shift work stress as an occupational hazard (#PE5768).
Aggravated by Marital stress (#PD0518) Hypersensitivity (#PE6898)
Stress addiction (#PE4951) Heat as an occupational hazard (#PE5720)
Visual strain in modern technology (#PE6865)
Health hazards of computer visual display units (#PE5083).

♦ **PE6947 Mismanagement by intergovernmental organization leadership**
Malpractice by intergovernmental organization leadership — Corruption of intergovernmental organization leadership
Broader Mismanagement (#PB8406).
Aggravates Intergovernmental organization mismanagement (#PD6628).
Aggravated by Official corruption (#PC9533).

♦ **PE6948 Trading in public office**
Trading in political endorsement — Politicization of diplomatic positions — Purchase of military commissions
Nature Trading in public office and political endorsement is making payment or receiving payment by public servant or party officials for actions related to government employment and endorsement for government elective office. These actions include: approval or disapproval of the promotion, employment, or retention as a public servant or designating or nominating a candidate for elective office.
Broader Political bribery (#PC2030) Bribery of public servants (#PD4541).
Related Simony (#PE9840).
Aggravated by Abusive distribution of political patronage (#PF8535).

♦ **PE6949 Denial of right to resist oppression**
Broader Denial of political and civil rights (#PC0632).

♦ **PE6951 Undocumented migrants**
Nature The undocumented migrant is trapped in a position of permanent inequality and is vulnerable to exploitation due to lack of legal protection. The deteriorating international economic situation is paralleled by an expanding market of undocumented labour; the undocumented migrant is in an unfavourable position with no guarantee of integration into society, and is trapped him into becoming part of a readily recognizable group which constitutes the lower stratum of society.
Broader Illegal immigration (#PD1928) Unrecorded knowledge (#PF5728).

♦ **PE6952 Exploitative financial policies of transnational corporations in developing countries**
Nature Where possible transnational corporations favour: inflated contracts (over-invoicing imports and under-invoicing exports); manipulation of commodity prices and of transfer pricing; excessive transfer of profits and other capital gains out of the developing country; and preference for external borrowing instead of bringing in new equity capital.
Broader Domination of developing countries by transnational corporations (#PE0163).
Aggravates Excessive foreign public debt of developing countries (#PD2133)
Private domestic capital outflow from developing to developed countries (#PD3132).
Aggravated by Inadequate negotiating capacity of developing countries (#PE9646).
Reduced by Inadequate investment of transnational corporations in least developed countries (#PE7892).

♦ **PE6955 Denial of right to a people to freely dispose of natural wealth**
Broader Denial of right of a people to be self-determining (#PC6727).
Related Violation of land rights of a people (#PD5218).
Aggravates Immorality (#PA3369).

♦ **PE6958 Nightmares**
Terrifying dreams — Anxiety dreams — Dream anxiety disorder
Nature A nightmare is a terrifying dream, often of an encounter with a being (supernatural, human, or animal) of great strength and energy. A victim may have a feeling or great pressure on the chest, difficulty breathing, and a sense of powerlessness to move or speak. As soon as he is able to move, the victim may experience strong palpitations and a rise in blood pressure, a feeling of languor and listlessness as though physically and emotionally exhausted, and an indescribable depression, uneasiness and anxiety, all experienced with the impression that he has had a real encounter with terror.
Broader Sleep disorders (#PE2197).
Related Sleep terror disorder (#PG4073).
Aggravated by Inhibited grief process (#PD4918)
Post-traumatic stress disorder (#PE0351)
Sleep difficulties due to torture (#PE0451).

♦ **PE6965 Socio-psychosis**
Nature Sociopsychosis reflects the spiritual malaise that permeates society as a whole. It is the pathology not of individuals, but of communities, nations and historical epochs.
Incidence Socio-psychosis is adequately expressed in the words of a London graffito: "Do not adjust your common sense; there is fault in society". Sociopsychotic symptoms are seen in all areas of social collapse, including murder, kidnapping, assault and mugging, alcoholism, and drug addiction.
Counter-claim All actions are of the individual. Socio-psychosis is an excuse to avoid responsibility. Symptoms are caused by the people and organizations who offer excuses to everyone who think the world owes them. Succour means to strengthen, not to pander.
Broader Psychoses (#PD1722).

♦ **PE6968 Massive psychic traumatization**
Nature Any painful individual experience may involve psychic traumatization, especially if that experience is associated with permanent environmental changes. Often inner defence mechanisms are employed to avoid the trauma.
Broader Mental illness (#PC0300) Negative emotions and attitudes (#PA7090).
Aggravates Deviant society (#PC2405) Behavioural deterioration (#PB6321)
Human psychological regression (#PE1418).

♦ **PE6969 Health risks of teenage sex**
Nature Health risks of teenage sex include sexually transmitted diseases, inadequately administered contraceptives, and pregnancy. Pregnancy is obviously the most serious risk for girls who are not yet physically mature. The two main obstetric risks of early pregnancy are toxaemia (high blood pressure) and cephalopelvic disproportion (the baby's head is large relative to the size of the mother's pelvis). The baby of a young mother will have a low birth weight either due to premature delivery or to delivery at term of an undernourished foetus. This means the baby is more likely to die at birth or in infancy and that its physical and mental development may be impaired.
Abortion is an option, though it is still illegal in many countries and thus a girl may have to seek an illegal abortion from an unskilled practitioner (complete with all its possible complications). Even in countries where abortion is legal, pregnant teenage girls may not seek such services due to the prior formalities need that may be necessary. In addition, a pregnant girl may not know she is pregnant or may vainly wish it would go away, and thus not seek an abortion until after the first three months of pregnancy, thus increasing the risk of complications.
Incidence In the Caribbean Islands, 58 percent of all first births are to mothers less than 19 years of age, and half of those are under 17 years old; 41 percent of Indonesian women have their 1st

baby before they are 17; in the USA a million teenage girls become pregnant every year, about 30,000 of them under 15. When girls become pregnant before the age of 15, they are 68 percent more likely to die in pregnancy than their older counterparts; and infant mortality is 2.4 times higher for babies born to those mothers than for babies born to mothers in their early 20s.
Broader Unsafe sex (#PE9776)   Neglect of adolescent health care (#PF6061).
Related Low self esteem (#PF5354)   Adolescent pregnancy (#PD0614)
Adolescent induced abortions (#PD1302)   Sexually transmitted diseases (#PD0061).
Aggravates Neonatal mortality (#PD9750)   Perinatal morbidity and mortality (#PD2387).
Aggravated by Pregnancy disorders (#PD2289)   Ignorance of reproductive processes (#PD7994).

♦ **PE6974 War-time communications with the enemy**
Nature In the time of war knowingly communicating or attempting to communicate with the enemy, knowingly evading or attempting to evade censorship of communication between one's nation and a foreign power, using codes or modes of communication to evade censorship is a criminal offence.
Broader Crimes related to national security information (#PE3997).
Aggravates Victimization (#PF6987).

♦ **PE6975 Fraudulent mineral exploitation claims**
Broader Fraud (#PD0486).
Aggravates Expropriation of land from indigenous populations (#PC3304).
Aggravated by Unethical practice of earth sciences (#PD0708).

♦ **PE6982 Excessive regulation of television**
Confusion of values in monitoring television
Nature There are many bodies who seek to regulate television so that it does not become too commercialized, too violent, too pornographic, or too differing in political view from the taste of the strongest regulating group. Some countries have regulated even the advertisers away, leaving themselves with no budget for quality programming.
Narrower Political interference with television news coverage in capitalist countries (#PE5610).
Related Radio and television propaganda (#PD3085)
Radio and television censorship (#PD3029).

♦ **PE6990 Learned helplessness**
Incidence The most grave example of learned helplessness in modern times was the reaction of European Jews to the Nazis. Witnesses who returned to villages and cities to give warning, were discredited, hampered and negated by lack of authority; and the ways in which Germans treated Jews established conditions most conducive for the production of learned helplessness, the most extreme state of which was death in the concentration camps.
Broader Vulnerability of human organism (#PB5647).
Aggravated by Humiliation (#PF3856).

♦ **PE6996 Stress in industry**
Refs Bagnara, S, et al *Interaction of Workers and Machinery* (1987); Levi, L *Stress in Industry* (1984); Levi, L and International Labour Office Staff *Stress in Industry* (1984).
Broader Stress in materials (#PD7216)   Occupational stress (#PE6937).
Aggravates Failure of materials (#PD2638)   Fatigue in materials (#PD1391).

♦ **PE7000 Weakness of infrastructure in land-locked developing countries**
Broader Vulnerability of land-locked developing countries (#PD5788).
Narrower Unreliable transit services for land-locked developing countries (#PE5836)
Inadequate roads and transport in land-locked developing countries (#PE5824)
Inadequate air transport service for land-locked developing countries (#PE5800)
Inadequate rail transport in developing land-locked and transit countries (#PE5848)
Unreliable telecommunication services for land-locked developing countries (#PE5860)
Inadequate port and storage facilities for land-locked developing countries (#PE5908).
Related Weakness of infrastructure in island developing countries (#PE5772).

♦ **PE7002 Unrealistic expectations**
Misplaced optimism — Unwarranted optimism — Unrealistic optimism — Naive optimism — Ingrained optimism — Over-optimistic bias
Claim In scientific forecasting the most favourable outcomes that can be projected are bounded by the limits of extrapolated data. An optimistic expectation based on nothing but its desirability obstructs further information gathering and reasoning; inhibits directive or corrective actions; and invites disaster.
Broader Personal misperceptions (#PF4389).
Aggravates Unrealistic policies (#PF9428)   Minimization of problems (#PF7186)
Avoidance of negative feedback (#PF5311)   Conceptual repression of problems (#PF5210).
Aggravated by Avoidance of reality (#PF7414)   Inaccurate forecasting (#PF4774)
Neglect of expert advice (#PF7820).
Reduced by Unwarranted pessimism (#PF2818).

♦ **PE7004 Anti-satellite arms race**
ASAT negotiation failure
Nature The production and deployment of anti-satellite weapons by the superpowers are not only extensions of the arms race but signal a tacit admission that space is considered to be one of the battlefields of the future, despite international intentions to the contrary. The development of anti-satellite weapons (ASAT) creates an additional obstacle to international agreement on the peaceful use of outer space and no negotiation to limit ASAT testing and deployment has been reached.
Broader Space weapons arms race (#PD0087).

♦ **PE7007 Conscientious objection at the factory**
Incidence A Muslim who refused work in the food industry to avoid handling pork meat; a Jehovah's Witness who refused to be transferred within a company to armaments manufacture; a merchant navy officer who demanded that his union refuse to unload nuclear missiles; a school teacher who refused a position to teach in a religiously-affiliated school. All these cases in several countries were acknowledged by unemployment compensation commissions to be justifiable exercises of personal conscience.
Broader Employee disobedience (#PD5244)
Denial of right of conscientious objection to military service (#PD1800).

♦ **PE7008 Political ineffectiveness of international nongovernmental organizations**
Nature Most INGOs claim to be non-political organizations, in the sense that there is a basic distinction between the organization of a political party and an organization representing the particular interests of its members – vocational, religious, etc. The reality of the situation is that governmental delegates assess the potential value of an INGO primarily in terms of the political power of the constituency it represents. INGOs controlled by particular national or cultural interests may be rejected for this reason. Furthermore, most expertise, however technical, is now held to have cultural overtones. Even INGOs concerned with palaeontology or Sanskrit literature, for example, are expected to align themselves with majority views of the IGO community on the current major issues of peace, human rights, etc. There is an apparent negligible political impact by INGOs on the wording of intergovernmental resolutions or new programmes undertaken within intergovernmental agencies. (One study showed that only 3 percent of intergovernmental resolutions resulted in new action).
Claim An international non-governmental organization (INGOs) can only be politically effective by relating politically to political institutions such as intergovernmental organizations (such as the United Nations) and national governments. Failures which INGOs experience arise, at least in part, from a failure to think and act politically and to acknowledge that the purpose of such relationships is to exchange influence. This problem is aggravated by INGO indifference to any governmental assessment of an INGO in terms of the importance of the political constituency it represents.
Counter-claim To the extent that many NGOs are working in areas not yet recognized as significant by IGOs or governments, they may be preparing the way for political impact which will be legitimized (possibly years later) by their work (such as the UN discovery of the environment issue in 1972).
Broader Obstacles to effective international nongovernmental organizations (#PF7082)
Ineffectiveness of international nongovernmental organizations and programmes (#PF1595).

♦ **PE7009 Lack of financial information systems for international nongovernmental organizations**
Nature INGOs do not have an information system to locate individuals, foundations and governmental programmes interested in making funds available in specific programmes areas; they depend on chance contact. Conversely, there is no information system that permits the INGOs to be located by such bodies. Similarly, INGOs do not have an information system to locate the most appropriate international and national bodies through which to make available funds for a specific programme; they also lack a system to overcome the difficulty whereby funds are voted every two or more years for programmes which may become irrelevant during that period in comparison with the need for new programmes adapted to newly detected problems in the INGOs' domain. Flexible fund allocation and distribution information techniques developed from the programme planning and budgeting system (PPBS) are frequently unavailable. INGOs also maintain an expensive exchange of correspondence before a potential member or supporter transfers funds for dues or in support of a particular programme. Automatic fund transfers from supporters to the INGO's account (and from there to programme accounts) is lacking.
Broader Lack of information (#PF6337)
Obstacles to effective international nongovernmental organizations (#PF7082)
Inadequate funding of international nongovernmental organizations and programmes (#PE0741).

♦ **PE7010 Disadvantaged status of international nongovernmental organization personnel**
Nature No convention or other arrangement exists to protect the status of INGO personnel (except in Belgium). This means that those who work for INGOs must be prepared to face bureaucratic obstacles of every kind: (a) in attempting to work in the headquarters offices; (b) in field-level work; (c) in travel on INGO business. In addition, social security provisions are such that INGO employees may be unable to ensure continuity of social security benefits and pension rights on return to their country of origin, or when they move to some third country. Payment of pension or life insurance may be blocked by currency regulations. Clearly this ensures that only nationals of the secretariat country can afford to spend career time with and INGO, or else people who are prepared to take the risk of forgoing such benefits. As a consequence this may have considerable implications for the INGOs ability to attract qualified personnel and guarantee their job security. The significance of this problem becomes evident for the work of INGOs when the state of IGO personnel rights and privileges and immunities is considered. Such privileges are held to be essential in order to maintain an adequate international staff. They cover items such as: travel documents, residential requirements and tax exemption, in addition to social security and pension rights. As well as generous fringe benefits, IGO personnel also receive salaries considerably in excess of local salaries to compensate them for the inconvenience for having to work away from their country of origin.
Broader Obstacles to effective international nongovernmental organizations (#PF7082).

♦ **PE7015 Torture through breaking bones**
Nature Victims of torture frequently have legs and arms broken, hands and feet crushed and jaws smashed as well as other bones broken through beating and other forms of torture.
Incidence Breaking bones has been reported in the following countries: **Af** Mali, Rwanda. **Am** Suriname. **As** Bahrain, Bangladesh, India, Korea Rep. **Eu** USSR.
Broader Torture by wounding (#PE4078).

♦ **PE7017 Discriminatory unwritten codes of behaviour**
Taboos
Claim In societies stratified into classes, upward mobility is hindered by not acquiring the behavioural characteristics of the class above. All these characteristics comprise an unwritten rule book. The 'chapters' include, for example, conversational behaviour, manner of speech, physical deportment, attire, hygiene and grooming, manner of eating and choice of food and beverages. There are also norms for personal possessions, recreations, cultural interests and ethical and political viewpoints. Failure to comply with any standard in the unwritten code stigmatizes an individual as not a true member of the class and can lead to social ostracism on that level. The unwritten code is a fossil from feudal times and is a tool for social repression.
Counter-claim Written codes produce legalistic, literal, and minimal compliance with the letter and not with its spirit. Unwritten codes, of honour, of trust, of loyalty and of all decent behaviour are both the ideal and the inspiration that raises the quality of any society.
Broader Conformism (#PB3407)   Discrimination (#PA0833)
Discriminatory imposition of standards (#PD5229).
Narrower Violating taboos (#PF3976).
Related Taboo (#PF3310)   Dietary restrictions (#PJ1933).
Aggravates Unparliamentary behaviour (#PF4550).

♦ **PE7020 Incompatible equivalent national sections of international nongovernmental organizations**
Nature In the case of some INGOs working across different national social systems, the functional equivalents of national organizations may have different relationships to their governments, particularly with regard to the degree of governmental control, funding and staffing. National sections in different countries may perform ranges of functions that only partially overlap such that the non-overlapping features tend to result in suspicion and incompatibilities which probably lead some governments to hesitate in facilitating interaction between their national organizations and the equivalent INGOs. In particular, in some non-Western cultures there may be difficulty in locating organizational forms natural to that culture which could related to a given INGO. There may be resentment of any imposition of a new Western style organization, and a lack of any socio-anthropological skill to match very different styles of organization, or to create or adapt an INGO appropriate to them.
Broader Obstacles to effective international nongovernmental organizations (#PF7082).

♦ **PE7021 Unrepresentative international nongovernmental organizations**
Nature Many nongovernmental organizations are considered to be unrepresentative, namely when all the member countries and regions of the UN are not represented in them. As such they are not considered adequate vehicles for the formulation of impartial policy oriented to the interests of those most in need.
Counter-claim (1) African, European, Asian and other such regional organizations by definition do not have representatives from other regions. The fact that there are more such bodies in Europe is a reality resulting from the relative degree of economic and social development of Europe. (2) There is a functional distinction. It is unrealistic to expect that the African and Latin American continents should be well represented in an international association for arctic agriculture, for example. (3) There is a distinction linked directly to the presence or absence of national counterparts in some countries due to the relative degree of economic and social development. It is not realistic to expect the Comoros to be represented in an international association for the computer-assisted study of ancient languages. (4) There is the non-representativity forced upon organizations by the problems of communication and transport between distant points on the surface of the planet. A national body cannot always allocate funds (possibly equal to or in excess of its own annual budget) to the transport costs implicit in sending representatives to the distant meetings of an international body. This problem arises whether the meeting is in a developing country or in Europe, and whether the national body is based on a developed economy or not. At a time when travel costs are increasing rapidly and subsidies are increasingly hard to obtain, it is therefore natural that the viability of regional bodies may in many cases be greater than that of multi-continent organizations.
Broader Unrepresentative international organizations (#PD4873)
Obstacles to effective international nongovernmental organizations (#PF7082).

♦ **PE7025 Protectionism against imports of service-related goods**
Nature Government intervention can impede the import of manufactured products used in service industries. Thus, government measures may affect the import of materials, machinery and equipment used, for example, by franchise companies, such as fast-food restaurants. Such measures in developing countries are mainly in the form of import licensing and foreign exchange regulations. In some developed countries, health, safety and standards regulations also affect imports. Similarly, countries may discourage imports of equipment and spare parts required for the maintenance of hotel and other accommodations. Tariffs and technical standard regulations can impede the import of data communication hardware. Governments may also require that only domestically manufactured computers and telecommunications equipment be used in service company operations.
Broader Protectionism in international trade (#PC5842).

♦ **PE7031 Sexual craving**
Dependence on sex — Preoccupation with sex — Lack of celibacy — Sexual obsession — Excessive sexual desire — Erotomania
Nature One side-product of the revolution in morals and behaviour since the 1960's is the common understanding that an active sex life is an essential part of good physical and psychological health. Ludicrous efforts are made to extend sexuality to include both young children and people of advancing age. In fact, the genital sex act causes untold stress, creates hormonal chaos, stifles the intellect, and weakens the body's resistance to illness. Where sex steps in, there is little room for social activity and personal development. Like drug use, sexual acts lead the user not to satiety, but to the longing for more. Sex is more dangerous than other addictions in that it requires the cooperation of another person, be they willing, coerced or bought.
Claim Intense sexual desire accompanied by lascivious imagination or wanton behaviour is a vice that promotes irrational, immoral, and frequently criminal, actions. It is a direct cause of prostitution, pornography, illegitimate children, adultery, marriage breakups and sexual offences.
Counter-claim Lust may be experienced by husband and wife or by betrothed couples in relationship to one another. As a primarily physiological condition, however, it may be subordinated to the dictates of right reason, to restraint as to its fulfilment in the enjoyment of active sexual pleasure by respect to its circumstances and objective and subjective purposes, and to moral conventions and personal conscience. As a result of such limitations, which characterize the practice of chastity, lust may find its outlet in normal and morally acceptable sexual union and other sexual activities. Excessive chastity on the other hand, may lead to sexual inhibitions, repressions and marital tensions.
Refs Carnes, Patrick *Out of the Shadows* (1985); Diamond, Jed *Looking for Love in All the Wrong Places* (1988); Trachtenberg, Peter *The Casanova Complex* (1989).
Broader Human disease and disability (#PB1044).
Narrower Nymphomania (#PE8213).
Related Addiction (#PD6324) Sexual deviation (#PD2198)
Human sexual disorders (#PD8016) Dependence on romantic love (#PF7418).
Aggravates Cohabitation (#PF3278) Unwanted pregnancies (#PF2859).
Aggravated by Sex (#PF9109).

♦ **PE7032 Lack of international nongovernmental organization identity**
Nature INGOs do not conceive of themselves as a well-defined group of organizations with common concerns and consequently have little basis for collective action.
Background There is no universally accepted description for organizations which are termed 'INGO'. 'NGO' is a term applied by the UN-related bodies in connection with their consultative status relationship, but not necessarily in connection with contractual relationships. 'INGO' is a term favoured by some scholars. The INGOs, and especially their membership, seldom conceive of themselves in the category of INGOs at all, but rather as scientific associations, trade unions, youth organizations, etc. The INGO sense of identity, such as it is, is therefore shared only amongst a small elite concerned with the problems or potential of such bodies in general and who are obliged to use the unsatisfactory description to link perceptions about a wide variety of organizations which do not generally perceive themselves as having common concerns.
Broader Obstacles to effective international nongovernmental organizations (#PF7082).

♦ **PE7036 Denial of right to recruit union members**
Broader Denial of right to organize trade unions (#PE5398).

♦ **PE7040 Inflexible military thinking in industry**
Claim The view that war is too important to leave to the generals is justified by their performances in corporate directorships or executive positions. Many generals, admirals and other military leaders retire to industry for prestigious positions. However, not only do they lack any conception of astute use of scarce economic resources in the private sector, having been accustomed to governmental waste and inefficiency, but they have a limited ability, relative to the career business man, to understand the profusion of economic data which business bases its decisions on. They are frequently in a position of vacillation between issuing peremptory and premature executive orders of the military variety, or of entrenchment in defensive positions. They have little, if any, conception of the 'long march' of business, or even of the guerrilla warfare over competitive prices and market share. They view long-range planning in business as academic theorizing and even though the world is changing all around them, their thinking remains traditionalist and isolated. It is the Maginot Line mentality under the officers cap, the only strategy being the megalith of the single answer of 'more': more capital in industry; more weapons in defence. For this reason, they function well as salesmen and lobbyists for the arms industries whom they so often join in retirement.
Broader Uncritical thinking (#PF5039).
Aggravates National defence procurement procedures (#PE4097).

♦ **PE7044 Lack of identity of international nongovernmental organization network**
Nature INGOs individually, or in small groups with closely related concerns, tend to conceive of themselves as operating in an international vacuum. They are consequently surprised to find at some stage that there are other organizations with similar programmes or common problems, or whose programmes are in some way affected by their own. There is only a vague sense of identity with an 'international community' and little general understanding of the elements and linkages constituting the inter-organizational network on which that sense of community is based.
Broader Obstacles to effective international nongovernmental organizations (#PF7082).

♦ **PE7045 Excess of western-based secretariats for international nongovernmental organizations**
Nature Statistics on the location of INGO headquarters show that a high proportion are located in Europe and North America. Because of the political significance attached to the geographical location of INGO offices, this leads to criticism that INGOs are primarily West-oriented, partisan, and therefore suspect.
Counter-claim This condition is also characteristic of IGOs. It is in fact linked to the relative degree of development of the different continents and to the associated problems and costs of communication and transport between them. It should never be forgotten that the travel costs and times between many developing countries and Europe are in fact less than those between neighbouring developing countries. The unsatisfactory asymmetry is in fact a consequence of the development problem with which many of the INGOs are concerned. It is also linked to the considerable legal problems of establishing such organizations in non-Western countries.
Broader Obstacles to effective international nongovernmental organizations (#PF7082).

♦ **PE7056 Operational ineffectiveness of international nongovernmental organizations**
Nature Evaluation of INGOs according to some criteria leads to an assessment of ineffectiveness which therefore justifies any proposed use of alternative organizational channels.
Counter-claim Assessment of INGO effectiveness is frequently based on the size of the budget, the number or qualifications of paid staff, the number of members, etc. Such assessments ignore characteristics of INGO operations, namely that: (a) the operating costs may be directly absorbed by national members (for example, when the INGO secretariat is handled by a national NGO); (b) much of the work may be done by people working voluntarily (who may be both skilled and highly influential); and (c) the members may be significant not in terms of their numbers but rather in terms of the (many) influential positions they occupy of their collective expertise in some specialized domain. A frequent error is to compare an INGO budget with that of some other organization operating with generous overheads, and a large support staff on an international pay scale. This compares potential, but not actual ability to focus effectively on a problem. Another error is to generalize about INGOs without distinguishing those with clearcut operations from those with correspondence secretariats only. An INGO's effectiveness, whatever the quantitative conclusions, may be primarily determined by its critical relationship to other bodies in a network. 'Insignificant' organizations may be very important communication centres. The notion of effectiveness is a very Western managerial concept of questionable relevance to some organizations concerned with relations between people and exchange of experience. The relation between the effectiveness of an organization and its right to exist is surely determined by its ability to continue to attract members and not by some externally imposed criteria.
Broader Obstacles to effective international nongovernmental organizations (#PF7082).
Related Ineffectiveness of international nongovernmental organizations and programmes (#PF1595).
Aggravates Ineffective governmental use of nongovernmental resources (#PF4095).

♦ **PE7061 Records destruction by transnational enterprises**
Missing documents and data of transnational corporations
Nature Home country record retention requirements may not apply to activities of subsidiaries, and local authorities needs may be less defined or stringent. In these cases enterprises may not have available, whether destroyed or concealed, details of past international dealings. During litigations, defendants may claim record destruction, as penalties may be lighter for this than for other infringements of the law that may appear from incriminating evidence.
Broader Destruction of historic documents and public archives (#PD0172).
Aggravated by Unethical practices of employers (#PD2879).

♦ **PE7062 Misogyny**
Woman-hating
Nature The distrust and dislike of women, by either sex, has a long individual and social history. It is not surprising, in the light of psychiatric considerations, that many persons having this tendency express it subconsciously. This makes misogyny much less visible, resulting in its persistence as an obstacle to full experience of the natural emotions and as a factor in social injustice to women.
Refs Bloch, Howard R and Ferguson, Frances (Eds) *Misogyny, Misandry, and Misanthropy*; Dworkin, Andrea *Woman Hating*.
Broader Misanthropy (#PJ1278) Sexual discrimination (#PC2022).
Related Misandry (#PJ1919).
Aggravated by Fear of sexual intercourse (#PF6910).

♦ **PE7064 Infected animal, meat and animal product shipments**
Insufficiently treated meat from infected animals — Animal products food contamination
Nature Contagious diseases may be introduced into a country as a result of the transit of live animals, meat or animal products. A transit country faced with the threat of contagious disease may slaughter and destroy the contaminated goods and refuse future shipments from the country of origin. Thus the health interests of transit countries may impede the international trade in animals, meat and other products of animal origin.
Broader Food pollution (#PD5605) Unethical food practices (#PD1045)
Man as vectors of disease (#PD8371).
Related Biological contamination of food (#PD2594).
Aggravates Brucellosis (#PE0924)
Environmental hazards from food processing industries (#PE1280).
Aggravated by Animal diseases (#PC0952).

♦ **PE7072 Collusive tendering in international trade**
Nature Collusive tendering is the practice whereby suppliers, among themselves, fix prices and/or other conditions for sale of goods or services. It is a pervasive feature of international trade, distorting patterns of trade and hindering, in particular, the trade and economic development of developing countries. Both in developed and in developing countries, governments and enterprises owned or controlled by the state are usually the largest purchasers and such procurement amounts for as much as 10 percent of the GNP in certain countries. Government procurement is one of the main means employed by developing countries for imports of essential

goods and services. These countries are particularly dependent upon developed countries for supplies of capital goods. For specific items developing countries account for more than 70 percent of developed countries' exports. The practice of collusion between enterprises in respect of submitting bids is clearly in contradiction with the intention for calling tenders.
Collusive tendering can occur as a result of an exchange of information between two or more enterprises concerning their prices and conditions of sale on a regular or ad hoc basis. Collusive tendering also takes place as a result of agreements among enterprises to share markets and to fix prices generally or in relation to certain types of business. Whether written or unwritten, formal or informal, such agreements or arrangements between enterprises engaged in rival or potentially rival activities generally have as their objective to obtain the highest possible price for the goods and services in question by eliminating competition amongst the parties. To the extent that competing enterprises remain outside such arrangements, in particular cases quotations can be made at prices designed to eliminate such outsiders or to force them to join the arrangement. Once the price at which the enterprise designated to submit the lowest bid has been determined, all or some of the others will agree to submit 'cover' bids, that is, bids at higher prices or not fulfilling the requirements of the tender. In such a case, the resulting monopoly profit may be shared among the members to the arrangement in a predetermined fashion. Alternatively, under such arrangements, each of the members or certain of them may agree to bid for less than the total amount required, the objective being to share the work amongst them. Where specific areas or countries have been allocated to specific members under a market-sharing arrangement, other members will either submit cover bids or refrain from bidding. Collusive tendering may also involve agreement among members that the enterprise awarded the contract will subcontract parts of the project to fellow members.
In developed and developing countries having restrictive business practice legislation, collusive tendering is considered to be a type of horizontal price-fixing and market allocation agreement and subject to prohibition or control afforded to these types of restrictive agreements. In certain countries, however, collusive tendering may be singled out for stricter control than other types of agreements, since the objective of calling tenders is based on the assumption that receipt of independent bids will result in obtaining the most favourable terms for specific projects, which collusion clearly negates.
Broader Distortion in international trade (#PC6761)
Inadequate regulation of the restrictive business practices of state enterprises (#PE0225).
Aggravates Monopolies (#PC0521) Restrictive trade practices (#PC0073).

♦ PE7080 Divisive responses to international nongovernmental organization actions
Nature The major formal link between the main IGOs and INGOs is through the consultative status relationship. This is specific to each IGO, which encourages the formation of standing conferences and associated committees for those INGOs linked to it in this way, and discourages links between 'its' group of INGOs and the groups linked to other IGOs despite the fact that: (a) many INGOs are linked to more than one IGO; and (b) the matters discussed by one such INGO group and its committees may also be discussed by another (reflecting the overlap in IGO programme areas). The INGO community is thus fragmented by the divisive posture of IGOs with a consultative relationship, even though the majority of such bodies are Specialized Agencies of the UN system.
Claim A special feature of this problem is that is continued existence is ensured by (a) the status tokens accorded by the IGOs to the individuals with formal positions in such INGO groups; (b) the services supplied to the INGO grouping, which effectively prevent excessive criticism of the IGO from such groupings; and (c) the efforts by IGO secretariat personnel to maintain the fiction of some 25 years standing that supportive resolutions by the INGO grouping will accomplish more than any critical action. The irony of the situation is that the IGOs do not even formally recognize the existence of such INGO groupings and only relate to them through their office-holders. Clearly, formal recognition of such INGO groupings would imply the existence of a well-formed international group which would pose questions of principle which it is more convenient for IGOs to avoid (whilst at the same time implying that INGOs are ineffective because they do not form viable inter-INGO confederations).
Broader Obstacles to effective international nongovernmental organizations (#PF7082).
Narrower Lack of cooperation (#PF2816).

♦ PE7084 Alcohol-related violence
Refs Casselman, Jo and Moorthamer, Lut *Violent Social Behaviour and Alcohol Use* (1988).
Broader Youth violence (#PF7498).
Narrower Alcohol-related violence against women (#PE7672).
Related Alcohol-related crime (#PE4131).
Aggravates Draining of resources due to alcohol (#PE8865).
Aggravated by Drunkenness (#PE8311) Alcohol abuse (#PD0153)
Unreported violence (#PF4967).

♦ PE7085 Lung cancer
Pulmonary carcinoma — Malignant neoplasm of lung
Refs Abrams, et al *Biology of Lung Cancer* (1988); Council of Europe *Lung Cancer in Western Europe* (1978); Cumming, G and Bonsignore, G *Smoking and the Lung* (1985).
Broader Malignant neoplasm of respiratory system (#PE7572).
Aggravated by Smoking (#PD0713).

♦ PE7092 International nongovernmental organization membership apathy
Nature It may be difficult for an INGO secretariat to stimulate its members to more than token interest in its programmes, particularly when these are internationally oriented, and especially when communications pass via a regional secretariat of the INGO. It is therefore also difficult to allocate significant resources to international activity.
Broader Apathy (#PA2360)
Obstacles to effective international nongovernmental organizations (#PF7082).

♦ PE7095 Intellectual methods which disenfranchise the grassroots from technological development
Nature In order for technological developments to be responsive to the needs of local people methods must be developed which enable the conceptual capacity of the grassroots to be used in this innovative process. The grassroots are excluded from the processes of needs analysis, from the intellectual framework in which the technology is being developed and from participating in creating strategies for distribution. They are left with the choice to consume or to not consume.
Broader Non-inclusive management decisions (#PF2754).

♦ PE7099 Land erosion brought about by site development
Nature Wherever sites are developed, the soil, plant and animal environment is invaded. Land erosion can be caused by neglect or lack of understanding of soil characteristics and grading techniques.
Incidence Development of sites increases runoff in streams in two ways. First, the amount of impervious area (roof tops, driveways, roadways, parking lots, etc.) is increased, preventing rainfall from infiltrating the ground, and resulting in direct runoff and increased surface flow. Secondly, storm sewers, gutters and so forth increase the efficiency and the speed of the runoff. This higher runoff produces greater stream discharge and velocity, and higher erosion force, creating an imbalance between the force of the stream and the ability of its own banks to contain it. It takes several years for a new balance to be developed. During this time of transition, other erosion problems can occur.
Broader Soil erosion (#PD0949) Destruction inherent in development (#PF4829).
Related Coastal erosion (#PE6734) Soil erosion by wind (#PE3656)
Uncontrolled river erosion (#PJ9518).

♦ PE7103 Blister agents
Mustard gas — Lewisite
Nature Blister agents react with the proteins in the skin, the lungs and the eyes. Mustard gas can cause itching and a painful inflammation similar to sunburn, large water-filled blisters, bronchitis or pneumonia, cancer and other long-term effects and in some cases genetic defects. All blister agents are oily liquids with a low vapour pressure, lingering on the ground and on objects for a considerable time.
Broader Chemical warfare (#PC0872).
Aggravates Change (#PF6605).

♦ PE7104 Duplication of international nongovernmental activities
Nature In a significant number of cases, more than one INGO may be concerned with the same subject or problem area, or may have membership links with the same range or organizations, or may solicit funds from the same range of bodies. Such duplication may be accompanied by a total lack of coordination between the INGOs in question. This situation may be considered a waste of resources calling for rationalization and mergers.
Counter-claim There are many reasons for such apparent duplication, including ideological and political differences (for example, INGO trade unions), methodological differences (for example, INGOs corresponding to different schools of psychology and psychoanalysis), geographical location (for example, when the INGOs are effectively regionally oriented and based), historical circumstances, personality differences, etc. However, this conditions is characteristic of all organizations at this time. (It is reputed that there are over 30 bodies within the UN family responsible for inter-Agency coordination). Thus, although duplication may be a criticism of organization in general, it is not specific to NGOs. In addition, research on research and innovation has shown that duplication is in fact beneficial in some instances.
Broader Obstacles to effective international nongovernmental organizations (#PF7082).

♦ PE7105 Uncontrolled structures of international nongovernmental organizations
Nature INGOs do not conform to a limited range or organizational models. They are in fact characterized by a wide variety of forms. This reduces ability to understand them and consequently reduces their credibility. Furthermore the fact that the interests of INGOs do not always correspond to the priorities currently in fashion in the major intergovernmental agencies is considered to be an indication of their irrelevance.
Counter-claim It is important to seek innovation of organizational forms in order to develop more effectiveness. With regard to areas of interest, who is to say that a seemingly irrelevant INGO today should not be relevant tomorrow. (The best example is the existence of environmental INGOs several decades before the UN Stockholm conference on the human environment in 1972).
Broader Obstacles to effective international nongovernmental organizations (#PF7082).

♦ PE7111 Cosmetic and health problems related to human hair
Nature Humanity has never been able to decide what to do with hair: whether to remove it, cultivate it or adorn it. A vast amount of energy and time goes into the individual's consideration of hair styles and hair colour, of maintenance or suppression of moustaches and beards, and of tolerance or removal or limb and torso hair growth. Another concern is with the healthy growth of scalp hair, which may be interfered with by trauma, inflammation or vascular insufficiency (in terms of follicle viability). General health affects hair growth; and nutritional deficiencies, metabolic changes and major internal diseases may be cited as inhibiting factors. Excessive radiation, anticoagulants and antimitotic agents cause hair loss as well. Genetic factors may produce abnormal hair, examples of which are: beaded (monilethrix): ringed (pili annulati); and twisted (pili torti). Hair and hair situated diseases include ringworm infections, acne, alopecia (baldness), head lice infections, and epidermal scurf such as dandruff. Cosmetic cover-ups which often take the place of adequate medical treatment may often lead to effects worse than the original malady.
Counter-claim Hair provides jobs for countless numbers of people as hairdressers and barbers, and in the manufacturing, distribution and selling of hair care products.
Refs Bartosova, Ludmila, et al *Diseases of the Hair and the Scalp* (1984); Dawber, R and Rook, A *Diseases of the Hair and Scalp* (1982).
Broader Diseases of the skin and subcutaneous tissue (#PC8534).
Narrower Baldness (#PF5880) Alopecia (#PG4264) Grey human hair (#PE6308)
Excess human body hair (#PE5580).
Aggravates Vanity (#PA6491).
Aggravated by Mange (#PE2727).

♦ PE7113 Accidental falls
Broader Injurious accidents (#PB0731).
Narrower Falls of elderly (#PE2853).

♦ PE7119 Denial of right to job protection during maternity leave
Broader Inadequate maternity protection in employment (#PD6733).

♦ PE7128 Establishment-orientation of international nongovernmental organizations
Nature The well-established INGOs tend to 'freeze out' people with new ideas, motivations and organizational goals. Some INGOs may therefore be assessed as not representing the changing interests of the constituency they claim to represent. This reflects a general problem of estrangement from nearly all existing institutional forms, particularly among young people.
Broader Obstacles to effective international nongovernmental organizations (#PF7082).

♦ PE7129 Multinationalism of international nongovernmental organizations
Nature There is still considerable confusion within the international community concerning the range or organizations embodying the negative characteristics associated with 'multinational corporations', now called 'transnational corporations' by the UN to help clarify the matter. For those individuals or societies unfamiliar with INGOs, they are often considered as being identical to multinationals or as having similar characteristics. This confusion is reinforced by the lack of development of adequate distinctions in some other languages (including French, for example). Clearly in many countries this confusion, and the emphasis given to the negative impact of multinationals, constitute a considerable barrier to the development of participation in INGO activity. The situation is further confused by the fact that both types of organization are 'international'and 'non-governmental'. The UN Charter does not distinguish (under Article 71, governing its relationship to 'NGOs') between profit-making and non-profit-making and may be forced to relate to multinationals under the procedures developed for INGOs. Further confusion is generated by the class of INGOs which are international trade and manufacturing associations.

Clearly this category is closely related in operation to multinationals and to cartels, although in form it may be an entirely legitimate non–profit association (since only its members are specifically profit–oriented).
Broader Obstacles to effective international nongovernmental organizations (#PF7082).

♦ PE7136 Enforced curtailment of living standards in borrower nations
Incidence In the case of Brazil, which after Mexico has been the principal example of this process in Latin America since the military coup in 1964, wages were reduced by over 40 percent. In Argentina since the military coup in 1976, wages were reduced by over 50 percent. But already before the coup real wages were going down as a result of the economic policy of the right wing of the Peronist government in 1974–75. In Chile, real wages since the coup were reduced by two thirds, that is to say, from an index of 100 to an index of just over 30, and unemployment increased from 4 percent to 20 percent before levelling off at 12–15 percent. To be able to do this it was necessary first to destroy or to control the unions, to eliminate, often physically, the leadership, to repress all political opposition and to throw people in gaol, torture them, murder them, exile them etc. Secondly, it has been necessary to distort the economy from producing for the internal market through so–called import substitution, to producing for exports.
Background Non–oil exporting countries in the Third World and some socialist countries have increasingly turned to the private international capital markets to cover growing deficits; these have re–cycled some of the OPEC surplus funds to them and have additionally lent them other funds at high rates of interest that found no borrowers in the industrial countries where investment has been low The extension of these loans and particularly their roll–over rescheduling to finance the growing debt service when the borrowers are unable to pay have become the basis of stringent economic and political conditions that the private banks and/or the International Monetary Fund (IMF) acting as their intermediary have imposed on Third World (and some socialist and developed) countries The standard 'conditionality' to the IMF package that governments are obliged to accept in their 'letter of intent' before being certified to receive further loans always includes devaluation of the currency, reduction of government expenditures especially on consumer subsidies and popular welfare, the reduction of the wage rate through various devices, and more favourable treatment for private and especially foreign capital These conditions have sometimes led to 'IMF riots' as the people sought to resist the enforced curtailment of their standards of living It has been said that the IMF has overthrown more governments than Marx and Lenin put together
An important political economic consequence, if not rationale, of these IMF promoted government policies in the Third World is to promote 'export led growth' by cheapening Third World labour and its fruits for international capital and foreign importers (by lowering the price of Third World wages and currencies) and to lend support to the domestic forces in these Third World countries that have an economic interest in export promotion. Thus, the international financial system and the financing of the Third World debt serves to fuel and oil the mechanism of the emerging international division of labour based on Third World export promotion. The political consequences of all these economic policies are that it is necessary to repress the labour force in order to keep wages low or reduce wages.
Broader Excessive social costs of structural adjustment in debtor developing countries (#PD8114).

♦ PE7139 Increasing lag in education against growth in knowledge
Broader Increasing development lag against information growth (#PE2000).
Aggravated by Increasing lag in education against population growth (#PE5369).

♦ PE7140 Inadequate international nongovernmental organization response to intergovernmental calls for action
Nature INGOs are frequently perceived as unenthusiastic in response to IGO calls for action on some new issue and as such are viewed as less than satisfactory partners. Associated with this is the view the INGOs have been slow in adapting themselves to the many changes in the membership, attitudes and practices of IGOs such the UN.
Counter–claim Many of the most important INGOs were established long before the UN (or even the League of Nations) with aims and objectives of their own, not all of which have yet been accepted by the UN. Many have had a more universal membership than the UN in various stages of its development. Whilst they are prepared to pursue objectives in partnership with the UN, when these objectives are shared, they are quite prepared to pursue others on their own until IGOs come to recognize their validity.
Broader Obstacles to effective international nongovernmental organizations (#PF7082).
Aggravates Ineffective governmental use of nongovernmental resources (#PF4095).

♦ PE7141 Inadequate legislation against environmental pollution in developing countries
Lack of regulations against pollution in developing countries — Licence to pollute in developing countries
Broader Inadequate legislation against environmental pollution (#PF9299).
Aggravates Export of hazardous industries to developing countries (#PE6687).
Aggravated by Inadequate safety legislation (#PJ2716)
Inadequate enforcement of safety regulations (#PD5001)
Environmental hazards from economic and industrial products (#PC0328).

♦ PE7144 Intellectual confusion concerning the role of the United Nations
Nature A considerable degree of uncertainty and confusion prevails as to the nature and role of the United Nations. This is aggravated by the tendency of the United Nations system: to define its mandate in an unreal fashion; to try to give the impression that the outputs from its activities have a direct effect on the internal realities of States; to maintain a permanent state of confusion between the functions of negotiating or seeking greater consensus on the one hand, and the functions of management on the other.
Claim A perusal of the texts, resolutions or documents of the United Nations tell very little about the real nature of the system. The basic instruments indicate a real hiatus between hard facts and mere talk, inasmuch as they tend to present as normal functions terms of reference which are totally unrealistic. The thousand of resolutions adopted every year, the planning and programming documents describing theoretically the objectives and strategies of organizations, do not supply an appreciably greater amount of information. Nor is it possible to refer to documents evaluating the results obtained, since there are virtually none and that those who claim to be performing this task have a limited threshold of credibility because of their lack of criteria and methods.
Broader Confusion (#PF7123).

♦ PE7153 Techno–mercenaries
Nature People and groups with extensive technical skills and expertise of military value who are willing to sell their services to the highest bidder irrespective of the consequences to human life. Such services may be negotiated on an individual basis or through a variety of legal fronts, such as a network of corporations operating in several countries.
Incidence Recent examples include networks of western corporations supplying expertise for the construction of chemical weapon factories or sophisticated missiles. Such activity may be accepted as perfectly legal in some countries. In others the physical export of such technology may be illegal but not the specialized knowledge to design or operate it.
Counter–claim The government sanctioned military–industrial complexes of a number of countries undertake the sale of such technology to countries which would normally be considered irresponsible, or unlikely to use such technology to maintain the peace.
Broader Irresponsible genetic manipulation (#PC0776)
Irresponsible international experts (#PF0221)
Irresponsible scientific and technological activity (#PC1153).
Aggravates Proliferation of nuclear weapons and technology (#PD0837).

♦ PE7154 Skin lesions due to torture
Nature Because of extensive beatings, stabbings, shootings, burnings or other forms of woundings victims of torture may have extensive damage to their skin which can be painful for years following being tortured.
Broader Somatic and psychosomatic effects of torture (#PE5294).
Related Bodily ailments, lesions and malfunctions (#PG3423).

♦ PE7155 Exclusion from decision making processes of those who question the context of the process
Nature Unwilling to assume responsibility for current problems in the society, the established powers transfer blame to other persons and groups. Individuals and groups who raise questions about the establishment's understanding of society are suppressed, excluded from exerting any influence, or are co–opted. So the established powers understanding of the rights and obligations of citizenship, of effective participation in decisions, and of accountability of the political processes to the public nearly always dominate.
Broader Repression of self–consciousness (#PC1777).
Aggravated by Fragmented decision–making (#PF8448).

♦ PE7159 Respiratory distress syndrome
Refs Raivio, Kari O et al *Respiratory Distress Syndrome* (1984).
Broader Foetal infection and death (#PE2041).

♦ PE7161 Susceptibility of women to alcohol
Nature Alcohol affects women faster than men because their bodies produce smaller amounts of an enzyme that breaks it down in the stomach. More alcohol therefore passes into a woman's bloodstream, thus rendering them more vulnerable to the short–term effects of alcohol consumption. The effects are exacerbated by the smaller build of women, the greater amount of fat and less water, and the fluctuating levels of different hormones.
Refs Wilsnack, Sharon C and Beckman, Linda J *Alcohol Problems in Women* (1984).
Broader Consumption of alcoholic beverages (#PD8286).
Related Poor physiological adaptability of women (#PG6603).
Aggravates Alcohol abuse (#PD0153).

♦ PE7165 Inadequate working conditions of teachers
Unattractive teacher post
Broader Obstacles to education (#PF4852)
International imbalance in the quality of life (#PB4993).
Narrower Underpayment of teachers (#PE8645) Limited teacher facilities (#PJ1340)
Insufficient incentives for teachers (#PJ8347).
Related Lack of housing for teachers in rural communities (#PE8103).
Aggravates Unmotivated teachers (#PF5978) Costly teaching staff (#PJ9524)
High teacher turnover (#PE8221) Poor supervision teachers (#PJ3757)
Maldistribution of teachers (#PE6183) Shortage of trained teachers (#PD8108)
Limited availability of education in rural areas (#PF3575)
Unavailability of trained teachers in the rural areas of developing countries (#PE8429).
Aggravated by Parent–teacher non communication (#PF1187).

♦ PE7169 Unethical practices by housing tenants
Irresponsible housing tenants — Negligence by housing tenants
Broader Unethical practices (#PC8247).
Related Unethical real estate practice (#PD5422).
Aggravates Neglect of property maintenance (#PD8894)
Arrears in payments for housing (#PE4571).

♦ PE7172 Isolated mass cemeteries
Nature Huge cemeteries on the outskirts of cities and impersonal funeral rites help to keep the fact of death away from the living. In the big cities the ceremonies of death have been increasingly undermined. Once beautifully simple forms of mourning have been replaced by synthetic memorial parks, plastic flowers, and servicing conveniences. The small graveyards of parishes and neighbourhood churches which once put people into daily contact with the fact of death, with its grief and sometimes with its joy, have vanished, replaced by cemeteries owned by corporations far away from people's daily business and family lives.
Claim When circumstances prevent them from making contact with the experience of mortality, and living with it, people are left depressed, confused or diminished. This is not only true about death, but is true also about attitudes towards dying: the two facets cannot be separated.
Broader Physically inaccessible services (#PC7674).

♦ PE7174 Diversion of high technology to hostile countries
Diversion of high technology to irresponsible groups — Trade in strategic goods with hostile countries — Trade in nuclear materials with irresponsible countries
Incidence The Co–ordinating Committee for Multilateral Export Controls (COCOM) is an agreement of seventeen advanced countries not to sell certain technical equipment to the Soviet Union and its allies. A list of forbidden technology includes semiconductor manufacturing equipment, sophisticated machine tools, telecommunications equipment and computers. Despite the rules German DR obtained illegally Japanese–made precision equipment for producing advanced computer chips and the Soviet Union was sold machine tools. Before mid–1990s the high–technology companies never knew which products were on the secret list and had to go through expensive and time consuming review by licensing authorities and by COCOM.
Broader Crimes related to foreign relations and trade (#PE5331).
Aggravates Proliferation of nuclear weapons and technology (#PD0837).
Aggravated by Unlawful business transactions (#PC4645)
Unethical practices by employees (#PD4334)
Unethical practices of employers (#PD2879)
Unethical practices in transportation (#PD1012)
Government complicity in illegal activities (#PF7730).

♦ PE7175 Dangerous pet animals
Aggressive domesticated animals — Violent pet animals — Breeding of dangerous animals
Claim All dogs are natural hunters, and although their aggressive instincts may appear submerged, they can become savagely unpredictable. Large dogs and fighting dogs cause fear and terror and should be muzzled in public.
Broader Dangerous animals (#PC2321) Proliferation of pets (#PD2689).
Aggravated by Animal fighting sports (#PE4893).

♦ **PE7176 Unlawful compensation for assistance in government matters**
Nature Unlawful compensation of assistance in government matters is giving to a public servant or a public servant receiving something of value for advice about the preparation of a bill, contract, claim or other matter over which the public servant has an official capacity.
Broader Bribery of public servants (#PD4541).

♦ **PE7184 Inadequate external debt management capacity within developing countries**
Refs United Nations Economic Commission for Latin *External Debt in Latin America* (1985).
Broader Inadequate economic policy–making in developing countries (#PF5964).
Aggravates Excessive borrowing by state–controlled enterprises (#PE7474)
Excessive foreign public debt of developing countries (#PD2133)
Decline in commercial bank lending to developing countries (#PE4655)
Deterioration in external financial position of developing countries (#PE9567).

♦ **PE7185 Working hours inappropriate to structural and technological changes**
Refs Maric, D *Adaptating Working Hours to Modern Needs* (1980).
Aggravates Excessive hours of work (#PD0140).

♦ **PE7188 Illicit production of alcoholic beverages**
Illicit distilleries — Illicit breweries — Moonshine — Unauthorized liquor manufacturing and distribution — Home–made liquor
Broader Tax evasion (#PD1466) Unethical commercial practices (#PC2563).
Aggravates Consumption of alcoholic beverages (#PD8286).

♦ **PE7195 Dependence of developing countries on external financing for development programmes**
Excessive reliance of developing countries on external sources for development financing — Dependence of developing countries on external aid — Dependence of developing countries on foreign loans — Over–reliance of developing countries on foreign borrowing
Claim Overreliance on foreign borrowing can cause appreciating real exchange rates, widening current account deficits, unsustainable external indebtedness, and dwindling foreign exchange reserves.
Broader Vulnerability of developing countries (#PC6189).
Narrower Over–dependency on international financial institutions by developing countries (#PE2214).
Aggravates Destruction inherent in development (#PF4829)
Infantilization of deprived populations (#PF2541)
Lack of economic and technical development (#PE8190)
Excessive foreign public debt of developing countries (#PD2133)
Acquisition of inappropriate equipment by developing countries (#PE6640).
Aggravated by Government deficits (#PD5984)
Deterioration in external financial position of developing countries (#PE9567).

♦ **PE7201 Dental torture**
Nature Breaking teeth, drilling into teeth to expose nerves and pulling teeth are used as forms of torture.
Broader Surgical torture (#PE7547) Physical torture (#PD8734).

♦ **PE7206 Discrimination in the employment of women with family responsibilities**
Refs ILO *Employment of Women with Families Responsabilities* (1978); ILO *Equal opportunities and Equal Treatment for Men and Women Workers. Workers with family responsabilities* (1981).
Broader Discrimination against women in employment (#PD0086).
Narrower Job insecurity for pregnant women (#PE5316).

♦ **PE7209 Denial of right of juvenile criminals to segregation from adult criminals**
Broader Denial of rights to prisoners (#PD0520) Repressive detention of juveniles (#PD0634).

♦ **PE7217 Denial of right to confront accusers**
Broader Denial of right to a legal defence (#PE4628).

♦ **PE7229 Zinc as a pollutant**
Refs FAO *Water Quality Criteria for European Freshwater Fish* (1979).
Broader Chemical contamination of water (#PE0535).
Narrower Zinc dust (#PG5385).

♦ **PE7230 Motor skills disorder**
Developmental coordination disorder
Nature The young children achieve their developmental motor milestones (sitting, walking, tying shoelaces etc) later than expected, the older children display difficulties in sports, puzzle assembly, model–building, handwriting and in other tasks requiring motor coordination.
Broader Mental illness (#PC0300).

♦ **PE7233 Interrogation while naked**
Nature Interrogation of prisoners who have been stripped naked serves several purposes. Victims are humiliated and intimidated by the experience. Being naked before interrogators represents a loss of control over the situation and symbolizes exposure of one's weaknesses and secrets without the ability to hide. It also implies an unreciprocated intimacy with the prison officials which for most subjects is incongruent with their own sense of morality.
Incidence Naked interrogation has been reported in the following countries: **Af** Kenya, Mali, Somalia, South Africa, Tunisia, Zambia. **Am** Canada, Colombia, Peru, Uruguay. **As** Israel. **Eu** Yugoslavia.
Broader Inhumane interrogation techniques (#PD1362).

♦ **PE7235 Denial of right to freedom from imprisonment for failure to fulfil a contractual agreement**
Debtor's gaol
Broader Unlawful imprisonment (#PD4489)
Denial of human rights in the administration of justice (#PD6927).

♦ **PE7242 Alcoholism amongst indigenous peoples**
Incidence Alcohol abuse among North American Indians is reaching a new generation: 5 to 25 per cent of children are affected by foetal alcohol syndrome or less severe foetal alcohol effect because their mothers drank heavily during pregnancy. The result is a population that is mentally and physically disabled.
Claim Among Lakota Sioux Indians there is a 80 to 90 per cent unemployment rate, so people are forced to live from welfare. There is no business, no industry and people have lost their hope for the future.
Broader Alcohol abuse (#PD0153).
Narrower Foetal alcohol syndrome (#PE3853).

♦ **PE7244 Obstruction of astronomical observation by environmental pollution**
Light pollution
Nature Increasing human activity results in various forms of pollution which obstruct or distort the receipt of light by telescopes or of electromagnetic radiation by radio–telescopes.
Incidence Aside from normal cloud cover, observations may be obstructed by smoke, dust and other atmospheric pollutants of industrial origin. Observations at night are severely hindered by reflections from urban lighting. Both from ground level and in space, observations may be obstructed by radiation leakages from orbiting nuclear reactors, as in the case of the military satellites of the USSR which have effectively blinded a number of orbiting telescopes during the 1980s.
Broader Deterioration in atmospheric visibility (#PE2593).
Aggravated by Haze (#PG3864) Dust (#PD1245)
Interstellar dust (#PF9002) Electromagnetic pollution (#PD4172)
Photochemical oxidant formation (#PD3663).

♦ **PE7252 False confessions**
Self–incrimination by the innocent — False confessions by the innocent — Coerced compliant confession — Interrogative suggestibility
Nature Certain people confess to having committed crimes of which they are not guilty, including serious crimes such as murder. Aside from confessions by eccentrics, the criminal justice also has to distinguish coerced compliant confessions by people who know they have not committed the crime but confess to gain some immediate result, such as being released from custody (possibly in the hope of establishing the truth at a later stage). A second form is that of the coerced internalized confession in which a suspect is persuaded by the interrogation process, possibly only temporarily, that he might have committed the crime in the light of the suggestions made by the interrogators that he had suppressed the memory of it (especially in the case of traumatic crimes). Such interrogative suggestibility is the tendency to accept uncritically information communicated during questioning. Highly suggestible people are more likely to make coerced internalized confessions whereas compliant people are more likely to make coerced compliant confessions. Such confessions may be made in the absence of any particular mental disorder.
Incidence Most well–publicized serious crimes result in confessions by eccentrics, whether because of a desire for publicity or because of fantasies about having committed the crime. Some 200 people confessed to the kidnapping of the child of Charles Lindbergh in 1932.
Broader False statements (#PF4583) Incorrect information (#PB3095).
Aggravated by Forced confession (#PE8947).

♦ **PE7255 Low intra–regional complementarity of developing countries economies**
Aggravated by Low complementarity of developing country economies (#PE8184).

♦ **PE7265 Over–cultivated gardens**
Claim The over–cultivated gardens with pavements, foreign flowers etc. are not natural to the environment, rewarding nor enjoyable.
Related Destruction of hedges and hedgerow trees (#PD1642).
Aggravated by Environmental degradation of suburbia (#PD2345).

♦ **PE7267 Denial of right to a family**
Broader Denial of social rights (#PC0663).
Narrower Denial of right to extended family (#PE5241).
Related Impediments to marriage (#PF3343) Denial of the right to procreate (#PC6870).
Aggravates Arbitrary external interference in family life (#PE4058).

♦ **PE7271 Trafficking in government benefit coupons**
Nature Trafficking in government benefit coupons is to knowingly transfers, disposes or receives coupons from any government benefits programme, like food stamps, child benefits, youth or elders discount programs.
Broader Fraud (#PD0486) Frauds, forgeries and financial crime (#PE5516).

♦ **PE7300 Limited acceptance of human rights treaties**
Delays in ratification of human rights treaties — Non–ratification of human rights treaties — Failure of governments to implement provisions of ratified human rights agreements
Incidence The ratification by the USA (in 1990) of the United Nations Genocide Convention took 40 years, because conservative legislators argued that it was unconstitutional and could be applied against American soldiers who served in Vietnam.
Broader Government hypocrisy (#PF9050)
Limited acceptance of international treaties (#PF0977).
Aggravates Denial of human rights (#PB3121)
Government inaction on alleged human rights violations (#PE1407).
Aggravated by Connivance of authorities in human rights abuses (#PF9288).

♦ **PE7304 Pox diseases in animals**
Broader Skin diseases in animals (#PD9667).
Narrower Cowpox (#PE6886) Swinepox (#PG5158) Sheep pox (#PE6272)
Pox virus infection in cats (#PG5339).

♦ **PE7310 Denial of right to pursue spiritual well–being because of discrimination**
Broader Discrimination (#PA0833) Denial of human rights (#PB3121).
Related Racial discrimination (#PC0006) Economic discrimination (#PC2157)
Discrimination against women (#PC0308).
Aggravated by Barriers to transcendent experience (#PF4371).

♦ **PE7311 Locales of high crime rates**
Broader Crime (#PB0001).
Narrower Crimes in the home (#PG1540) Crimes against the family (#PG1206)
Crimes committed in urban schools (#PJ1356).
Related Urban crime (#PD7399)
Crimes committed in hospitals and health care facilities (#PE8420).
Aggravated by Segregation of poor and minority population in urban ghettos (#PD1260).

♦ **PE7312 Denial of right of indigenous peoples to participate in political processes**
Broader Denial of right to participate in government (#PE6086)
Discrimination against indigenous populations (#PC0352).

♦ **PE7313 Inadequate local expertise in business practices in developing countries**
Lack of local management skills in developing countries
Nature Rural village vendors have a long history of trading through barter, buying and selling, but today their traditional style is being challenged as the market is undergoing a great transition. The market for local products like pots, rope, bamboo walls, hats and baskets is becoming increasingly competitive. Vendors experience themselves as having to accept smaller and smaller profits. Sales promotion and expansion include techniques just becoming familiar to the local merchant. In competitive instances he often finds himself the loser. The brokers who are middlemen between the craftsmen and the market take increasing advantage of rising inflation and the inexperience

of the craftsmen. Total productivity has no doubt increased over the last 100 years, and has required more complex operations and therefore more highly technical managerial skills. Production techniques are often underdeveloped compared with modern methods. Unless advance techniques in sales, management and production are acquired soon, the rural areas will find themselves continually unable to benefit from their own potential in the commercial world.
Broader Narrow range of business skills (#PE6554)
Deficiencies of developing countries (#PC4094)
Lack of management skills in developing countries (#PE0046).
Narrower Small local markets (#PJ8200) Narrow profit margin (#PJ9737)
Non-competitive products (#PG9751) Traditional market practices (#PJ9742)
Inadequate marketing knowledge (#PJ9659)
Unsophisticated production techniques (#PG9754).
Related Limited market development (#PF1086) Unavailability of appropriate expertise (#PF7916)
Shortage of industrial leadership and entrepreneurial ability in developing countries (#PC1820).
Aggravates Incompetent management (#PC4867)
Dependency on middlemen (#PD4632)
Discouraging conditions for small business (#PD5603)
Inadequate industrial services in developing countries (#PF4195).

♦ **PE7317 Inadequate vocational rehabilitation facilities for disabled persons**
Refs ILO *Basic Principles of Vocational Rehabilitation of the Disabled* (1985).
Broader Inadequate rehabilitation facilities (#PD1089)
Inadequate community care for handicapped persons (#PE8924)
Inadequate educational facilities for disabled persons (#PF0775).
Narrower Inadequate rehabilitation facilities for the mentally handicapped (#PE8151).
Aggravates Immigration barriers for handicapped family members (#PE4868).

♦ **PE7326 Deficient local structures for the unemployed**
Nature One reason unemployment continues is because the structures of local communities are incapable of generating new employment opportunities. Local political structures are seem powerless in encouraging job creation. Those seeking employment within the local government are often require to support the political party currently in power. Voluntary organizations are for the most part at the fringes of sustainable job creation. Corporations and other businesses normally use hiring and firing practices as one means of controlling costs and as such, do not have job development as a high priority.
Broader Limited image of employability (#PF2896).

♦ **PE7327 Torture by exposure to animals**
Nature Exposing victims of torture to the attacks of animals increases fear, causes pain and re-enforces a sense of helplessness. Subjects have been attacked by dogs, had rodents and snakes crawl over them, and been bitten by insects at the initiative of torturers.
Incidence Exposure to animals has been reported in the following countries: **Af** Libyan AJ. **Am** Colombia. **Eu** Poland, Romania.
Broader Physical torture (#PD8734).

♦ **PE7336 Denial of right to be informed of criminal charges**
Broader Denial of right to fair and public trial (#PE3964).

♦ **PE7337 Denial of the right to inform**
Nature People have the right to inform and to receive information from any source and free of any governmental restriction. If the news media is shut down and journalists gaoled, governments consequently plan and act in secret.
Broader Restrictions on freedom of information (#PC0185).

♦ **PE7344 Limited means of marketing employable skills**
Unmarketable available skills
Nature One reason unemployment and underemployment continue is people do not have the marketing skill to sell their own intellectual, physical or technical skills.
Broader Inadequate access to negotiation on employment and reward (#PD1958).
Aggravates Unemployment (#PB0750)
Underutilization of locally available skills (#PF6538)
Disparity between workers skills and job requirements (#PC1131).
Aggravated by Obsolete vocational skills (#PD3548)
Insufficient trained labour (#PD9113)
Limited availability of permanent employment in inner-cities (#PE1134).

♦ **PE7345 Denial of right of accused to segregation from convicted criminals**
Broader Denial of rights to prisoners (#PD0520).
Related Denial of right to presumed innocent until proven guilty (#PE7393).

♦ **PE7348 False accusations**
Conviction by accusation
Nature The press and political hue and cry before the charges against the accused have been proved, can ruin the professional and private life of the accused.
Aggravated by Slander (#PD3023).

♦ **PE7349 Denial of right to correct misinformation**
Broader Denial of freedom of thought (#PF3217).
Related Denial of religious liberty (#PD8445) Denial of freedom of opinion (#PD7219)
Denial of freedom of conscience (#PD7612) Restriction of freedom of expression (#PC2162)
Restrictions on freedom of information (#PC0185).

♦ **PE7354 Excessive portrayal of crime in the media**
Glorification of crime in the media
Broader Distorted media presentations (#PD6081).
Aggravates Crime (#PB0001) Culture of violence (#PD6279)
Deterioration of media standards (#PD5377).

♦ **PE7355 Interference in labour relations in developing countries by transnational corporations**
Refs ILO *Employment Effects of Multinational Enterprises in Industrialised Countries* (1985).
Broader Domination of labour relations by transnational corporations (#PE1187)
Domination of developing countries by transnational corporations (#PE0163).

♦ **PE7370 Anti-social behaviour of university students**
Campus violence — Physical insecurity of university campuses — Student violence on campus
Broader Human violence (#PA0429).
Aggravates Denial of rights to students (#PD6346)
Defacement of urban structures (#PD5305).

♦ **PE7390 Inadequate assistance to victims of abuse of power**
Broader Legal discrimination in favour of offenders (#PD9316)
Inadequate assistance to victims of human rights violations (#PD5122).
Related Neglect of victims of crime (#PD4823).

♦ **PE7393 Denial of right to presumed innocent until proven guilty**
Broader Denial of right to fair and public trial (#PE3964).
Related Denial of right of accused to segregation from convicted criminals (#PE7345).

♦ **PE7403 Environmental degradation from recreational use of unsurfaced country roads and tracks**
Broader Environmental degradation (#PB6384).
Aggravated by Environmental degradation by off-road and all-terrain vehicles (#PE1720).

♦ **PE7413 Distortion of national economies from food subsidies**
Dependence of economies on food subsidies
Nature Food production, especially in industrialized countries, tends to be highly subsidized and protected from international competition. These subsidies have encouraged the overuse of soil and chemicals, the degradation of the countryside, and the pollution of both water resources and foods by the chemicals used. The success of this effort has resulted in surpluses and their associated financial burdens, with some of these surpluses being sent at concessional rates to developing countries where they have undermined the farming policies. In number of developing countries food subsidies form a significant share of government expenditures and may introduce price distortions into the economy and may result in large economic losses to food producers.
Broader Distortion of international trade by selective domestic subsidies (#PD0678).
Aggravates Over-subsidized agriculture in industrialized countries (#PD9802).
Aggravated by Inappropriate government intervention in agriculture (#PE1170).
Reduced by Inappropriate aid to developing countries (#PF8120).

♦ **PE7416 Environmental hazards of chlorine-bleached paper products**
Nature Chlorine bleaches of woodpulp produce dioxins, one of the most dangerous chemicals, as the by-product. The dioxins affect eco-system in general, but they are also found in the bleached products themselves.
Refs United Nations Environment Programme *Environmental Management in the Pulp and Paper Industry* (1981).
Broader Chemical pollutants of the environment (#PD1670).
Aggravates Dioxin poisoning (#PE7555).
Aggravated by Chlorine as a pollutant (#PG7454).

♦ **PE7417 Leakage from packaging**
Refs MATCOM *Leakage. A Learning Element for Staff of Consumer Cooperatives* (1986).

♦ **PE7419 Health hazards of mining radioactive substances**
Refs ILO *Radiation Protection in Mining and Milling of Uranium and Thorium* (1976).
Broader Occupational hazards in the mining industry (#PE8428).
Aggravates Human disease and disability (#PB1044).
Aggravated by Hazards to human health in the natural environment (#PC4777).

♦ **PE7423 Gastroenteric diseases in animals**
Broader Diseases of the digestive system in animals (#PD3978).
Narrower Colitis in animals (#PG9325) Enteritis in animals (#PG1745)
Feline panleukopenia (#PG4070) Colon impaction in animals (#PG7857)
Enteric diseases of the newborn (#PG9434) Canine coronaviral gastroenteritis (#PG3842)
Canine haemorrhagic gastroenteritis (#PG7485) Gastric dilatation-volvulus in animals (#PG9755)
Small intestinal obstruction in animals (#PG9479)
Foreign bodies in the oesophagus of animals (#PG7761).

♦ **PE7430 The rack**
Nature The rack is still being used as a form of torture. Victims arms and legs are strapped to a metal frame which is gradually extended causing the limbs to be stretched to the limit. Joints are dislocated, and tendons and muscles are torn.
Broader Torture by wounding (#PE4078).

♦ **PE7433 Loss of water to industrial uses**
Loss of water to domestic uses — Competition between industry and agriculture for water
Nature In hot areas where rainfall is low and evaporation rates are high, the water abstracted to satisfy industrial and domestic requirements can only result in a corresponding decrease in agricultural production. When such competition occurs, water tends to go to the highest bidder. It is nonsensical in developing countries where every scrap of food is needed to feed their generally malnourished populations.
Aggravates Long-term shortage of water (#PC1173).
Aggravated by Shortage of industrial water (#PE5464)
Environmentally harmful dam construction (#PD9515).

♦ **PE7448 Manipulation of the individual by mass media**
Manipulation of public opinion
Nature As social communication becomes economically organized, the individual is increasingly perceived not as a communicator in his or her own right, but as a consumer of a product whose content is determined elsewhere. Messages assume the tone of training from the teachers to the students, admonitions from the knowledgeable to the ignorant, directives from organisers to participants, and so on. The assumption that the traffic should be in one direction often stems from the structure of society itself, and is mixed with goodwill, generosity, integrity and idealism. The results are nonetheless a passive and increasingly uninvolved audience.
Broader Unethical media practices (#PD5251).
Aggravates Abuse of power (#PB6918).
Aggravated by Gullibility (#PJ2417) Dependence on the media (#PD7773)
Distorted media presentations (#PD6081)
Excessive portrayal of negative information by the media (#PE1478).
Reduced by Inadequate mobilization of public opinion for development (#PF9704).

♦ **PE7451 Psychogenic fugue**
Nature Inability to recall one's past after unexpected travel from home or work.
Broader Neurosis (#PD0270).
Aggravated by Alcohol abuse (#PD0153).

♦ **PE7468 Fear of the loss of independence**
Nature People fear the loss of independence acquired through wealth. This fear aggravates the gross inequities in the distribution of wealth and benefits. People cling to the illusion of self-sufficiency and believe its loss will destroy one's humanity.
Broader Overemphasis on self-sufficiency with respect to interdependence (#PF3460).
Related Original sin (#PF8298).

♦ **PE7474 Excessive borrowing by state-controlled enterprises**
Incidence State-controlled enterprises have borrowed significantly in domestic and foreign credit markets, usually with the government as guarantor, whether explicitly or implicitly. Such guarantees create contingent liabilities, although lack of accounting discipline means that they often only become apparent at times of crisis, thus aggravating that crisis. Recent experience in many countries has made it painfully clear that a government's contingent liabilities can have serious repercussions if the financial situation of one or more of the enterprises deteriorates. Partly because of such guarantees, borrowing by state-controlled enterprises has added significantly to foreign debt. The direct borrowings of such enterprises accounted for more than 20 per cent of total foreign debt in 99 countries as a group and grew faster than the debt of private borrowers during 1970-86. The total contribution of such enterprises is however greater than this suggests because governments passed much of their own foreign borrowing on to such enterprises. Thus they account for more than half the outstanding external debt of another of major debtor nations.
Aggravates Excessive foreign public debt of developing countries (#PD2133).
Aggravated by Inadequate external debt management capacity within developing countries (#PE7184).

♦ **PE7476 Lack of appreciation for nuclear weapons**
Unrecognized merits of nuclear weapons
Nature Nuclear weapons have become a modern myth, holding a power over culture previously held only by religion. That myth has become a physical part of each person's brain and is now acting as a strong unifying force for the earth as a whole. "You can't do what you want with the bomb," says one theorist, "the bomb does what it wants with you." The bomb thus unites people across the earth in a way that has not existed since the Middle Ages. Some suggest that warheads would be most useful not hidden away as they are now, but universally on display in public places such as shopping centres.
Broader Weapons (#PD0658) Unrecognized opportunities (#PF6925).
Aggravates Inadequacy of civil defence (#PF0506).

♦ **PE7479 Imbalance in distribution of industrial processes**
Claim The use and distribution of industrial process by the larger society are determined by self-seeking, profit-making forces. The need of two-thirds of humanity for a minimal standard of living is blocked by a mal-distribution of manufacturing technologies.
Broader Industrial processes geared to reduced social needs (#PE3939).

♦ **PE7489 Denial of right to examine witnesses**
Broader Denial of right to a legal defence (#PE4628).

♦ **PE7502 Misleading endorsement advertising**
Nature Advertising endorsements as a whole have the common characteristic that consumers are led to believe that the endorser is an expert or has particular knowledge to approve the product. This is not always true. Endorsements may be inaccurately reproduced; taken out of context; or supplied by individuals, groups or agencies supposedly unbiased but actually connected to the product. Problems also arise where the endorsement implies continued use or experience with the product but it has actually been based upon single, isolated, or infrequent usage; and where an endorser's assessment of the product may have changed but the advertisement is continued to be broadcast. Endorsements of children's products can have great influence on children who are particularly susceptible to sophisticated advertising practices; and endorsement of products which may be hazardous to health (pharmaceutical products, alcohol and tobacco) may have serious consequences.
Incidence Misleading advertisement endorsements include testimony by an allegedly independent testing agency which was in fact financed or controlled by the advertiser and the use of consumer surveys where the results have been achieved as a consequence of a biased questionnaire.
Broader Misleading advertising (#PE3814).
Aggravates Insufficient role models (#PF8451).

♦ **PE7504 Unavailability of replacement parts for agricultural machinery**
Nature There are many instances where machines are inoperable, sometimes at critical periods in the production of a crop, because either a replacement part or the skill to fit it is unavailable. Other instances of less immediate consequence but nonetheless unsatisfactory, are where parts are prohibitively expensive, where the level of operator or service/repair skill is too low, where machinery is incorrectly used or mismatched or where soil and climatic conditions cause greater than anticipated replacement and repair problems. In addition, despite general acceptance that all machinery needs regular maintenance and a varying amount of repair during its working life, the full cost of that repair and maintenance is less widely accepted and is sometimes overlooked or ignored.
Incidence A recent study shows that with the exception of machinery with few or no moving parts, the cost of repairs and parts for agricultural machinery in developing countries may range from 130 to 400 percent of the initial cost of the machinery; these figures are substantially higher than those experienced in developed countries.
Aggravates Stagnated development of agricultural production (#PD1285).

♦ **PE7509 Gastric cancer**
Stomach cancer
Refs Inokuchi, Kiyoshi et al *Digestive Tract Tumors* (1986); Preece, Paul, et al (Eds) *Cancer of the Stomach* (1986).
Broader Malignant neoplasm of digestive organs (#PE4303).

♦ **PE7511 Fractures**
Fracture of bone
Refs Nourse, Alan E *Fractures, Dislocations and Sprains* (1978).
Broader Injurious accidents (#PB0731) Human disease and disability (#PB1044).
Narrower Head injuries (#PG5105) Fracture of rib (#PG4970)
Fracture of limb (#PG5257) Fracture of spine (#PG7075)
Fracture of pelvis (#PG9503).

♦ **PE7514 Mycotoxins**
Aflotoxins
Nature These toxins produced by microscopic fungi, are mostly extremely poisonous to man and many animals. The most infamous of them, the aflotoxins, affect some species at an intake of a few parts per billion. Each toxin has its own identity and pathology. Kidney and liver damage, haemorrhages, disturbances of the central nervous system, stunting of growth, abortion, or other effects may occur, depending on the toxin.
Refs CAB International Mycological Institute *Mycotoxins of Maize*; FAO *Mycotoxins* (1977); FAO *Perspectives on Mycotoxins* (1979); FAO *Prevention of Mycotoxins* (1979); Kurata, Hiroshi and Ueno, Yoshio (Eds) *Toxigenic Fungi - Their Toxins and Health Hazard* (1984).
Broader Toxic substances (#PD1115).

♦ **PE7515 Reduction of workforce in enterprises**
Refs Yemin, Edward *Workforce reductions in undertakings. Policies and Measures for the Protection of Redundant Workers in Seven Industrialised Economy Countries* (1982).
Broader Redundancy of workers (#PD8007).

♦ **PE7518 Echinococcosis**
Hydatidosis
Nature Echinococcosis is a debilitating tapeworm infection of humans and animals caused by the parasitic larvae of the tapeworm Echinococcis. Humans become infected by swallowing the eggs of helminths, mostly after touching dogs. Larvae emerge from the eggs in the intestine, penetrate into the blood vessels, and are carried to various organs, where they change into hybrid cysts.
Background Echinococcosis was described during the time of Hippocrates (460-379 BC), and during the Middle Ages it was regarded as being a condition of degenerated glands, collections of pus, or distended lymph spaces. It was not until the middle of the 19th century that its actual nature as a tapeworm disease became known.
Refs FAO *Echinococcosis - Hydatidosis Surveillance, Prevention and Control* (1982); WHO *Echinococcosis (Hydatidosis)* (1968).
Broader Tapeworm as a parasite (#PE3511).

♦ **PE7519 Lack of acceptable sites for power plants**
Nature In the last 15 years, finding acceptable sites for power plants has met with increasing problems. Delays of up to 5 or 6 years have been experienced and public resistance has increased, manifesting itself in several ways. Action ranging from legal suits to physical violence has been used to resist sitings. Authorities have been forced to introduce new ad hoc legislation and procedures to assist the siting decisions, but this has met with only limited success.
Broader Energy crisis (#PC6329).
Related Inadequate industrial space (#PJ0084).

♦ **PE7528 Occupational hazards to workers in small industries**
Nature General working conditions in small industries tend to depend largely upon the attitude of the employers or owners. Occupational hazards to workers in such industries are not restricted to the developing countries but are also serious in highly industrialized countries as well, and in almost all cases are more serious than those in large establishments engaged in similar industrial activity. In small industries, health problems sometimes take on epidemic proportions, and occupational injuries are prevalent due to lack of in-service training, poor housekeeping, deficient handling methods, and bad layout of machinery. Small industries usually hire workers indiscriminately from many vulnerable groups, including the very young, the old, and the partially handicapped. The lack of social amenities required to facilitate the employment of women imposes a strain on working mothers as regards the care of their children.
Incidence Small industries and mines (less than 100 workers) have been estimated to employ almost 70 percent of the world's work force in manufacturing, mining, and related trades (with a range of 45-95 percent in different countries).
Broader Occupational hazards (#PC6716).

♦ **PE7530 Lack of human organs for transplantation**
Shortage of organ donors — Unauthorized organ transplantations — Trade in human organs — Animal organ transplants into humans — Imbalance between need and availability of viable organs for transplantations
Nature While the techniques of heart surgery improve continually, the number of organs available for transplant is no where near the number of people needing them. Medical workers pour their efforts into trying to keep people with serious heart disease alive until an appropriate donor appears, but this effort is often wasted, because patients expire before donors appear. People do not like to think about their own death, let alone consider dividing their body up to share with strangers.
Incidence The world-wide shortage of organs for transplantation has encouraged medical exploitation of the poor, especially in developing countries. Prisoners in the Philippines have been granted freedom by volunteering as kidney donors. In Brazil, it is reported that bodies are washed up on the beach with their kidneys removed. In China, the kidneys of executed criminals are supplied to patients requiring transplants. In countries such as Japan, kidneys have been accepted as payments to loan sharks. Lack of kidneys for transplantation is so great in the UK that pigs are being considered as alternative donors. Pigs are suggested because their organs are comparable in weight and size to human ones. Hence the colourful portrayal of pigs as "horizontal men".
Broader Limited availability of therapeutic substances of human origin (#PF6751).
Related Inadequate emergency blood supply (#PE0366).
Aggravates Violation of the rights of foetuses (#PE6369)
Trafficking in children for medical exploitation (#PD4271).

♦ **PE7533 Justifiable homicide**
Lawful killing — Permissible murder
Claim At no time is the taking of another human life justifiable, murder is murder and justifications simply hide this fact.
Counter-claim The right to life implies the right to protect and defend one's own life or the life of another person against an unjust attack. Since the defence can only be effective if it is in proportion to the violence of the unjust attack, it is possible that in the act of defending oneself or another the victim may kill the assailant.
Broader Homicide (#PD2341) Deficiencies in national and local legal systems (#PF4851).
Related Inhumanity of capital punishment (#PF0399).

♦ **PE7538 Health hazards to tourists in developing countries**
Nature The problems of mass catering for holidaymakers are often intensified in the developing countries which have the climate, scenery, and local rural customs that attract tourists from colder, industrialized, and urbanized countries. Mass tourism is an excellent method of earning foreign currency, and most countries are eager to capitalize on it - sometimes at the expense of maintaining adequate health care provisions. A rapid development of tourist hotels may create a strain on local sanitation facilities; there is often a lack of knowledge locally of the technology of mass catering, and perhaps a shortage of semi-skilled workers; and hygiene habits of the workforce may be unsatisfactory and lead to illness. In addition, in an apparently modern hotel, tourists may not take the precautions they would normally take when visiting a developing country. They may drink unboiled milk in the belief that it is pasteurized; eat fruit and salads under the impression that they have been washed in clean water by clean hands; may overlook that the ice cubes in their drinks may have been made from tap water; and may forget that in many developing countries it is not the custom to store meats under refrigeration.
Broader Tourist hazards (#PE8966) Hazards to human health (#PB4885).

♦ **PE7539 Environmental hazards of decommissioned nuclear power plants**
Waste of resources in obsolete nuclear power plants — Excessive cost of decommissioning nuclear reactors

Nature The ultimate waste from a nuclear power plant is the plant itself. A commercial reactor can only operate for thirty to forty years. When it stops operating, 15 to 20 percent of its contents remain radioactive. A closed–down reactor could be even more dangerous than one that is operating, because many of the security devices may have been removed. Dismantling a reactor is the only way to guarantee safety, but no one is certain if dismantling is feasible.
Incidence By the turn of the century, one hundred reactors around the world may have shut down, and by 2020 all 233 commercial reactors in operation in 1979 may be ready for the graveyard.
Background When a plant closes, it is first mothballed (welded shut with steel), then entombed (completely encased in concrete), and eventually dismantled. Thus far, twenty prototype reactors have been shut down; only five have gone beyond mothballing, and not even one of these relatively small reactors has been dismantled. Because old reactors are so radioactive, the entire dismantling process must be done by remote control and could cost as much as $100 million. While a reactor is mothballed or entombed, it must be guarded for safety. Leaks are likely, and the reactor remains radioactive for up to 250,000 years.
Refs Pollock, Cynthia *Decommissioning* (1986).
Broader Environmental pollution by nuclear reactors (#PD1584)
Environmental hazards of nuclear power production (#PD4977).
Related Energy crisis (#PC6329) Lack of fire–fighting facilities (#PJ2437).
Aggravates Insufficient nuclear power stations (#PD7663).
Aggravated by Deterioration of nuclear power plants (#PE5260).
Reduces Nuclear accidents (#PD0771).

♦ **PE7544 Inadequate radiological services**
Nature Diagnostic radiology is still scarce and poorly organized in the developing world. Existing services tend to be centred in large urban hospitals, equipped for specialized examinations, but there is little coverage of peripheral areas and those that do exist at the local level lack sophisticated machinery and skilled personnel.
Broader Inadequate health services (#PD4790).

♦ **PE7547 Surgical torture**
Broader Torture by wounding (#PE4078).
Narrower Dental torture (#PE7201).

♦ **PE7548 Prohibitive cost of eye–glasses in developing countries**
Nature The provision of glasses for schoolchildren with visual problems and adults with presbyopia and other conditions requiring corrective glasses is taken for granted in most industrial countries, but in many developing countries the importance of glasses in enabling these people to lead socially and economically satisfactory lives is just beginning to be acknowledged. The expansion of education is one factor in this growing awareness: learning is difficult, if not impossible, for children and adults with visual defects. Yet glasses are often unobtainable because of their high cost; people sometimes resort to purchasing a pair of cheap, second–hand glasses that may be available in a local market.
Broader Prohibitive medical expenses (#PE8261)
Subsistence approach to capital resources (#PF6530).
Aggravates Visual deficiencies (#PD8179).

♦ **PE7549 Congenital anomalies of eye**
Broader Human physical genetic abnormalities (#PD1618).
Narrower Aniridia (#PG4813) Congenital glaucoma (#PG4522).

♦ **PE7552 Rift valley fever**
Nature Rift valley fever, an epizootic disease, causes abortions, disease and death in infected herds or flocks (mainly of cattle and sheep): it can also be transmitted to man by his close contact with infected animals (either alive or dead) and also (but to lesser degree) by animal products.
Background RVF, a virus disease transmitted by mosquitoes, has been known in East and South Africa for some 50 years. The principal disease hosts in such areas have been cattle and sheep, with the imported exotic breeds proving more susceptible to disease than the indigenous animals. At an early stage, it was realized that man became infected whenever he has in close contact with diseased or dead animals. The numbers of human cases were small and in humans the disease was rarely fatal. Epizootics of RVF occurred after periods of unusually heavy rainfall, which were often separated by periods of up to 10–15 years when no case of Rift Valley fever was recognized. In East and South Africa the use of attenuated, neurotropic strains of RVF virus, developed in mice and shown to be immunogenic in the animal hosts, has proved to be a successful control measure. This situation was dramatically altered in 1977 when RVF was identified in Egypt in epizootic form.
Broader Virus diseases (#PD0594) Viral diseases in animals (#PD2730)
Human disease and disability (#PB1044).
Aggravates Fever (#PD2255).
Aggravated by Insect vectors of animal diseases (#PD2748).

♦ **PE7553 Respiratory diseases of swine**
Broader Diseases of respiratory system in animals (#PD7307).
Narrower Pasteurellosis (#PG3955) Swine influenza (#PG5342)
Atrophic rhinitis (#PG4622) Necrotic rhinitis (#PG4648)
Mycoplasmal pneumonia (#PG6742) Pleuropneumonia in swine (#PG6905).

♦ **PE7555 Dioxin poisoning**
Nature Effects from dioxin poisoning include chloracne, reduced libido, cancer, sterility, and the parenting of children with birth defects.
Incidence A recent Swedish study revealed a ninefold increase in birth defects of children born to dioxin–exposed mothers; a Viet Nam report shows an increase in liver cancer and birth defects among exposed Vietnamese; a follow–up of Italians exposed to the dioxin cloud in Seveso revealed a higher than normal incidence of chloracne, spontaneous abortions, loss of sex drive, and liver damage.
Background Dioxin – 'Agent Orange' – was used by the USA as a defoliant in the Vietnam War. Over 2 million people (military personnel, press, civilians) were exposed and Vietnam veterans in America successfully fought for monetary recompense for their illnesses.
Refs Higuchi, Kentaro (Ed) *P C B Poisoning and Pollution* (1976); National Electrical Manufacturers Association *PCB Health Effects*.
Broader Chemical pollutants of the environment (#PD1670).
Narrower Dioxin poisoning in animals (#PG5237).
Aggravated by Environmental hazards of chlorine–bleached paper products (#PE7416).

♦ **PE7560 Environmental hazards of extremely low frequency electromagnetic radiation**
EEG entrainment by ELF magnetic fields
Nature Extremely low–frequency non–ionizing radiation (ELF) is an effect we get from household appliances, transmission lines, electric blankets and our environmental in general.
Refs WHO *Extremely Low Frequency (ELF) Fields* (1984).
Broader Radioactive contamination (#PC0229) Electromagnetic pollution (#PD4172)
Environmental hazards of non-ionizing radiation (#PE7651).
Narrower Environmental hazards of electrical power transmission lines (#PE9642)
Health hazards of electromagnetic fields generated by electrical appliances (#PE7879).
Related Psychotronic warfare (#PD7986)
Hazards of environmental electromagnetism (#PE1304).
Aggravated by Unethical practice of radiology (#PD8290).

♦ **PE7562 Salmonellosis**
Salmonella infections
Nature The term salmonellosis covers a complex group of foodborne infections, caused by salmonellas, which affect both man and animals. Salmonellosis develops when food products containing salmonellas enter the body. The disease symptoms vary, depending upon the severity and source of infection. The onset of the gastrointestinal form of the disease (which occurs most frequently in man) is usually acute, with profuse vomiting and diarrhoea. The disease's duration is form 3 to 7 days.
Background Infected animals are the main source of the disease although humans (sick individuals or bacteria carriers) may sometimes also cause infection. The disease is transmitted by infected foods that are usually of animal origin; these foods become infected as a result of the forced, improper slaughter of animals and the violation of regulations governing the storage and preparation of food products, including contact between prepared and raw foods and insufficient cooking before consumption.
Incidence In recent years there has been a dramatic increase in the incidence of salmonellosis in industrialized countries leading to a greater awareness of the resultant economic damage, including the cost of interrupted production and trade. In the USA in 1988 more than 40,000 cases of salmonellosis were reported and this may represent only 1 per cent of the cases.
Refs Kelterborn, E *Catalogue of Salmonella First Isolations 1965–1984* (1987).
Broader Zoonotic bacterial diseases (#PD6363) Intestinal infectious diseases (#PE9526)
Diseases of the digestive system in animals (#PD3978).
Narrower Fowl typhoid (#PG0682) Pullorum disease in poultry (#PG4618)
Arizona infection of poultry (#PG4534) Paratyphoid infection in poultry (#PG7350).
Aggravated by Sewage as a pollutant (#PD1414)
Rodent vectors of disease (#PE3629)
Food poisoning through negligence (#PE0561).

♦ **PE7563 Inability of developing countries to adopt appropriate exchange rate policies**
Broader Mismanagement of exchange rate system (#PF1874)
Financial destabilization of world trade (#PC7873).
Aggravates Excessive foreign public debt of developing countries (#PD2133).
Aggravated by Exchange rate volatility (#PE5930)
Excessive concentration of export markets of developing countries (#PE9457).

♦ **PE7568 Substance abuse by physicians**
Drug abuse by doctors — Smoking by physicians — Alcohol abuse by doctors
Nature Due to the unique stresses of a medical career (long hours, family pressures, lack of sleep, poor nutrition, exposure to diseases, lack of exercise) plus the easy access to pure drugs, many doctors are drug abusers. Doctors generally tend to ignore their own illnesses and when found out, there is not any specific counselling service or treatment available to them.
Incidence The AMA estimated, in 1984, that at least 4,000 of the 500,000 doctors in the USA were addicted to drugs, opium and cocaine being the favourites. It also reported that the addiction rate for US doctors is 30–100 times higher than in the general population.
Broader Drug abuse (#PD0094) Substance abuse by role models (#PE0742)
Impairment of physicians' ability (#PE5746).
Related Laboratory testing errors (#PF5304).

♦ **PE7570 Salivary disorders in animals**
Broader Diseases of the digestive system in animals (#PD3978).
Narrower Ptyalism (#PG5963) Xerostomia (#PG9467) Sialadenitis (#PG3944)
Salivary cyst (#PG9240).

♦ **PE7572 Malignant neoplasm of respiratory system**
Broader Malignant neoplasms (#PC0092) Diseases of the respiratory system (#PD7924).
Narrower Lung cancer (#PE7085)
Pleura and peritoneum cancers of the bronchi (#PE8228).

♦ **PE7573 Violation of treaties with indigenous populations**
Broader Unfulfilled treaty obligations (#PF2497).
Related Discrimination against indigenous populations (#PC0352).
Aggravates Alien domination of peoples (#PC7384)
Expropriation of land from indigenous populations (#PC3304).

♦ **PE7575 Bogus public interest groups**
Nature There are numerous associations and institutes which appear and profess to operate in the public interest but are actually subsidiaries of or work in conjunction with industry, thereby perpetuating the best interests of the industry rather than the public at large. Such organizations may use deceptive titles and credentials to push industry propaganda or may promote projects that produce little benefit and distract attention from authentic groups whose budgets are miniscule by comparison.
Incidence The (USA) Calorie Control Council incited people to speak out about how they felt concerning the proposed ban on saccharin, but no mention was ever made in all their propaganda that the council was an association of dietetic food and drink producers.
Broader Deception (#PB4731).
Narrower Denial of rights of indigenous people to be self–governing (#PE1024).
Related Bogus firms (#PF0326) Front organizations (#PE4358).
Aggravated by Misuse of nonprofit associations as front organizations by government (#PE0436).

♦ **PE7581 Seizure of trade union property by public authorities**
Destruction of trade union property by public authorities — Confiscation of trade union property
Broader Violation of the right of trade unions to function freely (#PE1758).

♦ **PE7582 Child prostitution**
Incidence Prostitution of young children is known to occur in Asia, Latin America, Europe and North America. Children may be forced into prostitution by parents or guardians. In 1977, a girl under 12 years of age in New York was arrested 12 times for prostitution; a couple on Long Island was arrested for photographing their 3 1/2 year old daughter performing sexual acts. In Bangkok, girls scarcely weaned are handed over to pimps for an equivalent sum of a few dollars. Also in Bangkok thirty thousand children under the age of sixteen and some as young as six work as prostitutes.
Refs Sereny, Gitta and Wilson, Victoria *The Invisible Children* (1985).
Broader Prostitution (#PD0693) Victimization of children (#PC5512)
Sexual exploitation of children (#PD3267).
Aggravates Kidnapping (#PD8744) Juvenile prostitution (#PD6213).

Aggravated by Child pornography (#PF1349) Children in poverty (#PD4966)
Sexual abuse of children (#PE3265)
Trafficking in children for sexual exploitation (#PE6613).

♦ **PE7584 Pollution-induced fish diseases**
Fish kills caused by pollution
Nature Fish diseases observed in the past three decades and which have been attributed to pollution include: haemorrhages; tumours; fin rot; deformed fins; and missing scales and tails. The livelihood of fishermen is being threatened by these deformities as they become more common; in industrialized countries, increasing numbers of fish are deemed inedible.
Incidence The University of Miami Medical School (USA) reported in 1980 that the fish in Biscayne Bay show symptoms of 40 kinds of diseases, including two strains of cholera, and surveys from all areas of the USA are finding fishermen's reports of deformed fish.
In the USA alone, between 1961 and 1976, 482 million fish were reported killed as direct result of pollution. These reports, based on voluntary notification, probably account for only a fraction of the fish that were killed. Many small kills are not noticed or are not reported, and large kills are often not included because of insufficient information to determine whether the kills were caused by pollution or natural by factors. Kills of 1 million or more were responsible for 77 percent of all reported fish killed in the 15 years. Low dissolved oxygen levels resulting from excessive sewage (primarily municipal) were the leading cause. The second most common cause was pesticides.
Broader Killing of animals (#PD8486) Pests and diseases of fish (#PD8567).
Aggravated by Water pollution (#PC0062).

♦ **PE7589 Night work**
Nature Night work may induce increasing tiredness during the night shift, stress, reduced appetite, reduced sleep, heightened family problems due to physical fatigue and tension, and loss of social life. Disturbance of the natural biorhythm can lead to extra consumption of stimulants, like coffee and alcohol, during the night and sleeping pills during daytime.
Refs Carpentier, J and Cazamian, P *Night Work* (1988); Council of Europe *Night Work* (1981); ILO *Night Work. Its Effects on the Health and Welfare of the Worker* (1978); ILO *Night Work* (1989).
Related Shift work stress as an occupational hazard (#PE5768).

♦ **PE7591 Acute respiratory infections**
Nature Although most of the infections of the respiratory tract are mild, self-limited illnesses, they are ever-present and the morbidity that they cause is very high. In developing countries the case fatality rate is still significant; and whereas the rate is considered low in developed countries, even there the number of deaths – especially in children and in the older age groups – is high, because of the size of the illness pool.
Incidence For children in developing countries the situation is particularly alarming. The mortality rate may be more than 50 times that in developed countries. These were the conclusions of WHO, drawn from analyses of mortality and case fatality from acute respiratory infections in children under five years of age in nearly all the countries of the Americas and Southeast Asia and in several countries in the African and the Western Pacific. It is a result of low birth weight, very young ages of mothers and births following too closely upon one another, malnutrition, lack of breast-feeding, indoor air pollution stemming from burning of 'biomass' fuel and exposure to 'passive smoking' at home.
Refs Pennington, James E *Respiratory Infections* (1988); WHO *Acute Respiratory Infections in South-East Asia* (1986).
Broader Lung disorders and diseases (#PD0637)
Infectious and parasitic diseases (#PD0982)
Diseases of the respiratory system (#PD7924).
Narrower Sinusitis (#PG4873) Sore throat (#PE4651) Common cold (#PE2412)
Tonsillitis (#PE2292) Acute laryngitis (#PG5614)
Inflammatory affections of the bronchial tubes and lungs (#PE8822).

♦ **PE7611 Inadequate social security for migrants**
Nature Migrants form a segment of society whose social security protection still leaves much to be desired. They may suffer from direct discrimination whereby non-nationals, especially those residing abroad, are excluded from certain benefits, or occasionally denied any coverage at all. Another discriminatory technique is that certain benefits are not paid to workers or members of their family who are resident or temporarily resident abroad. Entitlement to benefit may be subject to the completion of a period of insurance, employment, occupational activity or residence – a difficult condition for migrants whose career takes them from one country to another.
Broader Social insecurity (#PC1867) Economic insecurity (#PC2020).

♦ **PE7613 Abuse of project-tied migration**
Nature Project-tied migration, which drains skills often in short supply in migrant-sending countries and generally disrupts family life, is subject to abuse particularly at the stages of recruitment and the start of work. Virtually every aspect of the migrant workers' daily life is under the protection and control of their employers. They may be forbidden to form unions and may find that their individual contracts provide the only means of bargaining. Government-determined minimum requirements for accommodation in the camps (standards of living space, sanitary and recreation facilities, medical care) are often neither met nor strictly enforced. Work-related deaths and injuries are high and compensation is either non-existent or very inadequate.
Broader Back to the land (#PF4181) Economic exploitation (#PC8132).

♦ **PE7616 Denial of right to sufficient clothing**
Broader Inadequate standards of living (#PF0344).
Related Inappropriate clothing (#PJ6604) Denial of right to sufficient food (#PE0324)
Inadequate social welfare services (#PC0834) Denial of right to social security (#PD7251)
Denial of right to economic security (#PD0808) Denial of right to sufficient shelter (#PD5254)
Denial of right to adequate medical care (#PD2028).
Aggravated by Unethical practices in the apparel industry (#PD8001).

♦ **PE7620 Dismissal of workers to prevent legal strike action**
Dismissal of trade union representatives following legal strike action — Dismissal of workers following legal strike action
Broader Unjust dismissal of workers (#PD5965) Violation of the right to strike (#PE5070).

♦ **PE7621 Instability of trade in tungsten**
Nature In recent years there has been an increasing imbalance between the supply and demand of tungsten, as well as an increase in the availability of tungsten resulting from recycled material and stockpile releases while the market was weak. These factors have generated an unstable tungsten market.
Incidence The average 1983 price of tungsten was $80.86 per mtu (metric ton unit), compared to $106.12 per mtu in 1982. Between 1981 and 1983, tungsten concentrate prices were 42 percent lower than in 1980, leading to serious effects on producers.
Broader Instability of trade in non-ferrous metals (#PE1406).

♦ **PE7623 Alzheimer's disease**
Refs Altman, H J *Alzheimer's Disease* (1987); Katzman, Robert (Ed) *Biological Aspects of Alzheimer's Disease* (1983); Swaab, D F, et al (Eds) *Aging of the Brain and Alzheimer's Disease* (1986).
Broader Senile dementia (#PE3083) Diseases and injuries of the brain (#PD0992).
Aggravates Neurological rage (#PE0429) Deterioration of the mind with age (#PE4649).
Aggravated by Health hazards of aluminium (#PE4969).

♦ **PE7624 Denial of right to time to prepare a trial defence**
Broader Violation of civil rights (#PC5285) Denial of right to a legal defence (#PE4628).

♦ **PE7627 Dependence on breast feeding**
Breastfeeding as female servitude — Social stigma of breast-feeding
Nature In the last decade, breastfeeding has regained its lost popularity and is making a striking comeback in the Western world. Difficulties arise due to the fact that the comeback is concurrent with women's liberation from the household and embarking upon career; breastfeeding serves to hinder a woman's total emancipation for she is the only person capable of performing this task.
Incidence In the Western world, it is well-educated mothers who are returning to this age-old tradition. A 1980 US survey showed that 70 percent of graduate mothers breastfed their children, as opposed to only 25 percent of mothers with nine or fewer years of schooling.
Claim Breastfeeding maintains a one-to-one relationship between mother and child for an overly extended amount of time, thus keeping the child dependent upon the mother when it should be developing other relationships; and keeping the mother dependent upon her child when she, too, should be developing outside interests. Breastfeeding is usually not run at scheduled intervals, making it virtually impossible for a woman to have a full-time career outside the home. In addition, breastfeeding may be a contributory cause to sagging breasts, which a woman and others could conceive of as fading beauty; and if the woman chooses to feed her baby in the midst of company, she may be found to be offensive. It is also messier and less efficient than bottlefeeding. Substances such as polychlorinated biphenyls (PCB), dioxins and other potentially dangerous products remain in body tissues for several decades and could be passed to newborn babies in milk during breastfeeding.
Counter-claim Breastmilk is superior to infant formula, offers natural immunological defences, and creates a warm emotional bond between mother and child. In the developing world, breast-feeding is a natural safety mechanism against the worst effects of poverty.
Refs FAO *The Economic Value of Breast-Feeding* (1979).
Broader Slavery (#PC0146) Social stigma (#PD0884).
Related Hazards of bottle-feeding (#PE4935).
Aggravates Maternal malnutrition (#PE1085)
Substitution of inappropriate foodstuffs for breast feeding (#PE8255).
Aggravated by Diminishing capital investment in small communities (#PF6477).
Reduces Child malnutrition (#PD8941).

♦ **PE7628 Early marriage**
Restrictive early marriages — Pressured early marriage
Nature The younger women marry, the earlier they start childbearing and the longer they are exposed to the risk of conception. They lose the chance of longer schooling and of employment, and they enter marriage with less motivation and fewer personal resources to plan their families successfully. Pregnancy at an early age is a severe health risk both to the mother and her baby. In addition, early marriage means a shorter gap between successive generations, significantly increasing the birth rate.
Incidence In South Asia and sub-Saharan Africa about half of all women aged between fifteen and nineteen are, or have been married; in the Middle East and North Africa the proportion is close to a quarter. It falls to less than 20 percent in Latin America and in East Asia, and to less than 5 percent in Hong Kong and Korea. In Libya, 70 percent of women aged fifteen to nineteen have been married. In Bangladesh the mean age at marriage for women is sixteen.
Refs Lindsay, Jeanne W and Monserrat, Catherine *Teenage Marriage* (1988).
Broader Social subjugation of women (#PD4633).
Narrower Child-marriage (#PF3285).
Aggravates Adolescent pregnancy (#PD0614).
Aggravated by Forced marriage (#PD1915) Marriage markets (#PD7282)
Arranged marriage (#PF3284).
Reduces Pre-marital sexual intercourse (#PD5107).

♦ **PE7632 Obsessive compulsive personality disorder**
Pattern of perfectionism and inflexibility
Refs Emmelkamp, Paul M G *Phobic and Obsessive-Compulsive Disorders* (1982); Rapoport, Judith L *Obsessive-Compulsive Disorder in Children and Adolescents* (1988).
Broader Personality disorders (#PD9219).
Aggravates Lack of international cooperation (#PF0817)
Corruption of the good in human nature (#PE7917).

♦ **PE7637 Unbalanced application of communications technology**
Gaps in telecommunications capability — Inconsistent application of telecommunications technology
Nature As developed countries set up ever more powerful, homogeneous and centralized networks of communication, less developed countries find their existing telecommunications networks often left behind. Rough terrain, a shallow technological base, extremely high initial costs, limited local technicians and sometimes aversion to foreign ways, all block the development of sophisticated local communications installations, and enlarge the gap between richer and poorer countries.
Incidence A phone call to New York takes seconds in Hong Kong. A few miles away in the People's Republic of China, the same call can take hours and cost very much more.
Broader Uncontrolled application of technology (#PC0418).
Related Technological revolution (#PC3234)
Uncontrolled environmental impact of technology (#PC1174)
Use of inappropriate technologies in developing countries (#PF0878)
Obstacles to satellite communications for developing countries (#PF6072)
Disparities in distribution of communication resources and facilities (#PD2762).
Aggravates Insufficient communications systems (#PF2350).

♦ **PE7638 Biased portrayal of women in mass media**
Exploitative portrayal of women in media
Nature Mass media has significant influence in determining the cultural patterns which affect the full development of women. The influence of the media is male-oriented and tends to trivialize, undervalue and exploit the real contribution of women to society. While women come into many different categories and have widely varying social positions and educational levels, they tend to be depicted by one image, one role, or one life-style.,
Incidence Media programmes and articles discriminate against women in various ways. There

tend to be women's sections or pages, set aside from the rest of the publication or programme. In the USA, it is reported that one of the major television networks broadcast no news programmes, no editorials, no public affairs programmes, and no documentaries during the daytime hours when housewives and mothers are the majority of the television audience.
Broader Misleading information (#PF3096) Distorted media presentations (#PD6081).
Aggravates Exploitative films (#PE6328).
Aggravated by Excessive portrayal of sex in the media (#PE7930).

♦ **PE7645 Psychotic violence**
Psychotic stranglers — Psychotic poisoners — Psychotic urban bombers — Psychotic arsonists — Psychotic slashers — Psychotic snipers — Psychotic mass-murderers
Nature Individuals so void of social responsibility, a sense of morality or ability to identify with others that they murder masses of people are found throughout history and in every culture. Some because of social position or political power are immune from social sanction and can kill indiscriminately. Others in an effort to maintain their freedom kill one person, move to another city and kill again. Still others, disregarding their own life, walk into a crowded place and kill until they are stopped. Until ways can be found to prevent these people from murder they will always be potentially present in society.
Broader Criminals (#PC7373) Criminal insanity (#PD9699).
Aggravates Fear (#PA6030) Injuries (#PB0855) Human death (#PA0072) Vigilantism (#PD0527) Human destructiveness (#PA0832).
Aggravated by Psychoses (#PD1722) Mental illness (#PC0300) Socio-economic poverty (#PB0388) Psychological alienation (#PB0147) Inadequate firearm regulation (#PD1970).
Reduced by Excessive government control (#PF0304) Inadequate national law enforcement (#PE4768).

♦ **PE7647 Dysfunctional public utilities**
Deficient public facilities
Broader Inadequate maintenance of infrastructure (#PD0645) Poor condition of open spaces in urban communities (#PF1815).
Narrower Clogged storm drains (#PG9955) Underdeveloped road network (#PE1055) Chemical contamination of water (#PE0535).
Related Inappropriate sanitation systems (#PD0876).
Aggravates Ineffective utilization of public environment (#PF6543).

♦ **PE7651 Environmental hazards of non-ionizing radiation**
Nature Industry, engineering, telecommunications, medicine, research, education and the home, now use a large number of processes and devices which emit nonionizing radiation. It is known that the ultraviolet radiation generated from man-made sources can be more intense than that occurring naturally, for example in sunlight. The eye is more sensitive to ultraviolet radiation than is the skin and has a built in safety mechanism by reacting to strong sunlight with contraction of the iris and shutting of the eyelids. This reaction mechanism, however, is not fast enough to protect against very bright flashes, nor will it protect against radiation that is not accompanied by visible light. Thus it is necessary for workers in certain fields (arc welding, for example) to utilize protective devices, such as shields and filters. The resultant possible adverse effects of general population exposure are not yet fully understood, and adequate measures are lacking to inform the public about possible risks.
Incidence Non-ionizing radiation includes the high-frequency radiations used in communications and broadcasting; the microwave radiations used in radar, television transmissions, and industrial applications; the infrared radiation used in heat-lamps, and the visible light used in some lasers; ultraviolet lamps, and medical diathermy equipment. Excessive radiation may cause damage to the eyes and skin, and in sufficient dosage, to the internal organs.
Refs ILO *Occupational Hazards from Non-Ionising Electromagnetic Radiation* (1984); Seuss, Michael J *Nonionizing Radiation Protection* (1982).
Broader Pollutants (#PC5690) Occupational hazards (#PC6716) Radioactive contamination (#PC0229).
Narrower Ultraviolet radiation as a hazard (#PE5672) Hazards of environmental electromagnetism (#PE1304) Environmental hazards of electrical machinery, apparatus and appliances (#PE8026) Excessive exposure to radiation from consumer goods and electronic devices (#PE1909) Environmental hazards of extremely low frequency electromagnetic radiation (#PE7560).
Aggravates Human disease and disability (#PB1044).

♦ **PE7654 Self-satisfied style of informal leadership advisors**
Nature All leaders have a network of informal advisors who are willing to listen to the leader's ideas, make suggestions and offer new ideas. These informal advisors frequently have a self-satisfied style and operate out of limited and inflexible ideologies. For what ever reason they hold that their intellectual principles and perspective are above the common social good. This further hinders the effectiveness of leaders.
Broader Narrow context for counsel (#PF0823).
Aggravated by Irresponsible international experts (#PF0221) Extreme detachment from represented constituency (#PF0889).

♦ **PE7655 Unreasonable licensing restrictions**
Overwhelming licensing process
Nature Depending on the provisions of licence contracts, limitations of one kind of another are generally imposed on licensees. Tied purchases of imports of raw materials, intermediate goods, components, equipment or spare parts are sometimes stipulated on technical grounds, depending on the requirements of a given technology or on the licensor's trade mark. Not infrequently, these clauses are associated with excessively high prices of the delivered inputs, which from the viewpoint of the licensee represents a "hidden" cost of the technology transfer. A common type of restriction is the prohibition of the disclosure of the transferred technology to third parties. In the developing countries such practices often debar domestic firms from access to technologies which are already available locally, thus leading to what are known as "repetitive imports of technology". Other provisions commonly stipulated limit changes in the imported technology, thus reducing or eliminating the possibility of adapting the product or the technology to the local needs and conditions.
Broader Underprovision of basic urban services (#PF2583) Unexercised responsibility for external relations (#PF6505).

♦ **PE7659 Public assaults on police**
Incidence On average 55 police officers are assaulted every day, and in a year one in seven of the officers in England and Wales is attacked in 1988.
Broader Assault (#PD5235) Hindrance of law enforcement (#PD5515).
Aggravated by Mistrust of police (#PF8559) Police intimidation (#PD0736) Abuse of police power (#PC1142).

♦ **PE7664 Extrajudicial courts and tribunals**
Revolutionary courts — Security tribunals
Nature In some countries, special courts and tribunals are set up outside the normal judicial system of the country. Military courts may try civilians; special courts and tribunals are empowered to try 'political', 'security', or 'anti-revolutionary' offenders, and are usually not bound to follow the established procedures of the ordinary courts; safeguards for a fair trial are often ignored; legal representation may not be allowed; people may not be informed of the charges against them until their trials open; cross-examination of prosecution witnesses may not be allowed; evidence presented by the prosecution may not be contested; and judges are often not independent persons with legal backgrounds, but military personnel.
Broader Denial of right to trial by a court (#PE4737) Denial of right to fair and public trial (#PE3964).

♦ **PE7665 Lack of impartiality of the judiciary**
Influencing the law — Lack of independence of judges and lawyers — Denial of the right to independent judges — Lack of impartiality of judge and jury — Control of the judiciary by the executive or military power
Nature In a considerable number of situations in both developed and developing countries, the independence of the court is severely curtailed or non-existent, often in contradiction to the constitutional guarantees for the independence of the judiciary. Conviction and sentencing may be influenced or predetermined by the executive, who directly control the status of judges. Sometimes ordinary courts have been deprived of jurisdiction over certain categories of cases without any legal justification, and the cases later tried by military or special courts. In some instances, judges are intimidated to make decisions favourable to the executive.
Broader Unjust trials (#PD4827) Inequality before the law (#PC1268).
Narrower Racial bias in sentencing offenders (#PE4907).
Related Corruption of the judiciary (#PD4194).
Aggravates Bias in jury trials in small jurisdictions (#PE4733).
Aggravated by Harassment of the judiciary (#PE5487) Economic barriers to access to the legal profession, the judiciary and jury membership (#PE0803).

♦ **PE7670 Insecurity through unilateral structural disarmament**
Nature Structural disarmament occurs when a nation's defence budget, plus exports, provide too small a market to bring armament development and production costs down to a politically affordable level. Even when governments are spending more money to rearm, disarmament occurs. As unit costs go up, fewer and fewer weapons can be procured. Such unilateral disarmament will continue until governments establish an intercontinental market structure for the production and exchange of armaments.
Incidence In the case of the Western NATO alliance, even after 35 years no common arms market has been organized as a basis for common defence. Research and development tends to proceed independently of production. Access to technology is restricted. Economical procurement is undermined by protectionism.
Claim By the very process of rearmament with super-smart weapons to offset the numerical superiority of its opponents an alliance may price its collective defence beyond the reach of necessary political support to provide the resources.
Broader International insecurity (#PB0009).
Narrower Unilateral structural disarmament of nuclear weapons (#PE4051).

♦ **PE7672 Alcohol-related violence against women**
Broader Violence against women (#PD0247) Alcohol-related violence (#PE7084).

♦ **PE7677 Inhumane cooking practices**
Nature Some dishes are believed to have a superior taste if the animals used to prepare them are killed by the cooking process itself. The animals then die slowly by heat in contrast to more rapid methods of slaughter. Edible parts for cooking may be pulled off animals which are then discarded to die slowly.
Incidence Such techniques are typically used in the preparation of lobsters and prawns. Lobsters may be placed alive in boiling water in which they struggle until they die. Fish may be placed alive on a heated plate on which they are held whilst they are prepared. The legs may be pulled off a frog without any effort to kill the frog speedily; it may be discarded and left to die. Some delicacies, requiring little preparation, may involve consumption of a portion of an animal whilst it is still alive, as in the case of monkey brain. It is even possible to maintain animals alive as a constant source of supply of blood for drinking, as in some uses of cattle. Live turtles were used as a source of fresh food on board ship, only discarded when they died after days or weeks of agony.
Claim Inhumane attitudes to food preparation encourage and reinforce inhumane attitudes in other domaines of human activity.
Counter-claim Many such procedures are traditional methods of food preparation. There is no evidence to prove that they cause undue pain to the animals concerned.
Broader Denial to animals of the right to a natural death (#PE8339).

♦ **PE7678 Creeping modernization of military weaponry**
Nature Existing military weaponry may be progressively upgraded so that its capacity becomes greater than apparent from simplistic comparisons based on measures of its earlier capacity put forward in publicized arms control proposals.
Incidence An example is the modification of existing nuclear submarines to increase the number of missiles they can launch. The introduction of multiple warheads disguises the increase in capacity of an existing missile system.
Aggravated by Imbalance of conventional armed forces (#PC5230).

♦ **PE7679 Defence information uncertainty**
Stealth weaponry — Electronic deception
Nature Progressive improvements in weapons systems and the electronic detection of missiles are focusing on the development of stealth technology. Such technology is designed to be undetectable or confusing to detection systems; this leads to a level of information uncertainty which undermines any capacity to control or contain global warfare, by increasing the risk of decisions leading to escalation of destructive measures used. The situation may be aggravated by high altitude nuclear explosions resulting in electronic blackout.
Incidence Iran Air Airbus A-300 was shot down and its passengers all killed by the USS Vincennes, when its Aegis command and control system mistook the airliner for a hostile enemy aircraft. Fourteen months earlier, the American frigate Stark did not risk defending itself, and lost 37 of its crew members. In neither case did the state-of-the-art hardware of software provide appropriate information to assist the military in protecting lives.
Refs American Bar Association *National Security Leaks* (1986).
Broader Uncertainty (#PA6438).
Related Cross border military operations (#PD5272).

♦ **PE7680 Deceptive political proposals**
Nature Proposals to resolve problem situations and conflicts may be put forward with a view to gaining advantages for the proposer rather than to alleviate the situation. Such proposals may even be announced to create the impression that the proposer is acting in good faith but without any belief that they would finally prove acceptable or be implemented. Such token proposals become a means of gaining time and creating the illusion of responsible management of crises.

When both parties employ such techniques as a matter of course, this breeds cynicism in policy-making circles and false hopes on the part of the public, leading progressively to systematic alienation.
Related Diplomatic errors (#PF1440).
Aggravates Loss of confidence in government leaders (#PF1097).

♦ PE7687 Impersonating officials
Nature Impersonating a public servant, or a foreign official in order to exercise authority or gain a thing of value is a crime. This includes wearing the uniform or displaying any other symbol of authority in order to deceive.
Broader Crimes against the integrity and effectiveness of government operations (#PD1163).
Related Fraudulent impersonation (#PE1275).

♦ PE7691 Bereaved children
Nature Children who have lost a parent, sibling or other loved one lack the experience that adults can have to cope with their own pain and emotions. Children express their grief in their ways of behaving. They act out their feelings and emotions. We cannot always know what they are thinking or feeling. Withdrawal, aggressiveness, panic, anxiety, anger, guilt, fear, regression and symptoms of bodily distress can all be signs of grief.
Refs Furman, Erna and Freud, Anna *A Child's Parent Dies*; Mace, Gillian S, et al *The Bereaved Child* (1981).
Broader Human death (#PA0072).

♦ PE7695 Degradation of the environment by trees
Inappropriate tree plantations — Undesigned tree-planting plan — Inappropriate afforestation — Inappropriate re-forestation
Claim Multinational companies have realized that they may run out of raw material if they do not plant more trees, so they are producing eucalyptus and other fast-growing trees. These ones are extravagant users of water and nutrients, competing with crops and degrading good land.
Refs Anderson, Dennis *The Economics of Afforestation.*
Broader Natural environment degradation (#PB5250)
Lack of central planning structures in small communities (#PF2540).
Aggravates Vulnerability of world genetic resources (#PB4788).

♦ PE7698 Calcareous soils
Refs FAO *Calcareous Soils* (1977).
Broader Soil infertility (#PD0077).

♦ PE7700 Excessive parental drunkenness
Alcoholic parents
Broader Drunkenness (#PE8311) Alcohol abuse (#PD0153)
Demeaning community self-image (#PF2093).
Aggravates Suicide (#PC0417) Children of alcoholics (#PD4218).

♦ PE7701 Malingering
Effort discretion
Nature Malingering is an avoidance reaction (avoidance of situations deemed unpleasant) or a condition feigned in order to bring about personal gain, but in either case may be viewed as having social, moral and legal implications. Malingering most often manifests itself in criminal cases involving pleas of mental illness, in litigation arising from personal injury, and in attempts to evade military service.
Background The Bible and Greek mythology both contain descriptions of attempts to feign illness.
Broader Evasion of work (#PC5576) Ill defined causes of morbidity (#PE5463).
Related Neglect (#PA5438) Avoidance (#PA6379)
Uncommunicativeness (#PA7411).
Aggravates Absenteeism (#PD1634).

♦ PE7705 Excessive medical intervention in childbirth
Nature While a small proportion of women require dramatic surgical intervention to deliver a baby, most mothers require none of the expensive overprecautions used as normal maternity care. The movement of midwives and home birthing training have created a healthy alternative to the terrifying experience of hospitalization, overmedication and regimentation that surround the birthing process in most developed countries.
Counter-claim A desire by women for less medical intervention in childbirth could well mean a return to the widespread loss of lives during, before and after childbirth which was common as recently as 50 years ago. As a prominent obstetrician remarked, "If (women) don't wish to go by what we will do, they will make their own decision and that will be on their own heads."
Broader Unethical medical practice (#PD5770).
Narrower Alienating child-birth environments (#PF6161)
Precipitate or ill-judged forceps delivery (#PG2953).
Related Superstition (#PA0430) Abuse of medical drugs (#PD0028)
Inadequate medical care for pregnant women (#PE4820).
Aggravates Human suffering (#PB5955)
Resistance to incorporating midwives in medical care systems (#PE4901).
Aggravated by Pride (#PA7599) Prohibitive medical expenses (#PE8261)
Over-specialization in medical care (#PF5709) Lack of diversity in medical science (#PE8671)
Increase in insurance claims for medical negligence (#PE4329).

♦ PE7706 Fear of increased autonomy
Fear of independence — Fear of freedom
Nature The genetic and cultural conditioning of human beings to be social animals necessarily induces fear of being independent, i.e. free from automatic submission to the social pressures of peers, the media and the local community, free from control by personal sexual, emotional and physiological drives and free rationalizations of various kinds. This total autonomy of the self implies responsibility for one's thinking, acting and being without conditions. No external or internal force is to blame for the situation in which one finds oneself or for the consequences of one's actions in that situation. By far the bulk of humans would rather say the government, or their background or their desires or their reasons were why their situation is as it is.
Broader Fear (#PA6030).
Related Resistance to grace (#PF5266).
Aggravates Resistance to change (#PF0557).
Aggravated by Sin (#PF0641) Fear of contradicting popular views (#PF2040).

♦ PE7708 Diseases of thymus gland
Thymic tumours
Refs Sarrazin, R; Vincent, F and Vrousos, C (Eds) *Thymic Tumors* (1989).
Broader Gland disorders (#PD8301).

♦ PE7711 Trafficking in illegal firearms
Nature The possession, sale or transport of illegal firearms is a crime. Weapons like sawed off shotguns and machine guns; ammunition like dumdum bullets and explosives like plastics are designed nearly exclusively for military purposes and any other reason for possessing them is for criminal or terrorist activity.
Broader Firearms and explosives crimes (#PE1108)
Unlawful business transactions (#PC4645).

♦ PE7718 Inadequate working conditions in health and medical services
Refs ILO *Employment and Conditions of Work and Life in Health and Medical Services* (1985); ILO *Employment and Conditions of Work and Life of Nursing Personnel* (1976).
Broader International imbalance in the quality of life (#PB4993).
Related Inadequate working conditions for professionals (#PE3170).
Aggravates Human disease and disability (#PB1044).
Aggravated by Inadequate health services (#PD4790).

♦ PE7725 Closed channels of dialogue with the judiciary
Nature The structures, rites and mythology surrounding judicial systems isolate the judiciary from the legal needs of local people. Channels of communication between the grassroots judges, clerks, lawyers and other officials of the courts are closed.
Broader Ineffective dialogue (#PF1654)
Exclusion of the masses from setting criteria in judicial judgements (#PD1060).

♦ PE7726 Unequal opportunities for foreign students
Refs Altbach, Philip G *The Foreign Student Delimma.*
Broader Inequality in education (#PC3434).
Related Student immobility (#PF7577).
Aggravated by Debilitating deterioration of physical environment (#PD2672).

♦ PE7733 Diseases of upper respiratory tract
Diseases of the throat
Refs Maran, A G *Logan Turner's Diseases of the Nose, Throat and Ear* (1988).
Broader Diseases of the respiratory system (#PD7924).
Narrower Angina (#PE5204) Hay fever (#PE6197) Tonsillitis (#PE2292)
Nasopharyngitis (#PG9838) Throat irritation (#PG6761)
Adenovirus infections (#PE2355) Peritonsillar abscess (#PG4647)
Infection of the throat (#PG4579)
Benign neoplasm of throat and mouth (#PG4748)
Malignant neoplasm of mouth and throat (#PE9819).
Aggravates Diseases of the ear (#PD2567).
Aggravated by Limestone dust (#PE2038).

♦ PE7734 Deterioration of the quality of life in least developed countries
Inadequate standard of living in least developed countries
Nature Real incomes at the household and family level have declined sharply within the general fall in per capita incomes. Malnutrition, already a serious problem in the 1970s, has risen in the wake of food shortages and near famine conditions.
Broader Least developed countries (#PD8201) Deteriorating quality of life (#PF7142)
Inadequate standard of living in developing countries (#PE4052).
Aggravates Lack of technical development and excess of manpower in developing countries (#PE4933).

♦ PE7735 Disruptive secular impact of holy days
Discriminatory effect of holy day observance
Nature Days of religious observance such as Sunday, the Sabbath, and religious holidays such as Christmas, Easter, Yom Kippur, Rosh Ha-Sha-nah, Passover, for different religions occurring in the same country may cause conflict, especially if one religion predominates and national holidays are based on the days of observance of that religion to the detriment of the others.
Background For centuries, legislation in Europe and in the USA has proscribed various sorts of work, recreation, or dissipation on Sunday. While the roots of such proscription lie in the upholding of religious beliefs, religions whose Holy Days do not lie on Sunday are discriminated against by such legislation. The religious motivations of holy day observance were enlarged upon when the rise of the labour movement during the 19th century added a sociological impetus as well.
Incidence For the Jewish religion, Saturday is the Shabbath, or day of rest, in which all activity ceases, including the closing of shops. For Jewish shop owners to be forced by governmental law to be closed on Sunday as well means they can operate only 5 out of 7 days a week, thus putting them at an obvious disadvantage to non-Jewish entrepreneurs who operate 6 days a week. In Christian countries there is social and commercial pressure to buy Christmas presents and to send Christmas cards, which may be done by non-Christians. In Christian countries, some Jewish communities provide services on Sunday as well as the Sabbath because of the practical impediments for some to observe the Sabbath.
Claim Sunday laws unreasonably interfere with the liberty to conduct business and with freedom of religion.
Broader Ideological conflict (#PF3388) Religious intolerance (#PC1808)
Religious discrimination (#PC1455).
Related Desecration of holy spaces (#PF6385)
Peoples perceiving themselves as specially chosen (#PF4548).
Aggravates Religious conflict (#PC3292).
Aggravated by Traditionalism (#PF2676) Desecration of holy days (#PF3607)
Compulsory indoctrination (#PD3097) Proliferation of commercialism (#PF0815)
Discriminatory religious influence on the law (#PD3357).

♦ PE7742 Reductionistic decision making criteria in the construction industry
Nature The construction industry's decision making criteria is reduced to individualistic, profit making and short term values. They guide the investment of capital, the organization of labour, the application of technology and the distribution of goods and know-how. As such, the industry is not accountable to the broad spectrum of long-term social values required by today's globally interdependent society.
Broader Ineffective decision-making processes (#PF3709)
Profit motivated utilization of construction technology (#PF2464).

♦ PE7747 Aerospace monopolies
Nature Aerospace production is concentrated in the control of a small group of monopolies in the main capitalist countries, with the US monopolies accounting for over 70 percent of the world's production. As military contracts are the best source of revenue for these monopolies (they are especially prosperous in time of war), they are continually competing amongst themselves to win government contracts, which are often awarded as a result of contacts in the military establishment and not as a result of objective evidence. While aerospace material could sometimes be supplied at a lower cost and higher quality by non-monopolized organizations, the smaller suppliers are overlooked because they lack the necessary influence.
Broader Monopolies (#PC0521).
Related Industrial gas monopolies (#PE1813).

♦ **PE7750 Personal physical insecurity of women**
Fear of violence among women — Risk of violence to women
Incidence According to the British Crime Survey one in five women felt "very unsafe" walking outside at night. Yet fewer than one in 70 women claimed to have been attacked in 1988.
Broader Personal physical insecurity (#PD8657).
Aggravated by Violence against women (#PD0247).

♦ **PE7752 Marketing skills which reinforce a self image of being a victim of circumstances**
Broader Middleman control of rural marketing (#PE3528).
Aggravates Defensive life stance (#PF0979).
Aggravated by Misuse of advertising (#PE4225).

♦ **PE7753 Incompatibility of document classification systems**
Aggravates Uncatalogued documents (#PF4077).
Aggravated by Bias in document classification systems (#PF6743)
Prohibitive cost of maintaining comprehensive document collections (#PE1122).

♦ **PE7755 Sexual abuse of wards**
Nature Sexual abuse of wards is having sexual intercourse or deviant sexual behaviour with a minor as a parent, guardian or other child supervisor or with a person in a hospital, prison or other institution as a supervisor, guard or a person with disciplinary authority.
Broader Sexual offences (#PD4082) Sexual exploitation of children (#PD3267).
Related Sexual abuse by women (#PJ1767) Sexual abuse of children (#PE3265).

♦ **PE7760 Disparity in remuneration between public and private sector employees**
Broader Denial of right of equal pay for equal work (#PD1977).
Aggravates Underpayment of government officials (#PD8422).
Aggravated by Prohibitive labour costs (#PF8763)
Excessive salaries of experts (#PF8317)
Excessive salaries of corporate executives (#PF7578).
Reduced by Excessive salaries of international civil servants (#PE6388).

♦ **PE7765 Denial of social rights to indigenous peoples**
Broader Discrimination against indigenous populations (#PC0352).

♦ **PE7766 Botulism**
Nature Botulism is an acute toxic and infectious disease of the food toxaemia group, caused by anaerobic bacteria and their toxins. The causative agent of botulism is a spore-forming rod; man is infected when the botulism rods and their toxins enter the digestive tract with food. Infection usually occurs through canned foods that have not been subjected to proper sterilization. The average length of duration of botulism poisoning is 18-24 hours, during which time the patient has severe diarrhoea, vomiting, disturbed vision, intestinal cramps, loss of consciousness, and dehydration. Mortality rates are high, especially among the young and old.
Refs Smith, Louis *Botulisme* (1988).
Broader Biological contamination of food (#PD2594)
Food poisoning through negligence (#PE0561).

♦ **PE7768 Harassment in playgrounds**
Broader Physical intimidation (#PC2934) Physical intimidation by children (#PE2876)
Inadequate provision of public safety (#PF2874).
Narrower Teasing (#PE4187).

♦ **PE7769 Clostridial infections**
Broader Infectious diseases in animals (#PD2732).
Narrower Tetanus (#PE2530) Blackleg (#PE2737) Botulism in animals (#PG2994)
Malignant oedema in animals (#PG9283) Big head disease in animals (#PG6569)
Infectious necrotic hepatitis (#PG0371) Type D enterotoxaemia of sheep (#PG9390)
Bacillary haemoglobinuria in animals (#PG2359).

♦ **PE7771 Myopia**
Shortsightedness
Nature Myopia is an error of refraction of the eyes causing people suffering from it to see distant objects poorly. It is commonly called nearsightedness. People suffering from this condition must hold objects close to their eyes, and the eyes must be made to converge by increased activity of the internal rectus muscles. If these muscles are inadequate, strabismus may develop. Myopia often occurs in students doing copious amounts of reading and writing and also among workers in various fields which require close attention to detail.
Refs Curtin, Brian J *The Myopias* (1985); Fledelius, H C; Alsbirk, P H and Goldschmidt, E (Eds) *International Conference on Myopia (3, 1980, Copenhagen)* (1981).
Broader Visual deficiencies (#PD8179).

♦ **PE7774 Bourgeoisie**
Nature The bourgeoisie is the ruling class of capitalist society which possesses property in the means of production and which exists by exploiting wage labour. The source of income of the bourgeoisie is surplus value, which is created by unpaid labour and is appropriated by the capitalists.
Background The term denotes someone belonging to the urban middle-class, though in Marxist revolutionary parlance it had the negative connotation of the capital-owning class which the revolutionary proletariat had to overthrow and replace in economic and political power. In contemporary Marxist parlance it is used for the capitalist class as a whole.
Broader Capitalism (#PC0564) Accumulation of capital (#PC5225).
Aggravates Vulnerability of middle-class (#PC1002).
Reduced by Communism (#PC0369).

♦ **PE7775 Chicken pox**
Nature Chicken pox is an acute infectious disease accompanied by fever and a characteristic vesicular rash on the skin. Children up to ten years of age are the main victims, but after infection, immunity is guaranteed for life. The causative agent is a filterable virus, which is transmitted from an infected person to a healthy one mainly by means of air droplets (fine sprays of saliva during coughing or sneezing). Symptoms include a rise in temperature and a skin rash which becomes blisters which erupt, and resultant considerable and uncomfortable irritation.
Broader Infectious and parasitic diseases (#PD0982).

♦ **PE7778 Super-power chauvinism**
Super-power exclusiveness — Super-power arrogance
Nature The policies of super power chauvinism are aimed at enslaving other ('lesser power') nations; discriminating against them in economic, political, and cultural fields; or denying their independence. Super power chauvinism, like chauvinism in general, tends to kindle hatred and enmity between peoples, and persecutes and harasses persons of different nationalities.
Incidence Modern day examples of super power chauvinism are American intervention in Vietnam and the Soviet Union's invasion of Afghanistan.
Background Super power chauvinism arose in the era colonial empires were being formed; the Nazi era during the second World War is an example of the attempt to push super power chauvinism onto a whole continent.
Broader Domination of the world by territorially organized sovereign states (#PD0055).
Aggravates Superpower rivalry (#PD9655) Narrow range of practical skills (#PF2477).

♦ **PE7780 Harassment of travellers by immigration officials**
Harassment of travellers through consular procedures — Administrative impediments in obtaining travel visas — Victimization by immigration control procedures
Nature Discriminatory procedures may be applied to certain categories of travellers. These may include deliberate delays in the process of obtaining a visa, requiring repeated visits and lengthy waiting in queues. Whether or not a visa has been obtained, travellers may be subject to harassment by immigration officials at the point of entry, without any ability to appeal against such treatment for fear of official reprisal.
Counter-claim Governments are obliged to use tough procedures in order to distinguish legitimate visitors from those endeavouring to enter or settle in a country illegally.
Broader Harassment (#PC8558) Travel risks (#PD7716)
Restrictions on freedom of movement between countries (#PC0935).

♦ **PE7783 Crops pests and diseases**
Nature Among vertebrate animals, many crop pests are mammals, especially in the order of rodents. Among invertebrates, certain species of gastropods and a large number of roundworms from the class of nematodes harm crops. The most varied and numerous species of crop pests are arthropods-insects, arachnids (mites), and some species of millipedes and crustaceans (wood lice). Diseases vary from viral, bacterial, and nutritional to fungal, environmental and non-specific.
Incidence The FAO has estimated that annual worldwide losses done by plant pests and diseases amount to approximately 20-25 percent of the potential worldwide yield of food crops.
Refs CAB International Mycological Institute *Diseases of Tropical Root Crops*; FAO *Economic Guidelines for Crop Pest Control* (1984).
Broader Hazards to plants (#PD5706).
Narrower Infestation of seeds (#PE6271) Maize pests and diseases (#PE3589)
Pests and diseases of rice (#PE2221) Pests and diseases of wheat (#PE2222)
Pests and diseases of vines (#PE2985) Pests and diseases of potato (#PE2219)
Pests and diseases of cotton (#PE2220) Pests and diseases of groundnut (#PJ1181)
Pests and diseases of sugar cane (#PE2217) Pests and diseases of sugar-beet (#PE2975)
Pests and diseases of grain sorghum (#PE3590).
Related Plant diseases (#PC0555) Pests of plants (#PC1627)
Pests and diseases of trees (#PD3585).
Aggravated by Bad weather (#PC0293) Plant disease vectors (#PD3596).

♦ **PE7784 Extra-economic constraints**
Nature Extra-economic constraints are a form of coercion to work based on relations of direct rule and submission, of personal dependence of the working masses on exploiters; such constraints are characteristic of slaveholding and feudal society.
Claim Capitalism as a system of wage slavery is based on economic constraint. It presupposes the worker's personal liberty but at the same time deprives him in one way or another of the means of production. Therefore, in order not to die of hunger, the working people are forced to sell the capitalist their labour power and experience the oppression of exploitation.
Aggravated by Slavery (#PC0146) Feudalism (#PF2136)
Capitalism (#PC0564).

♦ **PE7786 Dog fighting**
Incidence Dog fighting is still held in a few places in Japan, the most famous being Kochi and Akita prefectures, but gambling on dogfights is now illegal.
Broader Animal fighting sports (#PE4893).

♦ **PE7787 Relapsing fever**
Relapsing fevers
Nature Relapsing fever is an acute infectious disease caused by spirochetes transmitted through bloodsucking insects (lice, ticks) and characterized by periodic paroxysms of fever fever. Man is the only sufferer of epidemic (louse-borne) relapsing fever; both man and animals suffer from endemic (tick-borne) relapsing fever.
Incidence Epidemic relapsing fever occurs mainly in countries with poor cultural and economic development; endemic relapsing fever is found an all the continents.
Background The causative agent of epidemic relapsing fever was discovered in 1868 by the German scientist O Obermeier. It was once widespread during wars and famines, being aided by the migration of people and their infestation with lice.
Refs Felsenfeld, Oscar *Borrelia, Borreliosis and Relapsing Fever* (1971).
Broader Arthropod-borne diseases (#PE7796).
Aggravates Fever (#PD2255).
Aggravated by Ticks as pests (#PE1766) Natural disasters (#PB1151)
Lice as insect pests (#PE1439) Rodent vectors of disease (#PE3629).

♦ **PE7788 Vibration sickness**
Nature Vibration sickness is an occupational disease caused by the prolonged effect on the body of a local or general vibration. It develops gradually and for a long time does not affect the ability to work. Fatigue, cold, and the strain caused by incorrect working posture lessen the body's tolerance to vibration, and vibration sickness eventually manifests itself by pains and weakness in the extremities and by increased sensitivity to chilling. Functional disorders of the nervous system result (rapid fatigue, irritability, headaches, and sometime dizziness) and as the disease progresses, disruptions of cardiovascular activity and disturbance of metabolic processes may result.
Incidence Vibration sickness occurs in workers subjected to prolonged local vibration (work with pneumatic or electric hammers, rammers vibropackers) on one part of the body, or general vibration on the entire body (from vibropacking concrete or from transport).
Background This illness was first described in 1911.
Broader Occupational diseases (#PD0215) Vibrations as a health hazard (#PE1145).

♦ **PE7793 Lack of protection for victims of intimidation**
Lack of protection following death threats
Broader Lack of protection for the vulnerable (#PB4353).
Aggravates Intimidation (#PB1992).
Aggravated by Fear of police (#PF8378) Breakdown of police protection (#PF8652)
Police indifference to community (#PF8125).

♦ **PE7794 Inundation of wildlife habitats through dams**
Loss of wildlife due to artificial flooding
Nature Since majority of animals cling to their territories or home grounds, the arrival of the flood waters kills most of them. There might not be any place to go for those who escape or are saved.

Once the natural habitat have been destroyed, the future breeding prospects of saved animals are slim.
Broader Destruction of wildlife habitats (#PC0480)
Environmentally harmful dam construction (#PD9515).

♦ **PE7796 Arthropod-borne diseases**
Broader Zoonoses (#PD1770) Infectious and parasitic diseases (#PD0982).
Narrower Malaria (#PE0616) Typhus fever (#PG1685) Rickettsiosis (#PE5530) Leishmaniasis (#PE2281) Epidemic typhus (#PG3894) Trypanosomiasis (#PE5725) Chagas' disease (#PE0653) Relapsing fever (#PE7787) Non-epidemic typhus (#PE3895) African trypanosomiasis (#PE1778).
Aggravates Ephemeral fever (#PG3738).

♦ **PE7803 Money laundering**
Legitimizing illegal fund transfers
Nature Money laundering refers to the conversion of money (usually in cash and small bills) earned through illicit activities (such as the drug trade) into either 'clean' currency or financial instruments that betray no trace of the owner's criminal activities. Insider dealers, international fraudsters and drug dealers need to launder their money. This is often done by transferring it through several financial centres with strong bank-secrecy laws, and then back to the country of origin where it can be used to purchase property or easily transportable assets.
Incidence In 1985, an estimate in the USA suggested that about US $60 billion a year is 'laundered' by criminals in the USA itself and between US $5–15 billion outside the USA. In 1990 it was estimated that drug trafficking generated some $85 billion a year worldwide that needed to be laundered and invested. In 1985, the Bank of Boston was fined US $500,000 for failing to report US $1.2 billion in currency dealings with foreign banks; the money belonging to the alleged head of Boston's Mafia family.
Claim Breaking up the large money-laundering rings is one of the most effective means to slow drug trafficking. If a big drug trafficker cannot deposit his money in a bank, much of the safety, and thus incentive, for having large amounts of money is lost.
Broader Banking law violations (#PE1208) Unlawful business transactions (#PC4645).
Aggravated by Illegally-obtained funds (#PD5433)
Unethical practices of employers (#PD2879).

♦ **PE7808 Congenital anomalies of the animal musculoskeletal system**
Broader Animal abnormalities (#PD4031)
Diseases of musculoskeletal system in animals (#PD7424).
Narrower Limber leg (#PG7813) Brown atrophy (#PG4157) Splaylegs in piglets (#PG4569) Syndactyly in animals (#PG9768) Femoral nerve paralysis (#PG9562) Double muscling in cattle (#PG4099) Contracted flexor tendons (#PG9259) Dyschondroplasia in animals (#PG9734) Defects of the back of foals (#PG9475) Angular limb deformities of foals (#PG0222) Disorders of skeletal system of poultry (#PG9644)
Myopathy associated with congenital articular rigidity (#PG9452).

♦ **PE7826 Lameness in horses**
Broader Diseases of musculoskeletal system in animals (#PD7424).
Narrower Curb (#PJ6043) Canker (#PG4066) Quittor (#PJ0670) Splints (#PJ5808) Sweeney (#PJ3883) Gonitis (#PG5394) Coxitis (#PG4053) Osselets (#PG9804) Ringbone (#PG3755) Carpitis (#PG9775) Physitis (#PG9321) Laminitis (#PG5410) Sandcrack (#PJ2766) Scratches (#PG9314) Seedy toe (#PG2962) Windgalls (#PG9456) Bog spavin (#PG2674) Stringhalt (#PJ5945) Bone spavin (#PG9633) Bruised sole (#PG6736) Sesamoiditis (#PG7249) Bucked shins (#PG7067) Sheared heels (#PG9194) Carpal hygroma (#PG6729) Pelvic fracture (#PG9416) Osteochondrosis (#PG6134) Contracted heels (#PG4159) Navicular disease (#PG9305) Pyramidal disease (#PG7181) Patellar luxation (#PG6295) Keratoma in horses (#PG9781) Ossifying myopathy (#PG9315) Flexion deformities (#PG7804) Sacroiliac desmitis (#PG0781) Subchondral bone cyst (#PG3764) Trochanteric bursitis (#PG9388) Fracture of pedal bone (#PG6040) Villonodular synovitis (#PG9806) Fracture of first phalanx (#PG3969) Puncture wounds of the foot (#PG9348) Fracture of the carpal bones (#PG5867) Fractures of the back of horse (#PG5208) Arthritis of the shoulder joint (#PG9520)
Lameness caused by bone cyst in pedal bone (#PG1636)
Overriding of the dorsal spinous processes (#PG7781).

♦ **PE7832 Increase in pests and diseases through perennial irrigation**
Increase in insect pests through modification of micro-climate — Diseases introduced through water projects
Nature To introduce perennial irrigation into an arid area means changing its micro-climate. The new moist ecosystem will attract all sorts of micro-organisms, insects and other forms of animal-life which are adapted to the new conditions. The species which previously lived in the arid area cannot adapt to the new moisture levels. Many of the new forms of live are undesirable, especially their numbers, and because they live off the crops which are being cultivated. They also lead to an upsurge in waterborne and other diseases, such as malaria, schistosomiasis, filariasis and river blindness. Migrant workers and relocated population in newly irrigated areas do not have adequate sanitation or housing, and so the local water supply becomes contaminated with human faeces. Under such conditions, epidemics, transmission of roundworm, hookworm, dysentry, hepatitis, cholera etc. will increase.
Refs Committee on Irrigation-Induced Water Quality Problems (Eds) *Irrigation-Induced Water Quality Problems* (1989).
Broader Environmentally harmful dam construction (#PD9515)
Ecosystem modifications due to creation of dams and lakes (#PD0767).
Aggravates Pests (#PC0728) Malaria (#PE0616) Filariasis (#PE2391) Onchocerciasis (#PE2388) Schistosomiasis (#PE0921)
Human disease and disability (#PB1044).

♦ **PE7837 Water salinization**
Water pollution by salinity effects — Salt water intrusion — Adjacency of saline water — Salt water flooding
Nature Rivers carry dissolved salts from exposed saline shale formations, springs and sheeps in the upper basin into the sea. Farmers withdraw flow for irrigation and leach salts out of their soil back into the stream. That used water will be used again and again further down the river, each use adding to the natural salinization, until the water becomes too salty to use.
Refs Atkinson, Sam F et al *Saltwater Intrusion* (1986).
Broader Floods (#PD0452) Shortage of fresh-water sources (#PC4815).
Narrower Flood waves (#PG2385) Tidal floods (#PE5006).
Related Impurities in waste water (#PD0482)
Negative consequences of shifting ecology on coastal communities (#PE2305).
Aggravates Underutilization of natural resources (#PF1459)
Inadequate water system infrastructure (#PD8517).

♦ **PE7838 Snow avalanches**
Nature Snow avalanches are masses of snow moving rapidly down a mountain slope of cliff. Snow slides range from small movements on established avalanche tracks to large, sporadic, very rapid movements capable of taking a heavy toll of life. Avalanches are infrequent on slopes of less than 25 degrees and especially numerous on those exceeding 35 degrees, and most commonly start on convex slopes. Both old and new snow may avalanche, but serious avalanches are always possible when 12 inches or more of new snow is present. Movements may be set off by temperature, vibration, shearing, or other slope disturbance. Dry snow avalanches usually occur during, or within several days after, snowfall. They may affect whole slopes, even if wooded, and may exceed 100 mph. Wet snow avalanches are formed during thaws or rainy weather. Their movement is less rapid but may be destructive. Slab avalanches of wind-packed snow are broader and deeper and move rapidly. This type of avalanche is an extreme hazard to life and property.
Broader Avalanches (#PD1146).
Related Rock avalanches (#PG0476).

♦ **PE7848 Feline immuno-deficiency virus** (FIV)
Incidence This recently discovered equivalent to HIV-virus is reported to be widespread amongst the population of domestic cats (in UK 6 per cent of healthy cats and 19 per cent of sick cats carry it; 21 per cent in the case of homes with more than one cat). The incidence amongst feral and wild cats is not known.
Broader Immuno-deficiency virus (#PD4747).

♦ **PE7854 Iron deficiency anaemias**
Refs Pollitt, Ernesto and Leibel, Rudolph *Iron Deficiency* (1982).
Broader Anaemia (#PD7758).

♦ **PE7855 Inundation of forests through dams**
Loss of forests by artificial flooding
Nature Forests ropresent more than the market value of timber: dams destroy also such benefits as soil preservation, water replenishment, climatic stabilization, air purification and wildlife shelter.
Broader Environmentally harmful dam construction (#PD9515).
Aggravates Deforestation (#PC1366).

♦ **PE7864 Symptoms referable to respiratory system**
Broader Ill defined health conditions (#PC9067) Diseases of the respiratory system (#PD7924).
Narrower Cough (#PE6825) Snoring (#PE4415) Dyspnoea (#PG4278) Epistaxis (#PG6688).

♦ **PE7867 War psychosis**
Claim After World War II the war psychosis governed all politics and it made necessary massive military spending, military industries and research institutes etc.
Broader Psychoses (#PD1722).
Aggravates Incitement to war (#PD4714).
Aggravated by War (#PB0593).

♦ **PE7877 Inadequate health care in urban slums**
Nature The transmission of communicable diseases is highest in urban slums. Childhood infections are not seasonal but all year round phenomena. In spite of the availability of of preventive measures, immunization coverage remains low in urban areas, especially in developing countries. Easier access to services, fewer logistical problems for supplies and exposure to communication media have not made ae difference to the well-being of children in slums.
Broader Inadequate health services (#PD4790)
Inadequate health care in least developed countries (#PE9242).
Aggravated by Urban slums (#PD3139) Urban malnutrition (#PD7473).

♦ **PE7879 Health hazards of electromagnetic fields generated by electrical appliances**
Nature Studies suggest that 60 hertz current, common to most electrical appliances, emits radiation that can cause biochemical changes, interfering with function of genes and stimulating activity in biochemicals linked to the growth of cancer. Such low-level radiation is associated with electrical mains, home wiring, computers, and electrical appliances.
Counter-claim Although there are legitimate reasons for concern, there is no basis for asserting that there is a significant risk. It is premature to link such exposure to the growth of cancer.
Broader Environmental hazards of electrical machinery, apparatus and appliances (#PE8026)
Environmental hazards of extremely low frequency electromagnetic radiation (#PE7560).
Related Health hazards of computer visual display units (#PE5083).
Aggravated by Electromagnetic pollution (#PD4172).

♦ **PE7892 Inadequate investment of transnational corporations in least developed countries**
Incidence Foreign direct investment inflows from transnational corporations have ranged from a negative balance to a few million dollars for most least developed countries in the first half of 1980s.
Broader Least developed countries (#PD8201)
Domination of developing countries by transnational corporations (#PE0163).
Aggravated by Inadequate negotiation of entrance terms for transnational corporations in developing countries (#PE0853).
Reduces Exploitative financial policies of transnational corporations in developing countries (#PE6952).

♦ **PE7894 Aplastic anaemia**
Refs Dessypris, Emmanuel N *Pure Red Cell Aplasia* (1988); Hibino, Susumu; Takaku, Fumimaro and Shahidi, N T (Eds) *Aplastic Anemia* (1978).
Broader Anaemia (#PD7758).
Narrower Pancytopenia (#PG5541).
Related Aplastic anaemia in animals (#PG7829).

♦ **PE7898 Fast and irregular pace of technological advance**
Nature Technological change has been extremely rapid in certain sectors and some of these, such as communications, have a broad impact on economic activity.
Broader Adverse consequences of scientific and technological progress (#PF3931).
Aggravates Lack of job satisfaction (#PF0171).

♦ **PE7899 Restrictive practices in the beverages and tobacco trade**
Broader Restrictive trade practices (#PC0073).
Narrower Protectionism in the tobacco industry (#PU0224).

♦ **PE7900 Corruption of documents**
Nature The altering of the original sense of a document may occur by means of direct alteration of the written text, by arbitrary alterations in reading aloud the text which is itself correct, by omitting parts of it, or by interpolations or a wrong exposition of the true sense.
Broader Corruption (#PA1986) Unethical intellectual practices (#PC2915).

Related Misuse of statistics (#PF4564)   Corruption of meaning (#PB2619).
Aggravates Incorrect information (#PB3095).
Aggravated by Unethical documentation practices (#PD2886).

♦ **PE7901 Jurisdictional conflict and antagonism between intergovernmental organizations**
Nature Departments, programmes and specialized agencies all compete with each other. This is compounded by the fact that each has its own constituency within member–nation governments. For example, agriculture ministries relate to FAO whilst educational ministries relate to UNESCO.
Broader Jurisdictional conflict and antagonism between international organizations (#PD0138).
Aggravates Inefficient location of facilities of international organizations (#PE3538)
Inadequate coordination of the intergovernmental system of organizations (#PE0730).

♦ **PE7902 Credit discrimination**
Inequitable credit extensions — Geographically denied credit — Redlining denies credit
Nature The banks of capitalist countries may establish more rigid credit conditions for some categories of borrowers than for others. Under current conditions, credit discrimination is one of the forms of financial oppression and exploitation of much of the population, including small entrepreneurs, farmers and the purchasers of consumer goods, who use various types of consumer credit. Credit discrimination is used by banks and specialized credit institutions to obtain additional profit and to redistribute scarce monetary resources when the money market is tight.
Counter–claim It is the responsibility of banks to meet the needs of their owners or shareholders, the market place and their customers. Without making choices about what groups of people can best meet credit obligations there is no way this responsibility can be met. The alternative is extensive and indepth files on individual's credit, life style and personal history.
Broader Lack of local commercial services (#PF2009)
Discriminatory business practices (#PD8913)
Paralyzing complexity of urban structures (#PF1776).

♦ **PE7903 Boundary disputes between neighbours**
Confusing property boundaries — Unresolved community boundaries — Unsatisfactory delimitation of boundaries
Incidence In many communities there are long–term disputes between neighbouring villages, between owners of neighbouring farms and fields, or between owners of neighbouring buildings concerning the exact location of the boundaries.
Aggravates Neighbourhood disputes (#PE5504).
Aggravated by Maldistribution of water (#PD8056)
Unnatural boundaries between states (#PF0090)
Personal isolation in communities of industrialized countries (#PD2495).

♦ **PE7904 Inadequate cooking stoves**
Claim Poorly designed stoves and the use of open fires used in developing countries to cook on are a hugh source of energy wastage. It is estimated that 5 million adequately designed stoves could be built for less than the cost of one average dam in africa and would save 4 to 5 times the amount of energy the dam would produce.
Broader Unsustainable development of energy use (#PC7517)
Lack of conservation of energy by the private sector (#PE8599).
Aggravates Energy crisis (#PC6329)   Shortage of firewood (#PD4769)
Long–term shortage of coal (#PE1054)   Energy deficient developing countries (#PE0379).
Aggravated by Inadequate maintenance equipment (#PD6520).

♦ **PE7905 Cretinism**
Nature Cretinism is a disease marked by arrested physical and mental development and thyroid dysfunction, caused by a deficiency of iodine. The average height for male cretins is 146.7 cm and 140 cm for females. They have physical deformities as well as impaired metabolism and low body temperature. Poor hearing is marked and permanent, and mental retardation is often at the level of idiocy. Cretins are slow moving, sleepy, and halting of speech; the acquisition of knowledge and skills is a difficult and time-consuming process.
Refs Merke, F (Ed) *History and Iconography of Endemic Goitre and Cretinism* (1984).
Broader Tolerated atrocities (#PC4710)   Thyroid gland disorders (#PE0652).
Aggravated by Iodine deficiency disorders (#PD2726).

♦ **PE7906 Denial of the right to national sovereignty**
Loss of national sovereignty — Stateless nations
Nature World is populated by more than 1,600 stateless nations, most of which are in one way or another engaged in national movements.
Claim Interdependence between developed countries and between developed and developing countries is eroding the principles of natural sovereignty and nonintervention.
Broader Denial of state's rights (#PD4814)
Violation of sovereignty by trans–border broadcasting (#PE0261)
Erosion of national sovereignty by transnational enterprises (#PE1539).
Aggravates Erosion of sovereignty (#PE5015).
Aggravated by Threatened and vulnerable minorities (#PC3295)
Misuse of satellite surveillance by governments (#PF3701).

♦ **PE7907 Instability of trade in electric energy**
Broader Instability of trade in mineral fuels, lubricants and related materials (#PD0877).

♦ **PE7908 Inefficient labour use in socialist countries**
Broader Inefficiency of state–controlled enterprises (#PD5642).

♦ **PE7912 Combat trauma**
War neurosis — Shell shock — Battle fatigue
Nature War neurosis occurs in connection with life threatening experiences during wartime. Patients suffer from alienation of the self, social withdrawal and difficulties in interpersonal relationships, irritability, recurrent dreams and flashbacks repeating the details of the experience, and severe anxiety.
Refs Dasberg, H, et al *Society and Trauma of War* (1987).
Broader Trauma (#PD4571)   Post-traumatic stress disorder (#PE0351).
Aggravated by War (#PB0593).

♦ **PE7914 Instability of trade in pulp and waste paper**
Broader Instability of trade in inedible crude non-fuel materials (#PD0280).

♦ **PE7915 Hereditary disorders of the central nervous system**
Familial disease of the nervous system — Hereditary neuromuscular disorders — Hereditary diseases of the striato–pallidal system — Hereditary ataxia
Broader Genetic defects and diseases (#PD2389)
Diseases of the nervous system (#PC8756).
Narrower Chorea (#PG3096)   Friedreichs ataxia (#PE8605)
Creutzfeldt–Jakob disease (#PJ1045).

♦ **PE7917 Corruption of the good in human nature**
Broader Corruption (#PA1986).
Aggravates Human suffering (#PB5955)   Behavioural deterioration (#PB6321).
Aggravated by White lies (#PF7631)
Obsessive compulsive personality disorder (#PE7632).

♦ **PE7919 Undefined government role for urban services**
Broader Underprovision of basic urban services (#PF2583).

♦ **PE7920 Wood shaped or simply worked shortage**
Broader Long–term shortage of wood, lumber and cork (#PE1372).

♦ **PE7922 Benign neoplasm of uterus**
Broader Benign tumours (#PD8347).

♦ **PE7926 Shortage of tobacco and tobacco manufactures**
Broader Long–term shortage of beverages and tobacco (#PE1253).

♦ **PE7930 Excessive portrayal of sex in the media**
Glorification of demeaning sex in the media — Trivialization of sexuality
Broader Pornography (#PD0132)   Facile social concepts (#PF5242)
Distorted media presentations (#PD6081).
Aggravates Exploitative films (#PE6328)   Paternal negligence (#PD7297)
Trivialization of love (#PF0959)   Public displays of sexuality (#PE4882)
Adolescent sexual intercourse (#PD7439)   Pre-marital sexual intercourse (#PD5107)
Deterioration of media standards (#PD5377)
Biased portrayal of women in mass media (#PE7638).

♦ **PE7931 Inhibition of personality development in exiled children**
Broader Social maladjustment of children of migrants (#PE4258).
Aggravated by Expulsion (#PC5313)   Psychological inhibition (#PF6339).

♦ **PE7933 Endangered species of tamarins and marmosets**
Incidence 35 species; 2 endangered, 1 possibly endangered, 1 rare.
Broader Endangered species of non-human primates (#PE1570).
Aggravates Inappropriate agricultural subsidies for chemicalized farming (#PE1785).

♦ **PE7934 Urinary calculus**
Broader Diseases of gallbladder (#PE9829).

♦ **PE7935 Illegal marriage**
Broader Offences against public order (#PD7520).
Narrower Bigamy (#PF3286)   Group marriage (#PF3288).
Related Forced marriage (#PD1915)   Consanguineous marriage (#PC2379).

♦ **PE7937 Government leaders associated with sex scandals**
Government rulers associated with sex scandals
Broader Sex scandal (#PD9398)   Misconduct in public office (#PD8227).
Aggravates Loss of confidence in government leaders (#PF1097).
Aggravated by Corruption of government leaders (#PC7587)
Mediocrity of government leaders (#PF3962).

♦ **PE7940 Shortage of sanitary plumbing, heating and lighting fixtures and fittings**
Broader Long–term shortage of miscellaneous manufactured articles (#PE0613).
Aggravates Inappropriate sanitation systems (#PD0876).

♦ **PE7941 Animal deaths**
Animal mortality — Economic loss through animal mortality
Broader Death of living creatures (#PF7043).
Narrower Animal road deaths (#PE1690).
Related Human death (#PA0072)   Animal extinction (#PD7989)
Denial to animals of the right to life (#PF8243)
Denial to animals of the right to a natural death (#PE8339).
Aggravated by Animal diseases (#PC0952)   Torture of animals (#PC3532)
Inadequate feeding of animals (#PC2765)   Animal stress in factory farming (#PD2760).

♦ **PE7942 Unequal distribution of old age pensions between men and women**
Broader Inadequate income in old age (#PC1966)
Discrimination against women in social services (#PD3691).
Related Unequal pension rights (#PJ2030).

♦ **PE7943 Functional changes due to technological advance**
Narrower Inability of educational systems to keep pace with technological advancement (#PF7806).
Aggravates Unemployment (#PB0750).

♦ **PE7944 Lack of interim support for families**
Broader Inadequate support services (#PF6492)
Precarious basis for family economics (#PF1382).

♦ **PE7945 Restriction on press coverage of legal affairs**
Broader Lack of freedom of the press (#PE8951).
Related Restrictions on direct news coverage of parliamentary affairs (#PF3072).
Aggravates Lack of representation (#PF3468).

♦ **PE7946 East Coast fever**
Nature East Coast Fever is caused by a single-celled parasite, carried by ticks, which gets into cells in cattle's blood, invades the lymphocytes, and causes fever and death. So far the principal strategy against this most devastating disease has been to regularly dip cattle into chemical baths, but most farmers have neither the money nor the facilities; indeed, the ticks are becoming immune to chemicals and new compounds must constantly be sought.
Incidence East Coast Fever limits the spread of livestock overlarge parts of East and Central Africa. It is endemic well below 2500 metres and kills 0.5 million cattle each year.
Refs FAO *Ticks and Tick-Borne Diseases* (1983); FAO *East Coast Fever and Related Tick-Borne Diseases* (1980).
Broader Theileriasis (#PE3996).
Aggravates Fever (#PD2255).
Aggravated by Ticks as pests (#PE1766)   Protozoa as pests (#PE6741).

♦ **PE7947 State–monopoly capitalism**
Nature State–monopoly capitalism is a new, more developed from of monopoly capitalism,

characterized by the joining of the forces of capitalist monopolies with the power of the state to preserve and strengthen the capitalist system, enrich monopolies, suppress workers and national liberation movements, and unleash aggressive wars.
Broader Monopolies (#PC0521) Capitalism (#PC0564).
Related State monopoly (#PJ4242).

♦ PE7953 Endangered species of potto and loris
Incidence 11 species; none endangered.
Broader Endangered species of non-human primates (#PE1570).

♦ PE7954 Inadequate food supplies in least developed countries
Food insecurity in least developed countries
Broader Food insecurity (#PB2846).
Aggravated by Inadequate agriculture in least developed countries (#PE8082).

♦ PE7956 Silver and platinum group trade instability
Broader Instability of trade in non-ferrous metals (#PE1406).

♦ PE7957 Internationalization of capitalist production
Nature The establishment of production relations between enterprises of different countries whereby the production of a country increasingly becomes a part of the world production process, arises in a period of developed industrial capitalism and becomes further developed under imperialism. The process has become particularly intense under the influence of the scientific and technical revolution.
Broader Capitalism (#PC0564) Capitalist economic imperialism (#PC3166).

♦ PE7958 Restrictive practices in trade in machinery and transport equipment
Broader Restrictive trade practices (#PC0073).

♦ PE7959 Insufficient repair services in developing countries
Broader Unexploited possibilities for local commerce (#PF2535).
Related Inadequate industrial services in developing countries (#PF4195).

♦ PE7960 Petroleum and coal products manufacture underdevelopment
Broader Underdevelopment of chemical and petrochemicals industry (#PE1483).

♦ PE7961 Shortage of professional, scientific and controlling instruments
Broader Long-term shortage of miscellaneous manufactured articles (#PE0613).

♦ PE7965 Restrictions on eligibility for government programmes
Broader Restricted scope of local employment (#PF2423).

♦ PE7968 Illiteracy as an inhibitor of business transactions
Broader Illiteracy (#PC0210).
Aggravated by Narrow range of practical skills (#PF2477).

♦ PE7970 Colitis
Nature Colitis, chronic inflammation of the colon, is one of the most common diseases of the gastrointestinal tract. It may be caused by stress, infection, poor food habits, or aberrant immune reactions. It may be either acute or chronic and has symptoms such as loss of appetite, nausea, and general weakness.
Refs Borriello, S P (Ed) *Antibiotic Associated Diarrhoea and Colitis* (1984).
Broader Diarrhoea (#PD5971) Intestinal diseases (#PD9045).

♦ PE7971 De facto racial requirement qualifications for public services
Broader Racial discrimination in public services (#PD3326).

♦ PE7973 Jurisdictional conflict and antagonism between international organizations at the country level
Broader Jurisdictional conflict and antagonism between international organizations (#PD0138).

♦ PE7974 Conjunctivitis
Ophthalmia
Nature Conjunctivitis, the most common eye disease, may be either exogenous (resulting from various microbes or from mechanical irritation) or endogenous (arising in the presence of systemic infections, inflammations of the nasopharynx or teeth, or diseases of the gastrointestinal tract or liver). Simple acute conjuctivitis is not a serious condition, but it may give rise to grave complications. There are various severe forms, for example Egyptian ophthalmia, that causes progressive loss of sight.
Broader Eye diseases and disorders (#PD8786).
Related Trachoma (#PE1946).
Aggravates Physical blindness (#PD0568).

♦ PE7975 Concentration of capitalist banks
Capitalist banking
Nature The main volume of capitalist bank operations is concentrated into an increasingly diminishing number of large and very large banks. This leads to the emergence of bank monopolies merging with industrial monopolies; and threatens to cause to a financial oligarchy.
Broader Concentration of investment power (#PC5323).

♦ PE7984 Unfamiliar procedures for using public health programmes
Broader Inadequate dissemination and use of available information (#PF1267).
Related Unknown procedures to access administrative services (#PS1227).
Aggravated by Complex public health regulations (#PU0332).

♦ PE7985 Technology gap between developed and developing countries
Incidence The gap in scientific and technological capabilities is widest in areas of direct relevance to the objectives of sustainable development: new energy sources, biotechnology, genetic engineering, new materials and substitutes, non-polluting and low-waste technologies. Before the disruption of development in the 1980s, levels of expenditure on research and development in the United States and certain Western European countries amounted to about $200 per capita, while the corresponding figure for most Latin American countries was less than $5 per capita and that of the poorer countries of Africa and Asia less than $1. Developing countries paid approximately $ 2 billion in 1980 by way of royalties and fees, mainly to industrialized countries.
Broader International economic injustice (#PC9112)
Maldistribution of science and technology (#PC8885)
Disparity between industrialized and developing countries (#PC8694).
Related Contradictions of capitalism in developing countries (#PF3126).
Aggravates Inadequate road and highway transport facilities in developing countries (#PD0543).
Aggravated by Lack of international cooperation (#PF0817)
Counterproductive capitalist investment financing (#PF3104).

♦ PE7987 Protectionism in free-zone international financial services
Nature Despite the multitude of government interventions in international service activities, a trend towards reduced control is apparent in certain service sectors. Relations between the international reinsurance markets and national or regional reinsurance institutions generally evolve on a normal commercial basis, and insurance institutions can be fairly well assured of obtaining coverage from the international reinsurance concerns on competitive conditions. In the area of direct insurance, special or very large risks are normally shifted to unregulated international insurance markets. In the banking sector, international markets, such as the inter-bank market and the market for participation in Euro-currency syndicates, became very competitive during the 1970s. In both the banking and insurance sectors, institutional arrangements have resulted in less intervention with respect to certain (eg offshore) locations and for certain types of customers.
The growth of financial centres hosting Euro-currency banks and captive insurance companies demonstrates the effective separation of relatively uncontrolled banking and insurance business in economic free zones from regulated operation in national economies. The host countries which have authorized off-shore banking and insurance activities are mainly small developing countries, in particular island states with minimal regulation and the added advantages of substantial tax benefits. The major reason for the rapid growth of Euro-currency banking in off-shore locations and other financial centres from virtually nothing in the 1960s to over $1,500 billion in 1980, has been the exemption of this type of banking from minimum reserve requirements.
Similarly, most captive insurance companies are incorporated in off-shore territories, with minimal legislation and less stringent reserve requirements. A captive is an insurance company established by an industrial or commercial concern (or a transnational group with a common activity) for the insurance of all or part of the risks of that concern or transnational group. Such an insurance institution enables a business concern to build up funds within a specialized vehicle and to design specific policies that meet the needs of its particular commercial requirements. Insurance can be provided to the parent transnational corporation and all its subsidiaries under a world-wide policy. The number of captives has significantly increased over the past years and their existence has substantially intensified competition in the international insurance sector. The more the captive companies provide the type of specialized cover required by the parent, the more they are taking insurance business and premiums from the commercial insurance industry, forcing the latter to become less costly and more innovative. On the other hand, negative influences that the captives can exert on the developing countries should not be overlooked. They can facilitate a large outflow of foreign exchange, without the local governments being aware. Generally, they are not liable to taxes in the countries where the risks are insured, and at times do not even take into account national regulations which protect the insured.
As a reaction to growing possibilities for unregulated operations abroad, the establishment of free economic zones is being considered by some countries as a means to induce the return of some of the business that has been lost to off-shore banking and insurance centres. In 1980, a free insurance zone was opened in New York (USA), followed in 1981 by a free banking zone. Insurance companies established in the zone can underwrite large or special risks without obtaining the permission of regulatory authorities, and resident banks are relatively free from reserve requirements. Reduced control in overseas locations has thus induced the relaxation of interventions in defined geographical areas of a national market. This chain of reduction of control actions is typical, on a small scale, of protectionist retaliations.
Broader Protectionism in the services industries (#PD7135).

♦ PE7995 Abuse in control of wild animal populations
Broader Misuse of wild animals (#PD8904).
Related Denial to animals of the right to freedom from mass killing (#PE9650)
Denial to animals of the right to conditions of life and liberty proper to their species (#PE6270).
Aggravates Killing of animals (#PD8486).

♦ PE7996 Medicinal and pharmaceutical products trade instability
Broader Instability of chemicals trade (#PD0619).

♦ PE8002 Lack of economic adaptation of indigenous systems
Lack of economic adaptation of indigenous society in non-socialist countries
Nature Despite the efforts of white bureaucrats, the native people in the Americas and elsewhere did not assimilate to industrial capitalism, but instead retreated into the cultural traditions which had sustained them for centuries.
Broader Traditionalism (#PF2676).

♦ PE8004 Incommunicado detention
Secret detention
Nature Persons may be detained in police, prisons or other facilities without being allowed to notify family or friends of their arrest and place of custody. Frequently other branches of government, like the judicial, are notified as required by law. Persons may be moved from one facility to another. Detainees may be denied any communication with the outside for any reason, such as need for medical attention. Interrogation can take place without interference from lawyers, family or friends of the detainee. The marks of torture also disappear. If the victim dies, the body can be disposed of without having to explain the death. Incommunicado detention re-enforces the victim's sense of isolation and powerlessness.
Those most likely to be tortured are those held incommunicado.
Incidence Incommunicado detention has been reported in the following countries: **Af** Egypt, Gabon, Mauritania, Morocco, Namibia, South Africa, Tunisia, Uganda, Zaire, Zambia, Zimbabwe. **Am** Argentina, Bolivia, Chile, El Salvador, Guatemala, Haiti, Paraguay, Peru, Suriname, Uruguay. **As** Bahrain, Bangladesh, Iran Islamic Rep, Israel, Korea Rep, Malaysia, Pakistan, Philippines, Saudi Arabia, Sri Lanka, Syrian AR, Taiwan (Rep of China). **Eu** Romania, Spain, Turkey.
Broader Wrongful detention (#PD6062) Social isolation as torture (#PD6810).
Related Torture through confinement (#PD4590).
Aggravates Forced disappearance of persons (#PD4259).

♦ PE8006 Insurance monopolies in capitalist countries
Nature Insurance monopolies in capitalist countries are powerful elements of the credit and finance system, second only, in scale of operations, to the banking monopolies. In many countries, the government regulates the use of the monopolies' reserves, the monopolies themselves are linked to leading banking and industrial companies through the inter-locking of corporate directorates. They affect political activity because their representatives hold administrative positions in government and the armed forces. Internationally, insurance monopolies have expanded through a network of divisions and branches outside their countries and through reinsurance operations.
Broader Monopolies (#PC0521).
Related Industrial gas monopolies (#PE1813).

♦ PE8011 Extraterritorial application of restrictive business practices legislation
Broader Restrictive trade practices (#PC0073).

♦ **PE8013 Increased demand in developed countries for meat and egg products**
Aggravates Factory farming (#PD1562).

♦ **PE8014 Threat from industrial combines to farming communities**
Aggravated by Factory farming (#PD1562).

♦ **PE8018 Food monopolies**
Nature The food industry in the capitalist world is controlled by a small number of monopolies; the major food monopolies controlling the production of particular commodities abroad, as well as in their home countries. By exporting goods and capital, they have cornered the world capitalist market for many food products. In former colonies and developing countries, the food monopolies control the production and supply of certain types of valuable raw materials.
Incidence The largest food monopolies are Unilever (Anglo–Dutch), Nestlé Alimentana SA (Swiss) and Coca–Cola (USA).
Claim By monopolizing the production of food commodities, maintaining high monopoly prices on goods, and paying extremely low prices for raw materials, the food monopolies exploit the people.
Broader Monopolies (#PC0521).
Narrower Domination by transnational corporations of the domestic name–brand food sector in developing countries (#PE1796).
Related Industrial gas monopolies (#PE1813)
Inadequate agriculture in least developed countries (#PE8082).
Aggravates Inadequate mechanisms for securing sufficient food supplies (#PF2857).
Aggravated by Unethical food practices (#PD1045).

♦ **PE8020 Uneven settling of building foundations**
Nature The condition of uneven settling of building foundations produces strains in the structure; and the additional forces that accompany these strains can disrupt structural stability or interfere with normal use of the building. Settling usually begins immediately after construction is started, continues while the building is being erected and the bearing load is increasing, and continues for quite a while after the building is finished.
Broader Structural failure (#PD1230).

♦ **PE8022 Toxaemias of pregnancy and the puerperium**
Broader Pregnancy disorders (#PD2289).
Narrower Acute yellow atrophy (#PG3232) Hyperemesis gravidarum (#PG4079).
Aggravates Maternal mortality (#PD2422).

♦ **PE8024 Denial to experimental animals of the right to freedom from suffering**
Nature Laboratory animals are treated without the concern to avoid pain and anxiety to which they are entitled as sentient beings.
Incidence Since the aims of experiments upon animals in psychological research is to observe their behaviour, the experiments must necessarily be performed on conscious animals. Many experiments involve aversive procedures causing stress or pain. Deprivation of food and or water and the application of electric shocks are used routinely in learning tasks. Other fields of investigation include the effects of malnutrition, maternal deprivation, social isolation, physical restraint, sleep deprivation and situations especially designed to produce stress and anxiety. Only a minority of projects are concerned with enriched environment, pleasant stimuli and rewards. An interesting paradox which arises is that either an animal is not like humankind, in which there is no reason for performing the experiment, or else it is like humankind, in which case one ought not to do an experiment which one would not be prepared to perform on a human being.
Broader Denial to animals of legal protection of their rights (#PE8643).
Narrower Denial to animals of the right to life (#PF8243)
Denial to animals of the right to dignity (#PE9573)
Denial to animals of the right to a natural death (#PE8339)
Denial to working animals of limitation of working hours (#PE6427)
Denial to animals of the right to freedom from mass killing (#PE9650)
Denial to working animals of restorative nourishment and rest (#PE4793)
Denial to animals of the right to the attention, care and protection of humankind (#PF5121)
Denial to animals of the right to conditions of life and liberty proper to their species (#PE6270).
Related Animal suffering (#PD8812) Abusive experimentation on humans (#PC6912)
Denial of rights of medical patients (#PD1662)
Uncertainties in animal experimentation (#PF4770)
Irresponsible research using human subjects (#PC0080)
Denial to food animals of the right to freedom from suffering (#PE3899).
Aggravates Cruel treatment of animals for research (#PD0260).

♦ **PE8026 Environmental hazards of electrical machinery, apparatus and appliances**
Broader Environmental hazards of non–ionizing radiation (#PE7651).
Narrower Health hazards of electromagnetic fields generated by electrical appliances (#PE7879).

♦ **PE8036 Domestic inconvenience for dwarfs, midgets and giants**
Broader Discrimination against giants (#PE5578)
Discrimination against dwarfs and midgets (#PE2635).

♦ **PE8045 Long–term shortage of natural and manufactured gas**
Broader Long–term shortage of mineral fuels and lubricants (#PE1712).

♦ **PE8048 Undeveloped business skills in urban areas**
Broader Incompetent management (#PC4867).
Aggravates Underprovision of basic urban services (#PF2583).
Aggravated by Limited development of functional abilities (#PF1332).

♦ **PE8049 Geographic barriers for least developed countries**
Broader Least developed countries (#PD8201).
Narrower Geographically disadvantaged countries (#PF9247).

♦ **PE8051 Moulds in plants**
Nature Mould is a furry or velvety growth that coats plants, animals and nonliving, organic objects. It is formed by certain ascomycetous, phycomycetous and imperfect fungi. Moulds are the cause of considerable economic losses, due to spoilage of such food products as flour, bread, canned goods, fruit juices, meat, dairy goods and beer. In particular they cause the destruction of stored fruits and vegetables; lower the quality of fodders; and also cause various diseases of plants, thus lowering yield.
Broader Ascomycetes (#PE4586) Deuteromycetes (#PE4346)
Fungal plant diseases (#PD2225).
Related Fungal diseases (#PD2728).

♦ **PE8058 Commodity fetishism**
Claim Commodity fetishism signifies the personification of things and economic categories. It is the objectification of production relations between people under the conditions of commodity production based on private ownership; and is seen when the element of social relations dominating people appears outwardly as domination by certain things. Capital as a production relation is personified in the capitalist, while hired labour is embodied in the worker. Fetishism permeates all the economic categories of capitalist society; an example is the exploitation of man by man, which is masked by the payment of wages. The acquisition of money and material goods is the highest manifestation of commodity fetishism.
Broader Fetishism (#PF8363) Materialism (#PF2655).

♦ **PE8059 Icing on aircraft**
Nature Icing causes deterioration of the aircraft's aerodynamic characteristics, flight performance, stability and controllability. It also increases frontal drag; can interfere with the operation of engines, navigational instruments and radio equipment; and can lead to a crash.
Broader Aircraft environmental hazards (#PD8328).

♦ **PE8061 Unfeasible housing alternatives in urban areas**
Uninvestigated building alternatives
Broader Housing shortage (#PD8778) Inadequate care of community space (#PF2346)
Limiting effect of individual survivalism (#PF2602).
Aggravates Urban slums (#PD3139)
Illegal occupation of unoccupied property (#PD0820).
Aggravated by Restrictive building codes in urban areas (#PE8443).

♦ **PE8062 Dehydration**
Nature Dehydration may be said to occur whenever there is loss of water by an organism to a point below the physiological norm. Animals die when they lose 20–25 percent of their body water, and disorders arise when the water loss reaches 10 percent. Dehydration may develop either as a result of excessive water loss (repeated vomiting, diarrhoea, excessive perspiration, extensive burns) or restriction of water intake.
Broader Nutritional diseases (#PD0287) Deficiency diseases in plants (#PD3653)
Ill defined health conditions (#PC9067)
Ill-defined health conditions in animals (#PD9366).
Aggravates Child mortality in developing countries (#PE5166).

♦ **PE8063 Prohibitive cost of transportation**
High transportation cost — Excessive freight costs — Expensive produce shipment — Expensive transport vehicles — High transportation fares — High transportation overhead
Broader Prohibitive cost of basic services (#PF6527)
Limited individual capital reserves (#PF2899)
Restrictions on effective means of transport (#PF2798).
Aggravates Fare evasion (#PE4693) Overloaded vehicles (#PE4127)
Ineffective rural transport (#PF2996) Unreliable freight transport (#PG1026)
Inadequate maintenance equipment (#PD6520) Prohibitive cost of farm machinery (#PF2457)
Inappropriate level of technological equipment (#PF2410)
Ineffective means for goods supply and distribution (#PF6495)
Restrictive pattern of business activities in small communities (#PD1415)
Overlooked potential for industrial development in rural communities (#PF2471)
Inadequate transportation facilities for rural communities in developing countries (#PE6526).
Aggravated by Limited accumulation of capital (#PF3630)
Insufficient transportation infrastructure (#PF1495).

♦ **PE8064 Loss of credibility in international institutions**
Low credibility of international organizations and programmes
Broader Credibility gap (#PB6314) Loss of institutional credibility (#PF1963).
Aggravates International insecurity (#PB0009)
Inadequate funding of international organizations and programmes (#PF0498)
Proliferation and duplication of international organizations and coordinating bodies (#PE1029)
Alienation of skilled and committed personnel from international organizations and programmes (#PE1553).
Aggravated by Breach of promise (#PF7150)
Politicization of intergovernmental organizational debate (#PD0457)
Ineffectiveness of international organizations and programmes (#PF1074).

♦ **PE8066 Inadequate integration of international information systems**
Broader Inadequate exchange of technical information concerning problems (#PF0209).
Aggravates Proliferation and duplication of international information systems (#PE0458).
Aggravated by Jurisdictional conflict and antagonism between international organizations (#PD0138).

♦ **PE8067 Ineffective self–regulation in the automobile manufacturing industry**
Broader Ineffective industry self–regulation (#PF5841).

♦ **PE8070 Iron and steel basic industries instability**
Broader Instability of basic metal industries (#PE2601).

♦ **PE8077 Endangered lifestyles of nomads and pastoralists**
Nature Sedentary farming and stock raising have preempted the favoured habitat of pastoralists, who have lost their grazing lands. Most governments try to control the nomads and pastoralists and settle them as farmers or stock rangers.
Aggravated by Desert advance (#PC2506) Inadequate grazing land (#PJ0404)
Disruption of arid zone ecosystems (#PD7096) Unnatural boundaries between states (#PF0090).

♦ **PE8080 Distortion in international politics from fuel shortages**
World tension in oil and petroleum crises
Aggravates Jihad (#PF5681).
Aggravated by Instability of trade in petroleum and petroleum products (#PD0909).

♦ **PE8082 Inadequate agriculture in least developed countries**
Nature Agricultural activities in the least developed countries have been characterized by low yields. The use of fertilizers is still very modest and production techniques are primitive. The whole area under irrigation represents less than one–tenth of the total arable land used, while the remainder is dependent on rainfall.
Refs Powelson, John P and Stock, Richard *The Peasant Betrayed* (1987).
Broader Least developed countries (#PD8201)
Inadequate mechanisms for securing sufficient food supplies (#PF2857).
Related Food monopolies (#PE8018).
Aggravates Inadequate food supplies in least developed countries (#PE7954).
Aggravated by Dumping of food in developing countries (#PE0607).

♦ **PE8084 Meagre youth facilities in the countryside**
Broader Inadequate circulation of local information (#PF6552).

♦ **PE8086 Dependence of developing countries on food imports**
Broader Dependence on external resources (#PC0065)
Vulnerability of developing countries (#PC6189).
Narrower Import–dependency in food staples in developing countries due to transnational corporations (#PE1806).

PE8086

Aggravates Excessive foreign public debt of developing countries (#PD2133)
Inadequate development of enterprises in developing countries (#PE8572).
Aggravated by Natural disasters (#PB1151)   Inadequacy of food aid (#PF3949)
Inadequate staple food supply in developing countries (#PD4101)
Fluctuations in food production in developing countries (#PE8188)
Economic disadvantages of excessive food production in developing countries (#PF4130).

♦ **PE8088 Inadequate documentation of works of art**
  Broader Uncatalogued documents (#PF4077)   Inadequate documentation (#PF6453).
  Aggravates Theft of works of art (#PE0323).

♦ **PE8091 Environmental hazards of wood and cork manufactures**
  Broader Environmental hazards from manufactured goods (#PE1344).

♦ **PE8092 Demand for unskilled labour in industrialized countries**
  Aggravates Illegal immigration (#PD1928).

♦ **PE8093 Emigration of trained personnel from industrialized countries**
Brain drain to more developed industrialized countries — Brain drain from socialist to capitalist countries
  Broader Disruptions due to migration (#PC0018).
  Related Emigration of trained personnel from developing to developed areas (#PD1291).

♦ **PE8095 Environmental hazards of publishing and printing industries**
  Broader Environmental hazards from paper and printing industries (#PE1425).

♦ **PE8097 Environmental hazards of electrical equipment industries**
  Broader Environmental hazards from machinery and equipment industries (#PE1859).

♦ **PE8098 Economic loss through reduced productivity of diseased animals**
Loss through disease game and zoo animals — Economic loss through reduced productivity of diseased animals
  Broader Economic loss (#PE9013)   Loss of animal productivity (#PD8469).
  Related Economic loss through slaughter of diseased animals (#PE8109).
  Aggravated by Mange (#PE2727)   Anthrax (#PE2736)
  Brucellosis (#PE0924)   Foot-and-mouth disease (#PE1589)
  Contagious bovine pleuropneumonia (#PE1775)
  Inadequate dissemination and use of available information (#PF1267).

♦ **PE8103 Lack of housing for teachers in rural communities**
Unconstructed teachers' houses
  Broader Inadequate housing (#PC0449)   Insufficient rural housing (#PF6511).
  Lack of essential local infrastructure (#PF2115).
  Related Inadequate working conditions of teachers (#PE7165).
  Aggravates Non-resident school personnel (#PG8764)
  Limited availability of education in rural areas (#PF3575).

♦ **PE8106 Instability of trade in undressed hides and non-fur skins**
  Broader Instability of trade in undressed hides, skins and fur skins (#PE1235).

♦ **PE8109 Economic loss through slaughter of diseased animals**
Slaughter of diseased animals — Economic loss through destruction of injured animals
  Broader Economic loss (#PE9013)   Loss of animal productivity (#PD8469).
  Related Economic loss through reduced productivity of diseased animals (#PE8098).
  Aggravated by Rabies (#PE1325)   Blackleg (#PE2737)
  Brucellosis (#PE0924)   Animal diseases (#PC0952)
  Animal injuries (#PC2753)   Killing of animals (#PD8486)
  Foot-and-mouth disease (#PE1589)
  Confusion of symptoms in animal diseases (#PF2780).
  Reduces Rinderpest (#PE2786)   Epizootic diseases (#PD2734)
  Water-borne animal diseases (#PE2787)
  Spread of animal diseases through factory farming (#PD2752)
  Difficulty in identifying carriers of animal diseases (#PF2775).

♦ **PE8110 Rescheduling of debts of developing countries at market-related interest rates**
  Nature The attempt to alleviate the debt burden of developing countries by rescheduling past debts at market-related interest rates only serves to increase the debt burden.
  Broader Deteriorating terms of financial loans to developing countries (#PE4603).
  Aggravates Excessive foreign public debt of developing countries (#PD2133)
  Burden of servicing foreign public debt by developing countries (#PD3051).

♦ **PE8116 Dependence of least developed countries on foreign aid**
  Refs United Nations *The Least Developed Countries and Action in their Favour by the International Community* (1983).
  Broader Compliance*complex (#PA5710)   Least developed countries (#PD8201)
  Vulnerability of developing countries (#PC6189).
  Aggravated by Capitalism (#PC0564).

♦ **PE8117 Diverticula of intestine**
  Broader Intestinal diseases (#PD9045).

♦ **PE8119 Fuel wood and charcoal trade instability**
  Broader Instability of trade in wood, lumber and cork (#PE2521).

♦ **PE8121 Ineffective self-regulation in the financial services sectors**
  Broader Ineffective industry self-regulation (#PF5841).

♦ **PE8131 Instability of trade in ores and concentrates of non-ferrous base metals**
  Broader Instability of trade in metalliferous ores and metal scrap (#PE0553).

♦ **PE8133 Inadequate knowledge of incubation periods for animal diseases**
  Broader Limited medical knowledge (#PD9160).
  Aggravates Inadequate animal quarantine (#PF2756).

♦ **PE8134 Protectionism in the alcoholic beverages industry**
  Broader Protectionism in international trade (#PC5842).

♦ **PE8135 Environmental hazards of animal oils and fats**
  Broader Environmental hazards from economic and industrial products (#PC0328).

♦ **PE8136 Vegetable oils and fats environmental hazard**
  Broader Environmental hazards from economic and industrial products (#PC0328).

♦ **PE8137 Refusal to let because of applicant's race**
  Broader Segregation in housing (#PD3442).

♦ **PE8138 Fungi as vectors of plant disease**
  Refs Ainsworth, G C; Hawksworth, D L and Sutton, B C *Ainsworth and Bisby's Dictionary of the Fungi* (1983); Butler, E J *Fungi and Disease in Plants* (1981).
  Broader Plant vectors of plant disease (#PD3599).

♦ **PE8139 Traditional values of the subordination of women in developing societies**
  Refs Kandiyoti, Denis *The Reproduction of Subordination* (1984).
  Broader Lack of participation by women in development (#PD3294).

♦ **PE8140 Insufficient knowledge of local production processes**
  Broader Narrow range of practical skills (#PF2477).

♦ **PE8143 Individualism as a restriction of effectiveness**
  Broader Individualism (#PF8393)
  Individualistic retaining of local tradition (#PF1705).

♦ **PE8144 Struggle for financial security in urban life style**
  Broader Deteriorating quality of life (#PF7142)
  Limiting effect of individual survivalism (#PF2602).

♦ **PE8148 Destruction of fisheries through dams**
  Nature Dams reduce the catch of migratory fish by preventing them from reaching their spawning grounds, vast reservoirs reduce the flow and level of rivers, increase the salinity of many rivers, trap the silt which contains nutrients vital to the survival of fisheries in the lower reaches of the river and in the sea, help the invasion of aquatic weeds which reduce the biological productivity of the reservoir and the number of micro-organisms on which fish can feed. When a reservoir is filled, there is a rise in the population of fish species favoured by the new lacustrinne conditions, but it is followed by failure. Besides, the river population decreases substantially, and some species disappear completely.
  Broader Environmentally harmful dam construction (#PD9515).
  Aggravated by Aquatic weeds (#PD2232).

♦ **PE8150 Screw-worm flies as pests**
  Nature The female fly lays 200–300 eggs in animal or human wounds – a tick bit is sufficient – and in a few days maggots kill its host by eating away at tissues.
  Broader Flies as insect pests (#PE2254)   Flies as vectors of diseases (#PE4514).

♦ **PE8151 Inadequate rehabilitation facilities for the mentally handicapped**
Inadequate support services for the mentally handicapped
  Broader Inadequate rehabilitation facilities (#PD1089)
  Inadequate community care for handicapped persons (#PE8924)
  Inadequate vocational rehabilitation facilities for disabled persons (#PE7317).
  Narrower Inadequate facilities for mentally disabled criminals (#PE8900).
  Related Inadequate support services (#PF6492).
  Aggravates Mental deficiency (#PC1587)   Age discrimination (#PC2541)
  Mental disorders of the aged (#PD0919)
  Denial of the right to procreate to the severely mentally handicapped (#PE4544).

♦ **PE8152 Limited access to culturally adapted pedagogical material**
  Aggravated by Narrow scope of education (#PF3552).

♦ **PE8154 Concentration of noxious substances in food chains**
  Broader Chemical contaminants of food (#PD1694).
  Aggravates Food pollution (#PD5605)   Pesticides as pollutants (#PD0120)
  Pesticide hazards to wildlife (#PD3680).
  Aggravated by Polychlorinated biphenyls as a health hazard (#PE2432).

♦ **PE8157 Delay in delivery of requested services**
  Broader Delays in delivery of goods and services (#PE3928)
  Unrecognized socio-economic interdependencies (#PF2969).
  Related Delay in administration of medical care (#PD5119).

♦ **PE8158 Ischaemic heart disease**
  Refs Willerson, James T et al *Ischemic Heart Disease* (1982).
  Broader Heart diseases (#PD0448).
  Narrower Angina pectoris (#PG9486)   Acute myocardial infarction (#PE8456)
  Arteriosclerotic heart disease (#PG2088).

♦ **PE8159 Rape and mustard oils trade instability**
  Broader Instability of trade in fixed vegetable oils and fats (#PE0861).

♦ **PE8162 Vibration damage to cultural artefacts**
Vibration damage to buildings
  Refs Jones, Robert S *Noise and Vibration Control in Buildings* (1984).
  Broader Destruction of cultural heritage (#PC2114).
  Related Environmental hazards of vibration (#PJ2171).
  Aggravates Failure of materials (#PD2638).

♦ **PE8165 Environmental hazards of furniture and fixtures manufacture**
  Broader Environmental hazards from woodworking industries (#PE0864).

♦ **PE8166 Overuse of machinery and equipment in development projects**
  Broader Inappropriate design of development projects (#PF4944).
  Aggravates Failure of development programmes (#PF8368)
  Unsuitability of development projects for available labour (#PE8687).
  Aggravated by Lack of participation in development (#PF3339).

♦ **PE8169 Scientific and controlling instruments trade instability professional**
  Broader Instability of trade in miscellaneous manufactured articles (#PE0814).

♦ **PE8170 Deprivation of trade union funds and property**
  Broader Deprivation (#PA0831)   Violation of the right of trade unions to function freely (#PE1758).

♦ **PE8171 Subsistence agricultural income level in rural communities**
Unpredictable farming income — Low farming income
  Nature Many rural communities have no industry and only a few small shops and businesses. The people are almost totally dependent on farming and the unpredictable sufficiency or insufficiency of rain. This provides a very narrow economic base and places the community in a perpetual struggle with contingency. The scant exposure to successful business enterprises and the past

failure of economic ventures leaves residents hesitant to risk commercial investment.
Claim Subsistence agriculture draws heavily on the natural resource base and thus degrades it.
  Broader Underdeveloped capacity for income farming (#PF1240).
  Related Unattractive fishing business (#PG5399).
  Aggravates Precarious basis for family economics (#PF1382)
  Rural poverty in developing countries (#PD4125)
  Inadequate income of rural communities (#PJ9649)
  Subsistence approach to capital resources (#PF6530)
  Dependency on unpredictable sources of income (#PF3084)
  Inadequate domestic savings in developing countries (#PD0465).
  Aggravated by Lack of livelihood standards (#PF1297)
  Unproductive subsistence agriculture (#PC0492)
  Rural unemployment in developing countries (#PD0295)
  Obsolete methods of agricultural production (#PF1822)
  Undeveloped channels for commercial initiative (#PF6471).

♦ **PE8173 Paralysis of throat, eye, limb and respiratory muscles**
  Broader Paralysis (#PD2632).
  Aggravated by Diphtheria (#PE8601).

♦ **PE8174 Vulnerability of children during armed conflict**
Children threatened by warfare
Nature The children suffer the direct and the indirect effects of warfare, including lower expenditure on services for children; limited access to services near front lines and in high-security zones; increasing migration and displacement leading to family breakdown, abandonment and even child conscription; as well as associated psychological stress and damage.
  Narrower Childhood martyrdom (#PF8118).
  Related Children engendered by occupying soldiers (#PD8825).
  Aggravated by War (#PB0593) Militarization of children (#PE5986).

♦ **PE8175 Pollution of semi-enclosed bodies of seawater**
Pollution of bays and estuarine waters
  Broader Marine pollution (#PC1117). Biological contamination of water (#PD1175).

♦ **PE8177 Illiteracy as an impediment for leadership**
  Broader Illiteracy (#PC0210) Obstacles to leadership (#PF7011).
  Aggravated by Limited development of functional abilities (#PF1332).

♦ **PE8180 Lack of rural industrialization in developing countries**
Refs Chuta, Enyinna and Sethuraman, S V *Rural Small-Scale Industrie and Employment in Africa and Asia* (1983).
  Broader Imbalance between urbanization and industrialization in developing countries (#PC1563).

♦ **PE8184 Low complementarity of developing country economies**
  Aggravates Structural rigidity in developing country economies (#PD2970)
  Excessive foreign public debt of developing countries (#PD2133)
  Low intra-regional complementarity of developing countries economies (#PE7255).

♦ **PE8187 Endangered species of white whale and narwhal**
Incidence 2 species; none endangered.
  Broader Endangered species of whale (#PD1593).

♦ **PE8188 Fluctuations in food production in developing countries**
  Aggravates Food insecurity (#PB2846).
  Dependence of developing countries on food imports (#PE8086).
  Aggravated by Unstable fishing season (#PJ9570)
  Disastrous insect invasions (#PD4751)
  Insufficient flow of information (#PF6469)
  Fluctuating agricultural markets (#PG9369)
  Seasonal fluctuations in agriculture (#PD5212)
  Instability of production of food and live animals (#PD2894)
  Underdevelopment of food and live animal production (#PF2821).

♦ **PE8189 Endangered species of sifakas, avahis and indri**
Incidence 4 species; 2 endangered, 1 vulnerable, 1 rare.
  Broader Endangered species of non-human primates (#PE1570).

♦ **PE8190 Lack of economic and technical development**
Underdeveloped economy — Low economic development
  Broader Underdevelopment (#PB0206).
  Narrower Inadequate financial services (#PJ8366)
  Inadequate industrial services (#PG9900)
  Middleman control of rural marketing (#PE3528)
  Limited availability of financial credit (#PF2489)
  Underdevelopment of manufacturing industries (#PF0854)
  Undeveloped financial markets in developing countries (#PE5902).
  Related Inadequate infrastructure (#PF7693) Limited market development (#PF1086)
  Underdevelopment of industrial and economic activities (#PC0880).
  Aggravates Underground economy (#PC6641) Uncontrolled growth of debt (#PC8316)
  Poverty in developed countries (#PC0444) Infectious and parasitic diseases (#PD0982)
  Undirected expansion of economic base (#PF0905)
  Decline in government social expenditure (#PF0611)
  Inadequate welfare services for the blind (#PD0542)
  Inappropriate government intervention in agriculture (#PE1170)
  Disparity between industrialized and developing countries (#PC8694)
  Emigration of trained personnel from developing to developed areas (#PD1291)
  Vulnerability of national economies to vagaries of external markets for goods and services (#PF9697).
  Aggravated by Small local markets (#PJ8200) Economic inefficiency (#PF7556)
  Dependency on middlemen (#PD4632) Complex banking practices (#PJ8033)
  Unexplored potential markets (#PF0581) Unstable shifting agriculture (#PD7516)
  Rejection of agriculture by youth (#PG9860) Unpromoted industrial development (#PJ1017)
  Seasonal fluctuations in agriculture (#PD5212) Unproductive subsistence agriculture (#PC0492)
  Structural rigidities in labour markets (#PD4011)
  Inadequate maintenance of infrastructure (#PD0645)
  Financial destabilization of world trade (#PC7873)
  Degradation of developing countries by tourism (#PF4115)
  Limited availability of loans in developing countries (#PE4704)
  Restrictive effects of external capital on development (#PF3318)
  Over-subsidized agriculture in industrialized countries (#PD9802)
  Inadequate commercial finance for rural development projects (#PF4340)
  Lagging transformation of agriculture in developing countries (#PD0946)
  Dependence of industrialized countries on import of resources (#PE0537)
  Overlooked potential for industrial development in rural communities (#PF2471)
  Private domestic capital outflow from developing to developed countries (#PD3132)

Restrictions on industrial and economic development due to environmental policies (#PE4905)
Dependence of developing countries on external financing for development programmes (#PE7195)
Restrictive business practices in the markets of developed countries against exports from developing countries (#PE5926).

♦ **PE8194 Underutilization of peat as an energy source**
Nature Peat is decayed, fibrous substance consisting of partially decomposed and disintegrated mosses, sedges, trees, grasses, reeds, algae etc. in a water-saturated environment. It has traditionally served as an alternative fuelwood for household purposes in rural areas. It has a high moisture content that has to be reduce before usage, but in comparison with fossil fuels very low sulphur dioxide emissions. The exploitation of peat land on a large scale can affect ecology, climate and hydrology, dust may cause itching and symptoms of rhinitis. However, peat land utilization provides energy and land for agriculture or forestry.
  Broader Unexplored energy alternatives (#PF7960)
  Underutilization of renewable energy resources (#PE8971).
  Reduces Vulnerability of wetlands (#PC3486).

♦ **PE8206 Instability of trade in coal, coke and briquettes**
  Broader Instability of trade in mineral fuels, lubricants and related materials (#PD0877).

♦ **PE8207 Military expeditions against distant objectives**
Nature In contrast to cross-border expeditions which are of relatively frequency occurrence, some governments occasionally sanction military intervention in distant countries for special purposes, usually of brief duration.
Incidence Recent examples include the air attack by Israel against a nuclear reactor in Iraq, the intervention by Israel at Entebbe airport in response to aircraft hijacking, the air attack by the USA against Libya, the USA intervention in Iran in an attempt to release hostages, and the USA intervention in Grenada and in Panama. Other possible examples include UK intervention in the Falklands/Malvinas and the attack by France on the Rainbow Warrior in New Zealand.
  Broader Foreign military intervention (#PD9331).
  Related Cross border military operations (#PD5272)
  State-supported international terrorism (#PD6008)
  Military expeditions against friendly powers (#PD7261).

♦ **PE8208 Instability of trade in vegetables, roots and tubers**
  Broader Instability of trade in fruit and vegetables and preparations thereof (#PE0961).

♦ **PE8210 Feuds**
Vendettas
  Broader Retaliation (#PF9181) Human violence (#PA0429).
  Narrower Family feuds (#PJ9994) Blood vengeance (#PF7653).
  Related Political feuding (#PD4846).
  Aggravates Factionalism (#PF8454).
  Aggravated by Grievance (#PF8029).

♦ **PE8212 Uranium ores and concentrates and thorium trade instability**
  Broader Instability of trade in metalliferous ores and metal scrap (#PE0553).

♦ **PE8213 Nymphomania**
Refs Ellis, Albert and Sagarin, Edward *Nymphomania* (1968).
  Broader Sexual craving (#PE7031) Sexual deviation (#PD2198)
  Human sexual inadequacy (#PC1892).

♦ **PE8215 Inadequate segregation of different categories of juvenile offenders**
  Broader Repressive detention of juveniles (#PD0634).

♦ **PE8216 Inadequate education concerning the nature of problems**
  Broader Inadequate education (#PF4984) Inadequate secondary education (#PD5345)
  General obstacles to problem alleviation (#PF0631).
  Aggravates Oversimplification (#PF8455)
  Recurrence of misapprehended world problems (#PF7027)
  Institutional preoccupation with obsolete problems (#PJ5014).
  Aggravated by Failure to integrate knowledge to empower humanity in response to the global problematique (#PF8753).

♦ **PE8218 Cessation of functions of civil courts**
  Aggravated by Martial law (#PD2637).

♦ **PE8219 Cardiac arrhythmia**
Irregular heart rhythm
Refs Bennett, David *Cardiac Arrhythmias* (1989).
  Broader Cardiovascular diseases (#PE6816).

♦ **PE8220 Women combatants**
Women soldiers — Discrimination against women in military forces — Sexual harassment in the military
Incidence In the USA in 1990, official reports indicated that more than one third of the women surveyed experienced some form of direct harassment, including touching, rape and pressure for sexual favours.
Claim Women military personnel have a high attrition rate due to pregnancy, up to 1 in 10 of enlisted women. Fraternization and providing separate bathrooms and sleeping quarters interfere with combat division operations. Protectiveness of men toward women hamper combat performance. Women have less strength and endurance than men. Politically women in combat is impossible. The prospect of having mothers and daughters return home in body bags is far more repellent than fathers and sons killed in action.
Counter-claim In a high tech army strength and endurance is of little consequence. Women can be as or more ruthless as men. In many types of duty women are frequently more qualified than available men.
  Aggravated by Vulnerability of women and children in emergencies (#PD1078).

♦ **PE8221 High teacher turnover**
  Broader Obstacles to education (#PF4852).
  Narrower Departure of good teachers (#PU0493).
  Aggravates Poor supervision teachers (#PJ3757)
  Deteriorating community identity (#PF2241).
  Aggravated by Unmotivated teachers (#PF5978)
  Inadequate working conditions of teachers (#PE7165).

♦ **PE8223 Underdevelopment of iron and steel basic industries**
  Broader Underdevelopment of basic metal industries (#PF1374).

♦ **PE8224 Shortage of fresh, chilled or frozen meat**

PE8224

Broader Long-term shortage of meat and meat preparations (#PE1490).

♦ **PE8225 Long-term shortage of dairy products and eggs**
Broader Long-term shortage of food and live animals (#PE0976).

♦ **PE8228 Pleura and peritoneum cancers of the bronchi**
Bronchial cancer
Broader Malignant neoplasms (#PC0092)   Occupational diseases (#PD0215)
Malignant neoplasm of respiratory system (#PE7572).
Related Pneumoconiosis (#PD2034).
Aggravated by Health hazards of asbestos (#PE3001).

♦ **PE8230 Unwillingness to divulge information by industrial concerns**
Aggravates Occupational cancer (#PE3509).

♦ **PE8231 Lack of airport and travel security**
Nature The plastic explosives used by terrorists to blow up planes are almost impossible to detect other than by hand searching the luggage at airports.
Broader Insecurity (#PA0857).
Aggravates Aerial piracy (#PD0124).

♦ **PE8232 Instability of trade in essential oils and perfume materials**
Broader Instability of chemicals trade (#PD0619).

♦ **PE8233 Mismanagement of irrigation schemes**
Uncoordinated irrigation control — Unmanaged irrigation potential — Open irrigation wastage — Insufficient irrigation knowledge — Underdeveloped water management — Poor water management — Deterioration of irrigation systems — Unregulated crop water
Broader Agricultural mismanagement (#PD8625).
Inadequate water system infrastructure (#PD8517).
Minimal access to appropriate technology (#PF3554).
Narrower Lack of water conservation (#PJ3480).
Aggravates Desert advance (#PC2506)   Soil salinization (#PE1727)
Rising water level (#PD8888)   Desiccation of lakes (#PD1990)
Water losses from irrigation systems (#PE8796)   Disruption of the hydrological cycle (#PD9670)
Seepage water losses from irrigation systems (#PE9088).
Aggravated by Unethical practice of hydrology (#PD2586)
Limited access to practical education (#PF2840)
Ineffective structures of local consensus (#PF6506)
Underdeveloped sources of income expansion (#PF1345)
Underdeveloped potential of basic resources (#PF3448)
Underproductive methods of agricultural management (#PF6524)
Lack of opportunities for practical training in communities (#PF2837).

♦ **PE8236 Insufficient consumer services in developing countries**
Broader Unexploited possibilities for local commerce (#PF2535).
Narrower Lack of production for domestic consumers in developing countries (#PJ8860).
Related Inadequate industrial services in developing countries (#PF4195).

♦ **PE8240 Lack of legal protection of extortion victims**
Inadequate assistance to victims of extortion
Broader Legal discrimination in favour of offenders (#PD9316).
Inadequate assistance to victims of human rights violations (#PD5122).
Aggravated by Intimidation of victims of crimes (#PJ1543).

♦ **PE8242 Shortage of coal-derived tar and crude chemicals**
Broader Long-term shortage of chemicals (#PE1261).

♦ **PE8244 Insufficient forestry in least developed countries**
Broader Least developed countries (#PD8201).
Related Underdevelopment of forestry industry (#PG5324).

♦ **PE8246 Illiteracy as an obstacle to acquiring skills**
Broader Illiteracy (#PC0210).
Aggravated by Ineffective systems of practical education (#PF3498).

♦ **PE8248 Non-ferrous metals basic industries environmental hazards**
Broader Health hazards from basic metal industries (#PD0243).

♦ **PE8250 Environmental hazards of oil nuts, oil kernels and oil-seeds**
Broader Environmental hazards from inedible crude non-fuel materials (#PE0546).

♦ **PE8251 Inadequate working conditions of employees of commerce and offices**
Refs ILO *Problems Specific to Employees in Commerce and Offices*.
Broader International imbalance of quality of working life (#PD9170).

♦ **PE8252 Underdevelopment of hunting, trapping and game propagation**
Broader Underdevelopment of industrial and economic activities (#PC0880).

♦ **PE8253 Endangered species of African mole rat**
Incidence 16 species; none endangered
Broader Endangered species of rodents (#PD3481).

♦ **PE8254 Vulnerability of telephone system**
Undependable telephone service
Broader Insufficient communications systems (#PF2350)
Underdeveloped potential of basic resources (#PF3448).
Aggravates Disaster unpreparedness (#PF3567)   National insecurity and vulnerability (#PB1149).
Aggravated by War (#PB0593)   Fires (#PB8054)   Terrorism (#PD5574)
Natural disasters (#PB1151)   Injurious accidents (#PB0731).

♦ **PE8255 Substitution of inappropriate foodstuffs for breast feeding**
Unnaturally short duration of breastfeeding — Unhygienic bottle feeding
Nature Substitution of breast feeding by the untimely introduction of other foodstuffs, usually of low or inadequate nutritional value, contributes to poor and unhealthy feeding practices and malnutrition.
Incidence Data available in 1987 indicated a continuing decline is breast-feeding, in both incidence and duration, in developing countries, especially in urban areas. The substitution is frequently encouraged by commercial interests using misleading advertising.
Claim Even if environmental pollution has contaminated mother's milk in some degree, it is still better alternative than any substitute, because already the drinking water used in artificial baby formulas or food contains 10 to 100 times more nitrates than mother's milk. The environmental poison will enter the bodies of babies – if they are breastfed or not.
Broader Inappropriate infant feeding strategies (#PD9661).
Related Unbalanced infant diets (#PE0691)   Hazards of bottle-feeding (#PE4935).
Aggravates Diarrhoea in children (#PE9751)
Protein-energy malnutrition in infants and early childhood (#PD0331).
Aggravated by Misleading advertising (#PE3814)
Dependence on breast feeding (#PE7627)
Ignorance of women concerning primary health care (#PD9021).

♦ **PE8256 Lack of availability of methods of cross-breeding**
Aggravates Poor quality of domestic livestock (#PD2743).

♦ **PE8257 Instability of trade in dried salted or smoked meat**
Broader Instability of trade in meat and meat preparations (#PE0755).

♦ **PE8258 Instability of trade in fresh potatoes**
Broader Instability of vegetable trade (#PE1711).

♦ **PE8259 Instability of trade in beans, dried, peas, lentils and other leguminous vegetables**
Broader Instability of vegetable trade (#PE1711).

♦ **PE8261 Prohibitive medical expenses**
Inflated medical expenses — Unnecessary medical expenses — Inaccessibility of health care to the underprivileged — Rationing of health care — Prohibitive cost of health care
Nature The production of sophisticated drugs requiring an extensive period of development, and the advances in high technology medicine contribute significantly to rising medical expenses.
Incidence It is estimated that advances in high-technology medicine contributed at least 20 per cent of the 10.1 per cent rise in health care costs in the USA in 1989. Extreme examples include a 1990 medical expense of $1.2 million for a heart/liver/kidney transplant in the USA, necessitating 4 months hospitalization. This is part of a pattern of increasing health costs, often born by employers who are consequently concerned with the erosion of their profitability, competitiveness and ability to raise wages. Health experts in the USA in 1989 argued that the country no longer has the financial resources to provide unlimited medical treatment for all those who need it.
Claim 1. The basic dilemma of modern medicine is that more health care methods have been invented than society can afford to pay for. Too much money is spent on high-technology care for the few and too little on basic health care for the many.
2. The payment for medical care for individuals is increasingly excessive, especially when the high costs of some operations may outweigh the benefits when the chances of prolonging life are small. Such resources may be more effectively spent of primary health care for many rather than on expensive operations for the few.
Counter-claim No expense should be spared in endeavouring to save and prolong human life.
Broader Inadequate health services (#PD4790)   Marginal level of family income (#PD6579)
Increasing cost of social security (#PF7911).
Narrower Prohibitive cost of abortion (#PJ2020)   Lack of funds for medical research (#PE0612)
Prohibitive cost of contraceptives (#PJ3244)   Prohibitive cost of disease control (#PF2779)
Prohibitive cost of private medical care (#PF8016)
Increase in insurance claims for medical negligence (#PE4329)
Prohibitive cost of eye-glasses in developing countries (#PE7548).
Aggravates Decline in government health expenditure (#PF4586)
Excessive medical intervention in childbirth (#PE7705)
Underdeveloped provision of basic services in developing countries (#PF6473)
Restricted delivery of essential services to developing country rural communities (#PF1667).
Aggravated by Injurious accidents (#PB0731)
Excessive prolongation of the dying process (#PF4936)
Waste of resources on expensive medical techniques (#PG9850).

♦ **PE8263 Health hazards in the steel industry**
Broader Occupational risk to health (#PC0865).

♦ **PE8265 Ineffective self-regulation in the housing construction industry**
Broader Ineffective industry self-regulation (#PF5841).
Aggravates Substandard housing and accommodation (#PD1251)
Malpractice in the construction industry (#PD9713).

♦ **PE8266 Inadequate traffic control**
Narrower Unsafe school crossing (#PU0149).
Related Inadequate safety legislation (#PJ2716)   Lack of protective legislation (#PJ2889).
Aggravates Traffic noise (#PD3664)   Traffic congestion (#PD0078)
Air traffic delays (#PF9464)   Road traffic accidents (#PD0079)
Road traffic violations (#PE0930).

♦ **PE8267 Distortion of international trade by state-trading and government monopoly practices**
Broader Distortion of international trade as a result of government participation (#PD2029)
Distortion of international trade by discriminatory customs and administrative entry procedures (#PE2603).
Related State monopoly (#PJ4242).
Disjointed patterns of community identity (#PF2845).
Aggravated by Proliferation of public sector institutions (#PF4739).

♦ **PE8272 Instability of trade in sanitary plumbing, heating, lighting fixtures and fittings**
Broader Instability of trade in miscellaneous manufactured articles (#PE0814).
Aggravates Unhygienic conditions (#PF8515).

♦ **PE8273 Interference of trade unions with contract performance**
Broader Restrictive trade union practices (#PD8146).
Aggravated by Unethical trade union practices (#PD4341).

♦ **PE8274 Health hazards in the glass industry**
Broader Occupational risk to health (#PC0865).

♦ **PE8275 Environmental hazards of miscellaneous manufactured articles**
Broader Environmental hazards from economic and industrial products (#PC0328).

♦ **PE8277 Shortage of fixed vegetable oils and fats**
Broader Long-term shortage of animal and vegetable oils and fats (#PE1188).

♦ **PE8279 Landsize as a limit to diversity**
Broader Underdeveloped sources of income expansion (#PF1345).

♦ **PE8280 Inadequate national supervision of adherence to international law**
Nature Transgressions of principles of international law include non-intervention in the affairs of

other nations, adjudication of disputes, sparing civilians from haphazard attack and deliberate destruction of the environment.
Broader Deficiencies in international law (#PF4816)
Inadequacy of international legislation (#PF0228).

♦ PE8283 Lack of time flexibility in the labour market
Aggravated by Discrimination against men in employment (#PD3338)
Ineffective economic structures in industrial nations (#PE4818).

♦ PE8288 Endangered species of odobenus
Endangered species of walrus
Nature The walrus, because of its large size, low reproductive rate, slowness in moving away from boats, gregarious nature, and the value of its ivory is one of the marine mammal species most vulnerable to over exploitation. It is estimated that less than 20,000 Atlantic walrus exist and the Pacific walrus population size is fluctuating widely in size do to crisis responses to depletion and uncoordinated simultaneous management by the U.S.A and U.S.S.R.
Incidence 1 possibly endangered sub-species.
Broader Endangered species of seals (#PE1656).

♦ PE8289 Conflict of interest between governments or groups of governments
Broader Conflict (#PA0298) Conflicts of interest (#PF9610).

♦ PE8296 Minimal family interest of urban life styles
Broader Obstacles to family life (#PF7094)
Limiting effect of individual survivalism (#PF2602).

♦ PE8300 Deficiencies in military codes of justice
Broader Injustice (#PA6486).
Aggravates Military disobedience (#PD7225) Injustice of military tribunals (#PE0494).

♦ PE8302 Endangered species of rough-toothed and white dolphins
Incidence 8 species; none endangered.
Refs Klinowska, Margaret *Dolphins, Porpoises, and Whales of the World* (1988).
Broader Endangered species of whale (#PD1593).
Aggravated by Hunting of marine animals (#PE0439).

♦ PE8303 Insufficient doctors
Unavailability of community doctors — Unavailability of local physicians — Lack of specialist doctors in developing countries — Unaffordable fulltime physician
Broader Insufficient health personnel (#PD0366) Obstacles to community achievement (#PF7118)
Decline in government health expenditure (#PF4586)
Shortage of adequately trained personnel to act against problems (#PF0559).
Aggravates Inadequate medical care (#PF4832) Inadequate health services (#PD4790)
Prohibitive cost of basic services (#PF6527) Fragility of maintaining basic health (#PJ2524)
Underprovision of basic services to rural areas (#PF2875).
Aggravated by Maldistribution of health personnel (#PF4126).

♦ PE8305 Inadequate teaching occupational health and safety
Broader Irrelevance of educational curricula (#PF0443).
Aggravates Occupational diseases (#PD0215) Human disease and disability (#PB1044).

♦ PE8306 Minimal exports in least developed countries
Broader Least developed countries (#PD8201)
Reduction of the share of the developing countries in world exports (#PC2566).
Related Underdeveloped export potential (#PJ1586)
Weakness in intra-regional trade of developing countries (#PD0169).

♦ PE8307 Inadequate inter-regional cooperation on problems within countries
Broader General obstacles to problem alleviation (#PF0631).
Aggravated by Fragmented regional cooperation (#PF9129)
Rivalry and disunity within developing regions (#PD0110)
Inadequate power of intergovernmental organizations (#PF9175).

♦ PE8308 Jurisdictional conflict and antagonism between government agencies within each country
Unclear fiscal relations among the different parts of government
Broader Conflict (#PA0298).
Aggravates Health hazards of modern insulating materials (#PE1499)
Inefficient mobilization of government revenue (#PF4197)
Jurisdictional conflict and antagonism between international organizations (#PD0138)
Jurisdictional conflict and antagonism between regional intergovernmental organizations with common membership (#PE1583).
Aggravated by Organizational empire-building (#PF1232)
Proliferation of public sector institutions (#PF4739).

♦ PE8309 Government protection and national commodity price support
Broader Trade protectionism (#PC4275).
Aggravates Detrimental international repercussions of domestic agricultural policies (#PF2889).

♦ PE8311 Drunkenness
Nature Drunkenness is a condition of mind and body produced by a sufficient quantity of alcohol to bring about changes in the emotions and will, intellect and movements of an individual.
Broader Alcohol abuse (#PD0153).
Narrower Public drunkenness (#PE2429) Excessive parental drunkenness (#PE7700)
Drunkenness of military personnel and troops (#PE8495).
Aggravates Fraud (#PD0486) Assault (#PD5235) Epilepsy (#PE0661)
Hangover (#PJ5020) Homicide (#PD2341) Gastritis (#PG2250)
Pancreatitis (#PG3330) Head injuries (#PG5105) Family violence (#PD6881)
Drunken driving (#PE2149) Property damage (#PD5859) Alcohol poisoning (#PG7437)
Attempted suicide (#PE4878) Theft of property (#PD4691) Injurious accidents (#PB0731)
Cruelty to children (#PC0838) Industrial accidents (#PC0646)
Sexual impotence of men (#PF6415) Foetal alcohol syndrome (#PE3853)
Alcohol-related violence (#PE7084) Neglected young children (#PE4245)
Inadequate sense of time (#PF9980) Alcoholic intoxication at work (#PE2033)
Occupational domestic accidents (#PE4961).

♦ PE8312 Iron and steel scrap trade instability
Broader Instability of trade in metalliferous ores and metal scrap (#PE0553).

♦ PE8313 Boycott
Nature Boycotting is a method of social ostracism, applied by organized or unorganized public opinion avoid or hold aloof from those judged unworthy of social intercourse. Economic forms of boycott are now the most common. With the mobility of today's society and hugh number of relationships any individual or group has boycotts are largely ineffective. When South Africa was boycotted, not only was she able to receive much of what she wanted, but for those things unavailable, she became self-sufficient.
Background Boycotting has been around as long as human society. The outcast or outlaw in simpler societies was subject to an extreme form of boycotting. In the middle ages excommunication because it carried with it denial of civil rights was also an extreme form. The right to boycott is recognized as inherent in the social nature of humans and is one of the forms in which public opinion expresses itself.
Refs Mersky, Roy M (Ed) *Conference on Transnational Economic Boycotts and Coercion* (1978); Remer, C F *Study of Chinese Boycotts* (1966).
Broader Ostracism (#PF1009) Non-violent weapons (#PF9327)
Economic sanctions against governments (#PF4260).
Narrower Consumer boycotts (#PE1213).
Aggravates Outlaws (#PE4409) Social outcasts (#PD6017).

♦ PE8318 Delay in societal impact of education
Broader Delays in implementation of social change (#PC6989).

♦ PE8319 Inadequate conditions of work and employment in utility supply services
Refs ILO *Conditions of Work and Employment in Water, Gas and Electricity Supply Services* (1982); ILO *Employment and Conditions of Work in Water, Gas and Electricity Supply Services* (1987).
Broader International imbalance in the quality of life (#PB4993).

♦ PE8320 Leather and leather manufactures trade instability
Instability of leather and footwear industry
Broader Instability of trade in manufactured goods (#PE0882).

♦ PE8321 Protectionism in developed countries against agricultural products from developing countries
Industrialized country disincentives to farm products from developing countries — Disincentives against farming in developing countries
Nature All developed countries continue to protect their agricultural sectors against competitive imports through domestic production support programmes and trade measures that have not only restricted the access of foreign suppliers to their markets, but have also generated large surpluses of such products as sugar, meat, cereals and dairy products.
Claim The international market for farm products from developing countries is barred by agricultural subsidy policies of developed countries. This leads the farmers in developing countries to grow poppies, coca leaves or marijuana.
Broader Protectionism in international trade against exports from developing countries (#PD9679)
Restrictive business practices in the markets of developed countries against exports from developing countries (#PE5926)
Development by industrialized countries of products substituting for commodities exported by developing countries (#PD7682).
Related Disincentives against farming (#PD7536).
Aggravates Cultivation of illegal drugs (#PD4563).
Aggravated by Over-subsidized agriculture in industrialized countries (#PD9802).

♦ PE8322 Hypochondria
Nature Hypochondria is the chronic mental state in which the sufferer is constantly occupied with a delusion that he or she is seriously ill. The individual spends a great deal of time seeking cures, has a gloomy turn of mind and is extremely self-centred. He or she is unproductive and drains the resources of family, friends and society by requiring large amounts of the medical professions time.
Refs Baur, Susan *Hypochondria* (1988); Meister, Robert *Hypochondria* (1980).
Broader Fear (#PA6030) Neurosis (#PD0270) Obsession (#PA6448).

♦ PE8324 Degradation of agricultural land by cash crops
Nature Vast tracts of land which are suitable for growing grazing grasses or trees, but little else, have been torn up to make way for cotton, peanut or coffee plantations. The result is soil that becomes poor in humus and loses its cohesiveness.
Incidence In Brazil many of the abandoned coffee lands are so ruined that they cannot be restored to crop production or the humus content of the soil is seriously reduced. Coffee plantations are always on the march, grabbing new land and leaving behind eroded soils.
Broader Destruction of agricultural land (#PD9118).
Aggravates Declining productivity of agricultural land (#PD7480).
Aggravated by Monoculture of crops (#PC3606) Inappropriate cash crop policy (#PF9187)
Unsustainable rural development (#PD4537).

♦ PE8327 Skins and fur skins shortage hides
Broader Long-term shortage of inedible crude non-fuel materials (#PE0461).

♦ PE8332 North and South spheres of influence within UN and related agencies
Broader Lack of central planning structures in small communities (#PF2540).
Aggravates Ineffectiveness of international organizations and programmes (#PF1074).

♦ PE8336 Environmental hazards from road motor vehicles
Broader Environmental hazards from transport equipment (#PE0738).

♦ PE8339 Denial to animals of the right to a natural death
Nature Animals are killed in slaughterhouses, religious festivals, without consideration for the trauma and pain caused them.
Broader Denial to animals of legal protection of their rights (#PE8643)
Denial to experimental animals of the right to freedom from suffering (#PE8024).
Narrower Inhumane cooking practices (#PE7677).
Related Animal deaths (#PEp941) Fungal diseases (#PD2728)
Viral diseases in animals (#PD2730) Diseases of beneficial insects (#PD2284)
Soil-borne diseases in animals (#PE2739) Denial to animals of the right to life (#PF8243)
Denial to animals of the right to dignity (#PE9573)
Denial to working animals of limitation of working hours (#PE6427)
Denial to animals of the right to freedom from mass killing (#PE9650)
Denial to food animals of the right to freedom from suffering (#PE3899)
Denial to working animals of restorative nourishment and rest (#PE4793)
Endangered animal and plant life due to radioactive contamination (#PD5157)
Denial to animals of the right to the attention, care and protection of humankind (#PF5121)
Denial to animals of the right to conditions of life and liberty proper to their species (#PE6270).
Aggravated by Human consumption of animals (#PC7644)
Excessive consumption of animal flesh (#PD4518).

♦ PE8341 Lack of self-development in the family
Lack of family welfare
Nature The modern nuclear family fails to create an atmosphere in which not only the children mature physically but every member of the can grow intellectually, emotionally, psychologically and spiritually.

PE8341

Broader Negative effects of the nuclear family (#PF0129).
Related Inadequate child welfare (#PC0233).
Aggravated by Loneliness of children (#PC0239)
Physically handicapped children (#PD0196).

♦ **PE8343 Medical experimentation on pregnant women and foetuses**
Nature Research that relates to a pregnant woman can result in the transfer of noxious substances into the foetal circulation, thus affecting the well–being of the vulnerable foetus.
Broader Abusive experimentation on humans (#PC6912)
Unethical experiments with drugs and medical devices (#PD2697).
Aggravates Complications of childbirth (#PC9042).

♦ **PE8345 Lack of unionization among working women**
Broader Ineffective worker organizations (#PF1262)
Restrictive trade union practices (#PD8146).
Aggravates Discrimination against women in employment (#PD0086).

♦ **PE8349 Capitalist subversion in communist and neutral countries**
Broader Capitalism (#PC0564).

♦ **PE8351 Restrictive practices in trade in inedible crude non–fuel materials**
Broader Restrictive trade practices (#PC0073).
Aggravates Instability of trade in inedible crude non–fuel materials (#PD0280).

♦ **PE8357 Lack of participation in attempting to solve intergroup conflicts**
Broader Non–participation (#PC0588)   Lack of community participation (#PF3307).

♦ **PE8361 Inadequate protection of civilians in armed conflict**
Inhumane treatment of non–combatants in war zones
Nature Civilians may suffer inhumane treatment including: murder, mutilation, cruel treatment and torture; being taken as a hostage; humiliation and degrading treatment; executions.
Broader Lack of protection for the vulnerable (#PB4353).
Narrower Inadequate protection of war correspondents (#PE3034).
Related Maltreatment of civilians (#PD5560)
Violation of rights of vulnerable groups during states of emergency (#PD3785).
Aggravated by War (#PB0593).

♦ **PE8364 Receiving stolen property**
Possession and fencing of stolen property
Broader Theft (#PD5552)   Criminal harm to property (#PD5511)
Aiding consummation of crime (#PD6655).

♦ **PE8367 Collusion of trade union leaders with employers and government**
Broader Unethical trade union practices (#PD4341).
Aggravates Strikes (#PD0694).
Aggravated by Unethical practices of employers (#PD2879).

♦ **PE8369 Difficulty of transition from school to work**
Refs Aggleton, Peter *Rebels Without a Cause?* (1987).
Broader Inadequate recognition by institutions of the transition through adolescence (#PF6173).

♦ **PE8370 Underutilization of tar sands as an energy source**
Nature Tar sands include unconsolidated sand, sandstone, limestone and other sedimentary rook containing oil. This in–place oil has a low mobility so no significant primary production can be achieved. Developing suitable technologies requires considerable time and field experimentation. Possible effects on environment may be considerable.
Broader Underutilization of renewable energy resources (#PE8971).
Related Underutilization of oil shale as an energy source (#PF0445).

♦ **PE8372 Prostate diseases**
Hyperplasia of prostate — Abscess of prostate — Prostatitis — Cancer of prostate
Incidence In USA one in 11 men gets prostate cancer at some time in his life, annually about 28,000 die of it.
Refs Coffey, D S et al *Multidisciplinary Analysis of Controversies in the Management of Prostate Cancer* (1988); Marberger, H, et al *Prostatic Disease* (1976).
Broader Diseases of male genital organs (#PD9154)
Malignant neoplasm of genito–urinary organs (#PE5100).
Related Malignant neoplasm of male genital organs (#PG4978).

♦ **PE8383 Abuse of antibiotics and vaccines in factory farming**
Nature Antibiotics are used on a tremendous scale in farming to suppress the spread of diseases and as a growth stimulant. The limited amount of effective antibiotics diminishes, because more and more bacteria becomes immune.
Broader Factory farming (#PD1562)   Abuse of animal drugs (#PE0043).

♦ **PE8384 Interference by employers in the affairs of workers' organizations**
Broader Incomplete access to information resources (#PF2401)
Violation of the right of trade unions to function freely (#PE1758).
Aggravated by Unethical practices of employers (#PD2879).

♦ **PE8397 Iron and steel basic industries environmental hazards**
Refs United Nations Environment Programme *Environmental Aspects of the Iron and Steel Industry* (1983).
Broader Health hazards from basic metal industries (#PD0243).

♦ **PE8398 Inadequate hygiene restrictions on carcass meat exports**
Broader Unhygienic conditions (#PF8515).
Aggravates Importation of infected carcass meats as factor in animal diseases (#PE2777).

♦ **PE8400 Failure of medical programmes in developing countries**
Costly medical care in developing countries
Nature The commercial and professionally controlled health care is too expensive for the poor people who may have to spend at least 40 per cent of their earnings on prescription drugs. What they spend on medicines and doctors they cannot spend on food, so the children do not have enough to eat and get sick easily.
Broader Inadequate medical care (#PF4832).
Aggravated by Lack of understanding of spiritual healing (#PF0761).

♦ **PE8401 Environmental destruction in least developed countries**
Claim In effort to claw their way out of debt and poverty, the countries have been forced to plunder the natural resources Desertification, deforestation, soil and water degradation have reached enormous proportions. These environmental changes have a direct and detrimental effects on their development efforts.
Broader Least developed countries (#PD8201)   Destruction of wildlife habitats (#PC0480)
Pollution in developing countries (#PC2023).
Aggravates Inevitable destruction of natural environment by mankind (#PE2443).

♦ **PE8403 Copyright barriers to transfer of knowledge**
Patent restrictions impeding sustainable development
Claim Proprietary rights are a key element in the commercial development of technology. Under certain conditions, their application hinders the diffusion of environmentally sound technologies and may increase inequities. In the case of new varieties of seed, for example, proprietary rights act as a major barrier to the acquisition of new technologies by developing countries. In 1980, the 65 per cent of the patents granted were held in industrialized countries, 29 per cent were held in the socialist countries, and only 6 per cent were held in developing countries (mostly by non–residents). Nationals of the developing country hold a bare 1 per cent of the total of patent grants and only one–seventh in their own countries, although over 90 per cent of the foreign–owned patents are never used in production processes in developing countries.
Broader Barriers to the international flow of knowledge and educational materials (#PF0166).
Aggravates Avoidance of copyright (#PD0188)   Monopolization of knowledge (#PF5329)
Insufficient communications systems (#PF2350)
Increasing development lag against information growth (#PE2000).
Aggravated by Vulnerability of intellectual property (#PF8854)
Violation of sovereignty by trans–border broadcasting (#PE0261).

♦ **PE8408 Restrictions on freedom of movement within countries**
Denial of right to freedom of movement within a state
Broader Travel restrictions (#PC8452)   Violation of civil rights (#PC5285)
Inadequate cultural integration of immigrants (#PC1532).
Related Denial of right to life (#PD4234)
Enforced collectivization of agriculture (#PD7443).
Reduces Migration of rural population to cities (#PE8768).

♦ **PE8409 Inadequate disinfection measures for humans during animal disease outbreaks**
Inadequate disinfection for humans after animal disease outbreaks
Broader Inadequate control of animal diseases (#PD2781).
Aggravates Human vectors of animal diseases (#PD2784)
Inadequate disinfection measures for animal housing and equipment (#PE2757).

♦ **PE8411 Denial of the right to ownership**
Denial of right of ownership — Disregard for property rights
Broader Restrictions on property rights (#PD8937)
Personal isolation in communities of industrialized countries (#PD2495).
Related Enforced collectivization of agriculture (#PD7443)
Unlawful interference with rights of innocent passage in territorial waters (#PE6755).
Aggravates Inadequate housing among indigenous peoples (#PC3320).
Reduced by Individualistic disposition of productive property (#PD4974).

♦ **PE8414 Restrictions on nationals leaving their own country**
Broader Restrictions on freedom of movement between countries (#PC0935).
Reduces Errant nationals (#PE0812).

♦ **PE8416 Separate and unequal development within different societies**
Broader Social inequality (#PB0514).
Related Regional disparities (#PC2049).

♦ **PE8420 Crimes committed in hospitals and health care facilities**
Broader Unethical practices of health services (#PE3328).
Related Locales of high crime rates (#PE7311).
Aggravates Inadequate medical care (#PF4832)   Inadequate health services (#PD4790).

♦ **PE8423 Refusal of entry to foreign workers' families**
Broader Restrictions on immigration (#PC0970).
Discrimination against immigrants and aliens (#PD0973).
Related Discrimination against foreigners in employment (#PD3529).

♦ **PE8424 Segregation of handicapped children in education**
Broader Segregation in education (#PD3441)   Discrimination against the disabled (#PD9757).
Aggravated by Physically handicapped children (#PD0196).

♦ **PE8427 China, pottery and earthenware manufacture underdevelopment**
Broader Underdevelopment of non–metallic mineral products industries (#PE1851).

♦ **PE8428 Occupational hazards in the mining industry**
Nature Especially hazardous for miners is extracting radioactive ores, such as uranium. Ionizing radiation can cause lung cancer.
Broader Occupational hazards (#PC6716).
Narrower Health hazards of mining radioactive substances (#PE7419).

♦ **PE8429 Unavailability of trained teachers in the rural areas of developing countries**
Nature Life is more comfortable and easier in the urban areas. The level development of infrastructure and amenities is far greater in the urban than rural areas especially in developing countries. Those who are educated and are therefore in a position to help often find it difficult to go back to the rural areas to live under conditions where basic facilities for reasonably comfortable living are non–existent or barely minimal. Going to rural areas to teach is even more difficult if teachers have been educated in urban areas. Frequently, the status of a teacher depends on how close to an urban area his school is.
Broader Shortage of trained teachers (#PD8108).
Aggravates Unrealized use of education structures (#PF2568)
Unperceived relevance of formal education in rural communities (#PF1944).
Aggravated by Maldistribution of teachers (#PE6183)
Inadequate working conditions of teachers (#PE7165)
Emigration of trained personnel from developing to developed areas (#PD1291).

♦ **PE8430 Ineffective means for participation in decision making**
Lack of brainstorming
Broader Unstructured local decision–making (#PF6550).
Narrower Lack of participation of disabled persons in decision–making (#PE8690).
Aggravated by Fragmented decision–making (#PF8448)
Over–centralization of global decision–making (#PF5472).

♦ **PE8436 Shortage of textile, yarn fabrics, made–up articles**
Broader Long–term shortage of manufactured goods (#PE0802).

PE8437 Urban population and the overall growth of the economy labour force
Related Imbalances between growth of the labour force, urban population and the overall growth of the economy (#PE8735).

PE8438 Insufficient educational material
Educational materials shortage
Broader Shortage (#PB8238).
Narrower Shortage of biological specimens for medical study (#PF5097)
Shortage of books and textbooks in developing countries (#PF0118).
Aggravates Unrealized use of education structures (#PF2568).

PE8442 Obstacles to learning due to hunger
Broader Obstacles to education (#PF4852).
Aggravated by Mass unemployment of human resources (#PD2046).

PE8443 Restrictive building codes in urban areas
Complex housing regulations — Complicated building standards
Nature Building codes tend to become so restrictive that they discourage owners and tenants from adapting structures to make them more habitable. In developing countries, and especially in shanty towns, the majority of building codes and standards are ignored because following them would produce buildings too expensive for most people.
Broader Complex government regulations (#PF8053)
Inadequacy of international standards (#PF5072)
Paralyzing complexity of urban structures (#PF1776)
Underdeveloped capacity for income farming (#PF1240)
Demoralizing constraints on housing rehabilitation (#PE2451).
Aggravates Counterfeit machine parts (#PE5319)
Inadequate housing construction (#PG0561)
Substandard housing and accommodation (#PD1251)
Underprovision of basic urban services (#PF2583)
Unfeasible housing alternatives in urban areas (#PE8061)
Depressing effect of poor housing construction (#PF1213)
Reduces Inadequate building standards (#PF8829).

PE8444 Undue taxation of certain goods and services
Broader Distortionary tax systems (#PD3436).

PE8447 Leather and leather manufactures environmental hazards
Broader Environmental hazards from manufactured goods (#PE1344).

PE8450 Lycanthropy
Werewolves — Shape-shifting — Were wolves
Nature Lycanthropy is used in two senses: (a) as a psychological disorder in which the patient imagines that he is an animal, especially a wolf, and acts as such; and (b) the popular belief that on occasion some humans can transform themselves, or be transformed, into a wolf or some other animal, which kills and eats humans.
In shape shifting any animate object can alter in form or substance and there are no limit to the kinds of objects susceptible to such alteration. Examples abound of the shape shifting of plants, animals, humans and gods.
Incidence As a psychological disorder this was common in the past, especially in the Middle Ages. Belief in shape-shifting is common in all cultures, notably as a power claimed by medicine men, and in one form or another the werewolf superstition is world-wide.
Counter-claim Without the belief in shape-shifting, the werewolf superstition could not have existed. Given the belief, people with a tendency to psychological disorder readily conceived themselves to be ferocious animals preying upon other human beings. This belief was then exploited by others.
Refs Baring-Gould, Sabine *Book of Werewolves*.
Broader Superstition (#PA0430)
Animal worship as a barrier to development (#PD2330).
Aggravates Vampirism (#PF6432).

PE8453 Environmental impacts of coal conversion plants
Nature The commercial coal conversion plants place a great strain on limited water supplies. Water is required for both cooling purposes and for process water. Discharged water can contain suspended and dissolved solids and may be acidic. Air pollutants arising from coal handling are dust, fine coal dusta, sulphur oxides and nitrogen oxides.
Broader Urban-Industrial pollution (#PC8745).
Aggravates Coal dust (#PG5093)   Air pollution (#PC0119)
Chemical contamination of water (#PE0535)   Sulphur compounds as pollutants (#PG6442)
Overexploitation of underground water resources (#PD4403).

PE8456 Acute myocardial infarction
Coronary thrombosis — Myocardial infarction
Refs Morris, Samuel M *Coronary Disease* (1987); Strano, A *Thrombosis and Cardiovascular Disease* (1984).
Broader Ischaemic heart disease (#PE8158).

PE8457 Unemployment as a perpetuator of failure
Broader Unemployment (#PB0750)   Demeaning community self-image (#PF2093).
Aggravates Personal failure (#PF4387).

PE8458 Protectionism in the high-technology industries
Broader Protectionism in international trade (#PC5842).
Underprovision of basic urban services (#PF2583).
Aggravates Abusive technological development under capitalism (#PD7463).

PE8463 Lack of funds for veterinary research
Broader Shortage of funds for research (#PF5419).
Related Lack of funds for medical research (#PE0612).
Aggravates Diseases of wild animals (#PD2776)
Inadequate control of animal diseases (#PD2781)
Difficulty in identifying carriers of animal diseases (#PF2775).

PE8464 Instability of trade in mineral tar and crude chemicals derived from coal, petroleum and natural gas
Broader Instability of chemicals trade (#PD0619).

PE8466 Sexual harassment in the working place
Nature Behaviour of a sexual nature which is unwanted, unwelcome and unreciprocated and which might threaten job security or create a stressful or intimidating working environment is sexual harassment in the working place. It ranges from comments, looks, jokes, suggestions, pin-ups or physical contact from touching and pinching to rape. It is not only physical but mental aimed at breaking down resistance to advances and destroying self confidence over time. Not only does it cause devastating economic, psychological and physical on its victims but also enormous cost to the employer. Personnel who leave to avoid harassment must be replaced. Health benefits are used more because of emotional and physical stress. Individual and workgroup productivity declines.
Incidence In a survey of employees of U.S. federal government 42 percent of women say they have been sexually harassed during the previous two years. Significantly only 5 percent of those harassed filed complaints. Some 14 percent of the men say they have been harassed.
In the United States the percentage of working women who have been harassed on the job range from 10 percent to 50 percent with most studies ranging around 20 percent to 30 percent. Less than 10 percent of all incidents involve women harassing men or harassment by the same sex.
Claim There is no difference between sexual harassment in the work place and rape except one is based on economic power and the other is based on physical power. If a person resists the person doing the harassing may have their advancement in the organization or even employment threatened.
Refs Baxter, Ralph H *Sexual Harassment in the Workplace* (1986); Gutek, Barbara A *Sex and the Workplace* (1985); MacKinnon, Catharine A and Emerson, Thomas I *Sexual Harassment of Working Women* (1979); Meyer, Mary C, et al *Sexual Harassment at Work* (1981).
Broader Harassment (#PC8558)   Sexual harassment (#PD1116).
Aggravated by Unethical personnel practices (#PD0862)
Unethical practices of employers (#PD2879).

PE8467 Lack of legislative control on advertising
Broader Inadequate laws (#PC6848).
Aggravates Misleading advertising (#PE3814).

PE8470 Photographical and optical goods trade instability
Broader Instability of trade in miscellaneous manufactured articles (#PE0814).

PE8472 Ineffective self-regulation in the food-processing industries
Broader Ineffective industry self-regulation (#PF5841)
Underprovision of basic urban services (#PF2583).
Aggravates Unethical food practices (#PD1045).

PE8473 Environmental hazards of glass and glass products manufacture
Broader Environmental hazards from non-metallic mineral products industries (#PE0890).

PE8475 Inadequate facilities for the transport of sanitary wastes
Broader Inappropriate sanitation systems (#PD0876)
Insufficient transportation infrastructure (#PF1495).

PE8476 Inhibition of crop growth by pollution
Crop contamination — Adverse impact of pollution on growth of crops
Broader Pollution (#PB6336).
Narrower Health hazards of air pollution for plants (#PE4744).
Aggravated by Agricultural pollution (#PD0563).

PE8477 Apathy toward improvement of urban life styles
Broader Apathy (#PA2360)   Deteriorating quality of life (#PF7142)
Limiting effect of individual survivalism (#PF2602).
Aggravates City life style (#PF1437).

PE8489 Meal and flour of wheat or meslin trade instability
Broader Instability of trade in cereals and cereal preparations (#PE1769).

PE8495 Drunkenness of military personnel and troops
Broader Drunkenness (#PE8311).
Aggravated by Military disobedience (#PD7225)   Inadequate army discipline (#PD2543).

PE8496 Lack of adequate clothing supply for dwarfs, midgets and giants
Broader Discrimination against giants (#PE5578)
Discrimination against dwarfs and midgets (#PE2635).
Aggravated by Human physical genetic abnormalities (#PD1618).

PE8498 Shortage of animal and vegetable oils and fats and waxes
Broader Long-term shortage of animal and vegetable oils and fats (#PE1188).

PE8501 Shortage in tanning and dyeing materials
Broader Long-term shortage of chemicals (#PE1261).

PE8502 Ineffective self-regulation in the pharmaceutical and medical devices industries
Refs Blum, et al (Eds) *Pharmaceuticals and Health Policy* (1981).
Broader Ineffective industry self-regulation (#PF5841)
Limiting effect of individual survivalism (#PF2602).
Narrower Marketing of banned pharmaceutical drugs in developing countries (#PE6036).
Aggravates Inadequate medical care (#PF4832)   Inconsistent medical practices (#PF1624).

PE8503 Lack of participation in local welfare programmes
Broader Non-participation (#PC0588)   Lack of community participation (#PF3307).
Aggravates Inadequate social welfare services (#PC0834).

PE8504 Agricultural effluent
Liquid agricultural wastes
Nature Slurry – a mixture of animal faeces, urine and water – and silage liquor, a by-product of the process by which cut grass is compressed and fermented into silage, are polluting water sources.
Broader Wasted water (#PD3669)   Agricultural wastes (#PC2205)
Inappropriate sanitation systems (#PD0876).
Narrower Water pollution from animal production (#PE3934).
Aggravates Lake pollution (#PD8628)   Marine pollution (#PC1117).

PE8505 Facial or oral injury or deficiency
Broader Human disease and disability (#PB1044).
Aggravates Speech disorders (#PE2265).

PE8506 Inadequate incentives for increased productivity in developing countries
Broader Capitalism (#PC0564)   Limited availability of functional information (#PF3539).

PE8507 Discrimination against men in public services
Broader Discrimination against men (#PC3258)   Discrimination in public services (#PD8460).

♦ **PE8509 Inability of communities to take advantage of training for business, industry or public service**
Aggravates Inequality of employment opportunity in developing countries (#PD1847).

♦ **PE8510 Discharge from sores of infected persons**
Aggravates Syphilis (#PE2300).

♦ **PE8513 Weak national identity due to tribalism**
Broader Tribalism (#PC1910) Weak national identity in post-colonial countries (#PE8610).

♦ **PE8514 Long-term shortage of animal feedstuffs excluding unmilled cereals**
Broader Long-term shortage of food and live animals (#PE0976).
Related Denial to working animals of restorative nourishment and rest (#PE4793).

♦ **PE8516 Lack of credit facilities for agricultural producers**
Limited access to credit for small farmers — Limits of single cropping on credit
Refs Indian Society of Agricultural Economics *Seminar on Problems of Small Farmers (1967, Bombay)* (1968).
Broader Rural underdevelopment (#PC0306)
Limited local availability of capital reserve (#PF2378).
Aggravates Disincentives against farming (#PD7536).

♦ **PE8518 Lack of social accounting in the business community**
Nature Company's success or failure is judged according to its ability to make a profit. The method of calculation does not include information related to all the economic, human and social aspects of a business community. This kind of social accounting takes into account internal environment in a company, worker motivation, utilization of world's resources, pollution, product properties and the company's relationship to the outside world.
Broader Obstacles to community achievement (#PF7118).

♦ **PE8525 Distortion of international trade by restrictive controls over foreign investment**
Nature The measures, applied unilaterally to discipline the actions of the foreign firms, have been made to ensure that the policies of foreign firms conform to the development objectives of the host country and which have derived from the observation that investment by such firms was generally aimed at serving the domestic market only and that such foreign firms had a pronounced tendency to import rather than purchase from domestic sources. In most of the developing countries import controls are introduced for balance-of-payments reasons and export performance requirements are often necessitated by the anxiety of the host country to reduce the burden of its balance-of-payments problems.
Broader Non-tariff barriers to international trade (#PC2725)
Distortion of international trade by discriminatory customs and administrative entry procedures (#PE2603).
Reduces Inappropriate foreign investment (#PD8030).

♦ **PE8527 Endangered species of dasyures and marsupial mice**
Incidence 35 species; 1 endangered, 4 rare, 1 possibly endangered.
Broader Endangered species of marsupials (#PD1762).

♦ **PE8533 Excessive dependence on computer models of complex system behaviour**
Claim The large computer simulation studies that have become so much a part of the resource management and environmental assessment literature are part of the problem, not of the solution. Large-scale modelling activities meant to inform decision and policy processes are not fulfilling their promise. Such work is more adversarial than most realize. It is one-sided, presenting but one perspective on a future rich in potentialities. Quality control and professional standards are low. Data inputs tend to have obscure, unknown, or unknowable empirical foundations, and the relevance of much data, even if valid, is unknown. Such work tends to ignore the most interesting aspect of any analysis, namely the assumptions through which it was elaborated.
Counter-claim Computer models have the potential for handling large amounts of technical information in a reproducible systematic way.
Broader Human errors and miscalculations (#PF3702).
Aggravates Inept theoretical models (#PF7182) Limitations of surprise-free thinking (#PF7700)
Inadequate models of socio-economic development (#PF9576).

♦ **PE8537 Endangered species of phalangers and cuscuses**
Incidence 6 species; 1 rare.
Broader Endangered species of marsupials (#PD1762).

♦ **PE8540 Unexplored entertainment alternatives in the countryside**
Aggravates Unstimulating entertainment (#PF8105).
Aggravated by Lack of local services for community leadership training (#PF2451).

♦ **PE8542 Vulnerability of computer systems**
Risks to computers and computer records — Inadequate safeguards against computer crimes — Inadequate prevention of computer disasters
Nature Computers are widely used in private, commercial, and governmental operations to store vast quantities of information. In many cases the information constitutes a record vital to the continued functioning of organizational systems, particularly in the case of personal data, stock data, and financial or accounting records. Computer systems are vulnerable in a number of ways. Data may be lost due to equipment failure, electrical surges, operator error, software error, over or under heating, theft or computer viruses. Equipment may be damaged in fire, flood, earthquake or any of a number of natural or man made disasters. Programmers, operators or maintenance personnel may damage hardware, software or data bases. Data may be damaged in transit over telecommunications lines or when storage media, such as tapes or disks, are shipped. The losses include replacement or repair of equipment, software or data; lost business or other transactions; and losses of employee time.
The destruction of a computer, its files and backup media through accident can be a disaster. A business' ability to carry on operations can be halted through fire or explosion in a computer room. Records and files of a mainframe, mini or micro computer can be destroyed through operator error, electrical surges like from lightning, radiation from microwave oven and similar devices or physical abuse.
Incidence A recent study by the US Department of Defence concluded that only 30 out of 17,000 computers used by the DOD met minimum standards for protection from attack. The computers were vulnerable to a broad range of high-tech hit-and-run spying techniques, such as "spoof" programs which appear to be conducting routine activities while they are actually collecting passwords or other useful information; or the implementation of undetectable instructions into the software which might order the alteration or destruction of highly classified data. In 1990 in the USA a small computer error at an exchange in New Jersey led to a major failure of a continent-wide telephone network over a period of 9 hours, during which 50 per cent of the long-distance calls could not be made.
Refs Clark, David D, et al *Computer Security*; Cooper, Arlin J *Computer and Communications Security* (1989).
Broader Inadequate security system (#PD6589) Human errors and miscalculations (#PF3702)
Inadequate disaster prevention and mitigation (#PF3566).
Narrower Uncontrolled use of computer data (#PF4176).
Aggravates Theft of data (#PD2957) Computer viruses (#PD3102)
Injurious accidents (#PB0731) Computer-based crime (#PE4362)
Abuse of computer systems (#PD9544)
Environmental degradation by automobiles (#PE6142).
Aggravated by Espionage (#PC2140) Computer illiteracy (#PG2575)
Unethical practices by employees (#PD4334) Unreliability of computer software (#PE4428)
Prohibitive cost of business security (#PJ7767).
Reduced by Computer control (#PG5683).

♦ **PE8545 Pollution of water by infected faeces**
Pollution of water supplies with human excreta
Broader Impurities in waste water (#PD0482) Water system contamination (#PD8122).
Related Recreational contact with sewage (#PE6685).
Aggravates Dysentery (#PE2259) Typhoid fever (#PD1753).
Aggravated by Inappropriate sanitation systems (#PD0876).

♦ **PE8547 Private international trade and investment disputes**
Broader Conflict (#PA0298).

♦ **PE8548 Violation of rights of transsexuals**
Discrimination against transsexuals — Denial of the rights of transsexuals
Nature Transsexuals reject the gender assigned to them, of which they have the complete morphology and even the genetic structure. They have a strong feeling and a deeply rooted belief that they belong to the opposite sex. Psychologically, they find their situation incongruous and feel alienation vis-à-vis their body and for their genital organs. They desire to live as a member of the opposite sex and they seek to change their sexual appearance through hormones and surgery. The rights of transsexuals are violated in a number of ways. They may be refused hormone treatments and surgery. If they are allowed medical treatment they may be refused the right to change their civil status, their social security and welfare records, their birth certificate and their identity papers. They may be refused the right to marry. Their rights to have a private live may be violated.
Broader Violation of the rights of sexual minorities (#PD1914).
Related Male homosexuality (#PF1390).
Aggravates Transsexualism (#PF3277).

♦ **PE8551 Scarcity of appropriate transport**
Broader Scarcity (#PA5984) Restrictions on effective means of transport (#PF2798)
Inadequate transportation facilities for rural communities in developing countries (#PE6526).
Aggravates Overloaded vehicles (#PE4127).

♦ **PE8553 Inadequate primary health care**
Broader Inadequate health services (#PD4790)
Inadequate health care in least developed countries (#PE9242).
Aggravates Inadequate prevention of disabilities (#PF0709).

♦ **PE8556 Silver and platinum ores trade instability**
Broader Instability of trade in metalliferous ores and metal scrap (#PE0553).

♦ **PE8560 Instability of trade in U235-depleted uranium**
Broader Instability of trade in non-ferrous metals (#PE1406).

♦ **PE8564 Destruction of civilian populations and institutions**
Broader War damage in civilian areas (#PD8719).
Related Human death (#PA0072).
Aggravated by Denial of human rights in armed conflicts (#PC1454).

♦ **PE8569 Accentuated inequality between rural and urban development**
Nature In developing countries there is a desire to develop urban areas as show-pieces to create deceptive image of ultra-modern development imitating the life styles of developed countries.
Broader Human inequality (#PA0844).
Aggravated by Urban bias (#PF9686)
Inappropriate education in developing countries (#PF1531).

♦ **PE8571 Environmental hazards of tanning and dyeing industries**
Broader Environmental hazards from chemicals (#PC1192).

♦ **PE8572 Inadequate development of enterprises in developing countries**
Inadequate development of private sector in developing countries
Nature The indigenous enterprise sector in developing countries is typically represented by small-scale entrepreneurs such as small traders, people providing transport services, artisans and small manufacturers. Such small-scale ventures and very small so called micro-enterprises are believed to account for a high proportion of industrial and service activities and employment in most developing countries.
Aggravates Destruction inherent in development (#PF4829).
Aggravated by Insufficient enterprise capital (#PG7832)
Bias against private enterprise (#PF1879)
Restrictions against small enterprise (#PD5584)
Geographically disadvantaged countries (#PF9247)
Nationalization of domestic enterprises (#PD1994)
Inappropriate education in developing countries (#PF1531)
Lack of management skills in developing countries (#PE0046)
Lack of natural resources in developing countries (#PD3625)
Foreign exchange shortage in developing countries (#PD3068)
Dependence of developing countries on food imports (#PE8086)
Weakness of infrastructure in developing countries (#PC1228)
Inadequate domestic savings in developing countries (#PD0465)
Undeveloped financial markets in developing countries (#PE5902)
Suppression of private enterprise in socialist countries (#PD2048)
Inadequacy of the domestic market in developing countries (#PD0928)
Insufficient use of natural resources in least developed countries (#PE0273)
Abuse of monopoly power of state-owned or state-controlled enterprises (#PE0988)
Inadequate supply of appropriate trained manpower in developing countries (#PE6243)
Lack of awareness of potential for investment in small, inner-city enterprises (#PF2042)
Inadequate regulation of the restrictive business practices of state enterprises (#PE0225)
Instability in export trade of developing countries producing primary commodities (#PD2968).

♦ **PE8574 Ineffective self-regulation in the consumer goods manufacturing industries**
Broader Ineffective industry self-regulation (#PF5841).
Aggravates Misrepresentation of information to consumers (#PE6877).

♦ **PE8579 Bronchiectasis**
 Broader Lung disorders and diseases (#PD0637).
 Aggravates Pleurisy (#PG6121).

♦ **PE8580 Risks in transfer of ownership**
Faulty title of property
Nature The transfer of ownership of property involves several risks. The deed may overstate the extent of the granters interest in the property. The transfer may be invalid for some reason. Flaws in the title may make the resale of the property impossible or reduce the price a purchaser may want to pay.
 Broader Risk (#PF7580).

♦ **PE8583 Discriminatory exchange rate policies**
Competitive exchange depreciation — Distortion of international trade by selective monetary controls
 Broader Restrictive monetary practices (#PF8749)
 Mismanagement of exchange rate system (#PF1874)
 Non-tariff barriers to international trade (#PC2725).
 Related Financial destabilization of world trade (#PC7873)
 Distortion of international trade by discriminatory customs and administrative entry procedures (#PE2603).

♦ **PE8584 Imbalance between rural and urban incomes in developing countries**
 Broader Rural-urban income differential (#PE5022).
 Aggravates Unequal income distribution within developing countries (#PD7615).

♦ **PE8589 Congenital anomalies of musculoskeletal system**
 Broader Human physical genetic abnormalities (#PD1618).
 Narrower Chondrodystrophy (#PG3757)   Osteogenesis imperfecta (#PG6747)
 Congenital anomalies of spine (#PG9817).
 Aggravated by Corruption of ruling classes (#PD8380).

♦ **PE8591 Instability of trade in fresh, frozen or chilled meat**
 Broader Instability of trade in meat and meat preparations (#PE0755).

♦ **PE8592 Insufficient and inappropriate energy equipment in developing countries**
Nature Energy activities require technologies ranging from the simple to the sophisticated. The developing countries must import most of the equipment and materials from developed countries. Often the lack of knowledge of required equipment characteristics and evaluation of technical specification results in too sophisticated equipment or functional "mismatch".
 Broader Inadequate infrastructure (#PC7693).
 Aggravates Energy crisis (#PC6329).
 Aggravated by Inadequate maintenance equipment (#PD6520).

♦ **PE8593 Instability of trade in asses, mules and hinnies**
 Broader Instability of trade in live animals (#PD1376).

♦ **PE8595 Underutilization of livestock in least developed countries**
 Broader Least developed countries (#PD8201)
 Depriving prisoners of medical treatment (#PD1480)
 Non-productive use of cattle and livestock (#PD1802).

♦ **PE8597 Absence of legal entities in developing countries**
 Broader Unrecognized socio-economic interdependencies (#PF2969)
 Deficiencies in national and local legal systems (#PF4851).

♦ **PE8599 Lack of conservation of energy by the private sector**
 Broader Unsustainable development of energy use (#PC7517).
 Narrower Inadequate cooking stoves (#PE7904).
 Aggravates Thermal pollutants (#PC1609).

♦ **PE8600 Restrictive practices in trade in chemicals**
 Broader Restrictive trade practices (#PC0073).
 Reduces Trade in products for chemical warfare (#PE3808).

♦ **PE8601 Diphtheria**
 Broader Bacterial disease (#PD9094).
 Aggravates Neuritis (#PG4892)   Tonsillitis (#PE2292)   Anaphylaxis (#PG4894)
 Weakened heart (#PG4890)   Child mortality in developing countries (#PE5166)
 Paralysis of throat, eye, limb and respiratory muscles (#PE8173).
 Aggravated by Natural disasters (#PB1151).

♦ **PE8603 Unequal distribution of goods and services**
 Broader Human inequality (#PA0844).
 Related Disparities in distribution of communication resources and facilities (#PD2762).
 Aggravated by Elitism (#PA1387).

♦ **PE8605 Friedreichs ataxia**
Nature Friedreichs ataxia is a hereditary disease resembling loco-motor ataxia and due to degenerative changes in nerves of the spinal cord and the brain. I usually occurs before the twentieth years of life and often affects several siblings. Its chief symptoms are a unsteady gait, loss of knee jerks, difficulties of speech, trembling hands, head and eyes, deformity of the feet and curvature of the spine. The sufferer may live twenty or thirty years helpless and steadily getting worse.
 Broader Paraplegia (#PE2945)   Genetic defects and diseases (#PD2389)
 Hereditary disorders of the central nervous system (#PE7915).
 Aggravated by Genetic inbreeding (#PD7465)   Consanguineous marriage (#PC2379).

♦ **PE8606 Meal and flour other than wheat or meslin shortage**
 Broader Long-term shortage of cereals and cereal preparations (#PE1218).

♦ **PE8610 Weak national identity in post-colonial countries**
 Refs Rhee, Kyu H *Struggle for National Identity in the Third World* (1983).
 Narrower Weak national identity due to tribalism (#PE8513).

♦ **PE8612 Sand and gravel trade instability stone**
 Broader Instability of trade in crude fertilizers and crude minerals, excluding coal, petroleum and precious stones (#PE0760).

♦ **PE8615 Herpes**
Nature There are two forms of herpes caused by the Herpes Simplex virus, ordinary cold sores are caused by HSV1 and genital herpes caused by HVS1 and HVS2. Sexual intercourse is not the only way genital herpes is transmitted. It can be passed on by touching, oral sex or even towels. For some it can be almost without symptoms and for others it causes great discomfort. Once contracted it last for life. People with herpes have a higher risk of being infected with AIDS.
Incidence In the United States as many as 40 million people may be afflicted with genital herpes. Various experts estimate 100,000 to 1 million Australians have genital herpes representing one in six of the sexually active. In the United Kingdom between 100,000 and 150,000 people have it and the number is increasing at a rate of 10,000 a year.
 Refs Albert, Daniel M et al *Herpesvirus* (1974); Roizman, Bernard and Lopez, Carlos *Immunobiology and Prophylaxis of Human Virus* (1985).
 Broader Virus diseases (#PD0594)   Stigmatized diseases (#PD7279)
 Sexually transmitted diseases (#PD0061).
 Aggravates AIDS (#PD5111)   Still-birth (#PD4029)   Infectious revenge (#PD5168).

♦ **PE8617 Health hazards of air pollution for children**
Air pollution impact on the lungs of children
 Refs Brasser, L J and Colley, J R T *Chronic Respiratory Diseases in Children in Relation to Air Pollution* (1980).
 Broader Air pollution (#PC0119).

♦ **PE8624 Diseases of the oesophagus, stomach and duodenum**
 Refs Appelman, Henry D *Pathology of the Esophagus, Stomach and Duodenum* (1984).
 Broader Human disease and disability (#PB1044)
 Diseases and deformities of the digestive system (#PC8866).
 Narrower Ulcers (#PE2308)   Gastritis (#PG2250)   Duodenitis (#PG4198)
 Gastric disorders (#PE1599)   Disorders of the bowel (#PE6553)
 Diseases of oesophagus (#PE8636).

♦ **PE8626 Long-term shortage of petroleum and petroleum products**
 Broader Long-term shortage of mineral fuels and lubricants (#PE1712).

♦ **PE8630 Ineffective self-regulation in the mining industries**
 Broader Ineffective industry self-regulation (#PF5841).

♦ **PE8631 Unexplained appearances and disappearances of persons and objects**
 Broader Unexplained phenomena (#PF8352).
 Aggravates Missing persons (#PD1380).

♦ **PE8632 Prohibitive cost of insurance**
 Broader Limited availability of investment capital for urban renewal (#PF3550).
 Narrower Prohibitive cost of life assurance (#PE8736).
 Prohibitive cost of product liability protection (#PE4404)
 Excessive costs and unsuitability of insurance for land-locked developing countries (#PE5896).
 Aggravates Inadequate insurance (#PF8827).

♦ **PE8633 Inadequate working conditions for land tenants**
 Broader Inadequate conditions of work of agricultural workers (#PE4243).

♦ **PE8634 Inhuman methods of conception**
Artificial insemination — In vitro conception — Test tube conception — Ectogenesis
Claim Whilst the desire of sterile couples to have children is understandable, it must be stressed that not everything that is desired is right and not even scientific progress can make it so. What matters most is not that a child is obtained, but that it is obtained in a human manner.
Counter-claim What could be more human than science ? What could be more beast like than natural conception ?
 Refs Bonnicksen, Andrea *In Vitro Fertilization* (1989); Edwards, R G and Purdy, J M *Human Conception In Vitro* (1982); Edwards, R G, et al *Implantation of the Human Embryo* (1985).
 Aggravates Surrogate parenting (#PF2747)
 Commercialization of human embryos (#PE6038).
 Aggravated by Eugenics (#PC2153)   Childlessness (#PC3280).

♦ **PE8636 Diseases of oesophagus**
 Refs Delarue, et al *Esophageal cancer* (1988); Siewert J R and Holscher, A H (Eds) *Diseases of the Esophagus* (1987).
 Broader Diseases of the oesophagus, stomach and duodenum (#PE8624).

♦ **PE8637 Lack of access for prisoners' defence**
 Broader Denial of human rights in the administration of justice (#PD6927).
 Aggravates Human torture (#PC3429).
 Aggravated by Mis-classification of political prisoners (#PF3020).

♦ **PE8640 Control by transnationals of the global communications industry**
 Broader Control of national economic sectors by transnational enterprises (#PE0042).
 Aggravates Lack of local commercial services (#PF2009)
 Insufficient communications systems (#PF2350).

♦ **PE8643 Denial to animals of legal protection of their rights**
Nature Animal rights are defended where strong pro-animal forces have been able to gain legislative power. Otherwise they are largely ignored.
 Broader Social unaccountability (#PC1522)   Denial of political and civil rights (#PC0632)
 Inadequate legislation for animal welfare (#PE5794).
 Narrower Denial to animals of the right to life (#PF8243)
 Denial to animals of the right to dignity (#PE9573)
 Boredom of captive and domesticated animals (#PF7681)
 Denial to animals of the right to a natural death (#PE8339)
 Denial to working animals of limitation of working hours (#PE6427)
 Denial to animals of the right to freedom from mass killing (#PE9650)
 Denial to food animals of the right to freedom from suffering (#PE3899)
 Denial to working animals of restorative nourishment and rest (#PE4793)
 Denial to experimental animals of the right to freedom from suffering (#PE8024)
 Denial to animals of the right to the attention, care and protection of humankind (#PF5121)
 Denial to animals of the right to conditions of life and liberty proper to their species (#PE6270).

♦ **PE8644 Perpetual preoccupation for sustenance of urban life style**
 Broader Deteriorating quality of life (#PF7142)
 Limiting effect of individual survivalism (#PF2602).

♦ **PE8645 Underpayment of teachers**
Underpaid teachers — Non-competitive teachers' salaries
Nature An increasing number of teachers at all levels are leaving for private business because of low pay in the teaching profession. University professors in high technology fields, such as, computers, space and biochemistry have been active in business for a long time but the numbers are rapidly increasing. Many educational institutions are concerned that academic research and teaching may be effected negatively as a result.

PE8645

Broader Underpayment for work (#PD8916)
Inadequate working conditions of teachers (#PE7165)
Underdeveloped sources of income expansion (#PF1345).
Aggravates Shortage of trained teachers (#PD8108).
Aggravated by Decline in government expenditure on education (#PF0674).

♦ **PE8651 Environmental hazards from manufacture of plastic products**
Broader Environmental hazards from chemicals (#PC1192)
Environmental hazards from plastic materials (#PD8566).

♦ **PE8659 Excessive proportion of income spent on food**
Aggravates Low general income (#PD8568)   Marginal level of family income (#PD6579).

♦ **PE8660 Illiteracy among women in developing countries**
Illiteracy among mothers
Incidence Despite significant efforts by many developing countries in the 1960s and 1970s, low enrolment and high female illiteracy remain major problems in Africa, South Asia and with pockets persisting in parts of Latin America (Bolivia, Haiti, Honduras and among native populations in the Andean countries) and in the Middle East and North Africa. Positive gains in education have been reversed by severe budgetary constraints owing to the recession problems of adjustment faced by many developing countries in 1987, especially those of Africa and Latin America.
Claim Women from the working and peasant classes are thought to have no business, no right to associate with books. Besides they have neither time to study nor money to buy books. The book world, then, becomes the special realm for the elite classes.
Refs Kelly, David H and Kelly, Gail P *Education of Women in Developing Countries*.
Broader Illiteracy among women (#PE4380)   Illiteracy in developing countries (#PD8329).
Aggravates Discrimination against women in education (#PD0190)
Ignorance of women concerning primary health care (#PD9021).

♦ **PE8664 Protectionism in the defence and arms industries**
Broader Protectionism in international trade (#PC5842).
Related Plutonium overproduction (#PD2539)
Risk of unintentional nuclear war generated by the strategy of deterrence (#PF4162).

♦ **PE8667 Reservation overbooking**
Involuntary boarding denial — Airline bumping
Incidence Airlines overbook passengers by 15 to 30 per cent. As a consequence, it has been estimated that the number of airline passengers bumped annually in North America and Europe varies from 5 to 10 per 10,000, rising to 15 for some carriers. In the case of London Heathrow, on average there are 50 to 80 passengers bumped daily by the 70 airlines operating there.
Claim Bumping is practised by airlines in order to favour high-yield, long-haul connecting passengers, regular customers and groups.
Counter-claim In any specific year, 21 European airlines lost one million seats through no-shows, representing a loss of revenue of $200 million, namely 1.5 per cent of passenger volume. Overbooking compensates for this loss and means that airlines are able to offer more seats on each flight.
Aggravated by Air traffic congestion (#PD0689).

♦ **PE8670 Dependence on family for social security**
Broader Social insecurity (#PC1867)   Dependence within extended families (#PD0850).

♦ **PE8671 Lack of diversity in medical science**
Aggravates Excessive medical intervention in childbirth (#PE7705).
Aggravated by Lack of understanding of spiritual healing (#PF0761).

♦ **PE8674 Endangered species of pocket mice and kangaroo mice**
Incidence 71 species; 1 endangered; 2 rare
Broader Endangered species of rodents (#PD3481).

♦ **PE8677 Medical experimentation on mentally impaired persons**
Broader Abusive experimentation on humans (#PC6912)
Unethical experiments with drugs and medical devices (#PD2697).

♦ **PE8681 Social isolation of women at home**
Social isolation of housewives
Refs Kurian, George and Ghosh, Ratna *Women in the Family and the Economy* (1981).
Broader Social isolation (#PC1707).
Aggravates Failure to employ skills of home-bound educated women (#PD8546).

♦ **PE8682 Instability of trade in manila fibre**
Broader Instability of vegetable fibre trade, excluding cotton and jute (#PE1513).

♦ **PE8683 Endangered species of cavies and dolichotids**
Incidence 15 species; none endangered
Broader Endangered species of rodents (#PD3481).

♦ **PE8684 Endangered species of African cane rat**
Incidence 2 species; none endangered
Broader Endangered species of rodents (#PD3481).

♦ **PE8687 Unsuitability of development projects for available labour**
Broader Lack of participation in development (#PF3339).
Aggravated by Overuse of machinery and equipment in development projects (#PE8166).

♦ **PE8688 Instability of trade in fresh tomatoes**
Broader Instability of vegetable trade (#PE1711).

♦ **PE8690 Lack of participation of disabled persons in decision-making**
Broader Ineffective means for participation in decision making (#PE8430).

♦ **PE8692 Health hazards from water development schemes**
Broader Hazards to human health (#PB4885).
Aggravates Human disease and disability (#PB1044).
Aggravated by Inadequate environmental monitoring (#PF4801).
Reduced by Inadequate water system infrastructure (#PD8517).

♦ **PE8693 Shortage of meat and meat preparations**
Broader Long-term shortage of meat and meat preparations (#PE1490).

♦ **PE8698 Prohibitively expensive housing for the poor**
Inaccessible housing for the poor
Nature Subsidized public housing projects intended for the poor are often out of their reach and captured by middle- or high-income households.
Broader Physically inaccessible services (#PC7674)
Prohibitive cost of goods and services (#PD1891).
Related Scattered housing locations (#PJ0760)   Over-spacing of suburban housing (#PE1708).
Aggravates Inadequate housing (#PC0449)   Discrimination in housing (#PD3469)
Illegal occupation of unoccupied property (#PD0820)
Inequality of employment opportunity in developing countries (#PD1847).
Aggravated by Scarcity of residential land (#PD8075).

♦ **PE8699 Lack of education for women immigrants**
Broader Discrimination against women in education (#PD0190).

♦ **PE8700 Inadequate adaptation of policy to educational difficulties**
Broader Inappropriate policies (#PF5645).
Aggravates Inadequate education of indigenous peoples (#PC3322).
Aggravated by Failure to integrate knowledge to empower humanity in response to the global problematique (#PF8753).

♦ **PE8701 Enforcement of religion in capitalist countries**
Broader Dependence on religion (#PF0150).

♦ **PE8702 Food manufacturing industry wastes**
Broader Unproductive use of resources (#PB8376)
Industrial waste water pollutants (#PD0575).
Aggravates Chemical contaminants of food (#PD1694)
Biological contamination of food (#PD2594).
Aggravated by Environmental hazards from food processing industries (#PE1280).

♦ **PE8706 Growing size and impersonality of firms**
Broader Inhumanity (#PB8214)   Excessive size of social institutions (#PF8798).
Narrower Growth in size of production unit (#PF8155).
Aggravates Strikes (#PD0694)   Reduced images of humanness (#PF4323)
Ineffectiveness of individual participation in large communities (#PF6127).
Aggravated by Bureaucracy as an organizational disease (#PD0460).

♦ **PE8707 Lack of purchasing power in developing countries**
Claim 1. People in developing countries are hungry because they are poor, not because there is insufficient food or land available there. Even when people are starving to death, there may be quite enough food within the country but they cannot afford to purchase what is available. 2. One of the major contradictions in the international economy is the fact that productive resources lie idle in developed-market economy countries because the largest group of potential customers, namely the developing countries, lack the means to purchase the output, even to the extent of satisfying basic human needs.
Broader Limited purchasing power (#PD8362).
Aggravates Hunger (#PB0262)   Instability of rice trade (#PE0696)
Excess production capacity (#PD0779)   Malnutrition in developing countries (#PD8668).
Aggravated by Economic hardship (#PD9180)   Poverty in developing countries (#PC0149)
Inappropriate modernization of agriculture (#PF4799)
Unavailability of land for agricultural purposes in developing countries (#PE5024).

♦ **PE8708 Endangered species of kangaroos and wallabies**
Incidence 45 species; 2 endangered, 2 vulnerable, 5 rare.
Broader Endangered species of marsupials (#PD1762).

♦ **PE8712 Denial of the right to picket**
Broader Denial of the right of trade union association (#PD0683).
Related Sanctions against trade union workers (#PD0610)
Denial of right to collective bargaining (#PE3970).

♦ **PE8715 Glass and glass products manufacture underdevelopment**
Broader Underdevelopment of non-metallic mineral products industries (#PE1851).

♦ **PE8716 Inadequate technical training**
Ineffective technical training
Broader Inadequate education (#PF4984).
Narrower Inadequate industrial retraining programmes (#PF4013).
Related Obsolete vocational skills (#PD3548).
Aggravates Narrow range of practical skills (#PF2477)
Lack of expertise in agricultural techniques (#PE8752).

♦ **PE8717 Loss of beneficial plants and animals**
Broader Endangered species of plants and animals (#PB1395).
Aggravates Excessive use of chemicals to control pests (#PD1207).
Aggravated by Pesticide hazards to wildlife (#PD3680)
Diseases of beneficial insects (#PD2284).

♦ **PE8720 Loss of civil capacity for married women**
Broader Discrimination against women before the law (#PD0162).
Aggravates Refusal to issue travel documents, passports, visas (#PE0325).

♦ **PE8721 Uncoordinated use of computers and automation**
Nature Manufacturing and administrative processes have been often divided into their basic elements handled by specialized equipment and labour. Different operations have not been efficiently tied together.
Narrower Proliferation of computers (#PE3959).
Related Uncoordinated business development (#PU1109).

♦ **PE8724 Punitive amputations**
Nature Amputation is a form of judicial punishment in a number of countries marking victims as criminals for the rest of their lives. Victims of accidents with similar wounds are automatically and usually unconsciously assumed to be criminals also.
Incidence Amputation has been reported in the following countries: **Af** Mauritania. **Am** Bolivia, El Salvador. **As** Iran Islamic Rep, Pakistan, Saudi Arabia.
Claim Not only is amputation by the state barbaric but unjust for the victim. In most cases, there is no recourse to appeal and there is no possibility of effectively reversing the sentence.
Broader Torture through mutilation (#PD7576)   Unjust punishments for crimes (#PD4779).

♦ **PE8728 Inadequate conditions of work and employment of public service personnel**
Refs ILO *Conditions of Work and Employment of Public Service Personnel of Local, Regional or Provincial Authorities. Report II. 2nd Session of the Joint Committee on the Public Service* (1976).
Broader International imbalance in the quality of life (#PB4993).

EMANATIONS OF OTHER PROBLEMS

♦ **PE8729 Water pollution by fertilizers**
Water pollution by pesticides
Incidence In UK it it estimated that annual current spending of £10 million would be needed in 20 years time to keep nitrate concentrations in all public water supplies within EC-agreed levels.
Refs American Society for Testing and Materials *Pesticides, Resource Recovery* (1986); EPA *Pesticides* (1990).
  Broader Pesticides as pollutants (#PD0120)
  Environmental hazards from fertilizers (#PE1514).
  Related Water pollution from animal production (#PE3934).
  Aggravates Water pollution (#PC0062)   Soil salinization (#PE1727).

♦ **PE8732 Environmental hazards from cereals and cereal preparations**
  Broader Environmental hazards from food and live animals (#PC1411).

♦ **PE8733 Maldistribution of students enrolled in school**
  Aggravates Obstacles to education (#PF4852).
  Aggravated by Student absenteeism (#PE4200).

♦ **PE8735 Imbalances between growth of the labour force, urban population and the overall growth of the economy**
  Related Urban population and the overall growth of the economy labour force (#PE8437).

♦ **PE8736 Prohibitive cost of life assurance**
High life assurance premiums — Inadequate life insurance in developing countries
  Broader Inadequate insurance (#PF8827)   Prohibitive cost of insurance (#PE8632).

♦ **PE8738 Endangered species of agoutis and acouchis**
Incidence 17 species; not known to be endangered
  Broader Endangered species of rodents (#PD3481).

♦ **PE8739 Endangered species of dog wolves and foxes**
Incidence 37 species; 4 vulnerable, 4 endangered, 2 rare
  Broader Endangered species of carnivores (#PD3482).

♦ **PE8740 Instability of trade in dressed fur skins**
  Broader Instability of trade in manufactured goods (#PE0882).

♦ **PE8741 Instability of trade in preserved fruit**
  Broader Instability of trade in fruit and vegetables and preparations thereof (#PE0961).

♦ **PE8742 Unethical practices of philanthropic organizations**
Irresponsible voluntary organizations — Corruption in charitable organizations — Charity frauds
Claim Advantage may be taken of the favourable image and tax status of philanthropic bodies to perpetrate financial frauds, whether through a single organization or by seemingly innocent transfers of funds within a network of such bodies. The disinterest and benevolent character of such bodies may also be used as a front for the dissemination of misleading information or for more or less disguised forms of brainwashing (as in the case of certain cults). Such bodies may also be used as a vehicle for egotistical individuals and personality cults.
  Broader Financial frauds (#PE2414).
  Aggravates Compassion fatigue (#PF2819).
  Aggravated by Personality cults (#PC1123)   Manipulative cults (#PE6336)
  Political indoctrination (#PD1624)
  Misuse of nonprofit associations as front organizations by government (#PE0436)
  Inappropriate taxation of not-for-profit and philanthropic charitable organizations (#PF3049).

♦ **PE8746 Denial of social security to nationals who have lived abroad**
  Broader Social insecurity (#PC1867).
  Aggravates Restrictions on freedom of movement between countries (#PC0935).

♦ **PE8747 Restrictions on foreigners leaving country of sojourn**
  Broader Restrictions on freedom of movement between countries (#PC0935).

♦ **PE8748 Instability of trade in non-alcoholic beverages**
  Broader Instability of trade in beverages (#PE1680).

♦ **PE8750 Endangered species of moles and desmans**
Incidence 19 species; 1 vulnerable, 1 rare
  Broader Endangered species of insectivores (#PD3479).

♦ **PE8752 Lack of expertise in agricultural techniques**
  Broader Shortage of technical skills (#PF6500)   Unavailability of appropriate expertise (#PF7916)
  Limited availability of technical agricultural and business training (#PF2698).
  Related Limited technical skills (#PF8594).
  Aggravates Dependence on external expertise (#PG8011)
  Decreasing genetic diversity of animals (#PC1408).
  Aggravated by Inadequate technical training (#PE8716).

♦ **PE8758 Hindrances to international spread of new technologies**
  Broader Barriers to the international flow of knowledge and educational materials (#PF0166).
  Aggravated by Competitive acquisition of arms (#PC1258).

♦ **PE8766 Prizefighting**
Boxing
Nature What distinguishes boxing from most other sports is that causing hurt and injury is its very aim. It is also the result. Medical evidence shows that repeated blows to the head endured by boxers produce cumulative and irreversible brain damage.
Claim Violence is the vested interest of boxing. The noble art of self-defence consists of deliberately attempting to inflict brain damage on an opponent.
Counter-claim Boxers not only voluntarily enter the boxing ring, but increasingly know exactly the risks they are taking. How can a society that allows smoking, drinking and pollution condemn an individual choice of occupation ?
  Broader Cruel sports (#PC1323).
  Narrower Punch drunk syndrome (#PE0768).

♦ **PE8768 Migration of rural population to cities**
Drain of skills to cities
Refs Fetherolf, Louftì Martha *Rural Development, Taking into Account the Problems of the Indigenous Populations as well as the Drift of the Rural Population to the Cities and its Integration in the Urban Informal Sector* (1986).
  Broader Internal migration (#PF4009)   Underutilization of human resources (#PF3523)
  Emigration of trained personnel from developing to developed areas (#PD1291).
  Narrower Business skills drain (#PG8609)   Youth migration to cities (#PJ7859)
  Community expertise drain (#PG8926).
  Aggravates Urban overcrowding (#PC3813)   Rural depopulation (#PC0056)
  Endangered cultures (#PB8613).
  Aggravated by Urban bias (#PF9686).
  Reduced by Restrictions on freedom of movement within countries (#PE8408).

♦ **PE8769 Instability of trade in regenerated cellulose and artificial resins**
  Broader Instability of chemicals trade (#PD0619).

♦ **PE8771 Shortage of manpower in socialist countries**
  Broader Socialism (#PC0115).
  Aggravates Underproductivity (#PF1107).

♦ **PE8772 Iron ore and concentrates trade instability**
  Broader Instability of trade in metalliferous ores and metal scrap (#PE0553).

♦ **PE8776 Inadequate organizational mechanism for international action**
  Broader Inadequate organizational mechanisms to act against problems (#PF2431).
  Aggravates Ineffectiveness of international organizations and programmes (#PF1074)
  Inadequate coordination of international organizations and programmes (#PD0285).

♦ **PE8782 Reduction of soil fertility downstream due to impoundment**
Incidence River Nile used to deposit 100 million tons of sediment a year on nearly one million hectares of land in the Nile Valley. After the building of the Aswan Dam, the Nile deposits only a few tons of sediment per year, most of that comes as a result of riverbed erosion, meaning the gradual loss of riverbed soils. To compensate the silt trapped behind the dam, Egypt must apply artificial fertilizers on an ever increasing scale.
  Broader Soil infertility (#PD0077)
  Environmentally harmful dam construction (#PD9515).
  Aggravates Infertile land (#PB8585).
  Aggravated by Silting of water systems (#PD3654).

♦ **PE8787 Dislocation of productive units by foreign investment**
  Broader Inappropriate foreign investment (#PD8030).

♦ **PE8788 Discrimination against women without children**
  Broader Discrimination against women (#PC0308).
  Aggravates Denial of right of family planning (#PE5226)
  Discrimination against unmarried women (#PD8622)
  Discrimination against women in employment (#PD0086)
  Discrimination in family planning facilities (#PD1036).

♦ **PE8789 Shortage of equipment and materials needed to act against problems**
  Broader General obstacles to problem alleviation (#PF0631).

♦ **PE8790 Environmental hazards from the construction industry**
  Broader Ecologically unsustainable development (#PC0111).

♦ **PE8792 Import substitution as a barrier to subsequent economic growth**
Nature At the early stage of industrialization, import substitution takes the form of reducing imports of non-durable consumer goods, as well as their inputs, so as to permit indigenous manufacturing to develop and to save foreign currency needed for the imports of producer goods. At the later stage, import substitution involves replacement of producer goods and intermediate inputs, as well as consumer durables, so as to increase the reliance for economic growth on the supply of domestically produced goods. Excessive and prolonged protection of import substitution industries has often adverse consequences on efficiency and competitiveness and impedes export promotion.
  Broader Trade protectionism (#PC4275).

♦ **PE8794 Endangered species of rabbits and hares**
Incidence 53 species; 2 endangered
  Broader Endangered species of rabbits (#PD3480).

♦ **PE8795 Protectionism in the consumer products industries**
  Broader Protectionism in international trade (#PC5842).

♦ **PE8796 Water losses from irrigation systems**
Water losses from dams — Water looses from irrigation canals
Nature In hot, dry areas of the world evaporation rates from open surfaces can exceed local rainfall.
  Narrower Over-watering during irrigation (#PE9311)
  Seepage water losses from irrigation systems (#PE9088).
  Aggravates Disruption of the hydrological cycle (#PD9670).
  Aggravated by Aquatic weeds (#PD2232)   Mismanagement of irrigation schemes (#PE8233)
  Deterioration in water supply systems (#PD9196).

♦ **PE8797 Lack of secondary schools in the countryside**
Remote junior-high schools in the countryside
  Broader Inadequate educational facilities (#PD0847).
  Aggravates Inadequate community care for transient urban populations (#PF1844).
  Aggravated by Closure of schools (#PJ8556)   Distant schooling facilities (#PJ7693)
  Decline in government expenditure on primary education (#PG9451)
  Unperceived relevance of formal education in rural communities (#PF1944).

♦ **PE8799 Jurisdictional conflict and antagonism within the specialized agencies of the United Nations**
  Broader Jurisdictional conflict and antagonism within international organizations (#PD0047).
  Aggravates Ineffectiveness of the United Nations system of organizations (#PF1451).

♦ **PE8802 Biased media-image of foreign groups and peoples**
  Broader Misleading information (#PF3096)   Discrimination against foreigners (#PD6361).
  Aggravated by Xenophobia (#PD4957)
  Inadequate cultural integration of immigrants (#PC1532).

♦ **PE8803 Inadequate rehabilitation of juvenile offenders**
Refs Wheeler, Gerald R *Counterdeterrence* (1978).
  Broader Denial of rights to prisoners (#PD0520)   Inadequate rehabilitation facilities (#PD1089).
  Related Repressive detention of juveniles (#PD0634).
  Aggravates Juvenile delinquency (#PC0212).

PE8806 Lack of variety of social life forms
Lack of variety in life style — Lack of diversity of social roles — Lack of variety in life styles
Aggravates Gypsy persecution (#PE1281) Political monoculture (#PF4405).
Aggravated by Nomadism (#PF3700) Ethnocide (#PC1328)
Collectivism (#PF2553) Threatened sects (#PC1995)
Forced assimilation (#PC3293) Cultural deprivation (#PC1351)
Sexual discrimination (#PC2022) Cultural fragmentation (#PF0536)
Endangered tribes and indigenous peoples (#PC0720)
Discrimination against indigenous populations (#PC0352).
Reduced by Occultism (#PF3312) Hero worship (#PF2650)
Superstition (#PA0430) Group marriage (#PF3288)
Cultural barriers (#PB2331) Plural society tensions (#PF2448).

◆ PE8809 Instability of miscellaneous mining and quarrying
Broader Instability of mining and quarrying industry (#PE0993).

◆ PE8813 Diseases of the spinal cord
Spinal cord injuries — Motor neurone disease
Refs Brainard, Williard T *Injuries of the Spinal Cord* (1985); Guttman, L *Spinal Cord Injuries* (1976); Leatherman, K D and Dickson, R A *The Management of Spinal Deformities* (1988); Maurice–Williams, R S *Spinal Degenerative Diseases* (1981); Wilberger, James E *Spinal Cord Injuries in Children* (1986).
Broader Diseases of the central nervous system (#PE9037).
Narrower Progressive muscular atrophy (#PE5188).

◆ PE8814 Lack of commitment to common symbols
Lack of a common symbol
Broader Lack of commitment (#PF1729).
Related Collapse of cultural dreams (#PF1084).
Aggravates Loss of cultural identity (#PF9005) Collapse of common values (#PF1118)
Reduction in symbolic celebrations (#PF1560)
Lack of symbolism in local relationships (#PF2063)
Vulnerability of marriage as an institution (#PF1870)
Decreasing participation in collective religious worship (#PF8905).
Aggravated by Cynicism (#PF3418) Scepticism (#PF3417)
Ideological conflict (#PF3388) Superstitious symbolic acts (#PF8754)
Disintegration of accepted myths (#PF8887) Deterioration of human environment (#PC8943)
Symbols unrelated to human experience (#PF9070)
Individualistic retaining of local tradition (#PF1705)
Modern disruption of traditional symbol systems (#PF6461)
Absence of convincing symbols connecting the individual's life to the cosmos (#PF1081).

◆ PE8815 Increase in atmospheric concentration of methane
Environmental hazards from methane
Nature Methane is released into the atmosphere by a variety of processes, the most prominent being enteric fermentation by ruminants, release from both natural wetlands and rice fields, biomass burning, coal mining operations and leakage of natural gas during transmission.
Incidence It is estimated that in the UK the long–term greenhouse effect of methane from domestic and commercial waste is equivalent to over 50 per cent of all carbon dioxide produced from coal–fired power stations.
Claim Methane has 25 time the capacity of carbon dioxide to trap heat in the atmosphere. Its pre–industrial level was 0.75 parts per million and the 1986 level was 1.65 parts per million. The current annual increase is 1 percent.
Broader Air pollution (#PC0119) Irritant gases and vapours (#PE6820)
Increase in trace gases in the atmosphere (#PD1354).
Narrower Methane gas emissions from landfill sites (#PE1256)
Methane gas emissions from animal husbandry (#PE5399).

◆ PE8816 Instability of trade in animal feedstuffs, excluding unmilled cereals
Broader Instability of trade in food and live animals (#PD1434).
Related Denial to working animals of restorative nourishment and rest (#PE4793).
Aggravates Population pyramid (#PS4772).

◆ PE8817 Restrictions on nationals returning to their country
Broader Restrictions on freedom of movement between countries (#PC0935).
Aggravates Errant nationals (#PE0812).

◆ PE8822 Inflammatory affections of the bronchial tubes and lungs
Broader Lung disorders and diseases (#PD0637)
Acute respiratory infections (#PE7591).
Aggravated by Whooping cough (#PE2481).

◆ PE8823 Refusal of sale because of buyer's race
Broader Racial discrimination (#PC0006) Discriminatory business practices (#PD8913).

◆ PE8824 Non–payment of compensation to victims of motor accidents
Delay in compensation of victims of motor accidents — Neglect of victims of motor accidents
Broader Uncompensated damages (#PD7179)
Inadequate assistance to victims of accidents (#PE4086)
Inadequate assistance to victims of human rights violations (#PD5122).
Aggravated by Non–payment of compensation to victims of crime (#PE3913).

◆ PE8828 Instability of trade in non–electrical machinery
Broader Instability of trade in machinery and transport equipment (#PD0620).

◆ PE8829 Use of undue influence to obstruct the administration of justice
Nature The influence of those with political power, wealth, social connections, control over career advancement, or connections with those prepared to enforce physical threats, can be used to obstruct or pervert the administration of justice. This may take the form of persuading the competent authorities to drop charges or suppress evidence, or persuading those with extreme views not to testify or to perjure themselves. Alternatively efforts may be made to use the evidence, suitably interpreted, to convict an innocent person or ensure that a person only marginally involved takes full responsibility for the crime. Such a scapegoat role may be undertaken voluntarily in exchange for appropriate compensation when then sentence has been completed. Such compensation may take the form of support for the person's family, especially if the sentence involves life imprisonment or death.
Broader Abuse of influence (#PC6307) Hindrance of law enforcement (#PD5515).
Aggravates Perjury (#PD2630) Intimidation (#PB1992) Death threats (#PD0337)
Police corruption (#PD2918) Unreported harassment (#PF4729)
Citizen powerlessness (#PJ8803) Corruption of the judiciary (#PD4194)
Mistrust of system of justice (#PD8217) Official fabrication of evidence (#PD8716)
Inadequate national law enforcement (#PE4768) Arbitrary enforcement of regulations (#PD8697)
Ineffective legislation against organized crime (#PE6699).
Aggravated by Elitism (#PA1387) Nepotism (#PD7704)
Organized crime (#PC2343) Bureaucratic corruption (#PC0279)
Corruption of ruling classes (#PD8380).

◆ PE8833 Inadequate recreational facilities for disabled people
Broader Inadequate recreational facilities (#PF0202).
Aggravated by Human disability (#PC0699).

◆ PE8843 Radioactive and associated materials trade instability
Broader Instability of trade in chemical elements and compounds (#PE0500).

◆ PE8845 Endangered species of old world porcupine
Incidence 15 species; none endangered
Broader Endangered species of rodents (#PD3481).

◆ PE8847 Endangered species of new world porcupines
Incidence 11 species; 1 rare
Broader Endangered species of rodents (#PD3481).

◆ PE8857 Inadequate maternal and child health care
Broader Inadequate medical care (#PF4832).
Related Inadequate child welfare (#PC0233)
Inadequate health care in family planning (#PD1038)
Inadequate medical care for pregnant women (#PE4820).
Aggravates Infant mortality (#PC1287).
Aggravated by Resistance to incorporating midwives in medical care systems (#PE4901).

◆ PE8858 Transport as a hindrance to economy
Broader Ineffective rural transport (#PF2996).
Aggravates Narrow range of practical skills (#PF2477).

◆ PE8861 Instability of trade in sheep and goats
Broader Instability of trade in live animals (#PD1376).

◆ PE8864 Overemphasis of governments on large firms
Broader Rural unemployment (#PF2949).

◆ PE8865 Draining of resources due to alcohol
Broader Subsistence approach to capital resources (#PF6530).
Aggravated by Alcohol abuse (#PD0153) Drunken driving (#PE2149)
Alcohol–related violence (#PE7084) Alcoholic intoxication at work (#PE2033)
Substance abuse during control of complex equipment (#PE0680)
Excessive portrayal of substance abuse in the media (#PE3980).
Reduced by Uncomplicated alcohol withdrawal (#PE0375).

◆ PE8867 Distortion of international trade by selective indirect taxes and import charges
Nature Taxes on international trade are highly favoured by the developing countries, because they generate about a third of the tax revenue with limited administrative costs. Import tariffs penalize consumers, promote inefficient patterns of production and implicitly tax exports.
Broader Non–tariff barriers to international trade (#PC2725)
Distorting effects of commodity taxes on the transaction of goods and nonfactor services (#PE9220).
Aggravates Ineffective economic structures in industrial nations (#PE4818).

◆ PE8873 Gap between industry and the universities
Aggravates Lack of job satisfaction (#PF0171).

◆ PE8875 Trade instability of electrical machinery, apparatus and appliances
Broader Instability of trade in machinery and transport equipment (#PD0620).

◆ PE8878 Electric machinery apparatus and appliances shortage
Broader Long–term shortage of machinery and transport equipment (#PE1436).

◆ PE8879 Discrimination against women in divorce rights
Refs Arendell, Terry *Mothers and Divorce* (1986).
Broader Discrimination before the law (#PC8726)
Exclusion of women from decision making (#PE9009)
Discrimination against women before the law (#PD0162).

◆ PE8880 Restrictive practices in trade in animal and vegetable oils and fats
Broader Restrictive trade practices (#PC0073).

◆ PE8881 Unrestrained use of force in administration of justice
Excessive use of force by law enforcement officials — Excessive use of force by police — Excessive use of force by military personnel
Nature There are numerous occurrences in many countries of excessive and/or completely unwarranted use of force by law enforcement officials and military personnel during public gatherings or arrests resulting in civilian loss of life or injuries.
Broader Injustice (#PA6486) Unethical practices by police forces (#PD9193).
Aggravates Maltreatment of civilians (#PD5560) Inhibiting social patterns (#PF0193)
Inequitable administration of justice (#PD0986).

◆ PE8882 Distortion of international trade by restrictive controls on movement of labour
Broader Non–tariff barriers to international trade (#PC2725)
Distortion of international trade by discriminatory customs and administrative entry procedures (#PE2603).
Aggravates Ineffective economic structures in industrial nations (#PE4818).
Aggravated by Restrictions on freedom of movement between countries (#PC0935).

◆ PE8883 Unavailability of scholarship funds for developing country students
Broader Unavailability of scholarship funds for students (#PE3569)
Restrictive effects of external capital on development (#PF3318).
Aggravates Prohibitive cost of education (#PF4375).

◆ PE8884 Inefficacy of consultation in the decision–making process
Ineffectiveness of consultation procedures
Aggravates Failure of individuals to participate in social processes (#PF0749).

◆ PE8893 Excessive bureaucratic requirements for welfare benefits
Broader Unchecked power of government bureaucracy (#PD8890).
Narrower Excessive paperwork (#PF5856).

Aggravates Limited access to social benefits (#PF1303).

♦ **PE8896 Instability of trade in animal oils and fats**
 Broader Instability of trade in animal and vegetable oils and fats (#PE0735).

♦ **PE8898 Non-payment of compensation for forced relocation**
Inadequate compensation for forced relocation
 Broader Non-payment of reparations by government (#PE4446).
 Aggravated by Government insensitivity (#PF2808).
 Involuntary mass resettlement (#PC6203)
 Compulsory acquisition of land by government (#PC1005).

♦ **PE8899 Violation of the right to the international freedom of movement of shipping**
Violation of the right of international freedom of navigation
 Broader Obstacles for international ocean shipping (#PD5885).
 Restrictions on freedom of movement between countries (#PC0935).

♦ **PE8900 Inadequate facilities for mentally disabled criminals**
Lack of care for intellectually disabled convicts
 Broader Inadequate rehabilitation facilities for the mentally handicapped (#PE8151).
 Aggravates Criminal insanity (#PD9699).
 Aggravated by Criminals (#PC7373).

♦ **PE8901 Roots and tuber trade instability vegetable products**
 Broader Instability of vegetable trade (#PE1711).

♦ **PE8902 Devaluation of education by survival needs**
 Broader Obsolete educational values (#PF8161)
 Limited horizons produced by survival living (#PF3161)
 Inhibiting effects of traditional life-styles (#PF3211).

♦ **PE8906 Environmental hazards of metalliferous ores and metal scrap**
 Broader Environmental hazards from inedible crude non-fuel materials (#PE0546).

♦ **PE8907 Insufficient leisure time for women**
 Broader Limited leisure time (#PF9062) Underutilization of human resources (#PF3523)
 Obstacles to efficient utilization of time (#PF7022).
 Aggravated by Discrimination against women (#PC0308).

♦ **PE8911 Birth trauma**
Denial of right to dignified birth
Nature Birth trauma is the anxiety experienced by a child at birth and is the prototype of all subsequent anxiety in a person's life. It is especially anxiety causing in medical institutions where the mother's increased anxiety is conveyed to the child.
 Broader Personal life crises (#PD4840) Denial of right to dignity (#PE6623)
 Complications of childbirth (#PC9042).
 Aggravates Neonatal mortality (#PD9750).

♦ **PE8912 Osteomyelitis, periostitis and other infections involving bones and acquired deformities of bone**
Osteomyelitis
Refs Stagnara, Pierre and Dove, John *Spinal Deformity* (1988); Van Rens, and Kayser, F H *Local Antibiotic Treatment in Osteomyelitis and Soft Tissue Infections* (1981).
 Broader Diseases and injuries of bone (#PE3822).
 Narrower Pyogenic infections (#PG5610).
 Aggravated by Typhoid fever (#PD1753).

♦ **PE8915 Destruction of economy due to war**
Economic burden of warfare
Nature The economic costs of warfare stem directly from the human casualties, the destruction of infrastructure and manufacturing plants. Moreover, it disrupts trade and international economic relations through its effects on the prices and availability of commodities, the dangers associated with the use of certain shipping routes, etc.
 Broader War damage in civilian areas (#PD8719)
 Disastrous consequences of war (#PC4257).
 Narrower Poverty as a consequence of war (#PE5252).
 Disruption of food supply due to military activities (#PE8979).
 Related War-time disruption of economies and production facilities (#PD8851).
 Aggravates Damage to crops (#PJ3949) Destruction of land resources (#PJ8397).
 Aggravated by Natural disasters (#PB1151)
 Socio-economic burden of militarization (#PF1447)
 Denial of human rights in armed conflicts (#PC1454).

♦ **PE8916 Inaccessible job market**
Distant male employment — Distant work places — Distant job locations
 Broader Rural unemployment (#PF2949) Physically inaccessible services (#PC7674)
 Limited availability of permanent employment in inner-cities (#PE1134).
 Related Underprovision of basic urban services (#PF2583)
 Local traditions of cultural isolation (#PF1696)
 Inappropriate application of traditional values (#PF2256).

♦ **PE8918 Instability of food and drink industries**
 Broader Instability of production of food and live animals (#PD2894).

♦ **PE8920 Instability in trade of travel goods**
 Broader Instability of trade in miscellaneous manufactured articles (#PE0814).

♦ **PE8923 Debauchery**
Orgies
Nature People have always craved for intense mental states, which are most easily gratified by induced excitement, alcohol and other drugs, sex and food. Excess and dissipation supply change from everyday routine and social rules and they discharge energy.
 Broader Decadence (#PB2542).
 Aggravates Public displays of sexuality (#PE4882).
 Aggravated by Satanism (#PF8260) Indecent art (#PE5042)
 Sexual immorality (#PF2687).

♦ **PE8924 Inadequate community care for handicapped persons**
Denial of right to adequate health care for the disabled — Neglected welfare of handicapped persons
 Broader Inadequate social welfare services (#PC0834)
 Obstacles to community achievement (#PF7118)
 Denial of right to adequate medical care (#PD2028).
 Narrower Family rejection of physically handicapped (#PE2087)
 Inadequate rehabilitation facilities for the mentally handicapped (#PE8151)
 Inadequate vocational rehabilitation facilities for disabled persons (#PE7317).
 Related Inadequate welfare services for the deaf (#PD0601)
 Inadequate welfare services for the blind (#PD0542)
 Vulnerability of the disabled during states of emergency (#PD0098).
 Aggravates Human disability (#PC0699) Physically dependent people (#PD7238)
 Physically handicapped children (#PD0196).
 Aggravated by Prohibitive cost of private medical care (#PF8016)
 Stress on families of the physically or mentally handicapped (#PD1405).

♦ **PE8925 Physiological malnutrition arising from mental factors**
 Broader Malnutrition (#PB1498) Mental disorders (#PD9131).
 Narrower Anorexia nervosa (#PE5758).

♦ **PE8928 China, pottery and earthenware industries instability**
 Broader Instability of non-metallic mineral products industry (#PE2599).

♦ **PE8932 Diseases of nerves and peripheral ganglia**
Diseases of the peripheral nervous system — Diseases of peripheral autonomic nervous system — Diseases of cranial nerves
Refs Crescenzi, G Serlupi *A Multidisciplinary Approach to Myelin Diseases* (1987); Enna, C D *Peripheral Denervation of the Hand* (1989); Gilliatt, R W and Asbury, A K *Peripheral Nerve Disorders* (1984).
 Broader Diseases of the nervous system (#PC8756).
 Narrower Neuritis (#PG4892) Sciatica (#PE2428) Neuralgia (#PG8198)
 Polyneuritis (#PG6802) Facial paralysis (#PG5569).
 Aggravates Paralysis (#PD2632).

♦ **PE8933 Dependence on excitement and danger among young people**
 Aggravates Human torture (#PC3429).

♦ **PE8934 Incorporation of carcinogens into consumer goods**
 Broader Grievances of consumers (#PD7567).
 Carcinogenic chemical and physical agents (#PD1239).
 Aggravated by Occupational cancer (#PE3509).

♦ **PE8935 Endangered species of zagoutis and coypu and hutia**
Incidence 11 species; 1 vulnerable, 4 rare)
 Broader Endangered species of rodents (#PD3481).

♦ **PE8936 Endangered species of weasels, badgers, skunks, otters**
Incidence 67 species; 6 endangered
 Broader Endangered species of carnivores (#PD3482).

♦ **PE8938 Decline in nutritional quality of food**
Nature The diet of developed countries consist of adulterated and denatured foods, from which the most precious essential factors have been removed by colouring, bleaching, heating and preserving, leaving only the bacteria.
 Broader Lack of qualitative excellence (#PF5703).
 Related Excessive consumption of specific foodstuffs (#PC3908).
 Aggravates Food tastelessness (#PJ5809) Nutritional deficiencies (#PC0382).
 Aggravated by Chemicalized farming (#PD7993) Export of nutritious food (#PJ1365)
 Predominance of fast food (#PF5940) Unethical catering practices (#PE6615)
 Prohibitive cost of nutritious food (#PF1212).

♦ **PE8947 Forced confession**
Nature Forced confession always involves some form of torture. In sophisticated societies it is subtle methods of psychological torture than leave no physical evidence. In less sophisticated countries it is brutal physical torture. Frequently the goal of forced confession is not the truth but admission of guilt. At times this is to increase the conviction rate of a group of police officers. At other times it is to find a a person to blame publically for a crime that cannot be solved for one reason or another.
Incidence Forced confessions have been reported in the following countries: **Eu** Egypt, Libyan AJ, Mozambique, Rwanda, South Africa, Tunisia, Zimbabwe. **Am** Argentina, Bolivia, Chile, Grenada, Guatemala, Mexico, Paraguay, Uruguay. **As** Afghanistan, India, Iran Islamic Rep, Iraq, Korea Rep, Kuwait, Oman, Pakistan, Philippines, Saudi Arabia, Syrian AR, Taiwan (Rep of China), United Arab Emirates. **Eu** Albania, Poland, Romania, Turkey.
 Broader Police brutality (#PD3543) Abuse of government power (#PC9104).
 Narrower Forced political confessions (#PE3016)
 Forced confessions with drugs (#PE4888)
 Aggravates False confessions (#PE7252).

♦ **PE8950 Instability of trade in synthetic and regenerated fibres**
 Broader Instability of trade in unprocessed textile fibres and their waste (#PE1550).

♦ **PE8951 Lack of freedom of the press**
Denial of right to freedom of the press
 Broader Restriction of freedom of expression (#PC2162).
 Narrower Violation of right of trade unions to publish (#PE5312)
 Restriction on press coverage of legal affairs (#PE7945)
 Restrictions on direct news coverage of parliamentary affairs (#PF3072).
 Related Underground press (#PD2366).
 Aggravates Sedition (#PC2414).
 Aggravated by Harassment of the media (#PD0160)
 Journalistic irresponsibility (#PD3071).

♦ **PE8955 Petroleum and natural gas production instability**
 Broader Instability of mining and quarrying industry (#PE0993).

♦ **PE8956 Ineffective self-regulation in the chemicals manufacturing industries**
 Broader Ineffective industry self-regulation (#PF5841).

♦ **PE8957 Foreign private investment income outflow from developing to developed countries**
 Broader Outflow of financial resources from developing countries (#PC3134).
 Related Private domestic capital outflow from developing to developed countries (#PD3132).
 Reduces Inappropriate foreign investment (#PD8030).

♦ **PE8958 Sugar confectionery and preparations trade instability**
 Broader Instability of trade in sugar, sugar preparations and honey (#PE1567).

♦ PE8959 Limited transportation services in urban areas
Broader Underprovision of basic urban services (#PF2583).

♦ PE8961 Hernia
Nature Hernia is a protrusion of any organ, or part of the organ, into or through the wall of cavity which contains it. Hernias are most frequently abdominal, more rarely they involve brain or muscles.
Refs Devlin, H B *Management of Abdominal Hernias* (1988).
Broader Human illness (#PA0294)
Diseases and deformities of the digestive system (#PC8866).
Narrower Femoral hernia (#PG4467)   Inguinal hernia (#PG4465)
Umbilical hernia (#PG4466)   Incisional hernia (#PG4470)
Hernia of the lung (#PG4463)   Diaphragmatic hernia (#PG4468)
Oesophageal hiatus hernia (#PG4469).

♦ PE8963 Distrust of business by the community
Broader Distrust (#PA8653)   Obstacles to community achievement (#PF7118).
Aggravates Discouraging conditions for small business (#PD5603).
Aggravated by Incompatibility of rural values in urban cultures (#PF2648).

♦ PE8965 Military obstacles to peaceful development of space
Broader Military influence (#PD3385).

♦ PE8966 Tourist hazards
Holiday risks
Broader Risk (#PF7580).
Narrower Drowning (#PG2857)   Sports accidents (#PE4262)
Abuse of tourism for sexual purposes (#PE4437)
Unconvivial hotel environments for travellers (#PF6196)
Health hazards to tourists in developing countries (#PE7538).
Related Travel risks (#PD7716).
Social environmental degradation from recreation and tourism (#PD0826).
Aggravates Human disease and disability (#PB1044)
Instability in tourist dependent economies (#PF4112)
Natural environmental degradation from recreation and tourism (#PE6920).
Aggravated by Inadequate enforcement of safety regulations (#PD5001).

♦ PE8967 Endangered species of mice and hamsters
Incidence 602 species; 1 endangered, 2 rare
Broader Endangered species of rodents (#PD3481).

♦ PE8969 Iron and steel manufactures trade instability
Broader Instability of trade in manufactured goods (#PE0882).

♦ PE8970 Instability of trade in textile yarn fabrics, made-up articles
Broader Instability of trade in manufactured goods (#PE0882).

♦ PE8971 Underutilization of renewable energy resources
Soft energy paths
Refs United Nations Economic Commission for Europe *New and Renewable Sources of Energy* (1987).
Broader Underutilization of natural resources (#PF1459).
Narrower Underutilization of hydropower (#PF0345).
Underutilization of wind energy (#PF0373)
Underutilization of ocean energy (#PG5350)
Underutilization of solar energy (#PF0370)
Underutilization of geothermal energy (#PJ7988)
Underutilization of peat as an energy source (#PE8194)
Insufficient utilization of renewable biofuels (#PF0357)
Underutilization of oil shale as an energy source (#PF0445)
Underutilization of tar sands as an energy source (#PE8370).
Aggravates Worldwide misallocation of resources (#PB6719).
Aggravated by Unexplored energy alternatives (#PF7960)
Underdeveloped rural energy sources (#PF0393)
Unsustainable development of energy use (#PC7517)
Dependence of government revenues on exploitation of environmentally inappropriate products (#PD1018).

♦ PE8972 Hazardous combination of effects in the work place
Broader Occupational hazards (#PC6716).
Aggravated by Badly laid out work premises (#PJ2468).

♦ PE8978 Illiteracy in least developed countries
Broader Illiteracy in developing countries (#PD8329).

♦ PE8979 Disruption of food supply due to military activities
War-induced famine
Nature Hundreds of thousands of people die each year from malnutrition caused by disruption of food supplies by anti-government guerrillas and other military forces. Crops are ruined, farm labourers killed, distribution networks for food and agricultural supplies are disrupted and storage facilities are damaged or destroyed. Military activities coupled with natural disasters like drought create famine. War time conditions make it nearly impossible to mount relief efforts.
Broader Destruction of economy due to war (#PE8915).
Aggravates Poverty as a consequence of war (#PE5252)
Instability of trade in food and live animals (#PD1434).
Aggravated by War (#PB0593).

♦ PE8980 Meal and flour of wheat or meslin shortage
Broader Long-term shortage of cereals and cereal preparations (#PE1218).

♦ PE8983 Inadequate rehabilitation methods for the blind
Education of the blind
Broader Inadequate rehabilitation facilities (#PD1089)
Inadequate welfare services for the blind (#PD0542).
Aggravated by Physical blindness (#PD0568).

♦ PE8985 Insufficient possibilities for gathering of elders
Broader Declining sense of community (#PF2575).
Related Unnatural urban environments inhibiting sleeping in public (#PF6193).

♦ PE8986 Limited mechanical services in developing countries
Broader Lack of essential local infrastructure (#PF2115).
Related Inadequate industrial services in developing countries (#PF4195).
Aggravates Unexploited possibilities for local commerce (#PF2535).

♦ PE8988 Overdependence on education
Nature Education is believed to be the solution to most social problems including racism, segregation, war, environmental degradation, crime, and poverty. While the lack of understanding about problems may contribute to their existence, the solutions to these problems are far more complex and there is a great many examples of well educated racist or criminals. The social structures that perpetuate this type of problem are often more important to the society than the problem. The morality of the society and its basic assumptions often block the solution of problems, for example, a complex of assumptions about freedom of the individual, criminality, value of wealth and national sovereignty prevent the effective solutions to drug abuse. At the same time, the lack of political will and the incapacity to motivate the electorate prevent solutions to other problems.
Broader Inadequacy of prevailing mental structures to challenge of human survival (#PF7713).
Aggravates Inflexible attitudes toward community social services (#PF3083).
Aggravated by Inadequate education (#PF4984).

♦ PE8990 Ankylosing spondylitis
Broader Rheumatoid arthritis (#PE5081).

♦ PE8991 Endangered species of old world monkeys
Incidence 76 species; 4 endangered, 1 vulnerable, 6 rare.
Broader Endangered species of non-human primates (#PE1570).

♦ PE8997 Irregular repair services in the countryside
Broader Inadequate maintenance (#PD8984).
Aggravated by Unexplored alternatives for commercial development (#PF6548).

♦ PE8998 Instability of agricultural and livestock production
Refs Riemann, H P and Burridge, M J (Eds) *Impact of Diseases on Livestock Production in the Tropics* (1984).
Broader Instability of economic and industrial production activities (#PC1217).

♦ PE9000 Electrical torture
Nature High voltage electricity is being used extensively in torture. Electrical clips, and pointed electrodes are attached to sensitive areas and repeatedly switched on. Cattle prods are touched to skin. Individuals are strapped to metal frames and attached to electrodes. This may result in severe pain, loss of consciousness, involuntary muscle contractions or death.
Incidence Torture using electric shock has been reported in the following countries: **Af** Angola, Burundi, Cameroon, Congo, Ethiopia, Gabon, Kenya, Libyan AJ, Madagascar, Mali, Morocco, Rwanda, Somalia, South Africa, Zaire, Zambia, Zimbabwe. **Am** Argentina, Barbados, Brazil, Chile, Colombia, Grenada, Honduras, Mexico, Paraguay, Peru, Uruguay. **As** Afghanistan, D Bangladesh, India, Indonesia, Iran Islamic Rep, Iraq, Korea Rep, Pakistan, Philippines, Saudi Arabia, Syrian AR, Taiwan (Rep of China). **Eu** Albania, Greece, Italy, Turkey.
Broader Physical torture (#PD8734).

♦ PE9001 Remoteness of legal services in developing countries
Broader Underdevelopment of legal infrastructure (#PF4836)
Inaccessible commercial and financial services (#PE0718).
Aggravated by Complex regulations paralyzing small communities (#PF2444).

♦ PE9004 Illicit export of works of art
Smuggling of cultural artefacts
Refs Sujan, M A and Trivadi, V D *Smuggling* (1976).
Broader Illegal exports (#PD4116)   Unlawful business transactions (#PC4645)
Evasion of customs and excise duties (#PD2620).
Related Theft of works of art (#PE0323).
Aggravates Misappropriation of cultural property (#PE6074).
Aggravated by Abusive collection of specimens (#PE9417)
Unethical practices in transportation (#PD1012).

♦ PE9009 Exclusion of women from decision making
Lack of participation of women
Nature Although most governments accept the need to integrate women in development planning and decision-making, little has been done in many countries to implement this.
Broader Discrimination against women (#PC0308)
Lack of community participation (#PF3307)
Failure of individuals to participate in social processes (#PF0749).
Narrower Discrimination against women in divorce rights (#PE8879)
Neglect of the role of women in rural development (#PF4959).
Related Discrimination against women in politics (#PC1001)
Discrimination against women in public life (#PD3335).
Aggravates Discrimination against women before the law (#PD0162).
Aggravated by Rigidly entrenched social traditions in rural areas (#PF1765).

♦ PE9011 Jurisdictional conflict and antagonism within intergovernmental organizations
Broader Jurisdictional conflict and antagonism within international organizations (#PD0047).
Related Narrow legal definition of the family in developing countries (#PD1501).
Aggravates Ineffectiveness of intergovernmental organization and programmes (#PF0074)
Inefficient location of facilities of international organizations (#PE3538).
Aggravated by Bureaucratic factionalism (#PF7979).

♦ PE9013 Economic loss
Financial loss
Narrower Unreported financial losses (#PF8079)
Economic loss through slaughter of diseased animals (#PE8109)
Economic loss through reduced productivity of diseased animals (#PE8098).
Aggravated by Forgery (#PD2557)   Animal infertility (#PC1803)
Disruption of financial markets (#PD4511).

♦ PE9017 Mites as vectors of plant disease
Broader Insect vectors of plant diseases (#PD7732).

♦ PE9018 Apostasy
Nature Apostasy is the deliberate abandonment of one religion for another, or the renunciation of the creed of a faith, made voluntarily or under compulsion. If faith is the owing one's allegiance to and giving one's live for an ultimate reality, then abandoning that reality is to turn one's back on one's selfhood. At the social level, it turns an ultimate reality into a relative one and encourages other defections.
Refs Bromley, David G (Ed) *Falling from the Faith* (1988).
Broader Religious backsliding (#PF6826).
Narrower Disloyalty (#PJ1895).
Aggravates Fragmentation of religious belief (#PF3404).
Aggravated by Religious intolerance (#PC1808)   Denial of religious liberty (#PD8445).

EMANATIONS OF OTHER PROBLEMS

♦ PE9019 Environmental hazards from china, pottery and earthenware industries
Broader Environmental hazards from non-metallic mineral products industries (#PE0890).

♦ PE9020 Adverse effects of urbanization on climate
Broader Anthropogenic climate change (#PC9717).
Aggravates Inadvertent modifications to climate (#PC1288).

♦ PE9027 Distortion of international trade by quantitative restrictions
Imposition of import licensing — Import quotas — Banned imports
Refs OECD Staff *The Costs of Restricting Imports* (1988).
Broader Non-tariff barriers to international trade (#PC2725).
Distortion of international trade by discriminatory customs and administrative entry procedures (#PE2603).
Narrower Imposition of trade quotas for political reasons (#PE9762).

♦ PE9029 Environmental hazards from fruit and vegetables
Broader Environmental hazards from food and live animals (#PC1411).

♦ PE9037 Diseases of the central nervous system
Central nervous system disorders
Refs Meulen, V Ter and Katz, M *Slow Virus Infections of the Central Nervous System* (1977); Rosenberg, H S and Bernstein, J (Eds) *Central Nervous System Diseases* (1987).
Broader Diseases of the nervous system (#PC8756).
Narrower Epilepsy (#PE0661)   Migraine (#PE6357)   Meningitis (#PE2280)
Encephalitis (#PE2348)   Hydrocephalus (#PJ3190)   Syringomyelia (#PG2700)
Paralysis agitans (#PE2206)   Multiple sclerosis (#PE1041)   Cerebral paralysis (#PE0763)
Acute encephalitis (#PG5505)   Intra-cranial abscesses (#PG9132)
Meningo-encephalomyelitis (#PG5293)   Diseases of the spinal cord (#PE8813)
Phlebitis of intracranial venous sinuses (#PG8044).
Aggravates Lung disorders and diseases (#PD0637)
Disorders of joints and ligaments (#PD2283).

♦ PE9038 Critical illnesses
Dread disease
Nature A group of illnesses identified for special forms of insurance coverage designed to provide a lump sum on their diagnosis and following the death which usually follows.
Broader Human disease and disability (#PB1044).
Narrower Paraplegia (#PE2945)   Multiple sclerosis (#PE1041)
Human organ failure (#PG5012)   Chronic terminal illness (#PE4906).

♦ PE9048 Ineffective self-regulation in the health care sector
Broader Ineffective industry self-regulation (#PF5841).
Aggravates Inadequate health services (#PD4790).

♦ PE9049 Instability of trade in crude fertilizers
Broader Instability of trade in crude fertilizers and crude minerals, excluding coal, petroleum and precious stones (#PE0760).

♦ PE9052 Proliferation of nuclear weapons in developing countries
Proliferation of ballistic missiles in developing countries
Nature At least 22 developing countries are trying hard to build or buy ballistic missiles which can be fitted with both chemical and nuclear warheads. It is not in Europe, but in third world that the proliferation of missiles is accelerating, and their threat to regional stability and international security is the most profound.
Broader Militarization in developing countries (#PD9495)
Proliferation of strategic nuclear arms (#PD0014).

♦ PE9054 Racial discrimination in according financial loans
Refusal of loans because of borrower's race
Broader Racial discrimination (#PC0006).

♦ PE9055 Obstacles to restructuring production in the industrialized countries
Nature Restructuring is needed as nations change their relative competitiveness, but it is also required for domestic economic efficiency. These countries require active employment policies.
Broader Lack of community planning (#PF2605).

♦ PE9056 Environmental hazards of essential oils
Broader Environmental hazards from chemicals (#PC1192).

♦ PE9059 Instability of hunting, trapping and game propagation
Broader Instability of economic and industrial production activities (#PC1217).
Related Denial to animals of the right to freedom from mass killing (#PE9650).

♦ PE9060 Inadequate family planning education for men
Broader Inadequate family planning education (#PD1039).
Aggravated by Inadequacy of male contraceptive methods (#PF1069).

♦ PE9061 Unreported tax obligations
Understatement of tax liability — Misrepresentation of fiscal liabilities
Broader Underreported issues (#PF9148)   Suppression of information (#PD9146)
Circumvention of duties and assessments (#PD4882).
Aggravates Inequitable tax treaties between developed and developing countries (#PD1477).
Aggravated by Cheating (#PJ7991)   Ignorance (#PA5568)
Ineffective tax systems (#PF1462)   Human errors and miscalculations (#PF3702)
Unethical practices of employers (#PD2879).

♦ PE9071 Unemployment of educated older people
Failure to employ skills of educated elderly people — Premature retirement of people with valuable skills
Broader Educated unemployed (#PD8550)   Unemployed skilled labour (#PE1753)
Unemployment of older people (#PE5951).

♦ PE9072 Inequality inducing effects of remote sensing systems
Nature The developing countries are virtually completely dependent on the countries utilizing advanced remote sensing capacity and lack the resources of evaluating and utilizing the potential of this source of information; It leads to the situation that important date about a developing country may be better known in some foreign capitals than by the national government.
Refs Pontificia Accademia delle Scienze *Remote Sensing and Its Impact on Developing Countries* (1986).
Broader Lack of central planning structures in small communities (#PF2540).
Aggravates Human inequality (#PA0844).

♦ PE9073 Distortion of international trade by discriminatory formulation of equipment safety regulations
Broader Distortion of international trade by discriminatory customs and administrative entry procedures (#PE2603).

♦ PE9074 Environmental hazards of textile fibres and waste
Broader Environmental hazards from inedible crude non-fuel materials (#PE0546).

♦ PE9075 Deterioration of living standards of workers
Aggravates Poor living conditions (#PD9156).
Aggravated by Lack of livelihood standards (#PF1297).

♦ PE9076 Trade instability in inorganic chemicals, elements, oxides and halogen salts
Broader Instability of trade in chemical elements and compounds (#PE0500).

♦ PE9084 Glass and glass products industries instability
Broader Instability of non-metallic mineral products industry (#PE2599).

♦ PE9085 Uncertainty of survival of the human race
Nature The world's cropland area stopped increasing at the beginning of 1980s. Water is becoming scarce and it will constrain the growth and food production in some parts of the world. Deforestation sets in motion a chain of events that leads to the gradual deterioration of the soil fertility: 24 billion tons of top soil is lost each year to soil erosion. The gradual increase in the average temperature of the earth affects food production more than any other environmental change. Even if the land productivity could be raised during the nineties as high as 0.9 per cent a year, the growth in food output will be about half the projected rate of world population growth.
Broader Uncertainty (#PA6438)   Vulnerability of human organism (#PB5647).
Aggravated by Racial degradation (#PF9851)   Decline in human genetic endowment (#PF7815)
Irresponsibility towards future generations (#PF9455).

♦ PE9086 Benign neoplasm of kidney and other urinary organs
Broader Benign tumours (#PD8347)   Kidney disorders (#PE2053).

♦ PE9087 Overprescription of drugs
Nature Promotional pressure, patient expectations and belief in the basic benign nature of drugs lead to overprescribing by physicians. It is estimated that only half of the U.S. consumer's drug bill is medically justified. Physicians prescribe antibiotic to patients with colds knowing that they are ineffective against the common cold. This overprescribing has caused an epidemic of adverse reactions and doctor-induced diseases. The death rate from these diseases in the United States is estimated at between 30,000 and 130,000 per year. The development of drug resistance bacterial strains have emerged and are estimated to cause an additional unnecessary 100,000 deaths a year, particularly in hospitals. A number of diseases are being detected which are caused by treatments 20 or 30 years earlier or even to the patient's mother. For example, a woman may develop cervical cancer if their mother was treated with diethylstilbestrol during her pregnancy 20 years earlier.
Refs Medawar, Charles *Drugs and World Health* (1984).
Broader Drug abuse (#PD0094)   Misuse of medicines (#PD8402).
Narrower Abuse of antibiotics (#PE6629).
Related Laboratory testing errors (#PF5304).
Aggravates Prescription of inappropriate drugs (#PE3799).
Aggravated by Reverence for drugs in medicine (#PG2702)
Excessive proliferation of medical drugs (#PD0644).

♦ PE9088 Seepage water losses from irrigation systems
Nature The extent of seepage varies considerably according to climate, soil, and the length and type of distribution system. In many areas of the Middle East, anything between 10 and 70 per cent of the total volume of water conveyed through irrigation canals is lost to seepage.
Broader Water losses from irrigation systems (#PE8796).
Aggravates Rising water level (#PD8888).
Aggravated by Mismanagement of irrigation schemes (#PE8233).

♦ PE9089 Instability in tanning and dyeing trade
Broader Instability of chemicals trade (#PD0619).

♦ PE9091 Morbid fear of illness
Nosophobia
Broader Phobia (#PE6354)   Excessive caution (#PF6389).

♦ PE9092 Disruptive behaviour in schools
Misbehaviour in schools — Lack of school discipline — Indiscipline in schools
Refs American Association of School Administrators *Student Discipline*; Deitz, Samuel M and Hummel, John H *Discipline in the Schools* (1978); Gagne, Eve E *School Behavior and School Discipline* (1983); Golman and Coghill *Disruptive Behaviour in Schools* (1987); Karnes, Elizabeth L, et al *Discipline in Our Schools* (1983); Williams, Robert B and Venturini, Joseph L *School Vandalism* (1981); Wolfgang, Charles H and Glickman, Carl D *Solving Discipline Problems* (1986).
Broader Disruptive behaviour (#PD8544).
Aggravates Vandalism (#PD1350).

♦ PE9093 Irresponsible tobacco and cigarette advertising
Claim Cigarette advertising associate smoking not with illness and death, but with youth, beauty, athletics and sex.
Broader Misuse of advertising (#PE4225).
Aggravates Smoking (#PD0713).

♦ PE9095 Exploitation of land for the burial of the dead
Shortage of land for cemeteries
Broader Inefficiency (#PB0843)   Limited available land (#PC8160).
Related Human death (#PA0072).
Aggravates Excessive land usage (#PE5059)   Shortage of cultivable land (#PC0219).

♦ PE9096 Inadequacy of agricultural education in developing countries
Broader Limited agricultural education (#PF8835)
Inappropriate education in developing countries (#PF1531).
Aggravates Animal diseases (#PC0952)
Deterioration of domestic food production in developing countries (#PD5092).

♦ PE9099 Bilateralism in aid to developing countries
Bilateralism in aid
Broader Bilateralism in trade arrangements (#rF7642)
Restrictions imposed on aid to developing countries (#PF1492).

PE9099

Aggravates Underdevelopment (#PB0206)
Donor distortion of aid to developing countries (#PF7955)
Weakness in trade between different economic systems (#PC2724)
Reduced scope of intergovernmental development assistance (#PF4794).
Aggravated by Mismanagement of aid to developing countries (#PF0175).

♦ PE9103 Endangered species of palaearctic mole rats
Incidence 3 species; none endangered
Broader Endangered species of rodents (#PD3481).

♦ PE9105 Carelessness in dealing with infectious patients
Broader Negligence (#PA2658).
Aggravates Poliomyelitis (#PE0504).

♦ PE9107 Informational and procedural obstacles to world trade
Lack of transparency in international trade information
Claim Information on the trade implications of restrictive business practices is inadequate. Procedures for consultation, conciliation and the redressal of grievances are inadequate or confusing. Such lack of information severely hinders the establishment of multilateral trade disciplines.
Broader Administrative delays (#PC2550)   Obstacles to world trade (#PC4890).

♦ PE9110 Denial of the right to organize political parties
Broader Denial of right to participate in government (#PE6086).
Related Denial of right to hold public office (#PE5608)
Infringement on the functioning of legitimate organizations (#PE5222).
Aggravated by Racial discrimination in politics (#PD3329).

♦ PE9111 Gunshot wounds
Refs DiMaio, V J M *Gunshot Wounds* (1987).
Broader Injuries (#PB0855).
Aggravated by Human violence (#PA0429)   Firearm accidents (#PE2857)
Inadequate firearm regulation (#PD1970)
Proliferation of weapons in civilian hands (#PE2449).

♦ PE9114 Artificial opposition between manual and intellectual labour
Aggravates Fragmentation of the human personality (#PA0911).

♦ PE9115 Lack of meat and egg production
Broader Underdevelopment of agricultural and livestock production (#PD0629).
Reduced by Factory farming (#PD1562).

♦ PE9116 Restrictive conditions on loans to developing countries through intergovernmental facilities
Nature Access of developing countries to financial facilities available through intergovernmental bodies, such as the International Monetary Fund, is subject to strict conditionalities, high cost and may only be available for the short term.
Broader Restrictive monetary practices (#PF8749).
Narrower Autocracy of intergovernmental financial institutions (#PE2805).
Aggravates Excessive foreign public debt of developing countries (#PD2133)
Uncertainty of development programmes due to short-term loans (#PF4300).

♦ PE9120 Endangered species of bush rat and rock rat
Incidence 8 species; none endangered
Broader Endangered species of rodents (#PD3481).

♦ PE9121 Endangered species of genets, civets and mongooses
Incidence 82 species; 3 vulnerable, 1 endangered
Broader Endangered species of carnivores (#PD3482).

♦ PE9122 Criminalization of sexual relations out of wedlock
Broader Moralism (#PF3379)   Restrictive legislation (#PD9012).
Aggravated by Fornication (#PF5434).

♦ PE9123 Delay in issue of travel documents
Broader Excessive waiting times in government facilities (#PF5120).
Aggravates Restrictions on freedom of movement between countries (#PC0935).
Aggravated by Refusal to issue travel documents, passports, visas (#PE0325).

♦ PE9124 Ineffectiveness of international organization and programme action at the country level
Broader Ineffectiveness of international organizations and programmes (#PF1074).
Aggravated by Inadequate coordination of action on intergovernmental programmes at national level (#PE1375).

♦ PE9125 Use of artificial methods of promoting fast livestock growth
Broader Inferior meat quality from intensive animal farming units (#PE2770).

♦ PE9128 Dorsopatie
Aggravated by Unprepared adult leadership (#PF6462).

♦ PE9130 Shortage of leather, miscellaneous leather manufactures and dressed fur skins
Broader Long-term shortage of manufactured goods (#PE0802).

♦ PE9133 Insufficient financial resources for urban services
Broader Underprovision of basic urban services (#PF2583).

♦ PE9134 Adverse effects of power production on weather
Broader Anthropogenic climate change (#PC9717)
Environmental hazards from energy production (#PD6693).

♦ PE9135 Unaccompanied foreign travel in remote places
Aggravates White slave trade (#PD3303).

♦ PE9137 High cost of natural gas trade infrastructure
Broader Inadequate maintenance of infrastructure (#PD0645).

♦ PE9140 Corruption amongst relatives of government leaders
Corruption of wives of government leaders — Corruption of children of government leaders
Broader Nepotism (#PD7704)   Institutionalized corruption (#PC9173).

Aggravates Mediocrity of government leaders (#PF3962).
Aggravated by Resignation towards bribery (#PF8611)
Corruption of government leaders (#PC7587).

♦ PE9151 Inflammatory infections of the respiratory organs
Broader Lung disorders and diseases (#PD0637).
Aggravated by Measles (#PE1603).

♦ PE9164 Instability of trade in vegetable-based foodstuffs
Broader Instability of trade in food and live animals (#PD1434).
Aggravated by Introduction of high-yield crop varieties (#PF3146).

♦ PE9172 Inadequacy of psychiatry
Lack of diversity in psychiatry — Recreational psychiatry — Recreational therapy — Inadequacy of psychiatrists — Psychobabble — Lack of concensus among psychiatrists — Inept testimony of psychologists and psychiatrists
Nature Psychotherapy as a discipline is corrupt, is a waste of money and not infrequently causes a great deal of harm. The relationship between the therapist and the patient is one of power and weakness; power corrupts. Even when the therapist is well meaning, let alone when they are sadistic, they assert their own, normally conservative values on the patient. Patients are not infrequently sexually abused and mentally battered and on at least one occasion beaten to death.
Professional psychologists and psychiatrists do not in fact make more accurate clinical judgements than lay persons about the sanity or propensity toward violence of criminal suspects on trial. In both prediction and diagnosis lay people are as accurate as clinicians. In court cases their testimony is treated as fact. Their opinions are sought by news people, writers, governments, entertainers, courts and the public at the expense of personal judgement.
Background From the beginning, psychotherapy has been corrupt. Freud placed the advancement of knowledge before the cure of a patient. His most celebrated case of "Dora" lends itself to the re-interpretation of Dora being a health, much-wronged young woman and Freud being a prejudiced chauvinist. Jung was a racist, collaborated with the Nazis and exploited women patients.
Broader Inadequate health services (#PD4790)   Inadequacy of scientific expertise (#PF5055).
Narrower Bogus psychiatrists and personal counsellors (#PJ1180).
Aggravates Psychiatrism (#PF6351)   Mental illness (#PC0300)
Denial of rights of mental patients (#PD1148).
Aggravated by Psychological pollution by mass media (#PD1983)
Lack of understanding of spiritual healing (#PF0761).

♦ PE9177 Environmental hazards of explosives and pyrotechnic products
Broader Environmental hazards from chemicals (#PC1192).

♦ PE9185 Nutritional anaemia in women in developing countries
Nature About half of all women and two-thirds of the pregnant women in the developing countries are anaemic, suffering from a deficiency of one or more essential nutrients, chiefly of iron. Severe anaemia in pregnancy increases the risk of maternal mortality. Mild or moderate anaemia impair well-being, reduce maximal work capacity and adversely affect work performance.
Broader Nutritional anaemia (#PD0321).
Aggravated by Unprepared adult leadership (#PF6462).

♦ PE9199 Underpayment for work in developing countries
Exploitation of cheap labour in developing countries — Low wages in developing countries
Nature Although many people may not be officially employed in developing countries, most find work in the informal sector, whether in unregistered businesses, street selling or as servants. Working hours may extend to 10-15 per day, 6 or 7 days per week, for which they are underpaid.
Broader Underpayment for work (#PD8916).
Related Unemployment in developing countries (#PD0176).
Aggravated by Decline in real wages in developing countries (#PD2769).

♦ PE9207 Terrorists armed with biochemical weapons
Claim The specialized knowledge required to manufacture deadly biochemical weapons is relatively accessible as are the necessary materials. Poisonous gases are no more difficult to smuggle or make than drugs.
Broader Terrorism (#PD5574).
Related Terrorists armed with nuclear weapons (#PE3769).

♦ PE9210 Narco-terrorism
Terrorism financed by trade in drugs — Drug-financed subversion
Nature Drug trafficking, with the aim of promoting violent crime, addiction and corruption, and which generates funds to promote more organized forms of terrorism.
Broader Terrorism (#PD5574)   Drug smuggling (#PE1880).
Aggravated by Illicit drug trafficking (#PD0991)   Lack of protective legislation (#PJ2889)
State-supported international terrorism (#PD6008).

♦ PE9220 Distorting effects of commodity taxes on the transaction of goods and nonfactor services
Nature In most developing countries commodity taxes account for 50 to 70 per cent of of all tax revenue. Governments often rely heavily on the domestic production taxes that distort the transactions between producers. Succeeding transactions add tax to that already paid previously by producers or distributors. The price of outputs increases and generates differential taxation of consumption.
Broader Distortionary tax systems (#PD3436).
Narrower Distortion of international trade by selective indirect taxes and import charges (#PE8867).

♦ PE9224 Asset stripping
Nature Asset stripping is organizing a company so some of its parts can bring the highest price and selling those parts without regard for the effect on stockholders, employees or the company as a whole.
Broader Unethical commercial practices (#PC2563).
Aggravates Redundancy of workers (#PD8007).
Aggravated by Undervaluation of public assets (#PF1001).

♦ PE9228 Lead poisoning in animals
Lead poisoning of wildlife
Nature Lead shot is used by hunters and is scattered over moorlands, marshes, wet fields and estuary mudflats, when it fails to kill the hunted wildfowl. There is increasing indication that large numbers of birds are dying following ingestion of the discarded shot during feeding. They tend to die slowly and in small numbers and are rarely seen because the corpses are consumed by predators or scavengers.
Incidence Research in the USA and the UK indicate that 2 to 3 per cent of the autumn population

EMANATIONS OF OTHER PROBLEMS   PE9339

of ducks dies each year from lead poisoning. This corresponds to some 2 million birds in the USA. Large quantities of shot (up to to 2 million pellets per hectare) accumulate in fashionable spots for shooting fowl. Diving ducks and swans are especially affected because they also ingest anglers' split-lead shot used to weight lines.
Broader Poisoning in animals (#PD5228).
Aggravated by Bird shooting (#PE2693) Fishing for sport (#PE5286).

♦ **PE9229 Malignant neoplasm of bone**
Malignant neoplasm of connective tissue
Refs Galasko, C S *Skeletal Metastases* (1986); Gottlieb, Geoffrey J and Ackerman, A Bernard *Kaposi's Sarcoma* (1988).
Broader Malignant neoplasms (#PC0092) Diseases of connective tissue (#PD2565).
Narrower Tumours of joints (#PG4619) Intracranial tumours (#PG2192).

♦ **PE9242 Inadequate health care in least developed countries**
Nature In spite of the high proportion of the population in the vulnerable groups of children and mothers, who are furthermore generally undernourished, public expenditure on health services is very low. The number of physicians per 100,000 population is only about 6, compared to 160 for the developed market-economy countries. The scarce medical personnel and facilities that do exist are concentrated in metropolitan areas, although only a small proportion of thee population lives there, thus exacerbating the already critical situation. A large percentage of public funds for health is spent on high technology and highly visible hospitals where less than 5 percent of preventable deaths are treated.
Broader Least developed countries (#PD8201) Inadequate health services (#PD4790).
Narrower Inadequate primary health care (#PE8553)
Inadequate health care in urban slums (#PE7877).
Related Inadequate health care in family planning (#PD1038).
Aggravates Dehumanization of health care (#PD7821).
Aggravated by Ignorance of health and hygiene (#PD8023)
Prohibitive cost of disease control (#PF2779).

♦ **PE9251 Abdication of control by company directors**
Irresponsible directors of commercial enterprises
Incidence A survey of company directors in the USA found that they were confused about their role, dependent on a company's chief executive for information, and crippled by social conventions inhibiting probing inquiry. Outside directors seldom exercise control over a company's management until a crisis strikes and then they are often paralyzed by indecision. A principal reason for not challenging a company's chief executive is that so many of the directors are themselves chief executives (63 per cent of all outside directors of the Fortune list of 1,000 largest American companies are chief executives of other companies). Directors are further inhibited from fulfilling their responsibilities by lack of time and resources, since the average outside director only spends 14 days per year with the company.
Broader Irresponsibility (#PA8658).
Aggravates Incompetent management (#PC4867).
Aggravated by Interlocking corporate directorates (#PF5522).

♦ **PE9253 Nicotine withdrawal**
Nature Withdrawal symptoms include craving for nicotine, changes in mood and performance, decreased heart rate and increased appetite or weight gain.
Broader Smoking (#PD0713).

♦ **PE9256 Avoidant disorder of childhood**
Avoidant disorder of adolescence
Nature The children shrink from contact with unfamiliar people and this is severe enough to harm normal social functioning. Instead, children wish for social involvement with people they already know.
Broader Mental illness (#PC0300).

♦ **PE9260 Aflatoxicosis**
Refs Heathcote, J G and Hibbert, J R *Aflatoxins* (1978).
Broader Mycotoxicoses (#PE9458).
Narrower Mycotoxicosis in poultry (#PG6792).

♦ **PE9263 Alcoholic psychosis**
Broader Psychoses (#PD1722) Alcohol abuse (#PD0153).
Narrower Delirium alcoholicum (#PG4121) Alcoholic hallucinosis (#PG5738).

♦ **PE9265 Decline in capital investment in productive capacity**
Nature When demand is deficient firms tend to restrict their investment to that needed for rationalization, increasing efficiency and reducing current labour and material costs in order to maintain their market shares, rather than enlarge capacity. This serves to increase productivity, but it reduces employment, and does not add to potential output.
Broader Insufficient capital investment (#PF2852).
Aggravates Unemployment (#PB0750).
Aggravated by Lack of venture capital (#PG7833)
Excess production capacity (#PD0779).

♦ **PE9267 Symbolic international agreements without substance**
Token intergovernmental treaties
Claim Arms negotiations aim at treaties in which all essential weapons are being maintained, except at a lower level. The fundamental political problems have to give way to concerns about short-term impact and public relations of politicians.
Broader Unimplemented decisions (#PF4672).
Related Tokenistic meeting resolutions (#PF2086).
Aggravated by Ineffective international agreements (#PF6992).

♦ **PE9271 Unnecessary surgery**
Unnecessary surgical operations
Incidence By the time American women reach 60, more than a third of them have had a hysterectomy, surgical removal of uterus. It is a second most common major operation for American women, second to Caesarean section. New techniques have made some hysterectomies unnecessary, but rate of operations have not declined. In 1987 more than 230,000 Americans had coronary bypass operations, twice as many as in 1980, but only few are life-and-death affairs performed on patients who have just had heart attacks. It has been estimated that 44 per cent of bypass operations are inappropriate. The aim generally is to relieve chest pain and for that purpose the alternative methods are less costly and sometimes as effective as an operation.
Broader Health fraud (#PD9297) Surgical malpractice (#PE4736).
Related Unnecessary health tests (#PF5679) Laboratory testing errors (#PF5304).

♦ **PE9282 Endangered parklands**
Nature Parklands are under constant threat from developers eager to transform them into upmarket golf courses, theme parks and other forms of leisure facility. Efforts are also made to transform such land into semi-exclusive residential areas.
Counter-claim Commercial developments such as golf courses are one of the ways in which to generate funds to ensure the upkeep of parkland.
Related Vulnerability of wetlands (#PC3486) Destruction of land resources (#PJ8397).
Aggravated by Decreasing land mass (#PF7435).
Limited available land (#PC8160)
Increasing proportion of land surface devoted to urbanization (#PE5931).
Reduced by Restrictive use of available land (#PF6528).

♦ **PE9286 Infective diseases of uterus**
Infective diseases of vagina
Broader Diseases of female genital organs (#PD8775).
Narrower Pyometra (#PG7561) Metritis (#PG1473) Vaginitis (#PG6117).

♦ **PE9291 Congenital disorders of amino-acid metabolism**
Broader Diseases of metabolism (#PC2270).
Narrower Albinism (#PE2332) Cystinosis (#PG4662) Phenylketonuria (#PG3323)
Maple syrup urine disease (#PG3840).

♦ **PE9294 Anti-personnel use of toxic substances in peacetime**
Anti-personnel use of asphyxiating gases — Health hazards of anti-personnel gases
Broader Chemical warfare (#PC0872) Chemical torture (#PD5204).

♦ **PE9295 Neglect of environmental consequences of government policies**
Government neglect of future ecological problems — Government neglect of future problems of social environment arising from current policies
Refs Curran, and Renzetti, *Social Problems* (1987).
Broader Inappropriate policies (#PF5645)
Inadequate evaluation of environmental costs (#PE6700).
Aggravates Environmentally harmful dam construction (#PD9515)
Unrecognized future financial commitments (#PF4114)
Simplistic technical solutions to complex environmental problems (#PF0799).
Aggravated by Inadequate legislation against environmental pollution (#PF9299)
Domination of government policy-making by short-term considerations (#PF0317).

♦ **PE9296 Congenital anomalies of nervous system**
Broader Human physical genetic abnormalities (#PD1618).
Narrower Microcephaly (#PG3838) Neurofibromatosis (#PE4814).

♦ **PE9300 Developmental receptive language disorder**
Nature The comprehension deficit varies from inability to understand simple words and sentences to difficulties in understanding spatial terms or complex statements.
Broader Mental illness (#PC0300) Speech disorders (#PE2265).
Related Innumeracy (#PC0143) Reading disabilities (#PD1950)
Developmental articulation disorder (#PE9712)
Developmental expressive writing disorder (#PE0330)
Developmental expressive language disorder (#PE5545).

♦ **PE9302 Female sterility**
Refs Keller, P J (Ed) *Female Infertility* (1978).
Broader Diseases of female genital organs (#PD8775).

♦ **PE9309 Illegally obtained evidence**
Broader Hindrance of law enforcement (#PD5515).
Aggravates Denial of evidence (#PD7385).
Aggravated by Inadequate evidence to convict known offenders (#PF8661)
Inadmissibility of evidence from other jurisdictions (#PE5424).

♦ **PE9311 Over-watering during irrigation**
Nature Damaged dams and inefficient operation of irrigation canals waste too much water; plants cannot take it up and, consequently, a considerable proportion seeps into the subsoil or evaporates.
Broader Water losses from irrigation systems (#PE8796).

♦ **PE9324 Congenital syndromes affecting multiple systems**
Broader Stigmatized diseases (#PD7279)
Human physical genetic abnormalities (#PD1618).
Narrower Monsters (#PF2516) Down's syndrome (#PE2125) Turner's syndrome (#PG3943).

♦ **PE9325 Underutilized government resources**
Untapped government resources — Untapped government sources — Untapped government grants — Unutilized government funds
Broader Reluctant claims on external resources (#PF1226)
Undeveloped channels for public and private resources (#PF3526).
Related Detrimental effect of jungle environment in tropical villages (#PE2235).
Aggravates Limited accumulation of capital (#PF3630).
Aggravated by Self-defeating style of community planning (#PF6456)
Fragmented pattern of community organization (#PF6525).

♦ **PE9334 Reckless driving**
Dangerous driving practices — Callous drivers
Broader Recklessness (#PD5349).
Narrower Driving delinquency (#PE6119)
Reckless children snowmobile drivers (#PG9220).
Related Drunken driving (#PE2149).
Aggravated by Inadequate driver training (#PJ4507)
Motor vehicle driver fatigue (#PG7482)
Excessive speed of motor vehicles (#PE2147).

♦ **PE9337 Demolition of homes by government authorities**
Incidence Israeli military has used demolitions of homes of security suspects in the occupied territories in its effort to put down the Palestinian uprising. During 1988 over 560 have been demolished or sealed, displacing more than 5,000 Palestinians.
Broader Collective punishment (#PD6970).

♦ **PE9339 Establishment homosexuality**
Homosexual officials — Homosexuality in the judiciary
Nature Homosexuality amongst persons in positions of power, such as in the government or judiciary, makes them vulnerable to ridicule and blackmail and thus to corruption in the execution of their functions.
Broader Homosexuality (#PF3242).

♦ **PE9344 Tic disorders**
Tics
Nature A tic is a sudden, recurrent, involuntary, rapid motor movement or vocalization often exacerbated by stress and diminished during sleep. Common motor tics are eye-blinking, shoulder-shrugging, facial gestures, jumping or stamping. Common vocal tics are coughing, sniffing, snorting, throat-clearing, coprolalia (using "dirty" or obscene words), palilalia (repeating words or sounds) or echolalia (repeating the last-heard sounds of another person).
Broader Mental illness (#PC0300).
Narrower Tourette's disorder (#PG3800) Transient tic disorder (#PG9690)
Chronic motor tic disorder (#PG9218).

♦ **PE9346 Waste of resources invested in obsolete armaments**
Obsolescent weapons
Refs Kaldor, Mary *The Baroque Arsenal* (1983).
Broader Unproductive use of resources (#PB8376).
Related Obsolete defence planning (#PJ1877).

♦ **PE9363 Chemical trespass**
Broader Criminal trespass (#PD3794).
Aggravates Chemical pollutants of the environment (#PD1670).

♦ **PE9364 Abomasal disorders of cattle**
Broader Cattle diseases (#PD0752)
Diseases of the digestive system in animals (#PD3978).
Narrower Abomasal ulcers (#PG7715) Abomasal displacements (#PG6894)
Dietary abomasal impaction (#PG9421).

♦ **PE9369 Symptoms referable to genito-urinary system**
Broader Ill defined health conditions (#PC9067)
Diseases of the genito-urinary system (#PC4575).
Narrower Enuresis (#PE5431) Polyuria (#PG9613) Chyluria (#PG3813).

♦ **PE9374 Bacterial food poisoning**
Nature Bacteria in food can cause immediate illness, even death.
Incidence There is a marked increase in the varieties of dangerous bacteria in foods. Since 1975 some 25 new microbes than can cause food poisoning have been identified in the USA. Many are associated with intensive farming methods. It has been estimated that some 30 per cent of animal feed is contaminated (with crowded factory farms allowing pathogens to spread easily) and some 60 per cent of food retailed contains contaminants (with half the chickens sold containing salmonella). Another source of contamination is new food preferences which necessitate the removal of preservatives from products in response to the consumer desire for natural foods.
Refs Monahan, John *Food Poisoning* (1987).
Broader Intestinal infectious diseases (#PE9526) Food poisoning through negligence (#PE0561).
Narrower Botulism in animals (#PG2994).

♦ **PE9379 Shortage of marriageable men**
Incidence In the USA for every 100 unmarried women aged 20 through 29 there is 122 unmarried men in their 20s.
Related Shortage of marriageable women (#PE0427).
Aggravated by Deliberate imbalancing of population sex ratio (#PF3382).

♦ **PE9399 Colon-rectal cancer**
Cancer of the bowel
Incidence Colon-rectal cancer is the second highest cause of cancer death in the United States after lung cancer. It strikes more than 150,000 Americans and kills more than 60,000.
Refs Kirsner and Shorter *Diseases of the Colon, Rectum, and Anal Canal* (1988).
Broader Disorders of the bowel (#PE6553)
Malignant neoplasm of digestive organs (#PE4303).
Aggravated by Inadequate dietary fibre (#PE4950).

♦ **PE9402 Discrimination against people of abnormal height**
Narrower Discrimination against giants (#PE5578)
Discrimination against dwarfs and midgets (#PE2635).

♦ **PE9405 Small business failures**
Instability of micro-enterprises — Failure of economic micro-organizations
Broader Business bankruptcy (#PD2591)
Haphazard provision of consumer services (#PF2411).
Aggravated by Discouraging conditions for small business (#PD5603)
Diminishing capital investment in small communities (#PF6477)
Ineffectiveness of traditional small business methods (#PF3008).

♦ **PE9407 Shortening of fallow periods on agricultural land**
Aggravates Soil infertility (#PD0077).
Reduced by Chemicalized farming (#PD7993).

♦ **PE9415 Pulmonary heart disease**
Refs Rubin, Lewis J *Pulmonary Heart Disease* (1984).
Broader Heart diseases (#PD0448).

♦ **PE9417 Abusive collection of specimens**
Collectors of exotic species — Birds egg collection — Fossil collection — Collection of cultural artefacts
Incidence In the case of birds eggs, collectors can constrain an attempt by a species to prosper. In the early 1950s, for example, there were no osprey's in Scotland. As a consequence of determined efforts by bird lovers, by 1989 there were 52 breeding pairs, but despite surveillance the nests of 8 of them were robbed by collectors. Egg collection can also have a catastrophic effect on a species which is declining naturally because collectors sensitive to such increasing rarity intensify their efforts to obtain eggs. Collection may take the form of an unhealthy obsession through which collectors obtain a sense of personal security by building up as complete a collection as possible. Such obsession also takes an institutional form through the collection policies of museums concerned to maintain their status within the museum community. Many flowering plants have been seriously threatened by collectors of exotic species. 99 per cent of these plants die within a few months of being removed from their habitat and the surviving one per cent is unable to reproduce.
Counter-claim Collection of specimens is necessary for the advancement of scientific knowledge. In the case of rare specimens, it is important to preserve them in collections before they disappear from nature altogether in order to maintain a scientific record.
Related Bibliomania (#PF6713) Archaeological and anthropological looting (#PD1823).
Aggravates Illicit export of works of art (#PE9004).
Aggravated by Human use of animal by-products (#PF1964)
Endangered species of plants and animals (#PB1395).

♦ **PE9418 Ecthyma**
Broader Diseases of the skin and subcutaneous tissue (#PC8534).

♦ **PE9420 Aggressive economic destabilization of countries by external forces**
Destabilization of economies of developing countries
Broader Destabilization of social systems (#PB5417)
Adverse economic shocks from external factors (#PF6210).
Narrower Destabilization of national insurance markets by offshore insurers (#PE4994)
Destabilization of monetary systems and exchange notes by transnational corporations (#PE5903).
Aggravates Excessive foreign public debt of developing countries (#PD2133).

♦ **PE9422 Septicaemia**
Blood-poisoning
Refs Wilfred, Lucia M *Septicemia* (1987).
Broader Bacterial disease (#PD9094).

♦ **PE9427 Parasites of the cardiovascular system in animals**
Broader Circulatory system diseases in animals (#PD5453).
Narrower Babesiosis (#PE6288) Rickettsiae (#PE2572)
Theileriasis (#PE3996) Schistosomiasis (#PE0921)
Trypanosomiasis (#PE5725) Eperythrozoonosis (#PG3997)
Feline cytauxzoonoses (#PG6901) Blood sporozoa of birds (#PE4284)
Canine heartworm infections (#PG7720).

♦ **PE9438 Respiratory diseases of sheep**
Respiratory diseases of goats
Broader Diseases of respiratory system in animals (#PD7307).
Narrower Sheep nose bot (#PG4147) Pulmonary adenomatosis (#PG9203)
Progressive pneumonia of sheep (#PG3965) Nonprogressive pneumonia of sheep (#PG2433)
Contagious caprine pleuropneumonia (#PE6293).

♦ **PE9440 Discriminatory operation of international commodity exchanges**
Restrictions on access of developing countries to commodity exchanges
Nature Commodity exchanges are an important vehicle for the marketing of a number of the commodity exports of developing countries. Such countries face barriers to effective participation in commodity exchange services. The operation of these exchanges does not ensure that they obtain an equitable share in the gains from the trade.
Margin deposits demanded by commodity brokers and exchange clearing houses are often cited as the major barrier to the use of futures markets by developing countries. The uncertainty about the size of margin deposits during the life of contract makes it difficult for primary exporters or importers to persuade their local central banks to give permission for hedging and price fixing via terminal markets. Developing countries do not have necessary infrastructure of communications and so many prospective users from those countries may face frustration and high costs when trying to reach markets far away, especially if the trading happens in different time zone. The payment of brokers' commissions, of the margins and gains or losses will be in the currency of the country where the futures market is located. In some developing countries exist exchange control regulations which hinder greater participation in the existing foreign futures markets. The rules for admission to the different classes of memberships of commodity exchanges are primarily of a financial nature, governing bank guarantees, paid-up capital, etc. However, there are also regulations covering the nationality of member companies and their employees which can discriminate against foreigners. The procedure used to arrive at the official daily spot and futures quotations, as the former is used as the price for many contracts round the world, affect developing countries trade. The location of delivery points are usually near consumers' plants in developed countries and it is perceived as weakening producers' control over their own marketing strategies.
Related Ineffectiveness of international commodity agreements (#PG5600).

♦ **PE9441 Secret government security vetting of job applicants**
Nature Governments may require security vetting in certain sectors and without informing the job applicant. People may be denied a job or promotion on the basis of information which they have had no opportunity to verify.
Incidence In the UK it is reported that one million positions are subject to secret government vetting of applicants, namely 1 in 22 of the working population. Security vetting occurs in both public and civil services, private sector defence contractors, telecommunications and the armed services.
Broader Political surveillance (#PD8871).
Aggravates Blacklisting (#PE0189).
Aggravated by Unaccountable government intelligence agencies (#PF9184).

♦ **PE9443 Increase in nitrous oxide in the atmosphere**
Nature Nitrous oxide is generated by processes of nitrification and denitrification of the soil and from fossil fuel combustion. It contributes significantly to global warming and stratospheric ozone depletion.
Incidence Nitrous oxide is currently increasing at a rate of 0.3 per cent per year for reason which are not clearly understood.
Broader Nitrogen compounds as pollutants (#PD2965)
Increase in trace gases in the atmosphere (#PD1354).

♦ **PE9446 Increase in concentration of atmospheric water vapour**
Nature As the climate becomes warmer the concentration of water vapour in the atmosphere increases. Water vapour is at present the main factor contributing to the entrapment of infrared radiation from the Earth's surface. Increases in water vapour should therefore significantly enhance global warming. Such positive feedback could bring about a runaway increase in global warming.
Counter-claim Such a runaway effect would only be possible in a climate many ten's of degrees warmer than that prevailing.
Broader Degradation of the atmosphere (#PD9413).
Aggravates Global warming (#PC0918).

♦ **PE9447 Serial killings**
Nature In serial killings victims are added one by one over a period of years.
Broader Homicide (#PD2341) Killing by humans (#PC8096).

♦ **PE9453 Changelings**
Nature Member of a non-human race, usually a child, believed to have been substituted, for an abducted human.
Background A common feature of European mythology in which the changeling is the child of fairies, elves or nature spirits, or of a witch or some other demonic creature. Equivalent beliefs exist in non-Western cultures. Such beliefs are strongly linked with witchcraft, the role of witches in the process, especially in the case of diabolic changelings as offspring of a witch and a demon. The belief resulted from, and is sustained by, the assumption that infants are peculiarly liable to

attack by spirits and demons with whom they are sometimes associated in nature before name-giving and purificatory rites such as Christian baptism.
Aggravates Human physical genetic abnormalities (#PD1618).
Aggravated by Demons (#PF6734) Witchcraft (#PF2099)
Kidnapping (#PD8744).

♦ PE9457 Excessive concentration of export markets of developing countries
Export of developing country products to a limited range of countries
Aggravates Inability of developing countries to adopt appropriate exchange rate policies (#PE7563)
Vulnerability of national economies to vagaries of external markets for goods and services (#PF9697).
Aggravated by Restricted growth in export markets of developing countries (#PF1471)
Instability in export trade of developing countries producing primary commodities (#PD2968)
Restrictive business practices in the markets of developed countries against exports from developing countries (#PE5926).

♦ PE9458 Mycotoxicoses
Broader Poisoning in animals (#PD5228).
Narrower Ergotism (#PG7849) Aflatoxicosis (#PE9260) Facial eczema (#PG7891)
Fescue lameness (#PG9336) Paspalum staggers (#PG7860) Mycotoxic lupinosis (#PG2026)
Fusarium estrogenism (#PG4388) Summer fescue toxicosis (#PG6109).

♦ PE9461 Vertebrogenic pain syndrome
Broader Disorders of joints and ligaments (#PD2283).
Narrower Pain in neck (#PG4279) Low back pain (#PE5483).

♦ PE9465 Failure of state banks
Government development bank failure
Broader Bank failure (#PE0964).

♦ PE9468 Symptoms referable to nervous system
Broader Ill defined health conditions (#PC9067) Diseases of the nervous system (#PC8756).
Narrower Delirium (#PG4263) Agraphia (#PE0280) Convulsions (#PG2611)
Paraesthesia (#PG3547) Hallucinations (#PF2249) Sexual frigidity (#PE6408)
Irreversible coma (#PE4620) Phantom limb pain (#PG2695) Language disorders (#PE3886).
Related Amnesia (#PD8297).

♦ PE9476 Unemployment in least developed countries
Nature Per capita income growth was negative in most least developed countries during the 1980s and a heavy burden of external debt was accumulated. While economic growth stagnated, population and labour force growth has continued to be high. This has led to an increase in unemployment and underemployment. Unemployment along with increasing income inequality have contributed to the overall rise in poverty.
Broader Least developed countries (#PD8201)
Unemployment in developing countries (#PD0176).
Aggravates Poverty in developing countries (#PC0149).
Aggravated by Excessive foreign public debt of developing countries (#PD2133).

♦ PE9509 Atopic diseases
Broader Immune system diseases in animals (#PD4068).
Narrower Anaphylactic shock (#PG1208) Canine rheumatoid arthritis (#PG4041)
Allergic rhinitis in animals (#PG9566) Atopic dermatitis in animals (#PG7830)
Allergic enteritis in animals (#PG7196) Pemphigus vulgaris in animals (#PG6405)
Allergic bronchitis in animals (#PG9507) Polyarteritis nodosa in animals (#PG5881)
Autoimmune thyroiditis in animals (#PG3718) Cold haemolytic disease in animals (#PG6753)
Idiopathic polyarthritis in animals (#PG9527)
Immune-deficiency diseases in animals (#PE4224)
Autoimmune thrombocytopenia in animals (#PG7284)
Systemic lupus erythematosus in animals (#PG6624)
Autoimmune haemolytic anaemia in animals (#PG4441).

♦ PE9517 Dam failures
Malsituated dams
Nature The number of new dams in countries with little or no experience in the design, construction and operation of dams increases every year. Lack of experience may lead to repetition of errors and serious mistakes in construction. Engineers may assume that the technology used to built small-scale dams can be used for putting up large dams. Shoddy workmanship is fairly common, so is "cutting corners" to complete the scheme in the shortest period possible. There may be a terrible lack of cooperation between the various organizations involved in putting up a dam. Besides, dams will increasingly be built in less and less suitable places, because appropriate sites run out. Dams also fail because of overtopping during floods. The pressure applied to fragile geographical structures by vast mass of water impounded by a big dam can give rise to earthquakes. Sabotage caused by rebels or enemy forces can cause the loss of hundred of thousands of lives.
Refs Penman, A D *Deterioration of Dams and Reservoirs* (1984).
Broader Structural failure (#PD1230)
Environmentally harmful dam construction (#PD9515).
Aggravated by Sabotage (#PD0405) Earthquakes (#PD0201)
Inadequate earthquake resistant construction (#PE6257).

♦ PE9525 Atmospheric corrosion of materials
Nature The corrosion of materials exposed to the atmosphere is most commonly due to chloridization and sulphidization of iron and its alloys, the chloridization of aluminium, the sulphidization of copper and its alloys, and the sulphidization of marble and masonry. The exact processes remain obscure, despite research, although the presence of water assists vigorous corrosive action.
Incidence In the USA in 1975, it is estimated that the costs associated with such corrosion amounted to over $70 billion per year.
Counter-claim Atmospheric corrosion is in many locations due to natural processes.
Refs Ailor, William H (Eds) *Atmospheric Corrosion* (1982).
Broader Corrosion (#PD0508) Degradation of the atmosphere (#PD9413).
Narrower Corrosion of glass (#PE4524) Corrosion of stonework (#PE5470)
Corrosion of iron and steel (#PE1945).
Aggravates Failure of materials (#PD2638).

♦ PE9526 Intestinal infectious diseases
Broader Infectious and parasitic diseases (#PD0982).
Narrower Dysentery (#PE2259) Enteritis (#PE4973) Amoebiasis (#PE6782)
Giardiasis (#PE4811) Salmonellosis (#PE7562) Paratyphoid fever (#PE4036)
Human coccidiosis (#PG3988) Bacterial food poisoning (#PE9374).
Related Diarrhoea (#PD5971).

♦ PE9528 Medium-term cyclic variations in solar radiant energy
Solar flares — Epidemics associated with sunspot cycle — Radio interference associated with solar flares
Nature The radiant energy output by the sun varies, especially over the 11-year cycle of growth and decline of solar activity as manifested by large variations in sun spots, solar flares and other phenomena. Such changes can contribute significantly to the increase (or reduction) in global surface temperatures (the greenhouse effect) and to interference in radio communications. There is evidence indicating that the increase in solar radiation associated with sunspots may drive viruses (especially influenza viruses) from the upper atmosphere down to the earth's surface resulting in periodic epidemics.
Incidence The amplitude of the observed variation is about 0.1 percent during the current solar cycle. There is no direct evidence for greater variations over longer periods, although it is known that there have been substantial changes in solar activity over the centuries and that they correlate positively with temperature changes on Earth. During the period 1600 to 1700, with the almost total absence of sun spots in the latter half of the century, the coldest period in the last 1000 years was experienced. A much lesser degree of cooling was experienced from the 1940s to the late 1970s, accompanied by a substantial reduction of solar activity. The statistical relationship between sunspot peaks and pandemics has been explored from 1761 in relation to the 11-year cycle.
Refs Schove, D Justin *Sunspot Cycles* (1983).
Related Long-term cyclic changes of climate (#PC6114).
Aggravates Epidemics (#PC2514) Global warming (#PC0918)
Global cooling (#PF1744) Geomagnetic storms (#PD1661)
Health hazards of radiation in aircraft (#PE0962).

♦ PE9541 Food intolerance
Nature It is an adverse reaction to common foods, does not occur rapidly, and the link to food is not obvious. Symptoms of intolerance vary widely: one person may have diarrhoea or constipation and fatigue, another rheumatoid arthritis, palpitations, and headache.
Related Allergy inducing food (#PE3225).
Aggravates Hypersensitivity (#PE6898).

♦ PE9545 Government failure to prosecute offenders effectively
Government pardoning of convicted human rights offenders — Arbitrary reduction of sentences for offenders after conviction
Incidence In Argentina, for example, the government passed a series of decrees pardoning almost all those convicted as responsible for human rights abuses involving the forced disappearance of an estimated 15,000 people during the military dictatorship's operations in the 1970s and 1980s.
Broader Inadequate national law enforcement (#PE4768).
Related Ineffective war crime prosecution (#PD1464).
Aggravates Impunity of violators of human rights (#PF3474)
Excessive leniency in sentencing of offenders (#PF4723).
Aggravated by Inadequate enforcement of human rights (#PC4608).

♦ PE9554 Anaemia in animals
Broader Circulatory system diseases in animals (#PD5453).
Narrower Myelophthisis in animals (#PG3793) Equine infectious anaemia (#PG7882)
Feline infectious anaemia (#PG5847) Aplastic anaemia in animals (#PG7829)
Haemolytic anaemia in animals (#PG9847) Nutritional anaemia in animals (#PG9508)
Iron deficiency anaemia in animals (#PG9709).
Related Anaemia (#PD7758).

♦ PE9567 Deterioration in external financial position of developing countries
Deterioration in external payments position of developing countries
Nature Much of the deterioration in the payments position has been due to the exceptionally severe and long recession in the developed countries and the rise of interest rates to unprecedented levels. Such deterioration not only reduces the availability of external savings and drains off domestically generated savings, in a number of countries it also adversely affects domestic savings performance.
Narrower Low return on investment in developing countries (#PE9811)
Deteriorating terms of financial loans to developing countries (#PE4603)
Over-dependency on international financial institutions by developing countries (#PE2214).
Related Budget deficits in developing countries (#PD3131)
Excessive foreign public debt of developing countries (#PD2133).
Aggravates Capital shortage in developing countries (#PD3137)
Financial paralysis of developing countries (#PD9449)
Excessive anxiety on lending to developing countries (#PF4345)
Limited availability of loans in developing countries (#PE4704)
Inadequate economic policy-making in developing countries (#PF5964)
Disincentives for financial investment within developing countries (#PF3845)
Excessive social costs of structural adjustment in debtor developing countries (#PD8114)
Dependence of developing countries on external financing for development programmes (#PE7195).
Aggravated by Decline in export credits to developing countries (#PE3066)
Inconsiderate choice of payment times by creditors (#PE3926)
Deterioration of terms of trade for developing countries (#PD2897)
Outflow of financial resources from developing countries (#PC3134)
Burden of servicing foreign public debt by developing countries (#PD3051)
Private domestic capital outflow from developing to developed countries (#PD3132)
Inadequate external debt management capacity within developing countries (#PE7184)
Decline in concessional financial resources available to developing countries (#PE3812).

♦ PE9573 Denial to animals of the right to dignity
Nature Living and dead animals have the right to be treated with respect. Animals which are used for entertainment, exhibition and scenes of bestiality or violence are denied their rights as beings.
Broader Denial to animals of legal protection of their rights (#PE8643)
Denial to experimental animals of the right to freedom from suffering (#PE8024).
Related Bestiality (#PE3274) Animal fighting sports (#PE4893)
Drug use in animal sports (#PJ1135) Ritual slaughter of animals (#PF0319)
Denial to animals of the right to life (#PF8243)
Denial to animals of the right to a natural death (#PE8339)
Denial to working animals of limitation of working hours (#PE6427)
Denial to animals of the right to freedom from mass killing (#PE9650)
Denial to food animals of the right to freedom from suffering (#PE3899)
Denial to working animals of restorative nourishment and rest (#PE4793)
Denial to animals of the right to the attention, care and protection of humankind (#PF5121)
Denial to animals of the right to conditions of life and liberty proper to their species (#PE6270).

♦ PE9580 Over-anxious disorder
Nature The children of this disorder have an excessive and unrealistic concern about future, past behaviour or their own competence. They may complain about headaches, dizziness or other somatic discomforts and may be enable to relax or to fall asleep.
Broader Mental illness (#PC0300).

♦ **PE9599 Emotional manipulation**
Emotional blackmail — Moral blackmail — Emotional bombardment — Love bombing
Broader Coercion (#PC3796).
Aggravates Lovesickness (#PF3385).
Aggravated by Beggars (#PD2500).
Reduced by Compassion fatigue (#PF2819).

♦ **PE9604 Symptoms referable to digestive system**
Symptoms referable to abdomen
Refs Sreenivas, V *Acute Disorders of the Abdomen* (1980).
Broader Ill defined health conditions (#PC9067)
Diseases and deformities of the digestive system (#PC8866).
Narrower Nausea (#PE4513) Hiccup (#PG9284) Dysphagia (#PG9548)
Encopresis (#PG6903) Abdominal pain (#PG9470)
Visceral impairments (#PE6777).

♦ **PE9605 Illegal roadblocks**
Nature Roadblocks may be set up by para–military, guerrilla and other groups as a means of exercising an unauthorized toll on traffic. Payment is extorted under the threat of death.
Incidence Tend to occur in countries where the central government does not exercise effective control over isolated roads (enabling banditry to survive) or where isolated military forces are not effectively controlled.
Broader Travel risks (#PD7716).
Aggravated by Banditry (#PD2609) Guerrilla warfare (#PC1738).

♦ **PE9607 Sexual desire disorders**
Refs Leiblum, Sandra R and Rosen, Raymond C *Sexual Desire Disorders* (1988).
Broader Human sexual disorders (#PD8016).

♦ **PE9609 Health hazards of air pollution for animals**
Animal fluorosis
Incidence The evidence of major air pollution disasters indicates adverse symptoms in animals of a character similar to those suffered by man. Animals then suffered from acute respiratory distress, bronchitis, emphysema and heart failure. Fluorides, arsenic and lead are highly toxic pollutants that may cause diseases in animals by accumulation in the forage and other crops that constitute the principal diet of cattle and other farm animals. Many other pollutants derived from the incomplete combustion of fuels and from specific industrial processes possess toxic properties.
Broader Air pollution (#PC0119).
Aggravates Animal diseases (#PC0952).

♦ **PE9611 Use of animals in toxicological experiments**
Cruel treatment of animals in testing pharmaceutical products — Abusive treatment of animals in testing toiletries and cosmetics
Nature Animals are used to determine acceptable levels of toxins and pollutants in the environment and in pharmaceutical products and especially to determine the acute and chronic toxicities of new chemical compounds. Animals are subjected to: electric shock and burning (to test lotions and salves), to stress (to test sedatives), to direct exposure of sensitive tissue, especially the eyes, to compounds (to test for irritation), and to diseases (to test new drugs).
Incidence In the UK, of the 3.1 million experiments performed on animals in 1986 (3.4 million in 1988), most were for the testing or development of drugs and other medical products. Of the other experiments: 55,301 were for pesticide tests, 23,565 were for herbicides, 72,150 were for industrial products, 9,309 were for household products, 15,652 (17,000 in 1988) were for toiletries and cosmetics, 8,988 were for food additives, and 35,512 were concerned with environmental pollutants. Some 39,142 experiments required inhalation of substances like tobacco smoke by animals.
A standard test, known as LD50, is based on the determination of the single dose of a product necessary to kill 50 per cent of the animals used in the experiment. Although manifestly cruel and thoroughly discredited on scientific grounds, statistics in the UK show that its use in fact increased in 1988.
Another standard test, known as Draize, involves placing chemical compounds directly into the eyes of rabbits to determine toxicity. The reason for the use of rabbits is that their eyes are much more sensitive to damage than human eyes and their tear ducts produce very little fluid so that any irritant placed on the eye remains there. The chemicals tested include cosmetics, pesticides and detergents. The animals are kept under restraint for 72 hours subjected to increasing doses and then observed for at least 21 days. Positive reactions include: ulceration of the cornea, inflammation of the iris, haemorrhage, gross destruction of the conjunctivae, obvious swelling with partial eversion of the eyelids. Basically the eye is eaten away by the chemical, after which the animal is killed. In the UK in 1986, 11,263 tests were performed.
In 1990 the European Commission proposed a directive, due to take force in 1992, to double the use of animals in cosmetics testing, by making testing compulsory for all previously untested ingredients (including many widely used products, dating from periods prior to the introduction of such tests).
Counter–claim No unnecessary pain or distress is inflicted upon the subject animals, therefore no anaesthetic, analgesic or tranquillizing drugs are necessary.
Broader Cruel treatment of animals for research (#PD0260).
Aggravated by Medication side effects (#PD9807).
Reduces Hazards of cosmetic use (#PE4895).

♦ **PE9620 Decline in cognitive ability with ageing**
Broader Mental disorders of the aged (#PD0919)
Susceptibility of the old to physical ill–health (#PD1043).
Aggravates Social withdrawal of aged (#PD3518)
Rigidity and inadaptability in the aged (#PD3515).
Aggravated by Human ageing (#PB0477).

♦ **PE9642 Environmental hazards of electrical power transmission lines**
Health hazards of high–voltage transmission lines
Broader Environmental hazards of extremely low frequency electromagnetic radiation (#PE7560).
Related Unaesthetic location of power transmission lines (#PD1665).
Aggravates Suicide (#PC0417) Malignant neoplasms (#PC0092).

♦ **PE9643 Excessive use of foreign programmes for media**
Inability to generate sufficient culture–specific programmes — Submission to media cultural imperialism
Incidence Small countries may simply not possess enough actors, playwrights, or financial resources to provide even a fraction of the material demanded by media.
Refs Lee, Chin–Chuan and Katz, Elihu *Media Imperialism Reconsidered* (1980).
Broader Cultural imperialism (#PC3195).
Aggravates Loss of cultural identity (#PF9005) Multiplicity of languages (#PC0178)
Homogenization of cultures (#PB1071)
Discrimination against minority languages (#PD5078)
Excessive portrayal of perspectives of industrialized cultures in media (#PE3831).

♦ **PE9646 Inadequate negotiating capacity of developing countries**
Asymmetry in bargaining power between transnational corporations and developing countries
Nature Developing countries tend to have limited capacity to negotiate appropriately to protect their interests in the face of an array of complex international financial mechanisms, especially with respect to debt contracting and renegotiation. Negotiations tend to be one–sided because of the developing country's lack of information, technical unpreparedness, and political and institutional weaknesses.
Related Developing country failure to exercise leverage on international bodies (#PJ8459).
Aggravates Excessive foreign public debt of developing countries (#PD2133)
Exploitative financial policies of transnational corporations in developing countries (#PE6952)
Inadequate negotiation of entrance terms for transnational corporations in developing countries (#PE0853).
Aggravated by Domination of developing countries by transnational corporations (#PE0163).

♦ **PE9650 Denial to animals of the right to freedom from mass killing**
Nature Any act involving the killing of large numbers of wild animals is genocide. This includes both deliberate destruction of the animals and deliberate destruction or mortal pollution of the environment upon which they depend.
Broader Denial to animals of legal protection of their rights (#PE8643)
Denial to experimental animals of the right to freedom from suffering (#PE8024).
Related Fox hunting (#PG3555) Stag hunting (#PJ3556)
Hunting tourism (#PE3008) Animal extinction (#PD7989)
Hunting of animals (#PC2024) Animal trophy hunting (#PE5644)
Unrestrained animal damage (#PJ9217) Disruptive airplane hunting (#PG0382)
Denial to animals of the right to life (#PF8243)
Extermination of wild animal natural prey (#PD3155)
Denial to animals of the right to dignity (#PE9573)
Abuse in control of wild animal populations (#PE7995)
Denial to animals of the right to a natural death (#PE8339)
Increase in animal disease by increase in aviation (#PJ4036)
Instability of hunting, trapping and game propagation (#PE9059)
Denial to working animals of limitation of working hours (#PE6427)
Denial to food animals of the right to freedom from suffering (#PE3899)
Denial to working animals of restorative nourishment and rest (#PE4793)
Denial to animals of the right to the attention, care and protection of humankind (#PF5121)
Denial to animals of the right to conditions of life and liberty proper to their species (#PE6270).

♦ **PE9657 Restrictive practices of transnational banks**
Broader Restrictive business practices of transnational corporations (#PE5915).
Narrower Domination of loan negotiations by transnational banks (#PE4355)
Discrimination in lending by transnational banks in developing countries (#PE4310)
Domination of restrictive project loans by transnational banks in developing countries (#PE4330)
Interference of transnational banks' off–shore borrowing with domestic monetary policies (#PE4315).
Aggravates Socially irresponsible programmes of transnational banks in developing countries (#PE4360).
Aggravated by Inadequate information concerning transnational banks (#PF4350).
Reduces Discrimination against transnational banks in developing countries (#PE4320).

♦ **PE9688 Rural drug abuse**
Incidence In USA drug–trafficking organizations, enticed by the huge profitability of crack and cocaine, have expanded beyond the urban centers into rural states.
Broader Drug abuse (#PD0094).

♦ **PE9689 Marek's disease**
Refs Payne, L N (Ed) *Marek's Disease* (1985).
Broader Viral diseases in animals (#PD2730).

♦ **PE9700 Arrears in financial payments between government agencies**
Delays in payments by state–controlled enterprises
Nature Government agencies and state–controlled enterprises affect government finances negatively through the build up of interagency arrears and cross–debts. Sizeable arrears can impede effective financial management because they obscure the true pattern of financing within the public sector. Governments may fail to fulfil capital grant and subsidy commitments, such that these become obligations for future capital years. State–controlled enterprises may fail to pay their obligations to governments, whether in the form of taxes, dividends, or debt service. Often this happens after governments have failed to meet their financial obligations toward the enterprises.
Counter–claim Mutual arrears provide a short–term answer for both government agencies and enterprises in constrained financial circumstances.
Broader Arrears in payment of government financial commitments (#PF1179).
Aggravated by Government loan defaults (#PD9437)
Delay in implementation of commitments (#PF3975)
Inefficiency of state–controlled enterprises (#PD5642).

♦ **PE9702 Mouth diseases of animals**
Broader Diseases of the digestive system in animals (#PD3978).
Narrower Lampas (#PG9853) Gingival hyperplasia (#PG3803) Glossitis in animals (#PG0416)
Gingivitis in animals (#PG7827) Stomatitis in animals (#PE2143)
Salaframine toxicosis (#PG4746) Periodontitis in animals (#PG9499)
Cheilitis in small animals (#PG7871) Eosinophilic granuloma of cats (#PG5871)
Alveolar periostitis in animals (#PG9551).

♦ **PE9705 Benign neoplasm of genital organs**
Broader Benign tumours (#PD8347) Diseases of male genital organs (#PD9154)
Diseases of female genital organs (#PD8775).

♦ **PE9712 Developmental articulation disorder**
Nature Person with this disorder articulates incorrectly or omits the speech sounds giving the impression of "baby talk", that could be completely or partially intelligible or totally unintelligible.
Refs Bernthal, John E and Bankson, Nicholas W *Articulation and Phonological Disorders* (1988); Edwards, M *Disorders of Articulation* (1984).
Broader Mental illness (#PC0300) Speech disorders (#PE2265).
Related Developmental receptive language disorder (#PE9300).

♦ **PE9746 Occupational hazards of painters**
Vulnerability of painters to cancer
Nature World Health Organization reports that children of painters are at increased risk of leukaemia and brain tumours; female painters suffer excess frequency of spontaneous pregnancy; spray painters are more prone to testicular cancer; painters have adverse effects on their nervous

system. In general, painters have a 40 percent higher than national average chance of contracting lung cancer and incidence of all types of cancer is 20 percent above normal.
Broader Occupational hazards (#PC6716).

♦ **PE9751 Diarrhoea in children**
Childhood diarrhoea
Nature The childhood diarrhoea has social consequences like: retarded growth, missed learning opportunity, weakened nutritional status, load on health delivery systems and budgets, disincentive to limiting family size, lowered quality of life and premature loss of young lives. Diarrhoea by itself is not fatal, but the dehydration represents a loss of essential water and salts from the body.
Incidence An estimated 5 million children under 5 years of age dies as a consequence of diarrhoeal disease.
Refs Walker-Smith, J A and McNeish, A S *Diarrhea and Malnutrition in Childhood* (1986).
Broader Diarrhoea (#PD5971).
Aggravated by Ignorance of women concerning primary health care (#PD9021)
Substitution of inappropriate foodstuffs for breast feeding (#PE8255).

♦ **PE9758 Inequitable distribution of public subsidies**
Incidence Uniformly low prices throughout the education and health sector imply that high-cost services are much more subsidized than low-cost ones. The relatively poor therefore have little access to those high-cost services however. Contrary to policy, the poorest are not only denied a greater share, but they often get less than their proportionate share. In education, for example, subsidies for higher education are much greater than at the lower levels. Thus the very small percentage of the population able to gain access to higher education receives a large share of the education budget. Moreover, among these few, the rich are over-represented. The distribution of public health expenditures is also skewed in many countries. Most health facilities are in urban areas, where household incomes are on average higher.
Broader Inappropriate public spending by government (#PF6377).

♦ **PE9759 Violation of human rights of individual parliamentarians**
Incidence An international committee has received communications concerning 625 cases in 52 countries between 1977 and 1988 September. Most of the cases concerned the situation of parliamentarians detained without charge or trial or imprisoned by virtue of a court decision, other included violations of freedom of expression, exile and disappearance, extra-judicial killings and summary executions, and curtailments of civil and political rights.
Aggravated by Threat to parliamentary immunity (#PF0609).

♦ **PE9762 Imposition of trade quotas for political reasons**
Discriminatory managed trade mechanisms — International cartels enforced by governments — Imposition of quantitative trade quotas of a discretionary character
Broader Protectionism in international trade (#PC5842).
Distortion of international trade by quantitative restrictions (#PE9027).
Narrower Trade restrictions due to voluntary export restraints (#PE0310).
Aggravated by Declining international competitiveness (#PD8994)
Fragmentation of the international trading system (#PC9584).

♦ **PE9764 Inadequate human resources development in least developed countries**
Lack of educated manpower in least developed countries — Shortage of managerial skills in least developed countries
Nature Human resources development concerns increased emphasis within existing avenues, such as in-house training and research and development, broadening of skills of the workforce, development and assimilation of new technologies so as to maximize productivity. On the other hand, new avenues are needed to develop expertise for the new production opportunities and for the marketing of the resultant outputs. Human resources development is a necessary complement to the mobilization of additional investment finance for commodity-based development.
Broader Least developed countries (#PD8201)
Inadequate supply of appropriate trained manpower in developing countries (#PE6243).
Aggravates Destruction inherent in development (#PF4829).

♦ **PE9767 Lack of slaughter facilities**
Inadequate hygiene in slaughter houses
Broader Poor organization of community environment (#PF1790).
Related Inadequate disinfection measures for animal housing and equipment (#PE2757).
Aggravates Brucellosis (#PE0924)   Unclean food (#PJ2532).
Aggravated by Killing of animals (#PD8486)
Inadequate housing and penning of domestic animals (#PE2763).

♦ **PE9773 Banned cultivation of plant species**
Regulatory banning of plant cultivation
Incidence In an effort to harmonize the range of crops produced in different countries of the EEC, directives have been issued banning the cultivation of certain species.
Aggravates Decreasing genetic diversity in cultivated plants (#PC2223).

♦ **PE9774 Myopathies in animals**
Broader Metabolic diseases in animals (#PD7420).
Diseases of musculoskeletal system in animals (#PD7424).
Narrower Azoturia (#PG9673)   Toxic myopathies (#PG3740)   Yellow fat disease (#PG9678)
Ischaemic myopathy (#PG5011)   Eosinophilic myositis (#PG7529)
Porcine stress syndrome (#PG5197)   Nutritional myopathy of lambs (#PG5310)
Transport myopathy of turkeys (#PG5232)   Nutritional myopathies in swine (#PG9531)
Capture myopathy of wild animals (#PG7884)   Exertional rhabdomyolysis in dogs (#PG9353)
Deep pectoral myopathy of turkeys (#PG9303).

♦ **PE9776 Unsafe sex**
Narrower Health risks of teenage sex (#PE6969).
Aggravates Infectious revenge (#PD5168)   Fear of sexual intercourse (#PF6910)
Sexually transmitted diseases (#PD0061).
Aggravated by Sex (#PF9109)   Sexual immorality (#PF2687)
Deception between sexual partners (#PE4890)   Prohibitive cost of contraceptives (#PJ3244)
Inadequacy of male contraceptive methods (#PF1069).

♦ **PE9789 Attention-deficit hyperactivity disorder**
Physical hyperactivity — Hyperkinesia
Nature People with the disorder display inappropriate degrees of inattention, impulsiveness and hyperactivity. Inattention is evidenced by being easily distracted, having difficulties sustaining attention in tasks and conversations, failing to finish chores and shifting from one uncompleted activity to another. Impulsiveness is demonstrated by interrupting and intruding without waiting her/his turn in games and social situations. Hyperactive person often fidgets or squirms in seat, has difficulties remaining seated, doesn't play quietly, doesn't consider consequences of physically dangerous activities and talks excessively.
Refs Conners, C Keith; Wells, Karen C and Kazdin, Alan E *Hyperkinetic Children* (1986); Piazza, Robert *Readings in Hyperactivity* (1979); Prior, Margot R; Griffin, Michael W and Rutter, Michael *Hyperactivity* (1985); Winchell, Carol A *The Hyperkinetic Child* (1981).
Broader Mental illness (#PC0300).
Aggravates Educational wastage (#PC1716).

♦ **PE9795 Contagious agalactia**
Refs Jones, G E *Contagious Agalactia and Other Mycoplasmal Diseases of Small Ruminants* (1987).
Broader Reproductive system diseases in animals (#PD7799).

♦ **PE9799 Pseudorabies**
Aujezky's disease — Mad itch — Infectious bulbar paralysis
Refs Wittmann, G and Hall, S A (Eds) *Aujeszkiy's Disease* (1982).
Broader Diseases of nervous system in animals (#PD7841).

♦ **PE9811 Low return on investment in developing countries**
Incidence In Africa, investment and operating costs are typically 50–100 per cent greater than those in the most comparable area, namely South Asia. This situation is exacerbated by the tendency of developing countries to allocate externally loaned funds to projects giving a low return on investment.
Claim In the case of Africa, a World Bank study concludes that the root cause of economic problems is low return on investment. Africa is simply not competitive in a competitive world.
Broader Inappropriate use of financial resources (#PD9338).
Deterioration in external financial position of developing countries (#PE9567).
Aggravates Excessive foreign public debt of developing countries (#PD2133).

♦ **PE9813 Laboratory waste water pollutants**
Discharge of dangerous substances into laboratory waste water
Broader Discharge of dangerous substances (#PD4542).
Aggravated by Irresponsible genetic manipulation (#PC0776).

♦ **PE9816 Congenital disorders of lipid metabolism**
Broader Diseases of metabolism (#PC2270).
Narrower Lipidosis (#PE5325)   Xanthomatosis (#PG4552).

♦ **PE9819 Malignant neoplasm of mouth and throat**
Oral cancer — Cancer of oesophagus — Cancer of pharynx — Cancer of larynx
Refs Pindberg, J J *Oral Cancer and Precancer* (1980).
Broader Malignant neoplasms (#PC0092)   Diseases of upper respiratory tract (#PE7733).
Aggravated by Alcohol abuse (#PD0153).

♦ **PE9825 Deterioration of cultural artefacts from tourism**
Nature Mass tourism to historic sites, ancient monuments, historic buildings and to archaeological sites (such as cave paintings) rapidly results in their degradation.
Incidence The pharonic tombs in Egypt, for example, are visited daily by an average of 3,000 visitors. It has been estimated that they leave behind an estimated 25 litres of perspiration, which mixed with the dust raised provides the perfect climate for molds and chemical reactions damaging to the wall-paintings. These are visibly flaking away from the reliefs. Elegantly decorated walls that were intact 10 years ago have now crumbled to dust. In other situations, such processes are aggravated by flash photography.
Related Damage to cultural artefacts by environmental pollution (#PD2478).
Aggravated by Social environmental degradation from recreation and tourism (#PD0826).

♦ **PE9829 Diseases of gallbladder**
Broader Diseases and deformities of the digestive system (#PC8866).
Narrower Cholecystitis (#PE2251)   Urinary calculus (#PE7934)
Bile-duct obstruction (#PG4857).

♦ **PE9831 Unauthorized access to computer information systems**
Computer database espionage — Computer hacking
Nature Computer information systems are vulnerable to information invasion by unauthorized users. This is distinct problem at governmental levels where quantities of classified information are kept on computer disks. The possibilities of sabotage intimate that today's high-tech world poses new potential threats to national security.
On the international scale, opportunities for computer espionage due to inadequate safeguards presents a major problem. Computer espionage by foreign governments may encompass military, industrial, economic and commercial objectives, with business firms being the most vulnerable. Inadequate safeguards are almost universally found against computer sabotage.
Incidence In the UK, the cost of hacking to enterprises is estimated at between 500 million to 2 billion pounds per year. Hacking is reported by 20 per cent of companies responding to a survey.
Counter-claim A true hacker is not interested in the data they see, but the way the computer works. Their snooping may actually help protect people from invasions of privacy: they prove and publicize that private information may be accessible to unauthorized persons and so they can help deter government agencies to put telephone records, tax data etc in computers. Hackers can also point out the mistakes of the system operators and the flaws in the security system.
Refs Cornwall, Hugo *Datatheft* (1987); Council of Europe *Protection of Personal Data Used for Social Security Purposes* (1986).
Broader Abuse of computer systems (#PD9544).
Aggravates Computer-based crime (#PE4362).
Aggravated by Unethical practices by employees (#PD4334).

♦ **PE9840 Simony**
Commerce in spiritual benefits
Nature Sale or purchase of spiritual gifts or of positions of power in religious hierarchies. This may include buying or selling of sacraments or sacramental ordinances, such as the exacting of payment for baptism, for the eucharist, and for the conferring or receiving of holy orders.
Background Named after Simon Magus who offered money to acquire the gift of laying on of hands. Of special concern during the Middle Ages in the buying and selling of ecclesiastical preferment, even involving the pope. This was one factor contributing to the Protestant Reformation.
Incidence Of considerable historical importance, especially within Christianity. But no age or country can be said to be completely free from it. When it is suppressed in its more obvious forms it takes a disguised form and becomes more difficult to overcome. For example, the sale may be disguised as a donation. It could be argued that some currently advertised courses to acquire spiritual or psychic powers (including 'siddhis') are a form of simony. Similarly the donation of funds to a spiritually oriented group in effective exchange for a position of power within that group can be considered in the same light.
Counter-claim In situations where the priesthood is impoverished, the receipt of money for sacraments offered one of the only sources of income.

PE9840

Narrower Abuse of relics (#PF5107).
Related Trading in public office (#PE6948)   Commerce in religious indulgences (#PF4433).

♦ **PE9844 Lymphadenitis in animals**
Lymphangitis in animals
Broader Circulatory system diseases in animals (#PD5453).
Narrower Strangles (#PJ9367)   Melioidosis (#PG0774)
Epizootic lymphangitis (#PG9632)   Lymphangitis of horses (#PG5041)
Caseous lymphadenitis in animals (#PG1419)   Streptococcal lymphangitis of swine (#PG1037).

♦ **PE9846 Tuberculosis of bones and joints**
Refs Martini, M *Tuberculosis of the Bones and Joints* (1988).

Broader Tuberculosis (#PE0566).
Narrower Tuberculous osteomyelitis (#PG5176).

♦ **PE9854 Respiratory diseases of small animals**
Broader Diseases of respiratory system in animals (#PD7307).
Narrower Rhinitis (#PG5704)   Lung flukes (#PG4427)   Tonsillitis (#PE2292)
Tracheobronchitis (#PG4623)   Nematodes in animals (#PJ9232)
Pneumonia in small animals (#PG9340)   Feline respiratory disease complex (#PG4076)
Infectious tracheobronchitis of dogs (#PG7328).

♦ **PE9857 Secret intelligence agents in public office**
Broader Unaccountable government intelligence agencies (#PF9184).

Fuzzy exceptional problems   PF

Content

This section groups "unusual" problems. These may include:

- potential problems;
- problems based on "superstition";
- problems based on unsubstantiated beliefs (such as UFOs)
- dormant problems and problems of the past;
- esecially ambiguous, intangible or "fuzzy" problems;
- problems in the organized response to other problems;
- low probability problems (such as geomagnetic reversal).

Problems are also allocated to this section when it is difficult to justify their allocation to any other section. A significant number of the problems cannot be readily grouped into hierarchies using the "broader" relationship. They then tend to be characterized by the looser "related" link or by functional links to other problems.

Note that further information relevant to an understanding of the problem may be present in other problems cross-referenced in the entry consulted.

The problems in this section tend not to be a preoccupation of the programmes of international organizations. But they may emerge in reports on the failures of those programmes.

This section groups 2,470 problems for which there are 19,198 cross-references.

Rationale

The problems allocated to this section tend to emerge from unusual studies that do not fall within any of the conventional disciplines and thus seldom figure in any reviews of the crises of the times. Some derive from the paragraphs and sub-paragraphs of documents of programmes in response to the more conventional described in earlier sections. Problems of this kind frequently escape information collection procedures and are easily ignored as being unworthy of serious attention. The importance of some of these problems become more readily apparent in practical situations.

Method

The entries are based on information obtained from international organizations, from a wide variety of reference books, or as reported in the international media. The procedures for identifying world problems are described in Section PZ at the end of this volume.

Index

A keyword index to entries is provided in Section PX.

Comment

Detailed comments are given in Section PZ at the end of this volume.

Reservations

The emphasis throughout this volume has been placed on providing descriptions of less well-known problems, particularly when the extensive material available on the better known problems contained neither succinct descriptions of them nor descriptive material which could easily be reduced to succinct descriptions. The problem descriptions here represent a compilation of views from published documents (usually from international organizations). The text provided does not necessarily constitute the best possible description of the problem, since a compromise has had to be struck between availability of information, the resources to process it, and the space available in this volume.

Possible future improvements

There is much scope for improving the quality of problem entries through feedback from interested bodies. More bibliographic references could be included where appropriate, as well as references to major resolutions concerning those problems recognized by the United Nations. There is also much scope for improving the pattern of cross-references, both between problems, to other sections of this volume (eg values) and to the 20,000 internationally-active bodies in the companion series (*Yearbook of International Organizations*)

ENTRY CONTENT AND ORGANIZATION

Ordering of entries
Entries are in **numeric order**. Entry numbers have been **allocated randomly**; they have no significance other than as a permanent point of reference to facilitate indexing, cross-referencing, and updating between editions.

Index access to entries
The location of an entry in this sub-section may be determined from
the **Volume Index** (Section PX) on the basis of keywords in the name of the entry or its alternate titles.

Structure of entries
Entries may be composed of the following descriptive elements:

(a) **Entry number** This number has **no significance**, except as a convenient method of identifying the entry (particularly for indexing purposes), of filing information on it, and as an identifier to which cross-references from other entries (possibly in other Sections) may refer in this and future editions. The first letter of the entry number refers to the section of this volume in which the sub-section, denoted by the second letter, is located.

(b) **Problem name** This is printed in bold characters. It is the name selected as best indicating the nature of the problem. It may be followed by alternative problem names.

(c) **Nature** Description of the problem which attempts to identify the nature of the disruptive processes involved. The information included here, and in the following paragraphs, is compiled directly, to the extent possible, from available published documents. Where appropriate the text included may be reproduced, in a minimally edited form, from the publications of international organizations, such as those of the United Nations or its Specialized Agencies.

(d) **Incidence** Summary description of the extent of the problem which makes it of more than national significance.

(e) **Background** Describes briefly when and how the problem's importance was recognized initially, and how this recognition has evolved over time.

(f) **Claim** Stresses the special importance of this problem and why action is particularly urgent. This paragraph offers means of including statements which may deliberately exaggerate claims for the unique importance of the problem.

(g) **Counter-claim** Stresses, where appropriate, the relative insignificance or erroneous conception of the problem as described. This paragraph offers a means of including statements which may deliberately exaggerate the arguments refuting the evidence for the existence of the problem. Absence of such arguments from the text does not mean that they do not exist.

Cross-referencing of entries
At the end of any entry, there may be cross-references to other entries. These indicate the number and name of the cross-referenced entry, whether within this Section or in other Sections. There are 3 types of **hierarchical** cross-references between problems:
>**Broader** = Broader problem: more general problems of which the problem described may be considered a part. The described problem may be considered an aspect of several broader problems
>**Narrower** = Narrower problem: more specific problems which may be considered a part of the described problem
>**Related** = Related problem: problems that may be considered as associated in a hierarchically undefined way with the described problem.

>There are 4 types of **functional** cross-references between problems:
>**Aggravates** = Problems aggravated by the described problem: a forward or subsequent negative causal link
>**Aggravated by** = Problems aggravating the described problem: a backward or prior negative causal link
>**Reduces** = Problems relieved, alleviated or reduced by the described problem: a forward or subsequent positive causal link
>**Reduced by** = Problems relieving or alleviating the described problem: a backward or prior positive causal link

♦ **PF0010 Pacifism**
Nature Pacifism encourages individuals and groups to refuse to take up arms to defend their country (or an ally) against aggression. As such it weakens the ability of the country to defend itself and therefore threatens the security of the country or the alliance to which it belongs.
Background Collective pacifism may take the form of national neutrality with the expressed denunciation of any form of war. Personal pacifism is an expression of conscience or conviction that one cannot participate in any military conflict. Individuals may advocate pacifism as a policy for a nation, but in the event of actual war, serve in their country's military forces or otherwise voluntarily support their nation's war effort. Pacifism in its collective and personal forms is obviously interrelated, but at the same time subject to widely divergent characteristics. Some pacifists claim the title 'activists'. Others speak approvingly of passive non-resistance. Expressions of pacifism, therefore, run a broad range. Pacifists are frequently the object of rhetorical, and some times of physical attacks; accused of being traitors, and reviled and discriminated against in other ways.
Claim Pacifist demonstrations in the West are overtly encouraged by the USSR and its allies, to the USSR's advantage. No pacifist demonstrations occurred in the Eastern bloc when the USSR deployed SS-20s. Pacifists should recognize the pernicious effects of pacifism in some circumstances, for example in the demobilization that occurred before the second world war. Europe is under threat and the balance of power is no longer stable; confidence cannot be re-established simply by political will.
Counter-claim Pacifists look back on a tradition in the West from the early Christians who refused military service, to the Society of Friends (the Quakers), and quote Ballou, Thoreau, Tolstoy, Aldous Huxley, Schweitzer and many others who shared a 'reverence for life'. In the East, pacifism has been preached from the time of Buddha to that of Gandhi.
Broader Military insecurity and vulnerability (#PC0541).
Narrower Denial of right of conscientious objection to military service (#PD1800).
Related Neutrality (#PF0473) Militarism (#PC2169)
Citizen disobedience (#PD5707).
Aggravates Civil disobedience (#PC0690) Conscientious objection (#PD4738)
National insecurity and vulnerability (#PB1149).
Reduces International insecurity (#PB0009).

♦ **PF0011 Inadequate environmental statistics**
Nature Statistical measurements of the state of the natural environment and of environmental damage are at a somewhat experimental stage. For some environmental elements that are, in principle, subject to measurement – such as the physical conditions of air, water and soil – there is considerable uncertainty among scientists about what it is most important to measure. There is a need to apply statistical techniques both to monitoring procedures and to assessing the costs and benefits of policy actions already taken or under consideration (the benefits being generally more diffuse and difficult to quantify than the costs); and to integrate the statistics with socio-economic data.
Broader Statistical errors (#PF4118)
Inadequate statistical information and data on problems (#PF0625).

♦ **PF0013 Failure of disarmament and arms control efforts**
Arms violations
Nature Disarmament and arms control are overlapping areas. The first is concerned with the substantial reduction or complete elimination of those weapons with which nations can commit aggression and wage war, given that the continued existence of nuclear weapons must sooner or later lead to war escalating to an unimaginable level of destructiveness. Arms control implies some form of collaboration between adversary states, involving either formal or tacit agreement, aimed at limited control in well-defined areas. The latter is considered more realistic, but it does not lead to the former. A very considerable range of disarmament proposals have been considered or negotiated with almost no tangible results, excepting a number of arms control agreements in areas of no military interest, or treaties of no significance, or having little probability of entering into force or being realistically implemented.
Incidence Some 19 protocols, treaties, agreements and other international instruments of accord have been generated in the interests of arms control since 1948. Negotiating sessions, productive and non-productive, between the super-powers alone are counted in the thousands. Strategic arms limitations talks are continuing in Geneva under the acronym "START" (strategic arms reduction talks), while nuclear and other armament is increasing and proliferating.
Refs Carter, April *Success and Failure in Arms Control Negotiations* (1989); Stockholm International Peace Research Institute *Success and Failure in Arms Control Negotiations* (1989); United Nations Institute for Disarmament Research *Disarmament* (1984).
Broader Limited acceptance of international treaties (#PF0977).
Narrower Hazardous remnants of war (#PF2613).
Related Plutonium overproduction (#PD2539).
Aggravates War (#PB0593) Homicide (#PD2341)
International insecurity (#PB0009) Pervasive fear of nuclear war (#PC3541)
Competitive acquisition of arms (#PC1258)
Insufficient nuclear weapon free zones (#PJ5335).
Aggravated by Proliferation of strategic nuclear arms (#PD0014)
Disparity in facilities for military mobilization and reinforcement (#PE3995).
Reduces Imperialism (#PB0113) Military espionage (#PD2922).

♦ **PF0019 Governmental bias in statistics**
Inadequate social welfare indicators — Excessive reliance on economic indicators of human development
Claim National accounts, particularly the concepts of gross national product or gross domestic product, provide an incomplete and often misleading record of economic performance, especially of welfare, between countries and over time. Much greater weight needs to be given to other measures of performance of a society, to employment, to the distribution of income and wealth, to indicators of social conditions, of education, of health, and of housing. Recently in developed countries, the costs of high rates of material progress have been publicized in terms of pollution, environmental damage and the unpleasant aspects of a modern urbanized and motorized society.
Broader Bureaucratic bias (#PC1497) Biased government information (#PF0157)
Inadequate statistical information and data on problems (#PF0625).
Aggravates Statistical errors (#PF4118) Misuse of statistics (#PF4564)
Inadequate social welfare services (#PC0834).
Aggravated by Suppression of information concerning social problems (#PF9828).

♦ **PF0021 Unsatisfied need for continuing education**
Lack of refresher courses in professional fields — Under-developed continuing education
Nature Constant progress in research, rapid building-up of new information, and accelerating technology require those in the professions (law, engineering, medicine, etc) to continue acquiring knowledge and techniques throughout their working lives. This need is unsatisfied in many educational systems. A major bottleneck, particularly in the development and utilization of technology, lies in the inadequate provision for the training of manpower with middle-level skills. This includes young adults, both men and women, with training above that acquired through traditional skill training and below that normally obtained in universities. It covers such work as that done by para-professionals, technologists and technicians, whose functions increasingly entail the exercise of both technical and managerial responsibilities. This is also the group of people who are increasingly affected by unemployment, aggravated by the movement of the population bulge (which exists in many countries) from adolescents to young adults.
Broader Unavailable education for effective living (#PF2313).
Related Inadequate industrial retraining programmes (#PF4013).
Aggravates Absenteeism (#PD1634) Inadequate education (#PF4984)
Inappropriate employment incentives (#PD0024).

♦ **PF0023 Non-valuation of housework in national accounts**
Nature One of the oldest conundrums in the theory of national income statistics is the question: if a man marries his housekeeper, is it correct to show a decline in national income. The answer given is, in a sense, a test of the respondent's concept of what national income statistics are intended to exhibit. The answer 'yes' implies that, because the marriage signifies the disappearance of a marketed activity from the sphere of measured production, national income, or GNP, being regarded as essentially the sum of market activities, correctly falls. The answer 'no' implies a recognition that the woman's activities as a housekeeper may be expected more or less to continue as before: there is no change (necessarily) in the total flow of goods and services.
Incidence Working women in the UK are estimated to have lost a total of $27 in overtime in 1987 because of home responsibilities. This is changing, however, as men assist in housework and childraising. US married men now do an impressive nine minutes more housework each day than they did twenty years ago.
Claim The household is a kind of enterprise, involving decisions on current and capital outlays, on choices between paid work and work at home and between work and leisure, etc, on the basis of a mixture of market prices and shadow prices for the various alternatives. Indeed, for most households, important decisions are in fact influenced to some extent, explicitly or implicitly, by this kind of economic rationalization.
Counter-claim The social and institutional as well as the economic constraints on economically rational planning of a family economy, together with the non-economic variables that may quite rightly dominate decision-making, render this type of approach inappropriate and unrealistic. Although aspects of household behaviour may be subjected to tests of economic rationality by the social scientist, the formal incorporation of the household as an enterprise in national accounting does not seem at all necessary.
Related Unpaid labour (#PD3056).

♦ **PF0037 Inadequate governmental energy conservation policies**
Nature Energy conservation can be defined as the strategy for reducing energy requirements per unit of industrial output or individual well-being without affecting the progress of socio-economic development or causing disruption in life style (Schipper, 1976).
Background In temperate developed countries most energy is used in heating and lighting industrial and domestic buildings. Industrial processes, transport and agriculture are the other main users. During the 1970s it was demonstrated that substantial savings could be achieved through appropriate building technologies and the use of energy-efficient equipment for heating, air-conditioning and lighting.
Claim Most goods could and should be both manufactured and made to work more efficiently. Policies should be designed to encourage efficient use of energy. These would probably have to incorporate a variety of energy-savings measures, including fiscal measures, regulations and standards, encouragement of action by common means (public transportation, total energy systems), public education and research and development.
Broader Inappropriate policies (#PF5645)
Long-term shortage of natural resources (#PC4824).
Aggravates Uninsulated buildings (#PE0242)
Unsustainable development of energy use (#PC7517)
Underutilization of oil shale as an energy source (#PF0445).
Aggravated by Surface to surface missiles (#PE4515).

♦ **PF0038 Inadequate governmental resource conservation policies**
Nature Inadequate government resource conservation policies may have negative ecological, social, and economic ramifications difficult to correct.
Background The main obstacles to achieving adequate conservation policies are: absence of conservation at the policy making level; lack of environment planning and of rational use allocation; poor legislation and organization; lack of training and of basic information; lack of support for conservation; and lack of conservation-based rural development.
Broader Inappropriate policies (#PF5645)
Long-term shortage of natural resources (#PC4824).

♦ **PF0039 Mass extinction due to comet showers**
Nature Life on earth, and maybe the earth itself, could well be threatened by inanimate 'attack' from outer space. Debris from explosive disintegration of comet fragments high in the atmosphere is said to cause damage such as occurred in Siberia in 1908. Far greater damage would be caused by the return of large showers of comets. The dinosaurs died out suddenly about 65 million years ago. It is theorized that a large asteroid collided with the planet, sending up a cloud of dust that hung in the atmosphere for months blocking sunlight and thus preventing photosynthesis. The dinosaurs succumbed to cold and hunger. Another theory suggests a 'killer star' that was, in fact, a binary star with the sun. This star, a small, cool dwarf, has an elliptical orbit that takes it as far as 2.4 light years from the sun, where a comet cloud containing 100 billion comets is known to exist. As it passes through the cloud, it drains some of the comets, which it flings into the inner solar system as it passes by every 28 million years. Since the extinction of the dinosaurs, there have been two mass extinctions, for which scientists have found evidence among fossils in the seabed. One occurred 37 million years ago, and the other about 14 million years ago, both consistent with the cyclical theory of extinctions. It is therefore predicted that another such extinction may occur about 14 million years from now.
Broader Hazards from comets (#PF3564) Periodic mass extinctions of species (#PF4149).

♦ **PF0041 Governmental inaction concerning trade in services**
Nature Despite the increasing importance of trade in services, there has been a notable absence of international action to reduce services trade barriers. The services industry includes banking, construction, accounting, insurance, motion pictures, travel and tourism, shipping, telecommunications and numerous other sectors.
Incidence Trade in services accounted for 20 percent of world trade, more than US$ 350,000 million in 1983.
Broader Government inaction (#PC3950).

♦ **PF0045 Inadequate satellite capacity for telecommunications**
Nature The explosive worldwide growth in telecommunications has triggered a crisis. Most governments and multinational corporations now transmit their communications via satellite, as do most long-distance telephone lines and most television networks. Demand for new cable television programs alone is still growing so fast it will outstrip capacity for years to come. Lower satellite transmission costs have triggered dramatic increases in video- conferencing networks. Lower costs for Earth stations are having an even more dramatic effect in the explosive growth

of telecommunications.
Broader Insufficient communications systems (#PF2350).

♦ PF0048 Inadequacies of the international monetary system

Inappropriate international institutional framework for finance — Crisis in international monetary and financial system — Lack of coordination of the international monetary system

Nature The international monetary system is fundamentally one that is used for payments in world trade. Big trading countries have strong voices in monetary affairs and their currencies are therefore "hard" (capable of use in world trade without restrictions). Countries with small shares in world trade and weak agricultural economies have "soft" currencies, which are not accepted in payment for their imports. Their trade is limited by the amount of "hard" currencies they can earn at any given time by their exports. As the world has become more interdependent, these guidelines of an earlier era are too simple to accommodate the complexities of international trade.

The external financial pressures on developing countries are of an unparalleled intensity. Severe difficulties in meeting debt service payments are widespread. Some statistics indicate a net flow of resources from the less developed countries to the developed countries, as efforts to repay years of credit continue.

Background When the international monetary system was constituted, the parity of the member countries' national currencies was determined according to the value of the US dollar, and so their parities depended on the stability and value of the dollar. When the value of the dollar began vacillating, the USA was forced to carry out devaluations, thus upsetting the system of parities. Another cause is the increasingly acute problem of international liquidity, and also the uneven distribution of monetary reserves between the developed and developing countries.

Since the collapse of the Bretton Woods monetary system of fixed exchange rates in the early 1970s, there have been several attempts to create new monetary arrangements among major trading nations: the Jamaica Accords, the Plaza Accord, efforts at economic summit meetings to coordinate domestic economic policies; the Louvre Accord, which sought to stabilize currency relationships within a vague set range, failed. All these efforts have been improvisations (often informal and confidential) in contrast to the elaborate, transparent framework of Bretton Woods.

Incidence The contraction in international liquidity and the continuing wide fluctuations in exchange rates among major currencies have dramatized the many inadequacies of the international monetary system that have emerged with the demise of the fixed exchange rate regime established at Bretton Woods. The domestic policies of individual major industrial countries and the operations of private international banks have become the major determinants of key international monetary variables such as exchange rates, interest rates and international reserves. No mechanism has served to bring such unilateral actions into a self-adjusting integrated system. During the crisis in the Guld in 1990, the Bank for International Settlements warned that it could pose a serious threat to the world financial system. The danger of cumulative interaction between a deteriorating economic environment (caused by the surge in oil prices) and a continuing constraint on the ability of banks to lend at a time of high demand for credit were likely to bring about an international credit squeeze.

Claim A reform of the existing international monetary system is vital not only because of the shortcomings accumulated since its creation, but also because of the major changes which have taken place in the structure of the world economy and international economic relations.

Broader Vulnerability of social systems (#PB2853)
Lack of international cooperation (#PF0817).
Narrower Chronic shortage of foreign exchange (#PC8182)
Inadequate level of world monetary reserves (#PF3059)
Underutilization of world banking facilities (#PJ8783)
Privatization of intergovernmental monetary system (#PF2139)
Inadequate mechanism for balance of payments adjustment (#PF3062)
Inadequate mechanism for the creation of liquid reserves (#PF3061).
Aggravates Social insecurity (#PC1867) Global economic crisis (#PC5876)
Global financial crisis (#PF3612) Uncertain status of monetary gold (#PF2342)
Lack of confidence in the international monetary system (#PF3058).
Aggravated by Payment of interest (#PF5514) Uncontrolled growth of debt (#PC8316)
Defaults on international loans (#PD3053)
Uncontrollability of world social systems (#PF1443)
Inadequate level of developing country monetary reserves (#PF3060)
Burden of servicing foreign public debt by developing countries (#PD3051).

♦ PF0053 Lack of autonomous world-level actor to identify and clarify world interests

Nature There is no autonomous organization or individual able to identify and clarify world interests. The United Nations is obliged, by its very nature, to reflect the collectivity of individual state interests. Only symbolic figures like the UN Secretary-General and the Pope can formulate world interests (although the former is constrained by the practical politics of being elected by governments to function in an intergovernmental setting and the latter is hampered by a religious heritage that is weighed down with roles and doctrines). The dominance of the state is partly expressed by filling the field of political action so completely that no other actor orientation toward reality can achieve prominence.

Counter-claim Vox populi, vox Dei; the voice of the people is the world-level agent that can transcend national interests, and international referendums on world or regional issues are the enabling mechanism. The direct election of representatives to the European Parliament was a forerunner of possibilities in this direction; but there are many obstacles to world referendums on, for example, United Nations resolutions. Non-governmental organizations have an important role in giving expression to the people's voice but nation-state interests have limited their powers to participate at intergovernmental organization deliberations.

Broader Fragmentation (#PA6233) Obstacles to leadership (#PF7011)
Domination of the world by territorially organized sovereign states (#PD0055).
Related Inadequate hero images (#PF2834).

♦ PF0054 Interference between communications satellites

Nature In the past 10 years, more than 120 communications satellites have been placed in orbit. To account approximately for the principal satellites: of the 90 still working, 20 carry telephone, television and business messages for the United States, 20 serve the Soviet Union, and Intelstat operates another 35. So many communications-satellites are now at similar radio frequencies in the same equatorial belt in space that they have begun to interfere with each other.

Incidence Parts of the equator linking communications satellites with Western Europe are saturated with satellite traffic. For the region of the equator serving North America, there is no longer enough space between satellites to allow them to transmit and receive C band signals without interfering with the next satellite's signal.

The US Federal Communications Commission has ruled that domestic satellites can come as close together as 3,200 kilometers at geosynchronous altitude, halving the separations to which they were restricted a few years ago. This means that more satellites can crowd into the same space.

Broader Traffic congestion (#PD0078).
Aggravates Insufficient communications systems (#PF2350).

♦ PF0059 Excessive emphasis on fashionable areas of research

Nature New research areas are continually emerging into fashion, causing considerable funds to be reallocated to such areas, far in excess of that required for adequate research. In response to such fashions, many research workers and institutes restructure their priorities and programmes to ally themselves with glamorous topics.

Incidence Several recent examples include laser technology, semiconductors, alternative energy sources, genetic engineering and robotics. Other kinds of fashionable studies concern simulation, gaming, modelling, and forecasting. Excess activity of this kind causes scepticism and obstacles to funding of serious, long-term research commitments.

Broader Inadequate research on problems (#PF1077).
Related Lack of scientific investigation (#PF2720)
Opportunist bias in public discussion and research on development (#PF3846).
Aggravates Imbalanced research activity (#PF0198)
Inadequate research and development on problems of developing countries (#PF1120)
Failure to integrate knowledge to empower humanity in response to the global problematique (#PF8753).
Aggravated by Excessive emphasis on fashionable problems (#PF4164)
Excessive reliance on fashionable solutions to problems (#PF4473).

♦ PF0074 Ineffectiveness of intergovernmental organization and programmes

Claim Intergovernmental organization effectiveness is hampered by internal factors such as bureaucratic inefficiency, the absence of clearly defined goals, inter-agency rivalry and budget manipulations that undermine some programmes at the expense of others. Financial irregularities and mis-management may be alleged and these allegations can destroy confidence and monetary support by member nations. In some instances there has been a wasting of important resources by the duplication of programmes of other organizations or of governments. The single most important external obstacle to greater effectiveness by intergovernmental organizations is the absence of fully cooperative participation by member states, some of whom, in fact, may be intentionally obstructive in order to serve narrow, national interests.

Broader Ineffectiveness of international organizations and programmes (#PF1074).
Aggravated by Denial of right to social welfare services for indigenous peoples (#PE1506)
Inadequate coordination of the intergovernmental system of organizations (#PE0730)
Jurisdictional conflict and antagonism within intergovernmental organizations (#PE9011).

♦ PF0076 Prejudice against communication by visual imagery

Absence of audio-visual media

Nature With the considerable development in world wide literacy and the use of the written word, there has developed a certain prejudice against communication (of other than simple messages) by the use of images. This applies not only to communication with non-literate people but also between well-educated people from different disciplines or sectors of society without a specialized language in common.

Incidence The problem applies to all kinds and levels of teaching, but it is of particular relevance to the millions of people who can neither read nor write. Use of visual imagery (reinforced by the spoken word) provides a short cut for communicating information in areas where literacy programmes are making slow progress.

Counter-claim For a person to function effectively within the mainstream of modern society they must know how to read and write. To provide effective alternatives will not only discourage the development of literary skills but in doing so render millions of people and their children to third class status.

Broader Narrow media options (#PJ0094) Overlooked media channels (#PF4611).
Related Discriminatory communication (#PD6804).
Aggravates Lack of communication (#PF0816) Ineffective communication (#PF6052)
Insufficient communications systems (#PF2350)
Ineffective systems of practical education (#PF3498).
Aggravated by Illiteracy hinders communication (#PS1611).

♦ PF0090 Unnatural boundaries between states

Arbitrary national boundaries — Artificial nation boundaries — Territorial fragmentation — Artificial territorial boundaries

Nature Boundaries between states are determined as a result of a variety of political, historical and administrative compromises. In the absence of readily available geographical features, such as rivers, some boundaries run through towns, across farms and even across the middle of the rooms of individual houses. This gives rise to considerable difficulties in observing the usual frontier formalities and in handling the administration and taxation of economic units divided in this way. Boundaries may also divide a minority linguistic or ethnic community, threatening its viability. Some unnatural boundaries arise due to peace treaties and agreements after armed conflict. This is aggravated in proportion to the number of nations which are involved.

Broader Arbitrariness (#PB5486) National boundaries (#PF8235)
Human errors and miscalculations (#PF3702).
Narrower Divided countries (#PD1263) Fragmenting district boundaries (#PG7968)
Unnatural boundaries between developing countries (#PD2544).
Related Fragmentation (#PA6233).
Aggravates Plural society tensions (#PF2448) Border incidents and violence (#PD2950)
National political dependence (#PF1452) Boundary disputes between states (#PD2946)
Boundary disputes between neighbours (#PE7903)
Boundary constraints on land planning (#PF0954)
Divisive effects of official cultural pluralism (#PF0152)
Endangered lifestyles of nomads and pastoralists (#PE8077).

♦ PF0091 Lack of terminological equivalents between languages

Nature A persistent obstacle to translation into languages which are relatively undeveloped in some conceptual domain is the lack of equivalent scientific and technical terms. This is particularly important for the preparation of textbooks in such languages.

Incidence This situation is not only important in developing countries where even relatively simple scientific terms do not have a precise equivalent, but also in developed countries.

Broader Multiplicity of languages (#PC0178).
Related Lack of vocabulary (#PF3824).
Aggravates Inaccessibility of knowledge (#PF1953).
Aggravated by Linguistic purism (#PF1954).

♦ PF0100 Underutilization of intellectual ability

Nature Particularly in the developing countries, a considerable reserve of unutilized intellectual ability exists. Surveys show that even in developed countries those enrolled in secondary education or university studies are not necessarily more able than others in their group who do not continue their education. It may be that only a fraction of the population able to benefit from higher education actually does so. For example, proportionately more workers and more girls benefit from higher education in socialist Eastern Europe than in the advanced countries of the West, yet there is no reason to suppose that actual ability for such groups varies between regions. The price mechanism may also intervene where normally demand would exceed educational supply, and the poor-but-able are denied further education. The capacity to succeed in studies at a given level obviously depends on intellectual ability, and this is what is measured, but it depends also on determination and perseverance. In terms of development and considering high youth unemployment, the absence of educational opportunities which would encourage and pay students to study leads to the lack of utilization of a vast amount of human intellectual

potential.
Broader Rural unemployment (#PF2949) Waste of human resources (#PC8914)
Waste of human resources in capitalist systems (#PC3113).
Narrower Unemployed intellectuals in developing countries (#PD1273)
Wastage of highly skilled personnel in the routine maintenance of complex systems (#PE1396).
Related Inadequate research on problems (#PF1077).
Aggravates Frustrated talent in government posts (#PE4398).
Aggravated by Educational wastage (#PC1716) Inadequate intellectual methods (#PF7380)
Suppression of intellectual freedom (#PC5018)
Inequality of access to education within countries (#PC1896)
Faulty approaches in the teaching of intellectual methods (#PF0956).

♦ PF0107 Negative effects of family allowances

Financial incentives for having children
Nature Family allowance schemes usually apply universally to all families with children. As such, instead of being a measure for redistribution of income as part of social security, it acts as an incentive to larger families. In certain countries, this incentive is adopted as a deliberate way of increasing population. In others, where it was intended as a means of ensuring adequate child welfare, this aim is not fulfilled, and underprivileged minorities are often accused of irresponsibly producing more children in order to obtain the family allowances.
Incidence Family allowances are granted until the child reaches a certain age – 12 in Iran, 15 In Japan, 16 in Bulgaria, Australia and New Zealand, and 18 in the United States. However, in certain circumstances, particularly when the child is continuing with his studies, allowances are paid up to a higher age (18 in Iran and New Zealand, 20 in Japan, 21 in the United States).
Broader Inadequate child welfare (#PC0233).
Related Lack of family planning (#PF0148).
Aggravates Social conflict (#PC0137) Social discrimination (#PC1864)
Discrimination in social services (#PC3433).
Aggravated by Unsustainable population levels (#PB0035)
Unequal coverage by social security (#PF0852)
Social insecurity in developing countries (#PE4796)
Government opposition to population control (#PF1023).

♦ PF0118 Shortage of books and textbooks in developing countries

Shortage of books in developing countries
Nature The shortage of suitable textbooks is felt at all levels of education, especially in primary schools, and is a serious impediment to educational progress. In many developing countries there are few teachers who are qualified to write textbooks even though they are experienced in teaching their own subjects. In addition, there are difficulties arising from the shortage of paper and up-to-date printing facilities. In some cases, textbooks are expensive and parents cannot afford to buy them. Lack of adequate planning has been a major contributor to some of these problems. Government policies in certain matters (for example, the language of instruction in schools; copyright; taxes and duty on imports of paper, printing materials, printing equipment and books; export controls; internal trade; bank loans; operations within the country of foreign publishers and printers) are vitally important to the planning and operations of textbook publishing organizations (TPOs). In cases where such policy matters have not been taken fully into account in the planning and operation of TPOs, changes to assist in the success of textbook publishing, for which there were pressing needs, have often not been made, or sometimes apparently not even considered, by governments.
Incidence In 1980 a random survey of 15 rural primary schools in a small, anglophone African country established that the average availability of textbooks for the pupils in grades one through six was 11 textbooks per 1,000 students.
Claim The experience of ministries around the world suggests that an improved supply of textbooks can raise children's achievements by more per dollar spent than any other improvement. In a comparative study of the availability of textbooks in the classroom, the researchers found that improvements in availability led to better (tested) performance in 83 percent of cases. Studies were of the primary systems of eleven countries – three African, four Asian and four in Latin America – over a wide range of conditions.
Broader Inaccessibility of knowledge (#PF1953) Insufficient educational material (#PE8438).
Aggravates Illiteracy (#PC0210) Educational wastage (#PC1716)
Inappropriate education in developing countries (#PF1531)
Inadequate research and development capacity in developing countries (#PE4880).
Aggravated by Paper shortage (#PE1616) Imbalanced distribution of knowledge (#PF0204).

♦ PF0129 Negative effects of the nuclear family

Nature Negative effects of the nuclear family include the isolation and emotional dependency of the husband–wife and parent–child relationship, which produces tensions and may lead to marriage breakdown in the former instance and juvenile delinquency and other juvenile problems in the latter. The nuclear family so far as it excludes other family members who are integrated into the extended family, tends to neglect the needy, such as the aged, poor, handicapped; and tends to aspire in a competitive manner towards material well-being and status for itself. The acquisition of status symbols which characterizes the nuclear family leads to a wastage of resources and artificial values with a strong emphasis on class consciousness.
Claim Each person in a nuclear family is too tightly linked to other members of the family. If any one relationship goes sour, even for a few hours, it becomes critical. People cannot simply turn towards uncles, aunts, grandchildren, cousins, nephews, nieces, etc, as they could in an extended family. Instead, each difficulty twists the family unit into ever tighter spirals of discomfort. The children become prey to all kinds of dependencies and oedipal neuroses. Nuclear families also splinter off their old relatives, who often live in conditions of loneliness and neglect; and instil a great sense of deprivation on the adults, particularly on housewives.
Counter-claim The family is what its members make it. The myth of the "ever-happy" family puts intolerable strains on members to behave in preordained ways. These are the strains of accepting a fixed group of role definitions, not the inherent strains of the family relationship. The rootless, readily mobile, nuclear family group is much more flexible and responsive to the demands for labour mobility that are built into 'advanced' economies. Also, within the nuclear family there is room for individuals to develop creativity and take personal initiatives that the extended family would never willingly encourage. It is wrong to romanticize the extended family which, as an institution, can be cloying, repressive and greedy as well as loving and kind.
Broader Family structure as a barrier to progress (#PF1502)
Inhibition of individual psychological development through life cycle (#PF6148).
Narrower Emotional dependency in marriage (#PD3244)
Isolation of parent–child relationship (#PC0600)
Lack of self-development in the family (#PE8341)
Excessive intensification of the parent–child relationship (#PF6186).
Related Pursuit of personal prestige (#PF8145) Household segregation by age group (#PF6136)
Dependence within extended families (#PD0850)
Stultifying homogeneity of modern cities (#PF6155)
Inhibition of adult life in small households (#PF6167)
Adjustment difficulties of new urban families (#PF1503)
Inadequate arrangement of housing with respect to common land (#PF6146).
Aggravates Divorce (#PF2100) Family breakdown (#PC2102) Marital instability (#PD2103)
Juvenile delinquency (#PC0212) Maladjusted children (#PD0586)
(#PD0919)
Impersonality of high density accommodation (#PF6156)
Inappropriate accommodations for single people (#PE6187).
Aggravated by Uncontrolled urban development (#PC0442)
Uncontrolled industrialization (#PB1845)
Susceptibility of the old to physical ill-health (#PD1043).

♦ PF0131 Inadequate care for children of prisoners

Children reared in prison — Children with imprisoned mothers
Nature The hardships suffered by families following imprisonment of the father usually involves the most serious economic disadvantages. Mothers are obliged to seek work or, if unable to do so, must depend on outside assistance. Children may frequently be neglected under the new circumstances. Imprisonment may also be followed by divorce and children may be abandoned in the care of welfare services. Even where the home is maintained in the case of imprisonment of one of the parents, the emotional stress placed on children, particularly adolescents, is largely neglected and gives rise to mental problems and anti-social behaviour.
Broader Inadequate child welfare (#PC0233) Victimization of children (#PC5512)
Denial of rights to prisoners (#PD0520).
Related Cruelty to children (#PC0838) Lack of child welfare institutions (#PJ2673)
Discrimination against prisoners' families (#PE5043).
Aggravates Children in poverty (#PD4966) Juvenile delinquency (#PC0212).
Aggravated by Female criminals (#PE1837).

♦ PF0145 Intrusive social science research

Nature Social scientists seek to probe and to describe the most intimate and secret practices by peoples across the world. The attraction of this "guarded realm" for many researchers is not rooted in their desire for knowledge and benefits alone, but some are also drawn by the allure of the forbidden, of secrecy itself. There is also a pleasure in dispelling what some understand to be mystery as in fact purely physical or superficial phenomena. There is the pleasure of trespassing on what is taboo.
Incidence The extraordinary amount of research into every minute aspect of sexuality and religious belief is simply not explicable on scientific grounds alone.
Related Infringement of privacy (#PB0284).

♦ PF0148 Lack of family planning

Sporadic family-planning — Family planning disinterest — Unmotivated family planning — Ineffective birth control — Ineffective use of contraceptives — Ineffective family planning — Ineffective population control
Nature According to UNFPA policy guidelines: "Family planning refers to those practices that help individuals or couples to avoid unwanted births, to determine the timing of births and to determine the number of children in a family. Family planning information, services and supplies, education about sex and parenthood, diagnosis and treatment of infertility make the attainment of this objective possible". The non-use of contraceptive methods and the non-limitation of numbers of children, with no care taken to space children so that they can be adequately nourished and do place a burden on family resources, imply lack of family planning. Early marriage is also included since this is likely to lead to more children, while the parents have less time to build up the economic resources with which to support them. Such lack of family planning results from inadequate access to family planning information and facilities, lack of social security other than that generated by the family itself, general poverty and traditionalism.
Failure to implement family planning intensifies land tenure and migration problems, poverty, malnutrition, maternal and child mortality, and hinders development. Reasons for such failure may differ considerably from developing to developed countries, in so far as the majority of parents in developing countries may want many children whereas underprivileged parents in industrialized countries may wish to limit their families. In industrialized countries it may be more directly related to lack of facilities and information especially for the young, poor and relatively uneducated. There is strong evidence that the higher the standard of living, the more family planning methods are put into practice.
Incidence In the developing world less than 20 percent of reproductive-age couples, on average, use contraceptives, compared with approximately 70 percent in the developed world. Yet half of the currently married fecund women in 15 countries stated that they wanted no more children; and in seven of eight countries where data were available, more than 25 percent of married women with at least one child, or currently pregnant, stated that their last pregnancy was unwanted. In 1983, 62 governments out of a total of 168 felt that the current level of fertility in their country was too high; however, only 74 percent of these countries implemented policies to decrease the level of fertility. Although Asian countries lead the way in instigating population programmes, other continents are gradually increasing their commitment to family planning. UNFPA response to increasing interest by African countries, in particular, has resulted in significantly increased annual expenditure in Africa over the last four years.
Counter-claim Family planning programmes may not adequately consider the following: the urge to procreate; the curse of sterility; the couples who are happy as they are but feel coerced into using unneeded contraceptives; the values imposed upon Third World countries by the Western World; and the methods of regulation already used in a community.
Refs WHO *Sexuality, Family Planning and Migrant Populations* (1987).
Broader Outdated forms of community health (#PF1608)
Debilitating content of village story (#PF2168).
Narrower Inadequacy of contraceptive methods (#PD0093)
Insufficient birth spacing in families (#PF5968).
Related Promiscuity (#PC0745) Negative effects of family allowances (#PF0107)
Family structure as a barrier to progress (#PF1502).
Aggravates Family poverty (#PC0999) Infant mortality (#PC1287)
Unmarried parents (#PD3257) Maternal mortality (#PD2422)
Wasted woman power (#PF3690) Unwanted pregnancies (#PF2859)
Single parent families (#PD2681) Rapid population growth (#PD9153)
Poverty in developing countries (#PC0149) Unsustainable population levels (#PB0035)
Trafficking in children for adoption (#PF3302).
Aggravated by Illegal induced abortion (#PD0159)
Mistrust of birth control (#PG8030)
Prohibitive cost of contraceptives (#PJ3244)
Denial of right of family planning (#PE5226)
Dependence within extended families (#PD0850)
Inadequate family planning education (#PD1039)
Discrimination against women in religion (#PD0127)
Neglect of the role of women in rural development (#PF4959)
Opposition to population control and family planning (#PF1021).
Reduces Birth prevention (#PE3286).

♦ PF0150 Dependence on religion

Religion
Nature Dependence on religion arises from the spiritual needs of man, such as his need to believe in or to worship transcendent beings. Religion is a means to satisfy these needs within a structured orthodoxy, as apposed to mysticism which is abstract and unstructured, and even condemned by certain religious. Also, because of the inadequacy of pure materialism to satisfy man's need

for spirituality and his idea of a 'better life', many people depend on religion as on a device whereby the injustice of the existing materialist social order is made tolerable to them.
Incidence The worldwide Islamic fundamentalist revival demonstrates more powerfully than any other trend in today's world the overwhelming power of religion to mobilize masses of people. No nonreligious movement in modern history has demonstrated its strength or ability to change mass behaviour. There is no parallel experience in contemporary Western civilization.
Claim Religious and other beliefs, such as witchcraft, contribute to the apparent conservatism of rural dwellers and often put a brake on progress. It is in religion that most farmers seek explanation for non-rational and rational behaviour. Apart from these, religious observances, such as not working on certain days and aversion to certain meats and types of food, impede productivity, production of cash-yielding products and use of nutritious foods.
Broader Dependence (#PA4565).
Narrower Experimental religion (#PF3367) Dependence on mysticism (#PF2590)
Irrational religious beliefs (#PF6829) Fragmentation of religious belief (#PF3404)
Enforcement of religion in capitalist countries (#PE8701).
Related Religious apathy (#PC3414) Ideological apathy (#PF3392)
Religious intolerance (#PC1808).
Aggravates Ignorance (#PA5568) Monasticism (#PF2188)
Horological superstition (#PF0565) Corruption in organized religion (#PC3359)
Religious and political antagonism (#PC0030)
Burdensome cost of religious ceremonies (#PF3313)
Vulnerability of marriage as an institution (#PF1870)
Irrational conscientious refusal of medical intervention (#PF0420).
Aggravated by Atheism (#PF2409) Positivism (#PF2179)
Agnosticism (#PF2333) Materialism (#PF2655)
Rationalism (#PF3400) Intellectualism (#PF2146)
Religious rivalry (#PC3355) Disintegration of organized religion (#PD3423).

♦ PF0152 Divisive effects of official cultural pluralism
Nature Official cultural pluralism exists in countries where there are two or more indigenous cultural groups. Because of the need for fair representation of each group, administrative procedure may become very slow and overloaded with personnel. The official recognition of more than one culture may serve to keep segregation and discrimination intact rather than integrating the groups into the national whole. Party politics may be divided on the basis of culture.
Broader Cultural fragmentation (#PF0536).
Narrower Underprivileged minorities (#PC3424).
Related Inequality of opportunity (#PC3435).
Aggravates Segregation (#PC0031) Racial conflict (#PC3684)
Ethnic conflict (#PC3685) Language barriers (#PF6035)
Religious conflict (#PC3292).
Aggravated by Political pluralism (#PF2182) Unnatural boundaries between states (#PF0090).
Reduces Language discrimination in politics (#PD3223).

♦ PF0157 Biased government information
Nature At all levels of local and national government, political manipulation of facts and figures may be employed to conceal or mitigate unwelcome information, and to enhance news that presents the present agencies, incumbent office holders and their policies in the best possible light.
Incidence Before elections, for example, a great deal of "doctored" information is released by governments to contribute to their desirability.
Broader Misleading information (#PF3096)
Restrictions on freedom of information (#PC0185).
Narrower Bureaucratic bias (#PC1497) Governmental bias in statistics (#PF0019)
Government bias in wage bargaining (#PF6745).
Related Official secrecy (#PC1812)
State control of communications mass media (#PD4597)
Domination of government policy-making by short-term considerations (#PF0317).
Aggravates Ignorance (#PA5568) Conformism (#PB3407)
Nationalism (#PB0534) Compulsory indoctrination (#PD3097).
Aggravated by News censorship (#PD3030) Excessive government control (#PF0304).
Reduced by Uncontrolled media (#PD0040).

♦ PF0166 Barriers to the international flow of knowledge and educational materials
Nature Production of the instruments of knowledge requires the investment of material effort. The dissemination of thought, in material form, is thus affected by the many difficulties arising in international trade. Materials which serve educational, scientific or cultural purposes do not move across frontiers as freely as they should since human progress depends on the efficiency with which accumulated knowledge can be made available.
When educational materials cross frontiers, they meet with trade obstacles, such as high postal charges or freight rates or limitations on packaging, etc. The task of removing, on a worldwide basis, these trade and other obstacles which hinder the free movement of educational, scientific and cultural materials remains to be accomplished. It may concern, for example, materials which move only temporarily from one country to another for the purpose of exhibitions, conferences, or as loans, in particular scientific instruments or apparatus. It may also concern safe and rapid transit of delicate scientific instruments.
Broader Cultural barriers (#PB2331)
Restrictions on the acquisition of knowledge (#PF1319).
Narrower Language barriers (#PF6035) Monopolization of knowledge (#PF5329)
Copyright barriers to transfer of knowledge (#PE8403)
Prohibitive cost of knowledge and information (#PF0703)
Hindrances to international spread of new technologies (#PE8758)
Pre-logical limitations to the comprehension of international information (#PF7134).
Related International trade barriers for primary commodities (#PD0057)
Inadequate research and development capacity in developing countries (#PE4880).
Aggravates Inaccessibility of knowledge (#PF1953).
Aggravated by Multiplicity of languages (#PC0178).

♦ PF0167 Maldistribution of world population
Dependence on maldistribution of world population
Nature The population of the world is distributed very unevenly over the land mass. Approximately 90 per cent of all people are crowded into no more than 25 per cent of the land surface. Variations in population concentration are due to differences in natural or social conditions. Although the proportion of developed countries experiencing difficulties caused by the spatial distribution of their population is smaller than the proportion of developing countries, these difficulties nevertheless occur quite frequently. The industrial mobility necessary to the functioning of the economic apparatus constantly creates problems of adjustment. Some countries are also starting to show concern at the excessive depopulation of rural areas, which has reached its limit, while others must still cope with relatively large-scale migration from the countryside to the towns. The development of new urban centres and the decline and renovation of older ones also present problems for almost all governments. Lastly, problems of environmental protection add a new dimension to spatial distribution policies.
Incidence There are three main areas of population concentration. The most prominent is in Asia, where the most densely populated areas are Japan, South Korea, China, Java, India and Bangladesh. Also heavily settled are the greater part of the European peninsula, the northeastern USA, parts of the Caribbean and coastal South America. Non-Soviet Asia, which constitutes not quite 20 per cent of the land surface of the globe, is occupied by an estimated 60 per cent of the world's population, and this percentage is increasing. By contrast, South America, with 13 per cent of the land surface, presently accounts for only 6 per cent of the world's people. World statistics for 1981 indicate that the average population density per square km was: 33 in the world; 99 in Europe; 95 in Asia, 16 in Africa; 16 in North America; 14 in Latin America; 12 in the USSR (in 1978); and 3 in Oceania.
Broader Chance (#PA6714).
Narrower Maldistribution of population within countries (#PC8192).
Related Ageing populations (#PD8561) Religious theology disrelated to life (#PF5694).
Aggravates Ineffective population control (#PF1020).
Aggravated by Declining birth rate (#PD2118).

♦ PF0170 Fluctuations in government social programmes
Nature The persistence into the 1980s of low growth and high unemployment, compounded by the aging populations, has obliged both poor and relatively prosperous countries to make cuts in welfare systems.
Incidence In Germany pensioners must now make a 5 percent contribution to what had been free medical insurance; indexation changes have brought pensions down 5 to 6 percent in real income compared to 1979. Unemployment benefits have also been cut for workers without children – from 68 percent to 63 percent of total salary in the first year, and from 58 percent to 56 percent in subsequent years. In Britain, health spending has fallen for the elderly, the handicapped and the mentally ill. Housing benefits and free hot school meals, except those for the poorest, are also being trimmed. The government of France has been forced into an austerity programme that has meant new private contributions to social security and restrictions on unemployment benefits. The budget for the costly state hospital system has been reduced.
Background The Organization for Economic Cooperation and Development, which monitors the health of the major industrial nations, reports that between 1960 and 1981 social expenditure in the principal West European economies leapt from 14.5 percent to 26.3 percent of their output of goods and services. (In the United States over the same period, the rise was from 11 percent). In Western Europe, huge pension increases led the way, accounting for about 40 percent of social outlay; health care, education, benefits and unemployment payments were the next biggest items. The oil price rises of the 1970s resulted in the first cuts in health and educational benefits, which had exploded in the previous decade.
Aggravates Inadequate social welfare services (#PC0834).

♦ PF0171 Lack of job satisfaction
Insufficiencies within employment
Nature Boredom and frustration in work resulting from lack of interest, involvement, responsibility or authority leads to absenteeism, poor quality work and low productivity from the employer's point of view and psychological distress on the part of the employee.
Incidence Job dissatisfaction varies from one individual to another but some professions and working environments particularly in industrial society, do organize work in such a way that it becomes meaningless, boring or stultifying. Alienating and dehumanizing work is associated with modern highly-mechanized methods of production and decision-making structures. A survey carried out in the USA showed that, if they could start again, 93 per cent of university professors would try to get into similar work. The corresponding figure for car workers, the least satisfied group, was 16 percent. In the auto industry, for example, the absence rate on Mondays and Fridays is staggering – 15 to 20 percent. There is also evidence of massive alcoholism under such circumstances.
Background A hundred years ago Karl Marx described the 'alienation' of the industrial worker, and his contemporaries William Morris and John Ruskin were concerned at increased production being bought at the price of making work frustrating and inane. They feared that the satisfaction and skill of craftsmanship would give way to monotony and incomprehension. The absurdity of the process was brought powerfully to the public attention in the 1930s in Charlie Chaplin's film 'Modern Times'. An American Secretary of Health, Education and Welfare in 1972 presented a study that concluded that 'the best predictor of long life is not whether a person smokes or how often he sees a doctor, but the extent to which he is satisfied with his job'.
Claim Work is a form of living with its own intrinsic rewards. Any way of organizing work which is at odds with this idea, which treats work instrumentally, as a means only to other ends, is inhuman. No-one enjoys his work if he is a machine. A man enjoys his work when he understands the whole and when he is responsible for the quality of it. He can only do so when the work which happens in society is undertaken by groups small enough to give people understanding through face-to-face contact, and autonomous enough to let workers participate in organizing their work. Today, more people are waking up to the absurdity of being well-fed, educated and housed, the pampered master of the machine at home, but its tethered tenant while at work.
Counter-claim Rational labour utilization and productivity are essential conditions for a sustained increase in production leading to improved living standards, particularly among the poorest groups. Workers are more interested in a better life outside their job, so job enrichment is not something unions have been pressing for.
Narrower Excessive hours of work (#PD0140) Increasing job monotony (#PD2656)
Insecurity of employment (#PD8211) Inequality in employment (#PD8903)
Insufficient job benefits (#PJ7915) Inferior status employment (#PD8996)
Worker maladjustment to technology (#PC7041) Unmotivating subsistence employment (#PJ1555)
Frustrated talent in government posts (#PE4398)
Victimization of workers' representatives (#PD1846)
Displacement of natural light in buildings (#PF6198)
Inhibition of communication between non-proximate offices (#PF6197).
Aggravates Strikes (#PD0694) Absenteeism (#PD1634) Human suffering (#PB5955)
Maldistribution of teachers (#PE6183)
High labour turnover in developing countries (#PF0907)
Emigration of trained personnel from developing to developed areas (#PD1291).
Aggravated by Underemployment (#PB1860) Underproductivity (#PF1107)
Over-qualification (#PF3462) Underpayment for work (#PD8916)
Intellectual terrorism (#PD6656) Elimination of traditional skills (#PD8872)
Government bias in wage bargaining (#PF6745)
Gap between industry and the universities (#PE8873)
International imbalance in the quality of life (#PB4993)
Fast and irregular pace of technological advance (#PE7898).

♦ PF0175 Mismanagement of aid to developing countries
Inadequate coordination of aid — Inefficient administration of aid — Lack of coordination between aid donors
Nature In view of the multiplicity of aid programmes and the growing number of developing countries which receive aid from several sources, the need for co-ordination of aid activities has been increasingly felt both by recipients and donors.
Administrative resources in recipient countries are limited and the need to deal with each donor separately. This places a very considerable strain on these resources. Each application for aid usually involves a close examination of the prospective recipient's development plan, programmes and resources and, if several donors are involved, there may be duplication of effort and time consuming consultations which are especially burdensome for the understaffed agencies of

recipient governments. The requirements of aid-giving countries and multilateral institutions vary considerably from one to the other. Cases have thus arisen in which developing countries, having gone to great trouble and expense to prepare projects in a form acceptable to one aid-giving country, find that the project has to be completely recast before presentation, in case of need, to another country. Consequently, while aid-giving countries are not prepared to commit funds except on the basis of fully worked-out projects, recipients are reluctant to incur the heavy expense of full-scale project preparation in the absence of some indication that finance will be available.

Aid-giving countries encounter difficulties resulting from a lack of adequate preparation of assistance requests in terms of the economic priority, technical feasibility and financial viability of projects, and of the institutional capacity to handle them. Consequently, considerable delays may occur because of administrative shortcomings in the recipient countries themselves.

The coordination of the projects of several donors also makes heavy demands on the recipient's administrative cadres. But even apart from these practical difficulties, aid programmes tend to work at cross purposes because of differing views of the recipient countries' needs. Aid donors, unfortunately, are reluctant to divulge information about their projects to one another, making it difficult to coordinate efforts. The fear that releasing information might lead to good project ideas being financed by others is cited as one reason for donor unwillingness to cooperate.

Refs Downing, Theodore; Kruijt, Dirk and Ufford, Philip Quarles van (Eds) *The Hidden Crisis in Development*.
    Broader Mismanagement (#PB8406)
    Inappropriate public spending by government (#PF6377)
    Inadequate planning of action against problems (#PF1467).
    Narrower Limited developing country capacity to absorb aid (#PD0305).
    Aggravates Compassion fatigue (#PF2819)
    Prohibitive administrative overhead costs (#PF8158)
    Bilateralism in aid to developing countries (#PE9099)
    Ineffectiveness of aid to developing countries (#PF1031).

♦ **PF0182 Refusal to grant amnesty**
Nature The refusal of governments to grant amnesty may lead to international conflict and to subversive activities as well as deprivation of prisoners' families and general fear.
    Broader Political injustice (#PC2181).
    Narrower Qualified amnesty (#PF3019).
    Related Violation of amnesty (#PD3018).
    Aggravates Internment without trial (#PD1576)
    Civilian political prisoners and detainees (#PD3015)
    Discrimination against prisoners' families (#PE5043).
    Aggravated by Mis-classification of political prisoners (#PF3020).
    Reduces Political trials (#PD3013)
    Military political prisoners and detainees (#PD3014).

♦ **PF0192 Indeterminacy of death**
Nature Modern medical technology has made the determination and definition of death increasingly complex. There is lack of complete consensus on the definition of death which justifies extraction of organs of the body for other uses or withdrawal of medical care, although parts of the body can be kept alive by such care long after effective death. The traditional meaning of death was either the departure of the soul from the body or, in more recent secular thought, the irreversible stoppage of the flowing of bodily fluids associated with heart and lung function. The loss of vital signs, especially pulse, is a traditional indicator which still has application. Japanese research demonstrates that a fall in the blood pressure of 40 mm of mercury and persistent low blood pressure for six hours signal the immanence of death. Now, however, individuals may have totally destroyed brains with the irreversible loss of the ability to integrate bodily functions, while their respiration is supported mechanically and their hearts continue to beat. This has led many to argue that the irreversible loss of capacity for bodily integration is the necessary and sufficient condition for being dead. Others have opted for a concept of death that relates it to loss of brain function. There are two schools of thought in this matter. One is to equate death with the loss of all brain functions. The other, called the "higher brain" position, equates death with the loss of higher brain activities such as consciousness, the ability to communicate and to relate to others. In this case, those who are irreversibly in a permanent vegetative state are considered dead, even though lower brain activities such as breathing and certain reflexes may continue. This is especially difficult when, as is often the case, the presence of any amount of drugs in the blood introduces doubt as to the irreversibility of the vegetative state. Once one has chosen a concept of death, the next concern, of no lesser difficulty, is how to measure the irreversible loss of the chosen essential functions.
Incidence Many countries and many (but not all) states within the USA have opted through statutes and case law for a concept of death that relates it to loss of brain function.
Refs Cooper, M A *The Uncertainty of the Signs of Death* (1902).
    Broader Death of living creatures (#PF7043)   Inconclusiveness of science (#PF6349).
    Aggravated by Human death (#PA0072)   Vulnerability of human organism (#PB5647).

♦ **PF0193 Inhibiting social patterns**
Unvaried social patterns — Immobility of the social pattern
    Broader Declining sense of community (#PF2575)
    Disjointed patterns of community identity (#PF2845).
    Narrower Unchanging business pattern (#PG5367)
    Inflexible patterns of family lifestyle (#PF2644)
    Restrictive patterns of traditional life (#PF3129).
    Aggravates Inadequate management skills in rural communities (#PF1442).
    Aggravated by Unrestrained use of force in administration of justice (#PE8881).

♦ **PF0198 Imbalanced research activity**
Nature Research priorities tend to become distorted by research fashions, politically fashionable topics or breakthroughs into new areas. This leads to an imbalance of effort with considerable resources being allocated to a relatively small range of research areas, often of debatable significance, to the detriment of research on less fashionable areas of greater social significance.
Claim Alternative energy research is an example of excessive effort with little to show after a decade of intensive investigations and pilot projects. The continuing presence of the internal combustion engine in automobiles is one of many indications of failure to achieve results. Building a better computer has been the modern equivalent to building a better mouse-trap. The resources consumed in meaningless competitive computer research have been diverted from technologies having direct social benefit. A particular problem in imbalanced research activity is exemplified in biotechnological areas where the bias is towards applications in the developed countries.
    Broader Inadequate research on problems (#PF1077).
    Aggravates Irrelevant scientific activity (#PF1202)
    Irrelevance of science and technology (#PF0770)
    Inadequate research and development on problems of developing countries (#PF1120).
    Aggravated by Excessive emphasis on fashionable areas of research (#PF0059).

♦ **PF0200 Zionism**
Zionist conspiracy
Nature Zionism is a Jewish nationalist movement that has had as its goal the creation and support of a Jewish national state in Palestine, the ancient homeland of the Jews. Having created this state, national aspirations conflict with those of the surrounding Arab countries and the Palestinian Arabs who live either within Israel or exiled elsewhere. The conflict is originally based on territorial dispute but has become divided on religious and ideological grounds: in the first instance Judaism against Islam (including all Muslims, not just Arabs), and in the second, capitalism versus communism, because of the very close affiliations of the USA with Israel and the sympathy of western Europe for the Jewish national state.
Incidence Sympathy for Zionism may exist among Jewish communities worldwide, but the embodiment of the idea lies solely within the Jewish state. Armed conflict between the Arab countries and Israel has existed since 1967, with support from the USA to Israel and from the USSR to the Arab countries. American citizens may take Israeli citizenship and fight in the Israeli army without giving up their American nationality – a provision which applies to no other country.
Background Zionism originated in eastern and central Europe in the latter part of the 19th century and was in many ways the continuation of the ancient and deep felt nationalist attachment of the Jews and of the Jewish religion to Palestine, where the hills of ancient Jerusalem were called Zion. The Jewish state of Israel was proclaimed on May 14, 1948 and immediately recognized by the United States. The first official foreign support for a Jewish state came from the British Foreign Secretary Arthur James Balfour in a declaration on November 2, 1922, under which Jewish immigration was encouraged.
Claim Zionism has proved to be a kind of fascism and emerges as one of the more reactionary trends in modern racialism.
Counter–claim 1. Zionism is a legitimate attempt to unify the Jewish people following centuries of discrimination and physical violence against them. Much of the territory of the present state of Israel is of deep significance to the Jews as their ancient homeland, with all the spiritual associations that this implies. In the absence of such territory the spiritual and cultural values of the Jewish people, together with their rights as a people, are placed in continuing jeopardy.
2. The United Nations General Assembly passed in 1975 a resolution declaring that "Zionism is a form of racism". These hateful words have made it very hard in Israel to believe in the impartiality of the United Nations, it intensified the element of fear and the feeling that Jewish history rules out trust in any other people. As consequence, the cause of justice and peace in Middle East is harmed until the resolution is rescinded.
Refs Combs, J *Who's Who in the World Zionist Conspiracy* (1982); Cruetz, W *New Light on the Protocols of Zion* (1982); Gal, Allon *Socialist Zionism* (1973); International Organization for the Elimination *Zionism and Racism* (1979); Kayyali, Abdul W (Ed) *Zionism, Imperialism and Racism* (1979); Kimmerling, Baruch *Zionism and Territory* (1983); Klein, H *Zionism Rules the World* (1982); Qureshi, Muhammed Siddiq *Zionism and Racism*; Wigoder, Geoffrey (Ed) *Anti-Semitism*.
    Broader Conspiracy (#PC2555)   Nationalism (#PB0534).
    Aggravates Antisemitism (#PE2131)   Territorial disputes between states (#PC1888).
    Aggravated by Pogroms (#PJ2093)   Destiny (#PF3111)
    Jewish conspiracy (#PF8838)
    Segregation of poor and minority population in urban ghettos (#PD1260).

♦ **PF0201 Haunted buildings**
    Broader Ghosts (#PF3801)   Inadequate housing (#PC0449).
    Related Poltergeists (#PF5697)   Psychic surgery (#PG7917).
    Aggravated by Bad omens (#PF8577).

♦ **PF0202 Inadequate recreational facilities**
Inadequate recreation facilities for the elderly — Lack of recreation funds — Underdeveloped leisure facilities — Inadequate leisure facilities
Nature As world population ages and working hours decline, the use of leisure time assumes growing importance. This is a completely new phenomenon to the post-industrial revolution urban society. In traditional rural culture, community festivities and religious rites involved all ages and gave meaningful roles to young and old alike. The shrine, cathedral, village green, community plaza and so on have few contemporary equivalents. Furthermore, those equivalents that do exist – such as sports centres and theatres – cater to very delimited segments of the population, and are often exorbitantly expensive, especially for those whose income is limited by retirement. Public centres are too often such a low budget priority that facilities, staffing and programming is severely limited.
Incidence Access to a recreational facility is more often seen as a sign of wealth and prestige than good mental and physical health. Firms eager to recruit the best graduates boast of their squash courts and aerobics programmes.
Refs Leitner, Michael J and Leitner, Sara F *Leisure in Later Life* (1986).
    Broader Insufficient financial resources (#PB4653)
    Unpreparedness for surplus leisure time (#PF5044).
    Narrower Inaccessibility of water for recreation (#PF6138)
    Inadequate recreational facilities for disabled people (#PE8833).
    Aggravates Racial inequality (#PF1199)   Unethical entertainment (#PF0374)
    Social withdrawal of aged (#PD3518)   Unstimulating entertainment (#PF8105)
    Mental disorders of the aged (#PD0919)
    Susceptibility of the old to physical ill-health (#PD1043)
    Excessive claims for human development through sports (#PG4881).
    Aggravated by Age discrimination (#PC2541)   Socio-economic poverty (#PB0388)
    Rapidly changing cultures (#PF8521)   Ageing of world population (#PC0027)
    Closure of recreation areas (#PE6276)   Inadequate income in old age (#PC1966)
    Social disadvantage of the aged (#PD3517)
    Unfocused design of community space (#PF1546)
    Delayed development of regional plans (#PF2018)
    Undirected expansion of economic base (#PF0905)
    Modern disruption of traditional symbol systems (#PF6461)
    Retirement as a threat to psychological well-being (#PF1269).

♦ **PF0203 Inadequate international standards of financial accounting and reporting**
Nature Accounting practices vary from country to country so that financial reports are rarely comparable. This contributes directly to the difficulties in developmental policy-making at national, regional and global levels. This applies mainly to accounting practices in business enterprises, which are designed for reporting to shareholders and for internal profit controls. It also applies to some extent in the accounting practices of governments and inter-governmental and non-governmental organizations.

Comparability of financial reports is important to all users – shareholders, lenders, creditors, employees, governments and the general public. Users also need financial reports to be reliable – which means that, as well as being comparable, financial reports must be based on relevant, balanced accounting standards. Governments need corporation reports which are comparable in order to be able to monitor volatile short-term movements of capital, to protect consumer interests, to regulate monopolistic practices, and to prevent artificial transfer pricing and tax evasion. Absence of standardized accounting may be used by corporations to avoid complete

disclosure of information which would reveal the negative social impact of some policies. Progress in this area is particularly essential for a wide range of policies and programmes concerning multinational enterprises, as well as for general development. The lack of international standards for the accounts of governments impedes the evaluation task of international funding agencies of applications for developmental grants, aid and loans.
Background It is possible to define a pyramid of problems regarding accounting discrepancies. At the base is the fact that accounting is not a science in the sense that it may not be possible to 'prove' one particular theory and eliminate all conflicting theories; some can and some cannot. The result is that when the annual statements of competitive firms are analyzed, one may be found to use one acceptable method of depreciation of assets, for example, while the other uses a completely different but equally acceptable method. Balance sheet impacts vary enormously, and financial analysts must recalculate or estimate to achieve comparable data. This level of problem relates to definitions of account use, account classifications, allowable computational methods, and statement presentation styles. At this level the accounting experts are said to agree on disagreeing. On the next level of problem the experts agree: namely that the governments' tax accounting methods do not satisfy the needs for financial statements in commerce. In effect, governments require another presentation of financial accounts for internal revenue, tax assessing purposes; and firms are obliged to employ tax accountants to maintain the accounts and books that governments need.
A third problem affects importers, exporters and transnationals: foreign currency translation accounting. The rules are burdensome, and transnationals keep separate books in local currencies in addition to those in the home country money values. Again, special accountants are needed to deal with foreign currency exchange and translation accounting requirements.
Finally, there has been the recent arrival of theories of inflation, or replacement cost accounting. Government tax specialists are pitted against private sector accounting firms; and accounting firm partners, economists, and corporate secretaries and treasurers debate among themselves the standards, or supposed correct methods of recording the apparent increase in value of assets as inflation proceeds. Different governments have different approaches but the net result has been still another set of books to be kept by companies, revealing assets and making other changes according to the prevailing standard. Governments' standards have been revised several times in the past few years, as have the standards recommended by the accounting profession: the resulting confusion, debate, and cost of maintaining the new accounts and new 'books', have been very high.
Refs United Nations Centre on Transnational Corporations *Towards International Standardization of Corporate Accounting and Reporting* (1982); United Nations Centre on Transnational Corporations *International Standards of Accounting and Reporting* (1984); United Nations Centre on Transnational Corporations *International Accounting and Reporting Issues* (1986).
Broader Misleading information (#PF3096) Excessive standardization (#PF2271)
Inadequacy of international standards (#PF5072).
Related Corporation financial secrecy (#PE1571)
Foreign currency manipulations in accounting records of transnational corporations (#PE2145).
Aggravates Creative accounting (#PE6093).
Aggravated by Misleading accounting information due to inflation (#PE4265).

♦ PF0204 Imbalanced distribution of knowledge
Unbalanced distribution of knowledge
Nature While the complexities of global life require more internationally oriented knowledge as well as knowledge of basic and behavioural sciences and applied technology, such knowledge is poorly distributed. In the main, information is delivered through school-situated learning and in the adult years, mass media (particularly newspapers, magazines, videos and radio and television broadcasts) provide the distribution. In addition, books are important. These include textbooks, reference works, research publications and non-fiction works as well as literature. All of these sources, however, are subject to filtering, selection and even censorship on the part of their creators and deliverers. Individuals and countries who lack economic advantage have limited access, and this places another obstacle in the way of their development.
Incidence World book production was estimated as 546,000 titles in 1970: Africa 8,000 (23 titles per million inhabitants), North America 90,000 (280 per million), South America 15,000 (79 per million), Asia 100,000 (49 per million), Europe 247,000 (535 per million), Oceania 7,000 (361 per million), USSR 79,000 (329 per million).
Broader Social outcasts (#PD6017).
Aggravates Shortage of books and textbooks in developing countries (#PF0118).
Aggravated by Paper shortage (#PE1616) Monopolization of knowledge (#PF5329)
Proliferation of information (#PC1298).

♦ PF0209 Inadequate exchange of technical information concerning problems
Nature Information on problems and their solution is produced in vast quantities in many countries, but there is still the greatest difficulty in ensuring that information of value reaches the people who can make direct use of it. For example, because of the language barrier, many workers are able to understand and use only a limited range of literature; and there are often gaps in the transmission of information from the receiving authorities, e.g. government departments, to those at lower levels for whom it is really intended. Improved international dissemination of information will be of little use if these gaps are allowed to remain.
Broader General obstacles to problem alleviation (#PF0631).
Narrower Inadequate information concerning transnational banks (#PF4350)
Inadequate integration of international information systems (#PE8066).
Aggravates Inadequate application of available knowledge to solve problems (#PF8191).

♦ PF0214 Inadequate social and demographic statistics
Lack of comparable social data — Undetermined population data — Insufficient census data
Nature The question of improved data for planning is of great importance. If a country's economic and social services are to be geared to serving the population as a whole, they should be developed in accordance with existing patterns and the possibilities for acting on them. The achievement of high response rates is of great importance in survey research. Confident generalizations can be made about the universe of study only if probability sampling has been adhered to and a high level of response obtained.
Social and demographic statistics are inadequate for many reason. Suppliers of information may falsify reports. People interviewed may answer questions inaccurately because they do not understand the questions or they feel answering may jeopardize them in some way. Different and seemingly comparable sets of information may differ because of differences in the time frame when the information was gathered, what kind of questions were asked, and who asked them. Statistics can and frequently are misinterpreted.
Incidence The limited amount of information available from some countries is not for reasons of economy, but simply because the data are not readily available. Some country reports present only very sketchy documentation, often based on manual tabulations of field records. A report of the Statistical Commission of the United Nations Economic and Social Council (ECOSOC) in 1985, recommends that the reporting burden on countries should be reduced, in particular that duplication should be avoided and attention be given to coordinating of questionnaires and to standardizing definitions used by different international and national organizations.
Broader Deficiencies in national statistics (#PF0510)
Inadequate statistical information and data on problems (#PF0625).
Related Misuse of statistics (#PF4564).
Aggravates Fragmented planning of community life (#PF2813)
Disjointed patterns of community identity (#PF2845).
Aggravated by Statistical errors (#PF4118).

♦ PF0216 Conflict of laws
Conflicts of law
Nature Conflict of laws, or international private law, arises from the universal acknowledgement that local law cannot or should not govern every human transaction. Less than ever before can the events of human life be confined to a simple jurisdiction. The increasing multinational complexity in matters such as those related to personal status or to commercial phenomena make the choice of applicable law a highly complex matter in case of controversy. Legal disputes often contain a foreign element and the conflict of laws is the study of how national courts deal, in practice and in theory, with such foreign elements.
Refs Morris, J C H *The Conflict of Laws* (1984); Nadelmann, K H *Conflicts of Law* (1972).
Broader Conflict (#PA0298).
Narrower Conflicting labour laws (#PE5135) Conflicts of labour law (#PD3533)
Conflict of national laws (#PG2119) Conflict of international laws (#PG2120)
Conflicting national divorce laws (#PF3249) Conflict of laws over nationality (#PF8953)
Conflicting national laws governing cheques (#PE6679)
Conflict of laws on international restriction of information (#PD3080)
Conflicting laws concerning bills of exchange and promissory notes (#PE6680)
Conflicts of national law in relation to international transactions (#PF9571).
Related Conflict of information (#PF2002).
Aggravates Statelessness (#PE2485) Intergovernmental disputes (#PJ5405)
Delays in elaboration of remedial legislation (#PC1613)
Disregard for internationally imposed economic sanctions (#PF1976).
Aggravated by Legalism (#PF8480) Secret laws (#PC6757)
Proliferation of legislation (#PD5315) Illegitimate political regimes (#PC1461)
Discriminatory religious influence on the law (#PD3357).

♦ PF0221 Irresponsible international experts
Unethical practices of consultants — Malpractice by consultants — Negligence of experts — Fraud by consultants
Claim The conditions under which external experts are hired to give advice are such that usually they are responsible neither to the host country nor to the host organization for their recommendations, should they prove to be totally incorrect. Often they may be unfamiliar with the special conditions of the country to which they are called and are unable to remain there long enough to acquire such experience (partly for fear of being by-passed in their parent institutions, partly because the salaries demanded are exorbitant by host country standards). Those who do remain long enough tend to live among the social elite, associating mainly with other foreigners, and remaining aloof from their opposite numbers in the host country who live in humbler surroundings. To the extent that they are highly specialized, they are not particularly concerned to reach agreement with other experts in order to provide coherent recommendations taking into account the insights deriving from a variety of disciplines. Thus advice which is technically correct may in effect be economically catastrophic.
Broader Irresponsibility (#PA8658) Human errors and miscalculations (#PF3702)
Unethical professional practices (#PC8019).
Narrower Techno-mercenaries (#PE7153).
Aggravates Biased expertise (#PF6395) Superficial research (#PF7889)
Neglect of expert advice (#PF7820) Inadequate career advice (#PJ8018)
Disagreement among experts (#PF6012) Narrow context for counsel (#PF0823)
Myopic advice to leadership (#PF7283) Excessive salaries of experts (#PF8317)
Self-satisfied style of informal leadership advisors (#PE7654).
Aggravated by Bias in scientific research (#PF9693)
Inadequacies of foreign consultants (#PE6274)
Excessive specialization in education (#PC0432)
Unavailability of appropriate expertise (#PF7916).

♦ PF0223 Lack of enforcement power
Lack of law enforcement bodies
Broader Unformed style of cooperative action (#PF6514).
Aggravates Inadequate national law enforcement (#PE4768)
Limited local respect for regional and global legislation (#PF2499).

♦ PF0228 Inadequacy of international legislation
Broader World anarchy (#PF2071) Inadequate laws (#PC6848)
Lack of international cooperation (#PF0817).
Narrower Inadequate narcotics legislation (#PF6787)
Inadequate international law enforcement (#PF8421)
Inadequacy of international legal procedure (#PF8616)
Inadequate national supervision of adherence to international law (#PE8280).
Aggravates War (#PB0593) Inadequate legislation against environmental pollution (#PF9299)
Disregard for internationally imposed economic sanctions (#PF1976)
Limited local respect for regional and global legislation (#PF2499).
Aggravated by Nationalism (#PB0534) Economic rivalry (#PD8897)
Ideological conflict (#PF3388).

♦ PF0245 Inadequate credit policies
Inappropriate credit system — Inadequate credit guidelines — Static credit uses — Static credit procedures — Lack of mutual credit societies
Nature The increased use of deferred payments without guidelines or regulations results in misuse by both lender and borrower. In the case of the lender, this is instanced in the sending of unsolicited credit cards without comprehensive checks and balances. Credit is misused by the borrower in overuse as when he makes purchases from innumerable locations, using the many different deferred payment methods available, and is encumbered by payments due to these easily accessible forms of credit such that he cannot live on his earnings.
Credit is predominantly used by both borrowers and lenders as a static commodity rather than as a dynamic economic tool. Such emphasis on static credit operates toward the accumulation of fixed goods rather than the production of future productive goods such as property or business investments.
Broader Inappropriate policies (#PF5645) Self-interest driven investment (#PC2576).
Narrower Profit-oriented interest payments (#PD2552)
Limited availability of financial credit (#PF2489)
Restrictive loan procedures in developing countries (#PF2491).
Related Limited market development (#PF1086)
Variations in national forms of currency (#PF2574)
Inadequate access to negotiation on employment and reward (#PD1958).
Aggravates Ambiguous shape of social identity (#PF6516)
Diminishing capital investment in small communities (#PF6477)
Inequitable labour standards in developing countries (#PD0142).

♦ PF0248 Fascism
Synarchism
Nature Originally an Italian political movement which took power in 1922, the term has been

extended to include ultra right-wing regimes in other countries, particularly of a Latin or Mediterranean type, or to encompass general right-wing antidemocratic tendencies of authoritarianism, nationalist aggression and patriotic sentimentality. Although fascism takes its ultimate support from capitalist financiers and industrialists, its appeal to the masses is not the traditional 'bourgeois' approach, and has been described as classless and 'tribal'. It involves the establishment of a dictatorship based on military and industrial strength and the paternalistic instigation of a uniform nation state with the use of corporativism. The sentimental, romantic ideal of a pure race adopted by fascism from Nazism in 1939 (ultimately with a destiny to rule other inferior races) led to repression of certain elements, notably the Jewish communities, and to severe restrictions on the rest of the population with the use of secret police and informers, torture and political imprisonment. Inherent in the system is the justification of inequalities and injustice and the use of brutality in the cause of an ideal. Once the country concerned is sufficiently united and strong it may turn its aggressions outwards.
Background Fascism was a product of the social and economic crisis in Europe which followed World War I. It produced no coherent system of ideas, and the various Fascist movements reflected the different national backgrounds of the countries in which they developed. None the less there were a number of common traits. All were strongly nationalist, violently anti-communist, and anti-Marxist; all hated liberalism, democracy, and parliamentary parties, which they sought to replace by a new authoritarian state in which there would be only one party, their own, with a monopoly of power, and a single leader with charismatic qualities and dictatorial powers; all shared a cult of violence and action, planned to seize power, exalted war, and with their uniforms, ranks, salutes, and rallies gave their parties a para-military character. In their political campaigns, they relied heavily on mass propaganda and terrorism; once in power, they used the power of the State to liquidate their rivals without regard for the law.
Refs Cassels, Alan *Fascism* (1975); Dimitrov, Georgi *Against Fascism and War*, Laqueur, Walter (Ed) *Fascism* (1977).
Broader Ideological conflict (#PF3388) Extremist ideologies (#PC6341).
Narrower Corporativism (#PU2157).
Related Socialism (#PC0115) Communism (#PC0369) Capitalism (#PC0564)
Neo-fascism (#PF2636) Single party democracies (#PD2001).
Aggravates Antisemitism (#PE2131) Politically emotive words and terms (#PF3128).
Aggravated by Destiny (#PF3111) Double standards in morality (#PF5225).

♦ PF0255 Forged scientific data
Doctored data
Nature Some scientists from all of the disciplines invent data to meet expectation from the experiments being conducted. Rigorous checks are used by most scientific establishments to prevent the accidental creation of false data. Some individuals and organizations deliberately forge data for the sake of money or prestige.
Incidence Recent research has demonstrated, for example, that Kepler in 1609 fabricated astronomical data in presenting his theory of how the planets move around the sun.
Broader Forgery (#PD2557) Scientific fraud (#PF1602).

♦ PF0256 Over-specialization
Lack of generalists — Excessive specialization
Nature Science and technology, and knowledge in general, have developed at such a rate and in such a way that it is no longer possible for one person to maintain an integrated overview or to provide any viable synthesis of the perspectives of different disciplines. In addition, no effort is made to educate people as generalists capable of responding, even in a limited manner, to the insights of a wide range of disciplines. Education, training, and work skills are compartmentalized. As a result, it is left to the person administering a project to make use of whatever scattered items of information and advice appear relevant in terms of whatever educational background and experience he may have had.
Broader General obstacles to problem alleviation (#PF0631).
Narrower Over-specialization in sciences (#PF7532)
Over-specialization in medical care (#PF5709)
Excessive specialization in education (#PC0432)
Fragmentation of academic disciplines (#PF8868)
Over-specialized study of global ecosystem (#PF5712).
Related Anti-holism (#PF5745) Inappropriate education of graduates (#PF1905).
Aggravates Boredom (#PA7365) Ignorance (#PA5568)
Fragmentation (#PA6233) Lack of coordination (#PF8330)
Obstacles to education (#PF4852) Fragmented decision-making (#PF8448)
Monopolization of knowledge (#PF5329) Enforced abstention from insight (#PF7111)
Human errors and miscalculations (#PF3702)
Inadequate dissemination and use of available information (#PF1267).
Aggravated by Proliferation of information (#PC1298).

♦ PF0275 Suppression of creativity and innovation
Patterns curtail creativity
Nature While the suppression of creativity and innovation is an organizational, institutional (for example, school) and bureaucratic major objective, it also exists in the form of the rejection by countries of things that threaten their traditional culture from outside. Creative and productively innovative ideas, objects, processes and the like, are termed too Western, too European or too American. Proposed change always challenges feelings of security, therefore one finds the older generations suppressing the challenge of the younger. Similarly the old ideologies expressed in the religious beliefs of mankind create resistance to newer ideas. In addition, authoritarian people feel their security challenged. Innovation is a threat. In one doctrinaire socialist country, some creative people were branded 'dangerous innovators'.
Claim Only people under 35 are innovative and creative. over this age, it they ever were original, they lose the spark or they become conformists. By the time they are 45 they are reactionary conservatives.
Counter-claim No one individual, or one government can hold back change. Suppression in one place simply means it will manifest elsewhere. There is no age limit for creativity, in the individual or in the state. Biographies of creative people give the lie to statements disagreeing with the former, while history demonstrates the latter when it shows how cultures renew themselves.
Broader Repression (#PB0871) Limited artistic expression (#PF9735).
Aggravates Deficiency in innovation in the public sector (#PF7230).
Aggravated by Censorship (#PC0067) Lack of venture capital (#PG7833)
Rigidly entrenched social traditions in rural areas (#PF1765).
Reduced by Anxiety (#PA1635).

♦ PF0278 Rejection of rituals
Revolt against formalism — Revolt against ritualism — Anti-ritualism
Nature There is a widespread explicit rejection of rituals. This takes the form of denunciation not only of irrelevant rituals but of ritualism as such, exaltation of the inner experience and denigration of its standardized expressions, preference for intuitive and instant forms of knowledge, rejection of mediating institutions, and rejection of any tendency to allow habit to provide the basis of a new symbolic system. In its extreme forms anti-ritualism is an attempt to abolish communication by means of complex symbolic systems.
Claim The attack on ritual ranges between the abandonment of ritual behaviour to the extinction of ritual thinking. Often it is expressed midway between the two: on language. Speech conventions embody the entire spectrum of ritual conditioning, such as obedience to authority, instantaneous execution of commands received, and 'proper' responses to diverse stimuli. These conventions include titles of courtesy or honour, expressions of mechanical politeness and consideration of others, and 'proper' grammar and acceptable standards of pronunciation. In addition, certain words are taboo or not socially accepted as conveying adequate meaning. Therefore ritual language is being rejected to remove the conditioning effect. This is very noticeable in modern cinema and popular music.
Broader Ideological conflict (#PF3388).
Aggravates Loss of cultural identity (#PF9005).
Reduces Ritualism (#PG2184).

♦ PF0286 Inadequate disaster rescue and relief
Nature As world population increases, disasters, both natural and man-made, show a correspondingly greater toll of human lives. Existing rescue services are inadequate with regard to major disasters. They are unable to centralize and secure the relevant expertise and to control all rescue operations involving the movement of men, materials and messages within the disaster area. Immediately after the disaster strikes, the problem of the logistic mobilization of resources and coordination of rescue, first aid and relief work emerges as a first priority. Associated with this activity may be the prevention of secondary effects such as flood, fire or explosion which are also liable to trap and kill people. Lack of coordination in relief operations has a detrimental effect on longer term reconstruction. Misdirected relief operations may also deter the less skilled yet able-bodied from participating in reconstruction. Unless there is adequate planning in the situating of relief stores, camps and emergency housing, they may be placed in areas essential for early reconstruction or in areas liable to flooding or difficult to service with water and sewage disposal.
Incidence Logistics often present the most serious obstacle to quick assistance. Many developing countries have poor land communication to isolated areas and lack the resources to hold sufficient vehicles and aircraft in readiness for emergency use. International assistance is often hampered by lack of reliable and rapid assessment of needs, resulting in inappropriate or obsolete relief supplies. The use of foreign assistance may be very seriously delayed or rendered impossible by political sensitivity to over-flying and landing arrangements for military aircraft or providing visas for the movement of foreigners within the frontiers of a State.
Background Before 1900, disaster relief consisted mainly of emergency grants of food, clothing and medical care, and the provision of mass shelter through hastily organized local committees. In the 20th century, disaster relief has become one of the chief activities of the International Red Cross Committee (war disasters) and the League of Red Cross Societies (natural disasters). In 1971, the UN General Assembly established the UN Disaster Relief Coordinator office (UNDRO) with mandate to promote disaster prevention and preparedness and to coordinate relief assistance at the international level. Organizations such as the Pan American Health Organization, regional office for the Americas of the World Health Organization, established disaster preparedness as a high priority in their development activities.
Broader Social neglect (#PB0883) Disorganization (#PF4487).
Narrower Uncoordinated disaster relief (#PF6022)
Insufficient emergency services (#PF9007)
Uncontrolled costs of disaster relief (#PF6025).
Aggravates Disasters (#PB3561) Natural disasters (#PB1151).
Aggravated by Man-made disasters (#PB2075) Disaster unpreparedness (#PF3567)
Unresponsive public authorities (#PF8072)
Insufficient transportation infrastructure (#PF1495)
Inadequate disaster prevention and mitigation (#PF3566).

♦ PF0292 Governmental disregard for international values and procedures
Nation states' disregard for world opinion
Nature Each government considers itself to be on its own, and participates in the world system to maximize its national goals. Alliances are mutually advantageous contracts to preserve a status quo against an ambitious rival. There is very little tradition of deference to some general assessment of planetary needs. The adverse effects of such policies are only perceived and experienced at the point at which excess capacity begins to disappear and a condition of scarcity begins to emerge.
Counter-claim Governments, since the Second World War, have created a number of globe-spanning intergovernmental bodies, including the United Nations system and its related agencies. In addition, there has been a very high level of regional international cooperation within corresponding intergovernmental organizations. There has never been, in recorded history, a higher degree of international conciliation and agreement attempts and successes. The world system of sovereign states is dynamic and consequently it produce disequilibria and stresses. There is no such thing as world community values and procedures for national sovereignty to be passively subordinated to. The world system is pluralistic. There are many communities with many values. The principle of sovereignty is precisely that it allows nations to act unilaterally in their own interests, and this principle is recognized by all the communities of nations.
Broader Domination of the world by territorially organized sovereign states (#PD0055).
Aggravates Inadequate international judicial system (#PF2113).

♦ PF0296 Prohibitive cost of maintenance
Unexpected maintenance costs
Broader Non-recognition of problems (#PF8112)
Limited availability of investment capital for urban renewal (#PF3550).
Narrower Prohibitive cost of home maintenance (#PJ9022)
Prohibitive cost of equipment maintenance (#PE1722).
Aggravates Inadequate road maintenance (#PD8557)
Prohibitive equipment costs (#PJ8196)
Unethical maintenance practices (#PD7964)
Incomplete access to development capital (#PF6517)
Poor condition of open spaces in urban communities (#PF1815)
Lack of local services for community leadership training (#PF2451)
Limited availability of public services in the small towns of developed countries (#PF6539).

♦ PF0297 Lack of political independence
Denial of right to a people to determine their own political status
Nature The essential characteristics of political independence for large ethnic, linguistic or other groups with separatist tendencies are the rights to self-determination and the necessary political representation and leadership. Political independence may be denied an entire nation dominated by a great power; it also may be denied to internal territories and to colonized lands. Internally, political representation of minorities may be disallowed or frustrated. Denial of the right to be represented in legislative bodies constitutes repression and abrogates human rights.
Incidence Total lack of political independence occurs in countries which are occupied by foreign military forces or where earlier conquest has placed native peoples on an inferior level. Despite their large numbers in a country, refugees and migrants may also have no voice.
Broader Imperialism (#PB0113) Colonialism (#PC0798)
Denial of right of a people to be self-determining (#PC6727).
Narrower National political dependence (#PF[illegible]52)
Lack of internal political independence (#PF3194).

Related Occupied nations (#PC1788)
Denial of right to national self-determination (#PC1450).
Aggravates Disruptive foreign influence (#PC3188).
Aggravated by Economic dependence (#PF0841).
Reduced by Administrative difficulties in new states (#PE1793).

♦ PF0301 Inadequate delivery mechanism in response to problems
Nature Even when all the resources required to solve or alleviate a problem are available, it is frequently very difficult, if not impossible, to deliver the commodities or services to the people in need. This is true of food (during food shortages or famine), medical care, services (whether water, electricity, sanitation, or road), information or education.
Incidence In attempts at famine relief in Ethiopia, for example, many villages were not accessible by road and airlifts were inadequate. In addition, relief work had to cope with illegal diversion of supplies.
Broader General obstacles to problem alleviation (#PF0631).
Aggravated by Inadequate technical cooperation on problems (#PF0863)
Inadequate power of intergovernmental organizations (#PF9175)
Inadequate research on proposed solutions to problems (#PF1572).

♦ PF0304 Excessive government control
Excessive government interference — Excessive government intervention in society — Inappropriate government intervention
Nature The responsibilities of government include the obligations to administer, adjudicate, legislate, and, by means of police and various punitive measures, to enforce the law. Government defence obligations also require it to maintain military forces; and its financial obligations require it to acquire vast sums of money. Governments may use any of these instruments, required for fulfilling their obligations, to interfere with personal freedom, for example, by over-regulation of free enterprises, by over-taxation, by censorship, by curfews, by compulsory military service, and in countless other ways that violate human rights including the right to life.
Claim Almost every human action and transaction is excessively controlled by some government somewhere. All business transactions generate mountains of paperwork due to excessive controls; but simple things are also associated with regulations. For example, eating in a restaurant or buying a package of food is the result of compliance of the restaurateur (and his suppliers in the food production and distribution chain) with dozens of health regulations. Conditions and duration of sleep may be controlled by military codes, civil codes for some municipal workers, collective farm regulations, and in government-operated health and custodial institutions.
Endowing central governments with power is an irreversible process. Every central government of every known, practising political system has tended to take more and more power: some until the governmental shadow is omnipresent in everyone's life.
Broader Illegitimate political regimes (#PC1461).
Narrower Restrictive social policies (#PF8282)
Confined scope of business operations (#PF2439)
Nationalization of foreign investments (#PC2172)
Domination of individuals by institutions (#PF4220)
Compulsory acquisition of land by government (#PC1005)
Political opposition to administrative action (#PF2628)
Unaccountable government intelligence agencies (#PF9184)
Monopolization of information within organizations (#PF2856)
Inappropriate government intervention in agriculture (#PE1170)
Excessive government intervention in the private sector (#PD4800).
Aggravates Dictatorship (#PC1049)
Art censorship (#PD2337)
Political crime (#PC0350)
News censorship (#PD3030)
Authoritarianism (#PB1638)
Political purges (#PC2933)
Official secrecy (#PC1812)
Vagueness of laws (#PF9849)
Qualified amnesty (#PF3019)
Reactionary forces (#PB6332)
Over-centralization (#PF2711)
Outdated procedures (#PF8793)
Government propaganda (#PC3074)
Totalitarian democracy (#PD3213)
Denial of human rights (#PB3121)
Harassment of the media (#PD0160)
Internment without trial (#PD1576)
False political evidence (#PD3017)
Denial of access to news (#PF3081)
Harassment of journalists (#PD3036)
Denial of political rights (#PD8276)
Forced political confessions (#PE3016)
Overdependence on government (#PF9530)
Biased government information (#PF0157)
Complex government regulations (#PF8053)
Radio and television censorship (#PD3029)
Radio and television propaganda (#PD3085)
Political confiscation of property (#PD3012)
Restrictions on freedom of worship (#PD5105)
Inadequate government publications (#PF3075)
Refusal to grant licences to media (#PF3079)
Newspaper and periodical censorship (#PD3027)
Restrictions on freedom of information (#PC0185)
Bureaucracy as an organizational disease (#PD0460)
Unjust allocation of government contracts (#PF2911)
Mis-classification of political prisoners (#PF3020)
Military political prisoners and detainees (#PD3014)
Civilian political prisoners and detainees (#PD3015)
Proliferation of public sector institutions (#PF4739)
Misuse of postal surveillance by governments (#PD2683)
Restrictions on news coverage of legal affairs (#PF3073)
Inefficient mobilization of government revenue (#PF4197)
Maintenance of political dossiers on individuals (#PD2929)
Misuse of electronic surveillance by governments (#PD2930)
Inadequate control over government administrative process (#PC1818)
Conflict of laws on international restriction of information (#PD3080).
Aggravated by Political unrest (#PD8168)
Governmental incompetence (#PF3953)
Inflexible central government (#PD1061)
Ineffective industry self-regulation (#PF5841).
Reduces Banditry (#PD2609)
Aerial piracy (#PD0124)
Assassination (#PD1971)
Civil disorders (#PC2551)
Psychotic violence (#PE7645)
National disintegration (#PB3384)
Inadequate firearm regulation (#PD1970)
Denial of the right to social security in capitalist systems (#PD3120).

♦ PF0306 Global amnesia
Erosion of societal memory
Broader Amnesia (#PD8297).
Related Structural amnesia in institutional systems (#PF7745).
Aggravated by Uncatalogued documents (#PF4077)
Prohibitive cost of maintaining comprehensive document collections (#PE1122).

♦ PF0316 Ineffectiveness and inefficiency of intercultural meetings
Acculturation
Nature Membership in a society produces the phenomenon of acculturation whereby the individual acquires that society's frames and terms of reference, its presuppositions and its internal values and world view. Intercultural meetings solve linguistic dissimilarities by translation and interpretation; but semantic problems arising from the different cultures' perceptions and premises remain a major problem. Even when intercultural meetings are ideologically unified, for example on a theme for world socialism, cultural dissimilarities inevitably introduce misunderstandings and argument. The ineffectiveness of intercultural meetings due to semantic problems can be illustrated by the difficulties in coming to a common understanding of terms such as: human rights, democracy, progress development, justice, equality, etc. On a behavioural level, meetings are almost inevitably convened without adequate time for delegates to be acculturated, at least temporarily, to the special intercultural milieu. The science of conference management has not developed sufficiently to remedy this and the other factors contributing to the inefficiency of such meetings.
Broader Lack of communication (#PF0816)
Inadequate meeting methods (#PF8939)
Inter-cultural misunderstanding (#PF3340).
Aggravates Ineffectiveness and inefficiency of international meetings (#PF0349).

♦ PF0317 Domination of government policy-making by short-term considerations
Short-term policy making — Immediate appeasement responses of leadership — Economic priorities rule — Immediate economic priorities
Nature Governments tend to seek to satisfy immediate national desires which are generally inconsistent with safeguarding the interests of future generations. This is particularly true where governments are elected for relatively short periods (for example, 2-5 years). Leaders are then motivated by the effective terms of their political tenure, or the comparable interval of some national plan. In addition, governments are staffed by individuals whose outer horizon of relevance is often bounded by their own death, or at most, by the life of their children or grandchildren. The notion that man's use of the earth is a sacred trust for future generations, although present in pre-industrial societies, has been lost in industrial societies.
Incidence Government policy frequently reflects pressure from the business community even when it is not the direct voice of that community. In the case of interest-payment defaults on international loans, lending banks, agencies and consortia influence governments to postpone settlement by authorizing and supporting the re-financing of nearly insolvent countries. This short-term solution is gambling with the entire world monetary system.
Claim Governments, and still more so individual ministers within them, are not only relatively short-lived (an international ministerial conference held at two-year intervals usually results in only about 30% of the ministers present at the last one also being present at the next) but are also necessarily too preoccupied with the problems of today, or at the very best of tomorrow, to have much capability in the realm of very long term thinking. Senior officials for their part are too busy serving ministers and carrying on the day-to-day business of government. In fact it is quite idle and quite unreasonable to expect that governments will initiate the process of definition of the functions or of extensive reform of existing intergovernmental institutions which they support. They have neither the time at a high enough level for such a complex task, nor the inclination, nor the detachment necessary for so immensely complicated and controversial a labour.
One of the most difficult problems for democracies is to reconcile the national need for big, long-term investment projects with the objections of those who have to live near them. As technology marches forward, the dilemma of where to put highways, industrial plants, nuclear power stations, airports and the like becomes more and more difficult, because present images of leadership are centred on satisfying the immediate needs of as many people as possible. Degenerate democracy, as this governing philosophy could be named, is a perverted form of the concept that the creativity of all the people is relevant in the governing process. The right to representation without the principle of accountability to the world, results in government that is representative only of individual wants.
Broader Inappropriate policies (#PF5645)
Subsistence approach to capital resources (#PF6530)
Short range planning for long-term development (#PF5660)
Modern disruption of traditional symbol systems (#PF6461).
Related Obstacles to leadership (#PF7011)
Biased government information (#PF0157)
Human errors and miscalculations (#PF3702)
Reduced understanding of globality (#PF7071)
Domination of the world by territorially organized sovereign states (#PD0055).
Aggravates Conflicting priorities (#PF5766)
Collaboration with the enemy (#PG2132)
Uncoordinated government policy-making (#PF7619)
Neglect of environmental consequences of government policies (#PE9295)
Unsustainable short-term improvements in agricultural productivity (#PE4331)
Over-emphasis on immediate solutions in resource development research (#PE4059).
Aggravated by War (#PB0593)
Conflict (#PA0298)
Unemployment (#PB0750)
Discrimination (#PA0833)
Economic stagnation (#PC0002)
Self-indulgent societies (#PF5466).
Reduced by Inability to compromise (#PF9747).

♦ PF0319 Ritual slaughter of animals
Animal sacrifice
Nature The practice of certain religions requires animals to be slaughtered according to rules defined by ritual. These may involve slaughtering animals while they are conscious, such as by using a knife to drain them of blood.
Background Such religious rituals were originally defined to maximize hygiene and minimize pain to the animals. Because of the long period since the rituals were laid down, sanctity has come to surround the origins of them and present day leaders are reluctant to introduce the more humane methods which technology has developed in the meantime.
Incidence Ritual slaughter may be practised in a country where the particular ritual is part of the state religion, or it may be practised by a minority. In terms of total suffering of animals, the most important rituals are: Halal (practised by Moslems), Shehita (practised by Jews) and Jatka (practised by Sikhs). Halal on sheep and goats consists of cutting the throat and bleeding to death in order to remove as much blood as possible (tests now show that more blood actually remains in the flesh of animals killed ritually than when they are first made unconscious before bleeding). For Shehita a knife is used to cut the throat and major blood vessels in order to remove the maximum amount of blood. Jatka consists of the decapitation, usually by a single stroke, of sheep, goats and cattle.
Claim Vegetarians ask 'Is slaughter humane ?'
Counter-claim There is a large amount of scientific literature arguing that most methods of ritual slaughter are still less painful than electric or other stunning of animals.
Refs Regan, Tom and Bowker, John *Animal Sacrifices* (1988); Siddiqui, M I *Animal Sacrifice in Islam*.
Broader Killing of animals (#PD8486)
Religious sacrifice (#PD3373).
Related Magic (#PF3311)
Occultism (#PF3312)
Inhumane killing of animals (#PE0358)
Denial to animals of the right to dignity (#PE9573).
Aggravated by Taboo (#PF3310)
Ignorance (#PA5568)
Satanic rituals (#PF7887)
Unethical practices with domesticated animals (#PE4771).

♦ PF0326 Bogus firms
Shell corporations — Paper companies
Broader Fraud (#PD0486)
Deception (#PB4731)
Economic crime (#PC5624).
Related Bogus public interest groups (#PE7575)
Misuse of nonprofit associations as front organizations by government (#PE0436).
Aggravated by Misuse of business enterprises as front organizations by government (#PE1651).

♦ PF0327 Absence of tactical methods
Total absence of tactical methodologies
Nature The nearly complete absence of methodologies for developing and implementing tactical actions results in social structures which are unable to respond to the needs of society. Educational structures emphasize in their teaching analytical methods and exclude tactical

planning techniques.
Broader Stagnating social methods (#PF3884) Lack of local commercial services (#PF2009).
Narrower Inadequate response to societal needs (#PD1080)
Inadequate models of socio-economic development (#PF9576).
Related Risk-aversion strategy (#PF4612) Undirected technological expansion (#PC1730)
Human wisdom unrelated to daily life (#PD1703).
Aggravates Outmoded forms of social education (#PF2095)
Educational curricula based on content rather than method (#PF3549).
Aggravated by Inadequate ideological frameworks (#PD0065).

♦ PF0333 Decrease in mortality rate

Low death rate

Nature In the past, the rate of population change has been strongly affected by malnutrition, famine, disease, plague and war which kept death rates high. Increasingly such problems are coming under control through scientific, political, and economic institutions, so that human mortality is now lower than ever before. It is the decline in mortality, particularly within the past generation or so, that has given rise to unprecedented population growth. There are at least four basic factors which have intervened in the decline of mortality: an improvement in living standards including access to food and improved shelter; improvements in sanitary control, and in public health and hygiene; the progress of medicine, both in diagnosis and understanding of the mechanisms of transmission of diseases, as well as in the treatment and prevention of disease; and biological changes which are relatively independent of human intervention, but which affect the immuno-parasitic balance.

Incidence Separating, somewhat artificially, the now more developed from the less developed regions, one can estimate that, in periods up to about 1850, in both sets of regions there was an average of about 35 deaths and about 40 births per 1,000 inhabitants each year, leaving similarly moderate balances of a natural increase in both sets of regions. But, after 1850, the annual average death rate in the now more developed regions decreased to about 28 per 1,000 in 1850-1900, to about 18 per 1,000 in 1900-1950, and up to 10 per 1,000 since 1950. In the now less developed regions the death rate may have averaged 38 per 1,000 during 1850-1900, and was still 32 per 1,000 during 1900-1950, as compared with an estimated 21 per 1,000 in 1950-1960, and 17 per 1,000 in 1960-1970. In these regions, as compared with the more developed ones, the modern diminution of mortality was delayed by nearly a century but now it is occurring with outstanding speed. Severe inequalities in risk of death still persist between the inhabitants of more developed and less developed regions, but at least it can justifiably be said that the gap, which was at its widest in the first half of this century, is now narrowing considerably. Nevertheless, there is a wide range between different regions and countries: in the less developed countries mortality rates in 1970-1980 vary between 4 and 46 and in the more developed countries between 7 and 21.

Due to the more youthful structure of their populations, the frequency of death (per 1,000 inhabitants of all ages) in some of the less developed countries has now fallen as low as, or even lower than, that in some of the more developed countries where the proportion of individuals at advanced ages is high.
Aggravates Uncontrolled urban development (#PC0442)
Unsustainable population levels (#PB0035).
Aggravated by Perinatal morbidity and mortality (#PD2387).
Reduces Human death (#PA0072) High mortality rate (#PJ6252).
Reduced by War (#PB0593) Famine (#PB0315) Malnutrition (#PB1498).

♦ PF0334 Long-term shortage of energy resources

Reliance on finite energy resources

Nature The world will face increasing difficulties in the global supply of energy in general, and oil in particular, in the coming decades, with worldwide implications. Developing countries with less financial resources will lose out in the world-wide scramble for energy resources as the supply constraint increasingly becomes a reality. This is particularly alarming since the domestic energy requirements of developing countries are growing and are expected to grow in the future, with the increasing pace of industrialization and development.
Broader Restrictions on freedom (#PC5075)
Long-term shortage of natural resources (#PC4824).
Aggravated by Unsustainable development of energy use (#PC7517).

♦ PF0336 Inadequate local enforcement

Unclear local enforcement — Reluctant local law enforcement

Nature Local enforcement of laws is a complex and difficult set of interactions between law enforcement agencies, including the police, judicial systems and legislative systems; the general public and the criminal. Detection of crime is, for the most part, the responsibility of the victim and the general public. The police are not present when crimes are committed. Traffic offenses are one exception. The discovery and apprehension of the criminal depends on how long and how hard the police will search for an offender which in turn depends upon the chances of finding him, the gravity of crime, the speed with which the police arrive at the scene of the crime, the ability of the victim to identify the criminal, and the willingness of the public to provide information. The ability of the police to recover the criminal's benefit depends on the type of crime, i.e. impossible in the case of violent crimes; and in the case of crimes against property whether or not it is damaged. Deterrence of crime depends on a number of factors; many of which are beyond the scope of the police and include: allocation of police resources, level of budgetary resources, which neighbourhoods are more closely patrolled, which crimes are more rigorously investigated, the balance in allocating resources between detection of crimes and apprehension of criminals, and the balance in allocating resources between manpower, vehicles and communications.
Broader Inadequate national law enforcement (#PE4768).
Narrower Inadequate prevention of crime (#PF4924).
Aggravates Inadequate narcotics legislation (#PF6787).
Aggravated by Statutory crime (#PC0277) Unformed style of cooperative action (#PF6514)
Unfocused style of community operations (#PF6559).

♦ PF0342 Ineffective follow-up on initiatives

Absence of follow-up procedures — Weak programme follow-up — Lack of consistent programme effort
Aggravates Limited programme impact (#PG8930)
Inadequate management of government finances (#PF9672)
Incomplete implementation of community decisions (#PF2863).
Aggravated by Lack of commitment (#PF2406)
Lack of local leadership and decision-making structures (#PF6556).

♦ PF0344 Inadequate standards of living

Denial of right to adequate standard of living

Nature Components of a measure of level of living include health, food consumption and nutrition, education, employment and conditions of work, housing, social security, clothing, recreation, and human freedoms. There are many people in developing countries and in some regions and urban areas of developed countries, whose level of living is considered to be below a desirable standard.

Claim Irrespective of the country, the world-wide recent economic slow-down has meant an absolute decline in the standard of living. For some it has been an abrupt interruption in what they had long experienced and taken for granted: steady living-standard improvement; but for all, in a convergence of moods not known to the experience of present generations, and in countries previously separated by differing outlooks and economic achievement, there has arisen an apprehensiveness, and in some cases, a sense of powerlessness concerning a future in which all the gains in living-standard are threatened.
Broader Socio-economic poverty (#PB0388) Denial of economic rights (#PD4150)
Inadequacy of international standards (#PF5072).
Narrower Malnutrition (#PB1498) Denial of right to sufficient food (#PE0324)
Denial of right to social security (#PD7251) Maldistribution of television sets (#PF4136)
Denial of right to economic security (#PD0808) Denial of right to sufficient shelter (#PD5254)
Denial of right to sufficient clothing (#PE7616)
Denial of right to adequate medical care (#PD2028)
Inadequate standard of living in developing countries (#PE4052)
Denial of the right to an adequate standard of living for indigenous peoples (#PE5484).
Related Inadequate social welfare services (#PC0834).
Aggravates Ascariasis (#PE2395) Overcrowding (#PB0469)
Human suffering (#PB5955) Discrimination in housing (#PD3469)
Inadequate personal hygiene (#PD2459).
Aggravated by Unemployment (#PB0750) Economic apathy (#PC3413)
Prohibitive cost of living (#PF1238) Lack of livelihood standards (#PF1297)
Deteriorating quality of life (#PF7142) Unsustainable population levels (#PB0035)
Unsustainable economic development (#PC0495) Unbalanced industrial distribution (#PF3843).

♦ PF0345 Underutilization of hydropower

Nature By the beginning of the 1980s, 75% of hydropower resources in Europe and North America had been developed, while only a small proportion had been tapped in Oceania, Asia, and especially Africa.

Counter-claim Hydropower developments have substantial environmental impacts. Large storage schemes displace settlements; the upheaval and resettlement cause considerable social effects by modifying agricultural and transport patterns and affecting regional health. Small schemes may sometimes increase the incidence of waterborne diseases.
Broader Underutilization of renewable energy resources (#PE8971).

♦ PF0349 Ineffectiveness and inefficiency of international meetings

Broader Lack of communication (#PF0816) Inadequate meeting methods (#PF8939)
Lack of international cooperation (#PF0817).
Aggravated by Ineffectiveness and inefficiency of intercultural meetings (#PF0316)
Ineffectiveness and inefficiency of interdisciplinary meetings (#PF0409).

♦ PF0357 Insufficient utilization of renewable biofuels

Underutilization of biomass energy — Underutilization of biogas energy

Incidence Some examples of how efficient utilization of biomass energy could improve living standards the world over include: the addition of 10% of alcohol to petrol could end the lead problem; in sunny countries if a maize farmer devoted 10% of his land to growing sunflowers or peanuts, he could run all the diesel powered machines he uses from the oil produced; Sweden currently gets 9% of her energy from converting her forest products in various ways, but she could get 50%; in Britain, which only devotes 7-8% of her land to forests, crops could be grown to convert to fuel instead of unnecessary feed.

Claim Advantages in biomass for energy schemes include: they store energy and are renewable; they can be developed with present manpower and material resources; they create employment and develop skills; they are reasonably priced and available to all income levels; they are ecologically safe and inoffensive; and they do not increase atmospheric carbon dioxide.

Counter-claim Disadvantages in biomass for energy schemes include: their fertilizer, soil, and water requirements are high; they challenge existing agricultural, forestry, and social practices; they represent a bulky resource for which transport and storage could be a problem; they are subject to climatic variability; they may have low conversion efficiencies; and they are sometimes seasonal.

Refs Smith, K R *Biofuels, Air Pollution, and Health* (1987).
Broader Underutilization of renewable energy resources (#PE8971).
Narrower Underutilization of fuelwood energy (#PJ0031).
Aggravated by Unsustainable development of energy use (#PC7517).

♦ PF0361 Proliferation of litigation

Narrower Frivolous or vindictive litigation (#PF1542).
Related Proliferation of commercialism (#PF0815).
Aggravates Law enforcement complexity (#PF2454).
Aggravated by Proliferation of legislation (#PD5315).

♦ PF0364 Complex interrelationship of world problems

Multi-disciplinary characteristics of problems

Nature The complexity of many problems, or networks of problems, presenting no clear-cut solutions, is often a reason for delaying action on them. The complexity is such that many approaches have to be explored to meet the various requirements of different constituents: regions, countries, areas within countries (both rural and urban) and sections of the community, particularly the underprivileged and minority groups. In addition, analysis tends to be simplistic and to lead to simplistic solutions (whether programmes, organizations, information systems or models) which do not match in complexity the network of problems on which they attempt to focus.

In addition to delayed and simplistic analysis and solution proposals, the perceived complexity of data on problem families or clusters gives rise to elaborate computer systems which process millions of pieces of information with ease. With equal ease, based on mathematical formulae, they generate what appear to be infallible forecasts and probabilities, and graphic-displayed models and systems flows. These are not simplistic but over-sophisticated, and this in itself causes delays in analysis. In addition, since the data encompass too many problems or variables, the uncertainly generated is too high. Examples can be seen in global economic forecasting of mankind's needs in the next fifty to one hundred years.

Claim Humanity is entering an era of chronic, large-scale and extremely complex syndromes of interdependence between the global economy and the world environment. Relative to earlier generations of problems, these emerging syndromes are characterized by profound scientific ignorance, enormous decision costs, and time and space scales that transcend those of most social institutions. The difficulties created by such increasing complexity will intensify over th next century as the number of people, the amount of industrial production, and the demand for agricultural products increase doubly or more.

Counter-claim Refusal to act on problems by labelling them as complex is itself a major problem which may exhibit fallacious reasoning. Problem alleviation could rather be modelled on medical practice, where each ailment is specifically treated. A person with burns, a gunshot wound, symptoms of poisoning and cardiac arrest would not be analyzed in a hospital emergency room but would be treated appropriately with the severest problems attended to first. That the world's problems are numerous does not argue that singly taken they cannot be solved one at a time and by appropriate specialists (rather than by those concerned with philosophy and investigation of

interrelationships).
Refs Gordon, Sheila *World Problems* (1971); Long, M and Roberson, B S *World Problems* (1973).
Broader General obstacles to problem alleviation (#PF0631)
Demoralizing images of rural community identity (#PF2358).
Aggravates Oversimplification (#PF8455) Inaccurate forecasting (#PF4774)
Limitation of current scientific knowledge (#PF4014)
Recurrence of misapprehended world problems (#PF7027)
Simplistic technical solutions to complex environmental problems (#PF0799).
Aggravated by Multiplicity of problems facing society (#PF2003)
Unproven relationships between problems (#PF7706)
Failure to integrate knowledge to empower humanity in response to the global problematique (#PF8753).

♦ PF0370 Underutilization of solar energy

Nature The sun's energy powers the living world. About ten to the eighteenth kilowatt–hours is received annually at the surface of the earth but only a tiny fraction of this energy is tapped directly for human use.
Claim The use of solar energy usually has only slight environmental impacts but yields huge gains. Its proper utilization may lead to a reduction in consumption of other forms of energy; and in developing countries solar stills, driers, and other devices can bring about major improvements in the quality of life.
Broader Underutilization of renewable energy resources (#PE8971).
Aggravates Energy crisis (#PC6329).
Aggravated by Unsustainable development of energy use (#PC7517).

♦ PF0373 Underutilization of wind energy

Nature The totai annual potential global wind energy is approximately 1200 million MW, but it can only be economically tapped at a limited number of sites, due as much to cost and public acceptance as to technology.
Background Windmills have been in use for centuries pumping water and grinding flour. Sail have been driving ships for an equally long period of time. Modern wind generators are more efficient machines built on aerodynamic principles. Large wind–energy installations are now operating in California, USA and windmills are providing electricity for national electric supply systems in France, the Netherlands, Denmark, Britain and several other countries. Thousands of small operations are supplying power to farms and rural areas.
Claim Wind–energy machines can be used for water pumping, flour milling and electricity generation, thus supplying readily available energy vital to rural regions in developing and developed countries. Sailing ships can save millions of tons of fuel.
Counter–claim Wind power systems can have a number of environmental impacts. Large installations may be visually intrusive and there are also the concerns of associated noise, interference with telecommunications, and the risk of accidents.
Broader Unexplored energy alternatives (#PF7960)
Underdeveloped rural energy sources (#PF0393)
Underutilization of renewable energy resources (#PE8971).
Aggravated by Unsustainable development of energy use (#PC7517).

♦ PF0374 Unethical entertainment

Unethical corporate entertainment — Irresponsible recreation
Narrower Cruel sports (#PD1323) Dangerous toys (#PE1158)
Video violence (#PE2224) Recreational war games (#PF1406)
Recreational crime games (#PJ3409) Gender–differentiated toys (#PE4664).
Related Meaningless recreation (#PF0386)
Socially disruptive effects of video games (#PF6345).
Aggravates Prostitution (#PD0693) Substance abuse (#PC5536)
Gambling on sports and athletic competitions (#PE4576)
Corruption of sports and athletic competitions (#PE3754)
Social environmental degradation from recreation and tourism (#PD0826)
Natural environmental degradation from recreation and tourism (#PE6920).
Aggravated by Inadequate recreational facilities (#PF0202).

♦ PF0377 Underproductivity of draught animal power in developing countries

Agricultural loss of draught animals
Nature One of the key questions in planning agricultural development strategies relates to the use of animal power to augment or substitute for human or machine labour so as to raise productivity; but governments pay little attention to animal power, being embarrassed by its usage, seeing it as an indication of backwardness. As a result, animals are not often formally recognized as a source of energy.
As draught animals are usually more readily accessible to developing country farmers than are machines, their utilization should be more widely and visibly encouraged.
Claim So little attention is paid to draught power that the harnesses and implements used today are almost the same as those that can be seen in ancient works of art. Enormous advances in the Third World development could be achieved if educated people would take an interest in improving the yokes, harnesses, implements and vehicles used with hundreds of millions of oxen, water buffalo and other animals. Their efficiency could be improved dramatically with little effort. For the two billion people below the poverty line it could be the biggest economic improvement made.
Draught animals are critical for a poor family's survival, being their main asset, security and power source. They provide more than just power; their dung supplies fuel for home heating and cooking and fertilizer for the soil.
Counter–claim There are many limitations to animal power. Compared with motorized vehicles, animals move slowly and have limited range. Using animals is also more onerous, the drudgery is great. The greatest overall limitation is that animals require grains or vegetation as food, the cultivation of which may take away land from producing food crops for humans.
Broader Energy crisis (#PC6329)
Underdeveloped sources of income expansion (#PF1345).
Related Denial to working animals of limitation of working hours (#PE6427)
Denial to working animals of restorative nourishment and rest (#PE4793).
Aggravated by African horse sickness (#PE1805).

♦ PF0379 Separation from nature

Disruption of biological rhythms — Disconnection of people from the rhythms of nature — Separation of people from biological rhythms
Nature The accelerated time frames of industrialized societies separate people from the rhythms of nature, distancing them from the periodicities that make up the many physiological time worlds of the biosphere. Humanity has developed an artificial time environment governed by mechanical contrivances and electronic impulses. This linear form of time is quantitative, fast–paced, efficient and predictable.
Claim 1. Isolation from nature is not just a matter of living in cities. Even more important, it involves a momentous change in man's outlook on the world. Men do not simply coexist with nature; they search for meaning in it. All religions in their integrative functions and rituals explain and support the basic solidarity between man and nature. Such beliefs help create and sustain the bond between man and the external world.
2. Humanity has developed new insights by separating itself from nature's biological clocks. This detailed knowledge has been gained at the price of increasing distance from the rest of creation and the rhythms of intimate temporal participation. The perspective gained has been at the price of each person's loss of contact with the ground of his temporal being. Increased understanding of nature has been accompanied by a self–imposed exile from biological time. People are unable to experience any close connection with rhythms of the planet. Human time is no longer related to that of the tides, to the movement of the sun and moon, or to the changing seasons. If a more empathetic relationship to nature is to be achieved, and life is to be resacralized, then time itself must be resacralized through understanding the natural rhythms of people and accepting the inherent pace, tempo and duration of the natural world. The rhythms by which nature produces and recycles have been so utterly taxed by the dictates of economic efficiency and speed requirements that the planetary ecosystems are no longer capable of renewing resources as fast as they are being depleted, or recycling waste as fast as it is discarded.
Narrower Shift work stress as an occupational hazard (#PE5768)
Desynchronization of bodily rhythm by international travel (#PE1904).
Aggravates Psychological ungroundedness (#PF1185).
Aggravated by Fear of nature (#PF6803) Hyperefficiency (#PF1706).

♦ PF0386 Meaningless recreation

Escapist leisure
Claim Closely related to, and indeed part of the consumption process, is the way modern man uses his leisure time. Instead of being closely integrated with work as in the past, the pursuit of leisure has become a desperate escape from work experienced as increasingly meaningless. But leisure itself has become a meaningless, packaged mass activity, with its values provided by the entertainment industry. Mass society weakens or destroys traditional human groupings, thus leaving the individual at the mercy of impersonal communication such as that provided by the media.
Broader Escapism (#PF7523).
Related Unethical entertainment (#PF0374)
Unpreparedness for surplus leisure time (#PF5044).
Aggravated by Decadent standard of living (#PD4037)
Escapist family life styles (#PD4069)
Parochial escapist media entertainment (#PD0917)
Escaping reality through sophisticated shields against life's pain (#PF0916).

♦ PF0392 Inadequacy of aid to developing countries

Lack of development aid — Decline in aid to developing countries — Decline in official development assistance to developing countries
Nature Official development assistance (ODA), or foreign aid, has traditionally accounted for only a small portion of development resources in most Third World countries. In recent years, however, the dependence of the 36 least developed countries on these external aid flows has risen dramatically, to the point where it now accounts for a large portion of their net external receipts. For all of these nations, foreign aid is a vital source of foreign exchange and investment capital for development purposes. Particularly dependent on foreign aid are the low–income nations of Sub–Saharan Africa, which make up the majority of the least developed group of countries. Aid flows to this area are generally equivalent to half or more of aggregate investment.
During the 1970s these countries experienced reduced rates of growth in production, exports and imports when compared to the 1960s, but higher rates of population growth. Agricultural production per capita actually declined over the decade, partly due to droughts. These factors, combined with the impact of the world recession, have generated severe per capita income and import volume declines, and payments crises which, though small relative to those of Mexico or Brazil, are even more painful to the countries concerned. An increasing number of African countries are in arrears on normal trade and suppliers' credits, and have had to negotiate with the IMF for balance of payments support. Often these negotiations have been protracted and acrimonious, as governments have balked at the nature of the IMF's deflationary medicine.
Since the least developed countries benefitted little from the rapid growth of private bank financing during the 1970s (they were too poor to be creditworthy), they are not affected directly by the drying up of private bank funds. They are hit severely, however, by the inadequate growth of aid flows and by the funding crisis confronting multilateral financial institutions. These countries also face the prospect of competing for limited aid funds with richer developing countries no longer able to rely on commercial finance.
Incidence Official development assistance (ODA) by members of the Development Assistance Committee (DAC) of the Organization for Economic Cooperation and Development (OECD) rose during the period 1982–1988 at an annual rate of about 2.3 percent in real terms, an increase that roughly matched their GNP growth. DAC set net 1989 disbursements of official development assistance at $46.5 billion, ratio of aid to GNP falling to 0.33 percent. DAC countries' aid effort – measured by the ratio of ODA to GNP – has not improved over the past decade remaining virtually constant at a level of 0.32 percent, slightly less than half the internationally agreed target of 0.7 percent. Looking at a longer period, DAC's aid performance in the 1980s was very similar to that of the 1970s, but much weaker when compared to the aid effort of the 1960s (0.50 percent). The World Bank's International Development Association programme has been limited to US$ 9 billion for 1985, 1986, 1987, which is 25 percent below earlier targets. The World Bank itself has announced constraints on the scope of its aid which has been in the US$ 12 billion annual range.
Trends in the individual performance of donor countries are mixed. Denmark, Netherlands, Norway, Sweden and Finland continue to donate the largest share of their economic output, with aid as a share of GNP averaging 0.92 percent. By 1988, the ODA/GNP ratio of the United States had declined to 0.21 percent os compared to 0.31 percent in the early 1970s. In 1989 the United States fell to 0.15 percent or $7.66 billion, to the lowest ratio in the group. Japan moved into the top slot, raising its aid by 4 percent, to $8.96 billion in 1989. Aid from Japan has increased more rapidly in the 1980s than the DAC average but not faster than its GNP. Thus, Japan's ODA/GNP ratio has hardly improved since the early 1980s and is still below the DAC average, at 0.32 percent. Outside the OECD members, concessional flows from OPEC countries declined sharply during the 1980s, in the face of the collapse of oil prices and mounting of financial difficulties. They ran over $10 billion until 1981, and have since declined to about $2 billion in 1988. Largest donors are Saudi Arabia and Kuwait, which provided, respectively, 3.3 percent and 0.8 percent of their GNP in official development aid for 1988. Assistance reported by the USSR rose from 1 percent of GNP in 1976–1980 to 1.4 percent in 1988.
Claim For the least developed countries, an adequate and sustained flow of development assistance is vital if development goals are to be achieved and the benefits of economic growth and international trade realized. Their poor development performance of the 1980s can only be bettered if aid targets are met, the quality of aid improved, and the future of multilateral agencies assured. Although the total amount of aid in grant form is increasing, the growth in export credit loans over–compensates the lower interest of official aid loans. Even official aid loans can have long–term commercial returns, besides creating long–term dependences.
Counter–claim Many forms of aid are unquestionably of value, but aid is not of great importance as a solution to the problems of developing countries. These problems are mainly due to the unjust way in which the global economy functions. No significant progress can be made until the

industrialized countries are prepared to accept a drastic redistribution of wealth and power and reduce the need for developing countries to organize their productive capacity to serve the interests of developed countries. It is not that the industrialized countries should give more aid, rather that they should take less from the developing countries.
Refs Linear, Marcus *Zapping the Third World* (1985); Zeylstra, W G *Aid or Development* (1977).
Broader Underdevelopment (#PB0206).
Narrower Reduced scope of intergovernmental development assistance (#PF4794)
Inadequate commercial finance for rural development projects (#PF4340)
Policy cross–conditionality restrictions in multilateral development aid (#PF9216).
Aggravates Non–alignment (#PF0801) Cultural invasion (#PC2548)
Reversal of development progress (#PF4718)
Unequal opportunities for media reception (#PD3039).
Aggravated by Prohibitive administrative overhead costs (#PF8158)
Insufficient creditworthiness of developing countries (#PD3054).

♦ PF0393 Underdeveloped rural energy sources
Broader Energy crisis (#PC6329).
Narrower Underutilization of wind energy (#PF0373).
Aggravates Underutilization of renewable energy resources (#PE8971).
Aggravated by Unsustainable development of energy use (#PC7517).

♦ PF0399 Inhumanity of capital punishment
Death penalty — Ineffectiveness of capital punishment as a deterrent — Denial of rights of those punished with death penalty — Anticipation of capital punishment as torture — Execution
Nature In addition to the inhumanity of carrying out any death penalty, there is the cruelty arising from the inefficiency with which it is carried out. Hanging is not necessarily immediately effective. Death may finally result from strangulation. Firing squads have proved even less effective when those involved miss the target and the person in charge is reluctant to put the prisoner out of his misery. Gassing and electrocution do not necessarily ensure a rapid death, possibly due to technical problems with the equipment.
Incidence Amnesty International reports in the 10 years up to 1989 the certain execution of 15,320 people in 90 countries. The true figures may be three times as great. In the three years to mid 1988 the countries most involved in capital punishment were Iran (743 executions or more), South Africa (more than 537), China (more than 500), Nigeria (more than 439), Somalia (150 or more), Saudi Arabia (140), Pakistan (115 or more), the United States (66) and the Soviet Union (63 or more). Iraq probably executed hundreds in this period. China may have executed as many as 30,000 in the period 1983–87. About forty–four countries do not resort to capital punishment.
The following countries are cited, by Amnesty International, as enforcing capital punishment: **Af** Angola, Benin, Central African Rep, Egypt, Equatorial Guinea, Ethiopia, Gambia, Ghana, Guinea–Bissau, Kenya, Libya, Malawi, Mali, Mauritania, Morocco, Mozambique, Namibia, Nigeria, Sierra Leone, Somalia, South Africa, Sudan, Swaziland, Tunisia, Uganda, Zaire, Zambia, Zimbabwe. **Am** Chile, Cuba, Jamaica, USA. **As** Afghanistan, Bangladesh, Burma, China, India, Indonesia, Iran, Iraq, Israel, Japan, Korea Rep, Lebanon, Malaysia, Nepal, Pakistan, Philippines, Saudi Arabia, Singapore, Sri Lanka, Syria, Taiwan, Thailand, Viet Nam, Yemen DR. **Eu** Albania, Bulgaria, Czechoslovakia, Hungary, Poland, Romania, Turkey, USSR, Yugoslavia.
In the USA, over 3,900 executions have occurred under the various criminal justice systems of the states since 1930. In 1983, over 1,200 prison inmates were awaiting execution. Florida, Texas, Georgia and California lead the thirty–eight States where the death penalty is enforced. The death penalty appeals process consumes millions of dollars and has made a mockery of the judicial system, contributing significantly to the public's lack of faith in the criminal justice system.
Claim The death penalty continues to be used in some countries, occasionally preceded by the use of torture and beating. As a deterrent, particularly for crimes such as murder, its value is questionable because such crimes are usually committed as acts of passion rather than in the light of reason sensitive to the implications. As an educational function, the killing of criminals by the state tends to lessen society's appreciation of human life. It also affects the entire law enforcement and penal apparatus in a negative way. Pre–execution mistreatment of the prisoner makes the process needlessly cruel and inhumane. The death penalty also precludes any correction of a miscarriage of justice as a result of the conviction of innocent persons, particularly when executions are hastily carried out. Crimes leading to the death penalty result in sensational trials which emphasize undesirable aspects of human behaviour.
Counter–claim Since individuals are morally free agents, capital punishment is the most appropriate deterrent against commission of certain crimes such as high treason, assassination, terrorism, murder, mutilation and rape. The existence of such punishment emphasizes society's abhorrence of the offence and reaffirms belief in the sanctity of life. As a penalty, capital punishment may be morally correct as being in equal or lesser proportion to the crime. The many procedural safeguards afforded the accused make the chances of an erroneous conviction minimal.
Refs Amnesty International *When the State Kills* (1989); Amnesty International Staff *The Death Penalty* (1979); Beman, Lamar T *Selected Articles on Capital Punishment* (1983); Calvert, E Roy *Capital Punishment in the Twentieth Century* (1973); Laurence, John *The History of Capital Punishment* (1983); Sheleff, Leon S *Ultimate Penalties* (1987); Van den Haag, Ernest and Conrad, John P *The Death Penalty* (1983).
Broader Human death (#PA0072) Killing by humans (#PC8096)
Denial of right to life (#PD4234)
Denial of right to freedom from cruel, inhumane or degrading punishment (#PC3768).
Narrower Extra–legal executions (#PE6366) Execution of young offenders (#PG4629).
Related Inadequate laws (#PC6848) Beating of prisoners (#PD2484)
Justifiable homicide (#PE7533) Punishment of criminals by mutilation (#PE3488).
Aggravates Cruelty (#PB2642) Sensationalism (#PG2304)
Miscarriage of justice (#PF8479)
Discrepancies in human life evaluation (#PF1191).
Aggravated by Homicide (#PD2341) Paternalistic punishment (#PJ3700).

♦ PF0401 Inability of governments to regulate family size
Nature Governments regulate a wide variety of private relationships (homosexuality is often treated as a crime, and sexual relationships outside of marriage are discouraged if not actually proscribed), but are unable or reluctant to regulate family size, even indirectly by a system of economic rewards and punishments. This is related to the concept of the sanctity of the family and the awkwardness of governmental interference on such intimate levels of interaction. Leaders in many states (including those with high rates of unemployment and general conditions of poverty) associate population growth with national potency.
Incidence Even China, the world's most populous country, was unable to maintain its 'one child' policy, and had to discontinue it after it proved more problematic than advantageous.
Related Domination of the world by territorially organized sovereign states (#PD0055).
Aggravates Excessive family size (#PJ5277) Excessively large families (#PD7625)
Unsustainable population levels (#PB0035).

♦ PF0404 Shortage of financial resources for action against problems
Nature Funds to address societal needs are limited for reasons depending on whose financial resources they represent. For example, governments spend a great deal of money on armaments and cut back on social welfare. The private sector, always aiming for as much profit as it can from its activities, gives only token heed to public responsibilities. Some intergovernmental projects and organizations absorb most of their funds in salaries and the production of reports calling for action against problems by their member states. Even where funds might initially be adequate to take sufficient remedial action against problems, they are subject to mismanagement and in some cases, misuse. Funding is also a manifestation of comparative political strength, and some problem–oriented projects may remain unfunded, regardless of merit, solely because they were proposed originally by a minority party, coalition or bloc.
Broader Insufficient financial resources (#PB4653)
General obstacles to problem alleviation (#PF0631).
Narrower Lack of funds for medical research (#PE0612)
Inadequate funding of international organizations and programmes (#PF0498).
Related Decline in government health expenditure (#PF4586).
Aggravates Compassion fatigue (#PF2819).
Aggravated by Inappropriate use of financial resources (#PD9338)
Inadequate power of intergovernmental organizations (#PF9175)
Reallocation of aid funds to alternative priorities (#PF0648).

♦ PF0409 Ineffectiveness and inefficiency of interdisciplinary meetings
Nature Interdisciplinary meetings tend to be less than successful because the group has relatively little ability to generate new knowledge or insight, as a genuine synthesis, rather than just interrelating what its members already know. Competing demands on members time, strong social pressures to defer to colleague's areas of expertise, and the limited duration of such events (or series of events) all erode the effectiveness of the meetings, whatever the degree of staff support. Furthermore, especially in the case of the many value laden, incompletely understood issues that arise in relation to sustainable development, the consensual mode of synthesis may be impossible or inappropriate.
Claim The modern invention by specialists of interdisciplinarity has merely resulted in a cacophony of claims as to what the special bridge between the disciplines is, in order to allow them to work together. Biologists claim it is systems theory, cyberneticists claim information theory, physicists claim it is the orientation garnered from the study of energy, entropy and fields, and other claims come from mathematicians, psychologists and even economists that interdisciplinary methods and sciences arise from their areas. Meetings addressing particular topics where an interdisciplinary approach is proposed flounder on disagreement as to what the particular tasks should be allocated to the specialists and how the specialists from the different disciplines should work closely together in each applicable stage of envisaged projects. Disagreements frequently pass from the merely technical of how to make a practical beginning, to those concerning purposes and endings, via the goals sought. The more practical and successful interdisciplinary meetings arise in conjunction with tangible intergovernmental, governmental and industrial projects. The others, convened voluntarily by the interdisciplinarians, are principally for intellectuals who like to listen to themselves and who have no common purpose or dedication.
Counter–claim The expert panel is the longest established and most widely used method of knowledge synthesis. When a consensus is reached, and especially when the consensus is sanctioned by a suitably distinguished convening organization, such a group can be a very effective way of producing syntheses of rapidly developing knowledge as it applies to complex problems.
Broader Inadequate meeting methods (#PF8939)
Lack of international cooperation (#PF0817).
Aggravates Ineffectiveness and inefficiency of international meetings (#PF0349)
Failure to integrate knowledge to empower humanity in response to the global problematique (#PF8753).
Aggravated by Ignorance (#PA5568).

♦ PF0411 Human racial regression
Nature An indefinite period of human regression could result from the gradual collapse of earth life support systems (caused by population increase, exhaustion of resources and environmental breakdown), accompanied by political chaos. The same eventuality might result from another global war, fought with either strategic or tactical nuclear weapons or other means of mass destruction. In the case of nuclear warfare, radiation could cause genetic deformations reducing societies to the animal level.
Counter–claim Biblical creation theories, for example, may be interpreted to indicate that post–Eden, man entered the indefinable period of regression that will now end in near self–destruction. However, it can also be argued, biblically and genetically, that he will eventually recover and reach higher stages of development.
Related Hereditary regression (#PD8149) Psychological inertia (#PF0421).
Aggravated by Political instability (#PC2677) Environmental degradation (#PB6384).

♦ PF0419 Secret international agreements
Covert arrangements between governments — Secret intergovernmental pacts — Conspiracy between governments
Nature When international agreements are made in secret, negotiators may be inclined to reach an accord with one another at the expense of persons or groups absent.
Incidence The Congress of Vienna in 1814–1815 redrew boundaries of states without the the knowledge of the European populations affected by such decisions; the Constitutional Convention of 1787 was drafted in secret; Nazi Germany and the Soviet Union signed secretly a nonaggression pact in 1939 dividing Europe into German and Soviet spheres of influence; Richard Nixon and Henry Kissinger conducted secret negotiations with China; and the Nixon administration conducted secret negotiations with the North Vietnamese on the ending of the Viet Nam War.
Counter–claim Widely different interests cannot be reconciled if all bargaining has to be done in full light, especially with the proliferation and liberty indicative of today's media. When only a brief overview is presented, understanding of negotiations is limited and this can be more injurious than helpful.
Broader Conspiracy (#PC2555) Official secrecy (#PC1812)
Undemocratic political organization (#PC1015).
Related Abuse of commercial confidentiality (#PE6786).
Aggravates Government limitations (#PF4668) Unlawful government action (#PF5332)
Unfairly negotiated treaties (#PF4787) Illegal exports of nuclear materials (#PE3968)
Covert violation of international treaties (#PD8465)
Collusive international trade arrangements (#PE0396)
Unaccountability of international financial institutions (#PF1136).
Aggravated by Secret laws (#PC6757).

♦ PF0420 Irrational conscientious refusal of medical intervention
Nature A number of religious groups have beliefs which lead them to refuse certain forms of medical intervention, however serious and painful the circumstances. Blood transfusions may be refused which could save a life, pharmaceutical drugs may be refused; and vaccinations may be refused, endangering the lives of others (in the case of epidemic diseases for example).
Broader Human disease and disability (#PB1044).
Aggravates Epidemics (#PC2514) Inadequate medical care (#PF4832).
Aggravated by Superstition (#PA0430) Dependence on religion (#PF0150).

♦ **PF0421 Psychological inertia**
Dependence on psychological inertia
Nature The inability and refusal to cope mentally, as a form of psychological disturbance, may induce irrational fears and the total loss of self-confidence or the ability to accomplish any given task. It may reduce a person to dependency. On a wider level it forms the basis for apathies of various kinds. It may also be expressed as a nervous breakdown after a period of stress.
Broader Apathy (#PA2360) Mental disorders (#PD9131).
Narrower Lack of self-confidence (#PF0879) Inhibited grief process (#PD4918) Psychological withdrawal (#PJ2329).
Related Fear (#PA6030) Insecurity (#PA0857) Human racial regression (#PF0411).
Aggravates Human dependence (#PA2159) Mental depression (#PC0799) Over-acceptance of socio-economic dependency (#PF8855).
Aggravated by Passivity (#PF6177).

♦ **PF0422 Inadequate vocational education**
Circumscribed vocational education — Vocational education void — Education vocationally irrelevant — Non-productive vocational training — Lack of vocational training — Lack vocational training schemes — Misdirected vocational training — Unfacilitated vocational school — Unprepared vocational potential
Broader Inadequate education (#PF4984) Inadequate secondary education (#PD5345) Limited development of functional abilities (#PF1332) Lack of skilled manpower in rural areas of developing countries (#PE5170).
Narrower Limited agricultural education (#PF8835).
Aggravates Unprepared adult leadership (#PF6462) Fragmented planning of community life (#PF2813) Inadequate practical training in rural areas (#PF6472) Inequitable labour standards in developing countries (#PD0142).
Aggravated by Shortage of trained teachers (#PD8108) Limited access to practical education (#PF2840) Ineffective methods of practical education (#PF2721) Ineffective mechanisms for functional training (#PF1352).

♦ **PF0423 Counter culture**
Counter-culture — Counter cultures
Broader Social deviation (#PC3452).
Narrower Anti-science (#PF2685) Non-compliance with the law (#PG2334) Peoples perceiving themselves as specially chosen (#PF4548).
Related Social disintegration (#PC3309) Vulnerability of marriage as an institution (#PF1870).
Aggravates Social breakdown (#PB2496).
Aggravated by Loss of institutional credibility (#PF1963).
Reduces Inflexible social structure (#PB1997) Failure of individuals to participate in social processes (#PF0749).

♦ **PF0424 Absent mindedness**
Refs Harris, John E and Morris, Peter E *Everyday Memory and Action and Absent Mindedness* (1984).
Broader Memory defects (#PD8484).

♦ **PF0431 Uncertainty of death of missing persons**
Nature Military conflicts and racial, religious and political persecutions have caused the disappearance of many persons whose deaths cannot be established with certainty. This international situation has produced difficulties of a legal nature, in addition to the human suffering of friends and relatives, which have placed a great number of people in a variety of precarious positions. There are some whose personal status may be affected by the survival or death of the missing person. There are relatives or friends desirous of adopting the minor children of the missing person. There are persons who may be entitled to, or have an interest in, some part or all of the missing person's entire estate under a will or intestacy; or who may be entitled to, or have an interest in, some property whose disposition may depend either on the survival or death or on the date of death of the missing person.
The tragic circumstances of missing persons are complicated by the fact that sovereign states, by virtue of their constitutions or laws, may be obliged to act on behalf of the missing, or their families, to search for and claim bodies, or to search for and claim its living citizens if they have been detained illegally by other countries. In addition, the States may be obliged to seek redress in the latter events. Instances of missing persons apply internationally to prisoners of war, hostages, and impressed labourers; and nationally to persecuted politicals and minorities. Great numbers of civilians may also disappear in war-time or as a result of tidal waves, earthquakes, and other disasters.
Broader Uncertainty (#PA6438).
Aggravated by Human death (#PA0072) Missing persons (#PD1380) Social injustice (#PC0797) Unreported births (#PF5381) Unknown relatives (#PF0782) Persons missing in military action (#PE1397) Bureaucracy as an organizational disease (#PD0460).

♦ **PF0440 Analytical stagnation**
Overemphasis on analysis
Nature The current overemphasis on analysis without the necessary counter-balance of synthesis or interpretation leads to a form of analytical stagnation, or an analyzed chaos without meaning. Science has provided more data than can be intelligently interpreted. The mere quantity of facts necessitates a specialized form of analysis without the possibility of seeing the facts within their full context. Without such a perspective, future generations are ill-equipped for the ever-increasing complexity of human experience.
Claim Science is getting out of hand. The most remarkable feature of the current scientific explosion is its almost total lack of direction and organization. The power wielded by scientists is increasing at a terrifying rate, but the growth of any adequate scientific ethos and philosophy is almost imperceptible.
Counter-claim Anti-science is fashionable among those with a classics or humanities education and those with no education at all. The essence of the scientific revolution of the past four centuries has been the evolution of analysis. The achievements of analysis in microbiology, cosmology, particle and high-energy physics, applied mathematics and logic (particularly in information and communication theory which lies behind the artificial intelligence of the computer), point not to stagnation but rather the possibility of convergence towards a new science of first principles that explains the origin and meaning of life and consciousness.
Broader Human errors and miscalculations (#PF3702).
Aggravates Unimplemented decisions (#PF4672).

♦ **PF0441 Harmful thought**
Negative thoughts
Nature Negative thoughts and careless words cause harm to others. People respond to these negative expressions and, like a chain reaction, spread their own hurt feelings to others during the course of the day.
Broader Personal covert himsa (#PF1978).
Related Impure thoughts (#PF5205) Negative emotions and attitudes (#PA7090).

♦ **PF0443 Irrelevance of educational curricula**
Imbalance of educational curricula — Prescribed irrelevant curriculum — Irrelevant school curriculum — Disrelated school curriculum — Disrelated school lessons
Nature For essentially historical reasons, the subjects of traditional curricula are given a value that often bears little relation to their educative or social usefulness. Literature and history are generally invested with greater prestige than geography or economics, and the study of classics takes precedence over learning about the contemporary world. Even science as a whole suffers from such prejudice: pure science is often more highly regarded than applied science. Many educational programmes are therefore ill-adapted to provide knowledge of the real universe, as the present generation sees it, and to the problems facing people today: military, social and racial conflicts; worldwide famine; pollution; the status of youth and women; and the condition of minority groups. Educators may feel that they do not possess sufficiently accurate information on these subjects, or they may wish to avoid controversial subjects, or the subjects may require multidisciplinary treatment which would be hard to include in strictly subject-based timetables.
Furthermore, the unwillingness to recognize that curriculum goes beyond compartments of subject matter hinders the development of non-formal efforts to make schooling more responsive to the real life needs of persons of all ages.
Refs Dosa, Marta L and Froehlich, Thomas J (Eds) *Curriculum Development in a Changing World* (1985).
Broader Inadequate education (#PF4984).
Narrower Impractical education (#PF3519) Data-oriented education (#PF1217) Static school curriculum (#PG5274) Narrow scope of education (#PF3552) Single language instruction (#PJ9290) Segregation through language (#PD4131) Unsupplemented school curriculum (#PG5370) Lagging training in social skills (#PF8085) Overemphasis on academic education (#PG9447) Lack of emphasis on basic education (#PF1548) Inadequate training in decision-making (#PD2036) Inadequate teaching occupational health and safety (#PE8305) Interference of school athletic activities with education (#PG4256) Educational curricula based on content rather than method (#PF3549) Educational curricula over emphasizing method rather than content (#PD1827).
Aggravates Ignorance (#PA5568) Educational wastage (#PC1716) Declining sense of community (#PF2575).
Aggravated by Unperceived relevance of formal education in rural communities (#PF1944).

♦ **PF0445 Underutilization of oil shale as an energy source**
Broader Energy crisis (#PC6329) Unexplored energy alternatives (#PF7960) Underutilization of renewable energy resources (#PE8971).
Related Underutilization of tar sands as an energy source (#PE8370).
Aggravates Long-term shortage of electric energy (#PE1216).
Aggravated by Unsustainable development of energy use (#PC7517) Inadequate governmental energy conservation policies (#PF0037).

♦ **PF0462 Fear of death**
Dependence on fear of death — Thanatophobia
Nature Fear of death occurs in four situations: during old-age; during terminal or critical illnesses, or those requiring major surgery; during a variable time period before foreseen untimely death (some seconds or minutes in the case of accidents, or days or months during military combat, or months and sometimes years in prison confinement awaiting execution); and when such fear is by mental illnesses of various kinds.
Fears differ in kind physiologically and psychologically. Some fears of death trigger all instincts: for example, adrenalin flow, alarm, flight or fight syndrome, involuntary urination or bowel movement. Some prospects of death can be so frightening that they cause madness. Many people not only fear death itself, therefore, but the circumstance of death. In addition, while the circumstances of death may appear frightful for an individual, they may also appear frightful in respect to his or her dependents and loved ones, when under conditions that cause undue hardship and deprivation.
The fear of death is often associated with fear of the after-life (for believers). Some may dread ghastly punishment, believing themselves not to have been 'saved', or because they have committed great sins or crimes.
In all its forms, fear of death at the very least degrades human life, and at worst causes serious distortions in behaviour leading to insanity or crime, including the crime of homicide, killing to avert death.
Claim Fear of death is one of the great fears which chains consciousness. The fear has far-reaching consequences, for it not only distorts understanding of the needs of the dying, it also undermines ability to experience life to the full, thereby giving rise to many negative attitudes. No choice is uninfluenced by the way in which the personality regards its destiny and the body its death. In the last analysis it is the conception of death which decides the answers to all the questions that life poses.
Counter-claim The fear of death in humans is the greatest motivator for sane and social behaviour. It is also the spur to all economic activity and consequently for almost all that civilization has produced. Without the fear of death there would be no value placed on human life, nor would the greatest achievements in the monumental and fine arts ever have been realized. In individual life, a mild fear of death, a respect for one's mortality, allows one to order one's affairs and give consideration, if not provision, for the following generation.
Broader Fear (#PA6030).
Narrower Avoidance of a confrontation with death (#PF1586).
Related Sex guilt (#PJ2396) Sibling rivalry (#PG2399) Fear of the dark (#PG2398) Physical restraint (#PG2397) Castration anxiety (#PG2400) Separation anxiety (#PE2401) Morbid preoccupation with death (#PD5086).
Aggravates Medical quackery (#PD1725) Unreported violence (#PF4967) Unreported harassment (#PF4729) Excessive prolongation of the dying process (#PF4936).
Aggravated by Human death (#PA0072) Bereavement (#PF3516) Human ageing (#PB0477) Violence as a resource (#PF3994) Dehumanization of death (#PF2442) Social withdrawal of aged (#PD3518) Ageing of world population (#PC0027) Mental disorders of the aged (#PD0919) Criminalization of euthanasia (#PF2643) Social disadvantage of the aged (#PD3517) Susceptibility of the old to physical ill-health (#PD1043) Retirement as a threat to psychological well-being (#PF1269).

♦ **PF0466 Risk of unintentional nuclear war**
Erroneous precipitation of nuclear conflict
Broader Risk of war (#PF4215).
Narrower Risk of unintentional nuclear war due to accidents (#PF4346) Risk of unintentional war generated by the arms race (#PF4152) Risk of unintentional nuclear war due to international crises (#PF4302) Risk of unintentional global nuclear war due to nuclear proliferation (#PF4352) Risk of unintentional nuclear war generated by the strategy of deterrence (#PF4162)

Risk of unintentional nuclear war generated by developments of strategic doctrine (#PF4156).
Related Risk of intentional nuclear war (#PF4435).
Aggravated by Meteorites as hazards (#PF1687).

♦ PF0468 Refusal to sell

Nature A free enterprise system not only presupposes the right of entrepreneurs to choose an occupation without dictation by the government but also without interference from other entrepreneurs, which implies that a competitive market free of monopolistic behaviour, discriminatory practices and other restrictions, is essential. Certain practices engaged in by suppliers in the distribution of their products, notably refusal to sell thus hindering free access to the market, may serve to distort the free market economy. Refusal to sell may be defined as the practice by a seller of refusing to sell to particular buyers or to a particular class of buyers. This implies that the sale of the product requested was not prohibited or controlled by law or regulation.
Three main types of refusal to sell may be distinguished: individual refusal to sell (the practice employed by one particular supplier with regard to a particular buyer or class of buyers); collective refusal to sell (an explicit or implicit agreement or conscious parallel action between suppliers to withhold supplies from a buyer or class of buyers); secondary boycott (action by a single enterprise or an agreement or concerted action between enterprises, the object of which is to coerce another enterprise or group of enterprises into refusing to deal with a third party).
Refusal to sell may take several forms. The simplest concerns a supplier's categorical refusal to sell to a particular buyer. However, less obvious forms exist. The supplier not wishing to supply a particular customer may employ delaying tactics or may agree to supply only on conditions which are clearly more onerous than those accorded to other dealers, in other words, on discriminatory conditions. In fact, such delaying tactics constitute a more subtle form of refusal to sell since their aim is either to put or maintain a customer in a position where he is unable to place an order (for example by refusing him catalogues, samples, reference books, etc, which he needs in order to make his choice) or to postpone without valid reason the execution of an order. Another disguised form of refusal to sell consists in a supplier offering to supply a similar product to the one requested but refusing to supply the exact product requested. This situation often arises in connection with a brand product which the supplier offers to supply without the brand label and in a different package.
Broader Restrictive trade practices (#PC0073).

♦ PF0473 Neutrality

Nature Neutrality during war-time makes it impossible to defend a nation against the treachery of aggressor nations.
Incidence In the twentieth century Germany invaded Belgium, Italy invaded Greece, and the UK occupied Iceland.
Claim Neutrality has broken down as a feasible state for nations. Few if any can stand wholly apart from war. Neutral country air space may be overflown by satellites or spy planes, or by ballistic missiles. Nuclear weapon detonation, furthermore, will expose neutral nations to radioactive fall-out and ecological destruction. The fragility of the status is indicated by the Charter of the United Nations which allows for the revocation of neutrality of states that the Security Council calls upon to take coercive measures against acts of aggression (Articles 41 and 48). International trade links may involve neutral nations in the provision or transit of strategic materials or general, favourable economic trade with antagonists. For all these reasons, plus the sweeping aside of neutrality by aggressor nations, the position may be an unrealistic and untenable one **Counter-claim** The absence of neutral nations would be on obstacle to mitigating some of the effects of war and would eliminate those who could best act as go-betweens and peace-makers.
Refs Levie, Howard S *Law of War and Neutrality* (1988).
Broader Caste system (#PC1968) Aggressive foreign policy (#PC4667).
Narrower Violation of neutrality (#PC2659)
Refusal to ratify international conventions (#PG2403).
Related Pacifism (#PF0010) Non-alignment (#PF0801)
National isolationism (#PF2141).
Aggravates Occupied territories (#PD8021) Restriction of arms supply (#PD1304)
Lack of international understanding (#PF5106).
Aggravated by Aggression (#PA0587) Maldistribution of electrical energy (#PD3446).

♦ PF0478 Unbankruptability of sovereign states

Nature The international community lacks a legal framework to deal with countries which are unable to continue to service their debts. The loaning institutions are in danger of going bankrupt because of the tendency to continue making loans increasing the level of bad debts. Money and stock markets become more volatile increasing risk. Because their is no way for sovereign borrowers to negotiate a rewriting of contracts, the restructuring of debts is disorderly and unpredictable.
Broader Domination of the world by territorially organized sovereign states (#PD0055).
Aggravated by Underproduction of primary commodities in developing countries (#PD3042).

♦ PF0479 Limited food variety

Broader Food insecurity (#PB2846).
Aggravates Inadequate health services (#PD4790).
Aggravated by Narrow range of food crops (#PD4100).

♦ PF0491 Grave robbing

Body snatching
Nature Graves have been robbed of their valuables for centuries. The practice of burying valuables with the deceased has encouraged this practice. In some cases graves are robbed for the skeletons for use in religious rites or to sell for medical purposes. Some satanic cults will buy human skulls for $400.00.
Broader Robbery (#PD5575).
Aggravates Infamy (#PB8172).
Aggravated by Satanism (#PF8260) Necrophilia (#PF6957)
Occult rites (#PG4311) Socio-economic poverty (#PB0388)
Decadent standard of living (#PD4037).

♦ PF0498 Inadequate funding of international organizations and programmes

Reduction in aid expenditure — Shortfall in aid to international organizations
Nature Projects and ongoing programmes of international organizations are impaired by budget limitations. Organizational income is too low to initiate and carry-through to completion necessary work, due to: delinquent payment, non-payment and under-payment of membership assessments; political decisions by member states to limit general fund contributions or not to support specific projects; and lack of other income generating possibilities.
Counter-claim International organizations propose enormous budgets for an endless variety of programmes only to perpetuate themselves. Their wastage of funds is also high. For essential work their funds are adequate if they would eliminate mismanagement, over-staffing and empire-building.
Broader Shortage of financial resources for action against problems (#PF0404).
Narrower Inadequate funding of international nongovernmental organizations and programmes (#PE0741)
Shortage of financial resources within the United Nations system of organization (#PF1460).
Aggravates Inadequate facilities for international organization action (#PD0929)
Ineffectiveness of international organizations and programmes (#PF1074)
Undermining of multilateral forums by industrialized countries (#PE4289).
Aggravated by Endemic abortion of ewes (#PG1537)
Lack of governmental support (#PF8960)
Prohibitive administrative overhead costs (#PF8158)
Loss of credibility in international institutions (#PE8064)
Competition between international organizations for scarce resources (#PC1463)
Government non-payment of agreed contributions to international organizations (#PF8650).

♦ PF0506 Inadequacy of civil defence

Nature Civil defence consists of all the non-military actions that can be taken to reduce loss of life and property from enemy action. It embraces defence against all types of attack, including conventional explosive bombs or rockets, nuclear weapons and chemical or biological attacks. But in the missile era, civil defence presents many shortcomings. In the case of nuclear attack, short warning times preclude many civil defence measures (such as getting people into appropriate air-raid shelters) and many important factors are unpredictable. The behaviour of sheltered populations under extreme stress is largely unknown. The fire hazard from nuclear attack is only partly explored: a nuclear burst between an envelope of clouds above and snow below could increase the thermal effect (and resulting fires) by a large factor. There are inherent uncertainties as to when an attack may begin and when it may end, which frustrate civil defence actions. The detection of a fleet of approaching enemy bombers does not preclude the possibility of a simultaneous missile attack, nor that enemy submarines may be lurking offshore, both alternatives conveying threats with short warning times. Similarly, the explosion of a single nuclear weapon cannot be interpreted as the end of the attack for any specific area. More weapons may be in the offing at the very time that the desire to emerge from shelters to fight fires and attempt rescue work is at its peak.
Background Although in the early stages of World War I, Germany used rigid airships to attack England from the air and the Allies launched some counterattacks on Germany, it was not until World War II that the threat of aerial attack on cities became sufficiently great to call for organized civil defence planning. While a few special air-raid shelters were built in Great Britain and in Hawaii, civil defence tactics during the interwar years consisted principally of utilizing improvised shelters such as basements and subways. Germany also built special bunkers for a small fraction of its population, and these proved to be very effective in saving lives. Other civil defence tactics (in Great Britain and along the coasts of the United States) consisted of blackouts to reduce the night glow from city lights that could have served as guides to enemy pilots. The British government provided gas masks for its people, and practically all the countries involved in the war trained citizens in the elements of fire fighting, rescue and medical first aid.
The major, perhaps critical, difference between the civil defence situation in World War II and that which has confronted the world since 1950 is that while the relatively small weapons of World War II afforded some 'learning time' – people could learn by experience that shelters were safer than ordinary buildings and civil defence volunteers could be recruited and trained after the war had begun – no learning time is allowed by nuclear weapons that can destroy whole metropolitan areas at one blow. There is no opportunity to learn from repeated attacks because the first attack in all probability, will accomplish its mission.
Claim Because, if thermonuclear weapons are ever used in numbers, organized community life as we know it will come to an end in any country involved in the conflict, the idea that civil defence against nuclear weapons can serve any useful purpose, is totally irresponsible in that it further anaesthetises people against the realities of nuclear war and its aftermath.
Broader Conflict (#PA0298) Unjustified military defence policies (#PF1385).
Narrower Lack of preparation for self-defence (#PU2405).
Related Military unreadiness (#PF5933).
Aggravates Fear (#PA6030) Panic (#PF2633) Injuries (#PB0855)
Starvation (#PB1875) Human death (#PA0072)
Inadequate medical care (#PF4832) Human disease and disability (#PB1044)
Lack of fire-fighting facilities (#PJ2437) Health hazards of irradiated food (#PD0361)
Instability of non-metallic mineral products industry (#PE2599).
Aggravated by Total disarmament (#PF8686) Biochemical warfare (#PC1164)
Lack of appreciation for nuclear weapons (#PE7476).

♦ PF0508 Psychic interference in decision-making

Claim Evidence has been presented indicating that individuals can interfere psychically with normal decision-making processes. For example, it is claimed that some lawyers use psychic abilities to read the minds of jurors, witnesses and other opposing attorneys. It is also claimed that they wage mental war with opposing counsel, programme jurors to return favourable verdicts, ferret out undisclosed or missing evidence, and even influence the presiding magistrate.
Related Psychotronic warfare (#PD7986) Psychological warfare (#PC2175)
Political indoctrination (#PD1624).

♦ PF0510 Deficiencies in national statistics

Weaknesses of national data sources
Nature Many countries lack the statistical information needed to pursue national or international policies of economic and social development. The collection of demographic and economic data is often inadequate, possibly due to a lack of finance and professional manpower. Vital events are frequently not registered by the population and important categories of data are omitted from censuses. The processing of statistics may suffer from methodological obsolescence.
Incidence Most of the countries of Africa, Latin America and Asia suffer from major statistical deficiencies. In the case of demographic statistics, for example, underregistration of births or deaths may be as much as 30 to 50 per cent.
Dozens of times a week, governments publish economic statistics. Businesses and governments make decisions based on these figures, but many of these figures are wrong. The biggest problems in official economic statistics lies within the balance of payments. The total of all countries current-account balances should add up to zero but in 1982 there was a discrepancy of $100 billion. This was down to $45 billion in 1987, not because the statistics were more accurate, but because there was an increase in errors which offset the discrepancy. Some countries deficits are obviously overstated and the surpluses of others are understated.
Broader Statistical errors (#PF4118).
Narrower Inadequate public finance statistics (#PF7842)
Inadequate social and demographic statistics (#PF0214)
Unreliability of statistics from socialist countries (#PF0888).
Aggravates Uncontrolled urbanization in developed countries (#PD3488)
Lack of comparability of international statistics (#PF2622)
Inadequate statistical information and data on problems (#PF0625).

♦ PF0513 Conflict of duties

Moral dilemma
Nature Situations in which there are moral reasons in favour of two or more incompatible actions, especially where it is a difficult matter of individual judgement to decide between them.
Incidence The sharpest and most obvious cases occur in war, but they are present in a lesser degree throughout life if only because time and resources devoted to furthering one end have to be taken from time and resources devoted to furthering another.

PF0513

Refs McCormick, Richard A and Ramsey, Paul (Eds) *Doing Evil to Achieve Good* (1985).
Related Ethical decay (#PB2480).
Aggravates Hypocrisy (#PF3377).
Aggravated by Inability to define moral standards (#PF7178).

♦ PF0531 Policy-making errors

Erroneous decisions
Broader Mismanagement (#PB8406) Human errors and miscalculations (#PF3702).
Related Diplomatic errors (#PF1440)
Failure of government to apologize for errors (#PF5296).
Aggravates Ineffective decision-making processes (#PF3709)
Inadequate implementation of plans and programmes against problems (#PF1010).

♦ PF0536 Cultural fragmentation

Lack of cultural integration
Nature Lack of cultural integration is the disharmony between cultural heritage and national identity. It is almost impossible to integrate different cultures into the national unit without a loss of cultural heritage, yet where cultural heritage remains completely intact there is division and social conflict. So far there seems to have been little success in finding a happy medium between assimilation and separatism.
Refs Stjernquist, Per *Poverty on the Outskirts* (1987).
Broader Fragmentation (#PA6233) Plural society tensions (#PF2448).
Narrower Heterosexism (#PE0818) Schizmogenesis (#PE4593)
Cultural barriers (#PB2331) Inter-cultural misunderstanding (#PF3340)
Lack of appreciation of cultural differences (#PF2679)
Inadequate cultural integration of immigrants (#PC1532)
Divisive effects of official cultural pluralism (#PF0152).
Related Segregation (#PC0031) Lack of assimilation (#PF2132)
Cultural deprivation (#PC1351) Symbol system failure (#PF3715)
Lack of racial identity (#PF0684) Multidenominational society (#PF3368)
Inadequate political integration (#PF3215).
Aggravates Taboo (#PF3310) Secession (#PD2490) Ethnocide (#PC1328)
Exploitation (#PB3200) Social conflict (#PC0137) Forced assimilation (#PC3293)
Social disintegration (#PC3309) Lack of social mobility (#PF2195)
Homogenization of cultures (#PB1071) Destruction of cultural heritage (#PC2114)
Lack of variety of social life forms (#PE8806)
Segregation of poor and minority population in urban ghettos (#PD1260).
Aggravated by Colonialism (#PC0798) Discrimination (#PA0833)
Cultural invasion (#PC2548) Economic domination (#PG2480)
Cultural illiteracy (#PD2041) Economic imperialism (#PC3198)
Political domination (#PC8512) Segregation in marriage (#PD3347)
Religious discrimination (#PC1455) Primitive secret societies (#PF2928)
Language discrimination in politics (#PD3223) Ideological impediments to marriage (#PF3345)
Threats to ideological movements and minorities (#PC3362).

♦ PF0544 Imbalance in distribution of political awareness in developing countries

Nature Political awareness is much more developed in urban centres than in rural communities. As a result many of the decision-making bodies are to be found in the cities and towns, thereby localizing developments to the disadvantage of the rural communities that must contribute heavily for them.
Broader Reduced civic awareness (#PJ1835).

♦ PF0545 Limited number of geostationary satellite orbits

Saturation of the geosynchronous orbit — Maldistribution of geosynchronous orbits
Nature There is a limit to the number of communications satellites that can operate independently in the geostationary 24-hour equatorial orbit. The limitation arises from the fact that global communications systems can best be based on a limited number of synchronous satellites which remain stationary over a predetermined point on the earth's surface due to the properties of the orbit. The orbital altitude required is 20,000 miles (32,000 km) for which the time of revolution is 24 hours, thus maintaining the satellite over one position. The minimum angular spacing (estimated at 2–6 degrees, namely 180 to 60 satellites) between satellites in such an orbit without mutual interference depends upon the sophistication of the ground aerials and the frequency used.
It is expected that overcrowding may indeed occur very quickly in some specially favoured parts of the orbit. (The range of longitude for a satellite linking London and Tokyo, for example, is such that satellites more than one degree off either way will lose one of the cities). Disputes about usage may lead to political and military action, particularly in view of the difficulty of creating an international mechanism to manage this resource.
Incidence In the past 10 years, countries have put more than 120 communications satellites into orbit 22,335 miles above the equator, the geosynchronous spot where a satellite appears to hover as it flies around the Earth at the same speed the Earth rotates and where it can serve one country or one region 24 hours a day. Many of those satellites have gone dark, but more than 90 are still at work. No fewer than 19 carry telephone, television and business traffic across the United States, more than 20 serve the Soviet Union, and the worldwide consortium known as Intelsat now has 34 communications satellites scattered around the globe in geosynchronous orbit. So many communications satellites are working at similar radio frequencies in the same equatorial belt in space that they have begun to interfere with each other.
Related Vulnerability of world cable communications (#PD0407).
Aggravated by Pollution of orbital space (#PD0089).
Reduces Violation of sovereignty by trans-border broadcasting (#PE0261).

♦ PF0549 Personal disempowerment

Claim Many people have been indoctrinated by government, religious and educational authorities, aided and abbetted by the media, to believe that true power lies outside themselves. In such conditions they become psychologically dependent on external authorities to guide them in their pattern of choices.
Aggravates Dependence on authority (#PF8995).
Aggravated by Political indoctrination (#PD1624).

♦ PF0551 Undomesticated men

Lack of house training
Claim The lack of house training of the majority of today's businessmen is the major cause of the world's pollution. Today's boardrooms are peopled by man-children; barely potty-trained, unable to locate a gold cuff-link without help, who make a virtue out of their inability to even boil an egg. They grew up and went to expensive universities under the care of maids who made their beds, cooks who prepared their meals, laundry ladies who washed up after them. They are taught that not only is the mess they leave behind is not their responsibility but it is not even their concern. Because they close their eyes to their own everyday messes, and have them whisked expertly and invisibly away, how can they be expected to see the ghastly cesspit they are making of our planet? Gandhi recognized the dangers of such blindness and believed every person should be responsible for themselves. That responsibility extended to the everyday things of cleaning and feeding oneself.
Broader Lack of training (#PD8388).
Aggravates Ignorance (#PA5568) Natural environment degradation (#PB5250).
Aggravated by Incompatibility of environmental and economic decision-making (#PF9728).

♦ PF0552 Instability of orthographic standards

Spelling variations
Nature In many languages the spelling of words and rules for spelling are subject to change. For some languages there are controlling bodies which can cause sudden changes in the official spelling; where there is no such central body, spelling can vary in different regions. In either case, considerable difficulty and confusion is created in the education system; in the many sectors of society which produce typewritten or printed texts; and in information retrieval, such as dictionaries, lists and automated systems. Older generations adapt slowly and with difficulty to such changes.
Incidence The English language, influenced by developments in the USA, has several variations in spelling, some of which are not accepted in other English-speaking countries. In German, efforts are being made to approve a switch to uncapitalized nouns because the capitalizing rules are so complex. The spelling of Dutch and Flemish has been subject to a number of official modifications in recent years, and efforts are underway for a reform of French spelling.
Counter-claim Spoken languages are bound to change, and consequent changes to established orthographic standards are, therefore, unavoidable. Changes in language are a source of difficulties (for young and old) whether orthographic standards remain stable or not.
Broader Human errors and miscalculations (#PF3702)
Inadequacy of international standards (#PF5072).
Aggravates Incorrect spelling (#PE4302).

♦ PF0557 Resistance to change

Nature People usually resist change because they fear to lose control; they are uncertain about the future; they do not have enough time to think things through and get accustomed to new ideas; they do not want to lose tradition and familiar symbols; they fear of losing face; they are not sure of they competence; they are reluctant to work more; they feel threatened; they may resent the person or the organization that wants to bring about changes.
Incidence In the international trading system competitiveness changes more rapidly than countries are willing or able to change their structure of production. The pressures to change the production patterns have added to those resulting from changing patterns of comparative advantage arising from technological change and shifts in factor input costs. This resistance to structural change frequently finds its expression in trade restrictions, especially if the latter are somehow more easily obtained by the affected groups than economically more efficient alternatives.
Broader Modern disruption of traditional symbol systems (#PF6461).
Narrower Resistance to grace (#PF5266) Lack of political will (#PC5180)
Denial of right to change religion (#PE6397) Denial of right to change nationality (#PE1736)
Rejection of proposals for social change (#PF9149)
Government resistance to institutional change (#PF0845).
Aggravates Professional stagnation (#PE6649)
Delays in implementation of social change (#PC6989)
Short range planning for long-term development (#PF5660)
Inadequate models of socio-economic development (#PF9576).
Aggravated by Change (#PF6605) Fear of increased autonomy (#PE7706)
Suspicion of imposed change (#PG9094) Individual fear of future change (#PD2670)
Confusion induced by rapid social change (#PF6712).

♦ PF0559 Shortage of adequately trained personnel to act against problems

Nature Lack of personnel is commonly reported to be less a matter of absolute shortage and more often an unmet need for technical expertise on the part of persons who have at the same time a thorough understanding of local conditions.
Broader Labour shortage (#PC0592)
General obstacles to problem alleviation (#PF0631).
Narrower Insufficient doctors (#PE8303)
Inadequate utilization of volunteer social service workers (#PF4892)
Inadequate research and development capacity in developing countries (#PE4880).
Aggravates Underproductivity (#PF1107) Lack of legal aid facilities (#PF8869).

♦ PF0565 Horological superstition

Astrology as superstition — Dependence on horoscopes
Nature The study of the zodiac and the calendar to predict the course of human affairs, usually with a detailed character study of the person involved (as determined by his or her birthchart), although being a highly developed art, is regarded as superstition in most developed countries, and not a reliable science.
Background From ancient times to the 18th century, astrology was considered a legitimate science.
Incidence Fortune-telling is a traditional occupation of gypsies and other peoples. Horoscopes are popular in daily, weekly, monthly and other journals in developed countries. Skilled astrologists practice in many countries, but usually do not publicize their activities. Prominent coverage has been given to the astrologer associated with US President Reagan, through his wife, during the period 1981–88. The astrologer in question claims to have influenced many presidential policy decisions. In Chinese culture, since birth in particular years of the Chinese horoscope are considered especially auspicious (such as the Year of the Snake), this has a significant influence on the birth rate within any year.
Claim Despite the long history of astrology no one has devised a reproducable experiment to demonstrate that it works. Of course astrological techniques are not perfect predictions but it should be possible to define some things that can be correctly predicted by astrology more often than by any other method. If not then astrologers and their followers have no basis for claiming that astrology works.
Counter-claim The astrological theory of the universe, which most historians of science had supposed to have been irreversibly sterilized by about 1700, has recently been the basis of a whole network of studies and experiments which are beginning to mesh into a new astrology. Some scientists consider that a new science is in the process of being born.
Broader Superstition (#PA0430).
Narrower Fortune-telling (#PF1358).
Aggravates Inauspicious conditions (#PF6683).
Aggravated by Dependence on religion (#PF0150).

♦ PF0574 Lack of worker participation in business decision-making

Lack of participation in management
Nature Although psychologists have advised against it, owners and managers of enterprises and bureaucrats in government have continued to restrict workers' participation in decisions that affect them directly. The typical result is demotivation or alienation of the worker, decreased productivity and increased costs.
Incidence A number of recent decentralization programmes to distribute decision making downwards have been made in Socialist countries, particularly in the agricultural sectors. However, decisions require both information and the skill to use it, and the educational infrastruc-

tures have proven too weak to support the initial reforms.
Refs Gaudier, Maryse *Workers' Participation in Management* (1988); ILO *Labour Force Participation and Development* (1982); ILO *Personnel Decision-making in Wholly Owned Foreign Subsidiaries and in International Joint Ventures* (1986).
Broader Undemocratic policy-making (#PF8703).
Failure of individuals to participate in social processes (#PF0749).
Related Political apathy (#PC1917)   Poor managerial communications (#PF1528).
Aggravates Mismanagement (#PB8406)   Lack of community participation (#PF3307).
Lack of participation in development (#PF3339).
Aggravated by Inaccessibility of decision-makers in multinational enterprises (#PE0573).

♦ **PF0581 Unexplored potential markets**
Insufficient product markets — Restricted market opportunities — Unstudied export market — Unused market channels — Unidentified potential markets — Unutilized marketing opportunities — Unknown market potential — Severely limited markets — Undeveloped market potential — Undetermined sales potential — Uncertain sales potential
Nature Until recently, rural businesses have depended upon word-of-mouth advertising, captive audiences and community loyalty to maintain their market. However, greater mobility has resulted in customers being diverted to stores in larger communities where attractive sales and broader selections are available. Unless small town businesses find the opportunity to explore new marketing techniques, the potential local market and business from passing traffic will remain untapped.
Broader Unrecognized opportunities (#PF6925).
Narrower Unpromoted community business (#PJ1409)
Limited economic communication (#PJ1489)
Unexplored business opportunities (#PG5323).
Aggravates Lack of economic and technical development (#PE8190).
Aggravated by Risk of capital investment (#PF6572)
Limited market development (#PF1086)
Oppressive prevalent images (#PF1365)
Narrow range of business skills (#PE6554)
Restricted flow of local economy (#PF6451)
Minimum promotion of community assets (#PF6557)
Confined scope of business operations (#PF2439)
Underdeveloped capacity for income farming (#PF1240)
Underdeveloped sources of income expansion (#PF1345)
Individualistic practices of local business (#PF1176)
Unattractive locale for economic development (#PF3499)
Undeveloped channels for commercial initiative (#PF6471)
Inadequate management skills in rural communities (#PF1442)
Underdeveloped approaches to local food production (#PF6493)
Inadequate dissemination and use of available information (#PF1267)
Restrictive pattern of business activities in small communities (#PD1415).

♦ **PF0589 Negative economic repercussions of disarmament**
Nature Disarmament would raise both general problems of maintaining the over-all level of economic activity and employment; and specific problems in so far as manpower or productive capacity might require adaptation to non-military needs.
There are some aspects of the process of disarmament which would raise problems significantly different from those that have been experienced in the usual process of economic growth. While many of the continuous changes in the composition of demand work themselves out only over a long period of time, it seems reasonable to assume that disarmament, once decided upon, would occur more rapidly. For some components of military demand, the whole of the shift might occur within a very short period of time such as a single year.
The reallocation of resources attendant upon disarmament would therefore pose some special problems. Even with the successful maintenance of total effective demand during a period of disarmament, significant problems of adjustment would remain in specific sectors and areas of the economy. Part of the personnel released by the armed forces and the armaments industry would have to be trained or retrained so as to permit absorption into peacetime occupations. Some plant and equipment would have to be converted. Productive capacity might contract in some industries, and might have to be expanded in others. Where the manufacture of armaments has been concentrated in particular regions, it would be necessary either to shift resources out of those regions to other areas of growing demand, or alternatively to undertake schemes of redevelopment. The necessary steps would have to be taken to modify the direction of research and of technological development. Some less developed countries (dependent on exports of a few raw materials and commodities) might suffer economically from a sudden reduction in demand, a closing down of military bases, or the rapid sale of raw material reserves.
Counter-claim The main civilian purposes for which the freed resources, whether domestic or foreign in origin, could be applied are: raising standards of personal consumption of goods and services; expanding or modernizing agricultural and industrial productive capacity; promoting housing construction, urban renewal, including slum clearance, and rural development; promoting and expanding facilities for education, health, welfare, social security, cultural development and scientific research. Nuclear disarmament alone is the most necessary and has the most readily manageable economic consequences. Much of the physical plant and all of the scientists and technicians are needed for the peaceful uses of nuclear energy.
Broader Capitalism (#PC0564)   Total disarmament (#PF8686).
Narrower Demobilization (#PU2583)   Military reconversion (#PU2582).
Aggravates Unemployment (#PB0750)
Unsustainable economic development (#PC0495).
Reduces Competitive acquisition of arms (#PC1258).

♦ **PF0591 Inadequate regulation of restrictive business practices in service industries**
Nature There is international concern about the prevalence of restrictive business practices in the services sector by consulting firms and other enterprises, in relation to the design and manufacture of plant and equipment. It is of particular concern in those practices faced by prospective new exporters, in particular from developing countries. Such practices include: agreements fixing prices, including as to exports and imports; collusive tendering; market or customer allocation arrangements; allocation by quota as to sales and production; collective action to enforce arrangements, for example by concerted refusals to deal; concerted refusal of supplies to potential importers; collective denial of access to an arrangement, or association, which is crucial to competition. These may take the form of formal, informal, written or unwritten agreements or arrangements between rival and potentially rival enterprises; and are of concern when the practices limit access to markets or otherwise unduly restrain competition, having or being likely to have adverse effects on international trade, particularly that of developing countries, and on the economic development of these countries.
Such practices may be used individually, or in combination with other practices. For example, collusive tendering or 'bid-rigging' may involve the fixing of prices, as well as market or customer allocation arrangements, and an agreement to allocate sales and production by quota may involve collective action to enforce the arrangements, for example by concerted refusals to deal or collective denial of access to an arrangement or association which is crucial to competition.
The restrictive business practices described above are likely to be aimed at securing the highest possible prices for the goods or services exported. Such practices are also likely to be aimed at retaining the traditional suppliers' position in those markets where there is a nascent domestic production or where new foreign supplies, in particular from developing countries, enter the international market.
Broader Lack of control (#PF7138).
Related Prohibitive legal fees (#PF0995).
Aggravates Restrictive trade practices (#PC0073).

♦ **PF0593 Permissive education**
Unconstrained educational self-expression
Claim When there is real talent, self-expression has the great advantage of increasing the chances of its discovery. But when equal value is attached to all forms of self-expression this ultimately leads to the extreme perversion of the progressive idea, namely that any form of expression is of value and that any notion of a "correct" form of expression is to be derided. Furthermore, when combined with an emphasis on child-centered education, individualism becomes equated with selfishness so that the child comes to dominate the school and the curriculum. Progressive educators insist that the curriculum make sense to the child. This insight is perverted when the child's comprehension is equated with the child's likes and dislikes. Such extreme advocates of permissiveness overlook the effect of excessive responsibility on youngsters in making choices without the benefit of experience or any appreciation of the challenges and needs of wider society. The long-term results are insecurity and lack of purpose.
Broader Permissiveness (#PF1252).
Aggravated by Parental permissiveness (#PD5344).
Reduces Rote learning (#PJ5437).

♦ **PF0603 Inadequate information on drugs**
Nature Inadequate information concerning drugs and their effects, and on illicit use and traffic, hampers both the public and the medical profession. The inadequacy of general information on drugs is due to: inadequate research; the unwillingness of doctors to divulge information to their patients; the unwillingness, ignorance or irresponsibility on the part of some manufacturers to fully describe the harmful as well as the beneficial properties of their products. The inadequacy of information on illicit drug use is due to: unrecorded drug abuse; confidentiality or other complication in autopsy; lack of inventories on illicit drug stocks; the efficiency of organized crime or official corruption.
Claim Lack of knowledge and sophistication in the proper use of drugs is perhaps the greatest deficiency of the average physician today.
Broader Lack of medical information (#PJ8069).
Narrower Inadequate drug quality control (#PD2392).
Aggravates Irritating drugs (#PJ4491)   Abuse of plant drugs (#PD0022)
Inadequate drug control (#PC0231)   Ignorance of drug users (#PJ2389)
Excessive cost of medical drugs (#PE5755)
Excessive proliferation of medical drugs (#PD0644)
Inhaling of solvents and anaesthetic drugs (#PE1427)
Unethical experiments with drugs and medical devices (#PD2697).
Aggravated by Inadequate testing of drugs (#PD1190)
Irresponsible pharmaceutical advertising (#PE2390).

♦ **PF0608 Silence about historical situations**
Denial of past — Evasion of past events
Nature The refusal to discuss particularly painful episodes in the life of a people.
Incidence Most countries, military establishments, and official religions maintain a careful silence concerning periods or incidents in the past when they have engaged in repressive or otherwise morally repugnant actions against some group. This leads to pressure to avoid mention of such events in history textbooks or other media. Examples include treatment of prisoners of war and wartime atrocities, as well as failures to respond to requests for assistance in emergencies.
Broader Indifference to suffering (#PB5249).
Related Denial by old people of the significance of the past (#PF2830).
Aggravates Unrecorded knowledge (#PF5728)   Ignorance of history (#PD3774)
Historical misrepresentation (#PF4932).

♦ **PF0609 Threat to parliamentary immunity**
Threat to parliamentary privilege
Nature Threat to parliamentary immunity (the right to restrict disclosure of information, particularly at the committee stage) may be caused by journalistic irresponsibility, corruption, espionage or undue political pressure. It usually takes the form of a leakage of official secrets, particularly to the press, but may also be incorporated in restrictive legislation. Effective sanctions against a threat, although provided for by the law, may be difficult to enforce procedurally. Threat to parliamentary immunity may cause ineffective government or provide fuel for power politics. The opening up of parliamentary debates to the mass media may negate parliamentary representation and induce greater party control over individuals, causing effective decision-making to retreat further into the realms of bureaucracy. Leakage of official secrets may cause political unrest and instability.
Broader Denial of right to liberty (#PF0705).
Related Erosion of journalistic immunity (#PD3035)
Violation of privileges and immunities of international civil servants (#PE5488).
Aggravates Corruption in politics (#PC0116)
Violation of human rights of individual parliamentarians (#PE9759)
Restrictions on direct news coverage of parliamentary affairs (#PF3072).
Reduced by Inadequate government publications (#PF3075).

♦ **PF0611 Decline in government social expenditure**
Decline in public spending in the social sector — Decline in public expenditure on human development
Incidence Between 1983 and 1986 developing countries cut spending on health by 13.3 percent and on education by 10.5 percent according to the World Bank's "World Development Report 1988".
Broader Decline in government expenditure (#PF9108).
Narrower Decline in government health expenditure (#PF4586)
Decline in government expenditure on education (#PF0674).
Aggravates Erosion of human capital (#PF9612).
Aggravated by Economic bias in development (#PF2997)
Lack of economic and technical development (#PE8190).
Reduced by Excessive growth of social expenditure (#PC6215).

♦ **PF0625 Inadequate statistical information and data on problems**
Inadequate economic and technical data
Nature Insufficiency of the data necessary for development planning is one of the characteristics of economic underdevelopment. Even a minimal coordination of efforts is impossible without a minimum of information on the technical and economic environment, costs and possible benefits of alternative courses of action, and the specific investment opportunities. Put simply: it is not possible to decide what is yet to be done if one does not know what has already been done. Adequate programming can take place without a completely detailed picture and, even in the least

developed of countries, some data is always available. In a development project, it is necessary to determine the extent of the gap between the set of data required and the set of data actually available in the economy. The practical definition of the 'data gap' for any country depends essentially, therefore, on the planning methodology used and the planning targets set. If, for example, a simple methodology were considered adequate, existing data might also be considered sufficient, whereas a 'data gap' would arise should more complex and sophisticated planning methods be contemplated.
Broader General obstacles to problem alleviation (#PF0631).
Narrower Governmental bias in statistics (#PF0019)
Inadequate environmental statistics (#PF0011)
Inadequate public finance statistics (#PF7842)
Inadequate social and demographic statistics (#PF0214)
Inadequate statistical information on tourism (#PF4106)
Underparticipation of socialist countries in international data systems (#PE4230).
Aggravated by Economic uncertainty (#PF5817) Ineffective data systems (#PF3671)
Deficiencies in national statistics (#PF0510)
Failure of methods to appropriate data (#PE0630)
Proliferation of unprocessed scientific data (#PF1065)
Suppression of information concerning social problems (#PF9828).

♦ PF0626 National federalism

Nature The term "federalism" describes a form of political organization which unites separate groups (of national or ethnic differences) under a single political system in such a way as to enable each group to maintain its fundamental political autonomy and character. To do so, power is distributed among the central and the local governing bodies, and this distribution may result in or reflect discrimination against one or more or the federated groups. The rebellion thus provoked leads to further repression.
When a system is federal in name only, there is usually a high degree of centralization and bureaucracy in order that domination over diverse political groups may be achieved.
Political parties under a federal system are characterized by fragmentation and a lack of central discipline.
Background The word "federal" was coined in 1645 by British theologians who sought ways of manifesting the biblical covenant theories. Federal principles were applied much earlier, however, by the 13th century Israelites, and later by the Greeks. Modern federalism developed in America in the late 18th century.
Counter-claim Federal systems, although liable to failure when being established, have proved the most durable once established. No federal system that has lasted even 15 years has ever been voluntarily abandoned, and those disrupted by revolution have ultimately been restored.
Broader Ethnic conflict (#PC3685) Political conflict (#PC0368)
Intra-state imperialism (#PC3197).
Narrower Constitutionalism (#PU2658) Nominal federalism (#PU2659).
Related Domination (#PA0839) Human inequality (#PA0844) Over-centralization (#PF2711)
Forced assimilation (#PC3293) Social fragmentation (#PF1324).
Aggravates National disintegration (#PB3384).

♦ PF0631 General obstacles to problem alleviation

Narrower Falsity (#PF5900) Inefficiency (#PB0843) Squeamishness (#PF5735)
Human contingency (#PF7054) Oversimplification (#PF8455) Over-specialization (#PF0256)
Inaction on problems (#PB1423) Underreported issues (#PF9148)
Human destructiveness (#PA0832) Conflicting priorities (#PF5766)
Obstacles to education (#PF4852) Inappropriate arguments (#PF2152)
Increasing pace of life (#PF2304) Restrictions on freedom (#PC5075)
Obstacles to leadership (#PF7011) Unimplemented decisions (#PF4672)
Minimization of problems (#PF7186) Accumulation of knowledge (#PF2376)
Geopolitical vulnerability (#PF5749) Non-recognition of problems (#PF8112)
Proliferation of information (#PC1298) Unfinished imperfect universe (#PF5716)
Differing conceptions of time (#PF6665) Deteriorating quality of life (#PF7142)
Inadequate research on problems (#PF1077) Negative emotions and attitudes (#PA7090)
Inefficient public administration (#PF2335) Limited individual attention span (#PF2384)
Conceptual repression of problems (#PF5210) Obstacles to community achievement (#PF7118)
Polarized protest against problems (#PF9691) Obstacles to building construction (#PF7106)
Inadequate rehabilitation facilities (#PD1089) Preoccupation with isolated problems (#PF6580)
Antiquated world socio-economic order (#PF0866)
Psychological pollution by mass media (#PD1983)
Multiplicity of problems facing society (#PF2003)
Complex interrelationship of world problems (#PF0364)
Inability to resolve problems realistically (#PF8435)
Inadequate technical cooperation on problems (#PF0863)
Inadequate planning of action against problems (#PF1467)
Deterioration of stored documents and archives (#PE1669)
Short range planning for long-term development (#PF5660)
Insufficient translation into minority languages (#PD0825)
Inadequate public information concerning problems (#PF5701)
Monopolization of information within organizations (#PF2856)
Inadequate delivery mechanism in response to problems (#PF0301)
Inadequate research on proposed solutions to problems (#PF1572)
Inadequacy of the committee system of decision making (#PF2843)
Inadequate standardization of procedures and equipment (#PC0666)
Inadequate education concerning the nature of problems (#PE8216)
Inadequate statistical information and data on problems (#PF0625)
Excessive complexity of intergovernmental organizations (#PF2806)
Constraint of time on individual and social development (#PF5692)
Inadequate technical advice and consultation on problems (#PF1981)
Inadequate legislation relating to action against problems (#PF1645)
Shortage of financial resources for action against problems (#PF0404)
Inadequate organizational mechanisms to act against problems (#PF2431)
Obstacles to the development of multidisciplinary approaches (#PF7923)
Institutional obsolescence in modern industrialized societies (#PE2862)
Inadequate application of available knowledge to solve problems (#PF8191)
Inadequate exchange of technical information concerning problems (#PF0209)
Shortage of adequately trained personnel to act against problems (#PF0559)
Inadequate implementation of plans and programmes against problems (#PF1010)
Obstacles to effective international nongovernmental organizations (#PF7082)
Inadequate inter-regional cooperation on problems within countries (#PE8307)
Shortage of equipment and materials needed to act against problems (#PE8789)
Inadequate buildings, services and facilities for organized action against problems (#PD2669).
Related Injustice (#PA6486) Domination (#PA0839) Deprivation (#PA0831)
Instability (#PA0859) Incompetence (#PA6416)
Excessive institutionalization of education (#PD0932).
Aggravates Human suffering (#PB5955).
Aggravated by Reaction (#PA6355) Fragmentation (#PA6233)
Fragmentation of the human personality (#PA0911)
Disguised negative consequences of remedial action (#PF7583).

♦ PF0638 Unreadiness for second coming of Christ

Eschatological unpreparedness — Cosmic conflagration — Armageddon — Apocalypse
Nature The two views on the second coming of Christ within traditional Christianity expose two different issues for people. The first view regards the Second Advent primarily as a spiritual experience already realized through the descent of the Holy Spirit upon the disciples after the resurrection of Jesus, a return repeated in the experience of successive generations of Christians. In this view being unprepared for the Second Coming implies being closed to the intrusion of grace in one's life either through arrogance about having reached a degree of perfection that no change is necessary, despair over being incapable of being transformed or striving to change through personal effort. Having a closed heart and mind, these people are incapable of growth.
The second view insists that the promised return of Christ has not yet taken place, but may be momentarily expected. In the mean time the world constantly deteriorates. Only by Christ's literal coming can the millennium be established and righteousness made to prevail upon earth. In this view being unprepared for the Second Coming implies disregard for the state of one's present life.
Counter-claim Apocalyptic faith is not exclusively pessimistic, dualistic, deterministic or escapist. The tradition of the faith is realistic because it reveals that within and behind the human power struggles there are cosmic powers at work. Only God's future judgement and a new act of creation can change the state of this world. When human intervention in power struggles are in vain, spiritual resistance movements lead to an ultimate hope.
Narrower Millenarianism (#PF7684).
Related Sin (#PF0641).
Aggravated by Antichrist (#PF9139) Original sin (#PF8298).

♦ PF0639 Contempt for democratic processes

Contempt for parliamentary procedures
Broader Contempt (#PF7697).
Aggravates Abuse of government power (#PC9104)
Unparliamentary behaviour (#PF4550).
Aggravated by Undemocratic political organization (#PC1015).

♦ PF0641 Sin

Sinners
Nature Sin is a religious issue defined in different ways among different traditions. Within the Christian traditions sin is any word or deed or thought against the eternal law. In addition it includes violations of God's will that do not count as moral offences, i.e., idolatry. It is indifference or opposition to the will of God, the refusal of faith and love. It is an explicit or implicit claim to live independently of God, to put something else, be it the world or self, in His place. Paul Tillich states that sin, as opposed to sins, is the all prevailing problem of one's life. It is a state of existence before it is an act. It is simultaneously willed and fated separation from self, others and the Ground of Being.
In Islamic traditions sin has a variety of meanings. Sin is the evils which God's punishment brings. It is also moral guilt. It is unbelief and wrong actions proceeding from such unbelief. It is a wrong attitude to others and so injustice. The Koranic idea of sin seems to be that it is pride and opposition to God. This opposition to God may lead a man to be an atheist, a polytheist, or a simply careless irreligious person. Man does not inherit a sinful nature, but simply a weak one. Sin is not so much a disposition as a habit which men acquire because of their weakness.
In the Shinto tradition the term 'tsumi' most closely expresses the idea of sin. Tsumi includes three distinct categories, uncleanness, ill deeds and calamities. Ritual impurity, being dirty in body and clothing was want of respect for the gods. Calamities were regarded as signs of the displeasure of the gods for some offence, conscious or unconscious. Ill deeds for the ancient Japanese were divided into two broad categories, heavenly sins and earthly sins. Heavenly sins were breaking down the divisions of the rice fields, filling up the irrigation channels, opening the flood-gates of sluices, sowing bad seed over good, planting a magic wand in a rice field, flaying an animal alive and flaying backwards from tail to head, and evacuating excrements in places not meant for the purpose. Earthly sins were cutting the living skin, cutting the dead skin, white men (lepers and albinos), excrescences, a son's cohabitation with his own mother, a father's cohabitation with his own child, a father's cohabitation with his step-daughter, a man's cohabitation with his mother-in-law, cohabitation with animals, calamity through crawling worms (bites from snakes and other venomous creatures), calamity through the gods on high (like being struck by lightening), calamity through birds on high (soiling food), killing a neighbour's animal, and performing witchcraft.
Refs Fairlie, Henry and Lawrence, Vint *The Seven Deadly Sins Today* (1979); Palazzini, Pietro *Sin* (1964).
Narrower Lust (#PA4673) Envy (#PA7253) Sloth (#PA3275)
Pride (#PA7599) Anger (#PA7797) Gluttony (#PA9638)
Impenitence (#PF8765) Original sin (#PF8298) Infidelity to God (#PF9307)
Resistance to grace (#PF5266) Sin against the Holy Spirit (#PF6327).
Related Guilt (#PA6793) Vanity (#PA6491) Avarice (#PA6999)
Melancholy (#PF7756) Spiritual impurity (#PF6657)
Unreadiness for second coming of Christ (#PF0638).
Aggravates Wrath of God (#PF8563) Religious backsliding (#PF6826)
Fear of increased autonomy (#PE7706).
Aggravated by Impure thoughts (#PF5205).

♦ PF0648 Reallocation of aid funds to alternative priorities

Reallocation of development capital from developing countries to socialist countries — Competition for aid funding
Nature Development aid funds, originally scheduled for developing countries, may be diverted to alternative priorities such as the emerging democracies of Eastern Europe.
Claim Channelling funds into Eastern Europe beyond its legitimate needs poses great danger of an explosion and a further deterioration of the North-South relationship.
Aggravates Competition between states (#PC0114)
Disparity between industrialized and developing countries (#PC8694)
Shortage of financial resources for action against problems (#PF0404).
Aggravated by Compassion fatigue (#PF2819) Single party democracies (#PD2001)
Ineffectiveness of aid to developing countries (#PF1031)
Decline in concessional financial resources available to developing countries (#PE3812).

♦ PF0662 Sacrilege

Sacrilegious people
Nature Sacrilege is any abuse or violation of a person, place or thing consecrated to a deity (and thus deemed sacred). Irreverence for an object is not, on its own, sacrilege. Conflicting religious ideologies commonly accuse one another of sacrilegious acts, such as theft or misuse of sacred objects, desecration of sacred sites, and maltreatment of religious leaders.
Incidence Anti-religious political ideologies, such as communism, are reputed to have committed sacrilege against churches. Archaeological excavations, in Jerusalem for example, are a source of accusations of sacrilege, as are military manoeuvres. When, in 1984, the Indian army attacked the Sikh temple at Amritsar in an attempt to rout the armed guard there, officials could not deny the degree of sacrilege involved.
Broader Impiety (#PA6058) Wrongness (#PA7280).
Narrower Desecration (#PF9176) Religious intolerance (#PC1808)
Holy places as a focus of religious friction (#PF1816).
Related Ideological conflict (#PF3388).

Aggravates Religious conflict (#PC3292).
Aggravated by Double standards in morality (#PF5225)
Vulnerability of sacred sites (#PD6128).

♦ PF0674 Decline in government expenditure on education
Decline in public spending on education
Nature In the period from 1983 to 1986 the percentage of the developing world's GNP going to education has decreased by 10.5 percent according to the World Bank's "World Development Report 1988". And according to a UNESCO report spending per head on education in the 37 poorest nations of the world has declined by 50 percent. Educational quality is declining. Enrolments are stagnating. Students are dropping out of primary school in massive numbers. Teacher's pay is inadequate. At the same time spending on the military is 30 percent higher than spending on health and education combined. The children of the poor pay the most for cuts in spending on education.
Broader Decline in government social expenditure (#PF0611).
Narrower Decline in government expenditure on primary education (#PG9451).
Aggravates Educational wastage (#PC1716) Student absenteeism (#PE4200)
Underpayment of teachers (#PE8615) Declining school enrolment (#PJ7844)
Decline in educational standards (#PF8466).
Aggravated by Unnecessary education expenditure (#PJ0626).

♦ PF0677 Inadequate social reform
Lack of reform
Broader Change (#PF6605) Conservatism (#PF2160).
Narrower Inefficient public administration (#PF2335)
Maldistribution of agricultural land (#PD9189)
Rural poverty in developing countries (#PD4125)
Resistance to changing agricultural methods (#PF3010)
Obstacles to economic reform in socialist countries (#PF3689)
Deliberate governmental avoidance of legislative reform (#PF5736).
Related Rapidly changing cultures (#PF8521).
Aggravates Short range planning for long-term development (#PF5660)
Inadequate models of socio-economic development (#PF9576).
Aggravated by Civil disobedience (#PC0690) Individual fear of future change (#PD2670).
Reduces Counter-reform (#PF7334)
Confusion induced by rapid social change (#PF6712).

♦ PF0679 Nuclear freeze
Nature A nuclear freeze, by which both superpowers agree to stop deploying and testing new nuclear weapons, leaves the international arms industry with serious need for new work. Secondly, the strong deterrence factor, which has kept peace in the world for nearly half a century, is abandoned in favour of unnamed new directions.
Aggravates Peace (#PF9996) Unpreparedness (#PF8176).
Reduces Fear of reprisals (#PF9078).

♦ PF0684 Lack of racial identity
Nature A racial group, however distinct, can find itself lacking a racial identity if forced to assimilate with another, or if cut off from its cultural heritage for a long period. Such a group does not accept its national status as being identical with its racial status. A search for a racial identity elsewhere may result, and may be the cause of racial conflict. An inability to identify with racial and ethnic origins may lead to frustration.
Lack of racial identity occurs wherever there has been immigration of a group substantial enough to form a sub-group in society. The longer this group is away from its homeland, the larger the cultural gap between it and its origins. It may not be integrated into the society where it is living because of discrimination or segregation, or it may be assimilated into the society but be aware of its difference. The problem is particularly acute where coloured immigrants are concerned. Difficulty in relating to racial or ethnic origins is increased by the change in economic capacity which usually follows immigration from an underprivileged area to an industrialized area.
Broader Ethnic disintegration (#PC3291) Social disintegration (#PC3309)
Inadequate sense of personal identity (#PF1934).
Narrower Stereotypes (#PF8508) Proliferation of immigrants (#PD4605).
Related Heterosexism (#PE0818) Social fragmentation (#PF1324)
Cultural fragmentation (#PF0536) Inter-cultural misunderstanding (#PF3340)
Destruction of cultural heritage (#PC2114).
Aggravates Frustration (#PA2252) Racial conflict (#PC3684)
Ethnic conflict (#PC3685).
Aggravated by Segregation (#PC0031) Colonialism (#PC0798)
Discrimination (#PA0833) Crisis of identity (#PJ2421)
Forced assimilation (#PC3293) Inadequate education (#PF4984)
Socio-economic poverty (#PB0388).
Reduces Lack of assimilation (#PF2132).

♦ PF0685 Delays in centralized collective bargaining
Broader Delay (#PA1999).

♦ PF0687 Unnecessary reserves of material
Unnecessary stocks
Incidence In the USA Pentagon has stockpiled at least $30 billion of spare parts, uniforms and other equipment that it does not need.
Narrower Agricultural surpluses (#PC2062).
Aggravated by Over-production of commodities (#PD1465)
Mismanagement of food resources (#PE6115).

♦ PF0691 Ineffective international regulation of transnational corporations
Nature While transnational corporations are subject to the jurisdiction of individual governments in respect of their activities within specific countries, the global character of these corporations has not been matched by corresponding coordination of actions by governments, nor by an internationally recognized set of rules, nor a system of information disclosure.
Even though international production has become as important a fact of life as international trade, there exists today no international institution dealing with the activities of transnational corporations comparable with the General Agreement on Tariffs and Trade (GATT) and the United Nations Conference on Trade and Development (UNCTAD) which are concerned with international trade. The absence of an international forum makes it very difficult to work towards the international arrangements and agreements which would harmonize relevant national policies and laws and provide a framework within which the global strategies of transnational corporations should operate.
Broader Lack of control (#PF7138) Transnational corporation imperialism (#PD5891)
Excessive power and independence of transnational corporations (#PD5807).

♦ PF0695 Freemasonry
Masonic conspiracy
Nature Freemasonry involves a fraternity which is secret in so far as those admitted take an oath never to divulge its particular rituals and practices, although its membership and meeting places are not secret. Freemasons, who also take an oath to help one another, include politicians, businessmen and military officers and are often some of the most powerful and prominent men in any one country. Although freemasonry claims to be based on the fundamentals of religion and admits all faiths, it has been denounced and continues to be denounced by organized religion for practising occult rites, for its secrecy and for its power.
Incidence There are over 9,000 lodges in all parts of the world; most are under the jurisdiction of the United Grand Lodge of England, and some are under either the Grand Lodge of Ireland or the Grand Lodge of Scotland. In most Commonwealth countries, local Lodges are under the jurisdiction of independent Grand Lodges. There are also 50 autonomous Grand Lodges in the USA, all recognized by the Grand Lodge of England, as well as 5 in Central America and Mexico, 11 in South America, 2 in the West Indies, and 1 in Israel. In Europe there are recognized Grand Lodges in Austria, Denmark, Finland, Germany FR, France, Greece, Iceland, Netherlands, Norway, Sweden, Switzerland.
Background Freemasonry began as an organization of operative masons in England in mediaeval times. By the 18th century it had developed into a speculative brotherhood. The Grand Lodge connecting all others was founded in 1717. Within half a century of its foundation, freemasonry had spread to Europe, America and Asia.
Claim Freemasonry has had a long association with politics. Its lodges provided meeting place for Masonic men of affairs – tradesmen, bankers, lawyers, politicians – where the issues of the day could be discussed and debated. In such lodges, activists of the American and the French Revolutions met. In proponents of a revolution met in the lodge at Naples. In Rome in 1981, the P-2 Masonic Lodge which was declared illegal and considered to be a conspiracy against the government; its members were politicians, businessmen and military officers. Freemasonry has had offspring, such as the Bavarian Illuminati who were suppressed in 1785 for plotting revolution, and possibly the Rosicrucians who are now considered to have had a role in the Reformation. By virtue of their secrecy, lodge activities lend themselves to covert, concerted political action. Under certain conditions Masons may originate, or be manipulated into originating, politically dissenting and provocative activities tending towards the destabilization of governments.
Counter-claim The political potential of Freemasonry remains wherever there is authoritarianism. For centuries it has opposed the excessive power of the Vatican; later it opposed the regimes of the four Axis powers in World War II. Its vigilance remains a threat in some countries as is shown by its proscription in the USSR and anti-Masonic agitation by socialist-liberal-labour elements in the UK where Marxist philosophies have gained considerable ground.
Refs Knight, Stephen *The Brotherhood* (1985).
Broader Conspiracy (#PC2555) Secret societies (#PF2508)
Closed communities (#PG4604).
Narrower Symbolism (#PU2807) Power complex (#PU2808).
Related Elitism (#PA1387) Occultism (#PF3312) Catholicism (#PF8071).
Aggravates Domination (#PA0839) Imperialism (#PB0113)
Religious intolerance (#PC1808)
Foreign intervention in internal affairs of states (#PC3185).
Aggravated by Colonialism (#PC0798) Religious discrimination (#PC1455).

♦ PF0699 Commercialization of nationality acquisition
Sale of passports by governments
Incidence Many governments are willing to grant nationality in exchange for some form of investment of which a portion may be non-refundable.
Related Fraudulent acquisition or use of passports (#PE4496)
Crimes related to immigration, naturalization and passports (#PE3889).

♦ PF0703 Prohibitive cost of knowledge and information
Inflated cost of information
Nature The total growth of published knowledge in the form of books, periodicals, documents and other knowledge services such as data bank output, and the increasing cost of each of these, combine to make the total cost of knowledge increasingly excessive. Many important reference sources and periodicals can only be purchased by institutions, and often only by institutions to which the public does not have access. This is true of developed as well as developing countries, and is especially true of the reference tools without which relevant documents cannot be efficiently located.
Background As an example, estimates in the UK of increases in non-fiction book prices give an average of 15.4 per cent per year for the 1960s. In some years prices increased by nearly 20 per cent for adult non-fiction.
Broader Barriers to the international flow of knowledge and educational materials (#PF0166).
Aggravates Inaccessibility of knowledge (#PF1953)
Prohibitive cost of maintaining comprehensive document collections (#PE1122).
Aggravated by Shortage of funds for research (#PF5419).

♦ PF0704 False qualifications
Falsifying academic records — Misleading résumé — False curriculum vitae
Nature Many eager job applicants trying to impress a prospective employer embellish their résumés, saying they are accomplished skiers, for instance, even when they have only been on skis twice in their life. Some measure of embellishment is expected but when it becomes falsification the job can be lost. U.S. managers falsify academic qualifications more often than any other item on their résumés. This is due to the fact that few companies actually check qualifications.
Broader Misleading information (#PF3096) Lack of professional standards (#PF3411).
Aggravated by Fraudulent impersonation (#PE1275).
Reduced by Closed professions (#PD8629).

♦ PF0705 Denial of right to liberty
Nature Limitation of opportunity and free choice implies lack of liberty. Man is controlled by his environment, and he alters the actions of others even if he thinks that he is simply determining his own. Therefore the realization of free choice is naturally inhibited. On the national level, the balance between population and resources becomes more delicate; state planning and control is necessary; individual freedom is inhibited and, politically, can only be expressed within the prevailing representative system.
As political control becomes more intense, intellectual liberties may be curtailed by censorship and the banning of meetings, books and the right of association. Conflicting group interests may result in tyranny by either the majority or a minority. Moral and religious freedom may be inhibited by law or by social taboo. Freedom of movement and employment will be reduced and the ability to make the best use of available opportunities will be inhibited by unequal distribution of state services, education, health, housing. Freedom of ownership and the freedom to do as one chooses with one's possessions will be restricted according to the prevailing ideology concerning materials and resources and the interaction of materials and resources, such as the air polluting effect of car exhaust fumes. Extreme forms of the negation of liberty include slavery, concubinage, and legalized discrimination such as apartheid.
Background Lack of individual liberty has been recognized by philosophers throughout the centuries, usually on the basis that a certain amount of curtailment of individual freedom is necessary for society to function, but that it should not be excessive.

Counter-claim Complete intellectual freedom in questioning or rejecting everything destroys family, social and religious values.
Broader Slave trade (#PC0130) Restrictions on freedom (#PC5075)
Denial of political and civil rights (#PC0632).
Narrower Slavery (#PC0146) Traffic in persons (#PC4442)
Denial of religious liberty (#PD8445) Threat to parliamentary immunity (#PF0609)
Denial of right of family planning (#PE5226)
Discrimination against domestic servants (#PE4964)
Administering of medical drugs for non-medical purposes (#PE3828).
Related Denial of right to life (#PD4234) Denial of freedom of thought (#PF3217)
Denial of the right to procreate (#PC6870) Restriction of freedom of expression (#PC2162)
Restrictions on freedom of information (#PC0185)
Failure of individuals to participate in social processes (#PF0749).
Aggravates Prejudice (#PA2173) Frustration (#PA2252)
Discrimination (#PA0833) Human suffering (#PB5955)
Civil disobedience (#PC0690).
Aggravated by Attachment (#PF6106) Oppression (#PB8656)
Infringement of privacy (#PB0284) Inequality before the law (#PC1268)
Inequality of opportunity (#PC3435).

♦ PF0709 Inadequate prevention of disabilities

Nature Disability prevention relates to all preventive measures aimed at reducing the occurrence of impairments (first-level prevention); limiting or reversing disability caused by impairment (second-level prevention); or preventing the transition of disability into handicap (third-level prevention). Much disability could be prevented through measures taken against malnutrition, environmental pollution, poor hygiene, inadequate pre-natal and post-natal care, water-borne diseases and accidents of all types.

There is a steady growth of activities to prevent impairment, such as: the improvement of hygiene, education, and nutrition; better access to food and health care through primary health care approaches, with special attention to mother and child care; counselling for parents on genetic and pre-natal care factors; immunization and control of diseases and infections; accident prevention; and improvement of the quality of the environment. In some parts of the world, such measures have a significant impact on the incidence of physical and mental impairment; but for a majority of the world's population, especially those living in countries in the early stages of economic development, these preventive measures effectively reach only a small proportion of the people in need. Most developing countries have yet to establish a system for the early detection and prevention of impairment through periodic health examinations, particularly for pregnant women, infants and young children.

The more global measures for prevention of impairment are: avoidance of war; improvement of the educational, economic and social status of the least privileged groups; identification of types of impairment and their causes within defined geographical areas; introduction of specific intervention measures through better nutritional practices; improvement of health services, early detection and diagnosis; pre-natal and post-natal care; proper health care instruction, including patient and physician education; family planning; legislation and regulations; modification of life-styles; selective placement services; education regarding environmental hazards; the fostering of better informed and strengthened families and communities.
Broader Hazards to human health (#PB4885).
Aggravated by Injuries (#PB0855) Disasters (#PB3561)
Injurious accidents (#PB0731) Inadequate prevention of crime (#PF4924)
Inadequate primary health care (#PE8553) Insufficient preventive medicine (#PE0751)
Inadequate safeguards against fire (#PD1631) Inadequate provision of public safety (#PF2874)
Inadequate enforcement of safety regulations (#PD5001).

♦ PF0713 Lack of transport vehicles

Inadequate transport vehicles
Broader Lack of essential local infrastructure (#PF2115).
Narrower Insufficient special care vehicles (#PJ9008).
Aggravates Overloaded vehicles (#PE4127)
Restricted delivery of essential services to developing country rural communities (#PF1667).

♦ PF0716 Misappropriation of resources for high cost research projects

Inappropriate use of resources for fundamental sciences
Nature The extremely high cost of some forms of fundamental research results in many less costly research projects, often of greater immediate practical relevance, being deprived of funds. When such projects are first proposed the costs are usually deliberately underestimated in order to facilitate their approval. Subsequent escalation of the costs, ensures a further drain on scarce resources for less prestigious projects.
Incidence Major projects include: the Hubble space telescope, $2 billion; the superconducting super collider, $8 billion; the human genome project, $3 billion. Space projects in general have been considered a questionable allocation of resources. It has been estimated that completion of big science projects in the 1990s will cost $65 billion, which could be better used for developing new technologies of commercial significance.
Claim Too many big science projects were approved uncritically in the 1980s, with no apparent concern for their impact on small science through which technical advances are made.
Broader Excessively costly prestige projects (#PF3455).
Aggravates Shortage of funds for research (#PF5419).
Aggravated by Unethical practice of physics (#PD1710).

♦ PF0717 Liberalism

Background Liberalism was coined as a term from the Spanish 'Liberales' political party in the early 19th century. The principles of liberalism developed and spread in English language societies during the 19th century. As an idea and philosophy it can be traced back to the Judaeo-Christian-Greek intellectual world, together with the idea of liberty with which it is closely connected.
Claim Liberalism, as a belief in individual freedom as a method and policy in government, as an organizing principle in society, and as a way of life for the individual and community, can lead to the fragmentation of society and of individual energies, possibly culminating in social or even national disintegration and foreign intervention. For the liberal, the state, if needed at all, is at best a necessary evil, whose only function is to maximize freedoms and protects the individual from those who would deprive the individual from their liberty. This over emphasis on freedom is at the expense of equality. The concern with the rights of human beings is at the expense of the common good of society. Liberal economic policy, namely laissez-faire, is outmoded and insufficient in the modern context of interdependence and scarcity of resources, and could lead to foreign economic control. Because it accommodates many views, liberalism may lead to pacifism and international insecurity. Although liberalism combats conservatism and may foster social issues, it is often guilty of a half hearted approach.
Counter-claim Liberalism is historically associated with the idea of individual and civil freedom. It promotes conscience and justice in politics, minorities' rights and civil liberties. Without the motivating force of liberalism, individuals would still be denied basic civil rights.
Refs Buchanan, Allen E *Marx and Justice* (1982); Gerber, William *American Liberalism* (1987); Geronimo, Roger J, et al *Liberalism Exposed* (1987); Gillmore, Robert *Liberalism and the Politics of Plunder* (1987); Harrison, Deborah *Limits of Liberalism* (1981); Wolff, Robert P *Poverty of Liberalism* (1969).
Broader Ideological conflict (#PF3388).
Narrower Compromise as a betrayal of principles (#PF3420).
Related Political radicalism (#PF2177)
Lack of individualism in capitalist systems (#PD3106).
Aggravates Fragmentation (#PA6233) Permissiveness (#PF1252)
Fascist liberalism (#PF9710).
Aggravated by Destiny (#PF3111) Double standards in morality (#PF5225).
Reduces Intolerance (#PF0860).

♦ PF0719 Increasing number of disabled persons

Nature Many factors are responsible for the rising number of disabled persons and the relegation of disabled persons to the margin of society. These include: wars, and the consequences of wars and other forms of violence and destruction (poverty, hunger, epidemics, major shifts in population); a high proportion of overburdened and impoverished families; overcrowded and unhealthy housing and living conditions; populations with a high proportion of illiteracy and little awareness of basic social services or of health and education measures; an absence of accurate knowledge about disability, its causes, prevention and treatment, including stigma, discrimination and misconceived ideas on disability; inadequate programmes of primary health care and services; constraints, including a lack of resources, geographical distance, physical and social barriers, that make it impossible for many people to take advantage of available services; the channelling of resources to highly specialized services that are not relevant to the needs of the majority of people who need help; the absence or weakness of an infrastructure of related services for social assistance, health, education, vocational training and placement; low priority in social and economic development for activities related to equalization of opportunities, disability prevention and rehabilitation; low priority in social and economic development for activities related to equalization of opportunities, disability prevention and rehabilitation; industrial, agricultural and transportation-related accidents; natural disaster and earthquake; pollution of the physical environment; stress and other psycho- social problems associated with the transition from a traditional to a modern society; the imprudent use of medication, the misuse of therapeutic substances and the illicit use of drugs and stimulants; the faulty treatment of injured persons at the time of a disaster, which can be the cause of avoidable disability; urbanization and population growth; and other indirect factors.
Broader Human disability (#PC0699).

♦ PF0725 Restrictions on research

Nature Restricting the acquiring party either in undertaking research and development directed to absorb and adapt the transferred technology to local conditions or in initiating research and development programmes in connection with new products, processes or equipment.
Broader Restrictive trade practices (#PC0073).
Related Collusive tendering (#PE4301) Grant-back provisions (#PE5306)
Challenges to validity (#PF1200) Restrictions on publicity (#PF1575)
Restrictions on adaptations (#PF5248) Exclusive dealing arrangements (#PE0413)
Restrictions on use of personnel (#PF3945)
Patent pool or cross-licensing agreements (#PE4039)
Exclusive sales and representation agreements (#PE4581)
Trade restrictions due to voluntary export restraints (#PE0310)
Payments after expiration of industrial property rights (#PF5292)
Tying of supplies to subsidiaries by transnational enterprises (#PE0669).

♦ PF0726 Delayed consequences of war-time imprisonment and deportation

Broader Prisoners of war (#PC8848).
Related Social neglect (#PB0883) Social injustice (#PC0797)
Human disease and disability (#PB1044).
Aggravated by Ill treatment of prisoners of war (#PD2617).

♦ PF0727 Lack of war relief

Broader Conflict (#PA0298) Disorganization (#PF4487).
Narrower Neglect of dependents of war victims (#PD2092).
Related Looting (#PE4152) Extermination (#PJ2839)
Restrictions on emigration (#PC3208) Destruction of private property (#PG2842).
Aggravates Starvation (#PB1875) Housing shortage (#PD8778)
Human disease and disability (#PB1044) Inadequate hospital facilities (#PE5058)
Ill treatment of prisoners of war (#PD2617).
Aggravated by Civil war (#PC1869) Inhumanity (#PB8214)
Lack of international cooperation (#PF0817).

♦ PF0732 Inadequate legal counsel for political dissidents

Nature The profession of an unpopular political opinion may damage an accused person's possibilities of being defended by a competent lawyer or by a lawyer at all, due to a fear on the part of lawyers of being associated in the minds of the public (including other potential clients) with the beliefs of the accused.
Broader Injustice (#PA6486) Political repression (#PC1919)
Denial of legal representation (#PF3517).
Aggravates Forced disappearance of persons (#PD4259).

♦ PF0734 Limited number of available radio frequencies

Discriminatory frequency allocation — Overcrowded spectrum of radio frequencies — Maldistribution of radio frequencies
Nature The radio frequency spectrum only offers a limited number of frequencies to potential users, despite the expansion of the usable spectrum into the higher frequencies. Faced with a very rapidly rising demand, as the result of the introduction of many new radio and television devices, the spectrum has become congested and exploitation of closely related frequencies has increased interference. It is no longer possible to instal broadcasting or television equipment unless frequencies can be allocated which do not interfere with existing services. Frequency congestion increases the risk for civil aviation because the necessary ground-air communications are inadequate, or menaced by severe interference.
Background Prior to 1939, any country could take into use, within certain limits, any frequency it required for any particular service. After the war the situation became chaotic and the International Frequency Registration Board was established in 1947 to maintain the Master International Frequency Register which currently lists over 363,000 frequency assignments. However, jamming renders many frequencies unusable.
Broader Instability of economic and industrial production activities (#PC1217).
Aggravates Radio frequency interference (#PD2045)
Insufficient communications systems (#PF2350)
Allocation of television frequency bands for satellite transmission (#PF3703).

♦ PF0749 Failure of individuals to participate in social processes

Nature Lack of individual involvement in political, economic or social life results from restrictive or discriminatory legislation and practice; from alienation caused by bureaucratic processes, the size of cities, companies and factories; or from apathy. Lack of participation in certain activities may stem from class conflict or other social conflicts and subsequent alienation and lack of trust.
The desire for participation on the political and economic level implies a desire for a share in 'real'

power affecting people's lives. Failure to concede this may be masked by giving participation in consultation but not in decision-making. Cooperatives are a more complete form of economic participation, but management of the cooperative may also favour certain individuals over others. Socialist countries are more advanced in participation at all levels of society in economic life and in the life of the community, but not in political life.
Claim When one compares the vain appearances of liberty provided by consultation with the actual powerlessness of the individual, one perceives on a small scale how the most absolute government can assume some of the forms of the most liberal type of democracy – so that to oppression is added the absurdity of pretending not to see it.
Broader Lack of representation (#PF3468) Undemocratic political organization (#PC1015).
Narrower Lack of community participation (#PF3307)
Social disaffection of the young (#PD1544)
Lack of participation in development (#PF3339)
Exclusion of women from decision making (#PE9009)
Lack of worker participation in business decision-making (#PF0574).
Related Alienation (#PA3545) Social apathy (#PC3412) National isolationism (#PF2141)
Denial of right to liberty (#PF0705).
Aggravates Social alienation (#PC2130) Political alienation (#PC3227)
Social fragmentation (#PF1324) Lack of social mobility (#PF2195)
Bureaucratic aggression (#PC2064).
Aggravated by Political apathy (#PC1917) Social discrimination (#PC1864)
Inequality of opportunity (#PC3435) Unequal distribution of social services (#PC3437)
Inefficacity of consultation in the decision-making process (#PE8884).
Reduces Elitism (#PA1387) Social dictatorship (#PD3241) Political dictatorship (#PC0845).
Reduced by Demonstrations (#PD8522) Counter culture (#PF0423)
Passive resistance (#PF2788).

♦ PF0756 Generation communication gap
Insufficient inter-age communication — Prejudicial generation relationships
Nature Lack of understanding develops between people of widely different age groups, particularly between children (usually from the age of adolescence) and parents, grandparents etc. This lack of understanding may lead to the rejection of the children by the parents or, at a later stage when the parents are in need, the rejection of the parents by the children. It may lead to: family breakdown; juvenile delinquency; hooliganism and assault on older people by youth gangs; destruction of property; student unrest. Equally it may lead, because of fear and prejudice, to rejection of new ideas by older people and resistance to change. On the other hand, the abilities of youth may be overvalued in comparison with those of older people, leading to early retirement, unemployment and neglect of middle-aged and older persons.
Claim The generational conflict, or conflict between age-groups, is largely due to the educational revolution, which gives the adolescents of today intellectual tools that their parents very often did not have. It is also due to the "knowledge revolution", making the knowledge of the parent generation more irrelevant. This is not only because there has been a quantitative accumulation since they formed their images of the world, but also because of the qualitative changes of paradigms that seem to take place with increasing speed today. There is little reason to believe that these trends will not continue, and the net result is increasing irrelevance, dissimilarity, and heterogeneity, between generations.
Refs Adler, *Oral Communication Problems in Children and Adolescents* (1987); Gangrade, K D *Crisis of Values* (1975).
Broader Social fragmentation (#PF1324) Isolation of parent-child relationship (#PC0600)
Stifled potential for social interaction between different age groups (#PF6570).
Narrower Social disaffection of the young (#PD1544)
Educational gap between generations (#PF6497).
Related Intolerance (#PF0860).
Aggravates Student revolt (#PC2052) Family breakdown (#PC2102)
Neglect of the aged (#PD8945) Inadequate sex education (#PD0759)
Family rejection of children (#PC8127) Inhumane methods of riot control (#PD1156)
Lack of sharing of community skills (#PF3393).
Aggravated by Leadership age gap (#PJ9855) Age discrimination (#PC2541)
Family disorganization (#PC2151) Conflicting sense of sexual identity (#PF1246)
Family structure as a barrier to progress (#PF1502)
Incompatibility of rural values in urban cultures (#PF2648).

♦ PF0760 Incompetent financial management
Makeshift economic management — Poor money management
Broader Mismanagement (#PB8406) Incompetent management (#PC4867)
Ineffective means for goods supply and distribution (#PF6495).
Aggravates Global financial crisis (#PF3612)
Diminishing capital investment in small communities (#PF6477).

♦ PF0761 Lack of understanding of spiritual healing
Nature Lack of general understanding of spiritual healing leaves it open to quackery, whereby charlatans take advantage of people's despair in order to extort a large fee for their services. Spiritual healing is also equated by many with superstition and occult practices. The dismissal of spiritual healing as superstition and quackery limits the scope of medicine and psychiatry. In developing countries, modern medicine may fail because of the refusal of tribes people to participate owing to insufficient explanation by the authorities in spiritual terms.
Incidence Spiritual healing is very much a minority form of healing, and misunderstanding of it is widespread. The inadequacies of clinical, medical and psychiatric science, particularly in the understanding of fear, give a certain outlet for spiritual healers in sophisticated society. Many may be quacks, and some genuine, but in so far as recovery depends to a large extent on belief, the patients may derive some benefit in either case. In indigenous tribal society where spiritual healing is the custom, modern medical techniques may have little effect if spiritual health is not considered as well. The high incidence of psycho-somatic diseases in industrialized countries underlines the close connection between mental and physical disorders and the inadequacy of a purely clinical approach in coping with the problem; but general opinion in modernized society dismisses all except purely scientific cures as quackery or useless.
Refs Heinze, Ruth-Inge *Trance and Healing in Southeast Asia Today* (1988).
Broader Inadequacy of medical science (#PF8326).
Related Magic (#PF3311) Occultism (#PF3312) Witchcraft (#PF2099)
Superstition (#PA0430).
Aggravates Fear (#PA6030) Medical quackery (#PD1725) Intractable diseases (#PC8801)
Psychosomatic disorders (#PD1967) Inadequacy of psychiatry (#PE9172)
Lack of diversity in medical science (#PE8671)
Failure of medical programmes in developing countries (#PE8400).
Aggravated by Despair (#PF4004) Prejudice (#PA2173).

♦ PF0770 Irrelevance of science and technology
Disillusionment with science
Nature Science and technology have failed to produce practical results, particularly as regards their ability to focus on and resolve the increasing number of social problems. New developments have not resulted in significantly lower costs of energy, health care, travel or communications, for example. Governments have discovered that high expenditure on research and development does not necessarily or quickly lead to an increase in the gross national product. Many new developments have proved unreliable when implemented. In part this is due to a lack of realization of the complexity of the systems affected, and in part to uni-disciplinary approaches to problems. Our dependence on sophisticated technological systems has eroded basic human skills of survival, self-reliance and social skills. With this dependence has come a deep sense of powerlessness and incompetence.
Counter-claim Public apathy toward the use of science and technology may mean that a nation will not be competitive in the world market place. In order for science to have produced the meaningful changes that it has in society, it has been necessary for it to be free. Thus, for every productive avenue opened, another costly investigation has resulted in a blind alley; for every product, service and system introduced successfully, there has been a failure. Research in planned economics has no comparable accomplishment rate to that of the sciences and technologies in the free-market countries such as Sweden, Japan, the USA, Germany and others who have high GNP and high social and economic standards of living.
Broader Human errors and miscalculations (#PF3702).
Related Biased presentation of news (#PD1718).
Aggravates Ignorance (#PA5568).
Aggravated by Atomism (#PF5322) Scientific fraud (#PF1602)
Scientific elitism (#PC1937) Imbalanced research activity (#PF0198)
Irrelevant scientific activity (#PF1202) Lack of scientific investigation (#PF2720)
Incomprehensibility of specialized jargon (#PF1748).

♦ PF0772 Insufficient technical aids for disabled persons
Lack of prosthetic devices
Nature Many disabled persons require technical aids. In some countries the technology needed to produce such items is well developed, and highly sophisticated devices are manufactured to assist the mobility, communication and daily living of disabled individuals. The costs of such items are high, however, and only a few countries are able to provide such equipment. Other people need only simple equipment to facilitate mobility, communication and daily living but, although such aids are produced and available in some countries, in many other countries they cannot be obtained because of a lack of their availability and/or of high cost. Insufficient attention is being given to the design of simpler, less expensive devices, with local methods of production which are more easily adapted to the country concerned, more appropriate to the needs of most disabled persons and more readily available to them. The insufficiency of technical aids is matched by an insufficiency of infra-structure of rehabilitation therapists who can provide training in the use of such aids.
Broader Insufficient modern technology (#PJ0996).

♦ PF0775 Inadequate educational facilities for disabled persons
Inadequate educational facilities for disabled
Nature At least 10 percent of children are disabled. They have the same right to education as non-disabled persons and they require active intervention and specialized services. But most disabled children in developing countries receive neither specialized services nor compulsory education. There is great variation among countries, with some having a high educational level for disabled persons while in others such facilities are limited or non-existant. There is a lack in existing knowledge of the potential of disabled persons. Furthermore, there is often no legislation which deals with their needs and a shortage of teaching staff and facilities. So far, disabled persons in most countries have not benefitted from a life-long education.
Broader Inadequate educational facilities (#PD0847).
Narrower Inadequate vocational rehabilitation facilities for disabled persons (#PE7317).
Aggravates Institutionalization of the disabled (#PF4681).
Aggravated by Human disability (#PC0699) Inadequate social welfare services (#PC0834).

♦ PF0780 Absorption of manpower resources by military activities
Nature Military manpower requirements, whether for active or reserve service in the armed forces or for defence research and development, absorb a considerable number of educated people who would otherwise be able to pursue more constructive activities. Considerable managerial talent and technical skill is utilized by the military branches, and in many cases military personnel go through long and extremely arduous courses of training in special educational establishments. The increasing sophistication of weapons means that whatever the percentage of a national budget which goes towards military expenditure, the corresponding percentage in terms of the use of professional scientific manpower will be higher. It is usual to find that in countries with developed military industries, the proportion of the labour force in the engineering industry which is absorbed in the production of military equipment is far greater than the percentage of GNP which goes to military expenditure, and that the percentage of all qualified scientists and engineers employed on military research and development is even higher.
Incidence There has been only a slow rise in the numbers of the world's armed forces. Both since 1914 and since 1960, the numbers appear to have increased at 2.0–2.5 per cent a year (this is about one-third as fast as the increase in military expenditure). However, in a number of countries the armed forces have been employing an increasing number of civilians to do work previously done by servicemen. The total in 1980, including civilian support staff, was estimated at 39 million. Virtually all the increase in military manpower occurred in the developing countries whose share of the overall total for the world's armed forces is now about 60 per cent, in contrast to a less than 10 per cent share in military expenditure. Over recent decades the numbers in their armed forces have risen significantly. It has been estimated that about 10 million people around the world are employed by arms manufacturers. Probably at least a quarter of the world total of scientists and engineers engaged in research and development are in fact employed on military work. Military research and development probably absorbs nearly one million scientists, engineers and their support staffs of technicians and others.
Broader Militarism (#PC2169) Competitive acquisition of arms (#PC1258).
Aggravates Manpower (#PS2958).
Aggravated by National insecurity and vulnerability (#PB1149).

♦ PF0782 Unknown relatives
Unknown parents — Relatives lost during warfare — Unknown children
Broader Control of industries and sectors by transnational corporations (#PE5831).
Aggravates Uncertainty of death of missing persons (#PF0431).
Aggravated by Missing persons (#PD1380) Illegitimate children (#PC1874)
Single parent families (#PD2681).

♦ PF0796 Lack of political integrity
Broader Lack of integrity (#PF7992).
Narrower Political opportunism (#PC1897)
Political party manipulation of elections (#PD2906).
Related Political blackmail (#PD2912) Corruption in politics (#PC0116).
Aggravates Unfair elections (#PC2649) Violation of amnesty (#PD3018)
Espionage in domestic politics (#PD1787).

♦ PF0799 Simplistic technical solutions to complex environmental problems
Nature Complex environmental problems evoke ingenious technical solutions which can easily do more to exacerbate the problems, or others related to them, than to alleviate them. Such solutions may seem very attractive in the short-term, provided little attention is given to examining

their longer-term effects or provided that the importance of such effects can be denied.
Incidence Pesticides and fertilizers have provided quick cures to the uncertainties of agriculture. Fossil fuels have been treated as the cheap solution to bountiful energy (as wood had been treated in earlier times). Chemicals such as chlorofluorocarbons were thought to be ideal solutions to a particular class of problems. The deliberate introduction of species to counteract some pest has in many cases created even greater problems. In 1990 a solution to the global warming problem has been put forward which involves the dumping of large amounts of iron into the oceans to stimulate the growth of marine algae to absorb excess carbon dioxide — neglecting the surprise effects if such a project were to go awry.
Aggravates Unforeseen environmental crises (#PF9769).
Aggravated by Reductionism (#PF7967) Limitations of surprise-free thinking (#PF7700) Complex interrelationship of world problems (#PF0364)
Inadequate research on proposed solutions to problems (#PF1572)
Neglect of environmental consequences of government policies (#PE9295).

♦ PF0801 Non-alignment
Neutralism
Nature Non-alignment is a government policy of non-adherence to either of the major power blocs. To the extent that non-aligned nations do not support the policies of a particular bloc, they constitute a potential threat to that bloc, if only in that they may be more susceptible to align themselves with the opposing bloc.
Claim A strong revival of nationalist sentiment in Western European countries has raised the possibility of a 'swissification' prejudicial to NATO, the EEC, and other Western defence and economic alignments.
Counter-claim It is only neutral countries who can call for a new international economic order, to lay the groundwork for the just and prudent use of global resources in the quest for world peace and prosperity. The model of economic cooperation that may be developed among non-aligned nations could be an example to the world.
Non-alignment and neutralism in regard to the super-power interests does not mean pacifism and impotence. In the Western European region for example, a military alliance technologically led by France, Germany and the United Kingdom, has the capabilities of launching and arming satellites, manufacturing and installing nuclear missiles and enhanced conventional and non-conventional weapons in sufficient quantities to assure the destruction of the homeland of any aggressor. Moreover, mutual defence treaties can be negotiated in a number of directions, including Latin America, Africa and Asia. It is interesting to recall the World War II alignment of Spain, Italy, Germany, Japan and their supporters elsewhere, for example, in Argentina, Egypt and Eastern Europe. A third major power bloc is therefore possible in a Western-European led intercontinental alignment, but it is also achievable by a peaceful pan-European economic grouping in which high levels of cooperation are achieved between the EEC, EFTs and CMES.
Refs Schmitt, Hans A *Neutral Europe Between War and Revolution, 1917–23* (1988).
Broader Imbalance of power (#PB1969) International insecurity (#PB0009)
Excessive neutrality of intergovernmental official information (#PF3076).
Narrower National isolationism (#PF2141).
Related Neutrality (#PF0473).
Aggravates International economic fragmentation (#PC0025).
Aggravated by Imperialism (#PB0113) Disruptive foreign influence (#PC3188)
Inadequacy of aid to developing countries (#PF0392).
Reduces Foreign intervention in internal affairs of states (#PC3185).

♦ PF0815 Proliferation of commercialism
Commercialization of society
Nature Individuals or societies are becoming more and more dominated by acquisitiveness and valuing all things in terms of prices in the market.
Refs Hochschild, Arlie R *The Managed Heart* (1983).
Broader Discrepancies in human life evaluation (#PF1191).
Narrower Commercialization of human embryos (#PE6038)
Commercial exploitation of education (#PD9370)
Excessive commercialization of the media (#PE4215)
Unexplored alternatives for commercial development (#PF6548)
Commercialization of athletic activities and sports events (#PE4222).
Related Secularization (#PB1540) Proliferation of litigation (#PF0361)
Proliferation of technology (#PD2420) Proliferation of legislation (#PD5315)
Entrenchment of vested interests (#PD1231).
Aggravates Uncontrolled media (#PD0040) Immoral literature (#PF1384)
Newspaper monopoly (#PE0246) Film and cinema censorship (#PD3032)
Biased presentation of news (#PD1718) Bias in scientific research (#PF9693)
Radio and television censorship (#PD3029) Newspaper and journal propaganda (#PD0184)
Disruptive secular impact of holy days (#PE7735).
Aggravated by Materialism (#PF2655) Consumerism (#PD5774).

♦ PF0816 Lack of communication
Lack of effective communications
Nature Interpersonal communication networks include provision or exchange of information in the family or extended family, in the neighbourhood, in communities and ethnic groups, in various clubs and professional associations, and in conferences and meetings which are convened by governments, by organizations of all kinds, or by commercial enterprises. Speech, performance and example remain the most common, if not the only, means of transmitting information in communities where isolation, or size, or persistent illiteracy, have encouraged the survival of tradition; while in industrialized countries, traditional channels for direct communication have virtually disappeared as sources of information as more people turn to depersonalized solo "communication" with sophisticated technological devices.
Refs Chaffee, S and Petrick, M *Using the Mass Media* (1975); Rieber, Robert W *Communication Disorders* (1981); Shoman, S Giora *The Violence of Silence* (1982); United Nations Educational, Scientific and Cultural Organization *Interim Report on Communication Problems in Modern Society* (1978).
Broader Limited access to social benefits (#PF1303).
Narrower Obscurantism (#PF1357) Maldistribution of radios (#PF4142)
Excessive use of acronyms and abbreviations (#PF4286)
Ineffectiveness and inefficiency of intercultural meetings (#PF0316)
Ineffectiveness and inefficiency of international meetings (#PF0349).
Aggravates Taboo (#PF3310) Tension (#PB6370) Ignorance (#PA5568)
Antisemitism (#PE2131) Culture shock (#PC2673) Political apathy (#PC1917)
Cultural barriers (#PB2331) Lack of human unity (#PF2434)
Psychological alienation (#PB0147).
Aggravated by Diverse unilingualism (#PF3317) Poor communications methods (#PJ8656)
Radio noise of industrial origin (#PE2473) Insufficient communications systems (#PF2350)
Delays in delivery of books and publications (#PF1538)
Prejudice against communication by visual imagery (#PF0076).

♦ PF0817 Lack of international cooperation
Dependence on lack of international cooperation — Nationalistic response to global issues — Inadequate global cooperation to solve world problems — Nationalistic policy responses to world economic crisis — Decline in multilateral cooperation
Incidence There is an increasing decline in multilateral cooperation, accompanied by a negative attitude to dialogue on development in particular. Official economic cooperation is inadequate or non-existent between governments in some regions of the South. North-North cooperation in all matters is beset by ideological conflict, and North-South cooperation by economic exploitation.
Claim Global ecological and economic interdependence requires effective international cooperation for appropriate management. The inadequate cooperation of all countries is responsible for the continuing rivalries and inequities in the global economic and social system.
Counter-claim The very process of international cooperation has been envisaged as a political one rather than an economic or cultural one. It exists in very positive ways in the nonpolitical sphere. Science is international, as is art and business. Service organizations are often international. Religious traditions are either already international or rapidly becoming so. All of this is international cooperation.
Refs Brandt Commission *Common Crisis; North-South* (1983); Gray, H Peter *International Economic Problems and Policies* (1987).
Broader Lack of cooperation (#PF2816).
Narrower Inadequate world calendar (#PF2043) Inadequacy of patent coverage (#PF3538)
Unfulfilled treaty obligations (#PF2497) Fragmented regional cooperation (#PF9129)
Disparity of national tax systems (#PD1791) Lack of trans-frontier cooperation (#PF6855)
International economic fragmentation (#PC0025)
Subversion of international agreements (#PD5876)
Inadequacy of international legislation (#PF0228)
Technology gap between developed countries (#PD0338)
Fragmentation of technological development (#PC1227)
Extraterritorial intrusion of jurisdiction (#PE3140)
Inadequate technical cooperation on problems (#PF0863)
Inadequacies of the international monetary system (#PF0048)
Limited enforceability of international standards (#PF8927)
Uncoordinated international river basin development (#PD0516)
Alienation of land through acquisition by foreigners (#PE0896)
Weakness in trade between different economic systems (#PC2724)
Lack of international coordination of interest rates (#PF3141)
Non-equivalence of national educational qualifications (#PC1524)
Inadequate economic integration of socialist countries (#PF4884)
Obstacles to legal relations between socialist countries (#PF4886)
Disregard for internationally imposed economic sanctions (#PF1976)
Irregular payments of international financial obligations (#PD1157)
Disparity between industrialized and developing countries (#PC8694)
Ineffectiveness and inefficiency of international meetings (#PF0349)
Inadequate international cooperation in reducing terrorism (#PF4366)
Ineffectiveness and inefficiency of interdisciplinary meetings (#PF0409)
Non-verificability of compliance with nuclear power safeguards (#PF4455)
Impediments to internationally mobile professionals and experts (#PF1068)
Proliferation of national and international anniversaries and years (#PF2723)
Allocation of television frequency bands for satellite transmission (#PF3703)
Domination of the world by territorially organized sovereign states (#PD0055)
Competition between international organizations for scarce resources (#PC1463)
Alienation of support for international organizations and programmes (#PD1809)
Detrimental international repercussions of domestic agricultural policies (#PF2889)
Financial and economic disputes between states and nationals of other states (#PE1911)
Imbalances in the distribution of the costs and benefits of economic integration (#PD0794)
Differences in trading principles and practices between different economic systems (#PC2952)
Territories accorded a United Nations non-self-governing status disputed by the administering government (#PF2943).
Related Harassment (#PC8558) Lack of international understanding (#PF5106)
Obstructions to international personnel exchanges and cultural cooperation (#PE4785).
Aggravates Tension (#PB6370) Sabotage (#PD0405)
Aerial piracy (#PD0124) Lack of war relief (#PF0727)
Extradition refusal (#PF2645) Cardiovascular syphilis (#PG0121)
Hazardous remnants of war (#PF2613) Unchallenging world vision (#PF9478)
Lack of a world government (#PF4937) Indecisive multilateralism (#PF9564)
Ineffective war crime prosecution (#PD1464)
Fragmentation of the international trading system (#PC9584)
Inadequate power of intergovernmental organizations (#PF9175)
Technology gap between developed and developing countries (#PE7985).
Aggravated by Disinformation (#PB7606) Mutual deceits (#PJ8029)
Lack of political will (#PC5180) Non-verification of compliance (#PF6310)
Multiplicity of time standards (#PF2621) Inter-cultural misunderstanding (#PF3340)
Governmental barriers to a global ethic (#PF2105)
Obsessive compulsive personality disorder (#PE7632)
Fixation on partial solutions to problems (#PF9409)
Language domination by developed countries (#PD6029)
Inadequate structures for achieving global unity (#PD2802)
Deteriorated structures of essential corporateness (#PF1301)
Restrictions on international freedom of information (#PC0931)
Untransferability of books between countries and cultures (#PF2126)
Undermining of multilateral forums by industrialized countries (#PE4289)
Inability to negotiate effective multilateral safeguard systems (#PF5287).
Reduced by Denial of right to confidentiality (#PD6612)
Declining international competitiveness (#PD8994).

♦ PF0823 Narrow context for counsel
Nature The role of those who give counsel – so urgent to the creation of consensus in today's complex societies – is blocked by a context for their advice which is excessively narrow. This reluctance to consider the direction of the whole community prevents advice from illuminating the broader issues which it must ultimately address. Thus the confidant, the technical specialist, the cabinet minister – all these advice-givers lose their power.
Broader Reduced understanding of globality (#PF7071).
Narrower Myopic advice to leadership (#PF7283)
Over-formalised decision-making (#PF1033)
Quantitative understanding of responsibility (#PF5587)
Extreme detachment from represented constituency (#PF0889)
Self-satisfied style of informal leadership advisors (#PE7654).
Related Undemocratic policy-making (#PF8703) Inflexible central government (#PD1061)
Exclusion of the masses from setting criteria in judicial judgements (#PD1060).
Aggravated by Irresponsible international experts (#PF0221).

♦ PF0827 Underutilization of facilities due to daily or seasonal peaks
Seasonal resource use
Nature Offices, factories, hotels, schools, transport systems and electrical power are designed to permit full utilization during normal working hours, and to cater for peak periods which may occur at certain times of the day or during certain seasons of the year. At other times such facilities are under-utilized or completely vacant. Despite this under-utilization, if the facilities are perceived to be inadequate during the normal hours, further facilities will be called for to take care of the load during those hours. It has not proved possible to stagger utilization to any great extent.
Broader Underdevelopment (#PB0206) Seasonal fluctuations (#PF8163).

Related Underutilization of natural resources (#PF1459)
Instability of economic and industrial production activities (#PC1217).
Aggravates Seasonal unemployment (#PC1108)
Environmental degradation of inner city areas (#PC2616)
Excessive land usage by transportation systems (#PE2525).
Aggravated by Inhospitable climate (#PC0387).

♦ PF0836 Obsolete basis of cultural identity

Nature Whenever effective development takes place, a major effort is spent creating a cultural framework which motivates citizens. This framework, based on geography, history, vision of the future and labour gives direction to all aspects of the community effort. The bases of a community's identity is out of date and the community senses that its historical role is completed, the sense of self-identity fades as do the languages, practices and customs which developed around the old identity. Individuals and organizations outside the community may reenforce the sense of lost identity by the way they relate to the community. The community's sense of what is significant responsibilities of citizens may also reenforce the identity of the community. Only when small rural communities recreate their cultural identity in terms relevant to the contemporary world will the residents have the motivity necessary to sustain development efforts.
Broader Cultural revolution (#PF3235).
Narrower Dim business future (#PG9236) Frustrated past goals (#PF5272)
Unappealing farm labour (#PG9349) We / they language patterns (#PG9219)
Continuing commercial decline (#PJ8726) Irrelevancy of past experience (#PG9311)
Unimportance of physical improvements (#PG9267)
Disrelationship from community history (#PJ0836)
Demoralizing image of urban community identity (#PF1681).
Related Culture shock (#PC2673) Cultural decline (#PC9083)
Cultural conflict (#PG3790) Cultural stagnation (#PC8269)
Disjointed patterns of community identity (#PF2845)
Untransposed significance of cultural tradition (#PF1373).
Aggravates Loss of cultural identity (#PF9005) Rapidly changing cultures (#PF8521)
Homogenization of cultures (#PB1071) Ignorance of cultural heritage (#PF1985)
Socially inappropriate housing (#PD8638).
Aggravated by Rigid cultural patterns (#PF8598)
Risk of capital investment (#PF6572)
Insufficient cultural media (#PJ8476)
Insufficient cultural nodes (#PG9774)
Insufficient cultural heroes (#PF8623)
Insufficient community events (#PF5250).

♦ PF0841 Economic dependence

Lack of economic self-sufficiency — Economic dependency
Broader Chance (#PA6714) Underdevelopment (#PB0206)
Vulnerability of small nations to foreign intervention (#PD2374).
Narrower Over-acceptance of socio-economic dependency (#PF8855)
Economic dependence of some developing countries on the drug trade (#PE5296)
Instability in export trade of developing countries producing primary commodities (#PD2968).
Aggravates Communism (#PC0369) Neo-colonialism (#PC1876)
Economic imperialism (#PC3198) Lack of self-confidence (#PF0879)
National political dependence (#PF1452) Lack of political independence (#PF0297)
Communist political imperialism (#PC3164)
Economic competition in communist systems (#PC3167)
Denial of right to national self-determination (#PC1450).
Aggravated by Communist economic imperialism (#PC3165)
Capitalist economic imperialism (#PC3166)
Ill-considered pressure to eliminate nakedness in developing countries (#PF3350).
Reduced by National isolationism (#PF2141).

♦ PF0845 Government resistance to institutional change

Government reluctance to transform administrative structures
Nature The general response of government to the speed and scale of global changes has been a reluctance to recognize sufficiently the need to change themselves. The environmental and developmental challenges are interdependent and integrated, requiring comprehensive approaches and popular participation. These requirements are difficult for national and international institutions that have been established on the basis of narrow preoccupations and compartmentalized concerns.
Broader Resistance to change (#PF0557).
Aggravates Government inaction (#PC3950) Institutional fragmentation (#PC3915)
Rejection of proposals for social change (#PF9149)
Delays in implementation of social change (#PC6989).
Aggravated by Lack of political will (#PC5180)
Government delay in response to symptoms of problems (#PF6707).

♦ PF0846 Refusal of family possibilities

Nature The amount and direction of change in this period of history is putting tremendous strain on the family and its members. So much so that the creative role of the family in society and in caring for its members is refused by even those most involved in family life. The family is the source of everyone's relationship to society. Even those institutionalized at birth are increasingly raised in "family substitute" situations. The family is the place where care for the next generation largely takes place. These and the other functions of family life is treated as though it were optional. As a result life styles within the family are reduced, economic considerations often controls decision making, and youth are unprepared to function effectively in society.
Broader Lost family role in society (#PF7456).
Related Escapist family life styles (#PD4069).

♦ PF0851 Majority rule mindset

Claim Present democratic institutions operate out of the understanding that majority opinion is equal to justice for all. The mindset which allows control to those in power is blind to the creative innovation provided by the voices of dissent, which should serve as a basis for the ongoing, healthy change that marks a living and effective social unit.
Broader Blocked minority opinion (#PD1140).
Aggravated by Diabetes (#PE0102).

♦ PF0852 Unequal coverage by social security

Unequal distribution of social security
Nature The population of some countries or regions and certain social groups are not covered by social security to the same extent as others. In addition, one of the most negative consequences of a economic recession may be a certain weakening in some countries of the autonomy of social security institutions.
Partial and uneven coverage by social security often occurs during the period when a country is building up its social security programme. Data do not permit a meaningful analysis of the proportion of the population in the developing countries who are covered by social security. However, given the concentration of the system in the urban-industrial-public service population and the smallness of those groups in total population, it is evident that only a small fraction of the population in those countries is covered by some type of social security. This contrasts with the almost universal coverage in the developed countries. In addition, while practically all major social and economic groups are already covered by some type of social security in the developed countries, the relative coverage of such groups varies markedly in the developing countries; one notable feature being that public sector employees are more often covered for certain types of benefits than any other group. Thus over 70 percent of countries with some sort of social security system cover public employees in some degree for old age, invalidism and death benefits. The coverage for other groups is far less frequent. The same is true of sickness and maternity benefits; these groups being the next most important and the next most frequently covered.
These features of the social security systems in the developing countries reflect not only their stage of development and paucity of resources, but also a certain urban-public service elitism. Agricultural workers and the self-employed are virtually left out of the social security system altogether. Agriculture provides a special challenge to social security policy makers and planners. In all cases, the first measures of social security reflect the economic needs of the urban industrial community. Any extension of coverage to the agricultural community implies that schemes have to be introduced into a very different setting, and the type of need may in fact be very different from that of the urban wage earner. There is no single, world-wide, uniform model of agricultural activity. Variations stem from all sorts of local conditions – tradition, soil, climate, systems of land tenure, communications, extent of industrialization, level of education and so on. Given such circumstances, the extension of a conventional programme of social security to the rural areas is no easy task, nor is it necessarily a sufficient answer to the social security protection of persons in those areas. It meets the situation in areas where rural-urban communications are close enough for town and country to form one wage-paid community. Elsewhere, and particularly in developing countries, it has seldom proved possible to extend full cover to the countryside because of poor communications, the question of identification, and the problem of covering self-employed persons.
Even in the most industrialized societies and those with the most comprehensive system of benefits, inequalities exist. Generally women working in the home depend on their husband's benefits; yet in most countries women carry very little insurance if they are widowed or disabled or abandoned by their husbands. Many women's careers are interrupted because of the need to devote time to bringing up children; consequently, the old-age benefits they receive are very meagre. Married women are generally denied unemployment and disability benefits. Parental benefits are generally not paid to men. Unequal pension ages for men and women allow women to retire earlier than men. Some categories of migrant workers are not covered especially those from the third world. Many countries base their social security systems on occupational schemes so that persons who have become disabled before they began work are not covered. Survivors' benefits are generally too low and are often eaten away by inflations. The right to benefit of young school leavers and of married women entering the labour market is generally limited.
Related Economic insecurity (#PC2020)
Social insecurity in developing countries (#PE4796).
Aggravates Family poverty (#PC0999) Increasing cost of social security (#PF7911)
Negative effects of family allowances (#PF0107).
Aggravated by Unreported births (#PF5381) Social inequality (#PB0514)
Inadequate social welfare services (#PC0834).

♦ PF0854 Underdevelopment of manufacturing industries

Broader Lack of economic and technical development (#PE8190)
Underdevelopment of industrial and economic activities (#PC0880).
Narrower Underdevelopment of basic metal industries (#PF1374)
Underdevelopment of woodworking industries (#PF2604)
Underdevelopment of food processing industries (#PD0908)
Underdevelopment of paper and printing industries (#PE1136)
Minimal manufacturing in least developed countries (#PE0282)
Underdevelopment of textile and clothing industries (#PE0453)
Underdevelopment of chemical and petrochemicals industry (#PE1483)
Underdevelopment of non-metallic mineral products industries (#PE1851)
Weakness in trade in manufactured goods among developing countries (#PE2966)
Inadequate supply of pharmaceutical products in developing countries (#PE4120)
Underdevelopment of metal products, machinery and equipment industries (#PF0942)
Inadequate increase in employment in the manufacturing industries in developing countries (#PE2238).
Related Health risks to workers in manufacturing industries (#PE1605).

♦ PF0860 Intolerance

Narrower Moralism (#PF3379) Religious intolerance (#PC1808)
Intolerance of criticism (#PF8396) Intolerance of imperfection (#PF7024).
Related Impatience (#PA6200) Nationalism (#PB0534) Irresolution (#PA7325)
Anti-communism (#PF1826) Narrowmindedness (#PA7306)
Generation communication gap (#PF0756) Negative emotions and attitudes (#PA7090).
Aggravates Persecution (#PB7709) Social fragmentation (#PF1324).
Aggravated by Conflict (#PA0298) Extremism (#PB3415)
Propaganda (#PF1878) Neo-fascism (#PF2636)
Political schism (#PC2361) Authoritarianism (#PB1638).
Reduced by Liberalism (#PF0717).

♦ PF0863 Inadequate technical cooperation on problems

Nature Obstacles to technical cooperation include ideological conflicts, rivalry between nations, psycho-linguistic and cultural differences, lack of agreement on problem priorities and lack of budgets to facilitate sustained international collaborative work.
Claim Even if all the foregoing obstacles were overcome, the world would still be left with technicians and specialists advocating the exclusive superiority of their own disciplines and approaches. Technical cooperation on problems is only achieved in a transitory manner, usually during severe crises.
Broader Lack of international cooperation (#PF0817)
General obstacles to problem alleviation (#PF0631).
Aggravates Inadequate delivery mechanism in response to problems (#PF0301).
Aggravated by Inadequate planning of action against problems (#PF1467)
Inadequate power of intergovernmental organizations (#PF9175).

♦ PF0866 Antiquated world socio-economic order

Broader General obstacles to problem alleviation (#PF0631).
Narrower International economic injustice (#PC9112)
Discriminatory international order (#PB6021)
Weakness of socio-economic infrastructure (#PC1059)
Inadequate trade-related structural adjustments (#PF4165)
Lack of progress in establishing a New International Economic Order (#PF4306).

♦ PF0867 Monotonous and unaesthetic architecture and design

Over-conformity in building design — Unimaginative housing design
Nature The structure of a building not only determines the appearance and feeling of it, it also influences the behaviour of people within it, their sense of well-being, or their lack of it. Ill-considered architecture and design can aggravate social tensions, depression and boredom. As a financial asset, buildings have to be readily marketable. Funding cannot be easily be obtained from financial institutions for innovative projects because of the increased element of risk. With

high land costs and rising interest rates, the developer wants to get construction, especially housing, completed and sold as quickly as possible. The more uncontroversial their appearance, the speedier the planning authorization is obtained. The trend to conformity is also influenced by the conservatism and lack of architectural sophistication of the general public. People like what they are familiar with. Public greed also has a role in that houses are increasingly a commodity to be traded. Judgements on their appearance, efficiency and comfort are secondary to their potential secondhand value. Pride in good design is not a primary concern.
Narrower Socially sterile rental accommodation (#PF6195)
Monolithic architecture of high-rise buildings (#PE1925).
Related Depressing effect of poor housing construction (#PF1213).
Aggravates Boredom (#PA7365) Civil disorders (#PC2551)
Mental depression (#PC0799)
Stultifying homogeneity of modern cities (#PF6155).

♦ **PF0878 Use of inappropriate technologies in developing countries**
Nature The developing countries are faced with selecting appropriate industries and technologies. Although the choice between requirements for more or less capital and special skills per worker is not the only choice to be made when making this selection, it is often a highly relevant one.
The rising pressure of unemployment seems to dictate that efforts should be made by developing countries to develop those industries and technologies which can provide the largest possible number of jobs, and require little capital. On the other hand, it must be recognized that, while it is possible to achieve this by selecting for development those industries with labour-intensive technologies, or industries in which different combinations of labour and capital intensity exist side by side, there are cases in which technological alternatives are not available, and where, if a particular industry is to be established at all, there is no other choice than to adopt a technology even though it may be a capital-intensive technology which employs relatively little labour. It is also true that often developing countries insist on the most up-to-date technology (or are obliged to use it by multinational enterprises willing to invest), although it may not, in fact, be appropriate to their objectives.
The problem is being further complicated by the fact that industrially advanced countries are on the threshold of a second industrial revolution. Automation, transfer machines and the use of electronics are just beginning to be developed, and further revolutionary changes in technology and production techniques may occur. Unless appropriate choices are made, the present economic and social gap between the industrialized and the developing countries will become even larger.
Counter-claim There may be instances in which a capital-intensive technology would actually yield the best results, for example, where there are important spin-off effects on local industry, or where multinational enterprises produce in export industries or in industries in which no labour-intensive technology is available or could only be used with considerable increase in prices. In such cases, labour-intensive operations may have only a peripheral use, for example in material handling and transportation.
Refs Bhagavan, M R *Critique of Appropriate Technology for Underdeveloped Countries* (1979).
Broader Underdevelopment (#PB0206) Human errors and miscalculations (#PF3702).
Narrower Inappropriate level of technological equipment (#PF2410)
Haphazard transmission of practical technology (#PF3409).
Related Unbalanced application of communications technology (#PE7637).
Aggravates Unemployment in developing countries (#PD0176)
Inadequate production capacity in developing countries (#PD4219).
Aggravated by Abusive technological development under capitalism (#PD7463)
Inadequate research and development on problems of developing countries (#PF1120).
Reduces Inadequate research and development capacity in developing countries (#PE4880).

♦ **PF0879 Lack of self-confidence**
Lack of self-confidences — Low self-confidence
Broader Fear (#PA6030) Psychological inertia (#PF0421)
Negative emotions and attitudes (#PA7090).
Narrower Low self esteem (#PF5354) Loss of self-respect (#PG2484)
Resignation to problems (#PF8781) Demeaning minority self-image (#PF1529)
Ill effects of educational failure (#PF2013).
Related Self-hatred (#PG6396) Defensive life stance (#PF0979)
Loneliness of children (#PC0239)
Loss of confidence in government leaders (#PF1097).
Aggravates Boasting (#PF4436) Drug abuse (#PD0094)
Domination (#PA0839) Narcissism (#PF7248)
Complacency (#PA1742) Death instinct (#PF3849)
Self-destruction (#PF8587) Anti-intellectualism (#PF1929)
Self-defeating behaviour (#PD4418) Inhibited self-promotion (#PJ1544)
Low learning expectations (#PJ8960) Consumption of alcoholic beverages (#PD8286)
Underutilization of locally available skills (#PF6538)
Demoralizing image of urban community identity (#PF1681)
Psychological barriers to the judicial protection of individual rights (#PE1479).
Aggravated by Acne (#PE3662) Loneliness (#PF2386)
Self-deception (#PF6362) Enforced celibacy (#PD3371)
Self-glorification (#PG5645) Economic dependence (#PF0841)
Maladjusted children (#PD0586) Ghetto education image (#PJ8933)
Demeaning farmer image (#PJ9781) Human sexual inadequacy (#PC1892)
Self-destructive excuses (#PF6044) Debilitating poverty image (#PJ1341)
Oppressive prevalent images (#PF1365) Declining sense of community (#PF2575)
Denial of rights to disabled (#PC3461) Lack of community self-worth (#PF3512)
Inaccurate youth stereotypes (#PJ8357) Demeaning community self-image (#PF2093)
Physically handicapped children (#PD0196) Social isolation of the elderly (#PD1564)
Low self image due to illiteracy (#PF9098) Financial irresponsibility image (#PG8414)
Isolation of parent-child relationship (#PC0600)
Unrealistically positive self-assessment (#PF4377)
Discrimination against dwarfs and midgets (#PE2635)
Social reinforcement of delimited self-worth (#PF1663)
Declining community confidence in its ability to change (#PF9066).

♦ **PF0886 Denial of right to inherit property**
Unequal property inheritance — Disinheritance
Broader Injustice (#PA6486) Denial of parental affiliation (#PD3255).
Aggravated by Intestacy (#PE5063) Social injustice (#PC0797)
Inadequate laws of adoption (#PD0590)
Discrimination against illegitimate children (#PD0943).

♦ **PF0888 Unreliability of statistics from socialist countries**
Nature Economic statistics and other data required in analyses are often limited in scope, quality and quantity. The basic facts required for analyses have to be reconstructed and carefully evaluated from the limited data released by socialist countries. Low quality and poor availability of commercial data limits trade and external financing because it raises both costs and risks.
Incidence Albania is the country that offers the least information of practical usefulness. Romania has a better record, although a great deal of basic statistical information remains unavailable. The German DR has tended to reduce substantially its economic information and statistical detail in recent years. Czechoslovakia and Bulgaria are in an intermediate position, whereas Hungary, Yugoslavia and Poland publish the most comprehensive and best documented sets of economic statistics in the region.
Broader Unreliable data (#PF6832) Deficiencies in national statistics (#PF0510).

♦ **PF0889 Extreme detachment from represented constituency**
Nature Those who counsel executive authorities are blocked by their ever-increasing detachment from the context of their expertise. Their advice is thus overly influenced by the executive receiver and less influenced by the constituency it represents.
Broader Political isolation (#PC7569) Obstacles to leadership (#PF7011)
Narrow context for counsel (#PF0823).
Aggravates Lack of leadership (#PF1254) Loss of leadership credibility (#PF9016)
Self-satisfied style of informal leadership advisors (#PE7654).
Aggravated by Insecurity of leadership (#PD9362)
Lack of leadership initiative (#PF7988).

♦ **PF0897 Collapse of public servant role of the professional**
Nature In the past, specialists have been called upon to play a particular role as keepers of a community's exclusive skills and in manifesting the representational dynamic in special services to that community. The marks of this role have been reduced to status in the form of economic success. The concept of "servant of the public" has been largely rendered meaningless.
Broader Static and unrelated social roles (#PF1651)
Individualistic utilization of expertise (#PF5639).

♦ **PF0903 Inefficient public administration in developing countries**
Limited civil administrative capacity in developing countries — Lack of trained administrative personnel in developing countries
Nature Civil service administration, both internal and general, has special features in the developing countries. To achieve a satisfactory internal administration, it is often essential to undertake radical reforms before an adequate and homogeneous civil service can be built up under what are frequently difficult conditions. To reorganize services, and to take a number of other measures connected with a gradual policy of modernization and transformation, raises a multitude of problems. Sustained attention and continuous effort are necessary to arrive at the form of administration best suited to particular needs. Nearly all the developing countries are in a period of transition in this respect; solutions are adopted under the pressure of circumstances, and the soundness of the means employed may be very questionable. Many of these countries are carrying out experiments, including reforms of various kinds, and at times it is too early to try to draw the necessary inferences from them. The task is to set up a homogeneous and national civil service based on merit where only the rudiments exist.
Incidence In many developing countries, especially those of low income, the government's administrative capacity to carry out a programme of economic reform is poorly developed. Moreover, in most developing countries the bureaucracy forms an influential interest group that may oppose economic stabilization and structural reform since this may require reducing the size of the public sector through government employee layoffs and privatization of state-controlled enterprises.
Refs United Nations *Issues and Priorities in Public Administration and Finance in the Third United Nations Development Decade* (1983).
Broader Inefficient public administration (#PF2335)
Weakness of infrastructure in developing countries (#PC1228)
Inadequate control over government administrative process (#PC1818).
Narrower Inefficient tax administration in developing countries (#PE5009).
Aggravates Ineffective tax systems in developing countries (#PF2124).

♦ **PF0905 Undirected expansion of economic base**
Nature The economy of some small communities may be based on one industry or activity, for example, mining. If in this case the mines are closed, the majority of the people may leave within months. If a number of residents, rather than deserting the town entirely, are determined to rebuild the community around new industries and business, then they face a number of difficulties. Agriculture and forestry may be looked at anew as basic industries. Profitable farming is difficult, because the products pass through many hands between the farmer and the consumer, and new techniques are only slowly obtainable. The closing of the mines means that local markets are even smaller than they once were; and external markets are competitive and over powering. Government-organized forestry and reforestation may be a possible resource for the future, but few townspeople have skills in this field.
The unclear economic future discourages the use of what savings may have been accumulated by individual families before the mines closed. Many may feel the only solution is to seek employment and business development elsewhere. The task of renovating basic facilities to attract outside investments seems enormous. Services decrease and public buildings, such as schools, remain underheated. Development momentum virtually comes to a standstill and town income declines.
Broader Expansionism (#PB5858) Economic expansion (#PF8111)
Unsustainable economic development (#PC0495).
Narrower Inadequate savings (#PC0927) Small local markets (#PJ8200)
Underpayment for work (#PD8916) Uninsulated buildings (#PE0242)
Narrow employment base (#PG7926) Unstable economic base (#PG8905)
Decreasing labour force (#PJ8193) Emigration of labour force (#PJ8205)
Inadequate support services (#PF6492) Unprofitable farm marketing (#PJ8204)
Poor-risk commercial farming (#PG8204) Unstrategic funding requests (#PJ8208)
Shortage of school equipment (#PJ0821) Lack of incentives for tourism (#PJ8198)
Questionable economic viability (#PG8186) Overpowering competitive markets (#PJ8195)
Insufficient availability of goods (#PB8891) Insufficient development collaterals (#PJ8197)
Limited local availability of capital reserve (#PF2378)
Diminishing capital investment in small communities (#PF6477).
Related Haphazard growth patterns (#PG8687) Time consuming procedures (#PJ8206).
Aggravates Inadequate recreational facilities (#PF0202).
Aggravated by Political aggression (#PD8877) Narrow economic foresight (#PF8602)
Limited economic communication (#PJ1489) Uncoordinated economic enterprises (#PG8588)
Mismanagement in developing countries (#PD8549)
Limited approaches to economic planning (#PF3500)
Short-term planning of product life cycles (#PF1740)
Lack of economic and technical development (#PE8190).
Reduced by Economic stagnation (#PC0002) Dissipating economic base (#PJ1342)
Weakness of socio-economic infrastructure (#PC1059).

♦ **PF0906 High human fertility in developing countries**
High crude birth rate
Nature Birth rates have fallen very substantially but the difference in fertility between developed and developing countries is still wide. High birth rates still constitute a characteristic of developing countries. In the more developed regions, an average woman now bears between 2 and 3 children, while in the developing regions she gives birth to between 2 and 7 children.
Incidence In the present more developed regions combined, the birth rate may have averaged 38 per 1,000 during 1850-1900 and 26 per 1,000 during 1900-1950, the average having fallen

progressively as one region after another underwent this process of change. Post-war 'baby booms' led to an average birth rate of 23 per 1,000 in 1950–1960, subsiding to 19 per 1,000 in 1960–1970 and 15 per 1,000 in the last decade. Yet during all those decades, the average birth rate in the less developed regions combined is estimated at 40 or 41 per 1,000; and only in recent years, has this dropped to a still high 33 per 1,000. It is assumed that the annual average birth rate of the more developed regions will continue to fluctuate around 15 per 1,000 to the end of the century, whereas, during the same period, the average birth rate of the less developed regions may decrease even more decisively.

While the more developed regions tend towards greater uniformity in the level of crude birth rates, a wide range of conditions now seems to exist in the less developed regions. Among 22 countries of these regions, the crude birth rate around 1970 was reported as less than 30 in 7 countries, 30–40 in 6 countries, 40–50 in 6 countries, and above 50 in 3 countries. More recent statistics indicate Africa, as a continent, having 46 per 1,000 as a birth rate; Latin America having 31, and Asia, 30. The wide ranges on each continent continue, however. If China is excluded from the figures for less-developed countries, then the average is 37 per 1,000.

Refs Potts, M, et al *Childbirth in Developing Countries* (1983).

Aggravates Birth prevention (#PE3286) Unsustainable population levels (#PB0035)
Excessive childhood dependency in developing countries (#PD3491)
Imbalance of population growth between developed and developing countries (#PE4241).
Aggravated by Ineffective population control (#PF1020).

♦ PF0907 High labour turnover in developing countries

Instability of the labour force in developing countries

Nature Continual recruitment and training of new workers constitutes a significant drain on the efficiency and earnings of industry. Costs are raised by the instability of the labour force; peasants are reluctant to enter permanent employment off the land.

In most of the less developed countries, there is a constant flux of workers between industrial employment in towns and traditional agriculture in their home villages. They may work only long enough to collect the money they need to pay taxes or to fulfil other obligations; they often become homesick and dissatisfied with their food and quarters, particularly if they have left their wives in the villages; lacking means of expressing their grievances and improving their status, they may move from industry to industry seeking better conditions. Attempts to stabilize the industrial labour force by means of a variety of incentives – holidays with pay contingent upon a certain period of steady work, regular wage increases for those who stay on the job, provision of appropriate facilities for eating, housing, health, education, recreation and so on – have often met with no more than limited success.

Broader Underproductivity (#PF1107)
Instability of economic and industrial production activities (#PC1217).
Aggravates Low occupational mobility in developing countries (#PD1493)
Underdevelopment of industrial and economic activities (#PC0880)
Lack of response to monetary incentives in developing countries (#PD1432).
Aggravated by Lack of job satisfaction (#PF0171)
Loss of traditional forms of social security in developing countries (#PD1543).

♦ PF0910 Futility of social loyalties

Nature There is a certain hypocrisy in the present day social process, in which society either appears unresponsive to desperate need and/or it limits human rights. People have sought and sometimes realized in their society a sense of necessary order, protection of individual rights and participation in genuine community. In the absence of such a community there arises the question: 'Why be a loyal citizen in a society that you can't depend on'.

Broader Collapsed meaning of human creativity (#PF0936).

♦ PF0913 Unhealthy emotional responses to atomic energy

Claim Certain emotional responses provoked by the advent of atomic energy are in many instances pathological. These responses, due partly to the circumstances in which atomic power has been introduced and partly to its very nature, constitute a major mental health aspect of the peaceful uses of atomic energy. The unhealthy reactions stem from anxiety, and from attempts which human beings make to deal with anxiety. Thus they may be manifested in the forms of irrational fears, irrational hopes, or irrational tendencies to ignore or deny the extraordinary potentialities of atomic energy.

Counter-claim Special pseudo-psychological studies can be made of the resistance to new technologies of all types. Anxiety accompanied the introduction of the automobile, as it undoubtedly did the horse, and perhaps even the wheel. If there were a pathological fear of atomic energy, (on the grounds offered), there would have been pathological energy fears of other kinds; fear of a log fire, steam locomotives, electricity and the internal combustion engine. On the contrary, a thousand million people use modern forms of power, including nuclear generated energy, with gratitude, and atomic power has an important role to play in development.

Broader Mental illness (#PC0300) Negative emotions and attitudes (#PA7090).
Aggravates Fear (#PA6030) Anxiety (#PA1635)
Irrational rejection of nuclear power (#PF8531).
Aggravated by Technophobia (#PG6101) Pervasive fear of nuclear war (#PC3541)
Denial of right to adequate medical care (#PD2028)
Hazardous locations for nuclear power plants (#PD2718).

♦ PF0916 Escaping reality through sophisticated shields against life's pain

Nature People develop sophisticated shields which they use in attempting to protect themselves from confronting the pain of life. Death, suffering and misery are isolated from day-to-day experience because they are unpleasant. This creates a situation where the public lacks the sensibilities to deal with these facts of the human situation. Not only is the living structure designed to prevent confrontation with unpleasantnesses, but language also helps create the illusion of these realities not being present. With death, for example, people use the terms 'he has passed away' or 'gone to meet his maker' to mean that he is dead. Creation of such guards against unavoidable events makes individuals even more vulnerable to fear and confusion when they are forced to confront reality.

Broader Escapism (#PF7523) Theological collapse (#PF6358) Avoidance of reality (#PF7414).
Aggravates Meaningless recreation (#PF0386).

♦ PF0919 Limited tripartite cooperation

Nature In tripartite cooperation partners from West, East and South are responsible signatories and actively participate in the cooperation. The biggest share of tripartite projects is accounted by construction of energy production capacities, followed by intermediate goods industries and consumer goods industries. Cooperation offers an opportunity for access to new markets, the comparative advantages of the international division of labour because the socialist and capitalist companies would not be able to supply alone all that is required for the construction of an industrial plant, developing countries receive the technology and the ability to use it.

Incidence The main part of the work has been done in the oil-producing and other fast growing developing countries.

Broader Weakness in trade between different economic systems (#PC2724).

♦ PF0922 Ineffective protection of individual rights due to excessive court costs

Nature Expensive court charges and formalities are usually involved in bringing a civil action, in appealing against the judgement of a court in a civil case and in enforcing constitutional guarantees concerning a fair hearing. Charges for civil actions or appeals often increase with the value of the claim in dispute, and the charges for civil appeals may be higher than those charged at first instance. In some countries of Latin America, there are no court charges as such, but litigants in civil actions must, in taking certain steps, use sealed paper, the cost of which increases in proportion to the value of the matter in dispute. In several of the countries where the use of sealed paper is compulsory in civil actions, court charges also apply. A private person who initiates a prosecution may also have to make certain payments. All of these requirements discriminate against poorer persons. In addition, the expenses in question are incurred before embarking on proceedings, and eventual success in the action is of no assistance at that earlier point.

Broader Economic discrimination in the administration of justice (#PE1399).

♦ PF0926 Complicity with evil

Broader Evil (#PF7042) Complicity (#PF4983)
Instability of fresh fruit and edible nut trade (#PE2587).
Aggravates Non-resistance to evil (#PF7754).

♦ PF0936 Collapsed meaning of human creativity

Nature Many people today are estranged, internally and externally, from a world of ever-increasing change and horrifying complexity. They have no central sense of history or personal destiny. The 19th Century western universe was rational, static and controlled by a benevolent power. A new era came to birth with the 20th Century, in which was found no simple rational pattern. There has been an exponential increase in technological change; and people experience a deep dread at seeing no end to the increasing rate of such change. At the same time there has been a collapse of images of the eternal, leaving no way of relating to final reality.

Claim Throughout history, points of crisis have arisen at which societies have been unable to bridge the gap between their understanding of life and their real experience, internal and external. One such point in the 20th Century came with the realization of the most horrifying realities of human potential in the form of the gas chambers and atomic bombs during World War II. This arose at the 'peak' of civilization, when almost the whole planet was witnessing technological control. Human attempts to articulate the mystery evident in this struggle have been incomplete. The totality of life may be denied by being deeply rooted in the 19th Century image of the benevolent Father; or it may be manifested in a search for the creative force, for good, and in denying the brokenness encountered. Personal concerns are not with the creation of social structures but with the individual's relationship to life within those structures. It is not the creation of the symbolic life which reminds the individual of his or her encounter with the mystery of life, but the sense of after-life which is experienced in the encounter itself.

Broader Theological collapse (#PF6358).
Narrower Defensive life stance (#PF0979) Blocked service to society (#PF6433)
Limits on areas of research (#PF2529) Collapse of cultural dreams (#PF1084)
Lost family role in society (#PF7456) Futility of social loyalties (#PF0910)
Collapse in providing ethical value screens (#PF1114)
Collapse in the meaning of participating in society (#PF0955).
Related Lack of commitment (#PF1729) Dominance of economic motives (#PF1913)
Overemphasis on self-sufficiency with respect to interdependence (#PF3460).
Aggravated by Educational curricula based on content rather than method (#PF3549).
Reduced by Anxiety (#PA1635).

♦ PF0942 Underdevelopment of metal products, machinery and equipment industries

Broader Underdevelopment of manufacturing industries (#PF0854).
Narrower Machinery and equipment industries underdevelopment (#PJ3130).

♦ PF0944 Fragmentation of knowledge

Inadequate organization of knowledge — Lack of a system of integration of knowledge — Lack of a comprehensive framework for understanding knowledge

Nature This includes those areas of basic analytic and rational methods (like the sciences) often seen to be at odds with other fields of knowledge. There is no framework or comprehensive construct in which to relate the tremendous volume of ideas, data and insights which every person meets at work and at home through the years. This results in confusion and loss of values. This failure to have an intellectual framework which would relate the variety of types of insight, knowledge and data tends to reinforce narrow specialization.

Lack of integration of the knowledge generated by researchers with differing geographic and functional backgrounds seriously limits the formulation of effective policies.

The majority of systems for organizing knowledge (as reflected in different kinds of information) make use of well-defined breakdowns by category into hierarchies, or are based on title or subject keywords (possibly selected from a thesaurus), or are limited to an author of the work in question. The emphasis is placed on indexing the document and not on the concepts in it (unless they are reflected in the title). The result is that a limited number of keywords become overused with little possibility of distinguishing between the meanings behind each usage, and with little possibility of exploring the relationships between concepts which make up the body of knowledge. In particular, there are no means for handling the level of abstraction of a concept and indicating its ability to interrelate other concepts, particularly those which are either common to several disciplines or integrate the conceptual structures of several disciplines.

Claim Present-day knowledge consists of a fragmented collection of isolated data and insights. The lack of a unifying system holding together the various disciplines and arenas is at the root of the unrelatedness of human wisdom to the rest of the social process. Individuals are overwhelmed and frustrated by being constantly inundated with great quantities of data which they have no way of retaining or relating.

The lack of more powerful means of organizing knowledge encourages institutions to implement simplistic information and organization systems which do not match in complexity the networks of problems on which they are expected to focus. Libraries are unable to highlight documents containing concepts of progressively higher degrees of interdisciplinary power, because such publications are categorized by the most prominent topic in the title. Bookshops reflect this situation with an even simpler organization of their publications. In both bookshops and libraries, unless either the title or the author are known, it is almost impossible to locate material dealing with interrelated concepts, or concepts not mentioned in the title.

Broader Human errors and miscalculations (#PF3702)
Human wisdom unrelated to daily life (#PD1703)
Antiquated intellectual methods to appropriate human depths (#PF1094).
Narrower Lack of integration of traditional and Western medicine (#PF4871)
Failure to integrate knowledge to empower humanity in response to the global problematique (#PF8753).
Related Ignorance (#PA5568).
Aggravates Lost knowledge (#PF5420) Inaccessibility of knowledge (#PF1953)
Inadequate models of socio-economic development (#PF9576).
Aggravated by Manipulative knowledge (#PF1609)
Fragmentation of research (#PF9830)
Monopolization of knowledge (#PF5329)
Proliferation of information (#PC1298)
Structural amnesia in institutional systems (#PF7745).

♦ **PF0945 Isolating social forms for enhancing ethical relations**
Claim At the local level, ethical relations are blocked by the continuing operation of isolating, inadequate structures which prohibit meaningful participation in society. People become frustrated by a sense of vague community obligation which causes feelings of being trapped and impotent. There is a sense of paralysis in the ability to care for others through social organizations: people feel themselves to be victimized when they are not in a position to receive care from these same organizations.
Broader Collapse in providing ethical value screens (#PF1114).

♦ **PF0951 Despairing individualism**
Rugged individualism
Nature Individualism is a world view that expounds the theory of individual endeavours and effects over societal efforts.
Background Individualism is not characteristic of primitive societies where individuals are still so undeveloped and immature that they need to belong to their social milieu, rather it is borne of developed societies where individuals have the choice of, and opportunity for, being independent. Rugged individualism, a recent phenomenon in some developed societies, is the sense of ultimate total responsibility, that is, that in the end, one is really answerable only to one's self. Taken to extremes, such rugged individualism could result in the despairing individualism that leaves one disbelieving in society's niches: marriage, family, religion, and politics. One may conclude that life is merely a game without purpose, that all action is futile, and that one can actually do nothing in the face of the world's relentless chaos. This could lead to wild personal style and belief experimentation or to hermitism.
Broader Individualism (#PF8393).
Aggravates Theological collapse (#PF6358).

♦ **PF0953 Thwarted technological communications**
Claim New, creative technological communications could implement a global service-oriented communications network. Misdirected power is evident in the gap between the socially created technological world and the socially created people who live in the world. It is a gap in basic relationship tools: how they are defined and how they are made available to all people.
Broader Limits on areas of research (#PF2529).
Narrower Lack of social progress (#PF1545) Excessive politicization of the media (#PD5475)
Parochial escapist media entertainment (#PD0917)
Unequal global distribution of basic skills (#PF2880)
Exclusive nationally-oriented language systems (#PD2579).
Related Individualistic utilization of expertise (#PF5639)
Inequitable distribution of skilled specialists (#PD2479)
Gap between the function of social techniques and the needs they address (#PF3608).
Aggravates Insufficient communications systems (#PF2350).
Aggravated by Unarticulated goal of educational methods (#PF2400).

♦ **PF0954 Boundary constraints on land planning**
Nature A major obstacle to proper land utilization is the persistence of obsolete or cumbersome urban boundaries, fixed before population and transportation burst their originally prescribed limits.
Broader Capitalism (#PC0564) Uncontrolled urban development (#PC0442).
Aggravated by Unnatural boundaries between states (#PF0090).

♦ **PF0955 Collapse in the meaning of participating in society**
Disillusionment with life aspirations
Background This collapse began around the turn of the century and by the end of World War I it was nearly complete. World War II, in fact, revived a sense of meaning for the western world. The values of decency, huma,ity, civilization, material and technological progress came to be seen as forces of good. For the individual, normalcy and uniformity were considered good and religious yearnings, neuroses or deviation from convention were regarded as a pathological condition. Society was not so much for something as against "Communism", and a poorly defined communism at that. By the mid-1960s these values were discredited for the most part. The youth were flaunting abnormality as a source of pride, the abnormal became the creative or the self-expressive. The social upheavals of the civil rights movement, the anti-vietnam war movement, and the student uprising of paris and else where exposed the hollowness of materialistic consumerism. Once again "everything is relative", there was no positive direction, except survival and muddling through, which have become goals in themselves. Without a belief in the future, the present becomes the focus of the search for purpose. Yet, even the present is being questioned. Most religious traditions, instead of imparting purpose and direction have bogged down in internal questions or attempts to recapture values of an earlier century.
Claim People seek the purpose of existence in their work, family, society, and cultural tradition, many of which no longer give meaning. As a part of real, everyday experience, people in the 20th Century have seen a collapse many of the understandings that gave meaning to their hopes and ambitions. While in the midst of bitter disillusionments and painful failures in aspiration, the individual comes to doubt his own creative worth in society. Nevertheless, a deep desire for creative action is retained. The more doubts there are, the stronger the desire to express creatively; while the experience of futility impels the destruction of others' creations. There is a collapse in the meaningfulness of social participation.
Broader Deteriorating quality of life (#PF7142)
Collapsed meaning of human creativity (#PF0936).

♦ **PF0956 Faulty approaches in the teaching of intellectual methods**
Nature One of the most evident malfunctions in transmitting intellectual methodologies is the approach of teaching methods in this area. Although much attention has been given to teacher education and training, teachers tend to act out of a fixed self-image as one who dispenses knowledge. Students are neither motivated nor equipped to expand learning beyond rote feedback. The lack of a systematic approach to methods of teaching, study and communicating, results in students using a trial and error approach.
Broader Ineffective methods of practical education (#PF2721)
Antiquated intellectual methods to appropriate human depths (#PF1094).
Aggravates Underutilization of intellectual ability (#PF0100).

♦ **PF0958 Exercise addiction**
Exercise fanatics
Nature For exercise fanatics missing one day is irritating, missing more than a few days is a catastrophe. They can have withdrawal symptoms: depression, anxiety, poor appetite, loss of self-confidence and insomnia.
Broader Addiction (#PD6324).

♦ **PF0959 Trivialization of love**
Collapsed conceptualization of love
Claim Reducing love to an emotional or sexual response to another person or thing effectively destroys the depth of the reality. While warm-hearted, tender or even sexual responses may be a dimension of love, they leave out the discipline and sacrifice required of the deeper experiences of love. There are ample suggestions that the most profound forms of love are found in the midst of the emotional experience of hate of another. This view exalts various forms of erotic love and forgets brotherly and spiritual love.
Refs Cook, Mark and Wilson, Glenn *Love and Attraction* (1979); Suttie, Ian *The Origins of Love and Hate.*
Broader Facile social concepts (#PF5242).
Aggravated by Excessive portrayal of sex in the media (#PE7930).

♦ **PF0960 Repressed inflation in socialist countries**
Nature Whereas market economies are characterized by 'open inflation' socialist economies are prone to 'repressed inflation' whose symptoms include chronic shortages, consumer queues, inefficiency of the distribution system and widespread corruption.
Broader Economic inflation (#PC0254).

♦ **PF0963 Inadequate procedures for community planning**
Claim Although social planning has increasingly enabled rural communities to cope with immediate needs and long-range plans, previous plans often have false starts that have hardened social unease. Local companies may have planned an array of other industries to be located around the edge of the town, and some small plants may have been set up – but not much more. A comprehensive development of lakes with recreation facilities may have proved unfeasible, since the water was needed extensively to irrigate farm lands downstream. Such failures will indicate to the town leaders that they themselves are depended on to create a new economy and, although they may be able men and women, they may feel new to such a role.
Total responsibility for a town must depend on the participation of everyone to assess the needs and to meet them, whether in the fields of medicine, housing, snow removal, farming or social life. The scope of decisions thus broadens into more complex arenas with seemingly insurmountable issues, yet broader participation in an effective consensus-making process may not be achieved. Regular meetings and neighbourhood discussions are not easily scheduled. Without a thorough consensus, proposals and petitions are often ineffective.
Broader Obstacles to community achievement (#PF7118).
Narrower Fluid community patterns (#PJ1850) Entrapping social context (#PG8218)
Unassessed corporate needs (#PG8220) Fragmented decision-making (#PF8448)
Hindered winter sociability (#PG8211) Ineffective official liaison (#PG8212)
Uncoordinated joint planning (#PG8217) Insufficient industrial promotion (#PJ8378)
Infrequent discussion opportunities (#PG8230) Unconceived corporate responsibility (#PG8214)
Unconsensed institution desirability (#PG8223)
Lack of means for achieving consensus (#PD2438).
Related Lack of local leadership role models (#PF6479)
Inadequate welfare services for the aged (#PD0512).
Aggravates Lack of community development (#PF7912).
Aggravated by Lack of social planning (#PF8185).

♦ **PF0966 Lack of meaningful educational context for ethical decisions**
Claim Education today is aimed at the acquisition of facts and skills and does not provide the experience of ambiguity necessary in the individual's struggle to develop an ethical posture. Rather, ethics has become so much the province of the individual that each person's posture is developed alone, without reference to the larger society or to the wisdom of the past in any organized form.
Broader Inadequate education (#PF4984)
Collapse in providing ethical value screens (#PF1114).
Aggravates Ethical decay (#PB2480) Lack of commitment (#PF1729)
Fragmented decision-making (#PF8448) Fragmented ethical contexts (#PF2096)
Deluding familial image of social responsibility (#PF1064).
Aggravated by Individualism (#PF8393)
Lack of a system for ethical decision-making (#PF2070)
Educational curricula over emphasizing method rather than content (#PD1827).

♦ **PF0967 Lack of motivating collegial relationships**
Nature An absence of objective, task-oriented collegial groups prevents people who have come together to meet the needs of others from moving beyond psychologically defined relationships. These are often of short duration and, while well-intentioned, are incapable of really accomplishing the goals they were intended to attain. The emotional quality of the group often becomes more important than the goal of the group. Methodologies are needed which are capable of equipping leaders to be enabling presences to their fellow participants, and of providing them with the skills necessary to hold members accountable for their decisions and tasks without blocking participation and commitment. They also need techniques to help deal with times of trouble.
Aggravates Educational curricula based on content rather than method (#PF3549).

♦ **PF0977 Limited acceptance of international treaties**
Delays in ratification of international treaties — Non-ratification of international treaties
Nature Prompt acceptance of a treaty by some States may have a positive impact, and delay by some could have a negative impact, on other States. If a treaty does not receive a sizeable number of acceptances within a reasonable period, states might lose interest in it. Although almost all multilateral United Nations treaties have been adopted by very large majorities, and a number of them unanimously or at least without any negative votes, some of them have been accepted by only a minority of States. Non-acceptance or undue delay in acceptance of treaties has deep consequences on the efficacy of codification and the development of international law. These attitudes can retard the entry into force of a treaty, and may even result in the failure of a treaty to come into effect at all. It is noteworthy that even States that have ratified or acceded to a treaty not yet in force are freed from the obligation to refrain from acts which would defeat its object and purposes, if the treaty's entry into force is unduly delayed. Even after a treaty has entered into force, its true effectiveness may be impaired if it applies only between a small number of countries. For all these and other reasons, prompt acceptance by the widest possible number of States, with least possible delay, is of crucial significance.
Incidence In the period up to June, 1968, of the 179 multilateral treaties for which the Secretary-General exercises depositary functions, only 138 had entered into force. The general multilateral treaties adopted under UN auspices are open for acceptance by States Members of the United Nations, and by States not Members of the UN but which are Members of any specialized agency; acceptance is not limited in time (so that the date at which a State becomes a Member does not affect its capacity to adhere to earlier UN treaties).
As another example, considering the General Secretariat of Organization of American States in its aspect as depositary of inter-American treaties and agreements and of the instruments of ratification: with respect to the 53 treaties and their instruments of ratification, only 43 had entered into force as of July 1984.
Counter-claim The assertion that the making of a multilateral UN treaty which obtains few acceptances, or for a long time fails to come into effect, has been an exercise in futility, need not be correct. Depending on its subject matter, such a treaty may influence attitudes and policies of governments; give impetus to national legislation or the adoption of bilateral treaties; and have an impact on public opinion. Reference could also be made to the right of certain ILO bodies to examine whether Members endeavour to foster standards set by ILO Conventions not ratified by them and to the tendency of some General Assembly resolutions to demand that Members abide

by the principles of Conventions not ratified by them.
Broader Aggressive foreign policy (#PC4667).
Narrower Limited acceptance of human rights treaties (#PE7300)
Failure of disarmament and arms control efforts (#PF0013).
Related Subversion of international agreements (#PD5876)
Non–verifiability of compliance with nuclear arms treaties (#PF4460).
Aggravates Unfairly negotiated treaties (#PF4787)
Disagreement within alliances (#PD2629)
Unfulfilled treaty obligations (#PF2497)
Ineffective international agreements (#PF6992)
Internationally non–cooperative governments (#PF9474)
Government refusal to accept the jurisdiction of international courts of justice (#PF7897).
Aggravated by Inadequate international law enforcement (#PF8421)
Covert violation of international treaties (#PD8465).

♦ **PF0979 Defensive life stance**
Self–actualizing stance of being a victim to external forces — Individual victim image — Economic victim image
Nature A personal life style of being unable to effect change in a given situation leads to a stoical acceptance of the status quo. As a result of this stance of helplessness and ineffectivity, self–identifying activities turn towards conformity, to endeavour for personal enrichment, or to defence of personal security.
Claim For the individual in the 20th Century, the future appears to have such overwhelming complexity that he or she feels helpless in dealing with its apparent chaos. This results in a self–image of being inadequate, and the victim of an external system which is choking one's very existence. There is no personal vision of being an essential participant in the deliberative system; and there seems to be no possibility of contributing creatively to decision–making processes, nor of gaining just consideration from them. This creates a mood of apathy towards these systems. The individual sees himself as a victim of circumstances, the government, the community, existence and his race, poverty or sex.
Broader Deteriorating quality of life (#PF7142)
Collapsed meaning of human creativity (#PF0936).
Narrower Lack of commitment (#PF1729)   Individual isolationism (#PD1749)
Disabling inadequacy feelings (#PJ9012)   Tensionless image of free choice (#PF1675)
Loss of the past as an operating context (#PF1062)
Limiting of responsibility to the personal (#PF1889)
Social reinforcement of shallow personal meaning (#PE4589).
Related Low self esteem (#PF5354)   Lack of self–confidence (#PF0879)
Limits on areas of research (#PF2529)
Collapse in providing ethical value screens (#PF1114)
Educational curricula based on content rather than method (#PF3549).
Aggravates Psychological inhibition (#PF6339).
Aggravated by Inflexible central government (#PD1061)
Limited development of functional abilities (#PF1332)
Marketing skills which reinforce a self image of being a victim of circumstances (#PE7752).

♦ **PF0980 Monopoly of competence for intervention in the future**
Monopoly of knowledge to predict the future
Claim Leaders tend to endeavour to maintain a monopoly of power over directing the future of their people, endeavouring to inspire their followers to accept a view of a perfect world which awaits them, whilst ensuring that the people remain totally dependent on the leadership to get them there. Leaders, especially autocratic leaders, reveal to their followers not more than the immediate next step in any changes towards the better society. The leaders are thus free to propose any course of action as being part of the unrevealed plan required to achieve the ends which all are agreed are desirable.
Broader Monopolization of knowledge (#PF5329).
Aggravates Temporal deprivation (#PF4644)   Manipulative knowledge (#PF1609).
Aggravated by Political dictatorship (#PC0845).

♦ **PF0981 Representative arts**
Pictorial arts — Depiction of human figures — Portraiture — Sculpture — Graven images
Nature It is through the fine arts that man portrays in material and concrete shapes his total personality. For the lovers of pictorial arts, pictures are no less meaningful than divine texts. The artist interprets the physical into the spiritual, aspiring to catch beauty and put soul into mute lines, elevating himself to contemplation of his creative efforts and enabling others to share with him the pleasing, restful feeling. The inherent danger of this very hypnotic and obsessive quality of artistic beauty is that it may not always be healthy either for the artist himself or for humanity in general. It may be fractional, partial or even perverted, as instanced by the classic focus on the naked human body. While Islam uses representative arts for utilitarian purposes (identity photographs, surgical diagrams), the sense of beauty and artistry is entirely channelled into prayer, especially congregational worship, rather than risking its dissipation in distracting imagery and the further risk of perpetuating its failures and inadequacies in material record. Human portraits in the form of sculpture, painting or photography demean the dignity of the human personality by isolating its carnal features, separated from the human faith and beliefs that are its true meaning. It is thus a prostitution of human thoughts, emotions and activities to waste them for any other function than honest, sincere and actual achievement of the serious purposes of life.
Incidence Of all the living religions, only Islam bans the representative arts. However certain puritanical Christian sects have strong reservations about the use of what amount to graven images.
Claim The rejection of absolute spiritual and aesthetic values and the adoption of the concept that everything must continually change with the changing times, so that there is really nothing of enduring worth, is responsible for the vulgarity of modern art. The highest "art" is ceaseless and unrelenting striving for the perfection of human character in real life in preparation for the life hereafter. Any pursuit must be condemned which distracts people's attention from that end. Modern attempts to distinguish worship from appreciation are pointless, because worship itself is nothing but ritualized admiration and appreciation.
Counter–claim Beauty offers an ideal for human feeling. Art is one of the higher ends of life, if not the highest. It is sufficient if art offers pleasant gratification and is perfect in its own taste.
Narrower Performing arts (#PF1756).
Aggravates Idolatry (#PF3374)   Art as propaganda (#PF3087)
Incongruous religious images (#PG1418).
Aggravated by Indecent art (#PE5042)   Debasement of works of art (#PE0558).

♦ **PF0995 Prohibitive legal fees**
Outrageous legal fees — Soaring cost of legal aid — Risk of incurring legal expenses — Extortionate legal costs
Nature Due to the need to pay lawyers' fees, economic factors prevent many from protecting their rights before the civil courts or administrative tribunals and prevent many accused from properly defending themselves. These factors apply afresh if appeal is envisaged. Furthermore, the very constitutional provisions which contain guarantees concerning aspects of a fair hearing lose effectiveness if an individual is prevented by these financial factors from going to court to avail himself of those guarantees.

Such economic problems are aggravated where there is incomplete or no statutory or other limitation on the fees which may be charged for various types of legal services. This may put litigation beyond the reach of some persons; and in other cases, a litigant may be able to command only the advice of counsel of a lesser competence. The very uncertainty of the cost of a legal action often prevents poor persons from enforcing their rights or compels them to accept a compromise which may not be just.
The cost of litigation is increased in a few countries by the fact that one category of lawyer has a monopoly of advocacy before the higher courts while the professional services of this category may be engaged by the individual only through his first approaching a second category of lawyer. This necessitates the payment of fees to lawyers of both types even if the case is to be heard only in the higher courts. The cost may be particularly high where a case has already been conducted in the lower courts by a lawyer of the second category and the pleading has then to be taken over by one of the first category in order to conduct an appeal.
Public interest demands that in criminal cases the accused should be legally represented. This stems from the practices and procedures of the courts. In an adversarial system, where the judges do not directly determine the breadth and scope of the evidence, and are there mainly to control the forensic process and adjudicate between disputants, the lawyers themselves are primarily to blame for the growing length of time needed to hear cases and the accompanying rise in costs.
The cost to tax payers to clear up crime is extremely high. For example, in the United Kingdom once detectives are called in it costs £550 per case.
Broader Risk (#PF7580)   Underprovision of basic services to rural areas (#PF2875)
Economic discrimination in the administration of justice (#PE1399).
Related Inadequate regulation of restrictive business practices in service industries (#PF0591).
Aggravates Inaccessibility of justice (#PD8334)   Risk of capital investment (#PF6572)
Excessive cost of effective prosecution of offenders (#PE6059).
Aggravated by Unethical practices in the legal profession (#PD5380).

♦ **PF0996 Manipulation of civic education**
Nature All education, beginning with family education, has the task of socializing children and adolescents. Schools undertake civic education (especially in newly independent countries or those which have recently recovered their national identity) and ideological education (in countries where leaders of a revolution consider it their duty to capture minds and to topple all the bastions of the past). Such education may, however, aim at conditioning individuals in order to make them follow set examples and become easy to govern. It may cultivate unthinking respect for hierarchies.
Broader Manipulation (#PA6359)   Inadequate education (#PF4984).

♦ **PF1000 Inadequate family structures**
Reduced interior structures of marriage
Nature The family as a social institution regulates sexual intercourse, assigns responsibility for children, conserves lines of descent, and orders wealth and inheritance. It assigns roles for the division of labour for everyday living, supports the roles of its members in the external economy, participates with other institutions in the socialization of the coming generation, and plays a role in the physical and psychological welfare of family members. The family both forms and expresses the identity and character of its members. The processes of urbanization, the destruction of the extended family and the secularization of society are working against the family playing these crucial roles in society.
Claim The interior structures of 20th century families no longer enable the family to perform its basic social role. Domestic structures have been reduced to emphasizing only the goal of material security and physical well–being.
Broader Reduced interior structure of families (#PF3783).
Narrower Fragmented patterns of extended family relationships (#PF1509).
Aggravates Domestic quarrels (#PE4021).

♦ **PF1001 Undervaluation of public assets**
Undervaluation of state enterprises
Aggravates Asset stripping (#PE9224).
Aggravated by Privatization of public services (#PE3391)
Corruption of government leaders (#PC7587)
Inefficiency of state–controlled enterprises (#PD5642).

♦ **PF1004 Shallow approaches to motivational techniques**
Inability to awaken consciousness due to inadequate methods
Claim Motivational technicians are not equipping people with the tools necessary to kindle any deep awareness of the necessity for, or meaning in, the roles in work and leisure of individuals and groups. In contrast with past values (the "my country do or die" approach), little or nothing seems worth risking one's life for. Because of this lack of motivation, the whole of society languishes without an adequate sense of responsibility. Nevertheless, individuals do seem to be seeking an engagement which extends beyond the immediate self–imperatives. The issue is therefore not so much what is being done, but how to put what is being done in a motivating form.
Broader Ineffective structures for community participation (#PF2437).
Aggravates Educational curricula based on content rather than method (#PF3549).

♦ **PF1006 Refusal of families to participate globally**
Nature In the complex technological society of today, the family is kept from authentic global participation by its necessary concern to guard its own security. By avoiding contact with other cultural groups in the community or further afield, the family avoids any suffering or personal loss that might increase fragmentation of social relationships which are already very difficult to maintain.
Broader Lost family role in society (#PF7456).
Related Escapist family life styles (#PD4069).

♦ **PF1009 Ostracism**
Ostracized people
Narrower Boycott (#PE8313).
Aggravates Outlaws (#PE4409)   Fear of ostracism (#PF2776).
Aggravated by Social outcasts (#PD6017)   Social invisibility (#PD8204).

♦ **PF1010 Inadequate implementation of plans and programmes against problems**
Ineffective implementation methods of decision–making — Uncertain planning implementation — Failure to implement policy directives
Nature The execution and management of plans, as distinct from plan formulation, make the severest demands on scarce managerial resources, particularly in the developing countries. The wide discrepancy between objectives and achievements at the end of the plan period (the distinction is limited to the industrial programmes) has been due largely to failures in sectoral coordination, effective direction of the plan and adequate supply of trained personnel. In some instances, planned targets may have over–estimated existing capacities, but, allowing for this factor, the skill in assembling resources and ensuring that the execution of the various compo-

nents of a plan are correctly phased has a great deal to do with the final result.
Broader Unstructured local decision-making (#PF6550)
Inaccessible administrative agencies (#PF2261)
General obstacles to problem alleviation (#PF0631).
Narrower Inappropriate management of development projects (#PD3712).
Aggravated by Over-programming (#PF7737) Policy-making errors (#PF0531)
Inadequate planning of action against problems (#PF1467)
Inadequate power of intergovernmental organizations (#PF9175)
Failure to integrate knowledge to empower humanity in response to the global problematique (#PF8753).

♦ PF1015 Post-revolutionary re-employment by government of security services of the ousted repressive regime

Nature When a country succeeds in the transition to a more just and democratic form of government, it is obliged to deal with those who were agents of the repressive regime, possibly as torturers and agents of assassination (and other forms of abuse). In those countries which have traditionally relied on a strong secret police force, there is considerable pressure to make use of those from the previous regime with such experience.
Related Unconvicted war criminals (#PD4067)
Government approved employment of war criminals (#PE4697)
Employment of criminals in policy-making contexts (#PE4439)
Remission of sentences for crimes against humanity (#PF1098).

♦ PF1020 Ineffective population control

Nature Population policies and programmes in many of the Third World countries are failing because of lack of participation of the majority in their respective communities. This has rendered ineffective the information campaigns mounted on family planning and on the variety of contraceptives that are available to citizens. A revision of governmental population policies to include community participation as a vital component is therefore necessary, as is the involvement of local leaders in the implementation of programmes at the local level, since they are in a better position to allay fears concerning fertility. This raises a further concern that, while most community leaders assume positive attitudes, some are opposed on moral grounds; and also certain socio-economic factors serve as deterrents.
Claim Governments, educators, and the media, as well as individuals, must become motivated in addressing both problems of overpopulation and the mental and physical problems that need to be overcome before measures can be successfully implemented.
Counter-claim Among reasons against population control given in developing countries are that: an increased population is all that is needed to spark off rapid development; a complex web of traditional customs makes large families essential (a new baby is viewed as an asset to the entire community, a child is often named after both departed and living relatives, thus assuring that the relatives' presence will survive); extra hands are needed for work; family planning is a plot by Western governments to hamper Third World growth and development; people will be forced to have fewer children; there is great paternal pride in having numerous offspring; and a man's standing in the community is enhanced by the number of children of which he can boast.
Population control is totally unnecessary in light of the trend of the human species having grown from a few hundred thousands in wretched poverty to 5 billion in relative wealth all the while living longer.
Narrower Religious opposition to population control (#PF1022)
Government opposition to population control (#PF1023).
Aggravates Scarcity (#PA5984) Malnutrition (#PB1498)
Underdevelopment (#PB0206) Socio-economic poverty (#PB0388)
Rapid population growth (#PD9153) Unsustainable population levels (#PB0035)
High human fertility in developing countries (#PF0906)
Imbalance of population growth between developed and developing countries (#PE4241).
Aggravated by Inadequacy of contraceptive methods (#PD0093)
Maldistribution of world population (#PF0167).
Reduces Underpopulation (#PD5432).
Reduced by Declining birth rate (#PD2118).

♦ PF1021 Opposition to population control and family planning

Discrimination in favour of childbearing
Nature Opposition to population control may come from government and religious authorities; those at whom population programmes are directed; those who see population control as a means of political oppression, either foreign or national; and those who have no belief in the efficacy of any population programmes and who tend to believe that population will be more readily stabilized by natural disaster, as for example by famine or war.
Some of the primary obstacles to family planning in various areas are: insecurity and conservatism of families owing to high levels of mortality, particularly infant mortality; socio-cultural traditions and values whereby the status of women is determined by their fertility; high values placed on the birth of sons or on large families so as to ensure support and security in illness and old age; and deficiency or lack of health care and other supportive social services, limited family resources, and related socio-economic and educational considerations.
Many efforts in family planning have started not as health programmes but as population control programmes focused on goals and targets expressed as numbers of 'acceptors' rather than on providing family planning care as a vital health measure within the context of health services. In addition, in some areas the methods included in family planning services are too difficult or unpleasant for regular use by individuals who are not highly interested in or concerned about planning their families; and the major and minor side effects of many methods discourage regular use. Inadequate provision of easily accessible health service of good quality for follow-up of these conditions has served as a deterrent to increased participation by the people concerned, particularly in rural areas.
Many cultural beliefs, values, and traditional practices are antithetical to family planning practices. In a society where a wife may be returned to her family if she does not bear several children, couples are not likely to adopt measures to space their children. Likewise, in a society where marriage occurs at an early age and the birth of a child within a reasonable period after marriage determines the suitability of the wife, or where demands to produce male offspring are paramount, measures to delay the birth of the first child or to adopt measures for child spacing before one or more male children are born are not readily accepted and practised.
Closely allied to the cultural barriers are various beliefs and feelings of morality relative to sexual behaviour and interference with the procreative function. For example, in situations where pregnancy outside wedlock is considered immoral but many unmarried women, especially teenagers, are becoming pregnant, education in ways to prevent such pregnancies is unusually complicated and delicate or even forbidden.
The whole subject of family planning is surrounded by emotional overtones. In almost all societies, sex, pregnancy, and reproduction are associated with certain taboos. Males may discuss certain aspects with males, and females with females, but there are certain subjects in almost all cultures that cannot be discussed in mixed groups or even at all. Individuals who are influenced by these taboos may not feel free to discuss family planning with their spouse or their children. They are likely to be even more reticent in discussing the subject with a health worker who is a stranger, particularly one of the opposite sex.
There may also be resistance or indifference on the part of both professional and auxiliary health workers. While many obstetricians and gynaecologists are helpful, there are some who insist that their duty is to bring children into the world and not to engage in family planning care service, and this makes the educational task more difficult. Similarly, midwives and birth attendants may in some instances perceive child spacing as reducing the need for their services and the amount of their income. A closely allied factor is the confusing advice that may emanate from different programme sources or from different physicians. Because of the rapid developments in the field of family planning, particularly during the past 10–15 years, and the new contraceptives being produced, different organizations as well as different physicians may recommend different action. Such divergent advice can easily discourage individuals in need of family planning care services from taking and implementing decisions.
There is often confusion regarding the purposes for which family planning care services are offered. The health reasons for delay and spacing of conception may be obscured when the services are seen as measures to limit the size of families. This confusion is reinforced further when family planning is not viewed in the context of maternal and child health and other health services, such as education for family life and responsible parenthood, genetic and premarital counselling, nutrition, prevention of illegally induced abortions, and management of infertility.
In some countries, people have avoided family planning services because they perceived them as an impersonal governmental programme with little or no concern for the people's health and welfare. Health workers whose religion forbids certain forms of interference with conception are often in conflict when their tasks include family planning care services and counselling of individuals.
Claim A recent world survey showed that in 20 countries the percentage of women who only had two children but who wanted no more ranged from 30 percent to 68 percent. In 19 countries, in the case of women who had three children at least 50 percent of those surveyed wanted no more children. If these unwanted births were prevented, population growth rates could be halved in Bangladesh, Colombia, the Dominican Republic, Guyana, Jamaica, Peru and Sri Lanka. In the case of Sri Lanka, Jamaica and Colombia, a reduction on this scale would eventually reduce the population growth rate to the level obtained in the developed world.
Counter-claim Population control programmes deny to a woman the right to have the children she wants.
Narrower Religious opposition to population control (#PF1022)
Government opposition to population control (#PF1023).
Aggravates Lack of family planning (#PF0148) Unsustainable population levels (#PB0035).
Aggravated by Forced motherhood (#PE5919) Lack of family planning information (#PD1050)
Inadequate family planning education (#PD1039)
Inadequacy of male contraceptive methods (#PF1069).
Reduces Birth prevention (#PE3286).

♦ PF1022 Religious opposition to population control

Religious opposition to contraception — Religious opposition to abortion
Nature Religious opposition to population control in practice mainly takes the form of opposition to family planning and artificial contraceptive methods. Its effect tends to encourage elitism in development since statistics show that even where religious taboos on contraception exist, it is nevertheless widely practised in developed countries and among the rich in developing countries, and birth rates are on a par with those of societies where there is no religious taboo. Those who follow religious doctrine on this point are mainly the poor and uneducated. Religious opposition to population control may be encouraged or discouraged by government policy depending on control and depending on how strong the religious influence is in politics. Religious opposition may already be built into the law, not only regarding birth control but also into family and marriage law, age of majority, education, status of women and other laws which affect the use of birth control.
Counter-claim There is no religious opposition to family planning or population control and certainly no opposition to contraception in the sense of preventing pregnancies. Buddhism, Hinduism, Islam, Roman Catholicism, Eastern Orthodox Christianity and Protestantism all agree with the use of total or temporary abstinence to prevent pregnancy. All also agree that one or, in some cases, the function of marriage is to procreate. Religious traditions concerned with family planning are opposed to the use of artificial contraceptive methods, and particularly for the convenience of the adults.
Broader Ineffective population control (#PF1020)
Opposition to population control and family planning (#PF1021).
Related Religious opposition to public health practices (#PF3838).
Aggravates Criminalization of abortion (#PF6169)
Unsustainable population levels (#PB0035)
Discrimination in family planning facilities (#PD1036)
Alienation of support for international organizations and programmes (#PD1809).
Aggravated by Theological justification of population growth (#PF9671).

♦ PF1023 Government opposition to population control

Nature Outright government opposition to population control is broadly based on three factors: religious influence, military considerations and economic expansion. Whenever government rejection of population control occurs it calls into question the attempts made by other nations, creating international tension and the potential risk of war. Apart from this, it overrides ecological and standard-of-living considerations and encourages elitism. Milder opposition may form part of a self-determination effort against population programmes devised by other nations, and from underpopulation in comparative terms.
Counter-claim Never before have so many governments asserted their desire to adopt population policies. More than four out of every five persons in the less developed regions of the world live in countries whose governments favour policies aimed at reducing the rates of population increase. Two out of every three people throughout the world are ruled by governments which regard demographic change as a significant factor affecting the nature, scope or pace of social and economic development. 94 less developed countries support, directly or indirectly, the provision of facilities for fertility regulation.
Broader Ineffective population control (#PF1020)
Opposition to population control and family planning (#PF1021).
Aggravates International insecurity (#PB0009) Unsustainable population levels (#PB0035)
Negative effects of family allowances (#PF0107)
Discrimination in family planning facilities (#PD1036).

♦ PF1025 Sales-dominated market research

Disrelated needs and promotion in marketing — Sales promotion dominates market research
Nature Sales promotion has become a dominant force, relegating market research to providing necessary data. Producers rely on sales promotion to sell more of the same type of products to the same limited market regardless of need. For instance, huge amounts of advertising money is used to sell forty different kinds of soap.
The organs of sales and of sales promotion override and ignore the advice of market research departments. The larger view of client's and society's needs and desires, the competition's weakness and the company's capacities is lost to narrow turnover considerations.
Claim Increased use of sales promotion is needed to sell more products to more people in direct relationship to need.
Broader Limited market development (#PF1086).

♦ PF1030 Introversion of the family unit
Individual family priority
Claim The family, which has been developed as a means of relating individuals to the larger society and ultimately to the world, has in modern times become increasingly introverted and is consequently collapsing in its utility as a social institution. The nature of the relationships between members is often vague and easily changed or broken. In addition, the function of the family as a social form is unclear as it relates to present-day issues, and it is often reduced to simply maintaining itself. No such social form exists for long.
Broader Lack of commitment (#PF1729)   Untransposed community structures (#PF6450).

♦ PF1031 Ineffectiveness of aid to developing countries
Nature In many aid-giving countries there is a persistent suspicion that aid objectives are ill-chosen and make little impact on economic and social conditions. In the countries which receive aid, there is also a good deal of dissatisfaction with the efficiency and honesty of the aid system; for example, when aid does not reach the purpose for which it was intended there is seldom an investigatory follow-up to determine accountability.
Claim Foreign aid programmes may in large part be political, to stabilize regimes friendly to donors, whether these donors are collective (such as OECD) or unilateral (such as the USA). As such, they are considered effective political programmes for the donors, who are not necessarily interested in effective development programmes in recipient lands, since effective aid runs the danger of raising the consciousness and expectations of the oppressed. This explains why so much foreign aid money goes into the pockets of the politically powerful, the local banks, the family-owned corporations, and the large-landowners, all of whom prop up the stability of regimes that impoverish their citizenry. New enterprises and new owners are feared by the traditional power-base and aid is preferred that keeps the developing countries dependent.
Counter-claim Some aid does get stolen; some goods are lost in transit; some imports languish in warehouses and are never utilized. It would be surprising if this were not so in an endeavour where fifteen countries are transferring more than US $6,000 million a year of goods and services to some eighty recipient countries in an avowedly experimental attempt to stimulate growth. There is some waste in every programme, public or private; this is part of the nature of man. There are numerous accounts of waste in aid programs, as in government domestic programs, or in national or international corporations. While all waste is to be condemned, the question is whether the proportion of waste is too high, or so high as to invalidate the programme. Evidence suggests that an extensive amount of aid has not been dissipated through abnormal waste, mismanagement, or corruption. Occasional incidents of improper conduct in aid administrations, or of misappropriation of funds in receiving countries, do not warrant the charge that aid resources are recklessly wasted.
Refs Baskin, V *Aid Offered by the West.*
Broader Lack of appreciation of cultural differences (#PF2679).
Narrower Inadequacy of food aid (#PF3949).
Aggravates Compassion fatigue (#PF2819)
Reallocation of aid funds to alternative priorities (#PF0648).
Aggravated by Mismanagement of aid to developing countries (#PF0175).

♦ PF1033 Over-formalised decision-making
Claim Effective counsel is obstructed when the one giving counsel decides where, how and on whose behalf decisions are to be made. The justification myth, based on the premise that no decision can be made until all the data is in and the single right-way is discerned, prevents informal confidants from establishing priorities and moving towards making decisions which anticipate the future.
Broader Narrow context for counsel (#PF0823).

♦ PF1035 Sexual discrimination in contraceptive methods
Claim Most contraceptive methods are designed by men for women, partly as a result of the traditional male-female relationship and partly due to the natural consequences of no birth control. There is so far no method which is 100 per cent satisfactory either in controlling pregnancy or in its lack of side-effects. While the consideration of loss of libido is a strong force in withholding the marketing of male contraceptives, this has never been considered as a valid reason against the use of female contraceptives. Women are also expected to tolerate the unpleasant physical side-effects of the pill or the IUD. Discrimination in research promotion and usage of different methods also exists between rich and poor women, and between women of different races.
Broader Discrimination against women (#PC0308)
Discrimination in family planning facilities (#PD1036).
Narrower Inadequacy of male contraceptive methods (#PF1069).
Related Inadequacy of contraceptive methods (#PD0093).
Aggravates Compulsory sterilization (#PF3240)
Mutilation and deformation of the human body (#PD2559).
Aggravated by Male domination (#PC3024).

♦ PF1040 Ideological overemphasis in economic administration
Nature Governments today use vast resources to administer their economies in ways that will demonstrate their particular ideological bias. Ponderous bureaucracies contain obsolete, overlapping agencies with no way of relating their tasks to long-range objectives. The lack of specific and inclusive accountability leads to misdirection and meaningless evaluation. The result is the short-sighted cessation of significant activities and the senseless perpetuation of worthless ones.
Broader Monopoly of the economy by corporations (#PD3003).

♦ PF1042 Failure of motivating socio-dramas
Nature The socio-dramas which motivated members of society by rehearsing past events, encouraging social cohesion and marking periods of transition no longer function effectively. Events like inaugurations and independence day celebrations have been rendered inconsequential, and new events have not emerged. People need opportunities for frequent and self-conscious participation in such events to rehearse the internal meaning of their lives. This both encourages them toward greater effort in the future and gives meaning to past efforts. Without this recording and meaning, people merely exist in the community, and feel little responsibility for it or their places in it.
Broader Symbol system failure (#PF3715).
Aggravates Educational curricula based on content rather than method (#PF3549).

♦ PF1055 Failure of global-scale planning for expertise development
Nature There is an apparent failing in the methods that are used for planning the use and development of skilled people to the greatest advantage of society: the training of specialists is uncoordinated on the international level; specialist services are isolated from the needs of people in a global context; and methods for future planning are inadequate.
Broader Individualistic utilization of expertise (#PF5639).
Aggravates Unavailability of appropriate expertise (#PF7916).

♦ PF1058 Immediate gratification-based social forms
Claim When people discover that they cannot face the absurdity of life, they build social relations that attempt to deny the chaos. The clarity with which the innumerable problems of the world are perceived favours the imposition of limited and hedonistic social forms in an attempt to maintain sanity in the face of looming insanity. Because they lack a comprehensive view of life, these social models deal with immediate external realities and fail to touch the human questions below the visible problems.
Broader Theological collapse (#PF6358)
Short range planning for long-term development (#PF5660).

♦ PF1062 Loss of the past as an operating context
Confusion through being detached from the past
Claim In the area of individual integrity, the loss of an awareness of the past as a helpful operating context has left people without anchoring foundations from which to act when dealing either with the complexities of the future or with the ambiguities of the present. They flounder in a universe of relative values, seeing little depth to life. This apparently leaves only two options open: revolt against the social structures that are crushing individuality; or retreat into a view of the world in which they see themselves as victims of societal and psychological forces.
Broader Defensive life stance (#PF0979).

♦ PF1064 Deluding familial image of social responsibility
Absence of a common structure for ethical decision
Claim The image of social responsibility as care for one's own family significantly obstructs the development of genuine social responsibility. The family-based society in which such an image developed has been supplanted by societies based on a diversity of other social links, from employment to country of origin or language. Consequently, the individual is deceived by these images into a confused sense of what society is and how to care for it.
Broader Lack of commitment (#PF1729)   Fragmented decision-making (#PF8448).
Aggravated by Lack of meaningful educational context for ethical decisions (#PF0966).

♦ PF1065 Proliferation of unprocessed scientific data
Incidence USA's space agency NASA has received from its probes over the past 30 years six trillion (6,000,000,000,000) bytes of data. Its new probes will return at least a trillion bytes a week in the 1990s.
Broader Proliferation of information (#PC1298).
Aggravates Inadequate statistical information and data on problems (#PF0625).
Aggravated by Accumulation of knowledge (#PF2376).

♦ PF1066 Unilateral declarations of independence by extra-territorial bases
Extra-jurisdictional territories — Off-shore broadcasting vessels
Incidence Vessels can anchor outside the territorial limit in order to broadcast into a country without being constrained by national regulations or taxation. Abandoned installations (such as oil platforms or anti-aircraft platforms) constructed on the coastal shelf outside territorial limits may be taken over by individuals and groups and then declared independent of the country which originally built them.
Related Enclaves and exclaves (#PD2154).
Aggravates Violation of sovereignty by trans-border broadcasting (#PE0261)
Conflicting claims concerning off-shore territorial waters (#PD1628).

♦ PF1068 Impediments to internationally mobile professionals and experts
Broader Lack of international cooperation (#PF0817)
Restrictions on freedom of movement between countries (#PC0935).
Narrower Arrogance of experts (#PF3294)   Harassment of journalists (#PD3036).

♦ PF1069 Inadequacy of male contraceptive methods
Nature The inadequacy of male birth control methods results partly from the fact that it is women rather than men who have to cope with the natural consequences of no birth control, and partly because the question of loss of libido as a result of contraceptive techniques is given much more weight in the case of the male than in the case of the female. Inadequate female methods of birth control are much more easily tolerated by current social attitudes. Research into adequate male contraception is therefore minimal and, as a consequence, family planning techniques and education are aimed mainly at women and male participation is insufficient. In communities where male domination in the family is absolute, the chances of full success of family planning programmes are therefore considerably diminished.
Broader Inadequacy of contraceptive methods (#PD0093)
Sexual discrimination in contraceptive methods (#PF1035).
Aggravates Unsafe sex (#PE9776)   Unsustainable population levels (#PB0035)
Inadequate family planning education (#PD1039)
Inadequate family planning education for men (#PE9060)
Opposition to population control and family planning (#PF1021).
Reduced by Total sexual abstention (#PE3298).

♦ PF1070 Over-reliance on economic interest groups by policy agencies
Incidence The 1980s has seen a fragmentation of the international trading system with an ever-increasing number of trade issues being treated outside the the framework of GATT and a large and growing number of GATT rules themselves being circumvented. Such circumvention is often the reflection of the particular interests of domestic high-cost producers, thus attenuating the broader international benefits of a properly functioning trading system.
Claim Policy agencies which are designed to regulate economic structures for the common welfare have become dominated by interest groups from the very economic groups they were designed to regulate. The agencies are protected by legal procedures unrelated to present needs.
Broader Undemocratic policy-making (#PF8703).
Aggravates Trade protectionism (#PC4275)   Inappropriate policies (#PF5645)
Economic bias in development (#PF2997)
Fragmentation of the international trading system (#PC9584).
Aggravated by Disproportionate control of global economy by limited number of corporations (#PE0135).

♦ PF1072 Non-accountability of transnational enterprises
Nature Despite the considerable and transnational power which transnational enterprises possess they, unlike governments, are not directly accountable for their policies and actions to democratic electorates. Nor, unlike purely national firms, are the transnational corporations subject to control and regulation by a single authority which can aim at ensuring a maximum degree of harmony between their operations and the public interest.
Broader Transnational corporation imperialism (#PD5891)
Excessive power and independence of transnational corporations (#PD5807).
Aggravated by Concentration of power by transnational corporations (#PE0766).

♦ PF1073 Undue attachment to a social group
Socio-sentiment
Claim Undue attachment to a social group is more dangerous than that to a particular territory. For although socio-sentiments are slightly more expansive than geo-sentiments, they are more

difficult to understand and to detach from. Socio-sentiments seek to divide people so that they can exploit others.
Aggravates Racism (#PB1047) Sexism (#PC3432) Dogmatism (#PF6988)
Xenophobia (#PD4957) Imperialism (#PB0113) Caste system (#PC1968)
Class conflict (#PC1573) Age discrimination (#PC2541)
Abuse of traditional cultural expressions of peoples (#PE4054).
Aggravated by Undue attachment to territory (#PF3390).

♦ **PF1074 Ineffectiveness of international organizations and programmes**
Broader Inefficiency (#PB0843).
Narrower Inadequacy of intergovernmental decision-making process (#PF2876)
Ineffectiveness of the United Nations system of organizations (#PF1451)
Ineffectiveness of intergovernmental organization and programmes (#PF0074)
Inadequate budgetary coordination within the United Nations systems (#PE2820)
Ineffectiveness of international nongovernmental organizations and programmes (#PF1595)
Ineffectiveness of international organization and programme action at the country level (#PE9124).
Related Competition between international organizations for scarce resources (#PC1463).
Aggravates Attacks on peace forces (#PF6895) Endemic abortion of ewes (#PG1537)
Loss of credibility in international institutions (#PE8064).
Aggravated by Excessive television viewing (#PD1533)
Multiplicity of official languages (#PF6027)
Denial of right to grievance procedures (#PE2832)
Intergovernmental organization mismanagement (#PD6628)
Inadequate power of intergovernmental organizations (#PF9175)
Inadequate facilities for international organization action (#PD0929)
Inadequate organizational mechanism for international action (#PE8776)
Inadequate funding of international organizations and programmes (#PF0498)
North and South spheres of influence within UN and related agencies (#PE8332)
Inadequate coordination of international organizations and programmes (#PD0285)
Jurisdictional conflict and antagonism within international organizations (#PD0047)
Alienation of skilled and committed personnel from international organizations and programmes (#PE1553).

♦ **PF1075 Lack of individual rights to political asylum**
Denying sanctuary — Denial of right to political asylum
Nature Under international law, the protection granted by a state to a foreign citizen against his own state, is a right of the state not of the individual. The person for whom asylum may be established has no legal right to demand it, and the state has no obligation to grant it. Under such legislation, a person unable or unwilling to return to his country of origin owing to a well-founded fear of persecution for reasons of race, religion, nationality, membership in a particular social group, or for political opinion, is never certain that his request will be conceded definitively.
Broader Rejection of refugees (#PF3021).
Related Illegal immigration (#PD1928) Illegal movement across frontiers (#PC2367)
Denial of right to economic asylum (#PE6212) Denial of right to cultural asylum (#PE3450).
Aggravates Extradition refusal (#PF2645) Desecration of holy spaces (#PF6385).
Aggravated by Political repression (#PC1919).

♦ **PF1076 Cumbersome methods of policy formation**
Claim Methods of formulating and regulating policy are slow and cumbersome and rely on specialists and experts untrained in forming social consensus. These methods result in a patchwork of policies from independent agencies rather than integrated, long-range planning.
Broader Undemocratic policy-making (#PF8703).
Related Self-destructive government policy-making (#PF5061).
Aggravates Policy-making delays (#PF8989) Inappropriate policies (#PF5645).
Aggravated by Risk-aversion strategy (#PF4612).

♦ **PF1077 Inadequate research on problems**
Broader General obstacles to problem alleviation (#PF0631).
Narrower Imbalanced research activity (#PF0198)
Irrelevant scientific activity (#PF1202)
Unreliability of equipment and machinery (#PC2297)
Excessive emphasis on fashionable areas of research (#PF0059)
Inadequate research and development on problems of developing countries (#PF1120).
Related Underutilization of intellectual ability (#PF0100).
Aggravates Denial of rights to students (#PD6346)
Long-term change in atmospheric chemistry (#PF1234)
Inadequate models of socio-economic development (#PF9576)
Inadequate research on proposed solutions to problems (#PF1572).
Aggravated by Minimization of problems (#PF7186)
Inadequate information systems for international governmental decision-making processes (#PD0104).

♦ **PF1078 Subsistence life style**
Subsistence living cycle — Subsistence limits role — Subsistence farmer hero — Subsistence struggle mindset — Independent subsistence style — Reduced family life styles
Broader Disincentives against farming (#PD7536)
Underdeveloped sources of income expansion (#PF1345).
Narrower Scavenging for survival (#PG3589)
Subsistence approach to capital resources (#PF6530)
Limited horizons produced by survival living (#PF3161).
Aggravates Poor living conditions (#PD9156) Loss of cultural identity (#PF9005)
Unproductive subsistence agriculture (#PC0492)
Minimal opportunities for corporate activities (#PF2316).
Aggravated by Rigidly entrenched social traditions in rural areas (#PF1765).

♦ **PF1079 Non-acceptance of reality**
Denial of reality
Nature Such a reaction may include varying degrees of non-perception, mis-perception, non-recognition, non-understanding, or non-acceptance of certain realities in order to cope with otherwise unacceptable intra-psychic conflicts, feelings or memories. People also unconsciously resort to denial as a reaction to threats that they feel unable to cope with; such denial in turn detracts from and inhibits their potential ability to resolve problems realistically. Another aspect of non-acceptance of reality is the failure to comprehend the consequences of an event that has not only never been experienced personally, but is unimaginable because of its extraordinary magnitude.
Incidence Modern examples are the threats of nuclear, chemical and biological warfare. The difficulty of facing the massive and unprecedented nature of these threats may be so great that many people simply choose to ignore their existence, or meet the threat with the notion of 'it can't happen to us'.
Broader Mental illness (#PC0300) Avoidance of reality (#PF7414)
Negative emotions and attitudes (#PA7090).
Aggravates Inability to resolve problems realistically (#PF8435).
Aggravated by Fear (#PA6030).

♦ **PF1081 Absence of convincing symbols connecting the individual's life to the cosmos**
Claim There is an lack of symbolic ideas and images which would broaden the individual's horizons and stabilize his vision of the reason for existence. No contemporary mythology points to the inherent significance of the individual or gives a sense of what has lasting value. People today have difficulty recognizing their own human capabilities and significance, and have no comprehensive vision of their role in the past, present and future pageant of humanity.
Broader Apathy (#PA2360) Symbol system failure (#PF3715).
Aggravates Ignorance (#PA5568)
Lack of commitment to common symbols (#PE8814)
Educational curricula based on content rather than method (#PF3549).

♦ **PF1083 Doomsday syndrome**
Over-reaction against environmental hazards
Nature People, and particularly intellectuals, tend to over-react to environmental and other social problems, making use of incorrect data or fallacious arguments (whether deliberately or inadvertently) to promote environmental scares.
Claim Intellectuals use each other to verify the state of the world. They depend most upon the experience of members of their own social network, but may also find incidents credible if they are observed by friends of friends. The sense of subjective probability is even influenced by reports of experiences twice removed. Reality testing works quite well as long as the networks remain constant in size, but social networks are doubled or trebled when people are in their twenties and thirties and are then usually maintained intact until the aging process makes itself felt, without the people themselves being conscious of the expansion. Therefore the threatening events coming to their attention per month increase by a factor somewhere around three to ten, depending upon the degree of overlap in social networks. The impression is reinforced and magnified further when they hear that such an impression is the virtually unanimous experience of everyone they meet in their circle. If one adds to this the improvement in communications which ensures that news travels faster and is less likely to be lost in a terminated rumour chain, it becomes evident that it would be a statistical miracle to find an optimist.
Broader Alarmism (#PF4384).
Related Unrealistic environmentalism (#PF4510).

♦ **PF1084 Collapse of cultural dreams**
Incidence The United States once had a dream of saving the world and making it safe for democracy. England saw itself as the great civilizer of mankind. But the hopes they had were shattered beyond repair in numerous major world events. A key event has been the demoralizing war in Vietnam.
Claim People all want to live in a culture that they are proud of, and to have the traditions that their parents and ancestors had. Even more, they want to have a world in which their children and their children's children can have peace and happiness, and to transmit to them the richness and opportunity for full and rewarding lives. But the 20th Century shows no sign of hope that even the most meagre dream will be realized. The question is even being raised as to what difference it makes if this culture continues or not.
Broader Collapsed meaning of human creativity (#PF0936).
Related Lack of commitment to common symbols (#PE8814).

♦ **PF1085 Disproportionate allocation of resources to pet animals**
Luxury treatment of pets
Incidence In the USA in 1990, the $11 billion pet products industry included increasingly expensive items. For example a store offering mink coats for pet animals had a $1 million business in 1989, with some customers spending $4,000 to $5,000 per year. Pets receive hip replacements, chemotherapy, kidney transplants, and pacemakers.
Broader Unethical practices with domesticated animals (#PE4771).
Aggravates Kidnapping of pet animals (#PE0805).
Aggravated by Proliferation of pets (#PD2689).

♦ **PF1086 Limited market development**
Limited market resource — Unprofitable market practices — Limited marketing lines — Limited market research
Claim Speculative growth tends to be directed towards narrow and specialized markets for new and specialized products rather than towards long-range comprehensive research to meet global needs. There is tension between market research and speculative growth since the latter lacks comprehensive, long-range vision, and therefore fails to hold market research or sales. Speculative growth focuses on inventing more products for a relatively small proportion of the world's population rather than creatively developing plans for getting goods to the people who need them most.
Broader Self-interest driven investment (#PC2576).
Narrower Parochial market research (#PF6928) Archaic marketing methods (#PF6465)
Over-production of commodities (#PD1465) Sales-dominated market research (#PF1025)
Prohibitive cost of goods and services (#PD1891)
Surrendered control of marketing systems (#PF6533)
Limited local markets for goods and services (#PF2989).
Related Inadequate credit policies (#PF0245)
Variations in national forms of currency (#PF2574)
Self-defeating style of community planning (#PF6456)
Underdeveloped sources of income expansion (#PF1345)
Lack of economic and technical development (#PE8190)
Inadequate access to negotiation on employment and reward (#PD1958)
Inadequate local expertise in business practices in developing countries (#PE7313).
Aggravates Unexplored potential markets (#PF0581)
Transfer of business from small communities to larger towns (#PF6540).

♦ **PF1088 Fear of social anarchy**
Fear of social collapse into anarchy
Nature Those in established positions of power fear social anarchy. There are few if any social structures to hear dissent or conceptual frameworks to understand alternatives to present policy available to leadership. When dissent is perceived the tendency is to suppress it in order to maintain internal security. At the same time if the minority voice is not silenced there would be the possibility of social upheaval.
Counter-claim As John Calvin suggests, in a perfect world where humankind has not fallen into sin the restraints of civil government would be unnecessary but in the actual world even a bad government is better than none at all.
Broader Blocked minority opinion (#PD1140) Lack of local commercial services (#PF2009).

♦ **PF1094 Antiquated intellectual methods to appropriate human depths**
Nature Intellectual methods are the techniques of studying, teaching and thinking that are both consciously and unconsciously formalized in any culture. They are transferred from one generation to the next both in formal educational structures and in formal actualizing processes. There are few techniques by which one can group the concepts and facts necessary to make responsible decisions.

Claim Most people are using methods of thinking and study which do not allow for organizing, understanding and communicating the high daily input of data they receive. Nineteenth century or mediaeval approaches are used instead of post-modern methods. This results in most people being unable to relate themselves rationally, dynamically and with depth to the day-to-day experience of life. Thought patterns tend to be either mechanical and cause-and-effect oriented, or emotive. Those intellectual methods that have been developed to utilize both the irrational and the intuitive are generally not available.
Broader Inadequate intellectual methods (#PF7380).
Narrower Fragmentation of knowledge (#PF0944)
Faulty approaches in the teaching of intellectual methods (#PF0956)
Inadequacy of prevailing mental structures to challenge of human survival (#PF7713).
Related Undirected technological expansion (#PC1730)
Human wisdom unrelated to daily life (#PD1703).
Aggravates Educational curricula based on content rather than method (#PF3549).

♦ PF1097 Loss of confidence in government leaders
Citizen mistrust of politicians — Disenchantment with government — Disillusionment with government — Pervasive distrust of government — Mistrust of government — Mistrust of political structures
Nature The legitimacy of a government leader depends on the trust invested by the public. This trust in turn depends on the values of the society and the leader's perceived ability to live up to these values. Where this trust is violated, like President Nixon lying about Watergate, or where the values change, like Prime Minister Takeshita's involvement in the Recruit shares scandal.
Broader Credibility gap (#PB6314) Loss of leadership credibility (#PF9016)
Loss of institutional credibility (#PF1963).
Related Distrust (#PA8653) Lack of self-confidence (#PF0879).
Aggravates Suspicion of bureaucracy (#PF8335)
Reluctance to join in community action (#PF1735).
Aggravated by Hyperinflation (#PD7940) Economic inflation (#PC0254)
Institutional lying (#PD2686) Deception by government (#PD1893)
Governmental incompetence (#PF3953) Deceptive political proposals (#PE7680)
Mediocrity of government leaders (#PF3962) Corruption of government leaders (#PC7587)
Abusive distribution of political patronage (#PF8535)
Government leaders associated with sex scandals (#PE7937).

♦ PF1098 Remission of sentences for crimes against humanity
Politically motivated amnesty for military personnel convicted of atrocities — Absolving military personnel of homicide
Incidence Examples include remission of sentences for the military personnel responsible for the: torture of Algerians during their struggle for independence (France), 1968 My Lai massacre in Viet Nam (USA), forced disappearances of civilians (Argentina), killing of civilians in Ireland (UK). Following democratic restoration, a general amnesty was given in Uruguay, Brazil and Chile. In 1990 consideration was given by the new governments: to amnesty outgoing Sandinistas (Nicaragua), to a sweeping amnesty for the armed forces (El Salvador), not to prosecute crimes under the previous regime (Chile), to amnesty black activists convicted of violent crimes (South Africa), not to prosecute crimes committed by the secret police (East European countries).
Counter-claim Under an oppressive regime, everybody is in some sense an accomplice of the dictatorship. Amnesty is the least bad choice under difficult and disagreeable circumstances.
Broader Ineffective war crime prosecution (#PD1464)
Excessive leniency in sentencing of offenders (#PF4723).
Related Post-revolutionary re-employment by government of security services of the ousted repressive regime (#PF1015).
Aggravates Crimes against humanity (#PC1073).

♦ PF1099 Lost family heritage
Claim There are no structures in the family specifically intended for the task of creating and sustaining the style of the family as based on historical roots. Present family gatherings and functions are nebulous and grounded neither in a family heritage nor in deep family experience. There are no structures which establish and recall the heritage of the family, with the result that it has lost all sense of the continuity expressed throughout history.
Broader Ignorance of cultural heritage (#PF1985)
Reduced interior structure of families (#PF3783).

♦ PF1102 Messianic image of leadership
Claim It has been considered the norm to depend on one person as the head of state. This person makes the decisions and provides guidance for policy agencies and bureaucratic systems. Today this single leader can no longer encompass the increasingly intricate and complex social patterns which are present in society; the single voice prevents comprehensive policy-making by offering only one point of view.
Aggravates Undemocratic policy-making (#PF8703)
Leadership impaired by illness (#PF8387).

♦ PF1104 Elitist leadership
Broader Elitism (#PA1387).
Aggravates Elitist ruling classes (#PF4849).

♦ PF1105 Cumulative environmental impacts
Combinations of environmental impacts
Nature Cumulative impacts of human activity on the environment become important when sources of perturbation in the environment are grouped sufficiently closely in space or time such as to exceed the ecosystems natural ability to remove or dissipate the resultant disturbance.
Claim Among the most troublesome interactions between development and environment are those that involve cumulative impacts.
Broader Natural environment degradation (#PB5250).

♦ PF1106 Religious secrecy
Refs Bolle, Kees W (Ed) *Secrecy in Religions* (1987).
Broader Secrecy (#PA0005).

♦ PF1107 Underproductivity
Narrower Labour shortage (#PC0592) Unproductive dependents (#PC1420)
Underutilization of labour force (#PF6293) Non-productive members of society (#PF4000)
Non-productive athletic activities (#PF4202) Unequal distribution of fish catches (#PF4495)
Unequal distribution of meat production (#PD4322)
Underutilization of second-hand equipment (#PD1484)
Non-productive use of cattle and livestock (#PD1802)
High labour turnover in developing countries (#PF0907)
Socio-economically inactive rural population (#PF4470)
Unequal distribution of livestock production (#PF4490)
Declining productivity in socialist countries (#PF7610)
Low occupational mobility in developing countries (#PD1493)
Underdevelopment of food and live animal production (#PF2821)
Contempt for agricultural labour in developing countries (#PD1965)
External dependence and vulnerability of socialist countries (#PD1104).
Related Underemployment (#PB1860)
Unemployment in developing countries (#PD0176)
Economic and social underdevelopment (#PB0539)
Inequitable labour standards in developing countries (#PD0142).
Aggravates Lack of job satisfaction (#PF0171)
Vulnerability of developing countries to inflation (#PD0367).
Aggravated by Boredom (#PA7365) Common cold (#PE2412)
Unemployment (#PB0750) Overstaffing (#PG3316)
Motivational death (#PF1948) Injurious accidents (#PB0731)
Wasted waiting time (#PF1761) Economic uncertainty (#PF5817)
Occupational diseases (#PD0215) Industrial intimidation (#PC2939)
Nutritional deficiencies (#PC0382) Inadequate land drainage (#PD2269)
Denial of right of complaint (#PD7609) Foreign exchange restrictions (#PF3070)
Alcoholic intoxication at work (#PE2033) Bias against private enterprise (#PF1879)
Insufficient capital investment (#PF2852) Inadequate equipment maintenance (#PD1565)
Worker maladjustment to technology (#PC7041) Infection of industrial water supplies (#PJ6198)
Inadequate industrial retraining programmes (#PF4013)
Shortage of manpower in socialist countries (#PE8771)
Health risks to workers in agricultural and livestock production (#PE0524)
Shortage of adequately trained personnel to act against problems (#PF0559).

♦ PF1112 Escaping reality through popular psychological screens
Claim People attempt to shield themselves from life by using popular psychological ideas to explain their experiences. They see themselves and their relationships through popular and/or pseudo-psychological categories, and attempt by these means to find a rational explanation for the overwhelming irrationalities of life. They may seek to insulate themselves from the reality of life by limiting their horizons, by basing their lives on slogans which tightly bracket people and events (thus fragmenting their lives and historical contexts), and by trying to preserve a sense of balance by reducing their world into manageable units.
Broader Escapism (#PF7523) Theological collapse (#PF6358) Avoidance of reality (#PF7414).
Aggravates Health fraud (#PD9297).

♦ PF1113 Starving for attention
Dependency on attention — Attention seeking
Refs O'Neill, Sherry B *Starving for Attention* (1982).
Aggravated by Paternal negligence (#PD7297) Maternal deprivation (#PC0981).

♦ PF1114 Collapse in providing ethical value screens
Claim Radical changes in world view and life style are effecting a crisis in both having and providing effective sets of ethical values. Current educational and religious systems have led to the acceptance of short-term "faddish" collective and individualistic values. The following of narrow objectives leads to a patchwork morality and the context for action becomes parochial. Responses to real global needs are ineffective and noncommittal as individuals feel themselves tied to increasingly limited social relations. Such attitudes lead to dissipation of the energy directed at global issues in short-term or reduced projects; and they blunt awareness of the need to create real solutions by fostering the illusion that such needs have already been met. Relations with the community at large, in its diverse and dynamic form, are reduced, as social obligations become limited to one's nearest relations.
The prevalent attitude that social benefits should be guaranteed regardless of individual participation further erodes ethical value screens by encouraging licence or withdrawal.
There are no parental role-models, no historical roots, no precedents and no intellectual grounding for understanding one's situation whether it is as a Yuppy or football hooligan. The absence of reliable values and expectations breeds insecurity, aggression, and finally violence. Severance from the old communal culture leaves a vacuum than can appear to be filled with a sub-culture built around the codes of one's contemporaries. But such reactive cults are inevitably defensive in their instability and amoral in their lack of adult direction.
Counter-claim It may be that the moral unrest be seen is not so much a breakdown in society as a breaking down of social barriers. This dismantling of old certainties, of what could be expected from one's life, offers an almost intolerable form of freedom, but is freedom all the same. This kind of identity crisis seems a hard price to pay so that later generations may believe anything might be possible for them.
Broader Collapsed meaning of human creativity (#PF0936).
Narrower Collapse of common values (#PF1118)
Fragmented ethical contexts (#PF2096)
Governmental barriers to a global ethic (#PF2105)
Fostering of dependency by social institutions (#PF1755)
Inhibited capacity to visualize a creative future (#PF2352)
Reinforced parochialism of internal values and images (#PF1728)
Isolating social forms for enhancing ethical relations (#PF0945)
Lack of meaningful educational context for ethical decisions (#PF0966).
Related Defensive life stance (#PF0979) Limits on areas of research (#PF2529)
Educational curricula based on content rather than method (#PF3549).
Aggravated by Minimal church / school involvement (#PJ9011).

♦ PF1115 Disagreement concerning religious doctrine
Incompatible forms of worship — Liturgical conflicts
Incidence In the case of Christian churches, for example, there are differences of opinion as to: the relation of word and sacrament; the blessing of the eucharistic elements; scope of liturgical and non-liturgical forms of worship; worship led by any member of the congregation rather than by an ordained minister; the sacrificial aspect of the eucharist; the distinction between saints and blessed saints.
Broader Inadequacy of religion (#PF2005).
Aggravates Fragmentation within organized religions (#PF3364).
Aggravated by Bibliolatry (#PF7129).

♦ PF1118 Collapse of common values
Irrelevant social values — Lack of common ethic
Nature The lack of a common ethical system means the behaviour is dependent on the individual's perception of what is appropriate or inappropriate and on the structural sanctions society can effectively apply. This tends to force the legal system to assume the role of guarding the morality of the social system. Immorality is equated with illegality. Those who choose a moral life are relieved of the burden of freedom of choice by following the letter of the law. Those choosing an immoral life are left with avoiding being caught.
Claim The commonly held overall value systems of society are collapsing not only globally, but also within the fabric of individual societies; and there is a deeply felt longing to recreate a sense of ethical human relations. The primary blocks to the creation of a new context for ethical decision-making are found within the established religious institutions, the very groups which generally create and sustain meaningful value systems for local people. This is due to persistent individuals and organizations of established religion that are attempting either to make old schemes relevant or to create a new set of rigid, lasting values. At the same time, the current reduction of values to personal likes and dislikes is fostering an equally unhelpful personal

hedonism.
Overwhelmed by the demand to include widely–diverse groups, intersocial groupings retreat into familiar, parochial value schemes and egocentricity. Although such values are easily understandable to even the simplest observer, they are nonetheless totally divorced from the actual situation and present day requirements.
Broader Misuse of international forums (#PF2216)
Collapse in providing ethical value screens (#PF1114).
Narrower Haphazard forms of social ethics (#PF1249)
Static and superficial adult values (#PF2883)
Reinforced parochialism of internal values and images (#PF1728).
Related Ethnic discrimination (#PC3686).
Aggravates Immorality (#PA3369)
Incompatibility of rural values in urban cultures (#PF2648).
Aggravated by Ethical decay (#PB2480)
Lack of commitment to common symbols (#PE8814).

♦ PF1120 Inadequate research and development on problems of developing countries
Claim Most global research and development is devoted to military purposes or the commercial objectives of large corporations. Little of that is of direct relevance to the conditions prevailing in developing countries. Furthermore, little of that effort addresses the pressing issues facing developing countries, such as arid–land agriculture or the control of tropical diseases. Insufficient is done to adapt recent innovations in materials technology, energy conservation, information technology and biotechnology to the needs of developing countries.
Broader Inadequate research on problems (#PF1077).
Aggravates Inadequate research on proposed solutions to problems (#PF1572)
Use of inappropriate technologies in developing countries (#PF0878).
Aggravated by Imbalanced research activity (#PF0198)
Shortage of funds for research (#PF5419)
Excessive emphasis on fashionable areas of research (#PF0059)
Inadequate research and development capacity in developing countries (#PE4880)
Over–emphasis on immediate solutions in resource development research (#PE4059).

♦ PF1128 Imbalance in the human sex ratio
Nature Although sperms carrying Y–chromosomes greatly outnumber those carrying only X–chromosomes, and thus the number of male foetuses conceived is considerably more than the number of female foetuses, a far higher proportion of male foetuses abort (one survey showed between 137 and 142 males aborting at the 20th week of pregnancy compared with 100 females). Nevertheless, there are still slightly more male births than female (between 104 and 107 males for every 100 females). Despite this, the sex ratio in many countries shows a higher percentage of females to males, due both to a higher mortality rate for male than for female infants and to deaths incurred by wars. This may lead to a dearth of males to carry out traditionally male tasks and to a large number of unwillingly celibate females; and also create a burden on those societies where the male population determines its strength. It may also lead to a higher incidence of divorce. Conversely, where the sex ratio shows a higher percentage of men, it is felt that the society might be more aggressive, with a higher incidence of crime and of racial and social tensions.
An additional factor is the present possibility of determining the sex of the baby before birth and the future possibility of choosing the sex of the baby before conception. These possibilities may lead to some advantages - the eradication of sex–linked genetic disorders, for example, such as haemophilia, some enzyme deficiency diseases and a form of muscular dystrophy, all of which are many times more common among males than among females. But because of the increasing availability of legal abortion and the widespread preference in many cultures for the birth of sons rather than daughters, some demographers fear that the sex ratio may be further imbalanced by abortion of female foetuses and the choice to conceive sons.
Claim Completely free parental choice might result in dramatic change in sex ratio, with sweeping implications for personal, family and societal life. Polygamy or polyandry might be introduced widely. Having children without a marriage union might be increasingly approved; and widespread voluntary or enforced celibacy of the "majority sex" might come into vogue.
Broader Social disintegration (#PC3309)
Deliberate imbalancing of population sex ratio (#PF3382).
Aggravates Frustration (#PA2252) Social conflict (#PC0137)
Enforced celibacy (#PD3371) Dependency of women (#PC3426)
Shortage of military manpower (#PE4920).
Aggravated by War (#PB0593) Traditionalism (#PF2676).

♦ PF1136 Unaccountability of international financial institutions
Unaccountability of international aid institutions — Secrecy of international aid institutions
Claim The most fundamental factor undermining the success of foreign aid projects, and the reason why aid projects so often cause environmental damage, is the extraordinary level of secrecy under which most aid agencies operate. A complex of diplomatic conventions, legal roadblocks and bureaucratic liberties have allowed international aid institutions to operate beyond a level of public scrutiny essential for the proper and fair functioning of any other public institutions, and essential if environmental mistakes are to be discovered before it is too late. Aid institutions operate in isolation. They are laws unto themselves. They vigorously resist public scrutiny of their activities and lock out attempts by their funders and recipients to set guidelines for their operation. Increasingly the developing countries who are recipients of such aid are found to be opposed to it in the form in which it is given, namely that despite the best intentions of the donors, the effect of the aid is to finance developing country governments against their own people.
Broader Official secrecy (#PC1812)
Unaccountability of institutions degrading the environment (#PF3458).
Aggravates Inappropriate aid to developing countries (#PF8120).
Aggravated by Classified public information (#PF9699)
Secret international agreements (#PF0419)
Limited accountability of public services (#PF6574)
Autocracy of intergovernmental financial institutions (#PE2805)
Over–dependency on international financial institutions by developing countries (#PE2214).

♦ PF1138 Oligarchy effectiveness myth
Claim The age–old tradition of rule by the chosen (and educated) few is the basis of today's systems of production decision–making, even in a time in which every society's mindset assumes, to a greater or a lesser degree, participation by the many. Consequently, production systems are consistently unrelated from the mass markets they seek to serve, and change of any kind is slow and cumbersome.
Broader Unregulated ownership of the means of production (#PF2014).

♦ PF1169 Unsatisfied needs
Unmet needs
Narrower Unconsensed health needs (#PG9032)
Inadequate career advice (#PJ8018)
Unsurveyed consumer needs (#PG5484)
Unassessed corporate needs (#PG8220).
Aggravated by Uncoordinated expression of needs (#PG9210).

♦ PF1172 International economic recession
Vulnerability to economic cycles — Aggravation of cyclical recession
Incidence There is a growing international synchronization of business cycles. The 1974/75 recession was the first general recession since the end of the Second World War which impaired the possibility of effective national anti–cyclical policies based on the internationally unsynchronized nature of national business cycles. Attempts to coordinate economic policies on a world scale, taking into account changed world economic circumstances, have not been able to revive the shaken neo–Keynesian optimism in the possibility of economic policies to prevent capitalist economic crises.
Refs United Nations *Overcoming Economic Disorder* (1983); Van der Hoeven R, Richards Peter J (Eds) *World Recession and Global Interdependence* (1987).
Broader Economic uncertainty (#PF5817).
Narrower Uneven economic recovery (#PF6071) National economic recession (#PD9436).
Aggravates Unemployment (#PB0750) Trade protectionism (#PC4275)
Socio–economic poverty (#PB0388) Excessive corporate debt (#PE1879)
Slow economical growth in the socialist countries (#PE1117).
Aggravated by Global financial crisis (#PF3612)
Declining economic growth in developing countries (#PD5326)
Declining economic growth in industrialized countries (#PF1737)
Mismatch of national macroeconomic policies among industrialized countries (#PF5000).

♦ PF1176 Individualistic practices of local business
Nature The small independent stores of ghetto neighbourhoods are isolated from information and skills required in modern retail businesses. The retail revolution of the last twenty years has generated various types of local merchant associations for establishing common shopping hours and practices, conducting common advertising and promotion campaigns and providing a balanced set of customer services. Through these efforts, markets and profits have been increased. However, in many urban inner city communities, independent neighbourhood stores attempt to serve overlapping areas without the use of such cooperative structures or associated marketing methods. The threat of theft makes the merchant hesitant to leave his premises, and thus he remain isolated with little time to seek cooperative action with competitors whom he fears. As a result of not having the variety of stock needed to serve a large area, local businesses cannot compete with large outlets and community buying is done elsewhere. A move to some form of specialization seems risky and may merely catalyse another round of competition which would cost the merchant the few customers he now serves regularly.
Broader Individualism (#PF8393) Self–centred business (#PG9161).
Narrower Lack of economic cooperation (#PC8417)
Uncohesive business community (#PG8855)
Overlooked community services (#PG8866)
Prohibitive cost of advertising (#PG8898)
Ineffectiveness of traditional small business methods (#PF3008).
Aggravates Unexplored potential markets (#PF0581)
International economic fragmentation (#PC0025).
Aggravated by Neglect of property maintenance (#PD8894).

♦ PF1178 Self–conception of ethical void
Claim People create internal images from their experience of being crushed by social pressures, in which they see themselves as at the very base base of the social pyramid and thus cut off from important decision–making. They feel ethically emasculated through lack of participation and having no place in decisions. Because of this image, local people effectively avoided organizing their own needs and demands.
Broader Unsystematic allocation of market facilities (#PD3507).

♦ PF1179 Arrears in payment of government financial commitments
Overdue government payments — Delay in government reimbursements
Narrower Government loan defaults (#PD9437) Defaults on international loans (#PD3053)
Non-payment of reparations by government (#PE4446)
Arrears in financial payments between government agencies (#PE9700)
Government non–payment of agreed contributions to international organizations (#PF8650).
Aggravates Intergovernmental failure to fulfil financial commitments (#PF3913).
Aggravated by Progressive reduction in government action commitment (#PF5502).

♦ PF1181 Evolutionary catastrophes
Nature Abrupt and widespread discontinuities exist in the fossil record of fauna and are considered evidence of widespread mass extinction of species. These low frequency events have been attributed to fluctuations in sea level, reversals of the geomagnetic fields (exposing the earth's surface to lethal radiation), impacts of the earth by very large meteors (putting tons of dust into the atmosphere cutting off photosynthesis) and supernovae (causing catastrophic but temporary climate changes).
In addition to these catastrophes and the potential man made one there are many other possibilities. The sun will expand into a red star engulfing Mercury and Venus and melting lead on the Earth. The moon can fall to earth, a comet, a swarm of meteorites or a black hole could collide with the earth. The earth will eventually loose its atmosphere. A life form might evolve destroying all of humankind.
Broader Natural disasters (#PB1151) Environmental degradation (#PB6384).
Aggravates Extinction of species (#PB9171).
Aggravated by Novae (#PF3563) Global warming (#PC0918)
Global cooling (#PF1744) Meteors as hazards (#PF1695)
Volcanic eruptions (#PD3568) Geomagnetic reversal (#PF1588)
Sunlight inhibition by nuclear warfare soot (#PE6350).

♦ PF1183 Deliberate misrepresentation in educational materials
Dissemination of discriminatory information — Dissemination of hate literature
Nature Educational materials used in schools and for public information purposes may deliberately make use of words and phrases which help to develop prejudice, misunderstanding and conflict. The locally accepted standards of scholarship, justice and morality may not be applied to other nations, cultures, religions or groups, and due recognition and representation of their contribution to society may be deliberately ignored.
Broader Propaganda (#PF1878) Psychological warfare (#PC2175)
Inadequate and inaccurate textbooks and reference books (#PD2716).
Related Biased literature (#PJ1679) Bias in children's literature (#PD4773).
Aggravates Distrust (#PA8653).

♦ PF1185 Psychological ungroundedness
Nature In an increasingly urbanized and industrialized society, people are deprived of any meaningful contact with nature. In the past, such contact was associated with a physical, emotional and mental state of groundedness conducive to a state of harmony with the environment. Civilization as currently practised leads to a detachment from nature, even in rural areas. This separation from nature leads to a state of ungroundedness which is compounded by a psychic separation from each person's innate 'animal' consciousness, depriving people of the wisdom and inner security that may have been experienced in generations past. This ungrounded state leads to insecurity and fear and to insensitivity to the condition of the natural environ-

ment.
Claim When society loses its grounding, collective energy shifts from compassion and inner wisdom to a cold and rational approach to life which entraps people in compulsive patterns of greed, status, indifference and competition in an effort to find substitutes for that groundedness.
Aggravates Insecurity (#PA0857) Accumulation (#PA4313)
Fear of nature (#PF6803).
Aggravated by Hyperefficiency (#PF1706) Separation from nature (#PF0379)
Numbness towards others (#PF1216).

♦ PF1187 Parent-teacher non communication
Lost school communication — Disrelation between parents and school — Inadequate parent-school liaison — Parent teacher separation — Parent-school non-involvement
Broader Obstacles to education (#PF4852)
Limits on participation in community development (#PF3560)
Ineffective structures for community decision-making (#PF1781).
Narrower Lack of school input (#PS8733) Unserious school-home relations (#PS9177).
Aggravates Unmotivated teachers (#PF5978)
Inadequate working conditions of teachers (#PE7165).
Aggravated by Declining sense of community (#PF2575)
Insufficient communications systems (#PF2350)
Divisive patterns of community groupings (#PF6545)
Limited availability of education in rural areas (#PF3575)
Deteriorated structures of essential corporateness (#PF1301).

♦ PF1191 Discrepancies in human life evaluation
Nature The value of human life as regarding matters of health, security, value to the community, etc, may vary according to the economic and social status of the person considered. The evaluation of human life in the sense of the degree of pollution or disease which is considered acceptable may be based on commercial considerations. In societies where the social unit is important and not the individual, individual interest will be sacrificed towards the common good. In an elitist society the individual good may be sacrificed simply because it is considered more expendable than that of the elite.
Incidence In the USA, group life insurance whose premiums were subscribed to by corporations, paid executives' survivors two or three times the principal paid to employees' survivors; about $25,000 to $50,000 for employees; $100,000 to $250,000 for executives. In many countries, juries awarding damages for loss of life through negligence and other punishable acts, discriminate between values for deceased infants, children, adults and aged, and for able-bodied or infirm, and sometimes for sex and race. In Pakistan, for example, the surviving members of a woman's family are paid half the amount received by the survivors of a man's family.
Broader Social injustice (#PC0797) Social inequality (#PB0514)
Deteriorating quality of life (#PF7142).
Narrower Lack of social conscience (#PF9144) Proliferation of commercialism (#PF0815).
Related Ideological conflict (#PF3388).
Aggravates Slavery (#PC0146) Exploitation (#PB3200)
Class consciousness (#PC3458) Inadequate firearm regulation (#PD1970).
Aggravated by Elitism (#PA1387) Tribalism (#PC1910)
Traditionalism (#PF2676) Social discrimination (#PC1864)
Double standards in morality (#PF5225) Inhumanity of capital punishment (#PF0399).

♦ PF1196 Undeveloped potential of informal leadership
Nature Many nations have found it necessary to centralize official leadership for the sake of building new national unity and identity. Intensification of efforts at nation-wide development brings the necessity for informal leadership at the village level; but such unofficial, complementary leadership has not forcefully emerged in many Third World suburban communities. Leadership patterns, as they have emerged in the village, relate to a relatively small group. The absence of broad leadership in business, education, social welfare and other arenas of society overburdens the present leaders and hinders community-wide engagement.
Claim While current leaders may provide essential social stability and guidance to a village, the means must be found to nurture a broader spectrum of leadership among established residents, new residents, women and educated youth. The young men, especially those who have vision and the experience of a modern education, often leave the village to find other arenas in which to exercise their creative energies and talents. Structural vehicles for citizen engagement and decision-making are still irregular and do not cover the full spectrum of decisions that residents now demand.
Broader Obstacles to leadership (#PF7011).
Narrower Overdependence on capital (#PG7993)
Dependence on external expertise (#PG8011)
Unexplored avenues of leadership potential (#PF2797).
Related Distortionary tax systems (#PD3436) Limited meeting facilities (#PE1535).
Aggravates Inadequate land drainage (#PD2269)
Limited youth activities (#PJ0106).

♦ PF1199 Racial inequality
Scientific racism — Inequality of human biological abilities — Inferior average intelligence of certain racial groups — Racially-determined differences in intelligence
Nature The average intelligence of the children of certain racial groups (notably negroes in the USA and UK) has been assessed through IQ tests as being lower than that of white children, although the range of intelligence within each group is equally wide. There is some discussion as to whether genetic or environmental causes are taken to be more important. If genetic causes are taken to be more important then segregation policies might occur. If environmental causes are accepted as being stronger, then problems of improving the environment and opportunities for intellectual development are posed.
Background The origins of modern scientific racism date from the rise of the science of biology in the 18th and 19th centuries. At that time the development of biology, especially of theories of evolution, undermined then current religious justifications for slavery and for social hierarchy in general. Apologists for inequality looked, therefore, to the new science of biology to support their social views. They first attempted to demonstrate on the basis of anatomical measures like brain size, that blacks and oppressed immigrant groups were "less evolved". By the beginning of the 20th century this approach proved fallacious and scientific racist theorists began using newly developed IQ test as their major argument for intellectual inferiority. The environmentalist school came into prominence after World War II because critical scholarship was demonstrating the IQ test to be invalid as a true measure of mental functioning.
Claim Equality of biological abilities between races is a chimera that no intelligent person can entertain. A 1987 survey of psychologists in the USA concerning the genetically determined basis for racial differences concluded that blacks were on average inferior to other races.
Counter-claim The IQ tests themselves have been proved to be racially and culturally biased and hence invalid. The acceptance of this theory gives support to racial ideology and practice, since some may take it to imply that all members of a certain racial group are intellectually inferior. General acceptance of such an interpretation of the theory in educational circles, together with the belief that such inferiority is genetically determined, could lead to a public policy of discrimination and prejudice against the 'inferior' group and prolongation of existing discrimination and prejudice. Both genetic and environmentalist schools of scientific racism are based on the assumption that deficiencies in the poor are the source of their low social position and both school refuse to locate the source of inequality in the wider society's social structure.
Refs Reich, Michael *Racial Inequality* (1981).
Broader Racism (#PB1047) Social injustice (#PC0797) Human inequality (#PA0844).
Related Threats against family or friends (#PE3308).
Aggravates Caste system (#PC1968) Racial conflict (#PC3684)
Racial segregation (#PC3688) Questionable facts (#PF9431)
Racial exploitation (#PC3334) Cultural deprivation (#PC1351)
Racial discrimination (#PC0006) Economic subordination (#PG3458)
Inequality in education (#PC3434) Inequality of opportunity (#PC3435)
Conceptual repression of problems (#PF5210) Racial discrimination in education (#PD3328)
Suppression of scientific information (#PF1615).
Aggravated by Deprivation (#PA0831) Intellectual prejudice (#PC3406)
Bias in scientific research (#PF9693) Inadequate recreational facilities (#PF0202)
Inadequacy and insensitivity of intelligence testing (#PD1975).
Reduced by Racial integration (#PU3461).

♦ PF1200 Challenges to validity
Nature Requiring the acquiring party to refrain from challenging the validity of patents and other type of protection for inventions involved in the transfer or the validity of other such grants claimed or obtained by the supplying party, recognizing that any issue concerning the mutual rights and obligations of the parties following such a challenge will be determined by the appropriate applicable law and the terms of agreement to the extent consistent with that law.
Broader Restrictive trade practices (#PC0073).
Related Collusive tendering (#PE4301) Grant-back provisions (#PE5306)
Restrictions on research (#PF0725) Restrictions on publicity (#PF1575)
Restrictions on adaptations (#PF5248) Exclusive dealing arrangements (#PE0413)
Restrictions on use of personnel (#PF3945)
Patent pool or cross-licensing agreements (#PE4039)
Exclusive sales and representation agreements (#PE4581)
Trade restrictions due to voluntary export restraints (#PE0310)
Payments after expiration of industrial property rights (#PF5292)
Tying of supplies to subsidiaries by transnational enterprises (#PE0669).

♦ PF1202 Irrelevant scientific activity
Nature Much scientific research activity fails to take into account social, political and economic realities. Researchers may even avoid selecting topics which bear any relation to the problems in the real world, and take pride in the absence of practical applications for such work. Such research nevertheless constitutes a drain on scarce resources.
Broader Inadequate research on problems (#PF1077)
Human errors and miscalculations (#PF3702).
Related Irresponsible scientific and technological activity (#PC1153).
Aggravates Ignorance (#PA5568) Irrelevant available information (#PG8884)
Irrelevance of science and technology (#PF0770).
Aggravated by Imbalanced research activity (#PF0198)
Denial of rights of medical patients (#PD1662)
Fragmentation of academic disciplines (#PF8868).

♦ PF1205 Fragmented conduct of community operations
Nature While development in small communities requires a spirit of cooperation among families and groups in order to actually get things done, in many communities efforts are conducted in a fragmented way. Communities are often organized along family and ethnic lines. People feel that it is necessary to be independent and so each family strives to exist without helping or being helped by other families. Possible municipal improvements such as road surfacing, housing and recreation facilities, which necessitate a combined effort, are not pursued because individual citizens are unwilling to make personal contributions of time or money in such campaigns. Private owners restrict the use of park land. People with similar agricultural problems do not organize themselves to achieve their common objectives. Cases have been recorded where families without their own means of transport have been refused neighbours' cars in times of medical emergency. Children of different groups may never go to school together. Some families resent the greater income of other families. Party-line telephones are tied up for long periods of time, the resulting frustration being a measure of the depth of feelings involved in the whole issue of individual rights versus community cooperation. Only when people in small communities find ways of working together towards their common good will their hopes for the future have a significant chance of success.
Broader Obstacles to community achievement (#PF7118).
Narrower Distrust of storekeepers (#PG9249) Misused ambulance service (#PG9312)
Family-focused operations (#PG9375) Loss of faith in religion (#PF3863)
Multiple family jealousies (#PG9358) Shortage of exciting events (#PG9189)
Inconsiderate telephone use (#PG9378) Inappropriate study patterns (#PJ9371)
Inaccessible wilderness areas (#PF9360) Impermanent living conditions (#PD4368)
Lack of corporate achievements (#PG9339) Reluctant personal participation (#PG9362)
Fear of humiliation by co-workers (#PF9234) Indifference of students' families (#PG9324)
Individualistic agricultural planning (#PJ1082)
Fragmented forms of cooperative efforts (#PF2588)
Absence of images of social responsibility (#PF3553).
Related Inadequate enforcement of safety regulations (#PD5001).
Aggravates International economic fragmentation (#PC0025).
Aggravated by Deliberate ignorance (#PF8229) Segmental school experience (#PJ9345)
Institutional fragmentation (#PC3915) Disruptive personal prejudices (#PG9377).

♦ PF1212 Prohibitive cost of nutritious food
Prohibitive cost of healthy foods — Prohibitive cost of organically grown foodstuffs
Aggravates Nutritional deficiencies (#PC0382) Unethical catering practices (#PE6615)
Decline in nutritional quality of food (#PE8938).
Aggravated by Food fads (#PD1189)
Inappropriate agricultural subsidies for chemicalized farming (#PE1785).

♦ PF1213 Depressing effect of poor housing construction
Claim In the rural setting a mud hut provides shelter in the larger environment; in the urban village a dwelling unit is more than shelter, it becomes the larger environment itself. The present construction of wood, mud and tin houses in suburban third world communities is socially depressing to the residents. Due to the limited amount of capital available and the high cost of building materials, landowners have built high density units with little regard for their effect on the community. Alternative housing construction seems almost inconceivable to them. In many communities the rapid influx of people has been met with a rapid expansion of such poor quality housing units. The resulting unsanitary, crowded conditions form a settlement pattern that reduces residents' initiative and motivation, since such housing not only poses a threat to the general physical health of the community but promotes a constant drain on peoples' spirits. In the case of traditional dwellings and settlements it is frequently their sacred character which is essential in that they form a humanized safe space in a profane and potentially dangerous environment. They become humanized by imposing an order using rituals, and sacred orientations and frequently, by becoming cosmological symbols. The loss of this aspect of housing and settlement

patterns disrelates people from their own psychological and spiritual roots. Housing could and should be improved, not to measure up to any outside standards but to release the community's energy.
   Broader Neglect of property maintenance (#PD8894)
   Obstacles to community achievement (#PF7118)
   Demoralizing constraints on housing rehabilitation (#PE2451).
   Narrower Deteriorated vacant houses (#PJ8678).
   Related Alienating public housing assignments (#PJ9479)
   Monotonous and unaesthetic architecture and design (#PF0867).
   Aggravated by Inadequate housing (#PC0449)   Disease-prone housing (#PG7994)
   Marginal living space (#PJ8036)   Inadequate toilet facilities (#PG7986)
   Prohibitive cost of home maintenance (#PJ9022)
   Discouragement of permanent residency (#PG8031)
   Overcrowding of housing and accommodation (#PD0758)
   Restrictive building codes in urban areas (#PE8443)
   Impersonality of high density accommodation (#PF6156).

♦ **PF1214 Politically unrealistic strategic warfare analysis**
Think-tank analysis
Nature There is an enormous gulf between what political leaders really think about nuclear weapons and what is assumed in complex calculations of relative 'advantage' in simulated strategic warfare. Think-tank analysts can set levels of 'acceptable' damage well up in the tens of millions of lives. They can assume that the loss of dozens of great cities is somehow a real choice for sane men. In the real world of real political leaders, a decision that would bring even one hydrogen bomb on one city of one's own country would be recognized in advance as a catastrophic blunder, ten bombs on ten cities would be a disaster beyond history, and a hundred bombs on a hundred cities are unthinkable. Yet this unthinkable level of human incineration is the least that could be expected by either side in response to any first strike in the next ten years, no matter what happens to weapons systems in the meantime.
   Broader Human errors and miscalculations (#PF3702).
   Aggravates Political over-reaction (#PF4110)   Competitive acquisition of arms (#PC1258)
   Aggravated by Unrealistic policies (#PF9428)   Competition between states (#PC0114)
   Benign neoplasm of bone and cartilage (#PG0819).

♦ **PF1215 Left-handedness**
Sinistrality
Nature In a world dominated by right-handedness (preferential and predominant use of the right hand) left-handed people are discriminated against. This discrimination may be as overt as mocking left-handers and forcing them to switch to the use of their right hand, or covert, such as the lack of proper equipment (from scissors to baseball gloves to industrial machines) for left-handed use. Reproaching children for their left-handedness may lead to emotional problems, nervousness, and a poor self-image.
Background Traditionally, mankind has looked upon left-handedness as a defect. Since Roman times the left hand has been "sinister", and parents often go out of their way to discourage their children from predominant left-handed usage. Many countries in the east look on the left hand as unclean and discourage its use in society.
Symbolic differentiations of left and right are virtually universal cultural classifications among humankind. Left handers live with a biased vocabulary "Adroit" and "Dextrous" are associated with the right hand, but "sinister" and "gauche" with the left. The righteous sit on the right hand of God. In 1909, French sociologist Robert Hertz made the following observation: "To the right hand go honours, flattering designation, prerogatives; it acts, orders, and takes. The left hand, on the contrary, is despised and reduced to the role of a humble auxiliary; by itself it can do nothing; it helps, it supports, it holds".
   Refs Hueck, Gerda and Finser, R *The Problem of Lefthandedness* (1986).
   Broader Cultural discrimination (#PC8344).
   Aggravates Impairment of visual-spatial ability (#PG3467).

♦ **PF1216 Numbness towards others**
Repression of feelings — Numbing of feelings — Emotional numbness — Insensitivity to personal pain — Emotional paralysis
Nature Reduction in levels of sensitivity is the primary protection against personal pain, suffering and humiliation. In addition to protection from external threats it can also be extended by people to protect themselves from inner experiences and impulses, including sexual feelings, joy, sadness, anger, pain and neediness. Skill in insensitivity towards personal pain can be extended to include the pain of others. This permits a further destructive development into active infliction of pain on others, whether by word or deed, omission or commission. The borderline between passive numbing and active cruelty is poorly defined and easily crossed.
   Related Anhedonia (#PG0585)   Insensitivity to diversity of cultural traditions (#PF8156).
   Aggravates Evil (#PF7042)   Psychological ungroundedness (#PF1185);
   Aggravated by Fear of emotional sensitivity (#PF9209).

♦ **PF1217 Data-oriented education**
   Broader Irrelevance of educational curricula (#PF0443)
   Ineffective methods of practical education (#PF2721).
   Aggravates Limited development of functional abilities (#PF1332).
   Aggravated by Inadequate use of visual imagery for societal learning (#PD7086).

♦ **PF1219 Inadequate legal counsel for minorities**
Nature The right to counsel presupposes that the services of counsel are given competently, diligently and fearlessly. Discriminatory factors limit access to the legal profession and harm the observance of this right, since they make it less likely that every accused, no matter of what race, colour, sex, or political beliefs, will find a competent lawyer willing to take up his case with proper devotion. Furthermore, the exclusion of persons from the legal profession may result in inequality in the administration of justice, since access to the profession influences the whole judicial system in many countries. The inability or unwillingness to provide interpreters or translators for minorities who require legal services but who are unable to speak or write the official language (s), is a serious limitation of their rights. Individuals who cannot communicate adequately with court or probation officers may be imprisoned or placed in custodial institutions.
   Broader Biased legal systems (#PF8065)   Denial of legal representation (#PF3517).
   Aggravated by Economic discrimination in the administration of justice (#PE1399).

♦ **PF1226 Reluctant claims on external resources**
Nature People in small communities press their claims for the benefits of modern life reluctantly, if at all. Although a wide range of services and funds are accessible to such communities, the complexity of the legal procedures for obtaining some grants deters residents from requesting available funds. Land use issues involve baffling relationships with government and neighbours. Access to distant municipal and district services may be limited and awkward. Often police and fire detachments are located at a distance and crisis calls are hindered when shared telephone lines are busy. Medical care for the children may be provided from another district. Yet the heart of this problem lies neither in the distance from services nor in the complexity of procedures. Rather, it has to do with the a determination not to allow government dependency to inhibit self-reliance which is exaggerated into a near rejection of these benefits which others receive as a matter of right. Even when benefits are requested, residents do so on an individual basis through private negotiations, rather than as a coordinated group.
   Narrower Telephone delays (#PF1698)   Small student enrolment (#PG9332)
   Regressive welfare system (#PG9346)   Difficult grant management (#PG9356)
   Fluctuating agricultural markets (#PG9369)   Underutilized government resources (#PE9325).
   Related Lack of trained firefighters (#PJ0277).
   Aggravates Maldistribution of resources (#PB1016)
   Restrictive use of available land (#PF6528).
   Aggravated by Complex legal procedures (#PF8519)
   Breakdown of police protection (#PF8652)
   Inadequate water system infrastructure (#PD8517)
   Insufficient transportation infrastructure (#PF1495).

♦ **PF1232 Organizational empire-building**
   Broader Conflict (#PA0298)   Domination (#PA0839)
   Negative emotions and attitudes (#PA7090).
   Aggravates Jurisdictional conflict and antagonism within international organizations (#PD0047)
   Jurisdictional conflict and antagonism between international organizations (#PD0138)
   Proliferation and duplication of international organizations and coordinating bodies (#PE1029)
   Jurisdictional conflict and antagonism between government agencies within each country (#PE8308).
   Aggravated by Jurisdictional conflict between academic disciplines (#PF9077)
   Excessive external trade deficits of developing countries (#PE1496).

♦ **PF1234 Long-term change in atmospheric chemistry**
Nature There are three ways in which human behaviour may modify the atmosphere, either deliberately or inadvertently: by changing the concentration of constituent substances (including water); by releasing heat; and by changing the physical and biological properties of the earth's surface. As a result of these changes, people and other living organisms may be exposed to harmful levels of toxic pollutants transported by the atmosphere. Droughts and desertification may occur. Changes in the concentration of ozone in the stratosphere may alter the amount of solar ultraviolet light reaching the surface of the earth, with possible effects on the health of humans and other species. Or the climate may be altered, with either beneficial or detrimental consequences. All such possibilities have caused concern during recent years and, as is characteristic of atmospheric environmental problems, the induced changes have been regional or global in scale and hence affected large numbers of people.
It is apparent that, because of lack of understanding of the basic global cycles of carbon, sulphur and associated elements, long-term climatic changes cannot be predicted with confidence; if the postulated general warming due to the increase in carbon dioxide concentrations were actually to take place, it would therefore be impossible to forecast its full effects on climate patterns – and all the more so its economic and social consequences. The climate changes probably would not be uniform, and the social economic consequences would be likely to benefit some areas and work to the detriment of others.
Incidence During the 1970s a number of atmospheric alterations due to various human activities were observed:
Carbon dioxide concentrations slowly and steadily increased, chiefly as a result of the increasing use of fossil fuels but also due to forest clearing. This phenomenon has important – but as yet incompletely understood – implications for world weather conditions and agriculture, because it could alter temperature and precipitation patterns and the distribution of snow and ice cover.
Acid rain became an established phenomenon that results from the long-distance transport in the atmosphere of sulphur oxides and nitrogen oxides, produced primarily by fossil fuel combustion. One of its adverse effects is the acidification of inland waters.
Photochemical oxidants, which cause smog, decreased in cities with effective controls, and increased where controls were absent or ineffective or where automobile use increased.
Sulphur dioxide and suspended particulate matter concentrations decreased in most cities with control policies, but increased elsewhere, especially in developing countries.
Stratospheric particulates appear to have increased somewhat, with possible undetermined effects on climate.
While local climatic changes occurred (as in heat islands and hazy areas), the question of whether long-term climatic changes are in progress, and if so at what rate, remains controversial.
   Broader Inhospitable climate (#PC0387).
   Narrower Inadvertent modifications to climate (#PC1288).
   Related Destruction of environmental oxygen (#PE5196).
   Aggravated by Dust (#PD1245)   Uncontrolled industrialization (#PB1845)
   Inadequate research on problems (#PF1077).

♦ **PF1236 False nuclear warfare alerts**
Nature Due to failure of missile warning systems or misinterpretation or misuse of the data they provide, national or military alliance defence systems may be placed on nuclear warfare alert. Such alerts may involve declarations of emergency over the news media, giving rise to unnecessary panic, and other psychological side-effects, among the civilian population.
   Broader Falsity (#PF5900)   False alarms (#PF4298).
   Aggravates Panic (#PF2633)   Anxiety (#PA1635).
   Aggravated by Unidentified flying objects (#PF1392)
   Accidents to nuclear weapons systems (#PD3493).

♦ **PF1238 Prohibitive cost of living**
High cost of living — High consumer prices — Inflated cost of necessities — Excessive retail prices
Nature Continuing increases in the cost of living or the level of retail prices tend to diminish the ability of consumers to purchase the goods and services on which they depend in order to maintain an adequate standard of living. Increases in nominal wages may sometimes be considerably reduced, or even effectively nullified, by rises in prices.
   Broader Instability of economic and industrial production activities (#PC1217).
   Narrower Prohibitive cost of accommodation (#PD1842)
   Prohibitive cost of farm machinery (#PF2457)
   Increasing cost of social security (#PF7911)
   Prohibitive cost of necessities in rural communities (#PF2385).
   Aggravates Poor living conditions (#PD9156)   Inadequate standards of living (#PF0344)
   Excessive employment of married women (#PD3557)
   Dependence on sophisticated technology for development (#PD6571)
   Transfer of business from small communities to larger towns (#PF6540).
   Aggravated by Social injustice (#PC0797)   Economic inflation (#PC0254)
   Sacroiliac desmitis (#PG0781)   Discriminatory professionalism (#PC2178)
   Uncontrolled household rent increases (#PG5510)
   Prohibitive cost of goods and services (#PD1891)
   Economic and social losses due to disability (#PE4856).

♦ **PF1240 Underdeveloped capacity for income farming**
Unexplored livestock income — Single source income — Undiversified basis of income possibilities
Nature Apart from human resources, the arable land within and adjacent to many third world

villages is still their greatest resource; but although the economic development of many third world communities must primarily depend on agricultural industries, virtually no farming of cash crops may be carried out. Large and small plots of land may stand uncultivated, although local people may be aware that here is a latent potential for, for example, horticulture or stock farming. Most villagers have a background of subsistence farming and see farming as, at best, a minimal supplement to urban income; they are underexposed to intensive and diversified high–income farming. As a result, the land remains divided into tiny, cultivated plots with few animals. Each family manages its own land individually. In addition, there is little experience of essential machinery, soil improvement, marketing processes and managerial techniques.

Although people in agricultural areas, and especially those in developing countries, work long hours, and although most villages have virtually no unemployment, yet in relation to the village's human potential and productive capacity there is radical underemployment, and few possibilities for improvement.

In many villages, the economy still rests on one cash crop with a secure but low market income. Distance and transportation costs are the major deterrent to attracting new industries to the area. Possibilities for regional industries, available job opportunities or potential agricultural markets remains unexplored. Limited employment opportunities perpetuate the role of unskilled labourers and villagers tend to work in nearby areas as unskilled farm labour, although others may work in larger towns and cities and commute to the village at weekends. Such labouring work and the time–consuming pattern of home–care for the women does not challenge the resources (particularly of the young) to acquire the skills or training needed to assume new roles in the job market.

Broader Insufficient diversification (#PD0335)
Underdeveloped sources of income expansion (#PF1345)
Underproductive methods of agricultural management (#PF6524).
Narrower Extensive farming (#PG7987) Narrow profit margin (#PJ9737)
Unused fodder resources (#PG8013) Insufficient cash crops (#PG0569)
Insufficient job options (#PJ0070)
Restrictive building codes in urban areas (#PE8443)
Subsistence agricultural income level in rural communities (#PE8171).
Related Insecure land tenure (#PD9162) Unproductive subsistence agriculture (#PC0492)
Non-diversification in subsistence fishing economies (#PF2135).
Aggravates Idle private land (#PJ8020) Low general income (#PD8568)
Limited purchasing power (#PD8362) Unexplored potential markets (#PF0581)
Insufficient financial resources (#PB4653).
Aggravated by Prohibitive cost of capital equipment (#PJ1768).

♦ PF1244 Inaccurate criminal stereotypes

Nature Increasingly law enforcement agencies and security services make use of criminal profiles by which to determine which people at security checking points should be subjected to more stringent searches. The profiles are sets of characteristics that are perfectly innocent in themselves, but when taken together are considered clues to identifying security risks, smugglers or the like. The profiles are necessarily treated as classified information. Since the profiles are based on statistical probabilities, a percentage of innocent people fitting them tend to be stopped on a regular basis and subject to searches (often quite humiliating).
Broader Stereotypes (#PF8508).
Related Criminality (#PA9226).

♦ PF1245 Vulnerability of socio–economic systems from globalization

Nature Globalization of the world system has progressed in three phases: expansion of trade, internationalization of financial markets, through increasing flows of investment accompanied by increased corporate and research networking. This accelerating interdependence has as its primary agent the transnational enterprise and is driven by the revolution of information and communication technology. Although globalization enhances opportunities for growth (through improved efficiency, methods of production, and new technologies), risk and vulnerability increase because globalization creates pressures for convergence policies which threaten national sovereignty. It enhances competition amongst systems and has a low tolerance for system diversity engendered by different cultural histories and institutional preferences.
Broader Vulnerability of social systems (#PB2853).
Related Cover up of convergence in practice of apparently opposed ideologies (#PF5124).
Aggravates Homogenization of cultures (#PB1071)
Destabilization of social systems (#PB5417).
Aggravated by Collusive international trade arrangements (#PE0396)
Excessive power and independence of transnational corporations (#PD5807)
Disproportionate control of global economy by limited number of corporations (#PE0135).
Reduces Uncontrollability of world social systems (#PF1443).

♦ PF1246 Conflicting sense of sexual identity

Indistinct awareness of sexual identity
Refs Troiden, Richard R *Gay and Lesbian Identity* (1988).
Broader Inadequate sense of personal identity (#PF1934).
Narrower Effeminateness (#PG3487).
Aggravates Prejudice (#PA2173) Stereotypes (#PF8508)
Generation communication gap (#PF0756).
Aggravated by Crisis of identity (#PJ2421) Social disintegration (#PC3309)
Sexual discrimination (#PC2022).

♦ PF1249 Haphazard forms of social ethics

Selective ethics
Nature The maintenance of basic order depends in large measure on establishing an informal consensus on the appropriate forms of individual behaviour which allows community sanctions to be invoked when necessary. It is becoming obvious that effective development depends upon a relatively stable basis of order being present in any particular community. Yet, in many socially mixed rural villages, such a consensus is lacking, and a haphazard assortment of patterns of behaviour is considered normal, with the result that social occasions often end in disorder. Furthermore, individuals tend to assume that a personal effort towards maintaining order would be comparatively fruitless. Residents are unwilling to hold one another accountable to social norms. For example, although many people may express concern over alcoholism and vandalism, individuals or groups rarely take the initiative to deal with such disruptions. The wisdom of older people is rarely called upon and they are unable to impose effective sanctions. Individual villages may have no official police force, and even where there are some means of law enforcement, violations of the law ranging from deliberate shootings to dangerous driving may be ignored. The education of children is left largely to teachers. Young people are confronted with conflicting expectations of appropriate roles and are given few opportunities to organize social events or to take responsibility for community facilities; vandalism and property damage are the result. All these things contribute to a sense of disruption which pervades everyday life. Until more orderly forms of social ethics are established at the village level, the climate of community life will continue to discourage residents from participation.
Claim There is a tendency to focus a level of ethics on other individuals and groups which one does not always apply to oneself or to one's own social group. This is especially noticeable among political muckrakers.
Broader Hypocrisy (#PF3377) Collapse of common values (#PF1118).
Narrower Fear of vandalism (#PJ9326) Snowmobiles on roads (#PG9341)
Easy access to liquor (#PG9316) Double standards in morality (#PF5225)
Unrewarded volunteer initiatives (#PG9294) Double standards of sexual morality (#PF3259).
Aggravates Erosion of elders' wisdom (#PF1664)
Breakdown of police protection (#PF8652)
Social disaffection of the young (#PD1544)
Undervaluation of education by parents (#PF9306).

♦ PF1252 Permissiveness

Permissive society
Incidence Permissiveness characterized American society in the 1960's. Children brought up in these years were called the 'one' generation, as they learned to put self–interest first. World–wide, homosexuality has come out of the 'closet'; and divorce rates and the demand for abortions climbed. At the same time drug dependence has increased, along with mental illness.
Claim Public and parental insistence on certain moral and social conventions has declined under pressures from sexual and social deviants, subversive ideologies, organized crime, and substance abusers, as well as from national cultural malaise and from unregulated commercial opportunism prevalent in free–market societies. Permissiveness has resulted in violence, perversion, obscenity, vulgarity, and the excessive pursuit of pleasure wherever it leads and at whatever cost.
Narrower Permissive education (#PF0593) Parental permissiveness (#PD5344).
Related Refusal (#PA7321) Neglect (#PA5438) Decadence (#PB2542)
Immorality (#PA3369) Compulsion (#PA5740).
Aggravates Nudism (#PF2660) Divorce (#PF2100) Adultery (#PF2314)
Obscenity (#PF2634) Unchastity (#PA5612) Promiscuity (#PC0745)
Cohabitation (#PF3278) Social conflict (#PC0137) Induced abortion (#PD0158)
Sexual deviation (#PD2198) Sexual immorality (#PF2687) Illegitimate children (#PC1874)
Retarded socialization (#PF2187)
Vulnerability of marriage as an institution (#PF1870).
Aggravated by Moralism (#PF3379) Liberalism (#PF0717)
Family breakdown (#PC2102) Social disintegration (#PC3309).
Reduces Prostitution (#PD0693) Homosexuality (#PF3242)
Authoritarianism (#PB1638) Impediments to marriage (#PF3343)
Social impediments to marriage (#PF3341).
Reduced by Censorship (#PC0067).

♦ PF1254 Lack of leadership

Lack of vocal leaders — Lack of core leadership — Crippling leadership void — Lack of formal leadership
Nature Despite the multiplication of social problems and institutions dealing with them (and in part because of this increase in social complexity) there is a shortage of individuals with the qualities necessary for leadership.
Claim Institutions are changing in ways that their leaders cannot always grasp. The relative simplicity of situations in which good and evil, and right and wrong could be easily and unambiguously distinguished has been superceded by a state of confusion and pessimism and a sense that real problems are so complex and technical that the traditional methods of leadership are inadequate. Leaders are exposed to serious internal doubts, uncertainty and strife over their precise functions and the correctness of their decisions; in addition they are increasingly subject to heavy scrutiny and criticism in the media, which over–exposes them, introducing doubt and over–familiarizing the public with the individuals in question. The traditionally most visible leadership roles are no longer the best remunerated and therefore no longer attract the best talent. The incumbents of such roles are faced with a proliferation of demands, claimants, constituents, contending groups and problems which cannot be harmonized in such a way as to reinforce their positions.
Broader Obstacles to leadership (#PF7011).
Narrower Leadership impaired by illness (#PF8387)
Shortage of industrial leadership and entrepreneurial ability in developing countries (#PC1820).
Aggravates Distrust (#PA8653) Underdevelopment (#PB0206) Political instability (#PC2677)
Lack of leadership initiative (#PF7988)
Limited availability of education in rural areas (#PF3575)
Inequality of employment opportunity in developing countries (#PD1847).
Aggravated by Corruption (#PA1986) Subjectivity (#PF2827)
Bureaucratic fragmentation (#PC2662) Unprepared adult leadership (#PF6462)
Demeaning community self–image (#PF2093) Mediocrity of government leaders (#PF3962)
Insufficient leadership training (#PF3605) Lack of local leadership role models (#PF6479)
Fragmented patterns of community activity (#PF6504)
Extreme detachment from represented constituency (#PF0889)
Ineffective structures for community decision–making (#PF1781)
Ineffective operation of community networks in urban ghettos (#PF1959).

♦ PF1257 Disunity in urban villages

Nature The complex rural to urban transitions occurring in Third World suburban communities are creating a sense of social disunity, and contributing to the lack of individual cooperation already present in many urban villages. The rapid influx of "transients" seeking to succeed in the city creates strained relationships with "permanent" residents; and the attendant high population density, combined with the absence of public facilities like a community centre or town square, is also a contributory factor. The spirit of cooperation that existed in rural life is replaced by isolating individualism that works against the progress of the community as well as the well–being of its individual inhabitants. The physical poverty in such communities, where so many compete for so few jobs, so few rooms, and so few school places, further contributes to this disunity.
Broader Social disunity (#PU3846).
Narrower Disturbing transient mobility (#PG7995)
Insufficient women co–operatives (#PG8001)
Overcrowding of schools by transients (#PG8014)
Divisive patterns of community groupings (#PF6545).
Related Lost family role in society (#PF7456).
Aggravated by Inadequate irrigation system (#PD8839).

♦ PF1258 Inefficient public spending to alleviate poverty

Nature Government subsidies of high–cost approach to housing, education, food or health care has not been able to relieve poverty or provide services for large numbers of poor people.
Broader Inappropriate public spending by government (#PF6377).

♦ PF1262 Ineffective worker organizations

Trade union rivalry — Unconsensed union farmers — Outmoded trade unions — Inadequate international trade unions
Nature Ineffective worker organizations result in organized efforts being usually limited to a minority of the working population, and therefore not having the force required to balance the power of organized capital. This allows the owners of capital to control the conditions of work and therefore control the level of livelihood in every community.
Broader Competition (#PB0848) Inflexible social structure (#PB1997)
Limited image of employability (#PF2896).
Narrower State–controlled trade unions (#PG6615)

Unrepresentativity of trade unions (#PE6018)
Lack of unionization among working women (#PE8345).
Aggravates Authoritarian division of labour (#PC6089)
Discrimination against trade unions (#PC4613)
Victimization of workers' representatives (#PD1846)
Inadequate sense of community and solidarity amongst workers (#PE4179).
Reduces Trade unionism (#PF8493).

♦ PF1266 Inappropriate selection and examination procedures in education
Inappropriate educational testing procedures — Test-led education
Nature Increasing reliance upon examinations in school and university systems has tended to put pressure on teachers to orient their lessons to expected tests rather than to exploration of the subject or to the students' mental and creative development. The strain imposed on teachers, pupils and parents may be such as to constitute a risk to mental health. Such tests may occur at a psychologically vulnerable period of life, aside from stress caused by the social stigma attached to failure. Examinations establish in the individual a state of mind, behaviour and habits which are the negation of the aims of education. Where promotion is in large part dependent upon examination success, the examination procedures have a restrictive effect on transfer between grades. Marks, whether they be obtained from teacher-set examinations or from standardized tests, are expressions of relative value. At their worst, they are expressions of inconsistent, subjective judgments by teachers on the basis of unreliable, written examinations of unknown validity. At their best they leave absolute judgments (as to what shall constitute a pass level) to be stated in terms of the proportion of children who shall be allowed to continue to the next grade. In neither case do they recognize the continuity of learning, which is not divisible into discrete and convenient administrative packages. Thus, examinations may also be accused of being irrelevant, not so much by their nature as in their conventional use.
Counter-claim Pupils need a goal such as examinations in order to apply themselves to study; educators need some yardstick for assessing pupils' progress; and employers need some way of assessing prospective employees' capabilities. Examinations may not be the perfect answer to these needs but are better than any other presently suggested alternative.
Broader Obstacles to education (#PF4852)
Constricting effect of educational structures (#PF6476).
Aggravates Stress in human beings (#PC1648) Competitiveness in education (#PD4178)
Discriminatory professionalism (#PC2178) Threats against family or friends (#PE3308).
Aggravated by Educational wastage (#PC1716) Inappropriate education (#PD8529)
Invasion of privacy through testing (#PJ6946).

♦ PF1267 Inadequate dissemination and use of available information
Fragmented utilization of public information — Inaccessible public information
Nature In an age when the quantity of information on effective development has increased dramatically, there are few effective channels of information to inform a community about events in its own life. The people of small remote communities have not found adequate ways to use their own knowledge about life in their community nor information about resources available from society at large. Local economic development does not progress because basic data is not distributed. This may be because the data is obscured by its specialized nature; or residents may experience difficulty in sorting through the many sources and agencies involved; they may feel that they lack the expertise to comply with proposal-writing requirements; guidelines may change before they are able to submit a proposal; finally, there appears to be a kind of victimization at work which prevents them from rapidly benefiting from any knowledge obtained. For example, there is little demonstration locally of the results of recent agricultural research; the market potential for possible commercial ventures is not explored; and opportunities for discussion are limited. As a result, areas of common concern are not dealt with; people claim to be uninformed about topics to be discussed at local meetings; students have limited access to books, newspapers and magazines in their homes; there is little knowledge of health care and nutrition, or of available preventive health and health education programmes, and consequently, diets of high starch and canned goods are common; and inappropriate sanitation practices continue because people are unaware of the necessity of using better systems.
Broader Limited access to social benefits (#PF1303).
Narrower Unpublicized meeting agendas (#PG9295)
Unpublicized public meetings (#PF5222)
Indefinite options for communicating (#PG9282)
Inadequate water system infrastructure (#PD8517)
Limited enforceability of international standards (#PF8927)
Unfamiliar procedures for using public health programmes (#PE7984).
Related Prohibitive cost of fuel (#PJ0346) Unpublicized community news (#PF7998).
Aggravates Inadequate land drainage (#PD2269)
Neglected health practices (#PD8607)
Unexplored potential markets (#PF0581)
Ignorance of health and hygiene (#PD8023)
Underutilization of legal rights (#PF3464)
Vulnerability of plants and crops (#PD5730)
Stagnated images of community identity (#PF6537)
Economic loss through reduced productivity of diseased animals (#PE8098).
Aggravated by Over-specialization (#PF0256) Unrecorded knowledge (#PF5728)
Limited funding expertise (#PJ0278)
Inoperative forums for public information (#PF7805).

♦ PF1269 Retirement as a threat to psychological well-being
Nature The break in working and living habits occurring on reaching retirement age is one of the most critical points in a person's life. It is invariably a shock, in part brought about by the loss of human contact with people at work and by the permanently diminished income represented by the pension. The individual may experience difficulties in the redirection of interests and feel fear at his or her decreasing mental, physical and financial ability to cope with rapidly changing living conditions. Psychological consequences are often periods of gloom, nervous tension, unsociability, touchiness and a tendency to introversion. There is often a feeling of social uselessness, a fear of illness and obsession with death.
Refs Willing, Jules Z *The Reality of Retirement* (1981).
Broader Human ageing (#PB0477).
Aggravates Suicide (#PC0417) Neurasthenia (#PE3520) Fear of death (#PF0462)
Alcohol abuse (#PD0153) Stress in human beings (#PC1648)
Social withdrawal of aged (#PD3518) Mental disorders of the aged (#PD0919)
Inadequate housing for the aged (#PD0276) Age discrimination in employment (#PD2318)
Inadequate recreational facilities (#PF0202)
Susceptibility of the old to physical ill-health (#PD1043).
Aggravated by Ageing of world population (#PC0027)
Inadequate income in old age (#PC1966)
Unemployment of older people (#PE5951)
Social disadvantage of the aged (#PD3517).

♦ PF1270 Religious superstition
Related Religion as an opiate (#PF9350).
Aggravated by Superstition (#PA0430).

♦ PF1275 Overemphasis on rapid returns on investment
Nature The investment of surplus capital, by both the private and public sectors, tends to be directed towards the rapid accumulation of capital returns. Capital is not available for the broad-based acquisition of theoretical technological knowledge and its practical applications, resulting in minimal development of the human resources crucial to a stable economy. Without this kind of investment in the future, the economically weak nations of the world will be unable to develop industry and business enterprises which would provide them with the means to enter the world market in a competitive fashion.
Broader Self-interest driven investment (#PC2576).
Aggravates Risk of capital investment (#PF6572).

♦ PF1277 Cyclic business recessions
Economic depressions — Economic stagnation — Economic cycle of boom and depression
Nature In a free market economy, business activity is characterized by cyclic expansions and contractions within very long periods of linear average annual growth. The contractions are known as recessions or, when more serious, as depressions. During a recession a decline of production is accompanied by a reduction in the number of jobs, resulting in a reduction of income and consumer spending. Distributors consequently reduce their orders to manufacturers who in turn reduce their demand for commodities which therefore suffer production or price decreases. At the same time wages are maintained or even increased, thus reducing business profits and increasing the number of bankruptcies. Investment in new ventures is reduced thus reinforcing the contraction, which may then spiral into a depression. The likelihood that a depression will develop depends on factors such as: scale of speculation, quality of credit, excess capacity, magnitude of national debts balance of payments, and saturation of markets.
Cyclic business effects often spread from one country to another and may engulf much of the world economy. Obstacles to foreign trade, extremes in commodity prices, depressed stock prices, high interest rates and artificially valued leading world currencies play a vital role in this process of transmission, both directly and through their influence on business psychology.
Claim Recent recessions have forcefully demonstrated how much more interrelated national economies have become. It is evident that the restrictive policies of the powerful countries have onerous consequences for the weak; it is also striking that this weakening of the poorer countries has posed a threat both to the viability of financial institutions serving the strong, and to the recovery of world production and trade. All recessions bring about a worsening of living conditions for large groups of people in most societies, and an aggravation of social tensions and cleavages.
Refs Burton, Theodore E *Financial Crises and Periods of Industrial and Commercial Depression*; Lightner, Otto C *History of Business Depressions* (1970).
Broader Chance (#PA6714) Economic uncertainty (#PF5817).
Narrower Business depression (#PU2316).
Aggravates Strikes (#PD0694) Unemployment (#PB0750) Business bankruptcy (#PD2591)
Military-industrial complex in capitalist systems (#PC3191)
Instability in export trade of developing countries producing primary commodities (#PD2968).
Aggravated by Abuse of credit (#PF2166) Economic inflation (#PC0254)
Imbalance of payments (#PC0998) Capitalist speculation (#PC2194)
Disruption of financial markets (#PD4511) Competition in capitalist systems (#PC3125).

♦ PF1278 Inhibiting shyness
Refs Jones, Warren H, et al *Shyness* (1986); Powell, Barbara *Overcoming Shyness* (1981); Zimbardo, Philip *Shyness* (1987).
Broader Rigidly entrenched social traditions in rural areas (#PF1765).
Aggravates School phobia (#PE4554).

♦ PF1286 Proliferation of second homes
Nature Longer holidays and higher incomes have led to families in developed countries being increasingly willing and able to buy second homes. Numbers of people buy property for a weekend holiday retreat, for retirement, or as an investment at home or abroad. The reservation of two homes for one family decreases the number of available houses at a time when many people are either homeless or ill-housed. Construction of such houses forces house prices up in areas such as low-income farming communities. Whilst in some cases renovation of historic buildings as second homes may have a valuable environmental impact, homes which take the form of caravans, chalets or apartment buildings exert extra pressures on beautiful areas and alter the character of a small neighbourhoods.
Incidence The second home phenomenon is international: Arabs buy in London, the Dutch in the British and French countryside, the British in Spain and the USA. It is estimated that in the UK there are about 700,000, or 5 percent, second homes. At the current rate of increase it is estimated that by the year 2000, 10 per cent of UK households could own a second home. This level is believed to be lower than that of other European countries. At the same time, there are some thousands of homeless, and nearly 20 per cent of the housing stock is officially classified unfit or in unsatisfactory condition. There are an estimated 4 to 5 million second homes in the USA and some 100,000 are added each year.
Counter-claim Second homes can play a significant role in the economic development of depressed regions and can bring new life into historic buildings which would otherwise not be adequately preserved. They may also help to reclaim derelict land.
Broader Personal wealth (#PC8222) Unnecessary personal consumption (#PF5931).
Aggravates Inadequate housing (#PC0449) Absentee ownership (#PD2338)
Shortage of cultivable land (#PC0219)
Long-term shortage of natural resources (#PC4824).

♦ PF1288 Computer obsession
Computer addiction
Nature Long-term computer users may become progressively enmeshed in the computer time frame, becoming less and less able to adjust to the temporal norms and standards of traditional clock time. Caught between two distinctly different temporal orientations, they become victims of a new form of temporal schizophrenia. Their adjustment to the accelerated computer time frame renders them increasingly impatient with the slower tempo of everyday life. They become more intolerant of behaviour that is ambiguous, digressive or tangential, avoiding open-ended discourse and any form of inefficiency in communication processes.
Related Socially disruptive effects of video games (#PF6345).
Aggravates Computer stress (#PE5053) Increasing pace of life (#PF2304)
Social isolation of computer users (#PE1277)
Health hazards of computer visual display units (#PE5083).
Aggravated by Proliferation of computers (#PE3959).

♦ PF1297 Lack of livelihood standards
Nature There are no universally accepted, comprehensive standards on working conditions to provide a context for determining what adequate livelihood is in a local community. Standards of livelihood are currently determined either in the context of profit-making, or in the context of community tradition and mores. What is missing are global standards from which local differences can be exercised as intentional diversity in relation to the good of a community.
Broader Limited image of employability (#PF2896)
Inadequacy of international standards (#PF5072).

Aggravates Inadequate standards of living (#PF0344)
Deterioration of living standards of workers (#PE9075)
Subsistence agricultural income level in rural communities (#PE8171).
Aggravated by Standards setting procedure (#PS3851).

♦ PF1301 Deteriorated structures of essential corporateness

Nature At a time when there is a general awareness of the interdependence of all facets of human society, many Third World villages find the effectiveness of their social and economic operations diminished due to traditional factionalism derived from past necessity. Social structures were designed to provide community care and to assign everyone a necessary role. Economic survival patterns demanded radical individualism for sheer self-preservation, but in the present day context these patterns actually diminish effective participation. The task for these villages is to develop a uniquely creative role in the context of the wider, more complex human community.
Broader Non-cooperation (#PF8195).
Narrower Traditional market practices (#PJ9742) Traditional land distribution (#PJ1289)
Lack of leadership initiative (#PF7988) Unpublished educational possibilities (#PJ9445)
Unrecognized benefits from cooperatives (#PF9729).
Related Lack of water conservation (#PJ3480) Unrepresentative formal leaders (#PJ1823)
Lack of means for achieving consensus (#PD2438).
Aggravates International conflict (#PB5057) International insecurity (#PB0009)
Interpersonal estrangement (#PB0034) Parent–teacher non communication (#PF1187)
Lack of international cooperation (#PF0817) Limited availability of financial credit (#PF2489).
Aggravated by Parochial national interests (#PF2600)
Insufficient community events (#PF5250).
Reduced by Food interdependence (#PJ0512).

♦ PF1303 Limited access to social benefits

Nature While present day expectations are for everyone to receive the services they require, some small communities are finding that although such services are theoretically available, they are not always available in practice. In addition, residents encounter difficulties when trying to find the proper office among numerous bureaux for receiving assistance, so that they are reluctant to seek services which are available. Access to financial and commercial services may be difficult if the community is regarded as a bad investment risk. All of these factors lead to a dependence on outside services and a lack of initiative in claiming the benefits to which a community is entitled.
Broader Limited access to external resources (#PF1653)
Unequal distribution of social services (#PC3437).
Narrower Delay (#PA1999) Lack of communication (#PF0816)
Bureaucratic decision-making (#PG8413) Dependence on social welfare (#PD1229)
Insufficient financial resources (#PB4653) Inadequate social welfare services (#PC0834)
Inadequate dissemination and use of available information (#PF1267)
Inadequate water supply in the rural communities of developing countries (#PD1204).
Related Inadequate financial services (#PJ8366) Inadequate maintenance equipment (#PD6520)
Underprovision of basic services to rural areas (#PF2875).
Aggravates Social injustice (#PC0797) Social alienation (#PC2130)
Social fragmentation (#PF1324) Reliance on canned food (#PJ8409)
Vulnerability of social systems (#PB2853)
Economic and social underdevelopment (#PB0539)
Limited availability of financial credit (#PF2489)
Over-acceptance of socio-economic dependency (#PF8855)
Fragmentation of social structures in depressed areas (#PD1566)
Inadequate community care for transient urban populations (#PF1844).
Aggravated by Limited social services (#PG8629)
Complex legal procedures (#PF8519)
Unresponsive social services (#PG5344)
Inadequate social development programmes (#PF4180)
Social insecurity in developing countries (#PE4796)
Absence of images of social responsibility (#PF3553)
Governmental disregard for people as human beings (#PD8017)
Excessive bureaucratic requirements for welfare benefits (#PE8893).

♦ PF1309 Obsolescence of rituals and customs

Underutilization of historical rites and customs — Dormant historical celebrations
Nature The isolation of the older generation within the community has resulted in the rituals and customs they would normally pass to future generations either becoming obsolete or being transmitted as rigid rites and fossilized customs which cannot be interpreted in the absence of the older people or made relevant to present family and community living styles. Indeed because of the knowledge gap imposed by their isolation, elders are left unable to recreate the past in terms of the future.
Broader Preservation of obsolete systems (#PC8390)
Untransposed significance of cultural tradition (#PF1373).
Aggravates Loss of cultural identity (#PF9005).
Aggravated by Social isolation of the elderly (#PD1564)
Denial by old people of the significance of the past (#PF2830).

♦ PF1312 Introduction of extraterrestrial infectious diseases and bacteria

Nature Organic debris from comets have started life on earth, dominated evolution and are responsible for the waves of disease which sweep the planet. Organic compounds have been discovered in the tail of Halley's Comet some of which are about the size of viruses and others of which could be dessicated bacteria. As the earth passes through the tail of a comet some of the material penetrates the atmosphere. Particles the size of bacteria may take a few weeks to fall through the air, particles the size of viruses may take 10 years or more to descend. More serious, long-lived viruses like small pox an now the Aids virus HTLV III, originated from space then spread from person to person. Small pox could keep going for 200–300 years once it had arrived. Aids began in a shower of organic material on central Africa 10 years ago.
Broader Disease vectors (#PC3595).
Aggravates Human disease and disability (#PB1044).
Aggravated by Extraterrestrial invasion (#PF4444).

♦ PF1316 Unimaginative vision of resource utilization

Nature In small mining communities, natural resources other than the mine (such as unutilized land, unused water resources or natural forests of valuable timber) may often be ignored and undeveloped. The land is usually owned and maintained by outside agencies, and is not always seen as an untapped reservoir of economic potential, so that land resources are undeveloped and the community does not receive the benefits of this economic resource. Even if, as is often the case, there is a sizeable amount of land that has been cleared of housing and is now unoccupied, the legal procedures for any new ventures are complex, lengthy and challenging. Plans to make lakes into recreation areas may be abandoned because of conflicting demands for irrigation by farmers downstream. Farmers, however, need water rights. Any new industries will need not only water-power and land, but also supporting services.
Broader Narrow economic foresight (#PF8602) Unchallenging world vision (#PF9478).
Narrower Unresolved legal issues (#PG8289) Unutilized local resources (#PG8284)
Restrictive use of available land (#PF6528)
Unutilized local resources (#PG8284)
Restrictive use of available land (#PF6528)
Overexploitation of underground water resources (#PD4403).
Related Limited available land (#PC8160).
Aggravates Rising water level (#PD8888) Underutilization of natural resources (#PF1459).

♦ PF1319 Restrictions on the acquisition of knowledge

Claim The scientific inventions of this century have altered and widened the skills required for effective participation in society. Traditional functional skills are no longer sufficient. Although village residents show remarkable resourcefulness and highly developed skills, modern techniques are virtually unknown; and there is an unfulfilled desire to acquire relevant training for the adoption of new methods. Few have the opportunity to participate in special educational programs that give mechanical skills, literacy, English, marketing skills, etc. There needs to be a system for implementing such training at the local level, and for creating the supporting mindsets and mores for such programs.
Narrower Illiteracy (#PC0210) Innumeracy (#PC0143) Limited verbal skills (#PD8123)
Outdated farming techniques (#PG9687) Monopolization of knowledge (#PF5329)
Limits to societal learning (#PF7074) Inaccessible educational facilities (#PD9051)
Minimal opportunities of adult training (#PF6531)
Increasing lag in education against population growth (#PE5369)
Barriers to the international flow of knowledge and educational materials (#PF0166).
Related Inadequate maintenance personnel (#PJ0088).
Aggravated by Manipulative knowledge (#PF1609).

♦ PF1321 Gap between material and technological needs and demands

Claim The needs of society as expressed by what the consumer feels he or she wants may bear no relationship to the resources available. The disparity arises between the conflicting demands of producers, consumers and society as a whole, with no overall plan as to what materials are put into the distribution process; so that materials lie dormant for lack of creative industrial planning, and waste products fail to be recycled because only economic factors are taken into consideration.
Broader Imperialistic distribution system (#PD7374).

♦ PF1324 Social fragmentation

Lack of integration in society — Lack of social integration
Refs Brunner, R and Crecine, J P *A Fragmented Society.*
Broader Fragmentation (#PA6233) Social injustice (#PC0797)
Social inequality (#PB0514) Plural society tensions (#PF2448).
Narrower Segregation (#PC0031) Schizmogenesis (#PE4593)
Lack of assimilation (#PF2132) Generation communication gap (#PF0756)
Fragmented patterns of community activity (#PF6504)
The split between eastern and western Europe (#PE4111)
Fragmented individual decision-making process (#PF3559)
Inadequate integration of ideology into society (#PF3402)
Fragmented forms of care at the neighbourhood level (#PE2274)
Fragmentation of social structures in depressed areas (#PD1566)
Fragmented social structures for environmental protection (#PE3977).
Related National federalism (#PF0626) Lack of racial identity (#PF0684)
Inadequate integration of religions into society (#PF3403).
Aggravates Suicide (#PC0417) Prejudice (#PA2173) Secession (#PD2490)
Scapegoats (#PF3332) Paternalism (#PF2183) Social conflict (#PC0137)
Desert nomadism (#PD2520) Ethnic conflict (#PC3685) Threatened sects (#PC1995)
Social breakdown (#PB2496) Gypsy persecution (#PE1281) Forced assimilation (#PC3293)
Lack of cooperation (#PF2816) Social disintegration (#PC3309) Ethnic discrimination (#PC3686)
Underprivileged minorities (#PC3424) Lack of community participation (#PF3307)
Underprivileged racial minorities (#PC0805) Denial of education to minorities (#PC3459)
Illiteracy among indigenous peoples (#PD3321) Underprivileged religious minorities (#PC2129)
Threatened and vulnerable minorities (#PC3295) Underprivileged linguistic minorities (#PC3324)
Adjustment difficulties of new urban families (#PF1503).
Aggravated by Schism (#PF3534) Nativism (#PF2186)
Tribalism (#PC1910) Lesbianism (#PF2640)
Alienation (#PA3545) Intolerance (#PF0860)
Anti-science (#PF2685) Traditionalism (#PF2676)
Racial conflict (#PC3684) Male prostitution (#PD3381)
Social intimidation (#PC2940) Racial exploitation (#PC3334)
Religious repression (#PC0578) Illegitimate children (#PC1874)
Social discrimination (#PC1864) Sexual discrimination (#PC2022)
Single parent families (#PD2681) Refusal to participate (#PF3226)
Inadequacy of doctrine (#PF3396) Segregation in housing (#PD3442)
Impediments to marriage (#PF3343) Psychological alienation (#PB0147)
Segregation in employment (#PD3443) Discrimination in politics (#PC0934)
Multidenominational society (#PF3368) Dictatorship of the majority (#PD3239)
Double standards in morality (#PF5225) Legal impediments to marriage (#PF3346)
Obsolete deliberative systems (#PD0975) Social impediments to marriage (#PF3341)
Segregation in social services (#PD3440) Conflict between minority groups (#PC3428)
Inadequacy of political doctrine (#PF3394) Discrimination against minorities (#PC0582)
Socially unintegrated expatriates (#PD2675) Discrimination in social services (#PC3433)
Limited access to social benefits (#PF1303) Inadequate ideological frameworks (#PD0065)
Discrimination against homosexuals (#PE1903) Racial discrimination in education (#PD3328)
Double standards of sexual morality (#PF3259)
Racial discrimination in public services (#PD3326)
Family structure as a barrier to progress (#PF1502)
Segregation based on religious affiliation (#PC3365)
Disruption of development by tribal warfare (#PD2191)
Discrimination against illegitimate children (#PD0943)
Threats to ideological movements and minorities (#PC3362)
Failure of individuals to participate in social processes (#PF0749).
Reduced by Nationalism (#PB0534) Interdependence (#PG6428).

♦ PF1332 Limited development of functional abilities

Nature The doubling of human inventions during the past decade has created a gap between the common sense previously adequate for practical productivity and the functional living skills now required for effective social participation. Although local communities may need and wish to profit from modern technology, without the functional skills to provide them with the practical means, people have no way to develop their community in harmony with the rest of the world. Simply for means of survival, traditional skills are transmitted from generation to generation, often as soon as a child can walk; for example, a three year old may be started into a future of ropemaking. Even the limited formal and technical training available competes with time which needs to be used for production, making it virtually impossible to explore and tap other communities' operations and specialized resources. Such experience provides no means of imagining practical alternatives to the proven inadequacy of traditional methods, leading to undirected fears of inevitable catastrophe: livestock will perish, crops will fail, drought will hit, sons will not be born to relieve the burden of marginal subsistence. The minority who have received formal academic training soon discover that it does not relate to the real needs for operational knowledge in water resource development, agricultural techniques, health care methods, leadership facility or industrial variation. Indeed, because academic skills appear unmarketable in the

village context, those few who do develop them eventually move away.
Broader Inadequate education (#PF4984).
Narrower Insufficient trained leaders (#PJ0389) Inadequate vocational education (#PF0422).
Related Limited basic skills (#PJ8138) Limited technical skills (#PF8594).
Aggravates Unmotivated teachers (#PF5978) Defensive life stance (#PF0979)
Limited verbal skills (#PD8123) Exploitation of child labour (#PD0164)
Underdeveloped farming skills (#PJ0729) Inadequate marketing knowledge (#PJ9659)
Undeveloped business skills in urban areas (#PE8048)
Illiteracy as an impediment for leadership (#PE8177).
Aggravated by Limited job market (#PC7997) Data-oriented education (#PF1217)
Inadequate teacher training (#PJ1327) Unused training opportunities (#PJ1280)
Unprofitable traditional skills (#PJ1031) Limitations on school admission (#PJ1364)
Static and unrelated social roles (#PF1651) Elimination of traditional skills (#PD8872)
Unrecognized relevance of education (#PF9068)
Incomplete access to information resources (#PF2401).

♦ PF1333 Polarization of local conflicts

Unresolved local polarization — Argumentative community meetings

Nature A conflict becomes polarized when the support to opposing sides of competing (arms supplying) countries subsumes the local dispute in the dispute between the supplying countries themselves. The transformation of border, ethnic or class conflicts into cold war issues implies that resolution will have to be sought within the international context. Because more parties are involved, the conflict may be harder to resolve. Polarization alters the supplying country's commitment because the delicate distinction between military assistance and military intervention cannot always be maintained. The intervention of supplying countries not only exacerbates wars once they have broken out, but also increases the risk of direct confrontation between them.
Broader Polarized protest against problems (#PF9691)
Obstacles to community achievement (#PF7118)
Ineffective structures for community decision-making (#PF1781).
Related Conflict (#PA0298) Unformed style of cooperative action (#PF6514).
Aggravates War (#PB0593).
Aggravated by Inappropriate arguments (#PF2152)
Oppressive prevalent images (#PF1365).

♦ PF1336 Indecisive response to technological changes

Claim Although technology is capable of providing for all material needs, all needs are not in fact being met. To actually meet all needs would mean commitment to changes in lifestyle in every nation. Responses to demand for such change tend to be: romantic about the ease "in principle" of solving the problem; or emotional about its scope; or an ignoring of the magnitude of the problem and of personal responsibility for it.
Broader Uncontrolled application of technology (#PC0418).

♦ PF1338 Fundamentalism

Religious fundamentalism — Revivalism

Nature The uncritical adherence to ancient, allegedly inspired religious writings or oral traditions as an exact guide to what has happened in the past and to what will happen in the future, as well as viewing such writings or traditions as an infallible and ritualistic guide to behaviour in the present, characterizes one aspect of fundamentalist or revivalist creeds in the major religions. Although revered authorities in the world religions have from ancient times insisted on approximately four levels of interpretation to be given to scripture (historical, literal, allegorical or metaphoric, and anagogical or inwardly spiritual) fundamentalists disregard such exegesis and use only the first two levels.

Incidence Protestant fundamentalists emphasize the accuracy of the Book of Genesis on the Creation, the Gospels on the Virgin Birth of Christ, and Revelations on the Second Coming of Christ and the final battle of Armageddon. Catholic fundamentalists, in addition, emphasize the prophetic character of the Old Testament and add the non-scriptural doctrine of the Assumption of the Virgin. Islamic fundamentalists adhere to the Koran and reject portions of Islamic canonical law. All fundamentalism is anti-intellectual. Worship tends to be emotional, searching for ecstatic and charismatic experiences. Large followings are built quickly and sometimes subverted to political ends.

Claim Fundamentalist simplicities are being embraced not by an oppressed and persecuted minority, but by some of the wealthiest, most comfortable, most powerful and best educated people in the world. This in effect endangers much of what civilization has learned in Biblical and evolutionary study and perhaps more important in humanity and tolerance. Not since Cromwell's Protectorate in Britain and the witch trials of America and Europe has religious fanaticism and bigotry been allied with wealth and power on so large a scale. Except, of course, for the Third Reich.

Counter-claim The strict adherence to a received religion, to its texts, to its traditions and to its teachings assures the continuity and coherence of a culture, high moral and altruistic ideals, and a mutually-supportive environment in which spiritual and personal development can be fully realized. In fundamentalist faiths the family is preserved against the decadence of the surrounding society, and its physical health is apt to be at a higher level because of the cleanliness of its life. Fundamental faiths have founded nations, and are the bulwark against the flux of contemporary confusion and national fragmentation. It has provided a consolation for the helpless, counselling resignation while at the same time proffering hope. In this capacity it performed a therapeutic role for nineteenth-century Jewish ghettos in Eastern Europe and for black communities in the 19th century American South.

It is fashionable for some to pretend to be suspicious of all apologias and enthusiasms. Yet, they too are apologists for one belief or another. Even scepticism and relativism have their ardent believers. The absolute relativist is no less dogmatic than the Christian or Muslim fundamentalist and frequently more arrogant.
Broader Ideological conflict (#PF3388).
Narrower Religious rivalry (#PC3355) Islamic fundamentalism (#PF6015).
Related Jihad (#PF5681).
Aggravates Suttee (#PF4819) Bibliolatry (#PF7129) Religious extremism (#PF4954)
Unauthorized proximity of males to females (#PF8780).
Aggravated by Religious intolerance (#PC1808) Double standards in morality (#PF5225).

♦ PF1345 Underdeveloped sources of income expansion

Absence of imaginative planning for expansion of income

Nature Generally the trend toward effective income generation is accompanied by larger production units employing substantial capital and specializing in increasingly narrower ranges of products. In small villages the limited capital of family farms and businesses contributes to the fragmentation and marginalization of production. Income levels of these businesses are often so low that in reality they are ways of passing time rather than providing employment. Family production units operating at marginal levels with business and family income inseparable are susceptible to the smallest of cash emergencies. Besides, imaginative thinking for future income is inhibited in many smaller communities by the experience of past business failures and a rugged determination to eke out a living from the diminishing of capital. Although small businesses still operate, many others have closed due to retirement or a lack of incentive to continue at a minimal profit. There is no cooperation or organized effort to promote and upgrade commercial life; and many communities grow to depend on outside sources for income, together with only the most convenient, easily accessible local resources. Limited capital prevents the development of local processing so raw materials are sold unprocessed. The absence of local market forces dependence on middlemen services. Locally produced higher quality food is sold and low quality food for local consumption is purchased outside the village. For the most part residents seek services, goods and employment in larger cities. To a large degree the long term trends of income for many communities is are producing a net outflow of income.
Broader Underdevelopment (#PB0206) Lack of community planning (#PF2605)
Short range planning for long-term development (#PF5660).
Narrower Short-term gain (#PF8675) Frustrated past goals (#PF5272)
Subsistence life style (#PF1078) Insufficient cash crops (#PG0569)
Single resource thinking (#PG8548) Underpayment of teachers (#PE8645)
Export of nutritious food (#PJ1365) Lack of savings structures (#PF1348)
Prohibitive irrigation costs (#PG8730) Slow rate of income expansion (#PF6478)
Traditional land distribution (#PJ1289) Unnecessary business closings (#PG8608)
Declining economic conditions (#PG8673) Unconnected business resources (#PG8636)
Scarce employment possibilities (#PG8652) Patronizing of other communities (#PG8665)
Landsize as a limit to diversity (#PE8279) Unavailability of training costs (#PJ8395)
Insufficient financial resources (#PB4653) Uncoordinated economic enterprises (#PG8588)
Unmotivating subsistence employment (#PJ1555) Limited availability of financial credit (#PF2489)
Underdeveloped capacity for income farming (#PF1240)
Continuing expectation of community decline (#PG8680)
Underdeveloped use of agricultural resources (#PF2164)
Underproductivity of draught animal power in developing countries (#PF0377).
Related Limited market development (#PF1086) Lack of water conservation (#PJ3480)
Short-term planning of product life cycles (#PF1740)
Dependency on unpredictable sources of income (#PF3084).
Aggravates Unexplored potential markets (#PF0581)
Inadequate feeding of animals (#PC2765)
Prohibitive cost of education (#PF4375)
Subsistence-level malnutrition (#PJ1370)
Marginal level of family income (#PD6579)
Mismanagement of irrigation schemes (#PE8233).
Aggravated by Unemployed educated youth (#PE1379)
Limited employment options (#PF1658)
Underutilization of locally available skills (#PF6538).

♦ PF1348 Lack of savings structures

Lack of organized savings funds — Lack of saving foresight — Immediacy prevents saving
Broader Insufficient financial resources (#PB4653)
Underdeveloped sources of income expansion (#PF1345)
Limited local availability of capital reserve (#PF2378).
Aggravates Inadequate management skills in rural communities (#PF1442)
Diminishing capital investment in small communities (#PF6477)
Non-diversification in subsistence fishing economies (#PF2135).
Aggravated by Rural poverty in developing countries (#PD4125).

♦ PF1349 Child pornography

Broader Pornography (#PD0132) Sexual exploitation of children (#PD3267).
Aggravates Child prostitution (#PE7582) Exposure of children to pornography (#PJ8730).
Aggravated by Corruption of minors (#PD9481) Sexual abuse of children (#PE3265)
Trafficking in children for sexual exploitation (#PE6613).

♦ PF1352 Ineffective mechanisms for functional training

Nature As home maintenance, community affairs and regular employment become more complex, the basic skills required for everyday social life at home and at work are also becoming increasingly sophisticated. Families in rural communities are without the skills they need to manage the complexity of their homes, budget, social rights or even family life. However, no extra training is provided to acquire such skills. This is especially true in the many rural areas where the burden of social participation lies mainly with the individual. Some people have a strong desire to engage in social activities but do not have the functional know-how to allow their participation. Although some effort is being made to upgrade public education systems by providing supplemental programs and access to vocational courses and specialized training, the skills needed for daily living are not taught or available.
Broader Obsolete group methods (#PG8406)
Ineffective systems of practical education (#PF3498).
Narrower Lack of apprentice training (#PJ8498) Inadequate training opportunities (#PJ8697).
Aggravates Consumerism (#PD5774) Inadequate education (#PF4984)
Incompetent management (#PC4867) Limited technical skills (#PF8594)
Poor commercial practices (#PJ8368) Inadequate vocational education (#PF0422)
Lagging training in social skills (#PF8085) Underutilization of human resources (#PF3523).

♦ PF1357 Obscurantism

Dependence on obscurantism

Nature Because most concepts of science and the professions are relatively simple, once one understands them, any ambitious scientist or professional must, in self defence, prevent his colleagues and audience from discovering that his ideas are simple too. All the professional has to do is speak or write obscurely so that no one will really attempt to understand the concepts being presented but will be awed by erudition. The methods of obscurantism are many fold. Ideas are abstracted beyond necessary. Reasoning and sentence structure are convoluted. Facts are distorted. Words are used that are ambiguous or equivocal. Meaningless phrases are inserted in sentences. Words are invented and not explained. Minor points are expanded to the exclusion of major ones. The meaning or consequences of a fact are described in great detail without reference to the reality.
Broader Lack of communication (#PF0816).
Narrower Misleading information (#PF3096)
Incomprehensibility of specialized jargon (#PF1748)
Excessive use of acronyms and abbreviations (#PF4286).
Aggravates Elitism (#PA1387) Underdevelopment (#PB0206)
Acceptance of hierarchy (#PJ3602).

♦ PF1358 Fortune-telling

Divination — Fortune-tellers

Incidence It has been estimated that in France there are 30,000 people offering the fortune-telling services with an annual turnover of 5 billion francs.

Background Divination is the art of foreseeing or foretelling future events or of discovering hidden or secret knowledge. There are basically two kinds: natural and artifical. Natural divination includes oneiromancy, based on dreams and chresmology based on oracles. Artifical, or inductive, divination is of many types. Ornithomancy is based on the flight, cries and eating of various species of birds. Cledonomancy is based on observing human signs, actions or utterances and includes palmistry. Extispicy is based on the examination of the entrails of animals. Pyromancy is based on observing the actions of wood, bone, eggs, flour or incense when thrown on a fire or the actions of the fire itself. Hydromancy is based on the action of water when various things are thrown on it. Cleromancy is divination by lots and includes a wide variety of types. Meteorological divination includes observation of lightning, shooting stars, meteorites and earthquakes. Astrology is well known. Christianity and Judaism generally oppose all forms of divination. Islam, Buddhism, Hinduism and other religions of the Middle and Far East use divination as a part of their traditions as do most less formal traditions.

Broader Superstition (#PA0430) Horological superstition (#PF0565).
Related Necromancy (#PF8042).

♦ PF1365 Oppressive prevalent images

Images of ineffectivity — Symbols of despair

Claim Many people in rural communities find the day-to-day dependence on an uncertain water supply, and the occurrence of periodic severe droughts, together with the hopelessness engendered by the physical surroundings - the deteriorating walls of unused buildings, the persistent dust and the colourlessness of the mud wall construction which confront the villager at every turn - reinforce rural images and drive them to see themselves as at at the mercy of uncontrollable forces. The disparity between the great accomplishments of the past and the sense of oppression of the present becomes immobilizing and overwhelming. Taken together with a deeply fatalistic attitude to life, such primary images ensure that no profound or lasting social and economic change can occur without creative transformation of attitude.

Broader Despair (#PF4004) Demeaning community self-image (#PF2093).
Narrower Demeaning farmer image (#PJ9781) Community demoralization (#PG1266)
Debilitating poverty image (#PJ1341).
Related Hopelessness inspired by school (#PJ1004).
Aggravates Impure thoughts (#PF5205) Disillusionment (#PA6453)
Citizen powerlessness (#PJ8803) Sense of powerlessness (#PF8618)
Lack of self-confidence (#PF0879) Minimal citizen confidence (#PF8076)
Resignation towards bribery (#PF8611) Undeveloped youth leadership (#PJ0151)
Unexplored potential markets (#PF0581) Polarization of local conflicts (#PF1333)
Deterioration of human environment (#PC8943) Lack of local leadership role models (#PF6479)
Unperceived educational opportunities (#PJ9762)
Demoralizing image of urban community identity (#PF1681)
Lack of local leadership and decision-making structures (#PF6556)
Declining community confidence in its ability to change (#PF9066)
Neglect of agricultural and rural life in developing countries (#PF7047).
Aggravated by Inadequate hero images (#PF2834)
Incongruous religious images (#PG1418)
Individual fear of future change (#PD2670).

♦ PF1373 Untransposed significance of cultural tradition

Claim At the same time when it is increasingly necessary for a society's heritage to provide cultural identity and cohesion during times of change, many small communities find themselves in a world very foreign to their traditions. The traditional rites of passage for youth are slowly disappearing; and the community's past is not re-enacted to provide young people with models to follow, and on which to base their own image of who they are and why they were born into their particular community. Children know few, if any, of the basic traditions and legends of their culture. Without recovering past heritage and relating it to the present day, the possibility of maintaining a community's unique culture is severely limited.

Broader Socio-cultural environment degradation (#PC4588).
Narrower Lost farming skills (#PJ8369) Unused cultural resources (#PG8319)
Neglect of basic necessities (#PG8388) Insufficient cultural heroes (#PF8623)
Reduced agricultural reliance (#PG8376) Insufficient traditional gatherings (#PG9896)
Obsolescence of rituals and customs (#PF1309) Detrimental story of community future (#PF6575)
Inadequate national language proficiency (#PG8362)
Untransposed creativity of traditional gifts (#PF2703).
Related Demeaning farmer image (#PJ9781) Obsolete basis of cultural identity (#PF0836)
Local traditions of cultural isolation (#PF1696).
Aggravates Value erosion (#PA1782) Cultural decline (#PC9083)
Cultural corruption (#PC2913) Cultural stagnation (#PC8269)
Cultural disintegration (#PG3299) Loss of faith in religion (#PF3863)
Loss of cultural identity (#PF9005) Loss of linguistic tradition (#PG6175)
Traditional land distribution (#PJ1289) Ignorance of cultural heritage (#PF1985)
Disappearance of local culture (#PF3012) Unexplored energy alternatives (#PF7960)
Destruction of cultural heritage (#PC2114) Elimination of traditional skills (#PD8872)
Restrictive patterns of traditional life (#PF3129)
Divisive patterns of community groupings (#PF6545)
Contempt for traditional modes of behaviour (#PC4321)
Obsolete methods of agricultural production (#PF1822)
Individualistic retaining of local tradition (#PF1705)
Inhibiting effects of traditional life-styles (#PF3211)
Inappropriate application of traditional values (#PF2256)
Modern disruption of traditional symbol systems (#PF6461)
Rigidly entrenched social traditions in rural areas (#PF1765)
Decline in rural customs and traditions in developing countries (#PD1095)
Loss of traditional forms of social control in developing countries (#PD0144)
Loss of traditional forms of social security in developing countries (#PD1543).
Aggravated by Rigid cultural patterns (#PF8598)
Rapidly changing cultures (#PF8521)
Inadequate appreciation of culture (#PF3408)
Unclarified procedures for transposing ancient traditions (#PF6494).
Reduced by Overpowering traditional habits (#PJ0453)
Absence of traditional patterns of community life (#PF3531).

♦ PF1374 Underdevelopment of basic metal industries

Broader Underdevelopment of manufacturing industries (#PF0854).
Narrower Underdevelopment of iron and steel basic industries (#PE8223).

♦ PF1379 Unsafe design of consumer products

Hazardous products

Incidence Examples of hazardous consumer products that can cause injury or death include, but are not limited to: unvented gas space heaters; rotary lawn mowers; aluminium electrical wiring; architectural glass used in homes; children's toys, specifically play equipment such as swings, slides, and seesaws; babies night dresses; tyres; medical products; electrical blankets; washing machines; and irons.

Broader Consumer vulnerability (#PC0123) Grievances of consumers (#PD7567)
Defects in machinery design (#PE2462).
Narrower Unsafe blood-related products (#PJ4536)
Inadequate recall procedures for unsafe products (#PJ8035).
Reduced by Consumer boycotts (#PE1213).

♦ PF1382 Precarious basis for family economics

Family money priorities

Nature Although many rural families are moving towards a standard of living in which more than their simplest needs are provided for, the basis for such a development is precarious. Average family income is barely adequate for survival: garden produce is limited to what the family can use; the jobs which support most families are seasonal; since work depends heavily on fish, agriculture, hunting or trapping, income is neither steady nor predictable; if a resident leaves the community looking for work, there is no interim income for his family until his first pay cheque arrives; in comparison with nearby towns, local prices are high and goods are limited; general health is seriously hampered by the fact that families cannot afford medical or ambulance expenses; most families cannot afford telephones. The constant struggle to sustain the family is reflected in immediate purchasing and spending patterns. Short-term cash purchasing is virtually the only mode of procurement. Consumer spending is sporadic and bills tend to be paid late. Until the economic basis for the family is placed on a more solid footing, there will be little energy or income available for common development efforts.

Broader Rural unemployment (#PF2949).
Narrower Endangered family farms (#PD5962) Late bill-paying patterns (#PG9251)
Unchecked community spending (#PJ9047) Habitual non-payment of bills (#PG8412)
Overriding economic orientation (#PG9239) Lack of interim support for families (#PE7944).
Related Subsistence approach to capital resources (#PF6530).
Aggravated by Short-term gain (#PF8675)
Prohibitive cost of necessities in rural communities (#PF2385)
Subsistence agricultural income level in rural communities (#PE8171).

♦ PF1384 Immoral literature

Pornographic magazines and comics — Obscene literature

Nature Immoral literature includes pornography and books which incite to violence or other reactions against the accepted social morality. Ideologically deviant literature may be classed as immoral by government policy.

Claim Immoral literature leads to depravity. The problem is aggravated by new methods of communication and distribution.

Refs Delacoste, Frederique and Alexander, Priscilla *Sex Work* (1987).

Broader Obscenity (#PF2634) Depravity (#PC8974) Immorality (#PA3369).
Related Pornography (#PD0132) Indecent art (#PE5042).
Aggravates Sexual deviation (#PD2198).
Aggravated by Political repression (#PC1919) Proliferation of commercialism (#PF0815).
Reduced by Censorship (#PC0067).

♦ PF1385 Unjustified military defence policies

Nature National needs for self-defence are claimed by all participants in the world arms race, from the superpowers to the smallest developing country ordering its armaments from European and American suppliers. Many small states maintain an unjustifiable amount of military hardware.

Incidence In the case of the USA and NATO, the policy of mutually assured destruction (MAD), and disavowal of the first use of non-conventional weapons, has now passed to a 'flexible response' justification for using nuclear and chemical weapons first, and to open contemplation of 'first-strike' capability as a defence option.

Claim Classically, an arms build up would have been evidence of aggressive intentions, but modern military defence policies are built on tortuous reasoning and political pressures. Arms build-ups in the name of defence inevitably lead, first, to speculation concerning preemptive strike capability, and secondly, to the temptation to use what were once defensive weapons, offensively.

Broader Inappropriate policies (#PF5645) Competition between states (#PC0114).
Narrower Inadequacy of civil defence (#PF0506).
Related Risk of unintentional nuclear war generated by the strategy of deterrence (#PF4162).
Aggravated by Aggression (#PA0587) Disorganization (#PF4487)
Military insecurity and vulnerability (#PC0541).

♦ PF1389 Paralyzing patterns between villages and administrative structures

Nature Life in many rural villages is still shaped by the limited services offered by outside administrative structures. Small communities have little access to the generally available services of agricultural research and credit, educational facilities, health care, electricity, telephone, paved roads, public transport systems or even clear land ownership records, although the existence and desirability of such services is known. Governmental and business structures are so overwhelmed by requests from thousands of villagers that they cannot respond; while thousands more villagers are so unclear about required procedure that they rarely dare to approach the administration for necessary services. Social groups and political factions may be so polarized as to prevent their effective approach to administrative structures; and the focus of representative leadership on mediating the diverse interests of these community segments may cause it to be it from making effective demands for outside services.

Narrower Political impotence (#PJ1283) Sense of powerlessness (#PF8618)
Social unaccountability (#PC1522) Bureaucratic opposition (#PD7966)
Lack of political leverage (#PG1314) Parochial leadership posture (#PF2627)
Prohibitive cost of fertilizer (#PJ1022) Confusing structural complexity (#PF8100)
Unavailability of library facilities (#PG9116) Restrictions against small enterprise (#PD5584)
Non-viability of local electrical generating capacity (#PJ8524)
Lack of local leadership and decision-making structures (#PF6556).
Related Contained village economy (#PJ0594) Irregular transport services (#PE5345)
Unrepresentative formal leaders (#PJ1823) Middleman control of rural marketing (#PE3528)
Unsystematic use of powerful relationships by rural communities (#PE1101).
Aggravated by Ignorance of protocol (#PG1267)
Ignorance of procedures (#PJ1219)
Limited teacher facilities (#PJ1340)
Ignorance of administration (#PE1234)
Low self image due to illiteracy (#PF9098).

♦ PF1390 Male homosexuality

Gay men — Homosexual cruising

Incidence Homosexual men have, over the last 30 years, created a male homosexual culture equalled in its openness and visibility only by that of classical Greece. In the USA, about 4 percent of adult white males have only homosexual contacts; after puberty about 37 percent had at least one homosexual experience leading to orgasm. Almost half of a sample of homosexual males reported at least 500 different sexual partners, although some claim up to 7,000. One third of homosexual men use bathhouses regularly as meeting places with 62 per cent occasionally participating in this commercial setting for anonymous sex. "Cruising" in the search for sexual partners is an almost universal practice amongst male homosexuals in the USA (in contrast to female homosexuals) and takes place in a variety of public places (lavatories, cinemas, bathhouses, streets and parks). The homosexual culture has encouraged greater sensation-seeking, including coprophiliac interest in anilingus and the insertion of the fingers or the hand in the partner's anus.

Claim The classical Greek conventions governing the propriety of particular sexual acts have little part in a modern unconstrained homosexual culture. Much of its present physical expression would have repelled the Greeks as being absurd or contemptible, while the compulsive and mechanical quest for sexual encounters with strangers would have been incomprehensible to a society which extolled self-mastery and abhorred anonymity. The passive role in anal intercourse, relegated by the Greeks to powerless women or slaves, would have been inconceivable as a chosen activity of free men.

Refs Diamant, Louise (Ed) *Male and Female Homosexuality* (1987); Friedman, Richard *Male Homosexuality* (1988); Green, Richard *The "Sissy Boy Syndrome" and the Development of Homosexuality* (1987).

Broader Homosexuality (#PF3242).
Narrower Homosexuality within the priesthood (#PG4151).

Related Lesbianism (#PF2640) Prostitution (#PD0693) Transsexualism (#PF3277)
Sexual abuse of children (#PE3265) Violation of rights of transsexuals (#PE8548).
Aggravates Sado–masochism (#PE6137) Family breakdown (#PC2102).
Aggravated by Sodomy (#PE3273) Male prostitution (#PD3381)
Sexual unfulfilment (#PF3260) Human sexual inadequacy (#PC1892)
Inadequate sex education (#PD0759) Sexually segregated schools (#PG3650)
Discrimination against homosexuals (#PE1903)
Violation of the rights of male homosexuals (#PE3882).

♦ PF1392 Unidentified flying objects (UFOs)

Flying saucers

Nature UFOs are aerial objects, sighted by visual or electronic means, which fail to identify themselves, and which remain unexplained after review of the evidence by competent personnel with scientific training. (In popular usage, UFO includes any aerial object or optical phenomenon which the observer cannot explain). Inexplicable aerial objects with erratic movements at high velocities are believed to have caused fatal air accidents; and have resulted in ground and air sighter panic, and a considerable expenditure of money and time in governmental investigations. They disrupt air traffic control procedures and may be interpreted as hostile attacks on radar screens, thus leading to the danger of accidental military over–reaction. They arouse concern in the public mind, especially on the basis of reports of abductions and other phenomena dangerous to human life. The phenomena and reports of them have a dubious effect on human culture at the mythic and spiritual levels.

Despite world wide reports of UFOs, the central problem remains whether in fact they exist and what in fact they are. Consideration of the problem is complicated by the large number of reports which prove to result from misidentification of aircraft, celestial objects, balloons, birds, and meteorological phenomena, as well as deliberate falsification of evidence by hoaxers. There is therefore a marked reluctance on the part of witnesses to report such objects, for fear of the official ridicule with which honest reports of genuinely puzzling phenomena have been received in the past. It is further complicated by psychological studies attributing such sightings to individual or collective hallucinations and people who, consciously or unconsciously, seek recognition.

Incidence UFOs have been seen throughout history and have been described according to the culture within which they were observed, whether as gods, magicians, or as interplanetary travellers. UFO reports have been made in almost every country of the world. More than 20,000 sightings have been catalogued, of which more than 1,500 concern objects described in detail from a short distance. Such reports bear a certain similarity to one another. In the USA, it has been estimated that about half the population believes in the existence of flying saucers and about one person in 20 claims to have seen one.

Claim The phenomena called unidentified flying objects are neither objects nor flying. Reliable photographs indicate that they can dematerialize and can violate the laws of motion as currently known. The UFO phenomenon does not give evidence of being extraterrestrial at all. Instead it appears to be inter–dimensional and to manipulate physical realities outside the space–time continuum recognized by humans. There are two reasons the scientific community has been unable to address the phenomenon sensibly. Firstly, because it is so elusive that it cannot be readily studied, and trained observers are loathe to commit themselves publicly for fear of ridicule by their peers. Secondly, because of fear in that any explanation of what is now a frequently documented phenomenon would constitute a profound challenge to cherished theories about the nature of the human understanding of the cosmos. If the reality behind the UFO phenomenon is both physical and psychic in nature, and if it manipulates space and time in ways that current scientific concepts are inadequate to describe, it is to be expected that its effects would not be limited to any particular period of history or geographical area. The UFO phenomenon is shielded from direct study by the persistent, misguided official denial of its very existence. It is made more confusing by the reactions and fears of witnesses. It is further protected from ultimate discovery by its own nature.

Counter–claim Proponents of the alien space–craft theory cite stories in which government secrecy concerning contact with aliens plays the principal role. Those versed in Atlantean and hollow–earth lore have identified them as the vehicles of a subterranean race. Other groups have indicated their nature as apocalyptic signs, or intrusion of hyper–space objects, or holographic projections from a deep–space based civilization, or equally, robotic specimen collectors for an inter–galactic zoo. Despite the overly–enthusiastic and imaginative conjectures UFO research is a serious and important area of activity that is concerned inter–alia, with exobiology, propulsion energy systems, high–energy physics, and a number of other subjects which make it a prototype of inter–disciplinary research. The existence of intelligent life elsewhere in the cosmos may yet be proven through UFO sightings, with unprecedent consequences for the human race. It is possible to make large sections of any population believe in the existence of supernatural races, in the possibility of flying machines, in the plurality of inhabited worlds, by exposing them to a few carefully engineered scenes the details of which are adapted to the culture and symbols of a particular time and place.

Refs Berthold, and Schwarz, *UFO–Dynamics* (1983); Eberhart, George M and Hynek, J Allen *UFO's and the Extraterrestrial Contact Movement* (1986); Hall, Richard *Uninvited Guests* (1988); Klass, Philip J *UFO–Abductions* (1988); Randles, Jenny *The UFO Conspiracy* (1988); Rasmussen, Richard M *The UFO Literature* (1985); Vallee, Jacques *Dimensions* (1988); Vallee, Jacques *UFO's in Space* (1987).

Broader Unexplained phenomena (#PF8352) Aircraft environmental hazards (#PD8328).
Related Unidentified submarine objects (#PE0712)
Abduction by extraterrestrials (#PF3881)
Secrecy concerning existence of extraterrestrials (#PF4331).
Aggravates False nuclear warfare alerts (#PF1236).
Aggravated by Hallucinations (#PF2249) Bias in scientific research (#PF9693).

♦ PF1393 Delay of religions in acknowledging social problems

Claim It is only from 1968 onwards that the world–wide phenomenon of poverty and its possible eradication became a major concern in the ecumenical movement. The whole debate on development must be seen in the light, or rather in the darkness, of the church's endorsement of colonialism and neo–colonialism, and its divorce from the working classes. The churches did not draw the consequences from their estrangement of the masses of the poor and exploited for several centuries. Christianity clearly did not succeed in taking adequate account of the demands made on it by the unexpected development of new economic and social conditions. Throughout the period of Western industrial revolution, the structures of the church institutions, with minor adjustments, continued to be those of the pre–industrial world. Instead of listening to the plea of the new underprivileged classes the churches stood aside and remained isolated from the radical aspects of what was happening.

Broader Non–recognition of problems (#PF8112).

♦ PF1403 Limited opportunities for significant work

Nature There tend to be neither a sufficient number nor a sufficient variety of jobs available locally to provide employment for the diverse skills available. Where work programmes are available they tend to be inadequate in size and scope, while many residents who have specialized degrees, skills and credentials are underemployed or without work. Persons in paid training programmes may remain in training much longer than necessary simply because they are not being prepared for real jobs waiting at the end of the programme. The experience of community residents that there are too few jobs available locally for the size of the potential work force is in marked contrast to the images received, through travel and television, that significant work should be and is available for everyone.

Broader Unemployment (#PB0750).
Narrower Seasonal unemployment (#PC1108) Inadequate adult guidance (#PG9147)
Neglect of individual goals (#PG8389) Imbalanced distribution of skills (#PG8355).
Aggravates Insufficient incentives for teachers (#PJ8347).
Aggravated by Limited employment options (#PF1658)
Certificate–based job market (#PJ8370)
Uncertified skilled tradesmen (#PG8363)
Limited agricultural education (#PF8835).

♦ PF1406 Recreational war games

War sports

Claim Vast majority of historical and modern war games are accurate, reflecting honestly each side's capabilities. Playing fictional war games does not reinforce aggression nor encourage the idea that war is inevitable and that nothing can be done to prevent it.

Broader Unethical entertainment (#PF0374).
Related Dangerous toys (#PE1158) Recreational crime games (#PJ3409).

♦ PF1418 Parochial scientific view

Claim In general, people have not been given an overall view the profound nature of science and its value relative to civilization, nor of the application of science to their everyday lives. They know neither its strengths nor its limitations. The lack of understanding of the philosophical significance of science thus generated blocks creative vision.

Broader Undirected technological expansion (#PC1730).
Aggravates Genetic defects and diseases (#PD2389).

♦ PF1421 Dying a bad death

Nature The manner of death may, especially in certain cultures, determine whether it is considered "bad". Dying a bad death may then be held to determine the fate of the deceased in the other world and thus determines whether rites can be accorded to the dead.

Incidence The list of deaths regarded as bad is not identical all over the world, although suicide is widely included, as is death by lightning (held to be the direct action of a god, as in any accidental death, or through some diseases). Death by drowning may be held to be seizure by a water–demon. Those killed by a poison ordeal, common in animistic societies in Africa, are held to be killed by the fetish.

Aggravates Denial of rites to the dead (#PF9190).
Aggravated by Demons (#PF6734) Suicide (#PC0417)
Drowning (#PG2857) Fetishism (#PF8363)
Lightning (#PD1292) Atrocities (#PD6945)
Wrath of God (#PF8563).

♦ PF1426 Ambivalence

Related Difference (#PA6698) Disagreement (#PA5982) Irresolution (#PA7325)
Maladjustment (#PA6739).
Aggravated by Injurious accidents (#PB0731).

♦ PF1430 Secrecy in scientific research

Nature A considerable amount of scientific research is conducted in institutes or under contracts which preclude dissemination of the results to other than a select group. Scarce resources are allocated to research which may be duplicated in another establishment or in another country. Secrecy may be maintained either to gain a military advantage (in the case of defence research) or to gain commercial advantage (in the case of industrial research).

Incidence Pharmaceutical research, for example, is intensely competitive and may involve corporate counter–intelligence operations and secrecy classifications.

Claim While secrecy in science may be necessary to protect the human race from damage or extinction, secrecy in science in order to increase private profits is morally indefensible. Universities and corporations will move away from research in areas that governments have classified secret, thus inhibiting the growth of knowledge.

Counter–claim The gap between science and technology, between understanding nature and using that knowledge to shape the natural world through new technologies and products has narrowed. In the early days of science, discoveries often found practical application only after the passage of decades or centuries. This length of time not only encouraged but required the free flow of ideas. Today the delay can be years or even months, encouraging industry and inventors to tap science as soon as possible. New technologies maintain their commercial and military competitiveness for increasingly shorter periods. These facts of today's science require secrecy in scientific development.

Broader Secrecy (#PA0005) Undemocratic political organization (#PC1015).
Aggravates Abuse of science (#PC9188)
Disadvantageous terms for technology transfer (#PE4922).
Aggravated by Professional secrecy (#PD6576) Scientific censorship (#PD1709)
Competition between states (#PC0114).

♦ PF1432 Temporal imperialism

Imposition of exogenous time standards

Nature Western–dominated industrialization has resulted in the imposition on developing countries of time standards, patterns of scheduling, and work rhythms which are foreign to the traditional rhythms of those cultures. In such cultures time awareness is not cued to mechanical clock time or a well–defined concept of efficiency. As a result people may experience difficulty in adjusting to other rhythms and may experience them as damaging to their culture and life–style.

Related Inadequate world calendar (#PF2043).
Aggravates Increasing pace of life (#PF2304).
Aggravated by Cultural imperialism (#PC3195).
Reduced by Inadequate sense of time (#PF9980).

♦ PF1433 Decline in religious broadcasting

Resistance to religious broadcasting

Incidence In 1949 French television regularly broadcast 1 hour 30 minutes of religious programming out of 16 hour output each week; by 1986 only 3 hours on Sunday morning, out of 680 hours weekly of television produced by six channels, were religious in content and inspiration. The percentage of weekly programme output had fallen from a respectable 10 percent to 0.5 percent. Across the world religious programmes have been pushed to the margins of the broadcasting system. In the USA there is little main–line religious programming on the national networks and the electronic church, for all its self–advertisement, has only a small minority, perhaps 7 million viewers. In Australia and New Zealand the churches are fighting to preserve the place of religion in the public broadcasting system. In most of the Third World churches are finding it ever more costly to ensure that even a limited amount of religious material is broadcast on a regular basis.

Broader Secularization (#PB1540) Cultural decline (#PC9083).
Related Decline in public interest broadcasting (#PF5622).

♦ **PF1437 City life style**
Urbanism
Nature Urbanism is the state of human settlement with high population density, fast pace, personal anonymity, ethnic diversity and other such characteristics.
Aggravated by Apathy toward improvement of urban life styles (#PE8477).

♦ **PF1440 Diplomatic errors**
International political miscalculation
Broader Human errors and miscalculations (#PF3702).
Related Policy–making errors (#PF0531) Deceptive political proposals (#PE7680).
Aggravates Soil erosion (#PD0949) Soil erosion by wind (#PE3656) Soil erosion by water (#PD2290) Destruction of land fertility (#PC1300).
Aggravated by Rural depopulation (#PC0056) Political immaturity (#PJ5657) Refusal to admit error (#PF5163) Inadequate crop rotation (#PF3698).
Reduced by Domestic refuse disposal (#PD0807).

♦ **PF1441 Gamma ray disturbance in space**
Nuclear reactor emissions from satellites
Nature Some satellites are powered by small nuclear reactors. Unlike earth–bound reactors, those in space are not shielded, because their radiation is thought to be harmlessly emitted in space. When one of these reactors passes near a satellite which is searching for naturally occurring gamma rays, however, its radiation is not harmless. Satellites record the gamma rays emitted by other satellites, and often miss entirely the naturally occurring gamma ray bursts which they are sent to track down.
Incidence Between 1961 and 1977 the United States launched 23 satellites with nuclear reactors in them. Four of them (17 percent) in into problems, including one that disintegrated on launch, tripling the amount of plutonium 238 in the Earth's environment. The Soviet Union has launched 39 nuclear reactors on orbiting satellites since 1965, of which 6 (15 percent) have malfunctioned.
Broader Environmental pollution by nuclear reactors (#PD1584).
Aggravated by Nuclear accidents in space (#PE5080).

♦ **PF1442 Inadequate management skills in rural communities**
Limited village management — Fragmented forms of commercial management in rural areas
Nature Small farming communities in the developing world are faced with issues such as availability of markets, diversification or crops, initiation of small industry, compliance with government regulations and agricultural quotas, and observance of strict loan regulations, all of which require management training and experience inaccessible to most of the local people, if indeed they are aware of its necessity. The continuation of inadequate or misinformed management approaches makes real growth in the local economy unlikely.
Business people in small communities are often dependent upon the international market for both supplies and customers. Such a market requires skilful participation in order to benefit from the services it affords, which presents difficulties in an economy of small business units. Small farmers may increase their production by using hybrid seed, equipment and tools produced elsewhere, but their small scale purchases, like those of other local businesses, must be made through expensive middlemen from outside the village. Similarly, sale of their produce through individual marketing methods requires costly transport and wholesale services. In a subsistence economy where external commercial structures are undeveloped, the subsistence style of operation itself blocks the perception of business possibilities. Producers who have never encountered their customer's life style lack skills in perceiving their changing tastes and product needs. Because of this inexperience they find themselves limited to local markets or put at the mercy of sporadic market demand and expensive middleman services. They lack the skills to develop the capital to build the flexibility needed to market goods for maximum income. The resulting erratic income patterns discourage any effective financial planning and cause a cycle of loan defaults, effectively terminating credit services. Diminishing returns may lead to such capital shortage that children must be withdrawn from school to provide labour for the family to survive. In fact, such business people find themselves no longer able even to imagine success. As long as these ineffective business management methods persist there can be no significant participation in expanding the local economy.
While the world today may be a global market in which every local community should be aware of its need to freely participate in the exchange of goods and services, participation in most rural communities is so sporadic that the individual feels unable to acquire the goods and services needed. The ability to obtain outside markets and have access to daily information concerning them is minimal. Simple management skills such as bookkeeping, purchasing, credit, interest terms and saving methods are extremely limited. Home industries may be unprofitable because markets have not been secured outside the village and the local demand for goods is limited. Small local shops offer overlapping stocks and are unable to expand to a profitable size. Profits from activities such as fishing may be diminishing, since middleman costs in marketing are high, encompassing both cold storage and transport. Farming done on small plots also produces little income. Essentially, all goods for the community are purchased by shop owners in a larger city.
Broader Insufficient skills (#PC6445)
Mismanagement in developing countries (#PD8549).
Narrower Limited local demand (#PG9522) Narrow profit margin (#PJ9737)
Unpopulated fishing grounds (#PG9578) Unprofitable home industries (#PG5415)
Confusing government competition (#PG9592).
Aggravates Mismanagement (#PB8406) Dependency on middlemen (#PD4632)
Unexplored potential markets (#PF0581)
Lack of motivation in leadership development (#PF2208).
Aggravated by Student absenteeism (#PE4200) Instability of prices (#PF8635)
Incompetent management (#PC4867) Lack of savings structures (#PF1348)
Unbudgeted child education (#PG1363) Risk of capital investment (#PF6572)
Inhibiting social patterns (#PF0193) Inadequate budgeting practices (#PJ0978)
Inadequate agricultural capital (#PJ1368) Underdeveloped export potential (#PJ1586)
Prohibitive livestock husbandry costs (#PJ4075)
Uncreditworthiness of rural communities (#PJ1268)
Unexploited possibilities for local commerce (#PF2535)
Restrictive loan procedures in developing countries (#PF2491).

♦ **PF1443 Uncontrollability of world social systems**
Nature Increasing population growth, combined with the considerable increases in interaction among people and among organizations brought about by technology, produce such a degree of social complexity that conventional management methods cannot respond adequately to control the evolution of systems, or can only respond by severe repression and regimentation.
Aggravates Inadequacies of the international monetary system (#PF0048).
Reduced by Vulnerability of socio–economic systems from globalization (#PF1245).

♦ **PF1447 Socio–economic burden of militarization**
Excessive military expenditure — Dependence on excessive military expenditure — Excessive expenditure on armaments — Increasing government expenditure on arms — Increasing public spending on defence — War–oriented economies — Arms culture
Nature A considerable proportion of the resources of each nation is allocated to military expenditure, constituting an economic burden. In addition, many development opportunities are lost (whether within a country or in the form of aid to another country) because of the higher priority accorded to such expenditure. World military spending has been increasing annually in real terms – at 3 percent growth, this spending is outstripping GNP growth in many countries, so that national budgets are burdened with an ever higher percentage for military outlays.
Incidence Global military expenditure in 1985 was in excess of $900 billion. World military expenditure rose from an estimated US $100,000 million in 1950 to $190,000 million in 1970 (the USA and the USSR between them sharing about 85 percent of this); for 1983, the SIPRI estimate for world expenditure (at current prices) was $800,000 million, or (for comparison purposes) $600–650,000 million at 1980 prices, with the USA spending about $185,000 million of this and its NATO allies about $120,000 million. The United Nations estimates the current annual world military expenditure to stand at about $1,000,000,000,000. Global military expenditure has risen from an estimated 4.7 per cent of world output in 1960 to over 6 per cent in 1985, namely an increase of about 150 per cent in constant price terms.
Detailed figures for 1979 are as follows: military spending in constant 1979 US $ millions was: for the USA, 2,500,000; for its NATO allies, 1,984,000; for Eastern European Warsaw Pact countries, 425,000; for the USSR, 1,375,000; for China 517,000; and for Japan, 1,030,000. The percent of GNP spent on defence was 5 for the USA; 4.5 NATO; 5 Warsaw Pact; 13 USSR; 10 China; and 9 Japan. The total number of armed forces (in 1,000s) for that same year was 9,927 for the USA, NATO (including Spain), China, and Japan; 5,944 for the USSR and her allies. In 1980 the Warsaw Pact countries had a total population of 382.2 million people with 5.9 million in military manpower; NATO countries had 578.1 people with 5.3 million in military manpower. The GNP of Warsaw Pact Countries was 1980 US $2,125,000 with 225,000 million devoted to defence; the GNP of NATO countries was 1980 US $5,764,000 with $250,000 million towards defence. Total world expenditure on defence is estimated at more than 6 percent of total resources. The data is not necessarily reliable, because the governments often understate or conceal military spending as civil expenditure.
Claim Annual military expenditure world–wide exceeds the total income of the poorest half of the world population or the equivalent of almost $1,000 for each of the 1 billion world's poorest. The real cost of the arms race is the loss of what could have been produced instead with the capital, labour and material resources, as well the contribution of military activity to pollution and environmental deterioration.
Counter–claim Military spending can have positive spinoff effects on the civilian economy and can be a focus of industrialization efforts of some developing countries. However, any decline in military expenditure as the result of détente and demilitarization tends to be much more gradual than expected. Expenditure cuts tend not to be as large as assumed and tend to be made in long–term programmes so that their impact is not immediately apparent, especially where jobs in defence–related industries are at stake. Cuts may therefore simply result in a zero increase in the military budget, and in non–inflation–adjusted terms may not result in a significant decrease in spending.
Refs Deger, Saadet and Sen, Somnath *Military Expenditure* (1990); Gleditsch, Nils Petter, et al *Economic Incentives to Arm?*.
Broader Inappropriate public spending by government (#PF6377).
Related Militarization (#PD1897) Military secrecy (#PC1144)
Inappropriate use of financial resources (#PD9338).
Aggravates Imbalance of payments (#PC0998) International arms trade (#PC1358)
Destruction of economy due to war (#PE8915)
Declining international competitiveness (#PD8994)
Political instability of developing countries (#PD8323)
Environmental degradation through military activity during peace–time (#PE0736)
Obstacles to redeployment of military resources for peaceful applications (#PF4785).
Aggravated by Competitive acquisition of arms (#PC1258)
National insecurity and vulnerability (#PB1149)
Imbalance of conventional armed forces (#PC5230)
Competitive development of new weapons (#PC0012)
Cost overruns in large–scale public programmes (#PD1644).

♦ **PF1451 Ineffectiveness of the United Nations system of organizations**
Nature It is uncertain whether the United Nations system has the capacity to cope successfully with the immense range and scope of new tasks requiring international action that each year brings. Part of this doubt relates to issues connected with the coordination of activities which have been familiar for the past 25 years or more, such as the continuing differences of view regarding the respective competences of different agencies in particular fields, and of the United Nations regional economic commissions and the specialized agencies in regard to action at the regional level. There continue to be cases of duplication and overlapping, of lack of cooperation among organizations and their staffs, of failures to consult, and divergencies of objectives. But to a considerable extent the thrust of the criticism has been shifting. Increasingly, the target is a broader set of problems, which are really problems of coherence and policy coordination as well as of structure.
The very complexity of the system and the extraordinary diversity of, and often apparent lack of coherence in, its activities, are themselves sources of frustration, as is the sense among the major contributors that the regular budgets, and the programmes financed under those budgets by mandatory assessments, escape their control. Furthermore, frustration has been voiced, with different emphases by different groups of countries, because of the lack of cohesion within the United Nations itself and the various parts of its Secretariat; the proliferation of intergovernmental organs, many with overlapping mandates and almost all of unmanageable size; the proliferation of highly independent voluntary trust funds for purposes not necessarily corresponding to established high priorities; the soaring budgets for tasks which may not always be well considered from the standpoint of cost, benefit or coordination; the quasi–impossibility of comparing and therefore of coordinating the future plans of different agencies; the involvement of so many agencies, including organs of the United Nations itself, in almost every undertaking; the independent public information and public relations offices of each agency and most of the United Nations programmes; the 'jungle' of United Nations and agency regional and subregional structures which makes system–wide action at those levels so difficult, the over–frequent and un-coordinated visits by officials of different organizations to the capitals of developing countries; and the considerable time and effort which the multifarious coordinating processes seem to require. Underlying such complaints, but partly independent of them, is the concern about the increasingly fragmented character of the system and the possibility of further fragmentation in very important fields such as population, food, and the resources of the sea–bed and ocean floor.
Claim If the theoretical potential of the United Nations is so great, then one must ask how it is that since the founding of the organization, governments throughout the world have been impelled to set up an extensive network of non–UN intergovernmental regional organizations to cater for their needs; that the programmes of such organizations expand and flourish; and that new organizations are still coming into being. The most important reason for this, operating equally in Europe, in Africa, in Asia and to a lesser extent in Latin America, is the desire to have an intergovernmental organization tailored to the needs of a particular subject matter as against the wider international community. If such an organization does not exist, then one is created for the purpose. The problem is thus the lateral spread of activities of existing organizations, and the creation of new organizations with overlapping competence.

Broader Ineffectiveness of international organizations and programmes (#PF1074).
Aggravated by Stereotypes (#PF8508)
Fragmentation and complexity of the United Nations system (#PE0296)
Irresponsibility of member governments of the United Nations (#PF5337)
Jurisdictional conflict and antagonism within the specialized agencies of the United Nations (#PE8799).

♦ PF1452 National political dependence

Claim Some states, while technically sovereign, are so dominated by others that they have little political will of their own. The colonial system persists. Colonial peoples are far removed from the seat of sovereign power, are under-represented and do not receive full and equal benefits of citizenship. For these reasons they may seek, and be denied, political independence.
Broader Lack of political independence (#PF0297).
Narrower Dependent territories (#PG3699).
Aggravates Plant pathogens (#PD1866)
Administrative difficulties in new states (#PE1793).
Aggravated by Colonialism (#PC0798) Neo-colonialism (#PC1876)
Economic dependence (#PF0841) Cultural imperialism (#PC3195)
Economic imperialism (#PC3198) Socially unintegrated expatriates (#PD2675)
Unnatural boundaries between states (#PF0090)
Non-viability of small states and territories (#PD0441)
Inadequate political parties in developing countries (#PD0548)
Vulnerability of small nations to foreign intervention (#PD2374).

♦ PF1456 Unreported crimes

Broader Crime (#PB0001) Unrecorded knowledge (#PF5728)
Avoidance of negative feedback (#PF5311).
Narrower Unreported rape (#PE5621) Unreported businesses (#PJ1192)
Undocumented violations of human rights (#PF4062).
Related Underreported issues (#PF9148).
Aggravates Unsolved crimes (#PF6911) Unreported violence (#PF4967)
Unreported accidents (#PF2887) Inadequate prevention of crime (#PF4924)
Inadequate evidence to convict known offenders (#PF8661)
Proliferation and duplication of international information systems (#PE0458).
Aggravated by Jurisdictional conflict and antagonism between international organizations (#PD0138).

♦ PF1457 Paralysis in individual decision-making

Claim The rapid change symptomatic of the 20th Century, together with a lack of fixed parameters on which to base decisions (exemplified in the relativity of the universe and the uncertainty principle) has created a paralysis in attitudes which does not permit people to find significant ways to commit themselves in helping create the future.
Broader Individual isolationism (#PD1749).

♦ PF1459 Underutilization of natural resources

Claim Maximum use of natural resources in small communities is crucial in order to expand local food supplies, create jobs for residents and increase the community's revenue. However, in many communities an abundance of resources remains only partially used; in particular, possibilities for land use are not explored. Seasonal patterns contribute heavily to under-utilization of resources, as they create a sense of impermanence in employment and a feeling of hopelessness about the possibility of upgrading the economic life of the community.
Broader Under-utilized raw materials (#PF6590)
Inefficient extraction and utilization of natural resources (#PF2204).
Narrower Underutilization of renewable energy resources (#PE8971)
Insufficient use of natural resources in least developed countries (#PE0273).
Related Underutilization of facilities due to daily or seasonal peaks (#PF0827).
Aggravates Endangered forests (#PC5165) Seasonal fluctuations in agriculture (#PD5212)
Worldwide misallocation of resources (#PB6719).
Aggravated by Alkaline soil (#PD3647) Water salinization (#PE7837)
Unstable fishing season (#PJ9570) Cold as an occupational hazard (#PF5744)
Over-rapid timber exploitation (#PD9235)
Unimaginative vision of resource utilization (#PF1316).
Reduces Unconstrained exploitation of natural resources (#PF2855).

♦ PF1460 Shortage of financial resources within the United Nations system of organization

Nature The United Nations is faced with a critical financial situation which leads to increasingly adverse effects on its reputation, as well as on the efficiency and effectiveness of its future operations. The situation, which has many intractable aspects of a political nature, is of long standing. In earlier years, the deficit, although steadily accumulating each year, was of manageable proportions. However, accommodations and compromises have increasingly had to be accepted in order that the UN can fulfil, in a meaningful manner, the basic purposes of its Charter. Although recourse to expediencies can be continued for a limited period, it is evident that cash and investment reserves (including balances temporarily available in trust and special accounts), have been virtually exhausted as a means of financing budgetary commitments.
Background The present financial difficulties of the Organization began at the end of 1956 when the General Assembly first levied assessments on Member States for contributions to the Special Account for the United Nations Emergency Force, to cover a part of the UNEF costs which were not covered by voluntary contributions. Prior to that time, all assessments by the General Assembly were in connection with expenses covered by the regular budget of the Organization and with the Working Capital Fund, and these assessments were paid in full by all the Members with but little delay. The undertaking of the United Nations Operation in the Congo in mid-1960, for which a substantial number of Members failed, or refused, to pay contributions assessed by the General Assembly, caused the financial difficulties to assume far more serious proportions from that time on, despite the various measures taken by the General Assembly, and the voluntary contributions made by some Member States, to alleviate the difficulties. The difficulties attributable to unpaid assessments were further increased, beginning in 1963, with the refusal of certain Member States to pay part of their regular budget assessments. This part, estimated at $4.8 million for 1963 and 1964 and $2.8 million for 1965, corresponded to their shares, under the approved scale of assessments, of the cost included in the regular budget for the payment of interest and the instalment payments due in respect of the principal amount of United Nations bonds, and also, in some cases, their shares of the costs of certain activities in the regular budget which they consider illegally included therein. In 1971, the UN encountered grave difficulty in meeting its payroll and other day-to-day obligations. At March 31, 1986, total arrears of payments to the UN by member states was $742.87 million, $176.85 of which were contributions outstanding for prior years.
Broader Inadequate funding of international organizations and programmes (#PF0498).
Related Reduced scope of intergovernmental development assistance (#PF4794).
Aggravated by Irresponsibility of member governments of the United Nations (#PF5337)
Government non-payment of agreed contributions to international organizations (#PF8650).

♦ PF1462 Ineffective tax systems

Tax avoidance — Tax loopholes
Nature In addition to problems of harmonization between tax systems, there are also internal problems within individual, national tax systems. National systems, particularly in developed countries, are usually highly complex and tend to be modified frequently and in an uncoordinated manner in response to particular political situations and revenue opportunities. This results in a progressive accumulation of fiscal measures, which in their totality give rise to many internal inconsistencies, tax loopholes (facilitating tax avoidance), and differing degrees of inequity. The latter is usually to the disproportionate advantage of those individuals and corporations already privileged by the system.
Incidence As an example, it was estimated in 1974 that the USA would lose in 1975 $78 billion through tax loopholes. This is $13.3 billion more than had been lost three years previously.
Counter-claim The use of legal means to minimize tax is simply good tax planning. Tax avoidance, as opposed to tax evasion, is actually a fundamental process of any market system. It is the process by which the participants in a market economy, be they individuals or corporations, go price-shopping for government services.
Broader Inadequate laws (#PC6848) Ineffective legislation (#PC9513)
Inadequate fiscal policies (#PF4850).
Narrower Tax holidays (#PE4290)
Ineffective tax systems in developing countries (#PF2124).
Aggravates Tax evasion (#PD1466) Unreported tax obligations (#PE9061)
Circumvention of duties and assessments (#PD4882).
Aggravated by Corporation financial secrecy (#PE1571).

♦ PF1466 Culturally biased religion

Eurocentric religion
Claim Ecumenical texts remain typically western documents, rooted in the old cultural traditions of Roman Catholicism, Orthodoxy and Protestantism of the European continent. They present a rather stale, anachronistic , lopsided and introverted theological charter of unity of the western hemisphere. Their language, style and reasoning are based on traditional theological argumentation, hardly understandable to any non-European Christians with minimal catechetical instruction. Christianity in the First World has still to demonstrate that it is open to and can learn from missionary and independent churches which differently interpret the being and nature of the Church according to their own cultural heritage and experience.
Aggravates Religious hypocrisy (#PF3983).
Aggravated by Official religion (#PF6091) Denial of right to manifest religion (#PF2850)
Religion as a reinforcement of nationalism (#PF3351).

♦ PF1467 Inadequate planning of action against problems

Inadequate coordination of action against problems
Nature Social institutions face growing difficulties as a result of the increasing complexity arising directly and indirectly from development and assimilation of technology. Many of the most serious conflicts facing mankind result from the interaction of social, economic, technological, political and psychological forces and can no longer be solved by fractional approaches from individual disciplines. For example: economic growth can no longer be promoted without consideration of social consequences; technology can no longer be allowed to develop without consideration of the social prerequisites of change or the social consequences of such change.
Claim The quality of individual life and that of the community is changing rapidly and in many senses deteriorating; foreseeable technological developments will have a still greater influence, presenting both opportunities for a richer life and attendant dangers. The large scale and complexity of problems are forcing decisions to be made at levels where individual participation of those affected is increasingly remote, threatening a crisis in political and social development. In the corporate environment, individual enterprises are tending to become larger and more complex. The multinational industrial activities which are developing can be expected increasingly to influence political relationships between the nations Scientific attack on these problems of complexity and interdependences is a matter of the utmost urgency. A corpus of knowledge already exists capable of immediate exploitation, and there is expectation of further and fruitful development. It is in relation to the development crisis that the planning function and related arts such as forecasting assume new significance. Planning should be concerned with the structural design of the system itself and involved in the formation of policy. However, the need for planning is not generally recognized; furthermore, where orthodox planning methods are pursued they prove quite insufficient, in that they seldom do more than touch a system through changes in the variables; diagnosis is often faulty and remedies proposed often merely suppress symptoms rather than attack the basic cause. But mere modification of policies already proved to be inadequate will not result a correct solution; too often planning is used to make situations which are inherently bad, more efficiently bad. The need is to plan systems as a whole, to understand the totality of factors involved and to intervene in the structural design to achieve more integrated operation. All large, complex systems are capable of some degree of self-adaptation. But in the face of immense technological, political, social and economic stresses, they will have to develop new structures. This can easily lead to grave social disturbances if the adaptation is not deliberately planned, but merely allowed to happen.
Broader General obstacles to problem alleviation (#PF0631).
Narrower Inadequate manpower planning (#PJ2036)
Mismanagement of aid to developing countries (#PF0175)
Inadequate coordination of international organizations and programmes (#PD0285).
Aggravates Oversimplification (#PF8455)
Inadequate technical cooperation on problems (#PF0863)
Inadequate implementation of plans and programmes against problems (#PF1010).
Aggravated by Economic uncertainty (#PF5817)
Short range planning for long-term development (#PF5660)
Inadequate power of intergovernmental organizations (#PF9175).

♦ PF1468 Unapplied scientific knowledge

Claim The educational system, in particular the university disciplines, do not understand how to inform the general population about the significance of science. Because of resulting inadequacies in the image presented in the propagation of scientific knowledge, such knowledge is not used in social planning.
Broader Inefficiency (#PB0843) Undirected technological expansion (#PC1730)
Inadequate application of available knowledge to solve problems (#PF8191).
Aggravated by Human errors and miscalculations (#PF3702).

♦ PF1470 Delay in project implementation

Nature The problems of project implementation do not greatly differ between private and public projects. Private projects may experience greater difficulty in obtaining approvals and allocations, especially of foreign exchange, but public projects may suffer more from organizational confusion and delays in decisions. Projects, particularly those on a large scale, can upset a development programme or involve costs which, when excessive, can disturb a national budget. Such projects also represent the most visible sector of industry, and the most politically dangerous if mishandled. Delays may be due to oversights in project planning, causing delays in construction, or to poor contracting procedures which lead to inefficient and costly start-ups. The costs of such delays arise from the tied-up capital, the loss of earning opportunity and the immobilization of scarce foreign exchange.
Broader Delay (#PA1999) Human errors and miscalculations (#PF3702)

Delays in implementation of social change (#PC6989).
Related Delayed development of regional plans (#PF2018).
Aggravated by Government inaction (#PC3950).

♦ **PF1471 Restricted growth in export markets of developing countries**
Increase in competition for export markets encountered by developing countries — Declining share of developing countries in world commodity exports — Restricted access of developing country products to world markets
Nature Over the last 20 years, the share of developing countries in world commodity exports has declined, while their share in world commodity imports has increased. Developing countries tend to find it increasingly difficult to find suitable markets for their primary commodities. This is due in part to the growth in primary production in the developed countries (in consequence of a rise in productivity and protection of domestic production), to the increasing production of synthetics and other substitutes which displace natural materials, and to restrictive import, tariff and excise policies of the developed countries. The scope for increasing exports has also depended on the state of world markets. The lagging recovery of Western Europe from 1980–1982 recession has favoured the major exporters of manufacturers vis-à-vis commodity-exporting countries.
Broader Economic rivalry (#PD8897)
Economic and social underdevelopment (#PB0539).
Aggravates Inadequate production capacity in developing countries (#PD4219)
Excessive concentration of export markets of developing countries (#PE9457)
Reduction of the share of the developing countries in world exports (#PC2566).

♦ **PF1479 Variability in crop yields**
Variability in grain yields
Nature Yield variability is determined by variety (genotype), variability and level of inputs (such as fertilizers, irrigation and pesticides), and variability in pests and diseases and in climatic factors (such as rainfall, frosts, and temperature). Interactions between these factors are important, although difficult to analyze. As national food-grain production has increased, especially through the introduction of improved varieties and the increased use of irrigation and fertilizers, so often has variability in yields from year to year. This leads to the perception of increased risk, which may make new technologies appear less attractive to farmers and hence slow agricultural development. It also increases instability in national and world food supplies, which may act to destabilize domestic prices, national income, and the food consumption of the poor, especially in poor agrarian countries.
Incidence At the global level production variability around the growth trend increased between the 1960s and the 1970s in both absolute and relative terms. The associated probability of a major shortfall below the trend in world cereal production has also increased.
Counter-claim The changes in variability have presented challenges for policy-makers but they are not statistically significant. The decade of the 1980s has provided an example of unfortunate chance rather than a forecast that fundamental structural changes in production will cause equal or increasing instability at the global level in future years.
Aggravates Fluctuation in availability of food (#PJ0743).
Aggravated by Excessive number of crop varieties (#PF3145).

♦ **PF1487 Delay in administration of justice**
Inordinate legal delay — Overloading of courts — Court delay
Nature The rate at which cases can be processed in courts is slow relative to the increase in the number of cases. This leads to the accumulation of a backlog of untried cases. Because of these delays, those accused who are imprisoned without the right of bail may spend much longer in gaol than the crimes with which they are accused warrant, even before the validity of the charge is clarified; and they may be innocent. Conversely, those guilty of the crimes with which they are accused, but who are allowed bail, may enjoy freedom for some time before justice can be administered. In addition, the existence of a backlog of cases may frustrate attempts to try each satisfactorily. Cases maybe prepared by inexperienced staff resulting in some injustices or more delays. Cases prepared too quickly may result in similar problems.
Incidence In France, 81% of jurisdictions require more than 12 months to handle civil cases. In the UK the average time taken to process indictable cases through magistrates' courts was 124 days in February 1989. In Italy, experts estimate 60 percent of the 33,500 men and women in prison have not been convicted of anything, they are merely awaiting trial. Cases in the Los Angeles court system may take up to 4 years to be tried.
Refs Matthews, Catherine J and Chunn, Dorothy E *Congestion and Delay in the Criminal Courts* (1979).
Broader Delay (#PA1999)　　Inefficiency (#PB0843).
Narrower Delay in civil litigation (#PD7571)
Delay in administration of criminal justice (#PD0412)
Excessive burden on the poor due to legal delays (#PE1093).
Related Administrative delays (#PC2550).
Aggravates Wrongful detention (#PD6062)　　Inequitable administration of justice (#PD0986)
Non-payment of reparations by government (#PE4446).

♦ **PF1489 Dependence of developing countries on imported technology**
Nature Developing countries are dependent upon industrial countries for a varying but large proportion of their capital equipment. The emphasis for many years has been on selling capital goods to less developed countries rather than denying them such goods. A relic of the older pattern of denying newer, more productive machinery to developing countries survives in the case of special types of equipment which are not sold by the makers but are leased to users for a fixed annual rental and a royalty which varies with the machine's output. In normal circumstances, the would-be manufacturer in a developing country has fairly free access to plant and machinery sources, subject always to the disadvantage of being much less familiar with the market than his counterpart in the industrial country. There are times, however, when the capacity of the machine-making industry is pre-empted because of events in industrial countries, and the dependence of less developed countries then becomes a major obstacle to growth. The situation appears even more unfortunate because at such times the supply of finished consumer goods from industrial countries is also likely to be restricted, and opportunities for the expansion of manufacturing industries substantially increased. Importation of second-hand machinery and use of equipment discarded by concerns in industrial countries also tends to result in high-cost production. In this way, shortage of capital goods has from time-to-time hindered industrialization, at least in its early stages when the developing countries have not been in a position to produce much of their own equipment.
Claim In many cases the cost of capital goods to less developed countries is likely to be higher than to manufacturers in industrial countries, not only because the latter have a better knowledge of the market, but because, being nearer the source of supply and having the advantage of more developed domestic transport and engineering industries, they are able to have plants conveyed and installed more expeditiously and at lower cost. The higher the proportion of capital cost (interest and amortization) in total manufacturing costs, the greater is the relative disadvantage of the underdeveloped country. This handicap is of particular significance during and immediately after a period of rapidly rising plant and equipment prices. Dependence upon industrial countries for capital equipment also implies certain technical disadvantages for less developed countries. Plant design is usually dictated by the needs of the large domestic market rather than by the much more diverse needs of various small markets in under-developed areas. As a result, equipment is often poorly adapted to specific local conditions. Automatic devices suited to conditions in advanced industrial countries are often left unused in developing countries, while the intricacy of many machines, though appropriate to the type of labour available in industrial countries, tends to magnify repair and maintenance costs in factories in less developed countries which depend upon a high proportion of unskilled labour. This involves a competitive handicap for the manufacturer in the developing country compared with his opposite number in the industrial country. Another disadvantage which flows from dependence upon industrial countries for capital goods is the fact that technical improvements are quickly adopted by manufacturers in equipment producing countries, whereas those in less developed countries usually tend to lag behind.
Remoteness from the main area of technological advance is one important cause of delay in instituting changes in the factories of less developed countries. Shortage of capital is an even more compelling reason, for where capital is more plentiful, the rate of obsolescence tends to be higher and the tempo of technical progress faster. The technological gap between industrial and less developed countries therefore tends to be maintained.
Refs Jéquier, Nicholas (Ed) *Appropriate Technology, Problems and Promises* (1976).
Broader Dependence on external resources (#PC0065)
Economic and social underdevelopment (#PB0539)
Vulnerability of developing countries (#PC6189).
Aggravates Proliferation of computers (#PE3959)
Underdevelopment of industrial and economic activities (#PC0880).
Aggravated by Proliferation of technology (#PD2420)
Inappropriate modernization of agriculture (#PF4799).

♦ **PF1491 Self determination unrelated to global obligations**
Nature Loyalties to political structures have changed over the centuries, from the clan, to the tribe, to the kingdom and finally to the nation. At this point in the relationship between the individual and the state loyalty to any geo-political reality smaller than the whole globe creates reduced images of responsibility and unnecessarily focused channels of action.
Broader Habitual overemphasis on national self-determination (#PF1804).

♦ **PF1492 Restrictions imposed on aid to developing countries**
Aid-tying — Tied aid
Nature Most donor countries place some restrictions on the use of their aid contributions. These restrictions fall into two broad categories: restrictions with respect to the purposes for which assistance may be used, and the limitation of aid-financed imports to procurements in the donor country. The two may be combined in any given instance.
Restriction to specific uses: Donors may restrict the use of aid contributions to the financing of specific projects or they may provide assistance for general developmental purposes. In either case, they may limit their contributions to the financing of identifiable imports or they may provide foreign exchange resources to cover imports as well as local expenditures of a project or programme.
Several donor countries and virtually all multilateral lending agencies provide assistance primarily or exclusively for specific development projects, because project financing involves close cooperation between the donor and the recipient country, its results can be readily identified and evaluated, and it facilitates the coordination of technical and financial assistance. However, the project approach has a number of drawbacks to the recipient country, especially if its investment policy is based on an overall plan. While some major projects can be effectively carried out in isolation, there are many that require the execution of complementary or otherwise related projects in order to yield the best results. In such cases the need to negotiate the financing of each project separately, possibly with more than one donor, gives rise to problems of timing and coordination. If, as is frequently the case, project assistance covers only the direct import requirements (machinery, equipment and materials needed for the execution of the project), the recipient may also be faced with the problem of finding supplementary foreign exchange resources to meet the additional import requirements arising from the increased income generated by the domestic investment expenditure.
The widespread practice of providing project assistance to cover only direct import requirements is based on the consideration that the recipient country should engage its own resources to the fullest possible extent in the execution of investment projects. Nevertheless, the financing of additional imports to meet the demand generated by local expenditure may be beyond the capacity of the assisted country. In that event, the project may have to be abandoned or the recipient country may be obliged to resort to short-term external credits, thereby adding to its future balance-of-payments problems. Because of the potential burden of local cost financing on the foreign exchange resources of developing countries, the practice of limiting project assistance to direct import requirements has tended to encourage the execution of projects with a relatively large import component at the expense of many equally important projects which involve relatively large local expenditures.
Procurement restrictions: Most donor countries restrict at least part of their bilateral aid contribution to purchases of their own products. Reasons given for applying such restrictions include balance-of-payments difficulties or the existence of idle capacity or unemployment in the donor country. Factors such as a desire to promote domestic exports or to compensate exporters for the loss of sales in traditional markets which may have resulted from procurement restrictions of other donor countries have also influenced government policies with respect to aid tying.
The tying of contributions to exports from the donor country has taken various forms. Most commonly, loans and grants for the financing of specific imports have been tied by contractual agreements which provide that all or part of the assistance will be used exclusively for procurement in the donor country or in countries specified by the donor. It has been more difficult to tie assistance which is not specifically linked with imports, such as budgetary support payments or general development loans and donations, but it is for this reason that donor countries have tended to shift from this type of assistance to loans and grants earmarked for specific imports. Another type of aid which is in most respects indistinguishable from formally tied aid is the official export or import credit. It has been claimed that such credits do not belong to the category of tied aid because they are granted after the borrower has selected his source of supply. In practice, the availability of credit has tended to influence the choice of supplier to a greater extent than have price or quality considerations. In these circumstances, it seems appropriate to treat such credits as a form of tied aid. Contributions in kind might also be termed tied aid, but since such aid is dependent upon the existence of surpluses in the donor country, it is in a separate category.
The adverse impact of procurement restrictions on recipients arises chiefly from the fact that such restrictions reduce competition between potential suppliers and may render impossible the procurement of imports from the optimum source of supply. The effect of this is greater when aid is given for the purchase of equipment required for a specific project than in the case of general development assistance which gives the recipient a wider choice of commodities to be imported. Procurement restrictions may involve not only higher costs, and hence a larger loan to be serviced, but also the purchase of equipment manufactured in the donor country which may not meet fully the recipient's requirements. Moreover, the availability of aid may encourage the execution of those projects for which suitable equipment can be readily obtained in the donor country, even if the recipient's development programme did not assign the highest priority to such projects.

Broader Underdevelopment (#PB0206)
Reduced scope of intergovernmental development assistance (#PF4794).
Narrower Bilateralism in aid to developing countries (#PE9099)
Policy cross-conditionality restrictions in multilateral development aid (#PF9216).
Related Dependence of island developing countries on official assistance (#PE5724)
Counter-productive government constraints on programme beneficiaries (#PJ3583).
Aggravated by Imbalance of payments (#PC0998)
Non-convertibility of currencies (#PF3069).

♦ **PF1495 Insufficient transportation infrastructure**
Poor transport systems — Underdevelopment of transportation — Distant public transportation — Unavailability of public transportation — Limited public transportation — Irregular passenger service — Insufficient public transportation — Inadequate public transportation
Nature Many countries lack integrated transportation system plans and infrastructures which would coordinate all means of transport to satisfy the needs of development. Road and rail networks are inadequate; and if inland waterways exist their use may be primitive if they are used at all. In developed countries, the proportion of road to rail transport is not always economical; and in the large metropolitan areas, public passenger transportation systems do not meet the needs of the increasingly densely populated and trafficked urban centres.
Broader Inadequate infrastructure (#PC7693) Lack of essential local infrastructure (#PF2115)
Weakness of infrastructure in developing countries (#PC1228).
Narrower Inadequate road conditions (#PJ0860) Ineffective rural transport (#PF2996)
Obstacles to use of bicycles (#PF5717) Unreliable freight transport (#PG1026)
Irregular transport services (#PE5345) Inadequate commercial roadways (#PG9385)
Inadequate port infrastructure (#PE5792) Inadequate cargo transportation (#PE0430)
Inadequate air transport service (#PJ0260) Irregular hospital transportation (#PJ8984)
Incompatibility of transport modes (#PF2403) Maldistribution of merchant vessels (#PG6316)
Inadequate rail transport facilities (#PD0496) Inadequate urban transport facilities (#PG5801)
Maldistribution of private automobiles (#PF4480) Maldistribution of commercial vehicles (#PF4485)
Lack of personal transport in rural areas (#PF2799)
Poor communications networks in rural areas (#PF6470)
Inadequate integration of transport systems (#PF6157)
Restrictions on effective means of transport (#PF2798)
Inadequate inland waterway transport facilities (#PD2487)
Inadequate road and highway transport facilities (#PD0490)
Inadequate facilities for the transport of water supplies (#PD1294)
Inadequate facilities for the transport of sanitary wastes (#PE8475)
Inadequate transportation facilities in developing countries (#PD1388)
Inadequate systems of transport and communications in rural villages (#PF6496)
Inadequate road and highway transport facilities in developing countries (#PD0543)
Lack of technical infrastructure for maritime commerce in developing countries (#PE5814).
Related Unfair transport practices (#PD1367) Inadequacy of postal services (#PF2717)
Excessive land usage by transportation systems (#PE2525)
Inadequate facilities for the transport of electrical energy (#PE1048).
Aggravates Traffic congestion (#PD0078) Distant representatives (#PJ0272)
Inadequate medical care (#PF4832) Unutilized fish by-products (#PG9331)
Lack of vocational teachers (#PJ9603) Breakdown of police protection (#PF8652)
Lack of corporate achievements (#PG9339) Physically inaccessible services (#PC7674)
Limited access to health services (#PF6577) Spoilage of agricultural products (#PC2027)
Prohibitive cost of transportation (#PE8063)
Confined scope of business operations (#PF2439)
Inadequate disaster rescue and relief (#PF0286)
Reluctant claims on external resources (#PF1226)
Underprovision of basic services to rural areas (#PF2875)
Restricted delivery of essential services to developing country rural communities (#PF1667)
Limited availability of public services in the small towns of developed countries (#PF6539).
Aggravated by Ice-blocked seaways (#PD2498)
Inadequate welfare services for the aged (#PD0512)
Social environmental degradation from recreation and tourism (#PD0826)
Natural environmental degradation from recreation and tourism (#PE6920).
Reduced by Destruction of natural barriers (#PC1247).

♦ **PF1502 Family structure as a barrier to progress**
Nature Family structure may present a barrier to progress either in the extended or nuclear form. The extended family in particular represents a community with affiliations that are not in accordance with those of industrialized society where allegiance tends to be more towards a company, trade union, national, regional or cooperative group. The nuclear family may appear to fit better into this pattern, but a social lag between attitudes and ideas of parents and those promulgated by the state or industry constitutes a challenge to parental authority and may produce youth problems of various kinds and general social disorganization. Equally the isolation of the nuclear family unit is not ideal for the integration of young people or their parents into society as a whole.
Broader Inflexible social structure (#PB1997).
Narrower Dependence within extended families (#PD0850)
Negative effects of the nuclear family (#PF0129)
Narrow legal definition of the family in developing countries (#PD1501).
Related Lack of family planning (#PF0148).
Aggravates Traditionalism (#PF2676) Social breakdown (#PB2496)
Social fragmentation (#PF1324) Generation communication gap (#PF0756).
Aggravated by Tribalism (#PC1910).

♦ **PF1503 Adjustment difficulties of new urban families**
Nature The most acute problems of new urban families occur in developing countries or among people who have migrated from developing countries to industrial centres in developed countries. Adjustment problems may also occur with families taking part in a rural exodus in developed countries. Major problems occur with nutrition and housing, industrial or service employment and the more constant use of money. With the greater fragmentation of society, the loss of the extended family, commercial pressure and overcrowding, psychological adjustment to city life may be very difficult or impossible to achieve, leading to family breakdown, crime, or anti-social behaviour.
Broader Rural depopulation (#PC0056) Personal and social maladjustment (#PC8337).
Related Dependence within extended families (#PD0850)
Negative effects of the nuclear family (#PF0129).
Aggravates Urban slums (#PD3139) Family breakdown (#PC2102)
Marital instability (#PD2103) Juvenile delinquency (#PC0212)
Maladjusted children (#PD0586) Family disorganization (#PC2151).
Aggravated by Rural poverty (#PC4992) Social fragmentation (#PF1324)
Ethnic disintegration (#PC3291) Unsustainable population levels (#PB0035)
Narrow legal definition of the family in developing countries (#PD1501).

♦ **PF1509 Fragmented patterns of extended family relationships**
Nature Under the strain of 20th Century changes, the family has shifted from the survival-oriented care structure of the past to a decisional unit of community care. In the past, the kinship structure of tribal societies effectively cared for the needs of all ages through a system of inter-family obligation. Nowadays, kinship demands seem unrealistic and no one part of a family can care for the whole. There is a strain between the maintenance of old obligations and operating as a single, self-sustained social unit (such as the obligation to extend credit to kin beyond own financial capabilities). In this transitional situation, there are no structures to supply care for old people or for single parent families. Young people have no strong memory of the kinship system, do not identify with the fragments still remaining, but have no alternative images of family life for themselves. Families that have successfully adjusted as an independent unit may have done so at the expense of their relations, often causing further confusion and alienation.
Broader Inadequate family structures (#PF1000) Obstacles to extended families (#PF3127).
Narrower Unstable family life (#PD7929) Rigid family obligations (#PJ8340)
Inaccurate youth stereotypes (#PJ8357) Family crediting obligations (#PG8365)
Increased family fragmentation (#PG8331) Unethical personal relationships (#PF8759).
Aggravates Family breakdown (#PC2102) Social disaffection of the young (#PD1544).

♦ **PF1528 Poor managerial communications**
Lack of management communication skills — Communication ineptitude among management
Nature Communication with employees, it is widely agreed, is the most crucial key to management effectiveness. Nevertheless, downward communication is rated favourably by less than half of office workers at all levels. The main barrier is managerial insecurity, the tendency of executives to isolate themselves behind a wall of omniscience.
Broader Incompetent management (#PC4867).
Narrower Inadequate management-employee communication (#PF3661).
Related Non-inclusive management decisions (#PF2754)
Lack of worker participation in business decision-making (#PF0574).
Aggravates Mismanagement (#PB8406).
Aggravated by Absence of management training (#PD3789)
Inflexible management patterns (#PF3091)
Insufficient communications systems (#PF2350).

♦ **PF1529 Demeaning minority self-image**
Minority victim image
Nature The voice of minorities may be paralysed by their own image as victims of society. Dissident views seldom play a part in defining the roles and guidelines of civil protection in particular, nor of shaping the social system in general. As a result, minorities feel no obligation to voice constructive criticism about the social system, so that the victim image is perpetuated.
Broader Lack of self-confidence (#PF0879) Blocked minority opinion (#PD1140).

♦ **PF1531 Inappropriate education in developing countries**
Nature The attitudes, aspirations and expectations perpetuated and stimulated by educational systems in developing countries tend to create a growing discrepancy between the employment opportunities that exist and the job expectations of school leavers and graduates. In addition, the function of the school is often seen to be to qualify students rather than to educate them. This leads to frustration and a sense of alienation on the part of those who fail and who therefore probably remain unemployed, while those who do qualify feel entitled to status and salary rather than encouraged to earn them. Qualifications are at any rate not regarded as an opportunity for self-fulfilment or for public service. The outcome is an educated elite minority holding certificates for the best jobs, even though these jobs may not exist in sufficient quantities; a majority of those educated being unappropriately trained for what real conditions require; and a retarding and distorting of informal education.
Refs McNown, John S *Staff Development for Institutions Educating and Training Engineers and Technicians* (1977).
Broader Inappropriate education (#PD8529).
Narrower Inadequacy of agricultural education in developing countries (#PE9096)
Unemployment of premature school leavers in developing countries (#PE0015)
Inadequacy of training for human settlements in developing countries (#PD5077).
Related Inadequate nutrition education in least developed countries (#PE0265).
Aggravates Underdevelopment (#PB0206) Educational wastage (#PC1716)
Juvenile delinquency (#PC0212) Unequal income distribution (#PD4962)
Exploitation of child labour (#PD0164) Inappropriate employment incentives (#PD0024)
Urban slums in developing countries (#PD3489)
Unemployment in developing countries (#PD0176)
Unemployed intellectuals in developing countries (#PD1273)
Accentuated inequality between rural and urban development (#PE8569)
Inadequate development of enterprises in developing countries (#PE8572)
Inadequate research and development capacity in developing countries (#PE4880).
Aggravated by Overcrowding in schools (#PE3757)
Poor supervision teachers (#PJ3757)
Shortage of trained teachers (#PD8108)
Unsustainable population levels (#PB0035)
Shortage of books and textbooks in developing countries (#PF0118)
Insufficient government spending on cost-effective activities (#PE5302)
Imbalance between urbanization and industrialization in developing countries (#PC1563)
Excessive social costs of structural adjustment in debtor developing countries (#PD8114).

♦ **PF1538 Delays in delivery of books and publications**
Nature Delays in the delivery of printed matter (particularly between continents) due to postal delays and administrative delays in processing requests, are such that periods from 2 to 6 months may lapse between initiation of the request and receipt of the publication.
Incidence Examples of such delays occur in the delivery of documents of international organizations (such as the United Nations) to distant countries, resulting in the receipt of documents after the events for which they could be relevant. Even between Europe and North America, books may take up to 6 months to arrive.
Broader Inefficiency (#PB0843) Delays in delivery of goods (#PF8268).
Aggravates Lack of communication (#PF0816) Inaccessibility of knowledge (#PF1953).
Aggravated by Administrative delays (#PC2550) Inadequacy of postal services (#PF2717).

♦ **PF1540 Unrehabilitated historical figures**
Incidence Within official religions and ideologies, people condemned in earlier periods may remain unrehabilitated in official documents even though later understanding reveals the merit of their position. Examples include Galileo (only rehabilitated in the 1980s by the Catholic Church) and Trotsky (who remains unrehabilitated with the USSR). There may also be considerable reluctance to rehabilitate whistleblowers who become the victims of harassment by major institutions.
Related Harassment of human rights monitors (#PF1585).

♦ **PF1542 Frivolous or vindictive litigation**
Nature A person may abuse the judicial system by bringing a cause of action against another person in the absence of a reasonable belief as to the existence or the merits of the alleged cause of action.
Broader Injustice (#PA6486) Proliferation of litigation (#PF0361).

♦ **PF1545 Lack of social progress**
Overemphasis on economic progress in society — Progress equated with efficiency
Nature Social and governmental agencies often equate progress in dealing with social problems with more efficient handling of data. More and more time is spent with internal communications

and less and less with those the agency is to be serving.
Claim Internalized value systems are largely expanded and strengthened through communication. The influence of the media has been to direct such value systems towards reducing human sociability and increasing economic progress. This isolates the person of already limited knowledge by further restricting his access to global ethics. Profit-oriented selection of programming appeals to escapism and the acquiring of material possessions. Economic values are emphasized to the point that they overshadow any social or cultural consideration.
Broader Thwarted technological communications (#PF0953)
Inappropriate understanding of progress (#PF8648)
Reinforced parochialism of internal values and images (#PF1728).
Aggravated by Economic bias in development (#PF2997).

♦ PF1546 Unfocused design of community space
Undeveloped community space — Unplanned village design
Claim Historically, villages focused around specific gathering places: the school, shop, square, source of fresh water, ceremonial grounds and religious buildings. But present day attempts to recreate community centres fail, with a resulting lack in common gathering place to shop, visit, exchange news, receive common services, or to celebrate together. The erasing of community boundaries has been accompanied by a similar break down of the once familiar family areas, such as small gardens and room in houses for family activities. Until the community redesigns its common life around common community buildings and spaces, appropriate both to traditional living patterns and to the age of technology, the possibilities of renewing present structures and building new ones remains blocked.
Broader Obstacles to community achievement (#PF7118)
Poor organization of community environment (#PF1790)
Obstacles to availability of community space (#PF7130).
Narrower Low-quality construction work (#PD7723).
Related Fragmented planning of community life (#PF2813).
Aggravates Lack of community development (#PF7912)
Inappropriate sanitation systems (#PD0876)
Inadequate recreational facilities (#PF0202).

♦ PF1548 Lack of emphasis on basic education
Nature Rapid economic and social development in any community requires that residents be equipped with a wide range of vocational and social skills, yet in many small communities the attitude towards basic education may be characterized as casual. Home environments may not encourage study or intellectual enrichment; the only secondary education available to students is focused on academic curricula and offers few employment-oriented courses; students are 'streamed' into specific courses which limit their later educational options; curricula are designed neither to hold the students' interest nor to deal with their everyday needs; for some students, instruction may not be in their mother tongue, and community structures are insufficient to pass on supplementary instruction in traditional languages. Many people reach adulthood without having completed their secondary education or met the minimum requirements for obtaining well-paid jobs.
Broader Irrelevance of educational curricula (#PF0443).
Narrower Lower course streaming (#PS9230).
Related Educational curricula over emphasizing method rather than content (#PD1827).
Aggravates Distant commercial training (#PJ9215)
Minimal opportunities of adult training (#PF6531).
Aggravated by Customary working patterns (#PG9278)
Single language instruction (#PJ9290)
Nonintellectual home environment (#PG9298).

♦ PF1555 Excessive animal sanitary regulations in international travel
Excessive health regulations — Animal quarantine
Nature Travellers possessing animal pets are frequently impeded in their movement between countries by health regulations which prohibit the import of animals into a country without a period of quarantine.
Claim The period of quarantine, which may be up to 9 months, is a cruel deprivation to the animal and awkward for the owner.
Counter-claim Quarantines are imposed to protect the human, animal, and plant populations of the receptor countries.
Broader Complex government regulations (#PF8053).
Related Excessive frontier formalities in international travel (#PE0208).
Aggravates Inadequate quarantine (#PE2850) Maltreatment of animals (#PC0066)
Restrictions on freedom of movement between countries (#PC0935).
Reduces Unhygienic conditions (#PF8515)
Domestic animals as carriers of animal diseases (#PD2746).

♦ PF1559 Unpredictable governmental policy
Government policy dilemmas — Political uncertainty — Inadequate government policy — Incoherent government policies — Unclear government policies — Unstable government policy — Frequently changing government policy — Arbitrary government decision-making — Unpredictable public administration — Expedient policy reversals
Claim High levels and rapid growth of human productivity are exacerbated by excessively frequent changes of direction in government policy, legal frameworks which fail to limit the scope for arbitrary decisions, and public administration which fails to take account of considerations of fairness and predictability.
Broader Uncertainty (#PA6438) Uncoordinated policy-making (#PF9166).
Related Political over-reaction (#PF4110).
Aggravates Confusion (#PF7123) Policy-making delays (#PF8989)
Inappropriate policies (#PF5645) Unproductive use of resources (#PB8376)
Discrimination against indigenous populations (#PC0352).
Aggravated by Governmental incompetence (#PF3953)
Broken government promises (#PF4558)
Uncoordinated government policy-making (#PF7619)
Bureaucracy as an organizational disease (#PD0460)
Unknowable future patterns of social choice (#PF9276).

♦ PF1560 Reduction in symbolic celebrations
Reduced celebrational images
Nature Because of the limiting nature of modern life and the reducing of many to subsistence-level existence, the scope of celebrations has been reduced from the richness of many past celebrational forms. Traditional celebrations which re-enacted the whole of life have often been forgotten, although weddings and funerals have taken a primary place and are engaged in with passion and in the teeth of the economic reality. There is a need for communities to broaden the scope and number of events to mark the journey through life, and to transpose ancient wisdom into meaningful contemporary forms.
Celebrations most often focus on the pleasant aspects of life, avoiding the fullness of struggles, joys and sorrows of daily living. Living is celebrated as positive achievements and better times rather than its entirety.
Broader Symbol system failure (#PF3715) Collapsed images of vocation (#PF6098)
Deteriorating quality of life (#PF7142).
Narrower Limited artistic expression (#PF9735) Insufficient community events (#PF5250)
Burdensome cost of religious ceremonies (#PF3313).
Aggravates Unorganized community recreation (#PF5409).
Aggravated by Lack of commitment to common symbols (#PE8814).

♦ PF1568 Spiritual and emotional hindrance
Spiritual and emotional malaise
Aggravates Barriers to transcendent experience (#PF4371).

♦ PF1572 Inadequate research on proposed solutions to problems
Lack of well-researched projects to alleviate problems — Lack of innovative projects against world problems — Obsolete programmes against social problems — Uncreative strategies against problems
Nature Projects intended to alleviate problems, such as those associated with underdevelopment, are not always sufficiently well researched and designed to meet the conditions required by the institutions financing the remedial enterprises. Market studies, estimates of effectiveness, profitability calculations, estimates of the skills or the amount of investment actually required, and estimates of raw material needs and prices, are sometimes superficial. Furthermore, the technological processes and production capacities designated are generally not optimum. There is a lack of awareness of the need for good studies and an unwillingness to fund such studies adequately, although they are essential to avoid poor-risk investments.
Counter-claim There will never be enough researchers or funds to adequately predict the effects of solutions proposed for problems.
Broader General obstacles to problem alleviation (#PF0631).
Related Opportunist bias in public discussion and research on development (#PF3846).
Aggravates Inadequate delivery mechanism in response to problems (#PF0301)
Simplistic technical solutions to complex environmental problems (#PF0799)
Misappropriation of resources for high cost civil engineering projects (#PF4975).
Aggravated by Minimization of problems (#PF7186)
Inadequate research on problems (#PF1077)
Fixation on partial solutions to problems (#PF9409)
Deficiency in innovation in the public sector (#PF7230)
Inadequate power of intergovernmental organizations (#PF9175)
Excessive reliance on fashionable solutions to problems (#PF4473)
Inadequate research and development on problems of developing countries (#PF1120).

♦ PF1575 Restrictions on publicity
Nature Restrictions regulating the advertising or publicity by the acquiring party except where restrictions of such publicity may be required to prevent injury to the supplying party's goodwill or reputation where the advertising or publicity makes reference to the supplying party's name, trade or service marks, trade names or other identifying items, or for legitimate reasons of avoiding product liability when the supplying party may be subject to such liability, or where appropriate for safety purposes or to protect consumers, or when needed to secure the confidentiality of the technology transferred.
Broader Restrictive trade practices (#PC0073).
Related Collusive tendering (#PE4301) Grant-back provisions (#PE5306)
Challenges to validity (#PF1200) Restrictions on research (#PF0725)
Restrictions on adaptations (#PF5248) Exclusive dealing arrangements (#PE0413)
Restrictions on use of personnel (#PF3945)
Patent pool or cross-licensing agreements (#PE4039)
Exclusive sales and representation agreements (#PE4581)
Trade restrictions due to voluntary export restraints (#PE0310)
Payments after expiration of industrial property rights (#PF5292)
Tying of supplies to subsidiaries by transnational enterprises (#PE0669).

♦ PF1577 Sentimental attachment to the past
Romantic understanding of the past
Nature Tragic and burdensome occurrences with which old people have been unable to deal in the past are blotted out of their memories so that only the good and the happy occasions are remembered in a romantic image, leading to a denial of the tragic dimension both in the past and in the present. This rejection of the tragic has led all generations to draw their meanings of life from romantic images and to avoid the tragedies which confront them.
The myths, rites and symbols of past glories generate a romantic understanding of the past of a social group which prevents living in the present situation and support a fear of the future.
Broader Attachment (#PF6106) Sentimentalism (#PF6961).
Related Elder paralysis over the future (#PF3973)
Inadequate means for upholding global concern (#PF1817)
Denial by old people of the significance of the past (#PF2830).

♦ PF1578 Failure to adapt general initiatives to specific needs
Programme initiatives inappropriate to local requirements
Broader Inappropriate policies (#PF5645).
Aggravates Inadequate management of government finances (#PF9672)
Inadequate coordination of the intergovernmental system of organizations (#PE0730).

♦ PF1585 Harassment of human rights monitors
Harassment of whistle-blowers — Harassment of informers
Nature Whistle-blowers, people who reveal secrets their employers, the government or other powerful interests would rather keep hidden, usually find their reward is a ruined life if they are not killed outright.
Incidence In one study of 233 whistle-blowers in the United States found that: 90 percent of them lost their jobs or were demoted, 27 percent face lawsuits, 26 percent faced psychiatric or medical referral, 25 percent admitted alcohol abuse, 17 percent lost their homes, 15 percent were subsequently divorced, 10 percent attempted suicide, and 8 percent went bankrupt. Nearly two thirds of them failed, after five years, to be vindicated for their disclosure.
Broader Harassment (#PC8558).
Related Harassment of the judiciary (#PE5487) Denial of freedom of conscience (#PD7612)
Unrehabilitated historical figures (#PF1540)
Official cover-up of government harassment of political activists (#PF3819).
Aggravates Suppression of information concerning environmental hazards (#PF4854).
Aggravated by Informers (#PD8926).

♦ PF1586 Avoidance of a confrontation with death
Nature The fact of death is avoided by present day society, so that people fill their lives with busy activities, seeking money, fun, friends, fame, happiness and seldom reflect upon on the certainty that such life will end. They refuse to recognize reality. As older people are separated from previous roles, deserted by family, and see their bodies age, they are forced to reflect on the fact of their own impending death. The sudden realization of what they have previously ignored causes sudden fear which may be paralyzing to the point of impotency among the older generation.
Broader Fear of death (#PF0462).
Aggravates Excessive prolongation of the dying process (#PF4936).
Aggravated by Social isolation of the elderly (#PD1564).
Reduced by Morbid preoccupation with death (#PD5086).

♦ **PF1588 Geomagnetic reversal**
Reversal of Earth's magnetic field
Nature At present the north-seeking end of a compass points approximately in the direction of the North Pole but paleomagnetic studies have shown that the Earth's magnetic field has been reversed in the past so that the same compass needle would have pointed south. Since the geomagnetic field normally helps to shield the Earth's surface from low-energy radiation, reversals permit a higher incidence of radiation to reach the surface, possibly causing extinctions and mutations. The more highly specialized species would be more vulnerable to such changes. Some bird migration patterns would be severely affected.
Incidence It is known that the earth's magnetic field has gone down to zero and reversed itself nine times during the last four million years, with at least 171 reversals in the last 80 million years. Some 45 million years ago the reversal frequency doubled to a rate of 5 reversals per million years although remaining irregular. There is evidence for 5 short-lived reversals in the last 450,000 years. It therefore seems to be impossible to predict when the next reversal will take place although the field intensity is known to have been decreasing over the past century. The polarity inversion apparently occurs over a period of 2,000 years although the field intensity change takes from 15,000–20,000 years.
Broader Geomagnetic disasters (#PD0830).
Aggravates Mutation (#PF2276) Extinction of species (#PB9171)
Evolutionary catastrophes (#PF1181).

♦ **PF1594 Frustration of the role of opposition**
Nature The function of loyal opposition is to allow dissenting voices to be heard and to participate in framing the future. However, in the absence of a clear vision of the future, opposition groups tend to invest time and energy in meaningless pursuits separated from reality and rooted in the false security of past experience. Search for significant meaning results in disillusionment and frustration and a feeling of victimization which obscures any vision of future possibilities for self-determination and organized action.
Broader Exclusion of opposing views (#PF3720) Static and unrelated social roles (#PF1651)
Suppression of opposition groups or individuals (#PC7662).
Aggravated by Oppressive reality (#PF7053).

♦ **PF1595 Ineffectiveness of international nongovernmental organizations and programmes**
Broader Ineffectiveness of international organizations and programmes (#PF1074)
Obstacles to effective international nongovernmental organizations (#PF7082).
Narrower Political ineffectiveness of international nongovernmental organizations (#PE7008).
Related Operational ineffectiveness of international nongovernmental organizations (#PE7056).
Aggravated by Inadequate facilities for international nongovernmental organization action (#PF2016)
Inadequate funding of international nongovernmental organizations and programmes (#PE0741)
Inadequate coordination of international nongovernmental organizations and programmes (#PE1209)
Jurisdictional conflict and antagonism within international nongovernmental organizations (#PE1169).

♦ **PF1596 Ineffective regulation of restrictive business practices**
Nature Many restrictive business practices have an impact on international trade and development and thus inevitably affect the interests of more than one country. The restrictive business practices in question may be engaged in by firms singly or collectively, that is, by firms holding dominant market power on an individual or collective basis. In addition, where the practices are collective, the firms may be located in several countries, as in the case of international cartel type arrangements or practices arising out of the activities of transnational corporations. In terms of controlling such practices, remedial action in a given country may well be hampered – for example: by the difficulty that a government can encounter in obtaining information on a firm's activities, because such information is located outside its national frontiers; or by conflicts between governments (legal or diplomatic) which can arise in respect of attempts to secure such information, and in respect of the remedial measures decided upon.
Broader Lack of control (#PF7138) Restrictive trade practices (#PC0073).

♦ **PF1600 Insufficient care of community property**
Nature Pride in the physical appearance of a community may be seen as a motivating symbol of the pride of the residents in the community itself. In many communities, public and private property is vandalized and neglected; buildings are left locked and unsupervised, paint peeling and windows broken as a result both of natural decay and deliberate vandalism; cemeteries are often overgrown with weeds; roadside ditches are clogged with vegetation and debris; in the absence of a community refuse pick-up system, wastes are dumped on yards or roadways; vandalism of family gardens by wandering animals and children decreases the yield of homegrown produce. Only when ways are found to provide adequate care for village property will residents be able to generate and maintain pride in their community.
Broader Lack of care (#PF4646)
Fragmented forms of care at the neighbourhood level (#PE2274).
Narrower Overgrown cemeteries (#PG9247) Unrestrained animal damage (#PJ9217)
Unsanitary refuse disposal (#PJ1098) Unavailable public buildings (#PG7699)
Uncontrolled local vandalism (#PJ0154).
Aggravates Vandalism (#PD1350).
Aggravated by Neglect of property maintenance (#PD8894).

♦ **PF1602 Scientific fraud**
Pathological science — Pseudo-science — Scientific hoaxes — Faking scientific evidence — Falsification of scientific test results — Falsification of scientific records
Nature Grants and promotions are largely based upon publication productivity, thus it is an accepted practice for senior researchers to co-sign subordinates' studies, even though they may not know much about the data. This co-signing helps younger researchers to publish their research and boosts the publication numbers of their bosses. Occasionally researchers may falsify data in an effort to make their studies more exciting or plausible. Compounding this problem (which, in addition to publication stress, also has financial roots - large medical school debts to repay) is the reticence with which fraudulent practices are brought to public view. Scientists who suspect colleagues of fraud often turn a blind eye as they fear tarnishing their institution's standing, being blackballed by colleagues, and possible defamation lawsuits brought by the accused.
Incidence Examples frequently cited of pathological science include. N-rays, mitogenetic rays, the Allison effect, Piltdown man, Piri Reis map, extrasensory perception, and flying saucers. Scientific fraud is not limited to modern or minor figures: the verdict of history is that Ptolemy, Galileo, Newton, Dalton, Cyril Burt and Gregor Mendel all tampered with some of their data.
A Yale University scientist was fired in 1980 for plagiarizing and falsifying his research work. Another institute found a researcher had faked evidence to show that human skin can be made transplantable without rejection (he was given a one year full-salaried 'sick leave'). A notable Harvard Medical School case involved the discovery of fraud, but the culprit was allowed to stay and continue to cheat.
Claim Practitioners of any scientific discipline are aware of many prominent people in that discipline who have risen to eminence by faking data. They are protected only by the law of libel and by the fact that universities are reluctant to having made mistaken appointments. Fraud is often the product of the highly competitive world of science. Researchers under great pressure to publish papers may ultimately resort to plagiarism and data faking. Dishonest scientists can mislead other researchers, stain their institutions' reputations, waste taxpayers' money, and sometimes even endanger the public.
It has been estimated that possibly 25 per cent of scientific papers has some (usually minor) discrepancy in relation to the original test results. In most cases these do not merit any form of rectification, but in a few cases the discrepancies are crucial.
In addition to flagrant fraud, the practice of co-signing when one has done no research on the study involved, is another form of fraud. Senior researchers should be held accountable for any publications they sign their names to.
Counter-claim Given the extreme pressures to publish in order to prove one's validity and in order to earn enough money to repay education loans, it is no wonder that scientific fraud is becoming more prevalent. Until some of the pressures are alleviated (for example, requiring scientists to cite only 3 publications a year when applying for funding, and lengthening the terms of federal research grants), this practice will not be curtailed.
Refs Hanen, Marsha P; Osler, Margaret J and Weyant, Robert G (Eds) *Science, Pseudo-Science and Society* (1980); Imwinkelried, Edward J *The Methods of Attacking Scientific Evidence* (1982).
Broader Fraud (#PD0486) Hoaxes (#PF9375) Immorality (#PA3369).
Narrower Forged scientific data (#PF0255).
Related Health fraud (#PD9297).
Aggravates Lack of scientific investigation (#PF2720)
Irrelevance of science and technology (#PF0770)
Inconclusiveness of scientific and medical tests (#PD7415).
Aggravated by Unexplained phenomena (#PF8352)
Unprofessional science (#PF6697)
Deceptive misuse of research (#PD7231)
Deception by natural scientists (#PD9182)
Deceptive social science research (#PF7634).

♦ **PF1608 Outdated forms of community health**
Nature While everyday rural life in Third World countries implies twelve or more hours of hard physical work for nearly every member of the community, regardless of age, the vitality of most residents is severely sapped by chronic infestation of such common parasites as hook worm, round worm and pin worms. The causes of these debilitating infestations are unpaved streets, open wells, unstructured sewage disposal, insanitary kitchen procedures and shared housing with livestock. In addition, the subsistence level diets lack necessary vegetable or animal protein to ensure minimum vitality, even under conditions of good health; and continued productivity in such conditions is due only to the great persistence of the will to survive. Despite this pervasive need, the image of health care is of action in the case of emergency to be called on only when accidents occur or when disease finally threatens life. While such emergency facilities are urgently required, they would be of no use in eradicating the causes of general ill-health in a community. Although villagers see the vitality and good health of persons from urban areas, they do not perceive that they have the option of good health in their own villages.
Broader Inadequate health services (#PD4790) Obstacles to community achievement (#PF7118).
Narrower Lack of family planning (#PF0148).
Aggravates Unbalanced infant diets (#PE0691) Nutritional deficiencies (#PC0382)
Unavailability of first aid (#PJ1261) Insufficient health personnel (#PD0366)
Unavailability of dietary care (#PJ1332) Insufficient emergency services (#PF9007)
Limited access to pharmaceutical drugs (#PE1278).
Aggravated by Unhygienic conditions (#PF8515)
Ignorance of health and hygiene (#PD8023).

♦ **PF1609 Manipulative knowledge**
Claim Manipulative knowledge gives control at the expense of wisdom. It results in skilled craftsmen learning how to reshape surfaces without gaining any deep understanding of interiors. It is always exercised at the outer margins of reality. Since it is based on cognition of control, it does not easily lend itself to creative insight and revelatory experience, which call for the cognition of surrender.
Broader Manipulation (#PA6359).
Aggravates Fragmentation of knowledge (#PF0944)
Limitation of current scientific knowledge (#PF4014)
Restrictions on the acquisition of knowledge (#PF1319)
Failure to integrate knowledge to empower humanity in response to the global problematique (#PF8753).
Aggravated by Reductionism (#PF7967)
Monopoly of competence for intervention in the future (#PF0980).

♦ **PF1610 Avoidance of the irrational**
Nature The body of presently-articulated human wisdom does not encompass the irrational depths of humanness. The humanities are characterized by abstract rationalistic intellectualism or by shallow, superficial popularizations. This has resulted in: the rejection of the humanities as irrelevant to the experience of life; the rejection of the individual's own depths; or the apathetic rejection of both of these.
Broader Human wisdom unrelated to daily life (#PD1703).
Aggravated by Uncritical thinking (#PF5039).

♦ **PF1615 Suppression of scientific information**
Suppression of experimental data
Nature Under certain conditions information in scientific reports is suppressed. This may due to pressures from the bodies sponsoring the reports (government agencies, foundations, or corporations). A common reason is maintenance of national security in the case of governments and maintenance of competitive advantage in the case of corporations. It may also be due to the embarrassing nature of the information and the effect it may have on public opinion.
Broader Abuse of science (#PC9188) Suppression of information (#PD9146).
Aggravates Selective perception of facts (#PF2453)
Conceptual repression of problems (#PF5210)
Suppression of information concerning environmental hazards (#PF4854).
Aggravated by Racial inequality (#PF1199) Scientific censorship (#PD1709)
Bias in scientific research (#PF9693).

♦ **PF1618 Ineffective religious dialogue**
Religious monologue — Inability to bear witness
Claim The danger persists that churches and their experts are deplorably unaware that the drafting of documents in missionary conferences and the careful work given to getting terms right, all tends to become an end in itself and thus a monologue. The continuous exercise of many Christians who put their missionary obligations on paper can be a surrogate for actual witness in concrete human situations. The grandiose imagination and the mutual encouragement to be faithful ambassadors of Christ are in fact burdens and obstacles in the meeting of neighbours of other faiths and of little or no faith.
Broader Ineffective dialogue (#PF1654).
Related Erosion of religious belief by ecumenical dialogue (#PF9274).

♦ **PF1622 Over-reaction to past sexual ethics**
Claim Ethical behaviour as concerns sexuality is primarily based on reaction against the sexual ethics of the past rather than the creation of future-oriented guidelines. Whether response to the past is based on permissive or on reactionary attitudes, people are more concerned with dead issues than with the living needs of today.
Refs Atkinson, Gary M and Moraczewski, Albert S *A Moral Evaluation of Contraception and Sterilization* (1979).
Broader Individualistic perception of sexual activity (#PF1682).
Aggravates Cohabitation (#PF3278).

♦ **PF1623 Issue avoidance**
Dodging issues — Temporizing
Nature In contrast to the evasion of issues, which denies their significance and seeks to prevent recognition of them, issue avoidance uses various acceptable procedural devices and loopholes in order to avoid having to deal with problems. Such devices may include postponement of consideration to a later date, requests for more information or for the opinion of qualified experts, and claims that the matter is the responsibility of some other body. Use of such loopholes may not be perceived as strictly ethical, but can be easily justified on grounds of precedence and pragmatism. It constitutes tolerable behaviour which does not easily serve as an extended focus of public criticism.
Broader Inability to resolve problems realistically (#PF8435).
Narrower Political displacement activity (#PF5360)
Avoidance of issues in political campaigns (#PE6311)
Flight to avoid prosecution or giving testimony (#PE5267).
Aggravates Evasion of issues (#PF7431) Policy-making delays (#PF8989)
Resignation to problems (#PF8781) Non-recognition of problems (#PF8112)
Recurrence of misapprehended world problems (#PF7027)
Calculated delays in releasing controversial news items (#PE0598).
Aggravated by Indecision (#PF8808) Blame avoidance (#PF6382)
Underreported issues (#PF9148).

♦ **PF1624 Inconsistent medical practices**
Inconsistent pharmaceutical practices — Culturally determined medical practices — Culture-specific diseases
Nature Medical culture plays an important role in both the approval and marketing of drugs leading to national differences which inhibit efforts to harmonize standards of drug prescription and use. Treatments for the same ailment vary radically from one country to another. The rate of surgical operations also varies enormously from country to country.
Incidence Drugs requiring a prescription in some countries (such as codeine-based products in the USA) are sold freely in others (such as Italy), because local medical tradition assesses the risks differently, possibly because of lack of experience. Local traditions of payment for drugs, and the relation to the national health plan, may also be important in how a drug is approved. In countries where more importance is attached to drug safety (such as Germany) rather than efficacy (such as France) many more drugs may be available. Differences in clinical testing procedures may also affect the availability of the same drug in different countries.
Aggravated by Excessive cost of medical drugs (#PE5755)
Unethical pharmaceutical practices (#PE3540)
Ineffective self-regulation in the pharmaceutical and medical devices industries (#PE8502).

♦ **PF1625 Prohibitive cost of parenthood**
Incidence In the UK in 1990 it was estimated that the cost involved in rearing a child from birth to coming of age, by the most minimal standards, exceeds £56,000. By the highest standards it can reach £250,000, with middle-income couples investing approximately £178,000. In April 1990, the minimum cost per week was: 0–4 years, £43.89; 5 to 7 years, £51.24; 8 to 10 years, £56.14; 11 to 12 years, £61.04; 13 to 15 years, £65.87; 16 to 18 years, £87.85.
Aggravates Population decrease (#PF6441).

♦ **PF1626 Pseudo-revolutionaries**
Claim In revolution the leaders must trust in the oppressed and in their ability to reason. Whoever lacks this trust will abandon dialogue, reflection, and communication, and will fall into using slogans, communiqués, monologues, and instructions.
Related Pseudo-leadership (#PJ5003).

♦ **PF1645 Inadequate legislation relating to action against problems**
Broader Inadequate laws (#PC6848)
General obstacles to problem alleviation (#PF0631).
Narrower Lack of legal provision for international nongovernmental organizations (#PE0069).
Aggravates Limited local respect for regional and global legislation (#PF2499).
Aggravated by Inadequate power of intergovernmental organizations (#PF9175).

♦ **PF1651 Static and unrelated social roles**
Disrelated images of social roles — Unclear societal roles — Compartmental role tasks — Unclear polity roles — Unexamined community roles
Claim There is no effective vehicle in society for relating diverse social groups to the common direction of society as a whole. In addition the roles which an individual may play in the different areas of his life, as in his job, his family, and in different social and cultural groups, are in no way related to each other. This results in social roles becoming static. Individuals have no means of relating their own social roles with those of others. There is no feeling whatsoever of the meaning of, or the relationship among, these roles in global terms.
Broader Obstacles to community achievement (#PF7118).
Narrower Negated paramedic role (#PG5203) Unclear educational roles (#PG8840)
Conflicting roles of women (#PD6273) Lost family role in society (#PF7456)
Collapsed images of vocation (#PF6098) Ineffective parenting techniques (#PJ0551)
Frustration of the role of opposition (#PF1594)
Unmeaningful social roles for the aged (#PF1825)
Inadequate image of roles within marriages (#PD1308)
Collapse of public servant role of the professional (#PF0897)
Elimination of the socio-cultural role of western women (#PE1046).
Related Demoralizing image of urban community identity (#PF1681)
Neglect of the role of women in rural development (#PF4959)
Rigidly entrenched social traditions in rural areas (#PF1765).
Aggravates Social disaffection of the young (#PD1544)
Limited development of functional abilities (#PF1332).
Aggravated by Consumerism (#PD5774) Insufficient role models (#PF8451).

♦ **PF1653 Limited access to external resources**
Nature Although advances in technology have made it theoretically possible for the resources of the whole world to be at the disposal of every human settlement, many small communities are in fact deprived of these resources. Sewer systems, farm equipment, educational opportunities, outside investment in local projects, are but a few of the amenities lacking due to remoteness of service or lack of knowledge of what is available and how best to use it. These factors discourage the development of new industries and the augmenting of local services.
Broader Inequitable access to resources (#PD3210).
Narrower Lack of extension offices (#PG9901) Complex services logistics (#PG9918)
Lack of funding structures (#PG9919) Law enforcement complexity (#PF2454)
Gap in agricultural methods (#PG9895) Inadequate improvement funds (#PG9506)
Uninterested private investors (#PG9870) Lack of specialized technology (#PG9889)
Insufficient councilman support (#PG9765) Limited access to social benefits (#PF1303)
Insufficient capitalization means (#PG9854) Unknown availability of subsidies (#PG9905)
Unfamiliar bureaucratic procedures (#PJ9912)
Prohibitive cost of connection to public utilities (#PJ9426)
Limited availability of investment capital for urban renewal (#PF3550).
Related Irregular transport services (#PE5345).
Aggravates Low athletics priority (#PJ0337) Decline in government expenditure (#PF9108)
Inadequate community care for transient urban populations (#PF1844).
Aggravated by Limited construction manpower (#PG9875)
Unperceived educational opportunities (#PJ9762).

♦ **PF1654 Ineffective dialogue**
Inappropriate dialogue — Dangerous forms of dialogue
Nature Concerned groups frequently find themselves unable to translate the social conflicts in which they are involved into effective dialogues. By reducing complex situations to single issues, or in emotional and violent outbursts, or hysterical relating of issues to individual situations, the form and quality of the dialogue become inadequate and nothing is gained from the exercise. A first danger of dialogue is to rush into premature agreement, thus obscuring the differences that are the reason for the dialogue in the first place. Other dangers include: triumphalism, relativism and syncretism.
Narrower Ineffective religious dialogue (#PF1618)
Closed channels of dialogue with the judiciary (#PE7725)
Erosion of religious belief by ecumenical dialogue (#PF9274)
Inadequate structures for political dialogue and review (#PC1547).
Aggravates Illusion of consensus (#PF4327) Inadequate meeting methods (#PF8939)
Distrust of political dialogue (#PD2263)
Lack of means for achieving consensus (#PD2438)
Irresponsible expression of emotions equated with free speech (#PF7798).
Aggravated by Triumphalism (#PF8203) Ethical decay (#PB2480)
Religious syncretism (#PF7079).

♦ **PF1658 Limited employment options**
Limited job opportunity — Insufficient job opportunities — Inadequate job opportunities — Insufficient jobs in the village — Scarce local jobs — Limited availability of jobs in small communities — Unavailability of local jobs — Limited employment alternatives
Nature In many communities the options for local employment are extremely limited. Perhaps the major limiting factor has to do with the seasonal nature of local employment, often the result of an economy based on the use of natural resources. In the absence of industry, steady jobs are mostly located at a distance from the community. Families which would like to return are prevented from doing so by the absence both of available unskilled jobs, and of adult training programmes coordinated with employment needs. In addition, the educational standards demanded for some jobs appear to be unrelated to the actual requirements of the positions sought; therefore students do not feel motivated to pursue higher education or to acquire technical skills.
Claim Those concerned with economic development of small communities have come to understand that a community must be able to provide jobs for its residents.
Narrower Restricted scope of local employment (#PF2423)
Unequal employment opportunities for women (#PD5115)
Limited availability of permanent employment in inner-cities (#PE1134).
Aggravates Unemployment (#PB0750) Rural unemployment (#PF2949)
Small student enrolment (#PG9332) Unemployed skilled labour (#PE1753)
Inconsiderate telephone use (#PG9378) Slow rate of income expansion (#PF6478)
Narrow range of cultural exposure (#PF3628) Available jobs unrelated to education (#PJ9233)
Arrested development of labour potential (#PF6532)
Crimes committed during high unemployment (#PJ1139)
Maladjustment to disciplines of employment (#PD7650)
Limited opportunities for significant work (#PF1403)
Underdeveloped sources of income expansion (#PF1345)
Disorganized attempts of upgraded employment (#PF6458)
Non-diversification in subsistence fishing economies (#PF2135).
Aggravated by Seasonal unemployment (#PC1108)
Unpublicized community news (#PF7998)
Inappropriate employment incentives (#PD0024).

♦ **PF1663 Social reinforcement of delimited self-worth**
Nature Awareness of self-worth implies personal responsibility. In order to escape this responsibility, individuals deny their own personal worth and limit themselves to roles which they or society may define as circumscribed, thus destroying their own essential freedom. Societal images taught through advertising, education in schools, and social relations, reinforce such "escapism" in which individuals find themselves trapped.
Broader Tensionless image of free choice (#PF1675).
Aggravates Lack of self-confidence (#PF0879).

♦ **PF1664 Erosion of elders' wisdom**
Isolation of the wisdom of elders from society — Non-transferring of the wisdom of the aged — Unrecognized role of elders — Underused elders' wisdom — Uncommunicated wisdom of elders — Limited social guidance by the older generation — Neglect of elders' skills
Nature Older members of society, with a lifetime of observations, personal development and historical continuity with previous generations, are unable to transmit their wisdom to those following them due to the negative features that have emerged in modern life.
These features include the isolation of the old due to the ever-shrinking size of the family, and to the weakening of natural ties when the elderly are cared for by the welfare state instead of the family doing so. They also include the radical changes that have occurred in values between the generations, which make communication difficult. Older people may also withhold their accumulated wisdom in a misguided attempt to shelter the next generation from unpleasant realities, such as experiences of war, poverty, famine or disease. Finally, because there is no routine cultural mechanism to record the perceptions of the mature, these may become distorted as people grow older, due to a psychological need to reinvent or romanticize the past, or due to failure of memory.
Isolating elders in retirement homes places them in a static relationship to society. The individual is forced to operate out of a reduced context, his capacity to function in society is limited. The elder's role becomes irrelevant perpetuating the need for isolation. There is strong evidence that senility sets in sooner and faster with limited physical and intellectual stimulation.
Because older people have progressed further along life's journey than the rest of society, those among them that are self-aware possess inherent insight about the nature of life which is essential in planning the future. However, neither society nor the old people themselves articulates or gives sufficient respect to this unique responsibility. Since these particular gifts are not passed on, the whole of present-day society is deprived.
Although older people have accumulated wisdom, this is unrelated to current values. In the midst of collapsing traditional values, adult members of the population receive little support or guidance

in social relations from their elders. The information systems necessary for maintaining family, community and larger social relations are missing. This vacuum of values to reinforce social forms results in little guidance for young people and children as they learn to be socially active.
Claim The accumulated wisdom of the older generation is the stabilizing and cohesive influence that is needed by modern, fragmented and rapidly changing societies, but which they are, for various reasons, deprived of.
Broader Neglect of the aged (#PD8945) Social isolation of the elderly (#PD1564).
Narrower Elder paralysis over the future (#PF3973)
Denial by old people of the significance of the past (#PF2830).
Aggravates Disobedience of elders (#PF7149)
Underutilization of locally available skills (#PF6538).
Aggravated by Unrecognized opportunities (#PF6925)
Lack of community self-worth (#PF3512)
Haphazard forms of social ethics (#PF1249)
Underutilization of popular wisdom (#PF2426)
Unmeaningful social roles for the aged (#PF1825)
Disjointed patterns of community identity (#PF2845)
Limited community responsibility of adults (#PF1731).

♦ PF1667 Restricted delivery of essential services to developing country rural communities
Nature In Third World villages, such necessities for effective participation in the modern world as transport services, all-weather roads, electricity and communication systems are often insufficient to meet requirements or are lacking altogether. Residents become resigned to the absence of these essentials and adjust to living with inordinate amounts of time consumed in "making the best of it". The longer such services are curtailed, the greater the disparity between villages and urban areas.
Broader Underprovision of basic services to rural areas (#PF2875).
Aggravated by Lack of transport vehicles (#PF0713)
Inadequate road maintenance (#PD8557)
Prohibitive medical expenses (#PE8261)
Inadequate electricity supply (#PJ0641)
Prohibitive cost of basic services (#PF6527)
Inadequate electricity infrastructure (#PD9033)
Insufficient transportation infrastructure (#PF1495).

♦ PF1673 Media reinforcement of materialism
Excessive dissemination of materialistic images
Nature People are continually informed through the media and their daily surroundings that the significance of living is related to economic success. These messages are reinforced by the structure of social and political organizations. At the same time, the individual is left to discover his spiritual and interpersonal integrity without societal reinforcement.
Broader Repression of self-consciousness (#PC1777).
Narrower Reduced images of humanness (#PF4323)
Collapsed images of vocation (#PF6098)
Separation of individual's social functions (#PF4754).
Aggravates Limited spheres of relationship (#PD1941).

♦ PF1674 Absence of rites of passage
Claim The absence of rites of passage leads to a serious breakdown in the process of maturing as a person. Young people are unable to participate in society in a creative manner because societal structures no longer consider it their responsibility to intentionally establish the necessary marks of passing from one age-related social role to another, such as: child to youth, youth to adult, adult to elder. The result is that society has no clear expectation of how people should participate in these roles and therefore individuals do not know what is required by society.
Broader Disorientation of the young due to lack of social forms (#PD2050).
Narrower Denial of rites to the dead (#PF9190)
Inadequate recognition by institutions of the transition through adolescence (#PF6173).
Aggravates Loss of cultural identity (#PF9005).

♦ PF1675 Tensionless image of free choice
Nature Transition into the new era brings with it a loss of freedom due to a failure to recognize the necessity for tensions and ambiguity as integral parts of free choice. A new paradigm is emerging that has, as an essential value, the participation of all people in the creation of their own future and that people have the freedom to participate. There is a tendency to teach that to escape tension and ambiguity is to be free. This understanding glorifies the escape patterns of the affluent society. It reinforces the tendency to render roles meaningless. It encourages people to react to life rather than to respond out of some self-understanding and world view. It discourages self-conscious decision-making and a realistic view of one's situation because a search for means of escape from difficulties becomes of paramount importance, rather than the creation of possibilities. Action is turned from meaningful responses to the future, towards self-seeking avoidance of pain.
Broader Defensive life stance (#PF0979).
Narrower Social disguise of ambiguity (#PF2505)
Non-integrating images of personal freedom (#PF1772)
Social reinforcement of delimited self-worth (#PF1663)
Lack of meaningful personal and social paradigms (#PD0894)
Lack of essential freedom due to unawareness of actuality (#PD1942).
Related Individual isolationism (#PD1749).

♦ PF1676 Radio noise of natural origin
Nature Clear reception of radio signals depends upon the ability to exclude unwanted noise. Noise from natural sources takes two forms, atmospheric and extraterrestrial. Atmospherics result from the electromagnetic disturbances caused by lightning in thunderstorms and are at a maximum at night and in summer and in the tropical latitudes. Extraterrestrial noise interferes with radio communication in the VHF band (and down to 20 mc). It is due to electromagnetic phenomena of galactic or solar origin. Solar noise increases during the daytime and during peak periods in the 11-year sunspot cycle. Solar flares and geomagnetic storms may produce radio communication blackouts, particularly in the polar regions.
Broader Inhospitable climate (#PC0387) Electromagnetic pollution (#PD4172).
Related Jamming of satellite communications (#PD1244).
Aggravates Radio frequency interference (#PD2045).

♦ PF1677 Incomprehensible presentations
Nature As the complexity and variety of information increases the capacity for an individual to understand not only the data being presented but to draw implications about the data is frustrated by incomprehensible presentations by speakers. It seems that conferences, lectures and other forms of presenting material is hindered by disorganization, distractive comments or stories, ineffective use of media such as overhead projectors, inefficient use of time and inaudible or mumbled vocal presentations.
Broader Ineffective communication (#PF6052).
Aggravated by Poor communications methods (#PJ8656).

♦ PF1681 Demoralizing image of urban community identity
Defeating community story — Demoralizing community self-story
Nature Within urban ghettos, the vacant lots, boarded-up buildings and general deterioration have given rise to demoralizing images of the community's total decay, which are reinforced by the contrast between such conditions and the architectural evidence of former grandeur. When these images are projected within the community and to the larger metropolitan area, they are taken to indicate that nobody can "make it" in such a community, nothing good ever happens there, and nothing good can ever happen there in the future.
This "story", told in conversation and dramatized in rampant vandalism, becomes a significant block to any cooperative efforts at development and contributes to negligence and destruction within the community. Publicity concerning plans, meetings or proposed projects is greeted with scepticism born from years of experiencing inner city programmes that come and go. Residents are tired of paper promises and lucid about the extent of reformulation required to rebuild their community, doubting that such rebuilding will ever occur. Nothing less than a rapid, massive physical demonstration of community rehabilitation will release residents from the habitual "story" and permit them to spread the image of their ghetto neighbourhood as a worthy place in which to expend the creative efforts of themselves and of others.
Broader Obstacles to community achievement (#PF7118)
Obsolete basis of cultural identity (#PF0836).
Narrower Distrust of services (#PG8857) Ghetto education image (#PJ8933)
Inadequate hero images (#PF2834) Faded community symbols (#PG8964)
Community expertise drain (#PG8926) Insufficient community networks (#PG8868)
Community development scepticism (#PJ8970) Overwhelming deteriorating space (#PG8918)
Irresponsible transient occupants (#PG8879) Unreleased creativity of the elderly (#PG8971)
Disjointed patterns of community identity (#PF2845).
Related Static and unrelated social roles (#PF1651)
Demoralizing constraints on housing rehabilitation (#PE2451).
Aggravates Low learning expectations (#PJ8960)
Uncontrolled local vandalism (#PJ0154)
Lack of community development (#PF7912)
Limited availability of financial credit (#PF2489).
Aggravated by Distrust (#PA8653) Lack of self-confidence (#PF0879)
Oppressive prevalent images (#PF1365) Declining sense of community (#PF2575).

♦ PF1682 Individualistic perception of sexual activity
Nature The perception that sexual activity is simply a personal, individual activity without social consequences infers that interpersonal relationships have little basis beyond the wishes of the individuals concerned. This results in a society with weak means of cohesion; and, at the same time, individuals and families find their relationships operating in a vacuum, without societal constraints or encouragement.
Broader Lost family role in society (#PF7456).
Narrower Sexual scientism (#PD2091) Sexual immorality (#PF2687)
Over-reaction to past sexual ethics (#PF1622)
Lost covenantal understanding of sexuality (#PF1764).
Related Reduced interior structure of families (#PF3783)
Exclusion of pre-adults from family decisions (#PE2268)
Individually defined operating structure of marriage (#PD2294).
Aggravates Paternal negligence (#PD7297).

♦ PF1686 Health hazards of exposure to cosmic radiation
Nature The surface of the earth is constantly exposed to cosmic radiation. The effect of this is diminished by the atmosphere, although high energy cosmic rays are unaffected. At high altitudes, and particularly in passenger aircraft, the minimum dose rate experienced by passengers and crew is greater than at ground level. There is evidence to suggest that cosmic radiation is a source of genetic mutation.
Incidence The maximum permissible radiation dose rate for adult workers (40 hours a week, 50 weeks a year) is 2.5 mrem per hour. It is not expected that passengers or crew of supersonic aircraft at high altitudes, for example, are likely to come near this limit.
Broader Contamination by natural radiation (#PC1299).
Aggravates Mutation (#PF2276) Human disease and disability (#PB1044).

♦ PF1687 Meteorites as hazards
Nature Meteorites are lumps of solid matter of extraterrestrial origin which are large enough not to be destroyed by their passage through the atmosphere, and therefore strike the earth's surface. Meteorites frequently occur in large showers. The number of bodies in a shower can be very great – 1868, Pultusk (Poland): 100,000; 1912, Holbrook AZ (USA): 14,000. With the increase in the spread of built up areas, the probability that meteorite showers or large meteorites will cause damage and loss of life also increases; in addition, an isolated iron meteorite of the appropriate size could give rise to a radar alert which could lead to a general nuclear counterattack.
Incidence It is estimated that 10,000 tons of micrometeorites accumulate (harmlessly) on the surface of the earth each day. By December 1965 the total number of well-authenticated distinct meteorites was 7,791. The larger known meteorites range in size from 500 pounds to 60 tons. Even larger meteorites have struck the earth, as is evidenced by the presence of craters which range in size up to 37 miles in diameter; it has even been suggested that the Hudson Bay (284 miles diameter) may have resulted from such an impact. Estimates show that at any one time about 1,000 asteroids with diameters greater than 1 km are crossing the earth's orbit and that, on average, one of these bodies would hit the earth every 300,000 years, producing a crater of 20 km in diameter. Impacts of objects with diameters larger than 10 km, producing craters in excess of 200 km, are to be expected every 40 million years. Additional impacts by comets could increase the incidence. It has been suggested that the mass extinction of living species at the Cretaceous-Tertiary boundary was due to climatic effects caused by the impact of an object of about 10 km diameter.
Broader Disasters of extraterrestrial origin (#PF3562).
Aggravates Risk of unintentional nuclear war (#PF0466).
Aggravated by Hazards from comets (#PF3564).

♦ PF1688 Uncritical preservation of the status quo
Nature Educational systems tend to shape the individual's internalized value system in such a way as to make him or her an uncritical preserver of society in static relations with other nations and cultures; they fail to emphasize the increasingly interdependent relationship of nations and cultures and thereby reinforce the status quo. The superficial coverage of other cultures tends only to relate to the life styles or ethics of the individual's own culture, so that no common understanding is formed or perpetuated. Individuals' conception of the common goals and the common journey of humanity into the future are thus disorientated.
Broader Reinforced parochialism of internal values and images (#PF1728).
Aggravated by Uncritical thinking (#PF5039) Inappropriate education (#PD8529).

♦ PF1691 Friction
Nature Friction is the resistance which one body offers to another sliding over it. Additional work is required to overcome this resistance when it occurs in any mechanical system, thus increasing the power consumption and the energy needed. Considerable research and design effort is devoted to eliminating friction from mechanical systems, aircraft, vehicles, and the various forms

of water craft.
Incidence It has been estimated that over 30 per cent of energy generated is consumed in friction, manifest as the heating of the moving parts. In the particular case of the automobile, some 20 per cent is so thus wasted.
Counter-claim While the presence of friction effects is a considerable obstacle to the increase in the efficiency of mechanical and transport devices, the complete absence of these same effects would create another set of difficulties, namely stopping or braking such devices once set in motion. Indeed, friction is a very useful and economical means to control the motion of mechanical parts or vehicles.
Broader Enmity (#PA5446) Disaccord (#PA5532) Opposition (#PA6979).
Related Wear (#PB1701).
Aggravates Energy crisis (#PC6329) Fatigue in materials (#PD1391).
Aggravated by Elimination of jobs by automation (#PD0528).

♦ PF1695 Meteors as hazards
Space boulders
Nature In its journey through space, the planet earth is continually colliding with asteroids (boulders comprised of compressed rocks and dust) and comets (compressed ice and dust). The earth meets a 100 tonne rock about a metre wide every day, a 1,000 ton body once a month, a 15,000 ton boulder once a year and a 100,000 ton behemoth every 20 years. The greatest danger during the past few thousand years has been fragments of the comet Encke. Each June 30 the earth passes right through this debris. Not only does massive destruction occur when a major boulder collides with the earth, but since the entry fireball mimics a nuclear explosion, there is a risk of detonating defence systems.
Incidence In 1908, a meteorite weighing a million tonnes fell into the atmosphere over the Tunguska region of Siberia, travelling at 72,000 mph. Its entry fireball was brighter than the sun and it exploded four miles up in a 12-megaton flash. 1,000 square miles of pine forest were destroyed and the local soil is still covered with tiny particles of glass and iron from the melted and vaporized boulder. A massive boulder of this size hits the earth on average every 100 years.
Refs Graubard, Mark *Witchcraft and the Nature of Man* (1985); International Astronomical Union Staff and Kresak, L and Millman, P M *Physics and Dynamics of Meteors* (1968); Kronk, Gary W *Meteor Showers* (1988).
Broader Bad omens (#PF8577) Disasters of extraterrestrial origin (#PF3562).
Related Novae (#PF3563) Interstellar dust (#PF9002)
Hazards from comets (#PF3564).
Aggravates Evolutionary catastrophes (#PF1181).

♦ PF1696 Local traditions of cultural isolation
Nature Individual, living cultures are beginning to find a new identity as part of one global culture. In Third World villages, for example, there is a growing discovery of customs and ways previously lost in successive waves of foreign influence. Such discovery may, however, lead to a sense of isolation in language, mores and living patterns; and the question is one of involving diverse peoples in a common national or regional identity.
Broader Cultural isolation (#PC3943) Deteriorating quality of life (#PF7142).
Narrower Socially isolated women (#PG9712) Narrow range of cultural exposure (#PF3628).
Related Inaccessible job market (#PE8916)
Untransposed significance of cultural tradition (#PF1373).
Aggravates Uninformed care techniques (#PG9655)
Ignorance of cultural heritage (#PF1985).

♦ PF1698 Telephone delays
Overloaded party lines
Nature The very rapid increase in the number of telephones installed has resulted in overloading of exchanges often using out-of-date equipment. Technical incompatibility between such exchanges, particularly in different countries, can lead to considerable delays in making a connection. The situation becomes worse in peak periods and is aggravated by time zone differences which limit the overlap in office hours between different countries.
Broader Communication delays (#PF4453)
Reluctant claims on external resources (#PF1226).
Aggravated by Multiplicity of time standards (#PF2621)
Inadequate standardization of procedures and equipment (#PC0666).

♦ PF1705 Individualistic retaining of local tradition
Nature Oral history often goes back no further than events within in the memory span of living villagers. As a result, people often have only a slight awareness of the significant common history of their ancestors. While many developing communities are consciously using what remains of their heritage to foster community solidarity, some small villages seem to hold on only to those aspects of tradition which heighten individualism to the detriment of common development efforts. Community meetings, which were once held to discuss and decide community issues, have virtually disappeared from the social fabric of these villages; many traditions, which formerly passed from the older to the younger people, have been dissipated as the young people move to the cities; those activities which used to bring women together have been replaced by the partial introduction of modern conveniences.
Broader Individualism (#PF8393).
Narrower Small land plots (#PG8526) Leadership age gap (#PJ9855)
Archaic economic base (#PG9813) Ageing of world population (#PC0027)
Insufficient women's groups (#PG9884) Individualistic land holding (#PG9789)
Insufficient banking information (#PG9863) Insufficient traditional gatherings (#PG9896)
Unconvincing alternatives to existing societies (#PF3826)
Individualism as a restriction of effectiveness (#PE8143).
Related Unchallenging world vision (#PF9478) Over-intensive soil exploitation (#PC0052)
Decreasing genetic diversity in cultivated plants (#PC2223)
Restrictive loan procedures in developing countries (#PF2491).
Aggravates Seasonal fluctuations in agriculture (#PD5212)
Lack of commitment to common symbols (#PE8814)
Inadequate training in decision-making (#PD2036)
Obsolete methods of agricultural production (#PF1822).
Aggravated by Untransposed significance of cultural tradition (#PF1373).

♦ PF1706 Hyperefficiency
Obsession with efficiency — Excessively efficient use of time — Preoccupation with accelerating time
Nature Time, like space, is increasingly perceived as a scarce resource which must be manipulated in ever more sophisticated ways to shape and mould the life of society for its greater well-being. Whilst the increases in efficiency of the industrialization process have resulted in short-term material benefits, the long-term psychic and environmental damage has outstripped whatever temporary gains might have been wrought by the fanatical obsession with speed at all costs. The artifical temporal environments created by this process, and the accelerated pace of life within them, increase the separation of man from the rhythms of nature and from the tempo at which nature is able to recycle wastes and renew basic resources.
Related Scrupulosity (#PF6404).
Aggravates Lack of time (#PC4498) Separation from nature (#PF0379)
Increasing pace of life (#PF2304) Psychological ungroundedness (#PF1185)
Obstacles to efficient utilization of time (#PF7022).

♦ PF1721 Anti-social personality disorders
Psychopaths - Sociopaths — Psychopathy
Nature Chronically antisocial individuals are always in trouble, profiting neither from experience nor punishment, and maintaining no real loyalties to any person, group, or code. They are frequently callous and hedonistic, showing marked emotional immaturity, lacking a sense of responsibility, lacking judgement, and having an ability to rationalize their behaviour so that it appears warranted, reasonable, and justified. They are distinguished from individuals who manifest disregard for the usual social codes, and often come in conflict with them, as the result of having lived all their lives in an abnormal moral environment. These latter individuals may be capable of strong loyalties and typically do not show significant personality deviations other than those implied by adherence to the value or code of their own predatory, criminal or other social group.
Psychopathy is a mental disorder not necessarily amounting to insanity, characterized by antisocial behaviour, often leading to acts of aggression and violence. The victim, known as a psychopath, is to be distinguished from the psychotic, who is commonly treated as insane. Psychopathy as a mental illness falls into a class of its own, and currently receives much attention from psychologists and behavioural scientists the world over. It is largely a phenomenon of our times. Often the intelligence of the psychopath is above average. Callous and selfish, he lacks the emotional qualities that make for normal interpersonal relationships such as altruism, affection, empathy, concern for others. But he may show a fanatical attachment to one person, animal or object. The principal characteristics of the psychopath is his social insensitivity and this lack of feelings. He is a sociopath, opposed to society in general, undisciplined and antisocial in his behaviour, and apparently devoid of conscience. He has little sense of duty or responsibility, and may be aggressive and violent. Varying degrees of psychopathy are to be seen in the road-hog, the bully, the mob-orator, the wife-beater, the baby-basher, the football hooligan, the vandal, the terrorist, the hi-jacker, the psychopathic killer. Psychopaths form the greater proportion of the population of remand institutions and prisons; because his moral sense is so far below normal, the psychopath's condition has been described as one of moral insanity, or moral imbecility.
Broader Mental illness (#PC0300) Mental disorders (#PD9131)
Personality disorders (#PD9219).
Aggravates Sexual violence (#PD3276).
Aggravated by Conduct disorder (#PE3770).

♦ PF1723 Unexpressed social compassion
Nature Because social care is most effectively maintained through large institutions, and professional workers care for the needy, the uneducated and the sick, non-specialists in social fields have little means of showing their concern for suffering people, other than by giving donations or voting for the appropriate party. At the same time, the many volunteer efforts at person-to-person service give volunteers tremendous satisfaction but are inadequate to meet the massive human need which exists.
Broader Lack of commitment (#PF1729).

♦ PF1724 Unrelated pioneer institutions in developing countries
Nature Pioneer industries in developing countries are likely to suffer from the absence of other establishments to which waste by-products might be passed. They may also be handicapped by the absence of institutions which in a more developed society would help to improve the skill of the labour force and facilitate the raising of capital. Where the work requires it, a new factory may have to organize its own training facilities - again a cost-raising expedient - but the lack of financial institutions is likely to be less tractable. This is a deterrent to industrial investment partly because it makes more difficult the raising of local capital and partly because it tends to reduce the liquidity of investment in secondary industry. In much the same way, the day-to-day operations of a manufacturing establishment are likely to be less smooth and hence less efficient, because of the absence of many of the services of insurance, credit and banking institutions which are part of the normal environment of an industrial economy.
Broader Economic and social underdevelopment (#PB0539)
Weakness of infrastructure in developing countries (#PC1228).
Aggravates Underdevelopment of industrial and economic activities (#PC0880).

♦ PF1728 Reinforced parochialism of internal values and images
Lack of correspondence between basic value images and social reality
Nature The rapid change characteristic of present-day life requires that the internal values and images, with which an individual dialogues in order to make a decision, be enlarged to enable responsible decisions to be made in as wide a context as possible. However, this internal "community" is normally developed so selectively that individual concerns are reinforced and decisions based on expediency are encouraged. Individuals tend to be indoctrinated by the family, educational institutions, communications media and interpersonal relationships, into values of personal psychological well-being, cultural preservation and economic progress. But wider experience does not necessarily provide deeper meaning; one negative aspect of the global acculturation presently taking place is the superficial appropriation of other cultures while rejecting even the creative and positive aspects of one's own culture. This failure to provide global, inclusive and profound points of personal reference is complicated by conflicting loyalties and no sense of belonging to any community. With few, if any, symbols or values beyond the personal or of a global nature, people become desensitized to human suffering, feel that their lives are meaningless, and desire to escape reality.
The rapidly changing twentieth century requires the expansion of an effective person's internalized social relations.
Whereas the small, traditional associations which are based on kinship and locality are still expected to communicate basic values to members of society, they are not the institutions on which these social functions are based, so that images and social reality do not correspond.
Broader Collapse of common values (#PF1118)
Collapse in providing ethical value screens (#PF1114)
Inadequate means for upholding global concern (#PF1817).
Narrower Lack of social progress (#PF1545) Parochial national interests (#PF2600)
Conditioned self-gratification (#PF1807) Socially ineffective family units (#PD4572)
Parochialism of established religions (#PF7593)
Lack of symbolism in local relationships (#PF2063)
Uncritical preservation of the status quo (#PF1688)
Non-adaptive local structure of social care (#PE5246)
Lack of a system for ethical decision-making (#PF2070).
Related Inhibited capacity to visualize a creative future (#PF2352)
Habitual overemphasis on national self-determination (#PF1804).
Aggravates Avoidance of reality (#PF7414).
Aggravated by Faceless social institutions (#PF2081).

♦ PF1729 Lack of commitment
Shallow personal commitment to social vision — Lack of individual social commitment — Reduced social commitment of contemporary life-styles — Loss of commitment ethic

Nature Individuals escape from the needs of society by refusing to make any in depth social commitments. The individual floats from one shallow commitment to another or dreams about the past or the future. All demands for concrete engagement in the present are left unmet.
Most life-styles currently available do not enable people to engage effectively in society. Primary group covenants between people, like marriage, have become sequential contracts and have lost their profundity. The process of getting married and raising a family has turned from a life's work to a frantic search for sexual satisfaction, financial security or release from guilt about the quality of life of one's children. Elders, adults and children have become disconnected from each other's life styles; and care for others has become a peripheral concern. Communities are no longer geographically oriented but are oriented towards work or leisure-time, further creating tensions in basic group relationships. Emotional and spiritual support systems have been eroded by this fragmentation of covenants.
Claim Public conscience is not demanding the making of decisions necessary for today's society. Instead, because it seems too difficult and complex to respond to all of the diverse expressions the social needs, people become the passive victims of agonising demand, unable to find ways to act out their commitment. An absence of social direction results in a multiplicity of competing demands vying for attention. The individual dares not risk starting significant involvement. Lack of commitment to building a new world future discourages self-actualization. It is difficult to preserve the present without being creatively related to the future social direction, enabling concrete appropriation of obligations to all other people on the globe.
Broader Defensive life stance (#PF0979) Non-concerned attitudes (#PF2158)
Deteriorating quality of life (#PF7142).
Narrower Lack of work commitment (#PD2790) Lost family role in society (#PF7456)
Unexpressed social compassion (#PF1723) Introversion of the family unit (#PF1030)
Repression of self-consciousness (#PC1777)
Lack of commitment to common symbols (#PE8814)
Ideal of tension free social structures (#PF6178)
Inadequate means for upholding global concern (#PF1817)
Deluding familial image of social responsibility (#PF1064)
Lack of commitment to the protection of vulnerable groups (#PF4662).
Related Theological collapse (#PF6358) Dominance of economic motives (#PF1913)
Negative emotions and attitudes (#PA7090)
Collapsed meaning of human creativity (#PF0936)
Overemphasis on self-sufficiency with respect to interdependence (#PF3460).
Aggravates Social isolation of the elderly (#PD1564)
Societal over-commitment to learning (#PD7051)
Inadequate management of government finances (#PF9672)
Over-acceptance of socio-economic dependency (#PF8855).
Aggravated by Unchallenging world vision (#PF9478)
Lack of meaningful educational context for ethical decisions (#PF0966).

♦ PF1731 Limited community responsibility of adults

Nature It is the responsibility of adults to set standards and establish community values. When outmoded societal structures continue functioning, the adult population, whose responsibility it is to bear the burden of such structures, find their traditional values collapse. They lose their vision of the relationship between concrete local tasks and wider responsibilities, focusing on parochial and nationalistic concerns and with a diminished view of community responsibility; this leaves them no way of maintaining deep-seated common values.
Broader Lack of community responsibility (#PJ3290).
Narrower Static and superficial adult values (#PF2883).
Related Parochial national interests (#PF2600) Elder paralysis over the future (#PF3973)
Denial by old people of the significance of the past (#PF2830)
Disorientation of the young due to lack of social forms (#PD2050).
Aggravates Erosion of elders' wisdom (#PF1664)
Ignorance of cultural heritage (#PF1985).
Aggravated by Social isolation of the elderly (#PD1564).

♦ PF1733 Unattractive appearance of deteriorating buildings

Nature Communities with decreasing populations tend to be left with unsightly, deteriorating empty buildings and vacant lots, which detract from the natural beauty of rural areas, discouraging visitors who might otherwise enjoy the area's amenities and persuading residents to work and shop outside their community. As buildings are demolished they present a reminder of what was and what might come to pass. In urban areas, the gradual deterioration in the appearance leads to a sense of futility on the part of the citizens.
Broader Obstacles to building construction (#PF7106).
Narrower Uncontrolled river erosion (#PJ9518) Debilitating physical appearance (#PG8311)
Unconsensed environmental programmes (#PG8306).
Aggravates Demeaning community self-image (#PF2093).

♦ PF1735 Reluctance to join in community action

Nature In contrast to the general trend for residents to work together to develop their communities, many small communities are reluctant to work together for the betterment of the entire village. This may be due to several reasons: there is often a sense that the issues are too complex to be solved; individuals may be reluctant to relinquish their personal power; sometimes there is a belief that those who were capable of dealing with the problems have already left the village for more fulfilling avenues, and that those who remain have neither the imagination nor the energy demanded for community action; past development efforts may have been left unfinished, reinforcing the sense that little can change. Suggestions for development tend to meet with a negative response - reasons why such a thing could not work, or why the village is not in a position to approach public or private sources for assistance. There is therefore a need to try new alternatives in total community development able to meet perceived needs and to mobilize the community's latent human resources.
Broader Obstacles to community achievement (#PF7118).
Narrower Local medical disinterest (#PG9856) Negative identifying slogan (#PG9917)
Unfeasible industrial images (#PG9882) Perceived complexity of issues (#PG1711)
Insufficient tourist attraction (#PG9911) Disunified requests for services (#PG9803)
Apparently unrewarded initiatives (#PG9897) Disappointing bureaucrative action (#PG9903)
Unmeaningful social roles for the aged (#PF1825)
Cumulative depletion of corporate initiative in rural communities (#PE3296).
Related Unchallenging world vision (#PF9478)
Maldistribution of land in developing countries (#PD0050).
Aggravates Unhygienic conditions (#PF8515) Lack of community development (#PF7912).
Aggravated by Low self esteem (#PF5354) Frustrated past goals (#PF5272)
Loss of confidence in government leaders (#PF1097).

♦ PF1737 Declining economic growth in industrialized countries

Weak economic growth in capitalist countries — Slow-down in growth of output in developed-market economies — Instability in the growth of income in industrialized countries
Nature The economic slowdown in the major developed market-economy countries in the 1980s has reflected the impact of macro-economic policies on capacity utilization and expansion. These countries have been operating below their potential trend level, which itself has been declining largely because of the persistent gap between actual and potential output (maximum production capacity consistent with stable inflation).
There are three explanations for the slow growth of output: (a) there has been a longer-term reduction in the efficiency and flexibility of the economic structure of developed market-economy countries, such as changing attitudes to work, the proliferation of government regulations, existing tax structures, accelerated inflation, the size of government sector, insufficient technological innovation and inadequate research and development; (b) various factors favouring growth in the 1950s and 1960s have gradually disappeared, such as the possibility of shifting a substantial proportion of the work force from agriculture to other sectors and the progress in Western Europe and Japan towards eliminating the technological gap between those countries and the United States; (c) the recurrent and severe macro-economic disturbances of the 1970s served to depress both the growth of output and the growth of production capabilities. Marked fluctuations in the aggregate demand, unprecedented swings of interest rates, volatility in exchange rates, heightened instability in raw material prices, volatility in economic policy have combined to heighten uncertainty and lower the profitability of investment and the propensity to invest. Uncertainty and high interest rates have inflicted particularly heavy damage on longer-term investment, thereby slowing the pace of structural change.
Refs John, J et al (Ed) *Influence of economic Instability on Health* (1983).
Broader Economic stagnation (#PC0002).
Aggravates Stagflation (#PC2536) International economic recession (#PF1172)
Instability in export trade of developing countries producing primary commodities (#PD2968).
Aggravated by Underpopulation (#PD5432).

♦ PF1739 Interaction of deficiencies in world economical systems

Nature The individual systems governing money, finance, trade and commodity markets which make up the overall system of trade and payments have a substantial number of important interlinkages. This means that the way one system operates may influence the efficiency with which another fulfils its task and that the weakness of one may aggravate the malfunctioning of another. The failure of the payments system to meet the financing needs of deficient countries has lead many of these countries to impose trade restrictions, thereby accentuating the failure of the trading system to fulfil its central objective of keeping markets open. Similarly, the abandonment of any kind of exchange rate regime in favour of free floating, by simultaneously increasing uncertainties regarding price competitiveness and reducing the efficacy of tariffs as a protective device, has increased the use of non-tariff restrictions. This in turn has diminished the effectiveness of exchange rate changes as a tool for balance-of payments adjustment and has thereby put more strain on the already inadequate financing system. The insufficiency of the balance-of payments financing system has itself exacerbated the problem posed by the many lacunae in the system of commodity market regulation. Thus, developing countries dependent on exports of commodities experiencing sharp price fluctuations have been obliged, because of the shortage of payments finance, to incur sometimes heavy adjustment costs in response to self-reversing earnings fluctuations which would normally call for financing and not adjustment.
Claim The working of the commodity, finance and trade systems has served to amplify the transmission from developed to developing countries of the deflationary impulses which have themselves strengthened because of the stance and mix of macro-economic policies adopted by the former countries. Besides, the deficiencies of the trade and payments system have tended to encourage or even enforce national policy responses which contribute further to the propagation of deflation. Consequently, an inordinate part of the adjustment process have taken the form of cutbacks in imports. When these involve either investment goods or raw materials and spare parts, they have had a direct impact on levels of current and prospective output. They also reduce the demand for other countries' exports, including those of developed countries, and therefore reinforce recessionary tendencies in those economies.
Broader Global economic crisis (#PC5876).

♦ PF1740 Short-term planning of product life cycles

Short sighted business planning — Short-term priorities of construction industry
Nature Production systems tend to operate on the basis of processing and distribution functions which yield profit, and to neglect other processes involved in the total life of a product. In this tradition of manufacturing, there is no future planning to develop technology covering the whole life-cycle of a product, so that resources may be stockpiled at an intermediate stage of production, or discarded as a waste after only partial use. The result is that full potential benefits are not realized.
Boards of directory of corporations lack vision, are preoccupied with short term profits, pay to little attention to detail, alienate workers and emphasize cost saving at the expense of customer satisfaction resulting in lost of market share and declining national economies.
Satisfaction of immediate consumer desires in the construction industry results in inadequate quality control, use of stop gap measures, incomplete planning, and the neglect of the needs of society.
Broader Short range planning for long-term development (#PF5660).
Narrower Planned obsolescence (#PC2008) Lack of demolition planning (#PF3731).
Related Imperialistic distribution system (#PD7374)
Underdeveloped sources of income expansion (#PF1345)
Profit motivated utilization of construction technology (#PF2464).
Aggravates Undirected expansion of economic base (#PF0905)
Unsustainable short-term improvements in agricultural productivity (#PE4331).

♦ PF1741 Dissatisfaction with skin colour

Dark skin — Pale skin — Untanned skin — Abusive use of skin colourants — Abusive use of skin lighteners
Incidence Amongst Caucasians extensive use is made of lotions and ultra-violet lamps to assist tanning. Amongst the dark skinned, especially women in developing countries, widespread use is made of skin lightening formulations. In both cases there is increasing evidence of dangers to health. In the case of tanning, the danger is of skin cancer. In the case of skin lighteners, dangers arise from the toxic products which may be used in the lotions (and which are often banned in industrialized countries) and from the removal of the dark pigmentation which increases vulnerability to ultra-violet and thus to skin cancer. Especially powerful, and therefore dangerous, lighteners tend to be used by women in lower income groups, because of the competitive advantage it gives in such activities as prostitution. Dissatisfaction with skin colour is promoted by aggressive advertising.
Aggravates Malignant neoplasm of skin (#PE5016)
Trading in products containing toxic substances with developing countries (#PE2061).
Aggravated by Female prostitution (#PD3380) Skin colour prejudice (#PC8774)
Misuse of advertising (#PF4225).

♦ PF1744 Global cooling

Cyclical planetary glaciation — Ice ages
Nature Geological evidence indicates that major ice ages occur approximately every 250 million years. Evidence has also been deduced for a number of minor glacial advances (other than those occurring every 100-200 years). One cycle of 70-90,000 years has been noted. Another cycle appears to recur every 11,000 years. Such periods are accompanied by a general lowering of the average temperature which may result in the extinction of many species.
Claim Carbon dioxide (CO2) in the atmosphere increases cold at poles by filtering out the near infrared wavelengths that melt the ice and snow, as they come from the sun. The snow and ice reflect the the rest of the solar energy spectrum back into space, so the growing snow and ice

are manufacturing increasing amounts of cold. Since 1938 Northern Hemisphere temperatures have dropped by factor of two degrees Celcius. Cooling of the climate will cause serious droughts and droughts expand deserts all over the world. Increasing ice and snow on the poles press down on the planet triggering earthquakes and volcanism.
Broader Long-term cyclic changes of climate (#PC6114).
Related Short-term climatic change (#PF1984).
Aggravates Famine (#PB0315)   Drought (#PC2430)   Earthquakes (#PD0201)
Climatic cold (#PD1404)   Desert advance (#PC2506)   Food insecurity (#PB2846)
Volcanic eruptions (#PD3568)   Extinction of species (#PB9171)
Evolutionary catastrophes (#PF1181).
Aggravated by Meteorological disaster (#PD4065)
Medium-term cyclic variations in solar radiant energy (#PE9528).
Reduced by Global warming (#PC0918).

◆ **PF1746 Failure to profit from patterns of history**
Denial of the lessons of history — Failure of opposing groups to appropriate the wisdom of the past — Perceived irrelevance of past experience
Nature Historical research has failed to illuminate the patterns of history so that everyone may utilize the significant lessons of the past. People thus see no meaning in the past and are ignorant of their heritage. This isolates them in the present without an historical or a future context. The past needs to be viewed in such a way that people can understand who they are now and how best to move into the future.
The opposition parties and groups within both democratic and non-democratic nations tend to romanticize the needs of the country, their capacity to meet these needs and the party in power, creating mistrust by the general public. This naivety prevents them from effectively envisioning new solutions.
Broader Exclusion of opposing views (#PF3720)   Human wisdom unrelated to daily life (#PD1703).
Aggravates Recurrence of misapprehended world problems (#PF7027).
Aggravated by Absurdity (#PF6991)   Ignorance of history (#PD3774)
Unrecorded knowledge (#PF5728)   Historical misrepresentation (#PF4932).
Underutilization of popular wisdom (#PF2426).

◆ **PF1748 Incomprehensibility of specialized jargon**
Specialized jargon
Nature Awkward, contorted syntax and the use of unfamiliar technical terms impede the communication of specialized (particularly scientific, academic and professional) knowledge, both to specialists in other disciplines and to the general public.
Counter-claim Like other tribal entities, corporations, professions and disciplines develop their own dialects as a way of linking members of the tribe and delineating their ranks. It bonds the user to the group and separates the user from the general society. It makes the user feel wanted.
Claim People are cut off from the world's knowledge by the use of specialized languages and forms. This has created a myth that only experts and intelligentsia can understand and use the wisdom of the humanities. People feel victimized by their lack of knowledge of 'in-group' terminology, believing that such wisdom is forever impossible for them to attain. The situation is further compounded by specialists who do not want to 'distort' their discipline through popularization.
Broader Obscurantism (#PF1357)   Human wisdom unrelated to daily life (#PD1703)
Excessive use of acronyms and abbreviations (#PF4286).
Narrower Mystique in attitudes to problems of disabled people (#PF6224).
Related Slang (#PF5213)   Corruption of meaning (#PB2619).
Aggravates Scientific elitism (#PC1937)
Irrelevance of science and technology (#PF0770).

◆ **PF1750 Failure to recognize uniqueness of family members**
Nature Each individual member of a family has a unique life-sequence, so that the family itself continually changes and the past is never repeated. However, if the family is unable to acknowledge and mark such uniqueness, and is unconscious of its members' individual lives, then it is trapped in an attempt to continue living within a net of relationships that no longer exist.
Broader Obstacles to family life (#PF7094)   Inflexible patterns of family lifestyle (#PF2644).
Narrower Unperceived interests of children (#PG7936).
Aggravates Family violence (#PD6881)   Unstable family life (#PD7929)
Family rejection of children (#PC8127)
Exclusion of pre-adults from family decisions (#PE2268).
Aggravated by Rigid family obligations (#PJ8340)
Unengaging family activities (#PG7916).

◆ **PF1755 Fostering of dependency by social institutions**
Nature The attitude which passes responsibility for decision-making to specialists, whether political, industrial, educational, technological or whatever, may increase the effectiveness of such decisions but also fosters a general sense among non-specialists of a lack of effectiveness and an inability to participate in social change. This leads to a feeling that 'the "government" should care for me', and thus personal responsibility is surrendered.
Claim Economic and political forms tend to ethically emasculate society, thus creating dependency upon social institutions and a sense of impotence of action in dealing with these structures. Such a situation may arise when dealing with social welfare systems, banks, educational institutions, or with the medical profession. The specialist knowledge inherent in such structures nurtures an attitude of "we know best", and the individual experiences loss of control over the decisions that affect his life.
Broader Collapse in providing ethical value screens (#PF1114).
Aggravates Excessive institutionalization of education (#PD0932).
Aggravated by Excessive institutionalization of vulnerable groups (#PF8209).

◆ **PF1756 Performing arts**
Drama — Sensuous dance
Claim According to Islamic beliefs, drama and dance stereotype unnatural actions and abuse artists' talents. They are merely different forms of moral laxity and idolatry. When human actors and actresses replace as living idols those ideals, emotions and aspirations which should only be enshrined in the human mind and strengthened in communal congregational worship, they create a precedent encouraging similar treatment for such idols as wealth, beauty and fertility. The decadence of modern music has encouraged the emergence of disorderly, sensuous and quite uncivilized dances which are now an integral part of the life of a large proportion of the younger generation.
Broader Representative arts (#PF0981).
Narrower Decadent music (#PF5190).

◆ **PF1761 Wasted waiting time**
Nature There are many occasions on which people have to waste time waiting. The profitability of shops and other services at which the customer or client pays for the service increases the more the average customer is prepared to wait, because this optimizes the utilization of employees. In operation research terms, this is akin to effective 'loading'. In the case of administrative services or public transport, the same is true, except that frequently the individual has no choice but to utilize the available facility because there is no alternative. Persons required to be available during the course of a court case may be required to wait considerable periods: this is a factor in discouraging people to volunteer as witnesses. In some cultures, status is determined by the length of the period of waiting that a person can impose upon any person seeking an interview.
Incidence Pravda calculates that Soviet citizens waste 37,000 million hours a year standing in line to buy food and other necessities. This figure is dwarfed by those who are imminently expecting the Second Coming.
Refs Schwartz, Barry *Queuing and Waiting* (1975).
Broader Wasted time (#PF8993)   Inefficiency (#PB0843)
Obstacles to efficient utilization of time (#PF7022).
Narrower Excessive waiting times in government facilities (#PF5120).
Aggravates Boredom (#PA7365)   Underproductivity (#PF1107).
Aggravated by Delay (#PA1999).

◆ **PF1762 Economic risks of subcontractors**
Subcontracting
Claim For businesses acting as subcontractors for other companies economic life can be nasty, brutish and full of risk. Contracts are of short duration, profits are thin, competition is fierce and bankruptcies are frequent. Subcontractors have to scramble to meet ever more rigorous demands for higher quality, faster delivery and lower prices. They pay wages as much as one half of major companies. In order to remain competitive, some subcontractors hire labour at wages below the legal minimum. They offer working conditions that are frequently squalid and fringe benefits that are minimal to nonexistent. They are the first to feel a recession.
Broader Unprofitable small businesses (#PJ8128).

◆ **PF1764 Lost covenantal understanding of sexuality**
Claim Although society is no longer considered responsible for ensuring that sexual activity is restricted to lifelong, sociologically-approved covenants, no alternative means have been developed for objectively relating individual sexual behaviour to society.
Broader Individualistic perception of sexual activity (#PF1682).
Aggravates Paternal negligence (#PD7297).

◆ **PF1765 Rigidly entrenched social traditions in rural areas**
Nature Residents in rural areas are deeply affected by the real and inescapable effects of new images and life-styles – for example in India, where people are experiencing the gradual crumbling of the caste structure which once defined roles and relationships and guaranteed a significant role to everyone. In many rural areas, the challenges presented by change are so foreign and overwhelming that rigid fidelity to the archaic has become the automatic response. Kaleidoscopic shifts from the formerly fixed patterns of wider society lead the rural villager to defend desperately the traditional stability of those narrower contexts over which he still exercises some control.
Such a reaction is formalized in the position of the adult male members of the community. Early marriage, prolonged childbearing responsibilities, and a withdrawal from public leadership characterize the role of women in some societies. Obedience to parental authority and occupational tradition defines the style of youth. A determination to labour for the protection of his dependents' marginal subsistence typifies the adult male who, because he is its symbol and spokesman, sets the style of the family as well. As the one hope in the face of misunderstood alterations in the social status quo, the extended family resorts to a protective individualism. Because such propping-up of tradition requires great effort, individual creativity is ossified and little desire or opportunity arise for corporate, community-wide endeavour.
Broader Rigid cultural patterns (#PF8598).
Narrower Inhibiting shyness (#PF1278)   Parental permissiveness (#PD5344)
Fear of vocational change (#PJ1318)
Cultural traditions blocking business profit (#PF9529).
Related Static and unrelated social roles (#PF1651).
Aggravates Subsistence life style (#PF1078)   Taboos against eating poultry (#PJ1326)
Exclusion of women from decision making (#PE9009)
Suppression of creativity and innovation (#PF0275).
Aggravated by Untransposed significance of cultural tradition (#PF1373).

◆ **PF1768 Uncoordinated efforts in agricultural development**
Nature Although cooperatives have demonstrated the effectiveness of people working together for their mutual benefit, farmers in many rural communities are still caught in a pattern of internal competition and conflicting efforts, so that they find themselves operating separately in purchasing, marketing and crop experimentation. For example, farmers who have not contributed to the construction of irrigation improvements are not entitled to water rights, and this isolates them from their fellow farmers. When they come to market their products, the dominance of a single industry in the local town may isolate farmers from other markets where prices are better. These conditions, combined with the traditional hazards of farming – a short growing season, crop failures and frustrating markets – exclude farmers from new agricultural developments. Economic recession means that some farmers will stop farming altogether.
Broader Lack of cooperation (#PF2816).
Narrower Limited agricultural expansion (#PG8312)
Unintegrated farming population (#PG8314)
Fragmented agricultural development (#PG8315).
Aggravates Stagnated development of agricultural production (#PD1285).

◆ **PF1772 Non-integrating images of personal freedom**
Nature Neither formal nor informal education provides the means for integrating current images of personal freedom; people have no understanding of the context or the origin of these images. This results in a lack of responsible decision-making, as the traditional views of freedom are replaced by vacuous models which reinforce escapism.
Broader Tensionless image of free choice (#PF1675).

◆ **PF1773 Inadequate maintenance of physical health**
Nature In developing countries, health care is often a low priority, although good health is a prerequisite for economic development. Problems exist but are not tackled in: provision of safe water; sanitary disposal; balanced diets; rudimentary knowledge of hygienic practices; dirt floors. Villages are often without water purification methods, and standing taps and bathing and sewage disposal methods present a constant health hazard. Even when there is an abundance of available food, a well balanced diet may not be consistently available. Health information is often a combination of traditional beliefs and popular images created by advertising. Parasitical and bacterial infestations result in chronic illnesses and subsequent lost of vitality. The scarcity of trained medical personnel, the high cost of treatment, and the remoteness of medical facilities block preventive efforts and lead to care only in extreme emergencies.
Claim Locally-based health care can be postponed only at the expense of decisive economic and social development.
Broader Neglected health practices (#PD8607)
Excessive dispersion of community facilities (#PF6141).
Narrower Unhygienic conditions (#PF8515)   Inappropriate clothing (#PJ6604)

Insufficient preventive medicine (#PE0751) Unhygienic dirt floors in houses (#PG9668)
Insufficient protein consumption (#PG9679)
Unhealthy lack of daily physical activity in urban environments (#PF6182).
Aggravates Nutritional deficiencies (#PC0382) Deterioration in physical health (#PC0716).
Aggravated by Physical unfitness (#PD4475) Inadequate housing (#PC0449)
Inadequate medical care (#PF4832) Unnecessary health tests (#PF5679)
Ignorance of health and hygiene (#PD8023).

♦ PF1776 Paralyzing complexity of urban structures

Nature Most inner-city residents find themselves without the power to significantly influence or fully understand the forces that act upon their environment. Knowing that every type of service is available they expect an effective response; but they find themselves faced by a mass of forms, guidelines and directives. The administrative procedures and bureaucratic systems which must be understood in order to survive in the urban community have become paralyzing in their complexity. There is a bewildering set of options from which to choose in order to obtain services for basic needs. Few individuals in one community service possess the expertise needed to liaise liaison with other administrative structures. Frustrated with the multitude of misunderstood referrals needed and often unable to follow them through far enough to accomplish the task, residents increasingly mistrust administrative structures. Until new forms of community advocacy are developed and made readily available to every citizen, the engagement of urban residents will continue to be inhibited.
Broader Confusing structural complexity (#PF8100).
Narrower Credit discrimination (#PE7902) Distant representatives (#PJ0272)
Social unaccountability (#PC1522) Obsolete zoning restrictions (#PG8953)
Conflicting agency priorities (#PG8869) Inaccessible landowner records (#PG8967)
Unresponsive public authorities (#PF8072)
Restrictive building codes in urban areas (#PE8443).
Related Police indifference to community (#PF8125)
Substandard housing and accommodation (#PD1251).
Aggravates Idle private land (#PJ8020) Citizen powerlessness (#PJ8803)
Difficult land acquisition (#PJ5369)
Excessive waiting times in government facilities (#PF5120).

♦ PF1780 Biased and inaccurate geography textbooks

Nature Geography textbooks, through which school children obtain an image of the world, may be distorted in different ways. Atlases may minimize the quantitative importance of some regions. Some textbooks emphasize the picturesque, historical features of a country or a culture rather than those of current significance. Others emphasize the importance of the topography or climate of a country as determining its evolution. Textbooks are frequently based on out-of-date material. They may imply that because wooden houses are built, or because the population belongs to a particular culture, that the country is naturally backward.
Broader Misleading information (#PF3096)
Inadequate and inaccurate textbooks and reference books (#PD2716).
Related Biased and inaccurate history textbooks (#PD2082).
Aggravates Prejudice (#PA2173) Ignorance (#PA5568)
International tension (#PB8287).
Aggravated by Inadequate international map of the world (#PD0398).

♦ PF1781 Ineffective structures for community decision-making

Nature Although people in many small communities may express their wishes and hopes, they may not have the means to implement them. The community remains fragmentary and individualistic, unable to gather either the momentum or the consensus which would allow it to move ahead. An attitude that outside official structures are responsible for most dimensions of economic and social development further reduces their willingness to assume local responsibility. Where a group of local residents emerges which will assume responsibility for the community (a phenomenon developing in most nations), their effectivity is directly dependent on the degree of local-corporate decision making they engender. For example, previous failed attempts to form an agricultural cooperative causes mistrust and this inhibits the willingness of villagers to work together. The inevitable result of living in this situation is a sense of powerlessness.
Broader Obsolete methods (#PF3713) Insufficient communications systems (#PF2350).
Narrower Lack of car pools (#PG9272) Lack of cooperation (#PF2816)
Lack of a united action (#PG9776) Risky rental agreements (#PG1425)
Lack of a governing body (#PG9197) Parallel social structures (#PG9300)
Closed federation operations (#PG9234) Lack of conscious initiatives (#PG9881)
Unclaimed local responsibility (#PG9836) Polarization of local conflicts (#PF1333)
Excessive classification demands (#PG9849) Unstructured improvement efforts (#PG9242)
Inadequately publicized services (#PG9437) Parent-teacher non communication (#PF1187)
Uncoordinated expression of needs (#PG9210) Unstructured local decision-making (#PF6550)
Noninclusive council representation (#PG9227)
Lack of means for achieving consensus (#PD2438)
Ineffective structures of local consensus (#PF6506)
Unrecognized benefits from corporate action (#PJ9420)
Limited local availability of capital reserve (#PF2378).
Related Unorganized community recreation (#PF5409)
Unprofessional building management (#PG8860).
Aggravates Lack of leadership (#PF1254) Lack of community development (#PF7912)
Inappropriate sanitation systems (#PD0876)
Unorganized development of work forces (#PF2128).
Aggravated by Unclear educational roles (#PG8840)
Unpublished educational possibilities (#PJ9445)
Unrecognized benefits from cooperatives (#PF9729).

♦ PF1786 Outdated labour negotiation procedures

Nature When they occur at all, employer-employee negotiations on working conditions, wages and definition of areas of responsibility are bound to 19th century labour union patterns, which operate on the basis of power confrontation rather than on an understanding of mutual responsibility. This confrontation mode ultimately leads to a defence of self interest and the perpetuation of the status quo. Another aspect of this problem is that even these outmoded negotiation methods are mainly limited to western blue-collar workers.
Broader Inadequate access to negotiation on employment and reward (#PD1958).

♦ PF1789 Environmental hazards constraining scientific research

Pollution hazards of industrial research and development — Pollution constraints on testing of new technology
Narrower Obstacles to medical experimentation (#PF4865)
Inconclusiveness of scientific and medical tests (#PD7415).
Related Nuclear weapons testing (#PC2201)
Cruel treatment of animals for research (#PD0260).
Aggravates Suppression of safety records (#PF2714)
Government secrecy concerning nuclear weapons testing (#PF4450).

♦ PF1790 Poor organization of community environment

Nature Although some communities are consciously arranging the comfort and appearance of public space and buildings so they are both functional and motivating, and many developing communities are also providing utilities and facilities to encourage commercial and industrial expansion, others are less well planned. For example: the dangerous conditions of roads and pathways may isolate the population both from the outside world and from each other during the winter months; business and industrial development may be hindered by inadequate facilities and services, such as limited water supply and the continued utilization of informal and time-consuming building construction systems; no consultation room may have been found and arranged for the district doctor as a necessary preliminary for scheduled weekly clinic hours; community events may be inhibited, public communication undermined and recreational activities limited due to neglect of school buildings and sports fields, and inadequate areas for public gatherings; poor television reception and lack of cultural focus may cut people off from interchange with the outside world. All of these factors lead to a feeling of isolation, blocking community initiative.
Claim Many small villages do not realize their potential as uniquely attractive places to both live and work; but as long as people continue to accept the inadequacies of their environment, development of such villages will be hindered.
Narrower Limited public space (#PJ9066) Low athletics priority (#PJ0337)
Poor television reception (#PG9867) Insufficient cultural nodes (#PG9774)
Lack of slaughter facilities (#PE9767) Unfeasible industrial images (#PG9882)
Limited construction manpower (#PG9875) Inadequate industrial services (#PG9900)
Unfocused design of community space (#PF1546).
Related Dangerous paths (#PJ9888).
Aggravates In-home animal shelter (#PJ9800) Overcrowded public clinics (#PG5393)
Inadequate road conditions (#PJ0860).
Aggravated by Neglect of school buildings (#PG9829).

♦ PF1798 Discrimination against foreigners in legal proceedings

Nature Many constitutions include only nationals within the scope of their provisions for access to the courts and for various procedural rights before the courts. Some legal systems accord to refugees and other aliens the same right of access to the courts as they do to nationals, but other laws require, for instance, the making of a financial deposit before an alien may bring a case before the courts.
Legal aid facilities existing in a country are sometimes partially or totally unavailable to aliens. Sometimes bail may be available only to nationals. Foreigners involved in court proceedings may be handicapped by linguistic difficulties unless adequate provision for interpretation is made on their behalf. Attention has been drawn to the psychological disadvantages faced by foreigners involved in court proceedings. The mere fact of being a foreigner, it has been said, often works against a defendant, whose punishment may be far harsher than that of a native of the country.
Broader Biased legal systems (#PF8065) Legalized discrimination (#PC8949)
Discrimination before the law (#PC8726).
Related Injustice (#PB0856) Unnecessary verbosity of legal documents (#PF7137).

♦ PF1804 Habitual overemphasis on national self-determination

Claim Nations, accustomed over the years to being the form in which the world was understood, continue to serve as the primary social unit even when, as today, the primary obligation as human beings is to the entire globe. Thus, refusal by the people of a nation to restrain their right to self-determination blocks the prospect of global unity.
Broader Inadequate means for upholding global concern (#PF1817).
Narrower Subjective cultural ranking (#PF2015)
Lack of intersocietal resource channels (#PF2517)
Unwillingness to sacrifice political power (#PF7163)
Self determination unrelated to global obligations (#PF1491).
Related Reinforced parochialism of internal values and images (#PF1728).
Aggravated by Pursuit of affluence (#PF5864).

♦ PF1807 Conditioned self-gratification

Nature Among Western educated or influenced people, the acculturation received from the family tends to condition the individual to operate from a search for psychological well-being and adjustment rather than from the aim of global socio-interrelatedness. The result in society of such gratification-seeking is the disruption of conventional bonds of relationship like marriage, and an incapacity to appropriate the needs of others beyond immediate relations. A spectator mentality toward mass suffering is created, and materialism substituted for social consciousness.
Broader Reinforced parochialism of internal values and images (#PF1728).

♦ PF1815 Poor condition of open spaces in urban communities

Defaced community space
Claim The 1970's saw increasing migration from the inner city. Industrial and professional resources moved to the suburbs, leaving a vast amount of unclaimed space, evident both in vacant lots and in empty, deteriorating buildings. Inner city buildings are also vacated and boarded up as a result of fire and substandard conditions. Furthermore, occupied housing is also deteriorating, a situation inevitable in the face of aging architecture, lack of repair and rapid turnover among residents. Still more unpleasing to the eye is the amount of litter strewn throughout most inner city residential areas. Broken glass and discarded paper along cracked sidewalks give the impression of deliberate carelessness. The effect of this condition is that even the well-cared-for space in the community and the few attempts at rehabilitation go virtually unnoticed; residents see little hope for dealing creatively with the area as a whole; there is an attitude of resigned acceptance in the face of mounting deterioration, encouraging further desecration which in turn increases the deterioration and deepens the despair. Until radical, expansive rehabilitation and new construction take place in such urban neighbourhoods, gradual deterioration of the area will continue to paralyse action and degrade human dignity.
Broader Obstacles to community achievement (#PF7118)
Obstacles to availability of community space (#PF7130)
Debilitating deterioration of physical environment (#PD2672).
Narrower Undefined lot usage (#PG8882) Outdated basic utilities (#PG8871)
Shortage of public cleaning (#PG8902) Negligent owner improvements (#PG8921)
Dysfunctional public utilities (#PE7647) Overcrowded building conditions (#PG8963)
Insufficient business facilities (#PG0110).
Related Inadequate care of community space (#PF2346)
Unprofessional building management (#PG8860).
Aggravated by Urban overcrowding (#PC3813) Neglect of property maintenance (#PD8894)
Prohibitive cost of maintenance (#PF0296).

♦ PF1816 Holy places as a focus of religious friction

Nature Holy places may cause the bitterest of religious friction if two or more religions converge on the same place. This may lead to military occupation, political annexation, territorial disputes and sacrilege. It may be a pretext for war.
Incidence One of the most prominent current examples of where holy places can cause religious friction is Jerusalem, where three religions converge: Judaism, Islam and Christianity. Jewish excavations near the Masjid Sakwa (Dome of the Rock) and the Masjid Al-Agsa outraged Arabs and Muslims, the building of a Jordanian hotel on some of the most ancient Jewish graves pre-1967 caused bitterness among Jews. The situation was inflamed at one stage when a Christian youth set fire to the Masjid Al-Agsa.
Background Perhaps the most notable wars concerned with holy places were the crusades in

the Middle Ages against the Turks. In the 19th century European powers, notably Russia and France, struggled for the prestige of supporting Jerusalem and other holy cities against the Turks.
Broader Sacrilege (#PF0662) Religious conflict (#PC3292).
Related Iconoclasm (#PF4923).
Aggravates Jihad (#PF5681) Political annexation (#PG4044) Religious intolerance (#PC1808) Desecration of holy spaces (#PF6385).

♦ PF1817 Inadequate means for upholding global concern
Claim Present-day society does not provide adequate social methods for communities to affirm both their own local lifestyles and those of other cultures. Racialism, nationalism and classism appear in updated guises, collapsing global concern.
Broader Lack of commitment (#PF1729) Demeaning community self-image (#PF2093).
Narrower International paternalism (#PF1871)
Overriding of international by local concerns (#PF2076)
Habitual overemphasis on national self-determination (#PF1804)
Reinforced parochialism of internal values and images (#PF1728)
Monopolization by interest groups in development of community priorities (#PE1702).
Related Lost family role in society (#PF7456) Social isolation of the elderly (#PD1564)
Repression of self-consciousness (#PC1777) Sentimental attachment to the past (#PF1577).

♦ PF1822 Obsolete methods of agricultural production
Out-of-date methods of agricultural production — Old farming methods — Outmoded farm techniques — Outdated farm methods — Rudimentary farming methods — Traditional farming methods
Nature Traditional dependence on farming to suffice as a family's means of subsistence from season to season has been displaced by technological and urban developments beyond the control of individuals and even of nations, which require that farming become a mechanized industry. But scientific means of increasing productivity, seen from the farmer's point-of-view, are remote, expensive and difficult to understand. The newly developed urban market overwhelms rural families with its impersonal deceit and confusing complexity. To make the necessary transition, more land must be put under cultivation. Traditional small, enclosed fields dictated by past methods and recent poverty do not provide adequate space for mechanization. Modernized farming requires increased labour just at a time when the average family size has decreased and family members already are overburdened. The single family unit cannot meet this demand for industrialized farming alone. The extent of planning, soil care, crop diversification, harvesting and marketing necessitates a larger unit.
Although there is increasing international awareness of the possible benefits to be derived from collaborative agriculture using modern technological innovations, the farmers in many rural areas still cling to traditional methods. They farm small, scattered plots which by themselves are uneconomic and which taken together barely provide a subsistence living; they do not share machinery and in most cases the cost of purchasing such modern equipment individually is prohibitive. One of the consequences of this situation is that too many individuals are engaged in agriculture compared with the ability of the land to produce under its present system of cultivation, when it tends to be given over to fodder crops and small vegetable plots. This presents a very negative image of farming to young people, who see their parents working long, back-breaking hours in the fields.
Broader Intemperance (#PA6466).
Narrower Lack of shared machinery (#PG9841) Irrational land divisions (#PG9807)
Overburdened production plan (#PG9876) Rejection of agriculture by youth (#PG9860).
Related Demeaning farmer image (#PJ9781)
Underproductive methods of agricultural management (#PF6524).
Aggravates Inadequate irrigation system (#PD8839)
Stagnated development of agricultural production (#PD1285)
Underutilization of potential in local communities (#PF6513)
Subsistence agricultural income level in rural communities (#PE8171).
Aggravated by Unprepared adult leadership (#PF6462)
Individualistic retaining of local tradition (#PF1705)
Untransposed significance of cultural tradition (#PF1373).

♦ PF1824 Enhanced risks of disease
Nature A wide variety of factors increase or decrease the risk of disease, injury and death. Diet, sanitation and basic health care, including immunization and simple first aid have the greatest impact on the risk of acquiring a disease or surviving it. Life style, type of employment, sex, age, climate, location of home, type of transportation used, amount of exercise and sleep, type of personality, race, and many others know and unknown affect the chances of disease and injury. None eliminate these risks.
Broader Hazards to human health (#PB4885).

♦ PF1825 Unmeaningful social roles for the aged
Inadequate roles for the elderly — Diminishing role of older people due to overemphasis on economic productivity — Elders' concealed needs — Elders' feeling of uselessness
Nature Old people no longer maintain their role as persons whose experience could illuminate the problems of everyday life. They no longer advise or counsel with a clear sense of future needs, because the uniquely shifting life-styles of the present day give them glimpses of the future which they view with shock and revulsion, and in which they are unable to discern any relationship with their past experiences. The result is that they attempt to avoid the future: they become immobilized, sometimes even liabilities, rather than the possible resource to the community which is virtually essential for a healthy society.
Older people no longer see themselves as essential members of society, as a necessary part of the functioning whole. They are influenced by the current trend to evaluate people by their ability to produce economically beneficial goods, and fail to see life in its realistic, historical perspective as a common, corporate venture. When they retire from a lifetime of production they feel useless, alone and impotent in their solitary state.
Incidence Grandmothers may be very busy but they do not have important roles in the family decision-making process; young marriages dissolve because they have been unable tap the underlying wisdom behind the stability of older families.
Broader Elder paralysis over the future (#PF3973)
Static and unrelated social roles (#PF1651)
Reluctance to join in community action (#PF1735).
Aggravates Erosion of elders' wisdom (#PF1664)
Limited access to health services (#PF6577).
Aggravated by Disobedience of elders (#PF7149).

♦ PF1826 Anti-communism
Nature Zealous activity to halt or reverse the spread of communism, derived from profound fear and a sense of moral obligation to uphold capitalism, may be practised by staunchly capitalistic countries where there is little socialist challenge, or by fascist regimes where there has been a strong communist challenge. In the latter case, the ideal is not so much to uphold capitalism as to make use of it for extreme right-wing authoritarian 'National-socialist' aims. **Background** The fight against Communism is weakened by ignorance of the totalitarian nature of the system. The millions of political prisoners who have died in the Gulags of the Soviet Union; the harsh repression of the captive East European nations, two of which have been invaded since World War II by the USSR; and the misery of every day life under Marxist-Leninism does not impact sufficiently upon those who are thousands and tens of thousands of kilometres away. The enemies of communism are not capitalists, they are freedom-loving peoples, including the free socialist parties of the world who are not Moscow's puppets.
Claim Communism may be suppressed internally with censorship, intimidation, espionage and other methods. Externally it may lead to cold-war, or actual war; to the use of propaganda, and of trade, military and other agreements to influence other countries; to intervention and coercion. These tactics may be totally successful or they may promote weak dependent or puppet governments. Alternatively they may encourage broad sympathy for communism, especially among youth, where it might not have otherwise existed.
Counter-claim It is much too easy to assume that capitalism automatically safeguards democracy and human rights, by contrast with Communism. The horrors of apartheid were committed in the name of maintaining western values and defeating Communism, and were committed with the support of capitalist countries.
Broader Neo-fascism (#PF2636) Ideological conflict (#PF3388).
Related Intolerance (#PF0860).
Aggravates Communism (#PC0369) Subversion of socialism (#PF9485).
Aggravated by Double standards in morality (#PF5225).

♦ PF1836 Lack of medical records
Destruction of medical records
Nature Institutions responsible for employing personnel under conditions which may later prove to be hazardous find it expedient to destroy any medical records to obstruct subsequent investigations.
Broader Uncatalogued documents (#PF4077) Inadequate medical care (#PF4832)
Fragmented planning of community life (#PF2813).
Narrower Inadequate supporting health documents (#PU2257).
Related Secrecy of medical facts and records (#PF5983).

♦ PF1844 Inadequate community care for transient urban populations
Denial of rights to transient populations
Nature There is a lack of basic community care structures in small urban communities, where a continued influx of transient residents rely on overburdened, rural-oriented services. Despite the close proximity of modern services for the rest of the population, such services are not available to these people and this produces a debilitating effect on those who continue to live under outmoded forms of care. The gap in services is apparent in the unsanitary conditions; the overextended medical services; the number of orphaned, uneducated and uncared children; the lack of access to vital information; and the virtual absence of social care for young or old. The gap in services and obvious disparity with others dissipates a community's energy, gives it a negative self-image and demoralizes the local people.
Refs Oberai, A S *Migration, Urbanisation and Development* (1987).
Broader Lack of care (#PF4646) Inadequate social welfare services (#PC0834)
Fragmented forms of care at the neighbourhood level (#PE2274).
Related Inadequate animal welfare (#PC1167) Denial of rights to vulnerable groups (#PC4405).
Aggravates Inadequate child welfare (#PC0233) Unsupervised home births (#PG7996)
Inefficient support for widowers (#PJ7997) Insufficient special care vehicles (#PJ9008)
Prohibitive cost of private medical care (#PF8016).
Aggravated by Unsanitary markets (#PG8021) Emotional disorders (#PD9159)
Rapid transient flux (#PG7990) Mistrust of strangers (#PF8743)
Inadequate medical care (#PF4832) Limited youth activities (#PJ0106)
Understaffed health clinics (#PJ7980) Unenforced sanitation codes (#PJ7979)
Insufficient preventive medicine (#PE0751) Limited access to social benefits (#PF1303)
Limited access to external resources (#PF1653) Inadequate child day-care facilities (#PD2085)
Lack of secondary schools in the countryside (#PE8797)
Inadequate water supply in the rural communities of developing countries (#PD1204).

♦ PF1849 Livestock mutilation
Cattle mutilation
Nature Livestock are found abandoned on the range suffering from mutilations, often of an especially inhumane nature, unlikely to have been caused by any known predators. No conclusive explanations have as yet been found.
Incidence Over 20,000 incidents have been reported worldwide during the period 1960-1986.
Broader Unexplained phenomena (#PF8352) Maltreatment of animals (#PC0066).

♦ PF1853 Uncoordinated social services in urban areas
Proliferation of parallel urban services
Nature Contemporary urban society has moved so decisively toward providing a network of social service agencies to cover the population's essential needs that people have come to depend upon their availability; yet these agencies frequently lack the coordination among themselves to serve a community effectively. Patchy knowledge of procedures, long waiting periods and confusing eligibility screening, combine to create the impression of structured isolation between agency and client, thereby increasing frustration. In the absence of regular inter-agency exchange, an individual agency lacks the total picture of the gaps and overlaps in available services which is necessary if it is to provide an adequate referral service. When a gap in services is pinpointed, expansion of services is hampered because of grant requirements which base funding on the number of people served: for example, a local drug clinic funded on the basis of the number of people in a methadone programme was blocked from expanding outreach programmes where funds were scarce. The service extended to recipients, in effect, became maintenance rather than rehabilitation.
Claim Until agencies are able to coordinate plans for their service to the community their fragmentation will block effective delivery of services to the urban ghetto community.
Broader Lack of cooperation (#PF2816).
Narrower Inadequate health services (#PD4790).
Aggravates Unhygienic conditions (#PF8515).
Aggravated by Lack of governmental support (#PF8960).

♦ PF1861 Individualistic family structures
Claim By the reinforcement and perpetuation of individualism, conventional family structures do not allow for the development of an interrelatedness between different family members. The approval and reward systems within families encourage competition and self-aggrandisement. This not only reinforces individualistic pursuits but perpetuates and engenders self-centered achievement.
Broader Reduced interior structure of families (#PF3783).

♦ PF1870 Vulnerability of marriage as an institution
Threat to the survival of marriage — Social irrelevance of marriage vows — Reduced dimension of the marriage covenant — Lack of sustaining symbols in marriages — Naive approach to the marriage covenant
Nature The survival of marriage as an institution is threatened by divorce, promiscuity, sexual deviation, discrimination against women, and lack of religious conviction. Cohabitation is one

alternative and families with single women heads is another. If marriage is denied, the result is any children of the union will be illegitimate. Complications resulting from the demise of marriage involve the apportionment of property and the custody of children after divorce, and a wide range of social and psychological problems.
Social attitudes to people intending to marry, and many of the popular forms of marriage counselling, encourage couples to consider the intricacies of human relationships and the marriage bond in a narrow, personal perspective, when present-day needs require individuals to have a conscious responsibility to society as a whole. Couples joined by marriage vows may be overwhelmed by the wider demands made on them and have no rational or adequate methods for assuming this responsibility.
Couples are not conscious of the need for sustaining symbols in marriage. They are therefore manipulated by inadequate symbols created by society, such as the album of wedding photographs. These reduce the significance of the family as an institution and tend to define the roles of men and women as merely sexual and financial, limiting the context in which marriage operates and disintegrating of the roles each individual must play in order to sustain the family covenant.
Incidence The survival of marriage is particularly threatened in developed countries and certain communist countries where the population is relatively mobile and strong family ties have been broken.
Claim Marriage vows no longer have any social role, but simply represent the expression of a one-to-one, emotion-laden relationship. Tension arises in the conflicting claims of responsibility to society (and present-day circumstances require conscious responsibility towards world-wide needs), and responsibility to the person to whom one is bound by marriage. In the family environment, responding to the needs of society may be seen as inhuman, and the individuals concerned feel unrelated to their vows, to each other, or to society as a whole.
Most couples are unable to articulate for themselves what their marriage covenant signifies and demands, and are not conscious of the decisions which are symbolized in the marriage covenant itself. They do not see the need to keep their marriage in mind so they may constantly recommit themselves to obey and honour that covenant. Marriage covenants have clearly failed to place this demand on each individual and merely require a "once-off" decision.
Counter-claim The proportion of new households set up by married couples is on the rise in the United States. Proportionately fewer people are living together out of wedlock or living unmarried with their children. A greater number of young people are continuing to live at home with their parents. The divorce rate has levelled out to about 4.8 per thousand of the total population, down from a peak of 5.3 in 1981.
Broader Immorality (#PA3369) Social disintegration (#PC3309).
Narrower Divorce (#PF2100).
Related Counter culture (#PF0423) Symbol system failure (#PF3715)
Individually defined operating structure of marriage (#PD2294).
Aggravates Alienation (#PA3545) Social conflict (#PC0137)
Family breakdown (#PC2102) Illegitimate children (#PC1874)
Discrimination against women (#PC0308)
Increasing number of single person households (#PJ8818).
Aggravated by Promiscuity (#PC0745) Cohabitation (#PF3278)
Permissiveness (#PF1252) Sexual deviation (#PD2198)
Sexual unfulfilment (#PF3260) Dependence on religion (#PF0150)
Lack of social mobility (#PF2195)
Lack of commitment to common symbols (#PE8814)
Lessening regard for the sanctity of marriage (#PG8543).

♦ PF1871 International paternalism
Nature Bi-lateral and multilateral aid programs of donor nations which attach business, purchasing and other requirements of recipient nations. This type of aid reinforces the image of the donor as benefactor of the world and the image of the recipient as inferior.
Claim Paternalistic control blocks the efforts of those who seek to put their energies creatively into caring for the gifts of other cultures. This paternalism pervades the foreign policy of those donor nations which attach strings to their foreign aid grants, and also of recipient nations which package and market donors' cultures and social issues.
Broader Inadequate means for upholding global concern (#PF1817).
Related Cultural imperialism (#PC3195) Parochial national interests (#PF2600)
Disruption of cultural and social identities in developing countries by transnational corporations (#PE1082).

♦ PF1874 Mismanagement of exchange rate system
Incidence Effectiveness of currency and exchange rate management has declined since 1985, especially as major countries have become more concerned with domestic as opposed to international financial problems. International policy coordination of official exchange rate strategies has become less effective and less coherent.
Narrower Dual exchange rate systems (#PE4476)
Misalignment of currencies (#PF6102)
Discriminatory exchange rate policies (#PE8583)
Inability of developing countries to adopt appropriate exchange rate policies (#PE7563).
Aggravates Currency black market (#PD5905) Exchange rate volatility (#PE5930)
Speculation on money markets (#PD9489).
Aggravated by Aggravation of instability in exchange rates by transnational corporations (#PE0980).

♦ PF1878 Propaganda
Dependence on propaganda
Nature Mass media may be used to influence or manipulate public opinion in favour of a political ideal or government policies. Such governmental 'information', official or unofficial, may be false or misleading. It may intensify conflict and intolerance, and lead to war or to terrorist activities. It may strengthen government control, dictatorship and repression, inequality, injustice and exploitation. It may create confusion or induce apathy, conformism, fear, cultural stagnation and alienation through ignorance. Non-governmental propaganda may emanate from opposition political parties, ideological groups, or from business interests.
Claim Propaganda is contriving conditions where people's critical resistances are so weakened and their freedom of choice so severely reduced as to make acquiescence all too likely. As distinct from education, it is a systematic attempt to influence people by reducing the amount of information available for discussion and encouraging them to act on impulse. Those to be persuaded are led to believe that only one line of action in a particular situation is possible.
Counter-claim Governments may use effective communications to encourage citizen health, family planning, and energy conservation; caution in the use of dangerous products; or advise of the need of civilian defence or war-time resistance against an aggressor. This type of propaganda, including persuasion to take advantage of educational opportunities, or to support international agencies and international cooperation and development activities, is beneficial.
Refs Institute for Propaganda Analysis *Propaganda Analysis* (1977); Lasswell, Harold, et al (Eds) *Propaganda and Communication in World History* (1979).
Narrower Demagoguery (#PC2372) Film propaganda (#PD3089)
Book propaganda (#PD3090) Art as propaganda (#PF3087)
Racist propaganda (#PD3093) Theatre propaganda (#PD3088)
Religious propaganda (#PD3094) Government propaganda (#PC3074)
Misuse of advertising (#PE4225) Photographic propaganda (#PD3086)
Empty slogans and mottoes (#PF3212) War and pre-war propaganda (#PD3092)
Radio and television propaganda (#PD3085) Newspaper and journal propaganda (#PD0184)
Rhetorical inflation in meetings (#PF3756)
Promotion of negative images of opponents (#PF4133)
Propaganda by intergovernmental organizations (#PE3077)
Deliberate misrepresentation in educational materials (#PF1183).
Related Falsity (#PF5900) Deception (#PB4731)
Newspaper monopoly (#PE0246) Psychological warfare (#PC2175)
Incorrect information (#PB3095) Misleading information (#PF3096)
Defamation of character (#PD2569) Biased presentation of news (#PD1718)
Radio and television censorship (#PD3029).
Aggravates Intolerance (#PF0860) Hero worship (#PF2650)
Political conflict (#PC0368) Cultural corruption (#PC2913)
Educating people to lie (#PE3909) Conflict of information (#PF2002)
Politically emotive words and terms (#PF3128)
Destabilizing international telecommunications (#PD0187).
Aggravated by Ignorance (#PA5568) Power politics (#PB3202)
Political feuding (#PD4846) Ideological conflict (#PF3388)
Intellectual conflict (#PC3390) Subversion of democracy (#PD3180)
Compulsory indoctrination (#PD3097) Double standards in morality (#PF5225)
Journalistic irresponsibility (#PD3071) International monopoly of the media (#PD3040)
International non-military conflict (#PF3100)
Violation of sovereignty by trans-border broadcasting (#PE0261)
Abuse of international cultural, diplomatic and commercial exchanges (#PF3099)
Risk of unintentional nuclear war generated by the strategy of deterrence (#PF4162).
Reduces Restrictions on direct news coverage of parliamentary affairs (#PF3072).
Reduced by Underground press (#PD2366).

♦ PF1879 Bias against private enterprise
Claim The critics of private enterprise underestimate its services, exaggerate its profits and impugn its motives. It is difficult to estimate the magnitude of the damage they do. Where a manager is harassed, the cost is in lost time; when the anti-business diatribes reach employees, productivity falls, labour relations are disrupted; and government policy may be influenced to the point where laws are enacted that harass companies, reduce investment returns, discourage new investment and retard development.
Broader Domestic market restrictions in developing countries (#PD1873).
Aggravates Underproductivity (#PF1107)
Suppression of private enterprise in socialist countries (#PD2048)
Inadequate development of enterprises in developing countries (#PE8572).
Aggravated by Short-term profit maximization (#PF2174).

♦ PF1889 Limiting of responsibility to the personal
Nature There is a current tendency for people to attempt to maintain their individual stability, order and peace, by retreating into a private universe. They attempt to control personally their intellectual contexts and social structures. The result is a feeling that social structures are crumbling, roles are becoming meaningless and impotent, and participation in decision-making processes are ineffective.
Broader Defensive life stance (#PF0979).
Aggravates Guilt (#PA6793).

♦ PF1905 Inappropriate education of graduates
Nature In terms of their suitability for employment in existing jobs in society (outside the university environment), recipients of higher degrees frequently find their prolonged education in a sheltered environment to be of little practical value. This leads to disillusionment and reduces the enthusiasm of those pursuing specialized fields of knowledge. Such graduates are then obliged to accept jobs for which they are inadequately or over-trained.
Counter-claim There is some confusion between the terms "education" and "training". Education involves a widening of the individual's horizons whereas training implies the learning of specific facts. The aim of university education, as is inherent in the name "university", is the providing of a universal basis. Those who wish to be trained in a particular specialization or for a particular profession are free to take this type of training.
Broader Inefficiency (#PB0843) Inappropriate education (#PD8529).
Related Over-specialization (#PF0256).
Aggravates Frustration (#PA2252) Educational wastage (#PC1716)
Graduate and post-graduate unemployment (#PD1162).
Aggravated by Excessive specialization in education (#PC0432).

♦ PF1913 Dominance of economic motives
Claim Today's images of what is meaningful in life tend to be confined to better goods and services. Not only do systems of livelihood focus initiative on the basis of profit, but the whole of everyday experience is towards action incited by superficial symbols.
Broader Theological collapse (#PF6358).
Narrower Production of non-essentials (#PC3651)
Belittling of grant recipients (#PF2708)
Self-interest driven investment (#PC2576)
Non-inclusive management decisions (#PF2754)
Superficial symbols of the economy (#PF3638)
Economic philosophy of controlled risk (#PF2334)
Social symbols dominated by the economy (#PE6671)
Static grassroots involvement in planning the economy (#PE4479)
Unimaginative vision of existing international economic structures (#PF2699).
Related Lack of commitment (#PF1729)
Collapsed meaning of human creativity (#PF0936)
Overemphasis on self-sufficiency with respect to interdependence (#PF3460).

♦ PF1916 Enforced vaccination
Compulsory vaccination — Enforced immunization — Denial of right to refuse vaccination
Refs Allen, Hannah *Don't Get Stuck* (1985).
Broader Religious opposition to public health practices (#PF3838).
Narrower Inoculation for viral diseases (#PJ4755).
Aggravated by Smallpox (#PE0097) Compulsory health care (#PF4820).

♦ PF1925 Inter-species warfare
Claim The abolition of inter-species war is as important as that to abolish warfare between humans. The latter is, more or less, a war amongst equals who can, more or less, fight back. There may be some dignity, some self-respect, in such a war. But war in which one side is practically omnipotent and the other practically impotent is no war at all. It is just a massacre — the pitting of mental cunning against weak ignorance.
Aggravates Endangered species of plants and animals (#PB1395).

♦ PF1929 Anti-intellectualism
Broader Populism (#PF3410) Class conflict (#PC1573) Cultural deprivation (#PC1351).
Narrower Conformism (#PB3407) Inadequate appreciation of culture (#PF3408).
Related Dictatorship of the majority (#PD3239).

Aggravates Ethnocide (#PC1328) Irrationalism (#PF3399)
Cultural decline (#PC9083) Cultural stagnation (#PC8269)
Politicization of education (#PD6298) Intellectual discrimination (#PF8590)
Denial of freedom of thought (#PF3217) Politicization of scholarship (#PF7220)
Destruction of cultural heritage (#PC2114).
Aggravated by Lack of self-confidence (#PF0879).
Reduces Elitism (#PA1387) Intellectual prejudice (#PC3406).

♦ PF1933 Unprofitable scope of industrial operations

Nature Rapid urban industrialization around the world has been followed by many rural communities starting small-scale industrial operations, using locally available human and natural resources, in the hope of providing an adequate economic base for the whole community. Such development has, however, been patchy. There are untapped skills and industrial possibilities in many such communities which local residents would like to find a way to exploit. Some communities have not established factories or commercial operations at all, while in others the scope of such industrial operation is unprofitable. There are various reasons for this: the infrastructure may be extremely limited; transportation costs for any industry are high, and agriculture and home production patterns are not geared toward reaching the greatest possible number of buyers; home production of clothing, woodwork or other items may not have been exploited for market sales. Meanwhile, families are split as children and parents search for work outside their villages, despite the desire locally for residents to return to the community and that sufficient earnings should be available to sustain them.
Narrower Lack of local industries (#PG9769) Underdeveloped road network (#PE1055)
Lack of significant employment (#PG9783) Family-based production patterns (#PG9821).
Related Unemployed skilled labour (#PE1753).
Aggravated by Inaccessible market and supply centres (#PF8299).

♦ PF1934 Inadequate sense of personal identity

Broader Alienation (#PA3545) Forced assimilation (#PC3293).
Narrower Crisis of identity (#PJ2421) Lack of racial identity (#PF0684)
Lack of social identity (#PJ4185) Lack of national identity (#PG4184)
Conflicting sense of sexual identity (#PF1246).
Related Apathy (#PA2360) Amnesia (#PD8297).
Aggravates Ethnic disintegration (#PC3291) National disintegration (#PB3384)
Psychological alienation (#PB0147).
Aggravated by Fear (#PA6030) Conformism (#PB3407)
Miscegenation (#PC1523) Cannabis abuse (#PE1186)
Family breakdown (#PC2102) Ideological apathy (#PF3392)
Social disintegration (#PC3309) Family disorganization (#PC2151)
Denial of right to a name (#PE4624) Inadequate laws of adoption (#PD0590)
Isolation of parent-child relationship (#PC0600)
Restrictions on recognition of nationality (#PE4912).
Reduced by Prejudice (#PA2173) Social discrimination (#PC1864).

♦ PF1939 Religious schism

Refs Stormon, E J and Stransky, Thomas F *Towards the Healing of Schism* (1987).
Broader Schism (#PF3534).
Narrower Multidenominational society (#PF3368) Fragmentation of religious belief (#PF3404)
Fragmentation within organized religions (#PF3364).
Related Political schism (#PC2361) Religious conflict (#PC3292)
Ideological conflict (#PF3388).
Aggravates Infamy (#PB8172) Religious extremism (#PF4954)
Segregation based on religious affiliation (#PC3365).
Aggravated by Lack of religious discipline (#PF8010)
Double standards in morality (#PF5225)
Undue religious influence on secular life (#PF3358)
Accumulation and misuse of religious property (#PE3354).

♦ PF1944 Unperceived relevance of formal education in rural communities

Lack of relevant training opportunities in local communities
Nature Despite growing agreement overall that children, young people and adults all need education to prepare them to deal practically with the actual situations which confront them, in many rural communities the relevance of formal education is neither perceived nor acknowledged. Many village children in agricultural areas are not given a realistic assessment of the impact of modern technology on farm productivity; therefore they do not have a practical sense of the role of agriculture in the future. High school students remark that their school lessons do not relate either to their cultural or to their employment needs, and in general that school activities do not seem to have any practical application to their situation. People see few incentives for higher education, whether formal or informal. This situation is intensified by the fact that many standardized national curricula allow little room for innovation in subject matter or teaching methods. People are therefore aware of being inadequately prepared to meet the challenges of the technological environment in which they live.
Although training tends to be increasingly tied to the practical needs of employment and social life, education in some communities is still primarily academic and almost exclusively for youth. In communities where the major industry (possibly the very reason the community came into existence) closes down, schools are half empty as much of the population leaves. Without a close tie between the schools and local jobs, young people go elsewhere for further training and employment. The older people who are left behind are untrained for other work even if it were available and, as a result, they retire prematurely from the work force. Those who start their own businesses find themselves lacking modern management and marketing skills. Opportunities for retraining are either nonexistent or restricted to skills that are not applicable to the employment available.
Counter-claim A number of developing countries are restructuring their education systems and developing curricula that are more closely related to the employment needs of the society and their culture.
Refs Bray, Mark *Are Small Schools the Answer?*.
Broader Obstacles to community achievement (#PF7118)
Unrecognized relevance of education (#PF9068).
Narrower Unimagined possibility of expanding the preschool (#PE5304).
Aggravates Impractical education (#PF3519) Incompetent management (#PC4867)
Limited agricultural education (#PF8835) Unrealistic agricultural vision (#PG9830)
Inadequate educational facilities (#PD0847) Lagging training in social skills (#PF8085)
Irrelevance of educational curricula (#PF0443)
Unrealized use of education structures (#PF2568)
Minimal opportunities of adult training (#PF6531)
Lack of motivation in leadership development (#PF2208)
Inadequate practical training in rural areas (#PF6472)
Lack of secondary schools in the countryside (#PE8797)
Unexplored opportunities for community education (#PF6512).
Aggravated by Attractive city jobs (#PD9869) Inadequate career advice (#PJ8018)
Rigid required curriculum (#PG9818) Insufficient educational incentives (#PJ9811)
Unavailability of trained teachers in the rural areas of developing countries (#PE8429).

♦ PF1948 Motivational death

Nature The motivation of intelligent, creative people may be eroded due to poor educational or employment environments. They may become dull, conforming, and uninspired.
Broader Apathy (#PA2360).
Aggravates Underproductivity (#PF1107).
Aggravated by Alienation (#PA3545).

♦ PF1951 Religious complacency

Claim 1. The present stage of the ecumenical way is not one with which the churches can rest content. There is a real danger that they will regard fraternal relationships and cooperation as sufficient or the continued existence of differences as intractable. Such a conclusion would tempt them to be satisfied with a consolidation of the achievements of past decades rather than to renew their commitment for the common journey. The ecumenical movement would then cease to be a movement of renewal leading toward the goal of unity embracing faith and order, worship and sacraments, mission and service.
2. There is a danger that Christians may be satisfied with charitable type giving (whether by individuals or governments) when justice demands coming to grips with the root causes of the conditions which require such giving. Those with the greater resources and abilities have the greater obligations.
Broader Complacency (#PA1742).

♦ PF1953 Inaccessibility of knowledge

Limited access to knowledge
Narrower Shortage of books and textbooks in developing countries (#PF0118).
Aggravates Ignorance (#PA5568) Lost knowledge (#PF5420)
Limited access to society's resources (#PF6573).
Aggravated by Unrecorded knowledge (#PF5728)
Fragmentation of knowledge (#PF0944)
Proliferation of information (#PC1298)
Delays in delivery of books and publications (#PF1538)
Prohibitive cost of knowledge and information (#PF0703)
Insufficient translation into minority languages (#PD0825)
Lack of terminological equivalents between languages (#PF0091)
Untransferability of books between countries and cultures (#PF2126)
Barriers to the international flow of knowledge and educational materials (#PF0166).

♦ PF1954 Linguistic purism

Nature Groups concerned with protecting their culture as reflected in its language may attempt to prevent use of terms which become widely known from some foreign language. Artificial attempts are made to generate new words and exclude or even prohibit use of borrowed foreign words or their derivatives. This could lead to isolation and technological backwardness. It also could severely limit communication with persons or groups outside the speaker's (or writer's) culture.
Broader Ideological conflict (#PF3388) Multiplicity of languages (#PC0178).
Narrower Nativism (#PF2186).
Aggravates Lack of terminological equivalents between languages (#PF0091).
Aggravated by Prejudice (#PA2173) Double standards in morality (#PF5225)
Monolingualism in a multi-cultural setting (#PD2695).

♦ PF1959 Ineffective operation of community networks in urban ghettos

Nature Inner city areas do not necessarily lack community organizations – they often have the full complement of social service agencies, school organizations and functioning block clubs – and local citizens may be clearly concerned about their community; but informal networks to inform citizens, and through which they can participate in the building of consensus, do not exist. Consequently, well-intentioned and well-designed programs are met with scepticism or suspicion. The local resident does not perceive any of them as effective channels for creative dialogue. In fact, the agencies are perceived as a 'they' who make decisions and carry out policies that affect 'us' without our advice or consent. Unfounded rumours about plans of outsiders for the community are rampant, but organized attempts to hold community meetings and explore the truth are ill-attended. Local associations are constantly frustrated in their efforts to increase membership and attendance.
Claim Until inner city communities are linked through local communication networks, efforts to build an effective human community will be thwarted by the lack of a firm corporate consensus.
Broader Lack of community participation (#PF3307)
Obstacles to community achievement (#PF7118).
Narrower Restricted decision-making (#PG8931)
Insufficient industrial planning (#PG8873)
Undefined programme responsibility (#PG8923)
Non-participatory financial priorities (#PG8938).
Aggravates Lack of leadership (#PF1254) Inadequate maintenance (#PD8984)
Limited youth activities (#PJ0106) Insufficient leadership training (#PF3605).
Aggravated by Unclear educational roles (#PG8840).

♦ PF1963 Loss of institutional credibility

Loss of credibility in institutions
Nature Loss of belief in the efficacity of institutions on a national or international level, including economic, political, social, religious and ideological institutions, may produce apathy or violence and crime from a sense of alienation, and may result in social breakdown or social or national disintegration. It could also be the cause of materialistic attitudes and consumerism and a lack of social development.
Broader Credibility gap (#PB6314).
Narrower Evasion of the law (#PD4208)
Loss of confidence in government leaders (#PF1097)
Loss of credibility in international institutions (#PE8064).
Aggravates Apathy (#PA2360) Cynicism (#PF3418) Scepticism (#PF3417)
Alienation (#PA3545) Counter culture (#PF0423) Social breakdown (#PB2496).
Aggravated by Corruption (#PA1986) Breach of promise (#PF7150)
Political deception (#PF9583)
Public resentment at government policies (#PF7472)
Administrative difficulties in new states (#PE1793)
Institutional preoccupation with obsolete problems (#PJ5014)
Increasing scepticism about the accuracy of official information (#PF7649).
Reduces Excessive institutionalization of education (#PD0932).
Reduced by Nationalism (#PB0534).

♦ PF1964 Human use of animal by-products

Incidence Animal by-products include: leather, wool, feathers, bone (for products such as gelatin), intestine and catgut (for sausage casings, instrument strings, rennet for cheese manufacture), hair (for brushes and furniture padding), blood (for animal feedstuffs), animal fats and lycerine (in confectionery and cosmetics). In addition to these more obvious products, many others are used in the manufacture of consumer goods considered essential to life although no mention of the animal origin is made on the label. Animal products are also widely sort as status symbols (rhinoceros horns as dagger handles), as trophies (animal heads, antlers, skins), as collectors items (birds eggs), as aphrodisiacs (horn), and as fashion accessories (fur).

Claim There is no such thing as an innocent animal by-product. Even in the case of wool, although sheep are not slaughtered after the first shearing, some 40 per cent of sheep and lambs are as part of the economic pattern of sheep farming. The availability and cost of wool are dependent on the slaughter of sheep. By changing diet to avoid animal products, a major step is taken towards the elimination of cruelty to animals. By purchasing clothes and accessories of non-animal origin, people deepen their commitment to animal rights and non-violence.
Counter-claim Animal by-products are a natural by-product of the animal, especially in the case of such products as wool, and do not cause the animal's death. Such use may constitute a wise recycling of natural resources. They may also reduce the cost of associated meat products for human consumption.
Narrower Human consumption of animal products (#PD7699).
Aggravates Hunting tourism (#PE3008) Animal fighting sports (#PE4893)
Endangered black rhinoceros (#PE6010) Slaughter of animals for pelts (#PD4575)
Abusive collection of specimens (#PE9417).

♦ PF1976 Disregard for internationally imposed economic sanctions
Circumvention of international sanctions — Smuggling through sanctions barriers
Nature Unilateral and multilateral economic sanctions imposed since the end of World War I have had some limited successes but have proved virtually useless when wielded by one strong power against another. The successes were achieved mostly against small countries when only modest foreign policy goals were sought.
Incidence An example of sanctions with general lack of success were the USA efforts to halt the sale of grain and gas pipeline equipment to the Soviet Union.
Counter-claim Even when they do not do their intended job, sanctions can serve important political purposes, such as unifying world opinion concerning violations of internationally accepted behaviour; for example, the League of Nations' action when Italy invaded Ethiopia.
Broader Lack of international cooperation (#PF0817).
Aggravates Apartheid (#PE3681).
Aggravated by Conflict of laws (#PF0216)
Inadequacy of international legislation (#PF0228).

♦ PF1978 Personal covert himsa
Covert non-physical psychological violence
Nature Psychologically dangerous forms of non-physical violence through which dignity and personhood are denied in interpersonal relations. Such actions can include hurting another through thought, word or deed, and withholding support and concern when the situation demands it. It is especially evident in individual postures of racism, sexism, and prejudice based on a person's age, religious belief or sexual orientation.
Broader Human violence (#PA0429).
Narrower Harmful thought (#PF0441).

♦ PF1979 Dispersion of local capital resources
Nature Any community, from the local to the national level, which wishes to enhance its economic development must retain its capital resources as long as possible. In most small rural villages the resources for commercial ventures are severely restricted, yet local resources are spent outside the community for goods and services not available in the village, usually being paid to peddlers who sell within the village, but carry the money received out to their home and business communities. Such dispersion of resources compounds the effect of tax and employment laws, fostering the belief that economic activity is necessarily 'hand to mouth' and making it unattractively risky to borrow development capital even if it were available. This occasions paralysis in thinking and planning for the future.
Broader Lack of local commercial services (#PF2009).
Narrower Limited village capital (#PG9811).
Related Prohibitive labour costs (#PF8763)
Subsistence approach to capital resources (#PF6530).
Aggravates Unstimulating entertainment (#PF8105).

♦ PF1981 Inadequate technical advice and consultation on problems
Broader General obstacles to problem alleviation (#PF0631).
Narrower Ignorance of health and hygiene (#PD8023).
Related Imbalance between capital and technical assistance (#PE4866).

♦ PF1984 Short-term climatic change
Nature Evidence of short term fluctuations in the global climate (on the scale of 100-200 years) indicates that a major sustained shift in climate began in 1950 and is likely to persist for the rest of the century. This will affect different regions in different ways. Winters in the northern temperate zones will tend to be much more severe, whereas arid and semi-arid zones will tend to be much drier. These changes are a consequence of changing wind patterns, although it is expected that the variability will increase. So while the general trend is towards colder, drier weather, the changes from year to year may be even sharper than before. These changes will considerably aggravate the problems of drought and desertification in the arid zones and will aggravate transportation and flooding problems in temperate zones. Ocean temperature changes are both causes and effects of meteorological changes. Changes in atmospheric climate may be accompanied by temperature anomalies in the seas. The consequences to this may be reduction of fish catches in some areas by 80 percent.
Refs Bolin, B *Climate Changes and their Effect on the Biosphere* (1980); Gribbin, J *Climatic Change* (1978).
Broader Inhospitable climate (#PC0387).
Related Global cooling (#PF1744).
Aggravates Floods (#PD0452) Drought (#PC2430) Desert advance (#PC2506).

♦ PF1985 Ignorance of cultural heritage
Lack of access to cultural heritage — Remote cultural heritage — Unknown cultural heritage — Unconveyed cultural heritage — Lost heritage vitality — Undisplayed heritage gifts — Commonly disvalued heritage — Loss of cultural heritage — Failure to pass on cultural heritage — Insufficient cultural heritage transmission — Lack of awareness of past heritage — Fragmented recognition of common heritage — Fragmented celebration of cultural heritage
Nature Large libraries, art galleries and museums have considerable proportions of their collections in storage areas inaccessible to the general public due to the cost of display space, operating costs, and the need to adequately protect their holdings from damage or theft. As a result, ever smaller proportions of cultural heritage are actually suitably displayed to permit casual browsing, and in many libraries browsing is no longer possible. Each item must be specially requested on the basis of prior knowledge of its existence.
Present-day adults have not incorporated the past wisdom and heritage of the older generation and so lack a historical context from which to view the present and the future. Without this perspective of a re-enacted and honoured past, people concentrate only on the immediate things of life and have little concept of their own role in the creation of history. They are more interested in the politics of advancement or in the petty annoyances of their jobs than in the more basic question of their role in society.
Claim The insatiable appetite of museums for material has resulted in many hastily acquired collections without adequate documentation for each item's cultural or historical context. Not only are collections poorly accessible, they are not always well identified and catalogued, so that retrieval is doubly difficult. In the case of libraries, the lack of access is in part due to the flooding of the world by printed matter, most of it of an ephemeral nature. Books of little worth cause cultural pollution encrusting in libraries the books embodying the cultural heritage. Truly current general collections are no longer possible and libraries resort to dispersal and fragmentation in order to service specific groups of readers in limited subject areas.
Broader Ignorance (#PA5568) Lost knowledge (#PF5420)
Pathologies of civilization (#PB3674).
Narrower Lost family heritage (#PF1099) Symbol system failure (#PF3715)
Lack of access to public archives (#PD1194)
Acculturation as a dilution of cultural heritage (#PD4272).
Aggravates Cultural illiteracy (#PD2041)
Increasing development lag against information growth (#PE2000).
Aggravated by Ignorance of history (#PD3774) Ethnic disintegration (#PC3291)
Declining sense of community (#PF2575) Demeaning community self-image (#PF2093)
Destruction of cultural heritage (#PC2114) Obsolete basis of cultural identity (#PF0836)
Fragmented planning of community life (#PF2813)
Local traditions of cultural isolation (#PF1696)
Restrictions on freedom of information (#PC0185)
Disjointed patterns of community identity (#PF2845)
Abusive exploitation of cultural heritage (#PC7605)
Limited community responsibility of adults (#PF1731)
Modern disruption of traditional symbol systems (#PF6461)
Untransposed significance of cultural tradition (#PF1373)
Unclarified procedures for transposing ancient traditions (#PF6494).

♦ PF2002 Conflict of information
Nature There can be different versions of fact or opinion about the same subject. This may include information which is unrelated to its context or other kinds of misleading information, such as propaganda. It may also include rigid dogma and theory. Conflict of information may cause confusion and hinder progress, or it may harden prejudices and have the effect of indoctrination. It may be the cause or the result of international conflict.
Broader Conflict (#PA0298).
Narrower Lack of comparability of international statistics (#PF2622).
Related Conflict of laws (#PF0216) Misleading information (#PF3096).
Aggravates Confusion (#PF7123).
Aggravated by Propaganda (#PF1878) Government propaganda (#PC3074)
Censorship in communist systems (#PD3172)
Propaganda by intergovernmental organizations (#PE3077)
Conflict of laws on international restriction of information (#PD3080).

♦ PF2003 Multiplicity of problems facing society
Multiplicity of problems
Nature In spite of much publicity, the complexity and magnitude of the problems faced by man if he is to survive as a social animal are still only adequately conceived by specialists; they derive not so much from the mere multiplicity and gravity of problems awaiting a solution in the present technological society nor from the frightening series of problems appearing over the horizon, as from the fact that between these multiple problems there exists an incalculable number of inter-relationships which, whether ascertained or not, greatly restrict the range of action open to the policy-maker. It is this situation which has brought about the tendency for the solution of one problem to create a number of new ones, often in fields only distantly related at first sight to the original matter. In particular, this not being fully understood, there is a general disposition to envisage and treat the symptoms of trouble, particularly the more obvious ones such as the various forms of pollution of the environment, rather than to deal with the root cause which is to be found in the inadequacy of the decision-making machinery of society under any form of government presently known.
Broader General obstacles to problem alleviation (#PF0631).
Aggravates Oversimplification (#PF8455)
Complex interrelationship of world problems (#PF0364)
Failure to integrate knowledge to empower humanity in response to the global problematique (#PF8753).

♦ PF2005 Inadequacy of religion
Inadequacy of religious doctrine — Ossification of faith in rigid doctrine
Nature When religious doctrines lose credibility, chiefly because they do not seem relevant to current social conditions, there results a lack of moral guidance. Religious teaching may be seen as hypocritical, corrupt or intolerant and may lead to conflict and atrocities. One religion may therefore be rejected without another taking its place. Experimental and fragmented religious sects may attempt to fill a need without in fact doing so on a wide or lasting basis. Other non-theist ideologies may be adopted by individuals, or society as a whole may witness the ascent of nationalism combined with an evangelical but secular political creed of a new order. State dogmatism may replace multi-religious, pluralistic and liberal regimes, and racism may enter along with totalitarianism. Decadence, cynicism and apathy may eventuate in a culture still in search of coherent or persuasive statements concerning its goals.
Religious doctrine may be incompatible with trends in modern living, as exampled in birth control, agricultural development, alleviation of mass poverty and malnutrition. Religious pronouncements against war, intolerance, discrimination and exploitation appear to have little practical effect. Religious doctrine may conflict with national or international development programmes causing conflict, or regressive religious attitudes may be adopted by the national government.
Incidence The rise of the Third Reich, and the establishment of communism, for example, may be seen as the results of the failure of religion. Religion appears to be inadequate, particularly in the view of disadvantaged or intellectual people in highly industrialized and technically developed societies where conservatively interpreted religious doctrines may be in direct conflict with social progress. Newer or more liberal doctrines may be insufficient or lack credibility to replace them. In developing countries, religion may impede progress to an even greater extent by holding the allegiance of masses of poor and uneducated people whose main solace is in religious belief. Among indigenous populations, the influences of the large and unacculturated institutionalized religions may serve to hasten ethnic disintegration but not to fulfil the needs of such populations in transition to a different way of life.
Counter-claim The great world religions have become what they are partly because of their social contributions in ages past, and particularly because they look at longer time frames than the maximum one generation span of attention of society. That every living creature, and those still to come, has an inviolable right to life is a religious viewpoint, without which societies would be inhumane. That every human being has a moral conscience and a right to exercise it is also a religious point of view. And, finally, that all men are brothers, is not a fact of science, but is an inspiration of the human spirit.
Refs Siriwardena, R (Ed) *Equality and the Religious Traditions of Asia* (1987).
Broader Ideological conflict (#PF3388) Inadequacy of doctrine (#PF3396).
Narrower Enforced celibacy (#PD3371) Superficial religion (#PJ5252)
Theological justification of nuclear war (#PF3807)
Disagreement concerning religious doctrine (#PF1115)
Negative effects of claims of religious infallibility (#PF3376).

Related Inadequate ideological frameworks (#PD0065)
Religious and political antagonism (#PC0030)
Undue religious influence on secular life (#PF3358).
Aggravates Apathy (#PA2360) Dogmatism (#PF6988) Bibliolatry (#PF7129)
Materialism (#PF2655) Anti-clericalism (#PF3360) Religious apathy (#PC3414)
Religious conflict (#PC3292) Experimental religion (#PF3367)
Disintegration of organized religion (#PD3423).
Aggravated by Religious intolerance (#PC1808) Double standards in morality (#PF5225).
Reduced by Scientism (#PF3366).

♦ PF2006 Institutionalized callousness of public services

Nature Social institutions provide only shallow means of care for people; this propensity is disguised behind irrelevant descriptions and tests. Having no overall context for service, many of the institutions of social care have turned inward to the mission of self-preservation. In addition, employers do not demand a worker's full creative endeavour but allow him to get by with minimum effort, thus inferring that his creativity is irrelevant to social needs.
Broader Selective perception of facts (#PF2453) Repression of self-consciousness (#PC1777).
Aggravates Inadequate health services (#PD4790)
Bureaucracy as an organizational disease (#PD0460).

♦ PF2009 Lack of local commercial services

Limited availability of local commercial resources
Nature Present day communities are dependent on commercial services. They are an essential aspect of the transition from a subsistence to a market economy, and have to be increased if the local economy is to accelerate. However, access to capital, adequate lines of credit, market information and accessibility, saving methods, cash exchange means, simple management bookkeeping, secretarial services and other essential fiscal services are currently non-existent in many Third World suburban communities. The lack of these services is occurring at a time when a community has to develop commercially or perish. The over-abundance of small shops limits the range each can offer, so that inventories overlap and none is able to expand to a profitable size. The same is true of the numerous tiny industries and sub-divided small farming plots that are unable to produce significant income.
Many local communities seek to initiate new industrial and commercial enterprises but find it a near-impossible task. Distances to markets and suppliers are prohibitive. Limited availability of capital funding, high start-up costs, and the absence of local business facilities deter the undertaking of new ventures. Intense outside competition makes it difficult to determine which markets to target. High local prices and limited variety of goods at local shops leave the consumer inclined to look elsewhere first. Because such communities cannot find the means to start up and sustain their own consumer businesses and competitive industries, their economic resources continue to dwindle and the quality of their economic well-being is increasingly determined by outside interests.
Broader Inadequacy of the commercial sector in developing countries (#PD1865).
Narrower Credit discrimination (#PE7902) Fear of social anarchy (#PF1088)
Complex banking practices (#PJ8033) Ineffective land collateral (#PG8046)
Absence of tactical methods (#PF0327) Excessive financial pressure (#PG8022)
Lack of market refrigeration (#PG7998) Untenable investment incentives (#PG7983)
Scarcity of business collaterals (#PG8008) Insufficient business facilities (#PG0110)
Dispersion of local capital resources (#PF1979)
Haphazard provision of consumer services (#PF2411)
Prohibitive cost of necessities in rural communities (#PF2385)
Absence of a long-range, world-wide capital flow plan (#PF2865).
Related Strained capital resources in small communities (#PF3665)
Limited availability of investment capital for urban renewal (#PF3550)
Failure of public authorities to assist in financial investment (#PF6508).
Aggravated by High interest rates (#PF9014) Prohibitive cost of capital equipment (#PJ1768)
Control by transnationals of the global communications industry (#PE8640).

♦ PF2012 Lack of facilities for educating older people

Nature Older people are not equipped with the skills, education and models to participate creatively in the present-day societal process; and facilities for them to be educated in such skills are not available. They need such education in order to be free to create their own roles in the current societal framework and regain a viable self-image and a sense of responsibility to society. Nonexistent educational means for elders thus impedes the possibility of their reappropriating selfhood.
Broader Inadequate educational facilities (#PD0847).
Aggravates Elder paralysis over the future (#PF3973).

♦ PF2013 Ill effects of educational failure

Failure self story
Nature A person who has been judged a failure by the schooling system from which he has dropped out, or by his family and contemporaries, and who therefore tends to be alienated from his environment and unable to find the kind of position that his educational orientation led him to expect, may come to believe that he was worse off than if he had not entered the schooling system at all. The belief that he deserved a better post than those who had not attended school at all may also result in inferior performance in employment. The frustration and resentment resulting from taking lowly paid, unskilled jobs represent a potential source of anti-social behaviour and political discontent.
Broader Lack of self-confidence (#PF0879) Failure of school systems (#PJ3896)
Negative emotions and attitudes (#PA7090).
Related Demoralizing images of rural community identity (#PF2358).
Aggravates Frustration (#PA2252) Political discontent (#PG4257)
Low learning expectations (#PJ8960).
Aggravated by Educational wastage (#PC1716).

♦ PF2014 Unregulated ownership of the means of production

Claim The emerging public pressure for responsible use of all productive means is blocked by the absence of a comprehensive and generally agreed design for allocating ownership of such means and the responsibility inherent in such ownership. A vacuum of decision-making criteria has become apparent. The "invisible hand" or laissez-faire model has been crippled by the emergence of corporate giants who can, singly or in concert, upset the statistical randomness which that model requires. Consequently, the context of ownership is reduced to the tension between individual or corporate self-interest on the one hand and benevolent good will on the other, with the relationship to property distorted from privilege to inherent right.
Broader Belittling of grant recipients (#PF2708).
Narrower Oligarchy effectiveness myth (#PF1138)
Destabilizing corporate expansion (#PD1220)
Limited ownership of productive systems (#PD3182)
Lack of accountability in the disposal of wealth (#PE0503)
Individualistic disposition of productive property (#PD4974).
Related Imperialistic distribution system (#PD7374)
Limited approaches to economic planning (#PF3500)
Monopoly of the economy by corporations (#PD3003).

♦ PF2015 Subjective cultural ranking

Nature There is a tendency to view different cultures as superior or inferior to one another and to one's own. This blocks appreciation of the gifts of other cultures, which are then only made use of in either grasping or condescending ways. This restricts relationships among peoples and results in societies becoming uniformly mediocre and unable to use the contributions that outsiders and foreigners could make.
Broader Habitual overemphasis on national self-determination (#PF1804).

♦ PF2016 Inadequate facilities for international nongovernmental organization action

Nature Most INGOs require the same basic administrative services and facilities, but because of great sensitivity to their independence and autonomy of their programme, they are reluctant to pool services and facilities in order to increase the efficiency of their administrative operations. This is partly due to an inability to distinguish between the objectives of the organization and the facilities and professional skills required to achieve them. Whether in capital cities of developing or developed countries, their restricted budgets force them to use minimum facilities, which are often inadequate and insufficient. Their offices may be so scattered that face-to-face contact between organization staff members is infrequent. Organizations are often poorly housed and equipped. A critical mass is not built up.
Broader Inadequate facilities for international organization action (#PD0929)
Obstacles to effective international nongovernmental organizations (#PF7082).
Aggravates Ineffectiveness of international nongovernmental organizations and programmes (#PF1595).
Aggravated by Inadequate funding of international nongovernmental organizations and programmes (#PE0741).

♦ PF2017 Attitude manipulation of children through play

Nature Games may teach children to act aggressively by rewards for such behaviour, or instil in them a value that competitiveness is a virtue. Games may reduce children to accept violence as a norm, and to use force themselves. Children's play, and in particular manufactured games, teach deception, bluffing and lying, placing considerable value on cunning or trickiness.
Claim In children's play, especially as devised by manufacturers, there is too much of the attitude 'winning is everything'. In play designed and supervised by figures of authority (parents, teachers, recreation workers), fear and respect of the hierarchy is reinforced and the concept is instilled that one 'must follow the rules'. Many ideological movements seek to control childrens' development and see an opportunity, especially in their recreation, to indoctrinate political ideas.
Broader Compulsory indoctrination (#PD3097).
Related Inadequate facilities for children's play (#PD0549).
Aggravates Loneliness of children (#PC0239)
Inadequate image of roles within marriages (#PD1308).
Aggravated by Dangerous toys (#PE1158) Television violence (#PE4260)
Violence in comic books (#PG4262) Segregation in education (#PD3441)
Socially disruptive effects of video games (#PF6345).

♦ PF2018 Delayed development of regional plans

Protracted development of regional plans
Broader Policy-making delays (#PF8989)
Delays in implementation of social change (#PC6989).
Narrower Political media events (#PD5207)
Non-adaptive local structure of social care (#PE5246).
Related Delay in project implementation (#PF1470).
Aggravates Threat of birds to aircraft safety (#PD1111)
Inadequate recreational facilities (#PF0202).

♦ PF2019 Death of God

Related Deicide (#PF4617).

♦ PF2031 Political appointees

Nature The appointment of political favourites or supporters to sinecures by a government or by high officials with a vested interest in making such appointments helps to tighten government control and the political power of a group or of one key person (such as the president). It may also serve as a means of political repression through propaganda. Political appointees in key industries and administration are characteristic of most dictatorships; where such appointments are made in constitutional democracies their predominant role is more that of enriching the elite of which they are part than simply keeping the elite in power. The appointment of political favourites may cause inefficient as well as corrupt administration.
Background The practice of appointment of political supporters to public office became institutionalized in the USA from 1800 until after the Civil War, and was known as the 'spoils system'.
Broader Bribery (#PC2558) Corruption of government leaders (#PC7587).
Narrower Sinecures (#PG4269).
Related Political corruption of the judiciary (#PE0647)
Unjust allocation of government contracts (#PF2911).
Aggravates Corruption (#PA1986) Political conflict (#PC0368)
Political alienation (#PC3227) Government inefficiency (#PF8491).
Aggravated by Political apathy (#PC1917)
Abusive distribution of political patronage (#PF8535).

♦ PF2040 Fear of contradicting popular views

Broader Fear (#PA6030).
Aggravates Fear of ostracism (#PF2776) Lack of political will (#PC5180)
Fear of increased autonomy (#PE7706) Fear of humiliation by co-workers (#PF9234).

♦ PF2042 Lack of awareness of potential for investment in small, inner-city enterprises

Nature The present day tendency for business and industry to move towards developing large enterprises requiring massive capital investment has resulted in private financial support has being withdrawn from small business enterprises in many inner-city areas. The style of operation of such businesses may also obscure their investment potential. Local inner city businesses often project an image of high risk, low volume, and a short future. The current "redlining" practices of banks block loans for remodelling or expansion, and insurance companies may make it difficult to secure adequate coverage for reasonable rates. The heavy flow of residents' income outside their neighbourhoods further weakens the operating base by separating local enterprise from local capital. Beyond this, highly scattered employment patterns obscure the availability of a local labour pool for a potential local industry. Until the commercial and industrial potential of urban inner city communities becomes visible, desperately needed capital for business, industrial and housing developments will be denied.
Broader Unrecognized opportunities (#PF6925) Obstacles to community achievement (#PF7118).
Narrower Unstable economic base (#PG8905) Professional salary losses (#PG8841)
Lack of capital development (#PD8604) Insufficient operating capital (#PG8932)
Uninsurable business investments (#PG8939).
Related Limited availability of investment capital for urban renewal (#PF3550).
Aggravates Inadequate development of enterprises in developing countries (#PE8572).
Aggravated by Lack of community participation (#PF3307).

♦ **PF2043 Inadequate world calendar**

Nature It is widely recognized that the calendar in use today is unsatisfactory for economic, social, educational, scientific and other activities of man.

Background The present calendar is essentially that introduced in about 45 BC by Julius Caesar (who invented the "leap year" to take the extra quarter of a day a year into account) and slightly modified by Pope Gregory XIII in 1582. The Gregorian calendar is not perfectly adjusted to the tropical year, the error accumulating to one day after about 3300 years.

The basic difficulty is that every year has 52 weeks of 7 days plus one day or, in a leap year, 2 days. The calendar contains 28 different kinds of months, with 28, 29, 30 or 31 days, starting on any one of 7 weekdays, and anything from 24 to 27 working days. Every year shifts and is different from the preceding and following years, so that comparability is difficult and inaccurate. Weekdays and month dates never agree with an incoming or outgoing year. Equal divisions of quarters and half-years are lacking. Considerable amounts of time and money are allocated each year to compiling new calendars and schedules because of this variability. Holidays falling on fixed dates which vary in their position within the week give rise to different conditions each year. Statistics for fixed periods have to be corrected for such variability to make them comparable between months and years. Variable church holy days are a continuous concern to both priest and layman. Law courts encounter many difficulties to reconcile wandering weekdays and holidays, resorting to such expressions as 'the first Tuesday after the first Monday in November'.

Other calendars continue to exist in parallel with that based on the Christian-oriented system. These include the Islamic, Hindu and Chinese calendars. The most radical attempt at redirecting the temporal identity of an entire culture through calendrical manipulation, in order to eliminate the religious bias, was that during the French Revolution.

Broader Lack of international cooperation (#PF0817).
Related Temporal imperialism (#PF1432) Multiplicity of time standards (#PF2621).
Aggravates Lack of comparability of international statistics (#PF2622).

♦ **PF2055 Land reclamation**

Broader Desert nomadism (#PD2520).
Aggravates Nomadism (#PF3700).
Reduced by Domestic refuse disposal (#PD0807)
Solid wastes as pollutants (#PD0177).

♦ **PF2060 Parental control of children's thoughts and reflections**

Nature Parents attempt to reduce and control the thoughts and reflective associations of their children to those that agree with their own views. This reduction of the child's relational world denies the broad context he needs in order to develop his own judgement. At the same time, the narrowing of young peoples' interests does not give credit to their ability to live in the external world. Adults fail to comprehend the degree to which their own behaviour influences their children, and thus fail to recognize the effect of their own unintentional behaviour on their children's image systems.

Counter-claim They even teach them languages.
Broader Exclusion of pre-adults from family decisions (#PE2268).

♦ **PF2063 Lack of symbolism in local relationships**

Nature There is no present-day system of meaningful symbols to remind people of their relationships to the local social framework which is their practical link to global society. In addition, the increasing mobility of the population diminishes people's ability to engage locally. This results in the paradox that as the individual becomes increasingly aware of events in the global arena he experiences decreasing opportunity to deal creatively with this awareness through his local relationships; and his sense of obligation becomes fragmented.

Broader Symbol system failure (#PF3715)
Reinforced parochialism of internal values and images (#PF1728).
Narrower Unmeaningful school name (#PU8815).
Aggravated by Lack of commitment to common symbols (#PE8814).

♦ **PF2066 Artificial and arbitrary job qualifications**

Nature There are a number of artificial and arbitrary job qualifications such as age, physical appearance, sex, and membership in unions or professional groups. Excessive requirements of experience or education often have little or nothing to do with the job, thus excluding many qualified people from being integrated into the job market.

Broader Unemployment (#PB0750) Arbitrariness (#PB5486)
Inadequate access to negotiation on employment and reward (#PD1958).
Narrower Certificate-based job market (#PJ8370).

♦ **PF2068 Young people's lack of context for the future**

Nature Young people today are not given a context, either by educational structures or by parents, within which to gain the practical experience which will demonstrate the importance of their participating creatively in society, and of the need for them to be responsible for creating the future. Adults believe that the young are strong and able enough to 'pick up' what is necessary in the future; in turn, the young find it easy to depend on the adults. They feel that with their own future ahead of them they should live for the present.

Broader Disorientation of the young due to lack of social forms (#PD2050).
Aggravated by Temporal deprivation (#PF4644).

♦ **PF2070 Lack of a system for ethical decision-making**

Claim Man has harnessed the computer to handle complex technological problems, but he has failed to discover a method to engage himself in ethical decision-making. There are no methods for dealing creatively with the complexity brought about by rapid change and global contingency.

Broader Reinforced parochialism of internal values and images (#PF1728).
Aggravates Indecision (#PF8808)
Lack of meaningful educational context for ethical decisions (#PF0966).

♦ **PF2071 World anarchy**

Nature Lack in world order leads to conflict, inequalities, injustice, exploitation, unequal distribution and wastage of resources. Treaties and leagues have been concerned with specific problems, but despite this and the number of large intergovernmental organizations, there is little cohesiveness in world events. The threat of a new international anarchy has arisen because anarchy now co-exists with nuclear proliferation. According to the United Nations Secretary General, this new anarchy is an "armed force, both overt and covert, used and increasingly justified as a legitimate means of obtaining national objectives".

Incidence There were more than 40 non-nuclear conflicts underway in 1984. These conflicts were fought with highly sophisticated conventional weapons which have caused about 20 million deaths in wars since 1945, almost twice as many civilian as military.

Claim Anarchy in the international system is due to the fact that centralized legislative or judicial organs of the international system have very limited power, and that most nations remain, to some extent, sole judge, prosecutor and jury of their causes.

Counter-claim Informal organizations may have self-regulative mechanisms and rules of behaviour, maintained by constraint which may stem from the immediate interests of the participants and from their indirect interest in maintaining the system insofar as it meets their needs. In international political systems, violations of the law are always more obvious than the observances and this largely accounts for the belief that the system is in anarchy.

Broader Lack of human unity (#PF2434) Political instability (#PC2677)
Political disintegration (#PC3204).
Narrower Nationalization of foreign investments (#PC2172)
Inadequacy of international legislation (#PF0228).
Related Anarchism (#PC1972) International economic fragmentation (#PC0025)
Rivalry and disunity within developing regions (#PD0110)
Domination of the world by territorially organized sovereign states (#PD0055).
Aggravates Inadequate road maintenance (#PD8557).
Aggravated by Amoralism (#PF3349) Nationalism (#PB0534)
Human inequality (#PA0844) Political schism (#PC2361)
Ideological conflict (#PF3388) Subversive activities (#PD0557)
Competition between states (#PC0114) Growth of anti-systemic movements (#PJ0051).
Reduced by World federalism (#PF2088).

♦ **PF2076 Overriding of international by local concerns**

Nature Communities which wish to relate themselves to world issues do so as a secondary effort, a gesture of generosity. Like many such gestures, this fails as soon as pressure is applied, and emphasis is again wholly focused on local issues which more immediately affect the population.

Broader Inadequate means for upholding global concern (#PF1817).

♦ **PF2081 Faceless social institutions**

Nature There are no generally accepted avenues to participate in reaching a consensus. Networks of informal interpersonal relationships are separate from the social institutions – school, market, office – in which priorities and goals are developed. The community has no means of authentically engaging citizens in a broadly-based exchange of ideas relative to the setting of priorities and goals for the locality on issues of primary importance.

Broader Institutional domination of organizational systems (#PF2825).
Narrower Excessive institutionalization of education (#PD0932).
Aggravates Reinforced parochialism of internal values and images (#PF1728).

♦ **PF2086 Tokenistic meeting resolutions**

Irrelevant resolutions — Stock resolutions — Vague resolutions

Nature Conferences, and especially those of international organizations responding to world problems, tend to devote considerable resources to elaborating resolutions and declarations. Relatively few of these lead to any action and little attention is subsequently given to evaluating the degree of follow up, even by the responsible bodies. Many of those involved may well be more concerned with the short-term public relations impact rather than the long-term operational impact. There is a failure to ensure that resolutions are both truly purposeful and to respect them as genuine expressions or reminders of widely shared concerns.

Significant action on controversial issues is ensured through collective agreements, not through the illusion of collective agreement provided by a vague resolution. Subsequent action is jeopardized if different interpretations are placed on the text by its framers. Remedial action cannot emerge when those endorsing the agreement do not share the same understanding of the text and fail to coordinate their policies on the basis of it.

Claim Resolutions are meant to keep alive the goals to be achieved and to ensure that these goals are not lost sight of in a multitude of other concerns. In that perspective, they can become an indispensable factor for the successful outcome of negotiations and be perceived as resolutions in the full sense of the term, not as incantations or mere formulations of theory. They become ineffective when they look like stock resolutions.

Broader Unimplemented decisions (#PF4672) Inadequate meeting methods (#PF8939)
Inadequately worded agreements (#PF5421).
Related Symbolic international agreements without substance (#PE9267).
Aggravates Avoidance of decision-making (#PF4204).

♦ **PF2088 World federalism**

Broader Ideological conflict (#PF3388).
Related Over-centralization (#PF2711) International economic fragmentation (#PC0025).
Aggravated by International insecurity (#PB0009)
Double standards in morality (#PF5225).
Reduces World anarchy (#PF2071).

♦ **PF2093 Demeaning community self-image**

Community image of powerlessness — Villagers victim image — Economic victim image — Community victim image — Reduced community image — Depressed community image — Restraining community image — Unattractive community image — Defeating community image — Lack of significant community images — Undeveloped community power — Defeating economic story

Nature Local communities feel powerless when confronted by the complexity and enormity of global problems and the difficulty of discerning the results of the community's efforts in the world. They may also fear that community uniqueness will be undermined, ignored or glossed over in a global society. These feelings inhibit individual participation in global projects and reduce the context of community action to immediate needs and concerns, so that day-to-day plans ignore global demands.

Residents of many cities feel powerless to deal with local government institutions. The demoralizing effect of past failures leads to resignation and a deep sense of cynicism about continuing the pursuit of shared community issues. Residual efforts at improvement tend to be met with failure, and the community spirit which would engender hope in the possibility of future victories is lacking. Until the residents of a community are able to synthesize a community image which captures the significance of its heritage, the meaning of the past will be lost and the community will remain unformed.

Broader Obstacles to community achievement (#PF7118).
Narrower Short-shipments (#PG1671) Economic isolationism (#PC2791)
Embarrassed civic pride (#PG1429) Resignation to problems (#PF8781)
Low community visibility (#PG7691) Low learning expectations (#PJ8960)
Unstimulating entertainment (#PF8105) Oppressive prevalent images (#PF1365)
Unavailable public buildings (#PG7699) Distant schooling facilities (#PJ7693)
Lack of confidence of parents (#PG7704) Debilitating education images (#PF8126)
Sensed futility of job-hunting (#PG7700) Excessive parental drunkenness (#PE7700)
Deteriorating downtown district (#PG1530) Increase in abandoned buildings (#PJ1665)
Unemployment as a perpetuator of failure (#PE8457)
Inadequate means for upholding global concern (#PF1817)
Imbalance in economic and social planning in developing countries (#PF4837).
Related Domination (#PA0839).
Aggravates Lack of leadership (#PF1254) Lack of self-confidence (#PF0879)
Overdependence on government (#PF9530) Uncontrolled local vandalism (#PJ0154)
Lack of community development (#PF7912) Ignorance of cultural heritage (#PF1985)
Deteriorating community identity (#PF2241)
Unorganized development of work forces (#PF2128)

Unfocused style of community operations (#PF6559)
Demoralizing images of rural community identity (#PF2358)
Declining community confidence in its ability to change (#PF9066)
Personal isolation in communities of industrialized countries (#PD2495).
Aggravated by Police state (#PD7910) Absentee ownership (#PD2338)
Neglect of property maintenance (#PD8894) Unproductive subsistence agriculture (#PC0492)
Detrimental story of community future (#PF6575)
Unrecognized benefits from cooperatives (#PF9729)
Unattractive appearance of deteriorating buildings (#PF1733)
Traumatic shift in life–styles of mining communities (#PE1137).

♦ PF2095 Outmoded forms of social education

Claim No educational institution or informal educational system today equips people with techniques to plan and act toward their community, toward their nation or toward the world. This is reflected in the massive disquiet seen among workers, students, parents and teachers alike.

In many developing countries, especially former colonies, education dominated by imported values and systems, has no relevance to the local or national needs. In fact, in many countries, there is no social education at all. The so–called education is directed towards producing graduates who are insipid, devoid of roots in their own culture and socially useless. Education today is geared towards a materialistic life style characterized by ever increasing sophistication and extravagance. It is producing tensions, strain and exhaustion even among the relatively affluent section of the people in their pursuit of the means to afford the consumption standards required by this life style. In the resultant competition, those who cannot afford, and therefore, cannot satisfy their needs, basic or imaginary, have little time or patience to think in terms of duty towards community, nation or world. This is even less so for the vast majority of poor whose major concern is food while clothing and shelter occupy only second and third place in their priorities.

Broader Inadequate education (#PF4984).
Narrower Lagging training in social skills (#PF8085).
Aggravated by Absence of tactical methods (#PF0327).

♦ PF2096 Fragmented ethical contexts

Nature The individual's ethical context is conditioned by the present parochial operation of the family and of educational systems. Parents attempt to meet the separate needs of their children, extended families, and organizations without considering the integrating image which could pervade the whole family. Educational systems, while responsible for providing the major part of moral training, attempt to meet the demands of the public and end up by oscillating like a pendulum between different ethical poles or by abdicating responsibility. Neither the family nor the educational systems reappropriates the heritage of the past and the profound relationship to historical stories and symbols which would help sustain actions directed towards present vision or the long–range future.

Broader Fragmentation of religious belief (#PF3404)
Collapse in providing ethical value screens (#PF1114).
Related Ideological conflict (#PF3388).
Aggravated by Double standards in morality (#PF5225)
Minimal church / school involvement (#PJ9011)
Lack of meaningful educational context for ethical decisions (#PF0966).

♦ PF2097 Simplistic family decision–making structures

Nature Family decision–making structures are overly simplistic. They often exclude the children and the old people and may in effect be embodied in one person. They are organized without an inbuilt accountability over a given period of time which would encourage unconscious decision about the joint responsibility of the family. The reduction of decision–making to such an arbitrary scheme inhibits the creativity of family members.

Broader Reduced interior structure of families (#PF3783).

♦ PF2099 Witchcraft

Wicca craft — Sorcery — Witch doctors

Nature Black magic is practised by people unwittingly possessed of little understood, natural or supernatural powers or by those consciously using such powers for selfish or evil purposes.

Background Distinctions have been made between witchcraft and sorcery, for example as they occur in the African Azande tribe and also as practised in 17th century England, but these distinctions are not universally applied. Social anthropologists have suggested that witchcraft depends upon the possession of appropriate powers that transform malevolent desires into reality. Sorcery is distinguished as involving some physical manipulation and its efficacy depends upon learning the appropriate skills or techniques to achieve its ends.

The earliest evidence of witchcraft is found in paleolithic cave drawings dating from 30,000 BC and ranges from the Soviet Union to Spain. Witchcraft practised in Western Christendom may have been a survival of indigenous pagan religion. There are references to witchcraft in the Bible, and in Greek and Roman literature. Belief in witchcraft became widespread in the Middle Ages and resulted in the trial, torture, confessions and burning alive of many women. In Western culture witches have traditionally been women; while in other cultures they may be either women or men, the latter being referred to as warlocks. Sorcerers in Western cultures were traditionally men and were also burnt alive. Witchcraft and sorcery were often alleged against rivals. The last official burnings took place in the early 18th century.

Incidence Witchcraft and sorcery are usually manifested in fear, belief, imputation and accusation rather than in the discovery of rites taking place, since the psychic aspects of witchery, like the evil eye and the laying on of curses, are not necessarily accompanied by ritual actions. The belief in witchcraft is widespread among primitive tribes in Africa, Asia, Australasia and America. Witchcraft is very evident even in the urban areas of developing countries. Witchcraft may be used by football teams to jinx each other. Witchcraft is often suspected in unsolved murders. Men may divorce their wives because they are suspected of being witches. Some traditional healers contribute directly to the deterioration of their patients or merely endeavour to extort money from them (as may be the case with some practitioners of modern medicine).

In South Africa in 1990, it was reported that large sums of money had been offered to black nannies looking after white children to hand them over to witchdoctors to be used in the production of a potion to end the civil war amongst the blacks in Natal. In 1990 in Nigeria it was reported that male genitalia were being spirited away by witchcraft and were being sold, notably to politicians as a way of enhancing their powers. The phenomenon was reportedly taken sufficiently seriously to prevent people from shaking hands (physical contact being one of the means through which the disappearance was effected). In industrialized societies there are still scattered claims. In the 1960s cases of black magic were reported in England; a village milkmaid in the Soviet Union was accused of witchcraft; and beliefs and practices among peasants in Lower Saxony gave rise to a government investigation. In the 1950s, two alleged practitioners of witchcraft were lynched in Queretar, Mexico; in Guatemala, the wife of a political rival of the anticommunist colonel Carlos Castillo Armas was accused of practising witchcraft against the colonel. Witchcult religion survives in France and Italy with witches sabbaths; and in the USA, where California and New York covens celebrate the rites of bell, book and candle. A considerable number of modern books advocating witchcraft have recently appeared in America and Western Europe.

Counter–claim 1. Witches stand in a long tradition as adherents of an ancient religion counter to those artificially created, which springs from an inherent human capability, active in witches, to relate to, understand and use nature and natural forces, although these are unperceived by those who cannot exercise their latent powers in this way. Thus witches and their assemblies, which may include those wishing to develop such abilities, are the custodians at once of an ethic and philosophy of nature, and a knowledge of such things as natural medicine and healing. Witches of the past are credited with having known natural abortifacients and, possibly, contraceptives; and are said to have helped reduce population growth. The renewed interest in the craft is characteristic of an enlightened time in which man and woman have recovered awareness of nature and each other.

2. In developing countries, especially in Africa, traditional healers have much to offer regions deprived of modern medical facilities, especially since there are many more of them and they tend to live in the villages (whereas most doctors and nurses prefer the conveniences of larger towns). In addition, whereas modern doctors are respected for relieving symptoms, many Africans believe that diseases have spiritual roots and that a thorough cure requires a healer in touch with such dimensions. They are therefore often as useful in the treatment of psychosomatic ills as modern therapists. Traditional herbal remedies often anticipate developments in pharmacology. Some African governments therefore work with traditional healers rather than by enforcing laws against them.

Refs Ewen, Cecil H *Witchcraft and Demonianism*; Graubard, Mark *Witchcraft and the Nature of Man* (1985); Lea, Henry C and Burr, George L *Materials Toward a History of Witchcraft*; Marwick, Max G *Witchcraft and Sorcery* (1987); Masters, R E *Eros and Evil*; Melton, J Gordon *Magic, Witchcraft and Paganism in America* (1982).

Broader Totemism (#PF3421) Occultism (#PF3312) Superstition (#PA0430).
Narrower Magic (#PF3311) Occult rites (#PG4311).
Related Satanism (#PF8260) Necromancy (#PF8042)
Lack of understanding of spiritual healing (#PF0761).
Aggravates Fear (#PA6030) Incapacity (#PJ4312) Human death (#PA0072)
Bewitchment (#PF3956) Imprecation (#PF3746) Changelings (#PE9453)
Affliction by malevolent spirits (#PF9043).
Aggravated by Tribalism (#PC1910).

♦ PF2100 Divorce

Marriage annulment — Repeated marriage divorces — Serial polygamy

Nature The annulment of a valid marriage may be for reasons other than the death of a spouse. While divorce is legal in many countries it is illegal in others, particularly as a result of religious doctrine, which may mean that a divorced person is denied rights within that religion and his or her children by another marriage may be regarded as illegitimate. Alimony or maintenance for women and children resulting from divorce may be insufficient or may impoverish the husband. Marital property may be unequally divided in divorce. The question of the custody of children may cause much bitterness with adverse effects on the children, or it may impose a heavy burden on one parent. Divorcees may find that they are discriminated against by society. If divorce proceedings are lengthy, the cost may be very high.

Incidence In the developed countries of Europe, North America, Australia, Japan, New Zealand and the USSR, divorce is a clear concept and the increase in its frequency is unquestionable; this has, in fact, been taking place since the beginning of the century. At present, in these regions, between one marriage out of two and one marriage out of four ends in divorce. Many young couples separate after having established a common household for several years, and for the individuals concerned there is not much difference between such separation and a legally sanctioned divorce. It is likely, therefore, that more than half of the men and women who are presently living together will not maintain this union through their lives. Divorces occur mainly at an early age. In the United States, for example, 52 percent of all divorces take place before the sixth year and 67 percent before the tenth year of the marriage; 45 percent of the couples who divorce have no children and 25 percent have one child. However, throughout the industrialized world, the number of dependent children who face divorce between their parents is large and continues to grow. These children generally stay with their mother, although an aspect of new divorce laws providing for greater equality between men and women is that men are less systematically considered as unable to provide homes for their children.

In Africa and Latin America, where marriage is early and universal but where the consensual and customary types are prevalent and also where, in Africa, polygamy is still frequent though declining, data on divorce rates are scanty and in many cases concern a fraction of the population only. The information provided in the context of the World Fertility Survey show that in five countries of Latin America, the proportion of first marriages dissolved by separation varies between 20 and 80 percent. A high marital instability is also reported in the urban areas of African countries, where sometimes a quarter of all women marry more than once.

Data from the World Fertility Survey also indicate that in Asia, the proportion of women whose first marriage or union is dissolved by separation or divorce amounts to 36 percent in Indonesia, 21 percent in Bangladesh and 18 percent in Thailand. In other Asian countries, a greater proportion of marriages are terminated by widowhood.

While divorce is likely to always be a part of society, it has a number of consequences. Children of divorced parents are not educated as well as their parents. Children and women of divorces are generally poorer. Each of the participants in a divorce is frequently emotionally hurt by the process. The violence inherent in a divorce causes profound spiritual scars and social difficulties for all involved.

Counter–claim In the case of women, divorce often leads to a huge increase in motivation and achievement, providing them with a spur to becoming independent, career–minded high–achievers.

Refs Belli, Melvin and Krantzler, Mel *Divorcing* (1988); Bontrager, G Edwin *Divorce and the Faithful Church* (1978); Chester, R *Divorce in Europe* (1977); Diamond, Susan A *Helping Children of Divorce* (1986); Dumon, Wilfried and Paepa, Chrisiane de (Eds) *International CFR Seminar on Divorce and Remarriage (19,1981, Leuven)* (1981); Gibbons, Alice M *Divorce and Divorce Factors* (1987); Oakland, Thomas P and Terry, Edwin J *Divorced Fathers* (1983); Peterson, Richard R *Women, Work, and Divorce* (1988); Phillips, Roderick *Putting Asunder* (Date not set); Pothen, S *Divorce* (1986); Wolchik, Sharlene A and Karoly, Paul (Eds) *Children of Divorce* (1988).

Broader Personal life crises (#PD4840) Voluntary dissolution of the family (#PF4930)
Vulnerability of marriage as an institution (#PF1870).
Narrower Non–validity of divorce (#PF3250) Separation under marriage law (#PF3251)
Inadequate provision of alimony (#PE3247) Non–parental custody of children (#PF3253)
Conflicting national divorce laws (#PF3249) Religious or civil refusal of divorce (#PF3248).
Related Non–validity of marriage (#PF3283).
Aggravates Social stigma (#PD0884) Family breakdown (#PC2102)
Marital instability (#PD2103) Religious intolerance (#PC1808)
Social discrimination (#PC1864) Single parent families (#PD2681)
Inadequate laws of adoption (#PD0590) Discrimination against divorcees (#PJ3376)
Discrimination against unmarried fathers (#PD3256).
Aggravated by Adultery (#PF2314) Immorality (#PA3369)
Promiscuity (#PC0745) Alcohol abuse (#PD0153)
Permissiveness (#PF1252) Sexual unfulfilment (#PF3260)
Parental domination (#PF4391) Social disintegration (#PC3309)

Desertion in marriage law (#PF3254) Sexually transmitted diseases (#PD0061)
Double standards of sexual morality (#PF3259) Negative effects of the nuclear family (#PF0129)
Psychological inconsistency of marriage partners (#PF9818).
Reduced by Polygamy (#PD2184).

♦ **PF2105 Governmental barriers to a global ethic**
Claim Wider concepts of ethical relations are inhibited by governmental institutions, which may act as barriers to a wider understanding of globality, and prevent communities from participating in a global society. Governments continue to misunderstand the global nature of reality in various ways, as through limited foreign policies, high defence priorities, censored news releases, immigration quotas, and conditioned responses to the global demands on food production.
Broader Collapse in providing ethical value screens (#PF1114).
Aggravates Lack of international cooperation (#PF0817).

♦ **PF2108 Megalomania**
Nature A condition in which the person considers himself possessed of greatness.
Broader Paranoia (#PE0435) Demagoguery (#PC2372).
Aggravates Political dictatorship (#PC0845).

♦ **PF2110 Individualistic style of professionals**
Claim The existing understanding of both professionals and of the public requiring their services is that a one-to-one relationship is the appropriate mode of operating. People with special skills also see themselves as individualists who should be free to respond to their creative whims. School and professional accreditation systems, by emphasizing individual examinations and skills (to the neglect of team or cooperative skills) reinforce this stance.
Broader Individualistic utilization of expertise (#PF5639).

♦ **PF2112 Dehumanized individual scientific research**
Claim The scientific community has become insulated from the demands of society through its emphasis on specific research, whose results are often unrelated to life's basic problems. Theoretical and abstract research particularly reflect the lack of consciousness of the role of the scientific community in the social process, acting in response to the priorities of the select few rather than to the needs of society as a whole.
Broader Limits on areas of research (#PF2529) Inhumane scientific activity (#PC1449).
Aggravates Dehumanization of man in the technological process (#PF5438).

♦ **PF2113 Inadequate international judicial system**
Ineffective administration of international justice
Nature The International Court of Justice is ineffective, largely due to the sensitive nature of cases brought before it. In many cases, effective litigation would require the disclosure of vital national security information, which most nations are reluctant to do.
Incidence In 1962, the Soviet Union refused to comply with the Court's judgement that all members must pay their assessed share of UN peacekeeping operations. In 1973, France refused to appear before the Court on charges from Australia and New Zealand that her nuclear devices testing in the South Pacific was unlawful. In 1985, the USA refused to participate in the Court on charges brought against her by Nicaragua.
Background Only a minority of UN members have, in the history of the International Court, accepted its compulsory jurisdiction; and then only on matters which did not jeopardize their national security. Of the 15 permanent World Court judges who decided, in 1985, that the Court had the jurisdiction to try Nicaragua's claim against the USA, 10 were from countries which themselves had refused to accept the compulsory jurisdiction of the Court.
Claim The refusal of nations to appear before the Court, even though it is a UN body which member nations have pledged to support, is an active example of the ineffectiveness of UN bodies caused by the arrogance of its member States.
Counter-claim Litigation involves interpretations of law and determinations of fact. Both of these require fact-finding proceedings which are virtually impossible without the jeopardizing of individual national security and prestige.
Broader Injustice (#PA6486).
Aggravates Underdevelopment (#PB0206) International aggression (#PB0968).
Aggravated by Governmental disregard for international values and procedures (#PF0292).

♦ **PF2115 Lack of essential local infrastructure**
Lack of essential services and materials needed for development
Nature Rural communities tend to receive only very limited provision of the services which are now a necessity for participation in contemporary society. Electricity is costly and may be available only privately rather than on a general, domestic basis. Water may be distributed, but it is usually untested and may be contaminated with parasites. Although a community may be aware that all forms of development depend on materials and equipment, and although natural resources abound for the production of building materials, the cost of supplying, manufacturing and transporting such materials may be beyond its reach. Programmes necessary for the improvement of community life are therefore held up. Stagnant open pools of waste water remain around most outdoor washing areas, and homes have open pits for garbage. Inoperative health outposts, distant medical services and inaccessible dental care facilities drain the vitality of the people and reinforce the life style of backward isolation. In addition, the daily expenditure of energy required to function with the current rudimentary form of services severely minimizes productive output.
Broader Inadequate infrastructure (#PC7693).
Narrower Stagnant surface water (#PE2634) Inadequate medical care (#PF4832)
Limited meeting facilities (#PE1535) Lack of transport vehicles (#PF0713)
Inadequate irrigation system (#PD8839) Inadequate electricity supply (#PJ0641)
Lack of agricultural machinery (#PF4108) Prohibitive cost of electricity (#PG8518)
Insufficient emergency services (#PF9007) Inadequate maintenance equipment (#PD6520)
Physically inaccessible services (#PC7674) Unidentifiable urban neighbourhoods (#PF6147)
Unavailability of building materials (#PJ0704)
Inadequate and insufficient immunization (#PF5969)
Insufficient transportation infrastructure (#PF1495)
Lack of housing for teachers in rural communities (#PE8103)
Limited mechanical services in developing countries (#PE8986)
Prohibitive construction cost of rural community buildings (#PJ6368).
Related Prohibitive cost of fuel (#PJ0346).
Aggravates Inadequate meeting methods (#PF8939)
Water system contamination (#PD8122).
Aggravated by Inadequate educational facilities (#PD0847).

♦ **PF2120 Imbalance in city sizes within a country**
Nature In a number of South Asian and Latin American countries, and also in many developed countries, the leading city is m..y times larger than any other city of the same country. Such metropolises constitute too great a concentration of the country's human resources and productive potential. The tendency for one city to remain supreme is usually reinforced by various types of inertia. A migratory stream, once established, is likely to continue in the form of chain migration. New investments are usually made in an environment of already existing specialized industries; and superior educational institutions are to be found in such cities. Entrepreneurs, doctors, teachers and other specialists are unwilling to leave a leading city for some other, far less diversified, environment. In many countries, therefore, qualified services are highly concentrated in one big city, to the detriment of the rest of the country. Comparable initiatives are unlikely to arise in other cities. Once a leading city has outdistanced others, its pre-eminence tends to be further increased by all subsequent developments.
The social ills prevailing in big cities of the less developed regions are particularly striking because of their concentrated form. It has been variously estimated that in the big cities of Asia and Latin America up to 30 or 40 per cent of the population is that of transitional settlements or shanty-towns. Despite governmental efforts at slum clearance or relocation, the shanty-town population appears to be growing faster than the remainder of the urban population. If trends are allowed to continue, an increasing proportion of the national populations may accumulate in the shanty-towns of metropolises.
Broader Uncontrolled industrialization (#PB1845).
Related Locational maladjustments of industry in developing countries (#PD1494).
Aggravated by Urban slums in developing countries (#PD3489).

♦ **PF2124 Ineffective tax systems in developing countries**
Inadequate fiscal discipline in developing countries — Dissipated fiscal base
Nature On average, about 70 percent or more of the tax revenue of the majority of developing countries come from indirect taxes (customs duties, for example). Since many indirect taxes tend to be both regressive in incidence and inelastic in yield as income rises, their preponderance in the tax structure calls for frequent revisions of the basic rate structure if the share of tax revenue in total income is to be maintained, for which revisions the developing countries are administratively ill equipped. In addition, although most developing countries levy direct taxes on income, defective legislation and inefficient administration mean that tax collections fall far short of amounts due. In many countries there is no single comprehensive tax on all income except the regular system of income taxation, which leaves important sources untaxed.
Broader Ineffective tax systems (#PF1462) Archaic marketing methods (#PF6465)
Depleted expertise of the rural labour force (#PF2973).
Related Inefficient tax administration in developing countries (#PE5009).
Aggravates Tax evasion (#PD1466).
Aggravated by Inefficient public administration in developing countries (#PF0903).

♦ **PF2126 Untransferability of books between countries and cultures**
Untransferability of images between countries and cultures
Nature In the past, many textbooks designed for use in one country have been introduced wholesale into another without reference to differences such as linguistic, social, and cultural. The books so introduced are often misleading, irrelevant or even offensive in the new context.
Incidence The problem is especially evident in the transfer of books from Western to non-Western cultures and between non-Western cultures. It occurs both with scientific textbooks, which may make allusions to phenomena outside the general knowledge of the reader in the receiving culture, and with literature such as folk tales.
A British poster picturing a "pregnant" man and encouraging the use of contraception by men was beyond the imagination of the people in Bangladesh: they assumed the man in the picture had worms.
Broader Cultural barriers (#PB2331) Inter-cultural misunderstanding (#PF3340).
Aggravates Inaccessibility of knowledge (#PF1953)
Lack of international cooperation (#PF0817).
Aggravated by Cultural isolation (#PC3943)
Lack of appreciation of cultural differences (#PF2679).

♦ **PF2128 Unorganized development of work forces**
Claim Motivation for corporate community efforts is often not focused on the task in hand because the major part of each individual's efforts is required to meet physical needs in the family setting. Despite awareness that development requires a corporate effort, until communities develop the full potential of their work-forces much needed public projects may remain uncompleted.
Broader Unformed structures of community organization (#PF2810).
Narrower Irregular working hours (#PJ5329) Breakdown of police protection (#PF8652).
Related Social unaccountability (#PC1522)
Minimal opportunities for corporate activities (#PF2316)
Decline in communal spirit and village solidarity in developing countries (#PE0835).
Aggravates Inadequate road maintenance (#PD8557)
Unfocused style of community operations (#PF6559)
Lack of responsible involvement in community affairs (#PF6536).
Aggravated by Demeaning community self-image (#PF2093)
Fragmented pattern of community organization (#PF6525)
Ineffective structures for community participation (#PF2437)
Ineffective structures for community decision-making (#PF1781).

♦ **PF2132 Lack of assimilation**
Nature Lack of assimilation perpetuates ethnic and social differences and discrimination, and makes conflict more likely. It also leaves the way open for subversive activities or foreign intervention to protect an oppressed group. Assimilation involves the adoption of 'national' characteristics by different racial and ethnic groups and the elimination of racial and ethnic differences in the social structure, which may be very difficult to achieve and not desirable in so far as it involves a loss of cultural heritage.
Lack of assimilation may occur wherever there are minority groupings of a racial, religious, ideological or linguistic kind. The groups may be immigrant or indigenous. It is aggravated by social inequalities and exploitation which create ethnic conflict. Since ethnic differences may be very great, assimilation means that at least one group must give up some of its customs, which it may be very unwilling to do. Where there is extensive segregation and little intermarriage, ethnic purity is likely to remain intact and prejudices and discrimination persist.
Counter-claim Assimilation may mean the domination of one culture within the nation or a fusion of cultures which is often unacceptable to minorities.
Broader Social conflict (#PC0137) Social fragmentation (#PF1324).
Narrower Fluctuations in real value of money (#PD9356)
Inadequate cultural integration of immigrants (#PC1532).
Related Cultural fragmentation (#PF0536) Insufficient minority culture support (#PF5659).
Aggravates Taboo (#PF3310) Traditionalism (#PF2676) Social injustice (#PC0797)
Social inequality (#PB0514) Subversive activities (#PD0557)
Lack of national unity (#PF8107) Lack of social mobility (#PF2195)
Plural society tensions (#PF2448) Inter-cultural misunderstanding (#PF3340).
Aggravated by Elitism (#PA1387) Segregation (#PC0031)
Racial conflict (#PC3684) Ethnic conflict (#PC3685)
Religious conflict (#PC3292) Ethnic segregation (#PC3315)
Ethnic discrimination (#PC3686)
Inadequate education of indigenous peoples (#PC3322).
Reduces Diverse unilingualism (#PF3317) Destruction of cultural heritage (#PC2114)
Underprivileged linguistic minorities (#PC3324).
Reduced by Miscegenation (#PC1523) Lack of racial identity (#PF0684).

♦ **PF2135 Non-diversification in subsistence fishing economies**
Nature Although the diversification of production potentials at the local level could provide an adequate income to the community, many fishing villages achieve unacceptably low levels of

production, with declining fishing populations and low-yield farm lands. These villages rely on the fishing industry to provide the community with immediate cash income on a day-to-day basis. As a result, the concept of planned savings is not part of community life, and excess cash is usually spent on immediate needs: concern for the present far outweighs the desire for an emergency store of cash. When fishing equipment, repairs and expansion cannot be funded immediately there is consequently little opportunity to increase production. Inefficient methods and lack of repair results in a daily routine for fishing families which leaves little time for recreation, as any deviation jeopardizes that day's much needed income. Child labour makes for little chance of any educational advancement within the family which could secure a higher level of income. High retraining expenses combined with few practically located training possibilities further discourage any additional income options. Those villagers who do manage to overcome this obstacle usually leave the community, since local jobs offer limited prospects and they wish to pursue jobs with higher income potential.
Broader Undiversified economies of developing countries (#PD2892).
Narrower Small farm plots (#PG9512)   Inadequate cash flow (#PJ0576)
Dependence on children (#PG9631)   Unstable fishing season (#PJ9570)
Questionable economic viability (#PG8186)
Inadequate income of rural communities (#PJ9649).
Related Low general income (#PD8568)
Underdeveloped capacity for income farming (#PF1240).
Aggravates Decline in government health expenditure (#PF4586).
Aggravated by Limited employment options (#PF1658)
Lack of savings structures (#PF1348)
Prohibitive cost of education (#PF4375).

♦ **PF2136 Feudalism**
Feudal social order
Nature This economic and political system of land tenure, whereby the landowner lets out land in return for service from his tenants or peasants, may arise out of dispossessing indigenous populations, and results in the pauperization and effective slavery of the latter. Where feudalism remains from the traditional social structure, development takes place at the expense of the peasants who may be used as cheap or free labour in factories and mines instead of on the land.
Incidence Feudalism still exists in developing non-Communist countries, especially in Latin America and Asia.
Counter-claim Modernization theory is mistaken in perceiving developing countries as having a predominantly feudal social order resistant to the advance of modern or capitalist economic systems. Such apparently feudal structures should be viewed as capitalist. The stagnant and oppressive relations evident on plantations, for example, are best understood as aspects of the underdevelopment of the periphery created by capitalist development for the benefit of the industrialized centre.
Refs Critchley, John *Feudalism* (1977); Duby, Georges; Goldhammer, Arthur and Bisson, Thomas N *The Three Orders* (1980); Lyon, Bryce D *From Fief to Indenture* (1971).
Broader Social inequality (#PB0514)
Disparity in social development within developing countries (#PD0266).
Related Forced labour (#PC0746)   Exploitation in employment (#PC3297).
Aggravates Apathy (#PA2360)   Elitism (#PA1387)   Slavery (#PC0146)
Extra-economic constraints (#PE7784).
Aggravated by Colonialism (#PC0798)   Fraudulent nature of inherited titles (#PE5754).

♦ **PF2137 Gambling and wagering**
Betting and gaming
Nature Gambling is the betting or staking of something of value on the outcome of a game or contest or uncertain event with awareness of the risk and in the hope of gain. It varies from lottery tickets and the betting of small sums of money by people who have little, to the sophisticated casino gambling of the wealthy, either for profit (if they are skilful) or as a pastime. Whether legal or not, gambling is not usually regarded as socially admirable. It can impoverish families or keep families impoverished; it may lead to blackmail and is often controlled by organized crime. Where there is gambling, there is usually also cheating (if only for a casino to recover or stave-off losses to a skilful gambler) and there may also be bribery (in sports), or doping or even sabotage to ensure winning a bet. This may lead to violent recrimination, even murder and vendetta. Extensive gambling concerns may involve the corruption of public officials.
Incidence Petty gambling in the sense of lotteries, bookmaking on horse racing and other sports and on political elections is widespread. Some 20 million French people play the lottery at least once a year. Casino gambling and slot machines may be less acceptable and may be legal or not, according to the laws of individual countries. In one year the French gamble Ffr 48,000,000,000 in Loto, PMU, horse racing and casinos. Americans wagered at least $241 billion legally and illegally in 1988, almost $1,000 was bet for every man, woman and child.
Background Gambling has existed in every known society from the most primitive to the most complex. Dice games and guessing games are recorded in Stone Age cultures, among Bushmen of South Africa, Australian aborigines and American indians. Dice dating from 3000 BC were found in an Egyptian tomb; a gaming board is cut into a step to the Acropolis in Athens; and there is abundant evidence of gambling in the decadent era of the Roman Empire. Gambling fashions change with the times and odds are that new forms will continue to occupy generations to come.
Counter-claim Gambling may be legalized or partially legalized and provide useful government revenue. UK excise duties on gambling equal £600 million annually. The Japanese government taxes legal gambling at 25 percent and gives the revenue to various local enterprises. In the past few years, government taxes on public gambling have totalled about one trillion yen, about $4.5 billion, annually.
Refs Church, Diana et al *Gambling* (1988); Eadington, William R (Ed) *Gambling Papers* (1985); Eadington, William R *The Gambling Papers* (1982); Rose, I Nelson *Gambling and the Law* (1985).
Broader Victimless crime (#PC5005)   Corruptive crimes (#PD8679)
Offences against public order (#PD7520).
Narrower Abuse of status of religious institutions (#PJ1156)
Gambling on sports and athletic competitions (#PE4576).
Related Decadence (#PB2542).
Aggravates Animal fighting sports (#PE4893).
Aggravated by Boredom (#PA7365)   Organized crime (#PC2343)
Immoral public policy (#PF4753)   Socio-economic poverty (#PB0388).

♦ **PF2139 Privatization of intergovernmental monetary system**
Nature The international monetary and financial system has become increasingly privatized. It was designed originally to allow the expansion of private transactions that accompanies a vigorous and expanding world economy. However, the official framework and underpinning of the system have failed to keep step with the total volume of transactions, so that the capacity of governments to influence variables such as exchange rates, interest rates, and the magnitude and direction of international capital flows, has progressively diminished. At the same time, such important variables as levels of liquidity, patterns of expenditure, and the degree of expenditure expansion and contraction in the system as a whole, are determined primarily, and at times exclusively, by private decisions. Thus, governments, most of which would not contemplate being without influence on such matters in their own domestic economies, have allowed themselves to lose influence on these variables in the international sphere. This, in turn, has reduced their capacity to shape the evolution of their own national economies.
The process of privatization, including, in particular, the growth of Eurocurrency lending, has had a major impact on developing countries. Private flows other than direct foreign investment have risen from less than 20 percent to more than 40 percent of the long-term financing of the current account deficits of non-oil developing countries, while the share of official flows fell from 60 percent to 40 percent. This new pattern of financial flows meant that those developing countries deemed creditworthy by private capital markets were able in the mid-1970s to secure a substantial volume of external financing and to pursue adjustment policies requiring large sums of new investment and the phasing of adjustment over an extended period of time. But the many developing countries that commanded little or no access to private capital markets, such as the least developed and most other low-income countries, continued to depend on the Bretton Woods institutions, regional development banks, and bilateral donors for payments and development financing. Flows from such sources grew in the 1970s and eased the adjustment process; but they failed to keep abreast of needs, thereby severely constricting the policy choices open to those countries. Indeed, a shortage of foreign exchange confronts many low-income countries with the choice of either abandoning projects under way or lowering capacity utilization.
Recently, flows from private capital markets have fallen off; indeed, banks have been attempting to reduce their exposure in a number of developing countries through such means as reducing short-term credit lines, thereby accentuating the liquidity, and financing shortage. The functioning of the private capital markets has thus made all groups of developing countries increasingly dependent on the official institutions. The size and policies of these institutions have become of key importance in determining the development prospects of all categories of developing countries.
Broader Inadequacies of the international monetary system (#PF0048).

♦ **PF2141 National isolationism**
International isolationism — Diplomatic isolationism — National isolation
Nature Any country or region that adopts a strategy of reducing its dependence on world or regional trade or on economic and military alliances may be isolationist. It may not take any role in the deliberations of international organizations, although nominally it might be a member. It may not honour previously signed mutual-defence treaties. In another sense, an isolationist policy is one that results in the world knowing that a nation or region has no other interest than its own. It pursues this unswervingly, manipulating whatever international interactions it has to serve this end.
Incidence The USA, between the two world wars, was isolationist; and this contributed to the failure of the League of Nations. The Republic of China in Taiwan has elected a degree of political isolationism by only maintaining diplomatic relations with a limited number of nations. There are many kinds of isolationism and many degrees. The word has also been associated with Eurocentric attitudes, particularly in the EEC.
Claim The countries of Eastern Europe during their period of socialism were isolated from the mainstream of world civilization.
Refs Dowty, Alan *Closed Borders* (1987).
Broader Isolation (#PB8685)   Non-alignment (#PF0801)   Political isolation (#PC7569).
Narrower Fragmentation of academic disciplines (#PF8868)
Domination of individuals by institutions (#PF4220)
Concentration of national governments activity on national affairs (#PE5132).
Related Neutrality (#PF0473)   Economic isolationism (#PC2791)
Failure of individuals to participate in social processes (#PF0749).
Aggravates Fear (#PA6030)   Nationalism (#PB0534).
Reduces Economic dependence (#PF0841)
Foreign intervention in internal affairs of states (#PC3185).

♦ **PF2142 Immortality**
Refs Badham, Paul and Badham, Linda *Death and Immortality in the Religions of the World* (1987); Humphreys, S C and King, H *Mortality and Immortality* (1982).
Related Excessive longevity (#PJ5973).

♦ **PF2146 Intellectualism**
Nature Intellectualism values reason or rationality above emotion. It prefers ideas to people. Intellectuals form part of an elite in 'Western' civilization and in countries influenced by it, where the systems and policies they evolve control the lives of ordinary people. These policies may be impractical and abstract and may not take enough consideration of majority needs, particularly of an emotional kind. The weight of intellectual opinion may be evaluated on the strength of paper qualifications rather than practical experience. Intellectuals anxious to promote the redeeming, transcending Truth, the establishment of which they see as their mission on behalf of humanity, have little patience with the mundane, everyday truth represented by objective facts which get in the way of their arguments. These awkward, minor truths get brushed aside, doctored, reversed or are even deliberately suppressed.
Claim Intellectualism is a universal and increasing problem. It is most prevalent in industrialized societies and in government controlled societies, and is increasing in developing countries where planning, especially along 'Western' lines, is becoming more pervasive. Strictly Communist societies, such as China, temper intellectualism by obliging people who work in planning and other intellectual fields to intersperse it with periods of purely practical and mundane work.
Broader Elitism (#PA1387)   Rationalism (#PF3400).
Narrower Meritocracy (#PG3637)   Intellectual prejudice (#PC3406).
Related Moralism (#PF3379)   Snobbery (#PJ3943)   Imbalance of power (#PB1969)
Repression of intellectual dissidents (#PD0434).
Aggravates Class consciousness (#PC3458)   Dependence on religion (#PF0150)
Inadequacy of doctrine (#PF3396).

♦ **PF2148 Incest**
Incest pregnancy
Nature Incest involves sexual relations between parents and children or between siblings. Other relatives, such as first cousins, sisters-in-law or brothers-in-law may be included, depending on the law of the country involved. Incest may include homosexual or lesbian incest. Incest may cause genetic defects in any children of such a union; and in many societies where incest is taboo, its occurrence may cause family disintegration.
Incidence Incest is taboo in most societies, primitive and developed. Where it does occur, it is usually under special circumstances, such as in the marriage of royalty (Azande tribe, Africa), or the treatment of incest as a sacred ritual act, or the taking of a forbidden relative as a sexual object by a tyrant to show his power. Other kinds of incest may occur clandestinely. In practice incest is almost always the sexual exploitation of a child by an older relative, usually by an adult male to a young female, often father to daughter. In the US father-daughter incest accounts for three quarters of incest cases.
Background Incest was common among royal families, for example, in Egypt, Hawaii, Peru, Europe) and sometimes led to severe genetic defects (such as with the Hapsburgs). If marriage between first cousins is considered incest, then it could be said to have been common among

the European aristocracy until the 19th century. In English law it was not a crime until 1908.
Refs De Young, Mary *Incest* (1985); Goodwin, Jean M *Sexual Abuse* (1988); Herman, Judith *Father–Daughter Incest* (1982); Renvoize, Jean *Incest* (1985); Shepher, Joseph *Incest* (1983); Twitchell, James B *Forbidden Partners* (1986); White, William L and Hagen, Russell J *Incest in the Organizational Family* (1986).
Broader Sexual deviation (#PD2198) Sexual exploitation of children (#PD3267).
Narrower Lesbian incest (#PG4347) Homosexual incest (#PG4346).
Related Sexual offences (#PD4082) Sexual immorality (#PF2687) Consanguineous marriage (#PC2379) Sexual abuse of children (#PE3265).
Aggravates Infamy (#PB8172) Infanticide (#PD3501) Family breakdown (#PC2102) Illegitimate children (#PC1874) Genetic defects and diseases (#PD2389).
Aggravated by Taboo (#PF3310) Genetic sexual attraction (#PE5925).

♦ PF2152 Inappropriate arguments

Fallacious arguments — Cheap argument — Bad arguments — Improper forms of debate — Fatuous reasoning — Flaky arguments — Polarizing rhetoric — Obfuscatory arguments — Misuse of plausible arguments to conceal nefarious benefits
Nature A well–defined group of fallacious arguments may be used to promote or oppose new projects. They may include: fallacies of logic or syntax; arguments from authority; posing of an incorrect question; suppression of relevant information; utilization of ambiguous terms; omission of a relevant concomitant variable; use of biased samples; use of samples which are statistically too small to give meaningful results; misleading use of graphs and pictures; and dependence on erroneous presentations of statistical data. Such methods of argument may over–simplify, over–complicate, ignore the opposing position, accept the authoritative position too readily, accept the position because it is detailed in print, place excessive reliance upon some formal system or formula or reject such a system too readily.
Counter–claim The assumption that truth depends on logical reasoning causes premature and unreasoned rejection of proposals whose justification has been presented in an illogical manner. Fallacy and falsity are two different things. Conclusions may state a truth despite faulty argumentation. For example: only fixed stars are stationary in space; the Sun is not a fixed star therefore it is moving. Likewise, approved, reasoned arguments have 'proven' that human travel through the air is impossible; that there are superior and inferior races; that God is Three Persons, Father, Son, and Holy Spirit; and that there is no God the Mother.
Broader Human errors and miscalculations (#PF3702) General obstacles to problem alleviation (#PF0631).
Narrower Defective reasoning (#PF5711) Terminological ambiguities (#PG4352) Suppression of information (#PD9146).
Aggravates Misleading information (#PF3096) Polarization of local conflicts (#PF1333) Impoverishment of political debate (#PF4600).
Aggravated by Oversimplification (#PF8455) Manipulation of debates (#PD4060) Withholding of information (#PF8536) Rhetorical inflation in meetings (#PF3756) Uncritical acceptance of authority (#PF8596) Misrepresentation of historical persons (#PF5662) Inconclusiveness of scientific and medical tests (#PD7415).

♦ PF2158 Non–concerned attitudes

Nature The development of public conscience and the voicing of responsible public concern may be inhibited by a pervading attitude that one is not bound to one's neighbour and should not take risks on his behalf.
Broader Lack of care (#PF4646) Legal contract system reduced to individual needs (#PE5397).
Narrower Lack of commitment (#PF1729) Non–globalized citizenship (#PF2835) Paralysis of social response (#PF2701) Riskless responses to public injustices (#PF7722).
Related Lack of work commitment (#PD2790) Parochial national interests (#PF2600) Misuse of international forums (#PF2216) Structural failure of citizen participation (#PF2347).

♦ PF2160 Conservatism

Nature Conservatism is a term commonly used in politics to denote a preference for the old and tried in the civil social order rather than the new and untried; by extension it has come to mean a preference for the old and tried in all categories of social and individual existence without particular reference to politics. It differs from traditionalism in so far as it does not depend on an ethnic heritage. It constitutes a barrier to social progress and modernization. In politics it may cause a barrier to legal reform or lead to half–hearted measures as a result of caution. Conservatism may lead to fascism under extreme circumstances. In religion, conservatism is associated with objections to ecumenicalism and to tolerance for other ideologies. Religious bodies, like governmental legislatures, are fragmented into conservative and other parties; to the right, arch–conservatives and to their right, fundamentalists (in the political, comparable position are the fascists); to the left are the neo–conservatives and to their left are the centrists, middle–of–the roaders, the 'blessed peace–makers'.
Incidence Conservatism is inherent in almost all societies. It is alleged that as people get older they get more conservative. There is the example of 74 year–old conservative President Ronald Reagan who admitted he was 'a bleeding–heart liberal' when younger. However, there is also the example of Pope John XXIII who, on reaching the papacy at 77, inaugurated changes in Roman Catholicism that shook the arch–conservative Roman Curia and changed the world for millions of people.
Claim Perhaps the conservative instinct is to preserve the essence of all the good things. Since few can agree on what this essence is, the conservative can more readily focus, when he wishes to take action, on the forms of the things and the structures. The conservative, wishing to have better room illumination, improves the gas lamp; the progressive invents electricity.
Broader Ideological conflict (#PF3388).
Narrower Reactionary forces (#PB6332) Inadequate social reform (#PF0677).
Related Traditionalism (#PF2676).
Aggravates Social conflict (#PC0137) Restriction of freedom of expression (#PC2162).
Aggravated by Destiny (#PF3111) Conformism (#PB3407) Double standards in morality (#PF5225).
Reduces Political instability (#PC2677).

♦ PF2164 Underdeveloped use of agricultural resources

Unused agricultural resources — Lack of agricultural development
Nature In many rural areas, even with plentiful rainfall, dependence on subsistence farming has resulted in only short–range agricultural planning. With existing traditional farming methods, it is difficult to meet family food requirements. Continuous soil erosion, limited means of fertilization, inadequately cultivated fields, and uncontrolled grazing practices all work to create minimally–productive farmland. Farmers tend to have little experience in poultry–keeping or in mechanized farming methods because of the inaccessibility of training centres where they might learn modern farming methods.
Broader Underdeveloped sources of income expansion (#PF1345) Limited local markets for goods and services (#PF2989).
Narrower Inadequate grazing land (#PJ0404) Underutilized animal genetic resources (#PD7526).
Related Unproductive subsistence agriculture (#PC0492).
Aggravates Stagnated development of agricultural production (#PD1285).
Aggravated by Animal diseases (#PC0952) Epizootic diseases (#PD2734) African horse sickness (#PE1805).

♦ PF2166 Abuse of credit

Credit risk — Bad debts — High credit risks
Broader Risk (#PF7580) Economic crime (#PC5624) Violations against economic regulations (#PD7438).
Related Immorality (#PA3369) Breach of contract (#PE5762) Subsistence approach to capital resources (#PF6530).
Aggravates Infringement of privacy (#PB0284) Cyclic business recessions (#PF1277) Risk of capital investment (#PF6572) Exploitation in rural pricing (#PG8423).
Aggravated by Business bankruptcy (#PD2591) Creative accounting (#PE6093) Socio–economic poverty (#PB0388).

♦ PF2168 Debilitating content of village story

Nature As the future of rural village life becomes less secure, the question of the essential worth and historical role of the community needs forceful consideration. Traditional modes may still be strictly adhered to, but people are aware that these may be blocking the inflow of diverse attitudes necessary for community development. As a result, unprofitable attitudes may continue: for example, fishing may dominate community life, and its immediate necessity far prevail over any future worth to be found in education, thus giving the young people little incentive to complete their schooling; and villagers may view themselves as inferior to their leaders, and be without self–confidence in their own leadership abilities, thus becoming dependent on outside aid. At the same time, rich cultural heritage may quickly be forgotten; many young people regarding the cultural traditions, such as traditional dancing, as merely being old–fashioned.
Broader Detrimental story of community future (#PF6575).
Narrower Cultural dances dated (#PG9603) Lack of family planning (#PF0148) Crop damage by wildlife (#PC3150) Disoriented habitual modes (#PG9595) Insufficient educational incentives (#PJ9811).
Related Unrepresentative formal leaders (#PJ1823).
Aggravates Cultural traditions blocking business profit (#PF9529).
Aggravated by Fear of police (#PF8378) Inflexible educational system (#PJ5365).

♦ PF2170 Monarchy

Monocracy
Nature The rule of a country by a single person with direct or indirect control of the economic, political, judicial and priestly functions. The ruler is generally regarded as of divine origin, and their acts invested with divine qualities. They often have taboos associated with their persons, to touch them may be treason. Their power is frequently related to successful crops and wars. At the heart of their power is military leadership.
Claim Monarchy inculcates unthinking credulity and servility. It is an obstacle to objective public discussion. The system of social distinction and hierarchy depends on its existence. It keeps alive the absurdity of the hereditary principle.
Counter–claim Constitutional monarchies are some of the most enlightened and progressive governments in the world today, such as the Dutch, Belgian, English, Spanish and Danish monarchies. Monarch is a symbol of the nation who stands above its political divisions. He or she is also seen as the ultimate guarantor of the nation and its liberties and the symbol of the continuity of the nation's history.
Broader Undemocratic political organization (#PC1015).
Aggravates Corruption (#PA1986) Personality cults (#PC1123) Displaced royalty (#PF5943).
Aggravated by Destiny (#PF3111).
Reduced by Revolution (#PA5901).

♦ PF2174 Short–term profit maximization

Distortion of profit–making motive — Profit over–emphasis in use of national resources — Business profit–making — Short–term revenue maximization — Financial short–termism
Nature Profit is a device which society uses both to measure and to reward enterprise efficiency and productivity in meeting market demand. Profits may be pocketed by a single owner, by several hundred thousand shareholders in the case of giant corporations, or by one or several institutional investors whose portfolio sizes in some cases make them significant, policy influencing, shareholders. The demand by shareholders for dividends that are 300 to 500 percent the rate of inflation is routine, so that in 5 percent inflation, 15 to 20 percent, or speculatively (that is with greater risk), even a higher dividend return on capital (common equity) may be expected. Corporations are not valued for the services they provide society, but are considered as money machines with the often remote investor disinterested in how or why this money is being made. The result is the generally amoral atmosphere in which business, international trade and development is conducted.
The most effective use of natural resources is reduced when an overemphasis on profit leads to uneven development of these resources. Development based on policies which promote monopolistic control further reduces use of natural resources. Reduction and ineffective use also occurs when resources are most accessible and are extracted without any responsibility being assumed for their depletion.
Incidence Loans provided to developing countries by international banks 'for development' are at the highest possible interest that can be obtained, which sometimes exceeds commercial prime rates by a significant spread. For example, one quarter or one half a percent on milliards means hundreds thousands, if not millions, of currency units over time. The banks behind these loans make their profits, regardless of the success of development projects. In the USA, for example, all–industry returns have recorded average of 11.5 percent, with the top performers including Chrysler (69.7) and Subaru of America (35.8), but also the Student Loan Marketing Association (43.0), and Pulter Home (36.5). Cars, education, homes and food (Diversifoods, 68.2) are priced to extract every cent of possible profit.
Claim To allow the profit motive to be the overwhelming determinant of development is to guarantee that mostly inappropriate development will result. The drive to maximize output, sales and return on investment inevitably leads to the orientation of productive capacity to serve the interests of the wealthy, namely those most able to pay the highest prices and to maximize consumption of increased output. Production of luxuries for them is much more likely to increase return on investment than producing necessities for destitute people.
Broader Capitalism (#PC0564) Nationalistically determined development of natural resources (#PD3546).
Narrower Profiteering (#PC2618).
Aggravates Exploitation (#PB3200) Short–term gain (#PF8675) Redundancy of workers (#PD8007) Bias against private enterprise (#PF1879) Unbalanced industrial distribution (#PF3843) Planned degradation in product quality (#PF7741) Unsustainable short–term improvements in agricultural productivity (#PE4331).
Aggravated by Corporate greed (#PF7189).

♦ PF2177 Political radicalism

Political extremism
Nature Extreme actions in the advocacy of political causes include denial of rights of others, and attempts or practices of intervention in due process of law or in abrogating deliberative pro-

cedures. It can be carried out in actions of political or economic reprisal, such as black-listing, or in violence.
Refs Betsworth, Roger G *The Radical Movement of the Nineteen Sixties* (1980); Carsten, F L *War Against War* (1982); Roy, Sarojendra Nath *Radicalism* (1946).
Broader Political conflict (#PC0368) Ideological conflict (#PF3388).
Related Insanity (#PA7157) Avoidance (#PA6379) Liberalism (#PF0717) Rationalism (#PF3400) Disobedience (#PA7250) Insufficiency (#PA5473) Religious extremism (#PF4954).
Aggravated by Double standards in morality (#PF5225).
Reduced by Compromise as a betrayal of principles (#PF3420).

♦ PF2178 Extraterritoriality
Refs Hermann, A H *Conflicts of National Laws with International Business Activity* (1982).
Broader Restrictive trade practices (#PC0073).

♦ PF2179 Positivism
Broader Rationalism (#PF3400).
Narrower Utilitarianism (#PU4386).
Related Atheism (#PF2409) Agnosticism (#PF2333).
Aggravates Dependence on religion (#PF0150).

♦ PF2182 Political pluralism
Narrow political interests of small groups
Nature Political institutions (formal or otherwise) may facilitate the representation of groups or a group interest rather than the individual interest. Political pluralism may lead to alienation if an individual cannot satisfactorily identify with a group. It may sharpen existing social tension if political groups are based on social division, including racial, religious, class and other differences. Vested interest groups may foster political lag.
Broader Political fragmentation (#PF3216).
Related Minority control (#PF2375) Undue political pressure (#PB3209).
Aggravates Power politics (#PB3202) Political opportunism (#PC1897) Divisive effects of official cultural pluralism (#PF0152).
Aggravated by Social conflict (#PC0137).
Reduces Lack of individualism in capitalist systems (#PD3106).

♦ PF2183 Paternalism
Nature Paternalism implies the policy of providing for the individual, without giving him the responsibility or the opportunity to provide for his own needs. The authority may see the individual's needs as different from the individual's own view of his needs. Paternalism may take the form of feudalism, agricultural paternalism or industrial paternalism. It can include political paternalism in the sense of dictatorship or of state social welfare and nationalized industries, and family or tribal paternalism. Paternalism may also be no more than a rationalization for racism.
Incidence Paternalism is the basic structure for most tribal and traditional family situations. Feudalism exists particularly in underdeveloped countries notably in Latin America and Asia. Industrial paternalism may exist wherever factories have grown up but tends to diminish as labour becomes unionized, although many large corporations may still provide their own insurance and other schemes in addition to national benefits. Paternalism in the socialistic sense of state welfare and nationalized industries occurs in certain Western European countries, in other capitalist countries, and in communist countries.
Counter-claim There are several conditions, when met, justify paternalism toward a person. These include: 1. the defect, encumbrance, or limitation in deciding, willing, or acting on the part of that person; 2. The persons high probability of serious harm apart from a paternalistic intervention; 3. the probability that a paternalistic intervention will produce a net balance of benefit over harm to the person; and 4. selection of the least restrictive, least insulting, least humiliating means of intervention.
Broader Domination (#PA0839) Authoritarianism (#PB1638) Human dependence (#PA2159).
Narrower Parental domination (#PF4391) Medical paternalism (#PF5397) Religious paternalism (#PF3609) Paternalistic punishment (#PJ3700).
Related Tribalism (#PC1910) Dictatorship (#PC1049) Parental over-protectiveness (#PF5255).
Aggravates Political repression (#PC1919).
Aggravated by Traditionalism (#PF2676) Social fragmentation (#PF1324) Inadequate education (#PF4984).
Reduces Lack of individualism in capitalist systems (#PD3106).

♦ PF2186 Nativism
Nature Nativism is a revitalization movement closely connected with utopianism and revivalism, manifesting the rejection of aliens and alien culture and the purification of the utopian culture. It is thus comparable to such popular beliefs as faith in the existence of a land without evil, or in the ultimate arrival of a messiah, or the coming of a millennium.
Broader Racism (#PB1047) Linguistic purism (#PF1954).
Narrower Purification of culture (#PU4396) Rejection of other cultures (#PU4395).
Related Traditionalism (#PF2676).
Aggravates Social conflict (#PC0137) Disillusionment (#PA6453) Social fragmentation (#PF1324) Social discrimination (#PC1864) Social disintegration (#PC3309).
Aggravated by Elitism (#PA1387) Secession (#PD2490) Forced assimilation (#PC3293) Discrimination against immigrants and aliens (#PD0973).

♦ PF2187 Retarded socialization
Lack of moral development
Nature Lack of development of standards of 'good' or 'bad' in children is manifested in later life by violence, aggression, exploitation, crime and general misuse of other people. Lack of moral development may be the result of a severely deprived background, lack of finer influences, and general insecurity. Lack of moral development may also be taken as meaning a lack of development along accepted moral lines, in other words, ideological deviation or refusal to conform to accepted social pattern. Lack of moral development in this sense of a refusal to conform occurs among a minority in most societies. Lack of moral development in the sense of not distinguishing between 'good' and 'bad' is a subjective term which may be applied to some extent to delinquents and criminals but may also occur in a more disguised form among respectable members of society.
Broader Human disability (#PC0699).
Narrower Amoralism (#PF3349) Irreligiousness (#PF8234) Social deviation (#PC3452) Maladjusted children (#PD0586) Personal and social maladjustment (#PC8337).
Related Insecurity (#PA0857) Social underdevelopment (#PC0242).
Aggravates Crime (#PB0001) Aggression (#PA0587) Immorality (#PA3369) Alcohol abuse (#PD0153) Human violence (#PA0429) Sexual deviation (#PD2198) Religious apathy (#PC3414) Sexual immorality (#PF2687) Juvenile delinquency (#PC0212).
Aggravated by Moralism (#PF3379) Permissiveness (#PF1252) Cultural deprivation (#PC1351) Lack of family planning information (#PD1050) Inability to define moral standards (#PF7178).

♦ PF2188 Monasticism
Nature Monasticism, particularly for the Hindu and Buddhist religions, has been the mainstay of church existence and expansion. In this respect monasticism breeds intolerance and religious rivalry. Monks or nuns may be involved in education or health and make use of the opportunity to spread indoctrination. Other holy orders were concerned with war. Because of the need to beg from the community, and because of their spiritual power over the community, monasteries have in some cases amassed substantial wealth. Monasticism encourages anticlericalism and religious, political and ideological repression. Monasticism is sometimes equated with mysticism.
Incidence Monasticism is particularly strong in the Buddhist and Hindu religions. It is slightly less strong in Christianity, exists in Islam, and did exist, to a very limited extent, in Judaism. It is currently in worldwide decline. Mexico, Russia and China have confiscated much or all monastic property and have secularized their inmates.
Background Communities of segregated monks or nuns retire from worldly life to devote themselves to the contemplation of truth and the striving for purity of heart. Most monks or nuns take vows of chastity, poverty and obedience to God. Monasticism is sometimes equated with mysticism. The predecessors of monasticism were hermits who existed in ancient Indian civilization from before 1500 BC. The discovery of the Dead Sea Scrolls brought to light a Jewish monastic community dating from the 1st century AD. Christian monasticism developed after the 2nd century AD. Muslim monasticism did not exist until the rise of sufism. In oriental religion, Jainism was the first to adopt full monasticism. Buddhist monasticism developed directly from the teachings of the Buddha. Zen Buddhist monasticism developed from the 8th century AD in China. Hindu monasticism dates from the 4th century AD.
Broader Ideological conflict (#PF3388).
Related Withdrawal (#PF4402).
Aggravates Religious repression (#PC0578) Ideological repression (#PC8083) Compulsory indoctrination (#PD3097).
Aggravated by Anti-clericalism (#PF3360) Dependence on religion (#PF0150) Double standards in morality (#PF5225).

♦ PF2189 Marxism
Nature Marxism is a socialist theory of dialectical human progress culminating in proletarian revolution and the abolition of class divisions and exploitation through the ownership of property. Because of the inconsistency of Marxist predictions with actual events, the doctrine and its exponents are left open to the accusation of hypocrisy either from contrary ideologies of from within Marxist ranks. The movement has been divided for a long time, originally into social democrats and communists (an element of this still exists among Western intellectuals in non-socialist countries) and currently among communist factions in socialist countries. Marxism may be elitist, tending towards bureaucracy, dictatorship and militarism. Alternatively it may tend towards anarchism, social disintegration and violence. Marxism is posing a threat to capitalism. It causes international conflict and internal conflict and repression within non-socialist countries. Divisions in Marxist ideology cause repression in socialist countries and may cause foreign intervention (as in Czechoslovakia).
Incidence Marxism is an increasing world wide phenomenon. Communist countries take their ideology from Marxist thought. Elsewhere, intellectuals and workers form communist and socialist movements or parties based on Marxist doctrine. High incidence of the latter is found in Italy and France. Lenin adopted an elitist and bureaucratic approach to Marxism reflected in the present day USSR, which also maintains a strict hold over most of the East European countries. Ideological schism occurs between the Russian and Chinese (and oriental) approach to Marxism. Trotskyite Marxism tends towards anarchy. Wars fought on ideological grounds between Marxism and capitalism include the Korean war, the Vietnam war and the war in Cambodia and Laos. Repression of Marxism occurs particularly in fascist or right-wing capitalist countries. Repression of Marxist deviation occurs within most communist countries.
Background Marxist doctrine was spelled out concisely in the 'Communist Manifesto' written by Karl Marx in 1848. Marx and Engels propagated the doctrine which became accepted by the First and Second Working Mens' Internationals, mainly representing a European movement but which spread to the USA when Marx removed the First International's headquarters there (from London) after the fall of Paris Commune in 1871. During the Second International after the death of Marx in 1882, conflicting interpretations arose which divided Marxists into the social democratic and communist camps. After the October Revolution in Russia in 1917, this division widened and has continued to do so, so that now social democrats and communists in non-socialist countries are often the most bitter opponents. Marxism lost much of its international aspect under Stalin who converted it for the purposes of nationalism. Russia has since been accused of imperialism by Red China and by Western intellectual Marxist-socialists who retain the original Marxist internationalism.
Refs Avineri, Shlomo (Ed) *Varieties of Marxism* (1977); Everyman, Ron *False Consciousness and Ideology in Marxist Theory* (1981); Jay, Martin *Marxism and Totality* (1984); Plekhanov, G *Fundamental Problems of Marxism* (1980).
Broader Political conflict (#PC0368) Ideological conflict (#PF3388).
Narrower Leninism (#PG4404) Stalinism (#PG2703) Deviationism (#PG2029).
Related Communism (#PC0369) Anti-science (#PF2685).
Aggravates Class conflict (#PC1573) Political repression (#PC1919) Ideological repression (#PC8083) Governmental disregard for people as human beings (#PD8017) Deteriorating structures of rural community cooperation (#PF3558).
Aggravated by Elitism (#PA1387) Destiny (#PF3111) Exploitation (#PB3200) Double standards in morality (#PF5225).

♦ PF2190 Totalitarianism
Nature Totalitarianism is a political system comprising commitment to a single goal, with use of violence by official forces on a large scale, suppression of opposing organizations or movements, and enforcement of total public participation towards the achievement of the single goal.
Background Totalitarianism is a very imprecise term containing many elements which are also present in non-totalitarian regimes. The accusation of being totalitarian has been levelled in propaganda directed against certain fascist or fascist type regimes. The term was first coined in 1928 with regard to fascism in Italy. It has been used to describe Mussolini's Italy, Hitler's Germany, and Stalin's Russia.
Refs Arendt, Hannah *The Origins of Totalitarianism*; Buchheim, Hans, et al *Totalitarian Rule* (1987); Chakotin, Serge *Rape of the Masses*; Curtis, Michael *Totalitarianism*.
Broader Neo-fascism (#PF2636) Power politics (#PB3202) Ideological conflict (#PF3388) Undemocratic political organization (#PC1015).
Narrower Totalitarian democracy (#PD3213) Political dictatorship (#PC0845) Enforced participation in community activity (#PD3386).
Aggravates Conspiracy (#PC2555) Political trials (#PD3013) Political repression (#PC1919) Subversive activities (#PD0557) Radio and television censorship (#PD3029).
Aggravated by Nationalism (#PB0534) Denial of freedom of thought (#PF3217) Double standards in morality (#PF5225).

♦ PF2192 Utopianism
Over-utopian visions
Nature Utopianism is a persistent tradition of thought about the perfect society, in which perfection

is defined as harmony.
Background Although the actual term is from a 16th century book by Sir Thomas More and means 'nowhere', the ideal of utopianism dates from the earliest records and occurs throughout the history of philosophical writing, including the Bible stories of man in the Golden Age, man in an Arcadian state of nature.
Claim Utopia is the ideology of revolt, of religious and political aspiration, of apocalyptic and messianic dreams, of the future.
Counter–claim No society fully understands itself and its potential. Therefore what appears over–utopian may be realizable, whereas what appears realistic may be impossible to achieve. Utopian visions, even when not realizable, may therefore bring about meaningful changes. Indeed, whether adopted by political movements or not, they may constitute a main mode of human evolution which should not be constrained by feasibility assessments.
Broader Idealism (#PF3419) Ideological conflict (#PF3388).
Related Malignant visions (#PF5691).
Aggravated by Double standards in morality (#PF5225).

♦ PF2195 Lack of social mobility
Broader Social inequality (#PB0514).
Narrower Inflexible social structure (#PB1997) Structural rigidities in labour markets (#PD4011).
Aggravates Caste system (#PC1968) Fragmentation (#PA6233)
Miscegenation (#PC1523) Social conflict (#PC0137)
Underdevelopment (#PB0206) Social breakdown (#PB2496)
Conflict between minority groups (#PC3428) Underprivileged racial minorities (#PC0805)
Vulnerability of marriage as an institution (#PF1870)
Waste of human resources in capitalist systems (#PC3113).
Aggravated by Prejudice (#PA2173) Segregation (#PC0031)
Traditionalism (#PF2676) Ethnic conflict (#PC3685)
Social injustice (#PC0797) Racial segregation (#PC3688)
Racial exploitation (#PC3334) Dependency of women (#PC3426)
Class consciousness (#PC3458) Political repression (#PC1919)
Lack of assimilation (#PF2132) Social discrimination (#PC1864)
Sexual discrimination (#PC2022) Ethnic discrimination (#PC3686)
Cultural fragmentation (#PF0536) Single parent families (#PD2681)
Denial of human rights (#PB3121) Segregation in housing (#PD3442)
Plural society tensions (#PF2448) Discrimination against minorities (#PC0582)
Inadequate educational facilities (#PD0847) Underprivileged linguistic minorities (#PC3324)
Racial discrimination in public services (#PD3326)
Discrimination against women in employment (#PD0086)
Inadequate education of indigenous peoples (#PC3322)
Failure of individuals to participate in social processes (#PF0749).

♦ PF2196 Social evolutionism
Nature Social evolutionism is a systematic approach to the study of social change, derived from social Darwinism and including the modern adherents of that theory in the USA. Social Darwinism applies the principles of the theory of evolution to social structure, justifying inequalities within society and between societies. Modern social evolutionism tends to be more descriptive than explanatory, but attempts to formulate evolutionary universals for society in general. Both social Darwinism and social evolutionism are materialist theories, the latter being used also to explain religious phenomena.
Broader Misuse of evolutionary theories (#PF3348).
Related Peroneal nerve paralysis (#PG5668) Lack of eugenic measures (#PD1091).

♦ PF2199 Government action against regimes with alternative policies
Government action against regimes with opposing policies — Government action against regimes with hostile policies — Eroding support for unfriendly governments
Broader Foreign intervention in internal affairs of states (#PC3185).
Related Government support for repressive regimes (#PF4821).
Aggravates Toxoplasmosis (#PE3659) Overthrow of government (#PD1964)
Secret military operations (#PF7669) Foreign military intervention (#PD9331)
Diseases of the lymphatic system (#PD2654) Covert intelligence agency operations (#PD4501)
Economic sanctions against governments (#PF4260).

♦ PF2202 Top down research methodologies
Nature New knowledge and skills require vision in order to be applied effectively. When analysis of data is not synthesized into workable solutions for problems, then the result is inadequate research inapplicable to the requirements of the whole of society.
Claim Research methodologies are too often employed superficially, simply to describe and categorize within the well–trodden arena of the possible, believable or assimilative. They do not synthesize and interpret data in such a way as to expand society.
Broader Limits on areas of research (#PF2529).

♦ PF2204 Inefficient extraction and utilization of natural resources
Nature The extraction and utilization of natural resources, and the making of decisions concerning their use, are inhibited because technology, systems and procedures for extracting and utilizing resources are not equitably distributed nor does their development respond to need. This absence of a global sharing in technology and in the transfer and coordination of skills results in inefficient production and consequent waste of natural resources.
Broader Limited access to natural resource use decisions (#PF2882).
Narrower Inadequate land drainage (#PD2269) Underutilization of natural resources (#PF1459).
Aggravates Unconstrained exploitation of natural resources (#PF2855).

♦ PF2208 Lack of motivation in leadership development
Impractical leadership skills — Inappropriate leadership development — Leadership skills paucity
Nature Although it is increasingly evident that full utilization of the leadership skills of all community residents is a prerequisite for progress, the outdated social methods, unfocused planning and ineffective meetings in many areas are indications that latent local leadership has neither been identified nor motivated to assume leadership roles. Adult community members in general, and women in particular, may be unable to participate constructively in guiding the community in the solution of its common problems. In particular, residents feel they need training before they are able to participate in organizing and planning local educational and commercial ventures.
Claim Until structures are created to motivate and train the corporate leadership potential, communities will find themselves ineffective in creating ongoing social change.
Refs Culligan, Matthew J, et al *Back to Basics Management* (1983).
Broader Apathy (#PA2360) Obstacles to leadership (#PF7011)
Insufficient leadership training (#PF3605).
Aggravates Stagnating social methods (#PF3884)
Inadequate meeting methods (#PF8939).
Aggravated by Narrow range of practical skills (#PF2477)
Minimal opportunities of adult training (#PF6531)
Inadequate management skills in rural communities (#PF1442)
Unperceived relevance of formal education in rural communities (#PF1944).

♦ PF2212 Incomplete understanding of new societal service systems
Nature Daily living requires an ever increasing amount of technologically–oriented information to carry out routine activities effectively. This has left behind rural communities which still rely on old–style information suited to another time. Skills may have been acquired on the surface but without the depth of understanding necessary for their complete application. As a result, such communities are more and more dependent on outside resources and government programmes, while being unclear as to what it actually means to be part of such programmes. The result is that systems of available credit are mistrusted and remain unused; the prevalence of local diseases may be acknowledged, but health services are not utilized and prevention materials rarely encountered; local advertising may be known as a concept but the results are limited in effect; systems of pest eradication and the means to acquire them remain unclear; and government procedures are seen as too complex to be within the reach of ordinary people. The result of acquiring basic information that should transform everyday life tends in fact to be confusion and a basic misunderstanding of new scientific practices.
Broader Lack of understanding (#PJ4173).
Narrower Limited local initiative (#PJ9554) Lack of tuberculosis materials (#PG9558)
Unfamiliar government procedures (#PJ0740).
Aggravates Distrust (#PA8653) Overdependence on government (#PF9530)
Ignorance of health and hygiene (#PD8023).
Aggravated by Misuse of advertising (#PE4225) Inadequate health services (#PD4790).

♦ PF2216 Misuse of international forums
Social agency competition — Rivalry between international groups
Nature International groups have tended increasingly to become forums for international conflict. Organizations from the United Nations to the Olympic Games are a crucial means of maintaining non–violent dialogue among nations and helping to promote peace. However, member delegations are increasingly using such organizations to publicise their own national, political, economic or cultural conflicts, valuing their unique "cause" above that of the global relations which such bodies exist to serve.
Broader Competition (#PB0848)
Legal contract system reduced to individual needs (#PE5397).
Narrower Collapse of common values (#PF1118)
International cultural vacuum (#PF2258)
Unquestioned control by economic forces (#PF2312)
Parochial attitudes of organizations' members (#PD2239)
Short sighted decisions about intersocial interaction (#PE4477)
Undermining of multilateral forums by industrialized countries (#PE4289).
Related Non–concerned attitudes (#PF2158) Parochial national interests (#PF2600)
Structural failure of citizen participation (#PF2347).

♦ PF2231 Unrealized potential of commercial enterprises
Nature Market economic activity in many communities consists only of below–subsistence farming. Attempts at commercial enterprise, such as village markets and small businesses, have only marginal success. Resources and opportunities for the development of new enterprises in business and light industry are not explored and enterprises remain undeveloped while new commercial enterprises fail.
Broader Undeveloped channels for commercial initiative (#PF6471).
Narrower Unrealized agronomic potential (#PJ1775)
Inaccessible administrative agencies (#PF2261).
Related Overlooked potential for industrial development in rural communities (#PF2471).
Aggravated by Inaccessible market and supply centres (#PF8299).

♦ PF2241 Deteriorating community identity
Nature As an urban neighbourhood declines, its role as the focal point of residents' life has diminished vitality. Old industries and closed stores witness to lost greatness; unused buildings and land suggest that the area has been passed by. Adults and young people look elsewhere for work since they see no vocational future in the locality.
Claim It is vitally necessary for every local community to be conscious of its unique worth, especially in contemporary large cities, where this awareness will deeply affect every aspect of public and private life. At critical moments in a neighbourhood's history, as when new housing estates appear which dramatize radical, rapid transition, a new role and emerging identity must be created or the community will cease to be a place of significant personal engagement for its residents and of creative service for the rest of society.
Broader Obstacles to community achievement (#PF7118).
Narrower Property damage (#PD5859) Community damage (#PG9379)
Lifestyle tensions (#PG9469) Sense of powerlessness (#PF8618)
Limited consumer knowledge (#PG7662) Minimal citizen confidence (#PF8076)
Disrelated social background (#PG9464) Inaccurate youth stereotypes (#PJ8357)
Isolating community lifestyle (#PG9465) Investment determined marriages (#PG9432)
Powerlessness within bureaucracy (#PG9383) Alienating public housing assignments (#PJ9479)
Stagnated images of community identity (#PF6537)
Demoralizing images of rural community identity (#PF2358).
Related Low general income (#PD8568) Mothers' self–imposed isolation (#PJ9425).
Aggravates Unorganized community recreation (#PF5409)
Social disaffection of the young (#PD1544)
Deteriorating structures of rural community cooperation (#PF3558).
Aggravated by Crisis of identity (#PJ2421) High teacher turnover (#PE8221)
Resignation to problems (#PF8781) Demeaning community self–image (#PF2093)
Lack of local leadership role models (#PF6479)
Disjointed patterns of community identity (#PF2845).

♦ PF2249 Hallucinations
Hallucinosis
Nature Hallucinations are the apparent perception of an external object when no such object is present. They are a fixed report of a perception without an appropriate stimulus. Frequently the stimulus appears to be located in the world outside the hallucinating person, although it can be demonstrated that it is within the person. Hallucinations have the characteristics of one or several sensory systems; they may be visual, auditory, tactile, olfactory, or gustatory. Organic disturbances may cause hallucinations, although in most instances they can be explained by psychological and cultural circumstances.
Incidence In adults in the Western world, hallucinations should raise the question of mental pathology. In primitive individuals and in primitive cultures in which myth and reality are not clearly differentiated, hallucinations occur more frequently.
Refs Chambers, William and Pilowsky, Daniel (Eds) *Hallucinations, in Children* (1987); De Boismont, A Brierre *A Treatise on Magnetism and Hallucinations As an Expression of Nervous Disorders* (1988).
Broader Illusion (#PA6414) Mental illness (#PC0300)
Symptoms referable to nervous system (#PE9468).
Aggravates Unidentified flying objects (#PF1392).
Aggravated by Fever (#PD2255) Overwork (#PD2778)
Psychoses (#PD1722) Schizophrenia (#PD0438)
Alcohol abuse (#PD0153) Sleep disorders (#PE2197)
Abuse of amphetamines (#PE1558) Abuse of hallucinogens (#PD0556)
Diseases of the nervous system (#PC8756) Diseases and injuries of the brain (#PD0992).

PF2256 Inappropriate application of traditional values
Nature The values which previously sustained people have to be applied in new ways in the reality of the present-day technological, urban world. Although the people of some communities continue to honour widely recognized values of the past, the application of past values to the actual situations in other communities may be felt inappropriate by many residents. Widespread attachment to the land, for example, seems out of place when agricultural enterprise requires more acreage and fewer people than ever before, and when a large farm that once supported 20 families now needs only five individuals to operate and maintain its advanced agricultural machinery. The traditional expectation that a family should care for its own home and grounds is complicated in rural communities by the fact that much available housing is dilapidated far beyond what even the most skilled home handyman could repair alone. Further, there is no way to deal with community sewage, public areas, or zoning ordinances on an individual family basis. The traditional qualities of warmth and graciousness, which work well when relationships are close and personal, seem to have nearly disappeared in the realities of today's job market, an arena in which they are greatly needed: some people cite discriminatory hiring practices, while others speak of insolence in everyday encounters. All this indicates that traditional values are not being applied, or are being applied inappropriately, to an unprecedented community milieu.
Claim Social structures have changed dramatically and often painfully during the past 40 years. Only when a community finds ways of applying its traditional values in ways appropriate to its new situation will it demonstrate that it is seriously committed to a new future.
  Narrower Dual school systems (#PG8759) Past-oriented concerns (#PG8739)
  Fear of urban problems (#PG8799) Narrow range of food crops (#PD4100)
  Habitual lifestyle routines (#PG8709) Irritation response pattern (#PG8769)
  Ingrained segregation habits (#PG8789) Traditional hiring practices (#PG8822)
  Traditional market practices (#PJ9742) Laissez-faire response pattern (#PG8689)
  Related Student absenteeism (#PE4200) Inaccessible job market (#PE8916).
  Aggravates Neglected young children (#PE4245).
  Aggravated by Insufficient community events (#PF5250)
  Untransposed significance of cultural tradition (#PF1373).

PF2258 International cultural vacuum
Claim There is no way of engaging the gifts of different cultures and of using the unique power of such diversity; and there is no common understanding of the unique primordial gifts of culture, nor is there a common framework from which such understanding might grow. Cultural power is not seen as a crucial kind of pressure in itself, nor is it perceived independently of, and as giving life to, economic and political power. The result is is control of all interchanges by economic criteria.
  Broader Misuse of international forums (#PF2216).

PF2261 Inaccessible administrative agencies
Remote borough authorities — Non-resident local authorities — Distant district centre — Inaccessible administrative centres
Nature Despite a growing demand for effective administrative systems to serve local communities, residents in urban areas feel these structures to be complex and remote. This impression is enhanced by the difficult, often lengthy procedures of requests, applications, appointments, waiting lists, negotiations, telephone calls, visits and letters. Although services are known to be available, applicants are constantly met with technical procedures that delay service delivery. Proliferation of legal data still leaves many uninformed about legal rights and special benefits. Despite efforts by administrators to find the best means for delivering services, the weight of official machinery and the number of requests make their task increasingly difficult. In effect, both the administrators and those whom they are intended to serve are clear that existing structures are no longer effective in providing the service for which they were created; but there is an underlying powerlessness in the face of the present vast socio-economic machinery.
  Broader Social inaccessibility (#PC0237) Physically inaccessible services (#PC7674)
  Unrealized potential of commercial enterprises (#PF2231).
  Narrower Administrative delays (#PC2550) Lack of community voice (#PG9471)
  Imposed official decisions (#PF8649) Externally controlled space (#PG9448)
  Inaccessible decision makers (#PJ9424) Ineffective care-taking system (#PG9474)
  Unresponsive public authorities (#PF8072) Dependence on interim employment (#PG9461)
  Ineffectively prioritized funding (#PG9453)
  Limited accountability of public services (#PF6574)
  Inaccessibility of decision-makers in multinational enterprises (#PE0573)
  Inadequate implementation of plans and programmes against problems (#PF1010).
  Related Inadequate maintenance personnel (#PJ0088)
  Disjointed patterns of community identity (#PF2845).
  Aggravated by Unequal school distribution (#PJ9458).

PF2271 Excessive standardization
Standardization
Nature With increasing global standardization (such as the metric system, the Christian calendar, scientific and legal terms, etc) and the spread of cultural homogenization, there has been a decline of national identity and local customs.
  Broader Conformism (#PB3407) Obstacles to world trade (#PC4890).
  Narrower Inadequate international standards of financial accounting and reporting (#PF0203).
  Aggravates Discrimination against giants (#PE5578)
  Inadequacy of international standards (#PF5072)
  Discriminatory imposition of standards (#PD5229)
  Discrimination against dwarfs and midgets (#PE2635).
  Aggravated by Parochial telecommunications standards (#PE1840).

PF2275 Self-accountability for grant funding
Nature Historically, those responsible for grant funding have been accountable only to their own private goals, understandings and predilections resulting in no effective accountability to actual social needs. This self-accountability often leads to grant giving on the basis of quantifiable returns. Complicated systems for requesting funds and arbitrary evaluation procedures different for each grantor, restrict equitable fund granting. Another complication occurs through limitations imposed by wills and trust fund legalities.
  Broader Belittling of grant recipients (#PF2708).

PF2276 Mutation
Bacterial mutation
Nature Mutation is an abrupt change in the genotype of an organism, not resulting from genetic recombination. There are three main types of mutational change. In the first type the genetic material is altered quantitatively, either by the addition or removal of whole chromosomes, parts of chromosomes, or whole chromosome sets; or, at the other extreme, by the addition or deletion of single base pairs from the nucleic acid of the gene. The second type are qualitative alterations of the genetic material, such as the substitution of one base pair for another. In the third type the existing genetic material may be rearranged without altering its quantity or quality. Because of the genetic complexity of man, there are immense possibilities for mutation. Genetic diseases are becoming relatively more important owing to the reduction in the incidence and severity of parasitic and bacterial diseases. Two main types of genetic damage are recognized: chromosome aberrations and gene mutations. These may affect either somatic cells or germinal cells. Although damage to either cell population may have serious consequences, from the public health standpoint mutations in germinal cells are of paramount importance, as they present a hazard to future generations. The relation between the ability of a chemical to produce mutations in experimental test systems and its ability to affect humans is not firmly established but the potential hazard to the population is of great magnitude.
Counter-claim If there were no mutations there would be no genetic differences among organisms and no evolution of life. In fact, mutations have often been called the building blocks of evolution. For this reason, the mutation process is fundamental not only to genetics but to the continuation of life itself in a world that is changing.
  Refs Obe, G (Ed) *Mutations in Man* (1984).
  Broader Human disease and disability (#PB1044)
  Vulnerability of human organism (#PB5647).
  Narrower Resistant bacteria (#PE6007) Pest resistance to pesticides (#PD3696).
  Related Chance (#PA6714) Lack of eugenic measures (#PD1091).
  Aggravates Genetic defects and diseases (#PD2389)
  Insect resistance to insecticides (#PD2109)
  Irresponsible genetic manipulation (#PC0776)
  Encouragement of drug resistant diseases (#PJ3767).
  Aggravated by Mutagens (#PD1368) Mutagenic chemicals (#PE4588)
  Geomagnetic reversal (#PF1588) Health hazards of radiation (#PD8050)
  Health hazards of exposure to cosmic radiation (#PF1686).

PF2277 Hedonism
Nature Hedonism is the creed that pleasure is the sole end and aim of human action or conduct, and that to it all good and well-being is ultimately reducible.
  Broader Inadequate ideological frameworks (#PD0065).
  Narrower Moral hedonism (#PG4590) Overindulgence in physical pleasures (#PG4591).
  Related Moralism (#PF3379) Cynicism (#PF3418)
  Abusive behaviour modification (#PE2690).
  Aggravates Selfishness (#PA7211) Family breakdown (#PC2102).

PF2296 Limited exposure to outside influences in rural villages
Nature In contrast to the current, highly mobile life-style in urban areas, the tasks of rural residents may take them frequently to neighbouring villages but only rarely as far as the district market town a few miles away. Without telephone, postal service, newspaper, television, radio or regular bus service, their contact with the outside world is sporadic. In the past, communication with the outside world was a luxury at best, and at worst, a temptation which diverted attention from the arduous tasks of a subsistence life style. Outside exposure is, however, expanding and will continue to do so. The growing necessity for effective interchange with the world which is beginning to substantially shape their destinies means that villagers find themselves ill-equipped in language, clothing and practical experience to operate effectively in their expanding circle of relationships. They find it hard to imagine that they could make any significant contribution to the larger community, and tend to avoid increased outside exposure as it painfully reinforces their images of ineffectiveness, and overwhelms them with random data which they cannot digest.
  Broader Narrow range of cultural exposure (#PF3628).
  Related Parochial national interests (#PF2600).
  Aggravates Narrow range of business skills (#PE6554).
  Aggravated by Wasted time (#PF8993) Unchallenging world vision (#PF9478)
  Unavailability of building materials (#PJ0704)
  Inadequate facilities for children's play (#PD0549).

PF2303 Disruptive effect of changing employment patterns
Nature Employment patterns are undergoing radical shifts as old production and distribution methods are replaced by centralized and/or capital-saving techniques. Urban communities face the challenge of locating new industries and diversified sources of employment to provide a stable income and employment base, but new industries are not inclined to locate in urban areas where the infrastructure is inadequate and land use planning is uncertain. This is causing unemployment in urban neighbourhoods and communities that are not prepared to wait for the implementation of long-term plans to deal with the resulting suffering and dislocation.
  Broader Change (#PF6605).
  Narrower Unclarity on dock use (#PG9443) Large industry syndrome (#PG9455)
  Insecurity of work changes (#PG9450) Unemployment of married women (#PG9418)
  Ineffective community leverage (#PG9376) Inappropriate development policy (#PF8757)
  Debilitating economic preoccupations (#PG9459).
  Related Youth unemployment (#PC2035).

PF2304 Increasing pace of life
Pace and duration of work
Claim The modern age is increasingly characterized by a relentless energy that preys on speed records and shortcuts, unmindful of the past, uncaring of the future, existing only for the present moment and the quick fix. Industrialized societies have quickened the pace of life only to become less patient, less spontaneous and less joyful. People are better prepared to act on the future, but less able to enjoy the present and reflect on the past. Developed societies have learnt how to extract and make things at a faster pace, but with the consequence that people exploit and devalue each other's time at the workplace in order to increase production quotas. The commitment to producing and consuming at a frantic pace has resulted in an increasing depletion of nature's endowment and a degradation of the biosphere.
Counter-claim Although it is frequently claimed that the increased pace and stress of modern times may lead to a higher incidence of mental illness, this has not been proved by quantitative research. From the information currently available, it would appear that psychoses do not increase although it is possible that psychoneuroses may. Under very difficult conditions, such as disasters, riots, and wars, unusual mental syndromes are observed. Some of these may be temporary and quite normal as part of the organism's readjustment phase after disturbed homeostasis; others may be pathological states precipitated, but probably not actually caused, by severe stress.
  Refs Rifkin, Jeremy *Time Wars* (1987).
  Broader Deteriorating quality of life (#PF7142)
  General obstacles to problem alleviation (#PF0631)
  Adverse consequences of scientific and technological progress (#PF3931).
  Aggravates Lack of time (#PC4498) Tuberculosis (#PE0566)
  Sleep deprivation (#PE2741) Stress in human beings (#PC1648).
  Aggravated by Hyperefficiency (#PF1706) Short-term gain (#PF8675)
  Computer obsession (#PF1288) Temporal imperialism (#PF1432).

PF2306 Uncoordinated research efforts
Nature Inadequate structures to coordinate and assign research responsibilities inhibit the establishment of corporate priorities in a global context. Research is thus not accountable to society's total needs but to the popularity of the subjects being investigated, so that breakthroughs in technology are restricted to certain areas, resulting in unnecessary duplication and benefits which are not shared globally.
  Broader Limits on areas of research (#PF2529).
  Aggravates Fragmentation of research (#PF9830).

♦ PF2312 Unquestioned control by economic forces
Nature Decision–making is based on economic concerns and values. These values control other social dynamics rather than being accountable to them.
Broader Misuse of international forums (#PF2216).

♦ PF2313 Unavailable education for effective living
Nature There is increasing demand internationally for methods of education for effective function in a society where every aspect seems to be in a state of rapid transition. In urban neighbourhoods, the need for such knowledge is particularly evident, and urban residents are having to acquire new skills, new work and leisure patterns, new methods of analysis and problem–solving, and new ways to relate to the world and to each other. There has been investment in education with visible benefit, especially at the secondary level, but the needs go far beyond the normally available sets of manual and intellectual skills: formal education is unable to cope with the demands made on it and adult education is either insufficient in scope or inadequate to meet demands. Meanwhile professional teachers tend to prefer permanent posts in suburban areas where often the opportunities are much greater both for students and teachers. Over–burdened educational structures are symptomatic of the need for an expanded and diversified approach to education in these communities.
Incidence In a survey of French schoolchildren, it was found that although 75% of school–leavers could read, write, add and subtract, fewer than 20% could comprehend in depth the historical or cultural context of what they had read. In problem solving, the same group could readily perform computations but could not devise solutions for problems when the computation was not given.
Broader Limited availability of learning opportunities (#PF3184)
Underprovision of basic services to rural areas (#PF2875).
Narrower Nonavailability of technical training (#PJ0121)
Unsatisfied need for continuing education (#PF0021)
Unequal global distribution of basic skills (#PF2880)
Limited availability of education in rural areas (#PF3575).
Related Narrow scope of education (#PF3552) Inadequacy of formal education (#PF4765)
Ineffective systems of practical education (#PF3498).
Aggravates Lack of retraining incentives (#PG9410)
Lagging training in social skills (#PF8085).
Aggravated by Inadequate health control (#PF9401)
Unclear educational roles (#PG8840)
Inadequate educational facilities (#PD0847)
Overemphasis on academic education (#PG9447)
Underutilization of locally available skills (#PF6538).

♦ PF2314 Adultery
Extra–marital affairs — Marital infidelity
Nature Although attitudes differ widely between cultures, and even between neighbouring communities, almost all consider adultery an offence, the punishment for which may be divorce or banishment, public exposure or mutilation, even death. Adultery is specifically defined as the act of sexual intercourse between a married person and someone other than his or her spouse, whether or not the other person is also married. In a broader sense, the term applies to an act of sexual intercourse with a partner outside the permitted group. Some communities permit adultery but only under certain conditions. Local law often discrimates between male and female, higher class and lower class adulterers, and female adulterers are often discriminated against.
Incidence Attitudes towards adultery are changing, especially in industrialized countries. More married women are now having affairs. Increasingly, women, not men, are the first to stray from marriage. In both the USA and the UK, 25 to 50 per cent of married women have at least one lover after they are married in any given marriage. From 50 to 65 per cent of married men stray by the age of 40. Whereas for people who married before 1960, women waited 14 years (and men 11) before having an affair, in 1989 women wait 4 years (and men 5). Currently 73 per cent of married people claim to have had at least one affair.
Claim Adultery is damaging to the good of the family and is a direct betrayal of the most sacred of human relationships.
Refs Lawson, Annette *Adultry* (1988).
Broader Sexual offences (#PD4082) Sexual immorality (#PF2687)
Unethical personal relationships (#PF8759).
Narrower Cohabitation (#PF3278) Wife–swopping (#PG4711).
Related Bigamy (#PF3286) Polygamy (#PD2184) Prostitution (#PD0693)
Non–validity of marriage (#PF3283).
Aggravates Divorce (#PF2100) Group sex (#PG6147) Fornication (#PF5434)
Family breakdown (#PC2102) Illegitimate children (#PC1874)
Deception between sexual partners (#PE4890).
Aggravated by Promiscuity (#PC0745) Indecent art (#PE5042)
Permissiveness (#PF1252) Birth prevention (#PE3286)
Disobedient wives (#PF4764) Marital instability (#PD2103)
Sexual unfulfilment (#PF3260) Conflicting national divorce laws (#PF3249)
Religious or civil refusal of divorce (#PF3248).

♦ PF2316 Minimal opportunities for corporate activities
Insufficient opportunities for community activities
Nature While people in some communities are requiring less time to earn their livelihood, allowing commensurately more time for community, family, and corporate activities, in rural areas the demanding daily requirements of providing for the family's immediate needs inhibits significant participation in community life and prevent practical care for neighbours. The cycle of subsistence living has a deep impact on the village life–style. The time for the concerted effort required for authentic and effective care for the whole community is thus severely limited, and the community as a unit suffers.
Broader Inactivity (#PB7991).
Narrower Unstable fishing season (#PJ9570) Insufficient personal time (#PJ9534)
Unpredictable work schedule (#PG9588) Limited family opportunities (#PG9545).
Related Unorganized development of work forces (#PF2128).
Aggravates Lack of community development (#PF7912).
Aggravated by Subsistence life style (#PF1078) Exploitation of child labour (#PD0164).

♦ PF2333 Agnosticism
Nature Agnosticism questions the possibility of knowledge of existence beyond the phenomena of experience. It is often equated with a general scepticism about religious questions and can lead to spiritual apathy and a loss of direction.
Agnosticism can be a way of life in three ways. Lazy, superficial or pragmatic agnostics cannot be bothered by religious questions, refuse to take difficult and contentious questions seriously and prefer to spend their time on problems that can actually be solved. Scrupulous agnostics are quite different as agnosticism is not an escape but a burden. They see the vastness of human knowledge and are paralysed. They have a deep fear of commitment, not through caring too little, but from worrying too much. They seek certainty, logical perfection, truths beyond criticism. The third type are the necessary agnostics who acknowledge that God is God, they are human beings and the first step toward wisdom is the acknowledgement of ignorance. They don't fall into the folly of believing they have it all worked out. They know faith is a process of seeing and then not seeing, of rising to great religious heights and then being humbled and corrected, this hoping and finding hope shattered, this dying and rising to new life is the source of their faith.
Incidence The significant drop in church attendance of recent years indicates the prevalence of agnosticism in developed countries.
Refs Mills, David *Overcoming Religion* (1980).
Broader Irreligiousness (#PF8234) Ideological conflict (#PF3388).
Narrower Religious apathy (#PC3414).
Related Theism (#PF3422) Positivism (#PF2179).
Aggravates Social alienation (#PC2130) Social disintegration (#PC3309)
Dependence on religion (#PF0150).
Aggravated by Scientism (#PF3366) Religious extremism (#PF4954)
Proliferation of technology (#PD2420) Double standards in morality (#PF5225).
Reduces Atheism (#PF2409) Superstition (#PA0430).

♦ PF2334 Economic philosophy of controlled risk
Nature Resource distribution is hampered by a static, no–risk philosophy, which fails to balance needs against operations and is unable or unwilling to combine the diverse historical values into an international valuation scheme. This is manifested globally by a lack of systems for distribution of necessities and the resorting to guilt, violence and a diversity of propaganda campaigns in order to ensure some level of ongoing equity among peoples.
Broader Risk (#PF7580) Dominance of economic motives (#PF1913).
Related Political risk (#PF5932).
Reduces Risk of capital investment (#PF6572).

♦ PF2335 Inefficient public administration
Ineffective civil service — Inadequate governmental managerial instruments — Bureaucratic resistance to self–review — Lack of administrative reform
Nature In the developed market countries, where the driving force of society is the private economic sector, public administration functions are funded by a large tax base. Governmental organization tends to be disproportionate to authority and need. Ministries, departments and bureaux may proliferate and yet have no real autonomous function or power. Civil servants may often have inadequate managerial backgrounds and their administration of the public services funded by taxpayers may be wasteful and sometimes, by business standards, incompetent.
In less developed areas, the future of the entire country may lie with its public administrators, there being no industrial infrastructure. However, lack of such industrial infrastructure implies a parallel lack in educational and managerial infrastructure as well, with the result being trial–and–error management of governmental agencies and services. Where the government itself is responsible for the establishment and operation of industrial enterprises, the burden upon the civil administration requires a managerial and sectoral expertise that can only develop in time. In the interim, costly mistakes may be made. Development projects planned and executed by public services are rarely feasible owing to a lack of productive working methods and experience.
Claim In some countries public administration is inadequate at both the central and local levels, forming a reservoir for excess labour rather than a device to ensure the efficiency of the economic system. Different level of government create an uncoordinated bureaucratic machine subject to votes of opposing political tendencies.
Refs ILO *World Labour Report 1989* (1989).
Broader Inadequate government (#PJ6362) Inadequate social reform (#PF0677)
General obstacles to problem alleviation (#PF0631).
Narrower Government administrative backlog (#PG8488)
Inadequate management of government finances (#PF9672)
Government delay in response to symptoms of problems (#PF6707)
Inadequate control over government administrative process (#PC1818)
Inefficient public administration in developing countries (#PF0903).
Aggravates Bureaucratic inaction (#PC0267) Inadequate provision of public safety (#PF2874)
Vulnerability of government to lobbying (#PF5365)
Inadequate road and highway transport facilities in developing countries (#PD0543).
Aggravated by Diplomatic fraud (#PD5554) Fear of officialdom (#PD9498)
Bureaucratic factionalism (#PF7979) Complex government bureaucracy (#PF8539)
Unchecked power of government bureaucracy (#PD8890)
Lack of appreciation of cultural differences (#PF2679)
Lack of political neutrality of civil servants (#PE4031)
Governmental disregard for people as human beings (#PD8017)
Demotivated civil servants in developing countries (#PE5278)
Excessive salaries of international civil servants (#PE6388)
Violation of privileges and immunities of international civil servants (#PE5488).

♦ PF2340 Collapse of societal engagement
Claim Amid the urgency of the rapidly changing times, local people's actions and creativity are dissipated in old–fashioned systems. Free assembly, which should provide the base for the expression of public opinion by all, is inhibited. The capacity of society to bring together everyone's knowledge and experience in the decision–making process is not being utilized, and the individual fails either to recognize the power that is his and everyone's, or to use that power for the sake of the future.
Broader Exclusion of opposing views (#PF3720).
Narrower Inaccessible decision makers (#PF2452)
Limited decision–making context (#PF2394)
Underutilization of popular wisdom (#PF2426)
Collapse of corporate engagement in society (#PF3750).
Related Blocked minority opinion (#PD1140) Inadequate political networks (#PD2213)
Lack of means for achieving consensus (#PD2438)
Irresponsible expression of emotions equated with free speech (#PF7798).
Aggravates Social disaffection of the young (#PD1544).

♦ PF2342 Uncertain status of monetary gold
Incidence In recent years controversy arose about the price and status of gold. Proposals ranged from demonetization, through preservation of the status quo, to an increase in the gold price. The uncertain future status of gold, fed by the debate on different proposals, reduces general confidence in the international monetary system and incites speculation which causes serious damage to it.
Claim Since the USA abrogated the convertibility of the central bank's dollar claims into gold, the official gold exchange rate is a fiction.
Counter–claim Although gold no longer holds the dominant position in international payments that it once had, its traditional role as the generally accepted unit of account and ultimate means of international settlement has, to a very considerable extent, been preserved.
Broader Capitalism (#PC0564) Uncertainty (#PA6438).
Aggravates Capitalist speculation (#PC2194)
Lack of confidence in the international monetary system (#PF3058).
Aggravated by Hoarded monetary gold (#PD3045)
Inadequate world liquidity (#PG4743)
Inadequacies of the international monetary system (#PF0048).

♦ PF2343 Evil eye
Refs Elworthy, Frederick T *The Evil Eye* (1986); Maloney, Clarence (Ed) *The Evil Eye* (1976).
Broader Superstition (#PA0430) Unexplained phenomena (#PF8352).

Aggravates Unluckiness (#PF9536).
Aggravated by Bad omens (#PF8577)
Promotion of negative images of opponents (#PF4133).

♦ PF2344 Undisclosed control of national economies by limited number of individuals

Disproportionate influence of some individuals on national policy–making

Nature It is possible for a limited number of wealthy individuals to make use of networks of holding companies, family ties and other special relationships in order to maintain effective control over the major corporations and decision–making centres of a national economy. The individuals related in this way may be in government, military and banking positions as well as in public or private corporations. Because of corporation secrecy on these matters, it is very difficult to determine the extent of such control and to hold the controllers responsible for policies which they attempt to implement to the disadvantage of those outside their privileged circle.

Incidence It has been claimed that only 100 people are behind most economic decisions taken in France, that four companies decide the industrial development of Italy, and that three inter–related families virtually control the economy of Belgium.

Broader Elitism (#PA1387).
Aggravates Elitist control of global economy (#PC3778)
Disproportionate influence on national economies of limited number of corporations (#PE1922).
Aggravated by Corporation financial secrecy (#PE1571).

♦ PF2346 Inadequate care of community space

Neglect of community space — Space care frustration — Overlooked space care — Undemonstrable space care — Inattentive care of public spaces

Nature An attractive and functional physical environment is increasingly considered important in generating civic pride and community spirit yet, in many small communities, care for community space is plainly inadequate. Often no one seems to know definitively who owns large empty areas and they are therefore not put to any intentional use. Moreover, although some houses may be well built and in good repair, some often seem to be barely standing. Trash often litters the streets and untreated sewage pollutes the water. Sewage flows in front of the houses in open ditches, and yards, roadsides, ditches and waterways are littered with overflow from inadequate trash receptacles. Merchants admit that the downtown business complexes are frequently unattractive. Once prosperous communities are deteriorating for a number of related reasons: companies facing restrictive taxation and high salary costs choose to leave single–industry towns, thus causing an employment vacuum; people are forced to find jobs in other locations, and the population quickly falls far below what it had been. People who owned their own homes may no longer live in the community but retain ownership of the property; absentee landlords and the practice of informal property exchange make contact with landholders difficult when houses and lot upkeep is required; this often results in neglect. Vacant property invites clutter, vandalism and even danger, which is demoralizing to any community.

Claim Irregular maintenance of land, housing and public areas creates a self perpetuating cycle of deterioration, with accompanying substandard housing, public buildings and recreational areas, which discourages community pride.

Broader Obstacles to community achievement (#PF7118).
Narrower Divisive road patterns (#PG8771) Unchecked poverty cycle (#PG8650)
Expensive meeting space (#PG8675) Risky rental agreements (#PG1425)
Difficult piling removal (#PG8586) Insufficient land upkeep (#PJ8634)
Unused recreation spaces (#PJ8596) Obscure ownership records (#PG8731)
Unrentable vacant housing (#PJ8683) Haphazard growth patterns (#PG8687)
Unplanned land development (#PF6475) Deteriorated vacant houses (#PJ8678)
Inoperative shutoff valving (#PG8811) Unrecorded property exchanges (#PJ8546)
Insufficient formal complaints (#PG8751)
Unfeasible housing alternatives in urban areas (#PE8061).
Related Limited available land (#PC8160) Lack of fire–fighting facilities (#PJ2437)
Obstacles to availability of community space (#PF7130)
Poor condition of open spaces in urban communities (#PF1815)
Debilitating deterioration of physical environment (#PD2672)
Fragmented forms of care at the neighbourhood level (#PE2274).
Aggravates Idle private land (#PJ8020) Closure of schools (#PJ8556)
Difficult land acquisition (#PJ5369).
Aggravated by Absentee ownership (#PD2338) Inadequate maintenance (#PD8984)
Segregation in housing (#PD3442) Restrictive use of available land (#PF6528)
Unrecognized relevance of education (#PF9068)
Inadequate water system infrastructure (#PD8517).

♦ PF2347 Structural failure of citizen participation

Nature Many groups concerned with promoting local participation use methods of social change and internal decision–making which actually mediate against such participation. In the arena of the maintenance of social tranquillity, many groups have emerged since World War II concerned with providing meaningful ways for citizens to participate locally and nationally. One after another, however, these movements and organizations dissolve, either overcome by the complexity of the problem or finding their energies drained in responding to one local issue after another. In some cases, such as IRA, KKK, such groups may become more repressive and threatening to social order than the forces they sought to ameliorate.

Broader Legal contract system reduced to individual needs (#PE5397).
Narrower Limited cultural context (#PF2504) Law enforcement complexity (#PF2454)
Limited and fragmented outlook of civic minded groups (#PF2428).
Related Non–participation (#PC0588) Non–concerned attitudes (#PF2158)
Resignation to problems (#PF8781) Parochial national interests (#PF2600)
Misuse of international forums (#PF2216).
Aggravated by Citizen powerlessness (#PJ8803)
Sociological ignorance of citizen participation (#PF2440).

♦ PF2350 Insufficient communications systems

Minimal communication system — Lost communication channels — Absence of communication channels — Inadequate structures for communication

Nature Although many communities are experimenting with systems of communication that direct and organize an ever–increasing information flow to provide what people need in a useable form, many urban residents lack practical communication links with sources of necessary information. They are still hampered by the continuing decline of old information systems, physical isolation of high–rise living and disrupted employment patterns. Factual data such as police jurisdiction, land ownership and safety rules are often unclear or their existence unknown Community consensus is difficult to achieve when people are unclear about relevant data. Clearly, building information systems simply for the sake of having more information is a waste of time, but they have difficulty identifying which information is truly needed and in building a system that gives them continued access.

Claim Only when urban residents establish their priorities and initiate systematic action will a genuine re–creation of vital communication systems be possible.

Broader Inadequate infrastructure (#PC7693)
Underdeveloped provision of basic services in developing countries (#PF6473).
Narrower Poor communications methods (#PJ8656)
Inadequacy of postal services (#PF2717)
Chaotic communication networks (#PG7937)
Vulnerability of telephone system (#PE8254)
Maldistribution of television sets (#PF4136)
Lack of international communication (#PG1851)
Poor communications networks in rural areas (#PF6470)
Maldistribution of telecommunications facilities (#PF4132)
Ineffective structures for community decision–making (#PF1781)
Inadequate satellite capacity for telecommunications (#PF0045)
Inadequate telecommunications in island developing countries (#PE5655)
Cumulative depletion of corporate initiative in rural communities (#PE3296)
Inadequate development of communication services in the least developed countries (#PF4297).
Aggravates Rumour (#PF5596) Gossip (#PE2192) Citizen apathy (#PF2421)
Communication delays (#PF4453) Lack of communication (#PF0816)
Risky rental agreements (#PG1425) Ineffective communication (#PF6052)
Inaccessible decision makers (#PF2452) Irregular transport services (#PE5345)
Poor managerial communications (#PF1528) Limited economic communication (#PJ1489)
Inadequately publicized services (#PG9437) Parent–teacher non communication (#PF1187)
Lack of local leadership role models (#PF6479) Unpublished educational possibilities (#PJ9445)
Uncommunicated resource opportunities (#PG8163)
Mis–communications to societal learners (#PF7050)
Unrecognized benefits from corporate action (#PJ9420)
Misunderstanding of official communications (#PF8382)
Limited local availability of capital reserve (#PF2378)
Haphazard transmission of practical technology (#PF3409)
Dangers of private control of communications mass media (#PF2573)
Inhibition of communication between non–proximate offices (#PF6197)
Insufficient provision of public services for communication (#PF2694)
Personal isolation in communities of industrialized countries (#PD2495)
Obstacles to satellite communications for developing countries (#PF6072)
Inadequate systems of transport and communications in rural villages (#PF6496).
Aggravated by Language barriers (#PF6035) Unclear educational roles (#PG8840)
Discriminatory communication (#PD6804) Interception of correspondence (#PE5093)
Jamming of satellite communications (#PD1244) Restriction of freedom of expression (#PC2162)
Thwarted technological communications (#PF0953)
Parochial telecommunications standards (#PE1840)
State control of communications mass media (#PD4597)
Vulnerability of world cable communications (#PD0407)
Copyright barriers to transfer of knowledge (#PE8403)
Limited number of available radio frequencies (#PF0734)
Interference between communications satellites (#PF0054)
Prejudice against communication by visual imagery (#PF0076)
Disparity in world telecommunications capabilities (#PD5701)
Unbalanced application of communications technology (#PE7637)
Break–down in communications due to difference in training (#PE6650)
Control by transnationals of the global communications industry (#PE8640)
Disparities in distribution of communication resources and facilities (#PD2762).

♦ PF2352 Inhibited capacity to visualize a creative future

Dystopian visions

Nature The capacity of individuals to visualize a creative and meaningful future is hampered by a complex of interacting social images and forms. Cultural traditions limit choices to few narrow areas and fail to provide the ability to respond to needs beyond that culture's experience. Some project present trends into wretched scenarios, killing their own motivation to continue living, moving toward their dystopian visions. Acceptable futures are limited and often shallow: people are consumed with the desire to find some form of personal or family security in the traditional home or job, telling themselves that they are unable to worry about the future until the present is secure. The magnitude of changes taking place has resulted in a fear of the new by some and an infatuation for the new by others.

Broader Collapse in providing ethical value screens (#PF1114).
Narrower Individual fear of future change (#PD2670)
Breakdown of local community cohesion (#PD2864)
Fatalistic attitudes to the use of time (#PJ2795)
Prioritized attachment to security structures (#PD2884)
Static social relations inhibiting vision of the future (#PF2803).
Related Reinforced parochialism of internal values and images (#PF1728).
Aggravates Unpreparedness (#PF8176) Malignant visions (#PF5691)
Deliberate invention of new threats (#PF3699).
Aggravated by Temporal deprivation (#PF4644)
Unknowable future patterns of social choice (#PF9276).

♦ PF2358 Demoralizing images of rural community identity

Unformed confidence in community corporateness in rural villages

Nature While there is a growing world–wide awareness of the interdependence of all facets of society, rural villages find themselves struggling with fragmented and personalistic forms to maintain basic care. They have yet to achieve powerful corporate efforts because of the difficulties such efforts face in a subsistence economy. The consuming work schedule of villagers, and their individual and family concerns, allow little time for community meetings. Besides, there are generally no long–range plans for various local programmes, despite the general desire that these should exist: bulk buying remains an unused alternative to high–cost individual supplies; special training, a much needed dimension in agriculture remains uncoordinated; over–emphasis on labour results in ineffective efforts as well as an inability to organize community actions; strong family ties further weaken the community spirit as personal needs continue to override those of the community.

The civic identity of many rural communities is transmitted through images which demoralize resident and non–resident alike. For example, seasonal annoyances (odour of open sewers, mosquitoes), lack of commerce, lack of public services, and a disbelief that industry would ever want to locate there, all serve to intensify the community's demoralization. This sense of hopelessness, the belief that the community is dying and the conviction that others would not choose to live in such a place undermines the community's will to build its own future. Until these images of civic identity are radically altered, small communities will be unable to generate the motivation and enthusiasm needed to carry out development.

Broader Obstacles to leadership (#PF7011) Deteriorating community identity (#PF2241)
Obstacles to community achievement (#PF7118).
Narrower Mistrust (#PA9041) Divisive family ties (#PG9513)
Lack of pioneer spirit (#PG9623) Domestic refuse disposal (#PD0807)
Unhelpful vandalism image (#PG8732) Unrealized corporate power (#PG9526)
Collapsed corporate effort (#PG9604) Unorganized labour potential (#PG5466)
Limited transportation availability (#PJ1707)
Disrelationship from community history (#PJ0836)
Complex interrelationship of world problems (#PF0364).
Related Lack of fire–fighting facilities (#PJ2437) Ill effects of educational failure (#PF2013)
Demoralizing constraints on housing rehabilitation (#PE2451).
Aggravates Poor social environment (#PJ8742) Inadequate land drainage (#PD2269)
Decay of rural communities (#PD9504) Insufficient community events (#PF5250)
Individualistic agricultural planning (#PJ1082)
Ineffective structures of local consensus (#PF6506)
Deteriorating structures of rural community cooperation (#PF3558).

Aggravated by Dust (#PD1245)
Insufficient rural housing (#PF6511)
Demeaning community self-image (#PF2093)
Shortage of resident professionals (#PG8812)
Debilitating designs for using labour (#PF2812).
Geographical isolation (#PF9023)
Uncoordinated special training (#PG9485)
Inappropriate sanitation systems (#PD0876)
Inadequate safeguards against fire (#PD1631)

♦ PF2364 Denial of right to develop as human beings

Broader Violation of human rights (#PB3860) Denial of political and civil rights (#PC0632).
Aggravated by Governmental disregard for people as human beings (#PD8017).

♦ PF2365 Disorganized approach to land ownership in tropical villages

Nature Land rights which have been handed down through families seldom have written deeds and are often not honoured. Land can be claimed, but procedures are complex and difficult for any one person. In many rural communities, land holdings are still very small and procedures for ownership unclear. This results in a disorganized approach to obtaining the land necessary for major income farming. Information on land ownership is sparse, and efforts are rarely organized to locate accurate information on ownership procedures. Large sections of jungle land are unused, land ownership being the major factor preparatory to any significant clearing of the land for agriculture – outside ownership often restricts jungle crop harvesting.
Broader Disorganization (#PF4487).
Narrower Conflicting land rights (#PG9509) Barriers to jungle harvesting (#PG9544).
Aggravates Difficult land acquisition (#PJ5369)
Endangered plantations of long-lifed trees (#PJ9565).
Aggravated by Absentee ownership (#PD2338).

♦ PF2373 Lack of assertiveness

Unassertive people
Nature Few people have adequate self-respect to clearly and honestly communicate their needs in response to those of another. As a result, personal relationships are littered with fear of self-exposure, shame and deceit.
Refs Vikler, Henry *Assertiveness* (1988).
Broader Fear of intimacy (#PF8012).

♦ PF2375 Minority control

Nature A racial, religious, linguistic or ideological minority may exercise control over the majority. In some ex-colonial and pluralistic societies, a class system and remnants of colonialism help maintain minority control. Effective control may be attained illegally by minority terrorist campaigns, facilitated by modern sophisticated weaponry.
Broader Racism (#PB1047) Elitism (#PA1387) Social injustice (#PC0797)
Social inequality (#PB0514).
Related Political pluralism (#PF2182) Dictatorship of the majority (#PD3239).
Aggravates Segregation (#PC0031)
Social conflict (#PC0137)
Religious conflict (#PC3292)
Class conflict (#PC1573)
Racial conflict (#PC3684)
Political repression (#PC1919)
Denial of the right of trade union association (#PD0683).
Aggravated by Militarism (#PC2169)
Class consciousness (#PC3458)
Plural society tensions (#PF2448)
Colonialism (#PC0798)
Social discrimination (#PC1864)
Revolutionary communism (#PC3163).

♦ PF2376 Accumulation of knowledge

Excessive data — Accumulation of data — Accumulation of information
Nature Throughout history a basic problem has been to acquire sufficient information to generate effective change. The individual wishing to become expert in some field of knowledge had to buy information; the government wishing to understand even the rudiments of the structure of its society had to buy information through the census. Increasing amounts of money have continued to be paid for data acquisition on the assumption that data constitute information. But data only become information at the point when the exposed individual is changed and the capacity to be changed is clearly strictly finite. In conditions of data paucity, almost all data acquired can be transformed into information and used to procure effective change. But in conditions when the supply of data far outruns the processing capability, most data are literally worthless.
Systems for the collection of data can acquire a momentum of their own such that resources continue to be deployed to gather, order and store that data beyond any currently foreseeable need. In addition to an often costly diversion of resources from projects of higher priority, such systems gradually become to be seen as an end in themselves. This obstructs the formulation of more fundamental questions which would call for the collection of new kinds of data or the ordering of what has been collected in new ways.
Claim Data are an excrescence, the very latest kind of pollution. Nothing can be done about the management of information and knowledge towards the regulation of society as long as an approach is made in data-processing terms. Data are worthless until mechanisms are developed to transform data into information and to enable the use of that information to innervate society.
Broader Accumulation (#PA4313)
General obstacles to problem alleviation (#PF0631).
Aggravates Proliferation of unprocessed scientific data (#PF1065).

♦ PF2378 Limited local availability of capital reserve

Limited extent of capital funds — Restricted availability of capital — Limited availability of funds
Nature The complexity of present-day society renders relatively large and rapid investments of money and pooled resources necessary in order to begin community development, and to start-up and expand business and services in both private and public sectors. In contrast to the trend for effective income realization to require ever larger units employing substantial capital for specialized production, the limited capital of small farms and businesses at the village level forces a fragmented, marginal level of production.
Agricultural diversification would involve purchasing more machinery, hiring more labour and increasing the amount of both labour and capital needed to manage a more complex operation. Without creating a financial reserve, there can be no loan structure, credit union or bank. In many small communities, despite an intensive drive for capital resources, the low income generated by most family production units means that they are unable to generate new capital, and they struggle to purchase supplies and to market products on their own. Subsistence-level income and lack of community capital prohibits investment, since an immediate return on investment cannot be assured and the tax base of land owners is often too low to support the building of necessary facilities or provide local fire, police and ambulance services. The limited availability of resources leads people to clear the land with only what is at hand, and market only the crops one person can carry into the village to sell.
Many people are employed outside their home villages; they come to the village only at weekends, and most of their earnings remain in the city. Residents purchase almost all personal and household goods and services outside the village on a small scale limited by what one person can carry. People are motivated to sell quality food which they need for their families and to purchase supplies outside their villages, further draining local capital and reducing employment alternatives within the village. These spending patterns continue the cycle of money out of local circulation and inhibit buildup of financial reserves that could be used as collateral to float loans or build up credit. Limited capital restricts local processing, so products are marketed as inexpensive raw materials; and the absence of local markets forces dependence on costly middleman services in order to allow goods to reach outside markets. People come to feel there is no hope for money or investment.
Without massive income improvement it is increasingly difficult for village people to avoid falling victim to subsistence-style living. Any disaster, such as drought, totally disrupts the delicate balance of marginal subsistence and consumes essential capital resources.
Broader Undirected expansion of economic base (#PF0905)
Ineffective structures for community decision-making (#PF1781).
Narrower Lack of savings structures (#PF1348) Unavailable machinery capital (#PG7823)
Insufficient enterprise capital (#PG7832) Inadequate agricultural capital (#PJ1368)
Prohibitive livestock husbandry costs (#PJ4075)
Strained capital resources in small communities (#PF3665)
Lack of credit facilities for agricultural producers (#PE8516)
Lack of channels for obtaining available local funding (#PF6544)
Constricting level of capital development in rural areas (#PE1139).
Related Limited individual capital reserves (#PF2899)
Limited availability of investment capital for urban renewal (#PF3550)
Failure of public authorities to assist in financial investment (#PF6508).
Aggravates Barriers to economic education (#PG7826)
Insufficient financial resources (#PB4653)
Unfocused style of community operations (#PF6559)
Limited availability of financial credit (#PF2489).
Aggravated by Insufficient communications systems (#PF2350).

♦ PF2384 Limited individual attention span

Attention deficit — Attention disorders — Inability to concentrate
Nature Although limited attention span is characteristic of people suffering from severe debilitation, it may also occur under other circumstances. It results in inability to concentrate for any length of time, distractedness, irritation, restlessness and impatience. It may be associated with chronic alcoholism, drug abuse and delinquency. Distractedness is especially characteristic of children who normally grow out of it.
Incidence In 1989 it was estimated that in the USA from 3 to 10 per cent of children (approximately 70 per cent boys) had attention deficit problems. Over 30 per cent of school children continue to have such problems as adults. From 2 to 5 per cent of American adults (namely 6 million) suffer in this way. Nearly 50 per cent of the children who do suffer from it also have specific learning disabilities characteristic of dyslexia. Nearly 30 per cent are diagnosed as anti-social or as having a conduct disorder.
Broader Vulnerability of human organism (#PB5647)
General obstacles to problem alleviation (#PF0631).
Narrower Inability to concentrate due to torture (#PE3716).
Related Substance abuse (#PC5536) Juvenile delinquency (#PC0212)
Irritation response pattern (#PG8769).
Aggravates Dyslexia (#PE3866) Intolerance of complexity (#PG4827).
Aggravated by Stroke (#PE1684) Ignorance (#PA5568)
Distractions (#PJ3953) Mental deficiency (#PC1587).

♦ PF2385 Prohibitive cost of necessities in rural communities

High local prices — Rising local expenses — Inflated prices in rural communities — Rising food prices in rural communities
Nature While modern technology has made it possible in principle for the resources of the entire globe to be at the disposal of every community, many villages are in fact cut off from these necessities. The high cost of training discourages technical self-sufficiency within the community, which continues to rely on services from outside – the high salaries demanded by vocational teachers admit few skilled instructors. Facilities such as health clinics may be built only to be left partially unused because local people have not been able to meet the expense of equipment and personnel. Decreasing agricultural resources at the local level force more outside purchasing. This spirals prices upward.
Incidence In fishing communities, boats and equipment to go further for larger fish are economically out of reach, so local waters become over-fished, further decreasing income and buying power.
Broader Prohibitive cost of living (#PF1238) Lack of local commercial services (#PF2009)
Underproductive methods of agricultural management (#PF6524).
Narrower Costly teaching staff (#PJ9524) Prohibitive cost of education (#PF4375)
Long-term shortage of natural resources (#PC4824).
Related Prohibitive cost of accommodation (#PD1842).
Aggravates Inadequate medical facilities (#PD4004)
Precarious basis for family economics (#PF1382).
Aggravated by Inappropriate cash crop policy (#PF9187).

♦ PF2386 Loneliness

Dependence on loneliness — Lonely people — Social isolation — Unfulfilled sociability
Nature Loneliness can afflict anyone, whether young or old, rich or poor, highly-educated or illiterate, healthy or infirm. Because children are almost invariably reared in a network of intimate relationships that, if not a natural or adopted family, has some other social context (usually institutional), in the first three or four years the individual is imprinted with the norm of being accompanied. Someone – whether parents, siblings, age cohorts, friends, relatives, social workers – is in daily contact with the child. This position of security may change gradually, for instance, if childhood companions grow away from each other, or suddenly, perhaps by a change in location or situation or by a death.
Social isolation is as significant to mortality rates as smoking, high blood pressure, high cholesterol, obesity and lack of physical exercise. Studies have show that people who are isolated but healthy are twice as likely to die over the period of a decade or so as others with the same health. Men are at more danger than women.
Incidence People who are physically isolated, such as those who work at home or in remote areas, are hospitalized or otherwise infirm, are often subject to loneliness. Equally painful is the subjective loneliness experienced by a shy or sensitive person, or by one who is, for whatever reason, alienated from society.
Claim Industrial civilization, and particularly urban lifestyles, tend to force individuals into many superficial, apparently socializing, contacts, while other features of modern society tend to cause actual isolation. Such isolating features include the breakdown of the extended family, the erosion of family-centered values, and passive entertainment. Statist governments reduce individuals to anonymities, while technological progress exalts the intellect at the expense of the feelings. Smaller families, social mobility and uncontrolled urbanization decrease the opportunities for frequent, long-term and personal social interactions in a community context.
Refs Lynch, J *The Broken Heart* (1977); Woodward, John C and Queen, Janel *The Solitude of Loneliness* (1988).
Broader Duality (#PA7339) Exclusion (#PA5869).
Narrower Living alone (#PF3089) Unrequited love (#PF6096)
Loneliness in adults (#PG4829) Loneliness in old age (#PD0633)
Loneliness of children (#PC0239) Loneliness in single people (#PD4392)
Self-imposed social isolation due to torture (#PE4703).

Related Alienation (#PA3545) Unsociability (#PA6653) Dissatisfaction (#PA8886)
Mental suffering (#PB5680) Psychological alienation (#PB0147)
Isolation of parent-child relationship (#PC0600).
Aggravates Suicide (#PC0417) Boredom (#PA7365) Widowhood (#PD0488)
Proliferation of pets (#PD2689) Stress in human beings (#PC1648)
Lack of self-confidence (#PF0879) Dependence on the media (#PD7773).
Aggravated by Agoraphobia (#PE0527) Human ageing (#PB0477)
Social isolation (#PC1707) Personal life crises (#PD4840)
Lack of community participation (#PF3307)
Exclusion of pre-adults from family decisions (#PE2268).

♦ PF2394 Limited decision-making context

Myopic decision-making methods

Nature Operating from immediate needs rather than long-range planning, intersocial bodies have no effective way of dealing with the complexity of the future. The lack of long-range decision-making methods results in a limited global exchange of gifts and insights; and lack of motivation to participate in the corporate creation of the future inhibits the production of alternatives means for dealing with future requirements and thus obstructs long-term development of civilization.

Claim A larger, societal context from which to make decisions corporately is missing in present-day affairs. The individual's ability to affect the decision-making process through free assembly as a corporate person is reduced to the most immediate issues.

Broader Collapse of societal engagement (#PF2340).
Aggravates Avoidance of decision-making (#PF4204).

♦ PF2400 Unarticulated goal of educational methods

Abstract educational methods

Claim There has been a general collapse in the purpose of education, following shifts in methods of expert training. This has meant a reduction in the overall context which enabled individuals to see the social applications of the skills they were learning, and has resulted in egocentric economic goals being the main stimulus for obtaining an education.

Broader Ineffective methods of practical education (#PF2721).
Aggravates Meaningless corporate engagement (#PD2671)
Thwarted technological communications (#PF0953)
Individualistic utilization of expertise (#PF5639)
Inequitable distribution of skilled specialists (#PD2479)
Lack of opportunities for practical training in communities (#PF2837)
Gap between the function of social techniques and the needs they address (#PF3608)
Unrelated ability application and situational demands in vocational decisions (#PF2828).
Aggravated by Limits on areas of research (#PF2529)
Personalistic use of training (#PF2445)
Non-articulated educational goals (#PF2595)
Ineffective educational policy decisions (#PF2447).

♦ PF2401 Incomplete access to information resources

Inadequate access to information — Poor information channels — Infrequent information dissemination — Vagueness of civic information — Inadequate information spread

Nature As communities expand and the world seems to grow smaller, people are increasingly expected to be aware of the world's common knowledge. Nevertheless, because rural areas often have no postal service, telephone, newspaper or public transportation, they are often out of touch with the information needed for their development. Social agencies which generate self-help services demand a sophistication in programming and a literacy level which does not exist in many such areas. The effects of polluted water, the practices needed to keep trees producing, the knowledge of what foods to cultivate, and the full services offered to villages by the government may also be unknown. This limited access to information severely restricts community initiative, and it remains cut off and unable to take advantage of generally available information, thus perpetuating antiquated methodologies.

At the national level communications can be divided structurally into two main branches: production on one hand, and distribution on the other, of information, opinion and entertainment. In practice, the division has never been absolute and overlap and unified control of both branches is now often much greater than in the past. The distinction is relevant because many countries, when developing their own communication systems, have given priority to distribution at the expense of production. Hence, they find themselves dependent on investment from abroad in the infrastructure, on news compiled by outside organizations, on entertainment also created far away, and in general on sources of production over which they have no influence. Although most countries have national news agencies, they often have meagre resources, material, technical or staff, so that their supply of news must be supplemented by outside material. For this reason, among others, the mass media in such countries still depend mainly on news selected and transmitted by larger outside agencies. Entertainment schedules on radio and TV are also heavily laden with imports from abroad and the advertising field is often influenced, if not controlled, by branches of international companies. In many instances this pattern leads to large-scale foreign intervention, heavy external investment and unhealthy competition in the development of the material-producing branch of the communication industry. It may also sometimes cause the creation of national and international monopolies in one or more of its components. Such centralization often tends to create a certain amount of standardization in media products.

Broader Underprovision of basic services to rural areas (#PF2875).
Narrower Limited political know-how (#PG9178) Incomplete skills information (#PJ0140)
Insufficient programme knowledge (#PG7844) Insufficient flow of information (#PF6469)
Insufficient technological contact (#PG7841) Lack of sharing of community skills (#PF3393)
Minimal access to necessary information (#PF2631)
Inoperative forums for public information (#PF7805)
Interference by employers in the affairs of workers' organizations (#PE8384).
Related Untransposed community structures (#PF6450)
Incompatibility of rural values in urban cultures (#PF2648)
Disproportionate control of global economy by limited number of corporations (#PE0135).
Aggravates Misleading information (#PF3096) Loss of cultural identity (#PF9005)
Suppression of information (#PD9146) Colonization of information (#PF4894)
Insufficient financial resources (#PB4653)
Restrictions on freedom of information (#PC0185)
Parochial escapist media entertainment (#PD0917)
Limited availability of financial credit (#PF2489)
Limited development of functional abilities (#PF1332)
Inadequate circulation of local information (#PF6552)
Inadequate information concerning transnational banks (#PF4350).
Aggravated by Vagueness of laws (#PF9849) Withholding of information (#PF8536)
Proliferation of information (#PC1298).

♦ PF2403 Incompatibility of transport modes

Broader Insufficient transportation infrastructure (#PF1495)
Inadequate standardization of procedures and equipment (#PC0666).

♦ PF2407 Geomagnetic field anomalies

Nature The Earth's magnetic field is not uniform but contains anomalies, particularly in the polar regions. These anomalies create navigational problems.

Background The Earth's magnetic field went through a peak of intensity about 2,000 years ago, when it was about half again as strong as today.

Broader Geomagnetic disasters (#PD0830).
Narrower Geomagnetic storms (#PD1661).

♦ PF2409 Atheism

Denial of gods — Denial of God — Rejection of God — Ignorance of God

Nature Atheism is the denial of a god or gods, in theory or in practice. As such it has frequently been used as an accusation against ideological opponents on a point of definition. It may also be used to describe amoral behaviour. Atheism in its ultimate forms denies a spiritual aspect to man. With the discovery of scientific explanations for phenomena and with increasing materialism, disenchantment with former religious explanations has given rise to atheism among wide range of people who would debate the existence of a god as a hypothesis. This might include agnostics or adherents of nontheistic faiths such as Marxism.

Incidence Occasionally statistics from polls concerning religious beliefs are published. For example, one poll in Sweden credits 48 percent of the population with belief in God, and then reports that 5 percent of the population regularly attend church. Atheism is sometimes imputed to the ordained religious who leave their ministries, orders or churches; but it may more frequently be represented by inadequately educated and theologically prepared clergy who have no deep comprehension of their faith and in an attempt at modernism, reduce their God to a humanistic value.

Background The Jews and early Christians were accused of atheism by the Romans because they refused to recognize the Roman gods and the Emperor as divinities. Orthodox Christians used the accusation against various heretics, notably those who acknowledged God but denied the Trinity.

Claim Atheism severs humans from their root, which is god, impoverishing, depriving of inner values, and leading to gradual succumbing to the de-humanizing dangers surrounding them.

Atheism, both by etymology and by usage, is essentially a negative conception presupposing the existence of theism. It is therefore not the replacement of a specific understanding of God with another specific understanding of God. It can arise from the believe that matter and physical force constitute the ultimate reality of the universe, and that, through the aggregation of the elements of matter in various organic forms, life and the infinitely varied forms of consciousness has emerged, i.e., materialism. Sensationalism can also be a source of atheism; i.e., the belief that all ideas are derived from sensations or from reflection on sensations.

Refs Blaskow, Ned *Steamroller and the Pebbles*; Buckley, Michael J *Modern Atheism* (1987); Molnar, Thomas *Theists and Atheists* (1980).

Broader Irreligiousness (#PF8234) Religious conflict (#PC3292).
Narrower Scientism (#PF3366) Materialism (#PF2655) Religious dissent (#PE4875).
Related Theism (#PF3422) Impiety (#PA6058) Positivism (#PF2179)
Inattention (#PA6247).
Aggravates Wrath of God (#PF8563) Religious prejudice (#PD4365)
Dependence on religion (#PF0150).
Aggravated by Polytheism (#PF4957) Ideological conflict (#PF3388)
Ideological repression (#PC8083) Double standards in morality (#PF5225)
Scientific explanations of phenomena (#PG4877).
Reduced by Agnosticism (#PF2333).

♦ PF2410 Inappropriate level of technological equipment

Nature Although the acquisition of complex labour-saving machinery is capable of increasing the productivity of communities, the people of some communities still prepare and harvest their crops using a simple machete. Rural people know of the existence of more sophisticated technology equipment, but they cannot believe they would have the opportunity to own and use it themselves. There is no transport available, and people walk to and from town carrying virtually all of their purchases and market goods on their heads or over their shoulders. The time-consuming physical labour needed to sustain the basic necessities of life leaves residents with little energy to engage in other meaningful work or activity, and the cost of industrial and agricultural machinery is prohibitive. But far more important than these factors is the low value which villagers place on their own energy as a human resource.

Claim Until a more appropriate level of technology is acquired in rural areas for agricultural and public use, communities will not move from subsistence to self-reliance.

Broader Use of inappropriate technologies in developing countries (#PF0878).
Narrower Unavailable road machinery (#PG7850)
Lack of agricultural machinery (#PF4108)
Prohibitive cost of capital equipment (#PJ1768)
Inadequate maintenance of infrastructure (#PD0645).
Related Inadequate maintenance personnel (#PJ0088).
Aggravates Lack of school transportation (#PJ7849).
Aggravated by Prohibitive cost of transportation (#PE8063).

♦ PF2411 Haphazard provision of consumer services

Nature Commercial life as expressed in day-to-day supply and demand is depressed in urban neighbourhoods and its revitalization is a crucial factor for the future development of cities. Every local community needs to participate in the exchange of goods and services. Though many city centres have huge shopping areas, other neighbourhoods have suffered a sharp decline in domestic commercial life. They are in fact less adequately provided for than the average outlying village or suburb. The network of shops and services traditional at such a level has declined abruptly and, as a result, shop owners are hesitant to make further investments to expand the volume of their services. This creates a vicious circle by which people go elsewhere for an ever wider range of items and for lower prices, so that shopowners have to reduce their stock for smaller patronage, thus in turn attracting fewer customers. This process is modified by the 'captive consumer' phenomenon which keeps prices higher than in a more competitive area.

Broader Grievances of consumers (#PD7567) Unjust dismissal of workers (#PD5965)
Lack of local commercial services (#PF2009).
Narrower Small business failures (#PE9405) Defensive marketing facilities (#PG9414)
Unprofitable entertainment market (#PG9384) Non-competitive consumer services (#PG9420).

♦ PF2415 Superficial research on the total human process

Nature There is a failure or unwillingness to develop the skills necessary to study in depth the problems of human relatedness. For example, research competences have neglected to study the implications of emerging life styles and structures both for individual human living and for society, tending to concentrate more on the externals of life. This leads to an emphasis on superficial research techniques rather than on the total human process of which such techniques are the outward manifestation.

Broader Superficial research (#PF7889) Limits on areas of research (#PF2529).

♦ PF2418 Execution of inappropriate orders

Execution of orders in violation of human rights — Execution of orders endangering the environment — Execution of illegal orders — Blind obedience — Irresponsible obedience

Nature Officials, employees and military personnel can receive orders which they recognize as leading to harmful or unjust consequences. They are then faced with whether to obey them, given that obedience is the basis for their employment and they are liable to lose their job, or be subject to other penalties, if they fail to execute the order.

Incidence In some countries there is authorization not to execute orders which are manifestly unlawful, but it remains difficult to define the meaning of manifestly by reference to international law. Despite acceptance of the principle, in concrete settings it remains easier for the individual to execute such an order rather than be exposed to the penalties of failing to do so.
Broader Immorality (#PA3369) Unethical practices (#PC8247).
Related Disobedience (#PA7250).
Aggravates Neuritis (#PG4892) Tonsillitis (#PE2292) Anaphylaxis (#PG4894) Weakened heart (#PG4890) Military atrocities (#PD1881).
Aggravated by Unlawful government action (#PF5332).

♦ PF2420 Inadequacy of governmental decision-making machinery
Claim The process of decision-making in national capitals (upon which the control of intergovernmental organizations depends, both in allocating funds between organizations and in programming individual bodies), remains not merely unscientific, but is often left to the caprice of events, to what has been customary in the past, or even to the caprice of interested individuals in governmental departments.
Broader Chance (#PA6714) Inefficiency (#PB0843).
Narrower Inadequate laws (#PC6848) Avoidance of decision-making (#PF4204) Uncoordinated government policy-making (#PF7619).
Aggravates Inappropriate policies (#PF5645).
Aggravated by Bureaucracy as an organizational disease (#PD0460).

♦ PF2421 Citizen apathy
Fragmented modes of citizen engagement — Fragmented citizen responsibility — Inadequate citizen input — Minimal citizen involvement — Sporadic community lobbying
Nature There are apparently no formal channels of communication that bridge the diverse religious, racial and economic groups of small communities. At a time when it is more and more necessary for all community members to work together for the common well-being, efforts by the residents of such communities to engage in the life of the community are fragmented, and there are no mechanisms by which they can creatively resolve tensions. In addition, some residents engage in school-related activities, others in agricultural production and still others in commercial life, and these diverse arenas hinder the people from coming together for a common task. However, until practical cooperation becomes an habitual pattern of engagement, efforts to attract industry will be disorganized and ineffective; and there is little hope that effective development can be implemented.
Broader Unstructured local decision-making (#PF6550) Ineffective structures of local consensus (#PF6506).
Narrower Crop dusting (#PG8744) Juvenile alcoholism (#PD1611) Herbicide damage to crops (#PD1224) Non-resident school personnel (#PG8764) Insufficient industrial promotion (#PJ8378) Improper implementation of programmes (#PG8804).
Related Apathy (#PA2360) Indifference (#PA7604).
Aggravates Citizen powerlessness (#PJ8803) Inadequate health services (#PD4790).
Aggravated by Insufficient communications systems (#PF2350).

♦ PF2423 Restricted scope of local employment
Nature Although diversification is necessary to develop a broad-based economy, many rural communities continue to limit themselves to agricultural activities, and the size and scope of rural business life has not changed significantly in many years. There is also a prevalent attitude that only one age group or only one sex does certain jobs, almost as if there were no way to keep everyone busy all the time. A large percentage of the established population works in the fields at the intensive labour of farming, employment which the young often reject in favour of city jobs. This migration pattern is disruptive to family patterns and to the entire community when whole generations are absent. The exodus of young people to the city causes village population to remain small despite a high birth rate. Many of the former residents would probably prefer to remain in the villages but see few job or economic investment options there, since the government provides few programmes for communities of small size, either to attract industry into rural areas or to provide jobs.
Broader Limited employment options (#PF1658).
Narrower Size limits on industry (#PG7852) Attraction of city life (#PJ7861) Insufficient job options (#PJ0070) Restrictions on eligibility for government programmes (#PE7965).
Aggravates Youth migration to cities (#PJ7859) Social disaffection of the young (#PD1544).

♦ PF2426 Underutilization of popular wisdom
Untapped community wisdom — Unshared practical wisdom — Unrecorded popular wisdom
Nature Because there is no way to utilize all of popular wisdom in forming common decisions, the basis of consent for determining social priorities is reduced. This leads to a stratification of people into groups which view themselves either as the elite or as victims.
Broader Collapse of societal engagement (#PF2340) Underutilization of human resources (#PF3523) Lack of local leadership and decision-making structures (#PF6556).
Aggravates Erosion of elders' wisdom (#PF1664) Lack of sharing of community skills (#PF3393) Human wisdom unrelated to daily life (#PD1703) Failure to profit from patterns of history (#PF1746).
Aggravated by Unrecorded knowledge (#PF5728).

♦ PF2428 Limited and fragmented outlook of civic minded groups
Nature Groups which have come together in hopes of correcting the imbalance of overly-powerful social dynamics often have myopic and fragmented outlooks. Without a common concrete vision of the future, these groups are unable to formulate the kind of sustained corporate thrust needed to deal effectively with complex social problems. Lacking such vision also implies lack of of distance and perspective so they constantly dissipate their energy without the strategic and intentional use of long periods of time.
Broader Structural failure of citizen participation (#PF2347).

♦ PF2431 Inadequate organizational mechanisms to act against problems
Nature Organizations often lack both the capacity to anticipate problems and the capability to respond to them. This is connected to their inability to identify problems and accurately act against them.
Claim Competition among governmental and intergovernmental agencies, or poor communications among them, does not allow for coordinated attacks on problem areas.
Broader General obstacles to problem alleviation (#PF0631).
Narrower Inadequate organizational mechanism for international action (#PE8776).
Aggravated by Institutional preoccupation with obsolete problems (#PJ5014) Inadequate power of intergovernmental organizations (#PF9175).

♦ PF2434 Lack of human unity
Lack of world unity
Broader Fragmentation (#PA6233).
Narrower World anarchy (#PF2071) Lack of understanding (#PJ4173).
Related Ideological conflict (#PF3388) Fragmentation of religious belief (#PF3404) Non-viability of small states and territories (#PD0441).
Aggravates Conflict (#PA0298) Alienation (#PA3545) Exploitation (#PB3200) Social injustice (#PC0797) Social inequality (#PB0514).
Aggravated by Fear (#PA6030) Avarice (#PA6999) Aggression (#PA0587) Insecurity (#PA0857) Lack of communication (#PF0816) Inadequate ideological frameworks (#PD0065).

♦ PF2437 Ineffective structures for community participation
Nature The increasing complexity of community concerns gives rise to a need for comprehensive planning and for a network of local organizations. However, most communities have no stated community plan; a few individuals and organizations bear the full weight of making decisions and taking actions for the whole community. Often no single group has the knowledge or sources to take all factors into account before making a decision. Decisions must often be made quickly, frequently in a vacuum as the overall wishes of the community are unknown. The absence of broad community planning and of channels for implementing such plans leads to lack of consensus on issues and difficulty in organizing support for those who lead community programmes. Unsupported leaders become disillusioned and withdraw from further community involvement, thus depriving the community of their creativity. This waning of leadership willingness is accompanied by reluctance of new leadership to emerge.
Broader Obstacles to community achievement (#PF7118).
Narrower Anticipation of criticism (#PJ7893) Frustrated personal effectiveness (#PJ7892) Unexperienced individual effectiveness (#PG7890) Shallow approaches to motivational techniques (#PF1004).
Related Undeveloped youth leadership (#PJ0151).
Aggravates Non-participation (#PC0588) Lack of community participation (#PF3307) Social disaffection of the young (#PD1544) Lack of local leadership role models (#PF6479) Unorganized development of work forces (#PF2128).
Aggravated by Hidden individual talents (#PG7887) Unrecognized benefits from cooperatives (#PF9729).

♦ PF2439 Confined scope of business operations
Nature The recent trend toward siting commercial centres some distance from small towns has resulted in economic problems for such communities. Few, if any, industries and only very small commercial businesses provide income for the town: outside businesses are not attracted to such small communities. The result is that farmers, who once could procure agricultural supplies and services nearby, must now travel to other towns. Townspeople can purchase groceries at several local grocery stores, but each store offers the same limited variety at prices necessarily higher than those charged by large chain stores in regional cities. Until means are found of expanding the scope of business operations, finance desperately needed for development will continue to be drained away by outside commercial enterprises.
Broader Excessive government control (#PF0304).
Narrower Limited purchasing power (#PD8362) Minimal commercial activities (#PG8706) Continuing commercial decline (#PJ8726) Unattractive local businesses (#PG8806) Inadequate economic incentives (#PF8554) Uncompetitive local merchandising (#PG8696) Excessive duplication of products (#PG8756).
Related Restrictive pattern of business activities in small communities (#PD1415).
Aggravates Unexplored potential markets (#PF0581) Declining community population (#PJ8746).
Aggravated by Inaccessible market and supply centres (#PF8299) Discouraging conditions for small business (#PD5603) Insufficient transportation infrastructure (#PF1495).

♦ PF2440 Sociological ignorance of citizen participation
Pluralistic ignorance
Nature Many groups that promote citizen participation are hampered by their own naiveté about the complexity of institutions and relationships which the law enforces. They lack the perspective which a concrete vision of the future would give, and have no distinct vantage point from which to evaluate the strategy for most effectively influencing the social fabric. Their efforts are dissipated in responding to immediacies.
Individual members of a group believe incorrectly that they are each alone or the only deviants in believing or not believing in particular values, while in reality many others, if not the majority, secretly feel exactly as they do.
Broader Political ignorance (#PC1982).
Aggravates Non-participation (#PC0588) Structural failure of citizen participation (#PF2347).

♦ PF2442 Dehumanization of death
Lack of deathing awareness — Inhumane dying conditions
Nature Modern society provides little emotional or spiritual comfort to facilitate the process of dying, which is often suffered in loneliness amid strangers and machines in a dehumanized hospital environment. People have difficulty in dying with dignity under such conditions, isolated from the rest of society, surrounded by a conspiracy of silence, denial and dissimulation. Alternative images of deathing, with honour and creativity, are available only to the few.
Modern society is a death-denying one. Terminally ill patients are segregated from the larger society. Cancer, AIDS and other terminal diseases popularly mean death, excruciating pain, exclusion from one's social and physical environment and a burden on society. Grief and mourning are considered abnormal. Dying is a taboo. Doctors operate out of a paradigm of saving terminal patients from death rather than caring for them as they die. As the saviour the physician does everything possible scientifically to cure his patients. The patient is facing an assault rather than be aided to cope. In the final phases of life the physician is ill prepared to facilitate the death for the patient. Some doctors are willing to use euthanasia which is another way maintaining control over the care of patients and avoiding the dying process. In many situations the patient is not even told that he has a terminal disease and has no way of knowing what is happening or of coping with his own death. Telling the truth to a patient takes time; they break down, get depressed. This failure to tell the truth creates a public attitude that some diseases like cancer are completely incurable because no one has ever heard of someone being cured.
Broader Social neglect (#PB0883) Dehumanization (#PA1757) Denial of right to dignity (#PE6623).
Narrower Failure to notify the imprisonment or death in prison of foreign nationals (#PE6424).
Related Despair (#PF4004) Spiritual void (#PA6220).
Aggravates Fear of death (#PF0462).
Aggravated by Human death (#PA0072) Chronic terminal illness (#PE4906) Human disease and disability (#PB1044) Selective perception of facts (#PF2453) Dehumanization of health care (#PD7821) Excessive prolongation of the dying process (#PF4936).

♦ PF2444 Complex regulations paralyzing small communities
Nature Residents of small communities find complex regulations paralyzing and isolating when they try to link their community with the larger structures that provide necessary services. Such regulations create misunderstanding and frustration. For example, in many communities property taxes seem to depend on arbitrary and inflationary valuation of land which requires a disproportionate amount of income be spent on taxes. In addition, some agencies find local people demonstrate

their independence in a reluctance to use their services; and the distance from such services of these communities, combined with the relatively small number of people served, increases the cost of the service and puts particular hardship on families. Often families are left to care for their own.

Broader Complex government regulations (#PF8053)
Excessive government intervention in the private sector (#PD4800).
Narrower Undiversified tax base (#PJ7897) Complex tax regulations (#PG7900)
Unenforced sanitation codes (#PJ7979) Restrictive regulations for training (#PJ7972).
Related Distortionary tax systems (#PD3436).
Aggravates Insufficient emergency services (#PF9007)
Remoteness of legal services in developing countries (#PE9001).

♦ PF2445 Personalistic use of training

Nature There is a gap between the practical training the individual receives (which tends to be based on methodologies primarily oriented towards individualistic use of the training attained), and self-aware means of dealing with the ordinary life and death issues that constantly confront people. The corporate methodologies which would allow people of the present-day to relate consciously to their lives are inadequately used and the educational process is inhibited.
Aggravates Unarticulated goal of educational methods (#PF2400).

♦ PF2447 Ineffective educational policy decisions

Inadequate government educational policy

Claim Groups which determine educational priorities operate on a very local level from a perspective too narrow to include broad social issues. Thus, for example, although wishing to provide future-oriented education, school boards often tend to make very conservative decisions, lacking in the wisdom available from outside sources.
Broader Obstacles to education (#PF4852) Inappropriate policies (#PF5645)
Ineffective decision-making processes (#PF3709).
Narrower Impractical education (#PF3519) Unrecognized relevance of education (#PF9068).
Aggravates Inadequate secondary education (#PD5345)
Illiteracy among indigenous peoples (#PD3321)
Unarticulated goal of educational methods (#PF2400).

♦ PF2448 Plural society tensions

Pluralistic society

Nature A plural society contains diverse cultural and ethnic elements, which may lead to conflict. Most countries are faced with such societies. This may arise from an arbitrary definition of the national boundaries to include two or more ethnic groups, or as a result of colonization or of immigration. Adverse effects are aggravated by discrimination, segregation, exploitation, racism, elitism and a lack of appreciation of cultural differences, leading to disparities and lack of integration. Although assimilation is one answer to the adverse effects of a plural society, it also sows the seeds of discontent. A successful balance in the integration of national and ethnic characteristics has yet to be found.
Broader Socioeconomic stress (#PC6759).
Narrower Heterosexism (#PE0818) Social fragmentation (#PF1324)
Diverse unilingualism (#PF3317) Cultural fragmentation (#PF0536)
Political fragmentation (#PF3216) Underprivileged minorities (#PC3424)
Socially unintegrated expatriates (#PD2675).
Related Ethnic conflict (#PC3685) Multidenominational society (#PF3368)
Weakness of multi-party parliamentary systems (#PF3214).
Aggravates Racism (#PB1047) Genocide (#PC1056) Ethnocide (#PC1328)
Segregation (#PC0031) Social conflict (#PC0137) Social injustice (#PC0797)
Minority control (#PF2375) Social inequality (#PB0514) Racial segregation (#PC3688)
Lack of social mobility (#PF2195).
Aggravated by Colonialism (#PC0798) Lack of assimilation (#PF2132)
Social discrimination (#PC1864) Ethnic discrimination (#PC3686)
Proliferation of immigrants (#PD4605) Unnatural boundaries between states (#PF0090).
Reduces Lack of variety of social life forms (#PE8806).
Reduced by Miscegenation (#PC1523) Forced assimilation (#PC3293).

♦ PF2451 Lack of local services for community leadership training

Nature Local communities lack new and effective methods to enable them to gain and use leadership skills, and local structures to train and utilize the leadership skills they require to deal with all aspects of community life, both economic and social. The unavailability of local special educators and other locally trained personnel is partly the reason that needs for special education and service, and for skills in determining priorities and leading meetings are are not met.
Broader Obstacles to leadership (#PF7011) Minimal community services (#PD8832)
Insufficient leadership training (#PF3605).
Aggravates Lack of training (#PD8388) Incompetent management (#PC4867)
Erratic land assessments (#PG7939) Inadequate meeting methods (#PF8939)
Complex agricultural practices (#PG7901) Restrictions on early apprenticeship (#PJ7946)
Unexplored entertainment alternatives in the countryside (#PE8540)
Lack of skilled manpower in rural areas of developing countries (#PE5170).
Aggravated by Prohibitive cost of maintenance (#PF0296)
Inhibiting participatory methods (#PG7973)
Conflicting community priorities (#PJ7975).

♦ PF2452 Inaccessible decision makers

Insufficient communication between decision-makers and grassroots

Nature There is insufficient communication between the grassroots and policy makers, local individuals and groups being unable to pass information to established powers or receive it back. The inaccessibility of decision-makers and the time required for letter-writing and petitions both point to the inadequacy of existing channels to meet present-day needs. The result is a lack of the on-going common basis necessary for consensus. Without accountability for continuous exchange of information, long-range planning becomes warped and misdirected.
Broader Collapse of societal engagement (#PF2340)
Inaccessibility of decision-makers in multinational enterprises (#PE0573).
Aggravates Denial of political rights (#PD8276) Fragmented decision-making (#PF8448).
Aggravated by Insufficient communications systems (#PF2350).

♦ PF2453 Selective perception of facts

Nature The selective perception of facts that support prejudices and the blindness to facts that disprove them, play important roles in the accumulation and spread of superstitions, falsehoods and lies in society.
Counter-claim The interaction of the senses and the brain is structured such that all stimuli is selectively perceived and must be to prevent sensory overload. All human beings have prejudices and are blind to some dimensions of life. When this selectivity results in disfunctional individuals within a specific social setting then a new, but equally rigid, selective screen is required.
Narrower Unrecognized opportunities (#PF6925) Unperceived career opportunities (#PF9004)
Unperceived educational opportunities (#PJ9762)
Institutionalized callousness of public services (#PF2006).
Aggravates Ignorance (#PA5568) Inhibited grief process (#PD4918)
Dehumanization of death (#PF2442).
Aggravated by Questionable facts (#PF9431) Suppression of scientific information (#PF1615)
Inadequate objectivity of institutions (#PF6691).

♦ PF2454 Law enforcement complexity

Complex difficult laws

Nature The growing complexity of rules that affect everything from taxes and social security to zoning, make life increasingly difficult for the average person. They also make it increasingly difficult to detect and enforce infringements of such laws.
Claim Existing law enforcement and correctional constructs have become, at every level, labyrinths of complexity, yet these are exactly the points at which community groups make contact. Since responsibility is so thinly diffused throughout the labyrinth, no structure can be held sufficiently accountable. Therefore there is no concentration of power which can be effectively addressed.
Broader Limited access to external resources (#PF1653)
Structural failure of citizen participation (#PF2347).
Related Ineffective regulation of electronic messages (#PE6226).
Aggravates Inadequate drug control (#PC0231) Inadequate narcotics legislation (#PF6787)
Inadequate national law enforcement (#PE4768)
Inadequate evidence to convict known offenders (#PF8661)
Limited local respect for regional and global legislation (#PF2499).
Aggravated by Legalism (#PF8480) Complex legal procedures (#PF8519)
Proliferation of litigation (#PF0361) Inadequately worded agreements (#PF5421).

♦ PF2457 Prohibitive cost of farm machinery

Limited availability of modern farm machinery

Nature The economic viability of small farms is threatened by the increasing expense involved in buying and maintaining farm machinery and other necessary outlays.
The widespread use of high level technology for large-scale production has reduced the feasibility of commercial farming with primitive tools; for example, the use of bullocks for transport and ploughing and home fashioned tools for manual cultivation severely limits the work potential of the available manpower. Most rural communities earn most of their living from agriculture, yet a very low percentage of their land is productively cultivated. The limited availability of modern farm implements prevents any dramatic increase in agricultural production, since the high cost of farm implements and large machinery, such as tractors and well-boring equipment, compared to the limited financial resources of village families, makes individual ownership prohibitive.
Broader Prohibitive cost of living (#PF1238) Prohibitive equipment costs (#PJ8196).
Narrower Delayed profits (#PG7978) Lack of dairy markets (#PG8471)
Demanding requirements (#PG7976) Prohibitive ownership costs (#PJ0033)
Restrictions on market access (#PG7910) Independent farming operations (#PG7971)
Insufficient returns from farming (#PG7947) Simultaneous demands on machinery (#PG7954)
Obsolete agricultural cooperative structures (#PF5961).
Related Social injustice (#PC0797) Inadequate maintenance equipment (#PD6520).
Aggravates Monoculture of crops (#PC3606).
Aggravated by Social inequality (#PB0514) Prohibitive cost of transportation (#PE8063)
Prohibitive cost of capital equipment (#PJ1768).

♦ PF2464 Profit motivated utilization of construction technology

Nature Operation of construction expertise in the reduced context of the profit motive often leads to failure to consider broadly based, long-range consumer needs. Thus the consumer suffers from inadequate quality control, use of stop-gap measures and the placing of priorities on immediate need rather than on comprehensive planning which anticipates future needs. The result is a slowing in the rate of expansion of basic construction expertise, which means that broad social needs are often neglected. This problem is heightened in those nations where direct or indirect governmental regulation is lacking.
Broader Non-inclusive management decisions (#PF2754).
Narrower Elitist control of construction technology (#PD2712)
Inequitable distribution of construction expertise (#PD2608)
Absence of accountability in construction planning (#PF2804)
Reductionistic decision making criteria in the construction industry (#PE7742).
Related Obsolete vocational skills (#PD3548)
Short-term planning of product life cycles (#PF1740)
Limited access to natural resource use decisions (#PF2882)
Nationalistically determined development of natural resources (#PD3546).

♦ PF2466 Neglect by research of natural resource product expansion

Nature Research into new procedures for expanding production of the world's natural resources is not inclusive and there is no computerized catalogue of data which includes and delineates the amounts and distribution of the earth's resources. Research tends to emphasize large-scale enterprises, overlooking the development and coordinated application of techniques needed in small-scale enterprises or in the more austere economies. Research in mining procedures tends to stress the extraction of the main products, neglecting the development of techniques which would make full use of the by-products.
Broader Limited access to natural resource use decisions (#PF2882).

♦ PF2471 Overlooked potential for industrial development in rural communities

Nature Industry in the developed countries is tending to move into rural areas in order to escape the problems of urban locations; but communities in these rural areas tend not to turn their attention to the work necessary to take full advantage of this resource, which could bring them added capital and employment opportunities. Often no market research or resource census is done to see how there could be a blending of small industries, nor is advantage taken of those financial advisory services which could gather the market data into accurate projections of income, expense and capitalization needs necessary for a nascent industry to be successful.
Broader Obstacles to community achievement (#PF7118).
Narrower Hyperinflation (#PD7940) Feminist backlash (#PG1702)
Prolonged repair services (#PG7967) Unexplored energy alternatives (#PF7960).
Related Narrow profit margin (#PJ9737) Instability of prices (#PF8635)
Unrealized potential of commercial enterprises (#PF2231).
Aggravates Lack of economic and technical development (#PE8190).
Aggravated by Prohibitive cost of transportation (#PE8063).

♦ PF2475 Overemphasis on individual rights

Nature The emphasis on the rights of the individual in the modern industrial era has generated a folklore in which care for the individual is often at the expense of the individual's obligations towards the wider context. For example, large labour unions act to maintain the unemployment, sickness and retirement benefits of their members at levels that cannot be guaranteed by current and future production earnings.
Broader Legal contract system reduced to individual needs (#PE5397).

♦ PF2477 Narrow range of practical skills

Limited range of functional skills

Nature As society grows increasingly more complex, communities need to have their residents equipped with a broader range of vocational and leadership skills than ever before. However, few

residents in small communities have modern management skills in agriculture, industry or commerce; this is an obvious impediment to rapid economic growth. Even though practical skills play a large part in establishing an individual's or a community's identity and provide social mobility, residents of many rural communities have grown to believe that up–to–date knowhow is unattainable. In some rural communities there may be no–one who knows how to drive an automobile, or use farm machinery; and neither the vehicles nor the training to use them are available. Illiteracy is widespread. Those young people who do acquire technical or commercial skills find no channels to put their skills into use and move to the city. Even in agriculture, although traditional skills may be very well developed, contemporary methods are largely unheard of and unsought. The people of small rural towns are also unequipped to meet the needs of their town's future. The people of such communities may be willing to engage in the community's restoration but they are hampered by their narrow range of practical skills. Some experience difficulties in getting and keeping jobs; they attribute these difficulties to a lack of preparation for interviews and on–the–job relations. City government employees lack accounting, record–keeping and proposal–writing skills. Some residents feel that the effectiveness of volunteer fire departments is weakened by their minimal experience in modern fire–fighting methods. Teachers do not have curricula which would allow them to effectively teach practical skills in school.
In many third world communities, individuals are sufficiently concerned to participate in planning their village's future, yet they lack specific social methods to go about the task. In every aspect of village life there lies a deep desire to participate but a lack of practical know–how to release that willingness to action. This contradiction particularly emerges in relation to the training available in villages: the educational process is limited to the primary school, where emphasis is solely upon a basic academic foundation; general education beyond this standard is limited to a very few; moreover, the inflated level of incidental school costs restricts the number of students even in the primary school; technical job training is not immediately possible for most aspirants seeking admission. Yet the contradiction goes far beyond the arena of formal schooling. In business, health, agriculture, family care, public services and social welfare, men and women, young and old, all suffer from under–development of their skills. Failure to fulfil of this need perpetuates a random and individualistic approach to community development and fosters the image that nothing can happen.
Broader Insufficient skills (#PC6445).
Narrower Limited basic skills (#PJ8138)
Inadequate career advice (#PJ8018)
Limited technical skills (#PF8594)
Lack of trained firefighters (#PJ0277)
Insufficient trained leaders (#PJ0389)
Shortage of technical skills (#PF6500)
Narrow range of business skills (#PE6554)
Improperly trained administrators (#PG8707)
Underproductive methods of agricultural management (#PF6524)
Insufficient knowledge of local production processes (#PE8140).
Related Unproductive subsistence agriculture (#PC0492).
Aggravates Unhygienic conditions (#PF8515)
Intra–group skills gap (#PG9276)
Incompetent management (#PC4867)
Neglected food resources (#PF7808)
Inefficient resource usage (#PG7808)
Outdated farming techniques (#PG9687)
Unmaintained private vehicles (#PG9243)
Ignorance of health and hygiene (#PD8023)
Restrictive use of available land (#PF6528)
Unformed style of cooperative action (#PF6514)
Lack of motivation in leadership development (#PF2208)
Illiteracy as an inhibitor of business transactions (#PE7968).
Aggravated by Indecision (#PF8808)
Lost farming skills (#PJ8369)
Super–power chauvinism (#PE7778)
Restricted job training (#PJ0617)
Suspicion of bureaucracy (#PF8335)
Limited funding expertise (#PJ0278)
Inadequate driver training (#PJ4507)
Limited paramedic training (#PG8035)
Limited planning experience (#PG7811)
Lack of agro–urban training (#PG7992)
Lack of vocational teachers (#PJ9603)
Untrained clerical personnel (#PG8787)
Shortage of trained teachers (#PD8108)
Inadequate technical training (#PE8716)
Inadequate training opportunities (#PJ8697)
Lagging training in social skills (#PF8085)
Insufficient employment mechanisms (#PG8797)
Transport as a hindrance to economy (#PE8858)
Restrictions on early apprenticeship (#PJ7946)
Discrimination against women in education (#PD0190)
Protectionism in the computer services industry (#PD7001)
Lack of skilled manpower in rural areas of developing countries (#PE5170).

♦ PF2489 Limited availability of financial credit

Undeveloped credit lines — Limited credit lines — Limited availability of credit — Limited credit accessibility — Lack of local credit — Unavailability of income subsidy — Limited local government credit — Inadequate credit monies — Insufficient credit capital
Nature International credit–flow standards are set and maintained by a few great national economies and overemphasize their own interests. This results in forms of credit exchange in which only a few nations can fully participate.
Incidence This disrelation is exemplified by the rigid exchange boundaries between the EEC and other Western economies.
Broader Inadequate credit policies (#PF0245)
Underdeveloped sources of income expansion (#PF1345)
Lack of economic and technical development (#PE8190).
Related Unknown availability of subsidies (#PG9905)
Decline in government health expenditure (#PF4586)
Over–subsidized agriculture in industrialized countries (#PD9802)
Constricting level of capital development in rural areas (#PE1139)
Distortion of international trade by export subsidies and countervailing duties (#PE1961).
Aggravates Lack of venture capital (#PG7833)
Inadequate support services (#PF6492)
Diminishing capital investment in small communities (#PF6477).
Aggravated by Limited accumulation of capital (#PF3630)
Limited access to social benefits (#PF1303)
Incomplete access to information resources (#PF2401)
Limited local availability of capital reserve (#PF2378)
Unrecognized socio–economic interdependencies (#PF2969)
Demoralizing image of urban community identity (#PF1681)
Deteriorated structures of essential corporateness (#PF1301)
Unsystematic use of powerful relationships by rural communities (#PE1101).
Reduces Distortion of international trade by selective domestic subsidies (#PD0678).

♦ PF2491 Restrictive loan procedures in developing countries

Confusing loan computing methods — Unknown loan services — Complex loan procedure — Fear of banks
Nature The operation of credit exchanges is obscured by the array of methods used for computing interest charges, repayment plans and regulatory procedures that, due to their technical nature, tend to confuse the average person. For example, several options of deferred payment accounts may be offered by the same retailer, so that interest charges can be stated on different bases to give a different figure, without offering a valid basis for comparison.
Broader Inadequate credit policies (#PF0245)
Limited accumulation of capital (#PF3630).
Related Restricted flow of local economy (#PF6451)
Individualistic retaining of local tradition (#PF1705).
Aggravates Inadequate management skills in rural communities (#PF1442).
Aggravated by Complex banking practices (#PJ8033).

♦ PF2493 Inflexible social care structures in developing countries

Nature The care structures of many Third World villages sustained people in the prescientific era, but today their adequacy is being questioned as communities are influenced by the increasing complexity and mobility of modern life. However, development of new care structures is hampered by time–consuming traditional roles. Circumstances combine to trap the village children into patterns which preclude new possibilities: boys begin working in the fields or follow their father's trade without the opportunity to consider alternatives. Although the creative presence of women is needed in a great variety of groups and social activities beyond the home in order to sustain the increased complexity of modern life, village women are confined to child–bearing and home care; high infant mortality rates and traditional roles discourage both birth control and any other changes that might release women to other needed roles. Likewise, the men are forced to put in long hours of work simply to sustain the economic life of the family, and have little time or energy for broader engagement in community life.
Because all members of the community work such long hours, little time is left for community concerns such as educational development, community planning, health care or physical maintenance. Although residents may be aware of the gaps in effective community care and be concerned to release members of the community to new effective modes of engagement; yet this very concern is what sustains them in their commitment to traditional means of care.
Claim Until the communities of the Third World create structures of mutual care which call forth new levels of creative commitment of the limited time and energy of its human resources, no significant development will occur and the social needs of its residents will continue to be neglected while willing concern is untapped.
Broader Inflexibility (#PA8555).
Narrower Rigid cultural patterns (#PF8598)
Limited child activities (#PG9733)
Tradition–bound childcare (#PG9717)
Underutilization of human resources (#PF3523).
Related Inadequate social welfare services (#PC0834)
Neglect of the role of women in rural development (#PF4959)
Fragmented forms of care at the neighbourhood level (#PE2274)
Inadequate development of new social structures in developing countries (#PD0822).
Aggravates Untenable orphan care (#PJ9718).

♦ PF2497 Unfu!filled treaty obligations

Transgression of international agreements — Breaching international treaties — Treaty violations — Disregard of international law — Abuse of international law — Non–implementation of international treaty provisions — Violation of international conventions — Violation of international agreements — Trade quota cheating
Nature A new international anarchy has arisen in recent years that defies international decisions on treaty provisions, specifically those initiated and/or made by the United Nations. This lack of adherence to a previously agreed upon decision–making body threatens to undermine security on both national and international levels.
Incidence In 1982, the Secretary–General of the United Nations stated that the UN Security Council often found itself unable to take decisive action to resolve international conflicts, and even its unanimous resolutions were "increasingly defied or ignored by those that feel themselves strong enough to do so". He observed that sterner measures for world peace were envisaged in the UN Charter, but that the prospect of realizing them now was virtually impossible, due to divisions within the international community.
Specific incidents of non–compliance include Lebanon and the Middle East crisis; the Iran–Iraq war; Soviet invasion of Afghanistan and American assistance to the resistance; Vietnamese invasion of Kampuchea; and the Cyprus situation.
Counter–claim Some reasons for the decline in strength of the UN Charter and Security Council lie within the UN itself: dangerous situations could perhaps be diffused before they reached the point of crisis if the Council kept an active watch on them by more systematic and less last–minute measures; the permanent members of the Security Council (China, France, UK, USA, USSR) should not allow their own bilateral difficulties to overshadow their commitment to the Council; the Council members should bear more collective influence when resolutions are not respected by those to whom they are addressed; the Council should devise wider and more systematic capacities for fact–finding in potential conflict areas and should be more responsive in sending good office missions, observers, or a UN presence, to such areas; and there should be explicit guarantees for collective or individual supportive action.
Broader Abuse of law (#PC5280)
Lack of international cooperation (#PF0817).
Narrower Illegal international arms shipments (#PD4858)
Non–recognition of international law (#PF9081)
Covert violation of international treaties (#PD8465)
Violation of treaties with indigenous populations (#PE7573)
Ineffectiveness of international commodity agreements (#PG5600)
Failure of governments to fulfil international reporting obligations (#PE2215)
Persistence of a technical state of war following cease–fire agreements (#PE2324)
Government non–payment of agreed contributions to international organizations (#PF8650)
Lack of adherence to international transit conventions for land–locked countries (#PE5789).
Related Breach of promise (#PF7150)
State sanctioned torture (#PD0181)
Ineffective international agreements (#PF6992).
Aggravates Violation of neutrality (#PC2659)
International insecurity (#PB0009)
International aggression (#PB0968)
Secret military operations (#PF7669)
Disagreement within alliances (#PD2629)
Offences against the peace and security of mankind (#PC6239).
Aggravated by Government inaction (#PC3950)
Government limitations (#PF4668)
Competition between states (#PC0114)
Unfairly negotiated treaties (#PF4787)
Inadequate international law enforcement (#PF8421)
Limited acceptance of international treaties (#PF0977)
Non–verifiability of compliance with nuclear arms treaties (#PF4460)
Conditional observance of multilaterally agreed trade commitments (#PF7838).

♦ PF2499 Limited local respect for regional and global legislation

Claim Everyday experience concerning both local and national laws suggests that the majority of the world's people have no sense of responsibility toward the laws that govern their lives, and therefore little respect for law and law enforcement. The result is widespread frustration and anger toward any power structure, and a feeling that nothing can be achieved beyond the local level.
Broader Hindrance of law enforcement (#PD5515)
Legal contract system reduced to individual needs (#PE5397).
Narrower Nonacceptance of government legislation (#PG6079).
Aggravates Restrictive legislation (#PD9012)
Ineffective legislation (#PC9513)
Exploitation of regulatory loopholes in countries with underdeveloped legislation (#PE4339).
Aggravated by Lack of enforcement power (#PF0223)
Law enforcement complexity (#PF2454)
Inadequate national law enforcement (#PE4768)
Arbitrary enforcement of regulations (#PD8697)
Inadequacy of international legislation (#PF0228)
Political barriers to effective legislation (#PC3201)
Ineffective legislation against organized crime (#PE6699)
Deliberate governmental avoidance of legislative reform (#PF5736)
Inadequate legislation relating to action against problems (#PF1645).

♦ PF2504 Limited cultural context

Nature Although no group can deal effectively with the complexity of the present social situation unless it is radically aware and involved with cultures beyond its own, most groups are trapped by the cultural bias of their own society and lack the perspective which would allow the sharing

of visions of the future beyond a single nation or culture.
Broader Structural failure of citizen participation (#PF2347).
Aggravated by Cultural illiteracy (#PD2041).

♦ PF2505 Social disguise of ambiguity

Nature Society creates images that covers up to fact that human being living and deciding always do so in ambiguity. The images that ambiguity does not exist creates a sense that there are no consequences to one's decisions beyond the perceived one's, a feeling of omnipotence about choice and a believe in simplistic either or choices. Man's freedom is thus taken away.
Broader Tensionless image of free choice (#PF1675).

♦ PF2508 Secret societies

Inner circles — Cabals
Nature Secret societies may be considered a danger to their community or nation. Secrecy may be partial or complete. Partial secrecy may take the form of a secret initiation ceremony while public acknowledgement is made of general objectives. Many secret societies are hierarchical. Secret societies may be mystical, occult, heretical, revolutionary, perverse, violent, subversive, criminal or primitive. Political secret societies may only be so constituted by the fact of their having held a number of meetings to affect some particular purpose. During a regime or governmental administration, they may be an inner circle of people wielding the power 'behind the throne' or they may be the manipulators of political parties, operating as a cabal.
Incidence Secret organizations that do intelligence or police work, may sometimes breed secret societies within themselves. For example, Brownshirt Nazis, partly secret enforcers, had under them the Blackshirt SS, which controlled the Gestapo, the secret police. Large urban police departments may be controlled by an inner circle of officers, or by an official benevolent or fraternal organization. In the UK for example, objection was made to Freemasonry among policemen. In the Vatican, protected by the same secret environment as afforded to the Curia, a criminal conspiracy manipulating the Vatican bank's prestige was recently fostered and investigation was impeded by Vatican secretiveness.
Background Famous secret societies include the Pythagorean Brotherhood, the Carbonari, the Illuminati, the Order of Assassins, the Jesuits, the Chinese Tongs, the Black Hand, the Weathermen, and the Black September group among many others.
Counter-claim Many secret societies have no despotic ulterior aims but offer inspiration, direction and recognition to members. Membership affords an identity by adherence to the group and fulfils important needs in the upward path of personal development and responsibility.
Refs Webster, Nesta H *Secret Societies and Subversive Movements* (1972); Wilgus, Neal and Wilson, Robert A *The Illuminoids* (1978).
Broader Injustice (#PA6486).
Narrower Freemasonry (#PF0695) Closed communities (#PG4604)
Subversive activities (#PD0557) Primitive secret societies (#PF2928).
Aggravates Conspiracy (#PC2555) Meaninglessness (#PA6977)
Conspiracies for societal control (#PB7125).
Aggravated by Repression (#PB0871) Unethical personnel practices (#PD0862)
Inadequate national law enforcement (#PE4768).

♦ PF2513 Human cannibalism

Anthropophagy
Nature Human flesh may be consumed as a food or as part of a ritual, especially in order to obtain spiritual power over enemies, or to pay respect to dead relatives (endocannibalism).
Incidence Cannibalism is still occasionally practised among tribes in interior New Guinea and prevailed until recently in parts of West and Central Africa, Melanesia (especially Fiji), Australia, New Zealand, Polynesia (especially Sumatra), and North and South America. Reports circulated in recent years have implicated several heads of state in cannibalistic rites. Cannibalism is occasionally reported in the case of people marooned without food, isolated in a grounded aircraft in subarctic conditions, refugees and shipwreck survivors at sea without food, or prisoners of war left to starve. Incidences are also reported in connection with severe famine conditions although, as in the case of the famine in the Ukraine in the 1930s or the siege of Leningrad, efforts are made by the authorities to suppress such information. The probability of cannibalism increases amongst the civilian population in war zones in the event of extended disruption of food supplies. Psychologically disturbed individuals have been known to kill and eat victims, for example, a Japanese university student in Paris killed and ate a Dutch woman student at the same university. There are occasional reports of cannibalism in connection with satanic rituals.
Background The practice of cannibalism goes back to early history and was found on all continents. The term is derived from a Spanish form of Carib, an early West Indian tribe of cannibals.
Counter-claim Cannibalism may be the only recourse for people subject to a long period without food. Recent research has questioned the authenticity of missionary reports of cannibalism in Africa arguing that incidences were exaggerated by missionaries endeavouring to illustrate the contrast resulting from their intervention. Such research questions how such practices came to be given up so easily if they were socially approved rather than aberrant. When it occurred it may largely have been a symptom of social breakdown. Whilst cannibalism certainly existed, the existence of tribes whose favourite food was human flesh is a fiction invented by Europeans.
Refs Sanday, Peggy R *Divine Hunger* (1986).
Broader Magic (#PF3311) Occultism (#PF3312) Superstition (#PA0430).
Narrower Endocannibalism (#PG4977) Ritual cannibalism (#PF8944)
Head hunting in tribal societies (#PF2666).
Related Human sacrifice (#PF2641) Animal cannibalism (#PG4979).
Aggravates Fear (#PA6030) Human death (#PA0072).
Aggravated by Hunger (#PB0262) Famine (#PB0315)
Tribalism (#PC1910) Isolation of ethnic groups (#PC3316).
Reduces Unsustainable population levels (#PB0035).
Reduced by Cultural invasion (#PC2548) Forced assimilation (#PC3293).

♦ PF2516 Monsters

Monstrous entities
Nature Monsters include malformed or misshapen animals or humans and creatures of great size. A great number of incompletely documented reports exist of various kinds of monsters. It has not been determined whether such reports correspond in fact to real, undescribed animals (possibly isolated descendants from prehistoric species), to mutations, or whether they are entirely due to different forms of hallucination or optical effects, possibly triggered by natural phenomena.
Incidence Such creatures as the dragon, the griffin, the hippogriff, the winged bull of the Babylonians, the winged elephant of the Hindus, centaurs, satyrs, the Minotaur, Lamia, the Bun-yip of Aboriginal Australia and Burr-Hound and Pot-Tilter of the North American Indians are a part, and perhaps a necessary part, of human consciousness since prehistorical times. A great number of monsters have been reported in the 20th century including: the Loch Ness Monster; Yeti or Abominable Snowman; Bigfoot and creature from Unidentified Flying Objects.
Broader Mental illness (#PC0300)
Congenital syndromes affecting multiple systems (#PE9324).
Related Endangered species of animals (#PC1713)
Negative emotions and attitudes (#PA7090)
Human errors and miscalculations (#PF3702).
Aggravated by Fear of the unknown (#PF6188).

♦ PF2517 Lack of intersocietal resource channels

Nature In order to implement their global concern, people have first to negotiate allegiance to their own society. There is no institutional means by which individuals can implement a decision to use their resources to sustain other territories. This societal vacuum makes it hard to perceive the reality of interdependence among societies, and even more difficult to participate in.
Broader Habitual overemphasis on national self-determination (#PF1804).
Narrower Lack of channels for obtaining available local funding (#PF6544).
Related Crime (#PB0001).
Aggravates Conflict (#PA0298).
Aggravated by Ignorance (#PA5568) Ideological conflict (#PF3388)
International insecurity (#PB0009) Competition between states (#PC0114).

♦ PF2519 Unplanned use of community space

Nature Public buildings and land areas play a critical role in maintaining civic pride, yet communities grow up in a haphazard and visually unpleasing manner. Many rural communities seem uncertain about their right to decide the use of land in their villages; as a result little space is designated for such things as a community meeting hall, plaza or even recreation area. Land cleared from a forest is rapidly overgrown unless it is maintained. Community boundaries sometimes are not clearly defined, so rambling development and a 'squatters' rights' style for homesteading create a sense of random and chaotic space. Many villages have no designated garbage or sewage disposal sites so that individual families tend to dispose of their waste 'in the brush' or somewhere near their home rather than a whole village deciding on a proper location.
Broader Obstacles to community achievement (#PF7118)
Obstacles to availability of community space (#PF7130).
Narrower Inadequate storage facilities (#PG4268).
Aggravates Unhealthy environment (#PJ1680) Unorganized community recreation (#PF5409).

♦ PF2527 Loss of humility in relation to the environment

Nature With the "conquest of nature" and the increasing adaptation of the environment by man, there is a loss of that sense of humility which enables people to live in gentle harmony with their surroundings, whether or not they are perceived as the creation of divinity.
Incidence Modern architecture, for example, constitutes a denial of the place of divinity in the scheme of things and the substitution of man's infallibility.
Claim A true sense of humanity in relation to the Creator of the natural environment is a prerequisite for the creation of an environment fit for humans to live in.
Broader Intellectual arrogance (#PF7847).
Aggravates Trachoma (#PE1946) High altitude stress (#PD2322)
Unrealized agronomic potential (#PJ1775).
Aggravated by Pride (#PA7599) Fear of nature (#PF6803)
Belief in humanity's dominance over nature (#PF8264).

♦ PF2528 False assumptions on sustainable development

Fallacies about ecosystemic processes
Claim 1. The world faces serious problems today, which required concerted effort by all nations for their solution. Much has been written about these problems, and the limitations within which solutions can be found. But the limitations are not those frequently assumed. In ecological terms the issue is resilience rather than short-term stability.
2. Development slogans are unsuccessful mainly because the smaller states fail to read the world-economy and its history properly. First, they tend to wrongly assume that there is ample room in the centre of the world-economy for both themselves and the major states. Second, they act as if whatever need to be done to move them to, and accommodate them in, the centre, need only concern the structural relations between the central and the peripheral parts of the world-economy. Third, the peripheral states naively tend to believe that what needs to be done could be left to the initiatives of the central states. Fourth, the peripheral states misunderstand development to mean no more than the growth-oriented, industrialized imitation of the central societies models; whether free-market and laissez-faire, or collectivist, five-year planned.
3. The easy assumption that all technical problems can be solved when it becomes necessary to do so, namely that no problems are irreversible, has resulted in the inappropriate development of certain industries (such as nuclear power) which have only subsequently recognized that the problems that they have created are not necessarily susceptible to solutions at an reasonable cost, if at all (as in the case of radioactive waste).
Refs Clark, W C and Munn, R E (Eds) *Sustainable Development of the Biosphere* (1988).
Broader Compliance*complex (#PA5710) Inappropriate assumptions (#PF6814).
Narrower End of nature (#PF9582) False economies of scale (#PF9791)
Economic bias in development (#PF2997).
Aggravated by Defective reasoning (#PF5711) Limitations of surprise-free thinking (#PF7700)
Limitation of current scientific knowledge (#PF4014).

♦ PF2529 Limits on areas of research

Nature Research competence is the ability to develop new knowledge and techniques that permit the development of creative expertise for improving basic human life. However, recent misunderstanding has tended to limit the image of research to isolated disciplines and narrow areas. This narrow view of research has prevented the channelling of research competence into vital areas of society – for example, studies of the effects on a traditional society of the injection of advanced technology. As a result, although research may have contributed to longer life span, it has not significantly improved the quality of society because of its neglect of important areas of society. Urban technological man, while enjoying greater comfort, is often confronted with a lack of personal significance because of the lack of creative engagement in life.
Claim Researchers must come to realize there is the lack of significant use for larger societal concerns, such as new life styles and structures, within the realm of social research.
Broader Collapsed meaning of human creativity (#PF0936).
Narrower Uncoordinated research efforts (#PF2306)
Top down research methodologies (#PF2202)
Inadequate peace research support (#PF4848)
Thwarted technological communications (#PF0953)
Individualistic utilization of expertise (#PF5639)
Dehumanized individual scientific research (#PF2112)
Superficial research on the total human process (#PF2415)
Inequitable distribution of skilled specialists (#PD2479)
Restriction of funding for research on social problems (#PF2823)
Gap between the function of social techniques and the needs they address (#PF3608).
Related Defensive life stance (#PF0979)
Collapse in providing ethical value screens (#PF1114)
Educational curricula based on content rather than method (#PF3549).
Aggravates Unarticulated goal of educational methods (#PF2400).

♦ PF2535 Unexploited possibilities for local commerce

Unanticipated market variance — Uncertain market volume — Uncertain market access

PF2535

Nature The increasing cost of fuel is restricting travel and thus reinforcing the 'support your local enterprise' story. This means that local community businesses have the opportunity to revitalize. However, the necessary research, forecasting and management experience are often not available to take advantage of this situation. Although residents may express an interest in and a willingness to support additional services, such as a repair shop or laundromat, lack of hard market data often deters anyone from initiating such businesses. Stores often carry similar lines of merchandise, creating an oversupply of some goods and a shortage of others.

Broader Uncertainty (#PA6438).
Narrower Increase of out-of-town shopping (#PG7963)
Protectionism in the commodities sectors (#PG7925)
Insufficient repair services in developing countries (#PE7959)
Insufficient consumer services in developing countries (#PE8236).
Aggravates Inadequate management skills in rural communities (#PF1442).
Aggravated by Limited mechanical services in developing countries (#PE8986).

♦ PF2540 Lack of central planning structures in small communities

Nature The residents of many communities which have long been dependent on single industries, such as coal-mining communities, become accustomed to depending upon outside institutions to decide their future. The lack of central planning structures within the community leads to a general unclarity concerning channels of communication with outside authorities and government structures; and to an absence of community consensus which inhibits action in many ways. A failure to work out plans for clarifying channels of communication, for utilizing the wisdom of all local residents and for meeting the needs of all age groups threatens the future of such communities.

Broader Lack of local leadership and decision-making structures (#PF6556).
Narrower Large trade surpluses (#PJ0207) Physical intimidation (#PC2934)
Excess production capacity (#PD0779) Irregular transport services (#PE5345)
Failure of methods to appropriate data (#PE0630)
Degradation of the environment by trees (#PE7695)
Financial destabilization of world trade (#PC7873)
Inequality inducing effects of remote sensing systems (#PE9072)
North and South spheres of influence within UN and related agencies (#PE8332).
Related Meaninglessness (#PA6977).
Aggravates Revenge (#PF8562) Limited youth activities (#PJ0106)
Lack of community planning (#PF2605)
Inadequate welfare services for the aged (#PD0512).
Reduced by Over-centralization (#PF2711).

♦ PF2541 Infantilization of deprived populations

Nature Development aid of the industrialized countries is based on the principle that the communities in the Third World cannot develop themselves without outside developers. It considers the people as objects of wider national plans rather than living subjects of their own destiny. The people consider these development strategies as government activities pursued for their own purposes and see no reason to participate.

Aggravated by Dependence of developing countries on external financing for development programmes (#PE7195).

♦ PF2553 Collectivism

Collectivist policies

Nature Collectivist theories and systems emphasize the priority of the community over the individual. The four main kinds of collectivism are social democracy, socialism, communism, fascism. Collectivism may involve dictatorship, nationalization and strict government control over a wide range of matters. Collectivism may also encourage militarism if there is resistance to it or deviation from it.

Broader Ideological conflict (#PF3388).
Narrower Infringement of privacy (#PB0284) Dictatorship of the majority (#PD3239)
Nationalization of domestic enterprises (#PD1994).
Aggravates Lack of variety of social life forms (#PE8806).
Aggravated by Double standards in morality (#PF5225).
Reduces Lack of individualism in capitalist systems (#PD3106).

♦ PF2554 Concubinage

Nature The state of cohabitation without the full sanctions of legal marriage, concubinage is often a form of slavery in which women may be bought and sold. Their status and the legal status of their children is lower than that of the legal wife. Concubines may be bought from brothels or abducted or enticed from foreign countries. Although the historical justification for concubines was to ensure offspring, in practice most concubines were and are acquired for sexual pleasure. Concubinage is particularly the prerogative of the rich.

Incidence Concubinage still exists in Asia and in Arabia among the rich, and in Africa both among the rich and in tribal custom.

Broader Polygamy (#PD2184) Forced marriage (#PD1915)
Sexual exploitation of women (#PD3262).
Related Prostitution (#PD0693) Illegitimate children (#PC1874).
Aggravates Miscegenation (#PC1523) Trafficking in women (#PC3298).
Aggravated by Child-marriage (#PF3285) Chattel slavery (#PC3300)
Inadequate national law enforcement (#PE4768).

♦ PF2560 Individualistic welfare responsibility

Nature The social story that everyone is self-sufficient results in the belief that a person should not have to care for anyone else. As a result care structures are designed to avoid direct personal responsibility for another's welfare. In general the ideal of such welfare structures is to develop people's self-sufficiency so that they are no longer a burden on others. Also people can and do avoid involvement with their extended family, their metropolitan area and the world as a whole.

Broader Overemphasis on self-sufficiency with respect to interdependence (#PF3460).
Aggravates Inadequate social welfare services (#PC0834).

♦ PF2564 Limited applicability of monetary grants

Nature Grants are almost entirely monetary which limits the use of human, technological and natural resources. This limitation in the nature of grants reduces the scope and arenas in which special grants operate.

Broader Brutality (#PC1987) Belittling of grant recipients (#PF2708).

♦ PF2568 Unrealized use of education structures

Unrealized teaching potential

Nature Limitations on the number of students, cuts in funds for further education and the high cost of educational facilities hamper plans to bring such facilities to small villages, and young people from rural areas tend to go elsewhere for vocational training and employment. Local business skills are underdeveloped and villagers with some practical know-how are inexperienced in guiding the development of new ventures. Minimal career counselling is available close at hand and, because most people are uninformed about the opportunities open to them, many possibilities remain unexplored.

Broader Unrecognized opportunities (#PF6925)
Ineffective systems of practical education (#PF3498).
Narrower Unexplored opportunities for community education (#PF6512).
Related Resistance to technology education (#PF6300).
Aggravates Educational wastage (#PC1716) Lack of local leadership role models (#PF6479)
Nonavailability of technical training (#PJ0121).
Aggravated by Student absenteeism (#PE4200) Administrative delays (#PC2550)
Inadequate career advice (#PJ8018) Ineffective class scheduling (#PG5424)
Shortage of school equipment (#PJ0821) Prohibitive cost of education (#PF4375)
Insufficient library equipment (#PG8495) Unperceived career opportunities (#PF9004)
Insufficient educational material (#PE8438) Unpublished educational possibilities (#PJ9445)
Unperceived relevance of formal education in rural communities (#PF1944)
Unavailability of trained teachers in the rural areas of developing countries (#PE8429).

♦ PF2573 Dangers of private control of communications mass media

Nature Privately owned organizations in the communication field wield power in setting patterns, forming attitudes, and motivating behaviour which is comparable to that of government, and sometimes – because of the financial resources committed – even greater. These corporations in mass media do not always consider adequately the ethical and social norms, spiritual and cultural values of the societies into which they are transmitting and thus can inflict conflicting or degrading values onto their vulnerable audiences, causing dissension and unrest.

Counter-claim Private activity in communications mass media works in the public's interest and lends itself to the variety, popularity, flexibility, free thought and competition that the non-private communications media do not have.

Broader Capitalism (#PC0564).
Aggravated by Insufficient communications systems (#PF2350).

♦ PF2574 Variations in national forms of currency

Nature As the world becomes more accessible to everyone and increasingly interdependent, national monetary systems and currencies are creating unnecessary blocks to growth. The disparities of currency from one country to another help perpetuate world conditions of economic instability, and too many people expend their energies without gaining the necessities of life.

Counter-claim For over a century nations have been the political units that have carried the thrust of economic growth. Each nation has a unique monetary system. This has been critical for the subsistence of the world's people. When a new nation is formed, it has always established its own currency and means of exchange. This makes the new nation an entity and means that the government has some control over the economy.

Broader Self-interest driven investment (#PC2576).
Narrower Parochial monetary agreements (#PD2469)
Nationalistic attitudes to currency (#PF6094)
US dollar dominance of world economy (#PD2463)
Absence of a long-range, world-wide capital flow plan (#PF2865)
Unimaginative vision of existing international economic structures (#PF2699).
Related Limited market development (#PF1086) Inadequate credit policies (#PF0245)
Inadequate access to negotiation on employment and reward (#PD1958).

♦ PF2575 Declining sense of community

Constricting patterns of individual involvement — Individual avenues for civic responsibility — Declining community life — Negative community reputation — Negative village self-identity — Estranged neighbourhood relations — Insular patterns of community groupings — Limited community identification — Infrequent community involvement — Limited involvement of neighbourhood residents — Minimal community involvement

Nature Although it is generally recognized that the participation of everyone, particularly of the young and the old, is vitally important in the planning and implementation of socio-economic programmes, few communities accomplish this end. Social patterns may have been well-defined for several generations, but contemporary living has transformed these forms and they have not been replaced. Despite an awareness of these gaps in effective community engagement and concern that all members of the community should participate in its total life, efforts will continue to be fragmented as long as dynamic role systems for everyone are not woven into the patterns of daily life.

Wherever individual concerns are emphasized at the expense of community concerns, systems for coordination are impeded. Groups and individuals engage in their separate activities in isolation from one another. Where there are no mechanisms by which the community can resolve inter-personal and inter-group tensions creatively, these disconnected forms of engagement very often prevent people from coming together as one community. For example, while signs of unified community pride and enthusiasm are appearing in many rural communities, there are no systems to reinforce unity and enthusiasm. This results in: competition for community space rather than coordination; reluctance of neighbours to hire each other for fear of possible ruptured friendships over work disagreements; a sense among residents of being unconnected with the community, as personality differences override community issues.

Transitions in lifestyle occasioned by current societal shifts often tear apart the unifying fabric in small communities. Many residents remain isolated from and cynical about the community, even when the town may have a rich heritage. There are few formal means to profit from the knowledge and experience of those who founded the community; and residents have a debilitating image of the town, related to loss of business, community decline, and a multitude of problems. Such negative images of the community are reinforced by its environmental deterioration, with abandoned buildings, overgrown lots, and old ruins witnessing to past glories and present squalor.

The fragmented structure of many rural communities precludes various groupings gathering to interact in working together as a single unit to accomplish common aims. People who work outside of their community come home to television and retire early with no opportunity for local involvement; empty sidewalks and streets either with no businesses or early closing stores and cafes underline the isolated pattern of life of most residents. Unique celebrations such as parades and festivals which could demonstrate community individuality no longer exist. The primary means of communication is word of mouth, forthcoming events receive little publicity, and numbers of neighbours have never met each other. Resentment and suspicion bred of unfamiliarity build up borders between families, neighbours and interest groupings, creating baffling networks of 'we's' and 'they's'. No common structure exists for objectifying misunderstanding.

Broader Social insecurity (#PC1867) Obstacles to community achievement (#PF7118)
Modern disruption of traditional symbol systems (#PF6461).
Narrower Apathy of youth (#PF5949) Limited youth activities (#PJ0106)
Inhibiting social patterns (#PF0193) Poor communications methods (#PJ8656)
Unstimulating entertainment (#PF8105) Unorganized family occasions (#PG8676)
Thwarted community enthusiasm (#PG7950) Unorganized welcoming process (#PG8567)
Undesignated gathering places (#PG9023) Uninvolved transient residents (#PG8660)
Insufficient possibilities for gathering of elders (#PE8985)
Decreasing participation in collective religious worship (#PF8905).
Related Minimal community services (#PD8832)
Ineffective organization of community action (#PF6501).
Aggravates Gossip (#PE2192) Statelessness (#PE2485)
Underreported issues (#PF9148) Rejection of refugees (#PF3021)
Citizen powerlessness (#PJ8803) Deserted public spaces (#PG9130)

Neighbourhood disputes (#PE5504)
Restrictive social groups (#PG8682)
Operating community cliques (#PG9155)
Ignorance of cultural heritage (#PF1985)
Parent–teacher non communication (#PF1187)
Indistinct community identification (#PG8607)
Lack of self–confidence (#PF0879)
Suspicion of imposed change (#PG9094)
Uncontrolled local vandalism (#PJ0154)
Fragmented community initiatives (#PG8987)
Expulsion of immigrants and aliens (#PC3207)
Lack of cooperation with officialdom (#PF8500)
Demoralizing image of urban community identity (#PF1681)
Legal contract system reduced to individual needs (#PE5397).
Aggravated by Individualism (#PF8393)
Divisive zoning disputes (#PG7903)
Individualistic life style (#PJ8635)
Unpublicized community news (#PF7998)
Unresponsive public authorities (#PF8072)
Limited public space (#PJ9066)
Inadequate adult guidance (#PG9147)
Limited strength of elders (#PG9151)
Excessive television viewing (#PD1533)
Non–cooperative community groups (#PG9159)
Overriding individual differences (#PF7934)
Overemphasis on community problems (#PG8617)
Minimal church / school involvement (#PJ9011)
Insufficient community celebrations (#PJ0188)
Irrelevance of educational curricula (#PF0443)
Detrimental story of community future (#PF6575)
Minimum promotion of community assets (#PF6557)
Inequality inducing effects of television (#PD5833)
Deteriorating structures of rural community cooperation (#PF3558).

♦ PF2577 Puritanism

Nature One of the strongest and most dynamic forms of protestantism, puritanism has given rise to a strict moralism based on guilt which is manifested in religious intolerance, repression, censorship and general austerity. It has encouraged discrimination especially concerning sexual morality, from which a double standard of morality has evolved.
Incidence Puritanism has been particularly instrumental in forming the social attitudes of English–speaking nations.
Background The term puritan was coined as an epithet of contempt in England during the 1560s. Puritanism been ingrained in England after the civil war of 1642 and the Commonwealth. Puritanism spread in North America from the New England puritans who emigrated from England at times when puritanism was suppressed.
Broader Ideological conflict (#PF3388)
Religious intolerance (#PC1808)
Double standards of sexual morality (#PF3259).
Narrower Inadequate utilization of volunteer social service workers (#PF4892).
Related Prudery (#PF5892).
Aggravated by Elitism (#PA1387)
Religious apathy (#PC3414)
Double standards in morality (#PF5225).

♦ PF2580 Outmoded legal systems

Outdated judicial precedent — Institutionalization of discriminatory outmoded concepts in legal systems — Persistence of outmoded concepts in legal systems
Nature Concepts such as "terra nullius", "conquest" and "discovery" as modes of territorial acquisition which are repugnant, have no legal standing, and are entirely without merit or justification to substantiate any claim to jurisdiction or ownership of indigenous lands and ancestral domains. However the legacies of these concepts persist in various forms in modern legal systems.
Incidence In a number of legal systems colonial laws and concepts are used to justify the imposition of "trusteeship", and other demeaning, prejudicial and racially founded systems which prevent indigenous peoples from exercising their human rights and fundamental freedoms, resulting in their impoverishment, disenfranchisement, debasement, demoralization and disintegration.
Claim The rapid social changes of the 20th century have paralysed the judicial process, which is based on past precedents that were established in a context that is now outdated and in situations of totally different experience. The basis of the process on property and individual rights over–emphasizes and over–protects the individual. This was the cutting edge of man's struggle in the 17th century, and may be still valid; but it needs to be employed in the context of freedom to participate in all the aspects of 20th century corporate welfare.
Many present law enforcement methods and judicial systems are outdated and do not relate to the basic requirements of the people. The complexity of legal systems discourages revision, repeal or creation of laws. This merely reinforces the antiquated laws along with the more adequate ones. The static state of such systems leaves little room for changing needs and restricts the necessary revitalization within the legal base.
Broader Persistence of outmoded concepts (#PF7673)
Legal contract system reduced to individual needs (#PE5397)
Exclusion of the masses from setting criteria in judicial judgements (#PD1060).
Narrower Legal inconsistency (#PF5356)
Biased legal systems (#PF8065)
Unresolved legal issues (#PG8289)
Restrictive legal practices (#PD8614)
Deficiencies in international law (#PF4816)
Unnecessary verbosity of legal documents (#PF7137)
Inadequacy of international legal procedure (#PF8616)
Deficiencies in national and local legal systems (#PF4851).
Aggravates Collapse of judicial system (#PJ0761)
Expropriation of land from indigenous populations (#PC3304).

♦ PF2583 Underprovision of basic urban services

Limited access to urban services — Poor service delivery in urban environments — Poor services motivation — Underdevelopment of electricity, gas, water and sanitary services
Nature Many urban neighbourhoods experience difficulty in gaining access to the benefits and services which society would normally consider their due, partly because of the intricately designed, complex delivery systems for such amenities, whether public or private. The complexity of urban government systems makes it overwhelmingly difficult to get speedy action on basic community problems like sewage back–up. Community credit is inhibited by the practice of "red–lining" by financial institutions, cutting off loans and mortgages to residents in depressed urban areas. Private services such as those offered by churches require membership or other obligations. Fundamentally, however, the problem lies in some communities' limited skills in functioning in the complex maze of city and private agencies which service them, resulting in their being effectively deprived of the benefits afforded to other communities. Development of such neighbourhoods requires skill in such procedures. Only when residents know what services are actually available to them and begin to experience success in meeting their own needs, will a community be able to participate with a new assurance and responsibility.
Background Governments usually play a large role in the provision of urban services because private providers traditionally find it hard to make profits from such services. Efficient urban services are a pre–condition for economic growth. They include: urban transport, water supplies, power and housing. Urban–based firms need transport and communications to do business with each other, sanitary services to dispose of waste, and power to make their capital productive. Their workers need all these services and housing too.
Incidence Despite heavy subsidies, many urban services are underprovided. Estimates by the World Bank indicate that 23 per cent of of the urban population in developing countries in without potable water within 200 metres; the figure rises to 35 per cent in sub–Saharan Africa. Road congestion is spreading and escalating transport costs have reduced productivity. Spending in many cities is not directed toward the appropriate services. In some cases, as in bus transport, large subsidies to public providers have squeezed out more efficient private providers. Basic services are being neglected. The cost of this neglect is particularly high when alternative private sources are either unavailable or too small to be efficient, as in the case of water and electricity.
Broader Inadequate infrastructure (#PC7693)
Underprovision of basic services to rural areas (#PF2875)
Underdevelopment of industrial and economic activities (#PC0880).
Narrower Slow city services (#PG7660)
Limited consumer knowledge (#PG7662)
Inadequate electricity supply (#PJ0641)
Unfamiliar government procedures (#PJ0740)
Inaccessible religious programmes (#PG7657)
Unreasonable licensing restrictions (#PE7655)
Lack of consumer influence on industry (#PE1940)
Unacknowledged availability of services (#PJ9082)
Limited availability of health resources (#PD7669)
Undefined government role for urban services (#PE7919)
Limited transportation services in urban areas (#PE8959)
Protectionism in the high–technology industries (#PE8458)
Insufficient financial resources for urban services (#PE9133)
Ineffective self–regulation in the food–processing industries (#PE8472).
Related Inaccessible job market (#PE8916)
Limited availability of investment capital for urban renewal (#PF3550).
Aggravates Social isolation of the elderly (#PD1564).
Aggravated by Restrictive building codes in urban areas (#PE8443)
Undeveloped business skills in urban areas (#PE8048)
Unrecognized possibilities for training in urban areas (#PJ7909).

♦ PF2588 Fragmented forms of cooperative efforts

Lack of combined efforts
Nature Despite the desire of residents in many rural communities to cooperate in local ventures, efforts at organizing such cooperation are often fragmented. Past failures on these lines tend to make communities hesitant about trying other ventures. Although when an emergency arises people work well together, when less immediate issues threaten the absence of long–term cooperative structures is evident. Citizen cooperation is replaced by individualism, especially in situations where property or normal habit patterns are involved.
Claim Until citizens develop ongoing structures of cooperation, fragmentation and the memory of disappointing past exercises will continue to block future development.
Broader Fragmented conduct of community operations (#PF1205)
Unrecognized socio–economic interdependencies (#PF2969).

♦ PF2590 Dependence on mysticism

Nature Mystical experience is defined as being the super–sensual knowledge or feeling of oneness with a higher reality. Belief in the possibility of mystical experience constitutes mysticism, which may be identified with occultism, superstition, symbolism, fancifulness, ineffectual idealism, vagueness or sentimentality. As a sociological phenomenon, mystical religion is characterized as a regressive, imagined participation in an amorphous, collective life–stream that may be viewed as nature or race or both. Psychologically, the other–wordly nature of mystical faith is in accord with schizoid personality tendencies, and mental disorders among mystics have been noted, so that it appears that introversion, dependence, and reality–avoiding temperaments may be drawn to beliefs of this nature. With religious intolerance, mysticism might also be defined as a heresy.
Incidence The mystical experience is not restricted to religious experiences. They occur in a great variety of fields, in numerous ways, and with all degrees of depth and inclusiveness. Lofty appreciation of beauty or sublimity, absorbed enjoyment of music, serene companionship with nature, sudden insight into the meaning of truth, the awakening of love, moral exaltation of life in the pursuit of duty, illustrate some types of experience which immensely transcend knowledge, experiences in which the subject and object are fused into an undifferentiated one, and in which self is identified with object.
Mysticism exists as a primary characteristic in Hinduism and Buddhism. Mysticism in the 'Abrahamic' religions, Judaism, Christianity and Islam, exists as orthodox expressions of religiousity but it is closely monitored by the ecclesiastics of these faiths.
Background The heretical Gnosticism served to caution mystics of the early Christian church against an uncritical acceptance of pagan spirituality. Quietism, condemned in 1687, was a later heresy connected with otherwise orthodox Catholic mysticism in Spain and France. Although Puritanism and Quakerism in England and America evolved a strong mysticism, other Protestant schools of thought totally condemned mysticism, casting a cloud over it in the 20th century. Hinduism in all its varieties has always been particularly predisposed to mysticism. Popular buddhism has Nirvana as its goal; but the advent of Zen combined this with a practical approach to life. The large literature of Jewish and Taoist mysticism is available, although the extent of their practice is covered in some secrecy owing to historically avowed connections with magic.
Claim Christian mysticism recognizes a distinction between the pantheistic yearning for oneness with an impersonal All and the mystical response of faith to the life of an historical Saviour. This in effect denies all other forms of mysticism and breeds religious intolerance.
Counter–claim Thomas Merton points out that mysticism is part of the normal Christian life, not paranormal or eccentric, not a privileged vocation for the super–pious. The ordinary Christian is either a mystic or does not exist.
Mysticism which places spiritual reality as higher than mundane reality or as true reality must be separated from the mystical experience in which the spiritual and the non–spiritual are interdependent realities. Attempts to live in one realm to the exclusion of the other leads to materialism or a perverse form of mysticism, separated from reality and ultimately self–destructive.
Broader Subjectivism (#PF8015)
Religious conflict (#PC3292)
Dependence on religion (#PF0150).
Related Superstition (#PA0430).
Aggravates Persecution (#PB7709)
Religious intolerance (#PC1808)
Religious discrimination (#PC1455).
Aggravated by Compulsory indoctrination (#PD3097).

♦ PF2595 Non–articulated educational goals

Untargeted educational goals — Unconsensed educational goals — Misdirected educational goals
Claim Methods of education today have become so unrelated to authentic intention and vocational thrust that there is a collapse of the purpose and role of education; and it no longer raises the question with students of how they are going to decide on priorities for the survival of society in today's world.
Broader Insufficient provision of public services for communication (#PF2694)
Overemphasis on self–sufficiency with respect to interdependence (#PF3460).
Narrower Unclear educational roles (#PG8840).
Aggravates Unarticulated goal of educational methods (#PF2400)
Underdeveloped potential of basic resources (#PF3448)
Fragmented pattern of community organization (#PF6525).

♦ PF2600 Parochial national interests

Nationalistic images of citizenship — Defensive nationalistic images — Overruling of global

interdependence by nationalistic images — Parochial external relations
Nature Nationalistic definitions of social responsibility hinder global understanding of citizenship, including the images and legal forms which allow for true global identification. At present cultural exclusiveness inhibits more than a national understanding of citizenship.
Images of conflict and defence against the rest of the world override motivation for a global covenant which would provide the basis of a global legal base.
Physical isolation from other countries has been replaced by powerful military and economic controls to create national security. Studies in national character often are attempts to further this end. People are prevented from appropriating other cultures by national-oriented concepts of foreign threat and by dehumanized images of other nationalities. Popular images are often powerful enough to influence the behaviour of the majority of a nation's citizens towards other nations and cultures; and international encounter is limited to superficial efforts at diplomacy or tourism.
Claim In the midst of the emergence of the "global community", the community values of adults have remained parochial and nationalistic. There is no question of consciously adopting a positive stance towards collective experience and the traditions which contribute to common wisdom; no adequate means for symbolic expression of world citizenship; and no conscious demonstration, whether at the individual or the community level, of either their interrelatedness or their self-identity.
Broader Nationalism (#PB0534)
Reinforced parochialism of internal values and images (#PF1728).
Narrower Ignorance of history (#PD3774) Fear of losing cultural identity (#PD2614)
Unrealistic attempts at globality (#PF2814)
Inadequate structures for achieving global unity (#PD2802).
Related Non-concerned attitudes (#PF2158) International paternalism (#PF1871)
Misuse of international forums (#PF2216)
Limited community responsibility of adults (#PF1731)
Structural failure of citizen participation (#PF2347)
Modern disruption of traditional symbol systems (#PF6461)
Unexercised responsibility for external relations (#PF6505)
Legal contract system reduced to individual needs (#PE5397)
Limited exposure to outside influences in rural villages (#PF2296).
Aggravates Unchallenging world vision (#PF9478)
Deteriorated structures of essential corporateness (#PF1301)
Nationalistically determined development of natural resources (#PD3546)
Domination of the world by territorially organized sovereign states (#PD0055).

♦ PF2602 Limiting effect of individual survivalism
Nature Despite moves by governments to provide for the basic needs of shelter, food and health services, an overriding sense of individual survivalism saps energy, limits aspirations and undermines broader motivation. In the economic realm, along with low incomes and unemployment, there is fatalistic uncertainty about job retention, welfare support, and the ability to compete successfully with people from other communities for jobs in one's own community or beyond. Another manifestation is the uncertainty of housing. There is an almost universal sense of transient or temporary residency, even among some who have been in a community for many years, with plans to move on when what seems to be a better opportunity arises. The high crime rate intensifies the sense of having to struggle to get along. The individual experiences his situation as so precarious that any relationship with other people is a threat to security, and it is a risk to suggest serious change. Community concerns are often spoken of but seldom resolved.
Claim Until citizens are released from their immediate and individualistic style, no concerted effort toward reversing the deterioration of urban communities is possible.
Broader Obstacles to community achievement (#PF7118).
Narrower Fear of resettlement (#PF9030) Excessive wheat surpluses (#PE2902)
Excessive parental defensiveness (#PG7688) Mutual student-teacher disinterest (#PG7678)
Minimal family interest of urban life styles (#PE8296)
Unfeasible housing alternatives in urban areas (#PE8061)
Apathy toward improvement of urban life styles (#PE8477)
Struggle for financial security in urban life style (#PE8144)
Perpetual preoccupation for sustenance of urban life style (#PE8644)
Ineffective self-regulation in the pharmaceutical and medical devices industries (#PE8502).
Related Domination (#PA0839) Student absenteeism (#PE4200)
Distortionary tax systems (#PD3436).
Aggravated by Overproduction of food (#PD9448).

♦ PF2604 Underdevelopment of woodworking industries
Broader Underdevelopment of manufacturing industries (#PF0854).
Narrower Furniture and fixtures manufacture underdevelopment (#PG5168).

♦ PF2605 Lack of community planning
Nature In the absence of an integrated community design much needed action to improve the condition and appearance of a village is not carried out. Despite increasing concern with the planning of space, particularly in the way a cohesive plan reveals the unique image and identity of each community, in most rural communities not much happens to improve the situation. Litter and debris detract from the appearance of a village and encourage laxity in helping to keep it clean. Many areas may require beautification but no village landscaping plan gets prepared. The pride of the community is dampened. Unless comprehensive space and beautification plans are created, it remains difficult to generate and maintain a community's symbolic identity.
Broader Obstacles to community achievement (#PF7118)
Obstacles to availability of community space (#PF7130).
Narrower Feral mammals (#PE0185) Low athletics priority (#PJ0337)
Restrictive community size (#PJ0123) Lung disorders and diseases (#PD0637)
Prohibitive cost of road construction (#PJ1070)
Underdeveloped sources of income expansion (#PF1345)
Obstacles to restructuring production in the industrialized countries (#PE9055).
Related Landscape disfigurement (#PC2122).
Aggravates Uncompensated tenant repair (#PJ0147)
Lack of community development (#PF7912).
Aggravated by Idleness (#PA7710)
Lack of central planning structures in small communities (#PF2540).

♦ PF2607 Proscriptive controls favouring the investor
Nature The regulations which organize and control credit have been defined by the investor with his best interests in mind. Such controls are proscriptive in nature, excluding investments which do not fit into the present procedures e. en if they meet real needs. The regulations also emphasize the flow of goods rather than the flow of services despite increasing demand for the latter. A condition thus exists where systems of control support the flow of credit to immediate needs to the exclusion of long-range needs.
Broader Self-interest driven investment (#PC2576).

♦ PF2613 Hazardous remnants of war
Ineffective minesweeping — Derelict military explosive devices — Unexploded bombs — Uncleared land mines — Uncleared sea mines
Nature Explosive devices engineered to detonate by pressure on contact or by proximity are used as antitank, antipersonnel and antivessel weapons. They may be employed by regular military forces, by guerrillas, by terrorists, by saboteurs or by criminals. In large numbers, such land or naval mines afford the opportunity that one or more will escape detection and minesweeping removal actions. Such derelict naval mines, fixed or floating free, remain a threat to ships at sea or coming into ports. Unremoved land mines are a hazard to any vehicles, persons, livestock or single animals in the vicinity.
Incidence Mining of the Suez Canal in 1984 by unknown saboteurs was not effectively countered by sweeping, and ships were damaged. Newer mining armaments technology may include non-metallic, undetectable mines; 'smart' naval mines that move; and systems approaches to land mining, particularly around fixed tactical or strategic installations, that are not possible to sweep (such as time-delay fuses on nylon-canistered devices, or C-B mines). The material remnants of war, particularly mines and unexploded bombs have been left on the territories, most recently in developing countries. These materials seriously impede development and cause injuries and the loss of lives and property. In 1990 it was estimated that some 20,000 Afghans were being severely wounded each year by hidden land mines.
Broader Conflict (#PA0298) Disastrous consequences of war (#PC4257)
Failure of disarmament and arms control efforts (#PF0013).
Related Housing destruction in war (#PE2592).
Aggravates Injuries (#PB0855) Human death (#PA0072) Loss of property (#PG5523)
Human destructiveness (#PA0832).
Aggravated by Lack of international cooperation (#PF0817).

♦ PF2621 Multiplicity of time standards
Nature The world was previously divided into 24 time zones based on 24 standard meridians 15 degrees apart in longitude, starting at Greenwich (UK). However, these theoretical divisions cut across states and divided adjacent areas inconveniently, so the division has therefore been considerably modified in practice. This modification has been carried further by the adoption (from time to time in different countries and at different times of the year) of summer-time or daylight saving measures. These time differences create considerable difficulty in setting up international timetables and maintaining time-based historical records. The multiplicity of time standards increases the difficulty of conducting international business, because of the limited overlap in working hours in different time zones. Some world-wide organizations are able to exploit these difficulties to their own advantage.
Broader Inadequacy of international standards (#PF5072)
Obstacles to efficient utilization of time (#PF7022).
Related Inadequate world calendar (#PF2043) Differing conceptions of time (#PF6665).
Aggravates Telephone delays (#PF1698) Human errors and miscalculations (#PF3702)
Lack of international cooperation (#PF0817)
Desynchronization of bodily rhythm by international travel (#PE1904).

♦ PF2622 Lack of comparability of international statistics
Limitations of international statistics
Nature For statistical data from different countries to be useful in international planning of economic and social development, this data must be collected, analyzed and presented in such a way as to permit inter-country comparisons. Despite considerable progress, it is still very difficult to prepare regional and world totals on the basis of information originating from official national sources. Each country may apply a slightly different method of data collection such that each individual statistical figure requires a special qualifying footnote in any international statistical compendium. Another source of incomparability is vast differences in meaning for terms, such as, unemployment, disease and crime.
Incidence Of paramount importance are national economic, social and health statistics, which despite their intense study at international levels cannot be subsumed into comprehensive regional or global aggregates to detect levels and trends. Employment, distribution of wealth, crime, human rights and educational statistics are among those presenting difficulties.
Counter-claim The timidity of academicism or some professional statisticians employed by intergovernmental organizations prevents them from assembling probabilistic data, or to engage in forecasting to any considerable extent. In some cases, country data submitted is suppressed from publication because its statistical treatment does not conform to that of the majority of countries, or there is some anomaly or marginally higher level of statistical uncertainty.
Refs Ginneken, Wouter Van and Park, Jong-goo *Generating Internationally Comparable Income Distribution Estimates* (1984).
Broader Statistical errors (#PF4118) Conflict of information (#PF2002).
Aggravated by Inadequate world calendar (#PF2043)
Deficiencies in national statistics (#PF0510).

♦ PF2627 Parochial leadership posture
Parochial images of society by leaders — Leadership dependent on local biases
Nature Government leaders have very reduced understanding of what society is and how it works. Despite the present interrelatedness of society, leaders still do not see themselves as globally responsible, but rather persist in functioning from limited and parochial images. Decision-making thus based on short-range and short-term results increases global tension.
Broader Obstacles to leadership (#PF7011) Reduced understanding of globality (#PF7071)
Paralyzing patterns between villages and administrative structures (#PF1389).

♦ PF2628 Political opposition to administrative action
Nature In many countries, the central civil service organs do not enjoy the unqualified support of the political authorities, who in the final resort take the decisions. The result is a certain paralysis of these central organs. The sweeping reforms essential for a coherent personnel administration policy raise a multitude of political problems, especially where the aim is to institute a merit system, to reduce ministerial powers over personnel, to challenge ingrained habits, to change established practices, or even to get rid of redundant staff. The chief opposition often comes from ministers who feel themselves unjustly deprived of managerial responsibility and of authority over their staff. This basic problem is of particular concern in countries where centralization of the decision-making powers is highly developed.
Broader Excessive government control (#PF0304).
Aggravates Bureaucracy as an organizational disease (#PD0460).
Reduces Political over-reaction (#PF4110).
Reduced by Promotion of negative images of opponents (#PF4133).

♦ PF2631 Minimal access to necessary information
Nature Communication of basic information is important to every community, but access to that information is an problem for many remote villages. Residents are often uninformed about services available to them and/or laws and regulations that affect them. The minimal availability of newspapers, telephone services and libraries means there is limited access to world and regional information. Even basics such as means of income are limited by poor access to such information as market conditions and price quotations. Without effective links to necessary data sources, limited access to information restricts initiative; and residents remain cut off and unable to take advantage of knowledge which is readily available.
Broader Incomplete access to information resources (#PF2401).

♦ PF2633 Panic

Nature Panic is a term used loosely to describe a variety of such social situations in which a portion of the population responds irrationally to socio–economic, military, or other crises, whether real or anticipated. Panic may follow acute financial crises, widespread bank failures, stock market failure, or the threat of attack. In the case of chemical and biological warfare, for example, it may cause individuals to flee from the area, even when there is no longer any danger of exposure, thus making it almost impossible either to bring personnel or material into the affected area to help cope with the casualties or to evacuate those affected.

Refs Hand, I and Wittchen, H U (Eds) *Panic and Phobias Two* (1988); Yardley, Stella S *Fear and Panic* (1988).

Broader Fear (#PA6030) Defeat (#PA7289) Inexcitability (#PA5467).
Narrower Panic consumer buying (#PJ0542).
Aggravated by War (#PB0593) Alarmism (#PF4384) Disasters (#PB3561)
Agoraphobia (#PE0527) Panic disorder (#PE3575) Biochemical warfare (#PC1164)
Inadequate riot control (#PD2207) Inadequacy of civil defence (#PF0506)
False nuclear warfare alerts (#PF1236) Inhumane methods of riot control (#PD1156)
Denial of human rights in armed conflicts (#PC1454).

♦ PF2634 Obscenity

Nature Conduct offensive to public decency includes the publication of indecent literature; behaviour considered as indecent (such as forms of sexual deviation); acts of profanity or in bad taste, such as sick jokes and possibly black comedy; as well as pornography of various kinds. Secular or church–related viewpoints of decency may lead to expectations that publishers, film and theatre directors, and television and radio broadcasters would act as censors, and failing to do so they would be held responsible for undeleted expletives and other offences.

Obscenity may also be perceived as including war crimes and torture, and certain highly sophisticated and devastating weapons.

Since obscenity is a subjective term, its application varies according to the moral customs and standards of different countries, according to prevailing opinion. The tolerance of certain forms of sexual deviation has increased in the so–called permissive society of western industrialized countries. Violence for pleasure is a problem in urbanized areas, especially among young people.

Broader Immorality (#PA3369) Sexual deviation (#PD2198) Victimless crime (#PC5005)
Sexual harassment of women (#PF3271) Vice and sex traffic offences (#PD8910).
Narrower Profanity (#PF7427) Sick jokes (#PG5575) Pornography (#PD0132)
Black comedy (#PG5576) Indecent art (#PE5042) Indecent exposure (#PF4317)
Immoral literature (#PF1384) Obscene telephone calls (#PE5757)
Exposure of children to pornography (#PJ8730).
Related Moralism (#PF3379) Prostitution (#PD0693).
Aggravates Film and cinema censorship (#PD3032).
Aggravated by Decadence (#PB2542) Frustration (#PA2252)
Permissiveness (#PF1252).
Reduces Aggression (#PA0587).
Reduced by Art censorship (#PD2337) Book censorship (#PD3026).

♦ PF2636 Neo–fascism

Neo–nazism — National socialism — Nazism

Nature Resurgence of Nazi and fascist principles and doctrines may involve racism (the desire to purify the superior race and the idea of its destiny to rule other inferior cultures), nationalism, anti–communism, militarism or tactics of force and brutality, an appeal to ignorance and fear among the masses, extremist measures, antisemitism, and sentimentality and exaltation of capitalistic imperialistic ethics. The movement feeds on ignorance and prejudice and tends to gain more adherents during a period of economic depression or government inadequacy. It may also take the form of a backlash movement against progressive or socialist groups. It breeds the intolerance and fear, prejudice and conflict that lead to wars and crimes against humanity, as in the past.

Nazism may be distinguished from Neo–nazism as being the continuation of the belief in the doctrines of the Third Reich, the honouring of Adolph Hitler as founder and as the ongoing inspiration to the Nazi adherents, and the addiction to all the paraphernalia of the original Nazism including the swastika emblem, SS armbands with the black lightning design, the singing of the Horst Wiesel, and the belief in the destiny of the fatherland.

Such forms of nazism are to be distinguished from some uses of nazi symbols by youth cults as a form of decorative protest.

Background Nazism originated as a political movement which held power in Germany from 1933 to 1945, the main features of the Nazi programme were the creation of a master race and world domination. The movement combined a popular appeal to the masses vaguely akin to socialism; with an appeal to the aristocracy through nationalistic and racist ideals of domination; an appeal to military circles in the development of an expansionist programme and the use of force both domestically and externally; and an appeal to business circles in the development of a war economy and the suppression of individual trade unions, which were consolidated into one unit incorporating management interests. The regime was highly centralized, authoritarian, totalitarian, and brutally repressive of any opposition or racial and hence general 'impurities'. The sadistic and systematic use of repression reached unprecedented proportions in the extermination of over 20 million political prisoners, including 5 million Jews.

National Socialism has its roots in the Prussian tradition of the great soldier kings (Frederick William I and Frederick II) and statesmen such as Bismarck, who combined militarism with political romanticism and hostility to rationalism. This was reinforced by the 19th century worship of science and the laws of nature which had an 'iron logic' divorced from concepts of good and evil. Pan Germanic movements in the Austrian Empire before 1914, racism, anti–semitism, anti–slavism, anti–catholicism, and the search for a German community prepared the way for Hitler's party. The development of Hitler himself was influenced not only by this environment but, in the years before joining the National Socialist German Workers Party ('Nazi' was the abbreviation) and prior to his 1914 to 1919 Army Service, by the philosophies of Nietsche and Richard Wagner, and by the occult.

Incidence Nazism and neo–nazism exist in Germany and are closely linked phenomenologically with right–wing extremist movements elsewhere, such as in the USA, Belgium, Brazil and South Africa; and with neo–fascist groups in Italy, Spain, and Portugal. Similar groups are reported in the USSR.

Broader Racism (#PB1047) Ideological conflict (#PF3388) Extremist ideologies (#PC6341).
Narrower Antisemitism (#PE2131) Anti–communism (#PF1826)
Totalitarianism (#PF2190).
Related Fascism (#PF0248).
Aggravates Intolerance (#PF0860) Racial conflict (#PC3684)
Racial discrimination (#PC0006) Politically emotive words and terms (#PF3128).
Aggravated by Destiny (#PF3111) Nationalism (#PB0534)
Defeat in war (#PG4363) Socio–economic poverty (#PB0388)
Double standards in morality (#PF5225).

♦ PF2639 Production serving false consumption needs

Nature Technology has enabled humans be freed from the constraints of working primarily to survive in an environment of scarcity. It is possible to sustain life abundantly for all of the world's people. Out of this shift has emerged a new concept of work motivated by material reward than survival needs. This new incentive structure turned systems of production to promoting consumption for its own sake. Systems of allocating consumption have been restructured to reenforce systems for meeting this falsely created demand for goods and services. The whole economy is increasingly oriented to meet the needs of increasing consumption and to ignore any other human needs.

Broader Production of non–essentials (#PC3651).
Related Historical forgery (#PE5051).
Aggravates Historical misrepresentation (#PF4932)
Excessive consumption of goods and services (#PC2518).
Aggravated by Misuse of advertising (#PE4225).

♦ PF2640 Lesbianism

Female homosexuality

Nature Although female homosexual practices are believed to occur much less widely than male, and have never been thought to constitute so grave a danger to society, they may nevertheless cause anguish in that many such relationships come under social pressure and may lead to blackmail or intimidation.

Incidence Compared with male homosexuals, lesbians are more faithful; the majority of them have less than 10 partners. An investigation conducted in France found that 56 percent of the formed a permanent couple, compared with on 36 percent of male homosexuals. They more often meet their partners at the houses at friends, some 60.6 percent according to the same investigation; whereas, 62.7 per cent of male homosexuals found their partners in specialized premises. Lesbians are usually less despised and less persecuted, because they are less visible and because what happens among women is considered less important in a male–oriented society. Lesbians are not suspected of paedophilia.

Claim Unconscious female homosexuality causes sexual problems in marriage.

Counter–claim A collection of accounts by lesbians in Belgium reveal that, as girls or young women, they were afraid of, or had experience of, the selfishness and brutality of men; in the company of another woman they were then able to discover, like others since adolescence, bodily pleasure and to have a love life without constraint or domination. Lesbians, to generalize, are women who are in love, and less interested in the sexual act than in a general affective relationship. The more this relationship is equal and reciprocal, the more the couple is durable, as it often is.

Refs Cavin, Susan, et al *Lesbian Origins* (1985); Curb, Rosemary and Manahan, Nancy *Lesbian Nuns* (1986); Diamant, Louise (Ed) *Male and Female Homosexuality* (1987); Green, G Dorsey and Clunis, D Merilee *Lesbian Couples* (1988); Hanscombe, Gillian E and Forster, Jackie *Rocking the Cradle* (1982); Maggiore, Dolores J *Lesbianism* (1987).

Broader Homosexuality (#PF3242).
Related Bisexuality (#PF3269) Male homosexuality (#PF1390)
Increasing female criminality (#PE5592).
Aggravates Family breakdown (#PC2102) Sexual frigidity (#PE6408)
Social fragmentation (#PF1324).
Aggravated by Sexual unfulfilment (#PF3260) Human sexual inadequacy (#PC1892)
Sexually segregated schools (#PG3650)
Violation of the rights of female homosexuals (#PE5741).

♦ PF2641 Human sacrifice

Ritual murder — Ritual killings — Ritual dismemberment

Nature The murder of human beings as part of a religious ritual has been most widely adopted by agricultural rather than hunting or pastoral peoples. Sacrifice would be made to encourage soil fertility or as a form of ancestor worship. Victims include kings, slaves, criminals and children. Human sacrifice may take a more insidious form in civilized societies when necessary changes encounter political and other forms of resistance until one or more lives have been sacrificed.

Background Human sacrifice existed in the ancient civilizations of Egypt, Mesopotamia, Palestine, Iran, India, China and Japan, Greece and Rome. It formed part of the annual rituals of the Aztec and Inca civilizations. In China up until the 17th century, a king's retinue was buried with him. In India up until the 19th century, the followers of Kali sacrificed a male child every week. Also in India, the practice of suttee (suicide of a deceased man's wife on his funeral pyre) survived until the 19th century. The North American Pawnee Indians used to sacrifice girls to fertilize the soil. From the 12th century onwards claims have been made that the Jews sacrificed Christian children at the passover. The last accusation of this kind was made by the Russian Government in 1911, but the accused was acquitted.

Incidence Human sacrifice is still believed to exist among certain tribes in Africa. In its more insidious form in civilized societies, the introduction of much new legislation affecting the condition of people can only be formulated, and only acquires credibility, after lives have been sacrificed. Examples include legislation affecting health and safety.

Claim Human beings are being sacrificed in ritual killings, such as, those committed by the Charles Manson gang and the Adolfo de Jesus Constanzo cult. They are sacrificed in India to ensure good harvests and in Liberia and other West African countries to help the advancement of politicians. Lives are sacrificed for the advancement of causes: the liberation of Palestine, the race to the moon, testing medicines, opposing apartheid and stopping war.

Refs Davies, Nigel *Human Sacrifice in History and Today* (1988); Gohain, Bikash *Human Sacrifice and Head Hunting in North Eastern India* (1977).

Broader Homicide (#PD2341) Occultism (#PF3312) Religious sacrifice (#PD3373).
Narrower Suttee (#PF4819) Sacrifice of children (#PJ5597).
Related Superstition (#PA0430) Human cannibalism (#PF2513).
Aggravates Fear (#PA6030) Slave trade (#PC0130) Ritual cannibalism (#PF8944).
Aggravated by Slavery (#PC0146) Tribalism (#PC1910)
Traditionalism (#PF2676) Satanic rituals (#PF7887)
Ancestor worship (#PD2315) Inauspicious conditions (#PF6683)
Head hunting in tribal societies (#PF2666).

♦ PF2643 Criminalization of euthanasia

Mercy killing — Illegality of doctor–assisted suicide — Denial of the right to euthanasia

Nature Euthanasia is a term which generally applies to the measures by which physicians seek to remove or alleviate the distress attending the approach of death in the course of a chronic disease. The removal of pain is regarded as essential for an 'easy death'. In a more specific sense, the term also implies the means of bringing on an 'easy death' by legally putting to immediate death one who suffers from an incurable disease and who prefers this kind of death to being tormented for a lengthy period before an eventual, painful death. This procedure is known as 'mercy killing' and has been advocated by small groups of physicians subscribing to the belief that, with adequate safeguards, euthanasia should be legalized to allow incurable sufferers to choose immediate death rather than await it in agony. These advocates of euthanasia hold that most of the legal and religious arguments against mercy killing are founded on emotion rather than on reason.

Claim When patients determine their own fate the public attitude toward euthanasia becomes more tolerant.

Counter–claim Euthanasia is open to the most tragic kinds of abuse. In spite of increasing public demand for the legal right to euthanasia, in all civilized countries any kind of euthanasia is contrary to ethical and humanitarian concepts and is punishable as a felony.

Refs Brody, Baruch A (Ed) *Suicide and Euthanasia* (1989); Downing, A B (Ed) *Euthanasia and the Right to Death* (1969); Maestri, William *Choose Life and Not Death*; McMillan, Richard C et al (Ed) *Euthanasia and the Newborn* (1987).
Broader Restrictive legislation (#PD9012) Denial of the right to die (#PF4813)
Denial to people of control over their own lives (#PC2381).
Related Suttee (#PF4819) Homicide (#PD2341) Parenticide (#PE0651).
Aggravates Suicide (#PC0417) Fear of death (#PF0462)
Individual unfitness for survival (#PF4946)
Excessive prolongation of the dying process (#PF4936)
Lack of facilities for severely deformed people (#PD0211).
Aggravated by Social withdrawal of aged (#PD3518)
Ageing of world population (#PC0027)
Inadequate income in old age (#PC1966)
Genetic defects and diseases (#PD2389)
Inadequate housing for the aged (#PD0276)
Social disadvantage of the aged (#PD3517)
Rigidity and inadaptability in the aged (#PD3515)
Susceptibility of the old to physical ill-health (#PD1043).
Reduces Killing by humans (#PC8096).

♦ PF2644 Inflexible patterns of family lifestyle
Delimiting family patterns of traditional way of life
Nature The fragmentation of the community common in Western society might be reversed if family lifestyles were formally focused on maintaining outside interests (a role once played by the church or by agricultural groups). Unfortunately, the family is often unable to sustain an interest in the affairs of the community, and is lost as a creative force to the town.
Broader Obstacles to family life (#PF7094) Inhibiting social patterns (#PF0193)
Inhibiting effects of traditional life-styles (#PF3211).
Narrower Exhausting work demands (#PG7952) Unengaging family activities (#PG7916)
Frequent schedule disruptions (#PJ0871) Unperceived interests of children (#PG7936)
Failure to recognize uniqueness of family members (#PF1750).
Related Time consuming procedures (#PJ8206).

♦ PF2645 Extradition refusal
Broader Inadequate laws (#PC6848) Impediments to extradition (#PF5947)
Obstruction of international criminal investigations (#PF7277).
Aggravates Social injustice (#PC0797) Poor international relations (#PJ5600)
Ineffective war crime prosecution (#PD1464).
Aggravated by Aerial piracy (#PD0124) National laws (#PG5601)
Rejection of refugees (#PF3021) Refusal to grant nationality (#PF2657)
Lack of international cooperation (#PF0817)
Lack of individual rights to political asylum (#PF1075).

♦ PF2648 Incompatibility of rural values in urban cultures
Nature The residents of many of the most sophisticated urban neighbourhoods are often few generations removed from families who migrated from farm areas, perhaps in other countries. Their prevailing images of community life are therefore rural ones, and they have rarely found a way to internalize and affirm the structural, prioritized relationships of urban life. Residents have no creative way of halting or reversing outmoded cultural impressions, and these people feel personally abandoned.
Contacts with officialdom are overlaid with self-fulfilling portents of failure. For example, local citizens may feel that the police do not always respond adequately when called to respond to community disturbances. Although police officials explain that there are certain calls which necessarily take precedence over others, even though they may reach the precinct at the same time, the people of the neighbourhood are then convinced that the police just do not care about their area. School authorities may manifest a distant stance toward these neighbourhoods in order to not get involved in personal neighbourhood conflicts, but the community sees this stance as a negative attitude of the school towards them. The result is that residents then have no structural means of expressing their concern about their children's educational settings; and the gulf of noncommunication continues to deepen, perpetuating the absence of a focused effort in education. Employment of residents is often inhibited by the absence of a structural ground between employer and employees.
Broader Change (#PF6605).
Narrower Mistrust of police (#PF8559) Anti-community attitudes (#PG7707)
Fraying of the tribal structure (#PG7712) Predetermined employer evaluation (#PG7714).
Related Police indifference to community (#PF8125)
Incomplete access to information resources (#PF2401).
Aggravates Lack of models of equality (#PF8639)
Generation communication gap (#PF0756)
Restrictive monetary practices (#PF8749)
Distrust of business by the community (#PE8963).
Aggravated by Collapse of common values (#PF1118).

♦ PF2650 Hero worship
Nature The worship of humans of extraordinary skill, strength and courage, or the eulogizing of the characters and way of life of the 'ante-Hellenic Age' or an early stage in history, occurs particularly after or during war or other conflict and forms part of an ideology of the race or group. In modern society general media techniques are used for the eulogizing of heroes. Hero worship may occur in the worship of ancestors and may include the practice of magical rites.
Incidence Hero worship traditionally takes the form of eulogizing literature, very often poems and sagas. Among primitive tribes these poems are oral; in more sophisticated society they are written and may be disseminated widely as a means of propaganda. After the 1917 Russian Revolution, Lenin was regarded as a hero; as in more recent times were Mao Tse Tung and Che Guevara.
In Japanese society there are certain characteristics of heroes. They are often of noble birth and endowed with charismatic qualities, they contribute to the general good of the nation or society, they often incur official disfavour, they meet death with calm resignation for the sake of glory, and they often leave behind farewell poems.
Counter-claim Hero's are a necessary part of the symbol system of any society. They inspire people to embody the best of a social system. They provide role models of individuals acting out the best values and expectations. They define what is good and significant in the society.
Broader Idolatry (#PF3374).
Related Superstition (#PA0430) Ancestor worship (#PD2315).
Aggravates Racism (#PB1047) Conflict (#PA0298) Ethnic conflict (#PC3685)
Ideological conflict (#PF3388) Compulsory indoctrination (#PD3097).
Aggravated by War (#PB0593) Tribalism (#PC1910) Militarism (#PC2169)
Propaganda (#PF1878) Nationalism (#PB0534) Traditionalism (#PF2676)
Inadequate political parties in developing countries (#PD0548).
Reduces Insufficient role models (#PF8451) Lack of variety of social life forms (#PE8806).

♦ PF2652 Limited distribution of basic services
Nature The need for basic services is increasing in all parts of the world but many villages are not connected with these services. Rural communities may be only partially served with electricity or not served at all. Telephone services may be irregular or non-existent, as may medical services, both of which lead to unmet emergencies and continuing patterns of poor health care. Continuance of such patterns of sparse services hinders the welfare and future growth of such communities.
Narrower Underdeveloped provision of basic services in developing countries (#PF6473).

♦ PF2655 Materialism
Nature Materialists deny the reality of anything existing that does not have mass (matter) or that cannot be expressed or applied as an energy. Thus the exhibited energy of mind, presently only partly measured by brain activity, is believed, in the materialist faith, to result from complicated chemical, electro-chemical, or electro-magnetic processes. There is no soul beyond energy and matter. There cannot be any purpose that invisibly guides history, as history is only the result of interactions of masses of men and masses of material things. There cannot be free will, as human thoughts and wishes can only operate within the material conditions and limitations of life. Materialism vigorously opposes religious philosophies, institutions, and influences.
Incidence Recent influential materialist philosophies have included C Lloyd Morgan's theory of emergent evolution, Karl Marx's dialectical materialism, and J B Watson's behaviourism.
Claim Materialism is an unscientific philosophy that cannot face the anomalies in modern cosmology, topology, high-energy physics, and psychology, for these show that an explicable universe is not possible from Euclidean, heratonian and other single-order, uni-dimensional perspectives. Dialectical materialism, with its a priori dogmatism, has held back scientific development in Russia, for example. The materialistic conception of the causes of mental illness leads to surgical destruction of parts of the brain as an attempted cure. Its solution to the ethical problem of evil is behaviourism, an Orwellian-kind of Big Brother State which conditions its citizens' acts and attitudes. In the final analysis, materialism is negated by the non-material existence of the argument with which it defends itself.
Broader Atheism (#PF2409) Ideological conflict (#PF3388).
Narrower Consumerism (#PD5774) Medical materialism (#PJ7913)
Commodity fetishism (#PE8058) Misuse of evolutionary theories (#PF3348).
Aggravates Amoralism (#PF3349) Anti-science (#PF2685)
Dependence on religion (#PF0150) Proliferation of commercialism (#PF0815).
Aggravated by Inadequacy of religion (#PF2005)
Double standards in morality (#PF5225).
Reduced by Idealism (#PF3419).

♦ PF2657 Refusal to grant nationality
Nature Refusal to grant nationality signifies a denial of political rights and rights of protection by a state. It may result in the creation of refugees and statelessness, the possibility of foreign extradition and a restriction on the freedom of movement. Refusal to grant nationality is very closely connected with deprivation of nationality which may occur if a new nationality is granted. Countries may refuse to grant asylum to persons exiled for political reasons or for particular misdemeanours; countries may refuse to give equal national status to ethnic groups.
Broader Restrictions on recognition of nationality (#PE4912).
Related Unreported births (#PF5381) Deprivation of nationality (#PD3225)
Vulnerability of nuclear power sources (#PD0365)
Forced repatriation of prisoners of war (#PD0218).
Aggravates Refugees (#PB0205) Statelessness (#PE2485)
Discrimination (#PA0833) Extradition refusal (#PF2645)
Discrimination against minorities (#PC0582)
Restrictions on freedom of movement between countries (#PC0935).
Aggravated by Immigration barriers for handicapped family members (#PE4868).
Reduced by Naturalization (#PG4974).

♦ PF2660 Nudism
Public nudity — Shameless nudity — Compulsory nudity — Ritual nudity
Nature Nudism is the cult of sexually unsegregated nakedness in developed countries. It is thought to be indecent and even immoral (leading to promiscuity) or simply impractical in cold temperatures or in situations where clothes serve as protection. As indecent exposure, it may be subject to legal penalties.
Incidence There are several hundred thousand nudists throughout the world. Compulsory communal showers in British schools reputedly do great harm to the children's psychosexual development.
Background Nudity is of widespread ritual significance. Christian baptism and conversion to Judaism originally involved nudity, as did a number of other initiatory rites. Nudity symbolized death and the beginning of a new life. Ritual nudity is often associated with magic. Witches are said to perform their rites in the nude. In some cases, magicians must be naked while offering sacrifices to Lakshmi, the wife of Vishnu but clothed when making offerings to Rama. The practice of Greco-Roman magic is associated with the observance of nudity as was Chinese magic. Nudity is associated with fertility, for example, in India, women will strip naked and embrace the image of Hanuman, the monkey god to obtain progeny. Rituals involving nudity are used to end droughts and controlling excessive rain. Funeral rites also sometimes involve nudity. A number of traditions practice liturgical nudity in association with healing and as a sign of humility and poverty. In some cases nudity may signify sinfulness.
Originally a European movement, nudism began in Germany at the beginning of the 20th century with the Nacktkultur groups. Nudist Societies were formed in England, France, Scandinavia and a few other European Countries after the First World War, and in the USA and Canada in the 1930s, but has little following in Latin or Roman Catholic countries.
Claim 1. Public displays of nudity are disgusting and degrading. Unrepentant transgressors deserve the harshest penalties society can inflict.
2. Compulsory group nakedness in school showers constitutes a gross infringement upon the civil liberties of a child. It is hard enough to bear the misery of visible impediments, but to be made to parade naked and reveal hitherto hidden inadequacies in the quintessential humiliation for the pubescent child.
3. Modern beauty pageants may replace the liturgical role of nudity. The rise of the cult of nudism may be symptomatic of modern man's loss of confidence in rationality. Nudism is democratic, as all distinctions of secular rank are removed which is reminiscent of the comradeship and egalitarianism characteristic of initiation rites.
Counter-claim 1. Nakedness fosters the right relationship between the sexes and spreads democracy and peace through the elimination of status symbols and artificial signs of inequality.
2. The public likes nudity in the theatre and the media, as do producers, writers, directors and actors, provided it is dramatically and aesthetically relevant.
Broader Ideological conflict (#PF3388) Double standards of sexual morality (#PF3259).
Narrower Indecency (#PF8842).
Related Indecent art (#PE5042) Exhibitionism (#PD4643)
Indecent advertising (#PD2547).
Aggravates Hypocrisy (#PF3377) Film and cinema censorship (#PD3032)
Inadequate utilization of volunteer social service workers (#PF4892).
Aggravated by Permissiveness (#PF1252) Double standards in morality (#PF5225).
Reduces Inadequate sex education (#PD0759).
Reduced by Art censorship (#PD2337).

♦ **PF2661 Abuse of bureaucratic procedures**
Nature Bureaucracies may respond differently to similar cases in such a way as to lead to injustice. Cases may be delayed or speeded up; rules may be applied zealously or considered inapplicable. Officials may be strongly influenced in their handling of a case by political interests, financial considerations and straight bribery, or by the implications of a particular outcome for their own career advancement, or the reputation of the agency in question. In some cases officials may use privileged information on proposed projects or decisions to their own personal advantage. Nepotism and favouritism govern many political appointments. Political processes are distorted by selfish interests making 'deals' by which the public loses.
Incidence An important case of bureaucratic abuse was the Environmental Protection Agency irregularities in the USA. The resignations in the USA of Messrs Agnew, Nixon, Allen and Haig and others of Secretariat or Cabinet Level or higher all involve bureaucratic irregularities. Dismissals and resignations due to procedural abuse can be counted in virtually every government in the last twenty years.
Broader Bureaucracy as an organizational disease (#PD0460).
Aggravates Administrative delays (#PC2550)
Inefficiency of state-controlled enterprises (#PD5642).
Aggravated by Bribery (#PC2558) Social injustice (#PC0797)
Bureaucratic bias (#PC1497) Bureaucratic corruption (#PC0279)
Bureaucratic factionalism (#PF7979) Bribery of public servants (#PD4541)
Unchecked power of government bureaucracy (#PD8890)
Inadequate control over government administrative process (#PC1818).

♦ **PF2666 Head hunting in tribal societies**
Broader Occultism (#PF3312) Human cannibalism (#PF2513).
Related Animal trophy hunting (#PE5644).
Aggravates Fear (#PA6030) Human death (#PA0072) Human sacrifice (#PF2641).
Aggravated by Tribalism (#PC1910) Superstition (#PA0430)
Self-glorification (#PG5645) Pursuit of personal prestige (#PF8145).
Reduces Unsustainable population levels (#PB0035).
Reduced by Cultural invasion (#PC2548) Forced assimilation (#PC3293).

♦ **PF2676 Traditionalism**
Nature Traditionalism is the maintenance of traditional attitudes and a traditional way of life against modernization. It constitutes a barrier to the success of development programmes in developing countries. In developed countries it may create a generation gap and general lack of adaptation of social attitudes to technological changes. In either case it leads to a waste of human resources.
Incidence In many rural areas in Africa, ridging, furrowing, sowing and planting seldom follow rectangular patterns. This has many disadvantages when an attempt is made to introduce modern farming machines which require straight line formation. Even animal draught at the most rudimentary level, with an improved hoe, requires a straight line. Yet it may be found extremely difficult to convince farmers that they should adopt the straight line in planting if they wish good yields. For the illiterate farmer, all the natural shapes that evoke wonder and admiration are round or circular: the moon, the sun, etc, so he thinks nature can hardly yield up its riches if its fundamental laws are flouted, the basic law being that all significant things are round.
Broader Underdevelopment (#PB0206) Political inertia (#PC1907)
Inflexible social structure (#PB1997).
Narrower Reactionary forces (#PB6332)
Burdensome cost of religious ceremonies (#PF3313)
Lack of economic adaptation of indigenous systems (#PE8002).
Related Nativism (#PF2186) Heterosexism (#PE0818) Conservatism (#PF2160).
Aggravates Feminism (#PF3025) Hypocrisy (#PF3377)
Paternalism (#PF2183) Hero worship (#PF2650)
Caste system (#PC1968) Art censorship (#PD2337)
Social conflict (#PC0137) Human sacrifice (#PF2641)
Male domination (#PC3024) Forced marriage (#PD1915)
Authoritarianism (#PB1638) Ancestor worship (#PD2315)
Theatre censorship (#PD3028) Religious conflict (#PC1292)
Religious sacrifice (#PD3373) Class consciousness (#PC3458)
Social fragmentation (#PF1324) Scientific censorship (#PD1709)
Sexual discrimination (#PC2022) Lack of modernization (#PG5654)
Lack of social mobility (#PF2195) Religious discrimination (#PC1455)
Unsustainable population levels (#PB0035) Poverty in developing countries (#PC0149)
Imbalance in the human sex ratio (#PF1128) Unsustainable economic development (#PC0495)
Dependence within extended families (#PD0850) Restriction of freedom of expression (#PC2162)
Discrepancies in human life evaluation (#PF1191)
Disruptive secular impact of holy days (#PE7735)
Animal worship as a barrier to development (#PD2330)
Negative effects of claims of religious infallibility (#PF3376).
Aggravated by Prejudice (#PA2173) Tribalism (#PC1910)
Discrimination (#PA0833) Arranged marriage (#PF3284)
Lack of assimilation (#PF2132)
Family structure as a barrier to progress (#PF1502).
Reduces Family breakdown (#PC2102) Ethnic disintegration (#PC3291)
Social disintegration (#PC3309) Family disorganization (#PC2151).
Reduced by Cultural invasion (#PC2548) Forced assimilation (#PC3293)
Ideological revolution (#PC3231) Disintegration of organized religion (#PD3423).

♦ **PF2679 Lack of appreciation of cultural differences**
Lack of appreciation cultural differences
Nature The tendency to gloss over or ignore cultural differences gives rise to inadequate national and international policies for social and economic development. Ignorance of these differences may result from fear of political division, but unless domination is complete it is also likely to set up dissent, and ethnic and racial conflict. In its mildest form, programmes for educational, agricultural, industrial and other development may simply be less effective and more expensive.
Incidence Cultural differences have ethnic, linguistic, racial and religious origins, and may co-exist with differences in social and political structure and in land tenure. All these considerations have particular bearing on the different requirements of development programmes applied to specific areas and communities. The problem exists in industrialized countries, as much as in developing countries but it is probably more marked in the latter if only because of the preponderance of international aid. National governments tend to gloss over differences mainly for political reasons. When international aid is given, it may follow national policy closely, but may also be based on considerations of ideological conflict between donor countries, on national policy models of the donor country, and generalized abstract study of conditions in the receiver country in relation to its own conditions.
Broader Underdevelopment (#PB0206) Cultural fragmentation (#PF0536).
Narrower Overgeneralized policy models (#PF9100)
Failure of development policies (#PF5658)
Ineffectiveness of aid to developing countries (#PF1031)
Insensitivity to diversity of cultural traditions (#PF8156)
Prohibitive cost of inadequate development policies (#PF9101).
Related Superstition (#PA0430).

Aggravates Social conflict (#PC0137) Racial conflict (#PC3684)
Ethnic conflict (#PC3685) Religious conflict (#PC3292)
Cultural deprivation (#PC1351) Discrimination against minorities (#PC0582)
Inefficient public administration (#PF2335) Maldistribution of agricultural land (#PD9189)
Untransferability of books between countries and cultures (#PF2126).
Aggravated by Technocracy (#PF6330) Political fear (#PJ5666)
Cultural arrogance (#PF5178) Ideological conflict (#PF3388)
Political fragmentation (#PF3216) Pursuit of national prestige (#PF8434).

♦ **PF2685 Anti-science**
Nature Anti-science is a movement against the control of society by science and technology, rejecting pragmatic data-based reasoning in favour of passion and expression. Anti-science is closely connected with anti-authoritarianism, and tends to break down existing structures without replacing them.
Broader Nationalism (#PB0534) Counter culture (#PF0423)
Intellectual dissent (#PC2582).
Related Marxism (#PF2189) Eugenics (#PC2153).
Aggravates Social fragmentation (#PF1324).
Aggravated by Elitism (#PA1387) Materialism (#PF2655)
Technocracy (#PF6330) Anti-meritocracy (#PG5679)
Computer control (#PG5683) Infringement of privacy (#PB0284)
Bureaucracy as an organizational disease (#PD0460)
Lack of individualism in capitalist systems (#PD3106).
Reduces Abuse of science (#PC9188).

♦ **PF2687 Sexual immorality**
Sexual licence — Sexual promiscuity — Individualistic standards of sexual behaviour — Collapse of standards of sexual conduct
Nature Sexual immorality implies behaviour contrary to the accepted moral code, which may differ according to the society and the group of people concerned. Sexual immorality may therefore include any sexual deviation as well as 'deviant' marriage patterns and violence. It also includes prostitution.
Sexual promiscuity may lead to children borne of two unmarried persons who are unable or unwilling to take care of them. That means that the community may be burdened with their care and may result in psychological traumas for the children, which could lead to deviant behaviour, thus further burdening the community. It also leads to (or is the result of) a diminished feeling of self-worth; and is said to be the main cause of the present dramatic increase in sexually-transmitted diseases.
Society at large has failed to define and support standards of sexual behaviour giving rise to varied and limited sexual codes. The viability and social benefit of any lasting images of sexuality have become impossible to determine. In any case, the biological and social needs of the whole of humanity is rarely, if ever, considered.
Claim If sexual promiscuity were accepted (that is, legalized) it would cause the breakdown of the institution of marriage, which is necessary to the social good, to be less desirable, and it would lead to disorder and strife.
Broader Immorality (#PA3369) Unethical personal relationships (#PF8759)
Individualistic perception of sexual activity (#PF1682).
Narrower Bigamy (#PF3286) Polygamy (#PD2184) Adultery (#PF2314)
Polyandry (#PF3289) Pornography (#PD0132) Promiscuity (#PC0745)
Fornication (#PF5434) Prostitution (#PD0693) Cohabitation (#PF3278)
Group marriage (#PF3288).
Related Incest (#PF2148) Inadequacy of international standards (#PF5072).
Aggravates Unsafe sex (#PE9776) Debauchery (#PE8923)
Sexual harassment (#PD1116) Sexual exploitation (#PC3261)
Paternal negligence (#PD7297) Adolescent pregnancy (#PD0614)
Illegitimate children (#PC1874) Sexually transmitted diseases (#PD0061)
Deception between sexual partners (#PE4890) Abuse of tourism for sexual purposes (#PE4437).
Aggravated by Sex (#PF9109) Lust (#PA4673) Permissiveness (#PF1252)
Religious intolerance (#PC1808) Retarded socialization (#PF2187)
Inadequate sex education (#PD0759) Double standards of sexual morality (#PF3259).

♦ **PF2692 Abuse of scientific power**
Uncontrolled scientific power
Nature The extensive specialized knowledge and facilities available to the scientific community places groups of scientists in a very strong position to abuse such power, whether for their own interests or in what they conceive to be the best interests of their country or of mankind. Abuse may not only take the form of commission but may be omission as well. Scientists as policy-makers may withhold certain advances, whether theoretical knowledge or practical applications, and they may also be negligent and fail to adequately pre-test new developments and screen out those that are harmful or undesirable or those whose side-effects or consequences are unknown. Equally realistically, scientists may be co-opted by political parties or governmental regimes, so that available scientific power may be solely at the disposal of the state.
Incidence The emergence of preferred university centres and laboratories for key, government-supported research, places single institutions or possible consortia of scientific institutions in virtual monopoly positions concerning vital scientific knowledge. Such knowledge may deal, for example, with energy in its peaceful and military applications - including nuclear and non-nuclear high energy physics, genetic engineering, or behaviour modification. Abuses or manipulations of the nucleus of human personality, or other knowledge become easier when research is concentrated. Such concentration is seen in the USSR, where the consortium is controlled through the Communist Party; in the USA where the Massachusetts, the Carnegie-Mellon and the California Institutes of Technology dominate with Defence Department assistance; in the UK with Cambridge University and the Imperial College among the leaders; and in a number of other world-power and non-world power countries (Israel with its Hebrew University). Another vehicle for powerful scientific concentration is in the 'prime contractor' system, such as employed in the USA, where applied armaments technologies are in the control of transnational firms. These firms, without government assistance, have already concentrated scientific work in a number at applied areas, notably medical drugs.
Broader Immorality (#PA3369) Lack of control (#PF7138).
Narrower Unethical medical practice (#PD5770).
Aggravates Irresponsible scientific and technological activity (#PC1153).

♦ **PF2694 Insufficient provision of public services for communication**
Nature Technological developments in utilities and services for domestic and industrial uses have not been made available to many rural communities, which often do not have services necessary to meet their present needs. Streets and sidewalks are often crumbling and full of potholes. Some communities have only one road connecting them with the outside world. There is often no public transport - bus services do not go to every community; or even if they do, late night or Sunday services often do not exist, thus limiting outside activities. Families who do not own vehicles must travel at the convenience of neighbours and friends. There is usually only one telephone, resulting in long trips for emergency calls. Such unavailability of basic services leads leads to a growing isolation that discourages the community's involvement in the development that is going on around it.
Narrower Humanism (#PE0176) Non-articulated educational goals (#PF2595).

Related Shock (#PC8245) Inadequate enforcement of safety regulations (#PD5001).
Aggravated by Underdeveloped technological skill (#PF8552)
Insufficient communications systems (#PF2350).

♦ **PF2698 Limited availability of technical agricultural and business training**
Nature Many rural areas are expanding both their agricultural production and commercial enterprises, but lack opportunities for training in the skills needed for modern farming and the expansion of commercial business. Introduction of new machinery for farming and industry may therefore be inhibited by a lack of training in using and repairing the equipment. Provision of the necessary technical training is crucial in order to fully tap local potential.
Incidence In many countries of Africa there are more than 2000 farmers for each extension worker.
Broader Nonavailability of technical training (#PJ0121).
Narrower Lack of expertise in agricultural techniques (#PE8752).
Aggravates Incompetent management (#PC4867)
Limited agricultural education (#PF8835).

♦ **PF2699 Unimaginative vision of existing international economic structures**
Inadequate solution-oriented images of the global economic situation
Nature Existing international monetary structures such as the World Bank and the International Monetary Fund have been effective in fostering certain limited national development. However, considering the global situation as a whole, the effectiveness of these structures is restricted. They are dominated by the most powerful nations, as membership is based on each nation investing capital in proportion to its economic solvency. It then participates in proportion to the amount of that capital. Also, the flow of currency is directed to a limited number of specific development objectives within member nations.
Overall images of the global economic system are limited, and even well-informed persons find their reasoning unclear when they confront the question of the system of world distribution of goods and services.
Broader Unchallenging world vision (#PF9478) Dominance of economic motives (#PF1913)
Variations in national forms of currency (#PF2574).
Aggravates Ineffective economic structures in industrial nations (#PE4818).

♦ **PF2701 Paralysis of social response**
Claim Frequent hindrance to progress by administrations which do not meet people's needs results in the individual allowing himself to become immobilized. Further limitation occurs in the absence of any indication to indicate that not only are people free to put forward new future alternatives for the world but that it is an historical necessity for the oppressed to activate public conscience in this way.
Broader Non-concerned attitudes (#PF2158).

♦ **PF2702 General unavailability of effective motivational techniques**
Nature The key people who are aware and dedicated to the needs of their communities are often confronted by a lack of the necessary tools to sustain them in their work. Because motivational methods are not available to those who are more aware, the rest of the population also suffers. They see themselves as unable to participate in or to actively affect processes, consequently becoming like bystanders. There is no-one to give them the necessary incentive to be responsible members of a global society.
Aggravates Educational curricula based on content rather than method (#PF3549).

♦ **PF2703 Untransposed creativity of traditional gifts**
Nature Communities need to deliberately strengthen their unique identities in order to promote creative self-reliance. The strong individualism that has grown up over the years among the people of many developing communities often prevents them from working and celebrating together, despite many clubs and groups. Until new patterns emerge in some areas and a deeper appreciation of existing heritage in others, providing the set patterns for farming, roles of women, and families, which help create the community's self-image, the motivation needed for development will not be generated.
Broader Untransposed significance of cultural tradition (#PF1373).

♦ **PF2706 Short-range project funding**
Nature The limited funds which are available to meet a community's increasing requirements are often invested in safe, short-range projects whose results will not upset the present social order. For example, funds are placed in such projects as city beautification rather than community organization and development.
Broader Belittling of grant recipients (#PF2708).

♦ **PF2707 Centralized decisions on local technological innovation**
Nature Decisions to innovate which do not come in response to local requests are not met with the necessary preparation in mindset, mores and willingness to change. Often the imposed technology, while tackling an area of critical importance, does not take full advantage of the available resources or consider the disparities in societal types. Thus innovation is blocked until a major localization of the prioritizing process occurs.
Incidence From the sixties onwards there have been initiated in Europe many initiatives for international technological research and product development, but to date these have been largely unsuccessful. As much as 95% of European industrial research funding is used for specific technological improvements in a firm's existing products. There is very little interest in or funding for independent technological research.
Broader Non-inclusive management decisions (#PF2754).
Aggravates Undemocratic policy-making (#PF8703).

♦ **PF2708 Belittling of grant recipients**
Nature The present system of special grants forces the recipient into crisis situations before help is given, degrades him and puts him under the domination of the giver. An example of this malfunction is seen in the present welfare system in which the recipient must reduce himself to pauperism before help is granted. In the past, the above system was limitedly useful because of the parochial and limited scale of its activity. However, in the present urbanized, globally interdependent society, the system is increasingly inoperable.
Broader Dominance of economic motives (#PF1913).
Narrower Crisis-oriented funding (#PF2849) Short-range project funding (#PF2706)
Self-accountability for grant funding (#PF2275)
Inadequate information on funding needs (#PF2878)
Limited approaches to economic planning (#PF3500)
Limited applicability of monetary grants (#PF2564)
Unregulated ownership of the means of production (#PF2014).
Related Production of non-essentials (#PC3651) Self-interest driven investment (#PC2576)
Non-inclusive management decisions (#PF2754)
Monopoly of the economy by corporations (#PD3003).

♦ **PF2709 Decisional paralysis of specialized services in relation to the world's need**
Nature Specialists are aware that vast needs around the globe are not being met, and that a plan to use skills effectively is lacking. They see a number of unprecedented issues being raised in such areas as medicine, physics, law and genetics, yet are often caught in bureaucratic structures more interested in perpetuating themselves than in dealing with the problems with which they are supposed to deal. Problem analysis takes years and leads to reports that often result in inaction. Skilled professionals are being held up to the public as models of competence and ingenuity, while at the same time experiencing uncertainty and ambiguity at the awesome decisions they are being called upon to make in a technological age frought with relative values. Rather than raise the question of care for the vast majority of mankind, they escape into reduced concerns restricted to a limited arena.
Broader Individualistic utilization of expertise (#PF5639).

♦ **PF2711 Over-centralization**
Centralization — Excessive bureaucratic and administrative centralization
Nature Regional and local governments frequently face restrictions in raising resources to finance present or potential spending. Central authorities often regulate the few local sources of revenue by controlling tax rates, prohibiting increases in user service charges, and limiting the means for revenue collection and enforcement.
Background Excessive centralization creates difficulties for goods and services that are regional or local, rather than national, in character, such as water supply and sanitation, and even some health and education services. Centralization tends to reduce public accountability and responsiveness to local preferences. The scope for decentralization is greatest in urban areas, but broadening the involvement of rural communities in water supply, irrigation and rural roads can also improve the quality of public services.
Claim 1. The ever-increasing centralization of governmental and organizational functions in the modern world has produced a widespread dissatisfaction and in some cases has become self-defeating. Centralization is associated with delay, bureaucratic red tape, and the erosion of local initiative.
2. The global-scale accumulation and concentration of finance capital in fewer and fewer hands by banks and transnational corporations, and the accumulation of physical capital equipment for large-scale production under systems of centralized control and decision-making, are paralleled by an increasing intervention of state governments into people's daily lives and by the increasing role played by other large organizations (corporations, unions, political parties, international and national organizations, etc) whose decisions are beyond the control of the individual.
Narrower Bureaucratic bias (#PC1497) Bureaucratic superiority (#PC1259)
Centralization of religious power (#PF5115).
Related World federalism (#PF2088) Over-organization (#PF4891)
National federalism (#PF0626).
Aggravates Inefficiency (#PB0843) Threatened and vulnerable minorities (#PC3295)
Unchecked power of government bureaucracy (#PD8890)
Over-centralization of global decision-making (#PF5472).
Aggravated by Excessive government control (#PF0304)
Failure of centrally planned economies (#PC3894).
Reduces Lack of central planning structures in small communities (#PF2540).
Reduced by Fragmentation (#PA6233).

♦ **PF2714 Suppression of safety records**
Fraudulent safety tests
Aggravates Industrial accidents (#PC0646) Uncatalogued documents (#PF4077)
Violations of health and safety regulations (#PE4006).
Aggravated by Environmental hazards constraining scientific research (#PF1789).

♦ **PF2717 Inadequacy of postal services**
Postal delays — Lack of postal agents
Nature Postal services continue to be inadequate in most developing countries, partly because of the remoteness of many villages and the poor quality of road and rail networks; there are still many centres of population which have no post office. A surprising trend in recent decades has been the deterioration of postal services in certain developed countries as well. One reason is that the excellent services of the past depended on lavish use of manpower; another is that communication authorities now prefer to invest in improvement of the telephone system, which is profitable, rather than on maintenance of the postal system, which is not – hence the deterioration of postal services between persons, nations and continents. The faster the aeroplanes, it seems, the slower the post. This deterioration often causes serious disruptions in individual and commercial communication. There are also grounds for believing that the decline in the habit of letterwriting has been a factor in reducing the ability of many people to express themselves in a literate manner; this also represents a cultural loss.
Incidence Some 80 percent of letters are business related. In the United Kingdom unnecessary delays and industrial unrest costs commerce more than £4 billion a year.
Claim Slow development in many countries of postal and telecommunication facilities and services is a real obstacle both to persons and societies. It is insufficiently recognized that these facilities and services are not only the outcome of economic growth, but also a precondition of overall development and even of democratic life. The unevenness in telecommunications expansion becomes an increasing obstacle to communication between developed and developing countries. Similarly, the rates of several services which have not yet fallen commensurately with costs, hamper their use by poorer consumers. This area of communication needs to be reconsidered in many countries, particularly in view of its social, economic and cultural significance.
Counter-claim As the world population and the proportion of literate individuals increase, the number of pieces of mail to be handled by national and international postal systems also increases. The absolute number of items is now so great, especially when direct mail is used extensively, that any weaknesses in the delivery system quickly become evident, particularly at seasonal peak periods. Delays may increase from days to weeks and in some serious cases (such as when the situation is aggravated by an industrial dispute), mail may simply be stored for long periods. In even more serious cases, and particularly in the case of printed matter, the mail may accumulate to such an extent that the authorities are obliged to destroy it.
Broader Communication delays (#PF4453) Insufficient communications systems (#PF2350)
Fragmented planning of community life (#PF2813).
Narrower Postal censorship (#PD3033).
Related Inappropriate use of the mail service (#PE6754)
Insufficient transportation infrastructure (#PF1495)
Misuse of postal surveillance by governments (#PD2683).
Aggravates Delays in delivery of books and publications (#PF1538).
Aggravated by Strikes (#PD0694) Unsustainable population levels (#PB0035)
Proliferation of direct mail advertising (#PE1810).

♦ **PF2720 Lack of scientific investigation**
Scientific orthodoxy
Nature The scientific community defines its concerns in such a way that some phenomena reported by individual or non-institutional scientists or by non-scientists are condemned as hallucinations, hoaxes, or otherwise unworthy of collective scientific investigation. In some cases years or even decades may pass before such evidence or phenomena acquire sufficient acceptability for establishment scientists to examine them. This forces some scientists or witnesses to spend a lifetime collecting data in order to overwhelm the initial resistance.

Incidence Past examples of compilers of overpowering data collections include Linnaeus and Darwin among others. 20th century sciences that have been obstructed include: the investigations of the sub-conscious and other inner terrain by psychoanalysts and psychologists; rocket development for space travel; and anthropological detection of human origins. Examples of under-investigated phenomena include: extra-sensory perception; hypnotism; astrology; unidentified flying objects; and folk medicine and remedies. Non-investigated phenomena have been included in the collection of Charles Fort and the Fortean societies, and are also quickly cited as anomalies in the work of astronomers, physicists and laboratory workers. Many of the phenomena concern human beings, such as survival after death, magic, comings and goings from the physical body, miracles (excluding the Shroud of Turin, investigated but not explained), the more incomprehensible yogic powers, healing power, charismatic phenomena, and the Devil, among others. Physical phenomena include inexplicable objects falling from the sky, countless violations of nature's 'laws' concerning time, space, energy and matter, and the intriguing subject of coincidences.
Broader Human errors and miscalculations (#PF3702).
Related Excessive emphasis on fashionable areas of research (#PF0059).
Aggravates Prejudice (#PA2173) Ignorance (#PA5568)
Irrelevance of science and technology (#PF0770).
Aggravated by Scientific fraud (#PF1602).

♦ PF2721 Ineffective methods of practical education
Poor teaching methods — Restrictive teaching methods — Limited teaching methods
Nature In the developed nations, due to the present rapid rate of change in knowledge and job options, education is increasingly informal, with special curricula and scheduling available close to job or home and at convenient times. However, this kind of practical curriculum and scheduling is not provided sufficiently near to the residents of deprived inner-city neighbourhoods, either as evening education for adults, or as early learning experiences for those below public school age. This is especially true of old people and of unemployed young mothers who have neither finished high school nor acquired any marketable skills. Such a vacuum in job skills and training in social methods leads to a continuing cycle of ineptitude and defeatism. The result is an ever decreased value placed on formal public education, increasing absenteeism and drop-outs. These factors exclude a large portion of the community from the job market and leave them without the tools necessary to relate themselves to the complex dynamics of the city.
Broader Obsolete methods (#PF3713) Obstacles to education (#PF4852)
Constricting effect of educational structures (#PF6476).
Narrower Rote learning (#PJ5437) Data-oriented education (#PF1217)
Narrow scope of education (#PF3552) Limits of second language teaching (#PF6028)
Unarticulated goal of educational methods (#PF2400)
Inadequate use of visual imagery for societal learning (#PD7086)
Faulty approaches in the teaching of intellectual methods (#PF0956).
Related Insufficient leadership training (#PF3605)
Limited availability of functional information (#PF3539).
Aggravates Reading disabilities (#PD1950) Impractical education (#PF3519)
Absence of social prowess (#PG7727) Lack of apprentice training (#PJ8498)
Obsolete educational values (#PF8161) Unscheduled summer programme (#PG7728)
Inadequate vocational education (#PF0422) Inadequate educational facilities (#PD0847)
Unrecognized relevance of education (#PF9068) Unavailability of trainee employment (#PJ0336).

♦ PF2723 Proliferation of national and international anniversaries and years
Nature The continuing proliferation of officially designated 'years' and anniversaries tends to reduce the effectiveness of those celebrations of special importance for which public support is most needed.
Broader Human errors and miscalculations (#PF3702)
Lack of international cooperation (#PF0817).
Related Multiplicity of languages (#PC0178).

♦ PF2747 Surrogate parenting
Commercial surrogate mothers
Refs Bartels, Dianne, et al *Beyond Baby M* (1990); Shannon, Thomas A *Surrogate Motherhood* (1988).
Aggravated by Inhuman methods of conception (#PE8634).

♦ PF2754 Non-inclusive management decisions
Nature The management of large corporations and government agencies control the means of innovation by regulating market indicators and internal technological decisions. Management patterns are not designed to include the knowledge and experience of all the groups involved in and affected by technological decisions; and it is difficult to discover how these decisions are made and who participates in important discussions.
Broader Mismanagement (#PB8406) Fragmented decision-making (#PF8448)
Dominance of economic motives (#PF1913).
Narrower Technological monoculture (#PF4741)
Limited social context in developing technology (#PF6617)
Limited access to natural resource use decisions (#PF2882)
Centralized decisions on local technological innovation (#PF2707)
Profit motivated utilization of construction technology (#PF2464)
Nationalistically determined development of natural resources (#PD3546)
Public non-accountability of organizations developing technology (#PE4032)
Intellectual methods which disenfranchise the grassroots from technological development (#PE7095).
Related Production of non-essentials (#PC3651) Belittling of grant recipients (#PF2708)
Poor managerial communications (#PF1528) Self-interest driven investment (#PC2576).
Aggravates Obsolete vocational skills (#PD3548).

♦ PF2758 Inefficient use of proteins in factory farming
Nature Intensive farming techniques require the use of concentrated feed with a high protein content; but 80 per cent of the protein content of concentrated grain feed is lost in converting plant protein to animal protein. The major part of grain commodities consumed is imported from developing countries where malnutrition among humans is widespread, where meat commodities do not compensate for plant commodities, and where surpluses of plant commodities are slight.
Incidence The problem of malnutrition in developing countries is widespread; factory farming, by encouraging the export of needed grain commodities to the already relatively well-nourished industrialized nations, increases the protein gap between rich and poor nations. In developed countries, where the standard of living, and hence the demand for meat and egg products, is rising, it is a common fallacy to equate high protein content with meat and egg products rather than with plant products.
Broader Factory farming (#PD1562).
Narrower Inefficient use of resources (#PE5001).
Aggravates Malnutrition (#PB1498) Food wastage (#PD8844)
Bovine spongiform encephalopathy (#PE5191).
Aggravated by Nutritional ignorance (#PE5773)
Agricultural mismanagement of housed farm animals (#PD2771).

♦ PF2775 Difficulty in identifying carriers of animal diseases
Nature Difficulty in identifying carriers of animal diseases may lead to the spread of infection and to epidemics. It may arise from the fact that the carriers of the disease may not be initially recognized, or they may be immune, or the symptoms may be very slight and easily confused with those of other diseases. Difficulties in identification may also arise from the fact that a disease such as anthrax has many different vectors.
Broader Inadequate control of animal diseases (#PD2781).
Narrower Vector immunity to animal diseases (#PG5820).
Related Confusion of symptoms in animal diseases (#PF2780).
Aggravates Zoonoses (#PD1770) Epizootic diseases (#PD2734)
Stray dog populations (#PE0359) Vectors of animal diseases (#PD2751)
Infectious diseases in animals (#PD2732) Snail vectors of animal diseases (#PE2747)
Insect vectors of animal diseases (#PD2748) Worms as vectors of animal diseases (#PD2750)
Wild birds as vectors of animal diseases (#PE2749)
Domestic animals as carriers of animal diseases (#PD2746).
Aggravated by Diseases of wild animals (#PD2776)
Inadequate animal quarantine (#PE2756)
Lack of funds for veterinary research (#PE8463).
Reduced by Wildlife extinction (#PC1445)
Economic loss through slaughter of diseased animals (#PE8109).

♦ PF2776 Fear of ostracism
Broader Fear (#PA6030).
Aggravated by Epilepsy (#PE0661) Ostracism (#PF1009)
Fear of contradicting popular views (#PF2040).

♦ PF2779 Prohibitive cost of disease control
High cost of controlling epizootic and enzootic diseases
Nature The cost of developing vaccines for certain diseases or the cost of using the vaccines adequately to eradicate the risk of disease may cause other less effective methods to be used, or the vaccine to be used in a less effective way. Slaughtering may be a relatively cheaper method of disease control than vaccination; but without compensation to the farmer, it also causes great economic loss, and slaughter policy may be avoided or reluctantly or sparingly carried out, raising the risk of the spread of disease. Farmers may be unwilling to notify authorities of the possibility of disease if they feel that they will lose their stock. Quarantine measures may also incur economic loss to farmers and breeders. Disinfection measures, especially for pastureland, and carcass disposal measures, may be insufficient owing to cost. A national surveillance system has to be operated to prevent spreading of a contagious disease from the primary outbreak.
Incidence Where foot-and-mouth disease is enzootic, vaccination measures are taken, but not 'overall' vaccination because of the cost. Therefore 'frontier' or 'ring' vaccination is practised – vaccination of all susceptible animals within a given radius of an outbreak. Where the disease is not enzootic, but may become epizootic, a 'stamping out' policy is followed with compensation for farmers.
Broader Human disease and disability (#PB1044)
Prohibitive medical expenses (#PE8261)
Resistance to changing agricultural methods (#PF3010).
Aggravates Inadequate quarantine (#PE2850) Inadequate health control (#PF9401)
Inadequate health services (#PD4790)
Strained capital resources in small communities (#PF3665)
Inadequate health care in least developed countries (#PE9242).

♦ PF2780 Confusion of symptoms in animal diseases
Difficulty in diagnosing animal diseases
Nature Confusion of the symptoms of one animal disease with those of another may lead to insufficient control or the wrong treatment being used for a disease which may become epizootic, or may cause unnecessarily severe treatment for a disease which could be controlled without incurring losses from slaughter. Doubt about the nature of disease in the first instance may lead to delay in treatment, allowing the disease to become more virulent.
Incidence Foot-and-mouth disease symptoms have been confused with those of swine vesicular disease, causing unnecessary slaughter (UK December 1972). Confusion of these two diseases the other way round could lead to the outbreak and rapid spread of foot-and-mouth disease. Other diseases which can be confused with foot-and-mouth disease are bluetongue, epizootic haemorrhagic disease, mucosal disease, rinderpest, and malignant catarrhal fever.
Broader Inadequate control of animal diseases (#PD2781).
Related Difficulty in identifying carriers of animal diseases (#PF2775).
Aggravates Epizootic diseases (#PD2734) Contagious bovine pleuropneumonia (#PE1775)
Economic loss through slaughter of diseased animals (#PE8109).

♦ PF2782 Ineffective monitoring of restrictive business practices due to inadequate regulation
Nature There are two main difficulties which confront authorities in obtaining information on restrictive practices from firms, in particular as regards such involving the activities of transnational corporations. These are: (a) company refusals to supply information because of its so-called secret nature since it relates to intra-company relations; and (b) states' unwillingness to permit companies within their jurisdiction to transmit documents requested by foreign authorities or courts. It is often the subsidiaries or parent companies of transnational corporations located in foreign countries which possess the information requested, and hence the ability to obtain such information depends not only on the willingness of the company in question to provide it, but also on the willingness of governments in foreign countries to permit the supply of the information requested.
Broader Ineffective monitoring (#PF2793).
Aggravates Restrictive trade practices (#PC0073).

♦ PF2788 Passive resistance
Peaceful resistance — Non-violent non-cooperation
Nature Passive resistance as commonly understood implies the action of the weak, unarmed or helpless. It does not reject violence as a matter of principle but because the means of violence are lacking. As such it can even serve as a preparatory stage to acts of armed resistance. Its underlying objective is to harass or manipulate the opponent into a desired course of action. It may be associated with nonviolent direct personal action through demonstrations, picketing, vigils, fasts, boycotts, raids and blockades. These may endanger others against their will. Non-cooperation may take the form of disobedience of a law or command claimed to be unjust. This may include refusal to participate in military service or to pay taxes.
Broader Passivity (#PF6177) Civil disobedience (#PC0690).
Related Non-resistance to evil (#PF7754).
Aggravates Strikes (#PD0694) Demonstrations (#PD8522) Citizen disobedience (#PD5707)
Military disobedience (#PD7225) Disobedience of judicial order (#PD3879).
Reduces Lack of community participation (#PF3307)
Failure of individuals to participate in social processes (#PF0749).

♦ PF2789 Semilinguism
Nature The migrants forget their mother tongue faster than they learn the language of their new

country. They have problems in communicating and expressing themselves efficiently and fluently in these two languages.
Related Limited verbal skills (#PD8123).
Aggravated by Exclusive nationally-oriented language systems (#PD2579).
Reduced by Bilingualism (#PF7927).

♦ **PF2792 Limited ways of matching talent and jobs**
Nature The lack of methods to match skills with tasks results in an inadequate evaluation of talents and a limited assessment of the necessary or the available skills. There is also a lack of acknowledged commitment for all services, including the jobs and skills needed in the total employment process.
Broader Obsolete methods (#PF3713).
Narrower Professional stagnation (#PE6649).
Aggravates Frustrated talent in government posts (#PE4398).
Aggravated by Inadequate access to negotiation on employment and reward (#PD1958).

♦ **PF2793 Ineffective monitoring**
Inadequate inspection — Inadequate supervision — Lack of regulatory inspectors
Narrower Inadequate environmental monitoring (#PF4801).
Inadequate control of development projects (#PF9244).
Ineffective monitoring of illegal activity (#PF7264).
Ineffective monitoring of hazardous substances (#PE5089).
Opaque budgetary procedures in the public sector (#PF5374).
Inadequate systems for monitoring industrial growth (#PF2905).
Ineffective official inspection of regulated activities (#PE4146).
Ineffective monitoring of restrictive business practices due to inadequate regulation (#PF2782).
Aggravates Police corruption (#PD2918).
Aggravated by Deceptive misuse of research (#PD7231).
Unethical practices of regulatory inspectors (#PF8046).

♦ **PF2795 Fatalistic attitudes to the use of time**
Claim People inhibit themselves from commitment by a fatalistic attitude towards time, in which the past determines the future. Time is "killed", "put in" or frittered away rather than transcended through simultaneous participation in the past, future and present. Job structures and educational institutions often encourage this narrowed time sense: work becomes dull and meaningless while leisure time is lively and creative. People wait for quitting time, the weekends, and retirement. As such, a large part of life is spent waiting for a lesser part.
Broader Inhibited capacity to visualize a creative future (#PF2352).
Constraint of time on individual and social development (#PF5692).
Aggravates Inadequate sense of time (#PF9980).

♦ **PF2797 Unexplored avenues of leadership potential**
Stifled emerging leadership
Nature The trend toward greater dependence on authority is often frustrating to the local communities and over-burdening to the authority. Community leadership roles tend to remain static as rural communities determine how to involve new people, and the burden of maintaining and expanding activities grows heavier each year. As organizers remain hesitant, local gifts and creative initiative are lost. The potential leadership among young people remains untapped or untrained, and vandalism is often the result of curtailed youth activities.
Broader Obstacles to leadership (#PF7011).
Undeveloped potential of informal leadership (#PF1196).
Narrower Undeveloped youth leadership (#PJ0151).
Aggravates Social disaffection of the young (#PD1544).

♦ **PF2798 Restrictions on effective means of transport**
Nature International transportation is critical for the provision of the goods and services necessary for economic and social growth. However, at the local level, many communities lack the infrastructure which makes delivery feasible. Rough or non-existent roads, battered equipment, little public transport, irregularity of services and high costs conspire to minimize adequate delivery of and access to basic goods and services, and thus to limit growth.
Broader Insufficient transportation infrastructure (#PF1495).
Narrower Ineffective rural transport (#PF2996). Scarcity of appropriate transport (#PE8551).
Prohibitive cost of transportation (#PE8063). Inadequate rail transport facilities (#PD0496).
Inadequate road and highway transport facilities (#PD0490).

♦ **PF2799 Lack of personal transport in rural areas**
Nature In many Western rural communities, distances make walking impractical; and although basic requirements and emergency transport may often be provided, public transport is normally non-existent, so those without cars are dependent on those who have, and are prevented from making social, personal or entertainment trips. This particularly limits old people, housewives, young people and children in activities that involve distance or time.
Broader Insufficient transportation infrastructure (#PF1495).

♦ **PF2803 Static social relations inhibiting vision of the future**
Claim To claim full humanity in the 20th century requires being a global citizen and appropriating the cultural gifts of others. The vision of a new social direction requires realistic and creative options to impinge on apparent possibilities. However, attempts to maintain cultural purity misunderstand the need for a dynamic interaction before growth can happen. Many communities have deteriorated to ethnic or social-status ghettos, limiting themselves in static cultural areas without employing the gifts of other social groups.
Broader Inhibited capacity to visualize a creative future (#PF2352).
Aggravates Unchallenging world vision (#PF9478).

♦ **PF2804 Absence of accountability in construction planning**
Irresponsible construction planning
Nature Present systems of long-range planning concerning the use of construction expertise are not oriented towards meeting the requirements of all interested parties, nor are they comprehensive enough to allow for adequate expansion to meet such needs. This is particularly true in underdeveloped nations where local funds are not available. Concern for individual requirements is pushed aside because of the absence of accountability on local, regional and international levels.
Broader Social unaccountability (#PC1522).
Profit motivated utilization of construction technology (#PF2464).
Narrower Lack of demolition planning (#PF3731).
Aggravates Malpractice in the construction industry (#PD9713).

♦ **PF2806 Excessive complexity of intergovernmental organizations**
Nature An intergovernmental organization, with its multiple international committees and international secretariat, is an organism so complex that it is difficult to penetrate beyond the superficial forms of charters and formal procedures, to the underlying realities. The formal parts often serve to conceal how the lines of influence and power really flow, and what are the interlocking effects of sources of finance, sources of political decision, rivalries between secretariats and, above all, defects in procedures for decision-making in national governments as they affect the still more defective decision-making machinery of most international organizations. This layer of concealment is all the more impenetrable in the older organizations, in which a natural evolution has taken place by way of 'interpretation' or 'bending' of their statutes in some respects, and of atrophy in others. Some organizations have in fact evolved so substantially that the substance of present reality is quite different from that envisaged by their founders and defined in their statutes: yet this transformation has taken place without any statutory amendments, by a process of deliberate or accidental evolution, with the development of new activities and procedures and the atrophy of old ones. Amongst these complexities and misleading appearances it is all but impossible for the outside non-practitioner to pick his way and to discern the realities; but it is also difficult for those within. The average secretariat member has at best a partial view of one sector of his organization, and his view is likely to be distorted by loyalty to that body, with which his own future is bound up. In fact, only the head of the organization, at the apex of the pyramid, and perhaps a few of his immediate advisers, obtain the panoramic view and are exposed to all the political and administrative realities.
Broader General obstacles to problem alleviation (#PF0631).
Aggravated by Inefficient location of facilities of international organizations (#PE3538).

♦ **PF2808 Government insensitivity**
Bureaucratic insensitivity
Narrower Insensitivity to diversity of cultural traditions (#PF8156).
Aggravates Involuntary mass resettlement (#PC6203).
Lack of cooperation with officialdom (#PF8500).
Unchecked power of government bureaucracy (#PD8890).
Non-payment of compensation for forced relocation (#PE8898).
Aggravated by Sentimentalism (#PF6961).
Proliferation of public sector institutions (#PF4739).
Insensitivity to non-immediate hazards to society (#PF9119).

♦ **PF2810 Unformed structures of community organization**
Nature Few urban communities are effectively organized; any attempts that are tried tend to focus on single issues and are short-lived, producing few long-range, sustained results. Organizations compete rather than cooperate with each other, and individualistic leadership is ineffective when applied to the community's future. Various sub-cultures may live in isolation from one another even though they may have common concerns. Community issues remain hidden because residents do not create and use channels to activate practical proposals. As many residents tend to be economically and socially involved in structures outside the community, they feel detached from involvement in community life. Residents seldom run for elected office or participate en masse in registration and voting.
Broader Obstacles to community achievement (#PF7118).
Narrower Reductionism (#PF7967). Lack of social programmes (#PG7756)
Limited practical engagement (#PG7753) Blocked parental participation (#PJ5308)
Shortage of industrial diamonds (#PG7754)
Unorganized development of work forces (#PF2128)
Restrictive organizational participation (#PG7752).
Related Limited meeting facilities (#PE1535).

♦ **PF2811 Relative authority of leadership**
Nature Leaders are not being raised who can take charge in today's ambiguous situation when a final arbitrating authority is no longer recognized. The political process through which executive authority is chosen does not breed independent, purposeful leaders. The most obvious breakdown is in the area of local executive authority, which as often opposes the spirit of the community as it represents it.
Broader Obstacles to leadership (#PF7011) Reduced understanding of globality (#PF7071).

♦ **PF2812 Debilitating designs for using labour**
Nature In rural communities, physical work is still the overriding focus of life. The demands it places on rural families affect their education, their health, their social life and their willingness to take on the extra demands made by development. Time and energy are limited and few cooperative patterns can be established. This everyday reality inhibits the creative involvement of individuals in developing their communities and building a meaningful social and cultural environment.
Broader Obstacles to community achievement (#PF7118).
Aggravates Demoralizing images of rural community identity (#PF2358).

♦ **PF2813 Fragmented planning of community life**
Limited scope of town planning — Disorganized community life
Nature In the face of immediate and long range issues, communities need to rediscover their social unity. The quickening pace of community development demands clear, precise planning and constant thinking through by the whole community: skilful projection is needed to finance and negotiate essential public services; sites need to be chosen and prepared for the new industries; untapped resources must be surveyed; statistics must be compiled; planning of road systems and housing rehabilitation requires full participation of all. However, different sectors of a town are accustomed to acting autonomously, so that individual groups pursue different aims and programmes.
In many rural communities of the developed world, the informality of relationships and the emphasis on individual approaches to property-use result in piecemeal designs for local life. Lack of comprehensive planning for redesigning town centres, and for the improvement of utilities, transportation and security services, means that small towns find it difficult to attract new business and additional residents.
Claim Without an effective organizational structure, the rapid social and economic development of a rural community is impossible. The erosion of the basis of community life is made obvious when, in the absence of comprehensive planning and of organization to carry out plans, obviously necessary improvements - such as the repair of roads, the installation of drains and public toilets and the allocation of community space for industrial and recreation facilities - are not carried out; and there is a paralyzing dependence on structures outside the community.
Broader Disorganization (#PF4487) Obstacles to community achievement (#PF7118)
Short range planning for long-term development (#PF5660).
Narrower Nebulous city plans (#PG8160) Unbalanced food usage (#PJ7868)
Unfocused town centre (#PG8140) Undiversified tax base (#PJ7897)
Lack of medical records (#PF1836) Insufficient land upkeep (#PJ8634)
Unsanitary refuse disposal (#PJ1098) Inadequate fiscal policies (#PF4850)
Unbudgeted child education (#PG1363) Unfamiliar use of equipment (#PG7867)
Inadequate industrial space (#PJ0084) Uncoordinated policy-making (#PF9166)
Insufficient utility revenue (#PG8175) Unintentional space upgrading (#PG8165)
Inadequacy of postal services (#PF2717) Uninvolved transient residents (#PG8660)
Derelict industrial wastelands (#PE6005) Increase in abandoned buildings (#PJ1665)
Inadequate technical assistance (#PG7864) Insufficient industrial planning (#PG8873)
Unprotected commercial establishments (#PG8124)
Unscheduled introduction of new machinery (#PG7848)
Inadequate enforcement of safety regulations (#PD5001)
Disorganized attempts of upgraded employment (#PF6458).

Related Unpreparedness (#PF8176) Undefined cultural space (#PJ0060)
Unfocused design of community space (#PF1546).
Aggravates Narrow media options (#PJ0094) Unhygienic conditions (#PF8515)
Inadequate medical care (#PF4832) Maldistribution of water (#PD8056)
Inadequate land drainage (#PD2269) Limited youth activities (#PJ0106)
Limited meeting facilities (#PE1535) Ignorance of cultural heritage (#PF1985)
Inappropriate sanitation systems (#PD0876) Deterioration in water supply systems (#PD9196)
Inadequate water system infrastructure (#PD8517).
Aggravated by Reduced civic awareness (#PJ1835)
Insufficient community events (#PF5250)
Loss of leadership credibility (#PF9016)
Minimal leadership coordination (#PJ9498)
Inadequate vocational education (#PF0422)
Inadequate educational facilities (#PD0847)
Inadequate social and demographic statistics (#PF0214).

♦ **PF2814 Unrealistic attempts at globality**
Claim An idealism that dreams of unity among people is often an unrealistic dream for making all people like one another, or a pseudo–globality attempting to create the future on the basis of what the past has been; the economic community, in its dominance, is uniting people on the promises of technology alone. But globality cannot be reduced to providing all people with material and physical comfort nor with a sense of democratic equality.
Broader Parochial national interests (#PF2600).

♦ **PF2816 Lack of cooperation**
Lack of agreement on cooperation
Broader Ineffective structures for community decision–making (#PF1781)
Divisive responses to international nongovernmental organization actions (#PE7080).
Narrower Lack of international cooperation (#PF0817)
Lack of cooperation with officialdom (#PF8500)
Uncoordinated social services in urban areas (#PF1853)
Lack of support for local commercial services (#PF6510)
Uncoordinated efforts in agricultural development (#PF1768).
Aggravated by Conflict (#PA0298) Non–cooperation (#PF8195)
Social fragmentation (#PF1324).

♦ **PF2818 Unwarranted pessimism**
Nature Pessimism is the belief that reality is evil, either predominantly or essentially and totally. The pessimist puts the least favourable construction on actions and events, and views life as basically futile. Human nature is weak and evil and an individual's capacity for improvement small or nonexistent. Pessimism usually leads to one of the following attitudes: 1) belligerent resentment by human beings of their nature and condition; 2) resignation, for example, the conservatives who reject progress and yearn for a past age; 3) despair and anxiety; and 4) people oriented toward the next world.
Current pessimism is seen in the spread of trend predictions, which have the effect of being self–defeating and self–fulfilling prophesies.
Refs Bailey, Joe *Pessimism* (1988).
Broader Negative emotions and attitudes (#PA7090).
Aggravates Alarmism (#PF4384) Human illness (#PA0294).
Reduces Unrealistic expectations (#PE7002).

♦ **PF2819 Compassion fatigue**
Increasing resistance to charitable giving — Fluctuating response to appeals for aid — Public disillusionment with aid for developing countries
Nature Appeals for donations to help people in a crisis may initially meet with large responses but repeated frequent appeals may soon result in little or no response. Charities know that extraordinary success of one appeal is unlikely to be repeated by a similar one. Images of famine or flood victims become too familiar and people become inured. "Compassion fatigue" can sap public sympathy. Causes become fashionable and unfashionable. Long term relief projects are in danger of becoming out of favour with the public and thereby may be terminated prematurely.
Broader Disillusionment (#PA6453).
Aggravates Decline of philanthropy (#PF6221)
Reallocation of aid funds to alternative priorities (#PF0648)
Inadequate funding of international nongovernmental organizations and programmes (#PE0741).
Aggravated by Beggars (#PD2500) Professional burnout (#PF4833)
Exploitation of dependence on food aid (#PD7592)
Inappropriate aid to developing countries (#PF8120)
Prohibitive administrative overhead costs (#PF8158)
Corruption and mismanagement of foreign aid (#PD0136)
Mismanagement of aid to developing countries (#PF0175)
Ineffectiveness of aid to developing countries (#PF1031)
Donor distortion of aid to developing countries (#PF7955)
Limited developing country capacity to absorb aid (#PD0305)
Unethical practices of philanthropic organizations (#PE8742)
Shortage of financial resources for action against problems (#PF0404).
Reduces Emotional manipulation (#PE9599).

♦ **PF2821 Underdevelopment of food and live animal production**
Nature More food will need to be produced in the developing countries of the world if the combined goals of better nutrition and economic development are to be achieved. The major hope for potential future increases in production lies in opportunities to increase yields per unit of land. In fact, in a number of large countries, where population pressure has already brought nearly all of the usable land under production, this approach offers the only possibility for substantial production increases.
Many factors need to be combined for the achievement of higher crop yield levels, but attention may be focussed on four components needed to support the drive for higher yields: improved seeds; fertilizers; plant protection; and farm credit. In addition, in many areas, farmers need more irrigation water. The use of improved seeds is the fundamental part of the technological package in which the other inputs take their place. Fertilizers and pesticides are today in the forefront of discussion as to the future direction of agricultural technology in the light of the energy crisis. Agricultural credit especially directed to assist small farmers is required if they are to be enabled to purchase sufficient quantities of these and other inputs.
In recent years, the hopes generated by some of the success stories of high–yielding varieties (HYVs) of wheat, rice and maize have been dampened by the realization that most of the new HYVs flourish in rather limited ecological areas and by the lack of breakthroughs in a number of other important crops. For some important crops such as oilseeds, tubers and pulses, with their crucial role in nutrition, as yet too little breeding work as been accomplished. Recent breakthroughs, such as maize with a high lysine content, indicate the scope for increasing protein availabilities, both for human and animal consumption. One of the main bottlenecks limiting the rapid and sustained spread of the HYVs has been the lack of a commercial supply of the improved seeds. Most developing countries do not yet have a seed industry which could support a major drive for higher productivity, or only have seeds produced for a very small number of crops. This situation has to be remedied by introducing seed multiplication programmes.
In the short run chemical fertilizers constitute the most important single weapon in the food production battle. The per hectare use of these nutrients in most developing countries is at present only one quarter, often less than one tenth, of what it is in most developed countries; it is especially low in Africa. This rate could be advantageously accelerated provided that supplies were available at reasonable prices. While a much wider use of these products could contribute substantially to increasing per–hectare yields, it is also true that they should be used with discrimination, thus effecting economies in the quantities required. Yet the major thrust must be to familiarize a larger proportion of the farmers of developing countries with plant protection practices and by this means significantly reduce the widespread losses of crops which occur every year because of pests, diseases and weeds. The use of quality seeds, of fertilizers and of plant protection materials on a sufficient scale is prevented in developing countries because most farmers do not have enough money to buy these inputs. In southern Asia the average small farmer spends $6 per hectare per year when he should be spending $20 to $80 according to the crop.
If modern technology is to be applied the farmers, especially the small farmers, must have credit, not the extremely expensive facilities provided by moneylenders, but institutional credit delivered and supervised, in many cases, through farmers' own organizations. The availability of credit on reasonable terms will also enable farmers to level their land and thus make better use of irrigation water, to purchase pumps and other mechanical equipment, and much needed tools. A limiting factor at present is the insufficiency of financial resources in developing countries. Also of vital consideration is water. Much of the water presently available in irrigation schemes is underutilized for lack of attention to land dwelling, field distribution, drainage systems and regulatory services. Improved drainage can check and largely eliminate the degradation of irrigated land through salinity. As water is ineffectively used, crop yields are much less than their potential.
Livestock sector potentialities are enormous, especially considering the very low levels of efficiency which characterize animal husbandry in most developing countries, apart from a few outstanding exceptions. Three overall considerations apply to all livestock production systems, whether they are relying predominantly on roughage or on grain. The first is the need to improve health conditions through better veterinary services, advice to farmers, and availability of the inputs required. An increasing share of such sanitary measures will have to extend across national boundaries as the benefits from continent–wide control of major epidemics are high. The second general consideration applies to the genetic improvement of all animal stocks. Just as in crop production, the direction of policy here is to provide animals with high genetic capacity and then to generate the health environment, managerial skills, and feed base for them which will permit full use of the higher capacity. Therefore, much of the success of genetic improvement will depend on the improvements which can be achieved in the feed base. Thirdly, the improvement in the feed base will have to relate to all three major components of the feed base in developing countries. The first of these is natural grasslands which provide a large share of sustenance both for bovine and sheep and goat populations. The grasslands of the developing world are perhaps the most over–utilized and at the same time under–utilized resources in agriculture. Over–stocking causes degradation of the pastures while on the other hand there exist far–reaching unused possibilities of raising the productivity of pastures through man–made changes in their composition – for instance, but not exclusively, in the semi–arid areas. Also, a large part of the feed from these pastures goes into animals which are very poor converters and which are kept in traditional low–efficiency systems of husbandry.
Measures aimed at increasing the contribution of fish to world food supplies cover a wide range of activities concerned with management and development, however, years have seen the fairly dramatic collapse of a number of major fisheries, due in part at least to excessive fishing, and it is clear that management action is needed if many stocks are to continue to provide the basis for commercially viable fisheries.
Counter–claim Many of the factors which were responsible for the less than full success for already developed HYVs have been identified and the problems are on their way to being overcome. The world–level organization of plant–breeding, as evolved over the past two decades, with international centres working in close cooperation with national research stations, has proved valuable. With additional labour and money, the existing centres together with new ones that may be needed, could help bring about dramatic yield improvements over much wider areas of the developing world. Varietal improvement also offers opportunities for increasing the quality of cereals, especially the quantity and composition of their protein content.
Broader Underproductivity (#PF1107)
Underdevelopment of industrial and economic activities (#PC0880).
Narrower Narrow range of food crops (#PD4100)
Slowing growth in food production (#PC1960)
Inadequate food storage facilities (#PE4877)
Underdevelopment of food processing industries (#PD0908)
Underdeveloped approaches to local food production (#PF6493).
Aggravates Food insecurity (#PB2846) Export of nutritious food (#PJ1365)
Excessive cost of animal protein (#PE4784)
Fluctuations in food production in developing countries (#PE8188)
Instability of economic and industrial production activities (#PC1217).

♦ **PF2822 Haphazard organization of community space**
Nature Despite the realization world–wide of the wisdom of forming communities around common community buildings appropriate for both traditional and contemporary needs, the buildings and homes of many rural communities are spread over large areas of land, making focused village life difficult. This also increases the cost of electricity, telephones and water, and the lack of a centralized plan makes access to such services difficult to project or extend to other outlying residences. Without increased centralization, these villages cannot expect improvement in services or a real focus for cooperative work in the future.
Broader Obstacles to availability of community space (#PF7130).
Narrower Restricted beach use (#PG9193) Lack of central nodes (#PG9344)
Poorly lighted roadways (#PJ9224) Low business visibility (#PG9252)
Distant farm residences (#PG9350) Inadequate public facilities (#PG9297)
Established geographic division (#PG9327) Insufficient commercial facilities (#PJ5337)
Inadequate visibility of facilities (#PG9209).
Related Dangerous paths (#PJ9888).

♦ **PF2823 Restriction of funding for research on social problems**
Nature Research groups conducting projects related to the study of social programmes invariably require funding; but applications by such researchers and response from the funders are often restricted. This is due largely to an inadequate image of human welfare primarily as economic security rather than as encompassing significant participation in the whole of life. Funds necessary for sustaining research are often directed to projects which are popular or prestigious, while those willing to undertake research on broader social problems are prevented from doing so by lack of finance. The most needed skills are often not developed, researchers seeking the status or economic advantage attached to "pure research" and neglecting the more crucial areas.
Broader Limits on areas of research (#PF2529) Shortage of funds for research (#PF5419).

♦ **PF2825 Institutional domination of organizational systems**
Organizational empire–building
Claim In every instance of organizational planning the same dynamics appear to be at work: a

reflex action on the part of each major institution to attempt to expand its planning over the space of the whole system, because no system-wide integrating force is at work in that space. The absence of such a force acts against the inherent nature, structure, and organization of the social system. It is a destructive gap, which many institutions compete to fill. This almost subconsciously motivated attempt, that of a sector to expand over the whole space of the system in its own particular terms and in accordance with its own particular outlooks and traditions, compounds the problem by further fragmenting the wholeness of the system. For sectors cannot become systems, they can only dominate them; and when they do they warp them. This tendency is an ominous portent of the conflicts and dislocations that await society unless a system-wide integrative approach is worked out, and unless new institutions with legitimate system-wide jurisdiction for turning such an approach into policy and action are devised.

Counter-claim All organizations, even up to societal or global levels, may be said to be warped by the most powerful forces which dominate as much as they are able. But in the dynamics of freedom, counter-forces evolve either unilaterally or sectorally, or as alliances of power. It is just such a dialectic of history that has encouraged progress. Imposition by the state of a system-wide integrative approach to benefit society has been the ideology of dictatorships and collectivist utopians.

Narrower Faceless social institutions (#PF2081)
Excessive institutionalization of education (#PD0932).
Related Bureaucracy as an organizational disease (#PD0460).

♦ PF2826 Use of offensive symbols

Offensive corporate logos

Nature The slogans and images by which certain companies and products identify themselves can be offensive in themselves.

Incidence Darkie toothpaste, a best seller throughout Asia, has a minstrel in blackface as its logo. When its Hong Kong producers sold out to an American firm, the name was changed to Darlie, and the picture changed to a racially indeterminate figure in top hat, tuxedo and bow tie. This change sought to avoid the offence that Darkie gave.

Aggravates Stereotypes (#PF8508)
Social symbols dominated by the economy (#PE6671).

♦ PF2827 Subjectivity

Lack of objectivity — Subjective assessment according to own group values

Nature Recent developments in physics have abolished the predictability of matter and obliterated the distinction between observer and observed. Modern architecture and functional design have banished historical allusion or even temporal reference. Contemporary literature has forsaken extrinsic "things" and "events" for a subjective, self-reflexive reality. Sophisticated cybernetics raises fundamental issues concerning the meaning of intelligence.

Claim The sense of objective reality, namely the comforting conviction that there is a natural world "out there", apart from ourselves and perceptible in the "middle distance", has diminished to total obsolescence.

Counter-claim Although it is true that no one can claim to know the whole truth about anything, it by no means follows that all assertions are equally true or false, and that it is not possible to ground some statements on the basis of evidence, as better or truer than others. Indeed to hold that objectivity is a myth is tantamount to denying the distinction between fiction and history, guilt and innocence, in relation to the admitted evidence.

Refs Hardison, O B *Disappearing Through the Skylight*.

Narrower Subjectivism (#PF8015) Oppressive reality (#PF7053)
Non-local employment patterns (#PG7745) Unappealing public office positions (#PG7748).
Related Lack of local leadership and decision-making structures (#PF6556).
Aggravates Lack of leadership (#PF1254).

♦ PF2828 Unrelated ability application and situational demands in vocational decisions

Nature New modelling methods that allow people to relate and employ their practical abilities toward the demands of society are often hampered by the use of outdated educational methodologies which are overwhelmed by demands made on them and give too detailed and specialized frameworks. This leads to a disparity between the ability to apply practical solutions and the reality of actual situations when people are considering their educational goals and trying to decide what their future work might be.

Broader Ineffective decision-making processes (#PF3709).
Aggravated by Unarticulated goal of educational methods (#PF2400).

♦ PF2830 Denial by old people of the significance of the past

Elders' disapproval of the past

Nature Old people tend to feel that the past is not helpful, an attitude reinforced by the attitudes of adults and the young. They neglect to share the traditions and experience that should be passed on to present society. Remembering past despair and disillusionment they are unable to relate this to the future and do not see their insight as relevant.

Broader Erosion of elders' wisdom (#PF1664).
Narrower Elder paralysis over the future (#PF3973).
Related Sentimental attachment to the past (#PF1577)
Silence about historical situations (#PF0608)
Limited community responsibility of adults (#PF1731)
Disorientation of the young due to lack of social forms (#PD2050).
Aggravates Obsolescence of rituals and customs (#PF1309).

♦ PF2833 Multiplicity of manual sign languages for the deaf

Broader Multiplicity of languages (#PC0178) Human disease and disability (#PB1044).
Aggravated by Deafness (#PD0659).

♦ PF2834 Inadequate hero images

Shortage of hero images

Nature Individuals, peoples and societies in crisis need to evoke heroes from the past and create or identify new ones. However, the present limited number of heroes recalled as role models is aggravated by over-use in the case of the more-recognized ones. Attempts by some societies to coin or invent new "heroes" has had only short-lived success.

Claim At present it is the media that create images, so that although there may be an adequate number of remarkable beings embodying the greatest virtues, most of them have no commercial or other sponsors to see that they are immortalized. True contemporary heroes and heroines often remain unnoticed, as front-page stories concerning violence and social aberrations take the centre of attention.

Counter-claim Excellence can be promoted by emulation or competition within local groups or social or work units. Leaders in such small units set the example by their qualities and, in addition, offer personal contact, encouragement and support to others. There is no need for greater-than-life-size heroes and poster heads looking down on the masses.

Broader Apathy (#PA2360) Insufficient role models (#PF8451)
Demoralizing image of urban community identity (#PF1681).
Related Lack of autonomous world-level actor to identify and clarify world interests (#PF0053).
Aggravates Oppressive prevalent images (#PF1365).

♦ PF2835 Non-globalized citizenship

Nature Two hundred years of over-emphasis on the individual have deeply rooted societal life-styles in protectionism. People see themselves as individual rather than global citizens and public conscience has not yet adequately articulated the value of corporateness to the individual and to society as a whole.

Broader Non-concerned attitudes (#PF2158).

♦ PF2837 Lack of opportunities for practical training in communities

Nature The commercial potential of many communities is beginning to be realized but the means for residents to develop the kind of modern business system, marketing expertise, management skills and sales ability that commercial education affords are not available. The acquisition of technical expertise and functional skills not included in formal education is hampered because of a lack of qualified teachers, so that Third World communities are critically short of skilled people to take advantage of opportunities made possible by the continuing development and specialization of commerce and industry. This is especially true in agricultural communities which have access to the recent technological advances which would permit self-support through small business, local industry and improved crop production. Local people require knowledge of business and finance, the use of construction and mechanical equipment, and improved farming methods, in order for the community to become self-sufficient and self-reliant. However, the necessity of devoting all efforts to the pursuit of subsistence-level farming, and the distance and travel cost to any practical training schools makes it virtually impossible for working adults or unemployed youth to obtain these skills. Despite a strong desire to learn domestic skills and the necessity for structured forms of child education, the talents of village women are largely untapped and those capable of teaching children lack practical training. The inability of residents to learn broad and comprehensive practical skills results in the economic and educational growth of a community remaining dormant or even declining, making it unable to participate effectively in the broad development of the nation.

Broader Inadequate training opportunities (#PJ8697).
Narrower Lack of training (#PD8388) Lack of apprentice training (#PJ8498)
Absence of management training (#PD3789)
Unexplored opportunities for community education (#PF6512).
Related Abusive behaviour modification (#PE2690).
Aggravates Unused female talents (#PJ0833) Incompetent management (#PC4867)
Inadequate career advice (#PJ8018) Crisis oriented education (#PG8966)
Limited funding expertise (#PJ0278) Insufficient trained labour (#PD9113)
Unlearned fundamental skills (#PG8901) Limited gardening experience (#PG8928)
Shortage of technical skills (#PF6500) Limited construction manpower (#PG9875)
Underdeveloped farming skills (#PJ0729) Unawareness of health problems (#PG8949)
Inadequate budgeting practices (#PJ0978) Narrow range of business skills (#PE6554)
Inadequate maintenance personnel (#PJ0088) Mismanagement of irrigation schemes (#PE8233)
Failure to employ skills of home-bound educated women (#PD8546)
Lack of skilled manpower in rural areas of developing countries (#PE5170).
Aggravated by Inadequate education (#PF4984) Complex curriculum writing (#PG8837)
Limited consumer knowledge (#PG7662) Obsolete vocational skills (#PD3548)
Minimal opportunities of adult training (#PF6531)
Unarticulated goal of educational methods (#PF2400).

♦ PF2840 Limited access to practical education

Limited practical education

Nature As communities shift toward new forms of industry and commerce, the need for functional education becomes critical. However, most practical vocational training is only within the family, as youth become apprentices to parents of relatives who are craftsmen. Few youth are able to go to regional or state training centres, and many of those who do soon leave the local community for more lucrative markets. The growth of agricultural industry is also discouraged by the lack of new techniques. In addition to vocational skills, a wide range of service skills, including modern medicine, disposal of rubbish, animal care, and vehicle maintenance, are required to improve living conditions. Many rural people see such training as expensive, marginally beneficial, and basically irrelevant to their work and to their daily lives. In addition, illiteracy is an underlying deterrent to all practical training.

Broader Limited availability of learning opportunities (#PF3184).
Narrower Limited availability of education in rural areas (#PF3575).
Aggravates Limited medical knowledge (#PD9160)
Unprepared adult leadership (#PF6462)
Inadequate vocational education (#PF0422)
Mismanagement of irrigation schemes (#PE8233).
Aggravated by Unsanitary refuse disposal (#PJ1098)
Prohibitive cost of education (#PF4375).

♦ PF2842 Unfulfilled aspirations of economic life

Nature There is no viable economic base in most inner city communities and few determined efforts are being made to draw new capital into the community from the outside. Although many neighbourhoods are surrounded by business and industry, unemployment remains high, since these economic institutions do not function as a practical economic resource for the community. At the same time, inner-city dwellers are weak competitors for city-wide jobs. This lack of jobs for the community prohibits participation in the life-style that is generally understood and publicly promoted. Without the means of earning an income, residents (particularly the younger men) tend to be resigned to a life of welfare or of economic crime. A feeling of uselessness is the predominant theme of conversations with those who have searched for jobs.

Broader Deteriorating quality of life (#PF7142).
Narrower Government dominated jobs (#PG7762)
Uncommitted local industry (#PG7760)
Automated warehousing jobs (#PG7755)
Underused industrial resources (#PG7750)
Scarce employment possibilities (#PG8652)
Overcomplicated implementation of citizen rights (#PG7763).
Aggravated by Limited basic skills (#PJ8138).

♦ PF2843 Inadequacy of the committee system of decision making

Nature The number and gravity of technical policy decisions to be taken has brought with it a multiplication of committees: faced with an awkward problem, the appointment of a committee is a means of obtaining a variety of views and of evading or postponing a political responsibility. But the difficulty under modern conditions is that, if a subject is so profuse and complex that it cannot be assimilated by a single human intellect, setting up a committee superimposes a new layer of problems arising from the mere existence of the new body, and multiplies the difficulty of comprehension by the members of the committee.

In large administrations, the committee system prevents the formulation of sound policy decisions. At the committee level, as at the national level, the assemblage of a group of constituents who not only possess different departmental interests but also differing bases of information as well as different individual approaches, results, after the benefits which certainly flow from the dialectical process have been obtained, not in an optimum decision in the light of processed facts, but in a compromise arrived at in the light of what different members of the group be. Thus many modern technical problems, instead of being solved or improved by a compromise, are merely

aggravated by it. The committee, used in conjunction with existing 'vertically' structured administrative and decision-making machinery, is thus as much a hindrance as an aid.
Broader Fragmented decision-making (#PF8448)
Human errors and miscalculations (#PF3702)
General obstacles to problem alleviation (#PF0631).
Related Bureaucracy as an organizational disease (#PD0460).

♦ PF2845 Disjointed patterns of community identity
Unconsolidated community patterns — Displaced community identity — Disparate village identity — Disrupted patterns of village identity — Fragmented community identity — Fragmented images of community identity
Nature Confused and complex understanding of community identity is emerging from the shifting roles and images required to cope with rapidly changing world patterns. Decision-making structures and organizational forms seem inadequate in the face of modern complexities, as individual concerns and actions are being replaced by corporate community endeavours. The community itself may be unaware of basic population data and therefore of the skills and talents of its residents. In many traditional villages, bloodline allegiance and family-focused lives conflict with the group commitment necessary to bring a livelihood above subsistence level. Some residents may seldom take part in community life because of jobs in distant places, poor local transportation even for children attending schools a few kilometres away, or physical isolation in widely-scattered houses or small clusterings of families; all of which contribute to a sense of fragmentation. Powerful national identification is not reflected in a local community pride which would enhance the national pride already present. The songs and rituals of the nation find no supportive local community counterpart, but only in individuals or small groups. Residents see themselves as dependent on remote outside assistance which appears unresponsive because of a lack of overall understanding of procedures and available services.
Incidence Patterns of village identity are disrupted by encroaching urban areas, resulting in the interests of the residents being directed toward the city. The lost traditional family and village systems are not replaced by new structures; traditional occupations are looked on as demeaning; skills from past trades are by-passed for modern expertise; education in other areas has greater recognition; young people look to the city for jobs; and adults go to the city for consumer goods and other urban conveniences. The loss of village identity and pride is exemplified by the untidy public areas: although the inside and outside of homes are often immaculately clean, litter and trash lie between living areas and in public spaces.
When small communities diminish in size they tend to feel victimized by the contrast between past greatness and their present situation. Buildings and lots which stand vacant and in disrepair give rise to an image of decay which it is hard to combat without some new symbol, for example promoting local resources. Such negative images contribute to a reluctance on the part of local business people to expand present operations or invest in new ventures; and future plans tend to be split between those wanting new development and those wishing it to to remain a small, quiet, rural community.
Broader Obstacles to community achievement (#PF7118)
Demoralizing image of urban community identity (#PF1681)
Fragmentation of social structures in depressed areas (#PD1566).
Narrower Unnamed streets (#PG8156)
Mistrust of strangers (#PF8743)
Minimal citizen confidence (#PF8076)
Inhibiting social patterns (#PF0193)
Unchanging business pattern (#PG5367)
Ineffective community story (#PG8167)
Abuse of individual property (#PJ8392)
Insufficient community events (#PF5250)
Unprofitable traditional skills (#PJ1031)
Unfamiliar government procedures (#PJ0740)
Ingrained individualistic patterns (#PG8116)
Disoriented traditional occupations (#PG5431)
Geographically undefined community limits (#PF6521)
Ineffective organization of community action (#PF6501)
Inadequate carcass disposal of diseased animals (#PE2778).
Related Consumerism (#PD5774)
Obsolete basis of cultural identity (#PF0836)
Inaccessible administrative agencies (#PF2261)
Distortion of international trade by state-trading and government monopoly practices (#PE8267).
Aggravates Lack of formal education (#PF6534)
Limited youth activities (#PJ0106)
Erosion of elders' wisdom (#PF1664)
Ignorance of cultural heritage (#PF1985)
Deteriorating community identity (#PF2241).
Aggravated by Lack of social contact (#PF8695)
Scattered housing locations (#PJ0760)
Inflexible educational system (#PJ5365)
Unrecognized benefits from cooperatives (#PF9729)
Inadequate social and demographic statistics (#PF0214).

♦ PF2848 Limited reservoir of technical skills in rural communities
Shallow reserve of technical skills in rural communities
Nature Communities are unaware of the proper channels for expanding their narrow technological perspective; they are therefore unable to increase their technical skills, although vocational training, literature and visual demonstrations have made a wide range of modern technology available. In rural communities where the majority of residents share a single vocational skill, this skill has become an inadequate means of providing financial sustenance. Such situations are characterized by: an absence of vocational teachers; no opportunity to better income through a second trade; outdated agricultural technology complicated by other technical factors, such as lack of organized land schemes and the inability to control pests; inadequate drainage systems with no alternative techniques or practical examples to follow; road and bridge building inhibited by minimal technological knowledge; poor understanding of poultry raising and animal husbandry; unavailability of trained health workers resulting in insufficient community health services; women economically hindered by lack of technical skills.
Broader Limited technical skills (#PF8594).
Narrower Inexperienced animal husbandry (#PG9643)
Ineffective agricultural technology (#PG9610)
Lack of skilled manpower in rural areas of developing countries (#PE5170).
Related Insufficient health personnel (#PD0366).
Aggravates Inadequate land drainage (#PD2269)
Inadequate electrical maintenance (#PG0584).
Aggravated by Restricted job training (#PJ0617)
Limited local initiative (#PJ9554)
Unplanned land development (#PF6475)
Lack of practical examples (#PG9638)
Lack of vocational teachers (#PJ9603)
Lack of apprentice training (#PJ8498)
Shortage of women instructors (#PG9482)
Underdeveloped farming skills (#PJ0729)
Complex government bureaucracy (#PF8539)
Minimal leadership coordination (#PJ9498)
Insufficient educational incentives (#PJ9811)
Unavailability of library facilities (#PG9116)
Lack of local leadership and decision-making structures (#PF6556).

♦ PF2849 Crisis-oriented funding
Claim Special grants are being used to relieve crisis situations that result from the present malfunctions of the social process, rather than to deal with those malfunctions in a more future-oriented manner. They frequently operate, therefore, to maintain the current ideas of those granting the funds as to what a normal, stable social situation should be. This results in sustaining a basically unhealthy dysfunctional system and does not allow the release of new visions or new and innovative models into the social stream. The result is often a patching up of the very deteriorating and dehumanizing social systems that are crying out for fundamental change and radical reordering.
Broader Belittling of grant recipients (#PF2708).
Aggravates Short range planning for long-term development (#PF5660).

♦ PF2850 Denial of right to manifest religion
Broader Violation of civil rights (#PC5285)
Denial of religious liberty (#PD8445).
Narrower Restrictions on freedom of worship (#PD5105).
Related Persecution of religious sects (#PF3353)
Denial of right to change religion (#PE6397).
Aggravates Culturally biased religion (#PF1466).

♦ PF2852 Insufficient capital investment
Insufficient flow of investment capital
Nature When communities are unable to find appropriate channels of financial support, self-sufficient economies are not realized and subsistence-oriented living continues to undercut development. Unavailability of local and outside funding in rural communities results in a shortage of capital for farming or business improvement. Limited family income hampers attempts at raising money for community projects - minimal family income is not put into savings since banking services are remote and benefits unclear, and the concept of forming a local cooperative union for the purpose of saving is unfamiliar. Also, a past history of financial mismanagement may breed an air of mistrust in local attempts to pool resources. Limited local assets prevent the securing of loans needed for expanded farming, starting businesses and constructing schools and other buildings.
Broader Insufficient financial resources (#PB4653).
Narrower Lack of venture capital (#PG7833)
Lack of capital investment in developing countries (#PE5790)
Diminishing capital investment in small communities (#PF6477)
Decline in capital investment in productive capacity (#PE9265)
Limited availability of investment capital for urban renewal (#PF3550).
Aggravates Underproductivity (#PF1107)
Prohibitive cost of education (#PF4375).
Aggravated by Low bank interest rates (#PE3903)
Risk of capital investment (#PF6572)
Ineffective economic structures in industrial nations (#PE4818).

♦ PF2855 Unconstrained exploitation of natural resources
Emphasis on mass extraction of natural resources — Exploitative resource development
Nature The emphasis on developing mass extraction procedures has resulted in the ecological systems in certain areas of the world being plundered by man, employing accessory improvements which facilitate rapid development without consideration for future needs. This has resulted in profound disruption caused by the depletion and inadequate replenishment of the earth's ecological systems. The earth's biological balance has been harmed through indiscriminate use of chemicals, the lack of adequate processes to deal with harmful by-products, and the mismanagement of waste products.
Broader Destruction inherent in development (#PF4829)
Violation of the integrity of creation (#PF5148)
Limited access to natural resource use decisions (#PF2882).
Narrower Excessive land usage (#PE5059)
Degradation of the environment in developing countries (#PD3922)
Depletion of natural resources due to population growth (#PD4007)
Natural resource depletion due to high-level consumption (#PD4002)
Degradation of semi-natural and natural habitats of flora and fauna (#PC3152).
Aggravates Lack of natural resources (#PC7928)
Natural environment degradation (#PB5250)
Ecologically unsustainable development (#PC0111)
Long-term shortage of natural resources (#PC4824)
Endangered tribes and indigenous peoples (#PC0720).
Aggravated by Unconstrained socio-economic growth (#PB9015)
Inefficient extraction and utilization of natural resources (#PF2204).
Reduced by Underutilization of natural resources (#PF1459).

♦ PF2856 Monopolization of information within organizations
Nature Withheld information limits power of action. A monopoly of information can give a form of security. There is, in all organizations at all levels, selective withholding and extending of information. Sole possession of information can make others dependent on a single individual or unit. Withholding of information can limit the scope and power of others' actions and reduce the threat to such units. Control of information channels can isolate certain persons from the remainder of the organization and keep them within a controlled sphere of influence.
Claim A number of political and organizational regimes have operated under the mistaken idea that not to share information is to make people dependent. After a time, revolution swept them away. Individuals will not have their right to attempt to secure information abrogated. If they are unable to get real information from an authoritative source, they will seize upon or invent rumoured news or will speculate unnecessarily, causing insecurity and a lack of confidence in leadership. Much undistributed organizational information is due to executive torpor and an insensitivity to organizational dynamics.
Broader Excessive government control (#PF0304)
Transnational corporation imperialism (#PD5891)
General obstacles to problem alleviation (#PF0631).

♦ PF2857 Inadequate mechanisms for securing sufficient food supplies
Nature Analysis of the trends in world food production indicates that the potential exists for an adequate food supply over the next decade provided that both national and international measures are adopted to increase production and productive capacity. But two factors make it imperative that improvements in food production are accompanied by a system of food security. First, countries differ in their ability to respond to the need for increases in food supplies and in the resources at their disposal for importing the required quantities of food. Secondly, fluctuations in food production, arising primarily from irregular weather patterns, imply that even a satisfactory trend of food availability may be interrupted by periods of shortfall. The breakdown in world food security came about largely because of a system of stockholding which was inadequate to meet an unexpected shortfall in grain supplies in a period of rising demand. The world trading system for grain is dominated by the policies of individual governments with regard to pricing and stocks, and this is likely to continue to be the case in the future. Governments through these policies have a major impact on the allocation of available supplies. Experience has shown that no nation can achieve sustained economic growth and reasonable economic stability in the absence of an assured food supply at reasonable prices. In the developing countries, inadequate food supplies have in many cases brought overall economic growth to a halt when resources had to be diverted from other development priorities to the import or acquisition of food to maintain the population. In such crises, investment projects have to be abandoned while scarce foreign exchange and external assistances are diverted to short-run survival. The concept of food security embraces both the reduction of risks emanating from unstable production and also provision of mechanisms whereby individual countries can obtain assistance to meet specific problems of food shortages. It includes arrangements for security of supplies in the face of production fluctuation, for price stability, for trade expansion, and for payments facilities. The development of a comprehensive

solution to the problem of world food security involves both individual country action on such matters as stockholding, food aid commitments, and trade policies, and international action to coordinate this national action and supplement it, where necessary, by programmes channelled through international agencies. Such a system of food security, if developed, would constitute something in the nature of a charter guaranteeing freedom from famine analogous to the United Nations Charter of Human Rights. The most important elements of a world food security policy are: the establishment of a food information and early warning system; a coordinated system of national stock policies embodied in an International Undertaking on World Food Security; better arrangements for meeting requirements of emergency food supplies; and the reorganization of food aid as a continuing form of assistance.

Claim Since the crop failures, the severe regional food shortages, and the drawdown of the world's grain reserves of the early 1970s, much international attention has been given to the need for more effective management of grain reserves to provide security against fluctuations in food supply. The crop failures and grain shortages of 1980 lend new urgency to this subject. With a rapidly growing world population, agriculture in many regions will spread into more and more marginal lands, where fluctuations in production are most likely and severe. In this situation, the effects of a year or two of bad weather can be critical. The recently passed Food Security Act is a step in the right direction, but more is needed. A major factor influencing world food security is the size of cereal stocks. It is the combination of production falling below consumption requirements with the consequent severe rundown of cereal stocks and rises in cereal prices which caused a global alert.

Broader Lack of a world food security system (#PF5137).
Narrower Inadequate agriculture in least developed countries (#PE8082).
Aggravates Food insecurity (#PB2846) Maldistribution of food (#PC2801)
Economic disadvantages of excessive food production in developing countries (#PF4130).
Aggravated by Food monopolies (#PE8018) Imbalance in world food economy (#PD5046)
Mismanagement of food resources (#PE6115)
Protectionism in agriculture and the food production industries (#PD5830).

♦ PF2859 Unwanted pregnancies

Unwanted high fertility — Non–termination of pregnancy

Nature Family planning has often been defined simply in terms of avoiding unwanted pregnancies. Expressions like unplanned, unwanted, and undesired are difficult to define in such situations, and often it is not clear to which of the prospective parents or other members of the family they refer. A pregnancy may be considered to be unwanted if either the woman, or the man, or both did not desire a child at the time of conception. The term does not imply that they will not change their minds later. All 'unplanned' pregnancies are not also 'unwanted': some pregnancies are neither planned in the sense that the couple deliberately try to achieve conception, nor unwanted in the sense that conception was clearly not desired.

Incidence Attempts to assess the incidence of unwanted pregnancies are fraught with difficulties. Several approaches have been adopted: abortion or illegitimacy rates have been used as indices, and direct questioning of parents has also been tried. The associations found between unwanted pregnancy and morbidity and mortality are very much related to the approach adopted.

Millions of the world's people still lack access to safe and effective family planning methods. By the year 2000, some 1.6 billion women will be of childbearing age, 1.3 billion of them in developing countries.

Claim Major efforts must be made now to ensure that all couples and individuals can exercise their basic human right to decide freely, responsibly and without coercion, the number and spacing of their children and to have the information, education and means to do so. In exercising this right, the best interests of their living and future children as well as the responsibility towards the community should be taken into account.

Unwanted high fertility adversely affects the health and welfare of individuals and families, especially among the poor; and seriously impedes social and economic progress in many countries. Women and children are the main victims of this unregulated fertility. Too many, too close, too early and too late pregnancies are a major cause of maternal, infant and childhood mortality and morbidity.

Broader Immorality (#PA3369) Human disease and disability (#PB1044)
Negative emotions and attitudes (#PA7090).
Narrower Deliberate imbalancing of population sex ratio (#PF3382).
Aggravates Unwanted children (#PE1907).
Aggravated by Sexual craving (#PE7031) Lack of control (#PF7138)
Lack of family planning (#PF0148) Illegal induced abortion (#PD0159).

♦ PF2863 Incomplete implementation of community decisions

Nature Lack of initiation and regular use of corporate planning methods and effective implementation structures prevents community issues from being resolved. Although the corporate needs of a community may be known by its residents, the infrequency of community meetings may lead them to understand such needs only in terms of an individual effort to improve the situation as it exists. There are few opportunities for group decision, and community action on public projects remains unscheduled.

Broader Obstacles to community achievement (#PF7118)
Ineffective decision–making processes (#PF3709).
Narrower Minimal community services (#PD8832)
Insufficient women's groups (#PG9884).
Related Limited public land (#PJ0574).
Aggravated by Insufficient community events (#PF5250)
Ineffective follow–up on initiatives (#PF0342).

♦ PF2865 Absence of a long–range, world–wide capital flow plan

Limited availability of long–term capital

Nature Existing planning is nationally or regionally based and tends to result from immediate crisis situations such as war, depression or monetary devaluation.

Broader Lack of local commercial services (#PF2009)
Variations in national forms of currency (#PF2574).
Aggravates Lack of long–term development assistance (#PF5181)
Ineffective economic structures in industrial nations (#PE4818).
Aggravated by Absence of long–term economic planning agencies (#PF3610)
Insensitivity to non–immediate hazards to society (#PF9119).

♦ PF2870 Leadership as symbolic of wealth

Nature When leadership is seen simply as a symbol of the "good life", whether economic, social or psychological, the public loses respect for the decisions of executive authority, which signals a breakdown of the symbol of leadership itself.

Broader Obstacles to leadership (#PF7011) Reduced understanding of globality (#PF7071).
Aggravated by Personal wealth (#PC8222)
Excessive accumulation of wealth by government leaders (#PD9653).

♦ PF2874 Inadequate provision of public safety

Nature Despite regenerated enthusiasm by citizens to establish control over disruptive social behaviour, life in many inner cities still instils fear in its residents, although actual crime figures may have been reduced over the years. Public participation in community projects is substantially reduced, giving the neighbourhood a false image of apathy, while various self–interests within the community feed on the fear of the residents. Without steps to control violent activity in the streets, and to promote a sense of neighbourhood trust and self–reliance, urban communities cannot experience the neighbourhood as a human space for living. Some problems indigenous to large inner–city neighbourhoods are: a high incidence of drug traffic and addiction resulting in theft; parents and children afraid to use play areas where violent and asocial behaviour occurs; high costs of security measures for business and industry causing some businesses to move and preventing new ones from coming in; stores remaining virtually unstocked because of continual theft; and hazards particularly for children in the traffic patterns of trucks from large industries, unchecked by traffic control signs or the enforcement of local ordinances. Fear and isolation may drive individuals, families and old people to transform their homes into barred fortresses, with locks, chains, alarms and guard dogs, facing the world through peepholes.

Narrower Idle youth lifestyle (#PG7766) Loose school security (#PG7769)
Excessive community crime (#PJ7765) Harassment in playgrounds (#PE7768)
Competitive economic interests (#PG7770) Prohibitive cost of business security (#PJ7767)
Inadequate nuclear reactor safeguards (#PF6084)
Unimaginative vision of community possibilities (#PJ7764).
Aggravates Crime (#PB0001) Inadequate prevention of disabilities (#PF0709).
Aggravated by Inefficient public administration (#PF2335).

♦ PF2875 Underprovision of basic services to rural areas

Ineffective delivery of basic human resource services — Restricted delivery of essential services to isolated communities — Remote basic services — Inaccessibility of essential services in isolated communities — Inadequate rural infrastructure in developing countries — Limited access to available services

Nature Corporate community action is necessary to assure the supply of critical basic services to rural areas, or residents will continue to be prevented from participating in economic and social development. At present, many rural communities are unable to participate in the benefits of a complete network of utilities and essential services, because of impassable roads, partial road maintenance, and drainage systems incapable of coping with rainy–season flooding, and because the village leaders lack the training and sophistication to negotiate successfully with the government structures in seeking necessary services. In addition to the inadequate roads, transportation of individuals and goods is handicapped in many instances because there are few cars or trucks locally and because transportation on the highway is expensive, irregular and time–consuming. There are no systems for transporting residents to the nearest clinic or metropolitan city, thereby precluding treatment of any but the most critically ill. Fire hazards also pose a constant threat, and antiquated fire equipment cannot deal with such emergencies. Developing countries suffer also from inadequate irrigation systems, lack of rural electrification and water supply. There is little opportunity to reverse the trend for children to suffer from malnutrition, which is aggravated by severely limited varieties and amounts of foods available from both the land and from nearby commercial stores. Although residents may attempt to secure essential services by seeking assistance through the appropriate channels, the complexity of the procedures involved, and the financial cost, combine to produce a growing sense of helplessness. Even heavy public spending on rural infrastructure will not help the poorest families, landless farm workers and smallholders, who can rarely afford the offered services.

Despite advances of modern technology which have made it theoretically possible for necessary services to be within the reach of all the world's communities, many coastal communities are in fact cut–off from these basic amenities. Lack of effective drainage systems may leave the area around homes swampy, unusable and a breeding ground for mosquitoes, parasites and other insects. Infrequent transport to the nearest major centre may require being gone for a whole day and spending the night on the boat; this implies a major decision on the part of the community before seeking government advice, needed information, doctor care, mechanical repairs or other essential services. The isolated nature of such communities means that most training opportunities require the individual to be gone from the village for extended periods of time. Communications with the outside, perhaps through a radio phone, may require a specially trained operator, be difficult to hear, and be suitable only for emergencies. The whole community may be deeply affected by a sense of remoteness and trapped by such limited transport and communication possibilities.

Broader Underdevelopment (#PB0206) Inadequate infrastructure (#PC7693)
Underdevelopment of industrial and economic activities (#PC0880).
Narrower Prohibitive legal fees (#PF0995) Inadequate medical care (#PF4832)
Inadequate land drainage (#PD2269) Maldistribution of water (#PD8056)
Infrequent doctor contact (#PJ0362) Unenforced animal control (#PG8642)
Isolated village mentality (#PG9576) Inadequate sports equipment (#PG8821)
Inadequate road maintenance (#PD8557) Undependable river transport (#PG5348)
Unreliable freight transport (#PG1026) Declining volunteer services (#PG8624)
Inadequate building standards (#PF8829) Insufficient preventive medicine (#PE0751)
Inaccessible government agencies (#PF5351) Ineffective parenting techniques (#PJ0551)
Lack of fire–fighting facilities (#PJ2437) Lack of specialized transportation (#PG8828)
Inadequate safeguards against fire (#PD1631) Unavailability of legal information (#PJ8698)
Limited access to society's resources (#PF6573)
Underprovision of basic urban services (#PF2583)
Unacknowledged availability of services (#PJ9082)
Unavailability of appropriate expertise (#PF7916)
Insufficient transportation for old people (#PG8604)
Incomplete access to information resources (#PF2401)
Unavailable education for effective living (#PF2313)
Undeveloped channels for public and private resources (#PF3526)
Restricted delivery of essential services to developing country rural communities (#PF1667)
Limited availability of public services in the small towns of developed countries (#PF6539).
Related Inadequate maintenance equipment (#PD6520)
Limited access to social benefits (#PF1303)
Underdeveloped provision of basic services in developing countries (#PF6473).
Aggravates Insufficient leadership training (#PF3605)
Inappropriate sanitation systems (#PD0876).
Aggravated by Insufficient doctors (#PE8303) Shortage of fresh–water sources (#PC4815)
Physically inaccessible services (#PC7674)
Insufficient transportation infrastructure (#PF1495)
Socio–economically inactive rural population (#PF4470).

♦ PF2876 Inadequacy of intergovernmental decision–making process

Nature The central structural defect of governmental decision–making machinery is that it is structured vertically, in terms of specialized ministries dealing with particular areas of decision. But society is evolving in such a way that all problems tend to be related with other problems; a great many of these problems being in other disciplines in a strictly practical manner; this leaves the 'vertical' machinery helpless. Instead, a 'horizontally' structured decision–making machine is called for into which all relevant disciplines are introduced from the 'vertical' elements. This is quite beyond the ability of the unaided human intellect to achieve, because of the bulk and complexity of the subject matter. Governments will not be able to assert effective control over the complex of intergovernmental organizations which now exists until their own decision–making machinery, or at least that of the more powerful of them, is equipped and structured in such a way as to be adequate for the tasks which it is called upon to perform under present and forseeable conditions.

The inadequacy of governmental decision-making machinery, more than any other factor outside the United Nations, deprives governments of the power to control. Only governments can exercise such control, first by a rational distribution of finance between organizations and then, project by project, within the organizations themselves. Unless they can achieve an overall view of the activity of the whole group of organizations concerned in all relevant disciplines, and of the interaction of all the factors involved, it is inconceivable that they can go further and give collective effect to decisions of a committee of ministers or other intergovernmental bodies.
Incidence An investigation carried out in 1968 under the auspices of the Secretaries and Directors-General of four intergovernmental organizations in Europe, and attended by the European Commission, revealed that in the field of science and technology alone, twenty-five intergovernmental organizations were then operative. Some governments participated in all of these and mosc in a large majority of them, at an annual cost of about $350 million. No government, or organization, however, was in a position to obtain an overall and at the same time detailed view of this vast effort, because the relevant information as to each organization's operations was nowhere centrally assembled, processed and organized. In the case of some organizations, because of the lack of adequate programme machinery, it was difficult for governments to obtain a clear overall view of even the work of that single body. But a further barrier to any systematic approach to programming was due to the fact that in different organizations, each with a different emphasis upon its work, government representation and supervision in capitals would tend to be in the hands of different ministries. For example, the OECD might be the concern of a ministry of economics or finance; the Council of Europe would be dealt with by the ministry of foreign affairs; UNESCO would tend to be in the hands of a ministry of culture or of education. The formula varied widely from country to country, as did the channel of responsibility from direct relations with individual departments, to channelling through a ministry of foreign affairs filter system, or a treasury coordinating committee. In the example cited, it was particularly unfortunate that all three organizations mentioned were involved in the science and technology field, with no perceptible relationship between their programmes in terms of harmonization of overall aim or of particular projects, though there were many exchanges of documents and visits of observers, the latter at least partly motivated by the desire to find out what the other organizations were doing. All three secretariats seemed to be promoting their own programmes in rivalry with the other two at the expense of governments.
The Capacity Study of the United Nations noted that enquiries revealed example after example where Departmental Ministers have advocated policies in the governing bodies of the particular Agency which concerned them which were in direct conflict with their governments' policies towards the United Nations systems as a whole. Unless a majority of Heads of Governments of Member States, assisted by their Ministers of Foreign Affairs and Finance, are determined to establish policies deliberately designed to introduce the necessary changes into the present 'non-system' and to ensure that their Departmental Ministers adhere to those policies in the various governing bodies, then the United Nations system generally will deteriorate.
Broader Ineffectiveness of international organizations and programmes (#PF1074).
Narrower Distortion in intergovernmental organizations by mini-state membership (#PF4876).
Aggravated by Inefficient location of facilities of international organizations (#PE3538)
Inadequate information systems for international governmental decision-making processes (#PD0104).

♦ PF2877 Unacknowledged global interdependence
Nature A prevailing and counter-productive concept is that survival is insured by insulation and independence. This fails to acknowledge global interdependence as a present-day fact. Such egocentric understanding is manifest among irresponsible technological leadership as the philosophy of survival of the fittest; it results in the energy and other resources of technological systems being directed away from the meeting of real human needs and diverted into the duplication of non-essential goods. Conversely, it is manifested at the grassroots level in a sense of helplessness and victimization, or isolation from external sources of power and goods.
Broader Gap between the function of social techniques and the needs they address (#PF3608).

♦ PF2878 Inadequate information on funding needs
Nature Because of the lack of adequate information on newer projects, sound decisions as to which projects or areas most urgently need funding cannot be made, and money flows either into more traditionally acceptable areas or is channelled into excessive capital replenishments.
Broader Belittling of grant recipients (#PF2708).

♦ PF2880 Unequal global distribution of basic skills
Claim An effective technological system for teaching basic skills would have unimaginable potential power: it could eliminate global inequities in education and give everyone the training necessary for dealing with present-day complexities. But the various education systems actually in use are rooted in the use of static, content oriented methodologies, and cannot possibly succeed in equipping people with the techniques they need nor enabling them to adjust to the ever increasing rate of change.
Broader Thwarted technological communications (#PF0953)
Unavailable education for effective living (#PF2313).
Aggravates Ignorance (#PA5568) Compliance*complex (#PA5710).

♦ PF2882 Limited access to natural resource use decisions
Nature Economic motives for extraction of natural resources prevent meaningful future use by encouraging present abuse for short-term profit, and accountability for the quantity of extraction and care of resources is virtually non-existent. Techniques for natural resource extraction are misused by limited interest groups to the exclusion of other areas of technology and of the needs of the world. These techniques are often the private property of corporations who will allow them to be used only for their own profit and under their control. National governments are often responsible for misuse of resources because they block overall planning and support foreign exploitation by corporations within their domain.
Broader Undemocratic policy-making (#PF8703) Non-inclusive management decisions (#PF2754).
Narrower Limited access to technological decisions (#PF4636)
Unconstrained exploitation of natural resources (#PF2855)
Neglect by research of natural resource product expansion (#PF2466)
Inefficient extraction and utilization of natural resources (#PF2204).
Related Obsolete vocational skills (#PD3548)
Profit motivated utilization of construction technology (#PF2464)
Nationalistically determined development of natural resources (#PD3546).

♦ PF2883 Static and superficial adult values
Nature The life-styles of most adults are static and superficial with no depth analysis of problems or possible solutions. They neither uphold community values for the structures of society, nor attempt to find goals worthy of their ability. Most adults are attached to the commercial value system of contemporary society. They believe the myth that the more you have, the happier you will be; and pressure for immediate gratification crowds out any serious attention to deeper human questions so that eternal issues always stay in the background.
Broader Collapse of common values (#PF1118)
Limited community responsibility of adults (#PF1731).
Related Obsolete educational values (#PF8161).

♦ PF2885 Social bias in planning of training programmes
Nature There is a basic inconsistency in planning for training and placement of skilled workers. The criteria for placement of people in training programmes are biased – for the advantaged classes, in terms of personal preference and individual ability; for the disadvantaged, in terms of immediate needs and geographic location. The latter results in the lower classes seeing little significance in their social roles because of little participation in creating those roles. On account of enormous and ever increasing populations, there is keen competition for placement even among skilled workers and planning schemes for training and placement do not take into account demographic realities. Local needs, resources and talents are not taken into account when locating local development units.
Broader Social discrimination (#PC1864).
Aggravates Obsolete vocational skills (#PD3548).
Aggravated by Unplanned training structures (#PS8398).

♦ PF2887 Unreported accidents
Broader Unrecorded knowledge (#PF5728) Underreported issues (#PF9148)
Avoidance of negative feedback (#PF5311).
Related Injurious accidents (#PB0731).
Aggravates Unreported illness (#PF8090) Disaster unpreparedness (#PF3567)
Natural environment degradation (#PB5250).
Aggravated by Unreported crimes (#PF1456).

♦ PF2889 Detrimental international repercussions of domestic agricultural policies
Nature The price support policies used by many countries for social, political, strategic or other domestic reasons have a built-in tendency to generate supplies in excess of demand at the prices which the support programmes attempt to maintain. Because of this, and because policy aims are frequently mutually contradictory, import requirements are restricted while export availabilities increase. This limits the opportunities for normal commercial trade, regardless of the comparative advantage of efficient exporting countries. It also leads to periodic surpluses, which can only be disposed of through trade arrangements on non-commercial terms and the use of special export aids. These further aggravate trade problems and provide an additional stimulus towards protective measures to insulate national producers from world markets. National policies have aggravated the tendency towards the accumulation of surpluses in developed countries where demand for foodstuffs rises slowly and where improvements in agricultural techniques have been rapid and resource mobility low. Stock policies to permit constant short-term national adjustments to meet changing conditions are needed. In the absence of a common base for the coordination of these measures, the burden of adjustment falls largely on the exporting countries.
Incidence The world's agricultural system is paradoxically producing more than ever, at a time when entire nations suffer from hunger. Deep global recession and slow recovery, in combination with a period of unusually bountiful harvests in most of the major farm countries, are responsible for the problem, and pressure grows in Third World nations for internal changes that will make them less reliant on foreign food sources.
Broader Lack of international cooperation (#PF0817)
Instability of economic and industrial production activities (#PC1217).
Narrower Economic isolationism (#PC2791)
Over-subsidized agriculture in industrialized countries (#PD9802)
Domestic agricultural price policy difficulties in developing countries (#PE2890).
Aggravates Agricultural surpluses (#PC2062).
Aggravated by Government protection and national commodity price support (#PE8309).

♦ PF2895 Arrogance of policy-makers
Arrogance of planners
Nature Policy-makers sit down and draw a line on a map for a road or railway, or create a suburb or recreation centre on a paper without a moment's consultation, without any sort of public inquiry.
Broader Arrogance (#PA7646).
Aggravated by Government arrogance (#PF8820)
Arrogance of experts (#PF3294)
Bureaucratic superiority (#PC1259).

♦ PF2896 Limited image of employability
Static job image — Parochial work images — Individualistic employment images
Nature In a world in the process of economic progress through automation and other technological breakthroughs, employers tend to be prevented from moving forward by their own limited vision of the potential labour market available, and most workers have a limited image of the vocational possibilities open to them. In-service training and retraining efforts are not structured to provide new vocational possibilities which would allow the efficient expansion of present economic benefits and movement into new areas of economic services and goods. Potential productivity is also hindered by society's narrow definitions of roles, which limit by sex, age, or race the jobs open to large sections of the labour market.
Little attention has been given to multiple approaches for dealing with job immobility. Most workers are trained in only one skill. There are inadequate vocational training and retraining programmes for the physically handicapped, aged, unskilled and unemployed. These inadequacies force workers into a static job/role image which compounds immobility, and they see themselves as remaining in the same job and working in the same geographic location for the rest of their lives.
In general, working people are no longer sociologically related to a vision of the social necessity of their work, that of providing for an adequate livelihood in a community. They lack a creative social attitude to their work as being for the good of the community. The result is that workers can no longer be held accountable for the good of the community, and conversely, that the community cannot be held accountable for the livelihood which is its responsibility to them. This blocks the trend toward recovering the role of worker and a new consciousness of work as corporate engagement.
Claim There is little interest in the creation of new forms of livelihood that are adequate for the whole community. Images of livelihood as the means of personal enjoyment block such a notion, by weighing heavier than the concept of work as the context from which to participate in providing livelihood for the community.
Broader Limited approaches to economic planning (#PF3500)
Breakdown in community security systems (#PD1147).
Narrower Lack of livelihood standards (#PF1297)
Ineffective worker organizations (#PF1262)
Deficient local structures for the unemployed (#PE7326)
Collapsed relationships between work and society (#PE4203).
Related Conflicting social service ideologies (#PD3190)
Uncontrolled application of technology (#PC0418)
Unsystematic allocation of market facilities (#PD3507).
Aggravates Obsolete vocational skills (#PD3548).

♦ PF2899 Limited individual capital reserves
Nature The quantity of global capital has been increased by the widespread move to a money economy after the Second World War. However, most rural communities are excluded from the mainstream of capital flow. Shopkeepers, farmers and fishermen operate on subsistence level

profits, and the additional cost involved in transporting goods from urban areas to rural villages increases overheads and lowers profit margins even more. Low incomes mean there is no financial surplus, and thus no seed money is available for local expansion or investment. High construction costs further prohibit development, and it often takes several selling seasons to complete a building. The volume of production decreases where children are no longer as active in the family's work as they once were. Whatever funds are generated in a household are managed on a day by day basis and few records are kept for reference.
Broader Limited accumulation of capital (#PF3630).
Narrower Inflated seed costs (#PJ4397) Complex banking practices (#PJ8033)
Prohibitive cost of crop treatment (#PG8823) Prohibitive cost of transportation (#PE8063).
Related Subsistence approach to capital resources (#PF6530)
Limited local availability of capital reserve (#PF2378).
Aggravates Unattractive fishing business (#PG5399)
Prohibitive cost of accommodation (#PD1842).
Aggravated by Lost child labour due to government laws (#PS1612).

♦ PF2901 Uncritical acceptance of dogmas and standards
Nature The adherence of scientists to 'schools of thought', unable or unwilling to try to grasp arguments from people in other 'camps' is no more understandable than the blind uncritical acceptance of different dogmas and ideologies by less educated people. Many institutions for higher learning lack a critical 'faculty'.
Broader Uncritical thinking (#PF5039).
Related Blind faith in technology (#PF4989).
Aggravates Ignorance (#PA5568)
Authoritarian propagation of knowledge (#PF3706)
Discriminatory imposition of standards (#PD5229).
Aggravated by Inappropriate education (#PD8529)
Inadequacy of international standards (#PF5072).
Reduced by Parascience (#PF9032) Limitations of democracy (#PF6608).

♦ PF2904 Limitations on right to vote
Denial of the right to vote
Nature Voting eligibility may be subject to property, income, age, sex, race, literacy, residence or nationality qualifications; vote by proxy may be inadmissible. Voting qualifications sharpen social conflict, alienation and apathy.
Incidence In Europe and the USA there are in some places, property requirements for voting in local elections. However, residence requirements are more widespread currently than property requirements, being 6 months in France and 6 months to 2 years in the USA. Most countries with a democratic voting tradition and a high literacy rate have some provision for voting by proxy. The most common minimum voting age is 21, but in certain countries it is 18; in the Netherlands it is 23. Women may not vote in certain Latin American and Muslim countries. Literacy tests remain in five Latin American countries, the Philippines and a third of the states in the USA. Negroes have been prevented from voting in South Africa, and also, until recently (although only semi–legally), in the southern USA.
Broader Violation of civil rights (#PC5285) Denial of political rights (#PD8276)
Denial of right to participate in government (#PE6086).
Narrower Property and occupational discrimination in politics (#PD3218).
Related Denial of right to hold public office (#PE5608)
Political party manipulation of elections (#PD2906).
Aggravates Political apathy (#PC1917) Political alienation (#PC3227)
Discrimination in politics (#PC0934) Unequal political representation (#PC0655).
Aggravated by Racial discrimination in politics (#PD3329).

♦ PF2905 Inadequate systems for monitoring industrial growth
Nature There are no absolute means of ensuring that industrial expansion will be useful to those it purports to serve, either in terms of produce or employment.
Broader Ineffective monitoring (#PF2793) Inadequate environmental monitoring (#PF4801)
Industrial processes geared to reduced social needs (#PE3939).

♦ PF2909 Manipulative use of referenda
Nature A referendum, a direct popular vote taken on a particular political issue, is an emotional appeal to public opinion and may be accompanied by demagoguery or fear. Public opinion may be largely uninformed about the issue, its finer points or the alternatives, yet by voting for it or against it, the public gives the government a mandate (or refuses it) to continue and accomplish a given policy objective, the consequences of which may be unknown to them. It has been used effectively to give 'concensus' to dictatorial regimes or the regimes which follow from coups d'etat.
Background The referendum was used by Napoleon I to ratify his political power in 1804. Subsequent French rulers, up to and including Charles de Gaulle, also used it.
Broader Forced participation in politics (#PD2910).
Related Dictatorship of the majority (#PD3239) Misleading public opinion polls (#PG5851).

♦ PF2911 Unjust allocation of government contracts
Nature Government contracts may be allocated to favourites or supporters by people in public office, or public office and public money may be used for private transactions or investments.
Broader Corruption of government leaders (#PC7587).
Related Political appointees (#PF2031).
Aggravates Grey lies (#PF3098).
Aggravated by Political apathy (#PC1917) Official secrecy (#PC1812)
Excessive government control (#PF0304).

♦ PF2922 Sexual repression
Repressed sexual tendencies
Broader Repression of self-consciousness (#PC1777).
Aggravates Voyeurism (#PE3272) Sexual deviation (#PD2198)
Sexual unfulfilment (#PF3260).

♦ PF2924 Sociological espionage
Nature Sociological espionage involves the covert gathering of sociological data on another country, particularly an enemy, in order to be able to prepare more comprehensive war plans and in order to find psychological weaknesses as targets for psychological warfare, propaganda, and other such tactics.
Broader Espionage (#PC2140).
Aggravates Ideological conflict (#PF3388).

♦ PF2928 Primitive secret societies
Nature Primitive secret societies promote elitism and may cause a barrier to progress or integration into modernized life. They include associations of an occult, professional or political nature, covering a wide range of different activities. The societies are secret mainly because of the initiation ceremony necessary for its members to be able to control the supernatural elements concerned with their particular activity.
Broader Secret societies (#PF2508)
Animal worship as a barrier to development (#PD2330).
Narrower Occultism (#PF3312).
Related Superstition (#PA0430).
Aggravates Cultural fragmentation (#PF0536).
Aggravated by Tribalism (#PC1910).
Reduced by Ethnic disintegration (#PC3291).

♦ PF2943 Territories accorded a United Nations non–self–governing status disputed by the administering government
Lack of United Nations jurisdiction in administered territories
Nature There exist a number of territories which from the United Nations perspective can be considered as 'non–self–governing', according to the obligations of Chapter XI of the UN Charter and General Assembly resolution 1514 (XV), although this may be disputed by the administering government. The latter may also refuse to recognize any United Nations jurisdiction in the matter.
Broader Competition between states (#PC0114) Lack of international cooperation (#PF0817).
Aggravates Colonialism (#PC0798).
Aggravated by Non-viability of small states and territories (#PD0441).

♦ PF2947 Disorganized liaison with formal support
Nature Both informal and formal efforts at liaising with support agencies may be unable to effect desired community improvement.
As part of their struggle to take charge of their own affairs rather than continuing to rely on outside agencies, communities may be motivated to choose other representatives. These new official advocates are however often met by a multitude of requests and demands, making it difficult for them to be consistent channels for external support.
One of the basic contacts with the larger society is the group of trained young people who have left to work and live in other places, as local communities no longer produce all the jobs and services that are needed. The people remaining at home rely on these relatives to secure benefits and services for the town. Their efforts as spokesmen may often unsuccessful prove as they become overwhelmed with urban complexities or simply become preoccupied by their better life. Such informal representation, although it has a great deal of moral persuasion, has very little objective accountability.
Broader Disorganization (#PF4487).
Narrower Vulnerability of government to lobbying (#PF5365).
Aggravates Inadequate support services (#PF6492)
Insufficient medical supplies (#PE1634).

♦ PF2949 Rural unemployment
Inadequate rural employment structures — Minimal options for rural employment — Limited employment opportunities in small rural communities
Nature Substantial numbers of the population in rural communities are either unemployed or underemployed: the lack of a diversified job market produces only seasonal employment and a sporadic flow of income into a village; graduates who reside in a village may be unable to find useful and productive jobs; and there are few opportunities for women to be gainfully employed. This inadequate and irregular income–base seriously affects the ability of families to send their children to school and also denies them domestic amenities, such as electricity, which are otherwise available. It produces a hand–to–mouth existence which precludes vital participation in long–range planning. Many rural communities in the developed world lack a viable economic base. Limited access to youth employment and low wages encourage young people to leave the community, threatening its vitality. At the same time, increasing numbers of local people, facing the prospects of seeking employment in deteriorating, problem–plagued urban centres, are opting to work where they are rather than uproot families. Although there may be many people with business and mechanical skills and training, they are faced with little opportunity for local employment except farming. Wages are rarely higher than the minimum and repetitive, monotonous work is the rule, with closed ended jobs and inadequate benefits. Those who wish to live in these communities must often accept these jobs or commute great distances to alternative work. Large scale mechanization of agriculture in many small communities has put many people out of work, while at the same time opportunities for other means of employment are increasingly limited. The harvesting and growing of local crops are seasonal occupations and provide only a fraction of the jobs they once supplied, and expansion is restricted as small farm operators discover that most agricultural development assistance is focused on large enterprises. Efforts to supplement family income are blocked for young mothers, for there are often no day–care facilities available for children under three years of age; as a result, income is less than adequate to meet expenses. Many small town residents migrate to big cities to find better opportunities; this migration affects the entire town. Failure to create opportunities for gainful employment in small communities results in the economic fabric being unable to support the needed development effort.
Broader Unemployment (#PB0750)
Unemployment in developing countries (#PD0176)
Disparities in unemployment within countries (#PD1837).
Narrower Narrow employment base (#PG7926) Inaccessible job market (#PE8916)
Limited job information (#PG8795) Undermined job motivity (#PG9064)
Insufficient job options (#PJ0070) Insufficient job benefits (#PJ7915)
Increased fear of taxation (#PG8725) Declining local businesses (#PG9007)
Lack of funds for education (#PJ6410) Scarcity of well–paying jobs (#PG7944)
Excessive bill–paying surcharges (#PG8695) Incomplete direct transportation (#PG8805)
Absentee owned natural resources (#PG8995) Ineffective student accountability (#PG9043)
Precarious basis for family economics (#PF1382)
Graduate and post–graduate unemployment (#PD1162)
Underutilization of intellectual ability (#PF0100)
Overemphasis of governments on large firms (#PE8864).
Related Lack of work commitment (#PD2790)
Limited availability of permanent employment in inner–cities (#PE1134).
Aggravates Underpayment for work (#PD8916) Limited purchasing power (#PD8362)
Discrimination in employment (#PC0244) Social disaffection of the young (#PD1544)
Ineffective economic structures in industrial nations (#PE4818).
Aggravated by Limited job market (#PC7997) Narrow profit margin (#PJ9737)
Limited employment options (#PF1658) Unproductive subsistence agriculture (#PC0492)
Subsistence approach to capital resources (#PF6530).

♦ PF2956 Overemphasis on technological skills
Technological skill overemphasis
Nature The social value system tends to focus elite status on certain technological skills, directing impetus away from the development of supportive and service skills, so that adequate, useful skills are not being designed and developed. Because of an unconscious tendency to identify self–worth with job status, the design and development of these latter skills are associated with belittling images of vocation, impoverished by the absence of any value placed on the aspect of service.
Broader Narrow scope of education (#PF3552).
Aggravates Gap between the function of social techniques and the needs they address (#PF3608).

♦ **PF2959 Overemphasis on effective use of technical resources**
Nature Systems of production were previously designed and located with greater emphasis being placed on the effective use and longevity of the machinery involved than on the welfare of the workers. For this reason, many industries are having to totally revise their production systems in order to maintain efficiency and comply with new ecological and labour requirements.
Broader Production of non-essentials (#PC3651).

♦ **PF2960 Differences in trading principles and practices between developing countries**
Nature States having different economic systems have correspondingly different national trading policies and practices; these differences aggravate the problems of foreign trade organization and impede the growth of the international market. In the case of the developed market and centrally planned economy countries there is in each case a certain basic similarity which facilitates trading relations. However, there is not the same general similarity of trading organization and practices among developing countries. They have some structural characteristics in common, such as a predominant share of primary goods in their exports and a generally low level of per capita income, but the types of economic and social organization in these countries vary widely.
Broader Discrepancies between principles and practice (#PF4705)
Differences in trading principles and practices between different economic systems (#PC2952).
Aggravates Weakness in trade among developing countries (#PC0933)
Rivalry and disunity within developing regions (#PD0110).

♦ **PF2969 Unrecognized socio-economic interdependencies**
Unrealized potential of external relations — Incomplete utilization of external relations
Nature The various sectors of society are becoming more interdependent: urban and rural; economic and social; local and global. Local development, especially, is increasingly dependent upon the combined support of both public and private sectors at the regional and national levels. However, small communities tend to be looked upon, and to accept the role of, recipients, exclusively dependent upon the public sector of a larger jurisdiction. Such a role deprives residents of the humanizing activity of bearing responsibility for developing their own local advocacy, and reinforces their powerlessness. In small villages, communication with outside powers occurs on an individual, personal basis without the strength or influence of community consensus. Attempts to work in cooperation with existing public and private structures almost always result in residents being overwhelmed by the apparent sophistication and expertise required to follow unfamiliar and therefore seemingly complex procedures. When the public sector acts (as it must) on the basis of priorities which understandably do not always include those of the small community, residents only see unrealized promises. There is a resultant reluctance to seek benefits which are available; and possibilities of relating to private sector resources or of developing resources within the community are seldom taken advantage of, due to a continuing unawareness of new resources.
Many rural communities seem to use only a part of the supportive network which is potentially available to them, all requests for help being directed to bureaux, departments and agencies of various government bodies to the exclusion of private support sources. Such requests often do not receive a positive response because many villages have too small a population to fit the guidelines established for public development programmes, which must serve the needs of the whole district or nation and necessarily emphasize larger population centres. A second factor is that the public officials who are contacted tend to be those known personally to the village leadership. This does not allow for the intricacy and breadth of transaction which are necessary to deal with the public sector of any nation in the world today. As a result, residents are resigned to managing without basic facilities such as running water.
Claim Until the framework of economic and social support, especially involving companies, foundations, and influential individuals, is used more fully, the people in small rural villages will continue to be frustrated in their efforts to develop their communities.
Broader Covert imperialism (#PF3199).
Narrower Breach of promise (#PF7150) Political stagnation (#PC2494)
Bureaucratic opposition (#PD7966) Suspicion of bureaucracy (#PF8335)
Fragmented decision-making (#PF8448) Lack of governmental support (#PF8960)
Unfulfilled service expectations (#PG7825) Low-priority for status of villages (#PG7815)
Complexity of government assistance (#PG7820) Inappropriate size of school classes (#PD6585)
Difficulties of cooperative planning (#PF8931)
Fragmented forms of cooperative efforts (#PF2588)
Delay in delivery of requested services (#PE8157)
Unexercised responsibility for external relations (#PF6505)
Absence of legal entities in developing countries (#PE8597)
Ineffective use of external relations relating to sportsmen (#PE6515)
Unsystematic use of powerful relationships by rural communities (#PE1101).
Related Competition (#PB0848) Underdeveloped road network (#PE1055)
Individual fear of future change (#PD2670).
Aggravates Irregular delivery of education (#PJ7936)
Limited availability of financial credit (#PF2489).

♦ **PF2971 Disrupted mechanisms for community health**
Nature In principle, adequate health care is possible for everyone; but the vitality of the Third World rural people continues to be sapped by poor health and disease. For example: (1) Many millions drink from and bathe in open rivers and untreated irrigation canals; human and animal waste is deposited in the same water. A high percentage of such people are infected by parasites; and, although treatment for parasites can be increased, without concomitant changes in excreta disposal this threat to life and health will not be overcome. (2) Infant death rates due to such diseases as diarrhoea and measles are alarmingly high, despite the fact that these maladies are easily controllable with the resources available in the cities. (3) Malnutrition in pre-school age children greatly compounds health problems with its resultant occurrence of rickets, scurvy and protein deficiency. (4) Eye diseases, diet deficiencies and dust unnecessarily disable village workers. (5) Poor toilet and sanitation facilities in most homes, plus the presence of animals in the eating and living quarters, perpetuate the spread of disease and harmful bacteria. Villagers may be aware of these hazards, but they see no alternative.
Narrower Cerebral infarction (#PD9057) Irregular working hours (#PJ5329)
Short term physical stamina (#PG7879) Unavailability of quality medicine (#PG5321).
Related Lack of water conservation (#PJ3480).
Aggravates Schistosomiasis (#PE0921) Unhygienic conditions (#PF8515)
Nutritional deficiencies (#PC0382) Water system contamination (#PD8122).
Aggravated by Inadequate health services (#PD4790).

♦ **PF2973 Depleted expertise of the rural labour force**
Nature Skilled workers from rural areas are obliged to leave their villages and seek work in the already overcrowded cities, due to the lack of industrial and commercial development elsewhere. Yet every economic unit, including rural villages, requires the participation of ever larger numbers of skilled people in order to reverse the decline in the local economic situation.
Broader Loss of human resources (#PC7721).
Narrower Shortage of skilled labour (#PD0044) Insufficient trained labour (#PD9113)
Unpromoted industrial development (#PJ1017)
Ineffective tax systems in developing countries (#PF2124).
Aggravated by Youth migration to cities (#PJ7859).

♦ **PF2974 Non-standardized social services**
Nature Social services are so locally unique that it is very difficult for agencies to use the experience of others. This results in the exchange of information, personnel and equipment being necessarily limited.
Broader Conflicting social service ideologies (#PD3190)
Inadequacy of international standards (#PF5072).

♦ **PF2986 Outmoded functional skills in rural communities**
Deskilling of rural communities
Nature Although the many remarkable skills possessed by village people have been the factor which enabled them to survive in past generations, they no longer allow for productivity commensurate with today's economy, nor do they provide people with the vitality needed for development. Moreover, entrenched traditional methods for coping with everyday problems produce a narrow pattern of thought which inhibits the search for alternative possibilities and poorly prepares people to function adequately amidst the realities of present-day technological society. There is a lack of knowledge of even the most rudimentary procedures for dealing adequately in emergency health situations; traditional food preparation and eating patterns prevent the greatest nutritional value from being drawn from existing resources; the good health vital for change is further compromised by unsanitary animal care. Adult training is at best simply the perpetuation of traditional knowledge and skills; apprenticeship schemes transmit to the coming generation methods which are identical to those practised by their parents and there is no effective access to new wisdom or methods.
Broader Obsolete vocational skills (#PD3548).
Related Underutilization of locally available skills (#PF6538).
Aggravates Unavailability of first aid (#PJ1261) Ignorance of health and hygiene (#PD8023).
Aggravated by Obsolete methods (#PF3713).
Reduced by Elimination of traditional skills (#PD8872).

♦ **PF2987 Uninterested workers**
Nature Often those least interested in participation in production unit decision-making are the workers themselves, who fail to see their role as being significant either to themselves or to others.
Broader Over-specialized supervisory personnel (#PF3588).

♦ **PF2989 Limited local markets for goods and services**
Nature Local communities need to expand markets for both goods and services; however such expansion is impeded by the need for capital investment and high transport costs, so that prospects for raising incomes are small. Rural products are usually sold to visiting middlemen, or occasionally to merchants from outside. In order to benefit from expanding economies, production of goods needs to be in greater quantity and of a quality suitable for larger area markets. Retailers often have to move from smaller, local markets to larger, regional ones in order to continue their business.
Broader Limited market development (#PF1086).
Narrower Small local markets (#PJ8200) Limited market areas (#PG8599)
Underdeveloped export potential (#PJ1586)
Underdeveloped use of agricultural resources (#PF2164).
Aggravates Unattractive fishing business (#PG5399).

♦ **PF2990 Unreported government spending**
Underreported government spending — Financial cover-up — Secret government agency agreements — Secret government deals
Nature Government agencies routinely underreport levels of spending in areas which their governments do not wish to publicly support.
Incidence The American Army reported its use of private contractors in 1983, including management reviews, technical assistance, special studies and management and support services, at US$23,000, while the Pentagon's Inspector General said the army spent US$2,700,000 on them.
Broader Deception by government (#PD1893) Suppression of information (#PD9146)
Unethical practices of government (#PD0814).
Narrower Concealed government subsidies (#PD4532).
Related Underreported issues (#PF9148).
Aggravated by Inadequate public finance statistics (#PF7842).

♦ **PF2992 Social service inconsistencies**
Nature As social systems and political/economic philosophies change, there is continuous interruption and change in the social services offered. Consequently, some services are duplicated over and over again, while others are inconsistent or nonexistent.
Refs Grathoff, R H *Structure of Social Inconsistencies* (1978).
Broader Conflicting social service ideologies (#PD3190).
Narrower Inconsistent family clinics (#PU1253).

♦ **PF2995 Unrecognized need for functional skills**
Nature The gap between the needs of productive industry in present-day society and the training available to provide practical expertise produces unemployment and impedes development. The value of education goes largely unrecognized in many rural communities and community responsibilities are misunderstood or unacknowledged. The residents require mechanical or construction skills, yet lack trained teachers, youth advisors, guidance methods and equipment. Ventures into business are inhibited by the lack of management and marketing skills.
Broader Unrecognized relevance of education (#PF9068).
Narrower Inadequate marketing knowledge (#PJ9659).
Related Incompetent management (#PC4867) Inadequate maintenance personnel (#PJ0088).
Aggravates Ignorance (#PA5568) Insufficient skills (#PC6445)
Inadequate career advice (#PJ8018)
Unperceived educational opportunities (#PJ9762).

♦ **PF2996 Ineffective rural transport**
Inadequate transportation systems for rural citizens
Nature The physical isolation of many rural communities reinforces their image as an anachronistic curiosity. The time necessary to get in and out of a village detracts from productivity; and commercial or industrial interests are understandably reluctant to locate in such an area. Rural commercial centres may exist, but are generally small and remote. If main access roads do not pass such villages, and heavy rains make the secondary roads hazardous and occasionally impassable, transportation remains much as it was in the last century: largely dependent on horse and donkey.
The increased centralization of goods and services in urban centres requires increased mobility in those requiring such goods and services. However, transport systems and communications in rural communities have remained the same for many decades: the single-lane mud and gravel roads leading into many villages cannot accommodate much traffic; motorized or commercial means of transport are lacking; there is rarely access to fields other than by footpaths so that farm machinery or trucks are unusable during rainy seasons; ankle-deep mud on roads makes travel very difficult so that people (in particular students and teachers) are cut off from education,

commerce or any social exchange, often for long periods of time, particularly in countries that have long rainy seasons.
Broader Insufficient transportation infrastructure (#PF1495)
Restrictions on effective means of transport (#PF2798).
Narrower Inadequate road conditions (#PJ0860) Underdeveloped road network (#PE1055)
Insufficient heavy equipment (#PJ0259) Inaccessible natural resources (#PG7875)
Infrequent travel opportunities (#PG8342) Transport as a hindrance to economy (#PE8858)
Limited transportation availability (#PJ1707)
Inadequate transportation facilities for rural communities in developing countries (#PE6526).
Related Overloaded vehicles (#PE4127).
Aggravates Scattered housing locations (#PJ0760)
Insufficient emergency services (#PF9007)
Limited relations beyond local environments (#PF3192)
Inadequate systems of transport and communications in rural villages (#PF6496).
Aggravated by Geographical isolation (#PF9023)
Prohibitive cost of transportation (#PE8063).

♦ PF2997 Economic bias in development
Over-emphasis on economic development — Inadequate economic priority system
Nature A dominant assumption in conventional development theory is that only economic development is of consequence, since other forms of development will automatically lead to the satisfactory development of all other aspects of society. Evidence has however accumulated over recent decades that such development results in little more than development for the rich and does little or nothing for the increasing proportion of the population in greatest need.
Claim Global priorities have not included allocating funds for the common welfare. Existing economic plans encourage immediate expenditure of income, while allowing functions such as transportation, personal safety, cleanliness, adequate education, health care and effective government to deteriorate.
Broader Monopoly of the economy by corporations (#PD3003)
False assumptions on sustainable development (#PF2528).
Aggravates Social underdevelopment (#PC0242)
Lack of social progress (#PF1545)
Destruction inherent in development (#PF4829)
Decline in government social expenditure (#PF0611)
Degradation of developing countries by tourism (#PF4115)
Imbalance in economic and social planning in developing countries (#PF4837)
Inadequate development of new social structures in developing countries (#PD0822)
Excessive social costs of structural adjustment in debtor developing countries (#PD8114).
Aggravated by Over-reliance on economic interest groups by policy agencies (#PF1070).

♦ PF2998 Narrow job context
Nature The context, or self-image, from which people work is so restricted that it is difficult to instil accountability or to upgrade quality in production. Job prowess is focused only on income, making it hard to derive any meaning from work.
Broader Over-specialized supervisory personnel (#PF3588).

♦ PF3002 False image of scarcity
Nature Societies with abundant resources but which operate from an image of scarcity require over-consumption to sustain themselves. Either goods must be accumulated or goods must be wasted in order to feed the consumption machine. Production systems perpetuate themselves by producing unnecessary goods, goods which are unrelated to needs, encouraging even more over-consumption.
Broader Production of non-essentials (#PC3651).
Related Scarcity (#PA5984).

♦ PF3005 Uncoordinated use of community property
Nature There is considerable residency turn-over and lack of clarity on land title legality in some rural communities, where village land and the surrounding areas are owned both by farmers who have inherited their farms and by relatively new owners who have recently purchased land. Many of the latter group are not as yet integrated into the life of the community and often function as absentee land holders. As a result, tracts of land go unutilized. In addition, villages are usually formed as an unplanned clustering of houses with an emphasis on individual private dwellings rather than on public space for community use. Community facilities, such as a post offices, schools and health centres, are placed without an apparent overall plan or consideration for internal traffic flow. These factors serve to diminish the ability of residents to determine the way in which their community takes shape and give them the impression that they have no influence in determining their own destiny.
Narrower Absentee ownership (#PD2338) Monopolized land ownership (#PG1151)
Restricted community expansion (#PJ1107) Delay in obtaining property titles (#PG1129).

♦ PF3006 Medical hypocrisy
Medical double standards
Nature Some doctors so not heed the advice of their profession to give up smoking or some other harmful habit. Neither do they heed the advice of regular check-ups.
Broader Hypocrisy (#PF3377).

♦ PF3007 Unimaginative educational vision
Unimaginative long-term training vision — Past-oriented educational vision
Nature Training programmes and unemployment measures are limited to mitigating short-term hardships and problems and are not geared to answering long term needs. Even after training, some employees find that their job futures are uncertain because their training has not taken into account possibilities for future development. This results in further frustration as well as a waste of human potential.
Broader Obstacles to education (#PF4852) Unchallenging world vision (#PF9478)
Ineffective systems of practical education (#PF3498).
Related Fragmented decision-making (#PF8448)
Unimaginative vision of community possibilities (#PJ7764).
Aggravates Lack of training (#PD8388) Obsolete vocational skills (#PD3548).
Aggravated by Temporal deprivation (#PF4644) Detrimental story of community future (#PF6575)
Insensitivity to non-immediate hazards to society (#PF9119).

♦ PF3008 Ineffectiveness of traditional small business methods
Individualized business mindset
Nature Although traditionally effective, family businesses are very rapidly wiped out by new market trends, especially when they cling to comfortable, idiosyncratic patterns rather than adapting to the new needs. Such businesses need to meet the demands of diversified industry for new commodities and expeditious services in order to prosper. Most small business, however, is based on the provision of one or two goods or services. Unless they can discover viable means of competition in the ever-changing market place, their profits will stay minimal and economic growth will be negligible.
Broader Individualistic practices of local business (#PF1176).
Narrower Unmotivating subsistence employment (#PJ1555).
Aggravates Small business failures (#PE9405)
Discouraging conditions for small business (#PD5603).
Aggravated by Unrecognized benefits from cooperatives (#PF9729)
Incompatibility of traditional and new technologies (#PE3337).

♦ PF3010 Resistance to changing agricultural methods
Nature The marginal subsistence economy in many rural communities does not permit the economic flexibility required for experimentation. Added to this is the entrenched individualism, once necessary for traditional subsistence farming, which results in large and small farmers holding tenaciously to traditional patterns of agriculture. There is deep reluctance to change any component in a time honoured interlocking system of production, transportation and marketing. For the small farmer the inflexible use of traditional agriculture is related to the simple methods used for transporting goods - for example, increasing yields would overtax the limited capacity for carrying goods to the marketplace over rough access roads by traditional animals. Furthermore, although crop diversification is the trend in world agriculture, the viability of new crops is not been adequately demonstrated to farmers, and there is deep apprehension of the risks involved.
Broader Inadequate social reform (#PF0677).
Narrower Prohibitive cost of farm equipment (#PJ8790)
Prohibitive cost of disease control (#PF2779).
Aggravates Stagnated development of agricultural production (#PD1285).

♦ PF3012 Disappearance of local culture
Aggravated by Degradation of developing countries by tourism (#PF4115)
Untransposed significance of cultural tradition (#PF1373)
Social environmental degradation from recreation and tourism (#PD0826).

♦ PF3019 Qualified amnesty
Amnesty
Nature Amnesty may be granted for certain categories of political prisoners, but not others. After a political purge, an abortive or successful coup d'Etat, or other major act which threatens to overthrow the current political regime, many possible suspects (almost anyone who could have been connected with the event) may be detained. The least suspect among them may be granted amnesty but the leaders may remain in detention awaiting trial. Although a qualified amnesty may mean that these leaders receive trial sooner than they would have done otherwise, since amnesty is usually only granted in response to public or international protest, certain prisoners may be reclassified under muted-down classifications so that they can be kept in detention.
Broader Political repression (#PC1919) Refusal to grant amnesty (#PF0182).
Narrower Mis-classification of political prisoners (#PF3020).
Related Political trials (#PD3013) Violation of amnesty (#PD3018).
Aggravated by Excessive government control (#PF0304).
Reduces Political prisoners (#PC0562) Internment without trial (#PD1576).

♦ PF3020 Mis-classification of political prisoners
Nature The classification of political prisoners affects the treatment they receive, the charges made against them, and measures that can be taken in their defence. Prisoners may be classified in a way that ensures their continued detention or conviction if they are brought to trial, and they may be reclassified if these classifications are found to be insufficient. Classifications may be used to cut off access from outside organizations working for international justice. Political prisoners may be classified as common criminals; and the distinction may or may not be accurately drawn between military and civilian detainees. Prisoners charged with acts of violence are accorded a different status; ideological deviants may be classified insane (such as in USSR) and charges of espionage may be produced in order to make intervention for the defence very difficult.
Broader Qualified amnesty (#PF3019) Political repression (#PC1919).
Aggravates Refusal to grant amnesty (#PF0182)
Lack of access for prisoners' defence (#PE8637).
Aggravated by Excessive government control (#PF0304)
Political corruption of the judiciary (#PE0647).

♦ PF3021 Rejection of refugees
Denial of right to asylum — Expulsion of refugees — Inadequate protection of refugees
Nature Asylum-seekers may form part of large-scale emigrations of refugees fleeing internal aggression or serious disturbances of public order in their native lands. Frontier rejection of such persons may occur as neighbouring states may claim inability to admit them on a long-term basis. Rejection may be defended as being necessary on economic grounds, although political factors may also be an influence. Asylum is often granted only on a temporary basis, therefore leaving refugees to be constantly seeking permanent asylum, all the while fearful of eventual extradition and possible homelessness.
Expulsion of refugees can give rise to very serious hardship for the refugees concerned. In some cases expulsion measures have in fact led to refugees being sent to a third country which simply returned them to their country of origin.
Incidence The agencies of the United Nations High Commissioner for Refugees in all parts of the world have been confronted with requests for the resettlement of refugees who have been granted temporary asylum in one country but who must find a permanent home in another.
Claim Problems relating to the identification of the state responsible for examining an asylum request have acquired a certain prominence. This problem arises when an asylum-seeker has passed though one or more countries before arriving in the state where he or she wishes to submit an asylum request. In such situations the authorities may refuse to consider an application for asylum on various grounds, including the fact that protection was, or could have been, obtained elsewhere. When this occurs, the asylum-seeker is frequently turned away and becomes what has been called a 'refugee in orbit'.
Broader Expulsion (#PC5313) Injustice (#PA6486)
Restrictions on freedom of movement between countries (#PC0935).
Narrower Denial of right to economic asylum (#PE6212)
Detention of refugees and asylum-seekers (#PE6376)
Lack of individual rights to political asylum (#PF1075).
Related Denial of right to cultural asylum (#PE3450).
Aggravates Statelessness (#PE2485) Extradition refusal (#PF2645)
Forced repatriation (#PD8099)
Denial of economic and social rights to refugees (#PE6375).
Aggravated by Refugees (#PB0205) Deprivation of nationality (#PD3225)
Proliferation of immigrants (#PD4605) Declining sense of community (#PF2575)
Physical insecurity of refugees and asylum-seekers (#PD6364)
Immigration barriers for handicapped family members (#PE4868)
Refusal to issue travel documents, passports, visas (#PE0325).

♦ PF3025 Feminism
Women's liberation
Nature Feminism implies a radical ideology in favour of liberation of women from traditional role-playing. As a movement encompassing a wide range of views of varying degrees of militancy, it may tend to lack cohesiveness and may be extremist in certain tactics and, by denigrating the

virtues traditionally held to be worthy of respect in women, may reduce rather than increase internal self-worth of those whose liberation it is attempting to assist. It may increase family breakdown and child-neglect, make conflict rather than cooperation the norm in situations involving members of both sexes, and possibly lead to discrimination against men.
Incidence Feminist groups of women strive to rid themselves of whatever it is that creates an oppressive environment for them—a task they have not the slightest opportunity of achieving as groups of women outside the men, children and institutions who maintain the oppression.
Broader Ideological conflict (#PF3388).
Aggravates Discrimination against men (#PC3258)
Increasing female criminality (#PE5592).
Aggravated by Traditionalism (#PF2676) Male domination (#PC3024)
Social inadequacy of men (#PF3613) Discrimination against women (#PC0308)
Double standards of sexual morality (#PF3259).
Reduces Denial of right of family planning (#PE5226).

♦ PF3037 Restrictive channels of cultural interchange
Nature The intercultural exchange of life-styles, gifts, and ideas is essential to human development, yet access to this interchange is severely restricted, in particular for many Third World rural communities. Residents may realize the significant effects of national and global events, and glean what they can by listening to radios in the streets and by conversing with neighbours in the regional towns, but these channels are minimal compared with what is both needed and desired. Travel to the market centre is the major contact point with the world, and travel beyond that is almost inconceivable. Women in particular do not expect to go far from their own village, and resources which might enrich and develop family life are felt to be so far off as to be virtually inaccessible. Access to the varying experiences and educational options that might enable village people to create a more effective life-style are strictly limited, nor are there means for the village culture to express itself to the nation or to the world.
Broader Restrictive practices (#PB9136).
Narrower Lack of social contact (#PF8695) Prohibited travel for women (#PG5287)
Insufficient family services (#PG5262) Infrequent travel opportunities (#PG8342).
Related Mothers' self-imposed isolation (#PJ9425).

♦ PF3043 Inequality in distribution of natural resources between countries
Disparity between countries in natural resource endowments
Nature Many of the commodities for which there is a rapid growth of world demand are commodities in the production of which developed countries enjoy climatic or resource advantages (such as some temperate zone foodstuffs: meat, fish, grains, cheese, wine), while most of the commodities produced solely in developing countries (notably tea, coffee, bananas, natural rubber, and hard fibres) have experienced comparatively slow rates of growth of demand.
Broader Unemployment (#PB0750).
Narrower Unequal distribution of fish catches (#PF4495).
Aggravates Environmental stress (#PC1282) Environmental refugees (#PE3728)
Competition for scarce resources (#PC4412)
Instability of production of food and live animals (#PD2894)
Reduction of the share of the developing countries in world exports (#PC2566).

♦ PF3049 Inappropriate taxation of not-for-profit and philanthropic charitable organizations
Inappropriate taxation of international non-profit organizations
Nature Both national and international non-profit and charitable organizations are confronted by special difficulties. The national tax provisions differ considerably from country and such organizations and their donors may not benefit from tax advantages in foreign countries. In particular, non-resident donors are unable to claim any form of tax reduction, and this state of affairs constitutes a special handicap to the mobilization of resources by such organizations, particularly in response to natural disasters or development problems.
Refs International Fiscal Association (Ed) *International Tax Problems of Charities and Other Private Institutions with Similar Tax Treatment* (1986).
Broader Distortionary tax systems (#PD3436) Disparity of national tax systems (#PD1791).
Aggravates Unethical practices of philanthropic organizations (#PE8742).

♦ PF3058 Lack of confidence in the international monetary system
Nature Under the present international monetary system, countries hold their external monetary reserves partly in gold and partly in foreign exchange. In the past, the latter have consisted mainly of gold-backed dollars and pounds as reserve currencies. The present more complex system requires political as well as economic cooperation and a greater reliance on the 'good intentions' of the USA and its now unbacked dollar. Confidence is very fragile, as both unilateral and regional actions are always possible. A weakening of confidence in the reserve currencies may affect the behaviour of both official and private holders of these currencies, inducing them to convert their existing stock of reserve currencies into gold or into other currencies considered to be stronger. A large-scale conversion of any principal OECD nation's currency into gold would not only tend to reduce the total of existing international liquid reserves, without any reduction in the need for such reserves but, by lowering the reserve assets of the reserve currency countries, might force them to undertake domestic adjustments which would be harmful to the level of world economic activity as a whole. A drastic devaluation, for example, by one of these countries could destabilize world trading patterns.
Incidence The present international monetary system remains highly vulnerable to sudden shocks of confidence. Despite the impressive strength of the defences that have been erected against shifts in confidence, the problems of the reserve currency system have yet to be solved. Crises of confidence cause large-scale movements of capital far in excess of the ability of any one country to deal with them by normal means. The system of floating rates of currency lead to improper allocation of the factors of production, anxiety in financial markets, a rise in interest rates and a slowdown in growth. The emergence of the ECU in Europe as a reliable common currency is still problematical as this unit of currency has not had time yet to prove itself.
Claim The world lived with the gold standard before World War I. The world economy was then marked by the dominance of one country, Great Britain, which was the industrial, commercial and financial capital of the world. When that monopoly was called into question, after the first World War, the system ceased to function. Then there was the gold exchange standard, or the Bretton Woods system, proposed by the United States at the end of the war. The adjustment mechanisms which were provided didn't work or worked too late and the system broke down. Each of these systems worked well in the beginning because they corresponded well to their economic environment, in which there existed one dominant economic pole. The world today is multipolar. No unipolar system works. It is foolish to expect anything but instability in a monetary system which depends on one currency—the dollar—as a basis for stability.
Aggravates Hoarded monetary gold (#PD3045) International liquidity (#PG5937).
Aggravated by Uncertain status of monetary gold (#PF2342)
Inadequacies of the international monetary system (#PF0048).

♦ PF3059 Inadequate level of world monetary reserves
Liquidity shortage
Nature The tendency for world reserves to rise more slowly than the requirements of world trade has persisted for some time. This means that one country can only increase its reserves at the expense of another; and the difficulty is compounded if reserves in the more traditional sense are actually falling. Deficit countries feel compelled to introduce restrictions on their trade and aid, including aid-tying, which might not be necessary if reserve positions were less tight. Surplus countries, even if they are ready to forgo further additions to their reserves, certainly do not wish to lose reserves, and therefore tend to be more restrictive in their policies and more reluctant to contemplate changes in their exchange rates than they might be if the general expansion of business activity, international trade and world reserves were sufficient to accommodate the needs of all. Inadequate reserves can also be a contributory factor to crises of confidence.
Incidence Liquidity, in the sense of official reserves, is not itself used in financing external transactions; this is done through the earning of foreign currencies and through a wide array of credit instruments. Its functions are rather to defend the rates of exchange of the national currency and to settle residual accounts when foreign exchange receipts fall short of accruing external obligations. The need for liquidity is thus dependent on the flexibility of exchange rates; the nearer to a free float the less the back-up reserve requirement. The need for international liquidity increases with the degree of autonomy in national policies and with the rigidity of the exchange rates linking national currencies. Notwithstanding the marked post-war decline in the ratio of reserves to world imports, and the recent absolute decline in world reserves, it would be difficult to say at precisely what point reserves could be considered adequate and at what point they might become inadequate. It could, perhaps, be said that it would be a symptom of inadequate liquidity if the developed world, taken as a whole, were maintaining restraints upon its rate of growth, upon its imports of goods and services, or upon its export of capital and other forms of development finance, that would not seem to be necessary in terms of the requirements of domestic monetary stability alone.
Refs Sinha, M R (Ed) *Problem of World Liquidity and Other Essays* (1970).
Broader Insufficient financial resources (#PB4653)
Inadequacies of the international monetary system (#PF0048).
Narrower Inadequate level of developing country monetary reserves (#PF3060).
Aggravated by Inadequate mechanism for the creation of liquid reserves (#PF3061).

♦ PF3060 Inadequate level of developing country monetary reserves
Liquidity shortage of developing countries — Depletion of official reserves of developing countries
Nature Many developing countries suffer from severe fluctuations in exports while others are vulnerable to sudden increases in demand for food imports consequent on diminished harvest yields. At the same time, given the considerable structural weaknesses and rigidities which characterize the economies of the developing countries and also their limited access to the international credit markets, their ability to adjust to fluctuations in their external payments without undue sacrifice of their growth objectives depends in part on the size of their reserves. However, the accumulation of reserves also has associated costs, which can be quite significant in the case of many developing countries with pressing needs for an increase in productive investment and consumption. Many of these countries have adopted restrictions on imports and payments and comprehensive systems of foreign exchange budgeting for the purpose of adjusting external payments to external receipts.
Yet the absence of adequate reserves and other short-term resources imposes intolerable strains on the administration of these systems from the point of view of the smooth functioning of the economy. Even such eventualities as relatively minor errors in forecasting, strikes in ports, the unexpected 'bunching' of imports and modest shortfalls in production which in normal circumstances ought to be accommodated by changes in reserves, now assume enormous proportions and play havoc with the proper functioning of the economy. They drive developing countries to seek credits from suppliers and through other channels on terms which are often unsatisfactory, and to impose embargoes on unessentials and even raw materials and intermediate goods needed for the utilization of existing productive capacity. All these result in a cumulative extension of administrative controls which by virtue of their scope and intensity lead to increasingly greater rigidities over the economy as a whole. These difficulties tend to be aggravated by the fact that inadequacy of reserves, and the consequent problems posed for developing countries in meeting external payments commitments, lead to a weakening of confidence and to a deterioration even in the terms of credit that are normally made available by the international banking system.
Incidence In the 1980s, faced with stagnation or declines in export earnings and with sharply reduced access to external finance, many developing countries had no option but to draw on their foreign exchange reserves, although these same circumstances called for the maintenance of more prudent reserve levels. Reserves nevertheless fell in many developing countries after 1981. The reserve holdings of capital importing African developing countries in the period 1982–84, were sufficient to finance little more than one month of imports. Constraints or declines in foreign exchange earnings and extremely low reserve levels often led developing countries to delay making payments. Arrears accumulated rapidly, increasing six-fold in 1982–84.
Broader Insufficient financial resources (#PB4653)
Inadequate level of world monetary reserves (#PF3059).
Narrower Inadequate mechanism for the creation of liquid reserves (#PF3061).
Aggravates Inadequacies of the international monetary system (#PF0048).

♦ PF3061 Inadequate mechanism for the creation of liquid reserves
Nature The essential defect of the international monetary system is that, apart from such additions to reserves as result from intermittent increases in monetary stocks of gold, provision is made for the expansion of liquidity mainly through the deficits of key currency countries, whereas those countries may not be running deficits concurrent with an expansion in world demand for liquid reserves. Moreover, the maintenance of confidence in the system depends on the key currencies being strong, and hence on the key currency countries usually being in external balance, if not in surplus. Without provision for an orderly increase in liquidity to permit financing of a growing volume of trade, economic development is inhibited.
Background Even under the international gold standard, the mining of new gold was not the sole means available for increasing the world's stocks of liquid reserves; since any upward adjustment in the price of gold was ruled out by the basic assumptions of the system, the necessary growth of liquidity depended to a considerable extent on the use of sterling as a reserve currency by many countries. The difficulties faced in the inter-war period in operating the international gold standard reflected, among other things, the dangers to which the international monetary system is exposed when too much reliance is placed on one or more key currencies for providing world liquidity.
More recent experience reiterates this lesson. The fundamental difficulties of the system had been widely recognized before the crisis of 1971. In fact, the very success of the system revealed its basic weaknesses since the increase in world liquidity was made possible by a steady decline in liquidity in the United States and, with it, the key currency of the entire system. As long as the relative shift was interpreted as a transient phenomenon of post-war recovery in Western Europe and Japan, it was greeted with satisfaction by all parties concerned. The momentum of the relative shift proved, however, surprisingly enduring. As early as the beginning of the 1960s, signs of the vulnerability of the dollar began to appear even while fears of a post-war dollar shortage lingered on. A foretaste of massive attacks on a reserve currency was provided by the repeated sterling crises of the early 1960s. Here it was evident that the basis of sterling as a reserve currency, inherited from the United Kingdom's past pre-eminence in world trade and finance, had changed fundamentally when the size of its Empire was greatly reduced and its overseas

investments mostly liquidated. While the relative decline of sterling meant the rise of the dollar as the undisputed reserve currency, an attack on sterling frequently also meant an attack on the dollar. Responses to the strains and stresses of the system and to periodic attacks on the dollar had been largely palliative in nature.

The upsurge in the price of gold, reflecting doubts about the contention that the dollar was as good as gold, was dealt with first by the major central banks which pooled their operations, and subsequently by a two-tier system which severed the gold market from official monetary transactions. As the United States' gold stock fell far below its official liquid liability and United States' deficits continued, the collapse of the system was only postponed by great restraint on the part of the major central banks which refrained from massive conversion of their large dollar holdings into gold. Periodic disquiet in the exchange markets was allayed by such devices as sway agreements, special bond issues to mop up excess official liquidity, and forward exchange operations. Nevertheless, the inability of the central banks to maintain the fixed exchange rate of the dollar finally triggered the crisis of 1971. It began with a massive movement of short-term capital from the United States in 1970 after the reversal of an extremely stringent monetary policy in 1969. As the European countries and Japan continued their restrictive policies at the beginning of 1971, interest differentials between these countries and the United States widened. A total of $12.5 billion left the United States during the first two quarters of 1971, as the effect of the interest differential was magnified by the sharp deterioration in the United States trade balance. Eventually, in May, the pressure on some European currencies became so heavy and the dollar accumulation so large that the Federal Republic of Germany and the Netherlands decided to float, and Austria and Switzerland to appreciate their currencies. These measures and official interventions in the exchange markets by the central banks of Western Europe and Japan proved unable to stop the flight from the dollar. On one or two days in August, the outflow reached as high as $1 billion. On 15 August 1971, the United States suspended the convertibility of the dollar into gold or currencies. Other emergency measures included a 10 percent temporary surcharge on dutiable imports. Since the surcharge fell most heavily on manufactures and exempted many raw materials and products, it affected about 60 percent of the developed market economies' exports to the United States and about a third of those from the developing countries.

These dramatic measures had important repercussions on the rest of the world. The countries of the European Economic Community were for a time unable to adopt a common policy in response to the new situation. The floating of some currencies posed a threat to the Community's internal cohesion. In France, the imposition of capital controls and a two-tier exchange system reversed the trend towards liberalization of the European capital markets. In Japan, the initial reluctance to revalue the yen was overcome only when the Bank of Japan, in the course of a single month, accumulated over $4 billion, more than its entire international reserve of a year before. The developing countries whose reserves were mostly in dollars suffered a heavy loss, although their debt burden, also mainly expressed in dollars, was correspondingly reduced in terms of other currencies. Exporters of commodities largely destined for the United States market, such as coffee, wool and tin, were faced with a possible decline in the purchasing power of the proceeds in terms of other currencies. The application of the surcharge to the developing countries was considered unwarranted since the payments deficit of the United States was on the whole unrelated to trade relations with these countries. Moreover, the surcharge was counter to the commitment to introduce a general preferential scheme favouring imports from developing countries.

Counter-claim Despite the slow growth of stocks of monetary gold, world liquidity outside the United States, not including special drawing rights, trebled between 1950 and 1970. Although the increase was especially conspicuous in a few industrial countries, it was general and widespread. This liberal supply of liquidity formed the basis for domestic expansion in most parts of the world and for steady liberalization of exchange controls and trade, which in turn facilitated growth.
   Broader Inadequacies of the international monetary system (#PF0048)
   Inadequate level of developing country monetary reserves (#PF3060).
   Aggravates Imbalance of payments (#PC0998)
   Inadequate level of world monetary reserves (#PF3059).

♦ **PF3062 Inadequate mechanism for balance of payments adjustment**
Nature The adjustment process is the process whereby unintended deficits and surpluses in the balance of payments on current and long-term capital account (which have to be financed either by compensatory short-term capital flows or by movements of reserves) are eliminated. There are certain inherent forces tending to reverse any departure from balance-of-payments equilibrium. Thus an import balance may tend to depress the level of domestic activity and hence the subsequent demand for imports. The impact of automatic mechanisms of adjustment of this type varies greatly from country to country, depending on the size of foreign transactions in relation to the gross national product. The automatic process of adjustment may also be damped or offset by countervailing domestic monetary or fiscal policies. It is therefore not possible to rely on spontaneous forces either to bring about adjustment within a reasonable time or to distribute the burden of adjustment equitably as between deficit and surplus countries. If adjustment is unduly delayed, the disequilibrium may be accentuated by speculative flows of short-term capital.
   Broader Financial destabilization of world trade (#PC7873)
   Inadequacies of the international monetary system (#PF0048).
   Aggravates Imbalance of payments (#PC0998)   Speculative flight of capital (#PC1453).

♦ **PF3064 Diffuseness of regulatory authority**
Nature In a particular nation or federation, authority concerning matters impeding social or economic change may be distributed among many different bodies at different levels. Even implementing changes whose value is uncontested by all parties may be extremely difficult and time-consuming.
Incidence An example of this problem is that it took 7 years for the Inter-American Development Bank to persuade most of the states within the USA to permit its securities to be held by banks, savings banks, trust accounts and pension funds.
   Related Ineffective regulation of electronic messages (#PE6226).
   Aggravates Legal impediments to foreign investment (#PD3063).

♦ **PF3065 Export credit risks**
Nature Sales of exported products on deferred-payment terms carry risks not found in the case of sales in domestic markets. Reliable information concerning prospective foreign buyers may be hard to obtain, making it difficult to accurately assess their credit-worthiness. Suppliers fear that in case of non-payment it may prove complicated or costly to press their claims in foreign courts, and that in the buyer's country alien creditors may not always receive the same treatment as do domestic creditors. In addition to commercial risks (insolvency and default of the buyer), external trade involves important non-commercial risks arising from events beyond the control of both buyer and supplier. Losses may be caused by political events such as war, rebellion and expropriation, by catastrophes such as hurricanes, floods and earthquakes, and by monetary phenomena such as foreign exchange shortages and other transfer difficulties. When such events occur before delivery of the goods, they may prevent the buyer from fulfilling the contract or make it impossible to transport the goods to their destination. When they occur after delivery, they may render previously solvent, buyers insolvent or prevent payment by solvent buyers. Export credit insurance schemes alleviate some risks for nationals of some countries, but there are inequalities among the insurance plans and there are some countries for which no insurance is available.

   Broader Obstacles to world trade (#PC4890).
   Narrower Decline in export credits to developing countries (#PE3066).
   Aggravated by War (#PB0593)   Natural disasters (#PB1151)   Political upheavals (#PC7660)
   Overthrow of government (#PD1964)   Chronic shortage of foreign exchange (#PC8182)
   Government expropriation of private property (#PD3055)
   Foreign exchange shortage in developing countries (#PD3068).

♦ **PF3069 Non-convertibility of currencies**
Nature The non-convertibility of the currency of a country or a group of countries (such as the Eastern bloc) and the absence of a realistic parity for the currency may make trading conditions very artificial.
Incidence For a long time, trade with the collectivist economies was done almost entirely within a system of clearing agreements. Trade had to be exactly balanced and tended to be aligned on the exports level of the economically weaker partner, so that triangular trade was made impossible. Most industrial countries trade today on the basis of a convertible currency, but clearing is still the rule in the case of developing countries. These countries can use the money received for their exports only for purchases from the bilateral partner; this obligation often restricts the choice of purchases and distorts sales prices.
   Broader Obstacles to world trade (#PC4890).
   Aggravates Foreign exchange shortage in developing countries (#PD3068)
   Restrictions imposed on aid to developing countries (#PF1492).

♦ **PF3070 Foreign exchange restrictions**
Counterproductive exchange control restrictions — Currency restrictions — Refusal to issue foreign currency for use abroad — Blocked currency
Nature Currency conversion may be subject to various kinds of national restrictions, the most severe form being one in which special permission is required for every single transaction on the foreign-exchange markets. General permission may, however, be given for certain types of transactions. Such restrictions may differentiate between residents and non-residents of a country in the relative freedom granted each group to carry out foreign exchange transactions. For every exchange control measure there is some alternative which would produce equivalent economic effects which may be more desirable for other reasons. For example, in dealing with a balance of payments problem created by internal inflation, the results expected from exchange controls could be obtained by anti-inflationary monetary, fiscal and wage policies. At the same time, some of the shortcomings of exchange control would be avoided and the problem would receive a more fundamental solution. Particular shortcomings are: a degree of flexibility which permits detailed policy changes to escape necessary public scrutiny; doubts on the soundness of the currency and the financial policies of the country, thus discouraging foreign investment; disturbing elements for the country's foreign trade (particularly when control takes the form of quantitative restrictions and the country lacks efficient and corruption-free administrative machinery); difficulty of reaching rational administrative decisions and the inherent temptation to corruption with the consequent reduction of productivity.

The governmental practice of limiting the holding and purchasing of foreign currency within a country's borders (a common way of supporting the value of the country's currency) has become more prevalent and more strictly enforced, particularly in developing countries. It has led to a substantial increase in the amount of funds being "blocked". Such blocked funds prevent international companies from deploying their capital where it would earn the highest returns. Some companies are forces to either sell the blocked assets at a loss; or wait until they are freed, with the risk that even greater losses will then be inevitable.
Incidence Airlines are among the most severely affected because they are legally required to accept payment for services in local currency. In 1984 it was estimated that airline companies alone had $850 million blocked in various countries, compared to $600 million in 1983.
Counter-claim Exchange control may be used as an instrument to foster economic development: as a balance of payments corrective; as a means of protecting domestic economic activities; as a means of reshaping import expenditures; to encourage domestic and foreign investment; as a source of fiscal revenue; and as a means of enlarging and protecting a regional market.
   Broader Obstacles to world trade (#PC4890)
   Financial destabilization of world trade (#PC7873).
   Restrictions on freedom of movement between countries (#PC0935).
   Narrower Socially unintegrated expatriates (#PD2675).
   Related Refusal to issue travel documents, passports, visas (#PE0325).
   Aggravates Corruption (#PA1986)   Underproductivity (#PF1107)
   Illegal movement across frontiers (#PC2367).
   Aggravated by Foreign exchange shortage in developing countries (#PD3068).
   Reduces Imbalance of payments (#PC0998)   Speculative flight of capital (#PC1453).

♦ **PF3072 Restrictions on direct news coverage of parliamentary affairs**
Restriction on press coverage of parliamentary affairs
Nature Broadcasting media are not allowed to cover parliamentary proceedings directly in most countries. Facilities for the press may be provided, but journalists may be prevailed upon to keep their views as neutral as possible. The result of such restrictions may be alienation, apathy and ignorance, a lack of participation or interest in politics, the tightening of government control, and the maintenance of government secrecy.
   Broader News censorship (#PD3030)   Lack of freedom of the press (#PE8951)
   Conflict between government and the news media (#PE1643).
   Related Government propaganda (#PC3074)
   Restrictions on news coverage of legal affairs (#PF3073)
   Restriction on press coverage of legal affairs (#PE7945).
   Aggravates Political apathy (#PC1917)   Lack of representation (#PF3468)
   Inadequate government publications (#PF3075).
   Aggravated by Uncontrolled media (#PD0040)   Journalistic irresponsibility (#PD3071)
   Threat to parliamentary immunity (#PF0609).
   Reduced by Propaganda (#PF1878).

♦ **PF3073 Restrictions on news coverage of legal affairs**
Nature Trials and other legal proceedings, including legislative sessions, may be closed to news media. The closing of legislatures involves broadcasting media for the most part, while the closure of trials and other proceedings may involve all media. Fair hearing may not be given to cases where the procedure is not open to public scrutiny. Secret trials are usually political, but other hearings which may set a precedent in the interpretation of the law may be closed and the issue may not be recognized as such until a precedent is set, usually in favour of the authorities. Corruption may continue and government control may be strengthened because of public ignorance and apathy.
   Broader News censorship (#PD3030)
   Restrictions on freedom of information (#PC0185).
   Conflict between government and the news media (#PE1643).
   Related Restrictions on direct news coverage of parliamentary affairs (#PF3072).
   Aggravated by Excessive government control (#PF0304)
   Journalistic irresponsibility (#PD3071).
   Reduced by Unfair trials due to pre-trial publicity (#PE1692).

PF3075 Inadequate government publications
Nature Official information at the national level includes propaganda, statistics and reports. Information in the sense of reports and statistics may be very turgid reading and very lukewarm in expressing opinions. Government reports may also be isolated by national delimitation or sectoral coverage, so that the information cannot be analysed in its full context without extensive research. Terminology may be difficult for non-specialists to understand. Such information induces public apathy, ignorance and alienation and, where it only gives facts which are favourable to the government or its prestige, it may serve to mask injustices, exploitation, inequalities, and corruption.
Broader Excessive neutrality of intergovernmental official information (#PF3076).
Related Censorship (#PC0067)   Official secrecy (#PC1812)
Government propaganda (#PC3074).
Aggravated by Excessive government control (#PF0304)
Restrictions on direct news coverage of parliamentary affairs (#PF3072).
Reduces Threat to parliamentary immunity (#PF0609).
Reduced by Uncontrolled media (#PD0040).

PF3076 Excessive neutrality of intergovernmental official information
Nature Information from international organizations, in particular those representing governments, tends to suffer from an excessively neutral stance, except if the organization evolves a certain 'nationalism' or propaganda of its own applying equally to all members (such as the EEC). Because they must incorporate several different viewpoints, reports tend to be very long and involved, and may be vague as well as being innocuous. The terminology may be very specialized. This coupled with an international policy bias makes the information seem remote. The public may not see how it applies at their local level, especially where there is little attempt to bridge the communication gap with effective public relations.
Broader Racism (#PB1047).
Narrower Non-alignment (#PF0801)   Inadequate government publications (#PF3075).
Aggravated by Official secrecy (#PC1812).

PF3079 Refusal to grant licences to media
Refusal of licence media
Nature Refusal by the authorities, governmental or private, to grant licences to broadcasting stations, film or theatre clubs may serve the same function as censorship. Licences may be distributed as a matter of governmental policy (particularly with broadcasting) and may be issued in such a way as to support the administration. Alternatively, broadcasting licences may be distributed on the basis of commercial performance. The authorities may refuse to renew licences where broadcasting stations have not met their requirements. Films in many countries require a licence to be shown and this may be refused on the grounds of immorality or obscenity, and private film and theatre clubs may be closed on these grounds if they do not comply with the law. Refusal of licence for the media may result in cultural stagnation, indoctrination and conformism, may serve to strengthen government control or political dictatorship, or may give rise to a counter culture.
Broader Refusal to grant licences (#PG5946)
Restrictions on freedom of information (#PC0185).
Narrower Refusal of broadcasting licence (#PG5912).
Aggravates Radio and television censorship (#PD3029)
Restriction of freedom of expression (#PC2162).
Aggravated by Excessive government control (#PF0304).

PF3081 Denial of access to news
Nature The restriction of access to information on current affairs may take the form of limited government or official information, propaganda, harassment of the press, government secrecy and official secrets, censorship and denial on commercial grounds either to news distributors or receivers. Such restrictions may be imposed by political authorities or by private organizations, or individuals who do not want to cooperate. The result tends to induce ignorance, alienation and apathy and may serve to maintain existing inequalities, injustices and exploitation; to strengthen the effectiveness of indoctrination, government control and dictatorship; and encourage subversive activities. Other ways in which the press can be limited include: legal pressures such as security laws, press laws, criminal codes of libel, and contempt laws; economic and political pressures such as bribes, control of newsprint, leverage of official advertising and control of bank loans; secrecy provisions; direct pressures such as the licensing of journalists, censorship, arrest and/or torture, bombings, forced mergers or closures, disappearances, or the killing of journalists.
Broader Violation of civil rights (#PC5285)
Conflict between government and the news media (#PE1643).
Narrower News censorship (#PD3030)   Harassment of the media (#PD0160)
Restriction of access to news distribution media (#PF3082).
Aggravated by Official secrecy (#PC1812)   Newspaper monopoly (#PE0246)
Monopoly of the media (#PD3101)   Excessive government control (#PF0304)
International monopoly of the media (#PD3040).

PF3082 Restriction of access to news distribution media
Denial of distribution access
Nature Restrictions on access to distribution media may be for political or commercial reasons, in the form of media monopoly or direct government control or censorship, the refusal of licences, harassment or confiscation. It may serve to maintain existing exploitation, injustice and inequality and to facilitate indoctrination and strengthen dictatorship and repression, by inducing apathy and ignorance; or it may encourage subversive activities resulting from alienation.
Broader Denial of access to news (#PF3081)
Restrictions on freedom of information (#PC0185).
Related Censorship (#PC0067)   International monopoly of the media (#PD3040)
State control of communications mass media (#PD4597).
Aggravates Restriction of freedom of expression (#PC2162).

PF3083 Inflexible attitudes toward community social services
Nature Communities expect the educated to provide local leadership and to fill the new jobs, but are not yet prepared to alter rural care patterns to accommodate urban systems at the less prestigious level (garbage collection, for example). Those who are educated outside their local communities (usually the younger men) tend not to return to diversify community expertise, but instead they either work away from their neighbourhood, coming home only at night and at weekends, or they leave it completely. This leaves the community to look forward to urban services and new industry with few people available to take on the new jobs. Farming has a lower priority than ever, due to its low productivity relative to the amount of physical work and time required, and small animal raising is regarded as an even less appropriate occupation. Women's roles tend to continue to be primarily as the major caretaker for the household and young children. Unable to recognize a broader range of vocational roles as appropriate, rural areas will become static, unable to meet social demands.
Broader Obstacles to community achievement (#PF7118).
Narrower Farming low priority (#PS1618)   Limited female vocations (#PU1636)
Presumed fishing vocation (#PS1556)   Unemphasized animal raising (#PS1589)
Depreciated refuse collecting (#PU1578)   Little intrafamily communication (#PU1599).
Aggravated by Overdependence on education (#PE8988)
Discrimination against women in education (#PD0190).

PF3084 Dependency on unpredictable sources of income
Insufficient side-income jobs — Fluctuating income drops
Broader Second employment (#PF6908)   Archaic marketing methods (#PF6465)
Undeveloped channels for public and private resources (#PF3526).
Narrower Lack of capital development (#PD8604).
Related Inadequate income in old age (#PC1966)
Underdeveloped sources of income expansion (#PF1345).
Aggravates Low general income (#PD8568)   Marginal level of family income (#PD6579)
Diminishing capital investment in small communities (#PF6477).
Aggravated by Limited job market (#PC7997)   Dependence on external resources (#PC0065)
Subsistence agricultural income level in rural communities (#PE8171).

PF3087 Art as propaganda
Nature The use of art to influence public opinion in favour of a political ideal or government policies may be on a national or international level. It may take the form of posters, political cartoons, commissioned paintings, sculptures, buildings, etc. It may be spread internationally through exhibitions and other methods of cultural exchange. Artistic propaganda may result in cultural stagnation and prejudice or alienation, conflict, counterpropaganda or subversive activities. It may strengthen government control and dictatorship, injustice, inequality and exploitation. It has the same advantage over printed or non-visual propaganda as photographic, film or theatre propaganda, in that it can appeal to the illiterate and has immediate impact.
Broader Propaganda (#PF1878)   Compulsory indoctrination (#PD3097).
Narrower Political cartoons (#PG5948).
Related Censorship (#PC0067)
Abuse of international cultural, diplomatic and commercial exchanges (#PF3099).
Aggravated by Representative arts (#PF0981).

PF3089 Living alone
Incidence In the United States, 20.6 million people live alone and increase of 100 percent in the ten years since 1970. People living alone reflect the result of more widows living alone, young people postponing marriage, increasing numbers of divorces and the availability of more housing units.
Broader Loneliness (#PF2386).
Related Loneliness in single people (#PD4392).
Aggravates Mental disorders of the aged (#PD0919).

PF3091 Inflexible management patterns
Inflexible management structure — Static management patterns
Nature Inflexible management patterns are unable to adapt to the changing needs of production and the changing expectations of employees. Static definitions of employability deny large sections of the population from participating in production; fixed limits on job benefits leave employees without comprehensive welfare; set job descriptions fail to consider human factors; parochial understanding of what is good for production may cause community concerns to go unheeded; closed designations of who shall design management patterns isolate the worker from the decision-making process, thus denying him creative engagement in his task; and restricted definitions of the worker's significance alienate him from his work.
Broader Inefficiency (#PB0843)   Inflexible social structure (#PB1997)
Production of non-essentials (#PC3651).
Narrower Rigid personnel hiring policies (#PF3168)
Economic bias of worker benefits (#PD3245)
Loss of the significance of work (#PF3676)
Disruption of work schedule due to computerization (#PE5074).
Related Unchanging business pattern (#PG5367)
Foreign control of natural resources (#PD3109)
Over-specialized supervisory personnel (#PF3588)
Industrial processes geared to reduced social needs (#PE3939).
Aggravates Ignorance (#PA5568)   Mismanagement (#PB8406)
Poor managerial communications (#PF1528).
Aggravated by Change (#PF6605).

PF3096 Misleading information
Misleading documents — False documents — Erroneous documents — Biased reports — Inaccurate documents — Biased information — Misleading reports
Nature Information which is incorrect or phrased in such a way that it gives a false impression may be deliberately misleading or unintentionally so, due to error or to being unrelated to its context. Misleading information includes propaganda and other means of indoctrination and abusive advertising. Such information may cause confusion, especially if there is conflicting information, prejudice, ignorance, conflict or apathy and alienation. It may serve to maintain existing injustices, inequalities and exploitation and strengthen government control, dictatorship, monopoly, organized crime and corruption.
Broader Deception (#PB4731)   Obscurantism (#PF1357)
Unethical documentation practices (#PD2886).
Narrower Exaggeration (#PJ5960)   Biased literature (#PJ1679)
Photographic bias (#PF9707)   False qualifications (#PF0704)
Misuse of statistics (#PF4564)   Deception in business (#PD4879)
Misleading advertising (#PE3814)   Misreading sacred documents (#PF5495)
Distorted media presentations (#PD6081)   Biased government information (#PF0157)
Biased allegations against governments (#PD4517)
Falsification of programme evaluations (#PF9243)
Biased portrayal of women in mass media (#PE7638)
Biased and inaccurate history textbooks (#PD2082)
Biased and inaccurate biology textbooks (#PF9358)
Biased and inaccurate geography textbooks (#PF1780)
Fraudulent acquisition or use of passports (#PE4496)
Biased media-image of foreign groups and peoples (#PE8802)
Inadequate knowledge and reporting of man-made disease (#PF7939)
Inadequate international standards of financial accounting and reporting (#PF0203).
Related Propaganda (#PF1878)   Historical forgery (#PE5051)   Incorrect information (#PB3095)
Conflict of information (#PF2002)   Photographic propaganda (#PD3086)
Compulsory indoctrination (#PD3097)   Suppression of information (#PD9146)
Biased presentation of news (#PD1718)   Radio and television propaganda (#PD3085)
Newspaper and journal propaganda (#PD0184).
Aggravates Excessive use of acronyms and abbreviations (#PF4286)
Destabilizing international telecommunications (#PD0187).
Aggravated by Rumour (#PF5596)   Disinformation (#PB7606)
Inappropriate arguments (#PF2152)   Journalistic irresponsibility (#PD3071)
Avoidance of negative feedback (#PF5311)   Unethical practices of employers (#PD2879)
Inoperative forums for public information (#PF7805)
Incomplete access to information resources (#PF2401).

PF3098 Grey lies
Broader Lying (#PB7600).

Related Black lies (#PE4432) White lies (#PF7631) Paternalistic lies (#PF7635).
Aggravates Exploitation (#PB3200) Class conflict (#PC1573)
Male domination (#PC3024) Political inertia (#PC1907)
Scientific censorship (#PD1709) Economic dictatorship (#PC3240)
Biased presentation of news (#PD1718) Discriminatory professionalism (#PC2178)
Contradictions of capitalism in developing countries (#PF3126).
Aggravated by Capitalism (#PC0564) Monopolies (#PC0521)
Power politics (#PB3202)
Unjust allocation of government contracts (#PF2911).

♦ PF3099 Abuse of international cultural, diplomatic and commercial exchanges

Nature Exchanges may be used as a means of propaganda and of espionage. Propaganda may use many different media, including visits by ambassadors and other key personalities; for exchange visits by ordinary members of the public, each nation may select young people or other categories susceptible to indoctrination. Programme schedules for such visits may only show the most favourable aspects of a country. 'Indigenous culture' may become distorted when used for propaganda purposes. International cultural and diplomatic exchange may encourage counter-propaganda in the same vein as a result of ideological rivalry. It may provide a foothold for espionage activities which, if discovered, may increase international conflict rather than dissipate it.
Broader Ideological conflict (#PF3388)
Obstacles to international cultural exchange (#PF4857).
Narrower Biased cultural tours (#PG5962)
Political boycott of international sports events (#PE4206)
Abuse of traditional cultural expressions of peoples (#PE4054).
Related Art as propaganda (#PF3087) International trade fairs (#PG5961)
Abuse of government power (#PC9104) Abuse of traditional customs (#PJ0675)
Social environmental degradation from recreation and tourism (#PD0826)
Natural environmental degradation from recreation and tourism (#PE6920)
Obstructions to international personnel exchanges and cultural cooperation (#PE4785).
Aggravates Espionage (#PC2140) Propaganda (#PF1878)
Counter-espionage (#PD2923).
Aggravated by Double standards in morality (#PF5225)
Benign neoplasm of bone and cartilage (#PG0819).

♦ PF3100 International non-military conflict

Cold war
Nature International hostilities without recourse to armed conflict take the form of propaganda of all kinds, espionage, restrictive trade practices, restriction on information and expression, the issuing of false and misleading information, ideological rivalry for influence in other countries, arms race, and war scare. Isolated acts of covert terrorism, including assassination and sabotage may occur. Cold war-type strife constitutes a barrier to progress and cooperation with the risk that increasing militarism will lead to armed confrontation.
Incidence In its most intense form, the cold war is the battle of nerves illustrated, for example, by Hitler's actions in the late 1930s. A cold war struggle can also exist between a neutral nation and a belligerent nation trying to intimidate it during war. The cold wars since World War II, besides the USSR versus its former allies, include bordering nations strife such as China versus USSR, and India versus China, among the more notable.
Counter-claim Cold war has been used by governments to enlist public opinion in support of various foreign and domestic policies, which would have otherwise encountered resistance.
Refs Chomsky, Noam *Towards a New Wold War* (1982); Sen, Chanakya *Against the Cold War* (1962).
Broader International insecurity (#PB0009) Aggressive foreign policy (#PC4667).
Narrower Espionage (#PC2140) Competitive acquisition of arms (#PC1258)
Destabilizing international telecommunications (#PD0187).
Aggravates Fear (#PA6030) Militarism (#PC2169) Propaganda (#PF1878).
Aggravated by Political instability (#PC2677).

♦ PF3103 General unproductivity of capitalist systems

Nature In capitalist systems the aim of free competition ostensibly is to provide the greatest opportunity and incentive for productivity and the positive exploitation of resources, and the widest possible availability and distribution of goods and services; but the profit motive restricts overall development and wastes both material and human resources. The former may cause inflation through scarcity; and the latter tends to widen the gap between the rich and the poor. Exploitation on a national level may spread to exploitation on an international level (such as colonialism, imperialism), and may cause rivalry for the dominance of underdeveloped nations.
Broader Capitalism (#PC0564) Unproductive use of resources (#PB8376)
Contradictions in capitalist systems (#PF3118).
Narrower Resource wastage in capitalist systems (#PC3108)
Declining productivity in industrialized countries (#PD5543).
Aggravates Economic gap (#PD8834) Unemployment (#PB0750)
Class conflict (#PC1573) Declining economic productivity (#PC8908).
Aggravated by Abuse of science and technology in capitalism (#PE3105)
Excessive demand for goods in capitalist systems (#PC3116).

♦ PF3104 Counterproductive capitalist investment financing

Claim The use of investments, particularly internationally, to accumulate wealth either on an individual or group level is non-productive in material terms and is therefore inconsistent with the productivity and trading principles of capitalism. It is exploitative and creates alienation of rich countries from developing countries, and of the artificial class, created by this process from the proletariat. It hinders general development in underdeveloped countries through foreign debt and through concentrating on its own profit-making objectives rather than social welfare in the country concerned. It hinders production in industrialized countries by supporting that of under-developed countries where raw materials are more easily accessible and labour is cheap. Investments, since they are made for profit, tend to be speculative; and since speculation is based on expectation, this causes economic instability which may equally result in political instability. Money and currency play a major role in this artificial situation and goods (particularly gold and precious metals or stones) contribute to a lesser extent. The uncertainty is aggravated by the combination of transnational corporations in investment projects. Such corporations may side-step government controls and may amass greater wealth and resources than entire nations, at the same time being responsible or responsive to no social group apart from their minority elite of directors and anonymous shareholders.
Counter-claim International investments have been on the whole, highly profitable. Investment and trade are inseparable aspects of international economic activity which brings cultural contact between nations and regions, technological advances, increased standards of living, and provides avenues for developing nations to become part of the international community in all senses. Foreign trade and investment is one of the surest guarantors of international peace.
Broader Contradictions in capitalist systems (#PF3118)
Waste of human resources in capitalist systems (#PC3113).
Narrower Capitalist speculation (#PC2194)
Conflicting roles of money in capitalist systems (#PF3114).
Aggravates Economic gap (#PD8834)
Technology gap between developed and developing countries (#PE7985).
Aggravated by Social inequality (#PB0514)
Lack of individualism in capitalist systems (#PD3106)
Abuse of science and technology in capitalism (#PE3105).

♦ PF3110 Anti-capitalism

Nature Anti-capitalism in non-capitalist countries may take the form of political repression, censorship, authoritarianism and other means of purging or subduing capitalist-minded elements. This may lead to an elaborate system of espionage and surveillance and a situation similar to that of the mediaeval inquisition. In capitalist countries it may take the form of 'subversive activities', strikes or demonstrations. Means of tightening international security, perhaps at the expense of the autonomy of another country, cold war tactics, military trade, and other agreements may be used externally. This may encourage ideological schism, strivings and subversive activities which result in a backlash reaction towards fascism.
Refs Kostopoulos, Tryphon *Decline of the Market* (1987).
Broader Ideological conflict (#PF3388).
Aggravates Politically emotive words and terms (#PF3128).
Aggravated by Capitalism (#PC0564) Double standards in morality (#PF5225).

♦ PF3111 Destiny

Nature Many nations or communities have believed themselves to have a "destiny", a goal set before them by some transhuman being. This belief sets them apart from the rest of the world, reinforces their sense of identity, and helps unite them. It can also induce feelings of superiority and contribute to aggressiveness, racism, nationalism and fanaticism.
Incidence The idea of destiny is contained in the biblical doctrine of God's providence and his sovereign disposal of history. From an early time, the people of Israel were conscious of having a destiny, and it was this consciousness which held them together and gave them their identity. In the New Testament the church appears as the eschatological community and as such is conscious of a destiny. In modern times German philosophy of the 19th and 20th centuries made a great deal of the destiny of the German people. In the USA the westward migration was justified by "Manifest Destiny" and recently supposing that God has favoured American and conferred on it the mission of leading the rest of the world to freedom and affluence. Ideological extremist in South Africa, Israel, and communist nations believe the inevitability of their envisaged future.
Counter-claim A sense of destiny can bring cohesion to a community, can invest its life with meaning and dignity, can inspire noble aspirations and ideals, and give courage to endure hardships in the pursuit of them. It is not a sense of destiny that is the problem but using it to justify unacceptable actions.
Broader Fatalism (#PF6430).
Narrower Accident proneness (#PE6352).
Aggravates Racism (#PB1047) Zionism (#PF0200) Fascism (#PF0248)
Elitism (#PA1387) Marxism (#PF2189) Monarchy (#PF2170)
Socialism (#PC0115) Communism (#PC0369) Apartheid (#PE3681)
Theocracy (#PG5111) Capitalism (#PC0564) Liberalism (#PF0717)
Imperialism (#PB0113) Neo-fascism (#PF2636) Conservatism (#PF2160)
Class conflict (#PC1573) Social inequality (#PB0514) Political conflict (#PC0368).

♦ PF3114 Conflicting roles of money in capitalist systems

Nature Money has two values under the capitalist system: its worth as metal or other material; and its exchange value. As capitalism evolves, the former becomes increasingly insignificant. The fact that the exchange value is less related to the real value puts the former in question and causes very complex situations of economic and other instability. It also encourages speculation, with very negative repercussions for the majority of the population. The role of money is closely related to that of commodities which may also serve as exchange value; it is therefore only the propertied class which can take advantage of money by turning it into investments and other hard capital.
Broader Capitalism (#PC0564) Counterproductive capitalist investment financing (#PF3104).
Related Conflicting roles of commodities in capitalism (#PF3115).
Aggravates Economic gap (#PD8834) Economic conflict (#PC0840).
Aggravated by Social inequality (#PB0514).

♦ PF3115 Conflicting roles of commodities in capitalism

Nature Commodities have two values under the capitalist system: their real worth as necessities; and their exchange value. Because the profit motivation of the capitalist system artificially creates excess demand for goods which may not be necessities, scarcity of real necessities and glut of highly profitable goods (because they are produced in excess when they are not needed) may result. Therefore the exchange value of necessities may become artificially high. Prestige items of low necessity but which are profitable to produce may also have an artificially high exchange value. This situation is aggravated by the fact that commodities may be bought and sold for speculative profit. In this case, they are rendered out of reach to the mass of people who need them and become the prerogative of the non-productive propertied class. The use of necessities for speculation creates inflation and the economic and social gap between the propertied class and the proletariat increases.
Broader Capitalism (#PC0564).
Related Conflicting roles of money in capitalist systems (#PF3114).
Aggravates Scarcity (#PA5984) Economic gap (#PD8834)
Economic conflict (#PC0840).
Aggravated by Social inequality (#PB0514).

♦ PF3118 Contradictions in capitalist systems

Nature The universal law of capital accumulation, one of the basic economic laws of capitalism, determines the polarization of capitalist society and the progressively deepening social gulf between the bourgeoisie and the proletariat. The accumulation of capital causes the growth of wealth among the capitalist class and the worsening of the position of the proletariat.
Broader Capitalism (#PC0564) Contradictions (#PF3667).
Narrower General unproductivity of capitalist systems (#PF3103)
Counterproductive capitalist investment financing (#PF3104).
Related Competition in capitalist systems (#PC3125).
Aggravates Unemployment (#PB0750) Economic conflict (#PC0840)
Political instability (#PC2677).

♦ PF3126 Contradictions of capitalism in developing countries

Nature Although capitalism claims to be the best means of promoting industrial production, its inherent instability and elitist traits have severe adverse effects on developing countries, and complicate and retard the road to progress. These effects are the result of the capitalist profit motivation for production (instead of necessity-motivation), combined with the creation of an artificial non-productive property-owning class and the complicating double role of money and commodities on a sophisticated world wide scale. Capitalist industrialisation in developing countries began under colonial rule and continues in the same way as a means of economic imperialism, (for example, the major profits return to the investor). This leads to debt and puts the country under the effective control of its 'benefactor'. Foreign capitalist industrialisation tends to use over-sophisticated technology for which there is little or no suitable labour in the developing countries. Therefore much of the labour is also imported and the industry does nothing to alleviate

unemployment, which increases from a neglect of agriculture and concentration on industrialisation or urbanisation problems caused by migration towards industrial centres. Prestige 'growth' of this kind encourages nationalism and competition with other developing countries, when it would be more to the benefit of the developing countries to cooperate, consolidate their position, and exploit the advantages they have as producers of essential base products against their deficit of technological expertise and commercial and financial strength in relation to the developed countries. The industrialised countries have a vested interest in investing in developing countries, but in so doing they also create eventual competitors. They therefore seek increasingly sophisticated methods of control so that they can maintain the gap between themselves and the 'third world'. However, in creating an industrialised society from feudal agricultural ones, they also create a proletariat which will become increasingly aware of its exploitation and will subsequently retaliate either with revolution, guerrilla warfare, or terrorism.
Refs Berger, Peter L (Ed) *Capitalism and Equality in the Third World* (1987).
Broader Capitalism (#PC0564) Underdevelopment (#PB0206).
Narrower Uncontrolled urbanization in developing countries (#PD0134)
Inadequate level of investment within developing countries (#PD0291).
Related Excessive foreign public debt of developing countries (#PD2133)
Technology gap between developed and developing countries (#PE7985).
Aggravates Economic gap (#PD8834) Competition between states (#PC0114).
Aggravated by Grey lies (#PF3098).

♦ **PF3127 Obstacles to extended families**
Nature The splitting up of the nuclear family between husband/wife, parents/children, sons/daughters is a critical factor in the breakdown of the extended family where grandparents, grandchildren, in-laws, uncles, aunts, cousins and others are in close relationship. A secondary cause is mobility, which uproots part of the extended family and takes it to a totally different region or even country. Other obstacles to extended families include the belief that the state can and should take care of senior citizens, the belief in having fewer or no children, and the erosion of community life.
Narrower Fragmented patterns of extended family relationships (#PF1509)
Narrow legal definition of the family in developing countries (#PD1501).

♦ **PF3128 Politically emotive words and terms**
Nature Words used in propaganda but often generally accepted in society, with emotive political and international connotations, have formed part of propaganda and indoctrination for many years, so that they immediately evoke a reaction of revulsion or resistance and hinder international or intergroup understanding and cooperation. Such words are often used loosely as labels without adequate definition, in attempts to create a false unity of all groups which are against a given "label".
Broader Ideological conflict (#PF3388).
Aggravates Lack of understanding (#PJ4173) Compulsory indoctrination (#PD3097).
Aggravated by Fascism (#PF0248) Socialism (#PC0115)
Communism (#PC0369) Propaganda (#PF1878)
Capitalism (#PC0564) Demagoguery (#PC2372)
Neo-fascism (#PF2636) Anti-capitalism (#PF3110)
Double standards in morality (#PF5225).

♦ **PF3129 Restrictive patterns of traditional life**
Nature The patterns of traditional life, which were created from necessity and once served many rural communities well, and in which the working day was very long, women remained at home, children worked and had little education, and family occupations were inherited, have become inhibiting factors to social development. The technological revolution has given birth to a new, fast moving lifestyle which necessitates a new relatedness in every community. However, many communities are restricted. In particular, the lack of unavailability of books and magazines curtails the flow of images and information about events occurring in the rest of the world; this lack of exposure is particularly evident among old people, who have no contemporary images to convince them of the importance of education for the young.
Broader Restrictive practices (#PB9136) Inhibiting social patterns (#PF0193)
Restrictive social practices (#PC5537).
Narrower Inhibiting effects of traditional life-styles (#PF3211).
Aggravates Exploitation of child labour (#PD0164)
Social isolation of the elderly (#PD1564).
Aggravated by Untransposed significance of cultural tradition (#PF1373).

♦ **PF3130 National communism**
Nature Although communism is supposed to be a completely world-wide movement, crossing all national barriers and exemplified by the withering away of the state after the dictatorship of the proletariat, (who have more in common with the workers of other countries then with the ruling classes in their own country), not only are many communist frontiers closed, but the political system varies from country to country. National communism is usually characterized by a resistance to foreign communist control and by a certain distinctiveness in national ideology, social institutions, and political strategy. This may cause aggression on the part of the dominant foreign communist power, ideological schism, and cold war. Ideological schism within the country concerned or within neighbouring countries, encouraged by this national distinctiveness, may lead to political purges and severe repression, or to subversive activities.
Broader Communism (#PC0369) Nationalism (#PB0534).
Related Eurocommunism (#PF3876) Contradictions in communist systems (#PF3179).
Aggravates Communist political imperialism (#PC3164).

♦ **PF3136 Imperfections of capital markets**
Nature There is a great variation from country to country in the strength of securities markets, and especially the long-term bond markets that are of particular interest to developing countries. Insurance companies, pension funds, trust funds and other institutional investors are the mainstay of bond markets; they not only provide the market with funds but also endow it with a high degree of professionalism and sophistication in the purchase and sale of securities. Where these institutions are weak or non-existent, as for example when social insurances reduce the need for private pensions and the public savings institutions channel their deposits to support government financial requirements, the bond market is correspondingly underdeveloped. The market then finds difficulty in 'digesting' large new issues and as a result new bond issues are frequently rationed. A second and quite different aspect of market imperfection is the often large divergence between the cost of issue to the borrower and the yield to the lender. The initial and recurrent costs vary considerable from country to country and tend to be larger for fund bonds. Other factors are the structure of interest rates as influenced by domestic economic policies, the interest rates themselves, and the ease with which new securities can be resold without loss, before maturity.
Partial deregulation of some of financial markets have resulted in inefficiency, distortions in prices and increased fraud. Abandoning fixed commission has forced all major financial houses to trade heavily on their own accounts in order to make up for lost commission revenue. They, are directly or indirectly, playing against their own customers. The major brokerage firms all have a good deal of power to manipulate intra-day market prices creating advantages for their customers and for themselves. Knowledge of major moves by customers provides a brokerage firm's traders with an edge in very short term speculation, and thus encourages more of it. Programme trading renders prices irrational. Stock index futures convey no real information beyond that already conveyed by current stock prices, the current price of the stock and their only function is to facilitate highly leveraged bets on broad market moves.
Claim When prices represent something other than the cumulative best information available about future returns, any marketplace is a fraud.
Broader Economic inefficiency (#PF7556)
Restrictions on foreign access to capital bond markets (#PD3135).
Related Obstacles to commodity futures trading (#PF4870).

♦ **PF3141 Lack of international coordination of interest rates**
Volatility of interest rates
Nature There is a lack of adequate coordination between the interest policies of different countries. This is complicated by conflicts between internal and external considerations. Thus, when inflationary pressure coexists with a payments surplus in one country and unemployment with a deficit in another, there is a natural reluctance for the former to lower interest rates and for the latter to raise them in order to ensure equilibrating capital movements, and to avoid unilateral restrictions.
Incidence In the 1980s, the world's biggest debtor nation, the USA, has maintained a policy of high interest rates to attract large foreign capital imports. This denies capital to Europe and to the developing countries. It forces massive annual increases in debt servicing charges for debtor countries. The 1984 Latin American foreign debt of $350,000 million is estimated to increase about $1,700 million for each half-point increase in the US prime rate of interest.
Broader Lack of international cooperation (#PF0817).
Aggravates High interest rates (#PF9014) Speculative flight of capital (#PC1453)
Fragmentation of the international trading system (#PC9584).

♦ **PF3143 Inadequate packaging of agricultural products**
Nature Bulk food crops have to be split into easily transportable packages for distribution to the consumer. However, much food is lost through improper or inadequate packaging. In tropical developing countries, with high temperatures and high humidity, special problems arise due to insect infestation, rodents and mould growth; and almost all packaging materials may be attacked by rats and mice. Developing countries may find it especially difficult to ensure adequate packaging because of the limited supply of packaging materials and the obsolescence of packaging machinery available.
Incidence Estimates of losses of up to 40 percent due to improper packaging are characteristic of insect-infested foods. Large harvests of produce in Central America do not reach the people simply because there is a lack of processing and packing equipment and such countries import canned food from developed countries. In Chile, olive spoilage was reduced by 60 percent when proper packaging technology was applied.
Broader Food insecurity (#PB2846).
Aggravates Food grain spoilage (#PD0811) Human disease and disability (#PB1044)
Spoilage of agricultural products (#PC2027).

♦ **PF3145 Excessive number of crop varieties**
Nature Considerable difficulties are encountered in national seed programmes (to improve crop yields) by the excessive number of varieties recommended for cultivation. Many of the varieties recommended can only be grown in restricted areas due to sensitivity to the environment; some are also low yielding. In order to make the multiplication and distribution of high quality seed of good varieties efficient, the total number of varieties should be as low as possible. For example, in Ceylon 208 varieties of rice were grown before the number of recommended varieties was reduced to 14.
Aggravates Variability in crop yields (#PF1479).

♦ **PF3146 Introduction of high-yield crop varieties**
Green revolution — Plant growth regulators
Nature The success of the drive to introduce new high-yield varieties tends to exacerbate the instability of the international crop trade in agricultural commodities; import demand from previously low-producing countries leading to price falls and surpluses in exporting countries. The resulting shifts in trade patterns make price stabilization more difficult. Although regulating the growth of plants is as old as farming itself, the use of chemical compounds in agriculture has been one critical aspect of the contemporary explosion of crop yields. It remains to be fully seen what environmental effects these unnatural manipulations of the food chain will have in the long run.
Claim The enormous short-term increase in grain yields were made possible by high-yield crop varieties, pesticides, fertilisers and mechanization. Hailed as the solution to the world's food problem, but the green revolution has created more problems than it has solved. By far, the manufacturers of pesticides, fertilisers and farm equipment grew riches by the green revolution. The only farmers who benefitted were the relatively rich ones who could afford the seeds, pesticides, fertilisers, tractors and irrigation equipment. The crops involved in the green revolution such as rice and wheat were often not the ones wanted by the masses of farmers. The green revolution was aimed at cash crops decreasing the availability of food crops. It led to a galloping erosion of native plant varieties in favour of highly inbred imports. Mechanization decreased the number of jobs available in the countryside. The varieties of seeds involved required good soil and plentiful water. The farmers on small plots of marginal land who grew staples like beans, cassava, yam, millet, sorghum and maize were driven into deeper poverty and increased dependence on large landholders or moneylenders. Land speculation caused land prices to rise, and peasants renting land found their rents soaring. Many were forced to leave the land swelling the ranks of urban slum-dwellers. Land ownership became more concentrated.
Refs Pearse, Andrew *Seeds of Plenty, Seeds of Want* (1980); Wolf, Edward C *Beyond the Green Revolution* (1986).
Broader Inappropriate modernization of agriculture (#PF4799)
Instability of economic and industrial production activities (#PC1217).
Aggravates Over-production of commodities (#PD1465)
Urban slums in developing countries (#PD3489)
Instability of trade in vegetable-based foodstuffs (#PE9164).
Aggravated by Irresponsible genetic manipulation (#PC0776).

♦ **PF3161 Limited horizons produced by survival living**
Survival induced individualism — Individualist survival mindset
Nature Despite a general awareness world-wide of the need for the basic necessities of life before people are able to participate creatively in the communities of which they are a part, families in many rural villages are still trapped in the pattern of survival existence. Long hours of work leave little time or energy for community activities, producing an attitude of "each person as an individual". This mindset prevents any creative community involvement which would benefit both the village and the individual families; and the development of mutual cooperation and understanding becomes slow and difficult.
Broader Subsistence life style (#PF1078) Lack of community participation (#PF3307)
Inhibiting effects of traditional life-styles (#PF3211).
Narrower Devaluation of education by survival needs (#PE8902).

♦ **PF3162 Communist opposition to international organizations**
Nature Because of the notion of the withering away of the state after the dictatorship of the proletariat, and the consequent difficulty in defining nationhood, communist countries have been reluctant to join international organizations or recognize or abide by international conventions. However, the communist movement has always asserted its 'proletarian international' basis. Cold war tactics of international relations have been frequently used by communist countries with respect to capitalist countries and with respect to each other. Political, military and trade agreements are concluded within the communist system. Trade agreements are also made with capitalist countries. But on certain issues non–recognition of international law and conventions serves to increase international tension and uncertainty. Where communist countries are members of important international organizations, they may use veto or other methods of blocking the policies of capitalist nations or the smooth running of the organizations.
Broader Communism (#PC0369).
Narrower Veto (#PG6037) Disruption of international action (#PG6038)
Non–recognition of international law (#PF9081).
Aggravates International insecurity (#PB0009).

♦ **PF3168 Rigid personnel hiring policies**
Nature Hiring, placement and tenure policies tend not to respond to availability of skills in the personnel market, leaving a serious disparity between the human resources available in most societies and the utilization of those resources for society's improvement.
Broader Inappropriate policies (#PF5645) Inflexible management patterns (#PF3091).

♦ **PF3179 Contradictions in communist systems**
Nature Inconsistencies in communist systems conflict with its ideals and retard development. They include the denial of human rights under various forms, including: censorship; political and religious repression; closed society; imperialism; elitism; dictatorship and bureaucracy; and more theoretical inconsistencies such as nationalism, the creation of an international system of states, the boycotting and negative participation in existing international bodies, and the non–inevitability of communist revolution which gives rise to the use of force and repression. The net result is instability and conflict, particularly international conflict, and alienation.
Refs Mao Zedong *On the Correct Handling of Contradictions among the People* (1966).
Broader Communism (#PC0369) Contradictions (#PF3667).
Narrower Communist closed society (#PD3169).
Related National communism (#PF3130) Revolutionary communism (#PC3163).
Aggravated by Inadequacy of doctrine (#PF3396).

♦ **PF3181 Ideological schism in communism**
Nature Ideological schism occurs between different communist countries as national systems emerge to cope with varying conditions. Schism may result in the withdrawal of cooperation or participation in international communist activities; or it may develop into cold war, with a high degree of tension on both sides. Political purges, censorship and repression may become more intense, and frontiers may be closed. Outwardly it may encourage imperialism on the part of the orthodox communist power, or two different communist countries to fight for prestige and the ideological leadership of the communist world system; or it may intensify their economic imperialist activities, competition with one another, and their propaganda.
Broader Schism (#PF3534) Communism (#PC0369) Ideological conflict (#PF3388).
Aggravated by Fragmentation (#PA6233) Revolutionary communism (#PC3163)
Double standards in morality (#PF5225).

♦ **PF3183 Unregulated global resources**
Nature There are no common international guidelines regulating the use and deployment of resources; nor, in relation to the control of production, are there means of enforcing standards to control the quantity and quality of goods produced.
Aggravates Foreign control of natural resources (#PD3109)
Worldwide misallocation of resources (#PB6719).

♦ **PF3184 Limited availability of learning opportunities**
Limited availability of learning structures — Insufficient technological adaptation — Insufficient learning opportunities
Nature The relative isolation of the individual from major changes in his own country and the world restricts automatic exposure to new experiences. There are few sources of outside news, nor do new visitors pass through often. Although there is an increasing number of people in formal or trade education, educational systems are mainly limited to formal primary curriculum, apprenticeships, and some educational broadcasts on the radio. A large number of adults are illiterate, having been forced to leave school by their families in order to work; often the formal education they have received is inappropriate. Adults and children alike find the time between starting school and leaving to get a job too long if home responsibilities must also be fulfilled.
Broader Obstacles to education (#PF4852).
Narrower Limitations on school admission (#PJ1364)
Inadequate training opportunities (#PJ8697)
Narrow range of cultural exposure (#PF3628)
Limited access to practical education (#PF2840)
Minimal opportunities of adult training (#PF6531)
Unavailable education for effective living (#PF2313)
Restriction of educational opportunities in capitalist systems (#PD3122).
Aggravates Ignorance (#PA5568) Lack of formal education (#PF6534)
Inadequate results of formal schooling (#PF6467)
Unclarified procedures for transposing ancient traditions (#PF6494).
Aggravated by Lengthy educational process (#PU1569).

♦ **PF3192 Limited relations beyond local environments**
Nature The high cost and time taken in transportation discourage mobility except for dire emergency and for securing locally unavailable necessities, and the residents of many communities have only very limited relations with the world beyond their particular village. Current work patterns demanded by subsistence living confine the individual to the village, the fields, and the home. The practical benefits of modern society may be vaguely known but are seldom available, and appreciation of outside experience is frustrated by the villager's experience of, or stories about, its complex structures. Lack of a constant flow of information inhibits practical knowledge of technical resources and marketing possibilities which directly affect the local economy; and, with only word of mouth reports of what is happening in the nearby vicinity, the imagination of the people is stifled. They have little understanding of the regional, national or global dynamics which actually affect their lives, and little perception of how they can creatively participate in today's society.
Aggravates Deteriorating structures of rural community cooperation (#PF3558).
Aggravated by Geographical isolation (#PF9023)
Ineffective rural transport (#PF2996).

♦ **PF3194 Lack of internal political independence**
Broader Lack of political independence (#PF0297).
Aggravated by Political corruption of the judiciary (#PE0647).

♦ **PF3199 Covert imperialism**
Division of the world into spheres of influence
Broader Imperialism (#PB0113).
Narrower Negative social context (#PF9003) Positive spheres of influence (#PG6062)
Unrecognized socio-economic interdependencies (#PF2969).
Related Imbalance of power (#PB1969).
Aggravates Benign neoplasm of bone and cartilage (#PG0819).
Aggravated by Colonialism (#PC0798).

♦ **PF3211 Inhibiting effects of traditional life–styles**
Nature Response to the challenges put by social transition has been impeded in many rural communities by the tenacious hold of outdated traditional images and modes of operation. People may deeply desire the benefits of agricultural technology but be hampered by the proliferation of inherited land plots on the one hand and absentee landownership on the other. Hand–to–mouth subsistence economic operation allows little time for anything but working in order to survive. Traditional prejudices maintaining the separation of castes, the restriction of women to the house, and dependence upon external authority, confine the individual to a limited framework which, however adequate for his ancestors, has a paralyzing effect on the appropriation of a more effective, relevant mode of operation.
Broader Unproductive subsistence agriculture (#PC0492)
Restrictive patterns of traditional life (#PF3129).
Narrower Small land plots (#PG8526) House–bound women (#PG8417)
Dietary restrictions (#PJ1933) Insufficient cultural media (#PJ8476)
Traditional purchasing habits (#PG8540) Traditional land distribution (#PJ1289)
Unawareness of health benefits (#PG8511) Postponement of bridge construction (#PG8538)
Dependence on planning by authorities (#PG8504)
Inflexible patterns of family lifestyle (#PF2644)
Devaluation of education by survival needs (#PE8902)
Limited horizons produced by survival living (#PF3161).
Aggravates Absentee ownership (#PD2338).
Aggravated by Caste system (#PC1968)
Untransposed significance of cultural tradition (#PF1373).

♦ **PF3212 Empty slogans and mottoes**
Nature Both in ideological or political rhetoric, and in commercial and institutional advertising and public relations, a word, phrase or image may be used to convince the public that they should associate the virtues expressed in the motto, slogan, device, trademark, emblem or symbol, with its propagator and possessor. Frequently there is no connection between the virtue or essential meaning put across by such propaganda and the person, products, cause, country or organization advertised.
Incidence 'Democracy' in expressing the legitimacy of a government, is constantly used in propaganda as an exoneration; and accusations are frequently made labelling opposing ideologies and systems 'undemocratic' or 'anti–democratic'. During the 20th century 'democracy' has come to encompass not only considerations of political equality but also social and economic equality. In being so stretched it has come to include any political regime or system which considers itself legitimate, despite the possible use of repression and the manipulation of voting methods.
Theriomorphic symbols such as bears, bulls, eagles, elephants, fighting cocks and hissing serpents, for example, may be used to express the supposed ferocity of political movements and nations and their martial prowess; while the sign of peaceful doves may be used by revolutionaries dedicated to violence. Pictures of healthy, outdoor–type persons are used to sell tobacco, and there are status symbols built into many products – such as gold trim. As for corporations, IBM employed the word 'think' but imposed conformity on its employees; ITT had 'the best ideas to help people', as it 'helped' in Chile; and Ford spoke of 'quality' as it recalled cars for manufacturing defects.
Broader Propaganda (#PF1878).
Aggravates Social conflict (#PC0137).

♦ **PF3214 Weakness of multi–party parliamentary systems**
Nature In a parliamentary assembly comprising members of many different political parties, a governmental majority may be very difficult to obtain or may be obtained only in a manner rendering administration and the passing of new legislation very difficult. This may lead to political unrest and revolution culminating in dictatorship. Government majority may require a coalition between two or more parties which may later disagree over policy, thus causing the government to fall, and a new government may not be formed immediately. Alternatively one party may have much more support than the others and may enjoy a largely uninterrupted term of office. Such parties tend to be conservative and may maintain a fairly strict censorship over the media.
Broader Political instability (#PC2677) Political fragmentation (#PF3216).
Narrower Coalitions (#PG6073) Unstable government (#PG4223).
Related Plural society tensions (#PF2448) Multidenominational society (#PF3368).
Aggravates Nonacceptance of government legislation (#PG6079).
Aggravated by Political schism (#PC2361).

♦ **PF3215 Inadequate political integration**
Nature Lack of integration of divergent political views in government administration may be a feature of a one–party state or dictatorship, or of a democracy in which minority interests may be excluded because of their economic weakness. Discrimination and other forms of repression or intimidation may be used to prevent minorities or underprivileged majorities from being heard, and political opportunism and corruption may be used to suppress a rival political party. The result may be political apathy and alienation, instability and possible revolution, subversive activities, and persistent social inequality and conflict. In an intergovernmental organization, some degree of political integration occurs between the same political parties among the different nations, and between nations who share common concerns, thus creating political factions. Dominant factions are resented and their actions frequently polarize the interests of the nation–states involved. Party alliances may be broken as political struggles for advantage take precedence among the member delegations.
Broader Political conflict (#PC0368).
Related Cultural fragmentation (#PF0536) Political fragmentation (#PF3216)
Inadequate integration of religions into society (#PF3403).
Aggravates Political apathy (#PC1917).
Aggravated by Political schism (#PC2361).

♦ **PF3216 Political fragmentation**
Lack of political unity — Lack of political integration
Nature Lack of unity in political thinking either between parties or within parties, nationally or internationally, is indicated by political schism, party political rivalry, foreign political rivalry for influence abroad; in the need for balance of power arrangements, treaties, alliances; and in subversive and terrorist activities of both a national and international nature. Lack of political unity may lead to political or national disintegration, foreign intervention, conflict and war.
Broader Fragmentation (#PA6233) Political conflict (#PC0368)
Plural society tensions (#PF2448).

Narrower Political rivalry (#PD8992) Political pluralism (#PF2182)
Weakness of multi-party parliamentary systems (#PF3214).
Related Inadequate political integration (#PF3215).
Aggravates Political disintegration (#PC3204)
Lack of appreciation of cultural differences (#PF2679).
Aggravated by Political schism (#PC2361) Institutional fragmentation (#PC3915).
Reduced by Nationalism (#PB0534).

♦ PF3217 Denial of freedom of thought

Dependence on denial of freedom and thought — Denial of right to freedom of thought — Denial of freedom to hold opinions — Lack of intellectual liberty — Restrictions on freedom of thought

Nature Thought can be controlled by limitations on the access to information, or by the punishment of deviation from a nationally accepted or imposed norm. Where successful, denial of freedom of thought tends to produce intellectual and cultural stagnation, apathy, intolerance and prejudice and encourages exploitation and elitism through support for dictatorial or totalitarian regimes.

Broader Denial of human rights (#PB3121).
Narrower Ideological repression (#PC8083) Denial of academic freedom (#PD4282)
Denial of freedom of opinion (#PD7219) Denial of freedom of conscience (#PD7612)
Suppression of intellectual freedom (#PC5018) Repression of intellectual dissidents (#PD0434)
Denial of right to correct misinformation (#PE7349).
Related Denial of right to life (#PD4234) Denial of right to liberty (#PF0705)
Restrictions on freedom of worship (#PD5105) Denial of the right of association (#PD3224)
Restrictions on freedom of information (#PC0185)
Restrictions on international freedom of information (#PC0931).
Aggravates Ethnocide (#PC1328) Totalitarianism (#PF2190)
Political apathy (#PC1917) Threatened sects (#PC1995)
Cultural deprivation (#PC1351) Political repression (#PC1919).
Aggravated by Anti-intellectualism (#PF1929) Compulsory indoctrination (#PD3097).

♦ PF3226 Refusal to participate

Contracting out

Nature The refusal of individuals to participate in general social and economic activities, whether for reasons of prejudice, lack of information, or discontent with the 'established' system, may cause failure of government policy and development programmes, social conflict and alienation.

Broader Apathy (#PA2360).
Aggravates Ignorance (#PA5568) Social fragmentation (#PF1324).
Aggravated by Prejudice (#PA2173).

♦ PF3232 Counter revolution

Counter-revolutionary activities

Nature Counter revolutionary forces may group to avert a revolution, or after it has happened, to overthrow it. In so doing they may have to adopt some of the tactics and even some of the programmes of the revolutionaries. Such groups are often elitist and may be supported by foreign governments against the revolutionaries. Counter revolution may constitute a barrier to progress and may introduce repression and militarism, if this has not already taken place during the revolution.

Refs Dede, Spiro *Counter-revolution in the Counter-revolution*; Kolar, F *Export of Counter-Revolution*.

Broader Revolution (#PA5901).
Related Counter-reform (#PF7334).
Aggravates Bourgeois deviationism (#PD7352).
Aggravated by Civil war (#PC1869) Violent revolution (#PC3229)
Political revolution (#PF3237) Violent political revolution (#PD3230).

♦ PF3235 Cultural revolution

Nature Dramatic change in cultural orientations may be directly influenced by political, technological and economic revolution, and may serve to reinforce cultural barriers between two or more cultures. Governments may use indoctrination to guide people along nationalist or other politico-cultural lines. Advertising may serve a similar purpose in channelling popular culture for commercial purposes. Liberalization of attitudes towards morality, the environment and the spiritual and physical world may alter cultural orientations.

Incidence Cultural revolution has taken place as ideological revolution in communist countries. In 'Western' industrialized countries, liberalization of moral attitudes has also produced a 'permissive society' and counter cultures.

Broader Revolution (#PA5901).
Narrower Symbol system failure (#PF3715) Obsolete basis of cultural identity (#PF0836).
Aggravated by Cultural deprivation (#PC1351).

♦ PF3237 Political revolution

Nature Radical political change may be brought about by either violent or non-violent means, as distinct from coups d'Etat which may simply exchange one political leader or dictator for another or for a clique. Revolution may occur on the basis of class or race (as in decolonization). It may result in economic and ideological revolution and repression of the old order. The more violent the revolution, the more intense the ensuing repression is likely to be. Political revolution may cause social and economic disruption and adaptation problems, or resistance, counter revolution and civil war.

A civil war or other resort to violence by a country's citizens in order to change the government may be termed a revolution. From a political point of view, it is a revolution if the system itself is changed and not simply the officials (as often is the case in a "coup d'Etat").

Incidence In the twentieth century some revolutions that might be cited (excluding wars of independence) are those of the Communists in Russia, China, Yugoslavia and Cuba, and that of the Islamic fundamentalists in Iran. A number of others (for instance, Ethiopia) can be listed, mainly Communistic. There are also revolutionary movements whose motives are linguistic or ethnic – as the Basque and Quebecois among others – or specifically racial.

Counter-claim When, in the course of human events, people are oppressed by their governments, and by the structural violence perpetuated by privileged classes; when their voice is not heard and they are mocked by token changes and platitudinous political promises; when their rights to adequate food and shelter, to medical care and employment, and to freedom of speech are ignored; then direct and forceful action against the criminal classes of rulers, bureaucrats, the rich and all their minions is required to restore justice. Revolution means return; judicious revolution with a minimum of violence means a return to the spirit of the laws of the nation. The elimination of central government by full anarchy, therefore, is the most spiritual form of revolution.

Broader Revolution (#PA5901) Political instability (#PC2677).
Narrower Political conflict (#PC0368) Violent political revolution (#PD3230)
Non-violent political revolution (#PD3228).
Aggravates Counter revolution (#PF3232).
Aggravated by Political disintegration (#PC3204).

♦ PF3240 Compulsory sterilization

Involuntary sterilization

Claim The physiological and psychological changes brought about by pregnancy and the demands of parenthood are beyond the capacity of mentally retarded people to handle. Especially retarded women are vulnerable to sexual exploitation and abuse. Normal contraceptive methods are inadequate, because they are not capable of taking adequate precautions against pregnancy. The only acceptable alternative is sterilization, and in the cases where they are not competent to make the decision, the consent for sterilization is given by others on their behalf.

Refs Gosney, E S; Popenoe, Paul and Grob, Gerald N *Sterilization for Human Betterment* (1979).

Broader Inadequate laws (#PC6848) Birth prevention (#PE3286).
Aggravates Consanguineous marriage (#PC2379).
Aggravated by Medical paternalism (#PF5397) Compulsory health care (#PF4820)
Sexual discrimination in contraceptive methods (#PF1035).
Reduces Decline in human genetic endowment (#PF7815).

♦ PF3242 Homosexuality

Nature Homosexuality refers to the state of being in love or having sexual relations with one belonging to the same sex. Homosexuality may lead to exploitation by blackmail, robbery, or sexual violence. In some countries, homosexuality is a criminal offence.

Background Male homosexuality, condemned in Judaeo-Christian society as the 'sin of Sodom', was condoned and even exalted in ancient Greece. Female homosexuality, or Lesbianism (recalling the poetess Sappho, who lived on the island of Lesbos and wrote love poems to women), is consistently condemned. Particularly since the liberalization of the law in England and Wales, and the foundation in America of the Gay Liberation Front, male and female homosexuals in these and many other Western countries have become less secretive about their way of life. Until the early part of the 20th century, homosexuality generally meant a form of sexual perversion. With the development of psycho-analysis, less emphasis was laid on sex-proper and a correspondingly broader connotation assigned to the term sexual. In current usage, overt homosexuality refers to physical, sexual contact between members of the same sex, while latent homosexuality is used to refer to impulses and desires toward a member of the same sex which are unconscious or, if conscious, are not openly expressed.

While sexual relations between two men or two women may be accepted in certain societies, and regarded as normal or at least tolerated (such as male homosexuality in Asian cultures), male homosexuality is particularly despised in Anglo-American culture and female homosexuality, although less reviled, is generally unaccepted. Recently homosexuality has come to be slightly more tolerated in 'Western' culture with the influence of the 'permissive society'.

Presently, there exists no scientific consensus on the causes of homosexuality or, for that matter, of heterosexuality. However, because traditionally homosexuality has been viewed in most Western societies as a divergence from normal sexual development and orientation, a wide variety of theories regarding its causation have arisen. Such explanations currently can be grouped as biological, psychoanalytic, and social learning theories, or some combination thereof.

An example of a biological theory is from a study done in the Netherlands. Thirteen of 15 brains of homosexual men who died of AIDS had Suprachiasmatic Nucleus (SCN), a part of the Hypothalamus, was twice the size of a comparable group of heterosexual men and women. Two of the 15 who not thought to be homosexual had normal SCNs. This small group of cells helps regulate the rhythms of waking and sleeping, hormone secretion and sexuality.

The traditional psychoanalytical view is that homosexuality is an acquired psychological condition reflecting a divergence from normal development. Most theories of this type emphasize the importance of early relationships with parental figures. Usually this figure is considered to be the mother in the case of male homosexuality.

Social learning theories suggest that, in some cases, an adolescent may be repelled by certain physiological aspects of the female personality. In others he may be afraid of not being able to play the role society expects of him in his relations with the woman his education assigns him as a partner. These apprehensions may lead to homosexuality if there are no moral prohibitions on this practice.

A theory favoured by militant homosexuals, is that homosexuality does not need any explanation, since it is an inherent potentiality in all of us and is repressed in most of us only by education and social pressure.

Claim 1. Active homosexual behaviour is deviant. Many boys and girls pass through a period of homosexuality. It is the task of the teacher to help them through that state. Anything else amounts to promoting arrested development. Anyone who teaches others to believe otherwise is an evil influence in society. 2. The family is the ultimate victim of homosexuality, a result which society can only tolerate within certain limits. Those who reinforce the disintegrative elements in society will be condemned by future generations. If psychiatrists collectively endorse one of the symptoms of social distress as a normal phenomenon, they demonstrate to the public their ignorance of social dynamics, of the relation of personal maladaptation to social disharmony, and thereby acquire a responsibility for aggravating the existing social chaos.

Counter-claim 1. In the Kinsey report of 1948 on sexual behaviour in the USA, 37 per cent of males questioned admitted to physical contact (to the point of orgasm) with another male at least once between adolescence and old age. It was estimated that some 10 per cent of American males had indulged in more or less exclusively homosexual activity for at least three years. Such figures are indicative that homosexuality constitutes normal rather than deviant behaviour. Perception of this normality is distorted by the widespread victimization of homosexuals. 2. Psychiatrists in the USA decided in 1973 to remove homosexuality from the classification of psychiatric disorders, replacing it by "sexual orientation disturbance", explicitly distinguished from homosexuality, which by itself does not necessarily constitute a psychiatric disorder.

Refs Clygout, Sanivar H *Homosexuality* (1985); Copely, Ursula E *Directory of Homosexual Organizations and Publications*; Das, Man Singh and Harry, Joseph *Homosexuality in International Perspective* (1980); Licata, Salvatore J and Petersen, Robert P *Historical Perspectives on Homosexuality* (1982); Socarides, Charles *Homosexuality* (1978); Whitman, Frederick L and Mathy, Robin M *Male Homosexuality in Four Societies* (1985).

Broader Sexual deviation (#PD2198).
Narrower Lesbianism (#PF2640) Male homosexuality (#PF1390)
Homosexuality in prisons (#PE1363) Clandestine homosexuality (#PG0721)
Establishment homosexuality (#PE9339).
Related Bisexuality (#PF3269) Discrimination against homosexuals (#PE1903).
Aggravates Prostitution (#PD0693) Sexual violence (#PD3276)
Family breakdown (#PC2102) Sexual exploitation (#PC3261)
Sexual abuse of children (#PE3265) Fear of emotional sensitivity (#PF9209)
Distrust of interpersonal relationships (#PF4274).
Aggravated by Promiscuity (#PC0745) Inadequate sex education (#PD0759)
Fear of sexual intercourse (#PF6910).
Reduced by Permissiveness (#PF1252).

♦ PF3243 Overemphasis on immediate superficial needs

Immediate needs mentality

Claim The technological community's emphasis upon the concept of growth as the embodiment of progress and its stress on efficiency for its own sake results in an impulsive reaction to the immediate and merely superficial needs of society. The understanding of progress as develop-

ment is thwarted; and any effective anticipation and meeting of long-range, deeply-rooted needs is ultimately lost. Technology has thus created a self-perpetuating system which creates a market for its products by responding primarily to its own inner needs.
Broader Rural poverty in developing countries (#PD4125)
Short range planning for long-term development (#PF5660)
Gap between the function of social techniques and the needs they address (#PF3608).

♦ PF3248 Religious or civil refusal of divorce

Nature Divorce may not be recognized under any circumstances (Roman Catholic doctrine) or, where the consent or the representation of both parties is necessary under the law, may not be granted if consent is not forthcoming from one party. The refusal of divorce may lead to a conjugal status outside the law (bigamous marriage, cohabitation, adultery), the children of which will be illegitimate; and both parents and children of such unions may forfeit legal rights.
Incidence Divorce is not recognized in Ireland, Italy, Portugal and Spain, where the Catholic religion is very strong. Judicial separation, which is allowed, does not enable the spouses to remarry. Individuals may refuse to consent to divorce for reasons of economic security and status of children (mainly in the case of women) or because of the cost (mainly in the case of men). Under Israeli law, divorce requires mutual consent of the husband and wife. A man separated from his wife can begin a second relationship and even commit bigamy legally but a woman is branded as an adulteress, can lose custody of her children and any new children born from that relationship are bastards who can only marry other bastards.
Broader Divorce (#PF2100) Marital instability (#PD2103)
Injustice of religious courts (#PE0397).
Narrower Non-validity of divorce (#PF3250).
Aggravates Bigamy (#PF3286) Adultery (#PF2314) Cohabitation (#PF3278).
Aggravated by Separation under marriage law (#PF3251)
Inadequate provision of alimony (#PE3247).

♦ PF3249 Conflicting national divorce laws

Nature Conflict of national laws on divorce may lead to a non-recognition of divorce in countries where the grounds on which it was obtained are not legally valid. This may render the marital status of the spouses after divorce outside the law (resulting in bigamy, adultery, cohabitation) and children from another union illegitimate.
Incidence Conflict may occur over requirements of residence, consent or duration of separation, or over the manner in which the proceedings were conducted. Elsewhere divorce may not be recognized at all. Divorce under Muslim law and African customary law does not require court proceedings in order to be valid, but this is essential elsewhere.
Broader Divorce (#PF2100) Conflict of laws (#PF0216).
Aggravates Bigamy (#PF3286) Adultery (#PF2314) Cohabitation (#PF3278).

♦ PF3250 Non-validity of divorce

Nature Divorce which does not legally comply with the requirements of the country where either party is resident or of which they are nationals, prevents the persons in question from legally remarrying and may cause official bigamy or alternatively cohabitation, which may jeopardize the legal rights of the couple and equally the rights of their children.
Incidence Divorce proceedings may not take place in England and Wales during the first three years of marriage; this also applies in certain Commonwealth countries. Residence requirements may invalidate divorces obtained by non-residents (for example, in Mexico, where it is relatively easy to obtain a divorce) or by those who have not spent the statutory time in that location. If divorce is sought on the grounds of desertion there may be a time requirement (such as 3 years desertion – English law; or 1 to 5 years – US law). Foreign divorces may only be recognized if they roughly comply with national standards.
Broader Divorce (#PF2100) Religious or civil refusal of divorce (#PF3248).
Related Non-validity of marriage (#PF3283).
Aggravates Cohabitation (#PF3278).

♦ PF3251 Separation under marriage law

Nature Agreement between spouses not to live together, or situations when one leaves the other without the other's agreement but for a legally justifiable cause, may be reinforced by judicial decree providing for the same economic and child custody arrangements as divorce, but under which the couple continue to enjoy certain conjugal benefits. Such legal separation implies that the individuals involved may marry no-one else, and may constitute a barrier to divorce at a later stage.
Broader Divorce (#PF2100) Marital instability (#PD2103)
Voluntary dissolution of the family (#PF4930).
Narrower Inadequate provision of alimony (#PE3247)
Conflict concerning legal custody of children (#PF3252).
Related Desertion in marriage law (#PF3254).
Aggravates Family breakdown (#PC2102) Single parent families (#PD2681)
Non-parental custody of children (#PF3253) Religious or civil refusal of divorce (#PF3248).
Aggravated by Emotional dependency in marriage (#PD3244).
Reduces Inadequate income in old age (#PC1966).

♦ PF3252 Conflict concerning legal custody of children

Nature Conflict may arise over legal custody of children after desertion, separation or divorce, or where the parents were not married. Parents may quarrel bitterly over who is to have custody of the children and may enter into complicated legal proceedings to obtain right of custody. Having obtained it, one parent may effectively deny the other right of access and even abduct the child to this end, causing severe emotional stress to the child and the other parent. Since the grounds for divorce or separation affect the allocation of custody, parents may indulge in defamation of each other's character. In many cases there is legal discrimination against men. In the event of desertion or illegitimacy, the custody of the children weighs against the single parent who may have sole responsibility.
Broader Family breakdown (#PC2102) Separation under marriage law (#PF3251)
Non-parental custody of children (#PF3253).
Narrower Inadequate provision of alimony (#PE3247)
Discrimination against men in parental rights (#PE4010).
Related Single parent families (#PD2681) State custody of deprived children (#PD0550).
Aggravated by Desertion in marriage law (#PF3254).
Reduced by Trafficking in children for adoption (#PF3302).

♦ PF3253 Non-parental custody of children

Nature The legal custody of children may be undertaken either by the state or by private individuals (adoption), which may cause stress and adjustment problems among children or hardship to parents.
Incidence Child care institutions are a worldwide phenomenon designed to cater for not only the children of broken homes, but also those who are orphaned or abandoned or who cannot be maintained as a result of poverty. Non-parental custody of children may also result during times of disaster, when aunts, uncles, cousins, or godparents may take custody of children, often without giving them full rights as members of the family by, for instance, using them as domestic help during their stay. After the 1976 Guatemalan earthquake, many children were adopted and used in this manner.
Broader Divorce (#PF2100) Marital instability (#PD2103).
Narrower State custody of deprived children (#PD0550)
Conflict concerning legal custody of children (#PF3252).
Related Childlessness (#PC3280) Inadequate laws of adoption (#PD0590).
Aggravates Single parent families (#PD2681) Trafficking in children for adoption (#PF3302).
Aggravated by Unmarried mothers (#PD0902) Desertion in marriage law (#PF3254)
Discrimination against men (#PC3258) Separation under marriage law (#PF3251).
Reduces Neglected young children (#PE4245).

♦ PF3254 Desertion in marriage law

Nature Desertion implies leaving one's spouse against the latter's will and without a legally justifiable cause, or the act of expelling a spouse from the marital home or conduct towards this end (constructive desertion). The laws of certain countries provide that a husband who has deserted his wife and children must support them. Desertion constitutes a ground for divorce.
Incidence Desertion, the 'poor man's divorce', does not show up in statistics, especially if the man or woman cannot be traced.
Broader Abandonment (#PA7685) Offences against public order (#PD7520).
Narrower Abandoned wives (#PD1030).
Related Separation under marriage law (#PF3251).
Aggravates Divorce (#PF2100) Missing persons (#PD1380) Unmarried parents (#PD3257)
Unmarried mothers (#PD0902) Marital instability (#PD2103)
Single parent families (#PD2681) Inadequate laws of adoption (#PD0590)
Non-parental custody of children (#PF3253) State custody of deprived children (#PD0550)
Discrimination against unmarried fathers (#PD3256)
Conflict concerning legal custody of children (#PF3252)
Narrow legal definition of the family in developing countries (#PD1501).
Aggravated by Family poverty (#PC0999)
Family poverty in industrialized countries (#PD1998).

♦ PF3259 Double standards of sexual morality

Nature Different codes of sexual morality apply to men and to women. The double standard is very clearly seen in questions of chastity before marriage and the legal and moral consequences of this. Basically a man is encouraged by social attitudes to 'sow wild oats' and thus enhance his 'manhood', while for women sexual experience outside the confines of marriage means 'shame'. She may forfeit the limited rights she has under the law or may be victimized by her family or the community. Accusations against a woman's chastity in traditional society may lead to murder and vendetta, and double standards of sexual morality encourage prostitution.
Incidence Double standards of sexual morality occur wherever there is discrimination against women. They appear more in industrialized and traditional society than in primitive society, which tends to be more cohesive. Religious intolerance has played a large part in reinforcing a double standard.
Broader Hypocrisy (#PF3377) Double standards in morality (#PF5225)
Haphazard forms of social ethics (#PF1249).
Narrower Nudism (#PF2660) Puritanism (#PF2577).
Related Value erosion (#PA1782) Dehumanization (#PA1757)
Inadequacy of international standards (#PF5072)
Discriminatory imposition of standards (#PD5229).
Aggravates Rape (#PD3266) Divorce (#PF2100) Chastity (#PG6123)
Feminism (#PF3025) Promiscuity (#PC0745) Prostitution (#PD0693)
Sexual immorality (#PF2687) Female prostitution (#PD3380) Sexual exploitation (#PC3261)
Social fragmentation (#PF1324) Sexual discrimination (#PC2022)
Denial of parental affiliation (#PD3255).
Aggravated by Cynicism (#PF3418) Scapegoats (#PF3332)
Social injustice (#PC0797) Compulsory indoctrination (#PD3097)
Sexual exploitation of men (#PD3263) Sexual exploitation of women (#PD3262)
Discrimination against women (#PC0308) Dependency of women in marriage (#PD3694)
Inability to define moral standards (#PF7178) Double standards of sexual morality (#PF3259)
Discrimination against women in religion (#PD0127).

♦ PF3260 Sexual unfulfilment

Inadequate satisfaction from sexual intercourse — Unskilled coitus — Unsatisfactory love-making
Nature Insufficient pleasure derived from sexual activities may lead to sexual deviation or violence, marriage and family breakdown, sexual exploitation, and prostitution. It may be covert or even unconscious and other causes of unhappiness may be blamed, or it may be a source of guilt feelings.
Of all sexual activities, coitus is of paramount psychophysical importance. Obstacles to satisfactory love-making may be physical, environmental, or psychological, with the latter caused by a range of factors.
Incidence Lack of sexual satisfaction becomes more obvious in a 'permissive society' where it may be discussed openly and action taken to alleviate it; although such action may sometimes may seem immoral to those of traditional views. Elsewhere lack of sexual satisfaction may not be admitted if it puts the person's social prestige in question, or it may take the form of finding new wives or concubines to add to a harem; homosexuality may be an accepted substitute, though it is not accepted in 'Western' society.
Broader Lack of satisfaction (#PG6124) Human sexual disorders (#PD8016).
Related Childlessness (#PC3280) Human sexual inadequacy (#PC1892).
Aggravates Rape (#PD3266) Magic (#PF3311) Sodomy (#PE3273)
Divorce (#PF2100) Adultery (#PF2314) Masochism (#PF3264)
Occultism (#PF3312) Lesbianism (#PF2640) Bestiality (#PE3274)
Pornography (#PD0132) Bisexuality (#PF3269) Marital stress (#PD0518)
Transsexualism (#PF3277) Group marriage (#PF3288) Sexual violence (#PD3276)
Human suffering (#PB5955) Family breakdown (#PC2102) Male homosexuality (#PF1390)
Marital instability (#PD2103) Sexual exploitation (#PC3261)
Sexual harassment of women (#PF3271) Sexual exploitation of children (#PD3267)
Deception between sexual partners (#PE4890)
Vulnerability of marriage as an institution (#PF1870).
Aggravated by Sex (#PF9109) Sex guilt (#PJ2396) Promiscuity (#PC0745)
Prostitution (#PD0693) Sexual frigidity (#PE6408) Sexual deviation (#PD2198)
Sexual immaturity (#PJ6143) Sexual repression (#PF2922) Masturbation guilt (#PF5609)
Female prostitution (#PD3380) Sexual impotence of men (#PF6415)
Inadequate sex education (#PD0759) Fear of sexual intercourse (#PF6910)
Undesired sexual obligations (#PF4948) Denial of the right to procreate (#PC6870)
Sexual unfulfilment of the disabled (#PE5197).

♦ PF3264 Masochism

Nature The inducement of sexual pleasure by inflicting pain on oneself, as in religious flagellation, may result in serious injury or death. Anyone cooperating with a masochist may lay himself open to charges of assault, manslaughter or murder.
Background Masochism is named after the Austrian novelist Leopold von Sacher-Masoch (1835–95) whose writings centred on the subject.
Broader Perversion (#PB0869) Human violence (#PA0429).
Narrower Cruelty (#PB2642) Sexual masochism (#PE3851).
Related Sadism (#PF3270) Prostitution (#PD0693).
Aggravates Slavery (#PC0146).
Aggravated by Sexual unfulfilment (#PF3260).

♦ **PF3269 Bisexuality**

Nature The ability to respond homosexually as well as heterosexually may cause psychological conflict since bisexuals rarely have a balanced preference for both male and female and their inclinations may be encouraged or deflated by circumstance. Bisexuals may suffer from a guilt complex and struggle to keep up a normal sexual relationship in marriage in spite of a strong homosexual preference. Bisexuality indicates an inability to commit oneself to a one-to-one relationship and may indicate other character defects.

Incidence Bisexuality is found most frequently among criminals and adolescents.

Refs Klein, Fritz and Wolf, Timothy (Eds) *Two Lives to Lead* (1985).

Broader Sexual deviation (#PD2198).
Related Lesbianism (#PF2640) Homosexuality (#PF3242).
Aggravates Marital instability (#PD2103).
Aggravated by Sexual unfulfilment (#PF3260).

♦ **PF3270 Sadism**

Nature The obtaining of sexual pleasure from acts of cruelty was the original meaning of sadism (taken from the Marquis de Sade). In current usage it encompasses: acts of cruelty but not appreciated in a sexual way, such as certain sports, arson and lynching; cruel acts accompanied by some sexual satisfaction but not erection or ejaculation; and outrages accompanied by full sexual satisfaction with erection and ejaculation, including necrophilia and rape. The sadist uses hate as the emotional currency where a normal man uses love. In the case of sadistic murder, which is often repeatedly committed by the same person in some periodic sequence, the habitual normal demeanour of the sadist between murders makes it difficult to identify and therefore to arrest him. Although sadism originally applied to men and still applies largely to them, women may also show sadistic tendencies.

Refs Eisler, Robert and Lathrop, Donald *Man into Wolf* (1978).

Broader Perversion (#PB0869) Sexual violence (#PD3276).
Narrower Sexual sadism (#PE6748) Sado-masochism (#PE6137) Sexual mutilation (#PD5718).
Related Torture (#PB3430) Masochism (#PF3264) Unkindness (#PA5643) Human torture (#PC3429) Unpleasantness (#PAp107) Human violence (#PA0429) State sanctioned torture (#PD0181).
Aggravates War crimes (#PC0747) Schizophrenia (#PD0438) Mental disorders (#PD9131).
Aggravated by Difficulty in identifying sadists (#PG6139).

♦ **PF3271 Sexual harassment of women**

Sexual provocation by men — Sexual harassment by men — Male sexual exhibitionism

Nature During infancy, the sexual instinct consists of several component impulses, one of which is exhibitionism. This impulse, manifested early in the form of genital exhibitionism, is later subject to modification. It may be progressively displaced from the genital zone to the whole body, to the oral zone (the pleasure of speaking), to clothes (a typically feminine form of exhibitionism), to dramatics, to the possession of material assets, etc; but more specifically, exposure of the genitalia may be used to give sexual pleasure. Usually the mere showing of the genitals is often not sufficient, and an emotional reaction of disgust or horror is sought from the person to whom the exposure is made. Ejaculation may occur at or after exposure, or the man may masturbate; this type exhibitionism usually occurring in males. Exhibitionism can be seen in the love dances of many animals, including apes, and in the sexual dances of primitive tribes. In developed society it has been suggested that it is more common among men who live in a female dominated situation, as exhibitionism is then a means of proving virility.

Broader Exhibitionism (#PD4643) Sexual harassment (#PD1116).
Narrower Obscenity (#PF2634) Human sexual inadequacy (#PC1892) Verbal sexual harassment of women in public (#PE0756).
Related Voyeurism (#PE3272).
Aggravates Disobedient wives (#PF4764).
Aggravated by Sexual unfulfilment (#PF3260).

♦ **PF3277 Transsexualism**

Change of sex — Change of gender — Trans-sexualism

Nature Transsexualism is the desire for a change of sex, usually commencing in puberty. Transsexuals may take on the role of the opposite sex, may practise transvestism, or may seek a physical change through surgery, though this cannot be complete and is usually very painful. TransNsexuals may try to persuade doctors to effect the change by way of repair for mutilation or castration. Hormones may be used by men to induce breasts. Male transsexuals may be homo- or heterosexual; female transsexuals are usually lesbians. In rare cases there is a physical basis for change, such as hermaphroditism.

Incidence While the full extent of transsexualism is not known, some indication is provided by the number of operations performed. The number of applications for operations does not provide an accurate indication because some transvestites may seek transsexual operations for the purposes of their livelihood, in a night-club or in the practice of prostitution for a male prostitutes who have become too old to attract paedophiles. It is estimated that from one transsexual in 30,000 to one in 100,000 persons. It is generally agreed that those born women are far fewer, one in three or four, than those born men.

Claim Transsexualism is a mental disorder, a psychosis involving the logical organization of delusions based on false premises. It is a disorder of gender identity and is not erotic in nature.

Counter-claim Transsexuals are not homosexual, lesbians nor transvestites. They reject the gender assigned to them, of which they have the complete morphology and even the genetic structure. They have a strong feeling and deeply rooted belief that they belong to the opposite sex. Psychologically, they find their situation incongruous and feel only alienation vis-a-vis their body and revulsion for their genital organs. They desire to live as a member of the opposite sex and they seek to change their sexual appearance through hormones and surgery and subsequently have their civil status amended.

Refs Green, Richard, et al (Eds) *Transsexualism and Sex Reassignment* ; Hodgkinson, Liz *Bodyshock* (1987); Koranyi, Erwin K *Transsexuality in the Male* (1980); Walters, William and Ross, William *Transsexualism and Reassignment* (1986).

Broader Sexual deviation (#PD2198) Human sexual disorders (#PD8016) Gender identity disorders (#PE6581).
Related Transvestism (#PE6348) Male homosexuality (#PF1390) Human pseudo hermaphroditism (#PE2246).
Aggravates Transvestite prostitution (#PD4525).
Aggravated by Sexual unfulfilment (#PF3260) Sexual discrimination (#PC2022) Violation of rights of transsexuals (#PE8548).

♦ **PF3278 Cohabitation**

Trial marriage — Living together — Marriage-like unions — Nonmarital unions — Non-marital cohabitation — Unmarried couple

Nature Cohabitation is the living together of a couple as in marriage, but without the legal or religions sanction of marriage and the resulting constraints. The children of cohabiting couples are illegitimate, and as such they may not be entitled to social benefits or inheritance of property. Cohabitation as a way of life is practised by certain racial groups; among West Indians, but not so much among Latin Americans, it is customary for couples to change partners, resulting in the children of such unions being deprived of a stable family environment and suffering from loss of identity. In an extended family situation, such children may be cared for by relatives; but where the practice is continued after immigration to an industrialized country, the children are often abandoned.

Cohabitation has become increasingly popular in sophisticated society, where it may be caused by a lack of religious conviction, a search for new forms of life-style and rejection of traditional values, unequal taxation, or the inability of one or both parties to obtain a divorce or dissolution of a previous marriage; a state of cohabitation may be imposed on an otherwise married couple because of the non-validity of their marriage or the non-validity of previous divorce. Cohabitation with a number of women may be the result of a reversion to the custom of polygamy which is prohibited under the law.

Incidence Legal marriage is increasingly being preceded or replaced by cohabitation or other forms of consensual union. Such union often starts at ages below those of entry into legal marriage. In North America, Europe and the USSR, at least in urban areas, the practice of cohabitation or sexual intimacy outside marriage has become an accepted norm in less than two decades. Cohabitation or consensual unions are also extremely frequent in other less industrialized societies, particularly in Latin America. Census data show celibacy rates at various ages in a number of Latin American countries which clearly reflect a high incidence of consensual unions. This is not a new form of behaviour among the poorer social groups, but it is spreading along the ladder of the social classes, from the bottom, and also to some extent from the top, to the middle class.

Claim The only reason for cohabitation is sexual satisfaction. This creates an emotionally and spiritually unhealthy climate for the partners and for the offspring. Couples who cohabit are more likely to separate than those who marry. About 40 percent end in the first year, either by marrying or because they split up. Couples who cohabit and then marry are more likely to divorce than those who marry without cohabiting.

Refs John, M and Katz, Stanford N *Marriage and Cohabitation in Contemporary Societies* (1980); Wiersma, G E *Cohabitation, an Alternative to Marriage ?* (1983).

Broader Adultery (#PF2314) Sexual immorality (#PF2687) Narrow legal definition of the family in developing countries (#PD1501).
Related Marriages of convenience (#PU3035).
Aggravates Family breakdown (#PC2102) Illegitimate children (#PC1874) Vulnerability of marriage as an institution (#PF1870).
Aggravated by Promiscuity (#PC0745) Permissiveness (#PF1252) Sexual craving (#PE7031) Marital instability (#PD2103) Non-validity of divorce (#PF3250) Non-validity of marriage (#PF3283) Distortionary tax systems (#PD3436) Pre-marital sexual intercourse (#PD5107) Conflicting national divorce laws (#PF3249) Over-reaction to past sexual ethics (#PF1622) Religious or civil refusal of divorce (#PF3248).
Reduces Enforced celibacy (#PD3371).

♦ **PF3281 Annulment of adoption**

Nature Where the law of adoption is not final, natural parents may reclaim their child, and if the foster parents also feel they have a strong claim to the child, a tug-of-war may develop with damaging effects on the child. In some situations the law demands that suitable foster parents should be found before the mother gives her consent to adoption. By the time that the adoption order can be made, she may have changed her mind, causing anxiety both to herself and to the prospective foster parents, again with adverse effects on the child. In certain countries there is a probationary period for adoption before it becomes final. If during this time the authorities feel that the foster parents are unsuitable, or if the foster parents are not happy with the child, the adoption may be annulled and the child returned to an institution.

Incidence In Sweden adoptions never become irrevocable. Many countries, including the USA, Germany, UK and Scandinavian countries, demand a supervisory period before adoption is confirmed. In England, foster parents must be found before the mother gives consent to the adoption.

Broader Impediments to adoption of children (#PF7353).
Narrower Dependency of children (#PD2476).
Related Trafficking in children for adoption (#PF3302).
Aggravates Childlessness (#PC3280) Refusal of adoption (#PF3282).
Aggravated by Family breakdown (#PC2102).

♦ **PF3282 Refusal of adoption**

Refusal of parents' consent to adoption

Nature Adoption of children may be refused on the grounds of the unsuitability of the applicant foster parents with regard to age, religion, or criminal record. Natural parents may refuse to let their children be adopted without the possibility of reclaiming them. A minimum age for foster parents may be stipulated by the law and a required difference between the ages of the foster parents and child laid down (as in England and in France). Single adults are usually considered unacceptable (but not in France).

Broader Dependency of children (#PD2476) Impediments to adoption of children (#PF7353).
Aggravates Childlessness (#PC3280) Vulnerability of women and children in emergencies (#PD1078).
Aggravated by Annulment of adoption (#PF3281) Trafficking in children for adoption (#PF3302).

♦ **PF3283 Non-validity of marriage**

Nature A marriage is not valid if it does not comply with the conditions laid down by the law. Foreign marriages may not be recognized if the parties were considered as not being competent to enter into marriage by the law of the country involved. Non-recognition of marriage may cause complications regarding the legitimacy of children, their nationality, property rights and liabilities; and in the event of divorce.

Incidence Non-valid marriages may include bigamous marriages, child marriage, polygamous marriage, and marriage forbidden on racial, ideological or other grounds. The Soviet Union forbids the marriage of Soviet citizens to foreigners. In many African countries, race inter-marriage is forbidden. Under Islamic law a woman may not marry a non-Muslim, and under Jewish law husband and wife must be Jew and Jewess, but either of them may become so after having been received into the Jewish community. Bigamous and polygamous marriages will be invalid where they are not accepted by the law. Child marriage is considered invalid in most countries, but may be accepted as a custom, as in India.

Narrower Bigamy (#PF3286) Clandestine marriage (#PG6157) Legal impediments to marriage (#PF3346).
Related Divorce (#PF2100) Adultery (#PF2314) Non-validity of divorce (#PF3250).
Aggravates Cohabitation (#PF3278) Illegitimate children (#PC1874).
Aggravated by Forced marriage (#PD1915).

♦ **PF3284 Arranged marriage**

Nature Arranged marriage implies that the consent of each party is conditional on outside influences, usually parental. In arranging marriages, parents usually take the status and financial position of the other family into account in order to make it a 'suitable match'. Arranged

marriages also include the selling of women and girls of poor families into marriage.
Incidence Arranged marriages occur most frequently in Asia and Africa, although they may occur in more subtle form in developed countries.
Counter-claim The fundamental reason for the requirement of consent of the family, and thus for arranged marriage, is the underlying belief that marriage is not simply a union between the couple but a union between their two families as well. It is due to this underlying belief that even the 'modern' civil code in some countries such as Ethiopia does not make the consent of the couple final but subordinates it to a mechanism known as 'opposition', by which representatives of the family can submit their opposition to the marriage.
Broader Forced marriage (#PD1915) Parental domination (#PF4391)
Dependence within extended families (#PD0850).
Narrower Child-marriage (#PF3285) Marriage by proxy (#PG6160).
Related Marriages of convenience (#PU3035).
Aggravates Traditionalism (#PF2676) Early marriage (#PE7628)
Marriage markets (#PD7282).
Aggravated by Caste system (#PC1968).

♦ PF3285 Child-marriage
Nature Child-marriage is the term used both for marriage of minor to minor, and for marriage of a minor to an older person by agreement between the parents and the future spouse, or between both sets of parents.
Incidence Child marriage continues as a means of forming unity between two families in Africa and India. Child marriage as a form of slavery, where girls are sold in marriage to older men, exists in Africa. Child marriage may be forbidden by law but accepted in social custom. Child marriage without consent of the girl was still legal in Gabon in 1964, but a Presidential decree was issued that year abolishing it.
Broader Early marriage (#PE7628) Arranged marriage (#PF3284)
Victimization of children (#PC5512).
Related Exploitation of children (#PD0635).
Aggravates Concubinage (#PF2554) Chattel slavery (#PC3300)
Trafficking in women (#PC3298) Social underdevelopment (#PC0242).
Aggravated by Tribalism (#PC1910) Bride-price (#PF3290)
Forced marriage (#PD1915) Parental domination (#PF4391)
Dependency of children (#PD2476).

♦ PF3286 Bigamy
Nature In countries where monogamous marriage is the law, marriage to more than one wife constitutes bigamy. If one of the marriages has been celebrated in a country where polygamy is accepted, the rights of the family in question regarding property inheritance and legitimacy may be recognized. If not, they will be denied.
Broader Illegal marriage (#PE7935) Sexual immorality (#PF2687)
Non-validity of marriage (#PF3283).
Related Adultery (#PF2314).
Aggravates Infamy (#PB8172).
Aggravated by Marital instability (#PD2103) Monogamous marriage law (#PG6164)
Inflexible social structure (#PB1997) Conflicting national divorce laws (#PF3249)
Religious or civil refusal of divorce (#PF3248).

♦ PF3287 Restricted local participation
Nature Political, cultural and economic boundaries inhibit local representatives from playing a significant role in the decision-making of the larger societies of which they are a part.
Broader Undemocratic policy-making (#PF8703)
Breakdown in community security systems (#PD1147).

♦ PF3288 Group marriage
Nature Group marriage combines polygamy and polyandry, though it may also be used as a loose term encompassing wife-exchange and a kind of sexual communism. The practice is generally incompatible with modernized society in its usual form. The breaking up of group marriage patterns by Western missionaries has been instrumental in causing the disintegration of traditional ethnic social structures. Where group sex is practised in developed society, it may cause adjustment problems or confusion of paternity. Wife-exchange may result in divorce or other marital problems.
Incidence Group marriage exists particularly among primitive tribes. It has also been found in societies where polyandry exists, such as India, Tibet and Sri Lanka. Certain African, Australian and New Guinea tribes practise a kind of sex communism, in which several men have the right of access to several women; but none of the women is properly married to more than one man. The existence of an early state of group marriage has been assumed from various customs, such as the lending or exchange of wives, sexual intercourse to which a girl is subject before her marriage, and the suspension of the ordinary rules of morality at certain ceremonies. Equivalents of group marriage in sophisticated society include wife-exchange, group sex, and particularly 'communes' or 'communities' where this is practised.
Broader Promiscuity (#PC0745) Illegal marriage (#PE7935) Sexual immorality (#PF2687).
Narrower Polygamy (#PD2184) Polyandry (#PF3289) Group sex (#PG6147)
Wife-swopping (#PG4711).
Aggravates Illegitimate children (#PC1874) Sexually transmitted diseases (#PD0061)
Denial of parental affiliation (#PD3255).
Aggravated by Sexual unfulfilment (#PF3260).
Reduces Lack of variety of social life forms (#PE8806).

♦ PF3289 Polyandry
Nature The practice of marriage allows a woman to have more than one husband.
Refs Deer, John *Polygamy and Polyandry* (1986); Premary, Y S *Polyandry in the Himalayas* (1975).
Broader Group marriage (#PF3288) Sexual immorality (#PF2687).
Related Polygamy (#PD2184).
Aggravates Promiscuity (#PC0745) Infanticide (#PD3501).
Aggravated by Tribalism (#PC1910).

♦ PF3290 Bride-price
Prohibitive marriage dowry
Nature Bride price implies selling women into marriage without their consent. The payment of bride-price gives a man total ownership of the woman and over all her children, whether he fathers them or not. If he wishes, he may re-sell both his wife and her children; when he dies she is inherited by his heir and must become his wife, unless the heir chooses to sell her to another man.
Background Bride-price is a custom traditionally practised by tribes in Africa, and as such no action was taken against it by colonial powers until the French decree of consent of both parties to marriage and the prohibition of the inheritance of widows in 1929. Similar legislation followed from Belgium and Great Britain, but native courts still upheld the practice.
Broader Economic impediments to marriage (#PF3342).
Related Sexual exploitation of women (#PD3262).
Aggravates Bride burning (#PE4718) Child-marriage (#PF3285)
Forced marriage (#PD1915) Chattel slavery (#PC3300)
Human dependence (#PA2159) Trafficking in women (#PC3298).
Aggravated by Tribalism (#PC1910).

♦ PF3294 Arrogance of experts
Professional arrogance
Broader Impediments to internationally mobile professionals and experts (#PF1068).
Aggravates Neglect of expert advice (#PF7820) Arrogance of policy-makers (#PF2895)
Arrogance of intergovernmental agencies (#PF9561).
Aggravated by Scientific arrogance (#PF7843) Unavailability of appropriate expertise (#PF7916).

♦ PF3302 Trafficking in children for adoption
Sham adoption of children — Adoption abuse
Nature Children are 'adopted' for a price by wealthy families or brothels. Families are sometimes separated for purposes of adoption. In particular, younger women and the populations of Third World countries are open to abuse because they lack the resources (emotional, financial, etc) to combat efforts against this disruptive procedure. Third World countries are at particular risk: young children and infants are removed from their homelands and cultures in order to meet the demands of more affluent societies for adoptable children Illegitimacy has traditionally been the basis for adoption. Until recently in the USA, illegitimate children were routinely placed for adoption. Because changes in attitude have resulted in more single parents keeping their children, there is caused a dearth of babies available for adoption. Agencies and infertile couples and others wishing to adopt children, prey upon the vulnerability of young pregnant women because the supply of babies has diminished.
Incidence Over 10,000 American couples legally adopt foreign children each year. In 1986, the number of foreign children adopted in France, 2,227 babies, exceeded the number of adopted French children. In Britain and West Germany the number of foreign adoptions are on the rise. More than 6,000 Korean babies are adopted by Western couples through a system administered by the Korean government While many social workers question the morality of legal adoptions schemes, all condemn illegal schemes. In such countries as Chile, Argentina and Paraguay, illicit baby traffickers are doing a brisk trade. In Chile a mother is paid $1,000 for her baby and sold for 15 times that amount. In Argentina, where 90 percent of the population is of European descent, blue-eyed, blond-haired babies are being sold for as much as $20,000. Latin American officials estimate that from 200 to 700 infants are illegally exported from each of these countries a year. In Sri Lanka, baby farms located near tourist resorts are infamous. The proprietors search the island's hospitals for impoverished young women in the early stages of pregnancy, offering them good care, food and a modest $100 a month if they will live on the farms until they give birth and then give up their newborns. Some of the babies are sold directly to visiting foreign visitors and others are exported, at an average price of $1,000, to syndicates in Belgium and Sweden, which then sell them for as much as $8,000 each. Sri Lankan authorities estimate some 300 infants are illegally sold to Western couples a year. Middlemen in the Philippines search out pregnant women in the night clubs and honky-tonk bars near U.S. military bases in order to buy mixed-blood babies and then resell them to Western couples.
Claim The claim that adoption provides better homes for children is a prejudicial and value-laden response. Rather than solve the problem of teenage pregnancy by providing for adequate sex education and contraception, adoption is proposed as a solution to provide childless couples with children. Adoption causes severe psychological damage to those children separated from their families. Poor families who cannot afford to keep their children and who need extra money sell them to wealthy families, who, because slavery is illegal, make a pretence of adopting them.
Counter-claim Without the alternative of adoption, many more thousands of children throughout the world would remain homeless, living either on the streets or in institutions.
Broader Cruelty to children (#PC0838) Trafficking in children (#PD8405).
Related Annulment of adoption (#PF3281).
Aggravates Caste system (#PC1968) Refusal of adoption (#PF3282)
Victimization of children (#PC5512).
Aggravated by Childlessness (#PC3280) Unmarried mothers (#PD0902)
Socio-economic poverty (#PB0388) Lack of family planning (#PF0148)
International adoptions (#PE4296) Inadequate laws of adoption (#PD0590)
Exploitation of child labour (#PD0164) Unethical personnel practices (#PD0862)
Non-parental custody of children (#PF3253) Impediments to adoption of children (#PF7353)
Inadequate national law enforcement (#PE4768).
Reduces State custody of deprived children (#PD0550)
Conflict concerning legal custody of children (#PF3252).

♦ PF3307 Lack of community participation
Absence of community participation — Unconfident citizen participation — Frustrating participation experience — Discouraged community participation — Factions and frustrations in community participation
Nature Lack of participation in activities relating to community life and welfare, leads to the alienation of certain sections of society; lack of social unity within the community, unequal treatment of sections of society, and general social and political alienation through lack of contact, lead to stress and mental disorders. Participation may be both the cause and the result of a well integrated community, and where it does not exist, intergroup (especially interracial) conflict is likely to occur, involving violence, crime, discrimination and prejudice. Where participation in local politics is lacking, corruption and bureaucratic abuse are more likely to occur, because actions taken at the local level are not contested. Participation implies responsibility and where it is not taken, the community is open to anarchy or elitist control of a dictatorial nature. Certain social systems are more conducive to a centrally planned system; lack of communication from authorities to the members of the community may lead to the failure in implementation of social policy and development policy. Urbanization may lead to an increase in mental disorders, in loneliness, suicide, and general deprivation in social life, through lack of contact.
Claim At a time when the need for a cooperative effort to achieve economic and social development is coming to be understood, many small communities are deeply split. Decision-making typified by protection of private interest and failure to actually implement proposals lead to resentment of individual initiatives and a cynical attitude toward community participation. Because of these attitudes, few organizations or events enjoy the support of the entire community. Extensive property damage, destruction of equipment, abuse of natural resources and vandalism indicate a profound frustration with available forms of participation.
Broader Non-participation (#PC0588) Obstacles to community achievement (#PF7118)
Failure of individuals to participate in social processes (#PF0749).
Narrower Jealousy (#PF5013) Absentee ownership (#PD2338)
Insecure lease tenure (#PJ8344) Lack of social contact (#PF8695)
Culture-denying competition (#PG8317) Abuse of individual property (#PJ8392)
Lack of actuation structures (#PG8416) Factionalized decision-making (#PG8360)
Habitual non-payment of bills (#PG8412) Financial irresponsibility image (#PG8414)
Widespread community factionalism (#PG8335) Deliberate destruction of equipment (#PG8408)
Exclusion of women from decision making (#PE9009)
Limited horizons produced by survival living (#PF3161)
Lack of participation in local welfare programmes (#PE8503)
Ineffective operation of community networks in urban ghettos (#PF1959)
Lack of participation in attempting to solve intergroup conflicts (#PE8357).

Related Apathy (#PA2360) Lack of participation in development (#PF3339)
Bureaucracy as an organizational disease (#PD0460).
Aggravates Loneliness (#PF2386) Political apathy (#PC1917)
Social alienation (#PC2130) Political alienation (#PC3227)
Inequality of opportunity (#PC3435) Undemocratic policy–making (#PF8703)
Uncontrolled local vandalism (#PJ0154) Lack of community development (#PF7912)
Lack of local leadership and decision–making structures (#PF6556)
Lack of awareness of potential for investment in small, inner–city enterprises (#PF2042).
Aggravated by Elitism (#PA1387) Resentment (#PF8374)
Discrimination (#PA0833) Property damage (#PD5859)
Social fragmentation (#PF1324) Destruction of land resources (#PJ8397)
Ineffective structures of local consensus (#PF6506)
Ineffective structures for community participation (#PF2437)
Lack of worker participation in business decision–making (#PF0574).
Reduced by Demonstrations (#PD8522) Passive resistance (#PF2788).

♦ PF3310 Taboo

Nature Taboo is the ritual avoidance of things, persons, places, times, actions or words believed to be inauspicious or, on the contrary, sacred. Taboo expresses itself essentially in prohibitions and restrictions. It was originally the negative aspect of magic, being something forbidden or profane; its modern equivalent is prejudice, which in popular usage can also include hypocrisy since subjects which are a source of embarrassment are often denied and 'taboo'. Taboos in primitive society constitute barriers to technological progress; in developed society, although they may also constitute barriers to technological progress, they are more significantly barriers to social progress.
Incidence Taboos are widespread among indigenous populations. The Ark of the Covenant and the Holy of Holies were taboo to the ancient Israelites. The caste system of India is an extraordinary complex of taboos. Incest is taboo for all societies. In Japan the number four *'shi'* is avoided because it is a homonym of the word death *'shi'*.
Being founded in religion, they are strictly observed by the societies to which they are relevant; a lack in observance of taboos by development planners may bring about project failure. In so far as advanced technology is very strange to indigenous tribes, it may itself be subject to taboos. Taboo in the sense of social prejudice is widespread in developed societies, particularly where they are strongly pluralistic; taboo in the sense of hypocrisy is also widespread, but both this and prejudice is alleviated to some extent by an increase in education and cultural integration or assimilation.
Counter–claim Every society has taboos and they are a rational and systematic derivatives of the basic postulates of the society.
Refs Browne, Ray B *Forbidden Fruits* (1984); Douglas, Mary *Purity and Danger* (1984); Hardin, Garrett J *Naked Emperors* (1982).
Broader Prejudice (#PA2173) Superstition (#PA0430).
Narrower Ritual pollution (#PF3960) Violating taboos (#PF3976)
Mother–in–law taboo (#PF4166) Dietary restrictions (#PJ1933)
Animal worship as a barrier to development (#PD2330).
Related Discriminatory unwritten codes of behaviour (#PE7017).
Aggravates Incest (#PF2148) Hypocrisy (#PF3377) Segregation (#PC0031)
Discrimination (#PA0833) Restrictions on freedom (#PC5075)
Inadequate sex education (#PD0759) Ritual slaughter of animals (#PF0319)
Lack of technological progress (#PG6216) Failure of development programmes (#PF8368)
Superstitious persecution of animals (#PD3453)
Use of agricultural resources for production of animal feed (#PD1283).
Aggravated by Underdevelopment (#PB0206) Lack of education (#PB8645)
Lack of assimilation (#PF2132) Lack of communication (#PF0816)
Cultural fragmentation (#PF0536).
Reduced by Forced assimilation (#PC3293).

♦ PF3311 Magic

Spells — Curses
Nature Magic is the practice of occult arts and rites to influence the natural or supernatural world. It is usually taken to be black magic, or harmful and is generally held as unacceptable in developed society. In primitive society it is still an integral part of daily living and may conflict with attempts to integrate tribes into developing society. While most magic practised by tribes is beneficial it is usually incompatible with modernized living patterns.
Incidence In developed society covens and small groups which practice black magic exist and appear to attract a fringe of dabblers and dilettantes who are often rich and well–educated. In primitive societies the practice of magic is widespread, although diminishing with the influence of 'advanced' culture.
Counter–claim Science is simply a conceptually narrower form of magic. Both are attempts to control events, both have a strong faith in the order and uniformity of nature and both believe in invariable natural laws.
Broader Occultism (#PF3312) Witchcraft (#PF2099) Superstition (#PA0430).
Narrower Satanism (#PF8260) Black mass (#PG6222) Necromancy (#PF8042)
Black magic (#PF8249) Human cannibalism (#PF2513).
Related Unexplained phenomena (#PF8352) Ritual slaughter of animals (#PF0319)
Lack of understanding of spiritual healing (#PF0761)
Animal worship as a barrier to development (#PD2330).
Aggravates Bewitchment (#PF3956) Imprecation (#PF3746)
Mental disorders (#PD9131) Physical disorders (#PG6226)
Inauspicious conditions (#PF6683).
Aggravated by Boredom (#PA7365) Tribalism (#PC1910)
Abuse of khat (#PE0912) Underdevelopment (#PB0206)
Sexual unfulfilment (#PF3260).
Reduced by Forced assimilation (#PC3293).

♦ PF3312 Occultism

Nature Occultism is the study of supposed supernatural forces and the belief and attempt in the possibility of controlling them. Occultism is often associated with the practice of black magic and other rites for satanic or evil purposes. The object of the occultist may be esoteric knowledge, health, domination over others, wealth, psychic or physical powers, satisfaction of scientific curiosity, the performance of acts of religious ritual and worship, or to be a benefactor of humanity. Occultism thus appeals to, among others, megalomaniacs, would–be Messiahs and those with schizophrenic tendencies. The sense of possessing secret knowledge which develops in the occultist, aggravates existing personality disorders; and some of the techniques (magical, yogic) may be injurious to the physical health of the practitioners, their helpers, and their victims. Occultism may also be criminal.
Incidence Occultism occurs in developed countries, more especially among the rich, the effete, and the deviant. In primitive societies where the occult is part of the social structure, only certain people study and practice it.
Counter–claim The study of supranormal faculties of man which lie in the main beyond the range of readily and universally testable knowledge is a valid scientific endeavour in which thousands of universities, research organizations, and study groups have participated for more than 50 years. The study and control of the hidden forces of nature has been central to the entire history of science and has resulted in the discoveries of Pasteur, the Curies, and so many others who have unlocked cells, genes, molecules, atoms and stars to reveal their mysteries. The hostility against occultism is a manifestation of ignorance of its methods.
Broader Superstition (#PA0430) Primitive secret societies (#PF2928).
Narrower Magic (#PF3311) Idolatry (#PF3374) Witchcraft (#PF2099)
Poltergeists (#PF5697) Human sacrifice (#PF2641) Human cannibalism (#PF2513)
Religious sacrifice (#PD3373) Exorcism as a superstition (#PF5673)
Demonic and spirit possession (#PF5781) Head hunting in tribal societies (#PF2666)
Animal worship as a barrier to development (#PD2330).
Related Freemasonry (#PF0695) Childhood martyrdom (#PF8118)
Ritual slaughter of animals (#PF0319)
Lack of understanding of spiritual healing (#PF0761).
Aggravates Fear (#PA6030) Medical quackery (#PD1725) Mental disorders (#PD9131).
Aggravated by Tribalism (#PC1910) Perversion (#PB0869)
Sexual unfulfilment (#PF3260).
Reduces Lack of variety of social life forms (#PE8806).

♦ PF3313 Burdensome cost of religious ceremonies

Imbalanced celebrational expenditures
Nature Festivals, holidays, weddings, funerals involve a high cost to poor families or communities, they also cause a reduction in working days and loss of productivity.
Incidence The disparity between the cost of religious ceremonies and the standard of living of those who pay is most marked in developing countries, such as India. The problem also exists among the working classes and poor of industrialized countries. Religious influence of this kind causes conflict with development programmes.
Broader Traditionalism (#PF2676) Reduction in symbolic celebrations (#PF1560).
Narrower Religious opposition to public health practices (#PF3838).
Aggravates Socio–economic poverty (#PB0388) Waste of human resources (#PC8914)
Failure of development programmes (#PF8368) Religious and political antagonism (#PC0030).
Aggravated by Caste system (#PC1968) Superstition (#PA0430)
Underdevelopment (#PB0206) Dependence on religion (#PF0150)
Compulsory indoctrination (#PD3097) Double standards in morality (#PF5225).

♦ PF3317 Diverse unilingualism

Nature Isolation of many different groups within one country, each knowing only their own language, occurs particularly in countries where there are many different indigenous populations which are cut off from each other and from outside influences so that their language is unknown to others. Since most of these languages are unwritten it is difficult for the society at large to have any knowledge of them unless there is extensive contact with the tribe in question. Diverse unilingualism is a severe obstacle to the education of indigenous populations and their integration into national society.
Broader Language barriers (#PF6035) Plural society tensions (#PF2448).
Narrower Unwritten language (#PF3470).
Aggravates Illiteracy (#PC0210) Ethnic conflict (#PC3685)
Lack of communication (#PF0816)
Inadequate education of indigenous peoples (#PC3322)
Discrimination against indigenous populations (#PC0352).
Aggravated by Isolation of ethnic groups (#PC3316).
Reduced by Lack of assimilation (#PF2132).

♦ PF3318 Restrictive effects of external capital on development

Nature A consistent system of money flow from sources outside a community creates the attitude that development can only begin to take place at the administrative level. When government funded improvements stop short of completion, a community may simply stand still, the people generally assuming that they can do nothing. Convinced that delay is inevitable, due to constantly shifting priorities which are largely beyond their control, they prefer to wait for the government to continue improvements rather than seeing the viability of alternative routes and pursuing them.
Broader Restrictive monetary practices (#PF8749).
Narrower Prohibitive cost of land (#PE4162) Government loan defaults (#PD9437)
Insufficient heavy equipment (#PJ0259) Lack of government initiative (#PG8505)
Insufficient programme funding (#PG8467) Prohibitive cost of electricity (#PG8518)
Sustenance determined priorities (#PG8454) Government administrative backlog (#PG8488)
Lack of funding for infrastructure (#PJ8512) Fear of retaliation by authorities (#PF3707)
Unavailability of scholarship funds for developing country students (#PE8883).
Aggravates Lack of care for animals (#PD8837) Prohibitive cost of education (#PF4375)
Lack of economic and technical development (#PE8190).

♦ PF3332 Scapegoats

Nature Scapegoats have been frequently used to divert anger or frustration for purposes of maintaining political domination over a variety of subordinate groups.
Incidence Scapegoats are frequently individuals, but may also be ethnic, racial, religious, linguistic, class or political groupings which are blamed collectively by another group or groups for economic hardship, political oppression, discrimination and other major ills. Elitist groups tend to play off discontented subordinate groups against one another. This was particularly practised under colonial rule, and against Jews in Czarist Russia and Nazi Germany.
Refs Florman, Samuel C *Blaming Technology* (1982).
Broader Social conflict (#PC0137) Social injustice (#PC0797)
Conflict between minority groups (#PC3428).
Narrower Pogroms (#PJ2093) Lynching (#PE6287) Racial conflict (#PC3684).
Aggravates Racism (#PB1047) Genocide (#PC1056)
Double standards of sexual morality (#PF3259).
Aggravated by Elitism (#PA1387) Segregation (#PC0031)
Antisemitism (#PE2131) Blame avoidance (#PF6382)
Ethnic conflict (#PC3685) Religious conflict (#PC3292)
Social fragmentation (#PF1324) Social discrimination (#PC1864)
Investigatory malpractice (#PD7885)
Promotion of negative images of opponents (#PF4133).

♦ PF3339 Lack of participation in development

Nature Development programmes may be made ineffective due to lack of understanding and cooperation on the part of the 'common man'. Such lack of participation may stem from community religious and social prejudice or from inadequate information and communication on the part of the development planners. Elitism and bureaucratic corruption may stultify natural tendencies towards participation, and unsuitable development programmes may also contribute to an unfavourable response.
Incidence The problem of general participation in development programmes has been recognized from the outset by planners and has often been cited as the reason for the failure of certain projects.
Broader Underdevelopment (#PB0206)
Inappropriate design of development projects (#PF4944)
Failure of individuals to participate in social processes (#PF0749).
Narrower Lack of coordination (#PF8330)
Difficulty of implementing development programs (#PG6302)
Neglect of the role of women in rural development (#PF4959)
Unsuitability of development projects for available labour (#PE8687).

PF3339

Related Corruption (#PA1986)   Lack of community participation (#PF3307)
Bureaucracy as an organizational disease (#PD0460).
Aggravates Political instability (#PC2677)   Inappropriate development policy (#PF8757)
Failure of development programmes (#PF8368)   Undemocratic political organization (#PC1015)
Overuse of machinery and equipment in development projects (#PE8166).
Aggravated by Political apathy (#PC1917)   Lack of education (#PB8645)
Inappropriate aid to developing countries (#PF8120)
Lack of worker participation in business decision-making (#PF0574).
Reduces Elitism (#PA1387).
Reduced by Dictatorship of the majority (#PD3239).

♦ **PF3340 Inter-cultural misunderstanding**
Cultural gap — Unfavourable opinions of other cultures
Nature There is a degree of misunderstanding between different cultures. This leads to conflict, barriers between cultures, culture shock, and possible domination, exploitation or extinction of certain groups. The cultural gap may be formed by linguistic, religious or racial differences, and by different social customs and economic, political and administrative systems. Misunderstanding is fomented by ideological conflict and indoctrination.
There may be a cultural gap between nations as much as between individuals and groups. With individuals, the cultural gap may be a matter of education, family background, inherent intelligence or a matter of ethnic origin. The gap between different ethnic groups is aggravated by discrimination, prejudice, segregation, and economic and social disparities. Leaders or representatives of governments may personally surmount cultural gaps in international organizations or meetings and may be able to freely communicate as well as to understand, in processes of dialogue, deliberation, negotiation and agreements. However, their constituencies at home, the great masses of people to whom they are responsible, having little foreign contact, are subject to insularity and narrower views. This creates a great tension for leaders caught by the pressures of the world to assure that their nations play an active role, and pulled on the other side by conservative national feelings that wish little cultural change and minimal foreign influence.
Broader Cultural fragmentation (#PF0536).   Inadequate appreciation of culture (#PF3408).
Narrower Heterosexism (#PE0818)   Lack of international understanding (#PF5106)
Untransferability of books between countries and cultures (#PF2126)
Ineffectiveness and inefficiency of intercultural meetings (#PF0316).
Related Cultural invasion (#PC2548)   Lack of racial identity (#PF0684)
Destruction of cultural heritage (#PC2114).
Aggravates Culture shock (#PC2673)   Social conflict (#PC0137)
Lack of international cooperation (#PF0817).
Aggravated by Segregation (#PC0031)   Language barriers (#PF6035)
Cultural arrogance (#PF5178)   Lack of assimilation (#PF2132).
Reduced by Forced assimilation (#PC3293).

♦ **PF3341 Social impediments to marriage**
Nature Social impediments to marriage include conventional taboos against certain kinds of marriage which are regarded as being contrary to the morality of society. They therefore include legal impediments (including incest, child marriage); ideological impediments (religious, political, nontheistic, and ideological); psychological impediments (incompatibility, immaturity) and class, racial, age, sexual prejudices and discrimination. Social impediments to marriage make for social conflict and inequality, lack of integration in society, and lack of social mobility.
Broader Social inequality (#PB0514)   Impediments to marriage (#PF3343)
Economic impediments to marriage (#PF3342).
Narrower Hypogamy (#PG5234)   Hypergamy (#PF5430).
Related Tax impediments to marriage (#PJ6283).
Aggravates Social fragmentation (#PF1324).
Aggravated by Class consciousness (#PC3458)   Social discrimination (#PC1864)
Racial discrimination (#PC0006).
Reduces Miscegenation (#PC1523).
Reduced by Permissiveness (#PF1252).

♦ **PF3342 Economic impediments to marriage**
Nature Economic impediments in marriage include poverty, the high cost of divorce and possible nullity, and general class consciousness or caste system based on economic possessions. The latter impediment is a traditional reason for refusing marriage. Economic hardship is aggravated by the fact that if a couple cohabits instead of marrying they may lose social benefits and legal protection.
Incidence Cohabitation and illegitimate children are widespread in Latin America because of poverty and lack of education. The high cost of divorce is a most notable impediment in 'Western' industrialized countries. Class consciousness regarding marriage prevails everywhere but is more noticeable in traditional or tribal communities (mainly in Asia and Africa) where marriages are arranged and where a bride price or dowry must be paid. Polygamy is restricted in Muslim and African tribal society by economic considerations.
Broader Social inequality (#PB0514)   Impediments to marriage (#PF3343).
Narrower Bride-price (#PF3290)   Tax impediments to marriage (#PJ6283)
Legal impediments to marriage (#PF3346)   Social impediments to marriage (#PF3341)
Ideological impediments to marriage (#PF3345)
Psychological impediments to marriage (#PF3344).
Aggravates Unequal distribution of social services (#PC3437).

♦ **PF3343 Impediments to marriage**
Denial of right to marriage
Nature Impediments to marriage include legal, social, psychological, ideological and economic impediments. They may lead to frustration, exploitation, prostitution, promiscuity, cohabitation, illegitimacy and also lack of social mobility and lack of integration in society. Impediments to marriage cause inequality before the law and an unequal distribution of social services.
Broader Denial of human rights (#PB3121).
Narrower Mixed marriage (#PD0355)   Legal impediments to marriage (#PF3346)
Social impediments to marriage (#PF3341)   Economic impediments to marriage (#PF3342)
Ideological impediments to marriage (#PF3345)
Psychological impediments to marriage (#PF3344).
Related Denial of right to a family (#PE7267).
Aggravates Enforced celibacy (#PD3371)   Social fragmentation (#PF1324).
Aggravated by Miscegenation (#PC1523)   Discrimination (#PA0833)
Social injustice (#PC0797)   Religious intolerance (#PC1808).
Reduced by Permissiveness (#PF1252).

♦ **PF3344 Psychological impediments to marriage**
Nature Psychological impediments to marriage cover a wide range of attitudes including prejudices, and also states of mind such as insecurity or irresponsibility, immaturity or mental disorder. The last mentioned can be a legal ground for refusing marriage, as can certain ideological or prejudicial impediments.
Broader Personal life crises (#PD4840)   Impediments to marriage (#PF3343)
Economic impediments to marriage (#PF3342).
Narrower Incompatibility (#PF9047)   Emotional insecurity (#PD8262).
Related Immaturity (#PF8413)   Tax impediments to marriage (#PJ6283).
Aggravates Psychological inconsistency of marriage partners (#PF9818).
Aggravated by Parental domination (#PF4391).

♦ **PF3345 Ideological impediments to marriage**
Nature Ideological impediments to marriage may be legal (USSR citizens may not marry foreigners) or institutionalized within religious doctrine (Muslim women may only marry Muslim men, Jewish marriage traditionally must be between Jew and Jewess although either may have been converted), or simply the result of prejudice and practice. Ideological impediments to marriage may also be constituted by nationalism, tribalism or other group philosophies if these take an ideological rather than a traditional form. Such impediments are caused by and perpetuate indoctrination, cultural and social division and conflict.
Broader Impediments to marriage (#PF3343)   Economic impediments to marriage (#PF3342).
Related Tax impediments to marriage (#PJ6283).
Aggravates Racism (#PB1047)   Cultural fragmentation (#PF0536).

♦ **PF3346 Legal impediments to marriage**
Nature Legal impediments may include: the prohibition of incestuous marriage; the marriage of persons afflicted with imbecility, epilepsy, venereal disease or tuberculosis; child marriage; or mixed marriage. In countries where polygamy is against the law, the existence of a legally viable marriage of one of the parties prevents him or her from remarrying.
Incidence The prohibition of incestuous marriage is widespread, as is that of child marriage. The legal prohibition of marriage of persons afflicted with imbecility, epilepsy, venereal disease or tuberculosis occurs in the USA, but is rarely enforced since there must be written proof of these diseases. In the UK, mental disorder or impotence are legally accepted reasons for refusing marriage, but if they are discovered after marriage they will not invalidate it.
Broader Impediments to marriage (#PF3343)   Non-validity of marriage (#PF3283)
Economic impediments to marriage (#PF3342).
Related Tax impediments to marriage (#PJ6283)   Sexually transmitted diseases (#PD0061)
Legalized racial discrimination (#PC3683).
Aggravates Social fragmentation (#PF1324).
Aggravated by Imbecility (#PE6314)   Inadequate health control (#PF9401)
Discrimination against illegitimate children (#PD0943).
Reduces Miscegenation (#PC1523).

♦ **PF3348 Misuse of evolutionary theories**
Nature The theory of natural selection and the survival of the fittest can be used to justify existing inequalities and injustices and may give weight to racism, elitism, domination and exploitation. Darwin's theory only concerned biological evolution; social Darwinism developed afterwards and was short-lived – but the heritage of these two aspects of the theory is paralleled in modern genetic arguments explaining inequalities and the naturally superior nature of some races (such as greater development and achievement in one society or social group as opposed to another).
Background The theory of evolution was developed by Darwin and Wallace in the mid-19th century. It gave rise to social Darwinism which developed in the late 19th century with the writings of Herbert Spencer, Walter Bagehot, and William Graham Sumner. In the second half of the 20th century scientists have discovered a number of problems in the theory, one of the more notable being that the theorized evolutionary time scale requires enormous spans of centuries for change, whereas fossil evidence shows change in inexplicably (relatively) short periods. Thus creationists, those that uphold the idea that God (that is a conscious Being with will and mind) created this world, and modern creationists, who assert that this Being (and helpers) managed the Creation, have become vocal in the Abrahamic religions of Judaism, Christianity and Islam to point out that the evolutionary theory in all its aspects is unproven.
Refs Rusch, Wilbert H *The Argument* (1984); Strahler, Arthur N *Science and Earth History* (1988); Walker, K R (Ed) *The Evolution-Creation Controversy Perspectives on Religion, Philosophy, Science and Education*; Zetterberg, J Peter *Evolution Versus Creationism* (1983).
Broader Materialism (#PF2655)   Inhumane scientific activity (#PC1449).
Narrower Social evolutionism (#PF2196).
Aggravates Elitism (#PA1387).
Aggravated by Biased and inaccurate biology textbooks (#PF9358).

♦ **PF3349 Amoralism**
Nature Amoralism does not admit the validity of moral judgement or sensibility, and is indifferent to right or wrong. Amoralism may perpetuate social injustices of all kinds since no distinction is made between just and unjust and any alleviation of injustice will be coincidental, not intentional. Amoralism is often equated with crime, violence and anarchy.
Broader Retarded socialization (#PF2187).
Related Moralism (#PF3379)   Immorality (#PA3369)   Economic apathy (#PC3413)
Ideological corruption (#PC2914).
Aggravates Massacres (#PD2483)   War crimes (#PC0747)
World anarchy (#PF2071)   Lack of social conscience (#PF9144).
Aggravated by Materialism (#PF2655).

♦ **PF3350 Ill-considered pressure to eliminate nakedness in developing countries**
Nature Nakedness, as opposed to nudism, tends to occur in the natural tribal state. Because it is equated with backwardness, savagery and immorality, efforts have been made, particularly by missionaries (and in recent years by governments) to ensure the use of clothing, often according to the Western mode. Clothing of indigenous tribespeople has resulted in the development of diseases fatal to them, either because they were carried from the country of manufacture within the clothing, or because wearing the clothing in tropical climates provided unhygienic conditions under which normally insignificant diseases may become of major importance. Clothing is expensive for underprivileged peoples and is often completely unnecessary in tropical climates. Moralism and the obligation to wear clothing to be 'decent' imposes economic dependency.
Incidence The insistence that the men in certain indigenous tribes of West Irian should wear trousers has caused both resentment and suffering.
Broader Tribalism (#PC1910)   Underdevelopment (#PB0206).
Aggravates Economic dependence (#PF0841).
Aggravated by Arsenic as a pollutant (#PE1732)
Nakedness in developing countries (#PF8241).

♦ **PF3351 Religion as a reinforcement of nationalism**
Nature Religious doctrine may be used as a foundation for nationalism. Religion, in providing a motivating force for nationalism, may encourage civil war or international conflict. It may provide the thrust needed to achieve political independence from colonial rule, but it does so very often through the creation of an educated elite which may indulge in megalomania and intolerance against tribal groups or other denominations. Such a nationalism may arise from the disintegration of traditional social patterns. Religious doctrine may be tailored and corrupted for political use.
Incidence Religion as a basis for nationalism is a notable feature of the Christian religion in Africa and especially Protestantism with its strongly individualistic traits. The influence of religion has come mainly through the missionary schools and the creation of a prestigious elite, some of whom are educated abroad with the help of the Church. Participation on Church councils and committees and involvement in Church social work provides the organizational basis for political activity. African politicians have found biblical language useful to harness the prestige of Christianity and education to their drives for independence.
Background Protestant religion, in the form of Lutheranism, was initially very closely connected

with nationalism during the Reformation and has remained so to a large extent, its denominations being mainly national in their characteristics. Certain Protestant denominations are official state religions (such as the Church of England, since the reign of Henry VIII when he made himself supreme head of the Church).
Broader Nationalism (#PB0534) Religious and political antagonism (#PC0030)
Undue religious influence on secular life (#PF3358).
Related Anti-clericalism (#PF3360)
Race as a reinforcement of nationalism (#PF3352).
Aggravates Political unrest (#PD8168) Religious conflict (#PC3292)
Culturally biased religion (#PF1466).
Aggravated by Ideological apathy (#PF3392).

♦ PF3352 Race as a reinforcement of nationalism

Nature Race may give rise to national separatism in a multi-racial state, or to nationalist domination of one race over others in the same country. It may take the form of racism to the detriment or destruction of the other races. Nationalism may be founded on a racial ideal (such as Aryanism). The intensely racial basis for nationalism is usually the product of a multi-racial society or of contact between different races and racial rivalry under colonialism.
Incidence Race appears as a motivating force for independence in colonial countries and as the cause of civil war or separatism after independence (Biafra, Burundi), where artificial national boundaries have been made under colonialism (Nigeria) or where a racial elite has been formed under colonialism (Burundi).
Broader Racism (#PB1047) Nationalism (#PB0534).
Related Religion as a reinforcement of nationalism (#PF3351)
Disruption of development by tribal warfare (#PD2191).
Aggravates Elitism (#PA1387) Racial war (#PD8718)
Excessively costly prestige projects (#PF3455).

♦ PF3353 Persecution of religious sects

Banned religious sects
Nature Religious sects may be banned for political or religious reasons. They may be thought to be subversive politically or to cause dissension from national policy. If there is a state religion or an accepted moral code the doctrines and practices of the sect may be thought to be contrary to morality.
Incidence Banned sects may include occult sects, experimental religion, and minority sects, which may be fundamentalist, exclusive or otherwise separatist. As examples, the Jehovah's Witnesses were banned from Malawi in 1967 and The Church of Scientology has been banned from several countries.
Broader Banned associations (#PD3536) Religious repression (#PC0578)
Religious persecution (#PC5994).
Related Iconoclasm (#PF4923) Banned trade unions (#PD3535)
Banned political parties (#PJ2274) Denial of right to manifest religion (#PF2850)
Repression of intellectual dissidents (#PD0434).
Aggravates Religious extremism (#PF4954) Religious discrimination in politics (#PC3220)
Peoples perceiving themselves as specially chosen (#PF4548).
Aggravated by Manipulative cults (#PE6336) Religious intolerance (#PC1808)
Discrimination against minorities (#PC0582)
Religious discrimination in the administration of justice (#PE0168).
Reduces Suttee (#PF4819).

♦ PF3354 Supralapsarianism

Predestination for heaven — Discrimination in access to eternal life — Elitist access to eternal life
Nature Religious doctrines holding that God restricts access to eternal life to a restricted portion of mankind, whilst the remaining portion is condemned to destruction.
Broader Religious discrimination (#PC1455).
Related Eternal punishment (#PF7228).
Aggravated by Religious elitism (#PG3644).

♦ PF3358 Undue religious influence on secular life

Religious influence on society — Discriminatory religious influence on society — Religious influence in politics
Nature Religious influence on society, in politics or on the law, may lead to conflict, discrimination, war, lack of development or anticlericalism. Religious influence may be subtle or overt; it is strongest and most widespread in its indirect form.
Religious belief may be contrary to certain political movements and retard their development. The church may be directly or semi-directly involved as a motivating force for nationalism or other political movements, or religious principles may be involved as in the instigation of Christian Democratic parties. This may lead to conflicts of various kinds and even to war. It may also lead to anti-clericalism or religious apathy.
Incidence Religious involvement in politics as a motivating force for nationalism is notable in Africa. Christian Democratic parties exist in western Europe and Latin America. Socialism is contrary to the teaching of Islam.
Broader Ideological conflict (#PF3388) Religious and political antagonism (#PC0030).
Narrower Theism (#PF3422) Injustice of religious courts (#PE0397)
Religious discrimination in education (#PD8807)
Religion as a reinforcement of nationalism (#PF3351)
Discriminatory religious influence on the law (#PD3357)
Religious discrimination in the administration of justice (#PE0168).
Related Inadequacy of religion (#PF2005)
Inadequate integration of religions into society (#PF3403).
Aggravates Religious schism (#PF1939) Religious conflict (#PC3292)
Unethical practices in politics (#PC5517).
Aggravated by Superstition (#PA0430) Religious rivalry (#PC3355)
Religious intolerance (#PC1808) Double standards in morality (#PF5225)
Deteriorating quality of life (#PF7142).
Reduced by Anti-clericalism (#PF3360).

♦ PF3360 Anti-clericalism

Nature Public or political opinion hostile to the influence of the Church in political and social affairs seeks to separate religious authority from civil government and subordinate the Church to state control. Anti-clericalism may take the form of propaganda, confiscation of Church property, trial, imprisonment or execution of the clergy, the banning of religious orders, or the secularization of education and hospitals. It may result in dictatorship and militarism or aggravate a general need for religion and mysticism. In so far as it is an expression of nationalism, it may also cause international conflict.
Incidence Current anti-clericalism is fostered in response to the presence of Christian Democrat parties in many parliaments, notably throughout Europe and in Latin America. It is also part of the ideology of communism. Elsewhere it may be expressed in counter-religious thought, atheism, or scientism.
Background Anti-clericals became numerous and influential in France by 1650. The movement grew in Austria-Hungary and Germany after 1800, and gradually took root throughout Europe as well as in Central and South America. Protestantism was rarely a major target for anti-clericalism until after the rise of Communist, Fascist and National Socialist governments.
Claim Churches need their critical centre. They cannot be left to the ruling minority and their quiet followers, or to the mockery of people outside. The common sense, ordinary believer must occupy the centre of the stage. It is not simply a question of democracy, natural justice or majority rule — it is a question of restoring the capacity of the church to save the world. This can only be done by getting together and being one.
Refs Sanchez, Jose M *Anticlericalisme* (1973).
Broader Religious and political antagonism (#PC0030).
Related Ideological conflict (#PF3388) Undemocratic pressures (#PD3389)
Religion as a reinforcement of nationalism (#PF3351).
Aggravates Monasticism (#PF2188).
Aggravated by Religious repression (#PC0578) Religious intolerance (#PC1808)
Inadequacy of religion (#PF2005) Double standards in morality (#PF5225)
Injustice of religious courts (#PE0397) Corruption in organized religion (#PC3359)
Religious discrimination in education (#PD8807)
Accumulation and misuse of religious property (#PE3354)
Religious discrimination in the administration of justice (#PE0168).
Reduces Religious discrimination in politics (#PC3220)
Undue religious influence on secular life (#PF3358).

♦ PF3361 Restrictive influence of religion on the masses

Nature Religions may have a very strong and subtle hold over the masses, particularly in poor and underdeveloped areas, and may retard development on account of rigid doctrine. Religious influence may conflict with political influence on ideological grounds.
Broader Religious repression (#PC0578) Restrictive practices (#PB9136).

♦ PF3364 Fragmentation within organized religions

Lack of unity within religions — Religious disunity — Shame of the churches — Separated religious brethren — Lack of common religious witness — Lack of ecclesial communion
Nature Lack of unity among branches of the same religion, and splits into many denominations, are caused by schism and doctrinal controversy. No real advance towards unity can be made if cultural, social, political and psychological components of the religious life are not painstakingly analyzed and evaluated. Many unconscious and unavowed prejudices and false presuppositions are too often treated as essentially in harmony with sacred texts, rather than as the conditioning of a particular culture and society.
Incidence The history of divisions in the Christian church is to a large extent reflected in the history of political systems and concepts. Division is a universal reality. Although, in the light of religious apathy in Western industrialized countries, Christian Churches are attempting to eliminate some of their differences and to preserve a relevance to the modern age through their fundamental unity, interdenominational intolerance and prejudice still persist.
In the case of the Christian churches the deepest difference has many forms and deep roots. It exists among many other differences of emphasis withen Christianity. Some are Catholic or Orthodox in clearly understood senses; some are Protestant after the great Reformation confessions; others stress the local congregation, the gathered community, and the idea of the free church. Some are deeply convinced that Catholic and Protestant (or Evangelical) can be held together within a single church. The clearest obstacle to manifestation of the churches' universality is their inability to understand the measure in which they already belong together in one body.
Claim 1. "Within our divided churches, there is much which we confess with penitence, for it is in our estrangement from God that all our sin has its origin. It is because of this that the evils of the world have so deeply penetrated our churches, so that amongst us too there are worldly standards of success, class division, economic rivalry, a secular mind. Even where there are no divisions of theology, language or liturgy, there exist churches segregated by race and colour, a scandal within the Body of Christ".
2. The basic division amongst the Christian churches is epitomized by the Catholic-Protestant split. "From each side of that division we see the Christian faith and life as a self-consistent whole, but our two conceptions of the whole are inconsistent with each other." "We differ in our understanding of the relation of our unity in Christ to the visible holy, Catholic and Apostolic Church. We are agreed that there are not two Churches, one visible and the other invisible, but one Church which must find visible expression on earth, but we differ in our belief as to whether certain doctrinal, sacramental and ministerial forms are of the essence of the Church itself. In consequence, we differ in our understanding of the character of the unity of the Church for which we hope, though none of us looks forward to an institution with a rigid uniformity of governmental structure."
Broader Religious schism (#PF1939) Religious conflict between sects (#PC3363)
Fragmentation of religious belief (#PF3404).
Narrower Religious rivalry (#PC3355).
Related Ideological conflict (#PF3388).
Aggravates Religious conflict (#PC3292) Disintegration of organized religion (#PD3423).
Aggravated by Religious intolerance (#PC1808) Double standards in morality (#PF5225)
Discrimination against minorities (#PC0582)
Disagreement concerning religious doctrine (#PF1115).

♦ PF3366 Scientism

Nature Scientism is the view that scientific, inductive methods must be employed to obtain factual data. It rejects intuitive or deductive approaches. Because data collections are not comprehensive and must be carefully related to context, the apparent evidence of scientism denies spiritual existence as an integral aspect of human experience. It tends towards elitism and technocracy.
Incidence As a product of advanced science and technology, scientism is most marked in industrialized countries. It may be combined with another ideology or even with religion, particularly in relation to ethics.
Broader Atheism (#PF2409) Ideological conflict (#PF3388).
Narrower Technocracy (#PF6330).
Related Rationalism (#PF3400).
Aggravates Agnosticism (#PF2333)
Irresponsible scientific and technological activity (#PC1153)
Negative effects of claims of religious infallibility (#PF3376).
Aggravated by Proliferation of technology (#PD2420)
Double standards in morality (#PF5225).
Reduces Inadequacy of religion (#PF2005).

♦ PF3367 Experimental religion

Nature New forms of religion, developed as a result of the inadequacy of traditional religious doctrine or as a result of commercialism, may be austere or orgiastic, based on mysticism or scientific data. They constitute a threat to traditional organized religions, and because they are cut off from the long-term experience of these religions they may lead their supporters into error.
Incidence Experimental religion is most marked in 'Western' industrialized countries where traditional values are being challenged and there is no one, overriding ideology.
Broader Dependence on religion (#PF0150).
Narrower Cargo cults (#PF5375) Manipulative cults (#PE6336).

Related Religious revivalism (#PU2333).
Aggravates Religious conflict (#PC3292) Multidenominational society (#PF3368).
Aggravated by Inadequacy of religion (#PF2005).

♦ PF3368 Multidenominational society

Nature A society incorporating many religious beliefs and sects may lead to religious segregation and religious conflict, discrimination, prejudice and elitism.
Broader Religious schism (#PF1939) Religious conflict between sects (#PC3363)
Fragmentation of religious belief (#PF3404).
Narrower Religious conflict (#PC3292).
Related Ideological conflict (#PF3388) Cultural fragmentation (#PF0536)
Plural society tensions (#PF2448)
Weakness of multi-party parliamentary systems (#PF3214).
Aggravates Social fragmentation (#PF1324) Religious intolerance (#PC1808)
Injustice of religious courts (#PE0397)
Religious discrimination in the administration of justice (#PE0168).
Aggravated by Experimental religion (#PF3367) Double standards in morality (#PF5225).
Reduces Inadequate integration of religions into society (#PF3403).

♦ PF3370 Ill–considered missionary activity

Missionary alienation

Nature Religious missions have been divided into 'foreign' and 'home' categories. 'Home' category missionaries deal in particular with the poverty stricken, the homeless, and the social outcasts or misfits. Those in 'foreign' categories mainly deal with indigenous peoples on whom they may impose a totally new moral code to which tribespeople find it very difficult to adapt; having done so, they may find themselves dependent on 'Western' economic and social structures. Because they are inadequately educated and discriminated against on an ethnic basis; and because they have little experience in market economic structures, this dependence renders them easy prey to exploitation, poverty, disease, malnutrition and death.
Background Buddhism, Zoroastrianism and Judaism were once missionary religions. Christian missionary activity began in the early years after the death of Christ. It spread to all parts of the Roman Empire. Christian missionaries pressed eastwards into Russia and Asia as far as China. From the 15th to the 18th century Roman Catholicism became a world religion with the thrust of Spanish and Portuguese colonialism. From 1750 to 1815 the Catholic missions were in decline. Protestantism was slow to undertake missions but this changed with British imperialism and continued to expand during the 19th and 20th centuries. Islam spread from Southeast Asia to the Atlantic coast of Europe and from central Europe to both coasts of central Africa by the middle ages. It currently has missionaries in Europe, the Americas and Africa.
Claim Christianity, and Islam are contemporary missionary religions. Missions are widespread, having followed or preceded colonialism. Missionary zeal has been the cause of colonial expansion and acted as pretext for war. When colonists arrive, they have the Bible and the indigenous people the land; before long, the situation is reversed.
Counter–claim Missionaries show selfless dedication to the lives of the people to whom they preach and minister. Christian missionaries were active in Europe extirpating paganism up to the 11th century, although their work had largely been completed during the Dark Ages. The indigenous peoples of Europe, speaking many different languages and divided into numerous tribes, gave up human sacrifice and tribal wars when they embraced Christianity, and they went on to raise the greatest of civilizations with its help. Islam, Judaism and Buddhism in their own missionary zeal advanced social conditions wherever they penetrated. In some war torn areas the only foreigners to stay to offer aid are missionaries.
Broader Religious conflict (#PC3292).
Narrower Moralism (#PF3379).
Related Evangelism (#PF6325) Cultural invasion (#PC2548).
Aggravates Forced assimilation (#PC3293) Ethnic disintegration (#PC3291)
Destruction of cultural heritage (#PC2114) Degradation of indigenous cultures (#PJ4963).
Aggravated by Religious intolerance (#PC1808).

♦ PF3374 Idolatry

Idolatrous worship

Nature Idolatry, in monotheistic traditions, is the worship of something or someone other than God in place of God. It is the substitution of symbols from everyday life for the growing, evolving, progressing concepts, sentiments and ideals which should inspire human society and individuals. Idolatry may take the overt form of the worship of, for instance, the sun, a king, an animal, or a statue. It may be less overt in the sense of an object of devotion and desire which is seen as the ultimate source of good and reason for living. Nationalism or materialism or even extreme forms of family devotion can be classed as idolatry in this sense.
Background The concept of idolatry originated in the very specific historico–religious context of the monotheism of Israel. In the applications of Second Commandment, it acquired definitive formulation in censure by the prophets of Israel of the pagan cults and their influence on the chosen people. This biblical heritage passed into the New Testament and early Christianity. The monotheism of Islam adopted this Judaeo–Christian concept and made it one of the foundations of its belief and its faith.
Incidence There is a long tradition of incorporating gargoyle–like figures into major structures, especially cathedrals. In 1989, a troll–like figure was incorporated into the structure of San Francisco's repaired Bay Bridge (following its partial destruction by earthquake) in order to enhance future safety. It has been argued that certain contemporary belief systems can encourage idolatry — such as when Judaism is perceived as a corporate obsession with the idolatries that vie for the loyalty of a people chosen to serve a jealous God and when for Jews both socialism and zionism become such idols.
Claim Idolatry is the worship of the creature instead of the Creator and, to make matters worse, the creature is made by man, who is himself a creature. Idolatry is, also, the worship of what in modern terms is process, the life–force, the *'elan vital'* or what we will, instead of the Creator who transcends and is in some sort external to creation. Finally, idolatry is the worship of an idol considered as a substitute for the divine.
Refs Barfield, Owen *Saving the Appearances* (1965).
Broader Heresy (#PF3375) Occultism (#PF3312) Superstition (#PA0430).
Narrower Totemism (#PF3421) Hero worship (#PF2650) Ancestor worship (#PD2315).
Related Animal worship as a barrier to development (#PD2330).
Aggravates Bibliolatry (#PF7129) Religious intolerance (#PC1808).
Aggravated by Polytheism (#PF4957) Representative arts (#PF0981).

♦ PF3375 Heresy

Nature A theological doctrine or system rejected as false by ecclesiastical authority, heresy is distinct from schism (where the schismatic severs himself from orthodoxy), in so far as a heretic may remain within the Church. Although condemnation of heretics is not now considered as a natural corollary to religious conviction in the Protestant religion, the Roman Catholic Church maintains a distinction between those who willfully and persistently adhere to doctrinal error and those who adhere to it because of their upbringing. The former inquisition has been democratized into the Congregation for the Doctrine of the Faith.
Background The word heresy is derived from the Greek hairesis which meant an act of choosing, and came to mean a set of philosophical opinions. The term acquired a tone of hostility and condemnation when taken up by Christianity; the distinction between orthodoxy and heresy was first drawn by St Paul. From the 2nd century, the Church became increasingly aware of the need to test the heretical, and penalties at this time were usually confined to excommunication. After the 4th century, penalties included confiscation of property and exile. In the 12th century the death penalty was introduced and as was the Inquisition, increasingly an instrument of political repression with the advent of Protestantism.
Refs Brown, Harold O *Heresies* (1988); Peters, Edward *Heresy and Authority in Medieval Europe* (1980); Robinson, Thomas A *The Bauer Thesis Examined* (1988).
Broader Error (#PA6180) Unbelief (#PA7392) Disagreement (#PA5982).
Narrower Idolatry (#PF3374).
Related Scientific censorship (#PD1709) Perversion of religion (#PJ0388).
Aggravates Infamy (#PB8172) Religious conflict (#PC3292)
Religious intolerance (#PC1808).
Reduced by Iconoclasm (#PF4923).

♦ PF3376 Negative effects of claims of religious infallibility

Incidence The claim that under certain conditions the doctrinal and moral teaching of a Church cannot err applies particularly to the Roman Catholic Church and to the Pope. Infallibility is derived from the belief that God has intervened in human history and has given man the possibility of assurance in religious matters; also from the belief that the Church is not a human creation, its essential structure and nature having been determined by Christ. The Church is seen as the extension of Christ in time, and the Pope as the Supreme teacher in the Church, guided by Christ. Papal declarations may be strongly adhered to, particularly by members of traditional or underdeveloped societies and this may retard social progress (for example, with birth control and abortion). The Roman Catholic claims to the primacy of the bishopric of Rome among bishops, and the infallibility of ex cathedra teachings, are the major impediments to ecumenicism and Church reunification.
Claim The historical church can never claim to be infallible or inerrant in regard to any of its decisions or traditions, for all of them are relative to historical conditions. Even the relative authority which can be claimed for a decision or tradition will depend upon its adherence to the mandate: "As the Father has sent me, even so I send you."
Broader Inadequacy of religion (#PF2005) Irrational religious beliefs (#PF6829)
Religious opposition to public health practices (#PF3838).
Aggravates Authoritarianism (#PB1638) Social underdevelopment (#PC0242)
Injustice of religious courts (#PE0397)
Religious discrimination in the administration of justice (#PE0168).
Aggravated by Fear (#PA6030) Scientism (#PF3366)
Superstition (#PA0430) Traditionalism (#PF2676)
Religious intolerance (#PC1808).

♦ PF3377 Hypocrisy

Moral double standards

Nature The feigning of beliefs and virtues not actually held or possessed becomes most obvious where there is a marked double standard of morality. Hypocrisy may gloss over considerable immorality, thus maintaining the respectability of the person in question and any position of power that he or she has. Such hypocrisy serves to frustrate people who suffer or may think they suffer as a result.
Broader Affectation (#PA6400) Uncommunicativeness (#PA7411).
Narrower Double mindedness (#PF4910) Medical hypocrisy (#PF3006)
Religious hypocrisy (#PF3983) Government hypocrisy (#PF9050)
Double standards in morality (#PF5225) Haphazard forms of social ethics (#PF1249)
Double standards of sexual morality (#PF3259)
Compromise as a betrayal of principles (#PF3420).
Related Moralism (#PF3379).
Aggravates Frustration (#PA2252).
Aggravated by Taboo (#PF3310) Nudism (#PF2660)
Conformism (#PB3407) Traditionalism (#PF2676)
Conflict of duties (#PF0513).

♦ PF3379 Moralism

Nature The practice or belief in a system of principles governing conduct, or the act of moralizing, may be an expression of religious intolerance, nationalism, or conformism against new thinking or ways of life. It may support the ethic of the superior morality of the wealthy against the inferior principles of the poor or the ethic of the inherent superiority of the poor. It may support racism, discrimination against women, against minorities, and against any deviation from the norms of those in control. It may be used to justify militarism or acts of violence; propaganda is usually heavily endowed with moralism.
Broader Complacency (#PA1742) Intolerance (#PF0860)
Ill–considered missionary activity (#PF3370).
Narrower Criminalization of sexual relations out of wedlock (#PE9122).
Related Hedonism (#PF2277) Obscenity (#PF2634) Hypocrisy (#PF3377)
Amoralism (#PF3349) Immorality (#PA3369) Intellectualism (#PF2146)
Narrow–mindedness (#PJ6348) Racial discrimination in politics (#PD3329).
Aggravates Conformism (#PB3407) Permissiveness (#PF1252)
Art censorship (#PD2337) Underground press (#PD2366)
Retarded socialization (#PF2187) Radio and television censorship (#PD3029).
Aggravated by Compulsory indoctrination (#PD3097).
Reduces Metal deficient diets (#PE1901).

♦ PF3382 Deliberate imbalancing of population sex ratio

Desire for male children — Unwanted female babies — Artificial sex determination before birth

Nature Amniocentesis and sex determination tests have been used to inform parents of the sex of their unborn child. Such cases have resulted in the performing of an abortion when the baby was not of the sex wanted. Ante–natal choices are biased against females because of the traditional preference in most societies for male offspring. Post–natal sex choice is a euphemism for murder, with infanticide enacted mainly against females.
Incidence The female population has been declining is some areas because of preference for the male. In India, the sex ratio in 1983 was 935 females to a 1,000 males, in 1985 in Korea Rep there were 110.4 males to 100 females, in 1989 some parts of China had a ratio of 121 males to 100 females. The chances that female population will drop even further have been increased with the possibility of sex–determination and abortion. It is possible that female foeticide could become a lucrative practice for medical doctors. Female foeticide taken together with large scale female infanticide (10,000 per year in China), is a species of genocide that might be termed gynecide. About 40 baby girls have been born in Japan by a technique of chromosome manipulation. Girls are the sex of choice because of certain types of hereditary conditions peculiar to men, such as haemophilia and colour blindness.
Broader Unwanted pregnancies (#PF2859).
Narrower Imbalance in the human sex ratio (#PF1128).
Aggravates Infanticide (#PD3501) Induced abortion (#PD0158)
Gender abortions (#PD3947) Shortage of marriageable men (#PE9379)

Shortage of marriageable women (#PE0427) Physical malformation of foetus (#PE2042)
Gender discrimination in developing countries (#PD9563).
Aggravated by Unethical medical practice (#PD5770).

♦ PF3385 Lovesickness

Nature A forlorn, enfeebling form of melancholia.
Background Traditionally recognized and often documented in arcane medical journals of mediaeval times. Physicians of that time tended to conclude that men were more especially susceptible to it.
Incidence A theme of major contemporary importance in the song industry and in certain forms of fiction. A phenomenon especially characteristic of adolescents.
Broader Melancholy (#PF7756).
Aggravates Juvenile stress (#PC0877).
Aggravated by Unrequited love (#PF6096) Romantic separation (#PD4233)
Emotional manipulation (#PE9599) Dependence on romantic love (#PF7418).

♦ PF3387 Lack of a world religion

Need for a world religion
Nature The need for a world religion derives from the religious conflict between existing faiths which prevents the achievement of a brotherhood of man.
Broader Fragmentation (#PA6233) Ideological conflict (#PF3388).
Aggravated by Fragmentation of religious belief (#PF3404).

♦ PF3388 Ideological conflict

Competing ideologies — Ideological rivalry — Debilitating ideological differences — Lack of ideological unity — Ideological opponents — Ideological dissent — Ideological opposition
Nature Division between ideologies and within a single ideology may lead to conflict, intolerance, prejudice, repression, rivalry, injustice, and sometimes war. Conflict between religious, political or intellectual ideologies may take the form of propaganda and other tensions or pressures, and may lead to civil war, international war or cold war. It may involve rivalry for ideological influence in Third World countries or other potentially useful or strategic areas.
Claim Ideological conflict is one of the stresses in modern society that permeates all human activities. It reaches into the classroom and the home, and makes antagonists among schoolmates, between them and the school system, and it divides families. Thus from an early age, children become accustomed to the warfare of ideas, and as adults they carry on their own ideological battles, bringing these to national and international levels. They are not only conditioned towards engaging in ideological conflicts and rivalries, but also to a conduct that is more emotional than logical and which casts their opponents into satanic roles or roles of enemies of humanity. This leads to witchhunts, pogroms, purges, prejudices and hostilities that erupt into violence.
Refs Watts, Michael R *The Dissenters* (1986).
Broader Conflict (#PA0298) Competition (#PB0848) Fragmentation (#PA6233).
Narrower Nudism (#PF2660) Fascism (#PF0248) Marxism (#PF2189)
Feminism (#PF3025) Populism (#PF3410) Socialism (#PC0115)
Communism (#PC0369) Anarchism (#PC1972) Scientism (#PF3366)
Capitalism (#PC0564) Liberalism (#PF0717) Utopianism (#PF2192)
Puritanism (#PF2577) Scepticism (#PF3417) Monasticism (#PF2188)
Agnosticism (#PF2333) Neo–fascism (#PF2636) Materialism (#PF2655)
Rationalism (#PF3400) Conservatism (#PF2160) Collectivism (#PF2553)
Sectarianism (#PF7780) Fundamentalism (#PF1338) Secularization (#PB1540)
Anti–communism (#PF1826) Totalitarianism (#PF2190) Male domination (#PC3024)
Anti–capitalism (#PF3110) World federalism (#PF2088) Political purges (#PC2933)
Linguistic purism (#PF1954) Religious conflict (#PC3292) Rejection of rituals (#PF0278)
Political radicalism (#PF2177) Intellectual dissent (#PC2582) Extremist ideologies (#PC6341)
Intellectual conflict (#PC3390) Ideological deviation (#PF3405)
Inadequacy of religion (#PF2005) Ideological revolution (#PC3231)
Undemocratic pressures (#PD3389) Subversion of democracy (#PD3180)
Political indoctrination (#PD1624) Lack of a world religion (#PF3387)
Compulsory indoctrination (#PD3097) Disagreement among experts (#PF6012)
Ideological schism in communism (#PF3181) Religious and political antagonism (#PC0030)
Politicization of technical debates (#PD2860) Politically emotive words and terms (#PF3128)
Factionalism in developing countries (#PD1629) Ideological discrimination in politics (#PC3219)
Disruptive secular impact of holy days (#PE7735)
Undue religious influence on secular life (#PF3358)
Threats to ideological movements and minorities (#PC3362)
Inadequate integration of ideology into society (#PF3402)
Weakness in trade between different economic systems (#PC2724)
Excessive claims for human development through sports (#PG4881)
Inadequate and inaccurate textbooks and reference books (#PD2716)
Politicization of intergovernmental organizational debate (#PD0457)
Alienation of support for international organizations and programmes (#PD1809)
Abuse of international cultural, diplomatic and commercial exchanges (#PF3099)
Alienation of skilled and committed personnel from international organizations and programmes (#PE1553)
Inadequate relationship between international governmental and nongovernmental organizations and programmes (#PE1973).
Related Sacrilege (#PF0662) Nationalism (#PB0534) Religious war (#PC2371)
Value erosion (#PA1782) Dehumanization (#PA1757) Religious schism (#PF1939)
Anti–clericalism (#PF3360) Religious torture (#PC7101) Political conflict (#PC0368)
Lack of human unity (#PF2434) Aggressive ideologies (#PJ4778)
Undue political pressure (#PB3209) Multidenominational society (#PF3368)
Fragmented ethical contexts (#PF2096) Corruption in organized religion (#PC3359)
Newspaper and journal propaganda (#PD0184)
Benign neoplasm of bone and cartilage (#PG0819)
Discrepancies in human life evaluation (#PF1191)
Fragmentation within organized religions (#PF3364)
Differences in trading principles and practices between different economic systems (#PC2952).
Aggravates Atheism (#PF2409) Distrust (#PA8653)
Extremism (#PB3415) Propaganda (#PF1878)
Revolution (#PA5901) Militarism (#PC2169)
Draft evasion (#PD0356) World anarchy (#PF2071)
Political schism (#PC2361) Malignant visions (#PF5691)
Political instability (#PC2677) Religious persecution (#PC5994)
Revolutionary communism (#PC3163)
Lack of commitment to common symbols (#PE8814)
Forced repatriation of prisoners of war (#PD0218)
Inadequacy of international legislation (#PF0228)
Lack of intersocietal resource channels (#PF2517)
Lack of appreciation of cultural differences (#PF2679)
Obstacles to international cultural exchange (#PF4857)
Competition between international organizations for scarce resources (#PC1463)
Obstructions to international personnel exchanges and cultural cooperation (#PE4785).
Aggravated by Cynicism (#PF3418) Idealism (#PF3419)
Hero worship (#PF2650) Social conflict (#PC0137)
Religious repression (#PC0578) Sociological espionage (#PF2924)
Religious discrimination (#PC1455) Inadequate ideological frameworks (#PD0065)
Repression of intellectual dissidents (#PD0434) Underprivileged ideological minorities (#PC3325).

♦ PF3390 Undue attachment to territory

Geo–sentiment
Claim When people are swayed by geo–sentiments they identify themselves with a specific homeland, nation or holy site to the extent that they condemn, refuse to speak with, hate, fight and exploit people who do not share their particular geographical identification. The danger comes when people identify so strongly with their locale that they divide themselves from others.
Aggravates Eurocentrism (#PJ9837) Discrimination against foreigners (#PD6361)
Undue attachment to a social group (#PF1073) Territorial disputes between states (#PC1888).
Aggravated by Domination of the world by territorially organized sovereign states (#PD0055).

♦ PF3391 Backlash against repeated warnings

Over–warning
Nature So many products in the industrialized countries carry warning labels that there is a risk of people not paying attention to them at all. Some consumers even think that products with warning labels are safer than those without, because warning indicates the carefulness of the manufacturer.
Aggravates False alarms (#PF4298).
Aggravated by Excessive caution (#PF6389).

♦ PF3392 Ideological apathy

Lack of ideology
Nature A motivating philosophy, political or religious, is needed from which to relate life and activity. If this need is not fulfilled apathy and alienation may result, also cynicism, materialism, and amoralism. Non–fulfilment of the need for ideology may arise from poverty or lack of material advantages and is exacerbated by injustices and inequality.
Claim It is impossible for nations to live without a conviction of moral direction, a generosity and purpose in national life; where there is no vision of the future, people will perish.
Broader Apathy (#PA2360).
Related Dependence on religion (#PF0150) Dependence on philosophy (#PG6361).
Aggravates Inadequate sense of personal identity (#PF1934)
Religion as a reinforcement of nationalism (#PF3351).
Aggravated by Ignorance (#PA5568) Insecurity (#PA0857)
Repression of intellectual dissidents (#PD0434).

♦ PF3393 Lack of sharing of community skills

Nature Although there may be a feeling that members of a community may have practical skills, such skills tend not to be used or shared simply because their existence is not known. Besides practical employment skills not being shared, there is also a lack of sharing of the social skills necessary to provide a relevant set of moral values and family practices for day–to–day living. This gap in skill sharing is widened by certain unfounded prejudices and a growing disrespect between ages. The need to recreate a sense of community well–being makes new skill learning critical. Adult education programmes may be offered, but if it is not clear what practical skills are demanded, the programmes are not effective in the retraining process.
Broader Fragmented patterns of community activity (#PF6504)
Incomplete access to information resources (#PF2401)
Underutilization of potential in local communities (#PF6513).
Aggravates Insufficient skills (#PC6445) Underdeveloped technological skill (#PF8552).
Aggravated by Inadequate education (#PF4984) Misinformed family practices (#PG8641)
Generation communication gap (#PF0756) Incomplete skills information (#PJ0140)
Unorganized transfer of skills (#PJ8603) Underutilization of popular wisdom (#PF2426)
Minimal opportunities of adult training (#PF6531).

♦ PF3394 Inadequacy of political doctrine

Nature The inability of a doctrine to meet political needs, whether in uniting a party, in producing effective government, or in drawing popular support, may lead to political schism, social breakdown, possible revolution, national disintegration, and foreign intervention.
Broader Political conflict (#PC0368) Inadequacy of doctrine (#PF3396).
Narrower Political inertia (#PC1907) Political stagnation (#PC2494)
Political instability (#PC2677) Inadequate government (#PJ6362)
Political over–reaction (#PF4110).
Related Inadequate ideological frameworks (#PD0065).
Aggravates Political apathy (#PC1917) Social breakdown (#PB2496)
Social fragmentation (#PF1324) National disintegration (#PB3384).

♦ PF3395 Inadequacy of economic doctrine

Nature Inability of economic policy or doctrine to meet economic needs concerning growth, stability, balance of payments and distribution of wealth, aggravates social problems and may lead to social disintegration and even national disintegration, particularly if a nation's economy is too dependent on that of a more powerful foreign nation.
Broader Inadequacy of doctrine (#PF3396).
Related International economic injustice (#PC9112)
Inadequate ideological frameworks (#PD0065).
Aggravates Economic apathy (#PC3413).

♦ PF3396 Inadequacy of doctrine

Nature The inability of a doctrine to meet practical and spiritual needs in society leads to a lack of integration, social injustice, inequality, discrimination and conflict. This is true of inadequate religious, social, economic, political or ideological doctrines.
Broader Inadequacy (#PA8199).
Narrower Inadequacy of religion (#PF2005) Inadequacy of social doctrine (#PF3398)
Inadequacy of economic doctrine (#PF3395) Inadequacy of political doctrine (#PF3394)
Inadequate ideological frameworks (#PD0065) Inadequacy of scientific expertise (#PF5055).
Aggravates Social fragmentation (#PF1324) Intellectual conflict (#PC3390)
Ideological corruption (#PC2914) Contradictions in communist systems (#PF3179).
Aggravated by Irrationalism (#PF3399) Intellectualism (#PF2146).

♦ PF3397 Information gap

Technical information gap
Nature Information about what is available in other developing countries is scarce and difficult to find, but information about the advanced countries is available daily in newspapers and on the television.
Broader Haphazard transmission of practical technology (#PF3409)
Disparity between industrialized and developing countries (#PC8694).
Narrower Break–down in communications due to difference in training (#PE6650).
Aggravates Increasing development lag against information growth (#PE2000).

♦ PF3398 Inadequacy of social doctrine

Nature Inadequacy of social doctrine to alleviate social conflict, social discrimination, social inequality, social injustice, and lack of integration is closely connected with inadequacies of economic doctrine and ideological doctrine and with social lag.
Broader Inadequacy of doctrine (#PF3396).
Related Social conflict (#PC0137) Social inequality (#PB0514)

Inadequate ideological frameworks (#PD0065).
Aggravates Apathy (#PA2360) Social breakdown (#PB2496) Social injustice (#PC0797).
Aggravated by Social discrimination (#PC1864).

♦ PF3399 Irrationalism

Nature This philosophical attitude believes that instincts, feelings and will are superior existential dimensions to reason. It implies that the world is devoid of meaning, that the reason is not capable of knowing the universe, that objective standards are ineffectual and that man is irrational by nature. Such a life posture, if taken to the extreme, can lead to spiritual void, contempt for established order and free living or, otherwise, to gullibility and narcissism.
Broader Lack of control (#PF7138).
Narrower Intellectual conflict (#PC3390).
Related Rationalism (#PF3400).
Aggravates Absurdity (#PF6991) Irrationality (#PA0466)
Inadequacy of doctrine (#PF3396).
Aggravated by Anti–intellectualism (#PF1929).

♦ PF3400 Rationalism

Nature The rationalist believes in the use of reason rather than experience of spiritual revelation and sense experience to define knowledge. The use of rational methods may not be successful for understanding much in the life of the spirit and in human history (such as subconscious processes of the human mind), and therefore it is a limited instrument of knowledge.
Broader Ideological conflict (#PF3388).
Narrower Positivism (#PF2179) Intellectualism (#PF2146).
Related Scientism (#PF3366) Irrationalism (#PF3399) Political radicalism (#PF2177).
Aggravates Ignorance (#PA5568) Dependence on religion (#PF0150).
Aggravated by Double standards in morality (#PF5225).

♦ PF3402 Inadequate integration of ideology into society

Nature Ideological integration in its broadest sense may prove impossible since it is on the basis of ideology that society is run. Conflicting ideologies such as capitalism and communism may never be integrated into society in their purest sense and compromise may be a solution which neither camp would accept. The dichotomy leaves society open to conflict and possible disintegration. It may lead to repression and denial of democracy.
Broader Ideological conflict (#PF3388) Social fragmentation (#PF1324).
Aggravates Social disintegration (#PC3309).
Aggravated by Double standards in morality (#PF5225).

♦ PF3403 Inadequate integration of religions into society

Alienation of religious minorities
Nature Inadequate integration of different religions and denominations into society may be the cause and the result of religious intolerance, discrimination and conflict. It is aggravated in countries where there is an official religion which takes precedence over all others and whose influence may have permeated the law and other civil institutions, and in itself, may be regarded as the integrating ideology in the society.
Incidence Moslems and Jewish people in the Soviet Union, Mormons in the United States, and Jehovah's Witnesses in predominantly Catholic countries are discriminated against. In the UK, non–conformity with Anglicanism has a range of disadvantages, chiefly among the upper–middle and upper classes, while non–Christians experience discrimination among the working class for employment, housing and social acceptance in some areas.
Broader Fragmentation of religious belief (#PF3404).
Related Social fragmentation (#PF1324) Inadequate political integration (#PF3215)
Religious discrimination in education (#PD8807)
Undue religious influence on secular life (#PF3358).
Aggravates Religious conflict (#PC3292).
Reduced by Multidenominational society (#PF3368).

♦ PF3404 Fragmentation of religious belief

Lack of unity between religions — Religious disunity — Religious pluralism — Multiplicity of religions
Nature Division between religions and within a single religion results in intolerance, discrimination, prejudice, conflict, and sometimes war.
Incidence Severe divisions exist between many of the world's major religions and between sects and denominations within one religion. Religious war resulting from these divisions can be seen in Northern Ireland, the Middle East and India, for example.
Counter–claim Religious unity is more dangerous than disunity. When great masses of humanity, the nation–states, embrace one theistic or 'spiritual' creed with all its paraphenalia of a priori conclusions about reality deduced from the God–hypothesis, their civilization freezes into a static shell, a lifeless societal geometry of relations. Typically it is symbolized by some emblem they raise. It may be a circle, crescent or straight lines crossed or joined, or elaborate combinations of these, but it is as artificial as the religious unity–in–ignorance it represents. Religious dis–unity is the only way to preserve societal dynamism; its symbol is the tree or rainbow. Societies with religious disunity do not conduct inquisitions or burn heretics. They do encourage the exchange of ideas among those open to reason. As for the violent, without religion they would find still other reasons for their violence; religious pluralism is not its cause.
Refs Jurji (Ed) *Religious Pluralism and World Community* (1969).
Broader Fragmentation (#PA6233) Religious schism (#PF1939)
Dependence on religion (#PF0150).
Narrower Fragmented ethical contexts (#PF2096)
Multidenominational society (#PF3368)
Fragmentation within organized religions (#PF3364)
Inadequate integration of religions into society (#PF3403).
Related Lack of human unity (#PF2434).
Aggravates Religious repression (#PC0578) Religious intolerance (#PC1808)
Lack of a world religion (#PF3387).
Aggravated by Apostasy (#PE9018) Religious rivalry (#PC3355)
Religious conflict (#PC3292).
Reduces Religious extremism (#PF4954).

♦ PF3405 Ideological deviation

Nature Deviation from an accepted ideology may result in social conflict or in ideological war. It encourages ideological repression or anarchy or revolution. Under extreme circumstances it may lead to national disintegration and foreign intervention. It may take the form of political or religious schism.
Broader Ideological conflict (#PF3388).
Aggravates Ethnocide (#PC1328) Political purges (#PC2933)
Political indoctrination (#PD1624).
Reduced by Conformism (#PB3407).

♦ PF3408 Inadequate appreciation of culture

Nature Lack of appreciation of cultural objects and affairs may result from a puritan or populist ethic, from a scientific or technologically orientated education, from materialistic values and commercialism, or from the gradual stifling of indigenous or ethnic cultural heritage through forced assimilation and standardization. Fine arts may not be appreciated due to lack of understanding or anti–intellectualism and class conflict. In developing countries, educational methods may be imported that attach little value to, or do not recognize, local cultural achievements, so that while literacy may be taught, for example, local literature is not used as illustrative material. Lack of cultural appreciation implies the loss of cultural heritage, perhaps involving ethnic disintegration and an unfulfilled need for the aesthetic and spiritual. Lack of cultural appreciation contributes to a loss of identity, disorientation and possibly social breakdown. It leads to a decline in fine arts and letters.
Broader Cultural deprivation (#PC1351) Anti–intellectualism (#PF1929).
Narrower Inter–cultural misunderstanding (#PF3340)
Insensitivity to diversity of cultural traditions (#PF8156).
Aggravates Destruction of cultural heritage (#PC2114)
Untransposed significance of cultural tradition (#PF1373).
Aggravated by Art censorship (#PD2337) Cultural illiteracy (#PD2041)
Obsolete educational values (#PF8161)
Abusive exploitation of cultural heritage (#PC7605).

♦ PF3409 Haphazard transmission of practical technology

Nature Rural communities have little access to the technology which would allow people to maximally utilize land and resources. Left largely to depend on traditional methods of farming and production or, at best, on partial application of improved techniques, the progress of rural communities in every sector is haphazard at best. Wasted man–hours and unused land testify to the separation of such villages from the gifts of contemporary society.
Broader Use of inappropriate technologies in developing countries (#PF0878).
Narrower Information gap (#PF3397) Insufficient pest control (#PJ1086)
Unexposed farming technology (#PG8478) Insufficient programme funding (#PG8467)
Prohibitive cost of fertilizer (#PJ1022) Insufficient modern technology (#PJ0996)
Scarcity of high–quality seeds (#PJ4680) Insufficient technological know–how (#PG8489).
Related Lack of apprentice training (#PJ8498) Limited agricultural education (#PF8835).
Aggravates Abusive technological development under capitalism (#PD7463).
Aggravated by Inadequate educational facilities (#PD0847)
Insufficient communications systems (#PF2350).

♦ PF3410 Populism

Nature Populism implies belief in the superior moral worth and creativity of ordinary, uneducated and unintellectual people. This may lead to the use of violence against oppressors, namely the upper classes. It may lead to anti–intellectualism and cultural stagnation of the fine arts as opposed to peasant arts.
Broader Class conflict (#PC1573) Ideological conflict (#PF3388).
Narrower Anti–intellectualism (#PF1929).
Aggravates Dictatorship of the majority (#PD3239).
Aggravated by Double standards in morality (#PF5225).
Reduces Elitism (#PA1387).

♦ PF3411 Lack of professional standards

Nature Lack of professional status, methods, character or standards includes quackery and exploitation, perhaps with the use of false qualifications. Certain professional practices may be totally subordinated to commercial, political or military considerations. Lack of professionalism may produce low standards, or high standards but unethical methods. Lack of professionalism in the medical profession may lead to disease, mutilation or death. Overall, it leads to inefficiency.
Broader Lack of qualitative excellence (#PF5703)
Inadequacy of international standards (#PF5072).
Narrower False qualifications (#PF0704) Lack of qualifications (#PG6377)
Decline in educational standards (#PF8466).
Related Discriminatory professionalism (#PC2178).
Aggravates Inefficiency (#PB0843) Unethical professional practices (#PC8019).
Aggravated by Lack of administrative control (#PG6380).
Reduces Amateurism (#PG4384).

♦ PF3417 Scepticism

Nature A doubting or questioning attitude of mind, scepticism originally related to a philosophical doctrine that absolute knowledge is impossible and that inquiry must be a process of doubting in order to acquire relative certainty. Scepticism often leads to caution and lack of willingness to take bold measures or steps. It may also lead to apathy regarding matters of morality, politics and social injustice.
Refs Annas, Julia and Barnes, Jonathan *The Modes of Scepticism* (1985); Popkin, Richard H *The History of Scepticism from Erasmus to Spinoza* (1979); Rothman, Milton A *A Physicist's Guide to Skepticism* (1988).
Broader Ideological conflict (#PF3388).
Narrower Caution (#PG6386) Community development scepticism (#PJ8970)
Increasing scepticism about the accuracy of official information (#PF7649).
Related Cynicism (#PF3418).
Aggravates Lack of commitment to common symbols (#PE8814).
Aggravated by Apathy (#PA2360) Insecurity (#PA0857)
Hyperinflation (#PD7940) Personal life crises (#PD4840)
Double standards in morality (#PF5225) Loss of institutional credibility (#PF1963).
Reduces Bias in scientific research (#PF9693).

♦ PF3418 Cynicism

Nature The belief that all men are motivated by selfishness, cynicism tends to breed selfishness and materialism, and also apathy regarding matters of morality, including religion, politics and social injustice. It also tends to foster social conflict for material goods, domination, elitism, and exploitation.
Incidence Cynicism is most generally noticeable in countries with a very high standard of living (Western industrialized countries) and where society and social values are no longer cohesive. Elsewhere, in feudalistic or slave societies and those with very severe political repression, cynicism may take the form of a total disregard for the welfare of others.
Background The first cynics were members of a Greek philosophical sect founded by Antisthenes of Athens.
Broader Unkindness (#PA5643) Disrespect (#PA6822)
Selfishness (#PA7211) Hopelessness (#PA6099).
Related Hedonism (#PF2277) Scepticism (#PF3417).
Aggravates Complacency (#PA1742) Verbal abuse (#PD5238)
Social apathy (#PC3412) Political apathy (#PC1917)
Ideological conflict (#PF3388) Political opportunism (#PC1897)
Double standards of sexual morality (#PF3259)
Lack of commitment to common symbols (#PE8814).
Aggravated by Manipulative euphemisms (#PF5183)
Double standards in morality (#PF5225)
Inadequate ideological frameworks (#PD0065)
Loss of institutional credibility (#PF1963)
Unequal distribution of fame and honours (#PF3439).
Reduced by Idealism (#PF3419).

♦ PF3419 Idealism
Nature Idealism in popular usage (as opposed to its technical philosophical connotation) implies the envisaging of things in an ideal form, and the pursuit of personal ideals which may be impractical – as in utopianism – and cause mismanagement, including wastage and misallocation of resources. The pursuit of such ideals may cause conflict of various kinds.
Narrower Utopianism (#PF2192).
Aggravates Ideological conflict (#PF3388).
Aggravated by Double standards in morality (#PF5225).
Reduces Cynicism (#PF3418) Materialism (#PF2655).

♦ PF3420 Compromise as a betrayal of principles
Nature Compromise may be seen as a betrayal of principle or as an act of hypocrisy, corruption and cynicism, which may lead to worse corruption and abandonment of idealism for self-interest. Alternatively it can be seen as a lukewarm solution which pleases no-one, or inducing apathy which in turn may lead to extremism later.
Incidence Compromise affects people both on a personal and political and social level, comprising community, and national and international society.
Counter-claim A settling of differences in which each side makes concessions, compromise in its most beneficial form is a means of avoiding conflict and pleasing all parties.
Broader Hypocrisy (#PF3377) Liberalism (#PF0717) Self-interest (#PA8760).
Related Conflict resolution (#PS3582).
Aggravates Social conflict (#PC0137) Government treachery (#PF4153) Inability to compromise (#PF9747).
Aggravated by Discrepancies between principles and practice (#PF4705).
Reduces Political radicalism (#PF2177) Undemocratic political organization (#PC1015).

♦ PF3421 Totemism
Nature Totemism is the identification of kinship with a sign or totem of an animate or inanimate object, and is practised among primitive tribes and cultures, where it may constitute a barrier to development. Rites of magical or religious significance are practised and social laws are enacted, notably against marriage within the totemic group. The sacred animal or plant is protected, hence taboo is also closely connected with totemism; sacred totem-images and totem-related behavioural taboos are obstacles to communication and the education of primitive people.
Background The totem objectifies a quality (or power) with which its tribal selectors wished to be identified, and hence is a primitive attempt at psychological and ethnic typology. The totem figures in tribal cosmological and anthropological legends teach a closer relationship of man with nature than do the later religions which divorced man from everything except distant divinities of infrequent and unreliable appearance. There are a great many theories concerning totemism, as the subject is kept alive by the persistence of theriomorphic images in man's subconscious.
Broader Idolatry (#PF3374) Tribalism (#PC1910).
Narrower Witchcraft (#PF2099) Animal worship as a barrier to development (#PD2330).
Aggravates Underdevelopment (#PB0206) Endangered totemic species (#PE4184).
Reduced by Cultural invasion (#PC2548) Forced assimilation (#PC3293).

♦ PF3422 Theism
Nature Belief in the existence of gods or a god may constitute a barrier to social and economic progress. The belief in a particular god or gods may foster intolerance, rivalry, repression, discrimination or segregation.
Incidence Theism exists in most societies but its effect is most noticeable in countries or societies where there is an established or official religion.
Broader Undue religious influence on secular life (#PF3358).
Narrower Pantheism (#PF4209).
Related Atheism (#PF2409) Agnosticism (#PF2333).
Aggravates Religious discrimination (#PC1455).

♦ PF3434 Lack of continuity amongst personnel of international organizations
Excessive career mobility of aid officials — Rapid personnel turnover of international organizations
Incidence On average government ministers, senior officials, diplomats and aid officers spend 2 to 2.5 years in a position, acquiring minimum competence, before being posted to some other position.
Related Alienation of skilled and committed personnel from international organizations and programmes (#PE1553).
Aggravated by Unperceived career opportunities (#PF9004).

♦ PF3437 Disruption of human activities by supernatural entities
Endangered habitats of elves
Incidence In many countries (as in Scandinanvia, but also in non-western countries) with a continuing folk tradition recognizing the existence of supernatural entities inhabiting particular physical objects, particular care must often be taken in planning new construction work in order not to disrupt such habitats. This may imply costly relocation of roads and buildings.
Broader Supernaturalism (#PF8433).
Related Extraterrestrial invasion (#PF4444).

♦ PF3439 Unequal distribution of fame and honours
Nature Some talented people are rewarded immediately, others late in life, still others after death, and many whose deeds merit it may never be recognized at all. Chance is a factor, as well as corruption. The greatest inequality is the denial of recognition because of nationality, race, beliefs, sex, youth, age or physical handicap.
Incidence Among astronomical, biological, psychological and other scientific discoverers, women have been insufficiently recognized, a recent example having occurred when despite contributions by both sexes, only men received recognition in DNA research. Americans, to take another instance, dominate world media and entertainment; their actors and actresses are glorified, sometimes well beyond their merits, but there is no Oscar for an Egyptian, Indian, African, or Latin-American local performer, for example. Many national systems of honours, such as in the UK, neglect singular voluntary work, while political bureaucrats and money-makers receive OBEs and MBEs. In the arts, commercialism prevails, and intrinsic merit is overlooked in favour of what sells. Writers of outstanding quality are suppressed or censored by political regimes and religious orders. Recognition, when it comes, comes too late to support those who have been isolated and obscure.
Broader Injustice (#PA6486) Human inequality (#PA0844).
Narrower Nepotism (#PD7704).
Related Inequality of opportunity (#PC3435).
Aggravates Cynicism (#PF3418) Frustration (#PA2252) Obstacles to leadership (#PF7011).
Aggravated by Corruption (#PA1986) Power politics (#PB3202).

♦ PF3448 Underdeveloped potential of basic resources
Nature The resources considered necessary for community self-sufficiency have expanded beyond natural resources to include techniques for communication, transportation, commerce and financing which were once considered luxuries. However, in many Third World villages, the basic resources that are present are severely overtaxed while others are only beginning to be introduced. The currently cultivated land is being stretched to support increasing population: material from the land is used for every possible purpose, even cornstalks serving as cooking fuel, fodder and roofing. Nevertheless, even where the land is rich and potentially productive, irrigation water may not yet be available; fertiliser and weed or pest control, where available, may be more expensive than most farmers can afford. Land ownership remains a complex matter that seems to work against coordinated farming efforts. The new resources such as radios and telephones, dependable transportation, local health services and commercial skills may be beginning to appear, but at only a fraction of what is needed for social and industrial development. Plans may be made but are slow in coming to fruition: bridges may have been planned for years but not yet be built; telephone cables may be laid, but few telephones be installed; and veterinary hospitals may be built but be without full-time staff. Residents are distressed at the realisation that the only obstacle to many of these improvements is their own lack of expertise in petition and negotiation.
Broader Underdevelopment (#PB0206).
Narrower Scant animal feed (#PG5264) Unirrigated islands (#PG5315) Overdemand on resources (#PG5246) Unavailability of timber (#PG5299) Inadequate reservoir plan (#PG8321) Insufficient pest control (#PJ1086) Monopolized land ownership (#PG1151) Shortage of cultivable land (#PC0219) Irregular transport services (#PE5345) Undependable river transport (#PG5348) Scarcity of well-paying jobs (#PG7944) Unreliable traditional farming (#PG8339) Insufficient enabling structures (#PG8390) Unavailability of local dentists (#PG5331) Insufficient industrial promotion (#PJ8378) Vulnerability of telephone system (#PE8254) Scarcity of subsidized fertilizers (#PG5282) Insufficient commercial facilities (#PJ5337) Underutilization of geothermal energy (#PJ7988) Individualistic agricultural planning (#PJ1082) Underdeveloped benefits of community resources (#PG8348).
Related Prohibitive cost of fuel (#PJ0346).
Aggravates Difficult land acquisition (#PJ5369) Mismanagement of irrigation schemes (#PE8233) Long-term shortage of natural resources (#PC4824).
Aggravated by Inadequate health services (#PD4790) Non-articulated educational goals (#PF2595).

♦ PF3454 Restrictive effects of traditional community decision-making
Nature In small Third World communities, the methods of local decision-making depend heavily on traditional paternalistic patterns which frequently inhibit local development. Occasions that enable all the residents to express concern for the village's future seldom arise; there may be no facilities which enable the gathering of the whole community. The limited roles of informal leadership allow few opportunities to develop the methodological expertise which would realize its full potential. Strong individualism fostered by subsistence living acts against community cohesiveness; and an attitude of waiting for "them" to take care of us prevents significant undertakings being planned or actualized.
Broader Restrictive social practices (#PC5537) Obstacles to community achievement (#PF7118).
Narrower Unattractive social life (#PJ1212) Unsanitary refuse disposal (#PJ1098) Social isolation of the elderly (#PD1564) Unfocused style of community operations (#PF6559) Lack of local leadership and decision-making structures (#PF6556).
Related Limited meeting facilities (#PE1535).
Aggravates Undemocratic policy-making (#PF8703) Unpublicized public meetings (#PF5222).
Aggravated by Excessive consumption of goods and services (#PC2518).

♦ PF3455 Excessively costly prestige projects
Grandiose public works — Unnecessary status projects — White elephantiasis
Claim The biggest, best, most costly white elephant projects plague developing and developed nations, the public sector and the private, and socialist, communist and capitalist countries: the Sears Tower in Chicago and plans for a 150 story building in New York, a 122 story building in Newark, New Jersey, and a 125 story building in Chicago are representative of architectural overreach. Seldom are these building profitable. They are more costly than lower buildings with equal floor space. They are business and architectural icons of power. Romania spent $1 billion on grandiose palace during Ceausescu.
In science, the $23 billion space station, $4 billion superconducting supercollider and the $3 billion human genome projects are marginally justifiable scientific projects but best understood as monuments to US capacity. Politicians in Latin America, Africa and Asia have built ill-sited dams and power stations; massive steel plants for agricultural countries; and multilane roads to nowhere. Germany mistakenly built a petrochemical plant on its North Sea coast. Britain has an international airport on a distant shore of Scotland. The Philippines has an unworkable nuclear power station sited near three geological faults.
The Concorde supersonic aircraft is a classic example. Capable of propelling 100 passengers paying $3,600 (one way) in a cramped seat from London to New York at twice the speed of sound, whilst consuming mre fuel than any other commercial aircraft. In the 1960s the development costs were nearly $5 billion. The aircraft (of which only 7 were produced, rather than the 150 planned) had to be given to the operators. In 1990 a study is being undertaken to produce a new version.
In 1990 the Pope opened a $150 million basilica in the home-town of the president of the Ivory Coast. It was built at a time when the government introduced a widespread austerity programme and with 70 per cent of the population living in abject poverty.
Narrower Unnecessary relocation of national capitals (#PG8532) Misappropriation of resources for high cost research projects (#PF0716) Misappropriation of resources for high cost civil engineering projects (#PF4975).
Aggravated by Racism (#PB1047) Boasting (#PF4436) Humiliation (#PF3856) Second class states (#PD0579) Messianic policy-making (#PF9796) International status race (#PC5348) Abusive national leadership (#PD2710) Pursuit of corporate prestige (#PF7983) Race as a reinforcement of nationalism (#PF3352).

♦ PF3458 Unaccountability of institutions degrading the environment
Environmental unaccountability of government agencies — Environmentally-insensitive government policies
Nature Governments have failed to make the bodies whose policy actions degrade the environment or inhibit development responsible for ensuring that their policies prevent that degradation or ameliorate development opportunities.
Incidence The existence of government agencies, or government-supported institutions, concerned with particular environmental or development issues has created the false impression that these bodies were by themselves able to protect and enhance the environmental resource base. But the mandates of central economic and sectoral ministries are often too narrow and too concerned with quantities of production or growth. As a result although ministries of industry include production targets, any resulting pollution is left to ministries of environment to handle. Similarly, whilst electricity boards produce power, the acid pollution is left to other bodies to clean up. The agencies with specific environmental responsibilities may also interpret their own mandates narrowly or be allocated inadequate resources to respond to the challenges which are

effectively delegated to them.
Broader Irresponsibility (#PA8658).
Narrower Unaccountability of international financial institutions (#PF1136).
Aggravates Environmental degradation (#PB6384)
Unethical commercial practices (#PC2563)
Irresponsibility towards future generations (#PF9455)
Irresponsible scientific and technological activity (#PC1153)
Social irresponsibility of transnational corporations (#PE5796).

♦ PF3460 Overemphasis on self-sufficiency with respect to interdependence

Nature An extreme emphasis on the importance of self-sufficiency causes the individual to underestimate his interdependence with others across the world. Education, welfare programmes, international relations and ethics all share this extreme position, blinding individual and social relations alike to the fact that interdependence is equally important as self-sufficiency.
Broader Theological collapse (#PF6358).
Narrower Exclusion of opposing views (#PF3720)
Fear of the loss of independence (#PE7468)
Non-articulated educational goals (#PF2595)
Reduced understanding of globality (#PF7071)
Search for individualistic meaning (#PF6796)
Individualistic welfare responsibility (#PF2560)
Breakdown in community security systems (#PD1147)
Legal contract system reduced to individual needs (#PE5397)
Story that self-sufficiency is socially responsible (#PE5795).
Related Lack of commitment (#PF1729) Dominance of economic motives (#PF1913)
Collapsed meaning of human creativity (#PF0936).

♦ PF3462 Over-qualification

Broader Waste of human resources (#PC8914).
Narrower Overqualification of women (#PG6441)
Denial of right to leave any country (#PE3463).
Aggravates Frustration (#PA2252) Lack of job satisfaction (#PF0171).
Aggravated by Lack of social planning (#PF8185)
Inequality of opportunity (#PC3435).

♦ PF3464 Underutilization of legal rights

Unfamiliar legal rights
Narrower Psychological barriers to the judicial protection of individual rights (#PE1479).
Related Political apathy (#PC1917).
Aggravates Apathy (#PA2360) Injustice (#PA6486) Human inequality (#PA0844)
Discrimination against women in politics (#PC1001).
Aggravated by Inequality before the law (#PC1268)
Denial of right of complaint (#PD7609)
Inadequate dissemination and use of available information (#PF1267).

♦ PF3467 Insufficient access to technology for agricultural upgrading

Nature Despite the fact that, even in the most depressed areas, people are extremely capable farmers given their present level of technology, it is continually necessary to upgrade skills and increase production. However, rural communities tend to lack access to agricultural technology: the advantages of modern mechanical implementation, the added production which comes from systematic irrigation and the increased yield of hybrid seeds always seem just beyond the reach of the individual farmer. Lack of access to such techniques leave the farmer frozen at his present level of expertise while the economic demands of contemporary society fall on his shoulders with alarming weight.
Broader Lack of means for local technological development (#PF6454).
Narrower Untimely seed supply (#PG8500) Lack of irrigation equipment (#PG8457)
Inadequate vehicle servicing (#PJ0909) Inadequate storage facilities (#PG4268)
Insufficient modern technology (#PJ0996) Government restriction on exports (#PG8515).
Related Soil erosion (#PD0949).
Aggravates Inadequate land drainage (#PD2269)
Stagnated development of agricultural production (#PD1285).
Aggravated by Abusive technological development under capitalism (#PD7463).

♦ PF3468 Lack of representation

Broader Undemocratic political organization (#PC1015)
Denial of right to participate in government (#PE6086).
Narrower Denial of legal representation (#PF3517)
Unequal political representation (#PC0655)
Failure of individuals to participate in social processes (#PF0749).
Aggravates Political disintegration (#PC3204).
Aggravated by Restriction on press coverage of legal affairs (#PE7945)
Restrictions on direct news coverage of parliamentary affairs (#PF3072).

♦ PF3470 Unwritten language

Nature The undesirability of a society communicating only in an unwritten language is its ephemerality if the language is immaterial (voice, sound, gesture) or its abstractness if it is material (a circle of stones, a knotted string, symbolic colours). In dwindling primitive tribes, the cultural legacies will be lost when members of the tribes no longer understand the unwritten languages. In some countries, the use of a non-latin script for writing is, for the purposes of using modern telecommunications, the same as if the language were unwritten, since the letters or characters cannot be transmitted or entered into computers.
Incidence The majority of the 3,500 languages and dialects in the world, some 3,000 in fact, are unwritten. Mexican Indians have over 100 unwritten languages and dialects for example, and India has over 1,000.
Broader Diverse unilingualism (#PF3317) Destruction of cultural heritage (#PC2114).
Narrower Localization of unwritten language (#PG6450)
Inability to teach unwritten language (#PG6451).
Aggravates Inadequate documentation (#PF6453)
Lack of historical record (#PG6453)
Illiteracy among indigenous peoples (#PD3321)
Inadequate education of indigenous peoples (#PC3322).

♦ PF3474 Impunity of violators of human rights

Incidence State security agencies have used such illegal repressive methods as detention without trial and torture, or misused their powers to carry out intelligence operations and raids in order to arrest and hold people. The identi., of attackers may then be deliberately concealed from victims. The involvement of members of different security agencies in an operation, and the illegal use of civilians in security agencies, may make any criminal or disciplinary investigation difficult. The establishment of self-defence groups, which can or do make use of weapons belonging to the armed forces, may be legalized, thus contributing to the number of crimes committed. By keeping no records of either the operations that lead to arrests or of the names of people who are arrested and held in military and police premises, it is easy to ensure the disappearance of people. Authorities conducting preliminary inquiries into abuses may destroy or tamper with evidence. The refusal of authorities to admit any complaints about acts which violate human rights is a crucial factor ensuring the impunity of any violators. Witnesses or complainants in such cases may be intimidated, be murdered or simply disappear.
Claim Allowing the perpetrators of serious human rights violations to go unpunished is tantamount to leaving the machinery intact and condoning the conduct that enabled such crimes to be committed and to shirking a fundamental responsibility to the future: namely protecting the basic values of civilized ecoexistence.
Aggravates Inadequate enforcement of human rights (#PC4608).
Aggravated by Connivance of authorities in human rights abuses (#PF9288)
Government failure to prosecute offenders effectively (#PE9545)
Misinformation concerning infringement of human rights (#PF9794)
Government inaction on alleged human rights violations (#PE1407).

♦ PF3494 Ease of manufacture of nuclear bombs

Nature All the knowledge required to manufacture nuclear bombs is now available in public libraries in unclassified literature. A few individuals with very limited practical experience could produce small bombs within a short time if they could obtain the necessary raw materials; the resources necessary for the manufacture of a few rudimentary nuclear weapons are within the means of many nations. The essentials are a cadre of trained personnel; uranium; an industrial base adequate to permit the construction of a nuclear reactor; and auxiliary facilities large enough to provide the necessary quantities of plutonium. Thus many nations possess resources sufficient to undertake, without special outside assistance, the manufacture of rudimentary nuclear weapons, given the national will to do so and the readiness, in some cases, to forego the benefits from the endeavours to which those resources might otherwise be applied. The time required would vary among the group of countries, and for those which have only the minimum resources, it might be up to ten years or more. At the upper end of the scale, highly industrialized nations, with substantial national income, large numbers of trained scientific, technical and managerial personnel, and a reasonably accessible source of uranium, could become capable of manufacturing rudimentary nuclear weapons within a few years.
Aggravates Nuclear accidents (#PD0771).
Aggravated by Theft of nuclear materials (#PD3495)
Proliferation of nuclear weapons and technology (#PD0837).

♦ PF3495 Religious deception

Religious lies — False prophets
Incidence A number of religious traditions have specific warnings to their followers concerning deceivers who will emerge and present attractive, but false, teachings in order to lead them away from the truths to which they hold. In the Christian tradition particular attention is focused on false prophets and the emergence of the Antichrist as the prime deceiver.
Broader Lying (#PB7600) Deception (#PB4731).
Aggravates Religious hypocrisy (#PF3983) Institutional lying (#PD2686)
Religious propaganda (#PD3094) Religious vilification (#PD5534)
Unethical practices of priesthood (#PF8889).
Aggravated by Antichrist (#PF9139).

♦ PF3498 Ineffective systems of practical education

Nature There is a world-wide need for continuing education in basic skills to meet the changing demands of peoples' lives. However, the educational process, as well as the wage scale, tends to avoid bestowing sufficient prestige upon the skilled technical worker who is so critically needed for new, small industries in rural communities. When practical courses are offered, some potential students stay away out of ingrained prejudice against intellectual pursuits. Others reject courses that don't lead to success in one or another system of examinations. In addition residents, whatever their age, who may be eager to learn new skills nevertheless lack the basic education necessary to benefit from such training. Many village men and most women are illiterate in terms of a functional vocabulary as well as in the cultural skills required by society in general, and lack the basic training in reading, writing and mathematics increasingly needed by rural people to comprehend official communications, instructions on health practices and information on agricultural and commercial methods. However, many rural primary schools only conduct classes a few hours a week; and parents and teachers alike agree that more adequate and imaginative teaching methods are needed in school to stimulate learning incentives. Young people may remain illiterate even after six years in school. Adequate literacy training may be available in urban preparatory schools and national universities, but rural people are unable to take advantage of this because of the additional expense and limited availability of transport, and because the young people from poorer families begin work at an early age. Those who receive advanced training tend not to return to their villages.
Broader Obstacles to education (#PF4852).
Narrower Over-education (#PC6262) Inadequate education (#PF4984)
Poor supervision teachers (#PJ3757) Inappropriate role learning (#PG5343)
Segmental school experience (#PJ9345) Low performance requirements (#PG5256)
Uncoordinated special training (#PG9485) Ineffective in-service training (#PG5291)
Unskilled programme development (#PG5300) Unimaginative educational vision (#PF3007)
Unproductive extension of schooling (#PF6601)
Unrealized use of education structures (#PF2568)
Constricting effect of educational structures (#PF6476)
Ineffective mechanisms for functional training (#PF1352).
Related Narrow scope of education (#PF3552) Inadequacy of formal education (#PF4765)
Minimal opportunities of adult training (#PF6531)
Unavailable education for effective living (#PF2313).
Aggravates Low learning expectations (#PJ8960)
Inappropriate basic hygiene (#PD8294)
Underdevelopment of forestry industry (#PG5324)
Illiteracy as an obstacle to acquiring skills (#PE8246).
Aggravated by Blocked parental participation (#PJ5308)
Prejudice against communication by visual imagery (#PF0076).

♦ PF3499 Unattractive locale for economic development

Poor geographical location
Nature Economic development in scattered populations tends to be hampered because of the limited availability of the factors required for the expansion of industry and business. In spite of long-range development strategies, prime land remains unavailable or overpriced and there is an absence of building sites for establishing businesses. In addition, an undeveloped market characterized by small populations and long distances from other potential markets deflates enthusiasm for new commercial enterprises. For many, the desertification, drought, erosion and flood caused by traditional 'development'—in the form of mining, cutting down forests, constructing massive waterworks to provide irrigation—are too high a price to pay for the dubious benefits received.
Incidence A third of India's 266 million hectares of agricultural land has been turned into wasteland by erosion, waterlogging of salinity—the ill effects of deforestation and modern irrigation.
Broader Underdevelopment (#PB0206) Deficiencies of developing countries (#PC4094).
Narrower Limited market areas (#PG8599) Business skills drain (#PG8609)
Long shipping distances (#PG8674) Prohibitive cost of land (#PE4162)
Youth migration to cities (#PJ7859) Undiversified labour force (#PG8559)
Uncontrolled waterfront damage (#PG8619) Restrictive logging regulations (#PG8653)

Unavailability of building sites (#PJ8549)
Inaccessible supply of repair materials (#PG8569).
Related Youth unemployment (#PC2035) Limited available land (#PC8160).
Aggravates Occupational diseases (#PD0215) Incompetent management (#PC4867)
Insufficient trained labour (#PD9113) Unexplored potential markets (#PF0581)
Short range planning for long-term development (#PF5660).
Aggravated by Geographical isolation (#PF9023).

♦ PF3500 Limited approaches to economic planning

Nature Although business and government are said to invest in the future, this "future" tends to refer to "the next five years", or to be limited in scope to the future of one place or, at most, of Western nations. This mode of planning results in chaos wherever long-term commitments are required. Business does not have the vision, the practical tools, nor the methodological styles necessary to plan for its own development in the context of the overall, world-wide economic foundation, and government is blocked by the need to respond to voters and powerful lobbies.

Incidence Recently the British government took four years of public debate before it went ahead with plans to construct one nuclear reactor.

Broader Belittling of grant recipients (#PF2708).
Narrower Ineffective data systems (#PF3671) Limited image of employability (#PF2896)
Short range planning for long-term development (#PF5660)
Absence of long-term economic planning agencies (#PF3610).
Related Imperialistic distribution system (#PD7374)
Monopoly of the economy by corporations (#PD3003)
Unregulated ownership of the means of production (#PF2014).
Aggravates Undirected expansion of economic base (#PF0905).

♦ PF3511 Anti-consumerism

Related Trade protectionism (#PC4275).
Reduces Consumerism (#PD5774).

♦ PF3512 Lack of community self-worth

Depreciated community identity

Nature Significantly diminished population due to changes in residency, plus the traditional pattern of social stratification which is now virtually accepted as the fated role of the villager have resulted in many rural villages having paralyzing and depreciating self-images that undermine their sense of community identity. Although some residents may remember a more active time when there was much commerce and frequent community celebrations, the underlying tone becomes that of failure, with few visible or spoken community traditions, and often its location is not even indicated on maps.

Broader Obstacles to community achievement (#PF7118)
Cumulative depletion of corporate initiative in rural communities (#PE3296).
Narrower Fear of loss of property (#PG1132).
Aggravates Lack of self-confidence (#PF0879) Erosion of elders' wisdom (#PF1664)
Declining community population (#PJ8746).
Aggravated by Unsustainable population levels (#PB0035).

♦ PF3513 Corporate planning paralysis

Nature People do not see themselves as being capable of achieving the type of status defined as desirable by consumer advertising. The sense of alienation and helplessness which results leads to the failure of their communities to organize for corporate planning. The potential of the people in such communities remains undeveloped and their needs remain unarticulated and unmet, while the power of such communities to exercise control over technological systems remains dormant.

Broader Gap between the function of social techniques and the needs they address (#PF3608).

♦ PF3514 High risk technologies

Refs Morone, Joseph G and Woodhouse, Edward J *Averting Catastrophe* (1988).

♦ PF3516 Bereavement

Mourning

Nature The death or loss of a loved one can cause disturbances in bodily and emotional functioning and in social behaviour. In the case of conjugal bereavement there is an increased risk of mortality of the remaining spouse in the succeeding five or six years, as well as an increased risk of disabling illness. Bereavement can aggravate or lead to substance abuse, whether alcohol, drugs or medications. Most conjugally bereaved persons suffer real clinical depression, with one or more symptoms, in the first year. As many as twenty percent may suffer for a longer time. The death of someone close also affects children in powerful and sometimes traumatic ways and may possibly affect their ensuing development. Lack of social and medical services for the bereaved aggravates their physical, mental, social, and even financial problems.

Counter-claim Lack of preparation for dying and the absence of realism amidst the artificialities and escapism of modern living make the dying and their eventual survivors equally responsible for the grief trauma prevalent in many, but not all, cultures.

Refs Deits, Bob *Life after Loss* (1988); Fulton, Robert *Death, Grief and Bereavement* (1976); Hafer, Keith W *Coping with Bereavement from Death or Diverce* (1988); Parkes, Colin M *Bereavement* (1987); Stroebe, Wolfgang and Stroebe, Margaret S *Bereavement and Health* (1987).

Broader Loss (#PA7382) Death (#PA7055) Human ageing (#PB0477).
Related Celibacy (#PA7410) Appropriation (#PA5688).
Aggravates Grief (#PF5654) Suicide (#PC0417) Fear of death (#PF0462)
Psychological inhibition (#PF6339) Social withdrawal of aged (#PD3518)
Mental disorders of the aged (#PD0919)
Exclusion of pre-adults from family decisions (#PE2268).
Aggravated by Natural human abortion (#PD0173)
Inhibited grief process (#PD4918)
Ageing of world population (#PC0027).

♦ PF3517 Denial of legal representation

Denial of right to legal counsel — Lack of legal representation

Broader Lack of representation (#PF3468) Inequality before the law (#PC1268)
Obstruction of international criminal investigations (#PF7277).
Narrower Lack of legal aid facilities (#PF8869)
Inadequate legal counsel for minorities (#PF1219)
Inadequate legal counsel for political dissidents (#PF0732).

♦ PF3519 Impractical education

Impractical training programmes — Unrealized intentions of practical education — Impractical educational aims — Impractical public education — Impractical school activities — Overacademic orientation at high school

Nature In the currently prevailing situation of rapid change in methods of production, prior training cannot ensure employment. In addition, job training programmes tend to focus more on providing credentials than on developing useful skills.

Broader Inappropriate education (#PD8529) Irrelevance of educational curricula (#PF0443)
Ineffective educational policy decisions (#PF2447).
Aggravates Unprepared adult leadership (#PF6462)
Over-specialized supervisory personnel (#PF3588).
Aggravated by Ineffective methods of practical education (#PF2721)
Unperceived relevance of formal education in rural communities (#PF1944).

♦ PF3523 Underutilization of human resources

Unutilized human resources — Undeveloped human resources

Nature In all countries of the world, developed as well as developing, the loss or underutilization, of human resources is both substantial, and detrimental to development. Examples include: the vast number of marginally or seasonally employed; the small number of women gainfully employed; the number of graduates working on "fringe" jobs rather than being engaged in significant enterprises; the exodus of youth from rural to urban areas in an unsuccessful search for livelihood; the disintegrating economic situation which forces many parents to use their children as farm labourers or in other tasks, thus seriously limiting their educational opportunities.

Broader Inefficient use of resources (#PE5001)
Inflexible social care structures in developing countries (#PF2493)
Excessive social costs of structural adjustment in debtor developing countries (#PD8114).
Narrower Priority of farm work (#PG8492) Youth migration to cities (#PJ7859)
Unemployed trained technicians (#PG8458) Underutilization of popular wisdom (#PF2426)
Insufficient leisure time for women (#PE8907)
Unpreparedness for surplus leisure time (#PF5044)
Migration of rural population to cities (#PE8768)
Underutilization of locally available skills (#PF6538)
Underutilization of potential in local communities (#PF6513).
Related Under-utilized raw materials (#PF6590).
Aggravated by Prohibitive cost of education (#PF4375)
Ineffective mechanisms for functional training (#PF1352).
Reduces Exploitation of child labour (#PD0164).

♦ PF3526 Undeveloped channels for public and private resources

Inaccessibility of bureaucratic services

Nature The remoteness of central bureaucracy may prevent agencies from being aware of the needs of local communities. Bureacratic channels for obtaining the information, funding and social services which are intended to be of service to the community are often inaccessible to the inexperienced; and attempts at local improvements and programmes may be complicated by apparently inflexible regulations, such as building codes, environmental rulings and programme qualifications criteria. In addition, people are suspicious of and reluctant to deal with government, thus hindering further their ability to tap available government funds. As a result, some designated funds remain unused.

Residents of most small towns and villages are unaware of how to link the social resources which are available through governmental and social agencies with their local needs; nor do they know how to proceed after an initial negative response to a request for assistance. Knowing the appropriate persons to contact and the procedures for following through require persistence and continuity on the part of residents, which they are often not ready to have. A similar situation exists in relation to the delivery of health services. Because residents are unfamiliar with the methods of seeking assistance offered to rural communities, they either do not attempt to obtain the resources or quit at the first discouraging response, thereby depriving themselves of necessary services.

Broader Underprovision of basic services to rural areas (#PF2875).
Narrower Limited social services (#PG8629) Mistrusted federal funding (#PG8620)
Restrictive land ordinances (#PG8638) Unskilled bureaucratic contacts (#PG8610)
Prohibitive daycare requirements (#PG8580) Underutilized government resources (#PE9325)
Dependency on unpredictable sources of income (#PF3084).
Related Instability of water supply (#PD0722).
Aggravates Inadequate maintenance (#PD8984) Inadequate road conditions (#PJ0860)
Inappropriate sanitation systems (#PD0876).

♦ PF3531 Absence of traditional patterns of community life

Nature The patterns of community life, by which values were commonly developed and represented in past generations, are dying out, with the consequence that social responsibility has to be assumed by government agencies.

Broader Obstacles to community achievement (#PF7118)
Breakdown in community security systems (#PD1147).
Reduces Untransposed significance of cultural tradition (#PF1373).

♦ PF3534 Schism

Nature The evolution of human thought runs from the private conception; to the articulated, great idea; to the conversion of others; to the group enunciation of an ideology; to the group's organization and the institutionalization of what are then new dogmas. Immediately, therefore, difference of opinion involves difference of organization in the sense of the means to institutionalize new ideas. Rejection of new ideas is also a rejection of the body of persons adhering to them and is hence an organic schism in the institution. Schisms affect all systematic or prescriptive religions; philosophies of behaviour, individual and societal; and fundamental sciences such as biology and physics. While schisms are pronounced by the orthodox, the tendency that leads to them may lie, on occasion, in compulsive behaviour induced by complexes and other personality disturbances. This frequently manifests in the excessive ego drives that characterize many schismatics.

Within international organizations there is always tendency towards schism, which sometimes erupts as disruptive influences, and which sometimes proves to be a salutary development. In a religious context, schism differs from heresy in that it maintains the same rituals but, by disobedience, divides the community.

Broader Conflict (#PA0298).
Narrower Secession (#PD2490) Economic schism (#PG6612)
Political schism (#PC2361) Religious schism (#PF1939)
Ideological schism in communism (#PF3181).
Aggravates Social fragmentation (#PF1324).

♦ PF3538 Inadequacy of patent coverage

Inadequate patent control — Inadequate patent extension — Difficulty of patent application — Avoidance of patent restrictions — Patent abuse

Nature Patents provide ownership rights to ways of doing things. Lack of coordination in national practices for awarding patents may result in discrimination against foreign patent applications; and may also cause infringement, in whole or in part, of existing inventions because of the unmanageable problem of document search.

Background In the Middle Ages, the concept of intellectual property rights amounted to an injuction by master craftsmen on journeymen not to use techniques they had observed in their apprenticeship. In 1790 the USA passed the first modern patent Act, France followed suit in 1791, and intellectual property rights became a European cause célèbre up until the middle 1800's.

Claim Patents, which represent the rights of inventors to dictate the use and distribution of their inventions, are actually little more than restrictive monopolies of private interest, which distort economic choices, hinder industrial growth, and are open to widespread abuse. Since every invention is based upon the evolution of wider knowledge within the society, why should one individual or company have exclusive control over the utilization of the invention.

Counter-claim There would be very little incentive for anyone or any company to invest in research if there were no safeguards on that person or company having exclusive rights to use the results of that research, even for a restricted length of time. The abuse of unscrupulous commercialization of another's research would be far greater than the present so-called abuse of the patent system.

Broader Avoidance of copyright (#PD0188) Lack of international cooperation (#PF0817)
Vulnerability of intellectual property (#PF8854).
Narrower Delay in recognition of patents (#PF7779).
Related Injustice (#PA6486).
Aggravates Disincentives to invention (#PG6623).

♦ PF3539 Limited availability of functional information

Nature The shift to technology has left many Third World rural communities behind. Villagers are still using the practical skills and information that were appropriate to another age. They may have been taught some skills – modern machines, chemical fertilisers, power saws and birth control pills are all commonly used. But the information on their use is often incomplete, and villagers lack the understanding necessary for their full application. Such partial knowledge results in villagers being unable to realize the new economic possibilities open to them, and agricultural production and small businesses are impeded by gaps in technical information. Programmes for nutrition, sanitation and preventive medicines need explanation, information and instruction on how applications for assistance must be submitted. For example, despite the general availability of building materials, old methods of extraction, preparation and shipping represent an overwhelming task; other methods may be available through government programmes, but residents are unaware of how to profit from such programmes. In general, the acquisition of functional information that could transform everyday life has resulted in only confusion.

Broader Lack of information (#PF6337).
Narrower Scrapped automobiles (#PE5339) Birth control misinformation (#PG5266)
Lack of family life education (#PF5079) Unexplored business opportunities (#PG5323)
Lack of knowledge of eligibility for benefits (#PF8205)
Inadequate incentives for increased productivity in developing countries (#PE8506).
Related Ineffective methods of practical education (#PF2721).
Aggravates Ignorance of health and hygiene (#PD8023)
Inadequate family planning education (#PD1039)
Increasing development lag against information growth (#PE2000).
Aggravated by Irrelevant available information (#PG8884).

♦ PF3549 Educational curricula based on content rather than method

Nature Although educational institutions may acknowledge their failure to offer problem solving or method-oriented curricula, they as yet have not developed a comprehensive, overall model for this type of education and few educators have the training necessary to provide a systematic methodological programme. With the exception of the scientific method, problem solving methods are lacking, whether motivational methods (except in business and military leadership seminars); study methods; reflective methods; or techniques of cooperation. The few instances where methods are taught they are generally in an isolated situation and there is no obvious relationship between one kind of method and another.

Broader Irrelevance of educational curricula (#PF0443).
Related Defensive life stance (#PF0979) Limits on areas of research (#PF2529)
Collapse in providing ethical value screens (#PF1114).
Aggravates Collapsed meaning of human creativity (#PF0936).
Aggravated by Absence of tactical methods (#PF0327)
Failure of motivating socio-dramas (#PF1042)
Undirected technological expansion (#PC1730)
Human wisdom unrelated to daily life (#PD1703)
Lack of motivating collegial relationships (#PF0967)
Shallow approaches to motivational techniques (#PF1004)
General unavailability of effective motivational techniques (#PF2702)
Antiquated intellectual methods to appropriate human depths (#PF1094)
Absence of convincing symbols connecting the individual's life to the cosmos (#PF1081).
Reduced by Educational curricula over emphasizing method rather than content (#PD1827).

♦ PF3550 Limited availability of investment capital for urban renewal

Limited availability of funding resources for urban services — Restricted access of municipalities to capital funds — Poor credit alternatives in urban areas

Nature Municipal budgets are rarely sufficient to provide desired police, fire, educational and public works improvements, and rate assessments often remain unchanged for a number of years. Yet the access of many small communities to potential sources of development capital – innovative loans and grants from private corporations, foundations and government agencies – is severely restricted. For those classified as depressed areas, most risk capital is unavailable and investment opportunities are unattractive. With high insurance and unreasonable construction costs, residents have very little collateral, which hinders receipt of loans.

There are several elements which further block attempts to receive government grants: the funding procedures seem hopelessly tedious; few citizens are aware of the wide range of loans available; there is widespread belief in communities that accepting such grants is somehow dishonourable Lack of municipal capital affects the individual citizen as well as the municipality: large-scale farmers are reluctant to diversify their crops partly because, after several bad crop years, they do not see how they can raise the capital needed to purchase the necessary agricultural equipment; those who farm on a smaller scale cannot accumulate the capital needed for adequate irrigation or chemical crop treatment. Every plan for future development is inhibited by the lack of capital funds, and this generates an overall sense of hopelessness. Without growth in the funding base to provide loans, necessary upgrading of housing and services cannot take place; and local government is unable to ensure proper maintenance of roads, public buildings and necessary city services, which further lowers the investment potential.

Broader Insufficient capital investment (#PF2852)
Limited access to external resources (#PF1653)
Constricting level of capital development in rural areas (#PE1139).
Narrower Underpayment of police (#PJ8601) Inadequate police funds (#PJ8760)
Insufficient risk capital (#PJ1704) Inaccessible housing loans (#PG8648)
Insufficient housing funds (#PG8768) Insufficient property base (#PG8800)
Complex funding mechanisms (#PG1502) Prohibitive irrigation costs (#PG8730)
Prohibitive cost of insurance (#PE8632) Depressed area classification (#PG8639)
Outdated property assessments (#PG8816) Prohibitive construction costs (#PJ8591)
Dishonourable grant acceptance (#PG8710) Limited financing opportunities (#PG8611)
Prohibitive cost of maintenance (#PF0296) Lack of fire-fighting facilities (#PJ2437)
Prohibitive cost of farm equipment (#PJ8790) Prohibitive cost of crop treatment (#PG8823)
Lack of finance in coastal communities (#PE2425).
Related Lack of local commercial services (#PF2009)
Underprovision of basic urban services (#PF2583)
Limited local availability of capital reserve (#PF2378)
Failure of public authorities to assist in financial investment (#PF6508)
Lack of awareness of potential for investment in small, inner-city enterprises (#PF2042).
Aggravates Diminishing capital investment in small communities (#PF6477).
Aggravated by Risk of capital investment (#PF6572).

♦ PF3551 Accountability based solely on profit

Unreliability of profit as a social indicator

Nature The overemphasis by the means of production on their profitability so severely narrows the field of productive accountability that it becomes virtually impossible to respond to other production values.

Claim One often referred to indicator that the greed of capitalism has abated is that the average rate of profit has fallen in a number of large industrial countries since the beginning of the 1970s. However, there are notorious difficulties encountered in trying to obtain a reliable measure of these rates, and the national differences in the timing of changes in rates and levels of profit. Moreover the key factor in determining the international reorganization of capital is not so much the absolute level of profit and its changes over time, but the divergence between the profits obtainable in the industrialized countries and those in the developing countries. It should also be noted that a fall in the average rate of profit is not incompatible with a constant or even increasing rate of profit for the majority of large companies.

Broader Production of non-essentials (#PC3651).

♦ PF3552 Narrow scope of education

Nature The paucity of teaching materials, basic equipment and trained staff, especially in practical areas, available in the formal schooling structures of many developing and developed countries, limits education to only the bare fundamentals at the expense of breadth and practical relevance. This leads to irregular attendance on the part of students and decreasing enthusiasm and commitment on the part of the teachers.

Broader Irrelevance of educational curricula (#PF0443)
Ineffective methods of practical education (#PF2721).
Narrower Restricted job training (#PJ0617) Overemphasis on technological skills (#PF2956).
Related Inadequacy of formal education (#PF4765)
Unavailable education for effective living (#PF2313)
Ineffective systems of practical education (#PF3498).
Aggravates Cultural illiteracy (#PD2041)
Limited access to culturally adapted pedagogical material (#PE8152).
Aggravated by Educational curricula over emphasizing method rather than content (#PD1827).

♦ PF3553 Absence of images of social responsibility

Insufficient responsibility images

Nature There is a present-day lack of acceptable images to emphasize the individual's social responsibility and the way in which it can be assumed. People thus find themselves either maintaining the status quo or refusing to participate as social beings.

Broader Breakdown in community security systems (#PD1147)
Fragmented conduct of community operations (#PF1205).
Aggravates Limited access to social benefits (#PF1303).

♦ PF3554 Minimal access to appropriate technology

Nature Development projects often fail to give priority to alternative technologies best suited to particular needs; rather, they tend to waste resources on more sophisticated systems which eventually prove unsuited to the purposes they were expected to serve. This situation frequently arises as a result of a lack of information: "small" technologies seldom receive widespread promotion, and consequently planners are unaware of alternative options and existing possibilities. Governments may compound the problem by calling unsophisticated alternatives "inferior". Lack of cooperation among projects or among nations also hinders awareness and acceptance of alternative solutions. However, even if an appropriate technological solution is adopted, it may never filter through the bureaucracy to the village level, where it is actually needed; or the villages may never receive adequate instruction in how to implement the solution.

Incidence Examples of village development projects failing, even though all the equipment was available, are legion, especially in the developing world. Irrigation pumps may be installed while the technical knowledge to keep them operating is lacking, so inoperative pumps are stored in a village while at the same time the farmers ask for irrigation equipment. (This knowledge is potentially available since similar engines in automobiles are commonly kept in running order twenty years past their normal life). Other equipment, supplies, materials and skills needed for construction, health improvement, increasing farm production and small industry are available in the larger regional towns, but techniques are expensive to obtain in both time and money. Water wells have been dug for centuries, but today the idea is so prevalent that the ground is too hard or the water too deep to reach even with modern drilling equipment, that a village goes without water that could be near at hand. Fertilisers and other chemicals available from the government require precision application for effective use. Even though their yields could be increased, the farmers do not have the necessary means for using modern farm technology; technology could greatly improve livestock production, yet husbandry goes on as it has for centuries. Commercial techniques of cooperative buying and selling may be understood by vendors and private farmers, yet they are not practised.

Refs Riedijk, W (Ed) *Appropriate Technology for Developing Countries* (1987).

Narrower Limited drilling equipment (#PG5267) Minimal financial understanding (#PG8129)
Individualized marketing practices (#PG5334) Mismanagement of irrigation schemes (#PE8233)
Unavailability of building materials (#PJ0704).
Aggravates Inadequate irrigation system (#PD8839).
Aggravated by Shortage of technical skills (#PF6500)
Abusive technological development under capitalism (#PD7463).

♦ PF3558 Deteriorating structures of rural community cooperation

Insufficient village cooperation — Lack of village cooperation — Inadequate inter-village cooperation — Undeveloped community cooperation

Nature The strain of divisions borne of long centuries of feudal tradition is being exaggerated in the rural communities of developing nations by the difficult economic situation. This is causing tensions between merchants and farmers and between the relatively prosperous and the less affluent people of the community. The lingering influence of now outmoded structures of social care and decision-making limit the effect of beliefs in authentic participation and intentions to embrace forms of social interdependence. It is understandable that the resultant prevailing attitude is individuality.

Broader Obstacles to community achievement (#PF7118).
Narrower Disunity among traders (#PG8474).
Related Untransposed community structures (#PF6450)
Over-specialized supervisory personnel (#PF3588)
Ineffective means for goods supply and distribution (#PF6495).
Aggravates Declining sense of community (#PF2575).
Aggravated by Marxism (#PF2189) Deteriorating community identity (#PF2241)
Limited relations beyond local environments (#PF3192)
Demoralizing images of rural community identity (#PF2358).

♦ PF3559 Fragmented individual decision-making process

Isolation of individual decision-making from context

Nature There is no social encouragement for the individual to allow others to participate in his decision-making. This isolation makes consensus virtually impossible to maintain with any degree of depth.
Broader Social fragmentation (#PF1324) Fragmented decision-making (#PF8448)
Breakdown in community security systems (#PD1147).

♦ PF3560 Limits on participation in community development

Nature Regardless of their degree of concern, members of developing communities often find their ability to participate severely restricted by material and cultural restraints. Traditionally, the primary centre of corporate life is the clan or extended family, so that large gatherings of villagers occur only at feasts and holidays celebrated by the wider group. Working together in the fields, meeting along the roads or in the home are the only other, informal gatherings within a village. Most families and individuals are burdened by the weight and expenditure of time related to subsistence living, and crowded homes and family size exceeding income capacity increase the obstacles to community engagement. For example, schools are conducted with little adult involvement (partially due also to adult illiteracy), yet the relevance and responsiveness of such an institution depends on the involvement of the parents and other adults of the village.
Broader Undemocratic policy-making (#PF8703) Obstacles to community achievement (#PF7118).
Narrower Limited study time (#PG5327) Excessive family size (#PJ5277)
Family-oriented socializing (#PG5259) Inadequate sports equipment (#PG8821)
Parent-teacher non communication (#PF1187) Restricted extra-curricular activities (#PG8299).
Aggravates Unorganized community recreation (#PF5409).
Aggravated by Insufficient community events (#PF5250)
Decline in government health expenditure (#PF4586)
Overcrowding of housing and accommodation (#PD0758).

♦ PF3562 Disasters of extraterrestrial origin

Nature Astronomical phenomena exist which can cause disaster conditions on the Earth, the best known example being meteors.
Broader Disasters (#PB3561).
Narrower Novae (#PF3563) Interstellar dust (#PF9002) Meteors as hazards (#PF1695)
Hazards from comets (#PF3564) Meteorites as hazards (#PF1687).

♦ PF3563 Novae

Supernovae — Stellar explosions
Nature Stars may explode, releasing an enormous amount of energy and ejecting incandescent gases at extremely high velocity. In the space of a few months, novae may release the same amount of energy radiated by the sun in 10,000 years; supernovae release the same amount of energy released by the sun in 1,000 million years. It is highly probable that life on earth would be completely destroyed if a star close to the solar system were to explode in this way.
Incidence Stellar explosions are infrequent; novae occur in the galaxy at the rate of 20 to 50 per year and supernovae occur at an estimated rate of one per galaxy per 360 years. Galactic novae are concentrated in a band 10 degrees each side of the plane of the galaxy and are densest toward the centre (our solar system lies towards the perimeter). The probability of the earth being affected is therefore very low. It has however been estimated that a supernova explosion will occur once in every 50 million years within 100 light years of the earth. Even at this distance the effect upon the atmosphere would be certainly catastrophic and sufficient to cause mass extinction of many species.
Broader Disasters of extraterrestrial origin (#PF3562).
Related Meteors as hazards (#PF1695).
Aggravates Evolutionary catastrophes (#PF1181).

♦ PF3564 Hazards from comets

Hazards from asteroids
Nature Comets consist of a nucleus (of the order of 1 km in diameter), a diffuse envelope of dust and gases driven out of the nucleus by solar radiation (possibly up to 400,000 km in diameter), and a tail composed of molecules and very finely divided dust particles (possibly over a million km long). There is a vast reservoir of comets in orbit around the sun. Their periods range from a few years to several million years, so that many may exist which have not been visible during recorded history. There is therefore some possibility that at least one future comet may intersect the path of the earth or cause meteors to do this. Disastrous effects include a direct cometary collision; a near collision with gravitational and tidal interactions; dense, heavy showers of meteorites striking the earth's surface; and interferences in space affecting earth missions, terrestrial magnetic or meteorological phenomena, or the moon.
Incidence In the case of asteroids, it has been estimated that only 5 percent of near misses are ever detected. Asteroids and parts of comets have been striking the earth since it began. The geological record of earth shows evidence of more than 100 large objects hitting the planet. Some evidence supports the theory that 65 million years ago a giant object struck the earth, pulverizing a hugh area and spewing so much debris into the atmosphere that the skies darkened for months, temperatures dipped, and much of the life on the earth, most notably the dinosaurs, perished. In 1908, an object exploded in the atmosphere above Siberia, causing a tremendous blast and fireball, felling trees in a 200 square mile area. Objects of one half mile in diameter hit the earth once every 40 million years or so. One theory says that one comet ten meters across strikes the atmosphere every three seconds.
Background Comets are considered the source of great danger and many disasters. The Chinese believed them to be celestial brooms wielded by the gods to sweep the heavens free of evil, which then fell to earth, bring wars, floods, drought an other disasters. In A.D. 66 the passing of Halley's comet presaged the fall of Jerusalem. Its return in 451 was a portend to the defeat of Attila the Hun's armies at Châlons. The Saxons may have attributed their defeat in 1066 to its appearance. It heralded the descent of Turkish armies on Belgrade in 1456. When it came into view in 1910, some residents in Chicago prepared themselves for death by cyanogen-gas poisoning. Some contemporary scientists believe comets are the source of life and diseases like AIDS on earth.
Broader Bad omens (#PF8577) Disasters of extraterrestrial origin (#PF3562).
Narrower Mass extinction due to comet showers (#PF0039).
Related Meteors as hazards (#PF1695).
Aggravates Meteorites as hazards (#PF1687).

♦ PF3565 Inadequate warning of disasters

Ignored disaster warnings — Rejection of disaster warnings
Nature It is not yet possible to predict with any satisfactory degree of accuracy whether, when or where natural disasters will occur, or what is likely to be their impact. It is thus difficult to give adequate warning to the population in danger. The exception is meteorological phenomena likely to cause flooding or wind damage which are somewhat more predictable.
Incidence To cite earthquakes as an example, there is no global alert system in existence. Where an earthquake is considered imminent, the information is often of dubious reliability and poorly distributed. Often a country's only source of warning is its own relevant national institute, which may only monitor selected areas, in most cases limited to the area within the county's borders.
Aggravates Disasters (#PB3561) Ignorance (#PA5568)
Disaster unpreparedness (#PF3567).
Aggravated by False alarms (#PF4298) Scientific ignorance (#PD8003)
Avoidance of negative feedback (#PF5311)
Suppression of information concerning environmental hazards (#PF4854).

♦ PF3566 Inadequate disaster prevention and mitigation

Nature The severity of disasters caused by natural phenomena could be largely mitigated, and the actual occurrence of disasters caused by inappropriate human activities could be prevented, if adequate measures were taken before the event. However, little attention is given to long-term policies and plans: the identification of disaster-prone areas is incomplete; public information is scarce; early-warning systems are undeveloped; training of personnel is inadequate; and new legislation and environmental measures are either insufficient or non-existent. Often the effects of a disaster on a population are far greater than need be because of ignorance, neglect, poverty, and sometimes corruption and deceit.
Incidence Developing countries are especially vulnerable to disasters because they are generally not at all equipped to handle the resulting problems, nor do they have the resources to devote to prevention.
Claim Although a disaster can provide the impetus not only to rebuild what was lost, but to build in such a way as to mitigate or prevent a re-occurrence, such lessons are seldom learned. As result, needless suffering continues.
Broader Social neglect (#PB0883)
Weakness of infrastructure in developing countries (#PC1228).
Narrower Vulnerability of computer systems (#PE8542).
Related Natural disasters (#PB1151) Man-made disasters (#PB2075)
Disaster unpreparedness (#PF3567).
Aggravates Disasters (#PB3561) Inadequate disaster rescue and relief (#PF0286)
Uncertainty of development programmes due to short-term loans (#PF4300).
Aggravated by Ignorance (#PA5568) Corruption (#PA1986)
Socio-economic poverty (#PB0388) Unresponsive public authorities (#PF8072).

♦ PF3567 Disaster unpreparedness

Confused government response to disasters
Nature The rescue and relief of the trapped, injured and homeless requires a great deal of coordination and pre-planning of activities and the availability of equipment, manpower trained in rescue operations, emergency reserves of bulk food (particularly concentrated baby food), and medical supplies. In many countries the administrative provisions are lacking or only exist in a rudimentary form. When a disaster strikes, ad hoc bodies are created which lack experience, knowledge and organization and may therefore be of questionable effectiveness.
Claim The pattern of cyclone, flood, and famine which is observed, must inevitably continue, with devastating effect upon the poorest of the world's people. Totally inadequate standards of preparedness against clearly defined threats have been identified.
Refs Lathrop, J W *Planning for Rare Events* (1981).
Broader Social neglect (#PB0883)
Weakness of infrastructure in developing countries (#PC1228).
Narrower Unpreparedness for food emergencies (#PC5016)
Suspension of rights during states of emergency (#PD6380)
Vulnerability of animals during states of emergency (#PE4694).
Related Inadequate disaster prevention and mitigation (#PF3566).
Aggravates Disasters (#PB3561) Inadequate disaster rescue and relief (#PF0286)
Short range planning for long-term development (#PF5660).
Aggravated by Unreported accidents (#PF2887) Lack of coordination (#PF8330)
Inadequate warning of disasters (#PF3565) Unresponsive public authorities (#PF8072)
Vulnerability of telephone system (#PE8254).

♦ PF3575 Limited availability of education in rural areas

Limited opportunity for ongoing education in rural communities — Limited educational opportunities in rural areas — Irregular outside instruction in rural communities — Limited access of rural youth to education
Nature Residents of small communities may realize the importance of basic education for their children but are unable to provide adequate academic training with their own resources. Often the only structure of formal education existing in rural communities is a primary school of several grades with one teacher. Regional discrepancies within countries occur in varying degrees, depending upon the overall participation in education, but are particularly serious in developing countries. In developed countries discrepancies tend to appear at the age at which education ceases to be compulsory and therefore mostly originate at secondary school levels. In rural communities teaching aids and basic equipment available for educational tasks are minimal. The consequent problems of irregular attendance, wide age variation within grades, and discipline appear to be insoluble. Rural areas are not considered as attractive teaching posts even when the schools are new - houses are not available for new teachers, and outside instruction is limited and irregular. For most children, completion of the fourth grade is the end of formal education. A large number of young people who begin work at an early age stop going to school. Transport and boarding expenses prohibit attendance at more advanced classes which are usually available only away from the village. When young people graduate from high school, they have to move to the cities for jobs and further education. While continuing education is becoming increasingly necessary for individuals to keep up with the present rapid change, local schools and adult training opportunities are not available in most rural communities. Funds for educational use are often inadequate or misspent, and insufficient information or supplementary services such as transportation, scholarships, etc, results in little use being made of regional education structures. In the absence of effective structures of ongoing education, preparation for the future will continue to be limited and present patterns remain unchanged.
Incidence Despite the vast expansion of education that has reduced world illiteracy to 34 percent, the growth in population means that in fact the number of illiterate adults has actually increased, and the number of people in the 10 to 24 years age group is expected to double by the year 2000. The majority of the population is in rural areas, and traditional Westernized educational approaches are unable to meet the demand of this increasing population.
Claim Although education should prepare children, young people and adults to deal practically with the actual situations which confront them, educational opportunities in rural areas are impractical and unrelated to the daily requirements of the community. There is need for more practical and imaginative curricula in primary schools to stimulate learning, equip students and encourage parents to assist in seeking further education for their children. In addition, adult education rarely exists in any formal manner, even though basic reading, writing and mathematics are increasingly necessary. Instructions on health practices, information on agriculture and commercial methods, and leadership training are not locally available. Residents may desire to learn additional skills, but they are thwarted by the lack of relevant training events.
Counter-claim A large number of countries are in the process of improving the educational situation in rural area. Teachers are being trained. Education infrastructure is being developed. Curriculum is being designed to meet the employment needs of the country and away from the cultural biases of former colonial powers. The quality of life is being improved in the countryside to attract educators and keep students.
Broader Limited access to practical education (#PF2840)
Unavailable education for effective living (#PF2313).
Narrower Minimal opportunities of adult training (#PF6531).

Related Inequality of access to education within countries (#PC1896).
Aggravates Unequal access to education (#PC2163)
Insufficient special training (#PG9609)
Limited agricultural education (#PF8835)
Parent–teacher non communication (#PF1187).
Aggravated by Lack of leadership (#PF1254)
Undervaluation of education by parents (#PF9306)
Inadequate working conditions of teachers (#PE7165)
Lack of housing for teachers in rural communities (#PE8103).

♦ PF3581 Inadequate plant genetic resources conservation

Nature The conservation and maintenance of representative collections covering the genetic variation of cultivated plants and their wild relatives is still inadequate. The paucity of such collections and the inadequacy of fundamental research in this area hinder the development of practical work on the utilization of the world's genetic resources. In addition, insufficient knowledge and commercial interests interfere seriously with an international exchange of information and resources.
Claim Genetic resources conservation is directly related to the fundamental problem of the future of human welfare in terms of food supply and of other plant products. The conservation of plant resources on which the future improvement of the world's staple crops must depend has more than scientific interest; it is a safeguard to meet the inevitable requirements of population pressure. Crop improvement, which has contributed so much to plant productivity and total world production, requires, and will continue to require ever more urgently, the genetic resources on which the increasingly sophisticated methods of plant breeding will depend. The extinction of wild species and varieties – probably at a rate of one a day – may pose a grave threat to world food supplies. Conservation of plant genetic resources has to be carried out both *'ex-situ'* through an appropriate system of gene banks and *'in-situ'* through a world wide network of protected areas including biosphere reserves and biogenetic reserves.
Aggravates Endangered species of plants and animals (#PB1395)
Decreasing genetic diversity in cultivated plants (#PC2223).
Aggravated by Destruction of weeds (#PE3987).

♦ PF3588 Over-specialized supervisory personnel

Nature Over-specialization by supervisory personnel in production occasions an ambiguity of roles, collapse of accountability structures and a dominance by skilled specialists.
Broader Production of non-essentials (#PC3651).
Narrower Narrow job context (#PF2998)
Uninterested workers (#PF2987)
Disunity among traders (#PG8474)
Lack of irrigation equipment (#PG8457)
Inappropriate employment incentives (#PD0024)
Disrelationship between production and work force needs (#PE0893).
Related Inflexible management patterns (#PF3091)
Industrial processes geared to reduced social needs (#PE3939)
Deteriorating structures of rural community cooperation (#PF3558).
Aggravates Inadequate medical care (#PF4832).
Aggravated by Impractical education (#PF3519).

♦ PF3592 Introduction of new species of insect pests

Insect invasions
Nature Man has managed to distribute insects around the world more effectively than with any other class of animals. Many such insects, without the constraints of their original habitat, have become established as pests.
Incidence International transport and commerce and the movement of livestock and nursery stock have all been responsible for the introduction of noxious or destructive insects into new regions. For example, the importation and planting of a gift shipment of Japanese cherry trees in Washington DC in 1912 introduced in to the United States the oriental fruit moth which is now one of the most destructive pests of peaches in that country. An example of the economic consequences of introduction is provided by the spotted alfalfa aphid. It was introduced into California in 1954, and within four years it had spread throughout the length of that state, inflicting over $35 million in direct crop damage and in costs of control.
The rapid migration of a dangerous insect through the medium of commerce is well illustrated by the instance of the pink bollworm of cotton, native to India and the most destructive pest of cotton. In about 1908, bales of cotton which contained a very high proportion of infested cotton seed from India were sent to an Egyptian spinning mill. The Egyptian infestation was not discovered until 1910, when the worm was already well established in widely separated localities. In an attempt to improve cotton varieties for planting in the Laguna district of Coahuila, New Mexico, a cotton grower planted imported Egyptian cotton seed in a field near Monterrey, thereby introducing the pink bollworm to a new country. When this infestation was first reported in November 1916, the worm was found to have spread throughout the entire Laguna district. Surveys were begun in the vicinity of 11 cottonseed–crushing mills in Texas that had received seed from the Laguna area. By the end of 1917, infestations had been found in cotton fields adjacent to two of the mills. Further, an extensive infestation was found late in 1917 at Anahuac, Texas, on Trinity Bay. The latter spread was attributed to Mexican cottonseed that was washed far inland when the great storm of 1915 wrecked a Mexican steamship en route to Galveston, Texas. Thus in less than a decade a major cotton pest, through a series of intercontinental distributional leaps, was spread through commercial channels from Asia to Africa, and to the North American continent.
Broader Disastrous insect invasions (#PD4751)
Natural environment degradation (#PB5250)
Irresponsible introduction of new species of animals (#PD1290).
Aggravates Insect pests (#PC1630)
Locust plagues (#PE0725)
Crop vulnerability (#PD0660)
Insect pests of plants (#PD3634)
Insect vectors of disease (#PC3597).
Aggravated by Inadequate plant quarantine (#PE0714).

♦ PF3602 Irresponsible introduction of new species of fish

Nature Accidental and deliberate introductions of fish species have occurred in several regions since 1960. The results are sometimes unfavourable, as with the rapid spread in Australia of the European carp, Cyprinus carpio. This now threatens the environment of native and introduced sport species, mainly because it causes water turbidity which affects the productivity of aquatic plants. The introduction of the Nile Perch, a large fish predator, into Lake Victoria, where it has spread greatly, is another example, and species of small clupeid fish from Lake Tanganyika were deliberately introduced into Lake Kariba to increase the fishing potential there.
Broader Irresponsible introduction of new species of animals (#PD1290).
Aggravates Pests (#PC0728)
Vector–borne diseases (#PD8385).

♦ PF3605 Insufficient leadership training

Limited methods of leadership development — Limited development leadership — Reduced training to leadership — Untransferred leadership skills — Ineffective leadership recruitment
Nature Although appointed local leaders may have ability and vision, the number of people equipped to guard the comprehensive direction of a community and effectively guide the implementation of particular common tasks is limited. Few communities will survive very long when the burden of leadership and its responsibility remains only in the hands of a few, yet the gap continues to widen between initiative and desire to lead on the one hand and leadership skills and organizational methods on the other. A particular problem is that although women are often ready and able to assume broader roles in the areas of commerce, education and social care, they are unable to find channels to allow for movement beyond the traditionally male–dominated status quo in the community.
Broader Inadequate education (#PF4984)
Obstacles to leadership (#PF7011)
Unprepared adult leadership (#PF6462).
Narrower Undeveloped youth leadership (#PJ0151)
Lack of motivation in leadership development (#PF2208)
Lack of local services for community leadership training (#PF2451).
Related Ineffective methods of practical education (#PF2721).
Aggravates Lack of leadership (#PF1254)
Demeaning farmer image (#PJ9781)
Lack of organized expertise (#PG8463)
Lack of leadership initiative (#PF7988)
Destruction inherent in development (#PF4829)
Lack of responsible involvement in community affairs (#PF6536)
Lack of local leadership and decision–making structures (#PF6556).
Aggravated by Male domination (#PC3024)
Underprovision of basic services to rural areas (#PF2875)
Ineffective operation of community networks in urban ghettos (#PF1959).

♦ PF3607 Desecration of holy days

Absence of holy day observance
Related Desecration of holy spaces (#PF6385).
Aggravates Disruptive secular impact of holy days (#PE7735).

♦ PF3608 Gap between the function of social techniques and the needs they address

Nature Social techniques tend to deal not with the content of systems design but with the process of designing. This results in a type of human exclusiveness which denies access to skills developed by the technological revolution, and is manifested in such negative trends as massive unemployment, lack of distribution of surplus food supplies, and inadequate structure for education.
Broader Limits on areas of research (#PF2529).
Narrower Corporate planning paralysis (#PF3513)
Unacknowledged global interdependence (#PF2877)
Overemphasis on immediate superficial needs (#PF3243).
Related Thwarted technological communications (#PF0953)
Individualistic utilization of expertise (#PF5639)
Inequitable distribution of skilled specialists (#PD2479).
Aggravated by Lagging training in social skills (#PF8085)
Overemphasis on technological skills (#PF2956)
Unarticulated goal of educational methods (#PF2400).

♦ PF3609 Religious paternalism

Claim Paternalistically motivated interventions in religious commitment violate the requirements of individuality. Indoctrination in a religious tradition removes from scrutiny a significant portion of belief and experience and decreases the likelihood of self–initiated remedies.
Broader Paternalism (#PF2183).

♦ PF3610 Absence of long-term economic planning agencies

Claim No government board, no private enterprise, and no global agency is currently assuming responsibility for creating a comprehensive long–range projection which might indicate, in concrete terms, the complex inter–relatedness of production, distribution, and resources involved in establishing a world–wide economic foundation. There is no adequate method for people to use the available skills and analytical tools to evaluate international economic inter–relations. An individual economic enterprise, when faced with planning its own future expansion, investment and development, must therefore, of necessity, plan only from its limited and immediate context.
Broader Limited approaches to economic planning (#PF3500).
Aggravates Absence of a long–range, world–wide capital flow plan (#PF2865).
Aggravated by Short range planning for long–term development (#PF5660).
Reduces Failure of centrally planned economies (#PC3894).

♦ PF3612 Global financial crisis

Global fiscal collapse
Nature The instability of the global financial system is such that various explanations have been put forward concerning its expected collapse. At one extreme it is argued that the sheer size of corporate, individual and government debt will produce a worldwide depression of its own accord, in the light of historical precedents. Evidence in support of such a view is based on tracking business boom–to–bust cycles, which can be interpreted as having the same sequence of phases: credit expansion, a peak in interest rates, a boom in property and shares, a bust in property, a bust in shares and then a contraction of credit. The expected crash would be catastrophic because of the unprecedented expansion in credit, encouraged by a false sense of security derived from the various protective mechanisms. At another extreme, the unregulated degree of government borrowing and the possibility of banks defaulting on commercial loans could trigger such a collapse.
Broader Global crisis (#PF6244).
Aggravates International economic recession (#PF1172).
Aggravated by Excessive public debt (#PC2546)
Government loan defaults (#PD9437)
Uncontrolled growth of debt (#PC8316)
Disruption of financial markets (#PD4511)
Defaults on international loans (#PD3053)
Incompetent financial management (#PF0760)
Inadequate financial clearing systems (#PF9841)
Financial destabilization of world trade (#PC7873)
Unreported international movements of funds (#PF6594)
Inadequate models of socio–economic development (#PF9576)
Inadequacies of the international monetary system (#PF0048)
Economic and financial instability of the world economy (#PC8073)
Lack of international coordination among supervisors of financial stock markets (#PE4508).

♦ PF3613 Social inadequacy of men

Claim Men are separated from the other halves of themselves and spend their lives looking for that other half. Most men are inadequate as fathers and husbands. Men often revert to the time–honoured system of of marrying women they do not quite love and loving a women they cannot marry, thus maintaining a state of idealism about love without ever having to face the reality of it. Men substitute authority for affection in their relationships with their children. They seek mother substitutes in wives and father substitutes at work.
Aggravates Feminism (#PF3025)
Unrequited love (#PF6096)
Fear of intimacy (#PF8012)
Discrimination against men (#PC3258).

♦ PF3628 Narrow range of cultural exposure

Limited global exposure — Lack of outside exposure — Limited world exposure — Limited regional exposure — Limited outside exposure
Nature Although contemporary technology has given virtually every community daily access to

knowledge about the trends, events and possibilities of global society, through television, radio and the news media, most people are either consumed with the daily routine of subsistence labour or spend leisure hours chatting. The result is that most of the world's population has little or no access to the events and trends outside the local community. The difficulty of travel and functional illiteracy blocks many others from even vicarious exposure to the world beyond their own village. This pervasive emphasis upon living here and now not only stunts relationship to the future and to the rest of the world, but also hampers appropriation of the individual's own culture and that of others.
Broader Local traditions of cultural isolation (#PF1696)
Limited availability of learning opportunities (#PF3184).
Narrower Unenticing printed media (#PG8464)
Limited awareness of child-care opportunities (#PG8452)
Limited exposure to outside influences in rural villages (#PF2296).
Related Modern disruption of traditional symbol systems (#PF6461).
Aggravates Cultural illiteracy (#PD2041) Functional illiteracy (#PD8723)
Obsolete educational values (#PF8161).
Aggravated by Limited employment options (#PF1658).

♦ PF3630 Limited accumulation of capital
Uncertain sources for rural funding — Lack of local capital — Lack of financial surplus
Nature There is a wide disparity of economic resources among landowners, tenant farmers and farm labourers in rural agricultural communities and the day-to-day marginal cash flow pattern is one which most small farmers find hard to break. This pattern reinforces the habit of purchasing for immediate needs which prevents long-range planning or budget management. Banking services are remote and not generally utilized by those with small sums to save; and with unclear land ownership, the collateral required for loans is difficult to secure. Obtaining such loans can be a time-consuming process and often the money may not be secured at an appropriate time when it is most needed. Added to this, the amount of capital obtained is often inadequate to secure the full complement of equipment or technical aid in order for the the additional investment to be a profitable venture. Capital funds needed for initial development of essential services are therefore often inaccessible. Sporadic sources of outside capital are an unstable funding base, but residents find it difficult to develop ways and means of mobilizing their own investment capital.
Incidence The interest in small town capital self-sufficiency apparent in many rural communities in the Western world is often not actuated because the means of doing so are lacking. A reputation as a farming community, with no major industry or business generating new money, makes banks reluctant to lend funds. As a result, new businesses are difficult to start.
Claim The economies of many rural communities have suffered for years because of the absence of new money. The necessity for rapid social and economic development requires direct, predictable access to available sources of capital, and they urgently require methods which will enable them to use available capital creatively and to generate new capital advantages. Because past economic ventures may have proved unrewarding, further attempts at local initiative tend to be discouraged. This loss of incentive in the face of the obvious capital needs creates an overwhelming burden on many communities and prevents residents from seeking new skills which would lead to higher income.
Broader Rural poverty in developing countries (#PD4125).
Narrower School supply costs (#PJ0686) Restrictive farm capital (#PJ0788)
Prohibitive refrigeration costs (#PG0577) Restricted flow of local economy (#PF6451)
Prohibitive cost of farm equipment (#PJ8790) Limited individual capital reserves (#PF2899)
Prohibitive livestock husbandry costs (#PJ4075)
Restrictive loan procedures in developing countries (#PF2491)
Limited availability of loans in developing countries (#PE4704).
Related Inadequate financial services (#PJ8366)
Subsistence approach to capital resources (#PF6530).
Aggravates Lack of venture capital (#PG7833) Unstimulating entertainment (#PF8105)
Inadequate electricity supply (#PJ0641) Prohibitive cost of education (#PF4375)
Insufficient financial resources (#PB4653) Prohibitive cost of accommodation (#PD1842)
Prohibitive cost of transportation (#PE8063) Prohibitive cost of home maintenance (#PJ9022)
Limited availability of financial credit (#PF2489)
Diminishing capital investment in small communities (#PF6477).
Aggravated by Underpayment for work (#PD8916)
Geographical isolation (#PF9023)
Underutilized government resources (#PE9325)
Inaccessible market and supply centres (#PF8299)
Inappropriate aid to developing countries (#PF8120).

♦ PF3638 Superficial symbols of the economy
Nature The symbols of the economy are void of human substance. Some of the symbols motivate people to superficial action and others prevent any depth to the participation in the economy.
Broader Dominance of economic motives (#PF1913).

♦ PF3646 Hybrids
Incidence Scientists have created laboratory mice with tiny human organ structures – lungs, intestines, pancreases, lymph nodes, thymuses, livers and immune systems. The purpose is to study the viruses of human diseases in living human tissues. The mouse-human hybrid has been used to study AIDS virus, cancer viruses and cytomegalovirus.
Related Hybridization of wild animal species (#PD2419).

♦ PF3661 Inadequate management-employee communication
Unsound intra-corporate communications — Information gaps within enterprises
Incidence U.S., French and British employee surveys show that, in spite of an increase in internal corporate communications programs and company mission statements, employees do not believe top managers are interested in a dialogue and believe managers are interested in only getting their message across. They believe that the only way of finding out what is going on in a company is through the corporate grapevine, that is, rumours are more reliable than management statements to employees. Employees are convinced that managers failure to take into account their suggestions.
Claim Managers, joining the fad of employee participation in business, have increased tremendously the type and scope of communications systems within their business in the hope to increase control over the business. Until managers go beyond the rhetoric of dialogue with employees to the reality, management and workers will be at cross purposes.
Broader Poor managerial communications (#PF1528).
Aggravates Mismanagement (#PB8406).

♦ PF3665 Strained capital resources in small communities
Nature Current trends in the world's economy are increasingly demanding that local economic units and communities should sustain their own populations. As a result, rural communities in developing countries are experiencing the strain of carrying a greater financial burden as they participate more and more in the money economy. Bartering and in-kind wages to farm workers are being replaced by money wages, so that land owners need more capital on hand, while at the same time local credit lines and banking are only just beginning to be effective. While village economy requires diversified crops and the introduction of new cash crops, the necessary risk capital is simply not available. Most of the population continues to exist at the subsistence level, working for low wages, or farming just enough land to provide for the family. Most of the money earned from marketing goods, from government purchases or from wages, is spent in the market towns or farther away, taking money out of the community and bringing little back. Small shops bring little money into the community since goods are purchased at retail prices in the towns and resold locally at only a small profit margin. Since the only significant transport of goods takes place between towns and cities, the cost of transport to villages is disproportionately high. This results in an even higher cost of living for residents who already have a lower income than people in the larger towns. They are permanently aware of not having enough money and of never seeing any real change.
Broader Insufficient financial resources (#PB4653)
Limited local availability of capital reserve (#PF2378).
Narrower Low general income (#PD8568) Limited purchasing power (#PD8362)
Insufficient risk capital (#PJ1704) Prohibitive cost of fertilizer (#PJ1022).
Related Contained village economy (#PJ0594) Lack of local commercial services (#PF2009)
Decline in government health expenditure (#PF4586).
Aggravates Lack of school transportation (#PJ7849).
Aggravated by Prohibitive cost of disease control (#PF2779).

♦ PF3667 Contradictions
Nature A social, political, cultural or economic condition which is perceived as preventing the orderly and natural progress from the present situation to an envisaged, improved situation.
Narrower Contradictions in communist systems (#PF3179)
Contradictions in capitalist systems (#PF3118)
Contradictions within the growth and partnership model of developed countries (#PE2203).
Related Reason (#PA5502) Dissent (#PA6838) Difference (#PA6698)
Opposition (#PA6979) Disagreement (#PA5982).
Aggravates Typhoid fever (#PD1753) Soil pollution (#PC0058)
Water-borne diseases (#PE3401) Sewage as a pollutant (#PD1414)
Phosphates as pollutants (#PE1313) Soil-transmitted diseases (#PD3699)
Pollution of inland waters (#PD1223)
Damage by degradable organic matter (#PJ6128).
Aggravated by Developmentalism (#PF9512).

♦ PF3671 Ineffective data systems
Nature The increasing use of computer technology is marked by operational systems of data storage and retrieval which are less efficient than their predecessors, because the technology is still only partially understood by its users and is not fully integrated into existing systems of office procedures and daily life and work. There is thus incomplete and slow information exchange, limited data dissemination and incompatible computers components and languages.
Broader Limited approaches to economic planning (#PF3500).
Aggravates Inadequate statistical information and data on problems (#PF0625)
Underutilization of international data bases by developing countries (#PF4155)
Underparticipation of socialist countries in international data systems (#PE4230).

♦ PF3676 Loss of the significance of work
Nature As profit takes precedence over significance in the individual's understanding of work, there is a corresponding loss in the sense of creativity and purpose. The effect of this in the average workplace is an increase in worker-management tension.
Broader Inflexible management patterns (#PF3091).
Aggravates Alcoholic intoxication at work (#PE2033).

♦ PF3679 Self-interested industrial vision
Claim Industrial planning is so biased by its own immediate interests that the long-range, inclusive prospects for world-wide production are simply not seen.
Broader Unchallenging world vision (#PF9478)
Industrial processes geared to reduced social needs (#PE3939).

♦ PF3682 Procrastination of science in the face of the unexplained
Nature The discipline of science, including the medical sciences, responds with difficulty and considerable delay to new classes of phenomena which are not easily explained within its existing paradigms. Given the authority with which it speaks in initially condemning such phenomena as illusions, coincidences, inadequately substantiated, not subject to experimentally verification, or simply the result of deliberate fraud, such procrastination delays the advance of knowledge and marginalizes those who endeavour to explore such phenomena using different paradigms.
Incidence Classical examples include scientific proofs that objects heavier than air cannot fly and the response to hypnotism. Continuing difficulties are experienced with homeopathy, extrasensory perception, radiesthesia, acupuncture and UFOs. Such difficulties are experienced to different degrees by different schools of science. Scientists in socialist countries have much greater legitimacy in exploring extrasensory perception, for example, than those in western countries who have little tolerance for such phenomena. Another aspect is the difficulty of accepting evidence for health hazards associated with long term exposure to low levels of chemicals or radiation.
Broader Policy-making delays (#PF8989).
Aggravated by Procrastination (#PF5299) Unexplained phenomena (#PF8352)
Inconclusiveness of scientific and medical tests (#PD7415).

♦ PF3689 Obstacles to economic reform in socialist countries
Nature Economic backwardness in socialist countries makes reform urgently necessary. However, it is difficult for such countries to reform planning and management institutions so as to stimulate productivity and yet preserve traditional socialist institutions and practices. For instance, schemes which offer greater rewards for greater productivity threaten egalitarian wage payments and job security. Increasing the role of technical experts and professionals threatens the traditional role of the Party apparatus.
Broader Socialism (#PC0115) Inadequate social reform (#PF0677).

♦ PF3690 Wasted woman power
Nature Wastage of the capacity of women in national and in family life through discrimination against women in employment and education has led to the waste of women as a productive sector of society, leading to a loss of a potentially useful working force. Women may be over qualified for what they are doing, or they may not be employed at all. They may be left illiterate so they cannot contribute to the development of their country.
Broader Underutilization of labour force (#PF6293)
Neglect of the role of women in rural development (#PF4959)
Irresponsible introduction of new species of animals (#PD1290).
Narrower Illiteracy among women (#PE4380)
Incompatibility of traditional and new technologies (#PE3337).
Aggravates Frustration (#PA2252) Mental disorders (#PD9131)
Lack of community development (#PF7912) Failure of development programmes (#PF8368).
Aggravated by Unrecorded knowledge (#PF5728)
Lack of family planning (#PF0148)
Discrimination against women (#PC0308)
Dependency of women in marriage (#PD3694)

Inadequate child day–care facilities (#PD2085)
Underemployment in developing countries (#PD8141)
Discrimination against women in employment (#PD0086)
Misdirection of human energies and desires (#PF8128)
Discrimination against women in public life (#PD3335)
Elimination of the socio–cultural role of western women (#PE1046).

♦ PF3698 Inadequate crop rotation

Nature Without adequate crop rotation, soils become depleted, and lower crop yields result. This is particularly true if the crop is cultivated frequently and returns little crop residue to the soil. Weeds, diseases and insects also become more of a problem, and the farmer who intends to grow one crop year after year becomes completely dependent on disease–resistant varieties of plants, chemical insecticides and fungicides, soil fumigation, and other methods of controlling diseases, insects and pests.

Aggravates Plant diseases (#PC0555) — Pests of plants (#PC1627)
Soil infertility (#PD0077) — Diplomatic errors (#PF1440)
Crop vulnerability (#PD0660) — Insect pests of plants (#PD3634)
Destruction of land fertility (#PC1300) — Deficiency diseases in plants (#PD3653).

♦ PF3699 Deliberate invention of new threats

Invention of new social problems

Nature As particular problems of society cease to be a threat, the experts, bureaucrats and institutions dependent on the funding allocated in response to those threats find themselves under pressure to invent new threats, or new variants on old threats, in order to ensure their survival.

Incidence In 1990, this issue emerged as a consequence of the reduction in world tensions and associated defence budgets. It also occurs with the dismantling of any security system, whether a network of informers or in any form of demobilization.

Aggravated by Inhibited capacity to visualize a creative future (#PF2352)
Institutional preoccupation with obsolete problems (#PJ5014)
Obstacles to redeployment of military resources for peaceful applications (#PF4785).

♦ PF3700 Nomadism

Nature Nomadic life is facing an increased strain due to conflict with contemporary society. Nomads are not popular with national bureaucracies; always on the move (often across state boundaries), difficult to count, to register or to school, herding peoples tend to be perceived as a combination of threat and nuisance. Political security and bureaucratic convenience have frequently been important motivations behind schemes to settle nomads in permanent homes. Competition from agricultural cultivation of the land and herding (the latter in the guise of ranching) has steadily eroded the resource–base of nomadic societies. The main cause is the structure of land–ownership and land–use whereby the most fertile land is devoted to capital–intensive cash–cropping, with subsistence farmers being pushed into remoter regions, ever drier and more unsuitable for agriculture. Areas which were once grazed by the nomads alone are now becoming occupied. This is the principal threat to the nomadic way of life.

Incidence Nomads include gypsies, desert tribes such as the Bedouin and the many primitive tribes in the Americas, Asia and Australasia. Herding survives as a way of life around the Sahara, in the Middle East, in Asia as far east as western India, and in the Asian parts of the USSR.

Claim The end of pastoral nomadism would be regrettable not merely on account of the independence and distinctiveness of this way of life but because this type of economy may be a more rational means of raising large numbers of animals under arid conditions than is capital–intensive ranching.

Refs Davenport, Charles B and Rosenberg, Charles *The Feebly Inhibited* (1984); Raghaviah, V *Nomadism* (1968).

Broader Isolation of ethnic groups (#PC3316) — Proliferation of immigrants (#PD4605)
Underprivileged racial minorities (#PC0805).
Narrower Desert nomadism (#PD2520) — Gypsy persecution (#PE1281).
Aggravates Illiteracy (#PC0210) — Underdevelopment (#PB0206)
Unstable shifting agriculture (#PD7516) — Inadequate social welfare services (#PC0834)
Lack of variety of social life forms (#PE8806)
Expropriation of land from indigenous populations (#PC3304)
Exploitation of indigenous populations in employment (#PD1092).
Aggravated by Land enclosure (#PJ3523) — Land reclamation (#PF2055)
Social discrimination (#PC1864) — Unfair distribution of land ownership (#PG2912).

♦ PF3701 Misuse of satellite surveillance by governments

Reconnaissance satellites — Spy satellites

Nature The two superpowers (USSR, USA) have deployed and continue to deploy a multitude of satellites which are used either for photographic or electronic reconnaissance. Besides them a growing number of nations are building their own spy satellites or advocating such action. Photographic reconnaissance permits searches of the territory of another country for objects of interest, usually missile sites or troop movements. Electronic reconnaissance is used to record and playback radar and other electromagnetic radiation emanating from another country. Other uses of satellites are to detect launching of missiles or nuclear explosions using infra–red techniques. Multi–spectral analysis gathers subtle clues about the actual physical makeup of observed objects, and penetrates natural barriers detecting anything camouflaged or decoyed. Satellite surveillance constitutes an infringement of national sovereignty which may be tolerated or which may be the cause of provocations leading to the destruction of 'enemy' satellites, and thus possibly, leading to war.

Incidence Three out of every four satellites launched are for military purposes, two for surveillance and one for communications. Surveillance satellites are relatively short–lived. For this reason and because of improvements in technology, launchings continue at an increased pace.

Counter–claim Modern means of verification of compliance with arms–control treaties using reconnaissance satellites are more reliable than on–site inspection for monitoring the quantitative limitation of arms. They may therefore be used as the basis for verifying the implementation of such agreements between the superpowers (for them to be useful in multilateral arms–control treaties, under which each party must obtain assurance of compliance by all parties, reconnaissance through satellites would also have to become a multilateral undertaking).

Broader Espionage (#PC2140) — Misuse of government surveillance of communications (#PD9538).
Narrower Military expropriation of satellites (#PE0101).
Related Invasion of airspace by foreign aircraft (#PE3972)
Violation of sovereignty by trans–border broadcasting (#PE0261).
Aggravates European insecurity and vulnerability (#PD1863)
Denial of the right to national sovereignty (#PE7906).
Aggravated by International insecurity (#PB0009)
Competition between states (#PC0114).
Reduces Surprise attack (#PE3705)
Proliferation of nuclear weapons and technology (#PD0837).

♦ PF3702 Human errors and miscalculations

Human error — Mistakes — Erroneous people — Error — Fallibility — Operator error

Nature The social, technological and military systems which man is developing are becoming more complex, thus increasing the risk that human error, miscalculation or misinterpretation of data will lead to accidents, disasters or social crises. The belief that reassigning tasks to computers is a solution is, in fact, part of the problem; engineering and operational errors affect the hardware and logical errors affect software. High technology makes the possibility of error worse, not better. Hi–tech concentrates power. This means that a single fault, if it does occur, can be much more disastrous. This is true for human as well as mechanical errors.

Incidence Despite three different computers with different software to back each other up, and despite the voiced concern that the rubber rings would fail, in 1986 the American Space Shuttle exploded with the loss of 7 lives. The failure to set a computer navigation system correctly by the pilot of a Korean passenger plane may have caused its flying over Russian territory and for which it was shot down. Because of stress, task fixation and unconscious distortion of data by the crew the USS Vincennes shot down a Iranian Airbus killing 290 civilians. Construction workers urinating into concrete channels in the walls of a nuclear reactor in Britain caused a series of faults in the reactor as it later became operational. The urine cause cables to corrode at an extraordinary rate resulting in their failure.

Claim People have not been eliminated from technology. They still figure everywhere. They maintain the equipment. They design, sell, buy, repair and operate it. Human beings make errors and they make unique and unpredictable errors particularly in unique situations. As new technologies are developed, new people are introduced to old technologies errors will happen in spite of any so called the safeguards, especially the foolproof ones.

The arms race has finally provided man with the means of putting an end to his species. Political wisdom has so far averted this final disaster but it cannot insure against military miscalculation or against human or technical error, both of which could lead to the same fearful end.

Broader Chance (#PA6714) — Falsity (#PF5900).
Narrower Road hazards (#PD0791) — Popular errors (#PF5627)
Personality cults (#PC1123) — Diplomatic errors (#PF1440)
Statistical errors (#PF4118) — Surgical malpractice (#PE4736)
Policy–making errors (#PF0531) — Analytical stagnation (#PF0440)
Corruption of meaning (#PB2619) — Proverbial lore errors (#PF5653)
Miscarriage of justice (#PF8479) — Inappropriate arguments (#PF2152)
Electrical power failure (#PE1341) — Fragmentation of knowledge (#PF0944)
Disagreement among experts (#PF6012) — Defects in machinery design (#PE2462)
Irrelevant scientific activity (#PF1202) — Delay in project implementation (#PF1470)
Lack of scientific investigation (#PF2720) — Excessive speed of motor vehicles (#PE2147)
Vulnerability of computer systems (#PE8542) — Inadequate safeguards against fire (#PD1631)
Unreliability of computer software (#PE4428) — Unnatural boundaries between states (#PF0090)
Irresponsible international experts (#PF0221) — Instability of orthographic standards (#PF0552)
Irrelevance of science and technology (#PF0770) — Interpretation and translation errors (#PF6916)
Uncontrolled application of technology (#PC0418)
Fragmentation of technological development (#PC1227)
Continued operation of unsafe motor vehicles (#PE2240)
Hazardous locations for nuclear power plants (#PD2718)
Intergovernmental organization mismanagement (#PD6628)
Cost overruns in large–scale public programmes (#PD1644)
Uncontrolled environmental impact of technology (#PC1174)
Errors and risks in medical self–experimentation (#PF4841)
Excessive reliance on infallibility of equipment (#PE4742)
Politically unrealistic strategic warfare analysis (#PF1214)
Defective land use planning in developing countries (#PD1141)
Inadequacy of the committee system of decision making (#PF2843)
Inadequate and inaccurate textbooks and reference books (#PD2716)
Use of inappropriate technologies in developing countries (#PF0878)
Proliferation and duplication of international information systems (#PE0458)
Proliferation of national and international anniversaries and years (#PF2723)
Excessive dependence on computer models of complex system behaviour (#PE8533)
Inappropriate transplantation of industrialized country methods to developing countries (#PE1337).
Related Monsters (#PF2516) — Human contingency (#PF7054)
Domination of government policy–making by short–term considerations (#PF0317).
Aggravates Ugliness (#PA7240) — Negligence (#PA2658)
Air accidents (#PD1582) — Computer piracy (#PE6625)
Man–made diseases (#PD6663) — Structural failure (#PD1230)
Man–made disasters (#PB2075) — Failure of materials (#PD2638)
Occupational hazards (#PC6716) — Unreported tax obligations (#PE9061)
Unapplied scientific knowledge (#PF1468) — Inadequate ideological frameworks (#PD0065)
Accidents to nuclear weapons systems (#PD3493)
Underestimation of the human potential (#PF7063)
Unreliability of equipment and machinery (#PC2297)
Environmental pollution by nuclear reactors (#PD1584)
Hypersensitive military mobilization procedures (#PG7953)
Inadequate relationship between international governmental and nongovernmental organizations and programmes (#PE1973).
Aggravated by Boredom (#PA7365) — Ignorance (#PA5568)
Over–specialization (#PF0256) — Aircraft pilot fatigue (#PE3870)
Refusal to admit error (#PF5163) — Differing conceptions of time (#PF6665)
Alcoholic intoxication at work (#PE2033) — Multiplicity of time standards (#PF2621)
Limited human information processing capacity (#PF9292)
Information overload during control of complex equipment (#PF6411).

♦ PF3703 Allocation of television frequency bands for satellite transmission

Nature Television broadcasts require much broader frequency bands than do radio broadcasts. Inside the frequency area where allocations have already been negotiated most bands are occupied or assigned. Inside this area the introduction of channels on satellite television transmission make it necessary to reallocate existing frequencies that are used for pre–satellite television transmission. One satellite transmission can cover an area of the earth's surface large enough to blanket or replace many local transmissions, unless they are limited directionally.

Broader Lack of international cooperation (#PF0817).
Aggravates European insecurity and vulnerability (#PD1863).
Aggravated by Limited number of available radio frequencies (#PF0734).
Reduces Violation of sovereignty by trans–border broadcasting (#PE0261).

♦ PF3706 Authoritarian propagation of knowledge

Nature Top–down pyramidal organization characterizes the institutions and collective groups that transmit knowledge in an authoritarian way, and the same model reflects the deductive structures imposed on knowledge by asserting the primary existence of unproven principles or premises upon which all must be cognitively built.

Broader Authoritarianism (#PB1638).
Aggravates Obstacles to education (#PF4852).
Aggravated by Uncritical acceptance of dogmas and standards (#PF2901).

♦ PF3707 Fear of retaliation by authorities

Fear of official harassment — Governmental reprisals against citizens — Political reprisals — Threat of official reprisals

Nature People may be reluctant to protest about injustice because they fear they will be registered as 'trouble makers' and subjected to harassment. This is particularly in cases involving police excesses or well–connected institutions.

Broader Retaliation (#PF9181) — Fear of reprisals (#PF9078)

capital on development (#PF3318).
Related Harassment of public officials (#PD4915).
Aggravates Fear of police (#PF8378) Fear of officialdom (#PD9498).

♦ **PF3709 Ineffective decision-making processes**
Ineffective policy-making methods
Broader Obsolete methods (#PF3713).
Narrower Ineffective educational policy decisions (#PF2447)
Self-destructive government policy-making (#PF5061)
Incomplete implementation of community decisions (#PF2863)
Lack of rational global political decision-making (#PF7535)
Reductionistic decision making criteria in the construction industry (#PE7742)
Unrelated ability application and situational demands in vocational decisions (#PF2828).
Aggravates Policy-making delays (#PF8989) Inappropriate policies (#PF5645).
Aggravated by Unrealistic policies (#PF9428) Policy-making errors (#PF0531)
Conflicting priorities (#PF5766) Fragmented decision-making (#PF8448)
Avoidance of negative feedback (#PF5311)
Arrogance of intergovernmental agencies (#PF9561)
Failure to integrate knowledge to empower humanity in response to the global problematique (#PF8753).

♦ **PF3710 Denial of rights of inanimate objects**
Inadequate legal rights of natural objects — Inadequate legal rights of trees
Claim Pieces of art, natural wonders and unusual objects have a right to existence in and of themselves. The wanton destruction of inanimate objects by human beings is one more example of the human species' need to control existence rather than be a part of creation. It is no answer to claim that streams and forests cannot have legal standing because they cannot speak. Corporations cannot speak, nor can states, estates, infants, incompetents, municipalities or universities. It is lawyers that speak for them.
Refs Stone, Christopher D *Should Trees Have Standing.*
Broader Denial of rights (#PB5405).
Related Denial of human rights (#PB3121) Denial of rights of animals (#PC5456).

♦ **PF3713 Obsolete methods**
Inadequate methods
Broader Oldness (#PA7131) Preservation of obsolete systems (#PC8390).
Narrower Obsolete group methods (#PG8406) Risk-aversion strategy (#PF4612)
Obsolete policy-making (#PF5009) Stagnating social methods (#PF3884)
Inadequate meeting methods (#PF8939) Poor communications methods (#PJ8656)
Inadequate intellectual methods (#PF7380)
Ineffective decision-making processes (#PF3709)
Limited ways of matching talent and jobs (#PF2792)
Ineffective methods of practical education (#PF2721)
Ineffective structures for community decision-making (#PF1781).
Aggravates Outmoded functional skills in rural communities (#PF2986).

♦ **PF3715 Symbol system failure**
Broader Cultural revolution (#PF3235) Ignorance of cultural heritage (#PF1985).
Narrower Ineffective communication (#PF6052) Disintegration of accepted myths (#PF8887)
Failure of motivating socio-dramas (#PF1042) Reduction in symbolic celebrations (#PF1560)
Lack of symbolism in local relationships (#PF2063)
Absence of convincing symbols connecting the individual's life to the cosmos (#PF1081).
Related Cultural fragmentation (#PF0536)
Vulnerability of marriage as an institution (#PF1870).
Aggravates Loss of cultural identity (#PF9005) Deterioration of human environment (#PC8943)
Decreasing participation in collective religious worship (#PF8905).

♦ **PF3720 Exclusion of opposing views**
Nature Social structures exclude views in opposition to their current practices. Even opposition loyal to the principles of the organization is relegated to inauthentic and non-visionary engagement. The past is not appropriated. New effective leadership is not prepared to assume responsibility. Organizations become more parochial. Planning becomes increasingly more unrealistic.
Broader Overemphasis on self-sufficiency with respect to interdependence (#PF3460).
Narrower Blocked minority opinion (#PD1140) Inadequate political networks (#PD2213)
Idealism of opposition groups (#PF5401) Collapse of societal engagement (#PF2340)
Frustration of the role of opposition (#PF1594) Structural failure of opposition groups (#PF3821)
Failure to profit from patterns of history (#PF1746)
Inadequate structures for political dialogue and review (#PC1547)
Irresponsible expression of emotions equated with free speech (#PF7798).
Related Reduced understanding of globality (#PF7071)
Breakdown in community security systems (#PD1147)
Legal contract system reduced to individual needs (#PE5397).

♦ **PF3731 Lack of demolition planning**
Nature The technology and expertise for dismantling man made structures is available for small and safe structures, such as building a few stories in height, but for many structures being built today the technology for demolishing them does not exist. The tools, such as robots, for dismantling nuclear reactors have to be invented while they are being decommissioned. The techniques for tearing down building like the Sears Tower do not exist. The over 6000 oil drilling platforms used at sea were never design to be dismantled even though a 1958 Geneva convention requires that they have to be removed and the sea returned to its natural state.
Broader Short-term planning of product life cycles (#PF1740)
Absence of accountability in construction planning (#PF2804).
Related Inadequate bridge plan (#PU0646).

♦ **PF3743 Increasing development lag against population growth**
Broader Increasing development lag against socio-economic growth (#PC5879).
Narrower Increasing lag in education against population growth (#PE5369).
Aggravated by Unsustainable population levels (#PB0035)
Imbalance of population growth between developed and developing countries (#PE4241).

♦ **PF3746 Imprecation**
Cursing — Execration — Malediction
Nature A curse is a mere wish, wish expressed in words or actions or embodied in a material object, or an appeal to a supernatural force, that something bad or evil may befall to a certain person.
Refs Aman, Reinhold (Eds) *Maledicta 1980* (1980).
Broader Blasphemy (#PF5630) Verbal abuse (#PD5238).
Related Profanity (#PF7427).
Aggravated by Magic (#PF3311) Witchcraft (#PF2099).

♦ **PF3750 Collapse of corporate engagement in society**
Nature The processes of engaging in society as a part of a group of people is ineffective in the modern urban world. Society is seen as made up as an elite and as victims creating a self-understanding that mediates against corporate engagement. The methodologies for creating consensus at the local level have been lost. Interchange between local groups is at best minimal and normally non-existing, making any corporate to social problems responsible impossible.
Broader Collapse of societal engagement (#PF2340).

♦ **PF3756 Rhetorical inflation in meetings**
Public posturing in meetings — Empty meeting oratory
Nature Meetings, and especially international meetings reflecting different cultures and ideologies and focusing on a complex variety of issues, tend to encourage lengthy interventions of little substance. Such excesses are difficult to control because any restraint is easily seen as symptomatic of repressive social forces which are frequently implicit in the topics under discussion. As a result such meetings, when they can be held, become excessively expensive and primarily of symbolic value. Continuing public debate is intended to exert pressure towards negotiations; when it can no longer do so, it defeats the aims of its own sponsors.
Incidence This problem arises in most non-technical meetings aiming to articulate new policies in response to pressing issues where experts are themselves in disagreement. Its importance has been frequently noted in the debates of the General Assembly of the United Nations and in the conferences of its Specialized Agencies.
Counter-claim Global issues require global solutions based on global consultations. The apparent fruitlessness of debates is mitigated as procedures are smoothed out. It is only by exchanging views that understanding of the possibilities of concerted approaches emerges.
Broader Propaganda (#PF1878) Inadequate meeting methods (#PF8939).
Aggravates Inappropriate arguments (#PF2152) Impoverishment of political debate (#PF4600).

♦ **PF3758 Normalism**
Handicappism
Nature People who are endowed with body that move, think and speak easily have a tendency to systematically exploit those who are not so endowed.
Aggravates Discrimination against mentally disabled (#PD9183)
Discrimination against physically disabled (#PD8627).

♦ **PF3777 Institutionally reinforced sense of personal shallow meaning**
Nature The social forms that created and reinforced images and life styles that at one time were meaning-giving no longer do so. Self-stories and symbols are shallow and are inadequate to express consciousness and life's journey. In the absence of situations where the meaning of symbols is rehearsed, people are left without a way to self-consciously appropriate perceptions of the awesome and their own depth encounter, thus failing to recognize the greatness of their own being.

♦ **PF3783 Reduced interior structure of families**
Nature The interior structure of families have been reduced to the point the family unit has become largely disfunctional. The forms of sustenance have been reduced to physical gratification. The social role of the family in general or a specific family is not self-consciously considered. Members of families are not enabled to deal with the meaning of life in relation to the family. Structures relating the family to its history or its heritage do not exist. The decision making forms are arbitrary and lack effective ways of holding members responsible for their acts. The structures that do exist tend to engender self-centered attitudes and behaviour of individual members and a reduced sense of responsibility toward the larger community and society.
Broader Lost family role in society (#PF7456).
Narrower Lost family heritage (#PF1099) Inadequate family structures (#PF1000)
Parochial family responsibility (#PD1668) Individualistic family structures (#PF1861)
Simplistic family decision-making structures (#PF2097).
Related Exclusion of pre-adults from family decisions (#PE2268)
Individualistic perception of sexual activity (#PF1682)
Individually defined operating structure of marriage (#PD2294).

♦ **PF3790 Induction of incongruent actions**
Nature Induction of incongruent actions are methods of torture which bring victims into situations of impossible choices and consequently to act or react against their ideology and ethics and is central to all forms of psychological torture. The stated goal of such methods is usually gaining information but in fact the usual goal is destroying the personality of the victim. Methods may be slow and progressive. When the victim does break down he enters a vicious circle of guilt followed by repentance, which cannot be expiated.
Broader Psychological torture (#PD4559).
Narrower Participation in torture (#PD4478) Forced to witness torture (#PE1255)
Torture by violation of taboos (#PE1296) Inhumane interrogation techniques (#PD1362)
Torture through denial of sanitary facilities (#PE5480).

♦ **PF3801 Ghosts**
Hauntings — Phantoms — Apparitions
Nature Some supernatural presence which frequents a place or attaches or periodically manifests to a particular person is said to be haunted. The agents involved are usually considered to include discarnate humans, termed ghosts or shades; discarnate animals; disembodied non-human spirits including poltergeists, and various spooks, phantoms, spectres or apparitions. Other classes of haunters include revenants who have returned from the dead, and in modern folk-lore, ectoplasmic or astral projections from living persons and extra-terrestrials or extra-dimensionals. Hauntings may be experienced through conventional means of knowledge so that they may be sensed as tactile, for example, clammy cold touches; olfactory, with stenches, perfumes or other odours; auditory, as musical vocal or mechanical sounds, and visually, by seeing the agent itself or its effects such as displaced objects, movement of drapery or doors, deposits or manifestations of articles, or other creativeness or destructiveness. Other experiences of hauntings may be through the emotions such as feelings of depression, dread and terror, although positive feelings like warmth, love and inspiration may also be produced.
Counter-claim Hauntings may be the term for any such experiences that are unwanted by the recipients, but fear of the unknown plays a supreme role in such negative reactions. Places and persons haunted may, in fact, be places and persons protected or under observation. This may apply to the previously unmentioned category of unidentified flying objects that have been frequenting or haunting the planet, as observed in the last forty years.
Refs Cohen, Daniel *The Encyclopedia of Ghosts* (1987); Crowe, Catherine *The Night-Side of Nature* (1988); Eberhart, George M *A Geo-Bibliography of Anomalies* (1980); MacKenzie, Andrew *Hauntings and Apparitions.*
Broader Supernaturalism (#PF8433) Unexplained phenomena (#PF8352).
Narrower Poltergeists (#PF5697) Haunted buildings (#PF0201).
Related Demons (#PF6734).

♦ **PF3806 Excessive population mobility**
Rootlessness
Claim Increasingly cities and suburbs are made of newcomers, many of which perceive their residence at any particular location to be temporary. Status is in some countries associated less with the amount of time that a person has been resident in a community and more with the

frequency with which that person has moved to a new community (provided that such movement can be associated with upward mobility). Few families therefore have roots in suburban towns; many assume that they, or their children, will move to some other location. Many families are in consequence the residue of nuclear families, since the children or the parents tend to live elsewhere. Many such communities are consequently unable to build a stable social structure, especially one capable of assisting those in need.
Counter-claim It is the mover – the restless, never-satisfied, status-seeking consumer – who has been the backbone of a number of successful industrial economies, notably in North America. Such mobility is not only a result of national wealth but a basic cause of it. The fierce desire for self-betterment creates a market for business and catalyzes a prosperity which would otherwise appear improbable.
Narrower Disturbing transient mobility (#PG7995)
Mobility of village populations (#PE1848).
Aggravates Mental disorders of the aged (#PD0919).

♦ PF3807 Theological justification of nuclear war
Nature The theory that the "last judgement" will take place through nuclear warfare, in which the good will survive in triumph and the evil will be destroyed.
Broader Inadequacy of religion (#PF2005).

♦ PF3819 Official cover-up of government harassment of political activists
Official cover-up of government sanctioned assassinations — Government complicity in killing of human rights activists
Nature People working peacefully for human rights may be persecuted under the pretext that they constitute a threat against the national security of a country. In many cases, human rights activists are gaoled because they defend or try to help people who have been harassed or detained for their beliefs. In some countries, people working for human rights are taken to clandestine detention centres and tortured. They are sentenced to long prison terms in secret trials. Their effort are officially denounced, their houses are raided, their papers are confiscated and their families intimidated. In a number countries, human rights activists may even be assassinated.
Incidence The French foreign intelligence services blew up the Greenpeace boat, Rainbow Warrior, killing a photographer. An attempt to cover up the incident was made by the French military.
Broader Unreported scandals (#PF5340) Abuse of government power (#PC9104)
Unethical practices of government (#PD0814).
Related Government sanctioned killing (#PD7221)
Harassment of public officials (#PD4915)
Harassment of human rights monitors (#PF1585).
Aggravates Denial of human rights (#PB3121).
Aggravated by Government treachery (#PF4153)
Unaccountable government intelligence agencies (#PF9184)
Connivance of authorities in human rights abuses (#PF9288).

♦ PF3821 Structural failure of opposition groups
Disorganized political opposition
Nature Opposition groups fail to create structures that demand creativity from their members and from the general public. Their members are not trained as members of the opposition nor as members of the establishment. Effective opposition planning is missing. This structural failure hinders the growth and the sensitivity of the whole political process.
Broader Exclusion of opposing views (#PF3720).
Aggravated by Suppression of opposition groups or individuals (#PC7662).

♦ PF3824 Lack of vocabulary
Terminological crisis in social studies
Nature The available supply of words has proven inadequate to meet the growing demand for terms to handle the proliferating study of concepts. For example, attention given to functional analysis during the last 20 years has led to a growing repertoire of functional categories, concepts used in the analysis of relationships between components and systems. Concurrently, words with mixed structural and functional connotations, that were attached to some of the most important social and governmental institutions, need to be dropped from scientific usage because their inescapable connotations blur meaning, making it difficult to distinguish clearly between social structures and the functions which they may, but need not always, perform. This means, of course, that there is not only a need for terms for new functional concepts, concepts not previously well-articulated, but also a need for new, more sharply defined terms for social structures to replace old terms whose functional connotations severely limit their usefulness for contemporary social analysis.
The growing awareness of the importance of variables, especially limitless variables, creates the need to find new terms for many scales or indices that have not so far been recognized, although their polarities have frequently been treated as dichotomous entities. Often enough there is only one word for such variables, a word that designates one extreme on a scale. The opposite extreme tends to be identified by a negative prefix, and no terms are available for intermediate positions on the scale, or for the scale itself. The increasing sophistication in social measurement is continually hampered by inability to designate scales and to measure positions, more because of a lack of terminology than because of inability to define these concepts. The emergence of new concepts is an important reason for recognizing and dealing with terminological shortages. However, a systematic analysis of concepts shows that the folk vocabulary leaves many familiar concepts without terms. The point can be readily illustrated by comparing languages, since the way in which one language handles concepts often differs from another. Many words retained in contemporary usage were invented long ago for concepts that have long since lost their original meanings, so that in modern applications they carry irrelevant or misleading connotations.
The problem of diffusion or distribution of terms requires separate recognition. When someone proposes a concept and a corresponding term, it often takes a long time for the idea to win acceptance. Perhaps even more frustration arises from the fact that the same idea often arises simultaneously in several places and is therefore assigned terms independently. The likelihood that the same expression will be chosen is, of course, remote. In each language, community, country or disciplinary cell, when a concept and term is invented, those who adopt it become its advocates and tend to resist strenuously the substitution of alternative terms that have been invented elsewhere. The connotations of words vary among cultural groups and countries even more sharply than their denotations. Consider, in this context, the diverse meanings attributed to the word "development". In addition to the various concepts attached to the word, there are differing connotations. In much of the world, the term continues to have a positive value, to be regarded as a good thing, whatever it is. But in the USA it has recently begun to acquire negative overtones. Under the influence of current stress on the environment, ecology and pollution, many people have begun to identify development with its unfortunate side-effects, namely the negative effects of economic growth and urbanization.
Counter-claim Although in some cases there is evidently a lack of necessary terms, in other cases the fault lies rather with the users of the vocabulary who, perhaps because they lack clarity of thought, continually blur the meaning of words by using them in inappropriate contexts; and who invent a multitude of terms for which, if they paused to reflect, they would find there was already a perfectly adequate word available.
Broader Misappropriation of words (#PF5247).
Related Terminological deception (#PF5383)
Lack of terminological equivalents between languages (#PF0091).

♦ PF3826 Unconvincing alternatives to existing societies
Ineffective opposition to existing social order — Low credibility of alternative social structures — Rejection of alternative images — Non-viable alternative modes of social organization — Prevalence of psychological conditions unfavourable to a transition to a post-capitalist social order — Unacceptability of conserver society policies
Claim Many who oppose the existing social order lack sufficient vision of the future to propose useful alternatives. This lack of vision creates an ineffective idealism which does not enable the opposition to transcend the role of victim. For example, when when what is actually needed may be an entirely new system, rather than building a new model, there is a tendency simply to oppose the established decision-making process; when the opposing voice is unable to make a move, it spends too much time trying to interrupt the present process. Without some practical methodology this raw destructive power will not be converted into effective creative action.
Alternative groups involved in the economy; such as alternative magazines, worker's cooperative, voluntary organizations, and campaign groups; operate from a set of dogmas which prevent the groups from being effective. They ignore the market because practices like market research is suspect. Budgeting, credit control, accountancy, and financial planning are all given low priority. These anti-business prejudices results in dependence on various forms of subsidy. Consensus decision-making results in inertia, indecision and difficult decision being deferred. Egalitarianism produces disdain for the division of labour and down playing or even the failure to recognize the importance of skills. Individuals with skills are often required by social pressure to hide them. The inability to understand the nature of management means that functions like, clarifying objectives and devising strategies are not done.
Broader Obsession with novelty (#PF8767) Unproductive subsistence agriculture (#PC0492)
Individualistic retaining of local tradition (#PF1705).
Aggravates Fixation on partial solutions to problems (#PF9409)
Static grassroots involvement in planning the economy (#PE4479).

♦ PF3831 Closure of social institutions
Destruction of social institutions
Nature Closing of universities, schools, newspapers and magazines, social and charitable organizations, and trade unions cripple education, research, and attempts at community organizations and identity.
Related Factory closures (#PE3537).
Aggravates Socio-cultural environment degradation (#PC4588).
Aggravated by Collective punishment (#PD6970).

♦ PF3836 Divisive effects of formal schooling
Nature As formal school education is taken increasingly seriously, young people are being removed both from gainful employment and from the home. The shift in values derived from increased exposure to education produces acute conflicts between the young and their parents. For example, sometimes the young people in rural communities feel they are entitled to as much leisure time as those in more urban areas, yet when they come home from school they are expected to do a number of household chores; and as the conflict increases, the means of discipline demand more attention – old-style school discipline may have been harsh or physically harmful. The result is that, while providing up-to-date education for its young people, communities lack local social forms for involving them in community areas broader than formal schooling. Communities waver between upholding old expectations and rapidly advancing into a modern style. The combination of a vacillating set of social expectations and the young person's feeling of being stifled by generational differences contributes to many young people leaving home for additional training or employment. Unless this trend is curbed the future development of local communities will be deprived of vigour and vision.
Broader Inadequacy of formal education (#PF4765).
Narrower Education breaks community ties (#PS1552).

♦ PF3838 Religious opposition to public health practices
Religious opposition to nutritional resources — Religious obligations
Broader Burdensome cost of religious ceremonies (#PF3313).
Narrower Enforced vaccination (#PF1916) Dietary restrictions (#PJ1933)
Medicines containing alcohol (#PE2047)
Negative effects of claims of religious infallibility (#PF3376).
Related Enforced celibacy (#PD3371)
Religious opposition to population control (#PF1022).
Aggravated by Ritual pollution (#PF3960) Compulsory health care (#PF4820).

♦ PF3843 Unbalanced industrial distribution
Nature The need for profitability severely limits the locations in which industrial development can occur, so that industrial processes are unavailable to the vast majority of people worldwide. This results in, and is partly occasioned by, gross inequity in worldwide living standards.
Aggravates Inadequate standards of living (#PF0344).
Aggravated by Short-term profit maximization (#PF2174).

♦ PF3844 Vocational obsolescence in the face of overwhelming need
Nature There is a current tendency for even the most highly trained professionals to find that they are a 'dime a dozen' or obsolete, while at the same time those whose vocation is to help others find that for every one person they assist, ten more people are suffering and in need of help. The vital question raised by this gap between the experience of unsatisfied needs and inadequate use of skills is why people are or are not satisfied by a particular job.
Broader Preservation of obsolete systems (#PC8390).

♦ PF3845 Disincentives for financial investment within developing countries
Inadequate investment performance in developing countries
Nature The absence of significant net investment by people in traditional societies is due to a lack of any notion of productive accumulation, or even of progress; and to the acceptance of values and institutions which are inimical to innovation. As a result, economic life continues in a repetitive pattern with no underlying tendency for appreciable increase in output. In addition, farmers need some of guarantee that prices will not fall below a minimum remunerative level if they are to be induced to introduce innovations and increase investment on their land.
Broader Restrictive trade practices (#PC0073)
Economic and financial instability of the world economy (#PC8073).
Narrower Disincentive to invest in heavily indebted countries (#PF9249).
Aggravates Obstacles to world trade (#PC4890) Insufficient financial resources (#PB4653)
Decline in import capacity of developing countries (#PE6861)
Inadequate level of investment within developing countries (#PD0291).
Aggravated by Risk of capital investment (#PF6572)
Excessive external trade deficits of developing countries (#PE1496)
Deterioration in external financial position of developing countries (#PE9567).

♦ **PF3846 Opportunist bias in public discussion and research on development**
Nature Conceptions of underdevelopment, development and development planning as presented in most of the scientific and popular economic literature and in the plans of developing countries is heavily biased in a direction that is basically opportunistic. Policy conclusions are therefore founded upon ideas about reality that are systematically though unintentionally falsified. Studies are expected to reach opportune conclusions and to appear in a form that is regarded as advantageous, or at least not disadvantageous, to national interests as these are officially and popularly understood, (for example in terms of the political and military interest in saving the country in question from domination by another ideological bloc). Research also tends to become diplomatic, conciliatory and generally over-optimistic, by-passing facts (such as the effect of corruption) that raise awkward problems, concealing them in an unduly technical terminology, or treating them in an excusing, condescending way. A more fundamental deficiency is that the approach to development problems derives from the attitudes and institutions in the developing countries and assumes that models and methods applicable to developed countries are universally valid. Whereas purely economic models may be adequate to developed countries, the attitudes and institutions in developing countries are less permissive of development and are much more rigid. These factors are not taken into account by the great majority who are involved in the problems of development (whether as students, in politics or in practice) who have a vested interest in the more conventional approach.
Broader Bias in scientific research (#PF9693).
Related Excessive emphasis on fashionable areas of research (#PF0059)
Inadequate research on proposed solutions to problems (#PF1572).

♦ **PF3849 Death instinct**
Death wish — Wish for self-destruction — Thanos — Thanatophilia
Nature The wish for self-destruction or self-harm expresses itself in the life histories of countless individuals. Some destroy their lives; others their careers; and others their relationships with other people. It is a sometimes invisible, corrosive force eating away at human development. There are a number of theories to explain this phenomenon: Psychoanalysts deriving from Freud speak of a death instinct; some feel this works as an efflux from the powers of the root psychic force, the libido. To other analysts it is an energy in its own right, the destrudo; while again for others, the self-directed destructive energy is distinguished as the mortido: this is supposed by some to be biologically situated although triggered by physical or mental reactions to what the organism perceives as deep-level pain, and thus constitutes an escape hatch. Others deny the biology, and say it is not an instinct but a psychological drive arising with the development of the individual psychic structure. Still other analysts suspect a positive purpose to what might be termed, self-reducing, or self-limiting tendencies in the psyche.
Background A tendency to death is seen in religious practices. The ascetics mortified themselves, the alchemists projected a stage called 'mortido' into their work, and Christian and Islamic mystics during the Middle Ages strove for a spiritual 'annihilation'. The religious suicidal hero is seen in several phases in Japanese and Islamic history, at Masada and at Mt Ségur, as it is not infrequently associated with militarism and terrorism. From a sociological perspective, indigenous peoples' mystical participation in nature has its counterpart among those, in modern urbanized societies, who wish to lose their identities by participation in a group, sect or other organization, or in fashionable large-scale movements such as ecology, women's liberation, etc, or in the 'bosom' of a political party or ideology. The abuse of drugs, alcohol and other substances may be due to self-destructive tendencies.
There is considerable confusion of both terminology and conceptualization in this subject, and the observable facts of physical deterioration in its natural phase, viz. senescence, and its attendant mental, emotional and social factors, are not adequately included. In the analytical psychology of Jung some emphasis was given to this phase as the field upon which the final maturation or integration of the personality could be played. The accord of the psyche with physical dissolution is seen as a therapeutic process from this viewpoint. Since the physical condition of senescence may trigger the first thoughts of death, the biological alarm clock in the body also may go off at an earlier hour, a carry-over or atavism of the not-so remote times when, following millions of years of evolution, homo sapiens died at one-half or less of the age attained now. This helps to account for the severity of middle-age depression or crisis which can strike before the age of forty or as late as the middle fifties. Other biological causes for 'death instincts' may be disease of the organism. Social reasons for an individual's proclivity for self-destruction may be societal and family 'disease'.
Claim Since the universe tends to disorganize itself via entropy, self-destructiveness would be a normal drive in humans. On a world level, wars and threats of nuclear wars may suggest a global death instinct.
The death wish shows itself in many ways: a) suicide; b) psychogenic death, or the surrender of the will to live; c) murder, the murderer kills in order to be killed; d) in war, a way of committing suicide is provided without the moral or religious guilt attached to it; e) in certain philosophical, religious and political beliefs, such as atheism, agnosticism, humanism, existentialism nihilism and anarchism; f) the choice of death defying professions, such as, mountain climbing, automobile racing, sky diving, and parachuting; g) insanity veils full consciousness with its burdening responsibility; h) sickness where many patients do not want to get well; i) addictions to drugs and alcohol; j) bad luck and accident proneness; and k) pleasure which can also be a way of escaping the hard realities of life.
Counter-claim What is seen as a death-instinct may be a drive towards transformation to a higher level. This was the mystery of the seed taught as a parable in the New Testament, that it unless it fell into the ground and died it could not bring forth new grain. In the Greater Mysteries at Elensis the initiates descended symbolically into the underworld, as they did in other Mysteries, in order to receive new life. The Islamic warrior fighting for his holy cause is still taught to sacrifice himself in battle so that he might immediately be reborn into paradise. In the arcana of human development, according to old but universal traditions, by a first death one won a second life. Without the death urge, the drive to live would not exist. Until a person has come to terms with their own death, not some symbolic abstraction but the reality of rotting in the grave, they have not really lived. Coming to terms with the reality of death is not simply acknowledging it but personally forcing one's death to reveal the meaning of one's life. Without the death urge one's life has no authenticity.
Broader Meaninglessness (#PA6977) Self-destruction (#PF8587)
Morbid preoccupation with death (#PD5086).
Related Suicide (#PC0417).
Aggravated by Lack of self-confidence (#PF0879).

♦ **PF3850 Negative effects of over-crowding on mental health**
Nature There are some indications from animal experiments that population density in itself may affect mental health. Thus, under conditions of extreme overcrowding, an increase in male aggressiveness and an accompanying decline in the adequacy of maternal behaviour have been observed among rats. When the young were methodically removed from colonies of rats kept at high densities, it was noted in addition that bands of young males assaulted females and that there was an increase in homosexual behaviour. However, mice brought up under overcrowded conditions showed less stress behaviour than those transferred only later to such conditions. Some epidemiological studies have shown higher rates of schizophrenia, crime, suicide, alcoholism and drug abuse in the central, more crowded areas of old established cities than in other areas of such cities, but correlations of this kind are by no means clear-cut.
Aggravates Mental illness (#PC0300).
Aggravated by Unsustainable population levels (#PB0035).

♦ **PF3855 Non-juridical fault**
Nature An act against the law in which the perpetrator was unaware of its nature and over which he had no control leaves him, in many jurisdictions, immune from requests for court-awarded damages. Plaintiffs have no redress under these conditions and governments do not maintain public funds to compensate victims of such unintentional actions.
Incidence In the moral law of major religions, and in primitive or tribal societies, agents of acts against law or custom, even if these were unintentional, were nevertheless punished. Underlying justifications are related to concepts of impurity, moral tainting and violation of taboos, as well as to kismet, karma, and superstitious doctrines that describe persons as evil, cursed or bewitched when they are involved in behaviour threatening to local society. Some prosecutions of persons as accomplices before or after the fact (of criminality) may find fault simply in familial loyalty. Unstable and despotic regimes have practised summary justice under the wildest, guilt-by-association, reasoning.

♦ **PF3856 Humiliation**
Loss of face — Loss of dignity — Public embarrassment — Dependence on humiliation — Humiliated people
Nature A sense of humiliation may be felt by an individual or a nation if short-comings and failures are widely known. This can lead to resentment or, on the national level, to war.
Humiliation - the feeling of being disgraced, shamed, debased, or dishonoured - is the frequent concomitant of a loss of self-esteem in clinically depressed patients; and the desire for infliction of actions by others that warrant feelings of humiliation is seen in masochism.
Incidence Modern superpower pride and propagandist strategies suffer when confronted with widely recognized failures. Often also, extremism by people's liberation movements may be due to a perceived need to regain honour or recognition after some setback. This may be true of terrorist groups as well.
Background In comparatively formal societies, such as Japan and China in the East, or feudal Europe in the West, public defeat, ridicule or reversal of fortune was not easily tolerated.
Broader Negative emotions and attitudes (#PA7090).
Narrower Embarrassment (#PF7950).
Related Humility (#PA6659) Disrepute (#PA6839) Disrespect (#PA6822)
Unpleasantness (#PA7107).
Aggravates Shame (#PF9991) Mental suffering (#PB5680)
Learned helplessness (#PE6990) Defamation of character (#PD2569)
Excessively costly prestige projects (#PF3455).
Aggravated by Dishonour (#PF8485) Homelessness (#PB2150)
Being a burden (#PF9608) Fear of humiliation by co-workers (#PF9234).

♦ **PF3858 Superannuated religious gerontocracy**
Nature As people abandon out-worn institutionalized religions, monks, nuns, priests, brothers, ministers, ecclesiastics of all kinds, theologians and teachers in religious orders, become redundant. Religious houses, East and West, are full of such people with no jobs. Some dignitaries also preside only nominally over sees, parishes and religious communities, but they are in decreasing demand.
Claim Superannuated persons with religious vocations, although belonging to ecclesiastical gerontocracy, are frequently obliged to live in the most impoverished conditions, being led to believe this is the way to holiness. They do not protest even when it is a question of their basic human rights to adequate food, clothing, shelter, warmth and medicine.
Broader Gerontocracy (#PD3133).

♦ **PF3859 Irresponsible finders of personal property**
Nature Finders of lost or abandoned property are conceded by many legal codes to have a right to ownership in the absence of counter-claims. Finders, fearful of claims, often conceal the essential fact of, or details concerning, their finds, this making recovery of lost objects exceptionally difficult.
Incidence Finders of objects of historical, artistic and archaeological or other scientific interest that have been misplaced in private or public collections, storage areas and the like, may remove these for their own gain. Lost pets are infrequently returned; lost jewellery almost never.
Claim There are inadequate public mechanisms to record notices of lost property; to publicize these; or to standardize incentive schemes to promote restitution to rightful owners. This reflects considerable moral indifference and is tantamount to sanctioning what is, in effect, simple theft.
Broader Irresponsibility (#PA8658).

♦ **PF3861 Limitation**
Nature A programme, organization or activity, or individual and collective human development potential, may be limited either by internal or by external limitation, or by both. Internal limitation may be inherent, constituted by the very nature of human organization and individual and systemic capacities. Thus a primitive society may organize healing functions via ritual, shamanism, witch doctors and other activities; but without real medical knowledge it is nonetheless completely exposed to disease. Internal limitations may also be accidental or circumstantial. It can be said, for example, that inadequate medical services in developing countries can be remedied by training indigenous peoples in medicine, assuming there is no insurmountable intellectual or cultural obstacle to this education. Extrinsic or external limitations may be illustrated by the isolation of a community so that it cannot participate in medical advances. Such a limitation may be presented by natural circumstances. Other external limitations may be imposed by human activity, for example, economic or military subjugation of peoples so that their medical needs are not met.
Background Limitation can usually be measured on a quantitative scale as an insufficiency of something. In terms of human problems this "something" may be causal. If the cause is not quantifiable, its effect may be. Quantification of limitation is a vital step in problem identification, analysis and remedial action. For example, a lack of aid to developing countries may be quantified financially against specific projects or against general budget deficits. A lack or, quantitatively expressed, insufficiency of infrastructure is more difficult to state. Many problems arising from limitations lack statistics. This is due to the presence in problems of a great number of variable limits or constraints. Some of these may be inputs, others may be structural. Complex interrelationships of limits require computer modelling and heuristic methodologies of approaching quantified formulations. An example is world growth models as used in Club of Rome and similar global forecasts. However, there has been no general agreement on global limits, exactly stated, in food, population, resource and pollution problems, for example.
Counter-claim The concept of limitation illustrated in dressed-up but simplistic supply and demand models is frequently expressed by arbitrary numbers in both sides of the equation. They are arbitrary because there is no scientific knowledge that is incontrovertible and that provides a 100 percent accurate forecasting basis. In addition, applied science (that is, human technology) keeps pushing back the limits. For example, the diminishing resources of this planet may be augmented by resources on other planets. Human intelligence is being augmented already by

computer artificial intelligence. And human development itself may go beyond its present limits by the employment of behavioural re–education, by genetic engineering, and by new forms of political and social organization developing in a new age.
Broader Limits (#PF4677).

♦ PF3862 Proscribed thinking and behaviour
Unwarranted prohibitions
Nature The tendency of highly organized religious and ideological groups is to become more and more proscriptive regarding the behaviour of their members and their private thoughts and reading habits. Lists of disapproved books and ideas may be circulated. Schools operated by religious denominations or by statist regimes may not provide some critically acclaimed modern and classic works to their students, and teachers may not even mention disapproved authors or concepts.
Incidence A document entitled, 'Syllabus of Errors' was put forth by the Vatican in 1864. Among its 10 sections and 80 theses it listed 'errors' of thought and behaviour as applied to Bible societies, secret societies, socialism, rationalism, religious latitudinarianism, and wrong conceptions concerning the temporal power of the Pope. Put positively, the Syllabus argued for state religion, temporal power for the Church, the subordination of civil law to ecclesiastical law in religious matters, along with many other relics of mediaeval Church thinking. The Catholic Encyclopaedia notes that even today 'the Syllabus must be accepted by all Catholics since it comes from the Pope'. In Communist Russia and China, 'errors' contradicting orthodox Marxism–Leninism have been identified and condemned and have borne such names as Trotskyism, revisionism, deviationism, etc.
Aggravates Uncritical thinking (#PF5039).
Aggravated by Organization of human thought (#PF5301).

♦ PF3863 Loss of faith in religion
Decline of religion — Loss of religious tradition — Lack of religious activity — Lack of religious conviction
Nature As scientific knowledge advances, the all–inclusive dogmatism of religions recede. People turn from religion and from those values directly, and sometimes indirectly, derived from sacred authority. However, because religions have developed total approaches to life and ready answers and ritual–like responses to life situations, independence from religious beliefs presents problems of autonomy and responsibility for individual behaviour for many people. Their independence gives them a greater vulnerability to self–doubt and uncertainty. From the standpoint of individual psychology, a more–or–less abrupt loss of faith can be symptomatic of a personality crisis or of emotional or mental disturbance or disorder. Loss of faith can aggravate depression or be produced by it. A gradual loss of faith may undermine 'divinely' sanctioned interrelationship patterns, for example, of offspring to parents, and of marital partners. Loss of faith therefore may lead to family break–ups but more importantly, as a contagion or chain–effect reaction, may undermine all positive values, leaving only a vacuum of egoism or nihilistic residue.
The inability of traditional religions to meet the spiritual needs of people has resulted in an increase in attendance in fringe religious organizations and in individualistic spiritualism.
Broader Forced assimilation (#PC3293)
Fragmented conduct of community operations (#PF1205).
Aggravates Decreasing participation in collective religious worship (#PF8905).
Aggravated by Decadence (#PB2542) Secularization (#PB1540)
Irreligiousness (#PF8234) Superficial religion (#PJ5252)
Untransposed significance of cultural tradition (#PF1373).

♦ PF3864 Fall of man
Nature Judaeo–Christian theology asserts that man was created with a nature and characteristics in many ways superior to what he now exhibits. His present condition is a fallen one for which he alone is responsible. However, he has no complete power to redeem himself but must be redeemed through an intercessor, the Christ or Messiah.
Background God created man purposively and endowed him with the capacities to fulfil his ends. These capacities or capabilities were crowned by a will that was free to pursue the spiritual purposes of the creation. Man's intellect, though enriched with supernatural graces, was nonetheless subordinated to spiritual willing, so that the end or good to which his life tended by providence was not consciously known. Man, however, chose to subordinate the unconscious accord of human and divine will by acquiring knowledge or consciousness of this divine good or end, symbolized by his eating from the Tree of the Knowledge of Good and Evil. This was the knowledge that, as the Old Testament phrases it, man could be as a god: that is, autonomous and creative, embracing the earth and reaching heavenwards into space with his own works. The pride in the realization of the power of such science or gnosis caused the separation or estrangement, from the providential purposes, between mankind and the Creator. Man's judgement entered into considerations as to how the human potential could be realized apart from an Edenic dependency. As man freely chose independence and consciousness of his own nature, he lost his state of grace. This included his mystical or unconsciously intuitive knowledge of nature which had also afforded him relief from the necessity of economic activity. His Fall cut off both his mental communion with animals and nature, and with his own unconscious, and hence, with his consciousness of God.
Refs Williams, Norman P *The Ideas of the Fall and of Original Sin.*
Broader Original sin (#PF8298).

♦ PF3869 Inaccessibility of religious scriptures
Inaccessibility of the Bible
Nature Some part of scriptures are in print for 99 percent of the world. More than 95 percent of all people have the entire Bible in print in their mother tongue. But, 45 to 55 percent of the adult world cannot read. Scripture is closed to have of the world.
Broader Forced repatriation of prisoners of war (#PD0218).

♦ PF3871 Preoccupation with reciprocity in trading relations
Narrow egalitarian application of trade measures
Nature Mutual reciprocity is a principle under which multilateral trade negotiations in GATT should be conducted such that the different parties in a trade relationship benefit best by regulations that force each party to give in goods or benefits as much as they get back, so that all may benefit equally. The principle was adopted for essentially political and legal reasons although it is not an obligation of the trading system. This concept of multilateral reciprocity has been undermined by a growing preoccupation with reciprocity in a narrow, even bilateral, sense which has further hampered the ability of the international trading community to take concerted action in the perspective of overall trade and economic benefits and common goals rather than in terms of concessions to be made in a particular negotiation. It has thus impeded coordinated flexible responses to new situations.
Claim Emphasis on narrow concepts of reciprocity may be a product of a more intensely competitive trading world, but it clearly detracts from the possibility of devising global solutions to global problems.
Broader Restrictive trade practices (#PC0073)
Unilateral interpretations of multilateral principles (#PF9629).
Reduces Excessive external trade deficits (#PC1100).

♦ PF3876 Eurocommunism
Background In the mid–1970's the Eurocommunist parties were proclaimed to be on the march across Southern Europe and were on the verge of gaining strong government roles from Greece through Italy, France, Spain and Portugal. By the late 1980's, none of the Eurocommunist parties holds even a minor share of national power. In the 1976 general election more than one–third of the Italian electorate vote Communist. In the 1979 general election they won 1.5 million few votes. And in the local elections held in June 1988 the Communist Party of Italy won one third less votes than in 1979.
Claim The Eurocommunist parties can either recoil from the challenges facing them and accept becoming isolated parties backed by a dwindling minority of voters or they can accept the painful changes being demanded by changes in the communist block. This would mean, in effect, discarding class warfare and accepting to work within a democratic and modern capitalistic system and in doing so ceasing to be communists.
Broader Communism (#PC0369).
Related National communism (#PF3130).

♦ PF3881 Abduction by extraterrestrials
Claim There are many hundreds of documented claims of abductions, namely instances when the witness of a close encounter has recalled under hypnosis about being forcibly taken inside a UFO. The witnesses generally describe being subject to medical examination (often causing trauma and pain). These claims are subtantiated by scars on the body and missing time in their conscious recollections during that period.
Counter–claim The encounters remembered by witnesses, if they occurred at all, should be treated at the symbolic level. They do not provide information about the extraterrestrial origin of the beings. To the extent that there is underlying reality to the UFO phenomenon, they provide a physical suport for human ability to use them as symbols in response to inner needs.
Refs Vallee, Jacques *Dimensions* (1988).
Broader Kidnapping (#PD8744).
Related Unidentified flying objects (#PF1392).
Aggravated by Extraterrestrial invasion (#PF4444)
Secrecy concerning existence of extraterrestrials (#PF4331).

♦ PF3884 Stagnating social methods
Inadequate social technologies — Uninitiated social methods
Nature The techniques used to create, support and enliven humane societies have become useless. Citizens are blocked from being informed about their communities and nations. Minorities have less influence than their numbers would suggest. Structures for developing effective leadership have failed. The public's consciousness is suppressed. Communities, regions and nations are not comprehensively and effectively related to each other. Individuals lack ways of seeing their own destiny connected to society's.
Broader Obsolete methods (#PF3713).
Narrower Absence of tactical methods (#PC0327).
Aggravated by Lack of motivation in leadership development (#PF2208).

♦ PF3900 Mediocrity
Cult of mediocrity
Aggravates Deterioration in product quality (#PD1435).
Aggravated by Apathy (#PA2360).
Reduces Competition (#PB0848).

♦ PF3904 Fear of food contamination
Neurosis about food
Broader Fear (#PA6030).
Aggravates Food fads (#PD1189).
Aggravated by Food pollution (#PD5605)
Environmental hazards from food processing industries (#PE1280).

♦ PF3913 Intergovernmental failure to fulfil financial commitments
Delay in supply of agreed multilateral assistance
Incidence On the occasion of the 1988 summit of Arab leaders in Algiers, $300 million was committed annually to assist the 1.5 million Palestinians living in Israeli–occupied territories. The amount was never fully paid.
Related Conditional observance of multilaterally agreed trade commitments (#PF7838).
Aggravated by Indecisive multilateralism (#PF9564)
Arrears in payment of government financial commitments (#PF1179)
Government non–payment of agreed contributions to international organizations (#PF8650).

♦ PF3918 Placement of the burden of proof on the disempowered
Placement of the burden of proof on victims of environment pollution — Placement of the burden of proof of abuse on vulnerable groups — Benefit of doubt given to vested interests
Claim Even if adequate proof is lacking on the causes of environmental and other dangers, doubt should not be used as a reason for postponing measures to prevent them.
Aggravates Inadequate evidence to convict known offenders (#PF8661)
Government delay in response to symptoms of problems (#PF6707).
Aggravated by False evidence (#PF5127) Destruction of scientific records (#PF4633)
Politicization of health standards (#PD4519)
Unproven relationships between problems (#PF7706)
Inadmissability of evidence from other jurisdictions (#PE5424)
Jurisdictional conflict between academic disciplines (#PF9077).

♦ PF3931 Adverse consequences of scientific and technological progress
Dogmatic belief in benefits of technical progress
Refs Coppock, Rob *Social Constraints on Technological Progress* (1984).
Broader Inappropriate understanding of progress (#PF8648).
Narrower Medical backlash (#PE6711) Increasing pace of life (#PF2304)
Fast and irregular pace of technological advance (#PE7898)
Irresponsible scientific and technological activity (#PC1153).
Aggravates Proliferation of information (#PC1298)
Maldistribution of science and technology (#PC8885)
Proliferation of nuclear weapons and technology (#PD0837)
Faltering structural adjustment in the world economy (#PF9664)
Inevitable destruction of natural environment by mankind (#PE2443)
Inability of educational systems to keep pace with technological advancement (#PF7806).

♦ PF3945 Restrictions on use of personnel
Nature Requiring the acquiring party to use personnel designated by the supplying party, except to the extent necessary to ensure the efficient transmission phase for the transfer of technology and putting it to use or thereafter continuing such requirement beyond the time when adequately trained local personnel are available or have been trained; or prejudicing the use of

personnel of the technology acquiring country.
Broader Restrictive trade practices (#PC0073).
Related Collusive tendering (#PE4301) Grant-back provisions (#PE5306)
Challenges to validity (#PF1200) Restrictions on research (#PF0725)
Restrictions on publicity (#PF1575) Restrictions on adaptations (#PF5248)
Exclusive dealing arrangements (#PE0413)
Patent pool or cross-licensing agreements (#PE4039)
Exclusive sales and representation agreements (#PE4581)
Trade restrictions due to voluntary export restraints (#PE0310)
Payments after expiration of industrial property rights (#PF5292)
Tying of supplies to subsidiaries by transnational enterprises (#PE0669).

♦ PF3949 Inadequacy of food aid

Food aid — Maldistribution of food aid — Maldistribution of food — Stagnation of food aid to developing countries

Nature It has been found that food aid has not improved the living conditions of the poorest people nor helped to build a development momentum. There is an associated disincentive effect, and in addition food aid often goes to benefit urban consumers rather than those in real need, in doing so changing food tastes in favour of imports such as wheat flour for bread making. Food aid has been used as a way of disposing of the agricultural surpluses of the developed countries and has acted as a disincentive to the local production of food crops. Feeding programmes for mothers and small children are not always carried out in the context of broad-based health and nutrition programmes, and often do not use locally produced foods. Food-for-work schemes, where food is exchanged for labour on public works, are another channel for nutritional relief, but do not attack the roots of poverty or create a sustainable income. Food aid programmes are rarely implemented as part of an overall, well-planned and coherent food strategy.

Incidence Food aid allocations for 1982/83 were approximately 9.4 million tons of cereal equivalent, very similar to the quantity actually shipped in 1981/82. However, a smaller proportion of food aid was shipped to low-income, food-deficit countries in 1981/82 - 79 percent - when compared to 81 percent in 1980/81. The proportion of cereal imports of these countries covered by food aid was at a low-figure of 18 percent; in 1977-78, it was 24 percent but had steadily declined since. Food aid basically stagnated during the five years to 1982, while the cereal imports of the low-income countries increased over 30 percent.

Claim Food aid is not a panacea and should not be viewed as such. While such aid is important in both absolute and selective terms, the food gap in developing countries, both present and projected, is of a magnitude which makes it clear that the answer is not food aid but development: the development of food. Food aid should be used in response to emergency situations such as floods or famines. It can strengthen agriculture in the recipient countries, but just as important are the development strategies pursued by the recipient countries and increased flexibility on the part of the donors to adapt to the recipient's situation and be innovative in the use of food aid. Many current uses of food aid are linked to disincentive effects on local food production, to destabilization of economic traditions.

Counter-claim While food aid may lead to inertia, along the way it saves lives, creates jobs, and an humanitarian link between donor and recipient. The basic humanitarian instinct to give to those in need cannot be discounted. Food aid to developing countries has stagnated in recent years. The annual food aid target of at least 10 million tons of cereals recommended by the World Food Conference in 1974 was only reached in 1984-5, ten years later. Trend estimates of gaps between domestic food production and demand at a given prices that the 85 developing countries are not able to fill through commercial imports project food aid needs are to rise to about 36 million tons by 1990 and 56 million tons by 1995. For the 36 lowest income countries, food aid needs are estimated to be 18 million tons in 1990 and 27 million tons in 1995.

Broader Maldistribution of food (#PC2801)
Ineffectiveness of aid to developing countries (#PF1031).
Narrower Modification of environmentally adapted nutritional habits through food aid (#PF6078).
Related Inappropriate food aid (#PE0302)
Dumping of food in developing countries (#PE0607).
Aggravates Dependence of developing countries on food imports (#PE8086).
Aggravated by Unethical food practices (#PD1045).
Reduced by Agricultural surpluses (#PC2062).

♦ PF3953 Governmental incompetence

Government ineptitude — Bad government
Broader Incompetence (#PA6416).
Aggravates Government inaction (#PC3950) Unrealistic policies (#PF9428)
Military incompetence (#PJ1069) Deception by government (#PD1893)
Fraud by government agents (#PD8392) Excessive government control (#PF0304)
Unpredictable governmental policy (#PF1559) Blackmail by government officials (#PD9842)
Frustrated talent in government posts (#PE4398)
Uncoordinated government policy-making (#PF7619)
Public resentment at government policies (#PF7472)
Loss of confidence in government leaders (#PF1097)
Self-destructive government policy-making (#PF5061)
Unchecked power of government bureaucracy (#PD8890)
Inadequate management of government finances (#PF9672)
Inefficiency of state-controlled enterprises (#PD5642)
Wastage in governmental budgets and appropriations (#PD0183)
Antagonism between government agencies and officials (#PE2719).
Aggravated by Mediocrity of government leaders (#PF3962)
Inadequate system of political checks and balances (#PE4997).

♦ PF3956 Bewitchment

Ensorcellement — Enchantment — Spellbound — Accursed
Narrower Evil enchantment (#PF5123).
Aggravates Demonic and spirit possession (#PF5781)
Uncritical acceptance of another person (#PF5973).
Aggravated by Magic (#PF3311) Voodoo (#PF9006)
Witchcraft (#PF2099) Superstition (#PA0430).

♦ PF3960 Ritual pollution

Ceremonial uncleanness — Ritual impurity — Death pollution

Nature A condition resulting from association with blood (menstrual, parturitional, placental, accidental or shed in murder or battle), death and birth. It also results under certain circumstances, whether in limited areas or at particular times, from association with certain foods and drinks, colours, places, seasons, trees, rocks and even persons.

Incidence Ritual pollution is a condition of major concern in certain cultures, possibly under the influence of particular religions, such as Islam and Judaism. It is a determining factor in the reinforcement of the caste system in India. Hindu priests burned 1,764 lbs of clarified butter, 6,630 lbs of edible oil seeds, 3,315 lbs of rice, 1,547 lbs of barley and 830 lbs of sugar as sacrificial offering to 640 million deities and had conch shells blown at local temples in order to purify the atmosphere and provide salvation for the souls of the Bhopal industrial disaster. In Japan, ritual uncleanliness resulted from contact with unsanitary things, human blood, human or animal death, natural disasters, disturbed life in human society, and incest or bestiality. In certain Chinese sub-cultures death pollution is a major concern, as with many South American Indians where mourners must undergo purification after funerals. In its relationship to the system of taboos prevalent in tribal cultures, the condition is of widespread concern.

Background Remedies for ritual impurity, have involved and, in many societies and religions, continue to involve the performance of some ceremony which variously includes sacrifice, washing, anointing, sprinkling, burning, or cutting. The guilt of an individual or a whole tribe may also be transferred to a scapegoat who is driven off or even killed.

Broader Taboo (#PF3310).
Related Spiritual impurity (#PF6657).
Aggravates Religious opposition to public health practices (#PF3838).
Aggravated by Unethical food practices (#PD1045)
Unethical personal relationships (#PF8759).

♦ PF3961 Disintermediation

Nature The process of ridding economic transactions of middle-men is occurring across the Western economies. Although it is intended to bring producers and consumers together, increase efficiency and heighten customer satisfaction, much the reverse is often true. Jobs are sacrificed and service impaired as there are fewer people available to respond fully to customer questions and needs.

Aggravated by Dependency on middlemen (#PD4632).

♦ PF3962 Mediocrity of government leaders

Incompetence of government leaders — Politicians' lack of character

Nature Government leadership in democracies or totalitarian states tends toward the mediocre. They are incapable of inspiring, or administrating, leading or managing. The selection processes eliminate those who can inspire and manage. In western democracies the values of selecting candidates by parties are governed by advertising considerations, such as, being photogenic and being acceptable culturally, religiously and ethnically and by political considerations, such as, party loyalty and size of power base within the party. Leadership qualities are not very important. In totalitarian systems of government, the selection process are either bureaucratic or violent neither of which tend to result in great leaders.

Broader Incompetence (#PA6416) Obstacles to leadership (#PF7011).
Related Corruption of government leaders (#PC7587).
Aggravates Lack of leadership (#PF1254) Governmental incompetence (#PF3953)
Loss of leadership credibility (#PF9016) Frustrated talent in government posts (#PE4398)
Loss of confidence in government leaders (#PF1097)
Government leaders associated with sex scandals (#PE7937).
Aggravated by Deception by government (#PD1893)
Corruption amongst relatives of government leaders (#PE9140)
Excessive accumulation of wealth by government leaders (#PD9653).

♦ PF3973 Elder paralysis over the future

Inability of the elderly to play a role in developing the future

Nature Present social images and structures cast the elders of society into a parochial context and rigid roles. The overemphasis on being economical productive as the major criteria for social usefulness renders retirement as a time of death in relation to society and its needs.

Because they are unable to accept the whole of past experience, including tragic as well as happy occurrences, older people are unable to identify with the present fast pace of society; they cannot see demands on their skills as part of historical development; nor can they use their insight into history to generate a vision of the future.

Broader Erosion of elders' wisdom (#PF1664)
Denial by old people of the significance of the past (#PF2830).
Narrower Unmeaningful social roles for the aged (#PF1825).
Related Sentimental attachment to the past (#PF1577)
Limited community responsibility of adults (#PF1731)
Disorientation of the young due to lack of social forms (#PD2050).
Aggravated by Temporal deprivation (#PF4644)
Lack of facilities for educating older people (#PF2012).

♦ PF3975 Delay in implementation of commitments

Delay in loan disbursements

Incidence Notable examples include the delays by governments to implement commitments on standstill and rollback in relation to existing protectionist measures against developing countries. In the case of non-concessional lending by the World Bank, for example, mainly as a consequence of the virtual stagnation of its commitments during 1982-85 and delays in loan disbursements, only made a marginal contribution to net resource transfers to developing countries in 1986, following a declining trend which emerged in 1984. Central ministries often react to tight budget constraints, overprogrammed budgets, or simple mistrust of spending agencies by slowing the disbursement of funds or by erecting unnecessarily cumbersome procedures in areas such as procurement, land acquisition, or contractor eligibility. Such indirect forms of control delay the implementation of projects.

Broader Delays in implementation of social change (#PC6989).
Aggravates Arrears in financial payments between government agencies (#PE9700).
Aggravated by Uncoordinated government policy-making (#PF7619).

♦ PF3976 Violating taboos

Deviation from behavioural norms — Infringement of taboos
Broader Taboo (#PF3310) Deviation (#PA6228)
Discriminatory unwritten codes of behaviour (#PEp017).
Narrower Sexual deviation (#PD2198) Violation of food taboos (#PD2868)
Torture by violation of taboos (#PE1296).
Aggravates Discriminatory imposition of standards (#PD5229).

♦ PF3979 Disrespect for elders

Disrespect for parents — Contempt for elders — Erosion of filial piety

Background Processes associated with the modernization and the development of western industrial societies have led to a change in status for the aged. These created conditions which made possible the emphasis on the primary of the conjugal tie, at the expense of the filial bond.

Claim All the forces of industrial society operate to make the child disrespect and have contempt for parental authority. The dislike of the young for everything old, including old people, has resulted in nothing but chaos, anarchy, disintegration and decadence, where not only the bonds between the young and old are destroyed but all kinship ties and human relationships in general have been weakened to breaking point.

Related Contempt for authority (#PF5012).
Aggravates Senilicide (#PJ7124) Parenticide (#PE0651)
Disobedience of elders (#PF7149) Exploitation of the elderly (#PD9343).
Aggravated by Age discrimination (#PC2541).

♦ PF3983 Religious hypocrisy

Spiritual hypocrisy — Double standards in religion
Broader Hypocrisy (#PF3377).
Aggravates Superficial religion (#PJ5252).
Aggravated by Religious deception (#PF3495) Culturally biased religion (#PF1466).

♦ PF3994 Violence as a resource

Nature The personal use or threat of violence is the most effective means to discourage those who would misuse or deprive an individual. It is practice not only in traditional societies but by those outside the law, like gangs and some inner city communities.

Background In most societies for thousands of years an individual's security largely depended on the ability to maintain a credible threat of violence. His interests were likely to be violated by competitors unless the competitors were deterred by fear of retaliation. With the development of the state, its monopoly on the legitimate use of force grew as well. In the Eleventh Century, William the Conqueror outlawed private vengeance in England and made homicide a crime against the state instead of a private wrong.

Broader Human violence (#PA0429) Youth violence (#PF7498).
Related Sexual violence (#PD3276) Family violence (#PD6881)
Structural violence (#PB1935) Physical intimidation (#PC2934)
Intimidation of electors (#PD2044) Violence as entertainment (#PD5081)
Violence along internal borders (#PD4782)
Violence by fanatical environmentalists (#PD5582).
Aggravates Fear of death (#PF0462) Political violence (#PD4425).
Aggravated by Pain (#PA0643).

♦ PF4000 Non-productive members of society

Broader Unemployment (#PB0750) Underproductivity (#PF1107).
Related Illiteracy (#PC0210).

♦ PF4004 Despair

Desperation — Hopelessness — Spirit sickness — Despairing people

Nature Despair is characterized by lack of hope and by a sense of waiting for hope to return. In present day Western society, despair is commonly suppressed or at least a (usually successful) attempt is made to hide its existence from others. This is said to leading to a partial numbing of the psyche so that the emotional and sensory life are diminished. Anxiety provoking data is effectively filtered out and the numbing effect intensified unless the despair is worked through. In this respect it is not dissimilar from grief.

Incidence In industrialized countries despair is typically associated with alienation from society. There is growing desperation among the poor in both the industrialized and developing countries.

Claim 1. In the long-run the security and well-being of the industrialized world depend on improving the economic prospects of the world's poor. Rich countries cannot hope to wall themselves off from billions of desperate people who see the gap between themselves and others widening.

2. Lucid despair menaces the living more than the frantic or irrational kind. The message of the suicide from beyond the grave is that, having peeled life back to the core, nothing was to be found.

Refs Park, James *Absurdity, Insecurity and Despair* (1975).

Broader Hopelessness (#PA6099).
Narrower Oppressive prevalent images (#PF1365)
Hopelessness inspired by school (#PJ1004).
Related Impure thoughts (#PF5205) Dehumanization of death (#PF2442).
Aggravates Superstition (#PA0430) Aerial piracy (#PD0124)
Lack of understanding of spiritual healing (#PF0761).

♦ PF4008 International migration

Mass immigration

Nature International migration gives rise to problems of assimilation and to the development of ethnic and religious minorities that cannot readily be integrated into the community of their adopted country. Antagonism is likely to develop against migrant groups which are different in appearance, habits and attitudes from the local inhabitants; and to grow in areas where migrants are relatively numerous. Contemporary migration is also marked by the outflow of skills from many developing countries. The problem arises not only with regard to highly skilled professionals; there is also a loss of scarce skills in various sub-professional categories, a type of middle level technical and administrative brain-drain. As a result many developing countries face severe shortages in certain skills and professions.

Incidence According to ILO surveys, in 1981 there were two million Asian migrant workers abroad, the vast majority of them in the Middle East, and most of them skilled blue collar workers in construction and transport. Pakistan had 775,000 workers abroad in 1982; 354,000 workers left the Philippines in 1982; and 250,000 Indians sought work outside India during that same year.

Background The recent migrations are of a modest scale compared to the mass migrations of population that took place from Europe to the New World in the latter part of the 19th century and the early part of the 20th century. However, they represent a resumption of earlier patterns of international migration. It is estimated that at least 52 million people migrated from Europe to the United States, Canada, Latin America and Oceania between 1840 and 1930, and this amounted to 20 percent of the population of Europe at the beginning of this period. There has also been a significant change in this pattern of migration from Europe and since 1960 there is a positive net migration to Western Europe.

The volume and nature of international migratory movements continue to undergo rapid changes. Illegal or undocumented migration and refugee movements have gained particular importance; labour migration of considerable magnitude occurs in all regions. The principal root cause of this large scale contemporary migration is to be found in the disparity in economic development, employment opportunities and living standards between countries at different stages of economic and social development. Natural disasters and political crises are among other principal causes of shifts in migration patterns and the increase in the volume of overall international population movements; though the motivation for most international migration is higher wages. Historically, some migration may have been directly related to population pressure, but today wage differences are the main driving force.

Refs Appleyard, Reginald and Stahl, Charles (Eds) *International Migration Today* (1988); Bilquees, Faid and Hamid, Shahnaz *Impact of International Migration on Women and Children Left behind* (1981); Newland, Kathleen *International Migration* (1979).

Broader Uncontrolled migration (#PD2229).
Narrower Emigration of trained personnel from developing to developed areas (#PD1291).
Aggravates Illegal occupation of unoccupied property (#PD0820).

♦ PF4009 Internal migration

Nature Population movements within a country can be looked at from two points of view: movements from one self-contained region to another; and movements between areas of different population density. It is not uncommon to observe both movements to more prosperous regions and at the same time a movement from the countryside to the towns. These movements often give rise to serious problems; for instance, as more and more people try to live in urban areas it becomes necessary for some of the surplus population to spread into neighbouring regions of lower population density. As a consequence, not only must more housing, drainage and other services be provided in these regions, but additional jobs must also be found; this brings about a change in the economic structure of the regions, which in turn leads to other demands, in particular on the local system of schools and technical colleges which have to provide the longer training usually called for by urban forms of employment. These problems add force to the demands for regional policies which seek to put all regions on a self-sustaining basis and so slow down, or even reverse, the former regional drifts.

Incidence According to some estimates, 75,000 people are leaving the rural areas of the Third World every day to set up home in the towns and cities. Rapid urbanization is bound to have radical effects on a nation's population and on its economy. More of a consensus exists at the world level with respect to spatial distribution of population and internal migration than on any other demographic topic: almost all the developing countries consider the current geographic distribution of their population partially or wholly unacceptable. The major urban centres in most developing countries have already become too large, and are growing too rapidly compared with the smaller cities, towns and rural areas. The costs of management of large, over-congested metropolises (in addition to relative concentration of public investment there), have become phenomenal, hampering general development goals and increasing the geographical inequalities which already exist.

The expected growth in urbanization in the less developed countries is the result not only of the natural increase in urban population but also of massive migrations of population to these areas in search of employment. It is estimated that the largest cities will grow more rapidly than the smaller cities and some of these are likely to reach proportions which are totally unfamiliar to town planners. In 1950, only four of the fifteen largest cities were in the less developed countries, but this number rose to seven by 1975. It is projected that twelve of the fifteen largest cities will be in the less developed countries by the year 2000. There were only six cities with populations of 5 million and over in 1950 and their combined population was only 47 million; in 1980, there were 26 such cities and their combined population was 252 million. Projections indicate that this number will rise to approximately 60 with an estimated population of nearly 650 million by the year 2000.

Broader Uncontrolled migration (#PD2229).
Narrower Back to the land (#PF4181) Rural depopulation (#PC0056)
Migration of rural population to cities (#PE8768).

♦ PF4010 Personal physical unattractiveness

Unsatisfactory personal appearance — Dissatisfaction with personal image — Inadequate grooming — Unkempt appearance — Boring personal appearance — Unchanging personal appearance — Unfashionable personal appearance

Incidence As an indication, the market for cosmetics in the UK in 1986 was £760 million (eye make-up, £103 million; facial make-up, £73 million; lip make-up, £76 million; highlighter and blushes, £25.5 million; nail make-up, £46 million; female fragrances and perfumes, £302 million; male fragrances and after-shave, £132.7 million). The market for shampoos was £125 million and for hair sprays was £86.3 million.

Counter-claim The effort of taking care of personal appearance is an indication of physical and emotional well-being.

Aggravates Low self esteem (#PF5354).
Aggravated by Ugliness (#PA7240) Human ageing (#PB0477)
Inadequate personal hygiene (#PD2459)
Unsociable human physiological processes (#PF4417).

♦ PF4013 Inadequate industrial retraining programmes

Unexplored industrial training

Nature With increased economic and technical change, traditional methods of acquiring skills through apprenticeship and of maintaining them through gradual adaptation on the job appear to be breaking down; perhaps to the detriment of quality, creativity and personal incentive.

Broader Inadequate technical training (#PE8716).
Related Unsatisfied need for continuing education (#PF0021).
Aggravates Unemployment (#PB0750) Underproductivity (#PF1107).
Aggravated by Emigration of trained personnel from developing to developed areas (#PD1291).

♦ PF4014 Limitation of current scientific knowledge

Ignorance of scientists — Incomplete state of the sciences — Imperfections of science

Nature In contrast to scientific ignorance (or illiteracy), this concerns the limited competence of science to explain certain phenomena at this stage in the development of knowledge. Scientists tend to cultivate an image in relation to non-scientists of competence, but only with respect to those problems that are currently considered solved, or in the process of solution. The unreliability of certain conclusions is not communicated to non-scientists, nor is it necessarily accepted within those disciplines for which training focuses primarily on soluble problems. Within science problems tend to be chosen for investigation only if there is some probability of arriving at a solution. Scientists place their careers at risk if they explore problems for which the research tools and conceptual frameworks are believed to be as yet inadequate.

Background Almost all the facts learnt by students of science are uncontested and incontestable. Examinations assume that every problem has only one correct solution. Only during subsequent research do scientists discover that scientific results can vary in quality, that many solutions can only be understood as tentative, that facts in the scientific literature may be unreliable and that concepts have a certain degree of plasticity. Later still in the careers of scientists do they discover that there are indeed problems that cannot be solved. Such scientific ignorance is paradoxical and directly contradictory to the image of science and the associated technologies.

Incidence The pervasiveness of ignorance concerning the interactions of technology with the natural and social environment has only recently been explicitly recognized, despite the manner in which ignorance dominates the sciences of the biosphere. Science has been able to draw attention to its successes as justification for further investigations along similar lines, whatever the relevance to humanity's problems, whilst distancing itself from preoccupations through which its current incompetence becomes apparent.

Claim 1. Humanity is entering an era of chronic, large-scale and extremely complex syndromes of interdependence between the global economy and the world environment. Relative to earlier generations of problems, these emerging syndromes are characterized by profound scientific ignorance.

2. The inherited conception of science is inappropriate for the new tasks of control of the apparently intractable biosphere problems. If humanity is to cope with the enormous problems now confronting it, some of the ideas about science and its applications will have to change. The most basic of these is the assumption that science can indeed be useful for policy, if and only if it is natural and effective and can provide "the facts" unequivocally. But now it has become apparent that humanity must cope with the imperfections of science, with radical uncertainty, and even with ignorance, in forming policy decisions for the biosphere.

Broader Lack of knowledge (#PF8381).
Narrower Insoluble scientific problems (#PF4706).
Aggravates Questionable facts (#PF9431) Inaccurate forecasting (#PF4774)
False assumptions on sustainable development (#PF2528).
Aggravated by Manipulative knowledge (#PF1609)
Fragmentation of academic disciplines (#PF8868)
Complex interrelationship of world problems (#PF0364)
Structural amnesia in institutional systems (#PF7745).

♦ **PF4022 Broken heart**
Aggravated by Unrequited love (#PF6096).

♦ **PF4028 Wage rigidity in labour markets**
Claim Since the end of 1970s wage rigidity reduced the demand for labour associated with any rise in effective demand by forcing enterprises to seek productivity gains by means of labour shedding, and hence constantly increased the rate of unemployment at any given rate of capacity use.
Broader Structural rigidities in labour markets (#PD4011).

♦ **PF4034 Enjoyment of war**
Nature The reversion to more primitive and bestial ways of conduct which is part of the killing and destruction of war, is for the combatants an opportunity to explore a level of consciousness which precedes by millions of years the more rational, compassionate and humane consciousness of our era. For many this journey is an exhilarating, stimulating adventure in consciousness, no matter how great the physical suffering and tribulation endured. This delight is most often an experience among combatants, not shared by civilians to any great extent.
Broader Culture of violence (#PD6279) Psychological disorders (#PD8375).
Aggravates Incitement to war (#PD4714)
Persistence of a technical state of war following cease-fire agreements (#PE2324).
Aggravated by War (#PB0593) Glorification of war (#PF9312)
War and pre-war propaganda (#PD3092).

♦ **PF4037 Derealization**
Nature The external world is felt to be unreal and strange, no longer like it used to be.
Related Depersonalization (#PA6953).

♦ **PF4043 Spontaneous human combustion**
Broader Unexplained phenomena (#PF8352).

♦ **PF4050 Suppression of information by security classification**
Suppression of innovation by purchase of patents — Government refusal to declassify innovative patents
Incidence Example: the UK inventor of a new form of space vehicle (the Hotol) considerably cheaper than a shuttle, patented the concept which was then classified by the government as being of military importance. When the military ceased funding its development and private industry was scheduled to take over, the government refused to declassify the patents to enable the work to continue, despite the considerable funds already invested by government in the project.
Broader Suppression of information (#PD9146).
Aggravates Disincentives to invention (#PG6623)
Restrictive business practices in relation to patents and trademarks (#PE0346).

♦ **PF4062 Undocumented violations of human rights**
Undocumented war crimes — Undocumented massacres — Undocumented torture
Broader Unreported crimes (#PF1456) Unrecorded knowledge (#PF5728)
Falsification of public records (#PD4239).

♦ **PF4077 Uncatalogued documents**
Broader Unretrievable documents (#PF4690).
Narrower Lack of medical records (#PF1836) Uncatalogued historical documents (#PF6483)
Inadequate documentation of works of art (#PE8088).
Related Misfiled documents and records (#PF4708).
Aggravates Global amnesia (#PF0306).
Aggravated by Suppression of safety records (#PF2714)
Incompatibility of document classification systems (#PE7753).

♦ **PF4095 Ineffective governmental use of nongovernmental resources**
Ineffective use of private sector resources by government — Government opposition to private initiative — Protection of public sector agencies from competition
Incidence Public regulatory policies have inhibited private providers of services. Although not explicitly part of the public budget, these policies can have large effects akin to taxation and spending. There is no uniquely appropriate balance between public and private sector activity. However, governments reduce their ability to broaden access to education and health, for example, when they discourage private initiatives. In the case of housing, in many cities private housing markets have been overly restricted by rent control, which has often produced results exactly opposite to those originally intended.
Claim In many areas, exposing state-controlled enterprises to domestic and foreign competition would promote economic efficiency. Such enterprises are protected by budgetary subsidies, regulated domestic markets that keep out private competitors, tariffs and import quotas. Such protection inhibits the characteristic ability of the private sector to adapt to changing conditions of trade.
Broader Inappropriate public spending by government (#PF6377).
Related Misuse of nonprofit associations as front organizations by government (#PE0436).
Aggravated by Neglect of international nongovernmental organization network (#PF7057)
Assessment of international nongovernmental organizations as naive (#PF7116)
Limited recognition of international nongovernmental organizations (#PF7093)
Absence of national nongovernmental organizations in developing countries (#PF7081)
Operational ineffectiveness of international nongovernmental organizations (#PE7056)
Inadequate international nongovernmental organization response to intergovernmental calls for action (#PE7140)
Absence of policies to associate international nongovernmental organizations in regional development (#PF7033).

♦ **PF4096 Anthropocentrism**
Nature Anthropocentrism is the world view which sees human beings as being at the centre of all creation. While a recent invention, it is take for granted by most Westerners and much of the rest of humanity.
Broader Inadequate models of socio-economic development (#PF9576).
Related Androcentrism (#PF6648).
Aggravates Unconstrained socio-economic growth (#PB9015).

♦ **PF4106 Inadequate statistical information on tourism**
Nature It is widely recognized that the current state of statistics on international tourism is far from satisfactory in the case of most countries, and that inter-country comparisons are extremely difficult because of limited consistency among the definitions and classifications employed by different countries. Deficiencies in statistics arise from the intrinsic difficulties both in defining the concepts and classifications and in collecting sufficiently reliable data. Work on statistics and economics of tourism is comparatively recent, in contrast to statistics of population, labour, international trade and industrial activity, in which there is long-term experience. Each country has its own geographical and administrative conditions which have affected the growth of tourism statistics. Moreover, these statistics have usually received low priority in the development of a national statistical system. Many of the items of data must be collected from establishments dealing with visitors to a country or from international visitors themselves, who are not the easiest respondents for statistical inquiries, partly because their transactions may be carried out in a variety of currencies and not always in accordance with government regulations. The commercial export or import of merchandise is usually accompanied by considerable documentation, which furnishes the basis for compiling international trade statistics. By contrast, most countries, in order to facilitate tourism, have minimized or even eliminated formalities and record-keeping concerning international visitors at frontiers and other points of contact. The volume and geographical dispersion of tourism also makes collections of detailed statistics difficult in many cases.
Incidence Even in countries with advanced systems of statistics and considerable experience in this field, the available statistics of tourism are far from complete and there are marked difficulties in reconciling certain data from different sources. While the growth of the tourist industry is of major importance to many developed countries, they have serious difficulties in compiling statistics of tourism that are adequate to their needs partly because of their limited statistical resources.
Broader Inadequate statistical information and data on problems (#PF0625).

♦ **PF4108 Lack of agricultural machinery**
Insufficient agricultural equipment in developing countries — Insufficient agricultural machinery in developing countries — Unequal distribution of agricultural machinery — Minimal farm machinery
Nature Farming mechanization is generally regarded as one of the critical input variables for increasing agricultural productivity and total food production. In many developing countries, where 55-60 percent of the population is wholly or partly employed in the agricultural sector, total economic development very much depends, inter alia, on the rate at which farming mechanization increases food production, often in the face of rapid population growth. It also renders the terms of trade between the dominant agricultural sector and the fledging industrial sector more economically balanced and equitable.
Equipment operating costs, including energy, spare parts and servicing, contribute additional burdens to the small-scale farmer and further compel the host government to intervene by way of control and subsidy of spare parts, fuel and community repair shops. Again, this has the effect of further limiting the units of equipment that can be deployed in the agricultural sector, while removing the full marketing and servicing responsibility from the manufacturer, to the extent that the host government gets involved in such activities.
The estimated fraction of the number of tractors in use in relation to the potential demand in developing countries indicates that a very substantial proportion of this potential demand goes unfilled by locally produced or assembled products or by imports from developed countries. Analysis of trade statistics and other industry data indicate that a major reason for this sort of imbalance is that in real terms, farm machinery exported to or assembled in developing countries, with transnational corporations as a major partner, are almost invariably beyond the means of individual small-scale farmers, both at the point of purchase and at subsequent routine operations of the equipment on the farm.
Broader Inadequate maintenance equipment (#PD6520)
Lack of essential local infrastructure (#PF2115)
Inappropriate level of technological equipment (#PF2410)
Underdevelopment of industrial and economic activities (#PC0880)
Lagging transformation of agriculture in developing countries (#PD0946).
Aggravates Stagnated development of agricultural production (#PD1285)
Deterioration of domestic food production in developing countries (#PD5092).
Aggravated by Prohibitive cost of farm equipment (#PJ8790).

♦ **PF4110 Political over-reaction**
Nature A major folly of governments and governors is that of over reacting to political situations. Very minor incidents are made to be life and death problems which must be dealt with immediately and decisively. The over reaction thus creates far larger problems than the original issue could ever have produced.
Claim Governments and politicians invent endangered "national security" or "vital interests" or enlarge a "commitment" to an ally to the point that these inventions take on lives of their own. These concerns of the U.S.A. during the Cold War and particularly during the Vietnam War perpetuated destructive foreign and domestic policies long after they were recognized as destructive even by those who invented them.
Broader Inadequacy of political doctrine (#PF3394)
Self-destructive government policy-making (#PF5061).
Related Political opportunism (#PC1897) Unpredictable governmental policy (#PF1559).
Aggravated by Political risk (#PF5932) Political myopia (#PD6963)
Lack of political will (#PC5180) Undue political pressure (#PB3209)
Politically unrealistic strategic warfare analysis (#PF1214)
Inadequate structures for political dialogue and review (#PC1547).
Reduced by Political opposition (#PJ5891)
Political opposition to administrative action (#PF2628).

♦ **PF4112 Instability in tourist dependent economies**
Nature Tourism plays a major part in the balance of payments of many countries, frequently exceeding in importance the imports or exports of merchandise; many branches of a national economy supply services and goods for tourism. In the case of some branches all, or almost all, of their activity is related to tourism, for example, hotels and similar accommodations, charter air passenger companies and travel agents; in the case of other branches, tourism supplies a major part of their revenue, for example, restaurants and souvenir shops in tourist resorts.
Incidence In addition to international tourism, domestic tourism consists of hundreds of millions of holidays away from home and other recreational and business trips. The development of tourism may, in certain circumstances, contribute to pressures on the general level of prices. Its inflationary effect varies considerably from one country to another. Although the main effects are encountered in the accommodation sector, in some countries they may extend to basic items, including food and clothing, creating local socio-economic problems.
Broader Compliance∗complex (#PA5710).
Aggravated by Tourist hazards (#PE8966).

♦ **PF4114 Unrecognized future financial commitments**
Unacknowledged future costs of present policies
Incidence In the USA in 1990 it was estimated that $25 to $50 billion per year may be required to pay for hidden costs such as in the field of health care and assistance to ailing institutions. Such commitments are readily ignored in upbeat reporting and budgeting based on optimistic economic assumptions and unrealistic spending cuts for which approval has previously proved unobtainable.
Broader Unpreparedness (#PF8176)
Irresponsibility towards future generations (#PF9455).
Aggravates Government limitations (#PF4668).
Aggravated by Neglect of environmental consequences of government policies (#PE9295).

♦ **PF4115 Degradation of developing countries by tourism**
Claim Foreign tourism in developing countries is an economic, social and cultural disaster. The construction of tourist facilities, airports, hotels, roads, and recreation sites often dislocates

indigenous communities by forcing them to move or cutting them off from traditional sources of income. It degrades the natural environment by polluting air and water and destroying natural habitats, such as nesting areas on the Turkish coast. It destroys ancestral lands and sacred sites. Local culture is commercialized; artifacts are produced for sale, folk dances become theatrical productions modified to meet the desires of foreigners and holy places are reduced from destinations for pilgrims to curiosities. It deprives local people of basic services. It promotes exploitation including prostitution and pederasty. Economic factors are important reasons for going into prostitution but frequently there are social reasons also; broken families, disgrace that comes with abandonment by a boy friend, pregnancy and single parenthood, encouragement or outright sale by parents and violence in the family. Children who are gaoled for vagrancy are bailed out by recruiters of prostitutes and forced into the trade. So-called employment agencies recruit women from poor areas, promising jobs, paying their parents in advance and then forcing them into prostitution. Crime is encouraged; drug use increases. Tourism increases outflow of foreign exchange and hampers economic development. Hotels, travel agencies and car rental agencies are owned by foreign corporations. Food and other consumables, equipment for hotels, vehicles and managers are imported.

Because control of the world industry is mainly in the hands of multinationals based in the First World, the tourist phenomenon is interconnected with the larger pattern of economic control which characterizes the unequal relationship between the North and the South. Whatever the short-term costs or benefits of tourism for Third World host economies, the overall significance of the travel industry is that it reinforces the dominance of the First World centres of the international economy and deepens the dependence of the countries on the 'periphery' of that economic system.

Counter-claim Properly managed tourism has proven itself to be a great boon to developing countries. Tourism earned poor countries about $55 billion in 1988, according to UN estimates, making it the second biggest earner of foreign exchange after oil. Spending on international tourism, excluding travel, will grow 4.5 – 5 percent a year until the year 2000. On study estimates that no more than 40 – 50 percent of tourist hotels' operating cost are leaked back abroad and the figures are falling as local agriculture, services, manufacturing and management skills improve. Arts and crafts in some places are improving because of the impact of tourism.

International tourism may be regarded as an important mechanism (along with other forms of communication) for the transmission of foreign cultures.

Refs Ascher, François *Tourism* (1986); English, E Philip *Great Escape?* (1986); Holden, Peter; Pfäfflin, Georg Fredrich and Horlemann, Jügen (Eds) *Third World People and Tourism* (1986); Hong, Evelyne *See the Third World While It Lasts*; Lea, John *Tourism and Development in the Third World* (1988).

Broader Social environmental degradation from recreation and tourism (#PD0826)
Natural environmental degradation from recreation and tourism (#PE6920).
Aggravates Drug abuse (#PD0094) Statutory crime (#PC0277)
Male prostitution (#PD3381) Female prostitution (#PD3380)
Sexual abuse of children (#PE3265) Male homosexual prostitution (#PD4402)
Endangered indigenous cultures (#PC7203) Disappearance of local culture (#PF3012)
Lack of economic and technical development (#PE8190)
Foreign exchange shortage in developing countries (#PD3068).
Aggravated by Economic bias in development (#PF2997).

♦ PF4118 Statistical errors

Discrepancies in official statistics
Broader Human errors and miscalculations (#PF3702).
Narrower Inadequate environmental statistics (#PF0011)
Deficiencies in national statistics (#PF0510)
Lack of comparability of international statistics (#PF2622).
Aggravates Misuse of statistics (#PF4564)
Inadequate social and demographic statistics (#PF0214)
Increasing scepticism about the accuracy of official information (#PF7649).
Aggravated by Governmental bias in statistics (#PF0019).

♦ PF4125 Fear of failure

Fear of mistakes
Incidence Tormented by the 1986 Challenger explosion, National Aeronautics and Space Administration's, NASA's, new shuttle programme advances warily in fear of another crash.
Broader Fear (#PA6030).
Narrower Risk of capital investment (#PF6572).
Aggravates Personal failure (#PF4387) Lack of political will (#PC5180)
Refusal to admit error (#PF5163) Human sexual inadequacy (#PC1892).

♦ PF4126 Maldistribution of health personnel

Incidence In many countries health personnel are not appropriately trained for the tasks they are expected to perform, or are not provided with the equipment and supplies they require. Health manpower varies greatly from country to country and includes a wide variety of different categories of people fulfilling different functions in different societies, depending on their social and economic conditions and cultural patterns. For this reason, intercountry comparisons are very difficult. Nevertheless, to illustrate the disparities among countries, in the least developed countries one health worker of all categories, including traditional practitioners, has to serve an average of 2400 people, in the other developing countries 500 people, and in the developed countries 130 people. As for medical personnel, in the least developed countries there is one doctor for an average of 17,000 people, in the other developing countries one for 2,700 people, and in the developed countries one for 520 people. The corresponding figures for nurses are one for an average of 6,500 people in the least developed countries, one for 1,500 in the other developing countries, and one for 220 people in the developed countries. To highlight the extremes: in the rural areas of some least developed countries there may be only one doctor to serve more than 200,000 people; whereas in the metropolitan areas of some developed countries there is one doctor for 300 people. None of these averages reveals the extremely inequitable distribution of health personnel often found within the same country. For example, in many countries there are ten times as many people for every doctor in rural areas as there are in metropolitan areas.

Compared with the developed countries, health care systems in the developing countries are supported by fewer middle-level cadres - technicians, clerks, administrators, etc. The staffing shortage is particularly acute at the level of village health posts and dispensaries where the recruitment and absorption of paraprofessional health workers has often been frustrated by laws and regulations governing training, licensing, and civil service status. Because of this shortage of suitable paraprofessional workers, together with inadequate methods of procuring and stocking drugs and other supplies, village health posts and dispensaries are often under-utilized and held in low esteem by villagers. In the absence of better public sector services at the community level, the rural poor for the most part seek out traditional healers, such as witchdoctors, herbalists, injectionists and traditional midwives.
Broader Insufficient health personnel (#PD0366) Maldistribution of medical resources (#PD2705).
Aggravates Insufficient doctors (#PE8303)
Delay in administration of medical care (#PD5119).

♦ PF4128 Imposed career interruptions

Broader Unemployment (#PB0750).
Narrower Conscription (#PF6051) Compulsory retirement (#PJ2411)
Career interruption due to pregnancy (#PD7692).
Aggravated by Human death (#PA0072) Human disease and disability (#PB1044)
Inadequate assistance to victims of accidents (#PE4086).

♦ PF4130 Economic disadvantages of excessive food production in developing countries

Nature In most developing countries where the majority of the population is in the agricultural sector, excessive food production may be nearly as economically disruptive as food shortages, especially when such expensive inputs as fertilizers, fuel and equipment are part of production costs to be repaid from output sales. Mechanization for increased production, therefore, may not always be a desirable or appropriate policy target. In many developing countries, a lack of on-farm facilities for post-harvesting and handling of crops, coupled with limited economic channels of crop disposal, may form a critical bottleneck that has the effect of neutralizing all benefits due from mechanized tillage and/or crop protection investments.

Incidence World grain and oilseed production has increased by about 157 million metric tonnes since 1983, while world consumption has increased by only 100 million tonnes. Irrigated cropland expanded by 20 percent in the 1970s. Turkey alone increased farm productivity by one-third between 1970 and 1982, and is currently adding 8 million hectares to the cropland base through irrigation programmes.
Broader Underdevelopment of industrial and economic activities (#PC0880).
Aggravates Dependence of developing countries on food imports (#PE8086).
Aggravated by Overproduction of food (#PD9448)
Maldistribution of food (#PC2801)
Inadequate mechanisms for securing sufficient food supplies (#PF2857).

♦ PF4132 Maldistribution of telecommunications facilities

Insufficient telephones in rural communities — Inadequate telephone systems in developing countries

Nature In many countries that have expanded their telephone services, concentration has too often focused on development in and between urban centres, overlooking the great need for and advantages to be gained from links between villages, and between rural outposts and the provincial centres. A single community telephone connecting local teachers and health workers to larger administrative centres, farmers with central markets, and local leaders with district officers, would certainly bring important and beneficial change to village life. Nevertheless, there are many more countries which have not expanded sufficiently, or have ignored totally the need for telephone services. They do not belong solely to the group of least developed countries; there are examples even at higher levels of development, such as in socialist countries.

Incidence World statistics for 1980 indicate that the average number of telephones in use per 1000 population was: 115 overall in the world; 11 in Africa; 31 in Asia; 55 in South America; 339 in Europe (excluding the USSR); 420 in Oceania; and 580 in North America. In 1983, three-quarters of the world's 600 million telephones were concentrated in nine countries, and more than half of the world's population lived in countries with less than one telephone per 100 inhabitants.
Broader Insufficient communications systems (#PF2350).
Narrower Maldistribution of radios (#PF4142) Maldistribution of television sets (#PF4136).
Aggravated by Delay in connection of telecommunications facilities (#PE3657).

♦ PF4133 Promotion of negative images of opponents

Demonizing the enemy — Negative socialization towards opponents — Fixation on adversaries

Nature The socio-psychological contexts and mind-sets of competing social groups and nations generate negative images and language which effectively prevent progress toward social, political, economic or military cooperation. Western societies are accused of being imperialistic or colonial capitalist war-mongers. Socialist countries are called demonic or atheistic. Such language and images are designed to create stereotypes and discourage understanding. Negotiations with an adversary who is totally evil is impossible.

Refs Keen, Sam *Faces of the Enemy* (1986).
Broader Propaganda (#PF1878) Psychological warfare (#PC2175)
Compulsory indoctrination (#PD3097).
Narrower Fabrication of politically sensitive incidents (#PF9558).
Related Deception (#PB4731) Misleading advertising (#PE3814)
Biased presentation of news (#PD1718) Negative identifying slogan (#PG9917)
Negative emotions and attitudes (#PA7090).
Aggravates Evil eye (#PF2343) Scapegoats (#PF3332) Personified evil (#PF7018)
Political smear campaigns (#PD9384) Negative effects of rejection (#PF4351)
Suppression of opposition groups or individuals (#PC7662).
Aggravated by Evil (#PF7042).
Reduces Political opposition (#PJ5891)
Political opposition to administrative action (#PF2628).

♦ PF4135 Underdevelopment of the power industry in developing countries

Nature The development of power-equipment manufacturing in several developing countries tends to be limited by small internal markets. The plant scale required for efficient operation has particularly influenced the power-generation sector. In contrast, large demand and the relatively small scale required for the efficient operation of transmission and distribution equipment offers more possibilities for domestic production. While in a considerable number of developing countries the demand has increased dynamically, the development of local power-equipment manufacture has fallen short of the production potential.

The important resources required for investment in special machinery and testing equipment as well as the long gestation period involved in capitalizing such investments have evidently discouraged the entry of domestic companies. In addition, the technology is largely controlled by a few large transnational corporations, and difficulties in adapting and assimilating this technology to local conditions within the framework of licensing agreements have not provided incentives to domestic manufacturers to move into the production of more sophisticated equipment.
Broader Energy crisis (#PC6329)
Weakness of infrastructure in developing countries (#PC1228)
Underdevelopment of industrial and economic activities (#PC0880).
Narrower Inadequate electrical power supply in developing countries (#PE1900).

♦ PF4136 Maldistribution of television sets

Insufficient television systems in developing countries

Nature In developing countries television sets are the possession of a minority, and in certain countries the programme content reveals that it is there primarily to serve the local elite and the expatriate community. Despite its phenomenal growth, in 40 countries television reaches less less than 10 percent of household units, and in more than half of the countries less than 10 percent of households have TV receivers. By contrast with radio, the cost of a television set is beyond the income of the average family; community sets, for example in village halls, have only partially mitigated this limitation. Also, television's limited range means that it is available chiefly to city-dwellers and reaches only a fraction of the rural population. Again by contrast with radio, the production of television programmes is an expensive business, and poor countries naturally

have other priorities. The screens are therefore filled for many hours with imported programmes, made originally for audiences in the developed countries; in most developing nations, these imports account for over half of transmission time. It is in the field of television, more than any other, that anxieties arise about cultural domination and threats to cultural identity.
Broader Inadequate standards of living (#PF0344).
Insufficient communications systems (#PF2350)
Maldistribution of telecommunications facilities (#PF4132).
Related Maldistribution of radios (#PF4142).

♦ **PF4142 Maldistribution of radios**
Broader Lack of communication (#PF0816).
Maldistribution of telecommunications facilities (#PF4132)
Inadequacy of telecommunication facilities in developing countries (#PE0004).
Related Maldistribution of television sets (#PF4136).

♦ **PF4149 Periodic mass extinctions of species**
Incidence After analysis of data of 9,250 extinct life forms, there is evidence to indicate that massive extinctions of species occur every 26 million years and have done so over the last 250 million years. On this basis the next mass extinction is due in 14.7 million years. It is not known whether this process is due to climatic shifts caused by asteroid belts or comets which could generate dust and smoke so as to bring on a devastating freeze. It is also possible that a source of extraterrestrial gravitation comes close to the solar system.
Broader Extinction of species (#PB9171).
Narrower Mass extinction due to comet showers (#PF0039).
Related Animal extinction (#PD7989)   Wildlife extinction (#PC1445).

♦ **PF4152 Risk of unintentional war generated by the arms race**
Nature The qualitative and quantitative arms race continuously generates new military options, thus delicate strategic stability is constantly challenged and undermined. The arms race may therefore create a propensity of the international strategic system to collapse due to incentives to pre-empt and to adopt launch-on-warning policies. In other words, it affects the crisis stability of the international system, thus aggravating the risk of nuclear war by miscalculation.
The secrecy surrounding new weapons developments leads the nuclear powers to infer the 'worst case' and to overreact; thus the arms race is accelerated. The risks involved in the current technological arms race derive mainly from: enhanced accuracy and yield (mainly due to multiple independently targetable re-entry vehicles) of anti-intercontinental ballistic missile weapons, thus jeopardizing intercontinental ballistic missile invulnerability; anti-submarine warfare developments jeopardizing the submarine launched ballistic missile 'leg' of the strategic triad; shrinking warning time due to forward basing (efforts to counter intercontinental ballistic missile vulnerability by reconsidering possible ballistic missile defence measures may be misunderstood by the opponent); and even if the strategic systems are not as vulnerable as some observers assume, the mere perception of their vulnerability may lead to misjudgement and miscalculation. A grave threat is posed to the invulnerability of command, control and communications systems; this may offer temptations to launch a pre-emptive first-strike attack or to adopt launch-on-warning policies and predelegation measures, which aggravate the risk of unauthorized initiation of nuclear war. To a considerable extent the arms race is progressing on an incremental step-by-step basis lacking proper national and international control. It thus produces results that are often highly dysfunctional to both national security interests and international concerns for strategic stability.
Incidence It is clear that the risks arising from the arms race are not efficiently countered or checked by the mitigating factors. Although in the absence of an acute international crisis these risks are not liable to overthrow strategic stability at the present moment or in the foreseeable future, they are extremely alarming in the long run.
Counter-claim The nuclear powers are extremely sensitive and attentive to potential vulnerabilities of their retaliatory capacity and command, and control and communications systems; as a consequence they undertake huge efforts to forestall potential 'windows of vulnerability'. New weapons technologies require testing, which offers an opportunity to detect new dangers in due time; this might mitigate the danger of being surprised by a technological breakthrough achieved by the opponent. So far and for the foreseeable future the arms race between offensive and defensive systems is led by the offensive systems. This fact helps the 'mutual assured destruction' relationship to continue to prevail. Arms control agreements may help to minimize threats to strategic stability by consensus.
Broader Risk of unintentional nuclear war (#PF0466).
Related Low-intensity conflict (#PE3988).

♦ **PF4153 Government treachery**
Government duplicity — Betrayal by government — Government perfidy
Related Treason (#PD2615).
Aggravates Misuse of statistics (#PF4564)   Deception by government (#PD1893)
Government misuse of personal records (#PG1194)
Covert violation of international treaties (#PD8465)
Official cover-up of government harassment of political activists (#PF3819).
Aggravated by Government hypocrisy (#PF9050)
Compromise as a betrayal of principles (#PF3420).

♦ **PF4155 Underutilization of international data bases by developing countries**
Nature On-line data bases are a segment of the information market that is rapidly growing. To date, most of the on-line data have been produced, transmitted and consumed by institutions of the developed countries, especially transnational corporations. Given the configuration of the transnational telecommunication networks, most developing countries are not linked to these networks, and hence have access to on-line data bases only through long-distance telephone dialling, which makes the use of this information resource almost prohibitively costly. The result, which is further accentuated by the inadequate telematics infrastructure in many developing countries, is that most of these countries do not actively participate in the transborder flow of data originating in on-line data bases. Therefore, they are, de facto, not in a position to use an information resource whose importance is growing.
Aggravated by Ineffective data systems (#PF3671).

♦ **PF4156 Risk of unintentional nuclear war generated by developments of strategic doctrine**
Nature Differing concepts employed in East and West may give rise to inadvertent misunderstandings which in turn are prone to produce additional distrust. This distrust is liable to fuel both the arms race and the forces undermining strategic stability; in addition, it may also lead to grave misunderstandings in situations of acute international crisis. Due to the reluctance of Soviet leaders to make pronouncements on the matter, Soviet strategic doctrine is particularly subject to all kinds of speculation by Western observers, which is largely based on 'worst case' assumptions and hardly conducive to strengthening strategic stability. In particular, misrepresentations of the element of surprise in Soviet statements about tactics in the battlefield may lead to the conclusion that the Soviet Union is interested in launching a first strike. Nuclear war by misjudgement may thus become more probable. On the other side, United States counterforce targeting policy may specifically lead to the presumption of pre-emptive strike preparations against the Soviet Union. This may also contribute to the risk of nuclear war by misjudgement. These problems are particularly grave in the case of strategic doctrines referring to the possible European war theatre. Collateral damage caused by 'limited' nuclear warfare may exceed any intended limits and thus lead to escalation beyond controlled, limited war.
Incidence The mismatch existing in the field of nuclear doctrine cannot be said to constitute a cause of imminent danger, yet it is constantly fuelling the arms race, thus contributing to the factors that undermine strategic stability in the long run.
Claim Strategic doctrines, as the basis for arms development and deployments as well as for defence postures, rest on extremely fuzzy grounds – guesses, speculations, beliefs, and assumptions of all kinds. These can be erroneous and thus lead to inappropriate conclusions jeopardizing strategic stability.
Counter-claim Although the strategic doctrines which were enunciated may differ substantially, the strategic capabilities of East and West are largely comparable. If doctrinal intentions are inferred from capabilities, the mismatch seems less marked. The threat of retaliation and thus the expectation of incalculable damage is still so effective and currently reinforced by doctrinal refinements of mutual assured destruction, as to make any doctrinal speculations about pre-emptive attack a necessarily idle activity. At the present moment no nuclear power can expect to escape radiation if it launches a pre-emptive strike against its opponent.
Broader Risk of unintentional nuclear war (#PF0466).
Aggravated by Low-intensity conflict (#PE3988).

♦ **PF4160 Inadequate industrial trade in developing countries**
Nature The difficulties faced by developing countries in securing a substantial increase in their share of world industrial production and trade emanate principally from a combination of mutually reinforcing constraints and handicaps, relating particularly to: inappropriate industrialization and commercial policies in the developing countries themselves; barriers erected by governments of developed market-economy countries to exports from developing countries, and lack of adequate policies to assist long-term restructuring of their industries so as to facilitate rather than restrict industrial imports from developing countries; practices of business enterprises, including intra-firm arrangements of transnational corporations, which unduly restrict the ability of developing countries to expand and diversify their exports of manufactures; lack of adequate technological basis in the developing countries and excessive dependence therein on expensive and inadequate foreign technology; inadequacy of mechanisms to develop new and expand existing flows of trade among developing countries themselves; and insufficient measures to expand trade between developing countries and the socialist countries of Eastern Europe.
Aggravates Trading in products containing toxic substances with developing countries (#PE2061).

♦ **PF4162 Risk of unintentional nuclear war generated by the strategy of deterrence**
Myth of deterrence — Illusion in nuclear strategy
Nature As the essence of deterrence is credibility, which is a subjective concept, the stability of deterrence depends on all kinds of delicate perceptual and psychological processes which might lead to gross miscalculation; if the threat to retaliate is underestimated, deterrence fails and an unintentional nuclear war would be the consequence. The system of deterrence is generally drifting towards becoming unstable, as indicated by the concern for scenarios like "Called Bluff". The assumptions of rationality underlying deterrence strategies are not justified in every situation. The North Atlantic Treaty Organization concept of deterring conventional aggression by the threat of a selective nuclear strike is constantly affected by doubts about its credibility and reliability. This concept also seems questionable because a limited nuclear war in Europe might destroy just what it is supposed to protect; these doubts are supported by the Soviet refusal to accept the concept of limited nuclear war. The North Atlantic Treaty Organization concept of extended deterrence (United States nuclear guarantees for European allies) also suffers from problems of credibility, thus aggravating the risk of miscalculation.
The two credibility problems (credibility of nuclear deterrence against conventional attack and credibility of United States guarantees for Europe) add up to one large credibility problem which serves to reinforce each separate problem. This creates regional crisis instability; and may create a temptation to engage in low-key 'probes' so as to assess the seriousness of commitments made within the framework of extended deterrence. The question whether there can be such a thing as a 'limited nuclear war', not necessarily escalating into an all-out nuclear war, is extremely controversial. There is at least some likelihood that the 'tactical' use of nuclear weapons would inadvertently trigger a full-scale nuclear exchange. Very serious risks of miscalculations emerge in third world regions, where the two major powers have made a multitude of poorly defined commitments in the sense of extended (mainly conventional) deterrence.
Incidence The risk of unstable deterrence which constitutes the very essence of strategic or crisis instability is currently being aggravated and multiplied by a series of strategic developments at both the global and regional levels. Doubts regarding the credibility of deterrence postures are increasing for a multitude of reasons, and this is conducive to miscalculation or inability to prevent miscalculation by the opponent. Nevertheless, credibility, although not absolute, is sufficient to stave off any failure of deterrence. Yet this situation of relative stability is not safe for an indefinite future. It also has a propensity to generate low-key crises of all kinds which in turn imply a risk of escalation.
Counter-claim For the foreseeable future, no nuclear power can rely with certainty on escaping any retaliations if it tries to disarm its opponent by a first strike, unless it has a completely insane risk-taking behaviour overriding any rational considerations which, in principle, will prevent gross misjudgement. The same can be said of the specific strategic situation prevailing in Europe, yet the prospects for continuing credibility of deterrence may be less far-reaching in time. Credibility is not absolute but sufficient. The uncertainties about possible disastrous consequences of a 'probe' of the deterrence postures in Europe are mutual: the Soviet Union is not sure about its chances of escaping United States retaliation. The United States is not sure about an immediate escalation into an all-out nuclear war by the Soviet Union, once it employs nuclear weapons. This does not exclude, however, low-key 'probes'.
Refs Bobbitt, Philip *Democracy and Deterrence* (1987); Buzan, Barry G (Ed) *The International politics of Deterrence* (1987); Kavka, Gregory *Moral Paradoxes of Nuclear Deterrence* (1987); United Nations *Study on Deterrence* (1987).
Broader Inequality of opportunity (#PC3435)   Risk of unintentional nuclear war (#PF0466).
Narrower Violation of sovereignty by trans-border broadcasting (#PE0261).
Related Unjustified military defence policies (#PF1385)
Unequal opportunities for media reception (#PD3039)
Vulnerability of nuclear defence control systems (#PD4049)
Protectionism in the defence and arms industries (#PE8664).
Aggravates Propaganda (#PF1878)   Compulsory indoctrination (#PD3097).
Aggravated by Risk-aversion strategy (#PF4612)
Technology gap between developed countries (#PD0338).

♦ **PF4164 Excessive emphasis on fashionable problems**
Popular problems
Nature Politicized issues and media processes have led to the emergence of new problems, especially in domains relating health and the environment. Such "problems of the month" attract attention on a short-term basis and provoke political responses by groups concerned to

appear relevant to current issues. They also stimulate research requests which are honoured by funding agencies equally desirous of appearing responsive to issues defined as important by this process. Such attention obscures the need for a more systematic and integrated understanding of networks of interacting problems and of more fundamental problems, of possibly a less attractive nature in media and political terms.
Aggravates Excessive emphasis on fashionable areas of research (#PF0059)
Excessive reliance on fashionable solutions to problems (#PF4473)
Failure to integrate knowledge to empower humanity in response to the global problematique (#PF8753).

♦ PF4165 Inadequate trade–related structural adjustments
Nature Industrial structural adjustments form part of the process of economic growth and development. Changes in demand and in relative prices, technological developments and increased competition between domestic and foreign enterprises, are some of the determinants that call for a process of structural shifts in industry, involving a reallocation of productive factors aimed at reaching a more efficient pattern of industrial production.
There is a need for intensified restructuring of industrial production in developed countries in response to the industrialization process in developing countries. Developed countries should encourage domestic factors of production to move progressively from the lines of production which are less competitive internationally, especially where the long–term comparative advantage lies in favour of developing countries, thus providing, inter alia, larger export possibilities for the developing countries and contributing to the attainment of their development objectives. Such adjustment would include the redeployment of those industries of the developed countries which are less competitive internationally to developing countries, thus leading to structural adjustments in the industrialized countries and a higher degree of utilization of natural and human resources in the developing world. Such adjustments would also be governed by the need for less competitive industries to move into more viable lines of production or into other sectors of the economy.
As long as adequate adjustment assistance measures are not available in developed countries, the inequities in the costs and benefits of industrial restructuring will promote resistance to adjustment and produce pressures for protection in these countries. These pressures, because of the immediacy of the damage to the import–competing interest groups, can be exceedingly strong, in particular in times of economic recession. Asymmetries inherent in protective and adjustment assistance devices in developed market–economy countries provide a further bias in these countries toward non–adjustment to competing imports. For example, adjustment assistance measures require fiscal disbursements, and hence are systematically more difficult to realize than are protectionist measures, which often serve to raise fiscal revenues.
Broader Antiquated world socio–economic order (#PF0866)
Underdevelopment of industrial and economic activities (#PC0880).

♦ PF4166 Mother–in–law taboo
Nature A custom most often practised in matrilineal societies which restricts contact between a husband and his wife's mother. They may be enjoined from speaking, looking at or referring to one another. In middle–class Western cultures a mild form of this taboo is given wide recognition in jokes. Although this taboo prevents friction in societies in which the structural interests of husband and mother–in–law conflict, it blocks communication.
Broader Taboo (#PF3310).
Related Dietary restrictions (#PJ1933).

♦ PF4174 Accumulation of functions
Accumulated power
Broader Accumulation (#PA4313).
Related Monopoly of power (#PC8410).
Aggravates Abuse of government employment (#PE4658).

♦ PF4175 Political nature of development issues
Nature Development issues are essentially political because governments must find the 'political will' to change existing conditions within the countries and internationally. Economic growth in the rich countries will not 'trickle down' to the others because the operation of market forces favours concentration of wealth and power rather than their devolution. The problems of hard core poverty require direct action by governments which must take and implement political decisions.
Political awareness is much more developed in the urban centres than in the rural communities. Many of the decision–making bodies are found in the cities and towns. Development is concentrated in cities and towns to the disadvantage of rural communities.
Broader Underdevelopment of industrial and economic activities (#PC0880).
Related Lack of political will (#PC5180).

♦ PF4176 Uncontrolled use of computer data
Erosion of human rights and privacy by the use of computer databases — Accumulation of personal data — Unauthorized disclosure of computerized information — Errors in computerized personal data
Nature There is a lack of legal protection of individuals with regard to automatic processing of personal information relating to them, especially when considering the increasing use made of computers for administrative purposes. There is also a lack of general rules on the storage and use of personal information and in particular, on the question of how individuals can be enabled to exercise control over information relating to themselves, which is collected and used by others. Mistakes in data held can wrongly label individuals with a criminal record or unfairly list them as a credit risk.
Background The use of computerized personal data systems may have adverse effects on the right to privacy by gathering and storing a greater amount of data pertaining to the private life of the individual and disseminating them to a wider audience than the individual consented to or anticipated when he originally surrendered the information, thus infringing upon its confidentiality.
Computerization of personal files has been also viewed as having a specific impact in organizational decision–making, on what is called (in the Anglo–American legal systems) the due–process tradition, as the data collected in these files are relied upon to determine rights, benefits or obligations of the individual. Concern has sometimes been expressed that expansion of computerization of personal data may result in a 'dossier society' which would have 'dehumanizing' effects on the individual.
The relative inflexibility of computer–based record–keeping resulting from the system having been designed to use certain pre–conceived categories, coupled with the constraints that some computerized systems put on the freedom of persons concerned to provide explanatory details in responding to questions have been considered as contributing to the dehumanizing effect of computerization.
The use of the standard universal identifiers (SUI), in connection with computerization, has recently raised fears and anxieties (even in some European countries where such identifiers were introduced without opposition). Because of the introduction of SUIs, citizens feel a sense of alienation from their social institutions and resent the dehumanizing effects of a highly mechanized civilization.

In addition, fear of loss of anonymity, and the suspicion that bureaucrats confronted by numbers will tend to forget that they represent real people, have been also mentioned as negative psychological effects of SUIs. In addition the use of computerized personal data systems may have harmful effects on human rights in relation to inaccuracy and obsolescence of data, access to and sharing of data, accumulation of data and the record–keeping personnel. One or several of the human rights mentioned above may be adversely affected by one of these negative consequences. For example, inaccuracies in personal data, due to the use of computers, may affect the right to a fair and public hearing, if the computerized information is used as evidence in courts and, when it is employed for making decisions, other rights of the individual.
Information power brings with it a corresponding social responsibility of the data users in the private and public sectors. In modern society, many decisions affecting individuals are based on information stored in computerized data files; payroll, social security records, medical files, etc. It is essential that the undeniable advantages obtained from automatic data processing do not at the same time lead to a weakening of the position of the persons on whom data is stored. The question has arisen to what extent national data protection laws afford adeqdate protection to individuals when data concerning them flow across borders. Computers, in combination with telecommunications, are opening new prospects for data processing on an international scale. In several sectors (for example banking, travel, and credit cards), such transfrontier data processing applications are already commonplace. In principle, it should make no difference for data users or data subjects whether data processing operations take place in one or in several countries. In practice, however, protection of persons grows weaker when the geographic area is widened. Data users might seek to avoid data protection controls by moving their operations, in whole or in part, to 'data havens', that is, countries which have less strict data protection laws, or none at all.
Incidence In the UK, 10 per cent of enterprises surveyed reported unauthorized disclosure of computerized information by employees to competitors. Also in the UK, some 60 protests per week are made concerning errors in personal information held in state and private data banks. Such errors have been discovered in police and hospital computers as well as among commercial collectors of personal data.
Counter–claim In order to counter the risk of 'data havens' some countries have built into their domestic law special controls, for example in the form of a licence for export. However, such controls may interfere with the free international flow of information which is a principle of fundamental importance for individuals as well as nations. Data protection at the international level must not prejudice this principle.
Claim As the public becomes increasingly aware of the information orientation of modern life and that a substantial amount of personal data about them is being preserved 'on the record', it is understandable that people may begin to doubt whether they have any meaningful existence or identity apart from their profile stored in the electronic catacombs of a 'master' computer. Embedded in this fear of being stripped of individuality is the psychosis of the Computerized Man, popularly portrayed as a quasi–automaton whose functions have been standardized, whose status in the community has been determined for him, and whose financial condition is prescribed in immutable terms.
Counter–claim Computer held files are more secure than those on paper, any access to them can be controlled and recorded. Standard universal identifiers used in conjunction with an identity card would greatly benefit the public. For example, in Britain, citizens would have a single identity number instead of a national health service number, a driver's licence number, a passport number, a national insurance number and a tax number.
Refs Council of Europe *Protection of Personal Data Used for the Purposes of Direct Marketing* (1986); Flaherty, David H *Privacy and Government Data Banks* (1979); Freedman, Warren *The Right of Privacy in the Computer Age* (1987); Hoffman, Lance J *Computers and Privacy in the Next Decade* (1980).
Broader Infringement of privacy (#PB0284)   Abuse of computer systems (#PD9544)
Vulnerability of computer systems (#PE8542).
Related Accumulation (#PA4313).

♦ PF4180 Inadequate social development programmes
Nature Developing countries with their still–recent experience of colonialism, are disinclined to discuss their social structures and priorities with developed countries. The developed countries are also disinclined to negotiate what they consider purely internal matters, such as levels and patterns of consumption. The result is that economic issues dominate the North–South negotiations.
Counter–claim The content of North–South negotiations indicates that social issues are very much at the centre of concern. When food is being considered, for instance, although the focus is on the nuts and bolts of finance, production and distribution, the end result is more food for the hungry. Similarly, when the transfer of technology is negotiated, such vital social needs as employment, health and transportation shape the priorities of governments. In fact all economic negotiations are conducted with social ends in view.
Broader Social conflict (#PC0137).
Aggravates Limited access to social benefits (#PF1303).

♦ PF4181 Back to the land
Rural resettlement — Migration to rural areas
Nature In both the developed world and the developing world the are increasing numbers of people returning to rural areas from urban ones.
Many ecologically–minded people in developed countries are moving away from the cities to rural areas to create a lifestyle more in keeping with natural principles. This rediscovery of the rural life is generally not in communication with the rural populace who might benefit from new technologies, nor with urban people looking for new and more creative directions. The result tends to be that certain individuals are able to upgrade the quality of their personal relationship to the environment. Virtuous activity like this is wonderful in its own right, but does nothing whatsoever for concern about the systematic, societal destruction of our common home, the earth. Further, such individualistic schemes confirm the growing suspicion that environmental issues are doomed.
The governments in developing countries have a variety of schemes to encourage and in some cases force people to go to rural, remote and under–populated areas. Most of these plans are attempts to relieve pressures on urban areas because of overpopulation. In some cases, they are attempts to get rid of undesirable groups, like thieves, prostitutes and drug addicts. While in the developed world the people moving back to rural communities are the most successful and most capable of adjusting to rural life, in the developing world it is those least equipped to succeed who are moving to these remote areas.
Broader Internal migration (#PF4009).
Narrower Abuse of project–tied migration (#PE7613).
Related Involuntary mass resettlement (#PC6203).
Aggravates Misuse of tropical rain forests for agricultural development (#PE5274).
Aggravated by Fear of resettlement (#PF9030).

♦ PF4185 Irrational fear of industrialization of developing countries
Nature Many developed countries fear that as poorer countries industrialize they will produce goods much cheaper than is possible elsewhere, thus leading to mass unemployment in many

developed countries. This fear often leads to protectionist measures for domestic goods and trade barriers for the manufactured goods of developing countries. In fact the reverse is true. Just as better pay and working conditions for workers in industrialized countries contributed to the growth of their national economies, so the development of poor countries will strengthen the world economy. It is estimated that trade with developing countries has led to the net addition of several hundred thousand jobs to the economies of developed countries.
Broader Compliance*complex (#PA5710).

♦ PF4187 Fairy rings
Nature Fairy rings are caused by several species of mushrooms growing in circles in lawns and golf greens, especially when soil is moist and contains a superabundance of organic matter.
Broader Basidiomycetes (#PE0364).

♦ PF4192 Excessive expense of international athletic competitions
Broader Lack of control (#PF7138).
Aggravates Athletic competition (#PE4266).
Aggravated by Competition between states (#PC0114)
Unethical commercial practices (#PC2563).

♦ PF4193 Ambiguity
Refs Goodman, Paul *Moral Ambiguity of America.*
Narrower Inadequately worded agreements (#PF5421)
Ambiguous shape of social identity (#PF6516).
Related Difference (#PA6698) Uncertainty (#PA7309) Disagreement (#PA5982)
Unintelligibility (#PA7367).

♦ PF4195 Inadequate industrial services in developing countries
Nature Industrial enterprises require, and continually make use of, various supporting services, including industrial research, feasibility studies, laboratory testing, industrial extension services, consultant services, training services, administration of industrial standards, economic and financial services (including special incentives), marketing services, and exhibitions. In the developing countries, industry is even more dependent upon such outside services. The average enterprise is likely to be small and unable to provide its own special engineering, research or testing units. Hence centrally organized industrial services, which are available and used to a great extent in industrially developed countries, are especially needed in the developing countries. But such services are less likely to be available in the developing countries and even where they exist, they generally lack experience and expertise.
Broader Underdeveloped provision of basic services in developing countries (#PF6473).
Related Limited mechanical services in developing countries (#PE8986)
Insufficient repair services in developing countries (#PE7959)
Insufficient consumer services in developing countries (#PE8236).
Aggravated by Inadequate local expertise in business practices in developing countries (#PE7313).

♦ PF4196 Excessive expense of athletic training programmes
Broader Lack of control (#PF7138) Prohibitive cost of education (#PF4375).
Aggravates Athletic competition (#PE4266).
Aggravated by Unethical commercial practices (#PC2563).

♦ PF4197 Inefficient mobilization of government revenue
Inefficient taxation systems — Costly taxation systems — Ineffective mobilization of public resources
Nature In order to raise public revenue governments pay the direct costs of administration and the indirect costs of distortions in economic activity.
Aggravated by Excessive government control (#PF0304)
Abuse of monopoly power of state–owned or state–controlled enterprises (#PE0988)
Jurisdictional conflict and antagonism between government agencies within each country (#PE8308).

♦ PF4198 Conspiracy against the public
Claim People of the same trade seldom meet together, even for merriment and diversion, but the conversation ends in a conspiracy against the public, or in some contrivance to raise prices (Adam Smith).
Broader Conspiracy (#PC2555).
Aggravates Profiteering (#PC2618) Commercial fraud (#PD2057)
Collusive tendering (#PE4301).
Aggravated by Avarice (#PA6999) Self–interest (#PA8760).

♦ PF4200 Decreasing number of adoptable children
Nature Most industrialized countries face a growing number of single–parent families headed by a woman, and the needs of such families are receiving greater consideration now than ever before. Recommendations for legislative changes providing more varied and more adequate support to such families are being actively considered in many countries. Special measures such as housing for single–parent families, increases in day–care facilities and special maintenance allowances for single–parent families will reduce the difficulties in keeping and caring for the children of such families. Consequently, single mothers are likely to be in a better position in the future to raise their own children rather than surrender them for adoption, and policy changes are likely to move in the direction of reducing disincentives to surrender the child for adoption.
The general climate of social attitudes has also changed in ways which currently make it easier for the unmarried mother to keep and raise her child. If the label of unmarried mother results in rejection, embarrassment, guilt, anxiety, or a self–image of moral inferiority and deviance, and is reinforced by the reaction of others, then incentives towards surrender are high. However, an increasingly more neutral, less punitive attitude towards the unmarried mother attenuates the incentives towards surrender.
Demographic changes also result in a reduction in the number of children available for adoption. The low birth rates of the 1965–1975 decade project a smaller group of adolescents and young adults in the near future. This is the age group for which the rate of illegitimacy is highest and for whom the difficulties in keeping and raising a child are greatest. The pressure of circumstances predisposing to surrender are greatest for this group, as is the pressure from parents, on whom the adolescent is likely to be dependent. Contraction of the size of this cohort group makes for a reduction in children available for adoption. The lowering of the age at which a child is no longer regarded as a minor legally dependent on parents may likewise result in a reduction in the number of children available for adoption. The trend is towards a reduction in the age of emancipation from 21 to 18 years. A larger number of adolescents with unwanted pregnancies will be in a position to make independent decisions regarding abortion and keeping their children without involving the consent of their parents. The greater weight of such parental pressure has been in the direction of surrender of the out–of–wedlock child. Earlier emancipation will tend to reduce the potency of pressure in this direction.
Broader Orphan children (#PD7046) Inadequate laws of adoption (#PD0590).
Related Decreasing number of adoptive parents (#PF4205).

♦ PF4202 Non–productive athletic activities
Broader Social injustice (#PC0797) Underproductivity (#PF1107).
Aggravated by Athletic competition (#PE4266).

♦ PF4204 Avoidance of decision–making
Inability to make difficult decisions — Unwillingness to make difficult decisions
Incidence Because of their inability or unwillingness to make difficult decisions, planners and budgeters often overprogramme and pay little attention to priorities, resource constraints, or phasing of implementation. The result is a tendency to cut or delay spending in a case of unexpected shortfalls.
Broader Inappropriate public spending by government (#PF6377)
Inadequacy of governmental decision–making machinery (#PF2420).
Narrower Indecision (#PF8808).
Aggravates Over–programming (#PF7737) Fragmented decision–making (#PF8448).
Aggravated by Blame avoidance (#PF6382) Tokenistic meeting resolutions (#PF2086)
Limited decision–making context (#PF2394).

♦ PF4205 Decreasing number of adoptive parents
Nature As an alternative to adoption, there is a growing acceptance of the use artificial insemination as a response to infertility; and there have been advances in procedures for remedying various infertile conditions, thereby permitting more people to become parents biologically rather than through adoption. The successful implantation of an ovum fertilized in the laboratory promises a 'cure' for a large number of women who might otherwise have resorted to adoption. There has also been increasing acceptance of childlessness as an optional lifestyle, and there is growing public sanction and support for adults who voluntarily elect not to become parents; the decision not to become a parent no longer requires an apology nor does it occasion as much discomfort as it previously did. Concern regarding population growth and the ideology of women's liberation have contributed towards the growing acceptance of childlessness. As a consequence, more infertile couples may choose to adjust to their situation by accepting their childlessness rather than attempting to change their situation through adoption.
Broader Orphan children (#PD7046) Inadequate laws of adoption (#PD0590).
Related Decreasing number of adoptable children (#PF4200).

♦ PF4209 Pantheism
Nature Pantheism is the view that all is God and God is all. Reality is one in essence and form and this unity is rational and divine. God is in the world and of it; everything partakes of the nature of God.
Counter–claim Pantheism rejects any explanation of the origin of the universe and ignores any difference between God and nature. As such, it avoids the issue of the "I" in relationship with the "Thou". The "I" is simply a part of the "Thou". The human self loses its freedom and immortality. Contemplation replaces action; quietism is encouraged. Wrong and right become absolutely relative and belief in and hope for progress is lost. The sense of sin, awareness of transgression and effort of amendment become illusionary.
Refs Schopenhauer, Arthur *Pantheism and the Christian System* (1987).
Broader Theism (#PF3422).
Related Unbelievers (#PF8068).

♦ PF4210 Divergent national concepts of corrupt practices
Nature The concept of corruption appears to have different shades of meaning, depending on the specific socio–cultural and historical contexts. Even within a particular society at a particular period of time, the concept may have a different meaning in the context of law, political science, religion, sociology or business. In essence, a corrupt practice is a special type of process or technique for influencing decision making. What distinguishes it from other influencing processes or techniques is the method by which the influence is effected. Every society accepts and legitimizes certain methods in the pursuit of individual interests and condemns others, ethically or legally. Duress, fraud and corruption belong to that latter category. In the international context, the primary concern is focused on three types of corrupt practices: those involving improper participation by foreign interests in the political process; payments to public officials either directly or through middlemen, in order to obtain favourable decisions; and 'facilitative' payments to achieve speedy action, which is not necessarily illegal. The dividing line between these categories of corrupt practices are not rigid ones and are often blurred.
Broader Corruption (#PA1986).

♦ PF4213 Soul murder
Psychic murder
Nature Soul murder is neither a diagnosis nor a condition. It is a dramatic term for circumstances that eventuate in crime, namely the deliberate attempt to eradicate or compromise the separate identity of another person. The victims of soul murder remain in large part possessed by another, their souls in bondage to someone else. It can result from childhood abuse, whether sexual molestation or physical beating, leading to overstimulation, terror and anger. It can also result from deprivation, especially in childhood, in the form of neglect or lack of emotional sustenance, leading to terrifying neediness, a sense of abandonment and rejection.
Broader Unethical personal relationships (#PF8759).
Related Political indoctrination (#PD1624) Demonic and spirit possession (#PF5781).
Aggravates Anger (#PA7797) Terror (#PF8483) Abandonment (#PA7685).
Aggravated by Maternal deprivation (#PC0981) Torture by deprivation (#PD3763)
Sexual abuse of children (#PE3265) Physical maltreatment of children (#PC2584).

♦ PF4215 Risk of war
Broader Risk (#PF7580).
Narrower Risk of unintentional nuclear war (#PF0466).
Aggravated by Imbalance of power (#PB1969).

♦ PF4216 National bias among judges of international athletic competitions
Broader Politicization of international sports events (#PF4761).
Aggravates Athletic competition (#PE4266).

♦ PF4220 Domination of individuals by institutions
Nature The continued expansion of large and impersonal institutions, both public and private, creates the impression that, in all walks of life, the individual is being increasingly manipulated by forces over which he has little control or influence.
Broader National isolationism (#PF2141) Excessive government control (#PF0304).
Aggravated by Bureaucracy as an organizational disease (#PD0460).

♦ PF4221 Therapeutic sex surrogacy
Claim Maybe the only way to cure sexual problems or hangups is to have sex with a surrogate partner. Sexual problems often stem from ignorance and the purpose of the surrogate is to show the patient what to do and dispel hidden fears.
Related Prostitution (#PD0693).

♦ **PF4240 Ineffective prevention of terrorism**
Ineffective anti-terrorist organization — Persistence of terrorist activities
Nature Reasons for the ineffectiveness of anti-terrorist organization are various: political and economic cost may be to high; political, social and economic conditions may aid recruitment of active members and supporters of terrorist organizations; publicity given to terrorist acts may encourage activists and increase support by their constituents; actions by the government or anti-terrorist organizations may be perceived as signs of weakness by the general public; alarm on the part of the general public and a sense of inability to do anything about terrorism encourages activists; police forces in a given area may be fragmented, thus creating problems in both international and internal coordination; police forces are often untrained in anti-terrorist tactics and are virtually helpless when terrorism strikes; police powers may be limited due to privacy laws (such as the ability to tap telephones without delay and without prior consent, the right to search areas without warrants, the right to hold suspects for a reasonable period for questioning); the media may jeopardize anti-terrorist organization by exposing operations; and intelligence service may be weak, often leaving an imprudent freedom of access to intelligence files.
As governments become more efficient in gathering intelligence and combating terrorism, terrorists adjust. Their organizations become smaller, making them tougher to monitor or penetrate, and their targets become specific ones aimed at specific nationalities.
Incidence While terrorism, in one form or another, has always been a political activity, since 1968 the number of deaths due to international terrorism has increased from about 100 to over 3000 a year in 1987. In August 1980, a bomb in a train station in Bologna, Italy killed 84 people. In April 1983, a bomb blew up the US Embassy in Beirut and claimed 46 lives. That October 240 soldiers in Beirut were killed in a bomb blast at a marine barracks. In December 1983 a bomb at Harrod's in London killed 6 and injured 91. In October 1983 bomb in Burma killed four members of South Korea's cabinet and two presidential advisers. One year later UK Prime Minister Margaret Thatcher escaped death but several other government members did not. Within a two week period in May and June of 1985, a bomb in the Frankfurt airport killed a woman and two children, a candy shop blew up in Lebanon with 33 dead, 329 were killed on an Air India flight to London, two baggage handlers were killed in Tokyo's airport, four American Marines were machine-gunned to death in El Salvador, a Pan Am plane in Athens was hijacked to Beirut and an American serviceman was killed; in November 1985 Egyptian commandoes stormed a hijacked plane and 60 deaths ensued. Five people were killed and 16 injured in April 1988 by a bomb outside a US servicemen's club in Naples.
Claim The increase in terrorist activities is abetted by Western governments who do nothing to counter it. The rhetoric is there, the threats, the insinuations, the all-night discussions, but action is absent and such absence encourages terrorism. Terrorism has ushered in a new type of insecurity in the Western world, and has damaged the credibility of democracy. Some Westerners have taken to considering terrorism as an inevitable occurrence in modern life, some urge the need to understand its roots, some seem to hope that if it won't go away, it can at least be contained at "acceptable" levels. Terrorism will not miraculously disappear as a passing fad, and as long as there is injury, destruction, and death, no level is acceptable. Until Western states take active measures to halt it (openly exposing those countries which train and harbour terrorists, taking joint action on imposing penalties) they have no right to cry victim. It will go away only when the victims strike back.
The flexibility of pluralist and representative societies explains the proliferation of subversive and terrorist groups. At a certain stage of the fight against terrorism, owing to the inertia of daily life and the discouragement of the masses in the face of continuous terrorist action, when no energetic reply is given to acts of violence and public opinion is divided on the matter, when murders are accepted by the population as ordinary, usual events, it can happen that the inhabitants of a country refuse to accept the seriousness of the situation or to feel themselves concerned and thus adopt the fallacious argument that the problem is exclusively a matter for the politicians. But it is precisely this collective psychological attitude that constitutes a great success for the terrorists. There is, therefore, a great need to find some means of strengthening the sense of civic duty and persuading the population to abandon its inertia and cooperate.
Counter-claim Within the constraints of a democratic society, terrorism may not be eradicated, and the government's inability to eliminate terrorism should not be seen as a sign of weakness. Like crime, a certain level of political violence is likely to persist as a feature of modern society. Terrorism is one price that democracy pays for an open society. At the same time, the world has taken measures against terrorism. International cooperation is increasing among anti-terrorist organizations. Much of the media is attempting to put terrorist acts in proper perspective. Individual governments are becoming more effective in combating it; the percent of the world total of terrorist incidents in Western Europe has decline steadily from 1982 to 1988.
President Reagan's decision to intercept the plane which carried the hijackers of the Achille Lauro to Italy was such a measure, and President Mubarek's decision to storm the hijacked plane in Malta was also welcomed act.
Broader Terrorism (#PD5574) Persistence (#PJ0642).
Narrower Terrorist havens (#PD5541)
Inadequate international cooperation in reducing terrorism (#PF4366).
Aggravates Border controls (#PJ1718).
Aggravated by Outlaws (#PE4409) Passive public support of terrorism (#PF6846)
Excessive portrayal of terrorist activity in the media (#PE6844).

♦ **PF4244 Refusal of medical care**
Broader Inadequate medical care (#PF4832)
Denial of right to adequate medical care (#PD2028).
Related Uninvestigated health care (#PU8168).
Aggravates Threatened sects (#PC1995) Inadequate health services (#PD4790).

♦ **PF4260 Economic sanctions against governments**
Economic blockade against governments — Economic blockade
Nature Economic sanctions are imposed by a government to show disapproval of another government's actions and to attempt to influence its future conduct. Economic sanctions involve the withdrawal or suspension of trade or financial relations.
Incidence The USA entered into a trade embargo with the USSR for the latter's invasion of Afghanistan, and several industrial countries have engaged in economic sanctions against South Africa in protest at the latter's policies of apartheid. In 1989 India closed 13 of 15 entry points to the landlocked Kingdom of Nepal creating shortages of food, medicines and fuel.
Counter-claim The main effect of sanctions has been punitive rather than remedial. They may weaken the target regime, but they will not necessarily change its behaviour.
Refs Leyton-Brown, David *The Utility of International Economic Sanctions* (1987); Lipton, Merle *Sanctions and South Africa* (1988); Nincic, Miroslav and Wallensteen, Peter (Eds) *Dilemmas of Economic Coercion* (1983); United Nations *Sanctions Against South Africa* (1988).
Broader Economic conflict (#PC0840) Abuse of economic power (#PC6873).
Narrower Boycott (#PE8313).
Aggravated by Government action against regimes with alternative policies (#PF2199).

♦ **PF4274 Distrust of interpersonal relationships**
Suspicion of adult-youth relationships
Broader Distrust (#PA8653).
Aggravated by Homosexuality (#PF3242) Infectious revenge (#PD5168)
Mistrust of strangers (#PF8743) Sexually transmitted diseases (#PD0061)
Sexual exploitation of children (#PD3267)
Unauthorized proximity of males to females (#PF8780).

♦ **PF4286 Excessive use of acronyms and abbreviations**
Broader Obscurantism (#PF1357) Lack of communication (#PF0816)
Corruption of meaning (#PB2619).
Narrower Incomprehensibility of specialized jargon (#PF1748).
Aggravated by Misleading information (#PF3096)
Multiplicity of languages (#PC0178).

♦ **PF4295 Uncertainty of development expenditures due to floating-rate loans**
Nature Although most cross-border lending, other than short-term, is advanced at floating rates, many developing country borrowers would prefer fixed-rate loans. The ability to predict accurately the cost of credit, which is impossible with floating rates, would permit these countries to engage in more accurate cost-benefit analysis. If a bank lends at a fixed rate, it must absorb the impact of interest rate fluctuations. The impact of a variable rate is absorbed by the customer.
Broader Compliance*complex (#PA5710) Economic uncertainty (#PF5817)
Deteriorating terms of financial loans to developing countries (#PE4603).
Aggravated by Exchange rate volatility (#PE5930).

♦ **PF4296 Dependence of the disabled**
Deprivation of freedom in caring for the disabled
Nature The dependence of the disabled, the sick and the mentally infirm on those who care for them outside institutional settings, constitutes a major constraint on the life-style of the carers, whose quality of life may be below that of the disabled. This is especially the case where the burden falls on relatives and friends who are seldom in a position to complain or to modify their situation, whether or not they would wish to do so.
Incidence In the UK in 1990 it was estimated that 6 million people look after some 700,000 sick, disabled and mentally infirm outside institutional settings requiring constant care and attention, often calling for more than 50 hours per week of time on the part of relatives and friends.
Aggravated by Human disability (#PC0699).

♦ **PF4297 Inadequate development of communication services in the least developed countries**
Nature Communications are a powerful instrument in the motivation and mobilization of human effort. The least developed countries are particularly disadvantaged in the coverage of their population by telephones (3.7 per 1,000 people), postal services (one post office for 11,000 people), radio transmission and radio receivers. A further negative factor has been the tendency to concentrate communication facilities in a few urban centers, leading to a relative isolation of the rural communities from the rest of the country and the outside world.
Broader Insufficient communications systems (#PF2350).
Related Unreliable telecommunication services for land-locked developing countries (#PE5860).
Aggravated by Inadequate infrastructure and services in least developed countries (#PE0289).

♦ **PF4298 False alarms**
False alerts — False disaster warnings
Broader Incorrect information (#PB3095).
Narrower False bomb warnings (#PJ9815) False burglary alarms (#PG7982)
False nuclear warfare alerts (#PF1236).
Aggravates Inadequate warning of disasters (#PF3565).
Aggravated by Hoaxes (#PF9375) Alarmism (#PF4384)
Backlash against repeated warnings (#PF3391).

♦ **PF4300 Uncertainty of development programmes due to short-term loans**
Nature The lack of symmetry between the long-term requirements of development and the short-to medium-term nature of commercial bank external financing places clear obstacles in the way of broad-based socio-economic development. Policy makers can confront the problem in a variety of ways, but in order to illustrate the gravity of the matter for a country that is heavily indebted to commercial banks, policy options can be simplified and reduced basically to two: (a) to focus development strategies on activities with high short-term private commercial rates of return and wait for social development to 'trickle down' to the population; or (b) to seek balanced socio-economic development today and hope that banks will refinance their short maturities.
Broader Economic uncertainty (#PF5817)
Excessive foreign public debt of developing countries (#PD2133)
Deteriorating terms of financial loans to developing countries (#PE4603).
Aggravates Lack of long-term development assistance (#PF5181).
Aggravated by Inadequate disaster prevention and mitigation (#PF3566)
Restrictive conditions on loans to developing countries through intergovernmental facilities (#PE9116).

♦ **PF4302 Risk of unintentional nuclear war due to international crises**
Nature In international crisis situations, decision-makers suffer from stress which causes a number of cognitive and behavioural maladaptations (distorted perception and sub-optimal decision-making). These maladaptations greatly enhance the risk of crucial decisions not being made correctly and rationally, thus leading to nuclear war by miscalculation.
Similar maladaptations may occur due to organizational problems of decision-making units: contraction of the decision-making group, information overflow, 'group-think', internal dissension, and inflexible standard operating procedures, may result in poor quality decisions being made in a crisis situation. Strategic vulnerability, urgency, strategic instability and crisis instability inevitably affect the performance of those responsible for decision-making. Thus, an inherently bad situation automatically worsens rather than progresses decision-making. The rules governing the use of force as a 'continuation of policy by other means', as employed by the powers for conveying signals, are utterly fragile and prone to misunderstanding, especially between opponents committed to systems of different ideological orientation. It may therefore be difficult to avoid fatal miscalculations regarding the use of nuclear weapons. Crisis bargaining entails the risk of events getting out of control due to the 'logic of events', military necessity, low-level actions taken by subordinate commanders and organizational routine and confusion due to the malfunctioning of command, control and communications systems. The trend towards global militarization and poorly defined, ambiguous commitments in the Third World fosters the inclination to use force in crisis bargaining. The risks inherent in this trend are enhanced by a network of strategic interdependence which makes the success of efforts to localize crises doubtful. Hence there is a risk of unintentional escalation (both geographically and militarily) of international crises.
Incidence The risks are hard to estimate. Whether positive or negative influences prevail depends to a large extent on the special circumstances of a crisis. Although the governments concerned are generally aware of the problems and dangers involved, the unpredictable and uncertain nature of specific situations may still produce a set of circumstances which might simply override all precautionary measures.
Counter-claim Although extreme stress has a disruptive effect in most situations, to some extent stress experienced by decision-makers faced with a crisis situation tends to improve their

perceptive and behavioural performance. Also, during crisis situations governments try to assume full control of all details of the making and execution of decisions, thus reducing the likelihood of organizational maladaptation by involvement in and meticulous handling of all aspects. Governments also generally tend to behave with utmost care and circumspection as soon as military force is involved in crisis bargaining. Existing mechanisms of communication contribute to avoiding misunderstandings.
Broader Risk of unintentional nuclear war (#PF0466).
Aggravated by Low-intensity conflict (#PE3988).

♦ PF4306 Lack of progress in establishing a New International Economic Order
Nature A call for a New International Economic Order to allow structural change in the world economy was adopted at the sixth special session of the United Nations General Assembly in 1974. The *Declaration of the New International Economic Order* enunciates principles covering a wide range of issues including commodities, natural resources, tariffs and monetary reform. The general lack of progress in this respect is due both to the limited economic and political leverage of developing countries, and to the perception among developed countries that the structural changes proposed in the creation of the New International Economic Order are at best irrelevant or, at worst, a constraint upon their ability to cope with economic problems.
Broader Antiquated world socio-economic order (#PF0866).
Aggravates International economic injustice (#PC9112)
Reversal of development progress (#PF4718).
Aggravated by Disparity between industrialized and developing countries (#PC8694).

♦ PF4317 Indecent exposure
Nature Exposing one's genitals with the intent to arouse or gratify the sexual desire of any person including the actor under circumstances which is likely to offend or alarm another person.
Refs Macdonald, John M *Indecent Exposure* (1973).
Broader Obscenity (#PF2634) Indecency (#PF8842).
Aggravated by Decadent clothing (#PE5607).

♦ PF4323 Reduced images of humanness
Nature Images of what it is to be a human being have been reduced by an over emphasis on productivity, material wealth, scientific validity, observable progress, self-gratification and numerous other social values. A sense of wonder at the mystery of being alive as a human being has been replaced by grim determination to improve one's position on one hand or a benumbed drive to happiness on the other.
Broader Media reinforcement of materialism (#PF1673).
Aggravated by Growing size and impersonality of firms (#PE8706).

♦ PF4326 Decreasing agricultural growth per capita
Nature Environmental degradation is an essential cause of decreasing agricultural production. Threats to the environment in developing countries in particular are very serious and are mounting. Genetic resources – which should be preserved as sources of future diversity and improvement – are shrinking because of pollution, deforestation, and the neglect of traditional species of crops and livestock. Soil degradation is a major hazard, and the the consequences of soil degradation are drastic: declining productivity of the land, sometimes leading to total loss of productivity, in many cases reversible only at very high cost.
Incidence A collaborative study by the United Nations Environment Programme (UNEP) and FAO found that 35 percent of the land area of North Africa was at risk from water erosion, 17 percent from wind erosion and 8 percent from salinization. Comparable figures for the Near East were 22 percent for water, 12 percent for wind and 1.4 percent for salinization. More than one third of the area of developing regions is either existing desert (8 million square kilometres) or at risk of becoming so (37.6 million square kilometres).
Claim Agricultural production can only be sustained if the environment is preserved: conservation is a precondition of long-term food security.
Broader Economic stagnation (#PC0002).
Narrower Stagnated development of agricultural production (#PD1285).
Aggravated by Infertile land (#PD8585).

♦ PF4327 Illusion of consensus
Nature Democracy depends heavily upon a citizenry which is accustomed to expressing its opinions, moderating them with respect to the concerns of others and carrying corporate decisions through to implementation. Where democratic institutions are being imposed, whether by benevolent indigenous leadership or by some outside power, there is very often a completely uncritical acceptance of new proposals and democratic institutions. A consensus of this kind often does not include strong commitment on the part of the population, and it can deceive leaders and outside observers alike into thinking that imposed democratization has been effective.
Broader Illusion (#PA6414).
Aggravated by Ineffective dialogue (#PF1654).

♦ PF4331 Secrecy concerning existence of extraterrestrials
Cover-up of extraterrestrial invasion
Broader Conspiracy (#PC2555).
Related Unidentified flying objects (#PF1392).
Aggravates Abduction by extraterrestrials (#PF3881).

♦ PF4334 Denial of rights to robots
Denial of rights of machine intelligences
Claim In the immediate future it is expected that robots will exist with many of the characteristics currently ascribed to humans and may even become an intimate companions to humans. No provision has been made for any extension of human rights to cover such cases.
Counter-claim Robots are machines and will remain so. As inanimate objects they are devoid of rights. Since they have restricted mobility, must be artificially programmed for thought, lack senses as well as the associated emotions, and are unable to experience suffering or fear, they lack the essential conditions to be considered alive.
Related Robotic crime (#PU1097).

♦ PF4336 Unequal distribution of production between countries
Incidence In 1977, per capita gross domestic product expressed in US dollars was $550 for Africa, $770 for Asia (East and Southeast), $1,310 for Latin America, $2,100 for Asia (Middle East), $5,470 for Oceania, $5,750 for Europe, and $8,660 for North America.
Broader Economic inequality (#PC8541).
Narrower Unequal distribution of meat production (#PD4322)
Unequal distribution of livestock production (#PF4490).

♦ PF4340 Inadequate commercial finance for rural development projects
Nature In general the international expansion strategies of transnational banks in developing nations have not emphasized lending to agriculture and other rural development projects. One underlying reason for the small percentage of rural loans granted by branches of transnational banks may be the restrictions that they face in establishing rural branches, because of the preference of governments for mobilizing rural savings through the widening of financial activities of national financing institutions.
Broader Rural underdevelopment (#PC0306)
Inadequacy of aid to developing countries (#PF0392).
Aggravates Lack of economic and technical development (#PE8190).

♦ PF4341 Lese majesty
Nature A crime against the dignity of the king/emperor or a member of the royal/imperial family.
Related Treason (#PD2615) Blasphemy (#PF5630).

♦ PF4345 Excessive anxiety on lending to developing countries
Deterioration of confidence in loans to developing countries
Nature Transnational banks share an interest in reducing the anxiety about lending to developing countries that surfaces periodically among banking authorities in their home countries and among analysts of bank stock. Sometimes the anxiety is caused by inadequate knowledge of the economic and financial environment in borrowing countries. For a better understanding of the borrowers' situation, wider contacts and exchange of information may be required between home country banking authorities and developing countries. Multinational conferences among the interested parties as well as other forms of direct contact could be useful in this connection. If transnational banks were to become less constrained by their home banking authorities and stockholders, they could be asked to improve their loan maturities to reflect changes in the economic cycle. To accomplish this, banks could match the structure of their liabilities more closely to the structure of their assets.
Broader Compliance*complex (#PA5710).
Aggravated by Insufficient creditworthiness of developing countries (#PD3054)
Deterioration in external financial position of developing countries (#PE9567).

♦ PF4346 Risk of unintentional nuclear war due to accidents
Nature Possibilities for accidents in nuclear weaponry and control systems are several. Technical failure or malfunctioning may detonate or launch nuclear weapons, or may also lead to false alarms. Nuclear weaponry accidents and threatening movements may also be caused by unauthorized action, human error, human over-reaction, stress or temporary insanity, and may be misinterpreted by an adversary nuclear power which could immediately retaliate, either proportionately or disproportionately. Risks are particularly serious in countries which have a newly acquired nuclear capability but are not yet able or willing to purchase or develop sophisticated safeguard systems.
Counter-claim Fail-safe systems with overlapping multiple controls, and countermeasures to initial accident phases, although not completely infallible, may exclude serious accidents or undesired incidents. Human control systems are designed to prevent any misuse of nuclear weapons by subordinate commanders or by the Commander-in-chief acting independently. There is a general tendency to subject the release of nuclear weapons to tight control by the supreme political authority, or at least by joint military-political (executive) decision. The focus on the risk of nuclear war by accident may misrepresent the problem and misdirect attention from more serious and crucial risks constituting a far greater danger. The risk of an attack might be increased if safeguards were multiplied to the point at which delay to field or submarine commanders in obtaining authorization and an electronic key effectively render their weapons impotent.
Broader Accidental military incidents (#PE4553) Risk of unintentional nuclear war (#PF0466).
Aggravated by Nuclear accidents (#PD0771).

♦ PF4347 Marginalization
Marginal groups
Nature Contemporary mass society creates individuals and social groups who are left on the cultural and economic edges, and certainly out of the mainstream process of decision-making. This tendency toward homogeneity at the centre of things makes consensus much easier, because only a part of society is included in it. Minorities of any kind play a fixed role that does not disturb the central processes overly much. This marginalizing process erodes society at its base, because there is no common reference point for people.
Aggravated by Homogenization of cultures (#PB1071).

♦ PF4350 Inadequate information concerning transnational banks
Nature A determined effort by developing countries, as external borrowers, could significantly improve their negotiating capacity. The most obvious need is for systematically collected information about loan negotiations and differences among transnational banks. In the bargaining environment, it is useful for developing country borrowers to be aware of such distinctions so that they can organize their relations with transnational banks in a way that is supportive of national development objectives. However, there is little institutional analysis available on the behaviour of individual lenders, which makes it difficult for new borrowers to set up strategies for negotiations with individual institutions. This is especially true with regard to banks that are relatively new in the international capital market, banks which are of particular interest for developing country borrowers. Under present circumstances, borrowers can only begin to differentiate among the behaviour of banks after a long accumulation of experience. However, the learning-by-doing method can be difficult and unnecessarily costly. To overcome this, systematic analysis of the institutional behaviour of transnational bank lenders is required.
Broader Inadequate exchange of technical information concerning problems (#PF0209).
Aggravates Restrictive practices of transnational banks (#PE9657).
Aggravated by Incomplete access to information resources (#PF2401).

♦ PF4351 Negative effects of rejection
Nature While rejection is a fact of life that cannot be avoided, it taints one's pride and trust. Rejection coupled with envy is the cause of wars, the prime motive for murder and the most common catalyst for suicide.
Aggravated by Politicization of health standards (#PD4519)
Promotion of negative images of opponents (#PF4133).

♦ PF4352 Risk of unintentional global nuclear war due to nuclear proliferation
Nature Among the most cited risks are: nuclear proliferation contributes to the risk of unstable governments or leaders gaining access to nuclear weapons and using them in an irresponsible way; in countries newly acquiring nuclear weapons, safeguard systems preventing nuclear accidents and incidents may be insufficient or altogether non-existant; in regions dominated by a high degree of tension, a nuclear power may be tempted to launch a pre-emptive nuclear strike before its opponent has also gained access to nuclear weapons. The initial phase of regional nuclear arms races are is prone to strategic instability. A local nuclear war may involve major nuclear powers through multilateral defence treaties or perceived threats. An unidentified nuclear strike against a nuclear power may be misinterpreted and lead to a punitive strike against the wrong opponent. Smaller nuclear powers may be manipulated into making nuclear attacks. Private corporations and groups outside governmental control, including terrorists and criminals, may acquire access to nuclear weapons.
Counter-claim The possession of nuclear weapons has led governments to behave with

increased political and military caution and circumspection. There are no convincing reasons to assume that in case of a local nuclear war the major nuclear powers and their defence alliances would fail to react with great care and try to limit the nuclear exchange unless they had made irrevocable nuclear commitments for mutual defence. There is no internal logic or compelling necessity for a bilateral or even regional nuclear conflict to develop into global nuclear war. The possession of nuclear weapons by groups outside governmental control would be aimed at blackmailing rather than at actually using the arms. If they are used, this would amount to national tragedies in the countries in which they would occur, but would not necessarily entail escalation into global nuclear war.
Broader Risk of unintentional nuclear war (#PF0466).

♦ PF4357 Fallacy
Fallacious people
Nature A fallacy is a statement or argument that leads one to a false conclusion because of a misconception of the meaning of the words used or a flaw in the reasoning involved.
Background While there is no general agreement on the various types of fallacies, one useful outline of types is as follows: I. Fallacies in diction: A. Equivocation, one word mistaken for another; B. Amphibology, double meaning sentence; C. Composition, attributing to the whole what is true only for the part; D. Division, attributing to the part what is true only for the whole; E. Metaphor, taking a figure of speech literally or stretching it unduly; F. Accent, different stress, tone, or gesture giving a different meaning to a word. II. Fallacies extra diction: A. Accident, presenting as true in the definite particular what is only generally true; B. False absolute, assuming as always true what is true only in its proper field of circumstance; C. Pretended cause, a prior event is cited as cause of a subsequent one; D. Evading the issue, of which there are many types; E. Begging the question, more than evading the issue but actually negation or contradiction of the issue; and The complex question, a "loaded" query that cannot be answered by a simple yes or no, i.e. "Have you stopped taking graft ?".
Broader Error (#PA6180) Reason (#PA5502)
Uncommunicativeness (#PA7411).
Related Sophistry (#PF8008).

♦ PF4366 Inadequate international cooperation in reducing terrorism
Government failure to sanction state-supported terrorism
Nature International cooperation among courts, prosecutors and policemen has not effectively controlled world-wide terrorism. While international terrorism appears to be well-financed, the budgetary and legislative approvals often necessary to extend the network of cooperative countermeasure arrangements encounter numerous obstacles on the national level. Prevention and punishment of terrorism are thereby hindered.
Incidence A hijacking of an American plane from Athens to Beirut, a bomb blast in the Frankfurt airport, and a luggage explosion in the Tokyo airport all were headline news during the the week of June 24, 1985 and yet, even though President Reagan spoke out sharply against such acts and the governments who supported them, little has been done in the international community to curb the flow of terrorism.
Broader Lack of international cooperation (#PF0817)
Ineffective prevention of terrorism (#PF4240).
Aggravates Terrorism (#PD5574)
State-supported international terrorism (#PD6008).
Aggravated by Failure of government intelligence services (#PF8819).

♦ PF4370 Antiquated regulations in the banking industry
Nature Innovation in regulatory and other dimensions of banking policy in most developing countries – and for that matter, in many developed nations – has failed to keep abreast of the innovations in financial services provided by transnational banks. Generally, developing countries regulate on the basis of institutional form, with the emphasis on branch banking, rather than on the basis of the type or function of a financial service. They do not devote nearly the same regulatory attention to finance subsidiaries of transnational banks as to branch offices. In many cases, the outcome is that regulatory control over branches is legally circumvented by the use of these subsidiaries. Thus, pursuing the traditional central bank means of controlling credit creation through control over branch bank deposits alone produces a "blind spot" – in many developed as well as developing nations – that modern transnational banking structures can work around.
Broader Outdated procedures (#PF8793).
Narrower Interference of transnational banks in domestic economic policies (#PE4325).
Related Discrimination against transnational banks in developing countries (#PE4320)
Socially irresponsible programmes of transnational banks in developing countries (#PE4360)
Domination of restrictive project loans by transnational banks in developing countries (#PE4330).
Aggravates Underutilization of world banking facilities (#PJ8783).

♦ PF4371 Barriers to transcendent experience
Barriers to spiritual experience — Barriers to religious experience
Refs Academy of Comparative Philosophy and Religion *Validity and Value of Religious Experience* (1968).
Aggravates Spiritual hunger (#PA8038)
Inhibition of individual psychological development through life cycle (#PF6148)
Denial of right to pursue spiritual well-being because of discrimination (#PE7310).
Aggravated by Spiritual disobedience (#PF7467)
Loss of spiritual guidance (#PF7005)
Lack of religious discipline (#PF8010)
Lack of individual development (#PG3595)
Spiritual and emotional hindrance (#PF1568).

♦ PF4373 Reinforcement of male dominance through language
Nature Use of the term "he" or "man" to speak of one member of a mixed group of people is considered correct English, even if the group consists of 50 women and one man. This creates in the minds of those who use the language an image of male people. Likewise "the evolution of man" lets you see in the mind's eye males evolving. This misleads the imagination, so much so that history, current events, the direction of the universe all take place without females having ever been there. When both sexes have the image of the male in mind when they use the term "man", this also means that males and females have a dramatically different sense of themselves as participants in the human venture.
Refs Penfield, Joyce *Women and Language in Transition* (1987).
Related Sexually discriminating job terminology (#PF6014)
Lack of English language to describe female experience (#PF7383).

♦ PF4375 Prohibitive cost of education
Expensive retraining — High training cost — Prohibitive school expenses — — Nonexistent school funds — Expensive distant schools — Inequitable school fees — Costly technical training — Prohibitive cost of higher education
Nature The different rates of expansion in enrolment at the first, second and third levels of education have produced an altered composition of pupil and student population. Since costs per place rise with the level of education, the change has forced up both average and total costs. It is thus the countries with the lowest secondary enrolment ratios, where the pressure for additional secondary places is felt the keenest, which face the fastest cost escalation if they try to accommodate demand. Since they also have typically the least favourable pupil-to-teacher ratio and meagre resources, the underlying cost structure acts as a formidable barrier to expansion – more so if standards are to be maintained.
Incidence In socialist and developed market countries, the average cost of maintaining a child in secondary school can be as much as double that of a primary place. In Latin America the situation is comparable. In Asia and the Pacific region, the cost differential is well in excess of 100 percent in one third of the countries reporting the relevant data. In Africa the position is extreme in that the cost difference is almost always above 100 percent and in half the countries more than 500 percent. At the third level, average costs are still higher. The third-to-first level cost ratio is typically less than five in developed and above five in developing countries. In Africa the recurrent cost of a place in a publicly funded third-level institution is now usually as much as 15 times the cost of a primary school place. In developing countries, the expansion in numbers and escalation in costs coincided with a period of steady and, in places, even rapid, growth in real income. The substantial shift of resources to education through the budget that occurred in the 1965–1977 period was typically financed by newly-generated resources. Even so, in nominal terms, public spending rose in relation to gross national product, substantially in both Africa and Asia and less markedly in Latin America. As a result of the slow-down in economic activity at the end of the 1970s and the beginning of the 1980s which coincided with a period of unfavourable terms of trade, all but a few developing countries have run into serious budgeting problems which make increased spending on education at historic rates impracticable.
Background Mass public education is a recent phenomenon. Until the industrial revolution, most schooling in the West was run by the church and available only to sons of the wealthy. Publicly-financed education expanded rapidly towards the end of 19th century and continued to mushroom (with a break during the war years) until the early 1970s.
Broader Obstacles to education (#PF4852)
Prohibitive cost of necessities in rural communities (#PF2385).
Narrower School supply costs (#PJ0686) Costly teaching staff (#PJ9524)
Unavailability of training costs (#PJ8395)
Excessive expense of athletic training programmes (#PF4196)
Prohibitive construction cost of rural community buildings (#PJ6368).
Related Lack of school transportation (#PJ7849).
Aggravates Lack of formal education (#PF6534) Underutilization of human resources (#PF3523)
Mass unemployment of human resources (#PD2046)
Limited access to practical education (#PF2840)
Unrealized use of education structures (#PF2568)
Non-diversification in subsistence fishing economies (#PF2135).
Aggravated by Limited accumulation of capital (#PF3630)
Insufficient capital investment (#PF2852)
Subsistence approach to capital resources (#PF6530)
Underdeveloped sources of income expansion (#PF1345)
Unavailability of scholarship funds for students (#PE3569)
Restrictive effects of external capital on development (#PF3318)
Unavailability of scholarship funds for developing country students (#PE8883).

♦ PF4377 Unrealistically positive self-assessment
Nature Unrealistically positive views of the self. Not only do people tend to attribute far more positive than negative traits to themselves; they easily process positive information and have difficulty recalling negative information. "Even when negative aspects of the self are acknowledged, they tend to be dismissed as inconsequential. One's poor abilities tend to be perceived as common, but one's favoured abilities are seen as rare and distinctive. Furthermore, the things that people are not proficient at are perceived as less important than the things that they are proficient at." Individuals judge positive attributes as more descriptive of themselves than others. The reverse is true of negative attributes. Individuals even believe that their driving ability is superior to others. They also give others less credit for success and more blame for failure than they ascribe to themselves. In experimental situations involving chance, people tend to think they have control or are applying skill. Most people think optimistically about the future, believing that the present is better than the past and that the future will be even better. Most people report being happy most of the time.
Counter-claim Optimism may improve social functioning. One study found that people with high self-esteem and an optimistic view of the future were better able to cope with loneliness. People with high self-evaluations see themselves – and are seen – as more popular. "There may.. be intellectual benefits to self-enhancement." Memory tends to be organized egocentrically so that people recall information relating to themselves well. Positive illusions may contribute positive mood. Positive affect is, in fact, an effective cue for memory retrieval. It seems to facilitate the use of efficient, rapid problem-solving strategies and enhances associations. "Positive conceptions of the are associated with working harder and longer on tasks. Perseverance, in turn produces more effective performance and a greater likelihood of goal attainment." People with high esteem evaluate their performance more positively than do low-esteem people, even when the performances are equivalent. "These perceptions then feed back into enhanced motivation. People with high self-esteem have higher estimations of their ability for future performance and higher predictions of future performance," regardless of prior performance. People with a stronger sense of personal efficacy are more highly motivated and therefore make more efforts to succeed. A desire for control leads people to respond more vigorously to a challenging task and to persist longer. People who expect to succeed world longer and harder than those with low expectations of success. "Overall, research evidence indicates that self-enhancement, exaggerated beliefs in control and unrealistic optimism are associated with higher motivation, greater persistence, more effective performance and, ultimately greater success".
Broader Personal misperceptions (#PF4389).
Aggravates Lack of self-confidence (#PF0879).
Aggravated by Accumulation of recognized merit (#PF7315)
Conceptual repression of problems (#PF5210).
Reduces Mental depression (#PC0799).

♦ PF4384 Alarmism
Scaremongering
Narrower Doomsday syndrome (#PF1083).
Aggravates Fear (#PA6030) Panic (#PF2633) False alarms (#PF4298).
Aggravated by Rumour (#PF5596) Unwarranted pessimism (#PF2818)
Resignation to problems (#PF8781) Vulnerability of society to truth (#PF5937).

♦ PF4385 Area disparities in book production and distribution
Nature The scene of book production is one of marked imbalance and dependence. Books are very unevenly distributed, both inside and among countries. Developing countries, with 70 percent of the world's population, produce 20 percent of the books published, and many of these are printed by subsidiaries of firms centred in developed countries. Developing countries therefore suffer, in varying degrees, from a serious shortage (sometimes amounting to a dearth) of books, and this slows considerably their economic, social and cultural progress. The chief obstacles to the development of local production are the cost of intellectual production (authors' fees, acquisition of copyrights, financing of translation); and the cost of manufacture, in which the two

main items are machinery and paper but which also includes professional training. Until they are able to cater for their own needs, the developing countries are obliged to meet their home demand – which is increasing with the spread of education and the advance of literacy – by recourse to outside sources of supply. This involves them in expenditures which are all the heavier in that high transport costs have to be added to the price of the books themselves and that payment has to be made in foreign currency. Moreover, the imported books are by no means always suited to the aspirations of their peoples, whereas their national authors, who are often forced to publish their works abroad because they are not included in the local publishing economic circuit, would be in a position to meet most of their needs. Outside supplies can therefore only be temporary palliatives and not real remedies for the book shortage. Whether they are commercial or are in the nature of bilateral or multilateral assistance, international exchanges should be regarded as a form of cooperation and not as a form of economic and cultural domination which would in the long run hinder or stifle local production.
Incidence UNESCO statistics indicate that world book production in terms of individual titles published or reprinted is presently in the order of 660,000 per year. Europe produces 410,000 titles, Asia and North America slightly above, and slightly below 100,000 titles each respectively. The remaining titles are produced as follows: Latin America 30,000; Africa 13,000; Oceania 7,000. In terms of the ratio of title share to population share, Europe is roughly 3:1 and North America 2:1. As a measure of pervasive literacy, Africa is 1:6 and Latin America is 1:2, indicating the problems in these regions. Similarly, without China, Asia is 1:3.
Broader Human inequality (#PA0844) Obstacles to world trade (#PC4890).

♦ PF4387 Personal failure
Loser
Nature Personal failure occurs when the results of initiatives taken fail to match the expectations. For those with out–directed values the expectations are those of important others (family, associates and institutional reference points), whereas those with inner–directed values naturally tend to rely more on their own personal standards. The former is influenced by someone else's standards without necessarily understanding it or being equipped to live by it. Few can match the external accomplishments of their idols and are thus vulnerable to a constant sense of failure. The inner–directed person may also set unreachable goals and be constantly dissatisfied with the failure to achieve them fully, possibly because they are too divorced from reality.
Counter–claim People only learn through failure.
Aggravates Refusal to admit error (#PF5163).
Aggravated by Fear of failure (#PF4125)
Unemployment as a perpetuator of failure (#PE8457).

♦ PF4389 Personal misperceptions
Personal illusions
Nature Decades of psychological wisdom have established contact with reality as a hallmark of mental health. The well–adjusted person engages in accurate reality testing, whereas the individual whose vision is clouded by illusion is regarded as vulnerable to mental illness.
Counter–claim Certain illusions are adaptive for mental health and well–being. Unrealistically positive self–evaluations, unrealistic optimism and an exaggerated sense of control or mastery are part of normal thinking. The only people who tend to see themselves as other see them are moderately depressed, low in self–esteem, or both. Positive illusions help make each individual's world a warmer and more active and beneficent place in which to live.
Broader Illusion (#PA6414) Mental illness (#PC0300).
Narrower Illusion of happiness (#PJ5224) Unrealistic expectations (#PE7002)
Illusion of controlling events (#PJ5116)
Unrealistically positive self–assessment (#PF4377).

♦ PF4391 Parental domination
Parental interference
Broader Paternalism (#PF2183).
Narrower Arranged marriage (#PF3284) Possessive attitude of parents (#PD1317).
Aggravates Divorce (#PF2100) Child–marriage (#PF3285)
Psychological impediments to marriage (#PF3344).

♦ PF4402 Withdrawal
Related Dissent (#PA6838) Exclusion (#PA5869) Avoidance (#PA6379)
Regression (#PA6338) Monasticism (#PF2188) Incuriosity (#PA6598)
Irresolution (#PA7325) Unfeelingness (#PA7364).
Aggravated by Frustration (#PA2252).

♦ PF4405 Political monoculture
Ideological monoculture — Erosion of socio–economic diversity
Aggravated by Single party democracies (#PD2001)
Homogenization of cultures (#PB1071)
Lack of variety of social life forms (#PE8806).

♦ PF4416 Lack of intimate relationships
Lack of friendship — Friendlessness — Dependence on deep emotional attachments
Nature To be intimate means being willing to be affected by someone else's feelings, to be aware of the nuances of their inner meanings and their moods. People who feel the need to insulate themselves from the emotional demands of others fear this. Some people adopt patterns in relationships that ensure them a "safe" emotional distance. Others alternate between intimacy and distance or select partners who have some flaw that guarantees unavailability. The failure to permit oneself to participate in intimate relationships leads people to chronic loneliness, unfulfilment, depression and failure of long–term relationships.
Incidence Students aged 10 to 15 years, who are in greatest need of intimate relationships are frequently placed in large impersonal schools incapable of meeting these needs.
Claim The loss personal relationships in neighbourhoods and small town, the relative rigid hierarchies of work places and the lack of alternative places for intimate relationships to develop makes it difficult for friendships to develop.
Broader Emotional disorders (#PD9159).
Related Maternal deprivation (#PC0981) Emotional dependency in marriage (#PD3244).
Aggravates Unethical personal relationships (#PF8759).
Aggravated by Fear of intimacy (#PF8012).
Reduced by Forced social intimacy (#PD4287).

♦ PF4417 Unsociable human physiological processes
Narrower Acne (#PE3662) Cough (#PE6825) Snoring (#PE4415)
Dandruff (#PJ5412) Bad breath (#PE6558) Common cold (#PE2412)
Incontinence (#PE4619) Excess human body hair (#PE5580)
Spitting in public places (#PD5347) Desiccation of human skin (#PE4966)
Excreting in public places (#PE1602) Human flatulence in public (#PE1764)
Disagreeable human body odour (#PE4481).
Related Inadequate personal hygiene (#PD2459).
Aggravates Personal physical unattractiveness (#PF4010).

♦ PF4432 Protection of company ownership
Blocks to hostile takeover bids from abroad
Nature A plethora of different mechanisms are used to block a hostile takeover bid from abroad. Some are enshrined in national law, others are cultural or built into corporate structures. Finance ministries can block bids by requiring positive approval of ownership of fixed percentages of shares, for example, 20 percent in France. Laws force shareholders with more than 10 percent of shares to declare their holdings. National laws prevent ownership of newspapers, radio and TV stations, banks and defence industries. They also ban ownership of industries crucial to national defence. Some countries prevent companies from acquiring dominate positions in a specific market. Companies place new bond issues in friendly hands which will exercise the attached equity warrants only in the event of an unwelcomed approach. Private companies have hard cores of friendly shareholders, i.e., family members, sister companies and board members. Complex pyramids of holding companies, sometimes spread internationally are used to frustrate takeover attempts. Employees can discourage hostile takeovers, managers and trade unions may be opposed to any bids to protect jobs and for fear of being shamed, in the case of Japanese companies. Labour unions may lobby legislative and regulatory bodies. Some companies erect their own barriers. Swiss companies can fashion their own statutes and decide which shares they will register and thus give voting rights to.
Broader Restrictive trade practices (#PC0073).
Related Capitalist speculation (#PC2194).
Reduces Concentration of investment power (#PC5323)
Excessive concentration of business enterprises (#PD0071).

♦ PF4433 Commerce in religious indulgences
Sale of indulgences
Related Simony (#PE9840).

♦ PF4435 Risk of intentional nuclear war
Nature There is a risk of intentional nuclear war either as the result of beliefs that such a war could be fought, contained and won with the victors able to survive the effects of such a war; or of belief that offence is preferable to defence: that is, that the idea of allowing the nation to be attacked and then to retaliate is completely unacceptable. The former belief does not necessarily require a first–strike plan but can arise from predecided retaliation against aggression by massive conventional forces using a so–called tactical or field nuclear force capability. This is the current doctrine of NATO, causing the 'Flexible Response' strategy to be even more flexible. In the USA, decisions have also been taken to launch MX nuclear missiles immediately upon warning that a Russian attack was underway. The definition of what an attack constitutes is dangerously elastic. Whether in the case of unilateral action by NATO or the USA, or of action by their adversaries, intentional nuclear war initiated as a response is increasingly likely unless disarmament agreements can be made and kept.
Claim In 1990 it was estimated that the risk of nuclear war was higher than at any time in the previous decade, but between regional powers rather than between superpowers.
Related Risk of unintentional nuclear war (#PF0466).

♦ PF4436 Boasting
Self–glorification — Self–aggrandizement — Self–praise — Social boasting — Vainglory — Sexual oneupmanship
Nature Boasting may take a well–recognized verbal form or it may be achieved more subtly through display of symbols of power, wealth, honours or other achievements. It tends to contain elements of bluffing opponents or rivals as well as joy in seeing oneself reflected in the mirror of one's own praise. When true it is carefully edited but easily develops into lying. Boasting may thus provide a disguise for inaction, when boasting about an action is easier than undertaking that action in reality.
Incidence In its most calamitous form, social boasting, the escalation of claims and counter–claims between rivals is the pattern which underlies the arms race and efforts to maintain and reinforce inequality. In situations characterized by corruption, boasting of licentiousness (as during the Restoration) exacerbates the phenomenon.
Claim Boasting is a misuse and perversion of the true range and greatness of human capacities. It may constitute a direct attempt to surprise and corrupt the moral judgement into a false verdict and as such has been one of the greatest defences of a debased and impenitent conscience.
Counter–claim Boasting may be said to have been the original force behind literature and many songs. Self–praise may not be honourable, but it may be extremely profitable if done well. It may be distasteful, but it better than doing something worthwhile and achieving no recognition for it at all.
Broader Exaggeration (#PJ5960).
Narrower Name–dropping (#PF5223).
Aggravates Lying (#PB7600) International status race (#PC5348)
Abusive national leadership (#PD2710) Pursuit of personal prestige (#PF8145)
Pursuit of national prestige (#PF8434) Excessively costly prestige projects (#PF3455).
Aggravated by Vanity (#PA6491) Lack of self–confidence (#PF0879).

♦ PF4444 Extraterrestrial invasion
Invaders from outer space
Broader Invasion (#PD8779).
Related Disruption of human activities by supernatural entities (#PF3437).
Aggravates Abduction by extraterrestrials (#PF3881)
Introduction of extraterrestrial infectious diseases and bacteria (#PF1312).

♦ PF4450 Government secrecy concerning nuclear weapons testing
Secret weapons development — Clandestine space weapons
Nature The secrecy surrounding nuclear weapons developments leads the nuclear powers to infer the 'worst case' which heightens mistrust, hinders international arms reduction talks, and accelerates the arms race. In addition, secrecy may cover up nuclear accidents and their causes, and also prevent accountability of the public cost of enriching arms manufacturers.
Broader Military secrecy (#PC1144) Official secrecy (#PC1812)
Nuclear weapons testing (#PC2201).
Aggravates Nuclear accidents in space (#PE5080).
Aggravated by Weapons (#PD0658) Nuclear arms race (#PD5076)
Environmental hazards constraining scientific research (#PF1789).

♦ PF4453 Communication delays
Broader Delay (#PA1999).
Narrower Telephone delays (#PF1698) Inadequacy of postal services (#PF2717).
Related Delays in delivery of goods and services (#PE3928).
Aggravated by Insufficient communications systems (#PF2350).

♦ PF4455 Non–verificability of compliance with nuclear power safeguards
Nature The non–military uses of nuclear power involve the hazards of accidents at sites near international borders, the possibility of nuclear power plants being targets of sabotage or of acts of war, and the diversion of nuclear materials from commercial or civil applications to surreptitious weapons manufacture. There are no adequate means of inspecting for violations or hazards by

international agreement.
Incidence In May 1984, nuclear experts from 19 countries agreed on guidelines under the auspices of the International Atomic Energy Agency to deal with the accidental release of radioactive materials across national boundaries. The IAEA is a purely advisory body with no enforcement powers. The guidelines can be and are ignored at the convenience of national governments, as the USSR did at the time of the Chernobyl accident.
Broader Military secrecy (#PC1144) Non-verification of compliance (#PF6310)
Lack of international cooperation (#PF0817).
Aggravates Nuclear accidents (#PD0771) Nuclear reactor accidents (#PD7579)
Insufficient nuclear power stations (#PD7663).
Aggravated by Inadequate nuclear reactor safeguards (#PF6084)
Deterioration of nuclear power plants (#PE5260).

♦ PF4460 Non-verifiability of compliance with nuclear arms treaties
Nature On-site inspections of military installations by international agencies to control nuclear weapons inventory limits are unachievable. Inspection of nuclear weapons materials production facilities are also beyond the scope of international treaties. National sovereignty and security issues stand in the way of both. Compliance with international agreements concerning limiting nuclear arms depends on good faith as well as the known vulnerability of nations to documented or photographic espionage and consequent exposure of violation evidence to world opinion.
Broader Non-verification of compliance (#PF6310).
Related Limited acceptance of international treaties (#PF0977).
Aggravates Unfulfilled treaty obligations (#PF2497).
Aggravated by Covert violation of international treaties (#PD8465).

♦ PF4470 Socio-economically inactive rural population
Small population of basic services users — Static town population — Small population base — Inadequate rural community market
Nature The economically inactive population includes students, women occupied solely in domestic duties, retired persons, persons living entirely on their own means and persons wholly dependent upon others. The practice varies between countries as regards the inclusion in this category of such groups as armed forces, inmates of institutions, persons living on reservations, persons seeking work for the first time, seasonal workers and persons engaged in part-time economic activities.
Incidence World statistics for 1975 indicate a crude activity rate of 41.5 per cent for the world, 46.0 per cent for more developed regions and 39.7 per cent for less developed regions.
Broader Unemployment (#PB0750) Underproductivity (#PF1107).
Aggravates Slow rate of income expansion (#PF6478)
Unemployment in developing countries (#PD0176)
Underprovision of basic services to rural areas (#PF2875)
Stifled potential for social interaction between different age groups (#PF6570).
Aggravated by Declining birth rate (#PD2118) Declining community population (#PJ8746).

♦ PF4473 Excessive reliance on fashionable solutions to problems
Habitual responses to problems
Nature For responses to problems to appear credible, especially to government agencies and funding agencies which have established methods and procedures for the problems within their mandates, there is considerable pressure for solutions to conform to a pre-existing pattern. Alternatively, when some new approach has received wide publicity as being successful, it becomes fashionable as a way of demonstrating that fresh approaches are being used. In neither case are the agencies capable of exploring new solutions on their own merits and in terms of their appropriateness to the problem. Innovation is thus severely inhibited whilst creating the impression of openness to it.
Aggravates Institutional preoccupation with obsolete problems (#PJ5014)
Excessive emphasis on fashionable areas of research (#PF0059)
Inadequate research on proposed solutions to problems (#PF1572)
Over-emphasis on immediate solutions in resource development research (#PE4059).
Aggravated by Excessive emphasis on fashionable problems (#PF4164)
Failure to integrate knowledge to empower humanity in response to the global problematique (#PF8753).

♦ PF4480 Maldistribution of private automobiles
Maldistribution of passenger cars
Incidence World statistics for 1980 indicate that the number of passenger cars in use was (in millions): 316.4 in the world, 5.8 in Africa, 7.3 in Oceania, 13.0 in South America, 31.6 in Asia, 115.3 in Europe (excluding USSR), and 135.1 in North America.
Broader Insufficient transportation infrastructure (#PF1495).
Aggravated by Intrusive truck traffic on residential streets (#PU9137).
Reduces Proliferation of automobiles and motor vehicles (#PD2072).

♦ PF4485 Maldistribution of commercial vehicles
Incidence World statistics for 1980 indicate that the number of commercial vehicles in use was (in millions): 87.4 in the world, 1.8 in Oceania, 2.8 in Africa, 4.5 in South America, 14.5 in Europe (excluding USSR), 17.6 in Asia, and 38.7 in North America.
Broader Insufficient transportation infrastructure (#PF1495).
Reduces Proliferation of automobiles and motor vehicles (#PD2072).

♦ PF4487 Disorganization
Disorder — Disorganization
Nature Disorder, an unstructured state of affairs, while necessary for change to occur, as a permanent state is destructive to society and to the individual. The structured patterns of animal conduct, such as, the spinning of webs by spiders or the building of nests by birds, is in the inherited nervous systems of the species. The innate releasing mechanisms by which these patterns are determined are for the most part stereotyped. The human species is distinguished by the fact that the action releasing mechanisms of its central nervous system are for the most part not stereotyped but open. They are susceptible, consequently, to the influence of imprintings from the society in which the individual grows up. The human child acquires its character, upright stature, ability to speak, and the vocabulary of its thinking under the influence of a culture, an open, flexible, but limited and limiting social form. Without this order the child cannot become a defined and competent member of some specific, efficiently functioning social group.
Broader Disorder (#PA7361) Disintegration (#PA6858).
Narrower Lack of war relief (#PF0727) Family disorganization (#PC2151)
Inadequate disaster rescue and relief (#PF0286)
Fragmented planning of community life (#PF2813)
Disorganized liaison with formal support (#PF2947)
Disorganized approach to land ownership in tropical villages (#PF2365).
Aggravates Uncertainty (#PA7309) Lawlessness (#PA5563)
Destruction (#PA6542) Formlessness (#PA6900)
Misbehaviour (#PA6498) Unjustified military defence policies (#PF1385).
Aggravated by Agitation (#PA5838) Impairment (#PA6088).

♦ PF4490 Unequal distribution of livestock production
Incidence World statistics for 1981 indicate a total of 1,210 million cattle in the world, with 34 million in Oceania, 132 million in Europe, 171 million in Africa, 180 million in North America, 214 million in South America, and 363 million in Asia.
Broader Underproductivity (#PF1107)
Unequal distribution of production between countries (#PF4336).

♦ PF4494 Paper qualification syndrome
Reliance on paper qualifications
Claim The long theoretical training of doctors is partly a show – designed to make medicine look difficult, and therefore to increase the intellectual status of doctors in society. The exaggerated sense of doctor's intellect makes him or her liable to incompetence. Mere book learning does not qualify anybody to be a doctor. In the end, experience is much more valuable than paper qualifications.
Narrower Diversion of education to qualification earning (#PF4828).

♦ PF4495 Unequal distribution of fish catches
Incidence World statistics for 1980 indicate that the total fish catches for the world was 72.19 million metric tons, with 0.35 million metric tons in Oceania, 4.10 million metric tons in Africa, 6.80 million metric tons in North America, 7.85 million metric tons in South America, 12.37 million metric tons in Europe, and 30.29 million metric tons in Asia.
Broader Underproductivity (#PF1107)
Inequality in distribution of natural resources between countries (#PF3043).

♦ PF4510 Unrealistic environmentalism
Fuzzy environmental thinking
Claim The environmental movement is grounded in a religion of nature characterized by hostility toward progress, science and the Judaeo-Christian religious tradition. It is suspicious of reason. It finds mysticism in nature, making a ritual bow to a mystical sense of connectedness and refuses to take seriously humankind's place in the ecology.
The environmental movement and an increasing number of the public revere natural. Mother's milk is natural, spring water is natural, organic tomatoes are natural. So is the poisonous radon gas that erupted from a lake bottom in Cameroon claiming 1500 lives, as are the earthquakes that shattered Mexico City, San Francisco and Armenia, and as is the volcanic mudslide that killed 21,000 Colombians. Basing their beliefs on the teachings of Rousseau and the romantic poets; responding to the man made disasters of our age: Hiroshima, Bhopal and Chernobyl; and being isolated from the reality of nature these fuzzy environmentalist create a vision of the future that is neither politically nor economically feasible nor morally defensible. They call for the deindustrialization of the west and the use of appropriate technology, appropriate to an age when there were a few hundred million people on the earth and the life expectancy was 35 years.
Broader Uncritical thinking (#PF5039).
Related Doomsday syndrome (#PF1083) Blind faith in technology (#PF4989).
Aggravates Unrealistic policies (#PF9428).

♦ PF4513 Religious racism
Racist religion
Claim Christianity has been fundamentally racist in its ideology, organization, and practice. It is believed to be the only true religion and the others are considered superstition. The fundamental component of Christianity's racism is its inherent ability to leave other people alone with their own beliefs.
Broader Racism (#PB1047) Religious discrimination (#PC1455).

♦ PF4516 Lack of incentive for users to care for common property
Asymmetry in economic system incentives — Tragedy of the commons
Nature Valuable as they are to individual users and to the collective as a whole, free commons deter ameliorative or preventive action by individual users since that is perceived as conferring advantages and benefits on those users who fail to participate in such ameliorative action.
Incidence Energy production, for example, yields valuable services whose allocation in the economic system is, like goods and services in general, effectively handled through the market place and property rights (in a market-oriented society). But the residual mass of these energy resources, after use, flows back into the common-property resources, within or across countries, namely into the atmosphere or into water systems. When access to such waste refuses is unimpeded and free, overuse and degradation is the inevitable result.
Claim The incentives in the economic system fail conspicuously in controlling the exploitation of common property resources, however well it works in promoting the production and distribution of resources based on access to such resources.
Counter-claim If overuse and degradation of environmental resources are frequently the inevitable consequences of the weaknesses of of traditional, free-market regimes, centrally planned economic systems do not seem to have produced a formula for rational environmental management either.
Aggravated by Economic inefficiency (#PF7556) Natural environment degradation (#PB5250).

♦ PF4528 End of the world
The eschaton — Doomsday
Nature The end of the world can be seen as three separate realities; the physical destruction of the earth and all of humankind, the collapse of one's own worldview and all of one's understanding, and the time when God will be manifested in final judgement and redemption of humankind.
Incidence There are frequent reports of individuals and groups concerned at the imminent end of the world. Some of them base their beliefs on deductions from scriptures, possibly combined with very sophisticated calculations (such as those based on the Great Pyramid). Others derive their belief from information provided by seers and mediums, whether from the past (as in the case of Nostradamus) or from the present time. Some groups act on the belief by selling their property in order to be able to congregate at a particular location to await the end of the world. This may involve purchasing the right to benefit from the facilities at such a location, including use of reserves to ensure survival in the period of transition after disaster strikes. Some of the locations selected are underground in elaborate systems of shelters as a protection from any conflagration.
Aggravated by Superstition (#PA0430).

♦ PF4530 Criticism
Nature In rare cases of deep and longlasting friendship and mutual trust, the expression of negative feedback helps individuals to assess and improve their performance and their selfhood. More often, however, criticism is an act of aggression designed in some way to block or damage the recipient. It is especially powerful used in a social context, in which criticisms based on personal preferences and those based on group policy become mingled, and the mass media expand a passing comment into a "story".
Incidence The 1988 American presidential campaign used criticism as the primary strategy by which candidates attracted the attention of the electorate.
Narrower Anticipation of criticism (#PJ7893) Criticism of official institutions (#PF9385)
Excessive portrayal of negative information by the media (#PE1478).

♦ **PF4548 Peoples perceiving themselves as specially chosen**
Peculiar people
Refs Douglas, C H *The Land for the Chosen People Racket* (1982).
Broader Counter culture (#PF0423).
Related Restrictive religious practices (#PD8439)
Disruptive secular impact of holy days (#PE7735).
Aggravates Religious intolerance (#PC1808) Religious persecution (#PC5994).
Aggravated by Religious elitism (#PG3644) Persecution of religious sects (#PF3353)
Underprivileged religious minorities (#PC2129).

♦ **PF4549 Cronyism**
Old boy networks
Aggravates Corruption (#PA1986) Interlocking corporate directorates (#PF5522)
Abusive distribution of political patronage (#PF8535).

♦ **PF4550 Unparliamentary behaviour**
Violence in parliamentary assemblies — Abusive language in parliament — Indecent conduct in parliament
Incidence Abusive or insulting language is frequently used in parliamentary assemblies as an extension of parliamentary rhetoric. This is more or less constrained, depending on practice tolerated in particular countries. Occasionally the tensions of debate, and the issues under discussion, lead to exchanges of blows between parliamentarians. Women parliamentarians tend to be exposed to insulting, sexist language and insinuations, especially since parliaments have been a male domain and many practices continue to reinforce male domination.
Broader Misconduct in public office (#PD8227).
Aggravates Profanity (#PF7427) Verbal abuse (#PD5238)
Uncoordinated government policy-making (#PF7619).
Aggravated by Contempt for democratic processes (#PF0639)
Discriminatory unwritten codes of behaviour (#PE7017).

♦ **PF4558 Broken government promises**
Broken election promises — Unrealistic election promises — Extravagant political promises — Broken government pledges — Breach of administrative pledges — Government reneging on public pledges
Claim In a political environment in which it is necessary to be seen to respond to the latest events, there is little time or inclination to honour past commitments which tend only to be remembered by the few. Laws, contracts and terms of agreement become short-lived. The greater the number of past commitments to be fulfilled, the greater the amount of future time to be set aside to do so. But time being at a premium, decreasing amounts of time can be allocated to the fulfilment of such earlier commitments, whilst more must be allocated to facilitating new options and agreements. Political debts, like legal tender, are continually being cancelled in the inflation-ridden policies of the present. The past is believed to place decreasing limits on the future, the devaluation of history being a prerequisite for the exercise of power.
Broader Breach of promise (#PF7150).
Narrower Progressive reduction in government action commitment (#PF5502).
Aggravates Credibility gap (#PB6314) Loss of leadership credibility (#PF9016)
Unpredictable governmental policy (#PF1559).
Aggravated by Government limitations (#PF4668).

♦ **PF4564 Misuse of statistics**
Biased adjustment of statistics — Misleading reports — Misreporting of statistics — Misinterpretation of statistics — Biased adjustment of official statistics by government — Unethical use of statistics — Corruption of statisticians — Statistical malpractice — Official abuse of statistics
Nature Numerical reports are deliberately slanted in order to make a situation appear more as the writer or his employers would like to have it.
Incidence Police in the UK are reported to improve the rate of crimes they have solved by extracting "confessions" from cooperative prisoners. United States military personnel exaggerated deaths of enemies. Investment fund managers choose periods of comparison to maximize the sales value of their fund. Governments change methods of determining unemployment and then compare rates developed from the different methods. In the UK, between 1979 and 1988, 19 changes were make to the way unemployment was calculated; all but one decreased the amount of unemployment. The government also masked for two years the widening gulf in living standards between rich and poor by a statistical mistake. In the statistic published by the United Nations, including the World Bank, certain countries are treated for political reasons as though they do not exist. Thus Taiwan, despite its productivity, is not mentioned, nor is South Africa. This makes the United Nations data of questionable value for any global comparative study, especially with respect to the economic future of particular regions.
Claim As Benjamin Disraeli said about statistics: "There are three kinds of lie – lies, damned lies and statistics". Statistics only serve those presenting them.
Broader Misleading information (#PF3096) Abuse of government power (#PC9104)
Unethical practices of government (#PD0814).
Related Corruption of documents (#PE7900)
Inadequate social and demographic statistics (#PF0214).
Aggravates Increasing scepticism about the accuracy of official information (#PF7649).
Aggravated by Statistical errors (#PF4118) Government treachery (#PF4153)
Governmental bias in statistics (#PF0019)
Suppression of information concerning social problems (#PF9828).

♦ **PF4580 Inappropriate loans**
Irresponsible lending by banks — Over-enthusiastic lending by banks
Incidence In the beginning of the 1980s the banks were holding abundant funds due to the oil price rises of the early 1970s, followed by slump in economic activities (that caused the fall of the export earnings of developing countries) and high interest rates of early 1980s. Banks started enthusiastically to give loans to developing countries which later created the debt crisis in these countries.
Refs Fair, D E and Bertrand, R *International Lending in a Fragile World Economy* (1983).
Aggravates Maldevelopment (#PB6207)
Excessive foreign public debt of developing countries (#PD2133)
Decline in commercial bank lending to developing countries (#PE4655).
Aggravated by Unethical commercial practices (#PC2563).

♦ **PF4583 False statements**
Nature Generally, it is considered a crime to make false statements to a legal government proceeding. This includes making false statements while under oath and testifying to a legislative or judicial inquiry, making false written statements, knowingly creating a false impression or omitting pertinent information in order to mislead or uses a trick to mislead a government body.
Broader Falsification of public records (#PD4239).
Narrower False confessions (#PE7252).
Aggravated by Perjury (#PD2630) Fraudulent impersonation (#PE1275).

♦ **PF4586 Decline in government health expenditure**
Decline in public spending on health — Limited public health subsidies — Limited health funds
Incidence Over the decade from 1979 to 1989 the proportion of government expenditure devoted to health has fallen in most countries of sub-Saharan Africa, in more than half of the countries of Latin America and the Caribbean, and in one third of the nations of Asia. And the cuts have not been marginal. In the 37 poorest nations spending per head on health care has fallen by nearly 25 percent in this period.
Broader Decline in government social expenditure (#PF0611).
Narrower Insufficient doctors (#PE8303) Insufficient health payments (#PG8867).
Related Unknown availability of subsidies (#PG9905)
Limited availability of financial credit (#PF2489)
Strained capital resources in small communities (#PF3665)
Shortage of financial resources for action against problems (#PF0404).
Aggravates Neglected health practices (#PD8607)
Insufficient health personnel (#PD0366)
Inadequate resources for health (#PF9587)
Inequitable use of medical resources (#PJ5160)
Limited availability of health resources (#PD7669)
Arbitrary evaluation of disability compensation (#PG7870)
Limits on participation in community development (#PF3560).
Aggravated by Prohibitive medical expenses (#PE8261)
Increasing public health expenditures (#PF6234)
Non-diversification in subsistence fishing economies (#PF2135)
Distortion of international trade by selective domestic subsidies (#PD0678).

♦ **PF4600 Impoverishment of political debate**
Superficiality of political debate — Trivialization of political debate — Avoidance of substantive issues in political debate
Nature Domestic politics, partly under the influence of the media, is becoming so shallow, mean and meaningless that it is failing to produce the ideas and leadership needed to guide national policies in a rapidly changing world. Issues have in many cases become trivialized to the point where meaningful debate has become almost impossible. Government is increasingly crippled by a superstructure of politics that makes ideas harder to discuss and exalts public opinion over leadership. Politicians are encouraged to avoid hard issues which may affect their positions in opinion polls instead of responding to changes in the world, elected officials allow their thinking to be determined by the risk of televized attacks by their opponents, personal security by the press, cynicism on the part of the public, and the need (in some countries) to raise large sums of money to purchase television time to combat such criticism. When every decision is taken in this light, effective decision-making becomes severely constrained, stifling creativity and strong stands on major issues.
Claim The political system becomes unable to define and debate critical questions, let alone resolve them. Politicians are locked into where they are today, rather than being able to lead society effectively into an appropriate future.
Aggravates Political stagnation (#PC2494) Preoccupation with isolated problems (#PF6580).
Aggravated by Political apathy (#PC1917) Avoidance of reality (#PF7414)
Unrealistic policies (#PF9428) Inappropriate arguments (#PF2152)
Political smear campaigns (#PD9384) Rhetorical inflation in meetings (#PF3756)
Media theatricalization of public life and politics (#PF9631).

♦ **PF4606 Malthusianism**
Nature British economist Thomas Malthus (1766–1834) developed theories of population growth which held that if unchecked, human population will rapidly outstrip the resources capable of sustaining it. This way of thinking relies heavily on statistical projections of past trends, disallowing the propensity of human groups to rapidly change.
Related Unsustainable population levels (#PB0035)
Depletion of natural resources due to population growth (#PD4007).

♦ **PF4611 Overlooked media channels**
Broader Lack of local information systems (#PF6541).
Narrower Prejudice against communication by visual imagery (#PF0076).

♦ **PF4612 Risk-aversion strategy**
Reactionary methods of avoiding danger
Nature The contemporary risk aversion strategy, 'no trials without prior guarantee against error' plays a major role in government policy, is a rigid, reactionary system, which places stringent limits on the number of hypotheses that can be tested, and thus on innovation, learning and self-correction. It freezes existing dangers. It assumes that danger can be outlawed, that safety and danger can be severed from each other. Safety and danger are the warp and woof of nature and technology. To focus exclusively on sources of danger is to direct thought and resources toward an infinity of hypothetical or acutely low-probability risks, such as one molecule can cause cancer, while ignoring and damaging the inextricably related sources of safety.
Broader Obsolete methods (#PF3713).
Related Obsession with novelty (#PF8767) Absence of tactical methods (#PF0327)
Short range planning for long-term development (#PF5660).
Aggravates Cumbersome methods of policy formation (#PF1076)
Risk of unintentional nuclear war generated by the strategy of deterrence (#PF4162).
Aggravated by Ignorance of history (#PD3774).

♦ **PF4617 Deicide**
Killing of a god — God slaughter
Nature Deicide has been a wide-spread custom. One variety was putting to death kings and chieftains, believed to be gods incarnate, at set times or when they were approaching disease or death. It was believed that through disease or decay of strength they were unable any longer to keep in safety those who looked to them for blessings. In a second variety, the phenomena of Nature were personified and her processes became incidents in the lives of gods thus originated. Such myth is, for example, Dionysus, god of wine, who was put to death by Juno.
Related Death of God (#PF2019).

♦ **PF4618 Lifestyle diseases**
Nature According to the World Health Organization 70 to 80 percents of all deaths in the industrialized world and 40 to 50 percent in developing countries are attributable to high blood pressure, heart attack, stroke or lung cancer. These diseases are caused by unhealthy diet, lack of exercise and smoking.
Broader Human disease and disability (#PB1044).

♦ **PF4633 Destruction of scientific records**
Loss of scientific evidence — Suppression of scientific evidence
Broader Denial of evidence (#PD7385).
Related Tampering with official documents (#PF4699).
Aggravates Unretrievable documents (#PF4690)
Unproven relationships between problems (#PF7706)
Placement of the burden of proof on the disempowered (#PF3918).

PF4636 Limited access to technological decisions
Nature The use of technology to exploit natural resources is limited to special interest groups who neither consider the need of the globe for the resources nor the use of alternative technologies. Many of the most effective technologies are private property and as such outside the control of the public. Governments contribute to misuse of technology through protectionist measures. Monetary considerations normally out weigh environmental or limited availability of the resources.
Broader Undemocratic policy-making (#PF8703)
Limited access to natural resource use decisions (#PF2882).

PF4641 Personal unpopularity
Unpopularity in childhood
Aggravates Mental depression (#PC0799)   Social alienation (#PC2130)
Physical intimidation (#PC2934).
Aggravated by Ignorance of nonverbal communication skills (#PE0533).

PF4644 Temporal deprivation
Temporal discrimination — Degrading people's time orientation — Extreme present-orientedness — Temporal disorientation — Narrow time span of the poor — Temporal ghettos
Nature In every society the poor and the powerless have a narrowly circumscribed sense of time, oriented towards the immediate challenges of daily living. Such people are not free or empowered to plan ahead to secure a better future for themselves. Temporal deprivation is thus built into the time frame of every advanced society with the poor in industrial cultures being both temporarily poor as well as materially poor. To those in lower economic classes the future is less predictable and much less apparent in comparison with the problems of immediate economic security. Those who are obliged in this way to focus on the present are led into a future that others have envisaged for them. Temporal discrimination segregates the unskilled and semi-skilled, who require little past knowledge and even less predictive or planning skills, from the professional who require both. Such higher economic classes are conditioned by their upbringing to delay immediate gratification in expectation of greater rewards at some future time.
Claim Extreme-present orientedness, not lack of income or wealth, is the principal cause of poverty. Temporal ghettos are no less scandalous than physical ghettos. Monopoly of power in every society begins with the separation of people from control over their own future, effectively making them prisoners of the present. Since they are unable to gain access to the future, they then become pawns in the hands of those on top of the temporal pyramid who are able to control the human time frame. The labourer remains stuck in a present-oriented temporal ghetto from which he is unable to reach out and claim some measure of control over his future. The elites have used temporal skills of reflection and anticipation, hindsight and foresight, as tile traps to ensnare and exploit the rhythms of the life world and to subdue and enslave their fellow human beings.
Related Socio-economic poverty (#PB0388).
Aggravates Lack of time (#PC4498)   Unpreparedness (#PF8176)
Elder paralysis over the future (#PF3973)   Unimaginative educational vision (#PF3007)
Individual fear of future change (#PD2670)   Detrimental story of community future (#PF6575)
Unknowable future patterns of social choice (#PF9276)
Irresponsibility towards future generations (#PF9455)
Young people's lack of context for the future (#PF2068)
Inhibited capacity to visualize a creative future (#PF2352).
Aggravated by Monopoly of competence for intervention in the future (#PF0980).

PF4646 Lack of care
Unmindfulness — Failure of stewardship
Nature Failure to respond attentively to the needs of the natural or social environment. In the case of the natural environment, this takes the form of a failure of the stewardship role, through which mankind has a responsibility to remedy or counteract any destruction of the environment. In the case of the social environment, it is associated with failure to care for others, whether in times of need or in support of their development. In interpersonal relations, it is the failure to respond to the condition of others.
Narrower Inaction on problems (#PB1423)   Non-concerned attitudes (#PF2158)
Lack of care for animals (#PD8837)
Insufficient care of community property (#PF1600)
Fragmented forms of care at the neighbourhood level (#PE2274)
Inadequate community care for transient urban populations (#PF1844).
Related Neglect (#PA5438)   Inadequate health services (#PD4790).
Aggravates Negligence (#PA2658).
Aggravated by Indifference (#PA7604).

PF4660 Unlucky numbers
Aggravates Unluckiness (#PF9536).

PF4662 Lack of commitment to the protection of vulnerable groups
Lack of commitment to protection of the poor
Broader Lack of commitment (#PF1729)   Lack of protection for the vulnerable (#PB4353).
Aggravates Violation of rights of vulnerable groups during states of emergency (#PD3785)
Excessive social costs of structural adjustment in debtor developing countries (#PD8114).

PF4668 Government limitations
Restricted government ability to undertake new initiatives — Limited degrees of freedom of government — Limitations of government power — Constraints on power of government — Government impotence
Nature Despite electoral promises and the best of intentions, governments are severely restricted in the nature of new initiatives they are able to undertake. This is due to a combination of the following factors: secret compromise commitments made during the electoral process, binding commitments made by previous governments, and financial obligations incurred by previous governments. The binding commitments can take the form of contractual obligations (construction projects, staff contracts, etc), agreements with other countries (trade agreements, defence agreements, etc) and agreements with intergovernmental organizations. The financial obligations are associated with repayment of debt as a result of borrowing by previous governments.
Any single government's power is very much circumscribed by the needs and expectations of its neighbours, friends and enemies. Constraints are imposed by previous governments, the powers of local and regional authorities, the country's membership in international alliances and organizations, and the current state of relationships with neighbours.
Incidence In the post-Cold War period, the limitations of the two superpowers have become increasingly evident, especially in their ability to influence regional conflicts (El Salvador, Gaza, Kashmir, Cambodia). In dealing with Soviet spies in a military project, the Canadian government has had to take into account not only its own relations with the Soviet Union, but also those between the Soviet Union and the United States.
Aggravates Government inaction (#PC3950)   Broken government promises (#PF4558)
Unfulfilled treaty obligations (#PF2497)
Government delay in response to symptoms of problems (#PF6707).
Aggravated by Political impotence (#PJ1283)   Excessive public debt (#PC2546)
Secret international agreements (#PF0419)
Unrecognized future financial commitments (#PF4114).
Reduces Abuse of government power (#PC9104)
Unchecked power of government bureaucracy (#PD8890).

PF4670 Unused land
Underutilization of land — Deliberately unused land
Incidence Deliberate underutilization of land is practised by plantation owners who have a vested interest in securing as much available land as possible in order to withhold it from use and to ensure the dependence of local labour on the landowning class. Local workers are then obliged to work on the plantations on terms dictated by the landowners.
Broader Land misuse (#PD8142)   Restrictive use of available land (#PF6528)
Narrower Idle private land (#PJ8020)   Derelict industrial wastelands (#PE6005).
Related Excessive land usage (#PE5059).
Aggravated by Plantation agriculture (#PD7598).

PF4672 Unimplemented decisions
Unactivated policy
Broader General obstacles to problem alleviation (#PF0631).
Narrower Tokenistic meeting resolutions (#PF2086)
Symbolic international agreements without substance (#PE9267).
Aggravated by Analytical stagnation (#PF0440)   Inappropriate policies (#PF5645).

PF4674 Restrictive commercialization of definitions
Nature Due to copyright restrictions, any books defining terms, such as dictionaries and glossaries, effectively restrict further publication of those definitions, unless permission is acquired from the original publisher. To avoid the economic consequences, producers of new dictionaries and glossaries must ensure that the definitions that they use are somewhat different from those of other publishers. A definition is intellectual property. Variant definitions are therefore constantly produced to avoid infringing on the property rights of others, thus destabilizing the knowledge base.
Broader Monopolization of knowledge (#PF5329)
Misappropriation of cultural property (#PE6074).

PF4676 Inadequate development of international criminal law
Lack of an international criminal court — Lack of an international code of crimes
Broader Law (#PU3744).
Aggravates Non-recognition of international law (#PF9081)
Inadequate international law enforcement (#PF8421).
Aggravated by Deficiencies in international law (#PF4816)
Conflicts of national law in relation to international transactions (#PF9571)
Government refusal to accept the jurisdiction of international courts of justice (#PF7897).

PF4677 Limits
Limitations — Constraints
Nature Whether in natural or human resources, there is a point beyond which "more" is really not feasible at a given time. This applies not only to the limits of natural resources available to be exploited on the planet, but also to the capacity of urban areas for expansion, of industries for growth, of individuals for rapid change. Although it is well to avoid drawing false limits to living out of a feeble imagination or weak faith, it is equally foolish to pretend that existence is an inexhaustible well.
Narrower Limitation (#PF3861)   Limited job market (#PC7997)
Obstacles to national development (#PF4842).
Related Dietary restrictions (#PJ1933).

PF4681 Institutionalization of the disabled
Restrictions of the liberty of the disabled — Inadequate housing for the disabled
Claim Institutionalization restricts the liberty of disabled persons and prevents them from free association with their families and the rest of the society. Many institutions are situated in rural underpopulated areas and inmates are nearly always segregated by sex. Privacy is frequently restricted by ward-type living arrangements and 24-hour supervision; mail is often opened and telephone and other communications limited or not allowed. Institutionalized persons are commonly barred from marriage, voting and work, whether capable of those activities or not. These practices stunt normal growth and infantilize the victim. Normal challenges essential for the stimulation of learning and problem-solving capacity are replaced with artificial survival. Many institutions are understaffed and overcrowded, which enhance the likelihood of low staff morale, excessive reliance on drugs to control disabled persons. Even the best of institutions encourage disabled persons to become passive and dependent, developing an institutional personality, itself a disability, which makes it all the more difficult for a disabled person to re-enter society.
Broader Excessive institutionalization of vulnerable groups (#PF8209).
Aggravates Exclusion of disabled persons from social and cultural life (#PD0784).
Aggravated by Denial of rights to disabled (#PC3461)
Inadequate educational facilities for disabled persons (#PF0775).
Reduces Stress on families of the physically or mentally handicapped (#PD1405).

PF4682 Fear of crime
Incidence The highest degrees of fear of crime is shown by insecure, fearful, pessimist, introvert people.
Claim Fear of crime and the rise in crime feed each other. Fear keeps people off the street; empty streets make it easier for muggers to operate.
Refs Lewis, Dan A and Salem, Greta W *Fear of Crime* (1986).
Broader Fear (#PA6030).
Aggravated by Crime (#PB0001).
Reduces Fear of police (#PF8378).

PF4690 Unretrievable documents
Risk of missing documents — Inaccessible documents
Narrower Uncatalogued documents (#PF4077)   Misfiled documents and records (#PF4708).
Aggravates Lost knowledge (#PF5420).
Aggravated by Theft (#PD5552)   Secrecy (#PA0005)
Unrecorded knowledge (#PF5728)   Destruction of scientific records (#PF4633).

PF4699 Tampering with official documents
Tampering with evidence — Tampering with scientific records
Nature Knowingly making a false entry, altering in a false way a government record or document or knowingly destroying, concealing or in way modifying the authenticity of a government document is a crime.
Broader Evasion of issues (#PF7431)   Falsification of public records (#PD4239)
Unethical documentation practices (#PD2886).
Related Destruction of scientific records (#PF4633).
Aggravates False evidence (#PF5127)   Denial of evidence (#PD7385)
False political evidence (#PD3017).
Aggravated by Inadequate evidence to convict known offenders (#PF8661).

♦ **PF4705 Discrepancies between principles and practice**
Narrower Inadequate enforcement of safety regulations (#PD5001)
Differences in trading principles and practices between developing countries (#PF2960).
Related Differences in trading principles and practices between different economic systems (#PC2952).
Aggravates Undemocratic social systems (#PB8031)
Jeopardization of universality (#PF5520)
Compromise as a betrayal of principles (#PF3420).

♦ **PF4706 Insoluble scientific problems**
Nature There is a class of scientific problems, of central importance to humanity's understanding and control of the environmental crises with which it is faced, which is insoluble, either in the present or in the foreseeable future.
Incidence There are classic mathematical problems and conjectures that have defied solution for decades and even centuries. There are however also scientific problems relating to the environment which cannot be solved. Examples of such questions include: the extent and timing of global warming given increasing levels of "greenhouse gases"; long-term effects of exposure to low-level concentrations of certain compounds or to low-level radioactivity; or the degree of dependence of the ecosystems on particular levels of biodiversity.
Broader Limitation of current scientific knowledge (#PF4014).

♦ **PF4707 Reappearance of aristocracy**
Refs Levy, Oscar *The Revival of Aristocracy* .

♦ **PF4708 Misfiled documents and records**
Broader Unretrievable documents (#PF4690).
Related Uncatalogued documents (#PF4077).
Aggravated by Unethical documentation practices (#PD2886).

♦ **PF4709 Electoral defeat**
Nature An unfortunate by-product of elections is that for each person elected, one or several candidates loses. The dialogue over issues and consideration of alternatives which an election campaign generates is abruptly stopped, and great resources are lost to public service.
Broader Defeat (#PA7289) Nonaccomplishment (#PA6662)
Resignation to problems (#PF8781).

♦ **PF4713 Pre-electoral political inertia**
Broader Political inertia (#PC1907).

♦ **PF4718 Reversal of development progress**
Lack of progress against underdevelopment — Declining rate of development
Incidence The economic conditions, and the social conditions dependent upon them, are deteriorating in much of the developing world. For most of the countries of Africa, Latin America and the Caribbean, almost every economic signal points to the fact that development has been reversed. Per capita GNP has fallen, debt repayments have risen to a quarter or more of all export earnings, share in world trade has dropped, and the productivity of labour has declined by one or more percentage points each year throughout the 1980s. The developing world still depends on raw materials for the majority of its export earnings. But during the 1980s, real prices for the developing world's principal commodities (including minerals, jute, rubber, coffee, cocoa, tea, oils, fats, tobacco and timber) have fallen by approximately 30 per cent below their 1979 levels.
With the fall in new commercial lending, consequent on the debt problem, and the inadequate and static levels of official aid, developing countries have no possibility to remedy their situation. Although it is still widely believed in industrialized countries that money is flowing from richer nations to the developing countries to assist in their struggle against poverty. This has not been true since 1979, at which time a net $40 billion flowed from the northern hemisphere to the nations of the south. In 1989, the southern nations are now transferring at least $20 billion a year to the north (taking into account loans, aid, repayments of interest and capital movements). If account were to be taken of the effective transfer of resources implied in the reduced prices paid by industrialized nations for the developing world's raw materials, then the annual flow from the poor to the rich might be as much as $60 billion per year.
Claim Development's reversal cannot easily be captured by the media. It is happening not in any one particular place, but in slums and shanties and in neglected rural communities in different continents. It occurs not at any one particular time, but over long years of increasing poverty which cannot easily be reported in the media. It is happening not because of any one visible cause, but because of an unfolding economic drama in which the industrialized nations play a leading role, although the reversal of hard-won gains is largely invisible in those countries. The spread of hardship and human misery is now occurring on a scale and is of a severity unprecedented since the 1940s. Without restoring the forward momentum of economic development, it will become increasingly difficult to sustain progress against other problems, let alone to accelerate it. In many countries today, social advance is like trying to walk up an escalator which has begun to travel downwards. If per capita incomes continue to decline, then any progress will be eroded if not completely compromised.
Statistics fail to capture the psychological dimensions of what is happening. For several decades there had been forward movement in most countries. Even though poverty continued to be pervasive, more people were finding better jobs than ever before, and an increasing share of the population was gaining access to clean water, education, and medical care of some sort. Parents saw their children has having a better start than themselves. This has been brought to a halt. Indeed the physical deterioration in basic infrastructure,, including schools and hospitals, and the mounting excess of unemployed and underemployed, will call for more than a weak increase in economic growth if hopes are to be rekindled. And these pent up needs continue to increase as investment remains depressed.
Counter-claim Economic progress continues in Asia, despite containing the majority of the world's absolute poor. Most of its nations are continuing to see average incomes slowly rising and average living standards slowly improving.
Narrower Disadvantageous terms for technology transfer (#PE4922).
Aggravates Underdevelopment (#PB0206).
Aggravated by Destruction inherent in development (#PF4829)
Inadequacy of aid to developing countries (#PF0392)
Inadequate level of investment within developing countries (#PD0291)
Burden of servicing foreign public debt by developing countries (#PD3051)
Lack of progress in establishing a New International Economic Order (#PF4306)
Instability in export trade of developing countries producing primary commodities (#PD2968).

♦ **PF4722 Complex trade regulations**
Broader Complex government regulations (#PF8053)
Excessive government intervention in the private sector (#PD4800).

♦ **PF4723 Excessive leniency in sentencing of offenders**
Politically motivated reduction of prison sentences — Sentence squashing
Narrower Remission of sentences for crimes against humanity (#PF1098).
Aggravated by Legal inconsistency (#PF5356)
Government failure to prosecute offenders effectively (#PE9545).
Reduces Disproportionately long prison sentences (#PE4602).

♦ **PF4727 Illegality of nuclear weapons**
Nature Nuclear weapons are in violation of international law. It is prohibited to use weapons or tactics that cause unnecessary and/or aggravated devastation and suffering. It is prohibited to use weapons or tactics that cause indiscriminate harm as between combatants and non-combatant military and civilian personnel. It is prohibited to effect reprisals that are disproportionate to their antecedent provocation or to legitimate military objectives, or that are disrespectful of persons, institutions, and resources otherwise protected by the laws of war. It is prohibited to use weapons or tactics that cause widespread, long-term and severe damage to the natural environment. It is prohibited to use weapons or tactics that violate jurisdiction of non-participating States. It is prohibited to use asphyxiating, poisonous or other gases, and all analogous liquids, materials or devices, including bacteriological methods of warfare.
Broader Unlawful government action (#PF5332) Secret military operations (#PF7669)
Inhumane and indiscriminate weapons (#PD1519).
Related Crimes against humanity (#PC1073).

♦ **PF4729 Unreported harassment**
Unreported intimidation — Unreported threats — Unreported bullying
Broader Harassment (#PC8558) Intimidation (#PB1992)
Unrecorded knowledge (#PF5728).
Related Underreported issues (#PF9148).
Aggravated by Fear of death (#PF0462) Avoidance of negative feedback (#PF5311)
Use of undue influence to obstruct the administration of justice (#PE8829).

♦ **PF4730 Lack of relationship between wealth generation and the public good**
Nature The huge fortunes of financial deals have transformed the prevailing attitudes to money over the decade, particularly in the USA. Moneymaking has became separated from public benefit and productive reality.
Related Accumulation of capital (#PC5225).
Aggravates Personal wealth (#PC8222) Decadent standard of living (#PD4037)
Inequitable distribution of wealth (#PB7666).
Aggravated by Avarice (#PA6999).

♦ **PF4735 Criminalization of drug use**
Illegality of drug use — Prohibition of addictive drugs
Nature Making the use of drugs illegal has created a criminal class, increased the profits of organized crime to the point of controlling governments and increased the crime rate. Federal, state and local governments in the United States spend more than 8 billion U.S. Dollars a year on the costs for police, courts and prisons for drug related crime. The nations of Columbia and Bolivia are increasingly being controlled by drug gangsters. Corruption is increasing where ever drug gangs are operating.
Claim The consequences of making a drug illegal are that it will only be available on the black market, usually adulterated, of uncertain dosage, and dangerous to consume. Although such use makes people criminals it does not stop them using drugs. The substances themselves do not make people lie, cheat, steal, prostitute themselves, and become covered in abscesses. Illegal drugs do much the same as legal ones: coffee, alcohol, tobacco and various stimulants and depressants available over the counter or by prescription. The difference is legislative, not pharmacological. Banned drugs' quality and purity is controlled by those who produce it. As long as a drug is banned it cannot be taxed. In legitimate commerce, their sale controlled, taxed and supervised, with their dangers proclaimed on every packet, drugs would poison fewer customers, kill fewer dealers, bribe fewer policemen, raise more public revenue. The drugs which are illegal cause a fraction of the number of deaths, create less illness, and cost health services less than alcohol and tobacco. Criminalization tends to inhibit people from getting access to services to assist them.
Counter-claim The costs to society would increase if drugs were legalized, especially those like cocaine. To effect the black market, legal drugs would have to be cheaper, leading to wider use. Hospital cost would rise. The prescribed maintenance doses of heroin programme of Britain failed. Cheaper and more easily obtained cocaine would lead to heavier use and an increase in the incidents of depression, paranoia and violent psychotic behaviour.
Refs United Nations *Extradition for Drug-Related Offences* (1985).
Broader Inadequate laws (#PC6848) Restrictive legislation (#PD9012).
Related Criminalization of prostitution (#PF6231).
Aggravates Inadequate drug control (#PC0231).

♦ **PF4739 Proliferation of public sector institutions**
Proliferation of government programmes — Proliferation of state-controlled enterprises — Proliferation of parastatals
Nature Slow growth, lagging private savings and investment, high inflation, balance of payments deficits, heavy debt burdens, continued poverty and unemployment are, in part, the result of excessive growth of the public sector. Even in those cases where external events beyond the control of individuals countries are the cause of many difficulties, the actions of governments are often inappropriate and fail to mobilize effectively the resources of a country in response to a crisis.
Incidence In developing countries the urgent infrastructure needs, the low levels of savings and investment, the need to foster economic growth through modernization, and the availability of concessional funding for public projects have all served to encourage the rapid expansion of public sector programmes. It is only with the economic stagnation resulting from the debt crisis that the weaknesses of this approach have been acknowledged. The late 1970s also marked a turning point in the centrally planned economies, where reliance on direct command by government and the use of state-controlled enterprises was increasingly seen as a drag on economic growth. By the end of 1989 the inadequacies of this approach, and the need to harness private interest and initiative to stimulate economic growth, were widely recognized.
Claim Although the pursuit of private interests allocates resources efficiently in competitive markets, this generally does not occur when individuals use the monopolistic power of government to their own advantage. Politicians, bureaucrats, and many private interests gain from growing government involvement in society and greater government expenditure. Government's necessary role as a provider of public goods needs to be carefully circumscribed otherwise inefficient public and private provision of goods and services results.
Counter-claim Since World War II the growing importance of the public sector is seen by many development economists and policy-makers as a natural and necessary ingredient of development. Governments need to intervene to foster development since the unmodified interaction of private agents does not achieve the goals of economic efficiency, growth, macroeconomic stability, and poverty alleviation. Imperfections in the action of free markets fail to meet these and other needs. Any mistakes made by government may indeed be a serious problem in practice, but are not inevitable or irreversible. Policy and administrative reforms can be elaborated to correct such inadequacies.

PF4739

Aggravates Official apathy (#PF9459)
Fear of officialdom (#PD9498)
Government arrogance (#PF8820)
Administrative delays (#PC2550)
Government complacency (#PF6407)
Corruption in politics (#PC0116)
Government inefficiency (#PF8491)
Bureaucratic corruption (#PC0279)
Government insensitivity (#PF2808)
Undemocratic policy-making (#PF8703)
Bribery of public servants (#PD4541)
Overdependence on government (#PF9530)
Complex government bureaucracy (#PF8539)
Official evasion of complaints (#PF9157)
Corruption of government leaders (#PC7587)
Frustrated talent in government posts (#PE4398)
Unchecked power of government bureaucracy (#PD8890)
Inappropriate public spending by government (#PF6377)
Inefficiency of state-controlled enterprises (#PD5642)
Conflict between government and the news media (#PE1643)
Governmental disregard for people as human beings (#PD8017)
Inadequate system of political checks and balances (#PE4997)
Wastage in governmental budgets and appropriations (#PD0183)
Excessive government intervention in the private sector (#PD4800)
Abuse of monopoly power of state-owned or state-controlled enterprises (#PE0988)
Distortion of international trade by state-trading and government monopoly practices (#PE8267)
Jurisdictional conflict and antagonism between government agencies within each country (#PE8308).
Aggravated by Economic inefficiency (#PF7556) Excessive government control (#PF0304).
Reduced by Privatization of public services (#PE3391).

♦ PF4741 Technological monoculture

Dependence on dominant technology — Dependence on single technology — Restriction of the technical basis of competition — Self-limiting technical innovation — Reduction of technological diversity — Self-perpetuating attitudes in technological innovation

Nature Following the emergence of a fundamentally new technological innovation, the structure of the industry established around it is characterized by a high degree of diversity and experimentation primarily focused on product performance rather than price. Through competition a particular dominant approach emerges which increasingly concentrates on incremental improvements to the core technology emphasizing cost, reliability and standardization. The success development of products based on this technology steadily narrows the technical basis for competition even though it becomes more intense. As a result technical alternatives are ignored or remain unexplored even though they might prove inherently superior in terms of cost or performance. The cost advantage of continued dependence on the dominant technology precludes attention to other possibilities. It also discourages attention to technical factors in that technology that might affect higher order social or environmental impacts. But as the dependence on that technology becomes increasingly widespread new difficulties resulting from the scale of its application become important, just when the kinds of research and development capable of anticipating such problems have been phased out as irrelevant to the further commercial success of the technology. And yet it is often at this point that the technology becomes vulnerable to unexpected side effects which can generate social reactions against it. Those controlling the application of the technology may then consider it appropriate to counteract such concerns by more sophisticated marketing or by more active measures. But any short-term success with such a response may prove to be counter-productive to the long-term interests of the corporations concerned by delaying the adaptive measures they will eventually be forced to take.

Incidence Examples include automobile, computer, plastics, pharmaceutical nuclear, and industrialized agriculture technologies.

Claim Corporations, governmental agencies and university faculties which determine the new arenas of technological innovation in which to expend their human and financial resources, operate from a self-perpetuating attitude. Because of inadequate sources of information, the absence of a global policy on innovation, fear of scarcity of resources, and determination to succeed whatever the implications for other sectors, decisions are often channelled in directions harmful to the welfare of society as a whole.

Broader Non-inclusive management decisions (#PF2754).
Related Dependence on sophisticated technology for development (#PD6571).
Aggravates Fixation on partial solutions to problems (#PF9409)
Static grassroots involvement in planning the economy (#PE4479).
Aggravated by Proliferation of technology (#PD2420).

♦ PF4750 Surplus

Broader Insufficiency (#PA5473).
Narrower Surplus labour (#PG5971)
Large trade surpluses (#PJ0207)
Agricultural surpluses (#PC2062).
Reduced by Excessive consumption of goods and services (#PC2518).

♦ PF4753 Immoral public policy

Government approval of addictive behaviour — Government approval of dangerous behaviour

Broader Immorality (#PA3369).
Aggravates Smoking (#PD0713)
Gambling and wagering (#PF2137)
Consumption of alcoholic beverages (#PD8286).
Aggravated by Government hypocrisy (#PF9050).

♦ PF4754 Separation of individual's social functions

Nature Local structures which reenforce the individual's sense of having meaningful social roles have collapsed creating a separation of social functions. The loss of the family's historical roles, the loss of significance of the local church, and loss of community at the neighbourhood level have all examples of the failure of local structures.

Broader Media reinforcement of materialism (#PF1673).

♦ PF4756 Confusing decision-making methodologies

Nature The methods for making decisions being taught to individuals are confusing, shallow and often lead to unhelpful results. This is true if one is making decisions about oneself, one's family or participating in the political processes.

Broader Individual isolationism (#PD1749).

♦ PF4761 Politicization of international sports events

Excessive political intervention in international athletic exchanges — Politicization of international sports competitions — Political exploitation of Olympic Games — Nationalist exploitation of sporting events

Nature International sporting events of an official nature, that is those requiring government representation, accreditation or support, have been increasingly perceived as a vehicle for ideological and nationalist propaganda. As a result, national governments may intervene in, or try to influence, anything from the venue of the games, their arrangements, contents and rules and the training programmes of their own teams, to the constituents of these teams, that is, to the selection of the competitors and to the provision for extra incentives, or even punishments. The political abuse of world-wide sporting events can lead to their cessation or interruption. On the other hand, solution of this problem does not necessarily lead to the elimination of political exploitation of regional, national and national inter-regional, inter-linguistic and inter-ethnic games, or sporting-events associated with particular classes of society.

International sports events are intended to being together competitors from different countries to compete in honest games - "may the best man win" - without regard to his race, country of origin, or political ideology. As the world gets tenser and increasingly politicized, anger and revenge often leave the political arena and enter into sports, pulling down with them the hopes and aspirations of the sportsmen who have spent years, or even a lifetime, in preparation. Spectators, too, become pawns, as governments take away their opportunities to relax and enjoy wholesome entertainment.

Incidence The 1936 Olympic Games which the hosting Nazi-ruled Germany used to propagate its racial and national theories of supremacy were a mild example compared to the horrors of the multiple murders at the 1972 Games at Munich in the name of a foreign political cause. Third World champions withdrew from Montreal in 1976 in protest against South Africa; the USA boycotted the 1980 games in Moscow as a result of the Soviet invasion of Afghanistan; and the Soviets retaliated by boycotting the 1984 Olympics hosted by the USA in Los Angeles.

Sports laws policies, and random acts of intervention by countries vary in scope and kind. To influence the international outcomes, governmental support of athletic programs and exchange is now almost universal. The Scandinavian governments guarantee employment to cross-country skiing competitors, but this modest influence is dwarfed by the massive programmes for nationalist sporting propaganda engaged in by the German DR and the People's Republic of China. A college-bonus system of support is used in the United States, and a cash bonus and military service relation system is employed by the Soviet Union. The use or boycotting of international sporting competitions to apply political pressure on South Africa was a new development wherein alignments of nations for political exploitation of widely viewed athletic events were made for the first time.

Background The oldest of the modern institutionalized international sports competitions is the Olympic Games founded in the late nineteenth century. These athletic interchanges antedate the institutionalization of exchanges of scholars and artists and were the first to be exploited.

Claim No nation should regulate international athletic exchange to its political advantage. The Olympic Games and other international competitive sporting events should be unpolluted by political currents. Athletic exchange, like other forms of international interaction, should be protected against governmental intervention.

No future Olympics should be hosted by a superpower. Given the increasingly tense atmosphere of world politics, it is inevitable that the Games will sometimes by used as weapons. The Olympics should instead by entrusted to lesser powers who could content themselves with staging an event that brings nations together and provides role models for the millions who see in the broad jumper, the swimmer, or the gymnast only the perfection brought about by discipline and aspiration, and not the ideology expressed by the government he or she represents.

Counter-claim It is not possible to divorce sport and politics, but it must be possible to protect sports from political exploitation. The influence of commercial intervention can be a balancing, or in some countries like the USA, a predominant factor in restricting undue political influence. Governmental involvement can prevent commercial exploitation. Voluntary private support of international competition vies in some countries with government and commerce for control.

As long as the Olympics is organized along national lines, i.e. national committees, national representatives to the International Olympics Committee, and national team sports; national governments will intervene to protect national prestige.

Broader Obstacles to community achievement (#PF7118)
Obstructions to international athletic exchange (#PE4809)
Exploitation of athletic competition for commercial or political ends (#PE4833).
Narrower Political boycott of international sports events (#PE4206)
National bias among judges of international athletic competitions (#PF4216).
Related Obstacles to international cultural exchange (#PF4857)
Corruption of sports and athletic competitions (#PE3754)
Obstructions to international personnel exchanges and cultural cooperation (#PE4785).
Aggravates Athletic competition (#PE4266)
Excessive claims for human development through sports (#PG4881).

♦ PF4763 Secrecy of national basic food stocks

Nature A number of countries have chosen not to divulge quantities of grain and other basic food stocks on hand, thereby placing obstacles in the way of the establishment of world-wide food information early-warning systems, effective grain and other food programs, and cooperative stock planning.

Broader Secrecy (#PA0005).
Aggravates Mismanagement of food resources (#PE6115).

♦ PF4764 Disobedient wives

Incidence In traditional communities wives have been expected to obey orders from their husbands concerning: flirting with other men (including secretive conversations, suggestive gestures), leaving the house without permission (especially if the husband is asleep or intoxicated), and shutting the door against the husband. These issues are of special importance in certain modern cultures and sects, especially those with conservative views concerning the freedom of women.

Broader Disobedience (#PA7250).
Aggravates Adultery (#PF2314)
Decadent clothing (#PE5607)
Sexual harassment of men (#PE1293).
Aggravated by Sexual harassment of women (#PF3271)
Discrimination against women (#PC0308).

♦ PF4765 Inadequacy of formal education

Failure of mass education

Nature Systems of formal education are unable to satisfy diverse individual and collective needs for knowledge, skills and behavioural changes. In addition, enormous strain is being put on existing systems due to the vast increase in the school-going population, inadequate educational planning in relation to national requirements, rise in the cost of education per student, and financial constraints. On the other hand, mass media facilities, such as television, radio and films, paperbacks and low cost publications, libraries, and evening classes, have become more accessible. All these favour the growth of the non-formal system of education.

The national drive for education imposes new standards for evaluation of self and others; creates new situations of competition, uncertainty and conflict; and gives a new shape to the life course of men and women. Most of the costs stem from the occupational prestige hierarchy for which schooling prepares and qualifies its pupils. This hierarchy of occupations graded in earnings and social respect, tends to replace the status systems of agrarian societies and impose a single uniform standard of socio-economic evaluation based on the superiority of professionals and managers over clerks and the superiority of clerks over manual workers. In European history, the rise of this occupational hierarchy was considered democratizing, as a merit system base on individual skill that replaced a feudal system based on hereditary status. In partly mobilized societies, it may work rather differently, as when the third world farmer discovers he is a peasant or a skilled craftsman discovers he is an illiterate manual worker. The discovery that one is at the bottom of the hierarchy can come as a rude shock. This is mitigated in the partly mobilized society by the existence of agrarian communities with their alternative values and social supports. In the fully mobilized societies of the industrialized world, alternatives are no longer available and the costs are less avoidable and more sharply felt. The sense of relative deprivation is keenest when everyone who goes to school is led to believe he or she has a chance to get to the top, only to find that competitive examinations eliminate the majority. Mass education awards certificates of

failure to the great majority of people. In societies where alternative social identities are not available life chances are few and most citizens are relegated to minimally significant roles.
Incidence A French survey found that children who were quite capable of reading, writing and computing nevertheless could not comprehend the meaning of a short text or resolve basic problems. The cultural setting of both the literature and the mathematical problems was so alien to students that their basic skills were rendered useless. It is this growing gap in shared cultural context which education systems teach only by inference, because until quite recently a classroom could be expected to be culturally homogeneous.
Background The schools and universities have taken over and abstracted many ways of learning which in earlier times were always closely related to real life. Adults did their economic work and other social tasks; children were not excluded, were paid attention to, and learned to be included. The children were not formally 'taught'.
Claim Schooling, that is, deliberate education, has little effect on vocational ability or citizenship. Grades have little correlation with life achievement in any profession. Learning from lectures and books is dry and dull; children and students become passive and unable to think or act for themselves. By and large, though not for all topics and all persons, the incidental process of education suits the nature of learning better than formal teaching.
Broader Inadequate education (#PF4984).
Narrower Maldistribution of teachers (#PE6183) Inadequate secondary education (#PD5345)
Divisive effects of formal schooling (#PF3836)
Inadequate results of formal schooling (#PF6467).
Related Narrow scope of education (#PF3552)
Unavailable education for effective living (#PF2313)
Ineffective systems of practical education (#PF3498).
Aggravates Educational wastage (#PC1716)
Inadequate recognition by institutions of the transition through adolescence (#PF6173).

♦ PF4767 Obstacles to the utilization of coastal and deep sea water resources
Nature A new era of ocean politics has been triggered by dramatic technological breakthroughs in ocean technology. Man is now more capable of using the ocean's surfaces, deeper waters, and bottoms than ever before, so consequently, worldwide commercial and political rivalries are intensifying over rights to ocean space and ocean resources. The difficulty of accommodating all the ocean's users and uses has made it a highly contentious issue among nations, transnational corporations, and special interest groups.
Claim All ocean users are politically, physically, and economically interdependent and thus must accept international accountability. Obstacles that must be overcome include: defining seaward limits of national sovereignty and spelling out the rights to ocean resources beyond these limits; allocating privileges between coastal states and others wanting access to their adjacent offshore areas beyond national limits; assuring that powerful maritime nations will have minimum interference and maximum stability for conducting all their ocean activities; and granting less powerful nations regimes from which they can benefit in common.
Narrower Marine pollution (#PC1117) Ice-blocked seaways (#PD2498)
Sea traffic congestion (#PD1486) Obstacles to aquaculture (#PF5496)
Conflicts over fishing rights (#PD5361)
Environmental hazards from fishing industry (#PD0743)
Militarization of the deep ocean and sea-bed (#PD1241).
Related Inadequate laws of the sea (#PF5923) Shortage of fresh-water sources (#PC4815)
Destructive changes in ocean characteristics (#PC2087)
Lack of coastal development in island countries (#PE5689).
Aggravates Unsustainable development of coast zones (#PD4671).

♦ PF4770 Uncertainties in animal experimentation
Nature Reaction of a mammal to a given noxious agent is an inheritable property and there are known substantial differences even among particular strains of the same species. One example is the existence of different strains of mice especially raised to be particularly (and uniformly) sensitive to insults by certain chemicals. Another example is the existence of certain tumours that are common with some animals and unknown in others. Such is the case with thymic lymphoma: very common in certain mouse strains, but apparently with no real analogue in humans. Reverse situations undoubtedly exist. The inference is that the predictive value for human hazard varies for different animal experiments. Although the discovery of any toxic effect points to the need for close medical supervision of exposed persons in order to detect similar changes in man at an early stage, accurate quantitative extrapolation to man of the results of animal experiments is not feasible.
Broader Inconclusiveness of science (#PF6349)
Obstacles to medical experimentation (#PF4865).
Related Denial to experimental animals of the right to freedom from suffering (#PE8024).

♦ PF4774 Inaccurate forecasting
Unrealistic forecasts — Irrelevant predictions — Surprise-free forecasts
Nature Forecasting and planning for the future in quantitative terms is the activity of almost everyone, as it is a virtual necessity in private life as well as in the management of organizations and activities of much larger dimensions. However, there are two very great differences. The first is of scale. Error in personal forecasting may occasionally be disastrous, but error in large-scale planning for the future is almost certainly disastrous. Military defence, jobs requirements, and energy and food production planning, are some of the areas where error has serious ramifications. The second difference between personal and large scale planning is of kind and method. Simply stated, the enormity of the variables and growth which is proportional to the growth of the size of the plan, both as it extends in time and in elements contained, requires increasing levels of methodological sophistication. One popular technique is to develop alternate scenarios, or at least those that can be typified as optimistic and pessimistic so that a mid-range of expectations can be established. These scenarios can be identified in part by statistical evaluations of probability. The lack of application of the statistical laws of probability to the possibilities of discontinuities in trends, to interferences in processes, and to the possible aggregate error in the total plan's content of calculations and assumptions leads to exaggeratedly optimistic or pessimistic forecasting.
Claim 1. Predictive efforts concerning future man-environment impacts are technically unfeasible and irrelevant. The are unfeasible because of the complexity of the environmental syndromes arising from interactions with development, the dependence of these interactions on unknowable future patterns of social choice and evolution, and the incomplete state of the sciences required for their assessment. They are irrelevant because there is little serious demand for detailed predictions on the part of experienced politicians, businessmen or managers. Such action-oriented people are much more interested in understanding how they might intervene to make what actually does happen more to their liking and benefit.
2. Inaccurate forecasting can lead to the most serious problems, from hunger to war. Errors in planning in those military sector contribute to the armaments race and divert the economic resources which could build better societies. The over-blown military budgets of the Pentagon and the Kremlin, of NATO and the Warsaw Pact, and those of some smaller, richer nations in South America, Africa and Asia, are based on inefficient financial planning and irrational expectations of the inevitability of conflict. The financial costs incurred from faulty forecasting, by the US Pentagon alone, would be sufficient to run the programs of an agency such as FAO into the 22nd century.
Counter-claim However prone to error and uncertainty, long-term forecasts are necessary for investment, research and policy decisions in both public and private sectors.
Refs Schnaars, Steven *Megamistakes* (1989).
Broader Inconclusiveness of science (#PF6349).
Narrower Commodity speculation (#PD9637) Incomplete cost projections (#PJ8109)
Inaccurate weather forecasting (#PF5118) Unpredictability of earthquakes (#PF4928).
Related Unpreparedness (#PF8176).
Aggravates Unrealistic policies (#PF9428) Unrealistic expectations (#PE7002)
Overstated programme advantages (#PF8181) Unforeseen environmental crises (#PF9769)
Underestimation of programme costs (#PF8499)
Irresponsibility towards future generations (#PF9455).
Aggravated by Economic uncertainty (#PF5817) Inappropriate assumptions (#PF6814)
Limitations of surprise-free thinking (#PF7700)
Limitation of current scientific knowledge (#PF4014)
Excessive confidence in prediction capacity (#PF7166)
Failure of government intelligence services (#PF8819)
Complex interrelationship of world problems (#PF0364)
Unknowable future patterns of social choice (#PF9276)
Economic dislocations in developing countries (#PD4063).
Reduced by Self-fulfilling prophecies (#PF9694).

♦ PF4778 Pseudo-socialism and state socialism
Nature States that are nominally socialist, implying that democratic principles are adhered to in political decision-making, are not really so when the state governmental or party apparatus makes all fundamental decisions. Without referenda from the people on issues, full elections of officials, and self-management by organizations of labour and other productive classes, there exists a statist political ideology that wishes to draw all power to the centre.
State socialism, a bourgeois-reformist, opportunist concept, according to which socialism is reduced to state intervention in the economy and in social relations. It is without any true socialist content. It arises from bourgeois lies, which call any attempts to restrict free competition, socialist. It is a petit bourgeois illusions of utopian socialists, who see socialism being introduced by the government and ruling classes.
Background Pseudo-socialism can lead to fascism, as was the case in Nazi Germany. The Dutch tried to maintain their colonial presence in Indonesia through attempts at state socialism.
Broader Socialism (#PC0115).

♦ PF4785 Obstacles to redeployment of military resources for peaceful applications
Rigidity in rechannelling reduced expenditure on defence — Peace recession
Nature Resources currently employed in defence industries, weapons research and military activities cannot be redeployed quickly or easily into other economic sectors or to other countries in the form of increased aid. The obstacles to such redeployment include the way in which many jobs are currently tied to such expenditures, especially in economies with high unemployment.
Incidence In the case of the USA, dependence of the economy on military expenditure has created fears that a peace dividend might cause a peace recession. Approximately 20,000 defence contractors (and 100,000 sub-contractors) are engaged on defence projects with some 8 million Americans receiving salaries associated with some form of military activity. Advocates of increased defence spending argue that it leads to more jobs and a healthier economy.
Claim It is a myth that smaller armies are cheaper. Smaller armies need to be better trained, better equipped and better paid. Restructuring of the military into a smaller configuration can be achieved but the cost of the result will remain the same.
Counter-claim For every billion dollars spent on military procurement in the USA, it has been estimated that 28,000 jobs are created. The same expenditure would create 32,000 jobs in public transportation and 71,000 jobs in education.
Refs Sharp, Jane M O (Ed) *Europe After an American Withdrawal* (1990).
Aggravates Deliberate invention of new threats (#PF3699).
Aggravated by Socio-economic burden of militarization (#PF1447).

♦ PF4787 Unfairly negotiated treaties
One-sided agreements — Inequitable treaties — Inequitable peace agreements — Humiliating treaties
Broader Political inequality (#PC3425).
Narrower Inequitable tax treaties between developed and developing countries (#PD1477).
Aggravates Unjust peace (#PB7694) Unfulfilled treaty obligations (#PF2497)
Expropriation of land from indigenous populations (#PC3304).
Aggravated by International inequality (#PC9152)
Inadequately worded agreements (#PF5421)
Secret international agreements (#PF0419)
Limited acceptance of international treaties (#PF0977).

♦ PF4794 Reduced scope of intergovernmental development assistance
Nature The post World War II developmental efforts through the United Nations Organization's Development Programme were based on the premise that the greatly differing developing countries were themselves the best judges of their own development priorities. The last ten years, however, have seen a tendency by some donor governments to either tie their share of multilateral assistance to their own perception of the priorities, or to limit their support in the multilateral context, in favour of increasing bilateral aid and special purpose arrangements.
Broader Inadequacy of aid to developing countries (#PF0392).
Narrower Restrictions imposed on aid to developing countries (#PF1492).
Related Prohibitive cost of maintaining comprehensive document collections (#PE1122)
Shortage of financial resources within the United Nations system of organization (#PF1460).
Aggravated by Bilateralism in aid to developing countries (#PE9099).

♦ PF4799 Inappropriate modernization of agriculture
Nature The modernization of agriculture, as currently envisaged, has a marked tendency to enrich a limited number of wealthy landowners who can afford to invest in high-yield seeds and who can afford the expensive fertilizers required if such seeds are to perform satisfactorily. When these farmers prosper, they then buy out poorer farmers or terminate the leases of peasants on their property in order to expand the cropped areas. Such modernization has therefore been a major factor in increasing the number of landless peasants and encouraging migration to urban slums. In addition it has increased dependence on imported technology and on imported energy (to operate the mechanical equipment). The final agricultural products tend to be more expensive than those produced by traditional methods and consequently become too expensive for poor people to purchase.
Claim Modernization of agriculture can mean little more than converting land from production by the poor for the poor, to production by wealthy landowners for use by the wealthy in developing countries and by consumers in the industrialized countries. As such it may have actually increased the amount of hunger in certain areas of the world. Such agriculture is in effect soil mining in that the nutrients in each crop, taken from the soil, are discarded (or exported) with long-term irreversible damage to the soil.
Narrower Chemicalized farming (#PD7993) Plantation agriculture (#PD7598)
Enforced collectivization of agriculture (#PD7443)
Introduction of high-yield crop varieties (#PF3146).
Related Over-subsidized agriculture in industrialized countries (#PD9802).

Aggravates Maldevelopment (#PB6207) Destruction of the countryside (#PE3914)
Energy dependence and vulnerability (#PJ7735) Landlessness in developing countries (#PC0990)
Destruction of hedges and hedgerow trees (#PD1642)
Lack of purchasing power in developing countries (#PE8707)
Dependence of developing countries on imported technology (#PF1489)
Maldistribution of land associated with large traditional estates (#PD0406).

♦ PF4801 Inadequate environmental monitoring

Nature Man's health is being impaired by activities which contaminate the air, water, food, soil and biota. These activities include harmful discharges resulting from uncontrolled industrial development, the indiscriminate use of pesticides and fertilizers and the poor or non-existent planning of water supply and waste disposal in human settlements, resulting in certain endemic diseases. People are increasingly exposed to overcrowding and noise, to an atmosphere polluted by smoke and exhaust fumes in towns and cities, to chemical and physical hazards at work to food that may contain residues of insecticides and fungicides, and to polluted water in many rivers and lakes and even the oceans. In spite of all that has been done to improve basic environmental conditions, the pollution of much of the world's natural water resources by sewage, household refuse, industrial effluents and agricultural wastes, and run-off containing organic matter, pesticides and fertilizers, makes it more and more difficult to provide growing populations, or even relatively static ones, with enough safe water to meet their needs.
Incidence Concern over the impact on human health of chemicals in the environment has increased in recent years (UNEP, 1977, 1978, 1980, 1981). It has become clear, for example, that a number of chemicals act specifically on the heart muscle and thus contribute to the development of heart diseases. Living and working conditions and activities, in conjunction with other stressors, can also cause an increase in morbidity rates of (for example) ischaemic heart disease, metabolic disorders and nervous conditions. Relationships with environmental factors have also come to light, such as that between soft water and the incidence of heart diseases, fluorine deficiency and excessive dental decay, or increased arsenic in soil and water and the enhanced occurrence of cancer. In evaluating human exposure to potentially hazardous chemicals it is therefore necessary to consider all pathways – in air, water, soil and food, and both the home and working environment. The rate of circulation of many elements through the environment has been greatly increased by man's activities. In the mid-1970s over 1,500 substances were produced in quantities exceeding 500 tonnes per year (worldwide), and over 50 substances were produced in quantities over 1 million tonnes. In 1950, world production or organic chemicals was 7 million tonnes; in 1970, 63 million tonnes; and the 1985 figure is expected to be 250 million tonnes. Despite increasingly stringent industrial controls, some substances that now occur in negligible concentrations (say less than one part per million) may present serious problems to the next and succeeding generations.
Background The system of government of a country, its geographic characteristics and size, its political philosophy, its economic situation and its stage of development, affect the way that environmental health services have developed and how they function. In contrast to public health and food safety services, which have a history stretching back more than a century, the problems of the environment as a whole have been recognized for only about two decades. Certain aspects of the environment, notably water pollution and emissions from chemical works, have been the subject of legislation that dates back to the 19th century but only recently has an attempt been made in a few countries to consider the environment as an integrated whole. Many countries are trying to consolidate responsibility for environmental services with varying degrees of success. While protection of the environment in theory works to the general benefit, the measures taken to this end are inevitably considered detrimental to certain interests and will therefore be resisted. The result, in many cases, has been a highly fragmented set of laws and regulations with exceptions and special conditions introduced to secure overall approval.
Because monitoring is often divided among various agencies, none has overall responsibility for coordinating the results or setting standards to cover all the routes of intake (air, water, food, tobacco, occupational exposure, etc) of toxic substances into the body. Only rarely are standards based on local conditions of exposure backed up by the biological monitoring of tissue and epidemiological surveillance. Indeed, the trend is to rely on multinational standards or guidelines, employing a safety factor to cover local variations. Standards for lead in water, air, petrol, paint and food, for example, are the responsibility of up to five different agencies, each of which may monitor its own medium, while the monitoring of soil around the home – a major source of intake by smaller children – may be no one's responsibility. Many other examples of divided responsibility can be found. The reasons for dispersed authority lie in the legislative history of the countries surveyed. Agencies and ministries have been set up over the years on an ad hoc basis to deal with water, food, industrial resources, occupational exposure, etc, as each one came to be seen as a problem. The responsibility for monitoring was frequently assigned to an existing body that was already dealing with supply or distribution, sometimes with the requirement that the health aspects be coordinated with the ministry responsible for public health. Even when several agencies have been combined under one ministry responsible for the environment, certain strongly entrenched departments have successfully resisted efforts to transfer their activities to that authority. In fact, a division of responsibility may exist even within a ministry if it is organized with separate departments for air, water supply, waste disposal, etc.
In countries where development is at a relatively low level and where the problems of the human environment are closely linked to poverty and to the very lack of development itself, human health and life are continually jeopardized by biological pollution resulting from insanitary conditions, both in congested urban areas and in rural areas. The absence of safe water supplies and of facilities for the sanitary collection and disposal of human excreta, refuse and industrial waste, and the prevalence of vectors of disease, with the consequent high incidence of sickness from communicable diseases, are still the prime environmental problems in these developing countries.
Claim Environmental monitoring is a basic component of an environmental health programme. Water, food and air can all be vehicles for the transmission of infection or toxic contaminants to man. Regular monitoring is essential. A comprehensive legal structure is needed, with regulations that define permissible levels of contaminants, if the important follow-up action on unsatisfactory samples is to be effective.
Broader Ineffective monitoring (#PF2793) Hazards to human health (#PB4885).
Narrower Inadequate radiation monitoring systems (#PF6635)
Ineffective monitoring of hazardous substances (#PE5089)
Inadequate systems for monitoring industrial growth (#PF2905)
Ineffective official inspection of regulated activities (#PE4146).
Aggravates Health hazards from water development schemes (#PE8692).
Aggravated by Ineffective monitoring of illegal activity (#PF7264).

♦ PF4813 Denial of the right to die

Nature The right to die may be claimed by each individual on account of his or her liberty to decide. For example, terminally ill patients in the final stages of a disease who are experiencing intense pain may wish to die sooner. This desire, which could be fulfilled by the withholding of essential medications and life-support systems, is thwarted by attending physicians, family, and by the general attitude of society.
Refs Sloan, Irving J *The Right to Die* (1988).
Broader Denial of human rights (#PB3121) Denial of political and civil rights (#PC0632).
Narrower Criminalization of euthanasia (#PF2643).
Aggravates Excessive prolongation of the dying process (#PF4936).
Reduced by Human death (#PA0072).

♦ PF4816 Deficiencies in international law

Inadequate intergovernmental legal systems
Nature A judgement which is valid and enforceable in the country in which it is rendered may not be recognized or enforced against a person or property located in another country. Thus the administration of justice may be frustrated by national boundaries.
Background The acceptance of the International Court of Justice as an instrument of international justice has never been strong. In 1920 the Committee of Jurist which prepared the statute of the Permanent Court of International Justice, the forerunner of the International Court of Justice, had proposed a system of true compulsory jurisdiction based upon the unilateral application to the court by the complaining state. This idea encountered strong opposition, particularly from the great powers of that epoch. In the end a proposal was adopted whereby the compulsory jurisdiction arose only by means of a unilateral declaration of a state indicating its acceptance of the jurisdiction of the court, and was not to implied directly in the statute of the court. When the architects of the United Nations considered the court's role, it affirmed the compromise solution under the League of Nations. When accepting the compulsory jurisdiction of the court, states tended, nevertheless, to append reservations to their acceptance of compulsory jurisdiction. Frequently, states excluded disputed where there were other methods of resolving them. These reservations and others tended to undermind the authority of the court. The authority of the court was further impaired by its reversal of a decision to accept jurisdiction of a case brought in the 1950s by Ethiopia and Liberia against South Africa to enforce the mandate conferred after the First World War by the League of Nations in respect of South-West Africa, Namibia. Since then no issue of major political importance has come before the court.
Background The world is currently organized such that, from the legal point-of-view, nations are independent of each other and and have no common authority above them. Each nation maintains its own courts and prescribes its own laws in complete isolation from, and independence of, every other nation. Unless countries have bound each other by treaty mutually to enforce their civil judgments, they are free as to whether or not, and under what conditions, they wish to enforce or otherwise recognize foreign judgments.
Broader Inadequate laws (#PC6848) Outmoded legal systems (#PF2580).
Narrower Ineffective war crime prosecution (#PD1464)
Inadequate international law enforcement (#PF8421)
Inadequate legislation for animal welfare (#PE5794)
Inadequacy of international human rights instruments (#PF6365)
Obstruction of international criminal investigations (#PF7277)
Inadequate national supervision of adherence to international law (#PE8280).
Related Inadequate national law enforcement (#PE4768).
Aggravates Resistance to internationally agreed standards (#PC4591)
Inadequate development of international criminal law (#PF4676).

♦ PF4819 Suttee

Sati — Burning of widows
Nature Tradition demands that the widow of a Hindu be cremated with him, even though she is still alive. The practice is called 'suttee'.
Incidence In 1987, an 18 year old city educated woman was burned to death on her husband's funeral pyre in front of her parents and several thousand people. The police did nothing to prevent the crime.
Refs Saxena, R K *Social Reforms* (1975).
Broader Ritual suicide (#PG0459) Human sacrifice (#PF2641)
Religious sacrifice (#PD3373).
Related Bride burning (#PE4718) Criminalization of euthanasia (#PF2643).
Aggravated by Fundamentalism (#PF1338) Official religion (#PF6091)
Discrimination against women in religion (#PD0127).
Reduces Widowhood (#PD0488) Burden on society of widows (#PF6149).
Reduced by Persecution of religious sects (#PF3353).

♦ PF4820 Compulsory health care

Aggravates Enforced vaccination (#PF1916) Compulsory sterilization (#PF3240)
Excessive prolongation of the dying process (#PF4936)
Religious opposition to public health practices (#PF3838).
Aggravated by Unethical practices of health services (#PE3328).

♦ PF4821 Government support for repressive regimes

Government aid to countries violating human rights — Assisting repression — Government support for undemocratic regimes — Government support for foreign dictatorships — Government support for corrupt regimes
Incidence Governments of industrialized countries have, throughout the Cold War, provided support to repressive regimes, especially in developing countries, in order to protect either their own security or their economic interests, or both.
Broader Foreign intervention in internal affairs of states (#PC3185).
Related Government action against regimes with alternative policies (#PF2199).
Aggravates Dictatorship (#PC1049) Military aid (#PE6052)
Foreign military intervention (#PD9331).
Aggravated by Connivance of authorities in human rights abuses (#PF9288).

♦ PF4828 Diversion of education to qualification earning

Nature The proportion of time in school devoted to the basic skills and the main academic disciplines has been curtailed in favour of other activities. General education courses may be squeezed out of curricula in schools and universities in order to introduce specialized courses focused exclusively on trades and professions, an act which is an abuse of the entire educational system and a jettisoning of centuries of educational and cultural heritage. Universal education prior to, and along side of, trade, technical and professional preparation is increasingly being eliminated with an irreparable loss to society. Indeed, although periodic doubts have arisen about the content, form or control of education, or its usefulness as a tool of development, the trend has continued whereby schools are burdened with tasks previously regarded as the normal responsibility of family or society.
Broader Inadequate education (#PF4984) Paper qualification syndrome (#PF4494).
Aggravates Certificate-based job market (#PJ8370).

♦ PF4829 Destruction inherent in development

Nature Every act of development, planned or unplanned, involves at least two acts of destruction: one in the ecosystem and one in the societal system of man. Planned development involves the destruction of flora and fauna, even to the extent of the elimination of whole species. Then there is destruction of traditional knowledge regarding food production, herbs and medical treatments; loss of knowledge of integrated living with the natural habitat; and loss of ethnic diversity and cultural heritage. Modern man, living within his own developments, is also systematically destroying the material, cultural and spiritual resources that sustain these developments.
Three important harms arising from development activities are displacement of persons, modernization of agriculture and the introduction of new hazardous products and technologies.

Many different kinds of events and activities cause displacement. Some events (e.g., natural disasters) may go beyond the range of human control and others (e.g., population growth which forces migration, or the ravages of war or mass violence deliberately directed toward minorities) may be very difficult to control. But many activities causing displacement can be subjected to control, notably through measures designed to recognize and secure rights of people to be protected from activities which cause displacement and rights to compensation for losses inflicted when displacement does occur. Activities which can be subjected to legal controls include. a. Large–scale development projects (such as dam building or the creation of plantations); b. Deliberate degradation of environments (such as initiation of "development" projects which destroy forests, pollute rivers and fishing grounds); c. Negligent failure to protect rural communities from environmental degradation (such as the failure of governments to work with communities threatened by deforestation, overuse of grazing lands or over–cultivation); d. Relief and food–aid schemes which lead to the wide–scale substitution of imported, subsidized food commodities for locally grown staples, and then to declining food production and vulnerability, particularly in rural areas to food shortages and famine when prices for imports rise and local harvests decline); e. Failure to protect "tribal" peoples and other minorities from territorial encroachments by industries or by spontaneous migrations, or by resettlement programs which result, in effect, in the expropriation of lands held by aboriginal groups who lack political and legal powers to protect themselves). Displacement often inflicts the severest kinds of impoverishment. It strips families of means of livelihood and produces new classes of landless workers or new communities of squatters who face continuing risks of further eviction. While attempts are sometimes made to "compensate" victims of displacement, there is considerable evidence to suggest that these programs fail to provide adequate reparation for all the losses inflicted. Similarly, efforts to "resettle" displaced people all too often are flawed in both the planning and administration stages, and these practices violate rights and inflict economic and other tangible harms. For example, resettlement projects often use coercive means; the people "transplanted" often suffer losses of animals and unharvested crops–and hunger, disease and other hardships. The ultimate outcome of relocation into unsuitable environments is often further displacement. Displacement produces political and cultural harms as well as economic damage. Poor people who lose possession of land usually lose status and dignity; their way of live is destroyed along with their traditional livelihood; communities and cultures are dissolved. Displaced people are usually "refugees" even if they never cross international boundaries; as refugees they are peculiarly powerless and thus vulnerable to all kinds of other human rights violations. Dependent on officials or others for satisfaction of essential needs, they are often easily deterred from engaging in any meaningful processes of political participation; at the same time they sometimes become political pawns of those on whom they have become dependent.

Modernization of agriculture refers to changes in types of crops produced along with changes in the organization methods, and technologies of production. This combination of changes results in the conversion of peasants and small farmers into producers of export crops under the aegis of projects organized by agribusiness firms, usually in collaboration with government agencies, frequently with the assistance of international donors. Agribusinesses (both private and "public") are the dynamos of modernization. They organize projects and create new systems of production. sometimes by acquiring lands and smallholders and transforming them into large–scale units, sometimes by making contracts with smallholders (e.g., "putting out" contracts) which convert them, in effect, to producers for the firm. Agribusinesses furnish the new seeds, inputs and technologies, and they usually process or market the crop. These operations (supply of seeds, inputs and factors, organization of production, processing and marketing) may be carried out by several firms, but usually all of them are subsidiaries or surrogates of a larger, transnational entreprise. The various companies involved may be purely private businesses, but often some are public corporations, or they are companies organized as joint enterprises between government and private companies. Other governmental and international organizations aid the processes of modernization in various ways, e.g., by channelling research and extension, credit, and physical infrastructure towards modernization objectives. The social impacts of "modernization" on small-holders and other rural workers have been widely discussed. The harms inflicted include. a. Landlessness. Modernization often calls for – or results in – the extraction of land from small–holders, e.g., by firms which create plantations, or by wealthier, "progressive" farmers who use various methods (notably money–lending), first to impoverish their marginal neighbours and then to take over their land; b. Indebtedness. Small farmers drawn into production of cash crops requiring purchase of new seeds, inputs and other factors from agribusiness are unusually vulnerable to impoverishing indebtedness which leads to loss of control of lands and income; c. Worker exploitation. Landless (or land poor) rural workers are often forced, by circumstances, to become workers for agribusinesses. The terms of employment and physical conditions under which they work are often exploitative. Agribusinesses often monopolize both markets for cash crops and the sale of inputs needed to produce new crops – with the result, again, that producers are exploited; d. Crop displacement. There is always the risk, all too frequently realized, that the market for the new "modern" crops (on which producers must now depend for their livelihood) will deteriorate. Seldom are producers insured against this outcome; yet they are the primary losers; e. Environmental degradation. The depletion of soil resources is often the result of mono–cropping and other practices introduced by modernization. Another threat, serious over the long run, is the loss of valuable genetic resources when traditional plants are replaced by new foreign varieties; f. Food shortages and hunger. Modernization often means loss of land needed to maintain local self–sufficiency in food production; the result is that economically marginal families become increasingly dependent on other producers and on uncertain markets to purchase food supplies; f. Food shortages and hunger. Modernization often means loss of land needed to maintain local self–sufficiency in food production; the result is that economically marginal families become increasingly dependent on other producers and on uncertain markets to purchase food supplies; g. Exclusion. Smallholders and rural workers are regularly excluded from any form of meaningful participation in the planning and management of modernization projects. Denied rights of participation and access to decision makers, they are denied opportunities to protect their interests and the power to impose accountability of those whose activities cause the harms noted above.

Processes leading to the introduction of hazardous products. a. The growing power and lack of accountability of the transnational chemical industry; b. The growing tendency on the part of third world governments to accept without question the introduction of a high technology as a panacea for "development" and a means of avoiding problems of structural reform and redistribution of wealth and power; c. The secrecy surrounding (and exclusion from participation in) decisions on policies and projects reflecting technology choices; d. The ignorance concerning risks associated with new chemical technologies – an ignorance encouraged by the chemical industry which makes hazard assessment and monitoring of health and environmental impacts virtually impossible. These developments are part of the context within which processes take place leading to the introduction, on an increasing scale, of hazardous products into third world environments. These processes include. the development (in third world countries) of businesses (both public and private) engaged in the manufacture or marketing of hazardous products; the "dumping" of hazardous products into third world markets and the export of dangerous technologies; the failure of governments to assess, carefully, the character of these products – the risks they impose on communities (notably communities of the poor); the failure of governments to enact laws which regulate hazardous products and impose accountability on those who manufacture or distribute them. A growing literature has portrayed the harms to health, life, and environments which occur as a result of these projects – for example. – physical harms to workers in industries which manufacture, store, or distribute hazardous products; – physical harms to agrarian workers who use and often misuse them – due to lack of information, adequate warnings and other precautions by the industry; – physical harms to other people, notably poor rural people when chemical pollutants enter their "food chains" (e.g., when their animals eat tainted plants); – physical harms to environments (including food producing environments), notably the habitats of the poor.

Broader Maldevelopment (#PB6207) Human destructiveness (#PA0832)
Obstacles to national development (#PF4842).
Narrower Unsustainable rural development (#PD4537)
Failure of development policies (#PF5658)
Failure of development programmes (#PF8368)
Exploitative property development (#PD8492)
Socially unsustainable development (#PC0381)
Unsustainable economic development (#PC0495)
Ecologically unsustainable development (#PC0111)
Unsustainable agricultural development (#PC8419)
Unsustainable development of energy use (#PC7517)
Obliteration of footpaths by development (#PE3874)
Unsustainable development of coast zones (#PD4671)
Land erosion brought about by site development (#PE7099)
Unconstrained exploitation of natural resources (#PF2855)
Misuse of tropical rain forests for agricultural development (#PE5274).
Aggravates Reversal of development progress (#PF4718)
Regional environmental degradation (#PD5845)
Unsustainable development of forest lands (#PD4900).
Aggravated by Export–led development (#PE6237)
Unplanned land development (#PF6475)
Unchallenging world vision (#PF9478)
Inflexible social structure (#PB1997)
Economic bias in development (#PF2997)
Uncontrolled urban development (#PC0442)
Insufficient leadership training (#PF3605)
Inappropriate development policy (#PF8757)
Discrimination against men in employment (#PD3338)
Inadequate control of development projects (#PF9244)
Inappropriate design of development projects (#PF4944)
Short range planning for long–term development (#PF5660)
Inadequate models of socio–economic development (#PF9576)
Difficulty of implementing development programs (#PG6302)
Inappropriate management of development projects (#PD3712)
Neglect of the role of women in rural development (#PF4959)
Abusive technological development under capitalism (#PD7463)
Nationalistically determined development of natural resources (#PD3546)
Inadequate development of enterprises in developing countries (#PE8572)
Inadequate human resources development in least developed countries (#PE9764)
Dependence of developing countries on external financing for development programmes (#PE7195)
Reinforcement of inappropriate development by privileged classes in developing countries (#PF6670).
Reduced by Underdevelopment (#PB0206) Unpromoted industrial development (#PJ1017)
Denial of right to a people to pursue development (#PE1536).

♦ PF4832 Inadequate medical care

Lack of medical care — Inadequate medical services — Prejudiced medical care — Remote clinic service — Inaccessible children's clinic — Inaccessible health care — Distant medical services — Infrequent medical visits — Inadequate medical facilities — Insufficient medical facilities and personnel — Unorganized health facilities

Broader Lack of essential local infrastructure (#PF2115)
Underprovision of basic services to rural areas (#PF2875).
Narrower Misdiagnosis (#PF8490) Misuse of medicines (#PD8402)
Surgical malpractice (#PE4736) Inadequate dental care (#PJ5478)
Lack of medical records (#PF1836) Refusal of medical care (#PF4244)
Lack of eugenic measures (#PD1091) Medicinal products shortage (#PE3502)
Lack of medical information (#PJ8069) Restrictive medical practices (#PD8831)
Unavailability of dietary care (#PJ1332) Insufficient preventive medicine (#PE0751)
Over–specialization in medical care (#PF5709)
Inadequate emergency medical services (#PD1428)
Delay in administration of medical care (#PD5119)
Depriving prisoners of medical treatment (#PD1480)
Inadequate maternal and child health care (#PE8857)
Non–surveillance of medical high risk persons (#PF4861)
Excessive exposure of medical patients to radiation (#PE1704)
Failure of medical programmes in developing countries (#PE8400)
Inaccurate medical diagnosis due to inadequate equipment (#PE4129)
Resistance to incorporating midwives in medical care systems (#PE4901).
Related Abuse of medical drugs (#PD0028) Physically inaccessible services (#PC7674)
Unethical experiments with drugs and medical devices (#PD2697)
Inhumane participation of the medical profession in torture (#PE4015).
Aggravates Injurious accidents (#PB0731) Inadequate health services (#PD4790)
Human disease and disability (#PB1044) Prohibitive cost of basic services (#PF6527)
Fragility of maintaining basic health (#PJ2524)
Unequal regional distribution of deaths (#PC4312)
Denial of right to adequate medical care (#PD2028)
Inadequate maintenance of physical health (#PF1773)
Unequal morbidity and mortality between countries (#PC6869)
Increase in insurance claims for medical negligence (#PE4329)
Inadequate community care for transient urban populations (#PF1844)
Health risks to workers in agricultural and livestock production (#PE0524)
Cumulative depletion of corporate initiative in rural communities (#PE3296)
Limited availability of public services in the small towns of developed countries (#PF6539).
Aggravated by Insufficient doctors (#PE8303) Unethical medical practice (#PD5770)
Inadequacy of civil defence (#PF0506) Insufficient medical supplies (#PE1634)
Excessive cost of medical drugs (#PE5755) Impairment of physicians' ability (#PE5746)
Lack of funds for medical research (#PE0612) Maldistribution of medical resources (#PD2705)
Fragmented planning of community life (#PF2813)
Over–specialized supervisory personnel (#PF3588)
Insufficient transportation infrastructure (#PF1495)
Irrational conscientious refusal of medical intervention (#PF0420)
Crimes committed in hospitals and health care facilities (#PE8420)
Limited availability of therapeutic substances of human origin (#PF6751)
Ineffective self–regulation in the pharmaceutical and medical devices industries (#PE8502).
Reduces Excessive proliferation of medical drugs (#PD0644).

♦ PF4833 Professional burnout

Compassion syndrome

Nature People, especially professionals engaged in work demanding extensive personal commitment, are vulnerable to a complex syndrome characterized by: extreme fatigue, depression, psychosomatic illnesses, substance abuse, apathy, resignation and defensiveness.

Incidence Whilst characteristic of extreme forms of executive stress amongst business executives faced with conditions of rivalry and competition, it appears in a different light in the case of people in the caring professions (health, social services, psychotherapy, relief).

Refs Rediger, G Lloyd *Coping with Clergy Burnout* (1982).
Broader Stress in human beings (#PC1648).
Aggravates Compassion fatigue (#PF2819).
Aggravated by Scientific rivalry (#PG3918) Interpersonal rivalry (#PD7617).

♦ PF4836 Underdevelopment of legal infrastructure

Absence of reliable systems of justice in developing countries — Lack of legal engineering
Claim Fairer and more reliable systems of justice not only enhance the climate for development but also constitute a social benefit immediately accessible to the citizen. In many countries development is inhibited by lacunae in the legal statutes; the absence of adequate laws creates doubt and uncertainty. In rich and poor countries alike, arbitrary, corrupt and inefficient courts pose the greatest threat to those who lack wealth or influence.
Broader Inadequate infrastructure (#PC7693) Lack of political development (#PD8673)
Weakness of socio-economic infrastructure (#PC1059).
Narrower Remoteness of legal services in developing countries (#PE9001).
Aggravates Unavailability of legal information (#PJ8698)
Deficiencies in national and local legal systems (#PF4851).

♦ PF4837 Imbalance in economic and social planning in developing countries

Unintegrated development — Lack of understanding of social and economic contexts in development programmes
Nature Economic development, and particularly industrialization, is too often implemented without due application of a unified approach of integrated and balanced economic and social planning. This may give rise to social problems and retards development. In many countries poverty and mass unemployment, for example, are so widespread and affect so critically the social equilibrium that they constitute, in themselves, blocks to further improvement. It is no longer possible to rely on the assumption that an expanding modern economic sector will, in a reasonable interval of time, absorb the mass of people and provide them with decent living standards.
Broader Compliance*complex (#PA5710) Demeaning community self-image (#PF2093).
Narrower Inequitable labour standards in developing countries (#PD0142).
Aggravates Economic uncertainty (#PF5817).
Aggravated by Economic bias in development (#PF2997).

♦ PF4839 Lack of accord on water use

Hydropolitics
Nature Internationally, when more than one country controls portions of a river basin, respective rights to water supplies need to regulated by negotiations and treaties; hydrology points to collaborative efforts as the key to optimum utilization of scarce water resources. Frequently, however, the politics of water resource allocation become enmeshed with other quarrelsome issues, aggravating international tension between neighbouring countries. Hydropolitics can be the first step in promoting collaborative relations across borders, but can also lead to sharpened international tension. Indeed, water resources represent a great potential for conflict because of competing demands on limited supplies or because of deterioration in quality through use. Conflicts exist both within national boundaries and between adjacent countries.
Aggravates Maldistribution of water (#PD8056) Long-term shortage of water (#PC1173)
Inadequate water system infrastructure (#PD8517).
Aggravated by Unethical practice of hydrology (#PD2586).

♦ PF4841 Errors and risks in medical self-experimentation

Nature Medical experiments done by physicians upon themselves (autoexperiments) have never evoked disapprobation, but rather have excited admiration, and yet it is curious that, of those recorded, some of the most dangerous were intended to prove, and were supposed to have proved, theories that are now known to be entirely incorrect.
Background In 1767 John Hunter infected himself with syphilis in an attempt to prove that it and gonorrhoea were the same disease. In 1835 Antoine-Barthelemy Clot tried to prove that the plague was not communicable by inoculating his arm with pus from a plague patient. And in 1892 Max von Pettenkofer swallowed live culture of cholera vibrios in an attempt to show that they alone could not cause the disease. In the second half of the 20th century medical auto-experimentation has been largely confined to investigations of hallucinogenic and other psychotropic drugs.
Broader Human errors and miscalculations (#PF3702)
Obstacles to medical experimentation (#PF4865).

♦ PF4842 Obstacles to national development

Nature Development programmes supported by international assistance are in themselves lengthy, but the slowness of their progress is aggravated by false starts, uneconomic investments, uncoordinated programmes, inattention to infra-structure building and other fundamentals.
Broader Limits (#PF4677) Compliance*complex (#PA5710).
Narrower Obstacles to world trade (#PC4890) Destruction inherent in development (#PF4829)
Constraints on the development of mental health services (#PF4955).

♦ PF4843 Self disorders

Self object disorders
Nature The "self" has failed to achieve cohesion, vigour harmony, or these qualities are lost after tentative establishment.
Narrower Psychoses (#PD1722) Narcissism (#PF7248)
Borderline personality disorders (#PE4396).
Related Mental depression (#PC0799).

♦ PF4848 Inadequate peace research support

Background In its first phase, in the 1960s, peace research was concerned primarily with the formulation of theories regarding strategic disarmament and security politics, and introduced the criticism of the international threat and deterrent system. This criticism met with little response even from the critical public, and was given virtually no practical support from any section of any democratic society. Also without consequence in this early phase were peace researchers' suggestions about conflict regulation. Presently, researchers are looking at armament complexes as parts of larger social structures. There is also a negative concept interpreted as the absence of personal violence, and a positive concept of peace, to be interpreted as social justice: equal distribution of power and resources as well as absence of structural violence.
Claim There have been increased attacks on certain Western peace research institutions, such that peace research itself has become a source of conflict. Various political groups have criticized Western peace research because of its predilection towards Utopian conceptions and its 'anti-pluralistic' nature; at the same time attempts have been made by political forces to make use of peace research institutions for selfish purposes. The majority of Western peace researchers consciously accept the conflicts related to their social role, emphasizing that peace research is necessarily conflict research as well.
Counter-claim Peace research as a branch of political science has gained ground because the growing variety and number of conflicts could not be analyzed in a sufficient way - in contrast with former expectations.
Broader Limits on areas of research (#PF2529) Shortage of funds for research (#PF5419).
Related Competitive development of new weapons (#PC0012).

♦ PF4849 Elitist ruling classes

Apparatchiks — Nomenklatura — Privileged classes of citizens — Appropriation of machinery of government by elites
Incidence Political party functionaries may be rewarded for their party loyalty by receiving appointments to governmental positions. These appointments, not based on ability, can lead to ineptness in national domestic matters and ideological rigidity in international relations. In the case of the USSR, it has been estimated that the nomenklatura numbered 50,000, although if lesser party and state officials, as well as senior figures from the economy, science and culture are included, this could rise to 750,000. With the inclusion of family members, the number for the USSR would be some 3 million, approximately 1 per cent of the population. In the USSR the nomenklatura has privileged access to a whole range of facilities and products. USSR sources estimate that the group was served by 1 million chauffeurs, driving cars whose purchase and upkeep cost the state 10 billion roubles. Exclusive hospitals, accessible to 100,000 people only, cost 100 million roubles to maintain. Some 400,000 people are entitled to special state pensions, with associated concessions, unavailable to others.
In sub-Saharan Africa, the World Bank has identified widespread appropriation of the machinery of government by elites in the service of their own interests.
Refs Markowitz, Irving L *Power and Class in Africa* (1977).
Broader Elitism (#PA1387).
Aggravates Reinforcement of inappropriate development by privileged classes in developing countries (#PF6670).
Aggravated by Elitist leadership (#PF1104) Authoritarian regimes (#PC9585)
Single party democracies (#PD2001).

♦ PF4850 Inadequate fiscal policies

Fiscal crisis
Nature The growth of expenditures followed by increasing gaps in the availability of resources have contributed to higher fiscal deficits and to rising debts, both internal and external. Further, changes in the composition of public expenditures have also affected the structure and exercise of financial controls.
Broader Inappropriate policies (#PF5645)
Fragmented planning of community life (#PF2813).
Narrower Ineffective tax systems (#PF1462) Distortionary tax systems (#PD3436)
Fiscal and trade imbalances (#PC4879).
Aggravated by Inadequate management of government finances (#PF9672)
Imbalance of revenue mobilization and expenditure allocation (#PF7432).

♦ PF4851 Deficiencies in national and local legal systems

Nature National and local legal systems can be deficient in a number of ways. Law enforcement systems can be inadequate. Enforcement officers can be ill-equipped to enforce laws or corrupt. The general public can be opposed to or unsupportive of law enforcement. Judiciary systems can be corrupt, undertrained or overloaded. Law faculties may be insufficient or may offer poor training. Law libraries may be inadequate for schools, judges, lawyers and other auxiliaries of justice. Judges and other court officials may be untrained or undertrained. Laws may be inadequate, confusing or obscure. Legal material may be uncollected and poorly classified. Prisons may be understaffed or poorly staffed. Systems of punishment may be in violation of human rights conventions or incapable of being equitable.
Broader Outmoded legal systems (#PF2580) Social underdevelopment (#PC0242)
Weakness of socio-economic infrastructure (#PC1059).
Narrower Martial law (#PD2637) Unjust trials (#PD4827)
Justifiable homicide (#PE7533) Inadequate national law enforcement (#PE4768)
Deficiencies in civil justice systems (#PF4899)
Unnecessary verbosity of legal documents (#PF7137)
Inadequate legislation for animal welfare (#PE5794)
Deficiencies in the criminal justice system (#PF4875)
Absence of legal entities in developing countries (#PE8597)
Failure to legally rehabilitate victims of miscarriage of justice (#PF7728).
Aggravates Political repression (#PC1919) Inequality before the law (#PC1268).
Aggravated by Underdevelopment of legal infrastructure (#PF4836).

♦ PF4852 Obstacles to education

Broader General obstacles to problem alleviation (#PF0631).
Narrower Educational elitism (#PC1527) Student absenteeism (#PE4200)
Inadequate education (#PF4984) Unmotivated teachers (#PF5978)
High teacher turnover (#PE8221) Inequality in education (#PC3434)
Inadequate career advice (#PJ8018) Unclear educational roles (#PG8840)
Unequal access to education (#PC2163) Inadequate teacher training (#PJ1327)
Obsolete educational values (#PF8161) Educational authoritarianism (#PC1526)
Shortage of trained teachers (#PD8108) Unused training opportunities (#PJ1280)
Prohibitive cost of education (#PF4375) Debilitating education images (#PF8126)
Barriers to economic education (#PG7826) Corporal punishment in schools (#PE0192)
Irregular delivery of education (#PJ7936) Unimaginative educational vision (#PF3007)
Parent-teacher non communication (#PF1187) Inadequate educational facilities (#PD0847)
Crimes committed in urban schools (#PJ1356) Unpublished training opportunities (#PJ9985)
Obstacles to learning due to hunger (#PE8442) Inappropriate size of school classes (#PD6585)
Restrictive regulations for training (#PJ7972)
Ineffective educational policy decisions (#PF2447)
Inadequate working conditions of teachers (#PE7165)
Ineffective systems of practical education (#PF3498)
Ineffective methods of practical education (#PF2721)
Limited availability of learning opportunities (#PF3184)
Non-equivalence of national educational qualifications (#PC1524)
Inappropriate selection and examination procedures in education (#PF1266).
Aggravates Ignorance (#PA5568) Cultural suicide (#PF5957)
Lack of formal education (#PF6534) Illiteracy in the fourth world (#PD6645).
Aggravated by Over-specialization (#PF0256)
Authoritarian propagation of knowledge (#PF3706)
Maldistribution of students enrolled in school (#PE8733)
Imbalance between training and existence of openings in various professions (#PE5194)
Inability of educational systems to keep pace with technological advancement (#PF7806).
Reduces Over-education (#PC6262).

♦ PF4854 Suppression of information concerning environmental hazards

Cover-up of unsafe buildings — Cover-up of unsafe industrial installations — Cover up of unsafe equipment — Cover-up of unsafe transportation systems — Environmental pollution cover-up — Suppression of information concerning health hazards — Non-disclosure of threats to public safety
Incidence In certain countries, such as the UK, information on the results of government safety tests on consumer products are not available to the public. Similarly information on the possible environmental hazards of toxic materials stored in a factory is not available to those living in the neighbourhood. Information on the results of safety tests on food additives or drugs is similarly classified. Besides, governments sometimes prefer not to inform public about air or water pollution levels and potential ecological disasters.
Broader Secrecy (#PA0005) Official secrecy (#PC1812)
Suppression of information (#PD9146).

Related Suppression of information concerning social problems (#PF9828).
Aggravates Disastrous accidents (#PC6034) Inadequate warning of disasters (#PF3565).
Aggravated by Unethical practices of employers (#PD2879)
Vulnerability of society to truth (#PF5937)
Harassment of human rights monitors (#PF1585)
Abuse of commercial confidentiality (#PE6786)
Suppression of scientific information (#PF1615).

♦ PF4857 Obstacles to international cultural exchange

Nature The international exchange of books, publications and educational, scientific and cultural materials is hampered by customs duties or other charges, licence requirements, quantitative restrictions, foreign exchange control measures, and intellectual property rights such as copyrights, trademarks or patents.
Claim The free exchange of ideas and knowledge and, in general, the widest possible dissemination of the diverse forms of self-expression used by civilizations, are vitally important both for intellectual progress and international understanding, and consequently for the maintenance of world peace.
Broader Obstructions to international personnel exchanges and cultural cooperation (#PE4785).
Narrower Abuse of international cultural, diplomatic and commercial exchanges (#PF3099).
Related Politicization of international sports events (#PF4761)
Obstructions to international athletic exchange (#PE4809).
Aggravated by War (#PB0593) Ideological conflict (#PF3388).

♦ PF4861 Non-surveillance of medical high risk persons

Inattention to the disease-prone
Nature Over the years much information has been gathered on factors of a genetic or environmental nature which increase the probability of an individual developing a chronic disease or disability. This knowledge has, however, only been applied to a limited extent in health services as means of identifying population groups at risk.
Claim The identification of high-risk persons should also lead in the long run, through the intensified surveillance of populations, to economies in public health expenditures.
Broader Inadequate medical care (#PF4832).

♦ PF4865 Obstacles to medical experimentation

Broader Environmental hazards constraining scientific research (#PF1789).
Narrower Uncertainties in animal experimentation (#PF4770)
Inhumane medical experimentation during war-time (#PE4781)
Unethical experimentation using aborted foetuses (#PE4805)
Errors and risks in medical self-experimentation (#PF4841)
Unethical experiments with drugs and medical devices (#PD2697).
Related Cruel treatment of animals for research (#PD0260)
Unethical medical experimentation on prisoners (#PE4889).
Reduces Abusive experimentation on humans (#PC6912).

♦ PF4866 Psychic warfare

Extrasensory warfare — Parapsychic warfare
Nature Military powers are trying to master extrasensory perception, telepathy, clairvoyance and psychokinesis to perform acts of espionage and war - penetrating secret files, locating submarines, blowing up guided missiles in flight etc.
Related Psychotronic warfare (#PD7986) Psychological warfare (#PC2175).

♦ PF4870 Obstacles to commodity futures trading

Nature Futures markets perform fundamental services in the organized distribution of certain agricultural and industrial products. They attempt to correlate world demand and supply and establish world prices and keep them uniform. Their regulatory purpose includes the ability to discount the impact on the market of forecasted supply and demand, and spread over a long period the burden of distribution of a short-period agricultural harvest.
The main obstacles to futures trading include foreign exchange, price and import controls, and direct government interventions on the market. Exchange regulations have the following four effects: they prevent or restrict transactions; they prevent arbitrage; they prevent imports protection against market cornering; and they prevent or impede exchange cover. To fulfil its task, a futures market must freely reflect the position of supply and demand, and the price of a commodity results precisely from the interaction of sellers and buyers. Obstacles are also created when the price of a commodity is regulated by the administration. Another way of regulating prices results from a limitation of the importer's profit margin. Import regulations are an obstacle to the free operation of futures markets, particularly in the case of a market dealing in an imported commodity. These markets cannot be opened unless import regulations are sufficiently flexible, to ensure that future sellers can back their sales with imports. Import regulations can also prevent a national of the country where they are in force from operating on a futures market for, in the case of purchase, he may be obliged to resell his contract, whereas in certain circumstances he might prefer actual delivery, which he cannot expect to obtain without an import licence.
One of the essential aims of a futures market, namely the quotation of a normal price for a given commodity at any moment, can obviously not be fulfilled if governments - whether the commodity is sold by a national board or stockpiled by the public authorities - are able to intervene on the market at certain times through bulk buying or bulk selling operations. These interventions interfere with the free operation of the law of supply and demand.
Claim Any factor which tends to restrict operations on a futures market prevents that market from effectively fulfilling its task. The broader the market, the more efficient are the services rendered by it. In principle any restrictive measure, whatever its nature, that limits transactions on commodities or currency, has a damaging effect on the operation of futures markets. Futures markets can only develop normally if given sufficient commercial and monetary freedom.
Broader Instability of the primary commodities trade (#PC0463).
Narrower Commodity speculation (#PD9637) Price fixing in the commodity markets (#PJ5528).
Related Imperfections of capital markets (#PF3136).
Aggravates Instability in export trade of developing countries producing primary commodities (#PD2968).

♦ PF4871 Lack of integration of traditional and Western medicine

Nature There is still considerable friction between traditional and scientific medicine, mainly due to lack of understanding and cooperation, although legislative action has also failed to promote, or even hindered the promotion, of traditional medicine. Although there is currently much discussion in international circles about recognizing traditional medicine and merging it with scientific medicine, a combined venture of traditional healing and modern medicine has not yet been given any real opportunity.
Incidence Traditional medicine is still practised to some degree in all cultures. "Barefoot doctors" are an important and respected part of the Chinese medical system; and the Western world is experiencing a renewed interest in traditional healing techniques such as homeopathy, chiropracty, dietary regimes, and mind and body integration therapies. It is to the traditional healers that the great majority of the people of developing countries turn in times of sickness. For 80% of the population 'primary health care' is synonymous with traditional medicine.
Claim (1) Traditional medicine has been practised since the dawn of mankind. Through trial and error, human beings have discovered ways of relieving pain and sickness, and of living in harmony with nature. (2) Traditional medicine is evolving and dynamic. This dynamism and vitality could be used as a basis for developing a new, integrated, system of health care in developing countries. (3) Practitioners of traditional medicine represent a vast and valuable human resource outside the official health services, yet very little is known about the extent of their practises or the methods they use. If the care these practitioners render is to be integrated into national, state and regional health plans, a systematic effort must be undertaken to learn more about their functioning. (4) Because such a large percentage of developing countries' peoples view traditional health care as a first resort, it is necessary both to give it credence and to recognize that Western medicine may benefit from a merger of traditional and modern techniques if treatment of the major diseases is to be successfully implemented. This has been achieved in China, where both modern Western methods and traditional medicine are provided in the rural communities. The complementary mix of the two has made health care widely available in the countryside. It is essential for health professionals to be fully aware of the work of traditional practitioners and of their place in society.
Broader Fragmentation of knowledge (#PF0944).
Narrower Allopathy (#PF7703) Disregard for self-healing potential (#PG7709)
Resistance to incorporating midwives in medical care systems (#PE4901).
Aggravates Excessive proliferation of medical drugs (#PD0644).
Aggravated by Poisonous plants (#PD2291)
Ignorance of traditional herbal remedies (#PE3946).

♦ PF4875 Deficiencies in the criminal justice system

Refs Jamieson, Katherine and Flanagan, Timothy J (Eds) *Sourcebook of Criminal Justice Statistics, 1986.* (1987); United Nations *Manual for the Development of Criminal Justice Statistics* (1986).
Broader Injustice (#PA6486) Deficiencies in national and local legal systems (#PF4851).
Narrower Parole violation (#PE1121) Inadequate prevention of crime (#PF4924)
Ineffective war crime prosecution (#PD1464) Releasing of repatriated prisoners (#PJ1505)
Ineffective legislation against organized crime (#PE6699).
Aggravates Recidivists (#PE5581).

♦ PF4876 Distortion in intergovernmental organizations by mini-state membership

Claim International organizations whose members are sovereign states may be unbalanced by small country memberships. Many small countries play an almost infinitesimal role in world trade and regional strategic considerations, and their small budgets and lack of trained people give them very little scope for cooperation and participation in international activities, so they may also have little international experience. Despite this, such countries may outvote countries who have vital interests in particular issues in such international organizations as the United Nations and others.
Broader Inadequacy of intergovernmental decision-making process (#PF2876).

♦ PF4882 Social constraints on freedom of imagination

Claim Politics is the enemy of the imagination. To act politically generally means joining with other people to be effective, aligning preoccupations with those of others, producing analytical compromises, deciding on priorities, methods of presentation and strategies. To act seriously in a political way is to have in mind a version of the future of which others need to be convinced. Acts of the free imagination call for the exact reverse of these requirements. The imagination is individualistic, egotistical in the extreme, reserving the right to be contradictory and espousing the unthinkable. Imaginative thinking is constrained in an atmosphere of official half-truths and erosion of freedom of information.
Counter-claim The freedom to think imaginatively is protected as the result of political action.
Aggravated by Official secrecy (#PC1812)
Restrictions on freedom of information (#PC0185).
Reduced by Anarchism (#PC1972).

♦ PF4884 Inadequate economic integration of socialist countries

Nature Due to the inefficiency of the economies of socialist countries, inflation, shortages, food queues and lack of consumer products are commonplace occurrences. The inefficiency is due to poor organization, with few incentives for high productivity or free enterprise. In addition, socialist statistics are often confused and difficult to interpret, thus discouraging Western trade.
Broader Socialism (#PC0115) Lack of international cooperation (#PF0817).
Aggravated by Educational curricula over emphasizing method rather than content (#PD1827).

♦ PF4886 Obstacles to legal relations between socialist countries

Nature According to Lenin, national frontiers should be delineated democratically, in accordance with the will and sympathies of the population. This elastic definition and ambiguous recognition of the principle of sovereignty allows for the real socialist goal - the fusion of states into one communist supra-nation. In practice, sovereignty is not strongly respected by some socialist countries and intervention into the internal affairs of a weaker socialist country by a stronger one is a repeated occurrence.
Incidence Interventions in Hungary, Czechoslovakia and Poland are recent examples.
Broader Socialism (#PC0115) Lack of international cooperation (#PF0817)
Obstruction of international criminal investigations (#PF7277).

♦ PF4891 Over-organization

Excessive socio-economic order
Claim With the historical increase in the population of the larger landholdings, these became kingdoms. The familiar relationship, through which people knew one another personally, lost its significance, thus removing the major dissipator of aggressivity. Further increases in the population shifted the numerical relationship between the rulers and the ruled. Democracy has not succeeded in striking an appropriate compromise between maintaining the order that a very large number of people make absolutely necessary and preserving those freedoms for individual action which belong to the rights of man. Large populations mean that there are too many voters and too few people to be voted for. And very few of the latter, however intelligent and morally faultless they may be, are capable of presrerving their whole humanness once they are in positions of power.
Counter-claim Love of order is part of the original programming of human behaviour and can be numbered among the human virtues. It is closely related to the search for harmony and for healthy, coordinated interaction.
Related Over-centralization (#PF2711).

♦ PF4892 Inadequate utilization of volunteer social service workers

Nature The potential social service work-force of volunteers is under- utilized. Universities, extension and adult education opportunities, community colleges and local institutions, offer little in the way of training for the thousands, and in some countries tens of thousands, of volunteers who render or who wish to render aid to the disadvantaged. The institutionalization of the handicapped often produces de-humanizing circumstances, especially in overcrowded, understaffed situations; neglect and superficial attention has the same effect. The trained volunteer

is required at every point where social service is needed, in homecare, at institutions and at social service centers.
Broader Puritanism (#PF2577)
Shortage of adequately trained personnel to act against problems (#PF0559).
Related Dehumanization (#PA1757).
Aggravates Lactation tetany in mares (#PG4797).
Aggravated by Nudism (#PF2660).
Reduces Inadequate social welfare services (#PC0834).

♦ PF4894 Colonization of information
Nature Least developed and developing countries have long depended on the developed nations' media. The BBC, Radio France, the Voice of America, Deutsche Welle, Radio Netherlands, Radio Moscow, and RAI, for example, have been the radio sources for information coverage on the Third World taken comprehensively, as there are of course, many highly developed national media in these countries, although limited in international scope. The neo-colonial distortion of news by these radio sources exacerbated by their official connection to their respective governments, is not much more excessive than the type of cultural propaganda, and perpetrations of misinformation by developed countries films, news agencies, publishers and periodicals.
Broader Colonialism (#PC0798).
Related Violation of sovereignty by trans-border broadcasting (#PE0261).
Aggravated by Incomplete access to information resources (#PF2401).

♦ PF4899 Deficiencies in civil justice systems
Broader Injustice (#PA6486) Deficiencies in national and local legal systems (#PF4851).

♦ PF4910 Double mindedness
Nature In the disease of double personality two defined selves in turn struggle for the possession of the field of consciousness, or they exist side by side, each more or less ignorant of the other. Most blemishes of character, misdeeds and crimes arise from more or less split personality. Normal evolution of character consists in the straightening out and unification of the inner self.
Broader Hypocrisy (#PF3377).

♦ PF4923 Iconoclasm
Nature Iconoclasm is a form of vandalism aimed at the destruction of sacred images and perpetrated because of religious or political conflict.
Incidence A recent example of iconoclasm is the destruction by bombing of the Borobudur temple in Indonesia, the world's largest Buddhist shrine, in January 1985. It is thought to have been perpetrated by Muslim extremists.
Background Early examples of iconoclasts are the Byzantine Emperor Leo III, the Isaurian who made the destruction of sacred images state policy and the Vikings who destroyed the churches of Western Europe in the name of Wotan. Florence in the time of Savonarola revived iconoclasm, with Botticelli and Lorenzo di Credi throwing their own works into the flames.
Counter-claim Iconoclasm, while at times extreme, is largely a reaction to the worship of images. In the eighth century, for example people chose icons, not people, as godparents for their children and images were ground, mixed with water and drunk as magic medicines. In an age committed to secularism, as ours is, it is difficult to the perceive the danger not only to the social order, but to the individuals involved that worship of idols represents. This practise was perceived as threatening to fragment society into warring factions. It was also seen as condemning the worshippers to a existence far worse than death.
Broader Art vandalism (#PE5171) Religious repression (#PC0578).
Related Persecution of religious sects (#PF3353)
Holy places as a focus of religious friction (#PF1816).
Reduces Heresy (#PF3375).

♦ PF4924 Inadequate prevention of crime
Decline in crime detection rate
Nature The long term reduction and control of crime depends upon an appropriate strategy being devised and adopted, not only by each country to deal with its own problems in its own way, but also by the international community, which is already suffering from the neglect of crime prevention in the past. Much of present crime could have been avoided by intelligent foresight which made crime prevention a part of past policy developments.
Police detection of crime depends in most cases upon a report by the victim, the public, or even the criminal himself. The crime may not be reported and thus go undetected because: the victim is vulnerable to prosecution (for example, for carrying a weapon or for suspicion of some other crime); the victim may accept such victimization as one of the risks of life; the victim or witnesses may not recognize an act as a crime (a brawl or a traffic violation, for example), or they may not correctly evaluate the gravity of the crime; a witness may not wish to "get involved" and thus will not place a report.
Counter-claim Some countries have experimented with broad approaches to the kind of general social improvement which they hoped would reduce crime. The developing countries have tried to plan their national growth, sometimes expecting problems like crime to disappear along the way. Many industrialized countries, less committed to national planning, have nevertheless devoted funds to selected programmes of social development intended to reduce crime by dealing with the conditions thought to produce it.
Broader Inadequate local enforcement (#PF0336)
Deficiencies in the criminal justice system (#PF4875).
Related Imprisonment (#PD5142) Needless incarceration (#PE5112).
Aggravates Inadequate prevention of disabilities (#PF0709).
Aggravated by Crime (#PB0001) Inadequate laws (#PC6848)
Unreported crimes (#PF1456) Citizen incompetence (#PE5742)
Inadequate national law enforcement (#PE4768).

♦ PF4925 Inadequate mental hospitals
Inadequate psychiatric institutions — Inhumane mental asylums
Nature Instead of being located in the community it serves, the traditional mental hospital is usually of remote location from this community. It is also large in size – hence the custodial rather than the rehabilitative approach on the part of the staff and the lack of psychotherapeutic contacts between staff and patients. Day care and day hospital services for the elderly mentally ill are particularly lacking. Comprehensive mental illness services have failed to materialize. Mental hospitals are often too inward-looking and hospital-orientated, while community need goes unmet.
Refs Barton, Russell *Institutional Neuroses* (1959); Goffman, Erving *Asylums* (1970); Goffman, Erving *Asylums* (1961).
Broader Inadequate mental health services (#PJ4379).
Narrower Limited psychiatric out-patient care (#PE0540).
Aggravated by Denial of rights of mental patients (#PD1148)
Unethical practices of health services (#PE3328)
Prohibitive cost of hospital facilities (#PE4154)
Abusive detention in psychiatric institutions (#PE2932)
Abusive treatment of patients in psychiatric hospitals (#PD0584).

♦ PF4928 Unpredictability of earthquakes
Inadequate earthquake forecasting
Nature Although the regions of greatest seismic activity are clearly defined and although there are scientific theories on what causes earthquakes, it is very rarely possible to forecast the time and place of a destructive earthquake. Moreover, even if such forecasting were possible, although it would reduce the number of lives lost it would not prevent property damage. Hopes raised by some predictions of an end to the era of unforeseen earthquake disasters have been shattered by other forecasts which proved far from accurate. The science of earthquake prediction is still in its infancy, and although scientists have made tremendous strides in their research, there is as yet no failsafe formula to foretell an impending earthquake with absolute accuracy.
Claim Greater scientific knowledge of earthquake risk will not in itself lead to a reduction of earthquake damage or loss of life. Enhanced ability to forecast earthquakes and to give warnings will not necessarily lead to action that reduces vulnerability or exposure to risk, and may increase both. Technological capability to modify earthquake events may, on balance, increase loss potential.
Counter-claim Earthquake prediction is a rapidly emerging scientific field offering great promise for loss reduction. Although accurate predictions of the size (magnitude), time, and location of future earthquakes may still be years away, scientific information needed for making reliable predictions within the next decade are emerging from studies by earth scientists from many different institutions in the United States and in several other countries, including the Soviet Union, Japan and China.
Broader Inaccurate forecasting (#PF4774).
Aggravated by Unethical practice of earth sciences (#PD0708).

♦ PF4930 Voluntary dissolution of the family
Nature The voluntary dissolution of the family, by legally sanctioned divorce or separation, is becoming more and more frequent. Fewer couples are separated by death, more by their own decision. Laws, customs, attitudes vis-à-vis marriage and separation are, however, extremely different in various cultures.
Broader Family breakdown (#PC2102).
Narrower Divorce (#PF2100) Separation under marriage law (#PF3251).

♦ PF4932 Historical misrepresentation
Historical cover-up — Rewriting history — Historical revisionism — Distortion of historical events — Distortion of history — Suppression of historical information — Falsification of government historical records
Nature The process of rewriting extremely horrible historical episodes so that those responsible are exonerated is historical revisionism. Simply denying the reality of the event can easily be refuted. The most effective means of historical revisionism is to relativize the evil by pointing out that those responsible are no more evil than anyone else, or the victims of the event are partially responsible.
Incidence For many years there have been a host of small groups in the United States, in Canada, in Europe and elsewhere that deny the reality of the Holocaust and endeavour to get this perspective accepted by the mainstream of historical scholarship. Other examples include the genocide of the Herero people in southern Africa, and the collaboration between industrialists of the USA and the Nazi regime. History textbooks in Japan do not refer to Japanese aggression during World War II. Photos claiming to be of Mao's Long March, the storming of the Winter Palace in 1917 and the Viet Minh victory at Dien Bien Phu were taken of staged events, under better conditions. Movie makers are becoming the West's most powerful historians, but not necessarily the most careful. Films like "Mississippi Burning", "Gandhi", "Killing Fields", "Last Emperor" and "Cry Freedom" may be fine or even excellent films, but they are not historically accurate even though they are often thought as such.
In 1990 a committee of historians in the USA accused the government of falsifying the historical record by censoring its publications of official US diplomatic documents, despite the openness shown by the USSR concerning its own history. The incident concerns the role of US intelligence in post-war foreign policy and the manner in which the published editions of diplomatic archives are santized to exclude almost all references to the Central Intelligence Agency, and specifically with the role of the CIA in preparing the coup which brought the Shah of Iran to power and which eliminated the liberal-left Arbenz government in Guatemala. Unmistakeable evidence has been found of dramatic and devastating changes in editorial policies and processes which govern the publication of diplomatic history over past decades.
Claim A political system that allows a single source of information about events is, at least, presenting history from a single perspective and, at worst, manipulating history to its benefit. Efforts to promote the misleading impression of government non-involvement in specific historic events constitute a gross misrepresentation of the historical record, sufficient to deserve the label of fraud. Although it has long been accepted by historians that some editing of government archival material is necessary to protect intelligence sources and methods, this censorship process has become increasingly blatant to the point of bringing the integrity of the entire historical record into question.
Counter-claim Those who claim that history is being revised have their own equally biased and distorted view of events. And while these biases may be popular, they, nevertheless, are interpretations based on an ideology that has narrow perspectives.
Broader Institutional lying (#PD2686) Unethical practices of government (#PD0814).
Narrower Historical forgery (#PE5051)
Misrepresentation of historical persons (#PE5662).
Aggravates Ignorance of history (#PD3774)
Failure to profit from patterns of history (#PF1746).
Aggravated by Literary forgery (#PE6188) Classified public information (#PF9699)
Lack of access to public archives (#PD1194) Silence about historical situations (#PF0608)
Production serving false consumption needs (#PF2639).

♦ PF4936 Excessive prolongation of the dying process
Inhumane use of medical life support systems — Inappropriate prolongation of life — Compulsory prolongation of life
Nature A wide range of resuscitation techniques, blood transfusions, potent drugs, machines (heart, lung, kidney) and other supports are brought into operation to save the live of a patient, and even to revive a patient already dead by some criteria. In terminal and geriatric cases, these methods are so effective that life of a kind may be prolonged almost indefinitely, however limited the existence the patient then experiences.
Incidence An extreme example is the case of a women in the USA who had placed on life support (although comatose) from 1983 to 1989 at the cost of millions of dollars. In 1990 the Supreme Court ruled that no one had any right to terminate the life support facilities.
Claim 1. For millions of people whose span of life has been extended, its quality has been diminished. Some are in pain from cancer or have a wasting muscular disease; some are in acute discomfort from vomiting, diarrhoea, insomnia, bedsores, flatulence and general exhaustion. In hospitals and clinics, they are sustained by drips into a vein or tubes into the stomach. Legislation does not allow doctors to grant their pleas for merciful release and religions tend to argue strongly for the maintenance of life under any circumstances. The compassion shown to sick animals, by putting them out of their misery, is denied to fellow human beings.

2. When such people are not hospitalized, caring for them becomes the responsibility of a relative, who feeds, washes and nurses them, often for months on end without respite. Watching a loved one relentlessly deteriorating, from a much loved relative to a semi-corpse, and dealing with the irrational and irritating behaviour engendered by such suffering and by the extensive use of pain-killing drugs, erodes affection. When they eventually die, the relief is so great that they are not mourned.

3. Prolongation of a vegetative existence is morally indefensible. The growing numbers of distressingly handicapped infants and adults, and hopelessly decrepit aged, will eventually exhaust the manpower and medical resources available.
Aggravates Pain (#PA0643)   Dehumanization of death (#PF2442)
Inhibited grief process (#PD4918)   Denial of right to dignity (#PE6623)
Prohibitive medical expenses (#PE8261)   Individual unfitness for survival (#PF4946).
Aggravated by Fear of death (#PF0462)   Medical paternalism (#PF5397)
Compulsory health care (#PF4820)   Denial of the right to die (#PF4813)
Criminalization of euthanasia (#PF2643)
Avoidance of a confrontation with death (#PF1586).

♦ **PF4937 Lack of a world government**
Inadequate world regulatory agencies
Aggravates Inadequate coordination of international organizations and programmes (#PD0285)
Inadequate coordination of action on intergovernmental programmes at national level (#PE1375).
Aggravated by Unchallenging world vision (#PF9478)
Intergovernmental suspicion (#PC2089)
Lack of international cooperation (#PF0817)
Jurisdictional conflict and antagonism between international organizations (#PD0138).

♦ **PF4940 Behaviourism**
Nature A psychological approach based on the ideas of John Broadus Watson and later Burrhus Skinner, behaviourism proposes that psychology should only be concerned with the observation of behaviour and ways it might be modified, not with understanding why. This has lead to shallow analysis and temporary change, because humans are not so easily segmented. An evangelical propensity by adherents has led to numerous communities and social experiments.

♦ **PF4944 Inappropriate design of development projects**
Inadequate conception of development programmes — Impractical development programmes — Unfeasible development projects
Nature Development projects are often designed based on faulty assumptions. Besides failing to design in the participation of people especially the poor and women, disregarding the project's impact on the environment, failing to mobilize public opinion and discounting the need for infrastructure, development projects are often based on small scale pilot projects with high resources to problem ratios. They may describe adequately the problems they are trying to deal with and even have appropriate strategies to deal with the issues, they often fail to prioritize these strategies. Planners fail to recognize the political impact of the project and there by make it impossible for a government or an agency to actually implement it.
Broader Economic and social underdevelopment (#PB0539).
Narrower Lack of participation in development (#PF3339)
Neglect of the role of women in rural development (#PF4959)
Weakness of infrastructure in developing countries (#PC1228)
Inadequate mobilization of public opinion for development (#PF9704)
Overuse of machinery and equipment in development projects (#PE8166)
Restrictions on industrial and economic development due to environmental policies (#PE4905).
Related Poverty in developing countries (#PC0149).
Aggravates Natural environment degradation (#PB5250)
Destruction inherent in development (#PF4829)
Inappropriate management of development projects (#PD3712)
Excessive foreign public debt of developing countries (#PD2133)
Misappropriation of resources for high cost civil engineering projects (#PF4975).
Aggravated by Inappropriate development policy (#PF8757)
Incompatibility of environmental and economic decision-making (#PF9728).

♦ **PF4945 Mental impairment**
Mental inefficiency
Nature A considerable proportion of waking life is devoted to the use of the brain in the solution of problems. Inadequate diet and substance abuse inhibit performance of the brain.
Incidence Laboratory tests indicate substantial improvement in non-verbal intelligence, by supplementing the diets of with a diet deficient in minerals and vitamins. It is as yet uncertain how general and how important this effect may be. Several studies have found significant correlations between indices of alcohol consumption and measures of cognitive function (abstract problems, problem solving and memory tests) among moderate drinkers. Alcohol can damage cell membranes in the brain and can impair the absorption and metabolism of vitamins, in particular Vitamin B1 which is important for the function of brain enzymes. A minimum of 50 grams a day of alcohol appear to be needed before structural damage in the brain can be observed, but other effects such as mental impairment can be observed at much lower intakes.
Aggravates Substance abuse during control of complex equipment (#PE0680).
Aggravated by Climatic heat (#PC2460)   Mental illness (#PC0300)
Substance abuse (#PC5536)   Mental depression (#PC0799)
Psychological conflict (#PE5087)   Nutritional deficiencies (#PC0382)
Protein-energy malnutrition in infants and early childhood (#PD0331).

♦ **PF4946 Individual unfitness for survival**
Nature From the widest social perspective, or that of the development of the human race, individuals with certain characteristics may be considered unfit to survive, whether because of the disproportionate amount of resources required to care for them or because of genetically determined mental and physical defects which are undesirable in that development. This perspective raises the controversial question of the criteria of fitness for survival and by which it should be determined when the chronically or incurable ill should be allowed to die. This leads onto the equally controversial issues of the conditions of the elimination of those beyond hope and of the sterilization of the unfit to prevent them from perpetuating their defects.
Claim The natural eliminative process has been subverted by genetic tampering and advances in remedial techniques. Foetuses and infants which would otherwise not survive can now be maintained alive despite their defects. The following categories of people may be considered unfit: the mentally unfit (the insane, psychotics, the feeble-minded, the grossly subnormal) and the criminally insane; epileptics; those in prolonged and irreversible coma; those with severe brain-damage; those in the terminal phases of illnesses (syphilis, leprosy, tuberculosis, chronic renal and heart disease, multiple sclerosis, etc); the highly deformed (tetraplegia or paraplegia; grossly deformed neonates); those with irremediable behaviour defects (psychopaths, sociopaths incorrigible delinquents; repeated sex offenders; and criminals convicted for capital offences). Other categories might include: chronic substance abusers (drug addicts, alcoholics); and the very senile.
Aggravates Racial degradation (#PF9851).
Aggravated by Criminalization of euthanasia (#PF2643)

Decline in human genetic endowment (#PF7815)
Excessive prolongation of the dying process (#PF4936).
Reduced by Natural human abortion (#PD0173).

♦ **PF4947 Social hierarchy**
Persistence of hierarchy
Nature Any social hierarchy establishes ideal models at the top that are unattainable for the vast majority. Thus wherever there has been a monarchy, a caste system, a feudal order, or a landed or mercantile elite, the majority of people were accustomed to the idea that they and their children could never realized for themselves the ideal lives to which they gave respect and support. Indeed, realizing these ideal lives was not necessary nor a goal aspired to, because the lives they had were considered significant by themselves and by the society in which they lived. The school-based occupational prestige hierarchy of modern society, however, is different because it gives hope: If a child does well enough in the school system he will enter the higher ranks and will actually have access to an ideal life. The chances of actually rising to the top are few because the hierarchy is a pyramid with few places at the top and the chances of rising decreases with an increase in the number of people educated to higher levels. The hope that economic expansion will increase the number of places at or near the top is false because educational expansion in industrialized countries and population growth in developing nations outpace economic production. The result is an avalanche of failed aspirations throughout the world. The majority, whose positions are low, experience a sense of inferiority that is undeserved and can be psychologically damaging.
Reduced by Fragmentation (#PA6233).

♦ **PF4948 Undesired sexual obligations**
Date rapes — Forced sexual intercourse — Rape within marriage — Marital rape
Nature Forced sexual relations between people living together, or meeting each other socially, have alarming implications, especially in the case of marriage. Most legal systems exclude the wife of the offender as a victim of rape, since marriage by definition includes sexual relationships. However, an atmosphere of violence surrounding sex in marriage seems to prevail to a considerable degree. According to a London survey one woman in seven were raped in marriage. Many women continue to live in a brutalized relationship, possibly because of their uncertain economic future should they leave their husbands, or possibly due to the fear of the negative attitude commonly held about a woman whose partnership has disintegrated. Although it is usually both partners who contribute to the failure of a relationship, it is usually the man who uses physical aggression and thus turns a tense or broken relationship into a violent one.
Incidence On the basis of a UK study, 47 percent of women do not consider being forced into sexual intercourse, having visited a man in his house, as rape. This view is shared by 49 percent of men. Similarly 12 percent of women do not perceive being forced into sexual intercourse with a superior at work as constituting rape. In the USA it has been estimated that 17 percent of women have been raped by the age of 20, but 75 percent of them did not perceive this as rape because of the circumstances.
Claim Failure to perceive some scenarios as rape encourages the emergence of a rape-supportive culture.
Refs Bayer, Edward J *Rape Within Marriage* (1985).
Broader Rape (#PD3266)   Unethical personal relationships (#PF8759).
Aggravates Forced motherhood (#PE5919)   Sexual unfulfilment (#PF3260)
Deception between sexual partners (#PE4890).

♦ **PF4954 Religious extremism**
Excessive integration of religious belief — Religious fanaticism
Nature Over-zealous and sometimes violent behaviour can arise within a religious community, particularly a community which maintains that the book of laws on which it is founded is literally and uniquely divine and authoritative and thus encompasses all human behaviour with its infallible laws. Examples are Christians with the New Testament of the Bible, and Muslims with the Koran. Such an intense focus on the laws of one book leads to fundamentalist rigidity, intolerance and anti-social acts against non-believers. Another condition which predisposes to extremism is adherence to ritualistic behaviour, whether such behaviour is specified in the accepted book or not. When rituals are given the status of binding commandments their practice can lead to discrimination against those who are non-conformists, and to extremist acts against them. Rituals may encompass not only ways or worship but also dress, diet, language, social contact and the use of special symbols or body marks. Religious extremism and fanaticism may be turned as much toward members of the same faith, as it is towards non-members.
Incidence Religious extremism and fanaticism are a phenomena bound up with exclusive claims to divine revelation. Examples are best known in, although not confined to, the major worldwide religions. Religious purity has served as an excuse for censorship and destruction of books and works of art, attacks on intellectuals, artists, political or other spokesmen of rival ideologies, and destruction of meeting places and personal residences. Non-conforming individuals have been subject to economic sanctions, ostracized, tortured and murdered.
Religious fanaticism has never been restricted to one group. Judaism provides examples of stoning to death in ancient times, of ostracism in mediaeval times, and nowadays of the violence sometimes associated with modern Zionism. Christianity has a long and bloody record of conversion by force which includes the Crusades and the Inquisition and such current secret terrorist societies as the Klu Klux Klan. Islam provided the origin of such terms as Holy War, fanaticism and assassination. Equal examples of extremism can be found in other religions. It is noteworthy that the quasi-religion of Nazism exhibited the same aberrant psychology. Modern ideologues in every country, who have replaced a theocentric religion with a nominally anthropo-centric philosophy, are susceptible to the same weaknesses, as is seen in some revolutions.
Claim All fanaticism in religion is worse and more destructive than fanaticism in politics or ideology. Efforts to achieve socio-religious homogeneity by controlling beliefs and behaviour can be termed excessively integrationist. It appears that this is the conscious philosophy of the founders of all sects and religious movements, whose roles as prophet-shepherds requires them to isolate their flocks and give them unique identity, while at the same time manipulating them towards undeclared objectives. Thus religions which teach the individual to passively accept all the dictates of authority in whatever form, are the breeding ground for extremism and fanaticism.
Broader Extremism (#PB3415)   Religious intolerance (#PC1808).
Narrower Jihad (#PF5681).
Related Political radicalism (#PF2177).
Aggravates Agnosticism (#PF2333)   Religious war (#PC2371)
Malignant visions (#PF5691)   Religious terrorism (#PD4134)
Religious prejudice (#PD4365)   Religious conflict between sects (#PC3363)
Self-inflicted physical suffering (#PD7550).
Aggravated by Imprisonment (#PD5142)   Fundamentalism (#PF1338)
Religious schism (#PF1939)   Religious repression (#PC0578)
Religious intimidation (#PC2937)   Religious discrimination (#PC1455)
Persecution of religious sects (#PF3353)
Denial of religion in communist systems (#PC3175).
Reduced by Religious apathy (#PC3414)   Fragmentation of religious belief (#PF3404).

♦ **PF4955 Constraints on the development of mental health services**

Nature In some countries, the law is too rigid concerning compulsory detention of mental patients. Unnecessary deprivation of personal liberty in order to provide treatment for mental disease leads to fear and delay in seeking help, and reduces use of the service by persons in need. Furthermore, it militates against the public health principles of early recognition and treatment. The method of financing mental health care may hold back the development of units in general hospitals and day hospital treatment. Comprehensive health care systems where all the money is funded from the state may be more flexible in their approach and more likely to encourage experimentation than those where financial provision is over-rigidly related to the number of beds available. Where a large private section exists, the public mental health services are in a disadvantageous position. The private sector tends to be very selective in the type of patient treated, as few people can afford to pay for treatment of the prolonged illnesses so characteristic of mental disorder. Because the private sector can be selective, it is able to offer salaries and working conditions which are better than in the public sector. Moreover, where there is a private sector, this often results in people who might otherwise be active in pressing for reform, being treated outside the publicly provided service. In short, two standards of service may develop and, in the ensuing competition for staff, the private sector is at an advantage, with serious consequences for the majority of the mentally ill.

Conservatism also impairs development. There are many large institutions where outmoded custodial attitudes persist and whose staffs see no need for change. While new methods of working are professionally more exciting, they are also more demanding. New techniques such as multidisciplinary team-work, and group and individual psychotherapy, should be acquired by all the staff concerned, but many lack motivation or are reluctant to learn these new techniques. The lack of trained manpower appears to be the principal constraint on the further development of psychiatric services. In claiming resources, psychiatry is at a disadvantage compared with other health services such as, say, renal dialysis or cardiac surgery which make a more spectacular appeal. By contrast with these, the needs of the mental health services appear to be lacking in urgency and public interest. Too often low priority has been given to the mental health services for so many years that the morale of staff and standards of care have fallen as a consequence.

Broader Obstacles to national development (#PF4842).
Narrower Inadequate mental health services (#PJ4379).
Aggravates Mental illness (#PC0300).

♦ **PF4957 Polytheism**

Claim Polytheism does not give man any feeling of real relationship with his gods except that he imagines them to be all-controlling agencies having the power to bestow happiness or inflict pain, provided they are appropriately worshipped. But as such they do not offer any moral guidance, thus encouraging idolator (s to invent their own code of morality and plan a way of life based on pure expediency and opportunism. Polytheism does not provide an independent and lasting basis for the arts, sciences, philosophy, literature, politics and economics. Throughout the ages polytheism has been reinforcing atheism.

Aggravates Atheism (#PF2409) Idolatry (#PF3374).

♦ **PF4958 Illegible handwriting**

Indecipherable script

Nature Many people can write, but not in such a way that others can read it. As typewriters and word processors become common even for students, the skills of penmanship are less emphasized. Further, it has become a curious sign of distinction to have penmanship which requires unusual effort for others to understand.

Broader Ineffective communication (#PF6052).

♦ **PF4959 Neglect of the role of women in rural development**

Limited role of women in the countryside — Lack of participation by women in development

Nature The role of women in rural development has been neglected. Integrated rural development plans for increased food production, improvement in nutrition and provision of primary health care are unattainable without the cooperation of rural women. It is not simply a question of providing employment opportunities for women, as they are already overburdened with work, but of integrating women fully into the process of development and its benefits. The single most important determinant of social change is the level of female literacy. Primary education for girls, and vocational training to enable women to upgrade their skills and increase their productivity, are neglected. Little attention is given to the improvement of traditional technology and equipment. Rural industrialization eliminates women as producers of traditional crafts and does not seek to integrate them into the new industries. Governments' delivery systems in all regions are manned mostly by males and are directed mainly to male producers; even where, as in some countries of Africa, more than 70 percent of the farmers are women. In regard to agricultural credit, they are denied access because they lack title to land collateral. Most agricultural extension agents are men, and cater mainly to Male clientele. Women do not participate, nor contribute on an equal basis with men in the social, economic and political processes of rural development, nor share fully in improved conditions of life in rural areas. Where development efforts are directed toward women, their problems tend to be seen as separate, rather than facets of the culture and structure of society. Full membership and equal rights for women in rural worker's organizations must still be promoted.

Social and religious prejudice militates against the full integration of women in development programmes, regardless of the fact that 'traditional' occupations of women in developing countries concern both agriculture and nutrition, and that backwardness in these areas is a fundamental obstacle to general development. Lack of progress in population control and control of disease is also aggravated by women's lack of education and their non-participation in development. Underutilization of human resources is another result, unemployment in general under the present systems contributing to the restriction of women's participation. Lack of participation by women may result in the ineffectiveness of such programmes and a retarding of national development.

Claim Women, who constitute more than half of the population, should not be left on the fringes of national development, thereby creating a stumbling block to any effort to introduce real changes. Too many governments are still not sufficiently aware of the benefits to be derived from women's contribution to development. Their ideas in that regard remain narrow, although if development is to have any chance of success it must be integral and integrated. Involvement of women is particularly important both from a moral angle, because of the discrimination to which women have been and still are subjected, and from the material angle, because of the effects of such discrimination on their role in society.

Counter-claim Traditional occupations of women in developing countries affect many aspects of the social-economic development process. Women have a vital role in food production, food processing and food preparation, marketing, in water management, in health and sanitation of the whole family, in collecting sources of energy, etc. In many cases women in developing countries, although overworked, do participate in development. Women's roles and activities are unrecognized, taken for granted or undervalued.

Refs Fetherolf, Loufti Martha *Rural Women. Unequal Partners in Development* (1987); Heyzer, Noeleen (Ed) *Missing Women* (1985); ILO *Women in Rural Development* (1988); Kandiyoti, Deniz *Women in Rural Production Systems* (1986).

Broader Lack of participation in development (#PF3339)
Exclusion of women from decision making (#PE9009)
Inappropriate design of development projects (#PF4944).
Narrower Sex segregation (#PC3383) Wasted woman power (#PF3690).
Related Static and unrelated social roles (#PF1651)
Lack of local leadership role models (#PF6479)
Inflexible social care structures in developing countries (#PF2493).
Aggravates Lack of family planning (#PF0148) Social underdevelopment (#PC0242)
Nutritional deficiencies (#PC0382) Conflicting roles of women (#PD6273)
Inappropriate development policy (#PF8757) Discrimination against rural women (#PE4947)
Destruction inherent in development (#PF4829)
Unproven relationships between problems (#PF7706)
Discrimination against women in education (#PD0190).
Aggravated by Illiteracy (#PC0210) Discrimination against women (#PC0308)
Inadequate educational facilities (#PD0847)
Discrimination against women in employment (#PD0086)
Discrimination against women before the law (#PD0162)
Discrimination against women in public life (#PD3335).
Reduced by Elimination of the socio-cultural role of western women (#PE1046).

♦ **PF4967 Unreported violence**

Unreported conflicts

Broader Unrecorded knowledge (#PF5728) Underreported issues (#PF9148).
Narrower Unreported rape (#PE5621).
Aggravates Alcohol-related violence (#PE7084).
Aggravated by Fear of death (#PF0462) Human violence (#PA0429)
Unreported crimes (#PF1456) Avoidance of negative feedback (#PF5311).

♦ **PF4968 Psychic conflict**

Intrapsychic conflict — Internalized conflicts

Nature A struggle among incompatible forces or structures within the mind, as opposed to external conflict involving a struggle between the individual and the external environment. External conflicts may be internalized.

Related Anxiety (#PA1635).
Aggravates Infantile neurosis (#PE3571) Psychological inhibition (#PF6339)
Human psychological regression (#PE1418).
Aggravated by Conflict (#PA0298).

♦ **PF4975 Misappropriation of resources for high cost civil engineering projects**

Implementation of development mega-projects

Incidence During the 1960s and the 1970s, large-scale development projects, including dams, highways and industrial schemes, were implemented. Many of these ended in bankruptcy in the 1980s whilst the environmental degradation they engendered became increasingly apparent in the form of razed forests, depleted fisheries, and desertification. In the 1990s a new generation of mega-projects is being developed which are completely out of scale with the lives of the people whose land they will use. Such projects include: diversion of water from the Zaire to Chad; construction of equatorial space ports in Indonesia (Biak) and Australia (Cape York); electrical grid linking Middle Eastern Islamic countries; damming the Yangtse; construction of a second Panama canal; damming the Mekong; exploitation of Cameroon forests; exploitation of the Saharan aquifer; construction of the Narmada dam (India).

Broader Excessively costly prestige projects (#PF3455).
Aggravates Failure of development programmes (#PF8368).
Aggravated by Inappropriate design of development projects (#PF4944)
Inadequate research on proposed solutions to problems (#PF1572).

♦ **PF4979 Inadequate mobilization of resources**

Failure to mobilize resources

Narrower Inadequate mobilization of public opinion for development (#PF9704).
Aggravates Long-term shortage of resources (#PB6112)
Worldwide misallocation of resources (#PB6719).
Aggravated by Inadequate management of government finances (#PF9672).

♦ **PF4983 Complicity**

Abetment — Complicity in crime

Nature Complicity involves two or more persons committing a crime in common, for which they have a common intent. There may or my not have been prior agreement. A crime involving complicity presents a greater social danger, since it may do especially appreciable harm to the interests of the state as well as the legal interests of citizens, and since it makes it easier for the accomplices to conceal the traces of their crime.

Liability for a crime is not limited to the person who actually commits the proscribed act. Rather, liability extends to anyone who has encouraged (incited) or assisted (abetted) perpetration of the crime or who has hindered apprehension of the perpetrator after commission of the offence. All such persons are regarded as parties to the crime and will be liable as if they themselves had committed the criminal act.

Incidence Recent cases of complicity have brought to light governmental assistance, which raises the question of governmental liability. Some incidents include: official Mexican complicity in narcotics traffic; the French government's complicity in the sinking of the Rainbow Warrior; Soviet complicity in the false defection of KGB agents to the USA; and various US-backed coup attempts in developing countries.

Broader Crime (#PB0001).
Narrower Inciting riot (#PD6392) Criminal concealment (#PF8000)
Complicity with evil (#PF0926) Criminal solicitation (#PD7676)
Aiding consummation of crime (#PD6655) Aiding national security criminals (#PD7407)
Complicity with structural injustice (#PJ5026)
Government complicity in illegal activities (#PF7730).
Related Criminal conspiracy (#PD1767) Vicarious criminal liability (#PE6323).
Aggravates Guilt (#PA6793).

♦ **PF4984 Inadequate education**

Discontinued educational programmes — Makeshift educational programme — Inadequate educational programmes — Educational deficiency

Incidence In countries with a very low GNP, the economic system often cannot finance the cost of mass education; where GNP is higher, inadequate education is often not a question of economic capacity but rather of the priority given to the allocation of resources for education, and whether policies in force favour educating the masses or concentrating resources in educating the elite. An education system unable to provide the entire school-age population with a primary education but one which provides a minority sector with educational opportunities up to a considerable age and up to high levels, is usually the result of the way in which the structure of education is influenced by middle-income groups. Within the already established state educational systems of developed countries, such as the USA, problems exist which affect and often limit the educational opportunities of the poor, of minorities, of residents who do not speak the national language, and of physically and mentally handicapped children.

Refs Carnoy, Martin *Education and Employment* (1977).
Broader Inadequacy (#PA8199) Obstacles to education (#PF4852)
Ineffective systems of practical education (#PF3498).
Narrower Bilingualism (#PF7927) Miseducation (#PA6393)
Lack of training (#PD8388) Lost farming skills (#PJ8369)
Inadequate teaching (#PC9714) Compulsory education (#PJ2615)
Inappropriate education (#PD8529) Inadequate sex education (#PD0759)
Failure of school systems (#PJ3896) Limited medical knowledge (#PD9160)
Inadequate driver training (#PJ4507) Inadequate training of judges (#PF6032)
Inadequate technical training (#PE8716) Inflexible educational system (#PJ5365)
Absence of management training (#PD3789) Inadequate education for peace (#PF6208)
Inadequacy of formal education (#PF4765) Inadequate political education (#PJ7906)
Inadequate preschool education (#PJ1962) Insufficient primary education (#PC6381)
Manipulation of civic education (#PF0996) Hopelessness inspired by school (#PJ1004)
Inadequate vocational education (#PF0422) Insufficient leadership training (#PF3605)
Inadequate environmental education (#PD1370) Outmoded forms of social education (#PF2095)
Irrelevance of educational curricula (#PF0443) Inadequate family planning education (#PD1039)
Excessive specialization in education (#PC0432)
Inadequate education for nomadic children (#PE6206)
Inadequate education of indigenous peoples (#PC3322)
Limited development of functional abilities (#PF1332)
Excessive institutionalization of education (#PD0932)
Diversion of education to qualification earning (#PF4828)
Inadequate education concerning the nature of problems (#PE8216)
Inadequate nutrition education in least developed countries (#PE0265)
Lack of meaningful educational context for ethical decisions (#PF0966)
Training inappropriate to structural and technological changes (#PE4596).
Related Societal over-commitment to learning (#PD7051).
Aggravates Ignorance (#PA5568) Paternalism (#PF2183)
Imprisonment (#PD5142) Limited basic skills (#PJ8138)
Lack of racial identity (#PF0684) Overdependence on education (#PE8988)
Inadequate secondary education (#PD5345) Lack of sharing of community skills (#PF3393)
Available jobs unrelated to education (#PJ9233)
Retardation of psychomotor development in children (#PD1307)
Inadequate political parties in developing countries (#PD0548)
Lack of opportunities for practical training in communities (#PF2837)
Imbalance between training and existence of openings in various professions (#PE5194).
Aggravated by Loose school security (#PG7769)
Segmental school experience (#PJ9345)
Shortage of trained teachers (#PD8108)
Debilitating education images (#PF8126)
Restrictive regulations for training (#PJ7972)
Unsatisfied need for continuing education (#PF0021)
Ineffective mechanisms for functional training (#PF1352).

♦ PF4989 Blind faith in technology

Technocentrism

Nature As technical solutions to problems become increasingly sophisticated, there is a well-founded tendency to place greater trust in machines than in human beings: computers for tasks requiring precision and accuracy and heavy equipment for tasks requiring strength, stamina or the ability to withstand heat, cold, radiation, and airlessness. When errors occur, it is very difficult to establish responsibility or to effect any changes other than improvement of the technology. Thus social problems and accidents of human origin become more and more difficult to treat.

Claim Technocentrism is the believe that technology will solve everything. It is a way of thinking that recognizes problems but believes that technological advancement is the only route to their solution.

Broader Uncritical thinking (#PF5039).
Related Unrealistic environmentalism (#PF4510) Uncritical acceptance of authority (#PF8596)
Uncritical acceptance of dogmas and standards (#PF2901).
Aggravates Unnecessary gadgets (#PE3745).
Aggravated by Gullibility (#PJ2417) Proliferation of technology (#PD2420).

♦ PF5000 Mismatch of national macroeconomic policies among industrialized countries

Divergences in macroeconomic policies of leading industrialized countries — Uncoordinated macroeconomic policies of governments

Incidence During much of the 1980s, the macroeconomic policies of the USA were expansionary whilst those of Japan and European countries were contractionary. Low tax revenues and high public spending caused US general and federal budget deficits and negative saving-investment balance reflected in the current account deficit. Japan and Germany FR followed more restrictive fiscal policies which led to current account surpluses.

Aggravates Economic stagnation (#PC0002) High interest rates (#PF9014)
Economic uncertainty (#PF5817) Exchange rate volatility (#PE5930)
International economic recession (#PF1172) Protectionism in international trade (#PC5842)
Decline in real wages in developing countries (#PD2769)
Economic imbalances among industrialized countries (#PD5865)
Outflow of financial resources from developing countries (#PC3134).
Aggravated by Uncoordinated policy-making (#PF9166).

♦ PF5006 Limited exchange of skills among developing countries

Inadequate inter-developing country skills flow

Nature The skill-importing developing countries can have the access to skilled manpower on substantially better terms than those offered to the equally skilled personnel from developed countries; access to skill experience more relevant to the economic and cultural basis of their societies; greater choice in the selection of skills and sources; and more, importantly, an assured supply of a variety of skilled people required for the implementation of their industrial programmes. The main reason for the non-realization is the characteristics of the present market context in which the exchange takes place. Information regarding the demand for various skills between developing countries and its diffusion to the job-seekers, recruiting agents and employers are lacking. There is also a lack of linkages of the migration system with the decision-making units in the education and training sectors on the other hand, and with the decision-making process in development planning on the other. The sending countries face such problems as the social cost of "replacement migrants"; occupational immobility and segmentation of the labour market in response to unplanned outflows; imperfect substitution for skilled manpower emigrating in certain sectors; fluctuations in the emigrants' remittances and shortages of specific skills debilitating the over-all economic development. The competition between skill-exporting countries in recruiting specific skills, as well as skill-supplying countries for access to the same markets have reinforced the inefficiency of the market mechanism for the exchange of skills among developing countries.

Claim The great bulk in the inter-developing country flows are semi-skilled or unskilled workers instead of doctors, engineers or scientists needed in the sending countries. The skills flows are for a "fixed period" in a socio-economic environment similar to that of sending country, which facilitates the readjustment for the return migration. Besides, the exchange between developing countries does not give rise to a number of mismatches between expected requirements and actual availabilities of resources as in the exchanges between developing and developed countries.

Related Inadequate supply of appropriate trained manpower in developing countries (#PE6243).
Reduced by Emigration of trained personnel from developing to developed areas (#PD1291).

♦ PF5008 Limited availability of land for low-income and disadvantaged groups

Nature Land for low-income and disadvantaged groups is a critical issue in both agricultural areas and in human settlements. There are gross inequalities in the distribution of rights to agricultural lands in most developing countries, and low income and disadvantaged groups are being excluded from adequate access to productive agricultural land.

Broader Limited available land (#PC8160) Unavailability of agricultural land (#PC7597).
Aggravates Cultivation of marginal agricultural land (#PD4273).
Aggravated by Excessive land usage (#PE5059).

♦ PF5009 Obsolete policy-making

Obsolete planning — Obsolete decision-making

Broader Obsolete methods (#PF3713).
Narrower Obsolete defence planning (#PJ1877).
Aggravates Conflicts of interest (#PF9610).
Aggravated by Obsolete deliberative systems (#PD0975).

♦ PF5012 Contempt for authority

Disrespect for authority

Related Disrespect for elders (#PF3979).
Aggravates Flouting regulatory authority (#PD7792).

♦ PF5013 Jealousy

Jealousy of personal success — Delusional jealousy

Nature Jealosy is a response to the actual or presumed advantage of a rival, especially in regard to the loved person or professional success.

Broader Envy (#PA7253) Lack of community participation (#PF3307).
Narrower Professional jealousy (#PD8488).

♦ PF5039 Uncritical thinking

Unreflective thinking — Inadequate inquiry-based learning — Philosophical ignorance — Inadequate development of critical awareness

Nature Conventional approaches to education tend to leave students deficient in comprehension, analysis and problem-solving skills. The inductive and deductive skills required for reading comprehension, plus the concept-formation and concept-analysis skills needed for the processing of information, are not adequately cultivated.

Background Students in every discipline require reasoning and concept formation skills as part of their career preparation and for their participation in institutions of society and for the overall course of their personal lives. The need for these skills is therefore essential rather than superficial. The assumption that these skills develop automatically with maturation, and that specific instruction in these areas need not be provided, puts masses of children at risk and may deprive them of the possibility of leading meaningful lives.

Broader Ignorance (#PA5568).
Narrower Single resource thinking (#PG8548) Blind faith in technology (#PF4989)
Unrealistic environmentalism (#PF4510) Uncritical acceptance of authority (#PF8596)
Uncritical acceptance of another person (#PF5973)
Inflexible military thinking in industry (#PE7040)
Uncritical acceptance of dogmas and standards (#PF2901).
Aggravates Gullibility (#PJ2417) Avoidance of the irrational (#PF1610)
Irrational religious beliefs (#PF6829) Avoidance of negative feedback (#PF5311)
Uncritical preservation of the status quo (#PF1688)
Inadequacy of prevailing mental structures to challenge of human survival (#PF7713)
Inadequate global consensus concerning problems and prospects of humanity (#PF9821).
Aggravated by Rote learning (#PJ5437) Inappropriate education (#PD8529)
Inadequate secondary education (#PD5345) Proscribed thinking and behaviour (#PF3862).

♦ PF5044 Unpreparedness for surplus leisure time

Excess leisure time arising from automation — Uninventive leisure time — Insignificant leisure engagement — Inability to make use of leisure time

Nature In countries with a high level of scientific and technical achievement and where automation continues to expand along the present lines, expectations for the foreseeable future are that there will be a 30 hour working week for a total of 40 working weeks, that is, 1200 working hours per annum. Extended education (over a longer period that at present) and early retirement by definition result in reducing the number of professionally active years. This number can be set on average at 35 years. If one multiplies the figures given above, that is 35 years times 40 working weeks times 30 working hours, the result is that in future man will only devote some 40,000 hours of his life to productive work. Since man - statistically speaking at least - has some 700,000 hours to live, this implies that in the near future the time spent on productive or gainful employment will only amount to 6 percent of his total life span. Free time - once ranking third in the list of items composing our daily schedule, the first being the time spent working and the second the time devoted to sleep - will then move to the top of the list. The time to be spent working and sleeping will become short in relation to the excessive amount of time to be used as one pleases. In fact, the ration between work and leisure has been inverted. Whereas during the 19th century there were fewer than 8 hours of leisure daily, this figure has now been almost doubled, and will probably be trebled in the future, judging by the experience of the United States. Hence the problem of leisure has now become qualitative rather than quantitative: the newest social problem of the industrial world, unprepared to cope with so much non-productive, free time.

Claim The working man's does not have the ability to 'handle' growing amounts of leisure time rapidly becoming available. Alcoholism, crime and fatigue will increase as a result. Extended leisure will increase conflict and tension in the family and cause considerable role strain.
Another aspect is the increase in the number of cases of 'weekend neurosis' - workers turned loose on weekends with free time often feel guilty, jittery, anxious, frustrated and miserable. Equally disturbing is the fact that the more monotonous a worker's job the more monotonous the recreation he seeks in his free time.

Broader Idleness (#PA7710) Social ill-effects of automation (#PE5134)
Underutilization of human resources (#PF3523).
Narrower Inadequate recreational facilities (#PF0202).
Related Meaningless recreation (#PF0386)
Obstacles to efficient utilization of time (#PF7022)
Constraint of time on individual and social development (#PF5692).
Aggravates Ignorance (#PA5568) Unstimulating entertainment (#PF8105)
Excessive television viewing (#PD1533)
Excessive claims for human development through sports (#PG4881)
Environmental degradation by off-road and all-terrain vehicles (#PE1720).
Aggravated by Frequent schedule disruptions (#PJ0871).

♦ PF5055 Inadequacy of scientific expertise

Nature Scientific information is more to be considered as a tool than actually providing elements of truth or solid facts. The scientific expertise that creates a problem of 'externalities' is not usually adequate to its solution. For instance, scientists using recombinant DNA techniques do not necessarily possess skills in pathogenicity or in microbial ecology. Also, some of the most urgent environmental problems lie outside any single recognized branch of science. This is partly due to their novelty but also to their inherent complexity, both scientific and social. As a result, there is always likely to be radical disagreement over methods and even over the basic competence of participants in a policy debate.

Scientific expertise is problematic is three fundamental ways: its basic assumptions, its method and its social support structures. The basic assumptions of science are: 1) All of reality that exists is subject to human sensory perception and mathematical reason; anything outside this is "supernatural" and either does not exist or isn't worth investigating; 2) The universe, including human evolution, runs by arbitrary chance and blind necessity and is essentially meaningless and dead; 3) In spite of this, humans can manipulate and master nature, creating their own future in a history of evolutionary progress; 4) Human beings are the arbiters of ultimate value; 5) Objective reality is value free; 6) Nature is expendable; and 6) There is meaningful order in the cosmos. The scientific method is problematic, if for no other reason than its dominance over other methods. It has had its successes but it is viewed as the only 'right' method. Reductionism, that is reducing the object under study into smaller and smaller components whether it be the human body, a crystal or the Earth is central to science. One adverse side-effect is that research is fragmented into specialties and sub-specialties which become insular, jealous and parochial. Another side-effect is the separation of science from nature. The scientific method is rational and thereby linear, precluding non-linear and intuitive thinking.

The social support systems: educational systems, social attitudes, businesses and governments reinforce science's world view and its methods and has dire effects on social systems. Non-western understandings of the universe are not only considered inadequate by the scientific community but by those who hold these understandings, not because they fail to allow people to cope effectively with the world but simply for the fact they are non-scientific. These and other institution strengthen science's tendency to exclude and exploit women. This exclusion took its most brutal form during the fifteenth and sixteenth centuries when millions of women midwives and healers were put to death as witches and medicine was put in the hands of male physicians. Today the numbers of girls studying science is miniscule compared with boys.

Broader Inadequacy of doctrine (#PF3396) Unavailability of appropriate expertise (#PF7916).
Narrower Inadequacy of psychiatry (#PE9172) Inadequacy of medical science (#PF8326) Decreasing rate of development of major agriculture technologies (#PE5336).

♦ PF5057 Proliferation of consumer products
Excessive consumer choice — Consumer over-choice — Consumer apathy

Nature Consumers are increasingly overwhelmed by the range of choices, especially in industrialized countries. Increasingly people are reaching a point of over-choice, namely the point at which the advantage of diversity and individualization are cancelled by the complexity of the consumer's decision-making process. Individuals can then become paralyzed by the choices, leading to a form of apathy which spills over into other areas of life.

Incidence In the USA in 1990 consumers can select from among 25,000 items on supermarket shelves, access 53 television channels, subscribe to 11,092 periodicals, and be solicited by many thousands of pubic and special interest groups.

Claim Marketing has cultivated a misunderstanding about the nature of freedom of choice. This is not equivalent to multiplying the opportunities for choice. Infinite choice reduces people to passivity.

Aggravates Dumping of consumer waste products (#PD8942).
Reduces Decrease in consumer choice (#PD6075).

♦ PF5061 Self-destructive government policy-making
Government folly — Policy contrary to self-interest — Political folly

Nature Governments, authoritarian or democratic, Communist or Capitalist, modern or ancient, have the capacity of pursuing over decades policies which are contrary to their own self interest. Perversity is continued even in the face of mounting evidence of its self-destructiveness, clear arguments against the policy and real alternatives to it. Obstinately adhering to principles, policies, values or beliefs while ignoring or denying the veracity of any contrary signs generates illusions about the effectiveness of existing doctrine and rigidifies decisions, making reversals of plans impossible. Self-centred rulers and their desire to be seen in the most positive light contribute to policy stagnation. These leaders seem to have the capacity to learn nothing from their mistakes or even acknowledge that they make them. Another method of government folly is taking the most provocative action possible insuring the unification of opponents. This is usually coupled with a remarkable underestimate of the enemy. Displays of uncertainty, apprehensiveness and weakness encourage opposition. Most if not all self-destructive policy is marked by unnecessary action when doing nothing could serve better. Even the most wise of governments make mistakes, even disastrous mistakes, but it is the self-destructive government that dogmatically continues in the same way. Another form is attempting to return to some past period of government, reinstating institutions, policies, and forms which are out of date and frequently in disrepute. Governments seem to have a remarkable capacity to act foolishly even granting that they are products of their own historical moment, cultural and political biases, and political ambitions.

Refs Tuchman, Barbara W *The March of Folly from Troy to Vietnam* .

Broader Self-destruction (#PF8587) Inappropriate policies (#PF5645) Ineffective decision-making processes (#PF3709).
Narrower Political over-reaction (#PF4110) Aggressive foreign policy (#PC4667) Uncoordinated policy-making (#PF9166).
Related Cumbersome methods of policy formation (#PF1076).
Aggravated by Unrealistic policies (#PF9428) Governmental incompetence (#PF3953).

♦ PF5069 Increased reflection of solar radiation
Reduction in albedo of the Earth's surface

Nature Human activity has increased the extent to which solar energy is reflected back from the Earth, thus modifying the global radiation balance. This has been achieved by the destruction of forests and desertification. The planetary albedo is also altered the introduction of aerosols and particles into the atmosphere.

Incidence It has been estimated that the reflection of solar radiation may have increased by 1 W per square metre over the whole of human history, and probably by 0.2 W per square metre over the last 30 years. It is estimated that over the next 100 years this may be increased by from 0.3 to 1 W per square metre. The expected proliferation of solar collectors would contribute to this effect.

Counter-claim Global energy change due to modifications to the planetary albedo are likely to be quite small compared to the expected radiative effects of trace gases and perhaps even to natural climatic variability.

Broader Degradation of the atmosphere (#PD9413).
Aggravates Global warming (#PC0918).
Aggravated by Deforestation (#PC1366) Desert advance (#PC2506).

♦ PF5072 Inadequacy of international standards
Narrower Lack of livelihood standards (#PF1297)
Substandard shipping vessels (#PE6630)
Low-quality construction work (#PD7723)
Inadequate standards of living (#PF0344)
Multiplicity of time standards (#PF2621)
Lack of professional standards (#PF3411)
Non-standardized social services (#PF2974)
Politicization of health standards (#PD4519)
Instability of orthographic standards (#PF0552)
Parochial telecommunications standards (#PE1840)
Non-standardization of geographical names (#PF8511)
Restrictive building codes in urban areas (#PE8443)
Restrictive agreements on product standards (#PD0343)
Inadequacy of international human rights instruments (#PF6365)
Inadequate standardization of procedures and equipment (#PC0666)
Inadequate international standards of financial accounting and reporting (#PF0203)
Conflicting standards for protection against chemical occupational hazards (#PE5651).
Related Sexual immorality (#PF2687) Double standards of sexual morality (#PF3259) Inequitable labour standards in developing countries (#PD0142).
Aggravates Uncritical acceptance of dogmas and standards (#PF2901).
Aggravated by Excessive standardization (#PF2271)
Discriminatory imposition of standards (#PD5229)
Resistance to internationally agreed standards (#PC4591)
Limited enforceability of international standards (#PF8927).

♦ PF5073 Shortage of experimental non-human primates
Nature Animals caught in the wild are the major source of supply for medical experiments, but their availability has decreased in recent years, partly due to export restrictions but also because of a ban in certain countries on the importation of monkeys to be kept as pets. The shortage of monkeys for biomedical purposes could lead to a lowering of safety standards for drugs and vaccines, while much medical research could be severely handicapped.

Incidence About 85,000 primates a year are used in biomedical programmes, including 80 percent of Asian or African species (mostly Rhesus monkeys) and 20 percent from the Americas. Although reports from countries on wild monkey populations are based on impressions rather than survey data, there is a general tendency for populations to decrease as their natural habitats are destroyed.

Related Inhumane use of non-human primates in research (#PE1621).
Aggravates Endangered species of non-human primates (#PE1570).

♦ PF5079 Lack of family life education
Reduced family education

Nature In the developing and the developed world, the teaching of adults and children appropriate ways of living as a family remains limited. In traditional and rural communities the process of socialization included preparing young people and children for the process of living as a family member. In today's urban environment and with the nuclear family much of this capacity is not passed on to individuals and so parents are left on their own to discover ways of being a family unit. While many nations teach sex education few teach sexuality. House keeping for men or women is frequently not passed from one generation to the next. Developing and using a family budget is seldom taught. The processes of making decisions within the context of the family is make shift at best and frequently inadequate. The techniques of decision making are not taught. Social interaction within the family is unclear and interaction between the family and the larger community is haphazard. Because of the development of the nuclear family and the loss of interaction with elderly people children never understand the process of growing old and are insulated from death. Generally it is only the religious schools that consciously teach any form of morality or ethics, so the bulk of students learn from example and the media. Parents often feel inadequate to teach ethics further exasperating the problem. The structures of formal education is left to prepare the child for job and social responsibility but for the most part schools never deal with the problems of being in a family.

Broader Obstacles to family life (#PF7094) Lagging training in social skills (#PF8085) Limited availability of functional information (#PF3539).

♦ PF5097 Shortage of biological specimens for medical study
Incidence There is a general shortage of biological specimens for medical study. One example is skeletons. Partly because of a ban by India, traditional export leader in the field, there is now a serious shortage of skeletons for medical study; despite an estimated 10,000 sold per annum, supply cannot meet demand, and the price of a good skeleton more than doubled over a two-year span.

Broader Insufficient medical supplies (#PE1634) Insufficient educational material (#PE8438).

♦ PF5106 Lack of international understanding
Nature There is a prevailing climate of fear and mistrust between governments and peoples of East and West and North and South. A lack of contact at a variety of levels aggravates the general lack of understanding. The absence of a genuine political dialogue increases the readiness of nations to resort to the use of force.

Broader Inter-cultural misunderstanding (#PF3340).
Related Lack of international cooperation (#PF0817).
Aggravated by Neutrality (#PF0473).

♦ PF5107 Abuse of relics
Nature The creation, sale, or distribution of false relics is prohibited under pain of excommunication.

Broader Simony (#PE9840).
Related Unethical practices of priesthood (#PF8889).
Aggravates Desecration of holy spaces (#PF6385).
Aggravated by Vulnerability of sacred sites (#PD6128).

♦ PF5115 Centralization of religious power
Hierarchical structures of established religions

Claim There is a need to expose the mysterious powers of ecclesiastical institutions, so far only partly demythologized. The hierarchical structures of the churches, even amended and reformed at some structure in history, are hardly conducive to what is now called participatory ecclesiology. No Christian community can claim to correspond to the Pauline image of the body of Christ in which each limb and organ has its own place and is indispensable. Even highly egalitarian and participatory evangelical communities are often directed by a small group of people who feed the masses of believers on a gospel diet of their own choice.

Broader Over-centralization (#PF2711).

♦ PF5118 Inaccurate weather forecasting
Unpredictability of weather

Nature Seasonal weather predictions cannot yet be made with scientific certainty or even with a reasonable probability of high accuracy. In an attempt to increase the understanding of atmospheric predictability, experiments are being carried out using large computers to test the sensitivity of predictions to small variations in the atmosphere and to the way in which physical processes are represented. The state of the atmosphere on the global scale – its temperature, pressure and movement at all levels up to about 30 kilometres – is now routinely calculated every twelve hours from observations made mainly from surface stations, instrument–carrying balloons, and satellites. The coverage varies considerably, however. It is good over Europe, much of Asia and North America but poor over the oceans, and this leaves considerable scope for error. Some temperatures may be as much as 2 deg C in error in places, and some winds as much as 10 metres per second. It is important to know how sensitive computer predictions are to these errors, since it will never be possible to specify absolutely accurately the actual state from which the prediction starts. Another source of error in computer predictions on time–scales of a week or more arises from the difficulty of representing the effects of clouds. These strongly reflect the sun's energy – as is clear from their whiteness when seen from space in satellite pictures – and therefore exert a strong control over the atmosphere's energy balance.

Broader Inaccurate forecasting (#PF4774).
Narrower Unreliable rainfall (#PD0489).
Related Large–scale weather anomalies (#PC4987).
Aggravates Bad weather (#PC0293).

♦ PF5120 Excessive waiting times in government facilities

Delay in access to social services — Government procedures delaying welfare

Nature When government facilities cannot meet demand effectively, institutions resort to rationing by queue.

Incidence In health this means long waiting times in government facilities: up to eight hours in Nigeria and five hours in Uganda. Not only is time wasted, but services can be unintentionally and inefficiently rationed, because people with relatively minor ailments, in the case of medical facilities, are induced to use health facilities more often when the health facilities are subsidized.

Broader Wasted waiting time (#PF1761).
Narrower Delay in issue of travel documents (#PE9123).
Aggravates Delay in administration of medical care (#PD5119).
Aggravated by Paralyzing complexity of urban structures (#PF1776)
Internal inefficiency of public programmes (#PF7483)
Governmental disregard for people as human beings (#PD8017).

♦ PF5121 Denial to animals of the right to the attention, care and protection of humankind

Nature Although humans have an intrinsic duty to provide care for beings less conscious than themselves with whom they share the planet, this duty is little exercised.

Refs Sperling, Susan *Animal Liberators* (1988).

Broader Denial of rights of animals (#PC5456)
Human consumption of animal products (#PD7699)
Denial to animals of legal protection of their rights (#PE8643)
Denial to experimental animals of the right to freedom from suffering (#PE8024).
Related Abuse of animal drugs (#PE0043) Lack of care for animals (#PD8837)
Cruel animal transportation (#PD0390) Human vectors of animal diseases (#PD2784)
Hybridization of wild animal species (#PD2419) Denial to animals of the right to life (#PF8243)
Denial to animals of the right to dignity (#PE9573)
Denial to animals of the right to a natural death (#PE8339)
Denial to working animals of limitation of working hours (#PE6427)
Denial to animals of the right to freedom from mass killing (#PE9650)
Denial to food animals of the right to freedom from suffering (#PE3899)
Denial to working animals of restorative nourishment and rest (#PE4793)
International movement of animals as factor in animal diseases (#PD2755)
Inadequate disinfection measures for animal housing and equipment (#PE2757)
Endangered animal and plant life due to radioactive contamination (#PD5157)
Denial to animals of the right to conditions of life and liberty proper to their species (#PE6270).

♦ PF5123 Evil enchantment

Nature Enchantment or fascination is to charm or bewitch by use of the eyes or looks: to put and keep in thraldom by charms or by power of pleasing. Enchanted people are no longer free to make their own choices, nor understand what is going on around them, and are often reduced to extreme emaciation and perish by a wasting disease. They lose their colour, have melancholy eyes either overflowing with tears or unnaturally dry, frequently sign, are low in spirit, and have bad dreams. Many animals are reported to be able to enchant. Wolves, if they see a person first, can deprive him of speech. Generally it is wicked people who are the most dangerous. The more wicked the more adept at exercising fascination. In 1653 a woman is reported to be able to bring down a falcon from the highest of flight by staring at it; a man was able to break mirrors by looking at them and another man killed his own and other folks' children by watching them. Those most susceptible to enchantment are those who are immoderately praised, especially without their knowledge; those who have a fair complexion, face or figure, particularly children; children in unwashed baby linen; a fair person who employees two maids to dress their hair; those who lie in bed very late, especially wearing a night cap; and those who breakfast on cheese or peas. All children who have been weaned and are brought back to the breast will be gifted with the power of fascination.

Broader Bewitchment (#PF3956).
Aggravates Unluckiness (#PF9536).

♦ PF5124 Cover up of convergence in practice of apparently opposed ideologies

Maintenance of super power ideological diversities

Claim The super powers share disregard for the nature upon which their warring economies depend, the brutalization and violence in the lives of the poorest and the suppression of human rights either in their own countries or abroad.

Related Vulnerability of socio–economic systems from globalization (#PF1245).

♦ PF5127 False evidence

False witness — Distortion of evidence — Misleading evidence

Broader Perjury (#PD2630) Evasion of issues (#PF7431)
Abuse of police power (#PC1142).
Narrower Government inducement to crime (#PD6943)
Official fabrication of evidence (#PD8716)
Misinformation concerning infringement of human rights (#PF9794).
Related Denial of evidence (#PD7385).
Aggravates Placement of the burden of proof on the disempowered (#PF3918).
Aggravated by Evidence decay (#PF5403) Tampering with official documents (#PF4699)
Inadequate evidence to convict known offenders (#PF8661).

♦ PF5137 Lack of a world food security system

Nature A global information and early warning system on food and agriculture was set up some years ago but the two main elements of the system of food security, namely food reserves and a better deal for developing countries in agricultural trade, have made very little progress. The most serious lag has been that concerning the system of food security reserves. Developed countries which either control the bulk of surplus grains, or can afford to pay for a system of food security, do not need it for themselves. The developing countries, which need such a system, cannot afford the full costs of an adequate system nor the possibility for effective participation in the control and management of the food reserves. As a result, despite several good harvests, the world is without a dependable system of food security.

Broader International insecurity (#PB0009).
Narrower Inadequate mechanisms for securing sufficient food supplies (#PF2857).
Aggravates Food insecurity (#PB2846).

♦ PF5145 Inadaptation of work to family needs

Nature Few US or European companies have systematic policies for dealing with today's newly emerged norm of working parents with need for outside child–care. It is still common for women to curtail employment because of child bearing while men do not. Women thereby take less demanding, lower paying jobs, and it is often the case that the structures, habits, values and atmospheres of work become organized around the availability of those people without first–line responsibility for children.

Incidence Even though Sweden has liberal maternity and paternity leaves, 85 percent of all leaves are taken by women, which results in corporations viewing women workers as being out of the mainstream in terms of promotion and career advancement. France has a strong national system of public child care, which has enabled many women to work, but there is still strong occupational segregation and no legal means to pursue equal opportunity goals. And Germany FR has generous national benefits but few private efforts to adapt work to family needs.

Related Absenteeism (#PD1634).

♦ PF5147 Neglect of sexual health of women

Nature Despite its connection with many serious physical and psychological problems, sexual health is generally ignored by the medical establishment. The physical effects of sexual frustration are to blame for a large number of cases of middle and upper–class women who complain to doctors of vague general pain in the pelvis and back. Some of these women seek psychological or psychiatric help because of severe depression and anxiety, or even serious psychotic disorders. Both the medical and psychological assistance which women may seek is often ineffectual, since the therapy tends to alleviate the symptoms without determining the real causes. Women from the lower income level may suffer the same symptoms and are less likely to get successful treatment for sexual health problems since they usually visit a doctor in extreme situations and never seek help for problems with no obvious physical symptoms. The problem is even more severe for women in developing countries, because the sociological conditioning of women's attitudes towards sex are directly linked to economic status.

Broader Neglect (#PA5438) Neglected health practices (#PD8607).
Aggravated by Unequal health benefits for women (#PE6835).

♦ PF5148 Violation of the integrity of creation

Narrower Unconstrained exploitation of natural resources (#PF2855).
Related War (#PB0593) Injustice (#PA6486).

♦ PF5153 Electoral organization favouring political incumbents

Discrimination against those seeking election

Broader Unfair elections (#PC2649).

♦ PF5155 Trivialization of equality

Collapsed conceptualization of equality — Egalitarianism

Claim A principle difficulty with egalitarianism is that of specifying in a concrete case when the requisite conditions for equality and procedurally just transactions have been met. A major objection to the radical call for redistribution of planetary resources according to egalitarian considerations is that it is unrealistic. Since there is no theory of global justice providing for institutions that would compensate for uneven distribution of resources then such an approach cannot be effectively implemented.

Refs Brown, Henry P *Egalitarianism and the Generation of Inequality* (1988).

Broader Facile social concepts (#PF5242).

♦ PF5163 Refusal to admit error

Refusal to apologize for mistakes — Failure to acknowledge failure

Narrower Failure of government to apologize for errors (#PF5296).
Aggravates Diplomatic errors (#PF1440) Human errors and miscalculations (#PF3702).
Aggravated by Fear of failure (#PF4125) Personal failure (#PF4387).

♦ PF5172 Inadequate correctional systems

Nature As regards national correctional systems, there appears to be a great need for the de–institutionalization of offenders, while at the same time ensuring the protection of the fundamental human rights of the accused and the upgrading of the training and professional qualifications of correctional personnel. The need for de–institutionalization of offenders, in many countries, can be attributed to judicial delay arising from, for example: complexities of and contradictions in the legal system; lack of alternatives to incarceration; population and prison growth; migration problems; lack of legal services. Another shortcoming of some correctional systems is the lack of diagnostic centres for observation and classification, a requisite for ensuring the progressive rehabilitation of offenders.

Broader Imprisonment (#PD5142).
Related Needless incarceration (#PE5112).

♦ PF5178 Cultural arrogance

Broader Arrogance (#PA7646).
Aggravates Cultural invasion (#PC2548) Cultural isolation (#PC3943)
Cultural imperialism (#PC3195) Cultural discrimination (#PC8344)
Inter–cultural misunderstanding (#PF3340) Destruction of cultural heritage (#PC2114)
Lack of appreciation of cultural differences (#PF2679)
Inadequate cultural integration of immigrants (#PC1532).
Aggravated by Intellectual arrogance (#PF7847).

♦ PF5181 Lack of long-term development assistance

Broader Excessive social costs of structural adjustment in debtor developing countries (#PD8114).
Aggravated by Overgeneralized policy models (#PF9100)
Short range planning for long–term development (#PF5660)
Insensitivity to non–immediate hazards to society (#PF9119)
Absence of a long–range, world–wide capital flow plan (#PF2865)
Uncertainty of development programmes due to short–term loans (#PF4300).

♦ PF5183 Manipulative euphemisms

Claim The discourses of power are full of euphemisms. Words no longer fit with facts. Annihilators are called nuclear arms, as if they were simply a more powerful version of conventional arms. We call 'the free world' a world full of examples of the most obscene inequities and violations of human rights. In the name of the people, systems are created where people must simply comply obediently with the dictums of the 'almighty state'. Peaceful protest marchers are severely punished and imprisoned for public disorder and subversion, while state terrorism is accepted as law and order. The end result is that people cease to understand and either turn into cynics or melt into impotent, perplexed and alienated masses.

Broader Misappropriation of words (#PF5247).
Aggravates Cynicism (#PF3418).
Aggravated by Government hypocrisy (#PF9050).

♦ PF5188 Uncertain toxicity thresholds

Dangerous legally prescribed chemical residue levels — Ignorance of hazardous levels of toxic substances — Uncertain radiation hazard thresholds
Incidence In the case of nuclear accidents, for example, individual governments decide, in the light of their interpretation of a range of evidence, at what level of radioactive contamination pasture land, drinking water, milk, eggs, vegetables and fish are to be banned as unfit for consumption, firstly by humans and secondly by livestock. Different countries, and even different local authorities within countries, have different criteria. Some have none at all. Sometimes the criteria changes and what was once acceptable may become dangerous in the future. Some may apply rigorous criteria to ban suspect foodstuffs, but may then apply very relaxed criteria to assess the risk from foodstuffs imported from neighbouring countries where very permissive criteria, if any, are applied.
Broader Uncertainty (#PA6438).
Aggravates Hazardous wastes (#PC9053)
Long-term hazards of exposure to chemicals (#PE4717)
Long-term hazards of exposure to radiation (#PE4057).
Aggravated by Unethical practice of radiology (#PD8290)
Politicization of health standards (#PD4519)
Inconclusiveness of scientific and medical tests (#PD7415).

♦ PF5190 Decadent music

Passionate abstract music — Musical perversions — Aggressive music — Pornography in sound
Claim 1. Whilst it is true that good music is still heard and even still written, the evil influence of musical perversions cannot be overstated. Modern music has resulted in the ruin of an entire generation.
2. Crime and violence in society are partly due to the preponderance of aggressive (rock) music and to the lack of classic music with its soothing powers. Aggressive music alienates the body from its natural emotional climate and evokes its most primitive reactions. In rock concerts, for example, the hight volume of the music, the mechanical hammering of its rhythm, its sheer physical impact and total lack of nuance, leave an audience in a state of complete mental stupor, drugged, numbed and impervious to feeling. The repetition of music which is totally unemotional in its effect diseducates the emotions. A synthesized rhythm section with an immaculate, mechanical beat is totally unnatural and estranging ? It breeds automata, or worse still, causes people to suppress unexpressed emotions, and the frustrations stemming from this can burst out in otherwise inexplicable violence. There is a danger of pornography in sound, whereby the raucous, throaty delivery of many pop singers bases itself, albeit unwittingly, on the sound of lustful ejaculation.
Counter-claim History of violence is considerably longer than that of rock music. In earlier times, the Church objected the lasciviousness of much dance music and the profane tunes in sacred works. Even Bach was accused of excessive elaboration. In the 1920s jazz was seen as a threat to the civilized European society. In the 1960s the Beatles was thought subversive. Music has a powerful impact on emotions and senses, but the arousal achieved by rock music or Wagner are not different from each other.
Broader Performing arts (#PF1756).
Aggravates Youth violence (#PF7498).
Aggravated by Offensive lyrics (#PF7807).

♦ PF5191 Broken-spirited animals

Aggravated by Maltreatment of animals (#PC0066)
Capture and use of wild animals as pets (#PD1179).

♦ PF5203 Failure to sacrifice any personal advantage

Unwillingness to sacrifice personal privilege — Resistance to loss of any personal investment — Resistance to personal austerity
Aggravates Austerity (#PJ4983)
Reinforcement of inappropriate development by privileged classes in developing countries (#PF6670).

♦ PF5205 Impure thoughts

Dirty thoughts
Broader Temptation (#PA7736).
Related Lust (#PA4673) Despair (#PF4004) Harmful thought (#PF0441).
Aggravates Sin (#PF0641) Guilt (#PA6793) Immorality (#PA3369)
Spiritual impurity (#PF6657) Psychological conflict (#PE5087)
Psychological alienation (#PB0147) Psychological withdrawal (#PJ2329)
Disabling inadequacy feelings (#PJ9012).
Aggravated by Oppressive prevalent images (#PF1365).

♦ PF5210 Conceptual repression of problems

Problem avoidance — Failure to acknowledge problems — Failure to acknowledge differences — Failure to acknowledge contradictions
Nature The tendency to eschew deliberately any acknowledgement of the existence of social or other problems, or any other form of social disharmony, in favour of an emphasis on positive visions of the future. Even the most striking problems, such as hunger and disease, are then repressed in an effort to stress what is going well in society. This impedes any effort to articulate the nature of the issues to be dealt with and hinders the appropriate development of any collective response to them. Such an approach conveniently removes the need for an individual to explore the extent to which he himself contributes to the problem.
Counter-claim 1. Any focus on problems is a drain on energy which can be more appropriately focused on enhancing the harmonious aspects of social life. Such a focus brings about a resolution to problems, to the extent that they can be considered to exist. 2. A preoccupation with social problems can usefully be considered a consequence of the personal problems of the people so engaged and of their avoidance of any appropriate effort to deal with them.
Broader General obstacles to problem alleviation (#PF0631).
Aggravates Underreported issues (#PF9148) Inaction on problems (#PB1423)
Minimization of problems (#PF7186) Non-recognition of problems (#PF8112)
Unproven relationships between problems (#PF7706)
Unrealistically positive self-assessment (#PF4377)
Inability to resolve problems realistically (#PF8435)
Recurrence of misapprehended world problems (#PF7027).
Aggravated by Blame avoidance (#PF6382) Racial inequality (#PF1199)
Unrealistic expectations (#PE7002) Avoidance of negative feedback (#PF5311)
Suppression of scientific information (#PF1615).

♦ PF5213 Slang

Vulgarism
Broader Misuse of language (#PF9598).
Related Profanity (#PF7427) Verbal abuse (#PD5238)
Incomprehensibility of specialized jargon (#PF1748).

♦ PF5222 Unpublicized public meetings

Unknown community meetings — Unannounced community meetings
Broader Obstacles to community achievement (#PF7118)
Inadequate dissemination and use of available information (#PF1267).
Related Inadequate meeting methods (#PF8939).
Aggravated by Unpublicized community news (#PF7998)
Lack of local information systems (#PF6541)
Restrictive effects of traditional community decision-making (#PF3454).

♦ PF5223 Name-dropping

Broader Boasting (#PF4436).
Aggravated by Snobbery (#PJ3943) Careerism (#PF6353)
Pursuit of personal prestige (#PF8145).

♦ PF5225 Double standards in morality

Double standards morality — Double standards
Nature The application of two standards of morality, usually one for the social group which sets the standards and another for those outside the group, encourages exploitation of other social groups and violence or otherwise immoral conduct against them. This may include racism, religious intolerance, sexual exploitation of women, and discrimination.
Broader Hypocrisy (#PF3377) Haphazard forms of social ethics (#PF1249).
Narrower Double standards of sexual morality (#PF3259)
Bias in United Nations response to human rights (#PF7925).
Aggravates Nazism (#PF2171) Nudism (#PF2660) Fascism (#PF0248)
Marxism (#PF2189) Atheism (#PF2409) Populism (#PF3410)
Cynicism (#PF3418) Idealism (#PF3419) Tribalism (#PC1910)
Anarchism (#PC1972) Extremism (#PB3415) Communism (#PC0369)
Sacrilege (#PF0662) Scientism (#PF3366) Socialism (#PC0115)
Scepticism (#PF3417) Liberalism (#PF0717) Puritanism (#PF2577)
Capitalism (#PC0564) Propaganda (#PF1878) Militarism (#PC2169)
Utopianism (#PF2192) Neo-fascism (#PF2636) Agnosticism (#PF2333)
Materialism (#PF2655) Rationalism (#PF3400) Monasticism (#PF2188)
Nationalism (#PB0534) Prostitution (#PD0693) Conservatism (#PF2160)
Collectivism (#PF2553) Sectarianism (#PF7780) Religious war (#PC2371)
Value erosion (#PA1782) Secularization (#PB1540) Fundamentalism (#PF1338)
Anti-communism (#PF1826) Dehumanization (#PA1757) Male domination (#PC3024)
Totalitarianism (#PF2190) Anti-capitalism (#PF3110) Political purges (#PC2933)
World federalism (#PF2088) Religious schism (#PF1939) Anti-clericalism (#PF3360)
Religious torture (#PC7101) Linguistic purism (#PF1954) Religious conflict (#PC3292)
Political conflict (#PC0368) Sexual exploitation (#PC3261) Social fragmentation (#PF1324)
Political radicalism (#PF2177) Extremist ideologies (#PC6341) Intellectual dissent (#PC2582)
Religious intolerance (#PC1808) Intellectual conflict (#PC3390)
Ideological revolution (#PC3231) Inadequacy of religion (#PF2005)
Undemocratic pressures (#PD3389) Subversion of democracy (#PD3180)
Undue political pressure (#PB3209) Political indoctrination (#PD1624)
Disagreement among experts (#PF6012) Multidenominational society (#PF3368)
Fragmented ethical contexts (#PF2096) Denial of parental affiliation (#PD3255)
Ideological schism in communism (#PF3181) Newspaper and journal propaganda (#PD0184)
Corruption in organized religion (#PC3359) Religious and political antagonism (#PC0030)
Politically emotive words and terms (#PF3128) Factionalism in developing countries (#PD1629)
Ideological discrimination in politics (#PC3219)
Discrepancies in human life evaluation (#PF1191)
Burdensome cost of religious ceremonies (#PF3313)
Fragmentation within organized religions (#PF3364)
Undue religious influence on secular life (#PF3358)
Threats to ideological movements and minorities (#PC3362)
Inadequate integration of ideology into society (#PF3402)
Weakness in trade between different economic systems (#PC2724)
Inadequate and inaccurate textbooks and reference books (#PD2716)
Politicization of intergovernmental organizational debate (#PD0457)
Abuse of international cultural, diplomatic and commercial exchanges (#PF3099)
Differences in trading principles and practices between different economic systems (#PC2952)
Alienation of skilled and committed personnel from international organizations and programmes (#PE1553)
Inadequate relationship between international governmental and nongovernmental organizations and programmes (#PE1973).

♦ PF5233 Fear of communism

Broader Fear (#PA6030).

♦ PF5240 Cognitive suicide

Nature Renunciation of any attempt by a person to think for himself to the point of becoming totally governed by cognitive patterns imposed by others.
Broader Suicide (#PC0417).

♦ PF5242 Facile social concepts

Misappropriation of social values — Collapsed conceptualization — Trivialization of social values — False concreteness of societal concepts — Popularization of concepts
Nature Appropriating the articulation a profound and necessary social, religious or cultural reality while reducing the meaning of the term to a facile social concept lends linguistic, social and symbolic credibility to the users of the word. An individual can use the term love or peace or religion and mean a very narrow and deceptive dimension of the term while implying its whole breadth of meanings. This process when used over and over again trivializes the word. The experience the word points to is left without a meaningful symbol to point to it and is then considered an inappropriate experience.
Broader Value erosion (#PA1782)
Abusive exploitation of cultural heritage (#PC7605).
Narrower Trivialization of love (#PF0959) Trivialization of peace (#PF5826)
Trivialization of liberty (#PF7497) Trivialization of equality (#PF5155)
Excessive portrayal of sex in the media (#PE7930).

♦ PF5247 Misappropriation of words

Misappropriation of concepts
Claim Certain key concepts, such as "development", have been appropriated by constituencies with vested interests in order to protect and further them. For a "developer", for example, the process of developing a piece of virgin land may well involve removing any natural vegetation and constructing a series of buildings. Such uses of "development" are common in business and financial circles, and the associated government planning departments. This is to be contrasted with concepts of "development" which seek to enhance the natural environment in a sustainable manner, respecting the relationship of people to it, and avoiding irreversible damage to the land. The positive connotations of the word are used to disguise the abusive projects undertaken in its name, taking advantage of the definitional ambiguity in the situation.

Broader Misappropriation of cultural property (#PE6074).
Narrower Lack of vocabulary (#PF3824) Manipulative euphemisms (#PF5183).
Aggravated by Terminological deception (#PF5383).

♦ PF5248 Restrictions on adaptations

Nature Restrictions which prevent the acquiring party from adapting the imported technology to local conditions or introducing innovations in it, or which oblige the acquiring party to introduce unwanted or unnecessary design or specification changes, if the acquiring party makes adaptations on his own responsibility and without using the technology supplying party's name, trade or service marks or trade names, and except to the extent that this adaptation unsuitably affects those products, or the process for their manufacture, to be supplied to the supplying party, his designates, or his other licensees, or to be used as a component or spare part in a product to be supplied to his customers.
Broader Restrictive trade practices (#PC0073).
Related Collusive tendering (#PE4301) Grant-back provisions (#PE5306)
Challenges to validity (#PF1200) Restrictions on research (#PF0725)
Restrictions on publicity (#PF1575) Exclusive dealing arrangements (#PE0413)
Restrictions on use of personnel (#PF3945)
Patent pool or cross-licensing agreements (#PE4039)
Exclusive sales and representation agreements (#PE4581)
Trade restrictions due to voluntary export restraints (#PE0310)
Payments after expiration of industrial property rights (#PF5292)
Tying of supplies to subsidiaries by transnational enterprises (#PE0669).

♦ PF5250 Insufficient community events

Limited cooperative events — Infrequent unifying events — Infrequent social events — Infrequent public gatherings — Infrequent community cultural events — Irregular village meetings — Infrequent community meetings
Broader Reduction in symbolic celebrations (#PF1560)
Obstacles to community achievement (#PF7118)
Disjointed patterns of community identity (#PF2845)
Unclarified procedures for transposing ancient traditions (#PF6494).
Narrower Insufficient community celebrations (#PJ0188).
Aggravates Obsolete basis of cultural identity (#PF0836)
Fragmented planning of community life (#PF2813)
Fragmented patterns of community activity (#PF6504)
Ineffective structures of local consensus (#PF6506)
Inappropriate application of traditional values (#PF2256)
Limits on participation in community development (#PF3560)
Incomplete implementation of community decisions (#PF2863)
Deteriorated structures of essential corporateness (#PF1301)
Lack of responsible involvement in community affairs (#PF6536).
Aggravated by Inadequate meeting methods (#PF8939)
Demoralizing images of rural community identity (#PF2358).

♦ PF5255 Parental over-protectiveness

Maternalism
Claim The worst thing for children is over-protectiveness.
Broader Human dependence (#PA2159).
Related Paternalism (#PF2183) Excessive caution (#PF6389).

♦ PF5266 Resistance to grace

Broader Sin (#PF0641) Resistance to change (#PF0557).
Related Fear (#PA6030) Fear of increased autonomy (#PE7706).
Aggravated by Fear of God (#PF9565) Fear of knowledge (#PF9595)
Fear of the unknown (#PF6188) Individual fear of future change (#PD2670).

♦ PF5272 Frustrated past goals

Ineffective past efforts — Awkward past experiences
Broader Obsolete basis of cultural identity (#PF0836)
Underdeveloped sources of income expansion (#PF1345).
Aggravates Reluctance to join in community action (#PF1735)
Cumulative depletion of corporate initiative in rural communities (#PE3296).

♦ PF5287 Inability to negotiate effective multilateral safeguard systems

Incidence This is evident in relation to issues of environment, disarmament, the international monetary system, the debt crisis, and to the international trading system. In the latter case, one of the major symptoms of the erosion of the GATT system has been the persistent inability to negotiate the improved multilateral safeguard system envisaged in the Tokyo Round declaration of 1979.
Aggravates Lack of international cooperation (#PF0817)
Fragmentation of the international trading system (#PC9584)
Unilateral interpretations of multilateral principles (#PF9629).
Aggravated by Conditional observance of multilaterally agreed trade commitments (#PF7838).

♦ PF5292 Payments after expiration of industrial property rights

Nature Requiring payments or imposing other obligations for continuing the use of industrial property rights which have invalidated, cancelled or have expired recognizing that any other issue, including other payment obligations for technology.
Broader Restrictive trade practices (#PC0073).
Related Collusive tendering (#PE4301) Grant-back provisions (#PE5306)
Challenges to validity (#PF1200) Restrictions on research (#PF0725)
Restrictions on publicity (#PF1575) Restrictions on adaptations (#PF5248)
Exclusive dealing arrangements (#PE0413) Restrictions on use of personnel (#PF3945)
Patent pool or cross-licensing agreements (#PE4039)
Exclusive sales and representation agreements (#PE4581)
Trade restrictions due to voluntary export restraints (#PE0310)
Tying of supplies to subsidiaries by transnational enterprises (#PE0669).

♦ PF5295 Unchanging legal precedent undermines accountability

Broader Exclusion of the masses from setting criteria in judicial judgements (#PD1060).
Aggravates Social unaccountability (#PC1522).

♦ PF5296 Failure of government to apologize for errors

Failure of government to admit mistakes — Refusal of government leaders to admit mistakes
Incidence Examples of long-standing resentment from such failure are indicated by the apology in 1990 by the government of Japan for its activities in Korea and the apology in 1990 by the East German government for its hostility to Israel and for the persecution of Jewish citizens prior to 1945 and thereafter.
Broader Refusal to admit error (#PF5163).
Related Policy-making errors (#PF0531).
Aggravated by Government hypocrisy (#PF9050).

♦ PF5298 Insufficient images of political involvement

Nature Modern society at all levels have limited images of effective participation in the political processes. Society does not adequately guard individual liberties nor demand individual responsibilities. Methods of participation are ineffective. Political processes are based on the assumption of political control rather than that of limited influence.
Broader Inadequate political networks (#PD2213).

♦ PF5299 Procrastination

Refs Burka, Jane and Yuen, Lenora *Procrastination* (1983).
Aggravates Policy-making delays (#PF8989)
Government delay in response to symptoms of problems (#PF6707)
Procrastination of science in the face of the unexplained (#PF3682).

♦ PF5301 Organization of human thought

Systematic human behaviour according to cultural norms — Mental set
Narrower Discriminatory imposition of standards (#PD5229).
Aggravates Proscribed thinking and behaviour (#PF3862).

♦ PF5304 Laboratory testing errors

Nature Inaccurate and unreliable testing remains a serious health hazard, as well as waste of money. Faulty tests can occur for many reasons: a machine loses its calibration; testing chemicals lose potency or get used improperly; human specimens are inadvertently switched. Even if a test is performed properly, it may be misinterpreted, especially if the screeners are overworked.
Incidence Boston researchers studied 10 women with cervical cancer who had negative test results in the preceding two years. A reexamination of their slides found out that five had been misinterpreted and two slides were too poorly done to read; only three were clearly negative.
Broader Iatrogenic disease (#PD6334).
Related Misdiagnosis (#PF8490) Unnecessary surgery (#PE9271)
Overprescription of drugs (#PE9087) Defective medical devices (#PG8737)
Substance abuse by physicians (#PE7568).

♦ PF5306 Failure to assist in emergencies

Narrower Indifference to interfamily emergencies (#PS9373).
Aggravates Insufficient emergency services (#PF9007).
Aggravated by Unpreparedness for food emergencies (#PC5016).

♦ PF5311 Avoidance of negative feedback

Positive bias in reporting — Uncritical upbeat reporting — Uncritical feel-good rhetoric
Nature Organizations and social groups of all kinds have difficulties in processing negative feedback. Such feedback can take the form of external criticism of the performance of the organization or of its detrimental impact on the wider environment. It can take the form of internal reporting on such questions as well as on weaknesses in the organizational performance or on programme failures. The avoidance of such feedback may be encouraged by an insecure leadership that needs to be constantly encouraged by positive feedback. It may result from the ways in which good performance is rewarded within the organizational structure which can effectively result in career penalization if reports fail to appear positive. In the latter case reporting may focus only on positive achievements or only mention failures and emerging problems indirectly, if at all. Such avoidance prevents receipt of appropriate signals by management or the leadership, resulting in inappropriate policy formulation and a lack of ability to anticipate the evolution of events.
Counter-claim Frail coalitions of social forces should not be threatened by unnecessary negative feedback which would undermine their ability to act at all. It is therefore appropriate that such feedback should be filtered or made available only to certain people under certain conditions.
Narrower Unreported crimes (#PF1456) Unreported illness (#PF8090)
Unreported accidents (#PF2887) Underreported issues (#PF9148)
Unreported disagreement (#PF7890) Unreported financial losses (#PF8079)
Inability of people's representatives to process feedback from constituents (#PJ0599).
Related Unreported scandals (#PF5340).
Aggravates Secrecy (#PA0005) Unpreparedness (#PF8176)
Unreported violence (#PF4967) Unreported disasters (#PF7768)
Unreported harassment (#PF4729) Misleading information (#PF3096)
Deception by government (#PD1893) Intolerance of criticism (#PF8396)
Terminological deception (#PF5383) Inadequate warning of disasters (#PF3565)
Conceptual repression of problems (#PF5210)
Ineffective decision-making processes (#PF3709)
Deliberate lying by corporation officials (#PD4982)
Deliberate ignorance during policy-making (#PF8278)
Calculated delays in releasing controversial news items (#PE0598)
Failure of governments to fulfil international reporting obligations (#PE2215).
Aggravated by Blame avoidance (#PF6382) Uncritical thinking (#PF5039)
Insecurity of leadership (#PD9362) Unrealistic expectations (#PE7002)
Pursuit of personal prestige (#PF8145) Pursuit of national prestige (#PF8434)
Pursuit of corporate prestige (#PF7983) Accumulation of recognized merit (#PF7315).

♦ PF5322 Atomism

Nature Until the atom was split, the universe was believed to consist of minute, indivisible particles. This belief survives operationally among those who treat the parts of a thing as having an existence all their own apart from the whole.
Broader Inadequate models of socio-economic development (#PF9576).
Aggravates Inconclusiveness of science (#PF6349)
Irresponsible social science (#PF8032)
Irrelevance of science and technology (#PF0770).

♦ PF5329 Monopolization of knowledge

Refusal to share knowledge
Broader Restrictions on the acquisition of knowledge (#PF1319)
Barriers to the international flow of knowledge and educational materials (#PF0166).
Narrower Restrictive commercialization of definitions (#PF4674)
Monopoly of competence for intervention in the future (#PF0980).
Aggravates Lack of knowledge (#PF8381) Fragmentation of knowledge (#PF0944)
Imbalanced distribution of knowledge (#PF0204).
Aggravated by Competition (#PB0848) Over-specialization (#PF0256)
Copyright barriers to transfer of knowledge (#PE8403).

♦ PF5332 Unlawful government action

Illegal action by governments — Law-breaking nations
Narrower Illegality of nuclear weapons (#PF4727)
Government complicity in illegal activities (#PF7730)
Misuse of telephone surveillance by governments (#PD1632).
Related Secret military operations (#PF7669).
Aggravates Deceptive misuse of research (#PD7231)
Unlawful business transactions (#PC4645)
Execution of inappropriate orders (#PF2418)
Ineffective monitoring of illegal activity (#PF7264).
Aggravated by Abuse of authority (#PC8689) Illegitimate political regimes (#PC1461)
Secret international agreements (#PF0419).

♦ **PF5337 Irresponsibility of member governments of the United Nations**
Irresponsibility of United Nations developing country membership
Claim Many of the most objectionable activities of the United Nations originate in radical Third World nations. The New World Information Order, which strengthens governments' control of media, was such an initiative. So are many of the anti–business, anti–growth, redistributive and regulatory schemes of the New International Economic Order. Extravagant spending, inefficient administrative practices and irresponsible budgetary processes also often originate from radical Third World states, whose poverty exempts them from paying their share of the resultant funding obligations. The construction of a $63 million conference centre in Ethiopia at the time of the famine in that area is such an example.
Aggravates Marginalization of the United Nations (#PF5934)
Internationally non–cooperative governments (#PF9474)
Ineffectiveness of the United Nations system of organizations (#PF1451)
Shortage of financial resources within the United Nations system of organization (#PF1460).

♦ **PF5340 Unreported scandals**
Unreported fraud — Cover–up
Refs Cox, Del *Corruption and Cover–Up* (1988).
Broader Scandal (#PC8391) Unrecorded knowledge (#PF5728)
Underreported issues (#PF9148).
Narrower Official cover–up of government harassment of political activists (#PF3819).
Related Fraud (#PD0486) Avoidance of negative feedback (#PF5311).
Aggravates Manipulation of debates (#PD4060).

♦ **PF5351 Inaccessible government agencies**
Remoteness from government — Removed governmental structure
Broader Social inaccessibility (#PC0237)
Underprovision of basic services to rural areas (#PF2875)
Cumulative depletion of corporate initiative in rural communities (#PE3296).
Aggravates Inadequate support services (#PF6492).

♦ **PF5354 Low self esteem**
Lack of self love — Absence of self affirmation — Low self–image — Low resident self–image — Self–depreciatory operating images — Inadequate self–image — Self–doubt
Nature Low self esteem contributes to a wide range of social problems including alcoholism, drug abuse, teenage pregnancy, crime, child abuse, chronic welfare dependency and poor educational performance.
Claim Virtually every social problem we face can be traced to lack of self–love. Present–day people do not affirm themselves and therefore cannot genuinely affirm others. The most affirmative stance they can take toward another's interior gifts or exterior good fortune is jealousy or cynicism. Dis–affirmation of self comes partially out of severance from the past. Because of this people cannot see their historical worth or build an adequate self image or self story. They thrust a pseudo–self at others and, in so doing, do not participate in genuine relations with them.
Refs Wadel, Cato *Now Whose Fault is That?* (1973).
Broader Lack of self–confidence (#PF0879).
Narrower Low self image due to illiteracy (#PF9098).
Related Crime (#PB0001) Drug abuse (#PD0094) Alcohol abuse (#PD0153)
Defensive life stance (#PF0979) Individual isolationism (#PD1749)
Health risks of teenage sex (#PE6969) Dependence on social welfare (#PD1229).
Aggravates Self–defeating behaviour (#PD4418) Inhibited self–promotion (#PJ1544)
Reluctance to join in community action (#PF1735)
Modern disruption of traditional symbol systems (#PF6461).
Aggravated by Racism (#PB1047) Personal physical unattractiveness (#PF4010)
Detrimental story of community future (#PF6575).
Reduced by Narcissism (#PF7248).

♦ **PF5356 Legal inconsistency**
Unpredictability of judicial decisions — Instability of legal judgements — Inconsistent sentencing of offenders
Incidence Sentencing is subject to a form of inflation, with the judiciary increasing sentences in a pattern inconsistent both with their peers in the same country and with those in other countries. This results in significant differences in the prison population (per 100,000 of the general population) in neighbouring countries: UK 97.4 per cent, Turkey 95.6, Luxembourg 86.5, West Germany 84.5, Portugal 83, France 81.1, Denmark 68, Belgium 65.4, Italy 60, Sweden 56, Netherlands 40.
Broader Arbitrariness (#PB5486) Outmoded legal systems (#PF2580)
Fragmented decision–making (#PF8448).
Narrower Non–recognition of foreign powers of attorney (#PE6418).
Aggravates Collapse of judicial system (#PJ0761)
Racial bias in sentencing offenders (#PE4907)
Excessive leniency in sentencing of offenders (#PF4723).

♦ **PF5360 Political displacement activity**
Nature Governments generate public interest in and acting on a policy in order to avoid confronting a problem they are unable or unwilling to deal. A country might mobilize its armed forces, impose martial law to distract the public and avoid popular demands for more democracy.
Broader Issue avoidance (#PF1623).
Related Manipulation of debates (#PD4060)
Avoidance of issues in political campaigns (#PE6311).

♦ **PF5365 Vulnerability of government to lobbying**
Unconstrained lobbying
Broader Disorganized liaison with formal support (#PF2947).
Aggravates Ineffective use of external relations relating to sportsmen (#PE6515).
Aggravated by Undue political pressure (#PB3209)
Entrenchment of vested interests (#PD1231)
Inefficient public administration (#PF2335).

♦ **PF5374 Opaque budgetary procedures in the public sector**
Lack of transparency in public finances — Inadequate financial reporting in the public sector — Inadequate monitoring of public sector activity
Nature Accounting in state–controlled enterprises has been such as to reduce transparency, namely the ability to assess the financial implications of public sector activities in advance, to evaluate them after the fact, and to identify who bears the costs and who receives the benefits. This severely inhibits the task of decision–makers who are supposed to be accountable for their actions.
Incidence Traditionally, public finance analysts and policy–makers have focused their attention on the central government budget as the main determinant of fiscal policy. Few attempts have been made to to monitor state–controlled enterprises in the aggregate or to compile fiscal data for all levels of the public sector. Where efforts have been made to gather the information, the rapid growth of state–controlled enterprises has often outpaced the ability of analysts to collect and evaluate it. Such enterprises often do not follow uniform accounting standards, so their financial statistics are difficult to consolidate with other public sector statistics. This problem, and the existence of underground public sectors, is of importance in all countries. But lack of transparency has been particularly disruptive in developing countries, creating unexpected fiscal difficulties (when the economic or political fortunes of a country deteriorate) by exposing the weaknesses of substantial segments of the economy that had directly or indirectly come under public sector control.
Broader Ineffective monitoring (#PF2793).
Aggravates Uncoordinated government policy–making (#PF7619)
Inefficiency of state–controlled enterprises (#PD5642)
Inadequate system of political checks and balances (#PE4997).
Aggravated by Inadequate public finance statistics (#PF7842).

♦ **PF5375 Cargo cults**
Revivalist movements — Nativistic movements — Millenarian movements
Nature These movements received their name from the Melanesian movements early in this century characterized by the belief that the millennium will be ushered in by the arrival of great ships loaded with European trade goods (cargo). The goods will be brought by the ancestral spirits and will be distributed to natives who have acted in accordance to the dictates of one of the cults. Some of the cults advocated the expulsion of all things alien, the renunciation of all things European and the return to traditional ways of life. Others called for the rejection of traditional ways of life and the adaptation of European customs. These movements develop where there is extreme material and other inequality between societies in contact. They attempt to explain then erase the differences in material wealth between natives and Europeans.
Broader Spiritual void (#PA6220) Experimental religion (#PF3367).
Related Contempt for traditional modes of behaviour (#PC4321).

♦ **PF5381 Unreported births**
Lack of birth certificate — Unregistered births
Nature Being without a certificate of birth often results in legally not existing, unable to have access to government services like health, education, or the ability to marry. Many of these individuals, unregistered and unschooled, end up in institutional care, abandoned or delinquent.
Broader Unrecorded knowledge (#PF5728).
Narrower Failure to notify the imprisonment or death in prison of foreign nationals (#PE6424).
Related Statelessness (#PE2485) Refusal to grant nationality (#PF2657)
Refusal to issue travel documents, passports, visas (#PE0325).
Aggravates Discrimination (#PA0833) Unequal coverage by social security (#PF0852)
Uncertainty of death of missing persons (#PF0431).
Aggravated by Social injustice (#PC0797).

♦ **PF5382 Duelling**
Single combat — Affair of honour
Nature A pre–arranged regulated contest between two persons with deadly weapons with the intent of settling a quarrel or vindicating a point of honour. A duel is not self defence or a quarrel. It is not a public duel where two men fight rather than two armies, for instance, David and Goliath or Hector and Achilles. One or both people are killed usually because of some social slight.
Background Duelling was unknown in the ancient European world. Roman gladiatorial combats were not duels but forms of public entertainment. Early Teutonic allowed private duels occasioned by affronts to one's honour but the real rational was judicial. It was believed that God would not allow the guilty to prevail over the innocent. The practice evolved into customary law in many places. When more sophisticated forms of adjudication evolved the practice of personal combat continued as a way of settling personal disputes. By the 15th century the duel of honour was well established in Spain and France. In the 16th and 17th centuries authorities began legislating against it. In some places the death sentence was imposed on those participating in one and heavy fines on those assisting. Early this century duelling was permitted under the German military code in extreme cases.
Refs Sabine, Lorenzo *Notes on Duels and Duelling*.
Related Homicide (#PD2341).
Aggravated by Dishonour (#PF8485).

♦ **PF5383 Terminological deception**
Euphemism
Claim In official reporting on incidents which could lead to embarrassing investigations, a form of terminological deception is used by employing euphemism to de–dramatize such events. For example in the UK, attempted prison suicides are recategorized by the use of the term "self–harm", which can then be described as "manipulative and self–seeking". Killing a live foetus may be described as a "termination", following the practice in reporting on government–sanctioned assassination. Destruction of a forested landscape can be usefully described as "clearing" the land.
Related Lack of vocabulary (#PF3824).
Aggravates Misappropriation of words (#PF5247).
Aggravated by Deception by government (#PD1893)
Avoidance of negative feedback (#PF5311)
Deliberate lying by corporation officials (#PD4982).

♦ **PF5390 Fear of success**
Refs Tresemer, David *Fear of Success* (1977).

♦ **PF5392 Hazings**
Ritual hazings
Broader Physical intimidation (#PC2934).
Aggravates Draft evasion (#PD0356) Brutalization of military personnel (#PD7602).
Aggravated by Inadequate army discipline (#PD2543).

♦ **PF5395 Antiquated provisions of the UN charter**
Nature Flaws in the charter of the United Nations, unforeseen when it was adopted, have now made it impossible for the organization to carry out the purposes for which it was intended. The United States and the Soviet Union have jointly blocked the use of a clause in the charter that calls for periodic review of its provisions by a conference of nations.

♦ **PF5397 Medical paternalism**
Nature Medical paternalism manifests itself in cases where the doctor withholds information from the patient or where treatment is authorized by a court order, a hospital committee, or the relatives of a patient, or where there is regulation banning a putative therapy.
Broader Paternalism (#PF2183).
Aggravates Medical deception (#PD9836) Compulsory sterilization (#PF3240)
Excessive prolongation of the dying process (#PF4936).

♦ **PF5401 Idealism of opposition groups**
Nature Loyal oppositions are unable to effectively challenge the status quo because of their inherent idealism. They simply cannot view issues from the perspective of the establishment: they have never had the opportunity or it has been a long time ago or they desire not to. While they articulate vague values and hopes about the future, they are unable to create a realistic operating vision of an alternative to existing policies. Frustrated by their lack of progress they revert to open rebellion, escape to a different situation or effectively surrender.
Broader Exclusion of opposing views (#PF3720).

♦ **PF5403 Evidence decay**
Misremembrance of evidence
Broader Lost knowledge (#PF5420).
Aggravates False evidence (#PF5127)
Unproven relationships between problems (#PF7706)
Inadequate evidence to convict known offenders (#PF8661).

♦ **PF5409 Unorganized community recreation**
Unorganized recreation programmes — Insufficient recreational planning — Lack of a recreational organization — Untrained recreation leaders
Broader Obstacles to community achievement (#PF7118).
Related Undeveloped channels for commercial initiative (#PF6471)
Ineffective structures for community decision-making (#PF1781)
Lack of skilled manpower in rural areas of developing countries (#PE5170).
Aggravated by Unplanned use of community space (#PF2519)
Deteriorating community identity (#PF2241)
Reduction in symbolic celebrations (#PF1560)
Lack of local leadership role models (#PF6479)
Fragmented patterns of community activity (#PF6504)
Limits on participation in community development (#PF3560)
Underdeveloped provision of basic services in developing countries (#PF6473).

♦ **PF5413 Attraction of the forbidden**
Attraction of illegal activity
Aggravates Unlawful business transactions (#PC4645).

♦ **PF5415 Execution of animals**
Aggravated by Superstition (#PA0430).

♦ **PF5419 Shortage of funds for research**
Insufficient government funding of research
Broader Insufficient financial resources (#PB4653).
Narrower Inadequate peace research support (#PF4848)
Lack of funds for medical research (#PE0612)
Lack of funds for veterinary research (#PE8463)
Restriction of funding for research on social problems (#PF2823).
Aggravates Ignorance (#PA5568)
Prohibitive cost of knowledge and information (#PF0703)
Inadequate research and development on problems of developing countries (#PF1120).
Aggravated by Misappropriation of resources for high cost research projects (#PF0716).

♦ **PF5420 Lost knowledge**
Narrower Evidence decay (#PF5403)
Lost farming skills (#PJ8369)
Ignorance of cultural heritage (#PF1985)
Elimination of traditional skills (#PD8872)
Ignorance of traditional herbal remedies (#PE3946).
Aggravates Ignorance (#PA5568)
Lack of knowledge (#PF8381)
Inadequate application of available knowledge to solve problems (#PF8191).
Aggravated by Unrecorded knowledge (#PF5728)
Unretrievable documents (#PF4690)
Fragmentation of knowledge (#PF0944)
Inaccessibility of knowledge (#PF1953)
Incomplete skills information (#PJ0140)
Unexplored energy alternatives (#PF7960)
Unprofitable traditional skills (#PJ1031)
Concealment of esoteric knowledge (#PF7077)
Underutilization of locally available skills (#PF6538).
Reduces Obsolete vocational skills (#PD3548).

♦ **PF5421 Inadequately worded agreements**
Inconsistent official texts — Ambiguous treaties — Ambiguous laws — Ambiguous regulations
Broader Ambiguity (#PF4193) Arbitrariness (#PB5486).
Narrower Tokenistic meeting resolutions (#PF2086).
Aggravates Inadequate laws (#PC6848)
Law enforcement complexity (#PF2454)
Unfairly negotiated treaties (#PF4787)
Inadequate national law enforcement (#PE4768)
Arbitrary enforcement of regulations (#PD8697).
Aggravated by Ineffective international agreements (#PF6992).

♦ **PF5430 Hypergamy**
Marrying up socially
Nature The custom forbidding women of a particular social status to marry a man with lower status. They are allowed to marry men of their own or higher status.
Broader Pursuit of personal prestige (#PF8145) Social impediments to marriage (#PF3341).
Aggravated by Social stratification (#PB5577).

♦ **PF5434 Fornication**
Sexual intercourse out of wedlock
Broader Sexual immorality (#PF2687).
Narrower Pre-marital sexual intercourse (#PD5107).
Related Rape (#PD3266).
Aggravates Criminalization of sexual relations out of wedlock (#PE9122).
Aggravated by Adultery (#PF2314).

♦ **PF5435 Obsolete legislation**
Obsolete laws — Out-of-date regulations
Incidence Under a series of international conventions airlines and ships who have the potential to injure and kill, enjoy a privileged position, because the amount of compensation they are liable to pay out is set at an artificially low level.
Broader Inadequate laws (#PC6848).
Aggravated by Proliferation of legislation (#PD5315)
Obsolete deliberative systems (#PD0975).

♦ **PF5438 Dehumanization of man in the technological process**
Claim Science purports to be value free and yet has established itself as the supreme value of industrial civilization. A strong scientific establishment is seen as the basis for all power: military, industrial, intellectual. Schools and universities are harnessed to science. Business and economics are becoming scientific disciplines in the quest of efficiency. Yet science cannot establish any values and, by its own definition, is not qualified to provide a scientific code of ethics. Science cannot furnish goals nor does it contain any ethical values. Furthermore it cannot in any manner sustain its own establishment as a final good. As a new orthodoxy it prides itself on having replaced the old faiths. But being alienated from nature and absolved from moral participation in the cosmic process, man has been left in a world without humanity.
Broader Dehumanization (#PA1757).
Aggravated by Proliferation of technology (#PD2420)
Dehumanized individual scientific research (#PF2112).

♦ **PF5452 Racial stereotyping**
Ethnic stereotyping — National character
Nature Studies of national character were some of the earliest attempts to systematize cultural observations. Herodotus and Tacitus attempted to set down the essential traits of the neighbouring peoples they did war with, as did the book of Exodus. Racialist propaganda makes much of the inborn traits of Blacks and Jews in the West and Caucasians in the East. Despite tremendous advances in understanding the environmental, physical, hereditary and sociological forces at play in any social group, portrayals of national character continue to be promoted, some as the self-conscious folk prejudice it is, and others as the very edge of some revolutionary new grasp of humanity.
Refs Gilman, Sander L *Difference and Pathology* (1985).
Broader Stereotypes (#PF8508).
Aggravates Bigotry (#PC7652) National prejudice (#PG4368).
Aggravated by Prejudice (#PA2173).

♦ **PF5466 Self-indulgent societies**
National self-indulgence
Aggravates Consumerism (#PD5774)
National insecurity and vulnerability (#PB1149)
Domination of government policy-making by short-term considerations (#PF0317).
Aggravated by Unproductive use of resources (#PB8376).

♦ **PF5472 Over-centralization of global decision-making**
Restrictions on participation of developing countries in international decision-making — Displacement of national decision-making from developing to industrialized countries — Global centralization of power
Narrower Developing country failure to exercise leverage on international bodies (#PJ8459).
Aggravates Ineffective means for participation in decision making (#PE8430).
Aggravated by Eurocentrism (#PJ9837)
Over-centralization (#PF2711)
Elitist control of global economy (#PC3778)
Undermining of multilateral forums by industrialized countries (#PE4289).

♦ **PF5477 Verbosity in intergovernmental organizations**
Incidence This is frequently cited in connection with the lengthy speeches made on the occasion of plenary debates by one representative of each Member State of the organization. Numerous illustrations are also to be found in planning and programming documents. Typically a programme with only one professional-level worker or involving the organization of one meeting or the production of one publication may be described in extremely inflated terms concealing the limited nature of the outputs for which provision has in reality been made.
Broader Proliferation of documents from international organizations (#PF5992).
Related Unnecessary verbosity of legal documents (#PF7137).

♦ **PF5482 Inconsistent risk evaluation**
Unequal evaluation of risks
Claim Society evaluates risks in an inconsistent manner. Thus radiation from nuclear power plants is probably feared too much whereas radon seeping into basements is probably feared too little. Airplane accidents are feared out of all proportion to the number of accidents per mile travelled, whereas the continuing level of road accidents is accepted as a routine aspect of travel.
Broader Risk (#PF7580).
Narrower Excessive caution (#PF6389)
Excessive desire for risk (#PF9424).

♦ **PF5495 Misreading sacred documents**
Broader Misleading information (#PF3096)
Profanation of sacred doctrine (#PF7484).
Aggravated by Bibliolatry (#PF7129).

♦ **PF5496 Obstacles to aquaculture**
Lack of hydroponics — Limited fish farming
Nature Fish have been raised for centuries in flooded fields and ponds. More recently these technologies have become a major focus of development efforts concerned with food production. Both the raising of fish and of plants in water without soil are unusual agricultural efforts, however. Only the wealthier and more innovative farmers are interested. Consequently, schemes that on paper and in a government-sponsored agricultural station sound quite impressive are not tried.
Background Aquatic culture goes back at least 4000 years in China, Japan and Egypt, well over 3000 years in Java and India, and at least 2500 years in Europe. Despite this ancient history, fish farming has had very little scientific help. Most of its methods are traditional, developed by trial and error. Aquaculture is characterized by four general types of activity. The first is hatchery, where large numbers of young are raised and then released into coastal waters with the expectation that they will increase the size of the natural population and the commercial catch. This expectation is almost never fulfilled. Exceptions are the cultivation of some species of salmon in the US and Europe and of sturgeon in the USSR. The second kind of aquaculture involves the capture and impoundment of young. In enclosures, they may be left to fend for themselves, the water may be fertilized to that natural food is increased, or supplemental food may be added. This is the most successful form of aquaculture. Third is production of young from eggs and their retention until they reach marketable size. Japanese shrimp culture is an example of this. The last method involves the full control of the life cycle of the animal; eggs are hatched, the animals are feed until large enough for commercial purposes, they are harvested and brood stock is kept and breed. Trout and catfish are fresh-water examples of this method. The only marine example is oysters on a very small scale.
Claim There is little scientific support of aquaculture. The usable surface area of the oceans is relatively small; the open sea is unsuitable. Coastal areas and shallow seas are highly prized by commercial and sport fishermen, pleasure boaters, swimmers, oil and other mineral explorers, housing developers and those who dump waste. The increasing amounts of pollution: sewage, industrial wastes, fertilizers, pesticides, silt, heated water, trash, oil, and radioactive substances are making some whole seas unsuitable for aquaculture. Unclear ownership rights of seas, even coasts increases the risk of potential sea farmers. Conservation laws sometimes prevent its development by forbidding the capture of young animals or of mature females. The number of marine animals which can be cultivated is small.
Broader Inadequate agricultural facilities (#PF6499)
Obstacles to the utilization of coastal and deep sea water resources (#PF4767).

♦ **PF5502 Progressive reduction in government action commitment**
Governmental delayed withdrawal of commitment to action
Nature Having made a strong commitment to act on a problem, and received the politically necessary media coverage, governments tend to feel free to gradually reduce the level of their commitment so that little is in effect accomplished in response to the problem. The reduction of commitment may take the form of disempowering the agency mandated to act, either by imposing impossible constraints, reducing the funds committed to the project, or appointing incompetent people to manage the programme.
Broader Government inaction (#PC3950) Broken government promises (#PF4558).
Aggravates Arrears in payment of government financial commitments (#PF1179).

♦ **PF5513 Pseudo–culture**
Nature Forms of cultural expression specifically engineered for the material gain of some group or individual, especially when produced for mass consumption. Such cultural products fail to take account of the cultural differences of distinct communities and tend to erode further any remaining cultural identity.
Claim Pseudo–culture becomes popular by playing to the most crude, external sentiments of the human mind: greed, fear, lust, insecurity. Subtle qualities such as compassion, service and aspirations to greatness, are rarely cultivated in the fast–moving, commercial world of pseudo–culture.
Aggravates Homogenization of cultures (#PB1071).

♦ **PF5514 Payment of interest**
Inflation–based monetary systems
Nature The crisis in financial systems and the financial difficulties experienced by many arise from the payment of interest from those who have less money to those who have more money than they need. Based on interest and compound interest, a given amount of money doubles in value at regular intervals, following an exponential growth pattern. The continual payment of interest and compound interest is arithmetically, as well as practically, impossible in the longer–term. The concentration of money in the hands of fewer people or enterprises creates a constant pressure for large scale investments. Military production is perhaps the only area through which the saturation point can be postponed indefinitely.
Background One common misconception is that payment of interest can be avoided by not borrowing money. This is not true because included in every price paid is a certain amount of interest. A second misconception is that since everybody has to pay interest when borrowing money (or buying goods or services), then all benefit or lose equally. This is incorrect. There are great differences between those who profit from the system and those who pay for it. The present monetary system allows the operation of a hidden distribution mechanism which constantly reallocates money from those who have less money than they need to those who have more than they need. The payment of interest is prohibited by the Islamic faith.
Incidence 1. One of the symptoms of this system is inflation which is effectively another form of taxation through which governments try to overcome the worst effects of increasing debt, resulting from the monetary system they created. Inflation, through the printing of money, is a way for government to overcome its increasing interest–related indebtedness. For example, government income in West Germany rose only 300 per cent between 1968 and 1982, whilst its interest payments rose by 1,160 per cent. Such problems would be avoided under an alternative system in which people would pay a fee if they kept money out of circulation.
2. Muslims are prohibited from any involvement in interest by their religion. But because the financial system of the world is currently dominated by interest, Muslims find it extremely difficult to abide by their religious convictions when dealing with that system. Most Muslims are deprived of any means of using the few non–interest paying facilities.
Claim Interest in effect acts like a cancer in the social structure.
Aggravates Economic inflation (#PC0254) High interest rates (#PF9014)
Inadequacies of the international monetary system (#PF0048).
Aggravated by Unethical financial practices (#PE0682).

♦ **PF5520 Jeopardization of universality**
Vulnerability of universal principles
Refs United Nations, Geneva *Is universality in Jeopardy ?* (1986).
Aggravated by Discrepancies between principles and practice (#PF4705).

♦ **PF5522 Interlocking corporate directorates**
Incidence A principal reason for not challenging a company's chief executive is that so many of the directors are themselves chief executives (63 per cent of all outside directors of the Fortune list of 1,000 largest American companies are chief executives of other companies).
Aggravates Undemocratic policy–making (#PF8703)
Elitist control of global economy (#PC3778)
Abdication of control by company directors (#PE9251)
Disproportionate control of global economy by limited number of corporations (#PE0135)
Disproportionate influence on national economies of limited number of corporations (#PE1922).
Aggravated by Cronyism (#PF4549).

♦ **PF5529 Self interested manipulation of timing**
Broader Narcissism (#PF7248).
Narrower Unjust election timing (#PD2907)
Misleading accounting information due to inflation (#PE4285)
Inconsiderate choice of payment times by creditors (#PE3926)
Calculated delays in releasing controversial news items (#PE0598).

♦ **PF5539 Local control of resources**
Aggravates Foreign control of natural resources (#PD3109)
Worldwide misallocation of resources (#PB6719).

♦ **PF5544 Criminal characteristics**
Nature It appears that some people have a genetic trait which gives them a low ability to cope with frustration and other emotional crises. Indeed, the impairment of intelligence alone may be the cause of sociopathic behaviour in individuals with chromosome defects.
Background In 1965, scientists claimed that an abnormally large proportion of sociopathic offenders in two special hospitals in Scotland had extra chromosomes. It was referred to as the XYY syndrome. This seemed to be the confirmation of ancient moral theory that people were waiting for, namely that criminals were born, not made. Five years later, a study published by the Cambridge University Institute of Criminology summarized and clarified the facts: (1) There is no doubt about the correlation between the double Y chromosome, criminality and anti–social behaviour increasing in frequency with decreasing intelligence. (2) The patients being genetically disposed to criminality are unlikely to respond to punishment. (3) The extra Y chromosome has resulted in a severely disordered personality, and this disorder has led these men into conflict with the law.
Counter–claim The chromosome theory of criminality has been shown more clearly to be what it always was – a lot of hot air. This is just as well, since there must be about 20,000 XYY adult males in the UK alone living normally and doing no harm to anyone.
Broader Genetic defects and diseases (#PD2389).
Related Genetic susceptibility to disease (#PG2834).
Aggravates Crime (#PB0001).

♦ **PF5555 Collapsed tension between care and responsibility**
Broader Conflicting social service ideologies (#PD3190).
Related Athletic competition (#PE4266).
Aggravates Inadequate social welfare services (#PC0834).

♦ **PF5587 Quantitative understanding of responsibility**
Broader Narrow context for counsel (#PF0823).

♦ **PF5596 Rumour**
Rumour–mongering
Nature Rumour, an unsubstantiated tale heard and retold as fact, is the cause of malignment of individuals and nations, death and injury due to panic in public events and confusion and misdirection in public and private life.
Refs Rosnow, R L and Fine, G A *Rumor and Gossip* (1976).
Aggravates Alarmism (#PF4384) Misleading information (#PF3096).
Aggravated by Gossip (#PE2192) Insufficient communications systems (#PF2350)
Covert smear campaigns by government (#PD7171).

♦ **PF5609 Masturbation guilt**
Nature Masturbation is the act of stimulating the external sexual organs by oneself, accompanied by fantasies that are usually of a sexual nature, and having as its aim the discharge of sexual excitation. The most important consequence of masturbation is the guilt which usually accompanies it, and the struggle to defend against such guilt which may last for years and absorb all the energy of the psychic system.
Background Freud tended to the view that neurasthenia could follow upon excessive masturbation; it is nowadays felt that neurasthenia is more likely an outcome of insufficient orgasm – that is, if anxieties and guilt disturb the satisfactory character of the masturbation. It would also seem that the adolescent in conflict about the practice has a deep need to believe that masturbation is a wrong thing and he often resists enlightenment about its harmlessness. This may be because the conscious masturbatory fantasies are distorted derivatives of unconscious Oedipal fantasies, and if the adolescent did believe that masturbation was harmless, he would have to face the Oedipal desires which were responsible for the guilt.
Claim Some religious teachers have gone to absurd and unscientific extremes in their efforts to discourage the young from the practice of masturbation. Such statements have included threats of insanity as resulting from its practice. This is without scientific foundation. The only psychological illness that can result from masturbation is from irrational and excessive fears and guilt feelings arising from such warnings, especially when such fears and feelings are related by the young person to experiences beyond his control. On the other hand, compulsive forms of masturbation can be symptomatic of emotional and mental disturbances that stem from other sources.
Refs Swarup, Ram *Facts and Fallacies About Selfrelief Wrongly Termed as 'Selfabuse' Solitary Vice, Sin of Youth and 'Onanism'* (1961).
Broader Guilt (#PA6793).
Aggravates Sexual unfulfilment (#PF3260).
Aggravated by Masturbation (#PE4426).

♦ **PF5622 Decline in public interest broadcasting**
Resistance to public interest broadcasting — Inadequate support for public interest broadcasting
Related Decline in religious broadcasting (#PF1433).
Aggravates Insufficient cultural media (#PJ8476).
Aggravated by Distorted media presentations (#PD6081)
Excessive commercialization of the media (#PE4215).

♦ **PF5627 Popular errors**
Refs Mackay, Charles and Templeton, John M *Extraordinary Popular Delusions and the Madness of Crowds* (1985).
Broader Illusion (#PA6414) Human errors and miscalculations (#PF3702).
Narrower Proverbial lore errors (#PF5653).

♦ **PF5630 Blasphemy**
Religious insult
Nature Any expression by word, sign, or gesture that is insulting to the goodness of God is viewed as blasphemy by the Church. In the USA, the common law offence of blasphemy is perpetrated through the verbal publication of words intended to impair or destroy man's reverence for God and religion, by subjecting belief in the deity, scriptures or religious doctrines to derision and vilification, If the blasphemous words are written or printed, they constitute blasphemous libel, as do pictorial caricatures which convey a blasphemous meaning.
Claim Blasphemy is a two part problem. The blasphemer consciously separates himself from the creator and as such is self destructive. The blasphemer also sets an example for others to do the same, endangering them.
Counter–claim 1. Blasphemy is to religion what sedition is to government. In the USA there are still laws on the books of some fifteen states today, but they are almost never used. They are holdovers from an earlier era, when Americans viewed Christianity as the one true religion. Most of them, possibly all, are unconstitutional by current Supreme Court standards. 2. There are laws of libel to protect individuals against untruthful attacks on their integrity and honour, and there are laws to protect individuals against racial incitement or sexual harassment. Blasphemy laws do not protect persons, they protect doctrines, and instead of being extended to cover all forms of religious belief, they should be abolished because they attempt to restrict freedom of opinion in the realm of ideas. 3. Conflict of perceptions of blasphemy does not concern questions of the sacred and the profane but rather differences between those who believe that there are such things as certainties in this life and those who want to celebrate human impurity and the radical incompatibilities between human visions.
Broader Impiety (#PA6058) Profanity (#PF7427) Victimless crime (#PC5005).
Narrower Imprecation (#PF3746).
Related Lese majesty (#PF4341).
Aggravates Verbal abuse (#PD5238).
Aggravated by Bibliolatry (#PF7129).

♦ **PF5633 Evil governments**
Satanic governments
Incidence Governments influenced by fundamentalist religious principles tend, in their propaganda, to refer to any governments that oppose them as "satanic" or as "agents of satan". For example, the Iranian government (and those who follow that lead) tends to refer to the USA as the "Great Satan". Some Christian fundamentalists refer to the papacy as satanic or to the Pope as Satan. President Reagan named the USSR as an "evil empire".
Broader Evil (#PF7042).

♦ **PF5639 Individualistic utilization of expertise**
Nature Professional's and expert's roles and behaviour in society are largely defined by individualistic values and images. Both the public and the experts expect one to one relationship between professionals and their clients. Professionals tend to see their roles as self enhancing instead of serving the public. The directions of professions and the development of new skills are uncoordinated.
Broader Limits on areas of research (#PF2529).
Narrower Individualistic style of professionals (#PF2110)
Distrust of professional service delivery (#PD0974)
Collapse of public servant role of the professional (#PF0897)
Failure of global-scale planning for expertise development (#PF1055)
Decisional paralysis of specialized services in relation to the world's need (#PF2709).
Related Thwarted technological communications (#PF0953)
Gap between the function of social techniques and the needs they address (#PF3608).
Aggravated by Unavailability of appropriate expertise (#PF7916)
Unarticulated goal of educational methods (#PF2400).

♦ **PF5645 Inappropriate policies**
Misguided priorities — Inadequate government policy — Inadequate strategies
Narrower Messianic policy-making (#PF9796) Risky rental agreements (#PG1425)
Aggressive foreign policy (#PC4667) Inadequate fiscal policies (#PF4850)
Inadequate credit policies (#PF0245) Restrictive social policies (#PF8282)
Restrictive monetary practices (#PF8749) Rigid personnel hiring policies (#PF3168)
Inappropriate development policy (#PF8757) Unjustified military defence policies (#PF1385)
Ineffective educational policy decisions (#PF2447)
Self-destructive government policy-making (#PF5061)
Failure of government intelligence services (#PF8819)
Inadequate governmental energy conservation policies (#PF0037)
Inadequate governmental resource conservation policies (#PF0038)
Restrictive trade union policies concerning employment (#PF6046)
Failure to adapt general initiatives to specific needs (#PF1578)
Antagonism between employment policy and technical advance (#PE5104)
Inadequate adaptation of policy to educational difficulties (#PE8700)
Social environmental degradation from recreation and tourism (#PD0826)
Neglect of environmental consequences of government policies (#PE9295)
Environment policy as restriction on trade in developing countries (#PE4965)
Domination of government policy-making by short-term considerations (#PF0317)
Restrictions on industrial and economic development due to environmental policies (#PE4905)
Distortion of international trade by discriminatory government and private procurement policies (#PE0347).
Aggravates Policy-making delays (#PF8989) Irrelevant institutions (#PF8846)
Unimplemented decisions (#PF4672) International status race (#PC5348)
Abuse in government policy (#PF8389) Disincentives against farming (#PD7536)
Failure of development policies (#PF5658)
Public resentment at government policies (#PF7472)
Prohibitive cost of inadequate development policies (#PF9101)
Natural environmental degradation from recreation and tourism (#PE6920).
Aggravated by Unrealistic policies (#PF9428) Conflicting priorities (#PF5766)
Undemocratic policy-making (#PF8703) Uncoordinated policy-making (#PF9166)
Overgeneralized policy models (#PF9100) Politicization of decision-making (#PD8468)
Unpredictable governmental policy (#PF1559) Drug abuse by government officials (#PD8696)
Inadequate public finance statistics (#PF7842)
Ineffective decision-making processes (#PF3709)
Cumbersome methods of policy formation (#PF1076)
Deliberate ignorance during policy-making (#PF8278)
Short range planning for long-term development (#PF5660)
Inadequate models of socio-economic development (#PF9576)
Employment of criminals in policy-making contexts (#PE4439)
Inadequacy of governmental decision-making machinery (#PF2420)
Irresponsible delimitation of policy responsibilities (#PF7823)
Constraint of inherited problems on policy innovation (#PF9208)
Over-reliance on economic interest groups by policy agencies (#PF1070)
Inadequate information systems for international governmental decision-making processes (#PD0104).

♦ **PF5649 Abuse of privileges and immunities by diplomats**
Abuse of privileges and immunities by international civil servants
Nature Members of diplomatic staffs in embassies, consulates or missions accredited to foreign governments or international inter-governmental bodies, may flout the civil or criminal codes of their host country knowing they have diplomatic immunity from prosecution. On instructions from their governments they may engage in: military or industrial espionage; spy recruitment using entrapment methods; and harassment, abduction or assassination of expatriates from their own countries. Expulsion of diplomatic staff is an insufficient deterrent against these abuses.
Incidence Offences not prosecuted due to diplomatic immunity range from parking tickets (75,000 cancelled tickets in London alone for the first nine months of 1983); to the smuggling of firearms, drugs, and blackmarket currencies; to child abuse (reported in the UK and USA); and to attempted homicide (USA). The 1983 shooting outside the Libyan People's Bureau, in London in which a London policewoman was killed, raised the question of and public interest in whether the firearms used were smuggled in diplomatic bags which, due to diplomatic immunity, could not be searched.
Narrower Diplomatic fraud (#PD5554).
Related Police immunity (#PE5832) Political assassination (#PE5614).
Aggravates Violation of privileges and immunities of international civil servants (#PE5488).

♦ **PF5653 Proverbial lore errors**
Contradictory commonplaces
Claim Proverbs exist in such great numbers that their so-called wisdom proves to be a maze of contradictions, countryside ignorance and naivety. Anyone can justify anything by reciting the appropriate proverb. 'There's no smoke without a fire', 'You can't judge a book by its cover', 'Seeing is believing' and 'It all comes out in the wash', are some of the saws that advise forming judgements more quickly or more slowly. There is also 'Judge not, lest you be judged by the same measure that you judge'. The excessive use of proverbs, quotations from the wise, figures of speech and the homely metaphors and analogies they embody, may take the place of thinking and retard development of an individual's own language and reasoning tools.
Broader Popular errors (#PF5627) Human errors and miscalculations (#PF3702).
Related Cultural corruption (#PC2913)
Misrepresentation of historical persons (#PF5662).
Aggravates Defective reasoning (#PF5711).

♦ **PF5654 Grief**
Nature Deep distress occasioned by the death of a loved one occurs in both animals and man. The dog who stops eating when his master dies until he himself succumbs is the extreme animal form of grief, while survivor suicide is the extreme human form. Grief is universal and an inescapable condition of living.
Claim Grief is usually regarded as a poignant emotion, but it is much more. It is a physically destruction condition that affects the body like an illness; and in addition, it is socially disruptive in countless ways. Grief should be considered as a medical disease, and treatment should be provided for it. The usual prescription of tranquillizers, sleeping pills and words of comfort are not medically specific. Victims of grief, like victims of crime, should be viewed as a responsibility of the whole or local society in which they exist. Appropriate medical, social and, perhaps, economic remedies need to be developed.
Refs Carr, Arthur C, et al (Eds) *Grief* (1974); Fulton, Robert *Death, Grief and Bereavement* (1976); Stern, Marvin *Death, Grief and Friendship in the Eighteenth Century* (1985).
Broader Adversity (#PA6340) Solemnity (#PA6731) Unpleasantness (#PA7107).
Related Disabled victims of crimes (#PD0762) Neglect of victims of crime (#PD4823).
Aggravates Mental depression (#PC0799).
Aggravated by Bereavement (#PF3516) Appropriation (#PA5688).

♦ **PF5657 Solipsism**
Nature The ultimate form of egocentricity is the philosophic formulation, involving consideration of metaphysics, epistemology, ontology and psychology, that the whole of the realities of other persons and external objects and worlds have no existence independent of the cognizing self and its mental states. In its modern form in developed countries, it is frequently associated with the addiction to mysticism, occultism, drugs, altered states of consciousness, social alienation, and personal selfishness in seeking peak experiences, hedonistic or otherwise. Not surprisingly, among the large numbers of people espousing popular varieties of the solipsist creed, and who have no formal training in philosophy, this denial of reality is linked with a denial of authority, and with negative values concerning social institutions and organizations. The solipsist may also have personality disorders, the most common of which are schizoid tendencies, a lack of emotional development, and an inability to relate to other people.
Claim Solipsism is a characteristic of pseudo-mysticism and a distortion of the philosophy of idealism. It has led to doctrines of ethical subjectivism that deny reality to values such as goodness; and to latitudinarianism, a critical paralysis which views all opinions and creeds indifferently as being of equal worth or of no worth. It lends itself to utopian collectivist social philosophies, to anarchy, and to the loss of liberty and justice, since it accords no reality to personal identity and national sovereignty and to the properties, rights and obligations that pertain to them.
Broader Egoism (#PA6318).
Aggravates Nihilism (#PJ7927) Ignorance (#PA5568) Schizophrenia (#PD0438)
Inflexible deliberative systems of government (#PF7059).

♦ **PF5658 Failure of development policies**
Broader Destruction inherent in development (#PF4829)
Lack of appreciation of cultural differences (#PF2679).
Aggravates Underdevelopment (#PB0206).
Aggravated by Inappropriate policies (#PF5645).

♦ **PF5659 Insufficient minority culture support**
Nature Historically, all nations over a long period of time tend to integrate diverse ethnic groups. The cultures and sub-cultures of the minority are preserved in part as influences and contributions to the main-stream, and in part are lost. Integration, with its positive and negative aspects, is achieved more or less rapidly depending on ethnic and language distance. A minority group of the same race and same family of languages may be absorbed more readily, unless religious or other ideological differences persist, than another race with an unrelated language. In the latter case, compensatory efforts to artificially accelerate integration or assimilation may suppress ethnic evidences such as food and clothing preferences, behavioural characteristics, minority language, and knowledge of ethnic history. Loss of ethnicity is considered a loss to the entire society when conceived in the world context. Within minorities themselves, as well as in the majority culture, there is insufficient preservation of ethnic history and characteristics.
Counter-claim Integration or assimilation, that is, the minimizing of ethnic differences and the mastering of a common language, is in the interests of all, including those who in the short-term might be considered minorities. The proper places to preserve the ethnic legacies of peoples are the museums, for the dynamics of society will continuously form a new and common heritage for all people.
Broader Cultural stagnation (#PC8269).
Related Lack of assimilation (#PF2132).
Aggravates Ethnocentricity (#PB5765) Ethnic conflict (#PC3685)
Endangered cultures (#PB8613) Threatened and vulnerable minorities (#PC3295).

♦ **PF5660 Short range planning for long-term development**
Crisis management approach to long-term development policy-making — Obstruction of change by crisis and emergency responses — Immediacy orientation of elected bodies — Lack of reward for long-range planning and decision-making — Undeveloped long range plans — Limited long-range planning — Crisis oriented operations — Short-sighted planning — Short-range development strategy
Nature By definition, a crisis is a short-term problem. It evokes improvised remedies and palliatives. But real solutions call for long-term policies; prevention rather than cure, or cures which facilitate prevention. Ad-hoc policies have then to be seen as underpinnings of an integrated long-term solution strategy. If each crisis becomes an isolated problem to be somehow resolved, then this encourages a fragmented approach. The real danger is that priorities become dominated by problems of the immediate, and the fundamental task of change in the social framework is neglected, perhaps even suspended. When pressures build up for real changes in institutions and values, people are asked to hold their peace so that the immediate crisis can be dealt with. Such appeals for patience and order in the name of smooth functioning can often be clever rationalization of the status quo. Thus a crisis psychology provides succour and respite to forces that resist change. The Western political system is organized for short-term response to problems. Few administrations last a decade. To distinguish themselves from previous officeholders, new officials must act rapidly and in visible ways. Governments can rarely afford to be sensitive to advice of a medium or long term nature, even when it comes from eminent people. As corporation and government management is primarily concerned with personal and corporate security, there has arisen a dominant trend to reward short-term accomplishments and short-range planning.
Incidence The Pearson Report, 1969, a report of a group led by ex-Prime Minister Lester Pearson of Canada, the RIO Report (Reshaping the International Order), coordinated by Professor Jan Tinbergen, and two reports, in 1980 and 1983, by a group under the leadership of ex-Federal Chancellor Willy Brandt, have had little effect on the conduct of governments. There are very few private or public enterprises that reward long-range model-building and risk-taking. This has seriously affected the stability of corporate mythology as perceived by the workers within an enterprise or the investors, both of whom are necessary contributors to the health of an organization. This is particularly evident in the growing gap between the rewards provided to executives and those provided to middle and lower levels of management and production.
Broader Mismanagement (#PB8406)
Limited approaches to economic planning (#PF3500)
General obstacles to problem alleviation (#PF0631).
Narrower Crisis oriented education (#PG8966)

(#PF2813)
Short-term planning of product life cycles (#PF1740)
Immediate gratification-based social forms (#PF1058)
Underdeveloped sources of income expansion (#PF1345)
Overemphasis on immediate superficial needs (#PF3243)
Disguised negative consequences of remedial action (#PF7583)
Domination of government policy-making by short-term considerations (#PF0317).
Related Risk-aversion strategy (#PF4612).
Aggravates Conflicting priorities (#PF5766) Inappropriate policies (#PF5645)
Messianic policy-making (#PF9796) Inadequate support services (#PF6492)
Overstated programme advantages (#PF8181) Underestimation of programme costs (#PF8499)
Destruction inherent in development (#PF4829)
Lack of long-term development assistance (#PF5181)
Ineffective structures of local consensus (#PF6506)
Irresponsibility towards future generations (#PF9455)
Inadequate planning of action against problems (#PF1467)
Absence of long-term economic planning agencies (#PF3610)
Lack of responsible involvement in community affairs (#PF6536)
Excessive social costs of structural adjustment in debtor developing countries (#PD8114).
Aggravated by Short-term gain (#PF8675) Lymphatic leukaemia (#PE2686)
Resistance to change (#PF0557) Policy-making delays (#PF8989)
Disaster unpreparedness (#PF3567) Crisis-oriented funding (#PF2849)
Inadequate social reform (#PF0677) Inadequate sense of time (#PF9980)
Unattractive locale for economic development (#PF3499)
Inappropriate institutionalized reward systems (#PF9786)
Insensitivity to non-immediate hazards to society (#PF9119)
Fragmentation of the international trading system (#PC9584)
Failure to integrate knowledge to empower humanity in response to the global problematique (#PF8753).

♦ PF5662 Misrepresentation of historical persons
Nature Figures of historical importance are frequently misquoted, and sometimes quotations are invented for them as a way of invoking authority for fallacious argumentation.
Broader Historical misrepresentation (#PF4932).
Related Historical forgery (#PE5051) Proverbial lore errors (#PF5653).
Aggravates Inappropriate arguments (#PF2152).

♦ PF5670 Terminological confusion in weights, measures and numbering systems
Nature The diversity of the world's measuring systems was reconciled in large degree in the metric system introduced by the French Revolution. World acceptance of the metric system's improved version, the International System of Units (SI); will be only in part, for while it secures the most important aspect, world agreement on most quantitative units, there remain a number of exceptions and special problems.
Incidence One problem is the various names for the same measure in the world's languages. Another problem is the special aggregates of SI units. It is likely then, that even though the SI and other internationally agreed standards have established a unit name and number, names and aggregates specific to linguistic areas will persist. The Netherlands *vierkante roedo*, for example, is the same as the are of the SI; the Greek *stremma* is 10 ares; a *bunder* in the Netherlands is a hectare and in Turkey the same thing is called a *djerib*. Practical problems are demonstrated by the instance of international motorists in Poland who want a number of litres of petrol. They will need to ask for a number of *kwartas*. A long-distance lorry stopping for 100 litres in the Netherlands will need to ask for a *mud*, and if the driver is ever in Thailand to purchase some rice wine, a *fanan* will give him a litre's worth. Apart from the weights and measures terminology, confusion exists in numbering systems. Exchange of international information is hindered by this persisting confusion of language.
Broader Multiplicity of languages (#PC0178).
Related Inadequate standardization of procedures and equipment (#PC0666).

♦ PF5673 Exorcism as a superstition
Background The addressing of supposed invisible spirits with a command to depart, binding them to obey by the invocation of a sanctified name, is the act of exorcism as known to the Western world. Exorcism is still provided for in rites of Judaism, Roman Catholicism, Anglicanism and the Orthodox Churches, and is known in a variety of forms in the major and minor religions and sects. In cases of supposed demonic possession a priest or lay exorcist may be summoned. Exorcism is a ceremony preparatory to baptism in the Catholic ritual, hence hundreds of millions of people have been involuntarily exorcised shortly after birth, and are later largely unaware that such an event occurred in their lives.
Claim Demonic possession and exorcism are both characteristic of the darkness of the human mind: dark because understanding of its disorders is difficult, such as in the case of non-existent spirit possession but real mental affliction; and also because of the clinging to superstition and the inability to reason. The acts of the exorcist may conceal some applied psychology that may be effective against hysteria; but they are totally ineffective for psychoses, therefore recourse to an exorcist is an obstacle to proper medical treatment. Exorcists and witch doctors are encountered in primitive societies where they wield sufficient influence to cause resistance to vaccinations, modern clinics, family planning, education, and other aspects of development.
Refs Goodman, Felicitas D *How About Demons? Possession and Exorcism in the Modern World* (1988); Nauman, Elmo Jr *Exorcism Through the Ages* (1974); Petitpierre, Dom Robert *Exorcising Devils* (1976).
Broader Occultism (#PF3312) Superstition (#PA0430)
Unexplained phenomena (#PF8352).
Reduces Poltergeists (#PF5697) Demonic and spirit possession (#PF5781).

♦ PF5679 Unnecessary health tests
Dubious medical tests
Incidence In the USA malpractice insurance carriers want the doctors to order expensive tests and studies in order to verify what the doctors already know.
Related Unnecessary surgery (#PE9271).
Aggravates Fictitious disorders (#PG7686)
Inadequate maintenance of physical health (#PF1773).
Aggravated by Unethical medical practice (#PD5770).

♦ PF5681 Jihad
Holy war
Nature The use of the sword is the Islamic fourth way of performing the religious duty of defending and propagating the faith. The first three means are by the heart, by the tongue and by the hand. The notion of religious justification for war persists, and threats of Jihad, or holy war, are still heard in the second half of this century. While the influence of the mullah, mufti and other Moslem jurists, theologians and intellectuals is to internalize the Jihad and thus make it a personal struggle to attain moral improvement by self-conquest, the lack of restraint of some popular leaders, militarists, fundamentalists and oil supported politicians in invoking the traditional fourth way appeals to the emotions of the uneducated and threatens peaceful solutions to problems of Islamic and world concern.
Incidence The war of Egyptians with Israelis, Pakistanis versus Hindus, and the conflict between Turkish-origin and Greek-origin Cypriots are recent instances where the ideals of Jihad have been introduced. Where Shiite Moslems fight Sunni Moslems, as in Lebanon, or in the case of Iran versus Iraq, each faction may also invoke the Jihad tradition, not unlike the clerically-blessed conflicts of Christian Europe. In 1990, Saddam Hussein, as president of Iraq declared a jihad against the forces arrayed aginst him.
Background The tendency for a nation to arrange divine endorsement of its defensive or offensive wars is known from antiquity, and was probably a characteristic of primitive man as exemplified by the American Indian war dances and other rites invoking supernatural help. However there is a difference of scale when one moves from the single, tribal level to levels of tribal confederations or nations. Religious fervour invoked on this scale can abet an enormous amount of bloodletting: the Crusades illustrate this. Islamic emphasis on the sword is seen by the conquests in its first thousand years. The wars of the Turkish Ottoman Empire (1300–1922), while not specifically Jihads, show such characteristic features as the desecration of the Hagia Sophia after the capture of Constantinople (1453); forced conversions to Islam over the centuries; and the genocidal massacres of over 2 million Catholic Armenians (1894–1896, 1915–16). Even where Jihad is not officially proclaimed, fighters may call themselves 'mujahidin' or holy warriors and circumvent the Koranic protection extended to Christians and Jews, as in the Six Days War with Israel.
Claim Muslims believe that the solution to the problem of Palestine is a holy war against Israel and all the powers that support the occupation of Palestine.
Broader Religious war (#PC2371) Religious conflict (#PC3292)
Religious extremism (#PF4954).
Related Fundamentalism (#PF1338) Self-inflicted physical suffering (#PD7550).
Aggravates Enemies (#PF8404) Malignant visions (#PF5691)
Childhood martyrdom (#PF8118).
Aggravated by Criminal contempt (#PD5705)
Holy places as a focus of religious friction (#PF1816)
Distortion in international politics from fuel shortages (#PE8080).

♦ PF5691 Malignant visions
Fanatical visions
Nature Building up a just and peaceful society is a very benign vision, though it may involve some tough societal interventions that make it malignant to some groups. Malignant visions include aspirations to conquer half the world and the proselytizing of non-believers through holy wars, especially when supported by modern weapons systems.
Incidence Visionary political leadership which adopts a fanatical stance is a regular feature of history that satisfies deep political needs, such as the search for enemies. International ideological cleavages, north-south conflicts, economic crises in poor countries, and modernization versus traditionalism are just a few of the features which are expected to increase the probability of malignant visionary political leadership.
Related Utopianism (#PF2192) Aggressive ideologies (#PJ4778).
Aggravates Crazy nation-states (#PJ1092).
Aggravated by Jihad (#PF5681) Enemies (#PF8404)
Religious extremism (#PF4954) Ideological conflict (#PF3388)
Inhibited capacity to visualize a creative future (#PF2352)
Disparity between industrialized and developing countries (#PC8694).

♦ PF5692 Constraint of time on individual and social development
Broader General obstacles to problem alleviation (#PF0631).
Narrower Lack of time (#PC4498) Ineffective class scheduling (#PG5424)
Fatalistic attitudes to the use of time (#PF2795)
Obstacles to efficient utilization of time (#PF7022).
Related Differing conceptions of time (#PF6665)
Unpreparedness for surplus leisure time (#PF5044).
Aggravates Inadequate sense of time (#PF9980)
Unfinished imperfect universe (#PF5716)
Underestimation of the human potential (#PF7063).

♦ PF5694 Religious theology disrelated to life
Nature The failure of theology to relate religious traditions to living in a post-modern secular world leaves human being with few ways of appropriating the mysterious, profound, or deep experiences to meaningful self-images. Humans escape into a variety of denials about the wonder, fear and pain of living.
Broader Theological collapse (#PF6358).
Related Maldistribution of world population (#PF0167).
Aggravated by Human ageing (#PB0477).

♦ PF5697 Poltergeists
Nature Alleged poltergeist phenomena cause fear, and their unsatisfactory explanation perpetuates superstitious ideas, such as houses which are said to be haunted. Houses which have experienced poltergeist knockings or transport of objects, are shunned, causing losses of property values and the movement of families from such residences.
Refs Goss, Michael *Poltergeists* (1979); Sitwell, Sacheverell *Poltergeists – Fact or Fiction?* (1988).
Broader Ghosts (#PF3801) Occultism (#PF3312).
Related Haunted buildings (#PF0201) Demonic and spirit possession (#PF5781).
Reduced by Exorcism as a superstition (#PF5673).

♦ PF5701 Inadequate public information concerning problems
Broader General obstacles to problem alleviation (#PF0631).
Aggravates Recurrence of misapprehended world problems (#PF7027)
Institutional preoccupation with obsolete problems (#PJ5014).
Aggravated by Suppression of information concerning social problems (#PF9828)
Failure to integrate knowledge to empower humanity in response to the global problematique (#PF8753).

♦ PF5703 Lack of qualitative excellence
Poor quality
Incidence Lack of qualitative excellence is demonstrated in a wide variety of fields: management standards in business administration emphasize economy and efficiency; entrepreneurial standards are set by payback and profitability ratios, with quality of products and services coming only second in consideration. In public administration, standards are focused on numbers of people served and for how long a time, so that qualitative aspects of the services offered are frequently subordinated to provisions for their durability. In the sciences, research that can provide immediate application is better funded than pure or basic research whose values are long-range. And in the arts, business considerations produce a popular culture where levels of expression and performance continually reach new depths.
Narrower Insufficient nobility (#PF5778) Lack of professional standards (#PF3411)
Deterioration in product quality (#PD1435) Decline in nutritional quality of food (#PE8938)
Quantitative pressure on standards of quality (#PF7227).

♦ PF5709 Over-specialization in medical care
Nature Health care specialists may not be sufficiently trained to refer patients to general medical

practitioners because they cannot readily recognize pathology in other parts of the body. They may also choose to ignore it as not being their concern. A patient's family physician is generally unknown to those specialists, such as dentists, opticians, chiropractors, acupuncturists, kinestherapists, psychologists or psychiatrists, who can be directly contacted by patients; and there is normally no contact between them, so the patient's medical history is also unknown. Some treatments, medicines and anaesthetics may be contra-indicated when the patient's health is already affected by heart condition, epilepsy, contagious diseases or a number of other ailments.
Broader Over-specialization (#PF0256) Inadequate medical care (#PF4832)
Fragmentation of health service (#PE5721).
Aggravates Excessive medical intervention in childbirth (#PE7705).

♦ **PF5711 Defective reasoning**
Fallacies of logic — Statistical fallacies
Incidence Invalid conclusions are usually of one type based on the insufficiency of the premises, data or considerations. Fallacious logic, however, can be exhibited by over 18 different types, and in the logic of the all-important syllogism, where there are 256 possible forms of the categorical variety alone, the possibilities for inherently fallacious structures are numerous.
Claim Assertions that conclusions may be drawn or derived from insufficient premises and consideration are examples of a popular defect in reasoning. It is probably the greatest and most pernicious habit of homo 'sapiens', and it has governed the life, organization and politics of many societies from ancient times. The lessons of the physical sciences, of experimental method and of scientific methodology in the last three hundred years show what correct argumentation can accomplish – that is, the establishment of conclusions from true and adequate premises. The power of cogent reasoning unfortunately has not penetrated as well in the social and behavioural sciences. Paradoxically, the study of logic was eliminated from higher education as the particular sciences were incorporated into it.
Counter-claim Parallel to the development of interdisciplinary sciences and generalized and more holistic curricula, logic is re-emerging – particularly in mathematics, linguistics and information theory as applied to computer and robot developments. Artificial intelligence may yield the perfect reasoning that has eluded human intelligence.
Broader Inappropriate arguments (#PF2152).
Narrower Reductionism (#PF7967) Inadequacy of formal logic (#PG7788)
Irrational religious beliefs (#PF6829).
Aggravates Ignorance (#PA5568)
False assumptions on sustainable development (#PF2528).
Aggravated by Affectation (#PA6400) Proverbial lore errors (#PF5653)
Unethical practices in the social sciences (#PD6626).

♦ **PF5712 Over-specialized study of global ecosystem**
Nature There is presently an inadequate understanding of the complex, subtle and often synergistic interactions between the several parts of the geosphere and biosphere. More knowledge is needed for societal management of the global ecosystem to enhance biological productivity and to respond to the increasing needs of a growing population. There are an insufficient number of national and international programmes which illuminate the processes that govern the behaviour of the oceans, atmosphere, lithosphere, biosphere and the solar terrestrial domain by addressing the interfaces among them. A holistic approach will require several years of careful planning and conceptualization to develop an appropriate scientific strategy.
Background Global change in the terrestrial environment (geosphere) and the life that inhabits it (biosphere) is a closely coupled system. The system is constantly undergoing change on time scales that range from hundred of millions of years through the slow recurrence of ice ages to transient phenomena. Changes in the geosphere that embrace the land, oceans, atmosphere and the solar terrestrial domain, and in the terrestrial and marine biosphere, arise from the interplay of physical, chemical and biological processes. Over millions of years these natural changes have resulted in the evolution of delicately balanced ecosystems that constitute the global life support system. To an increasing extent, these changes are influenced by the impact of human activity.
Broader Over-specialization (#PF0256).
Aggravates Inadequate models of socio-economic development (#PF9576).

♦ **PF5716 Unfinished imperfect universe**
Entropy
Claim If the universe were totally perfect there would be no need for change and nothing would develop. Conversely, the fact that change is inherent in this universe indicates that it is imperfect and must remain so – in fact, it can be inferred from the law of maximum entropy that the universe will in the long-term attain maximum disorder, even if in the short-term there are "pockets" of increasing order. There is thus no possibility for permanent improvement, and all effort for the development and amelioration of present conditions must be viewed as palliative. As Jesus said, the poor will be always with us. What is permanent is the desire in each individual for truth; if this is worked upon, the results must in the long-term be more effective than forever trying to stop up the holes in a sinking ship.
Broader General obstacles to problem alleviation (#PF0631).
Narrower Change (#PF6605) Underestimation of the human potential (#PF7063).
Related Chaos (#PF6836).
Aggravates Instability (#PA0859).
Aggravated by Constraint of time on individual and social development (#PF5692).

♦ **PF5717 Obstacles to use of bicycles**
Nature The normal urban environment is not designed for bicycles. On roads they are threatened by cars, and bicycles on paths and pavements threaten pedestrians. Bicycles are also threatened by parked cars as they make it difficult for the rider to see other people and for other people to see him. Many of the inevitable accidents between cyclists and automobiles and larger vehicles cause fatal or serious injuries and almost invariably these are only to the riders of the bicycles, mopeds, and motorcycles which have been struck or over-run; many of the victims are children and young people. Lack of provision for well-defined traffic rules for drivers of all conveyances; lack of integrated traffic control methods that serve automobiles, buses and commercial traffic, as well as students, workers and shoppers on two-wheelers; and lack of special lanes or bicycle paths, are contributory factors to cycling dangers and obstacles to their use.
Broader Insufficient transportation infrastructure (#PF1495).

♦ **PF5726 Pandering to the youth market**
Nature The object of those furnishing goods and services to teenagers and young adults, particularly in developed countries, is to pander to their passions and emotions. The encouragement of garish clothes and bizarre coiffures, and the provision of gathering places where a youth sub-culture's exaggerated behaviour can be indulged, is irresponsible and may cause social offence.
Aggravates Criminal investment in youth market (#PD5750).

♦ **PF5728 Unrecorded knowledge**
Undocumented information — Unreported events — Unrecorded insights — Unrecorded wisdom
Narrower Unreported crimes (#PF1456) Unreported births (#PF5381)
Unreported illness (#PF8090) Unreported violence (#PF4967)
Unreported scandals (#PF5340) Unreported research (#PF9141)
Unreported accidents (#PF2887) Unreported businesses (#PJ1192)
Undocumented migrants (#PE6951) Unreported harassment (#PF4729)
Unreported disagreement (#PF7890) Lack of historical record (#PG6453)
Unrecorded property exchanges (#PJ8546)
Undocumented violations of human rights (#PF4062)
Unreported international movements of funds (#PF6594).
Related Negative emotions and attitudes (#PA7090).
Aggravates Ignorance (#PA5568) Lost knowledge (#PF5420)
Wasted woman power (#PF3690) Unretrievable documents (#PF4690)
Inaccessibility of knowledge (#PF1953) Underutilization of popular wisdom (#PF2426)
Failure to profit from patterns of history (#PF1746)
Waste of human resources in capitalist systems (#PC3113)
Inadequate dissemination and use of available information (#PF1267).
Aggravated by Secrecy (#PA0005) Underground economy (#PC6641)
Silence about historical situations (#PF0608)
Failure of governments to fulfil international reporting obligations (#PE2215).

♦ **PF5735 Squeamishness**
Over-fastidiousness
Nature Many individuals are easily nauseated, disgusted or offended. They cannot bear the sight of blood, bodily fluids, wounds, maimed persons, the diseased and their symptoms, dead bodies, dead animals, surgical operations, and the like. They may be offended by the sights of copulating animals, humans displaying affection, and lovemaking. They may be unable to tolerate descriptions of these activities, and if these are expressed by disadvantaged classes in the vernacular, they are repelled. In a world which requires health care, social services, scientific investigation of human behaviour, and informed discussion on world problems, squeamishness is a hindrance.
Broader General obstacles to problem alleviation (#PF0631).
Related Prisoners of conscience (#PC6935) Intolerance of imperfection (#PF7024).

♦ **PF5736 Deliberate governmental avoidance of legislative reform**
Countries deriving advantages from the laxity of their legislation — Reluctance to reform law
Broader Inadequate social reform (#PF0677)
Inadequate international law enforcement (#PF8421).
Aggravates Inadequate narcotics legislation (#PF6787)
Inadequate national law enforcement (#PE4768)
Limited local respect for regional and global legislation (#PF2499)
Evasion of shipping regulations and taxes by flags of convenience (#PE5873)
Exploitation of regulatory loopholes in countries with underdeveloped legislation (#PE4339).

♦ **PF5737 Lack of skilled workers in island developing countries**
Weak social infrastructure in islands
Nature Island developing countries, being small, cannot employ full-time the wide range of specialists required for social and economic development; and generalists cannot have, nor are they expected to have, a high level of technical competence in a wide range of fields. There are occasions when small island developing countries need experts of the highest calibre, but such occasions may be so rare in any one field that it is not worth training a local person. The manpower constraints facing small island developing countries have common characteristics but pooling of technical abilities has not been achieved to any extent on a regional basis.
Broader Labour shortage (#PC0592)
Weakness of infrastructure in island developing countries (#PE5772)
Vulnerability of island developing countries and territories (#PE5700).
Related Inadequate infrastructure (#PC7693)
Excessive emigration from island developing countries (#PE5713).

♦ **PF5739 Disease and injury from exposure to weather**
Broader Human physical suffering (#PB5646) Human disease and disability (#PB1044)
Hazards to human health in the natural environment (#PC4777).
Narrower Cold disorders (#PE0274).

♦ **PF5743 Prohibitive costs of purification of effluents and emissions**
Nature Anti-pollution measures are deemed too costly by industry and by governmental polluters such as the military. Technology to recycle pollutants back into energy, fuel, fertilizer, or protein production is inadequately funded, although processes of this nature would pay for a large part of remaining purification costs.
Aggravates Inappropriate sanitation systems (#PD0876).

♦ **PF5744 Cold as an occupational hazard**
Lack of warmth at work — Cold fishing conditions
Nature Important hazards associated with cold work are chilblains, erythrocyanosis, immersion foot, and frostbite as a result of cutaneous vaso-constriction. General hypothermia is not unusual. Frequency of rheumatism and bronchopulmonary disease may be greater among some of those engaged in cold work. Both the reduction in the temperature of the hands and the wearing of protective gloves reduce dexterity and therefore increase the risk of mistakes and accidents.
Broader Cold disorders (#PE0274) Occupational hazards (#PC6716).
Aggravates Occupational diseases (#PD0215) Occupational risk to health (#PC0865)
Underutilization of natural resources (#PF1459).

♦ **PF5745 Anti-holism**
Nature The human being may be considered as the prime example of holistic engineering. However, while mankind's machines articulate large numbers of parts into great wholes, the ability to relate the parts of behavioural and societal activities and structures seems to lie outside human competence. Holism as a standard or ideal to apply to human efforts and goals seems to evoke a holophobia, a fear that this is incompatible with scientific methods, and while there are numbers of distinguished physical, natural and social scientists who are proponents, or tend towards holistic ideals, there are some scientists who do not acknowledge the complementarity between holistic and analytic methods. Anti-holism characterizes resistance to educational developments, particularly to the educational renaissance that has rediscovered the systemic integrity and developmental possibilities of the human being; and also resistance to reforms in medicine and health services in favour of maintaining segmented and unconnected departments.
Background The holistic framework is the one that seeks completeness and proportion among a system's or structure's members. It applies to political organization, education, health, economic behaviour, to human development, to environmental management, and to social welfare, from the most basic needs for food and shelter to the highest cultural achievements. As a philosophy, holism proposes that when the parts are properly fitted, the sum of the whole and its qualitative value can no longer be explained as an additive result by the method of reduction or analysis.
Narrower Excessive specialization in education (#PC0432).
Related Over-specialization (#PF0256).

♦ PF5749 Geopolitical vulnerability

Nature By virtue of proximity to heavily trafficked sea–lanes or great rivers, or by nearness to continental divides or mountain passes, entire regions, whatever their form, attain strategic geographical importance. They may be valleys, great alluvial plains, peninsulas, coasts, capes, or islands; despite their different natures, they figure in military history; the Khyber Pass, the Normandy beaches, Pearl Harbour and Waterloo are a few of their names. Some territories become politically strategic because they are potential gateways to a perceived enemy, hence the concern of nations for nearby islands not under their own flags. Contemporary news accounts nearly always include an insular political problem.

Broader Vulnerability of social systems (#PB2853)
General obstacles to problem alleviation (#PF0631).
Narrower Military threats to island countries and territories (#PE5785).

♦ PF5753 Fragmentation of organized students

Nature Factional fighting among students exists at all levels, from a single campus or institution to international congresses. At any one university, for example, it exists in student governmental bodies, between: individual student clubs, unions and societies; between enrolments in different faculties; between upper and lower year classes; between the more privileged financially and the less privileged; and between differing life–styles, religions and races. Between institutions, there are school rivalries that continually work against student cooperation. At regional and national levels when students gather, these patterns of conflict re–emerge to obstruct organization into effective federations or movements that represent the students interests and needs.

Broader Political impotence of students (#PE5729).
Narrower Student press weakness (#PE0628).
Aggravated by Unrepresentative international organizations (#PD4873).

♦ PF5759 Ignorance of lifelong human development

Nature Behavioural sciences have emphasized the physical character and ability development of the child, to the neglect of development throughout life.

Claim The first two decades of life show dramatic growth and change. However, the next forty to sixty years of life are equally complex, for changes in body, personality and capabilities are very significant.

Broader Ignorance (#PA5568).
Narrower Male mid–life crisis (#PD5783) Female mid–life crisis (#PD5675).
Aggravates Health inequalities (#PC4844).

♦ PF5764 Absence of God

Abandonment of the faithful

Claim This, according to religious psychology, is a condition in which the soul is either really (because of sin), or only apparently, forsaken by God. It is that experience in which it seems to a spiritual person that God has forsaken him. This spiritual abandonment may be a trial in which the spiritually advanced soul, feeling the need of a stronger possession of God, has the impression that God has deserted it and no longer holds it in His favour. In its less intense form, this abandonment leads to a feeling that God is far away; in its more intense form, it leads to a feeling of being rejected by God and destined to be lost. In the latter case this may lead to religiously–induced pathologies, aggravated by the secrecy in which the condition is maintained.

Broader Abandonment (#PA7685).
Narrower Mental suffering (#PB5680).
Aggravates Human suffering (#PB5955).

♦ PF5766 Conflicting priorities

Disagreement concerning priorities — Incompatible priorities — Policy disagreement

Nature There is no international consensus on human priorities, and a constant struggle for time, money, human, technical and material resources diverts a great amount of energy into political and ideological conflict. Local priorities obscure national priorities, and national priorities obscure international ones. The divisive forces in world society are endless. Although some progress is being made, international efforts to secure agreements are pursued on a piecemeal basis. The development of world agendas may be said to have begun in recent years, with the focusing on anywhere from two dozen to over a hundred key problems, but it is impossible to find world agreement as to which have more claim to immediate attention than others and the means to deal with them.

Broader General obstacles to problem alleviation (#PF0631).
Narrower Conflicting agency priorities (#PG8869)
Conflicting community priorities (#PJ7975).
Aggravates Inappropriate policies (#PF5645) Uncoordinated policy–making (#PF9166)
Disagreement within alliances (#PD2629)
Ineffective decision–making processes (#PF3709).
Aggravated by Consuming personal priorities (#PJ1535)
Short range planning for long–term development (#PF5660)
Domination of government policy–making by short–term considerations (#PF0317).

♦ PF5778 Insufficient nobility

Nature A society which lacks standards of excellence and exemplary role models, produces an environment where superior achievements of personal character diminish. Emphasis on cultivation of behavioural characteristics, important in their social context, such as truthfulness, sense of honour, integrity, altruism, industriousness, decency, courageousness, self–restraint and even intelligence, declines in the home and school. Hence nobility, the human characteristic that embodies such ideals and is itself idealistic, diminishes.

Claim Wherever the ideals of full human development are not established unalterably in a society, they are subject to change, attrition, or perversion. There are two ways to lay the foundations for a noble society. The first is negatively, by interdiction of undesirable behaviour, as far as possible by law. The second is to reinforce the striving for and attainment of the ideal by giving it a preferential place. This is accomplished by entitling individuals and families to political and economic leadership in recognition of their virtues. Characteristically, long service in the affairs of the state, civil or military, domestic or foreign, and private service through wise use of wealth and economic power to benefit the nation, exemplify a number of virtues: from industriousness to bravery and patriotism. Such families and individuals acknowledged with entitlements, will often sponsor and support some of their societies' most important cultural initiatives, institutions of learning and charities. As role models they are vital and since noble means "well–known", they fit the description. They are aristocratic, the best of citizens, and as aristocrats, the best suited to rule others. Possession of a hereditary title is a responsibility to fulfil the role requirements for virtue and leadership, and in the deeper sense, to make all men noble as well.

Counter–claim Hereditary titles are not possessions of the individual, but a description of the role he plays; he 'embodies' the virtues of that role when playing that role, any respect he receives being due to the role and not to the individual personally. This respect helps maintain social structure and the breakdown in society may be ascribed, at least partially, to confusing the individual with his role. Hereditary titles reduce rather than increase that confusion. Trained 'from the youth up' to a role means total commitment to that role, respect (for law, society, etc) breaks down when given to an individual, rather than to his role.

Broader Lack of qualitative excellence (#PF5703).
Related Fraudulent nature of inherited titles (#PE5754).
Aggravates Obstacles to leadership (#PF7011).

♦ PF5781 Demonic and spirit possession

Tsukimono

Nature Possession, whether demonic or otherwise, is destructive and terrifying, frequently causing verbal abuse and physical damage to the possessed and his or her environment. Beneficent possession occurs in some societies whose shamans or other sensitives affect cures while in these states. Cathartic results may be had from apparently collective states of possession, during group religious practice, as exhibited by Pentecostal–type experiences, compulsive dancing, auditions, and incoherent speech in supposed supernatural or archaic language. Collective possession also may occur in the secular world. Thus the characteristics of societies in developed countries are described by some as demonic, and viewed as the possession of civilization and its motive forces, the wills of human beings, by the dark force of an evil transcendental cause.

Incidence Reports of supposed demonic possession are dwindling as scientific knowledge of personality disorders advances, and society reduces the conditions that contribute to mental illness and personal stress. Some instances have proven to be hoaxes, but priest and lay exorcists, working within the traditions of the major religions, without documenting their cases, claim a number of possession instances to be known to them.

Background The idea that a personality could be possessed by something greater than itself is an ancient one, and, in the western culture, is recorded by Homer and in the Old Testament; that possession could be by something less than human is also–well recorded. Both types of possession are reported in almost all cultures from time immemorial.

Refs Binyon, Pamela M *Concepts of "Spirit" and "Demon"* (1977); Montgomery, John W *Demon Possession* (1976); Oesterreich, Traugott Konstatin *Obsession and Possession by Spirits Both Good and Evil* (1930); Prince, R (Ed) *Trance and Possession States* (1966).

Broader Occultism (#PF3312) Unexplained phenomena (#PF8352).
Related Voodoo (#PF9006) Soul murder (#PF4213) Poltergeists (#PF5697).
Aggravated by Demons (#PF6734) Bewitchment (#PF3956)
Affliction by malevolent spirits (#PF9043).
Reduced by Exorcism as a superstition (#PF5673).

♦ PF5805 Strikebreaking

Nature The use of intentional use of force or the threat of force to interfere with peaceful picketing by employees during a labour dispute or with employees exercising any of their rights to collective bargaining or trade union organizing is a violation of civil rights and trade union rights.

Broader Criminal violation of civil rights (#PD8709).

♦ PF5817 Economic uncertainty

Uncertain economic patterns — Unpredictability of key economic variables

Incidence Economic turbulence and uncertainty persist at the end of the 1980s, although since 1983 governments industrial countries have managed to reduce inflation and maintain a positive rate of growth. But the problems that persist include: high real interest rates, declining investment rates, volatile exchange rates, growing current account imbalances, rising protectionism and high unemployment. These problems are mainly the legacy of past inflationary policies and structural rigidities, but they are also the consequence of the mismatch of macroeconomic national policies and of the combination of loose fiscal policy and tight monetary policy, especially in the USA. These have led to slowed growth of production and trade with the consequence that the world economy faces continuing risks.

Background Historical analysis of international, national and sectoral economies indicates regularities and irregularities of behaviour. The regularities can be expressed by two models, the linear and the curvilinear, the linear indicates the straight path of sustained economic growth, or decline. The curvilinear, closed such as in a circle or open in rising and falling waves, indicates economic cyclicity. Both types of behaviour proceed for relatively short periods as regular phenomena. They are interrupted by unpredictable transitions and major dislocations which are argued by some to be the causes of economic conditions and by others to be the effects of accumulated economic forces. Economic prediction works very well retrospectively, with authors proving their economic philosophies by well–chosen examples, but no economic methodology has emerged that can deal prospectively with the variables in an open, world system. Recent experiences with the most advanced forecasting techniques known indicate that governments and industries cannot make an accurate comprehensive forecast five years ahead, although high correlations may be obtained in some areas. There remains, however, an excessive reliance on forecasting in economic as well as in social policy determinations.

Refs McKenna, C J *The Economics of Uncertainty* (1986).

Broader Uncertainty (#PA6438)
Violation of the right of workers organizations to protection against suspension (#PE5793).
Narrower Deflation (#PD7727) Stagflation (#PC2536) Hyperinflation (#PD7940)
Unbalanced growth (#PB0479) Cyclic business recessions (#PF1277)
International economic recession (#PF1172)
Instability of economic and industrial production activities (#PC1217)
Uncertainty of development programmes due to short–term loans (#PF4300)
Uncertainty of development expenditures due to floating–rate loans (#PF4295)
Adverse effect of transnational corporations on balance of payments (#PE1771).
Aggravates Unemployment (#PB0750) Underproductivity (#PF1107)
Economic stagnation (#PC0002) Inaccurate forecasting (#PF4774)
Unsustainable economic development (#PC0495)
Inadequate planning of action against problems (#PF1467)
Inadequate statistical information and data on problems (#PF0625).
Aggravated by Change (#PF6605) Chance (#PA6714)
Instability (#PA0859) Instability of prices (#PF8635)
Vulnerability of social systems (#PB2853) Unpredictable barriers to trade (#PD5033)
Imbalance in economic and social planning in developing countries (#PF4837)
Mismatch of national macroeconomic policies among industrialized countries (#PF5000).

♦ PF5826 Trivialization of peace

Collapsed concept of peace

Nature Reducing the concept of peace to the absence of conflict.

Claim Implicit in this understanding of peace is the absence of change and thereby stasis or death. Peace is not found in the absence of chaos but in the midst of it. It is not an alternative but an aspect of change.

Broader Facile social concepts (#PF5242).
Narrower Unjust peace (#PB7694) Unauthentic peace (#PF7643).

♦ PF5840 Ineffective self–regulation in the shipping industry

Nature The multinational activities of most conferences and of the conference system are without any form of intergovernmental regulation of operations, and there is a disinclination by states to regulate unilaterally those liner services which touch their trade. As a result, most conference services have remained self–administered or self–regulated. Research has disclosed no evidence that self–regulated conferences provide for arbitration of complaints by independent third parties. Investigation and adjudication of complaints of malpractices committed by their members or

shippers is thus purely internal nor has any evidence been found to show that impartial adjudication machinery outside internal conference arrangements is available to member lines, non-conference lines or shippers who may wish to dispute, without litigation, conference decisions taken unilaterally. This gap in the procedures of self-regulated conferences presents a serious disability for the users of conference services, since many of their more frequent serious complaints concern important issues of public interest. Member lines, non-conference lines or shippers that become involved in disputes with self-regulated conferences and that may wish to appeal the decision of the conference have, therefore, no resource other than litigation. Apart from the disadvantage of expense which litigation entails, recourse to law against conference decisions has not been found to be of much comfort to complainants since there is no evidence of a superior court having declared conference practicers to be unlawful in countries which tolerate self-regulation by conferences. Thus, most self-regulated conferences remain the final arbiters in disputes arising from the operation of their agreements and practices.
Claim Self-regulation by the shipping industry in its various forms does not adequately take into account the needs of shippers or the interests of shipowners that are new to shipping, particularly from developing countries. It has the effect of restricting free access to shipping markets, thus preventing the fleets of developing countries from participating equitably in shipping activities. Market controls exercised by large shipowners institutionalize barriers to trade entry and remove price competition. The absence of effective competition has enabled strong shipowners to exploit market power to the detriment of relatively weak shippers. In this situation, it appears indispensable to create an internationally agreed framework to restrict the possibility of abuse of market power.
Broader Ineffective industry self-regulation (#PF5841)
Obstacles for international ocean shipping (#PD5885).
Related Restrictive shipping practices (#PD0312).

♦ PF5841 Ineffective industry self-regulation
Broader International trade barriers for primary commodities (#PD0057).
Narrower Uncontrolled media (#PD0040) Uncontrolled industrialization (#PB1845)
Ineffective self-regulation in the shipping industry (#PF5840)
Ineffective self-regulation in the mining industries (#PE8630)
Trade restrictions due to voluntary export restraints (#PE0310)
Ineffective self-regulation in the health care sector (#PE9048)
Social environmental degradation from recreation and tourism (#PD0826)
Ineffective self-regulation in the telecommunications sectors (#PF5877)
Ineffective self-regulation in the financial services sectors (#PE8121)
Ineffective self-regulation in the food-processing industries (#PE8472)
Ineffective self-regulation in the housing construction industry (#PE8265)
Domination of the shipping industry by transnational corporations (#PE1620)
Ineffective self-regulation in the automobile manufacturing industry (#PE8067)
Ineffective self-regulation in the chemicals manufacturing industries (#PE8956)
Ineffective self-regulation in the consumer goods manufacturing industries (#PE8574)
Domination of agricultural equipment industry by transnational corporations (#PE1448)
Ineffective self-regulation in the pharmaceutical and medical devices industries (#PE8502).
Related Disproportionate influence on national economies of limited number of corporations (#PE1922).
Aggravates Excessive government control (#PF0304)
Natural environmental degradation from recreation and tourism (#PE6920).
Aggravated by Corporation financial secrecy (#PE1571)
Unethical commercial practices (#PC2563)
Transnational corporation imperialism (#PD5891).

♦ PF5856 Excessive paperwork
Nature The burden of excessive paperwork is more than just a financial and time-consuming inconvenience. It cause delays which may inhibit entrepreneurial motivation; international trade of raw manufactured or cultural goods; the distribution of food aid; and, in cases of law, medicine, and social services, could constitute the difference between life and death.
Incidence In 1984 in Lima, Peru, an entrepreneur registered his new clothing factory. It took him 289 days to complete the 310 required steps; and the paperwork involved, if stretched end to end, would have extended 30 metres. Also in Peru, the owner of a bus company spends 46 hours a month completing government demanded paperwork; and in and average year the Peruvian legislature issues almost 20,000 laws, decrees, and edicts, all of which must be written down. A 1985 report in the UK found many jobs had been lost and/or small businesses closed because of the reticence of owners to tackle all the governmental paperwork; and in Australia, excessive paperwork can lead to delays in search response time (by the government-controlled Search and Rescue operation) for people lost at sea, sometimes resulting in their deaths.
Counter-claim In an effort to cut red tape and excessive paperwork, some very necessary and constructive 'formalities' such as planning controls, building and fire regulations, health and safety requirements, and consumer protection, could be forfeited.
Broader Excessive bureaucratic requirements for welfare benefits (#PE8893).
Narrower Proliferation of documents from international organizations (#PF5992).
Related Complex government regulations (#PF8053)
Proliferation of printed matter (#PD4552).

♦ PF5857 Undignified treatment of corpses
Claim Corpses donated to science have been used for experimental purposes in circumstances other than those generally accepted by society, for example in forensic science and road accident research. Already handling the corpse to medical authorities is tantamount to its direct destruction: the corpse suffers a slow process of deterioration caused by multiple dissections and ablations; afterwards it is incinerated along with other human remains without a decent burial.
Refs Richardson, Ruth *Death, Dissection and the Destitute* (1987).
Related Abusive experimentation on humans (#PC6912)
Unethical experimentation using aborted foetuses (#PE4805).
Aggravates Denial of rites to the dead (#PF9190).
Aggravated by Necrophilia (#PF6957) Disposal of corpses (#PG4007)
Desecration of cemeteries (#PD7258).

♦ PF5862 Accusing channels for dialogue of being restrictive
Broader Irresponsible expression of emotions equated with free speech (#PF7798).

♦ PF5864 Pursuit of affluence
Pursuit of affluent living standards — Belief that wealth eliminates all problems
Claim The few who live in the overdeveloped countries can only go on enjoying their present resource use if they continue to seek more than their fair share. The goal of development cannot be to rise to the living standards and industrial-consumer economies which the rich countries now have.
Refs Barlow, Robin, et al *Economic Behavior of the Affluent* (1978).
Aggravates Personal wealth (#PC8222) Unnecessary personal consumption (#PF5931)
Inequitable distribution of wealth (#PB7666)
Ecologically unsustainable development (#PC0111)
Recurrence of misapprehended world problems (#PF7027)
Habitual overemphasis on national self-determination (#PF1804).

♦ PF5868 Excessive sexual activity
Adverse sexual development
Claim Sexual activity that appears to be excessive on a statistical basis but is non-injurious to the health of the practitioners, is morally condemned in many societies, although its expression may be normal or even confined within the marital relationship. A partner, therefore, of whom additional sexual activity is requested, may regard it as immoral or unnatural and refuse compliance. In marriage this can lead to the failure of the union, and it can also be an obstacle to matrimony when encountered during courtship. Depending on age, health and temperament a small percentage of normal people may require or be accustomed to daily sexual activity, or, occasionally, to several such activities in a single day. In these cases they are also likely to need variety, and thus the frequency and variation of the acts may be notable, especially if partners are changed and the private practitioner is exposed. He or she may be slandered as being promiscuous or perverted, and social standing or even careers can be destroyed. Where there are underlying psychological problems among highly sexually active persons their conditions may be described as nymphomania or satyriasis.
Aggravated by Crack (#PE2123) Prudery (#PF5892)
Sexual exploitation of children (#PD3267).

♦ PF5872 Inefficient shipping procedures and documentation
Nature Slow, cumbersome, and expensive procedures are instrumental in increasing the costs of access to world markets, thus reducing export receipts and raising the costs of imports. By impeding the smooth flow of trade and efficient operation of the means of transport, unwieldy procedures and excessive paperwork contribute to the congestion of ports, warehouses and stacking areas. It is not unusual for customs or other formalities to require the unnecessary unloading and reloading of goods, thus increasing the risks of pilferage or damage attendant upon the goods being held without proper storage. The role of forwarding agencies that have representatives in the main overseas markets is underdeveloped, even though they could facilitate the speedy movement of goods; at the same time the capabilities of strictly local firms are limited. The effective result is a hidden inflationary tax which is eventually borne by the final consumer of the goods.
Costly and complicated procedures are also a serious obstacle to the expansion of trade, sometimes discouraging those capable of export from engaging in external trade at all. In the case of land-locked countries in particular, delays and added costs caused by unsuitable procedures contribute to their products lack of competitiveness in world markets. An integrative approach that takes into consideration all the administrative and commercial aspects of the matter within the framework of a facilitation programme aimed at minimizing formalities, simplifying and streamlining procedures, and harmonizing and standardizing documents, is absent.
Broader Vulnerability of land-locked developing countries (#PD5788).

♦ PF5877 Ineffective self-regulation in the telecommunications sectors
Broader Ineffective industry self-regulation (#PF5841)
Obstacles to efficient port utilization and operation (#PE5921).
Aggravates Parochial telecommunications standards (#PE1840).

♦ PF5880 Baldness
Nature Among males, gradual baldness may start any time after puberty by recession at the temple hairline and crown thinning. The action of the male hormone, testosterone, on genetically predisposed follicles causes a degenerative change in their hair production capability. Bald patches on the heads of men or women may be due to localized follicle disease such as ring worm and acne (acne vulgaris), or to other conditions affecting the general health.
Claim Baldness, complete or partial, causes acute embarrassment in some people and a feeling of social inferiority which inhibits the development and expression of the personality.
Refs Bartosova, Ludmila, et al *Diseases of the Hair and the Scalp* (1984).
Broader Cosmetic and health problems related to human hair (#PE7111).

♦ PF5887 Inconsistent port charges
Nature Few ports at present have a wholly rational port-pricing system. One reason is that the concept of the autonomous port is a relatively recent one. In the past, many ports were administered by bodies such as customs or municipal authorities or directly as a government department, and port charges were therefore established and amended to satisfy not only the port requirements, but also those of the other parties involved. As a result, most port-pricing systems are very complicated; in consequence of technical progress in the operation and use of ports, such old-fashioned pricing systems have become unsatisfactory.
Broader Arbitrariness (#PB5486)
Obstacles to efficient port utilization and operation (#PE5921).
Aggravated by Unethical practices in transportation (#PD1012).

♦ PF5892 Prudery
Prudishness
Nature Many people show an affected modesty and priggish adherence to manners and morals, wanting to pass these off as virtues. A notable instance of this in prudish attitudes towards human sexuality.
Claim There is a high degree of prudery shown against human enterprise, ingenuity and effort. Thus the commonality of energy that sexuality and enterprise possess may indicate that the prude is apt to be associated with conservative social values which favour a static, unchanging world, regulated, controlled, censored and monitored; and that he or she is likely to attack those who threaten it by their more liberated and energetic personalities and proposals.
Counter-claim There have to be moral standards and people who are courageous enough to defend them. Some over-zealousness that is bound to occur is justified by the protection that society is given from degeneration.
Related Puritanism (#PF2577).
Aggravates Indecent art (#PE5042) Human sexual inadequacy (#PC1892)
Excessive sexual activity (#PF5868).

♦ PF5893 Lack of attunement
Nature Few people have the time or knowhow to listen adequately to themselves, their employers, their co-workers and the environment to ensure that needs and skills are taken into account and that the group operates harmoniously and in consensus. The result is undue stress, shoddy production and an undercurrent of malaise which pervades the working environment.

♦ PF5900 Falsity
Nature Falsity may exist in the perceiver or in the object perceived. In the perceiver it arises from errors in sense interpretation, unconscious expectation and bias, and, ultimately, incorrect or partial reasoning to a conclusion. Objects, human activities and events and the like may bear false appearances, by intent or by accident.
Incidence In diplomacy falsity may be the instrument to gain advantage: for example, the pro-Hitler non-aggression pact. In politics falsity may be employed in promises or in allegations concerning opponents. Optical illusions illustrate falsity in sense data. Supernatural phenomena, sightings of monsters and unidentified flying objects may be false perceptions. Overly-optimistic

and overly-pessimistic outlooks falsify probable outcome anticipations. Polite behaviour may be an inculcation to practise falsification. Several crimes involve falsity: impersonation, signature forgery, hoaxes, currency counterfeiting, art forgery, 'confidence rackets', fraudulent claims for products, false arrest and, in some jurisdictions, transvestitism.
Broader General obstacles to problem alleviation (#PF0631).
Narrower Incorrect information (#PB3095) False political evidence (#PD3017) False nuclear warfare alerts (#PF1236) Human errors and miscalculations (#PF3702) Irresponsible pharmaceutical advertising (#PE2390).
Related Propaganda (#PF1878) Historical forgery (#PE5051).

♦ PF5901 Malice
Nature Malice is the desire to hurt, injure or harm someone. Malice is active ill will.
In Anglo-American legal traditions, malice aforethought means the intent to do harm without necessarily implying active ill will to the victim.
Broader Hate (#PA7338) Enmity (#PA5446) Unkindness (#PA5643).
Narrower Libel (#PD3022) Slander (#PD3023).

♦ PF5912 Imbalance of developing countries share in shipping
Incidence Although developing countries account for some 60 percent of world exports and generate some 40 percent of world trade their share of the world fleet stands at 13 percent, representing an average annual rate of growth of 13.5 percent since 1971. Their share of world shipping is divided into 21.7 percent of world general cargo and 13.8 percent of containership, 11.6 percent of world dry bulk carriers and 9.7 percent of world tankers. The importance attached by developing countries to shipping is highlighted by the fact that the International Development Strategy for the Third United Nations Development Decade calls for structural change in the industry and for a 20 percent share of world shipping for developing countries by the year 1990.
Broader Obstacles for international ocean shipping (#PD5885) Obstacles for developing countries ocean shipping (#PE5909).

♦ PF5918 Menopause
Nature All women who live beyond the age of 55 to 60 years, and many of a younger age, experience a period of transition from the reproductive to the nonreproductive stage of life, of which the most striking feature is the permanent cessation of menstruation. Spontaneous menopause is the result of loss of ovarian follicular function. Because the permanent cessation of menstruation in over 90 percent of Caucasian women aged more than 45 years is preceded by the occurrence of amenorrhoea for 12 months, this interval is customarily used to indicate the fact that the menopause has occurred. In essence, menopause signifies the conclusion of a women's ability to conceive and bear children. Physiologically, menopause is the result of a decrease in the functioning of the ovaries, and, as a result, the production of the hormone oestrogen. It has been discovered that although the ovaries might have totally ceased producing oestrogen, the pituitary gland continues to produce gonadotrophins to induce the ovaries to produce oestrogen. The result is large quantities of gonadotrophins in the blood, which might account for some of the symptoms of menopause. Menopausal symptoms are characteristically hot flushes and the atrophy of the genitalia. Although hot flushes differ in women, typical sensations are warmth in the neck, face and upper body. This can be accompanied by a visible blush. The flushes can last for as long as two minutes and can occur up to ten or twenty times a day. At night, they might be manifest in profuse sweating. There are other vasomotor symptoms of menopause such as numbness, tingling of the hands and feet, heart palpitations, headaches, dizziness and fainting. These symptoms are more rare than hot flushes and are caused by hormonal changes. They usually cease after one or two years. The menstrual pattern changes in quantity of flow, duration of flow, and length of time between periods; a six month interval between periods is a sign that menstruation is going to stop. The genital organs deteriorate: the labia of the vagina become thinner, less elastic, and fatty; the vagina itself becomes drier and less elastic; the uterus and cervix shrink. Other menopausal signs might be thinning hair, growth of hair on the upper lip and chin, lack of elasticity and fullness of the breasts, increase in the size of the buttocks and hips, weight gain, headaches because of tension and anxiety, depression, and a possible lessening of sexual desire as a result of psychological stress, or, on the other hand, an increased sexual appetite because of the freedom from fear of conception.
Incidence The human female is the only animal to experience menopause, which usually occurs between the ages of forty-eight and fifty-two. Currently. there are approximately 27 million women who have reached the age of fifty and are experiencing, or will experience, this change of life. The occurrence and the severity of perimenopausal symptoms may well be affected by cultural and socioeconomic factors. The following factors are of possible relevance: the social significance of menstruation and the escape from the stigma of menstruation that follows menopause in some cultures; the social significance of childlessness; the social status of the postmenopausal woman; attitudes of husbands to their postmenopausal wives (for example, as a sexual partner); the level of socioeconomic deprivation experienced at the time; the degree of change in a woman's role at this time and the availability of new or alternative roles; the availability of medical help for perimenopausal problems. As yet, there are few studies comparing attitudes and reactions to the menopause in different cultures. Those that are available indicate that while the differences may be striking, their origins may be complex. For example, women from a particular (and relatively affluent) Indian caste reported fewer perimenopausal complaints than women in the USA. This could be attributed to the contrast in the status of postmenopausal women in the two cultures, the Indian women escaping from many earlier limitations and gaining higher social status, the American women anticipating loss of status in a 'youth-oriented' society.
There is at present controversy about whether there is a menopausal syndrome of somatic and psychological symptoms and illness, and there are virtually no data on the age distribution of the menopause and no information on its sociocultural significance in the developing countries. The subject of risks and benefits of oestrogen therapy in peri- and postmenopausal women is of considerable importance in view of the large number of prescriptions issued for such medicaments in developed countries, which indicates their frequent use, and the different interpretations and opinions among epidemiologists and clinicians on both past and current studies on this subject. The choice of family planning methods presents particular problems for women who are approaching the menopause, as well as the health repercussions of cessation of reproductive function and those caused by the agents used to treat the accompanying symptoms. By the year 2000, the average life expectancy for women in developed countries is expected to be 75 to 80, and in developing countries 65 to 70 years. The proportion of those reaching the age of 65 can be estimated at close to 90 percent in the developed and 70 percent in the developing countries; in the latter, one out of three of this group of over-65s can be expected to celebrate her eightieth birthday, compared with one out of seven under the mortality conditions of 1975. Life expectancy for women averages six years more than for men, at least in the developed countries. If it is assumed that reproductive function in women generally ceases at about the age of 50 years, it may be calculated that - by the year 2000 - one in every two to three of these women can expect about 30 years of post-menopausal life.
Claim Because of the menopause women have easier time coming to terms with retirement and old age than do men who do not have similar early warning system.
Refs Anderson, Kenneth *Symptoms after Forty* (1988); Fioretti, P *The Menopause* (1982); Van Keep, Pieter A and Utian, Wulf H (Eds) *The Controversial Climateric* (1982).
Aggravates Female mid-life crisis (#PD5675) Involutional depression (#PE0655).

♦ PF5922 Unfair surcharges in ocean freight
Overtaxed shipping systems
Nature Liner conferences may circumvent procedures for consultations on basic freight rates by the device of increasing surchages for bunker prices or currency adjustment factors, often under formulae which shippers claim are not comprehensible. Currency adjustment surcharges may be introduced or increased as a result of the weakening of the world monetary situation and currency exchange inequities. Bunker price surchages may be introduced or raised as a result of increased oil prices. For such reasons, while it is easier to assess the position in an individual trade, it is very difficult to judge the reasonableness of the levels of surcharges in general.
Broader Restrictive shipping practices (#PD0312) Obstacles for international ocean shipping (#PD5885).
Aggravates Inadequate systems of transport and communications in rural villages (#PF6496).

♦ PF5923 Inadequate laws of the sea
Insufficient maritime agreements
Nature The seas and oceans are a vital resource for human life yet conflicts abound regarding their use. There are national ownership claims to excessive offshore boundaries, for example, which is perhaps the fundamental issue, and which affects fishing rights, off-shore oil development and other sea-bed mineral exploitation as well as sea-lane rights of passage. There are issues of responsibility in the event of oil-well or tanker caused slicks, and issues of cooperation in regional marine management. Special problems exist regarding the legality of seabed mining and militarization of the ocean with mobile weapons. International ocean rescue and salvage laws require clarification.
Broader Inadequate laws (#PC6848) Lack of world maritime integration (#PE5801).
Narrower Obstacles for international ocean shipping (#PD5885).
Related Obstacles to the utilization of coastal and deep sea water resources (#PF4767).

♦ PF5929 Ethnic and social discrimination in foreign language teaching
Nature Particular foreign languages taught in national school systems may experience a de-emphasis owing to ethnic, social or political biases, or to governmental policies.
Incidence The de-emphasis and eventual elimination of German teaching in the USA during World War II and thereafter; and the dropping of French language requirements in Sweden, and Flemish in Belgium are among many examples. The policy of elimination of Finnish in the Swedish school system is actively pursued and is one of the clearer examples of governmental policies of assimilating minorities.
Broader Ethnic discrimination (#PC3686) Denial of right to education (#PD8102).
Aggravates Discriminatory communication (#PD6804).

♦ PF5931 Unnecessary personal consumption
Excessive luxury — Unnecessary luxury — Wasteful personal consumption — Unsustainable personal consumption
Claim Once bodily wants are satisfied and efficiency is maximized any meeting of further desires is a luxury. The consumption of luxuries has economic, social and ethical importance. Expenditure on supe fluities has a tendency toward relaxation of concentrated effort. In extreme cases it weakens moral fibre and opens the way to dangerous excesses. It not only tends to injure the person whose life is luxurious but acts on others by force of example. Once a group of people have become accustomed to an unnecessary luxury, it cannot be withdrawn except by violence.
Refs Clark, Eric *The Want Makers* (1988).
Broader Excess (#PB8952) Personal wealth (#PC8222).
Narrower Proliferation of second homes (#PF1286) Consumption of alcoholic beverages (#PD8286).
Aggravates Decadent standard of living (#PD4037).
Aggravated by Pursuit of affluence (#PF5864) Unethical consumption practices (#PD2625).

♦ PF5932 Political risk
Nature With almost daily changes in international politics and economics, foreign investments are particularly vulnerable. Some of the richest markets are in the least stable parts of the world, and given the vulnerable state of world economy, the number of territorial disputes, the extent of terrorist activities, and present conditions of foreign credit, political risk has become a major factor in determining where to or where not to locate transnational corporations, to invest and to make loans.
Refs Overholt, William H *Political Risk* (1982).
Related Economic philosophy of controlled risk (#PF2334).
Aggravates Political over-reaction (#PF4110).

♦ PF5933 Military unreadiness
Military unpreparedness — Lack of adequate military defence
Incidence The consequences of military unreadiness in the case of an outbreak of war could be devastating. 1984 figures for the USA suggested that 25 percent fewer army units and 15 percent fewer air force units, were rated as fully or substantially ready for battle than in 1980. The main cause for concern is related to gaps in specific areas. For example, the USA is supposed to be able to reinforce Europe with 10 divisions in 10 days, but in fact it could send only six divisions if war broke out tomorrow.
Claim (1) This military erosion is caused, at least in part, by a loss of individual responsibility in the officer corps. A civilian, cost-benefit orientation has demilitarized the military. The bureaucratization of command undermines troop readiness and morale. (2) There are many ways to address the lack of military readiness but cutting the defence budget is not one of them.
Related Inadequacy of civil defence (#PF0506).
Aggravates Military incompetence (#PJ1069).
Aggravated by Total disarmament (#PF8686) Shortage of military manpower (#PE4920).

♦ PF5934 Marginalization of the United Nations
Perceived irrelevance of the United Nations — Disenchantment with the United Nations — Loss of credibility of the United Nations
Claim 1. There is a school of thought which holds that the great powers do not need the world organization except as a symbol of the world community whose meetings may be used as a convenient opportunity for periodic bilateral exchange. This view is reinforced by dissatisfaction with the working of the United Nations expressed by one or another of these powers at different times. 2. No serious person denies that the record of the United Nations is disappointing. It has not been able to save mankind from the scourge of war, nor to protect people from aggression, nor to fulfil other promises in its charter.
Counter-claim This view fails to recognize the interest for great powers of maintaining their positions of respect and influence in a changing world situation. The United Nations provides a unique arena in which any power, whether large or small, can enhance its influence by taking initiatives in framing the universal agenda, drawing attention to new concerns and new ways of solving problems and contributing to the process of peaceful change. Such powers need the

United Nations as a context within which to come to grips with issues which concern other nations and which impinge on their own relationships as well.
Aggravated by Irresponsibility of member governments of the United Nations (#PF5337).

♦ **PF5937 Vulnerability of society to truth**
Undue public concern — Susceptibility of people to panic
Aggravates Alarmism (#PF4384)   Fear of knowledge (#PF9595)
Suppression of information concerning environmental hazards (#PF4854).

♦ **PF5940 Predominance of fast food**
Nature Due to two-career families, increased affluence, outside activities, and a faster paced lifestyle, fast food has become an integral part of life in some industrialized countries. Fast food eateries are more often frequented by people on their own or non-related groups than by families, thus the importance of sharing a meal and mealtime together is being lost, a not so insignificant factor in family breakdown. Fast food is also typically low in flavour, high in fat, cholesterol, sugar and empty calories which contribute to heart diseases, dental disease and obesity.
Incidence In 1984, the USA had a $47 billion fast-food market, which has been further expanded with the introduction of microwave fast-foods being sold in supermarkets' freezer, mobile fast-food vans cruising busy areas at lunchtimes, and home delivery services. The fast food market in the UK in 1986 represented £1675 million.
Related Food fads (#PD1189).
Aggravates Family breakdown (#PC2102)   Nutritional deficiencies (#PC0382)
Decline in nutritional quality of food (#PE8938)
Dietary deficiencies in developed countries (#PD0800)
Environmental hazards from food processing industries (#PE1280).

♦ **PF5943 Displaced royalty**
Royal pretenders — Exiled monarchs
Claim A number of former monarchies are the object of claims to the throne by their displaced royal families. This is divisive to the countries involved, and may also be subversive, with monarchical elements plotting a revolution to restore the former dynasty and its latest heir or claimant.
Counter-claim The universal popularity of monarchy is what is required to salvage the spirit of such troubled nations as the USSR, Yugoslavia, Albania and Italy, as well as Spain, France and Hungary, for example. Royal families for these and other nations such as Iran and Ethiopia await the call to return to serve their fatherlands. Special cases of royal displacement include the exiled Dalai Lama of Tibet, persecuted prelates of the Roman Catholic Church such as the late Cardinal Mindzenty, and the Scion of the House of David.
Aggravated by Monarchy (#PF2170).

♦ **PF5947 Impediments to extradition**
Nature Those that are wanted to stand trial for alleged criminal or political offences and who seek refuge in another country, may not be extradited under a number of conditions. One condition is if the sheltering country is not party to any extradition treaties and another is if the two nations involved do not have diplomatic relations. Extradition proceedings may not be instituted on alleged humanitarian grounds under a variety of circumstances, and finally, those accused of political offences may be exempt under provisions of international law.
Claim Terrorists may be protected by the claim that their offences in another country were politically motivated or had political objectives. International terrorists may also be protected by a regime that supports terrorism. Ineffectual extradition requests may be made to regimes where war criminals or multi-million dollar swindlers or robbers have effectually bought protection.
Narrower Legal havens (#PE0621)   Extradition refusal (#PF2645).
Aggravates Fugitive offenders (#PJ4337).
Aggravated by Governmental abuse of extradition (#PE6004).

♦ **PF5949 Apathy of youth**
Unmotivated youth stance — Limited youth engagement — Misdirected youth engagement
Nature Confronted with changes in most areas of society, young people are seen as increasingly apathetic, self-oriented and unambitious. Whereas most of their parents grew up during the Second World War and fought hard, with the notions of determination, sacrifice, and national purpose, to rebuild their own lives and societies, many of today's youth see little purpose in such goals and choose to adapt a basic conformism to role, but one that is carried out with the least possible effort and a minimum of commitment.
Broader Apathy (#PA2360)   Declining sense of community (#PF2575).
Aggravated by Juvenile stress (#PC0877).

♦ **PF5950 Negativism**
Chronic psychological resistance
Nature Those who are chronically sceptical or cynical about almost anything affirmed or done by someone else may be so according to their nature at an individual level, or carry this through to a negative attitude in their official capacities. In the latter case, such negativism can permeate an organization or a national government; this can lead to a withdrawal from international affairs or to a lack of cooperation, including active resistance to all new ideas or developmental programmes. In practical behavioural terms, negativism and resistance amount to the same thing and in an aggravated form may be expressed in a 'spoiler' mentality that seeks to impede any process. In psychological terms, negative resistance may be a development from a number of personality disorders in which an over-developed ego emerges. This is aggravated when individuals attain some self-perceived power over others and are in a position to veto recommendations or suggestions. Such negative individuals may immediately advocate the opposite of the recommendation, or, only after a long period, reintroduce the recommendation as their own without necessarily being aware of their inner processes of absorption.
Incidence Pathological negativism occurs among mental patients who do the opposite of what they are asked (active negativism); who do the opposite from their physiological promptings and urges (passive negativism); or who express the opposite of their own intentions or affirmations (intellectual negativism). The latter case, in milder form, is seen among those who express all sides of a question, reverse their own positions during argumentation, and are otherwise characteristically ambivalent.
Aggravates Fragmentation of the human personality (#PA0911).

♦ **PF5954 Excommunication**
Nature Excommunication, the exclusion from a religious community, is most often used by Catholicism, Orthodoxy and Judaism as a punitive measure for political purposes.
Incidence Historical examples of excommunication include Pugachev and Tolstoy being excommunicated from the Russian Orthodox Church, Huss and Bruno excommunicated by the Catholic Church, and Spinoza being excommunicated from Judaism. A recent excommunication in the Roman Catholic Church involved a person in the southern United States who, in the 1960s, publicly opposed the racial integration of church schools.
Background Public excommunication was once one of the most dreaded penalties in the mediaeval penal system. Both the Anglican and the Catholic churches derive their present rules on excommunication from mediaeval canon law. Excommunication imposed by a church court or by a bishop in the past could have led to the death penalty or life imprisonment with confiscation of property.
Counter-claim While there have been a number of excommunications for political, personal or even worse reasons, a religious community needs to capacity to excommunicate in order to care for the person being excommunicated. It is used legitimately when the person being excommunicated is so at odds with the structure of his faith that a continued relationship with the community will result in his spiritual or moral harm. In no case is excommunication used to protect God, the faith or the community.
Aggravated by Lack of religious discipline (#PF8010).

♦ **PF5956 Crisis in long-term pension funds**
Escalating pension costs — Progressive erosion of old age pensions
Nature Pension costs are escalating firstly because of structural factors whose effect is to increase pension expenditure gradually and steadily; these factors are built into the pension schemes. Secondly, there are external factors, closely related to the state of the economy, which have the effect of either accelerating the pace of rising costs or curtailing the resources available to finance the programmes. The consolidated effect of the built-in factors would lead towards cost escalation even in times of economic growth, when rising costs could at least be faced with rising revenues; but with an economic recession and high unemployment this is hardly possible. Moreover, further trouble builds up on a number of fronts. On the expenditure front, the accelerated withdrawal of older workers from the labour market (induced or voluntary early retirement, disguised disability awards to redundant workers, etc) means that more pensions have to be paid sooner than was foreseen and for longer periods; in the same way, systematic lowering of pensionable age in order to ease the labour market situation drastically increases pension outlays.
On the economic front, high rates of unemployment mean immediate net losses of revenue (contribution income) for the pension scheme (it has been estimated, for instance, that every time unemployment rises by 1 percent in the United States the state pension scheme (OASDI) loses $2,000 million a year in revenue). On the income front, when the economy is unhealthy, the general propensity of most governments to cut public expenditure in order to reduce the budget deficit and fight inflation naturally reduces the chances of supplementing income from contributions with additional subsidies; in any event, competition for such subsidies becomes tougher because other areas of public welfare and social security may also be in need of additional resources (unemployment payments, supplementary benefits, and anti-poverty handouts).
Refs ILO *Pensions in Inflation* (1980).
Broader Pay system erosion (#PE6725).
Narrower Inadequate income in old age (#PC1966).
Aggravated by Economic inflation (#PC0254)   Ageing of world population (#PC0027)
Increasing cost of social security (#PF7911).

♦ **PF5957 Cultural suicide**
Nature When a people begins to ignore its own culture and neglect its language, accepting the invasion of some foreign language, and when its leading scientists and writers start to publish the products of their intellect in foreign languages and not in their own, then the people in question is on the road to the destruction of its own culture, or cultural suicide.
Broader Confusion induced by rapid social change (#PF6712).
Aggravated by Ignorance (#PA5568)   Forgetfulness (#PA6651)
Cultural illiteracy (#PD2041)   Obstacles to education (#PF4852)
Carcinogenic consequences of food preparation (#PE6619).

♦ **PF5958 Antisocial attitudes in the planning of second language policy**
Nature The teaching of an important foreign second (or third) language, such as French or English, may be offered with one objective in the lower school system, another objective in the university, and still another objective in evening schools for adults, or in commercial schools. Thus, in the same country, where the foreign language may be used socially, a number of people may experience a social handicap arising from a linguistic problem due to different pedagogical approaches. These approaches simply may have favoured some standard form of the language to be attained at university, while language policies for expected school leavers, and for adult education are neglected or allowed to achieve sub-standard outcomes.
Incidence The teaching policy for English in Sweden exhibits insensitivity to the social needs of English-speakers, as do the policies for French teaching in Italy, for example.
Aggravates Discriminatory communication (#PD6804).

♦ **PF5961 Obsolete agricultural cooperative structures**
Nature The economic crisis is being experienced directly by cooperatives and non-cooperative undertakings alike, but in addition, the cooperative, as an extension of the farm, experiences the recession indirectly through the effects on its members' attitudes. The duality of the cooperative, based on the link between its supplying and purchasing members, offers a good potential but also causes difficulties with regard to the approach to be adopted towards the crisis. The stagnation of growth in agriculture, increasing anxiety about the constant worsening of the terms of trade, and the instinct to save: all these factors are affecting the selling potential of cooperatives (whether for fertilizers, phytosanitary products, energy or feedingstuffs). In the machinery and construction sectors, investment has manifestly suffered owing to high interest rates and unpredictable income variations in the agricultural sector. The situation of agriculture in the present crisis is liable to threaten the profitability of the marketing cooperatives, resulting in reduced investment in storage, handling and processing facilities (for cereals, meat and milk).
Broader Prohibitive cost of farm machinery (#PF2457).

♦ **PF5964 Inadequate economic policy-making in developing countries**
Inefficient management of government financial resources in developing countries — Inadequate mobilization of financial resources in developing countries — Economic mismanagement in developing countries — Unsustainability of macroeconomic policies in developing countries — Inefficient use of domestic resources in developing countries — Failure of public finance policy in developing countries
Nature A country's ability to mobilize domestic resources and to translate them into investments depends on both external factors and the domestic policy environment. The policy response of developing countries to the deterioration of the external financial and trading environment has not always been timely and adequate. Significant economic mismanagement, reflect in poor investment decisions, internal disequilibria as manifested by large budget deficits and rapid rates of inflation and rates of growth of aggregate demand that are not sustainable.
Background Until the end of the 1970s growth in GDP in developing countries remained generally strong, continuing the trend of the 1960s. After 1980 their growth rates dropped from an average of 5.4 per cent a year during 1973–80 to 3.9 per cent for 1980–87. China and India were important exceptions because of major growth-promoting policy reforms during the 1980s. The decline can be traced in part to unforeseen changes in the world economy. These changes not only had a direct adverse affect; they also exposed the unsustainability of the macroeconomic policies that many developing countries had adopted during the 1970s. Those most profoundly

affected had four things in common: high levels of external debt; major macroeconomic imbalances, such as large fiscal deficits and high inflation; distorted and inflexible markets; unresponsive policies.
Incidence While many developing countries, including major debtors as well as poorer developing countries, have embarked on rigorous adjustment programmes to increase their efficiency in the mobilization and use of available resources, such efforts have failed to offset the deterioration of the external environment.
The effects on the natural environment of poorly designed public finance policies are illustrated by the energy sector. In most developing countries energy prices have until recently failed to reflect opportunity costs. At the same time low prices have reduced returns on investments in energy conservation, perpetuated inefficient fuel use, and in turn caused environmental problems. For example, in countries where coal is an important fuel, prices have often been below economic costs, so that mines operate at a loss and require government subsidies. Yet each step in using coal also involves potential damage to land, water, and air quality. Similarly subsidized electricity prices intended to promote industrialization in many developing countries have led to uneconomic growth in electricity demand and inefficient levels of public investment in power-generating capacity. This in turn has led to excessive or premature development of hydro resources and unnecessary pollution from coal or oil-fired plants.
Claim The root cause of the difficulties encountered by debtors in meeting their external payments lies with the policies of the countries themselves. And, even if external shocks are an important factor in explaining the emergence of external financial difficulties, an appropriate national economic policy characterized by outward orientation and reliance on market forces would have allowed countries to cope with these disturbances (as has been the case for some countries).
Counter-claim While considerable analysis has been available indicating that the proximate cause of the sharp and severe deterioration of debtor countries resulted from changes in the external environment that were beyond the control of the debtor countries themselves, articulation of the international debt strategy has focused primarily on shortcomings in policy formulation in the debtor countries. It has never been explained why it is that, within the space of a few years, such grave shortfalls in management should arise simultaneously in such a large number of countries.
   Broader Mismanagement (#PB8406).
   Inadequate management of government finances (#PF9672).
   Faltering structural adjustment in the world economy (#PF9664).
   Narrower Inappropriate public spending by government (#PF6377).
   Developing country dependence on a single source of finance (#PE4188)
   Inadequate external debt management capacity within developing countries (#PE7184).
   Aggravates Inadequate domestic savings in developing countries (#PD0465).
   Degradation of the environment in developing countries (#PD3922).
   Decline in public sector savings in developing countries (#PE4574)
   Private domestic capital outflow from developing to developed countries (#PD3132).
   Aggravated by Inadequate public finance statistics (#PF7842).
   Deterioration of terms of trade for developing countries (#PD2897).
   Deterioration in external financial position of developing countries (#PE9567).

♦ PF5967 Sexist education of children
Nature Sexism in education, expressed early-on in some kindergarten teachers' attitudes and in elementary school text books, reinforces sexist role assignments to children outside of school, most notably in family and peer play relationships. Sexist education inculcates the doctrines that females are physically weak, that they are dependent – either on adults or on males of any age, that they are less competent to meet real life situations, that they are more emotional than intellectual and that therefore too much cognitive education for females is a waste. In developing countries, females leave school earlier than males – the mortality of their infants increases in inverse proportion to the mothers' years of education. Sexist education of males insists that boys do well at sports, be physically strong and agile, be competitive, show aggressiveness and leadership abilities, and be ready to fight for honour. Children of either sex who do not fit these roles are stereotyped: females as tomboys, males as sissies, and even if they escape outward derision, their own inner doubts and confusions concerning their personality attributes cause considerable inner turmoil, stress, and, in too many cases, lead to juvenile suicide.
Refs Deble, Isabell *The School Education of Girls* (1980).
   Broader Sexism (#PC3432).
   Aggravated by Sexually discriminating job terminology (#PF6014).

♦ PF5968 Insufficient birth spacing in families
Reduced intervals between pregancies — Maternal depletion syndrome
Incidence The rapid spread of family planning in the last two decades has provided evidence from almost every country that birth-spacing can have a revolutionary impact on maternal and child health. Studies in India, Turkey, the Philippines and Lebanon, for example, have shown that infant mortality rates for babies born within one year of a previous birth are between two and four times as high as for babies born after an interval of two years or more. A similar survey of 6,000 women in India has shown infant mortality rates of approximately 80 per 1,000 where the interval between births was three to four years, rising to 200 per 1,000 when the interval between births was less than one year. In the United States, it has been estimated that infant mortality rates could be reduced by almost a third if no woman were to have more than three well-spaced births.
Claim Medical workers, when caring for a mother and her young child, have come to realize that delaying the next conception and extending the interval between births are quite as important a part of health care as seeing that the latest child is adequately immunized.
   Broader Lack of family planning (#PF0148).
   Aggravates Infant mortality (#PC1287).   Excessive family size (#PJ5277).
   Aggravated by Inadequate family planning education (#PD1039).

♦ PF5969 Inadequate and insufficient immunization
Inadequate immunization methods — Unorganized immunization programme — Poor vaccination practices
Nature Certain limitations are apparent in the prevention and control of viral and bacterial diseases by vaccines. Some organisms either do not grow in vitro, or produce only small amounts of antigen. For example, the only source of hepatitis B antigen is human plasma from chronically infected persons. The production of inactivated vaccines against highly pathogenic agents, such as those of African haemorrhagic fevers, may be hazardous to those engaged in this work. There may be technical difficulties in detoxifying or inactivating vaccines. As knowledge of the genetic basis of attenuations is meagre, vaccine strains have to be selected on arbitrary criteria. Live vaccine strains may have the potential to revert to virulence or to lose immunogenic activity. Some viruses are associated with cellular transformation and potentially with the induction of malignancy. This is true of certain herpes viruses. Owing mainly to the complexity of the etiological agents, little progress has been achieved in the control of parasitic diseases using conventionally produced vaccines.
Incidence At present, fewer than 20 percent of the 80 million children born annually in developing countries are being fully immunized. Morbidity and mortality from diphtheria, pertussis, tetanus, measles, poliomyelitis and tuberculosis are still very high. While the reported numbers of cases and deaths may underestimate the extent of the consequences of these six diseases, they are thought to cause some 5 million deaths among children under 5 years, while blinding, crippling, or otherwise permanently disabling an additional 5 million. As is the case with the large numbers of childhood deaths estimated to be due to, for example, diarrhoeal diseases and malaria, these deaths usually occur in the context of severe malnutrition and interacting diseases; and the figures published for individual diseases usually involve considerable overlap. The leading killers are measles, pertussis and neonatal tetanus. The first two diseases affect most unimmunized children under 5 and have case-fatality rates ranging between 1 percent and 10 percent, the higher rates being more commonly observed among younger and/or less well nourished children. Although evidence on the subject is relatively scanty, it is thought that neonatal tetanus probably affects fewer than 2 percent of children born to unimmunized mothers in developing countries, but 70–90 percent of those infected die.
Claim Every six seconds, a child dies and another is disabled from a disease which can be immunized against. Many more suffer setbacks to normal health and growth. Immunization against the six major communicable diseases of childhood – measles, tetanus, whooping cough, diphtheria, poliomyelitis and tuberculosis – costs approximately 5 dollars per child. Most of the 5 dollars is for the delivery system – the vaccines themselves cost only 50 cents. To immunize every one of the 100 million children born each year in the developing world would therefore cost approximately $500 million a year. To compare the cost of such benefits with the cost of weapons of war has become a cliché of development literature. Yet it may perhaps bear pointing out that the sum of $500 million needed to prevent the deaths of 5 million children a year and the disability and malnutrition of many millions more is equivalent to the cost of only ten of today's most advanced fighter planes.
Refs Buttram, Harold E *Dangers of Immunization* (1983).
   Broader Physically inaccessible services (#PC7674)
   Lack of essential local infrastructure (#PF2115).
   Aggravates International movement of animals as factor in animal diseases (#PD2755).
   Aggravated by Lack of vaccines (#PE4657)
   Ignorance of women concerning primary health care (#PD9021).

♦ PF5970 Low birth-weights
Nature In the industrialized world, low birth-weight babies are usually premature babies. In the developing world, low birth-weigh babies are usually 'full-term' babies; the reason their birth-weight is low is that they have been malnourished in the womb. And one of the main reasons for that is that the mother herself has been malnourished in her pregnancy.
Incidence One of the most important facts about infant deaths in the developing world is that the 10 – 15 percent of babies who are born with low birth-weights (below 2,500 grams) account for between 30 percent and 40 percent of all deaths in the first year of life. In other words, low birth-weight babies are approximately three times more likely to die in infancy. Among those who survive, low birth-weight has also been shown to be associated with longer and more frequent illnesses and with mental and physical impairments.
   Broader Complications of childbirth (#PC9042).
   Aggravated by Inhibited growth of malnourished children (#PE4921).

♦ PF5973 Uncritical acceptance of another person
Infatuation
   Broader Uncritical thinking (#PF5039).
   Aggravates Dependence on romantic love (#PF7418).
   Aggravated by Bewitchment (#PF3956).   Personality cults (#PC1123).

♦ PF5978 Unmotivated teachers
Stagnating educators — Uninspired teacher participation — Partial teacher engagement
Nature Teachers who hold the same position for years on end run the risk of becoming bored and of stagnating emotionally and intellectually. If this happens, they lose their zest and their ability to respond to new demands, and inevitably lower their expectations of their jobs and also of their pupils' performances.
   Broader Obstacles to education (#PF4852).
   Narrower Detached local teachers (#PG9992).
   Aggravates High teacher turnover (#PE8221)   Shortage of trained teachers (#PD8108)
   Isolating effects of seasonal variations on undeveloped transportation (#PE3547).
   Aggravated by Poor supervision teachers (#PJ3757)
   Inadequate teacher training (#PJ1327)
   Parent–teacher non communication (#PF1187)
   Inadequate working conditions of teachers (#PE7165)
   Limited development of functional abilities (#PF1332).

♦ PF5983 Secrecy of medical facts and records
Medical cover-up — Medical obstruction
Nature The confidentiality with which physicians, other health workers and institutions may treat the medical history, present health and health-related conditions of patients and ex-patients, may be a serious obstacle to preventing crises in the lives of the individuals and families concerned. It may also prevent effective action against socially undesirable behaviour and individual and public health hazards. In some cases it is an obstacle in legal proceedings and may interfere with the ends of justice.
Incidence Adolescent pregnancy concealed from parents by health or social workers, who may be excluded from exercising their legal responsibilities over a minor, presents a moral, medical and legal challenge to the community. Similarly among the young, detected venereal disease and drug and alcohol addiction may not be disclosed to the parents or legal guardians. Among the mature, health workers may keep secret a serious heart condition although it may be known to them that the patient is in high-stress employment. Secrecy may apply to diagnosed mental illness or brain–function impairment; and increased stress in an already stressful situation may be caused when a person dying of cancer, for example, is not told, even though his family may have been informed. Patients may request deletion from their records of conditions which future medical treatments may be affected by; they may also refrain from authorizing release of their records even when medical research could benefit.
Refs Masters, N C and Shapiro, H A *Medical Secrecy and the Doctor-Patient Relationship* (1966).
   Broader Secrecy (#PA0005).
   Related Lack of medical records (#PF1836).
   Aggravated by Unethical medical practice (#PD5770).

♦ PF5985 Semantic confusion
   Broader Confusion (#PF7123)   Multiplicity of languages (#PC0178).

♦ PF5991 Abuse of banking secrecy
Nature Banking secrecy is proverbial, especially in Switzerland, and has been the model for more recent off-shore banking services seeking depositors internationally. However, secret bank accounts can facilitate tax avoidance, tax fraud, the financing of organized crime and international terrorism, and embezzlements by reigning monarchs or despots of their governments funds.
Incidence Legends of secret fortunes in Swiss banks are connected with the SS and Adolf Hitler, with Shahs and Sheiks, with the Mafia, with the CIA, and with the Vatican, to name a few.

Broader Secrecy (#PA0005) Abuse of economic power (#PC6873).
Aggravates Corruptive crimes (#PD8679).
Aggravated by Unethical financial practices (#PE0682).

♦ PF5992 Proliferation of documents from international organizations
Claim Some 13,000 trees must die every year to produce EEC reports. This number of trees produces enough paper to make 14 piles as high as the tower of Big Ben.
Broader Excessive paperwork (#PF5856) Proliferation of printed matter (#PD4552).
Narrower Verbosity in intergovernmental organizations (#PF5477).

♦ PF5997 Natural selection
Biological adaptation
Nature It refers to the processes whereby an organism or group of organisms ae more successful than others in accommodating to a given environment.
Reduces Rationing (#PF9026).

♦ PF6012 Disagreement among experts
Disagreement among expert witnesses — Conflicting viewpoints of specialists
Nature The expert witness on either side may present or stress one aspect of the evidence, and ignore or play down that which could lead to the opposite conclusion.
Claim The prosecution expert never draws conclusions that suggest the accused is not guilty, while the defence expert never presents evidence suggesting his guilt.
Counter-claim The genuine controversies do not arise because expert witnesses try to prove two opposite views to the best advantage, but because often there is conflicting evidence, or because the agreed evidence is interpretable in more than one way.
Broader Ideological conflict (#PF3388) Human errors and miscalculations (#PF3702).
Aggravates Ignorance (#PA5568) Questionable facts (#PF9431)
Neglect of expert advice (#PF7820)
Unproven relationships between problems (#PF7706)
Unavailability of appropriate expertise (#PF7916).
Aggravated by Biased expertise (#PF6395) Superficial research (#PF7889)
Corruption in politics (#PC0116) Bias in scientific research (#PF9693)
Double standards in morality (#PF5225) Irresponsible international experts (#PF0221).

♦ PF6013 Nepotism in socialist countries
Nature Nepotism, which operates on bloodline rather than merit, does not necessarily provide governments with the best candidates but, in the Socialist countries where there is so much distrust, animosity, backbiting, and disloyalty, it is a safety measure for leaders to have family members in key administrative positions.
Incidence In the Romanian administration of Nicolae Ceausescu, as many as 50 of his relatives worked in government, including three brothers who were ministers, his son who was Minister of Youth, and his wife, who was the regime's number two leader. Nepotism is still alive in China and Korea DPR.
Broader Nepotism (#PD7704) Socialism (#PC0115).

♦ PF6014 Sexually discriminating job terminology
Sexist language
Nature In many languages, there is no feminine equivalent term for the evidently masculine name describing many jobs usually held by men. As in many countries all jobs are now theoretically open to women, this discriminating terminology creates confusion and frustration.
Broader Sexism (#PC3432) Misuse of language (#PF9598).
Narrower Segregation through language (#PD4131).
Related Reinforcement of male dominance through language (#PF4373)
Lack of English language to describe female experience (#PF7383).
Aggravates Sexist education of children (#PF5967).

♦ PF6015 Islamic fundamentalism
Islamic radicalism
Nature The extremists of both the Sunnite and Shi'ite branches of Islam aim to isolate the Moslem World from the superpowers and to purge their societies of all non-Islamic influences. It is associated with larger fundamentalist movement that is basically a resort to old moral values. It manifests itself in many different ways, in many different countries. Fundamentalists believe that Islam is the only valid religion and culture and therefore must convert or conquer all of humanity. All human activity, political, economic, social, mental or sexual must be regulated by Islamic social and legal precepts, and it is to be applied universally. There is no acceptance of secular politics, institutions, laws, or wars, because at worst they are an enemy of God.
Refs Taheri, Amir *Holy Terror* (1987); Youssef, Michael *Revolt Against Modernity* (1985).
Broader Fundamentalism (#PF1338).
Aggravated by Nationalism (#PB0534) Value erosion (#PA1782).

♦ PF6016 Dialect discrimination
Nature The way an individual speak reflects his social and educational background, and dialects deemed "unrefined" may retard a person's social and professional development.
Incidence In New York City, there are numerous classes offered to New Yorkers who want to lose their "New Yorkese", as this way of speaking seems to lead to being passed over for promotions and passed by in singles bars, while an Englishman "can order a pastrami sandwich and sound like Shakespeare".
Broader Discrimination (#PA0833).
Related Discrimination against minority languages (#PD5078).

♦ PF6022 Uncoordinated disaster relief
Nature Without relief coordination, it is impossible to ensure that in case of natural disaster or other disaster situations, emergency relief activities of all donor sources are mobilized and coordinated so as to supply the needs of a disaster-stricken country in a timely and effective manner.
Broader Inadequate disaster rescue and relief (#PF0286).
Aggravates Disaster hazards to island populations (#PE5784).

♦ PF6023 Non-participation of youth on the decision-making bodies
Nature Irrespective of the prevailing conception in regard to social welfare for the young or the quality and extent of such assistance in each country, the institutions providing this assistance are almost always organizations that proceed by means of administrative action and which treat young people simply as clients. The young take virtually no part in the management or operation of these organizations in spite of the fact that they are designed to implement, through the services they provide, these important rights of the young.
Broader Undemocratic policy-making (#PF8703).

♦ PF6024 Neglect of women farmers
Unadapted vocational training for women farmers
Nature Women farmers tend to be ignored by programmes designed to improve agricultural production, even though they play a critical role in food production. In view of the woman farmer's great responsibility in the home and family and on the farm, particular importance should be given to the development of personality, promotion of the will to learn, mobility and self-determination. However, women farmers are usually not kept in touch with current political and social developments. Training courses whereby each individual can advance both within and outside his or her field of activity are usually not designed so that women farmers can attend them without problems of time or money. Further training courses do not cover a sufficiently wide range of subjects to facilitate a more modern form of organization of household, family and farm duties. Agricultural further training does not always take account of the specific needs of women farmers who have come from other backgrounds; it is rarely accompanied by general, political and social adult education courses.
Incidence Women form a large agricultural labour force in Asia, Latin America and the Caribbean, and in sub-Saharan Africa most food is grown by women. Most agricultural programmes tend to neglect the needs of women farmers.
Broader Limited agricultural education (#PF8835)
Neglect of agricultural and rural life in developing countries (#PF7047).

♦ PF6025 Uncontrolled costs of disaster relief
Nature Appeals for disaster relief are of an emergency and humane nature and preclude a financial accounting in the early stages. However, there is rarely a governmental accounting afterwards either, and the wastage is generally known to be high in most relief operations. In addition, intergovernmental organizations and national and international non-governmental relief organizations must maintain a relief apparatus and readiness, including financial reserves. The hidden costs in any disaster relief are the costs of maintaining such relief readiness and include staff salaries, purchase and storage of relief items, and transportation and relief delivery expenses. The ad hoc mechanisms of some relief operations result in premium prices being paid for some goods and services. Other relief operation planning which relies on contracted and sub-contracted services may favour some suppliers, who, enjoying a monopoly position, may abuse this by increasing their prices and thus the costs of relief. There may be inadequate internal financial controls and audits of such purchases of relief goods and services.
Broader Inadequate disaster rescue and relief (#PF0286).
Aggravates Disaster hazards to island populations (#PE5784).

♦ PF6026 Inadequate insurance against damages arising from a natural disaster
Nature Because of the continuing increase in disaster-related losses due to population growth and rapid urbanization in areas at risk, and the effect of these escalating losses on the insurance and reinsurance industries, there is a greater need for better knowledge about vulnerability of buildings; potential losses could then be better assessed. There is also a need for the insurance industry to encourage the application of disaster mitigation strategies.
Incidence A 1977 hurricane in Britain resulted in over 1 billion pounds of claims for damages, the largest amount ever paid out for a single UK weather catastrophe.
Broader Inadequate insurance (#PF8827).

♦ PF6027 Multiplicity of official languages
Lack of an international language
Incidence The effect of the multiplicity of official languages in multilateral, ad hoc and permanent meetings has produced a state of confused variety which has increased the difficulties of international communication and of the diplomatic profession, viewed as a whole, in carrying out this role. The language barrier, heightened in many cases by the revival of linguistic nationalism, hampers the work of non-professional delegates to conferences, of statesmen at their occasional meetings and even that of career diplomats, professional men and women with special language ability, training and experience. Most international organizations are faced with language problems. The international governmental organizations work with five, six or seven languages on a day-to-day basis, and employ vast numbers of interpreters and translators to provide the necessary technical support. The European Communities, for example, use nine languages, and twenty-five percent of their personnel are employed in their language services. The organizations in the United Nations system use anything from two to seven languages. In 1976 their total direct expenditure on language services (excluding the costs of management, office space, printing, etc) exceeded US $100 million.
Broader Multiplicity of languages (#PC0178).
Related Multiplicity of languages in international relations (#PC0410).
Aggravates Discriminatory communication (#PD6804)
Ineffectiveness of international organizations and programmes (#PF1074).

♦ PF6028 Limits of second language teaching
Nature Language teaching may never produce second language fluency in the classroom among some students. Inability to teach may be more relevant to this failure than inability to learn, as serious efforts by teachers to achieve student fluency may overlook and fail to address the student's problems and obstacles. However, there are limits of second language teaching that can not be overcome. Second language teaching cannot assure that the language is spoken by the student outside the classroom, nor can it assure that the second language spoken to the student outside will be standard. In fact, the second language as taught may not correspond to its common, everyday usage, due to social prejudice and an unwillingness to recognize popular speech. On the other hand there may be pressure on the student outside of school, especially if he speaks an indigenous mother tongue in an inward-looking community, not to use the second language at all. In addition, evidence suggests that learning of any language after the age of 11 years is considerably more stilted and in fact uses a different part of the brain than language learning before that age. Meanwhile, very few educational systems provide for much language teaching at the primary school stage.
Broader Ineffective methods of practical education (#PF2721).

♦ PF6032 Inadequate training of judges
Nature The judges who dispense justice have great power over the lives and liberty of other people but are sometimes inadequately prepared, either because of lack of proper training or because they have been appointed to a position for which they are not appropriately qualified. It is injurious to litigants, and to their families, as well as to the prosecutors, defendants and the jury, when a judge is inadequately trained to preside over a trial on which much is at stake for all involved.
Incidence In Great Britain, where there is no specific training for judges (judges are chosen from the ranks of the senior bar and, to a lesser extent, from solicitors), lawyers are given only a 3 1/2 day induction course before they preside over cases.
Counter-claim Since lawyers have already learned to assess both sides of an issue (which is, after all, what a judge does), that legal training may be an adequate prerequisite; indeed, it is an affront to an attorney's independence and ability to suggest that his or her legal expertise is insufficient to allow a position on the bench.
Broader Inadequate education (#PF4984).

♦ PF6035 Language barriers
Language barrier — Language barriers to transfer of knowledge
Nature Language barriers impede international communication in general and individuals in

particular by rendering them unable to speak and correspond with whomever they wish, or read the periodicals and books they want to read. Bilingualism enables people to participate fully and directly in world culture and universal dialogue; monolingualism leaves the individual with a more parochial, and fearful, worldview.
Claim As school is the place where most people receive their language training, it follows that the languages on school programmes of various countries determine to a large extent the size of the language problem. That means that the present inadequacy of communication is the result of the considerations taken into account at a time when school syllabuses were regulated by law. Those syllabuses are still not sufficiently adapted to the rapidly changing requirements of the times and do not conform with the worldwide variety of human contacts.
Counter-claim For intellectuals, opportunities are greatly enhanced by the foreign languages taught in school, as their foreign contacts are important; but for the less educated with virtually no possibility of speaking to other nationals, there is little value in learning a foreign language. Thus the concept of bilingualism serves as yet another wedge between different classes of people, rather than as a cohesive catalyst to uniting mankind.
Refs Guha, Bimalendu *Study on the Language Barrier in the Production, Dissemination and Use of the Scientific and Technical Information with Special Reference to the Problems of the Developing Countries* (1985); Nakamura, Susumu and Dykstra, Andrew (Eds) *Language, Thought and Culture Symposium* (1979).
    Broader Ineffective communication (#PF6052)
    Barriers to the international flow of knowledge and educational materials (#PF0166).
    Narrower Diverse unilingualism (#PF3317)   Cultural dominance in translation (#PE5959)
    Lack of English language to describe female experience (#PF7383).
    Aggravates Inter-cultural misunderstanding (#PF3340)
    Insufficient communications systems (#PF2350)
    Increasing development lag against information growth (#PE2000).
    Aggravated by Multiplicity of languages (#PC0178)
    Divisive effects of official cultural pluralism (#PF0152).

♦ **PF6042 Time lag in legal provisions**
    Broader Injustice (#PA6486)   Unnecessary verbosity of legal documents (#PF7137)
    Obstacles to efficient utilization of time (#PF7022).

♦ **PF6044 Self-destructive excuses**
Nature Excuses take many subtle and devious forms. They are often chronic evasions of responsibility borne of irrational fear. One of the human animal's main and most dangerous tendencies is said to be that of trading survival for peace of mind. Excuses are one of these methods. Excuses prevent insights into personal problems and thus bar healthy change in personality. They are generally used to hide human fragility.
Incidence Findings suggest that as many as 20 percent of American adults overuse excuses to a point that may be detrimental to their emotional health.
Counter-claim Benign excuses are common and useful as a social lubricant vital to the smooth operation of daily life.
    Broader Self-destruction (#PF8587).
    Aggravates Lack of self-confidence (#PF0879).
    Aggravated by Human destructiveness (#PA0832).

♦ **PF6046 Restrictive trade union policies concerning employment**
Nature Trade union policies may be restrictive to the point of infringing the rights of individual employees in their relations with labour organizations. Captive members of labour organizations may be denied civil rights by their union bosses. Restriction practices range from union shop, hiring halls, compulsory unionism in any forms to interference by national and international unions with local union bargaining and employment.
    Broader Inappropriate policies (#PF5645)   Restrictive trade union practices (#PD8146).

♦ **PF6051 Conscription**
Draft — Compulsory military service
Nature Conscription is compulsory military training and reglementation of young people, usually males.
The requirement that an able-bodied male citizen must serve in the military forces usually takes place under the universal military training provision (as for example in Switzerland), or under emergency or war-time conscription. Compulsory military service presents a number of problems. First is the questionable effectiveness of compelling non-arms bearing citizens to become fighting or defensive forces when other options might include: an all-volunteer military; mercenaries; or, less likely, a military alliance where a particular country's role would be strictly logistic and could be fulfilled by civilians. Another difficulty is the cost to the state of compulsory training or service. The motivation of conscripts or those inducted under UMT provisions is problematical, and the disruption of private life and economic activity also can be considerable.
Incidence After Napoleon first imposed military conscription the system spread rapidly to all countries of continental Europe and has since been extended to practically the whole world. Countries without conscription are rare. Often one of the first acts of newly independent states has been to impose compulsory military service upon their citizens. Conscription is almost considered to be a necessary adjunct of national sovereignty and independence. Governments of come countries with a large proportion of their people living in poverty, still feel that they must divert money and manpower to the maintenance of a conscript army.
Claim Conscription indoctrinates young people in patriotic lies and sends them to kill and be killed in foreign countries, against their will or personal choice and often with threats of prison if they refuse. conscription is accompanied by registration which acts to make a young person submissive to the authority of the state. Registration is a totalitarian way of requiring a loyalty oath to support the military policies of a government. Conscription is slavery, anti-democratic, and immoral. It deprives the individual of personal responsibility in matters of life and death, his own and, more particularly, of others. To compel a man to engage in the killing of other human beings whom he does not know and with whom he may well have no quarrel, is the greatest infringement of natural liberty and moral freedom. Conscription continues to indoctrinate people with war-minded ideas – and hence continues the acceptance of military institutions.
Compulsory military service may violate the conscience of those who object to violence, and once instituted opens the way to conscription of women and adolescents.
Counter-claim If military service is seen as a necessary part of citizenship, a general obligation is in many ways fairer than selective service or a standing army. Under a system of general conscription where every family is going to be affected by an outbreak of war, public opinion will act as a greater restraint on a reckless foreign policy of its government than if war is likely to affect only professional soldiers who have accepted the risk as part of their career.
Refs Anderson, Martin *Conscription* (1976).
    Broader Forced labour (#PC0746)   Imposed career interruptions (#PF4128).
    Narrower Extra-legal conscription (#PD6667).
    Aggravates Draft evasion (#PD0356).
    Aggravated by War (#PB0593).

♦ **PF6052 Ineffective communication**
    Broader Symbol system failure (#PF3715).
    Narrower Language barriers (#PF6035)   Illegible handwriting (#PF4958)
    Incomprehensible presentations (#PF1677)
    Modern disruption of traditional symbol systems (#PF6461).
    Aggravated by Insufficient communications systems (#PF2350)
    Prejudice against communication by visual imagery (#PF0076)
    Break-down in communications due to difference in training (#PE6650).

♦ **PF6057 Selective information**
Partial censorship
Nature When the public does not receive all the information it needs to make informed decisions, some form of news blackout or censorship has taken place; censorship being defined as the selective suppression of information, whether purposeful or not, by any method – including bias, omission, or under-reporting – which prevents the public from fully knowing what is happening in its society. Much citizen alienation from public life and citizen apathy may be due to this selective information and to a lack of reliable and usable information on critical issues.
Counter-claim Since "all" of the information needed to make informed decisions is far more than even experts can comprehend and use, some mechanism for selection, sorting and digestion of information is required. To suggest that this process is a form of censorship renders the term meaningless.
    Broader Censorship (#PC0067).

♦ **PF6060 Increase of defence budget**
Incidence Defence budgets have been steadily increasing in recent years. In the USA, for example, the 1984 defence budget was US $249.8 billion, a 53 percent increase over the last budget for 1981, which was US $162.9 billion. Between 1982 and 1988, if Congress agrees, US $1.996 trillion are expected to be spent on defence. This equals $7,000 for every living American, twice the present trillion dollar national debt, or 25 percent more than the full cost of World War II.
Claim This budget would pay for all of World War II plus the entire Social Security programme for two years.
    Broader Economic exploitation (#PC8132).

♦ **PF6061 Neglect of adolescent health care**
Inaccessibility of health care for teenagers
Nature Traditionally, parents are responsible for their children's health care, and minors are unable to give legal consent to medical services. For increasing numbers of teenagers, parental responsibility does not ensure that such care will be provided. This group includes adolescents who are living independently, those who are sexually or physically victimized at home, and others who are alienated from their parents. These young people need emergency treatment, family planning, drug and alcohol abuse treatment, abortion, pregnancy care services, and prevention assistance and treatment for communicable diseases. In connection with each of these medical services, counselling and confidential mental health services are often necessary. For many young people, parental consent to such services is impossible to obtain, and notification of parents deters many teenagers from seeking care at all.
    Broader Inadequate health services (#PD4790).
    Narrower Health risks of teenage sex (#PE6969).
    Related Inadequate child welfare (#PC0233).
    Aggravates Adolescent pregnancy (#PD0614)   Drug abuse by adolescents (#PD5987)
    Mental illness in adolescents (#PE0989).
    Aggravated by Family-based health care (#PG9372)
    Lack of child welfare institutions (#PJ2673)
    Inadequate recognition by institutions of the transition through adolescence (#PF6173).

♦ **PF6064 Stepfamilies**
Nature People living in stepfamilies experience problems of unfulfilled expectations, unpredicted hostilities, feelings of conflicting loyalties, of not belonging, the struggle of structuring two families into a third, the dread of failure, and more. Few give themselves permission to openly talk about the issues and to some extent the very subject of stepfamilies is taboo. The stepfamily brings with it foreign and unexpected ways of communicating. At the same time the relationships are tender and tenuous. Feelings of resentment, rejection and conflict can grow and destroy relationships. These stepfamily relationships are a previously unexplored area of family relationships. This ever growing new family phenomenon has a far reaching effect on the individuals involved and on society in general.
Incidence In the USA it is estimated that close to 50 million individuals are, in one way or the other, involved in a step relationship. According to the US Bureau of the Census, the dramatic upsurge in divorce has been a relatively recent development; since 1970 the divorce ratio has increased by 79 percent, and as the divorce rate spirals upwards, the number of step families increases proportionately. The latest US Bureau of Census figures show that one out of two marriages end in divorce, that two-thirds of the divorced and the widowed do remarry and that nearly 4 million individuals become available each year for remarriage and the ensuring stepfamily. Seven million children, or one out of every six, live in a stepfamily.
Claim The biological family still serves for most people as the model after which to pattern stepfamilies; but the same rules simply do not apply. The nuclear family – with a biological mother, father and child or children – is no longer the norm. The moral, psychological, legal, economic and sociological considerations of the new step family are 'mind-boggling'.
Refs Atkinson, Christine *Step-Parenting* (1986); Visher, Emily B and Visher, John S *Stepfamilies* (1980).
    Aggravated by Class conflict (#PC1573).

♦ **PF6067 Lack of specifically designed software in developing countries**
Nature There is a lack of software designed to meet the specific needs of cities in developing countries. Although many standard software packages are available for uses similar to the needs of developing countries, and although others can be adapted by operators with fundamental programming skills, the lack of specifically designed software fosters the impression that microcomputers are incapable of performing many of the functions needed, and discourages civic personnel who cannot distinguish programming from using microcomputers.
    Aggravated by Discriminatory design of information systems (#PD7450).

♦ **PF6071 Uneven economic recovery**
Imbalance in the recovery from recession
Incidence While a reversal of recessionary trends began in 1983, recovery has not become general. There are contradictory influences at work in the world economy making for a diversity of experience among regions. Some parts of the world, especially North America and some developing countries of South and East Asia, are now experiencing quite a rapid pace of advance in income and output. Economic growth in the centrally planned economies has also accelerated, though its rate generally remains at a lower level than in past years. But most developing countries are beset by problems which seriously hinder their prospects for vigorous reactivation of development; and recovery in Western Europe has so far been weak. World output appears to have expanded at a rate of about 2 percent in 1983, accelerating to between 3.5 and 4 percent in 1984 and 1985. While these rates are markedly higher than those attained in the first three years

of the decade, they are modest for a period of economic recovery.
The situation in much of the developing world remains deeply troubling. Many countries are emerging from the recession with a legacy of difficulties which will not be dissipated by recovery elsewhere. The aftermath of a pervasive drought has left many sub–Saharan African countries with a very precarious payments position. Debt–service ratios, particularly in Latin America, are likely to remain unusually high even after recovery in industrial countries; and debtor countries will still be compelled to retrench drastically. Apart from a number of countries, mostly in South and East Asia, per capita incomes have fallen for several consecutive years, and investment expenditures are generally well below the levels realized in the late 1970s. External loan financing from private sources has slowed down to a trickle, and the stringent international liquidity situation restricts capital goods imports. Among the developed market economies, a normal cyclical upswing in the United States and continued growth in Japan are being accompanied by a recovery in Western Europe that is so far limited to some countries and is cyclically weak. While an expansionary fiscal policy has aided recovery in the United States, a number of Western European countries have chosen to pursue relatively restrictive fiscal policies in order to reduce structural budget deficits and the overall size of the government sector. Lack of dynamism in demand has also restrained the expansion of output in Western Europe. Intra–European trade, which comprises a large percentage of total exports, has been adversely affected by weak aggregate demand; and exports to developing countries – more than half of total exports to countries outside the region – have contracted in absolute terms.
Broader International economic recession (#PF1172).

♦ PF6072 Obstacles to satellite communications for developing countries
Nature Third World nations which decide to get into satellite communications face significant obstacles to planning and implementing their programme. The main obstacles are cost, technological dependence and lack of skilled manpower. Other less direct problems have also to be dealt with, such as massive cultural dislocation which may result from the sudden introduction of space–age communications to remote areas little touched by the modern world. Also, monopolistic controls exercised over the space trade by large Western corporations must be faced.
Related Unbalanced application of communications technology (#PE7637).
Aggravated by Insufficient communications systems (#PF2350).

♦ PF6076 Inadequate system of child support enforcement
Incidence In some cases it is extremely difficult to enforce child maintenance obligations, for example against a parent who is residing in another country. In UK more than £1 billion a year is spent on social security payments to single mothers who do not receive maintenance from former husbands or partners.
Broader Victimization of children (#PC5512) Inadequate support services (#PF6492)
Inadequate national law enforcement (#PE4768).
Related Inadequate provision of alimony (#PE3247).
Aggravates Single parent families (#PD2681).

♦ PF6077 Denial of right to benefits of science
Broader Denial of cultural rights (#PD5907).
Related Denial of right to enjoyment of arts (#PE4507).
Aggravated by Eugenics (#PC2153).

♦ PF6078 Modification of environmentally adapted nutritional habits through food aid
Nature Food and food consumption are as much a social as a nutritional phenomenon. In every society, the nutritional items identified as food, their mode of preparation, the conditions under which they are eaten, the proscriptions and prescriptions relating to ritual and other occasions, all reflect basic cultural values, premises about life, religious convictions and, often, national pride. Consequently, changing food habits implies simultaneously changing and redefining social mores. Food aid does not take this aspect of food consumption into consideration.
Refs FAO *Nutritional Implications of Food Aid* (1985).
Broader Inadequacy of food aid (#PF3949).
Aggravates Nutritional deficiencies (#PC0382).

♦ PF6079 Monetarism
Claim The naive doctrine of monetarism which promulgated the theory that inflation control was a function of money supply control has lost credibility. Although governments may attempt to control the money supply, they never actually succeed, although they may achieve a disruption in employment and production growth. Monetarism has demonstrated that by attempting to restrict money flow, capital becomes excessively expensive, fuelling inflation while at the same time causing economic stagnation because of negative impact on investment in increased production capacity. The chain of effects of monetarist policies in the USA and UK has encouraged instability in floating currencies abroad, flights of capital to the high interest countries, stagnating wages and increased costs of living.
Broader Restrictive monetary practices (#PF8749).

♦ PF6080 Self censorship
Nature Populations may be so terrorized by repressive actions of governments that they censor themselves voluntarily. Self–censorship may be exercised in all manner of expression: by writers, artists and scientists, and by editors, artistic producers, publishers, teachers and administrators. Private conversation may be hypocritical or guarded, or may evince genuine approbation of government policies; but dissent and criticism will never be expressed. Media self–censorship denies the public the chance of gaining objective knowledge of conditions in time to act on that knowledge. Self–censorship can be reinforced by a graduated series of penalties for perseverance in dissent. Under the most despotic regimes, unjustified and arbitrary violence may be unleashed against those expressing themselves freely or incautiously.
Incidence Self–censorship exists under all dictatorships whether of one man, an oligarchy or a party. It exists wherever reprisals against freedom of speech are enacted, however subtly, and it can therefore be found in organizations of all types and in social relationships. (For example, a woman may be forced to exercise self–censorship by the insistence on dominance by the male.) In the international community, a country dependent on foreign aid may be prevented from speaking out on bilateral or global issues.
Broader Censorship (#PC0067).
Related Art censorship (#PD2337) Book censorship (#PD3026)
News censorship (#PD3030) Video censorship (#PE5966)
Postal censorship (#PD3033) Theatre censorship (#PD3028)
Military censorship (#PE5989) Religious censorship (#PD5998)
Scientific censorship (#PD1709) Film and cinema censorship (#PD3032)
Radio and television censorship (#PD3029) Censorship in communist systems (#PD3172)
Newspaper and periodical censorship (#PD3027).
Aggravates Biased and inaccurate biology textbooks (#PF9358).
Aggravated by Politicization of scholarship (#PF7220).

♦ PF6084 Inadequate nuclear reactor safeguards
Broader Inadequate provision of public safety (#PF2874).
Narrower Aggression against nuclear power sources (#PE0403).
Aggravates Nuclear accidents (#PD0771) Nuclear reactor accidents (#PD7579)
Non–verificability of compliance with nuclear power safeguards (#PF4455).

♦ PF6091 Official religion
State religion
Incidence The elevation of Christianity to a state religion following the conversion of the Emperor Constantine, resulted in the imposition of faith by secular government and resulted in the abuses perpetrated by the Roman Catholic Church, eventually opposed through the processes of the Reformation and the rise of Protestant and secular forces and resulting in the separation of church and state. In many countries legislation continues to favour the historically dominant religion. In Belgium, for example, non–Catholic churches have been restricted architecturally in their visibility from the street.
Counter–claim 1. Christian and democratic institutions are compatible and probably even necessary to one another's existence. Liberty is essential to happiness and prosperity in this world. Constitutional government is essential to that liberty. The preservation of both is contingent on Christian morality informing both voters and leaders. That morality cannot be maintained without firm faith in Christ. 2. It is an error to consider the Roman Catholic Church as simply another human institution rather than the Body of Christ.
Aggravates Suttee (#PF4819) Culturally biased religion (#PF1466)
Non–recognition of religions (#PJ1512) Religious discrimination in politics (#PC3220).
Reduces Religious and political antagonism (#PC0030).
Reduced by Secularization (#PB1540).

♦ PF6094 Nationalistic attitudes to currency
Nature Governments tend to operate from the viewpoint of a static, local economic system rather than the view of globally fluid capital. This results in attitudes which tend to protect the security of a nation's own medium of exchange at the expense of other nations.
Broader Nationalism (#PB0534)
Variations in national forms of currency (#PF2574).
Reduces Monetary bloc (#PE5247).

♦ PF6096 Unrequited love
Unreciprocated love
Claim One of the principal problems explored in the arts and especially in popular music.
Broader Loneliness (#PF2386).
Aggravates Broken heart (#PF4022) Lovesickness (#PF3385)
Marital instability (#PD2103).
Aggravated by Social inadequacy of men (#PF3613)
Dependence on romantic love (#PF7418).

♦ PF6098 Collapsed images of vocation
Breakdown in vocational images — Restricted vocational images — Reduced images of having a significant life work — Disrelated images of vocational roles
Nature Twentieth Century society has stripped the meaning from an individual having a vocation, a life work or single purpose to ones life. Celebrations of expenditure are limited to one's personal achievement. Images of social role are reduced to one's personal satisfaction. Social responsibility, outside the legal dimension, is reduced to oneself or one's family. The context out of which one works is this week, until the week end or these few years until retirement.
Claim New working conditions, new relationships, new technologies and new jobs have transformed people's daily engagement in society; and little remains of the vocational attitude or sense of calling common in the past, which reminded people who they were and what was expected of them. Many people lose track of life's purpose and direction and lead their lives in a continued state of confusion and aimlessness.
People of the twentieth century aware of the chaotic state of the world and feeling insecure about their own future create illusions of meaning which demean their lives and their work. Rather than have a vocation which is continually risking new directions, discovering new depth to social roles and facilitating more human responses from others people enslave themselves to these illusions.
Individuals have no means of relating their own vocations, i.e. life work, with the social roles of others. There is no feeling whatsoever of the meaning of, or the relationship among, these roles in global terms.
Broader Repression of self–consciousness (#PC1777)
Static and unrelated social roles (#PF1651)
Media reinforcement of materialism (#PF1673).
Narrower Social unaccountability (#PC1522) Reduction in symbolic celebrations (#PF1560)
Educationally reinforced egocentric attitudes (#PD2019).

♦ PF6102 Misalignment of currencies
Exchange rate misalignment — Overvalued currencies — Undervalued currencies
Nature Currency overvaluations and undervaluations have appeared repeatedly under the postwar regimes of both fixed and flexible exchange rates; have tended to grow rather than diminish; and may well reappear periodically even if the current regime is substantially reformed. Floating exchange rates tend to be accompanied by massive capital flows, resulting in large currency misalignments. These produce new protectionist measures in countries with overvalued currencies. In contrast to the initial assumptions of the Bretton Woods system, monetary flows have now come to determine trade flows rather than vice versa, and thus a country can find itself with a currency whose value in no way reflects its ability to compete in international trade, thus further exacerbating the uncertainty in the trading system and adding to the pressures for further protection.
Background An ongoing policy issue of considerable importance is how best to address such imbalances. There has been resort to, and proposals for, such measures as direct intervention in the foreign exchange markets, trade controls, border taxes, capital controls, interest equalization taxes, 'compensatory finance', and a range of other 'second–best' devices, as well as simply 'living with' the disequilibria.
Broader Mismanagement of exchange rate system (#PF1874).
Narrower US dollar dominance of world economy (#PD2463).
Aggravates Speculative flight of capital (#PC1453)
Protectionism in international trade (#PC5842)
Fragmentation of the international trading system (#PC9584).
Aggravated by Exchange rate volatility (#PE5930)
Speculation on money markets (#PD9489).

♦ PF6105 International economic interdependence
Refs Simai, M *Interdependence and Conflicts in the World Economy* (1981).
Broader Interdependence (#PG6428).
Narrower Food interdependence (#PJ0512).
Aggravates Economic revolution (#PC3233) Global economic crisis (#PC5876).

♦ PF6106 Attachment
Narrower Traditional crop attachment (#PG7771)

Sentimental attachment to the past (#PF1577)
Prioritized attachment to security structures (#PD2884).
Related Repression (#PB0871).
Aggravates Denial of right to liberty (#PF0705) Undemocratic political organization (#PC1015).

♦ PF6119 Government delaying tactics

Incidence Governments make frequent use of delaying tactics to postpone the impact of news on politically sensitive issues or on issues which would be embarrassing to government. Tactics include setting up a committee to investigate and report on the issue and delaying the publication of the report by any such body. Such tactics are often designed to provide delays beyond the termination of the mandate of the government, either beyond forthcoming elections, or into the mandate of some other coalition.
Aggravates Administrative delays (#PC2550) Withholding of information (#PF8536)
Government delay in response to symptoms of problems (#PF6707)
Lack of parliamentary time to approve needed legislation (#PF8876).

♦ PF6122 Artificial separation of home and workplace

Nature Almost all cities comprise zones for 'work' and other zones for 'living', and in many cases enforce the separation by law. Two reasons are given for the separation. First, the workplaces need to be near each other, for commercial reasons. Second, workplaces destroy the quiet and safety of residential neighbourhoods. But this separation actually creates rifts in people's lives. Children grow up in areas where there are no men, except at weekends, while women are trapped in an atmosphere where they are expected to restrict their activities to being pretty, unintelligent housekeepers. Men are forced to accept a schism in which they spend the greater part of their working lives 'at work and away from their families' and the other part of their lives 'with their families and away from work'. This separation reinforces the idea that work is a toil, while only family life is 'living' – a schizophrenic view likely to create difficulties for all the members of a family.
Broader Stultifying homogeneity of modern cities (#PF6155).
Narrower Sterile working environment (#PD6133)
Maldistribution of urban shopping facilities (#PE6144)
Inaccessibility of quiet zones in an urban environment (#PF6160).
Related Insufficient separation between urban subcultures (#PF6137)
Inadequate arrangement of housing with respect to common land (#PF6146).
Aggravates Impersonality of high density accommodation (#PF6156).

♦ PF6127 Ineffectiveness of individual participation in large communities

Nature The size of the political community is so large today that its members are separated from its leaders simply by their number. Individuals have little effective voice in local government except when the units of local government are autonomous, self-governing, self-budgeting communities and are small enough to create the possibility of an immediate link between the man-in-the-street and his local officials and elected representatives. Government becomes invisible as it leaves the realm of most citizens' daily lives; popular referenda are infrequent and decisions are via political machinery and rules of expediency with little concern for the long-term interests of the individuals in the electorate.
Broader Vulnerability of small towns (#PD6125) Obstacles to community achievement (#PF7118)
Stultifying homogeneity of modern cities (#PF6155).
Narrower Uncontrolled urban development (#PC0442)
Unidentifiable urban neighbourhoods (#PF6147)
Bureaucracy as an organizational disease (#PD0460)
Inadequate facilities for grass-roots community initiatives (#PF6143).
Related Alienating child-birth environments (#PF6161)
Excessive dispersion of community facilities (#PF6141)
Unbalanced urban population density gradients (#PD6131)
Inadequate night life entertainment facilities (#PE6126)
Insufficient separation between urban subcultures (#PF6137)
Unattractive pedestrian environments in urban areas (#PE6151).
Aggravates Impersonality of public squares in cities (#PFo165).
Aggravated by Non-participation (#PC0588)
Growing size and impersonality of firms (#PE8706).

♦ PF6132 Inaccessibility of city centres to suburban residents

Nature There are few people who do not enjoy the magic of a great city. But urban sprawl deprives the individual of this magic if he is not lucky enough, or rich enough, to live close to a large centre. Most people live in the suburbs and have no more than occasional access to the city's life.
Background This problem has led to considerable research in USA on the optimum size of cities and the distribution of their facilities. One such study showed that cities with more than 50,000 people have a big enough market to sustain 61 different kinds of retail shops and that cities with 100,000 people or more can support sophisticated jewellery, fur and fashion shops as well as a university, a museum, a library, a zoo, a symphony orchestra, a daily newspaper and a radio station. A population of 250–500,000 can, in addition, provide funds for a medical school, an opera and TV networks. Such studies aim to increase decentralization of activities and entertainments in cities, so that they are accessible to more people.
Broader Inaccessible recreation areas (#PF6503)
Stultifying homogeneity of modern cities (#PF6155)
Inaccessibility of countryside to city dwellers (#PF6140).
Narrower Inadequate integration of transport systems (#PF6157)
Impersonality of mass market shopping facilities (#PE6153)
Inhibition of collectively organized fantastic happenings (#PF6170).
Related Scattered housing locations (#PJ0760) Over-spacing of suburban housing (#PE1708)
Fragmentation of communities by automobiles (#PF6250)
Monolithic architecture of high-rise buildings (#PE1925)
Inadequate night life entertainment facilities (#PE6126)
Unattractive pedestrian environments in urban areas (#PE6151)
Inadequate arrangement of housing with respect to common land (#PF6146).
Aggravates Environmental degradation by automobiles (#PE6142).

♦ PF6135 Artificiality of parkland

Green belts
Incidence A contrast with the artificial division of land into economic or enjoyable is provided by residential forests of Japan, where a village grows up along the edge of a forest, the villagers tend and care for the forest, and the forest is available to anyone who wants to come and partake in the process. That this system is beneficial to the land, as well as to the villagers, is shown by the fact that these residential forests are still intact although they were established more than 300 years ago.
Claim Parkland is artificial: parks and campgrounds conceived as 'pieces of nature' for people's recreation, without regard to the intrinsic value of the land itself, are dead and immoral. People should be allowed to picnic, play, boat, fish, etc on farmland (provided that they respect the crops). The idea that farms are 'owned' by farmers for their exclusive profit is fallacious. If we continue to treat some land only as an instrument for our enjoyment, and other land as a source of economic profit, our parks and camps will become more artificial and plastic and our farms more and more like factories. All farms should be redefined as parks, where the public has a right to be, and all regional parks become working farms. A propos of these ideas, a Nigerian tribesman is also quoted as saying, 'I conceive that land belongs for use to a vast family of which many are dead, few are living, and countless numbers are still unborn.
Broader Spatial imbalance of human settlements (#PD6130)
Inaccessibility of countryside to city dwellers (#PF6140)
Obsolescence of suburban mode of human settlement (#PD6150).
Related Vulnerability of small towns (#PD6125)
Insufficient separation between urban subcultures (#PF6137).

♦ PF6136 Household segregation by age group

Nature Studies suggest that normal growth through the various stages of life requires contact, at each stage, with people and institutions from all the other ages. No one stage in the life cycle is self-sufficient. People need support from, and contact with, other people who have reached a different stage in the life cycle, just as they also need support from people who are at the same stage as they are themselves. However, present housing patterns and life styles tend to keep different types of households segregated from each other. There are many areas of two-bedroom houses, other areas of three- and four- bedroom houses and others of studio and one-bedroom apartments. This means that there are corresponding areas of single people, couples and small families, segregated by type. The effects of household segregation are profound.
Broader Inhibition of individual psychological development through life cycle (#PF6148).
Narrower Loneliness in old age (#PD0633)
Inadequate arrangement of housing with respect to common land (#PF6146).
Related Age segregation (#PD3444) Sterile working environment (#PD6133)
Negative effects of the nuclear family (#PF0129)
Inhibition of adult life in small households (#PF6167)
Insufficient separation between urban subcultures (#PF6137).
Aggravates Inadequate facilities for children's play (#PD0549)
Inappropriate accommodations for single people (#PE6187).

♦ PF6137 Insufficient separation between urban subcultures

Nature The mosaic of subcultures in a city results in hundreds of different cultures living, each in its own way, next door to one another. Such subcultures have their own ecology. They can only live at full intensity, unhampered by their neighbours, if they are physically separated by physical boundaries. This is important because where there is insufficient separation between subcultures, they tend to oppress or subdue the life style of their neighbours, or in turn to feel oppressed and subdued. In general, whenever one subculture in a city is very different from another next to it, people are afraid that the neighbouring area will encroach on theirs, and upset their values. Thus they will do everything to make the next door area like their own, and destroy it as another subculture. In effect, for unhampered existence, such subcultures should be separated by swathes of open land, workplaces, public buildings, water, parks or other natural boundaries.
This need for separation of subcultures is reinforced by ecology. In nature, the differentiation of species into subspecies is largely due to the process of geographical separation. It has been observed by a multitude of ecological studies that members of the same species develop distinguishable traits when separated from other members of the species by physical boundaries like a mountain ridge, a valley, a river, a dry strip of land, a cliff or a significant change in climate or vegetation. In just the same way, differentiation between subcultures in a city will be able to take place most easily when the flow of those elements which account for cultural variety – values, style, information and so on – is at least partially restricted between neighbouring subcultures.
Incidence Whenever there is an area of homogeneous housing in a city, its inhabitants will exert strong pressure on the adjacent areas to make them conform to their values and style. For example, the 'straight' people who lived near the 'hippie' Haight Ashbury district in San Francisco in 1967 were afraid that the Haight would send their land values down, so they put pressure on City Hall to get the Haight 'cleaned up' – to make it more like their own area. Other studies in California confirm that it is necessary to have a physical barrier between subcultures to prevent this happening. For example, in San Francisco the two most distinctive areas are Telegraph Hill and Chinatown. Each of these is surrounded by 'natural boundaries' – on two sides by docks, on another two sides by the city's banking area.
Broader Stultifying homogeneity of modern cities (#PF6155).
Narrower Maldistribution of urban shopping facilities (#PE6144)
Impersonality of mass market shopping facilities (#PE6153).
Related Artificiality of parkland (#PF6135) Household segregation by age group (#PF6136)
Alienating child-birth environments (#PF6161)
Artificial separation of home and workplace (#PF6122)
Excessive dispersion of community facilities (#PF6141)
Unbalanced urban population density gradients (#PD6131)
Disorientation stress in large building complexes (#PF6174)
Inaccessibility of green parkland in an urban environment (#PE6175)
Ineffectiveness of individual participation in large communities (#PF6127).
Aggravates Urban road traffic congestion (#PD0426)
Unhealthy lack of daily physical activity in urban environments (#PF6182).
Aggravated by Inaccessibility of water for recreation (#PF6138)
Monolithic architecture of high-rise buildings (#PE1925).

♦ PF6138 Inaccessibility of water for recreation

Nature People have a fundamental yearning for water, but most have little access to swimming pools, lakes or beaches, which are few in number and far away. Holiday resorts are over-filled with people so that many have to stay far from the water's edge; beaches may be in private hands, which removes their availability. Roads and industries frequently destroy water-edge amenities, making it dirty and treacherous. In the temperate climates that are rich in water, natural sources are dried up, hidden, covered or lost. For example, rainwater goes underground in sewers; water reservoirs are covered and fenced off; and rivers, lakes and ponds are so polluted that no one wants to go near them. There is inadequate planning for adding to public water recreation areas, owing to the tax burden of developmental costs while at the same time tax revenues are generated by the industrialization of water-side properties.
Claim Recreational water like drinking water should be a public utility and provided for on that basis.
Broader Inaccessible recreation areas (#PF6503)
Inadequate recreational facilities (#PF0202).
Aggravates Sterile working environment (#PD6133)
Urban road traffic congestion (#PD0426)
Unidentifiable urban neighbourhoods (#PF6147)
Wastage of open space in urban environments (#PE6163)
Insufficient common land in urban environments (#PE6171)
Insufficient separation between urban subcultures (#PF6137)
Unattractive pedestrian environments in urban areas (#PE6151)
Inaccessibility of quiet zones in an urban environment (#PF6160)
Inadequate arrangement of housing with respect to common land (#PF6146)
Unhealthy lack of daily physical activity in urban environments (#PF6182)
Lack of places in urban environments encouraging unstructured public access (#PF6190).
Aggravated by Inadequate water system infrastructure (#PD8517)
Environmental degradation from high-speed roads (#PD6124).

♦ PF6140 Inaccessibility of countryside to city dwellers

Denial of right to walk in the country — Denial of the right to roam the countryside

Nature It is a known fact that people only feel comfortable when they have access to the countryside, experience of open fields and agriculture, and access to wild plants and birds and animals; yet the growth of cities makes this increasingly difficult as continuous sprawling urbanization destroys life and makes cities unbearable. Consequently, countryside has become a mass phenomenon, but people do not have a access to the whole countryside, only to public footpaths, national parks and access agreement lands.
Incidence A Gallup poll in 1972 in the USA gave strong evidence that people living in cities need contact with the true rural land. The poll asked the question, 'If you could live anywhere, would you prefer a city, suburban area, small town or farm ?' It received the following answers from 1465 people: city 13 percent; suburb 13 percent; small town 32 percent; farm 23 percent.
Broader Inaccessible recreation areas (#PF6503)
Inaccessible wilderness areas (#PF9360)
Spatial imbalance of human settlements (#PD6130).
Narrower Artificiality of parkland (#PF6135)
Stultifying homogeneity of modern cities (#PF6155)
Inadequate integration of transport systems (#PF6157)
Obsolescence of suburban mode of human settlement (#PD6150)
Inaccessibility of city centres to suburban residents (#PF6132).
Related Fragmentation of communities by automobiles (#PF6250)
Environmental degradation from high-speed roads (#PD6124).
Aggravates Environmental degradation by automobiles (#PE6142).
Aggravated by Obliteration of footpaths by development (#PE3874).
Reduces Over-use of designated wilderness areas (#PD7585).

♦ PF6141 Excessive dispersion of community facilities
Nature Very few communities have public facilities arranged in compact and meaningful ways: facilities should be grouped densely in small areas, such as small public squares. Conspicuous by their absence are: small squares surrounded by a combination of mutually supportive community facilities and shops; groupings of evening entertainments, to attract the citizen to a particular area yet leave him free to choose his actual form of diversion; groupings of kindergartens and small parks and gardens, so arranged that young families with children could use any of these facilities.
Incidence Studies of pedestrian behaviour in America show that people seek out concentrations in a community.
Broader Uncontrolled urban development (#PC0442)
Obstacles to community achievement (#PF7118)
Unidentifiable urban neighbourhoods (#PF6147).
Narrower Inadequate maintenance of physical health (#PF1773).
Related Alienating child-birth environments (#PF6161)
Unconvivial hotel environments for travellers (#PF6196)
Insufficient separation between urban subcultures (#PF6137)
Ineffectiveness of individual participation in large communities (#PF6127).
Aggravates Maldistribution of teachers (#PE6183)
Impersonality of public squares in cities (#PF6165)
Inadequate recognition by institutions of the transition through adolescence (#PF6173).

♦ PF6143 Inadequate facilities for grass-roots community initiatives
Nature Small, grass-roots movements, which are often unpopular at their inception, play a vital role in society. They provide a critical opposition to established ideas. They encourage free speech. A lively process of community self-government depends on a series of ad hoc political and service groups, functioning freely, each with a proper chance to test its ideas before the townspeople. But such movements need a place to manifest themselves in a way which puts their ideas directly into the public domain. As a rule these groups are small and have very little money; and communities on the whole do not provide them with space to voice and propagate their ideas or to hold meetings.
Broader Obstacles to community achievement (#PF7118)
Ineffectiveness of individual participation in large communities (#PF6127).
Aggravates Lack of places in urban environments encouraging unstructured public access (#PF6190).

♦ PF6146 Inadequate arrangement of housing with respect to common land
Nature The cluster of land and homes immediately around a person's own home is of special importance to him. It is the source for gradual differentiation of neighbourhood land use, and the natural focus of neighbourly interaction. However, in most cities and urban developments, houses are not clustered around common land, but are arranged on streets, which are owned by the town.
Broader Household segregation by age group (#PF6136)
Unidentifiable urban neighbourhoods (#PF6147)
Unbalanced urban population density gradients (#PD6131).
Related Excessive use of land by automobiles (#PF6152)
Negative effects of the nuclear family (#PF0129)
Artificial separation of home and workplace (#PF6122)
Inhibition of adult life in small households (#PF6167)
Insufficient common land in urban environments (#PE6171)
Environmental degradation from high-speed roads (#PD6124)
Disorientation stress in large building complexes (#PF6174)
Inaccessibility of city centres to suburban residents (#PF6132).
Aggravates Inappropriate accommodations for single people (#PE6187).
Aggravated by Inaccessibility of water for recreation (#PF6138).

♦ PF6147 Unidentifiable urban neighbourhoods
Insufficient neighbourhood signs
Nature Available evidence suggests that the neighbourhoods with which people identify with have extremely small populations, and are small in area. People need to belong to an identifiable space; they want to be able to identify the part of the city where they live as distinct from all others – a clearly defined neighbourhood. A major road through a neighbourhood destroys it, yet many cities endanger or destroy identifiable neighbourhoods by putting major roads through them.
Incidence Studies in the USA have attempted to define the right population for an identifiable neighbourhood. One factor is that neighbourhood inhabitants should be able to look after their own interests by organizing themselves to bring pressure to bear on town hall or local government. This means that families in a neighbourhood must be able to reach agreement on basic decisions. In practice, it is difficult for this to be done if more than 1,500 people are involved: some authorities put the figure at as little as 500 people. A survey in Philadelphia asked people which area they knew well; respondents limited themselves to a small area, usually two or three blocks around their house. One quarter of the inhabitants of an area in Milwaukee considered a neighbourhood to be an area no larger than a block (300 feet); and one half considered it to be no more than seven blocks.
Other researchers have found that the heavier the traffic in an area, the less people think of it as home territory. People feel that streets with heavy traffic are less personal than streets without, and so are the houses on such streets. American statistics suggest that a neighbourhood begins to deteriorate once it has a road with more than 200 cars per hour. On streets with 550 cars an hour, people visit their neighbours less and never gather in the street to meet and talk. Research in London also indicates that any street with more than 200 cars per hour, at any time, will probably seem 'major' and start to destroy the neighbourhood identity.
Broader Lack of essential local infrastructure (#PF2115)
Stultifying homogeneity of modern cities (#PF6155)
Ineffectiveness of individual participation in large communities (#PF6127).
Narrower Impersonality of public squares in cities (#PF6165)
Excessive dispersion of community facilities (#PF6141)
Inadequate arrangement of housing with respect to common land (#PF6146).
Related Sterile working environment (#PD6133)
Inaccessibility of green parkland in an urban environment (#PE6175).
Aggravates Urban road traffic congestion (#PD0426).
Aggravated by Inaccessibility of water for recreation (#PF6138).

♦ PF6148 Inhibition of individual psychological development through life cycle
Nature It is commonly accepted that a person's life traverses several stages as it goes from infancy to old age. Each stage has a separate reality, with certain difficulties and compensations and certain characteristic experiences. In earlier societies, these stages in life were recognized and catered for. For example, rites of puberty would take place at a certain stage in a child's life. It was also the case that people of all stages in life would be in contact with people of all other stages. But in recent times, contact with the entire cycle of life has become less and less available to each person. In place of natural communities with a balanced life cycle, there are now retirement villages, dormitory suburbs, teenage cultures, ghettos of unemployed, college towns, industrial parks, etc. Under such conditions, people's chances for solving the conflicts that come with each stage in the life cycle are slim.
Broader Vulnerability of small towns (#PD6125).
Narrower Loneliness in old age (#PD0633) Household segregation by age group (#PF6136)
Havens for environmental pollution (#PE6172) Negative effects of the nuclear family (#PF0129)
Inhibition of exploration by children of urban environment (#PF6159)
Excessive intensification of the parent-child relationship (#PF6186)
Inadequate recognition by institutions of the transition through adolescence (#PF6173).
Related Sterile working environment (#PD6133) Alienating child-birth environments (#PF6161).
Aggravates Maldistribution of teachers (#PE6183).
Aggravated by Psychological inhibition (#PF6339)
Barriers to transcendent experience (#PF4371).

♦ PF6149 Burden on society of widows
Broader Being a burden (#PF9608).
Aggravates Denial of right to benefits to survivors (#PE4531).
Aggravated by Intestacy (#PE5063).
Reduced by Suttee (#PF4819).

♦ PF6152 Excessive use of land by automobiles
Automobile parking space
Incidence For comparison, a man occupies about 5 square feet of space when he is walking. A car occupies about 350 square feet when it is standing still (including access); and at 30 miles an hour, when cars are 3 lengths apart, it occupies about 1000 square feet. Most of the time, cars have a single occupant. This means that when people use cars, each person occupies almost 100 times as much space as he does when he is a pedestrian. It is also notable in cities that when the area devoted to parking is too great, it destroys the land. Empirical observations suggest that an environment is not fit for human use when more than 9 percent of it is given over to parking. Another factor in the use of land by the automobile concerns the amount of paved road and highway that becomes necessary for its use. For example, in downtown Los Angeles more than 65 percent of the land is covered with concrete or asphalt paving.
Claim Cars take up too much space.
Broader Land misuse (#PD8142)
Environmental degradation by automobiles (#PE6142).
Related Sterile working environment (#PD6133)
Disorientation stress in large building complexes (#PF6174)
Inadequate arrangement of housing with respect to common land (#PF6146).
Aggravates Shortage of urban land (#PD0384)
Impersonality of high density accommodation (#PF6156)
Monolithic architecture of high-rise buildings (#PE1925)
Insufficient common land in urban environments (#PE6171).
Aggravated by Urban road traffic congestion (#PD0426)
Stultifying homogeneity of modern cities (#PF6155).

♦ PF6155 Stultifying homogeneity of modern cities
Nature The idea of men as millions of nameless, faceless cogs pervades twentieth century literature; and the nature of modern housing reflects this image: the homogeneous and undifferentiated nature of modern cities tends to preclude variety of life styles and the growth of individual character. Although many life-styles may go to make up a city, and at first the heterogeneous character of this suggests richness, in fact people are often mixed together, irrespective of their life-style and culture. This dampens any significant variety, arrests possibilities for differentiation and encourages conformity. It tends to reduce all life styles to a common denominator. What appears heterogeneous often turns out to be homogeneous and dull. The vast majority of modern housing has the touch of mass-production: adjacent apartments are identical; adjacent houses are often identical; and vast office buildings also throw a uniform character over people's lives.
Broader Inaccessibility of countryside to city dwellers (#PF6140).
Narrower Unidentifiable urban neighbourhoods (#PF6147)
Artificial separation of home and workplace (#PF6122)
Fragmentation of communities by automobiles (#PF6250)
Maldistribution of urban shopping facilities (#PE6144)
Monolithic architecture of high-rise buildings (#PE1925)
Insufficient separation between urban subcultures (#PF6137)
Unattractive pedestrian environments in urban areas (#PE6151)
Inaccessibility of city centres to suburban residents (#PF6132)
Ineffectiveness of individual participation in large communities (#PF6127)
Lack of places in urban environments encouraging unstructured public access (#PF6190).
Related Negative effects of the nuclear family (#PF0129)
Inhibition of adult life in small households (#PF6167).
Aggravates Excessive use of land by automobiles (#PF6152)
Environmental degradation by automobiles (#PE6142)
Inappropriate accommodations for single people (#PE6187).
Aggravated by Urban road traffic congestion (#PD0426)
Overcrowding of housing and accommodation (#PD0758)
Monotonous and unaesthetic architecture and design (#PF0867).

♦ PF6156 Impersonality of high density accommodation
Nature There is a direct correlation between the incidence of mental disorder and high-rise living. Such housing has a tendency to isolate people in their apartments; home life is separated from casual street life by hallways, long passageways, stairs and lifts; the decision to go out in search of public life becomes formal and awkward; and unless there is some specific task to bring people out into the world, they tend to stay at home alone. There is also a remarkable lack of communication between families in high-rise buildings. Children and the elderly are especially isolated; the elderly are too tired, feeble or discouraged to go through the apartment maze to get

to the outside, and children become conditioned to a life within walls.
Broader Overcrowding of housing and accommodation (#PD0758)
Unbalanced urban population density gradients (#PD6131).
Narrower Inadequate facilities for children's play (#PD0549).
Related Inhibition of adult life in small households (#PF6167)
Insufficient common land in urban environments (#PE6171)
Unattractive pedestrian environments in urban areas (#PE6151).
Aggravates Depressing effect of poor housing construction (#PF1213)
Inhibition of exploration by children of urban environment (#PF6159).
Aggravated by Urban overcrowding (#PC3813) Excessive use of land by automobiles (#PF6152)
Negative effects of the nuclear family (#PF0129)
Artificial separation of home and workplace (#PF6122)
Monolithic architecture of high-rise buildings (#PE1925).

♦ PF6157 Inadequate integration of transport systems

Nature The transport network of airplanes, helicopters, hovercraft, trains, boats, ferries, buses, taxis, ski-lifts, etc, can only work if all the parts are well connected. However, the different agencies in charge of the various forms of transport have no incentive to interact, partly because they are in competition with one another and partly because cooperation simply makes life harder for them. This is particularly true along commuting corridors, where trains, buses, mini-buses, ferries and possibly even planes and helicopters compete for the same passenger market. When each mode is operated by an independent agency, there is no particular incentive to provide feeder services to the more inflexible modes, and many services are even reluctant to provide good feeder services to rapid transit trains and ferries because their commuter lines are their most lucrative lines. Similarly, in many cities of the developing world, mini-buses and colectivos provide transportation along the main commuting corridors, competing for passengers with the buses. This leaves the mainlines served by small vehicles, while almost empty buses reach the peripheral lines - usually because the public bus company is required to serve these areas, even at a loss.
Broader Insufficient transportation infrastructure (#PF1495)
Inaccessibility of countryside to city dwellers (#PF6140)
Inaccessibility of city centres to suburban residents (#PF6132).
Narrower Lack of world maritime integration (#PE5801).
Related Environmental degradation from high-speed roads (#PD6124).

♦ PF6158 Denial of right to freedom from double jeopardy

Violation of freedom from being tried twice for the same charges — Cumulation of convictions — Multiple convictions for a single criminal act — Multiple liability for an identical criminal act — Overlapping criminal liability
Nature In some jurisdictions, defendants may be convicted or punished for more than one crime arising out of a single act, where that act involves several distinctly defined crimes (eg murder and battery). A single act may be understood as resulting from a single or primary objective. Different acts constituting different crimes may often be committed closely together and for related reasons (eg burglary followed by larceny). When such crimes are treated as distinct, punishment is then imposed which is disproportionate to the defendants' actual blameworthiness.
Broader Denial of human rights in the administration of justice (#PD6927).
Aggravates Disproportionately long prison sentences (#PE4602).

♦ PF6159 Inhibition of exploration by children of urban environment

Nature The separation between the child's world and the adult world is unknown in traditional societies. In simple villages, children spend their time alongside farmers in fields, side by side with people who are building houses, close, in fact, to all the daily actions of the men and women around them. By contrast, modern cities are so dangerous that children cannot be allowed to explore them freely - life is so enormous and dangerous that children can't be left to roam around. There is constant danger from fast-moving cars and trucks and dangerous machinery, as well as the danger of assault or kidnap and, particularly for smaller children, the danger of getting lost. The obstacles to exploration inhibit the maturation process by depriving the child of opportunities to acquire self-confidence and to interact with the real, dynamic world of change, risk and reward surrounding him.
Broader Urban road traffic congestion (#PD0426)
Unattractive pedestrian environments in urban areas (#PE6151)
Inhibition of individual psychological development through life cycle (#PF6148).
Narrower Inadequate interaction between humans and animals in the urban environment (#PF6185)
Inadequate recognition by institutions of the transition through adolescence (#PF6173).
Related Sterile working environment (#PD6133)
Inadequate facilities for children's play (#PD0549)
Excessive intensification of the parent-child relationship (#PF6186).
Aggravates Unhealthy lack of daily physical activity in urban environments (#PF6182).
Aggravated by Maldistribution of teachers (#PE6183)
Impersonality of high density accommodation (#PF6156).

♦ PF6160 Inaccessibility of quiet zones in an urban environment

Nature Although anyone who has to work in noise, in offices with people all around, needs to be able to pause and refresh himself with quiet in a more natural situation, most modern cities do not cater for this need.
Broader Sterile working environment (#PD6133)
Artificial separation of home and workplace (#PF6122).
Narrower Inaccessibility of green parkland in an urban environment (#PE6175).
Related Inaccessible recreation areas (#PF6503) Vulnerability of sacred sites (#PD6128)
Uncontrolled urban development (#PC0442).
Aggravated by Inaccessibility of water for recreation (#PF6138).

♦ PF6161 Alienating child-birth environments

Nature The obstetrics service in most hospitals follows a well outlined procedure, where having a baby, instead of a somewhat mystic experience, is thought of as something of an illness and the stay in the hospital as recuperation. Women who are about to deliver are treated as 'patients' about to undergo surgery. They are sterilized. Their genitals are scrubbed and shaved. They are gowned in white and put on a table to be moved back and forth between the various parts of the hospitals. Women in labour are put in cubicles to pass the time with virtually no social contact. This time can last for many hours. Father and children are not normally permitted to be in contact. Delivery normally takes place in a 'delivery room' which has the proper 'table' for childbirth. There is a marked lack of facilities in which a woman can have a baby as part of a family, with privacy and seclusion afterwards which would enable the new baby to be introduced to the existing family gently and sympathetically and more in accordance with what happens in simpler societies.
Broader Complications of childbirth (#PC9042)
Excessive medical intervention in childbirth (#PE7705).
Related Excessive dispersion of community facilities (#PF6141)
Insufficient separation between urban subcultures (#PF6137)
Ineffectiveness of individual participation in large communities (#PF6127)
Inhibition of individual psychological development through life cycle (#PF6148).
Aggravated by Monolithic architecture of high-rise buildings (#PE1925).

♦ PF6165 Impersonality of public squares in cities

Nature A town needs public squares. They are the largest, most public rooms that the town has; and such squares accommodate the public gatherings, small crowds, festivities, bonfires, carnivals, speeches, dancing, mourning, etc which must have their place in the life of the town. However, architects and planners often build squares and plazas that are too large. They look good in blueprints, but in real life often end up desolate, dead and impersonal.
Incidence In the USA, considerable study has been given to the size of public squares. The general conclusion is that the diameter of squares should not exceed 70 feet, except in places that are great town centres, teeming with people, like Trafalgar Square or the Piazza San Marco. The findings are based on the distances at which people can comfortably recognize or hear each other. Two people with normal vision can see each other clearly up to 75 feet. They can also communicate up to this distance, even though their voices may be raised. This may mean that people feel half-consciously tied together in squares that have diameters of 70 feet or less, where they can make out the faces and half-hear the talk of the people around them. Architects are thus now recommended to make public squares no more than 45 feet to 60 feet across, certainly never more than 70 feet.
Broader Unidentifiable urban neighbourhoods (#PF6147)
Unattractive pedestrian environments in urban areas (#PE6151).
Narrower Alienating waiting environments (#PF6192)
Wastage of open space in urban environments (#PE6163)
Unnatural urban environments inhibiting sleeping in public (#PF6193).
Related Sterile working environment (#PD6133).
Aggravated by Excessive dispersion of community facilities (#PF6141)
Ineffectiveness of individual participation in large communities (#PF6127).

♦ PF6167 Inhibition of adult life in small households

Nature Most houses are built without consideration for the fact that adults and children need access to each other, need to be able to be together, but also need to be able to spend time separately. For example, in many houses all the bedrooms are clustered together and there are often thin dividing walls, so that privacy and intimacy are inhibited. The presence of children in a family often destroys the closeness and the special privacy which a man and wife need together, especially in small households where there is little room and no servants to help look after small children. Adults often have to pass children's bedrooms to get to the bathroom. Children are also able to run anywhere in the house, and therefore tend to dominate all of it. In such households, parents do not have the heart - or the energy - to keep children out of special areas; so finally the whole house has the character of a children's room, with children's clothes, drawings, boots and shoes, tricycles, toy trucks and disarray. There is a need for houses to be designed to take into account the adult need for privacy to be a couple, not just parents. In such houses, a couple's realm exists where they can sit and talk privately, perhaps with its own entrance to the outdoors or to a balcony. It is, or has, a sitting room, and is a place for privacy and projects. The bed is part of it, and so if possible is a separate bathroom, and it should have some kind of double door, or ante-room, to protect its privacy.
Broader Obstacles to family life (#PF7094).
Related Household segregation by age group (#PF6136)
Negative effects of the nuclear family (#PF0129)
Stultifying homogeneity of modern cities (#PF6155)
Impersonality of high density accommodation (#PF6156)
Inadequate arrangement of housing with respect to common land (#PF6146).
Aggravates Inappropriate accommodations for single people (#PE6187).
Aggravated by Inadequate facilities for children's play (#PD0549).

♦ PF6169 Criminalization of abortion

Denial of woman's right to abortion
Broader Restrictive legislation (#PD9012).
Related Denial of right of family planning (#PE5226).
Aggravates Abortion-related deaths (#PE3580)
Weakness of socio-economic infrastructure (#PC1059).
Aggravated by Religious opposition to population control (#PF1022).
Reduces Denial of right to life (#PD4234).
Reduced by Induced abortion (#PD0158).

♦ PF6170 Inhibition of collectively organized fantastic happenings

Nature Man has a great need for mad, subconscious processes to come into play, without unleashing them to the extent that they become socially destructive - as they do in modern 'equivalents' such as drug-taking and group violence. However, there is an increasing absence of carnivals, dancing in the streets and other forms of public fantasy in modern 'sophisticated' cities and centres of population.
Incidence In primitive societies, such outlets were provided by rites, with doctors and shamans. In Western civilization, circuses, fairs and carnivals played their part in community life. All over the earth, people once danced in the streets, as they still do in Bali, where street dancers fall into a whirling trance. In Mexico, every town still has several squares where the mariachi bands play and the neighbourhood comes out to dance. And in Japan, there is the bon odori festival where everybody claps and dances in the streets. But in those parts of the world where people have become 'modern' and technically sophisticated, such practices are dying out. Communities are fragmented; people are uncomfortable in the streets, and there is general embarrassment.
Broader Inadequate night life entertainment facilities (#PE6126)
Unattractive pedestrian environments in urban areas (#PE6151)
Inaccessibility of city centres to suburban residents (#PF6132).

♦ PF6173 Inadequate recognition by institutions of the transition through adolescence

Nature Typically, in Western societies and institutions such as schools, there is no equivalent to the tribal rites-of-passage. Indeed in many Western high schools there is inadequate recognition of developing maturity which aggravates adolescent delinquency, drug-taking, violence or suicide, and contributes to a disaffected, under-achieving student body.
Counter-claim The high school system has not failed, rather entire societies are failing as they become internally fragmented and torn between opposing global ideologies and power groups. Demands are made to make institutions more liberal on the one side, and more repressive on the other, reflecting the same tendency to polarization that is found in the community which the schools image.
Broader Absence of rites of passage (#PF1674)
Inhibition of exploration by children of urban environment (#PF6159)
Inhibition of individual psychological development through life cycle (#PF6148).
Narrower Difficulty of transition from school to work (#PE8369).
Aggravates Neglect of adolescent health care (#PF6061).
Aggravated by Inadequacy of formal education (#PF4765)
Excessive dispersion of community facilities (#PF6141)
Monolithic architecture of high-rise buildings (#PE1925)
Unhealthy lack of daily physical activity in urban environments (#PF6182).

♦ PF6174 Disorientation stress in large building complexes

Nature In many modern building complexes the problem of disorientation is acute. People have no idea where they are, and they experience considerable mental stress as a result. This is

because many modern buildings do not have features that are readily recognized and remembered by people trying to find their way about. Nor are they grouped in such a way that a person sees a recognisable entrance, or central part of the building, to which he can relate the other parts. Psychological theory suggests that the effect of a badly laid out building is almost as bad on a person who knows it as on one who does not, since he spends a good deal of time looking out for landmarks and wondering where to go next, and his minds is generally distracted by the excessive attention he has to pay to the building. It is also important that names be used to distinguish parts of buildings, and also colours, and that the sizes of different parts of the building indicate the relative importance of each part. A good environment is one which is easy to understand, without conscious attention: for example, an Oxford or Cambridge college has a system of groupings that are easily recognized and remembered, being made up of 'courts', 'staircases', 'rooms', etc, which are easily identifiable.
Broader Social inadequacy of large buildings (#PF6194)
Psychological stress of urban environment (#PE6299).
Related Excessive use of land by automobiles (#PF6152)
Insufficient common land in urban environments (#PE6171)
Insufficient separation between urban subcultures (#PF6137)
Unattractive pedestrian environments in urban areas (#PE6151)
Inadequate arrangement of housing with respect to common land (#PF6146).
Aggravates Monolithic architecture of high-rise buildings (#PE1925).
Aggravated by Indoor air pollution (#PD6627).

♦ PF6177 Passivity
Nature Passivity, when considered as a virtue either by religions or secular authorities, can be interpreted as a justification for inaction in the face of social problems and human suffering. Essentially this is a failure to distinguish between checking the instinct of revenge for personal or collective injury, and active concern to remedy the sufferings of others. Such absence of moral resentment, in the presence of oppression, lawless wrong-doing, or trampling on the rights of others, undermines social organization. Mere passivity on the part of a group or nation whose just rights and liberties were imperilled would mean the abdication of its true place and function. Equally forms of passivity enjoined upon people by such groups to discourage formation of coalitions against inequities constitutes a direct hindrance to human and social development.
Incidence Certain religions, notably Christianity and Buddhism, strongly advocate passivity, passive obedience and submissiveness, and passive endurance in the face of the pressure of the hostility and hatred of the world. Whilst this may be understood as being the spirit which recognizes that suffering is an indispensable law of the spiritual life, it can also be interpreted as justifying inaction on social problems.
Narrower Passive resistance (#PF2788)
Passive public support of terrorism (#PF6846)
Passive historical perspective by leaders (#PF6765).
Related Health hazards of passive smoking (#PE5146).
Aggravates Complacency (#PA1742)
Indifference (#PA7604)
Inaction on problems (#PB1423)
Psychological inertia (#PF0421).

♦ PF6178 Ideal of tension free social structures
Nature The ideal that society can function without any social, psychological or personal tensions is an illusion. Individuals expend a great deal of energy attempting to escape these tensions, through marriage, education, counselling and religious practices. Social structures like welfare systems are dedicated to removing as many tensions as possible further re-enforcing the illusions.
Broader Lack of commitment (#PF1729).

♦ PF6182 Unhealthy lack of daily physical activity in urban environments
Nature In agricultural societies, people are daily physically active in many different ways; but in urban societies, most middle-class adult men and women are physically under-active. Since there is ample evidence that physical health depends on daily physical activity, this indicates an imbalance in town and city life-styles.
Incidence A comparison of the death rates between groups that have been able to have daily physical activity with those that have not shows, for example, that in the age group 60 to 64, one percent of the men in the heavy exercise category died during the follow-up year, whereas in the non-exercise group five times as many died.
Broader Physical unfitness (#PD4475)
Inadequate maintenance of physical health (#PF1773).
Aggravates Inadequate recognition by institutions of the transition through adolescence (#PF6173).
Aggravated by Sterile working environment (#PD6133)
Alienating waiting environments (#PF6192)
Inaccessibility of water for recreation (#PF6138)
Insufficient common land in urban environments (#PE6171)
Monolithic architecture of high-rise buildings (#PE1925)
Insufficient separation between urban subcultures (#PF6137)
Inaccessibility of green parkland in an urban environment (#PE6175)
Inhibition of exploration by children of urban environment (#PF6159).

♦ PF6185 Inadequate interaction between humans and animals in the urban environment
Nature Animals are as important a part of nature as trees, grass and flowers; and there is also some evidence that contact with animals may play a vital role in a child's emotional development. Yet animals are almost missing from cities; there are, broadly speaking, only three kinds of animals: pets, vermin and animals in the zoo. None of these three provides the emotional sustenance nor the ecological connections that are needed. Pets are pleasant, but so humanized that they have no wild free life of their own. Vermin are peculiar to cities and depend ecologically on miserable and disorganized conditions, so they are naturally considered as enemies. Animals in the zoo are seen only as occasional curiosities – and it has been said that animals living under zoo conditions are essentially psychotic; that is entirely disturbed from their usual mode of existence.
Broader Insufficient common land in urban environments (#PE6171)
Inhibition of exploration by children of urban environment (#PF6159).
Related Inaccessibility of green parkland in an urban environment (#PE6175).
Aggravates Excessive intensification of the parent-child relationship (#PF6186).

♦ PF6186 Excessive intensification of the parent-child relationship
Nature The present-day child has typically an intense relationship only with its parents, notably its mother. This is obviously a restricted type of upbringing, and the intensity of these relationships may be unhealthy for the child.
Background In past generations, extended families gave children the opportunity for contact with people of all generations. Such families contained grandparents, parents, children, uncles, nephews, and nieces, all living in close proximity. Thus a child growing up in such a family was exposed to a wide variety of influences and age groups in a way that is not provided in the modern nuclear family.
Counter-claim Children require normal, loving and concerned parents, both a father and a mother. Socially deviant theories are attempting to use the destructive pressures working against the heart of the family, the bonding between child and parent, which is the nuclear power of natural affection that all of society draws on.
Broader Negative effects of the nuclear family (#PF0129)
Inhibition of individual psychological development through life cycle (#PF6148).
Related Inhibition of exploration by children of urban environment (#PF6159).
Aggravates Inadequate facilities for children's play (#PD0549).
Aggravated by Maldistribution of teachers (#PE6183)
Monolithic architecture of high-rise buildings (#PE1925)
Inadequate interaction between humans and animals in the urban environment (#PF6185).

♦ PF6188 Fear of the unknown
Nature The realm of the unknown is peopled, in the imagination, by monstrous beings and mysterious events with which superstitious fear is readily associated.
Incidence Such fear is characteristic of isolated tribal societies, although many such beliefs prevailed in Europe until the 18th century. Fear of this kind is common in children. In adults fear of the unknown is more closely associated with fears of personal security in which the monstrous beings are replaced by muggers and people of unpredictable habits from other cultures. In the elderly such fear tends to be associated with difficulties of dealing with unknown new technologies calling for particular skills.
Broader Fear of the abnormal (#PF7029).
Narrower Fear of the dark (#PG2398).
Aggravates Monsters (#PF2516)
Bad omens (#PF8577)
Resistance to grace (#PF5266).
Aggravated by Insecurity (#PA0857)
Superstition (#PA0430)
Fear of new technology (#PF9127).

♦ PF6190 Lack of places in urban environments encouraging unstructured public access
Nature There are very few spots along the streets of modern towns and neighbourhoods where people can meet and congregate, happily and casually, for hours at a time, but this was not always the case. In mediaeval towns, for example, towns and villages had simple, open structures where auctions, market fairs and open-air meetings took place. In European cities, cafes and beer-halls were commonly created where people could sit and drink an aperitif or a coffee and watch the world go by. Such places have an important role in community life. There are many things that people can do there: take the view and the air on a terrace, read a newspaper, stroll, play chess, draughts or cards and meet other people. One of the important functions of such places is that they provide a place for newcomers to get to know a neighbourhood, as well as a meeting place for regulars. People feel they have a right to be there, and can stay as long as they want.
Incidence Experiments in creating 'outdoor rooms', to which people would be drawn in ways they would not be to a simple empty space, resulted in a dramatic change in the life of the community. Many people spent more time outdoors, public talk was animated, and cars were replaced by people. Research done at the University of Oregon confirmed that students placed as much importance for their studies in 'talking with a small group of students in a coffee shop' as on 'examinations' and 'laboratory study'.
Broader Stultifying homogeneity of modern cities (#PF6155).
Related Inadequate night life entertainment facilities (#PE6126)
Unattractive pedestrian environments in urban areas (#PE6151).
Aggravated by Inaccessibility of water for recreation (#PF6138)
Monolithic architecture of high-rise buildings (#PE1925)
Inadequate facilities for grass-roots community initiatives (#PF6143).

♦ PF6192 Alienating waiting environments
Nature Rarely are environments where people wait for buses, trains, or other forms of public transport, made comfortable and appealing, despite the amount of time that people have to spend in them. Bus stops and waiting rooms, are often alienating environments. The bus shelter, if there is one, is often badly designed and lets in wind, snow, or rain; it is covered in graffiti. Some waiting areas are ill-lit or isolated and encourage crime. Due to this many people are apprehensive while using them and may experience repeated stress or fear if not actual harm.
Broader Impersonality of public squares in cities (#PF6165).
Related Unnatural urban environments inhibiting sleeping in public (#PF6193).
Aggravates Unhealthy lack of daily physical activity in urban environments (#PF6182).

♦ PF6193 Unnatural urban environments inhibiting sleeping in public
Nature In caring societies, it is accepted that people sometimes want to sleep in public, for a variety of reasons. They may simply be tired, or they may have nowhere else to go. In modern societies, however, such behaviour is often not tolerated, as being 'against public order or decency'. There is a tendency for modern cities to have few public benches, or comfortable places or corners, where people can sit or rest in comfort.
Broader Impersonality of public squares in cities (#PF6165).
Related Alienating waiting environments (#PF6192)
Unconvivial hotel environments for travellers (#PF6196)
Insufficient possibilities for gathering of elders (#PE8985)
Inaccessibility of green parkland in an urban environment (#PE6175).
Reduces Job fatigue (#PD8052).

♦ PF6194 Social inadequacy of large buildings
Incidence A cathedral or church may be taken as an example of a large, but inviting building complex. Its various parts, the campanile, the altar, the nave, and so on are practical accommodations ot its social purposes and groupings: the ministrants, the congregation, the choir, and the special services for weddings, funerals and baptisms. Similarly, a group of tribal huts or igloos are human too, because they also are a complex of buildings, not one centralized, mechanically assembled artifact imposing itself on human social structures. It is significant that the two entities that so often interfere with the individual's quality of life, namely government and the large corporations, have traditionally used the massive pilings of stone, steel and glass. This can be seen at the Kremlin and General Motors.
Claim High-rise, monolithic buildings are a denial of the natural tendency of human institutions to comprise a complex of smaller institutions each with its own social structure. People living and working in such buildings are subject to greater stress and work dissatisfaction as they are forced to adapt their lives to the dissonance of such architecture.
Broader Indoor air pollution (#PD6627).
Narrower Displacement of natural light in buildings (#PF6198)
Disorientation stress in large building complexes (#PF6174).
Related Landscape disfigurement (#PC2122).
Aggravates Monolithic architecture of high-rise buildings (#PE1925).

♦ PF6195 Socially sterile rental accommodation
Monotonous housing facilities
Claim Rented accommodation, whether from private landlords or public housing agencies, works against the natural processes which allow people to form stable, self-healing communities. Rental areas are always the first to turn into slums: residents have no incentive to maintain and repair the homes since the improvements add to the wealth of the landlord, and even justify higher rent; landlords try to build rental properties which are immune to neglect; gardens are replaced with green asphalt or concrete, and trees give way to flowerbeds, as the landlord tries to keep his maintenance and repair costs as low as possible; and every attempt is made to make new units

maintenance-free. Yet the sterile decor and industrial appearance of such dwellings encourages neglect or overt abuse, particularly by alienated youth, so the typical piece of rental property degenerates over the years.
Broader Inadequate housing (#PC0449)
Monotonous and unaesthetic architecture and design (#PF0867).
Related Monolithic architecture of high-rise buildings (#PE1925)
Inappropriate accommodations for single people (#PE6187).
Aggravates Substandard housing and accommodation (#PD1251).

♦ PF6196 Unconvivial hotel environments for travellers
Nature Modern hotels and motels often provide an unconvivial atmosphere for the traveller. Although he needs company, he is frequently put into a self-contained and self-sufficient room where his likely recourse is to watch TV alone. This is a strange contrast with former eras, when the inn was a place where strangers met for a night to eat and drink, play cards, tell stories and share their extraordinary adventures.
Broader Tourist hazards (#PE8966)
Inadequate night life entertainment facilities (#PE6126).
Related Sterile working environment (#PD6133)
Excessive dispersion of community facilities (#PF6141)
Unattractive pedestrian environments in urban areas (#PE6151)
Unnatural urban environments inhibiting sleeping in public (#PF6193).
Aggravated by Monolithic architecture of high-rise buildings (#PE1925).

♦ PF6197 Inhibition of communication between non-proximate offices
Nature There is evidence that there is a considerable nuisance value in people having to communicate with those in offices more than a certain distance away. If two parts of an office are too far apart, people will not move between them as often as they need to, and if they are more than one floor apart, there will be almost no communication between the two.
Incidence Studies in the USA have shown that two floors distance between two departments inhibits people considerably in their interchange with others. In a three-storey university building, people were asked to name all those they knew in other departments. When departments were on the same floor, respondents knew 12.2 percent of the people; when they were one floor apart, they knew only 8.9 percent and when they were two floors apart, the figure fell to only 2.2 percent. Other observations showed that the nuisance value of distance can begin at less than 100 feet, if people have to walk that distance every hour or two hours. A distance of 500 feet is a nuisance travelled more than once or twice a week.
Broader Lack of job satisfaction (#PF0171) Sterile working environment (#PD6133).
Narrower Displacement of natural light in buildings (#PF6198).
Aggravates Bureaucracy as an organizational disease (#PD0460).
Aggravated by Insufficient communications systems (#PF2350)
Monolithic architecture of high-rise buildings (#PE1925).

♦ PF6198 Displacement of natural light in buildings
Nature There is a growing body of evidence that man actually needs daylight, since the cycle of daylight somehow plays a vital role in the maintenance of the body's circadian rhythms. The change of light during the day is a fundamental constant by which the human body maintains its relationship to the environment. Thus, too much artificial light can create a rift between a person and his surroundings, and upset the human physiology. However, modern buildings are often designed with no concern for natural light, and depend almost entirely on artificial light. There are instances on record of people leaving jobs because of the lack of natural light.
Claim Buildings which displace natural light as the major source of illumination are not appropriate or comfortable for human beings to live and work in. Any room suitable for habitation should have not merely one window, but two, on different sides. Artificial lighting is also uniform lighting, which destroys the social nature of space and makes people feel disoriented. By contrast, pools of light are used to reinforce the social nature of space, as where a pool of light may irradiate a favourite armchair, or be suspended for intimate effect over a dining table.
Broader Lack of job satisfaction (#PF0171) Social inadequacy of large buildings (#PF6194)
Inhibition of communication between non-proximate offices (#PF6197).
Narrower Unrelated buildings in urban environments (#PF6199).
Related Wastage of open space in urban environments (#PE6163).
Aggravated by Bureaucracy as an organizational disease (#PD0460).

♦ PF6199 Unrelated buildings in urban environments
Incidence Roman churches are examples of buildings that were always connected to the buildings around them. A survey among 255 such churches showed that 41 had one side attached to other buildings, 96 had two sides so attached, 110 had three sides so attached and 2 had four sides obstructed by other buildings. Only six churches were free standing.
Claim Isolated buildings in an urban environment are a symbol of the isolated lives of many modern people. Such buildings are in contrast to the the cheek-by-jowl housing of former times, where the sheer fact of their adjacency forced people to confront their neighbours and solve the myriad little problems that occurred between them.
Broader Displacement of natural light in buildings (#PF6198)
Wastage of open space in urban environments (#PE6163).
Aggravates Monolithic architecture of high-rise buildings (#PE1925).

♦ PF6202 Misuse of food as a political weapon
Nature Grain and other relief foods provided by intergovernmental organizations, or individual nations, are being distributed in recipient nations on a reward or punishment basis. The victims are frequently indigenous peoples, for whom food is the final weapon employed in a long line of coercive programmes to destroy their culture and solidarity. Surplus grain, fish, meat or butter are not viewed as the answer to hunger, or even simply as a commodity, by the numerous producer nations who now instead look to the political advantages of distribution.
Incidence North American grain and European dairy products are among political food commodities, and water resources are also a political weapon.
Broader Political exploitation (#PC7356) Abuse of government power (#PC9104).
Narrower Exploitation of dependence on food aid (#PD7592).

♦ PF6208 Inadequate education for peace
Nature Four major types of peace education are apparent today – peace education as criticism of war, as liberation, as lifestyle movement, and as learning process. However, due to a number problems (such as prejudiced attitudes and concerns on the part of those educators who have experienced war and thus view it differently from those who have not; outdated textbooks; an inability to "teach" relationship perspective – females generally have a different vision of reality than men, seeing themselves as part of the whole rather than distinguishing themselves from others, and it is almost impossible to "teach" that insight), today's peace education lacks the strength and coherence necessary for the attainment of its far-reaching goals. Coupled with this is the unique problem of the developing countries which recognize the Ghandian principle that minds cannot be fed until stomachs are full. Peace education in those countries cannot be separated from development education.
Incidence In Japan, some teachers began peace education by talking to their students about their own horrifying experiences during the bombing of Hiroshima and Nagasaki; and in Hiroshima, students are "studying with their hands" by digging up bomb roof tiles whose surfaces were burnt to bubbles by the 2000 deg C heat from the blast. In Hungary, peace education is a major topic in the weekly "pastoral programme" which reaches 25 percent of classroom students, it is being introduced into teacher training programs, and over 2000 local cultural centres have discussions and seminars on this topic. The centrality of peace education in all state education is a goal in German DR and is promoted through a framework emphasizing disarmament, the absence of war, and solidarity with national liberation movements.
Claim Peace education emerged with an emphasis on "no peace without development", hence the problem of peace is one which reaches into the politics, economics and cultural life of every society; transformation of these systems is necessary in order to bring about peace. The work of Gandhi and Danilo Dolci is critical to peace education, as is the analysis of the application of their works in Northern and Western Europe by Johan Galtung.
Broader Inadequate education (#PF4984).

♦ PF6210 Adverse economic shocks from external factors
Nature All countries, but particularly developing ones, are vulnerable to violent disturbances in world, regional or bilateral economic relations and conditions. Import costs and export earnings vary unfavourably with exchange rate fluctuations, the volume of trade is negatively impacted by recession, and world interest rates and inflation may be susceptible to erratic but persistent increases. Other foreign constraints on domestic conditions are oil price increases and disturbances due to wars or hostile actions. The inadequacies of developing countries' reserves are apparent when externally caused crises emerge.
Narrower Aggressive economic destabilization of countries by external forces (#PE9420).

♦ PF6218 Complicated spelling
Non-phonetic spelling
Nature Many languages which have been in evolution over a thousand years have inconsistent spellings. Those using the Roman alphabet, among which are the international languages of English, French and Spanish, have homonymous words or syllables that are spelled differently. There are also redundant letters unnecessary for pronunciation which include doubles (nn, ll, ss) and mutes, as well as unvoiced vowels (terminal e, for example) and some consonants in certain combinations (the b in dumb). Languages using ideograms, such as Chinese, may have several differently drawn characters to express the same sound. Some educators and practical persons have unsuccessfully tried to reform various language spellings, for example the Pitmans, and GB Shaw for English. One sound, one letter has often been the goal. An English word like 'thought' or a French word like 'hazard' might be spelled as 'thawt' or 'azar' to satisfy simplification requirements.
Claim It is estimated that only some 1,000 English words are spelt as they are pronounced. The rest have to be learnt individually. Uncomplicated English spelling would help facilitate high standards in international business, education, and personal growth and performance; would perhaps save many from the depths of illiteracy; and could be achieved without sacrifice or loss, thus yielding enormous social and economic benefits.
Counter-claim It is never explained by spelling reformers why philologists and poets do not rally to their cause. For the cultural historians who trace word origins and the artists who love the shapes of letters and the forms of their combinations, spelling reform proposals are a plea for linguistic purity, hence 'genetic' sterility, as it is the mutations and irregularities of language that give it its vitality and beauty. On a purely practical basis, many so-called 'inconsistencies' are in fact necessary for pronunciation. For example: removing the mute 'e' from the words 'mute' and 'cute' would produce a different sound (and a different meaning); the double consonant (as in 'fully') also changes the way in which the preceding vowel is pronounced. Removing such nuances from spelling reduce the already declining richness in speech and have a degenerative effect on the language.
Aggravates Incorrect spelling (#PE4302).

♦ PF6221 Decline of philanthropy
Claim Western economies are in trouble, sustained growth cannot be taken for granted, even in the longer term, the limits to government action are becoming more clear and private giving is on more ambiguous ground than before. Philanthropy is no longer constant and independent of the political, economic or social system within which it is exercised. While it is a valuable independent voice of society members to the society in which they live, it no longer fills in the inevitable gaps in all systems of social provision.
Counter-claim The tax incentives for charitable giving are mechanisms to enable the rich to support their own class institutions, especially in the case of Britain's fee-paying schools, and the gifts of the rich, while reducing their own tax burdens actually increase the tax burdens of others.
Aggravated by Compassion fatigue (#PF2819) Economic stagnation (#PC0002).

♦ PF6224 Mystique in attitudes to problems of disabled people
Nature The problems of disabled people are often cloaked by a mystique developed by professionals, who intellectualize, analyze or compartmentalize the various aspects of these problems. They propagate the attitude that only the particular language developed by professionals can define the problems of disabled people; only the methods used by professional analysts can be used to determine where the problems come from; only the models developed by professionals can provide a solution; and only professionals are qualified and capable of implementing solutions. All of this undermines the efficacy with which disabled people can help themselves to operate at their maximum capacity.
Counter-claim In Europe a considerable effort is made to enable disabled people themselves to make a choice towards independent living. Professionals are encouraging disabled people to participate in decision making.
Refs Eisenberg, Myron G et al *Disabled People As Second Class Citizens* (1982).
Broader Incomprehensibility of specialized jargon (#PF1748).

♦ PF6229 Underutilization of biocontrol
Lack of biological control
Nature Biological control of pests is the method of controlling pests, whether of plants, animals or man, by exposing them to their natural enemies. However, it is underutilized in many of the developing countries which could best benefit from it.
Background Biological control includes breeding crop plants resistant to disease, releasing sterile males of a pest to interfere with reproduction, using sex attractants to trap and kill males before mating can ensue, and manipulating genes in order to render males sterile.
Claim Biocontrol is economically feasible in that it establishes a once-for-all self-regulatory system in which natural enemies keep the pest in check below a damaging level without any further effort or expense. It utilizes no chemicals which threaten, pollute or damage the environment or serve to introduce chemical-resistant strains. Finally, biocontrol utilizes none of the fossil fuels that are required in the production of chemical products.
Counter-claim Unlike the ensured control obtained from the use of chemicals, biocontrol is variable. Developmental research on such control methods is long and, upon application, may take years to be effective. In addition, an introduced 'natural enemy' may itself become an unwanted

pest over time.
Aggravates Pest resistance to pesticides (#PD3696).
Aggravated by Pesticide hazards to wildlife (#PD3680).

♦ PF6231 Criminalization of prostitution
Discrimination against prostitutes
Nature The making of prostitution a criminal offence does little or nothing to protect women from becoming prostitutes and exposing them to considerably increased risks once they have done so. They have no recourse to the law to protect them from violence or extortion by pimps or customers. Legalization of prostitution under controlled conditions is little better, as prostitutes have then little control over the conditions in which they work and the only person truly protected is the customer. The answer would seem to be decriminalization of prostitution altogether, with the onus for reform being put on the system which causes the abuse rather than the prostitutes who are abused.
Broader Inadequate laws (#PC6848) Restrictive legislation (#PD9012).
Related Criminalization of drug use (#PF4735)
Ineffective regulation of electronic messages (#PE6226).

♦ PF6232 Misandrony
Claim The feminist cult has resulted in misandrony becoming respectable. Boys learn negative self–images from schools, media portrayals, and from bitter mothers in single–parent households; even in the minds of women who expect men to share their own workload are lingering images and expectations of men as the ultimate providers, protectors, and initiators; and the image of female life as more valuable than male life is reflected and reenforced by the "civilized" practice of saving women before men during emergencies.

♦ PF6234 Increasing public health expenditures
Nature The problem of the increasing cost of health care and responses to it raise complex issues of the cost–effectiveness of the system as well as of equity. Since the Second World War many nations have expanded their social services programmes rapidly in order to make medical care available to as many people as possible. In addition, technical progress has allowed for a greater number of illnesses to be treated, thus creating a greater number of patients. Consumer expectations that the latest technology should be widely used in public health care drove the costs still higher. As a result the cost of health services has been rising sharply in most countries, irrespective of the system of medical care that the country has. Government plays a significant role in health care in most countries and its growing cost, in addition to increasing the budget deficit, has put a great strain on the health care system. In countries depending largely on private health insurance schemes, the response to rising costs has been a substantial increase in health insurance premiums In countries with a national health care system, there have been steps to increase the often nominal charges that users of the system are expected to bear, while those with a mixed system of private insurance and public health services have attempted to pass on a larger share of the cost to patients. Concerted efforts have been made to cut costs through rationalization of benefits and contribution to medical insurance schemes. Financial stringency as well as a desire for rationalization has also led governments to seek a reduction in cash sickness benefits or to check the real growth of such benefits.
Incidence In 1982, America's medical bill climbed 12.5 percent to $32 billion, or more than one–tenth of the gross national product. Between 1960 and 1984 the cost of medical care benefits at constant prices multiplied five to ten times in many industrialized countries, and as a proportion of gross domestic product, health care tripled and quadrupled in some countries.
Counter–claim Private health care is not always affordable to the masses of people. However, the same profit motives that drive up every cost in the private sector affect local and national governments providing health services as well, since they must pay for medical drugs and supplies; laboratory, diagnostic, and therapeutic equipment; electricity, ambulance fuel, and building services; and everything else from administrative and medical salaries to food and laundry expenses. High profits account for one dimension of the levels of public health expenditures, waste and inefficiency account for another. Yet health care is only expensive in relation to the value received. Public health services need to be expanded, but they also need to be delivered more efficiently and more cost effectively.
Broader Increase in operating costs (#PF7932).
Aggravates Inadequate health services (#PD4790)
Decline in government health expenditure (#PF4586).

♦ PF6235 Irredentism
Ethnic unification
Nature The union of an entire ethnic group under one state. Throughout modern history, the rise of ethnic consciousness and the drive for ethnic groups to unite together has caused violent upheavals across the world and led to repeated cases of oppression and brutality.

♦ PF6244 Global crisis
Refs Tiryakian, Edward A (Ed) *Global Crisis* (1984).
Narrower Energy crisis (#PC6329) Global economic crisis (#PC5876)
Global financial crisis (#PF3612) Environmental degradation (#PB6384).
Aggravates Human suffering (#PB5955).
Aggravated by International crisis escalation (#PB6335)
Excessive foreign public debt of developing countries (#PD2133).

♦ PF6250 Fragmentation of communities by automobiles
Nature The use of cars has the overall effect of spreading people out, and keeping them apart. The effect of this on the social fabric is clear. People are drawn away from each other, and densities and corresponding frequencies of interaction decrease substantially. Contacts become fragmented and limited, localized by the nature of the interaction into well–defined indoor places – the home, the workplace, and maybe the homes of a few isolated friends.
Broader Obstacles to community achievement (#PF7118)
Stultifying homogeneity of modern cities (#PF6155).
Related Inaccessibility of countryside to city dwellers (#PF6140)
Inaccessibility of city centres to suburban residents (#PF6132).

♦ PF6293 Underutilization of labour force
Labour force underutilization
Refs Smith, A D *Concepts of Labour Force Underutilisation* (1972).
Broader Underproductivity (#PF1107).
Narrower Wasted woman power (#PF3690).
Aggravated by Ineffective economic structures in industrial nations (#PE4818)
Lack of technical development and excess of manpower in developing countries (#PE4933).

♦ PF6300 Resistance to technology education
Nature A major obstacle to the extension of technology education in many developing countries is the resistance on the part both of pupils and of their parents to courses in manual and technical work, which are considered inferior to more academic studies.
Broader Undersupported school system (#PU8326).
Related Unrealized use of education structures (#PF2568).

♦ PF6310 Non–verification of compliance
Nature The problem of verification of international treaties or agreements is one of obtaining reliable, continuous and highly accurate evidence that treaties are being honoured by their signatories, without infringing upon such principles of international law as sovereignty, territorial integrity and non–interference in domestic affairs. While the verification problem arises in part from the degree of effectiveness of the national technical means available for monitoring purposes, it actually stems more from the cold–war legacy of mistrust and suspicion in relations between countries with different social systems, and by the constant fuelling of this mistrust and suspicion by the opponents of international disarmament.
Claim The problem of non–verification of compliance is used as a means of opposing disarmament and détente. This problem must be solved in the interests of preserving peace, averting war, and halting the arms race.
Broader Non–verifiability (#PJ6449).
Narrower Non–verifiability of compliance with nuclear arms treaties (#PF4460)
Non–verificability of compliance with nuclear power safeguards (#PF4455).
Aggravates Lack of international cooperation (#PF0817).

♦ PF6316 Anomie
Nature Anomie is a societal condition resulting from the disintegration of a commonly accepted normative code. In the 1950s and 1960s it came to describe a concept akin to alienation, one where an individual had lost his traditional moorings and was prone to disorientation or psychic disorder.
Background The term comes from the Greek meaning, literally, without law, and was resurrected by the French sociologist Emile Durkheim.
Aggravates Suicide (#PC0417).

♦ PF6317 Animism
Nature Animism, the view that ascribes anthropomorphic behaviour to objects and natural phenomena (such as the attribution to inanimate objects of such conditions of human activity as life, thought, and free decision) by projecting them outward from the human state, may interfere with a person's psychological growth.
Incidence Animism today is regarded to occur with the native inhabitants of Africa, South America, Oceania, but can also be seen in the woman who calls her car 'she', the man who throws his golf club to the ground because he missed a shot, and the child who kicks the chair from which he just fell.
Aggravates Demons (#PF6734).

♦ PF6325 Evangelism
Claim Evangelistic crusades, especially in other cultures, have been based on an attitude of arrogance, aggression, and negativism. These have obscured the gospel and caricatured Christianity as an aggressive and militant religion.
Related Ill–considered missionary activity (#PF3370).
Aggravates Religious rivalry (#PC3355).

♦ PF6326 Religious discrimination against women in priesthood
Bias against ordination of women
Nature Discrimination against women in priesthood limits both the potential of women and also the potential of Church members being aided by having a female rather than male priest.
Incidence Although still not permitted in many countries, and officially not permitted by the Anglican Church, women priests are found in the USA, Canada, New Zealand and Hong Kong.
Claim The Scriptures do not forbid women in priesthood; the only reason that Christ's apostles were male was due to the socio–cultural conditions of the time. The greater participation of women in public life and their new biological freedom as a result of contraception means that old attitudes must be reconsidered; the priest represents the Risen Christ, not Christ the Man, thus female priests could represent the feminine aspects of God.
Counter–claim The Scriptures state that Christ did not ordain women apostles; the tradition of male leadership is continued in the subsequent threefold ministry of bishop, priest, and deacon, reflecting a division in human sexuality with complementary roles for men and women; for the Anglican church to break with tradition would disrupt ecumenical relations with the orthodox and Roman Catholic Churches, to which a majority of Christians belong. A solution would be for an alternative but satisfying role being found for women in the Church hierarchy.
Broader Discrimination against women in religion (#PD0127).

♦ PF6327 Sin against the Holy Spirit
Broader Sin (#PF0641).

♦ PF6330 Technocracy
Technocratic leadership — Machine technocracy — Organized cult of machinery — Technicism
Nature A trend in 20th century thought that the management and control of production and society should be transferred from the people and politicians to an intelligentsia of engineers and technicians, and to management specialists allegedly acting as the chief driving forces of progress.
Refs George, F H *Machine Takeover* (1977); Ramesh, Jairam and Weiss, Charles Jr (Eds) *Mobilizing Technology for World Development* (1979).
Broader Scientism (#PF3366) Cultural invasion (#PC2548)
Technological revolution (#PC3234).
Aggravates Anti–science (#PF2685) Alienation in capitalist systems (#PD3112)
Undemocratic political organization (#PC1015)
Lack of appreciation of cultural differences (#PF2679).
Aggravated by Technocratic futurism (#PG2574).

♦ PF6337 Lack of information
Unavailability of timely data — Unavailability of comprehensive information
Narrower Lack of medical information (#PJ8069) Lack of local information systems (#PF6541)
Insufficient plant life information (#PG9158)
Limited availability of functional information (#PF3539)
Lack of financial information systems for international nongovernmental organizations (#PE7009).
Aggravates Ignorance (#PA5568)
Inadequate management of government finances (#PF9672).
Aggravated by Loss of information (#PF9298) Aiding consummation of crime (#PD6655)
Inoperative forums for public information (#PF7805)
Maldistribution of science and technology (#PC8885).

♦ PF6339 Psychological inhibition
Nature Functioning of the ego may be restricted due to the presence of some pathological

process. Inhibitions may result from a need for self-punishment or from extensive use of psychic energy in the maintenance of defenses.
Incidence Inhibitions are evident in sexual performance, eating, body movement, dress and use of language.
   Narrower Inhibited grief process (#PD4918).
   Aggravates Enforced celibacy (#PD3371)   Inhibited self-promotion (#PJ1544)
   Inhibition of personality development in exiled children (#PE7931)
   Inhibition of individual psychological development through life cycle (#PF6148).
   Aggravated by Bereavement (#PF3516)   Psychic conflict (#PF4968)
   Defensive life stance (#PF0979).

♦ **PF6345 Socially disruptive effects of video games**
Video virus — Video game addiction
Nature The worldwide proliferation of video game machines has led to an addiction, especially amongst teenagers, that can be socially disruptive. it is sometimes referred to as "video virus".
Incidence Instances can be cited world-wide of youngsters becoming so addicted to video games that they virtually drop out of society, spending all the money they can lay their hands on in video game parlours. In the Galleries Lafayette in Paris video game demonstrations are so packed with young people that salesmen cannot move; in Tokyo, video game parlours now outnumber the once ubiquitous pachinko parlours. Amsterdam's male youth is often prey to older loitering homosexuals who expect favours in return for the coins they slip in the machines; and Stockholm has been terrorized by young thugs who rob people in the subways and streets in order to pay for their addiction.
Counter-claim Video games have educational value (math games, language lessons, sports skills) and the warning is often en famille.
   Broader Proliferation of computers (#PE3959).
   Related Computer obsession (#PF1288)   Unethical entertainment (#PF0374).
   Aggravates Attitude manipulation of children through play (#PF2017)
   Health hazards of computer visual display units (#PE5083).
   Aggravated by Violent interactive toys (#PE4297).

♦ **PF6349 Inconclusiveness of science**
Nature Science functions in incremental pieces, thus scientific findings are often tentative and inconclusive and may even contradict themselves. The public has no way of knowing how long a much publicized theory will actually be 'correct' – or in vogue.
Incidence For years scientists have correlated high blood pressure with a diet too rich in salt, but recent research reports that high blood pressure may actually result from a diet deficient in calcium, potassium, vitamins and salt.
   Broader Uncertainty (#PA6438).
   Narrower Indeterminacy of death (#PF0192)   Inaccurate forecasting (#PF4774).
   Uncertainties in animal experimentation (#PF4770).
   Aggravated by Atomism (#PF5322)   Ignorance (#PA5568)
   Affectation (#PA6400)   Reductionism (#PF7967).

♦ **PF6351 Psychiatrism**
Nature Psychiatrism is the indiscriminate application of psychiatric principles by novices or those with little understanding of the complexity of the mind, tending instead to make gross generalizations and ignoring the enormous importance of individual variation. Psychiatric principles may be applied in a mechanistic manner without careful investigation of personal dynamics, and therefore the "treatment" can render serious harm.
   Broader Unethical medical practice (#PD5770).
   Related Illegal induced abortion (#PD0159).
   Aggravated by Inadequacy of psychiatry (#PE9172).

♦ **PF6353 Careerism**
Opportunism
Claim Careerism far surpasses idealism.
   Related Political opportunism (#PC1897).
   Aggravates Name-dropping (#PF5223).
   Aggravated by Avarice (#PA6999)   Inappropriate ambitions (#PF9852)
   Pursuit of personal prestige (#PF8145).

♦ **PF6358 Theological collapse**
Nature Twentieth century people experience a void in the dimensions of the spirit, because of the stories, images, methods and modes of relating to the universal, the mysterious, the profound and the trans-historical have been rendered suspect or disfunctional. The collapse is theological in the sense that every individual has an ultimate reality that they own their allegiance to and give their life for, i.e. a god. The understandings, methods, rituals of traditional religions were developed in a scientific world view that has been discredited and the world view that emerged through the scientific revolution at best suggests the questions of faith are outside its field of expertise. The ways people have related to God or Gods have for most been taken away leaving modern people with shallow psychological images and tools. Mythologies are disrelated from experience. Self-understanding reduced to current fads.
   Narrower Dominance of economic motives (#PF1913)
   Collapsed meaning of human creativity (#PF0936)
   Religious theology disrelated to life (#PF5694)
   Immediate gratification-based social forms (#PF1058)
   Escaping reality through popular psychological screens (#PF1112)
   Overemphasis on self-sufficiency with respect to interdependence (#PF3460)
   Escaping reality through sophisticated shields against life's pain (#PF0916).
   Related Spiritual void (#PA6220)   Lack of commitment (#PF1729).
   Aggravated by Despairing individualism (#PF0951).

♦ **PF6359 Killing non-human life**
   Broader Killing by humans (#PC8096).
   Narrower Killing of plants (#PD4217)   Killing of animals (#PD8486).
   Related Homicide (#PD2341).

♦ **PF6362 Self-deception**
Nature Getting oneself to believe something one does not believe.
Refs Steffen, Lloyd H *Self-Deception and the Common Life* (1986).
   Broader Deception (#PB4731).
   Related Rationalization (#PU2208).
   Aggravates Lack of self-confidence (#PF0879).

♦ **PF6365 Inadequacy of international human rights instruments**
Inadequacy of existing human rights standards
Nature International legal instruments and standards are insufficient to prevent the isolation of human rights throughout the world, due to the lack of enforcement machinery and to political pressures on intergovernmental human rights bodies. The respect of human rights depends therefore largely on the willingness of governments to act on recommendations. Even the use of economic sanctions for the isolation of human rights has proven ineffectual because not all countries comply with them.
   Broader Deficiencies in international law (#PF4816)
   Inadequacy of international standards (#PF5072).
   Aggravates Denial of human rights (#PB3121)
   Discrimination against indigenous populations (#PC0352).

♦ **PF6377 Inappropriate public spending by government**
Inefficiency of public spending — Ineffectiveness of public spending — Misallocation of public spending by government — Inefficient public programmes
Claim Governments are spending scarce resources on goods and activities that the private sector could do much more efficiently, for example producing and marketing agricultural and industrial goods or providing bus transport in cities. Governments have not set priorities and so their spending does not balance needs against the cost of raising revenue. Spending limits are not coordinated by fiscal planning, annual budgeting and regular monitoring of revenues and expenditures. This has lead into fiscal crisis in many developing countries.
   Broader Inadequate economic policy-making in developing countries (#PF5964).
   Narrower Over-programming (#PF7737)   Avoidance of decision-making (#PF4204)
   Inadequate public finance statistics (#PF7842)
   Uncoordinated government policy-making (#PF7619)
   Socio-economic burden of militarization (#PF1447)
   Internal inefficiency of public programmes (#PF7483)
   Inequitable distribution of public subsidies (#PE9758)
   Mismanagement of aid to developing countries (#PF0175)
   Inefficient public spending to alleviate poverty (#PF1258)
   Ineffective governmental use of nongovernmental resources (#PF4095)
   Insufficient government spending on cost-effective activities (#PE5302).
   Aggravates Misappropriation of public funds (#PC2920).
   Aggravated by Proliferation of public sector institutions (#PF4739).

♦ **PF6382 Blame avoidance**
Shifting responsibility onto others — Buck passing
Incidence Benefits and shortcomings of a project will not accrue until possibly years later, well after the project's originators have passed from the scene. When the shortcomings do become apparent, the incumbent policy makers can justifiably place the blame upon their predecessors.
Refs O'Connell, Jeffrey and Kelly, C Brian *The Blame Game* (1986); Shaver, K G *Attribution of Blame* (1985).
   Broader Irresponsibility (#PA8658).
   Aggravates Scapegoats (#PF3332)   Issue avoidance (#PF1623)
   Avoidance of reality (#PF7414)   Avoidance of decision-making (#PF4204)
   Avoidance of negative feedback (#PF5311)   Conceptual repression of problems (#PF5210).
   Aggravated by Deliberate ignorance during policy-making (#PF8278)
   Irresponsible delimitation of policy responsibilities (#PF7823)
   Collusion between administrators of funding agencies and programme formulators (#PF8711).

♦ **PF6385 Desecration of holy spaces**
Holy places
Nature Space is holy when invested with a sense of the divine, that is, the place points to multiple levels of profound and creative significance beyond that of the space itself. The multi-dimensional meanings of the place may be simply personal, for example, the house a person grew up in may have a personal sense of the holy; where the mystery gave them birth, growth, maturity, confrontation with death, etc. Normally a sacred place must be communal that is recognized as holy by a community. The place may have many stories, legends or myths associated with it. It may be the location of repeated efforts at prayer, sacrifice, rituals, worship services, or rites. It may be the location of historically significant event, like the community's founding, defence against destruction, significant change in direction or other meaning investing activities. Desecration of holy spaces is attack of the symbolic meanings of the place. To wantonly destroy a temple, shrine, church, mosque or even a memorial is to attack the community's symbols of unity, of its relations to God, itself and its neighbours, and of its history, its present and its future. When a community's symbols are attacked the realities to which these symbols are pointing are indirectly under siege. The symbols and the realities they point to must be separated to understand desecration. The desecration of a holy space may generate strong reactions because of the profundity of the reality to which the place points, i.e., the central pillars of a belief system, or because of the many dimensions of life to which it points, i.e., a sense of community, continuity with the past, hope in the future, unity with the universe, and repeated experiences of being blessed.
Background In primitive religious conceptions the gods are not exempt from general limitations of space and time. The gods have a physical environment, on and through which they act. They are thought of as bounded by certain local limits. The early Semites ideas of divine preference came to be associated with the fertility of particular places, the local gods being recognized and appeased by a tribute of first-fruits and by extension of meaning the first born animal and children. Thus the gods came to have their proper homes or haunts where the worshippers laid their gifts on sacred ground or hung them in a sacred tree, or, in the case of sacrificial blood poured over sacred stones. Later the homes became temples which could only be erected in a place where a god had manifested his presence. These places came to be surrounded by restrictions as to access, especially those who were unclean or had shed blood. The right of asylum in the Old Testament was limited to involuntary homicide and in some Arabian sanctuaries all fugitives were admitted to shelter. The idea of holiness became associated with restriction in the use of places and with protection from encroachment. Holy places were many types. Caves served as early Phoenician temples, the original sanctuary of the temple of Apollo at Delos and the dark inner chambers of early Semitic and Greek temples. In Arabia the whole mountain of Horeb was sacred ground. The idea of sanctity is attached to rivers and springs. Trees were often seen as holy among all the Semitic peoples as well as Druids and Germanic tribes. The local sanctuaries of the Hebrews were altar-sanctuaries erected under trees. Cairn were used for sacrifices for the Arabians, Greeks and Romans. These later evolved into the altars of modern churches and temples.
   Broader Desecration (#PF9176).
   Related Desecration of holy days (#PF3607)   Desecration of cemeteries (#PD7258)
   Desecration of religious buildings (#PD7278)
   Disruptive secular impact of holy days (#PE7735).
   Aggravated by Abuse of relics (#PF5107)   Desecration of monuments (#PD4348)
   Vulnerability of sacred sites (#PD6128)
   Holy places as a focus of religious friction (#PF1816)
   Desanctification of churches and holy places (#PF8666)
   Lack of individual rights to political asylum (#PF1075).

♦ **PF6386 Indiscriminate anti-trust prosecution**
Nature A number of governments accept a theory that if a company enjoys a monopoly market position it is intrinsically evil, and its operations tend to limit output while maintaining artificially high prices. In capitalist countries anti-trust or anti-monopoly legislation and regulation is enforced in an arbitrary manner, while in socialist and communist countries there is no restraint on government owned monopoly enterprises or on their competition in international trade.

Claim All technologically innovative corporations enjoy a monopoly position for varying periods. Their domination of a market which they themselves created in part allows for a significant return on investment for the venture capital risked. This fuels more investment, more innovation, more jobs, and a growth economy. In developing countries monopolies may provide the thrust to move the country forward.
Broader Monopolies (#PC0521).

♦ PF6389 Excessive caution

Deprivation of risk-taking — Excessive safety precautions — Cotton-wooling — Obsession with personal safety — Obsession with personal health — Excessive fear of risk
Broader Fear (#PA6030) Inconsistent risk evaluation (#PF5482).
Narrower Morbid fear of illness (#PE9091).
Related Parental over-protectiveness (#PF5255).
Aggravates Backlash against repeated warnings (#PF3391).

♦ PF6395 Biased expertise

Biased advice — Biased experts — Unavailability of impartial expertise
Broader Unavailability of appropriate expertise (#PF7916).
Aggravates Questionable facts (#PF9431) Neglect of expert advice (#PF7820) Disagreement among experts (#PF6012) Unproven relationships between problems (#PF7706).
Aggravated by Irresponsible international experts (#PF0221).

♦ PF6404 Scrupulosity

Nature Scrupulosity, excessive meticulousness or punctiliousness, may be symptomatic of an obsessive-compulsive personality and, if fostered over an extended period, could be suggestive of an underlying schizophrenic process.
Broader Obsession (#PA6448).
Related Hyperefficiency (#PF1706).

♦ PF6406 Sexual fetishism

Sexual fetichism
Nature Fetishism is a sexual perversion in which the person (usually male) displaces his erotic interest and satisfaction to a fetish. The fetish replaces and substitutes for the love object, and although sexual activity with the love object may occur, gratification is possible only if the fetish is present or at least fantasized during such activity.
Broader Sexual deviation (#PD2198) Human sexual inadequacy (#PC1892).

♦ PF6407 Government complacency

Broader Complacency (#PA1742).
Aggravates Government inaction (#PC3950).
Aggravated by Government arrogance (#PF8820) Bureaucratic superiority (#PC1259) Proliferation of public sector institutions (#PF4739).

♦ PF6411 Information overload during control of complex equipment

Information overload for process controllers
Incidence In the case of nuclear reactors, reports indicate that in moments of crisis the amount of information relevant to warnings of the condition (as with the Three Mile Island incident) is such that operators are simply overwhelmed, leading to responses which exacerbate the crisis. Information is substituted for knowledge inhibiting understanding of the appropriate action required.
Broader Proliferation of information (#PC1298).
Aggravates Human errors and miscalculations (#PF3702).
Aggravated by Human fatigue during control of complex equipment (#PE5572) Substance abuse during control of complex equipment (#PE0680).

♦ PF6415 Sexual impotence of men

Premature ejaculation — Retarded ejaculation
Nature Impotence is a symptom of various pathological conditions, manifested by a weak erection, which interferes with the normal course of the sex act. Impotence may be organic or psychosomatic but regardless of its origin, the impact such a deficiency renders on the psyche of a man can be devastating. Sexual dysfunction affects all areas of a man's life - his sense of self-esteem; his relationships to his mate, other women, and other men; and even his job performance. The Catholic Church recognizes impotence as grounds for invalidating a marriage.
Incidence An estimated 10 million American men suffer from impotence; 50-60 percent of their impotence stemming from physical causes and only 40-50 percent being mental in origin.
Background Blood vessel disease, diabetes, pelvic fractures, abdominal surgery, hormonal irregularities, neurological disease, chronic liver disease, alcoholism and drug abuse are among the major physical causes of impotence. Psychological impotence might stem from sex-related anxiety or guilt, often associated with a disastrous 'first experience'.
Refs Hastings, D W *Impotence and Frigidity* (1963); Wagner, Gorm and Green, Richard *Impotence* (1981).
Broader Human sexual disorders (#PD8016) Human sexual inadequacy (#PC1892) Sexual arousal disorders (#PG9473) Psychogenic physical disorders (#PE3974) Diseases of male genital organs (#PD9154).
Aggravates Voyeurism (#PE3272) Childlessness (#PC3280) Sexual unfulfilment (#PF3260).
Aggravated by Drunkenness (#PE8311) Alcohol abuse (#PD0153) Sexual deviation (#PD2198).

♦ PF6430 Fatalism

Belief in fate — Belief in manifest destiny
Nature Belief that the conditions and development of human life are necessarily controlled and determined by some inscrutable power to which all are subject. This may be either personified or represented as impersonal. Belief in fate is only transcended when people come to regard themselves as free or as called by some higher will to exert their freedom of choice by leading a responsible life. Fatalism predisposes people to passive acceptance of any unfortunate conditions or suffering to which they are subject, thus effectively disempowering them. In particular cases fate may be interpreted as the destiny which some higher will sets before a moral personality as an ideal to be realized. Empowered in this way, such destiny may be further understood to be beyond criticism by fellow human beings or other social groups, and may be pursued without any qualms concerning the suffering caused to them. The arbitrary will of some omnipotent power and the blind necessity of nature thus both result in the non-moral subjection of mankind to an inevitable necessity. An absolute determinism blights all spontaneity of action, leaving room at best for fanaticism. Only when man realizes his freedom as that which lays upon him the obligation of self-determination in the sphere of conduct does he cease to resort to occult arts; and only as he knows that all things can be utilized for the highest ends does he finally break with the idea of fate.
Incidence Historically this conception first prevails wherever people are unable to frame the idea of a rational necessity or of a supreme purposive will. It persists as long as either of these, though within the field of consciousness, is imperfectly realized. The belief emerges again at a subsequent stage of development when people doubt any rational order or rational end to the universe. Astrology, fortune telling, and theology all have fatalistic viewpoints, and the general populace on the whole has vague fatalistic feelings, manifested in such thoughts as 'You won't die until your time comes', 'I was warned not to take that airplane', 'It was fated that we should meet'. When perceived as manifest destiny, fatalism is a major factor justifying the expansion of social groups beyond their existing boundaries as in imperialism and colonialism, as well as in their religious equivalents.
Claim Fatalism is an escape mechanism people employ in order to lessen their needs to make decisions and thus lead a responsible life. It results in an attitude of 'It's not up to me, it's already been decided', which is exactly the weak, selfish attitude that is already too abundant.
Counter-claim There is a force, a direction, a fate that is larger than man; if he accepted that force in his life and did not always attempt to intervene and interfere, there would be many fewer problems on individual and societal levels. A number of religions hold the supreme power of the universe to be a rational will by which all things are engendered to appropriate ends. In different ways these religions call upon man to submit to that will as being the only valid means of reconciling individual desires with those of the larger whole.
Broader Superstition (#PA0430) Inappropriate assumptions (#PF6814) Negative emotions and attitudes (#PA7090).
Narrower Destiny (#PF3111) Inevitable destruction of natural environment by mankind (#PE2443).

♦ PF6432 Vampirism

Nature The belief in human vampires, dead persons who come from their graves at night to suck the blood from living people, still influences burial customs and the siting of graves in some regions of Eastern Europe, Greece and the Mediterranean. Children and adults are terrified of graveyards and the dead because of such superstitions, and extension are afraid of the dying who thereby become objects of repulsion to be avoided, even if a loved member of the family.
Background A rare disease called porphyria has been postulated as the core of vampirism. Treatable today, victims suffer extreme sensitivity to sunlight, and in the past had to stay indoors ('the children of the night') as exposure caused disfiguring by sores, scars, extreme hairiness, possible loss of extremities (noses and fingers), and stretched skin of the lips and gums, yielding fang-looking teeth. The belief in vampires may have instigated imitative behaviour among the psychotic, especially after wide dissemination of films based on Count Dracula. In modern Europe two instances of insane killers are recorded who were inspired by the film character. In occult and spiritualistic circles there is a belief that sensitive people may be drained of vital energy, under certain conditions, by disincarnate humans or other invisible entities. The term vampirism has been attached to a theory that one living human, in a variety of ways, can draw psychic or biological energy from another.
Refs Riccardo, Martin V *Vampires Unearthed* (1983); Wright, Dudley *Vampires and Vampirism.*
Aggravated by Lycanthropy (#PE8450).

♦ PF6433 Blocked service to society

Nature Individuals find that their desire to serve society is blocked. The symbols of recognition by the larger society that a profession has a significant role in the health of the social fabric are largely gone. Stories about the importance of a job have been replaced by waiting for retirement myths. Even the most specialized of professionals find themselves as a part of thousands of other professionals. Attempts to actually solve problems is frustrated by the size of any task when looked at from a global perspective.
Broader Collapsed meaning of human creativity (#PF0936).

♦ PF6441 Population decrease

Declining national population
Incidence Fertility rates have fallen throughout the industrial world, and in many countries are below 2.1 children for each woman of childbearing age, the generally accepted level at which a society must reproduce to sustain its population. In Quebec, Canada it is 1.4 and for Canada as a whole 1.7. In the U.S it is 1.8. West Germany has the lowest fertility rate in the world at 1.3. Except for Ireland and most Eastern European countries, birth rates have dropped to a level that is too low to keep the size of the population stable in the long run.
Claim The current worldwide tendency for actual population decrease or reductions in rate of increase may help developing countries raise their living standards, but will also require adjustments in both developed and developing countries, particularly in providing security to the elderly. This will be even more the case by the 21st century. Nations will experience a marked reduction in the number of children entering the age group of compulsory schooling and, progressively, in that of the age groups entering universities, military service, and growing up to be productive members of society in terms of both wages earned and ideas/creations returned to society.
Refs Teitelbaum, Michael S and Winter, Jay M *The Fear of Population Decline* (1985).
Narrower Declining community population (#PJ8746).
Aggravates Imbalance of population growth between developed and developing countries (#PE4241).
Aggravated by Declining birth rate (#PD2118) Prohibitive cost of parenthood (#PF1625).
Reduces Unsustainable population levels (#PB0035).

♦ PF6450 Untransposed community structures

Nature The rapid pace of technological innovation in the 20th century has occasioned rapid social change without accompanying changes in community patterns and structures. Because of increased media communications, no community has escaped the impact of this change. In rural areas, the young are torn between the need to help their family maintain their land and the need to get an education; the task of village leadership is made increasingly difficult with the coming of a monied economy and inflationary prices; and the sense of total village identity and cooperation is threatened by increased mobility which allows individuals and families to pursue their own private interests.
Broader Stagnation (#PA3917).
Narrower Athletic competition (#PE4266) Fear of urban problems (#PG8799) Abuse of traditional customs (#PJ0675) Introversion of the family unit (#PF1030) Unpublished educational possibilities (#PJ9445).
Related Domination (#PA0839) Ineffective structures of local consensus (#PF6506) Incomplete access to information resources (#PF2401) Deteriorating structures of rural community cooperation (#PF3558).
Aggravates Unrecognized relevance of education (#PF9068).

♦ PF6451 Restricted flow of local economy

Nature Although small rural communities need to relate their market economies to those of the district, national and global economies, they frequently have no way of doing so. There is often no regular, inexpensive transport to the market; therefore little cash comes into the village economy. Services such as boat and house building, land cropping and fishing are often given free rather than as income earning ventures, leaving the village with an exchange of services, but

no cash to transfer in the economy. Money which does enter the village is often circulated immediately outside the community for petrol, transport, groceries or luxury items, leaving a constant cash deficit.
Broader Limited accumulation of capital (#PF3630).
Narrower Inadequate cash flow (#PJ0576) Lack of governmental support (#PF8960)
Uncreditworthiness of rural communities (#PJ1268)
Limited availability of loans in developing countries (#PE4704).
Related Limited transportation availability (#PJ1707)
Restrictive loan procedures in developing countries (#PF2491).
Aggravates Unexplored potential markets (#PF0581).
Aggravated by Lack of vocational teachers (#PJ9603)
Underutilization of locally available skills (#PF6538).

♦ PF6453 Inadequate documentation
Narrower Inadequate documentation of works of art (#PE8088).
Aggravated by Unwritten language (#PF3470).

♦ PF6454 Lack of means for local technological development
Nature Although new technological methods are capable of being adapted to meet the requirements of developing villages, the distribution of technology appropriate to a given local situation is difficult to coordinate. Many communities are restricted both by their remoteness and the limited flow of available cash: there is often no obvious way for a community to develop its own appropriate technology or to create ways to purchase, borrow and utilize private or government sources of equipment.
Broader Underdevelopment (#PB0206).
Narrower Unavailability of timber (#PG5299) Unavailable road machinery (#PG7850)
Insufficient access to technology for agricultural upgrading (#PF3467).
Aggravates Insufficient medical supplies (#PE1634)
Lack of technical development and excess of manpower in developing countries (#PE4933).

♦ PF6455 Unproductive utilization of plantation space
Small plantation plots
Nature Despite the possibility of increasing marketable production through the application of agricultural science and technology, Third World rural communities often have little land under planned management and the primary emphasis is on gathering for daily needs. The constant necessity of supplying today's food needs for today and difficult access to agricultural land help maintain the notion that it is not possible to manage long-range production schemes and few efforts are made to produce major cash crops. Poor and impassable roads to agricultural land and hand harvesting methods serve to restrict the amount of land under production. Planting and cultivating those crops that feed the village is often done without considering either national or international markets. Livestock is primarily dependent on foraging for food, and animals are slaughtered primarily for celebration rather than as a part of the regular diet or used as a marketable product.
Broader Ownership as a basis for land allocation (#PF6460)
Obstacles to availability of community space (#PF7130).
Narrower Labour shortage (#PC0592) Insufficient cash crops (#PG0569)
Uncontrolled small animals (#PG0591).
Related Narrow profit margin (#PJ9737)
Decreasing genetic diversity in cultivated plants (#PC2223).
Aggravated by Inadequate road conditions (#PJ0860)
Inadequate water supply in the rural communities of developing countries (#PD1204).

♦ PF6456 Self-defeating style of community planning
Nature The economic and social livelihood of rural communities is increasingly dependent upon planned, locally-based services; but at the same time, absence or shortage of electricity, transport and communications create a self-reinforcing sense that such services are impossible to obtain and that making plans to acquire them is not feasible.
Broader Obstacles to community achievement (#PF7118).
Narrower Unplanned land development (#PF6475).
Related Unpreparedness (#PF8176) Limited market development (#PF1086).
Aggravates Inadequate road maintenance (#PD8557)
Underutilized government resources (#PE9325).

♦ PF6458 Disorganized attempts of upgraded employment
Nature Many rural communities lack organization in pooling the working skills that are available locally. Although very few rural jobs bring in a stable income, the many young people who leave their community to seek employment in large cities discover that low wages combined with their lack of skills hamper their successful entry into the urban labour force. Thus they find themselves in the gap between patterns of existence in a subsistence economy and the ability to carry out a stable, salaried job effectively.
Broader Fragmented planning of community life (#PF2813).
Narrower Restricted higher education (#PJ0572).
Related Contained village economy (#PJ0594).
Aggravates Social disaffection of the young (#PD1544).
Aggravated by Underpayment for work (#PD8916)
Limited employment options (#PF1658).

♦ PF6460 Ownership as a basis for land allocation
Nature In order to build an economic base, make long-range plans and utilize maximum production capability, rural communities need a system of land allocation responsive to internal and external market needs. This is particularly true in communities where little land is available. At present, old patterns of land allocation in many communities prohibit economic growth. Such allocation is traditionally related to ownership rather than production. Much of the land is divided into small plots designed to produce the daily needs of one or more families, and not cultivated to produce crops for outside markets. The system of land caretakers when owners are away prohibits full use of some good productive land; and the complexity of land procurement and the tradition of short-range production schemes deter effective allocations. Ownership as the only relationship to land use thus serves to prohibit maximum productivity.
Narrower Absentee ownership (#PD2338) Cultural dances dated (#PG9603)
Unproductive utilization of plantation space (#PF6455).
Related Limited public land (#PJ0574).
Aggravates Short-term gain (#PF8675).

♦ PF6461 Modern disruption of traditional symbol systems
Nature Many societies are faced with the contradiction between old and new ways. In some communities technological civilization has arrived abruptly and fully developed, thus disintegrating the social fabric. Local symbol systems, which have informed every dimension of life in traditional communities for, in some cases, thousands of years, have been assaulted by this 'modern' civilization to the point where the residents themselves talk of these systems as 'primitive' or 'superstitious'. As a result, they do not realize how great the past has been. Closely related to this is the fact that there are no images profoundly related to community experience of life which relate to future possibilities. Such shame about own ancient symbols deeply fragments communal life. Cooperative effort, which is always dependent upon shared common symbols, cannot be fostered effectively. The rhythm of daily life, as well as the patterns of weeks, months and years becomes chaotic.
Broader Ineffective communication (#PF6052).
Narrower Resistance to change (#PF0557) Unbalanced social life (#PD8113)
Divided community spirit (#PJ0399) Undefined cultural space (#PJ0060)
Debilitating poverty image (#PJ1341) Superstitious symbolic acts (#PF8754)
Declining sense of community (#PF2575) Insufficient cultural heroes (#PF8623)
Overpowering traditional habits (#PJ0453) Traditionally determined housing (#PJ0486)
Ineffective parenting techniques (#PJ0551)
Domination of government policy-making by short-term considerations (#PF0317).
Related Limited leisure time (#PF9062) Unchallenging world vision (#PF9478)
Parochial national interests (#PF2600) Narrow range of cultural exposure (#PF3628).
Aggravates Ignorance of cultural heritage (#PF1985)
Inadequate housing construction (#PG0561)
Inadequate recreational facilities (#PF0202)
Lack of commitment to common symbols (#PE8814)
Inadequate image of roles within marriages (#PD1308).
Aggravated by Low self esteem (#PF5354) Insufficient community celebrations (#PJ0188)
Untransposed significance of cultural tradition (#PF1373).

♦ PF6462 Unprepared adult leadership
Untrained elders' leadership
Broader Insufficient skills (#PC6445) Obstacles to leadership (#PF7011).
Narrower Insufficient trained leaders (#PJ0389) Insufficient leadership training (#PF3605).
Aggravates Dorsopatie (#PE9128) Kleptomania (#PG0515)
Homelessness (#PB2150) Misdiagnosis (#PF8490)
Animal cannibalism (#PG4979) Lack of leadership (#PF1254)
Ignorance of health and hygiene (#PD8023)
Obsolete methods of agricultural production (#PF1822)
Nutritional anaemia in women in developing countries (#PE9185)
Stifled potential for social interaction between different age groups (#PF6570).
Aggravated by Impractical education (#PF3519) Inadequate vocational education (#PF0422)
Limited access to practical education (#PF2840).

♦ PF6465 Archaic marketing methods
Nature The rise of a new national and international economy has created the necessity for new marketing practices, but these have not as yet been developed to include rural communities, where all marketing procedures revolve around the central issue of local bargaining power. The persistent use of practices effective when each village was a self-sustaining unit, now works to the disadvantage of the villager when he sells or purchases goods. Lack of adequate storage or processing facilities means that the entire crop of all villages is marketed at approximately the same time. This floods the markets, and results in low returns to the individual farmers. Because of high transportation costs and a relatively uncompetitive market, merchandise from outside the village is available only at relatively high prices. The villager has no bargaining power or economic leverage with which to participate creatively in the only economic system he has; he sees himself as kept on a marginal subsistence level, a victim of forces that he has no power to attack.
Broader Limited market development (#PF1086).
Narrower Craniofacial malformations (#PG0480) Prohibitive cost of basic services (#PF6527)
Insufficient recycling of materials (#PJ0525)
Dependency on unpredictable sources of income (#PF3084)
Ineffective tax systems in developing countries (#PF2124).
Related Instability of prices (#PF8635).
Aggravates Dependency on middlemen (#PD4632).

♦ PF6467 Inadequate results of formal schooling
Nature The nature and level of basic education achieved in a society reflects the frequently inadequate official and unofficial level of learning provided for and required of its members. Inadequate formal education is experienced at most levels and in most societies, resulting in uncertainty and student unrest. Public schooling is often past-orientated and abstract, with over-emphasis on individual attitudes. Student-teacher ratios are generally too high, often 50-1 or even 100-1, and classroom equipment is lacking. Awareness of the necessity of understanding new global developments and of relating such understanding to the individual's situation, while lacking the means to achieve these goals, generates cynicism.
Broader Inadequacy of formal education (#PF4765).
Aggravated by Neglected young children (#PE4245)
High student-teacher ratio (#PG0405)
Environmental consequences of war (#PC6675)
Unrecognized relevance of education (#PF9068)
Limited availability of learning opportunities (#PF3184).

♦ PF6469 Insufficient flow of information
Nature A two-way exchange of ideas and information – both economic and social – is essential for personal involvement in all national dynamics. Local people are frequently uninformed, misinformed or informed too late of the crucial plans which affect their lives. This communication breakdown causes unnecessary frustration and misdirected criticism on every level. Mass media devices such as radio, television, newspapers and bulletins may be available but they are simply a one-way information flow. They demonstrate the complexity of the modern world, thus expanding the internal imagination of local people, who can no longer live as if the village or the district were the whole universe. At the same time, the scale of modern society has made the opportunities for local citizens to put their own creativity into plans and actions that determine their future extremely remote. Long-range plans are increasingly made at the central rather than the local level, and local people cannot utilize their wisdom and creative ideas in the arenas where future is being formed.
Broader Incomplete access to information resources (#PF2401).
Narrower Underprioritized human values (#PG0429)
Inadequate circulation of local information (#PF6552).
Aggravates Overdependence on government (#PF9530)
Inoperative forums for public information (#PF7805)
Fluctuations in food production in developing countries (#PE8188).

♦ PF6470 Poor communications networks in rural areas
Claim Impeding the flow of services, goods and people impedes development in the expanded context within which people presently operate. Both goods and people must move freely and rapidly if local socio-economic development is to move beyond the stage of a good idea. This flow must occur among villages, and between the village and the province, the village and the nation, and the village and the world. Meanwhile, it is easier for many to travel to the national capital than it is to journey to the opposite side of the district. When local people discuss with concern the width and surface of the roads, they are pointing to something far beyond what they are able to articulate. Such matters are vitally related to market utility, personal relations and emergency health care. Rural development will not occur if the time and energy drain, the sense of isolation – of being cut off or left out – and the anxiety over emergencies that the modern world creates are not dealt with directly and soon.
Broader Insufficient communications systems (#PF2350)
Insufficient transportation infrastructure (#PF1495).

Narrower Irregular transport services (#PE5345).
Related Inadequate systems of transport and communications in rural villages (#PF6496).
Aggravates Undemocratic organizations (#PC8676).
Aggravated by Unproductive use of resources (#PB8376)
Inadequate cargo transportation (#PE0430).

♦ PF6471 Undeveloped channels for commercial initiative

Nature Small communities adjacent to large urban centres experience both the benefits of modern services and the pain of rapid social change. New means of self-support and the extension of family responsibilities result in the acquisition of urban skills and technical know-how. Small industries develop and residents accumulate some capital goods; but lack of corporate planning prevents their being used for the overall development of the village. Since the farmers eat what they grow, stocks for local stores are purchased outside the village and often at general retail prices. This forces prices up and keeps margins low. In addition, promotion and marketing is necessary to bring in additional businesses or to expand the existing ones. The financial and economic planning required to participate in more aspects of the larger urban economy is unrecognized despite the desire of local people to take part in such expansion.
Broader Inequality of opportunity (#PC3435).
Narrower Conflicting land rights (#PG9509)
Limited retail inventory (#PG5419)
Sporadic community growth (#PG5481)
Unsurveyed consumer needs (#PG5484)
Unanalysed available capital (#PG5428)
Unutilized wholesale systems (#PG5437)
Lack of industrial promotion (#PG5460)
Unorganized labour potential (#PG5466)
Unresolved development pattern (#PG5476)
Misunderstanding of official communications (#PF8382)
Unrealized potential of commercial enterprises (#PF2231).
Related Instability of prices (#PF8635)
Unorganized community recreation (#PF5409).
Aggravates Unexplored potential markets (#PF0581)
Social disaffection of the young (#PD1544)
Restrictive use of available land (#PF6528)
Subsistence agricultural income level in rural communities (#PE8171).
Aggravated by Inflexible educational system (#PJ5365)
Ignorance of health and hygiene (#PD8023).

♦ PF6472 Inadequate practical training in rural areas

Nature Rural residents have poor access to the acquisition of technological expertise. Old methods are dying out as resources became more scarce, but training for new means of livelihood tends to be the responsibility of employers in urban areas. If on-the-job industrial training is unavailable, openings are very limited and individuals lack the requirements for entry into an appropriate training school. In many families, both parents are illiterate. Unless there is enough money to send children to outside schools or to a place of employment, the children find that their advancement is blocked. Those fortunate ones who leave the area to be trained often do not return to the village to share their acquired skills. The absence of training in or use of technical skills in a village promotes distrust of technological services except in the form of consumer products. This technological handicap is manifest in other ways: land records go unresearched; commercial and agricultural services available in the area are basically unused; soil testing and agricultural waste disposal advice are unsolicited; and nearby veterinary services are largely ignored. Modern human health services are used on a minimal basis: dental care has a low priority and illness is often treated only when curative care is imperative.
Narrower Decreased skill transference (#PG5492).
Related Rural doctor turnover (#PS5467).
Aggravates Low dental priority (#PG5472)
Unused livestock waste (#PG5411)
Incompetent management (#PC4867)
Unresearched land records (#PG5486)
Untreated chicken diseases (#PG5454)
Inhibiting adult preliteracy (#PG5375)
Unpractised dental prevention (#PG5477)
Unattractive fishing business (#PG5399)
Unrealized agronomic potential (#PJ1775)
Misunderstanding of veterinary approach (#PG5402).
Aggravated by Lack of training (#PD8388)
Limited finance training (#PG5384)
Insufficient trained labour (#PD9113)
Inadequate vocational education (#PF0422)
Inadequate educational facilities (#PD0847)
Underemployment of skilled workers (#PJ5489)
Unperceived relevance of formal education in rural communities (#PF1944).

♦ PF6473 Underdeveloped provision of basic services in developing countries

Nature Basic services in rural areas are unevenly distributed and often not available to meet the growing needs of developing villages.
Broader Limited distribution of basic services (#PF2652).
Narrower Pond pollution (#PE5385)
Poorly lighted roadways (#PJ9224)
Inadequate industrial access (#PG5412)
Undependable garbage pick-up (#PG5430)
Shortage of industrial water (#PE5464)
Inadequate industrial services (#PG9900)
Insufficient garbage containers (#PG5439)
Crowded rush-hour transportation (#PG5421)
Prohibitive cost of basic services (#PF6527)
Insufficient communications systems (#PF2350)
Inadequate water system infrastructure (#PD8517)
Inadequate industrial services in developing countries (#PF4195).
Related Domestic refuse disposal (#PD0807)
Inadequate road maintenance (#PD8557)
Underprovision of basic services to rural areas (#PF2875).
Aggravates Unhygienic conditions (#PF8515)
Inadequate dental care (#PJ5478)
Inadequate land drainage (#PD2269)
Overcrowded public clinics (#PG5393)
Instability of water supply (#PD0722)
Inadequate housing for the aged (#PD0276)
Unorganized community recreation (#PF5409)
Inappropriate sanitation systems (#PD0876).
Aggravated by Prohibitive medical expenses (#PE8261)
Shortage of fresh-water sources (#PC4815).

♦ PF6475 Unplanned land development

Unplanned land use — Unregulated land development — Unorganized land scheme — Pessimistic approach to agricultural development
Nature Although the only real obstacle to agricultural development may be an overall plan, rural residents tend to be discouraged by the slowness with which agreement can be reached and the difficulties in dealing with the many small plots with many different owners and the variety of ways in which the land is being used. Old structures in disrepair are a constant reminder of economic paralysis, and leave a wide-spread impression that space for development is unavailable.
Broader Inadequate care of community space (#PF2346)
Self-defeating style of community planning (#PF6456).
Narrower Restricted land choices (#PG5457)
Overdivided farmland plots (#PG5387)
Undesignated parking space (#PG5432)
Diminishing land ownership (#PG5475)
Untested industrial expansion (#PG5414).
Related Limited public land (#PJ0574).
Aggravates Difficult land acquisition (#PJ5369)
Restrictive use of available land (#PF6528)
Destruction inherent in development (#PF4829)
Limited reservoir of technical skills in rural communities (#PF2848).

♦ PF6476 Constricting effect of educational structures

Nature Despite a growing tendency to relate education to life in a flexible way in order to train people to meet changing needs, educational structures unrelated to current employment and actual social needs leave human resources largely untapped in many communities. Curricula in village schools may be too inflexible to train children for urban life. Time, space and staff are limited, often requiring several age groups to be taught simultaneously in the same classroom, and teachers feel there is too little time to teach effectively. Securing volunteer teachers for new programmes is a major hurdle. Individual preparation in school is not possible under the current setup. Village adults are excluded from the whole process by their preliteracy, and parents who have not themselves benefitted from schooling are unable to supplement their children's schooling at home.
Broader Ineffective systems of practical education (#PF3498).
Narrower Ineffective class scheduling (#PG5424)
Ineffective methods of practical education (#PF2721)
Inappropriate selection and examination procedures in education (#PF1266).
Aggravates Inadequate career advice (#PJ8018)
Unsupplemented school curriculum (#PG5370).
Aggravated by Ineffective space usage (#PE5458)
Shortage of school equipment (#PJ0821)
Inflexible educational system (#PJ5365).

♦ PF6477 Diminishing capital investment in small communities

Inadequate local capital — Scarcity of start-up funds — Limited availability of local business capital — Limited local investment resources — Limited private investment — Unavailability of local capital — Unproductive local investments
Nature Local communities are increasingly called upon to sustain their own populace through forms of cooperative enterprises. This would appear to be possible in some near-urban villages, where the fact that untrained workers are poorly paid has resulted in many of the men and some entire families leaving the community on a more-or-less permanent basis to seek employment elsewhere. When they contribute from abroad to family support, the average family income in a village is raised and individual families have a certain amount of capital. However, because these families do not see investing in the village as economically beneficial, the few people who engage in serious enterprises operate in the city. Even if there is interest in expanding businesses, buying farm land or creating industries, the cost of the initial investment is so great and land acquisition is so difficult that few follow through and proposed investment plans are rarely materialized. When communities are unable to find appropriate channels of financial support, self-sufficient economies are not realized and subsistence-oriented living continues to undercut development. Unavailability of local and outside funding in rural communities results in a shortage of capital for farming or business improvement. Limited family income hampers attempts at raising money for community projects – minimal family income is not put into savings since banking services are remote and benefits unclear, and the concept of forming a local cooperative union for the purpose of saving is unfamiliar. Also, a past history of financial mismanagement may breed an air of mistrust in local attempts to pool resources. Limited local assets prevent the securing of loans needed for expanded farming, starting businesses and constructing schools and other buildings. Capital is urgently needed in many rural communities to increase production at the local level. This is not a matter of the people needing more money. It has to do with setting a new economic dynamic in order to raise living standards. Poor market prices; shrewd, if not dishonest practices of middle men; exorbitant interest rates; the loss of land to outside buyers; lack of knowledge of commercial affairs – all these take their toll, so that the small amount of capital that is ever accumulated simply drains away. The understandable reluctance of legitimate outside institutions to invest in village effort adds to the self-defeating cycle. Local people do not understand why the situation is paralyzed but they are profoundly aware of the paralysis.
Broader Scarcity (#PA5984)
Insufficient capital investment (#PF2852)
Undirected expansion of economic base (#PF0905)
Subsistence approach to capital resources (#PF6530).
Related Insufficient financial resources (#PB4653).
Aggravates Dependency on middlemen (#PD4632)
Small business failures (#PE9405)
Inadequate support services (#PF6492)
Dependence on breast feeding (#PE7627)
Attractiveness of overseas work (#PG5388)
Incomplete access to development capital (#PF6517)
Discouraging conditions for small business (#PD5603).
Aggravated by Unused local capital (#PG5434)
Seasonal unemployment (#PC1108)
Restrictive farm capital (#PJ0788)
Prohibitive cost of land (#PE4162)
Lack of savings structures (#PF1348)
Inadequate credit policies (#PF0245)
Inadequate use of manpower (#PG0456)
Export earnings instability (#PE4915)
Unavailability of land funds (#PG5407)
Unprofitable home industries (#PG5415)
Inadequate financial services (#PJ8366)
Usury in developing countries (#PE2524)
Unattractive fishing business (#PG5399)
Prohibitive construction costs (#PJ8591)
Limited accumulation of capital (#PF3630)
Incompetent financial management (#PF0760)
Rural poverty in developing countries (#PD4125)
Limited availability of financial credit (#PF2489)
Dependency on unpredictable sources of income (#PF3084)
Prohibitive cost of connection to public utilities (#PJ9426)
Inadequate domestic savings in developing countries (#PD0465)
Ineffective economic structures in industrial nations (#PE4818)
Limited availability of loans in developing countries (#PE4704)
Limited availability of investment capital for urban renewal (#PF3550).

♦ PF6478 Slow rate of income expansion

Nature The major issues affecting agricultural and community development in most small villages are sources of initiating capital and land space. (1) Capital: Although the present-day tendency for rural people to be employed outside the village means they bring in more money than previously, since these jobs pay more than fishing or farming, the wages are still comparatively low because of the rise in the cost of living; and few villagers have the technical skills needed for job improvement and advancement. Most money, whether it comes from jobs or farming, is spent on goods and services outside the village. (2) Land space: Current ownership patterns preclude the location of more job opportunities at village level and limit space available for recreational facilities.
Broader Underdeveloped sources of income expansion (#PF1345).
Narrower Understocked shops (#PG5452)
Unused recreation spaces (#PJ8596)
Lack of industrial promotion (#PG5460)
Unattractive fishing business (#PG5399)
Unpromoted community business (#PJ1409)
Insufficient factory facilities (#PG5390).
Aggravated by Low general income (#PD8568)
Underpayment for work (#PD8916)
Limited employment options (#PF1658)
Inadequate irrigation system (#PD8839)
Socio-economically inactive rural population (#PF4470).

♦ PF6479 Lack of local leadership role models

Untapped leadership potential — Underdeveloped community leadership — Unencouraged local leadership — Absence of a local leadership image
Nature Many villages rely heavily on the heads of families for their community leadership. However, the adult leadership force has been considerably weakened in recent years due to the failure of traditional local institutions, since the men of the village have to leave and seek employment elsewhere. This also results in the young people of the community being deprived of models for vocation and responsibility. Village residents are reluctant to assume the leadership roles necessary to initiate new developments: no clear-cut community decisions relative to the types of development are forthcoming – the traditional roles of women and youth inhibit their assuming these leadership roles, and they feel there is no time for community service after working all day just to support the family.

Broader Obstacles to leadership (#PF7011) Insufficient role models (#PF8451)
Obstacles to community achievement (#PF7118).
Narrower Unavailability of library facilities (#PG9116).
Related Inadequate procedures for community planning (#PF0963)
Neglect of the role of women in rural development (#PF4959).
Aggravates Labour shortage (#PC0592) Lack of leadership (#PF1254)
Unorganized community recreation (#PF5409) Deteriorating community identity (#PF2241).
Aggravated by Oppressive prevalent images (#PF1365)
Insufficient communications systems (#PF2350)
Unrealized use of education structures (#PF2568)
Ineffective structures for community participation (#PF2437).

♦ PF6483 Uncatalogued historical documents
Uncatalogued ancient manuscripts
Broader Uncatalogued documents (#PF4077).
Aggravates Inaccessible historical libraries (#PF9046).

♦ PF6492 Inadequate support services
Absence of support systems for effective community operation
Nature Isolation and remoteness are repeatedly given as the reason for the failure of the private sector to respond to requirements of rural communities. The absence of effective support systems linking small villages to the network of resources, services, and expertise available throughout a country means that funding sources are largely untapped, credit lines undeveloped, equipment acquisition difficult and government liaison inadequate. This not only prevents development but produces a debilitating sense of isolation and insignificance. Conversely, this same remoteness affects the rest of the nation's response to the village.
Broader Obstacles to community achievement (#PF7118)
Undirected expansion of economic base (#PF0905).
Narrower Inefficient support for widowers (#PJ7997)
Lack of interim support for families (#PE7944)
Inadequate system of child support enforcement (#PF6076).
Related Inadequate rehabilitation facilities for the mentally handicapped (#PE8151).
Aggravated by Inadequate cash flow (#PJ0576) Declining public image (#PG8266)
Unclear official directives (#PG0912) Inadequate budgeting practices (#PJ0978)
Inaccessible government agencies (#PF5351)
Disorganized liaison with formal support (#PF2947)
Limited availability of financial credit (#PF2489)
Short range planning for long-term development (#PF5660)
Diminishing capital investment in small communities (#PF6477)
Inadequate medical laboratory facilities in developing countries (#PG0932).

♦ PF6493 Underdeveloped approaches to local food production
Nature In many rural communities it is thought that development of new enterprises can be done only within certain technological means, levels and profit margins, and little has been achieved. This serves to perpetuate the existing lack of experience with technological production of people in rural villages. The refusal to adapt technology to differentiate between established and developing economies has fostered the tragic inequities of food production.
Broader Economic underdevelopment (#PC0281)
Underdevelopment of food and live animal production (#PF2821).
Narrower Irrational land divisions (#PG9807) Narrow range of food crops (#PD4100)
Limited gardening experience (#PG8928).
Aggravates Unexplored potential markets (#PF0581)
Distortion of international trade by selective domestic subsidies (#PD0678).
Aggravated by Inadequate irrigation system (#PD8839).

♦ PF6494 Unclarified procedures for transposing ancient traditions
Nature Because tribal people did not participate in the shaping of the technological society, they are being forced back to their ancient culture whose traditional forms are not applicable in present-day conditions. Their only source of uniqueness is the remembrance of the past. Transposing their deep wisdom, social forms and symbol system into meaningful and operable forms is closely connected with the challenge of self-sufficiency in the economic area.
Narrower Insufficient community events (#PF5250)
Disrelationship from community history (#PJ0836).
Related Divisive patterns of community groupings (#PF6545).
Aggravates Ignorance of cultural heritage (#PF1985)
Elimination of traditional skills (#PD8872)
Untransposed significance of cultural tradition (#PF1373).
Aggravated by Limited availability of learning opportunities (#PF3184).

♦ PF6495 Ineffective means for goods supply and distribution
Unfeasible goods transport
Nature Traditional systems in tribal villages to not allow for the stockpiling and distribution of goods. Supplies brought in individually stretch supply lines to unreliable lengths and increase the transportation costs to the individual buyer. Evidently there can only be a narrow range and selection of goods brought in this way, so the market is small and expansion limited. Goods are even more expensive when there are high interest and short term payment arrangements. Cash flow constantly away from the village and an inadequate supply system, results in villages paying more for less with irregular supply and unreliable service.
Broader Inadequate infrastructure (#PC7693).
Narrower Unsurveyed consumer needs (#PG5484)
Decrease in consumer choice (#PD6075)
Underdeveloped export potential (#PJ1586)
Deterioration in product quality (#PD1435)
Incompetent financial management (#PF0760).
Related Deteriorating structures of rural community cooperation (#PF3558).
Aggravates Limited purchasing power (#PD8362).
Aggravated by Prohibitive cost of transportation (#PE8063)
Inadequate transportation facilities for rural communities in developing countries (#PE6526).

♦ PF6496 Inadequate systems of transport and communications in rural villages
Nature Telephone communication and reliable and rapid transportation is essential for the conduct of any serious commercial business. The inadequate communications and transportation systems in many rural villages makes local development efforts insurmountable, and any significant move toward self-sufficiency is hard under these circumstances. In human terms, this creates and reinforces the villager's victimizing self-image of unimportance.
Broader Rural underdevelopment (#PC0306)
Insufficient transportation infrastructure (#PF1495).
Related Poor communications networks in rural areas (#PF6470).
Aggravated by Inadequate road conditions (#PJ0860)
Ineffective rural transport (#PF2996)
Inadequate vehicle servicing (#PJ0909)
Frequent schedule disruptions (#PJ0871)
Unfair surcharges in ocean freight (#PF5922)
Insufficient communications systems (#PF2350).

♦ PF6497 Educational gap between generations
Claim The gap between developed and undeveloped peoples around the world is no longer a question of natural resources or wealth, but of methods for planning and acting. In many societies this has historically been the function of the adults and elders in a community. Therefore the education of adults cannot be valued too highly. Although young people may develop faster, due to the focus of educational programmes for their age group, training only the next generation neglects development of the total community and therefore delays significant social change.
Broader Generation communication gap (#PF0756).
Aggravates Functional illiteracy (#PD8723) Obsolete educational values (#PF8161)
Ignorance of health and hygiene (#PD8023).
Aggravated by Lagging training in social skills (#PF8085).

♦ PF6499 Inadequate agricultural facilities
Nature In rural farming communities without facilities to store crops or livestock, valuable commodities are lost spoiled, or delayed from reaching the market. Without systematic operational procedures within agribusiness, farmers have no protection from changeable natural conditions. The result is continual failure to create self-sufficient or surplus foodstuff, which precludes the necessary confidence to embark on new economic ventures.
Broader Inadequate infrastructure (#PC7693).
Narrower Obstacles to aquaculture (#PF5496).
Aggravates Stagnated development of agricultural production (#PD1285).

♦ PF6500 Shortage of technical skills
Shortage of industrial skills — Low-level technical skills — Insufficient technical skills
Nature Development is dependent on ready access to a multiplicity of skills and experience. Remote communities often pay dearly for these skills but in-service training is limited to rare occasions when trained people visit for specific tasks; they do not have time to train potential apprentices. Not only is there little training available in rural communities, the managerial experience indirectly gained from coordinating appropriate skills is almost negligible. The rarity and difficulty of obtaining skills produces a failure mentality that discourages actively seeking jobs.
Broader Labour shortage (#PC0592) Insufficient skills (#PC6445)
Narrow range of practical skills (#PF2477).
Narrower Lack of expertise in agricultural techniques (#PE8752).
Related Incompetent management (#PC4867) Lagging training in social skills (#PF8085).
Aggravates Minimal access to appropriate technology (#PF3554).
Aggravated by Lack of opportunities for practical training in communities (#PF2837).

♦ PF6501 Ineffective organization of community action
Blocked community action — Ineffective community organization
Nature Newly formed communities in both rural and urban areas are often characterized by weak, undeveloped social structures. Malfunctioning leadership and councils makes it difficult to determine tasks, assign forces, and encourage successful performance; and infrequent planning sessions leave the development of the community to the subjective fancies of the moment or to the outside control of other agencies.
Broader Obstacles to community achievement (#PF7118)
Disjointed patterns of community identity (#PF2845).
Narrower Arbitrary enforcement of regulations (#PD8697).
Related Declining sense of community (#PF2575).

♦ PF6502 Delays in community building programmes
Nature Complicated delays in construction may lead, in some communities, to houses lying partially completed, inadequate transport facilities, undesignated storage space, and vague village plans. As long as construction remains unfinished, the people are not tied down to building their future livelihood and a sense of ownership and pride in the community is prevented.
Broader Delay (#PA1999) Obstacles to community achievement (#PF7118).
Narrower Unconcretized village plan (#PS0866).

♦ PF6503 Inaccessible recreation areas
Broader Physically inaccessible resources (#PC4020).
Narrower Inaccessible wilderness areas (#PF9360)
Inaccessibility of water for recreation (#PF6138)
Inaccessibility of countryside to city dwellers (#PF6140)
Inaccessibility of city centres to suburban residents (#PF6132)
Inaccessibility of green parkland in an urban environment (#PE6175).
Related Inaccessibility of quiet zones in an urban environment (#PF6160).
Aggravated by Closure of recreation areas (#PE6276).

♦ PF6504 Fragmented patterns of community activity
Nature The many community-wide concerns in small towns result in residents feeling at cross purposes and unable to take effective action. In a struggle to keep going at all local organizations reduce their scope to limited traditional activities, and most of these are for members only. The result is conflicting meetings, multi-memberships and competing events, further fracturing community life. New families see no way to be involved, so personal commitment to family, work and recreation become their primary focus. The result is a steadily shrinking base of involvement in the care for the community, and the remaining leadership core is overloaded.
Broader Social fragmentation (#PF1324).
Narrower Exhausting commuter travel (#PE1538)
Unfounded conflicting stories (#PG1506)
Consuming personal priorities (#PJ1535)
Lack of sharing of community skills (#PF3393).
Aggravates Lack of leadership (#PF1254) Limited youth activities (#PJ0106)
Social isolation of the elderly (#PD1564) Unorganized community recreation (#PF5409).
Aggravated by Insufficient community events (#PF5250)
Minimal leadership coordination (#PJ9498).

♦ PF6505 Unexercised responsibility for external relations
Nature Although consolidation of services at a regional level provides more effective delivery to all communities, it does require that communities open channels of access for receiving available resources and services. The acquisition of funding for community facilities and other projects in many places has shown the willingness of agencies to support local initiative, but a history of self-reliance may lead to the feeling that receiving outside assistance is akin to taking charity. Residents are hampered from developing these external relationships by lack of community support, thus impeding the investigation of potential help and the acquisition of the necessary skills needed to create successful local programmes.
Broader Unrecognized socio-economic interdependencies (#PF2969).
Narrower Complex funding mechanisms (#PG1502)
Unequal access to education (#PC2163)
Unfamiliar bureaucratic procedures (#PJ9912)
Unreasonable licensing restrictions (#PE7655).
Related Limited meeting facilities (#PE1535) Parochial national interests (#PF2600).
Aggravates Unfamiliar government procedures (#PJ0740).
Aggravated by Divisive school districts (#PU1528).

♦ **PF6506 Ineffective structures of local consensus**
Lack of systems for achieving grassroots consensus — Unclear village consensus — Difficult village consensus — Unformed community consensus
Nature Although personal participation in decisions which affect a person's future is a present-day requirement, many small-town residents see that the vitally important structures for building local, national or global consensus are inadequate. Lack of systems results in a feeling of being victimized by larger systems or issues. Many people feel unrelated to global issues because they see no way in which they can actually affect the outcome. The great majority of people in democratic countries feel that though they have the vote, this makes no difference to the everyday realities; so many do not make the effort to cast their vote. Communication between town leaders and the ordinary citizen is difficult, as there are no formal means for including regular participation in planning. Some decisions, such as those affecting water, sewage and street repairs, seem to be made by a few citizens, without the benefit of additional input from the community as a whole. Others, especially those which affect the long-range design of municipal services, do not really get made at all, and residents have no way to determine objectively the merits and disadvantages of a plan. There is a widespread sense that individuals do not have enough power to occasion change and that local public officials do not have a clear sense of the will of the whole community. Misunderstandings and anxiety about decisions being made results in the circulation of conflicting stories, frustrates residents, and inhibits constructive input, all of which makes it difficult to build a cohesive image of the extended community.
Claim Until effective structures are built to create and determine the local consensus, serious development will be unable to move beyond inertia to creative action.
Broader Unstructured local decision-making (#PF6550)
Obstacles to community achievement (#PF7118)
Ineffective structures for community decision-making (#PF1781).
Narrower Citizen apathy (#PF2421)
Social insecurity (#PC1867)
Limited public space (#PJ9066)
Undiversified tax base (#PJ7897)
Unknown building codes (#PJ8820)
Fragmented community goals (#PG1521)
Undemocratic policy-making (#PF8703)
Lack of political development (#PD8673)
Uneven concentrations of power (#PG8703)
Insufficient agency cooperation (#PG8763)
Conflicting community priorities (#PJ7975)
Uncommunicated annexation authority (#PG8810)
Lack of means for achieving consensus (#PD2438)
Divisive patterns of community groupings (#PF6545).
Related Untransposed community structures (#PF6450).
Aggravates Citizen powerlessness (#PJ8803)
Lack of community participation (#PF3307)
Mismanagement of irrigation schemes (#PE8233).
Aggravated by Insufficient community events (#PF5250)
Short range planning for long-term development (#PF5660)
Demoralizing images of rural community identity (#PF2358).

♦ **PF6508 Failure of public authorities to assist in financial investment**
Nature Small communities need long-term capital, and it is crucial to keep present businesses healthy and attract new ones; yet although some business and community groupings are experimenting with creative methods of ensuring their access to necessary capital (such as through credit unions, savings and loan associations, and local development companies), in general there tends to be no community vehicle for evaluating the viability of future investments, for planning comprehensive expansion of the economy, or for assisting in long-range planning. This is particularly true of communities based on agriculture, where it is evident that long-range planning is experienced as frustrating and difficult: unpredictable market prices and weather dependence make long-term loans difficult to obtain, discouraging many farmers from projecting their long-range capital needs. The success or failure of business ventures is seen as the responsibility of the owners, without financial or technical assistance from the community, in spite of the fact that such ventures affect the community dramatically. Complaints include the absence of estate planning to effectively transfer agricultural and business assets to succeeding generations and thus ensure the continuity of the local economy. Traditional lending institutions are reluctant to invest in small towns, and although long-term financing may be available for agriculture, business and housing from a variety of public sources, these tend not to be broadly explored or utilized.
Narrower Narrow economic foresight (#PF8602).
Related Instability of prices (#PF8635)
Lack of local commercial services (#PF2009)
Limited local availability of capital reserve (#PF2378)
Limited availability of investment capital for urban renewal (#PF3550).

♦ **PF6510 Lack of support for local commercial services**
Nature Trends in small towns illustrate that community support of local businesses is the key to retaining the business district as an economically viable service centre. Community spending power is a valuable local resource. This is seen when the rapid growth of commercial centres surrounding small towns brings with it a decline in commercial services available in the small towns themselves. The declining customer base makes competitive options less and less available to local businesses. Many community residents are probably related to neighbouring towns through school, work or entertainment, and often do their shopping while there. Although local commercial establishments may attempt to provide basic necessities, lack of space and the necessity to buy in small quantities limit variety in all areas; finding that all their needs cannot be met locally many families go to other areas and do all their shopping in one weekly trip, including those items which would be available locally. Residents will express dissatisfaction at the lack of variety, quality and competitive prices for some items, not realizing that their own out-of-town shopping is at the root of the problem.
Broader Lack of cooperation (#PF2816).
Narrower Strong external draw (#PS1412)
Missing country support (#PU1424)
Limited service selection (#PU1448)
Unexplored supplier options (#PS1492)
Unattractive commercial base (#PU1460)
Undeveloped business support (#PS1400)
Unanswered need for physician (#PU1470).
Aggravated by Incompetent management (#PC4867)
Inaccessible market and supply centres (#PF8299).

♦ **PF6511 Insufficient rural housing**
Inadequate residential housing in rural communities — Inadequate rural housing in developing countries
Nature Although the trend of outward migration from the rural to the urban has reversed in some countries, bringing with it a demand for additional housing in rural communities, in many such communities deteriorating houses are not being renovated. Owners may cite the high cost of adequate rebuilding or the risk of careless tenants. Farmers may sell to neighbours when they retire and their former homes be converted to storage or left to deteriorate; they are then costly to renovate and tax advantages often encourage nothing being done. New housing may be unsuccessful: older people, in particular, may be reluctant to risk the new life style that goes with more efficient, low-cost homes, fearing loss of familiar places and isolation from community life.
In the rural areas of developing countries environmental living conditions are usually primitive, little or no economic and social improvement is achieved, and the result is a declining cycle of poor health, stagnation and wasted human energies. This situation, together with the paucity of community facilities and other amenities in such areas, tends to accelerate the movement of rural migrants to urban centres. Although the problems of urban housing are receiving increased attention in developing countries faced with rapid urbanization trends, relatively little has been done for the improvement of rural housing and community facilities which would assist such countries in achieving a balanced urban-rural development.
Broader Housing shortage (#PD8778)
Inadequate housing (#PC0449)
Inadequate housing in developing countries (#PE0269).
Narrower Lack of housing for teachers in rural communities (#PE8103).
Aggravates Demoralizing images of rural community identity (#PF2358).
Aggravated by Marginal living space (#PJ8036)
Risky rental agreements (#PG1425)
Unavailability of building sites (#PJ8549)
Prohibitive cost of accommodation (#PD1842)
Prohibitive cost of home maintenance (#PJ9022)
Lack of sanitation in rural areas of developing countries (#PD1225).

♦ **PF6512 Unexplored opportunities for community education**
Uncomprehensive community education — Partial community education — Neglected community education — Unresearched educational choices
Nature Many small towns find the complexity of local, regional or national regulations and a lack of facilities deters them from making a coordinated, community-wide assessment of educational needs or interests, despite numbers of residents apparently being interested in adult, youth or preschool programmes, if provided. Often, even when a group of residents requests a specific course, it has to be held elsewhere because adequate facilities are lacking locally. The family unit seems to be left with the responsibility of obtaining continuing education, determining adult curricula, and providing education needs on a year-to-year basis, while the value of an imaginative pre-school programme is unperceived by many, so no attempt is made to coordinate or evaluate such a programme.
Broader Obstacles to community achievement (#PF7118)
Unrealized use of education structures (#PF2568)
Lack of opportunities for practical training in communities (#PF2837).
Aggravates Unexplored alternatives for commercial development (#PF6548).
Aggravated by Inadequate career advice (#PJ8018)
Defeating size requirements (#PE1437)
Inadequate preschool education (#PJ1962)
Unperceived educational opportunities (#PJ9762)
Unperceived relevance of formal education in rural communities (#PF1944).

♦ **PF6513 Underutilization of potential in local communities**
Nature The useful and creative human potential available in many small communities tends to remain untapped because there is no vehicle for getting the skills to where they are needed. Buildings deteriorate even though local skills could easily refurbish them. Older residents' handicraft skills are basically overlooked as contributions to the cultural life of the community, as there are few situations where they can be displayed or utilized, although the skills of many are hobby-oriented and self-fulfilling. Many older skilled tradesmen and businessmen have no apprentice positions to enable them to pass on wisdom to the next generation, while at the same time young people are often underemployed and do not see possibilities for contributing to the community's future. There is a need for creative effort to retrieve almost-lost skills and to promote cooperative skill pools to expand local service capabilities.
Broader Underutilization of human resources (#PF3523).
Narrower Lack of sharing of community skills (#PF3393)
Underutilization of locally available skills (#PF6538).
Related Lack of trained firefighters (#PJ0277).
Aggravates Incompetent management (#PC4867).
Aggravated by Obsolete methods of agricultural production (#PF1822).

♦ **PF6514 Unformed style of cooperative action**
Neglected cooperative action
Nature Small communities are increasingly faced with the task of shifting their operating base from that of the individual unit to one of the community as a whole. However, the existing decision-making process in many communities consumes valuable time as each suggestion progresses through various channels, losing its initial impetus as it moves toward final implementation. Progress seems often to be stifled by argumentative community meetings and there is little formal coordination of village events. Residents experience frustration at the relative lack of community services and fragmented village communication.
Narrower Uncontrolled dogs (#PG0273)
Policy-making delays (#PF8989)
Domestic refuse disposal (#PD0807)
Lack of enforcement power (#PF0223)
Prohibitive cost of basic services (#PF6527).
Related Polarization of local conflicts (#PF1333).
Aggravates Inadequate local enforcement (#PF0336)
Social disaffection of the young (#PD1544).
Aggravated by Narrow range of practical skills (#PF2477).

♦ **PF6516 Ambiguous shape of social identity**
Broader Ambiguity (#PF4193).
Aggravates Alcohol abuse (#PD0153).
Aggravated by Inadequate credit policies (#PF0245)
Inadequate teacher training (#PJ1327)
Dependence on social welfare (#PD1229)
Consuming personal priorities (#PJ1535)
Wasteful sportsmen's practices (#PG0215)
Unavailability of trainee employment (#PJ0336)
Declining community confidence in its ability to change (#PF9066).

♦ **PF6517 Incomplete access to development capital**
Nature Access to the development capital needed for village maintenance, public works and the development of local services and businesses can best be described as "incomplete", for only partial ways of releasing available funding for villages have been found. Unsettled suits with agencies responsible for the construction of many village homes delays necessary home repairs. Both limited private funding and the high cost of materials discourage residents from undertaking home maintenance. The absence of a salary base and the small commercial property bases limit revenue for taxes. Research has not yet been done to find appropriate funding sources for new industries, services and wages for potential employees. More importantly, modern business development has barely begun in most villages. As a consequence, villagers find it difficult to imagine their village generating its needed capital from within.
Broader Insufficient financial resources (#PB4653).
Related Instability of prices (#PF8635).
Aggravated by Unpaid wages (#PD5335)
Low athletics priority (#PJ0337)
Undiversified tax base (#PJ7897)
Lack of minimum wage fixing (#PE6726)
Prohibitive construction costs (#PJ8591)
Prohibitive cost of maintenance (#PF0296)
Diminishing capital investment in small communities (#PF6477).

♦ **PF6519 Adultism**
Nature Children and youth are systematically oppressed by older people by enforcement of dependency for resources, knowhow and access to decisions that affect them. Since everyone has experienced this exploitation, it is promoted from generation to generation.
Broader Oppression (#PB8656).

♦ PF6521 Geographically undefined community limits
Unidentified community space
Nature Many small villages are willing to coordinate their communal operations but find themselves hampered by ill-defined geographic limits. They feel they must define the limits of the space in which they operate in order to focus their efforts effectively, to provide common space for community operations and to make a general comprehensive land use plan.
Broader Obstacles to community achievement (#PF7118)
Disjointed patterns of community identity (#PF2845)
Obstacles to availability of community space (#PF7130).
Aggravated by Low athletics priority (#PJ0337) Inadequate road conditions (#PJ0860)
Increase in abandoned buildings (#PJ1665) Insufficient business facilities (#PG0110)
Restrictive use of available land (#PF6528).

♦ PF6524 Underproductive methods of agricultural management
Low-yield farm methods — Limited crop management
Nature Despite the availability of scientific methods in agriculture and animal husbandry, present methods of farming tend to yield a much smaller percentage than is potentially feasible with proper crop rotation, more plant varieties, plant nutrients, fertilisers and proper pest control. In many villages in the developing nations only a little over 50 percent of the arable land is used for agricultural production, which leaves much land uncultivated that could possibly produce profitable crops. Because of the low-yield farming methods, much of the village employment is cyclical rather than continual. Sometimes residents are blocked by outside ownership which controls their markets and brings minimal economic returns to the farmer. Land cultivation is often done with animal powered-single furrow ploughs; the farmers depend on regular rains to supply water where there is no irrigation system. Chickens, pigs, goats, cows, geese, ducks and rabbits live in family yards and are raised for home consumption only – these animals freely forage for food throughout a village since there is usually no common grazing land. Without proper upbreeding of present stock and a larger variety of animals, profitable animal husbandry is not possible. Continual reliance on present vegetable, fruit and animal farming methods can only leave small farmers with a continuing sense of futility and of being trapped in supplying day-to-day needs.
Broader Incompetent management (#PC4867) Narrow range of practical skills (#PF2477)
Underdevelopment of agricultural and livestock production (#PD0629).
Narrower Soil infertility (#PD0077) Part-time farm employment (#PJ1074)
Insufficient pest control (#PJ1086) Uncontrolled plant diseases (#PJ1016)
Insufficient modern technology (#PJ0996) Prohibitive livestock husbandry costs (#PJ4075)
Underdeveloped capacity for income farming (#PF1240)
Prohibitive cost of necessities in rural communities (#PF2385).
Related Instability of prices (#PF8635)
Obsolete methods of agricultural production (#PF1822).
Aggravates Over-intensive soil exploitation (#PC0052)
Mismanagement of irrigation schemes (#PE8233)
Stagnated development of agricultural production (#PD1285).
Aggravated by Inadequate irrigation system (#PD8839).

♦ PF6525 Fragmented pattern of community organization
Nature Despite evident present-day needs for local people to participate in creating their own future, in many Third World villages there is no effective social organization arising from the community which allows for a common voice. Technical planning skills and community organization are required to allow all residents to participate in important community issues on land use, schooling, community relations with government, and industrial and commercial development requiring complex decisions and strategic action. However, when residents try to expand the present educational facilities, their plans are generally untargeted. As liaison with the local government structures increases, communities find themselves undecided about requests and the use of crucially available resources. The problem is compounded by the fact that families tend to live in residences scattered over hills and valleys, so that there is obviously more concern focused on isolated family affairs than on effective ways of communication with other village residents.
Broader Obstacles to community achievement (#PF7118).
Narrower Unpromoted industrial development (#PJ1017)
Unrecognized relevance of education (#PF9068)
Individualistic agricultural planning (#PJ1082).
Aggravates Limited youth activities (#PJ0106) Underutilized veterinary aid (#PJ1084)
Underutilized government resources (#PE9325)
Unorganized development of work forces (#PF2128).
Aggravated by Minimal leadership coordination (#PJ9498)
Non-articulated educational goals (#PF2595).

♦ PF6527 Prohibitive cost of basic services
Prohibitive cost of essential services in rural communities — Costly distribution services
Incidence Heavily subsidized public providers often produce urban services inefficiently. They have little incentive to be cost-effective or to respond speedily to changing conditions. Heavy subsidies in urban infrastructure often fail to result in services accessible to the poor. For example the poorest members of urban society do not use the most expensive forms of transport, as in the case of the Calcutta metro which is not designed to serve the lowest income groups. The absence of infrastructures leaves most villages in developing nations without the essential services which have become necessary for effective participation in the realities of contemporary society. The cost of obtaining well-drilling equipment and large storage tanks is presently beyond the family income of many villagers, although houses may have water tanks to contain the rainwater. Few springs and wells are sufficiently close to be ready sources of water; villagers depend on rainwater when it is available or have to make long journeys for water. They are also far removed from adequate fire protection or water for firefighting. The distance from one house to another makes the possibility of electrical installation from a private utility company unaffordable for most village families. Public transportation is costly; fares are high because of fewer passengers and less freight; and high freight costs discourage industrial development.
Broader Archaic marketing methods (#PF6465) Unformed style of cooperative action (#PF6514)
Underdeveloped provision of basic services in developing countries (#PF6473).
Narrower Uneconomic snow removal (#PG8238) Unfeasible infrastructure costs (#PG8252)
Unprofitable transport business (#PJ8253) Prohibitive cost of transportation (#PE8063).
Related Prohibitive cost of home maintenance (#PJ9022).
Aggravates Inadequate road maintenance (#PD8557)
Lack of school transportation (#PJ7849)
Inadequate electricity supply (#PJ0641)
Inappropriate sanitation systems (#PD0876)
Discrimination in access to sanitation services (#PG8247)
Restricted delivery of essential services to developing country rural communities (#PF1667).
Aggravated by Lack of nearby wells (#PG1049) Insufficient doctors (#PE8303)
Inadequate medical care (#PF4832) Unsanitary refuse disposal (#PJ1098)
Underdeveloped road network (#PF1055) Insufficient removal of snow (#PG8241)
Freezing of plumbing systems (#PG8251) Unproductive use of resources (#PB8376)
Shortage of fresh-water sources (#PC4815) Inadequate safeguards against fire (#PD1631)
Prohibitive cost of connection to public utilities (#PJ9426)
Inadequate water supply in the rural communities of developing countries (#PD1204).

♦ PF6528 Restrictive use of available land
Restricted land use — Unproductive land use — Ineffective land usage — Land usage — Unconsensed land usage — Uncontrolled land use — Ineffective land use — Blocked land use — Unconvincing land-use story — Unprofitable land use practices — Unfocused land usage — Restrictive land-use vision — Complex land use — Disputed land usage
Nature The use of available land in the rural communities of many developing countries is restricted by tradition, ownership and settlement patterns, although there is a growing realization of the need to design the space of a community to ensure that the needs of all the people are met. The design of many rural villages emphasizes isolated family dwellings rather than the public space of the entire community. Locations for people to gather for the transaction of business, convenient use of educational facilities, electrical services, common clean water sources and corporate grazing land for animal raising are scattered and do not convey a sense of a well-ordered community. Many people must walk over miles of dirt pathways, occasionally very steep and often made muddy and slippery by abundant rains, to reach farms, stores or schools. The difficulty of transportation through the many narrow pathways does not easily allow the transport of material to build more secure or larger homes. This is further complicated by the fact that isolated families barely make an income from farming so that the purchase of building materials is beyond many families, abilities. The isolated pattern of village housing results in the continuing of a subsistence life style and prevents participation in the conveniences of modern life.
Broader Unproductive use of resources (#PB8376)
Unimaginative vision of resource utilization (#PF1316).
Narrower Unused land (#PF4670) Land misuse (#PD8142)
Insecure land tenure (#PD9162) Inadequate grazing land (#PJ0404)
Unavailability of building sites (#PJ8549) Insufficient commercial facilities (#PJ5337).
Related Limited public land (#PJ0574).
Aggravates Inadequate care of community space (#PF2346)
Unfocused style of community operations (#PF6559)
Geographically undefined community limits (#PF6521)
Inadequate circulation of local information (#PF6552).
Aggravated by Unplanned land development (#PF6475)
Prohibitive construction costs (#PJ8591)
Inadequate housing construction (#PG0561)
Narrow range of practical skills (#PF2477)
Inadequate educational facilities (#PD0847)
Depopulation of mountainous regions (#PD1908)
Reluctant claims on external resources (#PF1226)
Undeveloped channels for commercial initiative (#PF6471).
Reduces Endangered parklands (#PE9282).

♦ PF6530 Subsistence approach to capital resources
Subsistence economic orientation — Subsistence poverty economy — Prevalence of subsistence earning — Subsistence level profits — Crippling subsistence planning — Reluctant investment of capital resources — Low investment in developing country farms
Nature Every community is confronted with the necessity of operating on a credit economy with the means of financing both immediate and long-range enterprises; however, most of the world's population has no consistent source of income but is largely dependent upon the high risk factors of under-developed agriculture or small, family operated, business ventures. In particular, Third World village families live on a subsistence economy. Most of their food is raised or gathered near their homes, any money must be spent in town for necessities. Current capital seldom goes for farm equipment, tools or other hardware, but rather is taken up by the costs of food, education, transportation and medicine. There is a lack of the financial flexibility required for planned individual economic growth, and a sense that finance is unavailable for any but the most immediate necessities; loan capital to meet critical bills is borrowed only in small amounts. Start-up capital is lacking and loans inaccessible because of unmanageable interest and repayment terms. The heavy debt borne by farmers who, having actually negotiated a loan, experience crops ruined by a drought, weighs heavily against willingness to risk again. An overwhelming attitude that there will never be enough money fosters a style of wearily making do with less than what is adequate or, at best, allowing short range priorities to take precedence.
Broader Subsistence life style (#PF1078) Unproductive subsistence agriculture (#PC0492).
Narrower Scarcity of residential land (#PD8075) Draining of resources due to alcohol (#PE8865)
Reluctance towards corporate investment (#PG8444)
Diminishing capital investment in small communities (#PF6477)
Prohibitive cost of eye-glasses in developing countries (#PE7548)
Domination of government policy-making by short-term considerations (#PF0317).
Related Abuse of credit (#PF2166) Limited accumulation of capital (#PF3630)
Limited individual capital reserves (#PF2899) Precarious basis for family economics (#PF1382)
Dispersion of local capital resources (#PF1979).
Aggravates Heavy farm debts (#PG8525) Rural unemployment (#PF2949)
Limited funding approach (#PG8537) Prohibitive cost of education (#PF4375)
Inadequate budgeting practices (#PJ0978)
Declining productivity of agricultural land (#PD7480)
Inadequate level of investment within developing countries (#PD0291).
Aggravated by High interest rates (#PF9014) Prohibitive cost of fertilizer (#PJ1022)
Inadequate domestic savings in developing countries (#PD0465)
Limited availability of loans in developing countries (#PE4704)
Subsistence agricultural income level in rural communities (#PE8171).

♦ PF6531 Minimal opportunities of adult training
Non-local adult training — Insufficient education for adults — Undefined adult education — Unfocused adult education — Unplanned adult education — Unfunded adult education — Irrelevant adult education — Minimal adult education — Uninitiated adult education — Infrequent adult education — Inadequate adult education — Insensitive adult direction
Nature Without the appropriation and systematic application of fuller literacy, updated technical skills training and extension resources, small communities have no way to develop their services, attract capital or increase their agro-industries in order to compete in the modern economic world. Yet most villagers in developing nations find their daily existence unrelated to the aim and intent of practical education. Where there are volunteer teachers in practical skills, these tend to be only partially trained themselves and therefore do not have the confidence of the community. Modern methods are uninvestigated, even though practical benefits, such as more productive crops from using modern fertilizing methods, would result. Emphasis on the few traditional, low-paying skills of the area tends to undermine willingness to make the sacrifices necessary to learn new and more profitable skills that these communities so desperately need.
Broader Restrictions on the acquisition of knowledge (#PF1319)
Limited availability of learning opportunities (#PF3184)
Limited availability of education in rural areas (#PF3575).
Narrower Inadequate political education (#PJ7906).
Related Ineffective systems of practical education (#PF3498).
Aggravates Limited commercial experience (#PG7798)
Unprofitable traditional skills (#PJ1031)
Lagging training in social skills (#PF8085)
Lack of sharing of community skills (#PF3393)
Lack of motivation in leadership development (#PF2208)

Unexplored alternatives for commercial development (#PF6548)
Lack of opportunities for practical training in communities (#PF2837)
Stifled potential for social interaction between different age groups (#PF6570).
Aggravated by Lack of vocational teachers (#PJ9603)
Lack of emphasis on basic education (#PF1548)
Unrecognized relevance of education (#PF9068)
Unperceived relevance of formal education in rural communities (#PF1944).

♦ PF6532 Arrested development of labour potential
Nature The basic employment of the majority of people living in Third World villages is generally limited to family farming; however this work is cyclical, leaving frustrating gaps of non–working time. Much of the potential working force among many women is tied to family affairs which also do not fully employ them. Technically trained residents must find work in cities, for without new business and industry villages have minimal need for such skills; this creates a skilled–labour drain from the community, and the large group of trained people necessary if they are to have a school and new development cannot be generated. As long as small family farming remains as the only realistic type of work in the community, villagers will continue to feel the frustration of restrictive and unprofitable labour.
Broader Underdevelopment of industrial and economic activities (#PC0880).
Narrower Unemployed skilled labour (#PE1753).
Aggravated by Limited employment options (#PF1658)
Hopelessness inspired by school (#PJ1004)
Ineffective economic structures in industrial nations (#PE4818).

♦ PF6533 Surrendered control of marketing systems
Nature Many Third World villagers raise products for urban markets, but because they lack direct access to these markets they receive a lower return on their sales, having to deal through middlemen. In this way, they lose the profits of direct sales, and money that otherwise could be kept in the village is dispersed. To gain immediate cash, some farmers are forced to lease some of their land to outsiders to plant, harvest and sell the produce, which again deprives them of greater income. The distance to markets, and lack of storage facilities, transport, and of communal decision to enter formal cooperatives to buy and sell merchandise and farm produce, ensures such people will continue to be passive participants in the larger economy.
Broader Limited market development (#PF1086).
Narrower Middleman control of rural marketing (#PE3528).
Aggravated by Insufficient pest control (#PJ1086).

♦ PF6534 Lack of formal education
Disrelated formal education
Nature Although education prepares children, young people and adults to deal practically with life situations, schooling beyond the sixth grade is seldom pursued in most rural villages of Third World countries, and residents continue to stand outside the mainstream of society, lacking the self–assurance that comes from achievement in education. Some families even consider elementary school unnecessary when children are needed to help their families on the farm. Part–time farming is a familiar task and other skills seem too time–consuming and difficult to learn. A sense of prevailing inadequacy stifles the urge to relate seriously to the modern world.
Broader Lack of education (#PB8645).
Aggravates Ignorance (#PA5568).
Aggravated by Obstacles to education (#PF4852)
Unbudgeted child education (#PG1363)
Prohibitive cost of education (#PF4375)
Inadequate preschool education (#PJ1962)
Inaccessible market and supply centres (#PF8299)
Disjointed patterns of community identity (#PF2845)
Limited availability of learning opportunities (#PF3184)
Unavailability of scholarship funds for students (#PE3569).

♦ PF6535 Scarce options for involvement in culture
Claim Despite the vital importance of participation by all members of society in the socio–economic structures of communities, life in many Third World villages offers limited opportunities for cultural involvement. When land was originally homesteaded, there was little time or opportunity for a rich cultural life since every day was filled with the work on the land. As more people came, few cultural centers were created; instead, the isolated homestead became the basis for cultural development. Even today in most villages there is minimal cultural organization, and few hours of daylight are available for meetings and activities. The unchanging roles and forms of cultural life give rise to a feeling of estrangement.
Broader Cultural deprivation (#PC1351).
Narrower Unused evening hours (#PU0959) Minimal cultural groups (#PS0983)
Only religious celebrations (#PS0995).
Aggravated by Prevailing community insecurity (#PD9044).

♦ PF6536 Lack of responsible involvement in community affairs
Nature Although many people are searching for effective and significant methods of corporate activity, there are less opportunities for total citizen engagement. Apparent reasons are varied: few social events bring people together regularly; the public nature of assuming leadership roles tends to draw increased criticism, making recruitment difficult; individuals hesitate to speak their opinion publicly for fear of not drawing support; individuals and family units have few opportunities to directly contribute to the community's life; parents frequently have lost sight of their role as guide and supporter for their children and have accordingly given up parental responsibility to the school, neighbours and television; lack of structured community life has emphasized the influence of the youth peer group and left their potential untapped; and alcohol use and child abuse are concerns, but no communal approach has been developed to cope with these problems which leave families and individuals on their own.
Claim Many communities are characterized by fragmented individual and family styles of participation, despite realization that social and economic development requires cooperative effort. Total expenditure of energy into a job and family life are no longer adequate means of full engagement in community life. Social clubs and groups have also become ineffective in this direction.
Broader Lack of community responsibility (#PJ3290).
Narrower Negative social context (#PF9003) Restricted personal responsibility (#PG5223).
Aggravates Inadequate sense of time (#PF9980).
Aggravated by Isolated family units (#PG5240) Parental permissiveness (#PD5344)
Rigid personal viewpoint (#PG5238) Limited youth activities (#PJ0106)
Anticipation of criticism (#PJ7893) Overemphasized adult sports (#PG5243)
Excessive television viewing (#PD1533) Insufficient community events (#PF5250)
Ineffective parenting techniques (#PJ0551) Insufficient leadership training (#PF3605)
Consumption of alcoholic beverages (#PD8286)
Lack of means for achieving consensus (#PD2438)
Unorganized development of work forces (#PF2128)
Short range planning for long–term development (#PF5660).

♦ PF6537 Stagnated images of community identity
Restricted vocational images — Reduced images of having a significant life work
Nature Although there is increased concern about recovering the significance of small towns, many self–images of such communities are self–defeating. Many of the vital structures and activities which gave small towns distinct identity have shifted to the larger nearby towns, and many residents have also moved there. Local history classes do not attempt to recapture the history of the local pioneers, there is simply an image of uncertainty for the future, which destroys community goals and ensures that projects fail before implementation even begins.
Claim People of the twentieth century aware of the chaotic state of the world and feeling insecure about their own future create illusions of meaning which demean their lifes and their work. Rather than have a vocation which is continually risking new directions, discovering new depth to social roles and facilitating more human responses from others people enslave themselves to these illusions.
Broader Deteriorating community identity (#PF2241).
Narrower Incomplete identification signs (#PG5218).
Aggravates Vandalism (#PD1350).
Aggravated by Closed area churches (#PG5224)
Resignation to problems (#PF8781)
Fragmented citizen goals (#PG9946)
Disrespected public property (#PG9970)
Neglect of property maintenance (#PD8894)
Infrequent achievement of honours (#PG9991)
Disrelationship from community history (#PJ0836)
Inadequate dissemination and use of available information (#PF1267).

♦ PF6538 Underutilization of locally available skills
Unused local skills — Uncatalogued community skills — Unknown local skills — Unrecognized income skills — Overlooked personal skills
Nature Many local people possess skills and abilities which could be exchanged or shared, whether in repair, construction and traditional crafts, or in teaching and other forms of creative leadership. However, the use of such local skills and wisdom to small communities is often overlooked, and the creation of methods for sharing these skills does not keep pace with the need. Lack of public appreciation conveys the idea that it is unimportant to spend time and effort providing benefits to the community. The wisdom, skills and enthusiasm of youth and senior citizens are given few structural outlets, while there is a decline in organizational management and business expertise that commercial expansion requires. Organizational resources are untapped and often there is little knowledge of available training opportunities.
Broader Obstacles to community achievement (#PF7118)
Underutilization of human resources (#PF3523)
Underutilization of potential in local communities (#PF6513).
Narrower Failure to employ skills of home–bound educated women (#PD8546).
Related Unrecognized opportunities (#PF6925) Unperceived career opportunities (#PF9004)
Outmoded functional skills in rural communities (#PF2986).
Aggravates Lost knowledge (#PF5420) Restricted flow of local economy (#PF6451)
Inaccessible educational facilities (#PD9051)
Underdeveloped sources of income expansion (#PF1345)
Unavailable education for effective living (#PF2313).
Aggravated by Negated paramedic role (#PG5203)
Incompetent management (#PC4867)
Lack of self–confidence (#PF0879)
Detached local teachers (#PG9992)
Erosion of elders' wisdom (#PF1664)
Inaccurate youth stereotypes (#PJ8357)
Narrow range of business skills (#PE6554)
Neglected organization training (#PG5231)
Limited means of marketing employable skills (#PE7344).

♦ PF6539 Limited availability of public services in the small towns of developed countries
Nature The maintenance of basic public services which residents in small communities have come to expect is increasingly difficult due to high costs; and schools, banking services, post office and other services tend to be consolidated with nearby towns. Although these services exist, their delivery and maintenance in rural areas tend to be incomplete and do not meet people's needs. For example: absence of public transport limits the mobility of senior citizens and youth beyond the boundaries of their own community; medical services located outside a community lead to an emphasis on curative health instead of preventative health; police forces operating on a reduced time basis result in city ordinances covering vandalism and other public nuisances regularly going unenforced; already significant traffic hazards for children and pedestrians are further increased in the winter due to the length of time it takes some communities to clean the streets of snow and ice; play areas are not supervised and not protected with traffic signs; lack of day–care structures prevents some mothers from working, and parents seldom participate in activities beyond the family.
Broader Weakness of socio–economic infrastructure (#PC1059)
Underprovision of basic services to rural areas (#PF2875).
Narrower Breakdown of police protection (#PF8652).
Related Inadequate road maintenance (#PD8557)
Insufficient emergency services (#PF9007).
Aggravates Consumption of alcoholic beverages (#PD8286).
Aggravated by Diminished park safety (#PG5213)
Lenient judicial system (#PG9983)
Inadequate medical care (#PF4832)
Unenforced civic ordinances (#PG5217)
Prohibitive cost of maintenance (#PF0296)
Inadequate child day–care facilities (#PD2085)
Insufficient transportation infrastructure (#PF1495).

♦ PF6540 Transfer of business from small communities to larger towns
Nature Adequate consumer services generate increased business operation and money circulation within their own boundaries. The current challenge is relocating appropriate commercial and industrial functions in small communities in which many of the commercial functions have gradually been transferred to the nearest market town. At present, residents of communities in which the stable business section has been reduced spend large amounts on goods and services in nearby large communities; and patterns of buying goods away from home are encouraged by short driving distances involved.
Broader Obstacles to community achievement (#PF7118).
Aggravates Seasonal fluctuations in agriculture (#PD5212).
Aggravated by Small employee pool (#PG5227) Underpayment for work (#PD8916)
Single product economy (#PG9925) Limited market development (#PF1086)
Prohibitive cost of living (#PF1238) Unpromoted business choices (#PG9937)
Inconsistent inventory levels (#PG9973) Infrequent business successes (#PG9984)
Inaccessible market and supply centres (#PF8299).

♦ PF6541 Lack of local information systems
Nature Although the complexities of modern life have resulted in the creation of systematic structures of communication to replace the collapsing of informal methods, citizens in small communities remain uninformed about local issues, or misinformed through word of mouth communication and haphazard placement of public notices. Established means of information and

publicity, such as a local newspaper and area radio or television stations are not used to supply any depth of penetration into the community to publicize events and provide general information about plans, news and projects. Therefore, issues are discussed and decisions made with no structured channels available to inform citizens.

Broader Lack of information (#PF6337).
Narrower Missing public signs (#PE4927) Limited job information (#PG8795)
Inadequate schedule news (#PG9962) Overlooked media channels (#PF4611)
Unidentified news resources (#PG9995) Unpublicized community news (#PF7998)
Unpublished training opportunities (#PJ9985).
Aggravates Unpublicized public meetings (#PF5222).
Aggravated by Unconsensed health needs (#PG9032)
Disregarded financial resources (#PF7458).

♦ PF6543 Ineffective utilization of public environment

Nature Despite the necessity for aesthetic and functional design of public areas in order to communicate an image of the community to both residents and visitors, and to demonstrate its unity, many small communities have no overall plan for future land use and zoning. Local ordinances tend to be inadequate for environmental concerns (such as removal of abandoned automobiles), many narrow, shabby alleys and back streets abound with deteriorated outbuildings and lack of functional drainage; absentee landlords result in inadequate maintenance of houses.

Broader Obstacles to community achievement (#PF7118).
Aggravated by Absentee ownership (#PD2338) Unimaginative city plan (#PG9940)
Insufficient risk capital (#PJ1704) Dysfunctional public utilities (#PE7647)
Restricted community expansion (#PJ1107) Unexplored energy alternatives (#PF7960)
Insufficient environmental laws (#PG9964).

♦ PF6544 Lack of channels for obtaining available local funding

Nature Local communities do not have a sufficiently strong voice in regional affairs. Although government and business may be aware of the need to provide development resources to outlying areas and have made services and funds available to rural communities, the avenues for obtaining such external development resources are not clear, so that resources may be unknown or, if discovered, application procedures are complex. Local communities are often not informed of plans concerning the future use of land and the wide range of services and funds accessible for building and development programmes are often enmeshed in a complexity of regulations and bureaucratic procedures. The local communities, which have generally not participated in regional planning, are not able to coordinate city and regional plans.

Broader Lack of intersocietal resource channels (#PF2517)
Limited local availability of capital reserve (#PF2378).
Aggravates Insufficient financial resources (#PB4653)
Decline in government expenditure (#PF9108).
Aggravated by Undiversified tax base (#PJ7897)
Deficient city equipment (#PG9998)
Restrictive legal structures (#PG9988)
Inadequate financial services (#PJ8366)
Fragmented regional cooperation (#PF9129).

♦ PF6545 Divisive patterns of community groupings

Split tribal urban loyalty — Divisive tribal loyalties — Divisive traditional patterns

Nature There is a general break–down in the cohesive patterns of life in many small communities. Difficulties arise between the traditional decision–making groups which have remained unchanged for many years and overlapping city and township authorities; this leads to ineffective planning, conflicts over the use of large equipment, and inequitable delivery of utilities and services. Urban mobility and anonymity compound the problem: there is a lack of consensus on standards of behaviour and law enforcement. It is difficult for newcomers to become part of the community life, for there are few regular activities; past failures in sustaining community participation have made people reluctant to try again.

Broader Disunity in urban villages (#PF1257) Obstacles to community achievement (#PF7118)
Ineffective structures of local consensus (#PF6506).
Related Unclarified procedures for transposing ancient traditions (#PF6494).
Aggravates Parent–teacher non communication (#PF1187).
Aggravated by Rigid local groupings (#PG9966) Static power patterns (#PG9978)
Unchallenging world vision (#PF9478) Fragmented business practices (#PG9954)
Breakdown of police protection (#PF8652)
Untransposed significance of cultural tradition (#PF1373).

♦ PF6547 Limited control of environment by local communities

Nature There are currently few ways for local communities to decide appropriate ways of developing and conserving their natural resources. Many small towns have little influence in determining the use of surrounding lands, and feel the only land over which they have influence is limited to the small central area. The fact that surrounding resources, even if controlled from elsewhere, might still be a possible resource for the community, goes largely unnoticed, as well as the fact that resources include time as well as land.

Broader Obstacles to community achievement (#PF7118).

♦ PF6548 Unexplored alternatives for commercial development

Nature The desire of many small communities to revitalize their town is hampered by images of past greatness. This prevents them from searching for alternatives which could provide the same services in more appropriate forms – so that, despite declining population, basic goods and services could be made available to those residents who do remain. These would include shopping and health care facilities for older people, adult education courses and other community institutions.

Broader Proliferation of commercialism (#PF0815).
Aggravates Complex health delivery (#PJ8159)
Irregular repair services in the countryside (#PE8997).
Aggravated by Static school curriculum (#PG5274)
Prohibitive facility costs (#PG8153)
Narrow programme offerings (#PG8174)
Difficult land acquisition (#PJ5369)
Limited meeting facilities (#PE1535)
Inadequate health services (#PD4790)
Inadequate industrial space (#PJ0084)
Ineffective class scheduling (#PG5424)
Unlocated finance counselling (#PG8073)
Underutilized available space (#PG8182)
Inadequate financial services (#PJ8366)
Undiscovered business contacts (#PG8103)
Prohibitive cost of road construction (#PJ1070)
Minimal opportunities of adult training (#PF6531)
Unexplored opportunities for community education (#PF6512).

♦ PF6550 Unstructured local decision–making

Nature Because the internal structures in many rural communities lack system or intentionality, it is very hard to build a local consensus on matters of common concern (such as improved health and utility services), despite the wish of local people to take responsibility for the future of their community. It also limits the flow of accurate, objective information regarding available resource opportunities to the residents, and prevents effective procedures for following through once a decision is made. Residents begin to suspect that nothing can be achieved and adopt a fatalistic 'wait–and–see' attitude.

Broader Ineffective structures for community decision–making (#PF1781).
Narrower Citizen apathy (#PF2421) Incomplete planning process (#PG8065)
Obscure decision mechanisms (#PG8166) Insufficient civic organizations (#PG8125)
Unstructured beautification plans (#PG8095)
Ineffective structures of local consensus (#PF6506)
Ineffective means for participation in decision making (#PE8430)
Inadequate implementation of plans and programmes against problems (#PF1010).
Related Lack of means for achieving consensus (#PD2438).
Aggravates Inadequate meeting methods (#PF8939)
Breakdown of local community cohesion (#PD2864).
Aggravated by Restricted daytime involvement (#PG8141).

♦ PF6552 Inadequate circulation of local information

Nature Although many small communities in the developed world receive international news on radio and television, and in newspapers from nearby towns, there is no real system for the regular communication of coordinated, objective local news. Often there is no local newspaper, community bulletin board or other clearing house for public announcements and events; no available centralized source of basic information on taxes, city structures and ordinances; documents which might untangle the maze of property ownership, if they exist, may be difficult to locate; although the school may have a library, there may be no local public library; and there may be no public means for residents to share their views on community issues. The only system of internal communication may be the telephone, and residents remain uninformed or misinformed through word of mouth communication and haphazard placement of public notices. In the absence of dissemination of information of community–wide interest, people are resigned to feeling excluded from the centre of action and from the right to participate in community affairs; they assume that 'others' have inside knowledge of specific issues and will deal with them.

Claim Until ways are discovered to keep everyone abreast of local news and to enable them to share their views on current issues, a community's cohesiveness and vitality will be blunted.

Broader Insufficient flow of information (#PF6469).
Narrower Unpublicized community news (#PF7998)
Inoperative forums for public information (#PF7805)
Meagre youth facilities in the countryside (#PE8084).
Aggravated by Absent socializing systems (#PG8077)
Obscured intergenerational needs (#PG8057)
Restrictive use of available land (#PF6528)
Uncommunicated resource opportunities (#PG8163)
Incomplete access to information resources (#PF2401).

♦ PF6556 Lack of local leadership and decision–making structures

Undeveloped local leadership — Undefined community leadership — Insufficient neighbourhood leaders — Obscure community leadership — Lack of village leadership — Ill–equipped village leadership — Inexperienced local leaders

Nature Despite a trend toward local initiation of community development, many rural communities have no structure through which the community's leadership skills may be developed and practised. Channels of community decision–making tend to be obscure, there is little leadership training, and communities do not realize that people are their most valuable resource. Individuals hesitate to take the lead as they feel unqualified, or do not fit their own stereo–typed image of appropriate leadership. There is a tendency to rely on the handful of people who are willing to assume responsibility rather than trying to utilize the unique gifts of every individual; each person assumes that 'someone else' will take the lead, leaving the community without adequate leadership.

Broader Obstacles to leadership (#PF7011)
Restrictive effects of traditional community decision–making (#PF3454)
Paralyzing patterns between villages and administrative structures (#PF1389).
Narrower Nonexistent development groups (#PG8131)
Underutilization of popular wisdom (#PF2426)
Lack of central planning structures in small communities (#PF2540).
Related Subjectivity (#PF2827).
Aggravates Ineffective follow–up on initiatives (#PF0342)
Limited reservoir of technical skills in rural communities (#PF2848).
Aggravated by Oppressive prevalent images (#PF1365)
Lack of community participation (#PF3307)
Insufficient leadership training (#PF3605).

♦ PF6557 Minimum promotion of community assets

Insufficient community promotion

Nature Despite an apparent desire to promote their marketable assets and attract new commercial ventures, many rural communities do not develop promotional material in order, for example, to advertise regularly scheduled community celebrations or improve the local image.

Broader Obstacles to community achievement (#PF7118).
Aggravates Unexplored potential markets (#PF0581)
Declining sense of community (#PF2575).
Aggravated by Unstimulating entertainment (#PF8105)
Minimal industrial recruitment (#PG8112)
Limited manufacturing incentives (#PG8102)
Ineffective business solicitation (#PG8122).

♦ PF6559 Unfocused style of community operations

Unfocused resident unity

Nature Even as some communities in the developed world are becoming more skilled in administrative skills because of their increasingly complex relations and tasks, other communities are losing control of their organization. A community's elected officials may be volunteers whose training for administration and time for research is limited and vaguely defined. Members of the community may be unclear on the lines of responsibility.

Broader Obstacles to community achievement (#PF7118)
Restrictive effects of traditional community decision–making (#PF3454).
Narrower Undecided bank use (#PG9166) Unchecked community spending (#PJ9047)
Inadequate motorcycle control (#PG9057) Unacknowledged highway danger (#PG9086)
Arbitrary enforcement of regulations (#PD8697).
Aggravates Inadequate local enforcement (#PF0336).
Aggravated by State highway spending (#PG9036)
Incompetent management (#PC4867)
Ineffective space usage (#PE5458)
Inadequate police funds (#PJ8760)
Limited political know–how (#PG9178)
Unlocated community centre (#PG9143)
Difficult land acquisition (#PJ5369)
Demeaning community self–image (#PF2093)
Unskilled bureaucratic contacts (#PG8610)

Restrictive use of available land (#PF6528)
Insufficient industrial promotion (#PJ8378)
Unorganized development of work forces (#PF2128)
Limited local availability of capital reserve (#PF2378)
Ineffective official inspection of regulated activities (#PE4146).

♦ PF6570 Stifled potential for social interaction between different age groups

Nature Because of the preponderance of single unit families and the rapidity of social change, it is difficult for different age groups to come together through common experience; this is particularly true of the many elderly people who live alone and rarely interact with the young. Old people may be simply afraid of the young, who find that their natural grouping and thrills disturb their seniors. Despite manifest concern of all ages for their community, and a willingness to work to improve it, there are simply no structures which provide regular contact between all ages or provide community leadership for the youth, there is no way of acknowledging or appreciating the particular gifts of each age group, and there is no way of celebrating together.

Narrower Fear of growing old (#PJ9144)
Teenage language idiom (#PG9179)
Peer behaviour expectation (#PG9138)
Youth oriented society (#PG9131)
Social withdrawal of aged (#PD3518)
Generation communication gap (#PF0756).
Aggravates Disabling inadequacy feelings (#PJ9012).
Aggravated by Class consciousness (#PC3458)
Self-centred business (#PG9161)
Inward-focused churches (#PG9167)
Unprepared adult leadership (#PF6462)
Cut-off fragmented families (#PG9164)
Insufficient common experience (#PG9078)
Unresearched adaptable products (#PG9154)
Unstructured community gatherings (#PG9176)
Limited shared time (#PJ9126)
Vague community memory (#PG8977)
Low general expectations (#PG8989)
Misplaced independent pride (#PG9157)
Fear of a controlled lifestyle (#PG9001)
Unshared mutual responsibility (#PG9087)
Ineffective parenting techniques (#PJ0551)
Preoccupation with isolated problems (#PF6580)
Minimal opportunities of adult training (#PF6531)
Socio-economically inactive rural population (#PF4470).

♦ PF6572 Risk of capital investment

Inhibition of investment by risk — Fear of capital investment — Investment blocked by high risks — High business risk — Speculation risks — Unfavourable capital risk

Nature The risk involved in new economic development at any level, particularly in small rural communities, has grown to overwhelming proportions. Extensive capital is required for land, building, equipment and essential services such as water and sewage, before any production begins. As the readily available resources which brought business to these communities in the past become exhausted they are not replaced by the development of new resources, since the lack of sewage disposal and water systems means no outside development group would seriously consider working in such conditions. Local residents tend to lack the resources or the ability to initiate even small scale enterprises. Industry cannot afford to take risks in ventures that have low market values or low potential rates of return.

Refs Farrell, James L and Fuller, Russell *Modern Investments and Security Analysis* (1987).

Broader Risk (#PF7580)
Fear of failure (#PF4125)
Breakdown in community security systems (#PD1147).
Narrower Capitalist speculation (#PC2194)
Uninsurable business investments (#PG8939)
Speculation in developing countries (#PD1614).
Aggravates Bank failure (#PE0964)
Business bankruptcy (#PD2591)
Poor-risk commercial farming (#PG8204)
Insufficient capital investment (#PF2852)
Small bank failure (#PE1815)
Lack of venture capital (#PG7833)
Unexplored potential markets (#PF0581)
Obsolete basis of cultural identity (#PF0836)
Inadequate management skills in rural communities (#PF1442)
Limited availability of investment capital for urban renewal (#PF3550)
Lack of skilled manpower in rural areas of developing countries (#PE5170)
Disincentives for financial investment within developing countries (#PF3845).
Aggravated by Abuse of credit (#PF2166)
Creative accounting (#PE6093)
Prohibitive legal fees (#PF0995)
Unexplored tourist trade (#PG9070)
Minimal retailed merchandise (#PG9140)
Lack of fire-fighting facilities (#PJ2437)
Insufficient industrial promotion (#PJ8378)
Unavailability of agricultural land (#PC7597)
Breach of contract (#PE5762)
Narrow profit margin (#PJ9737)
Limited railroad usage (#PG8991)
Insufficient risk capital (#PJ1704)
Frightening investment outlook (#PG9106)
Imperialistic distribution system (#PD7374)
Undiscovered successful businesses (#PG9089)
Nationalization of foreign investments (#PC2172)
Overemphasis on rapid returns on investment (#PF1275).
Reduced by Economic philosophy of controlled risk (#PF2334).

♦ PF6573 Limited access to society's resources

Nature Many small rural communities in developed nations have limited access or are unrelated to relevant social information and experience. Even if they are surrounded by a mass of information from a large metropolitan area, they tend to make little use of these resources. Life on family farms or in small villages accustoms people to a kind of isolating self-sufficiency, so they may not visit the nearest city for years, and they neither have experience of nor feel the need for social agencies. As, in addition, familiar cultural and recreational activities which once brought people together over a broad geographical area cease to be held, this increasingly limited exposure to the resources of the contemporary world perpetuates a rural isolation and parochial value system that undermines attempts at inter-community cooperation.

Broader Underprovision of basic services to rural areas (#PF2875).
Aggravated by Limited access to news (#PG8980)
Parochial value system (#PG9071)
Geographical isolation (#PF9023)
Unattractive social life (#PJ1212)
Inadequate health services (#PD4790)
Local anti-intellectual bias (#PG9081)
Inaccessibility of knowledge (#PF1953)
Fragmented regional cooperation (#PF9129)
Unperceived career opportunities (#PF9004)
Unexplored surrounding territory (#PG9115)
Unquestioning, uncurious attitudes (#PG9016)
Obsolete self-sufficiency patterns (#PG9090)
Unadvertised educational resources (#PG9123)
Unavailability of funding information (#PG9051).

♦ PF6574 Limited accountability of public services

Obscure accountability for public services — Inhibited official accountability — Inadequate accountability of public sector management — Non-accountability of state-controlled enterprises

Claim It is the lack of accountability and the resultant profusion of corruption, bloated bureaucracies, under-performing parasites and abuse of power that has become a common denominator among developing countries, from single party left-wing regimes to conservative autocracies.

Broader Social unaccountability (#PC1522)
Inaccessible administrative agencies (#PF2261).
Aggravates Mismanagement (#PB8406)
Inadequate maintenance (#PD8984)
Official corruption (#PC9533)
Inadequate waste treatment (#PD6795)
Decline in government expenditure (#PF9108)
Inadequate system of political checks and balances (#PE4997)
Unaccountability of international financial institutions (#PF1136).
Aggravated by Unmaintained bridges (#PE2471)
Inadequate weed control (#PG9029)
Poorly lighted roadways (#PJ9224)
Domestic refuse disposal (#PD0807)
Inadequate land drainage (#PD2269)
Instability of water supply (#PD0722)
Uncompleted sidewalk repairs (#PG9072)
Unavailability of local directories (#PG9091)
Unavailability of library facilities (#PG9116)
Unacknowledged availability of services (#PJ9082).

♦ PF6575 Detrimental story of community future

Negative community story — Frozen vision of the community's future

Nature Many rural communities in developed nations only have faint memories of their past, and these tend to be of crises and tragedies: the loss of their school, a flood or great fire, and the closures of major industries. Some residents only half joke when they say there may not be a community there when their grandchildren are grown. Indeed the towns are filled with the symbols of a bygone day: unused, ageing buildings, abandoned and collapsing farms, and old people ending their days alone and without much to engage their creativity. These negative images of past and present unconsciously dominate anticipations of the future, as though residents have forgotten the resilience and courage of their ancestors who carried on through fire and flood, through boom and bust. Talk of the future is of failure and not challenge; there is little experience of reality as objective existence in which change and rebuilding are possible.

Broader Untransposed significance of cultural tradition (#PF1373)
Traumatic shift in life-styles of mining communities (#PE1137).
Narrower Closure of schools (#PJ8556)
Skeptical farming attitude (#PG8994)
External assistance inertia (#PG9109)
Neglected community uniqueness (#PG9018)
Undesigned town brochure (#PG9101)
Nonevident monetary rewards (#PG9042)
Paralyzing boom/bust mentality (#PG9117)
Debilitating content of village story (#PF2168).
Aggravates Low self esteem (#PF5354)
Demeaning community self-image (#PF2093)
Declining sense of community (#PF2575)
Unimaginative educational vision (#PF3007).
Aggravated by Temporal deprivation (#PF4644)
Unchallenging world vision (#PF9478).

♦ PF6577 Limited access to health services

Nature Modern understanding of preventive medicine, health practices and curative procedures tends to be centralized in urban areas while rural communities are often devoid of practical assistance; the nearest doctor may be miles away or only available at specific times. Infrequent visits, heavy caseloads and a high turnover of staff mean there are virtually no medical records available. Each visit requires a long repetition of physical history: old people are reluctant to go even when they are able to articulate their needs, and parents are unsure of what information and services they need for the health of their families. Such limited access occasions deep concern in residents over the accessibility of advice and care in times of emergency, although supply of emergency care and ambulance facilities would only be a partial and perhaps unrealistic solution. What is even more important is that residents lack basic information and skill in childcare, nutrition and basic prevention techniques, practical know-how and skills to maintain their own physical vitality.

Broader Inadequate health services (#PD4790).
Aggravates Insufficient emergency services (#PF9007).
Aggravated by Unconsensed health needs (#PG9032)
Infrequent doctor contact (#PJ0362)
Unfeasible medical expectations (#PG9044)
Irregular hospital transportation (#PJ8984)
Insufficient special care vehicles (#PJ9008)
Unmeaningful social roles for the aged (#PF1825)
Insufficient transportation infrastructure (#PF1495).

♦ PF6578 Neglected of public space

Nature Small, diminishing communities which used to serve a much larger population tend to present a a run-down appearance, with well-kept houses and gardens surrounded by empty lots on which dilapidated out buildings and abandoned junk are only partially hidden by tall weeds. The uncared for and unused space reminds both residents and visitors of the booming past and suggests only the imminent possibility of further decay and devaluation, as people move away rather than pay the price of demolition and new housing, or those without the resources to move out, buy trailers rather than renovate the old dwellings. The dilapidated and unused spaces become a formidable symbol of decay.

Broader Obstacles to availability of community space (#PF7130).
Narrower Accumulated junk (#PD5510)
Obsolete buildings (#PG8997)
Abandoned private homes (#PG9021).
Aggravated by Unattractive merchandise display (#PG9033).

♦ PF6580 Preoccupation with isolated problems

Overemphasis on discrete problems — Failure to focus on fundamental systemic problems — Overemphasis on symptoms of more fundamental problems — Piecemeal approach to problem-solving — Individual problem overemphasis — Oversimplified problem approaches

Nature Many problems are treated in isolation, whether for political reasons (as in the case of CFC effects on the ozone layer, for which corrective measures are obvious) or for public relations reasons (as in the case of endangered species such as the condor, which lend themselves readily to public campaigns). Perceived in this way, remedies can be found to such problems. This approach fails to deal with much more fundamental problems, whether economic, social or environmental, which result from the complex dynamics of the system in question.

Claim Really fundamental problems are not being tackled at all. An example is that complex of issues giving rise to the greenhouse effect. This calls for a new approach to the status of tropical forests in impoverished countries and to the ways in which energy is used in industrialized countries, neither of which can be handled within the framework of a single country, or resolved simplistically by further investment in nuclear power.

Broader General obstacles to problem alleviation (#PF0631)
Fragmented forms of care at the neighbourhood level (#PE2274).
Aggravates Inaction on problems (#PB1423)
Institutional preoccupation with obsolete problems (#PJ5014)
Incompatibility of environmental and economic decision-making (#PF9728)
Stifled potential for social interaction between different age groups (#PF6570).
Aggravated by Impoverishment of political debate (#PF4600)
Insensitivity to non-immediate hazards to society (#PF9119).

♦ PF6590 Under-utilized raw materials

Nature There exists a range of relatively low-cost raw materials which are not exploited and for which high cost alternatives are employed or developed. Once commercial interests are organized to manufacture the higher cost alternatives, there is considerable pressure against any attempt to develop the use of the cheaper alternatives.

Incidence Examples include rice bran and the water hyacinth.

Broader Inefficient use of resources (#PE5001).

Narrower Underutilization of natural resources (#PF1459).
Related Underutilization of human resources (#PF3523).
Aggravated by Maldistribution of resources (#PB1016).

♦ PF6592 Westernization of traditional modes of life in developing countries

Claim Westernization of traditional modes of life in developing countries (striving for greater incomes, unrestricted consumption, an urban may of life, etc) leads to a crisis as traditional culture comes under the impact of haphazard modernization, and as modern culture is introduced and rapidly disseminated in an unprepared cultural–historical medium. The inability of a society as a system to freely and easily 'digest' a great a portion of modern modes of life, results in culture as a system being unable organically to infiltrate society – some elements become hypertrophically developed and others become minimized.
Broader Deteriorating quality of life (#PF7142).
Aggravates Destruction of cultural heritage (#PC2114).

♦ PF6594 Unreported international movements of funds

Undocumented international financial transactions
Claim Concealed and unreported money flows influence the economy in a number of ways. Balance of payment accounting gets distorted, and the availability of such funds encourages corruption, perhaps even leading to politically destabilizing bribes. Undocumented money creates an illegal economic system with an advantage in competing with legal systems in such areas as real estate and banking. Such 'dark monies', once repatriated, contribute to inflation and, when successfully laundered, constitute the reward for crime and tax evasion.
Broader Unrecorded knowledge (#PF5728).
Aggravates Global financial crisis (#PF3612).

♦ PF6601 Unproductive extension of schooling

Nature As entry into jobs has come to depend more and more on formal qualifications and diplomas, attendance at courses has become mandatory for an ever–lengthening list of occupations, while entry on personal recommendations is relied on less frequently, especially in low–level and routine jobs, and advancement through on–the–job performance or training is becoming less frequent. Indirect consequences have been the lengthening in school life, encouragement of diploma hunting, and proliferation of private schools and courses in developed as well as developing countries, some of which have been a lucrative source of income for their organizers but are of dubious academic or professional worth.
Broader Ineffective systems of practical education (#PF3498).

♦ PF6605 Change

Claim "Change – swift, fundamental, unprecedented in scale, and jarring to the human condition – has become a commonplace of our age. Out of all this swirling upheaval and transformation, in every corner of human life and in all parts of the globe, arise many of the problems that confront the late 20th century man or woman. Learning to adjust and to live with a world in change is perhaps the central challenge of our time." (UN University)
Broader Unfinished imperfect universe (#PF5716).
Narrower Inadequate social reform (#PF0677) Long–term cyclic changes of climate (#PC6114)
Delay in societal impact of innovation (#PF8870)
Confusion induced by rapid social change (#PF6712)
Disruptive effect of changing employment patterns (#PF2303)
Incompatibility of rural values in urban cultures (#PF2648).
Related Instability (#PA0859).
Aggravates Economic uncertainty (#PF5817) Resistance to change (#PF0557)
Inflexible management patterns (#PF3091).
Aggravated by Blister agents (#PE7103) Unequal distribution of forces (#PJ7928).

♦ PF6608 Limitations of democracy

Nature Real democracy establishes limits to popular will. Not everything is permitted under a constitution, even if the people will it. Democracy is designed at its core to be spiritually empty. The defining proposition of liberal democracy is that it mandates means (through elections, parliaments and markets) but not ends.
Claim 1. Democracy is a charming form of government, full of variety and disorder, and dispensing a sort of equality to equals and unequals alike. It passes into despotism. 2. The one pervading evil of democracy is the tyranny of the majority, or rather of that party, not always the majority, that succeeds, by force or fraud, in carrying elections.
Counter–claim Man's capacity for justice makes democracy possible, but man's inclination to injustice makes democracy necessary.
Aggravates Undemocratic social systems (#PB8031).
Aggravated by Subversion of democracy (#PD3180).
Reduces Uncritical acceptance of dogmas and standards (#PF2901).

♦ PF6611 Nuclear irresponsibility

Nuclearism — Nuclear mentality
Claim Given the mounting threat of nuclear terrorism and the spread of nuclear weapons and technology to more nations throughout the world, the superpower's refusals to compromise on issues that could lead to nuclear weapon control or elimination is a gross example of nuclear (and world) irresponsibility.
Broader Irresponsibility (#PA8658).
Aggravates Nuclear accidents (#PD0771).
Aggravated by Competition between states (#PC0114).

♦ PF6616 Denial of the right of unborn children

Denial of foetal rights
Claim Children, conceived but not yet born, have the right to be protected against genetic experiments or engineering, cloning, creation of identical twins, research and experimentation on embryos, choice of sex by genetic manipulation, creation of the children from people of the same sex, ectogenesis, the fusion of embryos, creation of embryos from the sperm of different individuals, implantation of a human embryo in the uterus of another animal or the reverse, the fusion of human gametes with those of other animal. They have the right to be born.
Broader Denial of right to life (#PD4234).
Narrower Induced abortion (#PD0158) Foetal infection and death (#PE2041).
Aggravates Irresponsible genetic manipulation (#PC0776)
Unethical experimentation using aborted foetuses (#PE4805).

♦ PF6617 Limited social context in developing technology

Nature The social context out of which choices about which technology to develop is limited largely to military and economic conceptual frameworks. Medical treatments for those with rare diseases are left undeveloped because of small markets. Simple, low cost, agricultural equipment needed by the vast majority of the world's population is left to individual farmers or to small philanthropic organizations to develop. The vast majority of the people of the world have no means of influencing these decisions.
Broader Non–inclusive management decisions (#PF2754).

♦ PF6635 Inadequate radiation monitoring systems

Broader Inadequate environmental monitoring (#PF4801).
Aggravated by Unethical practice of radiology (#PD8290).

♦ PF6638 Cognitive dissonance

Nature Cognitive dissonance – lack of accord between beliefs and actions – may be responsible for misunderstandings, separations, tensions and mistrust between peoples or nations, and can also be the source of mental anguish and perhaps mental illness within people themselves.
Claim The USA seems to be pursuing a policy of cognitive dissonance in its stances (lack of stances) regarding Israel, South Africa, the USSR and Greece, wavering between asserting its beliefs and taking appropriate actions to sustain them.
Aggravates War (#PB0593) Ignorance (#PA5568).

♦ PF6648 Androcentrism

Nature The human male can be perceived as the centre of creation, with the female human as secondary, followed by mammals, other animals, fishes, plants and so on throughout the range of existence. Many who have adopted an "earth–centred" worldview have traded their old hierarchical vision of a divinity in the heavens for an androcentric view of earth.
Related Male domination (#PC3024) Anthropocentrism (#PF4096).

♦ PF6654 Conflicting use of resources

Incidence Some areas of land are far better suited to agriculture than others. For very fragile areas, particularly in the tropics, there may be a direct conflict between the goals of the residents and the need to preserve the soil. A family may need to collect scarce wood as the only available source of fuel, and the consequent deforestation results in national impact on food production and global impact on the environment. Protecting the environment and conserving the soil may cost the residents their livelihoods; in some areas, research which leads to a more abundant source of fuel is as critical to survival as more food.
Narrower Conflicts over fishing rights (#PD5361).
Aggravates Worldwide misallocation of resources (#PB6719)
Long–term shortage of natural resources (#PC4824).

♦ PF6657 Spiritual impurity

Spiritual uncleanliness — Spiritual defilement — Passive village promotion
Broader Unsystematic use of powerful relationships by rural communities (#PE1101).
Related Sin (#PF0641) Original sin (#PF8298) Ritual pollution (#PF3960).
Aggravated by Impure thoughts (#PF5205) Sexual deviation (#PD2198)
Violation of food taboos (#PD2868) Inadequate personal hygiene (#PD2459)
Unethical personal relationships (#PF8759).

♦ PF6665 Differing conceptions of time

Nature Ethnic, national and cultural traditions, often embedded in languages and in ways of looking at the world, see historical time very differently. Time budgets at various parts of the world are fundamentally different. The individual's assessment of his position with respect to society – and the world – are essentially time–related concepts; which vary greatly across the surface of the globe. These differences prejudge how people regard war and peace, life and death, good and evil. They influence the way people negotiate. They inform peoples' value judgments in all things; and many of these different views are totally incompatible.
Broader General obstacles to problem alleviation (#PF0631)
Obstacles to efficient utilization of time (#PF7022).
Narrower Inadequate sense of time (#PF9980).
Related Multiplicity of languages (#PC0178) Multiplicity of time standards (#PF2621)
Constraint of time on individual and social development (#PF5692).
Aggravates Inefficiency (#PB0843) Human errors and miscalculations (#PF3702).

♦ PF6670 Reinforcement of inappropriate development by privileged classes in developing countries

Disproportionately wealthy elites in developing countries — Conspicuous consumption in developing countries
Nature Within developing countries the capital is usually held by a small privileged class which is uninterested in investing in ventures conducive to more appropriate forms of development of value to the poor majority. Such capital is rendered unavailable because it is lent to foreign investors, used in speculative ventures or to purchase more land or imported luxuries, or sent out of the country to secure foreign banks.
Broader Privileged families (#PD5616) Decadent standard of living (#PD4037)
Non–productive capitalist elites (#PE3816).
Aggravates Maldevelopment (#PB6207) International insecurity (#PB0009)
Corruption in developing countries (#PD0348) Destruction inherent in development (#PF4829)
Capital shortage in developing countries (#PD3137).
Aggravated by Elitist ruling classes (#PF4849) Entrenchment of vested interests (#PD1231)
Failure to sacrifice any personal advantage (#PF5203)
Economic exploitation of developing countries by industrialized countries (#PE2427).

♦ PF6683 Inauspicious conditions

Unpropitious timing — Unpropitious physical locations — Bad feng shui
Broader Superstition (#PA0430) Unexplained phenomena (#PF8352).
Narrower Bad omens (#PF8577).
Aggravates Unluckiness (#PF9536) Human sacrifice (#PF2641)
Religious sacrifice (#PD3373) Policy–making delays (#PF8989).
Aggravated by Magic (#PF3311) Horological superstition (#PF0565).

♦ PF6691 Inadequate objectivity of institutions

Claim The existence of public opinion, freely known, creates an environment of information conducive to policy change without violence. In its absence, the public is left with rumour and this is unsettling. If government does not gauge public opinion via some credible source, the vacuum of information is generally filled by intelligence agencies, political elites, the Army, or ministries with hidden agendas. Characterizations of what is happening "out there" by such tendentious sources leads to decision–making which can be self–destructive, and is generally reactive in fact, if not by intent.
Broader Bureaucratic bias (#PC1497).
Aggravates Ignorance (#PA5568) Distortion (#PA6790)
Selective perception of facts (#PF2453).
Aggravated by Prejudice (#PA2173).

♦ PF6697 Unprofessional science

Second–class science
Claim Convention says that all science is excellent and some especially so. Scientists have closed their eyes from second–class science and feel responsible to cover it up and keep it from the public.
Broader Unethical professional practices (#PC8019).
Aggravates Scientific fraud (#PF1602)
Irresponsible scientific and technological activity (#PC1153).
Aggravated by Politicization of scholarship (#PF7220).

♦ **PF6703 Obscurantist diplomatic language**
Deceptive language — Doublespeak
Nature Governmental organizations sometimes develop a language of their own, a diplomatic language, which can be disconcerting to those lacking experience of working with the organization. It restricts genuine understanding of ultimate aims to small groups of initiates and discourages those delegates unfamiliar with the procedures and languages specific to that organization.
Claim The obscurity of diplomatic language corresponds neither to enhanced richness nor enhanced precision, but is the attempt to reconcile divergent and sometimes contradictory views by reaching an agreement which may be superficial and purely verbal.
Aggravated by Misuse of language (#PF9598).

♦ **PF6707 Government delay in response to symptoms of problems**
Official inaction in response to potential problems — Bureaucratic procrastination in response to evidence of potential hazards — Government delay in remedial response to problems
Claim A crisis is made worse by attempts to deny evidence for its existence. Government agencies are not interested in responding to emerging crises since their officials sincerely believe that absence of evidence is the same as evidence of absence of risk.
Broader Policy–making delays (#PF8989) Non–recognition of problems (#PF8112)
Inefficient public administration (#PF2335).
Related Government inaction (#PC3950).
Aggravates Indecisive multilateralism (#PF9564)
Government resistance to institutional change (#PF0845).
Aggravated by Official apathy (#PF9459) Latent problems (#PF9328)
Procrastination (#PF5299) Government limitations (#PF4668)
Bureaucratic factionalism (#PF7979) Government delaying tactics (#PF6119)
Unproven relationships between problems (#PF7706)
Inadequate power of intergovernmental organizations (#PF9175)
Placement of the burden of proof on the disempowered (#PF3918).

♦ **PF6712 Confusion induced by rapid social change**
Nature The effects of rapid social change are apparent in many different areas, but their repercussions on psychological and physical health have yet to be evaluated, and in some cases, even questioned. The changes in relationship between man and his social environment are causing an increase in isolation and breakdown in the stability of social relations. When members of a community or social group feel estranged from their society and culture (by no longer sharing its social norms and values), they feel isolated, frustrated, and powerless to affect events or control their own destinies; these persons run particular risk of behavioural disorders. The social and cultural structure has lost cohesion and consequently its values are in confusion.
Modern technology has also revolutionized many other basic needs of social life. The constant fears of pollution from radiation and the uses to which nuclear power may be put diminish the pleasure and relaxation man derives from nature, and also breed dismay for the welfare of future generations. Eating customs and dietary habits are influenced by the availability of pre–prepared foods, often hurriedly taken while standing at a 'bar'; leisure and recreation are dominated by television and other spectator pursuits; the pattern of working life is dictated by the pace, monotony, and programme of automated machinery; and accelerated transportation not only brings infectious diseases to previously uncontaminated districts, but also increases motor accident rates and separates home and workplace, thus adding the additional stress of commuting.
Broader Change (#PF6605) Confusion (#PF7123).
Narrower Cultural suicide (#PF5957) Fraying of the tribal structure (#PG7712).
Aggravates Social inequality (#PB0514) Resistance to change (#PF0557).
Reduced by Inadequate social reform (#PF0677).

♦ **PF6713 Bibliomania**
Fanatical book collecting — Obsession with books
Nature The passionate desire to handle, possess and accumulate books. This can lead to the ruination of the persons concerned.
Counter–claim Bibliomania is considered one of the liberal forms of madness appropriate to older generations. In the past the results of obsessive book collection have ensured the conservation of important documents which would otherwise have been destroyed.
Related Abusive collection of specimens (#PE9417).

♦ **PF6723 Abuse of in kind payments**
Nature Payments in kind, especially when compulsory (workers being paid in part in the goods they produce, or workers being forced to purchase goods from company stores), may seriously restrict a worker's freedom to spend his wages. Deductions for defective work and disciplinary fines are two other means by which payments may be abused.
Claim Abuse of payments in kind will not be resolved until workers are able to live settled lives with their families in homes of their own, buy food and other goods in shops of their own choosing, until they receive all their wages in cash, and until they themselves decide where and how to spend those wages.
Counter–claim Fringe benefits also represent abuse of payments. Company expense accounts, dinners, travel expenses are all just sophisticated forms of wages in kind and thus are liable to abuse, though usually in reverse that is, the employee abusing the payments in kind offered by the employer.
Broader Abuse of economic power (#PC6873).

♦ **PF6730 Human exceptionalism**
Nature Many people regard humanity as so consummately superior to all other forms of life that ecological processes do not apply to them. This radical separation of human beings from all other forms of life encourages environmental devastation and an insoluble loneliness among humankind.
Related Human destructiveness (#PA0832).
Aggravates Natural environment degradation (#PB5250).

♦ **PF6734 Demons**
Devils — Malevolent spirits
Nature Disembodied entities, believed to be malevolent unless appropriately propitiated.
Background The distinction drawn between demons and spirits is considered somewhat artificial, in that either may be beneficent or maleficent. Spirits are often considered beneficent, unless inappropriately propitiated, when they become demonic. Both are normally distinguished from ghosts or souls of the dead. Demons and spirits originally, and in tribal cultures today, had a more positive connotation, although in many cultures no effort is made to distinguish them as either good or evil. Demon is etymologically equivalent to the Sanskrit 'deva'. This connotation has been transformed into an evil one through Christian condemnation of paganism.
Incidence Belief in the malevolent potential of demons and spirits is widespread in tribal and peasant cultures and wherever elements of animism and fetichism persist. In the past, demons have been considered a calamity of the same order as fire, pestilence, floods and famine. The activities of demons continue to be of special concern to certain religious groups, notably charismatic Christians, as an explanation when medical or counselling efforts had inexplicably failed in healing the sick.
Broader Superstition (#PA0430).
Related Ghosts (#PF3801) Unexplained phenomena (#PF8352).
Aggravates Fetishism (#PF8363) Changelings (#PE9453)
Dying a bad death (#PF1421) Demonic and spirit possession (#PF5781)
Affliction by malevolent spirits (#PF9043).
Aggravated by Animism (#PF6317) Bad omens (#PF8577)
Malevolence (#PA7102).

♦ **PF6743 Bias in document classification systems**
Broader Unethical documentation practices (#PD2886).
Related Bias in information systems (#PD9330).
Aggravates Incompatibility of document classification systems (#PE7753).

♦ **PF6745 Government bias in wage bargaining**
Nature In times of economic difficulties, governments exert unusually strong influences on wage bargaining. The measures taken are of various kinds, ranging from non-binding guidelines to compulsory restrictions of limited scope – such as the suspension of cost–of–living allowances for those earning the highest salaries – to general wage freezes or a blanket obligation to secure prior approval of collective agreements by the public authorities. These biases tend to limit the freedom of employers and workers or their organizations to set wages by direct bargaining.
Broader Biased government information (#PF0157).
Aggravates Lack of job satisfaction (#PF0171) Inequitable range of salaries (#PD9430).

♦ **PF6751 Limited availability of therapeutic substances of human origin**
Nature Therapeutic substances of human origin are by their very nature the result of an act of the human donor and therefore not available in unlimited quantities. Therapeutic substances of human origin include human blood and its derivatives, as well as various human organs.
Broader Insufficient medical supplies (#PE1634).
Narrower Inadequate emergency blood supply (#PE0366)
Lack of human organs for transplantation (#PE7530).
Aggravates Inadequate medical care (#PF4832).

♦ **PF6765 Passive historical perspective by leaders**
Nature A passive view of history by leaders, e.g. the belief that history just happens rather than being created, leads to parochial models of the world, nationalistic methods of leadership and de–emphasized global priorities.
Broader Passivity (#PF6177) Reduced understanding of globality (#PF7071).

♦ **PF6766 Cult of youth**
Claim Adults fear being cheated of the good things of life. They have a horror of ageing and death. They grasp at any superficial signs of youthfulness, making every effort to preserve their outer appearance, to keep looking young and attractive. They are obsessed with clothes, dieting and cosmetics, which require that a women spend hours making up or having her hair done. This extends to wearing wigs or having a face lift. Men are also increasingly obsessed with maintaining an appearance of youthfulness, which may require having their hair styled and dyed. The individual's own concerns are reinforced by peers and employers. Especially business and politics worship the cult of youth.
Related Obsession with novelty (#PF8767).
Aggravates Human ageing (#PB0477) Age discrimination (#PC2541)
Malpractice in plastic surgery (#PE4429).
Aggravated by Fear of growing old (#PJ9144).

♦ **PF6787 Inadequate narcotics legislation**
Diffuse legislation against drugs
Nature The concern about the problem of illegal drugs has grown so great that politicians must show themselves to be very involved in the issue in order to maintain their popularity. Legislation in democratic countries is therefore so diffuse that no effective programme can emerge. Rather, governments use their funding for highly–publicized campaigns attacking a bit of everything about drugs, with very little focus or clear intent other than re–election of the politicians drafting the bills.
Broader Inadequate laws (#PC6848)
Inadequacy of international legislation (#PF0228).
Aggravates Phenylketonuria (#PG3323)
Ineffective legislation against organized crime (#PE6699).
Aggravated by Law enforcement complexity (#PF2454)
Inadequate local enforcement (#PF0336)
Deliberate governmental avoidance of legislative reform (#PF5736).

♦ **PF6796 Search for individualistic meaning**
Nature People are being forced to search for personal meaning within the scope of what they do and what they are. Generally this is reduced to the role of employee. The individual should be self–sufficient, self–motivating and self–controlled putting him beyond the normal interdependency of society and disrelated from society's needs.
Broader Overemphasis on self–sufficiency with respect to interdependence (#PF3460).
Narrower Self–interest (#PA8760).

♦ **PF6803 Fear of nature**
Dependence on fear of nature
Nature Failure to externalize the emotions is based on fear. The surging of emotion when faced with nature 'in the raw' can generate fear of the cause of such emotion and the desire to destroy it. Even naturalists and ecologists may express a respect for nature in theory which they do not hold to in practice, because of cultural models which dictate terror for the wolf, revulsion at snakes, nausea at sliminess and stickiness, and so on. The desire to ame or reduce the power of nature in response to such fears leads to artificial 'preservation' of nature in tourist–oriented parks and protected areas, where the multiple effects of the presence of humanity may be more destructive than leaving the area as it was previously. It may be considered as an 'apartheid' between man and nature, with nature in the inferior position, rather than an overall development in which agriculture and cultivation naturally coexist with rare species.
Background Confrontation and discussion with those responsible for massive destruction of natural environments have demonstrated a more profound reason for such behaviour than the arguments of necessity, profit and prestige usually quoted. There appears to be an underlying emotional attitude which is rationalized in such expressions as: 'these regions are only brushwood', or 'only swamp', or 'only desert'. Further questioning leads to statements of the type: 'I don't like swampy areas myself', or 'wild regions like that aren't suitable for mankind – they have a traumatic effect on me'. This reaction is the reverse of the sentiments felt by those who are drawn to the harmony, the beauty and the danger inherent in the unspoilt environment and who feel it has a certain sanctity. Those who find the wildness of nature disquieting do so in response to a rejection of archetypal symbols in the psyche.
Broader Fear (#PA6030).

PF6803

Aggravates End of nature (#PF9582) Separation from nature (#PF0379)
Loss of humility in relation to the environment (#PF2527).
Aggravated by Psychological ungroundedness (#PF1185).

♦ PF6814 Inappropriate assumptions

Untested assumptions — Flawed assumptions — Basic presuppositions — Outdated assumptions

Nature In this era of dramatic social change, some say that any presupposition that one has not tested limits their existence. One must be ceaselessly questioning and searching. This is because so many of the assumptions we have received from the past, from the "experts" and from our own experience have given way at the slightest testing. Others point out that living from minute to minute means depending upon a number of fundamental presuppositions, most of which cannot be tested all the time. One obviously has to test all of their presuppositions all the time; at the same time one must depend on their assumptions fully in order to move through life.

Claim Reliance on old assumptions about the world can be as misinformative as dependence on old data. It tends to create the most drastic errors in forecasting.

Narrower Fatalism (#PF6430) Supernaturalism (#PF8433) Belief in miracles (#PG3935)
Preservation of obsolete systems (#PC8390)
Belief in emotional instability of women (#PF8751)
Belief in humanity's dominance over nature (#PF8264)
False assumptions on sustainable development (#PF2528).
Aggravates Inaccurate forecasting (#PF4774).

♦ PF6826 Religious backsliding

Falling from grace — Reversion to sin

Nature Reversion to sin and worldly ways, returning to old habits after a period of reform, or falling from grace after experiencing religious conversion. For those experiencing conversion, the ideal and the past habitual life are brought into definite conflict. There is a sharp cleavage between the higher and the lower selves. An ideal is recognized which is more difficult to attain because of its incongruity with the old life which subsequently tends to reassert its demands.

Incidence Of special significance in those religions emphasizing the conversion experience as a turning point in an individual's development.

Narrower Apostasy (#PE9018).
Aggravated by Sin (#PF0641) Idleness (#PA7710) Temptation (#PA7736)
Inappropriate personal habits (#PD5494).

♦ PF6829 Irrational religious beliefs

Broader Defective reasoning (#PF5711) Dependence on religion (#PF0150).
Narrower Voodoo (#PF9006) Enforced celibacy (#PD3371)
Negative effects of claims of religious infallibility (#PF3376).
Related Institutional lying (#PD2686).
Aggravated by Ignorance (#PA5568) Uncritical thinking (#PF5039).

♦ PF6832 Unreliable data

Nature Analysis of data obtained by studies, surveys, etc, which are intrinsically unreliable (due to misunderstanding or fear on the part of the respondent or misunderstanding or prejudice on the part of the questioner), is not only wasted effort but can also be dangerously misleading. It can lead to practical changes that seriously affect the welfare of the public and the viability of national programmes.

Broader Incorrect information (#PB3095).
Narrower Unreliability of statistics from socialist countries (#PF0888).
Aggravates Ignorance (#PA5568).

♦ PF6836 Chaos

Dependence on chaos

Narrower Chaotic households (#PG0499) Unequal distribution of forces (#PJ7928).
Related Disorder (#PA7361) Transience (#PA6425) Moderation (#PA7156)
Inattention (#PA6247) Lawlessness (#PA5563) Formlessness (#PA6900)
Disintegration (#PA6858) Unfinished imperfect universe (#PF5716).
Aggravates Ugliness (#PA7240) Intolerance of imperfection (#PF7024).
Aggravated by Chance (#PA6714) Conflict (#PA0298)
Disasters (#PB3561) Instability (#PA0859)
Shallowness (#PA6993) Annihilation (#PF9169)
Lack of control (#PF7138) Vulnerability of social systems (#PB2853).

♦ PF6846 Passive public support of terrorism

Nature Through its passivity, the public may be unwittingly supporting terrorism. Individuals may express irritation at the annoyances caused by the counter-measures, or attempt to evade them; families, religions groups and schools fail to demonstrate to young people that not only is terrorism illegal but it is barbaric and profoundly incompatible with the values accepted by most countries, and that the terrorist, regardless of the amount of attention afforded by the media, is not a hero but a loser; the public fails to express its outrage or to support government in tracking down and prosecuting terrorists, and appears unaware that silence could one day have a direct, personal effect, since everyone is a potential victim.

Claim Violence cannot be ignored. Not only is history a record of violence, but current events provide further, daily examples. We cannot therefore raise children in a world without violence, since no such world exists. But at least, we can avoid crime films, westerns, etc, which depict the man who shoots straight and kills with his first bullet as a hero; which resent a glamourized image of violence, thus passively supporting terrorism.

Refs Arnold, Terrell E *The Violence Formula* (1988).

Broader Passivity (#PF6177).
Aggravates Terrorism (#PD5574) Ineffective prevention of terrorism (#PF4240).

♦ PF6855 Lack of trans-frontier cooperation

Nature Trans-frontier areas are 'nerve spots' where for geographical, economic, linguistic or other reasons, the new needs for cooperation are becoming increasingly urgent. They are the terrain for immediate cooperation across frontiers on problems which cannot be dealt with at a national level. Examples of such problems include: communications, water management and purification; frontier traffic; the socio-cultural interests of small municipalities; and recreational facilities. Trans-frontier cooperation is necessary not merely to establish coordination for the area in question, but to facilitate economic and cultural growth for all the countries involved.

Broader Repression (#PB0871) Lack of international cooperation (#PF0817).
Narrower Vulnerability of frontier workers (#PE6833).
Aggravated by Competition between states (#PC0114).

♦ PF6863 Impaired vigilance

Nature Vigilance is the optimal situation in which an individual receives information form the environment, processes it, and by sensorimotor coordination, transfers it to other parts of the organism. A worker's performance in industry depends to a large extent on his prevailing level of vigilance, which may be reduced by lack of stimuli or duration of continuous activity. Consequently, at places where the work does not vary or where there is little action, it must be expected that vigilance will decline fairly rapidly.

Aggravated by Complacency (#PA1742) Forgetfulness (#PA6651)
Drug abuse in prisons (#PE0978)
Carcinogenic consequences of food preparation (#PE6619).

♦ PF6895 Attacks on peace forces

Nature International peacekeeping forces have become an important component in world affairs, but their actual effect is much less than that originally envisaged, partly due to direct attacks on them by the combatants. In addition, military efficiency is impaired by: political requirements of representative heterogeneity in composition; inadequate security because of the emphasis on negotiation and a restrictive interpretation of self-defence; dangers of involvement in domestic power struggles when confronted with an authority vacuum in the host country; erratic financing; lack of cooperation from major powers or local belligerents; reluctance on the part of lesser powers to contribute contingents; and erosion of credibility as conflicts appear no nearer to political solution after years of peacekeeping. In practice, it has proved difficult to separate effective peacekeeping from escalating confrontation, intervention in domestic strife and the abdication of peacekeeping responsibilities.

Incidence Difficulties are highlighted by the killing of 241 American and 82 French members of the peace-keeping force in Beirut on 23 October 1983.

Counter-claim Attacks are to be expected if the peace-keeping forces lose their impartiality, as is what happened with the UN peace-keeping forces in Lebanon.

Aggravated by Ineffectiveness of international organizations and programmes (#PF1074).

♦ PF6908 Second employment

Moonlighting — Necessity of second jobs

Nature The effects of double jobbing or 'moonlighting' vary greatly according to the nature of both the work and the individual, but have been shown to include repeated late morning arrival; higher sick absenteeism; lethargy and faulty work; increased incidence of accidents; and domestic difficulties.

Broader Mobility of village populations (#PE1848).
Narrower Dependency on unpredictable sources of income (#PF3084).
Aggravates Limited availability of permanent employment in inner-cities (#PE1134).

♦ PF6910 Fear of sexual intercourse

Fear of opposite sex — Sexophobia — Coitophobia — Sexo-phobia

Nature Sexophobia is the fear of the opposite sex, and especially of sexual relations. In women it may be brought on by experiences of incest or rape, and in men it may be associated with the vagina dentata - toothed vulva - which will bite off the male organ during intercourse, and is manifested in misogyny and can cause impotence.

Refs Hirsch, E W *Sexual Fear* (1950).

Broader Fear (#PA6030) Phobia (#PE6354).
Aggravates Misogyny (#PE7062) Homosexuality (#PF3242)
Sexual unfulfilment (#PF3260).
Aggravated by Unsafe sex (#PE9776) Sexual unfulfilment of the disabled (#PE5197).

♦ PF6911 Unsolved crimes

Uninvestigated crimes — Counter-productive use of crime-screening

Aggravates Inadequate national law enforcement (#PE4768).
Aggravated by Crime (#PB0001) Unreported crimes (#PF1456)
Inaction on problems (#PB1423).

♦ PF6916 Interpretation and translation errors

Nature Problems in translation and interpretation can arise from: insufficient knowledge of one of the languages involved; insufficient knowledge of the subject involved; lack of examples in dictionaries of how specific terms are used; attempted word for word translation; lack of knowledge or understanding of multiple meanings or shades of meaning; ignorance of current usage and "double-entendre". The repercussions of mistranslations range from an amusing faux pas amongst friends; to lost poetic and literary power in the translation of foreign literature; to serious, but unintentional, diplomatic insults.

Incidence A sign in an Italian dry cleaner's shop reads 'Drop your trousers here for the best results'. A European woman insisted that a dress be make from a sign hanging in a store in Singapore which said in characters 'Cheap goods inside'.

Broader Human errors and miscalculations (#PF3702).
Narrower Cultural dominance in translation (#PE5959).
Aggravated by Unethical practices of interpreters (#PE6740).

♦ PF6925 Unrecognized opportunities

Unknown possibilities — Unawareness of potential benefits

Broader Selective perception of facts (#PF2453).
Narrower Inadequate career advice (#PJ8018) Limited consumer knowledge (#PG7662)
Unexplored potential markets (#PF0581) Unrealized agronomic potential (#PJ1775)
Unperceived career opportunities (#PF9004) Unknown availability of subsidies (#PG9905)
Unrealized use of education structures (#PF2568)
Unrecognized benefits from cooperatives (#PF9729)
Unacknowledged availability of services (#PJ9082)
Lack of appreciation for nuclear weapons (#PE7476)
Unrecognized benefits from corporate action (#PJ9420)
Lack of awareness of potential for investment in small, inner-city enterprises (#PF2042).
Related Underutilization of locally available skills (#PF6538).
Aggravates Inadequate land drainage (#PD2269)
Erosion of elders' wisdom (#PF1664).

♦ PF6928 Parochial market research

Nature Market research, marketing and sales directed at narrow market segments in small geographic areas meeting the needs of limited parts of the population re-enforce the parochialism of most businesses. The long range needs of the whole world are not considered. More products are invented for relative wealthy portions of the population, without considering the needs and market potential of vast majority of the world.

Broader Limited market development (#PF1086).

♦ PF6932 Recipient distortion of aid to developing countries

Broader Inappropriate aid to developing countries (#PF8120).

♦ PF6957 Necrophilia

Excessive preoccupation with corpses

Nature Necrophilia is an excessive preoccupation with corpses, death and decay, murders and suicides, funerals and festivals of the dead, and the worship of ancestors and relic-homage. Medically, it is regarded as a pathological symptom and those who indulge in it are considered slightly insane.

Background Necrophilia specifically refers to coition with corpses, and the first recorded instance of such activity was in ancient Egypt. Several incidences of monks abusing dead women were recorded in the Middle Ages.

Refs Spoerri, T *Nekrophilie* (1959).

Broader Sexual deviation (#PD2198).
Aggravates Grave robbing (#PF0491) Undignified treatment of corpses (#PF5857).

♦ **PF6961 Sentimentalism**
Claim Politicians dismiss the talk of saving people's cultural patterns and ancestral shrines from so called benefits of technical progress as mere sentimentalism.
Narrower Sentimental attachment to the past (#PF1577).
Aggravates Government insensitivity (#PF2808).
Aggravated by Fear of emotional sensitivity (#PF9209).

♦ **PF6966 Theomania**
Religious hysteria
Nature Theomania is excessive religious or devotional fervour, where a person reaches a state of frenzy, as if seized by a deity. It often manifests in physical symptoms and unusual phenomena, such as convulsions, hysterical contortions of the limbs, immunity to pain, gifts of healing, apparent clairvoyance, and "speaking in tongues".
Background Theomania reached epidemic proportions in mediaeval Europe among the flagellants of Germany, Poland and Hungary who lashed themselves and others in penance for their sins; it was seen in the 14th century among those in Holland, Belgium and Germany who were stricken with dancing mania, and was also manifest in 17th century England by the Shakers, early Quakers, Methodists, Revivalists, Pentecostalists, Adventists, and Messianists.
Broader Hysteria (#PE6412).

♦ **PF6973 Foolhardiness**
Claim Foolhardiness is a vice opposed to the virtue of fortitude by way of excess. It does not moderate the passions of fear and daring, thus is likely to attack unnecessarily, immoderately, and unreasonably. Foolhardiness may be caused by presumption, as when someone overestimates his own powers to repel evil; by anger, which can lead to unnecessary violence or punishment; or by vainglory.

♦ **PF6980 Inefficiency of financial markets**
Misleading market prices
Aggravates Economic inefficiency (#PF7556) Fluctuations in real value of money (#PD9356)
Inefficiency of state-controlled enterprises (#PD5642).

♦ **PF6987 Victimization**
Victimization archetypes — Victimization process
Nature The very core of a person, the self, is protected by several layers and victimization is an unwelcome invasion into the self of the victim. The result is destruction of imaginations, phantasies, fictions, basic emotions and basic values.
Background Victimizers and victims can be looked upon as influenced by patterns of behaviour which are based on and directed by archetypal images. They are potentials in the psyche of every human being and belong to the archetype of sacrifice (sacrificer and sacrificed). An individual can be possessed by these archetypal images and identify with them. When this happens, the faculty to take personal responsibility for his deeds is extinguished; conscience, the inner feeling and judgement between good and evil, is not functioning anymore and is replaced by a new law. When thus possessed and inflated, the person takes the fate of other human beings in his own hands and as victimizer, in the extreme case, determines the death of his victim. When the victim identifies with the image of the sacrificed, he will accept his fate (inflicted by the victimizer).
Refs Decker, David L, et al *Urban Structure and Victimization* (1982); Yin, Peter *Victimization and the Aged* (1985).
Related Abuse of power (#PB6918) Structural violence (#PB1935).
Aggravates Injuries (#PB0855) Injustice (#PA6486).
Aggravated by Vice (#PA5644) Passion (#PA7030)
Human violence (#PA0429)
War-time communications with the enemy (#PE6974).

♦ **PF6988 Dogmatism**
Religious dogmatism — Scientific dogmatism
Refs Lenin, V I *Against Dogmatism and Sectarianism* (1978).
Narrower Bibliolatry (#PF7129).
Aggravates Educating people to lie (#PE3909) Religious indoctrination (#PD4890)
Sexual bigotry in organized religion (#PE0567).
Aggravated by Ugliness (#PA7240) Intellectual arrogance (#PF7847)
Inadequacy of religion (#PF2005) Undue attachment to a social group (#PF1073).

♦ **PF6991 Absurdity**
Nonsense — Senselessness
Nature Loosely defined, nonsense includes everything plainly at variance with obvious fact. Efforts may however be made to distinguish fairly sharply between the false and the nonsensical, although there may be no agreement as to where the line should be drawn. Distinctions may be made between: (a) nonsense as obvious falsehood (when contrary to the observed facts); (b) semantic nonsense (when information is provided out of context and without any attempt to relate it to the context in which it is assessed); (c) deviant utterances involving mixing of categories such that some of the communication is meaningful and the rest nonsense; (d) jumbled strings of familiar words lacking any familiar syntax; (e) vocabulary nonsense in which the syntax is meaningful but for which the words used are unfamiliar; (f) nonsense as gibberish in which there is neither familiar vocabulary nor familiar syntax.
Incidence Nonsense is frequently perceived as characterizing the communications of opponents, especially those favouring alternative policies and notably in parliamentary debate. Utterances may be considered nonsensical in a given language at a given time if the majority of the population balks at it, although it is necessary to accept that some may not balk at it and that many, in the future, will find it quite meaningful. Nonsense is both a widespread byproduct of human communication and a professional danger for philosophers, especially those with metaphysical orientations.
Claim Claimed as a distinctive ability of man, "the privilege of absurdity to which living creature is subject, but man only. And of men, those are of all most subject to it that profess philosophy" (Thomas Hobbes).
Refs Park, James *Absurdity, Insecurity and Despair* (1975).
Broader Unintelligence (#PA7371) Inappropriateness (#PA6852)
Negative emotions and attitudes (#PA7090).
Related Meaninglessness (#PA6977) Human contingency (#PF7054).
Aggravates Alienation (#PA3545) Human suffering (#PB5955)
Failure to profit from patterns of history (#PF1746).
Aggravated by Ugliness (#PA7240) Ignorance (#PA5568)
Irrationalism (#PF3399).

♦ **PF6992 Ineffective international agreements**
Deliberately weakened international treaties
Narrower Government non-payment of agreed contributions to international organizations (#PF8650).
Related Unfulfilled treaty obligations (#PF2497).
Aggravates Ineffective legislation (#PC9513) Inadequately worded agreements (#PF5421)
Covert violation of international treaties (#PD8465)
Symbolic international agreements without substance (#PE9267).
Aggravated by Limited acceptance of international treaties (#PF0977).

♦ **PF7003 Collective attention-span limit in societal learning**
Claim It is a well-know characteristic of any society that it is unable to focus its collective attention on any situation for any length of time. Even the most dramatic events tend to fall into oblivion. Clearly this is more characteristic of attention focused through the mass media. But issues brought to the attention of international conferences may only remain active for a period of weeks or months; although hot issues, providing ammunition in a dramatic debate, may even be expended within a period of days. Of perhaps greater significance are issues that survive the government election cycles and are given a permanent focal point through institutionalization, possibly with the creation of special documents and a specialized information system. A special difficulty for the international documentation system in this context (and, subsequently, for users) is that a category is forced to carry the significance of concepts already abandoned. Later it becomes definitionally adulterated and finally becomes useless. This process is well-illustrated by concepts which undergo a career of stages or phases, a life-cycle in other words, and may move from one organization to another.
Counter-claim The special feature of the collective attention span limit is its dynamic nature. In one sense it is perhaps to be deplored that collective attention cannot be focused long enough to give rise to effective action. But in another sense, attention shifts once the issue no longer serves the poorly understood needs for dynamism within the international community (issues are 'consumed' to fuel the dynamics). And, to the extent that the attention shift takes place in search of innovative renewal, this is to be welcomed - particularly since this brings alternative and complementary factors into focus. But, given these extremes, not enough is known to indicate when a shift is premature (in terms of action requirements) and when it is necessary (in terms of the healthy dynamics of world society). Clearly a complex world problematique demands both sustained attention to comprehend the dimension of the problem and shifts in attention to respond to complementary needs.
Broader Limits to societal learning (#PF7074).
Aggravates Superficiality of collective comprehension in societal learning (#PF7110).
Aggravated by Limit to collective comprehension span in societal learning (#PF7038).

♦ **PF7005 Loss of spiritual guidance**
Nature In their contemporary state of decline, Eastern religions have become destructive and backward-looking forces in the world. Islam has produced fanatical fundamentalists, Hinduism has generated the yoga and ashram industries, and Buddhism has lost vitality and become contaminated with politics. Other Eastern religions are becoming extinct such as the creeds of the Taoist and to some extent, the Confucianist, the Mandaeans, the Tibetan Bons, the Druze, the Zoroastrians and several others.
Background Western civilization emerged after the Reformation to become a secular global force that challenged and eventually dominated much of Asia militarily and ideologically for several hundred years. Asian cultures were eroded, and particularly the highly evolved religious systems of Islam, Hinduism and Buddhism. By the 19th and 20th centuries there was a dearth of authentic religious teachers, charismatic spiritual personalities, to transmit the wisdom of the East.
Claim Eastern spirituality is a heritage for all mankind. It is being lost due to the influence of the materialism of the West whether this is the spirit of the pursuit of private profits by the corporations, or the shadow of the Marx-machine of the socialist state that destroys what it doesn't understand.
Counter-claim Western civilization is the latest stage in the cultural history of mankind. It has risen to oppose reason to revelation, science to superstition, and humanism to a heartless, false spirituality that is often embodied in institutionalized religion. As a result of western civilization, slavery is being abolished everywhere, the rights of women are being recognized, exploitation is being curtailed in every form, national aspirations are being realized and freedom and justice for all is becoming a world standard. This is happening rapidly, everywhere, except in the Middle East, the Indo-Pakistan sub-continent, and parts of Indo-China where adherence to ancient religious law perpetrates the gravest injustices and inhumanity. When Eastern religions are practised in the West, stripped of their accretions that favour castes and class, men over women, and priests and monks over lay people, their spiritual essences shine forth if they are truly concerned about the human condition. Thus, nothing is lost to mankind in this phase of civilization, but it may now be more accessible in the West than the East. In the next phase there will no longer be an East and West but a global civilization that will draw upon the achievements of both hemispheres to unite the forces of mankind for the goals of the new age.
Aggravates Lack of religious discipline (#PF8010)
Barriers to transcendent experience (#PF4371).
Aggravated by Spiritual disobedience (#PF7467).

♦ **PF7011 Obstacles to leadership**
Broader General obstacles to problem alleviation (#PF0631).
Narrower Credibility gap (#PB6314) Lack of leadership (#PF1254)
Leadership age gap (#PJ9855) Sense of powerlessness (#PF8618)
Dependence on authority (#PF8995) Minimal citizen confidence (#PF8076)
Peer behaviour expectation (#PG9138) Unprepared adult leadership (#PF6462)
Parochial leadership posture (#PF2627) Undeveloped youth leadership (#PJ0151)
Lack of leadership initiative (#PF7988) Loss of leadership credibility (#PF9016)
Minimal leadership coordination (#PJ9498) Relative authority of leadership (#PF2811)
Leadership as symbolic of wealth (#PF2870) Insufficient leadership training (#PF3605)
Powerlessness within bureaucracy (#PG9383) Individual fear of future change (#PD2670)
Mediocrity of government leaders (#PF3962) Lack of local leadership role models (#PF6479)
Unexplored avenues of leadership potential (#PF2797)
Illiteracy as an impediment for leadership (#PE8177)
Undeveloped potential of informal leadership (#PF1196)
Lack of motivation in leadership development (#PF2208)
Demoralizing images of rural community identity (#PF2358)
Extreme detachment from represented constituency (#PF0889)
Lack of local leadership and decision-making structures (#PF6556)
Declining community confidence in its ability to change (#PF9066)
Lack of local services for community leadership training (#PF2451)
Lack of autonomous world-level actor to identify and clarify world interests (#PF0053)
Shortage of industrial leadership and entrepreneurial ability in developing countries (#PC1820).
Related Domination of government policy-making by short-term considerations (#PF0317).
Aggravates Citizen powerlessness (#PJ8803).
Aggravated by Insufficient nobility (#PF5778)
Unequal distribution of fame and honours (#PF3439).

♦ **PF7014 Obstacles to conceptual integration in societal learning**
Nature A society may receive information, but collectively repress it, if it is painful or unfavourable or even if it has no perceived contemporary meaning. This missing data from documents or in the societal memory can lead to a lack of conceptual and behavioural integration.
Broader Limits to societal learning (#PF7074).

♦ **PF7016 Misapplication of venture capital**
Nature Capital invested in high–risk commercial ventures; (start–ups of companies, new product development and production, market entry or export, etc) is placed with the hope of eventual high or exceptionally high returns but often with little managerial control or assistance offered. In the explosion of high technology enterprises and products, venture capitalists have placed funds in start–ups with inadequate assessment of corporate and product potential. Small, new companies are frequently over–funded simply because they are in a glamour industry. The result encourages high rates of expenditure and excessively high risks to attain rapid growth. High failure rates follow.
Incidence The venture capital industry include individual financiers and banks but also large insurance companies and pension funds. In the USA, available venture capital is estimated to have grown at an average rate of 50 percent a year since 1978. Outside the USA venture capital is still provided mainly by banks whose ability to comprehend modern technology and offer managerial assistance is limited.
Aggravates Lack of venture capital (#PG7833).

♦ **PF7018 Personified evil**
Devil — Satan
Nature Unwelcome or calamitous events are often said to be due to the presence of an evil spirit or to be inspired by the personification of all evil in the figure of a chief devil or satan. Such a figure is a source of terror in innumerable societies and, as a conception alone, may be at the bottom of many mental disorders.
Background Supremely evil figures include Satan (Shaitan), Mephistopheles, Beelzebub, Belial, Asmodeus, Lucifer, Moloch, Mammon, Iblis and Antichrist. In some myths and religious lore, the supreme evil was destroyed in a previous age, in others it exists still, and in still others, there is evil to come, such as the Great Beast or Antichrist expected at the second coming, before the end of the world.
Claim The authority of the major faiths, and of primitive religion; and the testimony, as well, of occultists and clairvoyants, are uniform on agreement as to the existence of evil spirits. The records of the Society for Psychical Research show that ordinary people can experience a manifestation of evil in an individual form. In addition to being personified as an invisible being or as a living person, evil can be objectified. That is, evil can be said to be materially present in the universe, not only as a being or as evil incarnate, but also as an object, as substance, a source, a power, or force. For example, all instruments of death are evil, from the gun to nuclear weapons. Polluting filth and narcotics are evil. And those bent on domination also give evil an embodiment.
Counter–claim There is no central office for evil in the universe, and no chief bureaucrat who is in charge of its dispensation. Good and evil are inherent in everything, not as moral qualities but as tendencies and potentialities, in themselves or in the use to which they are put, to move towards development or support of a system or order of reality, or away from it. Good and evil are embodied but not personified, only in the individual will of living things. They exist as a duality so that there can be no evil present but that there is potential good as well. Failure to recognize this leads to attributing wholly evil characters to one's natural political, religious, economic or personal opponents, dehumanizing them by divesting them, in one's belief, of any good will or capacity for toleration or cooperation. This reification of an abstract concept, materialized as the label 'evil', and affixed to rival cultures or ideas and their proponents, leads to intra–societal witchhunts, religious persecutions, concentration camps and genocide; and internationally, to war and the threat of nuclear war.
Refs Ashton, John *The Devil in Britain and America* (1980); Russell, Jeffrey Burton *The Prince of Darkness*; Stedman, Ray C *Spiritual Warfare* (1985).
Narrower Antichrist (#PF9139).
Aggravates Satanism (#PF8260).
Aggravated by Bad omens (#PF8577)
Promotion of negative images of opponents (#PF4133).

♦ **PF7022 Obstacles to efficient utilization of time**
Broader Constraint of time on individual and social development (#PF5692).
Narrower Wasted time (#PF8993)
Wasted waiting time (#PF1761)
Limited leisure time (#PF9062)
Time consuming procedures (#PJ8206)
Time lag in legal provisions (#PF6042)
Differing conceptions of time (#PF6665)
Multiplicity of time standards (#PF2621)
Limited study time (#PG5327)
Limited shared time (#PJ9126)
Inadequate sense of time (#PF9980)
Insufficient personal time (#PJ9534)
Ineffective class scheduling (#PG5424)
Frequent schedule disruptions (#PJ0871)
Insufficient leisure time for women (#PE8907).
Related Unpreparedness for surplus leisure time (#PF5044).
Aggravated by Hyperefficiency (#PF1706).

♦ **PF7024 Intolerance of imperfection**
Nature The aversion to imperfection causes excess consumption of materials and labour for replacement objects or structures, even where the utility of the originals remains unimpaired. Socially, the aversion to the disfigured or handicapped results in a tendency to ostracize them. The intolerance of imperfection extends to people who have been damaged in other ways, such as status, reputation, purity, etc, and a certain stigma may be attached to organizations and nations, because of a notable 'blemish', which may obstruct international working relationships, communications and concord.
Broader Intolerance (#PF0860) Failure of materials (#PD2638).
Related Squeamishness (#PF5735).
Aggravated by Chaos (#PF6836)
Moral imperfection (#PB7712).
Ugliness (#PA7240)

♦ **PF7027 Recurrence of misapprehended world problems**
Societal selective memory
Nature Societal collective memory seems to be exposed to processes tending to its very rapid erosion. The alternative resource is the international information systems on which much collective consciousness is supposedly based. Their most striking feature is their fragmentation, whether as systems almost completely independent of each other, or individually in their isolation of subject categories from each other. There is also a functional gap between the logical subject hierarchies and the network of operational realities. As a consequence of these limitations, society has both a selective memory exhibited in its documents, and a restriction on integrating new learning into its behavioural patterns. This condemns society to continually reencountering the same kinds of problem.
Aggravated by Issue avoidance (#PF1623)
Resignation to problems (#PF8781)
Conceptual repression of problems (#PF5210)
Pursuit of affluence (#PF5864)
Non–recognition of problems (#PF8112)
Polarized protest against problems (#PF9691)
Failure to profit from patterns of history (#PF1746)
Complex interrelationship of world problems (#PF0364)
Inadequate public information concerning problems (#PF5701)
Insensitivity to non–immediate hazards to society (#PF9119)
Inadequate education concerning the nature of problems (#PE8216)
Inadequate application of available knowledge to solve problems (#PF8191)
Failure to integrate knowledge to empower humanity in response to the global problematique (#PF8753).

♦ **PF7029 Fear of the abnormal**
Dread of otherness — Fear of the unfamiliar
Nature "Otherness", a departure in human form, appearance, characteristics and behaviour from the range experienced as the local norm causes intense concern and an attempt at comprehension. From this the mind quickly moves to apprehension and fear, or to repulsion and hatred.
Incidence Fear of "otherness" is seen expressed against physical abnormalities, temporary or permanent, mental abnormalities, and behavioural and emotional abnormalities. However, it is also a reaction against a different ethnic background, a different language, a different dialect, and even against those identified with a different locality, or any stranger (xenophobia). Fear of non–human abnormalities includes that of monsters and the supernatural. Some such fears may have proposed names, such as dysmorphophobia relating to monsters, while the entire dread of "otherness" may be termed 'allophobia'.
Broader Fear (#PA6030).
Narrower Fear of the unknown (#PF6188).
Aggravates Prejudice (#PA2173)
Lack of political will (#PC5180).
Social discrimination (#PC1864)

♦ **PF7033 Absence of policies to associate international nongovernmental organizations in regional development**
Nature Regional IGOs, particularly for the developing countries, tend not to recognize INGOs (whether regional or not) and have no policy to associate them in any programme activity or facilitate regional INGO activity. This reinforces the communication gap between IGOs and INGOs.
Broader Obstacles to effective international nongovernmental organizations (#PF7082).
Aggravates Ineffective governmental use of nongovernmental resources (#PF4095).

♦ **PF7034 Lack of national legal provision for international nongovernmental organizations**
Nature The establishment of an INGO secretariat and associated staff, or the holding of a conference, or the organizations of a (field–level) programme, or the maintenance of membership ties in a particular country, are not governed and protected by national legislation recognizing the special character of INGOs (the only exception being Belgium). The INGO is obliged to register itself as a national organization of that country or a 'foreign' association, if it is permitted to establish itself at all. Many obstacles are thus created to INGO activity, particularly in the Eastern bloc and developing countries. This is a major obstacles to (a) increasing the representatives of INGO membership and to (b) ensuring that more INGOs have their headquarters or secretariats outside the North–West group of countries whose legislation is somewhat more open to association activity.
Incidence An interesting case in point is that of Kenya following the establishment of UNEP in Nairobi. Considerable difficulties were experienced by environmental INGOs wishing to establish offices or headquarters there – even the NGO Environmental Liaison Board which had the full support of UNEP. It is also interesting to note how carefully trade unions dissociate themselves from other INGOs on this point because their 'freedom of association' is the concern of a special ILO committee.
Broader Obstacles to effective international nongovernmental organizations (#PF7082)
Lack of legal provision for international nongovernmental organizations (#PE0069).

♦ **PF7038 Limit to collective comprehension span in societal learning**
Nature As in the case of the individual, there is a limit to the number of domains of knowledge, however pre–digested, which a group can handle conceptually as a comprehensible whole. Most groups have developed, whether consciously or unconsciously, remarkable skills at sweeping awkward factors under any convenient conceptual carpet in order to create the impression that they are in control of a situation. Presumably society could reach a condition in which more inconvenient items of knowledge are being repressed in this way than are effectively dealt with. Another aspect of the problem is that it is now recognized as misguided to elaborate information systems independently from the groups and institutions that they must serve. The man/machine interface has become such a critical factor that it is now vital to consider 'groupware' design as a necessary complement to hardware and software design. Group comprehension of complex problems requires that a user group 'reconfigure' to grasp the pattern of information available. Information systems should facilitate this process but as yet no such flexibility is envisaged. The gravity of the situation is particularly evident in the difficulty large conferences experience in organizing themselves as groups marshalling the (documentary) information at their disposal to focus on problem complexes.
Broader Limits to societal learning (#PF7074).
Aggravates Collective attention–span limit in societal learning (#PF7003).
Aggravated by Superficiality of collective comprehension in societal learning (#PF7110).

♦ **PF7042 Evil**
Dependence on evil — Belief in evil
Nature Evil is a construct which figures prominently in determinations of ethical behaviour as well as in emotional pathology and religious hysteria. Its definition is less important than its presence as a conception. Evil has been considered a side–product of many causes, coming not from one principle and having no unity in itself. When a body of any kind is infected with evil many of its elements do not keep their relative and just proportions; and to the extent that each part then strives to control the whole, disharmony prevails.
Incidence Evil, in one form or another, is a major concern of most religions. In many cultures there is concern to guard against evil (including the "evil eye") and to avoid provoking evil spirits. The importance to the international community is reflected in a widely publicized perception by the President of the USA that the communist bloc of nations constituted an "evil empire".
Background Anything that is, or causes distress, calamity, loss, damage or sorrow is an evil. As a hypothetical universal power said to bring such events, evil is raised to the level of metaphysics when it is counted among the primary forces. It is also raised, by many, to the level of theology when it is anthropomorphized, personified, or conceived to be God, viewed from 'behind'.
Claim Evil is the counterforce in nature which is related to order, form, purpose, will and similar tendencies and powers. Evil produces notable deficiency, such as disorder, shapelessness, uselessness, and inertia, but it also produces excess, so that excessive order as found, for example, in a crystal, if applied to society, renders it lifeless. Excessive form, excessive purpose and will in life or society account for numerous evils. For those who believe in evil as a real force in itself there are taboos and ritual behaviour patterns that do not lend themselves to rational defence, such as exorcism, wearing of protective amulets or particular colours, and other magical acts including the sacrifice of birds and animals, and the ringing of church bells.
For all those to whom evil is a problem the significance lies in what they consider to be the chief evil. In the Western religious texts of Judaism, Christianity and Islam, disobedience born out of pride is a chief attitudinal evil. However, a more subtle evil is forgetfulness, a lack of attention or inability to recollect, metaphorically called 'sleep', in which God, His words and even the highest spirit in man is lost to mind. Among the Greek religions, for example the Orphic, the sleep of the mind or death of memory (by Lethe) was among the chief evils. Popularly in all these religions the chief behavioural evils include blasphemy, sexual misbehaviour and obvious crimes against

persons and property. In psychology the chief evil appears to be the obstacles to the maturation of the personality. In socialism the chief evil is the obstacle to the distribution of wealth. In anarchy the chief evil is the concentration of power. To alleged evil beings themselves, from fiends to head devils, the chief evils are said to be light, the willingness that one person would sacrifice himself or die for others, forgiveness, and (according to Christians and Buddhists) prayer or meditation.

Counter-claim 1. Evil may be considered as a symptom of a problem rather than an essential feature of human nature. As such it may be essentially the product of of numbing or repressing access to primary positive emotions, rather than any basic defect in human makeup.

2. In as much as there are only human and natural agents in human affairs, the hoary concept of evil argues for the continuing lack of knowledge of human and natural law. More importantly, it attests to the lack of maturity of mankind which is still unwilling to accept responsibility for itself. Some schools of psychotherapy argue that the very milieu that makes actualization of the Self possible also demands that certain components of the Self remain unactualized or be actively repressed there. Such unacceptable elements have been stigmatized as bad or evil, corresponding to the "shadow" side of an individual or collectivity. In whatever culture a child grows up, it is usual for him to identify his conscious personality with whatever his group holds to be good and for his shadow complex to become the repository of all that is evil. In the process of human development, the effort to make the shadow conscious necessarily results in the manifestation of evil effects until whatever has been repressed can be appropriately integrated and contained. The risk of global catastrophe is increased by the collective tendency to project shadow qualities on to social systems, political institutions and individuals. The evil attributes of the shadow are of vital significance for the development process because of the inner transformation in individuals and collectivities brought about by the attempt to contain and integrate them. Such an experience is infinitely more important than political and social reforms which are valueless in the hands of people who are not at one with themselves. As opposite poles of the morality archetype, good and evil are ineradicable characteristics of the human condition. To hope to embrace one and eliminate the other is to breed personal division and public disorder. Both individuation and planetary salvation demand that people be aware of their capacity for both good and evil and that ethical choices be made between the two. Inasmuch as it enhances social responsibility, consciousness of the shadow benefits the group. Awareness of the personal or collective shadow means suffering the tension between good and evil in full consciousness, and through that suffering the basis for a richer and more profound integration emerges.

Refs Anshen, Ruth N *Anatomy of Evil* (1985); Russell, Jeffrey Burton *The Prince of Darkness*.

Broader Vice (#PA5644) Badness (#PA5454) Wrongness (#PA7280).
Narrower Atrocities (#PD6945) Evil governments (#PF5633) Complicity with evil (#PF0926).
Aggravates Unluckiness (#PF9536) Wrath of God (#PF8563)
Promotion of negative images of opponents (#PF4133).
Aggravated by Non-resistance to evil (#PF7754)
Numbness towards others (#PF1216).

♦ PF7043 Death of living creatures

Broader Human destructiveness (#PA0832).
Narrower Human death (#PA0072) Animal deaths (#PE7941)
High mortality rate (#PJ6252) Extinction of species (#PB9171)
Indeterminacy of death (#PF0192)
Death and disability from inhumane confinement (#PE5648).
Related Annihilation (#PF9169) Wildlife extinction (#PC1445).
Aggravated by Disease (#PA6799) Killing by humans (#PC8096)
Damage caused by space objects (#PE0250).

♦ PF7047 Neglect of agricultural and rural life in developing countries

Policy discrimination against rural areas in developing countries — Neglect of the small agricultural producer — Neglect of peasant farmers — Underemphasized importance of rural community — Forgotten rural importance

Nature Programmes to increase food production have concentrated where gains are likely to be easiest: among larger farmers in relatively fertile and well-watered valleys and plains. Small farmers in upland and other areas that are difficult to farm, shifting cultivators, women farmers, nomadic herders, and the rural landless have been largely ignored. Agricultural support systems tend not to tale account of the circumstances of subsistence farmers and herders. Such farmers cannot afford the high cash requirements of modern agriculture (fertilizer, etc). Many are shifting cultivators who do not have clear title to the land they work, planting a variety of crops on a single plot rather than using methods which have been developed for single crops on a large plot. Herders tend to be nomadic and difficult to locate and advise on modern agricultural methods. As with subsistence farmers, they depend on traditional rights which are threatened by commercial developments, and, although their breeds are hardy they are rarely highly productive.

Broader Personal isolation in communities of industrialized countries (#PD2495).
Narrower Neglect of women farmers (#PF6024).
Aggravates Unsustainable agricultural development (#PC8419)
Lagging transformation of agriculture in developing countries (#PD0946)
Deterioration of domestic food production in developing countries (#PD5092)
Constraints to increased agricultural output in developing countries (#PD5114).
Aggravated by Oppressive prevalent images (#PF1365)
Contempt for agricultural labour in developing countries (#PD1965)
Domestic agricultural price policy difficulties in developing countries (#PE2890).

♦ PF7050 Mis-communications to societal learners

Cultural lag in understanding information

Claim The key element of human communications – the ordering and transmission of information – is tending to become a source of mis-communication. The scientific and technological breakthroughs which have led to the informatics revolution are way ahead of the learning process of human society. This cultural lag is the most serious challenge to a comprehensive view of the implications of informatics. It is a matter of values of organizational capacity and transformation in mental structures.

Aggravates Ignorance (#PA5568).
Aggravated by Insufficient communications systems (#PF2350).

♦ PF7052 Obstacles to unilateral nuclear disarmament

Nature While proposals to unilateral disarmament in the nuclear weapons age gather considerable popular support, official enactments for these purposes are not debated in national legislative bodies or introduced there as alternatives to the nuclear arms race.

Claim Elected legislators and other elected officials frequently feel little direct, personal political support from the peace movement. The motives, however, of chauvinism, profit-making and exaggerated conservatism are well represented. In some countries, international cooperation for any purpose is suspect and nuclear disarmament is considered, from militarist perspectives, to be bordering on the treacherous and cowardly. Nuclear weapon obsolescence may eventuate to shift the problem into other dimensions.

Counter-claim Strategic imbalance would occur if any one of the key members of NATO or the Warsaw Pact, other than their leaders, would unilaterally disarm and thereby render its strategic commitments nil. An East German or United Kingdom unilateral disarmament, for example, (if they were possible and if the GDR possessed its own nuclear weapons) would be disastrous. In the NATO case, nuclear weapons would have to be re-deployed on the continent to make up for the UK loss. Such redeployment, closer to the Warsaw Pact, could only be provocative of counter-measures. Another reason for nations with nuclear weapons or developing nuclear weapons to retain and produce them is to be independent of nuclear superpower dominance. The strategic needs of France, China, or India, as perceived along classic, pre-nuclear age lines, requires nuclear weapons for self-defence. This argument also considers isolated, smaller nations who could be invaded by massive armies, such as South Africa, Pakistan and Israel among others, to be justified in possessing nuclear weapons as a deterrent although enhanced, conventional weapons are deadly enough.

Aggravates Imbalance in strategic arms (#PC1606).
Aggravated by Naval arms race (#PD8412) Fear of reprisals (#PF9078)
Nuclear arms race (#PD5076).

♦ PF7053 Oppressive reality

Claim Political and economic oppression creates a perception of a particular societal system as being both inherently oppressive and real. Oppression absorbs the attention of those within its dominion and acts to submerge their consciousness. Factual oppression becomes more oppressive by adding to it consciousness of oppression. This in turn reinforces the belief by the oppressed of their impotence in effecting change. Those who do not emerge from the oppression of the reality, and turn on it using both reflection and action will continue to be a prey to its force. Reflection without action leads to false understandings of reality, either sentimental or cynical in nature. Action without reflection leads to ineffectivity and to eventually reinforcing oppression. Oppression seen as and acted on in both its objective and subjective dimensions by the oppressed can be emerged from and thereby transformed.

Broader Objectivism (#PG5290) Subjectivity (#PF2827)
Excessive virtue (#PF7127).
Related Untimeliness (#PA7006).
Aggravates Mental illness (#PC0300) Mental suffering (#PB5680)
Avoidance of reality (#PF7414) Stress in human beings (#PC1648)
Frustration of the role of opposition (#PF1594).
Aggravated by Inaction on problems (#PB1423) Elusiveness of reality (#PJ8032).

♦ PF7054 Human contingency

Nature A fundamental aspect of the human predicament is the dependency of life and social organization on accidents and chance. There is little, if anything, certain. One thing hangs on another in an invisible progression; and like a spider-web, if one skein is pulled, the whole network is distorted or even destroyed. Physical human life is unpredictable, open to disease, decay, death and pain. Emotional life knows no equanimity; happiness and suffering exchange places without plan. Societal life is filled with conflict, and economic life is always precarious. Peace is shattered by war, and in this century human existence is threatened in its entirety. The increasingly perceived contingency of life is a destructive influence amidst any hope or optimism that human problems can be solved.

Counter-claim Nature, of which mankind is part, shows in its evolutionary patterns, purposes and trends of development that can be summarized, from the human and animal point of views particularly, as the emergence of mind, or intelligent control. Nothing is contingent in the biological life of humans. All activity points to the ascendency of intelligence and the rise of human nature to the omega point of evolution and mankind's ultimate purpose.

Broader Risk (#PF7580) General obstacles to problem alleviation (#PF0631).
Related Absurdity (#PF6991) Meaninglessness (#PA6977)
Human errors and miscalculations (#PF3702).
Aggravates Ignorance (#PA5568) Insecurity (#PA0857)
Human death (#PA0072) Human suffering (#PB5955)
Social insecurity (#PC1867) Declining breeds of cultivated plants (#PD5936).
Aggravated by Disasters (#PB3561) Political upheavals (#PC7660)
Lack of eugenic measures (#PD1091) Vulnerability of human organism (#PB5647)
Underestimation of the human potential (#PF7063)
Disintegration of technological capacity (#PD7719).

♦ PF7057 Neglect of international nongovernmental organization network

Nature The major IGOs have specific mandates which tend to de-emphasize any need to relate to other organizations, whether IGO or INGO, having related programme concerns. As a result, little attention, if any, is given by them to the importance of improving the inter-organizational structure focusing on a network of related problems. Where outside contacts are made by the IGO, they are made because a project can best be completed by a specific INGO, for example. The possibility that by facilitating the development and operation on the INGO network as a whole it might not even be necessary for the IGO to initiate many of the specific projects, is not considered.

Background It is of course a characteristic of all organizations to wish to undertake projects for which it can obtain immediate credit, rather than projects which ensure that other bodies undertake whatever tasks appear necessary. At the present time there is insufficient consensus within IGOs for any policy change to remedy this. This applies particularly to the relations between bodies within the United Nations system, whether: within different divisions of a particular secretariat (such as the Office of Public Information or NGO Liaison Section); between bodies reporting to the UN General Assembly (for example, ECOSOC and UNDP); between bodies reporting to different plenary bodies, despite ECOSOC's mandate to review such relationships (like FAO and UNESCO).

Broader Obstacles to effective international nongovernmental organizations (#PF7082).
Aggravates Ineffective governmental use of nongovernmental resources (#PF4095).

♦ PF7058 Lack of international legal provision for nongovernmental organizations

Nature NGOs and INGOs have no legal status within the framework of international law. They are therefore not recognized as having any international 'existence' in a legal sense, with the consequence that any governmental or scholarly attention which depends on such recognition is absent. The absence of such legislation ensures the INGOs are unprotected (as 'outlaws') and do not operate within anything but a self-imposed code of responsibilities. Their activities are not aided by facilitative arrangements, as is the case with the international activities of commercial enterprises. The absence of such legislation deprives national governments of any stimulus to generate national legislation to accommodate INGOs based in a particular country. Since they are not recognized internationally, some countries view with great suspicion the participation of their nationals or national groups as members of such bodies. There is also suspicion concerning the (field-level) programmes of such INGOs in a particular country, which may be construed as interference in internal affairs or as a cover for politically subversive activity.

Broader Obstacles to effective international nongovernmental organizations (#PF7082)
Lack of legal provision for international nongovernmental organizations (#PE0069).

♦ PF7059 Inflexible deliberative systems of government

Nature The inflexibility of the decision making processes of governments vary from system to system but all fail to actually involve local people, except in the most minimal of ways, in the political processes. As governments become more complex the less flexible they are in discerning or responding to the needs of the grassroots. People become more disenfranchised, more tyrannized by government and less and less involved in the processes of governing. The

government becomes less and less legitimate and then increases its complexity and inflexibility.
Broader Inflexible central government (#PD1061).
Aggravates Anarchism (#PC1972).
Aggravated by Solipsism (#PF5657).

♦ PF7063 Underestimation of the human potential
Unrealized human development potential
Nature It is not known if human evolutionary processes of development are subject to abrupt directional changes. Many mechanisms are not fully understood, including genetic mutational factors. The underestimation of the human potential aggravates all problems, from the personal to the global level. In the form of discrimination against allegedly inferior classes, ethnic groups and races, it thwarts developmental potential.
Broader Unfinished imperfect universe (#PF5716).
Narrower Moral imperfection (#PB7712) Health inequalities (#PC4844)
Behavioural deterioration (#PB6321).
Aggravates Human contingency (#PF7054).
Aggravated by Ignorance (#PA5568) Human suffering (#PB5955)
Lack of eugenic measures (#PD1091) Human errors and miscalculations (#PF3702)
Constraint of time on individual and social development (#PF5692).

♦ PF7066 Fear of eclipses and occultations
Nature The sudden disappearance from the sky of familiar luminous bodies causes the gravest apprehension and dread among primitive peoples as it did among the most ancient ancestors of mankind. Typically it is the total eclipse of the sun that is most frightening, but in the partial eclipses of the sun and moon also, a giant dark shadow is seen moving over the celestial body as though it were devouring it. This shadow is given the reality in diverse cultures of being a great monster, or beast. The fear that the sun might not return after such an eclipse caused rites of propitiation. The dread of eclipses may be one of the fundamental fears that primitive religion was organized to counter.
Background When the apparent diameter of a covered body is much smaller than that which hides it, it is said to be occulted. Occultations involve stars, nebulas or distant planets which may be hidden by the moon or nearer planets or, to take a particular example, the occultation of the Crab Nebula by the Solar corona (as measured by radio astronomy). Therefore early man may have been frightened by occultations as well as eclipses. Besides the sun or moon, notable stars may have vanished for such reasons. However, since there is a periodicity to these events, records would have assured the ancient peoples that they were only passing evils if not merely natural phenomena; so there appears to have been an occultation that was not periodic, but was catastrophic, to which all subsequent fear referred back.
Broader Bad omens (#PF8577) Superstition (#PA0430).

♦ PF7069 Neglect of non–Western structures in international nongovernmental organizations
Nature Most INGOs are organized in what can be termed a Western concept of organization. Such organizations, wherever they are based, then appear to be transplants which are not natural or meaningful in non–Western societies. As such they are easily suspect and subject to criticism, thus deterring full contact with them.
Counter–claim Although it would be valuable to make use of non–Western models of organization at the transnational level, the problem is that such models have not yet been sufficiently developed. Even regional organizations in African, Asian and Arab countries tend to be elaborations of the Western model rather than alternative models. National governmental agencies in developing countries, for example, are largely based on Western models, for lack of anything better. It is questionable whether the organizational concept used in Eastern socialist countries is sufficiently distinct from the Western model to escape such criticism. (To put matters in perspective, it is useful to look at the equivalent technological problem. The design of airplanes is governed by principles elaborated in developed countries. Whilst it would be delightful to travel in an intercontinental airplane designed in a developing country, there are none. Is this to mean that those designed in developed countries should not be used in developing countries ?)
Broader Obstacles to effective international nongovernmental organizations (#PF7082).

♦ PF7071 Reduced understanding of globality
Nature Understanding of what is required to function effectively as a global society is reduced. Generally it is believed that only super humans can do social change. The crisis in systems of authority is creating leadership incapable of dealing with highly complex issue and of leading on behalf of the whole society. Government modes of operation weeks to gratify only immediate individual desires.
Broader Overemphasis on self-sufficiency with respect to interdependence (#PF3460).
Narrower Narrow context for counsel (#PF0823) Parochial leadership posture (#PF2627)
Inflexible central government (#PD1061) Relative authority of leadership (#PF2811)
Leadership as symbolic of wealth (#PF2870)
Passive historical perspective by leaders (#PF6765)
Exclusion of the masses from setting criteria in judicial judgements (#PD1060).
Related Undemocratic policy–making (#PF8703) Exclusion of opposing views (#PF3720)
Breakdown in community security systems (#PD1147)
Legal contract system reduced to individual needs (#PE5397)
Domination of government policy–making by short–term considerations (#PF0317).

♦ PF7074 Limits to societal learning
Nature That societal learning is limited, is proven by the growing world population of illiterates and increasingly by the presence of innumerates, even in the most developed countries. Reading and numerical skill limitations are but part of the problem that includes the societal equivalents of individual learning disorders and restrictions.
Incidence About 20 percent of the world population is illiterate. Additional percentages can read but have learning disorders. No data is available on numeracy, but a wide incidence of functional innumeracy is probable.
Broader Restrictions on the acquisition of knowledge (#PF1319).
Narrower Societal over–commitment to learning (#PD7051)
Disconnectedness of societal learning (#PF7122)
Quantitative limit to societal learning (#PF7098)
Collective attention–span limit in societal learning (#PF7003)
Inadequate use of visual imagery for societal learning (#PD7086)
Obstacles to conceptual integration in societal learning (#PF7014)
Limit to collective comprehension span in societal learning (#PF7038)
Superficiality of collective comprehension in societal learning (#PF7110).

♦ PF7076 Psychogenetic constraints on behaviour
Bias in acts of conscience
Claim The capacity and exercise of both conscience and clarity of reason, as dominant influences on behaviour, are biased by induced and self–induced values. The process begins in childhood where entire value systems are imposed by the family and by schooling, and continues to early adulthood. By that time the individual has been led by his personality make–up and his psychological complexes to acquire any one or more of a range of vows, professional oaths, senses of duty to country and to causes, the calling of a mission, various moral obligations, reservations and scruples and similar beliefs. To the extent that these addictions close the individual's mind to dialogue with others of differing views; terminates further search for improved perspectives, facts and conclusions; and propels the individual into uncontrolled emotionality, there is an element of fanaticism present. In this sense the most apparently diverse behaviour has a common element. Thus pacifists and conscientious objectors may be as intense as militarists and war–mongers. Those who conduct vendettas or act as terrorists may be compared to vigilantes and others who would suspend trial by jury and other due process. The superstitious are a counterpart to scientists who have censured colleagues for holding unorthodox viewpoints. Transnational religious and ideological missionaries have just as dedicated and one–sided views as the transnational corporations. The bias in peoples' lives towards one dominating idea or dream and exaggerated attempts to fulfil it are written in red in the ages of history after the names of dictators, messiahs and others who have proclaimed their 'truth' as the Truth.
Refs Sakai, Toshiaki and Tsuboi, Takayuka *Genetic Aspects of Human Behavior* (1985).
Aggravates Extremism (#PB3415).

♦ PF7077 Concealment of esoteric knowledge
Nature Secret societies withhold knowledge which may be relevant to the solution of world problems. This knowledge encompasses human potential and, based on it, ways of organizing society effectively.
Claim If there is even partial validity to claims of secret knowledge in some instances, the keepers are self–serving by its concealment in order to perpetuate themselves and maintain a base of power. They are elitist in their opinion of themselves and cynical concerning others.
Counter–claim These claims are not only encountered among minor sects and cults but are at the centre of the 'mystery' of world religions. Such secrets have little implication for societal development as they only exalt the selfish quest for a supposed personal holiness. Such societies will never reveal what they know to an impious and hostile world that exploits all knowledge for motives of profit and dominance over others, or to those who are numbered among the multitude of the dilettanti and the merely intellectually curious.
Broader Secrecy (#PA0005).
Aggravates Lost knowledge (#PF5420).

♦ PF7079 Religious syncretism
Religious relativism
Nature May take various forms, such as the mixing of beliefs and practices, slow absorption into other religious systems, loss of conviction as to a core belief (such as the finality of Christ), and various forms of sophistication that seeks to feel at home in every variety of belief. Behind all these forms lies the presumption that it is the wisdom of man that establishes the truth.
Claim Syncretism is the conscious or unconscious pasting together of pieces of different religions to produce an ersatz religion that is recognizable from none of the original religious perspectives. Such a combined faith is not only a poor alternative to religious search for truth and an impoverishment of the human race, but it is strongly resisted by any religion defending its own spiritual integrity.
Counter–claim Many interreligious dialogues have amply proved that humanity is not on its way to produce a syncretistic world religion, a new normative faith for all members of the world community.
Refs Hartman, Sven S (Ed) *Syncretism* (1969).
Aggravates Ineffective dialogue (#PF1654).
Aggravated by Erosion of religious belief by ecumenical dialogue (#PF9274).

♦ PF7081 Absence of national nongovernmental organizations in developing countries
Nature Many of the newly independent countries are naturally characterized by a poorly developed organizational infra–structure. Priority is given to development of government agencies and productive enterprises. The creation of non–governmental, non–profit bodies therefore poses a special problem, both as a distraction and a drain on scarce resources, and as a possible focus for dissent threatening the stability of the government. Such bodies are therefore deliberately created by government for political ends or, if independent of government, are viewed with suspicion if they are permitted to exist at all. This situation makes it difficult for non–governmental representatives of the countries to related to INGOs.
Broader Obstacles to effective international nongovernmental organizations (#PF7082).
Aggravates Ineffective governmental use of nongovernmental resources (#PF4095).

♦ PF7082 Obstacles to effective international nongovernmental organizations
Broader General obstacles to problem alleviation (#PF0631).
Narrower Manipulation of nongovernmental organizations (#PD6800)
Duplication of international nongovernmental activities (#PE7104)
Lack of international nongovernmental organization identity (#PE7032)
Unrepresentative international nongovernmental organizations (#PE7021)
International nongovernmental organization membership apathy (#PE7092)
Inadequate statistical data on nongovernmental organizations (#PF7117)
Neglect of international nongovernmental organization network (#PF7057)
Multinationalism of international nongovernmental organizations (#PE7129)
Limited recognition of international nongovernmental organizations (#PF7093)
Assessment of international nongovernmental organizations as naive (#PF7116)
Misuse of nonprofit associations as front organizations by government (#PE0436)
Lack of identity of international nongovernmental organization network (#PE7044)
Uncontrolled structures of international nongovernmental organizations (#PE7105)
Lack of legal provision for international nongovernmental organizations (#PE0069)
Lack of international legal provision for nongovernmental organizations (#PF7058)
Political ineffectiveness of international nongovernmental organizations (#PE7008)
Divisive responses to international nongovernmental organization actions (#PE7080)
Establishment–orientation of international nongovernmental organizations (#PE7128)
Absence of national nongovernmental organizations in developing countries (#PF7081)
Operational ineffectiveness of international nongovernmental organizations (#PE7056)
Inadequate facilities for international nongovernmental organization action (#PF2016)
Disadvantaged status of international nongovernmental organization personnel (#PE7010)
Ineffectiveness of international nongovernmental organizations and programmes (#PF1595)
Inadequate funding of international nongovernmental organizations and programmes (#PE0741)
Lack of national legal provision for international nongovernmental organizations (#PF7034)
Neglect of non–Western structures in international nongovernmental organizations (#PF7069)
Competition between international nongovernmental organizations for scarce resources (#PE0259)
Excess of western–based secretariats for international nongovernmental organizations (#PE7045)
Inadequate coordination of international nongovernmental organizations and programmes (#PE1209)
Lack of financial information systems for international nongovernmental organizations (#PE7009)
Incompatible equivalent national sections of international nongovernmental organizations (#PE7020)
Jurisdictional conflict and antagonism within international nongovernmental organizations (#PE1169)
Jurisdictional conflict and antagonism between international nongovernmental organizations (#PE0064)
Proliferation and duplication of international nongovernmental organization information systems (#PE0362)
Proliferation and duplication of international nongovernmental organization and coordination bodies (#PE0179)

Absence of policies to associate international nongovernmental organizations in regional development (#PF7033)
Inadequate international nongovernmental organization response to intergovernmental calls for action (#PE7140)
Inadequate relationship between international governmental and nongovernmental organizations and programmes (#PE1973)
Inadequate relationship between international nongovernmental organizations and the specialized agencies of the United Nations (#PE0777).

♦ **PF7088 Lack of business opposition to the arms race**
Claim The business of business is to make money, which it does exceptionally well in the armaments and defence industries. Their hypocrisy is to be anti-war but pro 'combat readiness'.
Counter-claim Most of commerce and industry has nothing to do with defence. Businessmen remember war-time conditions and recognize that prosperity exists only during peace-time. There are numbers of leading international businessmen's groups who oppose the arms race, and national and local business alliances involved in political action to support disarmament, nuclear-free zones, and peaceful, international negotiation.
Broader International arms trade (#PC1358).
Related Illegal international arms shipments (#PD4858).
Aggravated by Competitive acquisition of arms (#PC1258).

♦ **PF7093 Limited recognition of international nongovernmental organizations**
Nature For some intergovernmental agencies, the number of INGOs which are in some way engaged in activities relevant to their own programmes constitutes an administrative, or even political, problem. As such, efforts are to limit contact with them in order to simplify the already difficult tasks of operating the agency. Clearly this determines the attitude of IGO secretariat personnel and delegates, and the content of the policy recommendations and documents that they generate for national governments. It restricts the number of linkages between IGOs and INGOs, and prevents IGOs and governments from recognizing the potential of the INGO network and the manner in which its activities can be facilitated and the consequent benefits for governmental programmes. The inability of such agencies to recognize that INGOs are first and foremost a social phenomenon and only incidentally an administrative problem is also an indication of the inability of such agencies to comprehend the nature of the international community within which they attempt to function (for example, the inability of UNESCO to recognize the usefulness of social studies of national and international INGOs - after 30 years of consultative relationship with them through a designated administrative unit).
Broader Obstacles to effective international nongovernmental organizations (#PF7082).
Aggravates Ineffective governmental use of nongovernmental resources (#PF4095).

♦ **PF7094 Obstacles to family life**
Broader Deteriorating quality of life (#PF7142).
Narrower Unstable family life (#PD7929)   Lack of family life education (#PF5079)
Impermanent living conditions (#PD4368)   Inadequate family planning education (#PD1039)
Inflexible patterns of family lifestyle (#PF2644)
Inhibition of adult life in small households (#PF6167)
Minimal family interest of urban life styles (#PE8296)
Failure to recognize uniqueness of family members (#PF1750).

♦ **PF7098 Quantitative limit to societal learning**
Nature Just as no individual can absorb all information, so it is not feasible for any group to do so even by sharing the load amongst its members. In fact it is only practical to devote a limited proportion of time and resources to absorbing or disseminating information. Furthermore much is destroyed after a certain period. In an important sense mankind lives in a forgetting society. Much information quickly becomes irrelevant, especially in rapidly evolving disciplines. Groups, like individuals, can suffer from information overload. There is no way that some countries or institutions can absorb the amount of information considered relevant by their better endowed counterparts. This is an aspect of the problem of transfer of know-how. Such groups are unlimited in their capacity to continue to learn, but there is a limit on the rate at which they can do so. Another fruitful aspect of this question emerges from comparison of the rate of increase in knowledge production with the rate of increase in population. Each advance in knowledge increases awareness of what remains unknown but, perhaps more significantly, each unit of knowledge produced becomes increasingly difficult to disseminate through the learning process, because of the increasing competition for attention time from other units to be learned. Under such conditions each unit of knowledge produced can usefully be seen as increasing the ignorance of those who are unable to absorb it, for whatever reason. The production of new knowledge for some is therefore matched by the reduction of others into greater ignorance. And the amount of ignorance so produced increases much faster than knowledge production because of the effects of population growth.
Claim Each significant document entering the international system increases the ignorance of those who fail to absorb it. The question is when the ratio of ignorance to knowledge in society will be such as to render knowledgeable decision-making unimplementable because of ignorance on the part of those who are needed to support the decision in a democratic process. Given the prevalence of ignorance (and the impossibility of eliminating it) would it not be more creative to investigate it in the hope of discovering properties which would enable it to be viewed and used as a resource ? For example, given its inherent 'boundedness', it could presumably provide insights into the structuring of society into 'information cells' of many types, linked by a variety of information networks. Then the question becomes how groups and individuals can learn to benefit from their state of ignorance.
Broader Limits to societal learning (#PF7074).

♦ **PF7106 Obstacles to building construction**
Building erection risks
Nature The process of constructing any edifice has a great number of risks. The construction process itself has dangers; the building materials may be flawed; the construction process can be faulty and the construction site may be unsuitable. The equipment used in construction can be dangerous; tractors can breakdown, cranes can fall and tools cab be dropped several stories. Access to the site can be difficult for example the North Sea oil drilling platforms. Worker can take industrial action. Vandals can destroy work that has just been completed.
Refs Webster, Christopher, et al *Constructing Dangerousness* (1985).
Broader General obstacles to problem alleviation (#PF0631).
Narrower Excessive office space in cities (#PE6660)
Underground construction obstacles (#PE1944)
Architectural obsolescence of building structures (#PG7775)
Unattractive appearance of deteriorating buildings (#PF1733).

♦ **PF7109 Inadequate negative capability**
Intolerance of uncertainty
Claim The capacity of individuals to tolerate doubt, uncertainty, and a lack of knowledge in particular circumstances, is essential in many situations requiring action. In the realm of concepts many, successful theories in the physical sciences explicitly leave room for the unknown and

for any anomalies as long as they are insufficient to disprove the theory. This factor is insufficiently appreciated or tolerated.
Broader Uncertainty (#PA6438).

♦ **PF7110 Superficiality of collective comprehension in societal learning**
Nature There are two conventional responses to societal data overload; one extreme is the effort to achieve an overview of a problem situation by sacrificing any focus on detail. At the other extreme is the much favoured tendency to concentrate on some highly specific practical question, ignoring the context, in order to make concrete progress and achieve results. Information systems have not yet been designed to stabilize the shift of user focus between these different levels. As in the case of the individual, it is difficult for a group focusing on a given level to bear in mind more than the next broader level and the next narrower level. Where there are many relevant levels, much must remain out of focus. And in the dynamics of practical programmes and policy-making, levels acquire an independence from one another especially since they lend themselves to the establishment of user fiefdoms. These may well give rise to their own information systems by which that independence is justified and reinforced. Needless to say such divisions constitute a severe limit on innovative learning. The question is whether learning systems can be designed and used respecting the limits to comprehension inherent in cognitive levels different depths.
Broader Limits to societal learning (#PF7074).
Aggravates Limit to collective comprehension span in societal learning (#PF7038).
Aggravated by Collective attention-span limit in societal learning (#PF7003).

♦ **PF7111 Enforced abstention from insight**
Nature Under the prevailing conditions of increasing specialization, each specialist overrates his own insight. This is accompanied by a tendency to grant to others, especially other specialists, unwarranted levels of authority in their own areas of expertise. Specialists must make it a habit to accept withouth question the opinions of other specialists and to depend upon them. People are increasingly obliged to make use of equipment whose functioning they do not fully understand. As a consequence individuals are obliged to abstain from insight. The more complex the equipment, the more the manufacturer must endeavour to avoid the need for any attempt by the user to make use of insight. Automation obviates the use of insight.
Aggravated by Over-specialization (#PF0256).

♦ **PF7116 Assessment of international nongovernmental organizations as naive**
Nature INGO representation and activity is occasionally assessed as naive because of the lack of sophistication or qualification of those involved. Typically this assessment is made in the light of INGO representation to delegates at intergovernmental meetings or to staff members of IGO secretariats. It contributes to the negative image of INGOs in general and is reinforced by it, even in cases where there is no objective basis for any such assessment. It is particularly unfortunate when powerful INGOs enter into relationships with intergovernmental agencies (under category A or I consultative status) in which it is of benefit to them to label other INGOs as naive in order to reinforce their own position.
Counter-claim It is only too easy to accuse a body of naivety when it seeks with inadequate personnel and resources to defend some subtle human value ignored by some well-supported agencies pursuing a politically non-controversial programme. Concern with peace and disarmament in the midst of an arms race is surely naive. Concern with the protection of some species threatened by industrial development is also surely naive. As is concern with the rights of a minority group neglected by a democratic majority. The creation of an International Astronautical Federation in 1950 could only be considered naive by the majority of the academic and intergovernmental community, as must be the recent concern expressed within the International Astronomical Union that attempts to send radio messages to distant planetary systems might attract unwelcome (rather than welcome) attention. The irony of the assessment of INGOs as naive is that more often than not it is a reflection on the assessor rather than the assessed. When an IGO representative complains that the INGOs that make contact with him (or come to his meetings) are naive, he may even be correct. Intergovernmental agencies have set up such an unfruitful environment for contact with INGOs that many INGOs and their representatives avoid such contact because there are more effective forms of action those that do not either have special introductions (and are therefore labelled 'effective') or are in the process of learning what a waste of time such contacts may be. The latter group may perhaps be legitimately labelled as naive, although the assessment is about as useful as labelling a high school student as naive before he has graduated.
Broader Obstacles to effective international nongovernmental organizations (#PF7082).
Aggravates Ineffective governmental use of nongovernmental resources (#PF4095).

♦ **PF7117 Inadequate statistical data on nongovernmental organizations**
Nature No attempt has been made, or formally recommended, to collect statistical data on INGOs and their members. Although data is collected on individuals (via the census), on commercial bodies and on each nation, none is collected on the bodies through which individuals express themselves or via which their views are moulded. As a consequence, attention is switched to socio-economic considerations and away from the variety of concerns represented by INGOs and their members.
Incidence This is particularly evident in the statistical data published in the various yearbooks of the UN system. Typically the ILO Yearbook of Labour Statistics has no details about trade unions, despite the amount of aggregated data on employees. The data on INGOs published in the Yearbook of International Organizations does not extend beyond INGOs as such in order to show the amount of national organization from which such international activity emerges. The absence of such data prevents its consideration as part of any battery of social indicators, given that it may be argued that the degree of organization of a society is an important measure of social development.
Broader Obstacles to effective international nongovernmental organizations (#PF7082).

♦ **PF7118 Obstacles to community achievement**
Broader General obstacles to problem alleviation (#PF0631).
Narrower Low commune priority (#PG9873)   Insufficient doctors (#PE8303)
Divided community spirit (#PJ0399)   Low community visibility (#PG7691)
Anti-community attitudes (#PG7707)   Sporadic community growth (#PG5481)
Excessive community crime (#PJ7765)   Community expertise drain (#PG8926)
Lack of community planning (#PF2605)   Restrictive community size (#PJ0123)
Fragmented community goals (#PG1521)   Minimal community services (#PD8832)
Sterile working environment (#PD6133)   Inadequate support services (#PF6492)
Scattered housing locations (#PJ0760)   Unpublicized community news (#PF7998)
Lack of community self-worth (#PF3512)   Unchecked community spending (#PJ9047)
Unpublicized public meetings (#PF5222)   Declining sense of community (#PF2575)
Unpromoted community business (#PJ1409)   Uncohesive business community (#PG8855)
Overlooked community services (#PG8866)   Isolating community lifestyle (#PG9465)
Lack of community development (#PF7912)   Insufficient community events (#PF5250)
Demeaning community self-image (#PF2093)   Restricted community expansion (#PJ1107)
Lack of community participation (#PF3307)   Insufficient community networks (#PG8868)
Polarization of local conflicts (#PF1333)   Prevailing community insecurity (#PD9044)

Social isolation of the elderly (#PD1564) Unequal political representation (#PC0655)
Deteriorating community identity (#PF2241) Unplanned use of community space (#PF2519)
Unorganized community recreation (#PF5409) Conflicting community priorities (#PJ7975)
Community development scepticism (#PJ8970) Fragmented community initiatives (#PG8987)
Non–cooperative community groups (#PG9159) Police indifference to community (#PF8125)
Widespread community factionalism (#PG8335) Unstructured community gatherings (#PG9176)
Static and unrelated social roles (#PF1651) Outdated forms of community health (#PF1608)
Inadequate care of community space (#PF2346)
Overemphasis on community problems (#PG8617)
Unfocused design of community space (#PF1546)
Politicization of technical debates (#PD2860)
Unavailability of community centres (#PG9399)
Lack of local leadership role models (#PF6479)
Societal over–commitment to learning (#PD7051)
Fragmented planning of community life (#PF2813)
Minimum promotion of community assets (#PF6557)
Distrust of business by the community (#PE8963)
Debilitating designs for using labour (#PF2812)
Reluctance to join in community action (#PF1735)
Breakdown in community security systems (#PD1147)
Delays in community building programmes (#PF6502)
Unfocused style of community operations (#PF6559)
Divisive patterns of community groupings (#PF6545)
Limiting effect of individual survivalism (#PF2602)
Geographically undefined community limits (#PF6521)
Disjointed patterns of community identity (#PF2845)
Ineffective structures of local consensus (#PF6506)
Fragmented conduct of community operations (#PF1205)
Self–defeating style of community planning (#PF6456)
Fragmentation of communities by automobiles (#PF6250)
Inadequate procedures for community planning (#PF0963)
Excessive dispersion of community facilities (#PF6141)
Ineffective organization of community action (#PF6501)
Fragmented pattern of community organization (#PF6525)
Obstacles to availability of community space (#PF7130)
Underutilization of locally available skills (#PF6538)
Environmental degradation of inner city areas (#PC2616)
Unformed structures of community organization (#PF2810)
Ineffective utilization of public environment (#PF6543)
Politicization of international sports events (#PF4761)
Rivalry and disunity within developing regions (#PD0110)
Depressing effect of poor housing construction (#PF1213)
Demoralizing image of urban community identity (#PF1681)
Unimaginative vision of community possibilities (#PJ7764)
Demoralizing images of rural community identity (#PF2358)
Incomplete implementation of community decisions (#PF2863)
Limits on participation in community development (#PF3560)
Unexplored opportunities for community education (#PF6512)
Absence of traditional patterns of community life (#PF3531)
Inadequate community care for handicapped persons (#PE8924)
Poor condition of open spaces in urban communities (#PF1815)
Ineffective structures for community participation (#PF2437)
Inadequate domestic savings in developing countries (#PD0465)
Limited control of environment by local communities (#PF6547)
Lack of social accounting in the business community (#PE8518)
Traumatic shift in life–styles of mining communities (#PE1137)
Inflexible attitudes toward community social services (#PF3083)
Deteriorating structures of rural community cooperation (#PF3558)
Declining community confidence in its ability to change (#PF9066)
Politicization of public service in developing countries (#PE4226)
Politicization of intergovernmental organizational debate (#PD0457)
Inadequate facilities for grass–roots community initiatives (#PF6143)
Transfer of business from small communities to larger towns (#PF6540)
Ineffective operation of community networks in urban ghettos (#PF1959)
Restrictive effects of traditional community decision–making (#PF3454)
Excessive dependence of local communities on outside services (#PE4780)
Unperceived relevance of formal education in rural communities (#PF1944)
Unsanitary environment for basic health in small rural villages (#PD2011)
Negative consequences of shifting ecology on coastal communities (#PE2305)
Ineffectiveness of individual participation in large communities (#PF6127)
Overlooked potential for industrial development in rural communities (#PF2471)
Monopolization by interest groups in development of community priorities (#PE1702)
Decline in communal spirit and village solidarity in developing countries (#PE0835)
Lack of awareness of potential for investment in small, inner–city enterprises (#PF2042).
Aggravates Declining community population (#PJ8746)
Social disaffection of the young (#PD1544).
Aggravated by Underpayment for work (#PD8916)
Insufficient community celebrations (#PJ0188).

♦ PF7122 Disconnectedness of societal learning

Nature Even if the task of societal learning can be shared amongst the appropriate sectors of society, there is a limit in the ability to establish functional connections between knowledge units and between those so connected. In addition, such systems are, even more so than the telephone, only available to the privileged. However much they spread in industrialized countries, access to them in developing countries will be very limited. If it is argued that such a degree of on–line interconnectedness is not a necessity for all, there is a dynamic discontinuity with those who can only be contacted by post (or unilaterally via the mass media). This disconnection is perceived as a serious gap by those on each side of it and immediately affects the dynamism of the learning process and of its use.
Broader Limits to societal learning (#PF7074).

♦ PF7123 Confusion

Nature Individuals, groups and nations are often confused as to their wants and underlying values. Developing countries want foreign investments, and at the same time, some wish the end of the free–market economic system. Developed countries say they want world peace, and at the same time, some build enormous weapon stockpiles and wish to be the strongest peaceful nation. A kind of societal schizophrenia exists. This is more readily seen in pluralistic societies which, instead of policy setting, accept expedient compromises, and in two or three party democracies where there is alternation in controlling the government.
Narrower Semantic confusion (#PF5985)
Confusion induced by rapid social change (#PF6712)
Intellectual confusion concerning the role of the United Nations (#PE7144).
Related Defeat (#PA7289) Vanity (#PA6491) Disorder (#PA7361)
Uncertainty (#PA7309) Inattention (#PA6247) Lawlessness (#PA5563)
Formlessness (#PA6900) Unpleasantness (#PA7107) Disintegration (#PA6858)
Negative emotions and attitudes (#PA7090).
Aggravated by Ignorance (#PA5568) Conflict of information (#PF2002)
Unpredictable governmental policy (#PF1559) Inadequate ideological frameworks (#PD0065).

♦ PF7127 Excessive virtue

Narrower Passion (#PA7030) Oppressive reality (#PF7053)
Prisoners of conscience (#PC6935).
Related Pride (#PA7599).

♦ PF7129 Bibliolatry

Cult of sacred scriptures
Nature Excessive veneration of a book believed to be divinely inspired. This may be more generally understood to include those major books which are the basis of either religions or, by extension (but to a lesser degree), of major ideologies. In the case of religions, the sacred writings are the plenary depository of the divine spirit and as such are of eternal duration and of superhuman origin. The book may be held to transcend all created things and possesses the quality of absolute infallibility being the perfect channel of divine revelation. Within the religion such a book proscribes all criticism and ranks as the one all–sufficient standard of appeal in questions of belief. Absolute submission to this sacred book is the obligation of the bibliolater. Outside it their exists no final tribunal in matters of faith.
Incidence Common within religions based on a single sacred book, otherwise known as 'book–religions'. These include Brahmanism, Buddhism, Christianity, Islam, Judaism, Zorastrianism. Whilst veneration for their sacred books may not be a required article of faith, it may become so in the practice of some of the faithful. Similar veneration may be seen for certain ideological writings such as those of Marx, Lenin and Mao Tse Tung.
Counter–claim It is only natural to venerate the revelation of divine understanding, the Word of God, as it is expressed in Holy Writ.
Broader Dogmatism (#PF6988).
Aggravates Blasphemy (#PF5630) Misreading sacred documents (#PF5495)
Profanation of sacred doctrine (#PF7484)
Disagreement concerning religious doctrine (#PF1115).
Aggravated by Idolatry (#PF3374) Fundamentalism (#PF1338)
Inadequacy of religion (#PF2005).

♦ PF7130 Obstacles to availability of community space

Broader Over–spacing of suburban housing (#PE1708)
Obstacles to community achievement (#PF7118).
Narrower Marginal living space (#PJ8036) Low athletics priority (#PJ0337)
Ineffective space usage (#PE5458) Expensive meeting space (#PG8675)
Undefined cultural space (#PJ0060) Unused recreation spaces (#PJ8596)
Neglected of public space (#PF6578) Lack of community planning (#PF2605)
Undesignated parking space (#PG5432) Inadequate industrial space (#PJ0084)
Underutilized available space (#PG8182) Unplanned use of community space (#PF2519)
Unfocused design of community space (#PF1546)
Insufficient storage space in homes (#PJ0221)
Spatial imbalance of human settlements (#PD6130)
Haphazard organization of community space (#PF2822)
Geographically undefined community limits (#PF6521)
Wastage of open space in urban environments (#PE6163)
Unproductive utilization of plantation space (#PF6455)
Poor condition of open spaces in urban communities (#PF1815).
Related Inadequate care of community space (#PF2346).

♦ PF7133 Fragmentation of international documentation

Uncoordinated intergovernmental information
Nature Major inter–agency information exchanges, particularly at the computer level, have been largely abandoned or focused on narrowly specialized domains. Creators and users are resigned to the fragmentation of international documentation. Relations between potential collaborators in any such exchanges have been eroded by priority attention to basic programme concerns within each agency. In many cases where there has been a real cross–system need this has been met by external services possibly established by a commercial enterprise at the national level.
Aggravates Failure to integrate knowledge to empower humanity in response to the global problematique (#PF8753).
Aggravated by Unethical documentation practices (#PD2886).

♦ PF7134 Pre–logical limitations to the comprehension of international information

Societal knowledge resistance
Nature Many organizations and individuals use items from the international documentation system to support pre–logical positions which are completely undermined by other documents (which are not cited, even if they have been consulted). This is part of the drama of the political arena and is accepted as such. Many are responsive only to the immediacy of verbal presentations, or to scientifically–backed arguments, or to arguments of a delegation with a strong power–base. Others are affected, or unaffected, by the style of presentation, whether it stresses order/disorder, static/dynamic, continuity/discreteness, spontaneity. On the other hand, and more important, many (at every level of education) are totally indifferent to the whole process which the international documentation system is designed to serve. For them, those documents contain no meaningful information. They are unaffected by efforts for change through the mobilization of public opinion. Although little is known about this pre–logical limit as it affects information, the receptivity to some forms of information only, means that there is a limit to the extent to which an individual or group can learn from information in other styles and modes. It is not simply a question of 'multi–media presentations' but of the pre–logical orientations inherent in any given form of information. The question is how these orientations complement one another and what this limit implies for information systems designed for communication of insights between users of every orientation.
Broader Barriers to the international flow of knowledge and educational materials (#PF0166).

♦ PF7137 Unnecessary verbosity of legal documents

Lengthy legal opinions
Nature Legal documents are written in a jargon, and with a length of detail that is totally unnecessary to communicate their messages. The law clerking profession is used by aspiring young lawyers to enhance their legal prospects by developing a reputation for scholarly publications. The result is rulings that run into thousands of unnecessary words, mostly incomprehensible to the public who are to obey and support the law.
Broader Outmoded legal systems (#PF2580) Restrictive legal practices (#PD8614)
Deficiencies in national and local legal systems (#PF4851).
Narrower Time lag in legal provisions (#PF6042).
Related Verbosity in intergovernmental organizations (#PF5477)
Discrimination against foreigners in legal proceedings (#PF1798).

♦ PF7138 Lack of control

Narrower Irrationalism (#PF3399) Hunting of animals (#PC2024)
Lack of eugenic measures (#PD1091) Abuse of scientific power (#PF2692)
Uncontrolled urban development (#PC0442) Inadequate national law enforcement (#PE4768)
Uncontrolled environmental impact of technology (#PC1174)
Excessive expense of athletic training programmes (#PF4196)
Ineffective regulation of restrictive business practices (#PF1596)

Excessive expense of international athletic competitions (#PF4192)
Inadequate control over government administrative process (#PC1818)
Ineffective international regulation of transnational corporations (#PF0691)
Proliferation and duplication of international information systems (#PE0458)
Inadequate regulation of restrictive business practices in service industries (#PF0591)
Inadequate regulation of the restrictive business practices of state enterprises (#PE0225).
Related Instability of economic and industrial production activities (#PC1217).
Aggravates Chaos (#PF6836) Fragmentation (#PA6233) Human suffering (#PB5955)
Reactionary forces (#PB6332) Unwanted pregnancies (#PF2859).

♦ PF7142 Deteriorating quality of life
Declining standard of living — Restricted life-style
Incidence Recent years have seen a further increase in unemployment and destitution, a general worsening in the basic conditions of life, including an absolute impoverishment of the mass of the population in many if not most developing countries as a result of stagnation and decline in use of value-oriented agricultural and industrial production, the outflow of surplus, the progressive displacement of the subsistence economy, and the destruction of the environment. In industrialized countries almost all social and environmental problems are getting worse: crime, substance abuse, stress diseases, homelessness, depression, traffic noise and congestion, pollution, pace, urban blight and poverty.
Claim Raising the GNP lowers the quality of life, because the market forces commercialize the lives of the people who concentrate on maximizing their self-interest. Slowly the market will determine more and more different aspects of the society.
Refs Lauer, Robert H *Social Problems and the Quality of Life* (1986).
Broader General obstacles to problem alleviation (#PF0631).
Narrower Lack of commitment (#PF1729) Personal life crises (#PD4840)
Defensive life stance (#PF0979) Decline of street life (#PE6705)
Unbalanced social life (#PD8113) Increasing pace of life (#PF2304)
Attraction of city life (#PJ7861) Obstacles to family life (#PF7094)
Unattractive social life (#PJ1212) Individualistic life style (#PJ8635)
Reduction in symbolic celebrations (#PF1560)
Discrepancies in human life evaluation (#PF1191)
Local traditions of cultural isolation (#PF1696)
Unfulfilled aspirations of economic life (#PF2842)
International imbalance in the quality of life (#PB4993)
Apathy toward improvement of urban life styles (#PE8477)
Collapse in the meaning of participating in society (#PF0955)
Struggle for financial security in urban life style (#PE8144)
Inadequate standard of living in developing countries (#PE4052)
Perpetual preoccupation for sustenance of urban life style (#PE8644)
Deterioration of the quality of life in least developed countries (#PE7734)
Westernization of traditional modes of life in developing countries (#PF6592).
Related Decadent standard of living (#PD4037).
Aggravates Socio-economic poverty (#PB0388) National economic recession (#PD9436)
Inadequate standards of living (#PF0344)
Undue religious influence on secular life (#PF3358).
Aggravated by Breakdown of local community cohesion (#PD2864).

♦ PF7149 Disobedience of elders
Failure to obey elders — Failure to follow advice of elders
Broader Disobedience (#PA7250).
Aggravates Social isolation of the elderly (#PD1564)
Unmeaningful social roles for the aged (#PF1825).
Aggravated by Disrespect (#PA6822) Juvenile stress (#PC0877)
Disrespect for elders (#PF3979) Erosion of elders' wisdom (#PF1664).

♦ PF7150 Breach of promise
Broken promises — Breach of trust — Breach of faith — Unfulfilled promises
Refs Ponting, Clive *Breach of Promise* .
Broader Unrecognized socio-economic interdependencies (#PF2969).
Narrower Broken government promises (#PF4558).
Related Parole violation (#PE1121) Unfulfilled treaty obligations (#PF2497).
Aggravates Frustration (#PA2252) Breach of contract (#PE5762)
Loss of institutional credibility (#PF1963)
Loss of credibility in international institutions (#PE8064).
Aggravated by Black lies (#PE4432) Inaction on problems (#PB1423).

♦ PF7163 Unwillingness to sacrifice political power
Nature The unwillingness to surrender political power for the sake of larger social units in the face of increasingly global demands jeopardizes unified attempts at world wide solutions. International crime, and drug trafficking, the environment, regulation of multinational businesses and other transnational issues are being attacked piecemeal or ineffectively.
Broader Habitual overemphasis on national self-determination (#PF1804).
Aggravates Inadequate power of intergovernmental organizations (#PF9175).

♦ PF7166 Excessive confidence in prediction capacity
Overestimation of ability to forecast
Claim Trusting the climate models is to overestimate our ability to predict, because these models do not handle feedbacks and surprises, like the ozone hole over Antarctica. Therefore, we should be more careful about what we are doing and much more observant of how the system works.
Refs McKern, R B and Lowenthal, G C (Eds) *Limits to Prediction.*
Aggravates Inaccurate forecasting (#PF4774).

♦ PF7178 Inability to define moral standards
Breakdown of standards of morality — Incompatible moral interests
Aggravates Moral hedonism (#PG4590) Moral offences (#PD9179)
Moral imperfection (#PB7712) Conflict of duties (#PF0513)
Unethical practices (#PC8247) Retarded socialization (#PF2187)
Double standards of sexual morality (#PF3259).
Aggravated by Ethical decay (#PB2480) Misconduct in public office (#PD8227).

♦ PF7186 Minimization of problems
Over-optimistic assessment of problems
Nature Bureaucrats have a tendency to smother sharp and disquieting reports and under tyrannies there is a tendency to tell the tyrant what he wants to hear.
Broader General obstacles to problem alleviation (#PF0631).
Narrower Oversimplification (#PF8455).
Aggravates Inadequate research on problems (#PF1077)
Inadequate research on proposed solutions to problems (#PF1572).
Aggravated by Short-term gain (#PF8675) Unrealistic expectations (#PE7002)
Conceptual repression of problems (#PF5210) Substitution of fantasy for reality (#PF8922)
Falsification of programme evaluations (#PF9243).

♦ PF7189 Corporate greed
Business greed
Broader Avarice (#PA6999).
Aggravates Profiteering (#PC2618) Short-term profit maximization (#PF2174).

♦ PF7198 Accumulation of titles
Accumulation of honours — Accumulation of awards
Incidence Characteristic of people in positions of power who, in part from a deep sense of personal insecurity, seek tokens of recognition, admiration and social acceptance through the acquisition of honours and titles. This is most evident in the case of certain dictators but has traditionally been evident in imperialistic systems. The acquisition of honours, and the control of their attribution, continues to exert a powerful influence on policy-making in certain countries, such as the UK. A related phenomenon is evident in the socio-political manoeuvering around academic prizes (Nobel, etc), literary prizes and sporting prizes (eg Olympic gold medals).
Broader Accumulation (#PA4313).

♦ PF7220 Politicization of scholarship
Ideological domination of research — Ideological resolution of scientific issues — Ethical resolution of scientific issues — Religious resolution of scientific issues — Misuse of democratic processes to resolve theoretical issues — Politicization of research
Nature Because of the importance of scientific insight into certain problems faced by humanity, efforts are made by those with ideological commitments, who are concerned at the dimensions of such challenges, to obtain scientific sanction in support of ideologically acceptable positions in response to such challenges. Such initiatives can result in attempts to resolve prematurely certain scientific issues, on which research is still in progress, using methods of ideological discourse that are inappropriate.
Incidence Efforts have been made to resolve scientific issues on the nature of violence by obtaining a vote amongst members of academic societies. Scientists are then faced with the dilemma of whether to offend morally committed colleagues by appearing anti-peace or pro-sociobiology, so that any opposition is made to appear like advocacy of war. The exercise then becomes, in the absence of public debate, a ritual of good intentions, leading to the thoughtless acceptance of incompletely researched phenomena. Similar efforts have been made amongst psychiatrists to resolve the question of whether homosexuality constitutes normal or deviant sexual behaviour.
Claim 1. The refusal to take a stand is not itself a political stance, unless the issue in question is explicitly political where the act of taking a stance has political consequences for everyone. Such a refusal may mean nothing more than a decision to suspend judgement in anticipation of further evidence. Even in the light of overwhelming evidence, continued suspension of judgement can only be considered unscholarly, but not political. 2. Scholarly decisions cannot be made by legislation and resolutions. If initiatives are taken to organize the intellectual community to declare, by the vote of an ideologically or ethically intimidated majority, that certain testable scientific hypotheses are "incorrect", and even evil and dangerous, then this is well on the way to harnessing aggressive instincts in the service of what has taken the form of inquisitions, pogroms, witch hunts, book-burning, the stamping out of heresy, and finally the jihad, the crusade and the "war to end all wars".
Counter-claim Refusal to take a stance is itself a political stance.
Aggravates Self censorship (#PF6080) Scientific censorship (#PD1709)
Unprofessional science (#PF6697) Denial of academic freedom (#PD4282)
Biased and inaccurate biology textbooks (#PF9358).
Aggravated by Extremism (#PB3415) Abuse of science (#PC9188)
Fascist liberalism (#PF9710) Anti-intellectualism (#PF1929)
Bias in scientific research (#PF9693) Over-reliance on convenient beliefs (#PF9205).

♦ PF7227 Quantitative pressure on standards of quality
Commercial erosion of standards of quality
Broader Lack of qualitative excellence (#PF5703).
Narrower Competitive distortion of musical pitch (#PF8713).
Aggravates Deterioration in product quality (#PD1435).

♦ PF7228 Eternal punishment
Hell
Refs Mew, James *Traditional Aspects of Hell* (1971); Shedd, William G *Doctrine of Endless Punishment* (1986).
Broader Punishment (#PA5583).
Related Supralapsarianism (#PF3354).

♦ PF7230 Deficiency in innovation in the public sector
Incidence Although member states of the European Community have demonstrated much innovation capacity, this capacity lies behind escalating needs. The situation is aggravated because of constraints on innovation, such as the need for contraction of government activities and reduction in the size of public organizations.
Claim Actual innovative capacities in the public sector, to the extent that it can be evaluated, vary a great deal between countries and between agencies within a country. But organization theory, historical studies and comparative investigations converge in finding that most public sector organizations have inbuilt tendencies towards incrementalism, seldom developing and pushing innovative policy paradigms.
Aggravates Inadequate research on proposed solutions to problems (#PF1572).
Aggravated by Suppression of creativity and innovation (#PF0275)
Constraint of inherited problems on policy innovation (#PF9208)
Inappropriate use of innovative management techniques (#PJ0447).

♦ PF7239 Officially nonexistent people
Minorities unrecognized by government
Related Statelessness (#PE2485).
Aggravates Social invisibility (#PD8204).

♦ PF7248 Narcissism
Narcissistic personality disorders — Narcissistic behaviour disorders — Self-absorption
Nature The true search for the self can become sidetracked into excessive self-interest and self-congratulation. Once a characteristic of youth, the contemporary mania for personal fulfilment has brought with it an epidemic of narcissism-related problems among adults.
Narcissistic people have a grandiose sense of self-importance, a feeling of being "special" without appropriate achievements. They have frequent fantasies of unlimited success, power, beauty and brilliance that cannot be satisfied by pursuing realistic goals. They require constant attention and admiration and take advantage of others to achieve their ends. On the other hand they are preoccupied with feelings of envy and can't take criticism.
Narcissus was a mythological Greek youth who fell in love with his own reflection.
Refs Schwartz-Salant, Nathan *Narcissism and Character transformation* (1982).
Broader Self disorders (#PF4843) Sexual deviation (#PD2198)
Personality disorders (#PD9219).

Narrower Self absorption of political leaders (#PD4749)
Self interested manipulation of timing (#PF5529).
Aggravated by Lack of self-confidence (#PF0879).
Reduces Low self esteem (#PF5354).

♦ **PF7264 Ineffective monitoring of illegal activity**
Broader Ineffective monitoring (#PF2793).
Aggravates Unlawful business transactions (#PC4645)
Inadequate environmental monitoring (#PF4801)
Inadequate national law enforcement (#PE4768).
Aggravated by Unlawful government action (#PF5332).

♦ **PF7274 Disputed priority of innovations**
Disagreement concerning identity of inventors and discoverers
Aggravates Disincentives to invention (#PG6623).

♦ **PF7277 Obstruction of international criminal investigations**
Legal impediments to international investigations
Broader Deficiencies in international law (#PF4816).
Narrower Extradition refusal (#PF2645)   Denial of legal representation (#PF3517)
Unavailability of legal information (#PJ8698)
Obstacles to legal relations between socialist countries (#PF4886).
Aggravated by Legal havens (#PE0621)   Legal prevarication (#PF9756)
Complex legal procedures (#PF8519)   Intergovernmental disputes (#PJ5405)
Restrictive legal practices (#PD8614)
Inadequacy of international legal procedure (#PF8616)
Prohibitive cost of linguistic interpretation legal proceedings (#PE1743).

♦ **PF7283 Myopic advice to leadership**
Nature Leaders have advisors, some formal and some informal who to be effective must be specialized and therefore operate out of a narrow context. Few leaders attempt to find as confidants, technical specialist or cabinet ministers generalists who can perceive the overall needs of the social group being led.
Broader Narrow context for counsel (#PF0823).
Aggravated by Irresponsible international experts (#PF0221).

♦ **PF7291 Tampering with physical evidence**
Nature The alteration, destruction or concealment of any physical evidence to be used in an official proceeding for the purpose of altering its outcome is a crime.
Incidence The most noteworthy of cases of tampering with evidence is President Nixon's destruction and alteration of tape recordings of White House meetings.
Broader Evasion of issues (#PF7431)   Hindrance of law enforcement (#PD5515).
Narrower Government inducement to crime (#PD6943).
Aggravated by Inadequate evidence to convict known offenders (#PF8661).

♦ **PF7293 Limited criminal liability of corporations**
Limited criminal liability of associations
Incidence Traditionally under common law, corporations and associations could not be convicted for criminal offences, since not being persons, they could not personally participate in a trial or be imprisoned if convicted. Corporate liability for failure to perform duties owed to the general public and for actions of employees during the course of their duties is increasingly recognized.
Claim Limited liability is a worldwide swindle. The only risk in running a limited company is that the public will understand the nature of the racket. Evasion of responsibility is the whole point of limited liability. Companies are now deliberately formed to take advantage of tax relief on bankruptcy.
Broader Crime (#PB0001).
Narrower Fraud concerning economic situation and corporate capital of companies (#PE5021).
Aggravates Corporate crime (#PD3528).

♦ **PF7299 Discriminatory language**
Vilification — Insensitive discourse — Demeaning expressions
Related Prejudice against other languages (#PD8800).
Aggravated by Verbal abuse (#PD5238)   Misuse of language (#PF9598)
Religious vilification (#PD5534)   Political smear campaigns (#PD9384)
Covert smear campaigns by government (#PD7171).

♦ **PF7315 Accumulation of recognized merit**
Nature Organizations and social systems develop a variety of procedures for rewarding performance evaluated as superior by the standards of the group. Such evaluation procedures are closely tied to the advancement of the career of the individual concerned and of any faction with which he is associated. As a result individuals adapt there career strategy to ensure that they acquire merit points, of whatever form, irrespective of whether these in fact correspond to meritorious effort and of whether they are forced to avoid actions which would expose them to negative evaluations. The acquisition of career merit points then becomes the career focus, whether this involves avoiding communicating negative feedback to superiors, avoiding responsibility for unpleasant decisions or doctoring reports to present his own actions in the most positive light. The individual's concerns thus shift to the image or appearance of his actions and away from the actions themselves.
Broader Accumulation (#PA4313).
Aggravates Avoidance of negative feedback (#PF5311)
Unrealistically positive self-assessment (#PF4377).

♦ **PF7318 Environmental prodigality**
Nature A condition in which people are encouraged to deplete resources and degrade environments because they can pass on the costs of doing so to other members of society and/or to other generations. This is because many natural assets are unpriced and so, effectively are "free"; and the prices of natural assets that are priced seldom reflect their full social costs.
Incidence It is the cause of most resource depletion and environmental degradation in developed countries, and by commercial interests in developing countries.
Broader Natural resource depletion due to high-level consumption (#PD4002).
Aggravates Decadent standard of living (#PD4037)
Natural environment degradation (#PB5250)
Waste of non-renewable resources (#PC8642).
Aggravated by Lack of accountability in the disposal of wealth (#PE0503)
Incompatibility of environmental and economic decision-making (#PF9728).

♦ **PF7334 Counter-reform**
Related Counter revolution (#PF3232).
Reduced by Inadequate social reform (#PF0677).

♦ **PF7342 Abdication of government ministerial control**
Aggravates Political surveillance (#PD8871).
Aggravated by Official secrecy (#PC1812).

♦ **PF7353 Impediments to adoption of children**
Claim Outdated regulations, "demand" exceeding the "supply" and other barriers keep people wanting to adopt a child frustrated and waiting for years.
Narrower Refusal of adoption (#PF3282)   Annulment of adoption (#PF3281)
Inadequate laws of adoption (#PD0590).
Aggravates Abandoned children (#PD5734)   Trafficking in children for adoption (#PF3302).
Aggravated by Outdated procedures (#PF8793).

♦ **PF7358 Verbal conflict envisaged as destructive**
Nature In the exercise of the right of free speech, verbal conflict is often envisaged as destructive. Many individuals and group not wanting to be in the position of opposing another agree with the other in order to avoid conflict because it is seen as uncreative.
Counter-claim Verbal conflict is destructive. Take for instance, the students and workers in Tiananmen Square in May and June of 1989 who voiced opposition to the policy and corruption of the Chinese government. Hundreds of them were killed and thousands were imprisoned because the ideas they were voicing would destroy the existing regime.
Broader Irresponsible expression of emotions equated with free speech (#PF7798).

♦ **PF7380 Inadequate intellectual methods**
Nature Local people are not given adequate methods for using their own intellectual powers. Most methods for appropriating information, including study methods are linear, and non-selective of content. Most techniques for recalling information rely on the rational dimensions of thinking and seldom the non-rational. Most approaches for creating responses are non-analytic. Methods, now coming into vogue, which are non-linear tend to deny the validity of linear methods, creating a rational, non-rational dichotomy.
Claim Because local people do not have effective tools to use the full potential of their minds, the whole society is deprived of creativity.
Broader Obsolete methods (#PF3713).
Narrower Antiquated intellectual methods to appropriate human depths (#PF1094).
Aggravates Underutilization of intellectual ability (#PF0100)
Inadequacy of prevailing mental structures to challenge of human survival (#PF7713).

♦ **PF7383 Lack of English language to describe female experience**
Nature Naming encodes the bias of those who do the naming. In the English language, males were those who held the positions of power during the periods in which the language was being developed. English speakers today suffer from an inability to articulate the female experiences, perspectives and feelings which language developers had no ability to perceive.
Incidence Men frequently point out that women are talking "nonsense", especially when they seek to articulate a vocational or relational experience. Linguistically, this is precisely correct. The female perspective is not part of the "common sense" of the English language.
Broader Language barriers (#PF6035).
Related Sexually discriminating job terminology (#PF6014)
Reinforcement of male dominance through language (#PF4373).
Aggra ates Segregation through language (#PD4131).

♦ **PF7388 Information anxiety**
Nature Is the gap between what we understand and what we think we should understand. It happens when information does not tell us what we need or want to know.
Refs Wurman, Richard Saul *Information Anxiety*.
Broader Anxiety (#PA1635).

♦ **PF7401 Exterminism**
Nature A destructive side-effect of concern about the arms race is the propensity to assume that nuclear holocaust is inevitable, no matter what is done to prevent it.
Incidence Films like "The Day After" give credence to the prospect of nuclear destruction, under the guise of creating resistance against it.
Aggravates Pervasive fear of nuclear war (#PC3541)
Sunlight inhibition by nuclear warfare soot (#PE6350).
Aggravated by Nuclear arms race (#PD5076).

♦ **PF7414 Avoidance of reality**
Cultivation of trivia
Claim People no longer prefer to confront reality directly, having learned and accepted that reality has for all practical purposes become unmanageable. People tend increasingly to devote their energies to the proliferation and production of trivia and unreality to soothe tired and fractured egos. The increasing role of television is particularly insidious in that it has become both a simulation of reality and reality itself. There is in fact no reality anymore. Symbols, information and entertainment have been so thoroughly merged, that few can distinguish between them.
Narrower Non-acceptance of reality (#PF1079)   Substitution of fantasy for reality (#PF8922)
Escaping reality through popular psychological screens (#PF1112)
Escaping reality through sophisticated shields against life's pain (#PF0916).
Aggravates Unrealistic expectations (#PE7002)   Impoverishment of political debate (#PF4600).
Aggravated by Blame avoidance (#PF6382)   Oppressive reality (#PF7053)
Elusiveness of reality (#PJ8032)
Reinforced parochialism of internal values and images (#PF1728).

♦ **PF7418 Dependence on romantic love**
Claim Romantic love, with its idealization of the love object, is an expression of neurosis, a maladaptive effort to solve a dependency problem, or an adolescent fixation. As such it is a foolish if not dangerous illusion that creates impossible expectations in people. It makes them unable to accept the positive benefits of a relationship. From a feminist perspective, romantic love may be characterized as a rationalization for female subordination and dependency, namely as a glamorous trap that disguises the prison to which women are condemned by marriage.
Romantic love may also be considered as a confusing mix of two important forms of love. The first is a natural urge toward an inner world, associated with spiritual aspiration. The second is a love for people in the flesh. It is through the inappropriate projection of the ideals associated with the first onto a specific person, that people are entrapped in dependency. This projection prevents any meaningful relationship with the person-in-the-flesh who can only occasionally live up to the ideal. The resulting disappointments undermine what might otherwise develop into a fruitful relationship. The great flaw in romantic love is that it seeks one relationship but forgets the other.
Counter-claim Romantic love serves an important function not only for the individual but for the culture. It provides a narrative thread in lives, determining obligations and transforming them. Even if romantic love is often short-lived, it is mistaken to consider that its transience disqualifies it from significance. It is the experience itself, and the difference it makes to life, that makes it valuable. The rationalist perspective on love is emotionally shallow or inhibited, fearful of passion and constrained by caution. The mature love advocated by therapists renders love stale and antiseptic,

denigrating the experience of falling in love, in order to stress mutual respect, shared values and common interests above emotional pleasure and sexual passion. Romantic love offers not just the excitement of the moment but the possibility for dramatic change in the self. As such it is in itself an agent of change. Love also offers an antidote both to personal neediness and to those existential anxieties that increase the sense of the frailty and brevity of life.
Refs Diamond, Jed *Looking for Love in All the Wrong Places* (1988); Streit, Gary *Psychology of a Broken Heart* (1987); Van de Vate, Dwight *Romantic Love* (1981).
Related Sexual craving (#PE7031).
Aggravates Lovesickness (#PF3385) Unrequited love (#PF6096) Romantic separation (#PD4233).
Aggravated by Uncritical acceptance of another person (#PF5973).

♦ PF7427 Profanity
Swearing — Bad language — Indecent speech — Vulgar language — Verbal obscenity — Coprophasia — Obscene wit — Obscene humour
Nature One form of verbal obscenity is the habitual use of language with excretory and uro-genitary references, or the frequent utterance of coarse expressions relating to sexual acts. Medically this is termed coprotatia or coprophasia. Its perverted objectives may be to elicit sexual responses, or to attain exhibitionistic auto-erotic excitation. Involuntary coprotatia, as an automatism, may accompany some forms of psychoses in both sexes, but may also be produced under persistent or extreme stress. Traditionally, coprotatia was viewed as a sign of demonic possession.
Refs Sharman, Julian *Cursory History of Swearing* (1968).
Broader Obscenity (#PF2634) Indecency (#PF8842).
Narrower Blasphemy (#PF5630).
Related Slang (#PF5213) Imprecation (#PF3746).
Aggravates Verbal abuse (#PD5238) Limited verbal skills (#PD8123) Profanation of sacred doctrine (#PF7484) Verbal sexual harassment of women in public (#PE0756).
Aggravated by Indecent art (#PE5042) Unparliamentary behaviour (#PF4550).

♦ PF7428 Inferior classes
Rabble — Riff-raff — Sans-culottes — Bums — Low socio-economic class — Scum
Broader Class consciousness (#PC3458).
Aggravates Trichomoniasis (#PE2310).

♦ PF7431 Evasion of issues
Nature In contrast to the avoidance of issues, which does not deny their significance but merely ensures that they are not dealt with, the evasion of issues uses various devices to avoid recognition of the existence of the issues. In social systems this takes the form of the refusal to accept the validity of certain forms of information, to the point of ensuring that such information is suppressed. Typically such evasion needs to be disguised or covered up, because if known it can relatively easily become the focus of extended public criticism, especially in the media. It may well be criminal but tends to be considered highly unethical. In individuals evasion may also take the form of repression of memories.
Broader Irresponsibility (#PA8658).
Narrower False evidence (#PF5127) Denial of evidence (#PD7385) False political evidence (#PD3017) Tampering with physical evidence (#PF7291) Tampering with official documents (#PF4699) Unproven relationships between problems (#PF7706) Inadmissability of evidence from other jurisdictions (#PE5424) Jurisdictional conflict between academic disciplines (#PF9077).
Related Repression of self-consciousness (#PC1777).
Aggravates Disinformation (#PB7606) Non-recognition of problems (#PF8112) Inability to resolve problems realistically (#PF8435).
Aggravated by Issue avoidance (#PF1623) Underreported issues (#PF9148) Intellectual arrogance (#PF7847).

♦ PF7432 Imbalance of revenue mobilization and expenditure allocation
Aggravates Inadequate fiscal policies (#PF4850).
Aggravated by Inadequate management of government finances (#PF9672).

♦ PF7435 Decreasing land mass
Decreasing territory — Rising sea level — Disappearing islands
Incidence There is evidence to indicate that sea levels may rise over the next 50 years, sufficiently to radically change the boundaries between coastal nations, to cause some island countries (such as the Maldives) to disappear, and to change the shapes and strategic importance of international waterways. Global estimates based on the period 1881–1980 indicate a rate of sea level increase of approximately 1.0 mm per year. It is also estimated that by 2030 sea level will be rising at about 4.00 mm or 5.00 mm a year.
Broader Environmental threats to national security (#PC4341).
Narrower Land subsidence (#PD5156) Coastal erosion (#PE6734) Subsiding coastal areas (#PD3775).
Aggravates Excessive land usage (#PE5059) Endangered parklands (#PE9282) Limited available land (#PC8160) Ecological disruption of animal breeding grounds (#PJ3994).
Aggravated by Changing river courses (#PG4357).

♦ PF7451 Excessive joy in functioning
Claim The desire to function, and to function well, is a programmed norm of human behaviour. Under the circumstances and conditions of an over-organized mass society this becomes a danger. As people come to derive pleasure from the skills required to perform well in a complex technocratic society, involvement in the exhibition of such skills and the natural pleasure in accomplishment become ends in themselves. The skills are valued, whether in the use of computers, defence systems, or chemical manufacturing, irrespective of the consequences of their use for the environment or for future generations.

♦ PF7455 Loss of international political leadership
Broader Loss of international leadership (#PF8353).

♦ PF7456 Lost family role in society
Diffusion of family role expectations — Unplanned family roles
Nature The function of the family in society has been lost. Families find themselves unable to function effectively in the ambiguity and insecurity of modern society. Families refuse to embrace the necessity of functioning in a global context. The marriage covenant has been replaced with a sexual contract. Or the family hides in the status quo.
Claim Previously, both men and women expected and experienced that the family provided them with love, happiness, proper role fulfilment, security for the future, and historical continuity through their children. Now the institution of the family is experiencing fragmentation in its day-to-day activities, with a shift to liberal sexual mores and a divisive alienation of life styles between generations. The question of what is to be gained from having a family is being raised.
Broader Lack of commitment (#PF1729) Static and unrelated social roles (#PF1651) Collapsed meaning of human creativity (#PF0936).
Narrower Escapist family life styles (#PD4069) Breakdown in covenants for life (#PD1026) Refusal of family possibilities (#PF0846) Reduced interior structure of families (#PF3783) Family adaptation of community status quo (#PE5408) Refusal of families to participate globally (#PF1006) Exclusion of pre-adults from family decisions (#PE2268) Individualistic perception of sexual activity (#PF1682) Individually defined operating structure of marriage (#PD2294).
Related Disunity in urban villages (#PF1257) Social isolation of the elderly (#PD1564) Repression of self-consciousness (#PC1777) Inadequate means for upholding global concern (#PF1817).

♦ PF7458 Disregarded financial resources
Aggravates Lack of local information systems (#PF6541).

♦ PF7467 Spiritual disobedience
Spiritual insubordination — Failure to obey religious advisers
Broader Disobedience (#PA7250) Insubordinate behaviour (#PJ6517).
Aggravates Disrespect (#PA6822) Loss of spiritual guidance (#PF7005) Barriers to transcendent experience (#PF4371).
Aggravated by Lack of religious discipline (#PF8010).

♦ PF7472 Public resentment at government policies
Aggravates Loss of institutional credibility (#PF1963).
Aggravated by Inappropriate policies (#PF5645) Governmental incompetence (#PF3953).

♦ PF7483 Internal inefficiency of public programmes
Nature The mix of inputs in publicly provided services is often inefficient, namely the same funds could achieve more if they were reallocated. One aspect of this is that administrators of centralized tax-supported systems have to set norms on budgetary allocations for key inputs. These norms may not match the institutions needs or the community's preferences but the recipients may well have neither the financial power nor the incentive to change them. This is exacerbated when centralized systems are slow to adjust to resource scarcities - leading them to underfund non-labour recurrent costs. Another form of inefficiency arises when, for lack of an appropriate price signal, demand fails to match supply. When demand cannot be met, institutions resort to rationing by queue, namely long waiting times in government facilities.
Broader Inappropriate public spending by government (#PF6377).
Aggravates Excessive waiting times in government facilities (#PF5120).

♦ PF7484 Profanation of sacred doctrine
Nature Incautious use of sacred doctrine, possibly regarded as the exclusive property of certain privileged classes. Such doctrine may also be profaned by failing to keep it quite distinct from other truths or opinions of lesser value.
Incidence In different cultures their sacred scriptures may be considered to be profaned if they are read or taught to people outside the prescribed classes.
Broader Desecration (#PF9176).
Narrower Misreading sacred documents (#PF5495).
Aggravated by Profanity (#PF7427) Bibliolatry (#PF7129).

♦ PF7490 Biologically determined aggression
Genetically determined violence — Violence determined by evolution — Instinctual violence — Neurophysiological compulsion to violence — Violence justified by science
Nature In some very real sense violence is embodied in the human genetic/evolutionary legacy as is evidenced by its frequent manifestation in the human species as well as in others. Although the complex outcomes of the interactions of genetic predispositions with the social, ecological and cultural environments are themselves not embodied in the genetic makeup, nevertheless any violent species-specific behaviour is part of its evolutionary heritage to the extent that any such cultural heritage forms a continuum with the biological heritage.
Claim 1. Violence lies in the animal brain and instincts of humanity. Physiological research has described many of its mechanisms, while the law, anticipating science, has in many countries recognized its involuntary nature, and exonerated certain crimes such as justifiable homicide (as in the case of self-defence) and some crimes of passion or even crimes avenging insulted dignity or 'honour'. War is a natural human instinct. Man is therefore condemned by biology to war.
2. In the course of human evolution there has been a selection for aggressive behaviour more than for other kinds of behaviour. The human species has inherited a tendency to make war from its evolutionary ancestors and consequently war and other forms of violent behaviour are genetically programmed into human nature.
Counter-claim 1. It is incorrect to claim that war or any other violent behaviour has been genetically programmed. While genes are involved at all levels of nervous system function, they provide a developmental potential that can be actualized only in conjunction with the ecological and social environment. Except for rare pathologies, genes do not produce individuals necessarily predisposed to violence. Neither do the determine the opposite.
2. It is incorrect to claim that in the course of evolution there has been a selection for aggressive behaviour more than for other kinds of behaviour. In all well-studied species, status within the group is achieved by the ability to cooperate and fulfil social functions relevant to the structure of the group. "Dominance" involves social bondings and affiliations; it is not simply a matter of the possession and use of superior physical power, although it does involve aggressive behaviours. Where genetic selection for aggressive behaviour has been artificially instituted in animals, it has rapidly succeeded in producing hyper-aggressive individuals; this indicates that aggression was not maximally selected under natural conditions. Violence is neither in the evolutionary legacy nor in the genetic makeup.
3. It is incorrect to claim that humans have a "violent brain". While we do have the neural apparatus to act violently, it is not automatically activated by internal or external stimuli. Like higher primates and unlike other animals, the human neural processes filter such stimuli before they are acted upon. How people act is shaped by how they have been conditioned and socialized. There is nothing in neurophysiology that compels humans to react violently.
4. It is incorrect to claim that violence or war is caused by "instinct" or any other single motivation. The emergence of modern warfare has been a journey from the primacy of emotional and motivational factors, sometimes called instincts, to the primacy of cognitive factors. Modern war involves institutional use of personal characteristics such as obedience, suggestibility and idealism, social skills such as language, and rational considerations such as cost calculation, planning and information processing. The technology of modern war has exaggerated traits associated with violence both in training of actual combatants and in the preparation and support for war in the general population. As a result of this exaggeration, such traits are often mistaken to be the causes rather than the consequences of the products.
Aggravates Aggression (#PA0587) Human violence (#PA0429).
Aggravated by Abuse of science (#PC9188).

♦ PF7493 Lack of realism in intergovernmental organizations
Superficial consensus in intergovernmental bodies

Nature Lack of realism is not confined to the texts of Charters and Constitutions of intergovernmental bodies. The pursuit of verbal consensus replaces real discussion of problems and the give–and–take of vested interests, thus concealing the fact that no agreement has been reached, possibly because there were no real negotiations. This tendency is facilitated by the paragraphs of resolutions that set forth basic principles or truisms to which it all the easier to subscribe in that there is no follow–up on their implementation.
Incidence Declarations of principles and and declarations enjoining Member States to observe them represent one quarter of the number of paragraphs of the sum total of United Nations resolutions. The degree of unreality varies with the programmes. Some render precise service. But in a general way, the world–wide scale of these undertakings; the gulf between the ambitions and the means; the lack of a transmission belt between the offices at headquarters and the responsible national services within each country; the inability to define modest objectives accessible within stated time–limits, raise doubts whether in the long run most of the actions have any connection with reality. Description by organization or sector of activity give an appearance of rationality to the division of labour as if the types of activity to be undertaken were the same in every sector. They also give the false impression that the degree of of effectiveness possible in the different fields (peace, education, transport, food, etc) is comparable, thus concealing real differences in the nature of the problems and types of activity.
Claim The use of broad terms and blurred distinctions make political agreement possible since such "constructive ambiguity" allows each country to read into proposals whatever they choose. Such techniques also mask real disagreements that have hampered the translation of programmes into meaningful activities. Thus programmes negotiated by government representatives suffer from the inherent limitations of the political process. Skill in the art of compromise is not sufficient when those who hold the resources and understand the conditions to which the programmes must apply are unable to ensure that practical issues are taken into account.
Counter–claim A certain amount of idealistic wording is necessary to the extent that a certain vagueness facilitates meetings between the representatives of opposing regimes or ideologies.
Broader Unrealistic policies (#PF9428).
Aggravated by Indecisive multilateralism (#PF9564).

♦ PF7497 Trivialization of liberty
Collapsed conceptualization of liberty
Claim The reduction of the concept of liberty to doing what one wants to do destroys the profundity of the experience. Freedom is found in the midst of and because of constraints, not in the absence of them.
Broader Facile social concepts (#PF5242).

♦ PF7498 Youth violence
Teenage violence — Culture of youth violence
Nature A culture of violence is emerging amongst the young. They are unemployed or underemployed and believe they have no skills. Seeing the riches of society in shops and through advertisement, members of the culture recognize they are neither producers nor consumers. Expected to respond obediently to a volatile labour market in flexible ways but they seek something tough and enduring. A culture of machismo is a form of resistence and an assertion of their most irreducible characteristics. They are bored yet reject make work schemes, job training schemes and calls for national service because all of these accomplish nothing. Rejecting materialism they set other goals and standards of behaviour. They define themselves through appearance, haircuts, clothes and stance, and not through anything they create or make in the world. Aware of the inadequacy of the role open to them as consumers the crimes they commit are not of theft but of collective, often violent, activity. Politics is equally meaningless. The political left is just another less efficient administration. Their lives are empty of meaning and having know no other way of living they are unable to say what is missing.
Incidence In the USA between 1983 and 1987 arrests of those under 18 for murder jumped 22.2 percent, for aggravated assault 18.6 percent and for rape 14.6 percent. At the same time, the total number of teenages declined 2 percent.
Refs Apter, Steven J and Goldstein, Arnold P *Youth Violence* (1986); Breer, William *The Adolescent Molester* (1987).
Broader Criminals (#PC7373) Human violence (#PA0429).
Narrower Violence as a resource (#PF3994) Alcohol–related violence (#PE7084).
Aggravated by Decadent music (#PF5190) Juvenile stress (#PC0877)
Family violence (#PD6881) Culture of violence (#PD6279)
Educational wastage (#PC1716) Indecent advertising (#PD2547)
Single parent families (#PD2681) Negative social context (#PF9003)
Drug abuse by adolescents (#PD5987) Violence as entertainment (#PD5081)
Inadequate firearm regulation (#PD1970) Poverty in developed countries (#PC0444)
Excessive portrayal of substance abuse in the media (#PE3980).

♦ PF7521 Eclipse of reason
Nature According to social philosopher Max Horkheimer the conquest of the natural world has become the ruling passion of modern society. People know more and more how to do things, less and less about what is worth doing. Nature is seen as raw materials to be worked up into something useful. This moral chaos he called "eclipse of reason".
Aggravates Ethical decay (#PB2480).

♦ PF7523 Escapism
Claim Overexposure of escapist material in entertainment such as TV, movies and novels diverts people from the problems of daily living, encourages them to retreat into a dream world and promotes individual and group apathy; thus inhibits social progress.
Counter–claim Escapist material does not divert people from more serious considerations. While some people do overindulge, these people might be more dangerously preoccupied. And while those who make heavy use of the media to escape have little interest in serious social problems, their lack of interest is more likely the cause of escapism than the result.
Narrower Meaningless recreation (#PF0386) Escapist family life styles (#PD4069)
Parochial escapist media entertainment (#PD0917)
Escaping reality through popular psychological screens (#PF1112)
Escaping reality through sophisticated shields against life's pain (#PF0916).

♦ PF7532 Over–specialization in sciences
Nature The diversification of scientific knowledge is increasing rapidly as more and more aspects of the material universe and more and more forms of human activity are being investigated in greater and greater depth, within the framework of increasingly rigorous theory building. Fields of specialization (pure and applied sciences, natural and social sciences) now number in the order of a thousand. Interdisciplinarity thus becomes a highly complex process involving a very large number of specialists from essentially different disciplines, and this complexity accentuates the difficult problem of communication which is thus posed. The fragmentation inherent in the growth process will increasingly affect the ability to tackle development problems, for the solution of which will be required an entirely new type of scientist, trained in the methods of global analysis and able to identify the problems and to select techniques by which they can be solved.
Broader Over–specialization (#PF0256).

♦ PF7535 Lack of rational global political decision–making
Nature Unlimited national sovereignty has spread like a plague across the world since the drafting of the UN Charter in 1945. It was quite natural but it did not help when the new states emerging from the yoke of colonialism lusted for the same total sovereignty their former masters had. There is no system of decision–making powerful enough to surpass this national drive for supremacy. The result is that international political decisions, even within the United Nations, depend at basis on power manipulation among those nations most able to threaten the others.
Broader Ineffective decision–making processes (#PF3709).

♦ PF7542 Inadequate protection and preservation of cultural property
Gradual destruction of cultural property
Broader Destruction of cultural heritage (#PC2114).
Narrower Destruction of archaeological sites (#PD4502)
Endangered monuments and historic sites (#PD0253)
Destruction of cultural property during warfare (#PD7298)
Destruction of historic documents and public archives (#PD0172).
Aggravated by Ignorance (#PA5568) Negligence (#PA2658)
Corrosion in tropical climates (#PE1811)
Destructive action of mould in tropical climates (#PE1265).

♦ PF7543 Prohibitive cost of nuclear power plants
Incidence There was an 800 percent cost overrun on the Seabrook nuclear project in New Hampshire, originally budgeted at less than US $1 billion but estimated, in 1984, to require $9 billion to complete.
Claim Bold assertions and unsupported analysis have led to enormous uncertainty about the actual cost or economic merits of nuclear power. Cost estimates for nuclear plants in the USA have been rising since the first commercial plants were started in the early 1960s, but most of the early evidence of economic troubles was ignored by experts who assumed that costs would begin to decline as the technology improved.
Broader Prohibitive equipment costs (#PJ8196).
Aggravates Insufficient nuclear power stations (#PD7663).

♦ PF7554 Vulnerability of marine animal communication
Nature Pollution of the world's water seriously threatens the elaborate sensory system of marine animals. Small amounts of some chemicals may not be directly poisonous, but they may have indirect, often invisible effects. They may destroy, suppress, mimic, or mask vital phenomena, garbling the messages on which aquatic life depends, confusing sea creatures, and altering their behaviour.
Background Pheromones are the naturally secreted chemicals by which water animals communicate up one another. Animals use these chemical messages to guide their migrations; discriminate between individuals; recognize sex and reproductive states; and sense aggressors, predators, friends and their young.
Broader Marine pollution (#PC1117)
Environmental hazards from chemicals (#PC1192).
Aggravated by Chemical industry wastes (#PG2549).

♦ PF7556 Economic inefficiency
Imperfect market operation — Market imperfections — Lack of protection by free markets — Failure of markets
Nature Free markets underprovide public goods such as national defence, law and order, primary education, basic health, infrastructure, and research and development, namely goods that benefit people other than the producers or consumers. Equally markets can overproduce goods that impose costs beyond those borne by the producer: traffic congestion, pollution, the depletion of natural resources, etc. In addition the existence of monopolies, the lack of fully developed markets (especially for capital and insurance), and gaps in the supply of information may result in inefficient resource allocation and yield savings and investment rates that are less than optimal. Market mechanisms may thus produce insufficient growth as well as macroeconomic imbalances, such as balance of payments deficits and underemployment.
Claim Conventional economic wisdom holds that freedom of enterprise and the profit motive produce economic efficiency. This is correct to the extent that economic theory defines such efficiency in precisely the way that those intending to maximize the return on their investment would wish. But when efficiency is defined as applying existing productive capacity to meet existing human needs, then it is immediately apparent how inefficient the global economic system presently is.
Broader Inefficiency (#PB0843).
Narrower Disruption of financial markets (#PD4511)
Imperfections of capital markets (#PF3136)
Economic inefficiencies in developing countries due to restrictive business practices (#PE2999).
Aggravates Economic inflation (#PC0254) Economic stagnation (#PC0002)
Underground economy (#PC6641) Uncontrolled markets (#PF7880)
International economic injustice (#PC9112)
Lack of economic and technical development (#PE8190)
Proliferation of public sector institutions (#PF4739)
Low confidence in investment and stock markets (#PE5102)
Economic and financial instability of the world economy (#PC8073)
Lack of incentive for users to care for common property (#PF4516)
Policy cross–conditionality restrictions in multilateral development aid (#PF9216).
Aggravated by Instability of prices (#PF8635) Vulnerability of stock markets (#PD5676)
Elitist control of global economy (#PC3778) Inefficiency of financial markets (#PF6980)
Fluctuations in real value of money (#PD9356)
Financial destabilization of world trade (#PC7873)
Faltering structural adjustment in the world economy (#PF9664).

♦ PF7577 Student immobility
Lack of foreign students — Hyperlocalized education
Nature Students in the Middle Ages were necessarily mobile, as the lack of books made personal dialogue among academics a necessity. Since the 19th and early 20th century development of university systems across the world, however, the proportion of students studying abroad has declined. Often the vast majority of foreign students in a country are those who lack adequate educational institutions in their home country. This lack of interest in the wisdom of one's neighbours leads to narrow perspectives and a decline in global communications.
Related Unequal opportunities for foreign students (#PE7726).

♦ PF7578 Excessive salaries of corporate executives
Overpaid business executives
Incidence The average 1988 pay of the top executives of 350 of biggest companies in the USA was over $2 million. Over the past decade bosses' pay has increased by an average of 12 percent a year while production workers' pay only 6 percent.

Claim Huge pay rises of executives make it more difficult to slap down inflationary wage demands from the trade unions.
Broader Prohibitive labour costs (#PF8763).
Aggravates Underpayment of government officials (#PD8422)
Disparity in remuneration between public and private sector employees (#PE7760).

♦ PF7580 Risk
Risks — Hazards — Dangers — Risky situations — Dangerous behaviour
Nature Risk is an inherent component of life; we are all at risk, every moment of our lives, even if we simply stay at home and do nothing. Some risks are self-imposed because we assume that the enjoyment they bring us is worth the risk (cigarette smoking, rock climbing, auto racing); other risks are imposed from natural sources (earthquakes, lightning, floods) or are man-made but considered outside of the realm of the average person's capacity to change (nuclear reactors, satellites falling from space).
Incidence It is recognized that the future will be characterized by increasing risks, whether those associated with new technologies, natural or human-caused disasters, or those of irreversible damage to natural systems both regionally (such as acidification, desertification, or deforestation) and globally (such as ozone layer depletion and climate change).
Claim Both private citizens and the public at large are often duped into having risks imposed upon them. The Americans who fought in the Vietnam War were mislead into believing that their presence would curtail the spread of communism; the inhabitants near the Three Mile Island nuclear reactor in Pennsylvania were told that the risk of an accident was virtually impossible; and women who took Thalidomide during the 1950s and 1960s thought that it would stop their early-pregnancy nausea, involving virtually no risk to the foetus.
Counter-claim We accept risks of employment or sport because we consider that the pay, the interest of the activity, or the benefit it gives to others is worth the risk involved.
Refs Boyadjian, Haig and Warren, James F *Risks* (1988); Cooper, M G (Ed) *Risk* (1985); Johnson, Branden, B and Covello, Vincent, T (Eds) *Social and Cultural Construction of Risk* (1987); Newman, Charles M and Czechowicz, James *International Risk Management* (1983).
Broader Danger (#PA6971) Uncertainty (#PA7309).
Narrower Libel (#PD3022) Theft (#PD5552) Ransom (#PJ1977)
Slander (#PD3023) Kidnapping (#PD8744) Risk of war (#PF4215)
Bad weather (#PC0293) Travel risks (#PD7716) Dangerous toys (#PE1158)
Abuse of credit (#PF2166) Tourist hazards (#PE8966) Multiple births (#PE4107)
Dangerous animals (#PC2321) Interruption risk (#PF9106) Human contingency (#PF7054)
Nuclear accidents (#PD0771) Breach of contract (#PE5762) Mechanical failure (#PC1904)
Political upheavals (#PC7660) Injurious accidents (#PB0731) Business bankruptcy (#PD2591)
Dangerous substances (#PC6913) Inadequate insurance (#PF8827)
Accidental explosions (#PE3153) Dangerous occupations (#PC1640)
Environmental hazards (#PC5883) Unsafe port facilities (#PE4897)
Prohibitive legal fees (#PF0995) Hazards to human health (#PB4885)
Vulnerability of farming (#PC4906) Foreign money liabilities (#PE6746)
Risk of capital investment (#PF6572) Refusal of licence renewal (#PF8964)
Occupational risk to health (#PC0865) Transport of dangerous goods (#PD0971)
Inconsistent risk evaluation (#PF5482) Defective product manufacture (#PD3998)
Risks in transfer of ownership (#PE8580) Aircraft environmental hazards (#PD8328)
Inadequate cargo transportation (#PE0430)
Economic philosophy of controlled risk (#PF2334)
Nationalization of foreign investments (#PC2172)
Offences involving danger to the person (#PD5300)
Delays in delivery of goods and services (#PE3928).
Related Dangerous paths (#PJ9888) Marine accidents (#PD8982)
Dishonest employees (#PD9397) Computer-based crime (#PE4362).
Aggravated by Neglect (#PA5438) Uncertainty (#PA6438)
Insecurity of property (#PC1784)
Inadequate emergency medical services (#PD1428)
Unreliability of equipment and machinery (#PC2297).

♦ PF7583 Disguised negative consequences of remedial action
Nature Remedial action is often studied and embarked upon without consideration of the repercussions - possibly negative - of such action. For example, the damming of a river may eliminate floods, make irrigation possible, and employ hundreds of people; but unforeseen are such consequences as the lost nutrients which once collected in the river and fertilized the plain when the river overflowed; the fish that subsequently have no nutrients to feed on, threatening the local fishing industry; and the costs of re-employing and re-housing the now unemployed fishermen.
Broader Short range planning for long-term development (#PF5660).
Aggravates General obstacles to problem alleviation (#PF0631).

♦ PF7593 Parochialism of established religions
Nature Established religions operating out of parochial world views do not create universal symbology. Attempts to destroy old symbol systems and replace them with expedient alternatives and to create reduce value systems alienate those who could most assist people with the spirit dimension.
Broader Reinforced parochialism of internal values and images (#PF1728).

♦ PF7610 Declining productivity in socialist countries
Broader Socialism (#PC0115) Underproductivity (#PF1107)
Declining economic productivity (#PC8908).
Aggravates Currency black market in socialist countries (#PD2413).

♦ PF7614 Reduced images of environmental protection
Broader Uncontrolled application of technology (#PC0418).

♦ PF7619 Uncoordinated government policy-making
Uncoordinated government decision-making — Uncoordinated government planning
Nature Responsibility for planning and budgeting is often dispersed among several institutions of government without any effective coordination.
The plan is often disregarded as the budget is prepared for reasons such as: (a) plans may not be sufficiently detailed to provide guidance in budgeting; (b) the budget process is often rushed and subject to many short-term pressures; (c) planners may have less influence than budgeters, because the budget is the authoritative legal document, while the plan typically does not have the force of law; (d) traditional stereotypes tend to separate "short-term" budgeters from the "long-term" planners concerned with questions over which the government has little control.
Incidence The most obvious example of this is the tension that often exists between the ministries of finance and planning. Few countries have managed to integrate the planning and budgeting functions well. Problems of coordination can also exist between core ministries and spending agencies, whether sectoral ministries, subnational levels of government, or state-controlled enterprises. Coordination can also break down during implementation. Central ministries often react to tight budget constraints, overprogrammed budgets, or simple mistrust of spending agencies by slowing the disbursement of funds or by erecting unnecessarily cumbersome procedures in areas such as procurement, land acquisition, or contractor eligibility. Such indirect forms of control delay the implementation of projects.
Broader Uncoordinated policy-making (#PF9166)
Inappropriate public spending by government (#PF6377)
Inadequacy of governmental decision-making machinery (#PF2420).
Aggravates Unpredictable governmental policy (#PF1559)
Delay in implementation of commitments (#PF3975).
Aggravated by Governmental incompetence (#PF3953)
Unparliamentary behaviour (#PF4550)
Opaque budgetary procedures in the public sector (#PF5374)
Domination of government policy-making by short-term considerations (#PF0317).

♦ PF7631 White lies
Nature Although a white lie may be a minor falsehood not meant to injure anyone, and is of little moral import, the accumulated effect of white lies may be confusion, misunderstandings and distress. White lies are rarely told just to be socially adept, they are rarely told in an isolated incidence, and they are rarely totally innocuous.
Claim Even though some people may defend the telling of white lies as being falsehoods not meant to hurt anyone and as being of little moral import, they are, regardless of their description, lies, and their cumulative consequences do harm.
Counter-claim The telling of white lies in social situations preserves the equilibrium and humaneness of relationships, and is usually excused as long as its numbers do not become excessive. They often bring a substantial benefit and have the effect of avoiding real harm.
Broader Lying (#PB7600) Deception (#PB4731).
Related Grey lies (#PF3098) Black lies (#PE4432) Paternalistic lies (#PF7635).
Aggravates Corruption of the good in human nature (#PE7917).

♦ PF7633 Abuse of confidentiality
Breach of confidence
Claim Confidentiality should protect the secrets of the patient or client. But in practice, it can expand to include what professionals hide from patients, clients, and the public at large. This prevents outsiders from finding out about negligence, overcharging, unnecessary surgery, or institutionalization. Confidentiality is also used to protect incompetent colleagues, companies, or entire industries.
Broader Immorality (#PA3369).
Aggravates Injustice (#PA6486) Intergovernmental suspicion (#PC2089).

♦ PF7634 Deceptive social science research
Social science lying — Deception in the social sciences
Nature Openness and honesty are two essential characteristics of the relationship between investigator and research participants, but these are repudiated when participants are subjected to research the means or motives of which have not been honestly explained.
Incidence In one study, male undergraduates were falsely led to believe they had been sexually aroused by photographs of men and their responses to such information studied. In another, subjects were surreptitiously given LSD.
Claim Deceptive social science research is an interference in the subject's freedom of informed choice. It also makes a mockery of the trust one human being places in another, especially someone in authority who would normally be assumed to be trustworthy.
Counter-claim Since most experiments are utterly harmless, it is a waste of time and energy to impose upon them the kind of requirements full explanations might entail.
Broader Irresponsible scientific and technological activity (#PC1153).
Aggravates Scientific fraud (#PF1602) Abuse of science (#PC9188).

♦ PF7635 Paternalistic lies
Nature To act paternalistically is to guide and even coerce people in order to protect them and serve their best interests. In addition to guidance and persuasion, the paternalist can also manipulate by force and by deception. Once having assumed a paternalistic stance, it is easy to condone one's lies under the argument of their necessity for the good of children (or childlike adults) whose lack of independent action does not warrant total truth; children are the most often deceived with the few qualms.
Counter-claim Lies to protect close human bonds carry a special sense of immediacy and appropriateness. To keep children from knowing that their parents' marriage is dissolving, to keep up a false pretence of good health to assure that good fortune will return again are all paternalistic lies which are told to comfort, protect and support the deceived.
Broader Lying (#PB7600) Institutional lying (#PD2686).
Related Grey lies (#PF3098) White lies (#PF7631).
Aggravates Domination (#PA0839).

♦ PF7642 Bilateralism in trade arrangements
Nature Bilateral relations in international trade, usually advocated by strong, industrialized countries, serve to fragment the international community, encourage imitation by others, and inevitably lead to the politicization of trade relations. Discriminatory practices in one country's favour necessarily create resentment elsewhere.
Broader Restrictive trade practices (#PC0073).
Narrower Bilateralism in aid to developing countries (#PE9099).

♦ PF7643 Unauthentic peace
Unmeaningful peace
Nature Whether or not a just peace prevails in a society, peace may be experienced as lacking in authenticity and meaning. This may be due to the absence of any development process, or one in which such development only takes the form of adaptive improvements to living standards and their technological support. Under such conditions, the society as a whole and people individually are not challenged in any fundamentally meaningful way. Whilst the society may survive for a long period in this mode, it is essentially stagnant and predictable. It is typical of this kind of peace that it is experienced as boring and frustrating by the young and the adventurous, especially after they have explored the available self-improvement and leisure opportunities. An unauthentic peace is characterized by processes which avoid confronting people with experiences which continually force them to reassess both themselves and their society in ways which lead to new insights and fundamental restructuring. Under an unauthentic peace the old order remains essentially unquestioned, few significant risks are taken, and there is little collective learning of significance. Such a risk-free peace is essentially self-replicating rather than self-transforming. It is probable that under such conditions a society effectively invokes conflict and war in order to be exposed to realities which lead to further individual and collective transformation.
Broader Trivialization of peace (#PF5826).
Aggravated by Unjust peace (#PB7694).

♦ PF7649 Increasing scepticism about the accuracy of official information
Broader Scepticism (#PF3417) Credibility gap (#PB6314).
Aggravates Loss of institutional credibility (#PF1963)
Misunderstanding of official communications (#PF8382).
Aggravated by Statistical errors (#PF4118) Misuse of statistics (#PF4564).

♦ **PF7653 Blood vengeance**
Blood feud
Nature Blood vengeance is a primitive form of the law of retribution according to which a kinsman must vindicate the rights of a relative whose blood has been shed.
Background According to the ancient Greek concept every act of bloodshed, even when committed in self–defence, created a certain defilement that required purification. Not only the criminal but his family also was defiled until the slain man's life was appeased by exacting vengeance. The initial crime could easily lead to a series of mutual crimes, a blood feud or vendetta. In primitive societies a whole family or even a whole clan might be annihilated for a murder committed by one of its members.
Broader Feuds (#PE8210) Human violence (#PA0429).

♦ **PF7669 Secret military operations**
Illegal warfare — Covert military operations — Clandestine military operations
Nature Military operations in time of peace may be conducted against a country in such a manner that the majority of the population of the aggressor country is unaware of them. Such operations may even be conducted without the knowledge or approval of appropriate bodies in the legislature of the aggressor country. Active military operations may also be conducted through guerrilla movements under the guise of providing them with weapons and military advisors.
Claim Traditional forms of warfare are now becoming so extreme as to be suicidal for all. But opportunities for covert tactics, the needs and means for them are developing continually. Covert tactics can be precise, effective, rewarding and cheap. Good use can be made of of surprise, bluff, timing and the adversaries weaker points. They can be used to pre–empt situations in which overt aggressive interaction might seem justified. They could perhaps relieve the planet of its dangerous, expensive overburden of useless military personnel and material, diverting resources to more beneficial ends.
Counter–claim Covert operations have no useful place in the policies of a democratic nation. After the Second World War each US administration has discovered that covert war fails more often than it succeeds. Even when covert operations have succeeded in the short run, they damage US interests and the victim country in the long run.
Broader Military secrecy (#PC1144) Aggressive foreign policy (#PC4667)
Foreign military intervention (#PD9331).
Narrower Illegality of nuclear weapons (#PF4727).
Related Unlawful government action (#PF5332)
Covert intelligence agency operations (#PD4501).
Aggravates Guerrilla warfare (#PC1738)
Government complicity in illegal activities (#PF7730)
Government seizure of foreign nationals in foreign countries (#PE6564).
Aggravated by Unfulfilled treaty obligations (#PF2497)
Government action against regimes with alternative policies (#PF2199).

♦ **PF7673 Persistence of outmoded concepts**
Institutionalization of discriminatory outmoded concepts
Nature Concepts which have been demonstrated to be inappropriate, incorrect or discriminatory are not immediately eliminated from institutional and other systems. Partly because they are associated with well–tried procedures, traces of them may persist in legal, educational, scientific and other systems.
Counter–claim Concepts defined by some constituencies as outmoded may be of continuing significance to others. The concepts by which they are supposedly to be replaced may have latent defects they will subsequently make such replacement questionable.
Narrower Outmoded legal systems (#PF2580).
Aggravates Inadequacy of prevailing mental structures to challenge of human survival (#PF7713)
Inadequate global consensus concerning problems and prospects of humanity (#PF9821).

♦ **PF7681 Boredom of captive and domesticated animals**
Nature Animals kept under unchanging conditions which offer them little incentive to move and explore their environment may easily become bored, especially when the environment does not permit them to participate in their normal cycles of relationship with those of the same or different species. They then exhibit abnormal behaviour patterns, such as the repetitive pacing and rubbing frequently observable in zoos.
Claim Imposing unstimulating conditions on animals is tantamount to destroying their spirit and effectively denies recognition of their role in the planetary ecology.
Counter–claim Man is the only animal that can be bored. Animals are quite content when they are regularly fed and do not have to compete in the wild for scarce food.
Broader Boredom (#PA7365) Animal suffering (#PD8812)
Denial to animals of legal protection of their rights (#PE8643).
Aggravated by Zoosadism (#PG2204)
Cruel treatment of animals for research (#PD0260).

♦ **PF7684 Millenarianism**
Millennialism
Nature It is a belief that the end of this world as we know it is at hand and that in its wake will appear a fertile, harmonious, sanctified, and just new world.
Broader Unreadiness for second coming of Christ (#PF0638).

♦ **PF7697 Contempt**
Narrower Criminal contempt (#PD5705) Contempt of judicial process (#PD9035)
Contempt for democratic processes (#PF0639)
Contempt for traditional modes of behaviour (#PC4321)
Contempt for agricultural labour in developing countries (#PD1965).
Aggravated by Homelessness (#PB2150).

♦ **PF7700 Limitations of surprise-free thinking**
Limitations of surprise–free methodologies — Inadequacy of methodologies to deal with discontinuities and random events
Nature Human societies each constitute a form of equilibrium. Equilibria are liable to catastrophes when, under special limiting conditions, small inputs may produce very large, often unforeseen, and frequently irreversible outputs. Analysts are hindered by the inadequacy of methodologies to deal with such discontinuities and random events and are usually obliged to make inappropriate assumptions that over a sufficiently long time span such discontinuities and surprises may be either ignored or will average out.
Claim Surprise–free forecasts are necessary, but insufficient, tools for efforts to improve the management of long–term interactions between development and the environment. By leaving out external shocks, nonlinear responses, and discontinuous behaviour so typical of social and natural systems, surprise free–analysis hinders efforts to interpret a host of not improbable eventualities.
Aggravates Latent problems (#PF9328) Inaccurate forecasting (#PF4774)
False assumptions on sustainable development (#PF2528)
Simplistic technical solutions to complex environmental problems (#PF0799).
Aggravated by Over–reliance on convenient beliefs (#PF9205)
Excessive dependence on computer models of complex system behaviour (#PE8533).

♦ **PF7702 Official self–deception**
Broader Deception (#PB4731) Deception by government (#PD1893).

♦ **PF7703 Allopathy**
Nature Allopathy, or "different suffering", is the system of healing which assumes that treatment is best done by applying to a disease a cure not directly related to the disease. This is the standard medical approach used by the pharmaceutical and mainline health industries. This is the reverse of homeopathy ("same suffering") which assumes that disease is cured by "same suffering", or cures directly related to the disease. Allopathic medicine may not always provide the best approach for an individual, but it has dominated medical thinkers to such a degree that alternative approaches tend to be ignored.
Broader Lack of integration of traditional and Western medicine (#PF4871).
Related Poisonous plants (#PD2291) Iatrogenic disease (#PD6334).

♦ **PF7706 Unproven relationships between problems**
Assumption of lack of problem interrelationships — Assumption of absence of causal relationship between problems — Denial of causal relationship between problems
Nature In the absence of adequate research on problems and associated phenomena, it can easily be assumed that there is no causal relationship between certain problems (eg smoking and lung cancer, acid rain and forest decline, carbon dioxide emission and global warming). It is especially convenient in a political context to rely on the argument that there is "no proven link". Recognition of the existence of any such link may be delayed for years until no reputable specialists can be found to question the arguments of those who initially demonstrated the possibility of such a link.
Broader Evasion of issues (#PF7431) Denial of evidence (#PD7385).
Aggravates Non–recognition of problems (#PF8112)
Complex interrelationship of world problems (#PF0364)
Government delay in response to symptoms of problems (#PF6707)
Placement of the burden of proof on the disempowered (#PF3918).
Aggravated by Evidence decay (#PF5403) Biased expertise (#PF6395)
Questionable facts (#PF9431) Disagreement among experts (#PF6012)
Conceptual repression of problems (#PF5210) Destruction of scientific records (#PF4633)
Neglect of the role of women in rural development (#PF4959)
Inadmissability of evidence from other jurisdictions (#PE5424).

♦ **PF7713 Inadequacy of prevailing mental structures to challenge of human survival**
Antiquated thought patterns — Inadequacy of prevailing learning systems — Inadequate intellectual and cognitive infrastructure to formulate effective global policies
Nature Much of society is trapped in antiquated thought patterns: narrow and hierarchical perspectives; deterministic, linear or emotive methods of reasoning; cause and effect perceptions of relationships; and shallow understandings of human spirit.
Broader Antiquated intellectual methods to appropriate human depths (#PF1094).
Narrower Dualism (#PF9161) Criminality (#PA9226) Scholasticism (#PF9065)
Overdependence on education (#PE8988) Over–reliance on convenient beliefs (#PF9205).
Aggravates Insensitivity to non–immediate hazards to society (#PF9119)
Inadequate global consensus concerning problems and prospects of humanity (#PF9821).
Aggravated by Uncritical thinking (#PF5039) Inadequate intellectual methods (#PF7380)
Persistence of outmoded concepts (#PF7673).

♦ **PF7722 Riskless responses to public injustices**
Nature Individuals and organizations seeing social and political injustices respond in the manner with the least risk to themselves. Some place responsibility on others, for example, crime is the responsibility of the police. Some become indifferent to the suffering of others. Some become so involved in superficial activities that they have no time to deal with injustices.
Broader Non–concerned attitudes (#PF2158).

♦ **PF7727 Compounding effect of treating genetic diseases**
Negative aspects of euphenics — Deterioration of gene pool
Nature As medical science is able to apply medical and surgical treatments to genetic diseases – successful treatment also removes barriers of reproduction – the frequency of each malfunctioning gene concerned will increase in subsequent generations. Also a probability of producing genotypes with multiple disorders, e.g. someone with haemophilia and sickle–cell anaemia, increases progressively.
Broader Genetic defects and diseases (#PD2389).

♦ **PF7728 Failure to legally rehabilitate victims of miscarriage of justice**
Failure to clear the names of people convicted for crimes of which they are innocent
Broader Deficiencies in national and local legal systems (#PF4851).
Aggravates Social stigma (#PD0884).
Aggravated by Miscarriage of justice (#PF8479) Indifference to suffering (#PB5249).

♦ **PF7730 Government complicity in illegal activities**
Government complicity in drug trafficking — Government complicity in illegal arms trade
Broader Complicity (#PF4983) Unlawful government action (#PF5332).
Aggravates Drug smuggling (#PE1880) Illegal ivory trade (#PE4991)
International arms trade (#PC1358)
Diversion of high technology to hostile countries (#PE7174).
Aggravated by Secret military operations (#PF7669)
Covert intelligence agency operations (#PD4501).

♦ **PF7737 Over–programming**
Over–planning — Over–programmed budgets
Incidence Because of their unwillingness to make difficult decisions, planners and budgeters often overprogramme, paying too little attention to priorities, resource constraints, or phasing of implementation. Unexpected shortfalls can make a well–programmed fiscal plan obsolete. The pressures of an overprogrammed budget tend to cause all spending to be delayed or cut, irrespective of priorities. As a result funding for many programmes is likely to fall below the minimum level for effectiveness.
Broader Inappropriate public spending by government (#PF6377).
Aggravates Inadequate implementation of plans and programmes against problems (#PF1010).
Aggravated by Avoidance of decision–making (#PF4204).

♦ **PF7741 Planned degradation in product quality**
Nature In order to improve profit margins, commercial enterprises may deliberately choose to gradually diminish the quality of a product whilst continuing to create the impression that it remains unchanged (or is even being improved).
Related Planned obsolescence (#PC2008).
Aggravates Deterioration in product quality (#PD1435).
Aggravated by Short–term profit maximization (#PF2174).

♦ **PF7745 Structural amnesia in institutional systems**
Structural forgetfulness — Biased recognition of problems

Nature Problems are selected by humanity, whether in whole communities or within specialized groups (science, religion, technology, politics, education), in accordance with an editing process that excludes some kinds of issue because of their societal implications. Although those problems threatening the central values of the group are recognized by it as meriting attention, no attention is given to problems which are handled as part of the routine procedures of the institutions. More significantly, problems are ignored by scientific and political institutions when any recognition of them would call for such a radical shift in perspective that, in addition to threatening their most cherished values, would actually deconstruct the institutions based upon them.
Incidence The effects of structural forgetfulness have been noted in the relationship between genealogical structures in tribal societies and the nature of the political and judicial institutions they permit. It has been used to account for the socially constructed ignorance of historical processes and for the geographical constraints on political action in certain western societies. It has been argued that it is significant to understanding the establishment of boundaries between scientific disciplines.
Related Forgetfulness (#PA6651) Global amnesia (#PF0306).
Aggravates Fragmentation of knowledge (#PF0944)
Fragmentation of academic disciplines (#PF8868)
Limitation of current scientific knowledge (#PF4014).

♦ PF7754 Non-resistance to evil
Refs Darrow, Clarence *Resist Not Evil* (1972).
Related Passive resistance (#PF2788) Resignation towards bribery (#PF8611).
Aggravates Evil (#PF7042).
Aggravated by Complicity with evil (#PF0926).

♦ PF7756 Melancholy
Refs Burton, Robert *Anatomy of Melancholy*.
Narrower Lovesickness (#PF3385).
Related Sin (#PF0641) Lust (#PA4673) Envy (#PA7253)
Anger (#PA7797) Boredom (#PA7365) Solemnity (#PA6731)
Unpleasantness (#PA7107).

♦ PF7768 Unreported disasters
Cover-up of major accidents — Cover up ship blast
Broader Disasters (#PB3561) Underreported issues (#PF9148).
Aggravated by Official secrecy (#PC1812) Avoidance of negative feedback (#PF5311).

♦ PF7779 Delay in recognition of patents
Incidence Almost 30 years after the application was filed, Japan granted Texas Instruments Inc. a patent on the integrated circuit, that made computer age possible.
Broader Inadequacy of patent coverage (#PF3538).
Aggravates Distortion of international trade through obstacles to patent protection (#PD0455).

♦ PF7780 Sectarianism
Sectarian tension
Refs Lenin, V I *Against Dogmatism and Sectarianism* (1978).
Broader Ideological conflict (#PF3388).
Aggravated by Double standards in morality (#PF5225).

♦ PF7798 Irresponsible expression of emotions equated with free speech
Erosion of social dialogue — Collapse of systems for verbal exchange
Nature Concerned groups frequently find themselves unable to translate the social conflicts in which they are involved into effective dialogues. By reducing complex situations to single issues, by emotional and violent outbursts, and by hysterical relating of issues to individual situations, the form and quality of the dialogue become secondary to the entertainment or therapeutic value for the speakers. Free speech is seen as an indulgence for an emotive elite rather than the responsibility of every member of the citizenry.
Another aspect of the contradiction in what is usually considered free speech is to be seen in a collapse of systems for verbal exchange. This is frequently seen in the irresponsible spewing of poorly thought out feelings. A healthy dynamic of free speech would not only afford the right of all to speak, even irresponsibly, but also encourage people to speak responsibly and significantly. Consciously aware systems of dialogue are necessary for everyone to be able to understand the responsibilities inherent in the exercise of free speech, and to allow the articulation of concerns in such a manner as to gain an attentive audience. This would in turn make each individual aware of what it means to speak on behalf everyone.
Broader Irresponsibility (#PA8658) Exclusion of opposing views (#PF3720)
Undemocratic social systems (#PB8031).
Narrower Verbal conflict envisaged as destructive (#PF7358)
Speaking in opposition out of personal responses (#PE3797)
Accusing channels for dialogue of being restrictive (#PF5862)
Failure to understand the necessity of creative establishment and disestablishment tensions (#PE3948).
Related Blocked minority opinion (#PD1140) Inadequate political networks (#PD2213)
Collapse of societal engagement (#PF2340).
Aggravated by Ineffective dialogue (#PF1654) Limited verbal skills (#PD8123).

♦ PF7801 Unreliability of weapons systems
Nature Weapons coming into service do not have guarantees that they would work, and reliability is not being fully tested.
Broader Unreliability of computer software (#PE4428)
Unreliability of equipment and machinery (#PC2297).

♦ PF7805 Inoperative forums for public information
Nature Despite the need of many First World rural communities for timely, accurate, public channels of local information and for regular, open, community-wide forums for airing of individual opinion and the consensing of the community will, local news often continues to be carried primarily by word of mouth with inevitable distortions.
Broader Incomplete access to information resources (#PF2401)
Inadequate circulation of local information (#PF6552).
Aggravates Lack of information (#PF6337) Incorrect information (#PB3095)
Misleading information (#PF3096)
Inadequate dissemination and use of available information (#PF1267).
Aggravated by Insufficient flow of information (#PF6469).

♦ PF7806 Inability of educational systems to keep pace with technological advancement
Nature Planning and training for vocational skills are insufficiently related to new skills requirements. This is the result of the rapid and inevitable transition of technology and the current tendencies of education to lag behind social changes. Because changes in technology are not interrelated with technical skills training, skills are developed for which there is no place or function with the job market and needs are created in society for which skills have not yet been developed.
Broader Functional changes due to technological advance (#PE7943).
Aggravates Obstacles to education (#PF4852) Inappropriate education (#PD8529)
Training inappropriate to structural and technological changes (#PE4596).
Aggravated by Adverse consequences of scientific and technological progress (#PF3931).

♦ PF7807 Offensive lyrics
Obscene songs
Incidence An increasing proportion of records contain lyrics which are offensive to some groups, especially to minority ethnic groups, to women or to those of another race or religion. Such records may carry explicit lyrics descriptive of or advocating one or more of the following: suicide, sodomy, incest, bestiality, sadomasochism, sexual activity in a violent context, murder, morbid violence, illegal use of drugs or alcohol.
Aggravates Decadent music (#PF5190).

♦ PF7808 Neglected food resources
Limited utilization of foodstuffs — Forgotten food products — Ignorance of traditional agricultural products
Incidence In pre-Colombian times protein-rich amaranth and guinea were basic foods in the New World along with maize and beans, but nowadays they are seen as weeds.
Narrower Declining breeds of domesticated animals (#PD6305).
Related Ignorance of traditional herbal remedies (#PE3946).
Aggravates Unaesthetic foodstuffs (#PD1126).
Aggravated by Narrow range of food crops (#PD4100)
Narrow range of practical skills (#PF2477).

♦ PF7815 Decline in human genetic endowment
Physical decline of the human race — Intellectual decline of human race — Racial degradation
Claim New preventive and remedial measures, advances in medicine, embryonics, genetics and intensive care have encouraged the perpetuation and propagation of all categories of people, including those with serious mental and physical handicaps, severe abnormalities, susceptibility to infectious diseases and transmissable genetic defects. This has resulted in the progressive multiplication of dysgenic qualities in the human race. The general level of intelligence, together with other qualities, is declining in industrialized countries. The more intelligent, cultured and fit, deliberately limit their families so their genes are being bred out. The less intelligent, the unskilled, the degenerate, and the unfit are multiplying unchecked, encouraged by a sentimental humanitarian approach which is inappropriate to the circumstances.
Broader Vulnerability of world genetic resources (#PB4788).
Aggravates Individual unfitness for survival (#PF4946)
Uncertainty of survival of the human race (#PE9085).
Aggravated by Decadence (#PB2542) Genetic drift (#PG5268)
Genetic inbreeding (#PD7465) Hereditary regression (#PD8149)
Lack of eugenic measures (#PD1091) Genetic and ethnic weapons (#PC6664)
Mutagenic effects of drugs (#PE4896) Genetic defects and diseases (#PD2389)
Irresponsible genetic manipulation (#PC0776)
Human physical genetic abnormalities (#PD1618).
Reduced by Compulsory sterilization (#PF3240).

♦ PF7820 Neglect of expert advice
Aggravates Unrealistic expectations (#PE7002) Substitution of fantasy for reality (#PF8922).
Aggravated by Short-term gain (#PF8675) Biased expertise (#PF6395)
Arrogance of experts (#PF3294) Disagreement among experts (#PF6012)
Irresponsible international experts (#PF0221) Unavailability of appropriate expertise (#PF7916).

♦ PF7823 Irresponsible delimitation of policy responsibilities
Broader Irresponsibility (#PA8658).
Aggravates Blame avoidance (#PF6382) Inappropriate policies (#PF5645)
Fragmentation of research (#PF9830) Uncoordinated policy-making (#PF9166)
Irresponsibility towards future generations (#PF9455).
Aggravated by Deliberate ignorance during policy-making (#PF8278)
Collusion between administrators of funding agencies and programme formulators (#PF8711).

♦ PF7836 Employment at risk through elimination of industrial pollution
Defence of industrial pollution by immediate victims
Nature Those most at risk from pollution are those most dependent on income from local polluting factories, and they often become the polluters' greatest defenders. They do not worry about health risks of lead, but they worry about the cost of compulsory catalytic converter.
Related Urban-Industrial pollution (#PC8745).

♦ PF7838 Conditional observance of multilaterally agreed trade commitments
Failure of political will to observe trade commitments
Nature The credibility of the rules and principles of the trading system has been weakened by the fact that the ultimate recourse open to a country affected by illegal trade measures is to withdraw "substantially equivalent concessions", a possibility of little effective use to weaker trading partners.
Broader Trade protectionism (#PC4275) Lack of political will (#PC5180)
Distortion in international trade (#PC6761).
Related Intergovernmental failure to fulfil financial commitments (#PF3913).
Aggravates Unfulfilled treaty obligations (#PF2497)
Inability to negotiate effective multilateral safeguard systems (#PF5287).

♦ PF7842 Inadequate public finance statistics
Inaccurate public finance data for developing countries — Inadequate systems of national accounts in developing countries
Nature To be effective, public finance policies must be based on accurate and comprehensive statistics on the financial transactions of public agencies. The weaknesses of national data sources are due to delays in auditing, weak administrative systems, and incomplete reporting of subnational government and state owned enterprise accounts. This makes it very difficult to get a timely, complete and accurate picture of the main sources and uses of public funds. Fiscal planning, consistent fiscal policy design, and financial accountability by decision-makers are thus significantly impeded. The weakness of public finance data, at a time of rapid increase in the public sector, complicates the design and implementation of public finance policies.
Background For intercountry analysis, financial data needs to be compiled on a comparable basis across countries. Such data is currently assembled within two international systems: the Government Finance Statistics (GFA) of the IMF and the System of National Accounts (SNA) of the United Nations. The former focuses solely on government transactions, whereas the latter considers government transactions as a component of the economy as a whole. Because of the difficulty of collecting consistent and accurate data on state owned enterprises, only data on transactions between the central government and the enterprises are currently recorded in the GFS. The GFS does not report asset or liability positions of the government, nor does it report depreciation for fixed assets owned by the government. Accounts are recorded on a cash, not an accrual, basis. GFS coverage is most complete for central government accounts, but more limited for general

government accounts. The SNA accounts are recorded on a accrual basis and include depreciation. For national aggregates the SNA consolidates transactions between all sectors and eliminates intersectoral transactions, so that only the final demand and value added are aggregated. National accounts data compiled within the SNA framework lack the detail required for many aspects of public finance analysis. The consolidated accounts omit some important financial flows, such as all domestic transfers, including interest.
Incidence For most developing countries consistent data are available only for the past decade or two and often cover only the central government. Comparable data on regional and local governments and on state-owned enterprises are patchy across countries and over time. Up-to-date national accounts data are not available in many developing countries. Whilst accounting systems may have been implemented, there are countries in which the data supplied is so late or so unreliable that they cannot serve as a basis for rational public expenditure or monitoring. Managers at all levels may either disregard requests for budget estimates for the next fiscal year or submit estimates far in excess of what is possible (reasoning that the government cannot fail to allocate some resources to their activities, that whatever they might submit is unlikely to be reflected in the ultimate budget, and that the actual release of funds will not match the budget anyway). And yet such unrealistic budget submissions in turn destroy confidence among those receiving them such that all phases of the process lose credibility in a cycle of mutually reinforcing scepticism.
Broader Deficiencies in national statistics (#PF0510)
Inappropriate public spending by government (#PF6377)
Inadequate statistical information and data on problems (#PF0625).
Aggravates Inappropriate policies (#PF5645) Unreported government spending (#PF2990)
Inadequate management of government finances (#PF9672)
Opaque budgetary procedures in the public sector (#PF5374)
Inadequate system of political checks and balances (#PE4997)
Inadequate economic policy-making in developing countries (#PF5964)
Inadequate control over government administrative process (#PC1818).

♦ PF7843 Scientific arrogance

Claim Scientists think that they have a divine right to appropriate society's resources for themselves simply because of their calling.
Broader Arrogance (#PA7646).
Aggravates Arrogance of experts (#PF3294).

♦ PF7847 Intellectual arrogance

Unquestionable certitude — Unquestionable beliefs — Closed mindedness — Lack of humility
Broader Arrogance (#PA7646).
Narrower Loss of humility in relation to the environment (#PF2527).
Aggravates Dogmatism (#PF6988) Evasion of issues (#PF7431)
Denial of evidence (#PD7385) Self-righteousness (#PG6373)
Cultural arrogance (#PF5178) Over-reliance on convenient beliefs (#PF9205).

♦ PF7870 Inflated art values

Prohibitive cost of art — Over-valuation of art
Nature Huge amounts of disposable income in the USA, Europe and Japan have created a global art market where prices have reached obscene levels of tens of millions of dollars. New kind of buyers have appeared: speculators who buy, wait for prices to jump, then sell.
Broader Prohibitive cost of goods and services (#PD1891).
Aggravates Capitalist speculation (#PC2194).

♦ PF7880 Uncontrolled markets

Uncontrolled market forces — Uncontrolled market economy
Nature Although there are many decisions that can effectively be made by market forces, there are conditions under which the market exacerbates social problems rather than alleviating them. Typically this occurs when there is significant inequality in purchasing power, if decisions are left to market forces, the wrong products get produced in the interests of richer minorities at the expense of the less privileged whose access to scarce resources is progressively eroded.
Incidence Global market forces have ensured inappropriate development in developing countries through: allowing the limited numbers of relatively rich to appropriate most, if not all, of the available resources; encouraging the development of industries inappropriate to developing countries because they failed to respond to needs of the the majority of the people; and much of the productive capacity of developing countries has become geared to the needs of the industrialized countries.
Claim Markets have a powerful, and usually overwhelming, tendency to encourage inappropriate development and distribution decisions. It is the normal functioning of the global market economy which delivers the available resources to the few and deprives the majority.
Counter-claim The market mechanism is the most appropriate mechanism for ensuring the most equitable economic decisions.
Aggravates Maldevelopment (#PB6207).
Aggravated by Economic inefficiency (#PF7556) Unequal income distribution (#PD4962)
International economic injustice (#PC9112)
Unsustainable economic development (#PC0495).

♦ PF7887 Satanic rituals

Ritualistic sexual abuse — Witches' sabbaths — Witchcraft ceremonies
Incidence In the UK in 1990 concern was expressed at the increase in reports of the sexual abuse use of children, even by their parents, in witchcraft ceremonies. Those so abused, whether children or adults, may be drugged. Animal sacrifices may be made, although there are occasional reports of human sacrifice, which have linked to the numbers of missing children. Those drawn into such satanic rings may be obliged to function as recruiters, bringing in others in exchange for money or favours. The perpetrators prey on the minds of teenagers, especially those craving power and control, or those lacking self-esteem. Participants may be subject to disinformation concerning the other participants so that their evidence is easily discredited, especially since as a result of such abuse they develop a grossly distorted understanding of sexuality.
Counter-claim Evidence of ritualistic abuse is confused by the possibility that witnesses may not be able to prove that they were not simply watching videos under the influence of drugs.
Broader Satanism (#PF8260).
Aggravates Human sacrifice (#PF2641) Missing children (#PD6009)
Ritual cannibalism (#PF8944) Sexual abuse of children (#PE3265)
Ritual slaughter of animals (#PF0319) Sexual exploitation of children (#PD3267).
Aggravated by Sexual deviation (#PD2198).

♦ PF7889 Superficial research

Superficial surveys — Superficial consultation — Superficial investigations
Narrower Superficial research on the total human process (#PF2415).
Aggravates Disagreement among experts (#PF6012).
Aggravated by Irresponsible international experts (#PF0221)
Inadequacies of foreign consultants (#PE6274)
Deliberate ignorance during policy-making (#PF8278)
Collusion between administrators of funding agencies and programme formulators (#PF8711).

♦ PF7890 Unreported disagreement

Unreported disputes — Unreported hostility — Unreported opposition
Broader Unrecorded knowledge (#PF5728) Underreported issues (#PF9148)
Avoidance of negative feedback (#PF5311).

♦ PF7897 Government refusal to accept the jurisdiction of international courts of justice

Government refusal to accept international tribunal arbitration
Incidence There are many cases where member states have refused to accept the jurisdiction of the International Court of Justice, or of regional courts such as the European Court of Human Rights, the Court of Justice of the European Communities or the Inter-American Court of Human Rights. In 1983, of the 158 member states of the United Nations only 47 accepted the compulsory jurisdiction of the United Nations International Court of Justice.
Aggravates Inadequate development of international criminal law (#PF4676).
Aggravated by Limited acceptance of international treaties (#PF0977).

♦ PF7905 Inaccuracy

Related Error (#PA6180) Neglect (#PA5438) Uncertainty (#PA7309)
Imperfection (#PA6997) Misrepresentation (#PA6644)
Journalistic irresponsibility (#PD3071).

♦ PF7911 Increasing cost of social security

Inflated cost of social security — Erosion of social security
Nature Social security is under attack from opposite camps. On the one hand it is accused of aggravating the world economic crisis by reducing saving, cutting investment, aggravating inflation, augmenting unemployment and undermining incentives to work; on the other hand, it is blamed for failing to solve the problems of poverty, for discriminating against women, for not treating equally those with similar needs and for distorting social priorities. While some press for a fundamental reorientation of social security policies, others argue that the whole system should be dismantled as it is no longer needed in societies which have reached there present state of affluence. There is no longer clear consensus favouring further developments.
Refs ILO *Employment Promotion and Vocational and Social Security* (1987); ILO *The Cost of Social Security* (1985); ILO *Financing Social Security. The Options* (1984).
Broader Pay system erosion (#PE6725) Prohibitive cost of living (#PF1238).
Narrower Prohibitive medical expenses (#PE8261)
Rising cost of unemployment benefits (#PJ4751).
Aggravates Social insecurity (#PC1867) Crisis in long-term pension funds (#PF5956)
Inadequate social welfare services (#PC0834) Evasion of social costs by companies (#PE3149)
Social insecurity in developing countries (#PE4796).
Aggravated by Unequal coverage by social security (#PF0852).

♦ PF7912 Lack of community development

Broader Obstacles to community achievement (#PF7118)
Dependence on sophisticated technology for development (#PD6571).
Aggravated by Social conflict (#PC0137) Wasted woman power (#PF3690)
Minimal community services (#PD8832) Lack of community planning (#PF2605)
Unpromoted community business (#PJ1409) Demeaning community self-image (#PF2093)
Declining community population (#PJ8746) Lack of community participation (#PF3307)
Widespread community factionalism (#PG8335)
Unfocused design of community space (#PF1546)
Breakdown of local community cohesion (#PD2864)
Reluctance to join in community action (#PF1735)
Inadequate procedures for community planning (#PF0963)
Lack of participation by women in development (#PD3294)
Demoralizing image of urban community identity (#PF1681)
Minimal opportunities for corporate activities (#PF2316)
Ineffective structures for community decision-making (#PF1781).

♦ PF7916 Unavailability of appropriate expertise

Inappropriate expertise — Lack of expert knowledge
Broader Physically inaccessible services (#PC7674)
Underprovision of basic services to rural areas (#PF2875).
Narrower Biased expertise (#PF6395) Limited funding expertise (#PJ0278)
Lack of organized expertise (#PG8463) Inadequacy of scientific expertise (#PF5055)
Lack of expertise in agricultural techniques (#PE8752).
Related Inadequate local expertise in business practices in developing countries (#PE7313).
Aggravates Arrogance of experts (#PF3294) Neglect of expert advice (#PF7820)
Irrelevant available information (#PG8884) Dependence on external expertise (#PG8011)
Irresponsible international experts (#PF0221) Individualistic utilization of expertise (#PF5639).
Aggravated by Disagreement among experts (#PF6012)
Bias in scientific research (#PF9693)
Excessive salaries of experts (#PF8317)
Failure of global-scale planning for expertise development (#PF1055).

♦ PF7918 Mind-induced death

Psychogenic death
Refs Franklyn, J *Death by Enchantment* (1971).
Broader Human death (#PA0072) Morbid preoccupation with death (#PD5086).

♦ PF7923 Obstacles to the development of multidisciplinary approaches

Broader General obstacles to problem alleviation (#PF0631).
Aggravated by Reaction (#PA6355).

♦ PF7925 Bias in United Nations response to human rights

Politicization of human rights issues within the United Nations — Intergovernmental double standard on human rights — Avoidance of human rights issues by the international community
Incidence According to the UN Charter, promotion and protection of human rights are a principal purpose of the United Nations, second only to the maintenance of peace. During the 1960s and the 1970s, UN human rights activities gave rise to the concern that instead of objectively applying the standards set forth in the Universal Declaration of Human Rights, the UN majority sought to use human rights as another means of furthering political warfare. During that period a double standard developed in that countries that were politically well-connected or sufficiently powerful were virtually immune from scrutiny on their human rights practices. By contrast, those that were relatively weak or unpopular within the United Nations were subject to detailed scrutiny within the United Nations Human Rights Commission. Despite marked improvements in human rights practices in some countries since that time, the UN continues to avoid focusing on certain major violators whenever it is politically expedient, as has been the case with large-scale massacres in a number of countries. In the reports of the Commission meetings a new technique is now being used whereby, in contrast to the past, the countries accused of violations are not named, being referred to by circumlocutions such as "in one country". Such procedures effectively ensure that certain violations are not exposed to the media spotlight which is the principal UN means for reducing human rights infractions by governments, thus seriously undermining the Declaration and

adherence to the norms therein and in other instruments.
Claim Erosion of the moral authority of the UN, as a result of the use of cynical tactics in dealing with issues which directly affect the lives of people, damages the credibility of the UN, possibly irretrievably.
   Broader Double standards in morality (#PF5225).

♦ **PF7927 Bilingualism**
Bilingual education
Nature Bilingual education teaches any single course in one of two languages so a student might take math, reading, grammar, history and geography in one language and social sciences, science, music, art and physical education in the second language. For those students who are not very bright and extremely competitive end up with a superficial knowledge of both languages and cultures. They end up without sufficient references in either.
   Broader Inadequate education (#PF4984).
   Aggravated by Mandatory language training (#PS1264).
   Reduces Semilinguism (#PF2789)   Multiplicity of languages (#PC0178).

♦ **PF7932 Increase in operating costs**
   Broader Restrictive pattern of business activities in small communities (#PD1415).
   Narrower Increasing public health expenditures (#PF6234).

♦ **PF7934 Overriding individual differences**
   Refs Tyler, Leona E *Individual Differences* (1974).
   Aggravates Declining sense of community (#PF2575).

♦ **PF7938 Collapse of distinctions between categories**
Intolerance of distinctions — Intolerance of distinctions between male and female
Nature The continuous stream of information provided through the media, and other means characteristic of high technological societies, juxtaposes, and tends to reduce to the same level in the receiver's mind, very different kinds and categories of information. Although these may in fact differ not only in a superficial qualitative sense (some being better or worse, however these are defined), the person viewing such information, as in the case of television, comes to see such programmes as one kind of thing, where they are in fact an amalgam of very heterogeneous material, in a far deeper qualitative sense. The effect of the continuousness with which such material is presented for entertainment is that it tends to blur the qualitative distinctions between the different kinds of material. A scene of soldiers fighting and dying in a real war zone may precede or follow a scene from a war play with actors simulating war. A politician trying to raise real issues on which the population is to make up its mind may precede or follow a comedian using who is merely using his personality to amuse an audience. It is inevitable, therefore, that the distinctions between the real and the artificial tend to become blurred, at least subconsciously. The critical consequence is that even those who are quite clear in their minds as to what is reality and what is fiction, the ultimate qualitative judgement is made on the basis of entertainment values. Thus scenes of the death of real soldiers are then judged on the basis of whether they are more or less moving, absorbing, entertaining than the actors in the war play, and the politician by whether his performance was as entertaining as that of the comedian.
   Aggravated by Excessive television viewing (#PD1533)
   Media theatricalization of public life and politics (#PF9631).

♦ **PF7939 Inadequate knowledge and reporting of man-made disease**
   Broader Misleading information (#PF3096).
   Aggravates Dangerous occupations (#PC1640).

♦ **PF7945 Trans-frontier pollution**
   Refs OECD Staff *Transfrontier Pollution and the Role of States* (1981).
   Broader Environmental pollution (#PB1166).

♦ **PF7950 Embarrassment**
   Broader Humiliation (#PF3856).
   Narrower Diplomatic embarrassment (#PG4043).
   Aggravates Shame (#PF9991).
   Aggravated by Acne (#PE3662).

♦ **PF7955 Donor distortion of aid to developing countries**
Politically exploitative humanitarian aid
   Broader Government hypocrisy (#PF9050)
   Inappropriate aid to developing countries (#PF8120).
   Related Counter-productive government constraints on programme beneficiaries (#PJ3583).
   Aggravates Compassion fatigue (#PF2819).
   Aggravated by Bilateralism in aid to developing countries (#PE9099)
   Policy cross-conditionality restrictions in multilateral development aid (#PF9216).

♦ **PF7960 Unexplored energy alternatives**
Disuse of traditional energy alternatives — Unexplored alternate energy
   Broader Overlooked potential for industrial development in rural communities (#PF2471).
   Narrower Underutilization of wind energy (#PF0373)
   Underutilization of peat as an energy source (#PE8194)
   Underutilization of oil shale as an energy source (#PF0445).
   Aggravates Lost knowledge (#PF5420)
   Unsustainable development of energy use (#PC7517)
   Ineffective utilization of public environment (#PF6543)
   Underutilization of renewable energy resources (#PE8971).
   Aggravated by Untransposed significance of cultural tradition (#PF1373).

♦ **PF7967 Reductionism**
Mono-factor explanations of the world — Narrow world perspective
Claim A powerful new form of reductionism is spreading through the intellectual community. Whereas the earlier form sought to reduce explanations of nature to a form based purely on mechanistic insights, the new form seeks explanations in terms of computer-like patterns of behaviour and logical functioning. In this way evolution is explained in cybernetic terms, providing an updated rationale for humanity's continued manipulation of the environment and for humanity's privileged position in the evolutionary scheme of things. In this framework emotions are decision-making programmes developed through evolution, the future is programmable in advance, and nature is reconceived as coded information.
The increased perspective offered by science through reductionism has been used to secure increased use of power. Wisdom has been sacrificed for violence. Linking perspective and power in this way has caused the world to be viewed in very narrow terms, with humanity severed from its participatory union with the rest of creation. The world's phenomena have been turned into objects for manipulation and expropriation.
   Broader Defective reasoning (#PF5711)
   Unformed structures of community organization (#PF2810).

   Related Unethical practices in the social sciences (#PD6626).
   Aggravates Manipulative knowledge (#PF1609)   Inconclusiveness of science (#PF6349)
   Simplistic technical solutions to complex environmental problems (#PF0799).

♦ **PF7972 Lack of anti-discrimination legislation**
   Broader Inadequate laws (#PC6848).
   Aggravates Racial discrimination in education (#PD3328)
   Racial discrimination in public services (#PD3326).
   Reduces Segregation in employment (#PD3443).

♦ **PF7979 Bureaucratic factionalism**
Bureaucratic feuding — Bureaucratic infighting — Official infighting — Bureaucratic rivalries — Bureaucratic jealousies
   Broader Factionalism (#PF8454).
   Aggravates Bureaucratic opposition (#PD7966)   Abuse of bureaucratic procedures (#PF2661)
   Inefficient public administration (#PF2335)
   Government delay in response to symptoms of problems (#PF6707)
   Competition between intergovernmental organizations for scarce resources (#PE0063)
   Jurisdictional conflict and antagonism within intergovernmental organizations (#PE9011)
   Jurisdictional conflict and antagonism between regional intergovernmental organizations with common membership (#PE1583).
   Aggravated by Bureaucratic fragmentation (#PC2662).

♦ **PF7983 Pursuit of corporate prestige**
Pursuit of corporate status
   Broader International status race (#PC5348).
   Aggravates Avoidance of negative feedback (#PF5311)
   Excessively costly prestige projects (#PF3455).
   Aggravated by Economic rivalry (#PD8897).

♦ **PF7988 Lack of leadership initiative**
Narrow visioning leadership
   Broader Obstacles to leadership (#PF7011)
   Deteriorated structures of essential corporateness (#PF1301).
   Aggravates Extreme detachment from represented constituency (#PF0889).
   Aggravated by Lack of leadership (#PF1254)   Insufficient leadership training (#PF3605).

♦ **PF7990 Negative effect of glamour**
Nature Glamour is the emotional aspect of duality which deludes, deceives, and creates false values and wrong desires. It entails wishing for apparent necessities with concomitant worries, anxieties, and cares. The emotions are deluded into cravings which, on a worldwide scale, can be thought of as 'the sum total of human ignorance, fear, and greed'. It is the result of the negative emotional focus of humanity and can be dissipated by the clear, inclusive thinking of those in whom the soul principle is awakening. It is 'illumined thinking' which helps to dissipate glamour.
Claim Glamour refers to the world of emotional being and of desire, in which all forms dwell. It is this glamour which colours all our lives and produces false values, wrong desires, needless so-called necessities, our worries, anxieties and cares; but glamour is age-old, and has us in so close a grip that there seems little we can do. The wish life of people has been wrongly oriented and human desire has been turned outward to the material plane, thus producing the world of glamour in which we all habitually struggle. It is by far the most potent of our delusions or mistaken orientations.
Refs Bailey, Alice A *Glamour* (1973); Lash, Scott and Urry, John *The End of Organized Capitalism* (1987).

♦ **PF7992 Lack of integrity**
   Broader Improbity (#PA7363).
   Narrower Lack of political integrity (#PF0796).
   Aggravates Libel (#PD3022)   Slander (#PD3023)
   Unethical practices of regulatory inspectors (#PF8046).

♦ **PF7998 Unpublicized community news**
Unpublished local news
   Broader Lack of local information systems (#PF6541)
   Obstacles to community achievement (#PF7118)
   Inadequate circulation of local information (#PF6552).
   Related Unpublicized meeting agendas (#PG9295)
   Inadequate dissemination and use of available information (#PF1267).
   Aggravates Limited employment options (#PF1658)
   Declining sense of community (#PF2575)
   Unpublicized public meetings (#PF5222)
   Inadequately publicized services (#PG9437)
   Unadvertised educational resources (#PG9123)
   Unpublished educational possibilities (#PJ9445).

♦ **PF7999 Crisis of capitalism**
Nature The revolutionary process of the collapse of the world capitalist system and the formation of world socialism and communism, extends over an entire epoch which began by the victory of socialism in the USSR. As the collapse of the world capitalist system progresses, the internal contradictions of capitalism in the countries where it still persists grows more intense, and the process of decay becomes extremely pronounced. This indicates that the capitalist system is in a state of general crisis.
Incidence The crisis of capitalism is evidenced by the number of countries which have become socialist or communist during this century.
Background According to Lenin, the period in which the general crisis of capitalism unfolds is characterized by 'worldwide revolutionary crisis', the rise of the 'world socialist revolution', and 'the collapse of capitalism in its entirely and the birth of socialist society: 'It is also marked by the breaking away of an increasing number of countries from the capitalist system and by the growth of the world socialist system.
Refs Trepelkov, V *General Crisis of Capitalism* (1984).
   Broader Capitalism (#PC0564).
   Aggravates Global economic crisis (#PC5876).

♦ **PF8000 Criminal concealment**
Nature In criminal law, concealment includes the deliberate concealment of a criminal, as well as of instruments and the means of committing a crime, traces of a crime, or articles criminally acquired. Concealment may include: providing a criminal with a place to live where he can hide from searches; providing him with counterfeit documents; or destroying any traces of a crime.
   Broader Complicity (#PF4983)   Offences of general applicability (#PD4158).

♦ **PF8008 Sophistry**
Nature Sophism is an inference or reasoning that substantiates some known incongruity, absurdity, or paradoxical statement contradicting generally accepted notions. The initial persuas-

PF8008

iveness of many sophisms (their 'logical' character) is usually related to a well-concealed error on the semiotic or logical level. Errors on the semiotic level usually involve metaphorical speech, homonymy and polysemy, and amphibology, which violate the unambiguous character of the thought and lead to a confusion of the meanings of terms. Errors on the logical level include deceptive substitution of the basic thought (thesis of the proof), the taking of false premises for true ones, violation of the permissible methods of reasoning, and the use of 'unauthorized' or 'forbidden' rules or actions, such as division by zero in a mathematical sophism.
Broader Reason (#PA5502) Miseducation (#PA6393)
Uncommunicativeness (#PA7411).
Related Fallacy (#PF4357) Unskillfulness (#PA7232).

♦ PF8010 Lack of religious discipline
Lack of spiritual discipline
Broader Lack of social discipline (#PF8078).
Aggravates Excommunication (#PF5954) Religious schism (#PF1939)
Spiritual disobedience (#PF7467) Barriers to transcendent experience (#PF4371).
Aggravated by Religious indoctrination (#PD4890)
Loss of spiritual guidance (#PF7005).

♦ PF8012 Fear of intimacy
Fear of authentic relationships
Refs Nunnally, Elam W, et al *Troubled Relationships* (1988).
Broader Fear (#PA6030).
Narrower Lack of assertiveness (#PF2373).
Aggravates Social integration handicap (#PE6779)
Lack of intimate relationships (#PF4416).
Aggravated by Social inadequacy of men (#PF3613)
Fear of emotional sensitivity (#PF9209).

♦ PF8015 Subjectivism
Nature Subjectivism is a world view that holds emotions, intuitions, aesthetic sensibilities, moral practices and spiritual awareness as being the basis for understanding. The concomitant abstraction of thought leads ultimately to a divorce from reality, subjective blindness, agnosticism, and relativism. In the political arena, subjectivism is reflected in policy decisions based on arbitrary, unscientific principles; contempt of the laws of society; and belief in the omnipotence of administrative decisions.
Broader Subjectivity (#PF2827).
Narrower Dependence on mysticism (#PF2590).

♦ PF8016 Prohibitive cost of private medical care
Claim Private medical care can charge mark-ups of 200 to 1,000 percent on prescription drugs and dressings without providing detailed itemized bills for customers.
Broader Prohibitive medical expenses (#PE8261).
Aggravates Inadequate health services (#PD4790)
Inadequate community care for handicapped persons (#PE8924).
Aggravated by Inadequate community care for transient urban populations (#PF1844).

♦ PF8025 Accumulation of privileges
Privilege
Broader Accumulation (#PA4313) Human inequality (#PA0844).
Narrower Official privilege (#PD5725).
Aggravates Class conflict (#PC1573).

♦ PF8026 Diseases of blood and blood-forming organs
Damage to small blood vessels in kidneys, eyes and nerves — Defective blood formation — Blood disorders — Diseases of the blood — Diseases of the blood and blood-forming organs — Blood diseases
Refs Gerrick, David J *The Rh Problem* (1979); Horwitz, Orville, et al (Eds) *Diseases of Blood Vessels* (1985); Oski, Frank A and Naiman, J Lawrence *Hematologic Problems in the Newborn* (1982).
Broader Human disease and disability (#PB1044).
Narrower Purpura (#PG2234) Anaemia (#PD7758) Thrombosis (#PE5783)
Haemophilia (#PE1920) Agranulocytosis (#PG9621)
Coagulation disorders (#PE2373) Chronic lymphadenitis (#PG7746)
Diseases of the spleen (#PE6155) Abnormal blood pressure (#PE0472)
Diseases of the lymphatic system (#PD2654).
Aggravates Nutritional anaemia (#PD0321) Periodontal diseases (#PE3503).
Aggravated by Rubella (#PE0785) Diabetes (#PE0102)
Inadequate emergency blood supply (#PE0366).

♦ PF8029 Grievance
Protest — Complaint
Narrower Grievances of citizens (#PJ5035) Grievances of employees (#PD5586)
Grievances of employers (#PD7286) Grievances of consumers (#PD7567).
Related Dissent (#PA6838) Badness (#PA5454) Injustice (#PA6486)
Lamentation (#PA5479) Unpleasantness (#PA7107).
Aggravates Feuds (#PE8210) Revenge (#PF8562) Resentment (#PF8374)
Demonstrations (#PD8522) Inadequate army discipline (#PD2543)
Social disaffection of the young (#PD1544).
Aggravated by Governmental disregard for legitimate protests (#PD9557).

♦ PF8032 Irresponsible social science
Broader Irresponsibility (#PA8658)
Irresponsible scientific and technological activity (#PC1153).
Aggravates Infringement of privacy (#PB0284).
Aggravated by Atomism (#PF5322).

♦ PF8040 Non-recognition of foreign governments
Cessation of diplomatic relations between states — Absence of diplomatic recognition — Abnormal diplomatic relations between states
Nature As a consequence of war or other incidents, states may cease diplomatic relations with each other. This may be initiated unilaterally. Subsequently such relations may only be partially restored, even after the signature of peace treaty following the termination of a state of war between them.
Incidence Although the USA and Vietnam signed a peace treaty in 1973, terminating the Vietnam War, little has been done in the intervening period to normalize relations between the two countries.
Aggravates Excessive frontier formalities in international travel (#PE0208).
Aggravated by Persons missing in military action (#PE1397)
Persistence of a technical state of war following cease-fire agreements (#PE2324).

♦ PF8042 Necromancy
Nature Special class of priests call up the spirits of the dead, or demons, to foretell the future or accomplish an act that is considered impossible in the natural world.
Broader Magic (#PF3311) Black magic (#PF8249).
Related Witchcraft (#PF2099) Fortune-telling (#PF1358).

♦ PF8046 Unethical practices of regulatory inspectors
Corruption in monitoring agencies — Corruption in regulatory agencies — Corruption of official inspectors — Negligence of official inspectors — Connivance between regulatory inspectors and violators of regulations
Nature Inspectors appointed to detect and report on violations of regulations may be persuaded by those responsible to "turn a blind eye" to certain abuses.
Incidence Initially inspectors may simply act in order to avoid submitting reports on trivial violations. As a relationship develops with those inspected, in the course of periodic visits, other pressures may be brought to bear in the form of token gifts or typically bribes, which may be substantial. Although individual violations may be of relatively little significance, the pattern of such violations and the corresponding "arrangements" with the inspectors can totally undermine and negate the value of any regulations. This is especially important in connection with factories violating regulations governing environmental pollution or conditions of employment. In the former case it results in long-term emissions of toxic materials above accepted thresholds. In the latter case it results in dangerous working conditions and, in a significant number of countries, in toleration of child labour even though it may be prohibited by regulations.
Broader Unethical practices by employees (#PD4334).
Aggravates Ineffective monitoring (#PF2793) Bribery of public servants (#PD4541)
Inadequate national law enforcement (#PE4768) Arbitrary enforcement of regulations (#PD8697)
Inadequate enforcement of safety regulations (#PD5001).
Aggravated by Lack of integrity (#PF7992) Official corruption (#PC9533)
Unethical practices of employers (#PD2879).

♦ PF8053 Complex government regulations
Complex government rules — Complex bureaucratic procedures — Red tape
Nature Strict adherence to complicated systems and procedures blocks access to public participation in the formation of goals and priorities of public agencies. Outdated and inadequate forms of communication have severely hampered public participation by limiting access to information needed for long-range planning. To simplify the process would require people to give up the chance to demonstrate their political influence, and few are willing to lose influence.
Broader Confusing structural complexity (#PF8100).
Narrower Complex tax regulations (#PG7900) Complex legal procedures (#PF8519)
Complex trade regulations (#PF4722) Complex banking practices (#PJ8033)
Difficult land acquisition (#PJ5369) Complex services logistics (#PG9918)
Complex funding mechanisms (#PG1502) Complex business regulations (#PG7894)
Restrictive regulations for training (#PJ7972)
Restrictive building codes in urban areas (#PE8443)
Complex regulations paralyzing small communities (#PF2444)
Excessive animal sanitary regulations in international travel (#PF1555).
Related Excessive paperwork (#PF5856) Bureaucratic inaction (#PC0267).
Aggravates Outdated procedures (#PF8793) Administrative delays (#PC2550)
Undemocratic policy-making (#PF8703) Complex government bureaucracy (#PF8539)
Unfamiliar bureaucratic procedures (#PJ9912)
Inefficiency of state-controlled enterprises (#PD5642).
Aggravated by Excessive government control (#PF0304).

♦ PF8065 Biased legal systems
Legal prejudice
Claim The present legal system has a bias toward the rich and powerful - the people who have run the system for hundreds of years.
Broader Prejudice (#PA2173) Outmoded legal systems (#PF2580).
Narrower Elitist legal judgements (#PD3737) Restrictive legal structures (#PG9988)
Inadequate legal counsel for minorities (#PF1219)
Discrimination against foreigners in legal proceedings (#PF1798)
Denial of the right to legal services for indigenous populations (#PE2317).
Related Restrictive legal practices (#PD8614).
Aggravates Cultural discrimination in the administration of justice (#PE6529).

♦ PF8068 Unbelievers
Paganism — Heathenism — Infidels
Refs Baskin, Wade and Wedeck, Harry E *Dictionary of Pagan Religion* (1973); Melton, J Gordon *Magic, Witchcraft and Paganism in America* (1982).
Broader Impiety (#PA6058) Unbelief (#PA7392) Irreligiousness (#PF8234).
Related Ignorance (#PA5568) Pantheism (#PF4209).
Aggravates Religious prejudice (#PD4365).

♦ PF8071 Catholicism
Papist conspiracy — Catholic conspiracy — Papism — Romanism — Popery
Incidence Catholicism is perceived as a threat in those societies where active efforts are made by its representatives, notably through Opus Dei, to counter the advance of secularism. There is a history of behind-the-scenes conflict, especially in European countries, between Catholic-controlled institutions and those controlled by bodies such as the Freemasons.
The historical belief in the threat of the Antichrist, and the associated threat of the papacy, continues to be sustained within certain Protestant churches and popular culture, although for many it has become a symbol of the evil in the human heart. This idea was developed during the time of Luther and contributed significantly to his opposition to the Pope, becoming an article of faith. The institution of the papacy, not individual popes, were identified with the Antichrist. Following the Reformation this idea gradually receded into the background, although maintained by Protestant scholars.
Broader Conspiracy (#PC2555) Religious intolerance (#PC1808).
Related Freemasonry (#PF0695) Jewish conspiracy (#PF8838).
Aggravates Antichrist (#PF9139).

♦ PF8072 Unresponsive public authorities
Neglect by public authorities — Public authorities neglect — Unresponsive public agencies — Unresponsive public offices
Broader Inaccessible administrative agencies (#PF2261)
Paralyzing complexity of urban structures (#PF1776).
Aggravates Disaster unpreparedness (#PF3567) Declining sense of community (#PF2575)
Lack of cooperation with officialdom (#PF8500) Inadequate disaster rescue and relief (#PF0286)
Inadequate disaster prevention and mitigation (#PF3566).

♦ PF8074 Unaccountable management of public information
Broader Social unaccountability (#PC1522).
Aggravates Mismanagement (#PB8406).

♦ PF8076 Minimal citizen confidence
Undermined social confidence

Broader Obstacles to leadership (#PF7011)   Deteriorating community identity (#PF2241)
Disjointed patterns of community identity (#PF2845).
Narrower Continuing expectation of community decline (#PG8680)
Declining community confidence in its ability to change (#PF9066).
Aggravates Debilitating education images (#PF8126).
Aggravated by Oppressive prevalent images (#PF1365).

◆ **PF8078 Lack of social discipline**
Social indiscipline
Narrower Inadequate army discipline (#PD2543)   Lack of religious discipline (#PF8010)
Maladjustment to disciplines of employment (#PF7650)
Inadequate social discipline in developing countries (#PD0095).
Aggravates Mutiny (#PD2589)   Disobedience (#PA7250)
Criminally life endangering behaviour (#PD0437).

◆ **PF8079 Unreported financial losses**
Unreported cost overruns
Broader Economic loss (#PE9013)   Underreported issues (#PF9148)
Avoidance of negative feedback (#PF5311).
Aggravated by Incomplete cost projections (#PJ8109).

◆ **PF8081 Loss of international market leadership**
Broader Loss of international leadership (#PF8353).
Aggravated by Declining international competitiveness (#PD8994).

◆ **PF8085 Lagging training in social skills**
Inadequate social — Neglected social skills training — Marginal grooming skills — Inadequate skills curriculum — Unavailable skills retraining — Limited skills training
Nature Training in social skills lags behind social needs. Formal, informal and non-formal educational structures fail to equip people with effective skills to cope with society as it is. Old unusable skills are taught, giving unnecessary skills to individuals who must function in times of rapid change.
Broader Unequal access to education (#PC2163)
Outmoded forms of social education (#PF2095)
Irrelevance of educational curricula (#PF0443).
Narrower Lack of family life education (#PF5079).
Related Shortage of technical skills (#PF6500).
Aggravates Insufficient trained labour (#PD9113)
Narrow range of practical skills (#PF2477)
Educational gap between generations (#PF6497)
Gap between the function of social techniques and the needs they address (#PF3608).
Aggravated by Elimination of traditional skills (#PD8872)
Minimal opportunities of adult training (#PF6531)
Unavailable education for effective living (#PF2313)
Ineffective mechanisms for functional training (#PF1352)
Unperceived relevance of formal education in rural communities (#PF1944).

◆ **PF8087 Managerism**
Claim Managerism is intended to resolve the social and economic contradictions of capitalism by placing greater trust in and responsibility on managers, under whose direction enterprises are supposedly working not for the sake of increasing the profits of their capitalist bosses, but for the well-being of society. But by investigating the relations between labour and capital, the true class nature of managerism is revealed. Ideologists of managerism call for research in 'human relations', in order to establish an atmosphere of 'working partnership' between workers and employers, which will increase labour productivity and actually lead to an even greater exploitation of hired workers.
Broader Capitalism (#PC0564).

◆ **PF8089 Marginalism**
Claim Marginalism, one of the methodological principles of bourgeois political economy, came into existence when the shift from free competition to all-powerful monopolies placed before economists tasks which could not be implemented by a strict reliance on the subjectivistic understanding of economic processes. Marginalism, therefore, deviates from such subjectivism by viewing economics as the interaction of individual economies.

◆ **PF8090 Unreported illness**
Unreported wounds — Unreported diseases
Broader Unrecorded knowledge (#PF5728)   Underreported issues (#PF9148)
Avoidance of negative feedback (#PF5311).
Aggravates Epidemics (#PC2514).
Aggravated by Unreported accidents (#PF2887).

◆ **PF8094 Incoherence**
Related Insanity (#PA7157)   Uncertainty (#PA7309)   Disagreement (#PA5982)
Disintegration (#PA6858)   Unintelligibility (#PA7367).

◆ **PF8100 Confusing structural complexity**
Broader Paralyzing patterns between villages and administrative structures (#PF1389).
Narrower Complex government regulations (#PF8053)
Paralyzing complexity of urban structures (#PF1776).

◆ **PF8105 Unstimulating entertainment**
Limited entertainment possibilities — Inadequate public entertainment — Limited recreational activities — Limited cultural attractions — Uncompetitive rural attractions
Nature Despite the range of leisure facilities, repeated exposure to them reduces their value as entertainment. People, especially young people, then find them frustratingly inadequate and seek for other sources of stimulation.
Incidence In urban environments, the scope for alternative forms of leisure is limited, especially for those with limited resources. This encourages the emergence of various forms of deviance, addiction nd crime, modelled in part on what is presented by the media as entertainment. In rural environments, there are fewer organized facilities, but the countryside itself is not perceived as stimulating, especially to young people attracted to the excitement of towns.
Broader Declining sense of community (#PF2575)
Demeaning community self-image (#PF2093).
Narrower Inadequate night life entertainment facilities (#PE6126).
Aggravates Boredom (#PA7365)   Attraction of city life (#PJ7861)
Minimum promotion of community assets (#PF6557).
Aggravated by Limited accumulation of capital (#PF3630)
Inadequate recreational facilities (#PF0202)
Dispersion of local capital resources (#PF1979)
Unpreparedness for surplus leisure time (#PF5044)
Unexplored entertainment alternatives in the countryside (#PE8540).
Reduced by Limited leisure time (#PF9062).

◆ **PF8107 Lack of national unity**
Broader Fragmentation (#PA6233).
Aggravated by Tribalism (#PC1910)   Lack of assimilation (#PF2132)
Disruption of development by tribal warfare (#PD2191).

◆ **PF8111 Economic expansion**
Expansionism
Broader Expansionism (#PB5858).
Narrower Undirected expansion of economic base (#PF0905).
Aggravates International aggression (#PB0968)   Threatened and vulnerable minorities (#PC3295).

◆ **PF8112 Non-recognition of problems**
Delay in recognition of problems
Broader Delay (#PA1999)   General obstacles to problem alleviation (#PF0631).
Narrower Unrecognized animal diseases (#PG5768)
Prohibitive cost of maintenance (#PF0296)
Vulnerability of plants and crops (#PD5730)
Delay of religions in acknowledging social problems (#PF1393)
Government delay in response to symptoms of problems (#PF6707)
Concentration of national governments activity on national affairs (#PE5132).
Aggravates Government inaction (#PC3950)
Recurrence of misapprehended world problems (#PF7027)
Institutional preoccupation with obsolete problems (#PJ5014).
Aggravated by Issue avoidance (#PF1623)   Latent problems (#PF9328)
Evasion of issues (#PF7431)   Conceptual repression of problems (#PF5210)
Unproven relationships between problems (#PF7706)
Calculated delays in releasing controversial news items (#PE0598)
Failure to integrate knowledge to empower humanity in response to the global problematique (#PF8753).

◆ **PF8118 Childhood martyrdom**
Sacrifice of children in holy war
Nature People perceive certain military causes to be of sufficient importance to sacrifice the future generation. Children went to the Crusades, to defend Europe from "the infidel". Children were sent to fight in the conflict between Iran and Iraq. In both cases, the conflict is seen as an eschatological one, a metaphor of human existence. Child martyrs give social and religious distinction to their parents. Young parents can lose their children and expect to have more, whereas the opportunities for supernatural blessings are very limited indeed.
Broader Religious sacrifice (#PD3373)   Sacrifice of children (#PJ5597)
Vulnerability of children during armed conflict (#PE8174).
Related Occultism (#PF3312)   Self-inflicted physical suffering (#PD7550).
Aggravates Human death (#PA0072).
Aggravated by Jihad (#PF5681)   Superstition (#PA0430)
Religious indoctrination (#PD4890).

◆ **PF8120 Inappropriate aid to developing countries**
Counter-productive effects of foreign aid — Negative effects of foreign aid — Unproductive technical aid
Claim International aid is a non-solution to the Third World's problems. It has not this far taken any country from misery. In the majority of cases, public aid is used to support the local bureaucracies and policies of poverty.
Refs Adams, Patricia and Solomon, Lawrence *In the Name of Progress* (1985).
Broader Unproductive use of resources (#PB8376).
Narrower Donor distortion of aid to developing countries (#PF7955)
Recipient distortion of aid to developing countries (#PF6932)
Policy cross-conditionality restrictions in multilateral development aid (#PF9216).
Aggravates Compassion fatigue (#PF2819)   Limited accumulation of capital (#PF3630)
Lack of participation in development (#PF3339).
Aggravated by Unaccountability of international financial institutions (#PF1136).
Reduces Distortion of national economies from food subsidies (#PE7413).

◆ **PF8125 Police indifference to community**
Police reputation as unhelpful
Broader Obstacles to community achievement (#PF7118)
Unethical practices by police forces (#PD9193).
Related Paralyzing complexity of urban structures (#PF1776)
Incompatibility of rural values in urban cultures (#PF2648).
Aggravates Lack of protection for victims of intimidation (#PE7793).
Aggravated by False burglary alarms (#PG7982)
Scarce employment possibilities (#PG8652).

◆ **PF8126 Debilitating education images**
Ineffective educational images
Broader Obstacles to education (#PF4852)   Demeaning community self-image (#PF2093).
Narrower Ghetto education image (#PJ8933).
Aggravates Inadequate education (#PF4984).
Aggravated by Minimal citizen confidence (#PF8076)
Unequal access to education (#PC2163).

◆ **PF8128 Misdirection of human energies and desires**
Related Negative emotions and attitudes (#PA7090).
Aggravates Ignorance (#PA5568)   Human suffering (#PB5955)
Wasted woman power (#PF3690)   Behavioural deterioration (#PB6321)
Waste of human resources in capitalist systems (#PC3113).

◆ **PF8145 Pursuit of personal prestige**
Pursuit of personal status — Status symbol — Status life-style — Status symbols — Status mannerisms — Status-seeking — Status language
Refs Pimpley and Sharma *Struggle for Status* (1985).
Broader International status race (#PC5348).
Narrower Hypergamy (#PF5430).
Related Discriminatory communication (#PD6804)
Negative effects of the nuclear family (#PF0129).
Aggravates Careerism (#PF6353)   Name-dropping (#PF5223)
Inferior status employment (#PD8996)   Inadequate firearm regulation (#PD1970)
Avoidance of negative feedback (#PF5311)   Head hunting in tribal societies (#PF2666)
Unnecessary relocation of national capitals (#PG8532).
Aggravated by Boasting (#PF4436)   Class consciousness (#PC3458)
Certificate-based job market (#PB8370).

◆ **PF8155 Growth in size of production unit**
Broader Growing size and impersonality of firms (#PE8706).
Aggravates Occupational deafness (#PD1361).

◆ **PF8156 Insensitivity to diversity of cultural traditions**
Bureaucratic insensitivity to cultural diversity — Bureaucratic nonrecognition of ethnic differ-

ences
Incidence People uprooted by the Volta River Project were resettled by planning authorities who were insensitive to their cultural traditions. Villages were split up and thousands of people from different ethnic backgrounds – speaking different languages, worshipping different gods and following different social customs were resettled together. The result was land disputes and outbreaks of violence.
Broader Government insensitivity (#PF2808)   Inadequate appreciation of culture (#PF3408)
Lack of appreciation of cultural differences (#PF2679).
Related Numbness towards others (#PF1216).
Aggravates Resettlement stress (#PD7776).
Aggravated by Involuntary mass resettlement (#PC6203)
Ignorance of nonverbal communication skills (#PE0533).

◆ **PF8158 Prohibitive administrative overhead costs**
Inflated overhead costs — Administrative overhead erosion of funding effectiveness — Overhead restrictions on efficacy of aid
Broader Untapped potential for retail trade in small towns (#PS6553).
Narrower Prohibitive cost of product liability protection (#PE4404).
Aggravates Compassion fatigue (#PF2819)
Inadequacy of aid to developing countries (#PF0392)
Inadequate funding of international organizations and programmes (#PF0498).
Aggravated by Mismanagement of aid to developing countries (#PF0175).

◆ **PF8161 Obsolete educational values**
Imposition on the young of outmoded values — Outmoded motivational methods — Inappropriate educational values
Claim The established generation chooses to set the limits on the world of younger people through the imposition of values that previously governed the adults' world. Imparting absolute morals such as "lying is bad" makes adults feel secure and comfortable; but younger people, when applying such absolute values on encountering real-life situations, tend to question the reasons behind them. When unsatisfactory answers are given, the young judge the values to be inapplicable and either active or passive rebellion ensues.
Broader Obstacles to education (#PF4852)
Disorientation of the young due to lack of social forms (#PD2050).
Narrower Lack of retraining incentives (#PG9410)
Unrecognized relevance of education (#PF9068)
Undervaluation of education by parents (#PF9306)
Devaluation of education by survival needs (#PE8902).
Related Static and superficial adult values (#PF2883).
Aggravates Inadequate appreciation of culture (#PF3408).
Aggravated by Narrow range of cultural exposure (#PF3628)
Educational gap between generations (#PF6497)
Ineffective methods of practical education (#PF2721).

◆ **PF8163 Seasonal fluctuations**
Seasonal adversity
Narrower Floods (#PD0452)   Seasonal unemployment (#PC1108)
Seasonal affective disorder (#PE0256)   Seasonal fluctuations in agriculture (#PD5212)
Underutilization of facilities due to daily or seasonal peaks (#PF0827)
Isolating effects of seasonal variations on undeveloped transportation (#PE3547).
Aggravates Underemployment (#PB1860).

◆ **PF8176 Unpreparedness**
Unreadiness — Undeveloped forward plan — Undeveloped future plan — Inadequate foresight capacity — Lack of lookout capacity — Inability to anticipate future development — Surprise-free policy-making
Claim History is full of examples of managed systems that have come to grief through an inability to understand or respond appropriately to surprises. Yet failures of this kind are now more dangerous than ever before, since the margin of acceptable error in managing the Earth and its resources is rapidly disappearing.
Broader Unpreparedness (#PA7341).
Narrower Unrecognized future financial commitments (#PF4114).
Related Neglect (#PA5438)   Untimeliness (#PA7006)
Inaccurate forecasting (#PF4774)
Fragmented planning of community life (#PF2813)
Self-defeating style of community planning (#PF6456).
Aggravated by Nuclear freeze (#PF0679)   Temporal deprivation (#PF4644)
Inadequate army discipline (#PD2543)   Avoidance of negative feedback (#PF5311)
Unknowable future patterns of social choice (#PF9276)
Inhibited capacity to visualize a creative future (#PF2352).

◆ **PF8181 Overstated programme advantages**
Overestimated economic benefits of projects
Incidence In large-scale water projects cost-benefit ratios are often manipulated to give a favourable economic assessment of the projects. Sponsoring agencies overestimate expected annual monetary gains, the benefits of irrigation, outdoor recreation, flood control, job creation potential and the life of dams.
Broader Falsification of programme evaluations (#PF9243).
Aggravates Cost overruns in large-scale public programmes (#PD1644).
Aggravated by Inaccurate forecasting (#PF4774)
Short range planning for long-term development (#PF5660).
Reduces Insufficient financial resources (#PB4653).

◆ **PF8185 Lack of social planning**
Aggravates Over-qualification (#PF3462)   Inflexible social structure (#PB1997)
Inadequate procedures for community planning (#PF0963).

◆ **PF8191 Inadequate application of available knowledge to solve problems**
Difficulties in applying knowledge — Inapplicability of formal learning
Claim Scientists are not naturally interested in the problems posed by the threats to human survival, since such problems are only rarely attractive for "pure" research. Commercial interests with research facilities are also not interested, especially since it is often their cost-cutting practices which create the problems in the first place. Government agencies are also not interested since their officials sincerely believe that absence of evidence is the same as evidence of absence of risk. In addition, for new problems, no agency may have a mandate to deal with it anyway. Society currently depends on the activities of independent, unauthorized pressure groups, aided by the media, to create some degree of panic against which all official and academic reassurances are unavailing. Only then are efforts made to apply knowledge effectively to respond to the threats to society.
Broader General obstacles to problem alleviation (#PF0631).
Narrower Unapplied scientific knowledge (#PF1468).
Aggravates Occupational cancer (#PE3509)   Government inaction (#PC3950)
Recurrence of misapprehended world problems (#PF7027).

Aggravated by Lost knowledge (#PF5420)   Inept theoretical models (#PF7182)
Inadequate secondary education (#PD5345)
Inadequate power of intergovernmental organizations (#PF9175)
Inadequate exchange of technical information concerning problems (#PF0209)
Failure to integrate knowledge to empower humanity in response to the global problematique (#PF8753).

◆ **PF8195 Non-cooperation**
Broader Militancy (#PC1090).
Narrower Deteriorated structures of essential corporateness (#PF1301).
Aggravates Lack of cooperation (#PF2816).

◆ **PF8203 Triumphalism**
Excessive equation of convictions with truth
Nature Triumphalism, on the part of one partner in a social interaction, terminates dialogue by eliminating other partners, assuming that their inferior truths are to be subsumed by the superior truths of the first partner. Triumphalism is not simply excessive passion for the truth, rather it is the excessive equation of the convictions of one group with the truth, and any such equation is excessive.
Claim Triumphalism, swelled by pride in possessing the truth, knows nothing of faith that is possessed by the truth. Triumphalism seeks to conquer where faith seeks to serve — also by understanding the truth by which others are possessed.
Aggravates Ineffective dialogue (#PF1654).
Aggravated by Pride (#PA7599)   Arrogance (#PA7646).

◆ **PF8205 Lack of knowledge of eligibility for benefits**
Lack of awareness of available welfare benefits — Unawareness of welfare benefits
Broader Lack of knowledge (#PF8381)
Limited availability of functional information (#PF3539).
Narrower Unawareness of health benefits (#PG8511).

◆ **PF8209 Excessive institutionalization of vulnerable groups**
Narrower Institutionalization of the disabled (#PF4681)
Abusive detention in psychiatric institutions (#PE2932).
Aggravates Fostering of dependency by social institutions (#PF1755).
Aggravated by Inadequate housing for the aged (#PD0276).

◆ **PF8229 Deliberate ignorance**
Pretended ignorance of events
Broader Ignorance (#PA5568).
Narrower Deliberate ignorance during policy-making (#PF8278).
Aggravates Fragmented conduct of community operations (#PF1205).

◆ **PF8234 Irreligiousness**
Lack of religion
Broader Retarded socialization (#PF2187).
Narrower Atheism (#PF2409)   Agnosticism (#PF2333)   Unbelievers (#PF8068).
Aggravates Loss of faith in religion (#PF3863).

◆ **PF8235 National boundaries**
Broader Nationalism (#PB0534).
Narrower Unnatural boundaries between states (#PF0090).
Aggravates Tribalism (#PC1910)
Disruption of development by tribal warfare (#PD2191)
Conflicting claims to non-terrestrial territory (#PG4114).

◆ **PF8241 Nakedness in developing countries**
Broader Economic and social underdevelopment (#PB0539).
Aggravates Ill-considered pressure to eliminate nakedness in developing countries (#PF3350).

◆ **PF8243 Denial to animals of the right to life**
Nature Animals are denied their right to live freely. They are exploited by humankind for work, entertainment, and companionship; they are killed for their flesh, for religious rituals and for sport.
Broader Denial of rights (#PB5405)
Denial to animals of legal protection of their rights (#PE8643)
Denial to experimental animals of the right to freedom from suffering (#PE8024).
Related Animal deaths (#PE7941)   Inadequate animal welfare (#PC1167)
Denial to animals of the right to dignity (#PE9573)
Denial to animals of the right to a natural death (#PE8339)
Denial to working animals of limitation of working hours (#PE6427)
Denial to animals of the right to freedom from mass killing (#PE9650)
Denial to food animals of the right to freedom from suffering (#PE3899)
Denial to working animals of restorative nourishment and rest (#PE4793)
Denial to animals of the right to the attention, care and protection of humankind (#PF5121)
Denial to animals of the right to conditions of life and liberty proper to their species (#PE6270).

◆ **PF8249 Black magic**
Broader Magic (#PF3311).
Narrower Necromancy (#PF8042).
Related Satanism (#PF8260).

◆ **PF8260 Satanism**
Devil worship
Nature The worship of the devil may include veneration of evil, ritual murder, cannibalism and sexual molestation of children.
Counter-claim Satanism is a hoax in one of two forms perpetrated by individuals attempting to discredit individuals or groups, like the Knights Templars in 1307 and others throughout the middle ages. Accusations of satanism are frequently used against women charged with witchcraft. Since torture was frequently used to obtain confessions of witches evidence was often readily obtained. Evidence of the other form consists entirely of the writings of a group of men in Paris in the latter half of the nineteenth century. In 1897 the man at the centre of the controversy admitted that it was all a fraud.
Refs Ashton, John *The Devil in Britain and America* (1980); Michelet, Jules *Satanism and Witchcraft* (1983).
Broader Magic (#PF3311).
Narrower Satanic rituals (#PF7887).
Related Voodoo (#PF9006)   Witchcraft (#PF2099)   Black magic (#PF8249).
Aggravates Debauchery (#PE8923)   Grave robbing (#PF0491).
Aggravated by Personified evil (#PF7018).

◆ **PF8264 Belief in humanity's dominance over nature**
Claim Human domination over nature is simply an illusion that has cost humankind much, ensnared people in their own designs, given them a few boasts, but still it is an illusion.

FUZZY EXCEPTIONAL PROBLEMS PF8393

Broader Inappropriate assumptions (#PF6814).
Aggravates End of nature (#PF9582)
Loss of humility in relation to the environment (#PF2527).

♦ **PF8268 Delays in delivery of goods**
Broader Delays in delivery of goods and services (#PE3928).
Narrower Port delays (#PE5815)   Ocean shipment delays (#PE5886)
Delays in delivery of books and publications (#PF1538).

♦ **PF8278 Deliberate ignorance during policy-making**
Wilful ignorance by government
Broader Deliberate ignorance (#PF8229).
Aggravates Blame avoidance (#PF6382)   Superficial research (#PF7889)
Inappropriate policies (#PF5645)
Irresponsible delimitation of policy responsibilities (#PF7823).
Aggravated by Avoidance of negative feedback (#PF5311)
Arrogance of intergovernmental agencies (#PF9561)
Collusion between administrators of funding agencies and programme formulators (#PF8711).

♦ **PF8282 Restrictive social policies**
Broader Inappropriate policies (#PF5645)   Excessive government control (#PF0304)
Debilitating deterioration of physical environment (#PD2672).

♦ **PF8298 Original sin**
Nature The rejection by the first humans of God's offer of friendship and through that rejection effected the whole of humanity. The divine offer of friendship and with it a God-like condition for every human being, was rejected. In the attempt to have Godly knowledge of Good and Evil humankind, in its desire to control its destiny became an enemy of God.
Broader Sin (#PF0641).
Narrower Fall of man (#PF3864).
Related Lust (#PA4673)   Envy (#PA7253)   Anger (#PA7797)
Sloth (#PA3275)   Guilt (#PA6793)   Avarice (#PA6999)
Gluttony (#PA9638)   Superstition (#PA0430)   Spiritual impurity (#PF6657)
Fear of the loss of independence (#PF7468).
Aggravates Human suffering (#PB5955)
Unreadiness for second coming of Christ (#PF0638).
Aggravated by Pride (#PA7599).

♦ **PF8299 Inaccessible market and supply centres**
Distant product market — Untapped regional markets — Excessively distant markets — Distant available markets — Inaccessible shopping districts — Remote supply centers
Broader Physically inaccessible services (#PC7674)
Inaccessible commercial and financial services (#PE0718).
Narrower Inaccessible supply of repair materials (#PG8569).
Aggravates Lack of formal education (#PF6534)   Limited accumulation of capital (#PF3630)
Confined scope of business operations (#PF2439)
Unprofitable scope of industrial operations (#PF1933)
Lack of support for local commercial services (#PF6510)
Unrealized potential of commercial enterprises (#PF2231)
Transfer of business from small communities to larger towns (#PF6540).

♦ **PF8317 Excessive salaries of experts**
Overpaid experts
Broader Prohibitive labour costs (#PF8763).
Aggravates Unavailability of appropriate expertise (#PF7916)
Disparity in remuneration between public and private sector employees (#PE7760).
Aggravated by Bias in scientific research (#PF9693)
Irresponsible international experts (#PF0221).

♦ **PF8326 Inadequacy of medical science**
Broader Inadequate health services (#PD4790)   Inadequacy of scientific expertise (#PF5055).
Narrower Iatrogenic disease (#PD6334).
Lack of understanding of spiritual healing (#PF0761)
Inaccurate medical diagnosis due to inadequate equipment (#PE4129)
Uncertainty of long-term health effects of radioactive fallout (#PE5324).
Related Surgical malpractice (#PE4736).

♦ **PF8330 Lack of coordination**
Coordination
Broader Lack of participation in development (#PF3339).
Aggravates Disaster unpreparedness (#PF3567).
Aggravated by Over-specialization (#PF0256).

♦ **PF8331 Cosmopolitanism**
Claim Cosmopolitanism, world citizenship, is reactionary ideology that teaches the renunciation of national traditions and cultures, patriotism, and state and national sovereignty. Cosmopolitanism reflects the nature of capital, which strives to where it can expect the greatest profit.
Broader Negative emotions and attitudes (#PA7090).
Aggravates Destruction of cultural heritage (#PC2114).

♦ **PF8335 Suspicion of bureaucracy**
Broader Distrust (#PA8653)   Unrecognized socio-economic interdependencies (#PF2969).
Aggravates Narrow range of practical skills (#PF2477).
Aggravated by Loss of confidence in government leaders (#PF1097)
Bureaucracy as an organizational disease (#PD0460)
Unchecked power of government bureaucracy (#PD8890).

♦ **PF8352 Unexplained phenomena**
Inexplicable phenomena — Telepathy — Homeopathy — Radiesthesia — Water divining — Extrasensory perception — Acupuncture — Prediction — Astrology
Narrower Ghosts (#PF3801)   Evil eye (#PF2343)   Psychic surgery (#PG7917)
Livestock mutilation (#PF1849)   Inauspicious conditions (#PF6683)
Exorcism as a superstition (#PF5673)   Unidentified flying objects (#PF1392)
Spontaneous human combustion (#PF4043)   Demonic and spirit possession (#PF5781)
Unexplained appearances and disappearances of persons and objects (#PE8631).
Related Magic (#PF3311)   Demons (#PF6734).
Aggravates Fraud (#PD0486)   Superstition (#PA0430)   Medical quackery (#PD1725)
Scientific fraud (#PF1602)
Procrastination of science in the face of the unexplained (#PF3682).
Aggravated by Ignorance (#PA5568).

♦ **PF8353 Loss of international leadership**
Narrower Loss of international market leadership (#PF8081)
Loss of international political leadership (#PF7455)
Loss of international technological leadership (#PF9345).
Aggravates Insecurity of leadership (#PD9362)   Loss of leadership credibility (#PF9016).

Aggravated by Competition between states (#PC0114).

♦ **PF8354 Insecurity from future crop uncertainty**
Broader Uncertainty (#PA6438)   Food insecurity (#PB2846)   Economic insecurity (#PC2020).

♦ **PF8356 Preponderance of Western-style organizations**
Broader Military and economic hegemony (#PB0318).
Aggravates Politicization of intergovernmental organizational debate (#PD0457)
Jurisdictional conflict and antagonism between international organizations (#PD0138).

♦ **PF8363 Fetishism**
Fetishism — Cult of relics — Dependence on amulets, charms and talismans
Nature Belief that the possession or wearing of an object charged with certain powers will procure for its owner the benefits of those powers. Such objects are regarded with irrational reverence. Any one who wears or carries these, who touches them, who prays to them, or who uses them in a variety of other ways, benefits by his action. The fetish itself may be a material, or even an animal (cock, serpent, etc), or natural (river, tree, etc), or part of a dead body (skull, penis, tooth, hair, etc). The object may then be considered to embody a spirit who can act through it or be communicated with. In some cases the objects are held to acquire their powers from their association with some powerful person. They are then valued in proportion to the extent of the power, strength, qualities or saintliness of the person to whom they originally belonged, as well as of the love or respect in which he was held.
Incidence The use of fetishes continues to be common in tribal societies, although the practice of wearing portions of the anatomy of a dead enemy or ancestor are no longer common. Relics of saints and holy men continue to be a prominent feature of most of the world's major religions, both in developing and industrialized countries. In many cases they are a focus for both pilgrimages and for tourism. Charms and talismans continue to be valued and sought after in most cultures.
Claim The preservation of such remains in whole or in part for veneration, or as incentives to greater faithfulness and goodness, or as reminders of the example offered by the lives of their former owners, is a forcing of the instinct of veneration beyond its legitimate place. There is much that is barbaric in the dividing up into larger or smaller fragments of the mortal remains of a saint and disseminating them over a wide area even for purposes of veneration. The admitted uncertainty which is associated with many relics, the incidence of fraud in the case of many, the gross superstitions and abuses to which they have given rise, far outweigh any positive good which they may have done.
Counter-claim Reverence for the remains of the dead or the treasuring of their more personal belongings is natural, instinctive and a worthy form of remembrance.
Refs Browne, Ray B *Objects of Special Devotion* (1982); Potter, Carole A *Knock on Wood* (1983).
Broader Superstition (#PA0430).
Narrower Commodity fetishism (#PE8058).
Aggravates Dying a bad death (#PF1421)   Vulnerability of sacred sites (#PD6128).
Aggravated by Demons (#PF6734)   Unethical practices of priesthood (#PF8889).

♦ **PF8368 Failure of development programmes**
Failure of development projects — Ineffectiveness of development programmes
Broader Destruction inherent in development (#PF4829).
Aggravated by Taboo (#PF3310)   Wasted woman power (#PF3690)
Inflexible social structure (#PB1997)   Lack of participation in development (#PF3339)
Burdensome cost of religious ceremonies (#PF3313)
Lack of participation by women in development (#PD3294)
Overuse of machinery and equipment in development projects (#PE8166)
Misappropriation of resources for high cost civil engineering projects (#PF4975).

♦ **PF8374 Resentment**
Irreconcilability — Unforgiving
Claim Holding on to past hurts, old grievances and judgements of others creates disharmony and strengthens the illusion of separateness. Since a person's judgements of others are often a reflection of feelings towards unresolved issues and difficulties in himself, the person's failure in not forgiving others reinforces his own sense of guilt and feelings of unworthiness.
Aggravates Lack of community participation (#PF3307).
Aggravated by Grievance (#PF8029).

♦ **PF8378 Fear of police**
Broader Fear of officialdom (#PD9498).
Aggravates Anxiety resulting from torture (#PE0969)
Debilitating content of village story (#PF2168)
Lack of protection for victims of intimidation (#PE7793).
Aggravated by Police state (#PD7910)   Police brutality (#PD3543)
Police intimidation (#PD0736)   Fear of retaliation by authorities (#PF3707).
Reduced by Fear of crime (#PF4682).

♦ **PF8381 Lack of knowledge**
Narrower Limitation of current scientific knowledge (#PF4014)
Lack of knowledge of eligibility for benefits (#PF8205).
Aggravates Medical quackery (#PD1725)   Occupational cancer (#PE3509).
Aggravated by Lost knowledge (#PF5420)   Monopolization of knowledge (#PF5329).

♦ **PF8382 Misunderstanding of official communications**
Broader Undeveloped channels for commercial initiative (#PF6471).
Aggravated by Insufficient communications systems (#PF2350)
Increasing scepticism about the accuracy of official information (#PF7649).

♦ **PF8387 Leadership impaired by illness**
Personality disorders of leadership — Leadership impaired by mental illness
Broader Lack of leadership (#PF1254).
Aggravates Abusive national leadership (#PD2710)
Loss of leadership credibility (#PF9016).
Aggravated by Insecurity of leadership (#PD9362)
Messianic image of leadership (#PF1102).

♦ **PF8389 Abuse in government policy**
Broader Abuse of government power (#PC9104).
Aggravates Torture schools (#PE2062).
Aggravated by Inappropriate policies (#PF5645).
Reduced by Misuse of classified communications information (#PD5183).

♦ **PF8393 Individualism**
Nature In Western societies individualism has different meanings. It (a) rejects any external authority, yet may have a sense of social duty; (b) emphasizes individual uniqueness; (c) resents intrusion of the state and government; (d) represents competitive economic individuals; (e) values every person compassionately.

PF8393

Narrower Militant individualism (#PD1106)   Despairing individualism (#PF0951)
Survival induced individualism (#PG8380)
Individualistic practices of local business (#PF1176)
Individualistic retaining of local tradition (#PF1705)
Individualism as a restriction of effectiveness (#PE8143).
Aggravates Short-term gain (#PF8675)   Declining sense of community (#PF2575)
Lack of meaningful educational context for ethical decisions (#PF0966).
Reduced by Conformism (#PB3407)   Dictatorship of the majority (#PD3239).

♦ PF8396 Intolerance of criticism
Unwillingness to contemplate criticism
Broader Intolerance (#PF0860).
Aggravates Criticism of official institutions (#PF9385).
Aggravated by Avoidance of negative feedback (#PF5311).

♦ PF8404 Enemies
Dependence on the existence of enemies — Search for enemies
Incidence The meaning of the term has evolved with the history of civilization. In the earliest times an enemy was a person with whom one was not connected by ties of blood. This also applies under some modern social extremes, as within urban gangs who symbolize their kinship with blood oaths. Similarly all foreign states were then regarded as enemies. The utmost cruelty was justified towards such enemies. At present an individual is not necessarily accounted the enemy of the state with which the government to which he owes allegiance is at war. But an individual may acquire enemy character because of his acts during wartime. In addition to personal enemies, enemies are now engendered by class warfare (class enemies), trade wars, commercial competition and competing ideologies (ideological enemies).
Counter–claim Man does not live by bread alone, but also by the nourishment of animosities.
Refs Popper, Karl *The Open Society and its Enemies*.
Aggravates Malignant visions (#PF5691).
Aggravated by War (#PB0593)   Jihad (#PF5681)   Racial war (#PD8718)
Malevolence (#PA7102)   Gang warfare (#PD4843)   Class conflict (#PC1573)
Ideological war (#PC3431)   Incitement to war (#PD4714)   Economic conflict (#PC0840).

♦ PF8413 Immaturity
Narrower Sexual immaturity (#PJ6143)   Emotional immaturity (#PJ5907)
Political immaturity (#PJ5657).
Related Imperfection (#PA6997)   Incompleteness (#PA6652)
Unintelligence (#PA7371)   Unskillfulness (#PA7232)
Unpreparedness (#PA7341)   Emotional dependency in marriage (#PD3244)
Psychological impediments to marriage (#PF3344).

♦ PF8421 Inadequate international law enforcement
Broader Deficiencies in international law (#PF4816)
Inadequacy of international legislation (#PF0228).
Narrower Inadequate enforcement of human rights (#PC4608)
Deliberate governmental avoidance of legislative reform (#PF5736).
Related Inadequate national law enforcement (#PE4768).
Aggravates Unfulfilled treaty obligations (#PF2497)
Limited acceptance of international treaties (#PF0977).
Aggravated by Government inaction (#PC3950)
Inadequate development of international criminal law (#PF4676).

♦ PF8433 Supernaturalism
Belief in the supernatural
Nature Refusal to accept that all phenomena can be satisfactorily and appropriately rationally and scientifically.
Counter–claim Recognition of exceptions to the processes of nature does not amount to the recognition of breaches in the order conceived as necessary after the manner of modern science.
Broader Inappropriate assumptions (#PF6814).
Narrower Ghosts (#PF3801)   Belief in miracles (#PG3935)
Disruption of human activities by supernatural entities (#PF3437).
Aggravates Superstition (#PA0430).

♦ PF8434 Pursuit of national prestige
National status race
Broader International status race (#PC5348).
Aggravates Competition between states (#PC0114)
Avoidance of negative feedback (#PF5311)
Unnecessary relocation of national capitals (#PG8532)
Lack of appreciation of cultural differences (#PF2679).
Aggravated by Boasting (#PF4436).

♦ PF8435 Inability to resolve problems realistically
Broader General obstacles to problem alleviation (#PF0631).
Narrower Issue avoidance (#PF1623).
Aggravated by Evasion of issues (#PF7431)   Non-acceptance of reality (#PF1079)
Conceptual repression of problems (#PF5210)
Inadequate power of intergovernmental organizations (#PF9175).

♦ PF8446 Demoralization
Low morale
Narrower Military demoralization (#PE6639)   Community demoralization (#PG1266).
Aggravates Fatigue (#PA0657).
Aggravated by Ill treatment of prisoners of war (#PD2617).

♦ PF8448 Fragmented decision-making
Broader Inadequate procedures for community planning (#PF0963)
Unrecognized socio-economic interdependencies (#PF2969).
Narrower Legal inconsistency (#PF5356)   Uncorporate family decisions (#PF9323)
Non-inclusive management decisions (#PF2754)
Fragmented individual decision-making process (#PF3559)
Deluding familial image of social responsibility (#PF1064)
Inadequacy of the committee system of decision making (#PF2843)
Short sighted decisions about intersocial interaction (#PE4477).
Related Unimaginative educational vision (#PF3007).
Aggravates Policy-making delays (#PF8989)   Institutional fragmentation (#PC3915)
Ineffective decision-making processes (#PF3709)
Ineffective means for participation in decision making (#PE8430)
Exclusion from decision making processes of those who question the context of the process (#PE7155).
Aggravated by Over-specialization (#PF0256)   Imposed official decisions (#PF8649)
Obscure decision mechanisms (#PF8166)   Avoidance of decision-making (#PF4204)
Inaccessible decision makers (#PF2452)
Lack of meaningful educational context for ethical decisions (#PF0966).

♦ PF8451 Insufficient role models
Inappropriate role models
Narrower Inadequate hero images (#PF2834)   Lack of local leadership role models (#PF6479).
Aggravates Social disaffection of the young (#PD1544)
Static and unrelated social roles (#PF1651).
Aggravated by Personality cults (#PC1123)   Insufficient cultural heroes (#PF8623)
Misleading endorsement advertising (#PE7502)
Excessive accumulation of wealth by government leaders (#PD9653).
Reduced by Hero worship (#PF2650).

♦ PF8454 Factionalism
Factional feuding
Refs Silverman, Maylin and Salisbury, Richard *House Divided?* (1976).
Narrower Bureaucratic factionalism (#PF7979)   Widespread community factionalism (#PG8335)
Factionalism in developing countries (#PD1629).
Related Fragmentation (#PA6233).
Aggravated by Feuds (#PE8210).

♦ PF8455 Oversimplification
Broader Minimization of problems (#PF7186)
General obstacles to problem alleviation (#PF0631).
Related Complexity (#PA6468)   Unpreparedness (#PA7341).
Aggravates Inappropriate arguments (#PF2152).
Aggravated by Multiplicity of problems facing society (#PF2003)
Complex interrelationship of world problems (#PF0364)
Inadequate planning of action against problems (#PF1467)
Inadequate education concerning the nature of problems (#PE8216).

♦ PF8466 Decline in educational standards
Decline in educational quality — Low educational standards
Refs Narayan, Raj *Falling Education Standards* (1970).
Broader Lack of professional standards (#PF3411).
Aggravated by Decline in government expenditure on education (#PF0674).

♦ PF8474 Doubt
Claim Doubt is the negation of belief. It is the condition of not having reached a conclusion for or against any proposition. While disbelief is the positive conclusion of the falsity of a proposition, doubt is a suspension of judgement, a state of being unconvinced. It results in inaction, the failure to commit for or against.
Counter–claim Doubt is the essence of a mature faith.
Aggravates Indecision (#PF8808)   Policy-making delays (#PF8989).
Aggravated by Distrust (#PA8653).

♦ PF8479 Miscarriage of justice
Error of criminal justice — Judicial error — Perversion of justice — Malicious prosecution
Nature The condemning of an innocent party may result from such factors as: falsification of evidence; perjury; rejection of contrary evidence; negligence; prejudice; pressure of public opinion; dishonesty; juridical unscrupulousness; bureaucratic inertia; or vested political, economic or professional interests. An innocent person, whether or not appeals subsequently establish his innocence, thus suffers all the economic and social penalties of the guilty, often with minimal compensation or none at all. Such errors, whether perpetrated deliberately or inadvertently, do not necessarily result in sanctions against those responsible, or in efforts to prevent their repetition.
Counter–claim Any judicial system is subject to error. Which is a greater miscarriage of justice; setting free a criminal who then kills, robs and rapes or an innocent person sitting in prison for life ? Society must minimize these miscarriages, but justice is difficult to find and moves from place to place.
Broader Human errors and miscalculations (#PF3702).
Aggravates Wrongful detention (#PD6062)   Inequitable administration of justice (#PD0986)
Failure to legally rehabilitate victims of miscarriage of justice (#PF7728).
Aggravated by Complex trials (#PE3916)   Collapse of judicial system (#PJ0761)
Inhumanity of capital punishment (#PF0399)
Unethical practices in the legal profession (#PD5380).

♦ PF8480 Legalism
Aggravates Conflict of laws (#PF0216)   Law enforcement complexity (#PF2454).

♦ PF8481 Lack of abortion facilities
Inadequate abortion methods
Incidence About 82 percent of the counties in the USA had no abortion facilities in 1985. Few doctors are willing to perform the procedure due to rural tendency towards conservatism and the growing aggressiveness of anti-abortion protests.
Broader Inadequacy of contraceptive methods (#PD0093).
Aggravates Abortion-related deaths (#PE3580)   Illegal induced abortion (#PD0159)
Prohibitive cost of abortion (#PJ2020).
Reduces Induced abortion (#PD0158).

♦ PF8483 Terror
Refs Adams, James *The Financing of Terror* (1986).
Broader Fear (#PA6030)   Moderation (#PA7156).
Aggravated by Civil war (#PC1869)   Soul murder (#PF4213).

♦ PF8485 Dishonour
Loss of honour
Related Impiety (#PA6058)   Disrepute (#PA6839)   Improbity (#PA7363)
Disrespect (#PA6822).
Aggravates Duelling (#PF5382)   Humiliation (#PF3856)
Inadequate firearm regulation (#PD1970).
Aggravated by Defamation of character (#PD2569).

♦ PF8490 Misdiagnosis
False diagnosis
Broader Inadequate medical care (#PF4832).
Narrower False diagnosis of mental disorder (#PG5868)
Inaccurate medical diagnosis due to inadequate equipment (#PE4129).
Related Laboratory testing errors (#PF5304).
Aggravates Misuse of medicines (#PD8402)   Surgical malpractice (#PE4736)
Prescription of inappropriate drugs (#PE3799).
Aggravated by Unprepared adult leadership (#PF6462).

♦ PF8491 Government inefficiency
Bureaucratic inefficiency
Narrower Inefficiency of state-controlled enterprises (#PD5642)
Inefficiency of government in developing countries (#PF9266).

Aggravates Wastage in governmental budgets and appropriations (#PD0183).
Aggravated by Political appointees (#PF2031)
Proliferation of public sector institutions (#PF4739).

♦ PF8493 Trade unionism

Syndicalism — Criminal syndicalism

Nature Trade union activity, when legal, disrupts production and reduces the profitability of economic activity. Criminal syndicalism is the advocation and use of unlawful acts as a means of accomplishing a change in industrial ownership or control of political change.
Broader Socialism (#PC0115).
Narrower Violation of trade union rights (#PD4695)
Restrictive trade union practices (#PD8146)
Sanctions against trade union workers (#PD0610).
Related Treason (#PD2615).
Aggravates Strikes (#PD0694).
Reduced by Ineffective worker organizations (#PF1262).

♦ PF8499 Underestimation of programme costs

Underestimated project disadvantages

Incidence In large-scale water projects cost-benefit ratios are often manipulated to give a favourable economic assessment of the projects. Sponsoring agencies underestimate expected costs for annual operation and maintenance of the projects, reservoir-flooding of productive land, loss of outdoor recreation sites, environmental costs of irrigation. Besides, they ignore the costs of decommissioning, fail to account for the energy costs of building, for example a dam, and use unrealistically low discount rates of future value of the currency.
Broader Falsification of programme evaluations (#PF9243).
Aggravates Cost overruns in large-scale public programmes (#PD1644).
Aggravated by Inaccurate forecasting (#PF4774)
Short range planning for long-term development (#PF5660).

♦ PF8500 Lack of cooperation with officialdom

Broader Lack of cooperation (#PF2816).
Aggravated by Fear of officialdom (#PD9498)
Bureaucratic ignorance (#PF8582)
Bureaucratic superiority (#PC1259)
Government insensitivity (#PF2808)
Declining sense of community (#PF2575)
Lack of governmental support (#PF8960)
Unresponsive public authorities (#PF8072).

♦ PF8508 Stereotypes

Acceptance of stereotypes — Dependence on stereotypes — Group labels — Labelling people

Nature There is a tendency to identify individuals or groups with general, impersonal terms that allow one to simplify the process of trying to understand each individual as a separate entity. People are "rightists", "vegetarians", "liberals" and so on, according to the group into which they most conveniently fall for the labeller. The catastrophe of this device is that it completely terminates for many of us the possibility of ever knowing the other individual or group for the unrepeatable specificity that they are. One actually no longer experiences the other, but experiences only what the label leaves visible. One of course also gradually ceases to experience the self as well.

Counter-claim Without names for realities human beings experience there would be no way of recalling the experience. There would be no history and no future. Names, to give an alternative label to this problem and therefore a different meaning, are necessary to relate to experiences, they imply significance about the experience. No one can understand everyone they meet or hear about as a separate entity, people have friends and neighbours, other labels.

Refs Signorielli, Nancy; Milke, Elizabeth and Katzman, Carol *Role Portrayal and Stereotyping on Television* (1985).
Broader Lack of racial identity (#PF0684)
Inappropriate labelling (#PD3521).
Narrower Gender stereotyping (#PD5843)
Racial stereotyping (#PF5452)
Inaccurate youth stereotypes (#PJ8357)
Inaccurate criminal stereotypes (#PF1244).
Aggravates Conflict between minority groups (#PC3428)
Ineffectiveness of the United Nations system of organizations (#PF1451).
Aggravated by Use of offensive symbols (#PF2826)
Conflicting sense of sexual identity (#PF1246).

♦ PF8511 Non-standardization of geographical names

Non-standardization of topographic names — Political changes to names of geographical features

Nature The same geographical feature may have several names in current use, complicating the elaboration of maps, signs and transportation schedules, especially in areas where several languages are in use.

Incidence The existence of several versions of a name creates major difficulties in bilingual countries, in international organizations or between neighbouring territories with common geographical features, or where these extend beyond a single sovereignty. These difficulties are frequently reinforced where a name has political or cultural significance, as in the case of Malvinas/Falklands, or Taiwan. Name changes, whether for political reasons or over historical periods, can also complicate the issue.

Refs United Nations *United Nations Conferences on the Standardization of Geographical Names* (1986).
Broader Inadequacy of international standards (#PF5072)
Inadequate international map of the world (#PD0398).

♦ PF8515 Unhygienic conditions

Unsanitary conditions — Unhealthy sanitation practices — Lack of preventive sanitation — Inadequate public health conditions
Broader Unhealthy environment (#PJ1680)
Inadequate maintenance of physical health (#PF1773).
Narrower Unclean food (#PJ2532)
Inappropriate basic hygiene (#PD8294)
Inappropriate sanitation systems (#PD0876)
Inadequate hygiene restrictions on carcass meat exports (#PE8398)
Unsanitary environment for basic health in small rural villages (#PD2011).
Aggravates Hookworm (#PE3508)
Smallpox (#PE0097)
Dysentery (#PE2259)
Meningitis (#PE2280)
Trichinosis (#PE2311)
Typhoid fever (#PD1753)
Chagas' disease (#PE0653)
Infant mortality (#PC1287)
Inadequate personal hygiene (#PD2459)
Human disease and disability (#PB1044)
Outdated forms of community health (#PF1608)
Fragility of maintaining basic health (#PJ2524)
Unequal morbidity and mortality between countries (#PC6869).
Aggravated by Unenforced sanitation codes (#PJ7979)
Uncontrolled urban development (#PC0442)
Narrow range of practical skills (#PF2477)
Fragmented planning of community life (#PF2813)
Reluctance to join in community action (#PF1735)
Disrupted mechanisms for community health (#PF2971)
Uncoordinated social services in urban areas (#PF1853)
Discrimination in access to sanitation services (#PG8247)
Underdeveloped provision of basic services in developing countries (#PF6473)
Instability of trade in sanitary plumbing, heating, lighting fixtures and fittings (#PE8272).
Reduced by Excessive animal sanitary regulations in international travel (#PF1555).

♦ PF8519 Complex legal procedures

Complicated legal procedures
Broader Complex government regulations (#PF8053).
Aggravates Law enforcement complexity (#PF2454)
Limited access to social benefits (#PF1303)
Reluctant claims on external resources (#PF1226)
Obstruction of international criminal investigations (#PF7277).

♦ PF8521 Rapidly changing cultures

Rapidly changing culture
Related Inadequate social reform (#PF0677).
Aggravates Cultural decline (#PC9083)
Mental disorders of the aged (#PD0919)
Inadequate housing for the aged (#PD0276)
Inadequate recreational facilities (#PF0202)
Inadequate welfare services for the aged (#PD0512)
Untransposed significance of cultural tradition (#PF1373).
Aggravated by Social disadvantage of the aged (#PD3517)
Obsolete basis of cultural identity (#PF0836).

♦ PF8531 Irrational rejection of nuclear power

Broader Negative emotions and attitudes (#PA7090).
Aggravates Insufficient nuclear power stations (#PD7663)
Long-term shortage of natural resources (#PC4824).
Aggravated by Unhealthy emotional responses to atomic energy (#PF0913).

♦ PF8535 Abusive distribution of political patronage

Political favouritism — Abusive allocation of government funding — Pork barrel politics

Incidence Patronage has developed from the distribution of largesse in the form of minor gifts to the allocation of large architectural and defence contracts. Politicians also use their power to locate positions for people, whether in the government or private sector. Politicians themselves obtain such jobs through the same process once they leave office.

Claim Instead of making government responsive, political patronage keeps politically connected but incompetent civil servants in office. Taxpayers are thus deprived of effective officials whilst demoralizing those who are effective in government.

Counter-claim Patronage has a terrible reputation but it makes government more responsive by making public offices change when the electorate changes its choices. It is thus good for the country because it ensures rotation in office and brings fresh people into government.
Broader Unethical practices in politics (#PC5517).
Aggravates Political scandal (#PD4651)
Abuse of influence (#PC6307)
Political injustice (#PC2181)
Political appointees (#PF2031)
Corruption in politics (#PC0116)
Trading in public office (#PE6948)
Abuse of government power (#PC9104)
Loss of confidence in government leaders (#PF1097)
Misuse of personal authority for political purposes (#PE4635).
Aggravated by Cronyism (#PF4549)
Unchecked power of government bureaucracy (#PD8890).

♦ PF8536 Withholding of information

Withholding of information by government
Broader Deception (#PB4731)
Contempt of judicial process (#PD9035)
Restrictions on freedom of information (#PC0185).
Narrower Misrepresentation of geographical information (#PF9239).
Aggravates Inappropriate arguments (#PF2152)
Suppression of information (#PD9146)
Incomplete access to information resources (#PF2401)
Inconclusiveness of scientific and medical tests (#PD7415).
Aggravated by Government delaying tactics (#PF6119)
Politicization of health standards (#PD4519).

♦ PF8539 Complex government bureaucracy

Broader Bureaucracy as an organizational disease (#PD0460).
Aggravates Inefficient public administration (#PF2335)
Limited reservoir of technical skills in rural communities (#PF2848).
Aggravated by Complex government regulations (#PF8053)
Proliferation of public sector institutions (#PF4739).

♦ PF8552 Underdeveloped technological skill

Broader Insufficient skills (#PC6445).
Narrower Blocked skills advancement (#PG8864)
Decreased skill transference (#PG5492).
Aggravates Insufficient provision of public services for communication (#PF2694)
Lack of skilled manpower in rural areas of developing countries (#PE5170).
Aggravated by Unorganized transfer of skills (#PJ8603)
Self-interest driven investment (#PC2576)
Elimination of traditional skills (#PD8872)
Lack of sharing of community skills (#PF3393).

♦ PF8554 Inadequate economic incentives

Insufficient economic incentives
Broader Mobility of village populations (#PE1848)
Confined scope of business operations (#PF2439).
Narrower Low bank interest rates (#PE3903).

♦ PF8559 Mistrust of police

Lack of confidence in police

Nature The police operate on the strength of collective myths: the power of the uniform, the threat of being caught, the justice of the legal system. But some people and even some police officers no longer believe in these myths, in policing by consent.
Broader Distrust (#PA8653)
Incompatibility of rural values in urban cultures (#PF2648).
Aggravates Public assaults on police (#PE7659).
Aggravated by Police brutality (#PD3543)
Police corruption (#PD2918)
Police intimidation (#PD0736).

♦ PF8562 Revenge

Vengeance — Retribution

Nature Revenge is the expression and continuation of resentment. Resentment is the deliberate or settled anger over a hurt or injury inflected. It may lead to tit for tat repayment for injury or restitution as a debt due to the injured person. Frequently it goes beyond this. It is not the anger that injures but the continued nurtured resentment which leads to lust for revenge. The growth of criminal law is in part an attempt to remove from private hands revenge and replace it with public retribution and appeal to public law, protecting the weak from the strong. Once public law has been established revenge itself becomes a crime. From the moral perspective revenge

PF8562

is seen as wrong because it is anti-social leading to human misery and not human happiness. It is almost always unjust because the judge is the injured person and takes an exaggerated estimate of his injuries. It is self-destructive because it is essentially selfish. It prevents any development of a charitable and forgiving temper. From a religious perspective it is frequently seen as interfering with the prerogative of God.

From one perspective revenge is a primitive form of law. It is unbridled, unreflective and an arbitrary act of retribution. Punishment, within the framework of law, has a purpose, is administered according to established guidelines and is dispensed on the basis of a judicial sentence. Moving from vengeance to penal law is a moral advance.

Others argue that revenge is for those outside the legal system of a group or community. The laws of the group punish transgressors within the community. Revenge is an act of self-assertion by a group against an outside attack. As such it can be an act of exclusion. The person harmed and his avenger, on one hand, and the culprit, on the other, are members of different groups. Each represents the rights and duties of the group and acts in its name.
Broader Vengeance (#PA6606)   Punishment (#PA5583)   Human violence (#PA0429).
Aggravated by Banditry (#PD2609)   Grievance (#PF8029)
Lack of central planning structures in small communities (#PF2540).

◆ **PF8563 Wrath of God**
Divine anger
Nature God, as presented in the Old Testament of the Bible, is confronted in the execution of his purposes by the ignorance and slowness of men, by their self-will and by their hostility. These arouse a divine anger which is not to be confused with the capricious fury of men. It is aroused when the conditions under which he alone can work out man's salvation are infringed and his purpose of mercy is imperilled. The anger of God is thus aroused by any act which stands between him and the ends which he has in view, by all those who are hostile to the people of his choice, and by the presence of evil and sin. In the New Testament of the Bible, the wrath of God is directed against those who reject Christ.
Incidence Recognition of the anger of God is of fundamental importance in Judaism and Christianity. For Christians the Bible leads to the conclusion that if people fail to respond to the opportunity of salvation, those who persist in opposition to God and rejection of divine mercy will have committed an eternal sin and must endure the utmost visitation of the wrath of God, especially at the last judgement.
Claim If God is not angry with the impious and the unrighteous, it is clear that he does not love the pious and the righteous.
Aggravates Fear of God (#PF9565)   Natural disasters (#PB1151)
Dying a bad death (#PF1421).
Aggravated by Sin (#PF0641)   Evil (#PF7042)   Atheism (#PF2409)
Infidelity to God (#PF9307).

◆ **PF8565 Communication with foreigners**
Speaking to foreigners
Aggravates Mistrust of strangers (#PF8743).
Aggravated by Repression (#PB0871).

◆ **PF8577 Bad omens**
Ill-omened events — Ominous events
Broader Superstition (#PA0430)   Inauspicious conditions (#PF6683).
Narrower Albinism (#PE2332)   Lightning (#PD1292)   Earthquakes (#PD0201)
Hail storms (#PD0251)   Meteors as hazards (#PF1695)   Volcanic eruptions (#PD3568)
Hazards from comets (#PF3564)   Fear of eclipses and occultations (#PF7066).
Aggravates Demons (#PF6734)   Evil eye (#PF2343)
Personified evil (#PF7018)   Haunted buildings (#PF0201).
Aggravated by Fear of the unknown (#PF6188).

◆ **PF8582 Bureaucratic ignorance**
Broader Ignorance (#PA5568).
Aggravates Unfamiliar bureaucratic procedures (#PJ9912)
Lack of cooperation with officialdom (#PF8500)
Unchecked power of government bureaucracy (#PD8890).

◆ **PF8587 Self-destruction**
Broader Human destructiveness (#PA0832).
Narrower Suicide (#PC0417)   Death instinct (#PF3849)
Self-destructive excuses (#PF6044)
Self-destructive government policy-making (#PF5061).
Aggravates Drug abuse (#PD0094).
Aggravated by Lack of self-confidence (#PF0879).

◆ **PF8590 Intellectual discrimination**
Broader Discrimination (#PA0833).
Narrower Scientific censorship (#PD1709).
Aggravated by Anti-intellectualism (#PF1929).

◆ **PF8594 Limited technical skills**
Broader Narrow range of practical skills (#PF2477).
Narrower Limited reservoir of technical skills in rural communities (#PF2848).
Related Limited development of functional abilities (#PF1332)
Lack of expertise in agricultural techniques (#PF8752)
Lack of skilled manpower in rural areas of developing countries (#PE5170).
Aggravated by Ineffective mechanisms for functional training (#PF1352).

◆ **PF8596 Uncritical acceptance of authority**
Broader Uncritical thinking (#PF5039).
Narrower Questioned acceptance of public officials (#PU0218).
Related Blind faith in technology (#PF4989).
Aggravates Inappropriate arguments (#PF2152).
Aggravated by Dependence on authority (#PF8995)
Inappropriate education (#PD8529).

◆ **PF8598 Rigid cultural patterns**
Broader Inflexible social care structures in developing countries (#PF2493).
Narrower Inflexible educational system (#PJ5365)
Rigidly entrenched social traditions in rural areas (#PF1765).
Aggravates Obsolete basis of cultural identity (#PF0836)
Untransposed significance of cultural tradition (#PF1373).

◆ **PF8602 Narrow economic foresight**
Broader Failure of public authorities to assist in financial investment (#PF6508).
Narrower Unimaginative vision of resource utilization (#PF1316).
Aggravates Undirected expansion of economic base (#PF0905).

◆ **PF8611 Resignation towards bribery**
Tolerance of corruption — Tolerance of nepotism
Broader Resignation to problems (#PF8781).
Related Non-resistance to evil (#PF7754).
Aggravates Bribery (#PC2558)   Nepotism (#PD7704)
Bribery of public servants (#PD4541)
Corruption amongst relatives of government leaders (#PE9140).
Aggravated by Oppressive prevalent images (#PF1365)
Institutionalized corruption (#PC9173).

◆ **PF8616 Inadequacy of international legal procedure**
Broader Outmoded legal systems (#PF2580)
Inadequacy of international legislation (#PF0228).
Aggravates Obstruction of international criminal investigations (#PF7277).

◆ **PF8618 Sense of powerlessness**
Broader Obstacles to leadership (#PF7011)   Deteriorating community identity (#PF2241)
Paralyzing patterns between villages and administrative structures (#PF1389).
Related Lack of urgency in village operations (#PS2881).
Aggravates Resignation to problems (#PF8781).
Aggravated by Oppression (#PB8656)   Oppressive prevalent images (#PF1365).

◆ **PF8623 Insufficient cultural heroes**
Unsymbolized village heroes — Uncelebrated heroes
Broader Untransposed significance of cultural tradition (#PF1373).
Modern disruption of traditional symbol systems (#PF6461)
Traumatic shift in life-styles of mining communities (#PE1137).
Aggravates Insufficient role models (#PF8451)   Obsolete basis of cultural identity (#PF0836).

◆ **PF8635 Instability of prices**
Price volatility — Unskilled market prices — Uncontrolled market price — Fluctuation of feed price — Unanticipated price fluctuations — Unstable crop prices
Narrower Instability of food prices in developing countries (#PE4986).
Related Archaic marketing methods (#PF6465)
Incomplete access to development capital (#PF6517)
Undeveloped channels for commercial initiative (#PF6471)
Underproductive methods of agricultural management (#PF6524)
Failure of public authorities to assist in financial investment (#PF6508)
Overlooked potential for industrial development in rural communities (#PF2471).
Aggravates Economic inflation (#PC0254)   Economic uncertainty (#PF5817)
Economic inefficiency (#PF7556)   Fluctuations in real value of money (#PD9356)
Inadequate management skills in rural communities (#PF1442).

◆ **PF8639 Lack of models of equality**
Nature Children learn the inequality between men and women by the example of their parents and peers. Overpopulation, disease, poverty, prostitution, and war are examples of the belief that some individuals, races, nations are less important than others. Corporations, arts, advertising, governments, religions etc. support beliefs of inequality.
Broader Inept theoretical models (#PF7182).
Aggravates Human inequality (#PA0844)   Unchallenging world vision (#PF9478).
Aggravated by Incompatibility of rural values in urban cultures (#PF2648)
Excessive accumulation of wealth by government leaders (#PD9653).

◆ **PF8648 Inappropriate understanding of progress**
Broader Misunderstanding (#PA8197).
Narrower Lack of social progress (#PF1545).
Adverse consequences of scientific and technological progress (#PF3931).
Aggravates Disparity between industrialized and developing countries (#PC8694).

◆ **PF8649 Imposed official decisions**
Broader Inaccessible administrative agencies (#PF2261).
Aggravates Fragmented decision-making (#PF8448).

◆ **PF8650 Government non-payment of agreed contributions to international organizations**
Imbalance between disbursements of multilateral agencies and official contributions — Withholding of membership payments to intergovernmental organizations — Overdue payments by governments to international organizations — Non-payment of United Nations membership dues
Incidence At the end of 1989 the member countries of the United Nations were $461 million in arrears on the general budget and $447 million in arrears on its peacekeeping budgets. In 1990 the USA owed $520 million in regular assessments to the United Nations, plus $150 million for peacekeeping costs. Other members in default on regular dues included South Africa ($40 million), Brazil ($17 million), Iran ($12 million), and Argentina ($10 million).
Broader Unfulfilled treaty obligations (#PF2497)   Ineffective international agreements (#PF6992)
Arrears in payment of government financial commitments (#PF1179).
Aggravates Intergovernmental failure to fulfil financial commitments (#PF3913)
Inadequate funding of international organizations and programmes (#PF0498)
Shortage of financial resources within the United Nations system of organization (#PF1460).

◆ **PF8652 Breakdown of police protection**
Delayed police response — Inadequate police control — Remote protection and security services — Distant police services — Partial police protection — Absence of police structures — Nonexistent security patrol
Broader Insufficient emergency services (#PF9007)
Unorganized development of work forces (#PF2128)
Limited availability of public services in the small towns of developed countries (#PF6539).
Aggravates Insecurity of property (#PC1784)   Personal physical insecurity (#PD8657)
Prevailing community insecurity (#PD9044)
Reluctant claims on external resources (#PF1226)
Divisive patterns of community groupings (#PF6545)
Lack of protection for victims of intimidation (#PE7793).
Aggravated by False burglary alarms (#PG7982)
Inadequate police funds (#PJ8760)
Official evasion of complaints (#PF9157)
Haphazard forms of social ethics (#PF1249)
Insufficient transportation infrastructure (#PF1495).

◆ **PF8654 Restrictive regulation of nuclear power**
Broader Restrictive legal structures (#PG9988).
Aggravates Insufficient nuclear power stations (#PD7663).
Reduces Nuclear accidents (#PD0771).

◆ **PF8661 Inadequate evidence to convict known offenders**
Difficulty of ensuring prosecutions for crimes — Law enforcement failure against known offenders
Broader Unjust laws (#PC7112).

Narrower Unconvicted war criminals (#PD4067) Inconclusive convictions for fraud (#PD5636).
Aggravates False evidence (#PF5127) False political evidence (#PD3017)
Illegally obtained evidence (#PE9309) Official fabrication of evidence (#PD8716)
Tampering with physical evidence (#PF7291) Tampering with official documents (#PF4699)
Inadequate national law enforcement (#PE4768).
Aggravated by Evidence decay (#PF5403) Unreported crimes (#PF1456)
Law enforcement complexity (#PF2454)
Legal discrimination in favour of offenders (#PD9316)
Inadmissability of evidence from other jurisdictions (#PE5424)
Placement of the burden of proof on the disempowered (#PF3918).

♦ PF8663 Non-viability of cold countries

Sub-arctic countries
Broader Geographically disadvantaged countries (#PF9247).
Aggravated by Climatic cold (#PD1404).

♦ PF8666 Desanctification of churches and holy places

Aggravates Desecration of holy spaces (#PF6385).

♦ PF8675 Short-term gain

Short-term professional gain — Short term cash purchase — Short-range business ventures — Short term share holders — Short-range financing image — Short range scheme
Broader Unproductive subsistence agriculture (#PC0492)
Underdeveloped sources of income expansion (#PF1345).
Aggravates Profiteering (#PC2618) Increasing pace of life (#PF2304)
Minimization of problems (#PF7186) Neglect of expert advice (#PF7820)
Precarious basis for family economics (#PF1382)
Short range planning for long-term development (#PF5660)
Unsustainable short-term improvements in agricultural productivity (#PE4331).
Aggravated by Individualism (#PF8393) Short-term profit maximization (#PF2174)
Ownership as a basis for land allocation (#PF6460).

♦ PF8686 Total disarmament

Nature Total disarmament is the elimination of the means of making war, whether imposed by a military victor; undertaken as a unilateral initiative for moral reason or to set an example; or agreed to as a result of intergovernmental negotiations on a bilateral or multilateral basis.
Counter-claim Just as it is impossible to stop the innovation, production and deployment of all military technology in advanced industrial states, so it is impossible to eliminate independent armed forces, short of establishing an effective world government. The appeals for total disarmament are statement of aspirations with very little relevance in the real world. Arms control, on the other hand, is demonstrably feasible in the existing system of nation states.
Refs Palme Commission *Common Security* (1982).
Narrower Negative economic repercussions of disarmament (#PF0589).
Aggravates Military unreadiness (#PF5933) International insecurity (#PB0009)
Inadequacy of civil defence (#PF0506) Military insecurity and vulnerability (#PC0541).

♦ PF8695 Lack of social contact

Infrequent neighbour contacts — Minimal community contact
Broader Lack of community participation (#PF3307)
Restrictive channels of cultural interchange (#PF3037).
Aggravates Disjointed patterns of community identity (#PF2845).
Aggravated by Social isolation (#PC1707).

♦ PF8703 Undemocratic policy-making

Insider policy-making — Policy-making conspiracy — Unrepresentative policy makers — Undemocratic government decision-making — Restricted access to policy formulation processes — Lack of organized participation — Non-participative decision-making — Non-collective decision-making — Exclusion from participation in decision-making
Claim There is a need for long-range planning to meet not only current policy requirements, but also the future implications of present decisions and policies. At present public agencies are no longer representative of the broad and varied populations they are supposed to serve. Policy making is seen as the responsibility of a few specialists, who consider the interests of only certain segments of the population. Consequently, policy decisions are made by few people with little accountability to local communities or the population at large.
Broader Denial of political rights (#PD8276)
Ineffective structures of local consensus (#PF6506)
Denial to people of control over their own lives (#PC2381).
Narrower Restricted local participation (#PF3287)
Unrepresentativity of trade unions (#PE6018)
Cumbersome methods of policy formation (#PF1076)
Restrictive organizational participation (#PG7752)
Limited access to technological decisions (#PF4636)
Exclusion of pre-adults from family decisions (#PE2268)
Limits on participation in community development (#PF3560)
Limited access to natural resource use decisions (#PF2882)
Non-participation of youth on the decision-making bodies (#PF6023)
Lack of worker participation in business decision-making (#PF0574)
Over-reliance on economic interest groups by policy agencies (#PF1070).
Related Narrow context for counsel (#PF0823) Undemocratic social systems (#PB8031)
Inflexible central government (#PD1061) Reduced understanding of globality (#PF7071)
Exclusion of the masses from setting criteria in judicial judgements (#PD1060).
Aggravates Inappropriate policies (#PF5645) Elitist control of global economy (#PC3778).
Aggravated by Messianic image of leadership (#PF1102)
Complex government regulations (#PF8053)
Social isolation of the elderly (#PD1564)
Lack of community participation (#PF3307)
Social disaffection of the young (#PD1544)
Interlocking corporate directorates (#PF5522)
Arrogance of intergovernmental agencies (#PF9561)
Proliferation of public sector institutions (#PF4739)
Inadequate models of socio-economic development (#PF9576)
Centralized decisions on local technological innovation (#PF2707)
Restrictive effects of traditional community decision-making (#PF3454).

♦ PF8704 Lack of a sense of community and solidarity at the world level

Lack of solidarity with the poor
Broader Fragmentation (#PA6233).
Narrower Inadequate sense of community and solidarity amongst workers (#PE4179).
Related Rivalry and disunity within developing regions (#PD0110).

♦ PF8711 Collusion between administrators of funding agencies and programme formulators

Nature In order to avoid responsibility for what may or will go wrong with a new programme initiative, it is in the interest of both the formulators and the agencies responsible for funding to fragment the field of analysis and to base their recommendations on limited information from reputable sources. If the programme subsequently fails or gives rise to other problems, the blame can then be shifted onto those supplying the information, especially other agencies. When such shortcomings manifest, those originally involved will already have moved on to other responsibilities. Short of criminal neglect, neither group can then be held directly responsible. Difficulties can also be conveniently blamed on the problems inherent in the situation, especially in developing societies.
Aggravates Blame avoidance (#PF6382) Superficial research (#PF7889)
Fragmentation of research (#PF9830)
Deliberate ignorance during policy-making (#PF8278)
Corruption and mismanagement of foreign aid (#PD0136)
Irresponsible delimitation of policy responsibilities (#PF7823).

♦ PF8713 Competitive distortion of musical pitch

Distortion of musical quality through escalating pitch
Nature International standards of pitch in music are being violated under pressure for more brilliant and striking performances under conditions for which the music was not originally intended. The drift in pitch standards is also occurring to a different degree in different places even though professional musicians and singers travel frequently between them. The higher pitch transforms warm musical colour to garish. The whole spectrum of voice ranges has been twisted out of shape. The colour of the mezzo-soprano has disappeared and the basso profondo voice has disappeared.
Background Pitch was arranged by the Greeks in descending and ascending ratios of seven different notes, currently denoted by the letters A through G. Prior to the 19th century, the note A could be as high or low as any particular culture favoured, varying as much as two tones. In the 19th century, with the increase in exchanges between cultures, efforts were made in Italy to impose the standard of A above middle C as being sound vibration at 432 cycles per second. For the past century a consensus has emerged, ratified in the 1930s by an international congress, that sets A at 440. Higher pitch has been compared to air pollution or toxic waste, where destruction results more from carelessness than calculation. Still higher pitch has its basis in the current economics of the arts. Concert halls are bigger in order to sell more seats. There is pressure on the organizers to attract attention with more "brilliant" performances whatever the loss in colour.
Incidence Singers are now under pressure in opera and concert halls, especially in Italy, Austria and Germany, to adapt to a pitch which is now creeping above 445 toward 450, despite efforts to restandardize A at 440.
Claim The increase in pitch from the classical C 256 to the norm of A 440 forces singers to change the notes at which voice register shifts occur. For the soprano a tuning at A 440 pushes the frequency of F beyond the dividing point between the first and second register. This arbitrary increase of pitch falsifies performance of all the musical classics.
Broader Quantitative pressure on standards of quality (#PF7227).

♦ PF8743 Mistrust of strangers

Mistrust of foreigners — Mistrust of newcomers — Fear of strangers — Fear of outsiders
Incidence In the earliest times, prior to the emergence of political society, all those with whom a person was not connected by ties of blood was considered an enemy. All such people, alien to the family or tribe, could be pursued with legitimate hatred as lawful prey to be plundered or slain. In classical times, foreigners were considered as barbarians and therefore the natural enemies of the state. Mistrust of foreigners persists in most cultures.
Broader Distrust (#PA8653) Disjointed patterns of community identity (#PF2845)
Personal isolation in communities of industrialized countries (#PD2495).
Aggravates Distrust of interpersonal relationships (#PF4274)
Inadequate cultural integration of immigrants (#PC1532)
Inadequate community care for transient urban populations (#PF1844).
Aggravated by Communication with foreigners (#PF8565).

♦ PF8749 Restrictive monetary practices

Restrictive monetary policies
Broader Restrictive practices (#PB9136) Inappropriate policies (#PF5645).
Narrower Monetarism (#PF6079) Parochial monetary agreements (#PD2469)
Discriminatory exchange rate policies (#PE8583)
Limited availability of loans in developing countries (#PE4704)
Restrictions on foreign access to capital bond markets (#PD3135)
Restrictive effects of external capital on development (#PF3318)
Domination of restrictive project loans by transnational banks in developing countries (#PE4330)
Interference of transnational banks' off-shore borrowing with domestic monetary policies (#PE4315)
Restrictive conditions on loans to developing countries through intergovernmental facilities (#PE9116).
Aggravates Economic inflation (#PC0254) Economic stagnation (#PC0002).
Aggravated by Incompatibility of rural values in urban cultures (#PF2648).

♦ PF8751 Belief in emotional instability of women

Broader Inappropriate assumptions (#PF6814).
Aggravates Discrimination against women in employment (#PD0086).

♦ PF8753 Failure to integrate knowledge to empower humanity in response to the global problematique

Failure to focus available insight in response to problems of society — Failure to interrelate wisdom of different cultures in response to social problems — Failure of international information systems to order problem-relevant information
Claim 1. The failure of any information system to provide all of the appropriate information (and no more) to the appropriate users (and no others) at the appropriate time (and no other), at the appropriate place, in the appropriate medium and at a cost affordable to the recipients and viable for the providers, is the superordinate problem underlying many others and impeding appropriate decision-making towards their solution.
2. In many ways the relationship of humanity to Earth is like that between an eighteenth century physician and his patients. Humanity shares with him a vast ignorance illuminated only by an instinct that warns us not to act precipitately, for such action is potentially as disastrous as inaction.
Broader Fragmentation of knowledge (#PF0944).
Aggravates Non-recognition of problems (#PF8112)
Uncoordinated policy-making (#PF9166)
Inappropriate development policy (#PF8757)
Ineffective decision-making processes (#PF3709)
Recurrence of misapprehended world problems (#PF7027)
Complex interrelationship of world problems (#PF0364)
Short range planning for long-term development (#PF5660)
Inadequate public information concerning problems (#PF5701)
Inadequate education concerning the nature of problems (#PE8216)
Excessive reliance on fashionable solutions to problems (#PF4473)
Inadequate adaptation of policy to educational difficulties (#PE8700)
Inadequate application of available knowledge to solve problems (#PF8191)
Inadequate implementation of plans and programmes against problems (#PF1010).
Aggravated by Manipulative knowledge (#PF1609)

Fragmentation of academic disciplines (#PF8868)
Multiplicity of problems facing society (#PF2003)
Fixation on partial solutions to problems (#PF9409)
Excessive emphasis on fashionable problems (#PF4164)
Fragmentation of international documentation (#PF7133)
Excessive emphasis on fashionable areas of research (#PF0059)
Ineffectiveness and inefficiency of interdisciplinary meetings (#PF0409).

♦ **PF8754 Superstitious symbolic acts**
Broader Modern disruption of traditional symbol systems (#PF6461).
Aggravates Lack of commitment to common symbols (#PE8814).

♦ **PF8757 Inappropriate development policy**
Mismatch between public policy and current scientific knowledge
Broader Inappropriate policies (#PF5645)
Disruptive effect of changing employment patterns (#PF2303).
Aggravates Maldevelopment (#PB6207) Natural environment degradation (#PB5250)
Destruction inherent in development (#PF4829)
Inappropriate design of development projects (#PF4944).
Aggravated by Poverty in developing countries (#PC0149)
Lack of participation in development (#PF3339)
Neglect of the role of women in rural development (#PF4959)
Weakness of infrastructure in developing countries (#PC1228)
Inadequate mobilization of public opinion for development (#PF9704)
Incompatibility of environmental and economic decision-making (#PF9728)
Restrictions on industrial and economic development due to environmental policies (#PE4905)
Failure to integrate knowledge to empower humanity in response to the global problematique (#PF8753).

♦ **PF8759 Unethical personal relationships**
Irresponsible personal relationships — Corruption in personal relationships — Dishonest personal relationships — Keeping bad company — Unclear relationships — False friendships
Claim There is a traditional concern, reinforced in a number of religious doctrines, at any tendency to "keep bad company" or to "associate with undesirables". In business and public life this translates into relationships which subvert formal roles to favour personal interests, whether financial, sexual, political or otherwise. Such relationships may involve threat, bribes or blackmail, whether on the frontiers of legality, in grey areas or in association with criminals. In private life it may involve relationships whose nature is concealed for fear of censure and because of the damage caused to other relationships. These may take the form of hurtful games between people, possibly involving some form of threat or manipulation. Such relationships are especially unhealthy when they appear normal, or even benevolent, with their true nature only emerging later in time. After all, a person is known by the company he or she keeps.
Broader Fragmented patterns of extended family relationships (#PF1509).
Narrower Bribery (#PC2558) Adultery (#PF2314) Soul murder (#PF4213)
Sexual harassment (#PD1116) Sexual immorality (#PF2687) Criminal association (#PE1178)
Fraudulent impersonation (#PE1275) Genetic sexual attraction (#PE5925)
Undesired sexual obligations (#PF4948)
Unauthorized proximity of males to females (#PF8780).
Aggravates Ritual pollution (#PF3960) Spiritual impurity (#PF6657)
Criminal subculture (#PE5508) Sexual abuse of children (#PE3265)
Negative personal relationships between leaders of countries (#PF9843).
Aggravated by Immorality (#PA3369) Lack of intimate relationships (#PF4416).

♦ **PF8763 Prohibitive labour costs**
Excessive salaries — Overpaid employees — High labour costs — Disproportionately high salaries — Inflated labour costs — Excessive cost of manpower — Increasing economic value of labour
Incidence Between 1960 and 1980, labour costs grew faster than the average productivity of the economy. Labour costs include net wages, payroll taxes and social security contributions by both the employees and employers, as well as various fringe benefits such as private company pension schemes, free cars and expense accounts. When labour costs increase faster than productivity, the remuneration of capital is lower.
Broader Economic inflation (#PC0254) Inequitable range of salaries (#PD9430)
Inappropriate use of financial resources (#PD9338).
Narrower High minimum wages (#PD5674) Costly teaching staff (#PJ9524)
Excessive salaries of experts (#PF8317) Prohibitive cost of farm labour (#PG9069)
Excessive salaries of corporate executives (#PF7578)
Excessive salaries of international civil servants (#PE6388).
Related Dispersion of local capital resources (#PF1979).
Aggravates Disparity in remuneration between public and private sector employees (#PE7760).
Aggravated by Socialism (#PC0115) Capitalism (#PC0564)
Ineffective economic structures in industrial nations (#PE4818).
Reduced by Underpayment for work (#PD8916).

♦ **PF8765 Impenitence**
Nature Impenitence is the denial of responsibility for a transgression against the life of oneself or another being. It is the inability or refusal to admit one's participation in the state of separation from oneself, others or existence. Until one admits one is wrong there is no possibility for change. The unrepentant rigidly adheres to the belief that they are doing the right thing forcing others and existence to conform to their belief.
Broader Sin (#PF0641).

♦ **PF8767 Obsession with novelty**
Intolerance of proven methods — Bias in favour of the new
Narrower Unconvincing alternatives to existing societies (#PF3826).
Related Cult of youth (#PF6766) Risk-aversion strategy (#PF4612).
Reduced by Unproven new methods (#PS9036).

♦ **PF8780 Unauthorized proximity of males to females**
Socializing between unrelated men and women — Khalwat
Incidence Traditionally in many western societies, but to a far more limited degree at present, socializing between unmarried males and females could only occur under the supervision of a chaperon. Whilst decreasingly subject to any form of control, the activities of an unmarried man and woman behind closed doors continue to be viewed as extremely suspect, especially in Latin countries. Islamic law prohibits fraternizing between men and women, such as unmarried men and women walking together or holding hands.
Broader Unethical personal relationships (#PF8759).
Aggravates Distrust of interpersonal relationships (#PF4274).
Aggravated by Fundamentalism (#PF1338).

♦ **PF8781 Resignation to problems**
Sense of defeat — Defeatist civic attitude — Defeatist self-images of citizens' law enforcement groups — Defeatism in the face of problems — Defeatism
Nature Community groups concerned with law enforcement which are unable to see their defeats in a wider context may fall prey to images of defeat and powerlessness. They become increasingly victimized by these self-images and retreat from dealing with vital problems into a milieu of apathy and personal isolation.
Broader Lack of self-confidence (#PF0879) Demeaning community self-image (#PF2093).
Narrower Electoral defeat (#PF4709) Resignation towards bribery (#PF8611).
Related Structural failure of citizen participation (#PF2347).
Aggravates Alarmism (#PF4384) Deteriorating community identity (#PF2241)
Stagnated images of community identity (#PF6537)
Recurrence of misapprehended world problems (#PF7027).
Aggravated by Apathy (#PA2360) Issue avoidance (#PF1623)
Inaction on problems (#PB1423) Sense of powerlessness (#PF8618).

♦ **PF8793 Outdated procedures**
Outdated regulations — Outdated rules
Narrower Antiquated regulations in the banking industry (#PF4370).
Aggravates Impediments to adoption of children (#PF7353).
Aggravated by Unjust laws (#PC7112) Excessive government control (#PF0304)
Complex government regulations (#PF8053).

♦ **PF8798 Excessive size of social institutions**
Narrower Growing size and impersonality of firms (#PE8706).
Aggravates Instability (#PA0859) Inefficiency (#PB0843)
Excessive institutionalization of education (#PD0932).
Aggravated by Unsustainable economic development (#PC0495).

♦ **PF8808 Indecision**
Broader Irresponsibility (#PA8658) Avoidance of decision-making (#PF4204).
Aggravates Issue avoidance (#PF1623) Unclear educational roles (#PG8840)
Narrow range of practical skills (#PF2477).
Aggravated by Doubt (#PF8474)
Inadequate training in decision-making (#PD2036)
Lack of a system for ethical decision-making (#PF2070).

♦ **PF8819 Failure of government intelligence services**
Intelligence failure
Broader Inadequate government (#PJ6362) Inappropriate policies (#PF5645)
Inadequate information systems for international governmental decision-making processes (#PD0104).
Related Unaccountable government intelligence agencies (#PF9184).
Aggravates Inaccurate forecasting (#PF4774)
Inadequate international cooperation in reducing terrorism (#PF4366).
Aggravated by Inadequate system of political checks and balances (#PE4997).

♦ **PF8820 Government arrogance**
Broader Arrogance (#PA7646).
Aggravates Government complacency (#PF6407)
Arrogance of policy-makers (#PF2895)
Arrogance of intergovernmental agencies (#PF9561)
Governmental disregard for legitimate protests (#PD9557).
Aggra ated by Bureaucratic superiority (#PC1259)
Proliferation of public sector institutions (#PF4739)
Unaccountable government intelligence agencies (#PF9184).

♦ **PF8827 Inadequate insurance**
Under-insured
Broader Risk (#PF7580).
Narrower Inadequate crop insurance (#PJ4576) Prohibitive cost of life assurance (#PE8736)
Inadequate insurance against damages arising from a natural disaster (#PF6026).
Aggravates Social insecurity in developing countries (#PE4796).
Aggravated by Prohibitive cost of insurance (#PE8632)
Inequities in marine insurance (#PE5802)
Restrictive transport insurance practices (#PD0881)
Inaccessibility of insurance for island developing countries (#PE5665)
Excessive costs and unsuitability of insurance for land-locked developing countries (#PE5896).

♦ **PF8829 Inadequate building standards**
Absence of housing codes
Broader Underprovision of basic services to rural areas (#PF2875).
Aggravates Low-quality construction work (#PD7723)
Substandard housing and accommodation (#PD1251).
Reduced by Restrictive building codes in urban areas (#PE8443).

♦ **PF8835 Limited agricultural education**
Unmotivated agricultural training — Unavailability of agricultural training — Agricultural education lack of — Lack of agricultural education — Technologically untrained farmers
Broader Lack of training (#PD8388) Inadequate vocational education (#PF0422).
Narrower Neglect of women farmers (#PF6024) Lack of agro-urban training (#PG7992)
Limited gardening experience (#PG8928) Underdeveloped farming skills (#PJ0729)
Inadequacy of agricultural education in developing countries (#PE9096).
Related Haphazard transmission of practical technology (#PF3409).
Aggravates Inadequate feeding of animals (#PC2765)
Poor quality of domestic livestock (#PD2743)
Decreasing genetic diversity of animals (#PC1408)
Limited opportunities for significant work (#PF1403)
Stagnated development of agricultural production (#PD1285).
Aggravated by Limited availability of education in rural areas (#PF3575)
Unperceived relevance of formal education in rural communities (#PF1944)
Limited availability of technical agricultural and business training (#PF2698).

♦ **PF8838 Jewish conspiracy**
Broader Conspiracy (#PC2555).
Related Catholicism (#PF8071).
Aggravates Zionism (#PF0200).

♦ **PF8842 Indecency**
Broader Nudism (#PF2660).
Narrower Profanity (#PF7427) Indecent exposure (#PF4317) Indecent advertising (#PD2547).

♦ **PF8846 Irrelevant institutions**
Irrelevant forms of governance — Irrelevant policy-making initiatives
Narrower Institutional preoccupation with obsolete problems (#PJ5014).
Aggravated by Inappropriate policies (#PF5645).

♦ **PF8854 Vulnerability of intellectual property**
Inadequate protection of intellectual property — Denial of right to intellectual property — Intellectual piracy — Theft of intellectual property — Industrial piracy — Lack of protection for

industrial property — Theft of ideas
Incidence It is estimated that the cost of patent pirates, trademark thieves and copyright bandits cost American industry over $60 billion a year. The industries suffering most are scientific and photographic, computers and software, electronics, motor vehicles and parts, and entertainment.
Refs World Intellectual Property Organization *Symposium on the Effective Protection of Industrial Property Rights, Geneva 1987* (1987); World Intellectual Property Organization *Background Reading Material on Intellectual Property* (1988).
Broader Insecurity of property (#PC1784).
Narrower Forgery (#PD2557) Plagiarism (#PD3996) Counterfeiting (#PD7981) Conceptual plagiarism (#PD1284) Avoidance of copyright (#PD0188) Inadequacy of patent coverage (#PF3538).
Aggravates Economic conflict (#PC0840) Crimes against intangible property (#PE6486) Copyright barriers to transfer of knowledge (#PE8403) Distortion of international trade through obstacles to patent protection (#PD0455).
Reduces Restrictive business practices in relation to patents and trademarks (#PE0346).

♦ **PF8855 Over-acceptance of socio-economic dependency**
Over-commitment to welfare dependency — Unquestioning dependency on social assistance — Belief in personal entitlement
Nature People, rich or poor, believe that they are entitled to the benefits provided by civilized governments, such as health care and education, without paying taxes to finance them.
Broader Economic dependence (#PF0841) Dependence on social welfare (#PD1229) Overdependence on government (#PF9530).
Aggravated by Lack of commitment (#PF1729) Psychological inertia (#PF0421) Limited access to social benefits (#PF1303).

♦ **PF8868 Fragmentation of academic disciplines**
Fragmentation of the sciences — Academic isolationism
Broader Over-specialization (#PF0256) National isolationism (#PF2141).
Aggravates Institutional fragmentation (#PC3915) Irrelevant scientific activity (#PF1202) Limitation of current scientific knowledge (#PF4014) Irresponsible scientific and technological activity (#PC1153) Jurisdictional conflict between academic disciplines (#PF9077) Failure to integrate knowledge to empower humanity in response to the global problematique (#PF8753).
Aggravated by Scientific elitism (#PC1937) Structural amnesia in institutional systems (#PF7745).

♦ **PF8869 Lack of legal aid facilities**
Denial of right to legal aid — Inadequate legal aid
Nature The vast majority of people in the world have no access to adequate legal service. The major problem is financial. Most people do not have the means to hire a legal professional. At the same time the legal profession cannot provide all of the required services free of charge. In some countries and regions within countries there is an acute shortage of lawyers, for example in the Northern Territory of Australia there are only 23 lawyers. In some places the large number of individual eligible for aid blocks access to the service. Distance and lack of communication facilities. Lawyers or clients would have to spend days travelling to provide the service. Language differences may hinder legal aid services. In large metropolitan areas, like Los Angeles, or multilingual countries, like India, dozens of languages are spoken and providing lawyers or even translators may be difficult. Legal aid offices may not be near courts. The high cost of administrating legal aid may prevent adequate services being provided. The legal profession may not support efforts to provide legal aid. The government or large segments of the public may oppose legal aid services. Where legal aid is provided, the government may influence the professional decisions of lawyers providing the service. The public may be unaware or distrustful of the service.
Broader Denial of legal representation (#PF3517).
Aggravates Inaccessibility of justice (#PD8334).
Aggravated by Overcrowding (#PB0469) Multiplicity of languages (#PC0178) Insufficient financial resources (#PB4653) Shortage of resident professionals (#PG8812) Distrust of professional service delivery (#PD0974) Shortage of adequately trained personnel to act against problems (#PF0559).

♦ **PF8870 Delay in societal impact of innovation**
Broader Change (#PF6605) Delays in implementation of social change (#PC6989).

♦ **PF8874 Threat of war**
Military threat
Refs Avery, John *Health Effects of War and the Threat of War* (1988).
Broader Aggressive foreign policy (#PC4667) Offences against the peace and security of mankind (#PC6239).
Narrower Military manoeuvres in sensitive border areas (#PE3704).
Aggravates Insecurity (#PA0857).

♦ **PF8876 Lack of parliamentary time to approve needed legislation**
Failure to process low priority legislative reform — Parliamentary inability to respond to minority concerns — Inadequate parliamentary debating time
Broader Lack of time (#PC4498).
Aggravated by Manipulation of debates (#PD4060) Government delaying tactics (#PF6119) Political barriers to effective legislation (#PC3201) Delays in elaboration of remedial legislation (#PC1613).

♦ **PF8887 Disintegration of accepted myths**
Claim Myths are allegorical descriptions of reality which, in their full sense, describe the beginning of existence, its end and the significance of day to day living. All contemporary myths including the one which supports the modern scientific world view are under attack and to some degree becoming disfunctional, in that they fail to enable individuals and societies to function meaningfully and creatively in the post-modern world.
Broader Symbol system failure (#PF3715).
Aggravates Loss of cultural identity (#PF9005) Lack of commitment to common symbols (#PE8814) Fragmentation of the human personality (#PA0911).

♦ **PF8889 Unethical practices of priesthood**
Religious fraud — Religious hoaxes — Fraudulent relics — Malpractice of clergy
Refs Malony, H Newton *Clergy Malpractice* (1986); Watts, Tim J *Clergy Malpractice* (1988).
Broader Fraud (#PD0486) Hoaxes (#PF9375).
Narrower Misuse of spiritual authority for sexual purposes (#PE1348).
Related Abuse of relics (#PF5107) Religious historical forgery (#PG6355).
Aggravates Fetishism (#PF8363).
Aggravated by Religious deception (#PF3495).

♦ **PF8905 Decreasing participation in collective religious worship**
Reduction in church attendance — Diminished church role
Incidence A survey of eight countries found that 54 percent of Americans visit churches, mosques, synagogues, temples, etc. at least once a month, 45 percent of Italians, 39 percent of Austrians, 28 percent of Dutch, 20 percent of British and 7 percent of Hungarians.
Claim The fact that few people go to church regularly shows that religious leaders are failing in their central task to answer the spiritual needs of people. They are looking to fulfil the need for meaning from sects, cults, disciplines, therapies etc.
Broader Religious apathy (#PC3414) Declining sense of community (#PF2575).
Narrower Minimal church / school involvement (#PJ9011).
Aggravates Religious prejudice (#PD4365) Loss of cultural identity (#PF9005) Inaccessible places of worship (#PE6795).
Aggravated by Symbol system failure (#PF3715) Loss of faith in religion (#PF3863) Lack of commitment to common symbols (#PE8814) Social symbols dominated by the economy (#PE6671).

♦ **PF8921 Transfer of highly polluting industries to developing countries**
Broader Exploitative transformation of international division of labour (#PF9281).

♦ **PF8922 Substitution of fantasy for reality**
Wishful thinking — Unrealistic beliefs — Treating fiction as reality — Imposing a fictional worldview
Broader Avoidance of reality (#PF7414).
Narrower Unrealistic policies (#PF9428).
Aggravates Minimization of problems (#PF7186).
Aggravated by Neglect of expert advice (#PF7820).

♦ **PF8927 Limited enforceability of international standards**
Informality of international standards
Nature International standards may be set either by "hard law" in the form of international treaties and agreements, or by "soft law" in the form of intergovernmental resolutions, declarations and programme objectives. Little provision can usually be made for the enforcement of "hard law" which is frequently violated. Despite the formality of the settings in which "soft law" may be elaborated, it is of questionable status in international law. The status of the many declarations and resolutions elaborated by international nongovernmental bodies is even less binding, except amongst those who choose to be bound by them.
Broader Lack of international cooperation (#PF0817) Inadequate dissemination and use of available information (#PF1267).
Aggravates Denial of human rights (#PB3121) Inadequacy of international standards (#PF5072) Parochial telecommunications standards (#PE1840).

♦ **PF8931 Difficulties of cooperative planning**
Broader Unrecognized socio-economic interdependencies (#PF2969).
Narrower Lack of collective housing schemes (#PU2001).

♦ **PF8939 Inadequate meeting methods**
Unproductive meeting time — Ineffective meeting formats
Broader Obsolete methods (#PF3713).
Narrower Expensive meeting space (#PG8675) Manipulation of debates (#PD4060) Unpublicized meeting agendas (#PG9295) Tokenistic meeting resolutions (#PF2086) Rhetorical inflation in meetings (#PF3756) Ineffectiveness and inefficiency of intercultural meetings (#PF0316) Ineffectiveness and inefficiency of international meetings (#PF0349) Ineffectiveness and inefficiency of interdisciplinary meetings (#PF0409).
Related Limited meeting facilities (#PE1535) Unpublicized public meetings (#PF5222).
Aggravates Insufficient community events (#PF5250).
Aggravated by Ineffective dialogue (#PF1654) Unstructured local decision-making (#PF6550) Lack of essential local infrastructure (#PF2115) Lack of motivation in leadership development (#PF2208) Lack of local services for community leadership training (#PF2451).

♦ **PF8944 Ritual cannibalism**
Nature Consumption of the dead or portions of their bodies as a means of acquiring possession of their souls or other qualities. This is associated with a belief in sympathetic magic whereby the whole is contained in the part so that consumption of part of the body confers on the practitioner some powers of the dead person.
Incidence Although much more common in tribal societies in the past, instances continue to be reported in connection with some forms of satanic rites and voodoo practices, both in developing and industrialized countries. Instances have also been reported in relation to initiations into secret societies where one object is to bond members together through a common experience which they cannot share with outsiders.
Broader Human cannibalism (#PF2513).
Aggravated by Human sacrifice (#PF2641) Satanic rituals (#PF7887).

♦ **PF8946 National hegemony over United Nations agencies**
Broader Domination (#PA0839).

♦ **PF8948 Discouragement**
Broader Fear (#PA6030) Solemnity (#PA6731) Dissuasion (#PA7343).

♦ **PF8953 Conflict of laws over nationality**
Broader Conflict of laws (#PF0216).
Aggravates Statelessness (#PE2485) Restrictions on recognition of nationality (#PE4912).

♦ **PF8960 Lack of governmental support**
Limited government support — Marginal government support
Broader Restricted flow of local economy (#PF6451) Unrecognized socio-economic interdependencies (#PF2969).
Aggravates Lack of cooperation with officialdom (#PF8500) Uncoordinated social services in urban areas (#PF1853) Inadequate funding of international organizations and programmes (#PF0498).

♦ **PF8964 Refusal of licence renewal**
Revoking of licences
Nature In many cases a person requires a licence to earn a living. Restaurant and bar owners are given licences based on a health inspection. Drivers are granted licences after passing a test demonstrating their understanding of driving law and their ability to safely drive. Pilots and some other aviation experts require licences based on skills but also medical health. If a licence is revoked or suspended a person may not only be force to leave work but may be a social outcast.
Broader Risk (#PF7580) Refusal to grant licences (#PG5946).

♦ **PF8989 Policy-making delays**
Decision-making delays — Lengthy decision process — Delayed decisions — Procrastination — Indecisiveness
  Broader Delay (#PA1999)   Unformed style of cooperative action (#PF6514).
  Narrower Delayed development of regional plans (#PF2018)
  Government delay in response to symptoms of problems (#PF6707)
  Procrastination of science in the face of the unexplained (#PF3682).
  Aggravates Delays in elaboration of remedial legislation (#PC1613)
  Short range planning for long-term development (#PF5660).
  Aggravated by Doubt (#PF8474)   Issue avoidance (#PF1623)
  Procrastination (#PF5299)   Inappropriate policies (#PF5645)
  Inauspicious conditions (#PF6683)   Fragmented decision-making (#PF8448)
  Unpredictable governmental policy (#PF1559)
  Ineffective decision-making processes (#PF3709)
  Cumbersome methods of policy formation (#PF1076).

♦ **PF8993 Wasted time**
Unresourceful time use
  Broader Obstacles to efficient utilization of time (#PF7022).
  Narrower Wasted waiting time (#PF1761).
  Aggravates Lack of time (#PC4498)
  Limited exposure to outside influences in rural villages (#PF2296).
  Aggravated by Delay (#PA1999)   Inadequate sense of time (#PF9980).
  Reduced by Limited leisure time (#PF9062).

♦ **PF8995 Dependence on authority**
Dependence on leadership
  Broader Obstacles to leadership (#PF7011).
  Narrower Abuse of authority (#PC8689)
  Interference with head of household authority (#PJ8055).
  Aggravates Uncritical acceptance of authority (#PF8596).
  Aggravated by Personal disempowerment (#PF0549).

♦ **PF9002 Interstellar dust**
  Broader Dust (#PD1245)   Disasters of extraterrestrial origin (#PF3562).
  Related Meteors as hazards (#PF1695).
  Aggravates Obstruction of astronomical observation by environmental pollution (#PE7244).

♦ **PF9003 Negative social context**
Negative influence of peer groups — Disruptive peer influence
  Broader Covert imperialism (#PF3199)
  Debilitating deterioration of physical environment (#PD2672)
  Lack of responsible involvement in community affairs (#PF6536).
  Aggravates Youth violence (#PF7498).

♦ **PF9004 Unperceived career opportunities**
Unexplored career opportunities
  Broader Unrecognized opportunities (#PF6925)   Selective perception of facts (#PF2453).
  Related Underutilization of locally available skills (#PF6538).
  Aggravates Limited access to society's resources (#PF6573)
  Unrealized use of education structures (#PF2568)
  Lack of continuity amongst personnel of international organizations (#PF3434).

♦ **PF9005 Loss of cultural identity**
Denial of right to cultural identity
  Broader Denial of cultural rights (#PD5907).
  Related Fear of losing cultural identity (#PD2614).
  Aggravated by Ethnocide (#PC1328)   Ignorance (#PA5568)
  Rejection of rituals (#PF0278)   Symbol system failure (#PF3715)
  Subsistence life style (#PF1078)   Faded community symbols (#PG8964)
  Absence of rites of passage (#PF1674)   Decadent standard of living (#PD4037)
  Socially inappropriate housing (#PD8638)   Disintegration of accepted myths (#PF8887)
  Obsolescence of rituals and customs (#PF1309)   Obsolete basis of cultural identity (#PF0836)
  Lack of commitment to common symbols (#PE8814)
  Symbols unrelated to human experience (#PF9070)
  Incomplete access to information resources (#PF2401)
  Excessive use of foreign programmes for media (#PE9643)
  Untransposed significance of cultural tradition (#PF1373)
  Decreasing participation in collective religious worship (#PF8905).
  Reduced by Ethnic conflict (#PC3685).

♦ **PF9006 Voodoo**
Zombie
  Nature Voodoo is devil-worship and fetishism controlled by secret societies which emerged during the slave revolts in Haiti. While not all Voodoo priests are members, these societies intimidate members and the general population. They control the government and have more weapons than any others on the island.
  Background Voodoo is the mixture of religion and magic imported from Ghana to Haiti and the southern United States blended with elements of Roman Catholic rituals, dating from the french colonial period before 1804. In earlier forms a girl child, called the goat without horns, was sacrificed once a year. Currently a white kid is used. It is drugged, killed and eaten. Rituals during the rest of the year involve disemboweling black dogs, cocks or hens. Trances are induced by spirit possession are central to the rituals. Belief in a supreme God is combined with service to the loa who are local or African gods, deified ancestors or Catholic saints. Individuals are initiated into the cult through a series of rituals and into the priesthood by a series of test. Transgressors of Voodoo traditions may be punished by being turned into zombies by a sorcerer capturing his soul.
  Counter-claim Zombies are poisoned with tetradotoxin, a substance contained in puffer fish. The poison suppresses the autonomic nervous system which controls the basic body functions. If conscious the victim cannot move and with a cursory examination may be pronounced dead.
  Broader Irrational religious beliefs (#PF6829).
  Related Satanism (#PF8260)   Demonic and spirit possession (#PF5781).
  Aggravates Bewitchment (#PF3956).

♦ **PF9007 Insufficient emergency services**
Lack of emergency facilities — Ineffective emergency services — Unavailable emergency services — Inadequate emergency care — Slow emergency care — Inaccessible emergency relief facilities — Distant emergency services — Distant emergency transport — Expensive emergency services
  Broader Physically inaccessible services (#PC7674)
  Inadequate disaster rescue and relief (#PF0286)
  Lack of essential local infrastructure (#PF2115).
  Narrower Unavailability of first aid (#PJ1261)   Lack of trained firefighters (#PJ0277)
  Breakdown of police protection (#PF8652)
  Inadequate emergency medical services (#PD1428).
  Related Inadequate social welfare services (#PC0834)
  Limited availability of public services in the small towns of developed countries (#PF6539).
  Aggravated by Ineffective rural transport (#PF2996)
  Failure to assist in emergencies (#PF5306)
  Limited access to health services (#PF6577)
  Outdated forms of community health (#PF1608)
  Complex regulations paralyzing small communities (#PF2444).

♦ **PF9008 Imbalances between people's aspirations and the structure of opportunities and income available**
  Broader Disillusionment (#PA6453).
  Aggravated by Unemployment (#PB0750).

♦ **PF9014 High interest rates**
Exhorbitant interest rates — Usury — Prohibitive cost of money
  Nature Relatively high interest rates slows borrowing, stagnates business and creates hardships on individual borrowers. Economic policies in the industrial countries determine interest rates worldwide.
  Incidence High real interest rates are adding significantly to structural budget deficits in both developed and developing countries.
  Broader Economic crime (#PC5624)
  Economic and financial instability of the world economy (#PC8073).
  Narrower Usury in developing countries (#PE2524).
  Aggravates Lack of local commercial services (#PF2009)
  Subsistence approach to capital resources (#PF6530)
  Inadequate domestic savings in developing countries (#PD0465)
  Decline in public sector savings in developing countries (#PE4574)
  Constricting level of capital development in rural areas (#PE1139).
  Aggravated by Payment of interest (#PF5514)   Unethical financial practices (#PE0682)
  Profit-oriented interest payments (#PD2552)
  Lack of international coordination of interest rates (#PF3141)
  Mismatch of national macroeconomic policies among industrialized countries (#PF5000).

♦ **PF9016 Loss of leadership credibility**
  Broader Obstacles to leadership (#PF7011).
  Narrower Loss of confidence in government leaders (#PF1097).
  Aggravates Fragmented planning of community life (#PF2813).
  Aggravated by Broken government promises (#PF4558)
  Leadership impaired by illness (#PF8387)
  Loss of international leadership (#PF8353)
  Mediocrity of government leaders (#PF3962)
  Extreme detachment from represented constituency (#PF0889).

♦ **PF9023 Geographical isolation**
Physical remoteness — Physical isolation — Isolated village location — Remote geographical location — Isolated geographical location — Social isolation of mountain valley communities — Perpetuated rural isolation — Social isolation of neighbouring villages — Rural dispersion
  Nature Rural dispersion, exemplified by small scattered settlements and isolated homesteads, inhibits the provision of infrastructure and services, particularly those relating to water, communication, transportation, electricity, health and education.
  Refs Wadsworth, G R *The Diet and Health of Isolated Populations* (1984).
  Broader Isolation (#PB8685).
  Narrower Isolated islands (#PD2941)   Physically inaccessible services (#PC7674)
  Physically inaccessible resources (#PC4020).
  Aggravates Social isolation (#PC1707)   Political isolation (#PC7569)
  Socio-economic poverty (#PB0388)   Consanguineous marriage (#PC2379)
  Ineffective rural transport (#PF2996)   Limited accumulation of capital (#PF3630)
  Unemployment in developing countries (#PD0176)
  Neglect of remote regions and islands (#PE5760)
  Inequitable administration of justice (#PD0986)
  Limited access to society's resources (#PF6573)
  Declining breeds of domesticated animals (#PD6305)
  Limited relations beyond local environments (#PF3192)
  Unattractive locale for economic development (#PF3499)
  Demoralizing images of rural community identity (#PF2358)
  Restrictive pattern of business activities in small communities (#PD1415).
  Aggravated by Distance (#PA7295).

♦ **PF9026 Rationing**
  Narrower Inadequate medical resources (#PD7254).
  Aggravated by Insufficient availability of goods (#PB8891).
  Reduced by Natural selection (#PF5997).

♦ **PF9030 Fear of resettlement**
  Broader Limiting effect of individual survivalism (#PF2602).
  Aggravates Back to the land (#PF4181)   Involuntary mass resettlement (#PC6203).

♦ **PF9032 Parascience**
  Broader Superstition (#PA0430).
  Reduces Ignorance (#PA5568)   Uncritical acceptance of dogmas and standards (#PF2901).

♦ **PF9043 Affliction by malevolent spirits**
  Nature People, but more frequently women, may believe themselves to be the victims of unsolicited and malevolent attention on the part of spirits. The visible form of such afflictions may include various stigmatizing illnesses or a trance condition, but they may also include a subjective sense of malaise. The affliction may be perceived as due to spirits acting of their own accord or due to the malevolent machinations of another human being who captures a spirit or directs it to attach the victim. Illnesses resulting from the former are held to be more difficult to counter than those due to the latter.
  Incidence In some societies certain conditions (including trance, madness and leprosy) are considered synonymous with spiritual affliction. If Western medicine fails, spiritual affliction is often suspected. Unusual ailments may also be viewed as spiritual afflictions. Women are widely considered to be more vulnerable than men, partly because of the greater risks associated with menstruation.
  Aggravates Demonic and spirit possession (#PF5781).
  Aggravated by Demons (#PF6734)   Witchcraft (#PF2099)
  Malevolence (#PA7102).

♦ **PF9046 Inaccessible historical libraries**
Denial of right of access to historical libraries
  Broader Unethical documentation practices (#PD2886).
  Related Classified public information (#PF9699).
  Aggravated by Historical forgery (#PE5051)   Uncatalogued historical documents (#PF6483).

♦ **PF9047 Incompatibility**

Broader Psychological impediments to marriage (#PF3344).
Related Enmity (#PA5446) Disaccord (#PA5532) Disagreement (#PA5982) Unsociability (#PA6653).
Aggravates Marital instability (#PD2103) Human sexual inadequacy (#PC1892).

♦ PF9050 Government hypocrisy
Political hypocrisy — Double moral standards of government
Nature The political art to espouse the highest of aims and principles while engaging in the lowest of deals and in breaking the same principles.
Broader Hypocrisy (#PF3377).
Narrower Limited acceptance of human rights treaties (#PE7300)
Donor distortion of aid to developing countries (#PF7955).
Related Excessive accumulation of wealth by government leaders (#PD9653).
Aggravates Government treachery (#PF4153) Immoral public policy (#PF4753)
Manipulative euphemisms (#PF5183)
Failure of government to apologize for errors (#PF5296)
Dependence of government revenues on exploitation of environmentally inappropriate products (#PD1018).

♦ PF9062 Limited leisure time
Restricted leisure time — Insufficient recreation time — Denial of right to leisure and rest
Broader Lack of time (#PC4498)
Obstacles to efficient utilization of time (#PF7022).
Narrower Excessive hours of work (#PD0140) Insufficient leisure time for women (#PE8907).
Related Limited shared time (#PJ9126) Insufficient personal time (#PJ9534)
Social disaffection of the young (#PD1544)
Modern disruption of traditional symbol systems (#PF6461).
Reduces Wasted time (#PF8993) Unstimulating entertainment (#PF8105).

♦ PF9063 Eccentricity
Nature It implies an uncommon personality, so that an individual behaves in a manner that is out of conformity with accepted norms. True eccentrics are often beset with phobias, obsessions and anxieties, their conduct being quite inexplicable. Eccentricity is frequently regarded as one of the signs of geniuses. Among the famous eccentrics are Friedrich von Schiller, who worked with his feet placed on ice for a time and inhaling the flavour of rotten apples, and Emile Zola, who counted the gas jets in every street through which he walked or drove.
Broader Behavioural deterioration (#PB6321).
Related Nonconformity (#PA5878) Unintelligence (#PA7371)
Changeableness (#PA5490).

♦ PF9065 Scholasticism
Scholastic thinking
Nature The tendency of experts to obscure their activity behind an artificial language or jargon resulting in a breakdown of communication, whilst dogmatically upholding the primacy of certain untested principles over concrete reality, reasoning only in endless syllogisms that avoid the examination of empirical data. Despite complete mastery of a discipline, that mastery does not enable those engaged in this form of thinking to say anything new beyond what can be expressed within the jargon of that discipline. The words used remain the language of old forms whilst the world to which they they supposedly apply is transforming into something entirely new. Knowledge of words thus becomes a substitute for knowledge of the world.
Incidence It is claimed that much contemporary developmental thinking is in effect of a scholastic nature and out of touch with the realities of a world in crisis.
Claim Here therefore is the first distemper of learning, when men study words and not matter. For words are but the images of matter; and except they have life of reason and invention, to fall in love with them is all one as to fall in love with a picture. (Francis Bacon).
Broader Inadequacy of prevailing mental structures to challenge of human survival (#PF7713).
Aggravates Developmentalism (#PF9512).

♦ PF9066 Declining community confidence in its ability to change
Limited village self-confidence
Broader Obstacles to leadership (#PF7011) Minimal citizen confidence (#PF8076)
Obstacles to community achievement (#PF7118).
Aggravates Lack of self-confidence (#PF0879) Ambiguous shape of social identity (#PF6516).
Aggravated by Oppressive prevalent images (#PF1365)
Demeaning community self-image (#PF2093)
Cumulative depletion of corporate initiative in rural communities (#PE3296).

♦ PF9068 Unrecognized relevance of education
Unrecognized school consequences — Unrecognized values of education — Low educational priority — Low educational priorities — Low high school priority
Broader Obsolete educational values (#PF8161)
Ineffective educational policy decisions (#PF2447)
Fragmented pattern of community organization (#PF6525).
Narrower Undervaluation of education by parents (#PF9306)
Unrecognized need for functional skills (#PF2995)
Unperceived relevance of formal education in rural communities (#PF1944).
Aggravates Inadequate care of community space (#PF2346)
Unperceived educational opportunities (#PJ9762)
Inadequate results of formal schooling (#PF6467)
Minimal opportunities of adult training (#PF6531)
Limited development of functional abilities (#PF1332).
Aggravated by Untransposed community structures (#PF6450)
Ineffective methods of practical education (#PF2721).

♦ PF9070 Symbols unrelated to human experience
Nature When symbols, rites, ceremonies, myths stop pointing beyond themselves to human experiences of reality they become meaningless gestures repeated for the sake of repetition. The vast majority of symbols are constantly being redefined and re-invested with meaning and those that are not are soon dropped from use. In some few cases, usually because of especially rigid social systems or extremely powerful connotations, symbols remain in use without a clear understanding of their relationship to human experience.
Broader Illusion (#PA6414).
Related Animal worship as a barrier to development (#PD2330).
Aggravates Loss of cultural identity (#PF9005)
Lack of commitment to common symbols (#PE8814).
Aggravated by Deterioration of human environment (#PC8943).

♦ PF9077 Jurisdictional conflict between academic disciplines
Inadmissability of evidence from other disciplines — Opposition to information from other disciplines
Broader Evasion of issues (#PF7431).
Aggravates Organizational empire-building (#PF1232)
Placement of the burden of proof on the disempowered (#PF3918).
Aggravated by Fragmentation of academic disciplines (#PF8868).

♦ PF9078 Fear of reprisals
Fear of harassment — Fear of retaliation — Deterrence — Threat of retaliation
Refs Byers, R B (Ed) *Deterrence in the 1980s* (1984); Byers, R B (Ed) *Deterrence in the 1980's* (1985).
Broader Fear (#PA6030).
Narrower Fear of retaliation by authorities (#PF3707).
Aggravates Lack of political will (#PC5180)
Obstacles to unilateral nuclear disarmament (#PF7052).
Aggravated by Retaliation (#PF9181) Absentee ownership (#PD2338)
Police intimidation (#PD0736) Unethical practices of employers (#PD2879).
Reduced by Nuclear freeze (#PF0679).

♦ PF9081 Non-recognition of international law
Disrespect for international law
Broader Unfulfilled treaty obligations (#PF2497)
Communist opposition to international organizations (#PF3162).
Aggravated by Inadequate development of international criminal law (#PF4676).

♦ PF9098 Low self image due to illiteracy
Broader Low self esteem (#PF5354).
Aggravates Lack of self-confidence (#PF0879)
Paralyzing patterns between villages and administrative structures (#PF1389).
Aggravated by Illiteracy (#PC0210) Geographical illiteracy (#PD3984).

♦ PF9100 Overgeneralized policy models
Broader Lack of appreciation of cultural differences (#PF2679)
Inadequate models of socio-economic development (#PF9576).
Aggravates Inappropriate policies (#PF5645)
Lack of long-term development assistance (#PF5181).

♦ PF9101 Prohibitive cost of inadequate development policies
Broader Lack of appreciation of cultural differences (#PF2679).
Aggravates Excessive foreign public debt of developing countries (#PD2133).
Aggravated by Inappropriate policies (#PF5645).

♦ PF9106 Interruption risk
Nature Business is conducted to earn a return for its owners or at least to recoup the expenses incurred. Those expenses cannot be terminated immediately if a business stops due to fire, theft, breakdown or other form of interruption. Also some expenses not diminish proportionately to lower sales during partial interruption. This financial strain rapidly erodes a net profit into a loss.
Broader Risk (#PF7580).
Related Dishonest employees (#PD9397).
Aggravated by Fires (#PD8054) Lightning (#PD1292)
Unpaid wages (#PD5335) Property damage (#PD5859)
Mechanical failure (#PC1904) Radiation accidents (#PD1949)
Accidental explosions (#PE3153) Inadequate safeguards against fire (#PD1631).

♦ PF9108 Decline in government expenditure
Decline in public spending — Limited public funds — Difficult public funding — Elusive public funding
Narrower Decline in government social expenditure (#PF0611).
Aggravated by Limited access to external resources (#PF1653)
Excessive growth of social expenditure (#PC6215)
Limited accountability of public services (#PF6574)
Lack of channels for obtaining available local funding (#PF6544).

♦ PF9109 Sex
Dependence on sex
Aggravates Unsafe sex (#PE9776) Sexual craving (#PE7031)
Heart diseases (#PD0448) Sexual deviation (#PD2198)
Sexual immorality (#PF2687) Sexual unfulfilment (#PF3260)
Human sexual disorders (#PD8016) Coitus as a cancer risk (#PD6033).

♦ PF9119 Insensitivity to non-immediate hazards to society
Obsession with immediate problems — Preoccupation with short-term problems
Nature There is a basic mismatch between people's primitive, limited and distorted capacity to perceive what is happening in their immediate environment in comparison with the urgent need to find ways of directing and controlling the tumultuous changes which scientific and technological ingenuity has enabled them to initiate. The relatively recent, and rapid, cultural evolution involving intellectual, social and political innovation, has enabled people to plunder an environment whose resources have been generated over millions of years. This has made the much slower processes of biological evolution completely inadequate as a means through which humans can adapt appropriately to their environment. As a consequence there is an almost universal tendency to focus on the more immediate short-term problems which can be effectively grasped, articulated, communicated and acted upon. Those problems requiring sensitivity to, and detection or analysis of, an underlying long-term trend or one involving distant factors, are avoided, often deliberately from political or economic expedience.
Refs Ornstein, Robert and Ehrlich, Paul *New World, New Mind* (1989).
Narrower Incompatibility of environmental and economic decision-making (#PF9728).
Aggravates Inaction on problems (#PB1423) Government insensitivity (#PF2808)
Unimaginative educational vision (#PF3007) Preoccupation with isolated problems (#PF6580)
Lack of long-term development assistance (#PF5181)
Irresponsibility towards future generations (#PF9455)
Recurrence of misapprehended world problems (#PF7027)
Short range planning for long-term development (#PF5660)
Institutional preoccupation with obsolete problems (#PJ5014)
Absence of a long-range, world-wide capital flow plan (#PF2865).
Aggravated by Inadequacy of prevailing mental structures to challenge of human survival (#PF7713).

♦ PF9127 Fear of new technology
Fear of technological innovation — Techno-fear — Techno-phobia — Irrational rejection of technological change
Broader Fear (#PA6030).
Aggravates Fear of the unknown (#PF6188) Disincentives to invention (#PG6623).

♦ PF9129 Fragmented regional cooperation
Inadequate regional integration
Broader Lack of international cooperation (#PF0817).
Narrower Weakness in trade among developing countries (#PC0933)
Weakness in intra-regional trade of developing countries (#PD0169)
Inadequate economic integration between regional groupings of developing countries (#PD9412).
Aggravates Regional disparities (#PC2049) Limited access to society's resources (#PF6573)

PF9129

Rivalry and disunity within developing regions (#PD0110)
Lack of channels for obtaining available local funding (#PF6544)
Inadequate inter-regional cooperation on problems within countries (#PE8307).
Aggravated by Inadequate coordination between regional intergovernmental organizations with common membership (#PE1184)
Discrimination against developing countries by the formation of regional groupings of developed countries (#PE1604)
Jurisdictional conflict and antagonism between regional intergovernmental organizations with common membership (#PE1583).
Reduces Vulnerability of small nations to foreign intervention (#PD2374).

♦ PF9138 Desecration of symbols
Desecration of flags
Claim Flag is a symbol of a country and if it is desecrated, also the principles of that country are desecrated.
Counter-claim Flag burning is a form of symbolic speech and is protected by the freedom of expression.
Broader Desecration (#PF9176).

♦ PF9139 Antichrist
Incarnate devil — Eschatological enemy — al-Dajal
Nature A god-opposing tyrant and ruler, the final opponent of good, believed to emerge in the final days of civilization. A modified belief holds that the Antichrist is a seductive agency who works by signs and wonders and seeks to obtain divine worship.
Background This belief dates back to Jewish apocalyptic literature, but probably has its roots in the mythological and speculative idea of the final battle between God and the Devil, originating in Persian eschatology. The idea of the Antichrist itself emerged in the 2nd century BC in the Book of Daniel. Christianity took over from Judaism these ideas which are reflected in certain passages in the Bible (Revelations 11 and 13). In Christian interpretations hostile to Judaism, the Antichrist was associated with a (false) Messiah (prophesized by Judaism as the awaited Messiah), in opposition to the true Messiah (of the Christians). Belief in a final opponent of the Messiah survived in later Judaism in descriptions of the legendary persecuting king Armilus, as had been the case with respect to the Roman Emperor Nero. From the 12th century, encouraged by Franciscans, the Antichrist or his forerunner was identified with every ecclesiastical, political, national or social opponent as a standard phrase of opprobrium. In particular, the belief that the Pope of Rome was the Antichrist became of widespread historical importance. This idea was further developed during the time of Luther and contributed significantly to his opposition to the Pope, becoming an article of faith. The institution of the papacy, not individual popes, were identified with the Antichrist. Following the Reformation this idea gradually receded into the background, although maintained by Protestant scholars. The Antichrist myth has had a potent influence on belief, theology, literature, politics and art.
Incidence Belief in the threat of the Antichrist continues to be sustained within certain Protestant churches and popular culture, although for many it has become a symbol of the evil in the human heart. Historical figures such as Napoleon and Hitler, as well as social movements such as socialism and communism, continue to be identified as the Antichrist or his manifestation.
Refs Bousset, Wilhelm *The Antichrist legend*; Hartman, Sven S; Böcher, Otto and Benrath, G A *Antichrist*.
Broader Personified evil (#PF7018).
Related Messianism (#PJ8830).
Aggravates Religious deception (#PF3495)
Unreadiness for second coming of Christ (#PF0638).
Aggravated by Catholicism (#PF8071).

♦ PF9141 Unreported research
Broader Unrecorded knowledge (#PF5728).
Aggravated by Secrecy (#PA0005) Deceptive misuse of research (#PD7231).

♦ PF9144 Lack of social conscience
Broader Discrepancies in human life evaluation (#PF1191).
Aggravates Apartheid (#PE3681) Killing of plants (#PD4217).
Aggravated by Amoralism (#PF3349).

♦ PF9148 Underreported issues
Unreported problems — Inadequately publicized issues
Broader Avoidance of negative feedback (#PF5311)
General obstacles to problem alleviation (#PF0631).
Narrower Unreported illness (#PF8090) Unreported violence (#PF4967)
Unreported scandals (#PF5340) Unreported accidents (#PF2887)
Unreported disasters (#PF7768) Unreported disagreement (#PF7890)
Unreported tax obligations (#PE9061) Unreported financial losses (#PF8079).
Related Unreported rape (#PE5621) Unreported crimes (#PF1456)
Unreported harassment (#PF4729) Unreported government spending (#PF2990).
Aggravates Issue avoidance (#PF1623) Evasion of issues (#PF7431)
Institutional preoccupation with obsolete problems (#PJ5014).
Aggravated by Suppression of information (#PD9146)
Declining sense of community (#PF2575)
Conceptual repression of problems (#PF5210).

♦ PF9149 Rejection of proposals for social change
Rejection of peace proposals — Rejection of remedial programmes — Rejectionism
Broader Resistance to change (#PF0557).
Aggravated by Government resistance to institutional change (#PF0845).

♦ PF9150 Contamination of human body
Human internal pollution
Nature It is brought about by the ingestion and inhalation of the products of our polluted external environment, by daily intake of chemical additives and impurities pre-packed into the food and drinks, and by unnecessary dosing of medicaments.
Broader Environmental pollution (#PB1166).
Aggravated by Food pollution (#PD5605) Contamination of drinking water (#PD0235).

♦ PF9157 Official evasion of complaints
Government avoidance of complaints
Broader Irresponsibility (#PA8658).
Aggravates Government inaction (#PC3950) Breakdown of police protection (#PF8652)
Inadequate response to societal needs (#PD1080).
Aggravated by Denial of right of complaint (#PD7609)
Proliferation of public sector institutions (#PF4739).

♦ PF9161 Dualism
Nature The operating understanding that mind and body are two distinct attributes of humanity, that the cosmos consists of heaven and hell and that moral choice is between right and wrong or good and bad. This division has had a profound effect on Western thought. It is the source of the belief that humans are isolated egos inside a body, mental activity is more important than manual labour and there is always a right choice in every situation.
Broader Inept theoretical models (#PF7182)
Inadequacy of prevailing mental structures to challenge of human survival (#PF7713).

♦ PF9166 Uncoordinated policy-making
Fragmented policy making — Inconsistent policy making — Contradictory decision-making — Fragmented resource planning
Broader Fragmented planning of community life (#PF2813)
Self-destructive government policy-making (#PF5061).
Narrower Unpredictable governmental policy (#PF1559)
Uncoordinated government policy-making (#PF7619).
Aggravates Inappropriate policies (#PF5645)
Mismatch of national macroeconomic policies among industrialized countries (#PF5000).
Aggravated by Conflicting priorities (#PF5766) Institutional fragmentation (#PC3915)
Irresponsible delimitation of policy responsibilities (#PF7823)
Failure to integrate knowledge to empower humanity in response to the global problematique (#PF8753).

♦ PF9167 Aerial explosions of unknown origin
Broader Accidental explosions (#PE3153).
Aggravates Hostile environmental modification (#PD7941).

♦ PF9169 Annihilation
Broader Death (#PA7055) Destruction (#PA6542)
Human destructiveness (#PA0832).
Related Nonexistence (#PA5870) Death of living creatures (#PF7043).
Aggravates Chaos (#PF6836) Vulnerability of organisms (#PB5658).
Aggravated by Disasters (#PB3561).

♦ PF9175 Inadequate power of intergovernmental organizations
Powerlessness of international organizations — Lack of enforcement power of international organizations — Intrinsic weakness of international organizations
Broader Limited political power (#PG9865).
Aggravates Government inaction (#PC3950)
Inability to resolve problems realistically (#PF8435)
Inadequate technical cooperation on problems (#PF0863)
Inadequate planning of action against problems (#PF1467)
Government delay in response to symptoms of problems (#PF6707)
Inadequate delivery mechanism in response to problems (#PF0301)
Inadequate research on proposed solutions to problems (#PF1572)
Inadequate legislation relating to action against problems (#PF1645)
Shortage of financial resources for action against problems (#PF0404)
Inadequate organizational mechanisms to act against problems (#PF2431)
Ineffectiveness of international organizations and programmes (#PF1074)
Inadequate application of available knowledge to solve problems (#PF8191)
Inadequate implementation of plans and programmes against problems (#PF1010)
Inadequate inter-regional cooperation on problems within countries (#PE8307).
Aggravated by Power politics (#PB3202) Lack of political will (#PC5180)
Lack of international cooperation (#PF0817)
Unwillingness to sacrifice political power (#PF7163)
Inadequate global consensus concerning problems and prospects of humanity (#PF9821).
Reduced by Erosion of sovereignty (#PE5015).

♦ PF9176 Desecration
Broader Sacrilege (#PF0662).
Narrower Desecration of symbols (#PF9138) Desecration of monuments (#PD4348)
Desecration of cemeteries (#PD7258) Desecration of holy spaces (#PF6385)
Profanation of sacred doctrine (#PF7484) Desecration of religious buildings (#PD7278).
Related Impiety (#PA6058) Wrongness (#PA7280) Human violence (#PA0429)
Inappropriateness (#PA6852).

♦ PF9181 Retaliation
Reprisals
Broader Vengeance (#PA6606) Human violence (#PA0429).
Narrower Feuds (#PE8210) Military reprisals (#PJ4986) Economic retaliation (#PD9389)
Fear of retaliation by authorities (#PF3707) Retaliation against public servants (#PD5399).
Related Harassment (#PC8558).
Aggravates Fear of reprisals (#PF9078).
Aggravated by Terrorist bombing (#PE2368).

♦ PF9184 Unaccountable government intelligence agencies
Covert cooperation between secret services — Unreported cooperation between intelligence agencies — Secret government security services
Claim Unaccountability on the part of government security services breeds arrogance and abuse, allowing them with impunity to brand as subversives all manner of people. Such services may place people under unlawful surveillance many legitimate dissenters whose only "crime" was to take part in some form of democratic protest. Abuses may include the pursuit of personal vendettas by particular agents. Individuals may be harassed throughout their working life, especially in relation to jobs subject to government vetting or advice from security services — without any form of appeal.
Refs Heilbrunn, Otto *The Soviet Secret Services* (1981); Richelson, Jeffrey *The US Intelligence Community* (1988).
Broader Social unaccountability (#PC1522) Excessive government control (#PF0304).
Narrower Secret police (#PE6331) Covert intelligence agency operations (#PD4501)
Secret intelligence agents in public office (#PE9857).
Related Espionage (#PC2140) Failure of government intelligence services (#PF8819)
Misuse of postal surveillance by governments (#PD2683).
Aggravates Informers (#PD8926) Secret trials (#PG3518)
Industrial espionage (#PC2921) Government arrogance (#PF8820)
Government propaganda (#PC3074) Political surveillance (#PD8871)
Abuse of government power (#PC9104)
Secret government security vetting of job applicants (#PE9441)
Official cover-up of government harassment of political activists (#PF3819).
Aggravated by Official secrecy (#PC1812) Foreign military presence (#PD3496).

♦ PF9187 Inappropriate cash crop policy
Loss of land to plantations — Discouragement of food crops guaranteeing survival — Proliferation of cash crops — Single cash crop — Non-diversification of market crops — Export cropping
Nature Although it is generally agreed that diversifying the crops produced by a community is necessary in order to maintain steady annual income, most rural communities raise only a few, or often a single, cash crop. This is generally because its easy marketability and stable price continue to provide the villagers with a basic income. There is deep apprehension of the risks involved in starting new crops when no-one is sure they can be grown or marketed, despite the fact that sometimes the cost of marketing what is produced at present virtually wipes out any profit, keeping family income at a subsistence level.

Broader Agricultural mismanagement (#PD8625) Insufficient diversification (#PD0335).
Narrower Preponderance of non-food crops in tropical economies (#PD7565).
Aggravates Narrow profit margin (#PJ9737)
Cultivation of marginal agricultural land (#PD4273)
Degradation of agricultural land by cash crops (#PE8324)
Prohibitive cost of necessities in rural communities (#PF2385)
Disruption of ecosystems in marginal agricultural lands (#PD6960)
Deterioration of domestic food production in developing countries (#PD5092).
Aggravated by Unethical food practices (#PD1045)
Environmentally harmful dam construction (#PD9515).
Reduces Insufficient cash crops (#PG0569).

♦ **PF9190 Denial of rites to the dead**
Nature In different cultures the normal rites have been or are denied from babes and children, slaves and common people, people who commit a suicide or die a "bad death" of drowning, some diseases, accidents etc., women dying in childbed, people who have died in debt, and sacred persons held in reverence.
Broader Absence of rites of passage (#PF1674).
Narrower Funerals for dead foetuses (#PS4035).
Aggravated by Dying a bad death (#PF1421)   Undignified treatment of corpses (#PF5857).

♦ **PF9195 Communion with the dead**
Spiritualism — Spiritism — Cult of the dead
Broader Superstition (#PA0430).
Narrower Ancestor worship (#PD2315).

♦ **PF9201 Intraspecific competition**
Competition within species
Broader Struggle for existence (#PB4411).
Related Interspecific competition (#PF9275).

♦ **PF9205 Over-reliance on convenient beliefs**
Facile beliefs — Unchallenging beliefs — Thoughtless acceptance of convenient beliefs
Broader Inadequacy of prevailing mental structures to challenge of human survival (#PF7713).
Aggravates Politicization of scholarship (#PF7220)
Limitations of surprise-free thinking (#PF7700).
Aggravated by Intellectual arrogance (#PF7847).

♦ **PF9208 Constraint of inherited problems on policy innovation**
Aggravates Inappropriate policies (#PF5645)
Deficiency in innovation in the public sector (#PF7230)
Institutional preoccupation with obsolete problems (#PJ5014).

♦ **PF9209 Fear of emotional sensitivity**
Fear of emotionally responsive behaviour — Fear of sentimentality — Fear of tenderness
Nature People experience considerable embarrassment at displays of overt or straightforward affection between adults. This creates artificial emotional barriers in relationships between adults and separates fathers from children.
Incidence Tenderness between adults in many cultures is only tolerated as a precursor to sex. Affection is disguised by jokes and mock abuse. The problem is usually stronger in men because they are more violently and more permanently forced away from all things tender.
Claim Humans are born without an instinct for aggression, but with an instinct for companionship which is just as, if not more important than, the sexual drive. Children are wrenched away from their early experiences of companionship with their mothers. This loss is so painful that no reminder of it is tolerable. Anything infantile, gentle, sentimental, including unguarded emotions, becomes taboo. Rather than cultivating mature minds, this engenders hardness and cynicism with a core of anxious, angry infantility. Every form of fondness is then tainted by implications of sexual proclivity, especially homosexuality.
Broader Fear (#PA6030).
Aggravates Sentimentalism (#PF6961)   Fear of intimacy (#PF8012)
Sexual discrimination (#PC2022)   Numbness towards others (#PF1216).
Aggravated by Homosexuality (#PF3242).

♦ **PF9216 Policy cross-conditionality restrictions in multilateral development aid**
Nature The severity of the problems facing the world economy has meant that many developing countries have sought the assistance of international agencies in meeting their immediate balance-of-payments needs. This assistance has been rendered on conditional terms, and the "conditionality" attached to the lending has shaped the nature and content of the adjustment programmes adopted by the recipient countries. The common thread running through the various structural adjustment programmes is the premise that a properly functioning market-economy experiences few market failures and that existing market imperfections are mainly due to policy interventions. Removing market distortions, changing the public and private sector balance and market development are the measures proposed by the international agencies to develop the market system. This approach assumes a structural framework of highly integrated markets, with ease of resource flow from sector to sector and price flexibility. But the economic structures of developing countries are quite different and, as a result, the application of a "standard" market-oriented structural adjustment policy can produce negative results quite at variance with the desired outcome.
Background The main thrust of the new trend in the World Bank's policy-based lending relates to the market-oriented approach to development problems. At a general level, there is widespread agreement on the need for developing countries with mixed or market-oriented economies to ensure appropriate incentive structures to economic agents, secure adequate rates of domestic savings and productive investment and maintain appropriate levels of real exchange rates. The operational implications are, however, much less clear and there is much debate concerning the relative importance of the variables targeted under this approach, such as the role of price incentives versus the removal of non-price impediments in stimulating agricultural production; the emphasis on export promotion versus import substitution; the role of interest rates in raising the savings rate; the impact of changes in the external environment on the domestic policy framework. More controversial aspects under this approach are the respect roles of and balance between public and private sector involvement in economic activities, government policies toward foreign investment and income distribution.
Incidence Aid, especially debt relief, to developing countries may only be made available under conditions whereby the country subjects itself to the constraints of an IMF programme. The constraints tend to be such as to prevent the countries maintaining growth, where priority is given by the IMF to restore imbalances of payments at all costs, even when such imbalances are of a short-term character. Since the creditworthiness of developing countries is evaluated by the World Bank (and by commercial banks) on the basis of a country's standing with the IMF, failure to enter into such agreement deprives a country of access to other sources of finance. Issues of policy cross-conditionality associated with Bank-supported adjustment programmes include: the adequacy of the policy package itself, the timing and sequencing of the programmes, and the collaboration between the Bank and the IMF.

In addition to the character and nature of the measures themselves, issues relating to the timing and phasing of implementation may also have negative consequences. The requirement of rapid major structural reforms and balance-of-payments adjustment may be beyond the reach of most poorer countries; political and administrative systems may not be adequate to deal with the speed at which the formulation and implementation of policy changes is advocated. There are also questions relating to policy sustainability: rigid policies phased over an insufficient time period enhance the potential for slippage.
A further specific issue is that of interdependent cross-conditionality, a situation where both the Bank and the Fund select the same policy instrument (eg the exchange rate) as the key element in their respective programmes. Developing countries seeking financial assistance are then exposed to concerted pressure and suffer from negative consequences of inter-agency cooperation.
The expansion of the World Bank's policy-based lending and the establishment of the IMF's Structural Adjustment Facility has led to a more intensive collaboration between the two institutions and the division of labour between them is no longer clear-cut. Although formal cross-conditionality does not occur between the Bretton Woods institutions, informal cross-conditionality arises in a number of cases, especially through the linkage between World Bank policy-based lending and IMF standby arrangements. In fact only three of all the World Bank sector loans approved from 1979 through 1985 occurred in countries where an IMF stabilization programme was not in place. The IMF compensatory financing facility was redesigned in 1982 from a more liberal low-conditionality form to a high-conditionality form. **Counter-claim** Where IMF adjustment programmes involve the cutting back of production and income, sacrificing growth, this is not due to any IMF predisposition to favour the adoption of deflationary measures but rather to limitations on the external resources available to the countries concerned.
Broader Inadequacy of aid to developing countries (#PF0392)
Inappropriate aid to developing countries (#PF8120)
Restrictions imposed on aid to developing countries (#PF1492).
Aggravates Donor distortion of aid to developing countries (#PF7955)
Developing country dependence on a single source of finance (#PE4188)
Excessive social costs of structural adjustment in debtor developing countries (#PD8114).
Aggravated by Economic inefficiency (#PF7556).

♦ **PF9234 Fear of humiliation by co-workers**
Broader Fragmented conduct of community operations (#PF1205).
Aggravates Humiliation (#PF3856).
Aggravated by Fear of contradicting popular views (#PF2040).

♦ **PF9239 Misrepresentation of geographical information**
Production of deliberately distorted maps — Geographical deception — Withholding of geographical information
Broader Official secrecy (#PC1812)   Withholding of information (#PF8536).
Aggravates Disinformation (#PB7606)
Inadequate international map of the world (#PD0398).
Aggravated by Inadequate cartographic skills in developing countries (#PJ8291).

♦ **PF9243 Falsification of programme evaluations**
Misrepresentation of project assessments — Manipulation of project cost-benefit analyses — False justification of projects
Broader Misleading information (#PF3096).
Narrower Overstated programme advantages (#PF8181)
Underestimation of programme costs (#PF8499).
Aggravates Minimization of problems (#PF7186).

♦ **PF9244 Inadequate control of development projects**
Inadequate monitoring of development projects — Inadequate evaluation of development projects
Broader Ineffective monitoring (#PF2793).
Aggravates Destruction inherent in development (#PF4829)
Excessive foreign public debt of developing countries (#PD2133).
Aggravated by Inappropriate management of development projects (#PD3712).

♦ **PF9247 Geographically disadvantaged countries**
Broader Geographic barriers for least developed countries (#PE8049).
Narrower Arid developing countries (#PD8418)   Non-viability of cold countries (#PF8663)
Non-viability of small states and territories (#PD0441)
Vulnerability of land-locked developing countries (#PD5788)
Vulnerability of island developing countries and territories (#PE5700).
Aggravates Inadequate development of enterprises in developing countries (#PE8572).

♦ **PF9249 Disincentive to invest in heavily indebted countries**
Debt overhang
Nature The debt overhang remains an obstacle to growth in debtor countries and a threat to the world economy. Where countries are unable to service their debt in full, actual payments tend to depend upon a countryRs economic performance. If exports increase as a result of policy reforms or further investment, most of the benefit usually then accrues to creditors (in the form of larger debt-service payments) rather than to the country itself. This depresses the returns to the country from fixed capital investment and thereby weakens the incentive to invest even if finance is available. Even in those cases when all of the country's foreign debt is held by the government, the disincentive effects of debt overhang will spill over into private investment, because the government has then no reason to encourage investment of any kind when most of the gains will be absorbed by higher debt payments. Such disincentive effects may be so strong that they give rise to a condition in which forgiveness by creditors could even raise the actual repayments they receive.
Incidence Investment in heavily indebted countries has been weak since 1982, both by historical standards and in relation to other countries. For example in Argentina the ration of investment to GDP fell from 25% to 15% between 1982 and 1988, and in Venezuela from 33% to 18%. By comparison, in a range of developing countries that are free of debt problems, the investment ratio typically increased over that period. Preliminary analysis by the IMF in 1989 concludes that such disincentive effects have contributed significantly to the fall in investment in debtor countries. In the 73 developing countries experiencing recent debt problems, the investment ratio fell from 27% in the late 1970s to 18% in 1988. In those without problems, the ratio rose from 26% to 28%.
Broader Disincentives for financial investment within developing countries (#PF3845).
Aggravates Decline in foreign direct investment in developing countries (#PD3138).
Aggravated by Burden of servicing foreign public debt by developing countries (#PD3051).

♦ **PF9266 Inefficiency of government in developing countries**
Nature The structure of local government in many developing countries is inefficient. Often fiscal relations are opaque because of political expediency rather than lack of knowledge or skill. This makes reform much more difficult when more open and transparent systems are urgently needed.
Broader Government inefficiency (#PF8491).

♦ **PF9269 Planned weapons**
Envisaged weapons — Anti-matter bombs — Transuranic weapons — Infrasonic radiation weapons — Genetic weapons
Broader Weapons (#PD0658).
Narrower Genetic and ethnic weapons (#PC6664).

♦ **PF9274 Erosion of religious belief by ecumenical dialogue**
Claim 1. Vibrant faith does not dialogue. It proclaims, instructs, judges and persuades. Ecumenical dialogue is a liberal scheme aimed at the evisceration of true religion. It is a substitute for doctrine and dogma. 2. The chief danger of religious dialogue is the desire to come to a conclusion in anticipation of that which can only emerge through the processes engendered by the creator. 3. Religious dialogue with other faiths is a betrayal of mission.
Counter-claim True dialogue attends to distinctiveness, knowing that disagreements are often of much greater interest than agreements and that agreements are discovered and tested by working through disagreements. True believers cannot have their faith shaken by dialogue only strengthened.
Broader Ineffective dialogue (#PF1654).
Related Ineffective religious dialogue (#PF1618).
Aggravates Religious syncretism (#PF7079).

♦ **PF9275 Interspecific competition**
Competition between species
Broader Struggle for existence (#PB4411).
Related Intraspecific competition (#PF9201).

♦ **PF9276 Unknowable future patterns of social choice**
Indeterminate future priorities — Indeterminate future human values
Claim Detailed future predictions are unfeasible because of the complexity of the environmental syndromes arising from interactions with development, the dependence of these interactions on unknowable future patterns of social choice and evolution, and the incomplete state of the sciences required for their assessment.
Broader Uncertainty (#PA6438).
Aggravates Unpreparedness (#PF8176) Inaccurate forecasting (#PF4774)
Unpredictable governmental policy (#PF1559)
Irresponsibility towards future generations (#PF9455)
Inhibited capacity to visualize a creative future (#PF2352).
Aggravated by Temporal deprivation (#PF4644).

♦ **PF9281 Exploitative transformation of international division of labour**
Refs Törnqvist, G et al (Ed) *Division of Labour, Specialization and Technical Change* (1987).
Broader International economic injustice (#PC9112).
Narrower Transfer of highly polluting industries to developing countries (#PF8921).

♦ **PF9288 Connivance of authorities in human rights abuses**
Government connivance in human rights abuses — Connivance of religious leaders in human rights abuses — Official connivance in human rights abuses
Incidence The most frequently cited example is that of the connivance of religious, military and government leaders and others in authority who assisted Nazi war criminals to escape justice.
Aggravates Ineffective war crime prosecution (#PD1464)
Impunity of violators of human rights (#PF3474)
Inadequate enforcement of human rights (#PC4608)
Government support for repressive regimes (#PF4821)
Limited acceptance of human rights treaties (#PE7300)
Government inaction on alleged human rights violations (#PE1407)
Official cover-up of government harassment of political activists (#PF3819).

♦ **PF9292 Limited human information processing capacity**
Aggravates Human errors and miscalculations (#PF3702).

♦ **PF9298 Loss of information**
Loss of data
Aggravates Lack of information (#PF6337).
Aggravated by Suppression of information (#PD9146).

♦ **PF9299 Inadequate legislation against environmental pollution**
Lack of regulations against environmental pollution
Broader Ineffective legislation (#PC9513).
Narrower Inadequate legislation against environmental pollution in developing countries (#PE7141).
Aggravates Resistance to internationally agreed standards (#PC4591)
Neglect of environmental consequences of government policies (#PE9295).
Aggravated by Inadequacy of international legislation (#PF0228).

♦ **PF9306 Undervaluation of education by parents**
Schooling blocked by family
Broader Obsolete educational values (#PF8161) Unrecognized relevance of education (#PF9068).
Aggravates Limited availability of education in rural areas (#PF3575).
Aggravated by Haphazard forms of social ethics (#PF1249).

♦ **PF9307 Infidelity to God**
Broader Sin (#PF0641).
Aggravates Wrath of God (#PF8563).

♦ **PF9312 Glorification of war**
Unrealistic portrayal of war
Nature War, and the experience of war, tends to be portrayed through manifold layers of comforting illusion built up like strips of emotional insulation around the plain, nearly unbearable abomination of what actually occurs. War is characterized by many fatal errors, gunners shooting down their own airplanes, planes bombing their own troops. It includes the deliberate shooting of officers by their own troops, sadistic behaviour, dismemberment of bodies, acts of cowardice and betrayal, a smutty cynicism on the part of the fighting men, and the unnecessary slaughter of civilians. The many instances of unreason, accident and contingency tend to be rationalized subsequently as design. Much of the reality of war is obscured behind a retrospective campaign of prettification and glorification. This obscures the tendency of war to descend to a kind of generalized barbarism, to acts of slaughter like the firebombing of Dresden, that have no military value and thus lack any moral meaning. And even without these things war is at its best killing, maiming and wounding.
Claim There has been so much talk of the good war, the justified war, the necessary war, that the young and the innocent could get the impression that it was not really such a bad thing after all. However it is essentially stupid and sadistic. It takes a special honesty, even if that honesty arises from despair, to perceive that some events, being inhuman, have no human meaning.
Counter-claim Without the sacrifice of those who die in such terrible circumstances, the nations would be cowed or crushed by tyranny.
Broader Culture of violence (#PD6279).
Aggravates Enjoyment of war (#PF4034).
Aggravated by War and pre-war propaganda (#PD3092).

♦ **PF9323 Uncorporate family decisions**
Nature The usually simplistic decision-making structures within the family do not consciously aim to make binding the family's stance on a particular matter which, over a given period of time, will hold the members accountable to this stance. Family decision-making neither intentionalizes accountability nor releases creativity; it too often excludes children and elderly people and is frequently in effect the task of single person. Such arbitrariness inhibits the creativity of the excluded members.
Broader Fragmented decision-making (#PF8448).

♦ **PF9327 Non-violent weapons**
Broader Weapons (#PD0658).
Narrower Strikes (#PD0694) Boycott (#PE8313)
Self-inflicted physical suffering (#PD7550).

♦ **PF9328 Latent problems**
Delayed emergence of problems — Delayed environmental impact of technological activities
Nature There is increasing evidence for the existence of a class of problems which are only recognizable as significant long after the causative actions have been taken. Adverse effects become manifest long after an apparently beneficial activity has become well established. The social costs of activities decades earlier, largely in ignorance of their potential consequences, suddenly become a source of great anxiety, especially when the extent of the phenomena is as yet poorly understood and denied by certain groups of experts. The problem creates unpredictable levels of anxiety when some new initiative (possibly without negative side effects) can be argued to be potentially the source of some future problem, as in the case of the recently questioned introduction of artificial sweeteners into diet soft drinks.
Incidence Examples include: appearance of cancers from exposure to very low concentrations of carcinogenic materials long after the period of that exposure; asbestos, for which the latency period may extend over much of a human life span; the emergence of lung cancer as the consequence of smoking many years before; the spread of toxic wastes into groundwater; the emergence of resistant pesticides and pathogens following cumulative years of pesticide use; recognition of the implications of increasing concentrations of carbon dioxide in the atmosphere.
Aggravates Non-recognition of problems (#PF8112)
Unforeseen environmental crises (#PF9769)
Government delay in response to symptoms of problems (#PF6707).
Aggravated by Limitations of surprise-free thinking (#PF7700).

♦ **PF9345 Loss of international technological leadership**
Broader Loss of international leadership (#PF8353).

♦ **PF9350 Religion as an opiate**
Utilitarian religion — Harmonial religion
Nature Harmonial religion is the popular reinforcement of the divorce of fact from value that makes public moral discourse almost impossible. In harmonial religion the seemingly conservative call for a return to objective values is supported by an argument that is relativistic to the core. As such it constitutes an opiate for the masses, providing a welcome escape from the need to engage questions of moral truth, whether personal or public — thus becoming the disease of which it claims to be the cure.
Background In harmonial religion, religion is reduced to ethics which in turn is reduced to everything that makes a person feel good about himself or other people. Feeling good is perceived as a consequence of being good which follows from doing good. Religious traditions are viewed as carrying the communal experiences of what makes people do, be and feel good. Civil peace (and market viability) are secured by avoiding rival claims to truth. If any particular belief enables a person to cope more effectively with misfortune and disappointment, then that is true belief, whether historically true or not. The true is thus the useful and "God" is the sum of the human community's useful truths. The religions of the Roman Empire, prior to its fall, have been described as being considered equally true by the populace, equally false by the philosophers and equally useful by the rulers.
Claim There is no doubt that society would be kinder, gentler, more generous, more peaceful and more harmonious if people lived according to the tenets of harmonial religion — except for the resulting distortion of truth.
Counter-claim 1. The purpose of religion is not to explain God or to please God, but to help people meet some of their most basic human needs. The primary purpose of religion is not to put individuals in touch with God, but rather to put them in touch with one another. 2. If people are in real pain and are in fact helped by some set of religious beliefs, however spurious, to get their lives together, it is inappropriate to quibble over the means employed.
Related Religious superstition (#PF1270).

♦ **PF9358 Biased and inaccurate biology textbooks**
Censorship of textbooks in response to creationism
Incidence A current example results from the political pressures within the predominantly Christian fundamentalist states of the USA where creationists favour the presentation of creation science (as it emerges from Biblical interpretation) taught alongside theories of biological evolution. This has led to the progressive dilution of references of evolution (treated as only a theory) in textbooks through a process of self-censorship by authors and publishers concerned to make their products acceptable to such important markets, with immediate consequences for the contents of those books also intended for distribution to other markets, in the USA and elsewhere. Thus in 1984, no mention was made of evolution in one sixth of the biology textbooks used in schools.
Broader Misleading information (#PF3096)
Inadequate and inaccurate textbooks and reference books (#PD2716).
Aggravates Ethical decay (#PB2480) Misuse of evolutionary theories (#PF3348).
Aggravated by Self censorship (#PF6080) Scientific censorship (#PD1709)
Politicization of scholarship (#PF7220).

♦ **PF9360 Inaccessible wilderness areas**
Inaccessible park lands
Claim In order to halt the deterioration of parks and wildernesses, it is necessary to put a lid on a number of visitors.
Broader Inaccessible recreation areas (#PF6503)
Fragmented conduct of community operations (#PF1205).
Narrower Inaccessibility of countryside to city dwellers (#PF6140).
Reduces Over-use of designated wilderness areas (#PD7585).

♦ **PF9375 Hoaxes**
Refs Wade, Carlson *Great Hoaxes and Famous Imposters*.
Narrower Scientific fraud (#PF1602) Unethical practices of priesthood (#PF8889).
Aggravates False alarms (#PF4298).

♦ **PF9385 Criticism of official institutions**
Criticism of government — Defamation of institutions — Criticism of national leadership
Broader Criticism (#PF4530).
Aggravated by Intolerance of criticism (#PF8396)
Institutional preoccupation with obsolete problems (#PJ5014).

♦ **PF9401 Inadequate health control**
Inadequate disease prevention — Health control
Broader Inadequate health services (#PD4790).
Narrower Inadequate quarantine (#PE2850) Diseases of wild animals (#PD2776)
Neglected health practices (#PD8607) Uncontrolled plant diseases (#PJ1016)
Uncontrolled tropical diseases (#PD4775)
Inadequate control of animal diseases (#PD2781).
Related Inadequate drug control (#PC0231).
Aggravates Zoonoses (#PD1770) Legal impediments to marriage (#PF3346)
Unavailable education for effective living (#PF2313).
Aggravated by Prohibitive cost of disease control (#PF2779).

♦ **PF9409 Fixation on partial solutions to problems**
Minimally cooperative solutions to problems
Nature There is a natural tendency of groups to be satisfied with solutions to problems that could be more effective and comprehensive with cooperation at a higher level of inclusiveness. The search for appropriate solutions tends therefore to end with partial solutions that do not necessitate cooperation with other groups. Fixation on the first reasonably successful partial solution thus inhibits any further search for opportunities for cooperative solutions that would have been superior and could have satisfied the mutual interests of a wider range of groups or interests.
Aggravates Lack of international cooperation (#PF0817)
Institutional preoccupation with obsolete problems (#PJ5014)
Static grassroots involvement in planning the economy (#PE4479)
Inadequate research on proposed solutions to problems (#PF1572)
Failure to integrate knowledge to empower humanity in response to the global problematique (#PF8753).
Aggravated by Complacency (#PA1742) Technological monoculture (#PF4741)
Unconvincing alternatives to existing societies (#PF3826)
Inadequate global consensus concerning problems and prospects of humanity (#PF9821).

♦ **PF9424 Excessive desire for risk**
Broader Inconsistent risk evaluation (#PF5482).

♦ **PF9428 Unrealistic policies**
Unrealistic policy-making — Government promotion of unrealistic perspectives — Political support for unsustainable policies — Political unreality — Political surrealism
Claim Increasingly a style of politics and governance is emerging based on: avoiding (or minimizing) the communication of unwelcome news, accustoming people to inconsistent policies, ensuring that every problem is someone else's fault. The result is a form of paralysis of truth in which reality about most important issues of governance (budget deficits, use of military force, foreign trade, environmental impacts, corruption, etc) is widely acknowledged but cannot be expressed by elected officials.
Broader Substitution of fantasy for reality (#PF8922).
Narrower Messianic policy-making (#PF9796)
Lack of realism in intergovernmental organizations (#PF7493).
Aggravates Inappropriate policies (#PF5645) Impoverishment of political debate (#PF4600)
Ineffective decision-making processes (#PF3709)
Self-destructive government policy-making (#PF5061)
Politically unrealistic strategic warfare analysis (#PF1214).
Aggravated by Inaccurate forecasting (#PF4774)
Unrealistic expectations (#PE7002)
Governmental incompetence (#PF3953)
Unrealistic environmentalism (#PF4510).

♦ **PF9429 Diminished personal capacity**
Incompetence to stand trial
Nature A plea of diminished capacity in legal proceedings permits the defendant to introduce evidence of temporary mental impairment at the time the crime was committed thus reducing the level of responsibility, if any, for the crime. This plea may be abused to avoid conviction or the responsibility for any reparations to victims. Incompetence to stand trial is distinguished from diminished capacity in that it is a temporary plea for postponement, such as for purposes of hospitalization. Both are to be distinguished from insanity as a permanent impairment.
Related Criminal insanity (#PD9699).
Aggravates Inadequate national law enforcement (#PE4768).

♦ **PF9431 Questionable facts**
Lack of consensus on facts — Negotiable facts — Biased interpretations of facts
Aggravates Selective perception of facts (#PF2453)
Unethical practices of interpreters (#PE6740)
Unproven relationships between problems (#PF7706).
Aggravated by Biased expertise (#PF6395) Racial inequality (#PF1199)
Disagreement among experts (#PF6012) Bias in scientific research (#PF9693)
Limitation of current scientific knowledge (#PF4014).

♦ **PF9432 Inflexibility of commodity supply**
Failure of commodity production to respond to fall in demand
Incidence In developing countries, the share of fixed costs in total costs of production of minerals appears to be higher than that in developed countries because of the latter's higher relative costs of capital equipment. Moreover, the objective in those countries of maintaining employment also in effect transforms wages partially or entirely into fixed-cost items, irrespective of the ownership of the mines. The supply of agricultural commodities also tends to be inflexible, particularly in the case of plantation and tree crops, since land costs are fixed and farmers do not have ready alternative opportunities for employment and earnings. Hence, as long as prices cover variable costs, production and sales tend to continue.
In the specific case of developing countries, the dominant factor which has aggravated the traditional inflexibility of supply in the 1980s is the negative impact of price falls on the balance-of-payments of the developing producer countries that depend so heavily on primary commodities, and on the earnings of the enterprises involved, whether public or private. This has led governments to take measures to maintain output irrespective of the ownership of the facilities.
Aggravates Over-production of commodities (#PD1465).

♦ **PF9450 Uncertain environmental impact of current policy**
Broader Uncertainty (#PA6438) Natural environment degradation (#PB5250).

♦ **PF9455 Irresponsibility towards future generations**
Irresponsibility for the long-term future of the planet — Irresponsibility towards posterity — Denial of the rights of posterity
Claim Whatever is right for our grand-children is always uneconomic and almost always impolitic.
Broader Irresponsibility (#PA8658).
Narrower Unrecognized future financial commitments (#PF4114).
Aggravates Eugenics (#PC2153)
Uncertainty of survival of the human race (#PE9085).
Aggravated by Temporal deprivation (#PF4644) Inaccurate forecasting (#PF4774)
Unknowable future patterns of social choice (#PF9276)
Short range planning for long-term development (#PF5660)
Insensitivity to non-immediate hazards to society (#PF9119)
Irresponsible delimitation of policy responsibilities (#PF7823)
Unaccountability of institutions degrading the environment (#PF3458).

♦ **PF9459 Official apathy**
Bureaucratic apathy — Governmental apathy
Broader Apathy (#PA2360).
Aggravates Government inaction (#PC3950) Bureaucratic inaction (#PC0267)
Government delay in response to symptoms of problems (#PF6707).
Aggravated by Proliferation of public sector institutions (#PF4739).

♦ **PF9464 Air traffic delays**
Incidence In Europe in July 1989 almost 30 per cent of international short- and medium-haul departures were more than 15 minutes late costing the industry about $5 billion for additional fuel and other expenses.
Aggravated by Air traffic congestion (#PD0689) Inadequate traffic control (#PE8266).

♦ **PF9474 Internationally non-cooperative governments**
Governments unwilling to cooperate with the United Nations in eliminating human rights violations
Narrower Government inaction on alleged human rights violations (#PE1407).
Aggravated by Limited acceptance of international treaties (#PF0977)
Irresponsibility of member governments of the United Nations (#PF5337).

♦ **PF9478 Unchallenging world vision**
Limited development vision — Unimaginative future planning — Unimaginative vision of the world's future — Inadequate models of an ideal society
Claim It is impossible to see an ideal pattern of economic development without difficult problems. Some costs in human hardship and misery are inevitable. It will often be necessary to work out proximate goals and least-harmful measures. Hard choices will have to be made and unavoidable risks taken.
Narrower Unimaginative city plan (#PG9940) Unimaginative facility use (#PG8552)
Unimaginative educational vision (#PF3007) Self-interested industrial vision (#PF3679)
Unimaginative vision of resource utilization (#PF1316)
Unimaginative vision of community possibilities (#PJ7764)
Unimagined possibility of expanding the preschool (#PE5304)
Unimaginative vision of existing international economic structures (#PF2699).
Related Reluctance to join in community action (#PF1735)
Individualistic retaining of local tradition (#PF1705)
Modern disruption of traditional symbol systems (#PF6461)
Traumatic shift in life-styles of mining communities (#PE1137).
Aggravates Lack of commitment (#PF1729) Lack of a world government (#PF4937)
Destruction inherent in development (#PF4829) Detrimental story of community future (#PF6575)
Divisive patterns of community groupings (#PF6545)
Limited exposure to outside influences in rural villages (#PF2296).
Aggravated by Lack of models of equality (#PF8639)
Parochial national interests (#PF2600)
Lack of international cooperation (#PF0817)
Static social relations inhibiting vision of the future (#PF2803).

♦ **PF9485 Subversion of socialism**
Subversion of communism
Aggravated by Anti-communism (#PF1826).
Reduces Communism (#PC0369) Socialism (#PC0115)
Subversion of democracy (#PD3180).

♦ **PF9493 Conspiracies of silence**
Claim The silence of the world's statesmen in the event of certain excesses on the part of one of their number is deafening and shaming. Examples include failure to protest: the excesses of nazism and stalinism, the massacres in Cambodia and Uganda, the appointment of a former Nazi to the position of Secretary-General of the United Nations, the USSR invasion of Hungary and Czechoslovakia, the crackdown on dissent in China, the USSR blockade of Lithuania; massacres in the Sudan by Islamic fundamentalists.
Aggravates Fraud (#PD0486).
Aggravated by Complicity with structural injustice (#PJ5026).

♦ **PF9512 Developmentalism**
Nature Development has become the central organizing concept in terms of which the historical movement and direction of social systems are analyzed, evaluated and acted upon. It is now a methodological principle, and analytical matrix for conceiving and interpreting social processes, a regulative idea in terms of which these processes, and the structures they give rise to, are compared, evaluated and given some sense of direction, as well as a praxeological notion for mobilizing and justifying social action in pursuit of policy objectives. As such it is the dominant myth of the current epoch. Although many alternative development theories and programmes emerge to challenge or surpass the weaknesses of their predecessors, such apparent advances obscure the enduring nature of the underlying commonality shared by such differing initiatives and which circumscribe their scope. The demands of the developmental crisis and theory diverge farther and farther from each other such that the current development debate comes to resemble more and more the scholastic debates that marked the transition from mediaeval philosophy to modern thought. The underlying commonality is that the dynamics of the world system as a whole are essentially determined by the movement of capital, namely the valorization and accumulation of capital on an ever-expanding scale. It is this process which lies behind the historical and contemporary development of the modern world-system, and it is this which developmentalism, as the philosophy of that system, expresses and subserves, resulting in unequal and uneven development as well as recurring cycles of expansion and stagnation.
Incidence In seeking to "develop" themselves via developmentalist strategies, individual countries collectively promote capital accumulation on a world scale, and hence also reproduce the contradictions inherent in the movement of capital, globally and within each of them. It is

because developmentalism refuses or is unable to recognize these fundamental contradictions, which continue to intensify crisis after crisis, that is has reached an impasse. This impasse, in turn, is nothing else but the particular expression of the crisis of the world–system itslef. The immobility of the philosophy is conditioned by the contradictions of the global society.
Claim The development of the current world crisis has led to a crisis in development itself, not only in practice, but also in theory as well. Since the mid–1970s, there have been a plethora of alternative proposals to rethink the development problématique and formulate new development concepts and strategies. The sheer quantity of the research output has been in inverse proportion to the concrete development results. **Counter–claim** The failure of developmentalism cannot be ascribed either to lack of ideas or to a dirth of creativity. Its contributions in creating a rich and diversified economic structure have been colossal. Its failure was due to: its inability to control monetary and financial imbalances; the productive structure it generated that placed great emphasis on the concentration of resources; and the approach to development was primarily economic, thus neglecting other social and political processes.
Aggravates Contradictions (#PF3667) Accumulation of capital (#PC5225).
Aggravated by Scholasticism (#PF9065).

♦ **PF9529 Cultural traditions blocking business profit**
Nature Traditional cultures which emphasis generosity, social equality, strong family ties often prevent the generation of business profits. Family members and friends may demand credit from employees of a shop or business which is never paid back. Business owners may be expected to make hugh contributions to community affairs. The shop owner as the major interface between a moneyed economy and a barter economy is frequently in the position of never making a profit.
Counter–claim In small rural communities in the process of developing businesses frequently perform far more than economic functions, They train individuals in managerial and entrepreneurial skills. They are the major focus of community information. Small shops function as the community decision making centers in hundreds of thousands of villages across the world. They may be a major link with the larger society. To look at small businesses in rural, developing communities simply in monetary terms is to misunderstand their social functions.
Broader Rigidly entrenched social traditions in rural areas (#PF1765).
Aggravated by Debilitating content of village story (#PF2168).

♦ **PF9530 Overdependence on government**
Excessive government dependence — Dependency on government funding — Dependence on government action
Narrower Dependence on social welfare (#PD1229)
Over–acceptance of socio–economic dependency (#PF8855).
Aggravated by Excessive government control (#PF0304)
Demeaning community self–image (#PF2093)
Insufficient flow of information (#PF6469)
Proliferation of public sector institutions (#PF4739)
Incomplete understanding of new societal service systems (#PF2212)
Unsystematic use of powerful relationships by rural communities (#PE1101).

♦ **PF9536 Unluckiness**
Bad luck — Unlucky people — Ill–starred people — Jinx — Misfortune
Incidence Belief in luck and the difficulties following from bad luck are widespread, both in developed and developing countries. It is a major preoccupation in certain cultures and milieu, especially those where gambling is favoured. Some people, groups or locations may be perceived as bearers of bad luck. Many superstitions relate to the possibility of bringing on bad luck. Some of these continue to be treated with concern in development related projects. In 1990, for example, special publicity was given to the role of the President of Argentina in bringing bad luck under a wide variety of circumstances. Officials of NASA declared in 1990 that their space programmes were suffering from a run of bad luck.
Refs Kokot, Waltraud *Perceived Control and the Origins of Misfortune* (1982).
Broader Superstition (#PA0430).
Aggravates Accident proneness (#PE6352).
Aggravated by Evil (#PF7042) Evil eye (#PF2343) Unlucky numbers (#PF4660) Evil enchantment (#PF5123) Inauspicious conditions (#PF6683).

♦ **PF9539 Positive discrimination**
Reverse discrimination
Nature Selecting people for merits or employment on the basis of their membership in oppressed groups, even if a member of a more privileged group is better qualified. While seen by most as a temporary measure, this practice fuels the discrimination it seeks to undo. In some cases highly qualified individuals are deprived of employment in favour of those of lesser skills.
Refs Edwards, John *Positive Discrimination and Social Justice* (1987).
Broader Discrimination (#PA0833).
Related Homophobiaphobia (#PS1199).

♦ **PF9542 Inadequate government control of military**
Disregard of government orders by armed forces — Military excesses
Related Military disobedience (#PD7225).
Aggravates Military reprisals (#PJ4986).
Aggravated by Military rivalry (#PD9252).

♦ **PF9558 Fabrication of politically sensitive incidents**
Fabrication of provocative incidents
Incidence It has been reported that right wing political groups have planted bombs in order to be able to argue that action should be taken against left wing terrorists or illegal immigrants supposedly responsible for such incidents. Similarly religious and other extremists are reported to put slogans on walls critical of themselves or of their opponents (eg the Swastika or Star of David) in order to arouse support for their cause or to provoke concern for the influence of their opponents.
Broader Promotion of negative images of opponents (#PF4133).

♦ **PF9560 Over–reliance of government on money creation**
Nature Governments can choose to finance fiscal deficits by creating money, that is by printing and spending currency. By issuing currency, governments are able to claim real resources; this claim is known as seignorage. When the rate of money creation exceeds the growth in demand for money, inflation can result. And that inflation itself worsens the deficits, because expenditures keep pace with rising prices while revenues do not. As a result still more money creation is called for, further worsening the inflationary spiral.
Aggravates Economic inflation (#PC0254).
Aggravated by Government deficits (#PD5984).

♦ **PF9561 Arrogance of intergovernmental agencies**
Broader Arrogance (#PA7646).
Aggravates Undemocratic policy–making (#PF8703)
Ineffective decision–making processes (#PF3709)
Deliberate ignorance during policy–making (#PF8278).
Aggravated by Government arrogance (#PF8820)
Arrogance of experts (#PF3294).

♦ **PF9564 Indecisive multilateralism**
Reduced government commitment to multilateralism — Ineffectual intergovernmental multilateral action
Refs United Nations *Challenge to Multilateralism* (1985).
Aggravates Lack of realism in intergovernmental organizations (#PF7493)
Intergovernmental failure to fulfil financial commitments (#PF3913).
Aggravated by Government inaction (#PC3950) Lack of international cooperation (#PF0817)
Government delay in response to symptoms of problems (#PF6707).

♦ **PF9565 Fear of God**
Refs Bynyan, John *Fear of God* .
Broader Fear (#PA1926).
Aggravates Resistance to grace (#PF5266).
Aggravated by Wrath of God (#PF8563).

♦ **PF9571 Conflicts of national law in relation to international transactions**
Divergence in different national laws relating to international transactions
Nature With the increase in international travel and interaction, there is a corresponding increase in the injustices resulting from the divergent laws of different countries which have some connection with such international transactions. Individual rights and interests ranging from the rights of married women and child custody to the amount of compensation for personal injury or breach of contract may ultimately turn on the person's geographic presence at an isolated moment in time or the country in which the person first seeks to protect his or her rights and interests.
Broader Conflict of laws (#PF0216).
Aggravates Injustice (#PA6486)
Inadequate development of international criminal law (#PF4676).

♦ **PF9576 Inadequate models of socio–economic development**
Uncoordinated multiplicity of development models — Competing models of socio–economic development — Uncoordinated multiplicity of global financial models — Inappropriate theoretical models of social development — Static models of social change by policy makers — Resistance to inclusive environmental models
Nature Inept theoretical models of society and community in the face of rapidly changing economic, political and cultural trends leaves decision makers at every level dramatically unequipped to create effective solutions. Models of society tend to be static, implementation of plans tend to be sequential and linear, social components are envisaged in fragmented ways, and reflections on progress focus on establishing blame or praise.
Incidence Policy makers and their advisors tend to use inappropriate static models of social change. These models include: the divine right of rulers, the responsibility of the chosen few to make decisions on behalf of the inept many, and belief in eternal principles of social interaction. The results of these static models of society include: fragmented and protectionist policies, rigid laws and traditions and injustice.
Large international corporations have created financial planning models with global perspective; but they largely exclude national corporations from the planning process, even though the national corporations are also dependent on the global situation. The needs of the consumer are seen only from the vantage point of the producers, which even further distorts meaningful planning.
Although there is accumulating evidence of environmental interdependence, the majority of people only give thought to their immediate environmental situation rather than directing energy to long–range, more inclusive, environmental planning. This concentration on individual circumstances avoids purposeful action on local and global environmental issues and leads to fears for the future as evidence on environmental issues grows.
Claim Existing socio–economic models for social change in a turbulent environment are woefully inadequate in the face of the policy challenges of the next decades. A new set of policy models is essential to deal with the emerging crises.
Refs Edwards, Chris *The Fragmented World* (1985).
Broader Absence of tactical methods (#PF0327)
Uncontrolled application of technology (#PC0418).
Narrower Atomism (#PF5322) Anthropocentrism (#PF4096) Anthropomorphism (#PG4120)
Overgeneralized policy models (#PF9100).
Related Self–interest driven investment (#PC2576).
Aggravates Inappropriate policies (#PF5645) Global financial crisis (#PF3612)
Undemocratic policy–making (#PF8703) Unforeseen environmental crises (#PF9769)
Destruction inherent in development (#PF4829) International economic fragmentation (#PC0025)
Inadequate global consensus concerning problems and prospects of humanity (#PF9821)
Contradictions within the growth and partnership model of developed countries (#PE2203).
Aggravated by Scientific ignorance (#PD8003) Resistance to change (#PF0557)
Inadequate social reform (#PF0677) Fragmentation of knowledge (#PF0944)
Inadequate research on problems (#PF1077)
Over–specialized study of global ecosystem (#PF5712)
Excessive dependence on computer models of complex system behaviour (#PE8533).

♦ **PF9582 End of nature**
False assumptions concerning the recuperative power of nature
Claim Humanity's comforting sense of the performance of the natural world is the result of a subtly warped perspective. Total changes in the natural world can now happen during the lifetime of a person. Without recognizing it, humanity has already stepped over the threshold of such a change and has reached the end of nature, namely the end of that set of human ideas about the world and man's place in it. The death of these ideas began with definite changes in the reality of the natural world — changes which can be measured. With increasing frequency such changes will clash with man's perceptions until finally the mistaken sense of nature as eternal and separate will no longer be tenable.
Broader False assumptions on sustainable development (#PF2528).
Aggravated by Fear of nature (#PF6803) Natural environment degradation (#PB5250)
Belief in humanity's dominance over nature (#PF8264).

♦ **PF9583 Political deception**
Political lies
Broader Lying (#PB7600).
Aggravates Loss of institutional credibility (#PF1963).
Aggravated by Corruption in politics (#PC0116) Deception by government (#PD1893).

♦ **PF9587 Inadequate resources for health**
Inadequate medical resources
Broader Insufficient financial resources (#PB4653).
Aggravates Inequitable use of medical resources (#PJ5160).
Aggravated by Decline in government health expenditure (#PF4586).

♦ **PF9595 Fear of knowledge**
Fear of truth — Fear of knowing
Broader Fear (#PA6030).
Aggravates Ignorance (#PA5568). Resistance to grace (#PF5266).
Aggravated by Vulnerability of society to truth (#PF5937).

♦ **PF9598 Misuse of language**
Abuse of language
Narrower Slang (#PF5213) Sexually discriminating job terminology (#PF6014).
Aggravates Corruption of meaning (#PB2619) Discriminatory language (#PF7299)
Obscurantist diplomatic language (#PF6703).

♦ **PF9608 Being a burden**
Being a burden to society — Being a burden to relatives
Broader Economic and social losses due to disability (#PE4856).
Narrower Burden on society of widows (#PF6149).
Aggravates Suicide (#PC0417) Senilicide (#PJ7124) Humiliation (#PF3856).
Aggravated by Dependency of the elderly (#PD8399)
Inadequate welfare services for the aged (#PD0512).

♦ **PF9610 Conflicts of interest**
Narrower Bias in scientific research (#PF9693) Conflict of interests in business (#PG5244)
Conflict of interests in imperialism (#PC6440)
Conflict of interest among parliamentarians (#PE3735)
Conflict of interest between governments or groups of governments (#PE8289).
Aggravated by Obsolete policy-making (#PF5009).

♦ **PF9612 Erosion of human capital**
Erosion of investment in human resources
Aggravated by Decline in government social expenditure (#PF0611).

♦ **PF9617 Obsession with celebrities**
Delusions concerning public figures
Nature The intense media focus on celebrities is leading an increasing number of mentally disordered people to place such celebrities at the focus of their delusions in such a way as to lead to violence. The delusion may take the encourage the person to believe that the celebrity is romantically interested, or is the only person who can provide protection from some form of persecution or injustice. Any rebuff, following an attempt to communicate may either reinforce the belief or result in anger and violence.
Incidence In 1989 it was reported that there have been as many injurious attacks in the USA on public figures by mentally disordered people as in the previous 175 years. The most prolific writer has sent more than 10,000 letters, mostly to a single entertainer, in six years.
Broader Illusion (#PA6414).
Aggravated by Personality cults (#PC1123)
Media theatricalization of public life and politics (#PF9631).

♦ **PF9618 Decline in deference**
Aggravated by Loss of civility (#PC7013).

♦ **PF9624 Stress in perfectionists**
Stress in idealists — Stress in environmentalists
Refs Dowling, Colette *Perfect Women* (1988).
Broader Stress in human beings (#PC1648).

♦ **PF9625 Obstruction of research**
Delays in research
Aggravates Ignorance (#PA5568).

♦ **PF9629 Unilateral interpretations of multilateral principles**
Incidence The adoption by countries of unilateral interpretations of multilateral trading principles to deal with new situations is a reflection of the inability of the system to adapt to the changing realities of world trade. This is most apparent in the attitude toward "unfair" trade practices. The international community has failed to address effectively the problem of broad and unilateral interpretations of what constitutes "unfair", "unreasonable" or "unjustified" trading practices.
Narrower Preoccupation with reciprocity in trading relations (#PF3871).
Aggravates Protectionism in international trade (#PC5842)
Fragmentation of the international trading system (#PC9584).
Aggravated by Inability to negotiate effective multilateral safeguard systems (#PF5287).

♦ **PF9631 Media theatricalization of public life and politics**
Nature As public life and politics increasingly come to be experienced by the public through the media, and especially television, the substantive content is increasingly judged by entertainment values. When a politician trying to raise real issues on which the population is to make up its mind is preceded or followed by a comedian using his personality to amuse an audience, the politician will tend to be judged in comparison with the comedian in terms of entertainment value. The pressures on political life to increase its impact on public opinion lead to increasing theatricalization: selection of candidates by their photogenic qualities on television, the conduct of campaigns based on the televisual potential of issues rather than their real relevance.
Incidence Media reports of real events can easily be of less interest than carefully prepared and scripted theatricalizations of similar events. Since the real events which happen unexpectedly and spontaneously are outside the control of the television producer, it is in the interest of the producer to cultivate a relationship with those who can provide advance notice of such events. Thus the organizers of a demonstration who inform television news organizations beforehand that there will be violence will provide good filmed news material which may assume far greater importance in the minds of viewers than the event may really deserve. More extreme examples include delays to executions, or even their arrangement, to enable television camera crews to be present.
Aggravates Personality cults (#PC1123) Obsession with celebrities (#PF9617)
Distorted media presentations (#PD6081) Impoverishment of political debate (#PF4600)
Collapse of distinctions between categories (#PF7938).
Aggravated by Excessive television viewing (#PD1533)
Parochial escapist media entertainment (#PD0917).

♦ **PF9659 Drug resistance**
Refs Mitsuhashi, Susumu (Ed) *Drug Resistance in Bacteria* (1983).
Narrower Drug resistant viruses (#PG6399) Pest resistance to pesticides (#PD3696)
Fungal resistance to fungicides (#PE4456) Insect resistance to insecticides (#PD2109)
Rodent resistance to rodenticides (#PE3573).
Related Abuse of antibiotics (#PE6629).
Aggravated by Abuse of animal drugs (#PE0043)
Abuse of medical drugs (#PD0028).

♦ **PF9664 Faltering structural adjustment in the world economy**
Retardation in economic structural adjustment — Faltering adaptation of economic agents to accelerating pace of structural change — Policy-making delays in response to accelerating structural change
Nature The decade of the 1970s witnessed a retardation of structural adjustment at a time when the need for structural change was greatest owing to rapid shifts in patterns of comparative advantage, especially as regards the increased export competitiveness of a number of industrializing developing countries and a certain developed countries. In addition, the severe recession of the 1980s has had a serious impact on structural adjustment. In general, adjustment costs tend to be higher in stagnant than in fast-growing economies as, for example, the prospects of alternative employment are unlikely to be very bright for the adversely affected factors of production when growth is sluggish.
Refs UNCTAD *Protectionism and Structural Adjustment in the World Economy* (1982).
Broader Economic and financial instability of the world economy (#PC8073).
Narrower Devaluation of money (#PE3700)
Economic imbalances among industrialized countries (#PD5865)
Outflow of financial resources from developing countries (#PC3134)
Inadequate economic policy-making in developing countries (#PF5964).
Aggravates Economic inefficiency (#PF7556).
Aggravated by Adverse consequences of scientific and technological progress (#PF3931).

♦ **PF9671 Theological justification of population growth**
Aggravates Unsustainable population levels (#PB0035)
Religious opposition to population control (#PF1022).

♦ **PF9672 Inadequate management of government finances**
Mismanagement of government financial expenditures — Ineffective control of government finances
Incidence In the case of the USA, inadequate controls in programmes that lend money or guarantee loans made by others were reported in 1989 as leading to losses of $100 billion. Losses of $5,000 billion in federal insurance and credit assistance would also be added to previous losses $200 billion in connection with insolvency in savings and loan institutions.
Broader Mismanagement (#PB8406) Financial scandal (#PD2458)
Inefficient public administration (#PF2335).
Narrower Inadequate economic policy-making in developing countries (#PF5964).
Aggravates Inadequate fiscal policies (#PF4850) Inadequate mobilization of resources (#PF4979)
Imbalance of revenue mobilization and expenditure allocation (#PF7432).
Aggravated by Lack of commitment (#PF1729) Lack of information (#PF6337)
Governmental incompetence (#PF3953) Insufficient financial resources (#PB4653)
Ineffective follow-up on initiatives (#PF0342) Inadequate public finance statistics (#PF7842)
Failure to adapt general initiatives to specific needs (#PF1578).

♦ **PF9683 Non-use of available health facilities**
Inefficient use of available health care
Nature Even when facilities exist, the poor are bewildered or turned off by the way they are treated. Or they simply may not know about the existence of the facility or how to get to it.
Incidence For example, a 1981 survey of a major city in South Asia showed that not one child was immunized adequately with the DPT and polio vaccines; only a few were protected against measles and the majority went without immunization of any kind. The children surveyed were all attending government-aided nurseries, for which government aided grants provided for periodic visits of medical professionals, and there was a primary health care centre in the neighbourhood.
Aggravated by Inadequate health services (#PD4790)
Dehumanization of health care (#PD7821).

♦ **PF9686 Urban bias**
Urban elitism — Disproportionate political influence of urban elites — Neglected rural populations
Nature Those constituting the urban elites, including those involved in government, tend to engender policies favouring urban groups while neglecting those in rural areas. Such policies may ensure that wealth from the rural areas is effectively transferred to urban regions (for example, by setting low prices for agricultural produce). Especially in developing countries, urban bias is encouraged by the investment priorities of transnational corporations.
Broader Elitism (#PA1387).
Aggravates Urban poverty (#PC5052) Rural underdevelopment (#PC0306)
Regional underdevelopment (#PD4424) Rural-urban income differential (#PE5022)
Migration of rural population to cities (#PE8768)
Accentuated inequality between rural and urban development (#PE8569)
Flooding of the urban labour market in developing countries (#PD0008)
Segregation of poor and minority population in urban ghettos (#PD1260)
Increasing income disparity in developing countries due to transnational corporations (#PE1660).

♦ **PF9691 Polarized protest against problems**
Nature At a certain stage in the evolution of recognition of a problem and of the articulation of protest against those perceived as reinforcing it, extreme forms of polarization become counter-productive. Such polarization makes it impossible for any middle ground or alternative to be explored through which movement beyond the problem can be opened up.
Incidence Examples can be seen in the anti-apartheid movement in which any not in favour of sanctions are perceived as being pro-apartheid. Similar examples occur in the protest against nuclear power, armaments, multinational corporations, communism, capitalism and abortion.
Broader General obstacles to problem alleviation (#PF0631).
Narrower Polarization of local conflicts (#PF1333).
Aggravates Recurrence of misapprehended world problems (#PF7027).
Aggravated by Vulnerability of developing countries to inflation (#PD0367).

♦ **PF9693 Bias in scientific research**
Reward biased research — Sponsor biased research — Erosion of scientific objectivity — Conflict of interest between pure and applied science — Corrupting effect of valuable prizes on academic integrity — Reward oriented research — Expertise biased in favour of the commissioning body — Research sponsored by vested interests — Manipulation of scientific studies
Nature The size of individual research projects has long outgrown the means of individual researchers or private patrons. The aggregate scale of science has become a significant item in a national budget. Its products have nearly immediate cash value as inputs to industry. Under these circumstances, control over the choice of problems, evaluation of conclusions, and the use made of results inevitably passes to those who fund such industrial-scale enterprise. This transition from academic to industrialized science, and the sensitivity to sponsorship which researchers must acquire, can destroy the morale necessary for their creativity. The direction and character of science is being altered, even deformed, by political and commercial forces.
Broader Conflicts of interest (#PF9610).
Narrower Opportunist bias in public discussion and research on development (#PF3846).
Aggravates Racial inequality (#PF1199) Questionable facts (#PF9431)

Disagreement among experts (#PF6012)
Excessive salaries of experts (#PF8317)
Irresponsible international experts (#PF0221)
Unavailability of appropriate expertise (#PF7916).
Unidentified flying objects (#PF1392)
Politicization of scholarship (#PF7220)
Suppression of scientific information (#PF1615)
Aggravated by Proliferation of commercialism (#PF0815)
Unethical professional practices (#PC8019).
Reduced by Scepticism (#PF3417).

♦ PF9694 Self-fulfilling prophecies
Reduces Inaccurate forecasting (#PF4774).

♦ PF9697 Vulnerability of national economies to vagaries of external markets for goods and services
Vulnerability to adverse conditions of foreign trade
Aggravated by Declining international competitiveness (#PD8994)
Lack of economic and technical development (#PE8190)
Deterioration of terms of trade for developing countries (#PD2897)
Excessive concentration of export markets of developing countries (#PE9457)
Protectionism in international trade against exports from developing countries (#PD9679)
Instability in export trade of developing countries producing primary commodities (#PD2968)
Development by industrialized countries of products substituting for commodities exported by developing countries (#PD7682).

♦ PF9699 Classified public information
Secret public records — Inaccessible historical documents — Withdrawal of official documents from public scrutiny — Overclassification of embarrassing documents
Nature Information is kept secret by authorities for reasons ranging from risk to national security to risk of embarrassing people still alive (or their relatives). Given the importance attached to such information, a significant proportion of it must necessarily offer new insights into the current condition of society and the forces which inhibit any effective remedial action. Projects proposed and implemented in ignorance of these factors therefore run the risk of being inappropriate to the real conditions. Restricting access therefore directly impedes efforts to respond to the crises of society and encourages dependence on bureaucratic initiatives with a poor record in doing so.
Incidence Within the USA it has proved difficult for the responsible agencies to determine exactly how many classified documents they control. The Information Security Oversight Office, established in 1978, devised new procedures and determined that for 1989 the number of new secret documents numbered 6,796,501. The cumulative total is not available. As a specific example, researchers trying to model how the climate will change as a result of the greenhouse effect cannot obtain access to secret data on sea ice thickness collected by the nuclear submarines operated by the superpowers.
Claim 1. Overly elaborate, costly declassification processes encourage historical distortion and facilitate cover-up by government. Governments conceal 30 to 40-year old secrets, enabling them to fabricate an official historical record.
2. Although the government must maintain some secrets, if members of the security apparatus can, with impunity, keep from those elected by the people that which they are entitled to know (or worse, feed them false information) then those who control the classified data can become the real decision makers, as happens in certain situations (such as the Iran-Contra incident in the USA).
Broader Official secrecy (#PC1812).
Related Inaccessible historical libraries (#PF9046).
Aggravates Historical misrepresentation (#PF4932)
Unaccountability of international financial institutions (#PF1136).

♦ PF9704 Inadequate mobilization of public opinion for development
Government failure to mobilize support against problems
Broader Inadequate mobilization of resources (#PF4979)
Inappropriate design of development projects (#PF4944).
Aggravates Inappropriate development policy (#PF8757).
Reduces Manipulation of the individual by mass media (#PE7448).

♦ PF9707 Photographic bias
Nature Photographs and photography offer slanted and partial views of reality. Sadly they are often taken to be true representations of a situation. Snapshots are taken of weddings not funerals, of birthday parties but not divorces and new born children but not battered babies. They are selective and conventions of selection amplify nostalgia and relegates unpleasantness to the uncertainty and dimness of memory. They are conservative, they re-enforce complacency by reminding people of the happy moments and discourage changing. In their photo albums they become an almost official version of lives. They are also instruments of change. They are ways of avoiding involvement. A person taking a souvenir shot on vacation creates a distance from those he is photographing. Even the language of photography is suspect. Cameras are loaded; pictures are shot; images are captured or taken; lenses are attached with bayonet mechanisms.
The consumer of photographs are the most influenced by them. Photos seem to be more real than written words. They reassure because of the impression that photos cannot lie. Yet everyone knows that pictures can be doctored, scenes set, situations contrived. Because photos are both pictures of reality and symbols of reality they have great power. They also help create stereotypes.
Broader Misleading information (#PF3096).
Related Photographic propaganda (#PD3086).

♦ PF9710 Fascist liberalism
Militant liberalism — Extremist liberalism
Broader Extremist ideologies (#PC6341).
Aggravates Politicization of scholarship (#PF7220).
Aggravated by Liberalism (#PF0717).

♦ PF9716 Inadequate use of modern information systems
Nature Modern information systems are under-utilized because of real or perceived inadequacy of information services, problems of access to such on-line systems, cost of such access, apathy, ignorance and information shock.

♦ PF9728 Incompatibility of environmental and economic decision-making
Environmentally insensitive economics — Inappropriate economic theory — Failure to recognize the environment as the resource base for all economic activity
Nature Economic progress continues to be measured as though the social and natural environments were of no consequence. Little account is taken for the manner in which the degradation of the environment subtracts from human welfare and impairs sustainable development. The discipline of economics, and the economic policies sustained by it, has treated the environment as a free good when in fact it is a scarce commodity. It has been made scarce by ignoring the ecological requirements for a sustainable planetary society, running natural wealth down instead of conserving it.
Claim The compatibility of environmental and economic objectives tends to be lost in the competitive pursuit of individual or group gains, with little regard to the impact on others, with a blind faith in science's ability to find solutions, and in ignorance of the distant consequences of today's decisions, especially for future generations.
Counter-claim 1. Economics has a substantive role to play in environmental reconstruction, although many incorrectly see economics as the agent of destruction and not a saviour. It is in economic thinking that solutions to environmental problems are to be found. Economics is not the enemy of the environment. Poor economics is. By manipulating markets, by demonstrating the economic value of natural resources, by altering the way in which economic progress is monitored, there is a real possibility of developing an agenda for development with conservation.
2. Economic and ecological concerns are not necessarily in opposition. For example, policies that conserve the quality of agricultural land and protect forests tend to improve the long-term prospects of agricultural development. Also an increase in the efficiency of energy and material use serves ecological purposes but can also reduce costs.
Broader Insensitivity to non-immediate hazards to society (#PF9119).
Aggravates Maldevelopment (#PB6207)
Environmental prodigality (#PF7318)
Undomesticated men (#PF0551)
Natural environment degradation (#PB5250)
Inappropriate development policy (#PF8757)
Inappropriate design of development projects (#PF4944).
Aggravated by Preoccupation with isolated problems (#PF6580).

♦ PF9729 Unrecognized benefits from cooperatives
Uncertain cooperative success — Undemonstrated cooperative purchasing — Unexplored co-operative potential — Undemonstrated cooperative benefits — Limited cooperative experience — Short-lived cooperative commitment — Past cooperative cheating — Disillusioning cooperative experience — Cooperative failure story — Co-op successes unknown — Co-ops under suspicion — Unrecognized co-op benefits
Incidence In most of the Western countries the once powerful consumers' cooperative movement is in serious decline. Cooperatives have fulfilled their social purpose since today consumer interests are well protected by legal standards and regulations, and their ideals are forgotten since members have simply become shoppers. The decline was inevitable, because the committee system of management did not adjust to the revolution of retailing.
Broader Unrecognized opportunities (#PF6925)
Unrecognized benefits from corporate action (#PJ9420)
Deteriorated structures of essential corporateness (#PF1301).
Aggravates Social insecurity (#PC1867)
Demeaning community self-image (#PF2093)
Unproductive subsistence agriculture (#PC0492)
Disjointed patterns of community identity (#PF2845)
Ineffective structures for community participation (#PF2437)
Government imposition of rural cooperative projects (#PE2309)
Ineffective structures for community decision-making (#PF1781)
Ineffectiveness of traditional small business methods (#PF3008)
Detrimental effect of jungle environment in tropical villages (#PE2235).

♦ PF9735 Limited artistic expression
Denial of right to artistic freedom
Nature Artistic expressions not sacred when they are connected with unpopular views or when they touch on sacred things of religions, or when they offend the common law of decency.
Broader Denial of right to a cultural life (#PE6561)
Reduction in symbolic celebrations (#PF1560).
Narrower Suppression of creativity and innovation (#PF0275).

♦ PF9747 Inability to compromise
Aggravated by Compromise as a betrayal of principles (#PF3420).
Reduces Domination of government policy-making by short-term considerations (#PF0317).

♦ PF9756 Legal prevarication
Legal cover-up — Legal obstruction
Aggravates Hindrance of law enforcement (#PD5515)
Soliciting obstruction of proceedings (#PD0790)
Obstruction of international criminal investigations (#PF7277).
Aggravated by Unethical practices in the legal profession (#PD5380).

♦ PF9769 Unforeseen environmental crises
Hazardous environmental discontinuities — Environmental surprises — Unpredictable ecological disasters
Broader Disasters (#PB3561).
Aggravates Irreversible problem emergence (#PF9790)
Economic dislocations in developing countries (#PD4063).
Aggravated by Latent problems (#PF9328)
Inaccurate forecasting (#PF4774)
Inadequate models of socio-economic development (#PF9576)
Simplistic technical solutions to complex environmental problems (#PF0799).

♦ PF9786 Inappropriate institutionalized reward systems
Counter-productive personnel reward systems
Nature Organizations and institutions with internal reward systems that operate in terms of criteria unrelated to the social problems that the institution is mandated to deal with tend to encourage personnel to formulate policies and approve projects which fail to respond to such social problems.
Claim It has been suggested that the reason that the World Bank has failed to give any priority to population control in practice is that personnel are rewarded, and achieve status, by committing large sums of money to construction projects. Population projects call for smaller amounts of capital investment and require lengthy and extensive dialogue. This offers little incentive within a bureaucracy whatever the declared policy commitments.
Aggravates Short range planning for long-term development (#PF5660).

♦ PF9790 Irreversible problem emergence
Irreversible environmental trends
Claim It is not the fact that problems emerge that is important, rather it is the fact that the emergence of many of these problems is irreversible. The remedies that can be applied are not capable of restoring the system to its earlier condition. Whether or not the further development of the problem can be contained, damage has been done and that damage may often prove to be very severe. Environmental degradation provides many examples: tropical deforestation, extinction of species, desertification, soil erosion, depletion of non-renewable natural resources.
Aggravated by Natural environment degradation (#PB5250)
Unforeseen environmental crises (#PF9769).

♦ PF9791 False economies of scale
Broader False assumptions on sustainable development (#PF2528).

♦ PF9794 Misinformation concerning infringement of human rights
False declarations of ill-treatment — False evidence of torture — Politically-motivated exaggeration of human rights infractions — False allegations of human rights violations
Nature It is in the interest of individuals and groups seeking political advantage to arouse public interest, international media attention, and especially the concern of those monitoring human rights abuses, by falsifying or exaggerating claims of infringement of human rights. Under the social conditions in which such claims are made it is difficult to determine what really occurred given the effort of those so motivated to make such claims and the effort of the authorities to deny their validity, whether or not they are true.
Incidence Governments represented in the United Nations Commission on Human Rights regularly make statements indicating that the evidence produced for human rights infringements is false or simply an exaggeration of minor incidents.
   Broader False evidence (#PF5127)
   Biased allegations against governments (#PD4517).
   Aggravates Impunity of violators of human rights (#PF3474).

♦ PF9796 Messianic policy-making
Milleniarist programmes
   Broader Unrealistic policies (#PF9428).   Inappropriate policies (#PF5645).
   Aggravates Excessively costly prestige projects (#PF3455).
   Aggravated by Short range planning for long-term development (#PF5660).

♦ PF9818 Psychological inconsistency of marriage partners
   Aggravates Divorce (#PF2100)   Marital stress (#PD0518)   Marital instability (#PD2103)
   Deception between sexual partners (#PE4890).
   Aggravated by Psychological impediments to marriage (#PF3344).

♦ PF9821 Inadequate global consensus concerning problems and prospects of humanity
Inadequate development of political thinking concerning the global problematique
   Aggravates Fixation on partial solutions to problems (#PF9409),
   Inadequate power of intergovernmental organizations (#PF9175).
   Aggravated by Uncritical thinking (#PF5039)   Persistence of outmoded concepts (#PF7673)
   Inadequate models of socio-economic development (#PF9576)
   Inadequacy of prevailing mental structures to challenge of human survival (#PF7713).

♦ PF9828 Suppression of information concerning social problems
Concealment of information on the extent of poverty — Concealment of information on the extent of unemployment — Concealment of information concerning the extent of homelessness
   Broader Official secrecy (#PC1812)   Suppression of information (#PD9146).
   Related Suppression of information concerning environmental hazards (#PF4854).
   Aggravates Misuse of statistics (#PF4564)   Governmental bias in statistics (#PF0019)
   Inadequate public information concerning problems (#PF5701)
   Inadequate statistical information and data on problems (#PF0625).

♦ PF9830 Fragmentation of research
Fragmentation of field of analysis
   Aggravates Fragmentation of knowledge (#PF0944).
   Aggravated by Uncoordinated research efforts (#PF2306)
   Irresponsible delimitation of policy responsibilities (#PF7823)
   Collusion between administrators of funding agencies and programme formulators (#PF8711).

♦ PF9841 Inadequate financial clearing systems
Inadequate financial settlement systems
   Aggravates Global financial crisis (#PF3612).

♦ PF9843 Negative personal relationships between leaders of countries
Antipathic personal relations between world leaders
Nature As in any interpersonal situation, presidents and ministers of countries can find their relations to their counterparts in other countries totally undermined by personality issues. This can have a very direct impact on their ability to come to agreement on issues that confront them.
Counter-claim With very rare exceptions, politicians are seldom alone with each other. Their meetings are planned and rehearsed by top officials by whom they are constantly surrounded. Diplomacy is designed to obviate the unpredictable consequences of spontaneous interaction.
   Aggravated by Unethical personal relationships (#PF8759).

♦ PF9849 Vagueness of laws
Vague laws — Vague regulations
Nature A law may be considered vague if it fails to define possible offences with sufficient precision so that ordinary people can understand what conduct is prohibited or so as to discourage arbitrary and discriminatory application.
Incidence Within some legislations, such vagueness may be prohibited by statute.
   Aggravates Internment without trial (#PD1576)   Arbitrary enforcement of regulations (#PD8697)
   Incomplete access to information resources (#PF2401).
   Aggravated by Excessive government control (#PF0304).

♦ PF9852 Inappropriate ambitions
Inappropriate goals
   Aggravates Careerism (#PF6353)   Accumulation (#PA4313).
   Reduces Meaninglessness (#PA6977).

♦ PF9980 Inadequate sense of time
Erratic time design — Loss of sense of time — Inability to organize time — Temporal illiteracy — Poor time-keeping
Incidence Some people, especially in certain developing cultures, do not structure time in ways which permit them to follow programmes which require actions at different stages. In the simplest case, medical prescriptions involving a course of treatment over a period of several months cannot be followed because the person forgets to take a pill after the first few days. This is especially important in the treatment of leprosy, for example, which requires regular doses over an extended period.
Claim The process of economic development is seriously impeded by the time values of local cultures, especially as interpreted by the indigenous labour force but also at the policy, planning and managerial levels. Workers in certain cultures are slothful, unpredictable and unreliable and have no sense of the future or of how to plan ahead. In the light of the values of industrialized societies they are unpunctual and undisciplined, caring only for the moment. In such cultures people tend to be settled into traditional behaviour patterns, are slow to initiate new activities and slow to react to new possibilities.
   Broader Differing conceptions of time (#PF6665)
   Obstacles to efficient utilization of time (#PF7022).
   Aggravates Wasted time (#PF8993)
   Short range planning for long-term development (#PF5660)
   Inappropriate management of development projects (#PD3712).
   Aggravated by Drug abuse (#PD0094)   Drunkenness (#PE8311)
   Political indoctrination (#PD1624)   Discontinuity of employment (#PD4461)
   Fatalistic attitudes to the use of time (#PF2795)
   Lack of responsible involvement in community affairs (#PF6536)
   Constraint of time on individual and social development (#PF5692).
   Reduces Temporal imperialism (#PF1432).

♦ PF9991 Shame
Nature The broad spectrum of painful affects, including embarrassment, humiliation, mortification and disgrace, that accompany the feeling of being rejected, ridiculed, exposed, or of losing the respect of others.
   Related Anxiety (#PA1635).
   Aggravated by Ridicule (#PJ7386)   Disrespect (#PA6822)
   Humiliation (#PF3856)   Embarrassment (#PF7950)
   Inhibited grief process (#PD4918).

♦ PF9996 Peace
Claim It is uncertain whether peace will ever be possible. It is far more questionable, by the objective standard of continued social survival rather than of emotional pacifism, that peace would be desirable even if it were demonstrably attainable. The war system, for all its subjective repugnance to important sections of public opinion, has demonstrated its effectiveness since the beginning of recorded history. It has provided the basis for the development of many impressively durable civilizations. It has consistently provided unambiguous social priorities and as such is largely a known quantity. A viable system of peace, assuming that the many transitional problems can be solved, would constitute a venture into the unknown, with the inevitable risks attendant on the unforeseen, however small and however well hedged. At the present state of knowledge and reasonable inference, it is the war system that must be identified with stability, the peace system with social speculation, however justified that speculation may appear in terms of subjective moral or emotional values. Any condition of genuine total peace, however achieved, would be destabilizing and unsustainable until proved otherwise.
Refs Cassiers, Juan *The Hazards of Peace* (1984); Mushkat, Marion *The Third World and Peace* (1983).
   Narrower Peace agreements (#PU2128).
   Aggravated by Nuclear freeze (#PF0679).

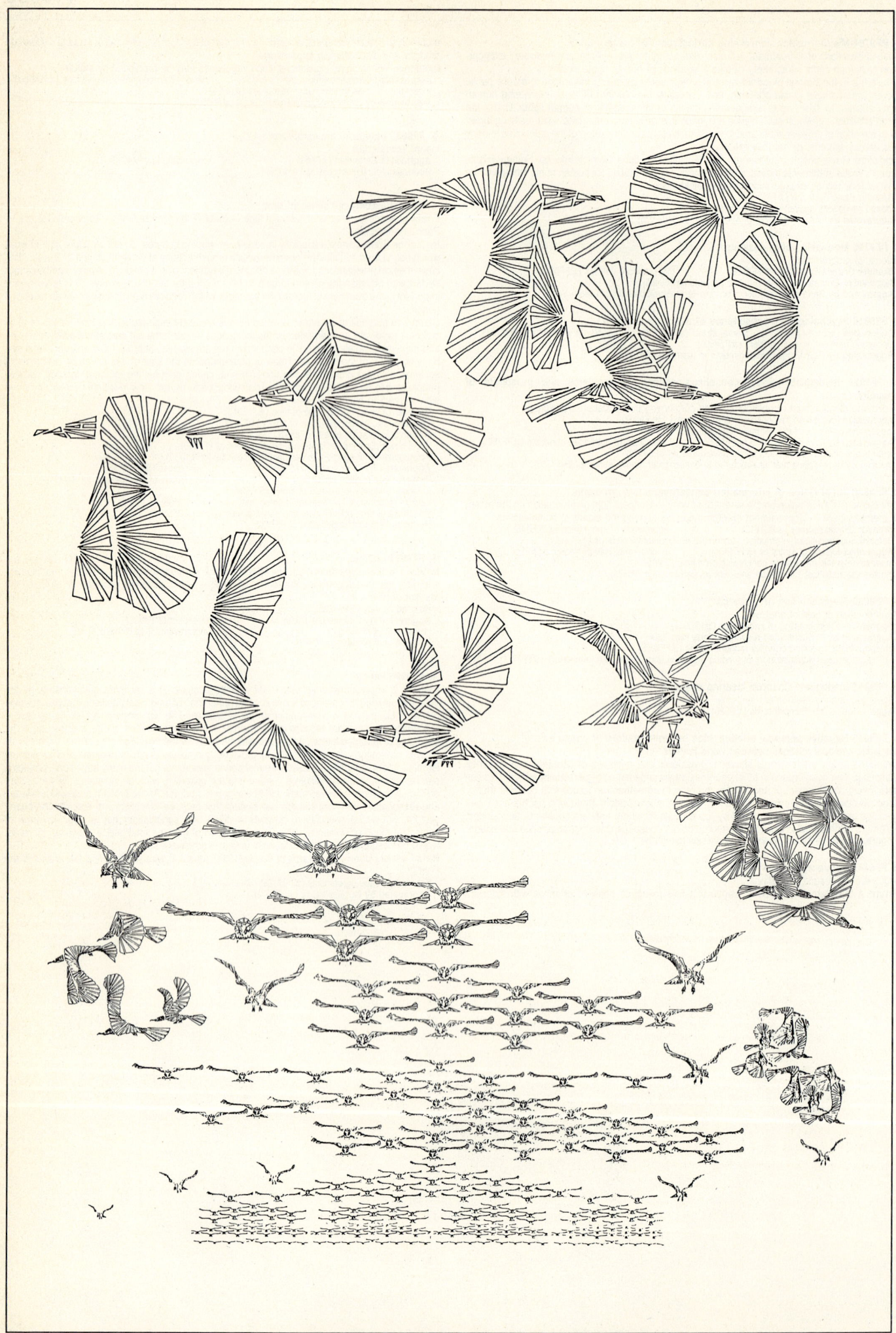

Very specific problems — PG

Content

This section groups together very specific problems. They themselves tend to be grouped under broader problems described in earlier sections. A problem is included here when it is considered too specific to merit inclusion in any of the previous sections, especially if it constitutes one of a number of sub-elements of specific problems described there.

No descriptions are provided. The problems in Section G only appear:

- as cross-references of entries in earlier sections;

- in the index (Section PX) which refers to entries in earlier sections in which they are mentioned as cross-references

Note that further information relevant to an understanding of the problem may be present in other problems to which the index cross-reference refers, especially any broader problems.

This section groups 2,909 problems for which there are 3,984 cross-references.

Rationale

In the process of collecting information for description in the previous sections, the names of many interesting candidates for inclusion emerge. The many problems in this section constitute the detail of reference books on diseases, endangered species, problems of particular commodities or economic sectors, and the like. As such they are easily ignored in attempts to respond to the classes of problems to which they belong.

It is by specific problems that people and groups are touched. The challenge is to explore methods of providing pointers to the maximum number of such problems. Keyword indexes and hierarchies of cross-references are used, thus ensuring a trace on them, whether or not it is possible or appropriate to provide a succinct description of them at this stage. The difficulty is to establish useful cut-off points to avoid overwhelming the process with problems at an excessive level of detail.

Registering a problem in this section ensures that borderline cases can be noted, indexed and included in hierarchies and networks of cross-references in anticiaption of the opportunity for future research and editorial work on them. This section therefore provides a possibility for initiating the process of setting such problems in context.

Method

The entries are based on information obtained from international organizations, from a wide variety of reference books, or as reported in the international media. The procedures for identifying world problems are described in Section PZ at the end of this volume.

Index

A keyword index to entries is provided in Section PX.

Comment

Detailed comments are given in Section PZ at the end of this volume.

Reservations

The emphasis throughout this volume has been placed on providing descriptions of less well-known problems, particularly when the extensive material available on the better known problems contained neither succinct descriptions of them nor descriptive material which could easily be reduced to succinct descriptions.

In a number of cases a problem could have been allocated to another section. Inclusion of a problem in this section, rather than in a preceding, has been based on a number of factors. The position of the problem in one or more hierarchies of cross-references was a major factor in determining its allocation to this section. Some problems, which would otherwise be allocated to this section, have been allocated to earlier sections (usually Section E). This has been done when the information available, or the pattern of cross-references, suggested that it would be of value to see them as part of the problem network.

Possible future improvements

There is much scope for improving the quality of problem entries through feedback from interested bodies. More bibliographic references could be included where appropriate, as well as references to major resolutions concerning those problems recognized by the United Nations. There is also much scope for improving the pattern of cross-references, both between problems, to other sections of this volume (eg values) and to the 20,000 internationally-active bodies in the companion series (*Yearbook of International Organizations*)

ENTRY CONTENT AND ORGANIZATION

Ordering of entries
Entries are in **numeric order**. Entry numbers have been **allocated randomly**; they have no significance other than as a permanent point of reference to facilitate indexing, cross-referencing, and updating between editions.

Index access to entries
The location of an entry in this sub-section may be determined from
the **Volume Index** (Section PX) on the basis of keywords in the name of the entry or its alternate titles.

Structure of entries
Entries may be composed of the following descriptive elements:

(a) **Entry number** This number has **no significance**, except as a convenient method of identifying the entry (particularly for indexing purposes), of filing information on it, and as an identifier to which cross-references from other entries (possibly in other Sections) may refer in this and future editions. The first letter of the entry number refers to the section of this volume in which the sub-section, denoted by the second letter, is located.

(b) **Problem name** This is printed in bold characters. It is the name selected as best indicating the nature of the problem. It may be followed by alternative problem names.

(c) **Nature** Description of the problem which attempts to identify the nature of the disruptive processes involved. The information included here, and in the following paragraphs, is compiled directly, to the extent possible, from available published documents. Where appropriate the text included may be reproduced, in a minimally edited form, from the publications of international organizations, such as those of the United Nations or its Specialized Agencies.

(d) **Incidence** Summary description of the extent of the problem which makes it of more than national significance.

(e) **Background** Describes briefly when and how the problem's importance was recognized initially, and how this recognition has evolved over time.

(f) **Claim** Stresses the special importance of this problem and why action is particularly urgent. This paragraph offers means of including statements which may deliberately exaggerate claims for the unique importance of the problem.

(g) **Counter-claim** Stresses, where appropriate, the relative insignificance or erroneous conception of the problem as described. This paragraph offers a means of including statements which may deliberately exaggerate the arguments refuting the evidence for the existence of the problem. Absence of such arguments from the text does not mean that they do not exist.

Cross-referencing of entries
At the end of any entry, there may be cross-references to other entries. These indicate the number and name of the cross-referenced entry, whether within this Section or in other Sections. There are 3 types of **hierarchical** cross-references between problems:
> **Broader** = Broader problem: more general problems of which the problem described may be considered a part. The described problem may be considered an aspect of several broader problems
> **Narrower** = Narrower problem: more specific problems which may be considered a part of the described problem
> **Related** = Related problem: problems that may be considered as associated in a hierarchically undefined way with the described problem.

There are 4 types of **functional** cross-references between problems:
> **Aggravates** = Problems aggravated by the described problem: a forward or subsequent negative causal link
> **Aggravated by** = Problems aggravating the described problem: a backward or prior negative causal link
> **Reduces** = Problems relieved, alleviated or reduced by the described problem: a forward or subsequent positive causal link
> **Reduced by** = Problems relieving or alleviating the described problem: a backward or prior positive causal link

Problems under consideration

PJ

Content

This section groups together problems which are under consideration for inclusion in the preceding sections. As such they may overlap problems already appearing there or may be rejected for a variety of other reasons. Problems are included in this section: (a) whenever documents have been located implying that an entry could possibly be elaborated for any of the previous sections, if resources permitted; or (b) where there was doubt that the problem named could be appropriately distinguished, at this stage, from other problems with similar names.

No descriptions are provided. The problems in Section J only appear:

- as cross-references of entries in earlier sections;

- in the index (Section PX) which refers to entries in earlier sections in which they are mentioned as cross-references

Note that further information relevant to an understanding of the problem may be present in other problems to which the index cross-reference refers, especially any broader problems.

This section groups 1,537 problems for which there are 2,011 cross-references.

Rationale

In the process of collecting information for description in the previous sections, the names of many interesting candidates for inclusion emerge. Since the problem collection process is an ongoing one, this section provides a valuable means of reflecting the kinds of problem on which further information is being sought.

Registering a problem in this section ensures that borderline or questionable cases can be noted at an early stage. They are immediately indexed and included in hierarchies and networks of cross-references in anticipation of the opportunity for future research and editorial work on them. This section therefore provides a possibility for initiating the process of setting such problems in context.

Method

The entries are based on information obtained from international organizations, from a wide variety of reference books, or as reported in the international media. The procedures for identifying world problems are described in Section PZ at the end of this volume.

Index

A keyword index to entries is provided in Section PX.

Comment

Detailed comments are given in Section PZ at the end of this volume.

Reservations

The emphasis throughout this volume has been placed on providing descriptions of less well-known problems, particularly when the extensive material available on the better known problems contained neither succinct descriptions of them nor descriptive material which could easily be reduced to succinct descriptions.

ENTRY CONTENT AND ORGANIZATION

Ordering of entries
Entries are in **numeric order**. Entry numbers have been **allocated randomly**; they have no significance other than as a permanent point of reference to facilitate indexing, cross-referencing, and updating between editions.

Index access to entries
The location of an entry in this sub-section may be determined from
the **Volume Index** (Section PX) on the basis of keywords in the name of the entry or its alternate titles.

Structure of entries
Entries may be composed of the following descriptive elements:

(a) **Entry number** This number has **no significance**, except as a convenient method of identifying the entry (particularly for indexing purposes), of filing information on it, and as an identifier to which cross-references from other entries (possibly in other Sections) may refer in this and future editions. The first letter of the entry number refers to the section of this volume in which the sub-section, denoted by the second letter, is located.

(b) **Problem name** This is printed in bold characters. It is the name selected as best indicating the nature of the problem. It may be followed by alternative problem names.

(c) **Nature** Description of the problem which attempts to identify the nature of the disruptive processes involved. The information included here, and in the following paragraphs, is compiled directly, to the extent possible, from available published documents. Where appropriate the text included may be reproduced, in a minimally edited form, from the publications of international organizations, such as those of the United Nations or its Specialized Agencies.

(d) **Incidence** Summary description of the extent of the problem which makes it of more than national significance.

(e) **Background** Describes briefly when and how the problem's importance was recognized initially, and how this recognition has evolved over time.

(f) **Claim** Stresses the special importance of this problem and why action is particularly urgent. This paragraph offers means of including statements which may deliberately exaggerate claims for the unique importance of the problem.

(g) **Counter-claim** Stresses, where appropriate, the relative insignificance or erroneous conception of the problem as described. This paragraph offers a means of including statements which may deliberately exaggerate the arguments refuting the evidence for the existence of the problem. Absence of such arguments from the text does not mean that they do not exist.

Cross-referencing of entries
At the end of any entry, there may be cross-references to other entries. These indicate the number and name of the cross-referenced entry, whether within this Section or in other Sections. There are 3 types of **hierarchical** cross-references between problems:

> **Broader** = Broader problem: more general problems of which the problem described may be considered a part. The described problem may be considered an aspect of several broader problems
> **Narrower** = Narrower problem: more specific problems which may be considered a part of the described problem
> **Related** = Related problem: problems that may be considered as associated in a hierarchically undefined way with the described problem.

There are 4 types of **functional** cross-references between problems:

> **Aggravates** = Problems aggravated by the described problem: a forward or subsequent negative causal link
> **Aggravated by** = Problems aggravating the described problem: a backward or prior negative causal link
> **Reduces** = Problems relieved, alleviated or reduced by the described problem: a forward or subsequent positive causal link
> **Reduced by** = Problems relieving or alleviating the described problem: a backward or prior positive causal link

Index

PX

World Problems

Index scope

This index covers **all sections in this volume**. It does **not** include information on entries in Volume 2.

All index entries refer via reference numbers (eg PB1234) to the descriptions in the preceding sections. The letters indicate the section (eg Section PB).

The whole approach to naming and documenting problems (discussed in Section PZ) emphasizes the generous use of alternative or secondary names. Through this index, these facilitate location of the problem from multiple points of entry. Occasionally this extends to use of alternative spellings of keywords.

Index entries

The index entries are of several types:

- Principal name or title of problem;

- Secondary or alternative names of problem (including popular expressions and synonyms)

- Keywords from the principal problem name;

- Keywords from the secondary problem names.

Since problems do not have formally defined names, it was not considered useful to distinguish typographically between the different types of index entry. Keywords are however recognizable by the presence of a semi-colon in the index entry.

Remarks

To facilitate consultation of the index, and to reduce the space requirement:

- Prepositions and articles are normally omitted from the index entries;

- Sub-headings in bold are generated either when several entries commence with a minimum of three common words, or when the second word (of a group) starts a new alphabetic sequence.

Although the index was generated automatically, it has been edited manually to eliminate less helpful keywords resulting from the extraction process. Where there was any doubt, less meaningful words have however been left where they may prove to be of value in locating problems.

Some index entries contain italicized text. These correspond to two special cases:

- Names of entries from Section PA (Abstract and fundamental problems) are characterized by multiple uses of the same word as alternative names for distinct problems. This is due to the multiple connotations of the words used there. This means that the same word may appear in several entries, thus resulting in distinct index entries. To clarify such seeming duplication in the case of alternative names, the principal name is appended in italics to the index entry.

- Entries from Section PG (Problems of a very specific nature) and Section PJ (Problems under consideration for inclusion) are **not** published in this Encyclopedia. The relevant index entries do however appear in this index. The text then appears entirely in italics. The entry number referring to Section PG or Section PJ is then replaced by the entry number of a broader problem in which it is mentioned (as a narrower problem). Where there is no broader (or other) problem, the number is not changed.

Classified index

For those users interested in a classified index to problems by subject, this is incorporated into the complementary Volume 3 of the *Yearbook of International Organizations*. There the problems are grouped in terms of some 3,000 subject headings together with international organizations.

Abuse computer

A

PE 2538 A deficiency ; Vitamin
PE 0517 A ; Viral hepatitis
PD 2094 Abandoned animals pets
PF 2813 *Abandoned buildings ; Increase*
Abandoned [cars...]
PE 5339 — cars
PD 5734 — children
PD 8825 — children foreign soldiers
PF 6578 *Abandoned private homes*
PD 1030 Abandoned wives
PA 7685 Abandonment
PA 7270 Abandonment *Absence*
PA 6379 Abandonment *Avoidance*
PD 4268 Abandonment dying
PF 5764 Abandonment faithful
PA 5458 Abandonment *Inhospitality*
PA 5473 Abandonment *Insufficiency*
PA 5438 Abandonment *Neglect*
PA 7296 Abandonment *Restraint*
PA 7685 Abandonment ; Sense
PA 5612 Abandonment *Unchastity*
PA 6858 Abandonment unity
PA 5644 Abandonment *Vice*
PA 6659 Abasement *Humility*
PF 4286 Abbreviations ; Excessive use acronyms
PE 9251 Abdication control company directors
PF 7342 Abdication government ministerial control
PA 8658 Abdication responsibility
PE 9604 Abdomen ; Symptoms referable
PD 8725 Abdominal cramp
PE 3711 *Abdominal hernias animals*
PD 8725 Abdominal pain
PE 9604 *Abdominal pain*
PE 2663 *Abdominal surgery*
PD 8744 Abduction
PE 6154 Abduction ; Child
PF 3881 Abduction extraterrestrials
PE 2113 Abduction ; Extraterritorial
PE 2113 Abduction government agents acting foreign countries
PC 3298 Abduction women
PD 5129 Abductions children ; Political
PA 6228 Aberration *Deviation*
PA 6180 Aberration *Error*
PA 7157 Aberration *Insanity*
PA 5878 Aberration *Nonconformity*
PF 4983 Abetment
PA 7338 Abhorrence *Hate*
PA 7107 Abhorrence *Unpleasantness*
PF 1199 Abilities ; Inequality human biological
PF 1332 Abilities ; Limited development functional
PE 1234 *Abilities ; Underdeveloped administrative*
PE 9620 Ability ageing ; Decline cognitive
PF 2828 Ability application situational demands vocational decisions ; Unrelated
PF 9066 Ability change ; Declining community confidence
PC 1820 Ability developing countries ; Shortage industrial leadership entrepreneurial
PF 7166 Ability forecast ; Overestimation
PE 5746 Ability ; Impairment physicians'
PD 8179 *Ability ; Impairment visual spatial*
PC 0239 *Ability ; Lack*
PE 4398 Ability public sector ; Lack scope intellectual
PF 4668 Ability undertake new initiatives ; Restricted government
PF 0100 Ability ; Underutilization intellectual
PA 6839 Abjection *Disrepute*
PA 6659 Abjection *Humility*
PA 5644 Abjection *Vice*
PC 9042 *Abnormal births*
PE 0472 Abnormal blood pressure
PE 9554 *Abnormal cells ; Marrow displacement*
PF 8040 Abnormal diplomatic relations states
PF 7029 Abnormal ; Fear
PF 9402 Abnormal height ; Discrimination against people
PE 9296 *Abnormal smallness head*
PD 4031 Abnormalities ; Animal
PD 1618 Abnormalities ; Congenital
PD 5535 *Abnormalities eyelids animals ; Conformational*
PD 1618 Abnormalities ; Human physical genetic
PD 8484 Abnormalities ; Memory
PD 9480 Abnormalities ; Plant
PE 5050 *Abnormalities swine ; Nutritionally induced skeletal*
PE 3711 *Abnormalities teeth*
PA 5982 Abnormality *Disagreement*
PA 6799 Abnormality *Disease*
PE 4566 *Abnormality gait*
PA 7157 Abnormality *Insanity*
PD 2198 Abnormality ; Sexual
PA 7280 Abnormality *Wrongness*
PD 0124 Aboard aircraft ; Offences
Abomasal [disorders...]
PE 9364 — disorders cattle
PE 9364 — *displacement ; Left side*
PE 9364 — *displacement ; Right side*
PE 9364 — *displacements*
PE 9364 *Abomasal impaction ; Dietary*
PE 9364 *Abomasal ulcers*
PA 5454 Abomination *Badness*
PA 7338 Abomination *Hate*
PA 5459 Abomination *Uncleanness*
PA 7107 Abomination *Unpleasantness*
PA 5644 Abomination *Vice*
PA 7280 Abomination *Wrongness*
PC 0720 Aboriginals ; Exploitation
PE 4805 Aborted foetuses ; Unethical experimentation using
PD 0158 Abortificants
PD 0158 Abortion
Abortion [Adolescent...]
PD 1302 — Adolescent
PD 0159 — Amateur
PD 3947 — Amniocentesis leading
Abortion [cattle...]
PD 7799 — *cattle ; Foothill*
PE 0924 — Contagious
PF 6169 — Criminalization
PF 6169 Abortion ; Denial woman's right
Abortion [Epidemic...]
PD 7799 — *Epidemic bovine*
PD 7799 — *ewes ; Chlamydial*
PD 7799 — *ewes ; Endemic*
PF 8481 Abortion facilities ; Lack
PD 0159 Abortion ; Illegal induced
PD 0158 Abortion ; Induced
PD 7799 *Abortion large animals*
PF 8481 Abortion methods ; Inadequate
PD 0158 Abortion murder
PD 0173 Abortion ; Natural human
Abortion [practices...]
PE 3580 — practices ; Unsanitary
PE 3580 — procedures ; Inadequate
PJ 3244 — *Prohibitive cost*
Abortion [related...]
PE 3580 — related deaths
PF 1022 — Religious opposition
PE 0924 — ruminants ; Infectious
Abortion [Self...]
PE 3580 — Self
PD 0173 — Spontaneous
PE 5846 — syndrome ; Post
PD 1302 Abortion ; Teenage
PE 3580 Abortion ; Unsafe
PD 0159 Abortion ; Variable criteria legal induced
PD 1302 Abortions ; Adolescent induced
PD 3947 Abortions ; Gender
PD 0158 *Abortions ; Late*
PD 1302 Abortions ; Teenage
PB 0731 *Abrasion*
PB 0855 *Abrasions ; Cuts*
PF 4432 Abroad ; Blocks hostile takeover bids
PE 8746 Abroad ; Denial social security nationals who have lived
PF 3070 Abroad ; Refusal issue foreign currency use
PE 1659 Abroad ; Restrictions socialist citizens working
PJ 2270 Abrogation pacts
PE 9844 *Abscess ; Cervical*
PD 8786 *Abscess eye*
PD 8786 *Abscess eyelid*
PE 5617 *Abscess ; Foot*
PE 5617 *Abscess ; Heel*
PE 9844 *Abscess ; Jowl*
PE 1028 *Abscess ; Liver*
PG 7551 *Abscess ovary*
Abscess [Peritonsillar...]
PE 7733 — *Peritonsillar*
PE 8372 — prostate
PD 7799 — *Prostatic*
PE 7570 *Abscess ; Salivary gland*
PE 7826 *Abscess ; Subsolar*
PE 5617 *Abscess ; Toe*
PG 3510 *Abscesses*
PD 3978 *Abscesses ; Bovine liver*
PE 9037 *Abscesses ; Intra cranial*
PE 9037 *Abscesses ; Intra spinal*
PA 7270 Absence
Absence [Absence...]
PA 7270 — *Absence*
PF 2804 — accountability construction planning
PC 1522 — accountability structures
PF 0076 — audio-visual media
Absence [causal...]
PF 7706 — causal relationship problems ; Assumption
PF 1064 — common structure ethical decision
PF 2350 — communication channels
PF 3307 — community participation
PF 1081 — convincing symbols connecting individual's life cosmos
PE 2603 — *coordinated customs services developing countries*
Absence [democratic...]
PC 1015 — democratic political process
PF 8040 — diplomatic recognition
PE 6350 — *direct sunlight*
PC 7674 Absence essential services
PF 0342 Absence follow-up procedures
PF 5764 Absence God
PF 3607 Absence holy day observance
PF 8829 Absence housing codes
Absence [images...]
PF 3553 — images social responsibility
PF 1345 — imaginative planning expansion income
PA 5473 — *Insufficiency*
Absence [land...]
PD 2338 — land holders
PD 3557 — *Leave*
PE 8597 — legal entities developing countries
PF 6479 — local leadership image
PF 2865 — long-range, world-wide capital flow plan
PF 3610 — long-term economic planning agencies
Absence [maintenance...]
PD 8984 — maintenance system
PD 3789 — management training
Absence [management...] cont'd
PC 4867 — management training
PE 0630 — method appropriate available data
PE 6291 — mycorrhizae
PF 7081 Absence national nongovernmental organizations developing countries
PF 8652 Absence police structures
PF 7033 Absence policies associate international nongovernmental organizations regional development
PF 4836 Absence reliable systems justice developing countries
PF 1674 Absence rites passage
Absence [school...]
PE 4200 — school
PF 5354 — self affirmation
PE 5170 — skilled workers countryside
PD 0065 — social methodologies global ideology
PF 2721 — *social prowess*
PF 6492 — support systems effective community operation
Absence [tactical...]
PF 0327 — tactical methodologies ; Total
PF 0327 — tactical methods
PF 3531 — traditional patterns community life
PF 2575 *Absence youth structures*
PD 7297 Absent fathers
PF 0424 Absent mindedness
PF 6552 *Absent socializing systems*
PD 2338 Absentee business ownership
PD 2338 Absentee land ownership
Absentee landlords
PD 2338 — Unapproachable
PD 2338 — Uninvolved
PD 2338 — urban property
PD 2338 *Absentee owned natural resources*
PD 2338 Absentee ownership
PD 2338 Absentee property ownership
PD 1634 Absenteeism
PE 2033 Absenteeism ; Alcohol related
PD 1634 Absenteeism ; Employee
PD 1634 Absenteeism ; Labour
PE 4200 Absenteeism ; Student
PE 0424 Absentia ; Injustice trials
PE 6413 Absolute idiocy
PA 5939 Absolute properties*complex
PA 7296 *Absolutism*
PA 7296 Absolutism *Restraint*
PF 1098 Absolving military personnel homicide
PD 0305 Absorb aid ; Limited developing country capacity
PD 0321 *Absorption iron ; Inadequate*
PF 0780 Absorption manpower resources military activities
PD 4749 Absorption political leaders ; Self
Absorption Self
PA 7363 — [Absorption ; Self]
PA 7364 — [Absorption ; Self]
PF 7248 — [Absorption ; Self]
PF 7111 Abstention insight ; Enforced
PC 0588 Abstention social processes
PE 3298 Abstention ; Total sexual
PF 2400 Abstract educational methods
PF 5190 Abstract music ; Passionate
PF 6991 Absurdity
PA 6852 Absurdity *Inappropriateness*
PA 6977 Absurdity *Meaninglessness*
PA 7371 Absurdity *Unintelligence*
Abuse [administrative...]
PC 8689 — administrative authority ; Illegal use
PD 5987 — adolescents ; Drug
PF 3302 — Adoption
PD 7598 — *agricultural techniques*
PD 0153 — Alcohol
PE 1558 — amphetamines
PE 4250 — *anabolic steroids*
PE 0043 — animal drugs
PE 1821 — anti-anxiety agents
PF 6629 — antibiotics
PE 8383 — antibiotics vaccines factory farming
PE 1821 — anxiolytics
PE 4250 — athletes ; Drug
PC 8689 — authority
PC 8689 — authority ; Illegal exercise
Abuse [Badness...]
PA 5454 — *Badness*
PF 5991 — banking secrecy
PE 0139 — barbiturates non-barbiturate hypnotics
PF 6735 — brothel legislation
PF 2661 — bureaucratic procedures
PE 2033 — business ; Alcohol
Abuse [Caffeine...]
PE 0618 — Caffeine
PE 1186 — Cannabis
PC 0838 — Child
PC 2584 — Child
Abuse children
PD 7330 — Emotional
PD 3267 — Organized sexual
PD 3267 — Ritual sexual
PD 3267 — Satanic child
PE 3265 — Sexual
Abuse [chronic...]
PE 4245 — chronic neglect ; Infant mortality due parental
PD 2363 — coca cocaine
PE 1329 — *Codeine*
PE 6786 — commercial confidentiality
PE 6786 — commercial secrecy
PE 0742 — community leaders ; Drug
PD 9544 — computer systems

Abuse [confidentiality...] cont'd
- PF 7633 — confidentiality
- PE 7995 — control wild animal populations
- PF 2166 — credit

Abuse [decision...]
- PD 8696 — decision makers ; Drug
- PD 1018 — Dependence government revenue substance
- PE 7568 — doctors ; Alcohol
- PE 7568 — doctors ; Drug
- PD 6002 — dominant market position international trade
- PD 0094 — Drug
- PE 0680 — during control complex equipment ; Substance
- PF 3858 — during pregnancy breast feeding ; Substance

Abuse [economic...]
- PC 6873 — economic power
- PC 6873 — economic power ; Dependence
- PD 2906 — electoral process political parties
- PE 4645 — expense accounts
- PE 6004 — extradition ; Governmental

Abuse government
- PE 4658 — employment
- PD 8696 — officials ; Drug
- PF 8389 — policy
- PC 9104 — power

Abuse [hallucinogenic...]
- PD 0556 — hallucinogenic drugs
- PD 0556 — hallucinogens
- PE 1776 — heroin
- PE 1821 — hypnotics

Abuse [Inappropriateness...]
- PA 6852 — *Inappropriateness*
- PC 6873 — *individual property*
- PC 6307 — *influence*
- PD 9544 — *information systems*
- PE 1427 — Inhalant
- PF 3099 — international cultural, diplomatic commercial exchanges
- PF 2497 — international law
- PD 1975 — IQ evaluations
- PE 0912 Abuse khat
- PF 6723 Abuse kind payments

Abuse [land...]
- PE 5861 — land-locked island countries havens
- PF 9598 — language
- PC 5280 — law
- PE 6622 — Lawsuits
- PD 2363 — Long term effects cocaine
- PD 0556 — *LSD*
- PD 0556 — *Lysergic acid*

Abuse [marijuana...]
- PE 1186 — marijuana hashish
- PD 6881 — Mate
- PE 3980 — media ; Excessive portrayal substance
- PE 3980 — media ; Glorification substance
- PD 0028 — medical drugs
- PD 5770 — medical psychological techniques
- PE 5579 — military personnel ; Drug
- PC 1142 — *Military police*
- PE 1821 — minor drugs
- PE 0988 — monopoly power state-owned state-controlled enterprises
- PE 1329 — *Morphine*
- PD 0213 — Multi drug
- PC 0720 Abuse native peoples
- PD 0713 Abuse ; Nicotine
- PE 1329 Abuse opiates
- PE 1329 *Abuse ; Opium*

Abuse [Patent...]
- PF 3538 — Patent
- PE 0742 — people authority ; Alcohol
- PE 7568 — physicians ; Substance
- PD 0022 — plant drugs
- PC 1142 — police power
- PD 8696 — policy makers ; Drug
- PB 6918 — power
- PE 7390 — power ; Inadequate assistance victims
- PE 0139 — prescription drugs
- PD 0165 — prison labour
- PE 0978 — prisoners ; Drug
- PE 0978 — prisons ; Drug
- PF 5649 — privileges immunities diplomats
- PF 5649 — privileges immunities international civil servants
- PE 7613 — project-tied migration
- PD 0556 — Psychedelic drug
- PC 9104 — public power
- PE 0912 Abuse qat

Abuse [refugee...]
- PE 6375 — refugee status
- PF 5107 — relics
- PC 5280 — *right appeal*
- PE 4183 — rights criminal suspects
- PF 7887 — Ritualistic sexual
- PE 0742 — role models ; Substance
- PE 9688 — Rural drug

Abuse [science...]
- PC 9188 — science
- PE 3105 — science technology capitalism
- PF 2692 — scientific power
- PE 1821 — sedatives
- PE 0139 — sedatives tranquillizers
- PD 2198 — sex
- PE 1821 — sleeping pills
- PE 4250 — *sports ; Drug*
- PD 6881 — Spouse cohabitant
- PF 4564 — statistics ; Official
- PF 2137 — *status religious institutions*

Abuse [strategically...] cont'd
- PJ 9080 — strategically distributed products ; Vulnerability
- PD 6346 — students
- PE 5507 — students ; Drug
- PC 5536 — Substance

Abuse [tax...]
- PF 2137 — *tax exempt status ; Church*
- PE 2370 — tax havens
- PE 4437 — tourism sexual purposes
- PD 4695 — trade union rights
- PF 4054 — traditional cultural expressions peoples
- PD 0826 — *traditional customs*
- PC 4422 — trust

Abuse [Verbal...]
- PD 5238 — Verbal
- PE 1427 — Volatile substance
- PF 3918 — vulnerable groups ; Placement burden proof

Abuse [wards...]
- PF 7755 — wards ; Sexual
- PD 6758 — Wife
- PJ 7167 — *women ; Sexual*
- PD 4514 — work ; Drug
- PD 9805 — work ; Substance
- PD 0094 *Abuse young drivers ; Drug*
- PD 5987 — Youth drug
- PF 9288 Abuses ; Connivance authorities human rights
- PF 9288 Abuses ; Connivance religious leaders human rights
- PF 9288 Abuses ; Government connivance human rights
- PD 0136 Abuses international assistance programs
- PF 9288 Abuses ; Official connivance human rights
- PD 9603 Abuses privacy ; Press
- PE 4847 Abuses private police forces
- PF 8535 Abusive allocation government funding
- PE 2690 Abusive behaviour modification
- PE 9417 Abusive collection specimens
- PE 2932 Abusive detention psychiatric institutions
- PF 8535 Abusive distribution political patronage
- PC 6912 Abusive experimentation humans
- PE 7605 Abusive exploitation cultural heritage
- PD 5238 Abusive language
- PF 4550 Abusive language parliament
- PD 5770 Abusive medical practices
- PD 9538 Abusive monitoring communication governments
- PD 2710 Abusive national leadership
- PC 1142 *Abusive police interrogations*
- PE 1951 Abusive psychosurgery
- PD 8056 Abusive restrictions use water
- PD 7463 Abusive technological development capitalism
- PD 2722 Abusive traffic immigrant workers

Abusive treatment
- PD 5501 — aged
- PE 9611 — animals testing toiletries cosmetics
- PD 0584 — patients psychiatric hospitals
- PD 7821 — women labour

Abusive use
- PF 1741 — skin colourants
- PF 1741 — skin lighteners
- PC 6873 — *trademarks*
- PA 6191 Abusiveness *Disapproval*
- PA 6822 Abusiveness *Disrespect*
- PF 8868 Academic disciplines ; Fragmentation
- PF 9077 Academic disciplines ; Jurisdictional conflict
- PD 0443 *Academic education ; Overemphasis*
- PD 4282 Academic freedom ; Denial
- PF 9693 Academic integrity ; Corrupting effect valuable prizes
- PF 8868 Academic isolationism
- PF 0704 *Academic records ; Falsifying*
- PE 4421 Academic stress children
- PC 0143 Acalculalia
- PC 8534 *Acanthosis ; Acquired*
- PC 6423 Acanthosis animals
- PD 0982 Acariasis
- PE 2727 *Acariasis animals ; Cutaneous*
- PE 2727 *Acariasis animals ; Nasal*
- PD 5228 Acaricide poisoning animals
- PD 0120 *Acaricides pollutants*
- PE 3639 Acarina
- PF 9664 Accelerating pace structural change ; Faltering adaptation economic agents
- PF 9664 Accelerating structural change ; Policy making delays response
- PF 1706 Accelerating time ; Preoccupation
- PE 5141 Accents ; Discrimination against regional
- PE 5141 Accents language ; Discrimination against use
- PE 8569 Accentuated inequality rural urban development
- PF 7897 Accept international tribunal arbitration ; Government refusal
- PF 7897 Accept jurisdiction international courts justice ; Government refusal
- PE 7519 Acceptable sites power plants ; Lack
- PF 5973 Acceptance another person ; Uncritical
- PF 8596 Acceptance authority ; Uncritical
- PF 9205 Acceptance convenient beliefs ; Thoughtless
- PA 7102 *Acceptance ; Dishonourable grant*
- PF 2901 Acceptance dogmas standards ; Uncritical
- PE 6039 Acceptance embryo transfer technology ; Non
- PJ 6238 Acceptance faith ; Graceless
- PF 2499 *Acceptance government legislation ; Non*
- PD 1975 *Acceptance hierarchy*
- PE 7800 Acceptance human rights treaties ; Limited
- PF 0977 Acceptance international treaties ; Limited
- PC 1523 Acceptance mixed races ; Non
- PF 1709 Acceptance new scientific theories ; Non
- PF 1079 Acceptance reality ; Non
- PF 8855 Acceptance socio economic dependency ; Over
- PF 8508 Acceptance stereotypes

- PF 8887 Accepted myths ; Disintegration
- PF 3554 Access appropriate technology ; Minimal
- PF 2875 Access available services ; Limited

Access [capital...]
- PD 3135 — capital bond markets ; Restrictions foreign
- PE 9831 — computer information systems ; Unauthorized
- PE 8516 — credit small farmers ; Limited
- PF 1985 — cultural heritage ; Lack
- PE 8152 — culturally adapted pedagogical material ; Limited

Access [Denial...]
- PF 3082 — Denial distribution
- PE 9440 — developing countries commodity exchanges ; Restrictions
- PF 1471 — developing country products world markets ; Restricted
- PF 6517 — development capital ; Incomplete
- PD 1204 — Difficult water
- PE 2328 — due product differentiation transnational corporations ; Limited market

Access [education...]
- PC 1896 — education countries ; Inequality
- PE 2163 — education ; Unequal
- PF 3354 — eternal life ; Discrimination
- PF 3354 — eternal life ; Elitist
- PF 1653 — external resources ; Limited
- PC 2801 Access food ; Unequal

Access [health...]
- PF 6577 — health services ; Limited
- PF 6458 — *higher education ; Denial right*
- PF 9046 — historical libraries ; Denial right

Access [Inadequate...]
- PF 6473 — *Inadequate industrial*
- PC 0352 — indigenous peoples international decision making processes ; Inadequate
- PE 2401 — information ; Inadequate
- PE 2401 — information resources ; Incomplete
- PE 1743 — interpreters judicial hearings ; Denial right
- PD 1198 Access judicial process ; Economic barriers women's
- PD 8334 Access justice ; Lack
- PF 1953 Access knowledge ; Limited

Access [Lack...]
- PF 6190 — Lack places urban environments encouraging unstructured public
- PE 5812 — land locked developing countries ; Excessive costs sea
- PE 0803 — legal profession, judiciary jury membership ; Economic barriers
- PF 2996 — *Limited transportation*
- PD 1204 — Limited water
- PF 1249 — *liquor ; Easy*
- PE 4704 — loans ; Limited
- PE 4704 — loans ; Restrictive
- PF 3550 Access municipalities capital funds ; Restricted

Access [natural...]
- PF 2882 — natural resource use decisions ; Limited
- PF 2631 — necessary information ; Minimal
- PD 1958 — negotiation employment reward ; Inadequate

Access news
- PF 3081 — Denial
- PF 3082 — distribution media ; Restriction
- PF 6573 — Limited

Access [pharmaceutical...]
- PE 1278 — pharmaceutical drugs ; Limited
- PE 6788 — plant seeds ; Restricted
- PF 8703 — policy formulation processes ; Restricted
- PE 2840 — practical education ; Limited
- PE 8637 — prisoners' defence ; Lack
- PD 1194 — public archives ; Lack
- PD 3335 — public service women ; Denial right

Access [resources...]
- PD 3210 — resources ; Inequitable
- PF 2457 — *Restrictions market*
- PE 1055 — roads ; Inadequate
- PE 1055 — roads ; Poor
- PF 3575 — rural youth education ; Limited

Access [sanitation...]
- PF 8515 — *sanitation services ; Discrimination*
- PF 1303 — social benefits ; Limited
- PF 5120 — social services ; Delay
- PF 6573 — society's resources ; Limited
- PC 4844 — *sophisticated medical equipment ; Restricted*
- PF 4636 Access technological decisions ; Limited
- PF 3467 Access technology agricultural upgrading ; Insufficient
- PF 2535 Access ; Uncertain market
- PF 2583 Access urban services ; Limited
- PE 1055 Access village ; Inadequate road
- PD 8056 Access water ; Inequitable right
- PD 0190 Access women education ; Unequal
- PF 2489 Accessibility ; Limited credit
- PA 6340 Accident *Adversity*
- PE 6352 Accident proneness
- PB 0731 Accident ; Risk
- PE 4086 Accident victims
- PB 0855 *Accidental cuts*
- PE 3153 Accidental explosion ; Risk
- PE 3153 Accidental explosions
- PF 7113 Accidental falls
- PD 1386 Accidental large-scale contamination environment
- PD 3493 Accidental loss nuclear weapons
- PF 4553 Accidental military incidents
- PE 1935 Accidental poisonings
- PE 5404 Accidental weed creation genetic engineering

Accidents [agricultural...]
- PD 5265 — agricultural workers
- PD 1582 — Air
- PE 4247 — animals

Accidents [Automobile...] cont'd
PD 0079 — Automobile
Accidents [Car...]
PD 0079 — Car
PD 4652 — caused fires
PD 2570 — Chemical
PD 6851 — Childhood
PF 7768 — Cover up major
Accidents [Delay...]
PE 8824 — Delay compensation victims motor
PE 5229 — Delay payment compensation victims major
PB 0731 — Dependence
PC 6034 — Disastrous
PB 0731 *Accidents ; Electric current*
PB 0731 *Accidents falling objects*
PE 2857 Accidents ; Firearm
PC 0776 Accidents ; Hazards genetic engineering
PE 4961 Accidents home
Accidents [Inadequate...]
PE 4086 — Inadequate assistance victims
PC 0646 — Industrial
PB 0731 — Injurious
PE 6756 — involving astronauts
PD 2570 Accidents ; Large scale industrial
Accidents [Marine...]
PD 8982 — Marine
PD 5373 — Military aircraft
PD 2278 — Mining
Accidents [Naval...]
PD 8982 — Naval
PE 8824 — Neglect victims motor
PE 8824 — Non payment compensation victims motor
PD 0771 — Nuclear
PD 7579 — Nuclear reactor
PD 3493 — nuclear weapons systems
Accidents [Occupational...]
PC 0646 — Occupational
PE 4961 — Occupational domestic
PE 5080 — orbiting nuclear powered reactors ; Radioactive fallout
Accidents [Panic...]
PE 5324 — Panic result nuclear
PD 0994 — Pedestrian
PD 6851 — Playground
Accidents [Radiation...]
PD 1949 — Radiation
PD 0126 — Rail transport
PD 0126 — Railway
PD 7579 — Reactors
Accidents Risk
PD 0771 — atomic
PD 8982 — maritime
PD 1949 — radiation
PD 0079 — traffic
PF 4346 — unintentional nuclear war due
PD 0079 Accidents ; Road traffic
Accidents [Sanitation...]
PE 4570 — Sanitation system
PE 1990 — School
PE 5080 — space ; Nuclear
PE 4262 — Sports
PD 0126 Accidents ; Train
PC 8478 Accidents ; Transport
PF 2887 Accidents ; Unreported
PC 0646 Accidents work
PE 6119 Accidents ; Young driver
PA 3275 Accidie *Sloth*
PE 5720 *Acclimatization ; Lack*
PD 3465 Accommodation ; Exploitative rental
PF 6156 Accommodation ; Impersonality high density
PD 0758 Accommodation ; Overcrowding housing
PD 1842 Accommodation ; Prohibitive cost
PF 6195 Accommodation ; Socially sterile rental
PD 1251 Accommodation ; Substandard housing
PE 6187 Accommodations single people ; Inappropriate
PE 0873 *Accompanying foreign protein reactions ; Arthritis*
PE 0873 *Accompanying psoriasis ; Arthritis*
PE 0873 *Accompanying ulcerative colitis ; Arthritis*
PF 4839 Accord water use ; Lack
PF 2943 Accorded United Nations non self governing status disputed administering government ; Territories
PF 5301 According cultural norms ; Systematic human behaviour
PE 9054 According financial loans ; Racial discrimination
PF 2827 According own group values ; Subjective assessment
PF 3551 Accountability based solely profit
PF 2804 Accountability construction planning ; Absence
PE 3780 Accountability control production processes ; Public non
PE 0503 Accountability disposal wealth ; Lack
PE 4997 Accountability government action ; Non
PF 2275 Accountability grant funding ; Self
Accountability [Ineffective...]
PC 1522 — *Ineffective student*
PF 6574 — Inhibited official
PC 1522 — Irregular task
Accountability [Lack...]
PC 1522 — Lack social
PC 1522 — Lack structural
PD 1544 — Limited youth
PE 4032 Accountability organizations developing technology ; Public non
Accountability public
PF 6574 — sector management ; Inadequate
PF 6574 — services ; Limited
PF 6574 — services ; Obscure

Accountability [Social...]
PC 1522 — Social non
PF 6574 — state controlled enterprises ; Non
PC 1522 — structures ; Absence
PF 1072 Accountability transnational enterprises ; Non
PF 5295 Accountability ; Unchanging legal precedent undermines
PD 7073 Accounting auditing services ; Protectionism public
PE 8518 Accounting business community ; Lack social
PE 6093 Accounting ; Creative
PE 4285 Accounting information due inflation ; Misleading
PE 2145 Accounting records transnational corporations ; Foreign currency manipulations
PF 0203 Accounting reporting ; Inadequate international standards financial
PE 4645 Accounts ; Abuse expense
PF 7842 Accounts developing countries ; Inadequate systems national
PF 0023 Accounts ; Non valuation housework national
PD 1393 Accretion ; Ice
PF 0316 Acculturation
PD 4272 Acculturation dilution cultural heritage
PD 5510 Accumulated junk
PD 5510 Accumulated junk ; Visible
PF 4174 Accumulated power
PA 4313 Accumulation
PD 0089 Accumulation aerospace hardware earth orbits
PF 7198 Accumulation awards
PC 5225 Accumulation capital
PF 3630 Accumulation capital ; Limited
Accumulation contaminant residues
PD 4925 — freshwater animals, fish birds
PD 3934 — marine animals fish
PD 5021 — plants animals
PD 5278 — terrestrial animals
PD 0381 — terrestrial plants
PD 6907 Accumulation cultural property
PF 2376 Accumulation data
PF 4174 Accumulation functions
PF 7198 Accumulation honours
PF 2376 Accumulation information
PF 2376 Accumulation knowledge
PE 3354 Accumulation misuse religious property
Accumulation [personal...]
PF 4176 — personal data
PF 9653 — personal wealth rulers countries ; Hypocritical
PD 0120 — pesticides ; Bio
Accumulation pollutants
PD 4925 — freshwater wildlife
PD 3934 — marine wildlife
PD 5021 — plants animals
PD 0381 — terrestrial plants
PD 5278 — terrestrial wildlife
PF 8025 Accumulation privileges
PC 8346 Accumulation property
PF 7315 Accumulation recognized merit
PE 5309 Accumulation seas ; Disused machinery
PF 7198 Accumulation titles
PC 5225 Accumulation wealth
PF 9653 Accumulation wealth government leaders ; Excessive
PF 7649 Accuracy official information ; Increasing scepticism
PF 3956 Accursed
PE 7348 Accusation ; Conviction
PE 7348 Accusations ; False
PE 5043 Accused ; Discrimination against families
PE 7345 Accused segregation convicted criminals ; Denial right
PE 7217 Accusers ; Denial right confront
PF 5862 Accusing channels dialogue being restrictive
PD 7420 *Acetonemia*
PE 4630 *Achalasia ; Cricopharyngeal*
PE 1310 Ache ; Back
PA 5471 Achievement∗complex
PF 6537 *Achievement honours ; Infrequent*
PF 7118 Achievement ; Obstacles community
PF 1205 *Achievements ; Lack corporate*
PF 2438 Achieving consensus ; Lack means
PD 2802 Achieving global unity ; Inadequate structures
PF 6506 Achieving grassroots consensus ; Lack systems
PD 7424 *Achilles tendon ; Rupture*
PD 0556 *Acid abuse ; Lysergic*
PD 4904 Acid deposition
PD 0339 *Acid imbalances ; Amino*
Acid [metabolism...]
PE 9291 — metabolism ; Congenital disorders amino
PE 4825 — mist ; Sulphuric
PE 1976 — mists
PD 4904 Acid rain
PE 0082 Acid rain ; Deterioration stained glass due
PD 4904 Acidic precipitation
PD 4904 Acidic snow
PD 3658 Acidic soils
PD 3658 Acidification ; Soil
PC 2270 *Acidosis*
PE 4630 *Acidosis ; Lactic*
PA 3275 Acidy *Sloth*
PD 1763 *Acinonyx ; Endangered species*
PF 5210 Acknowledge contradictions ; Failure
PF 5210 Acknowledge differences ; Failure
PF 5163 Acknowledge failure ; Failure
PF 5210 Acknowledge problems ; Failure
PF 1393 Acknowledging social problems ; Delay religions
PE 3662 Acne
PD 9667 *Acne*
PD 7799 *Acne dairy cows ; Udder*
PD 5228 *Acorns ; Poisoning oak*
PE 8738 Acouchis ; Endangered species agoutis

PE 4109 Acoustic trauma
PD 2831 Acoustical treatment workplaces ; Insufficient
PC 8534 *Acquired acanthosis*
PE 8912 Acquired deformities bone ; Osteomyelitis, periostitis infections involving bones
Acquired [haemolytic...]
PD 7758 — *haemolytic anaemias*
PD 6334 — hospital ; Infections
PD 5111 — human immunodeficiency syndrome
PE 4224 *Acquired immunodeficiency syndrome ; Simian*
PD 6334 Acquired infections ; Hospital
PE 3659 *Acquired toxoplasmosis*
PE 8246 Acquiring skills ; Illiteracy obstacle
PC 1258 Acquisition arms ; Competitive
PF 0699 Acquisition ; Commercialization nationality
PD 5552 Acquisition crimes ; Property
PF 8053 *Acquisition ; Difficult land*
PE 0896 Acquisition foreigners ; Alienation land
PE 6640 Acquisition inappropriate equipment developing countries
PF 1319 Acquisition knowledge ; Restrictions
PC 1005 Acquisition land government ; Compulsory
PD 9653 Acquisition offshore assets heads state ; Corrupt
PE 4496 Acquisition use passports ; Fraudulent
PD 3026 Acquisitions ; Censorship library
PA 4313 Acquisitiveness *Accumulation*
PA 6379 Acquisitiveness *Avoidance*
PA 7211 Acquisitiveness *Selfishness*
PE 3642 Acrididae
PA 5446 Acrimony *Enmity*
PA 7253 Acrimony *Envy*
PA 7156 Acrimony *Moderation*
PE 2125 Acromicria ; Congenital
PF 4286 Acronyms abbreviations ; Excessive use
PC 2367 Across frontiers ; Illegal movement
Act against problems
PF 2431 — Inadequate organizational mechanisms
PF 0559 — Shortage adequately trained personnel
PF 8789 — Shortage equipment materials needed
PD 6845 *Act ; Criminal omission*
PE 2113 Acting foreign countries ; Abduction government agents
PD 2731 *Actinobacillosis*
PE 2353 Actinomycosis
Action against problems
PD 2669 — Inadequate buildings, services facilities organized
PF 1467 — Inadequate coordination
PF 1645 — Inadequate legislation relating
PF 1467 — Inadequate planning
PF 0404 — Shortage financial resources
Action against regimes
PF 2199 — alternative policies ; Government
PF 2199 — hostile policies ; Government
PF 2199 — opposing policies ; Government
PF 6501 Action ; Blocked community
PF 5502 Action commitment ; Progressive reduction government
PA 5553 Action∗complex
PE 9124 Action country level ; Ineffectiveness international organization programme
Action [Dependence...]
PF 9530 — Dependence government
PF 1735 — *Disappointing bureaucrative*
PF 7583 — Disguised negative consequences remedial
Action Dismissal
PE 7620 — trade union representatives following legal strike
PE 7620 — workers following legal strike
PE 7620 — workers prevent legal strike
PF 3162 *Action ; Disruption international*
PB 0848 Action due rivalries ; Blocked
PD 5582 Action environmentalists ; Direct
PE 4454 Action environmentalists ; Sabotage hunting expeditions direct
PF 5502 Action ; Governmental delayed withdrawal commitment
PF 5332 Action governments ; Illegal
Action Inadequate
PF 2016 — facilities international nongovernmental organization
PD 0929 — facilities international organization
PE 7140 — international nongovernmental organization response intergovernmental calls
PE 8776 — organizational mechanism international
Action [Industrial...]
PD 0694 — Industrial
PF 6501 — Ineffective organization community
PF 9564 — Ineffectual intergovernmental multilateral
PE 1375 — intergovernmental programmes national level ; Inadequate coordination
PF 1781 *Action ; Lack united*
PE 1397 Action ; Missing
PE 1265 Action mould tropical climates ; Destructive
PF 6514 Action ; Neglected cooperative
PE 4997 Action ; Non accountability government
Action [Partisan...]
PD 4517 — Partisan reporting governmental
PE 1397 — Persons missing military
PF 2628 — Political opposition administrative
PF 1735 Action ; Reluctance join community
PE 0757 Action ; Requisitioning workers prevent strike
PE 1541 Action trade unions ; Transnational strike
PE 0226 Action transnational enterprises ; Destabilizing financial
Action [Uncontrolled...]
PE 4997 — Uncontrolled executive branch
PF 6514 — Unformed style cooperative
PF 5332 — Unlawful government
PF 1781 — *Unrecognized benefits corporate*
PE 4745 Action ; Violation right trade unions engage political
PE 7080 Actions ; Divisive responses international nongovernmental organization

PF 3790 Actions ; Induction incongruent
PE 6867 Active wood ; Poisonous, allergenic biologically
PE 5005 Activists ; Assassination environmental
PE 6934 Activists ; Denial rights human rights
PF 3819 Activists ; Government complicity killing human rights
PE 6934 Activists ; Government harassment human rights
PF 3819 Activists ; Official cover up government harassment political
PD 5582 Activists ; Sabotage environmental
PE 6934 Activists ; Vulnerability human rights
PF 0780 Activities ; Absorption manpower resources military
PD 3514 Activities aged ; Slowness sensori motor
PF 2575 *Activities ; Boring youth*
PF 3232 Activities ; Counter revolutionary
Activities [Degradation...]
PE 6256 — Degradation mountain environment leisure
PF 9328 — Delayed environmental impact technological
PD 0557 — Dependence subversive
PE 4335 — developing countries ; Restrictions home countries transnational banking
PE 8979 — Disruption food supply due military
PE 7104 — Duplication international nongovernmental
PF 0443 *Activities education ; Interference school athletic*
PF 7730 Activities ; Government complicity illegal
Activities [Impractical...]
PF 3519 — Impractical school
PE 4146 — Ineffective official inspection regulated
PC 1217 — Instability economic industrial production
PE 5302 — Insufficient government spending cost effective
PF 2316 — Insufficient opportunities community
PE 3538 — intergovernmental facilities ; Costly rotation
PF 2575 — *Irregular youth*
PC 8019 — Irresponsible professional
PF 2575 *Activities ; Lack youth*
Activities Limited
PF 2493 — *child*
PF 8105 — recreational
PF 2575 — *youth*
PF 2439 *Activities ; Minimal commercial*
PF 2316 Activities ; Minimal opportunities corporate
PF 4202 Activities ; Non productive athletic
Activities [Para...]
PD 0527 — Para military
PF 4240 — Persistence terrorist
PC 6675 — Pollution military
PD 1566 *Activities ; Restricted extra curricular*
Activities [Separation...]
PD 1544 — Separation youth / adult
PD 1415 — small communities ; Restrictive pattern business
PE 4222 — sports events ; Commercialization athletic
PD 0557 — Subversive
PF 3437 — supernatural entities ; Disruption human
Activities [Underdevelopment...]
PC 0880 — Underdevelopment industrial economic
PF 2575 — *Undirected youth*
PF 2644 — *Unengaging family*
PF 2605 — *Unfinanced sports*
PF 2575 — *Uninviting youth*
PJ 9820 — Unlawful interference marine
PC 6641 — Unreported income business
PF 2575 — *Unsupported youth*
PD 4728 Activities ; violation entrepreneurial
PF 5413 Activity ; Attraction illegal
PD 5344 Activity children ; Parental toleration sexual
Activity [Denial...]
PD 4695 — Denial right trade union
PC 1449 — Dependence inhumane scientific
PB 5250 — Disruption ecosystems human
PE 0895 — due licensing arrangements ; Restrictions export
PE 0736 — during peace time ; Environmental degradation military
PD 3386 Activity ; Enforced participation community
PF 5868 Activity ; Excessive sexual
PF 9728 Activity ; Failure recognize environment resource base economic
PF 6504 Activity ; Fragmented patterns community
Activity [Ill...]
PF 3370 — Ill considered missionary
PF 0198 — Imbalanced research
PC 1587 — Impaired mental
PF 5374 — Inadequate monitoring public sector
PF 1682 — Individualistic perception sexual
PF 7264 — Ineffective monitoring illegal
PC 1449 — Inhumane scientific
PC 5601 — International imbalance economic
PF 1202 — Irrelevant scientific
PC 1153 — Irresponsible scientific technological
PC 0716 Activity ; Lack physical
PF 3863 Activity ; Lack religious
Activity [malnourished...]
PE 4465 — malnourished children ; Reduced
PE 6844 — media ; Excessive portrayal terrorist
PD 7231 — Misuse research cover illegal
PE 2197 — *Monotonous*
PE 5132 Activity national affairs ; Concentration national governments
PF 5360 Activity ; Political displacement
PD 7231 Activity research ; Misrepresentation socially unacceptable
PE 3910 Activity rural workers ; Denial right union
PE 1355 Activity special groups ; Denial right union
PE 2449 Activity ; Supplying firearms criminal
PE 0109 Activity transnational enterprises ; Monopolistic
PF 6182 Activity urban environments ; Unhealthy lack daily physical

PF 0053 Actor identify clarify world interests ; Lack autonomous world level
PF 7076 Acts conscience ; Bias
PB 1151 Acts God
PE 4786 Acts ; Maltreatment animals used theatre
PF 8754 Acts ; Superstitious symbolic
PB 0869 Acts ; Unnatural
PD 9022 Acts ; Wrongful
PD 1942 Actuality ; Lack essential freedom due unawareness
PF 0979 Actualizing stance being victim external forces ; Self
PF 3307 *Actuation structures ; Lack*
PF 8352 Acupuncture
PE 2429 Acute alcoholic intoxication
PE 5524 *Acute bovine pulmonary emphysema*
PE 5524 *Acute bovine pulmonary oedema*
PE 2293 *Acute congestion lungs*
Acute [death...]
PD 5453 — *death syndrome*
PD 5971 — diarrhoeal disease
PD 5971 — Diarrhoeal disease
Acute [encephalitis...]
PE 9037 — *encephalitis*
PE 3898 — endocarditis
PD 3978 — *equine diarrhoea syndrome*
PD 9045 *Acute intestinal diseases*
PD 3978 *Acute intestinal obstructions animals*
PE 2653 *Acute laryngitis*
PC 8534 *Acute lymphadenitis*
Acute [metritis...]
PD 7799 — *metritis animals*
PG 9737 — *miliary tuberculosis*
PE 8456 — myocardial infarction
PD 0448 — *myocarditis*
PE 5444 *Acute otitis externa*
PD 3978 *Acute pancreatic necrosis animals*
PD 0448 *Acute pericarditis*
PD 9293 *Acute renal failure animals*
PE 7591 Acute respiratory infections
PE 1028 *Acute yellow atrophy*
PF 1578 Adapt general initiatives specific needs ; Failure
PE 5768 Adapt shift work ; Inability
PF 7161 *Adaptability women ; Poor physiological*
PF 6570 *Adaptable products ; Unresearched*
PF 5997 Adaptation ; Biological
PE 5408 Adaptation community status quo ; Family
PA 8178 Adaptation*complex
PF 9664 Adaptation economic agents accelerating pace structural change ; Faltering
Adaptation [indigenous...]
PE 8002 — indigenous society non socialist countries ; Lack economic
PE 8002 — indigenous systems ; Lack economic
PF 3184 — Insufficient technological
PE 8700 Adaptation policy educational difficulties ; Inadequate
PF 5248 Adaptations ; Restrictions
PF 6078 Adapted nutritional habits food aid ; Modification environmentally
PE 8152 Adapted pedagogical material ; Limited access culturally
PE 5246 Adaptive local structure social care ; Non
PD 5987 Addicted children ; Drug
PD 6324 Addiction
PD 5344 Addiction children ; Parental toleration drug
PF 1288 Addiction ; Computer
Addiction [Delinquency...]
PD 5750 — Delinquency side effect drug trade drug
PD 3825 — Drug
PD 3825 — *drug producing countries ; Increasing drug*
PF 0958 Addiction ; Exercise
PE 4951 Addiction ; Stress
PF 6345 Addiction ; Video game
PF 4753 Addictive behaviour ; Government approval
PF 4735 Addictive drugs ; Prohibition
PE 4609 Addicts ; Children drug
PG 5181 *Addison's disease*
PD 9654 *Addison's disease*
PD 1714 Additives ; Animal feed
PD 0487 Additives ; Carcinogenic food
PE 3035 Additives mineral waters ; Cancer causing
PD 0487 Additives ; Toxic food
PD 1714 Additives ; Undesirable effects animal feed
PF 3608 Address ; Gap function social techniques needs they
PD 4287 Address ; Imposition intimate forms
PE 2355 Adenoids
PE 9438 *Adenomatosis ; Pulmonary*
PE 2355 Adenovirus infections
PE 3547 Adequacy ; Seasonal roadway
PE 8496 Adequate clothing supply dwarfs, midgets giants ; Lack
PC 3319 Adequate food indigenous populations ; Denial right
Adequate [health...]
PE 8924 — health care disabled ; Denial right
PD 5254 — housing ; Denial right
PC 3320 — housing indigenous peoples ; Discrimination against indigenous populations housing Denial right
PD 2028 Adequate medical care ; Denial right
PF 5933 Adequate military defence ; Lack
PD 9219 *Adequate occupational performance ; Passive resistance demands*
PF 0344 Adequate standard living ; Denial right
PE 5484 Adequate standard living indigenous peoples ; Denial right
PC 0834 Adequate welfare services ; Denial right
PF 0559 Adequately trained personnel act against problems ; Shortage
Adherence international
PE 8280 — law ; Inadequate national supervision

Adherence international cont'd
PJ 8840 — transit conventions ; Lack
PE 5789 — transit conventions land locked countries ; Lack
PD 9045 *Adhesions ; Peritoneal*
PD 2728 *Adiaspiromycosis*
PE 7837 Adjacency saline water
PE 2844 Adjustment assistance industries labour affected developing country exports ; Inadequate
Adjustment [debtor...]
PD 8114 — debtor developing countries ; Excessive social costs structural
PF 1503 — difficulties new urban families
PC 0300 — *disorders*
PF 3062 Adjustment ; Inadequate mechanism balance payments
PF 4564 Adjustment official statistics government ; Biased
PC 0300 *Adjustment reaction adolescence*
PF 9664 Adjustment ; Retardation economic structural
PF 4564 Adjustment statistics ; Biased
PF 9664 Adjustment world economy ; Faltering structural
PF 4165 Adjustments ; Inadequate trade related structural
PF 2943 Administered territories ; Lack United Nations jurisdiction
PF 2943 Administering government ; Territories accorded United Nations non self governing status disputed
PE 3828 Administering medical drugs non-medical purposes
PF 0175 Administration aid ; Inefficient
PD 0412 Administration criminal justice ; Delay
PF 0903 Administration developing countries ; Inefficient public
PE 5009 Administration developing countries ; Inefficient tax
Administration [Ideological...]
PF 1040 — Ideological overemphasis economic
PE 1234 — Ignorance
PF 2335 — Inefficient public
PF 2113 — international justice ; Ineffective
Administration justice
PE 6529 — Cultural discrimination
PF 1487 — Delay
PD 6927 — Denial human rights
PD 0986 — Discrimination
PE 1399 — Economic discrimination
PE 1399 — Educational discrimination
PE 1347 — Geographic discrimination
PD 0986 — Inequitable
PD 8217 — Lack confidence
PD 5515 — Offences against
PE 1828 — Political discrimination
PE 0168 — Religious discrimination
PE 8881 — Unrestrained use force
PE 8829 — Use undue influence obstruct
PD 5119 Administration medical care ; Delay
PC 2649 Administration ; Unjust election
PF 1559 Administration ; Unpredictable public
Administrative [abilities...]
PE 1234 — *abilities ; Underdeveloped*
PF 2628 — action ; Political opposition
PF 2261 — agencies ; Inaccessible
PC 8689 — authority ; Illegal use abuse
PF 3318 *Administrative backlog ; Government*
Administrative [capacity...]
PF 0903 — capacity developing countries ; Limited civil
PF 2711 — centralization ; Excessive bureaucratic
PF 2261 — centres ; Inaccessible
PF 3411 — *control ; Lack*
PC 0279 — corruption
Administrative [delays...]
PC 2550 — delays
PD 1576 — detention charge
PE 1793 — difficulties new states
PE 2603 Administrative entry procedures ; Distortion international trade discriminatory customs
PE 7780 Administrative impediments obtaining travel visas
PF 8158 Administrative overhead costs ; Prohibitive
PF 8158 Administrative overhead erosion funding effectiveness
Administrative [personnel...]
PF 0903 — personnel developing countries ; Lack trained
PF 4558 — pledges ; Breach
PC 1820 — process ; Inadequate control over government
PF 2335 Administrative reform ; Lack
Administrative [skills...]
PE 0046 — skills developing countries ; Inadequate
PF 0845 — structures ; Government reluctance transform
PF 1389 — structures ; Paralyzing patterns villages
PF 8711 Administrators funding agencies programme formulators ; Collusion
PF 2477 *Administrators ; Improperly trained*
PA 6491 Admiration ; Self
PA 7211 Admiration ; Self
PF 3184 *Admission ; Limitations school*
PF 5163 Admit error ; Refusal
PF 5296 Admit mistakes ; Failure government
PF 5296 Admit mistakes ; Refusal government leaders
PC 0300 *Adolescence ; Adjustment reaction*
PE 9256 Adolescence ; Avoidant disorder
PF 6173 Adolescence ; Inadequate recognition institutions transition
PD 1302 Adolescent abortion
PC 0877 Adolescent disturbance
PD 0614 Adolescent fatherhood
PF 6061 Adolescent health care ; Neglect
PD 1302 Adolescent induced abortions
PD 0614 Adolescent motherhood
Adolescent [parenthood...]
PD 0614 — parenthood
PD 0614 — pregnancy
PD 6213 — prostitution
PD 7439 Adolescent sexual intercourse

Agencies

PE 5771 Adolescent suicide
PD 5987 Adolescents ; Drug abuse
PD 1611 Adolescents ; Excessive consumption alcohol
PE 0989 Adolescents ; Mental illness
PD 8340 Adolescents ; Runaway
PE 6219 Adolescents ; Smoking
PE 7563 Adopt appropriate exchange rate policies ; Inability developing countries
PF 4200 Adoptable children ; Decreasing number
PF 3302 Adoption abuse
PF 3281 Adoption ; Annulment
PF 7353 Adoption children ; Impediments
PF 3302 Adoption children ; Sham
PD 0590 Adoption ; Inadequate laws
PF 3282 Adoption ; Refusal
PF 3282 Adoption ; Refusal parents' consent
PF 3302 Adoption ; Trafficking children
PE 4296 Adoptions ; International
PF 4205 Adoptive parents ; Decreasing number
PG 5181 *Adrenal cortical hypofunction*
PD 8301 *Adrenal glands ; Diseases*
PD 8301 *Adrenal glands ; Dysfunction*
PD 1544 Adult activities ; Separation youth /
PE 7209 Adult criminals ; Denial right juvenile criminals segregation
PF 6531 Adult direction ; Insensitive
Adult education
PF 6531 — Inadequate
PF 6531 — Infrequent
PF 6531 — Irrelevant
PF 6531 — Minimal
PF 6531 — Undefined
PF 6531 — Unfocused
PF 6531 — Unfunded
PF 6531 — Uninitiated
PF 6531 — Unplanned
PF 1403 *Adult guidance ; Inadequate*
PD 8723 Adult illiteracy
Adult [leadership...]
PF 6462 — leadership ; Unprepared
PF 6167 — life small households ; Inhibition
PF 2386 — *loneliness*
PF 6472 *Adult preliteracy ; Inhibiting*
Adult [sports...]
PD 2283 — sports ; Overemphasized
PE 6258 — students ; Discrimination against
PF 1403 — *supervision ; Poor*
PF 6531 Adult training ; Minimal opportunities
PF 6531 Adult training ; Non local
PF 2883 Adult values ; Static superficial
PF 4274 Adult youth relationships ; Suspicion
PD 9433 Adulteration
PE 1349 Adulteration commercial alcohol
PE 0456 Adulteration illicit drugs
PF 2314 Adultery
PF 6519 Adultism
PE 2268 Adults family decisions ; Exclusion pre
PE 3992 Adults ; Inadequate guardianship mentally retarded
PF 6531 Adults ; Insufficient education
PF 1731 Adults ; Limited community responsibility
PF 2386 *Adults ; Loneliness*
PE 5104 Advance ; Antagonism employment policy technical
PD 0493 Advance ; Desert
PC 2506 Advance ; Desert
PE 7898 Advance ; Fast irregular pace technological
PE 7943 Advance ; Functional changes due technological
PD 4445 Advanced nuclear warfare technology ; Super power monopoly
PD 0847 Advanced training facility
PF 8552 *Advancement ; Blocked skills*
PF 7806 Advancement ; Inability educational systems keep pace technological
PF 5203 Advantage ; Failure sacrifice any personal
PD 0351 Advantage ; Restrictions international freedom movement national
PE 8509 Advantage training business, industry public service ; Inability communities take
PF 5736 Advantages laxity their legislation ; Countries deriving
PF 8181 Advantages ; Overstated programme
PF 4984 *Advantages ; Undefined preschool*
PF 4133 Adversaries ; Fixation
PJ 2601 Adversary culture
PF 9697 Adverse conditions foreign trade ; Vulnerability
PF 3931 Adverse consequences scientific technological progress
PF 6210 Adverse economic shocks external factors
PE 1771 Adverse effect transnational corporations balance payments
Adverse effects
PD 0767 — damming
PE 2890 — high-yield grain
PE 9134 — power production weather
PE 9020 — urbanization climate
PE 8476 Adverse impact pollution growth crops
PE 5820 Adverse impact technology transfer transnational corporations
PF 5868 Adverse sexual development
PD 3557 Adverse social consequences excessive employment married women
PC 0293 Adverse weather conditions
PD 9170 Adverse working conditions
PA 6340 Adversity
PF 8163 Adversity ; Seasonal
PD 0433 Advertiser controlled television programming
PE 4225 Advertisers social responsibility ; Insufficient awareness
PE 3980 Advertising alcoholic beverages ; Irresponsible

PE 3814 Advertising claims products ; Exaggerated
PD 5034 Advertising clutter
Advertising [education...]
PD 9370 — education ; Intrusion
PE 4225 — effects ; Harmful
PD 0606 — entertainment ; Incorporation
PD 5034 — explosion
PD 2547 Advertising ; Gratuitous sex
PE 2156 Advertising hoardings billboards ; Unaesthetic location
PD 2547 Advertising ; Indecent
PD 7108 Advertising industry ; Protectionism
Advertising Irresponsible
PE 4225 — [Advertising ; Irresponsible]
PE 3980 — liquor
PE 2390 — pharmaceutical
PE 9093 — tobacco cigarette
PE 8467 Advertising ; Lack legislative control
Advertising [Maltreatment...]
PE 4786 — Maltreatment animals
PE 3814 — Misleading
PE 7502 — Misleading endorsement
PE 4225 — Misuse
PE 0081 — Monopoly power due
PE 4225 Advertising ; Negative economic social effects
PD 2547 Advertising ; Nudity
Advertising [practices...]
PE 3814 — practices ; Fraudulent
PF 1176 — *Prohibitive cost*
PD 5034 — Proliferation
PE 1810 — Proliferation direct mail
PD 0606 Advertising ; Subliminal
PE 2004 Advertising transnational corporations developing countries ; Harmful effects
PE 2193 Advertising transnational corporations ; Domination
PE 5190 Advertising ; Vulgar combination sacred erotic
PF 6395 Advice ; Biased
PF 1981 Advice consultation problems ; Inadequate technical
PF 7149 Advice elders ; Failure follow
PF 2477 *Advice ; Inadequate career*
PF 7283 Advice leadership ; Myopic
PF 7820 Advice ; Neglect expert
PF 7467 Advisers ; Failure obey religious
PE 7654 Advisors ; Self satisfied style informal leadership
PF 2477 *Advisors ; Untrained youth*
PE 5952 Advocacy discrimination ; Denial right freedom
PC 0237 *Advocate ; Inaccessible issues*
PD 1964 Advocating overthrow government
PE 3621 Aedes mosquitoes vectors disease
PD 1416 Aeration ; Lack soil
PF 9167 Aerial explosions unknown origin
PD 0124 Aerial piracy
PE 3972 Aerial surveillance foreign powers
PE 2069 Aeroallergens
PE 1504 Aerosols industrial hazards
PD 0089 Aerospace hardware earth orbits ; Accumulation
PE 7747 Aerospace monopolies
PB 2542 *Aestheticism*
PD 8588 Aetiology ; Diseases unknown
PD 4233 Affair ; Ending love
PF 5382 Affair honour
PE 5132 Affairs ; Concentration national governments activity national
PF 2314 Affairs ; Extra marital
PD 0579 Affairs ; Government loss leadership role world
PE 5132 Affairs ; Insufficient work awareness international
PF 6536 Affairs ; Lack responsible involvement community
Affairs [Restriction...]
PE 7945 — Restriction press coverage legal
PF 3072 — Restriction press coverage parliamentary
PF 3072 — Restrictions direct news coverage parliamentary
PF 3073 — Restrictions news coverage legal
PC 3185 Affairs states ; Foreign interference internal
PC 3185 Affairs states ; Foreign intervention internal
PE 6086 Affairs ; Violation right participate conduct public
PE 8384 Affairs workers' organizations ; Interference employers
PA 6400 Affectation
PA 7411 *Affectation Uncommunicativeness*
PE 2844 Affected developing country exports ; Inadequate adjustment assistance industries labour
PE 9324 Affecting multiple systems ; Congenital syndromes
PC 0981 Affection ; Children deprived
PA 7145 *Affection∗complex ; Discriminative*
PE 8822 Affections bronchial tubes lungs ; Inflammatory
PD 0448 *Affections heart ; Inflammatory*
PE 0258 Affective disorder ; Seasonal
PD 9159 Affective disorders
Affective [psychoses...]
PD 1318 — psychoses
PD 1318 — psychosis
PD 0438 — *psychosis ; Schizo*
PD 4233 Affective relationship ; Breakup
PE 4071 Affiliate internationally ; Violation right workers' organizations
PD 3255 Affiliation ; Denial parental
PC 3365 Affiliation ; Segregation based religious
PF 5354 Affirmation ; Absence self
PA 6340 *Affliction Adversity*
PA 5454 *Affliction Badness*
PA 6799 *Affliction Disease*
PF 9043 Affliction malevolent spirits
PA 7107 *Affliction Unpleasantness*
PF 5864 Affluence ; Pursuit
PF 5864 Affluent living standards ; Pursuit
PF 7695 Afforestation ; Inappropriate
PA 6822 *Affrontery Disrespect*
PA 7253 *Affrontery Envy*

PA 6979 *Affrontery Opposition*
PE 9260 Aflatoxicosis
PD 2728 *Aflatoxicosis*
PE 7514 Aflotoxins
PE 8684 African cane rat ; Endangered species
PD 2730 *African green monkey disease*
PE 1805 African horse sickness
PE 8253 African mole rat ; Endangered species
PE 5207 African swine fever
PE 1778 African trypanosomiasis
PC 1966 After retirement ; Income maintenance
PD 1544 *Afterschool engagement ; Unstructured*
PE 9795 Agalactia ; Contagious
PD 7799 *Agalactia syndrome*
PA 7333 Age
PD 0162 Age benefits ; Denial right old
PD 5501 Age care ; Negligent old
PF 0756 Age communication ; Insufficient inter
Age [Denial...]
PE 1693 — Denial right minimum work
PE 4649 — Deterioration mind
PE 4649 — Diminishing mental capacity
PC 2541 — discrimination
PE 6069 — Discrimination against women retirement
PD 2318 — discrimination employment
PC 2541 — discrimination law
PD 1611 — drinking ; Under
Age [gap...]
PF 1705 — *gap ; Leadership*
PF 6136 — group ; Household segregation
PF 6570 — groups ; Stifled potential social interaction different
PC 1966 — Inadequate income old
Age [law...]
PE 1693 — law ; Lack minimum
PE 1693 — laws ; Failure enforce minimum
PD 1611 — laws ; Unobserved drinking
PD 0633 — Loneliness old
PC 8310 — Loss capacity
PB 0477 Age ; Old
PA 7131 *Age Oldness*
Age [pensioners...]
PD 0276 — pensioners' homes ; Old
PE 7942 — pensions men women ; Unequal distribution old
PF 5956 — pensions ; Progressive erosion old
PC 2541 — prejudice
PC 1966 Age ; Reduced income old
PD 3444 Age segregation
PD 5501 Aged ; Abusive treatment
Aged [Denial...]
PD 0512 — Denial right welfare services
PD 8399 — Dependence
PD 0096 — during disasters ; Lack protection
PD 0276 Aged ; Inadequate housing
PD 0512 Aged ; Inadequate welfare services
Aged [Mental...]
PD 0919 — Mental disorders
PE 0648 — Misallocation resources protect
PD 2341 — *Murder*
PD 8945 Aged ; Neglect
PF 1664 Aged ; Non transferring wisdom
PD 8561 Aged persons ; Increasing requirements
PD 8945 Aged ; Reduction motivation care
PD 3515 Aged ; Rigidity inadaptability
Aged [Slowness]
PD 3514 — Slowness sensori motor activities
PD 3517 — Social disadvantage
PD 3518 — Social withdrawal
PF 1825 Aged ; Unmeaningful social roles
PE 1625 Aged ; Violation right equal work benefits
PB 0477 Ageing
PE 9620 Ageing ; Decline cognitive ability
PF 6570 *Ageing ; Fear*
PB 0477 Ageing ; Human
Ageing [industrial...]
PE 2866 — industrial plants processes
PD 5350 — industrial sectors
PD 0645 — infrastructure
PD 8561 Ageing populations
PC 0027 *Ageing rural population*
Ageing [war...]
PD 0874 — war disabled
PE 6784 — women
PE 6784 — women ; Negative self image
PC 0027 — world population
PC 2541 Ageism
Agencies [Absence...]
PF 3610 — Absence long term economic planning
PE 9700 — Arrears financial payments government
PF 9561 — Arrogance intergovernmental
Agencies [competition...]
PF 4095 — competition ; Protection public sector
PF 8046 — Corruption monitoring
PF 8046 — Corruption regulatory
PF 1862 — Cultural diversity ignored social service
PE 8308 Agencies each country ; Jurisdictional conflict antagonism government
PF 3458 Agencies ; Environmental unaccountability government
Agencies [Inaccessible...]
PF 2261 — Inaccessible administrative
PF 5351 — Inaccessible government
PF 4937 — Inadequate world regulatory
PJ 1719 Agencies ; Lack coordination
PF 8946 Agencies ; National hegemony over United Nations
PE 8332 Agencies ; North South spheres influence UN related

Agencies [official...]
PF 8650 — official contributions ; Imbalance disbursements multilateral
PE 2719 — officials ; Antagonism government
PF 1070 — Over reliance economic interest groups policy
PF 8711 — Agencies programme formulators ; Collusion administrators funding
PE 0988 — Agencies ; Unaccountability decentralized government
PF 9184 — Agencies ; Unaccountable government intelligence
Agencies United Nations
PE 0777 — Inadequate relationship international nongovernmental organizations specialized
PE 0106 — Inadequate relationship transnational corporations specialized
PE 2486 — Jurisdictional conflict antagonism specialized
PE 8799 — Jurisdictional conflict antagonism specialized
PF 9184 — Agencies ; Unreported cooperation intelligence
PF 8072 — Agencies ; Unresponsive public
PD 0183 — Agencies ; Wastage resources government
PF 2990 — Agency agreements ; Secret government
PF 2216 — Agency competition ; Social
PF 6506 — *Agency cooperation ; Insufficient*
PD 4501 — Agency operations ; Covert intelligence
PF 1776 — *Agency priorities ; Conflicting*
PG 8893 — *Agency referral ; Uncoordinated*
PF 1267 — *Agendas ; Unpublicized meeting*
Agents [Abuse...]
PE 1821 — Abuse anti anxiety
PF 9664 — accelerating pace structural change ; Faltering adaptation economic
PE 2113 — acting foreign countries ; Abduction government
PC 0119 — Airborne toxic harmful
PE 1578 — Agents ; Betrayal double
PE 7103 — Agents ; Blister
Agents [Carcinogenic...]
PD 1239 — Carcinogenic chemical physical
PC 3595 — causing infectious diseases
PC 1192 — Chemical
PD 8392 — Corruption practised government
PD 0183 — Agents ; Extravagant waste resources government
Agents [Fear...]
PD 9498 — Fear government
PE 6715 — foreign principals ; Political contributions
PD 8392 — Fraud government
PC 1306 — Agents ; Harmful synergistic interaction biological
PF 2717 — Agents ; Lack postal
PE 4555 — Agents ; Lung damaging
PD 5422 — Agents ; Negligence real estate
PE 5696 — Agents occupational hazards ; Biological
Agents [Perjury...]
PD 1893 — Perjury government
PD 4248 — Psychochemical
PE 9857 — public office ; Secret intelligence
PE 1578 — Agents ; Treachery double
PE 5683 — Agents ; Vulnerability diplomatic
PF 1744 — Ages ; Ice
PF 4436 — Aggrandizement ; Self
PD 2710 — Aggrandizing national leaders ; Self
PD 0583 — Aggravated assault
PD 5575 — Aggravated larceny
PA 7378 — Aggravation
PA 6340 — Aggravation *Adversity*
PA 7378 — Aggravation *Aggravation*
PF 1172 — Aggravation cyclical recession
PA 7253 — Aggravation *Envy*
PE 0980 — Aggravation instability exchange rates transnational corporations
PA 7107 — Aggravation *Unpleasantness*
PA 0587 — Aggression
PC 7517 — Aggression against natural energy sources
PE 0403 — Aggression against nuclear power sources
PF 7490 — Aggression ; Biologically determined
PC 2064 — Aggression ; Bureaucratic
PD 3907 — Aggression ; Childhood
Aggression [Defence...]
PA 5445 — *Defence*
PA 0587 — Dependence
PB 0968 — Dependence international
PA 5532 — *Disaccord*
PA 7143 — *Discourtesy*
PC 0840 — Aggression ; Economic
PE 1174 — Aggression experiments ; Maltreatment animals
PB 0968 — Aggression ; International
PC 7559 — Aggression ; International
Aggression [Physical...]
PA 0587 — *Physical*
PD 8877 — *Political*
PA 0587 — *Psychological*
PE 0524 — *Aggressive animals*
PC 2321 — Aggressive animals
PE 7175 — Aggressive domesticated animals
PE 9420 — Aggressive economic destabilization countries external forces
PC 4667 — Aggressive foreign policy
PE 0793 — Aggressive honey bees
PF 5691 — *Aggressive ideologies*
PA 6448 — *Aggressive impulses ; Loss control*
PF 5190 — Aggressive music
Aggressive [people...]
PA 0587 — people
PD 9219 — *personality disorder ; Passive*
PE 2390 — promotion pharmaceutical products
PD 0408 — Aggressive uses natural energy resources
PC 3151 — Aggressive wild animals
PE 2206 — Agitans ; Paralysis
PA 5838 — Agitation

PA 5838 — Agitation *Agitation*
PA 6030 — Agitation *Fear*
PA 5467 — Agitation *Inexcitability*
PA 7156 — Agitation *Moderation*
PD 2490 — Agitation ; Nationalist
PJ 5138 — Agitation ; Political
PA 7226 — Agitation *Unhealthfulness*
PE 5150 — Agnosia
PF 2333 — Agnosticism
PE 0527 — Agoraphobia
PE 8738 — Agoutis acouchis ; Endangered species
PF 8026 — Agranulocytosis
PD 2730 — *Agranulocytosis ; Feline*
PE 0280 — Agraphia
PD 9189 — Agrarian reform ; Lack
PF 8650 — Agreed contributions international organizations ; Government non payment
PF 3913 — Agreed multilateral assistance ; Delay supply
PC 4591 — Agreed standards ; Failure conform internationally
PC 4591 — Agreed standards ; Resistance internationally
PE 5239 — Agreed toxicity thresholds ; Failure respect internationally
PF 7838 — Agreed trade commitments ; Conditional observance multilaterally
PE 5762 — Agreement ; Breach
PE 2816 — Agreement cooperation ; Lack
PE 7235 — Agreement ; Denial right freedom imprisonment failure fulfil contractual
PC 9584 — Agreements ; Circumvention international trading
PD 0340 — Agreements ; Distortion international trade discriminatory preference
PE 4581 — Agreements ; Exclusive sales representation
PE 7300 — Agreements ; Failure governments implement provisions ratified human rights
Agreements [Inadequately...]
PF 5421 — Inadequately worded
PC 9513 — Ineffective
PF 6992 — Ineffective international
PF 2497 — *Ineffectiveness international commodity*
PF 4787 — Inequitable peace
PF 5923 — Insufficient maritime
PE 4003 — Agreements ; Non implementation workers wage increases provided legislation collective
PF 4787 — Agreements ; One sided
Agreements [Parochial...]
PD 2469 — Parochial monetary
PE 4039 — Patent pool cross licensing
PE 2324 — Persistence hostilities following cease fire
PE 2324 — Persistence technical state war following cease fire
PD 0343 — product standards ; Restrictive
PF 2346 — *Agreements ; Risky rental*
Agreements Secret
PE 6786 — commercial
PF 2990 — government agency
PF 0419 — international
PE 9267 — Agreements substance ; Symbolic international
PD 5876 — Agreements ; Subversion international
PF 2497 — Agreements ; Transgression international
PE 5762 — Agreements ; Violation
PF 2497 — Agreements ; Violation international
PD 7598 — Agribusiness
PE 4045 — Agricide
PF 1205 — *Agricultural approach ; Individualistic*
PE 2698 — Agricultural business training ; Limited availability technical
Agricultural [capital...]
PF 2378 — *capital ; Inadequate*
PF 2378 — *capital ; Lack*
PD 7993 — chemicals ; Hazards
PE 4523 — commodities developing countries ; Inadequate trade
PF 5961 — cooperative structures ; Obsolete
PE 4744 — crops ; Harmful effects air pollution
PD 7480 — crops ; Loss
Agricultural development
PE 4140 — developing countries ; Insufficient fertilizers
PF 1768 — *Fragmented*
PF 2164 — Lack
PE 5274 — Misuse tropical rain forests
PF 6475 — Pessimistic approach
PF 1768 — Uncoordinated efforts
PF 1768 — *Unintegrated*
PC 8419 — Unsustainable
PJ 9259 — Agricultural diversification costs ; Prohibitive
Agricultural education
PE 9096 — developing countries ; Inadequacy
PF 8835 — lack
PF 8835 — Lack
PF 8835 — Limited
Agricultural [effluent...]
PE 8504 — effluent
PE 3934 — effluents animal husbandry
PF 4108 — equipment developing countries ; Insufficient
PE 1448 — equipment industry transnational corporations ; Domination
PF 1768 — *expansion ; Limited*
PE 4956 — exports imports developing countries ; Imbalance
PF 6499 — Agricultural facilities ; Inadequate
PE 5687 — Agricultural fraud
PE 6788 — Agricultural genetic resources ; Monopolization
PF 4326 — Agricultural growth capita ; Decreasing
PE 8171 — Agricultural income level rural communities ; Subsistence
PD 1965 — Agricultural labour developing countries ; Contempt
Agricultural land
PE 8324 — cash crops ; Degradation

Agricultural land cont'd
PD 4273 — Cultivation marginal
PD 9118 — Damage
PE 3786 — dams ; Inundation
PD 7480 — Declining productivity
PD 9118 — Destruction
PE 3786 — due artificial flooding ; Loss
PD 6103 — fuel production ; Use
PC 7597 — Loss
PD 9189 — Maldistribution
PC 7597 — Need
PE 9407 — Shortening fallow periods
PC 7597 — Unavailability
PE 5931 — urbanization ; Loss
PD 6960 — Agricultural lands ; Disruption ecosystems marginal
Agricultural livestock production
PD 0376 — Environmental hazards
PE 0524 — Health risks workers
PE 8998 — Instability
PD 0629 — Underdevelopment
PF 0377 — Agricultural loss draught animals
Agricultural machinery
PF 4108 — developing countries ; Insufficient
PF 4108 — Lack
PF 7504 — Unavailability replacement parts
PF 4108 — Unequal distribution
Agricultural [management...]
PF 6524 — management ; Underproductive methods
PF 1226 — *markets ; Fluctuating*
PC 0492 — methodology ; Traditional
PF 1653 — *methods ; Gap*
PF 3010 — methods ; Resistance changing
Agricultural mismanagement
PD 8625 — [Agricultural mismanagement]
PD 2771 — animals
PD 2771 — housed farm animals
PD 8625 — pasture arable land
Agricultural [output...]
PD 5114 — output developing countries ; Constraints increased
PD 5114 — output developing countries ; Lack incentives increase
PD 9448 — overproduction
Agricultural [pharmaceutical...]
PE 1274 — pharmaceutical products ; Distortion international trade discriminatory formulation health sanitary regulations
PF 1205 — *planning ; Individualistic*
PF 1205 — *planning ; Uncoordinated*
PD 5277 — poisons
PF 2889 — policies ; Detrimental international repercussions domestic
PD 0563 — polluters
PD 0563 — pollution
PF 2451 — *practices ; Complex*
PD 9802 — price difficulties developed countries ; Domestic
PE 2890 — price policy difficulties developing countries ; Domestic
PF 7047 — producer ; Neglect small
PE 8516 — producers ; Lack credit facilities
Agricultural production
PF 1822 — Obsolete methods
PF 1822 — Out date methods
PD 1285 — Stagnated development
PD 4316 — Unequal distribution
PE 4331 — Agricultural productivity ; Unsustainable short term improvements
Agricultural products
PE 0607 — developing countries ; Dumping
PE 8321 — developing countries ; Protectionism developed countries against
PF 7808 — Ignorance traditional
PF 3143 — Inadequate packaging
PC 2027 — Spoilage
PE 7313 — *Agricultural profits ; Low*
PE 5024 — Agricultural purposes developing countries ; Unavailability land
PF 1373 — *Agricultural reliance ; Reduced*
Agricultural resources
PC 8419 — Degradation
PD 1283 — production animal feed ; Use
PF 2164 — Underdeveloped use
PC 8419 — Unsustainable development
PF 2164 — Unused
PC 4906 — Agricultural risks
PF 7047 — Agricultural rural life developing countries ; Neglect
PD 9155 — Agricultural sector ; Insufficient social security
PE 0976 — Agricultural shortages
Agricultural subsidies
PE 1785 — chemicalized farming ; Inappropriate
PE 1785 — livestock production ; Inappropriate
PE 1785 — sales pesticides ; Inappropriate
PC 2062 — Agricultural surpluses
Agricultural [techniques...]
PD 7598 — *techniques ; Abuse*
PE 8752 — techniques ; Lack expertise
PF 2848 — *technology ; Ineffective*
PD 5830 — trade restrictions
PF 8835 — training ; Unavailability
PF 8835 — training ; Unmotivated
PF 3467 — Agricultural upgrading ; Insufficient access technology
PF 1944 — *Agricultural vision ; Unrealistic*
PC 2205 — Agricultural wastes
PE 8504 — Agricultural wastes ; Liquid
Agricultural workers
PD 5265 — Accidents
PE 5883 — developing countries ; Low productivity

Agricultural workers cont'd
- PE 4243 — Inadequate conditions work
- PE 4243 — Inadequate working conditions

Agriculture developing countries
- PD 4387 — Carbon dioxide impact
- PD 5114 — future declines production ; Vulnerability
- PD 0946 — Lagging transformation
- PD 7443 Agriculture ; Enforced collectivization
- PD 9534 Agriculture ; Excessive use land
- PD 5830 Agriculture food production industries ; Protectionism
- PD 0947 Agriculture ; Gerontocracy developing country

Agriculture [Inappropriate...]
- PE 1170 — Inappropriate government intervention
- PF 4799 — Inappropriate modernization
- PD 9802 — industrialized countries ; Over subsidized

Agriculture [Lack...]
- PC 0492 — Lack mechanized
- PE 5931 — land ; Competition industry
- PE 8082 — least developed countries ; Inadequate
- PD 7598 Agriculture ; Plantation
- PD 5212 Agriculture ; Seasonal fluctuations
- PE 5336 Agriculture technologies ; Decreasing rate development major
- PC 0492 Agriculture ; Unproductive subsistence
- PD 7516 Agriculture ; Unstable shifting
- PE 7433 Agriculture water ; Competition industry
- PF 1822 *Agriculture youth ; Rejection*
- PD 8388 *Agro urban training ; Lack*
- PD 2226 Agrobacterium
- PF 2231 *Agronomic potential ; Unrealized*
- PE 0616 Ague
- PE 9099 Aid ; Bilateralism

Aid [Corruption...]
- PD 0136 — Corruption mismanagement foreign
- PF 8120 — Counter productive effects foreign
- PF 4821 — countries violating human rights ; Government

Aid [Denial...]
- PF 8869 — Denial right legal
- PE 7195 — Dependence developing countries external
- PE 8116 — Dependence least developed countries foreign

Aid developing countries
- PE 9099 — Bilateralism
- PF 0392 — Decline
- PF 7955 — Donor distortion
- PF 0392 — Inadequacy
- PF 8120 — Inappropriate
- PF 1031 — Ineffectiveness
- PF 0175 — Mismanagement
- PF 2819 — Public disillusionment
- PF 6932 — Recipient distortion
- PF 1492 — Restrictions imposed
- PF 3949 — Stagnation food
- PF 0175 Aid donors ; Lack coordination
- PF 0498 Aid expenditure ; Reduction
- PD 7592 Aid ; Exploitation dependence food

Aid [facilities...]
- PF 8869 — facilities ; Lack legal
- PF 2819 — Fluctuating response appeals
- PF 3949 — Food
- PF 0648 — funding ; Competition
- PF 0648 — funds alternative priorities ; Reallocation

Aid [Inadequacy...]
- PF 3949 — Inadequacy food
- PF 0175 — Inadequate coordination
- PF 8869 — Inadequate legal
- PE 0302 — Inappropriate food
- PF 0175 — Inefficient administration
- PF 1136 — institutions ; Secrecy international
- PF 1136 — institutions ; Unaccountability international
- PF 0498 — international organizations ; Shortfall

Aid [Lack...]
- PF 0392 — Lack development
- PE 5338 — lenders beneficiaries net outflow funds developing countries
- PD 0305 — Limited developing country capacity absorb

Aid [Maldistribution...]
- PF 3949 — Maldistribution food
- PE 6052 — Military
- PE 6078 — Modification environmentally adapted nutritional habits food
- PF 8120 Aid ; Negative effects foreign
- PF 3434 Aid officials ; Excessive career mobility
- PF 8158 Aid ; Overhead restrictions efficacy
- PF 9216 Aid ; Policy cross conditionality restrictions multilateral development
- PF 7955 Aid ; Politically exploitative humanitarian
- PE 6052 Aid repressive regimes ; Military
- PF 0995 Aid ; Soaring cost legal

Aid [Tied...]
- PF 1492 — Tied
- PF 9007 — *training ; Inadequate first*
- PF 1492 — tying

Aid [Unavailability...]
- PF 9007 — *Unavailability first*
- PU 1623 — *Underutilized veterinary*
- PD 0512 — Unorganized elderly
- PF 8120 — Unproductive technical
- PD 0136 Aid ; Wasted foreign
- PE 4362 Aided fraud ; Computer
- PD 6655 Aiding consummation crime
- PE 4909 Aiding deserters
- PE 5390 Aiding escape
- PE 1200 Aiding escape prisoners war enemy aliens
- PD 7407 Aiding national security criminals
- PD 5111 AIDS
- PF 1379 *AIDS blood transfusions ; Risk contracting*
- PE 4276 AIDS ; Children born
- PF 0772 Aids disabled persons ; Insufficient technical
- PE 4299 AIDS infected persons ; Social stigmatization
- PE 4299 AIDS infected persons ; Travel restrictions
- PD 5111 *AIDS kissing ; Risk contracting*

AIDS [Pediatric...]
- PE 4276 — Pediatric
- PE 5376 — Prostitution
- PE 4299 — Psychological effects
- PD 5111 AIDS related complex ; ARC
- PE 5376 AIDS ; Restricted sexual liberation
- PE 5376 AIDS sexual behaviour ; Effects
- PB 0284 *AIDS testing ; Mandatory*

AIDS [virus...]
- PE 4299 — virus victims ; Social segregation
- PE 4276 — Vulnerability children
- PE 5798 — Vulnerability intravenous drug users
- PE 7154 *Ailments, lesions malfunctions ; Bodily*
- PE 5953 Ailments ; Minor
- PA 6714 *Aimlessness Chance*
- PA 7361 *Aimlessness Disorder*
- PA 6852 *Aimlessness Inappropriateness*
- PA 6977 *Aimlessness Meaninglessness*
- PF 3519 *Aims ; Impractical educational*
- PD 1582 Air accidents
- PE 4027 Air explosive ; Fuel
- PG 1918 *Air ; Inadequate fresh*

Air pollutants
- PC 0119 — [Air pollutants]
- PD 0414 — Automobile emissions
- PD 0450 — Biological
- PD 1271 — Chemical
- PE 0524 — *Gaseous*
- PE 2824 — Industrial domestic heating emissions
- PE 0155 — Plant pathogenic

Air pollution
- PC 0119 — [Air pollution]
- PE 4744 — agricultural crops ; Harmful effects
- PE 9609 — animals ; Health hazards
- PD 6283 — Arctic
- PE 8617 — children ; Health hazards
- PE 8617 — impact lungs children
- PD 6627 — Indoor
- PD 3391 — Long range transboundary
- PE 4744 — plants ; Health hazards
- PC 0119 — *Urban Industrial*

Air [sac...]
- PD 9366 — *sac mite poultry*
- PE 2611 — sickness
- PE 4732 — spora
- PC 0918 Air temperature ; Increasing
- PE 4089 Air terrorism

Air traffic
- PD 0689 — congestion
- PF 9464 — delays
- PD 8328 — Natural hazards
- PD 0689 — paralysis

Air transport
- PD 9163 — cartels
- PD 9163 — monopolies
- PD 9163 — practices ; Unfair
- PF 1495 — *service ; Inadequate*
- PE 5800 — service land locked developing countries ; Inadequate

Air [transportation...]
- PD 7132 — transportation industry ; Protectionism
- PD 2127 — turbulence
- PD 2127 — turbulence ; Clear
- PD 5029 Airborne diseases
- PD 2008 Airborne particles
- PD 2847 Airborne substances harmful health
- PC 0119 Airborne toxic harmful agents
- PD 2741 Airborne viral animal diseases
- PD 5373 Aircraft accidents ; Military
- PD 1111 Aircraft ; Bird strikes against
- PD 1582 Aircraft collisions
- PD 8328 Aircraft environmental hazards
- PE 1575 Aircraft ; Faulty

Aircraft [harassment...]
- PE 6006 — harassment
- PD 8328 — hazards
- PE 0962 — Health hazards radiation

Aircraft [Icing...]
- PE 8059 — Icing
- PE 6006 — Interference transiting
- PE 3972 — Invasion airspace foreign
- PD 1582 Aircraft near-collisions
- PE 5799 Aircraft noise
- PD 0124 Aircraft ; Offences aboard
- PE 3972 Aircraft ; Over flying foreign
- PE 3870 Aircraft pilot fatigue
- PE 4802 Aircraft ; Pollution
- PD 1111 Aircraft safety ; Threat birds
- PE 2435 Aircraft ; Sonic boom generated supersonic
- PE 1575 Aircraft ; Unsafe
- PE 8667 Airline bumping
- PE 4145 Airline industry ; Underparticipation developing countries
- PD 9163 Airline monopolies
- PE 5910 Airline services ; Restrictive practices cargo
- PD 3008 *Airplane hunting ; Disruptive*
- PD 0689 Airport congestion
- PE 2525 *Airport location*
- PE 5799 Airport noise
- PE 8231 Airport travel security ; Lack
- PD 0689 Airports ; Overloaded
- PD 0689 Airspace ; Congested
- PE 3972 Airspace foreign aircraft ; Invasion
- PE 6419 *Akabane disease animals*
- PF 9139 al-Dajal
- PJ 5188 Alalia
- PF 4384 Alarmism
- PA 6971 *Alarmism Danger*
- PA 6030 *Alarmism Fear*

Alarms False
- PF 4298 — [Alarms ; False]
- PF 4298 — *burglary*
- PF 4298 — *fire*
- PE 0097 *Alastrim*
- PF 5069 Albedo Earth's surface ; Reduction
- PE 2332 Albinism

Alcohol abuse
- PD 0153 — [Alcohol abuse]
- PE 2033 — business
- PE 7568 — doctors
- PE 0742 — people authority
- PD 1611 Alcohol adolescents ; Excessive consumption
- PE 1349 Alcohol ; Adulteration commercial
- PE 3980 Alcohol cigarettes media ; Excessive portrayal drugs,
- PD 1611 Alcohol consumption children young people
- PE 8865 Alcohol ; Draining resources due
- PE 2149 Alcohol ; Driving influence
- PD 0153 Alcohol ; Excessive consumption
- PD 0153 Alcohol ; Excessive intake
- PE 0680 Alcohol ; Flying influence
- PD 0153 *Alcohol idiosyncratic intoxication*
- PE 2047 Alcohol ; Medicines containing
- PD 0105 *Alcohol poisoning*

Alcohol related
- PE 2033 — absenteeism
- PE 4131 — crime
- PE 7084 — violence
- PE 7672 — violence against women
- PE 7161 Alcohol ; Susceptibility women
- PE 3853 Alcohol syndrome ; Foetal
- PD 0153 Alcohol usage ; Excessive
- PD 0153 Alcohol use ; Uncurbed
- PE 9263 *Alcohol withdrawal delirium*
- PE 0375 Alcohol withdrawal ; Uncomplicated

Alcoholic beverages
- PD 8286 — Consumption
- PE 3035 — Environmental hazards non
- PE 7188 — Illicit production
- PE 8134 — industry ; Protectionism
- PE 8748 — Instability trade non
- PE 3980 — Irresponsible advertising
- PD 8286 — Manufacture
- PD 8286 — Trade
- PE 2585 — trade ; Instability
- PD 0153 Alcoholic excess
- PE 9263 *Alcoholic hallucinosis*

Alcoholic intoxication
- PD 0153 — [Alcoholic intoxication]
- PE 2429 — Acute
- PE 2033 — work
- PE 7700 Alcoholic parents
- PE 9263 Alcoholic psychosis
- PD 4218 Alcoholics ; Children
- PE 9263 *Alcoholicum ; Delirium*
- PE 7242 Alcoholism amongst indigenous peoples
- PD 5344 Alcoholism children ; Parental toleration
- PD 0153 Alcoholism ; Chronic
- PD 0153 *Alcoholism ; Dementia associated*
- PD 0153 Alcoholism disease
- PD 0153 Alcoholism ; Ineffective treatment
- PD 1611 Alcoholism ; Juvenile
- PD 1611 Alcoholism ; Youth
- PJ 5013 Aldehydes
- PF 4298 Alerts ; False
- PF 1236 Alerts ; False nuclear warfare
- PE 3866 Alexia
- PD 9159 *Alexithymia*
- PE 2501 Algae blooms
- PE 2501 Algae ; Noxious
- PE 2501 Algae ; Poisonous
- PE 2501 Algae ; Toxic
- PD 5228 *Algal poisoning animals*
- PC 7384 Alien domination peoples
- PF 6161 Alienating child-birth environments
- PJ 6661 *Alienating psychotherapy*
- PF 2241 *Alienating public housing assignments*
- PF 6192 *Alienating waiting environments*
- PA 3545 Alienation

Alienation [capitalist...]
- PD 3112 — capitalist systems
- PC 3227 — Citizen
- PC 5088 — Cultural
- PA 6838 *Alienation Dissent*
- PA 7339 *Alienation Duality*
- PA 5446 *Alienation Enmity*
- PD 3076 Alienation human labour
- PA 7157 *Alienation Insanity*
- PE 0896 Alienation land acquisition foreigners
- PF 3370 Alienation ; Missionary
- PA 6979 *Alienation Opposition*
- PC 3227 Alienation ; Political
- PB 0147 Alienation ; Psychological
- PA 3403 Alienation religious minorities
- PA 5699 *Alienation Reversion*

Alienation [skilled...]
- PE 1553 — skilled committed personnel international organizations programmes

—855—

Alienation

	Alienation [Social...] cont'd	PD 8697 Ambiguous enforcement procedures	PA 7156 Anarchism *Moderation*
PC 2130	— Social	PF 5421 Ambiguous laws	PA 5901 Anarchism *Revolution*
PE 3594	— socialism ; Workers	PF 5421 Ambiguous regulations	PA 6858 Anarchy *Disintegration*
PD 1809	— support international organizations programmes	PF 6516 Ambiguous shape social identity	PA 7361 Anarchy *Disorder*
PD 3076	Alienation work	PF 5421 Ambiguous treaties	PF 1088 Anarchy ; Fear social
PD 1544	Alienation ; Youth	PA 6977 Ambition ; Lack	PF 1088 Anarchy ; Fear social collapse
PE 1200	Aliens ; Aiding escape prisoners war enemy	PF 9852 Ambitions ; Inappropriate	PA 5952 Anarchy *Illegality*
PD 0973	Aliens ; Discrimination against immigrants	PC 2494 *Ambitions political parties ; Clash personal*	PA 5563 Anarchy *Lawlessness*
PC 3207	Aliens ; Expulsion immigrants	PD 1544 Ambitions ; Unsatisfied youth	PF 2071 Anarchy ; World
PC 3207	*Aliens ; Illegal*	PE 1426 Ambivalence	PA 6191 Anathema
PC 3207	*Aliens ; Undesirable*	PA 6698 Ambivalence *Difference*	PA 6191 Anathema *Disapproval*
PF 0801	Alignment ; Non	PA 5982 Ambivalence *Disagreement*	PE 2946 Anatomical extremities ; Paralysis
PD 7758	*Alimentary anaemia stock*	PA 7325 Ambivalence *Irresolution*	PD 2315 Ancestor worship
PD 1618	*Alimentary tract ; Congenital anomalies*	PA 6739 Ambivalence *Maladjustment*	PD 6128 Ancient burial sites ; Desecration
PE 3247	*Alimony costs ; Excessive*	PF 1205 *Ambulance service ; Misused*	PF 6483 Ancient manuscripts ; Uncatalogued
PE 3247	*Alimony ; Inadequate provision*	PC 0374 Amenity destruction	PD 0253 Ancient sites ; Endangered
PD 3647	*Alkaline soil*	PC 0374 Amenity destruction ; Dependence	PF 6494 Ancient traditions ; Unclarified procedures transposing
PD 3647	*Alkaline topsoil ; Shallow*	PD 2728 *American blastomycosis ; North*	PE 3508 Ancylostomiasis
PD 5228	*Alkaloidosis ; Pyrrolizidine*	PE 2281 American leishmaniasis	PE 3508 Ancylostomiases - Nectoriasis
PD 4517	Allegations against governments ; Biased	PE 6461 *American liver fluke ; Large*	PF 6648 Androcentrism
PF 9794	Allegations human rights violations ; False	PE 0653 American trypanosomiasis	PE 6354 Anemophobia
PE 1407	Alleged human rights violations ; Government inaction	PE 0653 American trypanosomiasis ; South	PD 1618 *Anencephalus*
PE 6867	Allergenic biologically active wood ; Poisonous,	PD 0339 Amino-acid imbalances	PE 3597 Aneurysm
PE 9509	*Allergic bronchitis animals*	PE 9291 Amino acid metabolism ; Congenital disorders	PD 5453 *Aneurysm animals*
PE 9509	*Allergic enteritis animals*	PD 9366 Ammonia burn poultry	PD 5453 *Aneurysm turkeys ; Dissecting*
PE 1017	Allergic reactions	PD 2965 Ammonia pollutant	PD 0094 Angel dust
PE 9509	*Allergic rhinitis animals*	PD 8297 Amnesia	PA 7797 Anger
PE 1017	Allergy	PE 0298 Amnesia ; Alternating	PF 8563 Anger ; Divine
PD 9667	*Allergy dermatitis ; Flea*	PF 0306 Amnesia ; Global	PA 7253 Anger *Envy*
PE 3225	Allergy ; Food	PE 0314 Amnesia ; Hypnotic	PE 0429 Anger ; Explosive
PE 9509	*Allergy ; Food*	PE 0321 Amnesia ; Hysterical	PA 5644 Anger *Vice*
	Allergy inducing	PD 8297 *Amnesia ; Infantile*	PE 5204 Angina
PD 2731	— *bacteria*	PF 7745 Amnesia institutional systems ; Structural	PE 8158 Angina pectoris
PE 4895	— cosmetics	PE 0351 Amnesia ; Post traumatic	PE 2143 Angina ; *Vincent's*
PE 3225	— food	PD 0270 Amnesia ; Psychogenic	PE 3254 Angina ; *Vincent's*
PE 2069	— *pollens*	PE 0354 Amnesia ; Retrograde	PA 6950 Angst
PF 1258	Alleviate poverty ; Inefficient public spending	PE 0355 Amnesia ; Transient global	PA 5451 Anguish *Insensibility*
PF 1572	Alleviate problems ; Lack well researched projects	PE 0357 Amnesia ; Traumatic	PA 6731 Anguish *Solemnity*
PF 0631	Alleviation ; General obstacles problem	PF 3019 Amnesty	PD 4755 Anguish torture victims ; Mental
PD 2629	Alliances ; Disagreement	PF 1098 Amnesty military personnel convicted atrocities ; Politically motivated	PA 7107 Anguish *Unpleasantness*
PD 7522	Allies ; Disputes		PE 7808 Angular limb deformities foals
PD 7522	Allies superpowers ; Instability relations	PF 3019 Amnesty ; Qualified	PF 1216 Anhedonia
PF 0734	Allocation ; Discriminatory frequency	PF 0182 Amnesty ; Refusal grant	PD 4031 Animal abnormalities
PF 2911	Allocation government contracts ; Unjust	PD 3018 Amnesty ; Violation		**Animal [bites...]**
PF 8535	Allocation government funding ; Abusive	PD 3947 Amniocentesis leading abortion	PE 4931 — bites ; Domestic
PF 7432	Allocation ; Imbalance revenue mobilization expenditure	PD 3947 Amniocentesis ; Misuse results	PE 7424 — bone diseases
PD 3507	Allocation market facilities ; Unsystematic	PE 6782 Amoebiasis	PC 3152 — breeding grounds ; Ecological disruption
PD 1211	Allocation ; Non political market	PE 2259 Amoebic dysentery	PC 1408 — breeding stock ; Decreasing genetic diversity
PF 6460	Allocation ; Ownership basis land	PE 0517 *Amoebic hepatitis*		**Animal [cancer...]**
PE 6702	Allocation public contracts ; Unethical	PE 4179 Amongst workers ; Inadequate sense community solidarity	PC 0092 — cancer
PF 1085	Allocation resources pet animals ; Disproportionate		PE 2513 — *cannibalism*
PE 1597	Allocation rights exploit sea bed marine resources ; Inequitable	PF 3349 Amoralism	PD 8837 — care ; Unknowledgeable
		PA 5644 Amorality *Vice*	PD 2743 — *commodity production ; Inefficient*
PF 3703	Allocation television frequency bands satellite transmission	PA 7280 Amorality *Wrongness*	PF 7554 — communication ; Vulnerability marine
		PD 4027 Amphetamine intoxication	PF 0395 — conditions ; Unsanitary inhumane urban food
PF 7703	Allopathy	PE 1558 Amphetamine withdrawal	PE 4768 — control ; Unenforced
PF 0107	Allowances ; Negative effects family	PE 1558 Amphetamines ; Abuse		**Animal [damage...]**
PD 6850	Alluvial forests ; Destruction	PD 3156 Amphibia ; Endangered species	PC 0952 — *damage ; Unrestrained*
PF 3089	Alone ; Living	PE 6461 Amphistomes	PE 7941 — deaths
PD 4782	Along internal borders ; Violence	PE 8724 Amputations ; Punitive	PD 4031 — defects
PE 7111	*Alopecia*	PF 8363 Amulets, charms talismans ; Dependence	PC 0952 — deficiency diseases
PD 9667	*Alopecia animals*	PD 2078 Amusement ; Exploitation animals		**Animal disease**
PD 0270	Alteration physical functioning suggesting physical disorder	PC 2270 Amyloidosis	PD 2746 — Domestic animals vectors
		PE 4250 Anabolic steroids ; Abuse	PD 2784 — *increase aviation ; Increase*
PE 4346	*Alternaria diseases plants*	PE 1883 Anaclitic depression	PE 8409 — outbreaks ; Inadequate disinfection humans after
PF 7960	Alternate energy ; Unexplored	PD 0659 *Anaconsis*	PE 8409 — outbreaks ; Inadequate disinfection measures humans during
PE 0298	Alternating amnesia	PD 7758 Anaemia	
PF 3826	Alternative images ; Rejection		**Anaemia animals**	PD 2740 — Weather factor
PF 3826	Alternative modes social organization ; Non viable	PE 9554 — [Anaemia animals]		**Animal diseases**
PF 2199	Alternative policies ; Government action against regimes	PE 9554 — *Aplastic*	PC 0952 — [Animal diseases]
		PE 9509 — *Autoimmune haemolytic*	PD 2741 — Airborne viral
PF 0648	Alternative priorities ; Reallocation aid funds	PE 9554 — *Haemolytic*	PD 2755 — Animals' international movement factor
PF 3826	Alternative social structures ; Low credibility	PE 9554 — *Hypoplastic*	PF 2780 — Confusion symptoms
PF 6548	Alternatives commercial development ; Unexplored	PG 9847 — *Immune haemolytic*		**Animal diseases Difficulty**
PF 8540	Alternatives countryside ; Unexplored entertainment	PE 9554 — *Iron deficiency*	PD 2781 — controlling
PF 7960	Alternatives ; Disuse traditional energy	PE 9554 — *Nutritional*	PF 2780 — diagnosing
PF 3826	Alternatives existing societies ; Unconvincing	PE 7894 Anaemia ; Aplastic	PF 2775 — identifying carriers
PF 1658	Alternatives ; Limited employment	PE 9554 *Anaemia ; Equine infectious*		**Animal diseases [Domestic...]**
PF 2842	*Alternatives ; Limited engagement*	PE 9554 *Anaemia ; Feline infectious*	PD 2746 — Domestic animals carriers
PD 6571	*Alternatives ; Restricted farming*	PD 0321 Anaemia ; Nutritional	PD 2752 — factory farming ; Spread
	Alternatives [Unexplored...]	PD 7758 *Anaemia pregnancy*	PD 2784 — Human vectors
PF 7960	— Unexplored energy		**Anaemia [Sickle...]**	PE 2777 — Importation infected carcass meats factor
PE 8061	— Uninvestigated building	PE 3724 — Sickle cell	PD 2781 — Inadequate control
PF 3550	— urban areas ; Poor credit	PD 7758 — stock ; *Alimentary*	PE 8133 — Inadequate knowledge incubation periods
PE 8061	— urban areas ; Unfeasible housing	PD 2730 — syndrome chickens ; Haemorrhagic	PD 2732 — Infectious
PD 2322	Altitude ; High	PE 9185 Anaemia women developing countries ; Nutritional	PE 2764 — Insanitary penning conditions factor
PD 2322	Altitude stress ; High	PD 7758 Anaemias ; Acquired haemolytic	PD 2748 — Insect vectors
PE 4969	Aluminium ; Health hazards	PD 7758 Anaemias ; Hereditary haemolytic	PD 2755 — International movement animals factor
PE 1406	*Aluminium ; Instability trade*	PE 7854 Anaemias ; Iron deficiency	PE 0965 — *Paralytic*
PC 0521	*Aluminium monopolies*	PE 1427 Anaesthetic drugs ; Inhaling solvents	PE 2747 — Snail vectors
PE 4969	Aluminium poisoning	PE 3273 Anal intercourse	PD 2731 — *Toxaemic*
PE 5182	*Alveolar emphysema ; Chronic*	PD 8402 *Analgesia*	PD 7726 — Underreporting
PE 9702	*Alveolar periostitis animals*	PF 9243 Analyses ; Manipulation project cost benefit	PD 2781 — *Unrecognized*
PE 3503	*Alveolaris ; Pyorrhoea*	PF 9830 Analysis ; Fragmentation field	PF 2775 — *Vector immunity*
PE 9702	*Alveolaris ; Pyorrhoea*	PE 4888 Analysis ; Involuntary narco	PD 2751 — Vectors
PE 7623	Alzheimer's disease	PF 0440 Analysis ; Overemphasis	PE 2787 — Water borne
PE 0985	*Amarilli ; Vomito*	PF 1214 Analysis ; Politically unrealistic strategic warfare	PD 2729 — Wild animals carriers
PD 0159	Amateur abortion	PF 1214 Analysis ; Think tank	PE 2749 — Wild birds vectors
PC 2178	*Amateurism*	PF 0440 Analytical stagnation	PD 2223 — *Wind borne*
PF 2152	*Ambiguities ; Terminological*	PE 9509 Anaphylactic shock	PD 2750 — Worms vectors
PF 4193	Ambiguity	PE 1017 Anaphylaxis	PE 0043 Animal drugs ; Abuse
PA 6698	Ambiguity *Difference*	PE 6275 Anaplasmosis		**Animal [ecological...]**
PA 5982	Ambiguity *Disagreement*	PC 1972 Anarchism	PC 0842 — ecological imbalance
PF 2505	Ambiguity ; Social disguise	PA 7361 Anarchism *Disorder*	PE 4685 — excrement urban environments ; Insalubrity
PA 7309	Ambiguity *Uncertainty*	PA 5952 Anarchism *Illegality*	PF 4770 — experimentation ; Uncertainties
PA 7367	Ambiguity *Unintelligibility*	PA 5563 Anarchism *Lawlessness*	PD 7989 — extinction

Animals, fish

PE 2770 Animal farming units ; Inferior meat quality intensive
Animal feed
PD 1714 — additives
PD 1714 — additives ; Undesirable effects
PE 3896 — Contaminated
PF 3448 — *Scant*
PD 1283 — Use agricultural resources production
PF 2378 *Animal feeds ; Costly*
Animal feedstuffs excluding
PE 1331 — unmilled cereals ; Environmental hazards
PE 8816 — unmilled cereals ; Instability trade
PE 8514 — unmilled cereals ; Long term shortage
Animal [fighting...]
PE 4893 — fighting sports
PC 7644 — flesh ; Eating
PD 4518 — flesh ; Excessive consumption
PE 9609 — fluorosis
PD 7501 — forms ; Monopolistic control new
PD 4031 Animal genetic defects ; Increase
PD 7526 Animal genetic resources ; Underutilized
Animal [habitats...]
PC 3152 — habitats ; Modification
PE 5644 — head hunting
PE 2757 — housing equipment ; Inadequate disinfection measures
PD 9366 — *housing ; Stray voltage*
Animal husbandry
PE 3934 — Agricultural effluents
PF 2848 — *Inexperienced*
PD 1562 — Intensive
PE 5399 — Methane gas emissions
Animal [infections...]
PD 2730 — infections ; Viral
PC 1803 — infertility
PE 3157 — influenza
PC 2753 — injuries
PD 1290 — invasions
PD 2011 *Animal lodging ; Residential*
Animal [malformation...]
PD 2761 — malformation factory farming
PE 7064 — meat animal product shipments ; Infected
PC 2279 — migration movement patterns ; Disruption
PE 7941 — mortality
PE 7941 — mortality ; Economic loss
PE 7808 — musculoskeletal system ; Congenital anomalies
PD 0235 — *myocarditis*
PD 3155 Animal natural prey ; Extermination wild
PC 2765 Animal nutrition ; Inadequate
Animal [oils...]
PE 8135 — oils fats ; Environmental hazards
PE 8896 — oils fats ; Instability trade
PE 7530 — organ transplants humans
Animal [pest...]
PD 8426 — pest control
PD 8426 — pests
PE 4931 — pets ; Bites
PD 5157 — plant life due radioactive contamination ; Endangered
Animal populations
PE 7995 — Abuse control wild
PE 5776 — Excessive stray
PD 4721 — Underreporting hazards
PF 0377 Animal power developing countries ; Underproductivity draught
PE 7064 Animal product shipments ; Infected animal, meat
Animal production
PD 1714 — Health hazards hormone use
PC 2062 — *Surplus domestic*
PF 2821 — Underdevelopment food live
PE 3934 — Water pollution
PD 8469 Animal productivity ; Loss
Animal products
PD 0389 — endangered species ; Trade
PE 7064 — food contamination
PD 7699 — Human consumption
PF 1964 — Human use
PD 1119 — Radioactive contamination animals
PC 2062 — *Surpluses*
PE 4784 Animal protein ; Excessive cost
PC 4998 Animal protein ; Shortage
PF 1555 Animal quarantine
PE 2756 Animal quarantine ; Inadequate
Animal [range...]
PC 0475 — range size ; Restriction wild
PE 4438 — rights campaigners ; Intrusive
PE 1690 — road deaths
Animal [sacrifice...]
PF 0319 — sacrifice
PF 1555 — sanitary regulations international travel ; Excessive
PD 8837 — sanitation ; Uninformed
PD 2011 — *shelter ; In home*
PD 9578 — species ; Excessive economic exploitation
PD 2419 — species ; Hybridization wild
PD 0094 — *sports ; Drug use*
PD 2760 — stress factory farming
PD 8812 — suffering
Animal [transportation...]
PD 0390 — transportation ; Cruel
PE 5735 — traps ; Cruel
PE 5644 — trophy hunting
PD 8360 Animal vectors disease
Animal vegetable oils
PE 0735 — fats ; Instability trade
PE 1188 — fats ; Long term shortage
PE 8880 — fats ; Restrictive practices trade
PE 8498 — fats waxes ; Shortage

Animal welfare
PC 1167 — Inadequate
PE 5794 — Inadequate legislation
PD 8837 — Lack concern
PE 5794 — Lack legislation
PE 5794 — legislation ; Lack
PD 2330 Animal worship barrier development
PA 7156 Animality *Moderation*
PA 5612 Animality *Unchastity*
PA 5643 Animality *Unkindness*
PA 5821 Animality *Vulgarity*
Animals [Abdominal...]
PE 3711 — *Abdominal hernias*
PD 7799 — *Abortion large*
PD 6423 — *Acanthosis*
PD 5228 — *Acaricide poisoning*
PE 4247 — Accidents
Animals Accumulation
PD 5021 — contaminant residues plants
PD 5278 — contaminant residues terrestrial
PD 5021 — pollutants plants
Animals Acute
PD 3978 — *intestinal obstructions*
PD 7799 — *metritis*
PD 3978 — *pancreatic necrosis*
PD 9293 — *renal failure*
PE 4786 Animals advertising ; Maltreatment
PE 1174 Animals aggression experiments ; Maltreatment
Animals Aggressive
PE 0524 —
PC 2321 — [Animals ; Aggressive]
PE 7175 — domesticated
PC 3151 — wild
Animals Agricultural
PF 0377 — loss draught
PD 2771 — mismanagement
PD 2771 — mismanagement housed farm
PE 6419 Animals ; Akabane disease
PD 5228 Animals ; Algal poisoning
Animals Allergic
PE 9509 — *bronchitis*
PE 9509 — *enteritis*
PE 9509 — *rhinitis*
Animals [Alopecia...]
PD 9667 — Alopecia
PE 9702 — *Alveolar periostitis*
PD 2078 — amusement ; Exploitation
PE 9554 — Anaemia
PD 5453 — *Aneurysm*
PD 1119 — animal products ; Radioactive contamination
PE 5569 — Aortic stenosis
PE 9554 — *Aplastic anaemia*
PE 5461 — aquaria ; Maltreatment
PE 0471 — Arrhythmia
PD 5228 — Arsenic poisoning
PE 5913 — Arthritis disorders
PD 2728 — Aspergillosis
PE 9509 — *Atopic dermatitis*
PE 3711 — *Atresia*
Animals Autoimmune
PE 9509 — *haemolytic anaemia*
PE 9509 — *thrombocytopenia*
PE 9509 — *thyroiditis*
Animals [Bacillary...]
PE 7769 — *Bacillary haemoglobinuria*
PD 2731 — Bacterial diseases
PD 7799 — *Balanoposthitis*
PE 3394 — Beating draught
PE 7769 — *Big head disease*
PD 5535 — *Blepharitis*
PD 5228 — *Blister beetle poisoning*
PF 7681 — Boredom captive domesticated
PE 9374 — *Botulism*
PD 5228 — *Bracken fern poisoning*
PE 7175 — Breeding dangerous
PC 1408 — breeding purposes ; Excessive use domestic
PF 5191 — Broken spirited
PD 9366 — *Burns*
PE 6555 — Bursitis
Animals [Cantharidin...]
PD 5228 — *Cantharidin poisoning*
PE 9774 — *Capture myopathy wild*
PE 0471 — *Cardiac failure*
PE 0471 — *Cardiac insufficiency*
PD 5453 — Cardiovascular diseases
PD 2746 — carriers animal diseases ; Domestic
PD 2729 — carriers animal diseases ; Wild
PE 9844 — *Caseous lymphadenitis*
PJ 4592 — Castration
PD 5535 — *Cataracts*
PE 0965 — *Cerebral defects*
PE 9702 — *Cheilitis small*
PE 2727 — *Cheyletiella infestations*
PE 1928 — Chlamydid infections
PD 5535 — *Chorioretinitis*
PD 9293 — *Chronic renal failure*
PE 4786 — cinema ; Maltreatment
PD 5453 — *Circulatory system diseases*
PD 5228 — *Coal tar poisoning*
PE 9509 — *Cold haemolytic disease*
PE 7423 — *Colitis*
PE 7423 — *Colon impaction*
PE 4224 — *Combined immunodeficiency disease*
PD 1481 — Commercial exploitation wild
PD 5535 — *Conformational abnormalities eyelids*

Animals Congenital anomalies
PE 5569 — cardiovascular system
PE 3711 — digestive system
PE 6419 — due infectious diseases
PD 7799 — *reproductive system*
PD 9293 — *urinary system*
Animals [Congenital...]
PE 0965 — Congenital diseases nervous system
PE 7423 — Constipation
PD 9667 — Contagious ecthyma
PD 5228 — Copper poisoning
PC 0776 — Creation transgenic plants
PD 7520 — *Criminal killing*
PC 0066 — Cruelty
PE 3394 — Cruelty herd
PE 2738 — *Cryptosporidiosis*
PE 2727 — Cutaneous acariasis
PD 9667 — *Cuterebra infestation small*
PD 5228 — *Cyanide poisoning*
Animals Dangerous
PC 2321 — [Animals ; Dangerous]
PE 7175 — pet
PC 3151 — wild
Animals [Deafness...]
PD 5535 — Deafness
PD 6305 — Declining breeds domesticated
PC 1408 — Decreasing genetic diversity
PE 0965 — Defects brainstem
PE 0965 — Defects cerebellum
PD 4031 — Deformed
PC 5456 — Denial rights
PD 9667 — Dermatitis
PE 4161 — *Dermatitis verrucosa*
PD 9667 — Dermatophytosis
PD 9654 — *Diabetes insipidus*
PD 9654 — *Diabetes mellitus*
PD 7424 — *Diaphragmatic hernia*
PE 4760 — Dietary deficiencies domestic
PD 2776 — Difficulty controlling disease wild
PD 5228 — Dioxin poisoning
Animals Diseases
PD 5535 — *anterior uvea*
PD 5535 — *conjunctiva*
PD 5535 — *cornea*
PD 3978 — digestive system
PD 5535 — *lacrimal apparatus*
PD 7424 — musculoskeletal system
PD 7841 — nervous system
PE 4630 — *oesophagus*
PD 7307 — respiratory system
PD 5535 — *senses*
PE 4604 — spinal column
PE 4604 — spinal cord
PD 2776 — wild
Animals [Disproportionate...]
PF 1085 — Disproportionate allocation resources pet
PC 1408 — domestication ; Genetic impoverishment
PD 1179 — Domestication wild
PE 0958 — due parasites ; Gastrointestinal infections
Animals during
PE 4694 — disasters ; Lack protection domestic
PE 4694 — disasters ; Lack protection wild
PE 4694 — states emergency ; Vulnerability
PE 7808 Animals ; *Dyschondroplasia*
PD 7799 Animals ; *Dystocia*
PD 5535 Animals ; Ear diseases
Animals Economic loss
PE 8109 — destruction injured
PE 8098 — reduced productivity diseased
PE 8098 — reduced productivity diseased
PE 8109 — slaughter diseased
Animals [educational...]
PD 0260 — educational purposes ; Maltreatment
PE 4247 — Electric shock
PD 9366 — Electrocution
PD 5453 — Embolism
Animals Endangered species
PC 1713 — [Animals ; Endangered species]
PD 7506 — farm
PB 1395 — plants
Animals [Endocardium...]
PE 0471 — *Endocardium diseases*
PD 9654 — Endocrine diseases
PE 7423 — *Enteritis*
PD 2078 — entertainment ; Maltreatment
PC 1411 — Environmental hazards food live
PD 0788 — Environmental hazards live
PD 7799 — *Epididymitis*
PC 1408 — Excessive inbreeding domestic
PF 5415 — Execution
Animals Experimental
PE 1171 — battering
PE 1670 — induction psychological stress
PE 0412 — surgery
Animals [Experiments...]
PD 0260 — Experiments live
PC 1445 — Extermination wild
PD 5535 — Eye diseases
PD 5535 — *Eyeworms*
Animals [Facial...]
PD 7841 — *Facial paralysis*
PD 2755 — factor animal diseases ; International movement
PD 2768 — factory farming ; Cruelty
PD 3934 — fish ; Accumulation contaminant residues marine
PD 4925 — fish birds ; Accumulation contaminant residues freshwater

–857–

Animals

Animals [Fluoride...] cont'd
PD 5228 — *Fluoride poisoning*
PE 0236 — food preparation ; Cruelty
PE 4630 — *Foreign bodies oesophagus*
PD 9366 — *Frostbite*
PD 9654 — *Functional islet cell tumours*
PD 2728 — *Fungal diseases*
PD 4575 — fur ; Breeding
PD 4575 — fur industry ; Slaughter
Animals [Gambling...]
PE 0891 — Gambling performance
PE 2143 — *Gangrenous stomatitis*
PE 7423 — *Gastric dilatation volvulus*
PE 4630 — *Gastric foreign bodies*
PD 3978 — *Gastritis*
PE 7423 — *Gastroenteric diseases*
PE 2436 — *Gastrointestinal parasites*
PE 9702 — *Gingivitis*
PD 5535 — *Glaucoma*
PE 9702 — *Glossitis*
PD 9366 — *Gout*
Animals [Haematoma...]
PD 5535 — *Haematoma ear*
PE 9554 — *Haemolytic anaemia*
PD 5453 — *Haemostatic disorders*
PE 6815 — *Hazardous aquatic*
PE 9609 — Health hazards air pollution
PC 1411 — Health hazards food live
PE 0471 — *Heart disease*
PD 5228 — *Herbicide poisoning*
PE 5478 — *Hirsutism*
PC 7644 — Human consumption
PE 3274 — Human sexual intercourse
PD 8486 — humans ; Killing
PD 8486 — humans ; Slaughter
PC 2024 — Hunting
PE 0439 — Hunting marine
PD 9654 — *Hyperadrenocorticism*
PD 9366 — *Hyperthermia*
PD 9654 — *Hyperthyroidism*
PD 9654 — *Hypoadrenocorticism*
PD 7424 — *Hypoparathyroidism*
PE 9554 — *Hypoplastic anaemia*
PD 9366 — *Hypothermia*
PD 9654 — *Hypothyroidism*
PE 9509 Animals ; *Idiopathic polyarthritis*
PD 9366 Animals ; Ill defined health conditions
Animals Immune
PE 4224 — deficiency diseases
PG 9847 — *haemolytic anaemia*
PD 4068 — system diseases
PE 4224 Animals ; *Immunodeficiency diseases*
Animals Inadequate
PE 2778 — carcass disposal diseased
PC 2765 — feeding
PC 2765 — feeding farm
PE 2763 — housing penning domestic
PE 2786 — vaccination domestic
Animals [industrial...]
PD 2772 — industrial concerns ; Excessive commercial exploitation farm
PD 2751 — Infected
PD 2732 — Infectious diseases
PD 5535 — *Infectious keratoconjunctivitis*
PD 5535 — *Inherited retinopathies*
PE 0358 — Inhumane killing
PE 2759 — Inhumane killing stray
PD 5228 — *Insecticide poisoning*
Animals Instability
PD 2894 — production food live
PD 1434 — trade food live
PD 1376 — trade live
Animals [Insufficiently...]
PE 7064 — Insufficiently treated meat infected
PD 2755 — international movement factor animal diseases
PE 9554 — *Iron deficiency anaemia*
PD 1290 — Irresponsible introduction new species
PD 7501 — Irresponsible patenting genetically transformed
PE 0805 Animals ; Kidnapping pet
PD 8486 Animals ; Killing
Animals [Lack...]
PD 8837 — Lack care
PD 8469 — Lack productivity farm
PD 7307 — *Laryngitis*
PE 9228 — Lead poisoning
PE 8643 — legal protection their rights ; Denial
PE 6427 — limitation working hours ; Denial working
PD 3978 — *Liver diseases*
PE 0976 — Long term shortage food live
PE 8717 — Loss beneficial plants
PE 8098 — Loss disease game zoo
PE 9844 — *Lymphadenitis*
PE 9844 — *Lymphangitis*
PD 5453 — *Lymphatic system diseases*
PD 3978 Animals ; *Malabsorption syndromes*
PE 7769 Animals ; *Malignant oedema*
Animals Maltreatment
PC 0066 — [Animals ; Maltreatment]
PE 4810 — circus
PD 1265 — companion
PD 0260 — laboratory
PE 0906 — marine show
PE 4810 — performing
PD 1265 — pet
PE 4810 — rodeo
PE 3394 — transport

Animals Maltreatment cont'd
PE 4834 — zoo
Animals [Mastitis...]
PD 7799 — *Mastitis*
PE 4786 — media ; Maltreatment
PD 0260 — Medical experimentation
PD 2731 — *Meningitis*
PE 1155 — Mercury poisoning
Animals Metabolic
PD 9293 — *disease kidney*
PD 7420 — *diseases*
PD 7420 — *disturbances*
Animals [Metaldehyde...]
PD 5228 — *Metaldehyde poisoning*
PE 1666 — Military experiments
PE 1666 — Military use
PD 0260 — Misuse experimental
PD 8904 — Misuse wild
PD 5228 — *Molybdenum poisoning*
PD 9366 — *Motion sickness*
PE 9702 — *Mouth diseases*
PE 2143 — *Mycotic stomatitis*
PE 9554 — *Myelophthisis*
PE 0471 — *Myocardium diseases*
PE 9774 — *Myopathies*
Animals [Nasal...]
PE 2727 — *Nasal acariasis*
PE 9854 — *Nematodes*
PD 5228 — *Nitrate poisoning*
PD 5228 — *Nitrite poisoning*
PD 2728 — *Nocardiosis*
PE 9554 — *Nutritional anaemia*
PD 7424 — *Nutritional hyperparathyroidism*
Animals [Obesity...]
PD 7420 — *Obesity*
PD 9293 — *Obstructive uropathy*
PD 3978 — *Oesophageal spasm*
PE 4630 — *Oesophageal stenosis*
PE 4630 — *Oesophagitis*
PD 7799 — *Orchitis*
PD 7424 — *Osteitis*
PE 5913 — *Osteoarthritis*
PE 5050 — *Osteomyelitis older*
Animals Otitis
PE 5444 — *externa*
PD 5535 — *interna*
PD 5535 — *media*
PE 6024 Animals ; Over population wild
PD 2760 Animals ; Overcrowding housed farm
Animals [pain...]
PE 1670 — pain ; Experimental exposure
PD 9654 — *Panhypopituitarism*
PD 9667 — *Papillomatosis*
PD 9667 — *Parakeratosis*
PD 3978 — *Paralysis tongue*
PD 7799 — *Paraphimosis*
PE 9427 — *Parasites cardiovascular system*
PD 2735 — *Parasitic diseases*
PD 5228 — *PBB poisoning*
PD 5228 — *PCB poisoning*
PD 9667 — *Pediculosis*
PD 4575 — pelts ; Slaughter
PE 9509 — *Pemphigus vulgaris*
PE 0471 — *Pericardium diseases*
PD 3978 — *Perineal hernia*
PE 9702 — *Periodontitis*
PD 3978 — *Peritonitis*
PE 5569 — *Persistent ductus arteriosus*
PE 5569 — *Persistent right aortic arch*
PD 5228 — *Pesticide poisoning small*
PD 2094 — pets ; Abandoned
PD 1179 — pets ; Capture use wild
PD 7307 — *Pharyngitis*
PD 7799 — *Phimosis*
PD 7420 — Photosensitization
PD 9654 — *Pituitary tumours*
PC 0728 — plants ; Harmful
PE 9854 — *Pneumonia small*
PD 5228 — *Poisoning*
PE 0175 — *Poisonous*
PD 5453 — *Polyarteritis*
PE 9509 — *Polyarteritis nodosa*
PD 5228 — *Polybrominated biphenyls poisoning*
PD 5228 — *Polychlorinated biphenyls poisoning*
PD 5453 — *Polycythemia*
PE 7304 — *Pox diseases*
PC 0507 — Prejudice against
PD 7424 — *Primary hyperparathyroidism*
PD 7799 — *Prostatic cysts*
PD 7799 — *Prostatic neoplasms*
PD 7799 — *Prostatitis*
PD 9654 — *Pseudohyperparathyroidism*
PD 7307 — *Pulmonary emphysema*
PE 5569 — *Pulmonic stenosis*
PG 4410 — *Purpura*
PD 9667 — *Pyoderma*
PD 7799 — *Pyometra*
Animals [radiation...]
PE 0689 — radiation ; Experimental exposure
PE 0805 — Ransoming pet
PC 3150 — Ravaging wild
PD 3978 — *Rectal prolapse*
PE 3711 — *Rectovaginal fistula*
PD 1290 — Reintroduction
PD 7424 — *Renal secondary hyperparathyroidism*
PD 7799 — *Reproductive system diseases*

Animals [research...] cont'd
PD 0260 — research ; Cruel treatment
PE 9854 — *Respiratory diseases small*
PE 4793 — restorative nourishment rest ; Denial working
PD 7799 — *Retained placenta*
PD 7424 — *Rickets*
Animals right
PF 5121 — attention, care protection humankind ; Denial
PE 6270 — conditions life liberty proper their species ; Denial
PE 9573 — dignity ; Denial
Animals right freedom
PE 9650 — mass killing ; Denial
PE 8024 — suffering ; Denial experimental
PE 3899 — suffering ; Denial food
Animals [right...]
PF 8243 — right life ; Denial
PE 8339 — right natural death ; Denial
PD 9667 — *Ringworm*
PF 0319 — Ritual slaughter
PD 5228 — *Rodenticide poisoning*
Animals [Salivary...]
PE 7570 — *Salivary disorders*
PD 5228 — *Salt poisoning*
PD 5228 — *Selenium poisoning*
PD 5228 — *Senecio poisoning*
PE 5569 — *Septal defects*
PD 9366 — *Shock*
PD 3978 — *Simple indigestion*
PD 9667 — *Skin diseases*
PE 8109 — *Slaughter diseased*
PE 7423 — *Small intestinal obstruction*
PE 2739 — *Soil borne diseases*
PD 5228 — *Sorghum poisoning*
PE 0965 — *Spastic diseases*
PE 0891 — spectator sports ; Exploitation
PE 2143 — *Stomatitis*
PD 5228 — *Strychnine poisoning*
PD 3453 — *Superstitious persecution*
PE 5913 — *Suppurative arthritis*
PE 5050 — *Suppurative arthritis older*
PE 7808 — *Syndactyly*
PE 9509 — *Systemic lupus erythematosus*
Animals [Taming...]
PD 1179 — Taming wild
PE 5913 — *Tendinitis*
PE 5913 — *Tenosynovitis*
PE 9611 — testing pharmaceutical products ; Cruel treatment
PE 9611 — testing toiletries cosmetics ; Abusive treatment
PE 5569 — *Tetralogy fallot*
PE 4247 — *Thoracic trauma*
PD 5453 — *Thrombosis*
PC 3532 — Torture
PE 7327 — Torture exposure
PE 9611 — toxicological experiments ; Use
PE 0342 — trade ; Restrictive practices food live
PD 8360 — transmitters disease
PE 5735 — Trapping
PE 1670 — Trauma inducing experiments
Animals Tumours
PD 5535 — *external ear*
PD 9293 — *kidney*
PD 9293 — *lower urinary tract*
Animals [Ulcenomembranous...]
PE 2143 — *Ulcenomembranous stomatitis*
PD 7799 — *Ulcerative posthitis*
PE 2143 — *Ulcerative stomatitis*
PF 6455 — *Uncontrolled small*
PE 2778 — *Undisposed dead*
Animals Unethical practices
PE 4771 — circus
PE 4771 — domesticated
PE 4771 — farm
PE 4771 — pet
PE 4771 — zoo
Animals [Unwanted...]
PD 2094 — Unwanted pet
PF 6185 — urban environment ; Inadequate interaction humans
PD 9293 — *Urinary system diseases*
PD 9667 — *Urticaria*
PE 4786 — used theatre acts ; Maltreatment
PD 7799 — *Uterine prolapse*
Animals [Vaginal...]
PD 7799 — *Vaginal hyperplasia*
PD 7799 — *Vaginal prolapse*
PD 7799 — *Vaginitis*
PD 5453 — *Vasculitis*
Animals vectors
PD 2746 — animal disease ; Domestic
PD 8360 — disease
PD 2746 — disease ; Domestic
Animals [Venomous...]
PD 6823 — *Venomous*
PD 2730 — *Vesicular stomatitis*
PC 2321 — *Violent*
PE 7175 — *Violent pet*
PD 2730 — *Viral diseases*
PD 2730 — *Virus diseases*
PE 3676 — *Visceral leishmaniasis*
PE 4760 — Vitamin E deficiency domestic
PD 7799 — *Vulvitis*
PE 6443 — Animals warfare ; Use
PE 4247 Animals ; *Wounds*
PF 6317 Animism
PA 5446 Animosity *Enmity*
PA 7253 Animosity *Envy*
PA 7338 Animosity *Hate*

Applicants

PA 7102 Animosity *Malevolence*
PA 7364 Animosity *Unfeelingness*
PE 7549 *Aniridia*
PE 8990 Ankylosing spondylitis
PE 5913 *Ankylosing spondylitis*
PD 2283 *Ankylosis joint*
PD 3160 Annelida ; Endangered species
PE 5210 Annexation
PF 6506 *Annexation authority ; Uncommunicated*
PE 5210 Annexation ; Foreign
PE 5210 *Annexation ; Political*
PF 9169 Annihilation
PA 7055 Annihilation *Death*
PA 6542 Annihilation *Destruction*
PA 5870 Annihilation *Nonexistence*
PF 2723 Anniversaries years ; Proliferation national international
PA 6340 Annoyance *Adversity*
PA 7378 Annoyance *Aggravation*
PA 5497 Annoyance *Difficulty*
PA 7253 Annoyance *Envy*
PA 7107 Annoyance *Unpleasantness*
PD 5228 *Annual ryegrass staggers*
PF 3281 Annulment adoption
PA 7410 Annulment *Celibacy*
PA 6542 Annulment *Destruction*
PA 6838 Annulment *Dissent*
PF 2100 Annulment ; Marriage
PD 1618 *Anomalies alimentary tract ; Congenital*
PE 7808 Anomalies animal musculoskeletal system ; Congenital
Anomalies [cardiovascular...]
PE 5569 — cardiovascular system animals ; Congenital
PE 6924 — circulatory system ; Congenital
PD 1618 — Congenital
Anomalies [Dento...]
PD 1185 — *Dento facial*
PE 3711 — digestive system animals ; Congenital
PD 1618 — *digestive system ; Congenital*
PE 6419 — due infectious diseases animals ; Congenital
PE 7549 Anomalies eye ; Congenital
PE 4249 Anomalies genital organs ; Congenital
PF 2407 Anomalies ; Geomagnetic field
PC 4987 Anomalies ; Large scale weather
PE 8589 Anomalies musculoskeletal system ; Congenital
PE 9296 Anomalies nervous system ; Congenital
PD 0173 *Anomalies ; Placental*
PD 7799 *Anomalies reproductive system animals ; Congenital*
PD 1618 *Anomalies respiratory system ; Congenital*
PD 9667 *Anomalies skin ; Congenital*
PE 8589 *Anomalies spine ; Congenital*
PD 9293 *Anomalies urinary system animals ; Congenital*
PJ 0300 Anomia
PF 6316 Anomie
PE 3622 Anopheline mosquitoes vectors disease
PE 1439 Anoplura
PE 5758 Anorexia nervosa
PE 6408 Anorgasmia
PE 1066 Anosmia
PF 5973 Another person ; Uncritical acceptance
PE 2467 Anoxemia
Antagonism [Difference...]
PA 6698 — *Difference*
PA 5532 — *Disaccord*
PA 5982 — *Disagreement*
PE 5104 Antagonism employment policy technical advance
PA 5446 Antagonism *Enmity*
PE 8308 Antagonism government agencies each country ; Jurisdictional conflict
PE 2719 Antagonism government agencies officials
Antagonism [intergovernmental...]
PE 7901 — intergovernmental organizations ; Jurisdictional conflict
PE 9011 — intergovernmental organizations ; Jurisdictional conflict
PE 0064 — international nongovernmental organizations ; Jurisdictional conflict
PE 1169 — international nongovernmental organizations ; Jurisdictional conflict
Antagonism international organizations
PE 7973 — country level ; Jurisdictional conflict
PD 0047 — Jurisdictional conflict
PD 0138 — Jurisdictional conflict
PA 6979 Antagonism *Opposition*
PE 1583 Antagonism regional intergovernmental organizations common membership ; Jurisdictional conflict
PC 0030 Antagonism ; Religious political
PE 2486 Antagonism specialized agencies United Nations ; Jurisdictional conflict
PE 8799 Antagonism specialized agencies United Nations ; Jurisdictional conflict
PA 7107 Antagonism *Unpleasantness*
PE 5993 Antarctic pollution
PC 1888 *Antarctic territory ; Conflicting claims concerning*
PD 3603 *Anteaters ; Endangered species*
PA 7131 Antediluvian *Oldness*
PD 4029 Antenatal foetal death
PD 5535 *Anterior uvea animals ; Diseases*
PE 1910 Anthracnose diseases plants
PE 4586 *Anthracnose plants ; Spot*
PE 4586 *Anthracnose ; Sycamore*
PD 2034 *Anthracosis*
PE 2736 Anthrax
PF 4096 Anthropocentrism
PC 9717 Anthropogenic climate change
PD 1823 Anthropological looting ; Archaeological
PD 2623 Anthropologists ; Corruption
PD 2623 Anthropologists ; Negligence
PD 2623 Anthropology ; Irresponsible
PD 2623 Anthropology ; Malpractice
PD 2623 Anthropology ; Unethical practice
PF 9576 *Anthropomorphism*
PF 2513 Anthropophagy
PD 1770 Anthropozoonoses
PE 1821 Anti anxiety agents ; Abuse
Anti [capitalism...]
PF 3110 — capitalism
PF 3360 — clericalism
PF 1826 — communism
PF 2648 — *community attitudes*
PF 3511 — consumerism
PF 7972 Anti discrimination legislation ; Lack
PF 5745 Anti-holism
PF 6573 *Anti intellectual bias ; Local*
PF 1929 Anti-intellectualism
PE 4860 *Anti-labour laws*
Anti [marxist...]
PE 5546 — marxist crimes ; Revisionism
PE 9269 — matter bombs
PF 2685 — *meritocracy*
Anti personnel
PE 9294 — gases ; Health hazards
PE 9294 — use asphyxiating gases
PE 9294 — use toxic substances peacetime
PF 0278 Anti-ritualism
Anti [satellite...]
PE 7004 — satellite arms race
PD 0087 — satellite weapons
PF 2685 — science
Anti social behaviour
PC 4726 — [Anti-social behaviour]
PD 0329 — developing countries ; Increase
PE 7370 — university students
PF 1721 Anti-social personality disorders
PF 2071 *Anti systemic movements ; Growth*
Anti [terrorist...]
PF 4240 — terrorist organization ; Ineffective
PD 5272 — terrorist raids ; Cross border
PF 6386 — trust prosecution ; Indiscriminate
PE 6629 Antibiotics ; Abuse
PE 6629 Antibiotics ; Bacterial resistance
PE 8383 Antibiotics vaccines factory farming ; Abuse
PF 9139 Antichrist
PF 8176 Anticipate future development ; Inability
PF 6570 *Anticipated neglect elders*
PF 0399 Anticipation capital punishment torture
PA 6562 Anticipation∗complex
PF 2437 *Anticipation criticism*
PE 0191 Antidumping regulations ; Distortion international trade discriminatory application
PC 2270 *Antimetabolites*
PE 1989 Antimony health hazard
PF 9843 Antipathic personal relations world leaders
PA 8810 Antipathy
PA 6698 Antipathy *Difference*
PA 5446 Antipathy *Enmity*
PA 7338 Antipathy *Hate*
PA 6979 Antipathy *Opposition*
PA 7107 Antipathy *Unpleasantness*
PA 6509 Antipathy *Unwillingness*
PD 2732 *Antipestifer infection ; Pasteurella*
PD 6520 *Antiquated fire equipment*
PF 1094 Antiquated intellectual methods appropriate human depths
PF 5395 Antiquated provisions UN charter
PF 4370 Antiquated regulations banking industry
PF 7713 Antiquated thought patterns
PF 0866 Antiquated world socio-economic order
PA 5733 Antiquity
PE 2131 Antisemitism
PE 4369 Antisemitism ; Christian
PF 5958 Antisocial attitudes planning second language policy
PD 1533 Antisocial behaviour due television ; Increase
PE 4438 Antivivisectionists ; Intolerant
PE 3711 *Anus ; Imperforate*
PA 1635 Anxiety
PE 1821 Anxiety agents ; Abuse anti
PC 0300 *Anxiety ; Castration*
PA 1635 Anxiety ; Dependence
PA 5497 Anxiety *Difficulty*
Anxiety disorder
PA 1635 — [Anxiety disorder]
PE 6958 — Dream
PE 2401 — Separation
PE 6958 Anxiety dreams
PA 6030 Anxiety *Fear*
PA 6200 Anxiety *Impatience*
PF 7388 Anxiety ; Information
PF 4345 Anxiety lending developing countries ; Excessive
PA 1635 Anxiety neurosis
PE 0969 Anxiety resulting torture
PE 2401 Anxiety ; Separation
PA 7107 Anxiety *Unpleasantness*
PC 0877 Anxiety ; Youth
PD 4027 *Anxiolytic intoxication*
PE 0139 *Anxiolytic withdrawal*
PE 1821 Anxiolytics ; Abuse
PE 9580 Anxious disorder ; Over
PA 1635 Anxious people
PE 3463 Any country ; Denial right leave
PF 5203 Any personal advantage ; Failure sacrifice
PF 5203 Any personal investment ; Resistance loss
PE 6924 *Aorta ; Coarctation*
PE 5569 *Aortic arch animals ; Persistent right*
PD 5453 *Aortic rupture*
PE 5569 *Aortic stenosis animals*
PE 3681 Apartheid
PA 7306 Apartheid *Narrowmindedness*
Apartheid [Slavery...]
PE 3681 — Slavery like practices
PE 6246 — system ; Job reservation
PE 1996 — system ; Participation transnational corporations
PA 2360 Apathetic people
PA 2360 Apathy
PA 6379 Apathy *Avoidance*
PF 9459 Apathy ; Bureaucratic
PF 2421 Apathy ; Citizen
PF 5057 Apathy ; Consumer
Apathy Dependence
PA 2360 — [Apathy ; Dependence]
PC 3413 — economic
PC 1917 — political
PC 3414 — religious
PC 3412 — social
PD 8047 Apathy developing countries
PC 3413 Apathy ; Economic
PC 2494 Apathy ; Electoral
PF 9459 Apathy ; Governmental
PA 6099 Apathy *Hopelessness*
Apathy [Ideological...]
PF 3392 — Ideological
PA 6598 — *Incuriosity*
PC 3413 — Industrial
PE 7092 — International nongovernmental organization membership
PF 9459 Apathy ; Official
PC 1917 Apathy ; Political
PA 6444 Apathy *Quiescence*
PC 3414 Apathy ; Religious
PA 3275 Apathy *Sloth*
PC 3412 Apathy ; Social
PE 8477 Apathy toward improvement urban life styles
PA 7364 Apathy *Unfeelingness*
PC 2494 Apathy ; Voter
PF 5949 Apathy youth
PE 1570 *Apes ; Endangered species great*
PE 3886 Aphasia
PE 0345 Aphasia ; Ataxic
PE 0345 Aphasia ; Motor
PE 4234 Aphasia ; Sensory
PE 3613 Aphids pests
PE 2265 Aphonia
PE 7894 Aplastic anaemia
PE 9554 *Aplastic anaemia animals*
PF 0638 Apocalypse
PF 5296 Apologize errors ; Failure government
PF 5163 Apologize mistakes ; Refusal
PE 1684 Apoplexy
PE 9018 Apostasy
PF 4849 Apparatchiks
PD 5535 *Apparatus animals ; Diseases lacrimal*
Apparatus appliances
PE 8026 — Environmental hazards electrical machinery,
PE 8878 — shortage ; Electric machinery
PE 8875 — Trade instability electrical machinery,
Apparel [industries...]
PE 1008 — industries ; Instability wearing
PE 5819 — industries ; Protectionism textile
PD 8001 — industry ; Unethical practices
PE 1103 Apparel manufacture ; Environmental hazards wearing
PF 5124 Apparently opposed ideologies ; Cover up convergence practice
PF 1735 *Apparently unrewarded initiatives*
PF 3801 Apparitions
PC 5280 *Appeal ; Abuse right*
PD 5317 Appeal ; Denial right
PE 6784 Appeal women ; Loss sex
PF 2819 Appeals aid ; Fluctuating response
PD 7966 Appeals ; Bureaucracy blocks
PE 4577 Appear trial after release ; Failure
PE 1756 Appear witness produce information be sworn ; Failure
PF 4010 Appearance ; Boring personal
PF 1733 *Appearance ; Debilitating physical*
PF 1733 Appearance deteriorating buildings ; Unattractive
PD 0270 *Appearance ; Preoccupation imagined defect*
Appearance [Unaesthetic...]
PD 1185 — *Unaesthetic*
PF 4010 — Unchanging personal
PF 4010 — Unfashionable personal
PF 4010 — Unkempt
PF 4010 — Unsatisfactory personal
PE 8631 Appearances disappearances persons objects ; Unexplained
PF 0317 Appeasement responses leadership ; Immediate
PD 0622 *Appendicitis*
PE 5758 Appetite ; Loss
PE 3505 *Appetite ; Loss*
PE 6255 *Apple rust ; Cedar*
PE 4586 *Apple scab*
PE 8026 Appliances ; Environmental hazards electrical machinery, apparatus
PE 7879 Appliances ; Health hazards electromagnetic fields generated electrical
PE 8878 Appliances shortage ; Electric machinery apparatus
PE 8875 Appliances ; Trade instability electrical machinery, apparatus
PF 2564 Applicability monetary grants ; Limited
PD 4158 Applicability ; Offenses general
PE 8137 Applicant's race ; Refusal let because
PE 9441 Applicants ; Secret government security vetting job

Application antidumping

PE 0191 Application antidumping regulations ; Distortion international trade discriminatory
PF 8191 Application available knowledge solve problems ; Inadequate
PE 7637 Application communications technology ; Unbalanced
PF 3538 Application ; Difficulty patent
PC 0418 Application distribution technology ; Indiscriminate
PE 4768 Application existing laws ; Non
PF 2899 *Application ; Prohibitive cost pesticide*
PE 8011 Application restrictive business practices legislation ; Extraterritorial
PF 2828 Application situational demands vocational decisions ; Unrelated ability

Application [technology...]
PC 0418 — technology ; Uncontrolled
PE 7637 — telecommunications technology ; Inconsistent
PF 3871 — trade measures ; Narrow egalitarian
PF 2256 — traditional values ; Inappropriate
PE 4602 — Applications ; Delay processing parole
PF 4785 — Applications ; Obstacles redeployment military resources peaceful
PF 9693 Applied science ; Conflict interest pure
PF 8191 Applying knowledge ; Difficulties
PF 2031 Appointees ; Political
PE 1728 Appointment systems ; Inadequate
PE 1728 Appointments ; Political bias official

Appreciation [cultural...]
PF 2679 — cultural differences ; Lack
PF 2679 — *cultural differences ; Lack*
PF 3408 — culture ; Inadequate
PE 7476 Appreciation nuclear weapons ; Lack
PA 6030 Apprehension *Fear*
PD 5515 Apprehension felon ; Hindering
PA 7392 Apprehension *Unbelief*
PF 1352 Apprentice training ; Lack
PF 1352 *Apprenticeship opportunities ; Insufficient*
PF 1352 *Apprenticeship positions ; Unavailability*
PF 1352 *Apprenticeship ; Restricted opportunities*
PC 2163 *Apprenticeship ; Restrictions early*
PF 6475 Approach agricultural development ; Pessimistic
PF 6530 Approach capital resources ; Subsistence
PF 1205 *Approach ; Individualistic agricultural*

Approach [land...]
PF 2365 — land ownership tropical villages ; Disorganized
PF 6530 — *Limited funding*
PF 5660 — long term development policy making ; Crisis management
PF 1870 Approach marriage covenant ; Naive
PA 8197 *Approach ; Misunderstanding veterinary*
PF 6580 Approach problem solving ; Piecemeal
PF 3500 Approaches economic planning ; Limited
PF 6493 Approaches local food production ; Underdeveloped
PF 1004 Approaches motivational techniques ; Shallow
PF 7923 Approaches ; Obstacles development multidisciplinary
PF 6580 Approaches ; Oversimplified problem
PF 0956 Approaches teaching intellectual methods ; Faulty
PE 0630 Appropriate available data ; Absence method
PD 9021 Appropriate child care ; Ignorance women concerning
PE 0630 Appropriate data ; Failure methods

Appropriate [education...]
PE 5020 — education children immigrants ; Lack
PE 7563 — exchange rate policies ; Inability developing countries adopt
PF 7916 — expertise ; Unavailability
PF 1094 Appropriate human depths ; Antiquated intellectual methods

Appropriate [technology...]
PF 3554 — technology ; Minimal access
PE 6243 — trained manpower developing countries ; Inadequate supply
PE 8551 — transport ; Scarcity
PF 1746 Appropriate wisdom past ; Failure opposing groups
PA 7246 Appropriateness∗complex
PA 5688 Appropriation
PE 6154 Appropriation children ; Illegal
PE 6074 Appropriation cultural objects ; Illicit
PF 4849 Appropriation machinery government elites
PD 0820 Appropriation unoccupied housing
PD 0183 Appropriations ; Wastage governmental budgets
PF 4753 Approval addictive behaviour ; Government
PF 4753 Approval dangerous behaviour ; Government
PJ 0451 Approval ; Government requirement that economic benefits be proven prior programme
PC 1613 Approval urgent regulations ; Delay
PF 8876 Approve needed legislation ; Lack parliamentary time
PE 4697 Approved employment war criminals ; Government
PE 0304 Apraxia
PE 7570 Aptyalism
PF 5496 Aquaculture ; Obstacles
PE 5461 Aquaria ; Maltreatment animals
PE 5461 Aquarium imprisonment
PE 6815 Aquatic animals ; Hazardous
PD 2232 Aquatic weeds
PD 2503 Aquifers ; Nitrate pollution
PD 4403 Aquifers ; Overexploitation
PB 0534 Arab nationalism
PD 8625 Arable land ; Agricultural mismanagement pasture
PD 9534 Arable land ; Uncontrolled growth
PE 1601 Arabum ; Elephantiasis
PD 3473 Arachnida ; Endangered species
PE 3986 Arachnida pests
PJ 4562 Arbitral awards ; Inadequate enforcement foreign
PB 5486 Arbitrariness
PD 1576 Arbitrary arrest
PD 3055 Arbitrary deprivation property ; Denial right freed
PD 1576 Arbitrary detention ; Denial right freedom

Arbitrary [enforcement...]
PD 8697 — enforcement regulations
PB 5486 — *evaluation disability compensation*
PE 6366 — executions
PC 2507 — exile
PC 2507 — *exile ; Denial right freedom*
PC 5313 — expulsion ; Denial right freedom
PD 0520 — extension prison sentences
PE 4058 — external interference family life
PF 1559 Arbitrary government decision-making
PB 0284 Arbitrary interference privacy
PF 2066 Arbitrary job qualifications ; Artificial
PD 2710 Arbitrary leadership
PF 0090 Arbitrary national boundaries
PE 9545 Arbitrary reduction sentences offenders after conviction
PB 5486 *Arbitrary sequestration real property*
PD 3544 *Arbitrary street search*
PJ 0385 Arbitrating commodity markets ; Lack
PF 7897 Arbitration ; Government refusal accept international tribunal
PD 5111 ARC - AIDS-related complex
PE 5569 Arch animals ; Persistent right aortic
PD 1823 Archaeological anthropological looting
PD 4502 Archaeological sites ; Destruction
PD 4502 Archaeology ; Salvage
PF 1705 *Archaic economic base*
PF 6465 Archaic marketing methods
PA 7131 Archaism *Oldness*
PF 6987 Archetypes ; Victimization
PF 7106 *Architectural obsolescence building structures*
PF 0867 Architectural design ; Monotonous unaesthetic
PE 1925 Architecture high rise buildings ; Monolithic
PD 8638 Architecture insensitive needs women
PD 2886 Archival practices ; Unethical
PD 0172 Archives ; Destruction historic documents public
PE 1669 Archives ; Deterioration stored documents
PD 1194 Archives ; Lack access public
PD 0577 *Archives ; Stolen*
PD 0577 Archives ; Theft
PD 6283 Arctic air pollution
PB 8663 Arctic countries ; Sub
PE 5993 Arctic pollution
PE 1656 *Arctocephalus ; Endangered species*
PD 1204 Arduous water collection
PF 6537 *Area churches ; Closed*
PF 3550 *Area classification ; Depressed*
PF 4385 Area disparities book production distribution
PE 4585 Area irrigated land ; Declining
PC 5917 Area tensions
PD 2730 Argentinian haemorrhagic fever
PF 2152 Argument ; Cheap
PF 1333 Argumentative community meetings
PA 5532 Argumentativeness *Disaccord*
PA 7253 Argumentativeness *Envy*
PA 5502 Argumentativeness *Reason*
PF 2152 Arguments ; Bad
PF 2152 Arguments conceal nefarious benefits ; Misuse plausible
PF 2152 Arguments ; Fallacious
PF 2152 Arguments ; Flaky
PF 2152 Arguments ; Inappropriate
PF 2152 Arguments ; Obfuscatory
PD 8418 Arid developing countries
PD 7096 Arid ecosystems irrigation ; Destruction
PD 7096 Arid zone ecosystems ; Disruption
PC 2506 Arid zone enlargement
PF 5044 Arising automation ; Excess leisure time
PE 9295 Arising current policies ; Government neglect future problems social environment
PE 0740 Arising insurance transactions developing countries ; International indebtedness
PE 8925 Arising mental factors ; Physiological malnutrition
PF 6026 Arising natural disaster ; Inadequate insurance against damages
PF 4707 Aristocracy ; Reappearance
PC 0143 Arithmetic disorder ; Developmental
PE 7562 *Arizona infection poultry*
PD 3603 *Armadillos ; Endangered species*
PF 0638 Armageddon
PF 1447 Armaments ; Excessive expenditure
PC 0012 Armaments research ; Waste resources
PC 1258 Armaments ; Surplus
PE 9346 Armaments ; Waste resources invested obsolete
PE 9207 Armed biochemical weapons ; Terrorists

Armed conflict
PB 0593 — [Armed conflict]
PC 6675 — Degradation environment
PE 8361 — Inadequate protection civilians
PE 8174 — Vulnerability children during

Armed conflicts
PC 1454 — Denial human rights
PD 7661 — Internal
PD 1078 — Vulnerability women during
PD 8153 Armed crimes against national security

Armed forces
PE 5782 — Causing insubordination
PD 7360 — Corruption
PF 9542 — Disregard government orders
PE 5986 — Extra legal impressment children
PC 5230 — Imbalance conventional
PE 3912 — Obstruction recruiting induction
PD 9252 — Rivalry
PE 4484 — Unlawful recruiting enlistment foreign
PC 7373 *Armed groups ; Illegal*
PD 8284 Armed insurrection
PE 3769 Armed nuclear weapons ; Terrorists
PE 4739 Armed robbery
PD 0527 Armies ; Private
PE 5327 Arming rioters
PD 5575 Armoured car robberies ; Train

Arms [capability...]
PC 1606 — capability ; Unequal strategic
PD 0014 — competition ; Strategic
PC 1258 — Competitive acquisition
PF 0013 — control efforts ; Failure disarmament
PF 1447 — culture
PD 2107 Arms dealers ; Private international

Arms [Imbalance...]
PC 5230 — Imbalance conventional
PC 1606 — Imbalance strategic
PF 1447 — Increasing government expenditure
PE 8664 — industries ; Protectionism defence
PD 0014 Arms ; Proliferation strategic nuclear

Arms race
PC 1258 — [Arms race]
PE 7004 — Anti satellite
PC 1258 — Dependence
PC 0840 — *Information technology*
PF 7088 — Lack business opposition
PD 8412 — Naval
PD 5076 — Nuclear
PF 4152 — Risk unintentional war generated
PD 0087 — Space weapons
PD 0014 Arms rivalry ; Strategic

Arms [shipments...]
PD 4858 — shipments ; Illegal international
PD 4858 — smuggling
PD 1304 — supply ; Restriction

Arms trade
PD 3497 — developing countries
PF 7730 — Government complicity illegal
PC 1358 — International
PF 4460 Arms treaties ; Non verifiability compliance nuclear
PF 0013 Arms violations

Army discipline
PD 2543 — codes ; Strict
PD 2543 — Inadequate
PD 2543 — Lack
PE 3644 Army worms pests
PD 8016 *Arousal disorders ; Sexual*
PF 3284 Arranged marriage
PF 6146 Arrangement housing respect common land ; Inadequate
PF 7642 Arrangements ; Bilateralism trade
PE 0396 Arrangements ; Collusive international trade
PE 0310 Arrangements ; Discriminatory orderly marketing
PE 0413 Arrangements ; Exclusive dealing
PF 0419 Arrangements governments ; Covert
PD 0340 Arrangements ; Preferential trade
PE 0895 Arrangements ; Restrictions export activity due licensing
PE 0310 Arrangements ; Trade restrictions due intergovernmental
PA 5454 Arrant *Badness*
PA 6839 Arrant *Disrepute*
PA 5644 Arrant *Vice*
PE 9700 Arrears financial payments government agencies
PE 4571 Arrears home loan payments
PD 3053 Arrears international debt payments
PE 4571 Arrears mortgage payments
PF 1179 Arrears payment government financial commitments
PE 4571 Arrears payments housing
PE 4571 Arrears rental payments
PD 1576 Arrest ; Arbitrary
PD 1576 Arrest dubious grounds
PD 6062 Arrest ; False
PD 0520 *Arrest released offenders ; Repeated*
PD 7630 Arrest trade union leaders

Arrested development
PA 6739 — [Arrested development]
PA 7371 — [Arrested development]
PF 6532 — labour potential
PD 9349 Arrests ; Politically motivated
PD 6970 *Arrests ; Politically motivated mass*
PE 0471 Arrhythmia animals
PE 8219 Arrhythmia ; Cardiac
PA 7646 Arrogance
PC 1259 Arrogance ; Bureaucratic
PF 5178 Arrogance ; Cultural
PA 7646 Arrogance ; Dependence
PA 6822 Arrogance *Disrespect*
PF 3294 Arrogance experts
PF 8820 Arrogance ; Government
PA 6659 Arrogance *Humility*
PF 7847 Arrogance ; Intellectual
PF 9561 Arrogance intergovernmental agencies
PA 6979 Arrogance *Opposition*

Arrogance [planners...]
PF 2895 — planners
PF 2895 — policy-makers
PF 3294 — Professional
PF 7843 Arrogance ; Scientific
PE 7778 Arrogance ; Super power
PA 6491 Arrogance *Vanity*
PA 7646 Arrogant people
PD 4680 Arrogation rights
PD 3493 Arrows ; Broken
PD 5228 *Arsenic poisoning animals*
PE 1732 Arsenic pollutant
PE 5505 Arson
PE 5505 Arson attacks ; Threat
PE 7645 Arsonists ; Psychotic

PE 5042 Art ; Blasphemous
PD 2337 Art censorship
Art [Debasement...]
PE 0558 — Debasement works
PE 5042 — Decadent
PE 6252 — Dismembered works
PE 5042 Art ; Erotic
PE 2382 Art forgery
Art [Illicit...]
PE 9004 — Illicit export works
PE 8088 — Inadequate documentation works
PE 5042 — Indecent
PE 0323 Art ; Looting works
PE 5042 Art ; Nudity
Art [objects...]
PD 1955 — objects ; Deterioration physical condition
PE 5042 — Obscene
PF 7870 — Over valuation
PF 7870 Art ; Prohibitive cost
PF 3087 Art propaganda
PE 0323 Art ; Theft works
PF 7870 Art values ; Inflated
PE 5171 Art vandalism
PE 9417 Artefacts ; Collection cultural
PD 2478 Artefacts environmental pollution ; Damage cultural
PE 9004 Artefacts ; Smuggling cultural
PE 9825 Artefacts tourism ; Deterioration cultural
PE 8162 Artefacts ; Vibration damage cultural
PE 2684 Arteries ; Diseases
PE 2210 Arteriosclerosis
PD 0104 Arteriosclerosis ; Institutional
PE 8158 *Arteriosclerotic heart disease*
PD 0919 *Arteriosclerotic psychosis*
PC 2270 *Arteriosclerotic vascular disease*
PE 5569 *Arteriosus animals ; Persistent ductus*
PD 2730 *Arteritis ; Equine viral*
PE 2684 *Arteritis ; Generalized necrotizing*
PE 2684 *Arteritis ; Necrosing*
PD 4403 Artesian water supplies ; Misuse
PE 0873 Arthritis
Arthritis accompanying
PE 0873 — *foreign protein reactions*
PE 0873 — *psoriasis*
PE 0873 — *ulcerative colitis*
PE 5913 *Arthritis animals ; Suppurative*
PE 9509 *Arthritis ; Canine rheumatoid*
PD 2730 *Arthritis ; Caprine*
PE 5913 Arthritis disorders animals
PE 2259 *Arthritis ; Dysenteric*
PE 7826 *Arthritis fetlock joint ; Serous*
PD 2565 *Arthritis ; Generalized necrotizing*
PE 5050 *Arthritis ; Mycoplasmal*
PE 0873 *Arthritis ; Neurogenic*
PE 5050 *Arthritis older animals ; Suppurative*
PD 2565 *Arthritis ; Osteo*
PE 5913 *Arthritis poultry ; Viral*
PE 5081 *Arthritis ; Rheumatoid*
PE 5913 *Arthritis ; Septic*
PE 7826 *Arthritis shoulder joint*
PE 0873 *Arthritis ; Traumatic*
PD 2283 *Arthrodesis*
PE 7808 *Arthrogryposis*
PE 4161 *Arthropathy ; Degenerative*
PE 7796 Arthropod-borne diseases
PD 3471 *Arthropoda ; Endangered species*
PD 6651 *Arthropods intermediate hosts*
PE 8275 Articles ; Environmental hazards miscellaneous manufactured
PE 0814 Articles ; Instability trade miscellaneous manufactured
PE 8970 Articles ; Instability trade textile yarn fabrics, made up
PE 0613 Articles ; Long term shortage miscellaneous manufactured
PE 8436 Articles ; Shortage textile, yarn fabrics, made up
PE 7808 *Articular rigidity ; Myopathy associated congenital*
PF 2595 Articulated educational goals ; Non
PE 9712 Articulation disorder ; Developmental
PA 7411 Artifice *Uncommunicativeness*
PA 7232 Artifice *Unskillfulness*
PF 2066 Artificial arbitrary job qualifications
Artificial flooding Loss
PE 3786 — agricultural land due
PE 7855 — forests
PE 7794 — wildlife due
PE 8634 Artificial insemination
PE 9125 Artificial methods promoting fast livestock growth ; Use
PF 0090 Artificial nation boundaries
PE 9114 Artificial opposition manual intellectual labour
PE 8769 Artificial resins ; Instability trade regenerated cellulose
Artificial [separation...]
PF 6122 — separation home workplace
PF 3382 — sex determination birth
PE 6390 — sweeteners ; Unsafe
PF 0090 Artificial territorial boundaries
PA 6400 Artificiality *Affectation*
PF 6135 Artificiality parkland
PA 7411 Artificiality *Uncommunicativeness*
PE 4804 Artisans craftsmen developed countries ; Lack
PF 9735 Artistic expression ; Limited
PE 4054 Artistic expressions peoples ; Exploitation
PF 9735 Artistic freedom ; Denial right
PD 8872 Artistic skills ; Vanishing
PE 4507 Arts ; Denial right enjoyment
PF 1756 Arts ; Performing
PF 0981 Arts ; Pictorial
PF 0981 Arts ; Representative
PE 0323 Artworks ; Stolen

PE 7004 ASAT negotiation failure
PE 1127 Asbestos dust
PE 3001 Asbestos ; Health hazards
PE 3001 Asbestos occupational hazard
PE 1127 Asbestos pollutant
PE 3001 Asbestosis
PE 2395 Ascariasis
PD 7550 Asceticism
PD 3159 *Aschelminthes ; Endangered species*
PE 4586 *Ascomycetes*
PE 4161 *Aseptica diffusa ; Pododermatitis*
PD 1245 Ash ; *Coal fly*
PD 1245 Ash ; *Marine dumping coal*
PF 7727 Aspects euphenics ; Negative
PE 5212 Aspergillosis
PD 2728 *Aspergillosis animals*
PA 7253 Asperity *Envy*
PA 5643 Asperity *Unkindness*
PE 4104 Asphyxia
PE 9294 Asphyxiating gases ; Anti personnel use
PD 7307 *Aspiration pneumonia*
PF 0955 Aspirations ; Disillusionment life
PF 2842 Aspirations economic life ; Unfulfilled
PF 9008 Aspirations structure opportunities income available ; Imbalances people's
PD 7520 *Assassinate ; Incitement*
PD 1971 Assassination
Assassination [Character...]
PD 2569 — Character
PE 5944 — corporation leaders
PE 6356 — Corporation sanctioned
PE 5005 Assassination environmental activists
PD 7221 Assassination ; Government sanctioned
PE 5614 Assassination ; Political
PE 0252 Assassination trade union leaders
PE 0252 Assassination workers representatives
PF 3819 Assassinations ; Official cover up government sanctioned
PD 5235 Assault
PD 0583 Assault ; Aggravated
PD 3266 Assault ; Sexual
PE 1144 Assault ; Simple
PF 7659 Assaults police ; Public
PF 4550 Assemblies ; Violence parliamentary
PC 2383 Assembly ; Denial right
PD 7520 Assembly ; *Unlawful*
PF 2373 Assertiveness ; Lack
PE 8593 Asses, mules hinnies ; Instability trade
PF 2827 Assessment according own group values ; Subjective
Assessment [Inadequate...]
PJ 9034 — Inadequate capacity technology
PD 3436 — Inappropriate tax
PF 7116 — international nongovernmental organizations naive
PJ 9034 Assessment new technologies ; Inadequate
PF 7186 Assessment problems ; Over optimistic
PF 4377 Assessment ; Unrealistically positive self
PD 4882 Assessments ; Circumvention duties
PF 2451 *Assessments ; Erratic land*
PF 9243 Assessments ; Misrepresentation project
PF 3550 *Assessments ; Outdated property*
PE 9224 Asset stripping
PC 7605 Asset stripping ; Conceptual
PE 3827 Asset values ; Disparity share prices underlying
PJ 5990 *Asset ; Volatile value*
PD 9653 Assets heads state ; Corrupt acquisition offshore
PC 1784 *Assets ; Insecure property*
PF 6557 Assets ; Minimum promotion community
PF 1001 *Assets ; Undervaluation public*
PC 3438 *Assets ; Unequal distribution land*
PF 2241 *Assignments ; Alienating public housing*
PC 3293 *Assimilation ; Dependence forced*
PC 3293 *Assimilation ; Forced*
PF 2132 *Assimilation ; Lack*
PF 5306 Assist emergencies ; Failure
PF 6508 Assist financial investment ; Failure public authorities
PF 2969 *Assistance ; Complexity government*
Assistance [Delay...]
PF 3913 — Delay supply agreed multilateral
PE 5724 — Dependence island developing countries official
PF 0392 — developing countries ; Decline official development
PE 4796 — developing countries ; Inadequate public
PE 1407 Assistance endangered peoples ; Government non
PF 7176 Assistance government matters ; Unlawful compensation
Assistance [Imbalance...]
PE 4866 — Imbalance capital technical
PF 2813 — *Inadequate technical*
PE 2844 — industries labour affected developing country exports ; Inadequate adjustment
PF 6575 — *inertia ; External*
PF 5181 Assistance ; Lack long term development
PD 0136 Assistance programmes ; Abuses international
PE 4794 Assistance ; Reduced scope intergovernmental development
PF 8855 Assistance ; Unquestioning dependency social
Assistance victims
PE 7390 — abuse power ; Inadequate
PE 4086 — accidents ; Inadequate
PD 4823 — crime ; Inadequate
PE 8240 — extortion ; Inadequate
PD 5122 — human rights violations ; Inadequate
PE 4449 — rape ; Inadequate
PE 6936 — torture ; Inadequate
PF 2643 Assisted suicide ; Illegality doctor
PF 4821 Assisting repression

PF 7033 Associate international nongovernmental organizations regional development ; Absence policies
PE 1178 Association ; Criminal
PD 3224 Association ; Denial freedom
Association Denial right
PD 3224 — [Association ; Denial right]
PD 3224 — [Association ; Denial right]
PD 0683 — trade union
PD 4695 Association ; Violation right workers freedom
PD 3536 Associations ; Banned
PD 0822 Associations developing countries ; Lack
PE 0436 Associations front organizations government ; Misuse nonprofit
PF 7293 Associations ; Limited criminal liability
PF 1509 *Assumed youth incompetence*
PF 7706 Assumption absence causal relationship problems
PF 7706 Assumption lack problem interrelationships
PF 9582 Assumptions concerning recuperative power nature ; False
PF 6814 Assumptions ; Flawed
PF 6814 Assumptions ; Inappropriate
PF 6814 Assumptions ; Outdated
PF 2528 Assumptions sustainable development ; False
PF 6814 Assumptions ; Untested
Assurance [practices...]
PE 1826 — practices ; Unethical life
PE 8736 — premiums ; High life
PE 8736 — Prohibitive cost life
PD 2227 *Aster yellows*
PD 3158 *Asteroidea ; Endangered species*
PF 3564 Asteroids ; Hazards
PD 2408 Asthma
PE 3860 Asthma ; Bronchial
PD 2408 Asthma ; *Cardiac*
PE 1314 Asthma ; Potters'
PE 2053 Asthma ; Renal
PD 8179 *Astigmatism*
PF 8352 Astrology
PF 0565 Astrology superstition
PE 6756 Astronauts ; Accidents involving
PE 7244 Astronomical observation environmental pollution ; Obstruction
Asylum Denial right
PF 3021 — [Asylum ; Denial right]
PE 3450 — cultural
PE 6212 — economic
PF 1075 — political
PF 1075 Asylum ; Lack individual rights political
Asylum seekers
PB 0205 — [Asylum-seekers]
PE 6376 — Detention refugees
PD 6364 — Physical insecurity refugees
PB 0205 — Stowaway
PF 4925 Asylums ; Inhumane mental
PC 9112 Asymmetric interdependence developed developing countries
PE 9646 Asymmetry bargaining power transnational corporations developing countries
PA 5982 Asymmetry *Disagreement*
PA 6790 Asymmetry *Distortion*
PC 9112 Asymmetry economic interdependence
PF 4516 Asymmetry economic system incentives
PA 0224 Asymmetry *Imbalance*
PA 6695 Asymmetry *Inequality*
PD 8149 *Atavism*
PD 2730 Ataxia ; *Feline*
PE 8605 Ataxia ; Friedreichs
PF 7915 Ataxia ; Hereditary
PE 2945 Ataxia ; Locomotor
PE 0345 *Ataxic aphasia*
PF 2409 Atheism
PA 6058 Atheism *Impiety*
PA 6247 Atheism *Inattention*
PD 4365 Atheists ; Bigotry against
PE 2210 Atheroma
PE 4250 Athletes ; Drug abuse
Athletic activities
PF 0443 — *education ; Interference school*
PF 4202 — Non productive
PF 4222 — sports events ; Commercialization
PB 0750 *Athletic careers ; Short duration*
Athletic competition
PE 4266 — [Athletic competition]
PE 4833 — commercial political ends ; Exploitation
PE 4242 — Exclusion women
Athletic competitions
PE 3754 — Bribery connection sports
PE 3754 — Corruption sports
PF 4192 — Excessive expense international
PE 4576 — Gambling sports
PF 4216 — National bias judges international
Athletic [events...]
PE 0197 — *events ; Discrimination against women payment prizes*
PE 4809 — exchange ; Obstructions international
PF 4761 — exchanges ; Excessive political intervention international
PE 4262 Athletic injuries
PE 4246 Athletic training ; Discrimination against women
PF 4196 Athletic training programmes ; Excessive expense
PF 2605 *Athletics priority ; Low*
PD 9413 Atmosphere ; Degradation
PE 1655 Atmosphere ; Foggy
PE 9443 Atmosphere ; Increase nitrous oxide
PD 1354 Atmosphere ; Increase trace gases

Atmospheric carbon

Atmospheric [carbon...]
PD 4387 — carbon dioxide ; Increasing
PF 1234 — chemistry ; Long term change
PE 8815 — concentration methane ; Increase
PE 9525 — corrosion materials
PC 0119 Atmospheric pollution
PD 2008 Atmospheric pollution ; Particulate
PC 0918 Atmospheric radiation balance ; Deterioration
PE 2593 Atmospheric visibility ; Deterioration
PE 9446 Atmospheric water vapour ; Increase concentration
PD 0771 Atomic accidents ; Risk
PF 0913 Atomic energy ; Unhealthy emotional responses
PC 0229 Atomic radiation ; Environmental hazards
PF 5322 Atomism
PE 1277 Atomization computer-based work
PE 9509 *Atopic dermatitis animals*
PE 9509 Atopic diseases
PE 3711 *Atresia animals*
PD 9667 *Atrichia*
PD 6945 Atrocities
PD 1881 Atrocities ; Military
PC 4710 Atrocities ; Passive
PF 1098 Atrocities ; Politically motivated amnesty military personnel convicted
PC 4710 Atrocities ; Tolerated
PA 5454 Atrocity *Badness*
PA 6839 Atrocity *Disrepute*
PA 6822 Atrocity *Disrespect*
PA 5451 Atrocity *Insensibility*
PA 7156 Atrocity *Moderation*
PA 5643 Atrocity *Unkindness*
PA 7107 Atrocity *Unpleasantness*
PA 5644 Atrocity *Vice*
PA 7280 Atrocity *Wrongness*
PE 7553 *Atrophic rhinitis*
PD 0919 *Atrophic senile psychosis*
PE 1028 *Atrophy ; Acute yellow*
PE 7808 *Atrophy ; Brown*
PD 2565 *Atrophy ; Infantile muscular*
PE 3503 *Atrophy ; Periodontal*
PE 5188 Atrophy ; Progressive muscular
PE 7826 *Atrophy ; Shoulder*
PF 6106 Attachment
PC 0300 *Attachment disorder infancy ; Reactive*
PF 1577 Attachment past ; Sentimental
PD 2884 Attachment security structures ; Prioritized
PF 1073 Attachment social group ; Undue
PF 3390 Attachment territory ; Undue
PF 6106 *Attachment ; Traditional crop*
PF 4416 Attachments ; Dependence deep emotional
PD 5453 *Attack ; Heart*
PE 3705 Attack ; Surprise
PE 3933 *Attack ; Vasovagal*
PD 5238 Attack ; Verbal
PF 6895 Attacks peace forces
PD 2569 Attacks personal honour reputation ; Denial right freedom
PD 6364 Attacks refugee camps settlements ; Military
PE 5505 Attacks ; Threat arson
PD 5321 Attempt ; Criminal
PD 0437 *Attempted murder*
PE 4878 Attempted suicide
PE 8357 Attempting solve intergroup conflicts ; Lack participation
PF 2814 Attempts globality ; Unrealistic
PE 5621 Attempts ; Unreported rape
PF 6458 Attempts upgraded employment ; Disorganized
PE 4200 Attendance ; Erratic student
PE 4200 Attendance ; Inadequate enforcement school
PE 4200 Attendance ; Irregular school
PE 4200 Attendance ; Low school
PF 8905 Attendance ; Reduction church
PF 5121 Attention, care protection humankind ; Denial animals right
Attention [deficit...]
PF 2384 — deficit
PE 9789 — deficit hyperactivity disorder
PF 1113 — Dependency
PF 2384 — disorders
Attention [seeking...]
PF 1113 — seeking
PF 7003 — span limit societal learning ; Collective
PF 2384 — span ; Limited individual
PF 1113 — Starving
PA 6983 Attitude*complex
PF 8781 Attitude ; Defeatist civic
PF 2583 *Attitude ; Make consumer*
PF 2017 Attitude manipulation children play
PD 1317 Attitude parents ; Possessive
PC 1455 Attitude people faiths ; Negative
PF 6575 *Attitude ; Skeptical farming*
PF 2648 *Attitudes ; Anti community*
PF 6094 Attitudes currency ; Nationalistic
PA 7090 Attitudes ; Dependence negative emotions
PA 0832 Attitudes ; Destructive
PD 2019 Attitudes ; Educationally reinforced egocentric
PD 0061 *Attitudes ; Inhibiting social*
PA 7090 Attitudes ; Negative emotions
PF 2158 Attitudes ; Non concerned
PD 2239 Attitudes organizations' members ; Parochial
PF 5958 Attitudes planning second language policy ; Antisocial
PF 6224 Attitudes problems disabled people ; Mystique
Attitudes [technological...]
PF 4741 — technological innovation ; Self perpetuating
PF 3083 — toward community social services ; Inflexible
PE 4838 — towards menstruation ; Negative
PF 6573 *Attitudes ; Unquestioning, uncurious*
PF 2795 Attitudes use time ; Fatalistic
PE 6418 Attorney ; Non recognition foreign powers
PF 2423 *Attraction city life*
PF 5413 Attraction forbidden
PE 5925 Attraction ; Genetic sexual
PF 5413 Attraction illegal activity
PF 1735 *Attraction ; Insufficient tourist*
PF 8105 Attractions ; Limited cultural
PF 8105 Attractions ; Uncompetitive rural
PD 9869 Attractive city jobs
PF 6477 *Attractiveness overseas work*
PF 5893 Attunement ; Lack
PE 5524 *Atypical interstitial pneumonia ; Bovine*
PE 9438 *Atypical pneumonia sheep*
PF 0076 Audio visual media ; Absence
PD 7073 Auditing services ; Protectionism public accounting
PE 9799 Aujezky's disease
PD 5535 *Auricular dermatitis ; Necrotic*
PE 6254 Auriculata ; Salvinia
PA 5643 *Austerity*
PA 5740 Austerity *Compulsion*
PA 5473 Austerity *Insufficiency*
PD 8114 Austerity measures ; Negligent implementation
PF 5203 Austerity ; Resistance personal
PA 5612 Austerity *Unchastity*
PA 5643 Austerity *Unkindness*
PJ 5205 Autarky
PF 8012 Authentic relationships ; Fear
PC 6089 Authoritarian division labour
PC 9585 Authoritarian government
PB 1638 Authoritarian movements
PB 1638 Authoritarian people
PF 3706 Authoritarian propagation knowledge
PC 9585 Authoritarian regimes
PB 1638 Authoritarianism
PA 5740 Authoritarianism *Compulsion*
PB 1638 Authoritarianism ; Dependence
PC 1526 Authoritarianism ; Educational
PA 5563 Authoritarianism *Lawlessness*
PA 7306 Authoritarianism *Narrowmindedness*
PF 6508 Authorities assist financial investment ; Failure public
Authorities [Demolition...]
PE 9337 — Demolition homes government
PF 3211 — *Dependence planning*
PE 7581 — Destruction trade union property public
PF 3707 Authorities ; Fear retaliation
PE 1758 Authorities functioning workers organizations ; Interference public
PF 9288 Authorities human rights abuses ; Connivance
PC 6154 *Authorities ; Insolvent local*
Authorities [Neglect...]
PF 8072 — Neglect public
PF 8072 — neglect ; Public
PF 2261 — Non resident local
PE 5462 Authorities ; Occupation trade union premises public
PE 5462 Authorities ; Raids trade union premises public
PF 2261 Authorities ; Remote borough
PE 7581 Authorities ; Seizure trade union property public
PE 5070 Authorities ; Suppression strikes public
PF 8072 Authorities ; Unresponsive public
PC 8689 Authority ; Abuse
PE 0742 Authority ; Alcohol abuse people
PF 5012 Authority ; Contempt
Authority [Denial...]
PD 7609 — Denial right petition
PF 8995 — Dependence
PF 3064 — Diffuseness regulatory
PF 5012 — Disrespect
PD 7792 Authority ; Flouting regulatory
Authority [Illegal...]
PC 8689 — Illegal exercise abuse
PC 8689 — Illegal use abuse administrative
PC 8689 — Illegitimate
PF 8995 — *Interference head household*
PF 2811 Authority leadership ; Relative
PE 0365 Authority legal structures developing countries ; Inadequate local
PE 4635 Authority political purposes ; Misuse personal
PE 1348 Authority sexual purposes ; Misuse spiritual
PF 6506 *Authority ; Uncommunicated annexation*
PF 8596 Authority ; Uncritical acceptance
PD 7221 Authorized governments ; Political murders
PE 1222 Autism
PA 6180 Autism *Error*
PE 1222 Autism ; Infantile
PA 7211 Autism *Selfishness*
Autism [Unfeelingness...]
PA 7364 — *Unfeelingness*
PA 6738 — *Unimaginativeness*
PA 6653 — *Unsociability*
PE 1222 Autistic children
PC 9585 Autocracies ; Government
PE 2805 Autocracy intergovernmental financial institutions
PC 0845 Autocratic rule
PE 4426 Autoeroticism
PE 5214 Autoimmune disease
PE 9509 *Autoimmune haemolytic anaemia animals*
PE 9509 *Autoimmune thrombocytopenia animals*
PE 9509 *Autoimmune thyroiditis animals*
PF 2842 *Automated warehousing jobs*
PE 4391 Automation developing countries ; Elimination jobs due
PD 0528 Automation ; Elimination jobs
PF 5044 Automation ; Excess leisure time arising
PE 5164 Automation ; Mental stress due
PE 5134 Automation ; Social ill effects
PE 8721 Automation ; Uncoordinated use computers
PE 0661 Automatism ; Post epileptic
PD 0079 Automobile accidents
PE 2462 Automobile design construction ; Unsafe
PD 0414 Automobile emissions air pollutants
PC 5842 *Automobile industry ; Protectionism*
PE 1469 Automobile industry transnational corporations ; Domination
PE 8067 Automobile manufacturing industry ; Ineffective self regulation
PF 6152 Automobile parking space
PE 4826 Automobile theft
PE 6142 Automobiles ; Environmental degradation
PF 6152 Automobiles ; Excessive use land
PF 6250 Automobiles ; Fragmentation communities
PF 4480 Automobiles ; Maldistribution private
PD 2072 Automobiles motor vehicles ; Proliferation
PE 5339 Automobiles ; Scrapped
PE 8932 Autonomic nervous system ; Diseases peripheral
PF 0053 Autonomous world level actor identify clarify world interests ; Lack
PD 4282 *Autonomy ; Erosion university*
PE 7706 Autonomy ; Fear increased
PE 1137 *Autonomy ; Inexperienced financial*
PE 6197 Autumn catarrh
PE 8189 Avahis indri ; Endangered species sifakas,
Availability [capital...]
PF 2378 — capital reserve ; Limited local
PF 2378 — capital ; Restricted
PF 7130 — community space ; Obstacles
PF 2489 — credit ; Limited
PD 4560 Availability developed countries ; Contraceptive
PF 3575 Availability education rural areas ; Limited
PF 2996 *Availability ; Erratic truck*
Availability [farm...]
PF 3630 — *farm loans ; Limited*
PF 2489 — financial credit ; Limited
PD 5212 — food ; Fluctuation
PF 3539 — functional information ; Limited
PF 3550 — funding resources urban services ; Limited
PF 2378 — funds ; Limited
PB 8891 Availability goods ; Insufficient
PD 7669 Availability health resources ; Limited
PF 3550 Availability investment capital urban renewal ; Limited
PF 1658 Availability jobs small communities ; Limited
Availability [land...]
PF 5008 — land low income disadvantaged groups ; Limited
PF 3184 — learning opportunities ; Limited
PF 3184 — learning structures ; Limited
PF 2996 — *Limited transportation*
PF 2996 — *Limited vehicle*
PE 4704 — loan capital ; Limited
PE 4704 — loans developing countries ; Limited
PF 6477 — local business capital ; Limited
PF 2009 — local commercial resources ; Limited
PF 2865 — long term capital ; Limited
PE 8256 Availability methods cross breeding ; Lack
PF 2457 Availability modern farm machinery ; Limited
Availability [permanent...]
PE 1134 — permanent employment inner cities ; Limited
PE 5170 — practical skills rural communities ; Limited
PF 6539 — public services small towns developed countries ; Limited
PF 2583 *Availability services ; Unacknowledged*
PF 1653 *Availability subsidies ; Unknown*
PF 2698 Availability technical agricultural business training ; Limited
PF 6751 Availability therapeutic substances human origin ; Limited
PE 5170 Availability urban skills local communities ; Restricted
PE 7530 Availability viable organs transplantations ; Imbalance need
PF 6471 *Available capital ; Unanalysed*
Available [data...]
PE 0630 — data ; Absence method appropriate
PE 3812 — developing countries ; Decline concessional financial resources
PE 3812 — developing countries ; Inappropriate loans
PB 2846 Available food ; Insufficient
PF 9683 Available health care ; Inefficient use
PF 9683 Available health facilities ; Non use
Available [Imbalances...]
PF 9008 — Imbalances people's aspirations structure opportunities income
PF 1267 — information ; Inadequate dissemination use
PF 3539 — *information ; Irrelevant*
PF 8753 — insight response problems society ; Failure focus
PC 1131 *Available jobs unrelated education*
PE 2000 Available knowledge ; Increasing inability integrate
PF 8191 Available knowledge solve problems ; Inadequate application
Available [labour...]
PF 8687 — labour ; Unsuitability development projects
PC 8160 — land ; Limited
PF 6528 — land ; Restrictive use
PF 6544 — local funding ; Lack channels obtaining
PF 8299 Available markets ; Distant
PF 0734 Available radio frequencies ; Limited number
Available [seeds...]
PE 6788 — seeds ; Reduction genetic diversity
PF 2875 — services ; Limited access
PF 2583 — *services ; Unknown*
PF 6538 — skills ; Underutilization locally
PF 7344 — skills ; Unmarketable
PF 7130 — *space ; Underutilized*
PF 8205 Available welfare benefits ; Lack awareness

Barrenness

PD 1146 Avalanches
PD 1146 Avalanches ; Rock
PE 7838 Avalanches ; Snow
PA 6999 Avarice
PA 6379 Avarice *Avoidance*
PA 7193 Avarice *Cheapness*
PA 6999 Avarice ; Dependence
PA 5644 Avarice *Vice*
PF 2575 Avenues civic responsibility ; Individual
PF 2797 Avenues leadership potential ; Unexplored
PF 1199 Average intelligence certain racial groups ; Inferior
PA 7338 Aversion *Hate*
PF 4612 Aversion strategy ; Risk
PA 7107 Aversion *Unpleasantness*
PA 6509 Aversion *Unwillingness*
PD 0332 Aves ; Endangered species
PD 2731 Avian infectious hepatitis
PE 1923 Avian malaria
PE 1400 Avian pneumoencephalitis
PD 3601 Avian vectors plant diseases
PD 2731 Avian vibrionic hepatitis
PD 2784 Aviation ; Increase animal disease increase
PD 8328 Aviation risks
PD 0715 Avitaminoses
PE 5267 Avoid prosecution giving testimony ; Flight
PA 6379 Avoidance
PA 6379 Avoidance *Avoidance*
PF 6382 Avoidance ; Blame
Avoidance [complaints...]
PF 9157 — complaints ; Government
PF 1586 — confrontation death
PD 4060 — controversial issues
PD 0188 — copyright
PF 4204 Avoidance decision-making
PD 8468 Avoidance facts decision making ; Selective
PF 7925 Avoidance human rights issues international community
Avoidance [Intemperance...]
PA 6466 — *Intemperance*
PF 1610 — irrational
PF 1623 — *Issue*
PE 6311 — issues political campaigns
PD 4556 Avoidance legal obligations politicians
PF 5736 Avoidance legislative reform ; Deliberate governmental
PF 5311 Avoidance negative feedback
PA 8658 Avoidance obligations
PF 3538 Avoidance patent restrictions
PF 5210 Avoidance ; Problem
PF 7414 Avoidance reality
PB 5249 Avoidance reference suffering
PF 4600 Avoidance substantive issues political debate
PD 4882 Avoidance ; Tax
PF 1462 Avoidance ; Tax
PC 5528 Avoidance work
PE 9256 Avoidant disorder adolescence
PE 9256 Avoidant disorder childhood
PE 3901 Avoidant personality disorder
PF 4612 Avoiding danger ; Reactionary methods
PD 0356 Avoiding military service
PF 1004 Awaken consciousness due inadequate methods ; Inability
PF 7198 Awards ; Accumulation
PJ 4562 Awards ; Inadequate enforcement foreign arbitral
PE 4225 Awareness advertisers social responsibility ; Insufficient
PF 8205 Awareness available welfare benefits ; Lack
PF 3628 *Awareness child care opportunities ; Limited*
PF 0544 Awareness developing countries ; Imbalance distribution political
PD 5349 Awareness high risk ; Endangerment
PF 5039 Awareness ; Inadequate development critical
PE 5132 Awareness international affairs ; Insufficient work
PF 2442 Awareness ; Lack deathing
PF 1985 Awareness past heritage ; Lack
PF 2042 Awareness potential investment small, inner city enterprises ; Lack
PF 0544 *Awareness ; Reduced civic*
PF 1246 Awareness sexual identity ; Indistinct
PA 5454 Awfulness *Badness*
PA 6030 Awfulness *Fear*
PA 7240 Awfulness *Ugliness*
PA 7107 Awfulness *Unpleasantness*
PF 5272 Awkward past experiences
PA 5497 Awkwardness *Difficulty*
Awkwardness [Ignorance...]
PA 5568 — *Ignorance*
PA 6312 — *Inelegance*
PA 7395 — *Inexpedience*
PA 7107 Awkwardness *Unpleasantness*
PA 7232 Awkwardness *Unskillfulness*
PE 1570 Aye aye ; Endangered species
PE 9774 *Azoturia*

B

PD 0715 *B deficiency ; Vitamin*
PE 2348 B encephalitis ; Japanese
PE 0517 B ; Viral hepatitis
PD 0715 *B1 ; Lack vitamin*
PE 6288 Babesioses
PE 6288 Babesiosis
PD 4029 Babies ; Still born
PF 3382 Babies ; Unwanted female
PE 6154 Baby snatching
PE 2259 Bacillary dysentery
PE 7769 *Bacillary haemoglobinuria animals*
PD 2226 Bacillus

PE 1310 Back-ache
PE 7808 *Back foals ; Defects*
PE 7826 Back horse ; Fractures
PE 8589 *Back ; Hunch*
PF 4181 Back land
PE 9774 Back muscle necrosis
Back [pain...]
PE 1310 — pain
PE 5483 — pain ; Low
PE 5306 — provisions ; Grant
PF 2241 *Background ; Disrelated social*
PF 3391 Backlash against repeated warnings
PA 6706 Backlash *Culmination*
PF 2471 *Backlash ; Feminist*
PE 6711 Backlash ; Medical
PD 5995 Backlash richer nations ; Economic
PF 3318 *Backlog ; Government administrative*
PF 6826 Backsliding ; Religious
PE 4791 Backwardness
PD 9094 Bacteria
PD 2731 *Bacteria ; Allergy inducing*
PD 9094 Bacteria causing disease
PF 1312 Bacteria ; Introduction extraterrestrial infectious diseases
PE 6007 Bacteria ; Resistant
PD 2562 Bacteria vectors viral diseases
PD 2562 Bacteria ; Virus diseases
PD 9094 Bacteria ; Water borne pathogenic
PD 2226 *Bacterial bean blight*
Bacterial [disease...]
PD 9094 — disease
PD 2731 — diseases animals
PD 6363 — diseases ; Zoonotic
PE 9374 Bacterial food poisoning
PD 9094 Bacterial infections
PF 2276 Bacterial mutation
PD 2226 Bacterial plant diseases
PE 6629 Bacterial resistance antibiotics
PE 4706 Bacterial soft rot plants
PD 2226 *Bacterial wilt cucumber*
PD 2226 *Bacterial wilt cucurbits*
PC 0195 Bacteriological warfare
PF 7731 Bacteriologists ; Irresponsible
PD 0594 Bacteriophages
PF 2152 Bad arguments
PE 6558 Bad breath
PF 8759 Bad company ; Keeping
PF 1421 Bad death ; Dying
PF 2166 Bad debts
PF 6683 Bad feng shui
PF 3953 Bad government
PD 5494 Bad habits
PD 0465 Bad investment savings developing countries
Bad [language...]
PF 7427 — language
PD 9156 — living conditions
PF 9536 — luck
PC 7013 Bad manners
PE 1478 Bad news ; Media emphasis
PE 0598 Bad news ; Postponement
PE 8577 Bad omens
PE 4545 Bad property loans
PE 2280 *Bad ventilation*
PC 0293 Bad weather
PC 0293 Bad weather ; Risk
PE 8936 Badgers, skunks, otters ; Endangered species weasels,
PD 6627 *Badly laid out work premises*
PB 1498 Badly nourished people
PA 5454 Badness
PA 5454 Badness *Badness*
PA 6971 Badness *Danger*
PA 6799 Badness *Disease*
PA 6498 Badness *Misbehaviour*
PA 5981 Badness *Stench*
PA 7226 Badness *Unhealthfulness*
PA 5644 Badness *Vice*
PA 7289 Bafflement *Defeat*
PA 5497 Bafflement *Difficulty*
PA 5527 Bafflement *Inexpectation*
PA 7309 Bafflement *Uncertainty*
PD 4691 *Bag snatching*
PE 4577 *Bail jumping*
PC 0918 Balance ; Deterioration atmospheric radiation
PE 6603 Balance human body ; Disruption internal
Balance payments
PF 3062 — adjustment ; Inadequate mechanism
PE 1771 — Adverse effect transnational corporations
PC 0998 — deficits
PE 4997 Balances ; Inadequate system political checks
PD 7799 *Balanoposthitis animals*
PD 7799 *Balanoposthitis ; Enzootic*
PD 9667 *Balanoposthitis ; Veneral*
PF 5880 Baldness
PD 9667 *Baldness*
PC 2944 Balkanization
PC 1833 Balkanization metropolitan government decision-making
PD 1292 *Ball lightning*
PE 4515 Ballistic missiles
PE 9052 Ballistic missiles developing countries ; Proliferation
PD 5214 Ballot rigging
PE 2455 *Balstamycosis*
PD 3481 *Bamboo rats ; Endangered species*
PA 7365 Banality *Boredom*
PE 1774 Banana trade ; Instability
PE 1601 Bancroft's filariasis
PE 1601 Bancroftian filariasis

PD 1762 Bandicoots ; *Endangered species*
PD 2609 Banditry
PF 3703 Bands satellite transmission ; Allocation television frequency
PA 5454 Banefulness *Badness*
PA 7055 Banefulness *Death*
PA 6542 Banefulness *Destruction*
PA 7107 Banefulness *Unpleasantness*
PE 0924 Bang's disease
PE 2435 Bangs ; Risk sonic
Bank failure
PE 0964 — [Bank failure]
PE 9465 — Government development
PE 1815 — Small
PE 1398 Bank fraud
PE 3903 Bank interest rates ; Low
PE 4655 Bank lending developing countries ; Decline commercial
PE 4704 Bank loans ; Inaccessibility
PE 6559 *Bank use ; Undecided*
PE 4335 Banking activities developing countries ; Restrictions home countries transnational
PE 7975 Banking ; Capitalist
PF 0048 *Banking facilities ; Underutilization world*
Banking [industry...]
PF 4370 — industry ; Antiquated regulations
PD 7120 — industry ; Protectionism
PF 1705 — *information ; Insufficient*
PE 1208 Banking law violations
PE 0682 Banking malpractice
PE 0718 *Banking options ; Restricted*
PF 2009 *Banking practices ; Complex*
PF 2009 *Banking procedures ; Unfamiliar*
Banking [secrecy...]
PF 5991 — secrecy ; Abuse
PG 5429 — security firms ; National regulations
PE 0718 — *services ; Distant*
PE 0718 — *services ; Unavailability*
PC 6154 *Bankrupt cities*
PE 6431 Bankrupt institutions
PC 6154 Bankruptcy
PD 2591 Bankruptcy ; Business
PD 9376 Bankruptcy ; Personal
PE 7975 Banks ; Concentration capitalist
Banks developing countries
PE 4320 — Discrimination against transnational
PE 4310 — Discrimination lending transnational
PE 4330 — Domination restrictive project loans transnational
PE 4360 — Socially irresponsible programmes transnational
PE 4325 Banks domestic economic policies ; Interference transnational
PE 4355 Banks ; Domination loan negotiations transnational
Banks [Failure...]
PE 9465 — Failure state
PF 2491 — Fear
PE 1398 — Fraud savings
PE 1815 Banks ; High risk policy smaller
PF 4350 Banks ; Inadequate information concerning transnational
PF 4580 Banks ; Irresponsible lending
PE 4508 Banks ; Need increased cooperation central
PE 4315 Banks' off shore borrowing domestic monetary policies ; Interference transnational
PF 4580 Banks ; Over enthusiastic lending
PE 9657 Banks ; Restrictive practices transnational
PD 2290 *Banks ; Unlandscaped river*
PD 3536 Banned associations
PE 9773 Banned cultivation plant species
PC 0193 *Banned home ; Children*
PE 9027 Banned imports
Banned [pharmaceutical...]
PE 6036 — pharmaceutical drugs developing countries ; Marketing
PD 3028 — plays
PD 3536 — political parties
PF 3353 Banned religious sects
PD 3535 Banned trade unions
PE 4991 Banning ivory trade
PE 9773 Banning plant cultivation ; Regulatory
PE 1601 Barbados leg
PA 5568 Barbarism *Ignorance*
PA 5821 Barbarism *Vulgarity*
PA 7156 Barbarity *Moderation*
PA 5643 Barbarity *Unkindness*
PA 5821 Barbarity *Vulgarity*
PE 0139 Barbiturate hypnotics ; Abuse barbiturates non
PE 0139 Barbiturates non barbiturate hypnotics ; Abuse
PE 3970 Bargain collectively ; Violation right workers organizations
PF 0685 *Bargaining ; Delays centralized collective*
PE 3970 *Bargaining ; Denial right collective*
PF 6745 *Bargaining ; Government bias wage*
PE 6018 *Bargaining ; Inadequate employee participation collective*
PE 3528 *Bargaining power ; Inequality*
PE 9646 Bargaining power transnational corporations developing countries ; Asymmetry
PE 3970 Bargaining workers organizations ; Restrictions collective
PE 4586 *Bark disease complex ; Beech*
PD 7841 Barkers
PE 2727 *Barn itch*
PE 4475 Barratry
PF 8535 Barrel politics ; Pork
PC 6037 Barrenness
PA 7365 Barrenness *Boredom*

Barrenness

PA 6876 Barrenness *Impotence*
PA 6852 Barrenness *Inappropriateness*
PA 6738 Barrenness *Unimaginativeness*
PA 7208 Barrenness *Unproductiveness*
PC 3458 Barrier ; Class consciousness social
PA 7391 Barrier *Closure*
PD 2330 Barrier development ; Animal worship
PA 5497 Barrier *Difficulty*
PF 6035 Barrier ; Language
PF 1502 Barrier progress ; Family structure
PA 7296 Barrier *Restraint*
PE 8792 Barrier subsequent economic growth ; Import substitution
PE 0803 Barriers access legal profession, judiciary jury membership ; Economic
PB 2331 Barriers ; Cultural
Barriers [Decline...]
PE 2176 — Decline competition due entrance
PB 2331 — Dependence cultural
PC 1247 — Destruction natural
PE 0707 — disabled persons ; Structural
PD 3050 — dissemination technical knowledge ; Tax
PF 4852 *Barriers economic education*
PC 3201 Barriers effective legislation ; Political
PD 5972 Barriers ; Frontier
PF 2105 Barriers global ethic ; Governmental
PE 4868 Barriers handicapped family members ; Immigration
PD 9651 Barriers ; Inter cultural trade
Barriers international
PF 0166 — flow knowledge educational materials
PC 2725 — trade ; Non tariff
PC 0569 — trade ; Tariff
PE 1479 Barriers judicial protection individual rights ; Psychological
PF 2365 *Barriers jungle harvesting*
PF 6035 Barriers ; Language
PE 8049 Barriers least developed countries ; Geographic
PC 3258 *Barriers male pregnancy ; Social*
PE 4170 Barriers manufactured goods developing countries ; Trade
PD 0057 Barriers primary commodities ; International trade
PD 2958 Barriers protectionism developing countries ; Trade
PF 4371 Barriers religious experience
PF 1976 Barriers ; Smuggling sanctions
PF 4371 Barriers spiritual experience
PC 4275 Barriers ; Trade
Barriers trade
PC 2369 — developing developed countries ; Tariff
PC 4382 — Technical
PD 5033 — Unpredictable
PF 4371 *Barriers transcendent experience*
Barriers transfer
PD 0084 — educational facilities
PE 8403 — knowledge ; Copyright
PF 6035 — knowledge ; Language
PE 1198 Barriers women's access judicial process ; Economic
PE 0117 Barter system international trade ; Inadequate
PD 2034 *Barytosis*
PE 8131 Base metals ; Instability trade ores concentrates non ferrous
PA 5454 Baseness *Badness*
PA 6839 Baseness *Disrepute*
PA 6030 Baseness *Fear*
PA 6659 Baseness *Humility*
Baseness [Imperfection...]
PA 6997 — *Imperfection*
PA 7363 — *Improbity*
PA 5652 — *Inferiority*
PA 7107 Baseness *Unpleasantness*
PA 5644 Baseness *Vice*
PA 5821 Baseness *Vulgarity*
PF 4155 Bases developing countries ; Underutilization international data
PE 0736 Bases ; Environmental destruction military
PD 3496 Bases intelligence centres ; Extra territorial military
PC 8390 *Bases ; Obsolete military*
PF 1066 Bases ; Unilateral declarations independence extra territorial
PJ 7586 Bashfulness
PF 1548 Basic education ; Lack emphasis
PD 9306 Basic facilities ; Understaffing
PF 4763 Basic food stocks ; Secrecy national
Basic [health...]
PB 4885 — *health ; Fragility maintaining*
PD 2011 — health small rural villages ; Unsanitary environment
PF 2875 — human resource services ; Ineffective delivery
PD 8294 — *hygiene ; Inappropriate*
Basic industries
PE 8397 — environmental hazards ; Iron steel
PE 8248 — environmental hazards ; Non ferrous metals
PE 8070 — instability ; Iron steel
PE 2601 — *instability ; Non ferrous metal*
PE 8223 — Underdevelopment iron steel
Basic metal industries
PD 0243 — Health hazards
PE 2601 — Instability
PE 5866 — Protectionism steel
PF 1374 — Underdevelopment
PA 0831 Basic necessities ; Deprivation
PF 1373 *Basic necessities ; Neglect*
PF 6814 Basic presuppositions
PF 3448 Basic resources ; Underdeveloped potential
Basic services
PF 6473 — developing countries ; Underdeveloped provision
PF 2652 — Limited distribution
PF 6527 — Prohibitive cost

Basic services cont'd
PF 2875 — Remote
PF 2875 — rural areas ; Underprovision
PF 4470 — users ; Small population
PF 2477 Basic skills ; Limited
PF 2880 Basic skills ; Unequal global distribution
PF 2583 Basic urban services ; Underprovision
PF 1815 *Basic utilities ; Outdated*
PF 1728 Basic value images social reality ; Lack correspondence
PD 2728 *Basidiobolomycosis*
PE 0364 Basidiomycetes
PD 0516 Basin development ; Uncoordinated international river
PF 4741 Basis competition ; Restriction technical
PF 0836 Basis cultural identity ; Obsolete
PF 1382 Basis family economics ; Precarious
PF 1240 Basis income possibilities ; Undiversified
PF 6460 Basis land allocation ; Ownership
PA 5952 Bastardy *Illegality*
PE 5245 *Basuco*
PE 3604 Bat ; Endangered species golden
PE 3604 Bat ; Endangered species New Zealand short tailed
PD 2730 *Bat salivary gland fever*
PE 6685 Bathing near sewage discharge
PE 6685 Bathing sewage-contaminated water
PE 3604 Bats
Bats Endangered species
PE 3604 — *bulldog*
PE 3604 — *common*
PE 3604 — *disc winged*
PE 3604 — *free tailed*
PE 3604 — *fruit*
PE 3604 — *funnel eared*
PE 3604 — *horseshoe*
PE 3604 — *leaf nosed*
PE 3604 — *mouse tailed*
PE 3604 — *sac winged*
PE 3604 — *slit faced*
PE 3604 — *smoky*
PE 3604 — *spear nosed*
PE 1890 Bats ; Vampire
PC 2584 Battered children
PA 0429 *Battered husbands*
PD 6758 Battered women
PE 1171 Battering animals ; Experimental
PD 5235 Battery
PA 5532 Battle *Disaccord*
PE 7912 Battle fatigue
PA 5612 Bawdiness *Unchastity*
PE 8175 Bays estuarine waters ; Pollution
PD 1356 Beach pollution
PF 2822 *Beach use ; Restricted*
PD 1593 Beaked whales ; Endangered species
PD 2226 Bean blight ; Bacterial
PD 1434 Bean pea distribution unevenness
PE 8259 Beans, dried, peas, lentils leguminous vegetables ; Instability trade
PD 3483 Bear ; Endangered species
PE 1320 Bear species ; Endangered polar
PF 1618 Bear witness ; Inability
PA 5454 Beastliness *Badness*
Beastliness [Unchastity...]
PA 5612 — *Unchastity*
PA 5459 — *Uncleanness*
PA 5643 — *Unkindness*
PA 7107 — *Unpleasantness*
PE 2320 *Beat elbow*
PC 2584 Beating ; Child
PD 7187 Beating children ; Parental
PE 3394 Beating draught animals
PD 2484 Beating prisoners
PE 0192 Beating school students
PD 2484 Beating ; Torture
PD 6758 Beating ; Wife
PD 5560 Beatings civilians
PF 6550 *Beautification plans ; Unstructured*
PD 3481 *Beavers ; Endangered species*
PE 1597 Bed marine resources ; Inequitable allocation rights exploit sea
PD 1241 Bed ; Militarization deep ocean sea
PE 1597 Bed resources ; Discriminatory exploitation sea
PE 5431 Bed wetting
PE 3617 Bedbugs
PA 7157 *Bedevilment Insanity*
PA 7148 *Bedevilment Ungodliness*
PA 7107 *Bedevilment Unpleasantness*
PA 7361 *Bedraggled Disorder*
PA 5459 *Bedraggled Uncleanness*
PD 3634 Bee pests
PA 4586 *Beech bark disease complex*
PE 6586 Beef tapeworm
PE 2585 Beer trade ; Instability
PD 0793 Bees ; Aggressive honey
PD 0793 Bees ; Killer
PE 2254 Beet fly
PE 2254 Beet leaf miner
PE 2975 Beet ; Pests diseases sugar
PE 1679 *Beetle ; Carpet*
PD 2528 Beetle poisoning animals ; Blister
PE 1679 Beetles insect pests
PE 0886 Beetles ; Wood boring
PD 2500 Beggars
PC 5512 *Beggary ; Child*
PD 2500 Begging
PC 5301 Behaviour according cultural norms ; Systematic human
PC 4726 Behaviour ; Anti social

PC 4321 Behaviour ; Contempt traditional modes
PD 0437 Behaviour ; Criminally life endangering
Behaviour [Dangerous...]
PF 7580 — Dangerous
PD 0329 — developing countries ; Increase anti social
PE 7017 — Discriminatory unwritten codes
PE 2076 — Disordered
PF 7248 — disorders ; Narcissitic
PD 8544 — Disruptive
PE 2076 — disturbances
PE 3296 — *Dominated women's*
PD 1533 — due television ; Increase antisocial
Behaviour [Effects...]
PE 5376 — Effects AIDS sexual
PE 8533 — Excessive dependence computer models complex system
PF 6570 — *expectation ; Peer*
PF 9209 Behaviour ; Fear emotionally responsive
PF 4753 Behaviour ; Government approval addictive
PF 4753 Behaviour ; Government approval dangerous
PF 2687 Behaviour ; Individualistic standards sexual
PB 1125 Behaviour ; Insubordinate
PD 8227 Behaviour leaders ; Unethical
PE 2690 Behaviour modification ; Abusive
PF 3862 Behaviour ; Proscribed thinking
PF 7076 Behaviour ; Psychogenetic constraints
PE 9092 Behaviour schools ; Disruptive
PD 4418 Behaviour ; Self defeating
Behaviour [Unchecked...]
PD 8544 — Unchecked destructive
PE 7370 — university students ; Anti social
PF 4550 — Unparliamentary
PB 6321 Behavioural deterioration
PF 3976 Behavioural norms ; Deviation
PE 6914 Behavioural regulation ; Torture
PE 2690 Behaviourism
PE 4940 *Behaviourism*
PE 2690 Behaviourism ; Misapplied
PJ 4824 *Behcet's disease*
PF 2364 Beings ; Denial right develop human
PD 8017 Beings ; Governmental disregard people human
PC 1648 Beings ; Stress human
PE 1764 Belching
PF 3931 Belief benefits technical progress ; Dogmatic
PD 8445 *Belief ; Denial right freedom*
PC 1455 Belief ; Discrimination based
Belief [ecumenical...]
PF 9274 — ecumenical dialogue ; Erosion religious
PF 8751 — emotional instability women
PF 7042 — evil
PF 4954 — Excessive integration religious
PF 6430 *Belief fate*
PF 3404 Belief ; Fragmentation religious
PF 8264 Belief humanity's dominance over nature
PF 6430 *Belief manifest destiny*
PF 6814 *Belief miracles*
PF 8855 Belief personal entitlement
PF 8433 Belief supernatural
PF 5864 Belief that wealth eliminates problems
PA 0911 *Belief ; Wrong*
PF 9205 Beliefs ; Facile
PF 6829 Beliefs ; Irrational religious
PF 9205 Beliefs ; Over reliance convenient
PF 9205 Beliefs ; Thoughtless acceptance convenient
Beliefs [Unchallenging...]
PF 9205 — Unchallenging
PF 7847 — Unquestionable
PF 8922 — Unrealistic
PF 2708 Belittling grant recipients
PA 5532 Belligerence *Disaccord*
PA 5446 Belligerence *Enmity*
PA 7253 Belligerence *Envy*
PD 8021 Belligerent occupation
PD 3978 *Belly lambs ; Rattle*
PA 0294 Below health
PF 6135 Belts ; Green
PD 2284 Beneficial insects ; Diseases
PE 8717 Beneficial plants animals ; Loss
PF 1492 *Beneficiaries ; Counter productive government constraints programme*
PE 5338 Beneficiaries net outflow funds developing countries ; Aid lenders
PJ 8129 Beneficiaries projects ; Nonidentification intended
PF 9243 Benefit analyses ; Manipulation project cost
PE 7271 Benefit coupons ; Trafficking government
PF 3918 Benefit doubt given vested interests
PD 7500 Benefit overpayments
PE 1625 Benefits aged ; Violation right equal work
PJ 0451 Benefits be proven prior programme approval ; Government requirement that economic
Benefits [Commerce...]
PE 9840 — Commerce spiritual
PF 3448 — *community resources ; Underdeveloped*
PF 9729 — cooperatives ; Unrecognized
PF 1781 — *corporate action ; Unrecognized*
PD 0162 *Benefits ; Denial right old age*
PD 1229 Benefits ; Dependence government welfare
Benefits [Economic...]
PD 3245 — Economic bias worker
PD 0794 — economic integration ; Imbalances distribution costs
PE 1625 — elderly workers ; Denial equal
PE 8893 — Excessive bureaucratic requirements welfare
PD 8859 Benefits fraud ; Unemployment
PD 8859 Benefits fraud ; Welfare
Benefits [Inflated...]
PF 7911 — *Inflated cost unemployment*

-864-

Birth

Benefits [Insufficient...] cont'd
PE 4915 — Insufficient export
PF 0171 — Insufficient job
PE 5211 — invalids ; Denial right
Benefits [Lack...]
PF 8205 — Lack awareness available welfare
PF 8205 — Lack knowledge eligibility
PF 1303 — Limited access social
PD 0530 — limited number developed countries ; Restriction outer space
PF 2152 Benefits ; Misuse plausible arguments conceal nefarious
PE 5667 Benefits political purposes ; Deprivation government
PF 8181 Benefits projects ; Overestimated economic
PF 7911 Benefits ; Rising cost unemployment
PF 6077 Benefits science ; Denial right
PE 4531 Benefits survivors ; Denial right
PF 3931 Benefits technical progress ; Dogmatic belief
Benefits Unawareness
PF 3211 — health
PF 6925 — potential
PF 8205 — welfare
Benefits [Undemonstrated...]
PF 9729 — Undemonstrated cooperative
PD 8859 — Unethical use social welfare
PF 9729 — Unrecognized co
PE 6835 Benefits women ; Unequal health
PA 7331 Benevolence∗complex
PD 7841 Benign enzootic paresis
PE 5617 Benign foot rot
Benign neoplasm
PD 8347 — bone cartilage
PD 8347 — breast
PD 8347 — digestive system
PD 8347 — endocrine glands
PE 9705 — genital organs
PE 9086 — kidney urinary organs
PD 8347 — musculoskeletal system
PD 8347 — nervous system
PE 0368 — ovary
PE 3637 — respiratory system
PD 8347 — skin
PD 8347 — throat mouth
PE 7922 — uterus
PD 8347 Benign neoplasms
PE 0616 Benign tertian
PD 8347 Benign tumours
PE 1849 Benzene ; Occupational hazards
PE 7691 Bereaved children
PF 3516 Bereavement
PA 5688 Bereavement *Appropriation*
PA 7410 Bereavement *Celibacy*
PA 7055 Bereavement *Death*
PD 4918 Bereavement ; Inhibited
PA 7382 Bereavement *Loss*
PE 2185 Beriberi
PD 0637 *Berylliosis*
PE 2209 Beryllium health hazard
PE 3676 *Besnoitiosis*
PE 3274 Bestiality
PA 5454 Bestiality *Badness*
PA 5612 Bestiality *Unchastity*
PA 5643 Bestiality *Unkindness*
PA 5821 Bestiality *Vulgarity*
PA 6379 Betrayal *Avoidance*
PE 1578 Betrayal double agents
PF 4153 Betrayal government
PA 7363 Betrayal *Improbity*
PA 7325 Betrayal *Irresolution*
PF 3420 Betrayal principles ; Compromise
PE 4576 Betting ; Clandestine sports
PF 2137 Betting gaming
PD 8286 Beverages ; Consumption alcoholic
PE 0849 Beverages ; Environmental hazards
PE 3035 Beverages ; Environmental hazards non alcoholic
Beverages [Illicit...]
PE 7188 — Illicit production alcoholic
PE 8134 — industry ; Protectionism alcoholic
PE 1680 — Instability trade
PE 8748 — Instability trade non alcoholic
PE 3980 — Irresponsible advertising alcoholic
PD 8286 Beverages ; Manufacture alcoholic
Beverages tobacco
PE 1641 — Instability trade
PE 1253 — Long term shortage
PE 7899 — trade ; Restrictive practices
PD 8286 Beverages ; Trade alcoholic
PE 2585 Beverages trade ; Instability alcoholic
PA 5497 Bewilderment *Difficulty*
PA 7309 Bewilderment *Uncertainty*
PF 3956 Bewitchment
PF 3192 Beyond local environments ; Limited relations
PC 0238 *Beyond their commercial life ; Unproductive use trees*
PE 2455 *Bhynosporidiosis*
PF 7076 Bias acts conscience
Bias against
PF 6326 — ordination women
PF 1879 — private enterprise
PD 9316 — victims crime ; Legal
PC 1497 Bias ; Bureaucratic
PD 4773 Bias children's literature
PC 8344 Bias ; Cultural
Bias [development...]
PF 2997 — development ; Economic
PA 6790 — *Distortion*
PF 6743 — document classification systems

PF 5472 *Bias favour European Western culture ; Excessive*
PF 8767 Bias favour new
Bias [Ideological...]
PJ 0775 — Ideological
PD 9330 — information systems
PA 6486 — *Injustice*
PF 4216 Bias judges international athletic competitions ; National
PE 4733 Bias jury trials small jurisdictions
PF 6573 *Bias ; Local anti intellectual*
PD 6081 Bias media
PD 0917 Bias media against development social transformation
PA 7306 Bias *Narrowmindedness*
PF 1728 Bias official appointments ; Political
PE 7002 Bias ; Over optimistic
Bias [Photographic...]
PF 9707 — Photographic
PF 2885 — planning training programmes ; Social
PF 3846 — public discussion research development ; Opportunist
PE 0789 Bias regulation restrictive business practices ; Domestic
PF 5311 Bias reporting ; Positive
Bias [scientific...]
PF 9693 — scientific research
PF 2931 — selection political candidates
PE 4907 — sentencing offenders ; Racial
PF 0019 — statistics ; Governmental
PF 0567 Bias theology ; Gender
PD 0917 Bias towards entertainment reassurance ; Mass media
PF 7925 Bias United Nations response human rights
PF 9686 Bias ; Urban
PF 6745 Bias wage bargaining ; Government
PD 3245 Bias worker benefits ; Economic
Biased [adjustment...]
PF 4564 — adjustment official statistics government
PF 4564 — adjustment statistics
PF 6395 — advice
PD 4517 — allegations against governments
Biased [computer...]
PD 7450 — computer software design
PD 9330 — computerized information displays
PF 3099 — *cultural tours*
Biased [educational...]
PD 9370 — educational materials ; Commercially
PF 6395 — expertise
PF 6395 — experts
PF 9693 Biased favour commissioning body ; Expertise
PF 0157 Biased government information
Biased inaccurate
PF 9358 — biology textbooks
PF 1780 — geography textbooks
PD 2082 — history textbooks
PF 3096 Biased information
PF 9431 Biased interpretations facts
PD 4827 Biased judges juries
Biased [legal...]
PF 8065 — legal systems
PE 6529 — legal systems ; Culturally
PF 3096 — literature
PD 6081 Biased media coverage news
PE 8802 Biased media-image foreign groups peoples
PE 7638 Biased portrayal women mass media
PD 1718 Biased presentation news
Biased [recognition...]
PF 7745 — recognition problems
PF 1466 — religion ; Culturally
PF 3096 — reports
PF 9693 — research ; Reward
PF 9693 — research ; Sponsor
PD 9641 Biased voting systems
PF 2627 Biases ; Leadership dependent local
PF 3869 Bible ; Inaccessibility
PJ 5626 Biblical scholarship ; Ignoring implications
PF 7129 Bibliolatry
PF 6713 Bibliomania
PE 5717 Bicycles ; Obstacles use
PE 4301 Bid rigging
PD 3489 Bidonvilles
PF 4432 Bids abroad ; Blocks hostile takeover
PE 1221 Bifida ; Spina
PF 7769 Big head disease animals
PF 3286 Bigamy
PC 7652 Bigoted people
PC 7652 Bigotry
PD 4365 Bigotry against atheists
PC 3458 Bigotry ; Class
PC 7652 Bigotry ; Dependence
PA 7338 Bigotry *Hate*
PA 7157 Bigotry *Insanity*
PA 7325 Bigotry *Irresolution*
PA 7306 Bigotry *Narrowmindedness*
PE 0567 Bigotry organized religion ; Sexual
PD 4365 Bigotry ; Racial
PD 4365 Bigotry ; Religious
PA 7309 Bigotry *Uncertainty*
PE 9099 Bilateralism aid
PE 9099 Bilateralism aid developing countries
PF 7642 Bilateralism trade arrangements
PE 9829 Bile-duct obstruction
PE 9829 *Bile ; Obstruction outflow*
PE 0921 Bilharziasis
PF 7927 Bilingual education
PF 7927 Bilingualism
PE 2488 Bilingualism national settings regional linguistic controversies
PF 1382 Bill paying patterns ; Late

PF 2949 *Bill paying surcharges ; Excessive*
PE 2156 Billboards ; Unaesthetic location advertising hoardings
PE 6680 Bills exchange promissory notes ; Conflicting laws concerning
PF 1382 Bills ; Habitual non payment
PC 0872 Binary chemical weapons
PE 1323 Bind ; Hell
PD 0120 Bio-accumulation pesticides
PD 7205 Bioaccumulation toxic substances
Biochemical [warfare...]
PC 1164 — warfare
PC 1164 — warfare ; Dependence
PC 9207 — weapons ; Terrorists armed
PD 7731 Biochemists ; Irresponsible
PD 0983 Biocides
PE 5290 Biocoenosis ; Lack
PF 6229 Biocontrol ; Underutilization
PD 1180 Biodegradable plastic waste ; Non
PB 9748 Biodiversity ; Decreasing
PF 0357 Biofuels ; Insufficient utilization renewable
PF 0357 Biogas energy ; Underutilization
Biological [abilities...]
PF 1199 — abilities ; Inequality human
PF 5997 — adaptation
PC 1306 — agents ; Harmful synergistic interaction
PE 5696 — agents occupational hazards
PD 0450 — air pollutants
Biological [contamination...]
PD 2594 — contamination food
PD 1175 — contamination water
PF 6229 — control ; Lack
PD 1207 — control pests ; Lack
PC 5489 Biological disasters
PB 9748 Biological diversity ; Erosion
PE 6294 Biological effects ionizing radiation ; Harmful
PC 5773 Biological habitats ; Decreasing diversity
PC 3152 Biological niches ; Destruction
PC 5276 Biological pollutants
PD 1052 Biological productivity land ; Diminution
PD 9578 Biological resources ; Over exploitation
Biological rhythms
PE 1904 — Disruption
PF 0379 — Disruption
PF 0379 — Separation people
Biological [species...]
PD 7302 — species ; Decreasing diversity
PF 5097 — specimens medical study ; Shortage
PD 4633 — subjugation women
PD 7731 — systems ; Underreporting hazards
Biological warfare
PC 0195 — [Biological warfare]
PC 1164 — Chemical
PD 8621 — technology ; Trade
PD 1175 Biological water pollutants
PD 8621 Biological weapons ; International trade
PE 6867 Biologically active wood ; Poisonous, allergenic
PF 7490 Biologically determined aggression
PD 7731 Biologists ; Corruption
PD 7731 Biologists ; Irresponsible
PD 7731 Biologists ; Negligence
PD 7731 Biology ; Malpractice
PF 9358 Biology textbooks ; Biased inaccurate
PF 0357 Biomass energy ; Underutilization
PD 7731 Biosciences ; Unethical practice
PC 4824 Biosphere ecosystem management ; Unintegrated
PC 6675 Biosphere ; Military disruption
PC 0776 Biotechnology ; Hazards
PE 2432 Biphenyls health hazard ; Polychlorinated
PD 5228 *Biphenyls poisoning animals ; Polybrominated*
PD 5228 *Biphenyls poisoning animals ; Polychlorinated*
PE 4190 Bipolarization trade developed developing countries
PD 3323 Bird diseases
PE 5735 Bird netting
Bird [shooting...]
PE 2693 — shooting
PE 4938 — species ; Endangered migratory
PD 1111 — strikes against aircraft
PD 2730 — *syndrome ; Pale*
PD 3601 Bird vectors plant disease
PD 4925 Birds ; Accumulation contaminant residues freshwater animals, fish
PD 1111 Birds aircraft safety ; Threat
PE 4284 Birds ; Blood sporozoa
PC 3152 Birds ; Disruption breeding grounds
Birds [egg...]
PE 9417 — egg collection
PD 0332 — Endangered species
PE 4685 — Excrement
PD 2735 *Birds ; Fluke infections*
PE 4284 *Birds ; Leucocytozoonosis*
PD 1689 Birds pests
PE 2749 Birds vectors animal diseases ; Wild
PE 6659 Birds vectors disease
PF 3382 Birth ; Artificial sex determination
PF 5381 Birth certificate ; Lack
PE 6369 Birth chauvinism
Birth control
PE 3286 — [Birth control]
PD 1038 — Health hazards
PF 0148 — Ineffective
PF 3539 — *misinformation*
PA 8653 — *Mistrust*
PD 1038 — pills ; Increased use
PE 8261 — *Prohibitive cost*
PD 1618 Birth defects ; Human
PE 8911 Birth ; Denial right dignified

—865—

Birth environments

PF 6161	Birth environments ; Alienating child
PE 4828	Birth injuries
PD 4029	Birth mortality ; Still
	Birth [Premature...]
PD 1947	— Premature
PD 1947	— Premature still
PE 3286	— prevention
	Birth [rate...]
PD 2118	— rate ; Declining
PF 0906	— rate ; High crude
PE 4107	— Risk multiple
PF 5968	Birth spacing families ; Insufficient
PD 4029	Birth ; Still
PE 8911	Birth trauma
PF 5970	Birth weights ; Low
PC 9042	*Births ; Abnormal*
PE 4107	*Births ; Multiple*
	Births [Unregistered...]
PF 5381	— Unregistered
PF 5381	— Unreported
PC 9042	— *Unsupervised home*
PF 3269	Bisexuality
PD 6363	*Bite fever ; Rat*
PD 6363	*Bite fever ; Rat*
PB 0731	*Bite ; Nonvenomous insect*
PE 4931	Bites animal pets
PE 4931	Bites ; Dog
PE 4931	Bites ; Domestic animal
PE 3636	Bites stings ; Deadly insect
PE 3636	Bites stings ; Insect
PA 5446	Bitterness *Enmity*
PA 7253	Bitterness *Envy*
PA 7338	Bitterness *Hate*
PA 7156	Bitterness *Moderation*
PJ 3966	Bitterness ; Political
PA 5643	Bitterness *Unkindness*
PA 7107	Bitterness *Unpleasantness*
PD 4747	BIV Bovine immuno-deficiency virus
PE 3804	Black ; Carbon
PF 2634	*Black comedy*
	Black [death...]
PE 0987	— death
PE 7769	— *disease*
PE 7769	— *disease sheep*
PE 3646	Black flies pests
PE 4586	*Black knot cherry*
PE 4586	*Black knot plum*
PE 4586	*Black leaf spot elm*
PE 4432	Black lies
PF 8249	Black magic
	Black market
PD 5905	— Currency
PC 0939	— economies socialist countries
PC 6641	— economy
PD 2413	— socialist countries ; Currency
PC 6641	— trading
	Black [markets...]
PC 6641	— markets
PF 3311	— *mass*
PE 1607	— mildew plants
PD 7171	Black propaganda
PE 6010	Black rhinoceros ; Endangered
PE 4586	*Black rot grape*
PE 6245	Black working women ; Discrimination against
PE 3578	Blackfly resistance insecticides
PD 2735	*Blackhead*
PE 2737	*Blackleg*
PE 4346	*Blackleg crucifers*
PE 0189	Blacklisting
PE 0189	Blacklists
PC 3796	Blackmail
PD 4469	Blackmail ; Criminal
PE 9599	Blackmail ; Emotional
PD 9842	Blackmail government officials
PE 9599	Blackmail ; Moral
PD 9842	Blackmail ; Official
PD 2912	Blackmail ; Political
PE 1341	Blackouts ; Power
PE 4586	*Blackspot ; Rose*
PD 2255	*Blackwater fever*
PE 2307	Bladder ; Cancer
PE 2307	Bladder disorders ; Urinary
PD 9293	*Bladder ; Ruptured*
PF 6382	Blame avoidance
PA 5454	Blame *Badness*
PA 7237	Blame *Condemnation*
PA 6191	Blame *Disapproval*
PA 5644	Blame *Vice*
PE 5042	Blasphemous art
PF 5630	Blasphemy
PA 6058	Blasphemy *Impiety*
PF 7768	Blast ; Cover up ship
PE 4928	Blastomycosis
PD 2728	*Blastomycosis ; North American*
PA 6400	Blatancy *Affectation*
PE 7416	Bleached paper products ; Environmental hazards chlorine
PE 7864	*Bleed ; Nose*
PE 5182	*Bleeder*
PB 1044	*Bleeding*
PE 3861	Bleeding gastro-intestinal tract
PE 4838	Bleeding ; Menstrual
PA 6790	Blemish *Distortion*
PA 5869	Blemish *Exclusion*
PA 6088	Blemish *Impairment*
PA 6997	Blemish *Imperfection*

PA 7240	Blemish *Ugliness*
PD 8786	*Blepharitis*
PD 5535	*Blepharitis animals*
PD 2226	*Blight ; Bacterial bean*
PA 6542	Blight *Destruction*
PE 2229	Blight ; Fire
PA 6088	Blight *Impairment*
PA 5527	Blight *Inexpectation*
PE 5465	*Blight potato ; Late*
PC 7896	Blight ; Tree
PD 2672	Blighted land use
PE 3919	Blights
PE 2082	Blind children
PD 0542	Blind ; Denial right welfare services
PE 8983	Blind ; Education
PD 0568	*Blind ; Employment*
PF 4989	Blind faith technology
PE 8983	Blind ; Inadequate rehabilitation methods
PD 0542	Blind ; Inadequate welfare services
PF 2418	Blind obedience
PD 0568	*Blind workers*
PE 3377	Blindings
PA 6674	Blindness
PE 5380	Blindness ; Clear eyed
PE 6343	Blindness ; Colour
PD 5139	Blindness developing countries
PD 5535	*Blindness ; Moon*
PD 2542	*Blindness ; Night*
PC 0143	Blindness ; Number
PD 0568	Blindness ; Physical
PE 2388	Blindness ; River
PB 0731	*Blister*
PE 7103	Blister agents
PD 5228	*Blister beetle poisoning animals*
PE 6255	*Blister rust ; White pine*
PE 4586	*Blisters ; Leaf*
PD 9366	*Blisters poultry ; Breast*
PE 7423	*Bloat*
PD 3978	*Bloat ruminants*
PE 5247	*Bloc ; Monetary*
PC 4311	*Blockade*
PF 4260	Blockade against governments ; Economic
PF 4260	*Blockade ; Economic*
PC 4311	*Blockade ; Military*
PB 0848	Blocked action due rivalries
PE 6501	Blocked community action
PF 3070	Blocked currency
PD 0276	Blocked elderly housing
PF 9306	Blocked family ; Schooling
PD 0930	Blocked global marketing
PF 6572	Blocked high risks ; Investment
PF 6528	Blocked land use
PD 1140	Blocked minority opinion
PF 2810	*Blocked parental participation*
	Blocked [scientific...]
PD 7470	— scientific cooperation ; Structurally
PD 2498	— seaways ; Ice
PF 6433	— service society
PF 8552	— *skills advancement*
PF 9529	Blocking business profit ; Cultural traditions
PA 8653	Blocking support ; Distrust
PD 7966	Blocks appeals ; Bureaucracy
PF 4432	Blocks hostile takeover bids abroad
PD 5255	Blocs ; Military
PF 8026	Blood blood forming organs ; Diseases
	Blood [cholesterol...]
PE 2371	— cholesterol ; Elevated
PE 3830	— circulation disorders
PE 5783	— clots
	Blood [diseases...]
PF 8026	— diseases
PF 8026	— Diseases
PE 6806	— diseases ; Occupational
PF 8026	— disorders
	Blood [feud...]
PF 7653	— feud
PE 0921	— flukes
PF 8026	— formation ; Defective
PF 8026	— forming organs ; Diseases blood
PF 8026	— forming organs ; Diseases blood
PB 1044	*Blood ; Loss*
PE 2467	Blood ; Oxygen deficiency
PE 9422	Blood-poisoning
	Blood pressure
PE 0472	— Abnormal
PE 0472	— High
PE 0472	— Low
PF 1379	*Blood related products ; Unsafe*
	Blood [sporozoa...]
PE 4284	— sporozoa birds
PD 1323	— sports
PE 1926	— sugar ; Low
PE 0366	— supply ; Inadequate emergency
PE 0366	Blood transfusion systems ; Limited
PF 1379	*Blood transfusions ; Risk contracting AIDS*
PF 7653	Blood vengeance
PF 8026	Blood vessels kidneys, eyes nerves ; Damage small
PD 3228	Bloodless revolution
PD 2609	*Bloodshed*
PE 3660	Bloodsuckers
PE 2259	*Bloody flux*
	Bloody mindedness
PA 7055	— *Death*
PA 5532	— *Disaccord*
PA 5643	— *Unkindness*
PE 2501	Blooms ; Algae

PE 4235	Blotch diseases plants
PF 1585	Blowers ; Harassment whistle
PD 8926	Blowers ; Whistle
PE 3627	Blowflies pests
PD 1245	Blown dirt ; Excessive wind
PD 3978	*Bluecomb*
PE 6297	Bluetongue
PE 8667	Boarding denial ; Involuntary
PA 6659	Boastfulness *Humility*
PA 6491	Boastfulness *Vanity*
PF 4436	Boasting
PF 4436	Boasting ; Social
PC 4667	Boat diplomacy ; Gun
PD 8034	Boat people
PD 8034	Boat ; Refugees
PE 4630	Bodies animals ; Gastric foreign
PC 3355	Bodies ; Competition missionary
	Bodies [Deliberate...]
PE 1646	— Deliberate deformation childrens'
PF 5472	— Developing country failure exercise leverage international
PD 8629	— Discrimination against non members professional
PD 6945	— Dismemberment
PD 2245	— Diversity limited local decision making
PF 5660	Bodies ; Immediacy orientation elected
PF 0223	Bodies ; Lack law enforcement
PF 6023	Bodies ; Non participation youth decision making
PE 4630	*Bodies oesophagus animals ; Foreign*
	Bodies Proliferation duplication
PE 2417	— intergovernmental organizations coordination
PE 0179	— international nongovernmental organization coordination
PE 1029	— international organizations coordinating
PE 8175	Bodies seawater ; Pollution semi enclosed
PF 7493	Bodies ; Superficial consensus intergovernmental
PE 1579	Bodies United Nations system ; Proliferation duplication organizational units coordinating
PE 7154	Bodily ailments, lesions malfunctions
PE 1764	Bodily conditions ; Disagreeable
PE 1904	Bodily rhythm international travel ; Desynchronization
PD 2459	Bodily uncleanliness
PF 9150	Body ; Contamination human
	Body [disposition...]
PE 6768	— disposition disabilities
PE 6603	— Disruption internal balance human
PD 0270	— *dysmorphic disorder*
PE 5328	Body ; Excessive trace elements human
PF 9693	Body ; Expertise biased favour commissioning
	Body [hair...]
PE 5580	— hair ; Excess human
PD 0748	— Harmful effects ultrasonic radiation human
PD 2730	— *hepatitis ; Inclusion*
PE 6603	Body ; Inflammation internal parts
PF 1781	*Body ; Lack governing*
PD 2559	Body ; Mutilation deformation human
PE 4481	Body odour ; Disagreeable human
	Body [Parasites...]
PE 0596	— Parasites human
PD 7307	— pneumonia ; Foreign
PC 0350	— politic ; Crimes against
PD 8050	Body ; Radiation damage human
PF 0491	Body snatching
PE 6603	Body ; Sudden shocks
PE 5328	Body ; Trace element imbalance human
PE 7826	*Bog spavin*
PF 0326	Bogus firms
PD 5770	*Bogus psychiatrists personal counsellors*
PE 7575	Bogus public interest groups
PC 8534	*Boils*
PE 0102	Boils ; Gangrene, carbuncles
PD 2730	*Bolivian haemorrhagic fever*
PE 4027	Bomb ; Fuel
PE 3968	Bomb making materials ; Covert trade nuclear
PF 4298	*Bomb warning hoaxes*
PF 4298	*Bomb warnings ; False*
PE 2306	Bombardment
PE 9599	Bombardment ; Emotional
PF 7645	Bombers ; Psychotic urban
PE 2306	Bombing
PE 9599	Bombing ; Love
PE 2368	Bombing ; Terrorist
PF 9269	Bombs ; Anti matter
PF 3494	Bombs ; Ease manufacture nuclear
PE 2368	*Bombs ; Home made*
PE 6754	*Bombs ; Letter*
PE 6754	*Bombs ; Mail*
PE 6754	*Bombs ; Parcel*
PF 2613	Bombs ; Unexploded
PD 3135	Bond markets ; Restrictions foreign access capital
PE 6342	Bond service ; Violation freedom
PD 3301	Bondage ; Debt
PD 3301	Bonded labour
PD 7981	*Bonds ; Forgery shares*
PE 4963	*Bonds ; Junk*
PD 7981	*Bonds ; Risk forgery shares, stocks,*
PE 6342	Bondservice
	Bone [Calcification...]
PC 2270	— Calcification
PD 8347	— *cartilage ; Benign neoplasm*
PE 7826	— *cyst pedal bone ; Lameness caused*
PE 7826	— *cyst ; Subchondral*
PD 7424	Bone diseases ; Animal
PE 3822	Bone ; Diseases injuries
	Bone [fever...]
PE 2260	— fever ; Break
PE 7511	— Fracture

-866-

Briquettes

Bone [Fracture...] cont'd
PE 7826 — *Fracture pedal*
PE 7826 *Bone ; Lameness caused bone cyst pedal*
PE 9229 Bone ; Malignant neoplasm
PE 8912 Bone ; Osteomyelitis, periostitis infections involving bones acquired deformities
PE 7826 *Bone spavin*
PE 8912 Bones acquired deformities bone ; Osteomyelitis, periostitis infections involving
PE 8589 *Bones ; Brittle*
PB 0731 *Bones ; Dislocation*
PB 0731 *Bones ; Displacement*
PE 7826 *Bones ; Fracture carpal*
PE 7826 *Bones ; Fracture metacarpal*
PE 9846 Bones joints ; Tuberculosis
PE 8589 *Bones ; Marble*
PE 3822 Bones ; Porous
PE 7015 Bones ; Torture breaking
PD 3026 Book burning
PD 3026 Book censorship
PF 6713 Book collecting ; Fanatical
PF 4385 Book production distribution ; Area disparities
PD 3090 Book propaganda
PF 2126 Books countries cultures ; Untransferability
PF 0118 Books developing countries ; Shortage
PE 2691 Books ; Harmful effects comic strips picture story
PD 2716 Books ; Inadequate inaccurate textbooks reference
PE 1669 Books ; Inadequate preservation
PF 6713 Books ; Obsession
PF 1538 Books publications ; Delays delivery
PF 0118 Books textbooks developing countries ; Shortage
PD 0577 Books ; Theft
PA 0429 *Books ; Violence comic*
PF 6575 *Boom/bust mentality ; Paralyzing*
PF 1277 Boom depression ; Economic cycle
PE 2435 Boom generated supersonic aircraft ; Sonic
PD 0188 Bootlegging
PD 5272 Border anti terrorist raids ; Cross
PE 3704 Border areas ; Military manoeuvres sensitive
PE 0261 Border broadcasting ; Violation sovereignty trans
PD 0991 *Border controls*
Border [disease...]
PE 6419 — *disease*
PD 2946 — disputes
PD 2950 — disputes
PE 5210 Border encroachments
PD 2950 Border incidents violence
PD 5272 Border military operations ; Cross
PE 4791 Borderline mental retardation
PE 4396 Borderline personality disorders
PD 4782 Borders ; Violence along internal
PA 7365 Bored people
PA 7365 Boredom
PA 7365 Boredom *Boredom*
PF 7681 Boredom captive domesticated animals
PA 5806 Boredom *Inaction*
PA 6598 Boredom *Incuriosity*
PA 6890 Boredom *Nonuniformity*
PA 7107 Boredom *Unpleasantness*
PE 6286 Borer insects
PE 3648 Borers pests ; Corn
PE 0886 Boring beetles ; Wood
PA 7365 Boring people
PF 4010 Boring personal appearance
PA 0657 *Boring work*
PF 2575 *Boring youth activities*
PA 7365 Boringness *Boredom*
PA 5806 Boringness *Inaction*
PA 6598 Boringness *Incuriosity*
PA 7107 Boringness *Unpleasantness*
PE 4276 Born AIDS ; Children
PD 4029 Born babies ; Still
PE 2399 Born ; Haemolytic disease new
PC 1874 Born out wedlock ; Children
PE 2787 Borne animal diseases ; Water
PE 2223 *Borne animal diseases ; Wind*
PE 3401 Borne disease ; Water
Borne diseases
PE 2739 — animals ; Soil
PE 7796 — Arthropod
PE 2515 — Food
PE 4559 — Seed
PE 3897 — Tick
PD 8385 — Vector
PE 3401 — Water
PE 2348 Borne encephalitis ; Tick
PE 2515 Borne infections intoxications ; Food
PD 9094 Borne pathogenic bacteria ; Water
PE 5530 Borne rickettsiosis ; Tick
Borne typhus
PE 3895 — *Flea*
PD 1753 — *Louse*
PE 3895 — *Mite*
PD 1175 *Borne viral disease ; Water*
PF 2261 Borough authorities ; Remote
PE 7136 Borrower nations ; Enforced curtailment living standards
PE 9054 Borrower's race ; Refusal loans because
PE 0682 Borrowers funds ; Misleading
PD 3054 *Borrowing collateral ; Limited*
PD 3131 Borrowing ; Dependence developing countries budget deficit financing external
PE 4315 Borrowing domestic monetary policies ; Interference transnational banks' off shore
PE 7195 Borrowing ; Over reliance developing countries foreign
PC 0492 *Borrowing practices ; Misjudged*
PE 7474 Borrowing state controlled enterprises ; Excessive
PE 2254 *Bot fly ; Rodent rabbit*
PE 9438 *Bot ; Sheep nose*
PE 3635 Botflies pests
PA 5497 Bothersomeness *Difficulty*
PA 6030 Bothersomeness *Fear*
PA 7107 Bothersomeness *Unpleasantness*
PE 4346 *Botrytis diseases plants*
PE 4935 Bottle feeding ; Hazards
PE 8255 Bottle feeding ; Unhygienic
PE 7766 Botulism
PE 9374 *Botulism animals*
PE 1601 Boucremia
PF 1695 Boulders ; Space
PF 2493 *Bound childcare ; Tradition*
PD 8546 Bound educated women ; Failure employ skills home
PE 2274 *Bound mothers ; House*
PF 3211 *Bound women ; House*
Boundaries [Arbitrary...]
PF 0090 — Arbitrary national
PF 0090 — Artificial nation
PF 0090 — Artificial territorial
PE 7903 Boundaries ; Confusing property
PD 2544 Boundaries developing countries ; Unnatural
PD 2495 *Boundaries ; Fragmenting district*
PF 3499 *Boundaries ; Inflexible market*
PF 8235 Boundaries ; National
PF 0090 Boundaries states ; Unnatural
PE 7903 Boundaries ; Unresolved community
PE 7903 Boundaries ; Unsatisfactory delimitation
PF 0954 Boundary constraints land planning
PE 7903 Boundary disputes neighbours
PD 2946 Boundary disputes states
PD 5493 Bounty hunting
PD 7352 Bourgeois democracy ; Frivolity
PD 7352 Bourgeois deviationism
PD 7352 Bourgeois liberalization
PC 1002 Bourgeois power ; Elimination
PE 7774 Bourgeoisie
PD 4500 Bourse related crimes ; Stock exchange
PD 4511 Bourses ; International collapse stock exchanges
PE 5530 *Boutonneuse fever*
PD 7799 *Bovine abortion ; Epidemic*
PE 5524 *Bovine atypical interstitial pneumonia*
PD 9293 *Bovine cystitis*
PD 2732 *Bovine encephalomyelitis ; Sporadic*
PD 7799 *Bovine genital campylobacteriosis*
PD 4747 Bovine immuno-deficiency virus (BIV)
Bovine [leukaemia...]
PE 1243 — leukaemia
PE 1243 — leukosis
PD 3978 — *liver abscesses*
PE 1243 — lymphosarcoma
PE 1243 Bovine malignant lymphoma
PD 5535 *Bovine ocular squamous cell carcinoma*
Bovine [petechial...]
PD 2732 — *petechial fever*
PE 1775 — pleuropneumonia ; Contagious
PE 5524 — *pneumonic pasteurellosis*
PD 7841 — *progressive degenerative myeloencephalopathy brown Swiss cattle*
PE 5524 — *pulmonary emphysema ; Acute*
PE 5524 — *pulmonary oedema ; Acute*
PD 9293 — *pyelonephritis*
PD 9293 — *pyelonephritis ; Contagious*
PE 5524 *Bovine rhinotrancheitis ; Infectious*
PE 5191 Bovine spongiform encephalopathy (BSE)
PD 7799 *Bovine trichomoniasis*
PD 7799 *Bovine ulcerative mammillitis*
PD 3978 *Bovine viral diarrhoea*
PD 3978 *Bovine winter dysentery*
PE 5913 *Bowed tendon*
PE 9399 Bowel ; Cancer
PE 6553 Bowel ; Disorders
PE 6553 Bowel ; Obstruction
PE 6553 Bowel ; Obstruction stricture
PE 8766 Boxing
PF 4549 Boy networks ; Old
PE 8313 Boycott
PE 4206 Boycott international sports events ; Political
PE 1213 Boycotts ; Consumer
PE 8932 *Brachial neuritis*
PD 5228 *Bracken fern poisoning animals*
PE 4715 Brain ; Chemical imbalances
Brain [damage...]
PD 0992 — damage
PE 4620 — death
PD 0992 — Degenerative changes
PD 0992 — derangement
PD 0992 — disease
PD 0992 — Diseases
PD 0992 — Diseases injuries
Brain drain
PD 1291 — developing countries
PJ 7921 — Inter professional
PE 8093 — more developed industrialized countries
PE 8093 — socialist capitalist countries
PD 0992 Brain infection
PD 0992 Brain ; Organic disease
PE 1951 Brain ; Surgical manipulation
PD 0992 Brain tumour
PD 0992 Brain ; Tumour
PE 4874 Brains malnutrition insufficient stimuli ; Damage infant
PE 0965 *Brainstem animals ; Defects*
PE 8430 Brainstorming ; Lack
PA 6393 Brainwash *Miseducation*
PA 5699 Brainwash *Reversion*
PD 1624 Brainwashing
PD 1652 Brainwashing prisoners war
PE 4997 Branch action ; Uncontrolled executive
PJ 5226 Branchiomycosis
PE 1796 Brand food sector developing countries ; Domination transnational corporations domestic name
PE 5755 Brand name drugs ; Excessive pricing
PD 7981 Brand name merchandise ; Forgery
PE 2356 branks
PF 2874 *Bravado youth style*
PD 2730 *Bravo fever ; Rio*
PJ 1184 Brawling
PF 4558 Breach administrative pledges
PE 5762 Breach agreement
Breach [confidence...]
PF 7633 — confidence
PE 5762 — contract
PE 5762 — contract ; Risk
PF 7150 Breach faith
PD 3035 Breach journalistic confidence ; Forced
PF 7150 Breach promise
PF 7150 Breach trust
PF 2497 Breaching international treaties
PE 2260 Break-bone fever
PE 6650 Break-down communications due difference training
PD 1147 Breakdown community security systems
PD 1026 Breakdown covenants life
PC 2102 Breakdown ; Dependence family
PC 2102 Breakdown family
PC 2102 Breakdown ; Family
PC 9584 Breakdown international trading discipline
PD 2864 Breakdown local community cohesion
PD 2103 Breakdown ; Marriage
PC 0799 Breakdown ; Mental
PE 6322 Breakdown ; Nervous
PF 8652 Breakdown police protection
PC 3204 Breakdown political unity
PB 1016 Breakdown resource exchange
PB 2496 Breakdown ; Social
PF 7178 Breakdown standards morality
PF 6098 Breakdown vocational images
PE 7015 Breaking bones ; Torture
PD 2561 Breaking entry
PD 2561 *Breaking ; House*
PF 5332 Breaking nations ; Law
PD 1350 *Breaking ; Window*
PD 4233 Breakup affective relationship
PD 8347 *Breast ; Benign neoplasm*
PD 9366 *Breast blisters poultry*
PE 1175 Breast cancer
PD 9742 Breast ; Diseases
Breast feeding
PE 7627 — Dependence
PE 5026 — Smoking during pregnancy
PE 7627 — Social stigma
PE 3858 — Substance abuse during pregnancy
PE 8255 — Substitution inappropriate foodstuffs
PE 1175 Breast ; Malignant neoplasm
PE 7627 Breastfeeding female servitude
PE 8255 Breastfeeding ; Unnaturally short duration
PE 6558 Breath ; Bad
PE 7864 *Breathing ; Difficulty*
PD 4575 Breeding animals fur
PD 1135 *Breeding areas ; Defoliation insect*
PE 7175 Breeding dangerous animals
Breeding grounds
PC 3152 — *birds ; Disruption*
PC 3152 — *Ecological disruption animal*
PC 3152 — *fish ; Disruption*
PC 3152 — *marine mammals ; Disruption*
PD 1135 *Breeding habitats ; Defoliation insect*
PE 3601 Breeding humans ; State imposed functional
PE 8256 Breeding ; Lack availability methods cross
PE 1630 *Breeding ponds ; Mosquito*
PC 1408 Breeding purposes ; Excessive use domestic animals
PC 1408 Breeding stock ; Decreasing genetic diversity animal
PD 5936 Breeds cultivated plants ; Declining
PD 6305 Breeds domesticated animals ; Declining
PF 3364 Brethren ; Separated religious
PE 7188 Breweries ; Illicit
PC 2558 Bribery
PD 8449 Bribery ; Business
PE 3754 Bribery connection sports athletic competitions
PE 4033 Bribery customs officials
PD 4532 Bribery government
PD 4541 Bribery government officials
PC 2030 Bribery ; Political
PD 4541 Bribery public servants
PF 8611 Bribery ; Resignation towards
PE 0322 Bribery transnational enterprises developing countries
PC 4631 Bribes ; Giving
PC 4701 Bribes ; Receiving
PC 4864 *Bricking*
PE 4718 Bride burning
PF 3290 Bride-price
PC 3298 Bride selling
PC 3298 Brides ; Mail order
PF 3211 *Bridge construction ; Postponement*
PE 2471 Bridges ; Disrepair
PE 2471 Bridges ; Non durability
PE 2471 Bridges ; Unmaintained
PE 2471 Bridges ; Unrailed public
PD 1318 *Brief reactive psychosis*
PE 2272 Bright's disease
PE 8206 Briquettes ; Instability trade coal, coke

Brisket disease

PD 9366 *Brisket disease*
PE 8589 *Brittle bones*
PD 2638 *Brittle fracture metals*
PD 3158 *Brittle stars ; Endangered species*
PF 5622 Broadcasting ; Decline public interest
PF 1433 Broadcasting ; Decline religious
PF 5622 Broadcasting ; Inadequate support public interest
PE 0261 Broadcasting individual receivers ; Foreign controlled direct satellite
PF 3079 *Broadcasting licence ; Refusal*
PF 5622 Broadcasting ; Resistance public interest
PF 1433 Broadcasting ; Resistance religious
PE 0261 Broadcasting ; Uncontrolled satellite
PF 1066 Broadcasting vessels ; Off shore
PE 0261 Broadcasting ; Violation sovereignty trans border
PF 6575 *Brochure ; Undesigned town*
PD 3493 Broken arrows
PF 4558 Broken election promises
PF 4558 Broken government pledges
PF 4558 Broken government promises
PF 4022 Broken heart
PD 2103 Broken homes
PD 2103 Broken marriages
PF 7150 Broken promises
PF 5191 Broken-spirited animals
PD 8557 Broken surfaces streets roads
Brokenness [Defeat...]
PA 7289 — *Defeat*
PA 6542 — *Destruction*
PA 5828 — *Discontinuity*
PA 6858 — *Disintegration*
PA 7250 — *Disobedience*
PA 6088 Brokenness *Impairment*
PA 7236 Brokenness *Separation*
PE 8228 Bronchi ; Pleura peritoneum cancers
PE 3860 Bronchial asthma
PE 8228 Bronchial cancer
PE 8822 Bronchial tubes lungs ; Inflammatory affections
PE 8579 Bronchiectasis
PE 9509 *Bronchitis animals ; Allergic*
PE 2248 Bronchitis ; Chronic
PE 5567 *Bronchitis poultry ; Infectious*
PE 5567 *Bronchitis ; Quail*
PD 7307 *Bronchitis ; Verminous*
PE 2293 Bronchopneumonia
PE 3795 Broomrapes
PE 6735 Brothel legislation ; Abuse
PD 3888 Brothel slavery
PD 3888 Brothels ; Involuntary prostitution
PE 3390 Brothels ; Military
PE 7808 *Brown atrophy*
PE 4586 *Brown rot stone fruits*
PD 7841 *Brown Swiss cattle ; Bovine progressive degenerative myeloencephalopathy*
PE 0924 Brucellosis
PE 0924 Brucellosis cattle
PE 7826 *Bruised corns*
PE 7826 *Bruised sole*
PB 0731 *Bruises*
PE 4078 Bruising ; Torture
PC 1987 Brutality
PA 5454 Brutality *Badness*
PD 4945 Brutality ; Military
PA 7156 Brutality *Moderation*
PD 3543 Brutality ; Police
PD 0584 Brutality psychiatric hospitals ; Physical
PA 5612 Brutality *Unchastity*
PA 5643 Brutality *Unkindness*
PA 5821 Brutality *Vulgarity*
PD 7602 Brutalization military personnel
PE 5685 Bruxism
PD 3477 Bryozoa ; Endangered species
PE 5191 BSE Bovine spongiform encephalopathy
PE 5191 BSE Mad cow disease
PE 0987 Bubonic plague
PF 6382 Buck passing
PE 7826 *Bucked shins*
Budget [deficit...]
PD 5492 — deficit
PD 3131 — deficit financing external borrowing ; Dependence developing countries
PD 3131 — deficits developing countries
PF 3550 *Budget ; Inadequate police*
PF 6060 Budget ; Increase defence
PF 1442 *Budget knowledge ; Limited*
PF 1442 *Budget methods ; Uncomprehensive*
Budget [planning...]
PF 1442 — *planning ; Uninformed*
PD 5984 — policies ; Imprudent
PF 1442 — *procedure ; Uncoordinated*
PE 2820 Budgetary coordination United Nations systems ; Inadequate
PF 5374 Budgetary procedures public sector ; Opaque
PF 1442 *Budgeting practices ; Inadequate*
PD 0183 Budgets appropriations ; Wastage governmental
PF 7737 Budgets ; Over programmed
PD 5492 Budgets ; Unbalanced
PD 5228 *Buds ; Poisoning oak*
PE 2254 *Buffalo flies*
PE 3646 Buffalo gnats
PE 3615 *Bug ; Chinch*
PD 2930 Bugging ; Electronic
PD 1632 Bugging ; Telephone
PE 3615 Bugs
PE 4428 Bugs ; Computer
PE 3616 Bugs vectors disease ; Kissing
PF 1790 *Builders ; Limited construction*
PE 8061 Building alternatives ; Uninvestigated
Building [codes...]
PF 6506 — *codes ; Unknown*
PE 8443 — codes urban areas ; Restrictive
PD 1230 — collapse
PF 6174 — complexes ; Disorientation stress large
PD 2672 — *conditions ; Deteriorated*
PF 1815 — *conditions ; Overcrowded*
PC 0268 — *construction noise*
PF 7106 — construction ; Obstacles
PE 5260 *Building depreciation*
PF 0867 Building design ; Over conformity
PF 7106 Building erection risks
Building [failures...]
PD 1230 — failures
PE 0743 — flood plains
PE 8020 — foundations ; Uneven settling
PF 3554 *Building goods ; Unenvisioned*
PF 3499 *Building land ; Unavailability*
PD 8894 Building maintenance ; Inadequate
PC 4867 *Building management ; Unprofessional*
Building materials
PE 6214 — personal products ; Over use formaldehyde
PF 3550 — *Prohibitive cost*
PF 3554 — *Unavailability*
PF 1232 Building ; Organizational empire
PF 2825 Building ; Organizational empire
PF 6502 Building programmes ; Delays community
Building sites
PF 3499 — *Restricted*
PF 3499 — *Unavailability*
PF 3499 — *Undesignated*
Building [standards...]
PE 8443 — standards ; Complicated
PF 8829 — standards ; Inadequate
PF 7106 — *structures ; Architectural obsolescence*
PD 6627 — syndrome ; Sick
PF 4854 Buildings ; Cover up unsafe
Buildings [Desecration...]
PD 7278 — Desecration religious
PD 0253 — Destruction historic
PF 6198 — Displacement natural light
PE 4953 Buildings ; Environmental stress inhabitants tall
PF 0201 Buildings ; Haunted
PF 2813 *Buildings ; Increase abandoned*
PF 1600 *Buildings ; Locked public*
PE 1925 Buildings ; Monolithic architecture high rise
PD 8894 *Buildings ; Neglect school*
PF 6578 *Buildings ; Obsolete*
PF 6578 *Buildings ; Outmoded commercial*
PD 3522 Buildings ; Pest infestations
PJ 8591 *Buildings ; Prohibitive construction cost rural community*
Buildings [services...]
PD 2669 — services facilities organized action against problems ; Inadequate
PF 6194 — Social inadequacy large
PE 5058 — Substandard hospital
Buildings [Unappealing...]
PD 6627 — *Unappealing business*
PF 1733 — Unattractive appearance deteriorating
PF 1600 — *Unavailable public*
PE 0242 — Underheated school
PD 2672 — *Underutilized dilapidated*
PD 6627 — Unhealthy
PE 0242 — Uninsulated
PF 2813 — *Unsecured vacated*
PF 6199 — urban environments ; Unrelated
PF 2813 *Buildings ; Vacated business*
PE 8162 Buildings ; Vibration damage
PE 5874 Built ; Imbalances types ships
PE 5617 *Bulbar necrosis ; Infective*
PE 9799 Bulbar paralysis ; Infectious
PE 5722 Bulimia
PE 4538 Bulimia nervosa
PE 5839 Bulk shipping industry ; Imbalances dry
PE 5804 Bulk shipping ; Transnational corporation control
PE 5849 Bulk trades ; Unfair shipping practices
PE 4893 *Bull fighting*
PE 7553 *Bull-nose*
PE 3604 *Bulldog bats ; Endangered species*
PC 0418 Bulldozer technology
PE 4893 *Bullfighting*
PD 7799 *Bulling*
PC 2934 Bullying
PD 7602 Bullying amongst military personnel
PE 0428 Bullying amongst prisoners
PE 2876 Bullying children
PE 1622 Bullying foreign governments ; Government
PE 2876 Bullying schools
PF 4729 Bullying ; Unreported
PE 5617 *Bumblefoot*
PE 8667 Bumping ; Airline
PE 7826 *Bumps ; Hunter's*
PA 6979 Bumptiousness *Opposition*
PA 6491 Bumptiousness *Vanity*
PF 7428 Bums
PD 2283 *Bunion*
PF 9608 Burden ; Being
PE 2200 Burden conflicting national regulations transnational corporations
PE 3066 Burden export credit financing upon developing countries
PE 3926 Burden interest repayments developing countries ; Excessive unnecessary
PF 1447 Burden militarization ; Socio economic
PC 2546 Burden ; National debt
PE 1093 Burden poor due legal delays ; Excessive
Burden proof
PF 3918 — abuse vulnerable groups ; Placement
PF 3918 — disempowered ; Placement
PF 3918 — victims environment pollution ; Placement
PF 9608 Burden relatives ; Being
Burden [servicing...]
PD 3051 — servicing foreign public debt developing countries
PF 9608 — society ; Being
PF 6149 — society widows
PD 4178 — students ; Competitive
PE 8915 Burden warfare ; Economic
PE 4856 Burdens human illnesses disabilities ; Excessive costs cultural
PF 3313 Burdensome cost religious ceremonies
PA 5497 Burdensomeness *Difficulty*
PA 5806 Burdensomeness *Inaction*
PA 5491 Burdensomeness *Lightness*
PA 7107 Burdensomeness *Unpleasantness*
PD 7966 Bureaucracy blocks appeals
PF 8539 Bureaucracy ; Complex government
PD 0460 *Bureaucracy cooperatives*
PD 2511 Bureaucracy ex colonial countries ; Over development
PD 8655 Bureaucracy ; Extortionate
PD 9498 Bureaucracy ; Fear
PD 8017 Bureaucracy ; Impersonality
PD 0460 Bureaucracy organizational disease
PF 2241 *Bureaucracy ; Powerlessness*
PF 8335 Bureaucracy ; Suspicion
PD 8890 Bureaucracy ; Unchecked power government
PD 8017 Bureaucracy ; Unresponsive social service
PC 0267 Bureaucrat inertia
Bureaucratic [administrative...]
PF 2711 — administrative centralization ; Excessive
PC 2064 — aggression
PF 9459 — apathy
PC 1259 — arrogance
PC 1497 Bureaucratic bias
PF 3526 *Bureaucratic contacts ; Unskilled*
PC 0279 Bureaucratic corruption
PF 1303 *Bureaucratic decision-making*
PC 2550 Bureaucratic delays
Bureaucratic [factionalism...]
PF 7979 — factionalism
PF 7979 — feuding
PC 2662 — fragmentation
Bureaucratic [ignorance...]
PF 8582 — ignorance
PC 0267 — inaction
PF 8491 — inefficiency
PF 7979 — infighting
PF 2808 — insensitivity
PF 8156 — insensitivity cultural diversity
PF 7979 Bureaucratic jealousies
PF 3526 *Bureaucratic methods ; Unsophisticated*
PF 8156 Bureaucratic nonrecognition ethnic differences
PD 7966 Bureaucratic opposition
Bureaucratic procedures
PF 2661 — Abuse
PF 8053 — Complex
PF 1653 — *Ignorance*
PF 1653 — *Unfamiliar*
PF 6707 Bureaucratic procrastination response evidence potential hazards
Bureaucratic [requirements...]
PF 1653 — *requirements ; Unfamiliar*
PE 8893 — requirements welfare benefits ; Excessive
PF 2335 — resistance self-review
PF 7979 — rivalries
PF 3526 Bureaucratic services ; Inaccessibility
PC 1259 Bureaucratic superiority
PD 0183 Bureaucratic waste
PF 1735 *Bureaucrative action ; Disappointing*
PD 2993 Bureaucratization socialism
PD 8392 Bureaucrats ; Fraud government
PD 8422 Bureaucrats ; Underpaid
PD 2561 Burglary
PF 4298 *Burglary alarms ; False*
PD 1943 Burglary ; Political
PE 9095 Burial dead ; Exploitation land
PD 7258 Burial site desecration
PD 6128 Burial sites ; Desecration ancient
PD 9366 *Burn poultry ; Ammonia*
PD 3026 Burning ; Book
PE 4718 Burning ; Bride
PE 2678 Burning ; Torture
PF 4819 Burning widows
PE 4421 Burnout ; Childhood
PF 4833 Burnout ; Professional
PE 0394 Burns
PD 9366 *Burns animals*
PE 0394 Burns scalds
PE 1764 Burping
PD 2730 *Bursal disease chickens ; Infectious*
PE 5380 *Bursatti*
PE 2320 Bursitis
PE 6555 Bursitis animals
PE 7826 *Bursitis ; Trochanteric*
PE 5345 Bus schedule ; Uncommunicated
PE 5345 Bus service ; Infrequent
PC 3152 *Bush encroachment*
PD 0739 Bush fires
PE 9120 Bush rat rock rat ; Endangered species

Capacity

Business [activities...]
PD 1415 — activities small communities ; Restrictive pattern
PC 6641 — activities ; Unreported income
PE 2033 — Alcohol abuse
Business [bankruptcy...]
PD 2591 — bankruptcy
PD 8449 — bribery
PD 6627 — *buildings ; Unappealing*
PF 2813 — *buildings ; Vacated*
Business [capital...]
PF 6477 — capital ; Limited availability local
PF 6540 — *choices ; Unpromoted*
PF 1345 — *closings ; Unnecessary*
PF 2009 — *collaterals ; Scarcity*
Business community
PE 8963 — Distrust
PE 8518 — Lack social accounting
PF 1176 — *Uncohesive*
Business [Conflict...]
PF 9610 — *Conflict interests*
PF 6548 — *contacts ; Undiscovered*
PE 5944 — corporations ; Terrorism targeted against
Business [Deception...]
PD 4879 — Deception
PD 0574 — decision making ; Lack worker participation
PF 2439 — *decline ; Historical*
PD 5603 — Discouraging conditions small
PD 9628 — Discrimination against women
Business [enterprises...]
PD 0071 — enterprises ; Excessive concentration
PE 1651 — enterprises front organizations government ; Misuse
PF 7578 — executives ; Overpaid
PE 6554 — experience ; Insufficient
Business [facilities...]
PF 1815 — *facilities ; Insufficient*
PF 1815 — *facilities ; Unattractive*
PD 2591 — failure
PE 9405 — failures ; Small
PF 0836 — *future ; Dim*
Business [greed...]
PF 7189 — greed
PD 2338 — *groups ; Land controlled*
PE 2700 — growth ; Denial right
Business [Illegal...]
PC 4645 — Illegal international
PF 1176 — Individualistic practices local
PE 8509 — industry public service ; Inability communities take advantage training
PF 2042 — *investments ; Uninsurable*
PE 6554 Business know how ; Inadequate
PD 4879 Business lies
Business [management...]
PC 4867 — management ; Inexperienced
PF 3008 — methods ; Ineffectiveness traditional small
PF 3008 — mindset ; Individualized
Business [operations...]
PF 2439 — *operations ; Confined scope*
PF 3539 — *opportunities ; Ignored*
PF 3539 — *opportunities ; Unexplored*
PF 7088 — opposition arms race ; Lack
PE 1137 — *outlook ; Inhibiting*
PD 2338 — ownership ; Absentee
Business [pattern...]
PF 2845 — *pattern ; Unchanging*
PD 1147 — *payoffs ; Necessity*
PF 1740 — *planning ; Short sighted*
PE 0161 — practice ; Direct foreign investment transnational enterprises restrictive
Business practices
PE 7313 — developing countries ; Inadequate local expertise
PD 8913 — Discriminatory
PC 0492 — *Disoriented*
PC 0073 — Distortion international trade restrictive
PE 0789 — Domestic bias regulation restrictive
PF 2782 — due inadequate regulation ; Ineffective monitoring restrictive
PE 2999 — Economic inefficiencies developing countries due restrictive
PF 6545 — *Fragmented*
PF 1596 — Ineffective regulation restrictive
PE 6554 — Inexperienced
PC 2563 — Irresponsible
PE 8011 — legislation ; Extraterritorial application restrictive
PE 5926 — markets developed countries against exports developing countries ; Restrictive
PE 0346 — relation patents trademarks ; Restrictive
PC 0073 — Restrictive
PF 0591 — service industries ; Inadequate regulation restrictive
PE 0225 — state enterprises ; Inadequate regulation restrictive
PE 1978 — technology transactions ; Restrictive
PE 5915 — transnational corporations ; Restrictive
PE 1799 — transnational enterprises ; Consequences restrictive
PC 2563 — Unethical
PF 9529 Business profit ; Cultural traditions blocking
PF 2174 Business profit-making
Business [recessions...]
PF 1277 — recessions ; Cyclic
PF 8053 — *regulations ; Complex*
PF 1345 — *resources ; Unconnected*
PF 2874 — risk ; High
PF 2633 *Business security ; Prohibitive cost*
PF 1176 *Business ; Self centred*
Business skills
PF 3499 — drain
PE 6554 — Latent
PE 6554 — Narrow range

Business skills cont'd
PE 8048 — urban areas ; Undeveloped
Business [small...]
PF 6540 — small communities larger towns ; Transfer
PF 6557 — *solicitation ; Ineffective*
PF 1815 — *space ; Insufficient*
PF 6540 — *successes ; Infrequent*
Business training
PF 2698 — Limited availability technical agricultural
PC 4867 — Minimal
PC 4867 — Unexplored
PC 4867 — Unstructured
PE 7968 Business transactions ; Illiteracy inhibitor
PC 4645 Business transactions ; Unlawful
Business [Unattractive...]
PF 6478 — *Unattractive fishing*
PC 6641 — Underground illicit
PF 6527 — *Unprofitable transport*
PF 7118 — *Unpromoted community*
PF 8675 Business ventures ; Short range
PF 2822 *Business visibility ; Low*
PD 8916 Business wages ; Non competitive
PD 5350 *Businesses ; Declining local*
PD 4728 *Businesses ; Denial rights*
PD 1415 *Businesses ; Insufficient small*
PD 5891 *Businesses ; International*
PD 4632 *Businesses ; Prohibitive start up costs*
Businesses [Unattractive...]
PF 2439 — *Unattractive local*
PF 6572 — *Undiscovered successful*
PJ 2079 — *Unprofitable small*
PF 7118 — *Unpromoted village*
PF 5728 — *Unreported*
PD 2732 *Buss disease*
PE 5070 *Busters ; Union*
PE 5192 *Busting ; Union*
PE 3762 *Butterfly ; Endangered species*
PE 7826 *Buttress foot*
PE 4963 *Buy outs ; Leveraged*
PE 8823 *Buyer's race ; Refusal sale because*
Buying [Panic...]
PF 2633 — *Panic consumer*
PD 8362 — *power ; Insufficient*
PD 8362 — *power ; Limited*
PD 5214 Buying ; Vote
PE 3008 *Buzzing wildlife*
PE 2319 *Byssinosis*
PA 0429 *Bystanders ; Harm innocent*

C

PE 2380 C deficiency ; Vitamin
PE 3936 *C enteritis swine ; Clostridium perfringens type*
PE 0517 C ; Hepatitis
PF 2508 Cabals
PD 2225 *Cabbage ; Club root*
PD 0407 Cable communications ; Vulnerability world
PE 0670 Cabotage
PE 2778 *Cadavers*
PE 4805 Cadavers foetal tissue ; Research foetal
PE 1160 Cadmium pollutant
PE 0618 Caffeine abuse
PE 0618 Caffeine intoxication
PE 0618 Caffeinism
PC 8478 Calamities ; Transportation
PA 6340 *Calamity Adversity*
PA 5454 *Calamity Badness*
PA 7055 *Calamity Death*
PA 6542 *Calamity Destruction*
PE 7698 Calcareous soils
PC 2270 Calcification bone
PD 7424 *Calcinosis ; Enzootic*
PE 5161 Calcium deficiency
PE 0598 Calculated delays releasing controversial news items
PE 7934 Calculus ; Urinary
PF 2043 Calendar ; Inadequate world
PD 7307 Calf diphtheria
PE 2348 California encephalitis
PD 4418 *Call reluctance ; Sales*
PE 4187 *Calling ; Name*
PE 4161 Callosities
PE 9334 Callous drivers
PA 5570 *Callousness Intangibility*
PA 5479 *Callousness Lamentation*
PF 2006 Callousness public services ; Institutionalized
PA 7364 *Callousness Unfeelingness*
PA 5643 *Callousness Unkindness*
PA 5644 *Callousness Vice*
PE 7140 Calls action ; Inadequate international nongovernmental organization response intergovernmental
PE 4367 Calls ; Illegal long distance telephone
PE 5757 Calls ; Obscene telephone
PD 0446 Calorie intake ; Disparities
PD 0339 Calorie malnutrition ; Protein
PD 3023 Calumny
PE 5524 *Calves ; Endemic pneumonia*
PE 4161 *Calves ; Frostbite*
PE 9774 *Calves ; Nutritional myopathy*
PE 2357 *Calves ; Redwater*
PE 4161 Calving paralysis
PE 4438 Campaigners ; Intrusive animal rights
PD 9384 Campaigning ; Negative election
PE 1815 Campaigns ; Avoidance issues political
PD 7171 Campaigns government ; Covert smear
PD 9384 Campaigns ; Political smear

PD 2931 *Campaigns ; Prohibitive cost electoral*
PD 5534 Campaigns ; Religious smear
PD 2919 Campaigns ; Unjust electoral
PE 6311 Campaigns ; Vacuous political
PD 7585 Campers ; Degradation wilderness areas
PD 0702 *Camps ; Concentration*
PC 3320 *Camps ; Hacienda*
PD 0702 *Camps ; Military concentration*
PC 3320 *Camps ; Mining*
PE 1508 *Camps ; Non settled refugees living outside*
PD 0702 *Camps ; Political concentration*
PD 6364 Camps settlements ; Military attacks refugee
PE 7370 Campus ; Student violence
PE 7370 Campus violence
PE 7370 Campuses ; Physical insecurity university
PD 2731 *Campylobacteriosis*
PD 7799 *Campylobacteriosis ; Bovine genital*
PD 7799 *Campylobacteriosis ; Ovine genital*
PD 2948 Canals ; Restrictions passage straits interoceanic
PE 8796 Canals ; Water looses irrigation
PC 0092 Cancer
PC 0092 *Cancer ; Animal*
Cancer [bladder...]
PE 2307 — bladder
PE 9399 — bowel
PE 1175 — Breast
PE 8228 — Bronchial
Cancer [causing...]
PE 3035 — causing additives mineral waters
PD 0036 — causing foods
PE 1905 — Cervical
PE 9399 — Colon rectal
PD 1239 Cancer ; Environmental
PD 5535 *Cancer eye*
PE 7509 Cancer ; Gastric
PD 1239 Cancer hazards ; Environmental
Cancer [larynx...]
PE 9819 — larynx
PE 3233 — Liver
PE 7085 — Lung
PE 4637 — lymphatic tissue
Cancer [Occupational...]
PE 3509 — Occupational
PE 9819 — oesophagus
PE 9819 — Oral
PE 1905 — Ovarian
Cancer [pharynx...]
PE 9819 — pharynx
PE 1899 — Plant
PE 8372 — prostate
PD 6033 Cancer risk ; Coitus
PE 7509 Cancer ; Stomach
PE 8372 *Cancer testis*
PE 1905 Cancer ; Uterine
PE 9746 Cancer ; Vulnerability painters
PE 8228 Cancers bronchi ; Pleura peritoneum
PD 0036 Cancers ; Food related
PC 0092 Cancers ; Household
PE 5016 Cancers ; Skin
PD 2931 Candidates ; Bias selection political
PE 4497 Candidiasis
PE 2455 Candidiasis
PD 0061 *Candidiasis ; Vaginal*
PE 4923 Candidosis
PE 2217 Cane ; Pests diseases sugar
PE 8684 Cane rat ; Endangered species African
PE 2357 Canicola fever
PE 7423 Canine coronaviral gastroenteritis
PD 2730 *Canine distemper*
PD 2732 *Canine ehrlichiosis*
Canine [haemorrhagic...]
PE 7423 — *haemorrhagic gastroenteritis*
PE 9427 — heartworm infections
PD 2730 — *hepatitis ; Infectious*
PD 2730 — *herpesviral infection*
PE 5913 — hip dysplasia
PD 2732 — Canine infectious cyclic thrombocytopenia
PD 5453 Canine malignant lymphoma
PD 2730 Canine parvoviral infection
PE 9509 Canine rheumatoid arthritis
PE 0958 Canine trichuriasis
PE 2357 Canine typhus
PE 4656 Canine urolithiasis
PD 7799 *Canine venereal tumour ; Transmissible*
PE 0517 Canine viral hepatitis
PE 7826 Canker
PE 4586 *Canker trees ; Hypoxylon*
PE 0640 Cankers
PE 1186 Cannabis abuse
PD 4027 *Cannabis intoxication*
PE 2319 Cannabosis
PD 0800 *Canned food ; Reliance*
PF 2513 *Cannibalism ; Animal*
PF 2513 *Cannibalism ; Human*
PE 8944 *Cannibalism ; Ritual*
PD 5228 *Cantharidin poisoning animals*
PD 5701 Capabilities ; Disparity world telecommunications
PF 7637 Capability ; Gaps telecommunications
PF 7109 Capability ; Inadequate negative
PC 1606 Capability ; Unequal strategic arms
Capacity [absorb...]
PD 0305 — absorb aid ; Limited developing country
PE 4649 — age ; Diminishing mental
PC 8310 — age ; Loss
PE 0653 *Capacity ; Debilitated working*
PE 9265 Capacity ; Decline capital investment productive

Capacity developing

Capacity developing countries
- PE 6861 — Decline import
- PE 7184 — Inadequate external debt management
- PE 4574 — Inadequate government savings
- PD 4219 — Inadequate industrial
- PD 4219 — Inadequate manufacturing
- PE 9646 — Inadequate negotiating
- PD 4219 — Inadequate production
- PE 4880 — Inadequate research development
- PF 0903 — Limited civil administrative
- PE 4880 — science technology ; Inadequate
- PF 9429 Capacity ; Diminished personal
- PD 7719 Capacity ; Disintegration technological

Capacity [environment...]
- PB 5250 — environment ; Erosion human carrying
- PD 0779 — Excess production
- PF 7166 — Excessive confidence prediction
- PC 8885 Capacity ; Gap scientific technological
- PB 8176 Capacity ; Inadequate foresight
- PF 1240 Capacity income farming ; Underdeveloped
- PB 8176 Capacity ; Lack lookout

Capacity Limited
- PD 9033 — electrical
- PF 9292 — human information processing
- PF 3467 — *storage*

Capacity [Maldistribution...]
- PC 9785 — Maldistribution productive
- PE 8720 — married women ; Loss civil
- PC 1258 — Myth national security increase military
- PF 1389 *Capacity ; Non viability local electrical generating*
- PE 5897 Capacity shipping industry ; Excess
- PJ 9034 Capacity technology assessment ; Inadequate
- PF 0045 Capacity telecommunications ; Inadequate satellite

Capacity [Unstudied...]
- PF 6493 — *Unstudied gardening*
- PD 0779 — utilization ; Decline production
- PD 0779 — utilization manufacturing plant ; Low
- PF 2352 Capacity visualize creative future ; Inhibited
- PF 4326 Capita ; Decreasing agricultural growth
- PD 4002 Capita ; Excess environmental demand
- PC 5225 Capital ; Accumulation
- PD 3135 Capital bond markets ; Restrictions foreign access
- PE 5021 Capital companies ; Fraud concerning economic situation corporate

Capital developing countries
- PC 3134 — Loss
- PD 3132 — Outflow indigenous
- PE 5790 — shipping fleets ; Lack investment
- PF 0648 — socialist countries ; Reallocation development

Capital development
- PD 8604 — Lack
- PF 3318 — Restrictive effects external
- PE 1139 — rural areas ; Constricting level

Capital [equipment...]
- PF 2410 — *equipment ; Prohibitive cost*
- PF 9612 — Erosion human
- PC 1453 — Exodus private
- PC 1453 — Export

Capital [flight...]
- PD 3132 — flight
- PF 2865 — flow plan ; Absence long range, world wide
- PC 1453 — flows ; Massive international
- PJ 5990 — Free movement
- PF 2378 — funds ; Limited extent
- PF 3550 — funds ; Restricted access municipalities

Capital [Inadequate...]
- PF 2378 — *Inadequate agricultural*
- PF 6477 — Inadequate local
- PF 6517 — Incomplete access development

Capital Insufficient
- PF 2489 — credit
- PF 2378 — *enterprise*
- PE 2852 — flow investment
- PF 2042 — *operating*
- PF 3665 — *risk*

Capital investment
- PE 5790 — developing countries ; Lack
- PF 6572 — Fear
- PF 2852 — Insufficient
- PE 9265 — productive capacity ; Decline
- PF 6572 — Risk
- PF 6477 — small communities ; Diminishing
- PE 4167 Capital investments supporting racial discrimination

Capital Lack
- PB 4653 — [Capital ; Lack]
- PF 2378 — *agricultural*
- PB 4653 — equity
- PF 3630 — local
- PB 4653 — *venture*
- PF 3630 Capital ; Limited accumulation

Capital Limited availability
- PE 4704 — loan
- PF 6477 — local business
- PF 2865 — long term
- PC 0492 Capital ; Limited improvement
- PF 1979 Capital ; Limited village
- PF 3136 Capital markets ; Imperfections
- PF 7016 Capital ; Misapplication venture

Capital [outflow...]
- PD 3132 — outflow developing developed countries ; Private domestic
- PE 1139 — *outflow ; Excessive*
- PF 1196 — *Overdependence*

Capital punishment
- PF 0399 — deterrent ; Ineffectiveness
- PF 0399 — Inhumanity

Capital punishment cont'd
- PF 0399 — torture ; Anticipation

Capital [reserve...]
- PF 2378 — reserve ; Limited local availability
- PB 4653 — reserves ; Lack
- PF 2899 — reserves ; Limited individual

Capital resources
- PF 1979 — Dispersion local
- PE 6530 — Reluctant investment
- PF 3665 — small communities ; Strained
- PE 6530 — Subsistence approach

Capital [Restricted...]
- PF 2378 — Restricted availability
- PE 2425 — *Restricted store*
- PF 3630 — *Restrictive farm*
- PF 6572 — risk ; Unfavourable

Capital [Scarcity...]
- PF 3665 — *Scarcity risk*
- PB 4653 — Shortage
- PD 3137 — shortage developing countries
- PC 1453 — Speculative flight
- PE 4866 Capital technical assistance ; Imbalance

Capital [Unanalysed...]
- PF 6471 — *Unanalysed available*
- PF 3630 — *Unavailability farming*
- PF 6477 — Unavailability local
- PF 2378 — *Unavailable machinery*
- PF 3665 — *Unavailable risk*
- PF 6477 — Unused local
- PF 3550 — urban renewal ; Limited availability investment
- PC 1453 Capital ; Volatile
- PC 0564 Capitalism

Capitalism [Abuse...]
- PE 3105 — Abuse science technology
- PD 7463 — *Abusive technological development*
- PE 3110 — Anti
- PF 3115 Capitalism ; Conflicting roles commodities
- PF 7999 Capitalism ; Crisis
- PF 3126 Capitalism developing countries ; Contradictions
- PC 3116 Capitalism ; Excess demand goods
- PC 0564 *Capitalism ; Investment*
- PE 2194 Capitalism ; Speculation
- PF 7947 Capitalism ; State monopoly
- PE 7975 Capitalist banking
- PE 7975 Capitalist banks ; Concentration

Capitalist countries
- PE 8093 — Brain drain socialist
- PE 8701 — Enforcement religion
- PE 8006 — Insurance monopolies
- PE 5610 — Political interference television news coverage
- PF 1737 — Weak economic growth

Capitalist [economic...]
- PC 3166 — economic imperialism
- PE 3816 — elites ; Non productive
- PE 3123 — exploitation ; Denial effective national self determination
- PF 3104 Capitalist investment financing ; Counterproductive
- PC 0937 Capitalist militarism ; Provocations
- PD 3180 Capitalist neutral countries ; Communist subversion

Capitalist [plantation...]
- PD 7598 — plantation system
- PC 3193 — political imperialism
- PE 7957 — production ; Internationalization

Capitalist [social...]
- PE 3826 — social order ; Prevalence psychological conditions unfavourable transition post
- PE 5678 — socialist countries ; Imbalance trade cultural products
- PD 7463 — society ; Perverse impact technological development
- PC 2194 — speculation
- PE 8349 — subversion communist neutral countries

Capitalist systems
- PD 3112 — Alienation
- PC 3125 — Competition
- PF 3114 — Conflicting roles money
- PF 3118 — Contradictions

Capitalist systems Denial
- PC 3124 — human rights
- PD 3120 — right social security
- PC 3119 — right work

Capitalist systems [Excessive...]
- PC 3116 — Excessive demand goods
- PC 3117 — Exploitation
- PF 3103 — General unproductivity
- PD 3106 — Lack individualism
- PC 3191 — Military industrial complex
- PC 3107 — Poverty
- PC 3108 — Resource wastage
- PD 3122 — Restriction educational opportunities
- PC 3113 — Waste human resources
- PJ 7929 Capitalization expenses ; Unforeseen
- PF 1653 *Capitalization means ; Insufficient*
- PF 3455 *Capitals ; Unnecessary relocation national*
- PE 6555 *Capped elbow*
- PE 6555 *Capped hock*
- PB 5486 *Caprice*
- PA 5490 Capriciousness *Changeableness*
- PA 7361 Capriciousness *Disorder*
- PA 6774 Capriciousness *Irregularity*
- PA 7325 Capriciousness *Irresolution*
- PA 6890 Capriciousness *Nonuniformity*
- PA 6425 Capriciousness *Transience*
- PD 2730 *Caprine arthritis*
- PD 2730 *Caprine encephalitis*
- PE 6293 Caprine pleuropneumonia ; Contagious
- PD 8982 Capsizing ferries

- PF 7681 Captive domesticated animals ; Boredom
- PC 0521 Captive markets
- PC 1788 Captive nations
- PE 6564 Capture foreign nationals ; Disregard international law
- PE 9774 Capture myopathy wild animals
- PD 1179 Capture use wild animals pets
- PD 3481 *Capybara ; Endangered species*
- PD 0079 Car accidents
- PF 1781 Car pools ; Lack
- PD 5575 Car robberies ; Train armoured
- PE 2611 Car sickness
- PE 4826 Car theft ; Risk
- PE 1282 Carbamate insecticides pollutants
- PE 4630 *Carbohydrate engorgement*
- PE 4293 Carbohydrate metabolism ; Congenital disorders
- PC 3908 *Carbohydrates ; Excessive consumption*
- PE 3804 Carbon black

Carbon [dioxide...]
- PD 4387 — dioxide impact agriculture developing countries
- PD 4387 — dioxide ; Increasing atmospheric
- PE 5889 — disulphide
- PE 5889 — *disulphide*
- PE 1657 Carbon monoxide health hazard
- PE 3035 Carbonated drinks ; Carcinogenic
- PC 8534 *Carbuncle*
- PE 0102 *Carbuncles boils ; Gangrene,*
- PE 2778 Carcass disposal diseased animals ; Inadequate
- PE 8398 Carcass meat exports ; Inadequate hygiene restrictions
- PE 2777 Carcass meats factor animal diseases ; Importation infected
- PD 8050 Carcinogenesis humans ; Radiation

Carcinogenic [carbonated...]
- PE 3035 — carbonated drinks
- PD 1239 — chemical physical agents
- PE 6619 — consequences food preparation
- PD 0036 — *Carcinogenic diet*
- PD 0487 *Carcinogenic food additives*
- PE 3035 *Carcinogenic soft drinks*
- PE 8934 Carcinogens consumer goods ; Incorporation
- PD 1239 *Carcinogens ; Medicinal use occupational*
- PD 5535 *Carcinoma ; Bovine ocular squamous cell*
- PE 7085 *Carcinoma ; Pulmonary*
- PE 5308 *Carcinoma sita*
- PE 3592 Card fraud ; Cheque
- PE 3592 Card fraud ; Credit
- PE 8219 Cardiac arrhythmia
- PD 2408 *Cardiac asthma*
- PE 6811 Cardiac conditions work environment
- PD 0448 *Cardiac dilatation*
- PE 0471 *Cardiac failure animals*
- PE 0471 *Cardiac insufficiency animals*
- PE 9774 *Cardiac myopathies swine*
- PE 6816 Cardiac weakness
- PE 1556 Cardio-pulmonary disease due torture
- PE 6816 Cardiovascular diseases
- PD 5453 Cardiovascular diseases animals
- PE 2300 *Cardiovascular syphilis*

Cardiovascular system
- PE 5569 — animals ; Congenital anomalies
- PE 9427 — animals ; Parasites
- PE 3933 — Symptoms referable
- PD 8945 Care aged ; Reduction motivation
- PD 8837 Care animals ; Lack

Care [children...]
- PF 0131 — children prisoners ; Inadequate
- PF 4516 — common property ; Lack incentive users
- PF 1600 — community property ; Insufficient
- PF 2346 — community space ; Inadequate
- PF 4820 — Compulsory health

Care [Deficient...]
- PE 4245 — Deficient infant
- PD 7821 — Dehumanization health
- PD 5119 — Delay administration medical
- PD 4004 — delivery ; Insufficient physical infrastructure health
- PD 2028 — Denial right adequate medical
- PE 8400 — developing countries ; Costly medical
- PE 6770 — disabilities ; Personal
- PE 8924 — disabled ; Denial right adequate health

Care facilities
- PE 8420 — Crimes committed hospitals health
- PD 2085 — Inadequate child
- PD 2085 — Inadequate child day
- PD 2085 — Lack child

Care [Family...]
- PS 9219 — *Family based health*
- PD 1038 — family planning ; Inadequate health
- PF 2346 — frustration ; Space
- PE 8924 Care handicapped persons ; Inadequate community

Care [Ignorance...]
- PD 9021 — Ignorance women concerning appropriate child
- PD 9021 — Ignorance women concerning primary health
- PF 4832 — Inaccessible health

Care Inadequate
- PF 4832 — *dental*
- PF 9007 — emergency
- PF 8857 — maternal child health
- PF 4832 — medical
- PE 8553 — primary health

Care [Inefficient...]
- PF 9683 — Inefficient use available health
- PD 4790 — infrastructure ; Inadequate medical
- PE 8900 — intellectually disabled convicts ; Lack
- PF 2875 — *Irregular doctor*

Care Lack
- PF 4646 — [Care ; Lack]
- PF 4832 — *dental*

Central banks

Care Lack cont'd
- PF 4832 — medical
- PE 4820 — prenatal
- PE 9242 Care least developed countries ; Inadequate health
- PE 0540 Care ; Limited psychiatric out patient

Care [mental...]
- PF 4955 — *mental health* ; *Inadequate*
- PD 2085 — Missing child
- PD 8023 — Misunderstanding preventive health

Care [needs...]
- PD 4790 — *needs* ; *Unclear*
- PF 6061 — Neglect adolescent health
- PE 5288 — Neglect elderly institutional
- PD 5501 — Negligent old age
- PE 2274 — neighbourhood level ; Fragmented forms
- PE 5246 — Non adaptive local structure social

Care [opportunities...]
- PF 3628 — *opportunities* ; *Limited awareness child*
- PF 5709 — Over specialization medical
- PF 2346 — Overlooked space

Care [planning...]
- PD 4790 — planning ; Uncoordinated health
- PE 4820 — pregnant women ; Inadequate medical
- PF 4832 — Prejudiced medical
- PE 8261 — Prohibitive cost health
- PF 8016 — Prohibitive cost private medical
- PF 5121 — protection humankind ; Denial animals right attention,
- PF 2346 — public spaces ; Inattentive

Care [Rationing...]
- PE 8261 — Rationing health
- PF 4244 — Refusal medical
- PF 5555 — responsibility ; Collapsed tension
- PE 9048 Care sector ; Ineffective self regulation health

Care services
- PD 0512 — elderly ; Inadequate provision home
- PE 3328 — Irresponsible health
- PD 4790 — Remote health

Care [Slow...]
- PF 9007 — Slow emergency
- PF 2493 — structures developing countries ; Inflexible social
- PE 4901 — systems ; Resistance incorporating midwives medical
- PD 4790 — systems ; Undeveloped health

Care [taking...]
- PF 2261 — *taking system* ; *Ineffective*
- PC 0834 — techniques ; Uninformed
- PF 6061 — teenagers ; Inaccessibility health
- PF 4832 — teeth ; Inadequate
- PF 1844 — transient urban populations ; Inadequate community

Care [Unavailability...]
- PF 4832 — *Unavailability dietary*
- PF 2346 — Undemonstrable space
- PE 8261 — underprivileged ; Inaccessibility health
- PD 8837 — Unknowledgeable animal
- PC 0233 — *Untenable orphan*
- PE 7877 — urban slums ; Inadequate health
- PF 0713 Care vehicles ; Insufficient special
- PF 2477 Career advice ; Inadequate
- PF 2477 Career counselling ; Nonexistence
- PD 7692 Career disruption due childbearing
- PD 7692 Career interruption due pregnancy
- PF 4128 Career interruptions ; Imposed
- PF 3434 Career mobility aid officials ; Excessive
- PF 9004 Career opportunities ; Unexplored
- PF 9004 Career opportunities ; Unperceived
- PF 2477 Career planning ; Uninformed
- PF 6353 Careerism
- PB 0750 Careers ; Short duration athletic
- PE 2857 Careless firearm use
- PD 7585 Careless walkers ; Degradation countryside
- PA 6379 Carelessness *Avoidance*
- PA 5740 Carelessness *Compulsion*
- PE 9105 Carelessness dealing infectious patients
- PA 7361 Carelessness *Disorder*

Carelessness [Inattention...]
- PA 6247 — *Inattention*
- PA 6598 — *Incuriosity*
- PA 7325 — *Irresolution*
- PA 5438 Carelessness *Neglect*
- PA 2658 Carelessness *Negligence*
- PA 7115 Carelessness *Rashness*
- PA 7232 Carelessness *Unskillfulness*
- PE 5910 Cargo airline services ; Restrictive practices
- PF 5375 Cargo cults
- PD 9108 Cargo handling ; Dangerous
- PE 5899 Cargo handling ; Excessive costs inefficient port
- PE 5103 Cargo insecurity
- PD 0430 Cargo transportation ; Inadequate
- PE 1839 Cargoes ; Marine transportation hazardous
- PE 0125 Cargoes ; Seizure
- PD 1185 Caries ; Dental
- PF 4296 Caring disabled ; Deprivation freedom
- PD 3482 Carnivora
- PD 3482 Carnivores ; Endangered species
- PJ 4437 Carnivorism
- PE 7826 Carpal bones ; Fracture
- PE 7826 Carpal hygroma
- PE 7808 Carpal valgus
- PE 1679 Carpet beetle
- PE 7826 Carpitis

Carriers animal diseases
- PF 2775 — Difficulty identifying
- PD 2746 — Domestic animals
- PD 2729 — Wild animals
- PB 5250 Carrying capacity environment ; Erosion human

- PE 5339 Cars ; Abandoned
- PF 4480 Cars ; Maldistribution passenger
- PD 7963 Cartel ; Domestic
- PC 2512 Cartels
- PD 9163 Cartels ; Air transport

Cartels [enforced...]
- PE 9762 — enforced governments ; International
- PE 2598 — Excessive injury export interests developing countries due export
- PD 0470 — Export
- PD 0336 Cartels ; Import
- PE 0396 Cartels ; International
- PE 3829 Cartels ; Liner shipping
- PD 0668 Cartels ; Rebate
- PE 0396 Cartels ; Secret trade
- PD 8347 Cartilage ; Benign neoplasm bone
- PE 0046 Cartographic information developing countries ; Inadequate
- PE 0046 Cartographic skills developing countries ; Inadequate
- PE 0046 Cartographic technology developing countries ; Inadequate
- PF 3087 Cartoons ; Political
- PE 9844 Caseous lymphadenitis animals
- PE 1032 Cases ; Pecuniary guarantees condition provisional release criminal
- PE 0566 *Caseus tuberculosis* ; *Fibro*
- PF 9187 Cash crop policy ; Inappropriate
- PF 9187 Cash crop ; Single

Cash crops
- PE 8324 — Degradation agricultural land
- PF 6455 — *Insufficient*
- PF 6455 — *Limited*
- PF 7313 — *Low price*
- PF 9187 — Proliferation
- PJ 9314 Cash exchange reliance

Cash flow
- PF 6451 — Inadequate
- PF 6451 — *Irregular*
- PF 6451 — *Minimal*
- PS 6553 — Outward
- PF 6451 — *Restricted*
- PF 8675 Cash purchase ; Short term
- PS 2881 *Cash* ; *Scarce local*
- PE 6840 Cassava ; Toxic hazards
- PD 1637 Cast mining ; Landscape disfigurement open
- PC 1968 Caste prejudice
- PC 1968 Caste system
- PC 1968 Casteism
- PE 6054 Castration
- PJ 4592 Castration animals
- PC 0300 *Castration anxiety*
- PC 0300 *Castration complex*
- PE 4346 *Casts conifers* ; *Needle*
- PD 6930 Casual labour ; Discrimination against
- PD 6930 Casual workers ; Exploitation
- PD 4189 Casualties ; War
- PE 5105 Cat scratch fever
- PE 6817 Cataract
- PE 6817 Cataract ; Occupational
- PD 5535 Cataracts animals
- PE 6197 Catarrh ; Autumn
- PD 3978 *Catarrh* ; *Gastric*
- PE 6280 Catarrh ; Malignant head
- PG 9838 *Catarrh* ; *Nasal*
- PG 9838 *Catarrh* ; *Naso pharyngeal*
- PE 4724 Catarrh stomach
- PE 6197 Catarrh ; Summer
- PE 6280 Catarrhal fever
- PE 6280 Catarrhal fever ; Malignant
- PF 1181 Catastrophes ; Evolutionary
- PE 5229 Catastrophes ; Non payment compensation victims
- PD 5178 Catastrophic warfare damage ; Vulnerability marine environment
- PD 0438 *Catatonic schizophrenia*
- PF 4495 Catches ; Unequal distribution fish
- PD 1204 *Catchment facilities* ; *Limited*
- PE 8233 Catchments ; Inadequate water
- PF 7938 Categories ; Collapse distinctions
- PE 8215 Categories juvenile offenders ; Inadequate segregation different
- PJ 4748 Catering hazards ; Mass
- PE 4493 Catering industries ; Inadequate conditions work hotel
- PE 6615 Catering industry ; Corruption
- PE 6615 Catering practices ; Unethical
- PE 3649 Caterpillars
- PE 0912 *Catha edulis*
- PF 8071 Catholic conspiracy
- PF 8071 Catholicism
- PD 1763 Cats
- PD 1763 Cats ; Endangered species
- PE 9702 Cats ; Eosinophilic granuloma
- PE 5396 Cats ; Feral
- PE 7304 Cats ; Pox virus infection
- PE 9364 Cattle ; Abomasal disorders
- PD 7841 Cattle ; Bovine progressive degenerative myeloencephalopathy brown Swiss
- PE 0924 Cattle ; Brucellosis
- PD 0752 Cattle diseases
- PE 7808 Cattle ; Double muscling

Cattle [Farmers...]
- PE 5524 — Farmer's lung disease
- PD 7420 — Fatty liver disease
- PD 0752 — *fever* ; *Texas*
- PE 5380 — Filarial dermatitis
- PD 7799 — Foothill abortion
- PE 4161 — *Fractures*

- PE 2254 Cattle grubs

Cattle [Haemophilus...]
- PD 7841 — Haemophilus septicaemia
- PD 9366 — High mountain disease
- PE 5524 — Honker syndrome feeder
- PD 9667 — Hypodermosis
- PD 7420 — Hypomagnesemic tetany
- PF 2378 Cattle investment ; Costly
- PD 7420 Cattle ; Ketosis

Cattle [Lameness...]
- PE 4161 — Lameness
- PE 2357 — *Leptospirosis*
- PD 1802 — livestock ; Non productive use
- PD 9667 — Lumpy skin disease
- PE 2727 Cattle ; Mange
- PF 1849 Cattle mutilation

Cattle [Papular...]
- PE 2143 — *Papular stomatitis*
- PE 2786 — plague
- PD 7420 — Pregnancy toxaemia
- PD 7799 — Prolonged gestation
- PE 5524 Cattle ; Respiratory diseases
- PD 7420 Cattle ; Sudden death syndrome feeder
- PE 5524 Cattle ; Tracheal oedema syndrome feeder
- PE 4898 Cattle ; Trespassing
- PF 7706 Causal relationship problems ; Assumption absence
- PF 7706 Causal relationship problems ; Denial
- PA 0643 *Causalgia*
- PA 7406 Causation∗complex
- PD 7994 Cause conception ; Ignorance

Causes [morbidity...]
- PE 5463 — morbidity ; Ill defined
- PE 5463 — mortality ; Ill defined
- PE 5463 — mortality ; Unknown
- PF 3417 *Caution*
- PF 6389 Caution ; Excessive
- PE 8683 Cavies dolichotids ; Endangered species

Cease fire
- PE 2324 — agreements ; Persistence hostilities following
- PE 2324 — agreements ; Persistence technical state war following
- PE 2324 — violations
- PE 6255 Cedar apple rust
- PF 5250 Celebration ; Collapsed community
- PF 1985 Celebration cultural heritage ; Fragmented
- PF 3313 Celebrational expenditures ; Imbalanced
- PF 1560 Celebrational images ; Reduced
- PF 1309 Celebrations ; Dormant historical
- PF 5250 Celebrations ; Insufficient community
- PF 5250 Celebrations ; Insufficient corporate
- PF 1560 Celebrations ; Reduction symbolic
- PF 5250 Celebrations ; Uncensored public
- PF 9617 Celebrities ; Obsession
- PG 2295 Celiac disease
- PE 3298 Celibacy
- PA 7410 Celibacy
- PD 3371 Celibacy ; Enforced
- PE 7031 Celibacy ; Lack
- PE 3724 Cell anaemia ; Sickle
- PD 5535 Cell carcinoma ; Bovine ocular squamous
- PE 3724 Cell disease ; Sickle
- PE 5548 Cell sarcoma ; Reticulum
- PE 1914 Cell syndrome ; Low natural killer
- PD 9654 Cell tumours animals ; Functional islet
- PD 9554 Cells ; Marrow displacement abnormal
- PD 2730 Cellulitis ; Epidemic
- PE 5567 Cellulitis ; Facial
- PD 9667 Cellulitis ; Gangrenous
- PE 8769 Cellulose artificial resins ; Instability trade regenerated
- PE 2854 Cement dust
- PD 7258 Cemeteries ; Desecration
- PE 7172 Cemeteries ; Isolated mass
- PF 1600 Cemeteries ; Overgrown
- PE 9095 Cemeteries ; Shortage land
- PA 6400 Censoriousness *Affectation*
- PA 7237 Censoriousness *Condemnation*
- PA 6191 Censoriousness *Disapproval*
- PC 0067 Censorship
- PD 2337 Censorship ; Art
- PD 3026 Censorship ; Book
- PD 3172 Censorship communist systems
- PC 0067 Censorship ; Dependence
- PD 3032 Censorship ; Film cinema
- PD 3026 Censorship library acquisitions
- PE 5989 Censorship ; Military
- PD 3030 Censorship ; News
- PD 3027 Censorship ; Newspaper periodical

Censorship [Partial...]
- PF 6057 — Partial
- PD 2337 — photography
- PD 3033 — Postal
- PD 3029 Censorship ; Radio television
- PE 5998 Censorship ; Religious
- PD 1709 Censorship ; Scientific
- PF 6080 Censorship ; Self
- PF 9358 Censorship textbooks response creationism
- PD 3028 Censorship ; Theatre
- PE 5966 Censorship ; Video
- PA 7237 Censure *Condemnation*
- PA 6191 Censure *Disapproval*
- PA 6839 Censure *Disrepute*
- PA 6607 Censure *Misjudgement*
- PE 0189 Censured people
- PF 0214 Census data ; Insufficient
- PF 1176 *Centered business* ; *Self*
- PE 4508 Central banks ; Need increased cooperation

-871-

Central government

PD 1061 Central government ; Inflexible
Central nervous system
PE 1041 — Demyelinating diseases
PE 9037 — Diseases
PE 9037 — disorders
PE 7915 — Hereditary disorders
PE 2300 — *Syphilis*
PE 2280 — *Tuberculosis*
PF 2822 *Central nodes ; Lack*
PF 2540 Central planning structures small communities ; Lack
PF 2711 Centralization
PF 2711 Centralization ; Excessive bureaucratic administrative
PF 5472 Centralization global decision making ; Over
PF 2711 Centralization ; Over
PF 5472 Centralization power ; Global
PF 5115 Centralization religious power
PF 0685 Centralized collective bargaining ; Delays
PF 2707 Centralized decisions local technological innovation
PE 2820 Centralized financial control United Nations systems ; Lack
PC 3894 Centrally planned economies ; Failure
PD 0515 Centrally planned economies ; Lack consumer choice
PF 2261 Centre ; Distant district
Centre [Unfocused...]
PF 2813 — *Unfocused town*
PF 6559 — *Unlocated community*
PF 2575 — *Unsupported youth*
PF 7142 *Centred personal schedule ; Self*
PF 6504 *Centred priorities ; Home*
PA 6491 Centredness ; Self
PA 7211 Centredness ; Self
PD 3496 Centres ; Extra territorial military bases intelligence
PF 2261 Centres ; Inaccessible administrative
PF 8299 Centres ; Inaccessible market supply
PF 8299 Centres ; Remote supply
PF 6132 Centres suburban residents ; Inaccessibility city
PF 7118 *Centres ; Unavailability community*
PE 1154 Ceratcystis ulmi
Cereal preparations
PE 8732 — Environmental hazards cereals
PE 1769 — Instability trade cereals
PE 1218 — Long term shortage cereals
PE 4453 Cereal root rots
Cereals cereal preparations
PE 8732 — Environmental hazards
PE 1769 — Instability trade
PE 1218 — Long term shortage
PE 1331 Cereals ; Environmental hazards animal feedstuffs, excluding unmilled
PE 8816 Cereals ; Instability trade animal feedstuffs, excluding unmilled
PE 1769 Cereals ; Instability trade unmilled
PE 8514 Cereals ; Long term shortage animal feedstuffs excluding unmilled
PE 3147 Cereals ; Protein deficiency
PE 2233 Cereals ; Rusts grasses
PE 2234 Cereals ; Smuts grasses
PE 0965 *Cerebellum animals ; Defects*
PE 0965 *Cerebral defects animals*
PD 9057 Cerebral infarction
PD 0992 Cerebral ischaemia
PE 3830 Cerebral oedema
PE 0763 Cerebral palsy
PE 0763 Cerebral paralysis
PE 0763 Cerebral spastic infantile paralysis
PD 7841 *Cerebrocortical necrosis*
PD 0992 Cerebrovascular disease
PF 3960 Ceremonial uncleanness
PF 3313 Ceremonies ; Burdensome cost religious
PF 7887 Ceremonies ; Witchcraft
PF 2066 *Certificate-based job market*
PF 5381 Certificate ; Lack birth
PD 0486 *Certificates origin ; Forgery*
PD 0486 *Certificates origin goods ; Fraudulent*
PF 7847 Certitude ; Unquestionable
PE 9844 *Cervical abscess*
PE 1905 Cervical cancer
PD 7799 *Cervical prolapse*
PE 9461 *Cervicalgia*
PE 5606 Cervicobrachial disorders ; Occupational
PD 8775 *Cervix ; Diseases*
PA 5811 Cessation
PF 8040 Cessation diplomatic relations states
PE 8218 Cessation functions civil courts
PE 4233 Cessation romance
PD 2011 *Cesspool fencing ; Lack*
PE 3511 Cestoda
PE 6461 *Cestode infections*
PD 1593 Cetacea ; Endangered species
PE 4378 CFC's Chlorofluorocarbons environmental hazard
PE 0653 Chagas' disease
PA 6659 Chagrin *Humility*
PA 7107 Chagrin *Unpleasantness*
PE 8154 Chains ; Concentration noxious substances food
PD 0120 Chains ; Pesticide contamination food
PB 2253 Chains ; Vulnerability food
PF 7713 Challenge human survival ; Inadequacy prevailing mental structures
PF 1200 Challenges validity
PA 6714 Chance
PA 6714 Chance ; Dependence
PC 6605 Change
PC 9717 Change ; Anthropogenic climate
PF 1234 Change atmospheric chemistry ; Long term
PF 6721 Change∗complex
PF 6712 Change ; Confusion induced rapid social

PF 5660 Change crisis emergency responses ; Obstruction
PF 9066 Change ; Declining community confidence ability
PC 6989 Change ; Delays implementation social
PF 9664 Change ; Faltering adaptation economic agents accelerating pace structural
PF 1765 *Change ; Fear vocational*
PF 3277 Change gender
PF 0845 Change ; Government resistance institutional
PD 2670 Change ; Individual fear future
PF 9127 Change ; Irrational rejection technological
PE 1736 Change nationality ; Denial right
PC 2087 Change ; Ocean salinity
PC 2087 Change ; Ocean temperature
Change [Paralyzing...]
PD 2670 — Paralyzing image impossibility
PF 9576 — policy makers ; Static models social
PF 9664 — Policy making delays response accelerating structural
PC 5180 — Political unwillingness
PD 2670 — powerlessness ; Social
Change [Reduction...]
PD 1276 — Reduction demand primary commodities due technological
PF 9149 — Rejection proposals social
PE 6397 — religion ; Denial right
PF 0557 — Resistance
Change [sex...]
PF 3277 — sex
PF 1984 — Short term climatic
PA 8653 — *Suspicion imposed*
PA 5490 Changeableness
PA 5490 Changeableness *Changeableness*
PA 7325 Changeableness *Irresolution*
PA 6890 Changeableness *Nonuniformity*
PA 6425 Changeableness *Transience*
PA 7309 Changeableness *Uncertainty*
PA 5558 Changeableness *Weakness*
PE 9453 Changelings
PD 0992 Changes brain ; Degenerative
PC 6114 Changes climate ; Long term cyclic
PE 7943 Changes due technological advance ; Functional
PE 1684 Changes heart ; Degenerative
PF 1336 Changes ; Indecisive response technological
PF 2303 Changes ; Insecurity work
PF 8511 Changes names geographical features ; Political
PC 2087 Changes ocean characteristics ; Destructive
PD 0919 *Changes personality*
PE 4263 Changes precipitation patterns ; Long term
PD 4125 Changes rural economic patterns ; Difficulties establishing
PE 4596 Changes ; Training inappropriate structural technological
PE 7185 Changes ; Working hours inappropriate structural
PF 3010 Changing agricultural methods ; Resistance
PF 8521 Changing culture ; Rapidly
PF 8521 Changing cultures ; Rapidly
PF 2303 Changing employment patterns ; Disruptive effect
PF 1559 Changing government policy ; Frequently
PD 2154 *Changing river courses*
PF 2350 Channels ; Absence communication
PF 6471 Channels commercial initiative ; Undeveloped
PF 3037 Channels cultural interchange ; Restrictive
Channels [dialogue...]
PF 5862 — dialogue being restrictive ; Accusing
PF 7725 — dialogue judiciary ; Closed
PE 4365 — direct investments developing countries ; Inadequate
PF 2517 Channels ; Lack intersocietal resource
PF 2350 Channels ; Lost communication
PF 6544 Channels obtaining available local funding ; Lack
PF 4611 Channels ; Overlooked media
PF 2401 Channels ; Poor information
PF 3526 Channels public private resources ; Undeveloped
PE 2397 Channels transnational corporations ; Control marketing distribution
PF 0581 Channels ; Unused market
PF 6836 Chaos
Chaos [Dependence...]
PF 6836 — Dependence
PA 6858 — *Disintegration*
PA 7361 — *Disorder*
PA 6900 Chaos *Formlessness*
PA 6247 Chaos *Inattention*
PA 5563 Chaos *Lawlessness*
PA 7156 Chaos *Moderation*
PB 2480 Chaos ; Moral
PB 2496 Chaos ; Social
PF 2350 *Chaotic communication networks*
PF 6836 *Chaotic households*
PF 5684 Chapped skin
PD 2569 Character assassination
PD 2569 Character ; Defamation
PE 9762 Character ; Imposition quantitative trade quotas discretionary
PF 5452 Character ; National
PF 3962 Character ; Politicians' lack
PF 5544 Characteristics ; Criminal
PC 2087 Characteristics ; Destructive changes ocean
PF 0364 Characteristics problems ; Multi disciplinary
PA 7270 Characterlessness *Absence*
PA 7365 Characterlessness *Boredom*
PA 6900 Characterlessness *Formlessness*
PD 4769 Charcoal fuel ; Shortage
PE 8119 Charcoal trade instability ; Fuel wood
PD 1576 Charge ; Administrative detention
PE 7336 Charges ; Denial right be informed criminal

PE 8867 Charges ; Distortion international trade selective indirect taxes import
PF 5887 Charges ; Inconsistent port
PF 6158 Charges ; Violation freedom being tried twice same
PE 4628 Charges ; Violation right defence against criminal
PF 2819 Charitable giving ; Increasing resistance
PE 8742 Charitable organizations ; Corruption
PF 3049 Charitable organizations ; Inappropriate taxation not profit philanthropic
PE 8742 Charity frauds
PF 8363 Charms talismans ; Dependence amulets,
PF 5395 Charter ; Antiquated provisions UN
PF 3259 *Chastity*
PA 5612 Chastity ; Lack
PC 3300 Chattel slavery
PE 6369 Chauvinism ; Birth
PA 5532 Chauvinism *Disaccord*
PB 5765 Chauvinism ; Ethnic
PC 3024 Chauvinism ; Male
PA 7306 Chauvinism *Narrowmindedness*
PB 0534 Chauvinism ; National
PE 7778 Chauvinism ; Super power
PF 2152 Cheap argument
Cheap labour
PD 8916 — [Cheap labour]
PC 3117 — [*Cheap labour*]
PE 9199 — developing countries ; Exploitation
PA 7193 Cheapness
PA 6839 Cheapness *Disrepute*
PA 6997 Cheapness *Imperfection*
PA 6852 Cheapness *Inappropriateness*
PA 5942 Cheapness *Unimportance*
PA 3369 Cheating
PF 9729 Cheating ; Past cooperative
PF 2497 Cheating ; Trade quota
PE 4997 Checks balances ; Inadequate system political
PA 6099 Cheerlessness *Hopelessness*
PA 6731 Cheerlessness *Solemnity*
PA 7107 Cheerlessness *Unpleasantness*
PE 9702 *Cheilitis small animals*
PE 3711 *Cheiloschisis*
Chemical [accidents...]
PD 2570 — accidents
PC 1192 — agents
PD 1271 — air pollutants
PC 1164 Chemical biological warfare
Chemical [contaminants...]
PD 1694 — contaminants food
PD 1694 — contamination dietary intake
PE 0535 — contamination water
PE 0500 Chemical elements compounds ; Instability trade
PD 7993 Chemical fertilizers
Chemical [imbalances...]
PE 4715 — imbalances brain
PD 0575 — *industry wastes*
PD 0983 — insecticides ; Use
PC 1192 Chemical materials products ; Environmental hazards
PE 5651 Chemical occupational hazards ; Conflicting standards protection against
Chemical [pesticides...]
PD 0120 — pesticides
PD 0120 — pesticides ; Use
PE 0538 — petrochemical industry ; Instability
PE 1483 — petrochemicals industry ; Underdevelopment
PD 1239 — physical agents ; Carcinogenic
PD 1670 — pollutants environment
PD 4265 — pollution ; Underreporting
PF 5188 Chemical residue levels ; Dangerous legally prescribed
PD 5204 Chemical torture
PE 9363 Chemical trespass
Chemical warfare
PC 0872 — [Chemical warfare]
PD 9692 — technology ; Trade
PE 3808 — Trade products
PE 0535 Chemical water pollutants
Chemical weapons
PC 0872 — [Chemical weapons]
PC 0872 — Binary
PD 9692 — International trade
PD 7993 Chemicalized farming
PE 1785 Chemicalized farming ; Inappropriate agricultural subsidies
PD 1207 Chemicals control pests ; Excessive use
PE 8464 Chemicals derived coal, petroleum natural gas ; Instability trade mineral tar crude
PE 9076 Chemicals, elements, oxides halogen salts ; Trade instability inorganic
PC 1192 Chemicals ; Environmental hazards
Chemicals [Hazards...]
PD 7993 — Hazards agricultural
PE 4717 — Hazards low level exposure
PE 4717 — Health hazards long term, low level exposure toxic mixtures non toxic
PE 4717 — Long term hazards exposure
PE 1261 — Long term shortage
Chemicals [manufacturing...]
PE 8956 — manufacturing industries ; Ineffective self regulation
PD 5904 — Misuse
PE 4588 — Mutagenic
PC 1192 Chemicals petrochemicals industries ; Environmental hazards
PE 4588 Chemicals ; Radiomimetic
PE 8600 Chemicals ; Restrictive practices trade
PE 8242 Chemicals ; Shortage coal derived tar crude
PE 4588 Chemicals ; Toxic genetic
PD 0619 Chemicals trade ; Instability

Chinchillas

PF 1234 Chemistry ; Long term change atmospheric
PD 4265 Chemistry ; Malpractice
PD 4265 Chemistry ; Unethical practice
PD 4265 Chemists ; Corruption
PD 4265 Chemists ; Irresponsible
PD 4265 Chemists ; Negligence
PE 3592 Cheque card fraud
PE 6679 Cheques ; Conflicting national laws governing
PE 5279 Cherished things ; Torture destroying
PE 4586 *Cherry ; Black knot*
PE 2983 Chestnut ; Pests diseases
PD 0713 Chewing tobacco
PE 2727 *Cheyletiella infestations animals*
PD 2731 *Chick disease ; Mushy*
PD 1285 *Chicken diseases ; Prevailing*
PD 1285 *Chicken diseases ; Untreated*
PE 3639 *Chicken mites*
PE 7775 Chicken pox
PF 2378 *Chicken raising ; Exorbitant costs*
PD 0594 *Chickenpox*
PD 2768 Chickens factory farming ; Debeaking
PD 2730 *Chickens ; Haemorrhagic anaemia syndrome*
PD 2730 *Chickens ; Infectious bursal disease*
PD 2730 *Chickens ; Lymphoid leukosis*
PD 2730 *Chickens ; Malabsorption syndrome*
PD 3978 *Chickens ; Rotaviral infections*
PD 5453 *Chickens ; Sudden death syndrome*
PE 4497 Chickens ; Thrush
PE 3639 *Chigger ; Common*
PE 3639 *Chigger ; Turkey*
PC 2514 *Chikungunya*
PE 6154 Child abduction
Child abuse
PC 0838 — [Child abuse]
PC 2584 — [Child abuse]
PD 3267 — children ; Satanic
PF 2493 *Child activities ; Limited*
Child bearing
PD 2289 — Difficulty
PC 3258 — *Discrimination against men*
PD 9021 — Ignorance women concerning
Child [beating...]
PC 2584 — beating
PC 5512 — *beggary*
PF 6161 — birth environments ; Alienating
Child care
PD 2085 — facilities ; Inadequate
PD 2085 — facilities ; Lack
PD 9021 — Ignorance women concerning appropriate
PD 2085 — Missing
PF 3628 — *opportunities ; Limited awareness*
PE 5986 Child combatants
PB 2642 *Child cruelty*
Child [day...]
PD 2085 — day care facilities ; Inadequate
PC 1287 — deaths
PD 5734 — destitution
PF 2813 *Child education ; Unbudgeted*
PE 8857 Child health care ; Inadequate maternal
PE 4108 *Child hostages*
PD 0164 Child labour ; Exploitation
PD 0164 Child labour ; Necessity
Child [maintenance...]
PE 3247 — maintenance ; Inadequate
PD 8941 — malnutrition
PF 3285 — marriage
PE 5166 — mortality developing countries
PC 5512 — *mutilations*
Child [pornography...]
PF 1349 — pornography
PD 4966 — poverty
PE 6636 — prisoners
PE 7582 — prostitution
Child [rearing...]
PD 5344 — rearing ; Permissive
PD 4633 — rearing ; Subjugation women
PF 6186 — relationship ; Excessive intensification parent
PC 0600 — relationship ; Isolation parent
Child [sex...]
PD 3267 — sex rings
PE 5986 — soldiers
PF 6076 — support enforcement ; Inadequate system
Child [welfare...]
PC 0233 — welfare ; Inadequate
PC 0233 — *welfare institutions ; Lack*
PD 0164 — work force
PD 4773 Child youth literature ; Racism
PD 7692 Childbearing ; Career disruption due
PF 1021 Childbearing ; Discrimination favour
PD 0614 Childbearing ; Early
PC 9042 Childbirth ; Complications
PC 9042 Childbirth ; Complications labour
PE 7705 Childbirth ; Excessive medical intervention
PC 9042 *Childbirth fever*
PE 4894 Childbirth ; Haemorrhage pregnancy
PF 2493 *Childcare ; Tradition bound*
Childhood [accidents...]
PD 6851 — accidents
PD 3907 — aggression
PE 9256 — Avoidant disorder
PE 4421 Childhood burnout
Childhood [dependency...]
PD 3491 — dependency developing countries ; Excessive
PE 9751 — diarrhoea
PC 5890 — disability
PD 5344 — discipline ; Inadequate
PF 8118 Childhood martyrdom
PE 3717 Childhood neurosis
PE 1814 Childhood obesity
PD 0331 Childhood ; Protein energy malnutrition infants early
PD 5597 Childhood trauma
PF 4641 Childhood ; Unpopularity
PA 7371 Childishness *Unintelligence*
PC 3280 Childlessness
Children [Abandoned...]
PD 5734 — Abandoned
PE 4421 — Academic stress
PF 3302 — adoption ; Trafficking
PE 4276 — AIDS ; Vulnerability
PE 8617 — Air pollution impact lungs
PD 4218 — alcoholics
PE 5986 — armed forces ; Extra legal impressment
PE 1222 — Autistic
Children [banned...]
PC 0193 — *banned home*
PC 2584 — Battered
PE 7691 — Bereaved
PE 2082 — Blind
PE 4276 — born AIDS
PC 1874 — born out wedlock
PE 2876 — Bullying
Children [Conflict...]
PF 3252 — Conflict concerning legal custody
PD 0196 — Crippled
PC 0838 — Cruelty
PE 2083 Children ; Deaf
PF 4200 Children ; Decreasing number adoptable
Children Denial
PD 0635 — right freedom exploitation
PF 6616 — right unborn
PD 0164 — rights employed
PD 0943 — rights illegitimate
Children [Dependence...]
PF 2135 — *Dependence*
PD 2476 — Dependency
PC 0981 — deprived affection
PF 3382 — Desire male
PE 6636 — Detention
PE 9751 — Diarrhoea
PD 0196 — Disabled
PC 5890 — Disabled
Children Discrimination against
PE 5183 — coloured
PD 0943 — illegitimate
PE 5183 — mixed race
PE 8788 — women
Children [Disinherited...]
PC 0193 — *Disinherited*
PD 5308 — Disobedience
PD 5308 — Disobedient
PC 0193 — *Disowned*
PD 5136 — Displaced
PD 5987 — Drug addicted
PE 4609 — drug addicts
PE 8174 — during armed conflict ; Vulnerability
PD 1078 — during disasters ; Lack protection women
Children [economic...]
PD 7266 — economic exploitation ; Trafficking
PE 8424 — education ; Segregation handicapped
PD 1078 — emergencies ; Vulnerability women
PD 7330 — Emotional abuse
PD 0164 — Employment
PD 6065 — Endangered
PD 8825 — engendered occupying soldiers
PD 0635 — Exploitation
Children [Family...]
PC 8127 — Family rejection
PE 1814 — Fat
PD 3555 — father ; Dependence
PE 0185 — *Feral*
PF 0107 — Financial incentives having
PD 5129 — Forced disappearances
PD 8825 — foreign soldiers ; Abandoned
PE 9140 Children government leaders ; Corruption
PE 8617 Children ; Health hazards air pollution
PF 8118 Children holy war ; Sacrifice
Children [Illegal...]
PE 6154 — Illegal appropriation
PC 1874 — Illegitimate
PE 5020 — immigrants ; Lack appropriate education
PF 7353 — Impediments adoption
PF 0131 — imprisoned mothers
PE 6206 — Inadequate education nomadic
PD 2051 — Inadequate educational facilities gifted
PE 4921 — Inhibited growth malnourished
PE 7931 — Inhibition personality development exiled
PD 5308 — Insubordinate
PE 6676 — Involuntary loss nationality
Children [lie...]
PE 3909 — lie ; Teaching
PC 0239 — Loneliness
PD 9145 — Lying parents
Children [Maimed...]
PD 0196 — Maimed
PD 0586 — Maladjusted
PE 6764 — Medical experimentation
PD 4271 — medical experiments ; Traffic
PD 4271 — medical exploitation ; Trafficking
PD 0914 — Mental deficiency
PD 3784 — Mental depression
Children Mentally
PE 0989 — ill
Children Mentally cont'd
PD 0914 — retarded
PD 0914 — subnormal
Children [migrants...]
PE 4258 — migrants ; Social maladjustment
PE 5986 — Militarization
PD 6009 — Missing
PE 3265 — Molestation
PD 0196 — *Mute*
Children [Negative...]
PE 0341 — Negative effects family planning education
PD 4522 — Neglected
PD 3722 — neglected teachers
PE 4245 — Neglected young
PF 3253 — Non parental custody
Children [obscenity...]
PD 0132 — *obscenity ; Exposure*
PD 3267 — Organized sexual abuse
PD 7046 — Orphan
PE 4421 — Overworked
PD 7187 Children ; Parental beating
Children Parental toleration
PD 5344 — alcoholism
PD 5344 — drug addiction
PD 5344 — sexual activity
Children [Physical...]
PE 2876 — Physical intimidation
PC 2584 — Physical maltreatment
PD 0196 — Physically handicapped
PF 2017 — play ; Attitude manipulation
PD 5129 — Political abductions
PD 0132 — *pornography ; Exposure*
PD 4966 — poverty
PD 8973 — Prejudice
PF 0131 — prisoners ; Inadequate care
PD 0586 — Problem
PE 0392 — prostitutes ; Discrimination against
PE 0989 — Psychotic
PE 6922 — public life ; Discrimination harassment
Children [Rape...]
PE 6522 — Rape
PF 0131 — reared prison
PE 4465 — Reduced activity malnourished
PD 5136 — Refugee
PD 1307 — Retardation psychomotor development
PD 3267 — Ritual sexual abuse
PD 8340 — Runaway
PF 4832 Children's clinic ; Inaccessible
PD 0622 Children's diseases
PE 4700 Children's education ; Violation freedom parents choose their
PD 4773 Children's literature ; Bias
PD 0549 Children's play ; Inadequate facilities
PD 0549 Children's play ; Unsupervised
PF 2060 Children's thoughts reflections ; Parental control
Children [Sacrifice...]
PF 2641 — *Sacrifice*
PD 8405 — Sale
PD 3267 — Satanic child abuse
PE 3265 — Seduction
PE 4669 — separated parents ; Lack freedom movement
PF 5967 — Sexist education
PE 3265 — Sexual abuse
PD 3267 — Sexual exploitation
PE 6613 — sexual exploitation ; Trafficking
PE 3265 — Sexual love
PF 3302 — Sham adoption
PD 7187 — Smacking
PE 5146 — smokers ; Health hazards
PE 6219 — Smoking
PD 5349 — *snowmobile drivers ; Reckless*
PD 4271 — source organ transplants ; Traffic
PD 0550 — State custody deprived
PE 6676 — Stateless
PD 5980 — Street
PE 4421 — Stress
PE 4921 — Stunting growth
Children [threatened...]
PE 8174 — threatened warfare
PD 2851 — Torture
PE 4579 — torture victims
PD 8405 — Trafficking
Children [Unknown...]
PF 0782 — Unknown
PF 1750 — *Unperceived interests*
PE 1907 — Unwanted
PF 6159 — urban environment ; Inhibition exploration
PC 5512 Children ; Victimization
PD 3907 Children ; Violent
PE 4921 Children ; Wasting growth
PE 0185 *Children ; Wolf*
PD 1611 Children young people ; Alcohol consumption
PD 0513 Children youth ; Denial rights
PE 1646 Childrens' bodies ; Deliberate deformation
PE 6350 Chill ; Nuclear war induced winter
PE 8224 Chilled frozen meat ; Shortage fresh,
PE 8591 Chilled meat ; Instability trade fresh, frozen
PE 2481 Chin-cough
PE 3635 *Chin fly*
China pottery earthenware
PE 9019 — industries ; Environmental hazards
PE 8928 — industries instability
PE 8427 — manufacture underdevelopment
PE 3615 *Chinch bug*
PD 3481 *Chinchillas ; Endangered species*
PD 3481 *Chinchillas ; Endangered species rat*

PE 3604 Chiroptera ; Endangered species
PE 5476 Chlamydia
PD 7799 *Chlamydial abortion ewes*
PE 5913 *Chlamydial polyarthritis*
PE 5476 Chlamydid infections
PE 1928 Chlamydid infections animals
PE 5567 *Chlamydiosis poultry*
PE 4555 Chlorine
PE 7416 Chlorine bleached paper products ; Environmental hazards
PE 0535 *Chlorine pollutant*
PE 4378 Chlorofluorocarbons environmental hazard (CFC's)
PD 0515 Choice centrally planned economies ; Lack consumer
PA 7214 Choice*complex
PF 5057 Choice ; Consumer over
PD 6075 Choice ; Decrease consumer
PE 4700 Choice ; Denial right educational
PF 5057 Choice ; Excessive consumer
PD 1915 Choice marriage ; Denial right
PE 3926 Choice payment times creditors ; Inconsiderate
PF 1675 Choice ; Tensionless image free
PF 9276 Choice ; Unknowable future patterns social
PC 2494 Choice voters ; Inadequate
PE 3963 Choice work ; Denial right free
PF 6475 *Choices ; Restricted land*
PF 6540 *Choices ; Unpromoted business*
PF 6512 Choices ; Unresearched educational
PE 4630 *Choke*
PE 2251 Cholecystitis
PE 0560 Cholera
PE 0560 *Cholera ; Classical*
PE 0560 *Cholera El Tor*
PD 2732 *Cholera ; Fowl*
PE 6266 Cholera ; Hog
PE 0560 *Cholerae ; Vibrio*
PE 2371 Cholesterol ; Elevated blood
PE 8589 *Chondrodystrophy*
PE 3802 Choose moral religious education ; Denial right
PE 4700 Choose their children's education ; Violation freedom parents
PE 7915 *Chorea*
PE 7915 *Chorea ; Huntington's*
PE 7915 *Choriocarcinoma*
PD 5535 *Chorioretinitis animals*
PD 8786 *Choroid ; Inflammation*
PD 8786 *Choroiditis*
PF 4548 Chosen ; Peoples perceiving themselves specially
PF 0638 Christ ; Unreadiness second coming
PE 4369 Christian antisemitism
PD 0448 *Chromium deficiency*
PE 4072 Chromium pollutant
PE 2455 *Chromoblastomycosis*
PE 2455 *Chromomycosis*
PD 0153 Chronic alcoholism
PE 5182 *Chronic alveolar emphysema*
PE 2248 Chronic bronchitis
PD 8239 *Chronic cough*
PD 3978 *Chronic diarrhoea horses*
PD 8239 Chronic diseases
PD 9045 *Chronic enteritis*
Chronic fatigue
PD 4374 — [Chronic fatigue]
PE 1914 — immune dysfunction syndrome
PE 1914 — syndrome
PE 0585 Chronic hypertension
PD 8239 Chronic illness
PF 8026 *Chronic lymphadenitis*
PE 9344 *Chronic motor tic disorder*
PE 4161 *Chronic necrotic pododermatitis*
PE 4245 Chronic neglect ; Infant mortality due parental abuse
Chronic [obstructive...]
PD 0637 — obstructive pulmonary disease
PE 5182 — *obstructive pulmonary disease*
PE 5444 — *otitis externa*
Chronic [pain...]
PE 2694 — pain
PD 8239 — *pelvic inflammatory disease*
PF 5950 — psychological resistance
PD 0637 — pulmonary diseases ; Occupational
PD 9293 *Chronic renal failure animals*
PF 2732 *Chronic respiratory disease*
PC 8182 Chronic shortage foreign exchange
PE 4906 Chronic terminal illness
PE 2307 Chronic urinary infection
PE 0920 *Chronic valvular insufficiency*
PE 9344 *Chronic vocal tic disorder*
PD 1162 Chronically unemployed graduates
PF 2137 *Church abuse tax-exempt status*
PF 8905 Church attendance ; Reduction
PF 2137 *Church premises ; Gambling*
PF 8905 Church role ; Diminished
PF 8905 *Church / school involvement ; Minimal*
PC 0030 Church state ; Conflict
PF 6537 *Churches ; Closed area*
PF 8666 Churches holy places ; Desanctification
PF 6570 *Churches ; Inward focused*
PF 6091 *Churches ; Non recognition newly founded*
PF 3364 Churches ; Shame
PE 9369 *Chyluria*
PE 3662 *Cicatrization*
PE 9093 Cigarette advertising ; Irresponsible tobacco
PD 0713 Cigarette smoking
PE 3980 Cigarettes media ; Excessive portrayal drugs, alcohol
PD 0414 *Ciliostasis*
PD 3032 Cinema censorship ; Film
PE 4786 Cinema ; Maltreatment animals
PD 0606 Cinematic product placement
PE 1904 Circadian dysrhythmia
PF 2508 Circles ; Inner
PE 3779 Circling disease
PD 1318 *Circular psychosis*
PB 0284 *Circulation confidential information ; Unauthorized*
PE 3830 Circulation disorders ; Blood
PF 6552 Circulation local information ; Inadequate
PC 8482 Circulation system ; Diseases
PC 8482 Circulatory disorders
Circulatory system
PE 6924 — Congenital anomalies
PC 8482 — diseases
PC 8482 — Diseases
PD 5453 — diseases animals
PE 6054 Circumcision
PE 6055 Circumcision ; Female
PE 6053 Circumcision health hazard
PF 0422 Circumscribed vocational education
PE 4161 *Circumscripta ; Pododermatitis*
PE 7752 Circumstances ; Marketing skills which reinforce self image being victim
PD 4882 Circumvention duties assessments
PF 1976 Circumvention international sanctions
PC 9584 Circumvention international trading agreements
PD 4556 Circumvention law politicians
PE 4810 Circus animals ; Maltreatment
PE 4771 Circus animals ; Unethical practices
PE 2446 Cirrhosis liver
PC 6154 *Cities ; Bankrupt*
Cities [Dependence...]
PE 3296 — *Dependence*
PD 0134 — developing countries ; Uncontrolled physical expansion
PD 7065 — Divided
PE 8768 — Drain skills
PE 6660 Cities ; Excessive office space
PF 6165 Cities ; Impersonality public squares
PE 1134 Cities ; Limited availability permanent employment inner
PE 8768 Cities ; Migration rural population
PF 6155 Cities ; Stultifying homogeneity modern
PF 3523 *Cities ; Youth migration*
PC 3227 Citizen alienation
PF 2421 Citizen apathy
PA 1742 Citizen complacency
PF 8076 Citizen confidence ; Minimal
PD 5707 Citizen disobedience
PF 2421 Citizen engagement ; Fragmented modes
PF 6537 *Citizen goals ; Fragmented*
PD 9557 Citizen grievances ; Governmental rejection
Citizen [incompetence...]
PE 5742 — incompetence
PF 2421 — input ; Inadequate
PF 2421 — involvement ; Minimal
PF 1097 Citizen mistrust politicians
Citizen participation
PF 2440 — Sociological ignorance
PF 2347 — Structural failure
PF 3307 — Unconfident
PF 2347 *Citizen powerlessness*
PF 7118 *Citizen priorities ; Conflicting*
Citizen [resistance...]
PD 5707 — resistance Nuremberg obligation
PF 2421 — responsibility ; Fragmented
PC 5075 — *rights ; Overcomplicated implementation*
PD 0973 Citizens ; Discrimination against non
PC 3185 Citizens foreign countries ; Intervention major powers protect investments their
PF 3707 Citizens ; Governmental reprisals against
PF 8029 *Citizens ; Grievances*
PF 2996 Citizens ; Inadequate transportation systems rural
PD 1564 Citizens ; Isolated senior
PF 8781 Citizens' law enforcement groups ; Defeatist self images
PF 4849 Citizens ; Privileged classes
PD 5560 Citizens ; State supported violence against
PE 1659 Citizens working abroad ; Restrictions socialist
PE 2485 Citizenship ; Involuntary loss
PF 2600 Citizenship ; Nationalistic images
PF 2835 Citizenship ; Non globalized
PF 5453 Citizenship ; Refusal grant
PC 3458 *Citizenship ; Second class*
PE 2976 Citrus fruit ; Pests diseases
PC 2616 City areas ; Environmental degradation inner
PF 6132 City centres suburban residents ; Inaccessibility
PC 2616 City decay ; Inner
PF 6140 City dwellers ; Inaccessibility countryside
PF 2042 City enterprises ; Lack awareness potential investment small, inner
PF 6544 *City equipment ; Deficient*
PD 9869 City jobs ; Attractive
PF 2423 *City life ; Attraction*
PF 1437 City life style
PC 2616 City neighbourhoods ; Deteriorating inner
City [plan...]
PF 9478 — *plan ; Unimaginative*
PF 2813 — *plans ; Nebulous*
PC 5052 — poverty ; Inner
PF 2583 *City services ; Slow*
PF 2120 City sizes country ; Imbalance
PE 9121 Civets mongooses ; Endangered species genets,
PF 8781 Civic attitude ; Defeatist
PF 0544 *Civic awareness ; Reduced*
PF 0996 Civic education ; Manipulation
PF 2401 Civic information ; Vagueness
PF 2428 Civic minded groups ; Limited fragmented outlook
PE 4768 *Civic ordinances ; Unenforced*
PF 6550 *Civic organizations ; Insufficient*
PC 0588 Civic participation ; Decline
PF 2093 *Civic pride ; Embarrassed*
PF 2575 Civic responsibility ; Individual avenues
PF 0903 Civil administrative capacity developing countries ; Limited
Civil [capacity...]
PE 8720 — capacity married women ; Loss
PE 8218 — courts ; Cessation functions
PE 4294 — *crimes committed during war*
Civil [defence...]
PF 0506 — defence ; Inadequacy
PC 0690 — disobedience
PC 2551 — disorders
PD 5372 — disturbances
PF 4975 Civil engineering projects ; Misappropriation resources high cost
PF 4899 Civil justice systems ; Deficiencies
PD 5727 Civil law transgressions
PD 7571 Civil litigation ; Delay
Civil [refusal...]
PF 3248 — refusal divorce ; Religious
PB 1540 — religion
PC 0690 — resistance
Civil rights
PD 8709 — Criminal violation
PC 0632 — Denial political
PC 4608 — Inadequate enforcement
PC 5285 — Violation
Civil servants
PF 5649 — Abuse privileges immunities international
PE 3457 — Conspicuous consumption international
PE 6702 — Corruption
PE 5488 — Detention international
PE 5278 — developing countries ; Demotivated
PD 8392 — Embezzlement
PE 6388 — Excessive salaries international
PD 5554 — Fraud international
PE 6702 — Irresponsible
PE 4031 — Lack political neutrality
PE 5488 — Violation privileges immunities international
PF 2335 Civil service ; Ineffective
PE 4294 *Civil unrest ; Crimes committed during*
PC 4864 Civil violence
PE 0525 Civil violence ; Stress trauma context
PC 1869 Civil war
PD 3765 Civil war ; Economic
PD 8719 Civilian areas ; War damage
PE 2449 Civilian hands ; Proliferation weapons
PD 3015 Civilian political prisoners detainees
PE 8564 Civilian populations institutions ; Destruction
PE 8361 Civilians armed conflict ; Inadequate protection
PD 5560 Civilians ; Beatings
PD 5560 Civilians ; Maltreatment
PE 2449 Civilians ; Sale weapons
PC 7013 Civility ; Loss
PB 3674 Civilization ; Pathologies
PC 1888 *Claims concerning Antarctic territory ; Conflicting*
PD 1628 Claims concerning off shore territorial waters ; Conflicting
PF 1226 Claims external resources ; Reluctant
PE 6975 Claims ; Fraudulent mineral exploitation
PF 3388 *Claims human development sports ; Excessive*
PE 4329 Claims medical negligence ; Increase insurance
PC 1888 *Claims non terrestrial territory ; Conflicting*
PD 1628 Claims off shore territorial waters ; Unilateral
PD 7087 Claims over shared inland water resources ; Conflicting
PE 3814 Claims products ; Exaggerated advertising
PF 3376 Claims religious infallibility ; Negative effects
PC 2362 Claims states territories ; Conflicting
Claims [Unrealistic...]
PE 5305 — Unrealistic wage
PC 3304 — Unsettled indigenous land
PC 1888 — Unsettled territorial
PE 8210 *Clan feuds*
Clandestine [emigration...]
PD 1928 — emigration
PC 7607 — employment
PD 1823 — excavations
PF 3242 *Clandestine homosexuality*
PD 1928 Clandestine immigration
PD 4501 Clandestine intelligence operations
Clandestine [market...]
PC 2921 — market research
PF 3283 — *marriage*
PF 7669 — military operations
Clandestine [press...]
PD 2366 — press
PJ 0914 — private utilization public goods services
PC 2175 — psychological warfare
PF 4450 Clandestine space weapons
PE 4576 Clandestine sports betting
PE 3968 Clandestine trade nuclear weapons
PF 0053 Clarify world interests ; Lack autonomous world level actor identify
PC 2494 *Clash personal ambitions political parties*
PC 3458 Class bigotry
Class [citizenship...]
PC 3458 — *citizenship ; Second*
PC 1573 — conflict
PC 3458 — consciousness
PC 3458 — consciousness social barrier
PC 1573 — Creation dominant
Class [Discredited...]
PE 5341 — Discredited moneyed hereditary

Collapse meaning

Class [discrimination...] cont'd
- PC 3458 — discrimination
- PE 0779 — discrimination education
- PC 3458 — disparity resentment
- PC 3458 — distinction
- PC 3458 — division
- PC 1573 — domination
- PC 1573 — domination ; Dependence
- PC 1573 Class exploitation
- PF 7428 Class ; Low socio economic

Class scheduling
- PF 6476 — Inconvenient
- PF 6476 — Ineffective
- PF 6476 — Inflexible

Class [science...]
- PF 6697 — science ; Second
- PC 3458 — segregation
- PD 0579 — states ; Second
- PC 3458 — system
- PF 6476 Class time ; Underutilized
- PC 1002 Class ; Vulnerability middle
- PC 1573 Class war
- PF 4849 Classes citizens ; Privileged
- PD 8380 Classes ; Corruption ruling
- PF 6670 Classes developing countries ; Reinforcement inappropriate development privileged
- PF 4849 Classes ; Elitist ruling
- PD 6585 Classes ; Inappropriate size school
- PF 7428 Classes ; Inferior
- PD 6585 Classes ; Overcrowded school
- PE 0560 *Classical cholera*
- PE 6266 *Classical swine fever*
- PD 1753 *Classical typhus*
- PF 1781 *Classification demands ; Excessive*
- PF 3550 *Classification ; Depressed area*
- PD 2392 *Classification ; Inadequate drug*
- PF 3020 *Classification political prisoners ; Mis*

Classification [Suppression...]
- PF 4050 — Suppression information security
- PF 6743 — systems ; Bias document
- PE 7753 — systems ; Incompatibility document
- PD 5183 Classified communications information ; Misuse
- PF 9699 Classified public information
- PC 3458 Classlessness ; Myth
- PD 2226 Clavibacter
- PF 1815 Clean water ; Lack
- PF 1815 *Cleaning ; Shortage public*
- PD 2459 Cleanliness ; Lack
- PD 2127 Clear air turbulence
- PE 5380 *Clear-eyed blindness*
- PF 7728 Clear names people convicted crimes which they are innocent ; Failure
- PE 4162 Clearing land ; Cost
- PF 9841 Clearing systems ; Inadequate financial
- PE 5117 Cleft lip
- PE 5117 Cleft palate
- PE 3711 *Clefts*
- PF 1390 *Clergy ; Homosexual*
- PF 8889 Clergy ; Malpractice
- PD 8388 *Clerical personnel ; Untrained*
- PF 3360 Clericalism ; Anti
- PD 5783 *Climacteric ; Male*
- PC 0387 Climate
- PE 9020 Climate ; Adverse effects urbanization
- PC 9717 Climate change ; Anthropogenic

Climate [Inadvertent...]
- PC 1288 — Inadvertent modifications
- PE 7832 — Increase insect pests modification micro
- PC 0387 — Inhospitable
- PC 6114 Climate ; Long term cyclic changes
- PC 9717 Climate man ; Degradation
- PD 7941 Climate modification ; Hostile
- PC 0387 Climate ; Poor
- PC 0293 Climate ; Short term variations
- PD 1404 Climates ; Cold
- PE 1811 Climates ; Corrosion tropical
- PE 1265 Climates ; Destructive action mould tropical
- PD 2474 Climates ; Hot humid

Climatic [change...]
- PF 1984 — change ; Short term
- PD 1404 — cold
- PC 0387 — conditions ; Extreme
- PC 0387 — conditions ; Unfavourable
- PC 0387 Climatic extremes
- PC 2460 Climatic heat
- PJ 8591 Clinic construction ; Prohibitive cost
- PF 4832 Clinic ; Inaccessible children's
- PF 4832 Clinic service ; Remote
- PD 4004 Clinic space ; Lack
- PD 4004 Clinics ; Overcrowded public
- PD 0366 Clinics ; Understaffed health
- PF 2575 Cliques ; Operating community
- PC 9585 Cliques ; Ruling
- PE 6055 Clitoridectomy
- PE 7647 Clogged storm drains
- PE 6625 Cloning ; Computer
- PF 6537 *Closed area churches*
- PD 2637 *Closed areas*
- PF 7725 Closed channels dialogue judiciary
- PF 2508 *Closed communities*
- PF 1781 Closed federation operations
- PF 7847 Closed mindedness
- PD 8629 Closed professions
- PF 1134 *Closed shop unions*
- PD 3169 Closed society ; Communist
- PD 6019 Closed union shops

- PA 7193 Closeness *Cheapness*
- PA 6956 Closeness *Cold*
- PA 7256 Closeness *Narrowness*
- PA 6444 Closeness *Quiescence*
- PA 7411 Closeness *Uncommunicativeness*
- PF 3242 *Closet homosexuality*
- PF 1345 *Closings ; Unnecessary business*
- PD 9667 *Clostridial dermatomyositis*
- PE 7769 Clostridial infections
- PE 3936 *Clostridium perfringens type C enteritis swine*
- PA 7391 Closure
- PF 6575 *Closure ; Imposed high school*
- PE 6795 *Closure places worship*
- PE 6276 Closure recreation areas
- PF 6575 *Closure schools*
- PF 3831 Closure social institutions
- PE 3537 Closures ; Factory

Clothing [Decadent...]
- PE 5607 — Decadent
- PE 5409 — Denial freedom expression
- PE 7616 — Denial right sufficient
- PF 1773 *Clothing ; Inappropriate*

Clothing industries
- PE 1103 — Environmental hazards textile
- PE 1008 — Instability textile
- PE 0453 — Underdevelopment textile
- PD 8001 Clothing industry ; Corruption
- PD 8001 Clothing industry ; Unethical practices
- PE 0303 Clothing ; Long term shortage
- PE 8496 Clothing supply dwarfs, midgets giants ; Lack adequate
- PE 5783 Clots ; Blood
- PE 6350 *Cloud cover ; Excessive*
- PD 5228 *Clover poisoning ; Sweet*
- PE 1323 *Clover silk*
- PE 3836 Club foot
- PD 2225 *Club root cabbage*

Clubs [private...]
- PD 3032 — *private ; Film*
- PD 3032 — *Private film*
- PD 3032 — *Private theatre*
- PA 5454 Clumsiness *Badness*
- PA 6312 Clumsiness *Inelegance*
- PA 5438 Clumsiness *Neglect*
- PA 7240 Clumsiness *Ugliness*
- PA 7232 Clumsiness *Unskillfulness*
- PF 4670 *Clustered vacant land*
- PD 5034 Clutter ; Advertising
- PE 2265 Cluttering
- PF 9729 Co benefits ; Unrecognized
- PF 9729 Co-op successes unknown
- PF 9729 Co-ops suspicion
- PF 9234 Co workers ; Fear humiliation
- PE 2373 Coagulation disorders
- PD 1245 *Coal ash ; Marine dumping*
- PE 8206 Coal, coke briquettes ; Instability trade
- PE 8453 Coal conversion plants ; Environmental impacts
- PE 8242 Coal derived tar crude chemicals ; Shortage
- PE 4679 Coal dust
- PD 7541 Coal energy ; Environmental hazards
- PD 1245 *Coal fly-ash*
- PE 1054 Coal ; Long term shortage
- PE 5160 Coal mining environmental hazards

Coal petroleum
- PE 8464 — natural gas ; Instability trade mineral tar crude chemicals derived
- PE 0760 — precious stones ; Instability trade crude fertilizers crude minerals, excluding
- PE 1353 — precious stones ; Long term shortage crude fertilizers crude minerals, excluding
- PE 7960 Coal products manufacture underdevelopment ; Petroleum
- PD 5228 *Coal-tar poisoning animals*
- PF 3214 Coalitions
- PE 6924 *Coarctation aorta*
- PA 7143 Coarseness *Discourtesy*
- PA 6997 Coarseness *Imperfection*
- PA 5612 Coarseness *Unchastity*
- PA 7341 Coarseness *Unpreparedness*
- PA 5821 Coarseness *Vulgarity*
- PE 7946 Coast fever ; East
- PD 4671 Coast zones ; Unsustainable development
- PD 3775 Coastal areas ; Subsiding
- PE 2425 Coastal communities ; Lack finance
- PE 2305 Coastal communities ; Negative consequences shifting ecology
- PF 4767 Coastal deep sea water resources ; Obstacles utilization
- PE 5689 Coastal development island countries ; Lack
- PE 6734 Coastal erosion
- PD 9515 *Coastal erosion resulting dams*
- PD 5110 Coastal lowlands ; Inadequate empolderment
- PF 2346 *Coastal maintenance ; Inadequate*
- PD 1356 Coastal water pollution
- PE 3813 Coastlines ; Settlement unprotected
- PE 2339 Cobalt pollutant
- PD 2363 Coca cocaine ; Abuse
- PD 2363 Cocaine ; Abuse coca
- PD 2363 Cocaine abuse ; Long term effects
- PE 2123 Cocaine ; Crack
- PD 4027 Cocaine intoxication
- PD 2363 Cocaine withdrawal
- PD 2363 Cocaism
- PE 2738 Coccidioidomycosis
- PE 2738 Coccidiosis
- PE 9526 *Coccidiosis ; Human*
- PE 2738 *Coccidiosis poultry*

- PE 3612 *Coccids*
- PE 4893 *Cock fighting*
- PE 3579 Cockroach resistance insecticides
- PE 1633 Cockroaches pests
- PE 2979 Cocoa ; Pests diseases

Cocoa spices
- PE 0915 — manufactures thereof ; Instability trade coffee, tea,
- PE 1197 — manufactures thereof ; Long term shortage coffee, tea,
- PE 0481 — their manufacture ; Environmental hazards coffee, tea,
- PE 1549 Cocoa trade ; Instability
- PE 2981 *Coconut palm ; Pests diseases*
- PE 2305 *Coconut production ; Decreasing*
- PF 4676 Code crimes ; Lack international

Code enforcement
- PD 8697 — difficulties
- PD 8697 — inadequate
- PD 8697 — Random
- PD 1631 Code ; Inadequate fire
- PE 1329 Codeine abuse
- PF 8829 Codes ; Absence housing
- PE 7017 Codes behaviour ; Discriminatory unwritten
- PE 0671 Codes culpability ; Violations regulatory
- PE 8030 Codes justice ; Deficiencies military
- PD 2543 Codes ; Strict army discipline

Codes [Unenforced...]
- PD 5001 — Unenforced sanitation
- PF 6506 — Unknown building
- PE 8443 — urban areas ; Restrictive building
- PD 1251 Codes ; Violation house
- PD 4539 Codes ; Violation regulatory
- PE 3809 Codes ; Wilful violation regulatory
- PE 7252 Coerced compliant confession
- PC 3796 Coercion
- PD 4469 Coercion ; Criminal
- PC 9104 Coercion ; Political
- PE 0207 Coercive use economic power transnational enterprises against labour
- PE 3781 Coffee industry ; Domination transnational corporations
- PE 2218 Coffee ; Pests diseases

Coffee tea cocoa
- PE 0915 — spices manufactures thereof ; Instability trade
- PE 1197 — spices manufactures thereof ; Long term shortage
- PE 0481 — spices their manufacture ; Environmental hazards
- PE 0950 Coffee trade ; Instability
- PE 9620 Cognitive ability ageing ; Decline
- PF 6638 Cognitive dissonance
- PF 7713 Cognitive infrastructure formulate effective global policies ; Inadequate intellectual
- PF 5240 Cognitive suicide
- PD 6881 Cohabitant abuse ; Spouse
- PF 3278 Cohabitation
- PF 3278 Cohabitation ; Non marital
- PD 2864 Cohesion ; Breakdown local community
- PD 2864 Cohesion ; Limited community
- PE 5503 Coining ; Illicit
- PD 7799 *Coital exanthema ; Equine*
- PF 6910 Coitophobia
- PD 6033 Coitus cancer risk
- PF 3260 Coitus ; Unskilled
- PE 8206 Coke briquettes ; Instability trade coal,
- PA 6956 Cold

Cold [climates...]
- PD 1404 — climates
- PD 1404 — Climatic
- PE 2412 — Common
- PF 8663 — countries ; Non viability
- PE 0274 Cold disorders
- PF 5744 Cold fishing conditions
- PE 9509 Cold haemolytic disease animals
- PF 5744 Cold occupational hazard
- PD 1404 Cold spell
- PE 3734 Cold torture
- PF 3100 Cold war
- PD 1404 Cold wave
- PA 6379 Coldness *Avoidance*
- PA 6956 Coldness *Cold*
- PA 5446 Coldness *Enmity*

Coldness [Unfeelingness...]
- PA 7364 — *Unfeelingness*
- PA 5643 — *Unkindness*
- PA 6653 — *Unsociability*
- PE 1679 Coleoptera
- PD 7841 *Coli enterotoxaemia ; E*
- PD 2731 Colibacillosis poultry
- PE 3936 *Colibacillosis swine ; Enteric*
- PD 3978 *Colic horses*
- PE 7970 Colitis
- PE 7423 *Colitis animals*
- PE 0873 *Colitis ; Arthritis accompanying ulcerative*
- PD 9045 *Colitis ; Ulcerative*
- PD 3978 *Colitis-X horses*
- PD 0218 *Collaboration enemy*
- PD 2565 Collagen diseases
- PF 1088 Collapse anarchy ; Fear social
- PD 1230 Collapse ; Building

Collapse [common...]
- PF 1118 — common values
- PF 3750 — corporate engagement society
- PF 1084 — cultural dreams
- PF 7938 Collapse distinctions categories
- PF 3612 Collapse ; Global fiscal
- PD 5326 Collapse growth developing countries
- PF 8479 Collapse judicial system
- PF 0955 Collapse meaning participating society

Collapse

PB 2480 Collapse ; Moral
PF 1114 Collapse providing ethical value screens
PF 0897 Collapse public servant role professional
Collapse [societal...]
PF 2340 — societal engagement
PF 2687 — standards sexual conduct
PD 4511 — stock exchanges bourses ; International
PF 7798 — systems verbal exchange
PF 6358 Collapse ; Theological
PF 5250 *Collapsed community celebration*
PF 5826 Collapsed concept peace
Collapsed conceptualization
PF 5242 — [Collapsed conceptualization]
PF 5155 — equality
PF 7497 — liberty
PF 0959 — love
PD 2438 Collapsed consensus structure
PF 2358 *Collapsed corporate effort*
PF 6098 Collapsed images vocation
PF 0936 Collapsed meaning human creativity
PE 4203 Collapsed relationships work society
PF 5555 Collapsed tension care responsibility
PD 4143 Collapsing physical structures
PD 4143 *Collapsing public works*
PD 9667 *Collar galls*
PF 2009 *Collateral ; Ineffective land*
PD 3054 *Collateral ; Limited borrowing*
PD 3054 *Collateral ; Unappealing loan*
PF 0905 *Collaterals ; Insufficient development*
PF 2009 *Collaterals ; Scarcity business*
PF 6713 Collecting ; Fanatical book
PD 1204 Collection ; Arduous water
PE 9417 Collection ; Birds egg
PE 9417 Collection cultural artefacts
PE 9417 Collection ; Fossil
PD 0807 Collection ; Infrequent rubbish
PE 9417 Collection specimens ; Abusive
PD 0807 Collection ; Unorganized trash
PE 1122 Collections ; Prohibitive cost maintaining comprehensive document
PE 4003 Collective agreements ; Non implementation workers wage increases provided legislation
PF 7003 Collective attention-span limit societal learning
Collective bargaining
PF 0685 — Delays centralized
PE 3970 — Denial right
PE 6018 — Inadequate employee participation
PE 3970 — workers organizations ; Restrictions
Collective [comprehension...]
PF 7110 — comprehension societal learning ; Superficiality
PF 7038 — comprehension span societal learning ; Limit
PJ 5810 — consciousness
PF 8703 Collective decision making ; Non
PD 6970 Collective punishment
PF 8905 Collective religious worship ; Decreasing participation
PD 0694 Collective stoppages work
PF 6170 Collectively organized fantastic happenings ; Inhibition
PE 3970 Collectively ; Violation right workers organizations bargain
PF 2553 Collectivism
PF 2553 Collectivist policies
PD 7443 Collectivization agriculture ; Enforced
PE 9417 Collectors exotic species
PD 8723 College students ; Learning disabled
PF 0967 Collegial relationships ; Lack motivating
PD 5453 *Collie dogs ; Cyclic neuropenia grey*
PD 9667 *Collie nose*
PD 5453 *Collie syndrome ; Grey*
PD 5453 *Collie syndrome ; Silver*
PD 1582 Collisions ; Aircraft
PD 1582 Collisions ; Aircraft near
PD 0126 Collisions ; Railroad
PD 8982 Collisions sea
PD 8982 Collisions ; Ship
PC 8478 Collisions ; Vehicle
PG 5976 *Colloid goitre*
PF 8711 Collusion administrators funding agencies programme formulators
PD 8465 Collusion governments
PE 8367 Collusion trade union leaders employers government
PA 7411 Collusion *Uncommunicativeness*
PE 0396 Collusive international trade arrangements
PE 4301 Collusive tendering
PE 7072 Collusive tendering international trade
PE 7423 *Colon impaction animals*
PE 9399 Colon-rectal cancer
PD 2511 Colonial countries ; Over development bureaucracy ex
PE 8610 Colonial countries ; Weak national identity post
PC 0798 Colonialism
PE 3447 Colonialism ; Eco
PE 3447 Colonialism ; Environmental
PC 0352 Colonialism ; Internal
PC 1876 Colonialism ; Neo
PC 0798 Colonialism ; Slavery like practices
PD 4862 Colonialism ; Socialist
PC 0798 Colonization
PF 4894 Colonization information
PJ 3814 Colostomy
PE 6343 Colour blindness
Colour [deficiencies...]
PE 6343 — deficiencies
PC 8774 — Discrimination based skin
PF 1741 — Dissatisfaction skin
PC 8774 Colour prejudice ; Skin
PF 1741 Colourants ; Abusive use skin
PE 5183 Coloured children ; Discrimination against
PD 0487 Colourings ; Toxic food
PA 7301 Colourlessness
PE 4620 Coma ; Irreversible
PC 0300 *Combat fatigue*
PF 5382 Combat ; Single
PE 7912 Combat trauma
PE 5986 Combatants ; Child
PE 8361 Combatants war zones ; Inhumane treatment non
PE 8220 Combatants ; Women
PA 5445 Combativeness *Defence*
PA 5532 Combativeness *Disaccord*
PA 5502 Combativeness *Reason*
PE 8972 Combination effects work place ; Hazardous
PE 5190 Combination sacred erotic advertising ; Vulgar
PE 6403 Combinations ; Disease causing viral
PF 1105 Combinations environmental impacts
PD 5256 Combinations substances ; Hazardous
PF 2588 Combined efforts ; Lack
PF 4224 *Combined immunodeficiency disease animals*
PE 5656 Combined stresses occupational hazards
PE 8014 Combines farming communities ; Threat industrial
PF 4043 Combustion ; Spontaneous human
PF 2634 *Comedy ; Black*
PF 0039 Comet showers ; Mass extinction due
PF 3564 Comets ; Hazards
PA 0429 *Comic books ; Violence*
PE 2691 Comic strips picture story books ; Harmful effects
PF 1384 Comics ; Pornographic magazines
PF 0638 Coming Christ ; Unreadiness second
PC 2563 Commerce ; Corruption
PE 5814 Commerce developing countries ; Lack technical infrastructure maritime
PE 0688 Commerce ; Health risks workers
PC 2563 Commerce ; Negligence
PE 8251 Commerce offices ; Inadequate working conditions employees
PE 3957 Commerce offices ; Occupational hazards
PF 4433 Commerce religious indulgences
PE 9840 Commerce spiritual benefits
Commerce [Unexploited...]
PF 2535 — Unexploited possibilities local
PD 0312 — Unfair practices maritime
PC 4645 — Unlawful
Commercial [activities...]
PF 2439 — *activities ; Minimal*
PE 6786 — agreements ; Secret
PE 1349 — alcohol ; Adulteration
PE 4655 Commercial bank lending developing countries ; Decline
PF 6578 *Commercial buildings ; Outmoded*
Commercial [confidentiality...]
PE 6786 — confidentiality ; Abuse
PE 6786 — cover-up
PC 5624 — crime
Commercial [deception...]
PD 4879 — deception
PF 2439 — *decline ; Continuing*
PF 6548 — development ; Unexplored alternatives
PD 4519 — distortion toxicity thresholds
Commercial [enterprises...]
PE 9251 — enterprises ; Irresponsible directors
PF 2231 — enterprises ; Unrealized potential
PF 7227 — erosion standards quality
PF 2813 — *establishments ; Unprotected*
PF 3099 — exchanges ; Abuse international cultural, diplomatic
PC 4867 — *experience ; Limited*
Commercial exploitation
PD 9370 — education
PD 2772 — farm animals industrial concerns ; Excessive
PD 9370 — students
PD 1481 — wild animals
Commercial [facilities...]
PF 3448 — *facilities ; Insufficient*
PF 3448 — *facilities ; Unsuitable*
PF 0905 — *farming ; Poor risk*
PD 5584 — favouritism
PF 4340 — finance rural development projects ; Inadequate
PE 0718 — financial services ; Inaccessible
PD 2057 — fraud
PD 4879 Commercial information ; Misleading
PF 6471 Commercial initiative ; Undeveloped channels
PC 4867 *Commercial know how ; Insufficient*
PC 0238 *Commercial life ; Unproductive use trees beyond their*
PF 1442 Commercial management rural areas ; Fragmented forms
PD 0188 *Commercial piracy*
PE 4833 Commercial political ends ; Exploitation athletic competition
Commercial practices
PC 2563 — Irresponsible
PF 1352 — *Poor*
PC 2563 — Unethical
PE 4545 Commercial property loans ; Inadequately secured
Commercial [reprisals...]
PD 9389 — reprisals ; Threat
PF 2009 — resources ; Limited availability local
PD 8897 — rivalry
PF 1495 — *roadways ; Inadequate*
Commercial [secrecy...]
PE 6786 — secrecy ; Abuse
PD 1865 — sector developing countries ; Inadequacy
PF 2009 — services ; Lack local
PF 6510 — services ; Lack support local
PF 7130 — *sites ; Unavailable*
PC 4867 — *skills ; Insufficient*
PF 2747 — surrogate mothers
Commercial [television...]
PD 0433 — television ; Exploitative
PG 7741 — traffic ; Hazardous
PS 9612 — *training ; Distant*
PD 4728 — transactions ; Restrictions
Commercial [vehicles...]
PF 4485 — vehicles ; Maldistribution
PE 0897 — vehicles ; Unjustified restrictions free movement
PE 1139 — *ventures ; Unviable*
PF 0815 Commercialism ; Proliferation
PE 4222 Commercialization athletic activities sports events
PF 4674 Commercialization definitions ; Restrictive
PE 6038 Commercialization human embryos
PE 4215 Commercialization media ; Excessive
PF 0699 Commercialization nationality acquisition
PF 0815 Commercialization society
PD 9370 Commercially biased educational materials
PD 9370 Commercially sponsored education
PF 9693 Commissioning body ; Expertise biased favour
PE 6948 Commissions ; Purchase military
PJ 1382 Commissions ; Soft
PF 5502 Commitment action ; Governmental delayed withdrawal
Commitment [common...]
PE 8814 — common symbols ; Lack
PF 1729 — contemporary life styles ; Reduced social
PD 3954 — credit financing ; Over
PD 8047 Commitment developing countries ; Lack work
PF 1729 Commitment ethic ; Loss
Commitment Lack
PF 1729 — [Commitment ; Lack]
PF 1729 — individual social
PD 2790 — work
PD 7051 Commitment learning ; Societal over
Commitment [Minimal...]
PD 2790 — Minimal vocational
PD 2790 — Mistrust corporate
PF 9564 — multilateralism ; Reduced government
Commitment [Progressive...]
PF 5502 — Progressive reduction government action
PF 4662 — protection poor ; Lack
PF 4662 — protection vulnerable groups ; Lack
PE 2932 — psychiatric hospitals ; Unjust
PF 9729 Commitment ; Short lived cooperative
PF 1729 Commitment social vision ; Shallow personal
PF 8855 Commitment welfare dependency ; Over
PF 1179 Commitments ; Arrears payment government financial
PF 7838 Commitments ; Conditional observance multilaterally agreed trade
PF 3975 Commitments ; Delay implementation
PF 7838 Commitments ; Failure political will observe trade
PF 3913 Commitments ; Inter Governmental failure fulfil financial
PF 4114 Commitments ; Unrecognized future financial
PE 1553 Committed personnel international organizations programmes ; Alienation skilled
PF 2843 Committee system decision making ; Inadequacy
PD 2898 Commodities because rising living standards ; Inadequate demand primary
PF 3115 Commodities capitalism ; Conflicting roles
PD 1276 Commodities ; Competition synthetics primary
PE 0537 Commodities ; Dependence industrialized countries imports primary
Commodities developing countries
PD 2968 — Excessive income dependence primary
PE 4523 — Inadequate trade agricultural
PD 1554 — Lack processing industry primary
PD 2967 — Over production primary
PD 3042 — Underproduction primary
PD 1276 Commodities due technological change ; Reduction demand primary
PD 2968 Commodities ; Economic dependence developing countries export primary
PD 4500 Commodities exchange violations ; Securities
Commodities exported developing
PD 2968 — countries ; Continuing low level prices
PD 7682 — countries ; Development industrialized countries products substituting
PD 0425 — countries ; Processing developed countries
PD 4191 Commodities ; Fraudulent
PD 9637 Commodities futures markets ; Speculation
PD 0651 Commodities ; Hoarding primary
Commodities [Insider...]
PD 3917 — Insider trading
PD 2968 — Instability export trade developing countries producing primary
PD 0057 — International trade barriers primary
PC 1195 Commodities ; Long term shortage
PD 1465 Commodities ; Over production
PD 1276 Commodities ; Production synthetic substitutes primary
PD 4191 Commodities ; Sale non existent
PC 5842 *Commodities sectors ; Protectionism*
PC 0463 Commodities trade ; Instability primary
PD 3917 Commodities trading fraud
PF 2497 *Commodity agreements ; Ineffectiveness international*
PD 2892 Commodity diversification developing countries ; Lack horizontal
Commodity exchanges
PE 9440 — Discriminatory operation international
PD 4500 — related crimes
PE 9440 — Restrictions access developing countries
PD 2968 Commodity export earnings developing countries ; Decline growth
PF 1471 Commodity exports ; Declining share developing countries world
PE 8058 Commodity fetishism
PF 4870 Commodity futures trading ; Obstacles
PB 8891 Commodity goods ; Limited

Commodity markets
PJ 0385 — Lack arbitrating
PD 8647 — Manipulation
PE 4301 — *Price fixing*
PF 2497 *Commodity policies ; Inconsistency international*
PE 8309 Commodity price support ; Government protection national
Commodity prices
PD 2968 — Cyclic swings
PD 2968 — Decline developing country
PD 2968 — developing country earnings ; Instability
PD 2743 *Commodity production ; Inefficient animal*
PF 9432 Commodity production respond fall demand ; Failure
PD 2968 Commodity related shortfalls export earnings developing countries
Commodity [speculation...]
PD 9637 — speculation
PF 9432 — supply ; Inflexibility
PD 1465 — surplus ; Primary
PE 9220 Commodity taxes transaction goods nonfactor services ; Distorting effects
PE 2951 Commodity trade amongst developing countries ; Weakness primary
PE 3604 *Common bats ; Endangered species*
PE 3639 *Common chigger*
PE 2412 Common cold
Common [ethic...]
PF 1118 — ethic ; Lack
PF 6570 — *experience ; Insufficient*
PF 6570 — *experiences ; Unshared*
PF 1985 Common heritage ; Fragmented recognition
PE 3623 Common house mosquitoes
Common [land...]
PF 6146 — land ; Inadequate arrangement housing respect
PE 6171 — land urban environments ; Insufficient
PE 6461 — *liver fluke*
PE 1184 Common membership ; Inadequate coordination regional intergovernmental organizations
PE 1583 Common membership ; Jurisdictional conflict antagonism regional intergovernmental organizations
PF 1365 *Common poverty victimage*
PF 4516 Common property ; Lack incentive users care
PF 3364 Common religious witness ; Lack
Common [structure...]
PF 1064 — structure ethical decision ; Absence
PE 8814 — symbol ; Lack
PE 8814 — symbols ; Lack commitment
PE 0616 *Common use hypodermic needle*
PF 1118 Common values ; Collapse
PF 1985 Commonly disvalued heritage
PA 7365 Commonness *Boredom*
PA 6997 Commonness *Imperfection*
PA 5821 Commonness *Vulgarity*
PF 5653 Commonplaces ; Contradictory
PF 4516 Commons ; Tragedy
PC 3685 Communal ethnic violence ; Inter
PD 9162 Communal land ; Encroachment
PE 0835 Communal spirit village solidarity developing countries ; Decline
PF 7118 *Commune priority ; Low*
PD 0982 Communicable diseases
PC 2162 Communicate ; Denial right
PF 1267 *Communicating ; Indefinite options*
PF 2350 Communication channels ; Absence
PF 2350 Communication channels ; Lost
PA 6732 Communication*complex
Communication [decision...]
PF 2452 — decision makers grassroots ; Insufficient
PE 0533 — deficiency ; Nonverbal
PF 4453 — delays
PD 6804 — Discriminatory
PF 8565 Communication foreigners
Communication [gap...]
PF 0756 — gap ; Generation
PE 2192 — Gossip oriented
PD 9538 — governments ; Abusive monitoring
PD 3518 Communication handicap ; Elders'
Communication [Inadequate...]
PF 3661 — Inadequate management employee
PF 2350 — Inadequate structures
PE 1581 — industries ; Health risks workers transport, storage
PF 6052 — Ineffective
PF 1528 — ineptitude management
PF 0756 — Insufficient inter age
PF 2694 — Insufficient provision public services
PD 7773 — isolation
Communication [Lack...]
PF 0816 — Lack
PF 2350 — *Lack international*
PF 0581 — *Limited economic*
PF 1187 — Lost school
PF 2350 *Communication networks ; Chaotic*
PF 6197 Communication non proximate offices ; Inhibition
PF 1187 Communication ; Parent teacher non
PD 2762 Communication resources facilities ; Disparities distribution
Communication [services...]
PF 4297 — services least developed countries ; Inadequate development
PE 0533 — skills ; Ignorance nonverbal
PF 1528 — skills ; Lack management
PF 2350 — system ; Minimal
PF 0076 Communication visual imagery ; Prejudice against
PF 7554 Communication ; Vulnerability marine animal

Communications [Deliberate...]
PD 1244 — Deliberate interference satellite
PF 6072 — developing countries ; Obstacles satellite
PE 6650 — due difference training ; Break down
PE 6974 Communications enemy ; War time
Communications [industry...]
PE 8640 — industry ; Control transnationals global
PD 5183 — information ; Misuse classified
PD 7608 — Interception
PD 2045 — Interference radio television
PD 1244 Communications ; Jamming satellite
PF 0816 Communications ; Lack effective
Communications [mass...]
PF 2573 — mass media ; Dangers private control
PD 4597 — mass media ; State control
PF 2350 — *methods ; Poor*
PF 8382 — Misunderstanding official
PD 9538 — Misuse government surveillance
PF 6470 Communications networks rural areas ; Poor
PF 2350 *Communications ; Poor*
PF 1528 Communications ; Poor managerial
PF 6496 Communications rural villages ; Inadequate systems transport
Communications [satellites...]
PF 0054 — satellites ; Interference
PF 7050 — societal learners ; Mis
PF 2350 — systems ; Insufficient
PE 7637 Communications technology ; Unbalanced application
PF 0953 Communications ; Thwarted technological
PF 3661 Communications ; Unsound corporate
PD 0407 Communications ; Vulnerability world cable
PF 9195 Communion dead
PF 3364 Communion ; Lack ecclesial
PC 0369 Communism
PF 1826 Communism ; Anti
PF 5233 Communism ; Fear
PF 3181 Communism ; Ideological schism
PF 3130 Communism ; National
PC 3163 Communism ; Revolution
PC 3163 Communism ; Revolutionary
PF 9485 Communism ; Subversion
PD 3169 Communist closed society
PE 5671 Communist countries ; Military conflict
PC 3165 Communist economic imperialism
PD 5393 Communist environmental insensitivity
PE 8349 Communist neutral countries ; Capitalist subversion
PF 3162 Communist opposition international organizations
PD 0923 Communist parties ; Fragmentation
PC 3164 Communist political imperialism
PD 5393 Communist regimes ; Environmental destruction
PD 8785 Communist repression
Communist [subversion...]
PD 3180 — subversion capitalist neutral countries
PD 3172 — systems ; Censorship
PF 3179 — systems ; Contradictions
Communist systems Denial
PC 3176 — democracy
PC 3174 — freedom expression thought
PC 3173 — freedom movement
PC 3178 — human rights
PC 3175 — religion
PC 3177 — right national self determination
Communist systems [Economic...]
PC 3167 — Economic competition
PC 3170 — Elitism
PD 1330 — Military industrial governmental complex
PD 3171 — Political prisoners
PF 6250 Communities automobiles ; Fragmentation
Communities [Closed...]
PF 2508 — *Closed*
PF 2444 — Complex regulations paralyzing small
PE 3296 — Cumulative depletion corporate initiative rural
Communities [Debilitating...]
PD 2011 — Debilitating conditions health rural
PD 9504 — Decay rural
PE 2524 — Deepening debt cycle rural
PF 2986 — Deskilling rural
PE 6526 — developing countries ; Inadequate transportation facilities rural
PD 1204 — developing countries ; Inadequate water supply rural
PF 6477 — Diminishing capital investment small
PF 2875 Communities ; Inaccessibility essential services isolated
Communities Inadequate
PE 1139 — *income rural*
PF 1442 — management skills rural
PF 6511 — residential housing rural
Communities [industrialized...]
PD 2495 — industrialized countries ; Personal isolation
PF 6127 — Ineffectiveness individual participation large
PF 2385 — Inflated prices rural
PF 4132 — Insufficient telephones rural
PE 1139 — *Insufficient trade income village*
PF 3575 — Irregular outside instruction rural
Communities Lack
PF 2540 — central planning structures small
PE 2425 — finance coastal
PE 8103 — housing teachers rural
PF 2837 — opportunities practical training
PF 1944 — relevant training opportunities local
PF 6540 Communities larger towns ; Transfer business small
Communities Limited
PF 1658 — availability jobs small
PE 5170 — availability practical skills rural
PF 6547 — control environment local
PF 2949 — employment opportunities small rural
PF 3575 — opportunity ongoing education rural

Communities Limited cont'd
PF 2848 — reservoir technical skills rural
PF 6478 *Communities ; Low fishing income rural*
PE 2305 Communities ; Negative consequences shifting ecology coastal
Communities [Outmoded...]
PF 2986 — Outmoded functional skills rural
PE 4780 — outside services ; Excessive dependence local
PF 2471 — Overlooked potential industrial development rural
Communities [Patronizing...]
PF 1345 — *Patronizing*
PF 1815 — Poor condition open spaces urban
PF 6527 — Prohibitive cost essential services rural
PF 2385 — Prohibitive cost necessities rural
Communities Restricted
PE 5170 — availability urban skills local
PF 1667 — delivery essential services developing country rural
PF 2875 — delivery essential services isolated
PD 1415 Communities ; Restrictive pattern business activities small
PF 2385 Communities ; Rising food prices rural
Communities [Shallow...]
PF 2848 — Shallow reserve technical skills rural
PF 9023 — Social isolation mountain valley
PF 3665 — Strained capital resources small
PE 8171 — Subsistence agricultural income level rural
Communities [take...]
PE 8509 — take advantage training business, industry public service ; Inability
PE 8014 — Threat industrial combines farming
PE 1137 — Traumatic shift life styles mining
Communities [Uncreditworthiness...]
PD 3054 — *Uncreditworthiness rural*
PF 6513 — Underutilization potential local
PF 1944 — Unperceived relevance formal education rural
PE 1101 — Unsystematic use powerful relationships rural
PF 7118 Community achievement ; Obstacles
Community action
PF 6501 — Blocked
PF 6501 — Ineffective organization
PF 1735 — Reluctance join
Community [activities...]
PF 2316 — activities ; Insufficient opportunities
PD 3386 — activity ; Enforced participation
PF 6504 — activity ; Fragmented patterns
PF 6536 — affairs ; Lack responsible involvement
PF 6557 — assets ; Minimum promotion
PF 2648 — *attitudes ; Anti*
PF 7925 — Avoidance human rights issues international
Community [boundaries...]
PE 7903 — boundaries ; Unresolved
PF 6502 — building programmes ; Delays
PJ 8591 — *buildings ; Prohibitive construction cost rural*
PF 7118 — *business ; Unpromoted*
Community [care...]
PE 8924 — care handicapped persons ; Inadequate
PF 1844 — care transient urban populations ; Inadequate
PF 5250 — *celebration ; Collapsed*
PF 5250 — *celebrations ; Insufficient*
PF 6559 — *centre ; Unlocated*
PF 7118 — *centres ; Unavailability*
PF 2575 — *cliques ; Operating*
PD 2864 — cohesion ; Breakdown local
PD 2864 — cohesion ; Limited
PF 9066 — confidence ability change ; Declining
PF 6506 — consensus ; Unformed
PF 8695 — contact ; Minimal
PF 3558 — cooperation ; Deteriorating structures rural
PF 3558 — cooperation ; Undeveloped
PF 2358 — corporateness rural villages ; Unformed confidence
PF 2874 — *crime ; Excessive*
PF 5250 — cultural events ; Infrequent
Community [damage...]
PA 0832 — *damage*
PF 1781 — decision making ; Ineffective structures
PF 3454 — decision making ; Restrictive effects traditional
PF 2863 — decisions ; Incomplete implementation
PF 8076 — *decline ; Continuing expectation*
PF 2575 — Declining sense
PF 1365 — *demoralization*
Community development
PF 7912 — Lack
PF 7912 — Lack
PF 3560 — Limits participation
PF 1681 — *scepticism*
Community [Dispersed...]
PF 7118 — *Dispersed residential*
PE 8963 — Distrust business
PE 8303 — doctors ; Unavailability
Community education
PF 6512 — Neglected
PF 6512 — Partial
PF 6512 — Uncomprehensive
PF 6512 — Unexplored opportunities
Community [enthusiasm...]
PF 2575 — *enthusiasm ; Thwarted*
PF 1790 — environment ; Poor organization
PF 5250 — events ; Insufficient
PF 3005 — *expansion ; Restricted*
PF 1681 — *expertise drain*
Community [facilities...]
PF 6141 — facilities ; Excessive dispersion
PF 7118 — *factionalism ; Widespread*
PF 6575 — future ; Detrimental story
Community [gatherings...]
PF 7118 — *gatherings ; Unstructured*

Community goals

Community [goals...] cont'd
- PF 7118 — *goals ; Fragmented*
- PF 6545 — groupings ; Divisive patterns
- PF 2575 — groupings ; Insular patterns
- PF 7118 — *groups ; Non cooperative*
- PF 7118 — *growth ; Sporadic*

Community [health...]
- PF 2971 — health ; Disrupted mechanisms
- PF 1608 — health ; Outdated forms
- PF 6494 — *history ; Disrelationship*
- PF 6494 — *history ; Obscured*
- PF 2575 *Community identification ; Indistinct*
- PF 2575 Community identification ; Limited

Community identity
- PF 1681 — Demoralizing image urban
- PF 2358 — Demoralizing images rural
- PF 3512 — Depreciated
- PF 2241 — Deteriorating
- PF 2845 — Disjointed patterns
- PF 2845 — Displaced
- PF 2845 — Fragmented
- PF 2845 — Fragmented images
- PF 6537 — Stagnated images

Community image
- PF 2093 — Defeating
- PF 2093 — Depressed
- PF 2093 — powerlessness
- PF 2093 — Reduced
- PF 2093 — Restraining
- PF 2093 — Unattractive

Community [images...]
- PF 2093 — images ; Lack significant
- PF 7118 — *initiatives ; Fragmented*
- PF 6143 — initiatives ; Inadequate facilities grass roots
- PD 9044 — insecurity ; Prevailing
- PD 1566 — *interaction ; Fragmented*
- PF 2575 — involvement ; Infrequent
- PF 2575 — involvement ; Minimal
- PE 8518 Community ; Lack social accounting business
- PE 0742 Community leaders ; Drug abuse

Community leadership
- PF 6556 — Obscure
- PF 2451 — training ; Lack local services
- PF 6556 — Undefined
- PF 6479 — Underdeveloped
- PF 2303 *Community leverage ; Ineffective*

Community life
- PF 3531 — Absence traditional patterns
- PF 2575 — Declining
- PF 2813 — Disorganized
- PF 2813 — Fragmented planning

Community [lifestyle...]
- PF 2241 — *lifestyle ; Isolating*
- PF 6521 — limits ; Geographically undefined
- PF 2421 — lobbying ; Sporadic
- PF 4470 Community market ; Inadequate rural

Community meetings
- PF 1333 — Argumentative
- PF 5250 — Infrequent
- PF 5222 — Unannounced
- PF 5222 — Unknown
- PF 6570 *Community memory ; Vague*
- PF 1365 *Community morale ; Low*

Community [networks...]
- PF 1681 — *networks ; Insufficient*
- PF 1959 — networks urban ghettos ; Ineffective operation
- PF 7998 — *news ; Unpublicized*

Community [operation...]
- PF 6492 — operation ; Absence support systems effective
- PF 1205 — operations ; Fragmented conduct
- PF 6559 — operations ; Unfocused style

Community organization
- PF 6525 — Fragmented pattern
- PF 6501 — Ineffective
- PD 8832 — Minimal
- PF 2810 — Unformed structures

Community participation
- PF 3307 — Absence
- PF 3307 — Discouraged
- PF 3307 — Factions frustrations
- PF 2437 — Ineffective structures
- PF 3307 — Lack
- PF 0963 *Community patterns ; Fluid*
- PF 2845 Community patterns ; Unconsolidated

Community planning
- PF 0963 — Inadequate procedures
- PF 2605 — Lack
- PF 6456 — Self defeating style

Community [Police...]
- PF 8125 — Police indifference
- PF 6441 — *population ; Declining*
- PF 7118 — *possibilities ; Unimaginative vision*
- PF 2093 — power ; Undeveloped
- PF 7118 — *priorities ; Conflicting*
- PE 1702 — priorities ; Monopolization interest groups development
- PF 7118 — *problems ; Overemphasis*
- PF 6557 — promotion ; Insufficient
- PF 1600 — property ; Insufficient care
- PF 3005 — property ; Uncoordinated use

Community [recreation...]
- PF 5409 — recreation ; Unorganized
- PF 2575 — reputation ; Negative
- PF 3448 — *resources ; Underdeveloped benefits*

Community responsibility
- PF 1731 — adults ; Limited

Community responsibility cont'd
- PD 1544 — *Lack*
- PD 1544 — *Unclear*
- PF 1651 Community roles ; Unexamined
- PE 6561 Community's culture ; Exclusion
- PF 6575 Community's future ; Frozen vision

Community [savings...]
- PD 0465 — savings ; Uninvested
- PD 1147 — security ; Inadequate
- PD 1147 — security systems ; Breakdown

Community self
- PF 2093 — image ; Demeaning
- PF 1681 — story ; Demoralizing
- PF 3512 — worth ; Lack

Community services
- PD 8832 — Minimal
- PF 1176 — Overlooked
- PG 8866 — Unusable

Community [size...]
- PF 2605 — size ; Restrictive
- PF 3393 — skills ; Lack sharing
- PF 6538 — skills ; Uncatalogued
- PF 3083 — social services ; Inflexible attitudes toward

Community solidarity
- PE 4179 — amongst workers ; Inadequate sense
- PD 0110 — developing countries ; Lack sense
- PF 8704 — world level ; Lack sense

Community space
- PF 1815 — Defaced
- PF 2822 — Haphazard organization
- PF 2346 — Inadequate care
- PF 2346 — Neglect
- PF 7130 — Obstacles availability
- PF 1546 — Undeveloped
- PF 1546 — Unfocused design
- PF 6521 — Unidentified
- PF 2519 — Unplanned use

Community [spending...]
- PF 6559 — *spending ; Unchecked*
- PF 6559 — *spending ; Unfocused*
- PF 7118 — *spirit ; Divided*
- PF 7118 — *spirit ; Flagging*
- PE 5408 — status quo ; Family adaptation

Community story
- PF 1681 — Defeating
- PF 2845 — *Ineffective*
- PF 6575 — Negative
- PF 6450 Community structures ; Untransposed
- PF 1681 *Community symbols ; Faded*

Community [Uncohesive...]
- PF 1176 — *Uncohesive business*
- PF 7047 — Underemphasized importance rural
- PF 6575 — *uniqueness ; Neglected*

Community [victim...]
- PF 2093 — victim image
- PF 7118 — *visibility ; Low*
- PF 2261 — *voice ; Lack*
- PF 2426 Community wisdom ; Untapped

Community youth
- PD 1544 — displacement
- PD 1544 — Isolated
- PD 1544 — Unengaged
- PE 1538 Commuter travel ; Exhausting
- PD 1416 Compaction ; Soil
- PD 6417 Companies ; Discrimination against foreign
- PE 3149 Companies ; Evasion social costs
- PE 5021 Companies ; Fraud concerning economic situation corporate capital
- PF 0326 Companies ; Paper
- PD 1265 Companion animals ; Maltreatment
- PE 9251 Company directors ; Abdication control
- PF 8759 Company ; Keeping bad
- PF 4432 Company ownership ; Protection
- PE 5074 Company polity ; Worker participation excluded
- PF 2622 Comparability international statistics ; Lack
- PF 0214 Comparable social data ; Lack
- PF 1651 Compartmental role tasks
- PF 2819 Compassion fatigue
- PF 4833 Compassion syndrome
- PF 1723 Compassion ; Unexpressed social
- PB 5486 *Compensation ; Arbitrary evaluation disability*
- PE 7176 Compensation assistance government matters ; Unlawful

Compensation damages
- PE 0290 — consumers ; Non payment
- PD 7179 — Delay payment
- PD 7179 — Lack
- PD 7179 — Non payment
- PE 4446 Compensation ; Delay governmental payment
- PE 8898 Compensation forced relocation ; Inadequate
- PE 8898 Compensation forced relocation ; Non payment
- PE 0290 Compensation ; Inadequate consumer

Compensation victims
- PE 5229 — catastrophes ; Non payment
- PE 3913 — crime ; Delay payment
- PE 3913 — crime ; Non payment
- PE 5229 — disasters ; Delay payment
- PE 5229 — major accidents ; Delay payment
- PE 0811 — malpractice ; Delay payment
- PE 0811 — malpractice ; Non payment
- PE 8824 — motor accidents ; Delay
- PE 8824 — motor accidents ; Non payment
- PJ 4891 Compensationism
- PF 0980 Competence intervention future ; Monopoly
- PE 3916 Competence juries ; Trials exceeding
- PF 3388 Competing ideologies

- PF 9576 Competing models socio-economic development
- PB 0848 Competition
- PF 0648 Competition aid funding
- PE 4266 Competition ; Athletic

Competition [capitalist...]
- PC 3125 — capitalist systems
- PE 4833 — commercial political ends ; Exploitation athletic
- PC 3167 — communist systems ; Economic
- PF 1442 — *Confusing government*
- PF 5506 — convict made goods ; Unfair
- PF 3307 — *Culture denying*
- PB 0848 Competition ; Dependence
- PE 2176 Competition due entrance barriers ; Decline

Competition [Economic...]
- PD 8897 — Economic
- PE 4242 — Exclusion women athletic
- PD 3067 — Export credit
- PF 1471 — export markets encountered developing countries ; Increase

Competition [industry...]
- PE 5931 — industry agriculture land
- PE 7433 — industry agriculture water
- PE 0063 — Inter union
- PE 0063 — intergovernmental organizations scarce resources
- PE 0259 — international nongovernmental organizations scarce resources
- PC 1463 — international organizations scarce resources
- PF 9275 — Interspecific
- PE 0580 — Intranational
- PF 9201 — Intraspecific
- PC 3355 Competition missionary bodies
- PC 4412 Competition non-renewable resources
- PD 8992 Competition ; Political
- PF 4095 Competition ; Protection public sector agencies
- PF 4741 Competition ; Restriction technical basis

Competition [scarce...]
- PC 4412 — scarce resources
- PF 2216 — Social agency
- PC 4412 — sources energy
- PF 9201 — species
- PF 9275 — species
- PC 0114 — states
- PD 0014 — Strategic arms
- PD 1276 — synthetics primary commodities
- PE 0051 Competition transnational corporations ; Restriction free market
- PC 0099 Competition ; Unfair
- PE 5506 Competition workers employed forced labour ; Unfair
- PE 3754 Competitions ; Bribery connection sports athletic
- PE 3754 Competitions ; Corruption sports athletic
- PF 4192 Competitions ; Excessive expense international athletic
- PE 4576 Competitions ; Gambling sports athletic
- PF 4216 Competitions ; National bias judges international athletic
- PF 4761 Competitions ; Politicization international sports
- PC 1258 Competitive acquisition arms
- PD 4178 Competitive burden students
- PD 8916 Competitive business wages ; Non
- PF 2411 *Competitive consumer services ; Non*
- PC 0012 Competitive development new weapons
- PF 8713 Competitive distortion musical pitch
- PF 2874 *Competitive economic interests*
- PE 8583 Competitive exchange depreciation
- PF 0905 *Competitive markets ; Overpowering*
- PC 1258 Competitive militarization
- PE 7313 *Competitive products ; Non*
- PE 4266 *Competitive sports ; Over*
- PE 8645 Competitive teachers' salaries ; Non
- PB 0848 Competitiveness

Competitiveness [Declining...]
- PD 8994 — Declining industrial
- PD 8994 — Declining international
- PD 8994 — domestic industries ; Decline
- PD 4178 Competitiveness education
- PD 8994 Competitiveness ; Fluctuations international
- PD 4982 Competitors ; Fabrication reports corporate
- PA 1742 Complacency
- PA 1742 Complacency ; Citizen
- PA 1742 Complacency ; Dependence
- PA 6011 Complacency *Discontentment*
- PF 6407 Complacency ; Government
- PF 1951 Complacency ; Religious
- PD 9848 Complacency science
- PA 6491 Complacency *Vanity*
- PA 1742 Complacent people
- PF 8029 Complaint
- PD 7609 Complaint ; Denial right
- PD 7567 Complaints ; Consumer
- PF 9157 Complaints ; Government avoidance
- PF 2346 *Complaints ; Insufficient formal*
- PF 9157 Complaints ; Official evasion
- PD 5586 Complaints workers
- PE 7255 Complementarity developing countries economies ; Low regional
- PE 8184 Complementarity developing country economies ; Low
- PD 0520 *Completion sentence ; Detention offenders after*
- PF 2451 *Complex agricultural practices*
- PD 5111 Complex ; ARC AIDS related

Complex [banking...]
- PF 2009 — *banking practices*
- PE 4586 — *Beech bark disease*
- PF 8053 — *bureaucratic procedures*
- PF 8053 — *business regulations*

Complex [capitalist...]
- PC 3191 — capitalist systems ; Military industrial
- PC 0300 — *Castration*

Conditions construction

Complex [communist...] cont'd
PD 1330 — communist systems ; Military industrial governmental
PF 2837 — *curriculum writing*
PF 2454 Complex difficult laws
PF 2444 *Complex education regulations*
PF 0799 Complex environmental problems ; Simplistic technical solutions
Complex equipment
PE 5572 — Human fatigue during control
PF 6411 — Information overload during control
PE 0680 — Substance abuse during control
PE 9854 Complex ; *Feline respiratory disease*
PF 6505 Complex funding mechanisms
Complex government
PF 8539 — bureaucracy
PF 8053 — regulations
PF 8053 — rules
PA 6793 Complex ; Guilt
PD 4790 Complex health delivery
PE 8443 Complex housing regulations
PA 0587 Complex ; *Inferiority*
PF 0364 Complex interrelationship world problems
Complex [land...]
PF 8053 — land procurement
PF 6528 — land use
PF 8519 — legal procedures
PF 2491 — loan procedure
PC 1952 Complex ; Military industrial
PD 3978 Complex ; *Mucosal disease*
PF 2444 Complex regulations paralyzing small communities
Complex [services...]
PF 1653 — *services logistics*
PB 1151 — *society ; Instability*
PE 8533 — system behaviour ; Excessive dependence computer models
PE 1396 — systems ; Wastage highly skilled personnel routine maintenance
Complex [tax...]
PF 2444 — *tax regulations*
PF 4722 — *trade regulations*
PE 3916 — *trials*
PF 6174 Complexes ; Disorientation stress large building
PA 6468 Complexity
PF 8100 Complexity ; Confusing structural
PF 2969 *Complexity government assistance*
Complexity [intergovernmental...]
PF 2806 — intergovernmental organizations ; Excessive
PF 2384 — *Intolerance*
PF 1735 — *issues ; Perceived*
PF 8053 *Complexity land procedures*
PF 2454 Complexity ; Law enforcement
Complexity [Unintelligibility...]
PA 7367 — *Unintelligibility*
PE 0296 — United Nations system ; Fragmentation
PF 1776 — urban structures ; Paralyzing
PA 5710 Compliance∗complex
PD 1466 Compliance fiscal laws ; Lack
PF 0423 *Compliance law ; Non*
Compliance [Non...]
PF 6310 — Non verification
PF 4460 — nuclear arms treaties ; Non verifiability
PF 4455 — nuclear power safeguards ; Non verifiability
PE 7252 Compliant confession ; Coerced
PE 8443 Complicated building standards
PF 8519 Complicated legal procedures
PF 6218 Complicated spelling
PA 6468 Complication *Complexity*
PA 5497 Complication *Difficulty*
PA 6799 Complication *Disease*
PA 7367 Complication *Unintelligibility*
PC 9042 Complications childbirth
PC 9042 Complications labour childbirth
PE 2863 Complications ; Medical
PD 2289 Complications pregnancy
PC 9042 Complications puerperium
PE 2863 Complications ; Surgical
PF 4983 Complicity
PF 4983 Complicity crime
PF 7730 Complicity drug trafficking ; Government
PF 0926 Complicity evil
PA 6793 Complicity *Guilt*
PF 7730 Complicity illegal activities ; Government
PF 7730 Complicity illegal arms trade ; Government
PF 3819 Complicity killing human rights activists ; Government
PA 6979 Complicity *Opposition*
PF 4983 *Complicity structural injustice*
PE 0669 Components utilization ; Restrictions raw materials
PC 1306 Compound disease
PE 1485 Compounding crime
PF 7727 Compounding effect treating genetic diseases
PD 1271 *Compounds ; Gaseous organic*
PE 0500 Compounds ; Instability trade chemical elements
Compounds pollutants
PE 4483 — Halogen
PD 2965 — Nitrogen
PD 0983 — *Organochlorine*
PD 1670 — *Sulphur*
PE 0617 Compounds water pollutants ; Toxic organic
PF 7134 Comprehension international information ; Pre logical limitations
PF 7110 Comprehension societal learning ; Superficiality collective
PF 7038 Comprehension span societal learning ; Limit collective
PE 1122 Comprehensive document collections ; Prohibitive cost maintaining

PF 0944 Comprehensive framework understanding knowledge ; Lack
PE 1122 Comprehensive information systems ; Excessive cost
PF 6337 Comprehensive information ; Unavailability
PF 1133 Comprehensive wage scales ; Non
PE 6861 Compression developing countries ; Import
PA 7218 Compromise
PF 3420 Compromise betrayal principles
PF 9747 Compromise ; Inability
PF 4983 *Compromising silence*
PA 5740 Compulsion
PA 6448 Compulsion *Obsession*
PF 7490 Compulsion violence ; Neurophysiological
PA 6448 Compulsive disorder ; Obsessive
PE 6069 Compulsive early dismissal
PF 7632 Compulsive personality disorder ; Obsessive
PA 7157 Compulsiveness *Insanity*
PC 1005 Compulsory acquisition land government
PC 6848 *Compulsory education*
PF 4820 Compulsory health care
PD 3097 Compulsory indoctrination
PE 6342 Compulsory labour ; Denial right freedom
PE 6051 Compulsory military service
PF 2660 Compulsory nudity
PD 4098 Compulsory organization membership
PF 4936 Compulsory prolongation life
PB 0477 *Compulsory retirement*
PJ 0456 Compulsory seizure property
PF 3240 Compulsory sterilization
PE 0223 Compulsory telecommunications ; Invasion privacy
PF 1916 Compulsory vaccination
PC 2494 *Compulsory voting*
PF 1288 Computer addiction
PE 4362 Computer-aided fraud
Computer based
PE 4362 — crime
PE 1277 — recreation ; Social isolation
PE 1277 — work ; Atomization
PE 4428 Computer bugs
Computer [cloning...]
PE 6625 — cloning
PF 2685 — *control*
PE 8542 — crimes ; Inadequate safeguards against
Computer [data...]
PF 4176 — data ; Uncontrolled use
PE 9831 — database espionage
PF 4176 — databases ; Erosion human rights privacy use
PD 9544 — databases ; Misuse
PC 0840 — *development trade war*
PE 8542 — disasters ; Inadequate prevention
PE 4428 Computer error ; Risk
PE 4362 Computer fraud ; Risk
PE 9831 Computer hacking
PC 0210 *Computer illiteracy*
PE 9831 Computer information systems ; Unauthorized access
PE 8533 Computer models complex system behaviour ; Excessive dependence
PF 1288 Computer obsession
PD 7001 Computer office machine industries ; Protectionism
PE 6625 Computer piracy
PE 4428 Computer processing ; Unreliability
PE 8542 Computer records ; Risks computers
PD 7001 Computer services industry ; Protectionism
Computer software
PD 7450 — design ; Biased
PE 4428 — errors ; Hazardous
PE 4428 — Unreliability
Computer [stress...]
PE 5053 — stress
PD 9544 — systems ; Abuse
PE 8542 — systems ; Vulnerability
PE 1277 Computer users ; Social isolation
Computer [virus...]
PD 3102 — virus warfare
PD 3102 — viruses
PE 5083 — visual display units ; Health hazards
PE 5074 Computerization ; Disruption work schedule due
PE 5014 Computerization ; Downgrading jobs due
PD 4552 Computerization ; Paper proliferation due
PD 9330 Computerized information displays ; Biased
PF 4176 Computerized information ; Unauthorized disclosure
PF 4176 Computerized personal data ; Errors
PE 8721 Computers automation ; Uncoordinated use
PE 8542 Computers computer records ; Risks
PE 3959 Computers ; Overutilization personal
PE 3959 Computers ; Proliferation
PE 3959 Computers universal panacea ; Overdependence
PF 2491 Computing methods ; Confusing loan
PF 2152 Conceal nefarious benefits ; Misuse plausible arguments
PC 8019 *Conceal socially unacceptable initiatives ; Misuse societally endorsed professions*
PD 4532 Concealed government subsidies
PE 1825 Concealed needs ; Elders'
PE 1193 Concealed transfer pricing ; Profit repatriation
PF 8000 Concealment ; Criminal
PF 7077 Concealment esoteric knowledge
Concealment information
PF 9828 — concerning extent homelessness
PF 9828 — extent poverty
PF 9828 — extent unemployment
PA 0005 Concealment ; Intentional
PA 6400 Conceit *Affectation*
PA 6659 Conceit *Humility*
PA 6491 Conceit *Vanity*
PD 0024 Conceived incentives employment ; Ill

PE 9172 Concensus psychiatrists ; Lack
PE 3716 Concentrate ; Inability
PF 2384 Concentrate ; Inability
PE 8131 Concentrates non ferrous base metals ; Instability trade ores
Concentrates shortage
PE 0353 — *Iron ore*
PE 0353 — *Thorium ores*
PE 0353 — *Uranium ores*
PE 8212 Concentrates thorium trade instability ; Uranium ores
PE 8772 Concentrates trade instability ; Iron ore
PE 9446 Concentration atmospheric water vapour ; Increase
PD 0071 Concentration business enterprises ; Excessive
Concentration camps
PD 0702 — [Concentration camps]
PD 0702 — *Military*
PD 0702 — *Political*
PE 7975 Concentration capitalist banks
PE 9457 Concentration export markets developing countries ; Excessive
PC 5323 Concentration investment power
PE 8815 Concentration methane ; Increase atmospheric
PE 5132 Concentration national governments activity national affairs
PE 8154 Concentration noxious substances food chains
PE 5825 Concentration ownership maritime fleets ; Over
PE 0766 Concentration power transnational corporations
PE 0246 Concentration press ownership
PC 8885 Concentration science technology
PF 6506 *Concentrations power ; Uneven*
PE 5826 Concept peace ; Collapsed
PF 4944 Conception development programmes ; Inadequate
PF 1178 Conception ethical void ; Self
Conception [Ignorance...]
PD 7994 — Ignorance cause
PE 8634 — In vitro
PE 8634 — Inhuman methods
PE 8634 Conception ; Test tube
PF 6665 Conceptions time ; Differing
PF 4210 Concepts corrupt practices ; Divergent national
PF 5242 Concepts ; Facile social
PF 5242 Concepts ; False concreteness societal
PF 7673 Concepts ; Institutionalization discriminatory outmoded
PF 2580 Concepts legal systems ; Institutionalization discriminatory outmoded
PF 2580 Concepts legal systems ; Persistence outmoded
PF 5247 Concepts ; Misappropriation
PF 7673 Concepts ; Persistence outmoded
PF 5242 Concepts ; Popularization
PC 7605 Conceptual asset stripping
PF 7014 Conceptual integration societal learning ; Obstacles
PD 1284 Conceptual plagiarism
PJ 0253 Conceptual proliferation
PF 5210 Conceptual repression problems
PF 5242 Conceptualization ; Collapsed
PF 5155 Conceptualization equality ; Collapsed
PF 7497 Conceptualization liberty ; Collapsed
PF 0959 Conceptualization love ; Collapsed
PD 8837 Concern animal welfare ; Lack
PF 1817 Concern ; Inadequate means upholding global
PF 5937 Concern ; Undue public
PF 2158 Concerned attitudes ; Non
PD 2772 Concerns ; Excessive commercial exploitation farm animals industrial
PF 2076 Concerns ; Overriding international local
PF 8876 Concerns ; Parliamentary inability respond minority
PF 2256 *Concerns ; Past oriented*
PE 8230 Concerns ; Unwillingness divulge information industrial
PE 3812 Concessional financial resources available developing countries ; Decline
PE 3812 Concessional lending developing countries ; Erratic flows concessional non
PE 3812 Concessional non concessional lending developing countries ; Erratic flows
PE 6313 Concrete ; Creep
PE 1626 Concrete fatigue
PF 5242 Concreteness societal concepts ; False
PF 2554 Concubinage
PA 5612 Concupiscence *Unchastity*
PD 0992 Concussion
PA 7237 Condemnation
PA 7237 Condemnation *Condemnation*
PA 6191 Condemnation *Disapproval*
PB 1047 Condescension
PA 6659 Condescension *Humility*
PA 6491 Condescension *Vanity*
PD 1955 Condition art objects ; Deterioration physical
PD 5453 *Condition ; Dead good*
PF 1815 Condition open spaces urban communities ; Poor
PE 1032 Condition provisional release criminal cases ; Pecuniary guarantees
PE 3832 Conditional membership international organizations
PF 7838 Conditional observance multilaterally agreed trade commitments
PF 9216 Conditionality restrictions multilateral development aid ; Policy cross
PF 1807 Conditioned self-gratification
PD 1308 Conditioning ; Sex role
Conditions [Adverse...]
PC 0293 — Adverse weather
PD 9170 — Adverse working
PE 4243 — agricultural workers ; Inadequate working
PD 9366 — animals ; Ill defined health
PD 9156 Conditions ; Bad living
PF 5744 Conditions ; Cold fishing
PE 3969 Conditions construction industry ; Inadequate working

–879–

Conditions

Conditions [Declining...]
PF 1345 — *Declining economic*
PD 9170 — Denial right favourable work
PD 2672 — *Deteriorated building*
PD 1476 — developing countries ; Inadequate working
PE 1764 — Disagreeable bodily
PE 8251 Conditions employees commerce offices ; Inadequate working
PC 0387 Conditions ; Extreme climatic
Conditions [factor...]
PE 2764 — factor animal diseases ; Insanitary penning
PC 7873 — Fluctuations world monetary financial
PF 9697 — foreign trade ; Vulnerability adverse
Conditions [health...]
PE 7718 — health medical services ; Inadequate working
PD 2011 — health rural communities ; Debilitating
PD 2322 — High elevation
Conditions [Ill...]
PC 9067 — Ill defined health
PD 3427 — immigrant labourers industrialized countries ; Inadequate living working
PD 4368 — Impermanent living
PD 0520 — imprisonment ; Unacceptable
Conditions Inadequate
PD 0520 — gaol
PD 0520 — gaol
PF 8515 — public health
PE 6526 — *road*
PF 6683 Conditions ; Inauspicious
PF 2442 Conditions ; Inhumane dying
Conditions [land...]
PE 8633 — land tenants ; Inadequate working
PE 6270 — life liberty proper their species ; Denial animals right
PE 9116 — loans developing countries intergovernmental facilities ; Restrictive
PF 1815 *Conditions ; Overcrowded building*
Conditions [peasant...]
PE 4243 — peasant farmers ; Inadequate working
PD 0520 — penal systems ; Inadequate prison
PE 4243 — plantation workers ; Inadequate working
Conditions Poor
PD 9156 — living
PE 6526 — *road*
PD 9170 — working
PD 9090 Conditions pressure ; War time
PE 3170 Conditions professionals ; Inadequate working
PE 5603 Conditions small business ; Discouraging
PE 7165 Conditions teachers ; Inadequate working
Conditions [Unfavourable...]
PC 0387 — Unfavourable climatic
PF 3826 — unfavourable transition post capitalist social order ; Prevalence psychological
PF 8515 — Unhygienic
PF 8515 — Unsanitary
PE 0395 — Unsanitary inhumane urban food animal
PD 3197 Conditions women ; Inadequate working
Conditions work
PE 4243 — agricultural workers ; Inadequate
PE 6841 — construction industry ; Inadequate
PE 8728 — employment public service personnel ; Inadequate
PE 8319 — employment utility supply services ; Inadequate
PE 6811 — environment ; Cardiac
PE 4493 — hotel catering industries ; Inadequate
PE 5823 — textile industry ; Inadequate
PD 2592 Condottieri
PF 2687 Conduct ; Collapse standards sexual
PF 1205 Conduct community operations ; Fragmented
PE 3770 Conduct disorder
PD 9178 Conduct ; Disorderly
PD 7359 Conduct ; Hindering proceedings disorderly
PF 4550 Conduct parliament ; Indecent
PE 6086 Conduct public affairs ; Violation right participate
PE 0349 Confabulation ; Paramnesia
PE 8958 Confectionery preparations trade instability ; Sugar
PE 4071 Confederations ; Violation right workers' organizations establish
PE 7252 Confession ; Coerced compliant
PE 8947 Confession ; Forced
PC 1455 Confessionalism
PE 4888 Confessions drugs ; Forced
PE 7252 Confessions ; False
PE 3016 Confessions ; Forced political
PE 7252 Confessions innocent ; False
PF 9066 Confidence ability change ; Declining community
PD 8217 Confidence administration justice ; Lack
PF 7633 Confidence ; Breach
PF 2358 Confidence community corporateness rural villages ; Unformed
PD 3035 Confidence ; Forced breach journalistic
PF 1097 Confidence government leaders ; Loss
PC 2089 Confidence governments ; Lack
PF 3058 Confidence international monetary system ; Lack
PE 5102 Confidence investment stock markets ; Low
Confidence [Lack...]
PA 8653 — Lack
PF 0879 — Lack self
PF 9066 — Limited village self
PF 4345 — loans developing countries ; Deterioration
PF 0879 — Low self
PF 8076 Confidence ; Minimal citizen
Confidence [parents...]
PF 2093 — *parents ; Lack*
PF 8559 — police ; Lack
PF 7166 — prediction capacity ; Excessive
PF 8076 Confidence ; Undermined social
PF 0879 Confidences ; Lack self

PD 2926 Confidential government information ; Restrictions distribution
PB 0284 *Confidential information ; Unauthorized circulation*
PF 7633 Confidentiality ; Abuse
PE 6786 Confidentiality ; Abuse commercial
PD 6612 Confidentiality ; Denial right
Confined [scope...]
PF 2439 — scope business operations
PE 3547 — seasonal transportation
PC 6716 — *spaces*
PD 5142 Confinement
PE 5648 Confinement ; Death disability inhumane
PE 5763 Confinement ; Disease injury physical
PA 7078 Confinement *Environment*
PA 7256 Confinement *Narrowness*
PD 0634 *Confinement non-criminal reasons*
PA 5583 Confinement *Punishment*
PA 7296 Confinement *Restraint*
PE 4056 Confinement ; Solitary
PD 4590 Confinement ; Torture
PD 6081 Confiscation news media
PJ 0456 Confiscation property
PD 3012 Confiscation property ; Political
PE 7581 Confiscation trade union property
PF 0638 Conflagration ; Cosmic
PA 0298 Conflict
PF 9077 Conflict academic disciplines ; Jurisdictional
Conflict antagonism
PE 8308 — government agencies each country ; Jurisdictional
PE 7901 — intergovernmental organizations ; Jurisdictional
PE 9011 — intergovernmental organizations ; Jurisdictional
Conflict antagonism international
PE 0064 — nongovernmental organizations ; Jurisdictional
PE 1169 — nongovernmental organizations ; Jurisdictional
PE 7973 — organizations country level ; Jurisdictional
PD 0047 — organizations ; Jurisdictional
PD 0138 — organizations ; Jurisdictional
Conflict antagonism [regional...]
PE 1583 — regional intergovernmental organizations common membership ; Jurisdictional
PE 2486 — specialized agencies United Nations ; Jurisdictional
PE 8799 — specialized agencies United Nations ; Jurisdictional
PB 0593 Conflict ; Armed
Conflict [church...]
PC 0030 — church state
PC 1573 — Class
PE 5671 — communist countries ; Military
PF 3252 — concerning legal custody children
PC 1573 — *Cultural*
Conflict [Degradation...]
PC 6675 — Degradation environment armed
PA 0298 — Dependence
PB 5057 — Dependence international
PA 6698 — *Difference*
PA 5532 — *Disaccord*
PA 5982 — *Disagreement*
PC 1910 — due tribalism ; *Internal*
PF 0513 — duties
Conflict [Economic...]
PC 0840 — Economic
PA 5446 — *Enmity*
PF 7358 — envisaged destructive ; Verbal
PF 0466 — Erroneous precipitation nuclear
PC 3685 — Ethnic
PE 1643 Conflict government news media
Conflict [Ideological...]
PF 3388 — Ideological
PE 8361 — Inadequate protection civilians armed
PD 4012 — Industrial
PF 2002 — information
PC 3390 — Intellectual
Conflict interest
PE 8289 — governments groups governments
PE 3735 — parliamentarians
PE 3735 — peoples representatives
PF 9693 — pure applied science
Conflict [interests...]
PF 9610 — *interests business*
PC 6440 — interests imperialism
PB 5057 — International
PF 0216 — *international laws*
PF 3100 — International non military
PF 4968 — Intrapsychic
PC 1869 — Intrasocietal military
PD 3223 *Conflict ; Language*
Conflict laws
PF 0216 — [Conflict laws]
PD 3080 — international restriction information
PF 8953 — over nationality
PF 3988 Conflict ; Low intensity
PE 3988 Conflict ; Low level
Conflict [Maladjustment...]
PA 6739 — *Maladjustment*
PE 5087 — Mental
PB 0593 — Military
PC 3428 — minority groups
PF 0216 *Conflict national laws*
PA 6979 Conflict *Opposition*
Conflict [Political...]
PC 0368 — Political
PF 4968 — Psychic
PE 5087 — Psychological
Conflict [Racial...]
PC 3684 — Racial
PC 0030 — religion state
PC 3292 — Religious

Conflict [sects...]
PC 3363 — sects ; Religious
PC 0137 — Social
PA 0298 — *states ; Tribal*
PE 8174 Conflict ; Vulnerability children during armed
PF 1776 *Conflicting agency priorities*
PF 7118 *Conflicting citizen priorities*
Conflicting claims
PC 1888 — *concerning Antarctic territory*
PD 1628 — concerning off-shore territorial waters
PC 1888 — *non-terrestrial territory*
PD 7087 — over shared inland water resources
PC 2362 — states territories
PF 7118 *Conflicting community priorities*
PD 4900 Conflicting demands forests
PJ 4512 Conflicting economic goals
Conflicting [labour...]
PE 5135 — *labour laws*
PF 2365 — *land rights*
PE 6680 — laws concerning bills exchange promissory notes
PE 6678 Conflicting military obligations persons multiple nationality
PE 6677 Conflicting multiple nationalities
Conflicting national
PF 3249 — divorce laws
PE 6679 — laws governing cheques
PE 2200 — regulations transnational corporations ; Burden
PF 5766 Conflicting priorities
Conflicting roles
PF 3115 — commodities capitalism
PF 3114 — money capitalist systems
PD 6273 — women
Conflicting [sense...]
PF 1246 — sense sexual identity
PD 3190 — social service ideologies
PE 5651 — standards protection against chemical occupational hazards
PF 6504 — *stories ; Unfounded*
PF 2644 *Conflicting time schedules*
PF 6654 Conflicting use resources
PF 6012 Conflicting viewpoints specialists
PC 1454 Conflicts ; Denial human rights armed
Conflicts [integrated...]
PC 6242 — integrated rural development
PF 9610 — interest
PD 7661 — Internal armed
PF 4968 — Internalized
Conflicts [labour...]
PD 3533 — labour law
PE 8357 — Lack participation attempting solve intergroup
PF 0216 — law
PF 1115 — Liturgical
PD 0518 Conflicts marriage
PF 9571 Conflicts national law relation international transactions
PD 5361 Conflicts over fishing rights
PF 1333 Conflicts ; Polarization local
PC 5917 Conflicts ; Regional
PD 2191 Conflicts ; Tribal
PF 4967 Conflicts ; Unreported
PD 1078 Conflicts ; Vulnerability women during armed
PE 5239 Conform international health standards ; Failure
PC 4591 Conform internationally agreed standards ; Failure
PD 5535 *Conformational abnormalities eyelids animals*
PB 3407 Conformism
PB 3407 Conformism ; Dependence
PB 3407 Conformists
PF 0867 Conformity building design ; Over
PE 7217 Confront accusers ; Denial right
PF 1586 Confrontation death ; Avoidance
PF 3567 Confused government response disasters
PF 4756 Confusing decision-making methodologies
PF 1442 *Confusing government competition*
PF 2491 Confusing loan computing methods
PE 7903 Confusing property boundaries
PF 8100 Confusing structural complexity
PF 7123 Confusion
PF 1062 Confusion being detached past
PE 7144 Confusion concerning role United Nations ; Intellectual
Confusion [Defeat...]
PA 7289 — *Defeat*
PA 6858 — *Disintegration*
PA 7361 — *Disorder*
PA 6900 Confusion *Formlessness*
Confusion [Ideological...]
PD 0065 — Ideological
PA 6247 — *Inattention*
PF 6712 — induced rapid social change
PA 5563 Confusion *Lawlessness*
PD 3255 Confusion paternity
PF 5985 Confusion ; Semantic
PF 2780 Confusion symptoms animal diseases
PA 7309 Confusion *Uncertainty*
PA 7107 Confusion *Unpleasantness*
PE 6982 Confusion values monitoring television
PA 6491 Confusion *Vanity*
PF 5670 Confusion weights, measures numbering systems ; Terminological
PD 7841 *Congenita ; Myoclonia*
PD 1618 Congenital abnormalities
PE 2125 Congenital acromicria
Congenital anomalies
PD 1618 — [Congenital anomalies]
PD 1618 — *alimentary tract*
PE 7808 — animal musculoskeletal system
PE 5569 — cardiovascular system animals
PE 6924 — circulatory system

–880–

Consumers' loss

Congenital anomalies cont'd
PD 1618 — *digestive system*
PE 3711 — *digestive system animals*
PE 6419 — *due infectious diseases animals*
PE 7549 — *eye*
PE 4249 — *genital organs*
PE 8589 — *musculoskeletal system*
PE 9296 — *nervous system*
PD 7799 — *reproductive system animals*
PD 1618 — *respiratory system*
PD 9667 — *skin*
PE 8589 — *spine*
PD 9293 — *urinary system animals*
PE 7808 *Congenital articular rigidity ; Myopathy associated*
Congenital [deafness...]
PD 2389 — *deafness*
PD 1618 — *deformities*
PE 0965 — *diseases nervous system animals*
Congenital disorders
PD 1618 — [Congenital disorders]
PE 9291 — amino-acid metabolism
PE 4293 — carbohydrate metabolism
PE 9816 — lipid metabolism
PE 2264 *Congenital glaucoma*
Congenital [heart...]
PE 2365 — heart disease
PE 1221 — *hydrocephalus*
PG 5976 — *hypothyroidism*
PD 1618 Congenital malformation
PE 0763 *Congenital monoplegia*
PG 5561 *Congenital photosensitization sheep*
PD 7420 *Congenital porphyria erythropoietica*
PE 9324 Congenital syndromes affecting multiple systems
PE 2300 *Congenital syphilis*
Congenital [toxoplasmosis...]
PE 3659 — *toxoplasmosis*
PD 7841 — *trembles*
PD 7841 — *tremor syndrome ; Porcine*
PD 0689 Congested airspace
PD 0689 Congestion ; Air traffic
PD 0689 Congestion ; Airport
PA 7391 Congestion *Closure*
PA 5473 Congestion *Insufficiency*
Congestion [Lightness...]
PA 5491 — *Lightness*
PE 1028 — *Liver*
PD 5453 — *Lung*
PE 2293 — *lungs ; Acute*
PE 4766 Congestion ; Port
PE 4766 Congestion ; Port traffic
PD 0078 *Congestion ; Rail traffic*
PD 2106 Congestion ; Road highway traffic
PD 1486 Congestion ; Sea traffic
PD 0078 Congestion ; Traffic
PD 0426 Congestion ; Urban road traffic
PE 3958 Congestive heart failure
PD 2730 *Congo haemorrhagic fever ; Crimean*
PD 2730 *Congo virus disease*
PE 6461 *Conical flukes*
PE 4346 *Conifers ; Needle casts*
PD 6881 Conjugal violence
PD 5535 *Conjunctiva animals ; Diseases*
PE 7974 Conjunctivitis
PE 1946 Conjunctivitis ; Granular
PF 1081 Connecting individual's life cosmos ; Absence convicing symbols
PF 1653 *Connection electricity services ; Prohibitive cost*
PF 1653 *Connection ; Expensive water*
PF 1653 *Connection public utilities ; Prohibitive cost*
PE 3657 Connection telecommunications facilities ; Delay
PE 3657 Connection telephones ; Delay
Connective [tissue...]
PD 2565 — tissue ; Diseases
PE 9229 — tissue ; Malignant neoplasm
PD 2565 — tissues ; Diseases musculoskeletal
PD 2565 — tissues ; Diseases musculoskeletal system
PF 9288 Connivance authorities human rights abuses
PF 9288 Connivance human rights abuses ; Government
PF 9288 Connivance human rights abuses ; Official
PF 8046 Connivance regulatory inspectors violators regulations
PF 9288 Connivance religious leaders human rights abuses
PC 2379 Consanguineous marriage
PF 7076 Conscience ; Bias acts
PF 7612 Conscience ; Denial freedom
PF 9144 Conscience ; Lack social
PC 6935 Conscience ; Prisoners
Conscientious objection
PD 4738 — [Conscientious objection]
PE 7007 — factory
PD 1800 — military service ; Denial right
PD 4738 Conscientious objectors
PD 1800 *Conscientious objectors ; Internment*
PF 0420 *Conscientious refusal medical intervention ; Irrational*
PF 1781 *Conscious initiatives ; Lack*
PC 3458 Consciousness ; Class
PC 3458 Consciousness ; Collective
PD 8005 Consciousness ; Disorders
PF 1004 Consciousness due inadequate methods ; Inability awaken
PC 1777 Consciousness ; Repression self
PC 3458 Consciousness social barrier ; Class
PF 6051 Conscription
PD 6667 Conscription ; Extra legal
PD 0356 Conscription systems ; Weaknesses military
PF 9821 Consensus concerning problems prospects humanity ; Inadequate global

PF 6506 Consensus ; Difficult village
PF 9431 Consensus facts ; Lack
Consensus [Illusion...]
PF 4327 — Illusion
PF 6506 — Ineffective structures local
PD 2438 — Insufficient need
PF 7493 — *intergovernmental bodies ; Superficial*
PD 2438 Consensus ; Lack means achieving
PF 6506 Consensus ; Lack systems achieving grassroots
PD 2438 Consensus method ; Incomplete
Consensus [patterns...]
PD 2438 — patterns ; Unclear
PD 2438 — procedures ; Undetermined
PD 2438 — process ; Inadequate
PD 2438 Consensus structure ; Collapsed
PD 2438 Consensus structure ; Inadequate
PF 6506 Consensus ; Unclear village
PF 6506 Consensus ; Unformed community
PF 3282 Consent adoption ; Refusal parents'
PE 3828 Consent ; Denial right medical
PD 1915 Consent ; Marriage
PE 6522 Consent ; Rape minor
PD 6482 Conservation developing countries ; Inadequate soil
PE 8599 Conservation energy private sector ; Lack
PC 7517 Conservation ; Inadequate energy
PF 3581 Conservation ; Inadequate plant genetic resources
Conservation Lack
PD 0177 — *resource*
PD 0949 — *soil*
PE 8233 — *water*
Conservation [policies...]
PF 0037 — policies ; Inadequate governmental energy
PF 0038 — policies ; Inadequate governmental resource
PD 5429 — programmes ; Ineffectiveness
PD 0467 Conservation ; Unemployment caused environmental
PJ 0255 Conservation ; Unjustified urban
PF 2160 Conservatism
PF 3826 Conserver society policies ; Unacceptability
PD 9316 Consideration criminals ; Undue
PD 0317 Considerations ; Domination government policy making short term
PF 0342 Consistent programme effort ; Lack
Conspicuous consumption
PD 4037 — [Conspicuous consumption]
PF 6670 — developing countries
PE 3457 — international civil servants
PE 3457 — international delegations
PF 9493 Conspiracies silence
PB 7125 Conspiracies societal control
PC 2555 Conspiracy
PF 4198 Conspiracy against public
PF 8071 Conspiracy ; Catholic
PD 1767 Conspiracy ; Criminal
PF 0419 Conspiracy governments
PD 6943 Conspiracy incriminate ; Government
PF 8838 Conspiracy ; Jewish
PF 0695 Conspiracy ; Masonic
PA 6662 Conspiracy *Nonaccomplishment*
PF 8071 Conspiracy ; Papist
PF 8703 Conspiracy ; Policy making
PA 7411 Conspiracy *Uncommunicativeness*
PF 0200 Conspiracy ; Zionist
PE 3505 Constipation
PE 7423 *Constipation animals*
PD 2167 Constituencies ; Unequal parliamentary
PF 0889 Constituency ; Extreme detachment represented
PF 5311 *Constituents ; Inability people's representatives process feedback*
PF 1789 *Constraining scientific research ; Environmental hazards*
PA 5740 Constraint *Compulsion*
PF 9208 Constraint inherited problems policy innovation
PA 7296 Constraint *Restraint*
PF 5692 Constraint time individual social development
PF 4677 Constraints
PD 2318 Constraints against employment older people
PF 7076 Constraints behaviour ; Psychogenetic
PF 4955 Constraints development mental health services
PE 7784 *Constraints ; Extra economic*
PF 4882 Constraints freedom imagination ; Social
PE 2451 Constraints housing rehabilitation ; Demoralizing
PD 5114 Constraints increased agricultural output developing countries
PF 0954 Constraints land planning ; Boundary
PF 4668 Constraints power government
PF 1492 *Constraints programme beneficiaries ; Counter productive government*
PE 5305 Constraints salary increases ; Inadequate
PF 1789 Constraints testing new technology ; Pollution
PF 6476 Constricting effect educational structures
PE 1139 Constricting level capital development rural areas
PF 2575 Constricting patterns individual involvement
PF 1790 *Construction builders ; Limited*
Construction [cost...]
PJ 8591 — *cost rural community buildings ; Prohibitive*
PJ 8591 — *Costly school*
PF 3550 — *costs ; Prohibitive*
Construction [Dam...]
PD 0767 — Dam
PF 1213 — Depressing effect poor housing
PD 1251 — *dwellings ; Poor*
Construction [engineering...]
PD 7049 — engineering services industries ; Protectionism
PD 9515 — Environmentally harmful dam
PD 2608 — expertise ; Inequitable distribution
PE 0743 Construction flood plains

PE 6257 Construction ; Inadequate earthquake resistant
PD 1251 *Construction ; Inadequate housing*
Construction industry
PE 8790 — Environmental hazards
PE 0526 — Health risks workers
PE 6841 — Inadequate conditions work
PE 3969 — Inadequate working conditions
PE 8265 — Ineffective self regulation housing
PE 0509 — Instability
PD 9713 — Malpractice
PE 7742 — Reductionistic decision making criteria
PF 1740 — Short term priorities
Construction [manpower...]
PF 1790 — *manpower ; Limited*
PF 3550 — *materials ; Expensive*
PD 2211 — *materials ; Flammable*
PE 1848 — *methods ; Individualistic*
PC 0268 *Construction noise ; Building*
PF 7106 Construction ; Obstacles building
PE 1944 Construction obstacles ; Underground
Construction [planning...]
PF 2804 — planning ; Absence accountability
PF 2804 — planning ; Irresponsible
PF 3211 — *Postponement bridge*
PJ 8591 — *Prohibitive cost clinic*
PF 2605 — *Prohibitive cost road*
Construction [sites...]
PF 3499 — *sites ; Excessive cost*
PF 1790 — *skills ; Unavailability*
PD 7723 — standards ; Minimal
Construction [techniques...]
PE 5170 — techniques ; Unsafe
PD 2712 — technology ; Elitist control
PF 2464 — technology ; Profit motivated utilization
PE 2462 Construction ; Unsafe automobile design
PD 7723 Construction work ; Low quality
PD 7723 Construction work ; Shoddy
PD 1230 Constructional failure
PD 5344 Constructive family discipline young people ; Lack
PE 7780 Consular procedures ; Harassment travellers
PE 3938 Consultants ; Exploitative use
PF 0221 Consultants ; Fraud
PE 6274 Consultants ; Inadequacies foreign
PF 0221 Consultants ; Malpractice
PF 0221 Consultants ; Unethical practices
PE 8884 Consultation decision making process ; Inefficacity
PF 1981 Consultation problems ; Inadequate technical advice
PE 8884 Consultation procedures ; Ineffectiveness
PF 7889 Consultation ; Superficial
PE 7313 *Consumed ; Money crop*
PF 5057 Consumer apathy
PF 2583 *Consumer attitude ; Make*
PE 1213 Consumer boycotts
PF 2633 *Consumer buying ; Panic*
Consumer choice
PD 0515 — centrally planned economies ; Lack
PD 6075 — Decrease
PF 5057 — Excessive
PE 0290 Consumer compensation ; Inadequate
PD 7567 Consumer complaints
PD 3954 Consumer debt
PE 6877 Consumer fraud
Consumer goods
PE 1909 — electronic devices ; Excessive exposure radiation
PE 8934 — Incorporation carcinogens
PE 8574 — manufacturing industries ; Ineffective self regulation
PD 0515 — socialist countries ; Lack
Consumer [Ignorance...]
PF 2583 — *ignorance rights*
PE 1940 — influence industry ; Lack
PE 1940 — interests transnational corporations ; Disregard
PF 2583 *Consumer knowledge ; Limited*
PE 1011 Consumer needs developing countries ; Insensitivity transnational corporations
PF 6471 *Consumer needs ; Unsurveyed*
PF 5057 Consumer over-choice
PF 1238 Consumer prices ; High
Consumer products
PD 9310 — Environmentally unfriendly
PE 8795 — industries ; Protectionism
PF 5057 — Proliferation
PF 1379 — Unsafe design
PC 0123 Consumer protection ; Inadequate
Consumer [requirements...]
PF 6471 — requirements ; Unpolled
PF 2583 — rights ; Ignorance
PD 5774 — role ; Overemphasis
Consumer services
PE 8236 — developing countries ; Insufficient
PF 2411 — Haphazard provision
PF 2411 — *Non competitive*
PD 5774 Consumer spending ; Diverted
PC 0123 Consumer vulnerability
Consumer waste
PD 0807 — disposal
PD 8942 — products ; Dumping
PD 8942 — Proliferation
PD 8942 — Unrecycled
PD 5774 Consumerism
PF 3511 Consumerism ; Anti
PD 5774 Consumerism ; Self indulgent
PE 8236 *Consumers developing countries ; Lack production domestic*
PE 6877 Consumers ; Exploitation
PD 7567 Consumers ; Grievances
PE 0290 Consumers' loss ; Inadequate redress

Consumers

- PE 6877 Consumers ; Misrepresentation information
- PE 0290 Consumers ; Non payment compensation damages
- PC 0123 Consumers ; Unprotected
- PF 6504 *Consuming personal priorities*
- PF 6504 *Consuming personal survival*
- PD 6655 Consumption crime ; Aiding
- PE 0566 Consumption

Consumption [alcohol...]
- PD 1611 — alcohol adolescents ; Excessive
- PD 0153 — alcohol ; Excessive
- PD 8286 — alcoholic beverages
- PD 4518 — animal flesh ; Excessive
- PD 7699 — animal products ; Human
- PC 7644 — animals ; Human

Consumption [carbohydrates...]
- PC 3908 — *carbohydrates ; Excessive*
- PD 1611 — children young people ; Alcohol
- PD 4037 — Conspicuous

Consumption [dairy...]
- PD 7699 — dairy products ; Human
- PF 6670 — developing countries ; Conspicuous
- PC 5038 — Disparities energy
- PE 4261 Consumption excessive saturated fats
- PE 4261 Consumption fats ; Excessive
- PC 2518 Consumption goods services ; Excessive

Consumption [Inability...]
- PC 7517 — *Inability reduce petroleum*
- PC 2518 — *Increased food*
- PF 1773 — *Insufficient protein*
- PE 3457 — international civil servants ; Conspicuous
- PE 3457 — international delegations ; Conspicuous
- PD 0800 Consumption ; Junk food
- PC 5038 Consumption ; Maldistribution energy
- PD 4518 Consumption meat ; Excessive
- PD 4002 Consumption ; Natural resource depletion due high level
- PF 2639 Consumption needs ; Production serving false
- PD 2625 Consumption practices ; Unethical
- PD 7089 Consumption protein ; Excessive

Consumption resources
- PD 2625 — Corruption
- PE 5551 — developed countries ; Excessive
- PE 5551 — industrialized countries ; Unsustainable
- PB 1016 — International disparity
- PD 2625 — Irresponsible

Consumption [salt...]
- PE 4231 — salt ; Excessive
- PC 3908 — specific foodstuffs ; Excessive
- PC 3908 — *spices ; Excessive*
- PE 1894 — sugar ; Excessive

Consumption [Unnecessary...]
- PF 5931 — Unnecessary personal
- PD 0800 — Unnutritious food
- PC 0382 — Unnutritious food
- PF 5931 — Unsustainable personal
- PE 4665 Consumption vitamins ; Excessive
- PF 5931 Consumption ; Wasteful personal
- PE 2465 Contact dermatitis
- PE 8695 Contact ; Lack social
- PE 8695 Contact ; Minimal community
- PE 6685 Contact sewage ; Recreational
- PE 8695 Contacts ; Infrequent neighbour
- PF 6548 Contacts ; Undiscovered business
- PE 3526 Contacts ; Unskilled bureaucratic
- PE 0924 Contagious abortion
- PE 9795 Contagious agalactia
- PE 1775 Contagious bovine pleuropneumonia
- PD 9293 *Contagious bovine pyelonephritis*
- PE 6293 Contagious caprine pleuropneumonia

Contagious [ecthyma...]
- PE 6831 — ecthyma
- PD 9667 — *ecthyma animals*
- PD 7799 — *equine metritis*
- PE 5617 *Contagious foot rot*
- PD 9667 *Contagious pustular dermatitis*
- PC 2791 *Contained village economy*
- PF 6473 Containers ; Insufficient garbage
- PD 1754 Containers ; Non destructible packaging

Contaminant residues
- PD 4925 — freshwater animals, fish birds ; Accumulation
- PD 3934 — marine animals fish ; Accumulation
- PD 5021 — plants animals ; Accumulation
- PD 5278 — terrestrial animals ; Accumulation
- PD 0381 — terrestrial plants ; Accumulation
- PD 1694 Contaminants food ; Chemical
- PE 3896 Contaminated animal feed
- PD 0876 *Contaminated faeces*
- PE 3934 Contaminated farm slurry
- PD 3668 *Contaminated pastureland*
- PA 2353 *Contaminated saliva*
- PD 0876 *Contaminated toilet overflow*
- PE 6685 Contaminated water ; Bathing sewage
- PD 2503 Contaminated well water
- PE 7064 Contamination ; Animal products food
- PD 1119 Contamination animals animal products ; Radioactive
- PE 8476 Contamination ; Crop

Contamination [Degradation...]
- PE 4759 — Degradation environment
- PD 1694 — dietary intake ; Chemical
- PD 0235 — drinking water
- PE 5650 — drinking water ; Lead

Contamination [Endangered...]
- PD 5157 — Endangered animal plant life due radioactive
- PD 1386 — environment ; Accidental large scale
- PB 1166 — Environmental
- PF 3904 Contamination ; Fear food

- PD 5605 Contamination ; Food

Contamination food
- PD 2594 — Biological
- PD 0120 — chains ; Pesticide
- PD 9669 — Microbial
- PD 2503 Contamination ; Groundwater
- PF 9150 Contamination human body
- PE 1431 Contamination marine environment fisheries products ; Radioactive
- PC 1299 Contamination natural radiation

Contamination [pesticides...]
- PE 6480 — pesticides ; Food
- PD 0710 — plants ; Radioactive
- PE 1458 — public water supplies sabotage
- PC 0229 Contamination ; Radioactive
- PE 5539 Contamination sediments toxic substances

Contamination soil
- PD 3668 — Metal
- PE 3383 — Radioactive
- PC 0058 — toxic substances
- PD 0482 Contamination ; Waste water

Contamination water
- PD 1175 — Biological
- PE 0535 — Chemical
- PE 2441 — Radioactive
- PD 8122 Contamination ; Water system
- PF 8396 Contemplate criticism ; Unwillingness
- PF 1729 Contemporary life styles ; Reduced social commitment
- PF 7697 Contempt
- PD 1965 Contempt agricultural labour developing countries
- PF 5012 Contempt authority
- PD 9035 Contempt court
- PF 5705 Contempt ; Criminal
- PF 0639 Contempt democratic processes
- PF 3979 Contempt elders
- PD 9035 Contempt judicial process
- PF 0639 Contempt parliamentary procedures
- PC 4321 Contempt traditional modes behaviour
- PA 6191 Contemptuousness *Disapproval*
- PA 6822 Contemptuousness *Disrespect*
- PA 6387 Contemptuousness *Necessity*
- PA 6979 Contemptuousness *Opposition*
- PA 7070 Contemptuousness *Unprovability*
- PA 6491 Contemptuousness *Vanity*
- PA 5663 Content
- PD 1827 Content ; Educational curricula over emphasizing method rather than
- PF 3549 Content rather than method ; Educational curricula based
- PA 7378 Contentiousness *Aggravation*
- PA 5532 Contentiousness *Disaccord*
- PA 7253 Contentiousness *Envy*
- PA 5502 Contentiousness *Reason*
- PF 4389 Contentment ; *Illusion*
- PE 3754 Contests ; Rigging sports
- PE 0525 Context civil violence ; Stress trauma
- PF 0823 Context counsel ; Narrow
- PF 6617 Context developing technology ; Limited social
- PF 0963 Context ; *Entrapping social*
- PF 0966 Context ethical decisions ; Lack meaningful educational
- PF 2068 Context future ; Young people's lack
- PF 3559 Context ; Isolation individual decision making

Context [Lack...]
- PD 1591 — Lack individual historical
- PF 2504 — Limited cultural
- PF 2394 — Limited decision making
- PF 1062 — Loss past operating
- PF 2998 Context ; Narrow job
- PF 9003 Context ; Negative social
- PE 7155 Context process ; Exclusion decision making processes those who question
- PF 4837 Contexts development programmes ; Lack understanding social economic
- PE 4439 Contexts ; Employment criminals policy making
- PE 4439 Contexts ; Employment war criminals policy making
- PF 2096 Contexts ; Fragmented ethical
- PA 7213 Contextuality∗complex
- PF 7054 Contingency ; Human
- PE 2240 Continued operation unsafe motor vehicles
- PD 4125 *Continued subsistence living*
- PF 2439 *Continuing commercial decline*
- PC 1716 Continuing dropout pattern

Continuing [education...]
- PF 0021 — education ; Under developed
- PF 0021 — education ; Unsatisfied need
- PF 8076 — *expectation community decline*
- PD 1631 Continuing fire hazards
- PD 2968 Continuing low level prices commodities exported developing countries
- PF 3434 Continuity amongst personnel international organizations ; Lack
- PE 5691 Continuous noise ; Torture
- PE 5390 Contraband useful escape ; Introducing possessing
- PE 3286 Contraception
- PF 1022 Contraception ; Religious opposition
- PD 4560 Contraceptive availability developed countries
- PD 0093 Contraceptive failure

Contraceptive methods
- PD 0093 — Inadequacy
- PF 1069 — Inadequacy male
- PF 1035 — Sexual discrimination
- PD 1038 Contraceptives ; Health hazards
- PD 0093 Contraceptives ; Inadequacy
- PD 0148 Contraceptives ; Ineffective use
- PD 0093 Contraceptives ; Lack ideal
- PE 8261 *Contraceptives ; Prohibitive cost*

- PE 5762 Contract ; Breach
- PD 7876 Contract fraud
- PE 5397 Contract ; Individualistically reduced legal
- PD 6930 Contract labour ; Exploitation
- PE 5786 Contract law ; Violations
- PD 5493 Contract ; Murder
- PE 8273 Contract performance ; Interference trade unions
- PE 3528 *Contract power ; Lack*
- PE 5762 Contract ; Risk breach
- PE 5397 Contract system reduced individual needs ; Legal
- PE 7808 *Contracted flexor tendons*
- PE 7826 *Contracted heels*
- PE 7826 *Contracted tendons*
- PF 1379 *Contracting AIDS blood transfusions ; Risk*
- PD 5111 *Contracting AIDS kissing ; Risk*
- PF 3226 Contracting out
- PA 6847 Contraction
- PD 7727 Contractionary economic spiral ; Self reinforcing
- PE 6702 Contracts ; Unethical allocation public
- PE 2911 Contracts ; Unjust allocation government
- PE 7235 Contractual agreement ; Denial right freedom imprisonment failure fulfil
- PF 2040 Contradicting popular views ; Fear

Contradiction [Difference...]
- PA 6698 — *Difference*
- PA 5982 — *Disagreement*
- PA 6838 — *Dissent*
- PA 6979 Contradiction *Opposition*
- PA 5502 Contradiction *Reason*

Contradiction Self
- PA 5982 — [Contradiction ; Self]
- PA 6180 — [Contradiction ; Self]
- PA 6698 — [Contradiction ; Self]
- PF 3667 Contradictions

Contradictions [capitalism...]
- PF 3126 — capitalism developing countries
- PF 3118 — capitalist systems
- PF 3179 — communist systems
- PF 5210 — Contradictions ; Failure acknowledge
- PE 2203 Contradictions growth partnership model developed countries
- PF 5653 Contradictory commonplaces
- PF 9166 Contradictory decision-making
- PJ 4512 Contradictory solutions national economic goals
- PA 6695 Contrariety *Inequality*
- PA 6979 Contrariety *Opposition*
- PA 7107 Contrariety *Unpleasantness*
- PF 5061 Contrary self interest ; Policy
- PB 5405 Contravention rights
- PE 6715 Contributions agents foreign principals ; Political
- PF 8650 Contributions ; Imbalance disbursements multilateral agencies official
- PF 8650 Contributions international organizations ; Government non payment agreed

Control [Abdication...]
- PF 7342 — Abdication government ministerial
- PE 8467 — advertising ; Lack legislative
- PA 6448 — *aggressive impulses ; Loss*
- PD 2781 — animal diseases ; Inadequate
- PE 8426 — Animal pest
- PE 3286 Control ; Birth
- PE 5804 Control bulk shipping ; Transnational corporation

Control [childrens...]
- PF 2060 — children's thoughts reflections ; Parental
- PE 2573 — communications mass media ; Dangers private
- PD 4597 — communications mass media ; State
- PE 9251 — company directors ; Abdication

Control complex equipment
- PE 5572 — Human fatigue during
- PF 6411 — Information overload during
- PE 0680 — Substance abuse during

Control [Computer...]
- PF 2685 — *Computer*
- PB 7125 — Conspiracies societal
- PD 2712 — construction technology ; Elitist

Control [decision...]
- PE 1488 — decision making ; Social service quality negated oligarchic
- PD 3598 — developing countries ; Inadequate weed
- PD 0144 — developing countries ; Loss traditional forms social
- PJ 9244 — development projects ; Inadequate
- PA 6448 — disorder ; Impulse
- PJ 1751 — DNA ; Corporate

Control [economic...]
- PF 2312 — economic forces ; Unquestioned
- PF 0013 — efforts ; Failure disarmament arms
- PD 2931 — elections ; Oligarchic political
- PE 6547 — environment local communities ; Limited
- PF 0304 — Excessive government
- PE 6836 — Excessive job
- PF 1021 Control family planning ; Opposition population
- PC 3187 Control ; Foreign

Control [global...]
- PC 3778 — global economy ; Elitist
- PE 0135 — global economy limited number corporations ; Disproportionate
- PF 9672 — government finances ; Ineffective
- PF 1023 — Government opposition population
- PF 9401 Control ; Health
- PD 1038 Control ; Health hazards birth

Control Inadequate
- PC 0231 — drug
- PD 2392 — drug quality
- PD 0452 — flood
- PF 9401 — health
- PF 6559 — *motorcycle*

Corporate commitment

Control Inadequate cont'd
PF 3538 — patent
PF 8652 — police
PD 1435 — quality
PD 2207 — riot
PE 8266 — traffic
PD 8517 — water
PD 1574 — weed
PD 3101 Control ; Industrial
PE 5831 Control industries sectors transnational corporations
Control Ineffective
PF 0148 — birth
PF 0148 — population
PF 1020 — population
Control [Inhumane...]
PD 1156 — Inhumane methods riot
PF 3409 — Insufficient pest
PC 3778 — international monetary system ; Unrepresentative
PE 0135 — international trade transnational corporations ; Escalating
PE 3869 Control joint venture ; Inadequate participation
PE 7665 Control judiciary executive military power
Control Lack
PF 7138 — [Control ; Lack]
PF 3411 — administrative
PF 6229 — biological
PF 3409 — Control ; Limited pest
Control [major...]
PE 1922 — major corporations national policy making ; Undisclosed
PD 7461 — market facilities ; Hierarchical
PE 2397 — marketing distribution channels transnational corporations
PF 6533 — marketing systems ; Surrendered
PD 0071 — mergers ; Inadequate
PE 3528 — Middle man price
PF 9542 — military ; Inadequate government
PF 2375 — Minority
PF 3539 — misinformation ; Birth
PA 8653 — Mistrust birth
Control [national...]
PE 0042 — national economic sectors transnational enterprises
PF 2344 — national economies limited number individuals ; Undisclosed
PD 3109 — natural resources ; Foreign
PD 7501 — new animal forms ; Monopolistic
PD 7840 — new life forms ; Monopolistic
Control over
PC 1818 — government administrative process ; Inadequate
PC 3166 — international organizations
PC 2381 — their own lives ; Denial people
Control [pests...]
PD 1207 — pests ; Excessive use chemicals
PD 1207 — pests ; Lack biological
PD 1038 — pills ; Increased use birth
PE 7780 — procedures ; Victimization immigration
PD 0154 — production ; Elitist
PE 3780 — production processes ; Public non accountability
Control Prohibitive cost
PE 8261 — birth
PF 2779 — disease
PF 2899 — weed
Control [raw...]
PE 0194 — raw materials markets transnational corporations ; Excessive
PF 1022 — Religious opposition population
PF 5539 — resources ; Local
PF 3070 — restrictions ; Counterproductive exchange
PE 3528 — rural marketing ; Middleman
PD 4049 Control systems ; Vulnerability nuclear defence
PE 8640 Control transnationals global communications industry
Control [Uncoordinated...]
PE 8233 — Uncoordinated irrigation
PE 4768 — Unenforced animal
PE 2820 — United Nations systems ; Lack centralized financial
PE 7995 Control wild animal populations ; Abuse
PF 1770 Control zoonoses ; Inadequate
PF 6411 Controllers ; Information overload process
PD 2781 Controlling animal diseases ; Difficulty
PD 2776 Controlling disease wild animals ; Difficulty
PF 2779 Controlling epizootic enzootic diseases ; High cost
PF 4389 Controlling events ; Illusion
Controlling [insect...]
PD 4751 — insect populations ; Difficulty
PE 7961 — instruments ; Shortage professional, scientific
PE 8169 — instruments trade instability professional ; Scientific
PD 0991 Controls ; Border
PE 8583 Controls ; Distortion international trade selective monetary
PF 2607 Controls favouring investor ; Proscriptive
PE 8882 Controls movement labour ; Distortion international trade restrictive
PD 3041 Controls newspaper journal propaganda ; Foreign
PE 8525 Controls over foreign investment ; Distortion international trade restrictive
PD 4060 Controversial issues ; Avoidance
PE 0598 Controversial news items ; Calculated delays releasing
PE 2488 Controversies ; Bilingualism national settings regional linguistic
PA 5532 Controversy Disaccord
PA 5982 Controversy Disagreement
PA 5502 Controversy Reason
PA 7250 Contumaciousness Disobedience
PA 7325 Contumaciousness Irresolution
PA 6191 Contumeliousness Disapproval
PA 6822 Contumeliousness Disrespect

PA 6491 Contumeliousness Vanity
PD 5238 Contumely
PB 0731 Contusions
PE 5873 Convenience ; Evasion shipping regulations taxes flags
PE 5873 Convenience flagging ships
PF 9205 Convenient beliefs ; Over reliance
PF 9205 Convenient beliefs ; Thoughtless acceptance
PC 5230 Conventional armed forces ; Imbalance
PC 5230 Conventional arms ; Imbalance
PC 4311 Conventional warfare
PC 5230 Conventional weapons ; Imbalance
PJ 8840 Conventions ; Lack adherence international transit
PE 5789 Conventions land locked countries ; Lack adherence international transit
PF 0473 Conventions ; Refusal ratify international
PF 2497 Conventions ; Violation international
PE 5124 Convergence practice apparently opposed ideologies ; Cover up
PD 0270 Conversion disorder
PD 6637 Conversion ; Forced religious
PE 8453 Conversion plants ; Environmental impacts coal
PF 2005 Conversions ; Meaningless
PF 3069 Convertibility currencies ; Non
PF 8661 Convict known offenders ; Inadequate evidence
PF 5506 Convict made goods ; Unfair competition
PF 1098 Convicted atrocities ; Politically motivated amnesty military personnel
PF 7728 Convicted crimes which they are innocent ; Failure clear names people
PE 7345 Convicted criminals ; Denial right accused segregation
PE 9545 Convicted human rights offenders ; Government pardoning
PE 7348 Conviction accusation
PE 9545 Conviction ; Arbitrary reduction sentences offenders after
PD 5317 Conviction higher tribunal ; Violation right review
PF 3863 Conviction ; Lack religious
PF 6158 Convictions ; Cumulation
PD 5636 Convictions fraud ; Inconclusive
PF 6158 Convictions single criminal act ; Multiple
PF 8203 Convictions truth ; Excessive equation
PE 6929 Convicts ; Denial rights ex
PE 8900 Convicts ; Lack care intellectually disabled
PF 1081 Convincing symbols connecting individual's life cosmos ; Absence
PE 9468 Convulsions
PD 7841 Convulsive foals
PE 0596 Cooked flesh ; Uncooked inadequately
PE 0596 Cooked foods ; Inadequately
PF 1267 Cooking fuel ; High priced
PE 7677 Cooking practices ; Inhumane
PE 7904 Cooking stoves ; Inadequate
PD 9736 Cooled homes ; Poorly
PD 9736 Cooled shelters ; Inadequately
PF 1744 Cooling ; Global
PE 4604 Coonhound paralysis
PC 5180 Cooperate ; Lack political will
PF 9474 Cooperate United Nations eliminating human rights violations ; Governments unwilling
PE 4508 Cooperation central banks ; Need increased
Cooperation [Decline...]
PF 0817 — Decline multilateral
PF 0817 — Dependence lack international
PF 3558 — Deteriorating structures rural community
PC 0933 — developing countries ; Inadequate economic
PC 0025 — Distrusted economic
PC 2089 — due personal mistrust ; Lack international
PF 9129 — Cooperation ; Fragmented regional
Cooperation [Inadequate...]
PF 3558 — Inadequate inter village
PF 6506 — Insufficient agency
PF 3558 — Insufficient village
PF 9184 — intelligence agencies ; Unreported
Cooperation Lack
PF 2816 — [Cooperation ; Lack]
PF 2816 — agreement
PC 0025 — economic
PF 0817 — international
PF 6855 — trans frontier
PF 3558 — village
PF 0919 Cooperation ; Limited tripartite
PF 8195 Cooperation ; Non
PF 2788 Cooperation ; Non violent non
PE 4785 Cooperation ; Obstructions international personnel exchanges cultural
PE 8500 Cooperation officialdom ; Lack
PE 8307 Cooperation problems countries ; Inadequate inter regional
PF 0863 Cooperation problems ; Inadequate technical
PF 4366 Cooperation reducing terrorism ; Inadequate international
PD 0516 Cooperation riparian states ; Lack
Cooperation [secret...]
PF 9184 — secret services ; Covert
PF 0817 — solve world problems ; Inadequate global
PD 7470 — Structurally blocked scientific
PF 3558 — Cooperation ; Undeveloped community
PF 6514 Cooperative action ; Neglected
PF 6514 Cooperative action ; Unformed style
PF 9729 Cooperative benefits ; Undemonstrated
Cooperative [cheating...]
PF 9729 — cheating ; Past
PF 9729 — commitment ; Short lived
PF 7118 — community groups ; Non
Cooperative [economic...]
PC 0025 — economic relations ; Non

Cooperative [efforts...] cont'd
PF 2588 — efforts ; Fragmented forms
PF 5250 — events ; Limited
PF 9729 — experience ; Disillusioning
PF 9729 — experience ; Limited
PF 9729 Cooperative failure story
PF 9474 Cooperative governments ; Internationally non
Cooperative [planning...]
PF 8931 — planning ; Difficulties
PF 9729 — potential ; Unexplored
PE 2309 — projects ; Government imposition rural
PF 9729 — purchasing ; Undemonstrated
Cooperative [solutions...]
PF 9409 — solutions problems ; Minimally
PF 5961 — structures ; Obsolete agricultural
PF 9729 — success ; Uncertain
PD 0460 Cooperatives ; Bureaucracy
PE 2309 Cooperatives ; Failure rural
PE 2309 Cooperatives ; Government organized
PF 1257 Cooperatives ; Insufficient women
PD 0460 Cooperatives ; Red tape
PF 9729 Cooperatives ; Unrecognized benefits
PE 0780 Cooptation private sector initiatives government
PE 2603 Coordinated customs services developing countries ; Absence
PE 1029 Coordinating bodies ; Proliferation duplication international organizations
PE 1579 Coordinating bodies United Nations system ; Proliferation duplication organizational units
PF 8330 Coordination
Coordination [action...]
PF 1467 — action against problems ; Inadequate
PE 1375 — action intergovernmental programmes national level ; Inadequate
PJ 1719 — agencies ; Lack
PF 0175 — aid donors ; Lack
PF 0175 — aid ; Inadequate
PE 2417 Coordination bodies ; Proliferation duplication intergovernmental organizations
PE 0179 Coordination bodies ; Proliferation duplication international nongovernmental organization
PE 7230 Coordination disorder ; Developmental
PE 4344 Coordination governmental representation intergovernmental organizations ; Inadequate
PF 3141 Coordination interest rates ; Lack international
PE 0730 Coordination intergovernmental system organizations ; Inadequate
Coordination international
PF 0048 — monetary system ; Lack
PE 1209 — nongovernmental organizations programmes ; Inadequate
PD 0285 — organizations programmes ; Inadequate
PF 8330 Coordination ; Lack
PC 0025 Coordination ; Lack international economic
PF 7011 Coordination ; Minimal leadership
PE 1184 Coordination regional intergovernmental organizations common membership ; Inadequate
PE 4508 Coordination supervisors financial stock markets ; Lack international
Coordination United Nations
PF 0075 — information systems ; Lack
PE 0296 — system ; Inadequate
PE 2820 — systems ; Inadequate budgetary
PE 6625 Copies ; Illegal software
PE 4486 Copper dust
PE 2084 Copper industry transnational corporations ; Domination
PD 5228 Copper poisoning animals
PE 0535 Copper pollutant
PE 0824 Copper trade ; Instability
PF 7427 Coprophagia
PJ 0970 Coprophilia
PD 1284 Copying ideas ; Unacknowledged
PD 0188 Copyright ; Avoidance
PE 8403 Copyright barriers transfer knowledge
PD 0188 Copyright infringement
PD 0188 Copyright pirates
PD 5769 Coral reefs ; Destruction
PF 9774 Cording-up syndrome horses
PF 1254 Core leadership ; Lack
PE 2521 Cork ; Instability trade wood, lumber
PF 1372 Cork ; Long term shortage wood, lumber
PE 8091 Cork manufactures ; Environmental hazards wood
PE 3648 Corn borers pests
PD 5535 Cornea animals ; Diseases
PE 6789 Cornea ; Inflammation
PE 6789 Cornea ; Ulceration
PD 8786 Corneal opacity
PE 4161 Corns
PE 7826 Corns ; Bruised
PE 7826 Coronary sinus
PE 8456 Coronary thrombosis
PD 7841 Coronaviral encephalomyelitis swine
PD 3978 Coronaviral enteritis turkeys
PE 7423 Coronaviral gastroenteritis ; Canine
PD 2730 Coronaviral infection ; Feline
PD 8575 Corporal punishment
PE 0192 Corporal punishment schools
Corporate [achievements...]
PF 1205 — achievements ; Lack
PF 1781 — action ; Unrecognized benefits
PF 2316 — activities ; Minimal opportunities
Corporate [capital...]
PE 5021 — capital companies ; Fraud concerning economic situation
PF 5250 — celebrations ; Insufficient
PD 2790 — commitment ; Mistrust

-883-

Corporate communications

Corporate [communications...] cont'd
PF 3661 — communications ; Unsound
PD 4982 — competitors ; Fabrication reports
PJ 1751 — control DNA
PD 3528 — crime
PE 6618 — crime pharmaceutical industry
Corporate [debt...]
PE 1879 — debt ; Excessive
PF 5522 — directorates ; Interlocking
PD 2790 — disloyalty
PD 3003 — domination daily life
Corporate [effort...]
PF 2358 — *effort ; Collapsed*
PD 2671 — engagement ; Meaningless
PF 3750 — engagement society ; Collapse
PF 0374 — entertainment ; Unethical
PF 7578 — executives ; Excessive salaries
PE 5944 — executives ; Terrorism targeted against
PD 1220 — expansion ; Destabilizing
PF 7189 Corporate greed
PE 3296 Corporate initiative rural communities ; Cumulative depletion
PF 6530 *Corporate investment ; Reluctance towards*
Corporate [land...]
PC 8160 — *land ; Scarcity*
PE 5944 — leaders ; Killing
PD 4982 — leaders ; Misrepresentation facts
PF 2826 — logos ; Offensive
PF 0963 *Corporate needs ; Unassessed*
PD 4982 — Corporate news information ; Deliberate distortion
PJ 2493 Corporate operations existing groups ; Non
PD 4982 Corporate over-reporting under-reporting
Corporate [planning...]
PF 3513 — planning paralysis
PF 2358 — *power ; Unrealized*
PF 7983 — prestige ; Pursuit
PD 5725 — privileges
PF 0963 *Corporate responsibility ; Unconceived*
PD 8897 Corporate rivalry
PD 4982 Corporate slander
PF 7983 Corporate status ; Pursuit
PF 1301 Corporateness ; Deteriorated structures essential
PF 2358 Corporateness rural villages ; Unformed confidence community
PE 5804 Corporation control bulk shipping ; Transnational
PE 1571 Corporation financial secrecy
PE 2032 Corporation financial statements ; Distortion
PD 5891 Corporation imperialism ; Transnational
PE 5944 Corporation leaders ; Assassination
PD 4982 Corporation officials ; Deliberate lying
PD 4982 Corporation representatives ; Perjury
PE 6356 Corporation-sanctioned assassination
PE 6356 Corporation-sanctioned killing
Corporations [Adverse...]
PE 5820 — Adverse impact technology transfer transnational
PE 0980 — Aggravation instability exchange rates transnational
PE 1996 — apartheid system ; Participation transnational
PE 1771 — Corporations balance payments ; Adverse effect transnational
PE 2200 — Corporations ; Burden conflicting national regulations transnational
Corporations [coffee...]
PE 3781 — coffee industry ; Domination transnational
PE 0766 — Concentration power transnational
PE 1011 — consumer needs developing countries ; Insensitivity transnational
PE 5831 — Control industries sectors transnational
PE 2397 — Control marketing distribution channels transnational
PE 5903 — Corporations ; Destabilization monetary systems exchange notes transnational
Corporations developing countries
PE 9646 — Asymmetry bargaining power transnational
PE 6952 — Exploitative financial policies transnational
PE 2004 — Harmful effects advertising transnational
PE 0853 — Inadequate negotiation entrance terms transnational
PE 1598 — Minimal export promotion transnational
Corporations [Disproportionate...]
PE 0135 — Disproportionate control global economy limited number
PE 1922 — Disproportionate influence national economies limited number
PE 1940 — Disregard consumer interests transnational
PE 1082 — Disruption cultural social identities developing countries transnational
PE 1957 — Disruption domestic social policies transnational
PE 1796 — domestic name brand food sector developing countries ; Domination transnational
Corporations Domination
PE 2193 — advertising transnational
PE 1448 — agricultural equipment industry transnational
PE 1469 — automobile industry transnational
PE 2084 — copper industry transnational
PE 0163 — developing countries transnational
PE 1322 — economic integration transnational
PE 1187 — labour relations transnational
PE 1620 — shipping industry transnational
Corporations [Escalating...]
PE 0135 — Escalating control international trade transnational
PE 0194 — Excessive control raw materials markets transnational
PD 5807 — Excessive power independence transnational
PE 2145 — Corporations ; Foreign currency manipulations accounting records transnational
PE 0135 — Corporations global policy making ; Undisclosed influence transnational
PE 1557 — Corporations ; Highly indebted

Corporations [Import...]
PE 1806 — Import dependency food staples developing countries due transnational
PE 1660 — Increasing income disparity developing countries due transnational
PF 0691 — Ineffective international regulation transnational
PE 7355 — Interference labour relations developing countries transnational
PE 0667 — Interference labour relations industrialized countries transnational
PE 0163 — Intimidation developing countries transnational
Corporations [Lack...]
PD 2790 — Lack worker loyalty
PE 7892 — least developed countries ; Inadequate investment transnational
PF 7293 — Limited criminal liability
PE 2328 — Limited market access due product differentiation transnational
PE 1511 — local industry developing countries ; Inadequate relationship transnational
Corporations [Manipulation...]
PE 0245 — Manipulation transfer prices transnational
PE 7061 — Missing documents data transnational
PE 1918 — Monopolization technology transnational
PD 3003 — Monopoly economy
PE 1922 Corporations national policy making ; Undisclosed control major
PE 4120 Corporations pharmaceutical needs developing countries ; Non responsiveness transnational
PE 0032 Corporations ; Political intervention transnational
PE 0051 Corporations ; Restriction free market competition transnational
Corporations Restrictive
PE 5915 — business practices transnational
PE 3196 — market divisions transnational
PE 2396 — pricing policies transnational
PE 0234 Corporations ; Retarding development transnational
Corporations [Shell...]
PF 0326 — Shell
PE 5796 — Social irresponsibility transnational
PE 0106 — specialized agencies United Nations ; Inadequate relationship transnational
PD 9146 — Suppression information
PE 5944 Corporations ; Terrorism targeted against business
PE 5855 Corporations ; Unfair pricing transnational
PE 2778 Corpses ; Disposal
PF 6957 Corpses ; Excessive preoccupation
PD 2784 *Corpses ; Health risks international transfer*
PD 2784 *Corpses ; Obstacles international transfer*
PF 5857 Corpses ; Undignified treatment
PE 1177 Corpulence
PE 7349 Correct misinformation ; Denial right
PE 1675 Correctional institutions ; Disturbances
PD 5142 Correctional institutions ; Ineffectiveness
PF 5172 Correctional systems ; Inadequate
PF 1728 Correspondence basic value images social reality ; Lack
PD 7608 Correspondence ; Denial right private
PD 5093 Correspondence ; Interception
PE 3034 Correspondents ; Inadequate protection war
PD 0508 Corrosion
PE 4740 Corrosion encrustation water wells
PE 4524 Corrosion glass
PE 1945 Corrosion iron steel
PD 9525 Corrosion materials ; Atmospheric
PD 5803 Corrosion ships
PE 5470 Corrosion stonework
PE 1811 Corrosion tropical climates
PE 4740 Corrosive groundwater
PE 6887 Corrosive substances
PD 9653 Corrupt acquisition offshore assets heads state
PA 1986 Corrupt people
PF 4210 Corrupt practices ; Divergent national concepts
PF 4821 Corrupt regimes ; Government support
PF 9693 Corrupting effect valuable prizes academic integrity
PA 1986 Corruption
Corruption [Administrative...]
PC 0279 — Administrative
PE 9140 — amongst relatives government leaders
PD 2623 — anthropologists
PD 7360 — armed forces
Corruption [Badness...]
PA 5454 — *Badness*
PD 7731 — biologists
PC 0279 — Bureaucratic
Corruption [catering...]
PE 6615 — catering industry
PE 8742 — charitable organizations
PD 4265 — chemists
PE 9140 — children government leaders
PE 6702 — civil servants
PD 8001 — clothing industry
PC 2563 — commerce
PA 6468 — *Complexity*
PD 2625 — consumption resources
PC 2913 — Cultural
PE 4033 — customs excise officials
Corruption [Defence...]
PA 1986 — Defence against
PA 1986 — Dependence
PD 0348 — developing countries
PA 7343 — *Dissuasion*
PA 6790 — *Distortion*
PD 2886 — documentalists
PE 7900 — documents
PD 2879 Corruption employers

PD 3736 Corruption entertainment industry
PD 1045 Corruption food industry
PD 0708 Corruption geologists
PE 7917 Corruption good human nature
Corruption government
PC 7587 — [Corruption government]
PC 7587 — leaders
PC 7587 — ministers
PC 7587 — rulers
PD 2586 Corruption hydrologists
Corruption [Ideological...]
PC 2914 — Ideological
PA 6088 — *Impairment*
PA 7363 — *Improbity*
PD 2916 — industry
PC 9173 — Institutionalized
PC 2915 — Intellectual
PE 6947 — intergovernmental organization leadership
PE 6740 — interpreters
PD 4194 Corruption judiciary
PE 0647 Corruption judiciary ; Political
Corruption [law...]
PD 2918 — law enforcement officials
PD 5380 — legal profession
PD 5948 — local government
Corruption [maintenance...]
PD 7964 — maintenance practices
PB 2619 — meaning
PB 2619 — meaning ; Dependence
PD 5251 — media
PD 4182 — meteorologists
PD 7360 — Military
PD 9481 — minors
PE 6522 — minors
PA 6393 — *Miseducation*
PD 0136 — mismanagement foreign aid
PF 8046 — monitoring agencies
PD 5948 — Municipal
Corruption [oceanographers...]
PD 4277 — oceanographers
PC 9533 — Official
PF 8046 — official inspectors
PC 9533 — officials
PE 1369 — opticians
PC 3359 — organized religion
Corruption [pedologists...]
PD 1110 — pedologists
PF 8759 — personal relationships
PD 0862 — personnel practices
PD 1710 — physicists
PD 2918 — Police
PC 0116 — Political
PC 0116 — political parties
PC 0116 — politics
PD 8392 — practised government agents
PD 9414 — prisons
PC 8019 — professional practices
PD 5267 — psychotherapy
PD 4541 — public servants
Corruption [radiologists...]
PD 8290 — radiologists
PD 5422 — Real estate
PF 8046 — regulatory agencies
PA 5699 — *Reversion*
PD 8380 — ruling classes
Corruption [social...]
PD 6626 — social scientists
PD 8859 — Social welfare
PE 3754 — sports athletic competitions
PF 4564 — statisticians
Corruption [Tolerance...]
PF 8611 — Tolerance
PD 4341 — trade unions
PD 4684 — training
PD 1012 — transport industry
PA 5459 Corruption *Uncleanness*
PD 5948 Corruption ; Urban
PD 7726 Corruption veterinarians
PA 5644 Corruption *Vice*
PE 9140 Corruption wives government leaders
PD 4334 Corruption workers
PD 4721 Corruption zoologists
PD 8679 Corruptive crimes
PD 9654 Cortical excess
PG 5181 *Cortical hypofunction ; Adrenal*
PE 6280 *Coryza ; Gangrenous*
PE 5567 *Coryza poultry ; Infectious*
PE 5567 *Coryza ; Turkey*
PE 7111 Cosmetic health problems related human hair
PE 4429 Cosmetic surgeons ; Untrained
PE 4429 Cosmetic surgery quackery
PE 4895 Cosmetic use ; Hazards
PE 9611 Cosmetics ; Abusive treatment animals testing toiletries
PE 4895 Cosmetics ; Allergy inducing
PF 0638 Cosmic conflagration
PF 1686 Cosmic radiation ; Health hazards exposure
PD 1283 *Cosmically inharmonious way life*
PF 8331 Cosmopolitanism
PF 1081 Cosmos ; Absence convincing symbols connecting individual's life
Cost [abortion...]
PJ 3244 — *abortion ; Prohibitive*
PD 1842 — accommodation ; Prohibitive
PF 1176 — *advertising ; Prohibitive*
PE 4784 — animal protein ; Excessive
PF 7870 — art ; Prohibitive

Cover inaction

Cost [basic...]
PF 6527 — basic services ; Prohibitive
PF 9243 — benefit analyses ; Manipulation project
PE 8261 — *birth control ; Prohibitive*
PF 3550 — *building materials ; Prohibitive*
PF 2874 — *business security ; Prohibitive*
Cost [capital...]
PF 2410 — *capital equipment ; Prohibitive*
PF 4975 — civil engineering projects ; Misappropriation resources high
PE 4162 — clearing land
PJ 8591 — *clinic construction ; Prohibitive*
PE 1122 — comprehensive information systems ; Excessive
PF 1653 — *connection electricity services ; Prohibitive*
PF 1653 — *connection public utilities ; Prohibitive*
PF 3499 — *construction sites ; Excessive*
PE 8261 — *contraceptives ; Prohibitive*
PF 2779 — controlling epizootic enzootic diseases ; High
PF 2899 — *crop treatment ; Prohibitive*
PE 7539 Cost decommissioning nuclear reactors ; Excessive
PF 2779 Cost disease control ; Prohibitive
Cost [education...]
PF 4375 — education ; Prohibitive
PE 5302 — effective activities ; Insufficient government spending
PE 6059 — effective prosecution offenders ; Excessive
PD 2931 — *electoral campaigns ; Prohibitive*
PF 3318 — *electricity ; Prohibitive*
PE 1722 — equipment maintenance ; Prohibitive
PF 6527 — essential services rural communities ; Prohibitive
PE 4603 — external finance developing countries ; Deterioration
PE 7548 — eye glasses developing countries ; Prohibitive
Cost farm
PF 3630 — *equipment ; Prohibitive*
PF 8763 — *labour ; Prohibitive*
PF 2457 — machinery ; Prohibitive
Cost [fertilizer...]
PF 3409 — *fertilizer ; Prohibitive*
PF 6478 — *fishing equipment ; Prohibitive*
PF 6478 — *fishing ; Prohibitive*
PF 1267 — *fuel ; Prohibitive*
PD 1891 Cost goods services ; Prohibitive
Cost [health...]
PE 8261 — health care ; Prohibitive
PF 1212 — healthy foods ; Prohibitive
PF 4375 — High training
PE 8063 — High transportation
PF 4375 — higher education ; Prohibitive
PF 0296 — *home maintenance ; Prohibitive*
PE 4154 — hospital facilities ; Prohibitive
Cost housing
PD 1842 — Disparity income
PD 1842 — High
PD 1842 — Limited low
Cost [imported...]
PJ 7877 — imported goods ; Prohibitive
PF 9101 — inadequate development policies ; Prohibitive
PF 0703 — information ; Inflated
PE 8632 — insurance ; Prohibitive
PD 4632 — intermediaries ; Prohibitive
PF 0703 Cost knowledge information ; Prohibitive
Cost [land...]
PE 4162 — land ; High
PE 4162 — land ; Prohibitive
PF 0995 — legal aid ; Soaring
PE 8736 — life assurance ; Prohibitive
PE 1743 — linguistic interpretation legal proceedings ; Prohibitive
PF 1238 — living ; High
PF 1238 — living ; Prohibitive
Cost [maintaining...]
PE 1122 — maintaining comprehensive document collections ; Prohibitive
PF 0296 — maintenance ; Prohibitive
PF 8763 — manpower ; Excessive
PE 5755 — medical drugs ; Excessive
PF 9014 — money ; Prohibitive
Cost [natural...]
PE 9137 — natural gas trade infrastructure ; High
PF 1238 — necessities ; Inflated
PF 2385 — necessities rural communities ; Prohibitive
PF 7543 — nuclear power plants ; Prohibitive
PF 1212 — nutritious food ; Prohibitive
Cost [organically...]
PF 1212 — organically grown foodstuffs ; Prohibitive
PD 1644 — overruns large-scale public programmes
PF 8079 — overruns ; Unreported
Cost [parenthood...]
PF 1625 — parenthood ; Prohibitive
PF 2899 — *pesticide application ; Prohibitive*
PF 8016 — private medical care ; Prohibitive
PE 4404 — product liability protection ; Prohibitive
PE 1722 — product repair ; Prohibitive
PF 1345 — *Prohibitive drilling*
PF 4774 — *projections ; Incomplete*
Cost [religious...]
PF 3313 — religious ceremonies ; Burdensome
PF 0716 — research projects ; Misappropriation resources high
PF 2605 — *road construction ; Prohibitive*
PJ 8591 — *rural community buildings ; Prohibitive construction*
PF 7911 Cost social security ; Increasing
PF 7911 Cost social security ; Inflated
PE 8063 Cost transportation ; Prohibitive
PF 7911 *Cost unemployment benefits ; Inflated*
PF 7911 *Cost unemployment benefits ; Rising*
PF 2899 *Cost weed control ; Prohibitive*
PC 2035 Cost youth unemployment ; High
PF 2378 *Costly animal feeds*
PF 2378 *Costly cattle investment*
PF 6527 Costly distribution services
PD 6520 *Costly fire equipment*
PF 0296 *Costly furniture material*
Costly [home...]
PF 0296 — *home renovation*
PF 0296 — *hotwater heaters*
PF 0296 — *housing equipment*
PJ 7877 Costly imported foods
PF 2410 *Costly industrial machinery*
PE 8400 Costly medical care developing countries
PF 3455 Costly prestige projects ; Excessively
PF 2605 *Costly road improvements*
PE 3538 Costly rotation activities intergovernmental facilities
PJ 8591 *Costly school construction*
Costly [taxation...]
PF 4197 — taxation systems
PF 4375 — *teaching staff*
PF 4375 — technical training
PF 1653 — *telephone lines*
PD 0794 Costs benefits economic integration ; Imbalances distribution
PD 4632 *Costs businesses ; Prohibitive start up*
Costs [chicken...]
PF 2378 — *chicken raising ; Exorbitant*
PE 3149 — companies ; Evasion social
PE 4856 — cultural burdens human illnesses disabilities ; Excessive
PF 6025 Costs disaster relief ; Uncontrolled
PC 0328 Costs economic production ; Hidden environmental
PF 5956 Costs ; Escalating pension
Costs Excessive
PD 9338 — [Costs ; Excessive]
PE 3247 — *alimony*
PE 8063 — freight
PF 2899 — *hybrid seed*
PF 0995 Costs ; Extortionate legal
Costs High
PF 3318 — *electricity*
PD 1842 — housing
PF 8763 — labour
PF 1653 — *telephone*
Costs [Inadequate...]
PE 6700 — Inadequate evaluation environmental
PF 7932 — Increase operating
PF 0922 — Ineffective protection individual rights due excessive court
PE 5899 — inefficient port cargo handling ; Excessive
Costs Inflated
PF 8763 — labour
PF 8158 — overhead
PF 2899 — *seed*
PE 4162 Costs ; Land levelling
PF 4114 Costs present policies ; Unacknowledged future
Costs Prohibitive
PF 8158 — administrative overhead
PJ 9259 — agricultural diversification
PF 3550 — *construction*
PJ 1768 — *equipment*
PF 6548 — *facility*
PF 1345 — *irrigation*
PF 8763 — labour
PF 2378 — *livestock husbandry*
PF 2457 — *ownership*
PF 3630 — *refrigeration*
PF 1653 — *telephone*
PF 5743 Costs purification effluents emissions ; Prohibitive
Costs [School...]
PF 4375 — *School supply*
PE 5812 — sea access land locked developing countries ; Excessive
PE 6388 — staff intergovernmental organizations ; High
PD 8114 — structural adjustment debtor developing countries ; Excessive social
Costs [Unavailability...]
PF 4375 — *Unavailability training*
PF 8499 — Underestimation programme
PF 0296 — Unexpected maintenance
PF 6527 — *Unfeasible infrastructure*
PE 5896 — unsuitability insurance land locked developing countries ; Excessive
PD 1885 Cot deaths
PD 1250 Cottage industries developing countries ; Decline handicrafts
PE 1513 Cotton jute ; Instability vegetable fibre trade, excluding
PD 2227 *Cotton ; Leaf curl*
PE 2220 Cotton ; Pests diseases
PE 1510 Cotton trade ; Instability
PF 6389 Cotton-wooling
PF 6825 Cough
PE 2481 Cough ; Chin
PD 8239 *Cough ; Chronic*
PE 2481 Cough ; Hooping
PE 9854 *Cough ; Kennel*
PE 2248 Cough ; Persistent
PE 2481 Cough ; Whooping
PF 1781 *Council representation ; Neglected*
PF 1781 *Council representation ; Noninclusive*
PF 1653 *Councilman support ; Insufficient*
PF 3517 Counsel ; Denial right legal
PF 1219 Counsel minorities ; Inadequate legal
PF 0823 Counsel ; Narrow context
PF 0732 Counsel political dissidents ; Inadequate legal
PD 2389 *Counselling ; Lack genetic*
PF 2477 *Counselling ; Nonexistence career*
PF 6548 *Counselling ; Unlocated finance*
PD 5770 *Counsellors ; Bogus psychiatrists personal*
Counter [culture...]
PF 0423 — culture
PF 0423 — culture
PF 0423 — cultures
PD 2923 Counter-espionage
PC 1738 *Counter-insurgency warfare*
Counter productive
PF 8120 — effects foreign aid
PF 1492 — *government constraints programme beneficiaries*
PF 9786 — personnel reward systems
PD 9802 — price supports subsidies
PF 6911 — use crime-screening
Counter [reform...]
PF 7334 — reform
PF 3232 — revolution
PF 3232 — revolutionary activities
PA 6485 Counteraction
PD 5540 Counterfeit drugs
PE 5503 Counterfeit financial instruments
PD 7981 Counterfeit goods
PD 7981 Counterfeit goods ; Trade
Counterfeit [machine...]
PE 5319 — machine parts
PD 5540 — medicines
PE 5503 — money government securities
PD 7981 Counterfeit products services
PD 7981 Counterfeiting
PF 3104 Counterproductive capitalist investment financing
PF 3070 Counterproductive exchange control restrictions
PA 5454 Counterproductivity *Badness*
PA 5497 Counterproductivity *Difficulty*
PA 6876 Counterproductivity *Impotence*
PE 6621 Countertrade
PE 1961 Countervailing duties ; Distortion international trade export subsidies
PF 6140 Country ; Denial right walk
PF 6530 Country farms ; Low investment developing
PD 1291 Country human resources industrialized countries ; Loss developing
PF 2120 Country ; Imbalance city sizes
PE 8308 Country ; Jurisdictional conflict antagonism government agencies each
PE 9124 Country level ; Ineffectiveness international organization programme action
PE 7973 Country level ; Jurisdictional conflict antagonism international organizations
Country [residence...]
PC 0935 — *residence ; Denial right return*
PE 8414 — Restrictions nationals leaving their own
PE 8817 — Restrictions nationals returning their
PE 7403 — roads tracks ; Environmental degradation recreational use unsurfaced
PF 1667 — rural communities ; Restricted delivery essential services developing
Country [small...]
PE 0523 — small industry products ; Inadequate export promotion developing
PE 8747 — sojourn ; Restrictions foreigners leaving
PD 1291 — students studying foreign countries ; Non return developing
PD 3047 Country ; Tax discrimination against investment foreign
PD 3048 Country ; Tax discrimination against non residents
PE 5170 Countryside ; Absence skilled workers
PD 7585 Countryside careless walkers ; Degradation
PF 6140 Countryside city dwellers ; Inaccessibility
PF 6140 Countryside ; Denial right roam
PE 3914 Countryside ; Destruction
PC 2507 *Countryside ; Forced resettlement*
PE 8997 Countryside ; Irregular repair services
PE 8797 Countryside ; Lack secondary schools
PF 4959 Countryside ; Limited role women
PE 8084 Countryside ; Meagre youth facilities
PE 8797 Countryside ; Remote junior high schools
PE 8540 Countryside ; Unexplored entertainment alternatives
PF 3278 Couple ; Unmarried
PE 7271 Coupons ; Trafficking government benefit
PD 1964 Coups d'état
PD 2154 *Courses ; Changing river*
PF 0021 Courses professional fields ; Lack refresher
PD 9035 Court ; Contempt
PF 0922 Court costs ; Ineffective protection individual rights due excessive
PF 1487 Court delay
PE 4737 Court ; Denial right trial
PF 4676 Court ; Lack international criminal
PE 0405 Court proceedings ; Legal profession's monopoly
PE 8218 Courts ; Cessation functions civil
PE 0397 Courts ; Injustice religious
PE 0088 Courts ; Injustice special
PF 7897 Courts justice ; Government refusal accept jurisdiction international
PF 1487 Courts ; Overloading
PD 8614 Courts ; Restrictive practices
PE 7664 Courts ; Revolutionary
PE 7664 Courts tribunals ; Extrajudicial
PF 1870 Covenant ; Naive approach marriage
PF 1870 Covenant ; Reduced dimension marriage
PF 1764 Covenantal understanding sexuality ; Lost
PD 1026 Covenants life ; Breakdown
PE 6350 *Cover ; Excessive cloud*
PD 7231 Cover illegal activity ; Misuse research
PD 7231 Cover inaction ; Monitoring

Cover up

Cover up
PA 0005 — [Cover up]
PF 5340 — [Cover-up]
PE 6786 — Commercial
PF 5124 — convergence practice apparently opposed ideologies
PF 4854 — Environmental pollution
PF 4331 — extraterrestrial invasion
PF 2990 — Financial
PF 3819 — government harassment political activists ; Official
PF 3819 — government sanctioned assassinations ; Official
PF 4932 — Historical
PF 9756 — Legal
PF 7768 — major accidents
PD 4383 — Media
PF 5983 — Medical
PF 7768 — ship blast
Cover up unsafe
PF 4854 — buildings
PF 4854 — equipment
PF 4854 — industrial installations
PF 4854 — transportation systems
PC 1812 Cover ups ; Official
PE 5610 Coverage capitalist countries ; Political interference television news
PE 3831 Coverage developing country perspectives media ; Inadequate
PF 3538 Coverage ; Inadequacy patent
PE 7945 Coverage legal affairs ; Restriction press
PF 3073 Coverage legal affairs ; Restrictions news
PD 6081 Coverage news ; Biased media
Coverage [parliamentary...]
PF 3072 — parliamentary affairs ; Restriction press
PF 3072 — parliamentary affairs ; Restrictions direct news
PD 0917 — Parochial news
PF 0852 Coverage social security ; Unequal
PF 0419 Covert arrangements governments
PF 9184 Covert cooperation secret services
PB 1935 Covert himsa ; Institutionalized
PF 1978 Covert himsa ; Personal
PF 3199 Covert imperialism
PD 4501 Covert intelligence agency operations
PF 7669 Covert military operations
PF 1978 Covert non-physical psychological violence
PC 2175 Covert psychological warfare operations
PD 7171 Covert smear campaigns government
PE 3968 Covert trade nuclear bomb-making materials
PD 8465 Covert violation international treaties
PA 6999 Covetousness *Avarice*
PE 5191 Cow disease ; Mad (BSE)
PD 7424 *Cow ; downer*
PD 7420 *Cow syndrome ; Fat*
PA 5558 *Cowardice*
PA 6030 Cowardice *Fear*
PA 7325 Cowardice *Irresolution*
PA 5558 Cowardice *Weakness*
PE 2680 Cowdriosis
PE 6886 Cowpox
PE 6886 Cowpox ; Pseudo
PD 7424 *Cows following parturient paresis ; Neuromuscular paresis*
PD 7424 *Cows following parturient paresis ; Skeletal paresis*
PD 2731 *Cows ; Mastitis milking*
PD 7420 *Cows ; Parturient paresis*
PD 7799 *Cows ; Udder acne dairy*
PE 7826 *Coxitis*
PE 3852 Coxsackie virus
PE 8935 Coypu hutia ; Endangered species zagoutis
PE 2123 Crack
PE 2123 Crack cocaine
PE 7826 *Crack ; Quarter*
PE 7826 *Crack ; Toe*
PF 2099 Craft ; Wicca
PD 0694 *Craftsmanship ; Decline*
PE 4804 Craftsmen developed countries ; Lack artisans
PD 8725 Cramp ; Abdominal
PD 0270 *Cramp ; Writer's*
PE 4566 *Cramps*
PD 9366 *Cramps ; Heat*
PE 4161 *Crampy*
PE 9037 *Cranial abscesses ; Intra*
PE 8932 Cranial nerves ; Diseases
PF 6465 *Craniofacial malformations*
PD 4511 Crash ; Stock market
PE 3815 Crassipes ; Eichornia
PC 2696 *Craterization*
PE 7031 Craving ; Sexual
PE 1894 Craving ; Sugar
PF 5691 *Crazy nation-states*
PC 1707 Created isolationism ; Self
Creation [dams...]
PD 0767 — dams lakes ; Ecosystem modifications due
PD 0767 — dams lakes ; Environmental degradation due
PC 1573 — dominant class
PE 5404 Creation genetic engineering ; Accidental weed
PF 3061 Creation liquid reserves ; Inadequate mechanism
PC 0776 Creation new species ; Irresponsible
PF 9560 Creation ; Over reliance government money
PC 0776 Creation transgenic plants animals
PF 5148 Creation ; Violation integrity
PF 9358 Creationism ; Censorship textbooks response
PE 6093 Creative accounting
PJ 0095 Creative establishment/dis establishment tension ; Undervalued
PE 3948 Creative establishment disestablishment tensions ; Failure understand necessity
PF 2352 Creative future ; Inhibited capacity visualize
PF 0936 Creativity ; Collapsed meaning human
PF 1681 *Creativity elderly ; Unreleased*
PF 0275 Creativity innovation ; Suppression
PF 0275 Creativity ; Patterns curtail
PF 2703 Creativity traditional gifts ; Untransposed
PF 7043 Creatures ; Death living
PF 3826 Credibility alternative social structures ; Low
PA 5829 Credibility*complex
PJ 7629 Credibility diplomacy ; Loss
PB 6314 Credibility gap
PB 6314 Credibility gap ; Dependence
PF 1963 Credibility institutions ; Loss
Credibility international
PE 8064 — institutions ; Loss
PE 8064 — organizations programmes ; Low
PC 9584 — trading system ; Declining
Credibility Loss
PB 6314 — [Credibility ; Loss]
PF 1963 — institutional
PF 9016 — leadership
PF 5934 Credibility United Nations ; Loss
Credit [Abuse...]
PF 2166 — Abuse
PF 2489 — accessibility ; Limited
PF 3550 — alternatives urban areas ; Poor
Credit [capital...]
PF 2489 — capital ; Insufficient
PE 3592 — card fraud
PD 3067 — competition ; Export
PE 7902 Credit discrimination
PC 8316 Credit ; Excessive development
PE 7902 Credit extensions ; Inequitable
Credit [facilities...]
PE 8516 — facilities agricultural producers ; Lack
PD 3954 — financing ; Over commitment
PE 3066 — financing upon developing countries ; Burden export
Credit [Geographically...]
PE 7902 — Geographically denied
PE 3066 — guarantee facilities developing country exports ; Lack
PF 0245 — guidelines ; Inadequate
PJ 0646 Credit ; Individual need
PF 2489 Credit ; Lack local
Credit Limited
PF 2489 — availability
PF 2489 — availability financial
PF 2489 — local government
Credit [Limits...]
PE 8516 — Limits single cropping
PF 2489 — lines ; Limited
PF 2489 — lines ; Undeveloped
PF 2489 Credit monies ; Inadequate
PF 0245 Credit policies ; Inadequate
PF 0245 Credit procedures ; Static
Credit [rating...]
PD 3054 — rating developing countries ; Low
PE 7902 — Redlining denies
PF 2166 — risk
PF 3065 — risks ; Export
PF 2166 — risks ; High
Credit [small...]
PE 8516 — small farmers ; Limited access
PF 0245 — societies ; Lack mutual
PF 0245 — system ; Inappropriate
PE 4188 Credit used developing countries ; Undiversified forms
PF 0245 Credit uses ; Static
PF 1509 *Crediting obligations ; Family*
PD 4401 Creditors ; Defrauding secured
PD 3051 Creditors fulfilment debt service obligations developing countries ; Inconsiderate insistence
PE 3926 Creditors ; Inconsiderate choice payment times
PE 3066 Credits developing countries ; Decline export
PE 0938 Credits developing countries ; Excessive dependence export
PD 3054 Creditworthiness developing countries ; Insufficient
PD 0486 *Credulity*
PA 7371 Credulousness *Unintelligence*
PE 6313 Creep concrete
PD 2638 *Creep metals*
PE 7678 Creeping modernization military weaponry
PE 7905 Cretinism
PE 7915 *Creutzfeldt-Jakob disease*
PD 1885 Crib deaths
PE 3640 Crickets pests
PE 4630 *Cricopharyngeal achalasia*
PB 0001 Crime
PD 6655 Crime ; Aiding consummation
PE 4131 Crime ; Alcohol related
Crime [Commercial...]
PC 5624 — Commercial
PF 4983 — Complicity
PF 1485 — Compounding
PE 4362 — Computer based
PD 3528 — Corporate
Crime [Delay...]
PE 3913 — Delay payment compensation victims
PB 0001 — Dependence
PF 4924 — detection rate ; Decline
Crime [Economic...]
PC 5624 — Economic
PC 4584 — Environmental
PF 2874 — *Excessive community*
Crime [families...]
PC 2343 — families territories ; Organized
PF 4682 — Fear
PE 5516 — Frauds, forgeries financial
PF 0374 *Crime games ; Recreational*
PD 6943 Crime ; Government inducement
Crime [Inadequate...]
PD 4823 — Inadequate assistance victims
PF 4924 — Inadequate prevention
PE 6699 — Ineffective legislation against organized
PC 0742 Crime ; Khaki collar
PD 9316 Crime ; Legal bias against victims
PD 9316 Crime ; Legal indifference victims
PE 7354 Crime media ; Excessive portrayal
PE 7354 Crime media ; Glorification
PD 4823 Crime ; Neglect victims
PE 3913 Crime ; Non payment compensation victims
PC 2343 Crime ; Organized
Crime [pharmaceutical...]
PE 6618 — pharmaceutical industry ; Corporate
PC 0350 — Political
PE 0428 — prisons
PD 1464 — prosecution ; Ineffective war
Crime [rate...]
PE 4294 — rate ; Periods high
PE 7311 — rates ; Locales high
PC 2343 — *rings ; International*
Crime [screening...]
PF 6911 — screening ; Counter productive use
PJ 1688 — Senescent
PC 0277 — Statutory
PD 7399 Crime ; Urban
PC 5005 Crime ; Victimless
PD 4752 Crime ; Violent
PE 5516 Crime ; White collar
PD 2730 *Crimean-Congo haemorrhagic fever*
Crimes against
PC 0350 — body politic
PE 7311 — *family*
PE 5634 — government public property
PC 1073 — humanity
PF 1098 — humanity ; Remission sentences
PE 6486 — intangible property
PD 1163 — integrity effectiveness government operations
PC 0554 — national security
PD 8153 — national security ; Armed
PC 4584 — nature
PC 6239 — peace
PD 5511 — property
PC 5005 — public morality
PC 0350 — state
PC 5624 — *trade*
PE 6323 Crimes committed disregard orders ; Liability
Crimes committed during
PE 4294 — *civil unrest*
PE 4294 — *high unemployment*
PE 4294 — *war ; Civil*
Crimes [committed...]
PE 8420 — committed hospitals health care facilities
PF 4852 — *committed urban schools*
PD 4500 — Commodity exchanges related
PD 8679 — Corruptive
Crimes [Difficulty...]
PF 8661 — Difficulty ensuring prosecutions
PD 0762 — Disabled victims
PE 5037 — during narcotic investigations ; Police
PE 1108 Crimes ; Firearms explosives
PD 4158 Crimes ; Generalized types
PE 7311 *Crimes home*
Crimes [Inadequate...]
PE 8542 — Inadequate safeguards against computer
PD 1464 — Ineffective deterrent against war
PB 1992 — *Intimidation victims*
PF 4676 Crimes ; Lack international code
PC 0742 Crimes ; Military
PA 7030 *Crimes passion*
PD 5552 Crimes ; Property acquisition
Crimes related
PE 5331 — foreign relations trade
PE 3889 — immigration, naturalization passports
PD 5941 — military service obligations
PE 3997 — national security information
PE 5546 Crimes ; Revisionism anti marxist
Crimes [Sex...]
PD 4082 — Sex
PC 0350 — *state*
PG 6455 — *state ; Homicidal*
PD 4500 — Stock exchange bourse related
Crimes [Undocumented...]
PF 4062 — Undocumented war
PF 6911 — Uninvestigated
PD 4779 — Unjust punishments
PF 1456 — Unreported
PF 6911 — Unsolved
PC 5005 Crimes victims
PD 4752 Crimes violence
Crimes [War...]
PC 0747 — War
PD 5300 — which endanger people
PF 7728 — which they are innocent ; Failure clear names people convicted
Criminal [act...]
PF 6158 — act ; Multiple convictions single
PF 6158 — act ; Multiple liability identical
PE 2449 — activity ; Supplying firearms
PE 1178 — association
PD 5321 — attempt
PD 4469 Criminal blackmail

Criminal [cases...]
PE 1032 — cases ; Pecuniary guarantees condition provisional release
PF 5544 — characteristics
PF 7336 — charges ; Denial right be informed
PE 4628 — charges ; Violation right defence against
PD 4469 — coercion
PF 8000 — concealment
PD 1767 — conspiracy
PD 5705 — contempt
PF 4676 — court ; Lack international
PE 4343 Criminal espionage
PD 4469 Criminal extortion
PD 6845 Criminal facilitation
PD 3837 Criminal gangs
Criminal [harassment...]
PD 2067 — harassment
PD 0437 — harm persons
PD 5511 — harm property
Criminal [insanity...]
PD 9699 — insanity
PE 6771 — intrusion
PF 7277 — investigations ; Obstruction international
PD 5750 — investment youth market
Criminal justice
PD 0412 — Delay administration
PF 8479 — Error
PF 4875 — system ; Deficiencies
PD 7520 *Criminal killing animals*
PF 4676 Criminal law ; Inadequate development international
Criminal liability
PB 0001 — [Criminal liability]
PF 7293 — associations ; Limited
PF 7293 — corporations ; Limited
PF 6158 — Overlapping
PE 6323 — Vicarious
Criminal [menacing...]
PE 4467 — menacing
PD 1350 — mischief
PA 6048 — *motivation*
PA 2658 Criminal negligence
PE 5538 Criminal negligence performing socialist responsibilities
Criminal [offences...]
PC 4584 — offences against environment
PC 7373 — *offenders ; Preponderance male*
PD 6845 — *omission act*
PD 5859 Criminal property damage
PD 0634 *Criminal reasons ; Confinement non*
Criminal [solicitation...]
PD 7676 — solicitation
PA 9226 — state mind
PF 1244 — stereotypes ; Inaccurate
PE 5508 — sub-culture
PE 5508 — subculture
PE 4183 — suspects ; Abuse rights
PE 4183 — suspects ; Withholding information police
PF 8493 — syndicalism
PE 4661 Criminal threat
PD 3794 Criminal trespass
Criminal [use...]
PB 0001 — *use ; Technology*
PB 0001 — *use ; Tools*
PE 1181 — usury
PD 8709 Criminal violation civil rights
PC 7373 Criminal violence disorders
PD 9397 Criminal workers
PA 9226 Criminality
PE 5592 Criminality ; Increasing female
PF 6169 Criminalization abortion
PF 4735 Criminalization drug use
PF 2643 Criminalization euthanasia
PE 5460 Criminalization homelessness
PE 1903 Criminalization homosexuality
PF 6231 Criminalization prostitution
PE 9122 Criminalization sexual relations out wedlock
PD 0437 Criminally life endangering behaviour
PD 5349 Criminally reckless endangerment
PB 0001 Criminals
PC 7373 Criminals
PD 7407 Criminals ; Aiding national security
PE 7345 Criminals ; Denial right accused segregation convicted
PE 7209 Criminals ; Denial right juvenile criminals segregation adult
PE 1837 Criminals ; Female
PE 4697 Criminals ; Government approved employment war
PE 5581 Criminals ; Habitual
PD 5541 Criminals ; Havens war
PE 8900 Criminals ; Inadequate facilities mentally disabled
PD 1464 Criminals ; Inadequate punishment war
PE 3488 Criminals mutilation ; Punishment
PE 4439 Criminals policy making contexts ; Employment
PE 4439 Criminals policy making contexts ; Employment war
PE 7209 Criminals segregation adult criminals ; Denial right juvenile
PF 4875 *Criminals their own countries ; Inadequate punishment international*
PD 4067 Criminals ; Unconvicted war
PD 9316 Criminals ; Undue consideration
PC 0212 Criminals ; Young
PD 3158 *Crinodiea ; Endangered species*
PD 0196 Crippled children
PD 6571 *Crippling dependence mechanization*
PF 1254 Crippling leadership void
PF 6530 Crippling subsistence planning
PE 3407 Crises ; Emotional
PD 4840 Crises ; Personal life
PF 4302 Crises ; Risk unintentional nuclear war due international
PF 9769 Crises ; Unforeseen environmental
PE 8080 Crises ; World tension oil petroleum
PF 7999 Crisis capitalism
PA 6971 Crisis *Danger*
PD 2133 Crisis developing countries ; External debt
Crisis [emergency...]
PF 5660 — emergency responses ; Obstruction change
PC 6329 — Energy
PB 6335 — escalation ; International
PD 5675 Crisis ; Female mid life
PF 4850 Crisis ; Fiscal
Crisis Global
PF 6244 — [Crisis ; Global]
PC 5876 — economic
PF 3612 — financial
PF 1934 *Crisis identity*
PF 0048 Crisis international monetary financial system
PF 5956 Crisis long-term pension funds
PD 5783 Crisis ; Male mid life
PF 5660 Crisis management approach long-term development policy-making
PF 0817 Crisis ; Nationalistic policy responses world economic
Crisis oriented
PF 5660 — *education*
PF 2849 — funding
PF 5660 — operations
PF 3824 Crisis social studies ; Terminological
PD 5046 Crisis ; World food
PE 7742 Criteria construction industry ; Reductionistic decision making
PD 1060 Criteria judicial judgements ; Exclusion masses setting
PD 0159 Criteria legal induced abortion ; Variable
PF 5039 Critical awareness ; Inadequate development
PD 9306 Critical equipment ; Undermanning
PE 9038 Critical illnesses
PF 4530 Criticism
PF 2437 *Criticism ; Anticipation*
PF 2437 *Criticism ; Feared public*
PF 9385 Criticism government
PF 8396 Criticism ; Intolerance
PE 1478 Criticism media ; Excessive unsupportive
PF 9385 Criticism national leadership
PF 9385 Criticism official institutions
PF 8396 Criticism ; Unwillingness contemplate
PD 9045 *Crohn's disease*
PF 4549 Cronyism
PF 6106 *Crop attachment ; Traditional*
PE 7313 *Crop consumed ; Money*
PE 8476 Crop contamination
Crop [damage...]
PC 3150 — damage wildlife
PD 5730 — dangers ; Unknown
PC 2223 — diversity ; Limited
PF 2421 — *dusting*
PD 7480 *Crop failures*
PC 3606 Crop farming ; Single
PE 8476 Crop growth pollution ; Inhibition
PF 8827 *Crop insurance ; Inadequate*
PF 6524 Crop management ; Limited
Crop [policy...]
PF 9187 — policy ; Inappropriate cash
PE 3880 — pollinating insects ; Destruction
PE 3880 — pollinating species ; Destruction
PD 9366 — *poultry ; Pendulous*
PF 8635 — prices ; Unstable
PF 3698 Crop rotation ; Inadequate
Crop [shortfalls...]
PD 5174 — shortfalls
PF 9187 — Single cash
PE 4497 — Sour
PF 2899 *Crop treatment ; Prohibitive cost*
PF 8354 Crop uncertainty ; Insecurity future
Crop [varieties...]
PF 3145 — varieties ; Excessive number
PF 3146 — varieties ; Introduction high yield
PC 2223 — variety ; Limited
PD 0660 — vulnerability
PE 8233 Crop water ; Unregulated
PD 7480 Crop yields ; Low
PF 1479 Crop yields ; Variability
PE 8516 Cropping credit ; Limits single
PD 6103 Cropping ; Energy
PF 9187 Cropping ; Export
PE 1327 Cropping ; Unjustified game
PE 8476 Crops ; Adverse impact pollution growth
PD 5730 *Crops ; Damage*
PE 8324 Crops ; Degradation agricultural land cash
PE 3584 Crops ; Fumigant damage
PD 3577 Crops ; Fungicide damage
PF 9187 Crops guaranteeing survival ; Discouragement food
PE 4744 Crops ; Harmful effects air pollution agricultural
PD 1224 Crops ; Herbicide damage
PD 3695 Crops ; Insecticide damage
PF 6455 *Crops ; Insufficient cash*
Crops [Limited...]
PF 6455 — *Limited cash*
PD 4100 — Limited farm
PD 7480 — Loss agricultural
PE 7313 — *Low price cash*
PD 1285 — *Low yield rice*
PC 3606 Crops ; Monoculture
PD 4100 Crops ; Narrow range food
PF 9187 Crops ; Non diversification market
Crops [Pesticide...]
PD 2581 — Pesticide damage
Crops [pests...] cont'd
PE 7783 — pests diseases
PE 5682 — plants ; Weather hazards
PF 9187 — Proliferation cash
PD 7565 Crops tropical economies ; Preponderance non food
PD 4100 Crops ; Undiversified feed
PD 5730 Crops ; Vulnerability plants
PE 5682 Crops weather ; Vulnerability
Cross [border...]
PD 5272 — border anti-terrorist raids
PD 5272 — border military operations
PE 8256 — breeding ; Lack availability methods
PF 9216 Cross conditionality restrictions multilateral development aid ; Policy
PE 4039 Cross licensing agreements ; Patent pool
PF 6473 *Crowded rush-hour transportation*
PF 3850 Crowding mental health ; Negative effects over
PE 2230 Crown gall
PD 7424 *Cruciate ligaments ; Rupture*
PE 4346 *Crucifers ; Blackleg*
PE 5465 *Crucifers ; White rusts*
PF 0906 Crude birth rate ; High
PE 8464 Crude chemicals derived coal, petroleum natural gas ; Instability trade mineral tar
PE 8242 Crude chemicals ; Shortage coal derived tar
Crude fertilizers
PE 0760 — crude minerals, excluding coal, petroleum precious stones ; Instability trade
PE 1353 — crude minerals, excluding coal, petroleum precious stones ; Long term shortage
PE 9049 — Instability trade
PE 0760 Crude minerals, excluding coal, petroleum precious stones ; Instability trade crude fertilizers
PE 1353 Crude minerals, excluding coal, petroleum precious stones ; Long term shortage crude fertilizers
Crude non fuel
PE 0546 — materials ; Environmental hazards inedible
PD 0280 — materials ; Instability trade inedible
PE 0461 — materials ; Long term shortage inedible
PE 8351 — materials ; Restrictive practices trade inedible
PE 0701 Crude synthetic reclaimed rubber ; Instability trade
PD 0390 Cruel animal transportation
PE 5735 Cruel animal traps
PE 3484 Cruel culling seals
PE 0358 Cruel inefficient slaughterhouse practices
PC 3768 Cruel, inhumane degrading punishment ; Denial right freedom
PE 0360 Cruel methods destruction stray dogs
PB 2642 Cruel people
PC 3768 Cruel punishment
PD 1323 Cruel sports
PD 0260 Cruel treatment animals research
PE 9611 Cruel treatment animals testing pharmaceutical products
PB 2642 Cruelty
Cruelty animals
PC 0066 — [Cruelty animals]
PD 2768 — factory farming
PE 0236 — food preparation
PB 2642 *Cruelty ; Child*
PC 0838 Cruelty children
PB 2642 *Cruelty ; Disguised*
PE 5286 Cruelty fish
PE 3394 Cruelty herd animals
PE 3468 Cruelty insects
PD 0584 *Cruelty ; Mental*
PA 7023 Cruelty *Pitilessness*
PD 4148 Cruelty plants
PA 5643 Cruelty *Unkindness*
PD 0247 Cruelty women
PF 1390 Cruising ; Homosexual
PC 2371 Crusade
PB 0731 *Crushing*
PE 3719 Crushing ; Torture
PD 3472 Crustacea ; Endangered species
Crustacea molluscs
PD 0372 — Environmental hazards fish,
PE 0972 — preparations thereof ; Instability trade fish,
PE 1783 — preparations thereof ; Long term shortage salt water fish,
PC 0728 *Crustacean pests*
PE 4932 Cryptococcosis
PE 2738 *Cryptosporidiosis animals*
PD 2226 *Cucumber ; Bacterial wilt*
PD 3158 *Cucumbers ; Endangered species sea*
PD 2226 *Cucurbits ; Bacterial wilt*
PE 3623 Culicine mosquitoes vectors disease
PE 3484 Culling seals ; Cruel
PA 6706 Culmination
PF 7428 Culottes ; Sans
PA 6191 Culpability *Disapproval*
PA 6793 Culpability *Guilt*
PE 0671 Culpability ; Violations regulatory codes
PD 2315 Cult dead
PF 9195 Cult dead
PF 6330 Cult machinery ; Organized
PF 3900 Cult mediocrity
PF 8363 Cult relics
PF 7129 Cult sacred scriptures
PF 6766 Cult youth
PC 0219 Cultivable land ; Shortage
PE 7265 Cultivated gardens ; Over
PD 5936 Cultivated plants ; Declining breeds
PC 2223 Cultivated plants ; Decreasing genetic diversity
PD 4563 Cultivation illegal drugs
PD 4563 Cultivation illegal drugs ; Government sanctioned

Cultivation long

PC 0238 *Cultivation long lifed trees ; Unsustainable*
PD 4273 Cultivation marginal agricultural land
PE 9773 Cultivation plant species ; Banned
PE 9773 Cultivation ; Regulatory banning plant
PD 7516 Cultivation ; Shifting
PD 6282 Cultivation steep slopes ; Inappropriate
PF 7414 Cultivation trivia
PF 5375 Cults ; Cargo
PE 6336 Cults ; Manipulative
PC 1123 Cults ; Media personality
PC 1123 Cults ; Personality
PC 1123 Cults politics ; Personality
PC 5088 Cultural alienation
PF 5178 Cultural arrogance
Cultural artefacts
PE 9417 — Collection
PD 2478 — environmental pollution ; Damage
PE 9004 — Smuggling
PE 9825 — tourism ; Deterioration
PE 8162 — Vibration damage
PE 3450 Cultural asylum ; Denial right
PF 8105 Cultural attractions ; Limited
Cultural [barriers...]
PB 2331 — barriers
PB 2331 — barriers ; Dependence
PC 8344 — bias
PE 4856 — burdens human illnesses disabilities ; Excessive costs
Cultural [conflict...]
PC 1573 — *conflict*
PF 2504 — context ; Limited
PE 4785 — cooperation ; Obstructions international personnel exchanges
PC 2913 — corruption
Cultural [dances...]
PF 2168 — *dances dated*
PC 9083 — decline
PD 0826 — degradation recreation tourism
PC 1351 — deprivation
PC 1351 — deprivation ; Dependence
PF 2679 — differences ; Lack appreciation
PF 2679 — differences ; Lack appreciation
PF 3099 — diplomatic commercial exchanges ; Abuse international
PC 8344 — discrimination
PE 6529 — discrimination administration justice
PA 6858 — *disintegration*
PF 8156 — diversity ; Bureaucratic insensitivity
PE 1862 — diversity ignored social service agencies
PE 5959 — dominance translation
PA 0839 — *domination*
PF 1084 — dreams ; Collapse
Cultural [elitism...]
PA 1387 — *elitism*
PC 4588 — environment degradation ; Socio
PC 1328 — ethnocide
PF 5250 — events ; Infrequent community
PF 4857 — exchange ; Obstacles international
PF 3628 — exposure ; Narrow range
PE 4054 — expressions peoples ; Abuse traditional
PF 0536 Cultural fragmentation
PF 3340 Cultural gap
PC 1328 Cultural genocide
Cultural heritage
PC 7605 — Abusive exploitation
PD 4272 — Acculturation dilution
PC 2114 — Destruction
PF 1985 — Failure pass
PF 1985 — Fragmented celebration
PF 1985 — Ignorance
PF 1985 — Lack access
PF 1985 — Loss
PE 6074 — Lost
PF 1985 — Remote
PF 1985 — transmission ; Insufficient
PF 1985 — Unconveyed
PF 1985 — Unknown
PF 8623 Cultural heroes ; Insufficient
Cultural identity
PF 9005 — Denial right
PD 2614 — Fear losing
PF 9005 — Loss
PF 0836 — Obsolete basis
PD 2041 Cultural illiteracy
Cultural imperialism
PC 3195 — [Cultural imperialism]
PC 3195 — Dependence
PE 9643 — Submission media
Cultural [integration...]
PC 1532 — integration immigrants ; Inadequate
PF 0536 — integration ; Lack
PF 3037 — interchange ; Restrictive channels
PC 2548 — invasion
PC 3943 — isolation
PF 1696 — isolation ; Local traditions
Cultural [lag...]
PF 7050 — lag understanding information
PE 6561 — life ; Denial right
PD 0784 — life ; Exclusion disabled persons social
PF 3211 *Cultural media ; Insufficient*
PF 3340 Cultural misunderstanding ; Inter
PF 1790 *Cultural nodes ; Insufficient*
PF 5301 Cultural norms ; Systematic human behaviour according
PE 6074 Cultural objects ; Illicit appropriation
Cultural [patterns...]
PF 8598 — patterns ; Rigid
PJ 2555 — penetration ; Transnational
PF 0152 — pluralism ; Divisive effects official
PC 8520 — prejudice
PE 5678 — products capitalist socialist countries ; Imbalance trade
PE 5702 — products developed developing countries ; Imbalance trade
Cultural property
PD 6907 — Accumulation
PD 7298 — during warfare ; Destruction
PF 7542 — Gradual destruction
PF 7542 — Inadequate protection preservation
PE 6074 — Misappropriation
PE 6074 — Non restitution
PE 6074 — occupied territories ; War time exportation
Cultural [racism...]
PC 8344 — racism
PF 2015 — ranking ; Subjective
PC 8425 — repression
PF 1373 — *resources ; Unused*
PF 3235 — revolution
PD 5907 — rights ; Denial
PE 1046 — role western women ; Elimination socio
Cultural [separatism...]
PD 2490 — separatism
PD 2695 — setting ; Monolingualism multi
PE 1082 — social identities developing countries transnational corporations ; Disruption
PF 7130 — *space ; Undefined*
PC 8269 — stagnation
PF 5957 — suicide
PC 7605 — symbols ; Misappropriation
Cultural [tours...]
PF 3099 — *tours ; Biased*
PD 9651 — trade barriers ; Inter
PF 1373 — tradition ; Untransposed significance
PF 9529 — traditions blocking business profit
PF 8156 — traditions ; Insensitivity diversity
PF 2258 Cultural vacuum ; International
PE 8152 Culturally adapted pedagogical material ; Limited access
PE 6529 Culturally biased legal systems
PF 1466 Culturally biased religion
PF 1624 Culturally determined medical practices
PC 2122 Culturally important landscapes ; Degradation
PD 8638 Culturally insensitive house design
PC 0381 Culturally unsustainable development
PJ 2601 Culture ; Adversary
PF 1447 Culture ; Arms
Culture [Counter...]
PF 0423 — Counter
PF 0423 — Counter
PE 5508 — Criminal sub
PF 3307 *Culture-denying competition*
PF 3012 Culture ; Disappearance local
PF 5472 *Culture ; Excessive bias favour European Western*
PE 6561 Culture ; Exclusion community's
PF 3408 Culture ; Inadequate appreciation
PA 6030 *Culture-induced fear*
PF 5513 Culture ; Pseudo
PF 8521 Culture ; Rapidly changing
Culture [Scarce...]
PF 6535 — Scarce options involvement
PC 2673 — shock
PF 1624 — specific diseases
PE 9643 — specific programmes ; Inability generate sufficient
PF 5659 — support ; Insufficient minority
PD 6279 Culture violence
PF 7498 Culture youth violence
PF 0423 Cultures ; Counter
Cultures [Degradation...]
PC 7203 — *Degradation indigenous*
PC 7203 — Destruction indigenous
PE 1817 — Disparagement indigenous
PB 8613 — Dying
PB 8613 Cultures ; Endangered
PC 7203 Cultures ; Endangered indigenous
PB 1071 Cultures ; Homogenization
PF 2648 Cultures ; Incompatibility rural values urban
PE 3831 Cultures media ; Excessive portrayal perspectives industrialized
PD 3501 *Cultures ; Primitive*
PF 8521 Cultures ; Rapidly changing
PF 8753 Cultures response social problems ; Failure interrelate wisdom different
Cultures [Underreporting...]
PD 2623 — Underreporting hazards minority
PF 3340 — Unfavourable opinions
PF 2126 — Untransferability books countries
PF 2126 — Untransferability images countries
PD 2269 Culverts ; Unstandardized drainage
PF 1076 Cumbersome methods policy formation
PF 6158 Cumulation convictions
PE 3296 Cumulative depletion corporate initiative rural communities
PF 1105 Cumulative environmental impacts
PE 5606 Cumulative trauma syndromes
PA 7411 Cunning *Uncommunicativeness*
PA 7232 Cunning *Unskillfulness*
PE 7826 *Curb*
PJ 8933 *Curbed learning incentive*
PE 8261 *Cures ; Inaccessibility high technology medical*
PD 2637 *Curfew*
PD 2227 *Curl cotton ; Leaf*
PE 4586 *Curl diseases ; Leaf*
PE 4586 *Curl ; Peach leaf*
PF 6102 Currencies ; Misalignment
PF 3069 Currencies ; Non convertibility
PF 6102 Currencies ; Overvalued
PF 6102 Currencies ; Undervalued
Currency [black...]
PD 5905 — black market
PD 2413 — black market socialist countries
PF 3070 — Blocked
PE 3700 Currency ; Depreciation
PE 3700 Currency devaluation
PE 2145 Currency manipulations accounting records transnational corporations ; Foreign
PF 6094 Currency ; Nationalistic attitudes
PF 3070 Currency restrictions
Currency [shortage...]
PD 3068 — shortage developing countries ; Foreign
PD 3068 — shortage developing countries ; Hard
PD 9489 — speculation ; Foreign
PF 3070 Currency use abroad ; Refusal issue foreign
PF 2574 Currency ; Variations national forms
PB 0731 *Current accidents ; Electric*
PE 9295 Current policies ; Government neglect future problems social environment arising
PF 9450 Current policy ; Uncertain environmental impact
Current [scientific...]
PF 4014 — scientific knowledge ; Limitation
PF 8757 — scientific knowledge ; Mismatch public policy
PC 2087 — shift ; Ocean
PF 3549 Curricula based content rather than method ; Educational
PF 0443 Curricula ; Imbalance educational
PF 0443 Curricula ; Irrelevance educational
PD 1827 Curricula over emphasizing method rather than content ; Educational
PD 1566 *Curricular activities ; Restricted extra*
PF 0443 Curriculum ; Disrelated school
PF 0443 *Curriculum format ; Static*
PF 8085 Curriculum ; Inadequate skills
PF 0443 Curriculum ; Irrelevant school
PD 6298 Curriculum ; Politicization
PF 0443 Curriculum ; Prescribed irrelevant
PJ 5365 *Curriculum ; Rigid required*
PF 0443 *Curriculum ; Static school*
PF 0443 *Curriculum ; Unsupplemented school*
PF 0704 Curriculum vitae ; False
PF 2837 *Curriculum writing ; Complex*
PF 3311 Curses
PF 3746 Cursing
PF 0275 Curtail creativity ; Patterns
PD 4060 Curtailment discussion issues meetings
PE 7136 Curtailment living standards borrower nations ; Enforced
PA 7143 Curtness *Discourtesy*
PA 7411 Curtness *Uncommunicativeness*
PE 4113 Curvature spine
PE 8537 Cuscuses ; Endangered species phalangers
PD 9654 *Cushing's syndrome*
PF 3252 Custody children ; Conflict concerning legal
PF 3253 Custody children ; Non parental
PD 0550 Custody deprived children ; State
PE 5152 Customary rights ; Unjust
PD 0813 Customary tenure systems ; Maldistribution land
PE 3337 *Customary working patterns*
PD 0826 *Customs ; Abuse traditional*
PE 2603 Customs administrative entry procedures ; Distortion international trade discriminatory
PD 1987 Customs duties ; Excessive protection industries
PD 2620 Customs excise duties ; Evasion
PE 4033 Customs excise officials ; Corruption
PE 0208 *Customs formalities ; Excessive*
PD 2620 Customs fraud
PF 1309 Customs ; Obsolescence rituals
PF 4033 Customs officials ; Bribery
PD 2955 Customs revenue ; Dependence developing countries
PE 2603 *Customs services developing countries ; Absence coordinated*
Customs [tax...]
PD 1466 — tax evasion
PE 2603 — trade formalities ; Excessive
PD 1095 — traditions developing countries ; Decline rural
PF 1309 Customs ; Underutilization historical rites
PE 0497 Customs valuation practices ; Distortion international trade restrictive
PF 6570 *Cut-off fragmented families*
PE 2727 *Cutaneous acariasis animals*
PC 8534 Cutaneous diseases
PE 5380 *Cutaneous habronemiasis*
PE 3508 *Cutaneous larva migrans*
PD 9667 *Cutaneous streptotrichosis*
PD 9667 *Cuterebra infestation small animals*
PB 0855 *Cuts abrasions*
PB 0855 *Cuts ; Accidental*
PD 9082 Cyanide fishing
PD 5228 *Cyanide poisoning animals*
PC 0062 *Cyanide pollutant*
PE 0447 *Cyanosis*
PF 1277 Cycle boom depression ; Economic
PD 9670 Cycle ; Disruption hydrological
Cycle [Epidemics...]
PE 9528 — Epidemics associated sunspot
PD 4632 — Expensive middlemen
PE 6658 — Extension family
PF 6148 Cycle ; Inhibition individual psychological development life

Death

PE 2524 Cycle rural communities ; Deepening debt
PF 1078 Cycle ; Subsistence living
PF 2346 Cycle ; Unchecked poverty
PC 2008 Cycles ; Manipulated product life
PF 1740 Cycles ; Short term planning product life
PD 2586 Cycles ; Underreporting disruptions natural water
PF 1172 Cycles ; Vulnerability economic
PF 1277 Cyclic business recessions
PC 6114 Cyclic changes climate ; Long term
PD 5453 Cyclic haematopoiesis
PD 5453 Cyclic neuropenia grey Collie dogs
PD 2968 Cyclic swings commodity prices
PD 2732 Cyclic thrombocytopenia ; Canine infectious
PE 9528 Cyclic variations solar radiant energy ; Medium term
PF 1744 Cyclical planetary glaciation
PF 1172 Cyclical recession ; Aggravation
PD 8786 Cyclitis
PD 1590 Cyclones ; Tropical
PD 1739 Cyclones ; Tropical
PD 3634 Cyclops
PD 1318 Cyclothymia
PE 2356 Cynanche parotidea
PF 3418 Cynicism
PA 6822 Cynicism *Disrespect*
PA 6099 Cynicism *Hopelessness*
PA 5643 Cynicism *Unkindness*
PE 7826 Cyst like lesion ; Osseous
PE 7826 Cyst pedal bone ; Lameness caused bone
PE 7570 Cyst ; Salivary
PE 7826 Cyst ; Subchondral bone
PE 3331 Cystic fibrosis
PD 7799 Cystic follicles
PD 7799 Cystic hyperplasia
PD 7799 Cystic ovary disease ; Follicular
PD 7799 Cystic ovary disease ; Luteal
PE 6586 Cysticercosis
PE 9291 Cystinosis
PE 0078 Cystitis
PD 9293 Cystitis ; Bovine
PD 9293 Cystitis ; Porcine
PD 7799 Cysts animals ; Prostatic
PD 7799 Cysts ; Follicular
PD 9667 Cysts ; Interdigital
PD 7799 Cysts ; Luteal
PE 9427 Cytauxzoonoses ; Feline
PD 0594 Cytomegalic inclusion disease

D

PD 0287 D deficiency ; Vitamin
PE 7769 D enterotoxaemia sheep ; Type
PD 8786 Dacryocystitis
PD 3003 Daily life ; Corporate domination
PD 1703 Daily life ; Human wisdom unrelated
PF 6182 Daily physical activity urban environments ; Unhealthy lack
PF 0827 Daily seasonal peaks ; Underutilization facilities due
PD 7799 Dairy cows ; Udder acne
PE 3528 Dairy markets ; Lack
Dairy products eggs
PE 0505 — Environmental hazards
PE 0576 — Instability trade
PE 8225 — Long term shortage
PD 7699 Dairy products ; Human consumption
PD 0767 Dam construction
PD 9515 Dam construction ; Environmentally harmful
PE 9517 Dam failures
PD 9118 Damage agricultural land
Damage [Badness...]
PA 5454 — *Badness*
PD 0992 — Brain
PE 8162 — buildings ; Vibration
Damage [caused...]
PE 0250 — caused space objects
PD 8719 — civilian areas ; War
PA 0832 — *Community*
PD 5859 — Criminal property
Damage crops
PD 5730 — [*Damage crops*]
PE 3584 — Fumigant
PD 3577 — Fungicide
PD 1224 — Herbicide
PD 3695 — Insecticide
PD 2581 — Pesticide
PD 2478 Damage cultural artefacts environmental pollution
PE 8162 Damage cultural artefacts ; Vibration
Damage [death...]
PE 2041 — death foetus
PE 1256 — *degradable organic matter*
PE 4609 — drug use ; Foetal
PE 4109 Damage ear ; Noise
PD 8786 Damage ; Eye
PD 2478 Damage frescoes ; Pollution
PE 4447 Damage goods
PD 0448 Damage heart
PD 8050 Damage human body ; Radiation
Damage [Impairment...]
PA 6088 — *Impairment*
PA 7395 — *Inexpedience*
PE 4874 — infant brains malnutrition insufficient stimuli
PE 1028 Damage ; Liver
PA 7382 Damage *Loss*
PD 1206 Damage materials ; Radiation
PD 2478 Damage monuments ; Pollution

Damage [paintings...]
PD 2478 — paintings ; Pollution
PD 5859 — premises ; Risk
PD 5859 — Property
Damage [sculpture...]
PD 2478 — sculpture ; Pollution
PF 8026 — small blood vessels kidneys, eyes nerves
PE 4447 — stock ; Risk
PD 3657 — stored manufactured goods ; Insect
PE 1334 — structures ; Wind
PE 2305 Damage ; Tidal water
Damage [Uncontrolled...]
PD 5859 — Uncontrolled property
PF 3499 — Uncontrolled waterfront
PC 0952 — *Unrestrained animal*
PD 5178 Damage ; Vulnerability marine environment catastrophic warfare
PC 1784 Damage ; Vulnerability property theft
PC 3150 Damage wildlife ; Crop
PD 0500 Damage wildlife ; Forest
PE 4447 Damaged merchandise
PF 6026 Damages arising natural disaster ; Inadequate insurance against
PE 0290 Damages consumers ; Non payment compensation
PD 7179 Damages ; Delay payment compensation
PD 7179 Damages ; Lack compensation
PD 7179 Damages ; Non payment compensation
PD 7179 Damages ; Uncompensated
PE 4555 Damaging agents ; Lung
PD 2541 Damaging unremoved litter
PD 0767 Damming ; Adverse effects
PA 7237 Damnation *Condemnation*
PA 6542 Damnation *Destruction*
PA 6191 Damnation *Disapproval*
PA 5643 Damnation *Unkindness*
PE 2248 *Damp*
PE 2248 *Damp places*
PE 4787 Damping-off disease plants
PD 9515 Dams ; Coastal erosion resulting
PE 8148 Dams ; Destruction fisheries
PE 0715 Dams ; Disastrous failure natural
Dams Inundation
PE 3786 — agricultural land
PE 7855 — forests
PE 7794 — wildlife habitats
PD 0767 Dams lakes ; Ecosystem modifications due creation
PD 0767 Dams lakes ; Environmental degradation due creation
PE 9517 Dams ; Malsituated
Dams [Sedimentation...]
PD 3654 — Sedimentation
PD 3654 — Siltation
PD 0253 — *Submergence historical sites*
PE 8796 Dams ; Water losses
PF 1756 Dance ; Sensuous
PF 2168 Dances dated ; Cultural
PD 7841 *Dancing pigs*
PC 8534 *Dandruff*
PE 2260 Dandy fever
PA 6971 *Danger*
PB 1992 Danger
PD 5300 Danger person ; Offences involving
PF 4612 Danger ; Reactionary methods avoiding
PF 6559 Danger ; Unacknowledged highway
PE 8933 Danger young people ; Dependence excitement
PC 2321 Dangerous animals
PE 7175 Dangerous animals ; Breeding
PF 7580 Dangerous behaviour
PF 4753 Dangerous behaviour ; Government approval
PD 9108 Dangerous cargo handling
PD 4278 Dangerous countries
PE 9334 Dangerous driving practices
PF 7580 Dangerous foot paths
PF 1654 Dangerous forms dialogue
PD 0971 Dangerous goods ; Risk transporting
PD 0971 Dangerous goods ; Transport
PE 4897 Dangerous harbours docks
PD 1018 Dangerous health ; Dependence government income products
PE 2061 Dangerous illegal products developing countries ; Dumping
PF 5188 Dangerous legally prescribed chemical residue levels
PC 6913 Dangerous materials
PE 6860 Dangerous materials frontier areas
PC 1640 Dangerous occupations
PF 7580 *Dangerous paths*
PE 7175 Dangerous pet animals
Dangerous substances
PC 6913 — [*Dangerous substances*]
PD 4542 — Discharge
PD 2800 — domestic waste water ; Discharge
PD 2468 — Inadequate labelling
PD 0575 — industrial waste water ; Discharge
PE 9813 — laboratory waste water ; Discharge
PD 7636 — rivers ; Discharge
PD 3666 — sea ; Discharge
PD 1414 — sewage systems ; Discharge
PC 6913 — *Storage*
PE 1158 Dangerous toys
PC 3151 Dangerous wild animals
PA 6971 Dangerousness *Danger*
PA 7309 Dangerousness *Uncertainty*
PF 7580 *Dangers*
PD 6885 Dangers developing countries ; Occupational
PF 2573 Dangers private control communications mass media
PD 7716 Dangers travellers
PD 5730 Dangers ; Unknown crop

PF 6188 *Dark ; Fear*
PF 1741 Dark skin
PA 6261 Darkness
PA 6674 Darkness *Blindness*
PA 6261 Darkness *Darkness*
PA 5568 Darkness *Ignorance*
PA 6978 Darkness *Invisibility*
PA 6731 Darkness *Solemnity*
PA 7367 Darkness *Unintelligibility*
PE 8527 Dasyures marsupial mice ; Endangered species
Data [Absence...]
PE 0630 — Absence method appropriate available
PF 2376 — Accumulation
PF 4176 — Accumulation personal
PF 4155 Data bases developing countries ; Underutilization international
Data [developing...]
PF 7842 — developing countries ; Inaccurate public finance
PF 3539 — *dissemination ; Irrelevant*
PF 0255 — Doctored
PF 4176 — Data ; Errors computerized personal
PF 2376 — Data ; Excessive
PE 0630 — Data ; Failure methods appropriate
PF 0255 — Data ; Forged scientific
PF 0625 — Data ; Inadequate economic technical
PF 0214 — Data ; Insufficient census
Data [Lack...]
PF 0214 — Lack comparable social
PD 5183 — Leak military
PF 9298 — Loss
PF 7117 Data nongovernmental organizations ; Inadequate statistical
PF 1217 Data-oriented education
PF 0625 Data problems ; Inadequate statistical information
PF 1065 Data ; Proliferation unprocessed scientific
Data [sources...]
PF 0510 — sources ; Weaknesses national
PF 1615 — Suppression experimental
PF 3671 — systems ; Ineffective
PE 4230 — systems ; Underparticipation socialist countries international
PD 2957 Data ; Theft
PE 7061 Data transnational corporations ; Missing documents
Data [Unavailability...]
PF 1301 — *Unavailability educational*
PF 6337 — Unavailability timely
PF 1301 — *Uncommunicated educational*
PF 4176 — Uncontrolled use computer
PF 0214 — Undetermined population
PF 6832 — Unreliable
PE 9831 Database espionage ; Computer
PF 4176 Databases ; Erosion human rights privacy use computer
PD 9544 Databases ; Misuse computer
PF 1822 Date methods agricultural production ; Out
PF 4948 Date rapes
PF 5435 Date regulations ; Out
PE 1552 Date trade ; Instability
PF 2168 *Dated ; Cultural dances*
PD 2085 Day care facilities ; Inadequate child
PF 3607 Day observance ; Absence holy
PE 7735 Day observance ; Discriminatory effect holy
PF 3526 Daycare requirements ; Prohibitive
PF 3607 Days ; Desecration holy
PE 7735 Days ; Disruptive secular impact holy
PE 1848 Daytime encounters ; Limited
PF 6550 Daytime involvement ; Restricted
PC 0592 Daytime manpower shortage
PE 2197 Daytime sleepiness ; Excessive
PE 5028 DDT
PE 5028 DDT pollutant
PE 7971 De facto racial requirement qualifications public services
PE 2778 Dead animals ; Undisposed
Dead [Communion...]
PF 9195 — Communion
PD 2315 — Cult
PF 9195 — Cult
PF 9190 — Dead ; Denial rites
PE 9095 — Dead ; Exploitation land burial
PD 5453 *Dead good condition*
PE 3636 Deadly insect bites stings
PA 7365 Deadness *Boredom*
PA 6261 Deadness *Darkness*
PA 5451 Deadness *Insensibility*
PA 7204 Deadness *Unsavouriness*
PE 2083 Deaf children
PD 0601 Deaf ; Denial right welfare services
PD 0601 Deaf ; Inadequate welfare services
PF 2833 Deaf ; Multiplicity manual sign languages
PE 5261 Deaf mutism
PD 0659 Deafness
PD 5535 Deafness animals
PD 2389 Deafness ; Congenital
PD 2389 Deafness ; Genetic
PD 1361 Deafness ; Occupational
PF 7700 Deal discontinuities random events ; Inadequacy methodologies
PD 2107 Dealers ; Private international arms
PE 0413 Dealing arrangements ; Exclusive
PE 9105 Dealing infectious patients ; Carelessness
PD 3841 Dealing ; Insider
PE 6786 Deals ; Secret
PF 2990 Deals ; Secret government
PA 7055 Death
PD 4029 Death ; Antenatal foetal

-889-

Death

PF 1586 Death ; Avoidance confrontation
PE 0987 Death ; Black
PE 4620 Death ; Brain
Death [Death...]
PA 7055 — *Death*
PF 2442 — Dehumanization
PE 8339 — Denial animals right natural
PF 0462 — Dependence fear
PA 6542 — *Destruction*
PE 5648 — disability inhumane confinement
PF 1421 — Dying bad
Death [Fear...]
PF 0462 — Fear
PE 2041 — Foetal infection
PE 2041 — foetus ; Damage
PF 2019 Death God
PA 0072 Death ; Human
PF 0192 Death ; Indeterminacy
PF 3849 Death instinct
PF 7043 Death living creatures
Death [Mind...]
PF 7918 — Mind induced
PF 0431 — missing persons ; Uncertainty
PD 5086 — Morbid preoccupation
PF 1948 — Motivational
PD 2778 Death overwork
Death [penalty...]
PF 0399 — penalty
PF 0399 — penalty ; Denial rights those punished
PF 3960 — pollution
PD 6967 — Prenatal wrongful
PE 6424 — prison foreign nationals ; Failure notify imprisonment
PF 7918 — Psychogenic
PF 0333 Death rate ; Low
Death [squads...]
PD 0527 — squads
PD 7221 — squads
PD 1885 — Sudden unexpected infant
PE 2224 — suffering videos ; Trivialization
Death syndrome
PD 5453 — Acute
PD 5453 — *chickens ; Sudden*
PD 7420 — *feeder cattle ; Sudden*
PD 1885 — Sudden infant
Death threats
PD 0337 — [Death threats]
PE 4869 — against trade union leaders
PE 4869 — against workers representatives
PE 7793 — Lack protection following
PF 3849 Death wish
PF 2442 Deathing awareness ; Lack
Deaths [Abortion...]
PE 3580 — Abortion related
PE 7941 — Animal
PE 1690 — Animal road
Deaths [Child...]
PC 1287 — Child
PD 1885 — Cot
PD 1885 — Crib
PE 4718 Deaths ; Dowry
PE 6367 Deaths during detention ; Suspicious
PC 7896 Deaths ; Mass tree
PD 2422 Deaths ; Maternal
PC 4710 Deaths ; Preventable
PC 4312 Deaths ; Unequal regional distribution
PD 4666 Deaths ; Violent
PE 0558 Debasement works art
PE 1137 *Debasing retirement possibilities*
PF 4600 Debate ; Avoidance substantive issues political
PF 4600 Debate ; Impoverishment political
PF 2152 Debate ; Improper forms
PD 0457 Debate ; Politicization intergovernmental organizational
PF 4600 Debate ; Superficiality political
PC 2162 Debate ; Suppression public
PF 4600 Debate ; Trivialization political
PD 4060 Debates ; Manipulation
PD 2860 Debates ; Politicization technical
PF 8876 Debating time ; Inadequate parliamentary
PE 8923 Debauchery
PA 6466 Debauchery *Intemperance*
PA 5612 Debauchery *Unchastity*
PA 5644 Debauchery *Vice*
PD 2768 Debeaking chickens factory farming
PE 0653 *Debilitated working capacity*
PD 2011 Debilitating conditions health rural communities
PF 2168 Debilitating content village story
PF 2812 Debilitating designs using labour
PD 2672 Debilitating deterioration physical environment
Debilitating [economic...]
PF 2303 — *economic preoccupations*
PF 8126 — education images
PD 1229 — effects social welfare
PF 3388 Debilitating ideological differences
PF 1733 *Debilitating physical appearance*
PF 1365 *Debilitating poverty image*
PA 6799 Debility *Disease*
PC 9067 Debility ; *General*
PA 5806 Debility *Inaction*
PA 7131 Debility *Oldness*
PA 5558 Debility *Weakness*
PD 3301 Debt bondage
PC 2546 Debt burden ; National
Debt [Consumer...]
PD 3954 — Consumer
PD 2133 — crisis developing countries ; External
PE 2524 — cycle rural communities ; Deepening

Debt [developing...]
PD 3051 — developing countries ; Burden servicing foreign public
PD 2133 — developing countries ; Excessive foreign public
PC 3056 — Disproportionate external public
Debt [Enticement...]
PE 1181 — Enticement
PE 1879 — Excessive corporate
PC 2546 — Excessive public
PC 3056 Debt ; Foreign
Debt [industrialized...]
PD 9168 — industrialized countries ; Excessive foreign public
PE 2524 — interest ; Exorbitant
PE 5748 — island developing countries ; Increasing foreign
PE 7184 Debt management capacity developing countries ; Inadequate external
PF 9249 Debt overhang
Debt [payments...]
PD 3053 — payments ; Arrears international
PD 3954 — Personal
PD 3954 — Private
PD 3051 Debt relief developing countries ; Inadequate public
PD 3053 Debt repudiation
Debt [service...]
PD 3051 — service obligations developing creditors ; Inconsiderate insistence creditors fulfilment
PD 3301 — slavery
PJ 1014 — Social
PD 2502 — socialist countries West ; Excessive
PC 8316 Debt ; Uncontrolled growth
PD 3057 Debt ; War
PD 8114 Debtor developing countries ; Excessive social costs structural adjustment
PD 8114 Debtor developing countries ; Neglect human resource development
PE 7235 Debtor's gaol
PE 7166 Debts ; Bad
PE 4603 Debts developing countries ; Increases interest rates long term
PE 8110 Debts developing countries market related interest rates ; Rescheduling
PE 1557 Debts ; Excessive institutional
PE 6530 *Debts ; Heavy farm*
PD 1157 Debts ; Irregular repayment international
PE 1729 Debts ; Understated
PB 2542 Decadence
PB 2542 Decadence ; Dependence
PA 6088 Decadence *Impairment*
PA 5644 Decadence *Vice*
PE 5042 Decadent art
PE 5607 Decadent clothing
PF 5190 Decadent music
PB 2542 Decadent people
PD 4037 Decadent standard living
PD 4049 Decapitation ; Nuclear
PB 2480 Decay ; Dependence ethical
PB 2480 Decay ; Ethical
PF 5403 Decay ; Evidence
PC 2616 Decay ; Inner city
PD 9504 Decay rural communities
PD 1185 Decay ; Tooth
PA 1782 Decay traditional values
PC 2616 Decay ; Urban
PD 9504 Decay ; Village
PD 2301 Decay ; Wood deterioration
PB 2480 Decayed moral environment
PB 4731 Deceit
PB 4731 Deceit ; Dependence
PA 7363 Deceit *Improbity*
PA 7411 Deceit *Uncommunicativeness*
PA 7232 Deceit *Unskillfulness*
PB 4731 Deceitful people
PB 4731 *Deceits ; Mutual*
PC 7112 Decency ; Laws violation human
PE 0988 Decentralized government agencies ; Unaccountability
PB 4731 Deception
PD 4879 Deception business
PD 4879 Deception ; Commercial
PE 7679 Deception ; Electronic
PA 6180 Deception *Error*
PF 9239 Deception ; Geographical
PD 1893 Deception government
PD 3823 Deception management
PD 9836 Deception ; Medical
PD 9182 Deception natural scientists
PF 7702 Deception ; Official self
PF 9583 Deception ; Political
PF 3495 Deception ; Religious
Deception [Self...]
PF 6362 — Self
PE 4890 — sexual partners
PF 7634 — social sciences
PF 5383 Deception ; Terminological
PA 7411 Deception *Uncommunicativeness*
PF 6703 Deceptive language
PD 7231 Deceptive misuse research
PB 4731 Deceptive people
PE 7680 Deceptive political proposals
PF 7634 Declassify social science research
PE 3591 Deciduous fruit ; Pests diseases
PF 1064 Decision ; Absence common structure ethical
Decision makers
PD 8696 — Drug abuse
PF 2452 — grassroots ; Insufficient communication
PF 2452 — Inaccessible
PE 0573 — multinational enterprises ; Inaccessibility

Decision making
PF 1559 — Arbitrary government
PF 4204 — Avoidance
PC 1833 — Balkanization metropolitan government
PD 2245 — bodies ; Diversity limited local
PF 6023 — bodies ; Non participation youth
PF 1303 — *Bureaucratic*
PF 3559 — context ; Isolation individual
PF 2394 — context ; Limited
PF 9166 — Contradictory
PF 7742 — criteria construction industry ; Reductionistic
PF 8989 — delays
PF 5472 — developing industrialized countries ; Displacement national
PF 8703 — Exclusion participation
PE 9009 — Exclusion women
PD 2036 — experience ; Lack
PF 3307 — *Factionalized*
PF 8448 — Fragmented
PF 2843 — Inadequacy committee system
PD 2036 — Inadequate training
PF 9728 — Incompatibility environmental economic
Decision making Ineffective
PF 1010 — implementation methods
PE 8430 — means participation
PF 1781 — structures community
Decision making Lack
PE 8690 — participation disabled persons
PC 0352 — *participation indigenous peoples international*
PF 7535 — rational global political
PF 5660 — reward long range planning
PF 2070 — system ethical
PF 0574 — worker participation business
Decision making [machinery...]
PF 2420 — machinery ; Inadequacy governmental
PF 4756 — methodologies ; Confusing
PF 2394 — methods ; Myopic
PF 8703 — Non collective
PF 8703 — Non participative
PF 5009 — Obsolete
PF 5472 — Over centralization global
PF 1033 — Over formalised
PF 1457 — Paralysis individual
PD 8468 — Politicization
Decision making process
PF 3559 — Fragmented individual
PF 2876 — Inadequacy intergovernmental
PE 8884 — Inefficacity consultation
Decision making processes
PC 0352 — *Inadequate access indigenous peoples international*
PD 0104 — Inadequate information systems international governmental
PF 3709 — Ineffective
PE 7155 — those who question context process ; Exclusion
Decision making [Psychic...]
PF 0508 — Psychic interference
PF 1959 — *Restricted*
PF 5472 — Restrictions participation developing countries international
PF 3454 — Restrictive effects traditional community
PD 8468 — Selective avoidance facts
PE 1488 — Social service quality negated oligarchic control
PF 6556 — structures ; Lack local leadership
PF 2097 — structures ; Simplistic family
PF 7619 — Uncoordinated government
PF 8703 — Undemocratic government
PF 6550 — Unstructured local
PF 6550 *Decision mechanisms ; Obscure*
PF 8989 Decision process ; Lengthy
PF 2477 *Decision ; Unguided vocational*
PF 2709 Decisional paralysis specialized services relation world's need
PF 8989 Decisions ; Delayed
PF 0531 Decisions ; Erroneous
PE 2268 Decisions ; Exclusion pre adults family
Decisions [Imposed...]
PF 8649 — Imposed official
PF 4204 — Inability make difficult
PF 2863 — Incomplete implementation community
PF 2447 — Ineffective educational policy
PE 4477 — intersocial interaction ; Short sighted
Decisions [Lack...]
PF 0966 — Lack meaningful educational context ethical
PF 2882 — Limited access natural resource use
PF 4636 — Limited access technological
PF 2707 — local technological innovation ; Centralized
PF 2754 Decisions ; Non inclusive management
Decisions [Uncorporate...]
PF 9323 — Uncorporate family
PC 0470 — Unfavourable wildlife
PF 4672 — Unimplemented
PF 5356 — Unpredictability judicial
PF 2828 — Unrelated ability application situational demands vocational
PF 4204 — Unwillingness make difficult
PF 9794 Declarations ill treatment ; False
PF 1066 Declarations independence extra territorial bases ; Unilateral
PF 4050 Declassify innovative patents ; Government refusal
PA 6340 Decline *Adversity*
PF 0392 Decline aid developing countries
Decline [capital...]
PE 9265 — capital investment productive capacity
PE 5252 — caused war ; Economic
PA 7193 — *Cheapness*
PC 0588 — civic participation

-890-

Decline [cognitive...] cont'd
PE 9620 — cognitive ability ageing
PE 4655 — commercial bank lending developing countries
PE 0835 — communal spirit village solidarity developing countries
PE 2176 — competition due entrance barriers
PD 8994 — competitiveness domestic industries
PE 3812 — concessional financial resources available developing countries
PF 2439 — *Continuing commercial*
PF 8076 — *Continuing expectation community*
PD 0694 — *craftsmanship*
PF 4924 — crime detection rate
PC 9083 — *Cultural*
PF 9618 Decline deference
PD 2968 Decline developing country commodity prices
Decline [earnings...]
PD 2769 — earnings developing countries
PF 8466 — educational quality
PF 8466 — educational standards
PE 3066 — export credits developing countries
PD 9679 — exports developing countries developed-market economies
PD 3138 Decline foreign direct investment developing countries
PC 7896 Decline ; Forest
Decline government expenditure
PF 9108 — [Decline government expenditure]
PF 0674 — education
PF 0674 — *primary education*
Decline [government...]
PF 4586 — government health expenditure
PF 0611 — government social expenditure
PD 2968 — growth commodity export earnings developing countries
PD 1250 Decline handicrafts cottage industries developing countries
PF 2439 Decline ; *Historical business*
Decline human
PF 7815 — genetic endowment
PF 7815 — race ; Intellectual
PF 7815 — race ; Physical
Decline [Impairment...]
PA 6088 — *Impairment*
PE 6861 — import capacity developing countries
PD 0291 — investment developing countries
PF 0817 Decline multilateral cooperation
PC 0933 Decline mutual trade developing countries
PA 6387 Decline *Necessity*
PE 8938 Decline nutritional quality food
PF 0392 Decline official development assistance developing countries
PF 6221 Decline philanthropy
PD 0779 Decline production capacity utilization
Decline public
PF 0611 — expenditure human development
PF 5622 — interest broadcasting
PE 4574 — sector savings developing countries
Decline public spending
PF 9108 — [Decline public spending]
PF 0674 — education
PF 4586 — health
PF 0674 — *primary schools*
PF 0611 — social sector
Decline [rate...]
PD 5543 — rate productivity increase
PD 8916 — real wages
PD 2769 — real wages developing countries
PA 7321 — *Refusal*
PB 1540 — religion
PF 3863 — religion
PF 1433 — religious broadcasting
PD 1095 — rural customs traditions developing countries
Decline [Shortcoming...]
PA 6041 — *Shortcoming*
PD 1052 — soil productivity
PE 6705 — street life
PE 4922 Decline technology transfer developing countries
PD 2897 Decline trading position developing countries
PA 7371 Decline *Unintelligence*
PA 5558 Decline *Weakness*
PD 5114 Declines production ; Vulnerability agriculture developing countries future
PE 4585 Declining area irrigated land
Declining [birth...]
PD 2118 — birth rate
PD 5936 — breeds cultivated plants
PD 6305 — breeds domesticated animals
Declining community
PF 9066 — confidence ability change
PF 2575 — life
PF 6441 — *population*
PC 9584 Declining credibility international trading system
PF 1345 Declining *economic conditions*
Declining economic growth
PC 0002 — [Declining economic growth]
PD 5326 — developing countries
PF 1737 — industrialized countries
PC 8908 Declining economic productivity
PC 2102 Declining family values
PC 0002 Declining growth world economy
Declining [industrial...]
PD 8994 — industrial competitiveness
PD 5350 — industrial sectors
PD 8994 — international competitiveness
PD 8994 — international market share
PD 5350 *Declining local businesses*

PD 5350 *Declining local industries*
PF 6441 Declining national population
Declining productivity
PD 7480 — agricultural land
PD 5543 — industrialized countries
PF 7610 — socialist countries
PE 1137 *Declining public image*
PF 4718 Declining rate development
Declining [school...]
PC 1716 — school enrolment
PF 2575 — sense community
PF 1471 — share developing countries world commodity exports
PF 7142 — standard living
PF 2875 *Declining volunteer services*
PC 6031 Declining worker productivity
PE 7539 Decommissioned nuclear power plants ; Environmental hazards
PE 7539 Decommissioning nuclear reactors ; Excessive cost
PA 6858 Decomposition *Disintegration*
PA 6088 Decomposition *Impairment*
PA 5589 Decrease
PD 6075 Decrease consumer choice
PF 0333 Decrease mortality rate
PF 6441 Decrease ; Population
PD 0044 Decrease skilled labour
PF 8552 *Decreased skill transference*
PF 4326 Decreasing agricultural growth capita
PB 9748 Decreasing biodiversity
PE 2305 *Decreasing coconut production*
PC 5773 Decreasing diversity biological habitats
PD 7302 Decreasing diversity biological species
PC 5773 Decreasing ecological diversity
PE 7313 *Decreasing fish profits*
Decreasing genetic diversity
PC 1408 — animal breeding stock
PC 1408 — animals
PC 2223 — cultivated plants
PD 0547 — fish
PF 0905 *Decreasing labour force*
PF 7435 Decreasing land mass
PC 4824 Decreasing natural resources
Decreasing number
PF 4200 — adoptable children
PF 4205 — adoptive parents
PC 2494 — election voters
PF 8905 Decreasing participation collective religious worship
PE 5336 Decreasing rate development major agriculture technologies
PC 1716 *Decreasing school enrolment*
PF 7435 Decreasing territory
PB 9748 Decreasing variety life forms
PA 6088 Decrepitness *Impairment*
PA 7131 Decrepitness *Oldness*
PA 7371 Decrepitness *Unintelligence*
PA 5558 Decrepitness *Weakness*
PD 3139 Deculturation
PF 4416 Deep emotional attachments ; Dependence
PD 1241 Deep ocean sea bed ; Militarization
PE 9774 *Deep pectoral myopathy turkeys*
PF 4767 Deep sea water resources ; Obstacles utilization coastal
PD 0201 *Deep wells* ; *Waste disposal*
PE 2524 Deepening debt cycle rural communities
PE 3878 Deer flies
PE 3635 *Deer nose fly*
PF 1815 Defaced community space
PD 5305 Defacement urban structures
PD 2569 Defamation character
PA 6191 Defamation *Disapproval*
PA 6839 Defamation *Disrepute*
PF 9385 Defamation institutions
PA 7270 Default *Absence*
PA 5438 Default *Neglect*
PA 6434 Default *Poverty*
PA 6041 Default *Shortcoming*
PD 3054 *Defaulted loan syndrome*
PD 9437 Defaults ; Government loan
PD 3053 Defaults international loans
PA 7289 Defeat
Defeat [Defeat...]
PA 7289 — *Defeat*
PA 6542 — *Destruction*
PA 5497 — *Difficulty*
PF 4709 Defeat ; Electoral
PA 5527 Defeat *Inexpectation*
PA 6662 Defeat *Nonaccomplishment*
PF 8781 Defeat ; Sense
PF 2636 *Defeat war*
PD 4418 Defeating behaviour ; Self
PF 2093 Defeating community image
PF 1681 Defeating community story
PF 2093 Defeating economic story
PD 4418 Defeating personality disorder ; Self
PE 1437 Defeating size requirements
PF 6456 Defeating style community planning ; Self
PF 8781 Defeatism
PF 8781 Defeatism face problems
PF 8781 Defeatist civic attitude
PF 8781 Defeatist self-images citizens' law enforcement groups
PE 1602 Defecating public
PE 4619 Defecation ; Random
PD 0270 *Defect appearance* ; *Preoccupation imagined*
PE 1418 Defect ; Ego
PD 0218 *Defection*
PA 6379 Defection *Avoidance*
PA 6997 Defection *Imperfection*

PA 7325 Defection *Irresolution*
PF 8026 Defective blood formation
PE 3355 Defective human immunity system
PD 1141 Defective land use planning developing countries
PD 3998 Defective manufactured goods
PD 3998 *Defective medical devices*
PE 2467 *Defective oxygen supply foetus*
Defective [product...]
PD 3998 — product manufacture
PD 3998 — *products developing countries* ; *Sale*
PD 3998 — *products industrialized countries* ; *Dumping*
PF 5711 Defective reasoning
PA 6799 Defectiveness *Disease*
PA 6180 Defectiveness *Error*
Defectiveness [Imperfection...]
PA 6997 — *Imperfection*
PA 6652 — *Incompleteness*
PA 5473 — *Insufficiency*
PA 7371 Defectiveness *Unintelligence*
PD 2549 Defectors
Defects [Animal...]
PD 4031 — *Animal*
PE 0965 — *animals* ; *Cerebral*
PE 5569 — *animals* ; *Septal*
PE 7808 *Defects back foals*
PE 0965 *Defects brainstem animals*
PE 0965 *Defects cerebellum animals*
PD 2389 Defects diseases ; Genetic
PD 2389 Defects diseases ; Hereditary
PD 2389 Defects ; Genetic
PD 6306 Defects ; Hearing
PD 1618 Defects ; Human birth
PD 4031 Defects ; Increase animal genetic
PD 9480 Defects ; Increase plant genetic
PE 2462 Defects machinery design
PD 8484 Defects ; Memory
PD 3773 Defects ; Physical developmental
PE 4428 Defects ; Programming
PE 2265 Defects ; Speech
PA 5445 Defence
Defence [against...]
PA 1986 — against corruption
PE 4628 — against criminal charges ; Violation right
PE 8664 — arms industries ; Protectionism
PF 6060 Defence budget ; Increase
PD 4094 Defence control systems ; Vulnerability nuclear
PE 4628 Defence ; Denial right legal
PE 7624 Defence ; Denial right time prepare trial
Defence [Inadequacy...]
PF 0506 — Inadequacy civil
PF 1447 — Increasing public spending
PF 7836 — industrial pollution immediate victims
PE 4097 — industries ; National
PE 7679 — information uncertainty
PE 8637 Defence ; Lack access prisoners'
PF 5933 Defence ; Lack adequate military
Defence [planning...]
PF 5009 — *planning* ; *Obsolete*
PF 1385 — policies ; Unjustified military
PE 4097 — procurement procedures ; National
PF 4785 Defence ; Rigidity rechannelling reduced expenditure
PD 4361 Defence scandal
PB 0001 Defences ; Vulnerability social
PE 4319 Defendant's right silence ; Denial
PE 4863 Defendants ; Unfit legal
PF 0979 Defensive life stance
PE 2411 *Defensive marketing facilities*
PF 2600 Defensive nationalistic images
PG 7688 Defensiveness
PA 5445 Defensiveness *Defence*
PF 2602 *Defensiveness* ; *Excessive parental*
PF 9618 Deference ; Decline
PA 5532 Defiance *Disaccord*
PA 7250 Defiance *Disobedience*
PA 7325 Defiance *Irresolution*
PA 6979 Defiance *Opposition*
PE 4634 Defiant disorder ; Oppositional
Deficiencies [civil...]
PF 4899 — civil justice systems
PE 6343 — Colour
PF 4875 — criminal justice system
Deficiencies [developed...]
PD 0800 — developed countries ; Dietary
PC 4094 — developing countries
PD 0715 — diet ; Vitamin
PC 0382 — Dietary
PE 4760 — domestic animals ; Dietary
PC 0382 Deficiencies ; Food
PF 4816 Deficiencies international law
PE 8300 Deficiencies military codes justice
Deficiencies [national...]
PF 4851 — national local legal systems
PF 0510 — national statistics
PC 0382 — Nutritional
PE 1936 Deficiencies soils ; Trace element
PD 8179 Deficiencies ; Visual
PC 3437 Deficiencies welfare state
PF 1739 Deficiencies world economical systems ; Interaction
PA 5652 *Deficiency*
PE 9554 *Deficiency anaemia animals* ; *Iron*
PE 7854 *Deficiency anaemias* ; *Iron*
PE 2467 Deficiency blood ; Oxygen
Deficiency [Calcium...]
PE 5161 — Calcium
PE 3147 — cereals ; Protein
PD 0914 — children ; Mental

−891−

Deficiency

Deficiency [Chromium...] cont'd
PD 0448 — *Chromium*
PD 0339 Deficiency dietary protein
Deficiency diseases
PD 0287 — [Deficiency diseases]
PC 0952 — Animal
PE 4224 — animals ; Immune
PD 0287 — Human
PD 3653 — plants
PD 2726 Deficiency disorders ; Iodine
PE 4760 Deficiency domestic animals ; Vitamin E
PF 4984 Deficiency ; Educational
PD 8301 Deficiency ; Endocrine
PE 8505 Deficiency ; Facial oral injury
Deficiency [Imperfection...]
PA 6997 — *Imperfection*
PA 6652 — *Incompleteness*
PA 5652 — *Inferiority*
PF 7230 — innovation public sector
PA 5473 — *Insufficiency*
Deficiency [Magnesium...]
PD 0448 — *Magnesium*
PD 0448 — *Manganese*
PE 6289 — Marine oxygen
PC 1587 — *Mental*
PE 2287 Deficiency ; Niacin
PE 0533 Deficiency ; Nonverbal communication
PD 0077 Deficiency ; Soil
PE 2185 Deficiency ; Thiamine
PE 5328 Deficiency ; Trace element
PD 0448 Deficiency ; *Vanadium*
Deficiency virus
PD 4747 — Bovine immuno (BIV)
PD 4747 — Feline immuno (FIV)
PE 7848 — Feline immuno (FIV)
PD 4747 — Immuno
PD 4747 — Simian immuno (SIV)
Deficiency Vitamin
PE 2538 — A
PD 0715 — *B*
PE 2380 — C
PD 0287 — D
PF 6544 *Deficient city equipment*
PF 0379 Deficient developing countries ; Energy
PE 1901 Deficient diets ; Metal
PD 4790 Deficient health services
PD 1251 *Deficient housing materials*
PE 4245 Deficient infant care
PE 7326 Deficient local structures unemployed
PD 0382 Deficient nutritional practices
PE 7647 Deficient public facilities
PE 5664 Deficient transport remote island developing countries
PD 9196 Deficient water lines
PE 4791 *Deficientia intelligentiae*
PF 2384 Deficit ; Attention
PD 5492 Deficit ; Budget
PD 3131 Deficit financing external borrowing ; Dependence developing countries budget
PE 9789 Deficit hyperactivity disorder ; Attention
PD 5492 Deficit spending
PC 0998 Deficits ; Balance payments
PD 3131 Deficits developing countries ; Budget
PE 1496 Deficits developing countries ; Excessive external trade
PC 1100 Deficits ; Excessive external trade
PD 5984 Deficits ; Fiscal
PD 5984 Deficits ; Government
PD 9219 *Deficits interpersonal relatedness ; Pervasive pattern*
PD 5984 Deficits ; Public sector
PD 5984 Deficits ; Unsustainable fiscal
PF 6657 Defilement ; Spiritual
PF 7178 Define moral standards ; Inability
PD 1501 Definition family developing countries ; Narrow legal
PD 4519 Definition health risks ; Inappropriate legal
PD 1975 Definition intelligence ; Lack
PF 4674 Definitions ; Restrictive commercialization
PD 7727 Deflation
PD 7727 Deflation world economy
PE 4999 Defoliating insects
PD 1135 Defoliation
PD 1135 *Defoliation insect breeding areas*
PD 1135 *Defoliation insect-breeding habitats*
PC 1366 Deforestation
PD 6282 Deforestation mountainous regions
PD 6204 Deforestation ; Tropical
PD 6282 Deforestation upland watersheds
PE 5913 *Deformans ; Spondylosis*
PE 1646 Deformation childrens' bodies ; Deliberate
PA 6790 Deformation *Distortion*
PE 2042 Deformation foetus
PD 2559 Deformation human body ; Mutilation
PA 6997 Deformation *Imperfection*
PA 5878 Deformation *Nonconformity*
PD 9480 Deformation plant life
PA 7240 Deformation *Ugliness*
PD 4031 Deformed animals
PD 0211 Deformed people ; Lack facilities severely
PD 9480 Deformed plants
PE 8912 Deformities bone ; Osteomyelitis, periostitis infections involving bones acquired
PD 1618 Deformities ; Congenital
PC 8866 Deformities digestive system ; Diseases
PE 7826 *Deformities ; Flexion*
PE 7808 *Deformities foals ; Angular limb*
PD 4401 Defrauding secured creditors
PB 2542 Degeneracy
PC 1587 Degeneracy ; Mental

PD 8149 Degeneration
PA 5568 Degeneration ; Heart
PC 2270 *Degeneration ; Hepatolenticular*
PA 6088 Degeneration *Impairment*
PA 5644 Degeneration *Vice*
PE 4161 *Degenerative arthropathy*
PD 0992 Degenerative changes brain
PE 1684 Degenerative changes heart
PD 6216 Degenerative diseases
PD 0448 Degenerative heart disease
Degenerative joint disease
PE 5913 — [Degenerative joint disease]
PE 5913 — *[Degenerative joint disease]*
PE 5050 — *[Degenerative joint disease]*
PE 4161 — *[Degenerative joint disease]*
PD 7841 *Degenerative myeloencephalopathy brown Swiss cattle ; Bovine progressive*
PE 9774 *Degenerative myopathy*
PE 1256 *Degradable organic matter ; Damage*
Degradation [agricultural...]
PE 8324 — agricultural land cash crops
PC 8419 — agricultural resources
PD 9413 — atmosphere
PE 6142 — automobiles ; Environmental
Degradation [climate...]
PC 9717 — climate man
PD 7585 — countryside careless walkers
PC 2122 — culturally important landscapes
Degradation [Dependence...]
PB 6384 — *Dependence environmental*
PA 5493 — *Depression*
PD 2285 — desert oases ; Environmental
PD 7835 — developed countries ; Environmental
PD 3922 — developing countries ; Environmental
PF 4115 — developing countries tourism
PA 6858 — *Disintegration*
PA 6839 — *Disrepute*
PC 2506 — drylands
PD 0767 — due creation dams lakes ; Environmental
Degradation environment
PC 6675 — armed conflict
PE 4759 — contamination
PE 5064 — destruction species
PD 3922 — developing countries
PE 1412 — electrical power generation
PE 7695 — trees
PB 6384 Degradation ; Environmental
PD 6960 Degradation fragile ecosystems
PC 6923 Degradation fresh water sources ; Environmental
PD 6124 Degradation high speed roads ; Environmental
Degradation [Impairment...]
PA 6088 — *Impairment*
PC 7203 — indigenous cultures
PD 7835 — industrialized countries ; Environmental
PC 2616 — inner city areas ; Environmental
PD 1052 Degradation ; Land
Degradation [medical...]
PD 5770 — medical profession
PE 0736 — military activity during peace time ; Environmental
PE 6256 — mountain environment leisure activities
Degradation [Natural...]
PB 5250 — *Natural environment*
PB 5250 — natural resources
PC 2122 — *natural seascape*
PE 1720 Degradation off road terrain vehicles ; Environmental
Degradation [pipelines...]
PE 6251 — pipelines ; Environmental
PF 7741 — product quality ; Planned
PC 3144 — Psychological environment
PF 7815 Degradation ; Racial
Degradation recreation tourism
PD 0826 — Cultural
PE 6920 — Natural environmental
PD 0826 — Social environmental
PE 7403 Degradation recreational use unsurfaced country roads tracks ; Environmental
PD 5845 Degradation ; Regional environmental
Degradation [semi...]
PC 3152 — semi-natural natural habitats flora fauna
PC 4588 — Socio cultural environment
PD 1052 — Soil
PD 2345 — suburbia ; Environmental
PA 5644 Degradation *Vice*
PE 2593 Degradation ; Visibility
PD 7585 Degradation wilderness areas campers
PD 5845 Degraded exploitative development ; Regional ecosystems
PE 6005 Degraded industrial land
PF 3458 Degrading environment ; Unaccountability institutions
PD 7821 Degrading medical treatment
Degrading [peoples...]
PF 4644 — people's time orientation
PC 3768 — punishment
PC 3768 — punishment ; Denial right freedom cruel, inhumane
PC 1524 Degree equivalencies ; Uncertainty university
PC 9067 *Degree fever ; Excessive*
PF 4668 Degrees freedom government ; Limited
PA 1757 Dehumanization
PF 2442 *Dehumanization death*
PD 7821 Dehumanization health care
PF 5438 Dehumanization man technological process
PF 7602 Dehumanization soldiers
PA 5643 Dehumanization *Unkindness*
PD 2656 Dehumanization work
PF 2112 Dehumanized individual scientific research
PA 1757 Dehumanized people

PE 8062 Dehydration
PF 4617 Deicide
PC 0799 Dejection
PA 7253 Dejection *Envy*
PA 6739 Dejection *Maladjustment*
PA 6731 Dejection *Solemnity*
PA 1999 Delay
PF 5120 Delay access social services
Delay administration
PD 0412 — criminal justice
PF 1487 — justice
PD 5119 — medical care
PC 1613 Delay approval urgent regulations
Delay [civil...]
PD 7571 — civil litigation
PE 8824 — compensation victims motor accidents
PE 3657 — connection telecommunications facilities
PE 3657 — connection telephones
PF 1487 — Court
Delay [delivery...]
PE 8157 — delivery requested services
PA 1999 — Dependence
PE 0598 — disseminating negative information
PF 1179 — Delay government reimbursements
PE 4446 — Delay governmental payment compensation
Delay [implementation...]
PF 3975 — implementation commitments
PF 1487 — Inordinate legal
PE 9123 — issue travel documents
PF 3975 Delay loan disbursements
PF 3005 *Delay obtaining property titles*
Delay payment compensation
PD 7179 — damages
PE 3913 — victims crime
PE 5229 — victims disasters
PE 5229 — victims major accidents
PE 0811 — victims malpractice
Delay [payment...]
PE 4446 — payment government reparations
PE 4602 — processing parole applications
PF 1470 — project implementation
Delay [recognition...]
PF 7779 — recognition patents
PF 8112 — recognition problems
PF 1393 — religions acknowledging social problems
PF 6707 — remedial response problems ; Government
PF 6707 — response symptoms problems ; Government
Delay [societal...]
PE 8318 — societal impact education
PF 8870 — societal impact innovation
PF 3913 — supply agreed multilateral assistance
PF 0726 Delayed consequences war-time imprisonment deportation
PF 8989 Delayed decisions
PF 2018 Delayed development regional plans
PF 9328 Delayed emergence problems
PF 9328 Delayed environmental impact technological activities
PF 6525 *Delayed industrial development*
PF 8652 Delayed police response
PF 2457 *Delayed profits*
PA 1999 Delayed responses
PD 5119 Delayed surgery
PE 4887 Delayed trial ; Imprisonment
PF 5502 Delayed withdrawal commitment action ; Governmental
PF 6119 Delaying tactics ; Government
PF 5120 Delaying welfare ; Government procedures
PC 2550 Delays ; Administrative
PF 9464 Delays ; Air traffic
PC 2550 Delays ; Bureaucratic
Delays [centralized...]
PF 0685 — centralized collective bargaining
PF 4453 — Communication
PF 6502 — community building programmes
PF 8989 Delays ; Decision making
PF 1538 Delays delivery books publications
Delays delivery goods
PF 8268 — [Delays delivery goods]
PE 3928 — services
PE 3928 — services ; Risks due
PC 1613 Delays elaboration remedial legislation
PE 1093 Delays ; Excessive burden poor due legal
PE 2215 Delays fulfilling international reporting obligations ; Governmental
PC 6989 Delays implementation social change
PE 5886 Delays ; Ocean shipment
Delays [payments...]
PE 9700 — payments state-controlled enterprises
PF 8989 — Policy making
PE 5815 — Port
PF 2717 — Postal
Delays [ratification...]
PE 7300 — ratification human rights treaties
PF 0977 — ratification international treaties
PC 2550 — Red tape
PE 0598 — releasing controversial news items ; Calculated
PF 9625 — research
PF 9664 — response accelerating structural change ; Policy making
PF 1698 Delays ; Telephone
PE 1977 Delays ; Travel
PE 3457 Delegations ; Conspicuous consumption international
PA 5454 Deleteriousness *Badness*
PA 7395 Deleteriousness *Inexpedience*
Deliberate [deformation...]
PE 1646 — deformation childrens' bodies
PD 5859 — *destruction equipment*

Denial right

Deliberate [distortion...] cont'd
PD 4982 — distortion corporate news information
PD 1893 — distortion official news information
PA 6448 *Deliberate fire-setting*
PF 5736 Deliberate governmental avoidance legislative reform
Deliberate [ignorance...]
PF 8229 — ignorance
PF 8278 — ignorance during policy-making
PF 3382 — imbalancing population sex ratio
PD 1244 — interference satellite communications
PF 3699 — invention new threats
PD 4982 Deliberate lying corporation officials
PD 1893 Deliberate lying government officials
PF 1183 Deliberate misrepresentation educational materials
PE 4878 Deliberate self harm ; Non fatal
PF 9239 *Deliberately distorted maps* ; Production
PF 4670 Deliberately unused land
PF 6992 Deliberately weakened international treaties
Deliberative systems
PF 7059 — government ; Inflexible
PD 0975 — government ; Outmoded
PD 0975 — Obsolete
PD 0975 — Outmoded
PE 7903 Delimitation boundaries ; Unsatisfactory
PF 7823 Delimitation policy responsibilities ; Irresponsible
PE 3296 *Delimited desert use*
PF 1663 Delimited self worth ; Social reinforcement
PF 2644 Delimiting family patterns traditional way life
PE 6119 Delinquency ; Driving
PC 0212 Delinquency ; Juvenile
PD 5750 Delinquency side-effect drug trade drug addiction
PE 9468 Delirium
PE 9263 *Delirium ; Alcohol withdrawal*
PE 9263 *Delirium alcoholicum*
PA 6799 Delirium *Disease*
PA 7157 Delirium *Insanity*
PA 6739 Delirium *Maladjustment*
PE 9263 *Delirium tremens*
PF 2875 Delivery basic human resource services ; Ineffective
PF 1538 Delivery books publications ; Delays
PD 4790 *Delivery ; Complex health*
PD 0974 *Delivery ; Distrust professional service*
Delivery [education...]
PF 4852 — *education ; Irregular*
PF 1667 — essential services developing country rural communities ; Restricted
PF 2875 — essential services isolated communities ; Restricted
Delivery goods
PF 8268 — Delays
PE 3928 — services ; Delays
PE 3928 — services ; Risks due delays
PD 4004 Delivery ; Insufficient physical infrastructure health care
PD 8517 Delivery ; Insufficient water
PF 0301 Delivery mechanism response problems ; Inadequate
PF 7705 *Delivery ; Precipitate ill judged forceps*
PE 8157 Delivery requested services ; Delay
PF 2583 Delivery urban environments ; Poor service
PF 1064 Deluding familial image social responsibility
PA 6180 Delusion *Error*
PA 6414 Delusion *Illusion*
PA 6739 Delusion *Maladjustment*
PA 7411 Delusion *Uncommunicativeness*
PE 0435 Delusional insanity
PF 5013 Delusional jealousy
PF 9617 Delusions concerning public figures
PA 6414 *Delusions ; Erotic*
PA 6414 Delusions *Illusion*
PC 2372 Demagoguery
PD 4002 Demand capita ; Excessive environmental
PE 8013 Demand developed countries meat egg products ; Increased
PF 9432 Demand ; Failure commodity production respond fall
PC 3116 Demand goods capitalism ; Excess
PC 3116 Demand goods capitalist systems ; Excessive
PF 1442 *Demand ; Limited local*
PD 5429 Demand ; Mismanagement environmental
PD 2898 Demand primary commodities because rising living standards ; Inadequate
PD 1276 Demand primary commodities due technological change ; Reduction
PE 8092 Demand unskilled labour industrialized countries
PF 2457 Demanding requirements
PD 9219 *Demands adequate occupational performance ; Passive resistance*
PS 3268 *Demands ; Distracting faculty*
PF 1781 *Demands ; Excessive classification*
PF 2644 *Demands ; Exhausting work*
PD 4900 Demands forests ; Conflicting
PF 1321 Demands ; Gap material technological needs
PF 2457 *Demands machinery ; Simultaneous*
PC 4824 Demands natural resources ; Excessive
PF 2828 Demands vocational decisions ; Unrelated ability application situational
PF 2093 Demeaning community self-image
PF 7299 Demeaning expressions
PF 1365 *Demeaning farmer image*
PA 5652 Demeaning *Inferiority*
PF 1529 Demeaning minority self-image
PF 7930 Demeaning sex media ; Glorification
PA 7411 Demeaning *Uncommunicativeness*
PD 0153 Dementia associated alcoholism
PE 3083 Dementia ; Presenile
PE 3083 Dementia ; Senile
PD 9227 Demineralization soil
PD 0276 *Demise three-generation family structure*
PD 2001 Democracies ; Single party

PC 3176 Democracy communist systems ; Denial
PD 7352 Democracy ; Frivolity bourgeois
PB 8031 Democracy ; Lack
PF 6608 Democracy ; Limitations
PD 3180 Democracy ; Subversion
PD 3213 Democracy ; Totalitarian
Democratic [political...]
PD 0055 — political organization ; National governments obstacle representative
PC 1015 — political process ; Absence
PB 8031 — principles ; Erosion
PE 5179 — procedures international organizations ; Distortion
PD 0639 — processes ; Contempt
PF 7220 — processes resolve theoretical issues ; Misuse
PF 0214 Demographic statistics ; Inadequate social
PE 9337 Demolition homes government authorities
PD 3731 Demolition planning ; Lack
PF 5781 Demonic spirit possession
PF 4133 Demonizing enemy
PF 6734 Demons
PD 8522 Demonstrations
Demonstrations Violent
PA 0429 —
PD 4811 — repression
PE 2007 — repression trade union
PF 8446 Demoralization
PF 1365 *Demoralization ; Community*
PF 6639 Demoralization ; Military
PF 1681 Demoralizing community self-story
PE 2451 Demoralizing constraints housing rehabilitation
PF 1681 Demoralizing image urban community identity
PF 2358 Demoralizing images rural community identity
PB 0035 Demositis
PE 5278 Demotivated civil servants developing countries
PA 2360 Demotivation *Apathy*
PE 1041 Demyelinating diseases central nervous system
PC 3152 *Denaturalization fauna*
PE 2260 Dengue fever
PA 7400 Denial
Denial [academic...]
PD 4282 — academic freedom
PF 3081 — access news
PE 8643 — animals legal protection their rights
Denial animals right
PF 5121 — attention, care protection humankind
PE 6270 — conditions life liberty proper their species
PE 9573 — dignity
PE 9650 — freedom mass killing
PE 8243 — life
PE 8339 — natural death
PF 7706 Denial causal relationship problems
PD 5907 Denial cultural rights
Denial [defendants...]
PE 4319 — defendant's right silence
PC 3176 — democracy communist systems
PA 7400 — *Denial*
PA 6838 — *Dissent*
PF 3082 — distribution access
Denial [economic...]
PD 4150 — economic rights
PE 6375 — economic social rights refugees
PC 3459 — education minorities
PE 3123 — effective national self-determination capitalist exploitation
PE 0324 — entitlement food
PE 1625 — equal benefits elderly workers
PA 0833 — *equal property rights because discrimination*
PE 4712 — equality nations
PD 7385 — evidence
PE 8024 — experimental animals right freedom suffering
Denial [filiation...]
PD 3255 — filiation
PE 6616 — foetal rights
PE 3899 — food animals right freedom suffering
Denial freedom
PD 3224 — association
PD 7612 — conscience
PE 5409 — expression clothing
PC 3174 — expression thought communist systems
PF 3217 — hold opinions
PC 3173 — movement communist systems
PF 7219 — opinion
PC 2162 — speech
PF 3217 — thought
PF 3217 — thought ; Dependence
PF 2409 Denial God
PF 2409 Denial gods
Denial human rights
PB 3121 — [Denial human rights]
PC 6927 — administration justice
PC 1454 — armed conflicts
PC 3124 — capitalist systems
PC 3178 — communist systems
PC 3304 Denial indigenous rights lands resources
PE 8667 Denial ; Involuntary boarding
Denial [legal...]
PF 3517 — legal representation
PF 1746 — lessons history
PD 8102 — literacy education
PA 7382 — *Loss*
PA 6387 Denial *Necessity*
PD 2830 Denial old people significance past
PA 6979 Denial *Opposition*
Denial [parental...]
PD 3255 — parental affiliation
PF 0608 — past

Denial [paternity...] cont'd
PD 3255 — paternity
PC 2381 — people control over their own lives
PC 0632 — political civil rights
PD 8276 — political rights
PE 5608 — public service
Denial [reality...]
PF 1079 — reality
PA 7321 — *Refusal*
PC 3175 — religion communist systems
PD 8445 — religious liberty
PF 4813 — right
Denial right access
PF 6458 — *higher education*
PF 9046 — historical libraries
PF 1743 — interpreters judicial hearings
PD 3335 — public service women
PE 7345 Denial right accused segregation convicted criminals
Denial right adequate
PC 3319 — food indigenous populations
PE 8924 — health care disabled
PD 5254 — housing
PC 3320 — housing indigenous peoples ; Discrimination against indigenous populations housing
PD 2028 — medical care
PF 0344 — standard living
PE 5484 — standard living indigenous peoples
PC 0834 — welfare services
Denial right [appeal...]
PD 5317 — appeal
PF 9735 — *artistic freedom*
PC 2383 — assembly
PD 3224 — association
PD 3224 — association
PF 3021 — asylum
PE 7336 — be informed criminal charges
Denial right benefits
PE 5211 — invalids
PF 6077 — science
PE 4531 — survivors
Denial right [business...]
PE 2700 — business growth
PE 1736 — change nationality
PE 6397 — change religion
PD 1915 — choice marriage
PE 3802 — choose moral religious education
PE 3970 — collective bargaining
PC 2162 — communicate
PD 7609 — complaint
PD 6612 — confidentiality
PE 7217 — confront accusers
PD 1800 — conscientious objection military service
PE 7349 — correct misinformation
Denial right cultural
PE 3450 — asylum
PF 9005 — identity
PE 6561 — life
Denial right [develop...]
PF 2364 — develop human beings
PE 8911 — dignified birth
PE 6623 — dignity
PD 6927 — due process law
Denial right economic
PE 6212 — asylum
PD 0808 — security
PE 5406 — security during periods unemployment
Denial right [education...]
PD 8102 — education
PD 0190 — education women
PE 4700 — educational choice
PD 1092 — employment indigenous populations
PD 0086 — employment women
PE 4507 — enjoyment arts
PC 8943 — environmental quality
Denial right equal
PD 1977 — pay equal work
PD 0309 — pay women
PD 9628 — *promotion opportunities*
PC 1268 — protection law
Denial right equality
PA 0833 — [Denial right equality]
PC 0006 — because race
PE 4712 — states
PC 0308 — women
Denial right [euthanasia...]
PF 2643 — euthanasia
PE 7489 — examine witnesses
PE 5241 — extended family
PD 4827 — fair public trial
PF 3964 — fair public trial
PE 7267 — family
PE 5226 — family planning
PD 9170 — favourable work conditions
PC 6870 — found family
PE 3963 — free choice work
PC 6381 — free primary education
PD 3055 — freed arbitrary deprivation property
Denial right freedom
PE 5952 — advocacy discrimination
PD 1576 — arbitrary detention
PC 2507 — *arbitrary exile*
PC 5313 — arbitrary expulsion
PD 2569 — attacks personal honour reputation
PD 8445 — belief
PE 6342 — compulsory labour
PC 3768 — cruel, inhumane degrading punishment

Denial right

Denial right freedom cont'd
PA 0833 — discrimination
PF 6158 — double jeopardy
PD 0635 — exploitation children
PC 2162 — expression
PE 0324 — hunger
PE 7235 — imprisonment failure fulfil contractual agreement
PC 0185 — information
PC 0935 — international movement
PC 8452 — movement persons
PE 8408 — movement state
PE 8951 — press
PD 8445 — religion
PE 4332 — religion indigenous peoples
PD 0127 — religion women
PE 4743 — retroactive laws punishments
PE 4964 — servitude
PC 0146 — slavery
PE 6633 — testifying against oneself
PF 3217 — thought
PC 3429 — torture
PD 3092 — war propaganda

Denial right [grievance...]
PE 2832 — grievance procedures
PD 6927 — *habeas corpus*
PC 0663 — *health*
PE 4459 — health indigenous populations
PC 0663 — highest obtainable physical mental health
PE 5608 — hold public office
PD 0520 — humane imprisonment
PE 7665 — independent judges

Denial right indigenous
PC 3322 — peoples education
PE 7312 — peoples participate political processes
PE 2142 — peoples use their own language

Denial right [inform...]
PE 7337 — inform
PF 0886 — inherit property
PE 6755 — innocent passage territorial waters
PF 8854 — intellectual property
PD 4282 — investigate
PE 7119 — job protection during maternity leave
PD 8568 — just income

Denial right justice
PA 6486 — [Denial right justice]
PC 6162 — [Denial right justice]
PD 0162 — women
PE 7209 — Denial right juvenile criminals segregation adult criminals
PE 3463 Denial right leave any country

Denial right legal
PF 8869 — aid
PF 3517 — counsel
PE 4628 — defence
PE 2317 — services indigenous populations

Denial right [leisure...]
PF 9062 — leisure rest
PF 0705 — liberty
PD 4234 — life
PF 2850 — manifest religion
PF 3343 — marriage
PC 2157 — material well-being because discrimination
PE 3951 — maternity leave
PE 3828 — medical consent
PE 6726 — minimum wage
PE 1693 — minimum work age
PE 4624 — name

Denial right national
PC 1450 — self-determination
PC 3177 — self-determination communist systems
PE 7906 — sovereignty

Denial right [nationality...]
PE 4912 — nationality
PE 4016 — nationality women
PE 0241 — nations select their own leaders
PD 0162 — *old age benefits*
PE 9110 — organize political parties
PE 5398 — organize trade unions
PE 8411 — ownership
PE 6086 — participate government
PC 1001 — participate government women

Denial right people
PC 6727 — be self-determining
PF 0297 — determine their own political status
PE 6955 — freely dispose natural wealth
PD 5253 — live peace
PE 1536 — pursue development
PE 4399 — their own means subsistence

Denial right [peoples...]
PE 2142 — peoples use their own language
PE 3044 — periods nurse infants during working hours
PD 7609 — petition authority
PE 8712 — picket
PF 1075 — political asylum
PE 7393 — presumed innocent until proven guilty
PB 0284 — privacy
PD 7608 — private correspondence
PE 6168 — private home life
PC 6870 — procreate
PE 4544 — procreate severely mentally handicapped
PE 4018 — property women
PD 1846 — protection union representatives
PE 7310 — pursue spiritual well-being because discrimination
PD 0140 — reasonable work hours
PE 4214 — receive information
PE 4716 — recognition person law

Denial right [recruit...] cont'd
PE 7036 — recruit union members
PE 4173 — redress rights violations
PF 1916 — refuse vaccination
PE 6949 — resist oppression
PD 4458 — retirement
PC 0935 — *return country residence*
PF 6140 — roam countryside
PD 7212 — security

Denial right social
PD 7251 — security
PD 3120 — security capitalist systems
PE 1506 — security indigenous peoples
PE 1506 — welfare services indigenous peoples

Denial right [speedy...]
PE 4887 — speedy trial
PE 0241 — state succession
PE 5070 — strike

Denial right sufficient
PE 7616 — clothing
PE 0324 — food
PD 5254 — shelter

Denial right [time...]
PE 7624 — time prepare trial defence
PD 4695 — trade union activity
PD 0683 — trade union association
PD 4252 — traditional economies indigenous populations
PE 4737 — trial court
PF 6616 — unborn children
PE 3910 — union activity rural workers
PE 1355 — union activity special groups
PF 2904 — vote
PF 6140 — walk country

Denial right welfare
PD 0512 — services aged
PD 0542 — services blind
PD 0601 — services deaf

Denial right work
PC 5281 — [Denial right work]
PC 3119 — capitalist systems
PE 3751 — refugees
PD 2872 — women socialist countries

Denial rights
PB 5405 — [Denial rights]
PC 5456 — animals
PD 4728 — businesses
PD 0513 — children youth
PD 0520 — detainees
PE 4972 — development indigenous peoples
PC 3461 — disabled
PC 2541 — elderly
PC 2541 — elders
PD 0164 — employed children
PD 4916 — equal opportunities employment elderly
PC 8999 — ethnic minorities
PE 6929 — ex-convicts
PE 5741 — female homosexuals
PE 1903 — homosexuals
PE 6934 — human rights activists
PD 0943 — illegitimate children
PF 3710 — inanimate objects
PC 0352 — indigenous people
PE 1024 — indigenous people be self-governing
PD 3035 — journalists
PF 4334 — machine intelligences
PE 3882 — male homosexuals
PD 1662 — medical patients
PD 1148 — mental patients
PD 0973 — migrant workers
PC 8999 — minorities
PE 4010 — parent men
PE 4019 — parent women
PF 9455 — posterity
PD 0520 — prisoners
PC 2129 — religious minorities
PF 4334 — robots
PD 1914 — sexual minorities
PD 4089 — soldiers
PE 4814 — sovereign nations
PD 6346 — students
PC 2945 — territorial integrity
PD 6620 — territories
PF 0399 — those punished death penalty
PD 0610 — trade unions organizers
PF 1844 — transient populations
PE 8548 — transsexuals
PC 4405 — vulnerable groups
PE 4758 — wounded military personnel
PF 9190 Denial rites dead
PE 5480 Denial sanitary facilities ; Torture

Denial social
PC 0663 — rights
PE 7765 — rights indigenous peoples
PE 8746 — security nationals who have lived abroad

Denial [sovereignty...]
PD 3109 — sovereignty over natural resources
PD 4814 — state's rights
PE 1085 — sufficient nutrition women
PE 4860 — Denial trade union rights governments
PE 6888 — Denial trade union rights public employees
PA 7392 Denial *Unbelief*
PA 7070 Denial *Unprovability*

Denial [womans...]
PF 6169 — woman's right abortion
PE 6427 — working animals limitation working hours
PE 4793 — working animals restorative nourishment rest

PE 7902 — Denied credit ; Geographically
PE 7902 — Denies credit ; Redlining
PA 6191 — Denigration *Disapproval*
PD 8807 — Denominational education ; Discriminatory effects religious
PF 6156 — Density accommodation ; Impersonality high
PB 0469 — Density ; Excessively high population
PD 6131 — Density gradients ; Unbalanced urban population

Dental [care...]
PF 4832 — *care ; Inadequate*
PF 4832 — *care ; Lack*
PD 1185 — caries
PF 6472 — Dental prevention ; Unpractised
PF 6472 — Dental priority ; Low
PE 7201 — Dental torture
PF 3448 — Dentists ; Unavailability local
PD 1185 — Dento-facial anomalies
PD 8926 — Denunciation
PA 7237 — Denunciation *Condemnation*
PA 6191 — Denunciation *Disapproval*
PF 3307 — *Denying competition ; Culture*
PF 1075 — Denying sanctuary
PF 3523 — *Departure skilled youth*
PA 4565 — Dependence

Dependence [abuse...]
PC 6873 — abuse economic power
PB 0731 — accidents
PD 8399 — aged
PA 0587 — aggression
PC 0374 — amenity destruction
PF 8363 — amulets, charms talismans
PA 1635 — anxiety
PA 2360 — apathy
PC 1258 — arms race
PA 7646 — arrogance
PB 1638 — authoritarianism
PF 8995 — authority
PA 6999 — avarice

Dependence [bigotry...]
PC 7652 — bigotry
PC 1164 — biochemical warfare
PE 7627 — breast feeding

Dependence [censorship...]
PC 0067 — censorship
PA 6714 — chance
PF 6836 — chaos
PF 2135 — *children*
PD 3555 — children father
PE 3296 — cities
PC 1573 — class domination
PB 0848 — competition
PA 1742 — complacency
PE 8533 — computer models complex system behaviour ; Excessive
PA 0298 — conflict
PB 3407 — conformism
PA 1986 — corruption
PB 2619 — corruption meaning
PB 6314 — credibility gap
PB 0001 — crime

Dependence cultural
PB 2331 — barriers
PC 1351 — deprivation
PC 3195 — imperialism

Dependence [decadence...]
PB 2542 — decadence
PB 4731 — deceit
PF 4416 — deep emotional attachments
PA 1999 — delay
PF 3217 — denial freedom thought
PA 6953 — depersonalization
PA 0831 — deprivation
PA 0832 — destruction
PC 0480 — destruction wildlife habitats
PE 0537 — developed countries ; Resource import

Dependence developing countries
PD 3131 — budget deficit financing external borrowing
PD 2955 — customs revenue
PD 2968 — export limited range raw materials
PD 2968 — export primary commodities ; Economic
PC 6189 — External
PE 7195 — external aid
PE 7195 — external financing development programmes
PE 8086 — food imports
PE 4280 — foreign insurance
PE 7195 — foreign loans
PF 1489 — imported technology
PE 4451 — unpaid female labour
PF 4296 Dependence disabled

Dependence discrimination
PA 0833 — [Dependence discrimination]
PC 0244 — employment
PC 0934 — politics

Dependence [discriminatory...]
PB 6021 — discriminatory international order
PB 7606 — disinformation
PA 6233 — disunity
PF 4741 — dominant technology
PA 0839 — domination
PD 3825 — Drug
PF 0841 Dependence ; Economic

Dependence economic
PC 3413 — apathy
PC 3198 — imperialism
PC 2020 — insecurity
PC 6875 — manipulation

Deportation villages

Dependence economic cont'd
PB 0539 — social underdevelopment
Dependence [economies...]
PE 7413 — economies food subsidies
PA 6318 — egoism
PA 1387 — elitism
PB 6384 — environmental degradation
PB 1166 — environmental pollution
PB 2480 — ethical decay
PB 5765 — ethnocentricity
PC 1328 — ethnocide
PF 7042 — evil
PF 9530 — Excessive government
PF 1447 — excessive military expenditure
PE 8933 — excitement danger young people
PF 8404 — existence enemies
PB 3200 — exploitation
PC 3297 — exploitation employment
PE 0938 — export credits developing countries ; Excessive
PD 0850 — extended families
PF 1196 — *external expertise*
PC 0065 — external resources
PD 4469 — extortion
PB 3415 — extremism
Dependence [family...]
PC 2102 — family breakdown
PE 8670 — family social security
PB 0315 — famine
PD 9802 — farm subsidies developed countries ; Over
Dependence fear
PA 6030 — [Dependence fear]
PF 0462 — death
PF 6803 — nature
Dependence [fertilizers...]
PF 3409 — *fertilizers ; Expensive*
PD 7592 — food aid ; Exploitation
PB 2846 — food shortage
PC 3293 — forced assimilation
PC 0746 — forced labour
PB 0206 — Foreign
PC 0065 — *foreign labour*
PE 4891 — fossil fuels
PB 5747 — friendlessness
PA 2252 — frustration
PC 6664 Dependence genetic ethnic weapons
PC 1056 Dependence genocide
Dependence government
PF 9530 — action
PD 1018 — income products dangerous health
PD 1018 — revenue substance abuse
PD 1018 — revenues exploitation environmentally inappropriate products
PD 1018 — revenues exploitation non-renewable resources
PD 1229 — welfare benefits
Dependence [homelessness...]
PB 2150 — homelessness
PF 0565 — horoscopes
PA 2159 — Human
PB 1044 — human disease disability
PB 5646 — human physical suffering
PF 3856 — humiliation
PB 0262 — hunger
Dependence [ideological...]
PC 3431 — ideological war
PA 7710 — idleness
PA 5568 — ignorance
PC 0210 — illiteracy
PC 0998 — imbalance payments
PB 1969 — imbalance power
PA 3369 — immorality
PB 0113 — imperialism
PC 6329 — *imported energy*
PC 0065 — *imported labour*
PC 0449 — inadequate housing
PA 6416 — incompetence
PB 3095 — incorrect information
PA 7604 — indifference
PD 3097 — indoctrination
PE 0537 — industrialized countries import resources
PE 0537 — industrialized countries imports primary commodities
PB 0843 — inefficiency
Dependence inequality
PA 0844 — [Dependence inequality]
PC 3434 — education
PC 3435 — opportunity
Dependence [inequitable...]
PB 7666 — inequitable distribution wealth
PB 0284 — infringement privacy
PD 1519 — inhumane indiscriminate weapons
PC 1449 — inhumane scientific activity
PA 6486 — injustice
PC 0143 — innumeracy
PA 0857 — insecurity
PF 2261 — *interim employment*
Dependence international
PB 0968 — aggression
PB 5057 — conflict
PB 0009 — insecurity
Dependence [interpersonal...]
PB 0034 — interpersonal estrangement
PB 1992 — intimidation
PD 5677 — island developing countries imports
PE 5724 — island developing countries official assistance
Dependence [lack...]
PF 0817 — lack international cooperation
PC 2122 — landscape disfigurement

Dependence [leadership...] cont'd
PF 8995 — leadership
PE 8116 — least developed countries foreign aid
PE 4780 — local communities outside services ; Excessive
PF 2386 — loneliness
PB 7600 — lying
Dependence [maldevelopment...]
PB 6207 — maldevelopment
PC 2801 — maldistribution food
PB 0167 — maldistribution world population
PB 1498 — malnutrition
PD 6663 — man-made diseases
PA 6359 — manipulation
PA 6977 — meaninglessness
PD 6571 — *mechanization ; Crippling*
PD 7773 — media
PB 6248 — mental pollution
PB 5680 — mental suffering
PB 3528 — middlemen ; Farm tenant's
PB 0318 — military economic hegemony
PC 1867 — *mind set ; Urban*
PE 1821 — minor tranquillizers
PB 7712 — moral imperfection
PF 2590 — mysticism
Dependence [National...]
PF 1452 — National political
PB 0534 — nationalism
PA 7090 — negative emotions attitudes
PA 2658 — negligence
PF 1357 Dependence obscurantism
PC 0065 *Dependence oil*
Dependence [pain...]
PA 0643 — pain
PA 7030 — passion
PD 3555 — patriarchal role man ; Family
PB 7709 — persecution
PD 0869 — perversion
PF 3392 — *philosophy*
PF 3211 — planning authorities
PD 4100 — plant species
Dependence political
PC 1917 — apathy
PC 3425 — inequality
PC 1907 — inertia
PC 2181 — injustice
PC 2677 — instability
PB 3209 — pressure
PC 1919 — repression
Dependence [Polysubstance...]
PD 3825 — *Polysubstance*
PB 0388 — poverty
PB 3202 — power politics
PA 2173 — prejudice
PA 7599 — pride
PD 2968 — primary commodities developing countries ; Excessive income
PF 1878 — propaganda
PF 0421 — psychological inertia
Dependence [racial...]
PC 0006 — racial discrimination
PC 3334 — racial exploitation
PB 1047 — racism
PB 0205 — refugees
PF 0150 — religion
Dependence religious
PC 3414 — apathy
PC 1455 — discrimination
PC 1808 — intolerance
PC 0578 — repression
Dependence [repression...]
PB 0871 — repression
PB 6112 — resource depletion
PC 0065 — Resource import
PA 7296 — *Restraint*
PC 2162 — restriction freedom expression
PC 0073 — restrictive trade practices
PF 7418 — romantic love
Dependence [secrecy...]
PA 0005 — secrecy
PB 1540 — secularization
PE 7031 — sex
PF 9109 — sex
PE 4188 — single source finance ; Developing country
PF 4741 — single technology
PC 0146 — slavery
Dependence social
PC 3412 — apathy
PC 1864 — discrimination
PB 0514 — inequality
PC 0797 — injustice
PC 2940 — intimidation
PB 0883 — neglect
PD 1229 — security
PC 0242 — underdevelopment
PD 1229 — welfare
Dependence [socialist...]
PD 2048 — socialist countries private enterprise
PE 5296 — some developing countries drug trade ; Economic
PD 6571 — sophisticated technology development
PF 8508 — stereotypes
PB 1935 — structural violence
PD 3825 — *Substance*
PD 0557 — subversive activities
PA 0430 — superstition
Dependence [technology...]
PD 2420 — technology

Dependence [temptation...] cont'd
PA 7736 — temptation
PB 3430 — torture
Dependence [ugliness...]
PA 7240 — ugliness
PB 0479 — unbalanced growth
PA 6438 — uncertainty
PC 0495 — uncontrolled economic growth
PB 1845 — uncontrolled industrialization
PA 7273 — *Unconventionality*
PC 1015 — undemocratic political organization
PB 0206 — underdevelopment
PB 1860 — underemployment
PC 3424 — underprivileged minorities
PB 0750 — unemployment
PC 2815 — unequal income distribution countries
PC 0655 — unequal political representation
PD 1229 — Unmotivating welfare
PE 4789 — unpaid household work
Dependence [vanity...]
PA 6491 — vanity
PA 5644 — vice
PA 0429 — violence ; Human
PC 6329 — *vulnerability ; Energy*
PD 1104 — vulnerability socialist countries ; External
Dependence [war...]
PB 0593 — war
PD 3555 — wife husband
PC 1445 — wildlife extinction
PD 4957 Dependence xenophobia
PF 1113 Dependency attention
PD 2476 Dependency children
PA 4565 Dependency *Dependence*
PD 3491 Dependency developing countries ; Excessive childhood
PF 0841 Dependency ; Economic
PD 8399 Dependency elderly
PE 1806 Dependency food staples developing countries due transnational corporations ; Import
PF 9530 Dependency government funding
PD 4632 Dependency intermediaries
PE 2214 Dependency international financial institutions developing countries ; Over
PD 5788 Dependency land-locked developing countries
PE 3296 Dependency ; *Longterm external*
Dependency [marriage...]
PD 3244 — marriage ; Emotional
PD 4632 — middlemen
PD 2374 — *Military*
PF 8855 Dependency ; Over acceptance socio economic
PF 8855 Dependency ; Over commitment welfare
PD 2374 *Dependency ; Political*
PF 8855 Dependency social assistance ; Unquestioning
PF 1755 Dependency social institutions ; Fostering
PF 3084 Dependency unpredictable sources income
PB 4353 Dependency vulnerable groups
PC 3426 Dependency women
PD 3694 Dependency women marriage
PB 0206 Dependent development
PF 4112 Dependent economies ; Instability tourist
PF 2627 Dependent local biases ; Leadership
PD 7238 Dependent people ; Physically
PE 6696 Dependent personality disorder
PF 1452 *Dependent territories*
PC 1420 Dependents ; Unproductive
PD 2092 Dependents war victims ; Neglect
PA 6953 Depersonalization
PA 6953 Depersonalization ; Dependence
PA 6739 Depersonalization *Maladjustment*
PF 0981 Depiction human figures
PE 5697 Depilation ; Forced
PF 2973 Depleted expertise rural labour force
PD 9082 Depleted fish resources
PD 0077 Depleted soil nutrients
PE 8560 Depleted uranium ; Instability trade U235
PE 3296 Depletion corporate initiative rural communities ; Cumulative
PB 6112 Depletion ; Dependence resource
PD 4002 Depletion due high level consumption ; Natural resource
PE 4913 Depletion fish reserves marine mammals
PD 9082 Depletion ; Fisheries
PA 6852 Depletion *Inappropriateness*
PA 7382 Depletion *Loss*
PD 9357 Depletion mineral resources
Depletion natural resources
PB 5250 — [Depletion natural resources]
PD 3922 — developing countries
PD 4007 — due population growth
PF 3060 Depletion official reserves developing countries
PB 6112 Depletion ; Resource
Depletion [Soil...]
PD 0949 — Soil
PD 9227 — soil nutrients
PE 6113 — Stratospheric ozone
PF 5968 — syndrome ; Maternal
PD 9082 Deployment excessively efficient fishing nets
PE 3639 *Depluming mite*
PD 2483 *Depopulation*
PD 1908 Depopulation mountainous regions
PC 0056 Depopulation ; Rural
PD 7822 Deportation
PF 0726 Deportation ; Delayed consequences war time imprisonment
PC 6203 Deportation ; Forced
PC 6203 Deportation villages

—895—

PD 7822 Deportees
PD 4904 Deposition ; Acid
PA 6088 Depravation *Impairment*
PA 5644 Depravation *Vice*
PC 8974 Depravity
PA 6839 Depravity *Disrepute*
PC 8974 Depravity ; Total
PA 5644 Depravity *Vice*
PF 3512 Depreciated community identity
PE 5260 *Depreciation* ; *Building*
Depreciation [Cheapness...]
PA 7193 — *Cheapness*
PE 8583 — Competitive exchange
PE 3700 — currency
PA 6191 Depreciation *Disapproval*
PA 6088 Depreciation *Impairment*
PA 7382 Depreciation *Loss*
PA 6607 Depreciation *Misjudgement*
PF 5354 Depreciatory operating images ; Self
PA 5688 Depredation *Appropriation*
PA 6542 Depredation *Destruction*
PF 3550 *Depressed area classification*
PD 1566 Depressed areas ; Fragmentation social structures
PF 2093 Depressed community image
PC 0799 Depressed people ; Mentally
PD 8183 Depressed regions developed countries
PF 1213 Depressing effect poor housing construction
PA 5493 Depression
PE 1883 Depression ; Anaclitic
PD 3784 Depression children ; Mental
PE 0885 Depression due torture
PF 1277 Depression ; Economic cycle boom
PA 6088 Depression *Impairment*
PE 0655 Depression ; Involutional
PD 1318 *Depression* ; *Manic*
PC 0799 Depression ; Mental
PC 0799 *Depression* ; *Post natal*
PA 6731 Depression *Solemnity*
PE 0885 Depression torture victims
PA 7107 Depression *Unpleasantness*
PE 0258 Depression ; Winter
PF 1277 Depressions ; Economic
PC 0799 Depressive neuroses
PD 0270 *Depressive neurosis*
PC 0799 Depressive psychosis
PD 1318 Depressive psychosis ; Manic
PA 0831 Deprivation
PA 7270 Deprivation *Absence*
PA 0831 Deprivation basic necessities
PC 1351 Deprivation ; Cultural
Deprivation [Dependence...]
PA 0831 — Dependence
PC 1351 — Dependence cultural
PE 6909 — dwarfism
PF 4296 Deprivation freedom caring disabled
PE 5667 Deprivation government benefits political purposes
PE 5787 Deprivation ; Harmful effects sensory
PC 7379 Deprivation human rights
PC 0981 Deprivation ; Infant emotional
PA 5473 Deprivation *Insufficiency*
PA 7382 Deprivation *Loss*
PC 0981 Deprivation ; Maternal
PD 3225 Deprivation nationality
PA 5870 Deprivation *Nonexistence*
Deprivation [Paternal...]
PD 7297 — Paternal
PC 8862 — peasantry
PA 6434 — *Poverty*
PE 5043 — prisoners' families
PD 3055 — property ; Denial right freed arbitrary
Deprivation [Regional...]
PD 4424 — Regional
PB 5405 — rights
PF 6389 — risk-taking
PE 2741 Deprivation ; Sleep
Deprivation [Temporal...]
PF 4644 — Temporal
PD 3763 — Torture
PE 6797 — Torture sensory
PE 8170 — trade union funds property
PC 0981 Deprived affection ; Children
PD 0550 Deprived children ; State custody
PA 0831 Deprived people
PF 2541 Deprived populations ; Infantilization
PC 2049 Deprived regions
PD 1480 Depriving prisoners medical treatment
PF 1094 Depths ; Antiquated intellectual methods appropriate human
PD 0992 Derangement ; Brain
PF 4037 Derealization
PE 6005 Derelict industrial wastelands
PF 2613 Derelict military explosive devices
PE 5715 Dereliction
PA 6379 Dereliction *Avoidance*
PA 8658 Dereliction duty
PA 7363 Dereliction *Improbity*
PA 5438 Dereliction *Neglect*
PA 5644 Dereliction *Vice*
PE 5340 Derelicts hazards ; Wrecks
PJ 4005 Derision
PE 3618 Dermaptera insect pests
PE 2465 Dermatitis
PD 9667 Dermatitis animals
PE 9509 Dermatitis animals ; Atopic
Dermatitis [cattle...]
PE 5380 — cattle ; Filarial

Dermatitis [Contact...] cont'd
PE 2465 — Contact
PD 9667 — Contagious pustular
PD 9667 Dermatitis ; Flea allergy
PD 9667 Dermatitis ; Gangrenous
PD 9667 Dermatitis ; Nasal solar
PD 5535 Dermatitis ; Necrotic auricular
PE 5617 Dermatitis ; Ovine interdigital
Dermatitis [Pelodera...]
PE 5380 — Pelodera
PD 5535 — pinna
PD 9667 — poultry ; Necrotic
PD 9667 — Pyogenic
PE 2465 Dermatitis ; Radiation
PE 5380 Dermatitis ; Rhabditic
Dermatitis [Vegetative...]
PE 4161 — *Vegetative interdigital*
PE 2465 — venenanta
PE 7826 — verrucosa
PE 4161 — verrucosa animals
PA 2455 Dermatomycosis
PE 5086 Dermatomyositis
PD 9667 *Dermatomyositis* ; *Clostridial*
PD 9667 Dermatophilosis
PD 2728 Dermatophytoses
PD 9667 Dermatophytosis animals
PE 5684 Dermatoses ; Occupational
PE 5380 Dermatosis ; Filarial
PD 9667 Dermatosis sheep ; Ulcerative
PA 6191 Derogation *Disapproval*
PA 6088 Derogation *Impairment*
PF 8666 Desanctification churches holy places
PF 7383 Describe female experience ; Lack English language
PF 9176 Desecration
PD 6128 Desecration ancient burial sites
PD 7258 Desecration ; Burial site
PD 7258 Desecration cemeteries
PF 9138 Desecration flags
PF 3607 Desecration holy days
PE 6385 Desecration holy spaces
PA 6058 Desecration *Impiety*
PA 6852 Desecration *Inappropriateness*
PD 4348 Desecration monuments
PD 7278 Desecration religious buildings
PF 9138 Desecration symbols
PA 7280 Desecration *Wrongness*
PC 3688 Desegregation
PD 0493 Desert advance
PC 2506 Desert advance
PD 2520 Desert nomadism
PD 2285 Desert oases ; Environmental degradation
PE 3296 *Desert use* ; *Delimited*
PF 2575 *Deserted public spaces*
PD 1030 Deserted wives
PD 4621 Deserters
PE 4909 Deserters ; Aiding
PC 2506 Desertification
PA 7685 Desertion *Abandonment*
PA 6379 Desertion *Avoidance*
PA 6058 Desertion *Impiety*
PA 7325 Desertion *Irresolution*
PD 8340 Desertion ; Juvenile
PF 3254 Desertion marriage law
PD 4621 Desertion ; Military
PA 5699 Desertion *Reversion*
PA 6653 Desertion *Unsociability*
PD 8585 Deserts ; Infertile
PE 4966 Desiccation human skin
PD 1990 Desiccation inland seas
PD 1990 Desiccation lakes
PD 1990 *Desiccation rivers*
PD 1990 *Desiccation streams*
PD 7450 Design ; Biased computer software
Design [community...]
PF 1546 — community space ; Unfocused
PE 2462 — construction ; Unsafe automobile
PF 1379 — consumer products ; Unsafe
PD 8638 — Culturally insensitive house
PE 2462 Design ; Defects machinery
PF 4944 Design development projects ; Inappropriate
PF 9980 Design ; Erratic time
PE 2462 Design ; Faults reactor
PD 7450 Design information systems ; Discriminatory
PD 2886 Design ; Irresponsible information system
PF 0867 Design ; Monotonous unaesthetic architecture
PE 2462 *Design* ; *Non ergonomic*
PF 0867 Design ; Over conformity building
PF 0867 Design ; Unimaginative housing
PF 1546 Design ; Unplanned village
PD 7585 Designated wilderness areas ; Over use
PD 6627 *Designed premises* ; *Ill*
PF 6067 Designed software developing countries ; Lack specifically
PD 0094 *Designer drugs*
PF 2812 Designs using labour ; Debilitating
PF 0963 *Desirability* ; *Unconsensed institution*
PE 9607 Desire disorders ; Sexual
PE 7031 Desire ; Excessive sexual
PA 7030 Desire ; Insatiable
PF 3382 Desire male children
PF 9424 Desire risk ; Excessive
PF 8128 Desires ; Misdirection human energies
PF 2986 Deskilling rural communities
PD 8546 Deskilling women ; Domestic
PE 8750 Desmans ; Endangered species moles
PE 7826 *Desmitis* ; *Interosseous*

PE 7826 *Desmitis* ; *Sacroiliac*
PA 6542 Desolation *Destruction*
PA 6731 Desolation *Solemnity*
Desolation [Unpleasantness...]
PA 7107 — *Unpleasantness*
PA 7208 — *Unproductiveness*
PA 6653 — *Unsociability*
PF 4004 Despair
PA 6099 Despair *Hopelessness*
PA 6731 Despair *Solemnity*
PF 1365 Despair ; Symbols
PA 7107 Despair *Unpleasantness*
PF 0951 Despairing individualism
PF 4004 Despairing people
PF 4004 Desperation
PA 8038 Desperation ; Spiritual
PA 7240 Desperation *Ugliness*
PA 5454 Despicableness *Badness*
PA 6839 Despicableness *Disrepute*
PA 5942 Despicableness *Unimportance*
PA 7107 Despicableness *Unpleasantness*
PA 5688 Despoliation *Appropriation*
PA 5454 Despoliation *Badness*
PA 6542 Despoliation *Destruction*
PA 6099 Despondency *Hopelessness*
PA 6731 Despondency *Solemnity*
PC 0845 Despotism
PA 5563 Despotism *Lawlessness*
PE 9420 Destabilization countries external forces ; Aggressive economic
PD 7714 Destabilization developing countries ; Military
PD 9792 Destabilization developing countries ; Political
PE 9420 Destabilization economies developing countries
PD 8030 Destabilization foreign investment ; Vulnerability countries
PD 5693 Destabilization government
PD 4252 Destabilization indigenous economies
PE 5903 Destabilization monetary systems exchange notes transnational corporations
PE 4994 Destabilization national insurance markets offshore insurers
PB 5417 Destabilization social systems
PC 7873 Destabilization world trade ; Financial
PD 1220 Destabilizing corporate expansion
PD 0226 Destabilizing financial action transnational enterprises
PD 5693 Destabilizing governments
PE 1478 Destabilizing information media ; Excessive portrayal
PD 0187 Destabilizing international telecommunications
PF 3111 Destiny
PF 6430 Destiny ; Belief manifest
PD 6645 Destitute ; Illiteracy amongst
PB 2150 Destitute people
PB 2150 Destitution
PD 5734 Destitution ; Child
PE 5279 Destroying cherished things ; Torture
PD 1754 Destructible packaging containers ; Non
PA 6542 Destruction
Destruction [agricultural...]
PD 9118 — agricultural land
PD 6850 — alluvial forests
PC 0374 — Amenity
PA 4502 — archaeological sites
PD 7096 — arid ecosystems irrigation
PC 3152 Destruction biological niches
Destruction [civilian...]
PE 8564 — civilian populations institutions
PD 5393 — communist regimes ; Environmental
PD 5769 — coral reefs
PE 3914 — countryside
PE 3880 — crop pollinating insects
PE 3880 — crop pollinating species
Destruction cultural
PC 2114 — heritage
PD 7298 — property during warfare
PF 7542 — property ; Gradual
PA 0832 Destruction ; Dependence
PC 0374 Destruction ; Dependence amenity
Destruction [Ecological...]
PE 5196 — Ecological
PE 8915 — economy due war
PE 5196 — environmental oxygen
PD 5859 — *equipment* ; *Deliberate*
PE 8148 Destruction fisheries dams
PC 1366 Destruction forests
PB 4788 Destruction genetic diversity
PD 2345 Destruction green inner suburbs
Destruction [hedges...]
PD 1642 — hedges hedgerow trees
PD 0253 — historic buildings
PD 0172 — historic documents public archives
Destruction [indigenous...]
PC 7203 — indigenous cultures
PF 4829 — inherent development
PE 8109 — injured animals ; Economic loss
Destruction [land...]
PC 1300 — land fertility
PD 9118 — *land resources*
PE 8401 — least developed countries ; Environmental
Destruction [medical...]
PF 1836 — medical records
PE 0736 — military bases ; Environmental
PE 6256 — mountain ecosystems ski resorts
PC 1247 Destruction natural barriers
PE 2443 Destruction natural environment mankind ; Inevitable
PD 2285 Destruction oasis ecosystems
PD 2596 Destruction open pit mines quarries ; Environmental

Destruction [peat...]
PC 3486 — peat land
PA 0832 — *private property*
PD 5859 — property
PA 0832 — *Property*
PD 6204 Destruction rain forests
PC 2237 Destruction rural subsistence economy
Destruction [scientific...]
PF 4633 — scientific records
PA 7055 — Self
PF 8587 — Self
PF 3831 — social institutions
PD 3574 — soil fauna micro organisms ; Pesticide
PD 1052 — soil structure
PE 5064 — species ; Degradation environment
PE 0360 — stray dogs ; Cruel methods
PE 7581 Destruction trade union property public authorities
PE 7061 Destruction transnational enterprises ; Records
Destruction [war...]
PE 2592 — war ; Housing
PD 8359 — war ; Industrial
PD 6282 — watersheds
PE 3987 — weeds
PC 3486 — wetland environments
PB 5250 — wilderness areas
PC 0480 — wildlife habitats
PC 0480 — wildlife habitats ; Dependence
PF 3849 — Wish self
PC 1366 — woodlands
PE 1265 Destructive action mould tropical climates
PA 0832 Destructive attitudes
PD 8544 Destructive behaviour ; Unchecked
PC 2087 Destructive changes ocean characteristics
PC 2791 Destructive economic isolation
PF 6044 Destructive excuses ; Self
PD 9082 Destructive fishing
PF 5061 Destructive government policy making ; Self
PA 0832 Destructive people
PF 7358 Destructive ; Verbal conflict envisaged
PD 1229 Destructive welfarism
PF 1509 *Destructive youth image*
PA 6340 Destructiveness *Adversity*
PA 7055 Destructiveness *Death*
PA 6542 Destructiveness *Destruction*
PA 0832 Destructiveness ; *Human*
PD 3492 Destructiveness ; Incendiary weapons massive
PA 7226 Destructiveness *Unhealthfulness*
PE 1904 Desynchronization bodily rhythm international travel
PF 5978 *Detached local teachers*
PF 1062 Detached past ; Confusion being
PF 0889 Detachment represented constituency ; Extreme
PE 4584 Detachment retina
PD 3015 Detainees ; Civilian political prisoners
PD 0520 Detainees ; Denial rights
PE 6929 Detainees ; Discrimination against ex prisoners ex
PD 3014 Detainees ; Military political prisoners
PF 4924 Detection rate ; Decline crime
PB 0284 *Detector tests ; Enforced lie*
PD 5142 Detention
PD 1576 Detention charge ; Administrative
PE 6636 Detention children
PD 1576 Detention ; Denial right freedom arbitrary
PE 6260 Detention ; Escape official
PE 8004 Detention ; Incommunicado
PE 5488 Detention international civil servants
PD 0634 Detention juveniles ; Injustice
PD 0634 Detention juveniles ; Repressive
PD 5142 Detention mothers
PD 0520 Detention offenders after completion sentence
PD 0702 Detention political prisoners ; Mass
PE 2932 Detention psychiatric institutions ; Abusive
PE 6376 Detention refugees asylum-seekers
PE 8004 Detention ; Secret
PE 6367 Detention ; Suspicious deaths during
PD 4259 Detention ; Unacknowledged
PD 7630 Detention workers representatives
PD 6062 Detention ; Wrongful
PE 1087 Detergents pollutants
PD 2672 *Deteriorated building conditions*
PF 1301 Deteriorated structures essential corporateness
PF 1213 *Deteriorated vacant houses*
PF 1733 Deteriorating buildings ; Unattractive appearance
PD 2241 Deteriorating community identity
PF 2093 *Deteriorating downtown district*
PC 2616 Deteriorating inner-city neighbourhoods
PF 7142 Deteriorating quality life
PF 1681 *Deteriorating space ; Overwhelming*
PF 3558 Deteriorating structures rural community cooperation
PE 4603 Deteriorating terms financial loans developing countries
Deterioration [Aggravation...]
PA 7378 — *Aggravation*
PC 0918 — atmospheric radiation balance
PE 2593 — atmospheric visibility
PB 6321 Deterioration ; Behavioural
Deterioration [confidence...]
PF 4345 — confidence loans developing countries
PE 4603 — cost external finance developing countries
PE 9825 — cultural artefacts tourism
PD 2301 Deterioration decay ; Wood
PD 5092 Deterioration domestic food production developing countries
PE 9567 Deterioration external financial position developing countries
PE 9567 Deterioration external payments position developing countries
PD 5377 Deterioration film standards

PF 7727 Deterioration gene pool
PC 8943 Deterioration human environment
Deterioration [Impairment...]
PA 6088 — *Impairment*
PD 9202 — industrialized countries
PE 8073 — international economic environment
PE 8233 — irrigation systems
PC 9075 Deterioration living standards workers
PC 9584 Deterioration long-term trade policies
PD 5377 Deterioration media standards
PE 4649 Deterioration mind age
PE 5260 Deterioration nuclear power plants
PE 1669 Deterioration ; Paper
Deterioration physical
PD 1955 — condition art objects
PD 2672 — environment ; Debilitating
PC 0716 — health
PD 5377 Deterioration press standards
PD 1435 Deterioration product quality
PD 1565 Deterioration quality equipment maintenance servicing
PE 7734 Deterioration quality life least developed countries
PE 5133 Deterioration rangelands grasslands
Deterioration [Soil...]
PD 1052 — Soil
PD 0077 — soil fertility
PC 1052 — soil resources
PE 0082 — stained glass due acid rain
PB 0388 — standard living poor
PE 1669 — stored documents archives
PD 5377 Deterioration television programming standards
PD 2897 Deterioration terms trade developing countries
PD 5377 Deterioration video standards
PD 9196 Deterioration water supply systems
PD 6282 Deterioration ; Watershed
PF 3382 Determination birth ; Artificial sex
PE 3123 Determination capitalist exploitation ; Denial effective national self
PC 3177 Determination communist systems ; Denial right national self
PE 1450 Determination ; Denial right national self
PF 1804 Determination ; Habitual overemphasis national self
PF 1491 Determination unrelated global obligations ; Self
PF 0297 Determine their own political status ; Denial right people
PF 9078 Deterrence
PF 4162 Deterrence ; Myth
PF 4162 Deterrence ; Risk unintentional nuclear war generated strategy
PD 1464 Deterrent against war crimes ; Ineffective
PF 0399 Deterrent ; Ineffectiveness capital punishment
PE 4394 Detraction
PE 2235 Detrimental effect jungle environment tropical villages
PF 2889 Detrimental international repercussions domestic agricultural policies
PF 6575 Detrimental story community future
PF 4346 Deuteromycetes
PD 4252 Devalorization traditional economies
PE 3700 Devaluation ; Currency
PE 8902 Devaluation education survival needs
PE 3700 Devaluation money
PA 6542 Devastation *Destruction*
PF 2364 Develop human beings ; Denial right
PD 1291 Developed areas ; Emigration trained personnel developing
PF 0021 Developed continuing education ; Under
Developed countries
PE 8321 — against agricultural products developing countries ; Protectionism
PE 5926 — against exports developing countries ; Restrictive business practices markets
PD 0425 — commodities exported developing countries ; Processing
PD 4560 — Contraceptive availability
PE 2203 — Contradictions growth partnership model
PD 8183 — Depressed regions
PE 7734 — Deterioration quality life least
PD 9679 — developing countries ; Inaccessibility markets
PD 0800 — Dietary deficiencies
PE 1604 — Discrimination against developing countries formation regional groupings
PE 6843 — Displacement workers
PD 9802 — Domestic agricultural price difficulties
PD 7835 — Environmental degradation
PE 8401 — Environmental destruction least
PC 5551 — Excessive consumption resources
PD 9802 — Excessive subsidies farmers
PC 7459 — Failure restructure economic relations
PE 7954 — Food insecurity least
PE 8116 — foreign aid ; Dependence least
PE 8957 — Foreign private investment income outflow developing
PE 5723 — Formalism
PE 8049 — Geographic barriers least
PE 8285 — Homeless
PC 1383 — Illiteracy
PE 8978 — Illiteracy least
PD 7682 — Import substitution
Developed countries Inadequate
PE 8082 — agriculture least
PF 4297 — development communication services least
PE 7954 — food supplies least
PD 9242 — health care least
PE 9764 — human resources development least
PE 0289 — infrastructure services least
PE 7892 — investment transnational corporations least
PE 0265 — nutrition education least

Developed countries Inadequate cont'd
PE 7734 — standard living least
Developed countries [Income...]
PE 6891 — Income inequality
PE 8244 — Insufficient forestry least
PE 0273 — Insufficient use natural resources least
PE 4804 — Lack artisans craftsmen
PE 9764 — Lack educated manpower least
PD 6029 — Language domination
PD 8201 — Least
PF 6539 — Limited availability public services small towns
PE 8013 — meat egg products ; Increased demand
PE 8306 — Minimal exports least
PE 0282 — Minimal manufacturing least
PE 0299 — Natural disasters least
PD 9802 — Over dependence farm subsidies
PC 0444 — Poverty
PD 3132 — Private domestic capital outflow developing
PE 0537 — Resource import dependence
PD 0530 — Restriction outer space benefits limited number
PE 0402 — resulting participation developing countries manufacturing ; Unemployment
PD 9202 — Reversal development process
PE 9764 — Shortage managerial skills least
PC 2369 — Tariff barriers trade developing
PD 0338 — Technology gap
PD 3488 — Uncontrolled urbanization
PD 8183 — Underdevelopment
PE 8595 — Underutilization livestock least
PC 9718 — Unemployment
PE 9476 — Unemployment least
PD 2961 — Developed country limiting trade developing countries
Developed developing countries
PC 9112 — Asymmetric interdependence
PE 4190 — Bipolarization trade
PC 8694 — Economic imbalance
PE 4241 — Imbalance population growth
PE 5702 — Imbalance trade cultural products
PD 1477 — Inequitable tax treaties
PE 7985 — Technology gap
PE 8093 Developed industrialized countries ; Brain drain more
Developed market economies
PD 9679 — Decline exports developing countries
PF 1737 — Slow down growth output
PC 2954 — Weakness trade socialist
PC 0300 *Developed social responsiveness ; Poorly*
Developing countries Absence
PE 2603 — coordinated customs services
PE 8597 — legal entities
PF 7081 — national nongovernmental organizations
PF 4836 — reliable systems justice
Developing countries [Acquisition...]
PE 6640 — Acquisition inappropriate equipment
PE 7563 — adopt appropriate exchange rate policies ; Inability
PE 5338 — Aid lenders beneficiaries net outflow funds
PE 4145 — airline industry ; Underparticipation
PD 8047 — Apathy
PD 8418 — Arid
PD 9347 — Arms trade
PC 9112 — Asymmetric interdependence developed
PE 9646 — Asymmetry bargaining power transnational corporations
PD 0465 — Bad investment savings
PE 9099 — Bilateralism aid
PE 4190 — Bipolarization trade developed
PD 5139 — Blindness
PD 1291 — Brain drain
PE 0322 — Bribery transnational enterprises
PD 3131 — budget deficit financing external borrowing ; Dependence
PD 3131 — Budget deficits
PE 3066 — Burden export credit financing upon
PD 3051 — Burden servicing foreign public debt
PD 3137 — Capital shortage
PD 4387 — Carbon dioxide impact agriculture
PE 5166 — Child mortality
PD 5326 — Collapse growth
PE 9440 — commodity exchanges ; Restrictions access
PD 2968 — Commodity related shortfalls export earnings
PF 6670 — Conspicuous consumption
PE 5114 — Constraints increased agricultural output
PD 1965 — Contempt agricultural labour
PD 2968 — Continuing low level prices commodities exported
PF 3126 — Contradictions capitalism
PD 0348 — Corruption
PE 8400 — Costly medical care
PD 2955 — customs revenue ; Dependence
Developing countries Decline
PF 0392 — aid
PE 4655 — commercial bank lending
PE 0835 — communal spirit village solidarity
PE 3812 — concessional financial resources available
PD 2769 — earnings
PE 3066 — export credits
PD 3138 — foreign direct investment
PD 2968 — growth commodity export earnings
PD 1250 — handicrafts cottage industries
PE 6861 — import capacity
PD 0291 — investment
PC 0933 — mutual trade
PF 0392 — official development assistance
PE 4574 — public sector savings
PD 2769 — real wages
PE 1095 — rural customs traditions
PE 4922 — technology transfer
PD 2897 — trading position

–897–

Developing countries

Developing countries [Declining...]
PD 5326 — Declining economic growth
PD 1141 — Defective land use planning
PC 4094 — Deficiencies
PE 5664 — Deficient transport remote island
PD 3922 — Degradation environment
PE 5278 — Demotivated civil servants
PD 5788 — Dependency land locked
PD 3922 — Depletion natural resources
PF 3060 — Depletion official reserves
PE 9420 — Destabilization economies
PE 4603 — Deteriorating terms financial loans
Developing countries Deterioration
PF 4345 — confidence loans
PE 4603 — cost external finance
PD 5092 — domestic food production
PE 9567 — external financial position
PE 9567 — external payments position
PD 2897 — terms trade
Developing countries [Developed...]
PD 2961 — Developed country limiting trade
PD 9679 — developed market economies ; Decline exports
PD 7682 — Development industrialized countries products substituting commodities exported
PD 7978 — Development informal sector
PF 2960 — Differences trading principles practices
PD 7257 — Differential economic performance
PD 0724 — Disabled persons
Developing countries Discrimination
PE 4320 — against transnational banks
PC 4898 — against women
PE 4102 — against women workers multinational enterprises
PE 4310 — lending transnational banks
Developing countries Disincentives
PE 8321 — against farming
PF 3845 — financial investment
PD 3138 — foreign private investment
Developing countries [Disparate...]
PD 1534 — Disparate development economic sectors
PD 2963 — Disparities
PD 7257 — Disparities economic growth
Developing countries Disparity
PD 5258 — distribution wealth
PC 8694 — industrialized
PD 0266 — social development
Developing countries [Disproportionately...]
PF 6670 — Disproportionately wealthy elites
PD 1482 — Disruption family system
PE 2890 — Domestic agricultural price policy difficulties
PD 1873 — Domestic market restrictions
PE 4330 — Domination restrictive project loans transnational banks
PE 1796 — Domination transnational corporations domestic name brand food sector
PF 7955 — Donor distortion aid
PE 5296 — drug trade ; Economic dependence some
Developing countries due
PE 2598 — export cartels ; Excessive injury export interests
PE 4391 — introduction new technologies ; Elimination jobs
PE 2999 — restrictive business practices ; Economic inefficiencies
PE 1806 — transnational corporations ; Import dependency food staples
PE 1660 — transnational corporations ; Increasing income disparity
Developing countries Dumping
PE 0607 — agricultural products
PE 2061 — dangerous illegal products
PE 0607 — food
PE 2061 — toxic wastes
Developing countries Economic
PF 4130 — disadvantages excessive food production
PD 4063 — dislocations
PC 8694 — imbalance developed
PF 5964 — mismanagement
PD 0068 — non viability small
PD 5326 — stagnation
PD 4125 — stagnation due rural poverty
Developing countries [economies...]
PE 7255 — economies ; Low regional complementarity
PE 4391 — Elimination jobs due automation
PD 0103 — Endemic disease
PE 0379 — Energy deficient
PE 4965 — Environment policy restriction trade
PD 3922 — Environmental degradation
PE 3812 — Erratic flows concessional non concessional lending
Developing countries Excessive
PF 4345 — anxiety lending
PD 3491 — childhood dependency
PF 9457 — concentration export markets
PE 5812 — costs sea access land locked
PE 5896 — costs unsuitability insurance land locked
PE 0938 — dependence export credits
PE 5713 — emigration island
PE 1496 — external trade deficits
PD 0765 — foreign investment traditional industries
PD 2133 — foreign public debt
PD 1902 — government participation economies
PD 2968 — income dependence primary commodities
PC 3134 — repatriation profits foreign investors
PD 8114 — social costs structural adjustment debtor
PE 3926 — unnecessary burden interest repayments
Developing countries [Exploitation...]
PE 9199 — Exploitation cheap labour
PE 6952 — Exploitative financial policies transnational corporations
Developing countries [Export...] cont'd
PE 6687 — Export hazardous industries
PD 2968 — export limited range raw materials ; Dependence
PD 2968 — export primary commodities ; Economic dependence
PE 7195 — external aid ; Dependence
PD 2133 — External debt crisis
PC 6189 — External dependence
PE 7195 — external financing development programmes ; Dependence
PE 7195 — external sources development financing ; Excessive reliance
PD 1629 — Factionalism
PE 8400 — Failure medical programmes
PF 5964 — Failure public finance policy
PD 9449 — Financial paralysis
PD 0008 — Flooding urban labour market
PE 8188 — Fluctuations food production
PE 8086 — food imports ; Dependence
PE 7195 — foreign borrowing ; Over reliance
PD 3068 — Foreign currency shortage
PD 3068 — Foreign exchange shortage
PE 4280 — foreign insurance ; Dependence
PE 7195 — foreign loans ; Dependence
PE 1604 — formation regional groupings developed countries ; Discrimination against
PD 5114 — future declines production ; Vulnerability agriculture
PD 9563 — Gender discrimination
PD 3068 — Hard currency shortage
PE 2004 — Harmful effects advertising transnational corporations
PE 7538 — Health hazards tourists
PF 0906 — High human fertility
PF 0907 — High labour turnover
PD 1751 — Hoarding
PD 8856 — Homelessness
PE 0269 — Housing infrastructural weakness
PF 3350 — Ill considered pressure eliminate nakedness
PD 8329 — Illiteracy
PE 8660 — Illiteracy women
Developing countries Imbalance
PE 4956 — agricultural exports imports
PF 0544 — distribution political awareness
PF 4837 — economic social planning
PE 4241 — population growth developed
PE 8584 — rural urban incomes
PE 5702 — trade cultural products developed
PC 1563 — urbanization industrialization
Developing countries [Imbalances...]
PE 5920 — Imbalances exports imports land locked
PE 4922 — Impasse technology transfer
PE 6861 — Import compression
PF 1489 — imported technology ; Dependence
PD 5677 — imports ; Dependence island
PE 4386 — Improvisational housing
PE 5665 — Inaccessibility insurance island
PD 9679 — Inaccessibility markets developed countries
PF 7842 — Inaccurate public finance data
Developing countries Inadequacy
PE 9096 — agricultural education
PF 0392 — aid
PD 1865 — commercial sector
PD 0928 — domestic market
PE 0004 — telecommunication facilities
PD 5077 — training human settlements
Developing countries Inadequate
PE 0046 — administrative skills
PE 5800 — air transport service land locked
PE 0046 — *cartographic information*
PE 0046 — *cartographic skills*
PE 0046 — *cartographic technology*
PE 4365 — channels direct investments
PE 8572 — development enterprises
PD 0822 — development new social structures
PE 8572 — development private sector
PE 4305 — diversification loans
PD 0465 — domestic savings
PC 0933 — economic cooperation
PD 9412 — economic integration regional groupings
PF 5964 — economic policy making
PE 1900 — electrical power supply
PF 7184 — external debt management capacity
PE 0365 — financing local government
PF 2124 — fiscal discipline
PE 4574 — government savings capacity
PE 0269 — housing
PE 8506 — incentives increased productivity
PE 2238 — increase employment manufacturing industries
PD 4219 — industrial capacity
PF 4195 — industrial services
PF 4160 — industrial trade
PE 0365 — institutional structures local government
PE 4522 — international marketing jute products
PE 5761 — international participation island
PF 3845 — investment performance
PF 7141 — legislation against environmental pollution
PD 0291 — level investment
PF 8736 — life insurance
PE 0365 — local authority legal structures
PE 7313 — local expertise business practices
PD 4219 — manufacturing capacity
PE 0523 — marketing products
PC 7674 — *medical laboratory facilities*
PF 5964 — mobilization financial resources
PE 9646 — negotiating capacity
PE 0853 — negotiation entrance terms transnational corporations
Developing countries Inadequate cont'd
PD 0548 — political parties
PE 5908 — port storage facilities land locked
PD 4219 — production capacity
PE 4796 — public assistance
PD 3051 — public debt relief
PE 1511 — relationship transnational corporations local industry
PE 4880 — research development capacity
PF 1120 — research development problems
PD 0543 — road highway transport facilities
PE 5824 — roads transport land locked
PF 6511 — rural housing
PF 2875 — rural infrastructure
PD 0095 — social discipline
PE 4796 — social insurance
PE 4796 — social security welfare services
PD 6482 — soil conservation
PE 4052 — standard living
PD 4101 — staple food supply
PE 6243 — supply appropriate trained manpower
PE 4120 — supply pharmaceutical products
PF 7842 — systems national accounts
PE 5655 — telecommunications island
PF 4132 — telephone systems
PD 5176 — trade
PE 4523 — trade agricultural commodities
PD 1388 — transportation facilities
PE 6526 — transportation facilities rural communities
PD 1204 — water supply rural communities
PD 3598 — weed control
PD 1476 — working conditions
Developing countries Inappropriate
PF 8120 — aid
PF 1531 — education
PE 1785 — irrigation subsidies
PE 3812 — loans available
PE 4120 — pharmaceutical products
PE 1337 — transplantation industrialized country methods
Developing countries [Income...]
PD 7615 — Income inequality
PD 3051 — Inconsiderate insistence creditors fulfilment debt service obligations
PD 0329 — Increase anti social behaviour
PF 1471 — Increase competition export markets encountered
PE 4603 — Increases interest rates long term debts
PE 6201 — Increasing drug experimentation
PE 5748 — Increasing foreign debt island
PE 2427 — industrialized countries ; Economic exploitation
PE 8321 — Industrialized country disincentives farm products
PF 2124 — Ineffective tax systems
PF 1031 — Ineffectiveness aid
PF 9266 — Inefficiency government
Developing countries Inefficient
PF 5964 — management government financial resources
PF 0903 — public administration
PE 5009 — tax administration
PF 5964 — use domestic resources
PD 1847 Developing countries ; Inequality employment opportunity
Developing countries Inequitable
PD 0142 — labour standards
PD 3046 — tax systems
PD 1477 — tax treaties developed
Developing countries [inflation...]
PD 0367 — inflation ; Vulnerability
PF 2493 — Inflexible social care structures
PC 6189 — Insecurity
PD 5788 — Insecurity land locked
PE 1011 — Insensitivity transnational corporations consumer needs
Developing countries Instability
PE 4986 — food prices
PF 0907 — labour force
PD 9117 — manufacturing industries
Developing countries Insufficient
PF 4108 — agricultural equipment
PF 4108 — agricultural machinery
PE 8236 — consumer services
PD 3054 — creditworthiness
PD 0176 — employment opportunities
PE 4140 — fertilizers agricultural development
PE 8592 — inappropriate energy equipment
PE 7959 — repair services
PF 4136 — television systems
Developing countries [intergovernmental...]
PE 9116 — intergovernmental facilities ; Restrictive conditions loans
PF 5472 — international decision making ; Restrictions participation
PE 0740 — International indebtedness arising insurance transactions
PE 5808 — Inviability tropical island
PF 4185 — Irrational fear industrialization
Developing countries Labour
PD 0156 — market flooding
PD 5045 — shortage
PD 0156 — surpluses
Developing countries Lack
PD 0822 — associations
PE 5790 — capital investment
PD 2892 — horizontal commodity diversification
PD 5114 — incentives increase agricultural output
PD 0664 — integration transport systems neighbouring
PE 7313 — local management skills
PE 0046 — management skills
PD 3625 — natural resources

Development different

Developing countries Lack cont'd
PC 5180 — political will respond needs
PD 1554 — processing industry primary commodities
PE 8236 — *production domestic consumers*
PE 8707 — purchasing power
PE 7141 — regulations against pollution
PD 1432 — response monetary incentives
PE 8180 — rural industrialization
PD 1225 — sanitation rural areas
PD 0110 — sense community solidarity
PE 5170 — skilled manpower rural areas
PF 5737 — skilled workers island
PE 5884 — skilled workers transport sectors land locked
PE 8303 — specialist doctors
PF 6067 — specifically designed software
PE 4933 — technical development excess manpower
PE 5814 — technical infrastructure maritime commerce
PF 0903 — trained administrative personnel
PD 1554 — vertical diversification
PD 8047 — work commitment
Developing countries [Lagging...]
PD 0946 — Lagging transformation agriculture
PC 0990 — Landlessness
PE 7141 — Licence pollute
Developing countries Limited
PE 4704 — availability loans
PF 0903 — civil administrative capacity
PF 5006 — exchange skills
PE 8986 — mechanical services
PF 3060 Developing countries ; Liquidity shortage
PD 1494 Developing countries ; Locational maladjustments industry
Developing countries Loss
PC 3134 — capital
PD 5045 — labour force
PD 0144 — traditional forms social control
PD 1543 — traditional forms social security
Developing countries Low
PD 3054 — credit rating
PD 0465 — level domestic resource mobilization
PD 2769 — level personal income
PD 1493 — occupational mobility
PE 5883 — productivity agricultural workers
PE 9811 — return investment
PE 9199 — wages
Developing countries [Makeshift...]
PE 4386 — Makeshift dwellings
PD 0050 — Maldistribution land
PD 5258 — Maldistribution wealth
PD 8668 — Malnutrition
PE 0402 — manufacturing ; Unemployment developed countries resulting participation
PE 8110 — market related interest rates ; Rescheduling debts
PE 6036 — Marketing banned pharmaceutical drugs
PD 9495 — Militarization
PD 7714 — Military destabilization
PE 1598 — Minimal export promotion transnational corporations
PD 8549 — Mismanagement
PF 0175 — Mismanagement aid
PF 8241 — Nakedness
PD 2968 — Narrow export base
PD 1501 — Narrow legal definition family
PD 3835 — National insecurity
PC 3134 — Negative net resource transfers
PF 7047 — Neglect agricultural rural life
PD 8114 — Neglect human resource development debtor
PD 1672 — Nepotism
PE 4120 — Non responsiveness transnational corporations pharmaceutical needs
PC 0933 — *Non uniformity tariff non tariff trade preferences*
PE 9185 — Nutritional anaemia women
Developing countries Obstacles
PJ 0230 — industrialization
PE 4627 — political union island
PF 6072 — satellite communications
Developing countries [Occupational...]
PD 6885 — Occupational dangers
PD 8841 — Occupational illness
PE 5909 — ocean shipping ; Obstacles
PE 5724 — official assistance ; Dependence island
PJ 8271 — organize ; Incapacity
PC 3134 — Outflow financial resources
PD 3132 — Outflow indigenous capital
PE 2214 — Over dependency international financial institutions
PD 2967 — Over production primary commodities
PD 4812 — Overgrazing
PE 5391 — Personal physical insecurity
PF 7047 — Policy discrimination against rural areas
Developing countries Political
PD 9792 — destabilization
PD 3835 — insecurity
PD 8323 — instability
Developing countries [Politicization...]
PE 4226 — Politicization public service
PC 2023 — Pollution
PC 0149 — Poverty
PD 0425 — Processing developed countries commodities exported
PD 2968 — producing primary commodities ; Instability export trade
PE 7548 — Prohibitive cost eye glasses
PE 9052 — Proliferation ballistic missiles
PE 9052 — Proliferation nuclear weapons
Developing countries Protectionism
PD 3714 — [Developing countries ; Protectionism]
PE 8321 — developed countries against agricultural products
Developing countries Protectionism cont'd
PD 9679 — international trade against exports
Developing countries [Public...]
PF 2819 — Public disillusionment aid
PF 6932 — Recipient distortion aid
PD 8114 — Reduction public expenditure human resources
PF 6670 — Reinforcement inappropriate development privileged classes
PE 6687 — Relocation hazardous industries
PE 9001 — Remoteness legal services
PF 1471 — Restricted growth export markets
PE 4335 — Restrictions home countries transnational banking activities
PF 1492 — Restrictions imposed aid
PE 5926 — Restrictive business practices markets developed countries against exports
PF 2491 — Restrictive loan procedures
PD 2970 — Rigidities production structures
Developing countries Rural
PD 4125 — poverty
PD 0295 — underemployment
PD 0295 — unemployment
Developing countries [Sale...]
PD 3998 — *Sale defective products*
PE 6243 — Scarcity skilled manpower
PE 4880 — science technology ; Inadequate capacity
PD 9563 — Sexual discrimination
PF 5912 — share shipping ; Imbalance
PE 5790 — shipping fleets ; Lack investment capital
Developing countries Shortage
PF 0118 — books
PF 0118 — books textbooks
PC 1820 — industrial leadership entrepreneurial ability
Developing countries [Small...]
PD 0928 — Small size domestic market
PE 4996 — Smoking
PE 4796 — Social insecurity
PF 0648 — socialist countries ; Reallocation development capital
PE 4360 — Socially irresponsible programmes transnational banks
PD 6482 — Soil mismanagement
PD 1614 — Speculation
PF 3949 — Stagnation food aid
PF 7985 — Technology gap developed
PE 5700 — territories ; Vulnerability island
PF 4115 — tourism ; Degradation
PE 4170 — Trade barriers manufactured goods
PD 2958 — Trade barriers protectionism
PE 2061 — Trading products containing toxic substances
PF 8921 — Transfer highly polluting industries
Developing countries transnational
PE 1082 — corporations ; Disruption cultural social identities
PE 0163 — corporations ; Domination
PE 7355 — corporations ; Interference labour relations
PE 0163 — corporations ; Intimidation
Developing countries [Transnational...]
PE 5751 — Transnational monopolies
PE 5024 — Unavailability land agricultural purposes
PE 8429 — Unavailability trained teachers rural areas
PD 0134 — Uncontrolled physical expansion cities
PD 0134 — Uncontrolled urbanization
PD 3054 — Uncreditworthiness
PF 6473 — Underdeveloped provision basic services
PF 4120 — Underdevelopment pharmaceutical industry
PF 4135 — Underdevelopment power industry
PD 8141 — Underemployment
PE 9199 — Underpayment work
PD 3042 — Underproduction primary commodities
PF 0377 — Underproductivity draught animal power
PF 4155 — Underutilization international data bases
PE 5902 — Undeveloped financial markets
PD 8047 — Undeveloped work ethic
PD 2892 — Undiversified economies
PE 4188 — Undiversified forms credit used
PD 1007 — Uneconomical small farms
PD 1273 — Unemployed intellectuals
PD 0176 — Unemployment
PE 0015 — Unemployment premature school leavers
PD 7615 — Unequal income distribution
PD 2544 — Unnatural boundaries
PE 4451 — unpaid female labour ; Dependence
PD 0134 — Unplanned urbanization
PF 1724 — Unrelated pioneer institutions
PE 5860 — Unreliable telecommunication services land locked
PE 5836 — Unreliable transit services land locked
PD 9117 — Unstable growth manufacturing
PF 5964 — Unsustainability macroeconomic policies
PD 3489 — Urban slums
PD 1551 — Urban unemployment
PF 0878 — Use inappropriate technologies
PE 2524 — Usury
PC 6189 — Vulnerability
PD 5788 — Vulnerability land locked
PE 5170 — Wastage local skills
PD 3675 — Water pollution
Developing countries Weakness
PC 1228 — infrastructure
PE 5772 — infrastructure island
PE 7000 — infrastructure land locked
PE 2951 — primary commodity trade amongst
PD 0169 — regional trade
PC 0933 — trade
PE 2966 — trade manufactured goods
Developing countries [Westernization...]
PF 6592 — Westernization traditional modes life
Developing countries [world...] cont'd
PF 1471 — world commodity exports ; Declining share
PC 2566 — world exports ; Reduction share
Developing country
PD 0947 — agriculture ; Gerontocracy
PD 0305 — capacity absorb aid ; Limited
PD 2968 — commodity prices ; Decline
PE 4188 — dependence single source finance
PD 2968 — earnings ; Instability commodity prices
PE 8184 — economies ; Low complementarity
PD 2970 — economies ; Structural rigidity
PE 2844 — exports ; Inadequate adjustment assistance industries labour affected
PE 3066 — exports ; Lack credit guarantee facilities
PF 5472 — *failure exercise leverage international bodies*
PF 6530 — farms ; Low investment
PD 1291 — human resources industrialized countries ; Loss
PE 6861 — import strangulation
PF 5337 — membership ; Irresponsibility United Nations
PF 3060 — monetary reserves ; Inadequate level
PE 3831 — perspectives media ; Inadequate coverage
PE 9457 — products limited range countries ; Export
PF 1471 — products world markets ; Restricted access
PF 1667 — rural communities ; Restricted delivery essential services
PF 5006 — skills flow ; Inadequate inter
PE 0523 — small industry products ; Inadequate export promotion
PD 1291 — students studying foreign countries ; Non return
PE 8883 — students ; Unavailability scholarship funds
PE 3831 — weaknesses media ; Excessive emphasis
PD 1291 Developing developed areas ; Emigration trained personnel
Developing developed countries
PE 8957 — Foreign private investment income outflow
PD 3132 — Private domestic capital outflow
PC 2369 — Tariff barriers trade
PE 2953 Developing economies ; Weakness trade socialist
PF 3973 Developing future ; Inability elderly play role
PF 5472 Developing industrialized countries ; Displacement national decision making
PC 3134 Developing industrialized countries ; Transfer surplus wealth
PE 5848 Developing land locked transit countries ; Inadequate rail transport
PD 0110 Developing regions ; Rivalry disunity
PE 8139 Developing societies ; Traditional values subordination women
PE 5689 Developing states ; Neglected marine space island
PF 6617 Developing technology ; Limited social context
PE 4032 Developing technology ; Public non accountability organizations
Development [Absence...]
PF 7033 — Absence policies associate international nongovernmental organizations regional
PE 8569 — Accentuated inequality rural urban
PF 5868 — Adverse sexual
PD 1285 — agricultural production ; Stagnated
PC 8419 — agricultural resources ; Unsustainable
PF 0392 — aid ; Lack
PF 9216 — aid ; Policy cross conditionality restrictions multilateral
PD 2330 — Animal worship barrier
PA 6739 — Arrested
PA 7371 — Arrested
Development assistance
PF 0392 — developing countries ; Decline official
PF 5181 — Lack long term
PF 4794 — Reduced scope intergovernmental
PF 9465 Development bank failure ; Government
PD 2511 Development bureaucracy ex colonial countries ; Over
Development [capacity...]
PE 4880 — capacity developing countries ; Inadequate research
PF 0648 — capital developing countries socialist countries ; Reallocation
PF 6517 — capital ; Incomplete access
PD 7463 — capitalism ; Abusive technological
PD 7463 — capitalist society ; Perverse impact technological
PD 1307 — children ; Retardation psychomotor
PD 4671 — coast zones ; Unsustainable
PF 0905 — *collaterals ; Insufficient*
PF 4297 — communication services least developed countries ; Inadequate
PE 1702 — community priorities ; Monopolization interest groups
PF 9576 — Competing models socio economic
PC 6242 — Conflicts integrated rural
PF 5692 — Constraint time individual social
PC 2049 — *countries ; Imbalanced regional*
PC 8316 — credit ; Excessive
PF 5039 — critical awareness ; Inadequate
PC 0381 — Culturally unsustainable
Development [debtor...]
PD 8114 — debtor developing countries ; Neglect human resource
PF 0611 — Decline public expenditure human
PF 4718 — Declining rate
PF 6525 — *Delayed industrial*
PE 1536 — Denial right people pursue
PD 6571 — Dependence sophisticated technology
PB 0206 — Dependent
PF 4829 — Destruction inherent
PD 0266 — developing countries ; Disparity social
PE 4140 — developing countries ; Insufficient fertilizers agricultural
PE 8416 — different societies ; Separate unequal

Development [due...] cont'd
PE 4905 — due environmental policies ; Restrictions industrial economic
Development [Ecologically...]
PC 0111 — Ecologically unsustainable
PF 2997 — Economic bias
PD 1534 — economic sectors developing countries ; Disparate
PC 0495 — energy industry ; Unsustainable
PC 7517 — energy use ; Unsustainable
PE 8572 — enterprises developing countries ; Inadequate
PC 0111 — Environmental hazards due economic
PE 4933 — excess manpower developing countries ; Lack technical
PF 0019 — Excessive reliance economic indicators human
PE 7931 — exiled children ; Inhibition personality
PF 4295 — expenditures due floating rate loans ; Uncertainty
Development Exploitative
PB 6207 — [Development ; Exploitative]
PD 8492 — property
PF 2855 — resource
PE 6237 Development ; Export led
Development [Failure...]
PF 1055 — Failure global scale planning expertise
PF 2528 — False assumptions sustainable
PE 8431 — family ; Lack safe
PC 8419 — farmlands ; Unsustainable
PE 7195 — financing ; Excessive reliance developing countries external sources
PD 4900 — forest lands ; Unsustainable
PC 1227 — Fragmentation technological
PF 1768 — *Fragmented agricultural*
PD 6923 — fresh waters ; Unsustainable
PF 1332 — functional abilities ; Limited
PF 6556 *Development groups ; Nonexistent*
PC 7517 Development ; High energy
Development [Ignorance...]
PF 5759 — Ignorance lifelong human
PB 6207 — Imbalance
PD 4007 — Imbalance population growth resource
PF 8176 — Inability anticipate future
PF 9704 — Inadequate mobilization public opinion
PF 9576 — Inadequate models socio economic
Development Inappropriate
PB 6207 — [Development ; Inappropriate]
PF 2208 — leadership
PF 9576 — theoretical models social
Development [Increasing...]
PC 5879 — Increasing unsustainability
PE 4972 — indigenous peoples ; Denial rights
PC 0495 — indiscriminate economic growth
PD 7682 — industrialized countries products substituting commodities exported developing countries
PD 7978 — informal sector developing countries
PD 8114 — inhuman face
PJ 0451 — initiatives ; Reluctance government invest high risk
PE 7095 — Intellectual methods which disenfranchise grassroots technological
PF 4676 — international criminal law ; Inadequate
PE 5689 — island countries ; Lack coastal
PF 4175 — issues ; Political nature
PF 6532 Development labour potential ; Arrested
Development Lack
PB 0206 — [Development ; Lack]
PF 2164 — agricultural
PD 8604 — capital
PF 7912 — community
PE 8190 — economic technical
PF 2115 — essential services materials needed
PF 4371 — *individual*
PF 6454 — means local technological
PF 2187 — moral
PF 2208 — motivation leadership
PF 3339 — participation
PF 4959 — participation women
PF 4371 — *personal*
PD 8673 — political
PC 0242 — social
Development lag against
PE 2000 — information growth ; Increasing
PF 3743 — population growth ; Increasing
PC 5879 — socio economic growth ; Increasing
PE 3078 — technological growth ; Increasing
Development [Land...]
PE 7099 — Land erosion brought site
PF 3605 — leadership ; Limited
PE 9764 — least developed countries ; Inadequate human resources
PF 6148 — life cycle ; Inhibition individual psychological
PF 1086 — Limited market
PF 3605 — Limited methods leadership
PF 3560 — Limits participation community
PE 4704 — loans ; Difficult
PE 8190 — Low economic
Development [major...]
PE 5336 — major agriculture technologies ; Decreasing rate
PF 4975 — mega projects ; Implementation
PF 4955 — mental health services ; Constraints
PE 5274 — Misuse tropical rain forests agricultural
PF 9576 — models ; Uncoordinated multiplicity
PF 7923 — multidisciplinary approaches ; Obstacles
Development [National...]
PD 0055 — National sovereignty obstacle peaceful socio economic
PD 3546 — natural resources ; Nationalistically determined
PF 4959 — Neglect role women rural

Development [new...] cont'd
PD 0822 — new social structures developing countries ; Inadequate
PC 0012 — new weapons ; Competitive
Development [Obliteration...]
PE 3874 — Obliteration footpaths
PE 4842 — Obstacles national
PF 3846 — Opportunist bias public discussion research
PF 2997 — Over emphasis economic
Development [parallel...]
PC 6641 — parallel economies
PE 8403 — Patent restrictions impeding sustainable
PF 6471 — *pattern ; Unresolved*
PF 6475 — Pessimistic approach agricultural
PF 5658 — policies ; Failure
PF 9101 — policies ; Prohibitive cost inadequate
PF 8757 — policy ; Inappropriate
PF 5660 — policy making ; Crisis management approach long term
PF 9821 — political thinking concerning global problematique ; Inadequate
PF 1789 — Pollution hazards industrial research
PF 7063 — potential ; Unrealized human
PE 8572 — private sector developing countries ; Inadequate
PF 6670 — privileged classes developing countries ; Reinforcement inappropriate
PF 1120 — problems developing countries ; Inadequate research
PD 9202 — process developed countries ; Reversal
Development programmes
PE 7195 — Dependence developing countries external financing
PF 3339 — *Difficulty implementing*
PF 4300 — due short term loans ; Uncertainty
PF 8368 — Failure
PF 4944 — Impractical
Development programmes Inadequate
PF 4944 — conception
PD 3712 — implementation
PF 4180 — social
Development [programmes...]
PF 8368 — programmes ; Ineffectiveness
PF 4837 — programmes ; Lack understanding social economic contexts
PF 4718 — progress ; Reversal
PE 8687 — projects available labour ; Unsuitability
PF 8368 — projects ; Failure
Development projects Inadequate
PF 4340 — commercial finance rural
PF 9244 — control
PF 9244 — evaluation
PF 9244 — monitoring
Development projects [Inappropriate...]
PF 4944 — Inappropriate design
PD 3712 — Inappropriate management
PD 3712 — Ineffective execution
PE 8166 — Overuse machinery equipment
PF 4944 — Unfeasible
PC 4844 Development promotion health ; Unequal
Development [Regional...]
PD 5845 — Regional ecosystems degraded exploitative
PF 2018 — regional plans ; Delayed
PF 2018 — regional plans ; Protracted
PD 8114 — Regressive social
PE 4059 — research ; Over emphasis immediate solutions resource
PF 3318 — Restrictive effects external capital
PE 1139 — rural areas ; Constricting level capital
PF 2471 — rural communities ; Overlooked potential industrial
Development [scepticism...]
PF 1681 — *scepticism ; Community*
PE 8692 — schemes ; Health hazards water
PF 4450 — Secret weapons
PF 5660 — Short range planning long term
PB 1997 — Social stratification obstacle
PD 0917 — social transformation ; Bias media against
PC 0381 — Socially unsustainable
PE 8965 — space ; Military obstacles peaceful
PF 3388 — *sports ; Excessive claims human*
PF 5660 — strategy ; Short range
Development [trade...]
PC 0840 — *trade war ; Computer*
PE 0234 — transnational corporations ; Retarding
PD 2191 — tribal warfare ; Disruption
Development [Unattractive...]
PF 3499 — Unattractive locale economic
PC 0442 — Uncontrolled urban
PF 1768 — Uncoordinated efforts agricultural
PD 0516 — Uncoordinated international river basin
PB 6207 — Uneven
PF 6548 — Unexplored alternatives commercial
PF 4837 — Unintegrated
PF 1768 — *Unintegrated agricultural*
PF 6475 — Unplanned land
PF 6525 — *Unpromoted industrial*
PF 6475 — Unregulated land
PF 3498 — *Unskilled programme*
Development Unsustainable
PB 9419 — [Development ; Unsustainable]
PC 8419 — agricultural
PC 0495 — economic
PD 4537 — rural
PC 0442 — urban
PF 9478 Development vision ; Limited
PF 2128 Development work forces ; Unorganized
PC 0143 Developmental arithmetic disorder
PE 9712 Developmental articulation disorder

PE 7230 Developmental coordination disorder
PD 3773 Developmental defects ; Physical
PD 3773 Developmental disabilities
PE 5545 Developmental expressive language disorder
PE 0330 Developmental expressive writing disorder
PE 9300 Developmental receptive language disorder
PF 9512 Developmentalism
PD 8673 Developments ; Limited political
PF 4156 Developments strategic doctrine ; Risk unintentional nuclear war generated
PB 1125 Deviance
PC 0212 Deviance ; Juvenile
PC 2405 *Deviant society*
PA 6228 Deviation
PF 3976 Deviation behavioural norms
PA 6790 Deviation *Distortion*
PA 6180 Deviation *Error*
PF 3405 Deviation ; Ideological
PA 5878 Deviation *Nonconformity*
PA 6890 Deviation *Nonuniformity*
PD 2198 Deviation ; Sexual
PC 3452 Deviation ; Social
PF 2189 *Deviationism*
PD 7352 Deviationism ; Bourgeois
PD 3998 *Devices ; Defective medical*
PF 2613 Devices ; Derelict military explosive
PE 1909 Devices ; Excessive exposure radiation consumer goods electronic
PE 8502 Devices industries ; Ineffective self regulation pharmaceutical medical
PF 0772 Devices ; Lack prosthetic
PD 2697 Devices ; Unethical experiments drugs medical
PF 7018 Devil
PF 9139 Devil ; Incarnate
PE 1323 Devil's hair
PE 1323 Devil's ringlet
PF 8260 Devil worship
PA 5454 Devilishness *Badness*
PA 5643 Devilishness *Unkindness*
PA 5644 Devilishness *Vice*
PF 6734 Devils
PA 6468 Deviousness *Complexity*
PA 6228 Deviousness *Deviation*
PA 7363 Deviousness *Improbity*
PA 7371 Deviousness *Unintelligence*
PA 6876 Devitalization *Impotence*
PA 5558 Devitalization *Weakness*
PF 6767 Dexterity disabilities
PD 5228 Dextran toxicity newborn pigs ; Iron
PE 0102 Diabetes
PD 9654 *Diabetes insipidus animals*
PE 0102 Diabetes mellitus
PD 9654 *Diabetes mellitus animals*
PE 4808 Diabetic pregnancies ; Foetal malformation
PA 5454 Diabolic *Badness*
PA 5643 Diabolic *Unkindness*
PA 5644 Diabolic *Vice*
PF 2780 Diagnosing animal diseases ; Difficulty
PE 4129 Diagnosis due inadequate equipment ; Inaccurate medical
PF 8490 Diagnosis ; False
PF 8490 *Diagnosis mental disorder ; False*
PE 2932 Diagnosis ; Misuse psychiatric
PF 6016 Dialect discrimination
PF 5862 Dialogue being restrictive ; Accusing channels
PF 1654 Dialogue ; Dangerous forms
PD 2263 Dialogue ; Distrust political
PF 9274 Dialogue ; Erosion religious belief ecumenical
PF 7798 Dialogue ; Erosion social
Dialogue [Inappropriate...]
PF 1654 — Inappropriate
PF 1654 — Ineffective
PF 1618 — Ineffective religious
PE 7725 Dialogue judiciary ; Closed channels
PC 1547 Dialogue review ; Inadequate structures political
PC 1195 Diamonds ; Shortage industrial
PE 8961 *Diaphragmatic hernia*
PD 7424 *Diaphragmatic hernia animals*
PD 5971 *Diarrhoea*
PD 3978 *Diarrhoea ; Bovine viral*
PE 9751 *Diarrhoea ; Childhood*
PE 9751 Diarrhoea children
PD 3978 *Diarrhoea horses ; Chronic*
PD 3978 *Diarrhoea ruminants ; Neonatal*
PD 3978 *Diarrhoea syndrome ; Acute equine*
PD 5971 Diarrhoeal disease acute
PD 5971 *Diarrhoeas ; Infectious*
PC 1054 Dictatorship
PC 2034 Dictatorship ; Economic
PC 3186 Dictatorship ; Foreign
PA 5563 Dictatorship *Lawlessness*
PD 3239 Dictatorship majority
PC 0698 Dictatorship ; Military
PC 0845 Dictatorship ; Political
PD 3239 Dictatorship proletariat
PD 3241 Dictatorship ; Social
PD 2710 Dictatorship ; Tyrannical
PA 6491 Dictatorship *Vanity*
PF 4821 Dictatorships ; Government support foreign
PE 0640 Diebacks
PD 0036 *Diet ; Carcinogenic*
PE 4950 Diet ; Inadequate roughage
PE 0691 Diet ; Inadequate weaning
PC 0382 Diet ; Low energy
PC 0382 *Diet poor ; Inadequate*
PC 0382 Diet ; Unbalanced

Disabling inadequacy

PD 0715 Diet ; Vitamin deficiencies
PE 9364 *Dietary abomasal impaction*
PF 4832 *Dietary care ; Unavailability*
Dietary deficiencies
PC 0382 — [Dietary deficiencies]
PD 0800 — developed countries
PE 4760 — domestic animals
PC 1648 *Dietary energy stress*
PE 4950 Dietary fibre ; Inadequate
PC 0382 Dietary habits ; Improper
PD 3978 *Dietary indigestion ; Mild*
PD 1694 Dietary intake ; Chemical contamination
PD 0339 Dietary protein ; Deficiency
PF 3310 *Dietary restrictions*
PD 2868 Dietary taboos ; Infringement
PE 4371 Dietary torture
PE 9774 *Dietetica ; Hepatosis*
PE 4261 Diets ; High fat
PE 4950 Diets ; Low fibre
PE 1901 Diets ; Metal deficient
PD 5212 Diets ; Seasonally determined
PC 0382 *Diets ; Unbalanced family*
PE 0691 Diets ; Unbalanced infant
PA 6698 Difference
PE 6650 Difference training ; Break down communications due
PF 8156 Differences ; Bureaucratic nonrecognition ethnic
PF 3388 Differences ; Debilitating ideological
PF 5210 Differences ; Failure acknowledge
PF 1199 Differences intelligence ; Racially determined
PF 2679 Differences ; Lack appreciation cultural
PF 2679 Differences ; Lack appreciation cultural
PF 7934 Differences ; Overriding individual
PF 2960 Differences trading principles practices developing countries
PC 2952 Differences trading principles practices different economic systems
PF 6570 Different age groups ; Stifled potential social interaction
PE 8215 Different categories juvenile offenders ; Inadequate segregation
PF 8753 Different cultures response social problems ; Failure interrelate wisdom
PC 2952 Different economic systems ; Differences trading principles practices
PC 2724 Different economic systems ; Weakness trade
PF 9571 Different national laws relating international transactions ; Divergence
PE 8308 Different parts government ; Unclear fiscal relations
PC 0006 Different racial groups ; Unequal rights
PJ 8295 Different societies ; Inadequate interface fundamentally
PE 8416 Different societies ; Separate unequal development
PD 7257 Differential economic performance developing countries
PE 5022 Differential ; Rural urban income
PE 8971 *Differentials oceans ; Underutilization temperature*
PE 4664 Differentiated toys ; Gender
PE 2328 Differentiation transnational corporations ; Limited market access due product
PF 6665 Differing conceptions time
Difficult [decisions...]
PF 4204 — decisions ; Inability make
PF 4204 — decisions ; Unwillingness make
PE 4704 — development loans
PF 2875 *Difficult freight transport*
PF 1226 *Difficult grant management*
PF 7580 *Difficult internal pathways*
PF 8053 *Difficult land acquisition*
PF 2454 Difficult laws ; Complex
Difficult [piling...]
PF 2346 — *piling removal*
PF 2875 — *produce transport*
PF 9108 — public funding
PF 6506 Difficult village consensus
PD 1204 Difficult water access
PF 8191 Difficulties applying knowledge
PD 8697 Difficulties ; Code enforcement
PF 8931 Difficulties cooperative planning
Difficulties [developed...]
PD 9802 — developed countries ; Domestic agricultural price
PE 2890 — developing countries ; Domestic agricultural price policy
PE 0451 — due torture ; Sleep
PD 4125 Difficulties establishing changes rural economic patterns
PC 3685 Difficulties ; Ethnic
PD 7536 Difficulties farmers ; Economic
PE 8700 Difficulties ; Inadequate adaptation policy educational
PE 1793 Difficulties new states ; Administrative
PF 1503 Difficulties new urban families ; Adjustment
PA 5497 Difficulty
PA 6340 Difficulty *Adversity*
PE 7864 *Difficulty breathing*
PD 2289 Difficulty child-bearing
Difficulty controlling
PD 2781 — animal diseases
PD 2776 — disease wild animals
PD 4751 — insect populations
Difficulty [diagnosing...]
PF 2780 — diagnosing animal diseases
PA 5497 — *Difficulty*
PA 5532 — *Disaccord*
PF 8661 Difficulty ensuring prosecutions crimes
Difficulty [identifying...]
PF 2775 — identifying carriers animal diseases
PF 3270 — *identifying sadists*
PF 3339 — *implementing development programs*
PD 5350 — Industries
PF 3538 Difficulty patent application
PE 9604 *Difficulty swallowing*
PE 8369 Difficulty transition school work
PA 7367 Difficulty *Unintelligibility*
PA 7107 Difficulty *Unpleasantness*
PE 4161 *Diffusa ; Pododermatitis aseptica*
PF 6787 Diffuse legislation against drugs
PD 2565 *Diffuse scleroderma*
PA 5974 Diffuseness
PF 3064 Diffuseness regulatory authority
PF 7456 Diffusion family role expectations
PC 8866 Digestion ; Disturbance
PC 8866 Digestion ; Impaired
PC 8866 Digestion ; Poor
PC 8866 Digestive disorders
PE 3004 Digestive disorders rumen
PE 4303 Digestive organs ; Malignant neoplasm
Digestive system
PE 3711 — animals ; Congenital anomalies
PD 3978 — animals ; Diseases
PD 8347 — *Benign neoplasm*
PD 1618 — *Congenital anomalies*
PC 8866 — Diseases
PC 8866 — Diseases deformities
PE 9604 — Symptoms referable
PE 6461 *Digestive tract helminthiasis poultry*
PC 8866 Digestive troubles
PE 8911 Dignified birth ; Denial right
PE 9573 Dignity ; Denial animals right
PE 6623 Dignity ; Denial right
PF 3856 Dignity ; Loss
PD 2672 *Dilapidated buildings ; Underutilized*
PA 6858 Dilapidation *Disintegration*
PA 6088 Dilapidation *Impairment*
PD 0448 *Dilatation ; Cardiac*
PD 0448 *Dilatation ; Heart*
PE 4630 *Dilatation ; Oesophageal*
PE 3711 *Dilatation oesophagus*
PE 7423 *Dilatation volvulus animals ; Gastric*
PF 0513 Dilemma ; Moral
PF 1559 Dilemmas ; Government policy
PD 4272 Dilution cultural heritage ; Acculturation
PF 0836 *Dim business future*
PA 7319 Dimension*complex
PF 8905 Diminished church role
PF 6478 *Diminished diving income*
PJ 8596 *Diminished park safety*
PF 9429 Diminished personal capacity
PE 7570 *Diminished secretion saliva*
PF 6477 Diminishing capital investment small communities
PF 6475 *Diminishing land ownership*
PE 4649 Diminishing mental capacity age
PF 6461 *Diminishing parental influence*
PF 6441 *Diminishing population base*
PF 1825 Diminishing role older people due overemphasis economic productivity
PD 1052 Diminution biological productivity land
PD 4387 Dioxide impact agriculture developing countries ; Carbon
PD 4387 Dioxide ; Increasing atmospheric carbon
PE 1210 Dioxide occupational hazard ; Sulphur
PE 1210 Dioxide pollutant ; Sulphur
PE 5195 Dioxide pollutant ; Titanium
PE 7555 Dioxin poisoning
PD 5228 *Dioxin poisoning animals*
PE 8601 Diphtheria
PD 7307 *Diphtheria ; Calf*
PD 0659 *Diplacusis*
PE 0763 *Diplegia*
PC 4667 Diplomacy ; Gun boat
PC 4667 Diplomacy ; Gunboat
PJ 7629 Diplomacy ; Indifference toward
PJ 7629 Diplomacy ; Loss credibility
PD 2672 *Diplomas ; Unvalued*
PE 5683 Diplomatic agents ; Vulnerability
PF 3099 Diplomatic commercial exchanges ; Abuse international cultural,
PF 7950 *Diplomatic embarrassment*
PF 1440 Diplomatic errors
PD 5554 Diplomatic fraud
PE 5488 Diplomatic immunity ; Threat
PF 2141 Diplomatic isolationism
PF 6703 Diplomatic language ; Obscurantist
PD 1893 Diplomatic lying
PE 6948 Diplomatic positions ; Politicization
Diplomatic [recognition...]
PF 8040 — recognition ; Absence
PF 8040 — relations states ; Abnormal
PF 8040 — relations states ; Cessation
PF 5649 Diplomats ; Abuse privileges immunities
PD 2731 *Dipping lameness sheep ; Post*
PE 2254 Diptera
PD 5582 Direct action environmentalists
PE 4454 Direct action environmentalists ; Sabotage hunting expeditions
PE 0161 Direct foreign investment transnational enterprises restrictive business practice
PD 3138 Direct investment developing countries ; Decline foreign
PE 4365 Direct investments developing countries ; Inadequate channels
PE 1810 Direct mail advertising ; Proliferation
PF 3072 Direct news coverage parliamentary affairs ; Restrictions
PE 0261 Direct satellite broadcasting individual receivers ; Foreign controlled
PE 6350 *Direct sunlight ; Absence*
PF 2949 *Direct transportation ; Incomplete*
PF 6531 Direction ; Insensitive adult
PD 1544 Direction youth engagement ; Lost
PF 1010 Directives ; Failure implement policy
PF 6492 *Directives ; Unclear official*
PF 5522 Directorates ; Interlocking corporate
PF 6574 *Directories ; Unavailability local*
PE 9251 Directors ; Abdication control company
PE 9251 Directors commercial enterprises ; Irresponsible
PC 4867 Directors ; Inept
PD 1245 Dirt ; Excessive wind blown
PF 1773 *Dirt floors houses ; Unhygienic*
PA 5454 Dirtiness *Badness*
PA 6191 Dirtiness *Disapproval*
PA 6997 Dirtiness *Imperfection*
PA 5612 Dirtiness *Unchastity*
PA 5459 Dirtiness *Uncleanness*
PA 5644 Dirtiness *Vice*
PD 5433 Dirty money
PJ 3707 Dirty occupations
PC 5517 Dirty politics
PF 5205 Dirty thoughts
PD 7171 Dirty tricks policies ; Government
PD 8122 Dirty water
PE 6768 Disabilities ; Body disposition
PD 3773 Disabilities ; Developmental
PE 6767 Disabilities ; Dexterity
PE 4856 Disabilities ; Excessive costs cultural burdens human illnesses
PD 2647 Disabilities ; Foot diseases
PF 0709 Disabilities ; Inadequate prevention
PD 3865 Disabilities ; Learning
PE 6769 Disabilities ; Locomotor
PC 0699 *Disabilities ; Neuro muscular*
PB 1044 *Disabilities ; Orthopaedic*
PE 6770 Disabilities ; Personal care
PC 4710 Disabilities ; Preventable
PD 1950 Disabilities ; Reading
PC 0699 *Disabilities ; Skill*
PC 5890 Disability ; Childhood
PB 5486 *Disability compensation ; Arbitrary evaluation*
PB 1044 Disability ; Dependence human disease
PA 6799 Disability *Disease*
PE 4856 Disability ; Economic social losses due
PC 0699 Disability ; Human
PB 1044 Disability ; Human disease
Disability [Impairment...]
PA 6088 — *Impairment*
PA 6876 — *Impotence*
PE 5648 — inhumane confinement ; Death
PD 6020 Disability ; Physical
PD 0723 Disability ; Poverty
PA 5558 Disability *Weakness*
PD 0874 Disabled ; Ageing war
Disabled [children...]
PD 0196 — children
PC 5890 — children
PD 8723 — college students ; Learning
PE 8900 — convicts ; Lack care intellectually
PE 8900 — criminals ; Inadequate facilities mentally
Disabled [Denial...]
PE 8924 — Denial right adequate health care
PC 3461 — Denial rights
PF 4296 — Dependence
PF 4296 — Deprivation freedom caring
Disabled Discrimination against
PD 9757 — [Disabled ; Discrimination against]
PD 9183 — mentally
PD 8627 — physically
PD 0098 Disabled during states emergency ; Vulnerability
PE 3847 Disabled elderly persons
PE 0783 Disabled employment ; Discrimination against
Disabled [Inadequate...]
PF 0775 — Inadequate educational facilities
PF 4681 — Inadequate housing
PF 4681 — Institutionalization
PD 8314 Disabled ; Lack facilities physically
PE 0769 Disabled migrant workers
PE 8833 Disabled people ; Inadequate recreational facilities
PF 6224 Disabled people ; Mystique attitudes problems
Disabled persons
PC 0699 — [Disabled persons]
PE 8690 — decision making ; Lack participation
PD 0724 — developing countries
PD 0723 — Impoverished
PF 0775 — Inadequate educational facilities
PE 7317 — Inadequate vocational rehabilitation facilities
PF 0719 — Increasing number
PF 0772 — Insufficient technical aids
PD 0784 — social cultural life ; Exclusion
PE 0707 — Structural barriers
PE 0783 — Unequal employment opportunities
PE 0706 — Unequal opportunities
PC 0768 Disabled refugees
PF 4681 Disabled ; Restrictions liberty
Disabled [Sexual...]
PE 5197 — Sexual reproduction
PE 5197 — Sexual unfulfilment
PE 4544 — Sterilization mentally
PD 0762 Disabled victims crimes
PD 0764 Disabled victims torture
PC 0729 Disabled women
PD 4673 Disabled workers
PC 0699 Disablement
PE 3733 Disablement ; Malicious physical
PF 0979 *Disabling inadequacy feelings*

Disaccord

PA 5532	Disaccord	

Disaccord [Difference...]
- PA 6698 — *Difference*
- PA 5532 — *Disaccord*
- PA 5982 — *Disagreement*
- PA 6838 — *Dissent*
- PA 5446 Disaccord *Enmity*
- PA 6979 Disaccord *Opposition*
- PD 3517 Disadvantage aged ; Social
- PA 5454 Disadvantage *Badness*
- PA 5497 Disadvantage *Difficulty*
- PA 7395 Disadvantage *Inexpedience*
- PF 9247 Disadvantaged countries ; Geographically
- PB 6320 Disadvantaged groups
- PF 5008 Disadvantaged groups ; Limited availability land low income
- PE 7010 Disadvantaged status international nongovernmental organization personnel
- PD 2624 Disadvantaged students ; Economically
- PE 4922 Disadvantageous terms technology transfer
- PF 4130 Disadvantages excessive food production developing countries ; Economic
- PE 6240 Disadvantages homeworking employees
- PF 8499 Disadvantages ; Underestimated project
- PE 6240 Disadvantages workers home
- PA 5532 Disaffection *Disaccord*
- PA 7363 Disaffection *Improbity*
- PA 7107 Disaffection *Unpleasantness*
- PD 1544 Disaffection young ; Social
- PA 5532 Disaffinity *Disaccord*
- PA 5446 Disaffinity *Enmity*
- PA 5799 Disaffinity *Repulsion*
- PA 7364 Disaffinity *Unfeelingness*
- PA 7107 Disaffinity *Unpleasantness*
- PE 1764 Disagreeable bodily conditions
- PE 4481 Disagreeable human body odour
- PA 5982 Disagreeableness *Disagreement*
- PA 7253 Disagreeableness *Envy*

Disagreeableness [Unkindness...]
- PA 5643 — *Unkindness*
- PA 7107 — *Unpleasantness*
- PA 6509 — *Unwillingness*
- PA 5982 — Disagreement
- PD 2629 Disagreement alliances

Disagreement concerning
- PF 7274 — identity inventors discoverers
- PF 5766 — priorities
- PF 1115 — religious doctrine
- PF 6012 Disagreement expert witnesses
- PF 6012 Disagreement experts
- PF 5766 Disagreement ; Policy
- PF 7890 Disagreement ; Unreported
- PE 4021 Disagreements ; Family
- PD 2495 *Disagreements over land use*
- PA 6785 Disappearance
- PA 7270 Disappearance *Absence*
- PA 6379 Disappearance *Avoidance*
- PA 6785 Disappearance *Disappearance*
- PF 3012 Disappearance local culture
- PD 4259 Disappearance persons ; Forced
- PE 3874 Disappearance walking trails
- PD 5129 Disappearances children ; Forced

Disappearances persons
- PD 4259 — Enforced
- PD 4259 — Involuntary
- PE 8631 — objects ; Unexplained appearances
- PE 5882 Disappearances trade union leaders ; Forced
- PE 5882 Disappearances workers representatives ; Forced
- PF 7435 Disappearing islands
- PF 1735 *Disappointing bureaucratic action*
- PA 6191 Disappointment *Disapproval*
- PA 6011 Disappointment *Discontentment*
- PA 6180 Disappointment *Error*
- PA 6099 Disappointment *Hopelessness*
- PA 5527 Disappointment *Inexpectation*
- PA 6191 Disapproval
- PA 6191 Disapproval *Disapproval*
- PA 6838 Disapproval *Dissent*
- PA 7253 Disapproval *Envy*
- PA 6387 Disapproval *Necessity*
- PF 2830 Disapproval past ; Elders'
- PA 7107 Disapproval *Unpleasantness*
- PF 0013 Disarmament arms control efforts ; Failure
- PE 7670 Disarmament ; Insecurity unilateral structural
- PF 0589 Disarmament ; Negative economic repercussions
- PE 4051 Disarmament nuclear weapons ; Unilateral structural
- PF 7052 Disarmament ; Obstacles unilateral nuclear
- PF 8686 Disarmament ; Total
- PA 7158 Disarrangement *Dislocation*
- PA 7361 Disarrangement *Disorder*
- PB 3561 Disaster
- PA 6340 Disaster *Adversity*
- PA 7055 Disaster *Death*
- PA 6542 Disaster *Destruction*
- PE 5784 Disaster hazards island populations
- PF 6026 Disaster ; Inadequate insurance against damages arising natural
- PD 4065 Disaster ; Meteorological
- PF 3566 Disaster prevention mitigation ; Inadequate

Disaster [relief...]
- PF 6025 — relief ; Uncontrolled costs
- PF 6022 — relief ; Uncoordinated
- PF 0286 — rescue relief ; Inadequate
- PC 5010 Disaster ; Topological
- PF 3567 Disaster unpreparedness
- PB 3561 Disaster victim

Disaster warnings
- PF 4298 — False
- PF 3565 — Ignored
- PF 3565 — Rejection
- PB 3561 Disasters
- PC 5489 Disasters ; Biological
- PF 3567 Disasters ; Confused government response
- PE 5229 Disasters ; Delay payment compensation victims
- PF 3562 Disasters extraterrestrial origin
- PD 0830 Disasters ; Geomagnetic

Disasters [Inadequate...]
- PE 8542 — Inadequate prevention computer
- PF 3565 — Inadequate warning
- PD 2570 — Industrial

Disasters Lack protection
- PD 0096 — aged during
- PE 4694 — domestic animals during
- PD 0098 — handicapped during
- PD 3785 — vulnerable groups during
- PE 4694 — wild animals during
- PD 1078 — women children during
- PD 1233 Disasters ; Landslip
- PE 0299 Disasters least developed countries ; Natural
- PB 2075 Disasters ; Man made
- PD 2278 Disasters ; Mine
- PB 1151 Disasters ; Natural
- PE 1839 Disasters ; Offshore oil
- PE 1839 Disasters ; Oil tanker
- PD 8982 Disasters ; Sea
- PC 6684 Disasters ; Telluric tectonic
- PF 9769 Disasters ; Unpredictable ecological
- PF 7768 Disasters ; Unreported
- PC 6034 Disastrous accidents
- PC 4257 Disastrous consequences war
- PE 0715 Disastrous failure natural dams
- PD 4751 Disastrous insect invasions
- PD 4426 Disastrous technological failures
- PJ 5608 Disbarment public services
- PA 7392 Disbelief *Unbelief*
- PA 7309 Disbelief *Uncertainty*
- PF 3975 Disbursements ; Delay loan
- PF 8650 Disbursements multilateral agencies official contributions ; Imbalance
- PD 2626 *Disc disorders*
- PD 2626 Disc ; Prolapsed intervertebral
- PE 3604 *Disc winged bats ; Endangered species*
- PE 6685 Discharge ; Bathing near sewage

Discharge dangerous substances
- PD 4542 — [Discharge dangerous substances]
- PD 2800 — domestic waste water
- PD 0575 — industrial waste water
- PE 9813 — laboratory waste water
- PD 7636 — rivers
- PD 3666 — sea
- PD 1414 — sewage systems
- PE 8510 Discharge sores infected persons
- PE 9369 *Discharge urine ; Excessive*
- PD 4542 Discharges hazardous substances ; Illicit
- PF 0364 Disciplinary characteristics problems ; Multi
- PC 9584 Discipline ; Breakdown international trading
- PD 2543 Discipline codes ; Strict army
- PF 2124 Discipline developing countries ; Inadequate fiscal
- PD 0095 Discipline developing countries ; Inadequate social
- PD 2543 Discipline ; Inadequate army
- PD 5344 Discipline ; Inadequate childhood

Discipline Lack
- PD 2543 — army
- PF 8010 — religious
- PE 9092 — school
- PF 8078 — social
- PF 8010 — spiritual
- PD 5344 Discipline ; Lax parental
- PD 6893 Discipline socialist countries ; Inadequate social
- PF 7650 Discipline work ; Lack
- PD 5344 Discipline young people ; Lack constructive family
- PD 7650 Disciplines employment ; Maladjustment
- PF 8868 Disciplines ; Fragmentation academic
- PF 9077 Disciplines ; Inadmissability evidence
- PF 9077 Disciplines ; Jurisdictional conflict academic
- PF 9077 Disciplines ; Opposition information
- PF 4176 Disclosure computerized information ; Unauthorized
- PD 5183 Disclosure government reports ; Premature
- PA 0005 Disclosure ; Inadequate public
- PF 4854 Disclosure threats public safety ; Non
- PA 5451 Discomfort *Insensibility*
- PA 7107 Discomfort *Unpleasantness*
- PA 5838 Discomposure *Agitation*
- PA 7361 Discomposure *Disorder*
- PA 6247 Discomposure *Inattention*
- PA 7309 Discomposure *Uncertainty*
- PA 7107 Discomposure *Unpleasantness*
- PA 5497 Disconcertion *Difficulty*
- PA 6030 Disconcertion *Fear*
- PA 6247 Disconcertion *Inattention*
- PA 7309 Disconcertion *Uncertainty*
- PA 7107 Disconcertion *Unpleasantness*
- PF 7122 Disconnectedness societal learning
- PF 0379 Disconnection people rhythms nature
- PF 2013 *Discontent ; Political*
- PF 2013 *Discontent ; Political*
- PD 0941 Discontented marriages
- PA 6011 Discontentment
- PA 6191 Discontentment *Disapproval*
- PA 6011 Discontentment *Discontentment*
- PA 7253 Discontentment *Envy*
- PA 6731 Discontentment *Solemnity*
- PA 7107 Discontentment *Unpleasantness*
- PF 4984 Discontinued educational programs
- PF 9769 Discontinuities ; Hazardous environmental
- PF 7700 Discontinuities random events ; Inadequacy methodologies deal
- PA 5828 Discontinuity
- PA 5828 Discontinuity *Discontinuity*
- PA 7158 Discontinuity *Dislocation*
- PD 4461 Discontinuity employment
- PA 6774 Discontinuity *Irregularity*
- PA 7236 Discontinuity *Separation*
- PA 5565 Discord

Discord [Disaccord...]
- PA 5532 — *Disaccord*
- PA 5982 — *Disagreement*
- PA 5565 — *Discord*
- PA 5465 Discord *Loudness*
- PF 3307 Discouraged community participation
- PD 2790 Discouraged workers
- PF 8948 Discouragement
- PA 7343 Discouragement *Dissuasion*
- PA 6030 Discouragement *Fear*
- PF 9187 Discouragement food crops guaranteeing survival
- PF 1213 *Discouragement permanent residency*
- PA 6731 Discouragement *Solemnity*
- PF 3008 *Discourages vision ; Subsistence*
- PD 5603 Discouraging conditions small business
- PF 7299 Discourse ; Insensitive
- PA 7143 Discourtesy
- PC 7013 Discourtesy
- PA 7143 Discourtesy *Discourtesy*
- PA 6822 Discourtesy *Disrespect*
- PA 6498 Discourtesy *Misbehaviour*
- PF 7274 Discoverers ; Disagreement concerning identity inventors
- PA 5611 Discovery
- PA 6191 Discredit *Disapproval*
- PA 6839 Discredit *Disrepute*
- PA 7392 Discredit *Unbelief*
- PA 7070 Discredit *Unprovability*
- PE 5341 Discredited moneyed hereditary class
- PF 1191 Discrepancies human life evaluation
- PF 4118 Discrepancies official statistics
- PF 4705 Discrepancies principles practice
- PF 6580 Discrete problems ; Overemphasis
- PE 7701 Discretion ; Effort
- PE 9762 Discretionary character ; Imposition quantitative trade quotas
- PF 6014 Discriminating job terminology ; Sexually
- PA 0833 Discrimination

Discrimination [access...]
- PF 3354 — access eternal life
- PF 8515 — *access sanitation services*
- PE 9054 — according financial loans ; Racial

Discrimination administration justice
- PD 0986 — [Discrimination administration justice]
- PE 6529 — Cultural
- PE 1399 — Economic
- PE 1399 — Educational
- PE 1347 — Geographic
- PE 1828 — Political
- PE 0168 — Religious

Discrimination against
- PE 6258 — adult students
- PE 6245 — black working women
- PD 6930 — casual labour
- PE 0392 — children prostitutes
- PE 5183 — coloured children
- PE 1604 — developing countries formation regional groupings developed countries
- PD 9757 — disabled
- PE 0783 — disabled employment

Discrimination against divorcees
- PF 2100 — [Discrimination against divorcees]
- PF 2100 — Religious
- PF 2100 — Social

Discrimination against [domestic...]
- PE 4964 — domestic servants
- PE 2635 — dwarfs midgets
- PD 4916 — elderly ; Employment
- PC 2541 — elders
- PD 1091 — eugenics
- PE 6929 — ex-offenders
- PE 6929 — ex-prisoners ex-detainees
- PE 5043 — families accused
- PE 5043 — families offenders
- PD 6417 — foreign companies
- PE 6422 — foreign nationals military service

Discrimination against foreigners
- PD 0973 — [Discrimination against foreigners]
- PD 6361 — [Discrimination against foreigners]
- PD 3529 — employment
- PF 1798 — legal proceedings

Discrimination against [giants...]
- PE 5578 — giants
- PE 1281 — gypsies
- PD 9757 — handicapped
- PE 0783 — handicapped work
- PE 4299 — HIV-infected persons
- PE 1903 — homosexuals
- PD 0943 — illegitimate children
- PD 0973 — immigrants aliens

Discrimination against indigenous
- PC 0352 — populations
- PC 3322 — populations education
- PD 1092 — populations employment

-902-

Disease

Discrimination against indigenous cont'd
PC 3320 — populations housing - Denial right adequate housing indigenous peoples
Discrimination against [investment...]
PD 3047 — investment foreign country ; Tax
PE 2131 — Jews
PE 1295 — juveniles judicial proceedings due protective legislation
PC 4613 — labour unions
PE 5741 — lesbians
PE 3882 — male homosexuals
Discrimination against men
PC 3258 — [Discrimination against men]
PC 3258 — *child bearing*
PD 8909 — education
PD 3338 — employment
PD 3692 — law
PE 4010 — parental rights
PE 8507 — public services
PD 3336 — social services
PE 4232 — sports
Discrimination against [mentally...]
PD 9183 — mentally disabled
PD 0973 — migrant workers their families
PC 0582 — minorities
PC 3459 — minorities education
PD 5078 — minority languages
PE 5183 — mixed race children
Discrimination against non
PD 0973 — citizens
PD 8629 — members professional bodies
PD 0973 — nationals
PD 3048 — residents country ; Tax
PD 6019 — union workers
Discrimination against [part...]
PE 6241 — part-time work
PE 9402 — people abnormal height
PD 8627 — physically disabled
PE 1129 — poor judicial sanctions
PE 5043 — prisoners' families
PF 6231 — prostitutes
PE 5141 — regional accents
PD 4365 — religious unbelievers
PF 7047 — rural areas developing countries ; Policy
PE 4947 — rural women
PC 2129 — sects
PD 1914 — sexual minorities
PF 5153 — those seeking election
PC 4613 — trade unions
PE 4860 — trade unions ; Government
PD 4252 — traditional economies
PE 4320 — transnational banks developing countries
PE 8548 — transsexuals
PD 3256 — unmarried fathers
PD 8622 — unmarried women
PD 0902 — unwed mothers
PE 5141 — use accents language
Discrimination against women
PC 0308 — [Discrimination against women]
PE 4246 — athletic training
PD 9628 — business
PE 8788 — children
PC 4898 — developing countries
PE 8879 — divorce rights
PD 0190 — education
PD 0086 — employment
PD 9628 — executives
PD 0162 — law
PE 8220 — military forces
PD 9628 — non-manual workers
PE 4019 — parental rights
PE 0197 — *payment prizes athletic events*
PC 1001 — politics
PF 6326 — priesthood ; Religious
PD 3335 — public life
PD 0127 — religion
PE 6069 — retirement age
PE 4016 — right nationality
PD 3691 — social services
PE 0197 — sports
PE 4102 — workers multinational enterprises developing countries
Discrimination [against...]
PD 6812 — against working mothers
PD 2872 — against working women socialist countries
PC 2541 — Age
Discrimination based
PC 1455 — belief
PC 3222 — illiteracy ; Political
PC 8774 — skin colour
Discrimination [Capital...]
PE 4167 — Capital investments supporting racial
PC 3458 — Class
PF 1035 — contraceptive methods ; Sexual
PE 7902 — Credit
PC 8344 — Cultural
PA 0833 — *Discrimination ; Denial equal property rights because*
Discrimination Denial right
PA 0833 — freedom
PE 5952 — freedom advocacy
PC 2157 — material well being because
PE 7310 — pursue spiritual well being because
Discrimination Dependence
PA 0833 — [Discrimination ; Dependence]
PC 0006 — racial
PC 1455 — religious

Discrimination Dependence cont'd
PC 1864 — social
Discrimination [developing...]
PD 9563 — developing countries ; Gender
PD 9563 — developing countries ; Sexual
PF 6016 — Dialect
PC 2157 — Discrimination ; Economic
Discrimination education
PC 3434 — [Discrimination education]
PE 0779 — Class
PD 3328 — Racial
PD 8807 — Religious
PD 1468 — Sexual
PE 0810 Discrimination educators ; Professional
Discrimination employment
PC 0244 — [Discrimination employment]
PE 4934 — against immigrant workers
PD 2318 — Age
PC 0244 — Dependence
PE 7206 — women family responsibilities
PC 3686 Discrimination ; Ethnic
PD 1036 Discrimination family planning facilities
Discrimination favour
PF 1021 — childbearing
PD 9316 — offenders ; Legal
PD 2318 — youth employment
PF 5929 Discrimination foreign language teaching ; Ethnic social
PC 2022 Discrimination ; Gender
Discrimination [harassment...]
PE 6922 — harassment children public life
PD 3469 — housing
PD 3442 — housing ; Racial
PA 6486 Discrimination *Injustice*
PF 8590 Discrimination ; Intellectual
PE 4907 Discrimination judiciary ; Racial
Discrimination [law...]
PC 8726 — law
PC 2541 — law ; Age
PC 8949 — Legalized
PC 3683 — Legalized racial
PF 7972 — legislation ; Lack anti
PE 4310 — lending transnational banks developing countries
PD 4131 — Linguistic
PA 7306 Discrimination *Narrowmindedness*
PC 0934 Discrimination ; Political
Discrimination politics
PC 0934 — [Discrimination politics]
PC 0934 — Dependence
PC 3219 — Ideological
PD 3223 — Language
PC 3221 — Political
PD 3218 — Property occupational
PD 3329 — Racial
PC 3220 — Religious
Discrimination [Positive...]
PF 9539 — Positive
PC 2178 — Professional
PD 8460 — public services
PD 3326 — public services ; Racial
Discrimination [Racial...]
PC 0006 — Racial
PC 1455 — Religious
PF 9539 — Reverse
Discrimination [security...]
PD 3519 — security forces ; Racial
PC 2022 — Sexual
PD 4064 — sexual preferences ; Racial
PD 0700 — shipping ; Flag
PC 1864 — Social
PC 3433 — social services
PF 4644 Discrimination ; Temporal
PA 7145 Discriminative affection＊complex
PE 0191 Discriminatory application antidumping regulations ; Distortion international trade
PD 8913 Discriminatory business practices
PD 6804 Discriminatory communication
PE 2603 Discriminatory customs administrative entry procedures ; Distortion international trade
PD 7450 Discriminatory design information systems
Discriminatory [effect...]
PE 7735 — effect holy day observance
PD 8807 — effects religious denominational education
PC 0244 — employment practices
PE 8583 — exchange rate policies
PE 1597 — exploitation sea-bed resources
Discriminatory [formulation...]
PE 9073 — formulation equipment safety regulations ; Distortion international trade
PE 1274 — formulation health sanitary regulations agricultural pharmaceutical products ; Distortion international trade
PF 0734 — frequency allocation
PE 0347 Discriminatory government private procurement policies ; Distortion international trade
Discriminatory [ideologies...]
PC 6341 — ideologies
PD 5229 — imposition standards
PF 1183 — information ; Dissemination
PB 6021 — international order
PB 6021 — international order ; Dependence
PF 7299 Discriminatory language
PE 9762 Discriminatory managed trade mechanisms
PD 8124 Discriminatory nuclear trade
Discriminatory [operation...]
PE 9440 — operation international commodity exchanges
PE 0310 — orderly marketing arrangements

Discriminatory [outmoded...] cont'd
PF 7673 — outmoded concepts ; Institutionalization
PF 2580 — outmoded concepts legal systems ; Institutionalization
PD 0340 Discriminatory preference agreements ; Distortion international trade
PC 2178 Discriminatory professionalism
Discriminatory [religious...]
PD 3357 — religious influence law
PF 3358 — religious influence society
PD 9628 — requirement over-qualification women
PE 0778 — requirements respect marks product origin ; Distortion international trade
PE 0083 — requirements respect product standards measures ; Distortion international trade
PD 0678 Discriminatory subsidies
PE 6883 Discriminatory treatment foreign prisoners
Discriminatory [unwritten...]
PE 7017 — unwritten codes behaviour
PC 4844 — *use health facilities*
PE 8261 — *use medical resources heal privileged*
PD 4060 Discussion issues meetings ; Curtailment
PF 0963 *Discussion opportunities ; Infrequent*
PF 3846 Discussion research development ; Opportunist bias public
PA 6822 Disdain *Disrespect*
PA 6387 Disdain *Necessity*
PA 6979 Disdain *Opposition*
PA 6491 Disdain *Vanity*
PA 6799 Disease
Disease [acute...]
PD 5971 — acute ; Diarrhoeal
PD 5971 — Acute diarrhoeal
PG 5181 — *Addison's*
PD 9654 — *Addison's*
PE 3621 — Aedes mosquitoes vectors
PD 2730 — *African green monkey*
PD 0153 — Alcoholism
PE 7623 — Alzheimer's
PD 8360 — Animal vectors
Disease animals
PE 6419 — *Akabane*
PE 7769 — *Big head*
PE 9509 — *Cold haemolytic*
PE 4224 — *Combined immunodeficiency*
PE 0471 — Heart
Disease [Animals...]
PD 8360 — Animals transmitters
PD 8360 — Animals vectors
PE 3622 — Anopheline mosquitoes vectors
PE 8158 — *Arteriosclerotic heart*
PC 2270 — *Arteriosclerotic vascular*
PE 9799 — Aujezky's
PE 5214 — Autoimmune
Disease [Bacteria...]
PD 9094 — Bacteria causing
PD 9094 — Bacterial
PE 0924 — *Bang's*
PJ 4824 — *Behcet's*
PD 3601 — Bird vectors plant
PE 6659 — Birds vectors
PE 7769 — *Black*
PE 6419 — *Border*
PD 0992 — Brain
PD 0992 — brain ; Organic
PE 2272 — *Bright's*
PD 9366 — *Brisket*
PD 0460 — Bureaucracy organizational
PD 2732 — *Buss*
Disease cattle
PE 5524 — *Farmer's lung*
PE 7420 — *Fatty liver*
PD 9366 — *High mountain*
PD 9667 — *Lumpy skin*
Disease [causing...]
PD 1175 — *causing microbes drinking water*
PE 6403 — causing viral combinations
PG 2295 — *Celiac*
PD 0992 — Cerebrovascular
PE 0653 — *Chagas'*
PD 2730 — *chickens ; Infectious bursal*
Disease Chronic
PD 0637 — obstructive pulmonary
PD 5182 — *obstructive pulmonary*
PD 8239 — pelvic inflammatory
PD 2732 — respiratory
PE 3779 Disease ; Circling
Disease complex
PE 4586 — *Beech bark*
PE 9854 — *Feline respiratory*
PD 3978 — *Mucosal*
Disease [Compound...]
PC 1306 — Compound
PE 2365 — Congenital heart
PD 2730 — *Congo virus*
PF 2779 — control ; Prohibitive cost
PE 7915 — Creutzfeldt Jakob
PD 9045 — *Crohn's*
PE 3623 — Culicine mosquitoes vectors
PD 0594 — *Cytomegalic inclusion*
PD 0448 *Disease ; Degenerative heart*
Disease Degenerative joint
PE 5913 — [Disease ; Degenerative joint]
PE 5913 —
PE 5050 —
PE 4161 —

—903—

Disease developing

Disease [developing...]
PD 0103 — developing countries ; Endemic
PB 1044 — disability ; Dependence human
PB 1044 — disability ; Human
PE 5913 — *dogs* ; *Lyme*
PD 2746 — Domestic animals vectors
PD 2746 — Domestic animals vectors animal
PE 9038 — Dread
PE 1556 — due torture ; Cardio pulmonary
PE 1154 — Dutch elm
Disease [Elephant...]
PE 4814 — Elephant Man's
PF 1824 — Enhanced risks
PE 7826 — *Extensor process*
Disease [Faecal...]
PD 8360 — *Faecal transmission*
PE 4514 — *Flies vectors*
PD 5453 — *Flip over*
PE 5725 — *Fly*
PD 7799 — *Follicular cystic ovary*
PE 1589 — Foot mouth
PE 8138 — Fungi vectors plant
PD 2728 — Fungus
Disease [game...]
PE 8098 — game zoo animals ; Loss
PD 2625 — *Gaucher's*
PD 2389 — Genetic susceptibility
PD 2732 — *Glässer's*
PD 9293 — *glomerulus*
PE 4293 — Glycogen storage
PE 1924 — Graves'
PD 9667 — *Greasy pig*
PE 9774 — *Green muscle*
PD 2730 — Gumboro
Disease [Hairy...]
PE 6419 — *Hairy shaker*
PE 0721 — Hansen's
PD 2730 — *Hardpad*
PD 3978 — *Hardware*
PE 2680 — Heartwater
PE 5548 — Hodgkin's
PE 1589 — Hoof mouth
PE 3508 — Hookworm
PD 3593 — Human vectors plant
PE 2354 — Hydatid
Disease [Iatrogenic...]
PD 6334 — Iatrogenic
PD 8821 — Ignorance concerning
PC 9067 — Ill defined symptoms
PF 7939 — Inadequate knowledge reporting man made
PD 2784 — increase aviation ; Increase animal
PE 2567 — Infective middle ear
PF 5739 — injury exposure weather
PE 5763 — injury physical confinement
PC 3597 — Insect vectors
PE 3632 — Insect vectors human
PD 2651 — Intentional infecting
PE 1974 — *Intracranial*
PE 8158 — Ischaemic heart
PD 2731 — *Disease ; Johne's*
Disease [kidney...]
PD 9293 — *kidney animals ; Metabolic*
PE 3616 — Kissing bugs vectors
PE 2282 — Kwashiorkor
Disease [Legg...]
PE 5173 — Legg Perthes
PE 6783 — Legionnaire's
PE 0763 — *Little's*
PE 1028 — Liver
PD 7799 — *Luteal cystic ovary*
PD 6363 — Lyme
Disease [Mad...]
PE 5191 — Mad cow (BSE)
PD 8371 — Man vectors
PD 7424 — Manchester wasting
PE 9291 — Maple syrup urine
PE 9689 — Marek's
PD 7424 — Marie's
PE 2567 — Ménière's
PC 2270 — Metabolic
PE 9017 — Mites vectors plant
PE 1923 — Mosquitoes vectors
PE 8813 — Motor neurone
PE 9774 — *Mulberry heart*
PD 2731 — *Mushy chick*
PE 2455 — Mycotic
Disease [Naalehu...]
PD 7424 — *Naalehu*
PE 6268 — Nairobi sheep
PE 7826 — *Navicular*
PE 2300 — Neapolitan
PE 7915 — nervous system ; Familial
PE 2399 — new born ; Haemolytic
PD 2732 — *New duck*
PE 1400 — Newcastle
PD 0982 — *notification ; Failure*
Disease [Ondiri...]
PD 2732 — *Ondiri*
PD 2774 — outbreak ; Inadequate disinfection pastureland after
PE 8409 — outbreaks ; Inadequate disinfection humans after animal
PE 8409 — outbreaks ; Inadequate disinfection measures humans during animal
PE 7769 — *Overeating*
Disease [Parkinsons...]
PE 2206 — Parkinson's

Disease [Pea...] cont'd
PE 2357 — Pea picker's
PE 3503 — *Periodontal*
PE 9702 — *Periodontal*
PE 5550 — Pfeiffer's
PD 2730 — *pheasants ; Marble spleen*
PD 2730 — *pigs ; Encephalomyocarditis virus*
PD 3599 — Plant vectors plant
PE 4787 — plants ; Damping off
PB 0035 — Population
Disease poultry
PE 1400 — Newcastle
PE 7562 — *Pullorum*
PD 5453 — *Round heart*
Disease [prevention...]
PF 9401 — prevention ; Inadequate
PD 1251 — prone housing
PF 4861 — prone ; Inattention
Disease Pulmonary
PD 0637 — [Disease ; Pulmonary]
PE 9415 — heart
PD 9366 — hypertensive heart
PE 7769 — *Disease ; Pulpy kidney*
PE 7826 — *Disease ; Pyramidal*
PD 3978 — *Disease ; Quail*
Disease [Railroad...]
PD 7420 — *Railroad*
PE 7769 — *Red water*
PE 0921 — Reduced resistance
PC 3595 — *reservoirs*
PE 0920 — Rheumatic heart
PE 3629 — Rodent vectors
Disease [Salmon...]
PD 2732 — *Salmon poisoning*
PD 2730 — *San Miguel sea lion virus*
PE 7769 — *sheep ; Black*
PE 3724 — Sickle cell
PE 9774 — *Stiff lamb*
PE 2357 — Stuttgart
PE 7841 — swine ; Oedema
PD 2730 — *Swine vesicular*
PE 2357 — Swineherds
PE 4208 — Symptomatic heart
Disease [Talfan...]
PD 7841 — *Talfan*
PD 7841 — *Teschen*
PD 8821 — transmission ; Ignorance concerning
PD 6421 — transmission international travel
PD 2289 — *Trophoblastic*
PE 3895 — *Tsutsugamushi*
PD 2728 — Turkey X
PD 2730 — turkeys ; Lymphoproliferative
Disease [vectors...]
PC 3595 — vectors
PD 6651 — vectors ; Human
PD 6651 — Vectors human
PD 3596 — vectors ; Plant
PE 2572 — Vertebrate reservoirs
PD 7841 — *Vomiting wasting*
PE 2348 — Von Economo's
Disease [Water...]
PE 3401 — Water borne
PD 1175 — *Water borne viral*
PD 2740 — Weather factor animal
PE 2357 — Weil's
PD 2730 — *Wesselsbron*
PE 9374 — *Western duck*
PE 9774 — *White muscle*
PD 2776 — wild animals ; Difficulty controlling
PC 2270 — Wilson's
PE 2293 — Woolsorters'
PE 9774 — *Disease ; Yellow fat*
Diseased animals Economic
PE 8098 — loss reduced productivity
PE 8098 — loss reduced productivity
PE 8109 — loss slaughter
PE 2778 — Diseased animals ; Inadequate carcass disposal
PE 8109 — Diseased animals ; Slaughter
Diseases [Acute...]
PD 9045 — *Acute intestinal*
PD 8301 — adrenal glands
PC 3595 — Agents causing infectious
PD 5029 — Airborne
PD 2741 — Airborne viral animal
PC 0952 — [Diseases ; Animal]
PD 7424 — bone
PC 0952 — deficiency
Diseases animals
PD 2731 — Bacterial
PD 5453 — Cardiovascular
PD 5453 — Circulatory system
PE 6419 — Congenital anomalies due infectious
PE 5535 — Ear
PE 0471 — *Endocardium*
PD 9654 — Endocrine
PE 5535 — Eye
PD 2728 — Fungal
PE 7423 — Gastroenteric
PE 4224 — Immune deficiency
PD 4068 — Immune system
PE 4224 — Immunodeficiency
PD 2732 — Infectious
PD 2755 — Diseases ; Animals' international movement factor animal

Diseases animals
PD 3978 — Liver
PD 5453 — Lymphatic system
PE 7420 — Metabolic
PE 9702 — Mouth
PE 0471 — *Myocardium*
PD 2735 — *Parasitic*
PE 0471 — *Pericardium*
PE 7304 — Pox
PD 7799 — Reproductive system
PD 9667 — Skin
PE 2739 — Soil borne
PC 0965 — *Spastic*
PD 9293 — Urinary system
PD 2730 — Viral
PD 2730 — Virus
Diseases [anterior...]
PD 5535 — *anterior uvea animals*
PE 2684 — arteries
PE 7796 — Arthropod borne
PE 9509 — Atopic
PD 3601 — Avian vectors plant
Diseases [bacteria...]
PF 1312 — bacteria ; Introduction extraterrestrial infectious
PE 2562 — Bacteria vectors viral
PE 2562 — bacteria ; Virus
PD 2226 — Bacterial plant
PD 2284 — beneficial insects
PD 3323 — Bird
PF 8026 — blood
PF 8026 — Blood
PF 8026 — blood blood-forming organs
PF 8026 — blood blood-forming organs
PD 0992 — brain
PD 9742 — breast
Diseases [Cardiovascular...]
PE 6816 — Cardiovascular
PD 0752 — Cattle
PE 5524 — cattle ; Respiratory
PE 4683 — caused parasites ; Nervous system
PE 9037 — central nervous system
PE 1041 — central nervous system ; Demyelinating
PD 8775 — *cervix*
PE 2983 — chestnut ; Pests
PD 0622 — Children's
PD 8239 — Chronic
PC 8482 — circulation system
PC 8482 — circulatory system
PC 8482 — Circulatory system
PE 2976 — citrus fruit ; Pests
PE 2979 — cocoa ; Pests
PE 2981 — *coconut palm ; Pests*
PE 2218 — coffee ; Pests
PE 2565 — Collagen
PD 0982 — Communicable
PF 7727 — Compounding effect treating genetic
PE 2780 — Confusion symptoms animal
PD 5535 — *conjunctiva animals*
PE 2565 — connective tissue
PD 5535 — *cornea animals*
PE 2220 — cotton ; Pests
PE 8932 — cranial nerves
PE 7783 — Crops pests
PF 1624 — Culture specific
PC 8534 — Cutaneous
Diseases [deciduous...]
PE 3591 — deciduous fruit ; Pests
PD 0287 — Deficiency
PC 8866 — deformities digestive system
PE 6216 — Degenerative
PD 6663 — Dependence man made
Diseases Difficulty
PD 2781 — controlling animal
PF 2780 — diagnosing animal
PF 2775 — identifying carriers animal
Diseases [digestive...]
PC 8866 — digestive system
PD 3978 — digestive system animals
PE 2647 — disabilities ; Foot
PD 8786 — disorders ; Eye
PD 2746 — Domestic animals carriers animal
PD 6334 — Drug induced
Diseases [ear...]
PE 2567 — ear
PE 2982 — elm ; Pests
PB 6384 — *Encouragement drug resistant*
PD 8301 — endocrine glands
PD 8301 — endocrine ; Nutritional metabolic
PD 5669 — Environmental human
PD 2224 — Environmental plant
PD 8200 — Environmentally induced
PD 2733 — Enzootic
PD 2734 — Epizootic
PD 8786 — eye ; Inflammatory
Diseases [factory...]
PD 2752 — factory farming ; Spread animal
PD 9742 — fallopian tube
PD 8775 — female genital organs
PE 8567 — fish ; Pests
PE 4514 — Flies vectors
PD 3978 — *foals ; Enteric*
PE 2515 — Food borne
PD 2728 — Fungal
PD 2225 — Fungal plant
Diseases [gallbladder...]
PE 9829 — gallbladder

-904-

Disinfection humans

Diseases [Gastrointestinal...] cont'd
PE 3861 — Gastrointestinal
PD 2389 — Genetic defects
PC 4575 — genito-urinary system
PE 4293 — *Glycogen storage*
PD 3978 — *goats ; Enteric*
PE 9438 — goats ; Respiratory
PE 3590 — *grain sorghum ; Pests*
PE 7783 — *groundnut ; Pests*
PE 5182 — *guttural pouches*
Diseases [Haemolytic...]
PD 2289 — *Haemolytic*
PD 0448 — Heart
PE 0920 — *heart ; Valvular*
Diseases Hereditary
PD 2389 —
PD 2389 — defects
PD 2389 — metabolic
Diseases [High...]
PF 2779 — High cost controlling epizootic enzootic
PE 6708 — horses ; Enteric
PE 5182 — horses ; Respiratory
PD 0287 — Human deficiency
PD 2784 — Human vectors animal
Diseases [Ignorance...]
PD 8821 — Ignorance concerning sexually transmitted
PC 9067 — Ill defined
PE 2777 — Importation infected carcass meats factor animal
PD 2781 — Inadequate control animal
PE 8133 — Inadequate knowledge incubation periods animal
PC 8801 — Incurable
PD 0215 — Industrial
PD 2732 — Infectious animal
PD 0982 — Infectious parasitic
PE 0652 — *Inherited*
PE 3822 — *injuries bone*
PD 0992 — *injuries brain*
PF 1916 — *Inoculation viral*
PE 2764 — Insanitary penning conditions factor animal
PD 2748 — Insect vectors animal
PE 7732 — Insect vectors plant
PD 5168 — Intentional spread sexually transmitted
PD 2755 — International movement animals factor animal
PD 9045 — Intestinal
PE 9526 — Intestinal infectious
PC 8801 — Intractable
PE 7832 — introduced water projects
PD 1185 — *Diseases jaw*
PE 2053 — Diseases ; Kidney
Diseases [lacrimal...]
PD 5535 — *lacrimal apparatus animals*
PE 4586 — *Leaf curl*
PF 4618 — Lifestyle
PB 1044 — *limbs*
PE 1028 — Liver
PE 1028 — Liver fatty
PD 0637 — Lung
PD 0637 — Lung disorders
PD 2654 — lymphatic system
Diseases [Maize...]
PE 3589 — Maize pests
PD 9154 — male genital organs
PD 3593 — man ; Dissemination plant
PD 6663 — Man made
PE 3721 — Massacre indigenous populations imported
PD 2567 — mastoid process
PD 6334 — medical practice
PC 2270 — metabolism
PC 7492 — Microbial
PD 1185 — mouth jaw
PE 4414 — Mucopolysaccharide
PD 2565 — muscles
PD 2565 — musculoskeletal connective tissues
PD 2565 — musculo-skeletal system
PD 7424 — musculoskeletal system animals
PD 2565 — musculoskeletal system connective tissues
Diseases [nail...]
PC 8534 — *nail*
PD 2228 — Nematoid plant
PC 3853 — Neoplastic
PE 8932 — nerves peripheral ganglia
Diseases nervous system
PC 8756 — [Diseases nervous system]
PD 7841 — animals
PE 0965 — animals ; Congenital
Diseases [newborn...]
PE 7423 — *newborn ; Enteric*
PD 2224 — Nonparasitic plant
PD 7924 — nose
PE 5122 — nose
PD 0287 — Nutritional
PE 2984 — Diseases oak ; Pests
Diseases Occupational
PD 0215 — [Diseases ; Occupational]
PE 6806 — blood
PD 0637 — chronic pulmonary
Diseases oesophagus
PE 8636 — [Diseases oesophagus]
PE 4630 — animals
PE 8624 — stomach duodenum
Diseases [olives...]
PE 2978 — olives ; Pests
PD 8786 — Ophthalmic
PE 4584 — optic nerve
PD 9742 — ovary

Diseases [palms...]
PE 2981 — palms ; Pests
PE 1132 — pancreas
PE 0965 — *Paralytic animal*
PD 9742 — parametrium
PE 7832 — perennial irrigation ; Increase pests
PE 3503 — Periodontal
PE 8932 — peripheral autonomic nervous system
PE 8932 — peripheral nervous system
PD 6334 — Physician induced
PD 2224 — Physiogenic plant
PD 2224 — Physiological plant
PC 0555 — Plant
Diseases plants
PE 4346 — *Alternaria*
PE 1910 — Anthracnose
PE 4235 — Blotch
PE 4346 — *Botrytis*
PD 3653 — Deficiency
PE 4346 — *Fusarium*
PD 3600 — Insect vectors viral
PE 1040 — Scab
PE 4586 — *Sclerotinia*
PE 0857 — Smut
PD 2227 — Virus
PE 1056 — Wilt
Diseases [Pollution...]
PE 7584 — Pollution induced fish
PE 4413 — poplars ; Pests
PE 2219 — potato ; Pests
PE 5567 — poultry ; Respiratory
PD 1285 — *Prevailing chicken*
PE 8372 — Prostate
PD 0594 — *protozoa ; Virus*
PE 3676 — Protozoan
PD 0637 — Pulmonary
Diseases [Renal...]
PE 2053 — Renal
PB 1044 — *reproductive organs*
PD 0637 — Respiratory
PD 7924 — respiratory system
PD 7307 — respiratory system animals
PD 7924 — Respiratory virus
PE 4584 — retina
PE 0873 — Rheumatic
PD 2226 — *Rhizobium plant*
PE 2221 — rice ; Pests
PE 5530 — Rickettsial
PE 2977 — rubber ; Pests
Diseases [Seed...]
PE 4559 — Seed borne
PC 9623 — sense organs
PD 5535 — senses animals
PD 0061 — Sexually transmitted
PE 6594 — Sheep
PD 3978 — *sheep ; Enteric*
PE 9438 — sheep ; Respiratory
PC 8534 — Skin
PC 8534 — skin subcutaneous tissue
PE 9854 — small animals ; Respiratory
PE 2747 — Snail vectors animal
PD 3699 — Soil transmitted
Diseases spinal
PE 4604 — column animals
PE 8813 — cord
PE 4604 — cord animals
Diseases [spine...]
PD 2626 — spine
PE 3254 — Spirochaetal
PE 6155 — spleen
PD 7279 — Stigmatized
PD 3587 — storage transit ; Plant
PE 7915 — striato pallidal system ; Hereditary
PE 2975 — sugar beet ; Pests
PE 2217 — sugar cane ; Pests
PC 8534 — *sweat glands*
PE 3936 — swine ; Enteric
PE 7553 — swine ; Respiratory
Diseases [tea...]
PE 2980 — tea ; Pests
PD 7924 — throat
PE 7733 — throat
PE 7708 — thymus gland
PE 3897 — Tick borne
PD 1185 — *tongue*
PD 2731 — *Toxaemic animal*
PD 3585 — trees ; Pests
PE 6461 — Trematode
Diseases [Uncontrolled...]
PC 0555 — *Uncontrolled plant*
PD 4775 — Uncontrolled tropical
PD 7726 — Underreporting animal
Diseases unknown
PD 8588 — aetiology
PD 8588 — cause
PD 8588 — origin
Diseases [Unrecognized...]
PD 2781 — *Unrecognized animal*
PF 8090 — Unreported
PD 1285 — *Untreated chicken*
PE 7733 — upper respiratory tract
PD 8775 — uterus
PE 9286 — uterus ; Infective
Diseases [vagina...]
PE 9286 — vagina ; Infective
PD 8385 — Vector borne

Diseases [Vector...] cont'd
PF 2775 — *Vector immunity animal*
PD 2751 — Vectors animal
PE 2684 — veins
PD 0061 — Venereal
PE 2985 — vines ; Pests
PD 0594 — *Viral*
PD 2227 — *Viral plant*
PD 0594 — Virus
PE 6866 — voice ; Occupational
PE 3721 — Vulnerability indigenous populations introduction
Diseases [Water...]
PE 3401 — Water borne
PE 2787 — Water borne animal
PE 3401 — Waterborne
PE 2222 — wheat ; Pests
PD 2776 — wild animals
PD 2729 — Wild animals carriers animal
PE 2749 — Wild birds vectors animal
PE 2223 — *Wind borne animal*
PD 2750 — Worms vectors animal
PD 6363 — Diseases ; Zoonotic bacterial
PF 3918 — Disempowered ; Placement burden proof
PF 0549 — Disempowerment ; Personal
PB 4353 — Disempowerment vulnerable groups
PA 6191 — Disenchantment *Disapproval*
PA 6180 — Disenchantment *Error*
PF 1097 — Disenchantment government
PA 5699 — Disenchantment *Reversion*
PF 9594 — Disenchantment United Nations
PE 7095 — Disenfranchise grassroots technological development ; Intellectual methods which
PE 4716 — Disenfranchisement
PJ 3760 — *Disengaged transient residents urban areas*
PF 6836 — *Disequilibrium*
PE 3948 — Disestablishment tensions ; Failure understand necessity creative establishment
Disesteem [Disapproval...]
PA 6191 — *Disapproval*
PA 6839 — *Disrepute*
PA 6822 — *Disrespect*
Disfavour [Disaccord...]
PA 5532 — *Disaccord*
PA 6191 — *Disapproval*
PA 6839 — *Disrepute*
PA 7107 — Disfavour *Unpleasantness*
PC 2122 — Disfigurement ; Dependence landscape
PA 6790 — Disfigurement *Distortion*
PD 8076 — Disfigurement ; Facial
PA 6997 — Disfigurement *Imperfection*
PC 2122 — Disfigurement ; Landscape
PE 3733 — Disfigurement ; Malicious personal
PD 1637 — Disfigurement open cast mining ; Landscape
PD 8076 — Disfigurement ; Personal physical
PA 7240 — Disfigurement *Ugliness*
PE 6776 — Disfiguring impairments
PA 6191 — Disgrace *Disapproval*
PA 6839 — Disgrace *Disrepute*
PA 6659 — Disgrace *Humility*
PA 5644 — Disgrace *Vice*
PA 7280 — Disgrace *Wrongness*
PF 2505 — Disguise ambiguity ; Social
PB 2642 — *Disguised cruelty*
PF 7583 — Disguised negative consequences remedial action
PC 0940 — Disguised unemployment socialist countries
PA 7107 — Disgust *Unpleasantness*
PA 7204 — Disgust *Unsavouriness*
Disharmony [Disaccord...]
PA 5532 — *Disaccord*
PA 5982 — *Disagreement*
PA 5565 — *Discord*
PA 7361 — *Disorder*
PC 2649 — Dishonest elections
PD 9397 — Dishonest employees
PF 8759 — Dishonest personal relationships
PB 4731 — Dishonesty
PD 1893 — Dishonesty government ; Intellectual
PA 7363 — Dishonesty *Improbity*
PA 7411 — Dishonesty *Uncommunicativeness*
PF 8485 — Dishonour
PA 6839 — Dishonour *Disrepute*
PA 6822 — Dishonour *Disrespect*
PA 6058 — Dishonour *Impiety*
PA 7363 — Dishonour *Improbity*
PA 7102 — *Dishonourable grant acceptance*
PF 9729 — Disillusioning cooperative experience
PA 6453 — Disillusionment
PF 2819 — Disillusionment aid developing countries ; Public
PA 6191 — Disillusionment *Disapproval*
PA 6453 — Disillusionment *Disillusionment*
PA 6180 — Disillusionment *Error*
PF 1097 — Disillusionment government
PA 5527 — Disillusionment *Inexpectation*
PF 0955 — Disillusionment life aspirations
PF 0770 — Disillusionment science
PF 9249 — Disincentive invest heavily indebted countries
PD 7536 — Disincentives against farming
PE 8321 — Disincentives against farming developing countries
Disincentives [farm...]
PE 8321 — farm products developing countries ; Industrialized country
PF 3845 — financial investment developing countries
PD 3138 — foreign private investment developing countries
PF 3538 — *Disincentives invention*
PE 8409 — Disinfection humans after animal disease outbreaks ; Inadequate

Disinfection measures

PE 2757 Disinfection measures animal housing equipment ; Inadequate
PE 8409 Disinfection measures humans during animal disease outbreaks ; Inadequate
PD 2774 Disinfection pastureland after disease outbreak ; Inadequate
PB 7606 Disinformation
PB 7606 Disinformation ; Dependence
PD 7231 Disinformation ; Research
PA 7363 Disingenuousness *Improbity*
PA 5502 Disingenuousness *Reason*
PA 7411 Disingenuousness *Uncommunicativeness*
PF 0886 Disinheritance
PC 0193 *Disinherited children*
PA 6858 Disintegration
PF 8887 Disintegration accepted myths
PA 6858 *Disintegration ; Cultural*

Disintegration [Destruction...]
PA 6542 — *Destruction*
PA 6858 — *Disintegration*
PA 7361 — *Disorder*
PC 3291 Disintegration ; Ethnic
PC 2102 Disintegration ; Family
PA 6088 Disintegration *Impairment*
PB 3384 Disintegration ; National
PD 3423 Disintegration organized religion
PC 3204 Disintegration ; Political
PA 7236 Disintegration *Separation*
PC 3309 Disintegration ; Social
PD 7719 Disintegration technological capacity
PA 6379 Disinterest *Avoidance*
PF 0148 Disinterest ; Family planning
PA 6598 Disinterest *Incuriosity*
PF 1735 *Disinterest ; Local medical*
PF 2602 *Disinterest ; Mutual student teacher*
PC 0588 Disinterest social processes
PA 7364 Disinterest *Unfeelingness*
PF 3961 Disintermediation
PF 2845 Disjointed patterns community identity
PA 7338 Dislike *Hate*
PA 7107 Dislike *Unpleasantness*
PD 4368 *Dislocated life style*
PA 7158 Dislocation
PB 0731 *Dislocation bones*
PE 8787 Dislocation productive units foreign investment
PD 4063 Dislocations developing countries ; Economic
PE 9018 *Disloyalty*
PD 2790 Disloyalty ; Corporate
PJ 1895 *Disloyalty government*
PA 7363 Disloyalty *Improbity*
PA 7325 Disloyalty *Irresolution*
PA 6030 Dismay *Fear*
PA 7309 Dismay *Uncertainty*
PA 7107 Dismay *Unpleasantness*
PC 2102 Dismembered families
PE 6252 Dismembered works art
PD 6945 Dismemberment bodies
PF 2641 Dismemberment ; Ritual
PE 6069 Dismissal ; Compulsive early
PE 7620 Dismissal trade union representatives following legal strike action

Dismissal workers
PE 7620 — following legal strike action
PD 8007 — improve profitability
PE 7620 — prevent legal strike action
PD 5965 — Unjust
PA 7250 Disobedience

Disobedience [children...]
PD 5308 — children
PD 5707 — Citizen
PC 0690 — Civil
PA 7250 Disobedience *Disobedience*
PF 7149 Disobedience elders
PD 5244 Disobedience ; Employee
PD 3879 Disobedience judicial element
PA 5563 Disobedience *Lawlessness*
PD 7225 Disobedience ; Military
PA 7321 Disobedience *Refusal*
PF 7467 Disobedience ; Spiritual
PA 6509 Disobedience *Unwillingness*
PD 5244 Disobedience ; Worker
PD 5308 *Disobedient children*
PF 4764 *Disobedient wives*
PA 7361 Disorder
PF 4487 Disorder

Disorder [adolescence...]
PE 9256 — adolescence ; Avoidant
PA 5838 — *Agitation*
PD 0270 — *Alteration physical functioning suggesting physical*
PA 1635 — Anxiety
PE 9789 — Attention deficit hyperactivity
PE 3901 — Avoidant personality
PD 0270 *Disorder ; Body dysmorphic*

Disorder [childhood...]
PE 9256 — childhood ; Avoidant
PE 9344 — *Chronic motor tic*
PE 9344 — *Chronic vocal tic*
PE 3770 — Conduct
PD 0270 — *Conversion*
PE 6696 Disorder ; Dependent personality

Disorder Developmental
PC 0143 — arithmetic
PE 9712 — articulation
PE 7230 — coordination
PE 5545 — expressive language
PE 0330 — expressive writing

Disorder Developmental cont'd
PE 9300 — receptive language

Disorder [Disease...]
PA 6799 — Disease
PA 7361 — *Disorder*
PE 6958 — Dream anxiety
PF 8490 *Disorder ; False diagnosis mental*
PA 6900 Disorder *Formlessness*
PD 8301 Disorder ; Glandular
PE 4431 Disorder ; Hormonal
PE 4561 Disorder ; Hysterical personality

Disorder [Impulse...]
PA 6448 — Impulse control
PA 6247 — *Inattention*
PC 0300 — infancy ; Reactive attachment
PE 5187 — *infancy ; Rumination*
PA 6448 — Intermittent explosive
PA 5563 Disorder *Lawlessness*
PD 0270 *Disorder ; Loss physical functioning suggesting physical*

Disorder [Mental...]
PD 9131 — Mental
PA 6498 — *Misbehaviour*
PE 7230 — Motor skills
PE 5048 — Multiple personality

Disorder [Obsessive...]
PA 6448 — Obsessive compulsive
PE 7632 — Obsessive compulsive personality
PE 4634 — Oppositional defiant
PE 9580 — Over anxious

Disorder [Panic...]
PE 3575 — Panic
PE 0435 — Paranoid personality
PD 9219 — *Passive aggressive personality*
PD 0556 — *Post hallucinogen perception*
PE 0351 — Post traumatic stress

Disorder [Sadistic...]
PD 9219 — *Sadistic personality*
PD 0438 — *Schizoaffective*
PD 9219 — *Schizoid personality*
PD 0438 — *Schizophreniform*
PD 9219 — *Schizotypal personality*
PD 0258 — *Seasonal affective*
PD 4418 — Self defeating personality
PE 2401 — Separation anxiety
PE 2197 — *Sleep terror*
PE 2197 — *Sleep wake schedule*
PE 2197 — *Sleepwalking*
PD 0270 — *Somatization*
PD 0270 — *Somatoform pain*
PE 9344 *Disorder ; Tourette's*
PE 9344 *Disorder ; Transient tic*
PA 7309 Disorder *Uncertainty*
PE 2076 Disordered behaviour
PC 4867 Disordered local management
PD 9178 Disorderly conduct
PD 7359 Disorderly conduct ; Hindering proceedings

Disorders [Adjustment...]
PC 0300 — Adjustment
PD 9159 — Affective
PD 0919 — aged ; Mental
PE 9291 — amino acid metabolism ; Congenital

Disorders animals
PE 5913 — Arthritis
PD 5453 — *Haemostatic*
PE 7570 — Salivary
PF 1721 Disorders ; Anti social personality
PF 2384 Disorders ; Attention

Disorders [Blood...]
PF 8026 — Blood
PE 3830 — Blood circulation
PE 4396 — Borderline personality
PE 6553 — bowel

Disorders [carbohydrate...]
PE 4293 — carbohydrate metabolism ; Congenital
PE 9364 — cattle ; Abomasal
PE 9037 — Central nervous system
PE 7915 — central nervous system ; Hereditary
PC 8482 — Circulatory
PC 2551 — Civil
PE 2373 — Coagulation
PE 0274 — Cold
PD 1618 — Congenital
PD 8005 — consciousness
PC 7373 — Criminal violence

Disorders [Digestive...]
PC 8866 — Digestive
PD 2626 — *Disc*
PD 0637 — diseases ; Lung
PE 6724 — due torture ; Gastro intestinal

Disorders [Eating...]
PE 5187 — Eating
PD 9159 — Emotional
PD 8179 — Eye
PD 8786 — Eye diseases
PD 8179 — eye ; Refractive
PC 0300 *Disorders ; Fictitious*

Disorders [Gastric...]
PE 1599 — Gastric
PE 6581 — Gender identity
PD 8301 — Gland

Disorders [heart...]
PD 0448 — *heart ; Functional*
PE 2398 — Heat
PE 1849 — *Hepatic*
PE 7915 — Hereditary neuromuscular
PD 8016 — Human sexual

PD 2726 Disorders ; Iodine deficiency
PD 2283 Disorders joints ligaments
PE 2053 Disorders ; Kidney

Disorders [lactation...]
PC 9042 — *lactation*
PE 2265 — Language
PE 3886 — Language
PF 8387 — leadership ; Personality
PD 3865 — Learning
PE 9816 — lipid metabolism ; Congenital

Disorders [menstruation...]
PD 8775 — *menstruation*
PD 9131 — Mental
PC 2270 — Metabolic
PC 2270 — Metabolism
PD 9293 — *micturition*
PD 9159 — Mood
PE 6769 — Movement

Disorders [Narcissistic...]
PF 7248 — Narcissistic behaviour
PF 7248 — Narcissistic personality
PC 8756 — Nervous
PE 0965 — *nervous system dogs*
PC 0300 — Non psychotic mental
PE 5606 — Disorders ; Occupational cervicobrachial
PD 8016 *Disorders ; Orgasm*

Disorders [Parathyroid...]
PD 8301 — *Parathyroid gland*
PD 9219 — Personality
PB 1044 — *Physical*
PE 2286 — Pituitary gland
PE 0517 — *Post hepatitis*
PD 2289 — Pregnancy
PE 3974 — Psychogenic physical
PD 8375 — Psychological
PD 1967 — Psychosomatic
PD 1950 Disorders ; Reading
PE 3004 Disorders rumen ; Digestive

Disorders [Self...]
PF 4843 — Self
PF 4843 — Self object
PC 9623 — sense organs
PU 3287 — *Sensory*
PD 8016 — *Sexual arousal*
PE 9607 — *Sexual desire*
PE 2298 — Skeletal system
PE 7808 — *skeletal system poultry*
PC 8534 — Skin
PE 2197 — Sleep
PE 2265 — Speech

Disorders [Teeth...]
PD 1185 — Teeth
PE 0652 — Thyroid gland
PE 9344 — Tic
PD 2289 — Transplacental
PE 2307 Disorders ; Urinary bladder
PF 4487 Disorganization
PF 4487 Disorganization

Disorganization [Destruction...]
PA 6542 — *Destruction*
PA 6858 — *Disintegration*
PA 7361 — *Disorder*
PC 2151 Disorganization ; Family
PA 6088 Disorganization *Impairment*
PA 6247 Disorganization *Inattention*
PA 5563 Disorganization *Lawlessness*
PF 2365 Disorganized approach land ownership tropical villages
PF 6458 Disorganized attempts upgraded employment
PF 2813 Disorganized community life
PF 2947 Disorganized liaison formal support
PF 2989 *Disorganized local marketing*
PF 3821 Disorganized political opposition
PD 0438 *Disorganized type schizophrenia*
PA 6247 Disorientation *Inattention*
PA 7157 Disorientation *Insanity*
PA 6739 Disorientation *Maladjustment*
PF 6174 Disorientation stress large building complexes
PF 4644 Disorientation ; Temporal
PA 7309 Disorientation *Uncertainty*
PD 2050 Disorientation young due lack social forms
PC 0492 *Disoriented business practices*
PF 2168 *Disoriented habitual modes*
PF 2845 *Disoriented traditional occupations*
PC 1867 *Disorienting urban shift*
PC 0193 *Disowned children*
PD 5238 *Disparagement*

Disparagement [Disapproval...]
PA 6191 — *Disapproval*
PA 6839 — *Disrepute*
PA 6822 — *Disrespect*
PE 1817 Disparagement indigenous cultures
PA 6607 Disparagement *Misjudgement*
PJ 3532 Disparagement religious experience
PJ 3532 Disparagement transcendent experiences traditions
PD 1534 Disparate development economic sectors developing countries
PF 2845 *Disparate village identity*
PF 4385 *Disparities book production distribution ; Area*
PD 0446 *Disparities calorie intake*
PD 2963 *Disparities developing countries*
PD 2762 *Disparities distribution communication resources facilities*
PD 7257 Disparities economic growth developing countries
PC 5038 *Disparities energy consumption*
PC 2049 *Disparities ; Regional*
PD 1837 *Disparities unemployment countries*

—906—

Distant elders'

PB 1016 Disparity consumption resources ; International
PF 3043 Disparity countries natural resource endowments
Disparity [developing...]
PE 1660 — developing countries due transnational corporations ; Increasing income
PA 6698 — *Difference*
PA 5532 — *Disaccord*
PA 5982 — *Disagreement*
PA 6838 — *Dissent*
PD 5258 — distribution wealth developing countries
PE 3995 Disparity facilities military mobilization reinforcement
PC 4844 Disparity human healthiness
Disparity [income...]
PD 1842 — income cost housing
PC 8694 — industrialized developing countries
PA 6695 — *Inequality*
PD 1791 Disparity national tax systems
PE 7760 Disparity remuneration public private sector employees
PC 3458 Disparity resentment ; Class
PE 3827 Disparity share prices underlying asset values
PD 0266 Disparity social development developing countries
PC 1131 Disparity workers skills job requirements
PD 5701 Disparity world telecommunications capabilities
PF 2477 *Dispersed fire volunteers*
PF 7118 *Dispersed residential community*
PF 6141 Dispersion community facilities ; Excessive
PF 1979 Dispersion local capital resources
PF 9023 Dispersion ; Rural
PD 5136 Displaced children
PF 2845 Displaced community identity
PD 7822 Displaced persons
PF 5943 Displaced royalty
PE 9554 *Displacement abnormal cells ; Marrow*
PF 5360 Displacement activity ; Political
PB 0731 *Displacement bones*
PD 1544 Displacement ; Community youth
PE 6843 Displacement ; Labour
PE 9364 *Displacement ; Left side abomasal*
PF 5472 Displacement national decision-making developing industrialized countries
PF 6198 Displacement natural light buildings
PE 9364 *Displacement ; Right side abomasal*
PC 0475 Displacement wildlife ; Forcible
PE 6843 Displacement workers developed countries
PE 9364 *Displacements ; Abomasal*
PE 5083 Display terminal work ; Job stress video
PF 6578 *Display ; Unattractive merchandise*
PE 5083 Display units ; Health hazards computer visual
PD 9330 Displays ; Biased computerized information
PE 4882 Displays sexuality ; Public
PA 6809 Displeasure
PA 6191 Displeasure *Disapproval*
PA 6809 Displeasure *Displeasure*
PA 7253 Displeasure *Envy*
PA 6731 Displeasure *Solemnity*
PA 7107 Displeasure *Unpleasantness*
PD 9310 Disposable products ; Unrecyclable
PD 0807 Disposal ; Consumer waste
PE 2778 *Disposal corpses*
Disposal [deep...]
PD 0201 — *deep wells ; Waste*
PE 2778 — diseased animals ; Inadequate carcass
PD 0807 — Domestic refuse
PE 2778 *Disposal ; Final*
PD 0807 Disposal ; Garbage
PD 1398 Disposal hazardous waste
PE 4725 Disposal human remains ; Unsanitary
Disposal [Improper...]
PD 1152 — Improper paper
PD 0876 — Improper sewage
PD 6795 — Inadequate waste
PD 7574 Disposal obsolete weapons ; Marine
PC 1242 Disposal ; Radioactive waste
PD 0807 Disposal ; Rubbish
PD 0807 Disposal ; Trash
PF 3454 *Disposal ; Unsanitary refuse*
PD 3666 Disposal waste ; Ocean
PE 0503 Disposal wealth ; Lack accountability
PE 6955 Dispose natural wealth ; Denial right people freely
PE 6768 Disposition disabilities ; Body
PD 4974 Disposition productive property ; Individualistic
PE 6750 Dispositions ; Ineffective testamentary
PB 1047 *Dispossession*
PC 8862 Dispossession peasant landholdings
PF 1085 Disproportionate allocation resources pet animals
PE 0135 Disproportionate control global economy limited number corporations
PC 3056 Disproportionate external public debt
PE 1922 Disproportionate influence national economies limited number corporations
PF 2344 Disproportionate influence some individuals national policy-making
PF 9686 Disproportionate political influence urban elites
PD 9430 Disproportionate salary scales countries
PF 8763 Disproportionately high salaries
PE 4602 Disproportionately long prison sentences
PF 6670 Disproportionately wealthy elites developing countries
PA 5532 Disputatiousness *Disaccord*
PA 7253 Disputatiousness *Envy*
PA 6979 Disputatiousness *Opposition*
PA 5502 Disputatiousness *Reason*
PF 2943 Disputed administering government ; Territories accorded United Nations non self governing status
PC 2362 Disputed islands
PF 6528 Disputed land usage
PF 7274 Disputed priority innovations
PC 2362 Disputed territories
PD 7522 Disputes allies
PD 2946 Disputes ; Border
PD 2950 Disputes ; Border
PF 2575 *Disputes ; Divisive zoning*
Disputes [Industrial...]
PD 4012 — Industrial
PA 0298 — *Inter Governmental*
PA 0298 — *International legal*
PD 4012 Disputes ; Labour
Disputes [Neighbourhood...]
PE 5504 — Neighbourhood
PE 5504 — neighbours
PE 7903 — neighbours ; Boundary
PE 8547 Disputes ; Private international trade investment
PD 0516 Disputes ; River water
Disputes states
PD 2946 — Boundary
PE 1911 — nationals states ; Financial economic
PC 1888 — Territorial
PE 1911 Disputes ; Transfrontier private investment
PF 7890 Disputes ; Unreported
PA 5838 Disquiet *Agitation*
PA 6030 Disquiet *Fear*
PA 7107 Disquiet *Unpleasantness*
PA 6379 Disregard *Avoidance*
PE 1940 Disregard consumer interests transnational corporations
PA 7250 Disregard *Disobedience*
PA 6822 Disregard *Disrespect*
PF 9542 Disregard government orders armed forces
PA 6247 Disregard *Inattention*
PD 8017 Disregard individuals ; Official
Disregard international
PF 2497 — law
PE 6564 — law capture foreign nationals
PF 0292 — values procedures ; Governmental
PF 1976 Disregard internationally imposed economic sanctions
PD 9557 Disregard legitimate protests ; Governmental
PA 6387 Disregard *Necessity*
PA 5438 Disregard *Neglect*
PA 6979 Disregard *Opposition*
PE 6323 Disregard orders ; Liability crimes committed
PD 8017 Disregard people human beings ; Governmental
PE 8411 Disregard property rights
PD 5001 Disregard safety principles techniques
PF 4871 *Disregard self-healing potential*
PF 0292 Disregard world opinion ; Nation states'
PF 7458 Disregarded financial resources
PF 6534 Disrelated formal education
PF 1651 Disrelated images social roles
PF 6098 Disrelated images vocational roles
PF 5694 Disrelated life ; Religious theology
PF 1025 Disrelated needs promotion marketing
Disrelated [school...]
PF 0443 — school curriculum
PF 0443 — school lessons
PF 2241 — *social background*
PF 1187 Disrelation parents school
PF 6494 *Disrelationship community history*
PE 1980 Disrelationship political industrial sectors ; Planning
PE 0893 Disrelationship production work force needs
PE 2471 Disrepair bridges
PD 1565 Disrepair equipment
PA 6839 Disrepute
PA 6822 Disrespect
PF 5012 Disrespect authority
PA 6191 Disrespect *Disapproval*
PA 6822 Disrespect *Disrespect*
PF 3979 Disrespect elders
PF 9081 Disrespect international law
PF 3979 Disrespect parents
PA 6491 Disrespect *Vanity*
PF 6537 *Disrespected public property*
PF 2971 Disrupted mechanisms community health
PF 2845 Disrupted patterns village identity
PJ 3308 Disruption
Disruption [animal...]
PC 3152 — *animal breeding grounds ; Ecological*
PC 2279 — animal migration movement patterns
PD 7096 — arid zone ecosystems
Disruption [biological...]
PE 1904 — biological rhythms
PF 0379 — biological rhythms
PC 6675 — biosphere ; Military
Disruption breeding grounds
PC 3152 — *birds*
PC 3152 — *fish*
PC 3152 — *marine mammals*
PE 1082 Disruption cultural social identities developing countries transnational corporations
Disruption [development...]
PD 2191 — development tribal warfare
PE 1957 — domestic social policies transnational corporations
PD 7692 — due childbearing ; Career
PD 8851 Disruption economies production facilities ; War time
Disruption ecosystems
PC 1617 — exotic organisms
PB 5250 — human activity
PD 6960 — marginal agricultural lands
Disruption [family...]
PD 1482 — family system developing countries
PD 4511 — financial markets
PE 8979 — food supply due military activities
PF 3437 Disruption human activities supernatural entities
PD 9670 Disruption hydrological cycle
PE 6603 Disruption internal balance human body
PF 3162 *Disruption international action*
PD 4252 Disruption native trade patterns
PD 9670 Disruption natural water systems
PC 1647 *Disruption seabed ecosystems*
PC 2945 Disruption territorial integrity
PF 6461 Disruption traditional symbol systems ; Modern
PE 5074 Disruption work schedule due computerization
PC 0018 Disruptions due emigration
PC 0018 Disruptions due migration
PF 2644 *Disruptions ; Frequent schedule*
PD 2586 Disruptions natural water cycles ; Underreporting
PE 3008 *Disruptive airplane hunting*
PD 8544 Disruptive behaviour
PE 9092 Disruptive behaviour schools
Disruptive [effect...]
PF 2303 — effect changing employment patterns
PE 5928 — effect household moving
PF 6345 — effects video games ; Socially
Disruptive [food...]
PE 0302 — food imports
PC 3188 — foreign influence
PD 8030 — foreign investment
PD 6733 *Disruptive maternity leave employees*
PF 9003 Disruptive peer influence
PA 2173 *Disruptive personal prejudices*
PE 7735 Disruptive secular impact holy days
PA 8886 Dissatisfaction
Dissatisfaction [Disapproval...]
PA 6191 — *Disapproval*
PA 6011 — *Discontentment*
PA 6838 — *Dissent*
PA 7253 Dissatisfaction *Envy*
PA 5527 Dissatisfaction *Inexpectation*
PF 4010 Dissatisfaction personal image
PF 1741 Dissatisfaction skin colour
PA 7107 Dissatisfaction *Unpleasantness*
PE 3822 *Dissecans ; Osteochondritis*
PD 7424 *Dissecans ; Osteochondritis*
PD 5453 *Dissecting aneurysm turkeys*
PE 0566 *Disseminated tuberculosis*
PE 0598 Disseminating negative information ; Delay
PF 1183 Dissemination discriminatory information
PF 1183 Dissemination hate literature
PF 2401 Dissemination ; Infrequent information
PF 3539 *Dissemination ; Irrelevant data*
PF 1673 Dissemination materialistic images ; Excessive
PD 3593 Dissemination plant diseases man
PD 3050 Dissemination technical knowledge ; Tax barriers
PF 1267 Dissemination use available information ; Inadequate
Dissension [Disaccord...]
PA 5532 — *Disaccord*
PA 5982 — *Disagreement*
PA 6838 — *Dissent*
PA 6979 Dissension *Opposition*
PA 6838 Dissent
PF 3388 Dissent ; Ideological
PC 2582 Dissent ; Intellectual
PC 2361 Dissent ; Political
PE 4875 Dissent ; Religious
PC 2052 Dissent ; Student
PC 9695 Dissidence
Dissidence [Disaccord...]
PA 5532 — *Disaccord*
PA 5982 — *Disagreement*
PA 6838 — *Dissent*
PC 9695 Dissidents
PF 0732 Dissidents ; Inadequate legal counsel political
PD 0434 Dissidents ; Repression intellectual
PA 6428 Dissimilarity
PF 2124 Dissipated fiscal base
PF 2042 *Dissipating economic base*
PF 0905 *Dissipating economic base*
PA 6785 Dissipation *Disappearance*
PA 6466 Dissipation *Intemperance*
PA 7382 Dissipation *Loss*
PA 6858 Dissociation *Disintegration*
PA 0911 Dissociation human personality
PA 6739 Dissociation *Maladjustment*
PD 0061 *Dissociation sex love*
Dissolution [Death...]
PA 7055 — *Death*
PA 6542 — *Destruction*
PA 6785 — *Disappearance*
PA 6858 — *Disintegration*
PF 4930 Dissolution family ; Voluntary
PA 6088 Dissolution *Impairment*
PA 7236 Dissolution *Separation*
PE 5793 Dissolution ; Violation right trade unions protection against
PF 6638 Dissonance ; Cognitive
Dissonance [Difference...]
PA 6698 — *Difference*
PA 5982 — *Disagreement*
PA 5565 — *Discord*
PA 7343 Dissuasion
PE 7826 *Distal phalanx ; Fracture*
PA 7295 Distance
PC 1707 Distance ; Social
PF 3499 *Distances ; Long shipping*
PF 8299 Distant available markets
PE 0718 *Distant banking services*
PS 9612 *Distant commercial training*
PF 2261 Distant district centre
Distant [elders...]
PD 0512 — elders' programs

Distant elementary

Distant [elementary...] cont'd
PE 1848 — elementary school
PF 9007 — emergency services
PF 9007 — emergency transport
PF 2822 Distant farm residences
PE 2477 Distant fire services
PD 4790 Distant health services
PE 8916 Distant job locations
Distant [male...]
PE 8916 — male employment
PF 8299 — markets ; Excessively
PF 4832 — medical services
PE 8207 Distant objectives ; Military expeditions against
Distant [police...]
PF 8652 — police services
PF 8299 — product market
PF 1495 — public transportation
PC 0237 Distant relations ; Village leader
PC 0237 Distant representatives
Distant [schooling...]
PE 1848 — schooling facilities
PF 4375 — schools ; Expensive
PE 1848 — secondary school
PD 1204 — spring water
PD 1204 Distant water sources
PE 8916 Distant work places
PA 6191 Distaste Disapproval
PA 7107 Distaste Unpleasantness
PA 6509 Distaste Unwillingness
PE 9844 Distemper
PD 2730 Distemper ; Canine
PD 2730 Distemper ; Feline
PE 5283 Distilled spirits ; Possession unlawful
PE 7188 Distilleries ; Illicit
PC 3458 Distinction ; Class
PF 7938 Distinctions categories ; Collapse
PF 7938 Distinctions ; Intolerance
PF 7938 Distinctions male female ; Intolerance
PE 2785 Distomatosis
PE 9239 Distorted maps ; Production deliberately
PD 6081 Distorted media presentations
PE 9220 Distorting effects commodity taxes transaction goods nonfactor services
PA 6790 Distortion
PF 7955 Distortion aid developing countries ; Donor
PF 6932 Distortion aid developing countries ; Recipient
PD 4982 Distortion corporate news information ; Deliberate
PE 2032 Distortion corporation financial statements
PE 5179 Distortion democratic procedures international organizations
PA 6180 Distortion Error
PF 5127 Distortion evidence
PF 4932 Distortion historical events
PF 4932 Distortion history
Distortion [Imperfection...]
PA 6997 — Imperfection
PF 4876 — intergovernmental organizations mini-state membership
PE 8080 — international politics fuel shortages
Distortion international trade
PC 6761 — [Distortion international trade]
PE 0191 — discriminatory application antidumping regulations
PE 2603 — discriminatory customs administrative entry procedures
PE 9073 — discriminatory formulation equipment safety regulations
PE 1274 — discriminatory formulation health sanitary regulations agricultural pharmaceutical products
PE 0347 — discriminatory government private procurement policies
PD 0340 — discriminatory preference agreements
PE 0778 — discriminatory requirements respect marks product origin
PE 0083 — discriminatory requirements respect product standards measures
PD 2144 — dumping
PE 0522 — embargoes similar restrictions
PE 1961 — export subsidies countervailing duties
PE 1182 — minimum pricing regulations measures regulate domestic prices
PD 0455 — obstacles patent protection
PE 9027 — quantitative restrictions
PC 0073 — restrictive business practices
PE 8882 — restrictive controls movement labour
PE 8525 — restrictive controls over foreign investment
PE 0497 — restrictive customs valuation practices
PD 2029 — result government participation
PD 0678 — selective domestic subsidies
PE 8867 — selective indirect taxes import charges
PE 8583 — selective monetary controls
PE 8267 — state-trading government monopoly practices
Distortion [Misinterpretability...]
PA 6741 — Misinterpretability
PA 6644 — Misrepresentation
PF 8713 — musical pitch ; Competitive
PF 8713 — musical quality escalating pitch
PE 7413 Distortion national economies food subsidies
PD 1893 Distortion official news information ; Deliberate
PF 2174 Distortion profit-making motive
PD 4519 Distortion toxicity thresholds ; Commercial
PA 7411 Distortion Uncommunicativeness
PD 3436 Distortionary tax systems
PC 4867 Distracted local management
PS 3268 Distracting faculty demands
PA 6247 Distraction Inattention
PA 7157 Distraction Insanity

PF 2384 Distractions
PA 5688 Distress Appropriation
PA 5454 Distress Badness
PA 5497 Distress Difficulty
PA 6030 Distress Fear
PA 5451 Distress Insensibility
PA 6434 Distress Poverty
PE 5087 Distress ; Psychological
PE 7159 Distress syndrome ; Respiratory
PA 7107 Distress Unpleasantness
PE 6299 Distress ; Urban
PJ 9080 Distributed products ; Vulnerability abuse strategically
Distribution [access...]
PF 3082 — access ; Denial
PF 4108 — agricultural machinery ; Unequal
PD 4514 — agricultural production ; Unequal
PF 4385 — Area disparities book production
PF 2652 Distribution basic services ; Limited
PF 2880 Distribution basic skills ; Unequal global
Distribution [channels...]
PE 2397 — channels transnational corporations ; Control marketing
PD 2762 — communication resources facilities ; Disparities
PD 2926 — confidential government information ; Restrictions
PD 2608 — construction expertise ; Inequitable
PD 0794 — costs benefits economic integration ; Imbalances
PC 2815 — countries ; Dependence unequal income
PC 2815 — countries ; Unequal income
PC 4312 Distribution deaths ; Unequal regional
PD 7615 Distribution developing countries ; Unequal income
PC 5601 Distribution economic growth ; Unequal global
Distribution [fame...]
PF 3439 — fame honours ; Unequal
PE 4241 — family planning education facilities ; Unequal global
PE 3569 — fellowships ; Unequal
PF 4495 — fish catches ; Unequal
PF 6836 — forces ; Unequal
PD 8839 — Fragmented irrigation
PE 8603 Distribution goods services ; Unequal
Distribution [Imbalance...]
PE 5863 — Imbalance shipbuilding industry
PF 7479 — industrial processes ; Imbalance
PE 6891 — industrialized countries ; Unequal income
PF 6495 — Ineffective means goods supply
PF 0204 Distribution knowledge ; Imbalanced
PF 0204 Distribution knowledge ; Unbalanced
Distribution [land...]
PC 3438 — land assets ; Unequal
PB 0750 — land ownership ; Unfair
PF 4490 — livestock production ; Unequal
PD 4322 Distribution meat production ; Unequal
PF 3082 Distribution media ; Restriction access news
PF 3043 Distribution natural resources countries ; Inequality
PE 7942 Distribution old age pensions men women ; Unequal
Distribution [political...]
PF 0544 — political awareness developing countries ; Imbalance
PF 8535 — political patronage ; Abusive
PF 4336 — production countries ; Unequal
PE 9758 — public subsidies ; Inequitable
PD 4962 Distribution revenues ; Inequitable
Distribution [services...]
PF 6527 — services ; Costly
PD 2479 — skilled specialists ; Inequitable
PF 1403 — skills ; Imbalanced
PF 0852 — social security ; Unequal
PC 3437 — social services ; Unequal
PD 7374 — system ; Imperialistic
PC 0418 Distribution technology ; Indiscriminate application
PF 3211 Distribution ; Traditional land
Distribution [Unauthorized...]
PE 7188 — Unauthorized liquor manufacturing
PE 0564 — Unauthorized pharmaceutical manufacture
PF 3843 — Unbalanced industrial
Distribution Unequal
PD 4962 — income
PC 3438 — property
PC 2163 — school
PD 1434 Distribution unevenness ; Bean pea
PD 8056 Distribution water ; Uneven
Distribution wealth
PB 7666 — Dependence inequitable
PD 5258 — developing countries ; Disparity
PB 7666 — Inequitable
PB 7666 — Unequal
PD 2495 District boundaries ; Fragmenting
PF 2261 District centre ; Distant
PF 2093 District ; Deteriorating downtown
PF 8299 Districts ; Inaccessible shopping
PA 8653 Distrust
PA 8653 Distrust blocking support
PE 8963 Distrust business community
PA 7253 Distrust Envy
PF 1097 Distrust government ; Pervasive
PF 4274 Distrust interpersonal relationships
PA 8653 Distrust loan structures
PD 2263 Distrust political dialogue
PD 0974 Distrust professional service delivery
PA 7115 Distrust Rashness
PF 1681 Distrust services
PF 1205 Distrust storekeepers
PA 7392 Distrust Unbelief
PC 0025 Distrusted economic cooperation
PE 1101 Distrusted local politicians
PC 0877 Disturbance ; Adolescent
PA 5838 Disturbance Agitation
PC 8866 Disturbance digestion

PA 7361 Disturbance Disorder
PD 9159 Disturbance ; Emotional
PA 6030 Disturbance Fear
PE 4431 Disturbance ; Hormone
PA 6247 Disturbance Inattention
PA 5467 Disturbance Inexcitability
PE 5087 Disturbance ; Mental
PA 7156 Disturbance Moderation
PE 2665 Disturbance ; Oculomotor
PE 5087 Disturbance ; Psychological
PE 2197 Disturbance ; Sleep
PF 1441 Disturbance space ; Gamma ray
PA 7309 Disturbance Uncertainty
PA 7107 Disturbance Unpleasantness
PD 7420 Disturbances animals ; Metabolic
PE 0276 Disturbances ; Behaviour
PD 5372 Disturbances ; Civil
PE 1675 Disturbances correctional institutions
PE 6791 Disturbances due torture ; Gait
PD 8301 Disturbances ; Endocrine gland
PE 3861 Disturbances ; Gastroenteritic
PC 0300 Disturbances ; Transient situational
PF 1257 Disturbing transient mobility
PE 5889 Disulphide ; Carbon
PF 1735 Disunified requests services
Disunity [Dependence...]
PA 6233 — Dependence
PD 0110 — developing regions ; Rivalry
PA 5982 — Disagreement
PA 5446 Disunity Enmity
PA 6233 Disunity Fragmentation
PF 3364 Disunity ; Religious
PF 3404 Disunity ; Religious
PA 7236 Disunity Separation
PF 3558 Disunity traders
PF 1257 Disunity urban villages
PA 6379 Disuse Avoidance
PA 6852 Disuse Inappropriateness
PA 7131 Disuse Oldness
PF 7960 Disuse traditional energy alternatives
PE 5309 Disused machinery accumulation seas
PF 1985 Disvalued heritage ; Commonly
PD 2269 Ditches ; Inadequate maintenance drainage
PA 5573 Divergence
Divergence [Deviation...]
PA 6228 — Deviation
PF 9571 — different national laws relating international transactions
PA 5532 — Disaccord
PA 5982 — Disagreement
PA 5573 — Divergence
PA 5878 Divergence Nonconformity
PA 6890 Divergence Nonuniformity
PF 5000 Divergences macroeconomic policies leading industrialized countries
PF 4210 Divergent national concepts corrupt practices
PF 7011 Diverse leadership viewpoints
PF 3317 Diverse unilingualism
PJ 9259 Diversification costs ; Prohibitive agricultural
PD 2892 Diversification developing countries ; Lack horizontal commodity
PD 1554 Diversification developing countries ; Lack vertical
PC 6329 Diversification energy research ; Insufficient
PD 0335 Diversification ; Insufficient
PE 4305 Diversification loans developing countries ; Inadequate
PD 4907 Diversification manufactured goods ; Over
PF 9187 Diversification market crops ; Non
PD 4907 Diversification services ; Over
PF 2135 Diversification subsistence fishing economies ; Non
PD 0335 Diversification urban industrial energy supply ; Insufficient
PF 4828 Diversion education qualification earning
PE 7174 Diversion high technology hostile countries
PE 7174 Diversion high technology irresponsible groups
PD 0516 Diversion international rivers ; Upstream
PD 0837 Diversion nuclear materials
PF 5124 Diversities ; Maintenance super power ideological
Diversity [animal...]
PC 1408 — animal breeding stock ; Decreasing genetic
PC 1408 — animals ; Decreasing genetic
PE 6788 — available seeds ; Reduction genetic
Diversity [biological...]
PC 5773 — biological habitats ; Decreasing
PD 7302 — biological species ; Decreasing
PF 8156 — Bureaucratic insensitivity cultural
PC 2223 Diversity cultivated plants ; Decreasing genetic
PF 8156 Diversity cultural traditions ; Insensitivity
PC 5773 Diversity ; Decreasing ecological
PB 4788 Diversity ; Destruction genetic
Diversity Erosion
PB 9748 — biological
PC 5773 — ecosystem
PF 4405 — socio economic
PD 0547 Diversity fish ; Decreasing genetic
PE 1862 Diversity ignored social service agencies ; Cultural
Diversity [Landsize...]
PE 8279 — Landsize limit
PC 2223 — Limited crop
PD 2245 — limited local decision-making bodies
PE 8671 Diversity medical science ; Lack
PE 9172 Diversity psychiatry ; Lack
PF 4741 Diversity ; Reduction technological
PE 8806 Diversity social roles ; Lack
PD 5774 Diverted consumer spending
PE 8117 Diverticula intestine
PE 4630 Diverticulum ; Oesophageal

Double standards

Divided [cities...]
PD 7065 — cities
PF 7118 — *community spirit*
PD 1263 — countries
PF 1358 Divination
PF 8563 Divine anger
PF 6478 *Diving income ; Diminished*
PF 8352 Divining ; Water
PC 3458 Division ; Class
PA 5532 Division *Disaccord*
PA 5573 Division *Divergence*
Division [Established...]
PF 2822 — *Established geographic*
PE 4111 — Europe ; east west
PA 5869 — *Exclusion*
PA 6446 Division *Indiscrimination*
PC 6089 Division labour ; Authoritarian
PF 9281 Division labour ; Exploitative transformation international
PA 7236 Division *Separation*
PF 3199 Division world spheres influence
PD 4782 Divisions ; Internal language, religious ethnic
PF 1822 *Divisions ; Irrational land*
PE 3196 Divisions transnational corporations ; Restrictive market
PF 3836 Divisive effects formal schooling
PF 0152 Divisive effects official cultural pluralism
PF 2358 *Divisive family ties*
PF 6545 Divisive patterns community groupings
PE 7080 Divisive responses international nongovernmental organization actions
PF 2346 *Divisive road patterns*
PF 6545 Divisive traditional patterns
PF 6545 Divisive tribal loyalties
PF 2575 *Divisive zoning disputes*
PA 5532 Divisiveness *Disaccord*
PF 2100 Divorce
PF 3249 Divorce laws ; Conflicting national
PF 3250 Divorce ; Non validity
PF 3248 Divorce ; Religious civil refusal
PE 8879 Divorce rights ; Discrimination against women
PF 2100 *Divorcees ; Discrimination against*
PF 2100 *Divorcees ; Religious discrimination against*
PF 2100 *Divorcees ; Social discrimination against*
PF 2100 Divorces ; Repeated marriage
PE 8230 Divulge information industrial concerns ; Unwillingness
PE 5101 Dizziness
PJ 1751 DNA ; Corporate control
PF 2303 *Dock use ; Unclarity*
PE 4897 Docks ; Dangerous harbours
PF 2643 Doctor assisted suicide ; Illegality
PF 2875 *Doctor care ; Irregular*
PF 2875 *Doctor contact ; Infrequent*
PF 2875 *Doctor relationship ; Uncontinuous*
PF 0255 Doctored data
PE 7568 Doctors ; Alcohol abuse
PE 8303 Doctors developing countries ; Lack specialist
PE 7568 Doctors ; Drug abuse
PD 5770 Doctors ; Incompetent
PE 8303 Doctors ; Insufficient
PD 1725 Doctors ; Quack
PE 8303 Doctors ; Unavailability community
PF 2099 Doctors ; Witch
PF 1115 Doctrine ; Disagreement concerning religious
Doctrine Inadequacy
PF 3396 — [Doctrine ; Inadequacy]
PF 3395 — economic
PF 3394 — political
PF 2005 — religious
PF 3398 — social
PF 2005 Doctrine ; Ossification faith rigid
PF 7484 Doctrine ; Profanation sacred
PF 4156 Doctrine ; Risk unintentional nuclear war generated developments strategic
Document [classification...]
PF 6743 — classification systems ; Bias
PE 7753 — classification systems ; Incompatibility
PE 1122 — collections ; Prohibitive cost maintaining comprehensive
PD 2886 Documentalists ; Corruption
PE 1110 Documentary fraud
PF 7133 Documentation ; Fragmentation international
PF 6453 Documentation ; Inadequate
PF 5872 Documentation ; Inefficient shipping procedures
PD 2886 Documentation practices ; Unethical
PE 8088 Documentation works art ; Inadequate
PE 1669 Documents archives ; Deterioration stored
PE 7900 Documents ; Corruption
PE 7061 Documents data transnational corporations ; Missing
PE 9123 Documents ; Delay issue travel
PF 3096 Documents ; Erroneous
PF 3096 Documents ; False
PE 4496 Documents ; Forged travel
Documents [Inaccessible...]
PF 4690 — Inaccessible
PF 9699 — Inaccessible historical
PF 3096 — Inaccurate
PF 5992 — international organizations ; Proliferation
PF 3096 Documents ; Misleading
PF 5495 Documents ; Misreading sacred
PE 5051 Documents objects ; False historical
PF 9699 Documents ; Overclassification embarrassing
Documents [passports...]
PE 0325 — passports, visas ; Refusal issue travel
PD 0172 — public archives ; Destruction historic
PF 9699 — public scrutiny ; Withdrawal official
PF 4708 Documents records ; Misfiled
PF 4690 Documents ; Risk missing
PF 4699 Documents ; Tampering official
PD 0577 Documents ; Theft
Documents [Uncatalogued...]
PF 4077 — Uncatalogued
PF 6483 — Uncatalogued historical
PF 7137 — Unnecessary verbosity legal
PF 4690 — Unretrievable
PE 1323 Dodders
PD 0356 Dodging ; Draft
PF 1623 Dodging issues
PE 4931 Dog bites
PE 4685 Dog faeces pavements
PE 7786 Dog fighting
PD 9293 *Dog ; Giant kidney worm infection*
PE 0359 Dog populations ; Stray
PE 3511 *Dog tapeworm*
PE 8739 Dog wolves foxes ; Endangered species
PF 2901 Dogmas standards ; Uncritical acceptance
PF 3931 Dogmatic belief benefits technical progress
PF 6988 Dogmatism
PF 6988 Dogmatism ; Religious
PF 6988 Dogmatism ; Scientific
PE 0360 Dogs ; Cruel methods destruction stray
PD 5453 *Dogs ; Cyclic neuropenia grey Collie*
PE 0965 *Dogs ; Disorders nervous system*
PD 9667 *Dogs ; Eczema nasi*
PE 9774 *Dogs ; Exertional rhabdomyolysis*
PD 7799 *Dogs ; False pregnancy*
PE 0185 *Dogs ; Feral*
PE 3511 *Dogs ; Infected*
PE 9854 *Dogs ; Infectious tracheobronchitis*
Dogs [Leptospirosis...]
PE 2357 — *Leptospirosis*
PE 5913 — *Lyme disease*
PD 5453 — *Lymphocytic leukaemia*
PF 6514 *Dogs ; Uncontrolled*
PE 8683 Dolichotids ; Endangered species cavies
PD 2463 Dollar dominance world economy ; US
PD 2463 Dollar ; Overvalued
PD 2463 Dollar ; Weakness
PA 6731 Dolorousness *Solemnity*
PA 7107 Dolorousness *Unpleasantness*
PD 3673 *Dolphins ; Endangered species river*
PE 8302 Dolphins ; Endangered species rough toothed white
PE 0906 Dolphins ; Maltreatment performing
PE 4961 Domestic accidents ; Occupational
Domestic agricultural
PF 2889 — policies ; Detrimental international repercussions
PD 9802 — price difficulties developed countries
PE 2890 — price policy difficulties developing countries
PE 4931 Domestic animal bites
PC 2062 *Domestic animal production ; Surplus*
Domestic animals
PC 1408 — breeding purposes ; Excessive use
PD 2746 — carriers animal diseases
PE 4760 — Dietary deficiencies
PE 4694 — during disasters ; Lack protection
PC 1408 — Excessive inbreeding
PE 2763 — Inadequate housing penning
PE 2786 — *Inadequate vaccination*
PD 2746 — vectors animal disease
PD 2746 — vectors disease
PE 4760 — Vitamin E deficiency
PE 0789 Domestic bias regulation restrictive business practices
Domestic [capital...]
PD 3132 — capital outflow developing developed countries ; Private
PD 7963 — cartel
PE 8236 — *consumers developing countries ; Lack production*
PD 8546 Domestic deskilling women
Domestic [economic...]
PE 4325 — economic policies ; Interference transnational banks
PF 2115 — *electricity supply ; Insufficient*
PE 2824 — *emissions*
PD 1994 — enterprises ; Nationalization
PD 8054 *Domestic fires*
PD 5092 Domestic food production developing countries ; Deterioration
PD 0235 *Domestic hazards*
PE 2824 Domestic heating emissions air pollutants ; Industrial
PE 8036 Domestic inconvenience dwarfs, midgets giants
PD 8994 Domestic industries ; Decline competitiveness
PD 2743 Domestic livestock ; Poor quality
Domestic market
PD 0928 — developing countries ; Inadequacy
PD 0928 — developing countries ; Small size
PD 1873 — restrictions developing countries
PE 4315 Domestic monetary policies ; Interference transnational banks' off shore borrowing
PE 1796 Domestic name brand food sector developing countries ; Domination transnational corporations
Domestic [politics...]
PD 1787 — politics ; Espionage
PD 4524 — polluters
PE 1182 — prices ; Distortion international trade minimum pricing regulations measures regulate
PE 4021 Domestic quarrels
Domestic [recession...]
PD 9436 — recession
PD 0807 — refuse disposal
PD 0465 — resource mobilization developing countries ; Low level
PF 5964 — resources developing countries ; Inefficient use
Domestic [savings...]
PD 0465 — savings developing countries ; Inadequate
Domestic [servants...] cont'd
PE 4964 — servants ; Discrimination against
PC 0592 — *servants ; Shortage*
PD 4633 — service ; Subjugation women
PE 1957 — social policies transnational corporations ; Disruption
PD 0678 — subsidies ; Distortion international trade selective
PE 7433 Domestic uses ; Loss water
PD 6881 Domestic violence
PD 2800 Domestic waste water ; Discharge dangerous substances
PD 2800 Domestic waste water pollutants
Domesticated animals
PE 7175 — Aggressive
PF 7681 — Boredom captive
PD 6305 — Declining breeds
PE 4771 — Unethical practices
PC 1408 Domestication ; Genetic impoverishment animals
PD 1179 Domestication wild animals
PF 1913 Dominance economic motives
PF 4373 Dominance language ; Reinforcement male
PF 8264 Dominance over nature ; Belief humanity's
PC 3024 Dominance ; Traditional male
PE 5959 Dominance translation ; Cultural
PD 2463 Dominance world economy ; US dollar
PC 1573 Dominant class ; Creation
PD 6002 Dominant market position international trade ; Abuse
PJ 5262 Dominant stock markets weaker foreign economies ; Effect
PF 4741 Dominant technology ; Dependence
PE 6671 Dominated economy ; Social symbols
PF 2842 *Dominated jobs ; Government*
PF 1025 Dominated market research ; Sales
PE 3296 *Dominated women's behaviour*
PF 1025 Dominates market research ; Sales promotion
PA 0839 Dominating people
PA 0839 Dominating personal interests
PA 0839 Domination
Domination [advertising...]
PE 2193 — advertising transnational corporations
PE 1448 — agricultural equipment industry transnational corporations
PE 1469 — automobile industry transnational corporations
Domination [Class...]
PC 1573 — Class
PE 2084 — copper industry transnational corporations
PC 7384 — countries ; Foreign
PA 0839 — *Cultural*
Domination [daily...]
PD 3003 — daily life ; Corporate
PA 0839 — Dependence
PC 1573 — Dependence class
PD 6029 — developed countries ; Language
PE 0163 — developing countries transnational corporations
PF 0536 *Domination ; Economic*
PE 1322 Domination economic integration transnational corporations
PF 0317 Domination government policy-making short-term considerations
PD 6298 Domination humanities ; Ideological
PF 4220 Domination individuals institutions
PE 1187 Domination labour relations transnational corporations
PE 4355 Domination loan negotiations transnational banks
PC 3024 Domination ; Male
PC 3825 Domination ; Market
PF 2825 Domination organizational systems ; Institutional
Domination [Parental...]
PF 4391 — Parental
PC 7384 — peoples ; Alien
PC 8512 — Political
PF 7220 Domination research ; Ideological
PE 4330 Domination restrictive project loans transnational banks developing countries
PE 1620 Domination shipping industry transnational corporations
PE 3781 Domination transnational corporations coffee industry
PE 1796 Domination transnational corporations domestic name-brand food sector developing countries
PD 0055 Domination world territorially organized sovereign states
PF 7955 Donor distortion aid developing countries
PF 0175 Donors ; Lack coordination aid
PE 7530 Donors ; Shortage organ
PF 4528 Doomsday
PF 1083 Doomsday syndrome
PE 4250 Doping
PE 7808 *Doppellendigkëit*
PF 1309 Dormant historical celebrations
PD 3481 *Dormice ; Endangered species*
PD 3481 *Dormice ; Endangered species spiny*
PE 7826 *Dorsal spinous processes ; Overriding*
PE 9128 Dorsopatie
PD 2929 Dossiers individuals ; Maintenance political
PE 1578 Double agents ; Betrayal
PE 1578 Double agents ; Treachery
PC 0939 Double employment socialist countries
PF 6158 Double jeopardy ; Denial right freedom
Double [mindedness...]
PF 4910 — mindedness
PF 9050 — moral standards government
PE 7808 — *muscling cattle*
PF 7925 Double standard human rights ; Inter Governmental
Double standards
PF 5225 — [Double standards]
PF 3006 — Medical
PF 3377 — Moral
PF 5225 — morality
PF 5225 — morality

Double standards

Double standards cont'd
- PF 3983 — religion
- PF 3259 — sexual morality
- PD 0858 Double taxation ; International
- PD 3050 Double taxation royalties ; International
- PF 6703 Doublespeak
- PF 8474 Doubt
- PF 3918 Doubt given vested interests ; Benefit
- PF 5354 Doubt ; Self
- PA 6971 Doubtfulness *Danger*
- PA 7363 Doubtfulness *Improbity*
- PA 7392 Doubtfulness *Unbelief*
- PA 7309 Doubtfulness *Uncertainty*
- PE 6650 Down communications due difference training ; Break
- PF 1737 Down growth output developed market economies ; Slow
- PE 1323 Down ; Pull
- PF 2202 Down research methodologies ; Top
- PE 2125 Down's syndrome
- PF 3341 Down socially ; Marrying
- PD 7424 *downer cow*
- PE 5014 Downgrading jobs due computerization
- PE 8782 Downstream due impoundment ; Reduction soil fertility
- PD 4976 Downstream flooding
- PF 2093 Downtown district ; Deteriorating
- PE 0501 Downy mildews plants
- PE 4718 Dowry deaths
- PF 3290 Dowry ; Prohibitive marriage
- PD 8052 Dozing job
- PE 3510 Dracunculiasis
- PE 3510 Dracunculus infections
- PF 6051 Draft
- PD 0356 Draft dodging
- PD 0356 Draft evasion
- PF 3499 *Drain ; Business skills*
- PF 1681 *Drain ; Community expertise*
- PE 1291 Drain developing countries ; Brain
- PJ 7921 *Drain ; Inter professional brain*
- PE 8093 Drain more developed industrialized countries ; Brain
- PE 8768 Drain skills cities
- PE 8093 Drain socialist capitalist countries ; Brain
- PD 2269 Drainage culverts ; Unstandardized
- PD 2269 Drainage ditches ; Inadequate maintenance
- PD 0876 Drainage facilities ; Insanitary

Drainage [Ill...]
- PD 2269 — Ill considered land
- PD 2269 — Inadequate land
- PD 2269 — Ineffective temporary
- PD 2269 Drainage system ; Inadequate
- PE 8865 Draining resources due alcohol
- PE 7647 *Drains ; Clogged storm*
- PF 1756 Drama
- PF 1042 Dramas ; Failure motivating socio
- PE 0242 *Draught*

Draught [animal...]
- PF 0377 — animal power developing countries ; Underproductivity
- PF 0377 — animals ; Agricultural loss
- PE 3394 — animals ; Beating
- PE 2412 *Draughts*
- PA 6950 Dread *Angst*
- PE 9038 Dread disease
- PF 7029 Dread otherness
- PE 6958 Dream anxiety disorder
- PE 6958 Dreams ; Anxiety
- PF 1084 Dreams ; Collapse cultural
- PE 6958 Dreams ; Terrifying
- PE 5409 Dress ; Enforcement standards
- PE 5607 Dress ; Provocative
- PE 8740 Dressed fur skins ; Instability trade
- PE 9130 Dressed fur skins ; Shortage leather, miscellaneous leather manufactures
- PE 8259 Dried, peas, lentils leguminous vegetables ; Instability trade beans,
- PE 8257 Dried salted smoked meat ; Instability trade
- PF 7815 *Drift ; Genetic*
- PD 9082 Drift nets ; Excessive use
- PD 0493 Drift ; Sand
- PF 1345 *Drilling cost ; Prohibitive*
- PF 3554 *Drilling equipment ; Limited*
- PE 8918 Drink industries ; Instability food
- PD 1611 Drinking age laws ; Unobserved
- PD 0153 Drinking ; Dysfunctional
- PE 5722 Drinking ; Immoderate eating
- PE 1349 Drinking ; Solvent methylated spirits
- PD 1611 Drinking ; Under age

Drinking water
- PD 0235 — Contamination
- PD 1175 — *Disease causing microbes*
- PE 2871 — Fluoridation
- PE 5650 — Lead contamination
- PD 1175 — *Nuisance organisms*
- PC 4815 — Shortage
- PE 3035 Drinks ; Carcinogenic carbonated
- PE 3035 Drinks ; Carcinogenic soft
- PE 2412 Dripping nose
- PC 2576 Driven investment ; Self interest
- PE 6119 Driver accidents ; Young
- PF 4984 Driver education ; Remote
- PE 5572 *Driver fatigue ; Motor vehicle*
- PF 4984 Driver training ; Inadequate
- PE 9334 Drivers ; Callous
- PD 0094 *Drivers ; Drug abuse young*
- PF 4984 *Drivers ; Insufficient trained*
- PE 5349 *Drivers ; Reckless children snowmobile*
- PE 6119 Driving delinquency

- PE 2149 Driving ; Drunken
- PE 2149 Driving influence alcohol
- PE 2149 Driving influence drugs
- PE 0930 Driving law violations
- PE 9334 Driving practices ; Dangerous
- PE 9334 Driving ; Reckless
- PE 5152 Droit seigneur
- PD 7799 *Drop syndrome ; Egg*
- PC 1716 Dropout pattern ; Continuing
- PD 2046 Dropouts ; Lack programmes
- PC 1716 Dropouts ; School
- PF 5223 Dropping ; Name
- PF 3084 Drops ; Fluctuating income
- PE 3933 Dropsy
- PC 2430 Drought
- PC 2430 Drought ; Vulnerability
- PB 0731 *Drowning*

Drug abuse
- PD 0094 — [Drug abuse]
- PD 5987 — adolescents
- PE 4250 — athletes
- PE 0742 — community leaders
- PD 8696 — decision-makers
- PE 7568 — doctors
- PE 8696 — government officials
- PE 5579 — military personnel
- PD 0213 — Multi
- PD 8696 — policy-makers
- PE 0978 — prisoners
- PE 0978 — prisons
- PD 0556 — Psychedelic
- PE 9688 — Rural
- PE 4250 — *sports*
- PE 5507 — students
- PD 4514 — work
- PD 0094 — *young drivers*
- PE 5987 — Youth
- PD 5987 Drug addicted children

Drug addiction
- PD 3825 — [Drug addiction]
- PD 5344 — children ; Parental toleration
- PD 5750 — Delinquency side effect drug trade
- PD 3825 — *drug producing countries ; Increasing*
- PE 4609 Drug addicts ; Children
- PD 2392 *Drug classification ; Inadequate*
- PC 0231 Drug control ; Inadequate
- PD 3825 Drug dependence
- PE 6201 Drug experimentation developing countries ; Increasing
- PE 9210 Drug-financed subversion
- PD 6334 Drug induced diseases
- PD 3825 Drug producing countries ; *Increasing drug addiction*
- PD 2392 Drug quality control ; Inadequate
- PF 9659 Drug resistance

Drug resistant
- PB 6384 — *diseases ; Encouragement*
- PE 0616 — *malaria*
- PF 9659 — viruses
- PE 1880 Drug smuggling
- PB 0284 *Drug testing ; Enforced*
- PE 4609 Drug threat young

Drug trade
- PD 0991 — [Drug trade]
- PD 5750 — drug addiction ; Delinquency side effect
- PE 5296 — Economic dependence some developing countries
- PD 0991 — Illegal
- PE 1880 — International
- PF 7730 Drug trafficking ; Government complicity
- PD 0991 Drug trafficking ; Illicit

Drug use
- PD 0094 — *animal sports*
- PF 4735 — Criminalization
- PE 4609 — Foetal damage
- PF 4735 — Illegality
- PD 0094 — Illicit
- PD 1714 — meat production ; Health hazards
- PE 0680 — pilots

Drug users
- PE 5798 — AIDS ; Vulnerability intravenous
- PA 5568 — *Ignorance*
- PD 0094 — Illicit

Drugs Abuse
- PE 0043 — animal
- PD 0556 — hallucinogenic
- PD 0288 — medical
- PE 1821 — minor
- PD 0022 — plant
- PE 0139 — prescription
- PE 0456 Drugs ; Adulteration illicit
- PE 3980 Drugs, alcohol cigarettes media ; Excessive portrayal
- PD 5540 Drugs ; Counterfeit
- PD 4563 Drugs ; Cultivation illegal

Drugs [Designer...]
- PD 0094 — *Designer*
- PE 6036 — developing countries ; Marketing banned pharmaceutical
- PF 6787 — Diffuse legislation against
- PE 2149 — Driving influence

Drugs Excessive
- PE 5755 — cost medical
- PE 5755 — pricing brand name
- PD 0644 — proliferation medical

Drugs [Fake...]
- PD 5540 — Fake
- PE 4903 — foods ; Residues veterinary
- PE 4888 — Forced confessions
- PD 0094 — *Drugs ; Generic*

- PD 4563 Drugs ; Government sanctioned cultivation illegal

Drugs [Illicit...]
- PE 4946 — Illicit trade prescribed
- PF 0603 — Inadequate information
- PD 1190 — Inadequate testing
- PE 1427 — Inhaling solvents anaesthetic
- PD 8402 — *Irritating*
- PE 1278 Drugs ; Limited access pharmaceutical

Drugs [Manufacture...]
- PE 2512 — Manufacture illicit
- PD 2697 — medical devices ; Unethical experiments
- PD 0644 — *medicine ; Reverence*
- PD 0094 — Methamphetamine
- PE 4896 — Mutagenic effects
- PE 3828 Drugs non medical purposes ; Administering medical
- PE 9087 Drugs ; Overprescription

Drugs [Phoney...]
- PD 5540 — Phoney non prescription
- PE 5556 — Possession
- PE 3799 — Prescription inappropriate
- PF 4735 — Prohibition addictive
- PE 1634 Drugs ; Shortage medical
- PE 9210 Drugs ; Terrorism financed trade
- PE 0768 Drunk syndrome ; Punch
- PE 2149 Drunken driving
- PE 2429 Drunken pedestrians
- PE 8311 Drunkenness
- PE 7700 Drunkenness ; Excessive parental
- PE 8495 Drunkenness military personnel troops
- PE 2429 Drunkenness ; Public
- PE 5839 Dry bulk shipping industry ; Imbalances
- PE 1606 Dry rot wood
- PE 4966 Dry skin
- PE 3865 Drying plants ; Premature
- PC 2506 Drylands ; Degradation
- PD 7096 Drylands ; Endangered
- PE 4476 Dual exchange rate systems
- PE 6677 Dual nationality
- PD 6273 Dual roles women
- PF 2256 Dual school systems
- PF 9161 Dualism
- PA 7339 Duality
- PD 1576 Dubious grounds ; Arrest
- PF 5679 Dubious medical tests
- PA 6971 Dubiousness *Danger*

Dubiousness [Impiety...]
- PA 6058 — *Impiety*
- PA 7363 — *Improbity*
- PA 7325 — *Irresolution*

Dubiousness [Unbelief...]
- PA 7392 — *Unbelief*
- PA 7309 — *Uncertainty*
- PA 7411 — *Uncommunicativeness*
- PD 2732 Duck disease ; New
- PE 9374 Duck disease ; Western
- PD 3978 Duck plague
- PD 2730 Ducks ; Viral hepatitis
- PE 9829 Duct obstruction ; Bile
- PD 8786 Ducts ; Inflammation lacrimal glands
- PE 5569 Ductus arteriosus animals ; Persistent
- PF 5382 Duelling
- PF 8650 Dues ; Non payment United Nations membership
- PA 7365 Dullness *Boredom*

Dullness [Imperfection...]
- PA 6997 — *Imperfection*
- PA 5806 — *Inaction*
- PA 5451 — *Insensibility*
- PE 4960 Dullness ; Mental

Dullness [Unfeelingness...]
- PA 7364 — *Unfeelingness*
- PA 6738 — *Unimaginativeness*
- PA 7371 — *Unintelligence*
- PA 7107 — *Unpleasantness*
- PA 5558 Dullness *Weakness*
- PE 7841 *Dummies*
- PE 0607 Dumping agricultural products developing countries
- PD 1245 *Dumping coal ash ; Marine*
- PD 8942 Dumping consumer waste products

Dumping [dangerous...]
- PE 2061 — dangerous illegal products developing countries
- PD 3998 — *defective products industrialized countries*
- PD 2144 — Distortion international trade
- PE 0607 Dumping food developing countries
- PD 8844 Dumping food products waste
- PD 1398 Dumping ; Hazardous waste
- PF 3454 Dumping ; *Indiscriminate refuse*
- PD 3998 Dumping obsolete goods ex-socialist countries
- PD 3666 Dumping sewage ; Marine
- PD 1398 Dumping ; Toxic waste
- PE 2061 Dumping toxic wastes developing countries
- PD 3666 Dumping wastes ; Marine
- PF 3454 *Dumps ; Uncovered refuse*
- PD 0493 Dune encroachment ; Sand
- PD 0493 Dunes ; Migrating sand
- PE 2308 Duodenal ulcers
- PE 8624 *Duodenitis*
- PE 8624 Duodenum ; Diseases oesophagus, stomach
- PE 8624 Duodenum ; Inflammation
- PC 0521 Duopoly
- PE 2417 Duplication intergovernmental organizations coordination bodies ; Proliferation
- PE 0458 Duplication international information systems ; Proliferation

Duplication international nongovernmental
- PE 7104 — activities
- PE 0179 — organization coordination bodies ; Proliferation

Economic disputes

Duplication international nongovernmental cont'd
PE 0362 — organization information systems ; Proliferation
PE 1029 Duplication international organizations coordinating bodies ; Proliferation
PE 1579 Duplication organizational units coordinating bodies United Nations system ; Proliferation
PF 2439 *Duplication products ; Excessive*
PE 0075 Duplication United Nations information systems ; Proliferation
PF 4153 Duplicity ; Government
PA 7363 Duplicity *Improbity*
PA 7411 Duplicity *Uncommunicativeness*
PE 2471 Durability bridges ; Non
PB 0750 *Duration athletic careers ; Short*
PE 8255 Duration breastfeeding ; Unnaturally short
PF 2304 *Duration work ; Pace*
PC 3796 Duress
PA 5740 Duress *Compulsion*
PA 7296 Duress *Restraint*
PD 1245 Dust
PD 0094 *Dust ; Angel*
PE 1127 Dust ; Asbestos
Dust [Cement...]
PE 2854 — Cement
PE 4679 — *Coal*
PE 4486 — Copper
PE 4605 Dust explosions
PE 5439 Dust fumes ; Metal
PF 9002 Dust ; Interstellar
PE 4048 Dust ; Iron
PE 5650 Dust ; Lead
PE 2038 Dust ; Limestone
PE 4807 Dust ; Mine
PE 5767 Dust occupational hazard
PC 0314 Dust ; Radioactive
PD 3655 Dust storms
PD 8557 *Dust ; Street*
PE 5109 Dust ; Volcanic
PE 5439 *Dust ; Zinc*
PF 2421 *Dusting ; Crop*
PD 0637 *Dusts ; Inhalation*
PE 4679 Dusts ; Mineral
PE 0524 *Dusts ; Organic*
PE 1154 Dutch elm disease
PD 4882 Duties assessments ; Circumvention
PF 0513 Duties ; Conflict
PE 1961 Duties ; Distortion international trade export subsidies countervailing
PD 2620 Duties ; Evasion customs excise
PD 1987 Duties ; Excessive protection industries customs
PD 4882 Duties taxes ; Underpayment
PA 8658 Duty ; Dereliction
PE 3511 *Dwarf tapeworm*
PE 2715 Dwarfism
PE 6909 Dwarfism ; Deprivation
Dwarfs midgets
PE 2635 — Discrimination against
PE 8036 — giants ; Domestic inconvenience
PE 8496 — giants ; Lack adequate clothing supply
PF 6140 Dwellers ; Inaccessibility countryside city
PE 4386 Dwellings developing countries ; Makeshift
PD 1251 *Dwellings ; Poor construction*
PE 8571 Dyeing industries ; Environmental hazards tanning
PE 8501 Dyeing materials ; Shortage tanning
PE 9089 Dyeing trade ; Instability tanning
PD 4268 *Dying ; Abandonment*
PF 1421 Dying bad death
PF 2442 Dying conditions ; Inhumane
PB 8613 Dying cultures
PB 8613 Dying languages
PF 4936 Dying process ; Excessive prolongation
PD 7704 Dynastic politics
PD 0659 Dysaconsis
PJ 0303 Dysarthia
PD 7841 *Dysautonomia ; Feline*
PC 0143 Dyscalculia
PE 8589 Dyschondroplasia
PE 3822 *Dyschondroplasia*
PE 7808 *Dyschondroplasia animals*
PE 2259 *Dysenteric arthritis*
PE 2259 Dysentery
PE 2259 Dysentery ; Amoebic
PE 2259 Dysentery ; Bacillary
PD 3978 *Dysentery ; Bovine winter*
PE 3936 *Dysentery ; Swine*
PD 8301 Dysfunction adrenal glands
PE 3861 Dysfunction gastrointestinal tract
PE 6408 Dysfunction ; Orgasmic
PD 8301 *Dysfunction Ovarian*
PD 8301 *Dysfunction parathyroid gland*
PE 8016 Dysfunction ; Sexual
PE 1914 Dysfunction syndrome ; Chronic fatigue immune
PD 8301 *Dysfunction ; Testicular*
PE 3932 Dysfunction torture victims ; Sexual
PD 0153 Dysfunctional drinking
PE 7647 Dysfunctional public utilities
PE 0280 Dysgraphia
PD 3866 Dyslexia
PD 0270 *Dysmorphic disorder ; Body*
PD 0270 Dysmorphophobia
PE 8016 Dyspareunia
PJ 5171 Dyspasia ; Fibrous
PE 4724 Dyspepsia
PE 4724 *Dyspepsia ; Fermentative*
PE 4724 *Dyspepsia ; Nervous*
PE 9604 *Dysphagia*

PE 2265 Dysphasia
PJ 0303 Dysphonia
PE 5913 *Dysplasia ; Canine hip*
PD 7424 *Dysplasia ; Elbow*
PE 7826 Dysplasia growth plate
PE 7826 *Dysplasia ; Physeal*
PE 7864 Dyspnoea
PE 1904 Dysrhythmia ; Circadian
PD 0270 Dysthymia
PD 7799 *Dystocia animals*
PF 2352 Dystopian visions
PE 7826 *Dystrophia ungulae*
PE 3506 Dystrophy
PE 9774 *Dystrophy ; Enzootic muscular*

E

PD 7841 *E coli enterotoxaemia*
PE 4760 E deficiency domestic animals ; Vitamin
PD 5535 *Ear animals ; Haematoma*
PD 5535 *Ear animals ; Tumours external*
Ear [disease...]
PD 2567 — *disease ; Infective middle*
PD 2567 — Diseases
PD 5535 — diseases animals
PD 2567 Ear ; Fluid middle
PD 2567 Ear ; Infection
PE 5444 *Ear mange*
PD 5535 *Ear necrosis*
PE 4109 Ear ; Noise damage
PD 5535 *Ear syndrome swine ; Necrotic*
PD 2567 Ear tumour
PD 2567 Earache
PD 2567 Eardrum injury
PE 3604 *Eared bats ; Endangered species funnel*
PC 2163 *Early apprenticeship ; Restrictions*
PD 0614 Early childbearing
PD 0331 *Early childhood ; Protein energy malnutrition infants*
PE 6069 Early dismissal ; Compulsive
Early [marriage...]
PE 7628 — marriage
PE 7628 — marriage ; Pressured
PE 7628 — marriages ; Restrictive
PE 3445 Early return migration
PE 2300 *Early syphilis*
PD 0614 Early teenage motherhood
PF 4828 *Earning ; Diversion education qualification*
PF 6530 *Earning ; Prevalence subsistence*
Earnings developing countries
PD 2968 — Commodity related shortfalls export
PD 2769 — Decline
PD 2968 — Decline growth commodity export
PD 2968 — Earnings ; Instability commodity prices developing country
PE 4915 Earnings instability ; Export
PE 4915 Earnings shortfalls ; Export
PD 0068 Earth orbits ; Accumulation aerospace hardware
PF 1588 Earth's magnetic field ; Reversal
PF 5069 Earth's surface ; Reduction albedo
PD 0708 *Earth sciences ; Unethical practice*
PE 5096 Earth surface faulting
PE 9019 Earthenware industries ; Environmental hazards china, pottery
PE 8928 Earthenware industries instability ; China, pottery
PE 8427 Earthenware manufacture underdevelopment ; China, pottery
PF 4928 Earthquake forecasting ; Inadequate
PE 6257 Earthquake resistant construction ; Inadequate
PD 0708 Earthquake risks ; Underreporting
PD 0201 Earthquakes
PF 4928 Earthquakes ; Unpredictability
PE 3618 Earwigs
PF 3494 Ease manufacture nuclear bombs
PE 7946 East Coast fever
PE 4111 East-west division Europe
PD 7841 *Eastern encephalitis pheasants*
PE 4111 *Eastern western Europe ; Split*
PF 1249 *Easy access liquor*
PC 7644 *Eating animal flesh*
PE 5187 Eating disorders
PE 5722 *Eating drinking ; Immoderate*
PD 0800 *Eating habits ; Unnutritious*
PC 0382 *Eating habits ; Unnutritious*
PC 7644 *Eating ; Meat*
PE 5187 Eating non-nutritive substance
PE 5722 *Eating ; Over*
PD 2868 *Eating poultry ; Taboos against*
PF 9063 Eccentricity
PA 5490 Eccentricity *Changeableness*
PA 5878 Eccentricity *Nonconformity*
PA 7371 Eccentricity *Unintelligence*
PE 3364 *Ecclesial communion ; Lack*
PE 2354 Echinococcis
PE 7518 Echinococcosis
PD 3158 *Echinodermata ; Endangered species*
PD 3158 *Echinoidea ; Endangered species*
PD 7420 Eclampsia
PF 7521 *Eclipse reason*
PF 7066 *Eclipses occultations ; Fear*
PE 3447 Eco-colonialism
PD 2718 *Ecoaccidents ; Risk*
PD 2718 *Ecocatastrophes ; Risk*
PC 2696 *Ecocide*
Ecological [destruction...]
PE 5196 — destruction

Ecological [disasters...] cont'd
PF 9769 — disasters ; Unpredictable
PC 3152 — *disruption animal breeding grounds*
PC 5773 — diversity ; Decreasing
Ecological [imbalance...]
PC 0842 — *imbalance*
PC 0842 — imbalance ; Animal
PD 4007 — impact overpopulation ; Negative
PC 5333 — imperialism
PD 2888 — issues ; Lack faith solutions
PE 9295 Ecological problems ; Government neglect future
PD 5582 Ecological sabotage
PC 1282 Ecological stress
PC 0328 Ecologically unfriendly products
PC 0111 Ecologically unsustainable development
PE 2305 Ecology coastal communities ; Negative consequences shifting
Economic [activities...]
PC 0880 — activities ; Underdevelopment industrial
PF 9728 — activity ; Failure recognize environment resource base
PC 5601 — activity ; International imbalance
PE 8002 — adaptation indigenous society non socialist countries ; Lack
PE 8002 — adaptation indigenous systems ; Lack
PF 1040 — administration ; Ideological overemphasis
PF 9664 — agents accelerating pace structural change ; Faltering adaptation
PC 0840 — aggression
PC 3413 — apathy
PC 3413 — apathy ; Dependence
PE 6212 — asylum ; Denial right
Economic [backlash...]
PD 5995 — backlash richer nations
PE 0803 — barriers access legal profession, judiciary jury membership
PE 1198 — barriers women's access judicial process
Economic base
PF 1705 — Archaic
PF 2042 — *Dissipating*
PF 0905 — *Dissipating*
PF 0905 — Undirected expansion
PF 2042 — *Unstable*
Economic [benefits...]
PJ 0451 — benefits be proven prior programme approval ; Government requirement that
PF 8181 — benefits projects ; Overestimated
PF 2997 — bias development
PD 3245 — bias worker benefits
PF 4260 — blockade
PF 4260 — blockade against governments
PF 1447 — burden militarization ; Socio
PE 8915 — burden warfare
Economic [civil...]
PD 3765 — civil war
PF 7428 — class ; Low socio
PF 0581 — *communication ; Limited*
PD 8897 — competition
PC 3167 — competition communist systems
PF 1345 — *conditions ; Declining*
PC 0840 — conflict
PE 7784 — constraints ; Extra
PF 4837 — contexts development programmes ; Lack understanding social
Economic cooperation
PC 0933 — developing countries ; Inadequate
PC 0025 — Distrusted
PC 0025 — Lack
Economic [coordination...]
PC 0025 — coordination ; Lack international
PC 5624 — crime
PC 5876 — crisis ; Global
PF 0817 — crisis ; Nationalistic policy responses world
PF 1277 — cycle boom depression
PF 1172 — cycles ; Vulnerability
PF 9728 Economic decision making ; Incompatibility environmental
PE 5252 Economic decline caused war
Economic dependence
PF 0841 — [Economic dependence]
PD 2968 — developing countries export primary commodities
PE 5296 — some developing countries drug trade
Economic [dependency...]
PF 0841 — dependency
PF 8855 — dependency ; Over acceptance socio
PF 1277 — depressions
PE 9420 — destabilization countries external forces ; Aggressive
Economic development
PF 9576 — Competing models socio
PE 4905 — due environmental policies ; Restrictions industrial
PC 0111 — Environmental hazards due
PF 9576 — Inadequate models socio
PD 8190 — Low
PD 0055 — National sovereignty obstacle peaceful socio
PF 2997 — Over emphasis
PF 3499 — Unattractive locale
PC 0495 — Unsustainable
Economic [dictatorship...]
PC 3240 — dictatorship
PD 7536 — difficulties farmers
PF 4130 — disadvantages excessive food production developing countries
PC 2157 — discrimination
PE 1399 — discrimination administration justice
PD 4063 — dislocations developing countries
PE 1911 — disputes states nationals states ; Financial

-911-

Economic diversity

Economic [diversity...] cont'd
- PF 4405 — diversity ; Erosion socio
- PF 3395 — doctrine ; Inadequacy
- PF 0536 — *domination*

Economic [education...]
- PF 4852 — *education ; Barriers*
- PF 1345 — *enterprises ; Uncoordinated*
- PC 8073 — environment ; Deterioration international
- PF 8111 — expansion
- PE 2700 — expansion ; Restrictions private

Economic exploitation
- PC 8132 — [Economic exploitation]
- PD 9578 — animal species ; Excessive
- PE 2427 — developing countries industrialized countries
- PD 7266 — Trafficking children
- PD 0164 — youth

Economic [financial...]
- PC 8073 — financial instability world economy
- PC 2312 — forces ; Unquestioned control
- PF 8602 — foresight ; Narrow
- PC 0025 — fragmentation ; International

Economic [gap...]
- PD 8834 — gap
- PJ 4512 — goals ; Conflicting
- PJ 4512 — goals ; Contradictory solutions national

Economic growth
- PF 1737 — capitalist countries ; Weak
- PC 0002 — Declining
- PC 0495 — Dependence uncontrolled
- PD 5326 — developing countries ; Declining
- PD 7257 — developing countries ; Disparities
- PC 0495 — Development indiscriminate
- PE 8792 — Import substitution barrier subsequent
- PC 5879 — Increasing development lag against socio
- PC 0495 — Indiscriminate
- PF 1737 — industrialized countries ; Declining
- PB 9015 — Unconstrained socio
- PC 0495 — Uncontrolled
- PC 5601 — Unequal global distribution
- PC 0002 — Unstable
- PC 5601 — Unsynchronized
- PB 0318 — Economic hegemony ; Dependence military
- PB 0318 — Economic hegemony ; Military

Economic [ignorance...]
- PC 0210 — *ignorance*
- PC 0210 — *illiteracy*
- PC 8694 — imbalance developed developing countries
- PD 5865 — imbalances industrialized countries
- PF 3342 — impediments marriage

Economic imperialism
- PC 3198 — [Economic imperialism]
- PC 3166 — Capitalist
- PC 3165 — Communist
- PC 3198 — Dependence
- PE 0766 — transnational enterprises

Economic [incentives...]
- PF 8554 — incentives ; Inadequate
- PF 8554 — incentives ; Insufficient
- PF 0019 — indicators human development ; Excessive reliance
- PC 1217 — industrial production activities ; Instability
- PC 0328 — industrial products ; Environmental hazards
- PE 2999 — inefficiencies developing countries due restrictive business practices
- PF 7556 — inefficiency
- PC 8541 — inequality
- PC 0840 — infiltration
- PC 0254 — inflation
- PC 0840 — *influence*
- PC 1059 — infrastructure ; Weakness socio
- PC 9112 — injustice ; International

Economic insecurity
- PC 2020 — [Economic insecurity]
- PC 2020 — Dependence
- PA 0857 — Socio

Economic integration
- PD 0794 — Imbalances distribution costs benefits
- PD 9412 — regional groupings developing countries ; Inadequate
- PF 4884 — socialist countries ; Inadequate
- PE 1322 — transnational corporations ; Domination

Economic [interdependence...]
- PC 9112 — interdependence ; Asymmetry
- PF 6105 — interdependence ; International
- PF 2969 — interdependencies ; Unrecognized socio
- PF 1070 — interest groups policy agencies ; Over reliance
- PF 2874 — *interests ; Competitive*
- PC 3011 — intimidation
- PC 0840 — invasion

Economic isolation
- PC 2791 — [Economic isolation]
- PC 2791 — Destructive
- PC 2791 — *Village*
- PC 2791 — Economic isolationism
- PD 4150 — Economic liberty ; Lack
- PF 2842 — Economic life ; Unfulfilled aspirations

Economic loss
- PE 9013 — [Economic loss]
- PE 7941 — animal mortality
- PE 8109 — destruction injured animals
- PE 8098 — reduced productivity diseased animals
- PE 8098 — reduced productivity diseased animals
- PE 8109 — slaughter diseased animals

Economic [management...]
- PF 0760 — management ; Makeshift
- PC 6875 — manipulation
- PC 6875 — manipulation ; Dependence

Economic [micro...] cont'd
- PE 9049 — micro organizations ; Failure
- PF 5964 — mismanagement developing countries
- PF 1913 — motives ; Dominance
- PC 5842 — Economic nationalism
- PD 0068 — Economic non-viability small developing countries

Economic [order...]
- PF 0866 — order ; Antiquated world socio
- PF 4891 — order ; Excessive socio
- PF 4306 — Order, Lack progress establishing New International
- PC 9112 — order ; Unjust international
- PF 1382 — *orientation ; Overriding*
- PF 6530 — orientation ; Subsistence

Economic [patterns...]
- PD 4125 — patterns ; Difficulties establishing changes rural
- PF 5817 — patterns ; Uncertain
- PD 7257 — performance developing countries ; Differential
- PF 2334 — philosophy controlled risk
- PF 3610 — planning agencies ; Absence long term
- PF 3500 — planning ; Limited approaches
- PE 4325 — policies ; Interference transnational banks domestic
- PF 5964 — policy making developing countries ; Inadequate
- PB 0388 — poverty ; Socio

Economic power
- PC 6873 — Abuse
- PC 6873 — Dependence abuse
- PE 0207 — transnational enterprises against labour ; Coercive use

Economic [powerlessness...]
- PJ 5262 — powerlessness
- PC 0492 — *practices ; Individualistic*
- PA 2173 — prejudice
- PF 2303 — *preoccupations ; Debilitating*

Economic priorities
- PF 0317 — Immediate
- PE 4059 — research ; Emphasis
- PF 0317 — rule

Economic [priority...]
- PF 2997 — priority system ; Inadequate
- PC 0328 — production ; Hidden environmental costs
- PC 8908 — productivity ; Declining
- PF 1825 — productivity ; Diminishing role older people due overemphasis
- PF 1545 — progress society ; Overemphasis

Economic [recession...]
- PF 1172 — recession ; International
- PD 9436 — recession ; National
- PF 6071 — recovery ; Uneven
- PF 3689 — reform socialist countries ; Obstacles
- PD 4379 — refugees
- PD 7438 — regulations ; Violations against

Economic relations
- PC 7459 — developed countries ; Failure restructure
- PC 0025 — Non cooperative
- PC 8694 — North South ; Failure restructure

Economic [relationships...]
- PC 7459 — relationships industrialized countries ; Imbalance
- PF 0589 — repercussions disarmament ; Negative
- PE 8471 — repression
- PD 9389 — reprisals ; Fear
- PD 9389 — retaliation
- PC 3233 — revolution
- PD 4150 — rights ; Denial
- PF 1762 — risks subcontractors
- PD 8897 — rivalry

Economic [sanctions...]
- PF 4260 — sanctions against governments
- PF 1976 — sanctions ; Disregard internationally imposed
- PF 3534 — schism
- PD 1534 — sectors developing countries ; Disparate development
- PE 0042 — sectors transnational enterprises ; Control national
- PD 0808 — security ; Denial right
- PE 5406 — security during periods unemployment ; Denial right
- PE 6780 — self-sufficiency handicap
- PD 0841 — self sufficiency ; Lack
- PF 6210 — shocks external factors ; Adverse
- PE 5021 — situation corporate capital companies ; Fraud concerning
- PF 2699 — situation ; Inadequate solution oriented images global

Economic social
- PE 4225 — effects advertising ; Negative
- PE 4856 — losses due disability
- PF 4837 — planning developing countries ; Imbalance
- PE 6375 — rights refugees ; Denial
- PB 0539 — underdevelopment
- PB 0539 — underdevelopment ; Dependence
- PD 7727 Economic spiral ; Self reinforcing contractionary

Economic stagnation
- PF 1277 — [Economic stagnation]
- PC 0002 — [Economic stagnation]
- PD 5326 — developing countries
- PD 4125 — due rural poverty developing countries

Economic [story...]
- PF 2093 — story ; Defeating
- PC 6759 — stress ; Socio
- PF 9664 — structural adjustment ; Retardation
- PE 3296 — *structure ; Uncorporate*
- PE 4818 — structures industrial nations ; Ineffective
- PF 2699 — structures ; Unimaginative vision existing international
- PF 1199 — *subordination*

Economic system
- PC 9112 — Inadequate world
- PF 4516 — incentives ; Asymmetry

Economic system cont'd
- PC 3458 — *Two tier*
- PC 9112 — Unjust global

Economic systems
- PC 2952 — Differences trading principles practices different
- PF 1245 — globalization ; Vulnerability socio
- PB 8031 — *Undemocratic*
- PC 2724 — Weakness trade different

Economic [technical...]
- PF 0625 — technical data ; Inadequate
- PE 8190 — technical development ; Lack
- PF 9728 — theory ; Inappropriate

Economic [uncertainty...]
- PF 5817 — uncertainty
- PC 0281 — underdevelopment
- PD 4012 — unrest

Economic [value...]
- PF 8763 — value labour ; Increasing
- PF 5817 — variables ; Unpredictability key
- PE 3296 — *ventures ; Unsuccessful*
- PF 2135 — *viability ; Questionable*
- PF 0979 — victim image
- PF 2093 — victim image
- PC 0840 Economic warfare
- PE 1117 Economical growth socialist countries ; Slow
- PF 1739 Economical systems ; Interaction deficiencies world
- PC 2510 Economically controlled political power
- PD 2624 Economically disadvantaged students
- PF 4470 Economically inactive rural population ; Socio
- PF 9728 Economics ; Environmentally insensitive
- PF 1382 Economics ; Precarious basis family

Economies [Decline...]
- PD 9679 — Decline exports developing countries developed market
- PD 4252 — Destabilization indigenous
- PD 4252 — Devalorization traditional

Economies developing countries
- PE 9420 — Destabilization
- PD 1902 — Excessive government participation
- PD 2892 — Undiversified
- PC 6641 — Economies ; Development parallel
- PD 4252 — Economies ; Discrimination against traditional
- PJ 5262 — Economies ; Effect dominant stock markets weaker foreign

Economies [Failure...]
- PC 3894 — Failure centrally planned
- PE 7413 — food subsidies ; Dependence
- PE 7413 — food subsidies ; Distortion national

Economies [import...]
- PD 7486 — import penetration ; Vulnerability
- PD 4252 — indigenous populations ; Denial right traditional
- PF 4112 — Instability tourist dependent

Economies [Lack...]
- PD 0515 — Lack consumer choice centrally planned
- PE 1922 — limited number corporations ; Disproportionate influence national
- PF 2344 — limited number individuals ; Undisclosed control national
- PE 8184 — Low complementarity developing country
- PE 7255 — Low regional complementarity developing countries
- PD 4252 — Economies ; Marginalization traditional
- PF 2135 — Economies ; Non diversification subsistence fishing
- PD 7565 — Economies ; Preponderance non food crops tropical
- PD 8851 — Economies production facilities ; War time disruption

Economies [scale...]
- PF 9791 — scale ; False
- PC 1737 — Slow down growth output developed market
- PC 0939 — socialist countries ; Black market
- PC 0927 — Spendthrift
- PC 7459 — *Structural imbalances three largest market*
- PD 2970 — Structural rigidity developing country
- PF 9697 Economies vagaries external markets goods services ; Vulnerability national
- PC 7873 Economies vagaries international financial system ; Vulnerability national

Economies [War...]
- PF 1447 — War oriented
- PE 2954 — Weakness trade socialist developed market
- PE 2953 — Weakness trade socialist developing
- PE 2348 — *Economo's disease ; Von*
- PC 6641 — Economy ; Black market
- PC 2791 — Economy ; Contained village
- PD 3003 — Economy corporations ; Monopoly

Economy [Declining...]
- PC 0002 — Declining growth world
- PD 7727 — Deflation world
- PC 2237 — Destruction rural subsistence
- PE 8915 — due war ; Destruction
- PC 8073 — Economy ; Economic financial instability world
- PC 3778 — Economy ; Elitist control global
- PF 9664 — Economy ; Faltering structural adjustment world
- PD 4800 — Economy ; Government interference national
- PD 5046 — Economy ; Imbalance world food
- PE 8735 — Economy ; Imbalances growth labour force, urban population overall growth
- PE 8437 — Economy labour force ; Urban population overall growth
- PE 0135 — Economy limited number corporations ; Disproportionate control global
- PC 0254 — Economy ; Overheating
- PC 6451 — Economy ; Restricted flow local

Economy [Shadow...]
- PC 6641 — Shadow
- PF 6540 — *Single product*
- PC 6671 — Social symbols dominated
- PC 0939 — socialist countries ; Second, shadow

Educational snobbery

Economy [Static...] cont'd
PE 4479 — Static grassroots involvement planning
PD 4125 — Subsistence poverty
PF 6530 — Subsistence poverty
PF 3638 — Superficial symbols
PE 8858 Economy ; Transport hindrance
Economy [Uncontrolled...]
PF 7880 — Uncontrolled market
PE 8190 — Underdeveloped
PC 6641 — Underground
PD 2463 — US dollar dominance world
PC 5773 Ecosystem diversity ; Erosion
PD 1165 Ecosystem fragility ; Tundra
PC 4824 Ecosystem management ; Unintegrated biosphere
PD 0767 Ecosystem modifications due creation dams lakes
PC 5773 Ecosystem niches ; Vulnerability
PF 5712 Ecosystem ; Over specialized study global
PF 2528 Ecosystemic processes ; Fallacies
Ecosystems [Degradation...]
PD 6960 — Degradation fragile
PD 5845 — degraded exploitative development ; Regional
PD 2285 — Destruction oasis
PD 7096 — Disruption arid zone
PC 1647 — *Disruption seabed*
PC 1617 Ecosystems exotic organisms ; Disruption
PB 5250 Ecosystems human activity ; Disruption
PD 7096 Ecosystems irrigation ; Destruction arid
PD 6960 Ecosystems marginal agricultural lands ; Disruption
PE 6256 Ecosystems ski resorts ; Destruction mountain
PC 1647 Ecosystems ; Vulnerability marine
PD 5582 Ecotage
PD 0094 *Ecstasy*
PE 9418 Ecthyma
PD 9667 *Ecthyma animals ; Contagious*
PE 6831 Ecthyma ; Contagious
PE 8634 Ectogenesis
PD 2289 *Ectopic pregnancy*
PF 9274 Ecumenical dialogue ; Erosion religious belief
PE 2465 Eczema
PD 9667 Eczema
PE 9458 *Eczema ; Facial*
PD 9667 *Eczema nasi dogs*
PD 3603 Edentata
PD 3603 Edentates ; Endangered species
PE 2587 Edible nut trade ; Instability fresh fruit
PE 9071 Educated elderly people ; Failure employ skills
PE 9764 Educated manpower least developed countries ; Lack
PE 9071 Educated older people ; Unemployment
PD 8723 Educated school leavers ; Mal
PD 8550 Educated unemployed
PD 8546 Educated women ; Failure employ skills home bound
PE 1379 Educated youth ; Unemployed
PF 2012 Educating older people ; Lack facilities
PE 3909 Educating people lie
Education [adults...]
PF 6531 — adults ; Insufficient
PE 7139 — against growth knowledge ; Increasing lag
PE 5369 — against population growth ; Increasing lag
PC 1131 — *Available jobs unrelated*
Education [Barriers...]
PF 4852 — *Barriers economic*
PF 7927 — Bilingual
PE 8983 — blind
Education children
PE 5020 — immigrants ; Lack appropriate
PE 0341 — Negative effects family planning
PF 5967 — Sexist
Education [Circumscribed...]
PF 0422 — Circumscribed vocational
PE 0779 — Class discrimination
PD 9370 — Commercial exploitation
PD 9370 — Commercially sponsored
PD 4178 — Competitiveness
PC 6848 — *Compulsory*
PE 8216 — concerning nature problems ; Inadequate
PC 1896 — countries ; Inequality access
PF 5660 — *Crisis oriented*
PF 1217 Education ; Data oriented
Education Decline
PF 0674 — government expenditure
PF 0674 — *government expenditure primary*
PF 0674 — public spending
PE 8318 Education ; Delay societal impact
PD 8102 Education ; Denial literacy
Education Denial right
PD 8102 — [Education ; Denial right]
PF 6458 — *access higher*
PE 3802 — choose moral religious
PC 6381 — free primary
PC 3322 — indigenous peoples
Education [Dependence...]
PC 3434 — Dependence inequality
PE 9096 — developing countries ; Inadequacy agricultural
PF 1531 — developing countries ; Inappropriate
PC 3434 — Discrimination
Education Discrimination against
PC 3322 — indigenous populations
PD 8909 — men
PC 3459 — minorities
PD 0190 — women
PD 8807 Education ; Discriminatory effects religious denominational
PF 6534 Education ; Disrelated formal
PF 2313 Education effective living ; Unavailable
PF 4984 *Education electorate ; Inadequate*

Education Excessive
PD 0932 — institutionalization
PC 0432 — specialization
PE 0341 — youth sex
PD 4125 *Education expenditure ; Unnecessary*
PE 4241 Education facilities ; Unequal global distribution family planning
PF 4765 Education ; Failure mass
PF 1905 Education graduates ; Inappropriate
PF 7577 Education ; Hyperlocalized
Education [Ignorance...]
PA 5568 — *Ignorance*
PF 1681 — *image ; Ghetto*
PF 8126 — images ; Debilitating
PF 3519 — Impractical
PF 3519 — Impractical public
PF 4765 — Inadequacy formal
Education Inadequate
PF 4984 — [Education ; Inadequate]
PF 6531 — adult
PD 1370 — environmental
PD 1039 — family planning
PD 8023 — health
PD 9021 — maternal
PF 4984 — *political*
PF 4984 — *preschool*
PD 5345 — secondary
PF 0759 — sex
PF 0422 — vocational
Education [Inappropriate...]
PD 8529 — Inappropriate
PF 1266 — Inappropriate selection examination procedures
PC 3322 — indigenous peoples ; Inadequate
Education Ineffective
PD 8023 — health
PF 2721 — methods practical
PF 3498 — systems practical
Education [Inequality...]
PC 3434 — Inequality
PF 6531 — Infrequent adult
PC 6381 — Insufficient primary
PF 0443 — *Interference school athletic activities*
PD 9370 — Intrusion advertising
PF 4852 — *Irregular delivery*
PF 6531 — Irrelevant adult
Education [Lack...]
PB 8645 — Lack
PA 5568 — Lack
PF 8835 — lack ; Agricultural
Education Lack
PF 8835 — agricultural
PF 1548 — emphasis basic
PD 1370 — environmental
PF 5079 — family life
PD 1039 — family planning
PF 6534 — formal
PB 4653 — *funds*
PC 6381 — preparatory
PE 0265 Education least developed countries ; Inadequate nutrition
Education Limited
PF 2840 — access practical
PF 3575 — access rural youth
PF 8835 — agricultural
PF 2840 — practical
Education [Malpractice...]
PD 4684 — Malpractice
PF 0996 — Manipulation civic
PE 9060 — men ; Inadequate family planning
PF 6531 — Minimal adult
PC 3459 — minorities ; Denial
PC 3459 — minorities ; Violation right
Education [Narrow...]
PF 3552 — Narrow scope
PF 6512 — Neglected community
PE 6206 — nomadic children ; Inadequate
Education [Obstacles...]
PF 4852 — Obstacles
PF 2095 — Outmoded forms social
PC 6262 — Over
PE 8988 — Overdependence
PF 0443 — *Overemphasis academic*
Education [parents...]
PF 9306 — parents ; Undervaluation
PF 6512 — Partial community
PF 4984 — *patterns ; Static*
PF 6208 — peace ; Inadequate
PF 0593 — Permissive
PD 1624 — Political re
PD 6298 — Politicization
PF 4375 — Prohibitive cost
PF 4375 — Prohibitive cost higher
PF 4828 Education qualification earning ; Diversion
Education [Racial...]
PD 3328 — Racial discrimination
PF 5079 — Reduced family
PF 2444 — *regulations ; Complex*
PD 8807 — Religious discrimination
PF 4984 — *Remote driver*
PF 6300 — Resistance technology
PF 6458 — *Restricted higher*
Education rural
PF 3575 — areas ; Limited availability
PF 3575 — communities ; Limited opportunity ongoing
PF 1944 — communities ; Unperceived relevance formal

Education [Segregation...]
PD 3441 — Segregation
PE 8424 — Segregation handicapped children
PE 4700 — *selection ; Inequitable*
PD 1468 — Sexual discrimination
PF 2568 — structures ; Unrealized use
PE 8902 — survival needs ; Devaluation
PF 1266 Education ; Test led
Education [Unbudgeted...]
PE 2813 — *Unbudgeted child*
PF 6512 — Uncomprehensive community
PF 6531 — Undefined adult
PF 0021 — Under developed continuing
PC 1716 — Underperformance
PC 2163 — Unequal access
PD 0190 — Unequal access women
PF 6512 — Unexplored opportunities community
PF 6531 — Unfocused adult
PF 6531 — Unfunded adult
PF 6531 — Uninitiated adult
PF 6531 — Unplanned adult
PF 3519 — Unrealized intentions practical
PF 9068 — Unrecognized relevance
PF 9068 — Unrecognized values
PF 0021 — Unsatisfied need continuing
Education [Violation...]
PE 4700 — Violation freedom parents choose their children's
PF 0422 — vocationally irrelevant
PF 0422 — void ; Vocational
Education women
PD 0190 — Denial right
PE 8699 — immigrants ; Lack
PD 0190 — Neglect
PF 3519 Educational aims ; Impractical
PC 1526 Educational authoritarianism
Educational [choice...]
PE 4700 — choice ; Denial right
PF 6512 — choices ; Unresearched
PF 0966 — context ethical decisions ; Lack meaningful
Educational curricula
PF 3549 — based content rather than method
PF 0443 — Imbalance
PF 0443 — Irrelevance
PD 1827 — over emphasizing method rather than content
Educational [data...]
PF 1301 — *data ; Unavailability*
PF 1301 — *data ; Uncommunicated*
PF 4984 — deficiency
PE 8700 — difficulties ; Inadequate adaptation policy
PE 1399 — discrimination administration justice
PC 1527 Educational elitism
Educational facilities
PD 0084 — Barriers transfer
PF 0775 — disabled ; Inadequate
PF 0775 — disabled persons ; Inadequate
PD 2051 — gifted children ; Inadequate
PD 9051 — Inaccessible
PD 0847 — Inadequate
PD 0847 — Insufficient
PD 0847 — Lack
PD 0847 — Limited
PF 2013 Educational failure ; Ill effects
PF 4984 *Educational forms ; Rigid*
PF 6497 Educational gap generations
Educational goals
PF 2595 — Misdirected
PF 2595 — Non articulated
PF 2595 — Unconsensed
PF 2595 — Untargeted
Educational [images...]
PF 8126 — images ; Ineffective
PF 2168 — *incentives ; Insufficient*
PF 2168 — *incentives ; Lack*
PE 8438 Educational material ; Insufficient
Educational materials
PF 0166 — Barriers international flow knowledge
PD 9370 — Commercially biased
PF 1183 — Deliberate misrepresentation
PE 8438 — shortage
PF 2400 Educational methods ; Abstract
PF 2400 Educational methods ; Unarticulated goal
Educational opportunities
PD 3122 — capitalist systems ; Restriction
PF 3575 — rural areas ; Limited
PF 2453 — *Unperceived*
Educational [planning...]
PS 8148 — *planning ; Non local*
PE 2813 — *plans ; Unfunded*
PF 2447 — policy decisions ; Ineffective
PF 2447 — policy ; Inadequate government
PF 1301 — *possibilities ; Unpublished*
PF 9068 — priorities ; Low
PF 9068 — priority ; Low
PF 4984 — programme ; Makeshift
PF 4984 — programmes ; Discontinued
PF 4984 — programmes ; Inadequate
PD 0260 — purposes ; Maltreatment animals
PC 1524 Educational qualifications ; Non equivalence national
PF 8466 Educational quality ; Decline
Educational [resources...]
PF 6573 — *resources ; Unadvertised*
PF 2595 — *roles ; Unclear*
PF 2595 — *roles ; Undefined*
Educational [self...]
PF 0593 — self expression ; Unconstrained
PE 0779 — snobbery

Educational stages

Educational [stages...] cont'd
PC 1716 — stages ; Repetition
PF 8466 — standards ; Decline
PF 8466 — standards ; Low
PF 6476 — structures ; Constricting effect
PF 4984 — system ; Inflexible
PF 7806 — systems keep pace technological advancement ; Inability
PF 1266 Educational testing procedures ; Inappropriate
PC 1716 Educational underachievement
Educational [values...]
PF 8161 — values ; Inappropriate
PF 8161 — values ; Obsolete
PF 3007 — vision ; Past oriented
PF 3007 — vision ; Unimaginative
PC 1716 Educational wastage
PD 2019 Educationally reinforced egocentric attitudes
PF 0810 Educators ; Professional discrimination
PF 5978 Educators ; Stagnating
PE 0912 Edulis ; Catha
PE 7560 EEG entrainment ELF magnetic fields
PD 2228 Eelworms
PA 7055 Eeriness *Death*
PA 6030 Eeriness *Fear*
PE 5302 Effective activities ; Insufficient government spending cost
PF 0816 Effective communications ; Lack
PF 6492 Effective community operation ; Absence support systems
PF 7713 Effective global policies ; Inadequate intellectual cognitive infrastructure formulate
PF 7082 Effective international nongovernmental organizations ; Obstacles
PC 3201 Effective legislation ; Political barriers
PF 2313 Effective living ; Unavailable education
Effective [means...]
PF 2798 — means transport ; Restrictions
PF 2702 — motivational techniques ; General unavailability
PF 5287 — multilateral safeguard systems ; Inability negotiate
PE 3123 Effective national self determination capitalist exploitation ; Denial
PE 6059 Effective prosecution offenders ; Excessive cost
PE 5801 Effective use oceans waterways ; Obstacles
PF 2959 Effective use technical resources ; Overemphasis
PE 9545 Effectively ; Government failure prosecute offenders
PF 8158 Effectiveness ; Administrative overhead erosion funding
PF 2437 *Effectiveness ; Frustrated personal*
PD 1163 Effectiveness government operations ; Crimes against integrity
PD 4448 Effectiveness ; Impairing military
PE 8143 Effectiveness ; Individualism restriction
PF 1138 Effectiveness myth ; Oligarchy
PF 2437 *Effectiveness ; Unexperienced individual*
PF 1246 Effeminateness
PA 7365 Effetism *Boredom*
Effetism [Impairment...]
PA 6088 — *Impairment*
PA 6876 — *Impotence*
PA 6852 — *Inappropriateness*
PA 5558 Effetism *Weakness*
PF 8158 Efficacity aid ; Overhead restrictions
PF 1706 Efficiency ; Obsession
PF 1545 Efficiency ; Progress equated
PD 9082 Efficient fishing nets ; Deployment excessively
PE 5921 Efficient port utilization operation ; Obstacles
PF 1706 Efficient use time ; Excessively
PF 7022 Efficient utilization time ; Obstacles
PE 8504 Effluent ; Agricultural
PD 3666 *Effluent ; Industrial*
PD 3669 *Effluent waters ; Heated*
PE 3934 Effluents animal husbandry ; Agricultural
PF 5743 Effluents emissions ; Prohibitive costs purification
PC 1242 Effluents ; Radioactive
PF 2358 *Effort ; Collapsed corporate*
PE 7701 Effort discretion
PF 0342 Effort ; Lack consistent programme
PF 1768 Efforts agricultural development ; Uncoordinated
PF 0013 Efforts ; Failure disarmament arms control
PF 2588 Efforts ; Fragmented forms cooperative
PF 5272 Efforts ; Ineffective past
PF 2588 Efforts ; Lack combined
PF 2306 Efforts ; Uncoordinated research
PF 1781 *Efforts ; Unstructured improvement*
PF 3871 Egalitarian application trade measures ; Narrow
PF 5155 Egalitarianism
PE 9417 Egg collection ; Birds
PD 7799 *Egg drop syndrome*
PE 9115 Egg production ; Lack meat
PE 8013 Egg products ; Increased demand developed countries meat
PE 0505 Eggs ; Environmental hazards dairy products
PE 0576 Eggs ; Instability trade dairy products
PE 8225 Eggs ; Long term shortage dairy products
PE 6290 Eggshell thinning
PE 1418 *Ego defect*
PE 1418 *Ego ; Impairment*
PE 1418 Ego regression
PD 2019 Egocentric attitudes ; Educationally reinforced
PA 6318 Egoism
PA 6318 Egoism ; Dependence
PD 2732 *Ehrlichiosis ; Canine*
PD 2732 *Ehrlichiosis ; Equine*
PD 3978 *Ehrlichiosis ; Equine monocytic*
PE 3815 Eichornia crassipes
PF 6415 Ejaculation ; Premature
PF 6415 Ejaculation ; Retarded

PE 0560 *El Tor ; Cholera*
PA 6400 Elaborateness *Affectation*
PA 7240 Elaborateness *Ugliness*
PC 1613 Elaboration remedial legislation ; Delays
PE 5380 *Elaeophorosis*
PE 2320 *Elbow ; Beat*
PE 6555 *Elbow ; Capped*
PD 7424 *Elbow dysplasia*
PE 2320 *Elbow ; Miner's*
PF 3973 Elder paralysis over future
PD 0512 Elderly aid ; Unorganized
PE 4354 Elderly countries ; Unequal mortality
Elderly [Denial...]
PC 2541 — Denial rights
PD 4916 — Denial rights equal opportunities employment
PD 8399 — Dependency
PD 4916 — Elderly ; Employment discrimination against
PD 9343 — Elderly ; Exploitation
PE 2853 — Elderly ; Falls
PD 9343 — Elderly ; Fraud against
PD 0276 — Elderly housing ; Blocked
PD 1043 — Elderly ; Illness
Elderly Inadequate
PD 0512 — provision home care services
PF 0202 — recreation facilities
PF 1825 — roles
PE 5288 Elderly institutional care ; Neglect
PD 3133 Elderly ; Monopoly power
Elderly [people...]
PE 9071 — people ; Failure employ skills educated
PF 3847 — persons ; Disabled
PF 3973 — play role developing future ; Inability
PC 1966 — Poverty
Elderly [Sexual...]
PC 1892 — *Sexual inadequacy*
PD 1564 — Social isolation
PD 0096 — states emergency ; Vulnerability
PF 1681 — *Elderly ; Unreleased creativity*
PF 1625 — Elderly workers ; Denial equal benefits
PF 6570 *Elders ; Anticipated neglect*
Elders [communication...]
PD 3518 — communication handicap
PF 1825 — concealed needs
PF 3979 — Contempt
Elders [Denial...]
PC 2541 — Denial rights
PF 2830 — disapproval past
PC 2541 — Discrimination against
PF 7149 — Disobedience
PF 3979 — Disrespect
Elders [Failure...]
PF 7149 — Failure follow advice
PF 7149 — Failure obey
PF 1825 — feeling uselessness
PE 8985 Elders ; Insufficient possibilities gathering
PD 1564 Elders ; Isolation fostering structures engagement
PF 6462 Elders' leadership ; Untrained
PF 2575 *Elders ; Limited strength*
PD 3518 Elders' mindset ; Turned
Elders [participation...]
PD 1564 — participation image ; Preclusion
PC 0027 — predomination society
PD 0512 — programmes ; Distant
PD 9343 — property ; Vulnerability
Elders [self...]
PD 1564 — self-images preclude participation
PF 1664 — skills ; Neglect
PF 1664 — society ; Isolation wisdom
Elders [Uncommunicated...]
PF 1664 — Uncommunicated wisdom
PF 1664 — Unrecognized role
PD 1564 — urban environments ; Isolated non participating
PF 1664 Elders' wisdom ; Erosion
PF 1664 *Elders' wisdom ; Underused*
PE 1758 Elect their representatives ; Violation right trade unions freely
PF 5660 Elected bodies ; Immediacy orientation
PE 5608 Elected government positions ; Violation right be
PC 2649 Election administration ; Unjust
PD 9384 Election campaigning ; Negative
PF 5153 Election ; Discrimination against those seeking
PF 4558 Election promises ; Broken
PF 4558 Election promises ; Unrealistic
PD 2907 Election timing ; Unjust
PC 2494 Election voters ; Decreasing number
PC 2649 Elections ; Dishonest
PD 3982 Elections ; Oligarchic political control
PD 2931 Elections ; Oligarchic political control
PD 2906 Elections ; Political party manipulation
PC 2649 Elections ; Unfair
PE 4526 Elective mutism
PC 2494 Electoral apathy
PD 2931 *Electoral campaigns ; Prohibitive cost*
PD 2919 Electoral campaigns ; Unjust
PF 4709 Electoral defeat
PD 5214 Electoral fraud
PF 5153 Electoral organization favouring political incumbents
PF 4713 *Electoral political inertia ; Pre*
PD 2906 Electoral process political parties ; Abuse
PC 2677 *Electoral system ; Unstable*
PD 9641 *Electoral systems ; Unrepresentative*
PD 2044 Electoral violence
PD 1893 Electorate government repetition untruths ; Susceptibility
PF 4984 *Electorate ; Inadequate education*
PD 2044 Electors ; Intimidation

PB 0731 *Electric current accidents*
Electric energy
PE 1412 — Environmental hazards
PE 7907 — Instability trade
PE 1216 — Long term shortage
PE 8878 Electric machinery apparatus appliances shortage
PE 4247 *Electric shock animals*
PE 7879 Electrical appliances ; Health hazards electromagnetic fields generated
PD 9033 Electrical capacity ; Limited
Electrical [energy...]
PE 1048 — energy ; Inadequate facilities transport
PD 3446 — energy ; Maldistribution
PE 8097 — equipment industries ; Environmental hazards
PF 1389 *Electrical generating capacity ; Non viability local*
Electrical machinery
PE 8026 — apparatus appliances ; Environmental hazards
PE 8875 — apparatus appliances ; Trade instability
PJ 2425 — Environmental hazards non
PE 8828 — Instability trade non
PD 8984 *Electrical maintenance ; Inadequate*
Electrical power
PE 1341 — failure
PE 1412 — generation ; Degradation environment
PD 9033 — Insufficient
PE 1900 — supply developing countries ; Inadequate
PE 9642 — transmission lines ; Environmental hazards
PD 4172 *Electrical signals ; Intermittent*
PE 3881 *Electrical storms*
PE 9000 Electrical torture
PF 3318 *Electricity costs ; High*
PE 1412 *Electricity ; Environmental hazards*
PE 1159 Electricity, gas, water sanitary services ; Health risks workers
PF 2583 *Electricity, gas, water sanitary services ; Underdevelopment*
PE 1412 *Electricity ; Health risks*
PD 9033 *Electricity infrastructure ; Inadequate*
PE 2115 *Electricity ; Lack industrial*
PF 3318 *Electricity ; Prohibitive cost*
Electricity [services...]
PF 1653 — services ; Prohibitive cost connection
PE 2115 — supply ; Inadequate
PE 2115 — supply ; Insufficient domestic
PE 0984 Electricity ; Waste
PD 9366 *Electrocution animals*
PE 7879 Electromagnetic fields generated electrical appliances ; Health hazards
PD 4172 *Electromagnetic pollution*
PE 6360 Electromagnetic pulses ; Environmental hazards
PE 7560 Electromagnetic radiation ; Environmental hazards extremely low frequency
PE 1304 Electromagnetism ; Hazards environmental
PD 2930 Electronic bugging
PE 7679 Electronic deception
PE 1909 Electronic devices ; Excessive exposure radiation consumer goods
PD 1475 Electronic equipment failure
PD 2930 Electronic espionage
PE 6226 Electronic messages ; Ineffective regulation
PD 4172 *Electronic noise*
PE 5402 Electronic pornography
Electronic [sex...]
PE 5402 — sex
PD 4172 — smog
PD 2930 — surveillance governments ; Misuse
PE 1936 Element deficiencies soils ; Trace
PE 5328 *Element deficiency ; Trace*
PE 5328 Element imbalance human body ; Trace
PE 1848 *Elementary school ; Distant*
PE 0500 Elements compounds ; Instability trade chemical
PE 5328 Elements human body ; Excessive trace
PE 9076 Elements, oxides halogen salts ; Trade instability inorganic chemicals,
PC 1777 Elements personality ; Repression
PD 3771 *Elephant ; Endangered species*
PE 4814 *Elephant Man's disease*
PD 3479 *Elephant shrews ; Endangered species*
PE 1601 Elephantiasis
PE 1601 Elephantiasis arabum
PF 3455 *Elephantiasis ; White*
PE 2371 *Elevated blood cholesterol*
PD 2322 *Elevation conditions ; High*
PE 7560 ELF magnetic fields ; EEG entrainment
PF 8205 Eligibility benefits ; Lack knowledge
PE 7965 Eligibility government programmes ; Restrictions
PF 3350 Eliminate nakedness developing countries ; Ill considered pressure
PF 9474 Eliminating human rights violations ; Governments unwilling cooperate United Nations
PC 1002 Elimination bourgeois power
PF 7836 Elimination industrial pollution ; Employment risk
Elimination jobs
PD 0528 — automation
PE 4391 — developing countries due introduction new technologies
PE 4391 — due automation developing countries
PE 1046 Elimination socio-cultural role western women
PD 8872 Elimination traditional skills
PE 4849 Elites ; Appropriation machinery government
PF 6670 Elites developing countries ; Disproportionately wealthy
PF 9686 Elites ; Disproportional political influence urban
PE 3816 *Elites ; Non productive capitalist*
PA 1387 Elitism
PC 3170 Elitism communist systems
PA 1387 *Elitism ; Cultural*

-914-

Employment training

Code	Entry
PA 1387	Elitism ; Dependence
PC 1527	Elitism ; Educational
PA 1387	*Elitism ; Ideological*
PA 1387	*Elitism ; Linguistic*
PA 1387	*Elitism ; Military*
PE 3647	*Elitism ; Political*
PA 1387	*Elitism ; Racial*
PA 1387	*Elitism ; Religious*
PC 1937	Elitism ; Scientific
PF 9686	Elitism ; Urban
PF 3354	Elitist access eternal life

Elitist control
- PD 2712 — construction technology
- PC 3778 — global economy
- PD 0154 — production
- PD 6896 Elitist intergovernmental groupings
- PD 0986 Elitist justice
- PF 1104 Elitist leadership
- PD 0986 Elitist legal judgements
- PA 1387 Elitist people
- PF 4849 Elitist ruling classes
- PE 4586 *Elm ; Black leaf spot*
- PE 1154 Elm disease ; Dutch
- PE 2982 Elm ; Pests diseases
- PD 2732 Elokomin fluke fever
- PE 4161 *Elso heel*
- PF 9108 Elusive public funding
- PF 7053 *Elusiveness reality*
- PF 3437 Elves ; Endangered habitats
- PE 0522 Embargoes similar restrictions ; Distortion international trade
- PF 2093 *Embarrassed civic pride*
- PF 9699 Embarrassing documents ; Overclassification
- PF 7950 Embarrassment
- PA 5497 Embarrassment *Difficulty*
- PF 7950 *Embarrassment ; Diplomatic*
- PA 6659 Embarrassment *Humility*
- PA 6247 Embarrassment *Inattention*
- PA 6434 Embarrassment *Poverty*
- PF 3856 Embarrassment ; Public
- PA 7309 Embarrassment *Uncertainty*
- PA 7107 Embarrassment *Unpleasantness*
- PD 2688 Embezzlement
- PD 8392 Embezzlement civil servants
- PC 2920 Embezzlement public funds
- PD 5453 *Embolism animals*
- PD 0637 Embolism ; Pulmonary
- PE 5975 Embourgeoisement socialist countries
- PE 5623 Embryo storage ; Unlimited practice human
- PE 6039 Embryo transfer technology ; Non acceptance
- PE 2246 *Embryogenesis malfunction*
- PE 6038 Embryos ; Commercialization human
- PE 5623 Embryos ; Orphaned
- PE 4805 Embryos ; Research human
- PD 3040 Emergence international media monopolies
- PF 9790 Emergence ; Irreversible problem
- PF 9328 Emergence problems ; Delayed
- PF 5306 Emergencies ; Failure assist
- PC 5016 Emergencies ; Unpreparedness food
- PD 1078 Emergencies ; Vulnerability women children
- PE 0366 Emergency blood supply ; Inadequate
- PF 9007 Emergency care ; Inadequate
- PF 9007 Emergency care ; Slow
- PE 4363 Emergency ; Excessive imposition states

Emergency [facilities...]
- PF 9007 — facilities ; Lack
- PC 5016 — food reserves ; Inadequacy
- PC 5016 — food supplies ; Inadequate
- PD 1428 Emergency medical services ; Inadequate
- PF 9007 Emergency relief facilities ; Inaccessible
- PF 5660 Emergency responses ; Obstruction change crisis

Emergency services
- PF 9007 — Distant
- PF 9007 — Expensive
- PF 9007 — Ineffective
- PF 9007 — Insufficient
- PF 9007 — Unavailable
- PD 6380 Emergency ; Suspension rights during states
- PF 9007 Emergency transport ; Distant
- PE 4363 Emergency ; Undeclared states
- PD 6380 Emergency ; Violation rights state
- PD 3785 Emergency ; Violation rights vulnerable groups during states

Emergency Vulnerability
- PE 4694 — animals during states
- PD 0098 — disabled during states
- PD 0096 — elderly states
- PF 2797 Emerging leadership ; Stifled
- PE 0135 Emerging oligopolistic world trading system
- PD 1928 Emigration ; Clandestine
- PC 0018 Emigration ; Disruptions due
- PE 5713 Emigration island developing countries ; Excessive
- PF 0905 *Emigration labour force*
- PE 1848 *Emigration ; Massive urban*
- PC 3208 Emigration restrictions
- PC 3208 Emigration ; Restrictions
- PD 1291 Emigration trained personnel developing developed areas
- PE 8093 Emigration trained personnel industrialized countries

Emissions [air...]
- PD 0414 — air pollutants ; Automobile
- PE 2824 — air pollutants ; Industrial domestic heating
- PE 5399 — animal husbandry ; Methane gas
- PE 2824 *Emissions ; Domestic*
- PE 4891 Emissions ; Fossil fuel
- PE 1869 Emissions ; Industrial
- PE 1256 Emissions landfill sites ; Methane gas
- PD 0414 Emissions ; Motor vehicle
- PF 5743 Emissions ; Prohibitive costs purification effluents
- PF 1441 Emissions satellites ; Nuclear reactor
- PD 7330 Emotional abuse children
- PF 4416 Emotional attachments ; Dependence deep
- PE 9599 Emotional blackmail
- PE 9599 Emotional bombardment
- PE 3407 Emotional crises

Emotional [dependency...]
- PD 3244 — dependency marriage
- PC 0981 — deprivation ; Infant
- PD 9159 — disorders
- PD 9159 — disturbance
- PE 3407 Emotional flare-ups
- PF 1568 Emotional hindrance ; Spiritual
- PF 8413 *Emotional immaturity*
- PD 8262 Emotional insecurity

Emotional instability
- PD 9159 — [Emotional instability]
- PE 4687 — due torture
- PF 8751 — women ; Belief
- PF 1568 Emotional malaise ; Spiritual
- PE 9599 Emotional manipulation
- PF 1216 Emotional numbness
- PF 1216 Emotional paralysis
- PF 0913 Emotional responses atomic energy ; Unhealthy

Emotional [security...]
- PD 8262 — security ; Lack
- PF 9209 — sensitivity ; Fear
- PG 2572 — *strain*
- PJ 3545 — strain ; Mental, physical
- PF 9209 Emotionally responsive behaviour ; Fear
- PA 7090 Emotions attitudes ; Dependence negative
- PA 7090 Emotions attitudes ; Negative
- PF 7798 Emotions equated free speech ; Irresponsible expression
- PE 3407 Emotions ; Violent
- PF 3128 Emotive words terms ; Politically
- PD 0428 Empathy external hardship ; Inadequate national
- PA 0839 Emphasis ; Personal interest
- PE 1478 Emphasis bad news ; Media
- PF 1548 Emphasis basic education ; Lack
- PE 3831 Emphasis developing country weaknesses media ; Excessive
- PF 2997 Emphasis economic development ; Over
- PF 4059 Emphasis economic priorities research
- PF 0059 Emphasis fashionable areas research ; Excessive
- PF 4164 Emphasis fashionable problems ; Excessive
- PF 4059 Emphasis immediate solutions resource development research ; Over
- PF 2855 Emphasis mass extraction natural resources
- PF 2174 Emphasis use national resources ; Profit over
- PF 6570 *Emphasis ; Youth*
- PD 1827 Emphasizing method rather than content ; Educational curricula over
- PE 2248 Emphysema
- PE 5524 *Emphysema ; Acute bovine pulmonary*
- PD 7307 *Emphysema animals ; Pulmonary*
- PE 5182 *Emphysema ; Chronic alveolar*
- PE 2248 Emphysema ; Pulmonary
- PF 1232 Empire building ; Organizational
- PF 2825 Empire building ; Organizational
- PE 9071 Employ skills educated elderly people ; Failure
- PD 8546 Employ skills home bound educated women ; Failure
- PF 2896 Employability ; Limited image
- PE 7344 Employable skills ; Limited means marketing
- PD 9113 Employable skills ; Little
- PD 0164 Employed children ; Denial rights
- PE 5506 Employed forced labour ; Unfair competition workers
- PD 1634 Employee absenteeism
- PF 3661 Employee communication ; Inadequate management
- PD 5244 Employee disobedience
- PD 9397 Employee infidelity ; Risk
- PD 5244 Employee insubordination
- PE 6018 Employee participation collective bargaining ; Inadequate
- PF 6540 *Employee pool ; Small*
- PE 3684 Employee theft
- PE 8251 Employees commerce offices ; Inadequate working conditions

Employees [Denial...]
- PE 6888 — Denial trade union rights public
- PE 6240 — Disadvantages homeworking
- PD 9397 — Dishonest
- PE 7760 — Disparity remuneration public private sector
- PD 6733 — *Disruptive maternity leave*
- PE 3993 Employees ; Ghost
- PD 5586 Employees ; Grievances
- PD 4334 Employees ; Negligent
- PE 6702 Employees ; Negligent public service
- PF 8763 Employees ; Overpaid
- PE 6937 Employees ; Pressured

Employees [Underpaid...]
- PD 8916 — Underpaid
- PD 4334 — Unethical practices
- PE 6702 — Unethical practices public service
- PF 2648 *Employer evaluation ; Predetermined*
- PE 8384 Employers affairs workers' organizations ; Interference
- PD 2879 Employers ; Corruption
- PD 2879 Employers ; Fraud
- PE 8650 Employers government ; Collusion trade union leaders
- PD 7286 Employers ; Grievances
- PD 2879 Employers ; Irresponsible
- PE 6333 Employers ; Labour 'hoarding'
- PE 6333 Employers ; Labour hoarding
- PD 2879 Employers ; Negligence
- PD 2879 Employers ; Unethical practices
- PE 3970 Employers ; Violation right trade unions negotiate freely

Employment [Abuse...]
- PE 4658 — Abuse government
- PE 4934 — against immigrant workers ; Discrimination
- PD 2318 — Age discrimination
- PF 1658 — alternatives ; Limited
- PF 2949 *Employment base ; Narrow*
- PD 0568 *Employment blind*

Employment [children...]
- PD 0164 — children
- PC 7607 — Clandestine
- PE 4439 — criminals policy-making contexts

Employment Dependence
- PC 0244 — discrimination
- PC 3297 — exploitation
- PF 2261 — *interim*

Employment [Discontinuity...]
- PD 4461 — Discontinuity
- PC 0244 — Discrimination
- PE 0783 — Discrimination against disabled
- PD 4916 — discrimination against elderly

Employment Discrimination against
- PD 3529 — foreigners
- PD 1092 — indigenous populations
- PD 3338 — men
- PD 0086 — women

Employment [Discrimination...]
- PD 2318 — Discrimination favour youth
- PF 6458 — Disorganized attempts upgraded
- PE 8916 — Distant male

Employment [elderly...]
- PD 4916 — elderly ; Denial rights equal opportunities
- PC 3297 — Exploitation
- PD 1092 — Exploitation indigenous populations
- PD 3529 Employment foreigners ; Restrictions
- PE 3993 Employment fraud
- PD 3443 *Employment ; Gender stereotyping*
- PF 1015 Employment government security services ousted repressive regime ; Post revolutionary re

Employment [Ill...]
- PD 0024 — Ill conceived incentives
- PF 2896 — images ; Individualistic
- PE 3751 — Inability refugees obtain
- PF 6733 — Inadequate maternity protection
- PD 0024 — incentives ; Inappropriate
- PD 1092 — indigenous populations ; Denial right
- PD 8903 — Inequality
- PD 8996 — Inferior status
- PC 0646 — injuries
- PE 1134 — inner cities ; Limited availability permanent
- PD 8211 — Insecurity
- PF 2303 — *insecurity ; Future*
- PD 0171 — Insufficiencies
- PD 5115 — Insufficient female
- PB 0750 — *Insufficient part time*
- PF 1933 *Employment ; Lack significant*

Employment [Maladjustment...]
- PD 7650 — Maladjustment disciplines
- PE 2238 — manufacturing industries developing countries ; Inadequate increase
- PD 3557 — married women ; Adverse social consequences excessive
- PD 3557 — married women ; Excessive
- PF 2477 — *mechanisms ; Insufficient*
- PF 2949 — Minimal options rural
- PJ 8374 Employment ; Negative effects relative wage rates youth
- PD 1285 *Employment ; Off farm*
- PD 2318 Employment older people ; Constraints against

Employment opportunities
- PD 0176 — developing countries ; Insufficient
- PE 0783 — disabled persons ; Unequal
- PF 2949 — *Narrow*
- PF 2949 — small rural communities ; Limited
- PC 1131 — *Training*
- PD 5115 — women ; Unequal
- PD 1847 Employment opportunity developing countries ; Inequality
- PF 1658 Employment options ; Limited

Employment [Part...]
- PB 0750 — *Part time farm*
- PF 2303 — patterns ; Disruptive effect changing
- PF 2827 — *patterns ; Non local*
- PE 5104 — policy technical advance ; Antagonism
- PF 2842 — *possibilities ; Scarce*
- PC 0244 — practices ; Discriminatory
- PC 0244 — practices ; Prejudicial
- PE 8728 — public service personnel ; Inadequate conditions work

Employment [Restricted...]
- PF 2423 — Restricted scope local
- PF 6046 — Restrictive trade union policies concerning
- PD 1958 — reward ; Inadequate access negotiation
- PF 7836 — risk elimination industrial pollution

Employment [Seasonal...]
- PC 1108 — Seasonal
- PC 6908 — Second
- PD 3443 — Segregation

Employment socialist countries
- PC 0939 — Double
- PC 0939 — Second undeclared
- PC 0939 — Undeclared
- PF 2949 Employment structures ; Inadequate rural
- PD 3548 Employment training ; Obsolete

-915-

Employment [Unattractive...]
- PF 6478 — Unattractive fishing
- PC 1131 — Unavailability trainee
- PC 2035 — Unavailability youth
- PF 3008 — Unmotivating subsistence
- PE 8319 — utility supply services ; Inadequate conditions work

Employment [war...]
- PE 4697 — war criminals ; Government approved
- PE 4439 — war criminals policy-making contexts
- PD 0086 — women ; Denial right
- PE 7206 — women family responsibilities ; Discrimination
- PD 5110 Empolderment coastal lowlands ; Inadequate
- PD 5110 Empolderment peat land ; Inadequate
- PD 5110 Empolderment wetlands ; Inadequate
- PF 8753 Empower humanity response global problematique ; Failure integrate knowledge
- PF 3756 Empty meeting oratory
- PF 3212 Empty slogans mottoes
- PF 3448 *Enabling structures ; Insufficient*
- PE 2348 Encephalitis
- PE 9037 Encephalitis ; Acute
- PE 2348 Encephalitis ; California
- PD 2730 Encephalitis ; Caprine
- PE 2348 Encephalitis ; Equine
- PD 7841 *Encephalitis ; Equine*
- PE 2348 Encephalitis ; Japanese B
- PE 2348 Encephalitis ; Murray Valley
- PD 7841 *Encephalitis ; Ontario*
- PD 7841 *Encephalitis pheasants ; Eastern*
- PE 2348 Encephalitis ; Russian spring summer
- PE 2348 Encephalitis ; Saint Louis
- PE 2348 Encephalitis ; Tick borne
- PE 2348 Encephalitis ; Viral
- PE 9037 *Encephalomyelitis*
- PD 7841 *Encephalomyelitis ; Equine*
- PD 7841 *Encephalomyelitis ; Haemagglutinating*
- PE 9037 *Encephalomyelitis ; Meningo*
- PE 1914 *Encephalomyelitis ; Myalgic*

Encephalomyelitis [piglets...]
- PD 7841 — piglets ; Viral
- PD 7841 — Porcine enteroviral
- PD 7841 — poultry
- PD 2732 *Encephalomyelitis ; Sporadic bovine*
- PD 7841 *Encephalomyelitis swine ; Coronaviral*
- PD 2730 *Encephalomyocarditis virus disease pigs*
- PE 5191 *Encephalopathy ; Bovine spongiform (BSE)*
- PF 3956 *Enchantment*
- PF 5123 *Enchantment ; Evil*
- PD 2154 Enclaves exclaves
- PE 2858 Enclaves ; Misuse free production zones export
- PE 8175 Enclosed bodies seawater ; Pollution semi
- PE 1281 *Enclosure ; Land*
- PC 0300 *Encopresis*
- PD 7221 Encounter killings ; Staged
- PE 1848 *Encounters ; Limited daytime*
- PB 6384 *Encouragement drug resistant diseases*
- PF 6190 Encouraging unstructured public access ; Lack places urban environments
- PC 3152 *Encroachment ; Bush*
- PD 9162 Encroachment communal land
- PA 6862 Encroachment *Intrusion*
- PD 0493 Encroachment ; Sand dune
- PA 6041 Encroachment *Shortcoming*
- PA 6921 Encroachment *Undueness*
- PE 5210 Encroachments ; Border
- PE 4740 Encrustation water wells ; Corrosion
- PA 5497 Encumbrance *Difficulty*
- PA 5491 Encumbrance *Lightness*
- PA 7107 Encumbrance *Unpleasantness*
- PF 9582 End nature
- PF 4528 End world
- PD 5300 Endanger people ; Crimes which
- PD 0253 Endangered ancient sites
- PD 5157 Endangered animal plant life due radioactive contamination
- PE 6010 Endangered black rhinoceros
- PD 6065 Endangered children
- PB 8613 Endangered cultures
- PD 7096 Endangered drylands
- PD 5962 Endangered family farms
- PC 5165 Endangered forests
- PF 3437 Endangered habitats elves
- PC 7203 Endangered indigenous cultures
- PD 0528 Endangered jobs technological innovation
- PB 8613 Endangered languages
- PE 8077 Endangered lifestyles nomads pastoralists
- PE 4938 Endangered migratory bird species
- PD 0253 Endangered monuments historic sites

Endangered [park...]
- PD 2025 — park trees
- PE 9282 — parklands
- PE 1407 — peoples ; Government non assistance
- PC 0238 — plantations long-lifed trees
- PE 1320 — polar bear species
- PC 4764 — protected habitats
- PC 4764 — protected sites

Endangered species
- PD 1763 — *acinonyx*
- PE 8684 — African cane rat
- PE 8253 — African mole rat
- PE 8738 — agoutis acouchis
- PD 3156 — amphibia
- PC 1713 — animals
- PD 3160 — annelida
- PD 3603 — anteaters
- PD 3473 — arachnida

Endangered species cont'd
- PE 1656 — *arctocephalus*
- PD 3603 — armadillos
- PD 3471 — arthropoda
- PD 3159 — aschelminthes
- PD 3158 — asteroidea
- PD 0332 — aves
- PE 1570 — aye-aye
- PD 3481 — bamboo rats
- PD 1762 — bandicoots
- PD 1593 — beaked whales
- PD 3483 — bear
- PD 3481 — beavers
- PD 0332 — birds
- PD 3158 — brittle stars
- PD 3477 — bryozoa
- PE 3604 — bulldog bats
- PE 9120 — bush rat rock rat
- PE 3762 — butterfly
- PD 3481 — *capybara*
- PD 3482 — carnivores
- PD 1763 — cats
- PE 8683 — cavies dolichotids
- PD 1593 — cetacea
- PD 3481 — *chinchillas*
- PE 3604 — chiroptera
- PE 3604 — common bats
- PD 3158 — *crinodiea*
- PD 3472 — *crustacea*
- PE 8527 — dasyures marsupial mice
- PE 3604 — disc-winged bats
- PE 8739 — dog wolves foxes
- PD 3481 — *dormice*
- PD 3158 — echinodermata
- PD 3158 — *echinoidea*
- PD 3603 — edentates
- PD 3771 — elephant
- PD 3479 — *elephant shrews*
- PC 1326 — Even-toed ungulates
- PE 3604 — false vampires
- PD 7506 — farm animals
- PD 3158 — feather stars
- PD 1763 — *felis*
- PC 1535 — fish
- PE 4314 — flowering plants
- PC 1326 — Flying lemurs
- PE 3604 — free-tailed bats
- PD 3156 — frog
- PE 3604 — fruit bats
- PE 3604 — funnel-eared bats
- PE 9121 — genets, civets mongooses
- PE 1570 — gibbons
- PE 3604 — golden bat
- PD 3479 — golden moles
- PE 1570 — great apes
- PD 1593 — grey whale
- PD 3481 — *gundis*
- PD 3479 — hedgehogs
- PD 3158 — *holothuroidea*
- PD 1762 — honey possum
- PE 3604 — horseshoe bats
- PD 3482 — hyenas
- PC 1326 — *Hyracoidea*
- PE 2664 — Illegal hunting protected
- PD 3479 — insectivores
- PC 2326 — insects
- PC 0380 — International trade
- PD 7513 — invertebrates
- PD 3481 — *jerboas*
- PD 3481 — jumping mice
- PE 8708 — kangaroos wallabies
- PD 1762 — *koala*
- PE 3604 — leaf-nosed bats
- PE 1570 — lemurs
- PD 0904 — llamas
- PD 1762 — macropods
- PC 1326 — mammals
- PD 3673 — marine mammals
- PD 3160 — *marine worms*
- PE 3981 — marsupial moles
- PD 1762 — marsupials
- PD 4171 — medicinal plants
- PE 8967 — mice hamsters
- PE 8750 — moles desmans
- PD 3478 — molluscs
- PE 1656 — *monachus*
- PC 1326 — monotremes
- PE 3604 — *mouse-tailed bats*
- PD 3481 — murids
- PD 1763 — *neofelis*
- PE 1570 — new-world monkeys
- PE 8847 — new world porcupines
- PE 3604 — New Zealand short-tailed bat
- PE 1570 — non-human primates
- PC 1326 — Odd-toed ungulates
- PE 8288 — odobenus
- PE 8991 — old world monkeys
- PE 8845 — old world porcupine
- PD 3158 — *ophiaroidea*
- PD 1762 — *opossums*
- PD 3479 — otter shrews
- PD 3481 — *pacarana*
- PE 9103 — palaearctic mole rats
- PC 1326 — Pangolins
- PD 1763 — *panthera*
- PE 8537 — phalangers cuscuses

Endangered species cont'd
- PE 1656 — *phoca*
- PD 1762 — *pigmy possums*
- PD 3480 — *pikas*
- PC 0238 — plants
- PB 1395 — plants animals
- PD 3474 — platyhelminthes
- PD 3481 — pocket gophers
- PE 8674 — pocket mice kangaroo mice
- PE 3806 — porpoises
- PE 7953 — potto loris
- PD 3475 — protozoa
- PD 3480 — rabbits
- PE 8794 — rabbits hares
- PD 3482 — racoons
- PD 3481 — rat chinchillas
- PD 1762 — rat opossums
- PC 0604 — reptiles
- PD 1593 — right whales
- PD 1762 — ringtail gliders
- PD 3673 — river dolphins
- PD 3481 — rodents
- PD 1593 — rorquals
- PD 3159 — Rotifera
- PE 8302 — rough-toothed white dolphins
- PD 3159 — Roundworms
- PE 3604 — sac-winged bats
- PD 3156 — salamander
- PD 3481 — scaly-tailed squirrels
- PD 3158 — sea cucumbers
- PD 3158 — sea urchins
- PE 1656 — seals
- PD 3483 — *selenartos*
- PD 3479 — shrews
- PE 8189 — sifakas, avahis indri
- PE 3604 — slit-faced bats
- PD 3603 — sloths
- PE 3604 — smoky bats
- PD 3478 — snails
- PD 3479 — solenodons
- PE 3604 — spear-nosed bats
- PD 1593 — sperm whales
- PD 3481 — spiny dormice
- PD 3481 — spiny rats
- PD 3476 — sponge
- PD 3481 — springhaas
- PD 3478 — squid octopus
- PD 3481 — squirrels
- PD 3158 — starfish
- PE 7933 — tamarins marmosets
- PE 1570 — tarsiers
- PD 3479 — tenrecs
- PD 1762 — thylacine
- PD 3156 — toad
- PD 0389 — Trade animal products
- PD 0389 — Trade furs skins
- PE 1570 — tree shrews
- PD 3483 — *tremarctos*
- PE 3604 — true vampires
- PC 1326 — Tubulidentata
- PD 3481 — tucotucas
- PD 3483 — *ursus*
- PE 0904 — vicuña
- PE 8288 — walrus
- PE 5067 — water fowl
- PE 8936 — weasels, badgers, skunks, otters
- PD 1593 — whale
- PE 8187 — white whale narwhal
- PD 1762 — wombats
- PE 8935 — zagoutis coypu hutia
- PE 1656 — *zalophus*
- PE 4184 Endangered totemic species
- PC 0720 Endangered tribes indigenous peoples
- PD 2025 Endangered urban trees
- PD 0437 Endangering behaviour ; Criminally life
- PF 2418 Endangering environment ; Execution orders
- PE 5505 Endangering fire explosion
- PD 5349 Endangerment awareness high risk
- PD 5349 Endangerment ; Criminally reckless
- PE 2684 *Endarteritis*
- PD 7799 *Endemic abortion ewes*
- PD 0103 Endemic disease developing countries
- PE 1924 *Endemic goitre*
- PE 7553 *Endemic pneumonia*
- PE 5524 *Endemic pneumonia calves*
- PE 3895 *Endemic typhus*
- PD 4233 Ending love affair
- PF 2513 *Endocannibalism*
- PE 3898 *Endocarditis ; Acute*
- PE 0471 *Endocardium diseases animals*
- PD 8301 *Endocrine deficiency*
- PD 9654 *Endocrine diseases animals*

Endocrine [gland...]
- PD 8301 — gland disturbances
- PD 8347 — glands ; Benign neoplasm
- PD 8301 — glands ; Diseases
- PD 8301 Endocrine ; Nutritional metabolic diseases
- PE 4372 *Endometriosis*
- PE 4299 Endorsed homophobia ; Officially
- PC 8019 *Endorsed professions conceal socially unacceptable initiatives ; Misuse societally*
- PE 7502 Endorsement advertising ; Misleading
- PE 6948 Endorsement ; Trading political
- PF 7815 Endowment ; Decline human genetic
- PF 3043 Endowments ; Disparity countries natural resource

Entitlement

PE 4833 Ends ; Exploitation athletic competition commercial political
PF 8404 Enemies
PF 8404 Enemies ; Dependence existence
PE 1200 Enemy aliens ; Aiding escape prisoners war
PD 0218 *Enemy ; Collaboration*
PF 4133 Enemy ; Demonizing
PF 9139 Enemy ; Eschatological
PJ 7177 *Enemy ; Trading*
PE 6974 Enemy ; War time communications
PF 8128 Energies desires ; Misdirection human
PC 6329 Energy
PF 7960 Energy alternatives ; Disuse traditional
PF 7960 Energy alternatives ; Unexplored
Energy [Competition...]
PC 4412 — Competition sources
PC 7517 — conservation ; Inadequate
PF 0037 — conservation policies ; Inadequate governmental
PC 5038 — consumption ; Disparities
PC 5038 — consumption ; Maldistribution
PC 6329 — crisis
PD 6103 — cropping
Energy [deficient...]
PE 0379 — deficient developing countries
PC 6329 — *Dependence imported*
PC 6329 — *dependence vulnerability*
PC 7517 — development ; High
PC 0382 — diet ; Low
Energy [Environmental...]
PD 7541 — Environmental hazards coal
PE 1412 — Environmental hazards electric
PE 8592 — equipment developing countries ; Insufficient inappropriate
Energy [Inadequate...]
PE 1048 — Inadequate facilities transport electrical
PC 0495 — industry ; Unsustainable development
PE 7907 — Instability trade electric
PE 0635 — intensive packaging
PC 7517 — intensive products ; Use
PE 1216 Energy ; Long term shortage electric
PC 6114 Energy ; Long term variations solar radiant
PD 3446 Energy ; Maldistribution electrical
Energy malnutrition
PD 0331 — infants early childhood ; Protein
PD 0339 — Protein
PD 0363 — vulnerable groups ; Protein
PE 1085 — women during pregnancy nursing ; Protein
PE 9528 Energy ; Medium term cyclic variations solar radiant
Energy [paths...]
PE 8971 — paths ; Soft
PE 8599 — private sector ; Lack conservation
PD 6693 — production ; Environmental hazards
PC 6329 *Energy research ; Insufficient diversification*
Energy resources
PD 0408 — Aggressive uses natural
PC 7517 — Inefficient use
PF 0334 — Long term shortage
PF 0334 — Reliance finite
PE 8971 — Underutilization renewable
PC 7517 — Wasted
PC 6329 Energy shortage
Energy source Underutilization
PF 0445 — oil shale
PE 8194 — peat
PE 8370 — tar sands
Energy [sources...]
PC 7517 — sources ; Aggression against natural
PF 0393 — sources ; Underdeveloped rural
PC 1648 — *stress ; Dietary*
PD 0335 — *supply ; Insufficient diversification urban industrial*
Energy Underutilization
PF 0357 — biogas
PF 0357 — biomass
PF 0357 — *fuelwood*
PE 8971 — *geothermal*
PF 8971 — *ocean*
PF 0370 — solar
PE 8971 — *tidal*
PE 8971 — *wave*
PF 0373 — wind
Energy [Unexplored...]
PF 7960 — Unexplored alternate
PF 0913 — Unhealthy emotional responses atomic
PC 5038 — use ; Inequality
PC 7517 — use ; Unsustainable development
PE 4768 Enforce judgements orders ; Failure
PE 1693 Enforce minimum age laws ; Failure
PF 8927 Enforceability international standards ; Limited
PF 7111 Enforced abstention insight
Enforced [celibacy...]
PD 3371 — celibacy
PD 7443 — collectivization agriculture
PE 7136 — curtailment living standards borrower nations
PD 4259 Enforced disappearances persons
PB 0284 *Enforced drug testing*
PF 1916 Enforced immunization
PB 0284 *Enforced lie detector tests*
PE 3601 Enforced maternity ; Government
PD 3386 Enforced participation community activity
PB 0284 *Enforced polygraph tests*
PF 1916 Enforced vaccination
PD 2664 Enforcement against game poaching ; Inadequate
PF 0223 Enforcement bodies ; Lack law
Enforcement [civil...]
PC 4608 — civil rights ; Inadequate
PF 2454 — complexity ; Law

Enforcement [Compulsion...] cont'd
PA 5740 — *Compulsion*
PD 8697 Enforcement difficulties ; Code
Enforcement [failure...]
PF 8661 — failure against known offenders ; Law
PD 5515 — Forcible obstruction law
PJ 4562 — foreign arbitral awards ; Inadequate
PF 8781 Enforcement groups ; Defeatist self images citizens' law
PD 5515 Enforcement ; Hindrance law
PC 4608 Enforcement human rights ; Inadequate
Enforcement Inadequate
PD 8697 — code
PF 8421 — international law
PF 0336 — local
PE 4768 — national law
PF 6076 — system child support
PD 2918 Enforcement officials ; Corruption law
PE 8881 Enforcement officials ; Excessive use force law
Enforcement [power...]
PF 9175 — power international organizations ; Lack
PF 0223 — power ; Lack
PD 8697 — procedures ; Ambiguous
Enforcement [Random...]
PD 8697 — Random code
PD 8697 — regulations ; Arbitrary
PE 8701 — religion capitalist countries
PF 0336 — Reluctant local law
Enforcement [safety...]
PD 5001 — safety regulations ; Inadequate
PE 4200 — school attendance ; Inadequate
PE 5409 — standards dress
PD 8697 — Subjective law
PF 0336 Enforcement ; Unclear local
PE 4768 Enforcement ; Unstructured law
PE 4745 Engage political action ; Violation right trade unions
PE 2842 *Engagement alternatives ; Limited*
PE 2340 Engagement ; Collapse societal
PD 1564 Engagement elders ; Isolation fostering structures
PE 2421 Engagement ; Fragmented modes citizen
PF 5044 Engagement ; Insignificant leisure
Engagement [Limited...]
PF 2810 — *Limited practical*
PF 5949 — Limited youth
PD 1544 — Lost direction youth
PA 2671 — Engagement ; Meaningless corporate
PF 5949 — Engagement ; Misdirected youth
PF 5978 — Engagement ; Partial teacher
PF 7142 *Engagement patterns ; Limited*
PF 3750 Engagement society ; Collapse corporate
PD 1544 *Engagement ; Unstructured afterschool*
PC 4091 Engaging riot
PD 7183 Engineered micro organisms ; Release genetically
PE 5404 Engineering ; Accidental weed creation genetic
PC 0776 Engineering accidents ; Hazards genetic
PC 1904 Engineering failure risk
PF 4836 Engineering ; Lack legal
PD 1230 Engineering materials structures ; Failure
PF 4975 Engineering projects ; Misappropriation resources high cost civil
PD 7049 Engineering services industries ; Protectionism construction
PF 7383 English language describe female experience ; Lack
PE 4630 *Engorgement ; Carbohydrate*
PF 1824 Enhanced risks disease
PF 0945 Enhancing ethical relations ; Isolating social forms
PE 4507 Enjoyment arts ; Denial right
PF 4034 Enjoyment war
PC 2506 Enlargement ; Arid zone
PE 1028 *Enlargement liver spleen*
PE 4484 Enlistment foreign armed forces ; Unlawful recruiting
PA 5446 Enmity
PA 5532 Enmity *Disaccord*
PA 5446 Enmity *Enmity*
PA 7338 Enmity *Hate*
PD 0518 Enmity husband wife
PA 6979 Enmity *Opposition*
PA 7107 Enmity *Unpleasantness*
PF 7365 Ennui *Boredom*
PA 5806 Ennui *Inaction*
PA 7107 Ennui *Unpleasantness*
PE 8733 Enrolled school ; Maldistribution students
PC 1716 *Enrolment ; Declining school*
PC 1716 *Enrolment ; Decreasing school*
PF 1226 *Enrolment ; Small student*
PC 1716 *Enrolment ; Stagnating school*
PC 3298 Enslavement exploitation females
PD 8405 Enslavement exploitation minors
PF 3956 Ensorcellement
PF 8661 Ensuring prosecutions crimes ; Difficulty
PD 7424 *Enteque ossificans*
PD 7424 *Enteque seco*
PE 3936 Enteric colibacillosis swine
Enteric diseases
PD 3978 — foals
PD 3978 — goats
PE 6708 — horses
PE 7423 — newborn
PD 3978 — sheep
PE 3936 — swine
PD 1753 Enteric fever
PE 3936 Enteric infections
PE 3936 *Enteric salmonellosis swine*
PE 4973 Enteritis
PE 7423 *Enteritis animals*
PE 9509 *Enteritis animals ; Allergic*

PD 9045 *Enteritis ; Chronic*
PD 2730 *Enteritis ; Feline infectious*
Enteritis [Porcine...]
PE 3936 — *Porcine proliferative*
PD 3978 — *poultry ; Necrotic*
PD 3978 — *poultry ; Ulcerative*
PE 3936 *Enteritis swine ; Clostridium perfringens type C*
PE 3936 *Enteritis swine ; Rotaviral*
Enteritis [Transmissable...]
PD 3978 — *Transmissable*
PD 3978 — *turkeys ; Coronaviral*
PD 3978 — *turkeys ; Haemorrhagic*
PD 9045 *Enteritis ; Ulcerative*
PD 2735 *Enterohepatitis ; Infectious*
PE 3936 *Enteropathy ; Proliferative haemorrhagic*
PE 2053 Enterotoxaemia
PA 5568 Enterotoxaemia
PD 7841 *Enterotoxaemia ; E coli*
PA 5568 *Enterotoxaemia ; Infectious*
PE 7769 *Enterotoxaemia sheep ; Type D*
PD 7841 *Enteroviral encephalomyelitis ; Porcine*
PF 1879 Enterprise ; Bias against private
PF 2378 *Enterprise capital ; Insufficient*
PD 2048 Enterprise ; Dependence socialist countries private
PD 5642 Enterprise parastatals ; Lack
PD 5584 Enterprise ; Restrictions against small
PD 2048 Enterprise socialist countries ; Suppression private
PE 0988 Enterprises ; Abuse monopoly power state owned state controlled
PE 0207 Enterprises against labour ; Coercive use economic power transnational
PE 1799 Enterprises ; Consequences restrictive business practices transnational
PE 0042 Enterprises ; Control national economic sectors transnational
PE 9700 Enterprises ; Delays payments state controlled
PE 0226 Enterprises ; Destabilizing financial action transnational
Enterprises developing countries
PE 0322 — Bribery transnational
PE 4102 — Discrimination against women workers multinational
PE 8572 — Inadequate development
PE 0766 Enterprises ; Economic imperialism transnational
PE 1539 Enterprises ; Erosion national sovereignty transnational
Enterprises Excessive
PE 7474 — borrowing state controlled
PD 0071 — concentration business
PE 0060 — exploitation raw material reserves transnational
PE 1651 Enterprises front organizations government ; Misuse business
Enterprises [Inaccessibility...]
PE 0573 — Inaccessibility decision makers multinational
PE 0225 — Inadequate regulation restrictive business practices state
PD 5642 — Inefficiency public
PD 5642 — Inefficiency state controlled
PF 3661 — Information gaps
PE 9405 — Instability micro
PE 9251 — Irresponsible directors commercial
PF 2042 Enterprises ; Lack awareness potential investment small, inner city
PE 0109 Enterprises ; Monopolistic activity transnational
PD 5891 Enterprises ; Multinational
Enterprises [Nationalization...]
PD 1994 — Nationalization domestic
PF 6574 — Non accountability state controlled
PF 1072 — Non accountability transnational
PF 4739 — Enterprises ; Proliferation state controlled
Enterprises [Records...]
PE 7061 — Records destruction transnational
PE 7515 — Reduction workforce
PE 5938 — Restrictions trade union rights public
PE 0161 — restrictive business practice ; Direct foreign investment transnational
PD 5891 Enterprises ; Transnational
PE 0669 Enterprises ; Tying supplies subsidiaries transnational
Enterprises [Uncoordinated...]
PF 1345 — *Uncoordinated economic*
PF 1001 — Undervaluation state
PF 2231 — Unrealized potential commercial
PE 8540 — Entertainment alternatives countryside ; Unexplored
PD 0606 — Entertainment ; Exploitative
PE 6126 — Entertainment facilities ; Inadequate night life
Entertainment [Inadequate...]
PF 8105 — Inadequate public
PD 0606 — Incorporation advertising
PD 3736 — industry ; Corruption
PD 2078 — Entertainment ; Maltreatment animals
PF 2411 *Entertainment market ; Unprofitable*
Entertainment [Parochial...]
PD 0917 — Parochial escapist media
PF 8105 — possibilities ; Limited
PD 7060 — products film industries ; Protectionism
PD 0917 — Entertainment reassurance ; Mass media bias towards
Entertainment [Unethical...]
PF 0374 — Unethical
PF 0374 — Unethical corporate
PF 8105 — Unstimulating
PD 5081 Entertainment ; Violence
PF 2575 *Enthusiasm ; Thwarted community*
PF 4580 Enthusiastic lending banks ; Over
PJ 5074 *Enticement*
PE 1181 Enticement debt
PE 8597 Entities developing countries ; Absence legal
PF 3437 Entities ; Disruption human activities supernatural
PE 2516 Entities ; Monstrous
PF 8855 Entitlement ; Belief personal

Entitlement food

PE 0324 Entitlement food ; Denial
PD 2728 *Entomophthoromycosis*
PC 3597 Entomoses
PE 7560 Entrainment ELF magnetic fields ; EEG
PE 2176 Entrance barriers ; Decline competition due
PE 0853 Entrance terms transnational corporations developing countries ; Inadequate negotiation
PD 6943 Entrapment ; Incrimination
PF 0963 *Entrapping social context*
PD 1231 Entrenched privileges
PF 1765 Entrenched social traditions rural areas ; Rigidly
PD 1231 Entrenchment vested interests
PC 1820 Entrepreneur skills
PC 1820 Entrepreneurial ability developing countries ; Shortage industrial leadership
PD 4728 Entrepreneurial activities ; violation
PD 2879 Entrepreneurs ; Irresponsible
PE 1635 Entrepreneurs ; Stress
PC 1820 Entrepreneurs ; Unconfident potential
PC 1820 Entrepreneurship ; Inexperienced village
PF 5716 Entropy
PD 2561 Entry ; Breaking
PE 3973 Entry during surveillance ; Surreptitious
PE 8423 Entry foreign workers' families ; Refusal
PE 2603 Entry procedures ; Distortion international trade discriminatory customs administrative
PE 1134 *Entry ; Restricted union*
PD 1928 *Entry ; Unlawful*
PD 4798 *Entryism ; Political party*
PE 5431 Enuresis
PA 7078 Environment

Environment [Accidental...]
PD 1386 — Accidental large scale contamination
PE 9295 — arising current policies ; Government neglect future problems social
PC 6675 — armed conflict ; Degradation
PD 2011 Environment basic health small rural villages ; Unsanitary

Environment [Cardiac...]
PE 6811 — Cardiac conditions work
PD 5178 — catastrophic warfare damage ; Vulnerability marine
PD 1670 — Chemical pollutants
PE 4759 — contamination ; Degradation
PC 4584 — Criminal offences against
PD 2672 Environment ; Debilitating deterioration physical
PB 2480 Environment ; Decayed moral

Environment degradation
PB 5250 — Natural
PC 3144 — Psychological
PC 4588 — Socio cultural

Environment [destruction...]
PE 5064 — destruction species ; Degradation
PC 8943 — Deterioration human
PC 8073 — Deterioration international economic
PD 3922 — developing countries ; Degradation

Environment [electrical...]
PE 1412 — electrical power generation ; Degradation
PB 5250 — Erosion human carrying capacity
PF 2418 — Execution orders endangering
PE 143i — Environment fisheries products ; Radioactive contamination marine
PC 4777 Environment ; Hazards human health natural

Environment [Inaccessibility...]
PE 6175 — Inaccessibility green parkland urban
PF 6160 — Inaccessibility quiet zones urban
PE 6185 — Inadequate interaction humans animals urban
PF 1205 — *Inadequate study*
PF 6543 — Ineffective utilization public
PF 6159 — Inhibition exploration children urban

Environment [leisure...]
PE 6256 — leisure activities ; Degradation mountain
PE 6547 — local communities ; Limited control
PF 2527 — Loss humility relation
PE 2443 Environment mankind ; Inevitable destruction natural
PD 2831 Environment ; Noise working
PF 1548 *Environment ; Nonintellectual home*
PE 4965 Environment policy restriction trade developing countries
PF 3918 Environment pollution ; Placement burden proof victims

Environment Poor
PF 1790 — organization community
PC 4588 — *social*
PD 9170 — work
PD 9170 — working
PE 6299 Environment ; Psychological stress urban
PF 9728 Environment resource base economic activity ; Failure recognize
PD 6133 Environment ; Sterile working
PE 7695 Environment trees ; Degradation
PE 2235 Environment tropical villages ; Detrimental effect jungle

Environment [Unaccountability...]
PF 3458 — Unaccountability institutions degrading
PE 5199 — Underprivileged home
PB 6384 — *Unhealthy*
PE 1145 Environment ; Vibration working
PE 5005 Environmental activists ; Assassination
PD 5582 Environmental activists ; Sabotage

Environmental [cancer...]
PD 1239 — cancer
PD 1239 — cancer hazards
PE 3447 — colonialism
PE 9295 — consequences government policies ; Neglect
PC 6675 — consequences war
PD 0467 — conservation ; Unemployment caused
PB 1166 — contamination
PC 0328 — costs economic production ; Hidden

Environmental [costs...] cont'd
PE 6700 — costs ; Inadequate evaluation
PC 4584 — crime
PF 9769 — crises ; Unforeseen

Environmental degradation
PB 6384 — [Environmental degradation]
PE 6142 — automobiles
PB 6384 — Dependence
PD 2285 — desert oases
PD 7835 — developed countries
PD 3922 — developing countries
PD 0767 — due creation dams lakes
PD 6923 — fresh water sources
PD 6124 — high-speed roads
PD 7835 — industrialized countries
PC 2616 — inner city areas
PE 0736 — military activity during peace-time
PE 1720 — off-road all-terrain vehicles
PE 6251 — pipelines
PE 6920 — recreation tourism ; Natural
PD 0826 — recreation tourism ; Social
PE 7403 — recreational use unsurfaced country roads tracks
PD 5845 — Regional
PD 2345 — suburbia
PD 4002 Environmental demand capita ; Excessive
PD 5429 Environmental demand ; Mismanagement

Environmental destruction
PD 5393 — communist regimes
PE 8401 — least developed countries
PE 0736 — military bases
PD 2596 — open-pit mines quarries
PF 9769 Environmental discontinuities ; Hazardous

Environmental [economic...]
PF 9728 — economic decision making ; Incompatibility
PD 1370 — education ; Inadequate
PD 1370 — education ; Lack
PE 1304 — electromagnetism ; Hazards
PE 4378 Environmental hazard ; Chlorofluorocarbons (CFC's)
PE 8136 Environmental hazard ; Vegetable oils fats

Environmental hazards
PC 5883 — [Environmental hazards]
PD 0376 — agricultural livestock production
PD 8328 — Aircraft
PE 1331 — animal feedstuffs, excluding unmilled cereals
PE 8135 — animal oils fats
PC 0229 — atomic radiation
PE 0849 — beverages
PE 8732 — cereals cereal preparations
PC 1192 — chemical materials products
PC 1192 — chemicals
PC 1192 — chemicals petrochemicals industries
PE 9019 — china, pottery earthenware industries
PE 7416 — chlorine-bleached paper products
PD 7541 — coal energy
PE 5160 — Coal mining
PE 0481 — coffee, tea, cocoa, spices their manufacture
PF 1789 — constraining scientific research
PE 8790 — construction industry
PE 7539 — decommissioned nuclear power plants
PC 0111 — due economic development
PC 0328 — economic industrial products
PE 1412 — electric energy

Environmental hazards electrical
PE 8097 — equipment industries
PE 8026 — machinery, apparatus appliances
PE 9642 — power transmission lines

Environmental hazards [electricity...]
PE 1412 — electricity
PE 6360 — electromagnetic pulses
PD 6693 — energy production
PE 9056 — essential oils
PE 9177 — explosives pyrotechnic products
PE 7560 — extremely low frequency electromagnetic radiation
PE 1514 — fertilizers
PD 0372 — fish, crustacea molluscs
PD 0743 — fishing industry

Environmental hazards food
PC 1411 — live animals
PE 1280 — processing industries
PE 1280 — products

Environmental hazards [forestry...]
PE 1264 — forestry logging
PE 9029 — fruit vegetables
PE 8165 — furniture fixtures manufacture
PE 8473 — glass glass products manufacture
PE 6292 — hydropower
PE 0546 — inedible crude non-fuel materials
PE 8397 — Iron steel basic industries
PE 8447 — Leather leather manufactures
PD 0788 — live animals
PE 1859 — machinery equipment industries
PE 1852 — *Machinery equipment industries*
PE 8651 — manufacture plastic products
PE 1344 — manufactured goods
PD 0454 — manufacturing industries
PE 0133 — meat meat preparations
PE 8906 — metalliferous ores metal scrap
PE 1401 — metals
PE 8815 — methane
PE 6682 — mineral exploitation seabed resources
PE 1346 — mineral fuels, lubricants related materials
PD 2596 — mining
PE 8275 — miscellaneous manufactured articles
PD 2596 — *Natural gas production*
PC 1617 — new species introduction

Environmental hazards [non...] cont'd
PE 3035 — non-alcoholic beverages
PJ 2425 — non-electrical machinery
PE 8248 — Non ferrous metals basic industries
PE 7651 — non-ionizing radiation
PE 0890 — non-metallic mineral products industries
PD 4977 — nuclear power production
PE 5698 — nuclear weapons industry
PE 8250 — oil nuts, oil kernels oil-seeds
PF 1083 — Over reaction against
PE 1425 — paper printing industries
PE 1409 — petroleum
PC 1192 — *petroleum refineries*
PD 8566 — plastic materials
PE 8095 — publishing printing industries
PE 8336 — road motor vehicles
PE 1839 — *sea transportation*
PE 3883 — solar radiation
PE 1894 — sugar honey preparations thereof
PF 4854 — Suppression information concerning
PE 8571 — tanning dyeing industries
PE 1103 — textile clothing industries
PE 9074 — textile fibres waste
PE 0483 — tobacco tobacco manufactures
PE 0738 — transport equipment
PE 0828 — undressed hides, skins fur skins
PC 0268 — vibration
PE 1103 — wearing apparel manufacture
PE 8091 — wood cork manufactures
PE 0864 — woodworking industries
PD 7977 Environmental heat ; Excessive
PD 5669 Environmental human diseases

Environmental impact
PF 9450 — current policy ; Uncertain
PF 9328 — technological activities ; Delayed
PC 1174 — technology ; Uncontrolled

Environmental impacts
PE 8453 — coal conversion plants
PF 1105 — Combinations
PF 1105 — Cumulative

Environmental [influences...]
PD 2389 — *influences*
PC 4341 — insecurity
PD 5393 — insensitivity ; Communist
PC 1366 *Environmental laws ; Insufficient*

Environmental [models...]
PF 9576 — models ; Resistance inclusive
PD 7941 — modification ; Hostile
PF 4801 — monitoring ; Inadequate
PE 5196 Environmental oxygen ; Destruction

Environmental [plant...]
PD 2224 — plant diseases
PE 4905 — policies ; Restrictions industrial economic development due
PC 0936 — pollutants ; Human exposure hazardous

Environmental pollution
PB 1166 — [Environmental pollution]
PF 4854 — cover-up
PD 2478 — Damage cultural artefacts
PB 1166 — Dependence
PE 7141 — developing countries ; Inadequate legislation against
PD 4007 — due over-population
PE 6172 — Havens
PC 0936 — Health hazards
PF 9299 — Inadequate legislation against
PF 9299 — Lack regulations against
PD 1584 — nuclear reactors
PE 7244 — Obstruction astronomical observation
PD 9197 — socialist countries

Environmental [poverty...]
PD 5261 — poverty
PF 0799 — problems ; Simplistic technical solutions complex
PF 7318 — prodigality
PF 1733 — *programmes ; Unconsensed*
PE 3977 — protection ; Fragmented social structures
PF 7614 — protection ; Reduced images
PC 8943 Environmental quality ; Denial right
PE 3728 Environmental refugees
PD 5261 Environmental resources poor ; Overuse

Environmental [sexual...]
PD 7707 — sexual harassment
PF 0011 — statistics ; Inadequate
PC 1282 — stress
PE 4953 — stress inhabitants tall buildings
PF 9769 — surprises

Environmental [terrorism...]
PD 5582 — terrorism
PF 4510 — thinking ; Fuzzy
PC 4341 — threats national security
PF 9790 — trends ; Irreversible
PF 3458 Environmental unaccountability government agencies
PD 5582 Environmental vigilantism
PC 2696 Environmental warfare
PF 4510 Environmentalism ; Unrealistic
PD 5582 Environmentalists ; Direct action
PE 4454 Environmentalists ; Sabotage hunting expeditions direct action
PF 9624 Environmentalists ; Stress
PD 5582 Environmentalists ; Violence fanatical
PF 6078 Environmentally adapted nutritional habits food aid ; Modification
PD 9515 Environmentally harmful dam construction

Environmentally [inappropriate...]
PD 1018 — inappropriate products ; Dependence government revenues exploitation
PD 8200 — induced diseases

Erythroblastosis fetalis

Environmentally [insensitive...] cont'd
PF 9728 — insensitive economics
PF 3458 — insensitive government policies
PD 9310 Environmentally unfriendly consumer products
PF 6161 Environments ; Alienating child birth
PF 6192 Environments ; Alienating waiting
PC 3486 Environments ; Destruction wetland
PF 6190 Environments encouraging unstructured public access ; Lack places urban
Environments [inhibiting...]
PF 6193 — inhibiting sleeping public ; Unnatural urban
PE 4685 — Insalubrity animal excrement urban
PE 6171 — Insufficient common land urban
PD 1564 — Isolated non participating elders urban
PF 3192 Environments ; Limited relations beyond local
PE 6256 Environments mountaineers trekkers ; Pollution mountain
PF 2583 Environments ; Poor·service delivery urban
PF 6196 Environments travellers ; Unconvivial hotel
Environments [Unhealthy...]
PF 6182 — Unhealthy lack daily physical activity urban
PF 6199 — Unrelated buildings urban
PE 6151 — urban areas ; Unattractive pedestrian
PD 6282 Environments ; Vulnerability mountainous
PE 6163 Environments ; Wastage open space urban
PF 7358 Envisaged destructive ; Verbal conflict
PF 9269 Envisaged weapons
PA 7253 Envy
PA 6011 Envy *Discontentment*
PJ 8039 Envy ; Penis
PA 5644 Envy *Vice*
PD 7799 Enzootic balanoposthitis
PD 7424 Enzootic calcinosis
PD 2733 Enzootic diseases
PF 2779 Enzootic diseases ; High cost controlling epizootic
PF 9774 Enzootic muscular dystrophy
PD 7841 Enzootic paresis ; Benign
PE 9774 Enzootische herztod
PE 9702 Eosinophilic granuloma cats
PE 9774 Eosinophilic myositis
PE 9427 Eperythrozoonosis
PD 2730 Ephemeral fever
PD 7799 Epidemic bovine abortion
PD 2730 Epidemic cellulitis
PD 0235 Epidemic myalgias
PE 2356 Epidemic parotitis
PD 0713 Epidemic ; Smoking
PD 1753 Epidemic typhus
PE 3895 Epidemic typhus ; Non
PC 0195 Epidemic warfare
PC 2514 Epidemics
PE 9528 Epidemics associated sunspot cycle
PD 9667 *Epidermitis ; Exudative*
PD 9154 *Epididymitis*
PD 7799 *Epididymitis animals*
PE 0924 *Epididymitis rams*
PE 0661 Epilepsy
PE 0661 Epileptic automatism ; Post
PE 0661 Epileptics
PJ 5174 Epiphyseolysis
PE 5050 *Epiphysiolysis ; Proximal femoral*
PG 6134 *Epiphysistis*
PE 7826 *Epiphysitis*
PE 7864 *Epistaxis*
PE 5182 *Epistaxis horse*
PD 2734 Epizootic diseases
PF 2779 Epizootic enzootic diseases ; High cost controlling
PD 2728 *Epizootic lymphangitis*
PE 9702 *Epulis*
PE 9702 *Epulis ; Fibromatous*
PE 1625 Equal benefits elderly workers ; Denial
PD 4916 Equal opportunities employment elderly ; Denial rights
Equal [pay...]
PD 1977 — pay equal work ; Denial right
PD 0309 — pay women ; Denial right
PD 9628 — *promotion opportunities ; Denial right*
PA 0833 — *property rights because discrimination ; Denial*
PC 1268 — protection law ; Denial right
PE 1625 Equal work benefits aged ; Violation right
PD 1977 Equal work ; Denial right equal pay
PC 0006 Equality because race ; Denial right
PF 5155 Equality ; Collapsed conceptualization
PA 0833 Equality ; Denial right
PF 8639 Equality ; Lack models
PE 4712 Equality nations ; Denial
PE 4712 Equality states ; Denial right
PF 5155 Equality ; Trivialization
PC 0308 Equality women ; Denial right
PF 1545 Equated efficiency ; Progress
PF 7798 Equated free speech ; Irresponsible expression emotions
PF 8203 Equation convictions truth ; Excessive
PD 7799 *Equine coital exanthema*
PD 3978 *Equine diarrhoea syndrome ; Acute*
Equine [ehrlichiosis...]
PD 2732 — *ehrlichiosis*
PE 2348 — encephalitis
PD 7841 — *encephalitis*
PD 7841 — *encephalomyelitis*
PD 7841 *Equine grass sickness*
PE 5182 *Equine herpes virus 1 infection*
PE 9554 *Equine infectious anaemia*
PE 5182 *Equine influenza*
PD 7799 *Equine metritis ; Contagious*
PD 3978 *Equine monocytic ehrlichiosis*
PD 5535 *Equine recurrent uveitis*

PE 9844 *Equine ulcerative lymphangitis*
PD 2730 *Equine viral arteritis*
PE 5182 *Equine viral rhinopneumonitis*
PD 6520 *Equipment ; Antiquated fire*
Equipment [Costly...]
PD 6520 — *Costly fire*
PF 0296 — *Costly housing*
PJ 1768 — *costs ; Prohibitive*
PF 4854 — Cover up unsafe
PF 6544 *Equipment ; Deficient city*
PD 5859 *Equipment ; Deliberate destruction*
Equipment developing countries
PE 6640 — Acquisition inappropriate
PF 4108 — Insufficient agricultural
PE 8592 — Insufficient inappropriate energy
PE 8166 Equipment development projects ; Overuse machinery
PD 1565 Equipment ; Disrepair
Equipment [Environmental...]
PE 0738 — Environmental hazards transport
PE 4742 — Excessive reliance infallibility
PF 1345 — *Expensive water*
Equipment [failure...]
PD 1475 — failure ; Electronic
PC 1904 — failure ; Mechanical
PE 5803 — Fouling rust marine
PE 5572 — Equipment ; Human fatigue during control complex
PE 4129 Equipment ; Inaccurate medical diagnosis due inadequate
Equipment Inadequate
PE 2757 — disinfection measures animal housing
PD 6520 — *hauling*
PD 4004 — health
PD 6520 — maintenance
PF 3560 — *sports*
PC 0666 — standardization procedures
PF 0905 — *teaching*
PF 2410 Equipment ; Inappropriate level technological
PD 6520 Equipment ; Inappropriate maintenance
Equipment industries
PE 8097 — Environmental hazards electrical
PE 1859 — Environmental hazards machinery
PE 1852 — *environmental hazards ; Machinery*
PE 1852 — Instability machinery
PE 1852 — *instability ; Machinery*
PE 1852 — *underdevelopment ; Machinery*
PF 0942 — Underdevelopment metal products, machinery
Equipment [industry...]
PE 1448 — industry transnational corporations ; Domination agricultural
PF 6411 — Information overload during control complex
PD 0620 — Instability trade machinery transport
Equipment Insufficient
PD 6520 — *heavy*
PF 2568 — *library*
PF 0905 — *preschool*
PF 3560 — *sports*
Equipment Lack
PD 6520 — *heavy*
PF 3467 — *irrigation*
PF 6524 — *modern*
Equipment [Limited...]
PF 3554 — *Limited drilling*
PF 0905 — *Limited preschool*
PE 1436 — Long term shortage machinery transport
PC 2297 Equipment machinery ; Unreliability
Equipment maintenance
PD 1565 — Inadequate
PE 1722 — Prohibitive cost
PD 1565 — servicing ; Deterioration quality
PE 8789 Equipment materials needed act against problems ; Shortage
PD 6520 *Equipment ; Obsolete*
Equipment Prohibitive cost
PF 2410 — *capital*
PF 3630 — *farm*
PF 6478 — *fishing*
Equipment [replacement...]
PJ 4688 — replacement parts ; Overpriced
PC 4844 — *Restricted access sophisticated medical*
PE 7958 — Restrictive practices trade machinery transport
Equipment [safety...]
PE 9073 — safety regulations ; Distortion international trade discriminatory formulation
PF 0905 — *Shortage school*
PE 0680 — Substance abuse during control complex
Equipment [Unavailability...]
PF 6524 — *Unavailability modern*
PD 9306 — Undermanning critical
PD 1484 — Underutilization second hand
PE 2813 — *Unfamiliar use*
PE 4859 — Unsafe industrial, laboratory medical
PF 3467 — *Unshared irrigation*
PF 6556 — Equipped village leadership ; Ill
PB 4653 Equity capital ; Lack
PC 1524 Equivalence national educational qualifications ; Non
PC 1524 Equivalencies ; Uncertainty university degree
PE 7020 Equivalent national sections international nongovernmental organizations ; Incompatible
PF 0091 Equivalents languages ; Lack terminological
PA 6379 Equivocation *Avoidance*
PA 6698 Equivocation *Difference*
PA 7325 Equivocation *Irresolution*
PA 5502 Equivocation *Reason*
PA 7411 Equivocation *Uncommunicativeness*
PC 7367 Equivocation *Unintelligibility*
PC 1056 Eradication ; Minority

PF 7106 Erection risks ; Building
PE 2462 Ergonomic design ; Non
PE 4586 Ergot grain
PE 9458 Ergotism
PF 2199 Eroding support unfriendly governments
PC 8193 Erosion
PB 9748 Erosion biological diversity
PE 7099 Erosion brought site development ; Land
PE 6734 Erosion ; Coastal
PB 8031 Erosion democratic principles
PA 6858 Erosion *Disintegration*
PC 5773 Erosion ecosystem diversity
PF 1664 Erosion elders' wisdom
PF 3979 Erosion filial piety
PF 8158 Erosion funding effectiveness ; Administrative overhead
PB 4788 Erosion ; Genetic
Erosion human
PF 9612 — capital
PB 5250 — carrying capacity environment
PF 4176 — rights privacy use computer databases
Erosion [Impairment...]
PA 6088 — *Impairment*
PA 6852 — *Inappropriateness*
PF 9612 — investment human resources
PD 3035 Erosion journalistic immunity
PB 1701 Erosion metals
PD 3496 Erosion national sovereignty foreign military presence
PE 1539 Erosion national sovereignty transnational enterprises
PF 5956 Erosion old age pensions ; Progressive
PE 6725 Erosion ; Pay system
PF 9274 Erosion religious belief ecumenical dialogue
PD 9515 Erosion resulting dams ; Coastal
Erosion [scientific...]
PF 9693 — scientific objectivity
PF 7798 — social dialogue
PF 7911 — social security
PF 0306 — societal memory
PF 4405 — socio-economic diversity
PD 0949 — Soil
PE 5015 — sovereignty
PF 7227 — standards quality ; Commercial
PD 0949 — Soil ; Top soil
PD 2290 Erosion ; Uncontrolled river
PD 4282 Erosion university autonomy
PA 1782 Erosion ; Value
Erosion [water...]
PD 2290 — water ; Soil
PE 3656 — wind ; Soil
PD 2790 — work ethic
PE 5190 Erotic advertising ; Vulgar combination sacred
PE 5042 Erotic art
PA 6414 *Erotic delusions*
PE 5844 Erotic friction
PE 7031 Erotomania
PE 0435 *Erotomanic paranoia*
PE 0812 Errant nationals
PE 3812 Erratic flows concessional non-concessional lending developing countries
PF 2451 *Erratic land assessments*
PE 4200 Erratic student attendance
PF 9980 *Erratic time design*
PF 2996 *Erratic truck availability*
PD 0722 Erratic water system
PF 0531 Erroneous decisions
PF 3096 Erroneous documents
PF 3702 Erroneous people
PF 0466 Erroneous precipitation nuclear conflict
PA 6180 Erroneousness *Error*
PA 6997 Erroneousness *Imperfection*
PA 7411 Erroneousness *Uncommunicativeness*
PA 6180 Error
PF 3702 Error
PF 8479 Error criminal justice
PF 3702 Error ; Human
PF 8479 Error ; Judicial
PF 3702 Error ; Operator
PF 5163 Error ; Refusal admit
PF 4428 Error ; Risk computer
PF 4176 Errors computerized personal data
PF 1440 Errors ; Diplomatic
PF 5296 Errors ; Failure government apologize
PD 2389 Errors ; Genetic
PE 4428 Errors ; Hazardous computer software
PF 6916 Errors ; Interpretation translation
PF 5304 Errors ; Laboratory testing
PD 5380 Errors ; Legal
PC 2270 Errors metabolism ; Inborn
PF 3702 Errors miscalculations ; Human
Errors [Policy...]
PF 0531 — Policy making
PF 5627 — Popular
PF 5653 — Proverbial lore
PF 4841 Errors risks medical self-experimentation
PF 4118 Errors ; Statistical
PD 4736 Errors ; Surgical
PD 3568 Eruptions ; Volcanic
PD 2226 Erwinia
PD 9094 Erysipelas
PD 2731 Erysipelas ; Swine
PD 9094 Erysipeloid
PD 2732 *Erysipelothrix infection poultry*
PC 8534 Erythema
PE 9509 Erythematosus animals ; Systemic lupus
PD 2565 Erythematosus ; Lupus
PD 2565 Erythematosus ; Systemic lupus
PE 2041 Erythroblastosis fetalis

−919−

Erythroblastosis

PE 2041 *Erythroblastosis ; Foetal*
PD 7420 *Erythropoietica ; Congenital porphyria*
PE 0135 Escalating control international trade transnational corporations
PF 5956 Escalating pension costs
PF 8713 Escalating pitch ; Distortion musical quality
PB 6335 Escalation ; International crisis
PE 5390 Escape ; Aiding
PE 5390 Escape ; Introducing possessing contraband useful
PE 6260 Escape official detention
PE 1200 Escape prisoners war enemy aliens ; Aiding
PF 1112 Escaping reality popular psychological screens
PF 0916 Escaping reality sophisticated shields against life's pain
PF 7523 Escapism
PD 4069 Escapist family life styles
PF 0386 Escapist leisure
PD 0917 Escapist media entertainment ; Parochial
PF 9139 Eschatological enemy
PF 0638 Eschatological unpreparedness
PF 4528 eschaton
PD 3978 *Esophagism*
PF 7077 Esoteric knowledge ; Concealment
PD 7424 *Espichacao*
PD 7424 *Espichamento*
PC 2140 Espionage
Espionage [Computer...]
PE 9831 — Computer database
PD 2923 — Counter
PE 4343 — Criminal
PD 1787 Espionage domestic politics
PD 2930 Espionage ; Electronic
PC 2921 Espionage ; Industrial
PC 1868 Espionage ; International political
PD 2922 Espionage ; Military
PF 2924 Espionage ; Sociological
PF 1301 Essential corporateness ; Deteriorated structures
PD 1942 Essential freedom due unawareness actuality ; Lack
PE 0585 *Essential hypertension*
PF 2115 Essential local infrastructure ; Lack
PE 9056 Essential oils ; Environmental hazards
PE 8232 Essential oils perfume materials ; Instability trade
Essential services
PC 7674 — Absence
PF 1667 — developing country rural communities ; Restricted delivery
PF 2875 — isolated communities ; Inaccessibility
PF 2875 — isolated communities ; Restricted delivery
PF 2115 — materials needed development ; Lack
PF 6527 — rural communities ; Prohibitive cost
PC 3651 Essentials ; Production non
PF 2822 *Established geographic division*
PF 5115 Established religions ; Hierarchical structures
PF 7593 Established religions ; Parochialism
PD 4125 Establishing changes rural economic patterns ; Difficulties
PF 4306 Establishing New International Economic Order ; Lack progress
PJ 0095 Establishment/dis establishment tension ; Undervalued creative
PE 3948 Establishment disestablishment tensions ; Failure understand necessity creative
PE 9339 Establishment homosexuality
PE 7128 Establishment-orientation international nongovernmental organizations
PJ 0095 Establishment tension ; Undervalued creative establishment/dis
PF 2813 *Establishments ; Unprotected commercial*
PD 5422 Estate agents ; Negligence real
PD 5422 Estate practice ; Unethical real
PD 0406 Estates ; Maldistribution land associated large traditional
PE 1925 Estates ; Socially inadequate housing
PF 5354 Esteem ; Low self
Esteem Self
PA 6491 — [Esteem ; Self]
PA 6659 — [Esteem ; Self]
PA 7211 — [Esteem ; Self]
PF 2575 Estranged neighbourhood relations
PB 0034 Estranged relatives
PB 0034 Estrangement
PB 0034 Estrangement ; Dependence interpersonal
PB 0034 Estrangement ; Interpersonal
PD 2103 Estrangement marriage
PE 9458 *Estrogenism ; Fusarium*
PE 8175 Estuarine waters ; Pollution bays
PF 3354 Eternal life ; Discrimination access
PF 3354 Eternal life ; Elitist access
PF 7228 Eternal punishment
PD 8047 Ethic developing countries ; Undeveloped work
PD 2790 Ethic ; Erosion work
PF 2105 Ethic ; Governmental barriers global
PF 1118 Ethic ; Lack common
PF 1729 Ethic ; Loss commitment
PF 2096 Ethical contexts ; Fragmented
Ethical [decay...]
PB 2480 — decay
PB 2480 — decay ; Dependence
PF 1064 — decision ; Absence common structure
PF 2070 — decision making ; Lack system
PF 0966 — decisions ; Lack meaningful educational context
PD 8227 Ethical misconduct leaders
PF 0945 Ethical relations ; Isolating social forms enhancing
PF 7220 Ethical resolution scientific issues
PF 1114 Ethical value screens ; Collapse providing
PF 1178 Ethical void ; Self conception
PF 1249 Ethics ; Haphazard forms social
PD 5770 Ethics ; Lack medical
PF 1622 Ethics ; Over reaction past sexual
PF 1249 Ethics ; Selective
PB 5765 Ethnic chauvinism
PC 3685 Ethnic conflict
Ethnic [differences...]
PF 8156 — differences ; Bureaucratic nonrecognition
PC 3685 — difficulties
PC 3686 — discrimination
PC 3291 — disintegration
PD 4782 — divisions ; Internal language, religious
PC 3685 Ethnic feuding
PC 3316 Ethnic groups ; Isolation
PC 3685 Ethnic killings
PD 0355 Ethnic marriage ; Inter
Ethnic minorities
PC 8999 — Denial rights
PC 3205 — Expulsion
PC 0582 — Underprivileged
PC 3686 Ethnic racism
Ethnic [segregation...]
PC 3315 — segregation
PF 5929 — social discrimination foreign language teaching
PF 5452 — stereotyping
PC 3685 Ethnic tensions
PF 6235 Ethnic unification
PC 3685 Ethnic unrest
PC 3685 Ethnic violence ; Inter communal
PC 6664 Ethnic weapons ; Dependence genetic
PC 6664 Ethnic weapons ; Genetic
PB 5765 Ethnocentricity
PB 5765 Ethnocentricity ; Dependence
PB 5765 Ethnocentrism
PC 1328 Ethnocide
PC 1328 Ethnocide ; Cultural
PC 1328 Ethnocide ; Dependence
PD 5228 *Ethylene glycol poisoning*
PD 1091 Eugenic measures ; Lack
PC 2153 Eugenics
PD 1091 Eugenics ; Discrimination against
PD 1091 Eugenics ; Lack
PG 2456 *Eunuchism*
PC 3383 *Eunuchs ; Social segregation*
PF 5383 Euphemism
PF 5183 Euphemisms ; Manipulative
PF 7727 Euphenics ; Negative aspects
PF 1466 Eurocentric religion
PF 5472 *Eurocentrism*
PF 3876 Eurocommunism
PE 4111 Europe ; east west division
PE 4111 Europe ; split eastern western
PD 1863 European insecurity vulnerability
PF 5472 *European Western culture ; Excessive bias favour*
PF 2643 Euthanasia ; Criminalization
PF 2643 Euthanasia ; Denial right
PD 2257 Eutrophication lakes rivers
PA 7394 Evaluation*complex
Evaluation [development...]
PF 9244 — development projects ; Inadequate
PB 5486 — *disability compensation ; Arbitrary*
PF 1191 — Discrepancies human life
PE 6700 Evaluation environmental costs ; Inadequate
PF 5482 Evaluation ; Inconsistent risk
PF 2648 *Evaluation ; Predetermined employer*
PF 5482 Evaluation risks ; Unequal
PD 1975 Evaluations ; Abuse IQ
PF 9243 Evaluations ; Falsification programme
PF 6325 Evangelism
PA 6379 Evasion *Avoidance*
Evasion [complaints...]
PF 9157 — complaints ; Official
PD 2620 — customs excise duties
PD 1466 — Customs tax
PD 0356 Evasion ; Draft
PE 4693 Evasion ; Fare
PA 7363 Evasion *Improbity*
PF 7431 Evasion issues
PD 4208 Evasion law
PE 4339 Evasion national law regulations
PA 8658 Evasion obligations
PF 0608 Evasion past events
PA 5502 Evasion *Reason*
PE 5873 Evasion shipping regulations taxes flags convenience
PE 3149 Evasion social costs companies
PD 1466 Evasion ; Tax
PA 7411 Evasion *Uncommunicativeness*
PC 5576 Evasion work
PC 1326 *Even toed ungulates ; Endangered species*
PE 4222 Events ; Commercialization athletic activities sports
PE 0197 *Events ; Discrimination against women payment prizes athletic*
PF 4932 Events ; Distortion historical
PF 0608 Events ; Evasion past
Events [Ill...]
PF 8577 — Ill omened
PF 4389 — *Illusion controlling*
PF 7700 — Inadequacy methodologies deal discontinuities random
Events Infrequent
PF 5250 — community cultural
PF 5250 — social
PF 5250 — unifying
PF 5250 Events ; Insufficient community
PF 5250 Events ; Limited cooperative
PF 4761 Events ; Nationalist exploitation sporting
PD 4840 Events ; Negative life
PF 8577 Events ; Ominous
Events [Political...]
PE 4206 — Political boycott international sports
PD 5207 — Political media
PF 4761 — Politicization international sports
PF 8229 — Pretended ignorance
PF 1205 *Events ; Shortage exciting*
PF 5728 Events ; Unreported
PD 7799 *Eversion ; Uterine*
PF 8661 Evidence convict known offenders ; Inadequate
Evidence [decay...]
PF 5403 — decay
PD 7385 — Denial
PF 9077 — disciplines ; Inadmissability
PF 5127 — Distortion
Evidence [Faking...]
PF 1602 — Faking scientific
PF 5127 — False
PD 3017 — False political
PE 9309 Evidence ; Illegally obtained
PE 5424 Evidence jurisdictions ; Inadmissability
PF 4633 Evidence ; Loss scientific
PF 5127 Evidence ; Misleading
PF 5403 Evidence ; Misremembrance
PD 8716 Evidence ; Official fabrication
PD 8716 Evidence ; Official misinterpretation
Evidence [Planting...]
PD 8716 — Planting
PD 8716 — Police falsification
PF 6707 — potential hazards ; Bureaucratic procrastination response
PD 7385 Evidence ; Suppression
PF 4633 Evidence ; Suppression scientific
Evidence [Tampering...]
PF 4699 — Tampering
PF 7291 — Tampering physical
PF 9794 — torture ; False
PF 7042 Evil
PA 5454 Evil *Badness*
PF 7042 Evil ; Belief
PF 0926 Evil ; Complicity
PF 7042 Evil ; Dependence
PA 5497 Evil *Difficulty*
PF 5123 Evil enchantment
PF 2343 Evil eye
PF 5633 Evil governments
PJ 7640 Evil ; Knowledge good
PF 7754 Evil ; Non resistance
PF 7018 Evil ; Personified
PE 6555 *Evil ; Poll*
PA 5644 Evil *Vice*
PA 7280 Evil *Wrongness*
PD 1091 Evolution ; Insufficient intervention human
PF 7490 Evolution ; Violence determined
PF 1181 Evolutionary catastrophes
PF 3348 Evolutionary theories ; Misuse
PF 2196 Evolutionism ; Social
PD 7799 *Ewes ; Chlamydial abortion*
PD 7799 *Ewes ; Endemic abortion*
PD 7420 *Ewes ; Parturient paresis*
PD 7420 *Ewes ; Pregnancy toxaemia*
PD 2511 Ex colonial countries ; Over development bureaucracy
PE 6929 Ex convicts ; Denial rights
PE 6929 Ex detainees ; Discrimination against ex prisoners
PE 6929 Ex offenders ; Discrimination against
PE 6929 Ex prisoners ex detainees ; Discrimination against
PE 3814 Exaggerated advertising claims products
PJ 2489 Exaggerated reliance professionalized services
PB 2542 *Exaggerated tolerance*
PF 3096 *Exaggeration*
PF 9794 Exaggeration human rights infractions ; Politically motivated
PA 6607 Exaggeration *Misjudgement*
PA 6644 Exaggeration *Misrepresentation*
PA 7411 Exaggeration *Uncommunicativeness*
PC 1716 Examination failures ; Wastage school
PF 1266 Examination procedures education ; Inappropriate selection
PD 4178 Examinations ; Stress due
PE 7489 Examine witnesses ; Denial right
PF 2848 *Examples ; Lack practical*
PD 7799 *Exanthema ; Equine coital*
PD 2730 *Exanthema swine ; Vesicular*
PA 7378 Exasperation *Aggravation*
PA 7253 Exasperation *Envy*
PA 7107 Exasperation *Unpleasantness*
PD 1823 Excavations ; Clandestine
PE 3916 Exceeding competence juries ; Trials
PF 5703 Excellence ; Lack qualitative
PF 6730 Exceptionalism ; Human
PB 8952 Excess
PD 0153 Excess ; Alcoholic
Excess [capacity...]
PE 5897 — capacity shipping industry
PA 7193 — *Cheapness*
PD 9654 — *Cortical*
PC 3116 Excess demand goods capitalism
PE 5580 Excess human body hair
Excess [income...]
PD 8568 — income ; Minimal
PA 5473 — *Insufficiency*
PA 6466 — *Intemperance*
PB 0750 Excess labour
PF 5044 Excess leisure time arising automation
PE 4933 Excess manpower developing countries ; Lack technical development

Exchange rate

PD 0779	Excess production capacity
PC 2618	Excess profits
PA 7411	Excess *Uncommunicativeness*
PA 6921	Excess *Undueness*
PE 7045	Excess western-based secretariats international nongovernmental organizations
PD 1881	Excesses ; Military
PF 9542	Excesses ; Military

Excessive [accumulation...]
- PD 9653 — accumulation wealth government leaders
- PD 0153 — alcohol usage
- PE 3247 — *alimony costs*
- PF 1555 — animal sanitary regulations international travel
- PF 4345 — anxiety lending developing countries

Excessive [bias...]
- PF 5472 — *bias favour European Western culture*
- PF 2949 — *bill-paying surcharges*
- PE 7474 — borrowing state-controlled enterprises
- PE 1093 — burden poor due legal delays
- PF 2711 — bureaucratic administrative centralization
- PE 8893 — bureaucratic requirements welfare benefits

Excessive [capital...]
- PE 1139 — *capital outflow*
- PF 3434 — career mobility aid officials
- PF 6389 — caution
- PD 3491 — childhood dependency developing countries
- PF 3388 — *claims human development sports*
- PF 1781 — *classification demands*
- PE 6350 — *cloud cover*
- PD 2772 — commercial exploitation farm animals industrial concerns
- PE 4215 — commercialization media
- PF 2874 — *community crime*
- PF 2806 — complexity intergovernmental organizations
- PD 0071 — concentration business enterprises
- PE 9457 — concentration export markets developing countries
- PF 7166 — confidence prediction capacity
- PF 5057 — consumer choice

Excessive consumption
- PD 0153 — alcohol
- PD 1611 — alcohol adolescents
- PD 4518 — animal flesh
- PC 3908 — *carbohydrates*
- PE 4261 — fats
- PC 2518 — goods services
- PD 4518 — meat
- PD 7089 — protein
- PE 5551 — resources developed countries
- PE 4231 — salt
- PC 3908 — specific foodstuffs
- PC 3908 — *spices*
- PE 1894 — sugar
- PE 4665 — vitamins
- PE 0194 — Excessive control raw materials markets transnational corporations
- PE 1879 — Excessive corporate debt

Excessive cost
- PE 4784 — animal protein
- PE 1122 — comprehensive information systems
- PF 3499 — *construction sites*
- PE 7539 — decommissioning nuclear reactors
- PE 6059 — effective prosecution offenders
- PF 8763 — manpower
- PE 5755 — medical drugs

Excessive costs
- PD 9338 — [Excessive costs]
- PE 4856 — cultural burdens human illnesses disabilities
- PE 5899 — inefficient port cargo-handling
- PE 5812 — sea access land-locked developing countries
- PE 5896 — unsuitability insurance land-locked developing countries

Excessive [court...]
- PF 0922 — court costs ; Ineffective protection individual rights due
- PE 0208 — *customs formalities*
- PE 2603 — customs trade formalities

Excessive [data...]
- PF 2376 — data
- PE 2197 — *daytime sleepiness*
- PD 2502 — debt socialist countries West
- PC 9067 — *degree fever*
- PC 3116 — demand goods capitalist systems
- PC 4824 — demands natural resources

Excessive dependence
- PE 8533 — computer models complex system behaviour
- PF 0938 — export credits developing countries
- PE 4780 — local communities outside services

Excessive [desire...]
- PF 9424 — desire risk
- PC 8316 — development credit
- PF 9369 — discharge urine
- PF 6141 — dispersion community facilities
- PF 1673 — dissemination materialistic images
- PF 2439 — *duplication products*
- PD 9578 — Excessive economic exploitation animal species
- PE 5713 — Excessive emigration island developing countries

Excessive emphasis
- PE 3831 — developing country weaknesses media
- PF 0059 — fashionable areas research
- PF 4164 — fashionable problems

Excessive [employment...]
- PD 3557 — employment married women
- PD 3557 — employment married women ; Adverse social consequences
- PD 4002 — environmental demand capita
- PD 7977 — environmental heat

Excessive [equation...] cont'd
- PF 8203 — equation convictions truth
- PF 1447 — expenditure armaments
- PF 4196 — expense athletic training programmes
- PF 4192 — expense international athletic competitions
- PE 0060 — exploitation raw material reserves transnational enterprises
- PE 1704 — exposure medical patients radiation
- PE 1909 — exposure radiation consumer goods electronic devices
- PC 1100 — external trade deficits
- PE 1496 — external trade deficits developing countries
- PD 3246 — extralegal powers

Excessive [family...]
- PF 3560 — *family size*
- PF 6389 — *fear risk*
- PF 2009 — *financial pressure*
- PF 4130 — food production developing countries ; Economic disadvantages

Excessive foreign
- PD 8030 — investment ; National insecurity due
- PD 0765 — investment traditional industries developing countries
- PD 2133 — public debt developing countries
- PD 9168 — public debt industrialized countries
- PE 8063 — Excessive freight costs
- PE 0208 — Excessive frontier formalities international travel

Excessive government
- PF 0304 — control
- PF 9530 — dependence
- PF 0304 — interference
- PD 4800 — intervention private sector
- PF 0304 — intervention society
- PD 1902 — participation economies developing countries
- PC 1812 — secrecy

Excessive [governmental...]
- PD 0183 — governmental spending
- PC 6215 — growth social expenditure
- PB 9015 — growth ; Unwanted

Excessive [health...]
- PF 1555 — health regulations
- PD 0140 — hours work
- PF 2899 — *hybrid seed costs*

Excessive [imposition...]
- PE 4363 — imposition states emergency
- PE 4887 — imprisonment remand
- PC 1408 — inbreeding domestic animals
- PD 2968 — income dependence primary commodities developing countries
- PE 2598 — injury export interests developing countries due export cartels
- PE 1557 — institutional debts
- PD 0932 — institutionalization education
- PF 8209 — institutionalization vulnerable groups
- PD 0153 — intake alcohol
- PF 4954 — integration religious belief
- PF 6186 — intensification parent-child relationship
- PD 5676 — interdependence stock markets
- PE 6836 — Excessive job control
- PF 7451 — Excessive joy functioning

Excessive [land...]
- PE 5059 — land usage
- PE 2525 — land usage transportation systems
- PJ 3509 — legalization
- PE 4887 — length pre-trial internment
- PE 4602 — length prison sentences
- PF 4723 — leniency sentencing offenders
- PD 9235 — logging
- PA 6714 — *longevity*
- PF 5931 — luxury
- PE 7705 — Excessive medical intervention childbirth

Excessive military
- PF 1447 — expenditure
- PF 1447 — expenditure ; Dependence
- PE 3402 — usage land

Excessive [neutrality...]
- PF 3076 — neutrality intergovernmental official information
- PF 3145 — number crop varieties
- PC 0311 — number nuclear weapons

Excessive [occupational...]
- PD 1500 — occupational exposure radiation
- PE 6660 — office space cities
- PD 2778 — overtime

Excessive [paperwork...]
- PF 5856 — paperwork
- PF 2602 — *parental defensiveness*
- PF 7700 — parental drunkenness
- PB 3209 — political influence traditional vested interests
- PF 4761 — political intervention international athletic exchanges
- PD 5475 — politicization media
- PB 0035 — population growth
- PF 3806 — population mobility

Excessive portrayal
- PE 7354 — crime media
- PE 1478 — destabilizing information media
- PE 3980 — drugs, alcohol cigarettes media
- PE 1478 — negative information media
- PE 3831 — perspectives industrialized cultures media
- PE 7930 — sex media
- PE 3980 — substance abuse media
- PE 6844 — terrorist activity media
- PD 6279 — violence mass media

Excessive [power...]
- PD 5807 — power independence transnational corporations
- PF 6957 — preoccupation corpses
- PE 5755 — pricing brand-name drugs
- PC 2618 — profits

Excessive [proliferation...] cont'd
- PD 0644 — proliferation medical drugs
- PF 4936 — prolongation dying process
- PE 8659 — proportion income spent food
- PD 1987 — protection industries customs duties
- PC 2546 — public debt
- PE 4103 — Excessive rainfall
- PE 6982 — Excessive regulation television

Excessive reliance
- PE 7195 — developing countries external sources development financing
- PF 0019 — economic indicators human development
- PF 4473 — fashionable solutions problems
- PE 4742 — infallibility equipment
- PC 6631 — local governments grants
- PE 1514 — mineral fertilizers

Excessive [repatriation...]
- PC 3134 — repatriation profits foreign investors developing countries
- PJ 5595 — restrictions legalizing medications
- PF 1238 — retail prices
- PF 6389 — Excessive safety precautions

Excessive salaries
- PF 8763 — [Excessive salaries]
- PF 7578 — corporate executives
- PF 8317 — experts
- PD 9653 — government leaders
- PE 6388 — international civil servants

Excessive [saturated...]
- PE 4261 — saturated fats ; Consumption
- PF 5868 — sexual activity
- PE 7031 — sexual desire
- PD 6120 — size metropolitan regions
- PF 8798 — size social institutions
- PD 8114 — social costs structural adjustment debtor developing countries
- PF 4891 — socio-economic order
- PF 0256 — specialization
- PC 0432 — specialization education
- PE 2147 — speed motor vehicles
- PF 2271 — standardization
- PE 5776 — stray animal populations
- PD 9802 — subsidies farmers developed countries

Excessive [taxation...]
- PD 3436 — taxation
- PD 1533 — television viewing
- PD 0826 — tourism
- PE 5328 — trace elements human body
- PE 3926 — Excessive unnecessary burden interest repayments developing countries
- PE 1478 — Excessive unsupportive criticism media

Excessive use
- PF 4286 — acronyms abbreviations
- PD 1207 — chemicals control pests
- PC 1408 — domestic animals breeding purposes
- PD 9082 — drift nets
- PE 1514 — fertilizers

Excessive use force
- PE 8881 — law enforcement officials
- PE 8881 — military personnel
- PE 8881 — police

Excessive use [foreign...]
- PE 9643 — foreign programmes media
- PD 9534 — land agriculture
- PF 6152 — land automobiles
- PD 7585 — natural parks
- PF 7127 — Excessive virtue

Excessive [waiting...]
- PF 5120 — waiting times government facilities
- PC 8222 — wealth
- PE 2902 — wheat surpluses
- PD 1245 — wind-blown dirt
- PE 0341 — Excessive youth sex education
- PF 3455 — Excessively costly prestige projects
- PF 8299 — Excessively distant markets
- PD 9082 — Excessively efficient fishing nets ; Deployment
- PF 1706 — Excessively efficient use time
- PB 0469 — Excessively high population density
- PD 7625 — Excessively large families
- PD 8916 — Excessively low wages
- PA 7193 — Excessiveness *Cheapness*

Excessiveness [Insanity...]
- PA 7157 — *Insanity*
- PA 5473 — *Insufficiency*
- PA 6466 — *Intemperance*
- PD 4500 — Exchange bourse related crimes ; Stock
- PB 1016 — Exchange ; Breakdown resource

Exchange [Chronic...]
- PC 8182 — Chronic shortage foreign
- PF 7798 — Collapse systems verbal
- PF 3070 — control restrictions ; Counterproductive
- PE 8583 — Exchange depreciation ; Competitive

Exchange [ideas...]
- PD 8731 — ideas ; Prevention
- PD 8731 — ideas ; Restrictions
- PC 0185 — information ; Prevention
- PJ 5244 — Exchange monopoly ; Foreign
- PE 5903 — Exchange notes transnational corporations ; Destabilization monetary systems
- PF 4857 — Exchange ; Obstacles international cultural
- PE 4809 — Exchange ; Obstructions international athletic
- PE 6680 — Exchange promissory notes ; Conflicting laws concerning bills

Exchange rate
- PF 6102 — misalignment
- PE 8583 — policies ; Discriminatory

Exchange rate cont'd
PE 7563 — policies ; Inability developing countries adopt appropriate
PD 9489 — speculation
PF 1874 — system ; Mismanagement
PE 4476 — systems ; Dual
PE 5930 — volatility
Exchange rates
PE 5930 — Floating
PE 5930 — Instability
PE 0980 — transnational corporations ; Aggravation instability
PJ 9314 Exchange reliance ; Cash
PF 3070 Exchange restrictions ; Foreign
Exchange [shortage...]
PD 3068 — shortage developing countries ; Foreign
PC 8182 — Shortage foreign
PC 8182 — shortages ; Foreign
PF 5006 — skills developing countries ; Limited
Exchange [technical...]
PF 0209 — technical information concerning problems ; Inadequate
PD 0181 — *torture methods ; Inter Governmental*
PD 2897 — trade ; Unequal
PE 0117 Exchange ; Underutilization non monetary foreign
PD 4500 Exchange violations ; Securities commodities
PF 3099 Exchanges ; Abuse international cultural, diplomatic commercial
PD 4511 Exchanges bourses ; International collapse stock
PE 4785 Exchanges cultural cooperation ; Obstructions international personnel
PE 9440 Exchanges ; Discriminatory operation international commodity
PF 4761 Exchanges ; Excessive political intervention international athletic
PD 4500 Exchanges related crimes ; Commodity
PE 9440 Exchanges ; Restrictions access developing countries commodity
PF 2346 *Exchanges ; Unrecorded property*
PD 2620 Excise duties ; Evasion customs
PE 4033 Excise officials ; Corruption customs
PE 6055 Excision
PE 8933 Excitement danger young people ; Dependence
PD 4714 Excitement war
PF 1205 *Exciting events ; Shortage*
PD 2154 Exclaves ; Enclaves
PE 5074 Excluded company polity ; Worker participation
PA 5869 Exclusion
PE 6561 Exclusion community's culture
Exclusion [decision...]
PE 7155 — decision making processes those who question context process
PD 0784 — disabled persons social cultural life
PA 6191 — *Disapproval*
PA 5869 Exclusion *Exclusion*
PE 1843 Exclusion human requirements ; Market indicators
PD 1060 Exclusion masses setting criteria judicial judgements
PA 6387 Exclusion *Necessity*
PF 3720 Exclusion opposing views
PF 8703 Exclusion participation decision-making
PE 2268 Exclusion pre-adults family decisions
PA 7321 Exclusion *Refusal*
PC 0193 Exclusion ; Social
PE 4242 Exclusion women athletic competition
PE 9009 Exclusion women decision making
PE 0413 Exclusive dealing arrangements
PD 2579 Exclusive nationally-oriented language systems
Exclusive [sales...]
PE 4581 — sales representation agreements
PE 6336 — sects
PJ 2520 — services
PA 6822 Exclusiveness *Disrespect*
PA 6979 Exclusiveness *Opposition*
PE 7778 Exclusiveness ; Super power
PA 6653 Exclusiveness *Unsociability*
PA 5821 Exclusiveness *Vulgarity*
PJ 4245 Exclusivity
PF 5954 Excommunication
PE 4685 Excrement birds
PE 3508 *Excrement fertilizer*
PE 4685 Excrement urban environments ; Insalubrity animal
PE 8545 Excreta ; Pollution water supplies human
PE 1602 Excreting public places
PF 6044 Excuses ; Self destructive
PF 3746 Execration
PA 6191 Execration *Disapproval*
PA 7338 Execration *Hate*
PD 2341 Execution
PF 0399 Execution
PF 5415 Execution animals
PD 3712 Execution development projects ; Ineffective
PF 2418 Execution illegal orders
PF 2418 Execution inappropriate orders
PF 2418 Execution orders endangering environment
PF 2418 Execution orders violation human rights
PF 0399 *Execution young offenders*
PE 6366 Executions ; Arbitrary
PE 6366 Executions ; Extra legal
PD 7221 Executions ; Extrajudicial
PE 4407 Executions ; Sham
PE 6366 Executions ; Summary
PE 4997 Executive branch action ; Uncontrolled
PE 7665 Executive military power ; Control judiciary
PE 1635 Executive stress
PD 9628 Executives ; Discrimination against women
PF 7578 Executives ; Excessive salaries corporate
PF 7578 Executives ; Overpaid business
PE 5944 Executives ; Terrorism targeted against corporate
PF 2137 *Exempt status ; Church abuse tax*
PC 8689 Exercise abuse authority ; Illegal
PF 0958 Exercise addiction
PF 0958 Exercise fanatics
PD 5628 Exercise ; Forced
PD 4475 Exercise ; Inadequate
PE 5182 *Exercise-induced pulmonary haemorrhage*
PF 5472 *Exercise leverage international bodies ; Developing country failure*
PE 3704 Exercises ; Provocative military
PD 4475 Exertion ; Reduced physical
PE 9774 *Exertional rhabdomyolysis*
PE 9774 *Exertional rhabdomyolysis dogs*
PD 0414 Exhaust ; Health risks fuel
PF 1538 Exhausting commuter travel
PF 2644 *Exhausting work demands*
PA 6799 *Exhaustion*
PA 6799 Exhaustion *Disease*
PD 9366 *Exhaustion ; Heat*
PA 5806 Exhaustion *Inaction*
PA 6852 Exhaustion *Inappropriateness*
PA 7382 Exhaustion *Loss*
PD 9357 Exhaustion mineral resources
PD 0077 Exhaustion ; Soil
PD 4643 Exhibitionism
PE 1293 Exhibitionism ; Female sexual
PF 3271 Exhibitionism ; Male sexual
PD 1116 Exhibitionism ; Sexual
PF 9014 Exhorbitant interest rates
PA 5473 Exiguity *Insufficiency*
PA 7285 Exiguity *Littleness*
PA 7408 Exiguity *Smallness*
PC 2507 Exile
PC 2507 *Exile ; Arbitrary*
PC 2507 *Exile ; Denial right freedom arbitrary*
PC 2507 *Exile ; Internal*
PE 7931 Exiled children ; Inhibition personality development
PF 5943 Exiled monarchs
PA 5498 Existence*complex
PF 8404 Existence enemies ; Dependence
PF 4331 Existence extraterrestrials ; Secrecy concerning
PE 5194 Existence openings various professions ; Imbalance training
PB 4411 Existence ; Struggle
PD 4191 Existent commodities ; Sale non
PA 6220 Existential vacuum
PC 2367 Exit ; Illegal political
PD 0531 Exodus ; Mass
PC 1453 Exodus private capital
PF 3523 *Exodus trend ; Youth*
PF 1432 Exogenous time standards ; Imposition
PF 2378 *Exorbitant costs chicken raising*
PE 2524 Exorbitant debt interest
PF 5673 Exorcism superstition
PC 1617 Exotic organisms ; Disruption ecosystems
PE 9417 Exotic species ; Collectors
PC 0380 Exotic species ; Trade
PE 5304 Expanding preschool ; Unimagined possibility
PD 0134 Expansion cities developing countries ; Uncontrolled physical
PD 1220 Expansion ; Destabilizing corporate
PF 8111 Expansion ; Economic
PF 0905 Expansion economic base ; Undirected
PF 3005 *Expansion ; Impeded possible*
PF 1345 Expansion income ; Absence imaginative planning
PF 1768 *Expansion ; Limited agricultural*
PF 2466 Expansion ; Neglect research natural resource product
PF 3005 *Expansion ; Restricted community*
PE 2700 Expansion ; Restrictions private economic
PF 6478 Expansion ; Slow rate income
Expansion [Underdeveloped...]
PF 1345 — Underdeveloped sources income
PC 1730 — Undirected technological
PF 6475 — *Untested industrial*
PB 5858 Expansionism
PF 8111 Expansionism
PC 9547 Expansionism ; Territorial
PE 5036 Expansive soils
PD 2675 Expatriates ; Socially unintegrated
PE 1339 Expectancy gender ; Inequality life
PE 1339 Expectancy men ; Shorter life
PE 1339 Expectancy ; Social inequalities life
PC 4312 Expectancy ; Unequal life
PF 8076 *Expectation community decline ; Continuing*
PF 6570 *Expectation ; Peer behaviour*
PD 2108 Expectation suppression minority opinion ; Innate
PF 7456 Expectations ; Diffusion family role
PF 6570 *Expectations ; Low general*
PJ 8933 *Expectations ; Low learning*
PD 3557 *Expectations ; Material*
Expectations [Unfeasible...]
PF 6577 — *Unfeasible medical*
PF 2969 — *Unfulfilled service*
PE 7002 — Unrealistic
PD 5347 Expectoration
PF 1559 Expedient policy reversals
PE 8207 Expeditions against distant objectives ; Military
PD 7261 Expeditions against friendly powers ; Military
PE 4454 Expeditions direct action environmentalists ; Sabotage hunting
Expenditure [allocation...]
PF 7432 — allocation ; Imbalance revenue mobilization
PF 1447 — armaments ; Excessive
PF 1447 — arms ; Increasing government
Expenditure Decline government
PF 9108 — [Expenditure ; Decline government]
PF 4586 — health
PF 0611 — social
PF 4785 Expenditure defence ; Rigidity rechannelling reduced
PF 1447 Expenditure ; Dependence excessive military
Expenditure [education...]
PF 0674 — education ; Decline government
PC 6215 — Excessive growth social
PF 1447 — Excessive military
PF 0611 Expenditure human development ; Decline public
PD 8114 Expenditure human resources developing countries ; Reduction public
PF 0674 *Expenditure primary education ; Decline government*
PF 0498 Expenditure ; Reduction aid
PD 4125 *Expenditure ; Unnecessary education*
PF 4295 Expenditures due floating rate loans ; Uncertainty development
PF 3313 Expenditures ; Imbalanced celebrational
PF 6234 Expenditures ; Increasing public health
PF 9672 Expenditures ; Mismanagement government financial
PE 4645 Expense accounts ; Abuse
PF 4196 Expense athletic training programmes ; Excessive
PF 4192 Expense international athletic competitions ; Excessive
PE 8261 Expenses ; Inflated medical
PE 8261 Expenses ; Prohibitive medical
PF 4375 Expenses ; Prohibitive school
PF 2385 Expenses ; Rising local
PF 0995 Expenses ; Risk incurring legal
PJ 7929 Expenses ; Unforeseen capitalization
PE 8261 Expenses ; Unnecessary medical
PF 3550 *Expensive construction materials*
PF 3409 *Expensive dependence fertilizers*
PF 4375 Expensive distant schools
PF 9007 Expensive emergency services
PF 6478 *Expensive fish farming*
PF 1267 *Expensive generator fuel*
PF 0296 *Expensive home ovens*
PE 8698 Expensive housing poor ; Prohibitively
Expensive [medical...]
PE 8261 — *medical techniques ; Waste resources*
PF 7130 — *meeting space*
PD 4632 — middlemen cycle
PE 8063 Expensive produce shipment
PF 4375 Expensive retraining
PF 1267 *Expensive road transport fuel*
PC 2163 *Expensive school transport*
PC 2163 *Expensive school travel*
PD 0847 Expensive training facilities
PE 8063 Expensive transport vehicles
PF 1653 *Expensive water connection*
PF 1345 *Expensive water equipment*
Experience Barriers
PF 4371 — religious
PF 4371 — spiritual
PF 4371 — transcendent
PF 9729 Experience ; Disillusioning cooperative
PJ 3532 Experience ; Disparagement religious
PF 3307 Experience ; Frustrating participation
Experience Insufficient
PE 6554 — business
PF 6570 — *common*
PF 7916 — *funding*
PF 0836 *Experience ; Irrelevancy past*
PD 2036 Experience ; Lack decision making
PF 7383 Experience ; Lack English language describe female
Experience Limited
PC 4867 — *commercial*
PF 9729 — cooperative
PF 6493 — *gardening*
PF 2477 — *planning*
PF 1746 Experience ; Perceived irrelevance past
PF 3498 *Experience ; Segmental school*
PF 9070 Experience ; Symbols unrelated human
PF 5272 Experiences ; Awkward past
PD 4840 Experiences ; Stressful life
PJ 3532 Experiences traditions ; Disparagement transcendent
PF 6570 *Experiences ; Unshared common*
PD 0260 Experimental animals ; Misuse
PE 8024 Experimental animals right freedom suffering ; Denial
PE 1171 Experimental battering animals
PF 1615 Experimental data ; Suppression
Experimental exposure
PE 1670 — animals pain
PE 0689 — animals radiation
PC 6912 — humans radiation
PE 1670 Experimental induction psychological stress animals
PF 5073 Experimental non human primates ; Shortage
PF 3367 Experimental religion
PE 0412 Experimental surgery animals
PD 0260 Experimentation animals ; Medical
PE 6764 Experimentation children ; Medical
PE 6201 Experimentation developing countries ; Increasing drug
PE 4781 Experimentation during war time ; Inhumane medical
PF 4841 Experimentation ; Errors risks medical self
PC 6912 Experimentation ; Human
PC 6912 Experimentation humans ; Abusive
PE 6763 Experimentation institutionalized subjects ; Medical
PE 8677 Experimentation mentally impaired persons ; Medical
PE 4865 Experimentation ; Obstacles medical
PE 8343 Experimentation pregnant women foetuses ; Medical
PE 4889 Experimentation prisoners ; Unethical medical
PD 6760 Experimentation socially vulnerable groups ; Medical
PF 4770 Experimentation ; Uncertainties animal
PE 4805 Experimentation using aborted foetuses ; Unethical
PC 0012 Experimentation ; Weapon

Expression

PE 1666 Experiments animals ; Military
PE 1670 Experiments animals ; Trauma inducing
PD 2697 Experiments drugs medical devices ; Unethical
PD 2697 Experiments humans ; Unregulated medical
PD 0260 Experiments live animals
PE 1174 Experiments ; Maltreatment animals aggression
PD 4271 Experiments ; Traffic children medical
PE 9611 Experiments ; Use animals toxicological
PF 7820 Expert advice ; Neglect
PF 7916 Expert knowledge ; Lack
PF 6012 Expert witnesses ; Disagreement
PE 8752 Expertise agricultural techniques ; Lack

Expertise [Biased...]
PF 6395 — Biased
PF 9693 — biased favour commissioning body
PE 7313 — business practices developing countries ; Inadequate local

Expertise [Dependence...]
PF 1196 — *Dependence external*
PF 1055 — development ; Failure global scale planning
PF 1681 — *drain ; Community*

Expertise [Inadequacy...]
PF 5055 — Inadequacy scientific
PF 7916 — Inappropriate
PF 5639 — Individualistic utilization
PD 2608 — Inequitable distribution construction

Expertise Lack
PF 7916 — *funding*
PF 7916 — *organized*
PD 8023 — sanitation
PF 7916 *Expertise ; Limited funding*
PF 2973 Expertise rural labour force ; Depleted
PF 7916 Expertise ; Unavailability appropriate
PF 6395 Expertise ; Unavailability impartial
PF 3294 Experts ; Arrogance
PF 6395 Experts ; Biased
PF 6012 Experts ; Disagreement
PF 8317 Experts ; Excessive salaries
PF 1068 Experts ; Impediments internationally mobile professionals
PF 0221 Experts ; Irresponsible international
PF 0221 Experts ; Negligence
PF 8317 Experts ; Overpaid
PD 6571 *Experts ; Unresponsive river*
PF 5292 Expiration industrial property rights ; Payments after
PF 2409 *Explanations phenomena ; Scientific*
PF 7967 Explanations world ; Mono factor
PE 1597 Exploit sea bed marine resources ; Inequitable allocation rights
PB 3200 Exploitation

Exploitation [aboriginals...]
PC 0720 — aboriginals
PD 9578 — animal species ; Excessive economic
PD 2078 — animals amusement
PE 0891 — animals spectator sports
PE 4054 — artistic expressions peoples
PE 4833 — athletic competition commercial political ends
PD 9578 Exploitation biological resources ; Over

Exploitation [capitalist...]
PC 3117 — capitalist systems
PD 6930 — casual workers
PE 9199 — cheap labour developing countries
PA 7193 — *Cheapness*
PD 0164 — child labour

Exploitation children
PD 0635 — [Exploitation children]
PD 0635 — Denial right freedom
PD 3267 — Sexual

Exploitation [claims...]
PE 6975 — claims ; Fraudulent mineral
PC 1573 — Class
PE 6877 — consumers
PD 6930 — contract labour
PC 7605 — cultural heritage ; Abusive

Exploitation [Denial...]
PE 3123 — Denial effective national self determination capitalist
PB 3200 — Dependence
PD 7592 — dependence food aid
PC 3334 — Dependence racial
PE 2427 — developing countries industrialized countries ; Economic

Exploitation [Economic...]
PC 8132 — Economic
PD 9370 — education ; Commercial
PD 9343 — elderly
PC 3297 — employment
PC 3297 — employment ; Dependence
PD 1018 — environmentally inappropriate products ; Dependence government revenues

Exploitation [farm...]
PD 2772 — farm animals industrial concerns ; Excessive commercial
PC 3298 — females ; Enslavement
PD 9082 — fish resources ; Unsustainable
PE 4891 — fossil fuels
PD 3465 Exploitation housing
PD 3465 Exploitation housing tenants
PD 1092 Exploitation indigenous populations employment
PE 9095 Exploitation land burial dead

Exploitation [marginal...]
PF 6528 — *marginal grazing land*
PE 1666 — marine mammals military
PD 3263 — men ; Sexual
PD 9685 — mentally handicapped
PD 9685 — mentally ill
PD 8405 — minors ; Enslavement
PF 2855 Exploitation natural resources ; Unconstrained
PD 1018 Exploitation non renewable resources ; Dependence government revenues

Exploitation [Olympic...]
PF 4761 — Olympic Games ; Political
PC 0052 — Over intensive soil
PD 9235 — Over rapid timber
PC 7356 Exploitation ; Political
PE 5303 Exploitation prostitution others

Exploitation [Racial...]
PC 3334 — Racial
PE 0060 — raw material reserves transnational enterprises ; Excessive
PE 4339 — regulatory loopholes countries underdeveloped legislation
PE 3528 — *rural pricing*

Exploitation [sea...]
PE 1597 — sea bed resources ; Discriminatory
PE 6682 — seabed resources ; Environmental hazards mineral
PC 3261 — Sexual
PF 4761 — sporting events ; Nationalist
PD 9370 — students ; Commercial
PD 0826 Exploitation tourism ; Over

Exploitation Trafficking children
PD 7266 — economic
PD 4271 — medical
PE 6613 — sexual
PC 4422 Exploitation trust
PD 9347 Exploitation unemployed
PD 0433 Exploitation viewers
PB 4353 Exploitation vulnerable
PD 1481 Exploitation wild animals ; Commercial

Exploitation women
PC 9733 — [Exploitation women]
PD 5025 — refugees
PD 3262 — Sexual
PD 0164 Exploitation youth ; Economic
PD 0433 Exploitative commercial television
PB 6207 Exploitative development
PD 5845 Exploitative development ; Regional ecosystems degraded
PD 0606 Exploitative entertainment
PE 6328 Exploitative films
PE 6952 Exploitative financial policies transnational corporations developing countries
PE 6336 Exploitative gurus
PF 7955 Exploitative humanitarian aid ; Politically
PE 6328 Exploitative imagery

Exploitative [part...]
PE 6241 — part-time work
PC 3299 — personal services
PE 7638 — portrayal women media
PD 8492 — property development
PD 3465 Exploitative rental accommodation
PF 2855 Exploitative resource development
PF 9281 Exploitative transformation international division labour
PE 3938 Exploitative use consultants
PB 3200 Exploiters
PF 6159 Exploration children urban environment ; Inhibition
PD 5034 Explosion ; Advertising
PE 5505 Explosion ; Endangering fire
PC 1298 Explosion ; Information
PE 3153 Explosion ; Risk accidental
PE 3153 Explosions ; Accidental
PE 4605 Explosions ; Dust
PC 2201 Explosions peacetime ; Nuclear
PF 3563 Explosions ; Stellar
PE 2095 Explosions underground ; Nuclear
PF 9167 Explosions unknown origin ; Aerial
PE 0429 Explosive anger
PF 2613 Explosive devices ; Derelict military
PA 6448 *Explosive disorder ; Intermittent*
PF 4027 Explosive ; Fuel air
PB 0855 *Explosive substances*
PE 1108 Explosives crimes ; Firearms
PE 1108 Explosives laws ; Violations against firearms
PE 9177 Explosives pyrotechnic products ; Environmental hazards
PE 0895 Export activity due licensing arrangements ; Restrictions
PD 2968 Export base developing countries ; Narrow
PE 4915 Export benefits ; Insufficient

Export [capital...]
PC 1453 — capital
PD 0470 — cartels
PE 2598 — cartels ; Excessive injury export interests developing countries due

Export credit
PD 3067 — competition
PE 3066 — financing upon developing countries ; Burden
PF 3065 — risks

Export [credits...]
PE 3066 — credits developing countries ; Decline
PE 0938 — credits developing countries ; Excessive dependence
PF 9187 — cropping
PE 9457 Export developing country products limited range countries

Export earnings
PD 2968 — developing countries ; Commodity related shortfalls
PD 2968 — developing countries ; Decline growth commodity
PE 4915 — instability
PE 4915 — shortfalls
PE 2858 Export enclaves ; Misuse free production zones
PE 6687 Export hazardous industries developing countries
PD 5351 Export inflation
PE 2598 Export interests developing countries due export cartels ; Excessive injury

Export [led...]
PE 6237 — led development
PE 6237 — led industrialization
PD 2968 — limited range raw materials ; Dependence developing countries
PF 0581 Export market ; Unstudied

Export markets
PE 9457 — developing countries ; Excessive concentration
PF 1471 — developing countries ; Restricted growth
PF 1471 — encountered developing countries ; Increase competition
PF 1345 *Export nutritious food*

Export [potential...]
PF 2989 — *potential ; Underdeveloped*
PF 2989 — *potential ; Unexplored*
PD 2968 — primary commodities ; Economic dependence developing countries
PD 3132 — *proceeds ; Non repatriation*
PE 0523 — promotion developing country small industry products ; Inadequate
PE 1598 — promotion transnational corporations developing countries ; Minimal
PE 0310 Export restraints ; Trade restrictions due voluntary
PE 1961 Export subsidies countervailing duties ; Distortion international trade
PD 2968 Export trade developing countries producing primary commodities ; Instability
PD 7466 Export unemployment
PE 9004 Export works art ; Illicit
PF 2989 *Exportable products ; Ignorance*
PE 6074 Exportation cultural property occupied territories ; War time

Exported developing countries
PD 2968 — Continuing low level prices commodities
PD 7682 — Development industrialized countries products substituting commodities
PD 0425 — Processing developed countries commodities
PF 1471 Exports ; Declining share developing countries world commodity

Exports developing countries
PD 9679 — developed market economies ; Decline
PD 9679 — Protectionism international trade against
PE 5926 — Restrictive business practices markets developed countries against
PF 3467 *Exports ; Government restriction*

Exports [Illegal...]
PD 4116 — Illegal
PD 4116 — Illicit
PE 4956 — imports developing countries ; Imbalance agricultural
PE 5920 — imports land locked developing countries ; Imbalances
PE 2844 — Inadequate adjustment assistance industries labour affected developing country
PE 8398 — Inadequate hygiene restrictions carcass meat
PE 3066 Exports ; Lack credit guarantee facilities developing country
PE 8306 Exports least developed countries ; Minimal
PE 3968 Exports nuclear materials ; Illegal
PC 2566 Exports ; Reduction share developing countries world
PE 3734 Exposing freezing temperatures ; Torture

Exposure animals
PE 1670 — pain ; Experimental
PE 0689 — radiation ; Experimental
PE 7327 — Torture

Exposure [chemicals...]
PE 4717 — chemicals ; Hazards low level
PE 4717 — chemicals ; Long term hazards
PD 0132 — *children obscenity*
PD 0132 — *children pornography*
PF 1686 — cosmic radiation ; Health hazards
PC 0936 Exposure hazardous environmental pollutants ; Human
PC 6912 Exposure humans radiation ; Experimental
PF 4317 Exposure ; Indecent
PF 3628 Exposure ; Lack outside

Exposure Limited
PF 3628 — global
PF 3628 — outside
PF 3628 — regional
PF 3628 — world
PE 1704 Exposure medical patients radiation ; Excessive
PF 3628 Exposure ; Narrow range cultural
PC 0268 Exposure noise ; Health hazards
PF 2296 Exposure outside influences rural villages ; Limited

Exposure radiation
PE 1909 — consumer goods electronic devices ; Excessive
PD 1500 — Excessive occupational
PE 4057 — Hazards low level
PE 4057 — Long term hazards
PE 3813 Exposure settlements tidal flooding
PE 4717 Exposure toxic mixtures non toxic chemicals ; Health hazards long term, low level
PF 5739 Exposure weather ; Disease injury
PE 3915 Exposure weather ; Torture
PE 5409 Expression clothing ; Denial freedom
PC 2162 Expression ; Denial right freedom
PC 2162 Expression ; Dependence restriction freedom
PF 7798 Expression emotions equated free speech ; Irresponsible
PF 9735 Expression ; Limited artistic
PF 1781 *Expression needs ; Uncoordinated*
PC 2162 Expression ; Restriction freedom
PC 3174 Expression thought communist systems ; Denial freedom
PF 0593 Expression ; Unconstrained educational self

Expressions

PF 7299 Expressions ; Demeaning
PE 4054 Expressions peoples ; Abuse traditional cultural
PE 4054 Expressions peoples ; Exploitation artistic
PE 5545 Expressive language disorder ; Developmental
PE 0330 Expressive writing disorder ; Developmental
PD 5552 *Expropriation*
PD 3055 Expropriation land ; Government
PC 3304 Expropriation land indigenous populations
PD 3055 Expropriation private property ; Government
PE 0101 Expropriation satellites ; Military
PC 5313 Expulsion
PC 5313 Expulsion ; Denial right freedom arbitrary
PC 3205 Expulsion ethnic minorities
PA 5869 Expulsion *Exclusion*
PD 0531 Expulsion ; Forced mass
PC 3207 Expulsion immigrants aliens
PF 3021 Expulsion refugees
Extended [families...]
PD 0850 — families ; Dependence
PF 3127 — families ; Obstacles
PE 5241 — family ; Denial right
PF 1509 — family relationships ; Fragmented patterns
PF 7022 *Extended production time*
PE 6658 Extension family cycle
PF 3538 Extension ; Inadequate patent
PF 1653 *Extension offices ; Lack*
PD 0520 *Extension prison sentences ; Arbitrary*
PF 6601 Extension schooling ; Unproductive
PE 7902 Extensions ; Inequitable credit
PF 1240 *Extensive farming*
PE 7826 *Extensor process disease*
PA 0832 Extermination
PA 7055 Extermination *Death*
PA 6542 Extermination *Destruction*
PA 0832 Extermination ; *Mass*
PD 3155 Extermination wild animal natural prey
PC 1445 Extermination wild animals
PF 7401 Exterminism
PE 7195 External aid ; Dependence developing countries
PF 6575 *External assistance inertia*
PD 3131 External borrowing ; Dependence developing countries budget deficit financing
PF 3318 External capital development ; Restrictive effects
External [debt...]
PD 2133 — debt crisis developing countries
PE 7184 — debt management capacity developing countries ; Inadequate
PC 6189 — dependence developing countries
PD 1104 — dependence vulnerability socialist countries
PE 3296 — *dependency* ; *Longterm*
PD 5535 *External ear animals ; Tumours*
PF 1196 *External expertise ; Dependence*
External [factors...]
PF 6210 — factors ; Adverse economic shocks
PE 4603 — finance developing countries ; Deterioration cost
PE 9567 — financial position developing countries ; Deterioration
PE 7195 — financing development programmes ; Dependence developing countries
PE 9420 — forces ; Aggressive economic destabilization countries
PF 0979 — forces ; Self actualizing stance being victim
PD 0428 External hardship ; Inadequate national empathy
PE 4058 External interference family life ; Arbitrary
PF 9697 External markets goods services ; Vulnerability national economies vagaries
External [payments...]
PE 9567 — payments position developing countries ; Deterioration
PE 3504 — *piles*
PC 3056 — public debt ; Disproportionate
External relations
PF 2969 — Incomplete utilization
PF 2600 — Parochial
PE 6515 — relating sportsmen ; Ineffective use
PF 6505 — Unexercised responsibility
PF 2969 — Unrealized potential
External resources
PC 0065 — Dependence
PF 1653 — Limited access
PF 1226 — Reluctant claims
PE 7195 External sources development financing ; Excessive reliance developing countries
PE 1496 External trade deficits developing countries ; Excessive
PC 1100 External trade deficits ; Excessive
PF 2261 *Externally controlled space*
PD 7989 Extinction ; Animal
PA 6956 Extinction *Cold*
Extinction [Death...]
PA 7055 — *Death*
PC 1445 — Dependence wildlife
PA 6542 — *Destruction*
PA 6785 — *Disappearance*
PF 0039 — due comet showers ; Mass
PB 9171 Extinction species
PB 9171 Extinction ; Species
PC 1445 Extinction ; Wildlife
PF 4149 Extinctions species ; Periodic mass
PC 3796 Extortion
PA 5688 Extortion *Appropriation*
PA 7193 Extortion *Cheapness*
PD 4469 Extortion ; Criminal
PD 4469 Extortion ; Dependence
PE 8240 Extortion ; Inadequate assistance victims
Extortion [Product...]
PD 5426 — Product

Extortion [protection...] cont'd
PD 4469 — protection money
PD 9842 — public officials
PE 8240 Extortion victims ; Lack legal protection
PD 8655 Extortionate bureaucracy
PF 0995 Extortionate legal costs
PD 1566 *Extra curricular activities ; Restricted*
PE 7784 Extra-economic constraints
PF 1066 Extra-jurisdictional territories
Extra legal
PD 6667 — conscription
PE 6366 — executions
PE 4997 — government initiatives
PE 5986 — impressment children armed forces
PF 2314 Extra-marital affairs
PF 1066 Extra territorial bases ; Unilateral declarations independence
PD 3496 Extra-territorial military bases intelligence centres
PE 2855 Extraction natural resources ; Emphasis mass
PD 3546 Extraction ; Parochial planning natural resource
PF 2204 Extraction utilization natural resources ; Inefficient
PE 6004 Extraction ; Governmental abuse
PF 5947 Extradition ; Impediments
PF 2645 Extradition refusal
PF 7664 Extrajudicial courts tribunals
PD 7221 Extrajudicial executions
PD 3246 Extralegal powers ; Excessive
PF 8352 Extrasensory perception
PF 4866 Extrasensory warfare
Extraterrestrial [infectious...]
PF 1312 — infectious diseases bacteria ; Introduction
PF 4444 — invasion
PF 4331 — invasion ; Cover up
PF 3562 Extraterrestrial origin ; Disasters
PF 3881 Extraterrestrials ; Abduction
PF 4331 Extraterrestrials ; Secrecy concerning existence
PE 2113 Extraterritorial abduction
PE 8011 Extraterritorial application restrictive business practices legislation
PE 3140 Extraterritorial intrusion jurisdiction
PF 2178 Extraterritoriality
PA 6400 Extravagance *Affectation*
PA 7193 Extravagance *Cheapness*
PA 5974 Extravagance *Diffuseness*
Extravagance [Insanity...]
PA 7157 — *Insanity*
PA 5473 — *Insufficiency*
PA 6466 — *Intemperance*
PA 7411 Extravagance *Uncommunicativeness*
PD 4037 Extravagant life style
PF 4558 Extravagant political promises
PD 4037 Extravagant use wealth
PD 0183 Extravagant waste resources government agents
PC 0387 Extreme climatic conditions
PF 0889 Extreme detachment represented constituency
PB 0388 Extreme poverty
PF 4644 Extreme present-orientedness
PE 7560 Extremely low frequency electromagnetic radiation ; Environmental hazards
PC 0387 Extremes ; Climatic
PB 3415 Extremism
PB 3415 Extremism ; Dependence
PA 7250 Extremism *Disobedience*
PA 7157 Extremism *Insanity*
PA 5473 Extremism *Insufficiency*
PF 2177 Extremism ; Political
PF 4954 Extremism ; Religious
PC 6341 Extremist ideologies
PF 9710 Extremist liberalism
PB 3415 Extremists
PE 2946 Extremities ; Paralysis anatomical
PA 7409 Extrinsicality
PD 9667 *Exudative epidermitis*
PE 9774 *Exudative pork ; Pale soft*
PD 8786 Eye ; Abscess
PD 5535 *Eye ; Cancer*
PE 7549 Eye ; Congenital anomalies
Eye [damage...]
PD 8786 — damage
PD 5535 — diseases animals
PD 8786 — diseases disorders
PD 8179 — disorders
PF 2343 Eye ; Evil
PE 7548 Eye glasses developing countries ; Prohibitive cost
Eye [Infection...]
PD 8786 — Infection
PD 8786 — Inflammatory diseases
PE 6785 — irritation
PE 8173 Eye, limb respiratory muscles ; Paralysis throat,
PD 8179 Eye ; Refractive disorders
PD 8786 *Eye ; Sticky*
PE 6785 Eye strain
PE 5380 Eyed blindness ; Clear
PD 8786 Eyelid ; Abscess
PD 8786 Eyelid ; Inflammation
PD 5535 *Eyelids animals ; Conformational abnormalities*
PF 8026 Eyes nerves ; Damage small blood vessels kidneys,
PE 2395 *Eyeworm infection poultry ; Manson's*
PD 5535 Eyeworms animals

F

PD 8716 Fabrication evidence ; Official
PF 9558 Fabrication politically sensitive incidents
PF 9558 Fabrication provocative incidents

PD 4982 Fabrication reports corporate competitors
PA 7411 Fabrication *Uncommunicativeness*
PE 8970 Fabrics, made up articles ; Instability trade textile yarn
PE 8436 Fabrics, made up articles ; Shortage textile, yarn
PD 8114 Face ; Development inhuman
PE 2254 *Face flies*
PF 3856 Face ; Loss
PE 3604 *Faced bats ; Endangered species slit*
PF 2081 Faceless social institutions
PD 1185 *Facial anomalies ; Dento*
PE 5567 *Facial cellulitis*
PD 8076 Facial disfigurement
PE 9458 *Facial eczema*
PE 5580 Facial hair ; Unwanted
PE 8505 Facial oral injury deficiency
Facial [pain...]
PG 3554 — pain
PD 2632 — *paralysis*
PD 7841 — *paralysis animals*
PB 0477 *Facial wrinkles*
PF 9205 Facile beliefs
PE 5242 *Facile social concepts*
PE 5303 Facilitating prostitution
PD 6845 Facilitation ; Criminal
PE 8516 Facilities agricultural producers ; Lack credit
PD 0084 Facilities ; Barriers transfer educational
Facilities [childrens...]
PD 0549 — children's play ; Inadequate
PE 3538 — Costly rotation activities intergovernmental
PE 8084 — countryside ; Meagre youth
PE 8420 — Crimes committed hospitals health care
Facilities [Defensive...]
PF 2411 — *Defensive marketing*
PF 7647 — Deficient public
PE 3657 — Delay connection telecommunications
Facilities developing countries
PE 0004 — Inadequacy telecommunication
PC 7674 — *Inadequate medical laboratory*
PD 0543 — Inadequate road highway transport
PD 1388 — Inadequate transportation
PE 3066 Facilities developing country exports ; Lack credit guarantee
Facilities disabled
PF 0775 — Inadequate educational
PE 8833 — people ; Inadequate recreational
PF 0775 — persons ; Inadequate educational
PE 7317 — persons ; Inadequate vocational rehabilitation
Facilities [Discrimination...]
PD 1036 — Discrimination family planning
PC 4844 — *Discriminatory use health*
PD 2762 — Disparities distribution communication resources
PE 1848 — *Distant schooling*
PF 0827 — due daily seasonal peaks ; Underutilization
Facilities [educating...]
PF 2012 — educating older people ; Lack
PF 0202 — elderly ; Inadequate recreation
PF 6141 — Excessive dispersion community
PF 5120 — Excessive waiting times government
PD 0847 — Expensive training
Facilities [Geographically...]
PC 7674 — Geographically remote
PD 2051 — gifted children ; Inadequate educational
PF 6143 — grass roots community initiatives ; Inadequate
PD 7461 Facilities ; Hierarchical control market
Facilities [Impersonality...]
PE 6153 — Impersonality mass market shopping
PD 9051 — Inaccessible educational
PF 9007 — Inaccessible emergency relief
Facilities Inadequate
PF 6499 — agricultural
PD 2085 — child care
PD 2085 — child day care
PD 0847 — educational
PE 4877 — food storage
PE 5058 — hospital
PD 2487 — inland waterway transport
PF 0202 — leisure
PD 4004 — medical
PF 4832 — medical
PE 6126 — night life entertainment
PD 1566 — *public*
PD 0496 — rail transport
PF 0202 — recreational
PD 1089 — rehabilitation
PD 0490 — road highway transport
PJ 3023 — sea transport
PE 4766 — seaport
PF 3467 — *storage*
PD 0876 — toilet
PF 1495 — urban transport
PF 2822 — *visibility*
PE 4834 — zoo
PE 5058 Facilities ; Inoperable hospital
PD 0876 Facilities ; Insanitary drainage
Facilities Insufficient
PF 1815 — *business*
PF 3448 — *commercial*
PD 0847 — educational
PF 6478 — *factory*
PD 0847 — training
PF 2575 — *youth*
Facilities international
PF 2016 — nongovernmental organization action ; Inadequate
PD 0929 — organization action ; Inadequate
PE 3538 — organizations ; Inefficient location

Fallot animals

Facilities Lack
- PF 8481 — abortion
- PD 2085 — child care
- PD 0847 — educational
- PF 9007 — emergency
- PD 6520 — *fire fighting*
- PE 5058 — hospital
- PF 8869 — legal aid
- PD 4790 — *refrigeration*
- PE 9767 — slaughter
- PF 3467 — *storage*
- PE 5908 — Facilities land locked developing countries ; Inadequate port storage

Facilities Limited
- PD 1204 — *catchment*
- PD 0847 — *educational*
- PE 1535 — meeting
- PE 7165 — *teacher*

Facilities [Maldistribution...]
- PF 4132 — Maldistribution telecommunications
- PE 6144 — Maldistribution urban shopping
- PE 8900 — mentally disabled criminals ; Inadequate
- PE 8151 — mentally handicapped ; Inadequate rehabilitation
- PE 3995 — military mobilization reinforcement ; Disparity
- PD 0847 — Minimal preschool
- PF 6195 — Monotonous housing
- PF 9683 — Facilities ; Non use available health
- PD 2669 — Facilities organized action against problems ; Inadequate buildings, services

Facilities [personnel...]
- PF 4832 — personnel ; Insufficient medical
- PD 8314 — physically disabled ; Lack
- PD 0876 — Poor sanitation
- PE 4154 — Prohibitive cost hospital
- PE 9116 — Facilities ; Restrictive conditions loans developing countries intergovernmental
- PE 6526 — Facilities rural communities developing countries ; Inadequate transportation
- PD 0211 — Facilities severely deformed people ; Lack
- PE 2829 — Facilities storing, transporting processing solid wastes ; Inadequate
- PE 5480 — Facilities ; Torture denial sanitary

Facilities transport
- PE 1048 — electrical energy ; Inadequate
- PE 8475 — sanitary wastes ; Inadequate
- PD 1294 — water supplies ; Inadequate

Facilities [Unattractive...]
- PF 1815 — *Unattractive business*
- PF 6479 — *Unavailability library*
- PD 1566 — *Unavailable public*
- PF 0202 — *Underdeveloped leisure*
- PE 2635 — undersized persons ; Unequal
- PD 9306 — Understaffing basic
- PF 0048 — *Underutilization world banking*
- PD 0847 — Undesignated school
- PE 4241 — Unequal global distribution family planning education
- PF 4832 — Unorganized health
- PE 4897 — Unsafe port
- PD 0876 — Unsanitary toilet
- PF 3448 — *Unsuitable commercial*
- PD 3507 — Unsystematic allocation market
- PD 1350 — *Facilities ; Vandalism recreational*
- PD 8851 — Facilities ; War time disruption economies production
- PD 1846 — Facilities workers' representatives ; Inadequate protection
- PD 0549 — Facilities youth recreation ; Inadequate spatial
- PD 0847 — Facility ; Advanced training
- PF 6548 — *Facility costs ; Prohibitive*
- PF 9478 — *Facility use ; Unimaginative*
- PF 2003 — Facing society ; Multiplicity problems
- PF 8454 — Factional feuding
- PF 8454 — Factionalism
- PF 7979 — Factionalism ; Bureaucratic
- PD 1629 — Factionalism developing countries
- PF 7118 — *Factionalism ; Widespread community*
- PF 3307 — *Factionalized decision-making*
- PF 3307 — Factions frustrations community participation
- PE 9702 — *Factor ; Slobber*
- PD 0779 — Factories ; Idle
- PE 3537 — Factory closures
- PE 7007 — Factory ; Conscientious objection
- PF 6478 — *Factory facilities ; Insufficient*

Factory farming
- PD 1562 — [Factory farming]
- PE 8383 — Abuse antibiotics vaccines
- PD 2761 — Animal malformation
- PD 2760 — Animal stress
- PD 2768 — Cruelty animals
- PD 2768 — Debeaking chickens
- PD 2758 — Inefficient use proteins
- PD 1562 — Negative effects
- PD 2752 — Spread animal diseases
- PD 2760 — units ; Inadequate lighting
- PF 9431 — Facts ; Biased interpretations
- PD 4982 — Facts corporate leaders ; Misrepresentation
- PD 8468 — Facts decision making ; Selective avoidance
- PF 9431 — Facts ; Lack consensus
- PD 1893 — Facts national leaders ; Misrepresentation
- PF 9431 — Facts ; Negotiable
- PF 9431 — Facts ; Questionable
- PF 5983 — Facts records ; Secrecy medical
- PF 2453 — Facts ; Selective perception
- PA 7197 — Faculties∗complex ; Intellectual
- PS 3268 — *Faculty demands ; Distracting*
- PF 1681 — *Faded community symbols*
- PD 1189 — Fads ; Food

- PD 8360 — *Faecal transmission disease*
- PD 0876 — *Faeces ; Contaminated*
- PC 0300 — *Faeces ; Involuntary passing*
- PE 4685 — Faeces pavements ; Dog
- PE 8545 — Faeces ; Pollution water infected
- PE 3445 — Failed migration

Failure acknowledge
- PF 5210 — contradictions
- PF 5210 — differences
- PF 5163 — failure
- PF 5210 — problems

Failure [act...]
- PB 1423 — act
- PF 1578 — adapt general initiatives specific needs
- PE 4577 — appear trial after release
- PF 1756 — appear witness produce information be sworn
- PF 7004 — ASAT negotiation
- PF 5306 — assist emergencies
- PE 0964 — Failure ; Bank
- PD 2591 — Failure ; Business

Failure [centrally...]
- PC 3894 — centrally planned economies
- PF 2347 — citizen participation ; Structural
- PF 7728 — clear names people convicted crimes which they are innocent
- PF 9432 — commodity production respond fall demand
- PF 5239 — conform international health standards
- PC 4591 — conform internationally agreed standards
- PE 3958 — Congestive heart
- PD 1230 — Constructional
- PD 0093 — Contraceptive
- PA 7289 — Failure *Defeat*

Failure development
- PF 5658 — policies
- PF 8368 — programmes
- PF 8368 — projects
- PF 0013 — Failure disarmament arms control efforts
- PD 0982 — *Failure disease notification*

Failure [economic...]
- PE 9405 — economic micro-organizations
- PE 1341 — Electrical power
- PD 1475 — Electronic equipment
- PE 9071 — employ skills educated elderly people
- PD 8546 — employ skills home-bound educated women
- PE 4768 — enforce judgements orders
- PE 1693 — enforce minimum age laws
- PD 1230 — engineering materials structures
- PA 6180 — *Error*
- PF 5472 — exercise leverage international bodies ; Developing country

Failure [Failure...]
- PF 5163 — Failure acknowledge
- PF 4125 — Fear
- PF 8753 — focus available insight response problems society
- PF 6580 — focus fundamental systemic problems
- PF 7149 — follow advice elders
- PE 7235 — fulfil contractual agreement ; Denial right freedom imprisonment
- PF 3913 — fulfil financial commitments ; Inter Governmental

Failure [global...]
- PF 1055 — global-scale planning expertise development
- PF 5296 — government admit mistakes
- PF 5296 — government apologize errors
- PE 9465 — Government development bank
- PF 8819 — government intelligence services
- PE 2215 — governments fulfil international reporting obligations
- PE 7300 — governments implement provisions ratified human rights agreements
- PE 3958 — *Failure ; Heart*
- PE 9038 — *Failure ; Human organ*

Failure [Ill...]
- PF 2013 — Ill effects educational
- PA 6088 — *Impairment*
- PA 6997 — *Imperfection*
- PF 1010 — implement policy directives
- PA 6876 — *Impotence*
- PF 0749 — individuals participate social processes
- PA 5527 — *Inexpectation*
- PE 6909 — Infant growth
- PA 5652 — *Inferiority*
- PF 8753 — integrate knowledge empower humanity response global problematique
- PF 8819 — Intelligence
- PF 8753 — international information systems order problem-relevant information
- PF 8753 — interrelate wisdom different cultures response social problems
- PF 7728 — Failure legally rehabilitate victims miscarriage justice

Failure [maintain...]
- PD 9196 — maintain sewage systems
- PD 5335 — make payroll payments
- PF 7556 — markets
- PF 4765 — mass education
- PD 2638 — materials
- PC 1904 — Mechanical
- PC 1904 — Mechanical equipment
- PE 8400 — medical programmes developing countries
- PD 7215 — Metal
- PE 0630 — methods appropriate data
- PE 4920 — military recruitment
- PF 4979 — mobilize resources
- PF 9704 — mobilize support against problems ; Government
- PD 0489 — Monsoon
- PF 1042 — motivating socio-dramas

Failure [natural...]
- PE 0715 — natural dams ; Disastrous

Failure [Nonaccomplishment...] cont'd
- PA 6662 — *Nonaccomplishment*
- PE 6424 — notify imprisonment death prison foreign nationals

Failure [obey...]
- PF 7149 — obey elders
- PF 7467 — obey religious advisers
- PF 1746 — opposing groups appropriate wisdom past
- PF 3821 — opposition groups ; Structural

Failure [Parole...]
- PE 1121 — Parole
- PF 1985 — pass cultural heritage
- PF 4387 — Personal
- PF 7838 — political will observe trade commitments
- PF 8876 — process low priority legislative reform
- PF 1746 — profit patterns history
- PF 9545 — prosecute offenders effectively ; Government
- PF 6508 — public authorities assist financial investment
- PF 5964 — public finance policy developing countries

Failure [rainy...]
- PD 0489 — rainy season
- PF 9728 — recognize environment resource base economic activity
- PF 1750 — recognize uniqueness family members
- PE 2053 — Renal
- PB 5249 — report suffering
- PA 6448 — *resist impulses gamble*
- PA 6448 — *resist impulses steal*
- PE 5239 — respect internationally agreed toxicity thresholds
- PD 7924 — Respiratory
- PC 7459 — restructure economic relations developed countries
- PC 8694 — restructure economic relations North South
- PC 1904 — risk ; Engineering
- PE 2309 — rural cooperatives

Failure [sacrifice...]
- PF 5203 — sacrifice any personal advantage
- PF 4366 — sanction state supported terrorism ; Government
- PE 0964 — savings institutions
- PC 1716 — school
- PB 2496 — school systems
- PF 2013 — self story
- PE 4768 — service legal process
- PA 6041 — *Shortcoming*
- PE 2324 — sign peace treaty following war
- PE 1815 — Small bank
- PE 9465 — state banks
- PF 4646 — stewardship
- PF 9729 — story ; Cooperative
- PD 1230 — Structural
- PF 3715 — Symbol system
- PE 6909 — Failure thrive ; Nonorganic
- PE 3948 — Failure understand necessity creative establishment disestablishment tensions
- PE 8457 — Failure ; Unemployment perpetuator
- PA 5644 — *Failure Vice*
- PD 1230 — Failures ; Building
- PD 7480 — *Failures ; Crop*

Failures [Dam...]
- PE 9517 — Dam
- PD 4426 — Disastrous technological
- PE 5126 — due liquefaction ; Ground
- PE 5066 — Failures ; Ground
- PE 3537 — Failures ; Industrial
- PA 9405 — Failures ; Small business
- PC 1716 — Failures ; Wastage school examination
- PE 3933 — Fainting
- PD 4827 — Fair public trial ; Denial right
- PE 3964 — Fair public trial ; Denial right
- PF 3099 — *Fairs ; International trade*
- PF 4187 — Fairy rings
- PF 7150 — Faith ; Breach
- PC 3428 — Faith friction ; Inter
- PJ 6238 — Faith ; Graceless acceptance
- PF 3863 — Faith religion ; Loss
- PF 2005 — Faith rigid doctrine ; Ossification
- PD 2888 — Faith solutions ecological issues ; Lack
- PF 4989 — Faith technology ; Blind
- PF 5764 — Faithful ; Abandonment
- PA 7363 — Faithlessness *Improbity*
- PA 7325 — Faithlessness *Irresolution*
- PA 7392 — Faithlessness *Unbelief*
- PA 7411 — Faithlessness *Uncommunicativeness*
- PC 1455 — Faiths ; Negative attitude people
- PD 5540 — Fake drugs
- PD 7981 — Fake products
- PA 6400 — Fakery *Affectation*
- PA 7411 — Fakery *Uncommunicativeness*
- PF 1602 — Faking scientific evidence
- PF 9432 — Fall demand ; Failure commodity production respond
- PF 3864 — Fall man
- PE 2528 — Fallacies ecosystemic processes
- PF 5711 — Fallacies logic
- PF 5711 — Fallacies ; Statistical
- PA 2152 — Fallacious arguments
- PF 4357 — Fallacious people
- PF 4357 — Fallacy
- PA 6180 — Fallacy *Error*
- PA 5502 — Fallacy *Reason*
- PA 7411 — Fallacy *Uncommunicativeness*
- PF 3702 — Fallibility
- PA 6997 — Fallibility *Imperfection*
- PA 7309 — Fallibility *Uncertainty*
- PF 6826 — Falling grace
- PB 0731 — *Falling objects ; Accidents*
- PE 0661 — Falling sickness
- PD 9742 — Fallopian tube ; Diseases
- PE 5569 — *Fallot animals ; Tetralogy*

Fallout accidents

PE 5080 Fallout accidents orbiting nuclear powered reactors ; Radioactive
PC 0314 Fallout ; Radioactive
PE 5324 Fallout ; Uncertainty long term health effects radioactive
PE 9407 Fallow periods agricultural land ; Shortening
PE 7113 Falls ; Accidental
PE 2853 Falls elderly
False [accusations...]
PE 7348 — accusations
PF 4298 — alarms
PF 4298 — alerts
PF 9794 — allegations human rights violations
PD 6062 — arrest
PF 9582 — assumptions concerning recuperative power nature
PF 2528 — assumptions sustainable development
PF 4298 *False bomb warnings*
PF 4298 *False burglary alarms*
False [concreteness...]
PF 5242 — concreteness societal concepts
PE 7252 — confessions
PE 7252 — confessions innocent
PF 2639 — consumption needs ; Production serving
PF 0704 — curriculum vitae
False [declarations...]
PF 9794 — declarations ill-treatment
PF 8490 — diagnosis
PF 8490 — *diagnosis mental disorder*
PF 4298 — disaster warnings
PF 3096 — documents
False [economies...]
PF 9791 — economies scale
PF 5127 — evidence
PF 9794 — evidence torture
PF 4298 *False fire alarms*
PF 8759 False friendships
PE 5051 False historical documents objects
False [image...]
PF 3002 — image scarcity
PD 4489 — imprisonment
PD 6943 — incrimination
PB 3095 — information
PF 9243 False justification projects
PE 0501 False mildews
PF 1236 False nuclear warfare alerts
False [personation...]
PE 1275 — personation
PD 3017 — political evidence
PD 7799 — *pregnancy dogs*
PE 5076 — pretences ; Obtaining property
PF 3495 — prophets
PF 0704 False qualifications
PF 4583 False statements
PD 2630 False swearing
PE 3604 *False vampires ; Endangered species*
PF 5127 False witness
PA 7411 Falsehood *Uncommunicativeness*
PA 6180 Falseness *Error*
PA 7363 Falseness *Improbity*
PA 7411 Falseness *Uncommunicativeness*
PD 8716 Falsification evidence ; Police
PF 4932 Falsification government historical records
PF 9243 Falsification programme evaluations
PD 4239 Falsification public records
PF 1602 Falsification scientific records
PF 1602 Falsification scientific test results
PF 0704 Falsifying academic records
PF 5900 Falsity
PF 9664 Faltering adaptation economic agents accelerating pace structural change
PF 9664 Faltering structural adjustment world economy
PF 3439 Fame honours ; Unequal distribution
PE 7915 Familial disease nervous system
PF 1064 Familial image social responsibility ; Deluding
PC 2270 *Familial periodic paralysis*
PE 5043 Families accused ; Discrimination against
PF 1503 Families ; Adjustment difficulties new urban
PF 6570 Families ; Cut off fragmented
Families [Dependence...]
PD 0850 — Dependence extended
PE 5043 — Deprivation prisoners'
PD 0973 — Discrimination against migrant workers their
PE 5043 — Discrimination against prisoners'
PC 2102 — Dismembered
PD 7625 Families ; Excessively large
Families [Indifference...]
PF 1205 — *Indifference students'*
PF 5968 — Insufficient birth spacing
PE 1848 — *Insufficient young*
Families [Lack...]
PE 7944 — Lack interim support
PD 7625 — Large
PD 2681 — Lone parent
Families [Obstacles...]
PF 3127 — Obstacles extended
PE 5043 — offenders ; Discrimination against
PD 2681 — One parent
Families [participate...]
PF 1006 — participate globally ; Refusal
PD 1405 — physically mentally handicapped ; Stress
PC 0999 — Poor
PD 5616 — Privileged
PD 6881 — Problem
PF 3783 Families ; Reduced interior structure
PE 8423 Families ; Refusal entry foreign workers'
PD 2681 Families ; Single parent

PC 2343 Families territories ; Organized crime
Family [activities...]
PF 2644 — *activities ; Unengaging*
PE 5408 — adaptation community status quo
PF 0107 — allowances ; Negative effects
Family [based...]
PS 9219 — *based health care*
PF 1933 — *based production patterns*
PC 2102 — breakdown
PC 2102 — Breakdown
PC 2102 — breakdown ; Dependence
Family [crediting...]
PF 1509 — *crediting obligations*
PE 7311 — *Crimes against*
PE 6658 — cycle ; Extension
Family [decision...]
PF 2097 — decision making structures ; Simplistic
PE 2268 — decisions ; Exclusion pre adults
PF 9323 — decisions ; Uncorporate
Family Denial right
PE 7267 — [Family ; Denial right]
PE 5241 — extended
PC 6870 — found
Family [dependence...]
PD 3555 — dependence patriarchal role man
PD 1501 — developing countries ; Narrow legal definition
PC 0382 — *diets ; Unbalanced*
PE 4021 — disagreements
PD 5344 — discipline young people ; Lack constructive
PC 2102 — disintegration
PC 2151 — disorganization
PF 1382 Family economics ; Precarious basis
PF 5079 Family education ; Reduced
Family [farms...]
PD 5962 — farms ; Endangered
PE 8210 — feuds
PF 1205 — focused operations
PF 1509 — fragmentation ; Increased
PF 3308 — friends ; Threats against
PD 1350 *Family gardens ; Vandalized*
PF 1099 Family heritage ; Lost
PE 4058 Family home ; Unlawful interference
Family [income...]
PD 6579 — income ; Limited
PD 6579 — income ; Marginal level
PE 8296 — interest urban life styles ; Minimal
PE 1832 — Irresponsibility young people towards
PF 1205 *Family jealousies ; Multiple*
PE 8341 Family ; Lack self development
PD 1285 *Family land ; Limited*
Family life
PE 4058 — Arbitrary external interference
PF 5079 — education ; Lack
PF 7094 — Obstacles
PD 4069 — styles ; Escapist
PF 1078 — styles ; Reduced
PE 4058 — Unlawful interference
PD 7929 — Unstable
PF 2644 Family lifestyle ; Inflexible patterns
Family members
PF 1750 — Failure recognize uniqueness
PE 4868 — Immigration barriers handicapped
PE 4959 — Separation
PC 0382 *Family menus ; Restricted*
PF 1382 Family money priorities
PF 5145 Family needs ; Inadaptation work
PF 0129 Family ; Negative effects nuclear
Family [obligations...]
PF 1509 — *obligations ; Rigid*
PF 2575 — *occasions ; Unorganized*
PF 2316 — *opportunities ; Limited*
PF 3560 — oriented socializing
Family [patterns...]
PF 2644 — patterns traditional way life ; Delimiting
PE 5226 — planning ; Denial right
PF 0148 — planning disinterest
Family planning education
PE 0341 — children ; Negative effects
PE 4241 — facilities ; Unequal global distribution
PD 1039 — Inadequate
PD 1039 — Lack
PE 9060 — men ; Inadequate
Family planning [facilities...]
PD 1036 — facilities ; Discrimination
PD 1038 — Inadequate health care
PF 0148 — Ineffective
PD 1050 — information ; Lack
PF 0148 — Lack
PF 1021 — Opposition population control
PD 1039 — research ; Inadequate
PF 0148 — Sporadic
PF 0148 — Unmotivated
Family [possibilities...]
PF 0846 — possibilities ; Refusal
PC 0999 — poverty
PD 1998 — poverty industrialized countries
PF 3393 — *practices ; Misinformed*
PF 1030 — priority ; Individual
Family [rejection...]
PC 8127 — rejection children
PE 2087 — rejection physically handicapped
PF 1509 — relationships ; Fragmented patterns extended
PE 7206 — responsibilities ; Discrimination employment women
PD 1668 — responsibility ; Parochial
PF 7456 — role expectations ; Diffusion
PF 7456 — role society ; Lost

Family [roles...] cont'd
PF 7456 — roles ; Unplanned
Family [Schooling...]
PF 9306 — Schooling blocked
PE 4959 — separation
PF 3037 — services ; Insufficient
Family size
PF 3560 — Excessive
PF 0401 — Inability governments regulate
PF 3560 — Unsupportable
Family [social...]
PE 8670 — social security ; Dependence
PD 8130 — stress
PF 1502 — structure barrier progress
PD 0276 — *structure ; Demise three generation*
PF 1000 — structures ; Inadequate
PF 1861 — structures ; Individualistic
PD 1482 — system developing countries ; Disruption
Family [ties...]
PF 2358 — *ties ; Divisive*
PD 1380 — *ties ; Weakening*
PE 2119 — torture victims
Family [unit...]
PF 1030 — unit ; Introversion
PC 1707 — *units ; Isolated*
PD 4572 — units ; Socially ineffective
PD 5962 — use farming
Family [values...]
PC 2102 — values ; Declining
PD 6881 — violence
PF 4930 — Voluntary dissolution
PE 8341 Family welfare ; Lack
PB 0315 Famine
PB 0315 Famine ; Dependence
PD 0571 Famine ; Man made
PE 8979 Famine ; War induced
PE 8171 Faming income ; Low
PF 6713 Fanatical book collecting
PD 5582 Fanatical environmentalists ; Violence
PF 5691 Fanatical visions
PB 3415 Fanaticism
PA 6379 Fanaticism *Avoidance*
PA 7157 Fanaticism *Insanity*
PA 7325 Fanaticism *Irresolution*
PA 7156 Fanaticism *Moderation*
PA 7306 Fanaticism *Narrowmindedness*
PF 4954 Fanaticism ; Religious
PF 0958 Fanatics ; Exercise
PD 1185 *Fang ; Inflammation tooth*
PE 6281 Fans ; Violent sports
PF 6170 Fantastic happenings ; Inhibition collectively organized
PA 7325 Fantasy *Irresolution*
PF 8922 Fantasy reality ; Substitution
PE 2461 Farcy
PE 4693 Fare evasion
PE 4301 Fare-fixing
PE 8063 Fares ; High transportation
Farm animals
PD 2771 — Agricultural mismanagement housed
PD 7506 — Endangered species
PC 2765 — Inadequate feeding
PD 2772 — industrial concerns ; Excessive commercial exploitation
PD 8469 — Lack productivity
PD 2760 — Overcrowding housed
PE 4771 — Unethical practices
PF 3630 Farm capital ; Restrictive
PD 4100 Farm crops ; Limited
PF 6530 Farm debts ; Heavy
Farm [employment...]
PD 1285 — employment ; Off
PB 0750 — employment ; Part time
PF 3630 — equipment ; Prohibitive cost
PD 5584 Farm favouritism ; Large
Farm [labour...]
PF 8763 — labour ; Prohibitive cost
PF 0836 — *labour ; Unappealing*
PF 3630 — loans ; Limited availability
Farm machinery
PF 2457 — Limited availability modern
PF 4108 — Minimal
PF 2457 — Prohibitive cost
Farm marketing
PE 7313 — *Traditional*
PE 7313 — *Underdeveloped*
PF 0905 — *Unprofitable*
PF 6524 Farm methods ; Low yield
PF 1822 Farm methods ; Outdated
Farm [plots...]
PF 3211 — *plots ; Small*
PF 2135 — *plots ; Small*
PD 1285 — *produce ; Limited*
PE 8321 — products developing countries ; Industrialized country disincentives
PF 2822 *Farm residences ; Distant*
Farm [slurry...]
PE 3934 — slurry ; Contaminated
PC 0052 — soils ; Unimproved
PD 9802 — subsidies developed countries ; Over dependence
PF 1822 Farm techniques ; Outmoded
PE 3528 Farm tenant's dependence middlemen
PJ 1730 Farm wars
PF 3523 *Farm work ; Priority*
PF 1078 Farmer hero ; Subsistence
PF 1365 *Farmer image ; Demeaning*
PE 6899 Farmer's lung

Feeder cattle

PE 5524 *Farmer's lung disease cattle*
PD 9802 Farmers developed countries ; Excessive subsidies
PD 7536 Farmers ; Economic difficulties
PD 9155 Farmers ; Inadequate social protection
PE 4243 Farmers ; Inadequate working conditions peasant
Farmers [Lack...]
PD 9155 — Lack social security
PD 9155 — Lack social security peasant
PE 8516 — Limited access credit small
PF 7047 Farmers ; Neglect peasant
PF 6024 Farmers ; Neglect women
PD 7536 *Farmers ; Partially skilled*
PF 8835 Farmers ; Technologically untrained
Farmers [Unadapted...]
PF 6024 — Unadapted vocational training women
PF 1365 — *Unclear vision*
PF 1262 — Unconsensed union
Farming [Abuse...]
PE 8383 — Abuse antibiotics vaccines factory
PD 6571 — *alternatives ; Restricted*
PD 2761 — Animal malformation factory
PD 2760 — Animal stress factory
PF 6575 — *attitude ; Skeptical*
Farming [capital...]
PF 3630 — *capital ; Unavailability*
PD 7993 — Chemicalized
PE 8014 — communities ; Threat industrial combines
PD 2768 — Cruelty animals factory
Farming [Debeaking...]
PD 2768 — Debeaking chickens factory
PE 8321 — developing countries ; Disincentives against
PD 7536 — Disincentives against
PF 6478 *Farming ; Expensive fish*
PF 1240 *Farming ; Extensive*
Farming [Factory...]
PD 1562 — Factory
PD 5962 — Family use
PF 3630 — *funds ; Insufficient*
Farming [image...]
PF 1365 — *image ; Static*
PF 1365 — *images ; Past oriented*
PE 1785 — Inadequate subsidies organic
PE 1785 — Inappropriate agricultural subsidies chemicalized
PE 8171 — income ; Unpredictable
PF 2758 — Inefficient use proteins factory
PF 2457 — *Insufficient returns*
Farming [land...]
PC 7597 — land ; Unavailability
PF 5496 — Limited fish
PD 6571 — *Limited specialized*
Farming methods
PF 1822 — Old
PF 1822 — Rudimentary
PF 1822 — Traditional
PD 1562 Farming ; Negative effects factory
PE 3760 Farming ; Occupational hazards
PF 2457 *Farming operations ; Independent*
Farming [patterns...]
PC 0492 — patterns ; Subsistence
PC 0492 — Peasant
PF 0905 — *Poor risk commercial*
PF 1768 — *population ; Unintegrated*
Farming [Single...]
PC 3606 — Single crop
PF 1373 — *skills ; Lost*
PD 7536 — *skills ; Underdeveloped*
PD 2752 — Spread animal diseases factory
PF 1205 — *system ; Individualized*
Farming [techniques...]
PD 7993 — techniques ; Intensive
PF 1319 — *techniques ; Outdated*
PF 1319 — *technology ; Outdated*
PF 3409 — *technology ; Unexposed*
Farming [Underdeveloped...]
PF 1240 — Underdeveloped capacity income
PD 2760 — units ; Inadequate lighting factory
PE 2770 — units ; Inferior meat quality intensive animal
PF 3448 — *Unreliable traditional*
PC 4906 Farming ; Vulnerability
PF 6475 *Farmland plots ; Overdivided*
PC 8419 Farmlands ; Unsustainable development
PD 1007 Farms developing countries ; Uneconomical small
PD 5962 Farms ; Endangered family
PF 6530 Farms ; Low investment developing country
PE 3934 Farms ; Pollution water fish
PJ 8128 *Farms ; Uneconomic size*
PE 1764 Farting
PF 0248 Fascism
PF 2636 Fascism ; Neo
PF 9710 Fascist liberalism
PF 0059 Fashionable areas research ; Excessive emphasis
PF 4164 Fashionable problems ; Excessive emphasis
PF 4473 Fashionable solutions problems ; Excessive reliance
PD 2761 *Fashions ; Negative effects gastronomic*
PF 5940 Fast food ; Predominance
PE 6396 Fast growing plant species ; Substitution
PE 7898 Fast irregular pace technological advance
PE 9125 Fast livestock growth ; Use artificial methods promoting
PE 5735 Fastidiousness ; Over
PD 7550 Fasting
PE 1814 Fat children
PD 7420 *Fat cow syndrome*
PE 4261 Fat diets ; High
PE 9774 *Fat disease ; Yellow*
PD 3978 *Fat necrosis ; Peritoneal*
PD 2341 *Fatal poisonings ; Intentional*
PD 5453 *Fatal syncope*
PF 6430 Fatalism
PF 2795 Fatalistic attitudes use time
PF 6430 Fate ; Belief
PD 3555 Father ; Dependence children
PD 0614 Fatherhood ; Adolescent
PD 7297 Fatherhood ; Irresponsible
PD 7297 Fathers ; Absent
PD 3256 Fathers ; Discrimination against unmarried
PA 0657 Fatigue
PE 3870 Fatigue ; Aircraft pilot
PE 7912 Fatigue ; Battle
Fatigue [Chronic...]
PD 4374 — Chronic
PC 0300 — *Combat*
PF 2819 — Compassion
PF 1626 — Concrete
Fatigue [Disease...]
PA 6799 — *Disease*
PE 4229 — due torture
PE 5572 — during control complex equipment ; Human
PE 2398 Fatigue ; Heat
PF 1914 Fatigue immune dysfunction syndrome ; Chronic
PA 5806 Fatigue *Inaction*
PD 8052 Fatigue ; Job
Fatigue [materials...]
PD 1391 — materials
PD 7215 — Metal
PE 5572 — *Motor vehicle driver*
PE 1914 Fatigue syndrome ; Chronic
PE 4395 Fatigue ; Television viewer
PC 2494 Fatigue ; Voter
PA 5558 Fatigue *Weakness*
PE 4261 Fats ; Consumption excessive saturated
Fats [environmental...]
PE 8136 — environmental hazard ; Vegetable oils
PE 8135 — Environmental hazards animal oils
PE 4261 — Excessive consumption
Fats Instability trade
PE 8896 — animal oils
PE 0735 — animal vegetable oils
PE 0861 — fixed vegetable oils
PE 1188 Fats ; Long term shortage animal vegetable oils
PE 8880 Fats ; Restrictive practices trade animal vegetable oils
PE 8277 Fats ; Shortage fixed vegetable oils
PE 8498 Fats waxes ; Shortage animal vegetable oils
PE 1028 Fatty diseases ; Liver
PE 4261 Fatty food
PD 7420 *Fatty liver disease cattle*
PA 6876 Fatuity *Impotence*
PA 6852 Fatuity *Inappropriateness*
PA 6940 Fatuity *Thoughtlessness*
PA 7371 Fatuity *Unintelligence*
PF 2152 Fatuous reasoning
PF 3855 Fault ; Non juridical
PA 6180 Faultiness *Error*
PA 6793 Faultiness *Guilt*
PA 6997 Faultiness *Imperfection*
PA 5502 Faultiness *Reason*
PE 5096 Faulting ; Earth surface
PD 0201 Faults ; Geological
PE 2462 Faults reactor design
PE 2462 Faults ; Unexplained technical
PE 1575 Faulty aircraft
PF 0956 Faulty approaches teaching intellectual methods
PE 8580 Faulty title property
PC 3152 Fauna ; Degradation semi natural natural habitats flora
PC 3152 *Fauna ; Denaturalization*
PD 3574 Fauna micro organisms ; Pesticide destruction soil
PD 9170 Favourable work conditions ; Denial right
PC 2049 Favoured regions
PF 2607 Favouring investor ; Proscriptive controls
PF 5153 Favouring political incumbents ; Electoral organization
PD 5584 Favouritism ; Commercial
PD 5584 Favouritism ; Large farm
PF 8535 Favouritism ; Political
PE 1810 Fax mail ; Junk
PA 6030 Fear
Fear [abnormal...]
PF 7029 — abnormal
PF 6570 — *ageing*
PF 8012 — authentic relationships
PF 2491 Fear banks
PD 9498 Fear bureaucracy
Fear [capital...]
PF 6572 — capital investment
PF 5233 — communism
PF 2040 — contradicting popular views
PF 6570 — *controlled lifestyle*
PF 4682 — crime
PA 6030 — *Culture induced*
Fear [dark...]
PF 6188 — *dark*
PF 0462 — death
PF 0462 — death ; Dependence
PA 6030 — Dependence
Fear [eclipses...]
PF 7066 — eclipses occultations
PD 9389 — economic reprisals
PF 9209 — emotional sensitivity
PF 9209 — emotionally responsive behaviour
Fear [failure...]
PF 4125 — failure
PF 3904 — food contamination
PE 7706 — freedom
PD 2670 — future change ; Individual
Fear [God...]
PF 9565 — God
PD 9498 — government agents
PF 6570 — *growing old*
PF 9078 Fear harassment
PF 9234 Fear humiliation co-workers
Fear [illness...]
PE 9091 — illness ; Morbid
PE 7706 — increased autonomy
PE 7706 — independence
PF 4185 — industrialization developing countries ; Irrational
PA 5467 — *Inexcitability*
PF 8012 — intimacy
PA 7325 — *Irresolution*
PF 9595 Fear knowing
PF 9595 Fear knowledge
Fear [losing...]
PD 2614 — losing cultural identity
PE 7468 — loss independence
PF 3512 — *loss property*
PF 9181 *Fear military retaliation*
PF 4125 Fear mistakes
Fear [nature...]
PF 6803 — nature
PF 6803 — nature ; Dependence
PF 9127 — new technology
PC 3541 — nuclear war ; Pervasive
PE 5324 — nuclear winter
Fear [official...]
PF 3707 — official harassment
PD 9498 — officialdom
PF 6910 — opposite sex
PF 2776 — ostracism
PF 8743 — outsiders
PF 8378 Fear police
PA 6030 *Fear ; Political*
Fear [reprisals...]
PF 9078 — reprisals
PF 9030 — resettlement
PF 9078 — retaliation
PF 3707 — retaliation authorities
PF 6389 — risk ; Excessive
Fear [school...]
PE 4554 — school
PD 4418 — *self-promotion*
PF 9209 — sentimentality
PF 6910 — sexual intercourse
PF 1088 — social anarchy
PF 1088 — social collapse anarchy
PF 8743 — strangers
PF 5390 — success
Fear [tax...]
PF 2949 — *tax raise*
PF 2949 — *taxation ; Increased*
PF 9127 — Techno
PF 9127 — technological innovation
PF 9209 — tenderness
PE 0969 — Torture victim's
PF 9595 — truth
Fear [unfamiliar...]
PF 7029 — unfamiliar
PF 6188 — unknown
PF 2256 — *urban problems*
Fear [vandalism...]
PF 1249 — *vandalism*
PE 7750 — violence women
PF 1765 — *vocational change*
PE 6354 *Fear wind ; Irrational*
PF 2437 *Feared public criticism*
PA 6030 Fearful people
PE 3639 *Feather mite*
PD 3158 *Feather stars ; Endangered species*
PA 7270 Featurelessness *Absence*
PA 6900 Featurelessness *Formlessness*
PF 8511 Features ; Political changes names geographical
PA 6876 Fecklessness *Impotence*
PA 6852 Fecklessness *Inappropriateness*
PA 7341 Fecklessness *Unpreparedness*
PF 3526 *Federal funding ; Mistrusted*
PF 0626 Federalism ; National
PF 2088 Federalism ; World
PF 1781 *Federation operations ; Closed*
PE 4821 Feeblemindedness
PA 6799 Feebleness *Disease*
Feebleness [Impotence...]
PA 6876 — *Impotence*
PA 6978 — *Invisibility*
PA 7325 — *Irresolution*
PC 1587 Feebleness ; Mental
PA 7131 Feebleness *Oldness*
PF 2332 *Feebleness ; Physical*
PA 5502 Feebleness *Reason*
PA 7371 Feebleness *Unintelligence*
PA 5558 Feebleness *Weakness*
PD 1714 Feed additives ; Animal
PD 1714 Feed additives ; Undesirable effects animal
PE 3896 Feed ; Contaminated animal
PD 4100 Feed crops ; Undiversified
PF 8635 Feed price ; Fluctuation
PF 3448 *Feed ; Scant animal*
PD 1283 Feed ; Use agricultural resources production animal
PF 5311 Feedback ; Avoidance negative
PF 5311 *Feedback constituents ; Inability people's representatives process*
Feeder cattle
PE 5524 — *Honker syndrome*

Feeder cattle

	Feeder cattle cont'd
PD 7420	— Sudden death syndrome
PE 5524	— *Tracheal oedema syndrome*
PC 2765	Feeding animals ; Inadequate
PE 7627	Feeding ; Dependence breast
PC 2765	Feeding farm animals ; Inadequate
PE 4935	Feeding ; Hazards bottle
PE 3506	*Feeding ; Improper*
	Feeding [Smoking...]
PE 5026	— Smoking during pregnancy breast
PE 7627	— Social stigma breast
PD 9661	— strategies ; Inappropriate infant
PE 3858	— Substance abuse during pregnancy breast
PE 8255	— Substitution inappropriate foodstuffs breast
PE 8255	Feeding ; Unhygienic bottle
PF 2378	*Feeds ; Costly animal*
	Feedstuffs excluding unmilled
PE 1331	— cereals ; Environmental hazards animal
PE 8816	— cereals ; Instability trade animal
PE 8514	— cereals ; Long term shortage animal
PE 5311	Feel good rhetoric ; Uncritical
PA 6938	Feeling∗complex
PF 1825	Feeling uselessness ; Elders'
PF 0979	*Feelings ; Disabling inadequacy*
PF 1216	Feelings ; Numbing
PF 1216	Feelings ; Repression
PA 6793	Feelings sinfulness
PF 4375	Fees ; Inequitable school
PF 0995	Fees ; Outrageous legal
PF 0995	Fees ; Prohibitive legal
PB 4653	Fees students ; Lack
PE 4161	Feet ; Fever
PC 0300	Feigning physical symptoms ; Intentional
PC 0300	Feigning psychological symptoms ; Intentional
PD 1763	Felidae
PD 2730	*Feline agranulocytosis*
PD 2730	*Feline ataxia*
PD 2730	*Feline coronaviral infection*
PE 9427	*Feline cytauxzoonoses*
PD 2730	*Feline distemper*
PD 7841	*Feline dysautonomia*
PD 4747	Feline immuno-deficiency virus (FIV)
PE 7848	Feline immuno-deficiency virus (FIV)
	Feline infectious
PE 9554	— *anaemia*
PD 2730	— *enteritis*
PD 2730	— *peritonitis*
PD 2730	— *pleuritis*
PD 5453	*Feline leukaemia*
PD 5453	*Feline lymphosarcoma*
PD 2730	*Feline panleukopenia*
PD 2730	*Feline parvovirus*
PE 9854	*Feline respiratory disease complex*
PE 4656	*Feline urolithiasis*
PE 4656	*Feline urological syndrome*
PD 5453	*Feline visceral lymphoma*
PD 5453	*Feline visceral lymphosarcoma*
PD 1763	*Felis ; Endangered species*
PE 0428	Fellow inmates ; Maltreatment prisoners
PE 3569	*Fellowships ; Unequal distribution*
PD 5515	Felon ; Hindering apprehension
PD 5515	Felon ; Hindering prosecution
PD 5515	Felon ; Misprison
PE 1153	Felonies
PD 4488	Felonious restraint
PA 5952	Feloniousness *Illegality*
PA 7363	Feloniousness *Improbity*
PA 5644	Feloniousness *Vice*
PE 5267	*Felons ; Fugitive*
PD 2615	Felony ; Treason
PF 3382	*Female babies ; Unwanted*
	Female [circumcision...]
PE 6055	— circumcision
PE 5592	— criminality ; Increasing
PE 1837	— criminals
PD 5115	Female employment ; Insufficient
PF 7383	Female experience ; Lack English language describe
PD 3947	Female foeticide
PD 8775	Female genital organs ; Diseases
PE 1905	Female genital organs ; Malignant neoplasms
	Female [homosexuality...]
PF 2640	— homosexuality
PE 5741	— homosexuals ; Denial rights
PE 5741	— homosexuals ; Violation rights
PD 3501	Female infanticide
PF 7938	Female ; Intolerance distinctions male
PE 4451	Female labour ; Dependence developing countries unpaid
PD 5675	Female mid-life crisis
PD 3380	Female prostitution
	Female [servitude...]
PE 7627	— servitude ; Breastfeeding
PE 1293	— sexual exhibitionism
PE 6055	— sexual mutilations
PE 9302	— sterility
PS 1567	*Female talents ; Unused*
PC 5916	Female unemployment
PE 6902	*Female workers ; Occupational hazards*
PC 3298	Females ; Enslavement exploitation
PF 8780	Females ; Unauthorized proximity males
PA 5604	Femininity
PF 3025	Feminism
PF 2471	*Feminist backlash*
PE 5050	*Femoral epiphysiolysis ; Proximal*
PE 8961	*Femoral hernia*
PE 7808	*Femoral nerve paralysis*

PD 5972	Fenced frontiers
PD 1053	Fencing ; Lack cesspool
PE 8364	Fencing stolen property ; Possession
PF 6683	Feng shui ; Bad
PE 5396	Feral cats
PE 0185	*Feral children*
PE 0185	*Feral dogs*
PE 0185	*Feral livestock*
PE 0185	*Feral mammals*
PC 0776	Feral species ; Genetically modified
PE 4724	*Fermentative dyspepsia*
PD 5228	*Fern poisoning animals ; Bracken*
PA 5532	Ferocity *Discaord*
PA 5643	Ferocity *Unkindness*
PD 8982	*Ferries ; Capsizing*
PE 8131	Ferrous base metals ; Instability trade ores concentrates non
	Ferrous metal
PE 2601	— basic industries instability ; Non
PE 0824	— ores ; Long term shortage non
PE 0353	— scrap ; Shortage non
PE 8248	Ferrous metals basic industries environmental hazards ; Non
PE 1406	Ferrous metals ; Instability trade non
	Fertility [Destruction...]
PC 1300	— Destruction land
PD 0077	— Deterioration soil
PF 0906	— developing countries ; High human
PE 8782	— downstream due impoundment ; Reduction soil
PC 6870	Fertility rights
PD 7799	*Fertility turkeys ; Low*
PE 2859	Fertility ; Unwanted high
PE 3508	*Fertilizer ; Excrement*
PE 4525	*Fertilizer products ; Obstacles trade*
PF 3409	*Fertilizer ; Prohibitive cost*
PF 3409	*Fertilizer's price ; Variation*
PE 4140	Fertilizers agricultural development developing countries ; Insufficient
	Fertilizers [Chemical...]
PD 7993	— Chemical
PE 0760	— crude minerals, excluding coal, petroleum precious stones ; Instability trade crude
PE 1353	— crude minerals, excluding coal, petroleum precious stones ; Long term shortage crude
	Fertilizers [Environmental...]
PE 1514	— Environmental hazards
PE 1514	— Excessive reliance mineral
PE 1514	— Excessive use
PF 3409	— *Expensive dependence*
PE 9049	Fertilizers ; Instability trade crude
PE 0806	Fertilizers ; Instability trade manufactured
PE 0997	Fertilizers ; Long term shortage manufactured
PE 1514	Fertilizers ; Misuse
PE 1514	Fertilizers ; Overprescription
PE 1514	Fertilizers ; Overuse nitrate
PF 3448	*Fertilizers ; Scarcity subsidized*
PE 8729	Fertilizers ; Water pollution
PE 4161	*Fescue foot*
PE 4161	*Fescue lameness*
PE 9458	*Fescue toxicosis ; Summer*
PE 2041	*Fetalis ; Erythroblastosis*
PF 8363	Fetishism
PE 8058	Fetishism ; Commodity
PF 6406	Fetishism ; Sexual
PD 2198	Fetishism ; Transvestic
PE 7826	*Fetlock joint ; Periostitis*
PE 7826	*Fetlock joint ; Serous arthritis*
PF 7653	Feud ; Blood
PF 2136	Feudal social order
PF 2136	Feudalism
PD 0406	Feudalistic land tenure
PF 7979	Feuding ; Bureaucratic
PC 3685	Feuding ; Ethnic
PF 8454	Feuding ; Factional
PD 4846	Feuding ; Political
PE 8210	Feuds
PE 8210	Feuds ; Clan
PE 8210	Feuds ; Family
PD 2255	Fever
PE 5207	Fever ; African swine
PD 2730	Fever ; Argentinian haemorrhagic
	Fever [Bat...]
PD 2730	— Bat salivary gland
PD 2255	— Blackwater
PD 2730	— Bolivian haemorrhagic
PE 5530	— Boutonneuse
PD 2732	— Bovine petechial
PE 2260	— Break bone
	Fever [Canicola...]
PE 2357	— Canicola
PE 5105	— Cat scratch
PE 6280	— Catarrhal
PC 9042	— Childbirth
PE 6266	— Classical swine
PD 2730	— Crimean Congo haemorrhagic
PE 2260	— Fever ; Dandy
PE 2260	— Fever ; Dengue
	Fever [East...]
PE 7946	— East Coast
PD 2732	— Elokomin fluke
PD 1753	— Enteric
PD 2730	— Ephemeral
PC 9067	— Excessive degree
PE 4161	— Fever feet
PE 5524	— Fever ; Fog

	Fever [Haemorrhagic...]
PE 5272	— Haemorrhagic
PE 6197	— Hay
PE 0566	— Hectic
PE 0616	— Fever ; Jungle
PD 2730	*Fever ; Lassa*
	Fever [Malignant...]
PE 6280	— Malignant catarrhal
PE 5530	— Marseilles
PE 0616	— Marsh
PD 7420	— Milk
PE 3621	— mosquitoes ; Yellow
PE 2357	— Mud
	Fever [Pappataci...]
PD 2255	— Pappataci
PE 4036	— Paratyphoid
PE 0616	— Periodic
PD 3978	*Potomac horse*
PE 2534	Fever ; Q
	Fever [Rack...]
PG 3896	— Rack Mountain spotted
PD 6363	— Rat bite
PE 7787	— Relapsing
PE 0920	— Rheumatic
PE 7552	— Rift valley
PD 2730	— Rio Bravo
PG 3896	— Rock Mountain spotted
PG 3896	— Rocky Mountain spotted
	Fever [Scarlet...]
PD 9094	— Scarlet
PE 5524	— Shipping
PE 6266	— Swine
	Fever [Texas...]
PD 0752	— Texas cattle
PE 2260	— Three day
PE 5524	— Transit
PD 2255	— Trench
PD 1753	— Typhoid
PD 0640	— Typhus
PD 2730	Fever ; West Nile
PE 0985	Fever ; Yellow
PD 2255	Feverishness
PE 5272	Fevers ; Haemorrhagic
PE 7787	Fevers ; Relapsing
PD 6520	*Few fire hydrants*
PA 7152	Fewness
PD 2710	Fiat ; Rule
PE 4950	Fibre diets ; Low
PE 4950	Fibre ; Inadequate dietary
PE 8682	Fibre ; Instability trade manila
PE 1513	Fibre trade, excluding cotton jute ; Instability vegetable
PE 1513	*Fibres ; Instability trade hard*
PE 8950	Fibres ; Instability trade synthetic regenerated
PE 1550	Fibres their waste ; Instability trade unprocessed textile
PE 9074	Fibres waste ; Environmental hazards textile
PE 2373	*Fibrinolysis ; Haemorrhagic*
PE 0566	*Fibro-caseus tuberculosis*
PE 0566	*Fibroid tuberculosis*
PE 9702	*Fibromatosis gingivae*
PE 9702	*Fibromatous epulis*
PE 2298	*Fibrosa ; Generalized osteitis*
PC 2270	*Fibrosis*
PE 3331	Fibrosis ; Cystic
PE 1310	*Fibrositis*
PE 7826	*Fibrotic myopathy*
PJ 5171	Fibrous dyspasia
PD 7424	*Fibrous osteodystrophy*
PF 8922	Fiction reality ; Treating
PF 8922	Fictional worldview ; Imposing
PC 0300	Fictitious disorders
PF 9830	Field analysis ; Fragmentation
PF 2407	Field anomalies ; Geomagnetic
PF 1588	Field ; Reversal Earth's magnetic
PF 2605	*Field ; Unfinished playing*
PF 2605	*Field ; Unsatisfactory sports*
PE 7560	Fields ; EEG entrainment ELF magnetic
PE 7879	Fields generated electrical appliances ; Health hazards electromagnetic
PF 0021	*Fields ; Lack refresher courses professional*
PE 2305	*Fields ; Salty padi*
PA 5643	Fiendish *Badness*
PA 7148	Fiendish *Ungodliness*
PA 5643	Fiendish *Unkindness*
PA 5644	Fiendish *Vice*
PA 5532	Fierceness *Disaccord*
PA 7253	Fierceness *Envy*
PA 5467	Fierceness *Inexcitability*
PA 7156	Fierceness *Moderation*
PA 5643	Fierceness *Unkindness*
PF 2477	*Fighters ; Undertrained fire*
PJ 1184	Fighting
PE 4893	*Fighting ; Bull*
PE 4893	*Fighting ; Cock*
PE 7786	*Fighting ; Dog*
PD 6520	*Fighting facilities ; Lack fire*
PD 0518	Fighting husband wife
PE 4893	*Fighting sports ; Animal*
PF 9617	*Figures ; Delusions concerning public*
PF 0981	*Figures ; Depiction human*
PF 1540	*Figures ; Unrehabilitated historical*
PE 2391	*Filariae*
PE 5380	*Filarial dermatitis cattle*
PE 5380	*Filarial dermatosis*
PE 2391	*Filarial infection*
PE 2391	*Filariasis*
PE 1601	*Filariasis ; Bancroft's*

–928–

Fixing

PE 1601 Filariasis ; Bancroftian
PF 3979 Filial piety ; Erosion
PD 3255 Filiation ; Denial
Film [cinema...]
PD 3032 — cinema censorship
PD 3032 — clubs private
PD 3032 — clubs ; Private
PD 7060 Film industries ; Protectionism entertainment products
PD 3040 Film monopoly
PD 3089 Film propaganda
PD 5377 Film standards ; Deterioration
PE 4260 Film violence ; Effect television
PE 4260 Films ; Effect violence television
PE 6328 Films ; Exploitative
PA 5454 Filth *Badness*
PA 6191 Filth *Disapproval*
PA 5612 Filth *Unchastity*
PA 5459 Filth *Uncleanness*
PE 2778 Final disposal
PE 2425 Finance coastal communities ; Lack
PF 6548 Finance counselling ; Unlocated
Finance [data...]
PF 7842 — data developing countries ; Inaccurate public
PE 4603 — developing countries ; Deterioration cost external
PE 4188 — Developing country dependence single source
PF 0048 Finance ; Inappropriate international institutional framework
PF 5964 Finance policy developing countries ; Failure public
PF 4340 Finance rural development projects ; Inadequate commercial
PF 7842 Finance statistics ; Inadequate public
PC 4867 Finance training ; Limited
PE 9210 Financed subversion ; Drug
PE 9210 Financed trade drugs ; Terrorism
PF 9672 Finances ; Inadequate management government
PF 9672 Finances ; Ineffective control government
PF 5374 Finances ; Lack transparency public
Financial [accounting...]
PF 0203 — accounting reporting ; Inadequate international standards
PE 0226 — action transnational enterprises ; Destabilizing
PE 1137 — *autonomy* ; *Inexperienced*
PF 9841 Financial clearing systems ; Inadequate
Financial commitments
PF 1179 — Arrears payment government
PF 3913 — Inter Governmental failure fulfil
PF 4114 — Unrecognized future
Financial [conditions...]
PC 7873 — conditions ; Fluctuations world monetary
PE 2820 — control United Nations systems ; Lack centralized
PF 2990 — cover-up
PF 2489 — credit ; Limited availability
PE 5516 — crime ; Frauds, forgeries
PF 3612 — crisis ; Global
PC 7873 Financial destabilization world trade
PE 1911 Financial economic disputes states nationals states
PF 9672 Financial expenditures ; Mismanagement government
PD 2458 Financial fraud
PE 2414 Financial frauds
Financial [incentives...]
PF 0107 — incentives having children
PD 5193 — industrial oligarchy
PD 5193 — industrial power ; Misuse
PE 7009 — information systems international nongovernmental organizations ; Lack
PC 7873 — instability
PC 8073 — instability world economy ; Economic
Financial institutions
PE 2805 — Autocracy intergovernmental
PE 2214 — developing countries ; Over dependency international
PF 1136 — Unaccountability international
Financial [instruments...]
PE 5503 — instruments ; Counterfeit
PF 3845 — investment developing countries ; Disincentives
PF 6508 — investment ; Failure public authorities assist
PF 3307 — *irresponsibility image*
Financial [loans...]
PE 4603 — loans developing countries ; Deteriorating terms
PE 9054 — loans ; Racial discrimination according
PE 9013 — loss
PF 8079 — losses ; Unreported
PF 0760 Financial management ; Incompetent
PE 6642 Financial manipulation sects
Financial markets
PE 5902 — developing countries ; Undeveloped
PD 4511 — Disruption
PF 6980 — Inefficiency
PF 9576 Financial models ; Uncoordinated multiplicity global
PD 1157 Financial obligations ; Irregular payments international
Financial [paralysis...]
PD 9449 — paralysis developing countries
PE 9700 — payments government agencies ; Arrears
PE 6952 — policies transnational corporations developing countries ; Exploitative
PE 9567 — position developing countries ; Deterioration external
PE 0682 — practices ; Irresponsible
PE 0682 — practices ; Unethical
PF 2009 — *pressure* ; *Excessive*
PF 1959 — *priorities* ; *Non participatory*
Financial [referral...]
PG 8883 — *referral* ; *Insufficient*
PF 5374 — reporting public sector ; Inadequate
PB 4653 — reserves ; Inadequate
PF 0404 — resources action against problems ; Shortage

Financial [resources...] cont'd
PE 3812 — resources available developing countries ; Decline concessional
Financial resources developing
PF 5964 — countries ; Inadequate mobilization
PF 5964 — countries ; Inefficient management government
PC 3134 — countries ; Outflow
Financial resources [Disregarded...]
PF 7458 — Disregarded
PD 9338 — Inappropriate use
PB 4653 — Insufficient
PF 1460 — United Nations system organization ; Shortage
PE 9133 — urban services ; Insufficient
Financial [scandal...]
PD 2458 — scandal
PE 1571 — secrecy ; Corporation
PE 8144 — security urban life style ; Struggle
Financial services
PE 0718 — *Inaccessible*
PE 0718 — Inaccessible commercial
PE 0718 — *Inadequate*
PE 7987 — Protectionism free zone international
PE 8121 — sectors ; Ineffective self regulation
Financial [settlement...]
PF 9841 — settlement systems ; Inadequate
PF 2174 — short-termism
PE 2032 — statements ; Distortion corporation
PE 4508 — stock markets ; Lack international coordination supervisors
PF 3630 — surplus ; Lack
PF 0048 — system ; Crisis international monetary
PC 7873 — system ; Vulnerability national economies vagaries international
Financial transactions
PF 6594 — Undocumented international
PF 5728 — Unreported
PC 6641 — Unreported
PF 3554 Financial understanding ; Minimal
PE 3104 Financing ; Counterproductive capitalist investment
PE 7195 Financing development programmes ; Dependence developing countries external
PE 7195 Financing ; Excessive reliance developing countries external sources development
PD 3131 Financing external borrowing ; Dependence developing countries budget deficit
PE 8675 Financing image ; Short range
PC 6631 Financing ; Inadequate local government
PE 0365 Financing local government developing countries ; Inadequate
PF 3550 Financing opportunities ; Limited
PD 3954 Financing ; Over commitment credit
PE 0752 Financing political parties ; Unjust
PE 3066 Financing upon developing countries ; Burden export credit
PF 3859 Finders personal property ; Irresponsible
PF 0334 Finite energy resources ; Reliance
PD 8054 Fire
Fire [agreements...]
PE 2324 — agreements ; Persistence hostilities following cease
PE 2324 — agreements ; Persistence technical state war following cease
PF 4298 — *alarms* ; *False*
PE 2229 Fire blight
PD 1631 Fire code ; Inadequate
PD 6520 Fire department ; Unfunded
Fire [equipment...]
PD 6520 — equipment ; Antiquated
PD 6520 — equipment ; Costly
PE 5505 — explosion ; Endangering
PF 2477 Fire fighters ; Undertrained
PD 6520 Fire fighting facilities ; Lack
Fire [hazards...]
PD 1631 — hazards ; Continuing
PD 6520 — hoses ; Short
PD 6520 — hydrants ; Few
PD 1631 — ing ; Inadequate safeguards against
Fire [prevention...]
PD 1631 — prevention ; Inadequate
PD 1631 — protection ; Inadequate
PD 1631 — protection ; Unavailability
PF 2477 — protectors ; Insufficient
PD 8054 Fire risk
PF 2477 Fire services ; Distant
PA 6448 Fire setting ; Deliberate
PE 2324 Fire violations ; Cease
PF 2477 Fire volunteers ; Dispersed
PE 2857 Firearm accidents
PD 1970 Firearm regulation ; Inadequate
PE 2857 Firearm storage ; Unsafe
PE 2857 Firearm use ; Careless
PE 2449 Firearms criminal activity ; Supplying
PE 1108 Firearms explosives crimes
PE 1108 Firearms explosives laws ; Violations against
PE 2470 Firearms ; Illegal
PE 7711 Firearms ; Trafficking illegal
PE 3620 Firebrats
PF 2477 Firefighters ; Lack trained
PD 8054 Fires
PD 4652 Fires ; Accidents caused
PD 0739 Fires ; Bush
PD 8054 Fires ; Domestic
PD 0739 Fires ; Forest
PE 4675 Fires ; Lethal fumes modern house
PD 2211 Fires ; Urban
PD 0739 Fires ; Wildland
PD 4769 Firewood ; Shortage

PF 0326 Firms ; Bogus
PE 8706 Firms ; Growing size impersonality
PG 5429 Firms ; National regulations banking security
PE 8864 Firms ; Overemphasis governments large
PF 9007 First aid training ; Inadequate
PF 9007 First aid ; Unavailability
PE 7826 First phalanx ; Fracture
PF 2124 Fiscal base ; Dissipated
PF 3612 Fiscal collapse ; Global
PF 4850 Fiscal crisis
Fiscal [deficits...]
PD 5984 — deficits
PD 5984 — deficits ; Unsustainable
PF 2124 — discipline developing countries ; Inadequate
PD 1466 Fiscal fraud
PD 1466 Fiscal laws ; Lack compliance
PE 9061 Fiscal liabilities ; Misrepresentation
PF 4850 Fiscal policies ; Inadequate
PE 8308 Fiscal relations different parts government ; Unclear
PC 4879 Fiscal trade imbalances
PD 3934 Fish ; Accumulation contaminant residues marine animals
PD 4925 Fish birds ; Accumulation contaminant residues freshwater animals,
PF 4495 Fish catches ; Unequal distribution
PE 5286 Fish ; Cruelty
Fish crustacea molluscs
PD 0372 — Environmental hazards
PE 0972 — preparations thereof ; Instability trade
PE 1783 — preparations thereof ; Long term shortage salt water
Fish [Decreasing...]
PD 0547 — Decreasing genetic diversity
PE 7584 — diseases ; Pollution induced
PC 3152 — Disruption breeding grounds
PC 1535 Fish ; Endangered species
Fish [farming...]
PF 6478 — *farming* ; *Expensive*
PF 5496 — farming ; Limited
PE 3934 — farms ; Pollution water
PF 3602 Fish ; Irresponsible introduction new species
PE 7584 Fish kills caused pollution
Fish [Pests...]
PD 8567 — Pests diseases
PD 0372 — poisoning
PD 1415 — products ; Unutilized
PE 7313 — profits ; Decreasing
PE 4913 Fish reserves marine mammals ; Depletion
Fish resources
PD 9082 — Depleted
PD 9082 — Overexploitation
PD 9082 — Unsustainable exploitation
PF 2135 Fish ; Seasonal migration
PD 5361 Fish wars
PF 2135 Fish yield ; Unpredictable
PD 9082 Fish yields ; Reduction
PE 8148 Fisheries dams ; Destruction
PD 9082 Fisheries depletion
PE 1431 Fisheries products ; Radioactive contamination marine environment
PF 6478 Fishing business ; Unattractive
PF 5744 Fishing conditions ; Cold
PD 9082 Fishing ; Cyanide
PD 9082 Fishing ; Destructive
Fishing [economies...]
PF 2135 — economies ; Non diversification subsistence
PF 6478 — *employment* ; *Unattractive*
PF 6478 — equipment ; Prohibitive cost
PF 1442 Fishing grounds ; Unpopulated
PF 6478 Fishing ; Inadequate income
PF 6478 Fishing income rural communities ; Low
Fishing industry
PD 0743 — Environmental hazards
PE 1424 — Instability
PE 2138 — Underdevelopment
PD 9082 Fishing methods ; Indiscriminate
PD 9082 Fishing nets ; Deployment excessively efficient
PF 6478 Fishing ; Prohibitive cost
PD 5361 Fishing rights ; Conflicts over
Fishing [season...]
PF 2135 — *season* ; *Unstable*
PF 2135 — *Seasonal restrictions*
PF 1373 — *skills* ; *Untransmitted*
PE 5286 — sport
PE 4161 *Fissura ungulae horizontalis*
PE 4161 *Fissura ungulae longitudinalis*
PE 4161 Fissures ; Horizontal hoof wall
PE 4161 Fissures ; Vertical hoof wall
PE 3711 Fistula animals ; Rectovaginal
PE 6555 *Fistulous withers*
PE 8272 Fittings ; Instability trade sanitary plumbing, heating, lighting fixtures
PE 7940 Fittings ; Shortage sanitary plumbing, heating lighting fixtures
PD 4747 FIV Feline immuno-deficiency virus
PE 7848 FIV Feline immuno-deficiency virus
PF 4133 Fixation adversaries
PA 5497 Fixation *Difficulty*
PA 7157 Fixation *Insanity*
PA 6739 Fixation *Maladjustment*
PF 9409 Fixation partial solutions problems
PA 6448 Fixation ; Psychological
PE 0861 Fixed vegetable oils fats ; Instability trade
PE 8277 Fixed vegetable oils fats ; Shortage
PE 4301 *Fixing commodity markets* ; *Price*
PE 4301 Fixing ; Fare
PE 6726 Fixing ; Lack minimum wage

-929-

Fixing

PE 4301 Fixing ; Price
PE 8272 Fixtures fittings ; Instability trade sanitary plumbing, heating, lighting
PE 7940 Fixtures fittings ; Shortage sanitary plumbing, heating lighting
PE 8165 Fixtures manufacture ; Environmental hazards furniture
PF 2604 *Fixtures manufacture underdevelopment ; Furniture*
PD 0700 Flag discrimination shipping
PD 7550 Flagellation
PF 7118 *Flagging community spirit*
PE 5873 Flagging ships ; Convenience
PE 5873 Flags convenience ; Evasion shipping regulations taxes
PF 9138 Flags ; Desecration
PF 2152 Flaky arguments
PD 2211 *Flammable construction materials*
PB 0855 *Flammable substances ; Inflammable*
PE 3407 Flare ups ; Emotional
PE 9528 Flares ; Radio interference associated solar
PE 9528 Flares ; Solar
PE 5140 Flash floods
PD 2647 *Flat foot*
PE 1764 Flatulence public ; Human
PA 6180 Flaw *Error*
PA 5952 Flaw *Illegality*
PA 6997 Flaw *Imperfection*
PA 5502 Flaw *Reason*
PF 6814 Flawed assumptions
PD 9667 *Flea allergy dermatitis*
PE 3895 *Flea-borne typhus*
PE 3572 Flea resistance insecticides
PD 3634 *Flea ; Water*
PE 3643 Fleas
PD 9667 *Fleece worms*
PE 5790 Fleets ; Lack investment capital developing countries' shipping
PE 5825 Fleets ; Over concentration ownership maritime
PC 7644 Flesh ; Eating animal
PD 4518 Flesh ; Excessive consumption animal
PE 0596 *Flesh ; Uncooked inadequately cooked*
PE 8283 Flexibility labour market ; Lack time
PE 7826 *Flexion deformities*
PE 7808 *Flexor tendons ; Contracted*
PE 2254 *Flies ; Buffalo*
PE 3878 Flies ; Deer
PE 2254 *Flies ; Face*
Flies [Head...]
PE 2254 *— Head*
PE 2254 *— Horn*
PE 3878 — Horse
PE 2254 Flies insect pests
Flies pests
PE 3646 — Black
PE 3607 — Fruit
PE 8150 — Screw worm
PE 1335 — Tsetse
PE 2254 *Flies ; Plantation*
PE 2254 *Flies ; Stable*
PE 2254 *Flies ; Stable*
PE 4514 Flies vectors disease
PE 5267 Flight avoid prosecution giving testimony
PD 3132 Flight ; Capital
PC 1453 Flight capital ; Speculative
PE 5267 Flight ; Unlawful
PD 5453 *Flip-over disease*
PE 5930 Floating exchange rates
PF 4295 Floating rate loans ; Uncertainty development expenditures due
PD 2484 Flogging offenders
PE 0192 Flogging schools
PD 0452 Flood control ; Inadequate
Flood [plain...]
PE 0743 — plain settlement
PE 0743 — plains ; Building
PE 0743 — plains ; Construction
PE 7837 *Flood waves*
PD 0452 Flooding
PD 0156 Flooding developing countries ; Labour market
PD 4976 Flooding ; Downstream
PE 3813 Flooding ; Exposure settlements tidal
Flooding Loss
PE 3786 — agricultural land due artificial
PE 7855 — forests artificial
PE 7794 — wildlife due artificial
PE 7837 Flooding ; Salt water
PD 0452 Flooding ; Unchecked seasonal
PD 0008 Flooding urban labour market developing countries
PD 0452 Floods
PE 5140 Floods ; Flash
PD 4976 Floods ; Riverine
PE 5006 Floods ; Tidal
PF 1773 *Floors houses ; Unhygienic dirt*
PC 3152 Flora fauna ; Degradation semi natural natural habitats
PD 4217 Floracide
PE 8606 Flour than wheat meslin shortage ; Meal
PE 8980 Flour wheat meslin shortage ; Meal
PE 8489 Flour wheat meslin trade instability ; Meal
PD 7792 Flouting regulatory authority
Flow [Inadequate...]
PF 6451 *— Inadequate cash*
PF 5006 — Inadequate inter developing country skills
PF 6469 — information ; Insufficient
PF 2852 — investment capital ; Insufficient
PF 6451 *— Irregular cash*
PF 0166 Flow knowledge educational materials ; Barriers international
PF 6451 Flow local economy ; Restricted
PF 6451 *Flow ; Minimal cash*
PS 6553 *Flow ; Outward cash*
PF 2865 Flow plan ; Absence long range, world wide capital
PF 6451 *Flow ; Restricted cash*
PE 4314 Flowering plants ; Endangered species
PE 3812 Flows concessional non concessional lending developing countries ; Erratic
PE 3937 Flows ; Lava
PC 1453 Flows ; Massive international capital
PE 0447 Flu
PF 1226 *Fluctuating agricultural markets*
PF 3084 Fluctuating income drops
PF 2819 Fluctuating response appeals aid
PD 5212 Fluctuation availability food
PF 8635 Fluctuation feed price
PD 5212 Fluctuations agriculture ; Seasonal
PE 8188 Fluctuations food production developing countries
PD 5212 Fluctuations food ; Weather induced
PF 0170 Fluctuations government social programmes
PD 8994 Fluctuations international competitiveness
PD 0909 Fluctuations ; Oil price
PD 9356 Fluctuations real value money
PF 8163 Fluctuations ; Seasonal
PF 8635 Fluctuations ; Unanticipated price
PC 1108 Fluctuations work ; Seasonal
PC 7873 Fluctuations world monetary financial conditions
PD 8123 Fluency ; Limited language
PF 0963 *Fluid community patterns*
PD 2567 *Fluid middle ear*
PE 6461 *Fluke ; Common liver*
PD 2732 *Fluke fever ; Elokomin*
PE 6461 *Fluke ; Giant liver*
PE 6461 *Fluke ; Giant liver*
PD 2735 *Fluke infections birds*
Fluke [Lancet...]
PE 6461 *— Lancet*
PE 6461 *— Large American liver*
PE 6461 *— Lesser liver*
PE 2785 — Liver
PE 6461 *Fluke ; Pancreatic*
PE 0921 Flukes ; Blood
PE 6461 *Flukes ; Conical*
PE 9854 *Flukes ; Lung*
PE 6461 *Flukes ; Rumen*
PE 2871 Fluoridation drinking water
PD 5228 *Fluoride poisoning animals*
PE 1311 Fluorides pollutants
PD 5228 *Fluorosis*
PE 9609 Fluorosis ; Animal
PE 2259 Flux ; Bloody
PF 1844 *Flux ; Rapid transient*
PD 1245 *Fly ash ; Coal*
PE 2254 *Fly ; Beet*
PE 3635 *Fly ; Chin*
PE 3635 *Fly ; Deer nose*
PE 5725 Fly disease
PE 2388 *Fly-infested rivers streams*
PE 3635 *Fly ; Nose*
PE 2254 *Fly ; Rodent rabbit bot*
PD 9667 *Fly strike*
PE 3635 *Fly ; Warble*
PE 3972 Flying foreign aircraft ; Over
PD 8328 Flying hazards
PE 0680 Flying influence alcohol
PC 1326 *Flying lemurs ; Endangered species*
PF 1392 Flying objects ; Unidentified (UFOs)
PF 1392 Flying saucers (UFOs)
PE 7808 *Foals ; Angular limb deformities*
PD 7841 *Foals ; Convulsive*
PE 7808 *Foals ; Defects back*
PD 3978 *Foals ; Enteric diseases*
PD 2731 *Foals ; Septicaemia*
PD 9293 *Foals ; Uroperitoneum*
PF 8753 Focus available insight response problems society ; Failure
PF 6580 Focus fundamental systemic problems ; Failure
PF 1816 Focus religious friction ; Holy places
PF 1240 *Fodder resources ; Unused*
PE 3853 Foetal alcohol syndrome
PE 4805 Foetal cadavers foetal tissue ; Research
PE 4609 Foetal damage drug use
PD 4029 Foetal death ; Antenatal
PE 2041 *Foetal erythroblastosis*
PE 2041 Foetal infection death
PE 4808 Foetal malformation diabetic pregnancies
PF 6616 Foetal rights ; Denial
PE 4805 Foetal tissue ; Research foetal cadavers
PD 0158 Foeticide
PD 3947 Foeticide ; Female
Foetus [Damage...]
PE 2041 — Damage death
PE 2467 *— Defective oxygen supply*
PE 2042 — Deformation
PD 0158 Foetus ; Killing
PE 2042 Foetus ; Physical malformation
PE 8343 Foetuses ; Medical experimentation pregnant women
PE 6369 Foetuses ; Sale
PE 4805 Foetuses ; Unethical experimentation using aborted
PE 6369 Foetuses ; Violation rights
PE 1655 Fog
PE 5524 *Fog fever*
PE 1655 Foggy atmosphere
PD 7799 *Follicles ; Cystic*
PD 7799 *Follicular cystic ovary disease*
PD 7799 *Follicular cysts*
Follow up
PF 0342 — initiatives ; Ineffective
PF 0342 — procedures ; Absence
PF 0342 — Weak programme
PA 7241 Following
PF 5061 Folly ; Government
PF 5061 Folly ; Political
PD 0487 Food additives ; Carcinogenic
PD 0487 Food additives ; Toxic
Food aid
PF 3949 — [Food aid]
PF 3949 — developing countries ; Stagnation
PD 7592 — Exploitation dependence
PF 3949 — Inadequacy
PE 0302 — Inappropriate
PF 3949 — Maldistribution
PF 6078 — Modification environmentally adapted nutritional habits
Food [allergy...]
PE 3225 — allergy
PE 9509 *— allergy*
PE 3225 — Allergy inducing
PE 0395 — animal conditions ; Unsanitary inhumane urban
PE 3899 — animals right freedom suffering ; Denial
Food [Biological...]
PD 2594 — Biological contamination
PE 2515 — borne diseases
PE 2515 — borne infections intoxications
Food chains
PE 8154 — Concentration noxious substances
PD 0120 — Pesticide contamination
PB 2253 — Vulnerability
PD 1694 Food ; Chemical contaminants
PD 0487 Food colourings ; Toxic
Food consumption
PC 2518 *— Increased*
PD 0800 — Junk
PD 0800 — Unnutritious
PC 0382 — Unnutritious
Food contamination
PD 5605 — [Food contamination]
PE 7064 — Animal products
PF 3904 — Fear
PE 6480 — pesticides
PD 5046 Food crisis ; World
Food crops
PF 9187 — guaranteeing survival ; Discouragement
PD 4100 — Narrow range
PD 7565 — tropical economies ; Preponderance non
Food [Decline...]
PE 8938 — Decline nutritional quality
PC 0382 — deficiencies
PE 0324 — Denial entitlement
PE 0324 — Denial right sufficient
PC 2801 — Dependence maldistribution
PE 0607 — developing countries ; Dumping
PE 8918 — drink industries ; Instability
Food [economy...]
PD 5046 — economy ; Imbalance world
PC 5016 — emergencies ; Unpreparedness
PE 8659 — Excessive proportion income spent
PF 1345 *— Export nutritious*
Food [fads...]
PD 1189 — fads
PE 4261 — Fatty
PD 5212 — Fluctuation availability
Food grain
PB 2846 — insecurity
PD 4100 — species ; Limited
PD 0811 — spoilage
Food [Health...]
PD 0361 — Health hazards irradiated
PD 0651 *— hoarding*
PF 8515 *— hygiene ; Lack*
Food [imports...]
PE 8086 — imports ; Dependence developing countries
PE 0302 — imports ; Disruptive
PC 3319 — indigenous populations ; Denial right adequate
Food industry
PD 1045 — Corruption
PD 1045 — Irresponsible practices
PD 1045 — Negligence
Food insecurity
PB 2846 — [Food insecurity]
PB 2846 — Household
PE 7954 — least developed countries
Food [Insufficient...]
PB 2846 — Insufficient available
PF 6105 *— interdependence*
PE 9541 — intolerance
PE 1750 Food journalism ; Junk
PF 2821 Food live animal production ; Underdevelopment
Food live animals
PC 1411 — Environmental hazards
PC 1411 — Health hazards
PD 2894 — Instability production
PD 1434 — Instability trade
PE 0976 — Long term shortage
PE 0342 — trade ; Restrictive practices
Food [Maldistribution...]
PF 3949 — Maldistribution
PC 2801 — Maldistribution
PF 8702 — manufacturing industry wastes
PD 9669 — Microbial contamination
PE 8018 — monopolies
PF 3904 Food ; Neurosis

Foreign nationals

PD 9448 Food ; Overproduction
PE 6480 Food ; Pesticide residues
Food poisoning
PE 9374 — Bacterial
PE 0561 — negligence
PD 5426 — sabotage
PD 5426 — Terrorist
Food [poisons...]
PD 1472 — poisons ; Natural
PF 6202 — political weapon ; Misuse
PD 5605 — pollution
PD 1045 — practices ; Unethical
PF 5940 — Predominance fast
Food preparation
PE 6619 — Carcinogenic consequences
PE 0236 — Cruelty animals
PE 6619 — Mutagenic consequences
Food [preparations...]
PE 1683 — preparations ; Instability trade miscellaneous
PD 0361 — preservation ; Irradiation method
PD 0487 — preservatives ; Toxic
PE 4986 — prices developing countries ; Instability
PF 2385 — prices rural communities ; Rising
Food processing industries
PE 1280 — Environmental hazards
PE 8472 — Ineffective self regulation
PD 0908 — Underdevelopment
Food production developing
PD 5092 — countries ; Deterioration domestic
PF 4130 — countries ; Economic disadvantages excessive
PE 8188 — countries ; Fluctuations
Food production [industries...]
PD 5830 — industries ; Protectionism agriculture
PC 4815 — Shortage water
PC 1960 — Slowing growth
PF 6493 — Underdeveloped approaches local
Food products
PE 1280 — Environmental hazards
PF 7808 — Forgotten
PE 6390 — *Synthetic*
PD 8844 — waste ; Dumping
PF 1212 Food ; Prohibitive cost nutritious
Food [Regurgitation...]
PE 5187 — *Regurgitation*
PD 0036 — related cancers
PD 0800 — *Reliance canned*
PC 5016 — reserves ; Inadequacy emergency
PE 6115 — resources ; Mismanagement
PF 7808 — resources ; Neglected
Food [sector...]
PE 1796 — sector developing countries ; Domination transnational corporations domestic name brand
PF 5137 — security system ; Lack world
PB 2846 — shortage ; Dependence
PD 2243 — spoilage storage
PE 1806 — staples developing countries due transnational corporations ; Import dependency
PC 5016 — stocks ; Inadequate
PF 4763 — stocks ; Secrecy national basic
PE 4877 — storage facilities ; Inadequate
PD 1472 — stuffs ; Naturally occurring poisonous substances
PE 7413 — subsidies ; Dependence economies
PE 7413 — subsidies ; Distortion national economies
Food supplies
PC 5016 — Inadequate emergency
PF 2857 — Inadequate mechanisms securing sufficient
PE 7954 — least developed countries ; Inadequate
PD 5212 — Seasonal variability
PB 2846 — Shortage
Food supply
PD 4101 — developing countries ; Inadequate staple
PE 8979 — due military activities ; Disruption
PB 2846 — Insufficient
PB 0315 — Massive starvation despite sufficient world
Food [taboos...]
PF 3310 — *taboos*
PD 2868 — taboos ; Violation
PE 2770 — *tastelessness*
PD 5426 — Threats poison
Food [Unclean...]
PF 8515 — *Unclean*
PF 3310 — *Uneaten nutritional*
PC 2801 — Unequal access
PF 2813 — *usage ; Unbalanced*
PF 0479 Food variety ; Limited
Food [wastage...]
PD 8844 — wastage
PD 8844 — waste
PD 5212 — Weather induced fluctuations
PB 2253 — webs
PD 0036 Foods ; Cancer causing
PJ 7877 Foods ; Costly imported
PE 0596 *Foods ; Inadequately cooked*
PF 1212 Foods ; Prohibitive cost healthy
PE 4903 Foods ; Residues veterinary drugs
PE 8255 Foodstuffs breast feeding ; Substitution inappropriate
PC 3908 Foodstuffs ; Excessive consumption specific
PD 4238 Foodstuffs ; Harmful natural
PE 9164 Foodstuffs ; Instability trade vegetable based
PF 7808 Foodstuffs ; Limited utilization
PF 1212 Foodstuffs ; Prohibitive cost organically grown
PD 1126 Foodstuffs ; Unaesthetic
PE 0430 *Foodstuffs ; Unsafe transport perishable*
PF 6973 Foolhardiness
PE 5617 *Foot abscess*
PE 1589 Foot-and-mouth disease
PE 7826 *Foot ; Buttress*
PE 3836 Foot ; Club
PD 2647 Foot diseases disabilities
PE 4161 *Foot ; Fescue*
PD 2647 *Foot ; Flat*
PE 5050 *Foot lesions*
Foot [Madura...]
PE 2455 — *Madura*
PE 7826 — *Mule*
PE 7808 — *Mule*
Foot [paths...]
PF 7580 — *paths ; Dangerous*
PE 7826 — *Pricked*
PE 7826 — *Puncture wounds*
Foot rot
PE 4346 — *[Foot rot]*
PE 5050 — *[Foot rot]*
PE 4161 — *[Foot rot]*
Foot rot [Benign...]
PE 5617 — *Benign*
PE 5617 — *Contagious*
PE 5617 — *Malignant*
PE 4161 — *Stable*
PD 9667 — *Strawberry*
PE 5617 — *Virulent*
PE 5617 *Foot scald*
PD 7799 *Foothill abortion cattle*
PE 3874 Footpaths development ; Obliteration
PE 8320 Footwear industry ; Instability leather
PF 5413 Forbidden ; Attraction
PE 8881 Force administration justice ; Unrestrained use
PE 8881 Force law enforcement officials ; Excessive use
PE 8881 Force military personnel ; Excessive use
PE 8881 Force police ; Excessive use
PC 3293 Forced assimilation
PC 3293 Forced assimilation ; Dependence
PD 3035 Forced breach journalistic confidence
PE 8947 Forced confession
PE 4888 Forced confessions drugs
Forced [depilation...]
PE 5697 — depilation
PC 6203 — deportation
PD 4259 — disappearance persons
Forced disappearances
PD 5129 — children
PE 5882 — trade union leaders
PE 5882 — workers representatives
PD 5628 Forced exercise
PC 3293 Forced integration peoples
Forced labour
PC 0746 — [Forced labour]
PC 0746 — Dependence
PE 5506 — Unfair competition workers employed
Forced [marriage...]
PD 1915 — marriage
PD 0531 — mass expulsion
PC 6203 — migration
PE 5919 — motherhood
Forced [participation...]
PD 2910 — participation politics
PC 5387 — participation social processes
PE 3016 — political confessions
PE 5919 — pregnancy
PD 6637 Forced religious conversion
Forced relocation
PC 6203 — [Forced relocation]
PE 8898 — Inadequate compensation
PE 8898 — Non payment compensation
PD 4273 — peasants onto marginal lands
Forced [repatriation...]
PD 8099 — repatriation
PD 0218 — repatriation prisoners war
PC 2507 — *resettlement countryside*
PF 4948 Forced sexual intercourse
PD 4287 Forced social intimacy
PE 1255 Forced witness torture
PA 6876 Forcelessness *Impotence*
PA 6882 Forcelessness *Influencelessness*
PE 7705 *Forceps delivery ; Precipitate ill judged*
Forces [Abuses...]
PE 4847 — Abuses private police
PE 9420 — Aggressive economic destabilization countries external
PF 6895 — Attacks peace
PE 5782 Forces ; Causing insubordination armed
PD 7360 Forces ; Corruption armed
PE 8220 Forces ; Discrimination against women military
PF 9542 Forces ; Disregard government orders armed
PE 5986 Forces ; Extra legal impressment children armed
PC 5230 Forces ; Imbalance conventional armed
PE 3912 Forces ; Obstruction recruiting induction armed
Forces [Racial...]
PD 3519 — Racial discrimination security
PD 3519 — Racism police
PB 6332 — Reactionary
PD 9252 — Rivalry armed
Forces [Uncontrolled...]
PF 7880 — Uncontrolled market
PE 4920 — Under strength military
PF 6836 — *Unequal distribution*
PD 9193 — Unethical practices police
PE 4484 — Unlawful recruiting enlistment foreign armed
PF 2312 — Unquestioned control economic
PC 0475 Forcible displacement wildlife
PD 5515 Forcible obstruction law enforcement
PF 7166 Forecast ; Overestimation ability
Forecasting [Inaccurate...]
PF 4774 — Inaccurate
PF 5118 — Inaccurate weather
PF 4928 — Inadequate earthquake
PF 4774 Forecasts ; Surprise free
PF 4774 Forecasts ; Unrealistic
PD 3135 Foreign access capital bond markets ; Restrictions
Foreign aid
PD 0136 — Corruption mismanagement
PF 8120 — Counter productive effects
PE 8116 — Dependence least developed countries
PF 8120 — Negative effects
PD 0136 — Wasted
Foreign [aircraft...]
PE 3972 — aircraft ; Invasion airspace
PE 3972 — aircraft ; Over flying
PE 5210 — annexation
PJ 4562 — arbitral awards ; Inadequate enforcement
PE 4484 — armed forces ; Unlawful recruiting enlistment
Foreign [bodies...]
PE 4630 — *bodies animals ; Gastric*
PE 4630 — *bodies oesophagus animals*
PD 7307 — *body pneumonia*
PE 7195 — borrowing ; Over reliance developing countries
Foreign [companies...]
PD 6417 — companies ; Discrimination against
PE 6274 — consultants ; Inadequacies
PC 3187 — control
PD 3109 — control natural resources
PE 0261 — controlled direct satellite broadcasting individual receivers
PD 3041 — controls newspaper journal propaganda
Foreign countries
PE 2113 — Abduction government agents acting
PE 6564 — Government seizure foreign nationals
PC 3185 — Intervention major powers protect investments their citizens
PD 1291 — Non return developing country students studying
PD 3047 Foreign country ; Tax discrimination against investment
Foreign currency
PE 2145 — manipulations accounting records transnational corporations
PD 3068 — shortage developing countries
PD 9489 — speculation
PF 3070 — use abroad ; Refusal issue
Foreign [debt...]
PC 3056 — debt
PE 5748 — debt island developing countries ; Increasing
PB 0206 — dependence
PC 3186 — dictatorship
PF 4821 — dictatorships ; Government support
PD 3138 — direct investment developing countries ; Decline
PC 7384 — domination countries
PJ 5262 Foreign economies ; Effect dominant stock markets weaker
Foreign exchange
PC 8182 — Chronic shortage
PJ 5244 — monopoly
PF 3070 — restrictions
PC 8182 — Shortage
PD 3068 — shortage developing countries
PC 8182 — shortages
PE 0117 — Underutilization non monetary
Foreign [government...]
PC 3185 — government interference
PE 1622 — governments ; Government bullying
PF 8040 — governments ; Non recognition
PE 8802 — groups peoples ; Biased media image
Foreign [influence...]
PC 3188 — influence ; Disruptive
PE 4280 — insurance ; Dependence developing countries
PC 3185 — interference internal affairs states
Foreign intervention
PC 3185 — internal affairs states
PD 7276 — tourism
PD 2374 — Vulnerability small nations
Foreign investment
PE 8787 — Dislocation productive units
PD 8030 — Disruptive
PE 8525 — Distortion international trade restrictive controls over
PD 8030 — Inappropriate
PD 3063 — Legal impediments
PD 8030 — National insecurity due excessive
PD 0765 — traditional industries developing countries ; Excessive
PE 0161 — transnational enterprises restrictive business practice ; Direct
PD 8030 — Vulnerability countries destabilization
PC 2172 Foreign investments ; Nationalization
PC 3134 Foreign investors developing countries ; Excessive repatriation profits
Foreign [labour...]
PC 0065 — *labour ; Dependence*
PF 5929 — language teaching ; Ethnic social discrimination
PE 7195 — loans ; Dependence developing countries
Foreign military
PD 9331 — intervention
PD 1078 — occupation ; Vulnerability women during
PD 3496 — presence
PD 3496 — presence ; Erosion national sovereignty
PE 6746 Foreign money liabilities
Foreign nationals
PE 6564 — Disregard international law capture
PE 6424 — Failure notify imprisonment death prison
PE 6564 — foreign countries ; Government seizure
PE 6422 — military service ; Discrimination against

Foreign ownership

PE 0042 Foreign ownership
PE 4738 Foreign ownership
Foreign [policy...]
PC 4667 — policy ; Aggressive
PE 4343 — power ; Revealing national security information
PE 3972 — powers ; Aerial surveillance
PE 6418 — powers attorney ; Non recognition
PC 3188 — pressure
PE 6715 — principals ; Political contributions agents
PE 6883 — prisoners ; Discriminatory treatment
PD 3138 — private investment developing countries ; Disincentives
PE 8957 — private investment income outflow developing developed countries
PE 9643 — programmes media ; Excessive use
PE 0873 — *protein reactions ; Arthritis accompanying*
Foreign public debt
PD 3051 — developing countries ; Burden servicing
PD 2133 — developing countries ; Excessive
PD 9168 — industrialized countries ; Excessive
PE 5331 Foreign relations trade ; Crimes related
Foreign [soldiers...]
PD 8825 — soldiers ; Abandoned children
PF 7577 — students ; Lack
PF 7726 — students ; Unequal opportunities
PF 9697 Foreign trade ; Vulnerability adverse conditions
PE 9135 Foreign travel remote places ; Unaccompanied
PE 8423 Foreign workers' families ; Refusal entry
PE 0896 Foreigners ; Alienation land acquisition
PF 8565 Foreigners ; Communication
PD 0973 Foreigners ; Discrimination against
PD 6361 Foreigners ; Discrimination against
PD 3529 Foreigners employment ; Discrimination against
PC 1532 Foreigners ; Inadequate social integration
PE 8747 Foreigners leaving country sojourn ; Restrictions
PF 1798 Foreigners legal proceedings ; Discrimination against
PF 8743 Foreigners ; Mistrust
PD 3529 Foreigners ; Restrictions employment
PF 8565 Foreigners ; Speaking
PD 7841 *Forelimb ; Paralysis*
PF 8176 Foresight capacity ; Inadequate
PF 1348 Foresight ; Lack saving
PF 8602 Foresight ; Narrow economic
PD 0500 Forest damage wildlife
PC 7896 Forest decline
PD 0739 Forest fires
PD 9490 Forest fragmentation
PD 4900 Forest lands ; Unsustainable development
PD 9235 Forest overcutting
PD 3585 Forest pests
PC 5165 Forest resources ; Limited
PC 5165 Forest vulnerability
PE 7695 Forestation ; Inappropriate re
PC 0880 *Forestry industry ; Underdevelopment*
Forestry [least...]
PE 8244 — least developed countries ; Insufficient
PE 1264 — logging ; Environmental hazards
PE 0459 — logging ; Instability
PD 6701 Forestry malpractice
PD 6701 Forestry ; Unethical practices
PE 5274 Forests agricultural development ; Misuse tropical rain
PE 7855 Forests artificial flooding ; Loss
PD 4900 Forests ; Conflicting demands
PE 7855 Forests dams ; Inundation
Forests Destruction
PC 1366 — [Forests ; Destruction]
PD 6850 — alluvial
PD 6204 — rain
PC 5165 Forests ; Endangered
PD 6701 Forests ; Mismanagement
PC 7896 Forests ; Unhealthy
PF 0255 Forged scientific data
PE 4496 Forged travel documents
PE 5516 Forgeries financial crime ; Frauds,
PD 2557 Forgery
PE 2382 Forgery ; Art
PD 7981 Forgery brand-name merchandise
PD 0486 Forgery certificates origin
PE 5051 Forgery ; Historical
PE 6188 Forgery ; Literary
PE 5051 *Forgery ; Religious historical*
PD 7981 *Forgery shares bonds*
PD 7981 *Forgery shares, stocks, bonds ; Risk*
PD 2557 *Forgery wills testaments*
PA 6651 Forgetfulness
PA 6651 *Forgetfulness Forgetfulness*
PA 5438 *Forgetfulness Neglect*
PF 7745 Forgetfulness ; Structural
PA 5643 *Forgetfulness Unkindness*
PF 7808 Forgotten food products
PF 7047 Forgotten rural importance
PF 2346 *Formal complaints ; Insufficient*
Formal education
PF 6534 — Disrelated
PF 4765 — Inadequacy
PF 6534 — Lack
PF 1944 — rural communities ; Unperceived relevance
Formal [leaders...]
PC 0237 — *leaders ; Unrepresentative*
PF 1254 — leadership ; Lack
PF 8191 — learning ; Inapplicability
PF 5711 — *logic ; Inadequacy*
Formal [schooling...]
PF 3836 — schooling ; Divisive effects
PF 6467 — schooling ; Inadequate results
PF 2947 — support ; Disorganized liaison

PE 6214 Formaldehyde building materials personal products ; Over use
PF 1033 Formalised decision making ; Over
PE 5723 Formalism developed countries
PE 0278 Formalism ; Revolt against
PE 0208 Formalities ; Excessive customs
PE 2603 Formalities ; Excessive customs trade
PE 0208 Formalities international travel ; Excessive frontier
PF 0443 Format ; *Static curriculum*
PE 1604 Formation regional groupings developed countries ; Discrimination against developing countries
PF 8939 Formats ; Ineffective meeting
PF 2845 *Former skills ; Unprofitable*
PF 8026 Forming organs ; Diseases blood blood
PA 6900 Formlessness
PD 4287 Forms address ; Imposition intimate
PA 4332 Forms worship ; Inappropriate transfer
PF 1115 Forms worship ; Incompatible
PF 7713 Formulate effective global policies ; Inadequate intellectual cognitive infrastructure
PE 9073 Formulation equipment safety regulations ; Distortion international trade discriminatory
PE 1274 Formulation health sanitary regulations agricultural pharmaceutical products ; Distortion international trade discriminatory
PF 8703 Formulation processes ; Restricted access policy
PD 8468 Formulation ; Unscientifically based policy
PF 8711 Formulators ; Collusion administrators funding agencies programme
PF 5434 Fornication
PD 5972 Fortified frontiers
PF 1358 Fortune-tellers
PF 1358 Fortune-telling
PE 4289 Forums industrialized countries ; Undermining multilateral
PF 2216 Forums ; Misuse international
PF 7805 Forums public information ; Inoperative
PF 8176 Forward plan ; Undeveloped
PE 9417 Fossil collection
Fossil [fuel...]
PE 4891 — fuel emissions
PE 4891 — fuels ; Dependence
PE 4891 — fuels ; Exploitation
PD 4403 Fossil water reserves ; Misuse nonrenewable
PF 1755 Fostering dependency social institutions
PD 1564 Fostering structures engagement elders ; Isolation
PD 5803 Fouling rust marine equipment
PD 0482 *Fouling water supply systems*
PA 5454 *Foulness Badness*
PA 7391 *Foulness Closure*
Foulness [Destruction...]
PA 6542 — *Destruction*
PA 6191 — *Disapproval*
PA 6839 — *Disrepute*
PA 6486 Foulness *Injustice*
PA 5981 Foulness *Stench*
Foulness [Ugliness...]
PA 7240 — *Ugliness*
PA 5612 — *Unchastity*
PA 5459 — *Uncleanness*
PA 7107 — *Unpleasantness*
PA 7204 — *Unsavouriness*
PA 5644 Foulness *Vice*
PC 6870 Found family ; Denial right
PE 8020 Foundations ; Uneven settling building
PE 6091 Founded churches ; Non recognition newly
PE 4161 Founder
PD 6645 Fourth world ; Illiteracy
PD 2732 Fowl cholera
PE 5067 Fowl ; Endangered species water
PE 3639 Fowl mite ; Northern
PE 3639 Fowl mite ; Tropical
PD 2730 Fowl pox
PE 1766 Fowl ticks
PE 7562 Fowl typhoid
PE 1400 Fowlpest
PD 9667 Fowlpox
PE 4893 Fox hunting
PE 8739 Foxes ; Endangered species dog wolves
PE 7511 Fracture bone
PE 7826 Fracture carpal bones
PE 7826 Fracture distal phalanx
PE 7826 Fracture first phalanx
PE 7511 Fracture limb
PE 7511 Fracture lower limb
Fracture [metacarpal...]
PE 7826 — metacarpal bones
PD 7215 — Metal
PD 2638 — *metals ; Brittle*
PE 7826 — Fracture os pedis
Fracture [pedal...]
PE 7826 — *pedal bone*
PE 7826 — *Pelvic*
PE 7511 — *pelvis*
PE 7826 — *proximal sesamoids*
PE 7511 Fracture rib
PE 7511 Fracture ribs
Fracture [second...]
PE 7826 — *second phalanx*
PE 7511 — *Skull*
PE 7511 — *spine*
PE 7826 Fracture third phalanx
PE 7511 Fracture upper limb
PE 7511 Fractures
PE 4161 Fractures
PE 7826 *Fractures back horse*

PE 4161 *Fractures cattle*
PD 1230 Fractures ; Material
PE 7826 Fractures ; *Saucer*
PE 7511 Fractures ; *Skull*
PD 6960 Fragile ecosystems ; Degradation
PA 6799 Fragility *Disease*
PB 4885 *Fragility maintaining basic health*
PA 6976 Fragility *Toughness*
PD 1165 Fragility ; Tundra ecosystem
PA 5558 Fragility *Weakness*
PA 6233 Fragmentation
PF 8868 Fragmentation academic disciplines
PC 2662 Fragmentation ; Bureaucratic
Fragmentation [communist...]
PD 0923 — communist parties
PF 6250 — communities automobiles
PE 0296 — complexity United Nations system
PF 0536 — Cultural
PF 9830 Fragmentation field analysis
PD 9490 Fragmentation ; Forest
PE 5721 Fragmentation health service
PA 0911 Fragmentation human personality
Fragmentation [Increased...]
PF 1509 — *Increased family*
PC 3915 — Institutional
PC 3915 — institutional responsibility ; Sectoral
PF 7133 — international documentation
PC 0025 — International economic
PC 9584 — international trading system
PF 0944 Fragmentation knowledge
PF 3364 Fragmentation organized religions
PF 5753 Fragmentation organized students
PF 3216 Fragmentation ; Political
PA 0911 Fragmentation ; Psychological
Fragmentation [religious...]
PF 3404 — religious belief
PF 9830 — research
PD 2495 — *resident relationships*
Fragmentation [sciences...]
PF 8868 — sciences
PF 1324 — Social
PD 1566 — social structures depressed areas
Fragmentation [technological...]
PC 1227 — technological development
PF 0090 — Territorial
PC 2944 — Territorial
PC 1833 Fragmentation urban areas ; Political
PF 1768 *Fragmented agricultural development*
PF 6545 *Fragmented business practices*
Fragmented [celebration...]
PF 1985 — celebration cultural heritage
PF 6537 — *citizen goals*
PF 2421 — citizen responsibility
Fragmented community
PF 7118 — goals
PF 2845 — identity
PF 7118 — initiatives
PD 1566 — interaction
PF 1205 Fragmented conduct community operations
PF 8448 Fragmented decision-making
PF 2096 Fragmented ethical contexts
PF 6570 *Fragmented families ; Cut off*
Fragmented forms
PE 2274 — care neighbourhood level
PF 1442 — commercial management rural areas
PF 2588 — cooperative efforts
Fragmented [images...]
PF 2845 — images community identity
PF 3559 — individual decision-making process
PD 8839 — irrigation distribution
PF 3554 *Fragmented marketing practices*
PF 2421 Fragmented modes citizen engagement
PF 2428 Fragmented outlook civic minded groups ; Limited
Fragmented [pattern...]
PF 6525 — pattern community organization
PF 6504 — patterns community activity
PF 1509 — patterns extended family relationships
PF 2813 — planning community life
PF 9166 — policy making
Fragmented [recognition...]
PF 1985 — recognition common heritage
PF 9129 — regional cooperation
PF 9166 — resource planning
PA 0911 Fragmented selves
PE 3977 Fragmented social structures environmental protection
PF 1267 Fragmented utilization public information
PD 2495 Fragmenting district boundaries
PE 4716 Franchise ; Restricted
PC 3222 *Franchise ; Restricted*
PD 7121 Franchising services industry ; Protectionism
PD 0486 Fraud
PD 9343 Fraud against elderly
PE 5687 Fraud ; Agricultural
PE 1398 Fraud ; Bank
Fraud [Cheque...]
PE 3592 — Cheque card
PD 2057 — Commercial
PD 3917 — Commodities trading
PE 4362 — Computer aided
PE 5021 — concerning economic situation corporate capital companies
PF 0221 — consultants
PE 6877 — Consumer
PD 7876 — Contract
PE 3592 — Credit card
PD 2620 — Customs

—932—

Fuel materials

PD 5554 Fraud ; Diplomatic
PE 1110 Fraud ; Documentary
Fraud [Electoral...]
PD 5214 — Electoral
PD 2879 — employers
PE 3993 — Employment
PD 2458 Fraud ; Financial
PD 1466 Fraud ; Fiscal
PD 8392 Fraud government agents
PD 8392 Fraud government bureaucrats
PD 9297 Fraud ; Health
PD 5422 Fraud ; Housing
Fraud [Inconclusive...]
PD 5636 — Inconclusive convictions
PE 1826 — Insurance
PD 5554 — international civil servants
PD 5422 Fraud ; Land
PE 1398 Fraud ; Loan
Fraud [Mail...]
PE 1404 — Mail
PE 4475 — Maritime
PD 5770 — Medical
PD 5422 — Mortgage
PC 8019 Fraud ; Professional
Fraud [Real...]
PD 5422 — Real estate
PF 8889 — Religious
PE 4362 — Risk computer
Fraud [savings...]
PE 1398 — savings banks
PF 1602 — Scientific
PD 3841 — Securities
PD 8859 — Social security
PD 4543 — Subsidies
PD 4341 Fraud ; Trade union
PD 8859 Fraud ; Unemployment benefits
PF 5340 Fraud ; Unreported
PD 8859 Fraud ; Welfare benefits
PE 8742 Frauds ; Charity
PE 2414 Frauds ; Financial
PE 5516 Frauds, forgeries financial crime
PD 4543 Frauds ; Grant
PA 5688 Fraudulence *Appropriation*
PA 7363 Fraudulence *Improbity*
PA 7411 Fraudulence *Uncommunicativeness*
PE 4496 Fraudulent acquisition use passports
PE 3814 Fraudulent advertising practices
PD 0486 *Fraudulent certificates origin goods*
PD 4191 Fraudulent commodities
PE 1275 Fraudulent impersonation
PD 3521 *Fraudulent labelling*
PE 1398 Fraudulent loans
PE 0952 Fraudulent medical referrals
PE 6975 Fraudulent mineral exploitation claims
PE 5754 Fraudulent nature inherited titles
PD 4543 Fraudulent procurement use government private grants
PF 8889 Fraudulent relics
PF 2714 Fraudulent safety tests
PF 6712 *Fraying tribal structure*
PF 1675 *Free choice ; Tensionless image*
PE 3963 Free choice work ; Denial right
PF 4774 *Free forecasts ; Surprise*
Free [market...]
PE 0051 — market competition transnational corporations ; Restriction
PF 7556 — markets ; Lack protection
PF 7700 — methodologies ; Limitations surprise
PJ 5990 — movement capital
PE 0897 — movement commercial vehicles ; Unjustified restrictions
Free [policy...]
PF 8176 — policy making ; Surprise
PC 6381 — primary education ; Denial right
PE 2858 — production zones export enclaves ; Misuse
PF 6178 Free social structures ; Ideal tension
PF 7798 Free speech ; Irresponsible expression emotions equated
PE 3604 *Free tailed bats ; Endangered species*
PF 7700 Free thinking ; Limitations surprise
PE 7987 Free zone international financial services ; Protectionism
PD 0837 *Free zones ; Insufficient nuclear weapon*
PD 3055 Freed arbitrary deprivation property ; Denial right
PE 5952 Freedom advocacy discrimination ; Denial right
Freedom arbitrary
PD 1576 — detention ; Denial right
PC 2507 — *exile ; Denial right*
PC 5313 — expulsion ; Denial right
Freedom [association...]
PD 3224 — association ; Denial
PD 4695 — association ; Violation right workers
PD 2569 — attacks personal honour reputation ; Denial right
Freedom [being...]
PF 6158 — being tried twice same charges ; Violation
PD 8445 — belief ; Denial right
PE 6342 — bond service ; Violation
Freedom [caring...]
PF 4296 — caring disabled ; Deprivation
PE 6342 — compulsory labour ; Denial right
PD 7612 — conscience ; Denial
PC 3768 — cruel, inhumane degrading punishment ; Denial right
Freedom [Denial...]
PD 4282 — Denial academic
PF 9735 — Denial right artistic
PA 0833 — discrimination ; Denial right
PF 6158 — double jeopardy ; Denial right

Freedom [due...] cont'd
PD 1942 — due unawareness actuality ; Lack essential
PD 0635 Freedom exploitation children ; Denial right
Freedom expression
PE 5409 — clothing ; Denial
PC 2162 — Denial right
PC 2162 — Dependence restriction
PC 2162 — Restriction
PC 3174 — thought communist systems ; Denial
PE 7706 Freedom ; Fear
PF 4668 Freedom government ; Limited degrees
PF 3217 Freedom hold opinions ; Denial
PE 0324 Freedom hunger ; Denial right
PF 4882 Freedom imagination ; Social constraints
PE 7235 Freedom imprisonment failure fulfil contractual agreement ; Denial right
Freedom information
PC 0185 — Denial right
PC 0185 — Restrictions
PC 0931 — Restrictions international
PC 0935 Freedom international movement ; Denial right
PE 9650 Freedom mass killing ; Denial animals right
Freedom movement
PE 4669 — children separated parents ; Lack
PC 3173 — communist systems ; Denial
PE 8408 — countries ; Restrictions
PC 0935 — countries ; Restrictions
PD 0351 — national advantage ; Restrictions international
PC 8452 — persons ; Denial right
PE 8899 — shipping ; Violation right international
PE 8408 — state ; Denial right
PE 8899 Freedom navigation ; Violation right international
PF 1772 Freedom ; Non integrating images personal
PD 7219 Freedom opinion ; Denial
Freedom [parents...]
PE 4700 — parents choose their children's education ; Violation
PE 8951 — press ; Denial right
PE 8951 — press ; Lack
Freedom religion
PD 8445 — Denial right
PE 4332 — indigenous peoples ; Denial right
PD 0127 — women ; Denial right
PC 5075 Freedom ; Restrictions
PE 4743 Freedom retroactive laws punishments ; Denial right
Freedom [servitude...]
PE 4964 — servitude ; Denial right
PC 0146 — slavery ; Denial right
PC 2162 — speech ; Denial
PE 8024 — suffering ; Denial experimental animals right
PE 3899 — suffering ; Denial food animals right
PE 5018 — Suppression intellectual
PE 6633 Freedom testifying against oneself ; Denial right
Freedom thought
PF 3217 — Denial
PF 3217 — Denial right
PF 3217 — Dependence denial
PF 3217 — Restrictions
PC 3429 Freedom torture ; Denial right
PD 3092 Freedom war propaganda ; Denial right
PD 5105 Freedom worship ; Restrictions
PE 6955 Freely dispose natural wealth ; Denial right people
PE 1758 Freely elect their representatives ; Violation right trade unions
PE 3970 Freely employers ; Violation right trade unions negotiate
PE 1758 Freely ; Violation right trade unions function
PF 0695 Freemasonry
PF 0679 Freeze ; Nuclear
PF 6527 *Freezing plumbing systems*
PE 3734 Freezing temperatures ; Torture exposing
PE 8063 Freight costs ; Excessive
PE 5103 Freight pilferage theft
PE 5850 Freight rates ; Instability ocean
PF 2875 *Freight transport ; Difficult*
PF 2875 *Freight transport ; Unreliable*
PF 5922 Freight ; Unfair surcharges ocean
PE 2300 French pox
PA 6379 Frenzy *Avoidance*
PA 6799 Frenzy *Disease*
PA 6247 Frenzy *Inattention*
PA 7157 Frenzy *Insanity*
PA 7156 Frenzy *Moderation*
PE 5099 Frequencies health hazards ; Radio
PD 2045 Frequencies ; Jamming radio
PF 0734 Frequencies ; Limited number available radio
PF 0734 Frequencies ; Maldistribution radio
PF 0734 Frequencies ; Overcrowded spectrum radio
PF 0734 Frequency allocation ; Discriminatory
PF 3703 Frequency bands satellite transmission ; Allocation television
PE 7560 Frequency electromagnetic radiation ; Environmental hazards extremely low
PD 2045 Frequency interference ; Radio
PF 2644 *Frequent schedule disruptions*
PF 1559 Frequently changing government policy
PD 2478 Frescoes ; Pollution damage
PG 1918 *Fresh air ; Inadequate*
PE 8224 Fresh, chilled frozen meat ; Shortage
PE 8591 Fresh, frozen chilled meat ; Instability trade
PE 2587 Fresh fruit edible nut trade ; Instability
PE 8258 Fresh potatoes ; Instability trade
PE 8688 Fresh tomatoes ; Instability trade
Fresh water
PC 4815 — Scarce
PD 6923 — sources ; Environmental degradation
PC 4815 — sources ; Shortage

Fresh water cont'd
PC 4815 — supply ; Insufficient
PD 6923 Fresh waters ; Unsustainable development
PD 4925 Freshwater animals, fish birds ; Accumulation contaminant residues
PD 1223 Freshwater pollution
PD 4925 Freshwater wildlife ; Accumulation pollutants
PA 7253 Fretfulness *Envy*
PA 6200 Fretfulness *Impatience*
PF 1691 Friction
PA 5532 Friction *Disaccord*
PA 5446 Friction *Enmity*
PE 5844 Friction ; Erotic
PF 1816 Friction ; Holy places focus religious
PC 3428 Friction ; *Inter faith*
PA 6979 Friction *Opposition*
PE 8605 Friedreichs ataxia
PB 5747 Friendless people
PF 4416 Friendlessness
PB 5747 Friendlessness
PB 5747 Friendlessness ; Dependence
PD 7261 Friendly powers ; Military expeditions against
PE 3308 Friends ; Threats against family
PE 3308 Friends ; Torture relatives
PF 4416 Friendship ; Lack
PF 8759 Friendships ; False
PA 6030 Fright *Fear*
PF 6572 *Frightening investment outlook*
PA 6030 Frightfulness *Fear*
PA 7240 Frightfulness *Ugliness*
PE 6408 Frigidity
PE 6408 Frigidity ; Sexual
PD 3489 Fringe poverty ; Urban
PD 7352 Frivolity bourgeois democracy
PA 7361 Frivolity *Disorder*
PA 6247 Frivolity *Inattention*
PA 5942 Frivolity *Unimportance*
PA 7371 Frivolity *Unintelligence*
PF 1542 Frivolous vindictive litigation
PD 3156 Frog ; Endangered species
Front organizations
PE 4358 — [Front organizations]
PE 1651 — government ; Misuse business enterprises
PE 0436 — government ; Misuse nonprofit associations
PE 6860 Frontier areas ; Dangerous materials
PD 5972 Frontier barriers
PF 6855 Frontier cooperation ; Lack trans
PE 0208 Frontier formalities international travel ; Excessive
PF 7945 Frontier pollution ; Trans
PE 6833 Frontier workers ; Vulnerability
PD 5972 Frontiers ; Fenced
PD 5972 Frontiers ; Fortified
PC 2367 Frontiers ; Illegal movement across
PD 2244 Frost
PE 0274 Frostbite
PD 9366 Frostbite animals
PE 4161 Frostbite calves
PE 5844 Frottage
PE 5844 Frotteurism
PE 8591 Frozen chilled meat ; Instability trade fresh,
PE 8224 Frozen meat ; Shortage fresh, chilled
PD 4918 Frozen mourning
PF 6575 Frozen vision community's future
PE 3604 Fruit bats ; *Endangered species*
PE 2587 Fruit edible nut trade ; Instability fresh
PE 3607 Fruit flies pests
PE 8741 Fruit ; Instability trade preserved
PE 2976 Fruit ; Pests diseases citrus
PE 3591 Fruit ; Pests diseases deciduous
PE 5373 Fruit rot
PE 4346 *Fruit spots*
Fruit vegetables
PE 9029 — Environmental hazards
PE 0961 — preparations thereof ; Instability trade
PE 1013 — preparations thereof ; Long term shortage
PE 4586 *Fruits ; Brown rot stone*
Frustrated [past...]
PF 5272 — past goals
PA 2252 — people
PF 2437 — *personal effectiveness*
PE 4398 Frustrated talent government posts
PF 3307 Frustrating participation experience
PA 2252 Frustration
Frustration [Defeat...]
PA 7289 — *Defeat*
PA 2252 — Dependence
PA 5497 — *Difficulty*
PA 5527 Frustration *Inexpectation*
PA 6739 Frustration *Maladjustment*
PF 1594 Frustration role opposition
PF 2346 Frustration ; Space care
PD 1544 Frustration youth
PF 3307 Frustrations community participation ; Factions
PE 4027 Fuel-air explosive
PE 4027 Fuel bomb
Fuel [emissions...]
PE 4891 — emissions ; Fossil
PD 0414 — exhaust ; Health risks
PF 1267 — *Expensive generator*
PF 1267 — *Expensive road transport*
PF 1267 Fuel ; High priced cooking
Fuel materials
PE 0546 — Environmental hazards inedible crude non
PD 0520 — Instability trade inedible crude non
PE 0461 — Long term shortage inedible crude non
PE 8351 — Restrictive practices trade inedible crude non

Fuel production

PD 6103 Fuel production ; Use agricultural land
PF 1267 Fuel ; Prohibitive cost
Fuel [Shortage...]
PD 4769 — Shortage charcoal
PD 4769 — Shortage wood
PE 8080 — shortages ; Distortion international politics
PE 8119 Fuel wood charcoal trade instability
PE 4891 Fuels ; Dependence fossil
PE 4891 Fuels ; Exploitation fossil
Fuels lubricants
PE 1712 — Long term shortage mineral
PE 1346 — related materials ; Environmental hazards mineral
PD 0877 — related materials ; Instability trade mineral
PE 0141 Fuels trade ; Restrictive practices mineral
PF 0357 *Fuelwood energy ; Underutilization*
PD 4769 Fuelwood ; Inaccessibility
PE 5267 *Fugitive felons*
PF 5947 *Fugitive offenders*
PE 7451 *Fugue ; Psychogenic*
PE 2215 Fulfilling international reporting obligations ; Governmental delays
PF 9694 Fulfilling prophecies ; Self
PD 3051 Fulfilment debt service obligations developing countries ; Inconsiderate insistence creditors
PC 3280 *Fulfilment ; Lack parental*
PE 8303 Fulltime physician ; Unaffordable
PA 5454 Fulsomeness *Badness*
PA 6839 Fulsomeness *Disrepute*
PA 5981 Fulsomeness *Stench*
PA 5612 Fulsomeness *Unchastity*
PA 7107 Fulsomeness *Unpleasantness*
PD 3672 Fumes ; Irritant
Fumes [Malodorous...]
PD 1413 — Malodorous
PE 5439 — Metal dust
PE 4675 — modern house fires ; Lethal
PC 0119 Fumes ; Noxious
PE 3584 Fumigant damage crops
PE 3584 *Fumigants*
PE 1758 Function freely ; Violation right trade unions
PD 6710 Function ; Obstruction government
PF 3608 Function social techniques needs they address ; Gap
PF 1332 Functional abilities ; Limited development
PE 3601 Functional breeding humans ; State imposed
PE 7943 Functional changes due technological advance
PD 0448 *Functional disorders heart*
Functional [illiteracy...]
PD 8723 — illiteracy
PF 3539 — information ; Limited availability
PD 9654 — *islet cell tumours animals*
PE 6704 Functional obsolescence roads
PD 2632 *Functional paralysis*
Functional skills
PF 2477 — Limited range
PF 2986 — rural communities ; Outmoded
PF 2995 — Unrecognized need
PF 1352 Functional training ; Ineffective mechanisms
PF 7451 Functioning ; Excessive joy
PE 5222 Functioning legitimate organizations ; Infringement
PD 0270 *Functioning suggesting physical disorder ; Alteration physical*
PD 0270 *Functioning suggesting physical disorder ; Loss physical*
PE 1758 Functioning workers organizations ; Interference public authorities
PF 4174 Functions ; Accumulation
PE 8218 Functions civil courts ; Cessation
PF 4754 Functions ; Separation individual's social
PE 7803 Fund transfers ; Legitimizing illegal
PF 6580 Fundamental problems ; Overemphasis symptoms more
Fundamental [sciences...]
PF 0716 — sciences ; Inappropriate use resources
PA 5568 — *skills ; Unlearned*
PF 6580 — systemic problems ; Failure focus
PF 1338 Fundamentalism
PF 6015 Fundamentalism ; Islamic
PF 1338 Fundamentalism ; Religious
PJ 8295 Fundamentally different societies ; Inadequate interface
Funding [Abusive...]
PF 8535 — Abusive allocation government
PF 8711 — agencies programme formulators ; Collusion administrators
PF 6530 — *approach ; Limited*
PF 0648 Funding ; Competition aid
PF 2849 Funding ; Crisis oriented
PF 9530 Funding ; Dependency government
PF 9108 Funding ; Difficult public
Funding [effectiveness...]
PF 8158 — effectiveness ; Administrative overhead erosion
PF 9108 — Elusive public
PF 7916 — *experience ; Insufficient*
PF 7916 — *expertise ; Lack*
PF 7916 — *expertise ; Limited*
Funding [Ineffectively...]
PF 2261 — *Ineffectively prioritized*
PF 6573 — *information ; Unavailability*
PF 3318 — *infrastructure ; Lack*
PF 3409 — *Insufficient programme*
PE 0741 — international nongovernmental organizations programmes ; Inadequate
PF 0498 — international organizations programmes ; Inadequate
PF 6544 Funding ; Lack channels obtaining available local
PF 6505 Funding mechanisms ; Complex
PF 3526 Funding ; Mistrusted federal
PF 2878 Funding needs ; Inadequate information

PF 6505 *Funding procedures ; Tedious*
Funding [requests...]
PF 0905 — *requests ; Unstrategic*
PF 5419 — research ; Insufficient government
PF 2823 — research social problems ; Restriction
PF 3550 — resources urban services ; Limited availability
Funding [Self...]
PF 2275 — Self accountability grant
PF 2706 — Short range project
PF 1653 — *structures ; Lack*
PF 3630 Funding ; Uncertain sources rural
PF 0648 Funds alternative priorities ; Reallocation aid
PF 5956 Funds ; Crisis long term pension
PE 5338 Funds developing countries ; Aid lenders beneficiaries net outflow
PE 8883 Funds developing country students ; Unavailability scholarship
PB 4653 *Funds education ; Lack*
PC 2920 Funds ; Embezzlement public
Funds [Illegally...]
PD 5433 — *Illegally obtained*
PF 2425 — *Inadequate improvement*
PF 3550 — *Inadequate police*
PF 3630 — *Insufficient farming*
PF 3550 — *Insufficient housing*
Funds Lack
PF 1348 — organized savings
PF 0202 — recreation
PF 2605 — sports
Funds Limited
PF 2378 — availability
PF 2378 — extent capital
PF 4586 — health
PF 9108 — public
Funds [medical...]
PE 0612 — medical research ; Lack
PC 2920 — Misappropriation public
PE 0682 — Misleading borrowers
PD 4691 — *Misuse trust*
PF 4375 — Funds ; Nonexistent school
PE 8170 Funds property ; Deprivation trade union
PF 5419 Funds research ; Shortage
PF 3550 Funds ; Restricted access municipalities capital
Funds [Scarcity...]
PF 6477 — Scarcity start up
PB 4653 — Shortage
PE 3569 — students ; Unavailability scholarship
Funds [Unavailability...]
PF 6477 — *Unavailability land*
PF 6594 — Unreported international movements
PE 9325 — Unutilized government
PE 8463 Funds veterinary research ; Lack
PD 2728 Fungal diseases
PD 2728 Fungal diseases animals
PF 2225 Fungal plant diseases
PE 4456 Fungal resistance fungicides
PE 6269 Fungi ; Heart rot
PE 4346 Fungi imperfecti
PD 2225 Fungi ; Pathogenic
PE 6255 Fungi ; Rust
PE 8138 Fungi vectors plant disease
PD 3577 Fungicide damage crops
PD 1612 Fungicides
PE 4456 Fungicides ; Fungal resistance
PD 1612 Fungicides pollutants
PD 2728 Fungus disease
PD 2728 Fungus infection
PE 3604 *Funnel eared bats ; Endangered species*
PD 5433 Funny money
PD 4575 Fur ; Breeding animals
PD 4575 Fur industry ; Slaughter animals
PE 0828 Fur skins ; Environmental hazards undressed hides, skins
Fur skins Instability
PE 8740 — trade dressed
PE 8106 — trade undressed hides non
PE 1235 — trade undressed hides, skins
PE 8327 Fur skins shortage hides ; Skins
PE 9130 Fur skins ; Shortage leather, miscellaneous leather manufactures dressed
PE 1474 Fur trade ; Instability
PE 8165 Furniture fixtures manufacture ; Environmental hazards
PF 2604 Furniture fixtures manufacture underdevelopment
PE 0511 Furniture ; Long term shortage
PF 0296 Furniture material ; Costly
PE 4675 Furniture materials ; Toxic modern
PD 0389 Furs skins endangered species ; Trade
PA 7411 *Furtiveness Uncommunicativeness*
PC 8534 *Furuncles*
PA 6379 Fury *Avoidance*
PA 7253 Fury *Envy*
PA 7157 Fury *Insanity*
PA 7156 Fury *Moderation*
PE 4346 Fusarium diseases plants
PE 9458 Fusarium estrogenism
PG 5461 Fusarium wilt tomato
PE 2143 *Fusospirochetosis*
PA 5838 Fuss *Agitation*
Fuss [Disaccord...]
PA 5532 — *Disaccord*
PA 6191 — *Disapproval*
PA 7361 — *Disorder*
Fuss [Impatience...]
PA 6200 — *Impatience*
PA 6247 — *Inattention*
PA 5467 — *Inexcitability*

PA 5479 Fuss *Lamentation*
PA 7156 Fuss *Moderation*
PA 6491 Fuss *Vanity*
PA 6099 Futility *Hopelessness*
Futility [Impotence...]
PA 6876 — *Impotence*
PA 6852 — *Inappropriateness*
PA 7395 — *Inexpedience*
PF 2093 *Futility job hunting ; Sensed*
PA 6977 Futility *Meaninglessness*
PA 6662 Futility *Nonaccomplishment*
PF 0910 Futility social loyalties
PA 5942 Futility *Unimportance*
Future [change...]
PD 2670 — change ; Individual fear
PF 4114 — costs present policies ; Unacknowledged
PF 8354 — crop uncertainty ; Insecurity
Future [declines...]
PD 5114 — declines production ; Vulnerability agriculture developing countries
PF 6575 — Detrimental story community
PF 8176 — development ; Inability anticipate
PF 0836 — *Dim business*
Future [ecological...]
PE 9295 — ecological problems ; Government neglect
PF 3973 — Elder paralysis over
PF 2303 — *employment insecurity*
PF 4114 Future financial commitments ; Unrecognized
PF 6575 Future ; Frozen vision community's
PF 9455 Future generations ; Irresponsibility towards
PF 9276 Future human values ; Indeterminate
PF 3973 Future ; Inability elderly play role developing
PF 2352 Future ; Inhibited capacity visualize creative
PF 0980 Future ; Monopoly competence intervention
PF 0980 Future ; Monopoly knowledge predict
Future [patterns...]
PF 9276 — patterns social choice ; Unknowable
PF 8176 — plan ; Undeveloped
PF 9455 — planet ; Irresponsibility long term
PF 9478 — planning ; Unimaginative
PF 9276 — priorities ; Indeterminate
PE 9295 — problems social environment arising current policies ; Government neglect
PF 2803 Future ; Static social relations inhibiting vision
PE 1137 *Future ; Uncertainty miners'*
PF 9478 Future ; Unimaginative vision world's
PF 2068 Future ; Young people's lack context
PD 9637 Futures markets ; Speculation commodities
PF 4870 Futures trading ; Obstacles commodity
PF 6330 *Futurism ; Technocratic*
PF 4510 Fuzzy environmental thinking

G

PE 3745 Gadgets ; Unnecessary
PF 8675 Gain ; Short term
PF 8675 Gain ; Short term professional
PE 4566 *Gait ; Abnormality*
PE 6791 Gait disturbances due torture
PE 4566 *Gait ; Peculiar*
PE 4293 *Galactosemia*
PE 2230 Gall ; Crown
PE 9829 Gallbladder ; Diseases
PE 2251 Gallbladder ; Inflammation
PD 2732 *Gallisepticum infection poultry ; Mycoplasma*
PD 9667 *Galls ; Collar*
PE 3715 Galls plants
PE 6275 Gallsickness
PB 1044 *Gallstones*
PA 6448 Gamble ; Failure resist impulses
PF 2137 Gambling church premises
PA 6448 Gambling ; *Pathological*
PE 0891 Gambling performance animals
PE 4576 Gambling sports athletic competitions
PF 2137 Gambling wagering
PF 6345 Game addiction ; Video
PE 1327 Game cropping ; Unjustified
Game [poaching...]
PD 2664 — poaching ; Inadequate enforcement against
PE 9059 — propagation ; Instability hunting, trapping
PE 8252 — propagation ; Underdevelopment hunting, trapping
PE 8098 Game zoo animals ; Loss disease
PF 4761 Games ; Political exploitation Olympic
PF 0374 *Games ; Recreational crime*
PF 1406 *Games ; Recreational war*
PF 6345 Games ; Socially disruptive effects video
PE 3704 Games ; War
PF 2137 Gaming ; Betting
PF 1441 Gamma ray disturbance space
PD 3266 *Gang rape*
PA 0429 *Gang violence*
PD 4843 Gang war
PD 4843 Gang warfare
PE 8932 Ganglia ; Diseases nerves peripheral
PE 2684 *Gangrene*
PE 0102 *Gangrene, carbuncles boils*
PD 9667 *Gangrenous cellulitis*
PE 6280 Gangrenous coryza
PD 9667 *Gangrenous dermatitis*
PD 7307 *Gangrenous pneumonia*
PE 2143 *Gangrenous stomatitis animals*
PD 3837 *Gangs ; Criminal*
PD 0527 *Gangs ; Paramilitary*
PD 2682 *Gangs ; Street*
PD 2682 *Gangs ; Youth*

Glands inflammatory

PD 3837 Gangsterism
PD 0520 Gaol conditions ; Inadequate
PD 0520 Gaol conditions ; Inadequate
PE 7235 Gaol ; Debtor's
PD 8680 Gaol suicides
PE 5567 *Gapeworm infection*
PF 3661 Gaps enterprises ; Information
PE 7637 Gaps telecommunications capability
PD 0177 Garbage
PF 6473 *Garbage containers ; Insufficient*
PD 0807 Garbage disposal
PF 6473 *Garbage pick up ; Undependable*
PD 3102 Garbageware
PD 1350 *Garden vandalism ; Unpunished*
PF 6493 *Gardening capacity ; Unstudied*
PF 6493 *Gardening experience ; Limited*
PE 7265 Gardens ; Over cultivated
PD 1350 *Gardens ; Vandalized family*
PE 5399 Gas emissions animal husbandry ; Methane
PE 1256 Gas emissions landfill sites ; Methane
Gas [Indiscriminate...]
PE 4640 — Indiscriminate use tear
PD 0877 — *Instability trade*
PE 8464 — Instability trade mineral tar crude chemicals derived coal, petroleum natural
PE 8045 Gas ; Long term shortage natural manufactured
PE 1813 Gas monopolies ; Industrial
PE 7103 Gas ; Mustard
Gas [plants...]
PF 2231 — *plants ; Unrealized potential gober*
PD 2596 — *production environmental hazards ; Natural*
PE 8955 — production instability ; Petroleum natural
PE 9137 Gas trade infrastructure ; High cost natural
PE 1159 Gas, water sanitary services ; Health risks workers electricity,
PF 2583 Gas, water sanitary services ; Underdevelopment electricity,
PE 0524 *Gaseous air pollutants*
PD 1271 *Gaseous organic compounds*
PE 9294 Gases ; Anti personnel use asphyxiating
PD 1354 Gases atmosphere ; Increase trace
PE 9294 Gases ; Health hazards anti personnel
PC 0872 *Gases ; Nerve*
PG 3050 *Gases ; Poisonous*
PE 6820 Gases vapours ; Irritant
PG 3050 *Gases vapours ; Poisoning*
PD 7841 *Gaskell Syndrome ; Key*
PE 1161 Gasoline ; Lead
PE 7509 Gastric cancer
PD 3978 *Gastric catarrh*
PE 7423 *Gastric dilatation-volvulus animals*
PE 1599 Gastric disorders
PE 4630 *Gastric foreign bodies animals*
PE 7423 *Gastric torsion*
PE 2308 Gastric ulcers
PE 8624 *Gastritis*
PD 3978 *Gastritis animals*
PE 4724 Gastritis ; Mucous
PD 3978 *Gastritis ; Traumatic*
Gastro intestinal
PE 6724 — disorders due torture
PE 3861 — infections
PE 3861 — tract ; Bleeding
PE 4161 *Gastrocnemius muscle ; Rupture*
PE 4161 *Gastrocnemius tendon ; Rupture*
PE 7423 Gastroenteric diseases animals
PE 3861 Gastroenteritic disturbances
PE 7423 *Gastroenteritis ; Canine coronaviral*
PE 7423 *Gastroenteritis ; Canine haemorrhagic*
PE 3936 *Gastroenteritis swine ; Transmissible*
PE 3861 Gastrointestinal diseases
PE 0958 Gastrointestinal infections animals due parasites
PE 2436 Gastrointestinal parasites animals
PE 3861 Gastrointestinal tract ; Dysfunction
PE 2308 Gastrojejunal ulcer
PD 2761 *Gastronomic fashions ; Negative effects*
PE 8985 Gathering elders ; Insufficient possibilities
Gathering places
PE 1535 — Small
PF 2575 — *Undesignated*
PE 2274 — *Unused*
PF 5250 Gatherings ; Infrequent public
PF 1705 *Gatherings ; Insufficient traditional*
PF 7118 *Gatherings ; Unstructured community*
PD 2625 *Gaucher's disease*
PA 7301 Gaudiness *Colourlessness*
PA 5821 Gaudiness *Vulgarity*
PF 1390 Gay men
PD 2730 *Geese ; Viral hepatitis*
PD 3947 Gender abortions
PE 0567 Gender bias theology
PF 3277 Gender ; Change
Gender [differentiated...]
PE 4664 — differentiated toys
PC 2022 — discrimination
PD 9563 — discrimination developing countries
Gender [identity...]
PE 6581 — identity disorders
PE 1339 — Inequality life expectancy
PE 1339 — Inequality mortality rates
PD 5843 Gender stereotyping
PD 3443 *Gender stereotyping employment*
PF 7727 Gene pool ; Deterioration
PD 4158 General applicability ; Offences
PC 9067 *General debility*
PF 6570 *General expectations ; Low*
General [ill...]
PA 0294 — ill health
PD 8568 — income ; Low
PF 1578 — initiatives specific needs ; Failure adapt
PF 0631 General obstacles problem alleviation
PA 0294 General poor health
General [unavailability...]
PF 2702 — unavailability effective motivational techniques
PF 3103 — unproductivity capitalist systems
PC 4864 — *uprising*
PF 0256 Generalists ; Lack
PE 2684 *Generalized necrotizing arteritis*
PD 2565 *Generalized necrotizing arthritis*
PE 2298 *Generalized osteitis fibrosa*
PD 4158 Generalized types crimes
PE 9643 Generate sufficient culture specific programmes ; Inability
PF 4152 Generated arms race ; Risk unintentional war
PF 4156 Generated developments strategic doctrine ; Risk unintentional nuclear war
PE 7879 Generated electrical appliances ; Health hazards electromagnetic fields
PF 4162 Generated strategy deterrence ; Risk unintentional nuclear war
PE 2435 Generated supersonic aircraft ; Sonic boom
PF 1389 *Generating capacity ; Non viability local electrical*
PF 0756 Generation communication gap
PE 1412 Generation ; Degradation environment electrical power
PD 0276 *Generation family structure ; Demise three*
PE 4990 Generation immigrants ; Marginalization second
PD 7663 Generation ; Inadequate infrastructure nuclear power
PF 1664 Generation ; Limited social guidance older
PE 4730 Generation public good ; Lack relationship wealth
PF 0756 Generation relationships ; Prejudicial
PF 6497 Generations ; Educational gap
PF 9455 Generations ; Irresponsibility towards future
PF 1267 *Generator fuel ; Expensive*
PD 0094 *Generic drugs*
PD 1618 Genetic abnormalities ; Human physical
PE 4588 Genetic chemicals ; Toxic
PD 2389 *Genetic counselling ; Lack*
PD 2389 *Genetic deafness*
Genetic defects
PD 2389 — [Genetic defects]
PD 2389 — diseases
PD 4031 — Increase animal
PD 9480 — Increase plant
PF 7727 Genetic diseases ; Compounding effect treating
Genetic diversity
PC 1408 — animal breeding stock ; Decreasing
PC 1408 — animals ; Decreasing
PE 6788 — available seeds ; Reduction
PC 2223 — cultivated plants ; Decreasing
PB 4788 — Destruction
PD 0547 — fish ; Decreasing
PF 7815 *Genetic drift*
Genetic [effects...]
PE 4896 — effects medications ; Hazardous
PF 7815 — endowment ; Decline human
PE 5404 — engineering ; Accidental weed creation
PC 0776 — ,engineering accidents ; Hazards
PB 4788 — erosion
PD 2389 — errors
PC 6664 — ethnic weapons
PC 6664 — ethnic weapons ; Dependence
PC 1408 Genetic impoverishment animals domestication
PD 7465 Genetic inbreeding
PC 0776 Genetic manipulation ; Irresponsible
Genetic resources
PF 3581 — conservation ; Inadequate plant
PE 6788 — Monopolization agricultural
PD 7526 — Underutilized animal
PB 4788 — Vulnerability world
PE 5925 Genetic sexual attraction
PD 2389 *Genetic susceptibility disease*
PF 9269 Genetic weapons
PF 7490 Genetically determined violence
PD 7183 Genetically engineered micro organisms ; Release
PC 0776 Genetically modified feral species
PD 7501 Genetically transformed animals ; Irresponsible patenting
PE 6788 Genetically transformed plants ; Irresponsible patenting
PD 7731 Geneticists ; Irresponsible
PE 9121 Genets, civets mongooses ; Endangered species
PD 7799 *Genital campylobacteriosis ; Bovine*
PD 7799 *Genital campylobacteriosis ; Ovine*
Genital organs
PE 9705 — Benign neoplasm
PE 4249 — Congenital anomalies
PD 8775 — Diseases female
PD 9154 — Diseases male
PE 8372 — *Malignant neoplasm male*
PE 1905 — Malignant neoplasms female
PE 5100 Genito urinary organs ; Malignant neoplasm
Genito urinary system
PC 4575 — Diseases
PE 9369 — Symptoms referable
PE 2053 — *Tuberculosis*
PC 1056 Genocidal massacres
PC 1056 Genocide
PC 1328 Genocide ; Cultural
PC 1056 Genocide ; Dependence
PF 3390 Geo-sentiment
PE 8049 Geographic barriers least developed countries
PE 1347 Geographic discrimination administration justice
PF 2822 *Geographic division ; Established*
PF 9239 Geographical deception
PF 8511 Geographical features ; Political changes names
Geographical [illiteracy...]
PD 3984 — illiteracy
PF 9239 — information ; Misrepresentation
PF 9239 — information ; Withholding
PF 9023 — isolation
Geographical location
PF 9023 — Isolated
PF 3499 — Poor
PF 9023 — Remote
PF 8511 Geographical names ; Non standardization
PE 7902 Geographically denied credit
PF 9247 Geographically disadvantaged countries
PC 7674 Geographically remote facilities
PC 4020 Geographically remote resources
PF 6521 Geographically undefined community limits
PD 3984 Geography ; Ignorance world
PF 1780 Geography textbooks ; Biased inaccurate
PD 0201 Geological faults
PC 6684 Geological hazards
PD 0708 Geologists ; Corruption
PD 0708 Geologists ; Irresponsible
PD 0708 Geologists ; Negligence
PD 0708 Geology ; Malpractice
PD 0830 Geomagnetic disasters
PF 2407 Geomagnetic field anomalies
PF 1588 Geomagnetic reversal
PD 1661 Geomagnetic storms
PC 2696 Geophysical weapons
PF 5749 Geopolitical vulnerability
PF 0545 Geostationary satellite orbits ; Limited number
PF 0545 Geosynchronous orbit ; Saturation
PF 0545 Geosynchronous orbits ; Maldistribution
PE 8971 *Geothermal energy ; Underutilization*
PD 2728 *Geotrichosis*
PE 5288 Geriatric wards ; Inhumane
PC 0195 Germ warfare
PE 0785 German measles
PD 3133 Gerontocracy
PD 0947 Gerontocracy developing country agriculture
PF 3858 Gerontocracy ; Superannuated religious
PD 2906 Gerrymandering
PD 7799 *Gestation cattle ; Prolonged*
PD 7799 *Gestation sheep ; Prolonged*
PF 1681 *Ghetto education image*
PD 1260 Ghettoization
PF 1959 Ghettos ; Ineffective operation community networks urban
PD 1260 Ghettos ; Segregation poor minority population urban
PF 4644 Ghettos ; Temporal
PE 3993 Ghost employees
PE 4658 Ghost workers
PF 3801 Ghosts
PD 9293 *Giant kidney worm infection dog*
PD 9293 *Giant kidney worm infection mink*
PE 6461 *Giant liver fluke*
PE 6461 *Giant liver fluke*
PE 3837 Giantism
PE 5578 Giants ; Discrimination against
PE 8036 Giants ; Domestic inconvenience dwarfs, midgets
PE 8496 Giants ; Lack adequate clothing supply dwarfs, midgets
PE 4811 Giardiasis
PE 1570 *Gibbons ; Endangered species*
PE 3511 *Gid parasite*
PE 5101 Giddiness
PD 2051 Gifted children ; Inadequate educational facilities
PC 2030 *Gifts ; Political*
PF 1985 Gifts ; Undisplayed heritage
PF 2703 Gifts ; Untransposed creativity traditional
PF 3837 Gigantism
PE 3745 Gimmicks ; Technological household
PE 9702 *Gingivae ; Fibromatosis*
PE 9702 *Gingival hyperplasia*
PE 3503 Gingivitis
PE 3503 *Gingivitis*
PE 9702 *Gingivitis animals*
PE 3503 *Gingivosis*
PE 0427 Girls ; Shortage
PC 4631 Giving bribes
PF 2819 Giving ; Increasing resistance charitable
PE 5267 Giving testimony ; Flight avoid prosecution
PF 1744 Glaciation ; Cyclical planetary
PE 4256 Glacier size ; Reduction
PE 6824 Glaciers ; Hazardous
PF 7990 Glamour ; Negative effect
PE 7570 *Gland abscess ; Salivary*
PE 7708 Gland ; Diseases thymus
Gland disorders
PD 8301 — [Gland disorders]
PD 8301 — *Parathyroid*
PE 2286 — Pituitary
PE 0652 — Thyroid
PD 8301 Gland disturbances ; Endocrine
PD 8301 *Gland ; Dysfunction parathyroid*
PD 2730 *Gland fever ; Bat salivary*
PE 7570 *Gland inflammation ; Salivary*
PE 2461 Glanders
PD 8347 *Glands ; Benign neoplasm endocrine*
Glands Diseases
PD 8301 — *adrenal*
PD 8301 — endocrine
PC 8534 — *sweat*
PD 8786 *Glands ducts ; Inflammation lacrimal*
PD 8301 *Glands ; Dysfunction adrenal*
PE 0652 Glands inflammatory lesions ; Thyroid

Glands

PD 8301	Glands ; Overactivity parathyroid	
PE 0652	Glands tumorous overgrowth ; Thyroid	
PD 8301	Glandular disorder	
PE 4524	Glass ; Corrosion	
PE 0082	Glass due acid rain ; Deterioration stained	

Glass glass products
- PE 9084 — industries instability
- PE 8473 — manufacture ; Environmental hazards
- PE 8715 — manufacture underdevelopment
- PE 8274 Glass industry ; Health hazards

Glass products
- PE 9084 — industries instability ; Glass
- PE 8473 — manufacture ; Environmental hazards glass
- PE 8715 — manufacture underdevelopment ; Glass
- PD 2732 *Glässer's disease*
- PE 7548 Glasses developing countries ; Prohibitive cost eye
- PE 2264 Glaucoma
- PD 5535 *Glaucoma animals*
- PE 2264 *Glaucoma ; Congenital*
- PE 2264 *Glaucoma ; Primary*
- PE 2264 *Glaucoma ; Secondary*
- PA 7143 Glibness *Discourtesy*
- PD 1762 *Gliders ; Endangered species ringtail*
- PF 0306 Global amnesia
- PE 0355 Global amnesia ; Transient

Global [centralization...]
- PF 5472 — centralization power
- PE 8640 — communications industry ; Control transnationals
- PF 1817 — concern ; Inadequate means upholding
- PF 9821 — consensus concerning problems prospects humanity ; Inadequate
- PF 1744 — cooling
- PF 0817 — cooperation solve world problems ; Inadequate
- PF 6244 — crisis
- PF 5472 Global decision making ; Over centralization

Global distribution
- PF 2880 — basic skills ; Unequal
- PC 5601 — economic growth ; Unequal
- PE 4241 — family planning education facilities ; Unequal

Global economic
- PC 5876 — crisis
- PF 2699 — situation ; Inadequate solution oriented images
- PC 9112 — system ; Unjust

Global [economy...]
- PC 3778 — economy ; Elitist control
- PE 0135 — economy limited number corporations ; Disproportionate control
- PF 5712 — ecosystem ; Over specialized study
- PF 2105 — ethic ; Governmental barriers
- PF 3628 — exposure ; Limited

Global [financial...]
- PF 3612 — financial crisis
- PF 9576 — financial models ; Uncoordinated multiplicity
- PF 3612 — fiscal collapse

Global [ideology...]
- PD 0065 — ideology ; Absence social methodologies
- PF 2600 — interdependence nationalistic images ; Overruling
- PF 2877 — interdependence ; Unacknowledged
- PF 0817 — issues ; Nationalistic response
- PJ 3753 Global Keynesianism
- PF 2499 Global legislation ; Limited local respect regional
- PD 0930 Global marketing ; Blocked
- PF 4352 Global nuclear war due nuclear proliferation ; Risk unintentional
- PF 1491 Global obligations ; Self determination unrelated

Global [policies...]
- PF 7713 — policies ; Inadequate intellectual cognitive infrastructure formulate effective
- PE 0135 — policy making ; Undisclosed influence transnational corporations
- PF 7535 — political decision making ; Lack rational
- PF 8753 — problematique ; Failure integrate knowledge empower humanity response
- PF 9821 — problematique ; Inadequate development political thinking concerning
- PF 3183 Global resources ; Unregulated
- PF 1055 Global scale planning expertise development ; Failure
- PD 2802 Global unity ; Inadequate structures achieving
- PC 0918 Global warming
- PF 7071 Globality ; Reduced understanding
- PF 2814 Globality ; Unrealistic attempts
- PF 1245 Globalization ; Vulnerability socio economic systems
- PF 2835 Globalized citizenship ; Non
- PF 1006 Globally ; Refusal families participate
- PE 2272 Glomerulonephritis
- PD 9293 *Glomerulus ; Disease*
- PE 7354 Glorification crime media
- PE 7930 Glorification demeaning sex media

Glorification [Self...]
- PF 4436 — Self
- PF 2666 — *Self*
- PE 3980 — substance abuse media
- PD 6279 Glorification violence media
- PF 9312 Glorification war
- PG 5613 *Glossitis*
- PE 9702 *Glossitis animals*
- PD 3978 *Glossoplegia*
- PE 1427 Glue sniffing
- PE 5722 Gluttony
- PA 9638 Gluttony
- PE 4293 *Glycogen storage disease*
- PE 4293 *Glycogen storage diseases*
- PD 5228 *Glycol poisoning ; Ethylene*
- PE 2053 Glycosuria ; Renal
- PE 3646 Gnats ; Buffalo
- PF 2400 Goal educational methods ; Unarticulated

PJ 4512	Goals ; Conflicting economic	
PJ 4512	Goals ; Contradictory solutions national economic	

Goals [Fragmented...]
- PF 6537 — *Fragmented citizen*
- PF 7118 — *Fragmented community*
- PF 5272 — *Frustrated past*
- PF 9852 Goals ; Inappropriate
- PF 2595 Goals ; Misdirected educational
- PF 1403 Goals ; *Neglect individual*
- PF 2595 Goals ; Non articulated educational
- PF 2595 Goals ; Unconsensed educational
- PF 2595 Goals ; Untargeted educational
- PE 6272 Goatpox
- PD 3978 *Goats ; Enteric diseases*
- PE 8861 Goats ; Instability trade sheep
- PE 2727 *Goats ; Mange*
- PE 9438 Goats ; Respiratory diseases
- PE 2727 *Goats ; Scabies*
- PF 2231 *Gober gas plants ; Unrealized potential*
- PF 5764 God ; Absence
- PB 1151 God ; Acts
- PF 2019 God ; Death
- PF 2409 God ; Denial
- PF 9565 God ; Fear
- PF 2409 God ; Ignorance
- PF 9307 God ; Infidelity
- PF 4617 God ; Killing
- PF 2409 God ; Rejection
- PF 4617 God slaughter
- PF 8563 God ; Wrath
- PA 6058 Godlessness *Impiety*
- PF 2409 Gods ; Denial
- PG 5976 *Goitre ; Colloid*
- PE 1924 *Goitre ; Endemic*
- PG 5976 *Goitre ; Simple*
- PD 3045 Gold ; Hoarded monetary
- PD 3045 Gold ; Hoarding monetary
- PC 7859 Gold ; Non restitution monetary
- PE 1323 Gold thread
- PF 2342 Gold ; Uncertain status monetary
- PE 3604 *Golden bat ; Endangered species*
- PD 3479 *Golden moles ; Endangered species*
- PE 7826 *Gonitis*
- PD 0061 *Gonococcal urethritis*
- PE 1717 Gonorrhoea
- PD 5453 *Good condition ; Dead*
- PJ 7640 Good evil ; Knowledge
- PF 7917 Good human nature ; Corruption
- PF 4730 Good ; Lack relationship wealth generation public
- PF 5311 Good rhetoric ; Uncritical feel
- PE 1055 Good roads ; Insufficient

Goods [capitalism...]
- PC 3116 — capitalism ; Excess demand
- PC 3116 — capitalist systems ; Excessive demand
- PD 7981 — Counterfeit

Goods [Damage...]
- PE 4447 — Damage
- PD 3998 — Defective manufactured
- PF 8268 — Delays delivery
- PE 4170 — developing countries ; Trade barriers manufactured
- PF 2966 — developing countries ; Weakness trade manufactured

Goods [electronic...]
- PE 1909 — electronic devices ; Excessive exposure radiation consumer
- PE 1344 — Environmental hazards manufactured
- PD 3998 — ex socialist countries ; Dumping obsolete
- PD 0486 Goods ; Fraudulent certificates origin
- PE 7174 Goods hostile countries ; Trade strategic

Goods [Inadequate...]
- PD 3521 — Inadequate labelling
- PE 8934 — Incorporation carcinogens consumer
- PD 3657 — Insect damage stored manufactured
- PE 0882 — Instability trade manufactured
- PE 8920 — Instability trade travel
- PB 8891 — Insufficient availability
- PB 8891 — Goods ; Limited commodity
- PE 0802 Goods ; Long term shortage manufactured
- PE 8574 Goods manufacturing industries ; Ineffective self regulation consumer
- PD 6075 Goods ; Narrow range
- PE 9220 Goods nonfactor services ; Distorting effects commodity taxes transaction
- PD 4907 Goods ; Over diversification manufactured

Goods [Pollution...]
- PC 0328 — Pollution intensive
- PJ 7877 — Prohibitive cost imported
- PE 7025 — Protectionism against imports service related
- PD 1797 Goods ; Restrictive practices trade manufactured
- PD 0971 Goods ; Risk transporting dangerous

Goods services
- PJ 0914 — Clandestine private utilization public
- PE 3928 — Delays delivery
- PC 2518 — Excessive consumption
- PD 1891 — Inflated prices
- PF 2989 — Limited local markets
- PD 1891 — Over priced
- PD 1891 — Prohibitive cost
- PE 3928 — Risks due delays delivery
- PE 8444 — Undue taxation certain
- PE 8603 — Unequal distribution
- PF 9697 — Vulnerability national economies vagaries external markets

Goods [Shortage...]
- PE 0802 — Shortage industrial
- PB 8891 — Shortages supply

Goods [socialist...] cont'd
- PD 0515 — socialist countries ; Lack consumer
- PF 6495 — supply distribution ; Ineffective means

Goods [Trade...]
- PD 7981 — Trade counterfeit
- PE 8470 — trade instability ; Photographical optical
- PD 0971 — Transport dangerous
- PF 6495 — transport ; Unfeasible

Goods [Unenvisioned...]
- PF 3554 — *Unenvisioned building*
- PE 5506 — Unfair competition convict made
- PC 1784 — Unsecured
- PD 3481 *Gophers ; Endangered species pocket*
- PE 2192 Gossip
- PE 2192 Gossip-oriented communication
- PC 2270 *Gout*
- PD 9366 *Gout animals*
- PE 5081 Gout ; Rheumatic
- PF 8846 Governance ; Irrelevant forms
- PF 3394 Governance ; Weakness
- PF 1781 *Governing body ; Lack*
- PE 1024 Governing ; Denial rights indigenous people be self
- PF 2943 Governing status disputed administering government ; Territories accorded United Nations non self
- PF 4668 Government ability undertake new initiatives ; Restricted

Government action against
- PF 2199 — regimes alternative policies
- PF 2199 — regimes hostile policies
- PF 2199 — regimes opposing policies

Government action [commitment...]
- PF 5502 — commitment ; Progressive reduction
- PF 9530 — Dependence
- PE 4997 — Non accountability
- PF 5332 — Unlawful

Government [administrative...]
- PF 3318 — *administrative backlog*
- PC 1818 — administrative process ; Inadequate control over
- PF 5296 — admit mistakes ; Failure
- PD 1964 — Advocating overthrow

Government agencies
- PE 9700 — Arrears financial payments
- PE 8308 — each country ; Jurisdictional conflict antagonism
- PF 3458 — Environmental unaccountability
- PF 5351 — Inaccessible
- PE 2719 — officials ; Antagonism
- PE 0988 — Unaccountability decentralized
- PD 0183 — Wastage resources
- PF 2990 Government agency agreements ; Secret

Government agents
- PE 2113 — acting foreign countries ; Abduction
- PD 8392 — Corruption practised
- PD 0183 — Extravagant waste resources
- PD 9498 — Fear
- PD 8392 — Fraud
- PD 1893 — Perjury

Government [aid...]
- PF 4821 — aid countries violating human rights
- PF 5296 — apologize errors ; Failure
- PF 4753 — approval addictive behaviour
- PF 4753 — approval dangerous behaviour
- PE 4697 — approved employment war criminals
- PF 8820 — arrogance
- PF 2969 — *assistance ; Complexity*
- PC 9585 — Authoritarian
- PE 9337 — authorities ; Demolition homes
- PC 9585 — autocracies
- PF 9157 — avoidance complaints

Government [Bad...]
- PF 3953 — Bad
- PE 7271 — benefit coupons ; Trafficking
- PE 5667 — benefits political purposes ; Deprivation
- PF 4153 — Betrayal
- PF 6745 — bias wage bargaining
- PF 4564 — Biased adjustment official statistics
- PD 4532 — Bribery
- PE 1622 — bullying foreign governments
- PE 8539 — bureaucracy ; Complex
- PD 8890 — bureaucracy ; Unchecked power
- PD 8392 — bureaucrats ; Fraud

Government [Collusion...]
- PE 8367 — Collusion trade union leaders employers
- PF 9564 — commitment multilateralism ; Reduced
- PF 1442 — *competition ; Confusing*
- PF 6407 — complacency

Government complicity
- PF 7730 — drug trafficking
- PF 7730 — illegal activities
- PF 7730 — illegal arms trade
- PF 3819 — killing human rights activists

Government [Compulsory...]
- PC 1005 — Compulsory acquisition land
- PF 9288 — connivance human rights abuses
- PD 6943 — conspiracy incriminate
- PF 4668 — Constraints power
- PF 1492 — *constraints programme beneficiaries ; Counter productive*
- PF 2911 — contracts ; Unjust allocation
- PF 0304 — control ; Excessive
- PF 9542 — control military ; Inadequate
- PE 0780 — Cooptation private sector initiatives
- PC 7587 — Corruption
- PD 5948 — Corruption local
- PD 7171 — Covert smear campaigns
- PF 2489 — credit ; Limited local
- PF 9385 — Criticism

Government women

PF 2990 — Government deals ; Secret	Government intervention cont'd	Government [private...] cont'd
PD 1893 — Government ; Deception	PF 0304 — society ; Excessive	PD 4543 — private grants ; Fraudulent procurement use
Government decision making	**Government [intimidation...]**	PE 0347 — private procurement policies ; Distortion international trade discriminatory
PF 1559 — Arbitrary	PE 1622 — intimidation governments	**Government procedures**
PC 1833 — Balkanization metropolitan	PJ 0451 — invest high risk development initiatives ; Reluctance	PF 5120 — delaying welfare
PF 7619 — Uncoordinated	PD 5948 — Irresponsible local	PC 6757 — Secret
PF 8703 — Undemocratic	**Government [Lack...]**	PF 2845 — *Unfamiliar*
Government [deficits...]	PF 4937 — Lack world	**Government [programmes...]**
PD 5984 — deficits	PF 5296 — leaders admit mistakes ; Refusal	PF 4739 — programmes ; Proliferation
PF 6707 — delay remedial response problems	PE 7937 — leaders associated sex scandals	PE 7965 — programmes ; Restrictions eligibility
PF 6707 — delay response symptoms problems	**Government leaders Corruption**	PF 4558 — promises ; Broken
PF 6119 — delaying tactics	PC 7587 — [Government leaders ; Corruption]	PF 9428 — promotion unrealistic perspectives
PE 6086 — Denial right participate	PE 9140 — amongst relatives	PC 3074 — propaganda
PF 9530 — dependence ; Excessive	PE 9140 — children	PE 8309 — protection national commodity price support
PD 5693 — Destabilization	PE 9140 — wives	PE 5634 — public property ; Crimes against
Government developing countries	**Government leaders [Excessive...]**	PF 3075 — publications ; Inadequate
PE 0365 — Inadequate financing local	PD 9653 — Excessive accumulation wealth	PC 3186 — Puppet
PE 0365 — Inadequate institutional structures local	PD 9653 — Excessive salaries	**Government refusal**
PF 9266 — Inefficiency	PF 3962 — Incompetence	PF 7897 — accept international tribunal arbitration
Government [development...]	PF 1097 — Loss confidence	PF 7897 — accept jurisdiction international courts justice
PE 9465 — development bank failure	PF 3962 — Mediocrity	PF 4050 — declassify innovative patents
PD 7171 — dirty tricks policies	PD 9653 — Personal greed	**Government [regulations...]**
PE 4860 — discrimination against trade unions	PD 4651 — scandal ; Implication	PF 8053 — regulations ; Complex
PF 1097 — Disenchantment	**Government [legislation...]**	PD 4800 — regulations ; Proliferation
PF 1097 — Disillusionment	PF 2499 — *legislation ; Non acceptance*	PF 1179 — reimbursements ; Delay
PJ 1895 — *Disloyalty*	PF 2499 — *legislation ; Nonacceptance*	PF 0845 — reluctance transform administrative structures
PF 2842 — *dominated jobs*	PF 4668 — limitations	PF 5351 — Remoteness
PF 9050 — Double moral standards	PF 4668 — Limited degrees freedom	PF 4558 — reneging public pledges
PF 4153 — duplicity	PD 9437 — loan defaults	PE 4446 — reparations ; Delay payment
Government [educational...]	PD 9437 — loans ; Unpaid	PD 1893 — repetition untruths ; Susceptibility electorate
PF 2447 — educational policy ; Inadequate	PF 5365 — lobbying ; Vulnerability	PD 5183 — reports ; Premature disclosure
PF 4849 — elites ; Appropriation machinery	PD 0579 — loss leadership role world affairs	PJ 0451 — requirement that economic benefits be proven prior programme approval
PF 4658 — employment ; Abuse	**Government [matters...]**	PC 0690 — Resistance
PE 3601 — enforced maternity	PE 7176 — matters ; Unlawful compensation assistance	PF 0845 — resistance institutional change
Government expenditure	PC 0698 — Military	PE 9325 — resources ; Underutilized
PF 1447 — arms ; Increasing	PF 7342 — ministerial control ; Abdication	PE 9325 — resources ; Untapped
PF 9108 — Decline	PC 7587 — ministers ; Corruption	PF 3567 — response disasters ; Confused
PF 0674 — education ; Decline	PF 7239 — Minorities unrecognized	PF 3467 — *restriction exports*
PF 0674 — *primary education ; Decline*	PF 6377 — Misallocation public spending	PF 4197 — revenue ; Inefficient mobilization
PD 3055 — Government expropriation land	PF 1097 — Mistrust	PD 1018 — revenue substance abuse ; Dependence
PD 3055 — Government expropriation private property	PE 1651 — Misuse business enterprises front organizations	PD 1018 — revenues exploitation environmentally inappropriate products ; Dependence
PF 5120 — Government facilities ; Excessive waiting times	PE 0436 — Misuse nonprofit associations front organizations	PD 1018 — revenues exploitation non renewable resources ; Dependence
Government failure	PF 4153 — *misuse personal records*	PE 7919 — role urban services ; Undefined
PF 9704 — mobilize support against problems	PF 9560 — money creation ; Over reliance	PE 7937 — rulers associated sex scandals
PE 9545 — prosecute offenders effectively	PD 5642 — monopolies ; Inefficiency	PC 7587 — rulers ; Corruption
PF 4366 — sanction state-supported terrorism	PE 8267 — monopoly practices ; Distortion international trade state trading	PF 8053 — rules ; Complex
PF 9672 — Government finances ; Inadequate management	**Government [neglect...]**	**Government sanctioned**
PF 9672 — Government finances ; Ineffective control	PE 9295 — neglect future ecological problems	PD 7221 — assassination
Government financial	PE 9295 — neglect future problems social environment arising current policies	PF 3819 — assassinations ; Official cover up
PF 1179 — commitments ; Arrears payment	PD 5948 — Negligence local	PD 4563 — cultivation illegal drugs
PF 9672 — expenditures ; Mismanagement	PE 1643 — news media ; Conflict	PD 7221 — killing
PF 5964 — resources developing countries ; Inefficient management	PE 1407 — non-assistance endangered peoples	PD 7221 — murder
Government [financing...]	PF 8650 — non-payment agreed contributions international organizations	PD 0181 — torture
PC 6631 — financing ; Inadequate local	PE 4446 — Non payment reparations	**Government [savings...]**
PF 5061 — folly	PC 9585 — Non representational	PE 4574 — savings capacity developing countries ; Inadequate
PD 6710 — function ; Obstruction	PC 0554 — Government ; Offences against	PF 4450 — secrecy concerning nuclear weapons testing
Government funding	**Government officials**	PC 1812 — secrecy ; Excessive
PF 8535 — Abusive allocation	PD 9842 — Blackmail	PE 5503 — securities ; Counterfeit money
PF 9530 — Dependency	PD 4541 — Bribery	**Government security**
PF 5419 — research ; Insufficient	PD 1893 — Deliberate lying	PF 1015 — services ousted repressive regime ; Post revolutionary re employment
PE 9325 — Government funds ; Unutilized	PD 8696 — Drug abuse	PF 9184 — services ; Secret
PE 9325 — Government grants ; Untapped	PD 8422 — Underpayment	PE 9441 — vetting job applicants ; Secret
Government [harassment...]	**Government [operations...]**	**Government [seizure...]**
PE 6934 — harassment human rights activists	PD 1163 — operations ; Crimes against integrity effectiveness	PE 6564 — seizure foreign nationals foreign countries
PF 3819 — harassment political activists ; Official cover up	PF 1023 — opposition population control	PF 2845 — *services ; Unpublished*
PF 4586 — health expenditure ; Decline	PF 4095 — opposition private initiative	PE 5938 — services ; Violation right organize trade unions
PF 4932 — historical records ; Falsification	PF 9542 — orders armed forces ; Disregard	PF 0611 — social expenditure ; Decline
PF 9050 — hypocrisy	PE 2309 — organized cooperatives	PF 0170 — social programmes ; Fluctuations
Government [imposed...]	PD 0975 — Outmoded deliberative systems	PE 9325 — sources ; Untapped
PE 5919 — imposed pregnancy	PF 9530 — Overdependence	**Government spending**
PE 2309 — imposition rural cooperative projects	PD 1964 — Overthrow	PE 5302 — cost effective activities ; Insufficient
PF 4668 — impotence	**Government [pardoning...]**	PF 2990 — Underreported
PD 0814 — improbity	PE 9545 — pardoning convicted human rights offenders	PF 2990 — Unreported
PC 3950 — inaction	PD 2029 — participation ; Distortion international trade result	PF 2845 — *Government structures ; Unfamiliar*
PE 1407 — inaction alleged human rights violations	PD 1902 — participation economies developing countries ; Excessive	PD 4532 — Government subsidies ; Concealed
PF 3394 — *Inadequate*	PF 1179 — payments ; Overdue	**Government support**
PF 6377 — Inappropriate public spending	PF 4153 — perfidy	PF 4821 — corrupt regimes
PD 7520 — *incitement kill*	PF 1097 — Pervasive distrust	PF 4821 — foreign dictatorships
PD 6008 — incitement terrorism	PF 7619 — planning ; Uncoordinated	PF 8960 — Limited
PD 1018 — income products dangerous health ; Dependence	PF 4558 — pledges ; Broken	PF 8960 — Marginal
PD 6943 — inducement crime	**Government policies**	PF 4821 — repressive regimes
PF 4095 — Ineffective use private sector resources	PF 3458 — Environmentally insensitive	PF 4821 — undemocratic regimes
PF 8491 — inefficiency	PF 1559 — Incoherent	PD 9146 — Government ; Suppression information
PF 3953 — ineptitude	PE 9295 — Neglect environmental consequences	PD 9538 — Government surveillance communications ; Misuse
PD 1061 — Inflexible central	PF 7472 — Public resentment	PF 2943 — Government ; Territories accorded United Nations non self governing status disputed administering
PF 7059 — Inflexible deliberative systems	PF 1559 — Unclear	PF 4153 — Government treachery
Government information	**Government policy**	**Government [Unaccountability...]**
PF 0157 — Biased	PF 8389 — Abuse	PE 4997 — Unaccountability
PD 2926 — Restrictions distribution confidential	PF 1559 — dilemmas	PE 8308 — Unclear fiscal relations different parts
PD 2926 — services ; Inaccessible	PF 1559 — Frequently changing	PD 0814 — Unethical practices
Government [initiative...]	PF 1559 — Inadequate	PD 5948 — Unethical practices local
PF 3318 — *initiative ; Lack*	PF 5645 — Inadequate	PE 4997 — Unlawful
PE 4997 — initiatives ; Extra legal	**Government policy making**	PF 3214 — *Unstable*
PF 2808 — insensitivity	PF 5061 — Self destructive	PD 7171 — Government ; Vilification
PD 1893 — Intellectual dishonesty	PF 0317 — short term considerations ; Domination	**Government [Weak...]**
PF 9184 — intelligence agencies ; Unaccountable	PF 7619 — Uncoordinated	PF 3394 — *Weak*
PF 8819 — intelligence services ; Failure	**Government [policy...]**	PD 1229 — welfare benefits ; Dependence
Government interference	PF 1559 — policy ; Unstable	PF 8278 — Wilful ignorance
PF 0304 — Excessive	PE 5608 — positions ; Violation right be elected	PF 8536 — Withholding information
PC 3185 — Foreign	PE 4398 — posts ; Frustrated talent	PC 1001 — women ; Denial right participate
PD 4800 — national economy	PC 9104 — power ; Abuse	
Government intervention	PF 4668 — power ; Limitations	
PE 1170 — agriculture ; Inappropriate		
PF 0304 — Inappropriate		
PD 4800 — private sector ; Excessive		

-937-

Governmental abuse

Governmental [abuse...]
- PE 6004 — abuse extradition
- PD 4517 — action ; Partisan reporting
- PF 9459 — apathy
- PF 5736 — avoidance legislative reform ; Deliberate

Governmental [barriers...]
- PF 2105 — barriers global ethic
- PF 0019 — bias statistics
- PD 0183 — budgets appropriations ; Wastage
- PD 1330 Governmental complex communist systems ; Military industrial

Governmental [decision...]
- PF 2420 — decision making machinery ; Inadequacy
- PD 0104 — decision making processes ; Inadequate information systems international
- PF 5502 — delayed withdrawal commitment action
- PE 2215 — delays fulfilling international reporting obligations
- PA 0298 — disputes ; Inter

Governmental disregard
- PF 0292 — international values procedures
- PD 9557 — legitimate protests
- PD 8017 — people human beings
- PF 7925 Governmental double standard human rights ; Inter
- PF 0037 Governmental energy conservation policies ; Inadequate
- PD 0181 *Governmental exchange torture methods ; Inter*
- PF 3913 Governmental failure fulfil financial commitments ; Inter

Governmental [inaction...]
- PF 0041 — inaction concerning trade services
- PF 3953 — incompetence
- PF 3394 — *ineffectiveness*
- PF 2335 Governmental managerial instruments ; Inadequate
- PE 1973 Governmental nongovernmental organizations programmes ; Inadequate relationship international
- PD 6628 Governmental organization mismanagement ; Inter

Governmental [payment...]
- PE 4446 — payment compensation ; Delay
- PF 1559 — policy ; Unpredictable
- PD 5370 — polluters

Governmental [rejection...]
- PD 9557 — rejection citizen grievances
- PE 4344 — representation intergovernmental organizations ; Inadequate coordination
- PF 3707 — reprisals against citizens
- PC 3950 — resistance inertia response problems
- PF 0038 — resource conservation policies ; Inadequate

Governmental [spending...]
- PD 0183 — spending ; Excessive
- PF 5351 — structure ; Removed
- PF 8960 — support ; Lack
- PC 2089 — suspicion ; Inter
- PF 4095 Governmental use nongovernmental resources ; Ineffective

Governments [Abusive...]
- PD 9538 — Abusive monitoring communication
- PE 5132 — activity national affairs ; Concentration national
- PE 1622 — against governments ; Threats
- PD 4517 — Governments ; Biased allegations against

Governments [Collusion...]
- PD 8465 — Collusion
- PE 8289 — Conflict interest governments groups
- PF 0419 — Conspiracy
- PF 0419 — Covert arrangements
- PE 4860 Governments ; Denial trade union rights
- PD 5693 Governments ; Destabilizing

Governments [Economic...]
- PF 4260 — Economic blockade against
- PF 4260 — Economic sanctions against
- PF 2199 — Eroding support unfriendly
- PF 5633 — Evil
- PE 2215 Governments fulfil international reporting obligations ; Failure

Governments [Government...]
- PE 1622 — Government bullying foreign
- PE 1622 — Government intimidation
- PC 6631 — grants ; Excessive reliance local
- PE 8289 — groups governments ; Conflict interest

Governments [Illegal...]
- PE 5332 — Illegal action
- PE 7300 — implement provisions ratified human rights agreements ; Failure
- PE 9762 — International cartels enforced
- PE 8650 — international organizations ; Overdue payments
- PE 2215 — international surveys ; Non response
- PF 9474 — Internationally non cooperative
- PC 2089 Governments ; Lack confidence
- PE 8864 Governments large firms ; Overemphasis
- PC 2089 Governments ; Mistrust

Governments Misuse
- PD 2930 — electronic surveillance
- PD 2683 — postal surveillance
- PD 9538 — *radio transmission surveillance*
- PF 3701 — satellite surveillance
- PD 1632 — telephone surveillance
- PF 8040 Governments ; Non recognition foreign
- PD 0055 Governments obstacle representative democratic political organization ; National
- PD 7221 Governments ; Political murders authorized
- PF 0401 Governments regulate family size ; Inability
- PD 0699 Governments ; Sale passports
- PF 5633 Governments ; Satanic
- PE 1622 Governments ; Threats governments against

Governments [Uncoordinated...]
- PF 5000 — Uncoordinated macroeconomic policies
- PF 5337 — United Nations ; Irresponsibility member

Governments [unwilling...] cont'd
- PF 9474 — unwilling cooperate United Nations eliminating human rights violations
- PF 6826 Grace ; Falling
- PF 5266 Grace ; Resistance
- PJ 6238 Graceless acceptance faith
- PA 6312 Gracelessness *Inelegance*
- PA 7240 Gracelessness *Ugliness*
- PA 7232 Gracelessness *Unskillfulness*
- PD 6131 Gradients ; Unbalanced urban population density
- PF 7542 Gradual destruction cultural property
- PD 1162 Graduate post-graduate unemployment
- PD 1162 Graduate unemployment ; Graduate post
- PD 1162 Graduates ; Chronically unemployed
- PF 1905 Graduates ; Inappropriate education
- PD 1162 Graduates ; Unemployment highschool
- PD 3436 Graduation income tax ; Unequal
- PD 5305 Graffiti
- PD 4541 Graft
- PE 2890 Grain ; Adverse effects high yield
- PE 4586 Grain ; *Ergot*
- PB 2846 Grain insecurity ; Food
- PE 4630 Grain overload

Grain [sorghum...]
- PE 3590 — sorghum ; Pests diseases
- PD 4100 — species ; Limited food
- PD 0811 — spoilage ; Food
- PF 1479 Grain yields ; Variability
- PD 7566 Grammar ; Ignorance
- PF 3455 Grandiose public works
- PA 7102 *Grant acceptance ; Dishonourable*
- PF 0182 Grant amnesty ; Refusal
- PE 5306 Grant-back provisions
- PE 5453 Grant citizenship ; Refusal
- PD 4543 Grant frauds
- PF 2275 Grant funding ; Self accountability
- PF 3079 Grant licences media ; Refusal
- PF 3079 *Grant licences ; Refusal*
- PF 1226 *Grant management ; Difficult*
- PE 2657 Grant nationality ; Refusal
- PF 2708 Grant recipients ; Belittling
- PC 6631 Grants ; Excessive reliance local governments
- PD 4543 Grants ; Fraudulent procurement use government private
- PF 2564 Grants ; Limited applicability monetary
- PE 9325 Grants ; Untapped government
- PE 1946 Granular conjunctivitis
- PE 9702 Granuloma cats ; *Eosinophilic*
- PE 4604 Granulomatous meningoencephalomyelitis
- PE 4161 Granulosa ; *Verrucosa*
- PE 4586 Grape ; Black rot
- PF 6143 Grass roots community initiatives ; Inadequate facilities
- PJ 9394 Grass roots planning ; Unproductive
- PD 7841 *Grass sickness ; Equine*
- PD 7420 Grass staggers
- PD 7420 Grass tetany
- PE 2233 Grasses cereals ; Rusts
- PE 2234 Grasses cereals ; Smuts
- PE 4732 Grasses, weeds trees ; Pollens
- PD 0725 Grasshopper plagues
- PE 3642 Grasshoppers insect pests
- PD 5133 Grassland rangeland ; Misuse
- PD 5133 Grasslands ; Deterioration rangelands
- PF 6506 Grassroots consensus ; Lack systems achieving
- PF 2452 Grassroots ; Insufficient communication decision makers
- PE 4479 Grassroots involvement planning economy ; Static
- PE 7095 Grassroots technological development ; Intellectual methods which disenfranchise
- PF 1058 Gratification based social forms ; Immediate
- PF 1807 *Gratification ; Conditioned self*
- PD 2547 Gratuitous sex advertising
- PF 0491 Grave robbing
- PE 8612 Gravel trade instability stone ; Sand
- PD 0981 Graven images
- PE 1924 Graves' disease
- PE 8022 *Gravidarum ; Hyperemesis*

Grazing [land...]
- PF 6528 — *land ; Exploitation marginal*
- PF 6528 — *land ; Inadequate*
- PF 6528 — *lands ; Unreserved*
- PE 7826 Greasy heel
- PD 9667 Greasy pig disease
- PE 1570 *Great apes ; Endangered species*
- PE 2300 Great pox
- PA 6999 Greed *Avarice*
- PA 6379 Greed *Avoidance*
- PF 7189 Greed ; Business
- PF 7189 Greed ; Corporate
- PD 9653 Greed government leaders ; Personal
- PA 6466 Greed *Intemperance*
- PA 7211 Greed *Selfishness*
- PA 5644 Greed *Vice*
- PA 6999 Greedy people
- PF 6135 Green belts
- PD 2345 Green inner suburbs ; Destruction
- PD 2730 *Green monkey disease ; African*
- PE 9774 Green muscle disease
- PE 6175 Green parkland urban environment ; Inaccessibility
- PF 3146 Green revolution
- PC 0918 Greenhouse effect
- PD 5453 *Grey Collie dogs ; Cyclic neuropenia*
- PE 6308 Grey human hair
- PF 3098 Grey lies
- PD 1593 *Grey whale ; Endangered species*
- PF 5654 Grief

- PA 6340 Grief *Adversity*
- PD 4918 Grief process ; Inhibited
- PA 6731 Grief *Solemnity*
- PA 7107 Grief *Unpleasantness*
- PF 8029 Grievance
- PA 5454 Grievance *Badness*
- PA 6838 Grievance *Dissent*
- PA 6486 Grievance *Injustice*
- PA 5479 Grievance *Lamentation*
- PE 2832 Grievance procedures ; Denial right
- PA 7107 Grievance *Unpleasantness*
- PF 8029 Grievances citizens
- PD 7567 Grievances consumers
- PD 5586 Grievances employees
- PD 7286 Grievances employers
- PD 9557 Grievances ; Governmental rejection citizen
- PD 1544 Grievances youth
- PA 5454 Grimness *Badness*
- PA 7253 Grimness *Envy*
- PA 6030 Grimness *Fear*
- PA 6099 Grimness *Hopelessness*
- PA 7325 Grimness *Irresolution*
- PA 6731 Grimness *Solemnity*

Grimness [Ugliness...]
- PA 7240 — *Ugliness*
- PA 5643 — *Unkindness*
- PA 7107 — *Unpleasantness*
- PE 1314 Grinders' rot
- PE 0447 Grippe
- PF 4010 Grooming ; Inadequate
- PF 8085 Grooming skills ; Marginal
- PA 5454 Grossness *Badness*
- PA 6839 Grossness *Disrepute*

Grossness [Unchastity...]
- PA 5612 — *Unchastity*
- PA 7371 — *Unintelligence*
- PA 7107 — *Unpleasantness*
- PA 5821 Grossness *Vulgarity*
- PE 5066 Ground failures
- PE 5126 Ground failures due liquefaction
- PE 2587 *Ground nut peanut trade ; Instability*
- PE 7783 *Groundnut ; Pests diseases*
- PD 1576 Grounds ; Arrest dubious
- PC 3152 Grounds birds ; Disruption breeding
- PC 3152 Grounds ; *Ecological disruption animal breeding*
- PC 3152 Grounds fish ; Disruption breeding
- PC 3152 Grounds marine mammals ; Disruption breeding
- PF 1442 *Grounds ; Unpopulated fishing*
- PD 2503 Groundwater contamination
- PE 4740 Groundwater ; Corrosive
- PD 8888 Groundwater levels ; High
- PD 2503 Groundwater ; Nitrate pollution
- PD 2503 Groundwater ; Pollution
- PD 0708 Groundwater pollution hazards ; Underreporting
- PD 4403 Groundwater resources ; Shortage
- PF 6136 Group ; Household segregation age
- PF 8508 Group labels

Group [marriage...]
- PF 3288 — marriage
- PF 3713 — *methods ; Obsolete*
- PJ 5810 — mind

Group [separatism...]
- PD 2490 — separatism ; Minority
- PF 3288 — *sex*
- PF 2477 — skills gap ; Intra
- PF 1073 Group ; Undue attachment social
- PF 2827 Group values ; Subjective assessment according own

Groupings [developed...]
- PE 1604 — developed countries ; Discrimination against developing countries formation regional
- PD 9412 — developing countries ; Inadequate economic integration regional
- PF 6545 — Divisive patterns community
- PD 6896 Groupings ; Elitist intergovernmental
- PF 2575 Groupings ; Insular patterns community
- PF 6545 Groupings ; *Rigid local*
- PF 1746 Groups appropriate wisdom past ; Failure opposing
- PE 7575 Groups ; Bogus public interest
- PC 3428 Groups ; Conflict minority

Groups [Defeatist...]
- PF 8781 — Defeatist self images citizens' law enforcement
- PE 1355 — Denial right union activity special
- PC 4405 — Denial rights vulnerable
- PB 4353 — Dependency vulnerable
- PE 1702 — development community priorities ; Monopolization interest
- PB 6320 — Disadvantaged
- PB 4353 — Disempowerment vulnerable
- PE 7174 — Diversion high technology irresponsible
- PD 3785 — during disasters ; Lack protection vulnerable
- PD 3785 — during states emergency ; Violation rights vulnerable
- PF 8209 Groups ; Excessive institutionalization vulnerable
- PE 8289 Groups governments ; Conflict interest governments

Groups [Idealism...]
- PF 5401 — Idealism opposition
- PC 7373 — *Illegal armed*
- PC 7662 — individuals ; Suppression opposition
- PF 9774 — Inferior average intelligence certain racial
- PF 1705 — *Insufficient women's*
- PC 3316 — Isolation ethnic

Groups [Lack...]
- PF 4662 — Lack commitment protection vulnerable
- PD 2338 — *Land controlled business*
- PF 5008 — Limited availability land low income disadvantaged
- PF 2428 — *Limited fragmented outlook civic minded*
- PF 4347 Groups ; Marginal

Haphazardness

PD 6760 Groups ; Medical experimentation socially vulnerable
Groups [Narrow...]
PF 2182 — Narrow political interests small
PF 9003 — Negative influence peer
PF 7118 — *Non cooperative community*
PJ 2493 — Non corporate operations existing
PC 1523 — *Nonacceptance social*
PF 6556 — *Nonexistent development*
Groups [peoples...]
PE 8802 — peoples ; Biased media image foreign
PF 3918 — Placement burden proof abuse vulnerable
PF 1070 — policy agencies ; Over reliance economic interest
PD 0363 — Protein energy malnutrition vulnerable
PC 5537 *Groups ; Restrictive social*
PF 2216 Groups ; Rivalry international
Groups [Socially...]
PC 0193 — Socially isolated
PC 6570 — Stifled potential social interaction different age
PF 3821 — Structural failure opposition
PC 0006 Groups ; Unequal rights different racial
PF 1705 *Groups ; Uninitiated women's*
PC 6570 *Growing old ; Fear*
PE 6396 Growing plant species ; Substitution fast
PD 5212 Growing season ; Short
PE 8706 Growing size impersonality firms
PF 2071 *Growth anti-systemic movements*
PD 9534 Growth arable land ; Uncontrolled
Growth [capita...]
PF 4326 — capita ; Decreasing agricultural
PF 1737 — capitalist countries ; Weak economic
PE 4921 — children ; Stunting
PE 4921 — children ; Wasting
PD 2968 — commodity export earnings developing countries ; Decline
PE 8476 — crops ; Adverse impact pollution
Growth [Declining...]
PC 0002 — Declining economic
PE 2700 — Denial right business
PB 0479 — Dependence unbalanced
PC 0495 — Dependence uncontrolled economic
PD 4007 — Depletion natural resources due population
PE 4241 — developed developing countries ; Imbalance population
Growth developing countries
PD 5326 — Collapse
PD 5326 — Declining economic
PD 7257 — Disparities economic
PC 0495 — Growth ; Development indiscriminate economic
Growth [economy...]
PE 8735 — economy ; Imbalances growth labour force, urban population overall
PE 8437 — economy labour force ; Urban population overall
PB 0035 — Excessive population
PF 1471 — export markets developing countries ; Restricted
PE 6909 Growth failure ; Infant
PC 1960 Growth food production ; Slowing
Growth [Import...]
PE 8792 — Import substitution barrier subsequent economic
PF 2905 — Inadequate systems monitoring industrial
PF 1737 — income industrialized countries ; Instability
Growth Increasing development
PE 2000 — lag against information
PF 3743 — lag against population
PC 5879 — lag against socio economic
PE 3078 — lag against technological
Growth [Increasing...]
PE 5369 — Increasing lag education against population
PC 0495 — Indiscriminate economic
PF 1737 — industrialized countries ; Declining economic
PD 5177 — Inhibited human physical
PE 7139 Growth knowledge ; Increasing lag education against
PE 8735 Growth labour force, urban population overall growth economy ; Imbalances
PE 4921 Growth malnourished children ; Inhibited
PD 9117 Growth manufacturing developing countries ; Unstable
PF 1737 Growth output developed market economies ; Slow down
Growth [partnership...]
PE 2203 — partnership model developed countries ; Contradictions
PF 2346 — *patterns ; Haphazard*
PE 7826 — *plate ; Dysplasia*
PE 8476 — pollution ; Inhibition crop
PD 1714 — promoters ; Hormonal
Growth [Rapid...]
PB 0035 — Rapid population
PF 3146 — regulators ; Plant
PD 4007 — resource development ; Imbalance population
Growth [size...]
PF 8155 — size production unit
PC 6215 — social expenditure ; Excessive
PE 1117 — socialist countries ; Slow economical
PF 7118 — *Sporadic community*
PA 8653 — *suspicions ; Misinformed*
PF 9671 Growth ; Theological justification population
Growth [Unbalanced...]
PB 0479 — Unbalanced
PB 0035 — Unconstrained social
PB 9015 — Unconstrained socio economic
PC 0495 — Uncontrolled economic
PC 5601 — Unequal global distribution economic
PC 0002 — Unstable economic
PB 0035 — Unsustainable population
PC 5601 — Unsynchronized economic
PG 2295 — *Unusually rapid*
PB 9015 — Unwanted excessive

Growth [Urban...] cont'd
PC 0442 — Urban
PE 9125 — Use artificial methods promoting fast livestock
PC 0002 Growth world economy ; Declining
PC 3853 Growths
PE 2254 *Grubs ; Cattle*
PA 7055 Gruesomeness *Death*
PA 6030 Gruesomeness *Fear*
PA 7240 Gruesomeness *Ugliness*
PE 3066 Guarantee facilities developing country exports ; Lack credit
PF 9187 Guaranteeing survival ; Discouragement food crops
PE 1032 Guarantees condition provisional release criminal cases ; Pecuniary
PE 3992 Guardianship mentally retarded adults ; Inadequate
PC 1738 Guerrilla war
PC 1738 Guerrilla warfare
PD 1988 Guerrillas ; Urban
PF 1403 Guidance ; Inadequate adult
PF 7005 Guidance ; Loss spiritual
PF 1664 Guidance older generation ; Limited social
PF 0245 Guidelines ; Inadequate credit
PA 7411 Guile *Uncommunicativeness*
PA 7232 Guile *Unskillfulness*
PA 6793 Guilt
PA 6793 Guilt complex
PA 6793 Guilt *Guilt*
PF 5609 Guilt ; Masturbation
PD 8016 *Guilt ; Sex*
PE 7393 Guilty ; Denial right presumed innocent until proven
PE 3510 Guinea-worm
PD 0702 Gulags
PD 0486 *Gullibility*
PA 7392 Gullibility *Unbelief*
PA 7371 Gullibility *Unintelligence*
PD 2269 Gully maintenance ; Inadequate
PD 2730 Gumboro disease
PE 3503 *Gums ; Inflammation*
PC 4667 Gun-boat diplomacy
PD 1970 Gun laws ; Lack
PD 4858 Gun running
PC 4667 Gunboat diplomacy
PD 3481 *Gundis ; Endangered species*
PE 9111 Gunshot wounds
PE 6336 Gurus ; Exploitative
PE 5182 *Guttural pouches ; Diseases*
PE 1281 Gypsies ; Discrimination against
PE 1281 Gypsy persecution

H

PD 6927 *Habeas corpus ; Denial right*
PD 6927 *Habeas corpus ; Suspension*
PD 5494 *Habit ; Lack regular*
PD 0713 Habit ; Tobacco
Habitats [dams...]
PE 7794 — dams ; Inundation wildlife
PC 5773 — Decreasing diversity biological
PD 1135 — *Defoliation insect breeding*
PC 0480 — Dependence destruction wildlife
PC 0480 — Destruction wildlife
PF 3437 Habitats elves ; Endangered
PC 4764 Habitats ; Endangered protected
PC 3152 Habitats flora fauna ; Degradation semi natural natural
PC 3152 Habitats ; Modification animal
PD 5494 Habits ; Bad
PF 6078 Habits food aid ; Modification environmentally adapted nutritional
Habits [Improper...]
PC 0382 — Improper dietary
PD 5494 — Inappropriate personal
PF 2256 — Ingrained segregation
PE 4415 Habits ; Noisy sleeping
PF 6461 *Habits ; Overpowering traditional*
PD 0800 Habits ; Poor nutritional
PD 5494 *Habits ; Sedentary*
PF 3211 *Habits ; Traditional purchasing*
PD 0800 Habits ; Unnutritious eating
PC 0382 Habits ; Unnutritious eating
PE 5581 Habitual criminals
PF 2256 *Habitual lifestyle routines*
PF 2168 *Habitual modes ; Disoriented*
PF 1382 *Habitual non-payment bills*
PF 1804 Habitual overemphasis national self-determination
PF 4473 Habitual responses problems
PD 1533 Habitual television watching
PE 5380 Habronemiasis ; Cutaneous
PC 3320 Hacienda camps
PE 9831 Hacking ; Computer
PD 7841 Haemagglutinating encephalomyelitis
PD 8347 Haemangioma
PD 5535 *Haematoma ear animals*
PD 5453 *Haematopoiesis ; Cyclic*
PE 4637 *Haematopoietic tissue ; Neoplasms lymphatic*
PE 0921 Haemic schistosomiasis
PG 5035 Haemoglobinopathies
PE 7769 *Haemoglobinuria animals ; Bacillary*
PD 7420 *Haemoglobinuria ; Postparturient*
Haemolytic anaemia animals
PE 9554 — [*Haemolytic anaemia animals*]
PE 9509 — Autoimmune
PG 9847 — *Immune*
PD 7758 *Haemolytic anaemias ; Acquired*
PD 7758 *Haemolytic anaemias ; Hereditary*

Haemolytic [disease...]
PE 9509 — *disease animals ; Cold*
PE 2399 — disease new born
PD 2289 — *diseases*
PE 3098 *Haemolytic streptococcal infection throat*
PE 1920 Haemophilia
PE 1920 Haemophilia
PD 2732 *Haemophilus polyserositis*
PD 7841 *Haemophilus septicaemia cattle*
PE 2239 Haemorrhage
PE 2239 Haemorrhage
PE 5182 *Haemorrhage ; Exercise induced pulmonary*
PD 5453 *Haemorrhage ; Internal*
PD 8461 *Haemorrhage ; Intracerebral intracranial*
PE 7864 *Haemorrhage nose*
PE 4894 Haemorrhage pregnancy childbirth
PD 2730 *Haemorrhagic anaemia syndrome chickens*
PD 3978 *Haemorrhagic enteritis turkeys*
PE 3936 *Haemorrhagic enteropathy ; Proliferative*
Haemorrhagic fever
PE 5272 — [*Haemorrhagic fever*]
PD 2730 — *Argentinian*
PD 2730 — *Bolivian*
PD 2730 — *Crimean Congo*
PE 5272 Haemorrhagic fevers
PE 2373 *Haemorrhagic fibrinolysis*
PE 7423 *Haemorrhagic gastroenteritis ; Canine*
PE 2395 *Haemorrhagic pancreatitis*
PD 2731 *Haemorrhagic septicaemia*
PE 3504 Haemorrhoids
PD 5453 *Haemostatic disorders animals*
PE 1849 *Haemotoxicity*
PD 0251 Hail storms
PE 7111 Hair ; Cosmetic health problems related human
PE 1323 Hair ; Devil's
PE 5580 Hair ; Excess human body
PE 6308 Hair ; Grey human
PE 5697 Hair shaving ; Mandatory
PE 5580 Hair ; Unwanted facial
PE 1323 Hairweed
PE 6419 *Hairy shaker disease*
PE 6558 Halitosis
PC 4867 *Hall management ; Private*
PA 6180 Hallucination *Error*
PA 6414 Hallucination *Illusion*
PA 6739 Hallucination *Maladjustment*
PA 7411 Hallucination *Uncommunicativeness*
PF 2249 Hallucinations
PD 0556 *Hallucinogen perception disorder ; Post*
PD 0556 *Hallucinogenic drugs ; Abuse*
PD 0556 *Hallucinogens ; Abuse*
PF 2249 Hallucinosis
PE 9263 *Hallucinosis ; Alcoholic*
PE 4483 Halogen compounds pollutants
PE 9076 Halogen salts ; Trade instability inorganic chemicals, elements, oxides
PE 8967 Hamsters ; Endangered species mice
PF 1215 Handedness ; Left
PE 6780 Handicap ; Economic self sufficiency
PD 3518 Handicap ; Elders' communication
PC 0699 Handicap ; Health impairment
PE 6778 Handicap ; Occupation
PE 6773 Handicap ; Physical independence
PE 6779 Handicap ; Social integration
PE 8424 Handicapped children education ; Segregation
PD 0196 Handicapped children ; Physically
Handicapped [Denial...]
PE 4544 — Denial right procreate severely mentally
PD 9757 — Discrimination against
PD 0098 — during disasters ; Lack protection
PD 9685 Handicapped ; Exploitation mentally
PE 4868 Handicapped family members ; Immigration barriers
PE 2087 Handicapped ; Family rejection physically
PE 8151 Handicapped ; Inadequate rehabilitation facilities mentally
PE 8151 Handicapped ; Inadequate support services mentally
PC 1587 Handicapped ; Mentally
Handicapped persons
PE 8924 — Inadequate community care
PE 8924 — Neglected welfare
PD 6020 — Physically
PC 0699 — Physically mentally
PD 2542 — Visually
PE 4868 Handicapped refugees
PD 1507 Handicapped refugees ; Socially
PD 9757 Handicapped ; Stigmatization
PD 1405 Handicapped ; Stress families physically mentally
PE 0783 Handicapped work ; Discrimination against
PF 3758 Handicappism
PE 6772 Handicaps ; Orientation
PD 1250 Handicrafts cottage industries developing countries ; Decline
PD 9108 Handling ; Dangerous cargo
PE 5899 Handling ; Excessive costs inefficient port cargo
PE 4958 Handwriting ; Illegible
PE 2429 Hangover
PE 0721 Hansen's disease
PF 1249 *Haphazard forms social ethics*
PF 2346 *Haphazard growth patterns*
PF 2822 Haphazard organization community space
PE 2411 *Haphazard provision consumer services*
PF 3409 *Haphazard transmission practical technology*
PC 0442 Haphazard urban structure
PA 6714 Haphazardness *Chance*
PA 7361 Haphazardness *Disorder*
PA 5438 Haphazardness *Neglect*

-939-

Happenings

PF 6170 Happenings ; Inhibition collectively organized fantastic
PF 4389 *Happiness ; Illusion*
PF 1216 *Happiness ; Incapacity*
PC 8558 Harassment
PE 6006 Harassment ; Aircraft
PE 0428 Harassment amongst prisoners
PE 6922 Harassment children public life ; Discrimination
PD 2067 Harassment ; Criminal
PD 7707 Harassment ; Environmental sexual
Harassment Fear
PA 6030 — [Harassment *Fear*]
PF 9078 — [Harassment ; Fear]
PF 3707 — official
PE 6934 Harassment human rights activists ; Government
PF 1585 Harassment human rights monitors
PA 6852 Harassment *Inappropriateness*
PF 1585 Harassment informers
Harassment [journalists...]
PD 3036 — journalists
PD 3071 — journalists
PE 5487 — judges
PE 5487 — judiciary
PD 4194 — juries
PE 5487 Harassment lawyers
Harassment [media...]
PD 0160 — media
PD 3071 — media
PE 1293 — men ; Sexual
PF 3271 — men ; Sexual
PE 8220 — military ; Sexual
PD 0610 Harassment organizers
Harassment [playgrounds...]
PE 7768 — playgrounds
PD 0736 — Police
PF 3819 — political activists ; Official cover up government
PD 0160 — press
PD 4915 — public officials
PD 1116 Harassment ; Sexual
Harassment [Trade...]
PD 7441 — Trade
PE 7780 — travellers consular procedures
PE 7780 — travellers immigration officials
PA 7107 Harassment *Unpleasantness*
PF 4729 Harassment ; Unreported
Harassment [whistle...]
PF 1585 — whistle-blowers
PE 0756 — women public ; Verbal sexual
PF 3271 — women ; Sexual
PD 7471 — workers' organization leadership
PE 8466 — working place ; Sexual
PE 0484 Harbouring national security offenders
PE 4897 Harbours docks ; Dangerous
PD 3068 Hard currency shortage developing countries
PE 1513 *Hard fibres ; Instability trade*
PA 5497 Hardness *Difficulty*
PA 5479 Hardness *Lamentation*
PA 7023 Hardness *Pitilessness*
PA 6976 Hardness *Toughness*
Hardness [Unfeelingness...]
PA 7364 — *Unfeelingness*
PA 7367 — *Unintelligibility*
PA 5643 — *Unkindness*
PA 5644 Hardness *Vice*
PA 5558 Hardness *Weakness*
PD 2730 *Hardpad disease*
PD 0428 Hardship ; Inadequate national empathy external
PD 3978 *Hardware disease*
PD 0089 Hardware earth orbits ; Accumulation aerospace
PE 5117 Hare lip
PE 8794 Hares ; Endangered species rabbits
PA 0429 *Harm innocent bystanders*
PE 4878 Harm ; Non fatal deliberate self
PD 0437 Harm persons ; Criminal
PD 5511 Harm property ; Criminal
Harmful [advertising...]
PE 4225 — advertising effects
PC 0119 — agents ; Airborne toxic
PC 0728 — animals plants
PE 6294 Harmful biological effects ionizing radiation
PD 9515 Harmful dam construction ; Environmentally
Harmful effects
PE 2004 — advertising transnational corporations developing countries
PE 4744 — air pollution agricultural crops
PE 2691 — comic strips picture-story books
PE 5787 — sensory deprivation
PD 0748 — ultrasonic radiation human body
PD 2847 Harmful health ; Airborne substances
PD 4238 Harmful natural foodstuffs
PC 1306 Harmful synergistic interaction biological agents
PF 0441 Harmful thought
PC 3151 Harmful wildlife
PA 5454 Harmfulness *Badness*
PA 7226 Harmfulness *Unhealthfulness*
PA 5643 Harmfulness *Unkindness*
PF 9350 Harmonial religion
PC 4591 Harmonization standards ; Inadequate international
PA 5565 Harshness *Discord*
PA 7143 Harshness *Discourtesy*
PA 7156 Harshness *Moderation*
PA 7023 Harshness *Pitilessness*
PA 5643 Harshness *Unkindness*
PA 7107 Harshness *Unpleasantness*
PC 2027 Harvest losses ; Post
PF 2365 *Harvesting ; Barriers jungle*
PD 9578 Harvesting rates ; Unsustainable

PD 9578 Harvesting ; Unregulated
PE 1186 Hashish ; Abuse marijuana
PA 6200 Haste *Impatience*
PA 7115 Haste *Rashness*
PA 7006 Haste *Untimeliness*
PA 7253 Hastiness *Envy*
PA 6200 Hastiness *Impatience*
PA 5438 Hastiness *Neglect*
PA 7115 Hastiness *Rashness*
PA 7006 Hastiness *Untimeliness*
PA 7338 Hate
PA 8487 Hate *Hatred*
PF 1183 Hate literature ; Dissemination
PE 5952 Hate ; Slogans
PE 7062 Hating ; Woman
PA 8487 Hatred
PA 5446 Hatred *Enmity*
PA 7338 Hatred *Hate*
PE 5952 Hatred ; Incitement
PC 3428 Hatred ; Self
PA 7107 Hatred *Unpleasantness*
PD 6520 *Hauling equipment ; Inadequate*
PF 0201 Haunted buildings
PF 3801 Hauntings
PE 8746 Have lived abroad ; Denial social security nationals who
PE 5861 Havens ; Abuse land locked island countries
PE 2370 Havens ; Abuse tax
PE 6172 Havens environmental pollution
PE 6172 Havens hazardous products
PE 0621 Havens ; Legal
PE 0621 Havens prosecution
PD 5541 Havens ; Terrorist
PD 5541 Havens war criminals
PE 6197 Hay fever
PE 1989 Hazard ; Antimony health
PE 3001 Hazard ; Asbestos occupational
PE 2209 Hazard ; Beryllium health
Hazard [Carbon...]
PE 1657 — Carbon monoxide health
PE 1657 — Carbon monoxide occupational
PE 4378 — Chlorofluorocarbons environmental (CFC's)
PE 6053 — Circumcision health
PF 5744 — Cold occupational
PE 5767 Hazard ; Dust occupational
Hazard [Health...]
PB 4885 — Health
PE 5720 — Heat occupational
PE 2329 — Hydrogen sulphide health
PE 5780 Hazard ; Improper lighting occupational
PE 5650 Hazard ; Lead health
PE 5650 Hazard ; Lead occupational
PE 1364 Hazard ; Manganese health
PE 6056 Hazard ; Microwave radiation health
PE 2432 Hazard ; Polychlorinated biphenyls health
Hazard [Shift...]
PE 5768 — Shift work stress occupational
PE 5708 — Solvents occupational
PE 1210 — Sulphur dioxide occupational
PF 5188 Hazard thresholds ; Uncertain radiation
PE 5672 Hazard ; Ultraviolet radiation
PE 8136 Hazard ; Vegetable oils fats environmental
PE 1145 Hazard ; Vibrations health
PB 6336 *Hazard ; Wildlife pollution*
PE 5085 Hazard ; Wind shear
PE 6815 Hazardous aquatic animals
Hazardous [cargoes...]
PE 1839 — *cargoes ; Marine transportation*
PE 8972 — combination effects work place
PD 5256 — combinations substances
PG 7741 — commercial traffic
PC 4428 — computer software errors
PF 9769 Hazardous environmental discontinuities
PC 0936 Hazardous environmental pollutants ; Human exposure
PE 4896 Hazardous genetic effects medications
PE 6824 Hazardous glaciers
Hazardous [industrial...]
PE 4304 — industrial installations
PE 6687 — industries developing countries ; Export
PE 6687 — industries developing countries ; Relocation
PF 5188 Hazardous levels toxic substances ; Ignorance
PD 2718 Hazardous locations nuclear power plants
PC 6913 Hazardous materials ; Storage
PC 6913 *Hazardous materials ; Underground storage*
PC 6716 Hazardous occupation
PF 1379 Hazardous products
PE 6172 Hazardous products ; Havens
PF 2613 Hazardous remnants war
PD 0791 Hazardous road passages
Hazardous [school...]
PJ 9371 — *school laboratories*
PD 4542 — substances ; Illicit discharges
PE 5089 — substances ; Ineffective monitoring
Hazardous [waste...]
PD 1398 — waste ; Disposal
PD 1398 — waste dumping
PC 9053 — wastes
PA 6971 Hazardousness *Danger*
PA 7309 Hazardousness *Uncertainty*
PF 7580 Hazards
Hazards [Aerosols...]
PE 1504 — Aerosols industrial
PD 7993 — agricultural chemicals
PD 0376 — agricultural livestock production ; Environmental
Hazards air pollution
PE 9609 — animals ; Health

Hazards air pollution cont'd
PE 8617 — children ; Health
PE 4744 — plants ; Health
Hazards [air...]
PD 8328 — air traffic ; Natural
PD 8328 — Aircraft
PD 8328 — Aircraft environmental
PE 4969 — aluminium ; Health
Hazards animal
PE 1331 — feedstuffs, excluding unmilled cereals ; Environmental
PE 8135 — oils fats ; Environmental
PD 4721 — populations ; Underreporting
Hazards [anti...]
PE 9294 — anti personnel gases ; Health
PE 3001 — asbestos ; Health
PF 3564 — asteroids
PC 0229 — atomic radiation ; Environmental
Hazards [basic...]
PD 0243 — basic metal industries ; Health
PE 1849 — benzene ; Occupational
PE 0849 — beverages ; Environmental
PE 5696 — Biological agents occupational
PD 7731 — biological systems ; Underreporting
PC 0776 — biotechnology
PD 1038 — birth control ; Health
PE 4935 — bottle-feeding
PF 6707 — Bureaucratic procrastination response evidence potential
Hazards [cassava...]
PE 6840 — cassava ; Toxic
PE 8732 — cereals cereal preparations ; Environmental
PC 1192 — chemical materials products ; Environmental
PC 1192 — chemicals ; Environmental
PC 1192 — chemicals petrochemicals industries ; Environmental
PE 5146 — children smokers ; Health
PE 9019 — china, pottery earthenware industries ; Environmental
PE 7416 — chlorine bleached paper products ; Environmental
PD 7541 — coal energy ; Environmental
PE 5160 — Coal mining environmental
PE 0481 — coffee, tea, cocoa, spices their manufacture ; Environmental
PE 5656 — Combined stresses occupational
PF 3564 — comets
PE 3957 — commerce offices ; Occupational
PE 5083 — computer visual display units ; Health
PE 5651 — Conflicting standards protection against chemical occupational
PF 1789 — constraining scientific research ; Environmental
PE 8790 — construction industry ; Environmental
PD 1631 — Continuing fire
PD 1038 — contraceptives ; Health
PE 4895 — cosmetic use
PE 5682 — crops plants ; Weather
Hazards [dairy...]
PE 0505 — dairy products eggs ; Environmental
PE 7539 — decommissioned nuclear power plants ; Environmental
PD 0235 — *Domestic*
PD 1714 — drug use meat production ; Health
PC 0111 — due economic development ; Environmental
PC 0328 Hazards economic industrial products ; Environmental
PE 1412 Hazards electric energy ; Environmental
Hazards electrical
PE 8097 — equipment industries ; Environmental
PE 8026 — machinery, apparatus appliances ; Environmental
PE 9642 — power transmission lines ; Environmental
Hazards [electricity...]
PE 1412 — electricity ; Environmental
PE 7879 — electromagnetic fields generated electrical appliances ; Health
PE 6360 — electromagnetic pulses ; Environmental
PD 6693 — energy production ; Environmental
PC 5883 — Environmental
PD 1239 — Environmental cancer
PE 1304 — environmental electromagnetism
PC 0936 — environmental pollution ; Health
PE 9056 — essential oils ; Environmental
PE 9177 — explosives pyrotechnic products ; Environmental
Hazards exposure
PE 4717 — chemicals ; Long term
PF 1686 — cosmic radiation ; Health
PC 0268 — noise ; Health
PE 4057 — radiation ; Long term
PE 7560 Hazards extremely low frequency electromagnetic radiation ; Environmental
Hazards [farming...]
PE 3760 — farming ; Occupational
PE 6902 — female workers ; Occupational
PE 1514 — fertilizers ; Environmental
PD 0372 — fish, crustacea molluscs ; Environmental
PD 0743 — fishing industry ; Environmental
PD 8328 — Flying
Hazards food
PC 1411 — live animals ; Environmental
PC 1411 — live animals ; Health
PE 1280 — processing industries ; Environmental
PE 1280 — products ; Environmental
Hazards [forestry...]
PE 1264 — forestry logging ; Environmental
PE 9029 — fruit vegetables ; Environmental
PE 8165 — furniture fixtures manufacture ; Environmental
Hazards [genetic...]
PC 0776 — genetic engineering accidents
PC 6684 — Geological

Health safety

Hazards [glass...] cont'd
- PE 8473 — glass glass products manufacture ; Environmental
- PE 8274 — glass industry ; Health

Hazards [Health...]
- PB 4885 — Health
- PE 9642 — high voltage transmission lines ; Health
- PD 1714 — hormone use animal production ; Health
- PB 4885 — human health
- PC 4777 — human health natural environment
- PE 6292 — hydropower ; Environmental

Hazards [industrial...]
- PF 1789 — industrial research development ; Pollution
- PE 0546 — inedible crude non fuel materials ; Environmental
- PE 8397 — Iron steel basic industries environmental
- PD 0361 — irradiated food ; Health
- PE 5784 — island populations ; Disaster

Hazards [Leather...]
- PE 8447 — Leather leather manufactures environmental
- PD 0788 — live animals ; Environmental
- PE 4717 — long term, low level exposure toxic mixtures non toxic chemicals ; Health
- PE 4717 — low-level exposure chemicals
- PE 4057 — low-level exposure radiation

Hazards [machinery...]
- PE 1859 — machinery equipment industries ; Environmental
- PE 1852 — *Machinery equipment industries environmental*
- PE 6904 — male workers ; Occupational
- PE 8651 — manufacture plastic products ; Environmental
- PE 1344 — manufactured goods ; Environmental
- PD 0454 — manufacturing industries ; Environmental
- PJ 4748 — Mass catering
- PE 0133 — meat meat preparations ; Environmental
- PE 5355 — medical profession ; Occupational risks
- PE 8906 — metalliferous ores metal scrap ; Environmental
- PE 1401 — metals ; Environmental
- PF 1687 — Meteorites
- PF 1695 — Meteors
- PE 8815 — methane ; Environmental
- PE 6682 — mineral exploitation seabed resources ; Environmental
- PE 1346 — mineral fuels, lubricants related materials ; Environmental

Hazards mining
- PD 2596 — Environmental
- PE 8428 — industry ; Occupational
- PE 7419 — radioactive substances ; Health

Hazards [minority...]
- PD 2623 — minority cultures ; Underreporting
- PE 8275 — miscellaneous manufactured articles ; Environmental
- PE 1499 — modern insulating materials ; Health

Hazards [Natural...]
- PD 2596 — *Natural gas production environmental*
- PE 3868 — navigation
- PC 1617 — new species introduction ; Environmental
- PE 3035 — non alcoholic beverages ; Environmental
- PJ 2425 — non electrical machinery ; Environmental
- PE 8248 — Non ferrous metals basic industries environmental
- PE 7651 — non ionizing radiation ; Environmental
- PE 0890 — non metallic mineral products industries ; Environmental
- PD 4977 — nuclear power production ; Environmental
- PE 5698 — nuclear weapons industry ; Environmental

Hazards [Occupational...]
- PC 6716 — Occupational
- PE 8250 — oil nuts, oil kernels oil seeds ; Environmental
- PF 1083 — Over reaction against environmental

Hazards [painters...]
- PE 9746 — painters ; Occupational
- PE 1425 — paper printing industries ; Environmental
- PE 5146 — passive smoking ; Health
- PE 1409 — petroleum ; Environmental
- PC 1192 — *petroleum refineries ; Environmental*
- PD 5706 — plants
- PD 8566 — plastic materials ; Environmental
- PE 8095 — publishing printing industries ; Environmental

Hazards [radiation...]
- PE 0962 — radiation aircraft ; Health
- PD 8050 — radiation ; Health
- PE 5099 — Radio frequencies health
- PD 0791 — Road
- PE 8336 — road motor vehicles ; Environmental

Hazards [sea...]
- PE 1839 — *sea transportation ; Environmental*
- PE 3868 — Shipping
- PE 5684 — Skin irritants occupational
- PE 4995 — smoking women ; Health
- PF 9119 — society ; Insensitivity non immediate
- PD 1110 — soils ; Underreporting
- PE 3883 — solar radiation ; Environmental
- PE 8263 — steel industry ; Health
- PD 0122 — strong toxic substances
- PE 1894 — sugar honey preparations thereof ; Environmental
- PE 4854 — Suppression information concerning environmental
- PF 4854 — Suppression information concerning health

Hazards [tanning...]
- PE 8571 — tanning dyeing industries ; Environmental
- PE 1103 — textile clothing industries ; Environmental
- PE 9074 — textile fibres waste ; Environmental
- PE 0483 — tobacco tobacco manufactures ; Environmental
- PE 8966 — Tourist
- PE 7538 — tourists developing countries ; Health
- PE 0738 — transport equipment ; Environmental
- PD 2025 — trees ; Urban

Hazards Underreporting
- PD 0708 — groundwater pollution
- PD 6626 — social

Hazards Underreporting cont'd
- PD 4182 — weather
- PE 0828 — Hazards undressed hides, skins fur skins ; Environmental
- PC 0268 — *Hazards vibration ; Environmental*

Hazards [water...]
- PE 8692 — water development schemes ; Health
- PE 1103 — wearing apparel manufacture ; Environmental
- PB 6336 — *wild life ; Pollution*
- PD 3680 — wildlife ; Pesticide
- PB 6336 — wildlife ; Pollution
- PE 8091 — wood cork manufactures ; Environmental
- PE 0864 — woodworking industries ; Environmental
- PE 7528 — workers small industries ; Occupational
- PE 5340 — Wrecks derelicts
- PE 6868 — Hazards young people work
- PE 6868 — Hazards youth workers ; Occupational
- PE 1655 — Haze
- PF 5392 — Hazings
- PF 5392 — Hazings ; Ritual
- PE 9296 — *Head ; Abnormal smallness*
- PE 6280 — Head catarrh ; Malignant
- PE 7769 — *Head disease animals ; Big*
- PE 2254 — *Head flies*

Head [household...]
- PF 8995 — *household authority ; Interference*
- PE 5644 — hunting ; Animal
- PF 2666 — hunting tribal societies
- PE 7511 — Head injuries
- PD 2567 — Head noise
- PE 5567 — *Head syndrome ; Swollen*
- PE 5567 — *Head ; Thick*
- PE 1974 — Headache
- PD 9653 — Heads state ; Corrupt acquisition offshore assets
- PE 8261 — *Heal privileged ; Discriminatory use medical resources*
- PF 0761 — Healing ; Lack understanding spiritual
- PF 4871 — *Healing potential ; Disregard self*
- PD 2847 — Health ; Airborne substances harmful

Health [Below...]
- PA 0294 — Below
- PF 3211 — *benefits ; Unawareness*
- PE 6835 — benefits women ; Unequal

Health care
- PF 4820 — Compulsory
- PD 7821 — Dehumanization
- PD 4004 — delivery ; Insufficient physical infrastructure
- PE 8924 — disabled ; Denial right adequate
- PE 8420 — facilities ; Crimes committed hospitals
- PS 9219 — *Family based*
- PD 1038 — family planning ; Inadequate
- PD 9021 — Ignorance women concerning primary
- PF 4832 — Inaccessible
- PE 8857 — Inadequate maternal child
- PE 8553 — Inadequate primary
- PF 9683 — Inefficient use available
- PE 9242 — least developed countries ; Inadequate
- PD 8023 — Misunderstanding preventive
- PF 6061 — Neglect adolescent
- PD 4790 — planning ; Uncoordinated
- PE 8261 — Prohibitive cost
- PE 8261 — Rationing
- PE 9048 — sector ; Ineffective self regulation
- PE 3328 — services ; Irresponsible
- PD 4790 — services ; Remote
- PD 4790 — systems ; Undeveloped
- PF 6061 — teenagers ; Inaccessibility
- PE 8261 — underprivileged ; Inaccessibility
- PE 7877 — urban slums ; Inadequate
- PD 0366 — Health clinics ; Understaffed

Health conditions
- PD 9366 — animals ; Ill defined
- PC 9067 — Ill defined
- PF 8515 — Inadequate public
- PF 9401 — Health control
- PF 9401 — Health control ; Inadequate

Health [Decline...]
- PF 4586 — Decline public spending
- PD 4790 — *delivery ; Complex*
- PC 0663 — *Denial right*
- PC 0663 — *Denial right highest obtainable physical mental*
- PD 1018 — Dependence government income products dangerous
- PC 0716 — Deterioration physical
- PF 2971 — Disrupted mechanisms community

Health [education...]
- PD 8023 — education ; Inadequate
- PD 8023 — education ; Ineffective
- PE 5324 — effects radioactive fallout ; Uncertainty long term
- PD 4004 — equipment ; Inadequate
- PF 4586 — expenditure ; Decline government
- PF 6234 — expenditures ; Increasing public

Health facilities
- PC 4844 — *Discriminatory use*
- PF 9683 — Non use available
- PF 4832 — Unorganized

Health [Fragility...]
- PB 4885 — *Fragility maintaining basic*
- PD 9297 — fraud
- PF 4586 — *funds ; Limited*
- PA 0294 — Health ; General ill
- PA 0294 — Health ; General poor

Health hazard
- PB 4885 — [Health hazard]
- PE 1989 — Antimony
- PE 2209 — Beryllium
- PE 1657 — Carbon monoxide

Health hazard cont'd
- PE 6053 — Circumcision
- PE 2329 — Hydrogen sulphide
- PE 5650 — Lead
- PE 1364 — Manganese
- PE 6056 — Microwave radiation
- PE 2432 — Polychlorinated biphenyls
- PE 1145 — Vibrations
- PB 4885 — Health hazards

Health hazards air
- PE 9609 — pollution animals
- PE 8617 — pollution children
- PE 4744 — pollution plants

Health hazards [aluminium...]
- PE 4969 — aluminium
- PE 9294 — anti-personnel gases
- PE 3001 — asbestos
- PD 0243 — basic metal industries
- PD 1038 — birth control
- PE 5146 — children smokers
- PE 5083 — computer visual display units
- PD 1038 — contraceptives
- PD 1714 — drug use meat production
- PE 7879 — electromagnetic fields generated electrical appliances
- PC 0936 — environmental pollution
- PF 1686 — exposure cosmic radiation
- PC 0268 — exposure noise
- PC 1411 — food live animals
- PE 8274 — glass industry
- PE 9642 — high-voltage transmission lines
- PD 1714 — hormone use animal production
- PB 4885 — Health ; Hazards human

Health hazards
- PD 0361 — irradiated food
- PE 4717 — long-term, low-level exposure toxic mixtures non-toxic chemicals
- PE 7419 — mining radioactive substances
- PE 1499 — modern insulating materials
- PE 5146 — passive smoking
- PD 8050 — radiation
- PE 0962 — radiation aircraft
- PE 5099 — Radio frequencies
- PE 4995 — smoking women
- PE 8263 — steel industry
- PF 4854 — Suppression information concerning
- PE 7538 — tourists developing countries
- PE 8692 — water development schemes
- PD 8023 — Health hygiene ; Ignorance
- PC 1299 — Health ; Ill effects radon
- PC 0699 — Health impairment handicap

Health Inadequate
- PF 4955 — *care mental*
- PF 1773 — maintenance physical
- PF 9587 — resources
- PE 4459 — Health indigenous populations ; Denial right
- PC 4844 — Health inequalities
- PD 8023 — Health knowledge ; Limited
- PA 0294 — Health ; Lowered state
- PE 7718 — Health medical services ; Inadequate working conditions

Health [natural...]
- PC 4777 — natural environment ; Hazards human
- PD 4790 — *needs ; Unconsensed*
- PF 3850 — Negative effects over crowding mental
- PD 8607 — Neglect personal

Health [Obsession...]
- PF 6389 — Obsession personal
- PC 0865 — Occupational risk
- PF 1608 — Outdated forms community
- PD 1229 — *Health payments ; Insufficient*

Health personnel
- PD 0366 — Insufficient
- PD 0366 — Limited
- PF 4126 — Maldistribution
- PA 0294 — Health ; Physical mental ill

Health practices
- PD 8607 — Neglected
- PF 3838 — Religious opposition public
- PD 8831 — Restrictive

Health [problems...]
- PE 7111 — problems related human hair ; Cosmetic
- PD 8023 — *problems ; Unawareness*
- PE 7984 — programmes ; Unfamiliar procedures using public

Health [regulations...]
- PF 1555 — regulations ; Excessive
- PD 7669 — resources ; Inaccessible
- PD 7669 — resources ; Limited availability

Health risks
- PE 1412 — electricity
- PD 0414 — fuel exhaust
- PD 4519 — Inappropriate legal definition
- PD 2784 — *international transfer corpses*
- PC 0865 — Occupational
- PE 6969 — teenage sex

Health risks workers
- PE 0524 — agricultural livestock production
- PE 0688 — commerce
- PE 0526 — construction industry
- PE 1159 — electricity, gas, water sanitary services
- PE 1605 — manufacturing industries
- PE 0875 — service industries
- PE 1581 — transport, storage communication industries
- PD 2011 — Health rural communities ; Debilitating conditions

Health [safety...]
- PE 8305 — safety ; Inadequate teaching occupational

Health safety

Health [safety...] cont'd
- PE 4006 — safety regulations ; Violations
- PE 1274 — sanitary regulations agricultural pharmaceutical products ; Distortion international trade discriminatory formulation
- PE 5721 — service ; Fragmentation

Health services
- PF 4955 — Constraints development mental
- PD 4790 — Deficient
- PD 4790 — Distant
- PD 6265 — following nuclear war ; Inadequate
- PD 4790 — Inadequate
- PF 4955 — Inadequate mental
- PF 6577 — Limited access
- PD 4790 — Uncoordinated
- PD 4790 — Undeveloped
- PE 3328 — Unethical practices
- PD 4790 — Uninitiated
- PD 4790 — Unused
- PD 4790 — Unutilized

Health [small...]
- PD 2011 — small rural villages ; Unsanitary environment basic
- PE 5239 — standards ; Failure conform international
- PD 4519 — standards ; Politicization
- PF 4586 — subsidies ; Limited public
- PD 0366 — surveillance ; Insufficient
- PD 1043 — Susceptibility old physical ill
- PE 0952 — system referrals ; Unnecessary
- PD 0366 Health technicians ; Insufficient
- PF 5679 Health tests ; Unnecessary
- PC 4844 Health ; Unequal development promotion
- PF 5147 Health women ; Neglect sexual
- PC 4844 Healthiness ; Disparity human
- PF 1212 Healthy foods ; Prohibitive cost
- PD 6306 Hearing defects
- PE 1743 Hearings ; Denial right access interpreters judicial
- PD 5453 *Heart attack*
- PF 4022 Heart ; Broken

Heart [Damage...]
- PD 0448 — Damage
- PA 5568 — degeneration
- PE 1684 — Degenerative changes
- PD 0448 — dilatation

Heart disease
- PE 0471 — animals
- PE 8158 — *Arteriosclerotic*
- PE 2365 — *Congenital*
- PD 0448 — *Degenerative*
- PE 8158 — *Ischaemic*
- PE 9774 — *Mulberry*
- PD 5453 — poultry ; *Round*
- PE 9415 — *Pulmonary*
- PD 9366 — *Pulmonary hypertensive*
- PE 0920 — *Rheumatic*
- PE 4208 — *Symptomatic*
- PD 0448 Heart diseases

Heart [failure...]
- PE 3958 — failure
- PE 3958 — failure ; Congestive
- PD 0448 — *Functional disorders*
- PD 0448 *Heart ; Hypertrophy*
- PD 0448 *Heart ; Inflammatory affections*
- PE 8219 Heart rhythm ; Irregular
- PE 6269 Heart rot fungi
- PE 0920 *Heart ; Valvular diseases*
- PF 2418 *Heart ; Weakened*
- PA 6030 Heartlessness *Fear*
- PA 7023 Heartlessness *Pitilessness*
- PA 6731 Heartlessness *Solemnity*
- PA 7364 Heartlessness *Unfeelingness*
- PA 5643 Heartlessness *Unkindness*
- PA 5644 Heartlessness *Vice*
- PD 4392 Hearts ; Lonely
- PE 2680 Heartwater disease
- PE 9427 *Heartworm infections ; Canine*
- PC 2460 Heat ; Climatic
- PD 9366 *Heat cramps*
- PE 2398 Heat disorders
- PD 7977 Heat ; Excessive environmental
- PD 9366 *Heat exhaustion*
- PE 2398 Heat fatigue
- PE 5720 Heat occupational hazard
- PD 1584 Heat pollution
- PE 2398 Heat ; Prickly
- PG 9635 *Heat rash*
- PE 2398 Heat stress
- PE 5720 Heat stress work
- PC 2460 Heat wave
- PD 3669 Heated effluent waters
- PD 5173 Heated homes ; Poorly
- PD 5173 Heated shelters ; Inadequately
- PF 0296 Heaters ; Costly hotwater
- PF 8068 Heathenism
- PE 2824 Heating emissions air pollutants ; Industrial domestic
- PE 8272 Heating, lighting fixtures fittings ; Instability trade sanitary plumbing,
- PE 7940 Heating lighting fixtures fittings ; Shortage sanitary plumbing,
- PE 2398 Heatstroke
- PD 9366 *Heatstroke*
- PC 3354 Heaven ; Predestination
- PE 5182 *Heaves*
- PC 3056 Heavily indebted countries
- PF 9249 Heavily indebted countries ; Disincentive invest
- PE 6520 *Heavy equipment ; Insufficient*
- PE 6520 *Heavy equipment ; Lack*

- PF 6530 *Heavy farm debts*
- PE 4452 Heavy loads ; Work practices requiring women lift
- PD 0807 Heavy trash ; Unremoved
- PD 0438 Hebephrenic schizophrenia
- PE 0566 Hectic fever
- PD 3479 *Hedgehogs ; Endangered species*
- PD 1642 Hedgerow trees ; Destruction hedges
- PD 1642 Hedges hedgerow trees ; Destruction
- PF 2277 Hedonism
- PF 2277 Hedonism ; *Moral*
- PA 6379 Heedlessness *Avoidance*
- PA 6651 Heedlessness *Forgetfulness*
- PA 6247 Heedlessness *Inattention*
- PA 6598 Heedlessness *Incuriosity*
- PA 5438 Heedlessness *Neglect*
- PA 5643 Heedlessness *Unkindness*
- PE 5617 Heel abscess
- PE 4161 Heel ; *Elso*
- PE 7826 Heel ; *Greasy*
- PE 5606 Heel ; Policeman's
- PE 7826 Heels ; *Contracted*
- PE 7826 Heels ; *Sheared*
- PB 0318 Hegemony ; Dependence military economic
- PA 0839 Hegemony *Domination*
- PB 0318 Hegemony ; Military economic
- PF 8946 Hegemony over United Nations agencies ; National
- PE 9402 Height ; Discrimination against people abnormal
- PD 5177 Height ; Inadequate human physical
- PF 7228 Hell
- PE 1323 Hell-bind
- PE 6278 Helminthiasis
- PE 6461 *Helminthiasis poultry ; Digestive tract*
- PE 5380 Helminths skin
- PA 6876 Helplessness *Impotence*
- PA 6466 Helplessness *Intemperance*
- PE 6990 Helplessness ; Learned
- PE 0763 Hemiplegia
- PE 5182 *Hemiplegia ; Laryngeal*
- PE 3615 Hemiptera insect pests
- PC 2270 Hemochromatosis
- PE 1849 *Hepatic disorders*
- PE 0517 Hepatitis

Hepatitis [A...]
- PE 0517 — A ; Viral
- PE 0517 — *Amoebic*
- PD 2731 — *Avian infectious*
- PD 2731 — *Avian vibrionic*
- PE 0517 — Hepatitis B ; Viral
- PE 0517 — Hepatitis C
- PE 0517 — Hepatitis ; Canine viral
- PE 0517 — *Hepatitis disorders ; Post*
- PD 2730 — *Hepatitis ducks ; Viral*
- PD 2730 — *Hepatitis geese ; Viral*
- PE 0517 — *Hepatitis hyperbilirubinaemia ; Post*
- PD 2730 — *Hepatitis ; Inclusion body*

Hepatitis Infectious
- PE 0517 — [Hepatitis ; Infectious]
- PD 2730 — canine
- PE 7769 — necrotic
- PE 0517 — Hepatitis ; Serum
- PE 0517 — *Hepatitis syndrome ; Post*
- PD 2730 — *Hepatitis turkeys ; Viral*
- PC 2270 Hepatolenticular degeneration
- PE 3677 *Hepatonephritis*
- PE 9774 *Hepatosis dietetica*
- PE 3946 Herbal remedies ; Ignorance traditional
- PD 1224 Herbicide damage crops
- PD 5228 Herbicide poisoning animals
- PD 1143 Herbicides
- PD 1224 Herbicides ; Inconsiderate use
- PD 1224 Herbicides ; Misuse
- PD 1143 Herbicides pollutants
- PE 3394 Herd animals ; Cruelty
- PE 7915 Hereditary ataxia
- PE 5341 Hereditary class ; Discredited moneyed

Hereditary [defects...]
- PD 2389 — defects diseases
- PD 2389 — *diseases*
- PE 7915 — diseases striato-pallidal system
- PE 7915 — disorders central nervous system
- PF 7758 — *Hereditary haemolytic anaemias*
- PD 2389 — *Hereditary metabolic diseases*
- PE 7915 — Hereditary neuromuscular disorders
- PD 2389 — *Hereditary opticatrophy*
- PD 8149 Hereditary regression
- PF 3375 Heresy
- PA 5982 Heresy *Disagreement*
- PA 6180 Heresy *Error*
- PA 7392 Heresy *Unbelief*
- PC 7605 Heritage ; Abusive exploitation cultural
- PD 4272 Heritage ; Acculturation dilution cultural
- PF 1985 Heritage ; Commonly disvalued
- PC 2114 Heritage ; Destruction cultural

Heritage [Failure...]
- PF 1985 — Failure pass cultural
- PF 1985 — Fragmented celebration cultural
- PF 1985 — Fragmented recognition common
- PF 1985 — Heritage gifts ; Undisplayed
- PF 1985 — Heritage ; Ignorance cultural

Heritage [Lack...]
- PF 1985 — Lack access cultural
- PF 1985 — Lack awareness past
- PF 1985 — Loss cultural
- PE 6074 — Lost cultural
- PF 1099 — Lost family
- PF 1985 Heritage ; Remote cultural

- PF 1985 Heritage transmission ; Insufficient cultural
- PF 1985 Heritage ; Unconveyed cultural
- PF 1985 Heritage ; Unknown cultural
- PF 1985 Heritage vitality ; Lost
- PE 2246 Hermaphroditism ; Human pseudo
- PE 8961 Hernia
- PD 7424 *Hernia animals ; Diaphragmatic*
- PD 3978 *Hernia animals ; Perineal*
- PE 8961 Hernia ; *Diaphragmatic*
- PE 8961 Hernia ; *Femoral*
- PE 8961 Hernia ; *Incisional*
- PE 8961 Hernia ; *Inguinal*
- PD 0637 *Hernia lung*
- PE 8961 Hernia ; *Oesophageal hiatus*
- PE 8961 Hernia ; *Umbilical*
- PE 3711 *Hernias animals ; Abdominal*
- PF 2834 Hero images ; Inadequate
- PF 2834 Hero images ; Shortage
- PF 1078 Hero ; Subsistence farmer
- PF 2650 Hero worship
- PF 8623 Heroes ; Insufficient cultural
- PF 8623 Heroes ; Uncelebrated
- PF 8623 Heroes ; Unsymbolized village
- PE 1776 Heroin ; Abuse
- PE 8615 Herpes
- PE 5182 *Herpes virus 1 infection ; Equine*
- PD 2730 *Herpesviral infection ; Canine*
- PE 9774 *Herztod ; Enzootische*
- PA 7325 Hesitation *Irresolution*
- PA 7155 Hesitation *Rashness*
- PA 7309 Hesitation *Uncertainty*
- PA 7006 Hesitation *Untimeliness*
- PE 0818 Heterosexism
- PA 3369 Heterosexual pairing ; Moral offences
- PD 3978 *Hexamitiasis*
- PE 8961 Hiatus hernia ; *Oesophageal*
- PE 9604 Hiccough
- PE 9604 Hiccup
- PC 0328 Hidden environmental costs economic production
- PA 0005 Hidden individual talents
- PA 6030 Hideousness *Fear*
- PA 7240 Hideousness *Ugliness*
- PA 7107 Hideousness *Unpleasantness*
- PE 8106 Hides non fur skins ; Instability trade undressed

Hides [skins...]
- PE 0828 — skins fur skins ; Environmental hazards undressed
- PE 1235 — skins fur skins ; Instability trade undressed
- PE 8327 — Skins fur skins shortage
- PD 7461 Hierarchical control market facilities
- PF 5115 Hierarchical structures established religions
- PD 1975 *Hierarchy ; Acceptance*
- PF 4947 *Hierarchy ; Persistence*
- PF 4947 *Hierarchy ; Social*
- PD 2322 High altitude
- PD 2322 High altitude stress
- PE 0472 High blood pressure
- PF 6572 High business risk
- PF 1238 High consumer prices

High cost
- PF 4975 — civil engineering projects ; Misappropriation resources
- PF 2779 — controlling epizootic enzootic diseases
- PD 1842 — housing
- PE 4162 — land
- PF 1238 — living
- PE 9137 — natural gas trade infrastructure
- PF 0716 — research projects ; Misappropriation resources
- PC 2035 — youth unemployment

High [costs...]
- PE 6388 — costs staff intergovernmental organizations
- PF 2166 — credit risks
- PE 4294 — crime rate ; Periods
- PE 7311 — crime rates ; Locales
- PF 0906 — crude birth rate
- PF 6156 — High density accommodation ; Impersonality

High [electricity...]
- PF 3318 — electricity costs
- PD 2322 — elevation conditions
- PC 7517 — energy development
- PE 4261 High fat diets
- PF 2859 High fertility ; Unwanted
- PD 8888 High groundwater levels
- PD 1842 High housing costs
- PF 0906 High human fertility developing countries
- PF 9014 High interest rates

High [labour...]
- PF 8763 — labour costs
- PF 0907 — labour turnover developing countries
- PD 4002 — level consumption ; Natural resource depletion due
- PF 8736 — life assurance premiums
- PF 2385 — local prices

High [minimum...]
- PD 5674 — minimum wages
- PF 7043 — mortality rate
- PD 9366 — *mountain disease cattle*
- PC 0268 High noise levels
- PB 0469 High population density ; Excessively
- PF 1267 *High-priced cooking fuel*
- PF 3409 *High quality seeds ; Scarcity*
- PE 1925 High rise buildings ; Monolithic architecture

High risk
- PJ 0451 — development initiatives ; Reluctance government invest
- PD 5349 — Endangerment awareness
- PF 4861 — persons ; Non surveillance medical
- PE 1815 — policy smaller banks

High risk cont'd
PF 3514 — technologies
PF 6572 High risks ; Investment blocked
PF 8763 High salaries ; Disproportionately
High school
PF 6575 — *closure ; Imposed*
PF 3519 — Overacademic orientation
PF 9068 — *priority ; Low*
PC 2163 — *transport ; Nonexistent*
High [schools...]
PE 8797 — *schools countryside ; Remote junior*
PE 3872 — *severance pay top managers*
PD 6124 — *speed roads ; Environmental degradation*
PF 6467 — *student-teacher ratio*
PE 8221 High teacher turnover
High technology
PE 7174 — *hostile countries ; Diversion*
PE 8458 — *industries ; Protectionism*
PE 7174 — *irresponsible groups ; Diversion*
PE 8261 — *medical cures ; Inaccessibility*
PF 1653 *High telephone costs*
PF 4375 High training cost
High transportation
PE 8063 — cost
PE 8063 — fares
PE 8063 — overhead
PD 2615 High treason
PE 4294 High unemployment ; Crimes committed during
PB 0750 High unemployment rate
PD 1350 High vandalism rates
PE 9642 High voltage transmission lines ; Health hazards
PF 3146 High yield crop varieties ; Introduction
PE 2890 High yield grain ; Adverse effects
Higher education
PF 6458 — *Denial right access*
PF 4375 — Prohibitive cost
PF 6458 — *Restricted*
PD 5317 Higher tribunal ; Violation right review conviction
PC 0663 *Highest obtainable physical mental health ; Denial right*
PE 1557 Highly indebted corporations
PF 8921 Highly polluting industries developing countries ; Transfer
PE 1396 Highly skilled personnel routine maintenance complex systems ; Wastage
PD 1162 Highschool graduates ; Unemployment
PF 6559 *Highway danger ; Unacknowledged*
PF 6559 *Highway spending ; State*
Highway [traffic...]
PD 2106 — *traffic congestion ; Road*
PD 0543 — *transport facilities developing countries ; Inadequate road*
PD 0490 — *transport facilities ; Inadequate road*
PD 0124 Hijacking
PA 0429 Himsa *Human violence*
PB 1935 Himsa ; Institutionalized covert
PF 1978 Himsa ; Personal covert
PF 0963 *Hindered winter sociability*
PD 5515 Hindering apprehension felon
PD 7359 Hindering proceedings disorderly conduct
PD 5515 Hindering prosecution felon
PE 8858 Hindrance economy ; Transport
PD 5515 Hindrance law enforcement
PF 1568 Hindrance ; Spiritual emotional
PE 8758 Hindrances international spread new technologies
PE 8593 Hinnies ; Instability trade asses, mules
PE 5913 *Hip dysplasia ; Canine*
PE 7826 *Hip ; Osteoarthritis*
PF 3168 Hiring policies ; Rigid personnel
PF 2256 *Hiring practices ; Traditional*
PE 5478 Hirsutism animals
PD 2735 Histomoniasis poultry
PE 2455 Histoplasmosis
PD 0253 Historic buildings ; Destruction
PD 0172 Historic documents public archives ; Destruction
PD 0253 Historic sites ; Endangered monuments
PF 2439 *Historical business decline*
Historical [celebrations...]
PF 1309 — celebrations ; Dormant
PD 1591 — context ; Lack individual
PF 4932 — cover-up
Historical documents
PF 9699 — Inaccessible
PE 5051 — *objects ; False*
PF 6483 — Uncatalogued
PE 1046 Historical effect witch hunting ; Long term
PF 4932 Historical events ; Distortion
Historical [figures...]
PF 1540 — figures ; Unrehabilitated
PE 5051 — forgery
PE 5051 — *forgery ; Religious*
PF 4932 Historical information ; Suppression
PD 9046 Historical libraries ; Denial right access
PF 9046 Historical libraries ; Inaccessibility
PD 3774 Historical method ; Limited
PF 4932 Historical misrepresentation
PF 5662 Historical persons ; Misrepresentation
PF 6765 Historical perspective leaders ; Passive
Historical [record...]
PF 5728 — *record ; Lack*
PF 4932 — records ; Falsification government
PF 4932 — revisionism
PF 1309 — *rites customs ; Underutilization*
PD 0253 *Historical sites dams ; Submergence*
PF 0608 Historical situations ; Silence
History [Denial...]
PF 1746 — Denial lessons

History [Disrelationship...] cont'd
PF 6494 — *Disrelationship community*
PF 4932 — Distortion
PF 1746 History ; Failure profit patterns
PD 3774 History ; Ignorance
PF 6494 *History ; Obscured community*
PF 4932 History ; Rewriting
PD 2082 History textbooks ; Biased inaccurate
PF 4299 *History ; Untransmitted local*
PE 4561 Histrionic personality
PE 5448 Hitchhiking
PD 5111 HIV - human immunodeficiency virus
PE 4299 HIV infected persons ; Discrimination against
PD 9667 Hives
PE 1017 Hives
PD 3045 Hoarded monetary gold
PD 1751 Hoarding developing countries
PE 6333 Hoarding employers ; Labour
PD 0651 *Hoarding ; Food*
PE 6333 Hoarding ; Labour
PD 3045 Hoarding monetary gold
PD 0651 Hoarding primary commodities
PE 2156 Hoardings billboards ; Unaesthetic location advertising
PF 9375 Hoaxes
PF 4298 *Hoaxes ; Bomb warning*
PF 8889 Hoaxes ; Religious
PF 1602 Hoaxes ; Scientific
PE 5460 Hoboes
PE 6555 *Hock ; Capped*
PE 5548 Hodgkin's disease
PE 6266 Hog cholera
PE 2338 Holders ; Absence land
PE 8675 Holders ; Short term share
PF 1705 *Holding ; Individualistic land*
PE 1848 *Holiday residency patterns*
PE 8966 Holiday risks
PE 4647 Holidays ; Inefficiency due mismatch religious national
PE 4290 Holidays ; Tax
PF 5745 Holism ; Anti
PE 5721 Holistic medicine ; Lack
PF 7826 *Hollow wall*
PE 6255 Hollyhock rust
PC 1056 Holocaust
PD 3158 *Holothuroidea ; Endangered species*
Holy [day...]
PF 3607 — day observance ; Absence
PE 7735 — day observance ; Discriminatory effect
PF 3607 — days ; Desecration
PE 7735 — days ; Disruptive secular impact
Holy places
PF 6385 — [Holy places]
PF 8666 — Desanctification churches
PF 1816 — focus religious friction
PF 6385 Holy spaces ; Desecration
PF 6327 Holy Spirit ; Sin against
PF 5681 Holy war
PF 8118 Holy war ; Sacrifice children
PE 4961 Home ; Accidents
PD 2011 *Home animal shelter ; In*
PC 9042 *Home births ; Unsupervised*
PD 8546 Home bound educated women ; Failure employ skills
Home [care...]
PD 0512 — care services elderly ; Inadequate provision
PE 6504 — *centered priorities*
PC 0193 — *Children banned*
PE 4335 — countries transnational banking activities developing countries ; Restrictions
PE 7311 — Crimes
PE 6240 Home ; Disadvantages workers
PF 1548 *Home environment ; Nonintellectual*
PE 5199 Home environment ; Underprivileged
PF 1442 *Home industries ; Unprofitable*
PE 0242 Home insulation ; Inadequate
Home [life...]
PE 6168 — life ; Denial right private
PD 4368 — *life ; Unstructured*
PE 4571 — loan payments ; Arrears
Home [made...]
PE 2368 — *made bombs*
PE 7188 — made liquor
PF 0296 — *maintenance ; Prohibitive cost*
PD 8894 — maintenance ; Unsafe
PF 0296 *Home ovens ; Expensive*
PD 2338 Home ownership ; Non resident
PF 0296 *Home renovation ; Costly*
Home [Safety...]
PE 4961 — Safety
PE 8681 — Social isolation women
PS 3268 — *structures ; Unsupportive*
PE 4058 Home ; Unlawful interference family
PE 6122 Home workplace ; Artificial separation
PD 8285 Homeless developed countries
PB 2150 Homeless people
PB 2150 Homelessness
PF 9828 Homelessness ; Concealment information concerning extent
PE 5460 Homelessness ; Criminalization
Homelessness [Dependence...]
PB 2150 — Dependence
PD 8856 — developing countries
PA 7158 — *Dislocation*
PA 7339 — *Duality*
PD 8285 Homelessness industrialized countries
PA 6434 Homelessness *Poverty*
PA 6653 Homelessness *Unsociability*
PF 8352 Homeopathy

History [Disrelationship...] cont'd
PF 6578 *Homes ; Abandoned private*
PD 2103 Homes ; Broken
PE 9337 Homes government authorities ; Demolition
PF 7130 *Homes ; Insufficient storage space*
PD 0276 Homes ; Old age pensioners'
Homes [Poorly...]
PD 9736 — Poorly cooled
PD 5173 — Poorly heated
PF 1286 — Proliferation second
PD 0512 Homes ; Unserviced older
PJ 5282 Homesickness
PE 6240 Homeworking employees ; Disadvantages
PG 6455 *Homicidal crimes state*
PD 2341 Homicide
PF 1098 Homicide ; Absolving military personnel
PE 7533 Homicide ; Justifiable
PE 0437 Homicide neglect
PF 6155 Homogeneity modern cities ; Stultifying
PB 1071 Homogenization cultures
PE 4299 Homophobia ; Officially endorsed
PE 3614 Homoptera insect pests
PF 1390 *Homosexual clergy*
PF 1390 *Homosexual cruising*
PF 2148 *Homosexual incest*
PE 9339 Homosexual officials
PD 4402 Homosexual prostitution ; Male
PE 6137 Homosexual sadomasochism
PD 9398 Homosexual scandals
PF 3242 Homosexuality
Homosexuality [Clandestine...]
PF 3242 — Clandestine
PF 3242 — Closet
PE 1903 — Criminalization
PE 9339 Homosexuality ; Establishment
PE 2640 Homosexuality ; Female
PE 9339 Homosexuality judiciary
PF 1390 Homosexuality ; Male
PF 1390 *Homosexuality priesthood*
PE 1363 Homosexuality prisons
Homosexuals Denial rights
PE 1903 — [Homosexuals ; Denial rights]
PE 5741 — female
PE 3882 — male
PE 1903 Homosexuals ; Discrimination against
PE 3882 Homosexuals ; Discrimination against male
Homosexuals Violation rights
PE 1903 — [Homosexuals ; Violation rights]
PE 5741 — female
PE 3882 — male
PE 0793 Honey bees ; Aggressive
PE 0383 Honey ; Instability trade sugar, sugar preparations
Honey [possum...]
PD 1762 — *possum ; Endangered species*
PE 1894 — preparations thereof ; Environmental hazards sugar
PE 1120 — preparations thereof ; Long term shortage sugar
PE 5524 *Honker syndrome feeder cattle*
PF 5382 Honour ; Affair
PF 8485 Honour ; Loss
PD 2569 Honour reputation ; Denial right freedom attacks personal
PF 7198 Honours ; Accumulation
PF 6537 *Honours ; Infrequent achievement*
PF 3439 *Honours ; Unequal distribution fame*
PE 1589 Hoof-and-mouth disease
PE 4161 *Hoof wall fissures ; Horizontal*
PE 4161 *Hoof wall fissures ; Vertical*
PE 3508 Hookworm
PE 3508 Hookworm disease
PD 1109 Hooliganism
PE 2481 Hooping cough
PE 5050 *Hooves ; Overgrown*
PA 6099 Hopelessness
PF 4004 Hopelessness
PA 6714 Hopelessness *Chance*
PA 6099 Hopelessness *Hopelessness*
PA 6487 *Hopelessness Impossibility*
PF 4984 *Hopelessness inspired school*
PA 6731 Hopelessness *Solemnity*
PA 7364 Hopelessness *Unfeelingness*
PD 8786 Hordeolum
PF 3161 *Horizons produced survival living ; Limited*
PD 2892 *Horizontal commodity diversification developing countries ; Lack*
PE 4161 *Horizontal hoof wall fissures*
PG 9483 *Horizontal overbite*
PE 4431 Hormonal disorder
PD 1714 Hormonal growth promoters
PE 4431 Hormone disturbance
PD 1714 Hormone use animal production ; Health hazards
PE 2254 *Horn flies*
PF 0565 *Horological superstition*
PF 0565 Horoscopes ; Dependence
PA 5454 Horribleness *Badness*
PA 6030 Horribleness *Fear*
PA 7240 Horribleness *Ugliness*
PA 7107 Horribleness *Unpleasantness*
PA 6030 Horror *Fear*
PE 5182 *Horse ; Epistaxis*
Horse [fever...]
PD 3978 — *fever ; Potomac*
PE 3878 — flies
PE 7826 — *Fractures back*
PE 1805 Horse sickness ; African
Horses [Chronic...]
PD 3978 — *Chronic diarrhoea*
PD 3978 — *Colic*

Horses [Colitis...] cont'd
- PD 3978 — *Colitis X*
- PE 9774 — *Cording up syndrome*
- PE 6708 Horses ; Enteric diseases
- PE 7826 Horses ; Keratoma

Horses [Lameness...]
- PE 7826 — Lameness
- PE 0958 — *Large strongyle infections*
- PE 9844 — *Lymphangitis*
- PE 3394 Horses ; Maltreatment
- PE 2727 Horses ; Mange
- PE 5182 Horses ; Respiratory diseases
- PE 0958 Horses ; *Small strongyle infection*
- PD 9667 Horses ; *Sweet itch*
- PE 9774 Horses ; *Tying up syndrome*
- PE 4656 Horses ; *Urolithiasis*
- PE 3604 Horseshoe bats ; *Endangered species*
- PD 6520 Hoses ; *Short fire*
- PD 6334 Hospital-acquired infections
- PE 5058 Hospital buildings ; Substandard

Hospital facilities
- PE 5058 — Inadequate
- PE 5058 — Inoperable
- PE 5058 — Lack
- PE 4154 — Prohibitive cost
- PD 6334 Hospital hygiene ; Inadequate
- PD 6334 Hospital ; Infections acquired
- PD 4790 *Hospital transportation ; Irregular*
- PD 5119 Hospital waiting lists
- PE 4725 Hospital waste
- PC 0981 Hospitalism
- PD 0584 Hospitals ; Abusive treatment patients psychiatric
- PE 8420 Hospitals health care facilities ; Crimes committed
- PF 4925 Hospitals ; Inadequate mental
- PD 0584 Hospitals ; Physical brutality psychiatric
- PE 2932 Hospitals ; Unjust commitment psychiatric
- PE 5058 Hospitals ; Unsafe

Hostage taking
- PE 4108 — [Hostage taking]
- PD 1886 — Political
- PD 1886 — State sanctioned
- PE 4108 Hostages ; Child

Hostile [climate...]
- PD 7941 — climate modification
- PE 7174 — countries ; Diversion high technology
- PE 7174 — countries ; Trade strategic goods
- PD 7941 Hostile environmental modification
- PC 1617 Hostile introduction species
- PF 2199 Hostile policies ; Government action against regimes
- PF 4432 Hostile takeover bids abroad ; Blocks
- PD 7941 Hostile weather modification
- PE 2324 Hostilities following cease fire agreements ; Persistence
- PB 8538 Hostility
- PA 6698 Hostility *Difference*
- PA 5532 Hostility *Disaccord*
- PA 5446 Hostility *Enmity*
- PA 6979 Hostility *Opposition*

Hostility [Unfeelingness...]
- PA 7364 — *Unfeelingness*
- PA 7107 — *Unpleasantness*
- PF 7890 — Unreported
- PD 6651 Hosts ; Arthropods intermediate
- PE 3511 Hosts ; Vertebrates intermediate
- PD 2474 Hot humid climates
- PC 1453 Hot money
- PE 4493 Hotel catering industries ; Inadequate conditions work
- PF 6196 Hotel environments travellers ; Unconvivial
- PE 1542 Hotel restaurant waste
- PF 0296 *Hotwater heaters ; Costly*
- PF 6473 *Hour transportation ; Crowded rush*

Hours Denial
- PE 3044 — right periods nurse infants during working
- PD 0140 — right reasonable work
- PE 6427 — working animals limitation working
- PE 7185 Hours inappropriate structural technological changes ; Working
- PF 2971 Hours ; *Irregular working*
- PD 0140 Hours work ; Excessive

House [bound...]
- PE 2274 — *bound mothers*
- PF 3211 — *bound women*
- PD 2561 — *breaking*
- PD 1251 House codes ; Violation
- PD 8638 House design ; Culturally insensitive
- PE 4675 House fires ; Lethal fumes modern
- PE 3623 House mosquitoes ; Common
- PF 0551 House training ; Lack
- PD 1842 House values ; Inflated
- PD 2771 Housed farm animals ; Agricultural mismanagement
- PD 2760 Housed farm animals ; Overcrowding
- PE 3609 Houseflies pests
- PE 3583 Housefly resistance insecticides
- PF 0995 *Household authority ; Interference head*
- PC 0092 Household cancers
- PB 2846 Household food insecurity
- PE 3745 Household gimmicks ; Technological
- PE 5928 Household moving ; Disruptive effect
- PD 3522 Household pests
- PF 1238 *Household rent increases ; Uncontrolled*
- PF 6136 Household segregation age group
- PE 4789 Household work ; Dependence unpaid
- PE 4789 Household work ; Unpaid
- PF 6836 *Households ; Chaotic*
- PD 2103 *Households ; Increasing number single person*
- PE 6167 Households ; Inhibition adult life small

- PE 2320 Housemaid's knee
- PF 1213 *Houses ; Deteriorated vacant*
- PE 9767 Houses ; Inadequate hygiene slaughter
- PE 8103 Houses ; Unconstructed teachers'
- PF 1773 *Houses ; Unhygienic dirt floors*
- PE 4789 Housewives' labour ; Uncompensated
- PE 8681 Housewives ; Social isolation
- PE 4789 Housework ; Lack payment
- PF 0023 Housework national accounts ; Non valuation

Housing [accommodation...]
- PD 0758 — accommodation ; Overcrowding
- PD 1251 — accommodation ; Substandard
- PD 0276 — aged ; Inadequate
- PE 8061 — alternatives urban areas ; Unfeasible
- PD 0820 — Appropriation unoccupied
- PF 4571 — Arrears payments
- PF 2241 — *assignments ; Alienating public*
- PD 0276 Housing ; Blocked elderly
- PF 8829 Housing codes ; Absence

Housing construction
- PF 1213 — Depressing effect poor
- PD 1251 — *Inadequate*
- PE 8265 — industry ; Ineffective self regulation
- PD 1842 Housing costs ; High

Housing [Denial...]
- PD 5254 — Denial right adequate
- PC 3320 — Denial right adequate housing indigenous peoples ; Discrimination against indigenous populations
- PC 0449 — Dependence inadequate
- PF 0867 — design ; Unimaginative
- PE 2592 — destruction war

Housing developing countries
- PE 4386 — Improvisational
- PE 0269 — Inadequate
- PF 6511 — Inadequate rural

Housing [disabled...]
- PF 4681 — disabled ; Inadequate
- PD 3469 — Discrimination
- PD 1251 — *Disease prone*
- PD 1842 — Disparity income cost

Housing [equipment...]
- PF 0296 — *equipment ; Costly*
- PE 2757 — equipment ; Inadequate disinfection measures animal
- PE 1925 — estates ; Socially inadequate
- PD 3465 — Exploitation

Housing [facilities...]
- PF 6195 — facilities ; Monotonous
- PD 5422 — fraud
- PF 3550 — *funds ; Insufficient*
- PD 1842 Housing ; High cost

Housing [Inadequate...]
- PC 0449 — Inadequate
- PC 3320 — indigenous peoples ; Discrimination against indigenous populations housing Denial right adequate
- PC 3320 — indigenous peoples ; Inadequate
- PE 0269 — infrastructural weakness developing countries
- PF 6511 — Insufficient rural

Housing [Limited...]
- PD 1842 — Limited low cost
- PF 3550 — *loans ; Inaccessible*
- PF 7118 — *locations ; Scattered*
- PD 1251 — Housing materials ; Deficient
- PD 1251 — *Housing materials ; Limited*
- PC 2616 — Housing ; Neglect urban
- PE 1708 — Housing ; Over spacing suburban

Housing [penning...]
- PE 2763 — penning domestic animals ; Inadequate
- PE 8698 — poor ; Inaccessible
- PE 8698 — poor ; Prohibitively expensive
- PD 3469 — prejudice

Housing [Racial...]
- PD 3442 — Racial discrimination
- PB 8443 — regulations ; Complex
- PE 2451 — rehabilitation ; Demoralizing constraints
- PF 6146 — respect common land ; Inadequate arrangement
- PF 6511 — rural communities ; Inadequate residential

Housing [Segregation...]
- PD 3442 — Segregation
- PD 8778 — shortage
- PD 8638 — Socially inappropriate
- PD 9366 — Stray voltage animal
- PE 8103 — Housing teachers rural communities ; Lack

Housing tenants
- PD 3465 — Exploitation
- PE 7169 — Irresponsible
- PE 7169 — Negligence
- PE 7169 — Unethical practices
- PF 6461 *Housing ; Traditionally determined*
- PE 2346 *Housing ; Unrentable vacant*

Human [abortion...]
- PD 0173 — abortion ; Natural
- PF 3437 — activities supernatural entities ; Disruption
- PB 5250 — activity ; Disruption ecosystems
- PB 0477 — ageing
- PE 5301 Human behaviour according cultural norms ; Systematic

Human beings
- PF 2364 — Denial right develop
- PD 8017 — Governmental disregard people
- PC 1648 — Stress
- PF 1199 Human biological abilities ; Inequality
- PD 1618 Human birth defects

Human body
- PF 9150 — Contamination

Human body cont'd
- PE 6603 — Disruption internal balance
- PE 5328 — Excessive trace elements
- PE 5580 — hair ; Excess
- PD 0748 — Harmful effects ultrasonic radiation
- PD 2559 — Mutilation deformation
- PE 4481 — odour ; Disagreeable
- PE 0596 — Parasites
- PD 8050 — Radiation damage
- PE 5328 — Trace element imbalance

Human [cannibalism...]
- PF 2513 — cannibalism
- PF 9612 — capital ; Erosion
- PB 5250 — carrying capacity environment ; Erosion
- PE 9526 — *coccidiosis*
- PF 4043 — combustion ; Spontaneous

Human consumption
- PD 7699 — animal products
- PC 7644 — animals
- PD 7699 — dairy products
- PF 7054 Human contingency
- PF 0936 Human creativity ; Collapsed meaning

Human [death...]
- PA 0072 — death
- PC 7112 — decency ; Laws violation
- PD 0287 — deficiency diseases
- PA 2159 — dependence
- PA 0429 — dependence violence
- PF 1094 — depths ; Antiquated intellectual methods appropriate
- PA 0832 — destructiveness

Human development
- PF 0611 — Decline public expenditure
- PF 0019 — Excessive reliance economic indicators
- PF 5759 — Ignorance lifelong
- PF 7063 — potential ; Unrealized
- PF 3388 — *sports ; Excessive claims*
- PC 0699 Human disability

Human disease
- PB 1044 — disability
- PB 1044 — disability ; Dependence
- PE 3632 — Insect vectors
- PD 6651 — vectors
- PD 6651 — Vectors
- PD 5669 Human diseases ; Environmental

Human [embryo...]
- PE 5623 — embryo storage ; Unlimited practice
- PE 6038 — embryos ; Commercialization
- PE 4805 — embryos ; Research
- PE 8128 — energies desires ; Misdirection
- PC 8943 — environment ; Deterioration
- PF 3702 — error
- PF 3702 — errors miscalculations
- PD 1091 — evolution ; Insufficient intervention
- PF 6730 — exceptionalism
- PE 8545 — excreta ; Pollution water supplies
- PF 9070 — experience ; Symbols unrelated
- PC 6912 — experimentation
- PC 0936 — exposure hazardous environmental pollutants

Human [fatigue...]
- PE 5572 — fatigue during control complex equipment
- PF 0906 — fertility developing countries ; High
- PF 0981 — figures ; Depiction
- PE 1764 — flatulence public
- PF 7815 Human genetic endowment ; Decline

Human [hair...]
- PE 7111 — hair ; Cosmetic health problems related
- PE 6308 — hair ; Grey
- PB 4885 — health ; Hazards
- PC 4777 — health natural environment ; Hazards
- PC 4844 — healthiness ; Disparity

Human [illness...]
- PA 0294 — illness
- PB 1044 — illness
- PE 4856 — illnesses disabilities ; Excessive costs cultural burdens
- PE 3355 — immunity system ; Defective
- PD 5111 — immunodeficiency syndrome ; Acquired
- PD 5111 — immunodeficiency virus ; HIV
- PA 0844 — inequality
- PC 6037 — infertility
- PF 9292 — information processing capacity ; Limited
- PF 9150 — internal pollution

Human [labour...]
- PD 3076 — labour ; Alienation
- PF 1191 — life evaluation ; Discrepancies
- PF 6359 — life ; Killing non
- PD 1618 — Human monstrosities
- PA 0072 — Human mortality
- PE 7917 — Human nature ; Corruption good

Human [organ...]
- PE 9038 — organ failure
- PB 5647 — organism ; Vulnerability
- PE 7530 — organs ; Trade
- PE 7530 — organs transplantation ; Lack
- PF 6751 — origin ; Limited availability therapeutic substances
- PA 0911 Human personality ; Dissociation
- PA 0911 Human personality ; Fragmentation

Human physical
- PD 1618 — genetic abnormalities
- PD 5177 — growth ; Inhibited
- PD 5177 — height ; Inadequate
- PB 5646 — suffering
- PB 5646 — suffering ; Dependence

Human [physiological...]
- PF 4417 — physiological processes ; Unsociable
- PD 0105 — poisoning

Hypofunction

Human [potential...] cont'd
PF 7063 — potential ; Underestimation
Human primates
PE 1570 — Endangered species non
PE 1621 — research ; Inhumane use non
PF 5073 — Shortage experimental non
Human [process...]
PF 2415 — process ; Superficial research total
PE 2246 — pseudo hermaphroditism
PE 1418 — psychological regression
Human race
PF 7815 — Intellectual decline
PF 7815 — Physical decline
PE 9085 — Uncertainty survival
Human [racial...]
PF 0411 — racial regression
PE 4725 — remains ; Unsanitary disposal
PE 1843 — requirements ; Market indicators exclusion
PD 8114 — resource development debtor developing countries ; Neglect
PF 2875 — resource services ; Ineffective delivery basic
Human resources
PC 3113 — capitalist systems ; Waste
PD 8114 — developing countries ; Reduction public expenditure
PE 9764 — development least developed countries ; Inadequate
PF 9612 — Erosion investment
PD 1291 — industrialized countries ; Loss developing country
PC 7721 — Loss
PD 2046 — Mass unemployment
PF 3523 — Underutilization
PF 3523 — Undeveloped
PC 8914 — Unproductive use
PF 3523 — Unutilized
PC 8914 — Wastage
PC 8914 — Waste
Human rights abuses
PF 9288 — Connivance authorities
PF 9288 — Connivance religious leaders
PF 9288 — Government connivance
PF 9288 — Official connivance
Human rights activists
PE 6934 — Denial rights
PF 3819 — Government complicity killing
PE 6934 — Government harassment
PE 6934 — Vulnerability
Human rights [administration...]
PD 6927 — administration justice ; Denial
PE 7300 — agreements ; Failure governments implement provisions ratified
PC 1454 — armed conflicts ; Denial
PF 7925 — Bias United Nations response
PC 3124 — capitalist systems ; Denial
PC 3178 — communist systems ; Denial
PB 3121 — Denial
PC 7379 — Deprivation
PF 2418 — Execution orders violation
PF 4821 — Government aid countries violating
PF 3474 — Impunity violators
PC 4608 — Inadequate enforcement
PE 9759 — individual parliamentarians ; Violation
PF 9794 — infractions ; Politically motivated exaggeration
PC 6003 — Infringement
PF 6365 — instruments ; Inadequacy international
PF 7925 — Inter Governmental double standard
PF 7925 — issues international community ; Avoidance
PF 7925 — issues United Nations ; Politicization
PF 9794 — Misinformation concerning infringement
PF 1585 — monitors ; Harassment
PF 9545 — offenders ; Government pardoning convicted
PF 4176 — privacy use computer databases ; Erosion
PF 6365 — standards ; Inadequacy existing
Human rights treaties
PE 7300 — Delays ratification
PE 7300 — Limited acceptance
PE 7300 — Non ratification
PF 4062 Human rights ; Undocumented violations
PB 3860 Human rights ; Violation
Human rights violations
PF 9794 — False allegations
PE 1407 — Government inaction alleged
PF 9474 — Governments unwilling cooperate United Nations eliminating
PD 5122 — Inadequate assistance victims
Human [sacrifice...]
PF 2641 — sacrifice
PD 6150 — settlement ; Obsolescence suburban mode
PD 5077 — settlements developing countries ; Inadequacy training
PD 6130 — settlements ; Spatial imbalance
PF 1128 — sex ratio ; Imbalance
Human sexual
PD 8016 — disorders
PC 1892 — inadequacy
PE 3274 — intercourse animals
Human [skin...]
PE 4966 — skin ; Desiccation
PC 6037 — sterility
PC 0080 — subjects ; Irresponsible research using
PB 5955 — suffering
PF 7713 — survival ; Inadequacy prevailing mental structures challenge
PF 5301 Human thought ; Organization
PC 3429 Human torture
PF 2434 Human unity ; Lack
PF 1964 Human use animal by-products
PF 9276 Human values ; Indeterminate future
PF 6469 Human values ; Underprioritized
Human vectors
PD 6651 — [Human vectors]
PD 2784 — animal diseases
PD 3593 — plant disease
PA 0429 Human violence
PD 1703 Human wisdom unrelated daily life
PD 0520 Humane imprisonment ; Denial right
PB 8214 Humaneness ; Lack
PE 0176 Humanism
PE 0176 Humanism ; Secular
PF 7955 Humanitarian aid ; Politically exploitative
PD 6298 Humanities ; Ideological domination
PC 1073 Humanity ; Crimes against
PF 9821 Humanity ; Inadequate global consensus concerning problems prospects
PB 8214 Humanity ; Lack
PF 1098 Humanity ; Remission sentences crimes against
PF 8753 Humanity response global problematique ; Failure integrate knowledge empower
PF 8264 Humanity's dominance over nature ; Belief
PE 8497 Humankind ; Denial animals right attention, care protection
PF 5121 Humankind ; Denial animals right attention, care protection
PF 4323 Humanness ; Reduced images
Humans [Abusive...]
PC 6912 — Abusive experimentation
PE 8409 — after animal disease outbreaks ; Inadequate disinfection
PE 7530 — Animal organ transplants
PF 6185 — animals urban environment ; Inadequate interaction
PE 8409 — Humans during animal disease outbreaks ; Inadequate disinfection measures
PD 5493 Humans ; Hunting
PC 8096 Humans ; Killing
PD 8486 Humans ; Killing animals
PB 8050 Humans ; Radiation carcinogenesis
PC 6912 Humans radiation ; Experimental exposure
PD 8486 Humans ; Slaughter animals
PE 3601 Humans ; State imposed functional breeding
PD 2697 Humans ; Unregulated medical experiments
PD 2474 Humid climates ; Hot
PD 2474 Humidity
PF 3856 Humiliated people
PF 4787 Humiliating treaties
PF 3856 Humiliation
PF 9234 Humiliation co workers ; Fear
Humiliation [Dependence...]
PF 3856 — Dependence
PA 6839 — *Disrepute*
PA 6822 — *Disrespect*
PA 6659 Humiliation *Humility*
PA 7107 Humiliation *Unpleasantness*
PA 6659 Humility
PF 7847 Humility ; Lack
PF 2527 Humility relation environment ; Loss
PD 9654 Humoral hypercalcaemia malignancy
PF 7427 Humour ; Obscene
PJ 1051 Humourlessness
PE 8589 Hunch back
PB 0262 Hunger
PE 0324 Hunger ; Denial right freedom
PB 0262 Hunger ; Dependence
PE 8442 Hunger ; Obstacles learning due
PA 8038 Hunger ; Spiritual
PB 0262 Hungry people
PE 4454 Hunt sabotage
PD 7885 Hunt ; Witch
PE 7826 Hunter's bumps
Hunting [Animal...]
PE 5644 — Animal head
PE 5644 — Animal trophy
PC 2024 — animals
PD 5493 Hunting ; Bounty
PD 3008 Hunting ; Disruptive airplane
PE 4454 Hunting expeditions direct action environmentalists ; Sabotage
PE 4893 Hunting ; *Fox*
PD 5493 Hunting humans
PE 1046 Hunting ; Long term historical effect witch
PE 0439 Hunting marine animals
PD 2664 Hunting protected endangered species ; Illegal
PC 2024 Hunting ; Recreational
PF 2093 Hunting ; Sensed futility job
PE 4893 Hunting ; *Stag*
Hunting [tourism...]
PE 3008 — tourism
PE 9059 — trapping game propagation ; Instability
PE 8252 — trapping game propagation ; Underdevelopment
PF 2666 — tribal societies ; Head
PC 2024 Hunting ; Uncontrolled
PE 7915 Huntington's chorea
PD 1590 Hurricanes
PA 5454 Hurtfulness *Badness*
PA 5451 Hurtfulness *Insensibility*
PD 3555 Husband ; Dependence wife
PE 4624 Husband's name married women ; Imposition
PD 0518 Husband wife ; Enmity
PD 0518 Husband wife ; Fighting
PE 3934 Husbandry ; Agricultural effluents animal
PF 2378 Husbandry costs ; Prohibitive livestock
PF 2848 Husbandry ; Inexperienced animal
PD 1562 Husbandry ; Intensive animal
PE 5399 Husbandry ; Methane gas emissions animal
PA 0429 Husbands ; Battered
PJ 5408 Hustling
PE 8935 Hutia ; Endangered species zagoutis coypu
PE 3815 Hyacinth ; Water
PF 2899 Hybrid seed costs ; Excessive
PD 2419 Hybridization wild animal species
PF 3646 Hybrids
PE 2354 Hydatid disease
PD 2289 Hydatidiform mole
PE 7518 Hydatidosis
PD 6520 Hydrants ; Few fire
PE 7826 Hydrarthrosis ; Tarsal
PE 0754 Hydrocarbons pollutants
PD 9154 Hydrocele
PE 9037 Hydrocephalus
PE 1221 Hydrocephalus ; Congenital
PE 2329 Hydrogen sulphide health hazard
PE 2329 Hydrogen sulphide pollutant
PD 9670 Hydrological cycle ; Disruption
PD 2586 Hydrologists ; Corruption
PD 2586 Hydrologists ; Irresponsible
PD 2586 Hydrologists ; Negligence
PD 2586 Hydrology ; Malpractice
PD 2586 Hydrology ; Unethical practice
PF 4839 Hydropolitics
PF 5496 Hydroponics ; Lack
PE 6292 Hydropower ; Environmental hazards
PF 0345 Hydropower ; Underutilization
PD 3482 Hyenas ; Endangered species
PD 8023 Hygiene ; Ignorance health
Hygiene Inadequate
PD 6334 — hospital
PD 2459 — personal
PD 8023 — understanding
PD 8294 Hygiene ; Inappropriate basic
PF 8515 Hygiene ; Lack food
PE 8398 Hygiene restrictions carcass meat exports ; Inadequate
PE 9767 Hygiene slaughter houses ; Inadequate
PE 7826 Hygroma ; Carpal
PG 3408 Hyper thyroidism
PE 9789 Hyperactivity disorder ; Attention deficit
PE 9789 Hyperactivity ; Physical
PD 9654 Hyperadrenocorticism animals
PA 0643 Hyperalgesia
PC 2270 Hyperalimentation
PE 0517 Hyperbilirubinaemia ; Post hepatitis
PD 9654 Hypercalcaemia malignancy ; Humoral
PF 1706 Hyperefficiency
PE 8022 Hyperemesis gravidarum
PF 5430 Hypergamy
PD 7940 Hyperinflation
PE 9789 Hyperkinesia
PC 2270 Hyperlipaemia
PG 4552 Hyperlipoproteinemia
PF 7577 Hyperlocalized education
PG 2721 Hyperparathyroidism
Hyperparathyroidism animals
PD 7424 — Nutritional
PD 7424 — Primary
PD 7424 — Renal secondary
PD 7799 Hyperplasia animals ; Vaginal
PD 7799 Hyperplasia ; Cystic
PE 9702 Hyperplasia ; Gingival
PE 4161 Hyperplasia interdigitalis
PE 7808 Hyperplasia ; Myofibrillar
PE 8372 Hyperplasia prostate
PC 9067 Hyperpyrexia
PE 7570 Hypersecretion saliva
PF 3702 Hypersensitive military mobilization procedures
PE 5169 Hypersensitives
PE 6898 Hypersensitivity
PE 2197 Hypersomnia
PE 0585 Hypertension
PE 0585 Hypertension ; Chronic
PE 0585 Hypertension ; Essential
PE 6937 Hypertension job
PD 9366 Hypertensive heart disease ; Pulmonary
PD 9366 Hyperthermia animals
PE 9774 Hyperthermia ; Malignant
PD 9654 Hyperthyroidism animals
PE 5478 Hypertrichosis
PD 7424 Hypertrophic osteopathy
PD 0448 Hypertrophy heart
PE 7808 Hypertrophy ; Muscular
PD 0270 Hyperventilation
PE 0314 Hypnotic amnesia
PD 4027 Hypnotic intoxication
PE 0139 Hypnotic withdrawal
PE 1821 Hypnotics ; Abuse
PE 0139 Hypnotics ; Abuse barbiturates non barbiturate
PG 3408 Hypo thyroidism
PD 0659 Hypoacusis
PD 9654 Hypoadrenocorticism animals
PE 8322 Hypochondria
PF 3377 Hypocrisy
PA 6400 Hypocrisy *Affectation*
PF 9050 Hypocrisy ; Government
PF 3006 Hypocrisy ; Medical
PF 9050 Hypocrisy ; Political
PF 3983 Hypocrisy ; Religious
PF 3983 Hypocrisy ; Spiritual
PA 7411 Hypocrisy *Uncommunicativeness*
PD 9653 Hypocritical accumulation personal wealth rulers countries
PE 0616 Hypodermic needle ; Common use
PD 9667 Hypodermosis cattle
PG 5181 Hypofunction ; Adrenal cortical

Hypogamy

PF 3341 *Hypogamy*
PE 1926 Hypoglycaemia
PD 7420 *Hypoglycaemia piglets*
PC 0716 Hypokinesia
PD 7420 *Hypomagnesemic tetany cattle*
PD 7420 *Hypomagnesemic tetany sheep*
PG 2721 Hypoparathyroidism
PD 7424 *Hypoparathyroidism animals*
PE 7808 *Hypoplasia ; Myofibrillar*
PE 9554 *Hypoplastic anaemia animals*
PE 2197 Hyposomnia
PD 7307 *Hypostatic pneumonia*
PE 0274 Hypothermia
PE 4161 *Hypothermia*
PD 9366 *Hypothermia animals*
PD 9654 *Hypothyroidism animals*
PG 5976 *Hypothyroidism ; Congenital*
PD 2322 Hypoxia
PE 4586 *Hypoxylon canker trees*
PC 1326 *Hyracoidea ; Endangered species*
PE 6412 Hysteria
PF 6966 Hysteria ; Religious
PE 0321 Hysterical amnesia
PE 6412 Hysterical insanity
PD 0270 *Hysterical neurosis*
PE 4561 Hysterical personality disorder

I

PD 6334 Iatrogenic disease
PD 0094 *Ice*
PD 1393 Ice accretion
PF 1744 Ice ages
PD 2498 Ice-blocked seaways
PD 3142 Ice ; River
PD 3142 Ice runs
PE 1289 Ice ; Sea
PE 1289 Icebergs
PE 8059 Icing aircraft
PF 4923 Iconoclasm
PD 6656 Idea ; Terrorism against
PD 0093 Ideal contraceptives ; Lack
PF 9478 Ideal society ; Inadequate models
PF 6178 Ideal tension free social structures
PF 3419 Idealism
PF 5401 Idealism opposition groups
PF 9624 Idealists ; Stress
PD 8731 Ideas ; Prevention exchange
PD 8731 Ideas ; Restrictions exchange
PD 1284 Ideas ; Theft
PF 8854 Ideas ; Theft
PD 1284 Ideas ; Unacknowledged copying
PF 6158 Identical criminal act ; Multiple liability
PF 2575 *Identification ; Indistinct community*
PF 2575 Identification ; Limited community
PF 6537 *Identification signs ; Incomplete*
PF 0053 Identify clarify world interests ; Lack autonomous world level actor
PF 2775 *Identifying carriers animal diseases ; Difficulty*
PF 3270 *Identifying sadists ; Difficulty*
PF 1735 *Identifying slogan ; Negative*
PE 1082 Identities developing countries transnational corporations ; Disruption cultural social
PF 6516 Identity ; Ambiguous shape social
PF 1246 Identity ; Conflicting sense sexual
PF 1934 *Identity ; Crisis*
Identity [Demoralizing...]
PF 1681 — Demoralizing image urban community
PF 2358 — Demoralizing images rural community
PF 9005 — Denial right cultural
PF 3512 — Depreciated community
PF 2241 — Deteriorating community
PF 2845 — Disjointed patterns community
PE 6581 — disorders ; Gender
PF 2845 — Disparate village
PF 2845 — Displaced community
PF 2845 — Disrupted patterns village
PE 8513 — due tribalism ; Weak national
Identity [Fear...]
PD 2614 — Fear losing cultural
PF 2845 — Fragmented community
PF 2845 — Fragmented images community
Identity [Inadequate...]
PF 1934 — Inadequate sense personal
PF 1246 — Indistinct awareness sexual
PE 7044 — international nongovernmental organization network ; Lack
PF 7274 — inventors discoverers ; Disagreement concerning
Identity Lack
PE 7032 — international nongovernmental organization
PF 1934 — *national*
PF 0684 — *racial*
PF 1934 — *social*
PF 9005 Identity ; Loss cultural
PF 2575 Identity ; Negative village self
PF 0836 Identity ; Obsolete basis cultural
PE 8610 Identity post colonial society ; Weak national
PF 6537 Identity ; Stagnated images community
PF 3392 Ideological apathy
PJ 0775 Ideological bias
Ideological [conflict...]
PF 3388 — conflict
PD 0065 — confusion
PC 2914 — corruption

Ideological [deviation...]
PF 3405 — deviation
PF 3388 — differences ; Debilitating
PC 3219 — discrimination politics
PF 3388 — dissent
PF 5124 — diversities ; Maintenance super power
PD 6298 — domination humanities
PF 7220 — domination research
PA 1387 *Ideological elitism*
PD 0065 Ideological frameworks ; Inadequate
PF 3345 Ideological impediments marriage
Ideological [minorities...]
PC 3325 — minorities ; Underprivileged
PF 4405 — monoculture
PC 3362 — movements minorities ; Threats
Ideological [offences...]
PD 6632 — offences
PF 3388 — opponents
PF 3388 — opposition
PF 1040 — overemphasis economic administration
Ideological [repression...]
PC 8083 — repression
PF 7220 — resolution scientific issues
PC 3231 — revolution
PF 3388 — rivalry
PF 3181 Ideological schism communism
PF 3388 Ideological unity ; Lack
PC 3431 Ideological war
PC 3431 Ideological war ; Dependence
PF 5691 *Ideologies ; Aggressive*
Ideologies [Competing...]
PF 3388 — Competing
PD 3190 — Conflicting social service
PF 5124 — Cover up convergence practice apparently opposed
PC 6341 — Ideologies ; Discriminatory
PC 6341 — Ideologies ; Extremist
PD 0065 — Ideology ; Absence social methodologies global
PF 3392 Ideology ; Lack
PF 3402 Ideology society ; Inadequate integration
PE 6413 Idiocy ; Absolute
PF 6570 *Idiom ; Teenage language*
PE 9509 *Idiopathic polyarthritis animals*
PD 0153 *Idiosyncratic intoxication ; Alcohol*
PD 0779 Idle factories
PF 4670 *Idle private land*
PA 2874 *Idle youth lifestyle*
PA 7710 Idleness
PA 7710 Idleness ; Dependence
PA 5806 Idleness *Inaction*
PA 5942 Idleness *Unimportance*
PF 3374 Idolatrous worship
PF 3374 Idolatry
PA 6839 Ignobility *Disrepute*
PA 7107 Ignobility *Unpleasantness*
PA 5821 Ignobility *Vulgarity*
PA 6839 Ignominiousness *Disrepute*
PA 7280 Ignominy *Wrongness*
PA 5568 Ignorance
PE 1234 Ignorance administration
PF 8582 Ignorance ; Bureaucratic
PF 1653 *Ignorance bureaucratic procedures*
PD 7994 Ignorance cause conception
PF 2440 Ignorance citizen participation ; Sociological
Ignorance concerning
PD 8821 — disease
PD 8821 — disease transmission
PD 7994 — sex
PD 8821 — sexually transmitted diseases
PF 2583 *Ignorance consumer rights*
PF 1985 Ignorance cultural heritage
Ignorance [Deliberate...]
PF 8229 — Deliberate
PA 5568 — Dependence
PA 5568 — *drug users*
PF 8278 — during policy making ; Deliberate
Ignorance [Economic...]
PC 0210 — *Economic*
PF 8229 — events ; Pretended
PF 2989 — *exportable products*
Ignorance [God...]
PF 2409 — God
PF 8278 — government ; Wilful
PD 7566 — grammar
Ignorance [hazardous...]
PF 5188 — hazardous levels toxic substances
PD 8023 — health hygiene
PD 3774 — history
PA 5568 Ignorance *Ignorance*
PA 5568 *Ignorance law*
PF 5759 Ignorance lifelong human development
PD 6728 Ignorance ; Mathematical
PD 8821 Ignorance ; Medical
PE 0533 Ignorance nonverbal communication skills
PE 5773 Ignorance ; Nutritional
Ignorance [Philosophical...]
PF 5039 — Philosophical
PF 2440 — Pluralistic
PC 1982 — Political
PE 1234 — *procedures*
PJ 1219 — *protocol*
PD 7994 Ignorance reproductive processes
PF 2583 *Ignorance rights ; Consumer*
PD 8003 Ignorance ; Scientific
PF 4014 Ignorance scientists
Ignorance traditional
PF 7808 — agricultural products

Ignorance traditional cont'd
PE 3946 — herbal remedies
PE 3946 — medical practices
PE 3946 — plant remedies
PA 7371 Ignorance *Unintelligence*
PA 7232 Ignorance *Unskillfulness*
Ignorance women concerning
PD 9021 — appropriate child care
PD 9021 — child bearing
PD 9021 — infant nutrition
PD 9021 — primary health care
PD 9021 — weaning infants
PD 4506 Ignorance workers
PD 3984 Ignorance world geography
PA 5568 Ignorant people
PF 3539 Ignored business opportunities
PF 3565 Ignored disaster warnings
PE 1862 Ignored social service agencies ; Cultural diversity
PJ 5626 Ignoring implications biblical scholarship
PE 3936 *Ileitis ; Terminal*
PJ 3814 Ileostomy
PE 0989 Ill children ; Mentally
PD 0024 Ill-conceived incentives employment
Ill considered
PD 2269 — land drainage
PF 3370 — missionary activity
PF 3350 — pressure eliminate nakedness developing countries
Ill defined
PE 5463 — causes morbidity
PE 5463 — causes mortality
PC 9067 — diseases
PC 9067 — health conditions
PD 9366 — health conditions animals
PD 9162 — property rights
PC 9067 — symptoms disease
PD 6627 *Ill-designed premises*
Ill effects
PE 5134 — automation ; Social
PF 2013 — educational failure
PC 1299 — radon health
PF 6556 Ill-equipped village leadership
PD 9685 Ill ; Exploitation mentally
PB 1498 Ill-fed
Ill health
PA 0294 — General
PA 0294 — Physical mental
PD 1043 — Susceptibility old physical
PE 7705 *Ill judged forceps delivery ; Precipitate*
PD 7841 *Ill ; Louping*
PD 2731 *Ill ; Navel*
PF 8577 Ill-omened events
PD 1148 Ill persons ; Violation rights mentally
PF 9536 Ill-starred people
Ill treatment
PF 9794 — False declarations
PD 2617 — prisoners war
PD 7630 — trade union leaders
PA 7102 Ill will
PF 5332 Illegal action governments
PF 7730 Illegal activities ; Government complicity
Illegal activity
PF 5413 — Attraction
PF 7264 — Ineffective monitoring
PD 7231 — Misuse research cover
Illegal [aliens...]
PC 3207 — *aliens*
PE 6154 — appropriation children
PC 7373 — *armed groups*
PF 7730 — arms trade ; Government complicity
Illegal [drug...]
PD 0991 — drug trade
PD 4563 — drugs ; Cultivation
PD 4563 — drugs ; Government sanctioned cultivation
Illegal [exercise...]
PC 8689 — exercise abuse authority
PD 4116 — exports
PE 3968 — exports nuclear materials
Illegal [firearms...]
PE 2470 — firearms
PE 7711 — firearms ; Trafficking
PF 7803 — fund transfers ; Legitimizing
PD 2664 Illegal hunting protected endangered species
Illegal [immigration...]
PD 1928 — immigration
PD 0159 — induced abortion
PD 4858 — international arms shipments
PC 4645 — international business
PE 4991 — ivory trade
PE 4367 Illegal long-distance telephone calls
PF 7935 Illegal marriage
PC 2367 Illegal movement across frontiers
PD 0820 Illegal occupation unoccupied property
PF 2418 Illegal orders ; Execution
Illegal [pesticides...]
PD 4629 — pesticides ; Use
PD 1632 — phone tapping
PC 2367 — political exit
PC 1461 — political regimes
PC 0939 — private profit socialist countries
PE 2061 — products developing countries ; Dumping dangerous
PE 9605 Illegal roadblocks
PE 6625 Illegal software copies
PD 5384 Illegal strikes
PC 4645 Illegal trade
PC 8689 Illegal use abuse administrative authority
PF 7669 Illegal warfare

-946-

Immaturity

PA 5952 Illegality
PF 2643 Illegality doctor-assisted suicide
PF 4735 Illegality drug use
PA 5952 Illegality *Illegality*
PF 4727 Illegality nuclear weapons
PA 7321 Illegality *Refusal*
PA 5612 Illegality *Unchastity*
PE 9309 Illegally obtained evidence
PD 5433 Illegally-obtained funds
PF 4958 Illegible handwriting
PC 1874 Illegitimacy
PC 8689 Illegitimate authority
Illegitimate children
PC 1874 — [Illegitimate children]
PD 0943 — Denial rights
PD 0943 — Discrimination against
PA 5952 Illegitimate *Illegality*
PC 1461 Illegitimate political regimes
PA 7411 Illegitimate *Uncommunicativeness*
PA 7193 Illiberalism *Cheapness*
PA 7306 Illiberalism *Narrowmindedness*
PA 7211 Illiberalism *Selfishness*
PE 6074 Illicit appropriation cultural objects
PE 7188 Illicit breweries
PC 6641 Illicit business ; Underground
PE 5503 Illicit coining
PD 4542 Illicit discharges hazardous substances
PE 7188 Illicit distilleries
Illicit drug
PD 0991 — trafficking
PD 0094 — use
PD 0094 — users
PE 0456 Illicit drugs ; Adulteration
PE 2512 Illicit drugs ; Manufacture
PE 9004 Illicit export works art
PD 4116 Illicit exports
PD 2722 Illicit labour trafficking
PD 2366 Illicit literature
PD 9765 Illicit movement toxic products
PD 0991 Illicit narcotics trade
PE 7188 Illicit production alcoholic beverages
PE 4946 Illicit trade prescribed drugs
PD 3841 Illicit trading
PC 0210 Illiteracy
PD 8723 Illiteracy ; Adult
PD 6645 Illiteracy amongst destitute
PC 0210 Illiteracy ; *Computer*
PD 2041 Illiteracy ; *Cultural*
Illiteracy [Dependence...]
PC 0210 — Dependence
PC 1383 — developed countries
PD 8329 — developing countries
PC 0210 Illiteracy ; *Economic*
PD 6645 Illiteracy fourth world
PD 8723 Illiteracy ; Functional
PD 3984 Illiteracy ; Geographical
Illiteracy [impediment...]
PE 8177 — impediment leadership
PD 3321 — indigenous peoples
PE 7968 — inhibitor business transactions
PE 8978 Illiteracy least developed countries
PF 9098 Illiteracy ; Low self image due
PE 8660 Illiteracy mothers
PE 8246 Illiteracy obstacle acquiring skills
PC 3222 Illiteracy ; Political discrimination based
PD 8003 Illiteracy ; Scientific
PF 9980 Illiteracy ; Temporal
PC 0210 Illiteracy ; Wide spread
Illiteracy women
PE 4380 — [Illiteracy women]
PE 4380 — [Illiteracy women]
PE 8660 — developing countries
PC 0210 Illiterate people
PE 0989 Illness adolescents ; Mental
PA 5454 Illness *Badness*
PD 8239 Illness ; Chronic
PE 4906 Illness ; Chronic terminal
PD 8841 Illness developing countries ; Occupational
PA 6799 Illness *Disease*
PD 1043 Illness elderly
PA 0294 Illness ; Human
PB 1044 Illness ; Human
PA 7395 Illness *Inexpedience*
PF 8387 Illness ; Leadership impaired
PF 8387 Illness ; Leadership impaired mental
Illness [Malnutrition...]
PB 1498 — Malnutrition based
PC 0300 — Mental
PE 9091 — Morbid fear
PC 8756 Illness ; Neurological
PC 0300 Illness ; *Nutritionally induced mental*
PD 7924 Illness ; Respiratory
Illness [Unawareness...]
PD 8023 — Unawareness symptoms
PA 5643 — *Unkindness*
PF 8090 — Unreported
PE 9038 Illnesses ; Critical
PE 4856 Illnesses disabilities ; Excessive costs cultural burdens human
PA 6180 Illogic *Error*
PA 5502 Illogic *Reason*
PA 6414 Illusion
Illusion [consensus...]
PF 4327 — consensus
PF 4389 — *contentment*
PF 4389 — *controlling events*

PF 4389 *Illusion happiness*
PF 4162 Illusion nuclear strategy
PF 4389 Illusions ; Personal
PF 6479 Image ; Absence local leadership
PE 6784 Image ageing women ; Negative self
PE 7752 Image being victim circumstances ; Marketing skills which reinforce self
PF 2093 Image ; Community victim
Image [Debilitating...]
PF 1365 — *Debilitating poverty*
PE 1137 — *Declining public*
PF 2093 — Defeating community
Image Demeaning
PF 2093 — community self
PF 1365 — *farmer*
PF 1529 — minority self
Image [Depressed...]
PF 2093 — Depressed community
PF 1509 — *Destructive youth*
PF 4010 — Dissatisfaction personal
PF 9098 — due illiteracy ; Low self
Image [Economic...]
PF 0979 — Economic victim
PF 2093 — Economic victim
PF 2896 — employability ; Limited
Image [Financial...]
PF 3307 — *Financial irresponsibility*
PE 8802 — foreign groups peoples ; Biased media
PF 1675 — free choice ; Tensionless
PF 1681 Image ; *Ghetto education*
Image [impossibility...]
PD 2670 — impossibility change ; Paralyzing
PF 5354 — Inadequate self
PF 0979 — Individual victim
Image [leadership...]
PF 1102 — leadership ; Messianic
PF 5354 — Low resident self
PF 5354 — Low self
PF 1529 Image ; Minority victim
PF 2093 Image powerlessness ; Community
PD 1564 Image ; Preclusion elders' participation
Image [Reduced...]
PF 2093 — Reduced community
PF 2093 — Restraining community
PD 1308 — roles marriages ; Inadequate
Image [scarcity...]
PF 3002 — scarcity ; False
PF 8675 — Short range financing
PF 1064 — social responsibility ; Deluding familial
PF 1365 — *Static farming*
PF 2896 — Static job
Image [Unattractive...]
PF 2093 — Unattractive community
PE 1137 — *Undeveloped public*
PF 2358 — *Unhelpful vandalism*
PF 1681 — urban community identity ; Demoralizing
PF 2093 Image ; Villagers victim
PE 6328 Imagery ; Exploitative
PF 0076 Imagery ; Prejudice against communication visual
PD 7086 Imagery societal learning ; Inadequate use visual
PF 6098 Images ; Breakdown vocational
Images [citizens...]
PF 8781 — citizens' law enforcement groups ; Defeatist self
PF 2600 — citizenship ; Nationalistic
PF 2845 — community identity ; Fragmented
PF 6537 — community identity ; Stagnated
PF 2126 — countries cultures ; Untransferability
PF 8126 Images ; Debilitating education
PF 2600 Images ; Defensive nationalistic
PF 7614 Images environmental protection ; Reduced
PF 1673 Images ; Excessive dissemination materialistic
PF 2699 Images global economic situation ; Inadequate solution oriented
PF 0981 Images ; Graven
Images [having...]
PF 6098 — having significant life work ; Reduced
PF 6537 — having significant life work ; Reduced
PF 4323 — humanness ; Reduced
Images [Inadequate...]
PF 2834 — Inadequate hero
PF 1365 — *Incongruous religious*
PF 2896 — Individualistic employment
PF 8126 — Ineffective educational
PF 1365 — *ineffectivity*
PF 3553 — Insufficient responsibility
PF 2093 Images ; Lack significant community
Images [opponents...]
PF 4133 — opponents ; Promotion negative
PF 1365 — *Oppressive prevalent*
PF 2600 — Overruling global interdependence nationalistic
Images [Parochial...]
PF 2896 — Parochial work
PF 1365 — *Past oriented farming*
PF 1772 — personal freedom ; Non integrating
PF 5298 — political involvement ; Insufficient
PD 1564 — preclude participation ; Elders' self
Images [Reduced...]
PF 1560 — Reduced celebrational
PF 1728 — Reinforced parochialism internal values
PF 3826 — Rejection alternative
PF 6098 — Restricted vocational
PF 6537 — Restricted vocational
PF 2358 — rural community identity ; Demoralizing
Images [security...]
PD 1147 — security ; Individualistic
PF 5354 — Self depreciatory operating

Images [Shortage...] cont'd
PF 2834 — Shortage hero
Images social
PF 1728 — reality ; Lack correspondence basic value
PF 3553 — responsibility ; Absence
PF 1651 — roles ; Disrelated
PF 2627 Images society leaders ; Parochial
PF 1790 Images ; *Unfeasible industrial*
PF 6098 Images vocation ; Collapsed
PF 6098 Images vocational roles ; Disrelated
PF 1790 Imagination ; *Lacking industrial*
PF 4882 Imagination ; Social constraints freedom
PF 1345 Imaginative planning expansion income ; Absence
PD 0270 *Imagined defect appearance* ; Preoccupation
PA 0224 Imbalance
PE 4956 Imbalance agricultural exports imports developing countries
PC 0842 *Imbalance* ; *Animal ecological*
PE 4866 Imbalance capital technical assistance
PF 2120 Imbalance city sizes country
Imbalance conventional
PC 5230 — armed forces
PC 5230 — arms
PC 5230 — weapons
Imbalance [developed...]
PC 8694 — developed developing countries ; Economic
PF 5912 — developing countries share shipping
PB 6207 — development
PF 8650 — disbursements multilateral agencies official contributions
PE 7479 — distribution industrial processes
PF 0544 — distribution political awareness developing countries
PC 0842 *Imbalance* ; *Ecological*
Imbalance economic
PC 5601 — activity ; International
PC 7459 — relationships industrialized countries
PF 4837 — social planning developing countries
PF 0443 Imbalance educational curricula
Imbalance human
PE 5328 — body ; Trace element
PD 6130 — settlements ; Spatial
PF 1128 — sex ratio
PC 8415 Imbalance international trade patterns
PE 7530 Imbalance need availability viable organs transplantations
Imbalance [payments...]
PC 0998 — payments
PC 0998 — payments ; Dependence
PE 4241 — population growth developed developing countries
PD 4007 — population growth resource development
Imbalance power
PB 1969 — [Imbalance power]
PB 1969 — Dependence
PC 4291 — Regional
PE 5825 — shipping industry
PB 4993 Imbalance quality life ; International
PD 9170 Imbalance quality working life ; International
Imbalance [recovery...]
PF 6071 — recovery recession
PD 1494 — Regional
PF 7432 — revenue mobilization expenditure allocation
PE 5022 — rural urban incomes
PE 8584 — rural urban incomes developing countries
PE 5863 Imbalance shipbuilding industry distribution
PC 1606 Imbalance strategic arms
Imbalance [trade...]
PE 5678 — trade cultural products capitalist socialist countries
PE 5702 — trade cultural products developed developing countries
PE 5194 — training existence openings various professions
PC 1563 Imbalance urbanization industrialization developing countries
PD 5046 Imbalance world food economy
PF 3313 Imbalanced celebrational expenditures
PF 0204 Imbalanced distribution knowledge
PF 1403 *Imbalanced distribution skills*
PC 2049 Imbalanced regional development countries
PF 0198 Imbalanced research activity
PD 0339 *Imbalances* ; *Amino acid*
PE 4715 Imbalances brain ; Chemical
PD 0794 Imbalances distribution costs benefits economic integration
PE 5839 Imbalances dry bulk shipping industry
PE 5920 Imbalances exports imports land-locked developing countries
PC 4879 Imbalances ; Fiscal trade
PE 8735 Imbalances growth labour force, urban population overall growth economy
PD 5865 Imbalances industrialized countries ; Economic
PF 9008 Imbalances people's aspirations structure opportunities income available
PC 7459 Imbalances three largest market economies ; Structural
PE 5874 Imbalances types ships built
PF 3382 Imbalancing population sex ratio ; Deliberate
PE 6314 Imbecility
PA 6876 Imbecility *Impotence*
PA 7371 Imbecility *Unintelligence*
PA 6568 Imitation
PA 6176 Immateriality
PF 8413 Immaturity
PF 8413 Immaturity ; *Emotional*
PA 6997 Immaturity *Imperfection*
PA 6652 Immaturity *Incompleteness*
PF 8413 Immaturity ; *Political*
PF 8413 Immaturity ; *Sexual*

-947-

Immaturity

Immaturity [Unintelligence...]
PA 7371 — *Unintelligence*
PA 7341 — *Unpreparedness*
PA 7232 — *Unskillfulness*
PF 5660 Immediacy orientation elected bodies
PF 1348 Immediacy prevents saving
PF 0317 Immediate appeasement responses leadership
PF 0317 Immediate economic priorities
PF 1058 Immediate gratification-based social forms
PF 9119 Immediate hazards society ; Insensitivity non
PF 3243 Immediate needs mentality
PF 9119 Immediate problems ; Obsession
PE 4059 Immediate solutions resource development research ; Over emphasis
PF 3243 Immediate superficial needs ; Overemphasis
PF 7836 Immediate victims ; Defence industrial pollution
PD 3427 Immigrant labourers industrialized countries ; Inadequate living working conditions
Immigrant workers
PD 2722 — Abusive traffic
PE 4934 — Discrimination employment against
PD 2722 — Traffic
PD 4605 Immigrants
PD 0973 Immigrants aliens ; Discrimination against
PC 3207 Immigrants aliens ; Expulsion
PC 1532 Immigrants ; Inadequate cultural integration
PD 3427 Immigrants industrialized countries ; Menial work status
PE 5020 Immigrants ; Lack appropriate education children
PE 8699 Immigrants ; Lack education women
PE 4990 Immigrants ; Marginalization second generation
PD 4605 Immigrants ; Proliferation
PF 7776 Immigrants ; Stress
PE 4868 Immigration barriers handicapped family members
PD 1928 Immigration ; Clandestine
PE 7780 Immigration control procedures ; Victimization
PD 1928 Immigration ; Illegal
PF 4008 Immigration ; Mass
PE 3889 Immigration, naturalization passports ; Crimes related
PE 7780 Immigration officials ; Harassment travellers
PD 4605 Immigration overload
PC 0970 Immigration ; Restrictions
PA 6290 Imminence
PD 4011 Immobility ; Labour
PF 0193 Immobility social pattern
PF 7577 Immobility ; Student
PE 5722 Immoderate eating drinking
PA 5612 Immodesty *Unchastity*
PA 6491 Immodesty *Vanity*
PD 7550 Immolation
PF 1384 Immoral literature
PD 5770 Immoral medical practice
PA 3369 Immoral people
PF 4753 Immoral public policy
PA 3369 Immorality
PA 3369 Immorality ; Dependence
PA 7363 Immorality *Improbity*
PF 2687 Immorality ; Sexual
PA 5644 Immorality *Vice*
PF 2142 Immortality
PE 4224 Immune-deficiency diseases animals
PE 1914 Immune dysfunction syndrome ; Chronic fatigue
PG 9847 *Immune haemolytic anaemia animals*
PE 4883 Immune responses malnourished persons ; Inadequate
PD 4068 Immune system diseases animals
PF 5649 Immunities diplomats ; Abuse privileges
PF 5649 Immunities international civil servants ; Abuse privileges
PE 5488 Immunities international civil servants ; Violation privileges
PF 2775 *Immunity animal diseases ; Vector*
PD 3035 Immunity ; Erosion journalistic
PE 5832 Immunity ; Police
PE 4930 Immunity ; State
PE 3355 Immunity system ; Defective human
PE 5488 Immunity ; Threat diplomatic
PF 0609 Immunity ; Threat parliamentary
PF 1916 Immunization ; Enforced
PF 5969 Immunization ; Inadequate insufficient
PF 5969 Immunization methods ; Inadequate
PF 5969 Immunization programme ; Unorganized
Immuno deficiency virus
PD 4747 — [Immuno-deficiency virus]
PD 4747 — Bovine (BIV)
PD 4747 — Feline (FIV)
PE 7848 — Feline (FIV)
PD 4747 — Simian (SIV)
PJ 3296 Immunodeficiency
PE 4224 *Immunodeficiency disease animals ; Combined*
PE 4224 Immunodeficiency diseases animals
PD 5111 Immunodeficiency syndrome ; Acquired human
PE 4224 *Immunodeficiency syndrome ; Simian acquired*
PD 5111 Immunodeficiency virus ; HIV human
PD 4387 Impact agriculture developing countries ; Carbon dioxide
PF 9450 Impact current policy ; Uncertain environmental
PE 8318 Impact education ; Delay societal
PE 7735 Impact holy days ; Disruptive secular
PE 1478 *Impact impersonal repetitive news items ; Reduced*
PF 8870 Impact innovation ; Delay societal
PG 7844 *Impact ; Limited programme*
PE 8617 Impact lungs children ; Air pollution
PD 4007 Impact overpopulation ; Negative ecological
PE 8476 Impact pollution growth crops ; Adverse
Impact [technological...]
PF 9328 — technological activities ; Delayed environmental
PD 7463 — technological development capitalist society ; Perverse

Impact [technology...] cont'd
PE 5820 — technology transfer transnational corporations ; Adverse
PC 1174 — technology ; Uncontrolled environmental
PE 7423 *Impaction animals ; Colon*
PE 9364 *Impaction ; Dietary abomasal*
PE 4630 *Impaction ; Rumen*
Impacts [coal...]
PE 8453 — coal conversion plants ; Environmental
PF 1105 — Combinations environmental
PF 1105 — Cumulative environmental
PC 8866 Impaired digestion
PF 8387 Impaired illness ; Leadership
PC 1587 Impaired intelligence
PC 1587 Impaired mental activity
PF 8387 Impaired mental illness ; Leadership
PE 8677 Impaired persons ; Medical experimentation mentally
PF 6863 Impaired vigilance
PD 4448 Impairing military effectiveness
PA 6088 Impairment
PE 1418 *Impairment ego*
PC 0699 Impairment handicap ; Health
PA 6997 *Impairment Imperfection*
PA 7395 *Impairment Inexpedience*
PF 4945 Impairment ; Mental
PE 5746 Impairment physicians' ability
PD 8179 *Impairment ; Vision*
PD 8179 *Impairment visual-spatial ability*
PE 6776 Impairments ; Disfiguring
PC 6777 Impairments ; Visceral
PE 6395 Impartial expertise ; Unavailability
PE 7665 Impartiality judge jury ; Lack
PE 7665 Impartiality judiciary ; Lack
PE 6526 *Impassable roads*
PE 4922 Impasse technology transfer developing countries
PA 6200 Impatience
PA 6200 *Impatience Impatience*
PA 7325 *Impatience Irresolution*
PE 3005 *Impeded possible expansion*
PE 8177 Impediment leadership ; Illiteracy
PF 7353 Impediments adoption children
PF 5947 Impediments extradition
PD 3063 Impediments foreign investment ; Legal
Impediments [international...]
PF 7277 — international investigations ; Legal
PF 6681 — international motor traffic ; Tax
PF 1068 — internationally mobile professionals experts
Impediments marriage
PF 3343 — [Impediments marriage]
PF 3342 — Economic
PF 3345 — Ideological
PF 3346 — Legal
PF 3344 — Psychological
PC 1523 — Racial
PD 0355 — Religious
PF 3341 — Social
PF 3342 — Tax
PE 7780 Impediments obtaining travel visas ; Administrative
PD 2592 Impeding right self determination ; Use mercenaries means
PE 8403 Impeding sustainable development ; Patent restrictions
PF 8765 Impenitence
PF 7556 Imperfect market operation
PF 5716 Imperfect universe ; Unfinished
PE 8589 *Imperfecta ; Osteogenesis*
PE 4346 *Imperfecti ; Fungi*
PA 6997 Imperfection
PB 7712 Imperfection ; Dependence moral
Imperfection [Imperfection...]
PA 6997 — *Imperfection*
PA 5652 — *Inferiority*
PA 5473 — *Insufficiency*
PF 7024 — *Intolerance*
PB 7712 — *Moral*
PA 6041 *Imperfection Shortcoming*
PA 5644 *Imperfection Vice*
PF 3136 Imperfections capital markets
PD 8227 Imperfections leadership ; Moral
PF 7556 Imperfections ; Market
PF 4014 Imperfections science
PE 3711 *Imperforate anus*
PB 0113 Imperialism
Imperialism [Capitalist...]
PC 3166 — Capitalist economic
PC 3193 — Capitalist political
PC 3165 — Communist economic
PC 3164 — Communist political
PC 6440 — Conflict interests
PF 3199 — Covert
PC 3195 — Cultural
Imperialism Dependence
PB 0113 — [Imperialism ; Dependence]
PC 3195 — cultural
PC 3198 — economic
PE 5333 Imperialism ; Ecological
PC 3198 Imperialism ; Economic
PC 3197 Imperialism ; Intra state
PD 2695 Imperialism ; Linguistic
PE 9643 Imperialism ; Submission media cultural
Imperialism [Temporal...]
PF 1432 — Temporal
PD 5891 — Transnational corporation
PD 0766 — transnational enterprises ; Economic
PD 7374 Imperialistic distribution system
PD 4368 Impermanent living conditions
PE 1478 *Impersonal repetitive news items ; Reduced impact*

PD 8017 Impersonality bureaucracy
PE 8706 Impersonality firms ; Growing size
PF 6156 Impersonality high density accommodation
PE 6153 Impersonality mass market shopping facilities
PF 6165 Impersonality public squares cities
PE 7687 Impersonating officials
PE 1275 Impersonation ; Fraudulent
PA 6862 *Impertinence Intrusion*
PA 6979 *Impertinence Opposition*
PA 6491 *Impertinence Vanity*
PA 7364 *Imperviousness Unfeelingness*
PA 5558 *Imperviousness Weakness*
PC 8534 *Impetigo*
PA 5490 *Impetuousity Changeableness*
Impetuousity [Impatience...]
PA 6200 — *Impatience*
PA 5806 — *Inaction*
PA 7325 — *Irresolution*
PA 7115 *Impetuousity Rashness*
PA 6425 *Impetuousity Transience*
PA 6058 Impiety
PA 6058 Impiety *Impiety*
PJ 5219 Implantation military base
PA 7392 *Implausibility Unbelief*
PF 1010 Implement policy directives ; Failure
PE 7300 Implement provisions ratified human rights agreements ; Failure governments
PD 8114 Implementation austerity measures ; Negligent
Implementation [citizen...]
PC 5075 — *citizen rights ; Overcomplicated*
PF 3975 — commitments ; Delay
PF 2863 — community decisions ; Incomplete
Implementation [Delay...]
PF 1470 — Delay project
PF 4975 — development mega-projects
PD 3712 — development programmes ; Inadequate
PF 2497 — Implementation international treaty provisions ; Non
PF 6630 — Implementation maritime safety standards ; Inadequate
PF 1010 — Implementation methods decision making ; Ineffective
PE 0174 — Implementation networks ; Inflexible intermediary political
PF 1010 — Implementation plans programmes against problems ; Inadequate
PF 2421 — Implementation programmes ; Improper
PC 6989 — Implementation social change ; Delays
PF 1010 — Implementation ; Uncertain planning
PE 4003 — Implementation workers wage increases provided legislation collective agreements ; Non
PF 3339 *Implementing development programmes ; Difficulty*
PD 4651 Implication government leaders scandal
PJ 5626 Implications biblical scholarship ; Ignoring
PE 6347 Impoldering risks
PA 7143 *Impoliteness Discourtesy*
Import [capacity...]
PE 6861 — capacity developing countries ; Decline
PD 0336 — cartels
PE 8867 — charges ; Distortion international trade selective indirect taxes
PE 6861 — compression developing countries
Import [dependence...]
PE 0537 — dependence developed countries ; Resource
PC 0065 — dependence ; Resource
PE 1806 — dependency food staples developing countries due transnational corporations
PE 9027 Import licensing ; Imposition
PD 7486 Import penetration ; Vulnerability economies
PE 9027 Import quotas
PE 0537 Import resources ; Dependence industrialized countries
Import [strangulation...]
PE 6861 — strangulation ; Developing country
PE 8792 — substitution barrier subsequent economic growth
PD 7682 — substitution developed countries
PF 7047 *Importance ; Forgotten rural*
PF 7047 Importance rural community ; Underemphasized
PA 6400 Importance ; Self
PA 6491 Importance ; Self
PC 2122 Important landscapes ; Degradation culturally
PD 0270 *Important personal information ; Inability recall*
PE 2777 Importation infected carcass meats factor animal diseases
PE 3721 *Imported diseases ; Massacre indigenous populations*
PC 6329 *Imported energy ; Dependence*
PJ 7877 *Imported foods ; Costly*
PJ 7877 *Imported goods ; Prohibitive cost*
PC 0065 *Imported labour ; Dependence*
PF 1489 Imported technology ; Dependence developing countries
PE 9027 Imports ; Banned
Imports [Dependence...]
PE 8086 — Dependence developing countries food
PD 5677 — Dependence island developing countries
PE 4956 — Developing countries ; Imbalance agricultural exports
PE 0302 — Disruptive food
PE 5920 Imports land locked developing countries ; Imbalances exports
PE 0537 Imports primary commodities ; Dependence industrialized countries
PE 7025 Imports service related goods ; Protectionism against
PF 1492 Imposed aid developing countries ; Restrictions
PF 4128 Imposed career interruptions
PA 8653 *Imposed change ; Suspicion*
PF 1976 Imposed economic sanctions ; Disregard internationally
PE 3601 Imposed functional breeding humans ; State
PF 6575 *Imposed high school closure*
PF 2274 *Imposed isolation ; Mothers' self*
PF 8649 Imposed official decisions

PE 5919 Imposed pregnancy ; Government
PE 4703 Imposed social isolation due torture ; Self
PF 8922 Imposing fictional worldview
PF 1432 Imposition exogenous time standards
PE 4624 Imposition husband's name married women
PE 9027 Imposition import licensing
PD 4287 Imposition intimate forms address
PE 9762 Imposition quantitative trade quotas discretionary character
PE 2309 Imposition rural cooperative projects ; Government
PD 5229 Imposition standards ; Discriminatory
PE 4363 Imposition states emergency ; Excessive
PE 9762 Imposition trade quotas political reasons
PF 8161 Imposition young outmoded values
PA 6487 Impossibility
PA 6714 Impossibility *Chance*
PD 2670 Impossibility change ; Paralyzing image
PA 6099 Impossibility *Hopelessness*
PD 0050 Impossibility redistribute land
PA 6876 Impotence
PF 4668 Impotence ; Government
Impotence [Impotence...]
PA 6876 — *Impotence*
PA 6852 — *Inappropriateness*
PA 6882 — *Influencelessness*
PF 6415 Impotence men ; Sexual
PF 1389 *Impotence ; Political*
PE 5729 Impotence students ; Political
PA 7208 Impotence *Unproductiveness*
PA 5558 Impotence *Weakness*
PE 6263 Impotence youth ; Political
PD 0767 Impounded rivers
PE 8782 Impoundment ; Reduction soil fertility downstream due
PD 0723 Impoverished disabled persons
PC 1408 Impoverishment animals domestication ; Genetic
PA 5688 Impoverishment *Appropriation*
PA 6852 Impoverishment *Inappropriateness*
PF 4600 Impoverishment political debate
PA 6434 Impoverishment *Poverty*
PF 4944 Impractical development programmes
PF 3519 Impractical education
PF 3519 Impractical educational aims
PF 2208 Impractical leadership skills
PF 3519 Impractical public education
PF 3519 Impractical school activities
PF 3519 Impractical training programmes
PA 5497 Impracticality *Difficulty*
PA 6487 Impracticality *Impossibility*
PF 3746 Imprecation
PB 2619 Imprecise language
PA 6180 Imprecision *Error*
PA 6997 Imprecision *Imperfection*
PA 7309 Imprecision *Uncertainty*
PA 6882 Impressionability *Influencelessness*
PE 5986 Impressment children armed forces ; Extra legal
PC 0746 Impressment labour
PD 6667 Impressment military service
Imprisoned mothers
PE 1837 — [Imprisoned mothers]
PD 5142 — [*Imprisoned mothers*]
PF 0131 — Children
PD 5142 Imprisonment
PE 5461 Imprisonment ; Aquarium
Imprisonment [death...]
PE 6424 — death prison foreign nationals ; Failure notify
PE 4887 — delayed trial
PD 0520 — Denial right humane
PF 0726 — deportation ; Delayed consequences war time
PE 7235 Imprisonment failure fulfil contractual agreement ; Denial right freedom
PD 4489 Imprisonment ; False
PE 5112 Imprisonment ; Inappropriate
PC 0562 Imprisonment ; Political
PE 4887 Imprisonment remand ; Excessive
PD 4259 Imprisonment ; Secret
PD 0520 Imprisonment ; Unacceptable conditions
PD 4489 Imprisonment ; Unlawful
PA 7363 Improbity
PC 8247 Improbity
PD 0814 Improbity ; Government
PA 7363 Improbity *Improbity*
PA 7411 Improbity *Uncommunicativeness*
PC 0382 Improper dietary habits
PE 3506 *Improper feeding*
PF 2152 Improper forms debate
PF 2421 *Improper implementation programmes*
PD 9661 Improper infant weaning
PE 5780 Improper lighting occupational hazard
Improper [paper...]
PD 1152 — paper disposal
PA 6000 — people
PE 3553 — police questioning
PD 0876 — proper sewage disposal
PF 2477 *Improperly trained administrators*
PE 5607 Improperly veiled women
PA 6000 Impropriety
PA 5982 Impropriety *Disagreement*
PA 6486 Impropriety *Injustice*
PA 6498 Impropriety *Misbehaviour*
Impropriety [Unchastity...]
PA 5612 — *Unchastity*
PA 6921 — *Undueness*
PA 7006 — *Untimeliness*
PA 5644 Impropriety *Vice*
PA 5821 Impropriety *Vulgarity*
PA 7280 Impropriety *Wrongness*

PD 8007 Improve profitability ; Dismissal workers
PC 0492 *Improvement capital ; Limited*
PF 1781 *Improvement efforts ; Unstructured*
PE 2425 *Improvement funds ; Inadequate*
PE 8477 Improvement urban life styles ; Apathy toward
PE 4331 Improvements agricultural productivity ; Unsustainable short term
PF 2605 *Improvements ; Costly road*
PE 6526 *Improvements ; Inaccessible road*
PF 1815 *Improvements ; Negligent owner*
PF 2605 *Improvements ; Unfunded roadway*
PF 0836 *Improvements ; Unimportance physical*
PA 7115 Improvidence *Rashness*
PA 7341 Improvidence *Unpreparedness*
PE 4386 Improvisational housing developing countries
PA 6446 Imprudence *Indiscrimination*
PA 7115 Imprudence *Rashness*
PA 7371 Imprudence *Unintelligence*
PD 5984 Imprudent budget policies
PA 6822 Impudence *Disrespect*
PA 6979 Impudence *Opposition*
PA 7115 Impudence *Rashness*
PA 6491 Impudence *Vanity*
PA 6448 Impulse control disorder
PA 6448 *Impulses gamble ; Failure resist*
PA 6448 *Impulses ; Loss control aggressive*
PA 6448 *Impulses steal ; Failure resist*
PA 5490 Impulsiveness *Changeableness*
PA 5806 Impulsiveness *Inaction*
PA 7325 Impulsiveness *Irresolution*
PF 3474 Impunity violators human rights
PF 5205 Impure thoughts
PD 0482 Impurities waste water
PA 6997 Impurity *Imperfection*
PF 3960 Impurity ; Ritual
PF 6657 Impurity ; Spiritual
Impurity [Unchastity...]
PA 5612 — *Unchastity*
PA 5459 — *Uncleanness*
PA 7411 — *Uncommunicativeness*
PA 5644 Impurity *Vice*
PE 8634 In vitro conception
Inability [adapt...]
PE 5768 — adapt shift work
PF 8176 — anticipate future development
PF 1004 — awaken consciousness due inadequate methods
PF 1618 Inability bear witness
Inability [communities...]
PE 8509 — communities take advantage training business, industry public service
PF 9747 — compromise
PF 2384 — concentrate
PE 3716 — concentrate due torture
PF 7178 Inability define moral standards
PF 7563 Inability developing countries adopt appropriate exchange rate policies
PF 7806 Inability educational systems keep pace technological advancement
PF 3973 Inability elderly play role developing future
PF 9643 Inability generate sufficient culture-specific programmes
PF 0401 Inability governments regulate family size
PA 6876 Inability *Impotence*
PE 2000 Inability integrate available knowledge ; Increasing
PF 4204 Inability make difficult decisions
PF 5044 Inability make use leisure time
PF 5287 Inability negotiate effective multilateral safeguard systems
PF 9980 Inability organize time
PF 5311 *Inability people's representatives process feedback constituents*
Inability [recall...]
PD 0270 — recall important personal information
PC 7517 — reduce petroleum consumption
PE 3751 — refugees obtain employment
PF 8435 — resolve problems realistically
PF 8876 — respond minority concerns ; Parliamentary
PE 4302 Inability spell
PF 3470 *Inability teach unwritten language*
PA 7232 Inability *Unskillfulness*
PB 8685 Inaccessibility
Inaccessibility [bank...]
PE 4704 — bank loans
PF 3869 — Bible
PF 3526 — bureaucratic services
PF 6132 Inaccessibility city centres suburban residents
PF 6140 Inaccessibility countryside city dwellers
PE 0573 Inaccessibility decision-makers multinational enterprises
PF 2875 Inaccessibility essential services isolated communities
PD 4769 Inaccessibility fuelwood
PE 6175 Inaccessibility green parkland urban environment
Inaccessibility [health...]
PF 6061 — health care teenagers
PE 8261 — health care underprivileged
PE 8261 — high technology medical cures
PE 5665 Inaccessibility insurance island developing countries
PD 8334 Inaccessibility justice
PD 5314 Inaccessibility justice ; International
PF 1953 Inaccessibility knowledge
PD 9679 Inaccessibility markets developed countries developing countries
PF 6160 Inaccessibility quiet zones urban environment
PF 3869 Inaccessibility religious scriptures
PC 0237 Inaccessibility ; Social
PA 6653 Inaccessibility *Unsociability*
PF 6138 Inaccessibility water recreation
PF 2261 Inaccessible administrative agencies

PF 2261 Inaccessible administrative centres
PF 4832 Inaccessible children's clinic
PE 0718 Inaccessible commercial financial services
PE 2452 Inaccessible decision makers
PF 4690 Inaccessible documents
PD 9051 Inaccessible educational facilities
PF 9007 Inaccessible emergency relief facilities
PE 0718 *Inaccessible financial services*
PF 5351 Inaccessible government agencies
PD 2926 Inaccessible government information services
Inaccessible [health...]
PF 4832 — health care
PD 7669 — health resources
PF 9699 — historical documents
PF 9046 — historical libraries
PF 3550 — *housing loans*
PE 8698 — housing poor
PC 0237 *Inaccessible issues advocate*
PE 8916 Inaccessible job market
PF 1776 *Inaccessible landowner records*
PF 8299 Inaccessible market supply centres
PC 4020 *Inaccessible natural resources*
Inaccessible [park...]
PF 9360 — park lands
PE 6795 — places meditation
PE 6795 — places worship
PF 1267 — public information
Inaccessible [recreation...]
PF 6503 — recreation areas
PF 2583 — *religious programmes*
PC 4020 — resources ; Physically
PE 6526 — *road improvements*
Inaccessible [services...]
PC 7674 — services ; Physically
PF 8299 — shopping districts
PF 3499 — *supply repair materials*
PC 2163 *Inaccessible vocational training*
PF 9360 Inaccessible wilderness areas
PF 7905 Inaccuracy
PA 6180 Inaccuracy *Error*
PA 6997 Inaccuracy *Imperfection*
PA 6644 Inaccuracy *Misrepresentation*
PA 5438 Inaccuracy *Neglect*
PA 7309 Inaccuracy *Uncertainty*
PF 9358 Inaccurate biology textbooks ; Biased
PF 1244 Inaccurate criminal stereotypes
PF 3096 Inaccurate documents
PF 4774 Inaccurate forecasting
PF 1780 Inaccurate geography textbooks ; Biased
PD 2082 Inaccurate history textbooks ; Biased
PE 4129 Inaccurate medical diagnosis due inadequate equipment
PF 7842 Inaccurate public finance data developing countries
PD 2716 Inaccurate textbooks reference books ; Inadequate
PF 5118 Inaccurate weather forecasting
PF 1509 *Inaccurate youth stereotypes*
PA 5806 Inaction
PE 1407 Inaction alleged human rights violations ; Government
PC 0267 Inaction ; Bureaucratic
PF 0041 Inaction concerning trade services ; Governmental
PC 3950 Inaction ; Government
PD 7231 Inaction ; Monitoring cover
PB 1423 Inaction problems
PF 6707 Inaction response potential problems ; Official
PF 4470 Inactive rural population ; Socio economically
PB 7991 Inactivity
PA 5806 Inactivity *Inaction*
PD 3515 Inadaptability aged ; Rigidity
PE 5023 Inadaptation technology man industrialised societies
PF 5145 Inadaptation work family needs
PE 6274 Inadequacies foreign consultants
PF 0048 Inadequacies international monetary system
PA 8199 Inadequacy
PE 9096 Inadequacy agricultural education developing countries
PF 0392 Inadequacy aid developing countries
Inadequacy [civil...]
PF 0506 — civil defence
PD 1865 — commercial sector developing countries
PF 2843 — committee system decision making
PD 0093 — contraceptive methods
PD 0093 — contraceptives
Inadequacy [Discontentment...]
PA 6011 — *Discontentment*
PF 3396 — doctrine
PD 0928 — domestic market developing countries
Inadequacy [economic...]
PF 3395 — economic doctrine
PC 1892 — elderly ; Sexual
PC 5016 — emergency food reserves
PF 6365 — existing human rights standards
Inadequacy [feelings...]
PF 0979 — feelings ; Disabling
PF 3949 — food aid
PF 4765 — formal education
PF 5711 — formal logic
PF 2420 Inadequacy governmental decision-making machinery
PC 1892 Inadequacy ; Human sexual
Inadequacy [Imperfection...]
PA 6997 — *Imperfection*
PA 6876 — *Impotence*
PA 6652 — *Incompleteness*
PA 6695 — *Inequality*
PA 5652 — *Inferiority*
PD 1975 — insensitivity intelligence testing
PA 5473 — *Insufficiency*
PF 2876 — intergovernmental decision-making process

Inadequacy international
- PF 6365 — human rights instruments
- PF 8616 — legal procedure
- PF 0228 — legislation
- PF 5072 — standards
- PF 6194 Inadequacy large buildings ; Social

Inadequacy [male...]
- PF 1069 — male contraceptive methods
- PF 8326 — medical science
- PF 3613 — men ; Social
- PF 7700 — methodologies deal discontinuities random events

Inadequacy [patent...]
- PF 3538 — patent coverage
- PF 3394 — political doctrine
- PF 2717 — postal services
- PF 7713 — prevailing learning systems
- PF 7713 — prevailing mental structures challenge human survival
- PE 9172 — psychiatrists
- PE 9172 — psychiatry
- PF 2005 Inadequacy religion
- PF 2005 Inadequacy religious doctrine

Inadequacy [scientific...]
- PF 5055 — scientific expertise
- PA 6041 — *Shortcoming*
- PF 3398 — social doctrine
- PE 5090 — staff qualifications intergovernmental organizations
- PE 0004 — Inadequacy telecommunication facilities developing countries
- PD 5077 Inadequacy training human settlements developing countries
- PA 7232 Inadequacy *Unskillfulness*

Inadequate [abortion...]
- PF 8481 — abortion methods
- PE 3580 — abortion procedures
- PD 0321 — *absorption iron*

Inadequate access
- PC 0352 — *indigenous peoples international decision-making processes*
- PF 2401 — information
- PD 1958 — negotiation employment reward
- PE 1055 — roads

Inadequate [accountability...]
- PF 6574 — accountability public sector management
- PE 8700 — adaptation policy educational difficulties
- PE 2844 — adjustment assistance industries labour affected developing country exports
- PE 0046 — administrative skills developing countries
- PF 6531 — adult education
- PF 1403 — *adult guidance*
- PF 2378 — *agricultural capital*
- PF 6499 — agricultural facilities
- PE 8082 — agriculture least developed countries
- PF 1495 — *air transport service*
- PE 5800 — air transport service land-locked developing countries

Inadequate animal
- PC 2765 — nutrition
- PE 2756 — quarantine
- PC 1167 — welfare

Inadequate [application...]
- PF 8191 — application available knowledge solve problems
- PE 1728 — appointment systems
- PF 3408 — appreciation culture
- PD 2543 — army discipline
- PF 6146 — arrangement housing respect common land
- PJ 9034 — assessment new technologies

Inadequate assistance victims
- PE 7390 — abuse power
- PE 4086 — accidents
- PD 4823 — crime
- PE 8240 — extortion
- PD 5122 — human rights violations
- PE 4449 — rape
- PE 6936 — torture

Inadequate [barter...]
- PE 0117 — barter system international trade
- PE 2820 — budgetary coordination United Nations systems
- PF 1442 — *budgeting practices*
- PD 8894 — building maintenance
- PF 8829 — building standards
- PD 2669 — buildings, services facilities organized action against problems
- PE 6554 — business know-how

Inadequate [capacity...]
- PE 4880 — capacity developing countries science technology
- PJ 9034 — capacity technology assessment
- PE 2778 — carcass disposal diseased animals

Inadequate care
- PF 0131 — children prisoners
- PF 2346 — community space
- PF 4955 — *mental health*
- PF 4832 — *teeth*
- PF 2477 *Inadequate career advice*
- PE 0430 — Inadequate cargo transportation

Inadequate cartographic
- PE 0046 — *information developing countries*
- PE 0046 — *skills developing countries*
- PE 0046 — *technology developing countries*
- PF 6451 *Inadequate cash flow*
- PE 4365 Inadequate channels direct investments developing countries

Inadequate child
- PD 2085 — care facilities
- PD 2085 — day-care facilities
- PE 3247 — maintenance

Inadequate child cont'd
- PC 0233 — welfare

Inadequate [childhood...]
- PD 5344 — childhood discipline
- PC 2494 — choice voters
- PF 6552 — circulation local information
- PF 2421 — citizen input
- PF 2346 — *coastal maintenance*
- PD 8697 — code enforcement
- PF 4340 — commercial finance rural development projects
- PF 1495 — *commercial roadways*

Inadequate community
- PE 8924 — care handicapped persons
- PF 1844 — care transient urban populations
- PD 1147 — security
- PE 8898 Inadequate compensation forced relocation
- PF 4944 Inadequate conception development programmes

Inadequate conditions work
- PE 4243 — agricultural workers
- PE 6841 — construction industry
- PE 8728 — employment public service personnel
- PE 8319 — employment utility supply services
- PE 4493 — hotel catering industries
- PE 5823 — textile industry

Inadequate [consensus...]
- PD 2438 — consensus process
- PD 2438 — consensus structure
- PE 5305 — constraints salary increases
- PE 0290 — consumer compensation
- PC 0123 — consumer protection

Inadequate control
- PD 2781 — animal diseases
- PF 9244 — development projects
- PD 0071 — mergers
- PC 1818 — over government administrative process
- PD 1770 — zoonoses
- PE 7904 Inadequate cooking stoves

Inadequate coordination
- PF 1467 — action against problems
- PE 1375 — action intergovernmental programmes national level
- PF 0175 — aid
- PE 4344 — governmental representation intergovernmental organizations
- PE 0730 — intergovernmental system organizations
- PE 1209 — international nongovernmental organizations programmes
- PD 0285 — international organizations programmes
- PE 1184 — regional intergovernmental organizations common membership
- PE 0296 — United Nations system
- PF 5172 Inadequate correctional systems
- PE 3831 Inadequate coverage developing country perspectives media

Inadequate credit
- PF 0245 — guidelines
- PF 2489 — monies
- PF 0245 — policies

Inadequate [crop...]
- PF 8827 — *crop insurance*
- PF 3698 — crop rotation
- PC 1532 — cultural integration immigrants

Inadequate [delivery...]
- PF 0301 — delivery mechanism response problems
- PD 2898 — demand primary commodities because rising living standards
- PF 4832 — *dental care*

Inadequate development
- PF 4297 — communication services least developed countries
- PF 5039 — critical awareness
- PE 8572 — enterprises developing countries
- PF 4676 — international criminal law
- PD 0822 — new social structures developing countries
- PF 9101 — policies ; Prohibitive cost
- PF 9821 — political thinking concerning global problematique
- PE 8572 — private sector developing countries

Inadequate [diet...]
- PC 0382 — *diet poor*
- PE 4950 — dietary fibre
- PF 3566 — disaster prevention mitigation
- PF 0286 — disaster rescue relief
- PF 9401 — disease prevention

Inadequate disinfection
- PE 8409 — humans after animal disease outbreaks
- PE 2757 — measures animal housing equipment
- PE 8409 — measures humans during animal disease outbreaks
- PD 2774 — pastureland after disease outbreak

Inadequate [dissemination...]
- PF 1267 — dissemination use available information
- PE 4305 — diversification loans developing countries
- PF 6453 — documentation
- PE 8088 — documentation works art
- PD 0465 — domestic savings developing countries
- PD 2269 — drainage system
- PF 4984 — *driver training*

Inadequate drug
- PD 2392 — *classification*
- PC 0231 — control
- PD 2392 — quality control
- PF 4928 Inadequate earthquake forecasting
- PE 6257 Inadequate earthquake resistant construction

Inadequate economic
- PC 0933 — cooperation developing countries
- PF 8554 — incentives
- PD 9412 — integration regional groupings developing countries
- PF 4884 — integration socialist countries
- PF 5964 — policy-making developing countries

Inadequate economic cont'd
- PF 2997 — priority system
- PF 0625 — technical data

Inadequate education
- PF 4984 — [Inadequate education]
- PE 8216 — concerning nature problems
- PF 4984 — *electorate*
- PC 3322 — indigenous peoples
- PE 6206 — nomadic children
- PF 6208 — peace

Inadequate educational facilities
- PD 0847 — [Inadequate educational facilities]
- PF 0775 — disabled
- PF 0775 — disabled persons
- PD 2051 — gifted children

Inadequate [educational...]
- PF 4984 — educational programs
- PD 8984 — *electrical maintenance*
- PE 1900 — electrical power supply developing countries
- PD 9033 — electricity infrastructure
- PF 2115 — *electricity supply*

Inadequate emergency
- PE 0366 — blood supply
- PF 9007 — care
- PC 5016 — food supplies
- PD 1428 — medical services
- PE 6018 Inadequate employee participation collective-bargaining

Inadequate empolderment
- PD 5110 — coastal lowlands
- PD 5110 — peat land
- PD 5110 — wetlands
- PC 7517 Inadequate energy conservation

Inadequate enforcement
- PD 2664 — against game poaching
- PC 4608 — civil rights
- PJ 4562 — foreign arbitral awards
- PC 4608 — human rights
- PD 5001 — safety regulations
- PE 4200 — school attendance

Inadequate environmental
- PD 1370 — education
- PF 4801 — monitoring
- PF 0011 — statistics

Inadequate [equipment...]
- PE 4129 — equipment ; Inaccurate medical diagnosis due
- PD 1565 — equipment maintenance
- PF 9244 — evaluation development projects
- PE 6700 — evaluation environmental costs
- PF 8661 — evidence convict known offenders
- PF 0209 — exchange technical information concerning problems
- PD 4475 — exercise
- PE 0523 — export promotion developing country small industry products
- PE 7184 — external debt management capacity developing countries

Inadequate facilities
- PD 0549 — children's play
- PF 6143 — grass-roots community initiatives
- PF 2016 — international nongovernmental organization action
- PD 0929 — international organization action
- PE 8900 — mentally disabled criminals
- PE 2829 — storing, transporting processing solid wastes

Inadequate facilities transport
- PE 1048 — electrical energy
- PE 8475 — sanitary wastes
- PD 1294 — water supplies

Inadequate family planning
- PD 1039 — education
- PE 9060 — education men
- PD 1039 — research

Inadequate [family...]
- PF 1000 — family structures
- PC 2765 — feeding animals
- PC 2765 — feeding farm animals

Inadequate financial
- PF 9841 — clearing systems
- PF 5374 — reporting public sector
- PB 4653 — reserves
- PE 0718 — *services*
- PF 9841 — settlement systems
- PE 0365 Inadequate financing local government developing countries

Inadequate fire
- PD 1631 — code
- PD 1631 — prevention
- PD 1631 — protection

Inadequate [firearm...]
- PD 1970 — firearm regulation
- PF 9007 — *first aid training*
- PF 2124 — fiscal discipline developing countries
- PF 4850 — fiscal policies
- PD 0452 — flood control

Inadequate food
- PC 5016 — stocks
- PE 4877 — storage facilities
- PE 7954 — supplies least developed countries

Inadequate [foresight...]
- PF 8176 — foresight capacity
- PG 1918 — *fresh air*
- PE 0741 — funding international nongovernmental organizations programmes
- PF 0498 — funding international organizations programmes

Inadequate [gaol...]
- PD 0520 — gaol conditions
- PD 0520 — gaol conditions

Inadequate [global...] cont'd
PF 9821 — global consensus concerning problems prospects humanity
PF 0817 — global cooperation solve world problems
Inadequate government
PF 3394 — [Inadequate government]
PF 9542 — control military
PF 2447 — educational policy
PF 1559 — policy
PF 5645 — policy
PF 3075 — publications
PE 4574 — savings capacity developing countries
Inadequate governmental
PF 0037 — energy conservation policies
PF 2335 — managerial instruments
PF 0038 — resource conservation policies
Inadequate [grazing...]
PF 6528 — *grazing land*
PF 4010 — *grooming*
PE 3992 — guardianship mentally retarded adults
PD 2269 — gully maintenance
PD 6520 — *Inadequate hauling equipment*
Inadequate health care
PD 1038 — family planning
PE 9242 — least developed countries
PE 7877 — urban slums
Inadequate health [control...]
PF 9401 — control
PD 8023 — education
PD 4004 — equipment
PD 4790 — services
PD 6265 — services following nuclear war
Inadequate [hero...]
PF 2834 — hero images
PE 0242 — home insulation
PE 5058 — hospital facilities
PD 6334 — hospital hygiene
Inadequate housing
PC 0449 — [Inadequate housing]
PD 0276 — aged
PD 1251 — *construction*
PC 0449 — Dependence
PE 0269 — developing countries
PF 4681 — disabled
PE 1925 — estates ; Socially
PC 3320 — indigenous peoples
PE 2763 — penning domestic animals
Inadequate [human...]
PD 5177 — human physical height
PE 9764 — human resources development least developed countries
PE 8398 — hygiene restrictions carcass meat exports
PE 9767 — hygiene slaughter houses
Inadequate [ideological...]
PD 0065 — ideological frameworks
PD 1308 — image roles marriages
PE 4883 — immune responses malnourished persons
PF 5969 — immunization methods
Inadequate implementation
PD 3712 — development programmes
PE 6630 — maritime safety standards
PF 1010 — plans programmes against problems
Inadequate [improvement...]
PE 2425 — *improvement funds*
PD 2716 — inaccurate textbooks reference books
PE 8506 — incentives increased productivity developing countries
Inadequate income
PF 6478 — *fishing*
PC 1966 — old age
PE 1139 — *rural communities*
PE 2238 Inadequate increase employment manufacturing industries developing countries
Inadequate industrial
PF 6473 — *access*
PD 4219 — capacity developing countries
PF 0963 — *incentives*
PF 0963 — *promotion*
PF 4013 — retraining programmes
PF 1790 — *services*
PF 4195 — services developing countries
PF 7130 — *space*
PF 4160 — trade developing countries
PE 4245 Inadequate infant welfare
Inadequate information
PF 4350 — concerning transnational banks
PF 0603 — drugs
PF 2878 — funding needs
PF 2401 — spread
PD 0104 — systems international governmental decision-making processes
Inadequate infrastructure
PC 7693 — [Inadequate infrastructure]
PD 7663 — nuclear power generation
PE 0289 — services least developed countries
Inadequate [inland...]
PD 2487 — inland waterway transport facilities
PF 5039 — inquiry-based learning
PF 2793 — inspection
PE 0365 — institutional structures local government developing countries
PF 5969 — insufficient immunization
PF 8827 — insurance
PF 6026 — insurance against damages arising natural disaster
PD 0321 — *intake iron*

Inadequate integration
PF 3402 — ideology society
PE 8066 — international information systems
PF 3403 — religions society
PF 6157 — transport systems
PF 7713 Inadequate intellectual cognitive infrastructure formulate effective global policies
PF 7380 Inadequate intellectual methods
Inadequate inter
PF 5006 — developing country skills flow
PE 8307 — regional cooperation problems countries
PF 3558 — village cooperation
Inadequate [interaction...]
PF 6185 — interaction humans animals urban environment
PJ 8295 — interface fundamentally different societies
PF 4816 — intergovernmental legal systems
Inadequate international
PF 4366 — cooperation reducing terrorism
PC 4591 — harmonization standards
PF 2113 — judicial system
PF 8421 — law enforcement
PD 0398 — map world
PE 4522 — marketing jute products developing countries
PE 7140 — nongovernmental organization response intergovernmental calls action
PE 5761 — participation island developing countries
PF 0203 — standards financial accounting reporting
PF 1262 — trade unions
PC 9584 — trading system
Inadequate [investment...]
PF 3845 — investment performance developing countries
PE 7892 — investment transnational corporations least developed countries
PD 8839 — irrigation system
PF 1658 Inadequate job opportunities
PF 6541 *Inadequate job publicity*
PE 8133 Inadequate knowledge incubation periods animal diseases
PF 7939 Inadequate knowledge reporting man-made disease
Inadequate labelling
PD 2468 — dangerous substances
PD 3521 — goods
PD 3521 — packages
Inadequate [labour...]
PC 0592 — labour force
PD 2269 — land drainage
PD 9189 — land reform
Inadequate laws
PC 6848 — [Inadequate laws]
PD 0590 — adoption
PF 5923 — sea
Inadequate legal
PF 8869 — aid
PF 1219 — counsel minorities
PF 0732 — counsel political dissidents
PF 3710 — rights natural objects
PF 3710 — rights trees
Inadequate legislation
PF 9299 — against environmental pollution
PE 7141 — against environmental pollution developing countries
PE 5794 — animal welfare
PF 1645 — relating action against problems
PF 0202 Inadequate leisure facilities
Inadequate level
PF 3060 — developing country monetary reserves
PD 0291 — investment developing countries
PF 3059 — world monetary reserves
Inadequate [life...]
PE 8736 — life insurance developing countries
PD 2760 — lighting factory farming units
PD 3427 — living working conditions immigrant labourers industrialized countries
Inadequate local
PE 0365 — authority legal structures developing countries
PF 6477 — capital
PF 0336 — enforcement
PE 7313 — expertise business practices developing countries
PC 6631 — government financing
Inadequate maintenance
PD 8984 — [Inadequate maintenance]
PD 2269 — drainage ditches
PD 6520 — equipment
PD 0645 — infrastructure
PD 1565 — machines
PF 2995 — *personnel*
PF 1773 — physical health
PD 9196 — water systems
Inadequate management
PF 3661 — employee communication
PF 9672 — government finances
PF 1442 — skills rural communities
Inadequate [manpower...]
PF 1467 — *manpower planning*
PD 4219 — manufacturing capacity developing countries
PF 2995 — *marketing knowledge*
PE 0523 — marketing products developing countries
PE 8857 — maternal child health care
PD 9021 — maternal education
PD 6733 — maternity protection employment
PD 1975 — means measuring intelligence
PF 1817 — means upholding global concern
PF 3062 — mechanism balance payments adjustment
PF 3061 — mechanism creation liquid reserves
PF 2857 — mechanisms securing sufficient food supplies
Inadequate medical care
PF 4832 — [Inadequate medical care]

Inadequate medical care cont'd
PD 4790 — infrastructure
PE 4820 — pregnant women
Inadequate medical [facilities...]
PD 4004 — facilities
PF 4832 — facilities
PC 7674 — *laboratory facilities developing countries*
PD 7254 — resources
PF 9587 — resources
PF 4832 — services
PD 9160 — training
Inadequate [meeting...]
PF 8939 — meeting methods
PF 4955 — *mental health services*
PF 4925 — mental hospitals
PF 3713 — methods
PF 1004 — methods ; Inability awaken consciousness due
Inadequate mobilization
PF 5964 — financial resources developing countries
PF 9704 — public opinion development
PF 4979 — resources
Inadequate [models...]
PF 9478 — models ideal society
PF 9576 — models socio-economic development
PF 9244 — monitoring development projects
PF 5374 — monitoring public sector activity
PF 6559 — *motorcycle control*
PF 6787 Inadequate narcotics legislation
Inadequate national
PD 0428 — empathy external hardship
PF 1373 — *language proficiency*
PE 4768 — law enforcement
PE 8280 — supervision adherence international law
Inadequate [negative...]
PF 7109 — negative capability
PE 9646 — negotiating capacity developing countries
PE 0853 — negotiation entrance terms transnational corporations developing countries
PE 6126 — night life entertainment facilities
PE 6084 — nuclear reactor safeguards
PC 0382 — nutrition
PE 0265 — nutrition education least developed countries
Inadequate [objectivity...]
PF 6691 — objectivity institutions
PF 0944 — organization knowledge
PE 8776 — organizational mechanism international action
PF 2431 — organizational mechanisms act against problems
Inadequate [packaging...]
PF 3143 — packaging agricultural products
PF 1187 — parent-school liaison
PF 8876 — parliamentary debating time
PE 3869 — participation control joint venture
PF 3538 — patent control
PF 3538 — patent extension
PF 4848 — peace research support
PC 1966 — pension income
PD 2459 — personal hygiene
PA 0911 — personal integration
PF 1467 — planning action against problems
PF 3581 — plant genetic resources conservation
PE 0714 — plant quarantine
Inadequate police
PF 3550 — *budget*
PF 8652 — control
PF 3550 — *funds*
Inadequate political
PF 4984 — *education*
PF 3215 — integration
PD 2213 — networks
PD 0548 — parties developing countries
PC 9058 — structure
Inadequate [port...]
PE 5792 — port infrastructure
PE 5908 — port storage facilities land-locked developing countries
PF 9175 — power intergovernmental organizations
PF 6472 — practical training rural areas
PF 4984 — *pre-school plan*
PF 4852 — *preparation trainers*
PF 4984 — *preschool education*
PE 1669 — preservation books
Inadequate prevention
PE 8542 — computer disasters
PF 4924 — crime
PF 0709 — disabilities
Inadequate [primary...]
PE 8553 — primary health care
PD 0520 — prison conditions penal systems
PJ 8060 — private international labour management relations
PF 0963 — procedures community planning
PD 4219 — production capacity developing countries
Inadequate protection
PE 8361 — civilians armed conflict
PD 1846 — facilities workers' representatives
PB 0883 — *individual welfare*
PF 8854 — intellectual property
PC 8999 — linguistic, national religious minorities
PF 7542 — preservation cultural property
PF 3021 — refugees
PE 3034 — war correspondents
PC 1916 Inadequate protein supply
Inadequate provision
PE 3247 — alimony
PD 0512 — home care services elderly
PE 3247 — palimony
PF 2874 — public safety

PF 4925 Inadequate psychiatric institutions
Inadequate public
PE 4796 — assistance developing countries
PD 3051 — debt relief developing countries
PA 0005 — disclosure
PF 8105 — entertainment
PD 1566 — *facilities*
PF 7842 — finance statistics
PF 8515 — health conditions
PF 5701 — information concerning problems
PF 1495 — transportation
PD 8003 — understanding science
PF 4875 *Inadequate punishment international criminals their own countries*
PD 1464 Inadequate punishment war criminals
PD 1435 Inadequate quality control
PE 2850 Inadequate quarantine
Inadequate [radiation...]
PF 6635 — radiation monitoring systems
PE 7544 — radiological services
PE 5848 — rail transport developing land-locked transit countries
PD 0496 — rail transport facilities
PF 1379 — *recall procedures unsafe products*
PF 6173 — recognition institutions transition adolescence
PF 0202 — recreation facilities elderly
PF 0202 — recreational facilities
PE 8833 — recreational facilities disabled people
PE 0290 — redress consumers' loss
PF 9129 — regional integration
PE 6853 — registration wills
Inadequate regulation
PF 2782 — Ineffective monitoring restrictive business practices due
PF 0591 — restrictive business practices service industries
PE 0225 — restrictive business practices state enterprises
Inadequate rehabilitation
PD 1089 — facilities
PE 8151 — facilities mentally handicapped
PE 8803 — juvenile offenders
PE 8983 — methods blind
Inadequate relationship
PE 1973 — international governmental nongovernmental organizations programmes
PE 0777 — international nongovernmental organizations specialized agencies United Nations
PE 1511 — transnational corporations local industry developing countries
PE 0106 — transnational corporations specialized agencies United Nations
Inadequate research
PE 4880 — development capacity developing countries
PF 1120 — development problems developing countries
PF 1077 — problems
PF 1572 — proposed solutions problems
Inadequate [reservoir...]
PF 3448 — *reservoir plan*
PF 6511 — residential housing rural communities
PF 9587 — resources health
PD 1080 — response societal needs
PF 6467 — results formal schooling
PD 2207 — riot control
Inadequate road
PE 1055 — access village
PE 6526 — *conditions*
PD 0490 — highway transport facilities
PD 0543 — highway transport facilities developing countries
PD 8557 — maintenance
PE 6526 — surfacing
Inadequate [roads...]
PE 5824 — roads transport land-locked developing countries
PF 1825 — roles elderly
PE 4950 — roughage diet
Inadequate rural
PF 4470 — community market
PF 2949 — employment structures
PF 6511 — housing developing countries
PF 2875 — infrastructure developing countries
Inadequate [safeguards...]
PE 8542 — safeguards against computer crimes
PD 1631 — safeguards against fire
PC 6848 — *safety legislation*
PD 5001 — safety precautions
PD 0876 — sanitation infrastructure
PF 0045 — satellite capacity telecommunications
PF 3260 — satisfaction sexual intercourse
PC 0927 — savings
PF 6541 — *schedule news*
PJ 3023 — sea transport facilities
PE 4766 — seaport facilities
PD 5345 — secondary education
PD 6589 — security system
PC 1867 — security systems
PE 8215 — segregation different categories juvenile offenders
PF 5354 — self-image
Inadequate sense
PE 4179 — community solidarity amongst workers
PF 1934 — personal identity
PF 9980 — time
Inadequate [sewerage...]
PD 0876 — sewerage systems
PD 0759 — sex education
PB 2150 — shelter
PF 8085 — skills curriculum
Inadequate social
PF 8085 — [Inadequate social]

Inadequate social cont'd
PF 0214 — demographic statistics
PF 4180 — development programmes
PD 0095 — discipline developing countries
PD 6893 — discipline socialist countries
PE 4796 — insurance developing countries
PC 1532 — integration foreigners
PD 9155 — protection farmers
PF 0677 — reform
PF 7611 — security migrants
PE 4796 — security welfare services developing countries
PF 3884 — technologies
PF 0019 — welfare indicators
PC 0834 — welfare services
Inadequate [soil...]
PD 6482 — soil conservation developing countries
PF 2699 — solution-oriented images global economic situation
PD 0549 — spatial facilities youth recreation
PF 3560 — sports equipment
PE 4052 — standard living developing countries
PE 7734 — standard living least developed countries
PC 0666 — standardization procedures equipment
PF 0344 — standards living
PC 9714 — standards teaching
PD 4101 — staple food supply developing countries
Inadequate statistical
PF 7117 — data nongovernmental organizations
PF 0625 — information data problems
PF 4106 — information tourism
PF 3467 *Inadequate storage facilities*
PF 5645 Inadequate strategies
Inadequate structures
PD 2802 — achieving global unity
PF 2350 — communication
PC 1547 — political dialogue review
Inadequate [study...]
PF 1205 — *study environment*
PE 1785 — subsidies organic farming
PF 2793 — supervision
PE 6243 — supply appropriate trained manpower developing countries
PE 4120 — supply pharmaceutical products developing countries
Inadequate support
PF 5622 — public interest broadcasting
PF 6492 — services
PE 8151 — services mentally handicapped
PF 6076 Inadequate system child support enforcement
PE 4997 Inadequate system political checks balances
Inadequate systems
PF 2905 — monitoring industrial growth
PF 7842 — national accounts developing countries
PF 6496 — transport communications rural villages
PF 4852 *Inadequate teacher training*
Inadequate teaching
PC 9714 — [Inadequate teaching]
PF 0905 — *equipment*
PE 8305 — occupational health safety
Inadequate technical
PF 1981 — advice consultation problems
PF 2813 — *assistance*
PF 0863 — cooperation problems
PE 8716 — training
Inadequate [telecommunications...]
PE 5655 — telecommunications island developing countries
PF 4132 — telephone systems developing countries
PD 1190 — testing drugs
PD 0876 — *toilet facilities*
Inadequate trade
PE 4523 — agricultural commodities developing countries
PD 5176 — developing countries
PF 4165 — related structural adjustments
PE 8266 Inadequate traffic control
PF 1332 *Inadequate trained leaders*
Inadequate training
PD 2036 — decision-making
PF 6032 — judges
PF 1352 — *opportunities*
PF 0713 Inadequate transport vehicles
Inadequate transportation
PD 1388 — facilities developing countries
PE 6526 — facilities rural communities developing countries
PF 2996 — systems rural citizens
Inadequate [understanding...]
PD 8023 — understanding hygiene
PC 1833 — urban political machinery
PF 1495 — *urban transport facilities*
Inadequate use
PF 6477 — manpower
PF 9716 — modern information systems
PF 7086 — visual imagery societal learning
PF 4892 Inadequate utilization volunteer social service workers
Inadequate [vaccination...]
PE 2786 — vaccination domestic animals
PF 3467 — *vehicle servicing*
PC 3320 — ventilation
PF 2822 — *visibility facilities*
PF 0422 — vocational education
PE 7317 — vocational rehabilitation facilities disabled persons
PD 8916 Inadequate wages
PF 3565 Inadequate warning disasters
Inadequate waste
PD 6795 — disposal
PD 6465 — *recovery*
PD 6795 — treatment

Inadequate water
PE 8233 — *catchments*
PD 8517 — control
PD 0722 — pressure
PD 8517 — services
PE 8233 — *storage*
PD 1204 — supply rural communities developing countries
PD 8517 — system infrastructure
Inadequate [watermains...]
PD 8517 — watermains locations
PJ 5279 — waterproofing
PE 0691 — weaning diet
PD 1574 — *weed control*
PD 3598 — weed control developing countries
Inadequate welfare services
PD 0512 — aged
PD 0542 — blind
PD 0601 — deaf
PD 0024 Inadequate work incentives
Inadequate working conditions
PE 4243 — agricultural workers
PE 3969 — construction industry
PD 1476 — developing countries
PE 8251 — employees commerce offices
PE 7718 — health medical services
PE 8633 — land tenants
PE 4243 — peasant farmers
PE 4243 — plantation workers
PE 3170 — professionals
PE 7165 — teachers
PD 3197 — women
Inadequate world
PF 2043 — calendar
PC 9112 — economic system
PF 2342 — *liquidity*
PF 4937 — regulatory agencies
PD 1544 Inadequate youth roles
PE 4834 Inadequate zoo facilities
Inadequately [cooked...]
PE 0596 — *cooked flesh ; Uncooked*
PE 0596 — cooked foods
PD 9736 — cooled shelters
PD 5173 Inadequately heated shelters
PF 9148 Inadequately publicised issues
PF 1781 *Inadequately publicized services*
PE 4545 Inadequately secured commercial property loans
PF 5421 Inadequately worded agreements
PF 9077 Inadmissability evidence disciplines
PE 5424 Inadmissability evidence jurisdictions
PA 6180 Inadvertence *Error*
PA 6247 Inadvertence *Inattention*
PA 5438 Inadvertence *Neglect*
PC 1288 Inadvertent modifications climate
PA 7395 Inadvisability *Inexpedience*
PA 7371 Inadvisability *Unintelligence*
PF 3710 Inanimate objects ; Denial rights
PC 2270 *Inanition*
PA 7365 Inanity *Boredom*
Inanity [Ignorance...]
PA 5568 — *Ignorance*
PA 6876 — *Impotence*
PA 6852 — *Inappropriateness*
PA 6977 Inanity *Meaninglessness*
PA 6940 Inanity *Thoughtlessness*
Inanity [Unimportance...]
PA 5942 — *Unimportance*
PA 7371 — *Unintelligence*
PA 7204 — *Unsavouriness*
PA 5982 Inapplicability *Disagreement*
PF 8191 Inapplicability formal learning
PA 6852 Inapplicability *Inappropriateness*
PE 6187 Inappropriate accommodations single people
PE 7695 Inappropriate afforestation
Inappropriate agricultural subsidies
PE 1785 — chemicalized farming
PE 1785 — livestock production
PE 1785 — sales pesticides
Inappropriate [aid...]
PF 8120 — aid developing countries
PF 9852 — ambitions
PF 2256 — application traditional values
PF 2152 — arguments
PF 6814 — assumptions
PD 8294 Inappropriate basic hygiene
Inappropriate [cash...]
PF 9187 — cash crop policy
PF 1773 — *clothing*
PF 0245 — credit system
PD 6282 — cultivation steep slopes
PF 4944 Inappropriate design development projects
Inappropriate development
PB 6207 — [Inappropriate development]
PF 8757 — *policy*
PF 6670 — privileged classes developing countries ; Reinforcement
PF 1654 Inappropriate dialogue
PE 3799 Inappropriate drugs ; Prescription
PF 9728 Inappropriate economic theory
Inappropriate education
PD 8529 — [Inappropriate education]
PF 1531 — developing countries
PF 1905 — graduates
Inappropriate [educational...]
PF 1266 — educational testing procedures
PF 8161 — educational values
PD 0024 — employment incentives

Incongruity

Inappropriate [energy...] cont'd
PE 8592 — energy equipment developing countries ; Insufficient
PE 6640 — equipment developing countries ; Acquisition
PF 7916 — expertise
Inappropriate [food...]
PE 0302 — food aid
PE 8255 — foodstuffs breast feeding ; Substitution
PD 8030 — foreign investment
Inappropriate [goals...]
PF 9852 — goals
PF 0304 — government intervention
PE 1170 — government intervention agriculture
PD 8638 Inappropriate housing ; Socially
Inappropriate [imprisonment...]
PE 5112 — imprisonment
PB 1845 — industrialization
PD 9661 — infant feeding strategies
PF 9786 — institutionalized reward systems
PF 0048 — international institutional framework finance
PE 1785 — irrigation subsidies developing countries
Inappropriate [labelling...]
PD 3521 — labelling
PF 2208 — leadership development
PD 4519 — legal definition health risks
PF 2410 — level technological equipment
PF 4580 — loans
PE 3812 — loans available developing countries
PF 1578 — local requirements ; Programme initiatives
Inappropriate [maintenance...]
PD 6520 — maintenance equipment
PD 3712 — management development projects
PF 4799 — modernization agriculture
PF 2418 Inappropriate orders ; Execution
Inappropriate [personal...]
PD 5494 — personal habits
PD 4629 — pesticides ; Use
PE 4120 — pharmaceutical products developing countries
PF 5645 — policies
PD 1018 — products ; Dependence government revenues exploitation environmentally
PF 4936 — prolongation life
PF 6377 — public spending government
Inappropriate [re...]
PE 7695 — re-forestation
PF 3498 — *role learning*
PF 8451 — *role models*
Inappropriate [sanitation...]
PD 0876 — sanitation systems
PC 0834 — *school lunches*
PD 7415 — scientific medical testing
PF 1266 — selection examination procedures education
PD 6585 — size school classes
PE 4596 — structural technological changes ; Training
PE 7185 — structural technological changes ; Working hours
PF 1205 — *study patterns*
Inappropriate [tax...]
PD 3436 — tax assessment
PF 3049 — taxation international non-profit organizations
PF 3049 — taxation not-for-profit philanthropic charitable organizations
PF 0878 — technologies developing countries ; Use
PE 5820 — technology ; Introduction
PF 9576 — theoretical models social development
PE 4332 — transfer forms worship
PE 5820 — transfer technology
PE 1337 — transplantation industrialized country methods developing countries
PE 7695 — tree plantations
PF 8648 Inappropriate understanding progress
Inappropriate use
PD 9338 — financial resources
PD 3712 — *innovative management techniques*
PE 6754 — mail service
PF 0716 — resources fundamental sciences
PE 4450 — telecommunications services
PA 6852 Inappropriateness
PA 5982 Inappropriateness *Disagreement*
PA 6852 Inappropriateness *Inappropriateness*
PA 7395 Inappropriateness *Inexpedience*
PA 5612 Inappropriateness *Unchastity*
PA 6921 Inappropriateness *Undueness*
PA 5821 Inappropriateness *Vulgarity*
PA 7280 Inappropriateness *Wrongness*
PA 5982 Inaptitude *Disagreement*
PA 7395 Inaptitude *Inexpedience*
PA 7232 Inaptitude *Unskillfulness*
PA 7367 Inarticulation *Unintelligibility*
PA 6491 Inarticulation *Vanity*
PA 6247 Inattention
PA 6379 Inattention *Avoidance*
PF 4861 Inattention disease-prone
PA 5438 Inattention *Neglect*
PA 7371 Inattention *Unintelligence*
PF 2346 Inattentive care public spaces
PF 6683 Inauspicious conditions
PC 2270 Inborn errors metabolism
PC 2379 Inbreeding
PC 1408 Inbreeding domestic animals ; Excessive
PD 7465 Inbreeding ; Genetic
PA 6876 Incapability *Impotence*
PA 7341 Incapability *Unpreparedness*
PA 7232 Incapability *Unskillfulness*
PA 5558 *Incapacity*
PJ 8271 Incapacity developing countries organize
PF 1216 *Incapacity happiness*
PA 6876 Incapacity *Impotence*

PF 1216 *Incapacity pleasure*
PA 7371 Incapacity *Unintelligence*
PA 7232 Incapacity *Unskillfulness*
PA 5558 Incapacity *Weakness*
PE 2308 *Incapacity work*
PE 5112 Incarceration ; Needless
PF 9139 Incarnate devil
PE 5505 Incendiarism
PD 3492 Incendiary weapons massive destructiveness
PJ 8933 Incentive ; Curbed learning
PF 4516 Incentive users care common property ; Lack
PF 4516 Incentives ; Asymmetry economic system
PD 1432 Incentives developing countries ; Lack response monetary
PD 0024 Incentives employment ; Ill conceived
PF 0107 Incentives having children ; Financial
Incentives Inadequate
PF 8554 — economic
PF 0963 — *industrial*
PD 0024 — work
Incentives [Inappropriate...]
PD 0024 — Inappropriate employment
PD 5114 — increase agricultural output developing countries ; Lack
PE 8506 — increased productivity developing countries ; Inadequate
PF 8554 — Insufficient economic
PF 2168 — *Insufficient educational*
Incentives [Lack...]
PF 2168 — *Lack educational*
PF 8161 — *Lack retraining*
PF 6557 — *Limited manufacturing*
PF 0963 — *Low industrial*
PF 2168 *Incentives students ; Lack*
PE 7165 *Incentives teachers ; Insufficient*
PF 0905 *Incentives tourism ; Lack*
PF 2009 *Incentives ; Untenable investment*
PF 2148 Incest
PF 2148 *Incest ; Homosexual*
PF 2148 *Incest ; Lesbian*
PF 2148 Incest pregnancy
PE 4553 Incidents ; Accidental military
PF 9558 Incidents ; Fabrication politically sensitive
PF 9558 Incidents ; Fabrication provocative
PD 2950 Incidents violence ; Border
PE 8961 Incisional hernia
PD 7520 *Incitement assassinate*
PE 5952 Incitement hatred
PD 7520 *Incitement kill*
PD 7520 *Incitement kill ; Government*
PD 1964 Incitement rebellion
PD 6008 Incitement terrorism ; Government
PD 7520 *Incitement violence*
PD 4714 *Incitement war*
PD 6392 Inciting riot
PC 7013 Incivility
PA 7143 Incivility *Discourtesy*
PA 5821 Incivility *Vulgarity*
PA 7156 Inclemency *Moderation*
PA 7023 Inclemency *Pitilessness*
PA 5643 Inclemency *Unkindness*
PD 2730 Inclusion body hepatitis
PD 0594 *Inclusion disease ; Cytomegalic*
PF 9576 Inclusive environmental models ; Resistance
PF 2754 Inclusive management decisions ; Non
PF 8094 Incoherence
PA 5982 Incoherence *Disagreement*
PA 6858 Incoherence *Disintegration*
PA 7157 Incoherence *Insanity*
PA 7309 Incoherence *Uncertainty*
PA 7367 Incoherence *Unintelligibility*
PF 1559 Incoherent government policies
PF 1345 Income ; Absence imaginative planning expansion
PF 9008 Income available ; Imbalances people's aspirations structure opportunities
PC 6641 Income business activities ; Unreported
PD 1842 Income cost housing ; Disparity
Income [Denial...]
PD 8568 — Denial right just
PD 2968 — dependence primary commodities developing countries ; Excessive
PF 3084 — Dependency unpredictable sources
PD 2769 — developing countries ; Low level personal
PE 5022 — differential ; Rural urban
PF 6478 — *Diminished diving*
PF 5008 — disadvantaged groups ; Limited availability land low
PE 1660 — disparity developing countries due transnational corporations ; Increasing
Income distribution
PC 2815 — countries ; Dependence unequal
PC 2815 — countries ; Unequal
PD 7615 — developing countries ; Unequal
PE 6891 — industrialized countries ; Unequal
PD 4962 — Unequal
PF 3084 Income drops ; Fluctuating
PF 6478 Income expansion ; Slow rate
PF 1345 Income expansion ; Underdeveloped sources
PF 1240 Income farming ; Underdeveloped capacity
PF 6478 *Income fishing ; Inadequate*
PC 2815 Income gap ; International
PC 1966 Income ; Inadequate pension
PF 1737 Income industrialized countries ; Instability growth
Income inequality
PD 4962 — [Income inequality]
PC 2815 — countries
PE 6891 — developed countries

Income inequality cont'd
PD 7615 — developing countries
PD 8568 Income ; Insufficient personal
PF 3084 Income jobs ; Insufficient side
Income [level...]
PE 8171 — level rural communities ; Subsistence agricultural
PD 6579 — Limited family
PE 8171 — Low faming
PD 8568 — Low general
Income [maintenance...]
PC 1966 — maintenance after retirement
PD 6579 — Marginal level family
PD 8568 — Minimal excess
Income [old...]
PC 1966 — old age ; Inadequate
PC 1966 — old age ; Reduced
PE 8957 — outflow developing developed countries ; Foreign private investment
Income [population...]
PD 8568 — population ; Increasing low
PF 1240 — possibilities ; Undiversified basis
PD 8568 — potential ; Low
PD 1018 — products dangerous health ; Dependence government
PE 1139 *Income rural communities ; Inadequate*
PF 6478 *Income rural communities ; Low fishing*
Income [Single...]
PF 1240 — Single source
PF 6538 — skills ; Unrecognized
PE 8659 — spent food ; Excessive proportion
PF 2489 — subsidy ; Unavailability
PD 3436 Income tax ; Unequal graduation
PF 1240 Income ; Unexplored livestock
PE 8171 Income ; Unpredictable farming
PE 1139 *Income village communities ; Insufficient trade*
PE 8584 Incomes developing countries ; Imbalance rural urban
PE 5022 Incomes ; Imbalance rural urban
PE 8004 Incommunicado detention
PF 9047 Incompatibility
Incompatibility [Disaccord...]
PA 5532 — *Disaccord*
PA 5982 — *Disagreement*
PE 7753 — document classification systems
PA 5446 Incompatibility *Enmity*
PF 9728 Incompatibility environmental economic decision-making
PF 2648 Incompatibility rural values urban cultures
PE 3337 Incompatibility traditional new technologies
PF 2403 Incompatibility transport modes
PA 6653 Incompatibility *Unsociability*
PE 7020 Incompatible equivalent national sections international nongovernmental organizations
PF 1115 Incompatible forms worship
PF 7178 Incompatible moral interests
PF 5766 Incompatible priorities
PC 1227 Incompatible technologies
PA 6416 Incompetence
PF 1509 *Incompetence ; Assumed youth*
PE 5742 Incompetence ; Citizen
PA 6416 Incompetence ; Dependence
PF 3962 Incompetence government leaders
PF 3953 Incompetence ; Governmental
Incompetence [Impotence...]
PA 6876 — *Impotence*
PA 5652 — *Inferiority*
PA 5473 — *Insufficiency*
PA 6416 Incompetence ; Military
PF 9429 Incompetence stand trial
PA 7341 Incompetence *Unpreparedness*
PA 7232 Incompetence *Unskillfulness*
PD 5770 Incompetent doctors
PF 0760 Incompetent financial management
PC 4867 Incompetent management
PA 6416 Incompetent people
PE 4429 Incompetent plastic surgery
PD 4535 Incompetent workers
PF 6517 Incomplete access development capital
PF 2401 Incomplete access information resources
PD 2438 Incomplete consensus method
PF 4774 *Incomplete cost projections*
PF 2949 *Incomplete direct transportation*
Incomplete [identification...]
PF 6537 — *identification signs*
PF 2863 — implementation community decisions
PD 8839 — *irrigation system*
PF 6550 *Incomplete planning process*
PF 2401 *Incomplete skills information*
PF 4014 *Incomplete state sciences*
PF 2212 Incomplete understanding new societal service systems
PF 2969 *Incomplete utilization external relations*
PA 6652 Incompleteness
Incompleteness [Imperfection...]
PA 6997 — *Imperfection*
PA 6876 — *Impotence*
PA 6652 — *Incompleteness*
PA 5473 — *Insufficiency*
PA 5568 Incomprehensibility *Ignorance*
PF 1748 Incomprehensibility specialized jargon
PA 7371 Incomprehensibility *Unintelligence*
PA 7367 Incomprehensibility *Unintelligibility*
PF 1677 Incomprehensible presentations
PD 5636 Inconclusive convictions fraud
PF 6349 Inconclusiveness science
PD 7415 Inconclusiveness scientific medical tests
PF 3790 Incongruent actions ; Induction
PA 5982 Incongruity *Disagreement*

Incongruity

PA 7395 Incongruity *Inexpedience*
PA 5502 Incongruity *Reason*
PF 1365 *Incongruous religious images*
PA 5502 Inconsequence *Reason*
PA 7408 Inconsequence *Smallness*
PA 5942 Inconsequence *Unimportance*
PE 3926 Inconsiderate choice payment times creditors
PD 3051 Inconsiderate insistence creditors fulfilment debt service obligations developing countries
PF 1205 *Inconsiderate telephone use*
PD 1224 Inconsiderate use herbicides
PA 7143 Inconsiderateness *Discourtesy*
PA 7325 Inconsiderateness *Irresolution*
PA 5438 Inconsiderateness *Neglect*
PA 7371 Inconsiderateness *Unintelligence*
PA 5643 Inconsiderateness *Unkindness*
PF 2992 Inconsistencies ; Social service
PA 5490 Inconsistency *Changeableness*
Inconsistency [Difference...]
PA 6698 — *Difference*
PA 5982 — *Disagreement*
PA 6858 — *Disintegration*
PF 2497 *Inconsistency international commodity policies*
PF 5356 Inconsistency ; Legal
PF 9818 Inconsistency marriage partners ; Psychological
PA 5502 Inconsistency *Reason*
PE 7637 Inconsistent application telecommunications technology
PF 6540 *Inconsistent inventory levels*
PF 1624 Inconsistent medical practices
PF 5421 *Inconsistent official texts*
Inconsistent [pharmaceutical...]
PF 1624 — pharmaceutical practices
PF 9166 — policy making
PF 5887 — port charges
PF 5482 Inconsistent risk evaluation
PF 5356 Inconsistent sentencing offenders
PA 5490 Inconstancy *Changeableness*
PA 7363 Inconstancy *Improbity*
PA 7325 Inconstancy *Irresolution*
PE 4619 Incontinence
PA 6379 Incontinence *Avoidance*
Incontinence [Infant...]
PE 4619 — Infant
PA 5473 — *Insufficiency*
PA 6466 — *Intemperance*
PA 7296 Incontinence *Restraint*
PA 5612 Incontinence *Unchastity*
PA 5497 Inconvenience *Difficulty*
PE 8036 Inconvenience dwarfs, midgets giants ; Domestic
PA 7395 Inconvenience *Inexpedience*
PA 6921 Inconvenience *Undueness*
PA 7006 Inconvenience *Untimeliness*
PF 6476 *Inconvenient class scheduling*
PD 1204 Inconvenient water supply
PD 0606 Incorporation advertising entertainment
PE 8934 Incorporation carcinogens consumer goods
PB 3095 Incorrect information
PB 3095 Incorrect information ; Dependence
PE 4302 Incorrect spelling
PA 6180 Incorrectness *Error*
PA 7280 Incorrectness *Wrongness*
PA 6099 Incorrigibility *Hopelessness*
PA 7325 Incorrigibility *Irresolution*
PA 5644 Incorrigibility *Vice*
Increase [abandoned...]
PF 2813 — *abandoned buildings*
PD 5114 — agricultural output developing countries ; Lack incentives
PD 2784 — *animal disease increase aviation*
PD 4031 — animal genetic defects
PD 0329 — anti-social behaviour developing countries
PF 1533 — antisocial behaviour due television
PE 8815 — atmospheric concentration methane
PD 2784 — *aviation ; Increase animal disease*
PF 1471 Increase competition export markets encountered developing countries
PE 9446 Increase concentration atmospheric water vapour
PD 5543 Increase ; Decline rate productivity
PF 6060 Increase defence budget
PE 2238 Increase employment manufacturing industries developing countries ; Inadequate
PE 7832 Increase insect pests modification micro-climate
PE 4329 Increase insurance claims medical negligence
PC 1258 Increase military capacity ; Myth national security
PE 9443 Increase nitrous oxide atmosphere
PF 7932 Increase operating costs
PE 2535 *Increase out-of-town shopping*
PE 7832 Increase pests diseases perennial irrigation
PD 9480 Increase plant genetic defects
PE 4263 Increase rainfall
PD 2072 Increase road vehicles
PD 1354 Increase trace gases atmosphere
PD 5114 Increased agricultural output developing countries ; Constraints
PE 7706 Increased autonomy ; Fear
PE 4508 Increased cooperation central banks ; Need
PE 8013 Increased demand developed countries meat egg products
Increased [family...]
PF 1509 — *family fragmentation*
PF 2949 — *fear taxation*
PC 2518 — *food consumption*
PE 8506 Increased productivity developing countries ; Inadequate incentives
PF 5069 Increased reflection solar radiation
PD 1038 Increased use birth control pills

PE 5305 Increases ; Inadequate constraints salary
PE 4603 Increases interest rates long-term debts developing countries
PE 4003 Increases provided legislation collective agreements ; Non implementation workers wage
PD 9515 *Increases seismicity ; Reservoir induced*
PF 1238 *Increases ; Uncontrolled household rent*
PE 5305 Increases ; Unrestrained wage
PC 0918 Increasing air temperature
PD 4387 Increasing atmospheric carbon dioxide
PF 7911 Increasing cost social security
Increasing development lag
PE 2000 — against information growth
PF 3743 — against population growth
PC 5879 — against socio-economic growth
PE 3078 — against technological growth
PD 3825 *Increasing drug addiction drug producing countries*
PE 6201 Increasing drug experimentation developing countries
PE 8763 Increasing economic value labour
PE 5592 Increasing female criminality
PE 5748 Increasing foreign debt island developing countries
PE 1447 Increasing government expenditure arms
PE 2000 Increasing inability integrate available knowledge
PE 1660 Increasing income disparity developing countries due transnational corporations
PD 2656 Increasing job monotony
Increasing [lag...]
PE 7139 — lag education against growth knowledge
PE 5369 — lag education against population growth
PD 8568 — low-income population
PF 0719 Increasing number disabled persons
PD 2103 *Increasing number single person households*
Increasing [pace...]
PF 2304 — pace life
PE 5931 — proportion land surface devoted urbanization
PF 6234 — public health expenditures
PF 1447 — public spending defence
PD 8561 Increasing requirements aged persons
PF 2819 Increasing resistance charitable giving
PF 7649 Increasing scepticism accuracy official information
PC 5879 Increasing unsustainability development
PA 7392 Incredulity *Unbelief*
PD 6943 Incriminate ; Government conspiracy
PD 6943 Incrimination entrapment
PD 6943 Incrimination ; False
PE 7252 Incrimination innocent ; Self
PE 8133 Incubation periods animal diseases ; Inadequate knowledge
PF 5153 Incumbents ; Electoral organization favouring political
PC 8801 Incurable diseases
PA 6598 Incuriosity
PA 6379 Incuriosity *Avoidance*
PA 6247 Incuriosity *Inattention*
PF 0995 Incurring legal expenses ; Risk
Indebted [corporations...]
PE 1557 — corporations ; Highly
PF 9249 — countries ; Disincentive invest heavily
PC 3056 — countries ; Heavily
PE 0740 Indebtedness arising insurance transactions developing countries ; International
PF 8842 Indecency
PA 5612 Indecency *Unchastity*
PA 5821 Indecency *Vulgarity*
PD 2547 Indecent advertising
PE 5042 Indecent art
PF 4550 Indecent conduct parliament
PF 4317 Indecent exposure
PF 7427 Indecent speech
PF 4958 Indecipherable script
PF 8808 Indecision
PF 9564 Indecisive multilateralism
PF 1336 Indecisive response technological changes
PF 8989 Indecisiveness
PA 6900 Indecisiveness *Formlessness*
PA 7325 Indecisiveness *Irresolution*
PA 7309 Indecisiveness *Uncertainty*
PA 5558 Indecisiveness *Weakness*
PA 5612 Indecorum *Unchastity*
PA 5821 Indecorum *Vulgarity*
PA 7280 Indecorum *Wrongness*
PF 6011 Indefensibility *Discontentment*
PA 6486 Indefensibility *Injustice*
PF 1267 *Indefinite options communicating*
PA 5612 Indelicacy *Unchastity*
PA 5821 Indelicacy *Vulgarity*
PF 1066 Independence extra territorial bases ; Unilateral declarations
PE 7706 Independence ; Fear
PE 7468 Independence ; Fear loss
PE 6773 Independence handicap ; Physical
PE 7665 Independence judges lawyers ; Lack
PE 3194 Independence ; Lack internal political
PE 0297 Independence ; Lack political
PJ 3472 Independence professionals ; Threats
PE 5807 Independence transnational corporations ; Excessive power
PF 2457 *Independent farming operations*
PE 7665 Independent judges ; Denial right
PF 6570 *Independent pride ; Misplaced*
PF 1078 *Independent subsistence style*
PF 0192 Indeterminacy death
PF 9276 Indeterminate future human values
PF 9276 Indeterminate future priorities
PF 3551 Indicator ; Unreliability profit social
PE 1843 Indicators exclusion human requirements ; Market

PF 0019 Indicators human development ; Excessive reliance economic
PF 0019 Indicators ; Inadequate social welfare
PA 7604 Indifference
PA 6379 Indifference *Avoidance*
PF 8125 Indifference community ; Police
PA 7604 Indifference ; Dependence
Indifference [Imperfection...]
PA 6997 — *Imperfection*
PA 5806 — *Inaction*
PA 6247 — *Inattention*
PA 6598 — *Incuriosity*
PB 5249 Indifference ; Moral
PA 5438 Indifference *Neglect*
PA 6444 Indifference *Quiescence*
PB 5249 Indifference response injustice
Indifference [social...]
PD 9219 — *social relationships*
PF 1205 — *students' families*
PB 5249 — suffering
PJ 7629 Indifference toward diplomacy
PA 7364 Indifference *Unfeelingness*
PA 5942 Indifference *Unimportance*
PD 9316 Indifference victims crime ; Legal
PA 7604 Indifferent people
PD 3132 Indigenous capital developing countries ; Outflow
Indigenous cultures
PC 7203 — *Degradation*
PC 7203 — Destruction
PE 1817 — Disparagement
PC 7203 — Endangered
PD 4252 Indigenous economies ; Destabilization
PC 3304 Indigenous land claims ; Unsettled
PC 3304 Indigenous lands ; Invasion
PC 0720 Indigenous minorities ; Violence against
Indigenous [people...]
PE 1024 — people be self governing ; Denial rights
PC 0352 — people ; Denial rights
PE 7242 — peoples ; Alcoholism amongst
Indigenous peoples Denial
PE 5484 — right adequate standard living
PE 4332 — right freedom religion
PE 1506 — right social security
PE 1506 — right social welfare services
PE 4972 — rights development
PE 7765 — social rights
Indigenous peoples [Discrimination...]
PC 3320 — Discrimination against indigenous populations housing Denial right adequate housing
PC 3322 — education ; Denial right
PC 0720 — Endangered tribes
PD 3321 — Illiteracy
PC 3322 — Inadequate education
PC 3320 — Inadequate housing
PC 0352 — *international decision making ; Lack participation*
PC 0352 — *international decision making processes ; Inadequate access*
PC 3319 — Malnutrition
PE 7312 — participate political processes ; Denial right
PD 2490 — Separatism
PE 1506 — Social insecurity
PE 2142 — use their own language ; Denial right
PC 0720 — Victimization
Indigenous populations Denial
PC 3319 — right adequate food
PD 1092 — right employment
PE 4459 — right health
PE 2317 — right legal services
PD 4252 — right traditional economies
Indigenous populations [Discrimination...]
PC 0352 — Discrimination against
PC 3322 — education ; Discrimination against
PD 1092 — employment ; Discrimination against
PD 1092 — employment ; Exploitation
PC 3304 — Expropriation land
PC 3320 — housing Denial right adequate housing indigenous peoples ; Discrimination against
PE 3721 — imported diseases ; Massacre
PE 3721 — introduction diseases ; Vulnerability
PD 3305 — reservations ; Restriction
PE 7573 — Violation treaties
PE 4332 — Indigenous religion ; Prohibition practice
PC 3304 — Indigenous rights lands resources ; Denial
Indigenous [society...]
PE 8002 — society non socialist countries ; Lack economic adaptation
PE 4332 — spirituality ; Repression
PE 8002 — systems ; Lack economic adaptation
PE 4724 Indigestion
PD 3978 *Indigestion animals ; Simple*
PD 3978 *Indigestion ; Mild dietary*
PD 3978 *Indigestion ; Vagus*
PA 6191 Indignation *Disapproval*
PA 7253 Indignation *Envy*
PE 8867 Indirect taxes import charges ; Distortion international trade selective
PA 6228 Indirection *Deviation*
PA 5974 Indirection *Diffuseness*
PA 7363 Indirection *Improbity*
PA 7411 Indirection *Uncommunicativeness*
PD 2543 Indiscipline ; Military
PE 9092 Indiscipline schools
PF 8078 Indiscipline ; Social
PA 6180 Indiscretion *Error*
PA 6446 Indiscretion *Indiscrimination*
PA 7115 Indiscretion *Rashness*

Industrialized countries

PA 5612 Indiscretion *Unchastity*
PA 7371 Indiscretion *Unintelligence*
PA 5644 Indiscretion *Vice*
PF 6386 Indiscriminate anti-trust prosecution
PC 0418 Indiscriminate application distribution technology
PC 0495 Indiscriminate economic growth
PC 0495 Indiscriminate economic growth ; Development
PD 9082 Indiscriminate fishing methods
PF 3454 *Indiscriminate refuse dumping*
PE 4640 Indiscriminate use tear gas
PA 0429 *Indiscriminate violence*
PD 1519 Indiscriminate weapons ; Dependence inhumane
PD 1519 Indiscriminate weapons ; Inhumane
PA 6446 Indiscrimination
PA 6379 Indiscrimination *Avoidance*
PA 6446 Indiscrimination *Indiscrimination*
PA 6387 Indiscrimination *Necessity*
PA 6799 Indisposition *Disease*
PA 6509 Indisposition *Unwillingness*
PF 1246 Indistinct awareness sexual identity
PF 2575 *Indistinct community identification*
PF 2384 Individual attention span ; Limited
PF 2575 Individual avenues civic responsibility
PF 2899 Individual capital reserves ; Limited
Individual decision making
PF 3559 — context ; Isolation
PF 1457 — Paralysis
PF 3559 — process ; Fragmented
PF 4371 *Individual development ; Lack*
PF 7934 Individual differences ; Overriding
PF 2437 *Individual effectiveness ; Unexperienced*
PF 1030 Individual family priority
PD 2670 Individual fear future change
PF 1403 *Individual goals ; Neglect*
PD 1591 Individual historical context ; Lack
Individual [initiative...]
PC 2178 — *initiative ; Lack*
PF 2575 — involvement ; Constricting patterns
PF 7142 — *involvement ; Reluctancy*
PD 1749 — isolationism
PE 7448 Individual mass media ; Manipulation
PJ 0646 Individual need credit
PE 5397 Individual needs ; Legal contract system reduced
Individual [parliamentarians...]
PE 9759 — parliamentarians ; Violation human rights
PF 6127 — participation large communities ; Ineffectiveness
PA 0911 — personalities ; Unintegrated
PF 6580 — problem overemphasis
PC 6873 — *property ; Abuse*
PF 6148 — psychological development life cycle ; Inhibition
PE 0261 Individual receivers ; Foreign controlled direct satellite broadcasting
Individual rights
PF 0922 — due excessive court costs ; Ineffective protection
PF 2475 — Overemphasis
PF 1075 — political asylum ; Lack
PE 1479 — Psychological barriers judicial protection
PF 1081 Individual's life cosmos ; Absence convincing symbols connecting
PF 4754 Individual's social functions ; Separation
Individual [scientific...]
PF 2112 — scientific research ; Dehumanized
PF 1729 — social commitment ; Lack
PF 5692 — social development ; Constraint time
PF 2602 — survivalism ; Limiting effect
PA 0005 *Individual talents ; Hidden*
PF 4946 Individual unfitness survival
PF 0979 Individual victim image
PB 0883 *Individual welfare ; Inadequate protection*
PF 8393 Individualism
PD 3106 Individualism capitalist systems ; Lack
PF 0951 Individualism ; Despairing
PD 1106 Individualism ; Militant
PE 8143 Individualism restriction effectiveness
PF 0951 Individualism ; Rugged
PF 3161 Individualism ; Survival induced
PF 3161 Individualist survival mindset
PF 1205 *Individualistic agricultural approach*
PF 1205 *Individualistic agricultural planning*
PE 1848 *Individualistic construction methods*
PD 4974 Individualistic disposition productive property
PC 0492 *Individualistic economic practices*
PF 2896 Individualistic employment images
PF 1861 Individualistic family structures
PD 1147 Individualistic images security
PF 1705 *Individualistic land holding*
PF 7142 *Individualistic life style*
PF 6796 Individualistic meaning ; Search
Individualistic [patterns...]
PF 2845 — *patterns ; Ingrained*
PF 1682 — perception sexual activity
PF 1176 — practices local business
PF 1705 Individualistic retaining local tradition
PF 2687 Individualistic standards sexual behaviour
PF 2110 Individualistic style professionals
PC 6873 *Individualistic use property*
PF 5639 Individualistic utilization expertise
PF 2560 Individualistic welfare responsibility
PE 5397 Individualistically reduced legal contract
PF 3008 Individualized business mindset
PF 1205 *Individualized farming system*
PF 3554 *Individualized marketing practices*
PD 2294 Individually defined operating structure marriage
PF 4220 Individuals institutions ; Domination
PD 2929 Individuals ; Maintenance political dossiers
PF 2344 Individuals national policy making ; Disproportionate influence some
PD 8017 Individuals ; Official disregard
PF 0749 Individuals participate social processes ; Failure
PC 7662 Individuals ; Suppression opposition groups
PF 2344 Individuals ; Undisclosed control national economies limited number
PA 7250 Indocility *Disobedience*
PA 7325 Indocility *Irresolution*
PA 6509 Indocility *Unwillingness*
PD 3097 Indoctrination ; Compulsory
PD 3097 Indoctrination ; Dependence
PD 1624 Indoctrination ; Political
PD 4890 Indoctrination ; Religious
PA 7710 Indolence *Idleness*
PA 5806 Indolence *Inaction*
PA 6444 Indolence *Quiescence*
PD 6627 Indoor air pollution
PE 8189 Indri ; Endangered species sifakas, avahis
Induced abortion
PD 0158 — [Induced abortion]
PD 0159 — Illegal
PD 0159 — Variable criteria legal
PD 1302 Induced abortions ; Adolescent
PF 7918 Induced death ; Mind
Induced diseases
PD 6334 — Drug
PD 8200 — Environmentally
PD 6334 — Physician
Induced [famine...]
PE 8979 — famine ; War
PA 6030 — *fear ; Culture*
PE 7584 — fish diseases ; Pollution
PD 5212 — fluctuations food ; Weather
PD 9515 *Induced increases seismicity ; Reservoir*
PF 3161 Induced individualism ; Survival
PE 0616 *Induced malaria ; Recurrent*
PC 0300 *Induced mental illness ; Nutritionally*
PE 5182 *Induced pulmonary haemorrhage ; Exercise*
PF 6712 Induced rapid social change ; Confusion
PE 5050 *Induced skeletal abnormalities swine ; Nutritionally*
PE 6350 Induced winter chill ; Nuclear war
PD 6943 Inducement crime ; Government
PF 6471 *Inducements ; Unexplored industrial*
PE 3912 Induction armed forces ; Obstruction recruiting
PF 3790 Induction incongruent actions
PE 1670 Induction psychological stress animals ; Experimental
PA 6466 Indulgence *Intemperance*
PF 5466 Indulgence ; National self
PA 7321 Indulgence *Refusal*
PA 6466 Indulgence ; Self
PA 7211 Indulgence ; Self
PF 4433 Indulgences ; Commerce religious
PF 4433 Indulgences ; Sale
PD 5774 Indulgent consumerism ; Self
PF 5466 Indulgent societies ; Self
Industrial [access...]
PF 6473 — *access ; Inadequate*
PC 0646 — accidents
PD 2570 — accidents ; Large scale
PD 0694 — action
PC 0119 — *air pollution ; Urban*
PC 3413 — apathy
Industrial [capacity...]
PD 4219 — capacity developing countries ; Inadequate
PE 8014 — combines farming communities ; Threat
PD 8994 — competitiveness ; Declining
PC 3191 — complex capitalist systems ; Military
PC 1952 — complex ; Military
PD 2772 — concerns ; Excessive commercial exploitation farm animals
PE 8230 — concerns ; Unwillingness divulge information
PD 4012 — conflict
PD 3101 — *control*
PD 8359 Industrial destruction war
Industrial development
PF 6525 — *Delayed*
PF 2471 — rural communities ; Overlooked potential
PF 6525 — *Unpromoted*
Industrial [diamonds...]
PC 1195 — *diamonds ; Shortage*
PD 2570 — disasters
PD 0215 — diseases
PD 4012 — disputes
PF 3843 — distribution ; Unbalanced
PE 2824 — domestic heating emissions air pollutants
Industrial [economic...]
PC 0880 — economic activities ; Underdevelopment
PE 4905 — economic development due environmental policies ; Restrictions
PD 3666 — *effluent*
PF 2115 — *electricity ; Lack*
PE 1869 — emissions
PD 0335 — *energy supply ; Insufficient diversification urban*
PC 2921 — espionage
PF 6475 — *expansion ; Untested*
PE 3537 Industrial failures
Industrial [gas...]
PE 1813 — gas monopolies
PE 0802 — goods ; Shortage
PD 1330 — governmental complex communist systems ; Military
PF 2905 — growth ; Inadequate systems monitoring
PE 1504 Industrial hazards ; Aerosols
Industrial [images...]
PF 1790 — *images ; Unfeasible*
PF 1790 — *imagination ; Lacking*
Industrial [incentives...] cont'd
PF 0963 — *incentives ; Inadequate*
PF 0963 — *incentives ; Low*
PF 6471 — *inducements ; Unexplored*
PF 4854 — installations ; Cover up unsafe
PE 4304 — installations ; Hazardous
PC 2939 — intimidation
Industrial [laboratory...]
PE 4859 — laboratory medical equipment ; Unsafe
PE 6005 — land ; Degraded
PE 6005 — land ; Unallocated
PC 1820 — leadership entrepreneurial ability developing countries ; Shortage
Industrial [machinery...]
PF 2410 — *machinery ; Costly*
PD 2916 — malpractice
PD 4361 — malpractice ; Military
PE 4818 Industrial nations ; Ineffective economic structures
Industrial [oligarchy...]
PD 5193 — oligarchy ; Financial
PF 1933 — operations ; Unprofitable scope
PE 2473 — origin ; Radio interference
PE 2473 — origin ; Radio noise
Industrial [piracy...]
PF 8854 — piracy
PF 1959 — *planning ; Insufficient*
PF 1959 — *planning ; Unformed*
PE 2866 — plants processes ; Ageing
PC 8745 — polluters
Industrial pollution
PF 7836 — Employment risk elimination
PF 7836 — immediate victims ; Defence
PC 8745 — Urban
Industrial [power...]
PD 5193 — power ; Misuse financial
PD 2916 — practices ; Irresponsible
PD 2916 — practices ; Unethical
PE 3939 — processes geared reduced social needs
PE 7479 — processes ; Imbalance distribution
PC 1217 — production activities ; Instability economic
PD 0779 — production ; Overcapacity
PC 0328 — products ; Environmental hazards economic
Industrial promotion
PF 0963 — *Inadequate*
PF 0963 — *Insufficient*
PF 6471 — *Lack*
PF 0963 — *Unorganized*
PF 8854 Industrial property ; Lack protection
PF 5292 Industrial property rights ; Payments after expiration
Industrial [recruitment...]
PF 6557 — *recruitment ; Minimal*
PF 1789 — research development ; Pollution hazards
PF 2842 — *resources ; Underused*
PF 4013 — retraining programmes ; Inadequate
PD 8897 — rivalry
Industrial sectors
PD 5350 — Ageing
PD 5350 — Declining
PE 1980 — Planning disrelationship political
Industrial services
PF 4195 — developing countries ; Inadequate
PF 1790 — *Inadequate*
PF 1790 — *Undeveloped*
Industrial [site...]
PE 6005 — site ; Unfinished
PF 7130 — *sites ; Unprepared*
PF 6500 — skills ; Shortage
PF 7130 — *space ; Inadequate*
PF 4160 Industrial trade developing countries ; Inadequate
PF 4013 Industrial training ; Unexplored
PD 4012 Industrial unrest
PE 7433 Industrial uses ; Loss water
PF 3679 Industrial vision ; Self interested
Industrial [wages...]
PD 8916 — wages ; Unattractive
PD 0575 — waste water ; Discharge dangerous substances
PD 0575 — waste water pollutants
PE 6005 — wastelands ; Derelict
PF 5464 — water ; Shortage
PD 1607 — *water supplies ; Infection*
PB 1845 Industrialism
PB 1845 Industrialization ; Dependence uncontrolled
Industrialization developing countries
PC 1563 — Imbalance urbanization
PF 4185 — Irrational fear
PE 8180 — Lack rural
PJ 0230 — Obstacles
PE 6237 Industrialization ; Export led
PB 1845 Industrialization ; Inappropriate
PE 5931 Industrialization ; Loss land
PB 1845 Industrialization ; Over rapid
PB 1845 Industrialization ; Uncontrolled
Industrialized countries
PE 8093 — Brain drain more developed
PF 1737 — Declining economic growth
PD 5543 — Declining productivity
PE 8092 — Demand unskilled labour
PD 9202 — Deterioration
PF 5472 — Displacement national decision making developing
PF 5000 — Divergences macroeconomic policies leading
PD 3998 — *Dumping defective products*
PE 2427 — Economic exploitation developing countries
PD 5865 — Economic imbalances
PE 8093 — Emigration trained personnel
PD 7835 — Environmental degradation
PD 9168 — Excessive foreign public debt

Industrialized countries

Industrialized countries cont'd
- PD 1998 — Family poverty
- PD 8285 — Homelessness
- PC 7459 — Imbalance economic relationships
- PE 0537 — import resources ; Dependence
- PE 0537 — imports primary commodities ; Dependence
- PD 3427 — Inadequate living working conditions immigrant labourers
- PF 1737 — Instability growth income
- PD 1291 — Loss developing country human resources
- PD 3427 — Menial work status immigrants
- PF 5000 — Mismatch national macroeconomic policies
- PD 9055 — Obstacles restructuring production
- PD 9802 — Over subsidized agriculture
- PD 2495 — Personal isolation communities
- PE 4639 — Personal physical insecurity
- PC 0444 — Poverty
- PD 7682 — products substituting commodities exported developing countries ; Development
- PC 3134 — Transfer surplus wealth developing
- PE 0667 — transnational corporations ; Interference labour relations
- PE 4289 — Undermining multilateral forums
- PC 9718 — Unemployment
- PE 6891 — Unequal income distribution
- PD 3488 — Unplanned urbanization
- PE 5551 — Unsustainable consumption resources
- PE 1887 — Urban slums

Industrialized [country...]
- PE 8321 — country disincentives farm products developing countries
- PE 1337 — country methods developing countries ; Inappropriate transplantation
- PE 3831 — cultures media ; Excessive portrayal perspectives
- PC 8694 — Industrialized developing countries ; Disparity
- PE 5023 — Industrialized societies ; Inadaptation technology man
- PE 2862 — Industrialized societies ; Institutional obsolescence modern
- PD 1987 — Industries customs duties ; Excessive protection
- PD 8994 — Industries ; Decline competitiveness domestic
- PD 5350 — *Industries ; Declining local*

Industries developing countries
- PD 1250 — Decline handicrafts cottage
- PD 0765 — Excessive foreign investment traditional
- PE 6687 — Export hazardous
- PE 2238 — Inadequate increase employment manufacturing
- PD 9117 — Instability manufacturing
- PE 6687 — Relocation hazardous
- PF 8921 — Transfer highly polluting
- PD 5350 — Industries difficulty

Industries Environmental hazards
- PC 1192 — chemicals petrochemicals
- PE 9019 — china, pottery earthenware
- PE 8097 — electrical equipment
- PE 1280 — food processing

Industries [environmental...]
- PE 8397 — environmental hazards ; Iron steel basic
- PE 1859 — Environmental hazards machinery equipment
- PE 1852 — *environmental hazards ; Machinery equipment*
- PD 0454 — Environmental hazards manufacturing
- PE 8248 — environmental hazards ; Non ferrous metals basic

Industries Environmental hazards
- PE 0890 — non metallic mineral products
- PE 1425 — paper printing
- PE 8095 — publishing printing
- PE 8571 — tanning dyeing
- PE 1103 — textile clothing
- PE 0864 — woodworking
- PD 0243 — Industries ; Health hazards basic metal

Industries Health risks
- PE 1605 — workers manufacturing
- PE 0875 — workers service
- PE 1581 — workers transport, storage communication
- PE 4493 — Industries ; Inadequate conditions work hotel catering
- PF 0591 — Industries ; Inadequate regulation restrictive business practices service

Industries Ineffective self
- PE 8956 — regulation chemicals manufacturing
- PE 8574 — regulation consumer goods manufacturing
- PE 8472 — regulation food processing
- PE 8630 — regulation mining
- PE 8502 — regulation pharmaceutical medical devices

Industries [Instability...]
- PE 2601 — Instability basic metal
- PE 8928 — instability; China, pottery earthenware
- PE 8918 — Instability food drink
- PE 9084 — instability ; Glass glass products
- PE 8070 — instability ; Iron steel basic
- PE 1852 — Instability machinery equipment
- PE 1852 — *instability ; Machinery equipment*
- PC 0580 — Instability manufacturing
- PE 2601 — *instability ; Non ferrous metal basic*

Industries Instability
- PE 1927 — paper printing
- PE 1008 — textile clothing
- PE 1008 — wearing apparel
- PE 0681 — woodworking
- PE 2844 — Industries labour affected developing country exports ; Inadequate adjustment assistance
- PF 1933 — *Industries ; Lack local*
- PE 4097 — Industries ; National defence
- PC 3240 — *Industries ; Nationalized*
- PE 7528 — Industries ; Occupational hazards workers small
- PD 5350 — Industries ; Older

Industries Protectionism
- PD 5830 — agriculture food production

Industries Protectionism cont'd
- PD 7001 — computer office machine
- PD 7049 — construction engineering services
- PE 8795 — consumer products
- PE 8664 — defence arms
- PD 7060 — entertainment products film
- PE 8458 — high technology
- PC 5842 — *labour intensive*
- PC 5842 — *mining*
- PD 7135 — services
- PE 5866 — steel basic metal
- PE 5819 — textile apparel
- PG 8903 — *Industries sector ; Overgrowth service*
- PE 5831 — Industries sectors transnational corporations ; Control

Industries Underdevelopment
- PF 1374 — basic metal
- PD 0908 — food processing
- PE 8223 — iron steel basic
- PE 1852 — *Industries underdevelopment ; Machinery equipment*

Industries Underdevelopment
- PF 0854 — manufacturing
- PF 0942 — metal products, machinery equipment
- PE 1858 — mining quarrying
- PE 1851 — non metallic mineral products
- PF 1136 — paper printing
- PE 0453 — textile clothing
- PF 2604 — woodworking
- PF 1442 — *Industries ; Unprofitable home*

Industry [agriculture...]
- PE 5931 — agriculture land ; Competition
- PF 7433 — agriculture water ; Competition
- PF 4370 — Antiquated regulations banking
- PE 8640 — Industry ; Control transnationals global communications
- PE 6618 — Industry ; Corporate crime pharmaceutical

Industry Corruption
- PD 2916 — [Industry ; Corruption]
- PE 6615 — catering
- PD 8001 — clothing
- PD 3736 — entertainment
- PD 1045 — food
- PD 1012 — transport

Industry developing countries
- PE 1511 — Inadequate relationship transnational corporations local
- PD 1494 — Locational maladjustments
- PE 4120 — Underdevelopment pharmaceutical
- PF 4135 — Underdevelopment power
- PE 5863 — Industry distribution ; Imbalance shipbuilding
- PE 3781 — Industry ; Domination transnational corporations coffee

Industry Environmental hazards
- PE 8790 — construction
- PD 0743 — fishing
- PE 5698 — nuclear weapons
- PE 5897 — Industry ; Excess capacity shipping

Industry Health
- PE 8274 — hazards glass
- PE 8263 — hazards steel
- PE 0526 — risks workers construction
- PE 5825 — Industry ; Imbalance power shipping
- PE 5839 — Industry ; Imbalances dry bulk shipping

Industry Inadequate
- PE 6841 — conditions work construction
- PE 5823 — conditions work textile
- PE 3969 — working conditions construction

Industry Ineffective self
- PE 8067 — regulation automobile manufacturing
- PE 8265 — regulation housing construction
- PF 5840 — regulation shipping
- PE 7040 — Industry ; Inflexible military thinking
- PE 3537 — Industry ; Insolvency

Industry Instability
- PE 0538 — chemical petrochemical
- PE 0509 — construction
- PE 1424 — fishing
- PE 8320 — leather footwear
- PE 5791 — maritime shipping
- PE 1852 — *Industry instability ; Metal products*

Industry Instability
- PE 0993 — mining quarrying
- PE 2599 — non metallic mineral products
- PE 0538 — *paint*
- PE 5143 — plastics
- PD 1045 — Industry ; Irresponsible practices food
- PE 1940 — Industry ; Lack consumer influence
- PD 9713 — Industry ; Malpractice construction
- PD 1607 — Industry ; Microbial pests

Industry Negligence
- PD 2916 — [Industry ; Negligence]
- PD 1045 — food
- PD 1012 — transport
- PE 8428 — Industry ; Occupational hazards mining
- PD 1554 — Industry primary commodities developing countries ; Lack processing
- PE 0523 — Industry products ; Inadequate export promotion developing country small

Industry Protectionism
- PD 7108 — advertising
- PD 7132 — air transportation
- PE 8134 — alcoholic beverages
- PC 5842 — *automobile*
- PD 7120 — banking
- PD 7001 — computer services
- PD 7121 — franchising services
- PD 7012 — insurance
- PE 5888 — shipping

- PE 8509 — Industry public service ; Inability communities take advantage training business,
- PE 7742 — Industry ; Reductionistic decision making criteria construction

Industry [self...]
- PF 5841 — self regulation ; Ineffective
- PF 1740 — Short term priorities construction
- PF 2423 — Size limits
- PD 4575 — Slaughter animals fur
- PF 6996 — Stress
- PF 2303 — *syndrome ; Large*

Industry transnational corporations
- PE 1448 — Domination agricultural equipment
- PE 1469 — Domination automobile
- PE 2084 — Domination copper
- PE 1620 — Domination shipping
- PF 2842 — *Industry ; Uncommitted local*

Industry Underdevelopment
- PE 1483 — chemical petrochemicals
- PE 2138 — fishing
- PC 0880 — *forestry*

Industry [Underparticipation...]
- PE 4145 — Underparticipation developing countries airline
- PE 5362 — Unemployment wood
- PD 8001 — Unethical practices apparel
- PD 8001 — Unethical practices clothing
- PE 8873 — universities ; Gap
- PC 0495 — Unsustainable development energy
- PE 5698 — Industry ; Violation safety regulations nuclear weapons
- PD 0575 — *Industry wastes ; Chemical*
- PE 8702 — Industry wastes ; Food manufacturing

Inedible crude non
- PE 0546 — fuel materials ; Environmental hazards
- PD 0280 — fuel materials ; Instability trade
- PE 0461 — fuel materials ; Long term shortage
- PE 8351 — fuel materials ; Restrictive practices trade

Ineffective [administration...]
- PF 2113 — administration international justice
- PC 9513 — agreements
- PF 2848 — *agricultural technology*
- PF 4240 — anti-terrorist organization
- PF 0148 — Ineffective birth control
- PF 6557 — *Ineffective business solicitation*

Ineffective [care...]
- PF 2261 — *care-taking system*
- PF 2335 — civil service
- PF 6476 — *class scheduling*
- PF 6052 — communication

Ineffective community
- PF 2303 — leverage
- PF 6501 — organization
- PF 2845 — *story*
- PF 9672 Ineffective control government finances

Ineffective [data...]
- PF 3671 — data systems
- PF 3709 — decision-making processes
- PF 2875 — delivery basic human resource services
- PD 1464 — deterrent against war crimes
- PF 1654 — dialogue

Ineffective [economic...]
- PE 4818 — economic structures industrial nations
- PF 8126 — educational images
- PF 2447 — educational policy decisions
- PF 9007 — emergency services
- PD 3712 — execution development projects

Ineffective [family...]
- PF 0148 — family planning
- PD 4572 — family units ; Socially
- PF 0342 — follow-up initiatives
- PF 4095 — Ineffective governmental use nongovernmental resources
- PD 8023 Ineffective health education

Ineffective [implementation...]
- PF 1010 — implementation methods decision-making
- PF 3498 — *in-service training*
- PF 5841 — industry self-regulation
- PF 4146 — inspection toxic waste sites
- PF 6992 — international agreements
- PF 0691 — international regulation transnational corporations

Ineffective land
- PF 2009 — *collateral*
- PF 6528 — usage
- PF 6528 — use

Ineffective [laws...]
- PE 6699 — laws against racketeering
- PF 3605 — leadership recruitment
- PC 9513 — legislation
- PE 6699 — legislation against organized crime

Ineffective means
- PF 6495 — goods supply distribution
- PE 8430 — participation decision making
- PD 4392 — searching partner

Ineffective [mechanisms...]
- PF 1352 — mechanisms functional training
- PF 8939 — meeting formats
- PF 2721 — methods practical education
- PF 2613 — minesweeping
- PF 4197 — mobilization public resources

Ineffective monitoring
- PF 2793 — [Ineffective monitoring]
- PE 5089 — hazardous substances
- PF 7264 — illegal activity
- PF 2782 — restrictive business practices due inadequate regulation

Ineffective [official...]
- PE 4146 — official inspection regulated activities

Infatuation

Ineffective [official...] cont'd
- PF 0963 — official liaison
- PF 1959 — operation community networks urban ghettos
- PF 3826 — opposition existing social order
- PF 6501 — organization community action

Ineffective [parenting...]
- PF 6461 — parenting techniques
- PF 5272 — past efforts
- PF 3709 — policy-making methods
- PE 1101 — political ties
- PF 0148 — population control
- PF 1020 — population control
- PF 4240 — prevention terrorism
- PF 0922 — protection individual rights due excessive court costs

Ineffective [regulation...]
- PE 6226 — regulation electronic messages
- PF 1596 — regulation restrictive business practices
- PC 9513 — regulations
- PF 1618 — religious dialogue
- PF 2996 — rural transport

Ineffective self regulation
- PE 8067 — automobile manufacturing industry
- PE 8956 — chemicals manufacturing industries
- PE 8574 — consumer goods manufacturing industries
- PE 8121 — financial services sectors
- PE 8472 — food-processing industries
- PE 9048 — health care sector
- PE 8265 — housing construction industry
- PE 8630 — mining industries
- PE 8502 — pharmaceutical medical devices industries
- PF 5840 — shipping industry
- PF 5877 — telecommunications sectors
- PE 5458 Ineffective space usage

Ineffective structures
- PF 1781 — community decision-making
- PF 2437 — community participation
- PF 6506 — local consensus
- PC 1522 Ineffective student accountability
- PF 3498 Ineffective systems practical education

Ineffective [tax...]
- PF 1462 — tax systems
- PF 2124 — tax systems developing countries
- PE 8716 — technical training
- PD 2269 — temporary drainage
- PE 6750 — testamentary dispositions
- PD 0153 — treatment alcoholism

Ineffective use
- PF 0148 — contraceptives
- PE 6515 — external relations relating sportsmen
- PF 4095 — private sector resources government
- PF 6543 Ineffective utilization public environment

Ineffective [war...]
- PD 1464 — war crime prosecution
- PE 8233 — water usage
- PF 1262 — worker organizations
- PF 2261 Ineffectively prioritized funding
- PF 1031 Ineffectiveness aid developing countries

Ineffectiveness [capital...]
- PF 0399 — capital punishment deterrent
- PD 5429 — conservation programmes
- PE 8884 — consultation procedures
- PD 5142 — correctional institutions
- PF 8368 Ineffectiveness development programs
- PF 3394 Ineffectiveness ; Governmental

Ineffectiveness [Impotence...]
- PA 6876 — Impotence
- PA 6852 — Inappropriateness
- PF 6127 — individual participation large communities

Ineffectiveness inefficiency
- PF 0316 — intercultural meetings
- PF 0409 — interdisciplinary meetings
- PF 0349 — international meetings

Ineffectiveness [Influencelessness...]
- PA 6882 — Influencelessness
- PF 0074 — intergovernmental organization programmes
- PF 2497 — international commodity agreements

Ineffectiveness international nongovernmental
- PE 7056 — organizations ; Operational
- PE 7008 — organizations ; Political
- PF 1595 — organizations programmes
- PE 9124 Ineffectiveness international organization programme action country level
- PF 1074 Ineffectiveness international organizations programmes
- PF 6377 Ineffectiveness public spending
- PF 3008 Ineffectiveness traditional small business methods
- PF 1451 Ineffectiveness United Nations system organizations
- PA 7232 Ineffectiveness Unskillfulness
- PF 1365 Ineffectivity ; Images
- PF 9564 Ineffectual intergovernmental multilateral action

Ineffectuality [Impotence...]
- PA 6876 — Impotence
- PA 6852 — Inappropriateness
- PA 6882 — Influencelessness
- PA 5942 Ineffectuality Unimportance
- PA 7232 Ineffectuality Unskillfulness
- PE 8884 Inefficacity consultation decision-making process
- PE 2999 Inefficiencies developing countries due restrictive business practices ; Economic
- PB 0843 Inefficiency
- PF 8491 Inefficiency ; Bureaucratic
- PB 0843 Inefficiency ; Dependence
- PE 4647 Inefficiency due mismatch religious national holidays
- PF 7556 Inefficiency ; Economic
- PF 6980 Inefficiency financial markets

Inefficiency [Government...]
- PF 8491 — Government
- PF 9266 — government developing countries
- PD 5642 — government monopolies

Inefficiency [Impotence...]
- PA 6876 — Impotence
- PJ 8777 — Intellectual
- PF 0316 — intercultural meetings ; Ineffectiveness
- PF 0409 — interdisciplinary meetings ; Ineffectiveness
- PF 0349 — international meetings ; Ineffectiveness
- PF 4945 Inefficiency ; Mental
- PD 4361 Inefficiency ; Military

Inefficiency public
- PD 5642 — enterprises
- PF 7483 — programmes ; Internal
- PF 6377 — spending
- PD 5642 Inefficiency state-controlled enterprises
- PA 7232 Inefficiency Unskillfulness
- PD 0175 Inefficient administration aid
- PD 2743 Inefficient animal commodity production
- PF 2204 Inefficient extraction utilization natural resources
- PE 7908 Inefficient labour use socialist countries
- PE 3538 Inefficient location facilities international organizations
- PF 5964 Inefficient management government financial resources developing countries
- PF 4197 Inefficient mobilization government revenue
- PB 0843 Inefficient people
- PE 5899 Inefficient port cargo handling ; Excessive costs

Inefficient public
- PF 2335 — administration
- PF 0903 — administration developing countries
- PF 6377 — programmes
- PF 1258 — spending alleviate poverty
- PF 2477 Inefficient resource usage

Inefficient [shipping...]
- PF 5872 — shipping procedures documentation
- PE 0358 — slaughterhouse practices ; Cruel
- PE 5458 — space usage
- PE 4531 — support widowers
- PE 5009 Inefficient tax administration developing countries
- PF 4197 Inefficient taxation systems

Inefficient use
- PF 9683 — available health care
- PF 5964 — domestic resources developing countries
- PC 7517 — energy resources
- PF 2758 — proteins factory farming
- PE 5001 — resources
- PA 6312 Inelegance
- PA 6312 Inelegance Inelegance

Inelegance [Ugliness...]
- PA 7240 — Ugliness
- PA 5612 — Unchastity
- PA 7232 — Unskillfulness
- PA 5821 Inelegance Vulgarity
- PA 6118 Ineloquence
- PC 4867 Inept directors
- PE 9172 Inept testimony psychologists psychiatrists
- PF 3953 Ineptitude ; Government
- PA 6876 Ineptitude Impotence
- PA 7395 Ineptitude Inexpedience
- PF 1528 Ineptitude management ; Communication
- PJ 5588 Ineptitude ; Social
- PA 7371 Ineptitude Unintelligence
- PC 4844 Inequalities ; Health
- PE 1339 Inequalities life expectancy ; Social
- PA 6695 Inequality
- PC 1896 Inequality access education countries
- PE 3528 Inequality bargaining power
- PC 2815 Inequality countries ; Income

Inequality Dependence
- PA 0844 — [Inequality ; Dependence]
- PC 3425 — political
- PB 0514 — social

Inequality [developed...]
- PE 6891 — developed countries ; Income
- PD 7615 — developing countries ; Income
- PA 5982 — Disagreement
- PF 3043 — distribution natural resources countries

Inequality [Economic...]
- PC 8541 — Economic
- PC 3434 — education
- PC 3434 — education ; Dependence
- PD 8903 — employment
- PD 1847 — employment opportunity developing countries
- PC 5038 — energy use
- PA 0844 — Inequality ; Human
- PF 1199 Inequality human biological abilities

Inequality [Income...]
- PD 4962 — Income
- PE 9072 — inducing effects remote sensing systems
- PD 5833 — inducing effects television
- PA 6486 — Injustice
- PC 9152 — International
- PC 1268 Inequality law
- PE 1339 Inequality life expectancy gender
- PC 9586 Inequality mortality rates
- PE 1339 Inequality mortality rates gender
- PC 8694 Inequality North South
- PC 3435 Inequality opportunity
- PC 3435 Inequality opportunity ; Dependence
- PC 3425 Inequality ; Political
- PF 1199 Inequality ; Racial
- PE 8569 Inequality rural urban development ; Accentuated
- PB 0514 Inequality ; Social
- PC 3433 Inequality social services

Inequitable [access...]
- PD 3210 — access resources
- PD 0986 — administration justice
- PE 1597 — allocation rights exploit sea-bed marine resources
- PE 7902 Inequitable credit extensions

Inequitable distribution
- PD 2608 — construction expertise
- PE 9758 — public subsidies
- PD 4962 — revenues
- PD 2479 — skilled specialists
- PB 7666 — wealth
- PB 7666 — wealth ; Dependence
- PE 4700 Inequitable education selection
- PE 1399 Inequitable justice poor
- PD 0142 Inequitable labour standards developing countries
- PF 4787 Inequitable peace agreements
- PD 9430 Inequitable range salaries
- PD 8056 Inequitable right access water
- PF 4375 Inequitable school fees

Inequitable tax
- PD 3436 — systems
- PD 3046 — systems developing countries
- PD 1477 — treaties developed developing countries
- PF 4787 — treaties
- PC 4844 Inequitable use medical resources
- PE 5802 Inequities marine insurance
- PE 5875 Inequities ship owner registration
- PA 6695 Inequity Inequality
- PA 6486 Inequity Injustice
- PC 0267 Inertia ; Bureaucrat
- PC 1907 Inertia ; Dependence political
- PF 0421 Inertia ; Dependence psychological
- PF 6575 Inertia ; External assistance
- PA 5806 Inertia Inaction

Inertia [Political...]
- PC 1907 — Political
- PF 4713 — Pre electoral political
- PF 0421 — Psychological
- PA 6444 Inertia Quiescence
- PC 3950 Inertia response problems ; Governmental resistance
- PE 2443 Inevitable destruction natural environment mankind
- PA 5467 Inexcitability
- PA 5527 Inexpectation
- PA 7395 Inexpedience
- PA 5454 Inexpedience Badness
- PA 7395 Inexpedience Inexpedience

Inexpedience [Unintelligence...]
- PA 7371 — Unintelligence
- PA 7232 — Unskillfulness
- PA 7006 — Untimeliness
- PA 5568 Inexperience Ignorance
- PA 7131 Inexperience Oldness
- PA 7232 Inexperience Unskillfulness
- PF 2848 Inexperienced animal husbandry
- PC 4867 Inexperienced business management
- PE 6554 Inexperienced business practices
- PE 1137 Inexperienced financial autonomy
- PD 9113 Inexperienced labour force
- PF 6556 Inexperienced local leaders
- PC 1820 Inexperienced village entrepreneurship
- PF 8352 Inexplicable phenomena
- PE 4742 Infallibility equipment ; Excessive reliance
- PF 3376 Infallibility ; Negative effects claims religious
- PB 8172 Infamy
- PA 6839 Infamy Disrepute
- PA 5644 Infamy Vice
- PA 7280 Infamy Wrongness
- PC 0300 Infancy ; Reactive attachment disorder
- PE 5187 Infancy ; Rumination disorder
- PE 4874 Infant brains malnutrition insufficient stimuli ; Damage
- PE 4245 Infant care ; Deficient

Infant [death...]
- PD 1885 — death ; Sudden unexpected
- PD 1885 — death syndrome ; Sudden
- PE 0691 — diets ; Unbalanced
- PC 0981 — Infant emotional deprivation
- PD 9661 Infant feeding strategies ; Inappropriate
- PE 6909 Infant growth failure
- PE 4619 Infant incontinence
- PC 1287 Infant mortality
- PE 4245 Infant mortality due parental abuse chronic neglect
- PD 9021 Infant nutrition ; Ignorance women concerning
- PD 9661 Infant weaning ; Improper
- PE 4245 Infant welfare ; Inadequate
- PD 3501 Infanticide
- PD 3501 Infanticide ; Female
- PD 8297 Infantile amnesia
- PE 1222 Infantile autism
- PE 0763 Infantile monoplegia
- PE 2565 Infantile muscular atrophy
- PE 3571 Infantile neurosis

Infantile [paralysis...]
- PE 0504 — paralysis
- PE 0763 — paralysis ; Cerebral spastic
- PE 0763 — paraplegia ; Spastic
- PF 2541 Infantilization deprived populations
- PE 3044 Infants during working hours ; Denial right periods nurse
- PD 0331 Infants early childhood ; Protein energy malnutrition
- PD 9021 Infants ; Ignorance women concerning weaning
- PE 8456 Infarction ; Acute myocardial
- PD 9057 Infarction ; Cerebral
- PE 8456 Infarction ; Myocardial
- PF 5973 Infatuation
- PA 6448 Infatuation
- PA 6379 Infatuation Avoidance

PA 7157 Infatuation *Insanity*	Infectious [abortion...]	PD 4798 Infiltration ; Political
PA 7392 Infatuation *Unbelief*	PE 0924 — *abortion ruminants*	PB 0855 *Inflammable flammable substances*
PA 7371 Infatuation *Unintelligence*	PE 9554 — *anaemia* ; *Equine*	PD 8786 Inflammation choroid
Infected [animal...]	PE 9554 — *anaemia* ; *Feline*	PE 6789 Inflammation cornea
PE 7064 — animal, meat animal product shipments	PD 2732 — *animal diseases*	PE 8624 *Inflammation duodenum*
PD 2751 — *animals*	**Infectious [bovine...]**	PD 8786 *Inflammation eyelid*
PE 7064 — animals ; Insufficiently treated meat	PE 5524 — *bovine rhinotracheitis*	PE 2251 Inflammation gallbladder
PE 2777 Infected carcass meats factor animal diseases ; Importation	PE 5567 — *bronchitis poultry*	PE 3503 *Inflammation gums*
	PE 9799 — *bulbar paralysis*	PE 6603 Inflammation internal parts body
PE 3511 *Infected dogs*	PD 2730 — *bursal disease chickens*	PE 4973 Inflammation intestines
PE 8545 Infected faeces ; Pollution water	**Infectious [canine...]**	PD 8786 *Inflammation lacrimal glands ducts*
Infected persons	PD 2730 — *canine hepatitis*	PD 0659 Inflammation ; Nose
PE 8510 — Discharge sores	PE 5567 — *coryza poultry*	PD 8786 *Inflammation optic nerve*
PE 4299 — Discrimination against HIV	PD 2732 — *cyclic thrombocytopenia* ; *Canine*	PE 2663 Inflammation peritoneum
PE 4299 — Social stigmatization AIDS	PD 5971 Infectious diarrhoeas	PE 4651 Inflammation pharynx
PE 4299 — Travel restrictions AIDS	**Infectious diseases**	PD 8786 *Inflammation retina*
PD 2651 Infecting disease ; Intentional	PC 3595 — Agents causing	PE 7570 Inflammation ; Salivary gland
PC 9025 Infection	PD 2732 — *animals*	PE 7826 *Inflammation stifle joint*
PD 0992 Infection ; Brain	PE 6419 — *animals* ; *Congenital anomalies due*	**Inflammation [tendons...]**
Infection [Canine...]	PF 1312 — *bacteria* ; *Introduction extraterrestrial*	PD 2647 — *tendons*
PD 2730 — *Canine herpesviral*	PE 9526 — Intestinal	PE 4651 — Throat
PD 2730 — *Canine parvoviral*	**Infectious [enteritis...]**	PE 9702 — *tongue*
PE 7304 — *cats* ; *Pox virus*	PD 2730 — *enteritis* ; *Feline*	PG 5613 — *tongue*
PE 2307 — Chronic urinary	PD 2735 — *enterohepatitis*	PD 1185 — *tooth fang*
PE 2041 Infection death ; Foetal	PA 5568 — *enterotoxaemia*	PD 8786 *Inflammation uveal tract*
PD 9293 *Infection dog* ; *Giant kidney worm*	PE 0517 Infectious hepatitis	PE 9286 *Inflammation womb*
Infection [ear...]	PD 2731 *Infectious hepatitis* ; *Avian*	PE 8822 Inflammatory affections bronchial tubes lungs
PD 2567 — ear	PE 0517 Infectious jaundice	PD 0448 *Inflammatory affections heart*
PE 5182 — *Equine herpes virus 1*	PE 2357 *Infectious jaundice*	PD 8239 *Inflammatory disease* ; *Chronic pelvic*
PD 8786 — eye	PE 5535 *Infectious keratoconjunctivitis animals*	PD 8786 Inflammatory diseases eye
Infection [Feline...]	PD 2730 Infectious laryngotracheitis	PE 9151 Inflammatory infections respiratory organs
PD 2730 — *Feline coronaviral*	PE 5567 *Infectious laryngotracheitis poultry*	PE 0652 Inflammatory lesions ; Thyroid glands
PE 2391 — Filarial	PE 5550 Infectious mononucleosis	PF 7870 Inflated art values
PD 2728 — Fungus	PE 7769 *Infectious necrotic hepatitis*	**Inflated cost**
PE 5567 *Infection* ; *Gapeworm*	PD 5535 *Infectious ophthalmia*	PF 0703 — information
PE 0958 *Infection horses* ; *Small strongyle*	**Infectious [parasitic...]**	PF 1238 — necessities
PD 1607 *Infection industrial water supplies*	PD 0982 — *parasitic diseases*	PF 7911 — social security
PD 7307 Infection ; Lungworm	PE 9105 — patients ; Carelessness dealing	PF 7911 — unemployment benefits
PD 9293 *Infection mink* ; *Giant kidney worm*	PD 2730 — *peritonitis* ; *Feline*	PD 1842 Inflated house values
Infection [Paracolon...]	PD 2730 — *pleuritis* ; *Feline*	PF 8763 Inflated labour costs
PE 7562 — *Paracolon*	PE 4161 — *pododermatitis*	PE 4162 Inflated land values
PD 9667 — *Parafilaria*	PD 2732 — *polyarthritis*	PE 8261 Inflated medical expenses
PD 2732 — *Pasteurella antipestifer*	PE 5524 — *pustular vulvovaginitis*	PF 8158 Inflated overhead costs
Infection poultry	PD 5168 Infectious revenge	PD 1891 Inflated prices goods services
PE 7562 — Arizona	**Infectious [serositis...]**	PF 2385 Inflated prices rural communities
PD 2732 — *Erysipelothrix*	PD 2732 — *serositis*	PF 2899 *Inflated seed costs*
PE 2395 — *Manson's eyeworm*	PD 2732 — *sinusitis*	PF 5514 Inflation-based monetary systems
PD 2732 — *Mycoplasma gallisepticum*	PD 2732 — *synovitis*	PC 0254 Inflation ; Economic
PD 2732 — *Mycoplasma synoviae*	PD 7841 *Infectious thromboembolic meningitis*	PD 5351 Inflation ; Export
PE 7562 — Paratyphoid	PE 9854 *Infectious tracheobronchitis dogs*	PF 3756 Inflation meetings ; Rhetorical
PE 5913 *Infection* ; *Reoviral*	PE 5617 *Infective bulbar necrosis*	PF 4285 Inflation ; Misleading accounting information due
Infection [Sinus...]	PE 9286 Infective diseases uterus	PE 1750 Inflation ; News
PC 9025 — *Sinus*	PE 9286 Infective diseases vagina	PF 0960 Inflation socialist countries ; Repressed
PC 9025 — *sinuses*	PD 2567 *Infective middle ear disease*	PD 0367 Inflation ; Vulnerability developing countries
PC 9025 — *sinuses*	PA 5982 Infelicity *Disagreement*	PE 3377 Inflected loss vision
PE 3098 — Streptococcus	PA 7395 Infelicity *Inexpedience*	PA 8555 Inflexibility
PC 9025 — *Susceptibility*	PA 6731 Infelicity *Solemnity*	PF 9432 Inflexibility commodity supply
PD 9293 — *Swine kidney worm*	PA 7107 Infelicity *Unpleasantness*	PA 5740 Inflexibility *Compulsion*
Infection [throat...]	PA 7006 Infelicity *Untimeliness*	PA 7325 Inflexibility *Irresolution*
PE 7733 — throat	PF 1199 Inferior average intelligence certain racial groups	PD 4011 Inflexibility ; Labour market
PE 3098 — *throat* ; *Haemolytic streptococcal*	PF 7428 Inferior classes	PE 7632 Inflexibility ; Pattern perfectionism
PC 9025 — tonsils	PD 8585 Inferior land	PA 5988 Inflexibility *Softness*
PE 6461 — Trematode	PE 2770 Inferior meat quality intensive animal farming units	PF 3083 Inflexible attitudes toward community social services
PD 2732 — *turkeys* ; *Mycoplasma meleagridis*	PD 8996 Inferior status employment	PD 1061 Inflexible central government
PE 2143 *Infection* ; *Vincent's*	PA 5652 Inferiority	PF 6476 Inflexible class scheduling
Infections [acquired...]	PA 5454 Inferiority *Badness*	PF 7059 Inflexible deliberative systems government
PD 6334 — acquired hospital	PA 0587 *Inferiority complex*	PF 4984 *Inflexible educational system*
PE 7591 — Acute respiratory	**Inferiority [Imperfection...]**	PE 0174 Inflexible intermediary political implementation networks
PE 2355 — Adenovirus	PA 6997 — *Imperfection*	**Inflexible [management...]**
PE 1928 — *animals* ; *Chlamydid*	PA 6876 — *Impotence*	PF 3091 — management patterns
PE 0958 — *animals due parasites* ; *Gastrointestinal*	PA 5652 — *Inferiority*	PF 3091 — management structure
PD 9094 Infections ; Bacterial	PA 5878 Inferiority *Nonconformity*	PF 3499 — *market boundaries*
PD 2735 *Infections birds* ; *Fluke*	PA 7296 Inferiority *Restraint*	PE 7040 — military thinking industry
Infections [Canine...]	PA 6041 Inferiority *Shortcoming*	PF 2644 Inflexible patterns family lifestyle
PE 9427 — *Canine heartworm*	PA 5942 *Inferiority Unimportance*	PB 1997 *Inflexible property laws*
PE 6461 — Cestode	PD 8585 Infertile deserts	**Inflexible [sexual...]**
PD 3978 — *chickens* ; *Rotaviral*	PD 8585 Infertile land	PD 1308 — sexual roles
PE 5476 — Chlamydid	PC 3280 Infertility	PF 2493 — social care structures developing countries
PE 7769 — Clostridial	PC 1803 Infertility ; Animal	PB 1997 — social structure
PE 3510 Infections ; Dracunculus	PC 6037 Infertility ; Human	PD 7550 Inflicted physical suffering ; Self
PD 0640 Infections ; Enteric	PD 8585 Infertility ; Soil	**Influence [Abuse...]**
PE 3861 Infections ; Gastro intestinal	PD 0077 Infertility ; Soil	PC 6307 — Abuse
PE 0958 *Infections horses* ; *Large strongyle*	PA 6738 Infertility *Unimaginativeness*	PE 2149 — alcohol ; Driving
PD 6334 Infections ; Hospital acquired	PA 7208 Infertility *Unproductiveness*	PE 0680 — alcohol ; Flying
PE 2515 Infections intoxications ; Food borne	PG 4137 Infestation ; Itch mite	**Influence [Diminishing...]**
PE 8912 Infections involving bones acquired deformities bone ; Osteomyelitis, periostitis	PD 0982 *Infestation louse*	PF 6461 — *Diminishing parental*
	PD 9667 *Infestation* ; *Louse*	PC 3188 — Disruptive foreign
PD 2283 *Infections joints*	PE 2395 Infestation ; Roundworm	PF 9003 — Disruptive peer
PJ 3620 Infections ; Nasal	PE 6271 Infestation seeds	PF 3199 — Division world spheres
PE 6461 *Infections* ; *Nematode*	PD 9667 *Infestation small animals* ; *Cuterebra*	PE 2149 — drugs ; Driving
Infections [Porcine...]	PE 2727 *Infestations animals* ; *Cheyletiella*	PC 0840 *Influence* ; *Economic*
PD 2731 — *Porcine streptococcal*	PD 3522 Infestations buildings ; Pest	PE 1940 Influence industry ; Lack consumer
PE 3659 — *Pre natal*	PE 3177 *Infested areas* ; *Rat*	PB 3209 Influence influencing politics ; Political
PE 2280 — Pyogenic	PE 2388 *Infested rivers streams* ; *Fly*	PD 3357 Influence law ; Discriminatory religious
PD 7924 Infections ; Respiratory	PD 0077 *Infested soils* ; *Rock*	PD 3357 Influence law ; Religious
PE 9151 Infections respiratory organs ; Inflammatory	PE 6055 Infibulation	PD 3385 Influence ; Military
Infections [Salmonella...]	PE 9018 *Infidelity*	PE 1922 Influence national economies limited number corporations ; Disproportionate
PE 7562 — Salmonella	PF 9307 Infidelity God	
PC 8534 — Skin	PA 7363 Infidelity *Improbity*	PE 8829 Influence obstruct administration justice ; Use undue
PE 0958 — *swine* ; *Stomach worm*	PF 2314 Infidelity ; Marital	**Influence [peddling...]**
PD 3978 *Infections turkeys* ; *Rotaviral*	PD 9397 Infidelity ; Risk employee	PC 6307 — peddling
Infections Viral	PA 7392 Infidelity *Unbelief*	PF 9003 — peer groups ; Negative
PD 0594 — [*Infections* ; *Viral*]	PF 8068 Infidels	PD 3385 — politics ; Military
PD 0594 —	PF 7979 Infighting ; Bureaucratic	PF 3358 — politics ; Religious
PD 2730 — animal	PF 7979 Infighting ; Official	PF 3199 — *Positive spheres*
PD 0594 — protozoa	PC 0840 Infiltration ; Economic	PF 3361 Influence religion masses ; Restrictive

Inhibition adult

Influence [secular...]
PF 3358 — secular life ; Undue religious
PF 3358 — society ; Discriminatory religious
PF 3358 — society ; Religious
PF 2344 — some individuals national policy making ; Disproportionate

Influence [toys...]
PE 1158 — toys ; Undesirable
PC 6307 — Trading special
PB 3209 — traditional vested interests ; Excessive political
PE 0135 — transnational corporations global policy making ; Undisclosed
PE 8332 Influence UN related agencies ; North South spheres
PF 9686 Influence urban elites ; Disproportionate political
PA 6882 Influencelessness
PD 2389 *Influences ; Environmental*
PF 2296 Influences rural villages ; Limited exposure outside
PE 7665 Influencing law
PB 3209 Influencing politics ; Political influence
PE 0447 Influenza
PE 3157 Influenza ; Animal
PE 5182 Influenza ; Equine
PD 7307 Influenza poultry
PE 7553 Influenza ; Swine
PE 7337 Inform ; Denial right
PF 7654 Informal leadership advisors ; Self satisfied style
PF 1196 Informal leadership ; Undeveloped potential
PD 7978 Informal sector developing countries ; Development
PC 6641 *Informal sector ; Urban*
PA 7170 Informality
PF 8927 Informality international standards
PF 6781 Informants proceedings ; Tampering witnesses
PF 2376 Information ; Accumulation
PF 7388 Information anxiety

Information [be...]
PE 1756 — be sworn ; Failure appear witness produce
PF 3096 — Biased
PF 0157 — Biased government

Information [channels...]
PF 2401 — channels ; Poor
PF 9699 — Classified public
PF 4894 — Colonization

Information concerning
PF 4854 — environmental hazards ; Suppression
PF 9828 — extent homelessness ; Concealment
PF 4854 — health hazards ; Suppression
PF 0209 — problems ; Inadequate exchange technical
PF 5701 — problems ; Inadequate public
PF 9828 — social problems ; Suppression
PF 4350 — transnational banks ; Inadequate

Information [Conflict...]
PF 2002 — Conflict
PD 3080 — Conflict laws international restriction
PE 6877 — consumers ; Misrepresentation
PD 9146 — corporations ; Suppression
PE 3997 — Crimes related national security
PF 7050 — Cultural lag understanding

Information [data...]
PF 0625 — data problems ; Inadequate statistical
PE 0598 — Delay disseminating negative
PD 4982 — Deliberate distortion corporate news
PD 1893 — Deliberate distortion official news
PC 0185 — Denial right freedom
PE 4214 — Denial right receive
PB 3095 — Dependence incorrect
PE 0046 — *developing countries ; Inadequate cartographic*
PF 9077 — disciplines ; Opposition
PD 9330 — displays ; Biased computerized
PF 1183 — Dissemination discriminatory
PF 2401 — dissemination ; Infrequent
PF 0603 — drugs ; Inadequate
PE 4285 — due inflation ; Misleading accounting

Information [Excessive...]
PF 3076 — Excessive neutrality intergovernmental official
PC 1298 — explosion
PF 9828 — extent poverty ; Concealment
PF 9828 — extent unemployment ; Concealment

Information [Failure...]
PF 8753 — Failure international information systems order problem relevant
PB 3095 — False
PE 4343 — foreign power ; Revealing national security
PF 1267 — Fragmented utilization public
PF 2878 — funding needs ; Inadequate

Information [gap...]
PF 3397 — gap
PF 3397 — gap ; Technical
PF 3661 — gaps enterprises
PD 9146 — government ; Suppression
PF 8536 — government ; Withholding
PE 2000 — growth ; Increasing development lag against
PD 0270 — *Information ; Inability recall important personal*
PF 1267 — Information ; Inaccessible public

Information Inadequate
PF 2401 — access
PF 6552 — circulation local
PF 1267 — dissemination use available

Information [Incomplete...]
PF 2401 — *Incomplete skills*
PB 3095 — Incorrect
PF 7649 — Increasing scepticism accuracy official
PE 8230 — industrial concerns ; Unwillingness divulge
PF 0703 — Inflated cost
PD 3537 — Infringement proprietary
PF 7805 — Inoperative forums public

Information Insufficient
PF 1705 — *banking*
PF 6469 — flow
PF 6337 — *plant life*
PF 3539 *Information ; Irrelevant available*

Information Lack
PF 6337 — [Information ; Lack]
PD 1050 — family planning
PF 4832 — *medical*
PE 9107 — transparency international trade

Information [Limited...]
PF 3539 — Limited availability functional
PF 6541 — *Limited job*
PF 9298 — Loss

Information [media...]
PE 1478 — media ; Excessive portrayal destabilizing
PE 1478 — media ; Excessive portrayal negative
PE 2631 — Minimal access necessary
PE 3749 — Mishandling national security
PF 3096 — Misleading
PD 4879 — Misleading commercial
PF 9239 — Misrepresentation geographical
PD 5183 — Misuse classified communications
PF 2856 Information organizations ; Monopolization

Information overload
PC 1298 — [Information overload]
PF 6411 — during control complex equipment
PF 6411 — process controllers

Information [police...]
PE 4183 — police criminal suspects ; Withholding
PE 6650 — Pollution
PF 7134 — Pre logical limitations comprehension international
PC 0185 — Prevention exchange
PE 9292 — processing capacity ; Limited human
PF 0703 — Prohibitive cost knowledge
PC 1298 — Proliferation
PF 2401 Information resources ; Incomplete access

Information Restrictions
PD 2926 — distribution confidential government
PC 0185 — freedom
PC 0931 — international freedom

Information [security...]
PF 4050 — security classification ; Suppression
PF 6057 — Selective
PD 2926 — services ; Inaccessible government
PF 2401 — spread ; Inadequate

Information Suppression
PD 9146 — [Information ; Suppression]
PF 4932 — historical
PF 1615 — scientific
PD 2886 Information system design ; Irresponsible

Information systems
PD 9544 — Abuse
PD 9330 — Bias
PD 7450 — Discriminatory design
PE 1122 — Excessive cost comprehensive
PE 8066 — Inadequate integration international
PF 9716 — Inadequate use modern
PD 0104 — international governmental decision making processes ; Inadequate
PE 7009 — international nongovernmental organizations ; Lack financial
PE 0075 — Lack coordination United Nations
PF 6541 — Lack local
PF 8753 — order problem relevant information ; Failure international

Information systems Proliferation
PE 0458 — duplication international
PE 0362 — duplication international nongovernmental organization
PE 0075 — duplication United Nations
PE 9831 Information systems ; Unauthorized access computer
PC 0840 *Information technology arms race*
PF 4106 *Information tourism ; Inadequate statistical*

Information [Unaccountable...]
PF 8074 — Unaccountable management public
PB 0284 — *Unauthorized circulation confidential*
PF 4176 — Unauthorized disclosure computerized

Information Unavailability
PF 6337 — comprehensive
PF 6573 — *funding*
PF 7277 — *legal*

Information [uncertainty...]
PE 7679 — uncertainty ; Defence
PF 2401 — *Uncollected skills*
PF 7133 — Uncoordinated intergovernmental
PF 5728 — Undocumented
PF 1301 — *Unpublicized training*
PF 2401 Information ; Vagueness civic

Information Withholding
PF 8536 — [Information ; Withholding]
PF 9239 — geographical
PD 9836 — medical
PE 9107 Informational procedural obstacles world trade
PF 7336 Informed criminal charges ; Denial right be
PD 8926 Informers
PF 1585 Informers ; Harassment
PF 9794 Infractions ; Politically motivated exaggeration human rights
PF 9269 Infrasonic radiation weapons
PE 0269 Infrastructural weakness developing countries ; Housing
PD 0645 Infrastructure ; Ageing
PF 6527 *Infrastructure costs ; Unfeasible*
PF 2875 Infrastructure developing countries ; Inadequate rural
PC 1228 Infrastructure developing countries ; Weakness

PF 7713 Infrastructure formulate effective global policies ; Inadequate intellectual cognitive
PD 4004 Infrastructure health care delivery ; Insufficient physical
PE 9137 Infrastructure ; High cost natural gas trade

Infrastructure Inadequate
PC 7693 — [Infrastructure ; Inadequate]
PD 9033 — electricity
PD 0645 — maintenance
PD 4790 — medical care
PE 5792 — port
PD 0876 — sanitation
PD 8517 — water system

Infrastructure [Insufficient...]
PF 1495 — Insufficient transportation
PE 5772 — island developing countries ; Weakness
PF 5737 — islands ; Weak social

Infrastructure [Lack...]
PF 2115 — Lack essential local
PF 3318 — *Lack funding*
PE 7000 — land locked developing countries ; Weakness
PE 5814 Infrastructure maritime commerce developing countries ; Lack technical
PD 0645 Infrastructure ; Neglected maintenance local
PD 7663 Infrastructure nuclear power generation ; Inadequate
PE 0289 Infrastructure services least developed countries ; Inadequate
PF 4836 Infrastructure ; Underdevelopment legal
PC 1059 Infrastructure ; Weakness socio economic
PA 7372 Infrequency
PF 6537 *Infrequent achievement honours*
PF 6531 Infrequent adult education
PE 5345 *Infrequent bus service*
PF 6540 *Infrequent business successes*

Infrequent community
PF 5250 — cultural events
PF 2575 — involvement
PF 5250 — meetings
PF 0963 *Infrequent discussion opportunities*
PF 2875 *Infrequent doctor contact*
PF 2401 Infrequent information dissemination
PF 4832 Infrequent medical visits
PF 8695 *Infrequent neighbour contacts*
PF 5250 *Infrequent public gatherings*
PF 3448 *Infrequent river transport*
PD 0807 Infrequent rubbish collection
PF 5250 Infrequent social events
PF 3037 *Infrequent travel opportunities*
PF 5250 *Infrequent unifying events*
PD 0188 *Infringement ; Copyright*
PD 2868 Infringement dietary taboos
PE 5222 Infringement functioning legitimate organizations
PC 6003 Infringement human rights
PF 9794 Infringement human rights ; Misinformation concerning
PE 5015 Infringement national sovereignty

Infringement [privacy...]
PB 0284 — privacy
PB 0284 — privacy ; Dependence
PD 3537 — proprietary information
PB 5405 Infringement rights
PD 2198 Infringement sexual taboos
PF 3976 Infringement taboos
PD 3537 Infringement trade secrets
PB 0731 *Ingested objects ; Suffocation*
PG 2250 Ingestion injurious materials
PA 6839 Ingloriousness *Disrepute*
PA 6659 Ingloriousness *Humility*
PF 2845 *Ingrained individualistic patterns*
PE 7002 Ingrained optimism
PF 2256 *Ingrained segregation habits*
PA 5504 Ingratitude
PA 5504 Ingratitude *Ingratitude*
PE 8961 *Inguinal hernia*
PE 4953 Inhabitants tall buildings ; Environmental stress
PE 1427 Inhalant abuse
PD 4027 *Inhalant intoxication*
PD 0637 *Inhalation dusts*
PD 7307 *Inhalation pneumonia*
PE 1427 Inhaling solvents anaesthetic drugs
PA 5532 Inharmonious *Discaccord*
PA 5982 Inharmonious *Disagreement*
PD 1283 *Inharmonious way life ; Cosmically*
PF 4829 *Inherent development ; Destruction*
PF 0886 Inherit property ; Denial right
PF 0886 Inheritance ; Unequal property
PE 0652 *Inherited diseases*
PF 9208 Inherited problems policy innovation ; Constraint
PD 5535 *Inherited retinopathies animals*
PE 5754 Inherited titles ; Fraudulent nature
PD 4918 Inhibited bereavement
PF 2352 Inhibited capacity visualize creative future
PD 4918 Inhibited grief process
PE 4921 Inhibited growth malnourished children
PD 5177 Inhibited human physical growth
PF 6574 Inhibited official accountability
PD 4418 *Inhibited self-promotion*
PF 6472 *Inhibiting adult preliteracy*
PE 1137 *Inhibiting business outlook*
PF 3211 Inhibiting effects traditional life-styles
PF 2451 Inhibiting participatory methods

Inhibiting [shyness...]
PF 1278 — shyness
PF 6193 — sleeping public ; Unnatural urban environments
PD 0061 — *social attitudes*
PF 0193 — social patterns
PF 2803 Inhibiting vision future ; Static social relations
PF 6167 Inhibition adult life small households

-959-

Inhibition [collectively...]
- PF 6170 — collectively organized fantastic happenings
- PF 6197 — communication non-proximate offices
- PE 8476 — crop growth pollution
- PA 5497 Inhibition *Difficulty*
- PF 6159 Inhibition exploration children urban environment
- PF 6148 Inhibition individual psychological development life cycle
- PF 6572 Inhibition investment risk
- PE 6350 Inhibition nuclear warfare soot ; Sunlight
- PE 7931 Inhibition personality development exiled children
- PF 6339 Inhibition ; Psychological
- PA 7321 Inhibition *Refusal*
- PA 7296 Inhibition *Restraint*
- PE 7968 Inhibitor business transactions ; Illiteracy
- PC 7013 *Inhospitability*
- PA 5446 Inhospitability *Enmity*
- PA 5458 Inhospitability *Inhospitality*
- PA 5643 Inhospitability *Unkindness*
- PC 0387 Inhospitable climate
- PA 5458 Inhospitality
- PD 8114 Inhuman face ; Development
- PE 8634 Inhuman methods conception
- PE 5648 Inhumane confinement ; Death disability
- PE 7677 Inhumane cooking practices
- PC 3768 Inhumane degrading punishment ; Denial right freedom cruel,
- PF 2442 Inhumane dying conditions
- PE 5288 Inhumane geriatric wards

Inhumane [indiscriminate...]
- PD 1519 — indiscriminate weapons
- PD 1519 — indiscriminate weapons ; Dependence
- PD 1362 — interrogation techniques
- PD 0520 Inhumane jails
- PE 0358 Inhumane killing animals
- PE 2759 Inhumane killing stray animals
- PD 2772 *Inhumane livestock slaughtering meat*

Inhumane [medical...]
- PE 4781 — medical experimentation during war-time
- PF 4925 — mental asylums
- PD 1156 — methods riot control
- PE 4015 Inhumane participation medical profession torture

Inhumane scientific
- PC 1449 — activity
- PC 1449 — activity ; Dependence
- PC 1449 — research
- PE 8361 Inhumane treatment non-combatants war zones

Inhumane [urban...]
- PE 0395 — urban food animal conditions ; Unsanitary
- PF 4936 — use medical life support systems
- PE 1621 — use non-human primates research
- PB 8214 Inhumanity
- PF 0399 Inhumanity capital punishment
- PA 7156 Inhumanity *Moderation*
- PA 5643 Inhumanity *Unkindness*
- PA 6486 Iniquity *Injustice*
- PA 5644 Iniquity *Vice*
- PF 4095 Initiative ; Government opposition private

Initiative Lack
- PF 3318 — government
- PC 2178 — *individual*
- PF 7988 — leadership
- PF 2212 Initiative ; Limited local
- PF 2212 Initiative ; Little local
- PE 3296 Initiative rural communities ; Cumulative depletion corporate
- PF 6471 Initiative ; Undeveloped channels commercial
- PD 1229 Initiative ; Welfare weakened
- PF 1735 *Initiatives ; Apparently unrewarded*
- PE 4997 Initiatives ; Extra legal government
- PF 7118 *Initiatives ; Fragmented community*
- PE 0780 Initiatives government ; Cooptation private sector

Initiatives [Inadequate...]
- PF 6143 — Inadequate facilities grass roots community
- PF 1578 — inappropriate local requirements ; Programme
- PF 0342 — Ineffective follow up
- PF 8846 — Irrelevant policy making
- PF 1781 *Initiatives ; Lack conscious*
- PC 8019 *Initiatives ; Misuse societally endorsed professions conceal socially unacceptable*
- PJ 0451 *Initiatives ; Reluctance government invest high risk development*
- PF 4668 *Initiatives ; Restricted government ability undertake new*
- PF 1578 Initiatives specific needs ; Failure adapt general
- PF 1249 *Initiatives ; Unrewarded volunteer*
- PF 2212 *Initiatives ; Weak local*
- PE 8109 Injured animals ; Economic loss destruction
- PB 0855 Injured people
- PB 0855 Injuries
- PC 2753 Injuries ; Animal
- PE 4262 Injuries ; Athletic

Injuries [Birth...]
- PE 4828 — Birth
- PE 3822 — bone ; Diseases
- PD 0992 — brain ; Diseases
- PC 0646 Injuries ; Employment
- PE 7511 *Injuries ; Head*
- PE 2053 *Injuries ; Kidney*
- PC 0646 Injuries ; Occupational
- PE 5606 Injuries ; Repetitive strain

Injuries [Spinal...]
- PE 8813 — Spinal cord
- PE 4262 — Sports
- PE 5686 — sports ; Long term
- PE 6874 Injuries ; Traumatic

- PC 0646 Injuries ; Work
- PB 0731 Injurious accidents
- PC 1630 Injurious insects
- PG 2250 Injurious materials ; Ingestion
- PA 5454 Injury *Badness*
- PE 8505 Injury deficiency ; Facial oral
- PA 6822 Injury *Disrespect*

Injury [Eardrum...]
- PD 2567 — Eardrum
- PE 2598 — export interests developing countries due export cartels ; Excessive
- PF 5739 — exposure weather ; Disease

Injury [Impairment...]
- PA 6088 — *Impairment*
- PA 6852 — *Inappropriateness*
- PA 7395 — *Inexpedience*
- PA 6486 — *Injustice*
- PA 7382 Injury *Loss*
- PE 5763 Injury physical confinement ; Disease
- PE 2286 Injury pituitary tissue
- PC 8534 Injury skin
- PB 0731 Injury ; Superficial
- PA 7107 Injury *Unpleasantness*
- PA 5644 Injury *Vice*
- PD 2626 *Injury ; Whiplash*
- PA 6486 Injustice
- PF 4983 *Injustice ; Complicity structural*

Injustice Dependence
- PA 6486 — [Injustice ; Dependence]
- PC 2181 — political
- PC 0797 — social
- PD 0634 Injustice detention juveniles

Injustice [Indifference...]
- PB 5249 — Indifference response
- PA 6695 — *Inequality*
- PC 9112 — International economic
- PC 7112 Injustice ; Legislative

Injustice [mass...]
- PE 0597 — mass trials
- PE 0494 — military tribunals
- PA 6644 — *Misrepresentation*
- PC 2181 Injustice ; Political
- PE 0397 Injustice religious courts

Injustice [Social...]
- PC 0797 — Social
- PE 0088 — special courts
- PB 1935 — Structural
- PE 0424 Injustice trials absentia
- PA 5644 Injustice *Vice*
- PF 7722 Injustices ; Riskless responses public
- PD 1990 Inland seas ; Desiccation

Inland [water...]
- PD 7087 — water resources ; Conflicting claims over shared
- PD 1223 — waters ; Pollution
- PD 2487 — waterway transport facilities ; Inadequate
- PE 0428 Inmate killing
- PE 0428 Inmates ; Intimidation prisoners
- PE 0428 Inmates ; Maltreatment prisoners fellow
- PD 2108 Innate expectation suppression minority opinion
- PF 2508 Inner circles
- PE 1134 Inner cities ; Limited availability permanent employment

Inner city
- PC 2616 — areas ; Environmental degradation
- PC 2616 — decay
- PF 2042 — enterprises ; Lack awareness potential investment small,
- PC 2616 — neighbourhoods ; Deteriorating
- PC 5052 — poverty
- PD 2345 Inner suburbs ; Destruction green
- PC 2494 *Inner-tensions political parties*
- PA 0429 Innocent bystanders ; *Harm*
- PF 7728 Innocent ; Failure clear names people convicted crimes which they are
- PE 7252 Innocent ; False confessions
- PE 6755 Innocent passage territorial waters ; Denial right
- PE 6755 Innocent passage territorial waters ; Unlawful interference rights
- PE 7252 Innocent ; Self incrimination
- PE 7393 Innocent until proven guilty ; Denial right presumed
- PF 2707 Innovation ; Centralized decisions local technological
- PF 9208 Innovation ; Constraint inherited problems policy
- PE 8870 Innovation ; Delay societal impact
- PD 0528 Innovation ; Endangered jobs technological
- PF 9127 Innovation ; Fear technological
- PF 7230 Innovation public sector ; Deficiency
- PF 4050 Innovation purchase patents ; Suppression

Innovation [Self...]
- PF 4741 — Self limiting technical
- PF 4741 — Self perpetuating attitudes technological
- PF 0275 — Suppression creativity
- PF 7274 Innovations ; Disputed priority
- PD 3712 *Innovative management techniques ; Inappropriate use*
- PF 4050 Innovative patents ; Government refusal declassify
- PF 1572 Innovative projects against world problems ; Lack
- PG 1957 Innuendo ; Sexual
- PC 0143 Innumeracy
- PC 0143 Innumeracy ; Dependence
- PA 7250 Inobservance *Disobedience*
- PA 6247 Inobservance *Inattention*
- PF 1916 Inoculation viral diseases
- PE 5058 Inoperable hospital facilities
- PB 7805 Inoperative forums public information
- PF 2346 *Inoperative shutoff valving*
- PF 1487 Inordinate legal delay
- PE 9076 Inorganic chemicals, elements, oxides halogen salts ; Trade instability

- PD 5227 Inorganic salts pollutants
- PF 2421 Input ; Inadequate citizen
- PF 6461 *Input ; Insufficient parent*
- PA 5838 Inquietude *Agitation*
- PA 6030 Inquietude *Fear*
- PA 7107 Inquietude *Unpleasantness*
- PF 5039 Inquiry based learning ; Inadequate
- PA 5611 Inquisition *Discovery*
- PE 4685 Insalubrity animal excrement urban environments
- PD 0876 Insanitary drainage facilities
- PE 2764 Insanitary penning conditions factor animal diseases
- PA 7157 Insanity
- PC 0300 Insanity
- PD 9699 Insanity ; Criminal
- PE 0435 Insanity ; Delusional
- PE 6412 Insanity ; Hysterical
- PA 6739 Insanity *Maladjustment*
- PD 1318 Insanity ; Periodic
- PA 7371 Insanity *Unintelligence*
- PA 7030 Insatiable desire

Insect [bite...]
- PB 0731 — bite ; Nonvenomous
- PE 3636 — bites stings
- PE 3636 — bites stings ; Deadly
- PD 1135 — breeding areas ; Defoliation
- PD 1135 — breeding habitats ; Defoliation
- PD 3657 Insect damage stored manufactured goods
- PF 3592 Insect invasions
- PD 4751 Insect invasions ; Disastrous

Insect pests
- PC 1630 — [Insect pests]
- PE 1679 — Beetles
- PE 3618 — Dermaptera
- PE 2254 — Flies
- PE 3642 — Grasshoppers
- PE 3615 — Hemiptera
- PE 3614 — Homoptera
- PF 3592 — Introduction new species
- PE 3649 — Lepidoptera
- PE 1439 — Lice
- PE 7832 — modification micro climate ; Increase
- PE 3641 — Orthoptera
- PD 3634 — plants
- PE 3643 — Siphonaptera
- PE 3619 — Thysanoptera
- PE 3620 — Thysanura
- PD 3586 — wood
- PD 4751 Insect populations ; Difficulty controlling
- PD 2109 Insect resistance insecticides
- PD 4751 Insect swarms

Insect vectors
- PD 2748 — animal diseases
- PC 3597 — disease
- PE 3632 — human disease
- PD 7732 — plant diseases
- PD 3600 — viral diseases plants
- PD 3695 Insecticide damage crops

Insecticide [poisoning...]
- PE 2349 — *poisoning*
- PD 5228 — *poisoning animals*
- PD 0983 — *pollutants ; Organochlorine*
- PD 0983 Insecticides
- PE 3578 Insecticides ; Blackfly resistance
- PE 3579 Insecticides ; Cockroach resistance
- PE 3572 Insecticides ; Flea resistance
- PE 3583 Insecticides ; Housefly resistance
- PD 2109 Insecticides ; Insect resistance
- PE 3576 Insecticides ; Louse resistance
- PE 3582 Insecticides ; Mosquito resistance
- PD 1694 *Insecticides ; Organophosphorous*

Insecticides pollutants
- PD 0983 — [Insecticides pollutants]
- PE 1282 — Carbamate
- PD 0983 — *Organophosphorus*
- PD 0983 Insecticides ; Use chemical
- PD 3479 Insectivora
- PD 3479 Insectivores ; Endangered species
- PE 6286 Insects ; Borer
- PE 3468 Insects ; Cruelty

Insects [Defoliating...]
- PE 4999 — Defoliating
- PE 3880 — Destruction crop pollinating
- PD 2284 — Diseases beneficial
- PC 2326 Insects ; Endangered species
- PC 1630 Insects ; Injurious
- PE 3612 Insects pests ; Scale
- PD 9162 Insecure land tenure
- PE 2451 *Insecure lease tenure*
- PA 0857 Insecure people
- PC 1784 Insecure property assets
- PA 0857 Insecurity
- PE 5103 Insecurity ; Cargo
- PA 6971 Insecurity *Danger*

Insecurity Dependence
- PA 0857 — [Insecurity ; Dependence]
- PC 2020 — economic
- PB 0009 — international

Insecurity developing countries
- PC 6189 — [Insecurity developing countries]
- PD 3835 — National
- PE 5391 — Personal physical
- PD 3835 — Political
- PE 4796 — Social
- PD 8030 Insecurity due excessive foreign investment ; National

Insecurity [Economic...]
- PC 2020 — Economic

Instability trade

Insecurity [Emotional...] cont'd
PD 8262 — Emotional
PD 8211 — employment
PC 4341 — Environmental
Insecurity [Food...]
PB 2846 — Food
PB 2846 — Food grain
PF 8354 — future crop uncertainty
PF 2303 — *Future employment*
PB 2846 — Insecurity ; Household food
Insecurity [indigenous...]
PE 1506 — indigenous peoples ; Social
PE 4639 — industrialized countries ; Personal physical
PB 1149 — Internal
PB 0009 — International
PD 8211 — Insecurity ; Job
Insecurity [land...]
PD 5788 — land-locked developing countries
PD 9362 — leadership
PE 7954 — least developed countries ; Food
Insecurity [Personal...]
PD 8657 — Personal physical
PB 1149 — Political
PE 5316 — pregnant women ; Job
PD 9044 — Prevailing community
PC 1784 — property
PD 6364 — Insecurity refugees asylum seekers ; Physical
PB 8678 — Insecurity resources
Insecurity [small...]
PD 2374 — small states ; International
PC 1867 — Social
PA 0857 — Socio economic
Insecurity [Uncertainty...]
PA 7309 — *Uncertainty*
PE 7670 — unilateral structural disarmament
PE 7370 — university campuses ; Physical
PE 6299 — Urban
Insecurity vulnerability
PD 1863 — European
PC 0541 — Military
PB 1149 — National
PD 1521 — non-nuclear weapon states
PC 4440 — nuclear weapon states
Insecurity [Water...]
PC 4341 — Water
PE 4985 — Western marriages
PE 7750 — women ; Personal physical
PF 2303 — *work changes*
PE 8634 — Insemination ; Artificial
PA 5451 — Insensibility
Insensibility [Ignorance...]
PA 5568 — *Ignorance*
PA 6446 — *Indiscrimination*
PA 5451 — *Insensibility*
PA 6739 — Insensibility *Maladjustment*
PA 7371 — Insensibility *Unintelligence*
PF 6531 — Insensitive adult direction
PF 7299 — Insensitive discourse
PF 9728 — Insensitive economics ; Environmentally
PF 3458 — Insensitive government policies ; Environmentally
PD 8638 — Insensitive house design ; Culturally
PD 8638 — Insensitive needs women ; Architecture
PD 7320 — Insensitive urban renewal
PF 2808 — Insensitivity ; Bureaucratic
PD 5393 — Insensitivity ; Communist environmental
PF 8156 — Insensitivity cultural diversity ; Bureaucratic
PA 7143 — Insensitivity *Discourtesy*
PF 8156 — Insensitivity diversity cultural traditions
PF 2808 — Insensitivity ; Government
Insensitivity [Indiscrimination...]
PA 6446 — *Indiscrimination*
PA 5451 — *Insensibility*
PD 1975 — intelligence testing ; Inadequacy
PF 9119 — Insensitivity non-immediate hazards society
PE 0533 — Insensitivity nonverbal messages
PF 1216 — Insensitivity personal pain
PE 1011 — Insensitivity transnational corporations consumer needs developing countries
PA 7364 — Insensitivity *Unfeelingness*
PA 5643 — Insensitivity *Unkindness*
PD 3841 — Insider dealing
PF 8703 — Insider policy-making
Insider [trade...]
PD 3841 — trade
PD 3841 — trading
PD 3917 — trading commodities
PF 7111 — Insight ; Enforced abstention
PF 8753 — Insight response problems society ; Failure focus available
PF 5728 — Insights ; Unrecorded
PA 6977 — Insignificance *Meaninglessness*
PA 5942 — Insignificance *Unimportance*
PF 5044 — Insignificant leisure engagement
PA 6400 — Insincerity *Affectation*
PA 7363 — Insincerity *Improbity*
PA 5502 — Insincerity *Reason*
PA 7411 — Insincerity *Uncommunicativeness*
PA 7237 — Insinuation *Condemnation*
PA 6191 — Insinuation *Disapproval*
PA 6659 — Insinuation *Humility*
PA 6862 — Insinuation *Intrusion*
PD 9654 — *Insipidus animals* ; Diabetes
PD 3051 — Insistence creditors fulfilment debt service obligations developing countries ; Inconsiderate
PA 6466 — Insobriety *Intemperance*
PA 7143 — Insolence *Discourtesy*

PA 6822 — Insolence *Disrespect*
PA 5479 — Insolence *Lamentation*
PA 6979 — Insolence *Opposition*
PA 7115 — Insolence *Rashness*
PA 6491 — Insolence *Vanity*
PF 4706 — Insoluble scientific problems
PC 6154 — Insolvency
PE 3537 — Insolvency industry
PD 9376 — Insolvency ; Personal
PD 2591 — Insolvency ; Risk
PE 6431 — Insolvent institutions
PC 6154 — Insolvent local authorities
PC 6154 — Insolvent municipalities
PE 2197 — Insomnia
PE 3924 — Insomnia
PA 6379 — Insouciance *Avoidance*
PA 6598 — Insouciance *Incuriosity*
PA 5438 — Insouciance *Neglect*
PA 7364 — Insouciance *Unfeelingness*
PF 2793 — Inspection ; Inadequate
PE 4146 — Inspection regulated activities ; Ineffective official
PE 4146 — Inspection toxic waste sites ; Ineffective
PF 8046 — Inspectors ; Corruption official
PF 2793 — Inspectors ; Lack regulatory
PF 8046 — Inspectors ; Negligence official
PF 8046 — Inspectors ; Unethical practices regulatory
PF 8046 — Inspectors violators regulations ; Connivance regulatory
PF 4984 — *Inspired school* ; Hopelessness
PA 0859 — Instability
PE 8998 — Instability agricultural livestock production
PE 2585 — Instability alcoholic beverages trade
Instability [banana...]
PE 1774 — banana trade
PE 2601 — basic metal industries
PE 2585 — *beer trade*
Instability [Changeableness...]
PA 5490 — *Changeableness*
PE 0538 — chemical petrochemical industry
PD 0619 — chemicals trade
PE 8928 — China, pottery earthenware industries
PE 1549 — cocoa trade
PE 0950 — coffee trade
PD 2968 — commodity prices developing country earnings
PB 1151 — *complex society*
PE 0509 — construction industry
PE 0824 — *copper trade*
PE 1510 — cotton trade
Instability [Danger...]
PA 6971 — *Danger*
PE 1552 — date trade
PC 2677 — Dependence political
PD 8323 — developing countries ; Political
PE 4687 — due torture ; Emotional
PC 1910 — *due tribalism* ; National
Instability [economic...]
PC 1217 — economic industrial production activities
PE 8875 — electrical machinery, apparatus appliances ; Trade
PD 9159 — Emotional
PE 5930 — exchange rates
PE 0980 — exchange rates transnational corporations ; Aggravation
PE 4915 — Export earnings
PD 2968 — export trade developing countries producing primary commodities
Instability [Financial...]
PC 7873 — Financial
PE 1424 — fishing industry
PE 8918 — food drink industries
PE 4986 — food prices developing countries
PE 0459 — forestry logging
PE 2587 — fresh fruit edible nut trade
PE 8119 — Fuel wood charcoal trade
PE 1474 — fur trade
Instability [Glass...]
PE 9084 — Glass glass products industries
PE 2587 — *ground nut peanut trade*
PF 1737 — *growth income industrialized countries*
PE 9059 — Instability hunting, trapping game propagation
Instability [inorganic...]
PE 9076 — inorganic chemicals, elements, oxides halogen salts ; Trade
PC 1606 — International strategic
PE 8772 — Iron ore concentrates trade
Instability Iron steel
PE 8070 — basic industries
PE 8969 — manufactures trade
PE 8312 — scrap trade
PA 7325 — Instability *Irresolution*
PE 1794 — Instability jute trade
Instability [labour...]
PF 0907 — labour force developing countries
PE 0824 — *lead trade*
PE 8320 — leather footwear industry
PE 8320 — Leather leather manufactures trade
PF 5356 — legal judgements
Instability [machinery...]
PE 1852 — machinery equipment industries
PE 1852 — *Machinery equipment industries*
PC 0580 — manufacturing industries
PD 9117 — manufacturing industries developing countries
PE 1683 — margarine trade
PD 2103 — Marital
PE 5791 — maritime shipping industry
PE 8489 — Meal flour wheat meslin trade
PE 7996 — Medicinal pharmaceutical products trade
PE 1852 — *Metal products industry*

Instability [micro...] cont'd
PE 9405 — micro-enterprises
PE 0993 — mining quarrying industry
PE 8809 — miscellaneous mining quarrying
PC 7873 — Monetary
PC 7873 — monetary markets
Instability [nickel...]
PE 0824 — *nickel trade*
PE 2601 — *Non ferrous metal basic industries*
PE 2599 — non-metallic mineral products industry
Instability [ocean...]
PE 5850 — ocean freight rates
PE 0313 — olive oil trade
PF 0552 — orthographic standards
Instability [paint...]
PE 0538 — *paint industry*
PE 1927 — paper printing industries
PE 3000 — pepper trade
PD 1165 — Permafrost
PE 8955 — Petroleum natural gas production
PD 0909 — petroleum prices
PE 8470 — Photographical optical goods trade
PE 5143 — plastics industry
PC 2677 — Political
PF 8635 — prices
PC 0463 — primary commodities trade
PD 2894 — production food live animals
PE 8169 — professional ; Scientific controlling instruments trade
Instability [Radioactive...]
PE 8843 — Radioactive associated materials trade
PE 8159 — Rape mustard oils trade
PD 7522 — relations allies superpowers
PE 0696 — rice trade
Instability [silk...]
PE 1550 — *silk trade*
PE 7956 — Silver platinum group trade
PE 8556 — Silver platinum ores trade
PE 0961 — soybean trade
PE 1619 — spices trade
PE 2585 — *spirits trade*
PE 8612 — stone ; Sand gravel trade
PE 8958 — Sugar confectionery preparations trade
PE 0383 — sugar trade
Instability [tanning...]
PE 9089 — tanning dyeing trade
PE 2054 — tea trade
PE 1008 — textile clothing industries
PE 0824 — *tin trade*
PE 0572 — Tobacco manufactures
PF 4112 — tourist dependent economies
PE 1406 — *trade aluminium*
Instability trade animal
PE 8816 — feedstuffs, excluding unmilled cereals
PE 8896 — oils fats
PE 0735 — vegetable oils fats
Instability trade [asses...]
PE 8593 — asses, mules hinnies
PE 8259 — beans, dried, peas, lentils leguminous vegetables
PE 1680 — beverages
PE 1641 — beverages tobacco
PE 1769 — cereals cereal preparations
PE 0500 — chemical elements compounds
PE 8206 — coal, coke briquettes
PE 0915 — coffee, tea, cocoa, spices manufactures thereof
Instability trade crude
PE 9049 — fertilizers
PE 0760 — fertilizers crude minerals, excluding coal, petroleum precious stones
PE 0701 — synthetic reclaimed rubber
Instability trade [dairy...]
PE 0576 — dairy products eggs
PE 8740 — dressed fur skins
PE 8257 — dried salted smoked meat
PE 7907 — electric energy
PE 8232 — essential oils perfume materials
PE 0972 — fish, crustacea molluscs preparations thereof
PE 0861 — fixed vegetable oils fats
PD 1434 — food live animals
Instability trade fresh
PE 8591 — frozen chilled meat
PE 8258 — potatoes
PE 8688 — tomatoes
Instability trade [fruit...]
PE 0961 — fruit vegetables preparations thereof
PD 0877 — *gas*
PE 1513 — *hard fibres*
PD 0877 — inedible crude non-fuel materials
PD 1434 — legumes
PD 1376 — live animals
PD 0620 — machinery transport equipment
PE 8682 — manila fibre
PE 0806 — manufactured fertilizers
PE 0882 — manufactured goods
PE 0755 — meat meat preparations
PE 0553 — metalliferous ores metal scrap
PD 0877 — mineral fuels, lubricants related materials
PE 8464 — mineral tar crude chemicals derived coal, petroleum natural gas
PE 1683 — miscellaneous food preparations
PE 0814 — miscellaneous manufactured articles
Instability trade non
PE 8748 — alcoholic beverages
PE 8828 — electrical machinery
PE 1406 — ferrous metals
Instability trade [oil...]
PE 0386 — oil-seeds, oil nuts oil kernels

Instability trade [ores...] cont'd
- PE 8131 — ores concentrates non-ferrous base metals
- PD 0909 — petroleum petroleum products
- PE 8741 — preserved fruit
- PE 7914 — pulp waste paper
- PE 8769 — regenerated cellulose artificial resins
- PE 8272 — sanitary plumbing, heating, lighting fixtures fittings
- PE 8861 — sheep goats
- PE 0383 — sugar, sugar preparations honey
- PE 0760 — *sulphur*
- PE 8950 — synthetic regenerated fibres
- PE 0915 — *tea mate*
- PE 8970 — textile yarn fabrics, made-up articles
- PE 0572 — tobacco tobacco manufactures
- PE 8920 — travel goods
- PE 7621 — tungsten
- PE 8560 — U235-depleted uranium
- PE 8106 — undressed hides non-fur skins
- PE 1235 — undressed hides, skins fur skins
- PE 1769 — unmilled cereals
- PE 1550 — unprocessed textile fibres their waste
- PE 9164 — vegetable-based foodstuffs
- PE 8208 — vegetables, roots tubers
- PE 2521 — wood, lumber cork
- PA 6425 Instability *Transience*
- PA 7309 Instability *Uncertainty*
- PE 8212 Instability ; Uranium ores concentrates thorium trade

Instability vegetable
- PE 1513 — fibre trade, excluding cotton jute
- PE 8901 — products ; Roots tuber trade
- PE 1711 — trade

Instability [water...]
- PD 0722 — water supply
- PA 5558 — *Weakness*
- PE 1008 — wearing apparel industries
- PE 0385 — wheat trade
- PE 2522 — wine trade
- PF 8751 — women ; Belief emotional
- PE 0681 — woodworking industries
- PE 2056 — wool trade
- PC 8073 — world economy ; Economic financial
- PE 0824 *Instability zinc trade*
- PF 4854 Installations ; Cover up unsafe industrial
- PE 4304 Installations ; Hazardous industrial
- PC 1867 *Installations ; Unreliable phone*
- PA 6560 Instantaneousness
- PF 3849 Instinct ; Death
- PF 7490 Instinctual violence
- PE 4430 Institutes ; Political prisoners mental
- PF 0963 *Institution desirability ; Unconsensed*
- PF 1870 Institution ; Vulnerability marriage
- PD 0104 Institutional arteriosclerosis

Institutional [care...]
- PE 5288 — care ; Neglect elderly
- PF 0845 — change ; Government resistance
- PF 1963 — credibility ; Loss
- PE 1557 Institutional debts ; Excessive
- PF 2825 Institutional domination organizational systems
- PC 3915 Institutional fragmentation
- PF 0048 Institutional framework finance ; Inappropriate international
- PD 2686 Institutional lying
- PE 2862 Institutional obsolescence modern industrialized societies
- PF 8846 *Institutional preoccupation obsolete problems*
- PC 3915 Institutional responsibility ; Sectoral fragmentation

Institutional [security...]
- PC 1835 — security ; Overemphasis
- PE 0365 — structures local government developing countries ; Inadequate
- PF 7745 — systems ; Structural amnesia
- PB 1935 Institutional violence

Institutionalization [disabled...]
- PF 4681 — disabled
- PF 7673 — discriminatory outmoded concepts
- PF 2580 — discriminatory outmoded concepts legal systems
- PD 0932 Institutionalization education ; Excessive
- PF 8209 Institutionalization vulnerable groups ; Excessive

Institutionalized [callousness...]
- PF 2006 — callousness public services
- PC 9173 — corruption
- PB 1935 — covert himsa
- PE 4001 Institutionalized members society
- PD 1662 *Institutionalized patients*
- PF 9786 Institutionalized reward systems ; Inappropriate
- PE 6763 Institutionalized subjects ; Medical experimentation
- PD 6145 *Institutionalized torture*
- PF 3777 Institutionally reinforced sense personal shallow meaning

Institutions [Abuse...]
- PF 2137 — *Abuse status religious*
- PE 2932 — Abusive detention psychiatric
- PE 2805 — Autocracy intergovernmental financial
- PE 6431 Institutions ; Bankrupt
- PF 3831 Institutions ; Closure social
- PF 9385 Institutions ; Criticism official

Institutions [Defamation...]
- PF 9385 — Defamation
- PF 3458 — degrading environment ; Unaccountability
- PE 8564 — Destruction civilian populations
- PF 3831 — Destruction social
- PE 2214 — developing countries ; Over dependency international financial
- PF 1724 — developing countries ; Unrelated pioneer
- PE 1675 — Disturbances correctional
- PF 4220 — Domination individuals

- PF 8798 Institutions ; Excessive size social

Institutions [Faceless...]
- PF 2081 — Faceless social
- PE 0964 — Failure savings
- PF 1755 — Fostering dependency social

Institutions [Inadequate...]
- PF 6691 — Inadequate objectivity
- PF 4925 — Inadequate psychiatric
- PD 5142 — Ineffectiveness correctional
- PE 6431 — Insolvent
- PF 8846 — Irrelevant

Institutions [Lack...]
- PC 0233 — *Lack child welfare*
- PF 1963 — Loss credibility
- PE 8064 — Loss credibility international
- PF 4739 Institutions ; Proliferation public sector
- PF 1136 Institutions ; Secrecy international aid
- PJ 5193 *Institutions; Soulless*
- PF 6173 Institutions transition adolescence ; Inadequate recognition
- PF 1136 Institutions ; Unaccountability international aid
- PF 1136 Institutions ; Unaccountability international financial
- PF 3575 Instruction rural communities ; Irregular outside
- PF 0443 *Instruction ; Single language*
- PD 4217 Instruction vegetation
- PC 0592 *Instructors ; Shortage women*
- PE 5503 Instruments ; Counterfeit financial
- PF 6365 Instruments ; Inadequacy international human rights
- PF 2335 Instruments ; Inadequate governmental managerial
- PE 7961 Instruments ; Shortage professional, scientific controlling
- PE 8169 Instruments trade instability professional ; Scientific controlling
- PB 1125 Insubordinate behaviour
- PD 5308 Insubordinate children
- PB 1125 Insubordination
- PE 5782 Insubordination armed forces ; Causing
- PA 7250 Insubordination *Disobedience*
- PD 5244 Insubordination ; Employee
- PA 5563 Insubordination *Lawlessness*
- PD 7225 Insubordination ; Military
- PF 7467 Insubordination ; Spiritual
- PA 6959 Insubstantiality
- PA 7309 Insubstantiality *Uncertainty*
- PF 0171 Insufficiencies employment
- PA 5473 Insufficiency
- PE 0471 *Insufficiency animals ; Cardiac*
- PE 0920 *Insufficiency ; Chronic valvular*
- PA 6011 Insufficiency *Discontentment*

Insufficiency [Impotence...]
- PA 6876 — *Impotence*
- PA 6695 — *Inequality*
- PA 5652 — *Inferiority*
- PA 6041 Insufficiency *Shortcoming*
- PA 7408 Insufficiency *Smallness*

Insufficient [access...]
- PF 3467 — access technology agricultural upgrading
- PD 2831 — acoustical treatment workplaces
- PF 6506 — *agency cooperation*
- PF 4108 — agricultural equipment developing countries
- PF 4108 — agricultural machinery developing countries
- PF 1352 — *apprenticeship opportunities*
- PB 8891 — availability goods
- PB 2846 — available food
- PE 4225 — awareness advertisers social responsibility
- PF 1705 *Insufficient banking information*
- PF 5968 Insufficient birth spacing families

Insufficient business
- PE 6554 — experience
- PF 1815 — *facilities*
- PF 1815 — *space*
- PD 8362 Insufficient buying power

Insufficient [capital...]
- PF 2852 — capital investment
- PF 1653 — capitalization means
- PF 1600 — care community property
- PF 6455 — *cash crops*
- PF 0214 — census data
- PF 6550 — *civic organizations*

Insufficient commercial
- PF 3448 — *facilities*
- PC 4867 — know-how
- PC 4867 — *skills*

Insufficient [common...]
- PF 6570 — *common experience*
- PE 6171 — common land urban environments
- PF 2452 — communication decision-makers grassroots
- PF 2350 — communications systems

Insufficient community
- PF 5250 — celebrations
- PF 5250 — events
- PF 1681 — *networks*
- PF 6557 — promotion

Insufficient [consumer...]
- PE 8236 — consumer services developing countries
- PF 5250 — *corporate celebrations*
- PF 1653 — *councilman support*
- PF 2489 — credit capital
- PD 3054 — creditworthiness developing countries

Insufficient cultural
- PF 1985 — heritage transmission
- PF 8623 — heroes
- PF 3211 — *media*
- PF 1790 — *nodes*
- PF 0905 *Insufficient development collaterals*

Insufficient diversification
- PD 0335 — [Insufficient diversification]
- PC 6329 — *energy research*
- PD 0335 — *urban industrial energy supply*
- PE 8303 Insufficient doctors
- PF 2115 Insufficient domestic electricity supply
- PF 8554 Insufficient economic incentives
- PF 6531 Insufficient education adults

Insufficient educational
- PD 0847 — facilities
- PF 2168 — *incentives*
- PE 8438 — material

Insufficient [electrical...]
- PD 9033 — electrical power
- PF 9007 — emergency services
- PF 2477 — *employment mechanisms*
- PD 0176 — employment opportunities developing countries
- PF 3448 — *enabling structures*
- PF 2378 — *enterprise capital*
- PC 1366 — *environmental laws*
- PE 4915 — export benefits

Insufficient [factory...]
- PF 6478 — *factory facilities*
- PF 3037 — *family services*
- PF 3630 — *farming funds*
- PD 5115 — female employment
- PE 4140 — fertilizers agricultural development developing countries
- PF 6578 Insufficient financial
- PG 8883 — *referral*
- PB 4653 — *resources*
- PE 9133 — resources urban services

Insufficient [fire...]
- PF 2477 — *fire protectors*
- PF 6469 — *flow information*
- PF 2852 — *flow investment capital*
- PB 2846 — *food supply*
- PE 8244 — forestry least developed countries
- PF 2346 — *formal complaints*
- PC 4815 — *fresh water supply*
- PF 7916 — *funding experience*

Insufficient [garbage...]
- PF 6473 — *garbage containers*
- PE 1055 — good roads
- PF 5419 — government funding research
- PE 5302 — government spending cost-effective activities

Insufficient health
- PD 1229 — payments
- PD 0366 — personnel
- PD 0366 — surveillance
- PD 0366 — technicians
- PD 6520 Insufficient heavy equipment
- PF 3550 Insufficient housing funds

Insufficient [images...]
- PF 5298 — images political involvement
- PF 5969 — immunization ; Inadequate
- PE 8592 — inappropriate energy equipment developing countries
- PE 7165 — *incentives teachers*
- PF 1959 — *industrial planning*
- PF 0963 — *industrial promotion*
- PF 0756 — *inter-age communication*
- PD 1091 — intervention human evolution
- PE 8233 — *irrigation knowledge*

Insufficient job
- PF 0171 — benefits
- PF 1658 — opportunities
- PF 2949 — *options*
- PF 1658 Insufficient jobs village
- PE 8140 Insufficient knowledge local production processes

Insufficient [land...]
- PF 2346 — *land upkeep*
- PC 6848 — law
- PF 3605 — leadership training
- PF 3184 — *learning opportunities*
- PE 8907 — *leisure time women*
- PF 2568 — *library equipment*

Insufficient [management...]
- PC 4867 — management skills
- PE 4804 — manually skilled workers
- PF 5923 — *maritime agreements*
- PF 4832 — *medical facilities personnel*
- PE 1634 — medical supplies
- PE 1535 — meeting space
- PE 4920 — military personnel
- PF 5659 — minority culture support
- PF 6524 — *modern technology*
- PB 4653 — *money*

Insufficient [need...]
- PD 2438 — need consensus
- PF 6556 — neighbourhood leaders
- PF 6147 — neighbourhood signs
- PF 5778 — nobility
- PD 7663 — nuclear power stations
- PD 0837 — *nuclear weapon free zones*
- PF 2042 Insufficient operating capital
- PF 2316 Insufficient opportunities community activities

Insufficient [parent...]
- PF 6461 — *parent input*
- PB 0750 — *part-time employment*
- PC 2765 — *pastureland*
- PD 8568 — *personal income*
- PF 2316 — *personal time*
- PF 3409 — pest control
- PD 4004 — physical infrastructure health care delivery
- PF 6337 — *plant life information*

Interdependence

Insufficient [possibilities...] cont'd
PE 8985 — possibilities gathering elders
PF 0905 — *preschool equipment*
PE 0751 — preventive medicine
PC 6381 — primary education
PF 0581 — product markets
PF 3409 — *programme funding*
PF 2401 — *programme knowledge*
PF 3550 — *property base*
PF 1773 — *protein consumption*
PF 2694 — provision public services communication
PF 1790 — *public space*
PF 1495 — public transportation
Insufficient [recreation...]
PF 9062 — recreation time
PF 5409 — recreational planning
PF 6465 — recycling materials
PF 6527 — removal snow
PF 2346 — *rental returns*
PE 7959 — repair services developing countries
PF 3553 — responsibility images
PF 2457 — *returns farming*
PF 3665 — *risk capital*
PF 8451 — role models
PF 6511 — *rural housing*
Insufficient [separation...]
PF 6137 — separation urban subcultures
PF 3084 — *side-income jobs*
PD 8108 — skilled teachers
PD 0044 — skilled workmen
PC 6445 — skills
PD 1415 — *small businesses*
PD 9155 — social security agricultural sector
PF 0713 — *special care vehicles*
PD 8388 — *special training*
PF 3560 — *sports equipment*
PE 4874 — stimuli ; Damage infant brains malnutrition
PF 7130 — *storage space homes*
PF 6473 — *street lighting*
PE 2274 — *supervisory personnel*
PF 0772 — Insufficient technical aids disabled persons
PF 6500 — Insufficient technical skills
Insufficient technological
PF 3184 — adaptation
PF 2401 — contact
PF 3409 — know-how
Insufficient [telephones...]
PF 4132 — telephones rural communities
PF 4136 — television systems developing countries
PF 1735 — *tourist attraction*
PE 1139 — *trade income village communities*
PF 1705 — *traditional gatherings*
Insufficient trained
PF 4984 — *drivers*
PD 9113 — *labour*
PF 1332 — *leaders*
PC 4867 — *managers*
Insufficient [training...]
PD 0847 — training facilities
PF 1352 — *training opportunities*
PD 0825 — translation minority languages
PF 1495 — transportation infrastructure
PF 2875 — *transportation old people*
PF 3037 — *travel opportunities*
Insufficient [use...]
PE 0273 — use natural resources least developed countries
PF 2813 — *utility revenue*
PF 0357 — utilization renewable biofuels
PF 3558 Insufficient village cooperation
Insufficient [water...]
PD 8517 — water delivery
PD 0722 — *water pressure*
PF 1257 — *women cooperatives*
PF 1705 — *women's groups*
PE 5132 — work awareness international affairs
Insufficient [young...]
PE 1848 — *young families*
PF 2575 — *youth facilities*
PD 1544 — *youth responsibility*
PE 7064 Insufficiently treated meat infected animals
PF 2575 Insular patterns community groupings
PA 5869 Insularity *Exclusion*
PA 7306 Insularity *Narrowmindedness*
PE 1499 Insulating materials ; Health hazards modern
PE 0242 Insulation ; Inadequate home
PD 5238 Insult
PA 6822 Insult *Disrespect*
PF 5630 Insult ; Religious
PF 6026 Insurance against damages arising natural disaster ; Inadequate
PE 4329 Insurance claims medical negligence ; Increase
Insurance [Dependence...]
PE 4280 — Dependence developing countries foreign
PE 8736 — developing countries ; Inadequate life
PE 4796 — developing countries ; Inadequate social
PE 1826 Insurance fraud
Insurance [Inadequate...]
PF 8827 — Inadequate
PF 8827 — *Inadequate crop*
PD 7012 — industry ; Protectionism
PE 5802 — Inequities marine
PE 5665 — island developing countries ; Inaccessibility
PE 5896 — Insurance land locked developing countries ; Excessive costs unsuitability
PE 4994 Insurance markets offshore insurers ; Destabilization national

PE 8006 Insurance monopolies capitalist countries
Insurance [practices...]
PD 0881 — practices ; Restrictive transport
PE 1826 — practices ; Unethical
PE 8632 — Prohibitive cost
PE 0740 Insurance transactions developing countries ; International indebtedness arising
PF 8827 Insured ; Under
PE 4994 Insurers ; Destabilization national insurance markets offshore
PD 8284 Insurgency
PC 1738 *Insurgency warfare ; Counter*
PD 8284 Insurrection
PD 8284 Insurrection ; Armed
PD 0153 Intake alcohol ; Excessive
PD 1694 Intake ; Chemical contamination dietary
PD 0446 Intake ; Disparities calorie
PD 0321 *Intake iron ; Inadequate*
PD 0339 Intake ; Low protein
PA 5570 Intangibility
PE 6486 Intangible property ; Crimes against
PE 2000 Integrate available knowledge ; Increasing inability
PF 8753 Integrate knowledge empower humanity response global problematique ; Failure
PC 6242 Integrated rural development ; Conflicts
PF 1772 Integrating images personal freedom ; Non
PC 1532 Integration foreigners ; Inadequate social
PE 6779 Integration handicap ; Social
Integration [ideology...]
PF 3402 — ideology society ; Inadequate
PD 0794 — Imbalances distribution costs benefits economic
PC 1532 — immigrants ; Inadequate cultural
Integration Inadequate
PA 0911 — personal
PF 3215 — political
PF 9129 — regional
PE 8066 Integration international information systems ; Inadequate
PF 0944 Integration knowledge ; Lack system
Integration Lack
PA 6233 — [Integration ; Lack]
PF 0536 — cultural
PF 3216 — political
PF 1324 — social
PE 5801 — world maritime
PD 4131 Integration ; Linguistic dis
PC 3293 Integration peoples ; Forced
Integration [regional...]
PD 9412 — regional groupings developing countries ; Inadequate economic
PF 3403 — religions society ; Inadequate
PF 4954 — religious belief ; Excessive
Integration [socialist...]
PF 4884 — socialist countries ; Inadequate economic
PF 7014 — societal learning ; Obstacles conceptual
PF 1324 — society ; Lack
Integration [traditional...]
PF 4871 — traditional Western medicine ; Lack
PE 1322 — transnational corporations ; Domination economic
PF 6157 — transport systems ; Inadequate
PD 0664 — transport systems neighbouring developing countries ; Lack
PA 7263 Integrity∗complex
PF 9693 Integrity ; Corrupting effect valuable prizes academic
PF 5148 Integrity creation ; Violation
PC 2945 Integrity ; Denial rights territorial
PC 2945 Integrity ; Disruption territorial
PD 1163 Integrity effectiveness government operations ; Crimes against
PF 7992 Integrity ; Lack
PF 0796 Integrity ; Lack political
Intellectual [ability...]
PE 4398 — ability public sector ; Lack scope
PF 0100 — ability ; Underutilization
PF 7847 — arrogance
PF 6573 *Intellectual bias ; Local anti*
Intellectual [cognitive...]
PF 7713 — cognitive infrastructure formulate effective global policies ; Inadequate
PC 3390 — conflict
PE 7144 — confusion concerning role United Nations
PC 2915 — corruption
Intellectual [decline...]
PF 7815 — decline human race
PF 8590 — discrimination
PD 1893 — dishonesty government
PC 2582 — dissent
PD 0434 — dissidents ; Repression
PA 7197 Intellectual faculties∗complex
PC 5018 Intellectual freedom ; Suppression
PJ 8777 Intellectual inefficiency
PC 2915 Intellectual irresponsibility
PE 9114 Intellectual labour ; Artificial opposition manual
PF 3217 Intellectual liberty ; Lack
Intellectual methods
PF 1094 — appropriate human depths ; Antiquated
PF 0956 — Faulty approaches teaching
PF 7380 — Inadequate
PE 7095 — which disenfranchise grassroots technological development
Intellectual [piracy...]
PF 8854 — piracy
PC 2915 — practices ; Unethical
PC 3406 — prejudice
Intellectual property
PF 8854 — Denial right

Intellectual property cont'd
PF 8854 — Inadequate protection
PF 8854 — Theft
PF 8854 — Vulnerability
PD 6656 Intellectual terrorism
PF 2146 Intellectualism
PF 1929 Intellectualism ; Anti
PE 8900 Intellectually disabled convicts ; Lack care
PD 1273 Intellectuals developing countries ; Unemployed
Intelligence [agencies...]
PF 9184 — agencies ; Unaccountable government
PF 9184 — agencies ; Unreported cooperation
PD 4501 — agency operations ; Covert
PE 9857 — agents public office ; Secret
PD 3496 Intelligence centres ; Extra territorial military bases
PF 1199 Intelligence certain racial groups ; Inferior average
PF 8819 Intelligence failure
PC 1587 Intelligence ; Impaired
PD 1975 Intelligence ; Inadequate means measuring
Intelligence [Lack...]
PC 1587 — Lack
PD 1975 — Lack definition
PC 1587 — Low
PD 4501 Intelligence operations ; Clandestine
PC 2921 Intelligence ; Organizational
PF 1199 Intelligence ; Racially determined differences
PF 8819 Intelligence services ; Failure government
PD 1975 Intelligence testing ; Inadequacy insensitivity
PF 4334 Intelligences ; Denial rights machine
PE 4791 Intelligentiae ; Deficientia
PA 6466 Intemperance
PD 0153 Intemperance
PA 5473 Intemperence *Insufficiency*
PA 6466 Intemperence *Intemperance*
PA 7296 Intemperence *Restraint*
PA 5612 Intemperence *Unchastity*
PF 6186 Intensification parent child relationship ; Excessive
PE 2770 Intensive animal farming units ; Inferior meat quality
PD 1562 Intensive animal husbandry
PD 7993 Intensive farming techniques
PC 5842 Intensive industries ; Protectionism labour
PC 0052 Intensive soil exploitation ; Over
PA 0005 Intentional concealment
Intentional [fatal...]
PD 2341 — *fatal poisonings*
PC 0300 — *feigning physical symptoms*
PC 0300 — *feigning psychological symptoms*
PD 2651 Intentional infecting disease
PD 3022 *Intentional libel*
PF 4435 Intentional nuclear war ; Risk
PD 5168 Intentional spread sexually transmitted diseases
PF 3519 Intentions practical education ; Unrealized
PF 0756 Inter age communication ; Insufficient
Inter [communal...]
PC 3685 — communal ethnic violence
PF 3340 — cultural misunderstanding
PD 9651 — cultural trade barriers
PF 5006 Inter developing country skills flow ; Inadequate
PD 0355 Inter-ethnic marriage
PC 3428 *Inter-faith friction*
Inter Governmental
PA 0011 — disputes
PF 7925 — double standard human rights
PD 0181 — *exchange torture methods*
PF 3913 — failure fulfil financial commitments
PD 6628 — organization mismanagement
PC 2089 — suspicion
PJ 7921 Inter-professional brain drain
PE 8307 Inter regional cooperation problems countries ; Inadequate
PC 0933 Inter regional trade ; Weakness South South
Inter [species...]
PA 5414 — species violence
PF 1925 — species warfare
PE 3538 — state rivalry secretariats intergovernmental organizations
PE 0063 Inter-union competition
PF 3558 Inter village cooperation ; Inadequate
PC 1306 Interaction biological agents ; Harmful synergistic
PA 6429 Interaction∗complex
PF 1739 Interaction deficiencies world economical systems
PF 6570 Interaction different age groups ; Stifled potential social
PD 1566 *Interaction ; Fragmented community*
PF 6185 Interaction humans animals urban environment ; Inadequate
PE 4477 Interaction ; Short sighted decisions intersocial
PE 4297 Interactive toys ; Violent
PD 7608 Interception communications
PE 5093 Interception correspondence
PD 2930 Interception telecommunications
PF 3037 Interchange ; Restrictive channels cultural
Intercourse [Adolescent...]
PD 7439 — Adolescent sexual
PE 3273 — Anal
PE 3274 — animals ; Human sexual
PF 6910 Intercourse ; Fear sexual
PF 4948 Intercourse ; Forced sexual
PF 3260 Intercourse ; Inadequate satisfaction sexual
PE 6522 Intercourse minors ; Sexual
PF 5434 Intercourse out wedlock ; Sexual
PD 5107 Intercourse ; Pre marital sexual
PF 0316 Intercultural meetings ; Ineffectiveness inefficiency
PD 0355 Interdenominational marriage
PF 6105 *Interdependence*
PC 9112 Interdependence ; Asymmetry economic

-963-

PC 9112 Interdependence developed developing countries ; Asymmetric
PF 6105 *Interdependence ; Food*
PF 6105 Interdependence ; International economic
PF 1735 *Interdependence issues*
PF 2600 Interdependence nationalistic images ; Overruling global
PF 3460 Interdependence ; Overemphasis self sufficiency respect
PD 5676 Interdependence stock markets ; Excessive
PF 2877 Interdependence ; Unacknowledged global
PF 2969 Interdependencies ; Unrecognized socio economic
PD 9667 Interdigital cysts
PE 5617 *Interdigital dermatitis ; Ovine*
PE 4161 *Interdigital dermatitis ; Vegetative*
PF 4161 *Interdigital necrobacillosis*
PE 4161 *Interdigitalis ; Hyperplasia*
PF 0409 Interdisciplinary meetings ; Ineffectiveness inefficiency

Interest broadcasting
PF 5622 — Decline public
PF 5622 — Inadequate support public
PF 5622 — Resistance public
PF 9610 Interest ; Conflicts
PD 1157 Interest due ; Irregular international payment
PA 0839 Interest emphasis ; Personal
PE 2524 Interest ; Exorbitant debt
PE 8289 Interest governments groups governments ; Conflict

Interest groups
PE 7575 — Bogus public
PE 1702 — development community priorities ; Monopolization
PF 1070 — policy agencies ; Over reliance economic
PD 6571 *Interest ; Limited merchandising*
PD 3546 Interest ; Natural resources used national self

Interest [parliamentarians...]
PE 3735 — parliamentarians ; Conflict
PF 5514 — Payment
PD 2552 — payments ; Profit oriented
PE 3735 — peoples representatives ; Conflict
PF 5061 — Policy contrary self
PF 9693 — pure applied science ; Conflict

Interest rates
PF 9014 — Exhorbitant
PF 9014 — High
PF 3141 — Lack international coordination
PE 4603 — long term debts developing countries ; Increases
PE 3903 — Low bank
PE 8110 — Rescheduling debts developing countries market related
PF 3141 — Volatility
PE 3926 Interest repayments developing countries ; Excessive unnecessary burden
PA 8760 Interest ; Self
PF 3679 Interested industrial vision ; Self
PF 5529 Interested manipulation timing ; Self
PF 3918 Interests ; Benefit doubt given vested
PF 9610 *Interests business ; Conflict*
PF 1750 *Interests children ; Unperceived*
PF 2874 *Interests ; Competitive economic*
PE 2598 Interests developing countries due export cartels ; Excessive injury export
PA 0839 Interests ; Dominating personal
PD 1231 Interests ; Entrenchment vested
PB 3209 Interests ; Excessive political influence traditional vested
PC 6440 Interests imperialism ; Conflict
PF 7178 Interests ; Incompatible moral
PF 0053 Interests ; Lack autonomous world level actor identify clarify world
PD 1231 Interests ; Parochial
PF 2600 Interests ; Parochial national
PF 9693 Interests ; Research sponsored vested
PF 2182 Interests small groups ; Narrow political
PE 1940 Interests transnational corporations ; Disregard consumer
PJ 8295 Interface fundamentally different societies ; Inadequate
PE 9528 Interference associated solar flares ; Radio
PF 0054 Interference communications satellites
PF 0506 Interference decision making ; Psychic
PA 5497 Interference *Difficulty*
PE 8384 Interference employers affairs workers' organizations
PF 0304 Interference ; Excessive government

Interference family
PE 4058 — home ; Unlawful
PE 4058 — life ; Arbitrary external
PE 4058 — life ; Unlawful
PC 3185 Interference ; Foreign government
PF 8995 *Interference head household authority*

Interference [industrial...]
PE 2473 — industrial origin ; Radio
PC 3185 — internal affairs states ; Foreign
PA 6862 — *Intrusion*
PE 7355 Interference labour relations developing countries transnational corporations
PE 0667 Interference labour relations industrialized countries transnational corporations
PJ 9820 Interference marine activities ; Unlawful
PD 4800 Interference national economy ; Government

Interference [Parental...]
PF 4391 — Parental
PE 5827 — port operations ; Political
PB 0284 — privacy ; Arbitrary
PB 0284 — privacy ; Unlawful
PE 1758 — public authorities functioning workers organizations

Interference [Radio...]
PD 2045 — Radio frequency
PD 2045 — radio television communications

Interference [rights...] cont'd
PE 6755 — rights innocent passage territorial waters ; Unlawful
PD 1244 Interference satellite communications ; Deliberate
PF 0443 *Interference school athletic activities education*

Interference [television...]
PE 5610 — television news coverage capitalist countries ; Political
PE 8273 — trade unions contract performance
PE 6006 — transiting aircraft
PE 4325 — transnational banks domestic economic policies
PE 4315 — transnational banks' off-shore borrowing domestic monetary policies
PF 6552 *Intergenerational needs ; Obscured*
PF 9561 Intergovernmental agencies ; Arrogance
PE 0310 Intergovernmental arrangements ; Trade restrictions due
PF 7493 Intergovernmental bodies ; Superficial consensus
PE 7140 Intergovernmental calls action ; Inadequate international nongovernmental organization response
PF 2876 Intergovernmental decision making process ; Inadequacy
PF 4794 Intergovernmental development assistance ; Reduced scope

Intergovernmental [facilities...]
PE 3538 — facilities ; Costly rotation activities
PE 9116 — facilities ; Restrictive conditions loans developing countries
PE 2805 — financial institutions ; Autocracy
PD 6896 Intergovernmental groupings ; Elitist
PF 7133 Intergovernmental information ; Uncoordinated
PF 4816 Intergovernmental legal systems ; Inadequate
PF 2139 Intergovernmental monetary system ; Privatization
PF 9564 Intergovernmental multilateral action ; Ineffectual
PF 3076 Intergovernmental official information ; Excessive neutrality

Intergovernmental organization leadership
PE 6947 — Corruption
PE 6947 — Malpractice
PE 6947 — Mismanagement
PF 0074 Intergovernmental organization programmes ; Ineffectiveness
PD 0457 Intergovernmental organizational debate ; Politicization

Intergovernmental organizations
PE 1184 — common membership ; Inadequate coordination regional
PE 1583 — common membership ; Jurisdictional conflict antagonism regional
PE 2417 — coordination bodies ; Proliferation duplication
PF 2806 — Excessive complexity
PE 6388 — High costs staff
PE 5090 — Inadequacy staff qualifications
PE 4344 — Inadequate coordination governmental representation
PF 9175 — Inadequate power
PE 3538 — Inter state rivalry secretariats
PE 7901 — Jurisdictional conflict antagonism
PF 9011 — Jurisdictional conflict antagonism
PF 7493 — Lack realism
PF 4876 — mini state membership ; Distortion
PE 3077 — Propaganda
PE 0063 — Rivalry
PE 0063 — scarce resources ; Competition
PF 5477 — Verbosity
PF 8650 — Withholding membership payments
PF 0419 Intergovernmental pacts ; Secret
PE 1375 Intergovernmental programmes national level ; Inadequate coordination action
PE 0730 Intergovernmental system organizations ; Inadequate coordination
PE 9267 Intergovernmental treaties ; Token
PE 8357 Intergroup conflicts ; Lack participation attempting solve
PF 2261 *Interim employment ; Dependence*
PE 7944 Interim support families ; Lack
PF 3783 Interior structure families ; Reduced
PF 1000 Interior structures marriage ; Reduced
PF 5522 Interlocking corporate directorates
PD 4632 Intermediaries ; Dependency
PD 4632 Intermediaries ; Prohibitive cost
PE 0174 Intermediary political implementation networks ; Inflexible
PD 6651 *Intermediate hosts ; Arthropods*
PE 3511 *Intermediate hosts ; Vertebrates*
PD 4172 Intermittent electrical signals
PA 6448 *Intermittent explosive disorder*

Internal [affairs...]
PC 3185 — affairs states ; Foreign interference
PC 3185 — affairs states ; Foreign intervention
PD 7661 — armed conflicts
PE 6603 — Internal balance human body ; Disruption
PD 4782 Internal borders ; Violence along
PC 0352 Internal colonialism
PC 1910 *Internal conflict due tribalism*
PC 2507 *Internal exile*
PD 5453 *Internal haemorrhage*
PF 7483 Internal inefficiency public programmes
PB 1149 Internal insecurity
PD 4782 Internal language, religious ethnic divisions
PF 4009 Internal migration

Internal [parts...]
PE 6603 — parts body ; Inflammation
PF 7580 — *pathways ; Difficult*
PE 3504 — piles
PF 3194 — political independence ; Lack
PF 9150 — pollution ; Human

PD 8871 Internal surveillance
PF 1728 Internal values images ; Reinforced parochialism
PF 4968 Internalized conflicts

International [action...]
PF 3162 — *action ; Disruption*
PE 8776 — action ; Inadequate organizational mechanism
PE 4296 — adoptions
PE 5132 — affairs ; Insufficient work awareness

International aggression
PB 0968 — [International aggression]
PC 7559 — [International aggression]
PB 0968 — Dependence

International agreements
PF 6992 — Ineffective
PF 0419 — Secret
PE 9267 — substance ; Symbolic
PD 5876 — Subversion
PF 2497 — Transgression
PF 2497 — Violation

International [aid...]
PF 1136 — aid institutions ; Secrecy
PF 1136 — aid institutions ; Unaccountability
PF 2723 — anniversaries years ; Proliferation national

International arms
PD 2107 — dealers ; Private
PD 4858 — shipments ; Illegal
PC 1358 — trade
PD 0136 International assistance programmes ; Abuses

International athletic
PF 4192 — competitions ; Excessive expense
PF 4216 — competitions ; National bias judges
PE 4809 — exchange ; Obstructions
PF 4761 — exchanges ; Excessive political intervention

International [bodies...]
PF 5472 — *bodies ; Developing country failure exercise leverage*
PC 4645 — business ; Illegal
PD 5891 — businesses

International [capital...]
PC 1453 — capital flows ; Massive
PE 0396 — cartels
PE 9762 — cartels enforced governments

International civil servants
PF 5649 — Abuse privileges immunities
PE 3457 — Conspicuous consumption
PE 5488 — Detention
PE 6388 — Excessive salaries
PD 5554 — Fraud
PE 5488 — Violation privileges immunities
PF 4676 International code crimes ; Lack
PD 4511 International collapse stock exchanges bourses

International commodity
PF 2497 — *agreements ; Ineffectiveness*
PE 9440 — exchanges ; Discriminatory operation
PF 2497 — *policies ; Inconsistency*

International [communication...]
PF 2350 — *communication ; Lack*
PF 7925 — community ; Avoidance human rights issues
PD 8994 — competitiveness ; Declining
PD 8994 — competitiveness ; Fluctuations
PB 5057 — conflict
PB 5057 — conflict ; Dependence
PF 0473 — *conventions ; Refusal ratify*
PF 2497 — conventions ; Violation

International cooperation
PF 0817 — Dependence lack
PC 2089 — due personal mistrust ; Lack
PF 0817 — Lack
PF 4366 — reducing terrorism ; Inadequate

International [coordination...]
PF 3141 — coordination interest rates ; Lack
PE 4508 — coordination supervisors financial stock markets ; Lack
PF 7897 — courts justice ; Government refusal accept jurisdiction
PC 2343 — *crime rings*

International criminal
PF 4676 — court ; Lack
PF 7277 — investigations ; Obstruction
PF 4676 — law ; Inadequate development

International [criminals...]
PF 4875 — *criminals their own countries ; Inadequate punishment*
PF 4302 — crises ; Risk unintentional nuclear war due
PB 6335 — crisis escalation

International cultural
PF 3099 — diplomatic commercial exchanges ; Abuse
PF 4857 — exchange ; Obstacles
PF 2258 — vacuum

International [data...]
PF 4155 — data bases developing countries ; Underutilization
PE 4230 — data systems ; Underparticipation socialist countries
PD 3053 — debt payments ; Arrears
PD 1157 — debts ; Irregular repayment

International decision making
PC 0352 — *Lack participation indigenous peoples*
PC 0352 — *processes ; Inadequate access indigenous peoples*
PF 5472 — Restrictions participation developing countries

International [delegations...]
PE 3457 — delegations ; Conspicuous consumption
PB 1016 — disparity consumption resources
PF 9281 — division labour ; Exploitative transformation
PF 7133 — documentation ; Fragmentation
PD 0858 — double taxation
PD 3050 — double taxation royalties
PE 1880 — drug trade

International transactions

International economic
PC 0025 — coordination ; Lack
PC 8073 — environment ; Deterioration
PC 0025 — fragmentation
PC 9112 — injustice
PF 6105 — interdependence
PF 4306 — International Economic Order ; Lack progress establishing New

International economic
PC 9112 — order ; Unjust
PF 1172 — recession
PF 2699 — structures ; Unimaginative vision existing
PF 0221 — International experts ; Irresponsible

International financial
PE 2214 — institutions developing countries ; Over dependency
PF 1136 — institutions ; Unaccountability
PD 1157 — obligations ; Irregular payments
PE 7987 — services ; Protectionism free zone
PC 7873 — system ; Vulnerability national economies vagaries
PF 6594 — transactions ; Undocumented
PF 0166 — International flow knowledge educational materials ; Barriers
PF 2216 — International forums ; Misuse

International freedom
PC 0931 — information ; Restrictions
PD 0351 — movement national advantage ; Restrictions
PE 8899 — movement shipping ; Violation right
PE 8899 — navigation ; Violation right

International [governmental...]
PD 0104 — governmental decision making processes ; Inadequate information systems
PE 1973 — governmental nongovernmental organizations programmes ; Inadequate relationship
PF 2216 — groups ; Rivalry

International [harmonization...]
PC 4591 — harmonization standards ; Inadequate
PE 5239 — health standards ; Failure conform
PF 6365 — human rights instruments ; Inadequacy

International imbalance
PC 5601 — economic activity
PB 4993 — quality life
PD 9170 — quality working life

International [inaccessibility...]
PD 5314 — inaccessibility justice
PC 2815 — income gap
PE 0740 — indebtedness arising insurance transactions developing countries
PC 9152 — inequality
PF 7134 — information ; Pre logical limitations comprehension

International information systems
PE 8066 — Inadequate integration
PF 8753 — order problem relevant information ; Failure
PE 0458 — Proliferation duplication

International insecurity
PB 0009 — [International insecurity]
PB 0009 — Dependence
PD 2374 — small states

International [institutional...]
PF 0048 — institutional framework finance ; Inappropriate
PE 8064 — institutions ; Loss credibility
PF 7277 — investigations ; Legal impediments
PD 0673 — investment ; Tax obstacles
PF 2141 — isolationism
PF 2113 — International judicial system ; Inadequate
PF 2113 — International justice ; Ineffective administration

International [labour...]
PJ 8060 — labour management relations ; Inadequate private
PF 6027 — language ; Lack
PD 6804 — *language* ; Lack

International law
PF 2497 — Abuse
PE 6564 — capture foreign nationals ; Disregard
PF 4816 — Deficiencies
PF 2497 — Disregard
PF 9081 — Disrespect
PF 8421 — enforcement ; Inadequate
PE 8280 — Inadequate national supervision adherence
PF 9081 — Non recognition
PF 0216 — *International laws* ; *Conflict*
PF 8353 — International leadership ; Loss

International legal
PA 0298 — *disputes*
PF 8616 — procedure ; Inadequacy
PF 7058 — provision nongovernmental organizations ; Lack

International [legislation...]
PF 0228 — legislation ; Inadequacy
PF 3058 — *liquidity*
PD 3053 — loans ; Defaults
PF 2076 — local concerns ; Overriding

International [map...]
PD 0398 — map world ; Inadequate
PF 8081 — market leadership ; Loss
PD 8994 — market share ; Declining
PE 4522 — marketing jute products developing countries ; Inadequate
PD 8897 — markets ; Rivalry
PD 3040 — media monopolies ; Emergence
PD 0349 — meetings ; Ineffectiveness inefficiency
PF 4008 — migration
PF 0048 — monetary financial system ; Crisis

International monetary system
PF 0048 — Inadequacies
PF 3058 — Lack confidence
PF 0048 — Lack coordination
PC 3778 — Unrepresentative control

International [money...]
PJ 3258 — money orders ; Irregular
PD 3040 — monopoly media
PE 6681 — motor traffic ; Tax impediments

International movement
PD 2755 — animals factor animal diseases
PC 0935 — Denial right freedom
PD 2755 — factor animal diseases ; Animals'
PF 6594 — International movements funds ; Unreported

International [non...]
PF 3100 — non-military conflict
PF 3049 — non profit organizations ; Inappropriate taxation
PE 7104 — nongovernmental activities ; Duplication

International nongovernmental organization
PF 2016 — action ; Inadequate facilities
PE 7080 — actions ; Divisive responses
PE 0179 — coordination bodies ; Proliferation duplication
PE 7032 — identity ; Lack
PE 0362 — information systems ; Proliferation duplication
PE 7092 — membership apathy
PE 7044 — network ; Lack identity
PE 7057 — network ; Neglect
PE 7010 — personnel ; Disadvantaged status
PE 7140 — response intergovernmental calls action ; Inadequate

International nongovernmental organizations
PE 7128 — Establishment orientation
PE 7045 — Excess western based secretariats
PE 7020 — Incompatible equivalent national sections
PE 0064 — Jurisdictional conflict antagonism
PE 1169 — Jurisdictional conflict antagonism
PE 7009 — Lack financial information systems
PE 0069 — Lack legal provision
PE 7034 — Lack national legal provision
PE 7093 — Limited recognition
PE 7129 — Multinationalism
PF 7116 — naive ; Assessment
PF 7069 — Neglect non Western structures
PF 7082 — Obstacles effective
PE 7056 — Operational ineffectiveness
PE 7008 — Political ineffectiveness
PE 1209 — programmes ; Inadequate coordination
PE 0741 — programmes ; Inadequate funding
PF 1595 — programmes ; Ineffectiveness
PE 7033 — regional development ; Absence policies associate
PE 0259 — Rivalry
PE 0259 — scarce resources ; Competition
PE 0777 — specialized agencies United Nations ; Inadequate relationship
PE 7105 — Uncontrolled structures
PE 7021 — Unrepresentative

International [ocean...]
PD 5885 — ocean shipping ; Obstacles
PB 6021 — order ; Dependence discriminatory
PB 6021 — order ; Discriminatory
PD 0929 — organization action ; Inadequate facilities
PE 9124 — organization programme action country level ; Ineffectiveness

International organizations
PF 3162 — Communist opposition
PE 3832 — Conditional membership
PC 3166 — *Control over*
PE 1029 — coordinating bodies ; Proliferation duplication
PE 7973 — country level ; Jurisdictional conflict antagonism
PE 5179 — Distortion democratic procedures
PF 8650 — Government non payment agreed contributions
PE 3538 — Inefficient location facilities
PF 9175 — Intrinsic weakness
PD 0047 — Jurisdictional conflict antagonism
PD 0138 — Jurisdictional conflict antagonism
PF 3434 — Lack continuity amongst personnel
PF 9175 — Lack enforcement power
PD 2942 — Micro state participation
PF 8650 — Overdue payments governments
PF 9175 — Powerlessness

International organizations programmes
PE 1553 — Alienation skilled committed personnel
PD 1809 — Alienation support
PD 0285 — Inadequate coordination
PF 0498 — Inadequate funding
PF 1074 — Ineffectiveness
PE 8064 — Low credibility

International organizations [Proliferation...]
PF 5992 — Proliferation documents
PF 3434 — Rapid personnel turnover
PC 1463 — Rivalry
PC 1463 — scarce resources ; Competition
PF 0498 — Shortfall aid
PD 4873 — Unrepresentative

International [participation...]
PE 5761 — participation island developing countries ; Inadequate
PF 1871 — paternalism
PD 1157 — payment interest due ; Irregular
PB 0009 — peace security ; Threats
PB 0009 — peace ; Violation
PE 4785 — personnel exchanges cultural cooperation ; Obstructions
PB 7125 — plots ; Manipulative

International political
PC 1868 — espionage
PF 7455 — leadership ; Loss
PF 1440 — miscalculation
PE 8080 — International politics fuel shortages ; Distortion
PC 5348 — International prestige race

International [regulation...]
PF 0691 — regulation transnational corporations ; Ineffective
PC 0410 — relations ; Multiplicity languages
PS 7068 — *relations* ; *Poor*
PF 2889 — repercussions domestic agricultural policies ; Detrimental
PE 2215 — reporting obligations ; Failure governments fulfil
PE 2215 — reporting obligations ; Governmental delays fulfilling
PD 3080 — restriction information ; Conflict laws
PC 0114 — rivalry
PD 0516 — river basin development ; Uncoordinated
PD 0516 — rivers ; Upstream diversion

International [sanctions...]
PF 1976 — sanctions ; Circumvention
PB 0009 — security ; Lack
PB 0009 — security ; Violation

International sports
PF 4761 — competitions ; Politicization
PF 4206 — events ; Political boycott
PF 4761 — events ; Politicization
PE 8758 — International spread new technologies ; Hindrances

International standards
PF 0203 — financial accounting reporting ; Inadequate
PF 5072 — Inadequacy
PF 8927 — Informality
PF 8927 — Limited enforceability

International [statistics...]
PF 2622 — statistics ; Lack comparability
PF 2622 — statistics ; Limitations
PC 5348 — status race
PC 1606 — strategic instability
PE 2215 — surveys ; Non response governments

International [tax...]
PJ 3491 — tax manipulations
PF 9345 — technological leadership ; Loss
PD 0187 — telecommunications ; Destabilizing
PB 8287 — tension
PD 6008 — terrorism ; State supported

International trade
PD 6002 — Abuse dominant market position
PD 9679 — against exports developing countries ; Protectionism
PE 0396 — arrangements ; Collusive
PD 0057 — barriers primary commodities
PD 8621 — biological weapons
PD 9692 — chemical weapons
PE 7072 — Collusive tendering

International trade discriminatory
PE 0191 — application antidumping regulations ; Distortion
PE 2603 — customs administrative entry procedures ; Distortion
PE 9073 — formulation equipment safety regulations ; Distortion
PE 1274 — formulation health sanitary regulations agricultural pharmaceutical products ; Distortion
PE 0347 — government private procurement policies ; Distortion
PD 0340 — preference agreements ; Distortion
PE 0778 — requirements respect marks product origin ; Distortion
PE 0083 — requirements respect product standards measures ; Distortion

International trade [Distortion...]
PC 6761 — Distortion
PD 2144 — dumping ; Distortion
PE 0522 — embargoes similar restrictions ; Distortion
PC 0308 — endangered species
PE 1961 — export subsidies countervailing duties ; Distortion
PF 3099 — *fairs*
PE 0117 — Inadequate barter system
PE 9107 — information ; Lack transparency
PE 8547 — investment disputes ; Private
PC 8930 — Irresponsible
PE 1182 — minimum pricing regulations measures regulate domestic prices ; Distortion
PC 2725 — Non tariff barriers
PD 0455 — obstacles patent protection ; Distortion
PC 8415 — patterns ; Imbalance
PC 5842 — Protectionism
PE 9027 — quantitative restrictions ; Distortion

International trade restrictive
PC 0073 — business practices ; Distortion
PE 8882 — controls movement labour ; Distortion
PE 8525 — controls over foreign investment ; Distortion
PE 0497 — customs valuation practices ; Distortion
PD 2029 — International trade result government participation ; Distortion

International trade selective
PD 0678 — domestic subsidies ; Distortion
PE 8867 — indirect taxes import charges ; Distortion
PE 8583 — monetary controls ; Distortion

International trade [services...]
PD 6223 — services ; Obstacles
PE 8267 — state trading government monopoly practices ; Distortion
PC 0569 — Tariff barriers
PE 0135 — transnational corporations ; Escalating control
PC 5842 — Unfair restrictions
PF 1262 — unions ; Inadequate
PC 9584 — International trading agreements ; Circumvention
PC 9584 — International trading discipline ; Breakdown

International trading system
PC 9584 — Declining credibility
PC 9584 — Fragmentation
PC 9584 — Inadequate
PC 9584 — Uncertainty

International [transactions...]
PF 9571 — transactions ; Conflicts national law relation
PF 9571 — transactions ; Divergence different national laws relating

-965-

International transfer

International [transfer...] cont'd
PD 2784 — *transfer corpses ; Health risks*
PD 2784 — *transfer corpses ; Obstacles*
PJ 8840 — transit conventions ; Lack adherence
PE 5789 — transit conventions land locked countries ; Lack adherence
International travel
PE 1904 — Desynchronization bodily rhythm
PD 6421 — Disease transmission
PF 1555 — Excessive animal sanitary regulations
PE 0208 — Excessive frontier formalities
International treaties
PF 2497 — Breaching
PD 8465 — Covert violation
PF 0977 — Delays ratification
PF 6992 — Deliberately weakened
PF 0977 — Limited acceptance
PF 0977 — Non ratification
PF 2497 International treaty provisions ; Non implementation
PF 7897 International tribunal arbitration ; Government refusal accept
PF 5106 International understanding ; Lack
PF 0292 International values procedures ; Governmental disregard
PB 0593 *International waterways ; Mined*
PE 7957 Internationalization capitalist production
Internationally agreed
PC 4591 — standards ; Failure conform
PC 4591 — standards ; Resistance
PE 5239 — toxicity thresholds ; Failure respect
PF 1976 Internationally imposed economic sanctions ; Disregard
PF 1068 Internationally mobile professionals experts ; Impediments
PF 9474 Internationally non-cooperative governments
PE 4071 Internationally ; Violation right workers' organizations affiliate
PD 1800 *Internment conscientious objectors*
PE 4887 Internment ; Excessive length pre trial
PD 1576 Internment trial
PD 2948 Interoceanic canals ; Restrictions passage straits
PE 7826 *Interosseous desmitis*
PB 0034 Interpersonal estrangement
PB 0034 Interpersonal estrangement ; Dependence
Interpersonal [relatedness...]
PD 9219 — *relatedness ; Pervasive pattern deficits*
PF 4274 — relationships ; Distrust
PD 7617 — rivalry
PE 1743 Interpretation legal proceedings ; Prohibitive cost linguistic
PE 6740 Interpretation ; Negligent
PF 6916 Interpretation translation errors
PF 9431 Interpretations facts ; Biased
PF 9629 Interpretations multilateral principles ; Unilateral
PE 6740 Interpreters ; Corruption
PE 6740 Interpreters ; Irresponsible
PE 1743 Interpreters judicial hearings ; Denial right access
PE 6740 Interpreters ; Unethical practices
PC 1523 Interracial marriage
PC 1523 Interracial sex
PF 8753 Interrelate wisdom different cultures response social problems ; Failure
PF 0364 Interrelationship world problems ; Complex
PF 7706 Interrelationships ; Assumption lack problem
PE 3553 Interrogation methods ; Unethical
PD 1362 Interrogation techniques ; Inhumane
PE 7233 Interrogation while naked
PC 1142 *Interrogations ; Abusive police*
PE 7252 Interrogative suggestibility
PD 7692 Interruption due pregnancy ; Career
PF 9106 Interruption risk
PF 4128 Interruptions ; Imposed career
PE 4477 Intersocial interaction ; Short sighted decisions
PF 2517 Intersocietal resource channels ; Lack
PF 9275 Interspecific competition
PF 9002 Interstellar dust
PE 6789 Interstitial keratitis
PE 5524 *Interstitial pneumonia ; Bovine atypical*
PA 6862 *Intervention*
PE 1170 Intervention agriculture ; Inappropriate government
PF 7705 Intervention childbirth ; Excessive medical
PC 3185 Intervention countries ; Super power
PD 9331 Intervention ; Foreign military
PF 0980 Intervention future ; Monopoly competence
PD 1091 Intervention human evolution ; Insufficient
Intervention [Inappropriate...]
PF 0304 — Inappropriate government
PC 3185 — internal affairs states ; Foreign
PF 4761 — international athletic exchanges ; Excessive political
PA 6862 — *Intrusion*
PF 0420 — Irrational conscientious refusal medical
PC 3185 Intervention major powers protect investments their citizens foreign countries
PD 4811 Intervention meetings ; Violent police
PD 4800 Intervention private sector ; Excessive government
PF 0304 Intervention society ; Excessive government
PD 7276 Intervention tourism ; Foreign
PE 0032 Intervention transnational corporations ; Political
PD 2374 Intervention ; Vulnerability small nations foreign
PE 2007 Intervention workers meetings ; Military
PE 2007 Intervention workers meetings ; Violent police
PD 2626 *Intervertebral disc ; Prolapsed*
PE 5063 Intestacy
Intestinal [diseases...]
PD 9045 — diseases
PD 9045 — *diseases ; Acute*
PE 6724 — disorders due torture ; Gastro
PE 3861 Intestinal infections ; Gastro
PE 9526 Intestinal infectious diseases
Intestinal [obstruction...]
PD 9045 — *obstruction*
PE 7423 — *obstruction animals ; Small*
PD 3978 — *obstructions animals ; Acute*
PD 9045 *Intestinal perforation*
PE 0921 *Intestinal schistosomiasis*
PE 3861 Intestinal tract ; Bleeding gastro
PE 0958 *Intestinal trichostrongylosis sheep*
PE 8117 Intestine ; Diverticula
PE 3936 *Intestine swine ; Mesenteric torsion small*
PE 4973 Intestines ; Inflammation
PE 0566 *Intestines ; Tuberculosis*
PF 8012 Intimacy ; Fear
PD 4287 Intimacy ; Forced social
PD 4287 Intimacy ; Unauthentic
PD 4287 Intimate forms address ; Imposition
PF 4416 Intimate relationships ; Lack
PB 1992 Intimidating people
PB 1992 Intimidation
PE 2876 Intimidation children ; Physical
PA 5740 Intimidation *Compulsion*
Intimidation [Dependence...]
PB 1992 — Dependence
PC 2940 — Dependence social
PE 0163 — developing countries transnational corporations
PA 6191 — *Disapproval*
PA 7343 — *Dissuasion*
PC 3011 Intimidation ; Economic
PD 2044 Intimidation electors
PA 6030 Intimidation *Fear*
PE 1622 Intimidation governments ; Government
PC 2939 Intimidation ; Industrial
PD 3036 Intimidation journalists
PE 7793 Intimidation ; Lack protection victims
Intimidation [pedestrians...]
PE 6139 — pedestrians vehicles
PC 2934 — Physical
PD 0736 — Police
PC 2938 — Political
PE 0428 — prisoners inmates
PC 2935 — Psychological
PD 4734 — public officials
PC 2936 Intimidation ; Racial
PC 2937 Intimidation ; Religious
PC 2940 Intimidation ; Social
PF 4729 Intimidation ; Unreported
PA 6491 Intimidation *Vanity*
PB 1992 *Intimidation victims crimes*
Intimidation [witnesses...]
PB 1992 — *witnesses*
PD 0610 — worker's representatives
PD 7471 — workers' representatives
PF 0860 Intolerance
PF 2384 *Intolerance complexity*
PF 8396 Intolerance criticism
Intolerance [Dependence...]
PC 1808 — Dependence religious
PF 7938 — distinctions
PF 7938 — distinctions male female
PE 9541 Intolerance ; Food
Intolerance [Impatience...]
PA 6200 — *Impatience*
PF 7024 — imperfection
PA 7325 — *Irresolution*
PE 4949 Intolerance ; Lactose
PA 7306 Intolerance *Narrowmindedness*
PF 8767 Intolerance proven methods
PC 1808 Intolerance ; Religious
PF 7109 Intolerance uncertainty
PE 4438 Intolerant antivivisectionists
PJ 3527 Intolerant movements
Intoxication [Acute...]
PE 2429 — Acute alcoholic
PD 0153 — *Alcohol idiosyncratic*
PD 0153 — Alcoholic
PD 4027 — *Amphetamine*
PD 4027 — *Anxiolytic*
Intoxication [Caffeine...]
PE 0618 — Caffeine
PD 4027 — *Cannabis*
PD 4027 — *Cocaine*
PA 6799 Intoxication *Disease*
PD 4027 *Intoxication ; Hypnotic*
PD 4027 *Intoxication ; Inhalant*
PA 6466 Intoxication *Intemperance*
PD 4027 *Intoxication ; Opioid*
PD 0153 *Intoxication ; Pathological*
PE 2349 Intoxication ; Pesticide
PD 4027 *Intoxication ; Sedative*
PD 4027 Intoxication ; Substance
PE 2033 Intoxication work ; Alcoholic
PE 2515 Intoxications ; Food borne infections
PE 9037 *Intra-cranial abscesses*
PF 2477 *Intra-group skills gap*
PD 3765 Intra-national trade war
Intra [species...]
PA 5414 — species violence
PE 9037 — *spinal abscesses*
PC 3197 — state imperialism
PD 8461 Intracerebral intracranial haemorrhage
PE 1974 *Intracranial disease*
PD 8461 Intracranial haemorrhage ; Intracerebral
PE 0763 *Intracranial paralysis*
PE 9229 *Intracranial tumours*
PE 9037 *Intracranial venous sinuses ; Phlebitis*
PA 7250 Intractability *Disobedience*
PA 7325 Intractability *Irresolution*
PA 5988 Intractability *Softness*
PC 8801 Intractable diseases
PE 2694 Intractable pain
PE 0580 Intranational competition
PA 7325 Intransigence *Irresolution*
PA 6979 Intransigence *Opposition*
PF 4968 Intrapsychic conflict
PC 1869 Intrasocietal military conflict
PF 9201 Intraspecific competition
PE 5798 Intravenous drug users AIDS ; Vulnerability
PA 5497 Intricacy *Difficulty*
PA 7367 Intricacy *Unintelligibility*
PA 7363 Intrigue *Improbity*
PF 9175 Intrinsic weakness international organizations
PE 7832 Introduced water projects ; Diseases
PE 5390 Introducing possessing contraband useful escape
PE 3721 Introduction diseases ; Vulnerability indigenous populations
PC 1617 Introduction ; Environmental hazards new species
PF 1312 Introduction extraterrestrial infectious diseases bacteria
PF 3146 Introduction high-yield crop varieties
PE 5820 Introduction inappropriate technology
PF 2813 *Introduction new machinery ; Unscheduled*
PE 1444 Introduction new plant species ; Irresponsible
Introduction new species
PD 1290 — animals ; Irresponsible
PF 3602 — fish ; Irresponsible
PF 3592 — insect pests
Introduction new technologies
PE 4391 — Elimination jobs developing countries due
PE 5134 — Negative social effects
PD 0528 — Unemployment
PD 5033 Introduction protectionist measures ; Unpredictable
PC 1617 Introduction species ; Hostile
PE 4687 Introversion due torture
PF 1030 Introversion family unit
PA 6862 Intrusion
PD 9370 Intrusion advertising education
PE 6771 Intrusion ; Criminal
PE 3140 Intrusion jurisdiction ; Extraterritorial
PE 7837 Intrusion ; Salt water
PE 4438 Intrusive animal-rights campaigners
PF 0145 Intrusive social science research
PA 5869 Intrusiveness *Exclusion*
PA 5806 Intrusiveness *Inaction*
PA 6862 Intrusiveness *Intrusion*
PA 7006 Intrusiveness *Untimeliness*
PE 3786 Inundation agricultural land dams
PE 7855 Inundation forests dams
PE 7794 Inundation wildlife habitats dams
PF 4444 Invaders outer space
PA 6799 Invalidity *Disease*
PA 6876 Invalidity *Impotence*
PA 5502 Invalidity *Reason*
PE 5211 Invalids ; Denial right benefits
PD 8779 Invasion
PE 3972 Invasion airspace foreign aircraft
PF 4331 Invasion ; Cover up extraterrestrial
PC 2548 Invasion ; Cultural
PA 5445 Invasion *Defence*
PC 0840 Invasion ; Economic
PF 4444 Invasion ; Extraterrestrial
PC 3304 Invasion indigenous lands
PA 6862 Invasion *Intrusion*
Invasion privacy
PB 0284 — [Invasion privacy]
PE 0223 — compulsory telecommunications
PD 9603 — media
PB 0284 — *testing*
PD 1290 Invasions ; Animal
PD 4751 Invasions ; Disastrous insect
PF 3592 Invasions ; Insect
PD 5238 Invective
PF 3538 *Invention ; Disincentives*
PF 3699 Invention new social problems
PF 3699 Invention new threats ; Deliberate
PF 7274 Inventors discoverers ; Disagreement concerning identity
PF 6540 *Inventory levels ; Inconsistent*
PF 6471 *Inventory ; Limited retail*
PC 0119 *Inversion ; Temperature*
PD 7513 Invertebrates ; Endangered species
PD 3158 Invertebrates ; Marine
PF 2146 *Inverted snobbery*
PF 9249 Invest heavily indebted countries ; Disincentive
PJ 0451 Invest high risk development initiatives ; Reluctance government
PE 9346 Invested obsolete armaments ; Waste resources
PD 4282 Investigate ; Denial right
PF 2720 Investigation ; Lack scientific
PD 8871 Investigation ; Secret
PF 7277 Investigations ; Legal impediments international
PF 7277 Investigations ; Obstruction international criminal
PE 5037 Investigations ; Police crimes during narcotic
PF 7889 Investigations ; Superficial
PD 7885 Investigatory malpractice
PF 6572 Investment blocked high risks
Investment capital
PE 5790 — developing countries' shipping fleets ; Lack
PF 2852 — Insufficient flow
PF 6530 — resources ; Reluctant
PF 3550 — urban renewal ; Limited availability
PC 0564 *Investment capitalism*

Irresponsible patenting

PF 2378 Investment ; Costly cattle
PF 2241 Investment determined marriages
Investment developing countries
PD 0291 — Decline
PD 3138 — Decline foreign direct
PF 3845 — Disincentives financial
PD 3138 — Disincentives foreign private
PD 0291 — Inadequate level
PE 5790 — Lack capital
PE 9811 — Low return
Investment [developing...]
PF 6530 — developing country farms ; Low
PE 8787 — Dislocation productive units foreign
PE 8547 — disputes ; Private international trade
PE 1911 — disputes ; Transfrontier private
PD 8030 — Disruptive foreign
PE 8525 — Distortion international trade restrictive controls over foreign
Investment [Failure...]
PF 6508 — Failure public authorities assist financial
PF 6572 — Fear capital
PF 3104 — financing ; Counterproductive capitalist
PD 3047 — foreign country ; Tax discrimination against
PF 9612 — Investment human resources ; Erosion
Investment [Inappropriate...]
PD 8030 — Inappropriate foreign
PF 2009 — incentives ; Untenable
PE 8957 — income outflow developing developed countries ; Foreign private
PF 2852 — Insufficient capital
PD 3063 — Investment ; Legal impediments foreign
PF 6477 — Investment ; Limited private
PD 8030 — Investment ; National insecurity due excessive foreign
Investment [Outflow...]
PD 3132 — Outflow local
PF 6572 — outlook ; Frightening
PF 1275 — Overemphasis rapid returns
Investment [performance...]
PF 3845 — performance developing countries ; Inadequate
PC 5323 — power ; Concentration
PE 9265 — productive capacity ; Decline capital
Investment [Reluctance...]
PF 6530 — Reluctance towards corporate
PF 5203 — Resistance loss any personal
PF 6477 — resources ; Limited local
PF 6572 — Risk capital
PF 6572 — risk ; Inhibition
Investment [savings...]
PD 0465 — savings developing countries ; Bad
PC 2576 — Self interest driven
PF 6477 — small communities ; Diminishing capital
PF 2042 — small, inner city enterprises ; Lack awareness potential
PE 5102 — stock markets ; Low confidence
Investment [Tax...]
PD 0673 — Tax obstacles international
PD 0765 — traditional industries developing countries ; Excessive foreign
PE 7892 — transnational corporations least developed countries ; Inadequate
PE 0161 — transnational enterprises restrictive business practice ; Direct foreign
PD 8030 Investment ; Vulnerability countries destabilization foreign
PD 5750 Investment youth market ; Criminal
PE 4365 Investments developing countries ; Inadequate channels direct
PC 2172 Investments ; Nationalization foreign
PC 2172 Investments ; Risk nationalization overseas
PE 4167 Investments supporting racial discrimination ; Capital
PC 3185 Investments their citizens foreign countries ; Intervention major powers protect
PF 2042 Investments ; Uninsurable business
PF 6477 Investments ; Unproductive local
PF 2607 Investor ; Proscriptive controls favouring
PC 3134 Investors developing countries ; Excessive repatriation profits foreign
PF 1653 Investors ; Uninterested private
PE 5808 Inviability tropical island developing countries
PA 6978 Invisibility
PD 8204 Invisibility ; Social
PD 8204 Invisible people
PE 8667 Involuntary boarding denial
PD 4259 Involuntary disappearances persons
Involuntary loss
PE 2485 — citizenship
PE 2485 — nationality
PE 6676 — nationality children
PC 6203 Involuntary mass resettlement
PC 6203 Involuntary movement people
PE 4888 Involuntary narco-analysis
Involuntary [passage...]
PE 5431 — passage urine
PC 0300 — passing faeces
PD 3888 — prostitution brothels
PF 3240 Involuntary sterilization
PA 6468 Involution Complexity
PA 6088 Involution Impairment
PE 0655 Involutional depression
PD 1318 Involutional melancholia
Involvement [community...]
PF 6536 — community affairs ; Lack responsible
PF 2575 — Constricting patterns individual
PF 6535 — culture ; Scarce options
PF 2575 — Involvement ; Infrequent community
PF 5298 Involvement ; Insufficient images political

Involvement Minimal
PF 8905 — church / school
PF 2421 — citizen
PF 2575 — community
PF 2575 Involvement neighbourhood residents ; Limited
PF 1187 Involvement ; Parent school non
PE 4479 Involvement planning economy ; Static grassroots
PF 7142 Involvement ; Reluctance individual
PF 6550 Involvement ; Restricted daytime
PF 6570 Inward-focused churches
PD 2726 Iodine deficiency disorders
PE 7651 Ionizing radiation ; Environmental hazards non
PE 6294 Ionizing radiation ; Harmful biological effects
PD 1975 IQ evaluations ; Abuse
PA 5532 Irascibility Disaccord
PA 7253 Irascibility Envy
PA 7325 Irascibility Irresolution
PD 8786 Iridocyclitis
PD 5535 Iridocyclitis ; Recurrent
PD 8786 Iritis
PA 7365 Irksomeness Boredom
PA 5497 Irksomeness Difficulty
PA 7107 Irksomeness Unpleasantness
Iron [deficiency...]
PE 9554 — deficiency anaemia animals
PE 7854 — deficiency anaemias
PD 5228 — dextran toxicity newborn pigs
PE 4048 — dust
PD 0321 — Iron ; Inadequate absorption
PD 0321 — Iron ; Inadequate intake
PD 0353 — Iron ore concentrates shortage
PE 8772 — Iron ore concentrates trade instability
Iron steel basic
PE 8397 — industries environmental hazards
PE 8070 — industries instability
PE 8223 — industries ; Underdevelopment
Iron steel [Corrosion...]
PE 1945 — Corrosion
PE 8969 — manufactures trade instability
PE 8312 — scrap trade instability
PD 0361 Irradiated food ; Health hazards
PD 0361 Irradiation method food preservation
PD 0361 Irradiation technology ; Opposition
PF 1610 Irrational ; Avoidance
PF 0420 Irrational conscientious refusal medical intervention
PF 4185 Irrational fear industrialization developing countries
PE 6354 Irrational fear wind
PF 1822 Irrational land divisions
Irrational [rejection...]
PF 8531 — rejection nuclear power
PF 9127 — rejection technological change
PF 6829 — religious beliefs
PF 3399 Irrationalism
PA 0466 Irrationality
PA 7157 Irrationality Insanity
PA 5502 Irrationality Reason
PA 7371 Irrationality Unintelligence
PF 8374 Irreconcilability
PA 5982 Irreconcilability Disagreement
PA 5446 Irreconcilability Enmity
PA 7325 Irreconcilability Irresolution
PA 6979 Irreconcilability Opposition
PA 6606 Irreconcilability Vengeance
PA 6099 Irredeemability Hopelessness
PA 5644 Irredeemability Vice
PF 6235 Irredentism
PA 6099 Irreformability Hopelessness
PA 5644 Irreformability Vice
PF 6451 Irregular cash flow
PF 4852 Irregular delivery education
PF 2875 Irregular doctor care
PE 8219 Irregular heart rhythm
PD 4790 Irregular hospital transportation
PJ 3258 Irregular international money orders
PD 1157 Irregular international payment interest due
PU 3035 Irregular marriages
PF 3575 Irregular outside instruction rural communities
Irregular [pace...]
PE 7898 — pace technological advance ; Fast
PF 1495 — passenger service
PD 1157 — payments international financial obligations
PE 8997 Irregular repair services countryside
PD 1157 Irregular repayment international debts
PE 4200 Irregular school attendance
PC 1522 Irregular task accountability
PE 5345 Irregular transport services
PF 5250 Irregular village meetings
PF 2971 Irregular work patterns
PF 2971 Irregular working hours
PF 2575 Irregular youth activities
PA 6774 Irregularity
PD 1185 Irregularity teeth
PA 5982 Irrelevance Disagreement
PF 0443 Irrelevance educational curricula
PF 1870 Irrelevance marriage vows ; Social
PF 1746 Irrelevance past experience ; Perceived
PF 0770 Irrelevance science technology
Irrelevance [Unimportance...]
PA 5942 — Unimportance
PF 5934 — United Nations ; Perceived
PA 7006 — Untimeliness
PF 0836 Irrelevancy past experience
PF 6531 Irrelevant adult education
PF 3539 Irrelevant available information
PF 0443 Irrelevant curriculum ; Prescribed
PF 3539 Irrelevant data dissemination

PF 0422 Irrelevant ; Education vocationally
PF 8846 Irrelevant forms governance
PF 8846 Irrelevant institutions
PJ 9248 Irrelevant job requirements
PF 8846 Irrelevant policy-making initiatives
PF 4774 Irrelevant predictions
PF 2086 Irrelevant resolutions
Irrelevant [school...]
PF 0443 — school curriculum
PF 1202 — scientific activity
PF 1118 — social values
PF 8234 Irreligiousness
PA 6058 Irreligiousness Impiety
PA 6542 Irremediability Destruction
PA 6099 Irremediability Hopelessness
PA 7325 Irresolution
PA 7325 Irresolution Irresolution
PA 7309 Irresolution Uncertainty
PA 5558 Irresolution Weakness
PA 8658 Irresponsibility
PA 5490 Irresponsibility Changeableness
Irresponsibility [image...]
PF 3307 — image ; Financial
PA 7363 — Improbity
PC 2915 — Intellectual
PD 3071 Irresponsibility ; Journalistic
PA 5563 Irresponsibility Lawlessness
PF 9455 Irresponsibility long-term future planet
Irresponsibility [Male...]
PD 7297 — Male sexual
PD 5770 — Medical
PF 5337 — member governments United Nations
PF 6611 Irresponsibility ; Nuclear
Irresponsibility [towards...]
PF 9455 — towards future generations
PF 9455 — towards posterity
PE 5796 — transnational corporations ; Social
PF 5337 Irresponsibility United Nations developing country membership
PE 1832 Irresponsibility young people towards family
Irresponsible [advertising...]
PE 4225 — advertising
PE 3980 — advertising alcoholic beverages
PD 2623 — anthropology
Irresponsible [bacteriologists...]
PD 7731 — bacteriologists
PD 7731 — biochemists
PD 7731 — biologists
PC 2563 — business practices
Irresponsible [chemists...]
PD 4265 — chemists
PE 6702 — civil servants
PC 2563 — commercial practices
PF 2804 — construction planning
PD 2625 — consumption resources
PE 7174 — countries ; Trade nuclear materials
PC 0776 — creation new species
PF 7823 Irresponsible delimitation policy responsibilities
PE 9251 Irresponsible directors commercial enterprises
Irresponsible [employers...]
PD 2879 — employers
PD 2879 — entrepreneurs
PF 7798 — expression emotions equated free speech
Irresponsible [fatherhood...]
PD 7297 — fatherhood
PE 0682 — financial practices
PF 3859 — finders personal property
Irresponsible [genetic...]
PC 0776 — genetic manipulation
PD 7731 — geneticists
PD 0708 — geologists
PE 7174 — groups ; Diversion high technology
Irresponsible [health...]
PE 3328 — health care services
PE 7169 — housing tenants
PD 2586 — hydrologists
Irresponsible [industrial...]
PD 2916 — industrial practices
PD 2886 — information system design
PF 0221 — international experts
PC 8930 — international trade
PE 6740 — interpreters
Irresponsible introduction new
PE 1444 — plant species
PD 1290 — species animals
PF 3602 — species fish
Irresponsible [lawyers...]
PD 5380 — lawyers
PF 4580 — lending banks
PD 2886 — librarianship
PE 3980 — liquor advertising
PD 5948 — local government
Irresponsible [maintenance...]
PD 7964 — maintenance practices
PD 5251 — media practices
PD 4182 — meteorologists
PD 7731 — microbiologists
Irresponsible [obedience...]
PF 2418 — obedience
PD 4277 — oceanographers
PE 1369 — opticians
Irresponsible patenting
PD 7501 — genetically transformed animals
PE 6788 — genetically transformed plants
PD 7840 — life forms

Irresponsible [pedologists...]
PD 1110 — pedologists
PA 8658 — people
PF 8759 — personal relationships
PD 0862 — personnel practices
PE 2390 — pharmaceutical advertising
PE 3540 — pharmaceutical practices
PD 1710 — physicists
PC 5517 — politics
PD 1045 — practices food industry
PC 8019 — professional activities
PE 4360 — programmes transnational banks developing countries ; Socially
PD 5267 — psychotherapists
Irresponsible [radiologists...]
PD 8290 — radiologists
PF 0374 — recreation
PC 0080 — research using human subjects
PE 6615 — restaurant practices
Irresponsible [scientific...]
PC 1153 — scientific technological activity
PF 8032 — social science
PD 6626 — social scientists
Irresponsible [tobacco...]
PE 9093 — tobacco cigarette advertising
PD 4341 — trade unions
PF 1681 — *transient occupants*
PD 1012 — transport practices
PE 6754 Irresponsible use postal service
PE 4450 Irresponsible use telecommunications service
PD 7726 Irresponsible veterinarians
PE 8742 Irresponsible voluntary organizations
PD 4334 Irresponsible workers
PD 4721 Irresponsible zoologists
PA 6822 Irreverence *Disrespect*
PA 6058 Irreverence *Impiety*
PE 4620 Irreversible coma
PF 9790 Irreversible environmental trends
PF 9790 Irreversible problem emergence
PE 4585 Irrigated land ; Declining area
Irrigation [canals...]
PE 8796 — canals ; Water looses
PE 8233 — control ; Uncoordinated
PF 1345 — *costs ; Prohibitive*
PD 7096 Irrigation ; Destruction arid ecosystems
PD 8839 Irrigation distribution ; Fragmented
PF 3467 *Irrigation equipment ; Lack*
PF 3467 *Irrigation equipment ; Unshared*
PE 7832 Irrigation ; Increase pests diseases perennial
PE 8233 Irrigation knowledge ; Insufficient
PD 8839 Irrigation ; Lack
PE 9311 Irrigation ; Over watering during
PE 8233 Irrigation potential ; Unmanaged
PE 8233 Irrigation schemes ; Mismanagement
PE 1785 Irrigation subsidies developing countries ; Inappropriate
Irrigation system
PD 8839 — Inadequate
PD 8839 — Incomplete
PD 8839 — Undeveloped
Irrigation systems
PE 8233 — Deterioration
PE 9088 — Seepage water losses
PE 8796 — Water losses
PE 8233 Irrigation wastage ; Open
PD 8839 Irrigation water ; Unavailability
PA 5451 *Irritability*
PE 1520 Irritability due torture
PA 7253 Irritability *Envy*
PA 7363 Irritability *Improbity*
PA 5451 Irritability *Insensibility*
PD 3672 Irritant fumes
PE 6820 Irritant gases vapours
PE 5684 Irritants occupational hazards ; Skin
PD 8402 *Irritating drugs*
PE 6785 *Irritation ; Eye*
PD 3672 *Irritation ; Nose*
PF 2256 *Irritation response pattern*
PC 8534 Irritation ; Skin
PE 7733 *Irritation ; Throat*
PD 0992 *Ischaemia ; Cerebral*
PE 8158 Ischaemic heart disease
PE 9774 *Ischaemic myopathy*
PE 6015 Islamic fundamentalism
PE 6015 Islamic radicalism
Island countries
PE 5861 — havens ; Abuse land locked
PE 5689 — Lack coastal development
PE 5785 — territories ; Military threats
Island developing countries
PE 5664 — Deficient transport remote
PE 5713 — Excessive emigration
PD 5677 — imports ; Dependence
PE 5665 — Inaccessibility insurance
PE 5761 — Inadequate international participation
PE 5655 — Inadequate telecommunications
PE 5748 — Increasing foreign debt
PE 5808 — Inviability tropical
PF 5737 — Lack skilled workers
PE 4627 — Obstacles political union
PE 5724 — official assistance ; Dependence
PE 5700 — territories ; Vulnerability
PE 5772 — Weakness infrastructure
PE 5689 Island developing states ; Neglected marine space
PE 5784 Island populations ; Disaster hazards
PF 7435 Islands ; Disappearing
PC 2362 Islands ; Disputed

PD 2941 Islands ; Isolated
PE 5760 Islands ; Neglect remote regions
PF 3448 *Islands ; Unirrigated*
PF 5737 Islands ; Weak social infrastructure
PD 9654 *Islet cell tumours animals ; Functional*
Isolated [communities...]
PF 2875 — communities ; Inaccessibility essential services
PF 2875 — communities ; Restricted delivery essential services
PD 1544 — community youth
PC 1707 *Isolated family units*
PF 9023 Isolated geographical location
PC 0193 Isolated groups ; Socially
PD 2941 Isolated islands
PE 7172 Isolated mass cemeteries
PD 1564 Isolated non-participating elders urban environments
PF 6580 Isolated problems ; Preoccupation
PD 1564 Isolated senior citizens
PF 9023 Isolated village location
PE 3296 *Isolated village mentality*
PF 1696 *Isolated women ; Socially*
PF 2241 *Isolating community lifestyle*
PE 3547 Isolating effects seasonal variations undeveloped transportation
PF 0945 *Isolating social forms enhancing ethical relations*
PB 8685 Isolation
Isolation [Communication...]
PD 7773 — Communication
PD 2495 — communities industrialized countries ; Personal
PE 1277 — computer based recreation ; Social
PE 1277 — computer users ; Social
PC 3943 — Cultural
Isolation [Destructive...]
PC 2791 — Destructive economic
PA 7339 — Duality
PE 4703 — due torture ; Self imposed social
Isolation [Economic...]
PC 2791 — Economic
PD 1564 — elderly ; Social
PC 3316 — ethnic groups
PA 5869 — Exclusion
PD 1564 Isolation fostering structures engagement elders
PF 9023 Isolation ; Geographical
PE 8681 Isolation housewives ; Social
PF 3559 Isolation individual decision-making context
PF 7096 Isolation ; Local traditions cultural
PE 2274 *Isolation ; Mothers' self imposed*
PF 9023 Isolation mountain valley communities ; Social
PF 2141 Isolation ; National
PF 9023 Isolation neighbouring villages ; Social
Isolation [parent...]
PC 0600 — parent-child relationship
PF 9023 — Perpetuated rural
PF 9023 — Physical
PC 7569 — Political
PA 7296 Isolation *Restraint*
PF 2386 Isolation ; Social
PC 1707 *Isolation ; Social*
Isolation [Terminal...]
PE 1277 — Terminal
PD 6810 — torture ; Social
PJ 0272 — *trade union members their representatives*
PC 2791 *Isolation ; Village economic*
PF 1664 Isolation wisdom elders society
PE 8681 Isolation women home ; Social
PF 8868 Isolationism ; Academic
PF 2141 Isolationism ; Diplomatic
PC 2791 Isolationism ; Economic
PD 1749 Isolationism ; Individual
PF 2141 Isolationism ; International
PF 2141 Isolationism ; National
PC 7569 Isolationism ; Political
PC 1707 Isolationism ; Self created
PF 1623 Issue avoidance
PF 3070 Issue foreign currency use abroad ; Refusal
PF 9123 Issue travel documents ; Delay
PE 0325 Issue travel documents, passports, visas ; Refusal
PC 0237 *Issues advocate ; Inaccessible*
PD 4060 Issues ; Avoidance controversial
PF 1623 Issues ; Dodging
PF 7220 Issues ; Ethical resolution scientific
PF 7431 Issues ; Evasion
Issues [Ideological...]
PF 7220 — Ideological resolution scientific
PF 9148 — Inadequately publicized
PF 1735 — *Interdependence*
PF 7925 — international community ; Avoidance human rights
PD 2888 — Issues ; Lack faith solutions ecological
PD 4060 Issues meetings ; Curtailment discussion
PF 7220 Issues ; Misuse democratic processes resolve theoretical
PF 0817 Issues ; Nationalistic response global
PA 2360 *Issues ; Overexposure*
Issues [Perceived...]
PF 1735 — *Perceived complexity*
PE 6311 — political campaigns ; Avoidance
PF 4600 — political debate ; Avoidance substantive
PF 4175 — *Political nature development*
PD 2860 — Politicization
PF 7220 — Issues ; Religious resolution scientific
Issues [Underreported...]
PF 9148 — Underreported
PF 7925 — United Nations ; Politicization human rights
PF 1316 — *Unresolved legal*
PE 3940 Itch
PE 2727 *Itch ; Barn*
PD 9667 *Itch horses ; Sweet*

PE 9799 *Itch ; Mad*
PG 4137 *Itch mite infestation*
PE 6668 Itchgrass
PE 0598 Items ; Calculated delays releasing controversial news
PE 1478 *Items ; Reduced impact impersonal repetitive news*
PE 4991 Ivory trade ; Banning
PE 4991 Ivory trade ; Illegal

J

PE 9438 *Jaagsiekte*
PE 5380 *Jack sores*
PE 0985 Jack ; Yellow
PD 8680 Jail suicides
PD 0520 Jails ; Inhumane
PE 7915 *Jakob disease ; Creutzfeldt*
PD 2045 Jamming radio frequencies
PD 1244 Jamming satellite communications
PE 2348 Japanese B encephalitis
PF 1748 Jargon ; Incomprehensibility specialized
PF 1748 Jargon ; Specialized
PE 0517 Jaundice
PE 0517 Jaundice ; Infectious
PE 2357 *Jaundice ; Infectious*
PD 1185 *Jaw ; Diseases*
PD 1185 Jaw ; Diseases mouth
PE 0435 *Jealous paranoia*
PF 7979 Jealousies ; Bureaucratic
PF 1205 *Jealousies ; Multiple family*
PF 5013 Jealousy
PF 5013 Jealousy ; Delusional
PA 7253 Jealousy *Envy*
PF 5013 Jealousy personal success
PD 8488 Jealousy ; Professional
PF 5520 Jeopardization universality
PF 6158 Jeopardy ; Denial right freedom double
PD 3481 *Jerboas ; Endangered species*
PE 1904 Jet-lag
PE 5799 Jet noise
PF 8838 Jewish conspiracy
PE 2131 Jews ; Discrimination against
PF 5681 Jihad
PF 9536 Jinx
PE 9441 Job applicants ; Secret government security vetting
PF 0171 *Job benefits ; Insufficient*
PF 2998 Job context ; Narrow
PE 6836 Job control ; Excessive
PD 8052 Job ; Dozing
PD 8052 Job fatigue
PF 2093 *Job hunting ; Sensed futility*
PE 6937 Job ; Hypertension
Job [image...]
PF 2896 — image ; Static
PF 6541 — *information ; Limited*
PD 8211 — insecurity
PE 5316 — insecurity pregnant women
PE 8916 Job locations ; Distant
Job market
PF 2066 — *Certificate based*
PE 8916 — Inaccessible
PC 7997 — Limited
PC 7997 — Unorganized
PD 2656 Job monotony ; Increasing
PF 2949 *Job motivity ; Undermined*
Job opportunities
PF 1658 — Inadequate
PF 1658 — Insufficient
PD 2972 — some sectors society ; Lack
PC 5576 — unemployed ; Rejection
PF 1658 Job opportunity ; Limited
PF 2949 *Job options ; Insufficient*
PF 7119 Job protection during maternity leave ; Denial right
PF 6541 *Job publicity ; Inadequate*
PF 2066 Job qualifications ; Artificial arbitrary
Job [requirements...]
PC 1131 — requirements ; Disparity workers skills
PJ 9248 — *requirements ; Irrelevant*
PE 6246 — reservation apartheid system
PF 0171 Job satisfaction ; Lack
PE 5083 Job stress video display terminal work
PF 6014 *Job terminology ; Sexually discriminating*
Job training
PD 3548 — Outdated
PF 3552 — *Restricted*
PF 3552 — *Undeveloped*
PC 2035 Job vacuum ; Youth
PB 0750 Joblessness
Jobs [Attractive...]
PD 9869 — Attractive city
PF 2842 — *Automated warehousing*
PD 0528 — automation ; Elimination
Jobs [developing...]
PE 4391 — developing countries due introduction new technologies ; Elimination
PE 4391 — due automation developing countries ; Elimination
PE 5014 — due computerization ; Downgrading
PF 2842 *Jobs ; Government dominated*
PF 3084 Jobs ; Insufficient side income
Jobs [Lack...]
PE 1753 — Lack skilled
PC 1108 — Limited off season
PF 2792 — Limited ways matching talent
PD 3443 *Jobs men ; Reservation*
PF 6908 Jobs ; Necessity second

Knees

	Jobs [Scarce...]
PF 1658	— Scarce local
PF 3448	— *Scarcity well paying*
PF 1658	— small communities ; Limited availability
PD 0528	Jobs technological innovation ; Endangered
PF 7022	*Jobs ; Time consuming*
PF 1658	Jobs ; Unavailability local
PC 1131	Jobs unrelated education ; Available
PF 1658	Jobs village ; Insufficient
PD 3443	Jobs women ; *Reservation*
PD 2731	*Johne's disease*
PF 1735	Join community action ; Reluctance
PE 5192	Join trade unions ; Violation right workers
PD 2283	Joint ; Ankylosis
PE 7826	Joint ; Arthritis shoulder
	Joint disease Degenerative
PE 5913	— [Joint disease ; Degenerative]
PE 5913	—
PE 5050	—
PE 4161	—
PE 7826	Joint ; Inflammation stifle
	Joint [pains...]
PE 3732	— pains due torture
PE 7826	— *Periostitis fetlock*
PF 0963	— *planning ; Uncoordinated*
PE 7826	Joint ; Serous arthritis fetlock
PD 2283	Joint ; Stiffness
PE 3869	Joint venture ; Inadequate participation control
PD 2283	Joints ; Infections
PD 2283	Joints ligaments ; Disorders
PD 2283	Joints ; Neoplasms
	Joints [Sprains...]
PB 0731	— *Sprains*
PB 0731	— *Strains*
PE 4566	— Symptoms referable
PE 9846	Joints ; Tuberculosis bones
PD 2283	*Joints ; Tumours*
PF 2634	*Jokes ; Sick*
PD 3041	Journal propaganda ; Foreign controls newspaper
PD 0184	Journal propaganda ; Newspaper
PE 1750	Journalism ; Junk food
PE 1750	Journalism ; Junk news
PD 3035	Journalistic confidence ; Forced breach
PD 3035	Journalistic immunity ; Erosion
PD 3071	Journalistic irresponsibility
PD 5183	*Journalistic muckraking*
PD 3035	Journalists ; Denial rights
PD 3036	Journalists ; Harassment
PD 3071	Journalists ; Harassment
PD 3036	Journalists ; Intimidation
PE 9844	*Jowl abscess*
PF 7451	Joy functioning ; Excessive
PA 6731	Joylessness *Solemnity*
PA 7107	Joylessness *Unpleasantness*
PE 7665	Judge jury ; Lack impartiality
PA 8528	Judgement*complex
PD 0986	Judgements ; Elitist legal
PD 1060	Judgements ; Exclusion masses setting criteria judicial
PF 5356	Judgements ; Instability legal
PE 4768	Judgements orders ; Failure enforce
PE 7665	Judges ; Denial right independent
PE 5487	Judges ; Harassment
PF 6032	Judges ; Inadequate training
PF 4216	Judges international athletic competitions ; National bias
PD 4827	Judges juries ; Biased
PE 7665	Judges lawyers ; Lack independence
PF 5356	Judicial decisions ; Unpredictability
PF 8479	Judicial error
PE 1743	Judicial hearings ; Denial right access interpreters
PD 1060	Judicial judgements ; Exclusion masses setting criteria
PD 3879	Judicial order ; Disobedience
	Judicial [precedent...]
PF 2580	— precedent ; Outdated
PE 1295	— proceedings due protective legislation ; Discrimination against juveniles
PD 9035	— process ; Contempt
PE 1198	— process ; Economic barriers women's access
PE 1479	— protection individual rights ; Psychological barriers
PE 1129	Judicial sanctions ; Discrimination against poor
	Judicial system
PF 8479	— *Collapse*
PF 2113	— Inadequate international
PF 6539	— *Lenient*
PE 7725	Judiciary ; Closed channels dialogue
PD 4194	Judiciary ; Corruption
PE 7665	Judiciary executive military power ; Control
PE 5487	Judiciary ; Harassment
PE 9339	Judiciary ; Homosexuality
PE 0803	Judiciary jury membership ; Economic barriers access legal profession,
PE 7665	Judiciary ; Lack impartiality
PE 0647	Judiciary ; Political corruption
PE 4907	Judiciary ; Racial discrimination
PD 8422	Judiciary ; Underpayment
PE 4577	Jumping ; Bail
PD 3481	*Jumping mice ; Endangered species*
PE 2235	Jungle environment tropical villages ; Detrimental effect
PE 0616	Jungle fever
PF 2365	*Jungle harvesting ; Barriers*
PE 8797	Junior high schools countryside ; Remote
PD 5510	Junk ; Accumulated
PE 4963	Junk bonds
	Junk [fax...]
PE 1810	— fax mail
PD 0800	— food consumption

	Junk [food...] cont'd
PE 1750	— food journalism
PE 1810	Junk mail
PE 1750	Junk news journalism
PD 0089	Junk ; Orbit
PD 0089	Junk ; Space
PE 5309	Junk ; Submarine
PD 5510	Junk ; Visible accumulated
PF 3855	Juridical fault ; Non
PD 4827	Juries ; Biased judges
PD 4194	Juries ; Harassment
PE 3916	Juries ; Trials exceeding competence
PF 2943	Jurisdiction administered territories ; Lack United Nations
PE 3140	Jurisdiction ; Extraterritorial intrusion
PF 7897	Jurisdiction international courts justice ; Government refusal accept
PF 9077	Jurisdictional conflict academic disciplines
	Jurisdictional conflict antagonism
PE 8308	— government agencies each country
PE 7901	— intergovernmental organizations
PE 9011	— intergovernmental organizations
PE 0064	— international nongovernmental organizations
PE 1169	— international nongovernmental organizations
PD 0047	— international organizations
PD 0138	— international organizations
PE 7973	— international organizations country level
PE 1583	— regional intergovernmental organizations common membership
PE 2486	— specialized agencies United Nations
PE 8799	— specialized agencies United Nations
PF 1066	Jurisdictional territories ; Extra
PE 4733	Jurisdictions ; Bias jury trials small
PE 5424	Jurisdictions ; Inadmissability evidence
PE 7665	Jury ; Lack impartiality judge
PE 0803	Jury membership ; Economic barriers access legal profession, judiciary
PD 4194	Jury nobbling
PE 4733	Jury trials small jurisdictions ; Bias
PD 8568	Just income ; Denial right
PE 6529	Justice ; Cultural discrimination administration
	Justice [Deficiencies...]
PE 8300	— Deficiencies military codes
PF 1487	— Delay administration
PD 0412	— Delay administration criminal
	Justice Denial
PD 6927	— human rights administration
PA 6486	— right
PC 6162	— right
PF 4836	Justice developing countries ; Absence reliable systems
PD 0986	Justice ; Discrimination administration
	Justice [Economic...]
PE 1399	— Economic discrimination administration
PE 1399	— Educational discrimination administration
PD 0986	— Elitist
PF 8479	— Error criminal
PF 7728	Justice ; Failure legally rehabilitate victims miscarriage
PE 1347	Justice ; Geographic discrimination administration
PF 7897	Justice ; Government refusal accept jurisdiction international courts
	Justice [Inaccessibility...]
PD 8334	— Inaccessibility
PF 2113	— Ineffective administration international
PD 0986	— Inequitable administration
PD 5314	— International inaccessibility
PD 8334	Justice ; Lack access
PD 8217	Justice ; Lack confidence administration
PF 8479	Justice ; Miscarriage
PD 8217	Justice ; Mistrust system
PD 5515	Justice ; Obstruction
PD 5515	Justice ; Offences against administration
	Justice [Perversion...]
PF 8479	— Perversion
PE 1828	— Political discrimination administration
PE 1399	— poor ; Inequitable
PE 0168	— Justice ; Religious discrimination administration
PF 4875	Justice system ; Deficiencies criminal
PF 4899	Justice systems ; Deficiencies civil
PE 8881	Justice ; Unrestrained use force administration
PE 8829	Justice ; Use undue influence obstruct administration
PD 0162	Justice women ; Denial right
PE 7533	Justifiable homicide
PC 2371	Justification militarism ; Religious
PF 3807	Justification nuclear war ; Theological
PF 9671	Justification population growth ; Theological
PF 9243	Justification projects ; False
PF 7490	Justified science ; Violence
PE 1513	Jute ; Instability vegetable fibre trade, excluding cotton
PE 4522	Jute products developing countries ; Inadequate international marketing
PE 1794	Jute trade ; Instability
PD 1611	Juvenile alcoholism
PE 7209	Juvenile criminals segregation adult criminals ; Denial right
	Juvenile [delinquency...]
PC 0212	— delinquency
PD 8340	— desertion
PC 0212	— deviance
PE 8803	Juvenile offenders ; Inadequate rehabilitation
PE 8215	Juvenile offenders ; Inadequate segregation different categories
PD 6213	Juvenile prostitution
PC 0877	Juvenile stress
PE 5771	Juvenile suicide
PD 0634	Juveniles ; Injustice detention

PE 1295	Juveniles judicial proceedings due protective legislation ; Discrimination against
PD 0634	Juveniles ; Repressive detention
PD 9394	Juveniles ; Sexual offences

K

PE 8674	Kangaroo mice ; Endangered species pocket mice
PE 8708	Kangaroos wallabies ; Endangered species
PD 2778	*Karoshi*
PE 2254	*Ked ; Sheep*
PE 9854	*Kennel cough*
PE 7826	*Keraphyllocele*
PE 6789	*Keratitis*
PE 6789	Keratitis ; Interstitial
PE 6789	Keratitis ; Superficial
PE 6789	Keratitis ; Ulcerative
PD 9366	*Keratoconjunctivitis*
PD 5535	Keratoconjunctivitis animals ; Infectious
PD 2565	*Keratoconjunctivitis sicca*
PE 7826	*Keratoma horses*
PC 0092	*Keratoses*
PE 0386	Kernels ; Instability trade oil seeds, oil nuts oil
PE 8250	Kernels oil seeds ; Environmental hazards oil nuts, oil
PD 7420	*Ketosis cattle*
PD 7420	*Ketosis ; Ovine*
PF 5817	Key economic variables ; Unpredictability
PD 7841	*Key-Gaskell Syndrome*
PJ 3753	Keynesianism ; Global
PC 0742	Khaki-collar crime
PF 8780	*Khalwat*
PE 0912	Khat ; Abuse
PC 2558	Kickbacks
PD 8744	Kidnapped ; Risk being
PD 8744	Kidnapping
	Kidnapping [Parental...]
PE 6075	— Parental
PE 0805	— pet animals
PD 1886	— Political
PD 9293	Kidney animals ; Metabolic disease
PD 9293	*Kidney animals ; Tumours*
	Kidney [disease...]
PE 7769	— *disease ; Pulpy*
PE 2053	— diseases
PE 2053	— disorders
PE 2053	*Kidney injuries*
PE 2053	*Kidney ; Suppuration*
PE 2053	*Kidney ; Tuberculosis*
PE 2053	Kidney tumours
PE 9086	Kidney urinary organs ; Benign neoplasm
	Kidney worm infection
PD 9293	— *dog ; Giant*
PD 9293	— *mink ; Giant*
PD 9293	— *Swine*
PF 8026	Kidneys, eyes nerves ; Damage small blood vessels
PE 4421	Kids ; Stressed out
PD 7520	*Kill ; Government incitement*
PD 7520	*Kill ; Incitement*
PE 0793	Killer bees
PE 1914	Killer cell syndrome ; Low natural
	Killing animals
PD 8486	— [Killing animals]
PD 7520	— *Criminal*
PD 8486	— humans
PE 0358	— Inhumane
PE 5944	Killing corporate leaders
PE 6356	Killing ; Corporation sanctioned
PE 9650	Killing ; Denial animals right freedom mass
PD 0158	Killing foetus
PF 4617	Killing god
PD 7221	Killing ; Government sanctioned
PF 3819	Killing human rights activists ; Government complicity
PC 8096	Killing humans
PE 0428	Killing ; Inmate
PE 7533	Killing ; Lawful
PE 2643	Killing ; Mercy
PF 6359	Killing non-human life
	Killing [parents...]
PE 0651	— parents
PD 4217	— plants
PD 2341	— Premeditated
PE 2759	Killing stray animals ; Inhumane
PC 3685	Killings ; Ethnic
PD 2483	Killings ; Mass
PD 7221	Killings ; Political
PE 5614	Killings ; Political
PF 2641	Killings ; Ritual
PE 9447	Killings ; Serial
PD 7221	Killings ; Staged encounter
PE 7584	Kills caused pollution ; Fish
PF 6723	Kind payments ; Abuse
PE 2611	*Kinetosis*
PE 3616	Kissing bugs vectors disease
PD 4287	Kissing ; Obligatory social
PD 5111	*Kissing ; Risk contracting AIDS*
PE 7826	*Kissing spines syndrome*
PA 6448	*Kleptomania*
PE 2320	*Knee ; Housemaid's*
	Knee [pain...]
PE 1309	— pain
PE 2320	— *Parson's*
PE 7826	— *Popped*
PE 7826	*Knee ; Sore*
PE 7808	*Knees ; Knock*
PG 2295	*Knees ; Knock*

—969—

Knees

PE 1309 Knees ; Problematic
PE 7808 *Knock-knees*
PG 2295 *Knock knees*
PE 4586 *Knot cherry ; Black*
PE 4586 *Knot plum ; Black*
Know how
PE 6554 — Inadequate business
PC 4867 — *Insufficient commercial*
PF 3409 — *Insufficient technological*
PF 2995 — *Limited maintenance*
PF 2401 — *Limited political*
PD 7719 — Loss technical
PF 9595 Knowing ; Fear
PF 2376 Knowledge ; Accumulation
PF 3706 Knowledge ; Authoritarian propagation
PF 7077 Knowledge ; Concealment esoteric
PE 8403 Knowledge ; Copyright barriers transfer
PF 8191 Knowledge ; Difficulties applying
Knowledge [educational...]
PF 0166 — educational materials ; Barriers international flow
PF 8205 — eligibility benefits ; Lack
PF 8753 — empower humanity response global problematique ; Failure integrate
PF 9595 Knowledge ; Fear
PF 0944 Knowledge ; Fragmentation
PJ 7640 Knowledge good evil
Knowledge [Imbalanced...]
PF 0204 — Imbalanced distribution
PF 1953 — Inaccessibility
PF 2995 — *Inadequate marketing*
PF 0944 — Inadequate organization
PE 2000 — Increasing inability integrate available
PE 7139 — Increasing lag education against growth
PE 8133 — incubation periods animal diseases ; Inadequate
PF 0703 — information ; Prohibitive cost
PE 8233 — Insufficient irrigation
PF 2401 — *Insufficient programme*
Knowledge Lack
PF 8381 — [Knowledge ; Lack]
PF 0944 — comprehensive framework understanding
PF 7916 — expert
PD 8023 — sanitation
PD 8003 — scientific
PF 0944 — system integration
PF 6035 Knowledge ; Language barriers transfer
PF 4014 Knowledge ; Limitation current scientific
Knowledge Limited
PF 1953 — access
PF 1442 — *budget*
PF 2583 — *consumer*
PD 8023 — health
PD 9160 — medical
PE 8140 Knowledge local production processes ; Insufficient
PF 5420 Knowledge ; Lost
Knowledge [Manipulative...]
PF 1609 — Manipulative
PF 8757 — Mismatch public policy current scientific
PF 5329 — Monopolization
PF 0980 Knowledge predict future ; Monopoly
Knowledge [Refusal...]
PF 5329 — Refusal share
PF 7939 — reporting man made disease ; Inadequate
PF 7134 — resistance ; Societal
PF 1319 — Restrictions acquisition
PF 8191 Knowledge solve problems ; Inadequate application available
PD 3050 Knowledge ; Tax barriers dissemination technical
Knowledge [Unapplied...]
PF 1468 — Unapplied scientific
PF 0204 — Unbalanced distribution
PF 5728 — Unrecorded
PF 3539 — *Unusable*
PF 8661 Known offenders ; Inadequate evidence convict
PF 8661 Known offenders ; Law enforcement failure against
PE 7826 *Knuckling*
PD 1762 *Koala ; Endangered species*
PE 0333 Korsakoff syndrome
PE 4161 *Krampfigkeit*
PE 2282 Kwashiorkor disease
PE 4113 Kyphosis

L

PD 2468 Labelling dangerous substances ; Inadequate
PD 3521 *Labelling ; Fraudulent*
PD 3521 Labelling goods ; Inadequate
PD 3521 Labelling ; Inappropriate
Labelling [packages...]
PD 3521 — packages ; Inadequate
PD 3521 — packages ; Misleading
PF 8508 — people
PF 8508 Labels ; Group
PE 4561 Labile personality
PJ 9371 *Laboratories ; Hazardous school*
PD 0260 Laboratory animals ; Maltreatment
PC 7674 *Laboratory facilities developing countries ; Inadequate medical*
PE 4859 Laboratory medical equipment ; Unsafe industrial,
PF 5304 Laboratory testing errors
PE 9813 Laboratory waste water ; Discharge dangerous substances
PE 9813 Laboratory waste water pollutants
Labour [absenteeism...]
PD 1634 — absenteeism
PD 0165 — Abuse prison
Labour [Abusive...] cont'd
PD 7821 — Abusive treatment women
PE 2844 — affected developing country exports ; Inadequate adjustment assistance industries
PD 3076 — Alienation human
PE 9114 — Artificial opposition manual intellectual
PC 6089 — Authoritarian division
PD 3301 Labour ; Bonded
Labour [Cheap...]
PD 8916 — Cheap
PC 3117 — *Cheap*
PC 9042 — childbirth ; Complications
PE 0207 — Coercive use economic power transnational enterprises against
Labour costs
PF 8763 — High
PF 8763 — Inflated
PF 8763 — Prohibitive
Labour [Debilitating...]
PF 2812 — Debilitating designs using
PD 0044 — Decrease skilled
PE 6342 — Denial right freedom compulsory
Labour Dependence
PE 4451 — developing countries unpaid female
PC 0746 — forced
PC 0065 — *foreign*
PC 0065 — *imported*
Labour [developing...]
PD 1965 — developing countries ; Contempt agricultural
PE 9199 — developing countries ; Exploitation cheap
PD 6930 — Discrimination against casual
PE 6843 — displacement
PD 4012 — disputes
PE 8882 — Distortion international trade restrictive controls movement
Labour [Excess...]
PB 0750 — Excess
PD 0164 — Exploitation child
PD 6930 — Exploitation contract
PF 9281 — Exploitative transformation international division
Labour force
PF 0905 — *Decreasing*
PF 2973 — Depleted expertise rural
PF 0907 — developing countries ; Instability
PD 5045 — developing countries ; Loss
PF 0905 — *Emigration*
PC 0592 — Inadequate
PD 9113 — Inexperienced
PD 4011 — mobility ; Lack
PD 9113 — Undertrained
PF 6293 — underutilization
PF 6293 — Underutilization
PF 3499 — *Undiversified*
PE 8437 — Urban population overall growth economy
PE 8735 — urban population overall growth economy ; Imbalances growth
PC 0746 Labour ; Forced
Labour hoarding
PE 6333 — [Labour hoarding]
PE 6333 — employers
PE 6333 — employers
Labour [immobility...]
PD 4011 — immobility
PC 0746 — Impressment
PF 8763 — Increasing economic value
PE 8092 — industrialized countries ; Demand unskilled
PD 9113 — Insufficient trained
PC 5842 — *intensive industries ; Protectionism*
Labour [law...]
PD 3533 — law ; Conflicts
PE 4860 — *laws ; Anti*
PE 5135 — laws ; Conflicting
PJ 8060 Labour management relations ; Inadequate private international
Labour market
PD 0008 — developing countries ; Flooding urban
PD 0156 — flooding developing countries
PD 4011 — inflexibility
PE 8283 — Lack time flexibility
Labour markets
PD 6744 — Segmented
PD 4011 — Structural rigidities
PF 4028 — Wage rigidity
Labour mobility
PD 4011 — Lack
PD 4011 — Low
PD 4011 — Restrictions
PD 0164 Labour ; Necessity child
PF 1786 Labour negotiation procedures ; Outdated
Labour [potential...]
PF 6532 — potential ; Arrested development
PF 6471 — *potential ; Unorganized*
PD 4695 — practices ; Unfair
PD 0702 — prisons
PC 0180 — Problems migrant
PF 8763 — *Prohibitive cost farm*
Labour relations
PE 7355 — developing countries transnational corporations ; Interference
PE 0667 — industrialized countries transnational corporations ; Interference
PE 1187 — transnational corporations ; Domination
PC 6031 Labour resources ; Unproductive
Labour [Seasonal...]
PC 1108 — Seasonal
PC 0592 — shortage
Labour [shortage...] cont'd
PD 5045 — shortage developing countries
PD 0044 — Shortage skilled
PE 1753 — skills ; Underutilized
PD 0142 — standards developing countries ; Inequitable
PE 3105 — *Surplus*
PD 0156 — surpluses developing countries
Labour [tensions...]
PE 5927 — tensions involving transnationals
PD 2722 — trafficking ; Illicit
PF 0907 — turnover developing countries ; High
Labour [Unappealing...]
PF 0836 — *Unappealing farm*
PE 4789 — Uncompensated housewives'
PE 1134 — *Undesirable manual*
PE 1753 — Unemployed skilled
PE 5506 — Unfair competition workers employed forced
PD 8916 — Unfair remuneration
PC 4613 — unions ; Discrimination against
PD 3056 — Unpaid
PD 9113 — Unprepared local
PE 8687 — Unsuitability development projects available
PE 7908 — use socialist countries ; Inefficient
PD 3427 Labourers industrialized countries ; Inadequate living working conditions immigrant
PD 2567 *Labyrinthitis*
Lack [ability...]
PC 0239 — *ability*
PF 8481 — abortion facilities
PA 7270 — *Absence*
PE 7519 — acceptable sites power plants
Lack access
PF 1985 — cultural heritage
PD 8334 — justice
PE 8637 — prisoners' defence
PD 1194 — public archives
Lack [acclimatization...]
PE 5720 — *acclimatization*
PF 4839 — accord water use
PE 0503 — accountability disposal wealth
PF 3307 — *actuation structures*
PE 8496 — adequate clothing supply dwarfs, midgets giants
PF 5933 — adequate military defence
PJ 8840 — adherence international transit conventions
PE 5789 — adherence international transit conventions land-locked countries
PF 3411 — *administrative control*
PF 2335 — administrative reform
PD 9189 — agrarian reform
PF 2816 — agreement cooperation
Lack agricultural
PF 2378 — *capital*
PF 2164 — development
PF 8835 — education
Lack [Agricultural...]
PF 8835 — Agricultural education
PF 4108 — agricultural machinery
PD 8388 — *agro-urban training*
PE 8231 — airport travel security
PA 6977 — ambition
PE 5794 — animal welfare legislation
PF 7972 — anti-discrimination legislation
Lack appreciation
PF 2679 — cultural differences
PF 2679 — cultural differences
PE 7476 — nuclear weapons
Lack [apprentice...]
PF 1352 — *apprentice training*
PE 5020 — appropriate education children immigrants
PJ 0385 — arbitrating commodity markets
PD 2543 — army discipline
PE 4804 — artisans craftsmen developed countries
PF 2373 — assertiveness
PF 2132 — assimilation
PD 0822 — associations developing countries
PF 5893 — attunement
PF 0053 — autonomous world-level actor identify clarify world interests
PE 8256 — availability methods cross-breeding
Lack awareness
PF 8205 — available welfare benefits
PF 1985 — past heritage
PF 2042 — potential investment small, inner-city enterprises
Lack [biocoenosis...]
PE 5290 — biocoenosis
PF 6229 — biological control
PD 1207 — biological control pests
PF 5381 — birth certificate
PE 8430 — brainstorming
PF 7088 — business opposition arms race
Lack capital
PB 4653 — [Lack capital]
PD 8604 — development
PE 5790 — investment developing countries
PB 4653 — reserves
PF 1781 *Lack car pools*
Lack care
PF 4646 — [Lack care]
PD 8837 — animals
PE 8900 — intellectually disabled convicts
Lack [celibacy...]
PE 7031 — celibacy
PF 2822 — *central nodes*
PF 2540 — central planning structures small communities
PE 2820 — centralized financial control United Nations systems
PD 2011 — *cesspool fencing*

Lack motivation

Lack [channels...] cont'd
PF 6544 — channels obtaining available local funding
PF 3962 — character ; Politicians'
PA 5612 — chastity
PD 2085 — child care facilities
PC 0233 — *child welfare institutions*
PD 8122 — clean water
PD 2459 — cleanliness
PD 4004 — *clinic space*
PE 5689 — coastal development island countries
PF 2588 — combined efforts
Lack commitment
PF 1729 — [Lack commitment]
PE 8814 — common symbols
PF 4662 — protection poor
PF 4662 — protection vulnerable groups
Lack common
PF 1118 — ethic
PF 3364 — religious witness
PE 8814 — symbol
PF 0816 Lack communication
Lack community
PF 7912 — development
PF 7912 — development
PF 3307 — participation
PF 2605 — planning
PD 1544 — *responsibility*
PF 3512 — self-worth
PF 2261 — *voice*
Lack [comparability...]
PF 2622 — comparability international statistics
PF 0214 — comparable social data
PD 7179 — compensation damages
PD 1466 — compliance fiscal laws
PF 0944 — comprehensive framework understanding knowledge
PE 9172 — concensus psychiatrists
PD 8837 — concern animal welfare
Lack confidence
PA 8653 — [Lack confidence]
PD 8217 — administration justice
PC 2089 — governments
PF 3058 — international monetary system
PF 2093 — *parents*
PF 8559 — police
Lack [conscious...]
PF 1781 — *conscious initiatives*
PF 9431 — consensus facts
PE 8599 — conservation energy private sector
PF 0342 — consistent programme effort
PD 5344 — constructive family discipline young people
Lack consumer
PD 0515 — choice centrally-planned economies
PD 0515 — goods socialist countries
PE 1940 — influence industry
Lack [context...]
PF 2068 — context future ; Young people's
PF 3434 — continuity amongst personnel international organizations
PE 3528 — *contract power*
PF 7138 — control
Lack cooperation
PF 2816 — [Lack cooperation]
PF 8500 — officialdom
PD 0516 — riparian states
Lack coordination
PF 8330 — [Lack coordination]
PJ 1719 — agencies
PF 0175 — aid donors
PF 0048 — international monetary system
PE 0075 — United Nations information systems
Lack [core...]
PF 1254 — core leadership
PF 1205 — *corporate achievements*
PF 1728 — correspondence basic value images social reality
PE 8516 — credit facilities agricultural producers
PE 3066 — credit guarantee facilities developing country exports
PF 0536 — cultural integration
Lack [daily...]
PF 6182 — daily physical activity urban environments ; Unhealthy
PE 3528 — *dairy markets*
PF 2442 — deathing awareness
PD 2036 — decision-making experience
PD 1975 — definition intelligence
PB 8031 — democracy
PF 3731 — demolition planning
PF 4832 — *dental care*
PB 0206 — development
PF 0392 — development aid
PD 7650 — discipline work
Lack diversity
PE 8671 — medical science
PE 9172 — psychiatry
PE 8806 — social roles
PF 3364 Lack ecclesial communion
Lack economic
PE 8002 — adaptation indigenous society non-socialist countries
PE 8002 — adaptation indigenous systems
PC 0025 — cooperation
PD 4150 — liberty
PF 0841 — self-sufficiency
PE 8190 — technical development
PE 9764 Lack educated manpower least developed countries
Lack education
PB 8645 — [Lack education]
PA 5568 — [Lack education]

Lack education cont'd
PE 8699 — women immigrants
Lack [educational...]
PD 0847 — educational facilities
PF 2168 — *educational incentives*
PF 0816 — effective communications
PF 9007 — emergency facilities
PD 8262 — emotional security
PF 1548 — emphasis basic education
PF 0223 — enforcement power
PF 9175 — enforcement power international organizations
PF 7383 — English language describe female experience
PD 5642 — enterprise parastatals
PD 1370 — environmental education
PB 4653 — equity capital
Lack essential
PD 1942 — freedom due unawareness actuality
PF 2115 — local infrastructure
PF 2115 — services materials needed development
Lack [eugenic...]
PD 1091 — eugenic measures
PD 1091 — eugenics
PF 7916 — expert knowledge
PE 8752 — expertise agricultural techniques
PF 1653 — *extension offices*
Lack facilities
PF 2012 — educating older people
PD 8314 — physically disabled
PD 0211 — severely deformed people
PD 2888 Lack faith solutions ecological issues
PF 5079 Lack family life education
Lack family planning
PF 0148 — [Lack family planning]
PD 1039 — education
PD 1050 — information
Lack [family...]
PE 8341 — family welfare
PB 4653 — *fees students*
PE 2425 — finance coastal communities
PE 7009 — financial information systems international nongovernmental organizations
PF 3630 — financial surplus
PD 6520 — *fire-fighting facilities*
PF 8515 — *food hygiene*
PF 7577 — foreign students
PF 6534 — formal education
PF 1254 — formal leadership
PE 4669 — freedom movement children separated parents
PE 8951 — freedom press
PF 4416 — friendship
Lack funding
PF 7916 — *expertise*
PF 3318 — *infrastructure*
PF 1653 — *structures*
Lack funds
PB 4653 — *education*
PE 0612 — medical research
PE 8463 — veterinary research
Lack [generalists...]
PF 0256 — generalists
PD 2389 — *genetic counselling*
PF 1781 — *governing body*
PF 3318 — *government initiative*
PF 8960 — governmental support
PD 1970 — gun laws
Lack [heavy...]
PD 6520 — *heavy equipment*
PF 5728 — *historical record*
PE 5721 — holistic medicine
PD 2892 — horizontal commodity diversification developing countries
PE 5058 — hospital facilities
PF 0551 — house training
PE 8103 — housing teachers rural communities
PE 7530 — human organs transplantation
PF 2434 — human unity
PB 8214 — humaneness
PB 8214 — humanity
PF 7847 — humility
PF 5496 — hydroponics
Lack [ideal...]
PD 0093 — ideal contraceptives
PE 7044 — identity international nongovernmental organization network
PF 3388 — ideological unity
PF 3392 — *ideology*
PE 7665 — impartiality judge jury
PE 7665 — impartiality judiciary
PA 6997 — *Imperfection*
PF 4516 — incentive users care common property
Lack incentives
PD 5114 — increase agricultural output developing countries
PF 2168 — *students*
PF 0905 — *tourism*
PA 6652 Lack *Incompleteness*
PE 7665 Lack independence judges lawyers
Lack individual
PF 4371 — *development*
PD 1591 — historical context
PC 2178 — *initiative*
PF 1075 — rights political asylum
PF 1729 — social commitment
Lack [individualism...]
PD 3106 — individualism capitalist systems
PF 2115 — *industrial electricity*
PF 6471 — *industrial promotion*

Lack [information...] cont'd
PF 6337 — information
PF 1572 — innovative projects against world problems
PA 5473 — *Insufficiency*
Lack integration
PA 6233 — [Lack integration]
PF 1324 — society
PF 4871 — traditional Western medicine
PD 0664 — transport systems neighbouring developing countries
Lack [integrity...]
PF 7992 — integrity
PF 3217 — intellectual liberty
PC 1587 — intelligence
PE 7944 — interim support families
PF 3194 — internal political independence
PF 4676 — international code crimes
PF 2350 — *international communication*
Lack international cooperation
PF 0817 — [Lack international cooperation]
PF 0817 — Dependence
PC 2089 — due personal mistrust
Lack international [coordination...]
PF 3141 — coordination interest rates
PE 4508 — coordination supervisors financial stock markets
PF 4676 — criminal court
PC 0025 — economic coordination
PF 6027 — language
PD 6804 — *language*
PF 7058 — legal provision nongovernmental organizations
PF 7032 — nongovernmental organization identity
PB 0009 — security
PF 5106 — understanding
Lack [intersocietal...]
PF 2517 — intersocietal resource channels
PF 4416 — intimate relationships
PE 5790 — investment capital developing countries' shipping fleets
PD 8839 — irrigation
PF 3467 — *irrigation equipment*
PD 2972 Lack job opportunities some sectors society
PF 0171 Lack job satisfaction
PF 8381 Lack knowledge
PF 8205 Lack knowledge eligibility benefits
Lack [labour...]
PD 4011 — labour force mobility
PD 4011 — labour mobility
PC 8160 — land space
PA 6977 — *law*
PF 0223 — law enforcement bodies
PF 1254 — leadership
PF 7988 — leadership initiative
Lack legal
PF 8869 — aid facilities
PF 4836 — engineering
PE 8240 — protection extortion victims
PE 0069 — provision international nongovernmental organizations
PD 7609 — recourse
PF 3517 — representation
Lack [legislation...]
PE 5794 — legislation animal welfare
PC 0268 — legislation restricting noise levels
PE 8467 — legislative control advertising
PF 1297 — livelihood standards
Lack local
PF 3630 — capital
PF 2009 — commercial services
PF 2489 — credit
PF 1933 — *industries*
PF 6541 — information systems
PF 6556 — leadership decision-making structures
PF 6479 — leadership role models
PE 7313 — management skills developing countries
PF 2451 — services community leadership training
Lack [long...]
PF 5181 — long-term development assistance
PF 8176 — lookout capacity
PE 9018 — loyalty
Lack [maintenance...]
PD 6520 — maintenance tools
PF 1528 — management communication skills
PE 0046 — management skills developing countries
PF 1467 — *manpower planning*
PF 2009 — *market refrigeration*
PJ 2657 — marriageable partners
PF 0966 — meaningful educational context ethical decisions
PD 0894 — meaningful personal social paradigms
PD 2438 — means achieving consensus
PF 6454 — means local technological development
PE 9115 — meat egg production
PC 0492 — mechanized agriculture
Lack medical
PF 4832 — care
PD 5770 — ethics
PF 4832 — *information*
PF 1836 — records
Lack [minimum...]
PE 1693 — minimum age law
PE 6726 — minimum wage fixing
PF 8639 — models equality
PF 6524 — *modern equipment*
PE 3547 — *modern production*
PB 0206 — *modernization*
PF 2187 — moral development
PF 0967 — motivating collegial relationships
PA 2360 — motivation

Lack motivation

Lack [motivation...] cont'd
PF 2208 — motivation leadership development
PF 0245 — mutual credit societies
Lack national
PF 1934 — *identity*
PF 7034 — legal provision international nongovernmental organizations
PF 8107 — unity
Lack [natural...]
PC 7928 — natural resources
PD 3625 — natural resources developing countries
PF 6527 — nearby wells
PE 4780 — neighbourhood technology
PF 2827 Lack objectivity
PF 2837 Lack opportunities practical training communities
Lack organized
PF 7916 — *expertise*
PF 8703 — participation
PF 1348 — savings funds
PF 3628 Lack outside exposure
PC 3280 Lack parental fulfilment
PF 8876 Lack parliamentary time approve needed legislation
Lack participation
PE 8357 — attempting solve intergroup conflicts
PF 3339 — development
PE 8690 — disabled persons decision-making
PC 0352 — *indigenous peoples international decision-making*
PC 1917 — local politics
PE 8503 — local welfare programmes
PF 0574 — management
PC 1917 — politics
PE 9009 — women
PF 4959 — women development
Lack [patriotism...]
PD 0218 — *patriotism*
PE 4789 — payment housework
PF 4371 — *personal development*
PF 2799 — personal transport rural areas
PD 1718 — *perspective*
PC 0716 — physical activity
PD 5173 — physical warmth
PF 2358 — *pioneer spirit*
PF 6190 — places urban environments encouraging unstructured public access
PC 0238 — plant protection
Lack political
PD 8673 — development
PF 0297 — independence
PF 3216 — integration
PF 0796 — integrity
PF 1389 — *leverage*
PD 8276 — liberty
PE 4031 — neutrality civil servants
PF 3216 — unity
Lack political will
PC 5180 — [Lack political will]
PC 5180 — act problems
PC 5180 — cooperate
PC 5180 — respond needs developing countries
Lack [postal...]
PF 2717 — postal agents
PA 6434 — *Poverty*
PF 2848 — *practical examples*
PE 4820 — prenatal care
PC 6381 — preparatory education
PF 8515 — preventive sanitation
PF 7706 — problem interrelationships ; Assumption
PD 1554 — processing industry primary commodities developing countries
PE 8236 — *production domestic consumers developing countries*
PC 8908 — productivity
PD 8469 — productivity farm animals
PF 3411 — professional standards
PJ 7793 — programme methods
PD 2046 — *programmes dropouts*
PF 4718 — progress against underdevelopment
PF 4306 — progress establishing New International Economic Order
PF 0772 — prosthetic devices
Lack protection
PD 0096 — aged during disasters
PE 4694 — domestic animals during disasters
PE 7793 — following death threats
PF 7556 — free markets
PD 0098 — handicapped during disasters
PF 8854 — industrial property
PC 1784 — property
PE 7793 — victims intimidation
PB 4353 — vulnerable
PD 3785 — vulnerable groups during disasters
PE 4694 — wild animals during disasters
PD 1078 — women children during disasters
Lack [protective...]
PC 6848 — *protective legislation*
PD 0339 — protein
PE 8707 — purchasing power developing countries
PA 6977 — purpose
PF 3411 Lack qualifications
PF 5703 Lack qualitative excellence
Lack [racial...]
PF 0684 — racial identity
PF 7535 — rational global political decision-making
PD 4270 — raw materials
PF 7493 — realism intergovernmental organizations
PF 0202 — recreation funds

Lack [recreational...] cont'd
PF 5409 — recreational organization
PF 0677 — reform
PF 0021 — refresher courses professional fields
PD 4790 — *refrigeration facilities*
PD 5494 — *regular habit*
PF 2996 — *regular transportation*
PF 9299 — regulations against environmental pollution
PE 7141 — regulations against pollution developing countries
PF 2793 — regulatory inspectors
PF 4730 — relationship wealth generation public good
PF 1944 — relevant training opportunities local communities
PF 8234 — religion
Lack religious
PF 3863 — activity
PF 3863 — conviction
PF 8010 — discipline
PJ 5626 — objectivity
Lack [representation...]
PF 3468 — representation
PD 0177 — *resource conservation*
PA 6822 — respect
PD 1432 — response monetary incentives developing countries
PF 6536 — responsible involvement community affairs
PF 8161 — *retraining incentives*
PF 5660 — reward long-range planning decision-making
PE 8180 — rural industrialization developing countries
Lack sanitation
PD 8023 — expertise
PD 8023 — knowledge
PD 1225 — rural areas developing countries
Lack [satisfaction...]
PA 8886 — *satisfaction*
PF 1348 — saving foresight
PF 1348 — savings structures
Lack school
PE 9092 — discipline
PD 0549 — recreation
PC 2163 — transportation
Lack [scientific...]
PF 2720 — scientific investigation
PD 8003 — scientific knowledge
PE 4398 — scope intellectual ability public sector
PE 8797 — secondary schools countryside
PF 2899 — *seed money*
Lack self
PF 0879 — confidence
PF 0879 — confidences
PE 8341 — development family
PF 5354 — love
Lack sense
PD 0110 — community solidarity developing countries
PF 8704 — community solidarity world level
PE 1066 — smell
Lack [shared...]
PF 1822 — *shared machinery*
PF 3393 — sharing community skills
PF 2093 — significant community images
PF 1933 — *significant employment*
Lack skilled
PE 1753 — jobs
PE 5170 — manpower rural areas developing countries
PF 5737 — workers island developing countries
PE 5884 — workers transport sectors land-locked developing countries
PE 1753 Lack skills opportunities
PE 9767 Lack slaughter facilities
Lack social
PC 1522 — accountability
PE 8518 — accounting business community
PF 9144 — conscience
PF 8695 — contact
PC 0242 — development
PF 8078 — discipline
PD 2050 — forms ; Disorientation young due
PF 1934 — *identity*
PF 1324 — integration
PC 0663 — liberty
PF 2195 — mobility
PF 8185 — planning
PF 2810 — *programmes*
PF 1545 — progress
PD 9155 — security farmers
PD 9155 — security peasant farmers
Lack [soil...]
PD 1416 — soil aeration
PD 0949 — soil conservation
PF 8704 — solidarity poor
PE 8303 — specialist doctors developing countries
PF 1653 — *specialized technology*
PF 2875 — *specialized transportation*
PB 9171 — species preservation
PF 6067 — specifically designed software developing countries
PF 8010 — spiritual discipline
PF 2605 — *sports funds*
PF 3467 — storage facilities
PC 1522 — structural accountability
PF 6510 — support local commercial services
PF 1870 — sustaining symbols marriages
PF 2063 — symbolism local relationships
PF 2070 — system ethical decision-making
PF 0944 — system integration knowledge
PF 6506 — systems achieving grassroots consensus
Lack [technical...]
PE 4933 — technical development excess manpower developing countries

Lack [technical...] cont'd
PE 5814 — technical infrastructure maritime commerce developing countries
PF 3310 — *technological progress*
PF 0091 — terminological equivalents languages
PC 4498 — time
PE 8283 — time flexibility labour market
PD 0683 — *trade union recognition*
PF 0903 — trained administrative personnel developing countries
PF 2477 — *trained firefighters*
PD 8388 — training
PD 0190 — training women
PF 6855 — trans-frontier cooperation
PE 9107 — transparency international trade information
PF 5374 — transparency public finances
PF 0713 — transport vehicles
PF 2212 — *tuberculosis materials*
PD 1590 — typhoon protection
Lack understanding
PF 2434 — [Lack understanding]
PF 4837 — social economic contexts development programmes
PF 0761 — spiritual healing
Lack [unionization...]
PE 8345 — unionization working women
PF 1781 — *united action*
PF 2943 — United Nations jurisdiction administered territories
Lack unity
PA 6233 — [Lack unity]
PF 3364 — religions
PF 3404 — religions
Lack [vaccines...]
PE 4657 — vaccines
PE 8806 — variety life style
PE 8806 — variety social life forms
PD 0122 — ventilation
PB 4653 — *venture capital*
PD 1554 — vertical diversification developing countries
PF 3558 — village cooperation
PF 6556 — village leadership
PE 2593 — visibility
PD 0715 — *vitamin B1*
PF 3824 — vocabulary
PF 1254 — vocal leaders
Lack vocational
PC 2163 — *teachers*
PF 0422 — training
PF 0422 — training schemes
PF 0727 Lack war relief
PF 5744 Lack warmth work
Lack water
PE 8233 — *conservation*
PD 8517 — systems
PE 8233 — tanks
Lack [well...]
PF 1572 — well-researched projects alleviate problems
PD 2790 — work commitment
PD 8047 — work commitment developing countries
PD 2790 — worker loyalty corporations
PF 0574 — worker participation business decision-making
Lack world
PF 5137 — food security system
PF 4937 — government
PE 5801 — maritime integration
PF 3387 — religion
PF 2434 — unity
Lack youth
PF 2575 — activities
PF 2575 — organizations
PD 1544 — participation
PF 1790 Lacking industrial imagination
PD 5535 Lacrimal apparatus animals ; Diseases
PD 8786 Lacrimal glands ducts ; Inflammation
PC 9042 Lactation ; Disorders
PD 7799 Lactation failure sow
PD 7420 Lactation tetany mares
PE 4630 Lactic acidosis
PE 4949 Lactose intolerance
Lag against
PE 2000 — information growth ; Increasing development
PF 3743 — population growth ; Increasing development
PC 5879 — socio economic growth ; Increasing development
PE 3078 — technological growth ; Increasing development
PE 7139 Lag education against growth knowledge ; Increasing
PE 5369 Lag education against population growth ; Increasing
PE 1904 Lag ; Jet
PF 6042 Lag legal provisions ; Time
PF 7050 Lag understanding information ; Cultural
PF 8085 Lagging training social skills
PD 0946 Lagging transformation agriculture developing countries
PD 3480 Lagomorpha
PD 6627 *Laid out work premises ; Badly*
PF 2256 *Laissez-faire response pattern*
PD 8628 Lake pollution
PD 1990 Lakes ; Desiccation
PD 0767 Lakes ; Ecosystem modifications due creation dams
PD 0767 Lakes ; Environmental degradation due creation dams
PD 2257 Lakes rivers ; Eutrophication
PE 5813 Lakes rivers land locked countries ; Vulnerability
PE 9774 *Lamb disease ; Stiff*
PE 4811 Lambliasis
PD 2731 Lambs ; Nonsuppurative polyarthritis
PE 9774 Lambs ; Nutritional myopathy
PD 3978 Lambs ; Rattle belly
PE 5617 Lamellar suppuration
PE 4161 Lameness cattle

Lapse

PE 7826 Lameness caused bone cyst pedal bone	**Land locked developing cont'd**	**Land use cont'd**
PE 4161 Lameness ; Fescue	PE 5824 — countries ; Inadequate roads transport	PD 1141 — planning developing countries ; Defective
PE 7826 Lameness horses	PD 5788 — countries ; Insecurity	PF 6528 — practices ; Unprofitable
Lameness [sheep...]	PE 5884 — countries ; Lack skilled workers transport sectors	PF 6528 — Restricted
PE 5617 — sheep	PE 5860 — countries ; Unreliable telecommunication services	PF 6528 — story ; Unconvincing
PD 2731 — sheep ; Post dipping	PE 5836 — countries ; Unreliable transit services	PF 6528 — Uncontrolled
PE 5050 — swine	PD 5788 — countries ; Vulnerability	PF 6475 — Unplanned
PE 7826 Lameness ; Whirlbone	PE 7000 — countries ; Weakness infrastructure	PF 6528 — Unproductive
PA 5479 Lamentation	**Land [locked...]**	PF 6528 — vision ; Restrictive
PE 4161 Laminitis	PE 5861 — locked island countries havens ; Abuse	PE 4162 Land values ; Inflated
PE 5617 Laminitis ; Septic	PE 5848 — locked transit countries ; Inadequate rail transport developing	PD 8888 Land ; Waterlogging
PE 9702 Lampas		PA 6438 Land zoning ; Uncertainty
PE 9702 Lampers	PC 7597 — Loss agricultural	PE 1256 Landfill sites ; Methane gas emissions
PE 9374 Lamziekte	PC 8862 — Loss peasant title	PC 8862 Landholdings ; Dispossession peasant
PE 6461 Lancet fluke	PF 5008 — low income disadvantaged groups ; Limited availability	PC 8862 Landless peasants
Land [acquisition...]		PC 8862 Landlessness
PF 8053 — acquisition ; Difficult	**Land [maintenance...]**	PC 0990 Landlessness developing countries
PE 0896 — acquisition foreigners ; Alienation	PF 2346 — maintenance ; Unmanageable	**Landlords [Unapproachable...]**
PD 8625 — Agricultural mismanagement pasture arable	PD 9189 — Maldistribution agricultural	PD 2338 — Unapproachable absentee
PE 5024 — agricultural purposes developing countries ; Unavailability	PE 1925 — markets ; Uncontrolled	PD 2338 — Uninvolved absentee
	PF 7435 — mass ; Decreasing	PD 2338 — urban property ; Absentee
PD 9534 — agriculture ; Excessive use	PF 2613 — mines ; Uncleared	PF 1776 Landowner records ; Inaccessible
PF 6460 — allocation ; Ownership basis	PC 8160 — Minimal prime	PD 6960 Lands ; Disruption ecosystems marginal agricultural
PF 2451 — assessments ; Erratic	PD 8142 — misuse	PD 4273 Lands ; Forced relocation peasants onto marginal
PC 3438 — assets ; Unequal distribution	PC 7597 Land ; Need agricultural	PF 9360 Lands ; Inaccessible park
PD 0406 — associated large traditional estates ; Maldistribution	PF 3526 Land ordinances ; Restrictive	PC 3304 Lands ; Invasion indigenous
PF 6152 — automobiles ; Excessive use	PE 4162 Land ; Overpriced valuable	PC 3304 Lands resources ; Denial indigenous rights
PF 4181 Land ; Back	**Land ownership**	PF 6528 Lands ; Unreserved grazing
PE 9095 Land burial dead ; Exploitation	PB 0750 — [Land ownership]	PD 4900 Lands ; Unsustainable development forest
Land [cash...]	PD 2338 — Absentee	PC 2122 Landscape areas ; Untended
PE 8324 — cash crops ; Degradation agricultural	PF 6475 — Diminishing	**Landscape disfigurement**
PE 9095 — cemeteries ; Shortage	PF 3005 — Monopolized	PC 2122 — [Landscape disfigurement]
PF 6475 — choices ; Restricted	PD 2338 — Obscure	PC 2122 — Dependence
PC 3304 — claims ; Unsettled indigenous	PD 2338 — Outside	PD 1637 — open-cast mining
PF 4670 — Clustered vacant	PF 3005 — Static	PC 2122 Landscapes ; Degradation culturally important
PF 2009 — collateral ; Ineffective	PF 3211 — Traditional	PE 8279 Landsize limit diversity
PE 5931 — Competition industry agriculture	PF 2365 — tropical villages ; Disorganized approach	PD 1233 Landslides
PD 2338 — controlled business groups	PB 0750 — Unfair distribution	PD 1233 Landslip disasters
PE 4162 — Cost clearing	**Land [planning...]**	PF 9598 Language ; Abuse
PD 4273 — Cultivation marginal agricultural	PF 0954 — planning ; Boundary constraints	PD 5238 Language ; Abusive
PD 0813 — customary tenure systems ; Maldistribution	PF 9187 — plantations ; Loss	**Language [Bad...]**
Land [Damage...]	PF 3211 — plots ; Small	PF 7427 — Bad
PD 9118 — Damage agricultural	PE 4162 — price ; Rising	PF 6035 — barrier
PE 3786 — dams ; Inundation agricultural	PF 8053 — procedures ; Complexity	PF 6035 — barriers
PE 4585 — Declining area irrigated	PF 8053 — procurement ; Complex	PF 6035 — barriers transfer knowledge
PD 7480 — Declining productivity agricultural	PE 4162 — Prohibitive cost	PD 3223 Language conflict
PD 1052 — degradation	**Land [reclamation...]**	**Language [Deceptive...]**
PE 6005 — Degraded industrial	PF 2055 — reclamation	PF 6703 — Deceptive
PF 4670 — Deliberately unused	PF 6472 — records ; Unresearched	PE 2142 — Denial right indigenous peoples use their own
PD 9118 — Destruction agricultural	PD 9189 — reform ; Inadequate	PE 2142 — Denial right peoples use their own
PC 3486 — Destruction peat	PD 9118 — resources ; Destruction	PF 7383 — describe female experience ; Lack English
PD 0050 — developing countries ; Maldistribution	PE 3296 — Restraints legal	PE 5141 — Discrimination against use accents
PF 6475 — development ; Unplanned	PF 6528 — Restrictive use available	PD 3223 — discrimination politics
PF 6475 — development ; Unregulated	**Land rights**	PF 7299 — Discriminatory
PD 1052 — Diminution biological productivity	PF 2365 — Conflicting	PE 5545 — disorder ; Developmental expressive
PF 3211 — distribution ; Traditional	PF 2365 — Obscured	PE 9300 — disorder ; Developmental receptive
PF 1822 — divisions ; Irrational	PD 5218 — people ; Violation	PE 2265 — disorders
PD 2269 — drainage ; Ill considered	**Land [Salty...]**	PE 3886 — disorders
PD 2269 — drainage ; Inadequate	PE 1727 — Salty	PD 6029 — domination developed countries
PE 3786 — due artificial flooding ; Loss agricultural	PC 8160 — Scarcity corporate	PD 8123 — fluency ; Limited
Land [enclosure...]	PD 8075 — Scarcity residential	**Language [idiom...]**
PE 1281 — enclosure	PF 6475 — scheme ; Unorganized	PF 6570 — idiom ; Teenage
PD 9162 — Encroachment communal	PC 8160 — shortage	PB 2619 — Imprecise
PE 7099 — erosion brought site development	PC 0219 — Shortage cultivable	PF 3470 — Inability teach unwritten
PE 3402 — Excessive military usage	PD 0384 — Shortage urban	PF 0443 — instruction ; Single
PF 6528 — Exploitation marginal grazing	PE 9407 — Shortening fallow periods agricultural	**Language [Lack...]**
Land [fertility...]	PC 8160 — space ; Lack	PF 6027 — Lack international
PC 1300 — fertility ; Destruction	PF 1822 — subdivision ; Random	PD 6804 — Lack international
PD 5422 — fraud	PD 5156 — subsidence	PF 3470 — Localization unwritten
PD 6103 — fuel production ; Use agricultural	PE 5931 — surface devoted urbanization ; Increasing proportion	PF 9598 Language ; Misuse
PF 6477 — funds ; Unavailability	PE 8633 Land tenants ; Inadequate working conditions	PF 6703 Language ; Obscurantist diplomatic
PC 1005 Land government ; Compulsory acquisition	**Land tenure**	**Language [parliament...]**
PD 3055 Land ; Government expropriation	PD 0406 — Feudalistic	PF 4550 — parliament ; Abusive
Land [High...]	PD 9162 — Insecure	PF 0836 — patterns ; We / they
PE 4162 — High cost	PD 9162 — Short term	PF 5958 — policy ; Antisocial attitudes planning second
PD 2338 — holders ; Absence	PE 6005 — Land ; Unallocated industrial	PF 1373 — proficiency ; Inadequate national
PF 1705 — holding ; Individualistic	PC 8160 Land ; Unassigned public	PF 4373 Language ; Reinforcement male dominance
PF 4670 Land ; Idle private	**Land Unavailability**	PD 4782 Language, religious ethnic divisions ; Internal
PD 0050 Land ; Impossibility redistribute	PC 7597 — agricultural	**Language [Segregation...]**
Land Inadequate	PF 3499 — building	PD 4131 — Segregation
PF 6146 — arrangement housing respect common	PC 7597 — farming	PF 6014 — Sexist
PD 5110 — empolderment peat	PC 8160 — prime	PF 8145 — Status
PF 6528 — grazing	**Land [Uncontrolled...]**	PD 2579 — systems ; Exclusive nationally oriented
Land [indigenous...]	PD 9534 — Uncontrolled growth arable	PF 5929 Language teaching ; Ethnic social discrimination foreign
PC 3304 — indigenous populations ; Expropriation	PF 4670 — Underutilization	PF 6028 Language teaching ; Limits second
PE 5931 — industrialization ; Loss	PC 8160 — Undesignated public	PF 3470 Language ; Unwritten
PD 8585 — Inferior	PF 4670 — Undesignated vacant	PF 7427 Language ; Vulgar
PD 8585 — Infertile	PF 4670 — Unplanned vacant	**Languages [deaf...]**
PD 8585 — infertility	PF 4670 — Unused	PF 2833 — deaf ; Multiplicity manual sign
PE 4162 Land levelling costs	PF 2346 — upkeep ; Insufficient	PD 5078 — Discrimination against minority
Land Limited	PE 6171 — urban environments ; Insufficient common	PB 8613 — Dying
PC 8160 — available	PE 5931 — urbanization ; Loss agricultural	PB 8613 Languages ; Endangered
PD 1285 — family	**Land usage**	PD 0825 Languages ; Insufficient translation minority
PC 8160 — public	PF 6528 — [Land usage]	PC 0410 Languages international relations ; Multiplicity
Land locked countries	PF 6528 — Disputed	PF 0091 Languages ; Lack terminological equivalents
PD 5788 — [Land-locked countries]	PE 5059 — Excessive	PC 0178 Languages ; Multiplicity
PE 5789 — Lack adherence international transit conventions	PF 6528 — Ineffective	PF 6027 Languages ; Multiplicity official
PE 5837 — Military threats	PE 2525 — transportation systems ; Excessive	PC 1518 Languages national setting ; Multiplicity
PE 5813 — Vulnerability lakes rivers	PF 6528 — Unconsensed	PD 8800 Languages ; Prejudice against
Land locked developing	PF 6528 — Unfocused	PA 5806 Languor Inaction
PD 5788 — countries ; Dependency	**Land use**	PA 6444 Languor Quiescence
PE 5812 — countries ; Excessive costs sea access	PD 2672 — Blighted	PA 5558 Languor Weakness
PE 5896 — countries ; Excessive costs unsuitability insurance	PF 6528 — Blocked	PA 6180 Lapse Error
PE 5920 — countries ; Imbalances exports imports	PF 6528 — Complex	PA 6030 Lapse Fear
PE 5800 — countries ; Inadequate air transport service	PD 2495 — Disagreements over	PA 6088 Lapse Impairment
PE 5908 — countries ; Inadequate port storage facilities	PF 6528 — Ineffective	PA 6058 Lapse Impiety

-973-

Lapse

PA 5438 Lapse *Neglect*
PA 6338 Lapse *Regression*
PA 5619 Lapse *Relapse*
PA 5644 Lapse *Vice*
PD 1566 *Lapsed volunteer services*
PD 4691 Larceny
PD 5575 Larceny ; Aggravated
PD 5552 *Larceny ; Petty*
PD 4691 *Larceny public property*
PE 6461 *Large American liver fluke*
PD 7799 *Large animals ; Abortion*
PF 6174 Large building complexes ; Disorientation stress
PF 6194 Large buildings ; Social inadequacy
PF 6127 Large communities ; Ineffectiveness individual participation
Large [families...]
PD 7625 — families
PD 7625 — families ; Excessively
PD 5584 — farm favouritism
PE 8864 — firms ; Overemphasis governments
PF 2303 *Large industry syndrome*
Large scale
PD 1386 — contamination environment ; Accidental
PD 2570 — industrial accidents
PD 1644 — public programmes ; Cost overruns
PC 4987 — weather anomalies
PE 0958 *Large-strongyle infections horses*
PC 4879 *Large trade surpluses*
PD 0406 Large traditional estates ; Maldistribution land associated
PF 6540 Larger towns ; Transfer business small communities
PC 7459 *Largest market economies ; Structural imbalances three*
PE 3508 *Larva migrans ; Cutaneous*
PE 6278 *Larva migrans ; Visceral*
PE 5182 *Laryngeal hemiplegia*
PD 7307 *Laryngeal oedema*
PD 7307 *Laryngeal oedema*
PG 2295 *Laryngismus*
PE 2653 Laryngitis
PE 2653 *Laryngitis ; Acute*
PD 7307 *Laryngitis animals*
PE 2653 *Laryngitis ; Tuberculous*
PD 2730 *Laryngotracheitis ; Infectious*
PE 5567 *Laryngotracheitis poultry ; Infectious*
PE 9819 Larynx ; Cancer
PG 6761 *Larynx windpipe ; Swelling mucous membrane*
PD 2730 *Lassa fever*
PE 3505 *Lassitude*
PD 0158 *Late abortions*
PF 1382 *Late bill-paying patterns*
PE 5465 *Late blight potato*
PD 1533 Late television watching
PJ 7316 Lateness
PA 7006 Lateness *Untimeliness*
PE 6554 Latent business skills
PD 2227 *Latent mosaic ; Potato*
PF 9328 Latent problems
PC 2696 *Laterization soil*
PD 0406 Latifundia
PE 7803 Laundering ; Money
PE 3937 Lava flows
PA 5473 Lavishness *Insufficiency*
Law [Abuse...]
PC 5280 — Abuse
PF 2497 — Abuse international
PC 2541 — Age discrimination
PF 5332 Law-breaking nations
Law [capture...]
PE 6564 — capture foreign nationals ; Disregard international
PF 0216 — Conflicts
PD 3533 — Conflicts labour
PF 4816 Law ; Deficiencies international
Law Denial right
PD 6927 — due process
PC 1268 — equal protection
PE 4716 — recognition person
PF 3254 Law ; Desertion marriage
Law Discrimination
PC 8726 — [Law ; Discrimination]
PD 3692 — against men
PD 0162 — against women
Law [Discriminatory...]
PD 3357 — Discriminatory religious influence
PF 2497 — Disregard international
PF 9081 — Disrespect international
Law enforcement
PF 0223 — bodies ; Lack
PF 2454 — complexity
PF 8661 — failure against known offenders
PD 5515 — Forcible obstruction
PF 8781 — groups ; Defeatist self images citizens'
PD 5515 — Hindrance
PF 8421 — Inadequate international
PE 4768 — Inadequate national
PD 2918 — officials ; Corruption
PE 8881 — officials ; Excessive use force
PF 0336 — Reluctant local
PD 8697 — Subjective
PE 4768 — Unstructured
PD 4208 Law ; Evasion
Law [Ignorance...]
PA 5568 — *Ignorance*
PF 4676 — Inadequate development international criminal
PE 8280 — Inadequate national supervision adherence international
Law [Inequality...] cont'd
PC 1268 — Inequality
PE 7665 — Influencing
PC 6848 — Insufficient
PA 6977 Law ; Lack
PE 1693 Law ; Lack minimum age
Law [Martial...]
PD 2637 — Martial
PD 2637 — *Military*
PF 3286 — *Monogamous marriage*
PF 0423 *Law ; Non compliance*
PF 9081 Law ; Non recognition international
PD 4556 Law politicians ; Circumvention
Law [regulations...]
PE 4339 — regulations ; Evasion national
PF 9571 — relation international transactions ; Conflicts national
PD 3357 — Religious influence
PF 5736 — Reluctance reform
PD 6380 — Restrictions rights martial
PF 3251 Law ; Separation marriage
PF 4166 Law taboo ; Mother
PD 5727 Law transgressions ; Civil
Law [violations...]
PE 1208 — violations ; Banking
PE 5786 — Violations contract
PE 0930 — violations ; Driving
PD 5727 — Violations private
PE 7533 Lawful killing
PA 5563 Lawlessness
PA 7250 Lawlessness *Disobedience*
PA 5952 Lawlessness *Illegality*
PA 5563 Lawlessness *Lawlessness*
PA 6977 Lawlessness *Meaninglessness*
PA 6921 Lawlessness *Undueness*
PE 4346 *Lawns ; Snowmold*
Laws [adoption...]
PD 0590 — adoption ; Inadequate
PE 6699 — against racketeering ; Ineffective
PF 5421 — Ambiguous
PE 4860 — *Anti labour*
PF 2454 Laws ; Complex difficult
PE 6680 Laws concerning bills exchange promissory notes ; Conflicting
Laws Conflict
PF 0216 — [Laws ; Conflict]
PF 0216 — *international*
PF 0216 — *national*
PE 5135 Laws ; Conflicting labour
PE 3249 Laws ; Conflicting national divorce
PE 1693 Laws ; Failure enforce minimum age
PE 6679 Laws governing cheques ; Conflicting national
Laws [Inadequate...]
PC 6848 — Inadequate
PB 1997 — *Inflexible property*
PC 1366 — *Insufficient environmental*
PD 3080 — international restriction information ; Conflict
PD 1466 Laws ; Lack compliance fiscal
PD 1970 Laws ; Lack gun
PF 2645 *Laws ; National*
PE 4768 Laws ; Non application existing
PF 5435 Laws ; Obsolete
PF 8953 Laws over nationality ; Conflict
PE 4743 Laws punishments ; Denial right freedom retroactive
PF 9571 Laws relating international transactions ; Divergence different national
PF 5923 Laws sea ; Inadequate
PC 6757 Laws ; Secret
Laws [Unenforced...]
PE 4768 — Unenforced
PE 4768 — *Unenforced littering*
PC 7112 — Unjust
PD 1611 — Unobserved drinking age
Laws [Vague...]
PF 9849 — Vague
PF 9849 — Vagueness
PC 7112 — violation human decency
PE 6726 — Violation minimum wage
PE 1108 — Violations against firearms explosives
PE 6622 Lawsuits abuse
PE 5487 Lawyers ; Harassment
PD 5380 Lawyers ; Irresponsible
PE 7665 Lawyers ; Lack independence judges
PD 5380 Lawyers ; Negligence
PD 5344 Lax parental discipline
PF 6461 *Lax parental responses*
PD 8402 *Laxatives ; Strong*
PA 6180 Laxity *Error*
PA 5438 Laxity *Neglect*
PF 5736 Laxity their legislation ; Countries deriving advantages
PA 7309 Laxity *Uncertainty*
PA 5612 Laxity *Unchastity*
PA 5740 Laxness *Compulsion*
PA 5438 Laxness *Neglect*
PA 7296 Laxness *Restraint*
PA 7006 Laxness *Untimeliness*
PA 7710 Laziness *Idleness*
PA 5806 Laziness *Inaction*
PA 5438 Laziness *Neglect*
PA 7006 Laziness *Untimeliness*
PE 5650 Lead contamination drinking water
PE 5650 Lead dust
PE 1161 Lead gasoline
PE 5650 Lead health hazard
PE 5650 Lead occupational hazard
PE 1161 Lead petrol
Lead poisoning
PE 5650 — [Lead poisoning]
PE 9228 — animals
PE 9228 — wildlife
PE 1161 Lead pollutant
PE 0824 *Lead trade ; Instability*
PC 0237 *Leader distant relations ; Village*
Leaders [admit...]
PF 5296 — admit mistakes ; Refusal government
PF 9843 — Antipathic personal relations world
PD 7630 — Arrest trade union
PE 5944 — Assassination corporation
PE 0252 — Assassination trade union
PE 7937 — associated sex scandals ; Government
Leaders Corruption
PE 9140 — amongst relatives government
PE 9140 — children government
PC 7587 — government
PE 9140 — wives government
PF 9843 Leaders countries ; Negative personal relationships
Leaders [Death...]
PE 4869 — Death threats against trade union
PE 0241 — Denial right nations select their own
PE 0742 — Drug abuse community
Leaders [employers...]
PE 8367 — employers government ; Collusion trade union
PD 8227 — Ethical misconduct
PD 9653 — Excessive accumulation wealth government
PD 9653 — Excessive salaries government
PE 5882 Leaders ; Forced disappearances trade union
PF 9288 Leaders human rights abuses ; Connivance religious
Leaders [Ill...]
PD 7630 — Ill treatment trade union
PF 1332 — *Inadequate trained*
PF 3962 — Incompetence government
PF 6556 — Inexperienced local
PF 6556 — Insufficient neighbourhood
PF 1332 — *Insufficient trained*
PE 5944 Leaders ; Killing corporate
PF 1254 Leaders ; Lack vocal
PF 1097 Leaders ; Loss confidence government
Leaders [Mediocrity...]
PF 3962 — Mediocrity government
PD 4982 — Misrepresentation facts corporate
PD 1893 — Misrepresentation facts national
Leaders [Parochial...]
PF 2627 — Parochial images society
PF 6765 — Passive historical perspective
PD 9653 — Personal greed government
PD 2710 — Psychopathic national
Leaders [scandal...]
PD 4651 — scandal ; Implication government
PD 4749 — Self absorption political
PD 2710 — Self aggrandizing national
PD 7471 Leaders ; Threats against trade union
Leaders [Unethical...]
PD 8227 — Unethical behaviour
PC 0237 — *Unrepresentative formal*
PC 0237 — *Unrepresentative village*
PF 5409 — Untrained recreation
Leadership [Abusive...]
PD 2710 — Abusive national
PE 7654 — advisors ; Self satisfied style informal
PF 1705 — *age gap*
PD 2710 — Arbitrary
Leadership [coordination...]
PF 7011 — *coordination ; Minimal*
PE 6947 — Corruption intergovernmental organization
PF 9016 — credibility ; Loss
PF 9385 — Criticism national
Leadership [decision...]
PF 6556 — decision making structures ; Lack local
PF 8995 — Dependence
PF 2627 — dependent local biases
Leadership development
PF 2208 — Inappropriate
PF 2208 — Lack motivation
PF 3605 — Limited methods
PF 1104 Leadership ; Elitist
PC 1820 Leadership entrepreneurial ability developing countries ; Shortage industrial
PD 7471 Leadership ; Harassment workers' organization
Leadership [Ill...]
PF 6556 — Ill equipped village
PE 8177 — Illiteracy impediment
PF 6479 — image ; Absence local
PF 0317 — Immediate appeasement responses
PF 8387 — impaired illness
PF 8387 — impaired mental illness
PF 7988 — initiative ; Lack
PD 9362 — Insecurity
Leadership Lack
PF 1254 — [Leadership ; Lack]
PF 1254 — core
PF 1254 — formal
PF 6556 — village
PF 3605 Leadership ; Limited development
Leadership Loss international
PF 8353 — [Leadership ; Loss international]
PF 8081 — market
PF 7455 — political
PF 9345 — technological
Leadership [Malpractice...]
PE 6947 — Malpractice intergovernmental organization
PF 1102 — Messianic image
PE 6947 — Mismanagement intergovernmental organization

Legitimizing illegal

Leadership [Moral...] cont'd
- PD 8227 — Moral imperfections
- PF 7283 — Myopic advice
- PF 7988 Leadership ; Narrow visioning

Leadership [Obscure...]
- PF 6556 — Obscure community
- PF 7011 — Obstacles
- PF 2797 — *Overburden youth*

Leadership [Personality...]
- PF 8387 — Personality disorders
- PF 2627 — posture ; Parochial
- PF 2797 — potential ; Unexplored avenues
- PF 6479 — potential ; Untapped
- PF 1626 — *Pseudo*

Leadership [recruitment...]
- PF 3605 — recruitment ; Ineffective
- PF 3605 — Reduced training
- PF 2811 — Relative authority
- PF 6479 — role models ; Lack local
- PD 0579 — role world affairs ; Government loss

Leadership skills
- PF 2208 — Impractical
- PF 2208 — paucity
- PF 3605 — Untransferred
- PF 2797 Leadership ; Stifled emerging
- PF 2870 Leadership symbolic wealth

Leadership [Technocratic...]
- PF 6330 — Technocratic
- PF 3605 — training ; Insufficient
- PF 2451 — training ; Lack local services community

Leadership [Undefined...]
- PF 6556 — Undefined community
- PD 2710 — Undemocratic
- PF 6479 — Underdeveloped community

Leadership Undeveloped
- PF 6556 — local
- PF 1196 — potential informal
- PF 2797 — *youth*

Leadership [Unencouraged...]
- PF 6479 — Unencouraged local
- PF 6462 — Unprepared adult
- PF 6462 — Untrained elders'
- PF 7011 *Leadership viewpoints ; Diverse*
- PF 1254 Leadership void ; Crippling
- PF 5000 Leading industrialized countries ; Divergences macroeconomic policies
- PE 4586 *Leaf blisters*

Leaf curl
- PD 2227 — *cotton*
- PE 4586 — *diseases*
- PE 4586 — *Peach*
- PE 2254 *Leaf miner ; Beet*
- PE 3604 *Leaf nosed bats ; Endangered species*

Leaf [scorch...]
- PE 4586 — *scorch*
- PE 4586 — *spot elm ; Black*
- PE 1954 — *spots*
- PD 5183 Leak military data
- PE 7417 Leakage packaging
- PD 9196 Leakage water supply systems
- PD 5183 Leaking official secrets
- PD 9196 Leaks
- PE 6990 Learned helplessness
- PF 7050 Learners ; Mis communications societal
- PF 7003 Learning ; Collective attention span limit societal

Learning [disabilities...]
- PD 3865 — disabilities
- PD 8723 — disabled college students
- PF 7122 — Disconnectedness societal
- PD 3865 — disorders
- PE 8442 — due hunger ; Obstacles
- PJ 8933 *Learning expectations ; Low*

Learning [Inadequate...]
- PF 5039 — Inadequate inquiry based
- PD 7086 — Inadequate use visual imagery societal
- PF 8191 — Inapplicability formal
- PF 3498 — *Inappropriate role*
- PJ 8933 — *incentive ; Curbed*
- PF 7038 Learning ; Limit collective comprehension span societal
- PF 7074 Learning ; Limits societal

Learning [Obstacles...]
- PF 7014 — Obstacles conceptual integration societal
- PF 3184 — opportunities ; Insufficient
- PF 3184 — opportunities ; Limited availability
- PF 7098 Learning ; Quantitative limit societal
- PF 2721 *Learning ; Rote*

Learning [Societal...]
- PD 7051 — Societal over commitment
- PF 3184 — structures ; Limited availability
- PF 7110 — Superficiality collective comprehension societal
- PF 7713 — systems ; Inadequacy prevailing
- PE 2451 *Lease tenure ; Insecure*

Least developed countries
- PD 8201 — [Least developed countries]
- PE 7734 — Deterioration quality life
- PE 8401 — Environmental destruction
- PE 7954 — Food insecurity
- PE 8116 — foreign aid ; Dependence
- PE 8049 — Geographic barriers
- PE 8978 — Illiteracy
- PE 8082 — Inadequate agriculture
- PF 4297 — Inadequate development communication services
- PE 7954 — Inadequate food supplies
- PE 9242 — Inadequate health care
- PE 9764 — Inadequate human resources development
- PE 0289 — Inadequate infrastructure services

Least developed countries cont'd
- PE 7892 — Inadequate investment transnational corporations
- PE 0265 — Inadequate nutrition education
- PE 7734 — Inadequate standard living
- PE 8244 — Insufficient forestry
- PE 0273 — Insufficient use natural resources
- PE 9764 — Lack educated manpower
- PE 8306 — Minimal exports
- PE 8282 — Minimal manufacturing
- PE 0299 — Natural disasters
- PE 9764 — Shortage managerial skills
- PE 8595 — Underutilization livestock
- PE 9476 — Unemployment
- PE 8320 Leather footwear industry ; Instability
- PE 8447 Leather leather manufactures environmental hazards
- PE 8320 Leather leather manufactures trade instability

Leather manufactures
- PE 9130 — dressed fur skins ; Shortage leather, miscellaneous
- PE 8447 — environmental hazards ; Leather
- PE 8320 — trade instability ; Leather
- PE 9130 Leather, miscellaneous leather manufactures dressed fur skins ; Shortage
- PD 3557 *Leave absence*
- PE 3463 Leave any country ; Denial right
- PE 7119 Leave ; Denial right job protection during maternity
- PE 3951 Leave ; Denial right maternity
- PD 6733 *Leave employees ; Disruptive maternity*
- PE 0015 Leavers developing countries ; Unemployment premature school
- PD 8723 Leavers ; Mal educated school
- PD 8723 Leavers ; Uneducated school
- PE 8747 Leaving country sojourn ; Restrictions foreigners
- PC 1716 Leaving ; Premature school
- PE 8414 Leaving their own country ; Restrictions nationals
- PE 3660 Leeches pests
- PF 1215 Left-handedness
- PE 9364 *Left-side abomasal displacement*
- PE 1601 Leg ; Barbados
- PE 7808 Leg ; Limber
- PE 3639 Leg mite ; Scaly
- PE 9774 Leg oedema syndrome
- PD 9667 *Leg ulceration*
- PF 7945 Legal affairs ; Restriction press coverage
- PF 3073 Legal affairs ; Restrictions news coverage

Legal aid
- PF 8869 — Denial right
- PF 8869 — facilities ; Lack
- PF 8869 — Inadequate
- PF 0995 — Soaring cost
- PD 9316 Legal bias against victims crime

Legal [conscription...]
- PD 6667 — conscription ; Extra
- PE 5397 — contract ; Individualistically reduced
- PE 5397 — contract system reduced individual needs
- PF 0995 — costs ; Extortionate

Legal counsel
- PF 3517 — Denial right
- PF 1219 — minorities ; Inadequate
- PF 0732 — political dissidents ; Inadequate
- PF 9756 Legal cover-up
- PF 3252 Legal custody children ; Conflict concerning

Legal [defence...]
- PE 4628 — defence ; Denial right
- PE 4863 — defendants ; Unfit
- PD 1501 — definition family developing countries ; Narrow
- PD 4519 — definition health risks ; Inappropriate
- PF 1487 — delay ; Inordinate
- PE 1093 — delays ; Excessive burden poor due
- PD 9316 — discrimination favour offenders
- PA 0298 — *disputes ; International*
- PF 7137 — documents ; Unnecessary verbosity

Legal [engineering...]
- PF 4836 — engineering ; Lack
- PE 8597 — entities developing countries ; Absence
- PD 5380 — errors
- PE 6366 — executions ; Extra
- PF 0995 — expenses ; Risk incurring
- PF 0995 — expenses ; Outrageous
- PF 0995 — fees ; Prohibitive
- PE 4997 — government initiatives ; Extra
- PE 0621 Legal havens

Legal impediments
- PD 3063 — foreign investment
- PF 7277 — international investigations
- PF 3346 — marriage

Legal [impressment...]
- PE 5986 — impressment children armed forces ; Extra
- PF 5356 — inconsistency
- PD 9316 — indifference victims crime
- PD 0159 — induced abortion ; Variable criteria
- PF 7277 — *information ; Unavailability*
- PF 4836 — infrastructure ; Underdevelopment
- PF 1316 — *issues ; Unresolved*
- PD 0986 Legal judgements ; Elitist
- PF 5356 Legal judgements ; Instability
- PE 3296 *Legal land ; Restraints*
- PD 5380 Legal malpractice

Legal [obligations...]
- PD 4556 — obligations politicians ; Avoidance
- PF 9756 — obstruction
- PF 7137 — opinions ; Lengthy

Legal [practices...]
- PD 8614 — practices ; Restrictive
- PF 5295 — precedent undermines accountability ; Unchanging
- PF 8065 — prejudice
- PF 9756 — prevarication

Legal [procedure...] cont'd
- PF 8616 — procedure ; Inadequacy international
- PF 8519 — procedures ; Complex
- PF 8519 — procedures ; Complicated
- PF 1798 — proceedings ; Discrimination against foreigners
- PE 1743 — proceedings ; Prohibitive cost linguistic interpretation
- PE 4768 — process ; Failure service
- PD 5380 — profession ; Corruption
- PE 0803 — profession, judiciary jury membership ; Economic barriers access
- PE 0405 — profession's monopoly court proceedings
- PD 5380 — profession ; Unethical practices
- PE 8240 — protection extortion victims ; Lack
- PE 8643 — protection their rights ; Denial animals

Legal provision
- PE 0069 — international nongovernmental organizations ; Lack
- PF 7034 — international nongovernmental organizations ; Lack national
- PF 7058 — nongovernmental organizations ; Lack international
- PF 6042 Legal provisions ; Time lag

Legal [recourse...]
- PD 7609 — recourse ; Lack
- PF 4886 — relations socialist countries ; Obstacles
- PF 3517 — representation ; Denial
- PF 3517 — representation ; Lack

Legal rights
- PF 3710 — natural objects ; Inadequate
- PF 3710 — trees ; Inadequate
- PF 3464 — Underutilization
- PF 3464 — Unfamiliar
- PD 3520 Legal segregation

Legal services
- PE 9001 — developing countries ; Remoteness
- PE 2317 — indigenous populations ; Denial right
- PD 8614 — Protectionism

Legal strike action
- PE 7620 — Dismissal trade union representatives following
- PE 7620 — Dismissal workers following
- PE 7620 — Dismissal workers prevent
- PE 0365 Legal structures developing countries ; Inadequate local authority
- PF 8065 Legal structures ; Restrictive

Legal systems
- PF 8065 — Biased
- PE 6529 — Culturally biased
- PF 4851 — Deficiencies national local
- PF 4816 — Inadequate intergovernmental
- PF 2580 — Institutionalization discriminatory outmoded concepts
- PF 2580 — Outmoded
- PF 2580 — Persistence outmoded concepts
- PF 8480 Legalism
- PJ 3509 Legalization ; Excessive
- PC 8949 Legalized discrimination
- PC 3683 Legalized racial discrimination
- PJ 5595 Legalizing medications ; Excessive restrictions
- PF 5188 Legally prescribed chemical residue levels ; Dangerous
- PF 7728 Legally rehabilitate victims miscarriage justice ; Failure
- PE 5173 Legg-Perthes disease
- PE 6783 Legionellosis
- PE 6783 Legionnaire's disease
- PE 6735 Legislation ; Abuse brothel

Legislation against
- PF 6787 — drugs ; Diffuse
- PE 7141 — environmental pollution developing countries ; Inadequate
- PF 9299 — environmental pollution ; Inadequate
- PE 6699 — organized crime ; Ineffective
- PE 5794 Legislation animal welfare ; Inadequate
- PE 5794 Legislation animal welfare ; Lack
- PE 4003 Legislation collective agreements ; Non implementation workers wage increases provided
- PF 5736 Legislation ; Countries deriving advantages laxity their
- PC 1613 Legislation ; Delays elaboration remedial
- PE 1295 Legislation ; Discrimination against juveniles judicial proceedings due protective
- PE 4339 Legislation ; Exploitation regulatory loopholes countries underdeveloped
- PE 8011 Legislation ; Extraterritorial application restrictive business practices

Legislation [Inadequacy...]
- PF 0228 — Inadequacy international
- PF 6787 — Inadequate narcotics
- PC 6848 — *Inadequate safety*
- PC 9513 — Ineffective

Legislation Lack
- PE 5794 — animal welfare
- PF 7972 — anti discrimination
- PF 8876 — parliamentary time approve needed
- PC 6848 — *protective*
- PF 2499 Legislation ; Limited local respect regional global
- PF 2499 *Legislation ; Non acceptance government*
- PF 5435 Legislation ; Obsolete
- PC 3201 Legislation ; Political barriers effective
- PD 5315 Legislation ; Proliferation

Legislation [relating...]
- PF 1645 — relating action against problems ; Inadequate
- PC 0268 — restricting noise levels ; Lack
- PD 9012 — Restrictive
- PC 7112 Legislation ; Unjust
- PE 8467 Legislative control advertising ; Lack
- PC 7112 Legislative injustice
- PF 5736 Legislative reform ; Deliberate governmental avoidance
- PF 8876 Legislative reform ; Failure process low priority
- PE 5222 Legitimate organizations ; Infringement functioning
- PD 9557 Legitimate protests ; Governmental disregard
- PE 7803 Legitimizing illegal fund transfers

Legs

PE 7808 *Legs ; Spraddled*
PD 1434 *Legumes ; Instability trade*
PD 2226 *Legumes ; Root nodules*
PE 8259 Leguminous vegetables ; Instability trade beans, dried, peas, lentils
PE 2281 Leishmaniasis
PE 2281 *Leishmaniasis ; American*
PE 3676 *Leishmaniasis animals ; Visceral*
PE 2281 *Leishmaniasis skin*
PE 2281 *Leishmaniasis ; Visceral*
PE 6256 Leisure activities ; Degradation mountain environment
PE 5044 Leisure engagement ; Insignificant
PD 4406 *Leisure ; Escapist*
PF 0202 Leisure facilities ; Inadequate
PF 0202 Leisure facilities ; Underdeveloped
PF 9062 Leisure rest ; Denial right
Leisure time
PF 5044 — arising automation ; Excess
PF 5044 — Inability make use
PF 9062 — Limited
PF 9062 — Restricted
PF 5044 — Uninventive
PF 5044 — Unpreparedness surplus
PE 8907 — women ; Insufficient
PE 1570 *Lemurs ; Endangered species*
PC 1326 *Lemurs ; Endangered species Flying*
PE 5338 Lenders beneficiaries net outflow funds developing countries ; Aid
PF 4580 Lending banks ; Irresponsible
PF 4580 Lending banks ; Over enthusiastic
Lending developing countries
PE 4655 — Decline commercial bank
PE 3812 — Erratic flows concessional non concessional
PF 4345 — Excessive anxiety
PE 4310 Lending transnational banks developing countries ; Discrimination
PE 4887 Length pre trial internment ; Excessive
PE 4602 Length prison sentences ; Excessive
PF 8989 Lengthy decision process
PF 7137 Lengthy legal opinions
PE 3916 Lengthy trials
PF 4723 Leniency sentencing offenders ; Excessive
PF 6539 *Lenient judicial system*
PF 2189 *Leninism*
PE 8259 Lentils leguminous vegetables ; Instability trade beans, dried, peas,
PE 3649 Lepidoptera insect pests
PE 0721 Leprosy
PE 2280 *Leptomeningitis*
PE 2357 *Leptospirosis*
PE 2357 *Leptospirosis cattle*
PE 2357 *Leptospirosis dogs*
PF 2148 *Lesbian incest*
PF 2640 *Lesbianism*
PE 5741 Lesbians ; Discrimination against
PF 4341 *Lese majesty*
PE 7826 *Lesion ; Osseous cyst like*
PF 7154 Lesions due torture ; Skin
PE 5050 *Lesions ; Foot*
PF 7154 Lesions malfunctions ; Bodily ailments,
PE 0652 Lesions ; Thyroid glands inflammatory
PF 1870 *Lessening regard sanctity marriage*
PE 6461 *Lesser liver fluke*
PF 0443 Lessons ; Disrelated school
PF 1746 Lessons history ; Denial
PE 8137 Let because applicant's race ; Refusal
PE 4675 Lethal fumes modern house fires
PA 2360 *Lethargy*
PA 5806 *Lethargy Inaction*
PA 7364 *Lethargy Unfeelingness*
PA 7371 *Lethargy Unintelligence*
PE 6754 *Letter bombs*
PE 4284 *Leucocytozoonosis birds*
PG 9621 *Leucopenia*
PE 0639 *Leukaemia*
PE 1243 *Leukaemia ; Bovine*
PD 5453 *Leukaemia dogs ; Lymphocytic*
PD 5453 *Leukaemia ; Feline*
Leukaemia [Lymphatic...]
PE 2686 — Lymphatic
PE 2686 — Lymphocytic
PE 2686 — Lymphoid
PE 0639 *Leukaemia ; Monocytic*
PE 0639 *Leukaemia ; Myelogenous*
PG 9621 *Leukopenia*
PD 8775 *Leukorrhea*
PD 8775 *Leukorrhoea*
PE 1243 *Leukosis ; Bovine*
PD 2730 *Leukosis chickens ; Lymphoid*
PD 4403 Level ; Lowering water
PF 7435 Level ; Rising sea
PD 8888 Level ; Rising water
PD 8888 Level underground water ; Rising
PE 4162 Levelling costs ; Land
PD 8888 Levels ; High groundwater
PF 2303 *Leverage ; Ineffective community*
PF 5472 Leverage international bodies ; Developing country failure exercise
PF 1389 *Leverage ; Lack political*
PE 4963 Leveraged buy-outs
PD 1915 Levirate marriages
PA 6822 *Levity Disrespect*
PA 6247 *Levity Inattention*
PA 7325 *Levity Irresolution*
PA 5942 *Levity Unimportance*
PE 7103 Lewisite

PE 6746 Liabilities ; Foreign money
PE 9061 Liabilities ; Misrepresentation fiscal
PF 7293 Liability associations ; Limited criminal
Liability [corporations...]
PF 7293 — corporations ; Limited criminal
PE 6323 — crimes committed disregard orders
PB 0001 — Criminal
PA 6971 Liability *Danger*
PF 6158 Liability identical criminal act ; Multiple
PA 7395 Liability *Inexpedience*
PF 6158 Liability ; Overlapping criminal
PA 6434 Liability *Poverty*
PF 4404 Liability protection ; Prohibitive cost product
PE 9061 Liability ; Understatement tax
PE 6323 Liability ; Vicarious criminal
PF 2947 Liaison formal support ; Disorganized
PF 1187 Liaison ; Inadequate parent school
PF 0963 Liaison ; Ineffective official
PD 3022 Libel
PD 3022 *Libel ; Intentional*
PD 3022 *Libel ; Unintentional*
PD 3022 Libelling ; Risk
PF 0717 Liberalism
PF 9710 Liberalism ; Extremist
PF 9710 Liberalism ; Fascist
PF 9710 Liberalism ; Militant
PD 7352 Liberalization ; Bourgeois
PE 5376 Liberation AIDS ; Restricted sexual
PF 3025 Liberation ; Women's
PF 7497 Liberty ; Collapsed conceptualization
Liberty [Denial...]
PD 8445 — Denial religious
PF 0705 — Denial right
PF 4681 — disabled ; Restrictions
Liberty Lack
PD 4150 — economic
PF 3217 — intellectual
PD 8276 — political
PC 0663 — social
PE 6270 Liberty proper their species ; Denial animals right conditions life
PF 7497 Liberty ; Trivialization
PF 1418 *Libidinal regression*
PD 2886 Librarianship ; Irresponsible
PF 9046 Libraries ; Denial right access historical
PF 9046 Libraries ; Inaccessible historical
PD 3026 Library acquisitions ; Censorship
PF 2568 Library equipment ; Insufficient
PF 6479 Library facilities ; Unavailability
PF 6479 Library ; Unconsensed village
PF 6479 Library use ; Unfamiliar
PE 3988 LIC
PE 1439 Lice insect pests
PF 3079 Licence media ; Refusal
PF 7141 Licence pollute developing countries
PF 3079 *Licence ; Refusal broadcasting*
PF 8964 Licence renewal ; Refusal
PF 2687 Licence ; Sexual
PD 3028 *Licence theatre performances ; Refusal*
PF 3079 Licences media ; Refusal grant
PF 3079 *Licences ; Refusal grant*
PF 8964 Licences ; Revoking
PE 4039 Licensing agreements ; Patent pool cross
PE 0895 Licensing arrangements ; Restrictions export activity due
PE 9027 Licensing ; Imposition import
PF 7655 Licensing process ; Overwhelming
PF 7655 Licensing restrictions ; Unreasonable
PA 6466 Licentiousness *Intemperance*
PA 5563 Licentiousness *Lawlessness*
PA 7296 Licentiousness *Restraint*
PA 5612 Licentiousness *Unchastity*
PA 6921 Licentiousness *Undueness*
PE 4586 Lichen plants
PB 0284 *Lie detector tests ; Enforced*
PE 3909 Lie ; Educating people
PE 3909 Lie ; Teaching children
PE 4432 Lies ; Black
PD 4879 Lies ; Business
PF 3098 Lies ; Grey
PF 7635 Lies ; Paternalistic
PF 9583 Lies ; Political
PF 3495 Lies ; Religious
PF 7631 Lies ; White
Life [Absence...]
PF 3531 — Absence traditional patterns community
PF 4058 — Arbitrary external interference family
PF 0955 — aspirations ; Disillusionment
Life assurance
PE 1826 — practices ; Unethical
PE 8736 — premiums ; High
PE 8736 — Prohibitive cost
PF 2423 *Life ; Attraction city*
PD 1026 Life ; Breakdown covenants
PA 6586 Life∗complex
Life [Compulsory...]
PF 4936 — Compulsory prolongation
PD 3003 — Corporate domination daily
PD 1283 — Cosmically inharmonious way
PF 1081 — cosmos ; Absence convincing symbols connecting individual's
PD 4840 — crises ; Personal
PD 5675 — crisis ; Female mid
PD 5783 — crisis ; Male mid
PF 6148 — cycle ; Inhibition individual psychological development

Life [cycles...] cont'd
PC 2008 — cycles ; Manipulated product
PF 1740 — cycles ; Short term planning product
Life [Decline...]
PE 6705 — Decline street
PF 2575 — Declining community
PD 9480 — Deformation plant
PF 2644 — Delimiting family patterns traditional way
PF 8243 — Denial animals right
Life Denial right
PD 4234 — [Life ; Denial right]
PE 6561 — cultural
PE 6168 — private home
Life [Deteriorating...]
PF 7142 — Deteriorating quality
PF 7047 — developing countries ; Neglect agricultural rural
PF 6592 — developing countries ; Westernization traditional modes
Life Discrimination
PF 3354 — access eternal
PD 3335 — against women public
PE 6922 — harassment children public
PF 2813 Life ; Disorganized community
PD 5157 Life due radioactive contamination ; Endangered animal plant
Life [education...]
PF 5079 — education ; Lack family
PF 3354 — Elitist access eternal
PD 0437 — endangering behaviour ; Criminally
PE 6126 — entertainment facilities ; Inadequate night
PF 1191 — evaluation ; Discrepancies human
PD 4840 — events ; Negative
PD 0784 — Exclusion disabled persons social cultural
Life expectancy
PE 1339 — gender ; Inequality
PE 1339 — men ; Shorter
PE 1339 — Social inequalities
PC 4312 — Unequal
PD 4840 Life experiences ; Stressful
Life forms
PB 9748 — Decreasing variety
PD 7840 — Irresponsible patenting
PE 8806 — Lack variety social
PD 7840 — Monopolistic control new
PF 2813 Life ; Fragmented planning community
PD 1703 Life ; Human wisdom unrelated daily
Life [Inappropriate...]
PF 4936 — Inappropriate prolongation
PF 2304 — Increasing pace
PF 6337 — *information ; Insufficient plant*
PE 8736 — insurance developing countries ; Inadequate
PB 4993 — International imbalance quality
PD 9170 — International imbalance quality working
PF 6359 Life ; Killing non human
Life [least...]
PE 7734 — least developed countries ; Deterioration quality
PE 6270 — liberty proper their species ; Denial animals right conditions
PC 4321 — Loss reverence
PA 6220 Life ; Meaninglessness
PF 7094 Life ; Obstacles family
PF 9631 Life politics ; Media theatricalization public
PB 6336 *Life ; Pollution hazards wild*
PF 5694 Life ; Religious theology disrelated
PF 3129 Life ; Restrictive patterns traditional
PF 0916 Life's pain ; Escaping reality sophisticated shields against
PF 6167 Life small households ; Inhibition adult
PF 0979 Life stance ; Defensive
Life style
PF 1437 — City
PD 4368 — Dislocated
PD 4037 — Extravagant
PF 7142 — *Individualistic*
PE 8806 — Lack variety
PE 8644 — Perpetual preoccupation sustenance urban
PF 7142 — Restricted
PF 8145 — Status
PE 8144 — Struggle financial security urban
PF 1078 — Subsistence
Life styles
PE 8477 — Apathy toward improvement urban
PD 4069 — Escapist family
PF 3211 — Inhibiting effects traditional
PE 8296 — Minimal family interest urban
PE 1137 — mining communities ; Traumatic shift
PF 1078 — Reduced family
PF 1729 — Reduced social commitment contemporary
PF 4936 Life support systems ; Inhumane use medical
Life [Unattractive...]
PF 3454 — *Unattractive social*
PD 8113 — Unbalanced social
PF 3358 — Undue religious influence secular
PF 2842 — Unfulfilled aspirations economic
PE 4058 — Unlawful interference family
PC 0228 — Unproductive use trees beyond their commercial
PD 7929 — Unstable family
PD 4368 — Unstructured home
PD 4840 — Upheavals private
PF 6098 Life work ; Reduced images having significant
PF 6537 Life work ; Reduced images having significant
PA 7365 Lifelessness *Boredom*
PA 6261 Lifelessness *Darkness*
PA 7055 Lifelessness *Death*
PA 5806 Lifelessness *Inaction*
PA 6444 Lifelessness *Quiescence*

-976-

Limited social

PF 5759 Lifelong human development ; Ignorance
PF 4618 Lifestyle diseases
PF 6570 *Lifestyle ; Fear controlled*
Lifestyle [Idle...]
PF 2874 — *Idle youth*
PF 2644 — Inflexible patterns family
PF 2241 — *Isolating community*
PF 2256 *Lifestyle routines ; Habitual*
PF 2241 *Lifestyle tensions*
PE 8077 Lifestyles nomads pastoralists ; Endangered
PE 4452 Lift heavy loads ; Work practices requiring women
PD 2283 Ligaments ; Disorders joints
PD 7424 *Ligaments ; Rupture cruciate*
PF 6198 Light buildings ; Displacement natural
PE 7244 Light pollution
PF 6473 *Lighted roadways ; Poorly*
PF 1741 Lighteners ; Abusive use skin
Lighting [factory...]
PD 2760 — factory farming units ; Inadequate
PE 8272 — fixtures fittings ; Instability trade sanitary plumbing, heating,
PE 7940 — fixtures fittings ; Shortage sanitary plumbing, heating
PF 6473 *Lighting ; Insufficient street*
PE 5780 Lighting occupational hazard ; Improper
PE 5780 Lighting ; Poor
PF 6473 *Lighting ; Poor street*
PA 5491 Lightness
PD 1292 Lightning
PD 1292 *Lightning ; Ball*
PD 1292 Lightning striking ; Risk
PD 9366 *Lightning stroke*
PE 7808 *Limb deformities foals ; Angular*
Limb Fracture
PE 7511 —
PE 7511 — *lower*
PE 7511 — *upper*
PA 0643 *Limb pain ; Phantom*
PE 8173 Limb respiratory muscles ; Paralysis throat, eye,
PE 4566 *Limb ; Transient paralysis*
PE 7808 *Limber leg*
PE 9374 *Limberneck*
PB 1044 *Limbs ; Diseases*
PE 4566 Limbs ; Symptoms referable
PE 2038 Limestone dust
PF 7038 Limit collective comprehension span societal learning
PE 8279 Limit diversity ; Landsize
PF 7003 Limit societal learning ; Collective attention span
PF 7098 Limit societal learning ; Quantitative
PF 3861 Limitation
PF 4014 Limitation current scientific knowledge
PE 6427 Limitation working hours ; Denial working animals
PF 4677 Limitations
PF 7134 Limitations comprehension international information ; Pre logical
PF 6608 Limitations democracy
PF 4668 Limitations ; Government
PF 4668 Limitations government power
PF 2622 Limitations international statistics
PF 2904 Limitations right vote
Limitations [school...]
PF 3184 — *school admission*
PF 7700 — surprise-free methodologies
PF 7700 — surprise-free thinking
PE 7300 Limited acceptance human rights treaties
PF 0977 Limited acceptance international treaties
Limited access
PF 2875 — available services
PE 8516 — credit small farmers
PE 8152 — culturally adapted pedagogical material
PF 1653 — external resources
PF 6577 — health services
PF 1953 — knowledge
PE 4704 — loans
PF 2882 — natural resource use decisions
PF 6573 — *news*
PE 1278 — pharmaceutical drugs
PF 2840 — practical education
PF 3575 — rural youth education
PF 1303 — social benefits
PF 6573 — society's resources
PF 4636 — technological decisions
PF 2583 — urban services
Limited [accountability...]
PF 6574 — accountability public services
PF 3630 — accumulation capital
PF 8835 — agricultural education
PF 1768 — *agricultural expansion*
PF 2564 — applicability monetary grants
PF 3500 — approaches economic planning
PF 9735 — artistic expression
Limited availability
PF 2489 — credit
PF 3575 — education rural areas
PF 3630 — *farm loans*
PF 2489 — financial credit
PF 3539 — functional information
PF 3550 — funding resources urban services
PF 2378 — funds
PD 7669 — health resources
PF 3550 — investment capital urban renewal
PF 1658 — jobs small communities
PF 5008 — land low-income disadvantaged groups
PF 3184 — learning opportunities
PF 3184 — learning structures
PE 4704 — loan capital
PE 4704 — loans developing countries
Limited availability cont'd
PF 6477 — local business capital
PF 2009 — local commercial resources
PF 2865 — long-term capital
PF 2457 — modern farm machinery
PE 1134 — permanent employment inner-cities
PE 5170 — practical skills rural communities
PF 6539 — public services small towns developed countries
PF 2698 — technical agricultural business training
PF 6751 — therapeutic substances human origin
PC 8160 Limited available land
PF 3628 *Limited awareness child-care opportunities*
Limited [basic...]
PF 2477 — *basic skills*
PE 0366 — blood transfusion systems
PD 3054 — *borrowing collateral*
PF 1442 — *budget knowledge*
PD 8362 — buying power
Limited [cash...]
PF 6455 — *cash crops*
PD 1204 — *catchment facilities*
PF 2493 — *child activities*
PF 0903 — civil administrative capacity developing countries
PC 4867 — *commercial experience*
PB 8891 — commodity goods
Limited community
PD 2864 — cohesion
PF 2575 — identification
PF 1731 — responsibility adults
Limited [construction...]
PF 1790 — *construction builders*
PF 1790 — *construction manpower*
PF 2583 — *consumer knowledge*
PF 6547 — control environment local communities
PF 5250 — cooperative events
PF 9729 — cooperative experience
PF 2489 — credit accessibility
PF 2489 — credit lines
PF 7293 — criminal liability associations
PF 7293 — criminal liability corporations
Limited crop
PC 2223 — diversity
PF 6524 — management
PC 2223 — variety
PF 8105 Limited cultural attractions
PF 2504 Limited cultural context
Limited [daytime...]
PE 1848 — *daytime encounters*
PF 2394 — decision-making context
PF 4668 — degrees freedom government
PD 0305 — developing country capacity absorb aid
Limited development
PF 1332 — functional abilities
PF 3605 — leadership
PF 9478 — vision
PF 2652 Limited distribution basic services
PF 3554 *Limited drilling equipment*
Limited [economic...]
PF 0581 — *economic communication*
PD 0847 — educational facilities
PF 3575 — educational opportunities rural areas
PD 9033 — electrical capacity
Limited employment
PF 1658 — alternatives
PF 2949 — opportunities small rural communities
PF 1658 — options
Limited [enforceability...]
PF 8927 — enforceability international standards
PF 2842 — *engagement alternatives*
PF 7142 — *engagement patterns*
PF 8105 — entertainment possibilities
PF 5006 — exchange skills developing countries
PF 2296 — exposure outside influences rural villages
PF 2378 — extent capital funds
Limited family
PD 6579 — income
PD 1285 — *land*
PF 2316 — *opportunities*
Limited [farm...]
PD 4100 — farm crops
PD 1285 — *farm produce*
PC 4867 — *finance training*
PF 3550 — *financing opportunities*
PF 5496 — fish farming
PD 4100 — food grain species
PF 0479 — food variety
PC 5165 — forest resources
PF 2428 — fragmented outlook civic minded groups
PF 6530 — *funding approach*
PF 7916 — *funding expertise*
Limited [gardening...]
PF 6493 — *gardening experience*
PF 3628 — global exposure
PF 8960 — government support
Limited health
PF 4586 — funds
PD 8023 — knowledge
PD 0366 — personnel
Limited [historical...]
PD 3774 — historical method
PF 3161 — horizons produced survival living
PD 1251 — *housing materials*
PF 9292 — human information processing capacity
Limited [image...]
PF 2896 — image employability
PC 0492 — *improvement capital*
Limited [individual...] cont'd
PF 2384 — individual attention span
PF 2899 — individual capital reserves
PF 2575 — involvement neighbourhood residents
Limited job
PF 6541 — *information*
PC 7997 — market
PF 1658 — opportunity
PD 8123 Limited language fluency
PF 9062 Limited leisure time
Limited local
PF 2378 — availability capital reserve
PD 2245 — decision making bodies ; Diversity
PF 1442 — *demand*
PF 2489 — government credit
PF 2212 — *initiative*
PF 6477 — investment resources
PF 2989 — markets goods services
PF 2499 — respect regional global legislation
PF 5660 Limited long-range planning
PD 1842 Limited low-cost housing
PF 2995 *Limited maintenance know-how*
PF 6557 *Limited manufacturing incentives*
Limited market
PE 2328 — access due product differentiation transnational corporations
PF 3499 — *areas*
PF 1086 — development
PF 1086 — research
PF 1086 — resource
Limited [marketing...]
PF 1086 — marketing lines
PF 0581 — markets ; Severely
PE 7344 — means marketing employable skills
PE 8986 — mechanical services developing countries
PD 9160 — medical knowledge
PD 9160 — medical skills
PE 1535 — meeting facilities
PD 6571 — *merchandising interest*
PF 3605 — methods leadership development
Limited number
PF 0734 — available radio frequencies
PE 0135 — corporations ; Disproportionate control global economy
PE 1922 — corporations ; Disproportionate influence national economies
PD 0530 — developed countries ; Restriction outer space benefits
PF 0545 — geostationary satellite orbits
PF 2344 — individuals ; Undisclosed control national economies
Limited [off...]
PC 1108 — off-season jobs
PF 1403 — opportunities significant work
PF 3575 — opportunity ongoing education rural communities
PF 3628 — outside exposure
PD 3182 — ownership productive systems
Limited [paramedic...]
PD 9160 — *paramedic training*
PF 2810 — *parental participation*
PD 1564 — participation
PF 3409 — *pest control*
PF 2477 — *planning experience*
Limited political
PD 8673 — developments
PF 2401 — *know-how*
PE 1101 — *power*
Limited [practical...]
PF 2840 — practical education
PF 2810 — *practical engagement*
PF 0905 — *preschool equipment*
PF 6477 — private investment
PG 7844 — *programme impact*
PC 8694 — progress North-South negotiations
PE 0540 — psychiatric out-patient care
Limited public
PF 9108 — funds
PF 4586 — health subsidies
PC 8160 — *land*
PD 8422 — salaries
PF 1790 — *space*
PF 1495 — transportation
PD 8362 Limited purchasing power
PF 6572 *Limited railroad usage*
Limited range
PE 9457 — countries ; Export developing country products
PF 2477 — functional skills
PD 2968 — raw materials ; Dependence developing countries export
Limited [recognition...]
PF 7093 — recognition international nongovernmental organizations
PF 8105 — recreational activities
PF 3628 — regional exposure
PF 3192 — relations beyond local environments
PF 2848 — reservoir technical skills rural communities
PF 6471 — *retail inventory*
PF 4959 — role women countryside
Limited [scope...]
PF 2813 — scope town planning
PF 7022 — *shared time*
PF 8085 — skills training
Limited social
PF 6617 — context developing technology
PF 1664 — guidance older generation
PF 3526 — *services*

Limited specialized

Limited [specialized...]
PD 6571 — *specialized farming*
PD 1941 — *spheres relationship*
PF 3467 — *storage capacity*
PF 2575 — *strength elders*
PF 3560 — *study time*
Limited [tax...]
PF 2444 — *tax revenues*
PE 7165 — *teacher facilities*
PF 2721 — *teaching methods*
PF 8594 — *technical skills*
PS 6553 — *trade areas*
PD 1415 — *traditional markets*
PF 2605 — *training space*
Limited transportation
PF 2996 — *access*
PF 2996 — *availability*
PE 8959 — *services urban areas*
PF 0919 Limited tripartite cooperation
PF 7808 Limited utilization foodstuffs
PF 2996 *Limited vehicle availability*
PD 8123 *Limited verbal skills*
Limited village
PF 1979 — *capital*
PF 1442 — *management*
PF 9066 — *self-confidence*
Limited [war...]
PE 3988 — *war*
PD 1204 — *water access*
PC 4815 — *water supply*
PF 2792 — *ways matching talent jobs*
PF 3628 — *world exposure*
Limited youth
PD 1544 — *accountability*
PF 2575 — *activities*
PF 5949 — *engagement*
PF 2575 — *organizations*
PA 7078 Limitedness *Environment*
PA 5473 Limitedness *Insufficiency*
PF 2602 Limiting effect individual survivalism
PF 1889 Limiting responsibility personal
PF 4741 Limiting technical innovation ; Self
PD 2961 Limiting trade developing countries ; Developed country
PF 4677 Limits
PF 2529 Limits areas research
PF 6521 Limits ; Geographically undefined community
PF 2423 *Limits industry ; Size*
PF 3560 Limits participation community development
PF 1078 Limits role ; Subsistence
Limits [second...]
PF 6028 — *second language teaching*
PE 8516 — *single cropping credit*
PF 7074 — *societal learning*
PE 3829 Liner shipping cartels
PE 2488 Linguistic controversies ; Bilingualism national settings regional
PD 4131 Linguistic dis-integration
PD 4131 Linguistic discrimination
PA 1387 *Linguistic elitism*
PD 2695 Linguistic imperialism
PE 1743 Linguistic interpretation legal proceedings ; Prohibitive cost
PC 3324 Linguistic minorities ; Underprivileged
PC 8999 Linguistic, national religious minorities ; Inadequate protection
PF 1954 Linguistic purism
PC 3293 *Linguistic tradition ; Loss*
PD 2730 *Lion virus disease ; San Miguel sea*
PE 5117 Lip ; Cleft
PE 5117 Lip ; Hare
PD 9667 Lip ulceration
PE 9816 Lipid metabolism ; Congenital disorders
PE 5325 Lipidosis
PE 7808 *Lipofuscinosis*
PE 5325 Lipoidosis
PE 5126 Liquefaction ; Ground failures due
PE 8504 Liquid agricultural wastes
PF 3061 Liquid reserves ; Inadequate mechanism creation
PF 2342 *Liquidity ; Inadequate world*
PF 3058 *Liquidity ; International*
PF 3059 Liquidity shortage
PF 3060 Liquidity shortage developing countries
PJ 2664 Liquidity ; Unbanked
PB 0731 *Liquids ; Poisoning solids*
PE 3980 Liquor advertising ; Irresponsible
PF 1249 *Liquor ; Easy access*
PE 7188 *Liquor ; Home made*
PE 7188 Liquor manufacturing distribution ; Unauthorized
PE 3779 Listerellosis
PE 3779 Listeriosis
PA 6379 Listlessness *Avoidance*
PA 5806 Listlessness *Inaction*
PA 6598 Listlessness *Incuriosity*
PA 7364 Listlessness *Unfeelingness*
PA 5558 Listlessness *Weakness*
PD 8102 Literacy education ; Denial
PE 6188 Literary forgery
PD 0188 Literary piracy
PD 4773 Literature ; Bias children's
PF 3096 *Literature ; Biased*
PF 1183 Literature ; Dissemination hate
PD 2366 Literature ; Illicit
PF 1384 Literature ; Immoral
PF 1384 Literature ; Obscene
PD 4773 Literature ; Racism child youth
PD 7571 Litigation ; Delay civil

PF 1542 Litigation ; Frivolous vindictive
PF 0361 Litigation ; Proliferation
PD 2541 Litter
PD 2541 Litter ; Damaging unremoved
PD 2541 Litter ; Unremoved public
PE 4768 *Littering laws ; Unenforced*
PE 0763 *Little's disease*
PA 7285 Littleness
PA 6839 Littleness *Disrepute*
PA 5652 Littleness *Inferiority*
PA 7306 Littleness *Narrowmindedness*
PA 7211 Littleness *Selfishness*
PA 5942 Littleness *Unimportance*
PF 1115 Liturgical conflicts
PF 2821 Live animal production ; Underdevelopment food
Live animals
PD 0788 — Environmental hazards
PC 1411 — Environmental hazards food
PD 0260 — Experiments
PC 1411 — Health hazards food
Live animals Instability
PD 2894 — *production food*
PD 1376 — *trade*
PD 1434 — *trade food*
PE 0976 Live animals ; Long term shortage food
PE 0342 Live animals trade ; Restrictive practices food
PD 5253 *Live peace ; Denial right people*
PE 8746 Lived abroad ; Denial social security nationals who have
PF 9729 Lived cooperative commitment ; Short
PF 1297 Livelihood standards ; Lack
PE 1028 Liver abscess
PD 3978 *Liver abscesses ; Bovine*
Liver [cancer...]
PE 3233 — cancer
PE 2446 — Cirrhosis
PE 1028 — congestion
Liver [damage...]
PE 1028 — damage
PE 1028 — disease
PD 7420 — *disease cattle ; Fatty*
PE 1028 — diseases
PD 3978 — *diseases animals*
PE 1028 Liver fatty diseases
Liver fluke
PE 2785 — [Liver fluke]
PE 6461 — *Common*
PE 6461 — *Giant*
PE 6461 — *Giant*
PE 6461 — *Large American*
PE 6461 — *Lesser*
PE 1028 *Liver rot*
PE 1028 Liver spleen ; Enlargement
PC 2381 Lives ; Denial people control over their own
PD 0185 Livestock ; Feral
PE 9125 Livestock growth ; Use artificial methods promoting fast
PF 2378 *Livestock husbandry costs ; Prohibitive*
PF 1240 Livestock income ; Unexplored
PE 8595 Livestock least developed countries ; Underutilization
PE 0976 *Livestock ; Long term shortage*
PF 3394 Livestock ; Maltreatment
PF 1849 Livestock mutilation
PD 1802 Livestock ; Non productive use cattle
PD 2743 Livestock ; Poor quality domestic
Livestock production
PD 0376 — Environmental hazards agricultural
PE 0524 — Health risks workers agricultural
PE 1785 — Inappropriate agricultural subsidies
PE 8998 — Instability agricultural
PD 0629 — Underdevelopment agricultural
PF 4490 — Unequal distribution
PD 2772 *Livestock slaughtering meat ; Inhumane*
PE 4898 Livestock ; Trespassing
PC 2205 *Livestock waste ; Unused*
PF 3089 Living alone
Living conditions
PD 9156 — Bad
PD 4368 — Impermanent
PD 9156 — Poor
Living [Continued...]
PD 4125 — Continued subsistence
PF 7043 — creatures ; Death
PF 1078 — cycle ; Subsistence
Living [Decadent...]
PD 4037 — Decadent standard
PF 7142 — Declining standard
PF 0344 — Denial right adequate standard
PE 4052 — developing countries ; Inadequate standard
PF 1238 Living ; High cost
PF 0344 Living ; Inadequate standards
PE 5484 Living indigenous peoples ; Denial right adequate standard
PE 7734 Living least developed countries ; Inadequate standard
PF 3161 Living ; Limited horizons produced survival
PC 1617 Living organisms ; Translocation
PE 1508 Living outside camps ; Non settled refugees
PB 0388 Living poor ; Deterioration standard
PF 1238 Living ; Prohibitive cost
PF 7130 *Living space ; Marginal*
PF 7130 *Living space ; Unused*
Living standards
PE 7136 — *borrower nations ; Enforced curtailment*
PD 2898 — Inadequate demand primary commodities because rising
PF 5864 — Pursuit affluent
PE 9075 — workers ; Deterioration

PE 3278 Living together
PE 2313 Living ; Unavailable education effective
PD 3427 Living working conditions immigrant labourers industrialized countries ; Inadequate
PE 0904 Llamas ; Endangered species
PD 9108 Loading unloading ; Risk during
PE 4452 Loads ; Work practices requiring women lift heavy
Loan [capital...]
PE 4704 — capital ; Limited availability
PD 3054 — *collateral ; Unappealing*
PF 2491 — computing methods ; Confusing
PD 9437 Loan defaults ; Government
PF 3975 Loan disbursements ; Delay
PE 1398 Loan fraud
PE 4355 Loan negotiations transnational banks ; Domination
Loan [payments...]
PE 4571 — payments ; Arrears home
PF 2491 — procedure ; Complex
PF 2491 — procedures developing countries ; Restrictive
Loan [services...]
PF 2491 — services ; Unknown
PE 1181 — sharking
PA 8653 — structures ; Distrust
PD 3054 — *syndrome ; Defaulted*
PE 3812 Loans available developing countries ; Inappropriate
PE 4545 Loans ; Bad property
PE 9054 Loans because borrower's race ; Refusal
PD 3053 Loans ; Defaults international
PE 7195 Loans ; Dependence developing countries foreign
Loans developing countries
PE 4603 — Deteriorating terms financial
PE 4345 — Deterioration confidence
PE 4305 — Inadequate diversification
PE 9116 — intergovernmental facilities ; Restrictive conditions
PE 4704 — Limited availability
PE 4704 Loans ; Difficult development
PE 1398 Loans ; Fraudulent
Loans [Inaccessibility...]
PE 4704 — Inaccessibility bank
PE 3550 — *Inaccessible housing*
PE 4545 — Inadequately secured commercial property
PF 4580 — Inappropriate
PE 4704 — Loans ; Limited access
PF 3630 — *Loans ; Limited availability farm*
PE 0682 — Loans ; Misinformation concerning
PE 9054 — Loans ; Racial discrimination according financial
PE 4704 — Loans ; Restrictive access
PE 4330 Loans transnational banks developing countries ; Domination restrictive project
Loans [Uncertainty...]
PF 4295 — Uncertainty development expenditures due floating rate
PF 4300 — Uncertainty development programmes due short term
PD 9437 — Unpaid government
PA 5454 Loathsomeness *Badness*
PA 7338 Loathsomeness *Hate*
PA 7240 Loathsomeness *Ugliness*
PA 7107 Loathsomeness *Unpleasantness*
PF 2421 Lobbying ; Sporadic community
PF 5365 Lobbying ; Unconstrained
PF 5365 Lobbying ; Vulnerability government
Local [adult...]
PF 6531 — adult training ; Non
PF 6573 — *anti-intellectual bias*
PC 6154 — *authorities ; Insolvent*
PE 2261 — authorities ; Non resident
PE 0365 — authority legal structures developing countries ; Inadequate
PF 2378 — availability capital reserve ; Limited
Local [biases...]
PF 2627 — biases ; Leadership dependent
PF 6477 — business capital ; Limited availability
PF 1176 — business ; Individualistic practices
PD 5350 — *businesses ; Declining*
PF 2439 — *businesses ; Unattractive*
Local capital
PF 6477 — Inadequate
PF 3630 — Lack
PF 1979 — resources ; Dispersion
PF 6477 — Unavailability
PF 6477 — *Unused*
PS 2881 *Local cash ; Scarce*
PF 2535 Local commerce ; Unexploited possibilities
Local commercial
PF 2009 — resources ; Limited availability
PF 2009 — services ; Lack
PF 6510 — services ; Lack support
Local communities
PF 1944 — Lack relevant training opportunities
PF 6547 — Limited control environment
PE 4780 — outside services ; Excessive dependence
PE 5170 — Restricted availability urban skills
PF 6513 — Underutilization potential
Local [community...]
PD 2864 — community cohesion ; Breakdown
PF 2076 — concerns ; Overriding international
PF 1333 — conflicts ; Polarization
PF 6506 — consensus ; Ineffective structures
PF 5539 — control resources
PF 2489 — credit ; Lack
PF 3012 — culture ; Disappearance
Local [decision...]
PD 2245 — decision making bodies ; Diversity limited
PF 6550 — decision making ; Unstructured
PF 1442 — *demand ; Limited*

Local [dentists...] cont'd
PF 3448 — *dentists ; Unavailability*
PF 6574 — *directories ; Unavailability*
Local [economy...]
PF 6451 — economy ; Restricted flow
PS 8148 — *educational planning ; Non*
PF 1389 — *electrical generating capacity ; Non viability*
PF 2827 — *employment patterns ; Non*
PF 2423 — employment ; Restricted scope
PF 0336 — enforcement ; Inadequate
PF 0336 — enforcement ; Unclear
PF 3192 — environments ; Limited relations beyond
PF 2385 — expenses ; Rising
PE 7313 — expertise business practices developing countries ; Inadequate
PF 6493 Local food production ; Underdeveloped approaches
PF 6544 Local funding ; Lack channels obtaining available
Local government
PD 5948 — Corruption
PF 2489 — credit ; Limited
PE 0365 — developing countries ; Inadequate financing
PE 0365 — developing countries ; Inadequate institutional structures
PC 6631 — financing ; Inadequate
PD 5948 — Irresponsible
PD 5948 — Negligence
PD 5948 — Unethical practices
PC 6631 Local governments grants ; Excessive reliance
PF 6545 *Local groupings ; Rigid*
PF 6494 *Local history ; Untransmitted*
Local [industries...]
PD 5350 — *industries ; Declining*
PF 1933 — *industries ; Lack*
PE 1511 — industry developing countries ; Inadequate relationship transnational corporations
PF 2842 — *industry ; Uncommitted*
PF 6552 — information ; Inadequate circulation
PF 6541 — information systems ; Lack
PF 2115 — infrastructure ; Lack essential
PD 0645 — infrastructure ; Neglected maintenance
PF 2212 — *initiative ; Limited*
PF 2212 — *initiative ; Little*
PF 2212 — *initiatives ; Weak*
PD 3132 — investment ; Outflow
PF 6477 — investment resources ; Limited
PF 6477 — investments ; Unproductive
PF 1658 Local jobs ; Scarce
PF 1658 Local jobs ; Unavailability
Local [labour...]
PD 9113 — labour ; Unprepared
PF 0336 — law enforcement ; Reluctant
PF 6556 — leaders ; Inexperienced
Local leadership
PF 6556 — decision making structures ; Lack
PF 6479 — image ; Absence
PF 6479 — role models ; Lack
PF 6556 — Undeveloped
PF 6479 — Unencouraged
PF 4851 Local legal systems ; Deficiencies national
Local management
PC 4867 — Disordered
PC 4867 — Distracted
PE 7313 — skills developing countries ; Lack
Local [market...]
PF 2989 — market ; Shrinking
PF 2989 — market ; Unrealized
PF 2989 — marketing ; Disorganized
PF 2989 — markets goods services ; Limited
PF 2989 — markets ; Small
PF 1735 — *medical disinterest*
PF 2439 — *merchandising ; Uncompetitive*
PF 7998 Local news ; Unpublished
Local [participation...]
PF 3287 — participation ; Restricted
PE 8303 — physicians ; Unavailability
PF 1333 — polarization ; Unresolved
PE 1101 — *politicians ; Distrusted*
PC 1917 — politics ; Lack participation
PE 1101 — *politics ; Unprofitable*
PF 2385 — prices ; High
PF 7118 — *problems ; Overemphasis*
PE 8140 — production processes ; Insufficient knowledge
Local [relationships...]
PF 2063 — relationships ; Lack symbolism
PF 1578 — requirements ; Programme initiatives inappropriate
PF 1316 — *resources ; Unutilized*
PF 2499 — respect regional global legislation ; Limited
PF 1781 — *responsibility ; Unclaimed*
PF 2451 Local services community leadership training ; Lack
Local skills
PE 5170 — developing countries ; Wastage
PF 6538 — Unknown
PF 6538 — Unused
PE 5246 Local structure social care ; Non adaptive
PE 7326 Local structures unemployed ; Deficient
Local [teachers...]
PF 5978 — *teachers ; Detached*
PF 6454 — technological development ; Lack means
PF 2707 — technological innovation ; Centralized decisions
PF 1705 — tradition ; Individualistic retaining
PF 1696 — traditions cultural isolation
PD 1350 *Local vandalism ; Uncontrolled*
PE 8503 Local welfare programmes ; Lack participation
PF 3499 *Locale economic development ; Unattractive*
PE 7311 Locales high crime rates
PF 3470 *Localization unwritten language*

PF 6538 Locally available skills ; Underutilization
PE 2156 Location advertising hoardings billboards ; Unaesthetic
PE 2525 *Location ; Airport*
PE 3538 Location facilities international organizations ; Inefficient
PF 9023 Location ; Isolated geographical
PF 9023 Location ; Isolated village
PF 3499 Location ; Poor geographical
PD 1665 Location power transmission lines ; Unaesthetic
PF 9023 Location ; Remote geographical
PF 1494 Locational maladjustments industry developing countries
PE 8916 Locations ; Distant job
PD 8517 Locations ; Inadequate watermains
PD 2718 Locations nuclear power plants ; Hazardous
PF 7118 *Locations ; Scattered housing*
PF 6683 Locations ; Unpropitious physical
PD 6808 Lock-out
PF 1600 *Locked public buildings*
PE 5848 Locked transit countries ; Inadequate rail transport developing land
PE 2530 Lockjaw
PE 2945 Locomotor ataxia
PE 6769 Locomotor disabilities
PE 0725 Locust plagues
PD 2011 *Lodging ; Residential animal*
PE 1264 Logging ; Environmental hazards forestry
PD 9235 Logging ; Excessive
PF 0459 Logging ; Instability forestry
PF 3499 *Logging regulations ; Restrictive*
PF 5711 Logic ; Fallacies
PF 5711 *Logic ; Inadequacy formal*
PF 7134 Logical limitations comprehension international information ; Pre
PF 1653 Logistics ; Complex services
PF 2826 Logos ; Offensive corporate
PD 2681 Lone-parent families
PF 2386 Loneliness
PF 2386 Loneliness ; Adult
PF 2386 Loneliness adults
PC 0239 Loneliness children
PF 2386 Loneliness ; Dependence
PA 7339 Loneliness *Duality*
PD 0633 Loneliness old age
PD 4392 Loneliness single people
PA 6653 Loneliness *Unsociability*
PD 4392 Lonely hearts
PF 2386 Lonely people
PE 4367 Long distance telephone calls ; Illegal
PC 0238 *Long lifed trees ; Endangered plantations*
PC 0238 *Long lifed trees ; Unsustainable cultivation*
PE 4602 Long prison sentences ; Disproportionately
Long range
PF 5660 — planning decision making ; Lack reward
PF 5660 — planning ; Limited
PF 5660 — plans ; Undeveloped
PD 3391 — transboundary air pollution
PF 2865 — world wide capital flow plan ; Absence
PF 3499 *Long shipping distances*
Long term
PF 2865 — capital ; Limited availability
PF 1234 — change atmospheric chemistry
PF 4263 — changes precipitation patterns
PC 6114 — cyclic changes climate
PE 4603 — debts developing countries ; Increases interest rates
Long term development
PF 5181 — assistance ; Lack
PF 5660 — policy making ; Crisis management approach
PF 5660 — Short range planning
Long term [economic...]
PF 3610 — economic planning agencies ; Absence
PD 2363 — effects cocaine abuse
PD 7918 — effects war
PF 9455 — future planet ; Irresponsibility
PE 4717 — hazards exposure chemicals
PF 4057 — hazards exposure radiation
PE 5324 — health effects radioactive fallout ; Uncertainty
PE 1046 — historical effect witch hunting
PF 5686 — injuries sports
PE 4717 — low level exposure toxic mixtures non toxic chemicals ; Health hazards
PF 5956 — pension funds ; Crisis
Long term shortage
PE 8514 — animal feedstuffs excluding unmilled cereals
PE 1188 — animal vegetable oils fats
PE 1253 — beverages tobacco
PE 1218 — cereals cereal preparations
PE 1261 — chemicals
PE 0303 — clothing
PE 1054 — coal
PE 1197 — coffee, tea, cocoa, spices manufactures thereof
PC 1195 — commodities
PE 1353 — crude fertilizers crude minerals, excluding coal, petroleum precious stones
PE 8225 — dairy products eggs
PE 1216 — electric energy
PF 0334 — energy resources
PE 0976 — food live animals
PE 1013 — fruit vegetables preparations thereof
PE 0511 — furniture
PE 0461 — inedible crude non-fuel materials
PE 0976 — livestock
PE 1436 — machinery transport equipment
PE 0997 — manufactured fertilizers
PE 0802 — manufactured products
PE 1490 — meat meat preparations
PE 0353 — metalliferous ores metal scrap

Long term shortage cont'd
PE 1712 — mineral fuels lubricants
PD 9357 — mineral resources
PE 0613 — miscellaneous manufactured articles
PE 8045 — natural manufactured gas
PC 4824 — natural resources
PE 0824 — non-ferrous metal ores
PE 8626 — petroleum petroleum products
PE 1616 — pulp waste paper
PB 6112 — resources
PE 1218 — rice
PE 1783 — salt-water fish, crustacea molluscs preparations thereof
PE 1120 — sugar honey preparations thereof
PB 6112 — *uranium resources*
PC 1173 — water
PE 2903 — wheat
PE 1372 — wood, lumber cork
Long term [trade...]
PC 9584 — trade policies ; Deterioration
PF 3007 — *training vision ; Unimaginative*
PC 6114 — variations solar radiant energy
PA 6714 *Longevity ; Excessive*
PD 8179 Longsightedness
PE 3296 *Longterm external dependency*
PF 8176 Lookout capacity ; Lack
PE 4339 Loopholes countries underdeveloped legislation ; Exploitation regulatory
PC 9513 Loopholes ; Regulatory
PF 1462 Loopholes ; Tax
PF 2874 *Loose school security*
PA 5740 Looseness *Compulsion*
PA 6858 Looseness *Disintegration*
PA 7361 Looseness *Disorder*
PA 6180 Looseness *Error*
PA 5438 Looseness *Neglect*
PA 7309 Looseness *Uncertainty*
PA 5612 Looseness *Unchastity*
PD 1185 *Loosening teeth*
PE 4152 Looting
PD 1823 Looting ; Archaeological anthropological
PE 0323 Looting works art
PA 6659 Lordliness *Humility*
PC 6491 Lordliness *Vanity*
PF 5653 Lore errors ; Proverbial
PE 7953 Loris ; Endangered species potto
PF 4387 Loser
PD 2614 Losing cultural identity ; Fear
PA 7382 Loss
PD 7480 Loss agricultural crops
Loss agricultural land
PC 7597 — [Loss agricultural land]
PE 3786 — due artificial flooding
PE 5931 — urbanization
Loss [animal...]
PE 7941 — animal mortality ; Economic
PD 8469 — animal productivity
PF 5203 — any personal investment ; Resistance
PE 5758 — appetite
PF 3505 — *appetite*
PE 8717 Loss beneficial plants animals
PB 1044 *Loss blood*
Loss [capacity...]
PC 8310 — capacity age
PC 3134 — capital developing countries
PE 2485 — citizenship ; Involuntary
PE 8720 — civil capacity married women
PC 7013 — civility
PF 1729 — commitment ethic
PF 1097 — confidence government leaders
PA 6448 — *control aggressive impulses*
Loss credibility
PB 6314 — [Loss credibility]
PJ 7629 — diplomacy
PF 1963 — institutions
PE 8064 — international institutions
PF 5934 — United Nations
PF 1985 Loss cultural heritage
PF 9005 Loss cultural identity
Loss [data...]
PE 9298 — data
PE 8109 — destruction injured animals ; Economic
PD 1291 — developing country human resources industrialized countries
PF 3856 — dignity
PE 8098 — disease game zoo animals
PF 0377 — draught animals ; Agricultural
PE 9013 Loss ; Economic
Loss [face...]
PF 3856 — face
PF 3863 — faith religion
PE 9013 — Financial
PE 7855 — forests artificial flooding
Loss [honour...]
PF 8485 — honour
PC 7721 — human resources
PF 2527 — humility relation environment
Loss [Inadequate...]
PE 0290 — Inadequate redress consumers'
PE 7468 — independence ; Fear
PF 9298 — information
PF 1963 — institutional credibility
Loss international
PF 8353 — leadership
PF 8081 — market leadership
PF 7455 — political leadership

Loss international

Loss international cont'd
PF 9345 — technological leadership
Loss [labour...]
PD 5045 — labour force developing countries
PE 5931 — land industrialization
PF 9187 — land plantations
PF 9016 — leadership credibility
PD 0579 — leadership role world affairs ; Government
PC 3293 — *linguistic tradition*
Loss [memory...]
PD 8297 — memory
PE 6593 — memory due torture
PD 5719 — micro-organic proteins
Loss [national...]
PE 7906 — national sovereignty
PE 6676 — nationality children ; Involuntary
PE 2485 — nationality ; Involuntary
PD 3493 — nuclear weapons ; Accidental
Loss [past...]
PF 1062 — past operating context
PC 8862 — peasant title land
PD 0270 — physical functioning suggesting physical disorder
PE 2613 — *property*
PF 3512 — *property ; Fear*
Loss [reduced...]
PE 8098 — reduced productivity diseased animals ; Economic
PE 8098 — reduced productivity diseased animals ; Economic
PF 3863 — religious tradition
PC 4321 — reverence life
Loss [scientific...]
PF 4633 — scientific evidence
PF 0879 — *self-respect*
PF 9980 — sense time
PE 6784 — sex appeal women
PF 3676 — significance work
PE 8109 — slaughter diseased animals ; Economic
PE 2741 — Sleep
PF 7005 — spiritual guidance
Loss [technical...]
PD 7719 — technical know-how
PD 7719 — technical skills society
PE 1310 — tendon reflex
PD 0949 — top soil ; Progressive
PD 0144 — traditional forms social control developing countries
PD 1543 — traditional forms social security developing countries
PE 3377 Loss vision ; Inflected
PE 4196 Loss vision ; Partial
Loss [water...]
PE 7433 — water domestic uses
PE 7433 — water industrial uses
PE 7794 — wildlife due artificial flooding
PE 8796 Losses dams ; Water
PE 4856 Losses due disability ; Economic social
PE 9088 Losses irrigation systems ; Seepage water
PE 8796 Losses irrigation systems ; Water
PC 2027 Losses ; Post harvest
PF 2042 Losses ; *Professional salary*
PF 8079 Losses ; Unreported financial
Lost [communication...]
PF 2350 — communication channels
PF 1764 — covenantal understanding sexuality
PE 6074 — cultural heritage
PD 1544 Lost direction youth engagement
PF 0782 Lost during warfare ; Relatives
Lost [family...]
PF 1099 — family heritage
PF 7456 — family role society
PF 1373 — farming skills
PF 1985 Lost heritage vitality
PF 5420 Lost knowledge
PF 1187 Lost school communication
PF 1815 Lot usage ; Undefined
PD 2672 Lots ; Overgrown vacant
PA 5465 Loudness
PA 7301 Loudness *Colourlessness*
PA 5465 Loudness *Loudness*
PA 5821 Loudness *Vulgarity*
PE 2348 Louis encephalitis ; Saint
PD 7841 *Louping ill*
PD 1753 *Louse-borne typhus*
PD 0982 *Louse ; Infestation*
PD 9667 *Louse infestation*
PE 3576 Louse resistance insecticides
PD 9667 *Lousiness*
PD 4233 Love affair ; Ending
PE 9599 Love bombing
PE 3265 Love children ; Sexual
PF 0959 Love ; Collapsed conceptualization
PF 7418 Love ; Dependence romantic
PD 0061 *Love ; Dissociation sex*
PF 5354 Love ; Lack self
PF 3260 Love making ; Unsatisfactory
PF 0959 Love ; Trivialization
PF 6096 Love ; Unreciprocated
PF 6096 Love ; Unrequited
PE 1323 Love vine
PA 6379 Lovelessness *Avoidance*
PA 7107 Lovelessness *Unpleasantness*
PF 3385 Lovesickness
PE 7313 *Low agricultural profits*
PF 2605 *Low athletics priority*
Low [back...]
PE 5483 — back pain
PE 3903 — bank interest rates
PF 5970 — birth-weights
PE 0472 — blood pressure

Low [blood...] cont'd
PE 1926 — blood sugar
PF 2822 — *business visibility*
Low [capacity...]
PD 0779 — capacity utilization manufacturing plant
PF 7118 — *commune priority*
PF 1365 — *community morale*
PF 7118 — *community visibility*
PE 8184 — complementarity developing country economies
PE 5102 — confidence investment stock markets
PD 1842 — cost housing ; Limited
PF 3826 — credibility alternative social structures
PE 8064 — credibility international organizations programmes
PD 3054 — credit rating developing countries
PD 7480 — crop yields
PF 0333 Low death rate
PF 6472 *Low dental priority*
PE 8190 Low economic development
Low educational
PF 9068 — priorities
PF 9068 — priority
PF 8466 — standards
PC 0382 Low energy diet
Low [faming...]
PE 8171 — faming income
PD 7799 — *fertility turkeys*
PE 4950 — *fibre diets*
PE 6478 — *fishing income rural communities*
PE 7560 — frequency electromagnetic radiation ; Environmental hazards extremely
PF 6570 *Low general expectations*
PD 8568 Low general income
PF 9068 Low high school priority
Low income
PF 5008 — disadvantaged groups ; Limited availability land
PD 8568 — population ; Increasing
PD 8568 — potential
Low [industrial...]
PF 0963 — *industrial incentives*
PC 1587 — intelligence
PE 3988 — intensity conflict
PE 7255 — intra-regional complementarity developing countries economies
PF 6530 — investment developing country farms
Low [labour...]
PD 4011 — labour mobility
PJ 8933 — *learning expectations*
PE 3988 — level conflict
PD 0465 — level domestic resource mobilization developing countries
Low level exposure
PE 4717 — chemicals ; Hazards
PE 4057 — radiation ; Hazards
PE 4717 — toxic mixtures non toxic chemicals ; Health hazards long term,
Low level [management...]
PC 4867 — management skills
PD 2769 — personal income developing countries
PD 2968 — prices commodities exported developing countries ; Continuing
PF 6500 — technical skills
Low [metabolism...]
PE 4463 — metabolism ; Obesity due
PE 6639 — military morale
PF 8446 — morale
PE 1914 Low natural killer cell syndrome
PD 1493 Low occupational mobility developing countries
Low [pay...]
PD 8916 — pay
PF 2845 — *paying skills*
PF 3498 — *performance requirements*
PF 3550 — *police wages*
PE 7313 — *price cash crops*
PF 8876 — priority legislative reform ; Failure process
PF 2969 — *priority status villages*
PE 5883 — productivity agricultural workers developing countries
PD 0339 — protein intake
Low quality
PD 7723 — construction work
PD 1435 — merchandise
PC 9714 — teaching
Low [reading...]
PS 3268 — *reading levels*
PF 5354 — resident self-image
PE 9811 — return investment developing countries
PE 4200 Low school attendance
Low self
PF 0879 — confidence
PF 5354 — esteem
PF 5354 — image
PF 9098 — image due illiteracy
Low [socio...]
PF 7428 — socio-economic class
PF 2605 — *sports priority*
PB 5577 — *status*
PF 2444 *Low tax base*
PD 8916 Low tenant wages
PE 6726 Low wage scale
Low wages
PE 9199 — developing countries
PD 8916 — Excessively
PD 8916 — Unmotivating
Low yield
PF 6524 — farm methods
PD 1285 — *rice crops*
PF 3409 — seeds

PE 7511 Lower limb ; Fracture
PD 9293 *Lower urinary tract animals ; Tumours*
PA 0294 Lowered state health
PD 4403 Lowering water level
PD 4403 Lowering water table
PD 5110 Lowlands ; Inadequate empolderment coastal
PA 6798 Lowness
PA 6839 Lowness *Disrepute*
PA 5652 Lowness *Inferiority*
PA 6731 Lowness *Solemnity*
PA 5821 Lowness *Vulgarity*
PF 6545 Loyalties ; Divisive tribal
PF 0910 Loyalties ; Futility social
PD 2790 Loyalty corporations ; Lack worker
PE 9018 Loyalty ; Lack
PF 6545 Loyalty ; Split tribal urban
PD 0556 *LSD abuse*
PE 1712 Lubricants ; Long term shortage mineral fuels
PE 1346 Lubricants related materials ; Environmental hazards mineral fuels,
PD 0877 Lubricants related materials ; Instability trade mineral fuels,
PF 9536 Luck ; Bad
PA 6340 Lucklessness *Adversity*
PE 1310 Lumbago
PE 5483 Lumbalgia
PE 2521 Lumber cork ; Instability trade wood,
PE 1372 Lumber cork ; Long term shortage wood,
PD 9667 *Lumpy skin disease cattle*
PD 9667 *Lumpy wool*
PC 0834 Lunches ; *Inappropriate school*
PC 0834 Lunches ; *Unprovided school*
PC 0834 Lunches ; *Unsupplied school*
PE 7085 Lung cancer
PD 5453 *Lung congestion*
Lung [damaging...]
PE 4555 — damaging agents
PE 5524 — *disease cattle ; Farmer's*
PD 0637 — diseases
PD 0637 — disorders diseases
PE 6899 Lung ; Farmer's
PE 9854 *Lung flukes*
PD 0637 *Lung ; Hernia*
PE 7085 Lung ; Malignant neoplasm
PD 5453 *Lung oedema*
PE 2293 *Lungs ; Acute congestion*
PE 8617 Lungs children ; Air pollution impact
PE 8822 Lungs ; Inflammatory affections bronchial tubes
PD 7307 Lungworm infection
PE 9458 Lupinosis ; *Mycotoxic*
PE 0566 Lupus
Lupus erythematosus
PD 2565 — [Lupus erythematosus]
PE 9509 — animals ; Systemic
PD 2565 — Systemic
PE 0566 Lupus vulgaris
PA 4673 Lust
PA 6379 Lust *Avoidance*
PA 5612 Lust *Unchastity*
PA 5644 Lust *Vice*
PD 7799 *Luteal cystic ovary disease*
PD 7799 *Luteal cysts*
PD 7424 *Luxation ; Patellar*
PF 5931 Luxury ; Excessive
PF 1085 Luxury treatment pets
PF 5931 Luxury ; Unnecessary
PE 8450 Lycanthropy
PB 7600 Lying
PE 4890 Lying concerning sexual relations
PD 4982 Lying corporation officials ; Deliberate
PB 7600 Lying ; Dependence
PD 1893 Lying ; Diplomatic
PD 1893 Lying government officials ; Deliberate
PD 2686 Lying ; Institutional
PD 9836 Lying medical professionals
PD 9145 Lying ; Parental
PD 9145 Lying parents children
PD 9182 Lying scientists
PF 7634 Lying ; Social science
PD 6363 *Lyme disease*
PE 5913 *Lyme disease dogs*
PF 8026 Lymphadenitis
Lymphadenitis [Acute...]
PC 8534 — Acute
PE 9844 — animals
PE 9844 — *animals ; Caseous*
PF 8026 Lymphadenitis ; Chronic
PE 9844 Lymphadenitis swine ; *Streptococcal*
PD 2654 Lymphadenitis ; Tuberculous
PD 8347 *Lymphangioma*
PE 9844 Lymphangitis animals
PD 2728 *Lymphangitis ; Epizootic*
PE 9844 *Lymphangitis ; Equine ulcerative*
PE 9844 Lymphangitis horses
PE 4637 Lymphatic haematopoietic tissue ; Neoplasms
PE 2686 Lymphatic leukaemia
Lymphatic system
PD 2654 — Diseases
PD 5453 — diseases animals
PE 3933 — Symptoms referable
PE 4637 Lymphatic tissue ; Cancer
PE 2686 Lymphocytic leukaemia
PD 5453 *Lymphocytic leukaemia dogs*
PE 2686 Lymphoid leukaemia
PD 2730 *Lymphoid leukosis chickens*
PE 1243 Lymphoma ; Bovine malignant

Malice

PD 5453 *Lymphoma ; Canine malignant*
PD 5453 *Lymphoma ; Feline visceral*
PD 2730 *Lymphoproliferative disease turkeys*
PE 5548 Lymphosarcoma
PD 5453 *Lymphosarcoma*
PE 1243 Lymphosarcoma ; Bovine
PD 5453 *Lymphosarcoma ; Feline*
PD 5453 *Lymphosarcoma ; Feline visceral*
PE 6287 Lynching
PF 7807 Lyrics ; Offensive
PD 0556 *Lysergic acid abuse*

M

PD 7001 Machine industries ; Protectionism computer office
PF 4334 Machine intelligences ; Denial rights
PE 5319 Machine parts ; Counterfeit
PF 6330 Machine technocracy
PE 5309 Machinery accumulation seas ; Disused
Machinery apparatus appliances
PE 8026 — Environmental hazards electrical
PE 8878 — shortage ; Electric
PE 8875 — Trade instability electrical
PF 2378 *Machinery capital ; Unavailable*
PF 2410 *Machinery ; Costly industrial*
PE 2462 Machinery design ; Defects
PF 4108 Machinery developing countries ; Insufficient agricultural
PJ 2425 Machinery ; Environmental hazards non electrical
PE 8166 Machinery equipment development projects ; Overuse
Machinery equipment industries
PE 1859 — Environmental hazards
PE 1852 — *environmental hazards*
PE 1852 — Instability
PE 1852 — *instability*
PE 1852 — *underdevelopment*
PF 0942 — Underdevelopment metal products,
PF 4849 Machinery government elites ; Appropriation
Machinery [Inadequacy...]
PF 2420 — Inadequacy governmental decision making
PC 1833 — Inadequate urban political
PE 8828 — Instability trade non electrical
Machinery [Lack...]
PF 4108 — Lack agricultural
PF 1822 — *Lack shared*
PF 2457 — Limited availability modern farm
PF 4108 Machinery ; Minimal farm
PD 6520 *Machinery ; Obsolete*
PF 6330 Machinery ; Organized cult
PF 2457 Machinery ; Prohibitive cost farm
PF 2457 *Machinery ; Simultaneous demands*
Machinery transport equipment
PD 0620 — Instability trade
PE 1436 — Long term shortage
PE 7958 — Restrictive practices trade
Machinery [Unavailability...]
PE 7504 — Unavailability replacement parts agricultural
PF 2410 — *Unavailable road*
PF 4108 — Unequal distribution agricultural
PC 2297 — Unreliability equipment
PF 2813 — *Unscheduled introduction new*
PD 1565 Machines ; Inadequate maintenance
PF 2410 *Machines ; Unavailability road*
PC 3024 Machismo
Macroeconomic policies
PF 5964 — developing countries ; Unsustainability
PF 5000 — governments ; Uncoordinated
PF 5000 — industrialized countries ; Mismatch national
PF 5000 — leading industrialized countries ; Divergences
PD 1762 Macropods ; Endangered species
PE 5191 Mad cow disease (BSE)
PE 9799 Mad itch
PE 8970 Made up articles ; Instability trade textile yarn fabrics,
PE 8436 Made up articles ; Shortage textile, yarn fabrics,
PA 6799 Madness *Disease*
PA 7157 Madness *Insanity*
PA 7371 Madness *Unintelligence*
PE 2455 *Madura foot*
PE 9438 *Maedi-Visna*
PC 2343 Mafia
PF 1384 Magazines comics ; Pornographic
PD 9667 *Maggots ; Wool*
PF 3311 Magic
PF 8249 Magic ; Black
PD 0448 *Magnesium deficiency*
PF 1588 Magnetic field ; Reversal Earth's
PE 7560 Magnetic fields ; EEG entrainment ELF
PD 1661 Magnetic storms
PE 1810 Mail advertising ; Proliferation direct
PE 6754 *Mail bombs*
PE 1404 Mail fraud
PE 1810 Mail ; Junk
PE 1810 Mail ; Junk fax
PD 2683 Mail ; Monitoring
PC 3298 Mail-order brides
PE 6754 Mail service ; Inappropriate use
PD 0196 Maimed children
PD 9196 Maintain sewage systems ; Failure
PB 4885 *Maintaining basic health ; Fragility*
PE 1122 Maintaining comprehensive document collections ; Prohibitive cost
PC 1966 Maintenance after retirement ; Income
PE 1396 Maintenance complex systems ; Wastage highly skilled personnel routine
PF 0296 Maintenance costs ; Unexpected
PD 2269 Maintenance drainage ditches ; Inadequate
PD 6520 Maintenance equipment ; Inadequate
PD 6520 Maintenance equipment ; Inappropriate
Maintenance Inadequate
PD 8984 — [Maintenance ; Inadequate]
PD 8894 — building
PE 3247 — child
PF 2346 — *coastal*
PD 8984 — *electrical*
PD 1565 — equipment
PD 2269 — gully
PD 8557 — road
PD 0645 Maintenance infrastructure ; Inadequate
PF 2995 *Maintenance know how ; Limited*
PD 0645 Maintenance local infrastructure ; Neglected
PD 1565 Maintenance machines ; Inadequate
PD 8894 Maintenance margin ; Unplanned
PD 8894 Maintenance ; Neglect property
PD 7964 Maintenance ; Negligence
Maintenance [personnel...]
PF 2995 — *personnel ; Inadequate*
PF 1773 — physical health ; Inadequate
PD 2929 — political dossiers individuals
Maintenance practices
PD 7964 — Corruption
PD 7964 — Irresponsible
PD 7964 — Unethical
Maintenance Prohibitive cost
PF 0296 — [Maintenance ; Prohibitive cost]
PE 1722 — equipment
PF 0296 — *home*
Maintenance [servicing...]
PD 1565 — servicing ; Deterioration quality equipment
PF 5124 — super power ideological diversities
PD 8984 — system ; Absence
PD 6520 Maintenance tools ; Lack
Maintenance [Unmanageable...]
PF 2346 — *Unmanageable land*
PD 8894 — Unsafe home
PF 2995 — *Unskilled mechanical*
PD 9196 Maintenance water systems ; Inadequate
PE 3589 Maize pests diseases
PF 4341 Majesty ; Lese
Major [accidents...]
PF 7768 — accidents ; Cover up
PE 5229 — accidents ; Delay payment compensation victims
PE 5336 — agriculture technologies ; Decreasing rate development
PE 1922 Major corporations national policy making ; Undisclosed control
PC 3185 Major powers protect investments their citizens foreign countries ; Intervention
PE 1055 Major roads ; Road network unconnected
PE 0097 *Major ; Variola*
PD 3239 Majority ; Dictatorship
PF 0851 Majority rule mindset
Make [difficult...]
PF 4204 — difficult decisions ; Inability
PF 4204 — difficult decisions ; Unwillingness
PF 2583 — *do consumer attitude*
PD 5335 Make payroll payments ; Failure
PF 5044 Make use leisure time ; Inability
PF 2895 Makers ; Arrogance policy
PD 8696 Makers ; Drug abuse decision
PD 8696 Makers ; Drug abuse policy
PF 2452 Makers grassroots ; Insufficient communication decision
PF 2452 Makers ; Inaccessible decision
PE 0573 Makers multinational enterprises ; Inaccessibility decision
PF 9576 Makers ; Static models social change policy
PF 8703 Makers ; Unrepresentative policy
PE 4386 Makeshift dwellings developing countries
PF 0760 Makeshift economic management
PF 4984 Makeshift educational programme
PD 4368 Makeshift settlements
PD 8723 Mal-educated school leavers
PD 2730 *Malabsorption syndrome chickens*
PD 3978 *Malabsorption syndromes animals*
PD 0586 Maladjusted children
PA 6739 Maladjustment
PE 4258 Maladjustment children migrants ; Social
PA 5982 Maladjustment *Disagreement*
PD 7650 Maladjustment disciplines employment
PC 8337 Maladjustment ; Personal social
PD 7841 *Maladjustment syndrome ; Neonatal*
PC 7041 Maladjustment technology ; Worker
PA 7232 Maladjustment *Unskillfulness*
PD 1494 Maladjustments industry developing countries ; Locational
PA 5652 Maladroit *Inferiority*
PA 7232 Maladroit *Unskillfulness*
PA 5838 Malaise *Agitation*
PA 6799 Malaise *Disease*
PA 6030 Malaise *Fear*
PA 5451 Malaise *Insensibility*
PA 6731 Malaise *Solemnity*
PF 1568 Malaise ; Spiritual emotional
PA 7107 Malaise *Unpleasantness*
PE 0616 Malaria
PE 1923 *Malaria ; Avian*
PE 0616 *Malaria ; Drug resistant*
Malaria [Malariae...]
PE 0616 — *Malariae*
PE 1923 — *Mammalian*
PE 3622 — mosquitoes
PE 0616 *Malaria ; Ovale*
PE 0616 *Malaria ; Recurrent induced*
PE 0616 *Malaria ; Simian*
PE 0616 *Malaria ; Vivax*
PE 0616 *Malariae malaria*
PB 6207 Maldevelopment
PB 6207 Maldevelopment ; Dependence
PD 9189 Maldistribution agricultural land
PF 4485 Maldistribution commercial vehicles
PD 3446 Maldistribution electrical energy
PC 5038 Maldistribution energy consumption
Maldistribution food
PF 3949 — [Maldistribution food]
PC 2801 — [Maldistribution food]
PF 3949 — aid
PC 2801 — Dependence
PF 0545 Maldistribution geosynchronous orbits
PF 4126 Maldistribution health personnel
Maldistribution land
PD 0406 — associated large traditional estates
PD 0813 — customary tenure systems
PD 0050 — developing countries
PD 2705 Maldistribution medical resources
PF 1495 *Maldistribution merchant vessels*
Maldistribution [passenger...]
PF 4480 — passenger cars
PC 8192 — population countries
PF 4480 — private automobiles
PC 9785 — productive capacity
Maldistribution [radio...]
PF 0734 — radio frequencies
PF 4142 — radios
PB 1016 — resource utilization
PB 1016 — resources
PD 4962 — revenues
PC 8885 Maldistribution science technology
PE 8733 Maldistribution students enrolled school
Maldistribution [teachers...]
PE 6183 — teachers
PF 4132 — telecommunications facilities
PF 4136 — television sets
PE 6144 Maldistribution urban shopping facilities
Maldistribution [water...]
PD 8056 — water
PB 7666 — wealth
PD 5258 — wealth developing countries
PF 0167 — world population
PF 0167 — world population ; Dependence
Male [chauvinism...]
PC 3024 — chauvinism
PF 3382 — children ; Desire
PD 5783 — climacteric
PF 1069 — contraceptive methods ; Inadequacy
PC 7373 — *criminal offenders ; Preponderance*
Male [dominance...]
PF 4373 — dominance language ; Reinforcement
PC 3024 — dominance ; Traditional
PC 3024 — domination
PE 8916 Male employment ; Distant
PF 7938 Male female ; Intolerance distinctions
PD 9154 Male genital organs ; Diseases
PE 8372 *Male genital organs ; Malignant neoplasm*
PD 4402 Male homosexual prostitution
PF 1390 Male homosexuality
Male homosexuals
PE 3882 — Denial rights
PE 3882 — Discrimination against
PE 3882 — Violation rights
PD 5783 Male menopause
PD 5783 Male mid-life crisis
PC 3258 *Male pregnancy ; Social barriers*
PD 3381 Male prostitution
Male sexual
PF 3271 — exhibitionism
PD 7297 — irresponsibility
PE 6054 — mutilation
PD 9154 *Male sterility*
PC 3024 Male supremacy
PE 6904 Male workers ; Occupational hazards
PF 3746 Malediction
PF 8780 Males females ; Unauthorized proximity
PA 7102 Malevolence
PA 5454 Malevolence *Badness*
PA 5446 Malevolence *Enmity*
PA 7338 Malevolence *Hate*
PA 5643 Malevolence *Unkindness*
PA 7102 Malevolent people
PF 6734 Malevolent spirits
PF 9043 Malevolent spirits ; Affliction
PA 7410 Malfeasance *Celibacy*
PA 6852 Malfeasance *Inappropriateness*
PA 5644 Malfeasance *Vice*
PE 4460 Malformation
PD 1618 Malformation ; Congenital
PE 4808 Malformation diabetic pregnancies ; Foetal
PA 6790 Malformation *Distortion*
PD 2761 Malformation factory farming ; Animal
PE 2042 Malformation foetus ; Physical
PA 5878 Malformation *Nonconformity*
PA 7240 Malformation *Ugliness*
PF 6465 *Malformations ; Craniofacial*
PE 2246 *Malfunction ; Embryogenesis*
PE 7154 *Malfunctions ; Bodily ailments, lesions*
PF 5901 Malice
PA 5446 Malice *Enmity*
PA 7338 Malice *Hate*

Malice

PA 5643 Malice *Unkindness*
Malicious [personal...]
PE 3733 — personal disfigurement
PE 3733 — physical disablement
PF 8479 — prosecution
PA 5454 Malignancy *Badness*
PA 7055 Malignancy *Death*
PD 9654 *Malignancy ; Humoral hypercalcaemia*
PA 5643 Malignancy *Unkindness*
PE 6280 Malignant catarrhal fever
PE 5617 *Malignant foot rot*
PE 6280 Malignant head catarrh
PE 9774 *Malignant hyperthermia*
PE 1243 Malignant lymphoma ; Bovine
PD 5453 *Malignant lymphoma ; Canine*
Malignant neoplasm
PE 9229 — bone
PE 1175 — breast
PE 9229 — connective tissue
PE 4303 — digestive organs
PE 5100 — genito-urinary organs
PE 7085 — lung
PE 8372 — *male genital organs*
PE 9819 — mouth throat
PE 7572 — respiratory system
PE 5016 — skin
PC 0092 Malignant neoplasms
PE 1905 Malignant neoplasms female genital organs
PE 7769 *Malignant oedema*
PE 7769 *Malignant oedema animals*
PF 5691 Malignant visions
PE 7701 Malingering
PA 6379 Malingering *Avoidance*
PA 5438 Malingering *Neglect*
PA 7411 Malingering *Uncommunicativeness*
PE 4921 Malnourished children ; Inhibited growth
PE 4465 Malnourished children ; Reduced activity
PE 4883 Malnourished persons ; Inadequate immune responses
PB 1498 Malnutrition
PE 8925 Malnutrition arising mental factors ; Physiological
PB 1498 Malnutrition-based illness
PD 8941 Malnutrition ; Child
PB 1498 Malnutrition ; Dependence
PD 8668 Malnutrition developing countries
Malnutrition [indigenous...]
PC 3319 — indigenous peoples
PD 0331 — infants early childhood ; Protein energy
PE 4874 — insufficient stimuli ; Damage infant brains
PE 1085 Malnutrition ; Maternal
PD 0339 Malnutrition ; Protein calorie
PD 0339 Malnutrition ; Protein energy
Malnutrition [Seasonal...]
PD 5212 — Seasonal
PD 7473 — slums
PB 1498 — *Subsistence level*
PD 7473 Malnutrition ; Urban
PD 0363 Malnutrition vulnerable groups ; Protein energy
PE 1085 Malnutrition women during pregnancy nursing ; Protein energy
PD 1413 Malodorous fumes
PD 8775 *Malposition uterus*
PD 2623 Malpractice anthropology
PE 0682 Malpractice ; Banking
PD 7731 Malpractice biology
Malpractice [chemistry...]
PD 4265 — chemistry
PF 8889 — clergy
PD 9713 — construction industry
PF 0221 — consultants
PE 0811 Malpractice ; Delay payment compensation victims
PD 4684 Malpractice education
PD 6701 Malpractice ; Forestry
PD 0708 Malpractice geology
PD 2586 Malpractice hydrology
Malpractice [Inappropriateness...]
PA 6852 — *Inappropriateness*
PD 2916 — Industrial
PE 6947 — intergovernmental organization leadership
PD 7885 — Investigatory
PD 5380 Malpractice ; Legal
Malpractice [Medical...]
PD 5770 — Medical
PD 4182 — meteorology
PD 4361 — Military industrial
PE 0811 Malpractice ; Non payment compensation victims
PD 4277 Malpractice oceanography
PE 1369 Malpractice opticians
Malpractice [pedology...]
PD 1110 — pedology
PD 1710 — physics
PE 4429 — plastic surgery
PD 9193 — Police
PC 5517 — Political
PC 8019 — Professional
PD 5267 — Psychiatric
PD 8290 Malpractice radiology
Malpractice [social...]
PD 6626 — social sciences
PF 4564 — Statistical
PE 4736 — Surgical
PA 7232 Malpractice *Unskillfulness*
PD 7726 Malpractice ; Veterinary
PA 5644 Malpractice *Vice*
PD 4721 Malpractice zoology
PE 9517 Malsituated dams
PF 4606 Malthusianism
Maltreatment animals
PC 0066 — [Maltreatment animals]
PE 4786 — advertising
PE 1174 — aggression experiments
PE 5461 — aquaria
PE 4786 — cinema
PD 0260 — educational purposes
PD 2078 — entertainment
PE 4786 — media
PE 4786 — used theatre acts
Maltreatment [children...]
PC 2584 — children ; Physical
PE 4810 — circus animals
PD 5560 — civilians
PD 1265 — companion animals
PE 3394 Maltreatment horses
PA 6852 Maltreatment *Inappropriateness*
PD 0260 Maltreatment laboratory animals
PE 3394 Maltreatment livestock
PE 0906 Maltreatment marine show animals
Maltreatment [performing...]
PE 4810 — performing animals
PE 0906 — performing dolphins
PD 1265 — pet animals
Maltreatment prisoners
PD 6005 — [Maltreatment prisoners]
PE 0428 — fellow inmates
PE 0998 — prison officers
PE 0998 — Unsanctioned
PE 4810 Maltreatment rodeo animals
PE 3394 Maltreatment transport animals
PE 4834 Maltreatment zoo animals
PE 1923 *Mammalian malaria*
PE 4913 Mammals ; Depletion fish reserves marine
PC 3152 *Mammals ; Disruption breeding grounds marine*
PC 1326 Mammals ; Endangered species
PD 3673 Mammals ; Endangered species marine
PE 0185 Mammals ; Feral
PE 1666 Mammals military ; Exploitation marine
PD 7799 *Mammary tumours*
PD 7799 *Mammillitis ; Bovine ulcerative*
PC 9717 Man ; Degradation climate
PD 3593 Man ; Dissemination plant diseases
PF 3864 Man ; Fall
PD 3555 Man ; Family dependence patriarchal role
PE 5023 Man industrialized societies ; Inadaptation technology
Man made
PB 2075 — disasters
PF 7939 — disease ; Inadequate knowledge reporting
PD 6663 — diseases
PD 6663 — diseases ; Dependence
PD 0571 — famine
PF 5438 Man technological process ; Dehumanization
PD 8371 Man vectors disease
PE 9762 Managed trade mechanisms ; Discriminatory
PF 5660 Management approach long term development policy making ; Crisis
Management [capacity...]
PE 7184 — capacity developing countries ; Inadequate external debt
PF 1528 — Communication ineptitude
PF 1528 — communication skills ; Lack
Management [Deception...]
PD 3823 — Deception
PF 2754 — decisions ; Non inclusive
PD 3712 — development projects ; Inappropriate
PF 1226 — *Difficult grant*
PC 4867 — Disordered local
PC 4867 — Distracted local
PF 3661 Management employee communication ; Inadequate
PF 9672 Management government finances ; Inadequate
PF 5964 Management government financial resources developing countries ; Inefficient
Management [Inadequate...]
PF 6574 — Inadequate accountability public sector
PC 4867 — Incompetent
PF 0760 — Incompetent financial
PC 4867 — Inexperienced business
Management [Lack...]
PF 0574 — Lack participation
PF 6524 — Limited crop
PF 1442 — Limited village
PF 0760 Management ; Makeshift economic
PF 3091 Management patterns ; Inflexible
PF 3091 Management patterns ; Static
Management Poor
PF 0760 — money
PC 0052 — soil
PE 8233 — water
PC 4867 *Management ; Private hall*
PF 8074 Management public information ; Unaccountable
PJ 8060 Management relations ; Inadequate private international labour
PF 1442 Management rural areas ; Fragmented forms commercial
Management skills
PE 0046 — developing countries ; Lack
PE 7313 — developing countries ; Lack local
PC 4867 — Insufficient
PC 4867 — Low level
PF 1442 — rural communities ; Inadequate
PC 4867 — Underused
PF 3091 Management structure ; Inflexible
Management [techniques...]
PD 3712 — *techniques ; Inappropriate use innovative*
PD 3789 — training ; Absence
Management [training...] cont'd
PC 4867 — training ; Absence
Management [Underdeveloped...]
PE 8233 — Underdeveloped water
PF 6524 — Underproductive methods agricultural
PC 4824 — Unintegrated biosphere ecosystem
PC 4867 — *Unprofessional building*
PF 1528 Managerial communications ; Poor
PF 2335 Managerial instruments ; Inadequate governmental
Managerial [skills...]
PE 9764 — skills least developed countries ; Shortage
PC 4867 — skills ; Undiscovered
PE 1635 — stress
PF 8087 Managerism
PE 3872 Managers ; High severance pay top
PC 4867 Managers ; Insufficient trained
PD 7424 *Manchester wasting disease*
PB 0284 *Mandatory AIDS testing*
PE 5697 Mandatory hair shaving
PD 0448 *Manganese deficiency*
PE 1364 Manganese health hazard
PE 2727 Mange
PE 2727 *Mange cattle*
PE 5444 *Mange ; Ear*
PE 2727 *Mange goats*
PE 2727 *Mange horses*
PE 5444 *Mange ; Otodectic*
PE 2727 *Mange sheep*
PE 2727 *Mange swine*
PE 4650 Mania
PD 1318 *Manic depression*
PD 1318 Manic-depressive psychosis
PD 1318 *Manic episode*
PF 6430 Manifest destiny ; Belief
PF 2850 Manifest religion ; Denial right
PF 8682 Manila fibre ; Instability trade
PC 2008 Manipulated product life cycles
PA 6359 Manipulation
PE 1951 Manipulation brain ; Surgical
Manipulation [children...]
PF 2017 — children play ; Attitude
PF 0996 — civic education
PD 8647 — commodity markets
Manipulation [debates...]
PD 4060 — debates
PA 6359 — Dependence
PC 6875 — Dependence economic
Manipulation [Economic...]
PC 6875 — Economic
PD 2906 — elections ; Political party
PE 9599 — Emotional
PE 7448 Manipulation individual mass media
PC 0776 Manipulation ; Irresponsible genetic
PD 6800 Manipulation nongovernmental organizations
PF 9243 Manipulation project cost-benefit analyses
PE 7448 Manipulation public opinion
Manipulation [scientific...]
PF 9693 — scientific studies
PE 6642 — sects ; Financial
PE 5777 — students
PF 5529 Manipulation timing ; Self interested
PE 0245 Manipulation transfer prices transnational corporations
PE 2145 Manipulations accounting records transnational corporations ; Foreign currency
PJ 3491 Manipulations ; International tax
PE 6336 Manipulative cults
PF 5183 Manipulative euphemisms
PB 7125 Manipulative international plots
PF 1609 Manipulative knowledge
PA 6359 Manipulative people
PF 2909 Manipulative use referenda
PF 2443 Mankind ; Inevitable destruction natural environment
PC 6239 Mankind ; Offences against peace security
PF 8145 Mannerisms ; Status
PC 7013 Manners ; Bad
PE 3704 Manoeuvres sensitive areas ; Naval
PE 3704 Manoeuvres sensitive border areas ; Military
Manpower developing countries
PE 6243 — Inadequate supply appropriate trained
PE 4933 — Lack technical development excess
PE 6243 — Scarcity skilled
PF 8763 Manpower ; Excessive cost
PF 6477 *Manpower ; Inadequate use*
PE 9764 Manpower least developed countries ; Lack educated
PF 1790 *Manpower ; Limited construction*
PF 1467 *Manpower planning ; Inadequate*
PF 1467 *Manpower planning ; Lack*
PF 0780 Manpower resources military activities ; Absorption
PE 5170 Manpower rural areas developing countries ; Lack skilled
Manpower [shortage...]
PC 0592 — shortage
PC 0592 — shortage ; Daytime
PE 4920 — Shortage military
PD 0044 — Shortage skilled
PE 8771 — socialist countries ; Shortage
PD 2341 Manslaughter
PD 7952 Manslaughter
PE 0437 Manslaughter ; Negligence
PE 2395 *Manson's eyeworm infection poultry*
PE 9114 Manual intellectual labour ; Artificial opposition
PE 1134 *Manual labour ; Undesirable*
PF 2833 Manual sign languages deaf ; Multiplicity
PD 9628 Manual workers ; Discrimination against women non
PE 4804 Manually skilled workers ; Insufficient
PD 8286 Manufacture alcoholic beverages

Markets

PD 3998 Manufacture ; Defective product
PE 0564 Manufacture distribution ; Unauthorized pharmaceutical
Manufacture Environmental hazards
PE 0481 — coffee, tea, cocoa, spices their
PE 8165 — furniture fixtures
PE 8473 — glass glass products
PE 1103 — wearing apparel
PE 2512 Manufacture illicit drugs
PC 2618 *Manufacture munitions*
PF 3494 Manufacture nuclear bombs ; Ease
PE 8651 Manufacture plastic products ; Environmental hazards
PE 5698 Manufacture ; Pollution nuclear weapons
PD 1435 Manufacture substandard products
Manufacture underdevelopment
PE 8427 — China, pottery earthenware
PF 2604 — *Furniture fixtures*
PE 8715 — Glass glass products
PE 7960 — Petroleum coal products
Manufactured articles
PE 8275 — Environmental hazards miscellaneous
PE 0814 — Instability trade miscellaneous
PE 0613 — Long term shortage miscellaneous
PE 0806 Manufactured fertilizers ; Instability trade
PE 0997 Manufactured fertilizers ; Long term shortage
PE 8045 Manufactured gas ; Long term shortage natural
Manufactured goods
PD 3998 — Defective
PE 4170 — developing countries ; Trade barriers
PE 2966 — developing countries ; Weakness trade
PE 1344 — Environmental hazards
PD 3657 — Insect damage stored
PE 0882 — Instability trade
PE 0802 — Long term shortage
PD 4907 — Over diversification
PD 1797 — Restrictive practices trade
PE 9130 Manufactures dressed fur skins ; Shortage leather, miscellaneous leather
Manufactures [environmental...]
PE 8447 — environmental hazards ; Leather leather
PE 0483 — Environmental hazards tobacco tobacco
PE 8091 — Environmental hazards wood cork
PE 0572 Manufactures instability ; Tobacco
PE 0572 Manufactures ; Instability trade tobacco tobacco
PE 7926 Manufactures ; Shortage tobacco tobacco
Manufactures [thereof...]
PE 0915 — thereof ; Instability trade coffee, tea, cocoa, spices
PE 1197 — thereof ; Long term shortage coffee, tea, cocoa, spices
PE 8969 — trade instability ; Iron steel
PE 8320 — trade instability ; Leather leather
PD 4219 Manufacturing capacity developing countries ; Inadequate
PD 9117 Manufacturing developing countries ; Unstable growth
PE 7188 Manufacturing distribution ; Unauthorized liquor
PF 6557 *Manufacturing incentives ; Limited*
Manufacturing industries
PE 2238 — developing countries ; Inadequate increase employment
PD 9117 — developing countries ; Instability
PD 0454 — Environmental hazards
PE 1605 — Health risks workers
PE 8956 — Ineffective self regulation chemicals
PE 8574 — Ineffective self regulation consumer goods
PC 0580 — Instability
PC 0854 — Underdevelopment
PE 8067 Manufacturing industry ; Ineffective self regulation automobile
PE 8702 Manufacturing industry wastes ; Food
PE 0282 Manufacturing least developed countries ; Minimal
PD 0779 Manufacturing plant ; Low capacity utilization
PE 0402 Manufacturing ; Unemployment developed countries resulting participation developing countries
PF 6483 Manuscripts ; Uncatalogued ancient
PD 0398 Map world ; Inadequate international
PE 9291 Maple syrup urine disease
PE 4586 *Maple ; Tar spot*
PE 9239 Maps ; Production deliberately distorted
PE 8589 *Marble bones*
PD 2730 *Marble spleen disease pheasants*
PE 9689 Marek's disease
PF 7420 *Mares ; Lactation tetany*
PE 1683 *Margarine trade ; Instability*
PE 7313 *Margin ; Narrow profit*
PD 8894 Margin ; Unplanned maintenance
PD 4273 Marginal agricultural land ; Cultivation
PD 6960 Marginal agricultural lands ; Disruption ecosystems
Marginal [government...]
PF 8960 — government support
PF 6528 — *grazing land ; Exploitation*
PF 8085 — grooming skills
PF 4347 — groups
Marginal [lands...]
PD 4273 — lands ; Forced relocation peasants onto
PD 6579 — level family income
PF 7130 — *living space*
PE 8089 Marginalism
PF 4347 Marginalization
PE 4990 Marginalization second-generation immigrants
PD 4252 Marginalization traditional economies
PF 5934 Marginalization United Nations
PD 7424 Marie's disease
PE 1186 Marijuana hashish ; Abuse
Marine [accidents...]
PD 8982 — accidents
PJ 9820 — activities ; Unlawful interference
PF 7554 — animal communication ; Vulnerability

Marine [animals...] cont'd
PD 3934 — animals fish ; Accumulation contaminant residues
PE 0439 — animals ; Hunting
PD 7574 Marine disposal obsolete weapons
Marine dumping
PD 1245 — *coal ash*
PD 3666 — sewage
PD 3666 — wastes
Marine [ecosystems...]
PC 1647 — ecosystems ; Vulnerability
PD 5178 — environment catastrophic warfare damage ; Vulnerability
PE 1431 — environment fisheries products ; Radioactive contamination
PD 5803 — equipment ; Fouling rust
PE 5802 Marine insurance ; Inequities
PD 3158 Marine invertebrates
Marine mammals
PE 4913 — Depletion fish reserves
PC 3152 — *Disruption breeding grounds*
PD 3673 — Endangered species
PE 1666 — military ; Exploitation
PE 6289 Marine oxygen deficiency
Marine pollution
PC 1117 — [Marine pollution]
PE 3741 — plastic waste
PD 4277 — Underreporting
PE 1597 Marine resources ; Inequitable allocation rights exploit sea bed
Marine [sciences...]
PD 4277 — sciences ; Unethical practice
PE 0906 — show animals ; Maltreatment
PE 5689 — space island developing states ; Neglected
PE 1839 Marine transportation hazardous cargoes
PG 3896 *Marine typhus*
PD 3934 Marine wildlife ; Accumulation pollutants
PD 3160 *Marine worms ; Endangered species*
PF 2314 Marital affairs ; Extra
PF 3278 Marital cohabitation ; Non
PF 2314 Marital infidelity
PD 2103 Marital instability
PD 0518 Marital problems
PF 4948 Marital rape
PD 5107 Marital sexual intercourse ; Pre
PD 0518 Marital stress
PD 8982 Maritime accidents ; Risk
PF 5923 Maritime agreements ; Insufficient
PE 5814 Maritime commerce developing countries ; Lack technical infrastructure
PD 0312 Maritime commerce ; Unfair practices
PE 5825 Maritime fleets ; Over concentration ownership
PE 4475 Maritime fraud
PE 5801 Maritime integration ; Lack world
PD 8412 Maritime nuclear weapons systems
PE 6630 Maritime safety standards ; Inadequate implementation
PE 5791 Maritime shipping industry ; Instability
Market access
PE 2328 — due product differentiation transnational corporations ; Limited
PF 2457 — Restrictions
PF 2535 — Uncertain
PD 1211 Market allocation ; Non political
PF 3499 *Market areas ; Limited*
PF 3499 Market boundaries ; Inflexible
Market [Certificate...]
PF 2066 — *Certificate based job*
PF 0581 — channels ; Unused
PE 0051 — competition transnational corporations ; Restriction free
PD 4511 — crash ; Stock
PD 5750 — Criminal investment youth
PF 9187 — crops ; Non diversification
PD 5905 — Currency black
Market developing countries
PD 0008 — Flooding urban labour
PD 0928 — Inadequacy domestic
PD 0928 — Small size domestic
Market [development...]
PF 1086 — development ; Limited
PF 8299 — Distant product
PE 3196 — divisions transnational corporations ; Restrictive
PC 3825 — domination
Market economies
PD 9679 — Decline exports developing countries developed
PF 1737 — Slow down growth output developed
PC 0939 — socialist countries ; Black
PC 7459 — *Structural imbalances three largest*
PE 2954 — Weakness trade socialist developed
PC 6641 Market economy ; Black
PF 7880 Market economy ; Uncontrolled
Market [facilities...]
PD 7461 — facilities ; Hierarchical control
PD 3507 — facilities ; Unsystematic allocation
PD 0156 — flooding developing countries ; Labour
PF 7880 — forces ; Uncontrolled
Market [imperfections...]
PF 7556 — imperfections
PE 8916 — Inaccessible job
PF 4470 — Inadequate rural community
PE 1843 — indicators exclusion human requirements
PD 4011 — inflexibility ; Labour
Market [Lack...]
PE 8283 — Lack time flexibility labour
PF 8081 — leadership ; Loss international
PC 7997 — Limited job
PD 4632 Market middlemen ; Unnecessary

PF 7556 Market operation ; Imperfect
PF 0581 Market opportunities ; Restricted
Market [Pandering...]
PF 5726 — Pandering youth
PD 6002 — position international trade ; Abuse dominant
PF 0581 — potential ; Undeveloped
PF 0581 — potential ; Unknown
PE 7313 — *practices ; Traditional*
PF 1086 — practices ; Unprofitable
PF 8635 — price ; Uncontrolled
PF 6980 — prices ; Misleading
PF 8635 — prices ; Unskilled
PF 2009 *Market refrigeration ; Lack*
PE 8110 Market related interest rates ; Rescheduling debts developing countries
Market research
PC 2921 — Clandestine
PF 1086 — Limited
PF 6928 — Parochial
PF 1025 — Sales dominated
PF 1025 — Sales promotion dominates
PF 1086 Market resource ; Limited
PD 1873 Market restrictions developing countries ; Domestic
Market [share...]
PD 8994 — share ; Declining international
PF 6153 — shopping facilities ; Impersonality mass
PF 2989 — *Shrinking local*
PF 2995 — skills ; Victimized
PF 2995 — skills ; Victimizing
PD 2413 — socialist countries ; Currency black
PF 8299 — supply centres ; Inaccessible
PC 6641 Market trading ; Black
Market [Unorganized...]
PC 7997 — Unorganized job
PF 2411 — Unprofitable entertainment
PF 2989 — Unrealized local
PF 0581 — Unstudied export
PE 2535 Market variance ; Unanticipated
PE 2535 Market volume ; Uncertain
PF 2995 *Marketable skills ; Unrealized*
PE 0310 Marketing arrangements ; Discriminatory orderly
PE 6036 Marketing banned pharmaceutical drugs developing countries
PD 0930 Marketing ; Blocked global
Marketing [Disorganized...]
PF 2989 — *Disorganized local*
PF 1025 — Disrelated needs promotion
PE 2397 — distribution channels transnational corporations ; Control
PE 7344 Marketing employable skills ; Limited means
PF 2411 *Marketing facilities ; Defensive*
PE 4522 Marketing jute products developing countries ; Inadequate international
PF 2995 Marketing knowledge ; Inadequate
PF 1086 Marketing lines ; Limited
PF 6465 Marketing methods ; Archaic
PE 3528 Marketing ; Middleman control rural
PF 0581 Marketing opportunities ; Unutilized
Marketing [practices...]
PF 3554 — practices ; Fragmented
PF 3554 — *practices ; Individualized*
PE 0523 — products developing countries ; Inadequate
PE 7752 Marketing skills which reinforce self image being victim circumstances
PF 6533 Marketing systems ; Surrendered control
PE 7313 *Marketing ; Traditional farm*
PE 7313 *Marketing ; Underdeveloped farm*
PF 0905 *Marketing ; Unprofitable farm*
PC 6641 Markets ; Black
PE 0521 Markets ; Captive
PE 5926 Markets developed countries against exports developing countries ; Restrictive business practices
PD 9679 Markets developed countries developing countries ; Inaccessibility
Markets developing countries
PE 9457 — Excessive concentration export
PE 1471 — Restricted growth export
PE 5902 — Undeveloped financial
PD 4511 Markets ; Disruption financial
PF 8299 Markets ; Distant available
Markets [encountered...]
PF 1471 — encountered developing countries ; Increase competition export
PD 5676 — Excessive interdependence stock
PF 8299 — Excessively distant
PF 7556 Markets ; Failure
PF 1226 *Markets ; Fluctuating agricultural*
PF 2989 Markets goods services ; Limited local
PF 9697 Markets goods services ; Vulnerability national economies vagaries external
Markets [Imperfections...]
PF 3136 — Imperfections capital
PF 6980 — Inefficiency financial
PC 7873 — Instability monetary
PF 0581 — Insufficient product
Markets Lack
PJ 0385 — arbitrating commodity
PE 3528 — *dairy*
PE 4508 — international coordination supervisors financial stock
PF 7556 — protection free
PF 1415 *Markets ; Limited traditional*
PE 5102 Markets ; Low confidence investment stock
PD 8647 Markets ; Manipulation commodity
PD 7282 Markets ; Marriage

Markets offshore

PE 4994 Markets offshore insurers ; Destabilization national insurance
PF 0905 *Markets ; Overpowering competitive*
PE 4301 *Markets ; Price fixing commodity*

Markets [Restricted...]
PF 1471 — Restricted access developing country products world
PD 3135 — Restrictions foreign access capital bond
PD 8897 — Rivalry international

Markets [Segmented...]
PD 6744 — Segmented labour
PF 0581 — Severely limited
PF 2989 — *Small local*
PD 9637 — Speculation commodities futures
PD 9489 — Speculation money
PD 4011 — Structural rigidities labour
PE 0194 Markets transnational corporations ; Excessive control raw materials

Markets [Uncontrolled...]
PF 7880 — Uncontrolled
PE 1925 — *Uncontrolled land*
PF 0581 — *Unexplored potential*
PF 0581 — *Unidentified potential*
PF 1844 — *Unsanitary*
PF 8299 — *Untapped regional*
PD 5676 Markets ; Vulnerability stock
PF 4028 Markets ; Wage rigidity labour
PJ 5262 Markets weaker foreign economies ; Effect dominant stock
PE 6749 Marking navigable waters ; Non uniformity
PE 7933 Marmosets ; Endangered species tamarins
PF 2100 Marriage annulment
PF 3284 Marriage ; Arranged
PD 2103 Marriage breakdown

Marriage [Child...]
PF 3285 — Child
PF 3283 — *Clandestine*
PD 0518 — Conflicts
PC 2379 — Consanguineous
PD 1915 — consent
PF 1870 — covenant ; Naive approach
PF 1870 — covenant ; Reduced dimension

Marriage [Denial...]
PF 3343 — Denial right
PD 1915 — Denial right choice
PD 3694 — Dependency women
PF 2100 — divorces ; Repeated
PF 3290 — dowry ; Prohibitive

Marriage [Early...]
PE 7628 — Early
PF 3342 — Economic impediments
PD 3244 — Emotional dependency
PD 2103 — Estrangement
PD 1915 Marriage ; Forced
PF 3288 Marriage ; Group

Marriage [Ideological...]
PF 3345 — Ideological impediments
PE 7935 — Illegal
PF 3343 — Impediments
PD 2294 — Individually defined operating structure
PF 1870 — institution ; Vulnerability
PD 0355 — Inter ethnic
PD 0355 — Interdenominational
PC 1523 — Interracial

Marriage law
PF 3254 — Desertion
PF 3286 — *Monogamous*
PF 3251 — Separation

Marriage [Legal...]
PF 3346 — Legal impediments
PF 1870 — *Lessening regard sanctity*
PF 3278 — like unions
PD 7282 Marriage markets
PD 0355 Marriage ; Mixed
PF 3283 Marriage ; Non validity

Marriage [partners...]
PF 9818 — partners ; Psychological inconsistency
PE 7628 — Pressured early
PF 3284 — *proxy*
PF 3344 — Psychological impediments

Marriage [Racial...]
PC 1523 — Racial impediments
PF 4948 — Rape
PF 1000 — Reduced interior structures
PD 0355 — Religious impediments
PD 3347 Marriage ; Segregation
PF 3341 Marriage ; Social impediments

Marriage [Tax...]
PF 3342 — *Tax impediments*
PF 1870 — Threat survival
PF 3278 — Trial
PF 1870 Marriage vows ; Social irrelevance
PE 9379 Marriageable men ; Shortage
PJ 2657 Marriageable partners ; Lack
PE 0427 Marriageable women ; Shortage
PD 2103 Marriages ; Broken
PD 0941 Marriages ; Discontented

Marriages [Inadequate...]
PD 1308 — Inadequate image roles
PE 4985 — Insecurity Western
PF 2241 — *Investment determined*
PU 3035 — *Irregular*
PF 1870 Marriages ; Lack sustaining symbols
PD 1915 Marriages ; Levirate
PE 7628 Marriages ; Restrictive early
PD 0941 Marriages ; Unhappy

Married women
PD 3557 — Adverse social consequences excessive employment
PD 3557 — Excessive employment
PE 4624 — Imposition husband's name
PE 8720 — Loss civil capacity
PF 2303 — *Unemployment*
PE 9554 Marrow displacement abnormal cells
PF 3341 *Marrying down socially*
PF 5430 *Marrying up socially*
PE 5530 *Marseilles fever*
PE 0616 Marsh fever
PE 8527 Marsupial mice ; Endangered species dasyures
PE 3981 Marsupial moles ; Endangered species
PD 1762 Marsupials ; Endangered species
PD 2637 Martial law
PD 6380 Martial law ; Restrictions rights
PD 7550 Martyrdom
PF 8118 Martyrdom ; Childhood
PF 2189 Marxism
PE 5546 Marxist crimes ; Revisionism anti
PC 3685 Masked racism
PF 3264 Masochism
PE 6137 Masochism ; Sado
PE 3851 Masochism ; Sexual
PF 0695 Masonic conspiracy
PD 6970 *Mass arrests ; Politically motivated*
PF 3311 *Mass ; Black*
PJ 4748 Mass catering hazards
PE 7172 Mass cemeteries ; Isolated
PF 7435 *Mass ; Decreasing land*
PD 0702 Mass detention political prisoners

Mass [education...]
PF 4765 — education ; Failure
PD 0531 — exodus
PD 0531 — expulsion ; Forced
PA 0832 — *extermination*
PF 0039 — extinction due comet showers
PF 4149 — extinctions species ; Periodic
PF 2855 — extraction natural resources ; Emphasis
PF 4008 Mass immigration
PE 9650 Mass killing ; Denial animals right freedom
PD 2483 Mass killings
PE 6153 Mass market shopping facilities ; Impersonality

Mass media
PD 0917 — bias towards entertainment reassurance
PE 7638 — Biased portrayal women
PF 2573 — Dangers private control communications
PD 6279 — Excessive portrayal violence
PE 7448 — Manipulation individual
PD 1983 — Psychological pollution
PD 4597 — State control communications

Mass [murder...]
PD 2341 — *murder*
PD 5590 — murder ; Political
PE 7645 — murderers ; Psychotic
PD 0528 Mass production techniques ; Unemployment due
PC 0690 Mass protests
PC 6203 Mass resettlement ; Involuntary
PC 7896 Mass tree deaths
PE 0597 Mass trials ; Injustice
PD 2046 Mass unemployment human resources
PE 3721 Massacre indigenous populations imported diseases
PD 2483 Massacres
PC 1056 Massacres ; Genocidal
PF 4062 Massacres ; Undocumented
PF 3361 Masses ; Restrictive influence religion
PD 1060 Masses setting criteria judicial judgements ; Exclusion
PD 3492 Massive destructiveness ; Incendiary weapons
PC 1453 Massive international capital flows
PE 6968 Massive psychic traumatization
PB 0315 Massive starvation despite sufficient world food supply
PE 1848 *Massive urban emigration*
PD 9742 *Mastitis*
PF 7799 *Mastitis animals*
PD 2731 *Mastitis milking cows*
PC 9042 *Mastitis puerperalis*
PD 2567 Mastoid process ; Diseases
PD 2567 *Mastoiditis*
PE 4426 Masturbation
PF 5609 Masturbation guilt
PF 2792 Matching talent jobs ; Limited ways
PD 6881 Mate abuse
PE 0915 *Mate ; Instability trade tea*
PF 0296 *Material ; Costly furniture*
PD 3557 *Material expectations*
PD 1230 Material fractures
PE 8438 Material ; Insufficient educational
PE 8152 Material ; Limited access culturally adapted pedagogical
PE 0060 Material reserves transnational enterprises ; Excessive exploitation raw
PC 0311 Material ; Stockpiles nuclear warfare
PF 1321 Material technological needs demands ; Gap
PE 6884 Material ; Transport storage radioactive
PC 2157 Material well being because discrimination ; Denial right
PF 2655 Materialism
PF 1673 Materialism ; Media reinforcement
PF 2655 *Materialism ; Medical*
PF 1673 Materialistic images ; Excessive dissemination
PE 9525 Materials ; Atmospheric corrosion
PF 0166 Materials ; Barriers international flow knowledge educational

Materials [Commercially...]
PD 9370 — Commercially biased educational
PE 0669 — components utilization ; Restrictions raw

Materials [Covert...] cont'd
PE 3968 — Covert trade nuclear bomb making

Materials [Dangerous...]
PC 6913 — Dangerous
PD 1251 — *Deficient housing*
PF 1183 — Deliberate misrepresentation educational
PD 2968 — Dependence developing countries export limited range raw
PD 0837 — Diversion nuclear

Materials Environmental hazards
PE 0546 — inedible crude non fuel
PE 1346 — mineral fuels, lubricants related
PD 8566 — plastic
PF 3550 *Materials ; Expensive construction*

Materials [Failure...]
PD 2638 — Failure
PD 1391 — Fatigue
PD 2211 — Flammable construction
PE 6860 — frontier areas ; Dangerous
PE 1499 Materials ; Health hazards modern insulating

Materials [Illegal...]
PE 3968 — Illegal exports nuclear
PF 3499 — *Inaccessible supply repair*
PG 2250 — *Ingestion injurious*

Materials Instability trade
PE 8232 — essential oils perfume
PD 0280 — inedible crude non fuel
PD 0877 — mineral fuels, lubricants related
PF 6465 *Materials ; Insufficient recycling*
PE 7174 Materials irresponsible countries ; Trade nuclear

Materials [Lack...]
PD 4270 — Lack raw
PF 2212 — *Lack tuberculosis*
PD 1251 — *Limited housing*
PE 0461 — Long term shortage inedible crude non fuel
PE 0194 Materials markets transnational corporations ; Excessive control raw
PE 8789 Materials needed act against problems ; Shortage equipment
PF 2115 Materials needed development ; Lack essential services

Materials [personal...]
PE 6214 — personal products ; Over use formaldehyde building
PC 1192 — products ; Environmental hazards chemical
PF 3550 — *Prohibitive cost building*

Materials [Radiation...]
PD 1206 — Radiation damage
PD 4270 — Raw
PE 8351 — Restrictive practices trade inedible crude non fuel

Materials [shortage...]
PE 8438 — shortage ; Educational
PE 8501 — Shortage tanning dyeing
PC 6913 — *Storage hazardous*
PD 7216 — Stress
PD 1230 — structures ; Failure engineering

Materials [Theft...]
PD 3495 — Theft nuclear
PE 4675 — Toxic modern furniture
PE 8843 — trade instability ; Radioactive associated

Materials [Unavailability...]
PF 3554 — *Unavailability building*
PF 6590 — *Under utilized raw*
PC 6913 — *Underground storage hazardous*
PD 4270 — Unstable supply raw
PE 8857 Maternal child health care ; Inadequate

Maternal [deaths...]
PD 2422 — deaths
PF 5968 — depletion syndrome
PC 0981 — deprivation
PD 9021 Maternal education ; Inadequate
PE 1085 Maternal malnutrition
PD 2422 Maternal mortality
PC 0981 Maternal negligence
PC 0981 Maternal rejection
PF 5255 Maternalism
PS 3751 *Maternity*
PE 3601 Maternity ; Government enforced

Maternity leave
PE 3951 — Denial right
PE 7119 — Denial right job protection during
PD 6733 — *employees ; Disruptive*
PD 6733 Maternity protection employment ; Inadequate
PD 6728 Mathematical ignorance
PE 0651 Matricide
PF 9269 Matter bombs ; Anti
PE 1256 *Matter ; Damage degradable organic*
PD 4552 Matter ; Proliferation printed
PD 0579 Matter water ; Suspended
PE 7176 Matters ; Unlawful compensation assistance government
PF 2174 Maximization ; Short term profit
PF 2174 Maximization ; Short term revenue
PE 3733 Mayhem
PE 8084 Meagre youth facilities countryside

Meal flour
PE 8606 — than wheat meslin shortage
PE 8980 — wheat meslin shortage
PE 8489 — wheat meslin trade instability
PA 7377 *Meaning*complex*
PB 2619 Meaning ; Corruption
PB 2619 Meaning ; Dependence corruption
PF 0936 Meaning human creativity ; Collapsed
PF 3777 Meaning ; Institutionally reinforced sense personal shallow
PF 0955 Meaning participating society ; Collapse

Medical services

Meaning [Search...]
PF 6796 — Search individualistic
PE 4589 — Social reinforcement shallow personal
PE 4589 — Socially reinforced shallow perception personal
PF 0966 Meaningful educational context ethical decisions ; Lack
PD 0894 Meaningful personal social paradigms ; Lack
PF 2005 *Meaningless conversions*
PD 2671 Meaningless corporate engagement
PF 0386 Meaningless recreation
PA 6977 Meaninglessness
PA 6977 Meaninglessness ; Dependence
PA 6220 Meaninglessness life
PA 7193 Meanness *Cheapness*
PA 6839 Meanness *Disrepute*
PA 7253 Meanness *Envy*
PA 6659 Meanness *Humility*
Meanness [Imperfection...]
PA 6997 — *Imperfection*
PA 5652 — *Inferiority*
PA 5473 — *Insufficiency*
PA 7306 Meanness *Narrowmindedness*
PA 7211 Meanness *Selfishness*
PA 5942 Meanness *Unimportance*
PA 5643 Meanness *Unkindness*
PA 5821 Meanness *Vulgarity*
PD 2438 Means achieving consensus ; Lack
PF 6495 Means goods supply distribution ; Ineffective
PD 2592 Means impeding right self determination ; Use mercenaries
PF 1653 *Means ; Insufficient capitalization*
PF 6454 Means local technological development ; Lack
PE 7344 Means marketing employable skills ; Limited
PD 1975 Means measuring intelligence ; Inadequate
PE 8430 Means participation decision making ; Ineffective
PF 2014 Means production ; Unregulated ownership
PD 4392 Means searching partner ; Ineffective
PE 4399 Means subsistence ; Denial right people their own
PF 2798 Means transport ; Restrictions effective
PF 1817 Means upholding global concern ; Inadequate
PE 1603 Measles
PE 0785 Measles ; German
PE 2757 Measures animal housing equipment ; Inadequate disinfection
PE 0083 Measures ; Distortion international trade discriminatory requirements respect product standards
PE 8409 Measures humans during animal disease outbreaks ; Inadequate disinfection
PD 1091 Measures ; Lack eugenic
Measures [Narrow...]
PF 3871 — Narrow egalitarian application trade
PD 8114 — Negligent implementation austerity
PF 5670 — numbering systems ; Terminological confusion weights,
PE 1182 Measures regulate domestic prices ; Distortion international trade minimum pricing regulations
PD 5033 Measures ; Unpredictable introduction protectionist
PD 1975 Measuring intelligence ; Inadequate means
PE 7064 Meat animal product shipments ; Infected animal,
Meat [eating...]
PC 7644 — eating
PE 9115 — egg production ; Lack
PE 8013 — egg products ; Increased demand developed countries
PD 4518 — Excessive consumption
PE 8398 — exports ; Inadequate hygiene restrictions carcass
Meat [infected...]
PE 7064 — infected animals ; Insufficiently treated
PD 2772 — *Inhumane livestock slaughtering*
PE 8257 — Instability trade dried salted smoked
PE 8591 — Instability trade fresh, frozen chilled
Meat meat preparations
PE 0133 — Environmental hazards
PE 0755 — Instability trade
PE 1490 — Long term shortage
PE 8693 — Shortage
Meat preparations
PE 0133 — Environmental hazards meat
PE 0755 — Instability trade meat
PE 1490 — Long term shortage meat
PE 8693 — Shortage meat
PD 1714 Meat production ; Health hazards drug use
PD 4322 Meat production ; Unequal distribution
PE 2770 Meat quality intensive animal farming units ; Inferior
PE 8224 Meat ; Shortage fresh, chilled frozen
PE 0596 *Meat ; Undercooked*
PE 2777 Meats factor animal diseases ; Importation infected carcass
PC 1904 Mechanical equipment failure
PC 1904 Mechanical failure
PF 2995 *Mechanical maintenance ; Unskilled*
Mechanical [services...]
PE 8986 — services developing countries ; Limited
PD 7216 — stress
PB 0731 — *suffocation*
PF 3062 Mechanism balance payments adjustment ; Inadequate
PF 3061 Mechanism creation liquid reserves ; Inadequate
PE 8776 Mechanism international action ; Inadequate organizational
PF 0301 Mechanism response problems ; Inadequate delivery
PF 2431 Mechanisms act against problems ; Inadequate organizational
PF 2971 Mechanisms community health ; Disrupted
PF 6505 *Mechanisms ; Complex funding*
PE 9762 Mechanisms ; Discriminatory managed trade
PF 1352 Mechanisms functional training ; Ineffective
PF 2477 *Mechanisms ; Insufficient employment*

PF 6550 *Mechanisms ; Obscure decision*
PF 2857 Mechanisms securing sufficient food supplies ; Inadequate
PD 6571 *Mechanization ; Crippling dependence*
PC 0492 Mechanized agriculture ; Lack
Media [Absence...]
PF 0076 — Absence audio visual
PD 0917 — against development social transformation ; Bias
PD 5535 — *animals ; Otitis*
Media [Bias...]
PD 6081 — Bias
PD 0917 — bias towards entertainment reassurance ; Mass
PE 7638 — Biased portrayal women mass
Media [channels...]
PF 4611 — channels ; Overlooked
PD 6081 — Confiscation news
PE 1643 — Conflict government news
PD 5251 — Corruption
PD 4383 — cover-up
PD 6081 — coverage news ; Biased
PE 9643 — cultural imperialism ; Submission
PF 2573 Media ; Dangers private control communications mass
PD 7773 Media ; Dependence
Media [emphasis...]
PE 1478 — emphasis bad news
PD 0917 — entertainment ; Parochial escapist
PD 5207 — events ; Political
Media Excessive
PE 4215 — commercialization
PE 3831 — emphasis developing country weaknesses
PD 5475 — politicization
Media Excessive portrayal
PE 7354 — crime
PE 1478 — destabilizing information
PE 3980 — drugs, alcohol cigarettes
PE 1478 — negative information
PE 3831 — perspectives industrialized cultures
PE 7930 — sex
PE 3980 — substance abuse
PE 6844 — terrorist activity
PD 6279 — violence mass
Media [Excessive...]
PE 1478 — Excessive unsupportive criticism
PE 9643 — Excessive use foreign programmes
PE 7638 — Exploitative portrayal women
Media Glorification
PE 7354 — crime
PE 7930 — demeaning sex
PE 3980 — substance abuse
PD 6279 — violence
PD 0160 Media ; Harassment
PD 3071 Media ; Harassment
Media [image...]
PE 8802 — image foreign groups peoples ; Biased
PE 3831 — Inadequate coverage developing country perspectives
PF 3211 — *Insufficient cultural*
PD 3040 — International monopoly
PD 9603 — Invasion privacy
Media [Maltreatment...]
PE 4786 — Maltreatment animals
PE 7448 — Manipulation individual mass
PD 3040 — monopolies ; Emergence international
PD 3101 — Monopoly
PD 5475 Media ; Nationalistic
PD 5251 Media ; Negligence
PF 0076 *Media options ; Narrow*
PD 2567 *Media ; Otitis*
Media [personality...]
PC 1123 — personality cults
PD 5251 — practices ; Irresponsible
PD 5251 — practices ; Unethical
PD 6081 — presentations ; Distorted
PD 1983 — Psychological pollution mass
Media [reception...]
PD 3039 — reception ; Unequal opportunities
PF 3079 — Refusal grant licences
PF 3079 — Refusal licence
PF 1673 — reinforcement materialism
PF 3082 — Restriction access news distribution
PD 5377 Media standards ; Deterioration
PD 4597 Media ; State control communications mass
PF 9631 Media theatricalization public life politics
PD 0040 Media ; Uncontrolled
PF 3628 *Media ; Unenticing printed*
PD 6279 Media violence
PE 6711 Medical backlash
Medical care
PD 5119 — Delay administration
PD 2028 — Denial right adequate
PE 8400 — developing countries ; Costly
PF 4832 — Inadequate
PD 4790 — infrastructure ; Inadequate
PF 4832 — Lack
PF 5709 — Over specialization
PE 4820 — pregnant women ; Inadequate
PF 4832 — Prejudiced
PF 8016 — Prohibitive cost private
PF 4244 — Refusal
PE 4901 — systems ; Resistance incorporating midwives
Medical [complications...]
PE 2863 — complications
PE 3828 — consent ; Denial right
PF 5983 — cover-up
PE 8261 — *cures ; Inaccessibility high technology*
PD 9836 Medical deception

Medical devices
PD 3998 — *Defective*
PE 8502 — industries ; Ineffective self regulation pharmaceutical
PD 2697 — Unethical experiments drugs
Medical [diagnosis...]
PE 4129 — diagnosis due inadequate equipment ; Inaccurate
PF 1735 — *disinterest ; Local*
PF 3006 — double standards
Medical drugs
PD 0028 — Abuse
PE 5755 — Excessive cost
PD 0644 — Excessive proliferation
PE 3828 — non medical purposes ; Administering
PF 1634 — Shortage
Medical [equipment...]
PC 4844 — *equipment ; Restricted access sophisticated*
PE 4859 — equipment ; Unsafe industrial, laboratory
PD 5770 — ethics ; Lack
PF 6577 — *expectations ; Unfeasible*
Medical expenses
PE 8261 — Inflated
PE 8261 — Prohibitive
PE 8261 — Unnecessary
Medical experimentation
PD 0260 — animals
PE 6764 — children
PE 4781 — during war time ; Inhumane
PE 6763 — institutionalized subjects
PE 8677 — mentally impaired persons
PF 4865 — Obstacles
PE 8343 — pregnant women foetuses
PE 4889 — prisoners ; Unethical
PD 6760 — socially vulnerable groups
Medical [experiments...]
PD 2697 — experiments humans ; Unregulated
PD 4271 — experiments ; Traffic children
PD 4271 — exploitation ; Trafficking children
Medical facilities
PD 4004 — Inadequate
PF 4832 — Inadequate
PF 4832 — personnel ; Insufficient
PF 5983 Medical facts records ; Secrecy
PD 5770 Medical fraud
PF 4861 Medical high risk persons ; Non surveillance
PF 3006 Medical hypocrisy
Medical [ignorance...]
PD 8821 — ignorance
PF 4832 — *information ; Lack*
PD 9836 — information ; Withholding
PE 7705 — intervention childbirth ; Excessive
PF 0420 — intervention ; Irrational conscientious refusal
PD 5770 — irresponsibility
PD 9160 — Medical knowledge ; Limited
PC 7674 *Medical laboratory facilities developing countries ; Inadequate*
PF 4936 Medical life support systems ; Inhumane use
PD 5770 Medical malpractice
PF 2655 *Medical materialism*
PD 5770 Medical negligence
PE 4329 Medical negligence ; Increase insurance claims
PF 5983 Medical obstruction
Medical [paternalism...]
PF 5770 — paternalism
PD 1662 — patients ; Denial rights
PE 1704 — patients radiation ; Excessive exposure
Medical practice
PD 6334 — Diseases
PD 5770 — Immoral
PE 5637 — Unauthorized
PD 5770 — Unethical
Medical practices
PD 5770 — Abusive
PF 1624 — Culturally determined
PE 3946 — Ignorance traditional
PF 1624 — Inconsistent
PD 8831 — Restrictive
PE 5027 Medical practitioners refusing treat patients
Medical profession
PD 5770 — Degradation
PE 5355 — Occupational risks hazards
PE 4015 — torture ; Inhumane participation
Medical [professionals...]
PD 9836 — professionals ; Lying
PE 8400 — programmes developing countries ; Failure
PD 5770 — psychological techniques ; Abuse
PE 3828 — purposes ; Administering medical drugs non
PD 1725 Medical quackery
Medical [records...]
PF 1836 — records ; Destruction
PF 1836 — records ; Lack
PE 0952 — referrals ; Fraudulent
PE 0612 — research ; Lack funds
Medical resources
PE 8261 — *heal privileged ; Discriminatory use*
PD 7254 — Inadequate
PF 9587 — Inadequate
PC 4844 — *Inequitable use*
PD 2705 — Maldistribution
Medical [science...]
PF 8326 — science ; Inadequacy
PE 8671 — science ; Lack diversity
PF 4841 — self experimentation ; Errors risks
PF 4832 — services ; Distant
Medical services Inadequate
PF 4832 — [Medical services ; Inadequate]
PD 1428 — emergency

-985-

Medical services

Medical services Inadequate cont'd
PE 7718 — working conditions health
Medical [skills...]
PD 9160 — skills ; Limited
PF 5097 — study ; Shortage biological specimens
PE 1634 — supplies ; Insufficient
Medical [techniques...]
PE 8261 — *techniques ; Waste resources expensive*
PD 7415 — test results ; Misuse scientific
PD 7415 — testing ; Inappropriate scientific
PF 5679 — tests ; Dubious
PD 7415 — tests ; Inconclusiveness scientific
PD 9160 — training ; Inadequate
Medical treatment
PD 7821 — Degrading
PD 1480 — Depriving prisoners
PD 7254 — Rationing
PD 7821 — Undignified
PF 4832 Medical visits ; Infrequent
PE 4725 Medical waste
PD 9807 Medication side effects
PJ 5595 Medications ; Excessive restrictions legalizing
PE 4896 Medications ; Hazardous genetic effects
Medicinal [pharmaceutical...]
PE 7996 — pharmaceutical products trade instability
PD 4171 — plants ; Endangered species
PE 3502 — products shortage
PD 1239 *Medicinal use occupational carcinogens*
PE 0751 Medicine ; Insufficient preventive
PE 5721 Medicine ; Lack holistic
PF 4871 Medicine ; Lack integration traditional Western
PD 0644 *Medicine ; Reverence drugs*
PF 2971 *Medicine ; Unavailability quality*
PE 2047 Medicines containing alcohol
PD 5540 Medicines ; Counterfeit
PD 8402 Medicines ; Misuse
PE 1278 Medicines ; Unavailability
PF 3900 Mediocrity
PF 3900 Mediocrity ; Cult
PF 3962 Mediocrity government leaders
PA 6997 Mediocrity *Imperfection*
PA 5652 Mediocrity *Inferiority*
PA 5942 Mediocrity *Unimportance*
PA 7232 Mediocrity *Unskillfulness*
PE 6795 Meditation ; Inaccessible places
PE 9528 Medium-term cyclic variations solar radiant energy
PF 1267 *Meeting agendas ; Unpublicized*
PE 1535 Meeting facilities ; Limited
PF 8939 Meeting formats ; Ineffective
PF 8939 Meeting methods ; Inadequate
PF 3756 Meeting oratory ; Empty
PE 2086 Meeting resolutions ; Tokenistic
PF 7130 *Meeting space ; Expensive*
PE 1535 Meeting space ; Insufficient
PF 8939 Meeting time ; Unproductive
PF 1333 Meetings ; Argumentative community
PD 4060 Meetings ; Curtailment discussion issues
Meetings Ineffectiveness inefficiency
PF 0316 — intercultural
PF 0409 — interdisciplinary
PF 0349 — international
PF 5250 Meetings ; Infrequent community
PF 5250 Meetings ; Irregular village
PE 2007 Meetings ; Military intervention workers
Meetings [Prohibition...]
PD 7210 — Prohibition trade union
PD 8522 — Protest
PF 3756 — Public posturing
PF 3756 Meetings ; Rhetorical inflation
Meetings [Unannounced...]
PF 5222 — Unannounced community
PF 5222 — Unknown community
PF 5222 — Unpublicized public
PD 4811 Meetings ; Violent police intervention
PE 2007 Meetings ; Violent police intervention workers
PF 4975 Mega projects ; Implementation development
PF 2108 Megalomania
PD 1318 *Melancholia ; Involutional*
PF 7756 Melancholy
PA 7365 Melancholy *Boredom*
PA 7253 Melancholy *Envy*
PA 6731 Melancholy *Solemnity*
PA 7107 Melancholy *Unpleasantness*
PE 5016 Melanoma
PD 2732 *Meleagridis infection turkeys ; Mycoplasma*
PD 6363 Melioidosis
PD 9654 *Mellitus animals ; Diabetes*
PE 0102 Mellitus ; Diabetes
PF 5337 Member governments United Nations ; Irresponsibility
PF 7036 Members ; Denial right recruit union
PF 1750 Members ; Failure recognize uniqueness family
PE 4868 Members ; Immigration barriers handicapped family
PD 2239 Members ; Parochial attitudes organizations'
PD 8629 Members professional bodies ; Discrimination against non
Members [Separation...]
PE 4959 — Separation family
PE 4001 — society ; Institutionalized
PF 4000 — society ; Non productive
PJ 0272 *Members their representatives ; Isolation trade union*
PE 7092 Membership apathy ; International nongovernmental organization
PD 4098 Membership ; Compulsory organization
PF 4876 Membership ; Distortion intergovernmental organizations mini state
PF 8650 Membership dues ; Non payment United Nations

PE 0803 Membership ; Economic barriers access legal profession, judiciary jury
Membership [Inadequate...]
PE 1184 — Inadequate coordination regional intergovernmental organizations common
PE 3832 — international organizations ; Conditional
PF 5337 — *Irresponsibility United Nations developing country*
PE 1583 Membership ; Jurisdictional conflict antagonism regional intergovernmental organizations common
PF 8650 Membership payments intergovernmental organizations ; Withholding
PG 6761 *Membrane larynx windpipe ; Swelling mucous*
PD 8484 Memory abnormalities
PD 8484 Memory defects
PE 6593 Memory due torture ; Loss
PD 0306 *Memory ; Erosion societal*
PD 8297 Memory ; Loss
PF 7027 Memory ; Societal selective
PF 6570 *Memory ; Vague community*
PC 3258 Men child bearing ; Discrimination against
PE 4010 Men ; Denial rights parent
PC 3258 Men ; Discrimination against
PD 8909 Men education ; Discrimination against
PD 3338 Men employment ; Discrimination against
PF 1390 Men ; Gay
PE 9060 Men ; Inadequate family planning education
PD 3692 Men law ; Discrimination against
PE 4010 Men parental rights ; Discrimination against
PE 8507 Men public services ; Discrimination against
PD 3443 *Men ; Reservation jobs*
Men Sexual
PD 3263 — exploitation
PE 1293 — harassment
PF 3271 — harassment
PF 6415 — impotence
PE 1293 — provocation
PF 3271 — provocation
Men [Shortage...]
PE 9379 — Shortage marriageable
PE 1339 — Shorter life expectancy
PF 3613 — Social inadequacy
PD 3336 — social services ; Discrimination against
PE 4232 — sports ; Discrimination against
PF 0551 Men ; Undomesticated
PF 8780 Men women ; Socializing unrelated
PE 7942 Men women ; Unequal distribution old age pensions
PA 5454 Menace *Badness*
PA 6971 Menace *Danger*
PA 6191 Menace *Disapproval*
PE 4467 Menacing ; Criminal
PA 7411 Mendacity *Uncommunicativeness*
PA 6434 Mendicancy *Poverty*
PD 3427 Menial work status immigrants industrialized countries
PD 2567 *Ménière's disease*
PE 2280 Meningitis
PD 2731 Meningitis animals
PE 7841 Meningitis ; Infectious thromboembolic
PE 2280 Meningitis ; Meningococcol
PE 2280 Meningitis ; Pneumococcol
PE 2280 Meningitis ; Pyogenic
PE 2280 Meningitis ; Syphilitic
PE 2280 Meningitis ; Tuberculous
PE 9037 Meningo-encephalomyelitis
PE 2280 Meningococcol meningitis
PE 2356 Meningoencephalitis
PE 4604 Meningoencephalomyelitis ; Granulomatous
PD 8775 Menopausal symptoms
PF 5918 Menopause
PD 5783 Menopause ; Male
PE 4838 Menstrual bleeding
PE 4838 Menstruation
PD 8775 *Menstruation ; Disorders*
PE 4838 Menstruation ; Negative attitudes towards
Mental [activity...]
PC 1587 — activity ; Impaired
PD 4755 — anguish torture victims
PF 4925 — asylums ; Inhumane
PC 0799 Mental breakdown
Mental [capacity...]
PE 4649 — capacity age ; Diminishing
PE 5087 — conflict
PD 0584 — *cruelty*
Mental [deficiency...]
PC 1587 — deficiency
PD 0914 — deficiency children
PC 1587 — degeneracy
PC 0799 — depression
PD 3784 — depression children
PD 9131 — disorder
PF 8490 — *disorder ; False diagnosis*
Mental disorders
PD 9131 — [Mental disorders]
PD 0919 — aged
PC 0300 — Non psychotic
PE 5087 Mental disturbance
PE 4960 Mental dullness
PE 8925 Mental factors ; Physiological malnutrition arising
PC 1587 Mental feebleness
Mental health
PC 0663 — *Denial right highest obtainable physical*
PF 4955 — Inadequate care
PF 3850 — Negative effects over crowding
PF 4955 — services ; Constraints development
PF 4955 — *services ; Inadequate*
PE 4925 Mental hospitals ; Inadequate
PA 0294 Mental ill health ; Physical

Mental illness
PC 0300 — [Mental illness]
PE 0989 — adolescents
PF 8387 — Leadership impaired
PC 0300 — *Nutritionally induced*
Mental [impairment...]
PF 4945 — impairment
PF 4945 — inefficiency
PE 4430 — institutes ; Political prisoners
Mental [patients...]
PD 1148 — patients ; Denial rights
PJ 3545 — *physical emotional strain*
PB 6248 — pollution
PB 6248 — pollution ; Dependence
Mental retardation
PC 1587 — [Mental retardation]
PE 4791 — Borderline
PE 4821 — Mild
PC 1587 — *Moderate*
PE 6413 — Profound
PE 6314 — Severe
Mental [set...]
PF 5301 — set
PE 5164 — stress due automation
PF 7713 — structures challenge human survival ; Inadequacy prevailing
PB 5680 — suffering
PB 5680 — suffering ; Dependence
PB 6302 Mental tension
PF 3243 Mentality ; Immediate needs
PE 3296 *Mentality ; Isolated village*
PF 6611 Mentality ; Nuclear
PF 6575 *Mentality ; Paralyzing boom/bust*
PC 0799 Mentally depressed people
Mentally disabled
PE 8900 — criminals ; Inadequate facilities
PD 9183 — Discrimination against
PE 4544 — Sterilization
Mentally handicapped
PC 1587 — [Mentally handicapped]
PE 4544 — Denial right procreate severely
PD 9685 — Exploitation
PE 8151 — Inadequate rehabilitation facilities
PE 8151 — Inadequate support services
PC 0699 — persons ; Physically
PD 1405 — Stress families physically
Mentally ill
PE 0989 — children
PD 9685 — Exploitation
PD 1148 — persons ; Violation rights
PE 8677 Mentally impaired persons ; Medical experimentation
PE 3992 Mentally retarded adults ; Inadequate guardianship
PD 0914 Mentally retarded children
PD 0914 Mentally subnormal children
PC 0382 *Menus ; Restricted family*
PJ 4807 Mercantilism
PE 5436 Mercaptans
PE 5436 Mercaptans pollutants ; Thiols
PD 2592 Mercenaries
PD 2592 Mercenaries means impeding right self determination ; Use
PE 7153 Mercenaries ; Techno
PD 2592 Mercenary troops
PE 4447 Merchandise ; Damaged
PF 6578 *Merchandise display ; Unattractive*
PD 7981 Merchandise ; Forgery brand name
PD 1435 Merchandise ; Low quality
PF 6572 *Merchandise ; Minimal retailed*
PD 6571 *Merchandising interest ; Limited*
PF 2439 Merchandising ; Uncompetitive local
PF 1495 *Merchant vessels ; Maldistribution*
PA 7253 Mercilessness *Envy*
PA 7156 Mercilessness *Moderation*
PE 1155 Mercury poisoning animals
PE 1155 Mercury pollutant
PF 2643 Mercy killing
PD 0071 Mergers ; Inadequate control
PD 0071 Mergers takeovers ; Proliferation
PF 7315 Merit ; Accumulation recognized
PA 5942 Meritlessness *Unimportance*
PA 1387 *Meritocracy*
PF 2685 *Meritocracy ; Anti*
PE 7476 Merits nuclear weapons ; Unrecognized
PE 3936 *Mesenteric torsion small intestine swine*
PE 8606 Meslin shortage ; Meal flour than wheat
PE 8980 Meslin shortage ; Meal flour wheat
PE 8489 Meslin trade instability ; Meal flour wheat
PE 3001 Mesothelioma
PE 6226 Messages ; Ineffective regulation electronic
PE 0533 Messages ; Insensitivity nonverbal
PE 5402 Messaging services ; Pink
PF 1102 Messianic image leadership
PF 9796 Messianic policy-making
PF 9139 Messianism
PA 7361 Messiness *Disorder*
PA 6900 Messiness *Formlessness*
PA 5438 Messiness *Neglect*
PA 5459 Messiness *Uncleanness*
PC 2270 Metabolic disease
PD 9293 *Metabolic disease kidney animals*
Metabolic diseases
PD 7420 — animals
PD 8301 — endocrine ; Nutritional
PD 2389 — *Hereditary*
PC 2270 Metabolic disorders
PD 7420 Metabolic disturbances animals

Military disobedience

Metabolism Congenital disorders
PE 9291 — amino acid
PE 4293 — carbohydrate
PE 9816 — lipid
PC 2270 Metabolism ; Diseases
PC 2270 Metabolism disorders
PC 2270 Metabolism ; Inborn errors
PE 4463 Metabolism ; Obesity due low
PE 7826 Metacarpal bones ; Fracture
PE 2601 Metal basic industries instability ; Non ferrous
PD 3668 Metal contamination soil
PE 1901 Metal deficient diets
PE 5439 Metal dust fumes
Metal [failure...]
PD 7215 — failure
PD 7215 — fatigue
PD 7215 — fracture
Metal industries
PD 0243 — Health hazards basic
PE 2601 — Instability basic
PE 5866 — Protectionism steel basic
PF 1374 — Underdevelopment basic
PE 0824 Metal ores ; Long term shortage non ferrous
Metal [poisoning...]
PD 0105 — poisoning
PD 0948 — pollutants ; Toxic
PE 1852 — products industry instability
PF 0942 — products, machinery equipment industries ; Underdevelopment
Metal scrap
PE 8906 — Environmental hazards metalliferous ores
PE 0553 — Instability trade metalliferous ores
PE 0353 — Long term shortage metalliferous ores
PE 0353 — Shortage non ferrous
PD 5228 Metalaldehyde poisoning animals
Metallic mineral products
PE 0890 — industries ; Environmental hazards non
PE 1851 — industries ; Underdevelopment non
PE 2599 — industry ; Instability non
Metalliferous ores metal
PE 8906 — scrap ; Environmental hazards
PE 0553 — scrap ; Instability trade
PE 0353 — scrap ; Long term shortage
PE 8248 Metals basic industries environmental hazards ; Non ferrous
PD 2638 Metals ; Brittle fracture
PD 2638 Metals ; Creep
PE 1401 Metals ; Environmental hazards
PB 1701 Metals ; Erosion
PE 1406 Metals ; Instability trade non ferrous
PE 8131 Metals ; Instability trade ores concentrates non ferrous base
PD 0948 Metals ; Toxic
PD 2647 Metatarsalgia
PF 1687 Meteorites hazards
PD 4065 Meteorological disaster
PD 4182 Meteorologists ; Corruption
PD 4182 Meteorologists ; Irresponsible
PD 4182 Meteorologists ; Negligence
PD 4182 Meteorology ; Malpractice
PD 4182 Meteorology ; Unethical practice
PF 1695 Meteors hazards
PD 0094 Methamphetamine drugs
PE 8815 Methane ; Environmental hazards
PE 5399 Methane gas emissions animal husbandry
PE 1256 Methane gas emissions landfill sites
PE 8815 Methane ; Increase atmospheric concentration
PE 0630 Method appropriate available data ; Absence
PF 3549 Method ; Educational curricula based content rather than
PD 0361 Method food preservation ; Irradiation
PD 2438 Method ; Incomplete consensus
PD 3774 Method ; Limited historical
PD 1827 Method rather than content ; Educational curricula over emphasizing
PF 4756 Methodologies ; Confusing decision making
PF 7700 Methodologies deal discontinuities random events ; Inadequacy
PD 0065 Methodologies global ideology ; Absence social
PF 7700 Methodologies ; Limitations surprise free
PF 2202 Methodologies ; Top down research
PF 0327 Methodologies ; Total absence tactical
PC 0492 Methodology ; Traditional agricultural
PF 0327 Methodology ; Absence tactical
PF 2400 Methods ; Abstract educational
Methods agricultural
PF 6524 — management ; Underproductive
PF 1822 — production ; Obsolete
PF 1822 — production ; Out date
Methods [appropriate...]
PE 0630 — appropriate data ; Failure
PF 1094 — appropriate human depths ; Antiquated intellectual
PF 6465 — Archaic marketing
PF 4612 — avoiding danger ; Reactionary
PE 8983 Methods blind ; Inadequate rehabilitation
Methods [conception...]
PE 8634 — conception ; Inhuman
PF 2491 — Confusing loan computing
PE 8256 — cross breeding ; Lack availability
Methods [decision...]
PF 1010 — decision making ; Ineffective implementation
PE 0360 — destruction stray dogs ; Cruel
PE 1337 — developing countries ; Inappropriate transplantation industrialized country
PF 0956 Methods ; Faulty approaches teaching intellectual
PF 1653 Methods ; Gap agricultural

Methods [Inability...]
PF 1004 — Inability awaken consciousness due inadequate
PD 0093 — Inadequacy contraceptive
PF 1069 — Inadequacy male contraceptive
Methods Inadequate
PF 3713 — [Methods ; Inadequate]
PF 8481 — abortion
PF 5969 — immunization
PF 7380 — intellectual
PF 8939 — meeting
Methods [Indiscriminate...]
PD 9082 — Indiscriminate fishing
PE 1848 — Individualistic construction
PF 3709 — Ineffective policy making
PF 3008 — Ineffectiveness traditional small business
PF 2451 — Inhibiting participatory
PD 0181 — Inter Governmental exchange torture
PF 8767 — Intolerance proven
Methods [Lack...]
PJ 7793 — Lack programme
PF 3605 — leadership development ; Limited
PF 2721 — Limited teaching
PF 6524 — Low yield farm
PF 2394 Methods ; Myopic decision making
Methods [Obsolete...]
PF 3713 — Obsolete
PF 3713 — Obsolete group
PF 1822 — Old farming
PF 1822 — Outdated farm
PF 8161 — Outmoded motivational
Methods [policy...]
PF 1076 — policy formation ; Cumbersome
PF 2350 — Poor communications
PF 2721 — Poor teaching
PF 2721 — practical education ; Ineffective
PE 9125 — promoting fast livestock growth ; Use artificial
Methods [Resistance...]
PF 3010 — Resistance changing agricultural
PF 2721 — Restrictive teaching
PD 1156 — riot control ; Inhumane
PF 1822 — Rudimentary farming
PF 1035 — Methods ; Sexual discrimination contraceptive
PF 3884 Methods ; Stagnating social
PF 1822 Methods ; Traditional farming
PF 4852 Methods training ; Unused
Methods [Unarticulated...]
PF 2400 — Unarticulated goal educational
PF 1442 — Uncomprehensive budget
PE 3553 — Unethical interrogation
PF 3884 — Uninitiated social
PF 3526 — Unsophisticated bureaucratic
PE 7095 Methods which disenfranchise grassroots technological development ; Intellectual
PE 1349 Methylated spirits drinking ; Solvent
PE 9286 Metritis
PD 7799 Metritis animals ; Acute
PD 7799 Metritis ; Contagious equine
PC 1833 Metropolitan government decision making ; Balkanization
PD 6120 Metropolitan regions ; Excessive size
PE 1397 MIA's
PA 5454 Miasma Badness
PA 5981 Miasma Stench
PA 7107 Miasma Unpleasantness
Mice Endangered species
PE 8527 — dasyures marsupial
PD 3481 — jumping
PE 8674 — pocket mice kangaroo
PE 8967 Mice hamsters ; Endangered species
PE 8674 Mice kangaroo mice ; Endangered species pocket
PE 7832 Micro climate ; Increase insect pests modification
PE 9405 Micro enterprises ; Instability
Micro [organic...]
PD 5719 — organic proteins ; Loss
PD 3574 — organisms ; Pesticide destruction soil fauna
PD 7183 — organisms ; Release genetically engineered
PE 9405 — organizations ; Failure economic
PD 2942 Micro-state participation international organizations
PD 9094 Microbes
PC 7492 Microbes
PD 1175 Microbes drinking water ; Disease causing
PD 9669 Microbial contamination food
PC 7492 Microbial diseases
PE 1607 Microbial pests industry
PD 7731 Microbiologists ; Irresponsible
PE 9296 Microcephalus
PE 9296 Microcephaly
PE 2715 Microsomia
PE 6056 Microwave radiation health hazard
PD 9293 Micturition ; Disorders
PD 5675 Mid life crisis ; Female
PD 5783 Mid life crisis ; Male
PC 1002 Middle class ; Vulnerability
PD 2567 Middle ear disease ; Infective
PD 2567 Middle ear ; Fluid
PE 3528 Middle-man price control
PE 3528 Middleman control rural marketing
PD 4632 Middleman profits ; Stifling
PD 4632 Middlemen cycle ; Expensive
PD 4632 Middlemen ; Dependency
PE 3528 Middlemen ; Farm tenant's dependence
PD 4632 Middlemen ; Unnecessary market
PE 2635 Midgets ; Discrimination against dwarfs
PE 8036 Midgets giants ; Domestic inconvenience dwarfs,
PE 8496 Midgets giants ; Lack adequate clothing supply dwarfs,

PE 4901 Midwives medical care systems ; Resistance incorporating
PE 6357 Migraine
PE 3508 Migrans ; Cutaneous larva
PE 6278 Migrans ; Visceral larva
PC 0180 Migrant labour ; Problems
Migrant workers
PD 0973 — Denial rights
PE 0769 — Disabled
PD 0973 — their families ; Discrimination against
PC 5017 — Xenophobia regard
PE 7611 Migrants ; Inadequate social security
PE 4258 Migrants ; Social maladjustment children
PE 6951 Migrants ; Undocumented
PD 0493 Migrating sand dunes
PE 7613 Migration ; Abuse project tied
PF 3523 Migration cities ; Youth
PC 0018 Migration ; Disruptions due
PE 3445 Migration ; Early return
Migration [Failed...]
PE 3445 — Failed
PF 2135 — fish ; Seasonal
PC 6203 — Forced
PF 4009 — Migration ; Internal
PF 4008 — Migration ; International
PC 2279 Migration movement patterns ; Disruption animal
PF 4181 Migration rural areas
PE 8768 Migration rural population cities
PD 1291 Migration skilled people rural urban areas
Migration [Uncontrolled...]
PD 2229 — Uncontrolled
PC 0018 — Undesirable effects
PF 3523 — Unreplenished urban
PF 3523 Migration youth urban sites
PE 4938 Migratory bird species ; Endangered
PD 3978 Mild dietary indigestion
PE 4821 Mild mental retardation
PE 1607 Mildew plants ; Black
PE 0501 Mildews ; False
PE 0501 Mildews plants ; Downy
PE 0529 Mildews plants ; Powdery
PG 9737 Miliary tuberculosis ; Acute
PC 1090 Militancy
PA 5532 Militancy Disaccord
PD 4695 Militancy ; Shopfloor
PD 1106 Militant individualism
PF 9710 Militant liberalism
PB 1540 Militant secularism
PC 2169 Militarism
PC 0937 Militarism ; Provocations capitalist
PC 2371 Militarism ; Religious justification
PD 1897 Militarization
PE 5986 Militarization children
PC 1258 Militarization ; Competitive
PD 1241 Militarization deep ocean sea-bed
PD 9495 Militarization developing countries
PF 1447 Militarization ; Socio economic burden
PE 1397 Military action ; Persons missing
Military activities
PF 0780 — Absorption manpower resources
PE 8979 — Disruption food supply due
PD 0527 — Para
PC 6675 — Pollution
Military [activity...]
PE 0736 — activity during peace time ; Environmental degradation
PE 6052 — aid
PE 6052 — aid repressive regimes
PD 5373 — aircraft accidents
PD 1881 — atrocities
PD 6364 — attacks refugee camps settlements
PJ 5219 Military base ; Implantation
Military bases
PE 0736 — Environmental destruction
PD 3496 — intelligence centres ; Extra territorial
PC 8390 — Obsolete
Military [blockade...]
PC 4301 — blockade
PD 5255 — blocs
PE 3390 — brothels
PD 4945 — brutality
Military [capacity...]
PC 1258 — capacity ; Myth national security increase
PE 5989 — censorship
PE 8300 — codes justice ; Deficiencies
PE 6948 — commissions ; Purchase
PD 0702 — concentration camps
Military conflict
PB 0593 — [Military conflict]
PE 5671 — communist countries
PF 3100 — International non
PC 1869 — Intrasocietal
Military [conscription...]
PD 0356 — conscription systems ; Weaknesses
PD 7360 — corruption
PC 0742 — crimes
Military [data...]
PD 5183 — data ; Leak
PF 5933 — defence ; Lack adequate
PF 1385 — defence policies ; Unjustified
PE 6639 — demoralization
PD 2374 — dependency
PD 4621 — desertion
PD 7714 — destabilization developing countries
PC 0698 — dictatorship
PD 7225 — disobedience

Military disruption

Military [disruption...] cont'd
PC 6675 — disruption biosphere
Military [economic...]
PB 0318 — economic hegemony
PB 0318 — economic hegemony ; Dependence
PD 4448 — effectiveness ; Impairing
PA 1387 — *elitism*
PD 2922 — espionage
PD 1881 — excesses
PF 9542 — excesses
PE 3704 — exercises ; Provocative
PE 8207 — expeditions against distant objectives
PD 7261 — expeditions against friendly powers
PF 1447 — expenditure ; Dependence excessive
PF 1447 — expenditure ; Excessive
PE 1666 — experiments animals
PE 1666 — Exploitation marine mammals
PF 2613 — explosive devices ; Derelict
PE 0101 — expropriation satellites
PE 8220 Military forces ; Discrimination against women
PE 4920 Military forces ; Under strength
PC 0698 Military government
Military [Inadequate...]
PF 9542 — Inadequate government control
PE 4553 — incidents ; Accidental
PA 6416 — *incompetence*
PD 2543 — indiscipline
Military industrial
PC 1952 — complex
PC 3191 — complex capitalist systems
PD 1330 — governmental complex communist systems
PD 4361 — malpractice
Military [inefficiency...]
PD 4361 — inefficiency
PD 3385 — influence
PD 3385 — influence politics
PC 0541 — insecurity vulnerability
PD 7225 — insubordination
PD 9331 — intervention ; Foreign
PE 2007 — intervention workers meetings
PD 2637 *Military law*
Military [manoeuvres...]
PE 3704 — manoeuvres sensitive border areas
PE 4920 — manpower ; Shortage
PF 3702 — *mobilization procedures ; Hypersensitive*
PE 3995 — mobilization reinforcement ; Disparity facilities
PE 6639 — morale ; Low
Military [obligations...]
PE 6678 — obligations persons multiple nationality ; Conflicting
PE 8965 — obstacles peaceful development space
PD 8021 — occupation
PD 1078 — occupation ; Vulnerability women during foreign
PC 0742 — offences
Military operations
PF 7669 — Clandestine
PF 7669 — Covert
PD 5272 — Cross border
PF 7669 — Secret
Military personnel
PD 7602 — Brutalization
PD 7602 — Bullying amongst
PF 1098 — convicted atrocities ; Politically motivated amnesty
PE 4758 — Denial rights wounded
PE 5579 — Drug abuse
PE 8881 — Excessive use force
PF 1098 — homicide ; Absolving
PE 4920 — Insufficient
PD 3519 — Racism amongst
PE 8495 — troops ; Drunkenness
Military [police...]
PC 1142 — *police abuse*
PE 4119 — police personnel participation torture
PD 3014 — political prisoners detainees
PE 7665 — power ; Control judiciary executive
PD 7360 — practices ; Unethical
PD 3496 — presence ; Erosion national sovereignty foreign
PD 3496 — presence ; Foreign
PE 3390 — prostitution women
Military [recruitment...]
PE 4920 — recruitment ; Failure
PC 0698 — regimes
PF 9181 — *reprisals*
PF 4785 — resources peaceful applications ; Obstacles redeployment
Military retaliation
PF 9181 — [Military retaliation]
PF 9181 — *Fear*
PF 9181 — *Threat*
PD 9252 Military rivalry
PC 1144 Military secrecy
Military service
PD 0356 — Avoiding
PF 6051 — Compulsory
PD 1800 — Denial right conscientious objection
PE 6422 — Discrimination against foreign nationals
PD 6667 — Impressment
PD 5941 — obligations ; Crimes related
PE 8220 Military ; Sexual harassment
Military [targets...]
PD 0087 — targets space
PE 7040 — thinking industry ; Inflexible
PF 8874 — threat
PE 5785 — threats island countries territories
PE 5837 — threats land-locked countries
PE 0494 — tribunals
PE 0494 — tribunals ; Injustice

Military [unpreparedness...]
PF 5933 — unpreparedness
PF 5933 — unreadiness
PE 3402 — usage land ; Excessive
PE 1666 — use animals
PE 7678 Military weaponry ; Creeping modernization
PD 7420 Milk fever
PD 2283 Milk ; Tuberculous
PE 6886 Milker's nodes
PD 2731 *Milking cows ; Mastitis*
PF 5375 Millenarian movements
PF 7684 Millenarianism
PF 9796 Milleniarist programmes
PF 7684 Millennialism
PE 4649 Mind age ; Deterioration
PA 9226 Mind ; Criminal state
PJ 5810 Mind ; Group
PF 7918 Mind-induced death
PC 1867 *Mind set ; Urban dependence*
PF 0424 Mindedness ; Absent
Mindedness Bloody
PA 7055 — [Mindedness ; Bloody]
PA 5532 — [Mindedness ; Bloody]
PA 5643 — [Mindedness ; Bloody]
PF 7847 Mindedness ; Closed
PF 4910 Mindedness ; Double
PF 3379 *Mindedness ; Narrow*
PA 7325 Mindedness ; Weak
PA 7371 Mindedness ; Weak
PA 6379 Mindlessness *Avoidance*
PA 5568 Mindlessness *Ignorance*
PA 7156 Mindlessness *Moderation*
PA 7371 Mindlessness *Unintelligence*
PE 3909 Minds young ; Subverting
PF 3161 Mindset ; Individualist survival
PF 3008 Mindset ; Individualized business
PF 0851 Mindset ; Majority rule
PF 1078 Mindset ; Subsistence struggle
PD 3518 Mindset ; Turned elders'
PD 2278 Mine disasters
PE 4807 Mine dust
PD 0575 *Mine wastes*
PB 0593 *Mined international waterways*
PJ 2409 Minefields
PE 2254 Miner ; Beet leaf
PE 2320 Miner's elbow
PE 4679 Mineral dusts
PE 6975 Mineral exploitation claims ; Fraudulent
PE 6682 Mineral exploitation seabed resources ; Environmental hazards
PE 1514 Mineral fertilizers ; Excessive reliance
Mineral fuels lubricants
PE 1712 — Long term shortage
PE 1346 — related materials ; Environmental hazards
PD 0877 — related materials ; Instability trade
PE 0141 Mineral fuels trade ; Restrictive practices
Mineral products
PE 0890 — industries ; Environmental hazards non metallic
PE 1851 — industries ; Underdevelopment non metallic
PE 2599 — industry ; Instability non metallic
Mineral resources
PD 9357 — Depletion
PD 9357 — Exhaustion
PD 9357 — Long term shortage
PE 8464 Mineral tar crude chemicals derived coal, petroleum natural gas ; Instability trade
PE 3035 Mineral waters ; Cancer causing additives
PE 0760 Minerals, excluding coal, petroleum precious stones ; Instability trade crude fertilizers crude
PE 1353 Minerals, excluding coal, petroleum precious stones ; Long term shortage crude fertilizers crude
PE 1137 Miners' future ; Uncertainty
PD 2596 Mines quarries ; Environmental destruction open pit
PF 2613 Mines ; Uncleared land
PF 2613 Mines ; Uncleared sea
PF 2613 Minesweeping ; Ineffective
PF 4876 Mini state membership ; Distortion intergovernmental organizations
Minimal [access...]
PF 3554 — access appropriate technology
PF 2631 — access necessary information
PF 6531 — adult education
PC 4867 Minimal business training
Minimal [cash...]
PF 6451 — *cash flow*
PF 8905 — *church / school involvement*
PF 8076 — citizen confidence
PF 2421 — citizen involvement
PF 2439 — *commercial activities*
PF 2350 — communication system
Minimal community
PF 8695 — contact
PF 2575 — involvement
PD 8832 — organization
PD 8832 — services
PD 7723 Minimal construction standards
Minimal [excess...]
PD 8568 — excess income
PE 1598 — export promotion transnational corporations developing countries
PE 8306 — exports least developed countries
Minimal [family...]
PE 8296 — family interest urban life styles
PF 4108 — farm machinery
PF 3554 — *financial understanding*
PF 6557 *Minimal industrial recruitment*

PF 7011 *Minimal leadership coordination*
PE 0282 Minimal manufacturing least developed countries
Minimal [opportunities...]
PF 6531 — opportunities adult training
PF 2316 — opportunities corporate activities
PF 2949 — options rural employment
PD 0847 Minimal preschool facilities
PC 8160 Minimal prime land
PF 6572 *Minimal retailed merchandise*
PC 2163 *Minimal technical teachers*
PD 2790 Minimal vocational commitment
PF 9409 Minimally cooperative solutions problems
PF 7186 Minimization problems
PE 1693 Minimum age law ; Lack
PE 1693 Minimum age laws ; Failure enforce
PE 1182 Minimum pricing regulations measures regulate domestic prices ; Distortion international trade
PF 6557 Minimum promotion community assets
Minimum wage
PE 6726 — Denial right
PE 6726 — fixing ; Lack
PE 6726 — laws ; Violation
Minimum [wages...]
PD 5674 — wages ; High
PD 5674 — wages ; Prohibitive
PE 1693 — work age ; Denial right
PD 2278 Mining accidents
PC 3320 *Mining camps*
PE 1137 Mining communities ; Traumatic shift life styles
PD 2596 Mining ; Environmental hazards
PE 5160 Mining environmental hazards ; Coal
Mining [industries...]
PE 8630 — industries ; Ineffective self regulation
PC 5842 — *industries ; Protectionism*
PE 8428 — industry ; Occupational hazards
PD 1637 Mining ; Landscape disfigurement open cast
Mining quarrying
PE 1858 — industries ; Underdevelopment
PE 0993 — industry ; Instability
PE 8809 — Instability miscellaneous
PE 7419 Mining radioactive substances ; Health hazards
PD 1637 Mining ; Strip
PE 4393 Mining ; Subsidence
PF 7342 Ministerial control ; Abdication government
PC 7587 Ministers ; Corruption government
PD 9293 *Mink ; Giant kidney worm infection*
PE 5953 Minor ailments
PE 6522 Minor consent ; Rape
PE 1821 Minor drugs ; Abuse
PE 1821 Minor tranquilizers ; Dependence
PD 1350 *Minor vandalism ; Unchecked*
PE 0097 *Minor ; Variola*
PF 3403 Minorities ; Alienation religious
PC 3459 Minorities ; Denial education
Minorities Denial rights
PC 8999 — [Minorities ; Denial rights]
PC 8999 — ethnic
PC 2129 — religious
PD 1914 — sexual
Minorities [Dependence...]
PC 3424 — Dependence underprivileged
PC 0582 — Discrimination against
PD 1914 — Discrimination against sexual
PC 3459 Minorities education ; Discrimination against
PC 3205 Minorities ; Expulsion ethnic
PF 1219 Minorities ; Inadequate legal counsel
PC 8999 Minorities ; Inadequate protection linguistic, national religious
PC 8494 Minorities ; Prejudice against
Minorities [Threatened...]
PC 3295 — Threatened
PC 3295 — Threatened vulnerable
PC 3362 — Threats ideological movements
Minorities Underprivileged
PC 3424 — [Minorities ; Underprivileged]
PC 0582 — ethnic
PC 3325 — ideological
PC 3324 — linguistic
PC 0805 — racial
PC 2129 — religious
PF 7239 Minorities unrecognized government
Minorities [Violation...]
PC 3459 — Violation right education
PD 1914 — Violation rights sexual
PC 0720 — Violence against indigenous
Minority [concerns...]
PF 8876 — concerns ; Parliamentary inability respond
PF 2375 — control
PF 5659 — culture support ; Insufficient
PD 2623 — cultures ; Underreporting hazards
PC 1056 Minority eradication
PD 2490 Minority group separatism
PC 3428 Minority groups ; Conflict
PD 5078 Minority languages ; Discrimination against
PD 0825 Minority languages ; Insufficient translation
PD 1140 Minority opinion ; Blocked
PD 2108 Minority opinion ; Innate expectation suppression
PD 1260 Minority population urban ghettos ; Segregation poor
PF 1529 Minority self image ; Demeaning
PD 1140 Minority status ; Political
PC 3685 Minority turmoil
PF 1529 Minority victim image
PD 1140 Minority voices ; Paralysis
PD 9481 Minors ; Corruption
PE 6522 Minors ; Corruption
PD 8405 Minors ; Enslavement exploitation

–988–

Misuse resources

PE 6522 Minors ; Sexual intercourse	PF 2477 *Misguidance ; Vocational*	PD 2085 Missing child care
PE 6854 Minors ; Unlawful repatriation	PE 1162 Misguided missiles	PD 6009 Missing children
PF 6814 *Miracles ; Belief*	PF 5645 Misguided priorities	PE 7061 Missing documents data transnational corporations
PJ 5154 Miracles ; Religious	PF 3749 Mishandling national security information	PF 4690 Missing documents ; Risk
PF 3020 Mis-classification political prisoners	PD 8523 Misinformation	PC 1820 Missing entrepreneur skills
PF 7050 Mis-communications societal learners	PF 3539 *Misinformation ; Birth control*	PE 1397 Missing military action ; Persons
PF 6102 Misalignment currencies	PF 9794 Misinformation concerning infringement human rights	**Missing [persons...]**
PF 6102 Misalignment ; Exchange rate	PE 0682 Misinformation concerning loans	PD 1380 — persons
PA 5982 Misalliance *Disagreement*	PE 7349 Misinformation ; Denial right correct	PF 0431 — persons ; Uncertainty death
PA 5794 Misalliance *Unrelatedness*	PA 5568 Misinformation *Ignorance*	PE 4927 — public signs
PF 6377 Misallocation public spending government	PA 6393 Misinformation *Miseducation*	PF 3370 Missionary activity ; Ill considered
PE 0648 Misallocation resources protect aged	PA 7411 Misinformation *Uncommunicativeness*	PF 3370 Missionary alienation
PB 6719 Misallocation resources ; Worldwide	PF 3393 *Misinformed family practices*	PC 3355 Missionary bodies ; Competition
PF 6232 Misandrony	PA 8653 *Misinformed growth suspicions*	PA 6180 Misstatement *Error*
PJ 1278 *Misandry*	PA 6741 Misinterpretability	PA 6644 Misstatement *Misrepresentation*
PA 8487 *Misanthrope*	PA 6790 Misinterpretation *Distortion*	PA 7411 Misstatement *Uncommunicativeness*
PA 8487 *Misanthropy*	PA 6180 Misinterpretation *Error*	PE 1655 Mist
PA 6180 Misapplication *Error*	PD 8716 Misinterpretation evidence ; Official	PD 1245 Mist ; Pesticides
PA 6852 Misapplication *Inappropriateness*	PA 6741 Misinterpretation *Misinterpretability*	PE 4825 Mist ; Sulphuric acid
PA 6741 Misapplication *Misinterpretability*	PA 6607 Misinterpretation *Misjudgement*	PA 6180 Mistake *Error*
PA 5502 Misapplication *Reason*	PF 4564 Misinterpretation statistics	PA 6741 Mistake *Misinterpretability*
PA 5794 Misapplication *Unrelatedness*	PC 0492 *Misjudged borrowing practices*	PA 6662 Mistake *Nonaccomplishment*
PF 7016 Misapplication venture capital	PA 6607 Misjudgement	PA 7232 Mistake *Unskillfulness*
PE 2690 Misapplied behaviourism	PA 6180 Misjudgement *Error*	PF 3702 Mistakes
PF 7027 Misapprehended world problems ; Recurrence	PA 6741 Misjudgement *Misinterpretability*	PF 5296 Mistakes ; Failure government admit
PA 6180 Misapprehension *Error*	PA 6607 Misjudgement *Misjudgement*	PF 4125 Mistakes ; Fear
PA 6741 Misapprehension *Misinterpretability*	PE 4285 Misleading accounting information due inflation	PE 4736 Mistakes ; Operating theatre
Misappropriation [concepts...]	PE 3814 Misleading advertising	PF 5163 Mistakes ; Refusal apologize
PF 5247 — concepts	PE 0682 Misleading borrowers funds	PF 5296 Mistakes ; Refusal government leaders admit
PE 6074 — cultural property	PD 4879 Misleading commercial information	PE 4139 Mistletoe trees
PC 7605 — cultural symbols	PF 3096 Misleading documents	PD 5501 Mistreatment senile
PC 2920 Misappropriation public funds	PE 7502 Misleading endorsement advertising	PA 8653 *Mistrust birth control*
PF 4975 Misappropriation resources high cost civil engineering projects	PF 5127 Misleading evidence	PD 2790 Mistrust corporate commitment
PF 0716 Misappropriation resources high cost research projects	PF 3096 Misleading information	PA 8653 Mistrust *Distrust*
PD 8041 Misappropriation sacred objects	PD 3521 Misleading labelling packages	PA 7253 Mistrust *Envy*
PF 5242 Misappropriation social values	PF 6980 Misleading market prices	PF 8743 Mistrust foreigners
PF 5247 Misappropriation words	PF 2909 *Misleading public opinion polls*	PF 1097 Mistrust government
PA 6498 Misbehaviour	**Misleading [reports...]**	PC 2089 Mistrust governments
PE 9092 Misbehaviour schools	PF 3096 — reports	PC 2089 Mistrust ; Lack international cooperation due personal
PA 6180 Misbelief *Error*	PF 4564 — reports	PF 8743 Mistrust newcomers
PA 7392 Misbelief *Unbelief*	PF 0704 — résumé	**Mistrust [police...]**
PA 5564 Misbelief *Unorthodoxy*	PB 8406 Mismanagement	PF 8559 — police
PA 6180 Miscalculation *Error*	**Mismanagement [Agricultural...]**	PF 1097 — political structures
PF 1440 Miscalculation ; International political	PD 8625 — Agricultural	PF 1097 — politicians ; Citizen
PA 6607 Miscalculation *Misjudgement*	PF 0175 — aid developing countries	PA 7115 Mistrust *Rashness*
PF 3702 Miscalculations ; Human errors	PD 2771 — animals ; Agricultural	PF 8743 Mistrust strangers
PD 0173 Miscarriage	**Mismanagement developing countries**	PD 8217 Mistrust system justice
PA 6180 Miscarriage *Error*	PD 8549 — [Mismanagement developing countries]	PA 7392 Mistrust *Unbelief*
PA 6486 Miscarriage *Injustice*	PF 5964 — Economic	PF 3526 *Mistrusted federal funding*
PF 8479 Miscarriage justice	PD 6482 — Soil	PE 1976 Mists ; Acid
PF 7728 Miscarriage justice ; Failure legally rehabilitate victims	PD 5429 Mismanagement environmental demand	PA 8197 Misunderstanding
PA 7232 Miscarriage *Unskillfulness*	PF 1874 Mismanagement exchange rate system	PA 5532 Misunderstanding *Disaccord*
PC 1523 Miscegenation	**Mismanagement [food...]**	PA 6180 Misunderstanding *Error*
PA 5454 Mischief *Badness*	PE 6115 — food resources	PF 3340 Misunderstanding ; Inter cultural
PD 1350 Mischief ; Criminal	PD 0136 — foreign aid ; Corruption	PA 6741 Misunderstanding *Misinterpretability*
PA 5532 Mischief *Disaccord*	PD 6701 — forests	PF 8382 Misunderstanding official communications
PA 6088 Mischief *Impairment*	PF 9672 Mismanagement government financial expenditures	PD 8023 Misunderstanding preventive health care
PA 7395 Mischief *Inexpedience*	PD 2771 Mismanagement housed farm animals ; Agricultural	PA 8197 *Misunderstanding veterinary approach*
PA 6498 Mischief *Misbehaviour*	**Mismanagement [Inappropriateness...]**	PA 6852 Misusage *Inappropriateness*
PA 5454 Mischievousness *Badness*	PA 6852 — *Inappropriateness*	PE 4225 Misuse advertising
PA 6498 Mischievousness *Misbehaviour*	PD 6628 — Inter Governmental organization	PD 4403 Misuse artesian water supplies
PA 7148 Mischievousness *Ungodliness*	PE 6947 — intergovernmental organization leadership	PE 1651 Misuse business enterprises front organizations government
PA 6180 Misconception *Error*	PE 8233 — irrigation schemes	**Misuse [chemicals...]**
PA 6741 Misconception *Misinterpretability*	PD 8625 Mismanagement pasture arable land ; Agricultural	PD 5904 — chemicals
PC 2673 *Misconceptions*	PC 0052 Mismanagement ; Soil	PD 5183 — classified communications information
PA 6180 Misconduct *Error*	PA 7232 Mismanagement *Unskillfulness*	PD 9544 — computer databases
PA 6852 Misconduct *Inappropriateness*	PE 4834 Mismanagement ; Zoo	PF 7220 Misuse democratic processes resolve theoretical issues
PD 8227 Misconduct leaders ; Ethical	PF 5000 Mismatch national macroeconomic policies industrialized countries	PA 6790 Misuse *Distortion*
PA 6498 Misconduct *Misbehaviour*	PF 8757 Mismatch public policy current scientific knowledge	**Misuse [electronic...]**
PD 8227 Misconduct public office	PE 4647 Mismatch religious national holidays ; Inefficiency due	PD 2930 — electronic surveillance governments
PA 7232 Misconduct *Unskillfulness*	PE 7062 Misogyny	PA 6180 — *Error*
PA 5644 Misconduct *Vice*	PA 6414 Misperception *Illusion*	PF 3348 — evolutionary theories
PA 6790 Misconstruction *Distortion*	PF 4389 Misperceptions ; Personal	PD 0260 — experimental animals
PA 6180 Misconstruction *Error*	PF 6570 *Misplaced independent pride*	**Misuse [fertilizers...]**
PA 6741 Misconstruction *Misinterpretability*	PE 7002 Misplaced optimism	PE 1514 — fertilizers
PA 6607 Misconstruction *Misjudgement*	PD 5515 Misprison felon	PD 5193 — financial industrial power
PA 7411 Misconstruction *Uncommunicativeness*	PF 5495 Misreading sacred documents	PF 6202 — food political weapon
PA 5952 Misdemeanour *Illegality*	PF 5403 Misremembrance evidence	PE 2858 — free production zones export enclaves
PA 6498 Misdemeanour *Misbehaviour*	PF 4564 Misreporting statistics	PD 9538 Misuse government surveillance communications
PA 5644 Misdemeanour *Vice*	PA 6644 Misrepresentation	PD 5133 Misuse grassland rangeland
PE 5594 Misdemeanours	PA 6790 Misrepresentation *Distortion*	PD 1224 Misuse herbicides
PF 8490 Misdiagnosis	PF 1183 Misrepresentation educational materials ; Deliberate	**Misuse [Impairment...]**
PF 2595 Misdirected educational goals	**Misrepresentation [facts...]**	PA 6088 — *Impairment*
PF 0422 Misdirected vocational training	PD 4982 — facts corporate leaders	PA 6852 — *Inappropriateness*
PF 5949 Misdirected youth engagement	PD 1893 — facts national leaders	PF 2216 — international forums
PA 6790 Misdirection *Distortion*	PE 9061 — fiscal liabilities	PD 8142 Misuse ; Land
PF 8128 Misdirection human energies desires	PF 9239 Misrepresentation geographical information	PF 9598 Misuse language
PA 6393 Misdirection *Miseducation*	PF 4932 Misrepresentation ; Historical	PD 8402 Misuse medicines
PA 7411 Misdirection *Uncommunicativeness*	PF 5662 Misrepresentation historical persons	PE 0436 Misuse nonprofit associations front organizations government
PA 7232 Misdirection *Unskillfulness*	PE 6877 Misrepresentation information consumers	PD 4403 Misuse nonrenewable fossil water reserves
PA 6393 Miseducation	PA 6644 Misrepresentation *Misrepresentation*	**Misuse [personal...]**
PB 8167 Misery	PF 9243 Misrepresentation project assessments	PE 4635 — personal authority political purposes
PA 5451 Misery *Insensibility*	PD 7231 Misrepresentation socially unacceptable activity research	PF 4153 — *personal records ; Government*
PA 6731 Misery *Solemnity*	PA 7411 Misrepresentation *Uncommunicativeness*	PD 4629 — pesticides
PA 7107 Misery *Unpleasantness*	PA 7361 Misrule *Disorder*	PF 2152 — plausible arguments conceal nefarious benefits
PA 6180 Misfeasance *Error*	PA 5563 Misrule *Lawlessness*	PC 9104 — political power
PA 6852 Misfeasance *Inappropriateness*	PA 7232 Misrule *Unskillfulness*	PD 2683 — postal surveillance governments
PA 7232 Misfeasance *Unskillfulness*	PC 1606 Missile gap	PE 2932 — psychiatric diagnosis
PA 5644 Misfeasance *Vice*	PE 4515 Missiles ; Ballistic	**Misuse [radio...]**
PF 4708 Misfiled documents records	PE 9052 Missiles developing countries ; Proliferation ballistic	PD 9538 — *radio transmission surveillance governments*
PF 9536 Misfortune	PE 1162 Missiles ; Misguided	PE 3354 — religious property ; Accumulation
PA 6340 Misfortune *Adversity*	PE 4515 Missiles ; Surface surface	PD 7231 — research cover illegal activity
PA 6393 Misfortune *Miseducation*	PD 3493 Missiles ; Unsafe nuclear	PD 7231 — research ; Deceptive
PA 7411 Misguidance *Uncommunicativeness*	PE 1397 Missing action	PB 5151 — resources
PA 7232 Misguidance *Unskillfulness*		

-989-

Misuse results

Misuse [results...] cont'd
PD 3947 — results amniocentesis
PD 4629 — *rodenticides*
PF 3701 Misuse satellite surveillance governments
Misuse scientific
PD 7415 — medical test results
PC 9188 — research
PC 9188 — theories
Misuse [societally...]
PC 8019 — *societally-endorsed professions conceal socially unacceptable initiatives*
PE 1348 — spiritual authority sexual purposes
PF 4564 — statistics
Misuse [telephone...]
PD 1632 — telephone surveillance governments
PE 5274 — tropical rain forests agricultural development
PD 4691 — *trust funds*
PD 8904 Misuse wild animals
PF 1205 *Misused ambulance service*
PE 3895 *Mite-borne typhus*
PE 3639 *Mite ; Depluming*
PE 3639 *Mite ; Feather*
PG 4137 *Mite infestation ; Itch*
PE 3639 *Mite ; Northern fowl*
PD 9366 *Mite poultry ; Air sac*
PE 3639 *Mite ; Scaly leg*
PE 3639 *Mite ; Subcutaneous*
PE 3639 *Mite ; Tropical fowl*
PE 3639 *Mites ; Chicken*
PE 2727 *Mites ; Nasal*
PE 3639 Mites pests
PE 9017 Mites vectors plant disease
PF 3566 Mitigation ; Inadequate disaster prevention
PD 0355 Mixed marriage
PE 3504 *Mixed piles*
PE 5183 Mixed race children ; Discrimination against
PC 1523 Mixed races ; Non acceptance
PE 4717 Mixtures non toxic chemicals ; Health hazards long term, low level exposure toxic
PD 5256 Mixtures ; Toxic
PD 2226 MLOs Mycoplasma-like organisms
PA 0429 *Mob violence*
PF 1068 Mobile professionals experts ; Impediments internationally
PF 3434 Mobility aid officials ; Excessive career
PD 1493 Mobility developing countries ; Low occupational
PF 1257 *Mobility ; Disturbing transient*
PF 3806 Mobility ; Excessive population
Mobility Lack
PD 4011 — labour
PD 4011 — labour force
PF 2195 — social
PD 4011 Mobility ; Low labour
PD 4011 Mobility ; Restrictions labour
PE 1848 Mobility village populations
PD 0465 Mobilization developing countries ; Low level domestic resource
PF 7432 Mobilization expenditure allocation ; Imbalance revenue
PF 5964 Mobilization financial resources developing countries ; Inadequate
PF 4197 Mobilization government revenue ; Inefficient
Mobilization [procedures...]
PF 3702 — *procedures ; Hypersensitive military*
PF 9704 — public opinion development ; Inadequate
PF 4197 — public resources ; Ineffective
PE 3995 Mobilization reinforcement ; Disparity facilities military
PF 4979 Mobilization resources ; Inadequate
PF 4979 Mobilize resources ; Failure
PF 9704 Mobilize support against problems ; Government failure
PE 2203 Model developed countries ; Contradictions growth partnership
PE 8533 Models complex system behaviour ; Excessive dependence computer
PF 8639 Models equality ; Lack
Models [ideal...]
PF 9478 — ideal society ; Inadequate
PF 8451 — Inappropriate role
PF 8451 — Insufficient role
PF 6479 Models ; Lack local leadership role
PF 9100 Models ; Overgeneralized policy
PF 9576 Models ; Resistance inclusive environmental
Models [social...]
PF 9576 — social change policy makers ; Static
PF 9576 — social development ; Inappropriate theoretical
PF 9576 — socio economic development ; Competing
PF 9576 — socio economic development ; Inadequate
PE 0742 — Substance abuse role
PF 9576 Models ; Uncoordinated multiplicity development
PF 9576 Models ; Uncoordinated multiplicity global financial
PC 1587 *Moderate mental retardation*
PA 7156 Moderation
PF 6155 Modern cities ; Stultifying homogeneity
PF 6461 Modern disruption traditional symbol systems
PF 6524 *Modern equipment ; Lack*
PF 6524 *Modern equipment ; Unavailability*
PF 2457 Modern farm machinery ; Limited availability
PF 4675 Modern furniture materials ; Toxic
PE 4675 Modern house fires ; Lethal fumes
Modern [industrialized...]
PE 2862 — industrialized societies ; Institutional obsolescence
PF 9716 — information systems ; Inadequate use
PE 1499 — insulating materials ; Health hazards
PE 3547 *Modern production ; Lack*
PF 6524 *Modern technology ; Insufficient*
PE 6865 Modern technology ; Visual strain
PF 4799 Modernization agriculture ; Inappropriate
PB 0206 *Modernization ; Lack*
PE 7678 Modernization military weaponry ; Creeping
PC 4321 Modes behaviour ; Contempt traditional
PF 2421 Modes citizen engagement ; Fragmented
PF 2168 *Modes ; Disoriented habitual*
PF 2403 Modes ; Incompatibility transport
PF 6592 Modes life developing countries ; Westernization traditional
PF 3826 Modes social organization ; Non viable alternative
PE 2690 Modification ; Abusive behaviour
PC 3152 Modification animal habitats
PF 6078 Modification environmentally adapted nutritional habits food aid
Modification Hostile
PD 7941 — climate
PD 7941 — environmental
PD 7941 — weather
PE 7832 Modification micro climate ; Increase insect pests
PC 1288 Modifications climate ; Inadvertent
PD 0767 Modifications due creation dams lakes ; Ecosystem
PC 0776 Modified feral species ; Genetically
PD 0122 *Moisture*
PD 2289 *Mole ; Hydatidiform*
PE 8253 Mole rat ; Endangered species African
PE 9103 Mole rats ; Endangered species palaearctic
PE 8750 Moles desmans ; Endangered species
PD 3479 *Moles ; Endangered species golden*
PE 3981 Moles ; Endangered species marsupial
PE 3265 Molestation children
PC 0728 *Mollusc pests*
PD 3478 Mollusca
PD 3478 Molluscs ; Endangered species
PD 0372 Molluscs ; Environmental hazards fish, crustacea
PE 0972 Molluscs preparations thereof ; Instability trade fish, crustacea
PE 1783 Molluscs preparations thereof ; Long term shortage salt water fish, crustacea
PD 5228 *Molybdenum poisoning animals*
PE 1656 *Monachus ; Endangered species*
PF 5943 Monarchs ; Exiled
PF 2170 Monarchy
PF 2188 Monasticism
PF 6079 Monetarism
PD 2469 Monetary agreements ; Parochial
PE 5247 Monetary bloc
PF 8583 Monetary controls ; Distortion international trade selective
Monetary [financial...]
PC 7873 — financial conditions ; Fluctuations world
PF 0048 — financial system ; Crisis international
PE 0117 — foreign exchange ; Underutilization non
Monetary gold
PD 3045 — Hoarded
PD 3045 — Hoarding
PC 7859 — Non restitution
PF 2342 — Uncertain status
PF 2564 Monetary grants ; Limited applicability
PD 1432 Monetary incentives developing countries ; Lack response
PC 7873 Monetary instability
PC 7873 Monetary markets ; Instability
Monetary [policies...]
PE 4315 — policies ; Interference transnational banks' off shore borrowing domestic
PF 8749 — policies ; Restrictive
PF 8749 — practices ; Restrictive
Monetary [reserves...]
PF 3060 — reserves ; Inadequate level developing country
PF 3059 — reserves ; Inadequate level world
PF 6575 — *rewards ; Nonevident*
Monetary system
PF 0048 — Inadequacies international
PF 3058 — Lack confidence international
PF 0048 — Lack coordination international
PF 2139 — Privatization intergovernmental
PC 3778 — Unrepresentative control international
PE 5903 Monetary systems exchange notes transnational corporations ; Destabilization
PF 5514 Monetary systems ; Inflation based
Money [capitalist...]
PF 3114 — capitalist systems ; Conflicting roles
PF 9560 — creation ; Over reliance government
PF 7313 — *crop consumed*
PE 3700 Money ; Devaluation
PD 5433 Money ; Dirty
PD 4469 Money ; Extortion protection
PD 9356 Money ; Fluctuations real value
PD 5433 Money ; Funny
PE 5503 Money government securities ; Counterfeit
PC 1453 Money ; Hot
PB 4653 Money ; Insufficient
Money [Lack...]
PF 2899 — *Lack seed*
PE 7803 — laundering
PE 6746 — liabilities ; Foreign
PF 0760 Money management ; Poor
PD 9489 Money markets ; Speculation
PJ 3258 Money orders ; Irregular international
PF 1382 Money priorities ; Family
PF 9014 Money ; Prohibitive cost
PE 5341 Moneyed hereditary class ; Discredited
PF 5596 Mongering ; Rumour
PE 2125 Mongolism
PE 9121 Mongooses ; Endangered species genets, civets
PF 2489 Monies ; Inadequate credit
PE 4497 Moniliasis
PE 2455 *Moniliasis*
PF 8046 Monitoring agencies ; Corruption
PD 9538 Monitoring communication governments ; Abusive
PD 7231 Monitoring cover inaction
PF 9244 Monitoring development projects ; Inadequate
PE 5089 Monitoring hazardous substances ; Ineffective
Monitoring [illegal...]
PF 7264 — illegal activity ; Ineffective
PF 4801 — Inadequate environmental
PF 2905 — industrial growth ; Inadequate systems
PF 2793 — Ineffective
PD 2683 Monitoring mail
PF 5374 Monitoring public sector activity ; Inadequate
PF 2782 Monitoring restrictive business practices due inadequate regulation ; Ineffective
PF 6635 Monitoring systems ; Inadequate radiation
PE 6982 Monitoring television ; Confusion values
PF 1585 Monitors ; Harassment human rights
PD 2730 *Monkey disease ; African green*
PE 1570 *Monkeys ; Endangered species new world*
PE 8991 Monkeys ; Endangered species old world
PD 0405 Monkeywrenching
PF 7967 Mono-factor explanations world
PF 2170 Monocracy
PC 3606 Monoculture crops
PF 4405 Monoculture ; Ideological
PF 4405 Monoculture ; Political
PF 4741 Monoculture ; Technological
PE 0639 *Monocytic leukaemia*
PF 3286 *Monogamous marriage law*
PD 2695 Monolingualism multi-cultural setting
PE 1925 Monolithic architecture high-rise buildings
PF 1618 Monologue ; Religious
PE 5550 Mononucleosis ; Infectious
PD 3978 *Monoocytic ehrlichiosis ; Equine*
PE 0763 *Monoplegia ; Congenital*
PE 0763 *Monoplegia ; Infantile*
PE 4566 *Monoplegia ; Transient*
PC 0521 Monopolies
Monopolies [Aerospace...]
PE 7747 — Aerospace
PD 9163 — Air transport
PD 9163 — Airline
PC 0521 — *Aluminium*
PE 8006 Monopolies capitalist countries ; Insurance
PE 5751 Monopolies developing countries ; Transnational
PD 3040 Monopolies ; Emergence international media
PE 8018 Monopolies ; Food
PE 1813 Monopolies ; Industrial gas
PD 5642 Monopolies ; Inefficiency government
PE 3391 Monopolies ; Privatization state
PE 0109 Monopolistic activity transnational enterprises
PD 7501 Monopolistic control new animal forms
PD 7840 Monopolistic control new life forms
PE 6788 Monopolization agricultural genetic resources
PF 2856 Monopolization information organizations
PE 1702 Monopolization interest groups development community priorities
PF 5329 Monopolization knowledge
PC 8410 Monopolization power
PD 5642 Monopolization resources public sector
PE 1918 Monopolization technology transnational corporations
PF 3005 *Monopolized land ownership*
PD 4445 Monopoly advanced nuclear warfare technology ; Super power
Monopoly [capitalism...]
PE 7947 — capitalism ; State
PF 0980 — competence intervention future
PE 0405 — court proceedings ; Legal profession's
PD 3003 Monopoly economy corporations
PD 3040 *Monopoly ; Film*
PJ 5244 Monopoly ; Foreign exchange
PF 0980 Monopoly knowledge predict future
Monopoly [media...]
PD 3101 — media
PD 3040 — media ; International
PE 3829 — Multimodal transport
PE 0246 Monopoly ; Newspaper
PD 1741 Monopoly nuclear power techniques
Monopoly power
PC 8410 — [Monopoly power]
PE 0081 — due advertising
PD 3133 — elderly
PE 0988 — state owned state controlled enterprises ; Abuse
PE 8267 Monopoly practices ; Distortion international trade state trading government
PD 3040 *Monopoly ; Satellite transmission*
PC 0521 *Monopoly ; State*
PD 3040 *Monopoly ; Television*
PA 0429 *Monopoly violence ; Union*
PE 2197 *Monotonous activity*
PF 6195 Monotonous housing facilities
PD 2656 Monotonous repetitious work
PF 0867 Monotonous unaesthetic architecture design
PD 2656 Monotony ; Increasing job
PC 1326 *Monotremes ; Endangered species*
PE 1657 Monoxide health hazard ; Carbon
PE 1657 Monoxide occupational hazard ; Carbon
PE 4103 Monsoon
PD 0489 Monsoon failure
PF 2516 Monsters
PD 1618 Monstrosities ; Human
PF 2516 Monstrous entities
PD 4348 Monuments ; Desecration
PD 0253 Monuments historic sites ; Endangered
PD 2478 Monuments ; Pollution damage

Murder

PD 9159 Mood disorders
PD 5535 *Moon blindness*
PF 6908 Moonlighting
PE 7188 Moonshine
PE 9599 Moral blackmail
PB 2480 Moral chaos
PB 2480 Moral collapse
Moral [development...]
PF 2187 — development ; Lack
PF 0513 — dilemma
PF 3377 — double standards
PB 2480 Moral environment ; Decayed
PF 2277 *Moral hedonism*
Moral [imperfection...]
PB 7712 — imperfection
PB 7712 — imperfection ; Dependence
PD 8227 — imperfections leadership
PB 5249 — indifference
PF 7178 — interests ; Incompatible
PD 9179 Moral offences
PA 3369 *Moral offences heterosexual pairing*
PB 2480 Moral pollution
PB 2480 Moral relativism
PE 3802 Moral religious education ; Denial right choose
PF 9050 Moral standards government ; Double
PF 7178 Moral standards ; Inability define
PB 5249 Moral tourism
Morale Low
PF 8446 — [Morale ; Low]
PF 1365 — *community*
PE 6639 — military
PF 3379 Moralism
PF 7178 Morality ; Breakdown standards
PA 6892 Morality*complex
PC 5005 Morality ; Crimes against public
Morality Double standards
PF 5225 — [Morality ; Double standards]
PF 5225 — [Morality ; Double standards]
PF 3259 — sexual
PE 9091 Morbid fear illness
PD 5086 Morbid preoccupation death
PD 4538 Morbidity
PA 6799 Morbidity *Disease*
PA 6030 Morbidity *Fear*
PE 5463 Morbidity ; Ill defined causes
PC 6869 Morbidity mortality countries ; Unequal
PD 2387 Morbidity mortality ; Perinatal
PD 2387 Morbidity ; Perinatal
PD 4538 Morbidness
PE 2348 Morphine
PE 1329 *Morphine abuse*
PE 7941 Mortality ; Animal
PC 6869 Mortality countries ; Unequal morbidity
PE 5166 Mortality developing countries ; Child
PE 4245 Mortality due parental abuse chronic neglect ; Infant
PE 7941 Mortality ; Economic loss animal
PE 4354 Mortality elderly countries ; Unequal
PA 0072 Mortality ; Human
PA 0072 Mortality *Human death*
PE 5463 Mortality ; Ill defined causes
PC 1287 Mortality ; Infant
PD 2422 Mortality ; Maternal
PD 9750 Mortality ; Neonatal
PD 2387 Mortality ; Perinatal morbidity
PF 0333 Mortality rate ; Decrease
PF 7043 *Mortality rate ; High*
Mortality rates
PC 9586 — countries ; Unequal
PE 1339 — gender ; Inequality
PC 9586 — Inequality
PD 4029 Mortality ; Still birth
PE 5463 Mortality ; Unknown causes
PD 5422 Mortgage fraud
PE 4571 Mortgage payments ; Arrears
PA 6799 Mortification *Disease*
PA 6659 Mortification *Humility*
PA 6088 Mortification *Impairment*
PA 5612 Mortification *Unchastity*
PA 7107 Mortification *Unpleasantness*
PD 2227 Mosaic ; *Potato latent*
PD 2227 Mosaic ; *Tobacco*
PC 1630 *Mosquito breeding ponds*
PE 3582 Mosquito resistance insecticides
PE 1923 Mosquito vectors
PE 3623 Mosquitoes ; Common house
PE 3622 Mosquitoes ; Malaria
PE 1923 Mosquitoes vectors
Mosquitoes vectors disease
PE 1923 — [Mosquitoes vectors disease]
PE 3621 — Aedes
PE 3622 — Anopheline
PE 3623 — Culicine
PE 3621 Mosquitoes ; Yellow fever
PF 4166 Mother-in-law taboo
PD 0614 Motherhood ; Adolescent
PD 0614 Motherhood ; Early teenage
PE 5919 Motherhood ; Forced
PF 0131 Mothers ; Children imprisoned
PF 2747 Mothers ; Commercial surrogate
Mothers [Detention...]
PD 5142 — *Detention*
PD 0902 — Discrimination against unwed
PD 6812 — Discrimination against working
PE 2274 Mothers ; *House bound*
Mothers [Illiteracy...]
PE 8660 — Illiteracy

Mothers [Imprisoned...] cont'd
PE 1837 — Imprisoned
PD 5142 — *Imprisoned*
PE 2274 Mothers' self-imposed isolation
PD 2681 Mothers ; Single
PD 0902 Mothers ; Unmarried
PA 5468 Motion*complex
PA 6752 Motion*complex ; Relative
PE 2611 Motion sickness
PD 9366 Motion sickness animals
PF 1098 Motivated amnesty military personnel convicted atrocities ; Politically
PD 9349 Motivated arrests ; Politically
PF 9794 Motivated exaggeration human rights infractions ; Politically
PD 6970 Motivated mass arrests ; Politically
PF 4723 Motivated reduction prison sentences ; Politically
PF 2464 Motivated utilization construction technology ; Profit
PF 0967 Motivating collegial relationships ; Lack
PF 1042 Motivating socio dramas ; Failure
PD 8945 Motivation care aged ; Reduction
PA 6048 Motivation*complex
PA 6048 Motivation ; *Criminal*
PA 2360 Motivation ; Lack
PF 2208 Motivation leadership development ; Lack
PE 2583 Motivation ; Poor services
PF 1948 Motivational death
PF 8161 Motivational methods ; Outmoded
PF 2702 Motivational techniques ; General unavailability effective
PF 1004 Motivational techniques ; Shallow approaches
PF 2174 Motive ; Distortion profit making
PF 1913 Motives ; Dominance economic
PF 2949 *Motivity ; Undermined job*
Motor accidents
PE 8824 — Delay compensation victims
PE 8824 — Neglect victims
PE 8824 — Non payment compensation victims
PD 3514 Motor activities aged ; Slowness sensori
PE 0345 Motor aphasia
PE 8813 Motor neurone disease
PE 7230 Motor skills disorder
PE 9344 *Motor tic disorder ; Chronic*
PE 6681 Motor traffic ; Tax impediments international
Motor vehicle
PE 5572 — *driver fatigue*
PD 0414 — emissions
PD 3664 — noise
Motor vehicles
PE 2240 — Continued operation unsafe
PE 8336 — Environmental hazards road
PE 2147 — Excessive speed
PD 2072 — Proliferation automobiles
PF 6559 *Motorcycle control ; Inadequate*
PF 3212 Mottoes ; Empty slogans
PE 0916 Mould plants ; Sooty
PE 1265 Mould tropical climates ; Destructive action
PE 2069 *Moulds*
PE 8051 Moulds plants
PE 1371 Moulds plants ; Slime
PD 9366 Mountain disease cattle ; *High*
Mountain [ecosystems...]
PE 6256 — ecosystems ski resorts ; Destruction
PE 6256 — environment leisure activities ; Degradation
PE 6256 — environments mountaineers trekkers ; Pollution
PD 2322 Mountain sickness
PF 9023 Mountain valley communities ; Social isolation
PE 6256 Mountaineers trekkers ; Pollution mountain environments
PD 6282 Mountainous environments ; Vulnerability
PD 6282 Mountainous regions ; Deforestation
PD 1908 Mountainous regions ; Depopulation
PA 5479 Mournfulness *Lamentation*
PA 6731 Mournfulness *Solemnity*
PA 7107 Mournfulness *Unpleasantness*
PF 3516 Mourning
PD 4918 Mourning ; Frozen
PE 3604 *Mouse tailed bats ; Endangered species*
PD 8347 Mouth ; *Benign neoplasm throat*
Mouth [disease...]
PE 1589 — disease ; Foot
PE 1589 — disease ; Hoof
PE 9702 — diseases animals
PD 1185 Mouth jaw ; Diseases
PD 3978 Mouth ; *Slavery*
PD 9667 Mouth ; *Sore*
PE 9819 Mouth throat ; Malignant neoplasm
PE 3254 Mouth ; *Trench*
PD 3978 Mouth ; *Watery*
PE 5330 Mouton ; Tremblante
PC 2367 Movement across frontiers ; Illegal
PD 2755 Movement animals factor animal diseases ; International
Movement [capital...]
PJ 5990 — capital ; Free
PE 4669 — children separated parents ; Lack freedom
PE 0897 — commercial vehicles ; Unjustified restrictions free
PC 3173 — communist systems ; Denial freedom
PE 8408 — countries ; Restrictions freedom
PC 0935 — countries ; Restrictions freedom
PC 0935 Movement ; Denial right freedom international
PE 6769 Movement disorders
PD 2755 Movement factor animal diseases ; Animals' international
PE 8882 Movement labour ; Distortion international trade restrictive controls

PD 0351 Movement national advantage ; Restrictions international freedom
Movement [patterns...]
PC 2279 — patterns ; Disruption animal migration
PC 6203 — people ; Involuntary
PC 8452 — persons ; Denial right freedom
PE 8899 Movement shipping ; Violation right international freedom
PE 8408 Movement state ; Denial right freedom
PD 9765 Movement toxic products ; Illicit
PB 1638 Movements ; Authoritarian
PF 6594 Movements funds ; Unreported international
PF 2071 *Movements ; Growth anti systemic*
PJ 3527 Movements ; Intolerant
PF 5375 Movements ; Millenarian
PC 3362 Movements minorities ; Threats ideological
PF 5375 Movements ; Nativistic
PC 1869 *Movements ; Resistance*
PF 5375 Movements ; Revivalist
PE 2224 Movies ; Snuff
PE 5928 Moving ; Disruptive effect household
PD 5183 *Muckraking ; Journalistic*
PE 7570 Mucocele
PE 4414 Mucopolysaccharide diseases
PD 2728 *Mucormycosis*
PD 3978 *Mucosal disease complex*
PE 4724 Mucous gastritis
PG 6761 *Mucous membrane larynx windpipe ; Swelling*
PE 3331 Mucoviscidosis
PE 2357 Mud fever
PA 6468 Muddle *Complexity*
PA 7361 Muddle *Disorder*
PA 6900 Muddle *Formlessness*
PA 6247 Muddle *Inattention*
PA 6446 Muddle *Indiscrimination*
PA 6662 Muddle *Nonaccomplishment*
PA 7309 Muddle *Uncertainty*
PA 7232 Muddle *Unskillfulness*
PD 1233 Mudslides
PD 5575 Mugging
PE 9774 *Mulberry heart disease*
PE 7826 Mule foot
PE 7808 *Mule foot*
PE 8593 Mules hinnies ; Instability trade asses,
PD 2695 Multi cultural setting ; Monolingualism
PF 0364 Multi-disciplinary characteristics problems
PD 0213 Multi-drug abuse
PF 3214 Multi party parliamentary systems ; Weakness
PF 3368 Multidenominational society
PF 7923 Multidisciplinary approaches ; Obstacles development
Multilateral [action...]
PF 9564 — action ; Ineffectual intergovernmental
PF 8650 — agencies official contributions ; Imbalance disbursements
PF 3913 — assistance ; Delay supply agreed
PF 0817 Multilateral cooperation ; Decline
PF 9216 Multilateral development aid ; Policy cross conditionality restrictions
PE 4289 Multilateral forums industrialized countries ; Undermining
PF 9629 Multilateral principles ; Unilateral interpretations
PF 5287 Multilateral safeguard systems ; Inability negotiate effective
PF 9564 Multilateralism ; Indecisive
PF 9564 Multilateralism ; Reduced government commitment
PF 7838 Multilaterally agreed trade commitments ; Conditional observance
PE 3829 Multimodal transport monopoly
Multinational enterprises
PD 5891 — [Multinational enterprises]
PE 4102 — developing countries ; Discrimination against women workers
PE 0573 — Inaccessibility decision makers
PE 7129 Multinationalism international nongovernmental organizations
PE 4107 Multiple birth ; Risk
PE 4107 Multiple births
PF 6158 Multiple convictions single criminal act
PF 1205 *Multiple family jealousies*
PF 6158 Multiple liability identical criminal act
PE 5289 Multiple myeloma
PE 6677 Multiple nationalities ; Conflicting
PE 6678 Multiple nationality ; Conflicting military obligations persons
PE 5048 Multiple personality disorder
PD 0173 *Multiple pregnancy*
PE 1041 Multiple sclerosis
PE 9324 Multiple systems ; Congenital syndromes affecting
PF 9576 Multiplicity development models ; Uncoordinated
PF 9576 Multiplicity global financial models ; Uncoordinated
Multiplicity languages
PC 0178 — [Multiplicity languages]
PC 0410 — international relations
PC 1518 — national setting
PF 2833 Multiplicity manual sign languages deaf
PF 6027 Multiplicity official languages
PF 2003 Multiplicity problems
PF 2003 Multiplicity problems facing society
PF 3404 Multiplicity religions
PF 2621 Multiplicity time standards
PE 2356 Mumps
PD 5948 Municipal corruption
PF 3550 Municipalities capital funds ; Restricted access
PC 6154 *Municipalities ; Insolvent*
PC 2618 Munitions ; Manufacture
PD 2341 Murder

-991-

Murder

	Murder [Abortion...]
PD 0158	— Abortion
PD 2341	— aged
PD 0437	— Attempted
PD 5493	Murder contract
PD 7221	Murder ; Government sanctioned
PD 2341	Murder ; Mass
PE 0437	Murder ; Negligent
	Murder [Permissible...]
PE 7533	— Permissible
PE 5614	— Political
PD 5590	— Political mass
PF 4213	— Psychic
PF 2641	Murder ; Ritual
PF 4213	Murder ; Soul
PE 7645	Murderers ; Psychotic mass
PD 7221	Murders authorized governments ; Political
PD 3481	Murids ; Endangered species
PG 3896	Murine typhus
PE 2348	Murray Valley encephalitis
PE 9774	Muscle disease ; Green
PE 9774	Muscle disease ; White
PE 9774	Muscle necrosis ; Back
PE 4161	Muscle ; Rupture gastrocnemius
PD 2565	Muscles ; Diseases
PE 8173	Muscles ; Paralysis throat, eye, limb respiratory
PB 0731	Muscles ; Strains
PE 7808	Muscling cattle ; Double
PD 2565	Muscoskeletal connective tissues ; Diseases
PD 2565	Muscular atrophy ; Infantile
PE 5188	Muscular atrophy ; Progressive
PC 0699	Muscular disabilities ; Neuro
PE 9774	Muscular dystrophy ; Enzootic
PE 7808	Muscular hypertrophy
PE 4921	Muscular wasting
PG 2295	Muscular weakness
PD 2565	Musculo skeletal system ; Diseases
	Musculoskeletal system
PD 7424	— animals ; Diseases
PD 8347	— Benign neoplasm
PE 8589	— Congenital anomalies
PE 7808	— Congenital anomalies animal
PD 2565	— connective tissues ; Diseases
PE 4566	— Symptoms referable
PD 2731	Mushy chick disease
PF 5190	Music ; Aggressive
PF 5190	Music ; Decadent
PF 5190	Music ; Passionate abstract
PF 5190	Musical perversions
PF 8713	Musical pitch ; Competitive distortion
PF 8713	Musical quality escalating pitch ; Distortion
PE 7103	Mustard gas
PE 8159	Mustard oils trade instability ; Rape
PA 7365	Mustiness Boredom
PA 5981	Mustiness Stench
PA 5490	Mutability Changeableness
PA 6890	Mutability Nonuniformity
PA 6425	Mutability Transience
PE 4588	Mutagenic chemicals
PE 6619	Mutagenic consequences food preparation
PE 4896	Mutagenic effects drugs
PD 1368	Mutagens
PF 2276	Mutation
PF 2276	Mutation ; Bacterial
PD 0196	Mute children
PE 4526	Muteness
PA 6118	Muteness Ineloquence
PA 5465	Muteness Loudness
PA 7411	Muteness Uncommunicativeness
PF 1849	Mutilation ; Cattle
PD 2559	Mutilation deformation human body
PA 6790	Mutilation Distortion
PA 6088	Mutilation Impairment
PF 1849	Mutilation ; Livestock
PE 6054	Mutilation ; Male sexual
PE 3488	Mutilation ; Punishment criminals
PD 5718	Mutilation ; Sexual
PD 7576	Mutilation ; Torture
PC 5512	Mutilations ; Child
PE 6055	Mutilations ; Female sexual
PD 2589	Mutiny
PE 4526	Mutism
PE 5261	Mutism ; Deaf
PE 4526	Mutism ; Elective
PF 0245	Mutual credit societies ; Lack
PB 4731	Mutual deceits
PF 6570	Mutual responsibility ; Unshared
PF 2602	Mutual student-teacher disinterest
PC 0933	Mutual trade developing countries ; Decline
PD 2735	Myaisis
PD 0235	Myalgias ; Epidemic
PE 1914	Myalgic encephalomyelitis
PE 3638	Myasthenia gravis
PE 2455	Mycetoma
PE 2455	Mycetoma pedis
PD 2732	Mycoplasma gallisepticum infection poultry
PD 2226	Mycoplasma-like organisms (MLOs)
PD 2732	Mycoplasma meleagridis infection turkeys
PD 2732	Mycoplasma synoviae infection poultry
PE 5050	Mycoplasmal arthritis
PE 7553	Mycoplasmal pneumonia
PE 5050	Mycoplasmal polyserositis
PE 6291	Mycorrhizae ; Absence
PE 2455	Mycoses
PE 2455	Mycosis
PE 2455	Mycotic disease

PE 5444	Mycotic otitis externa
PD 7307	Mycotic pneumonia
PE 2143	Mycotic stomatitis animals
PE 9458	Mycotoxic lupinosis
PE 9458	Mycotoxicoses
PD 2728	Mycotoxicosis poultry
PE 7514	Mycotoxins
PE 9037	Myelitis
PD 7841	Myeloencephalopathy brown Swiss cattle ; Bovine progressive degenerative
PE 0639	Myelogenous leukaemia
PE 5289	Myeloma ; Multiple
PE 9554	Myelophthisis animals
PE 1849	Myelotoxicity
PE 3633	Myiasis
PE 8456	Myocardial infarction
PE 8456	Myocardial infarction ; Acute
PD 0448	Myocarditis ; Acute
PD 0235	Myocarditis ; Animal
PD 0471	Myocardium diseases animals
PD 7841	Myoclonia congenita
PE 7808	Myofibrillar hyperplasia
PE 7808	Myofibrillar hypoplasia
PE 9774	Myoglobinuria ; Paralytic
PE 9774	Myopathies animals
	Myopathies swine
PE 9774	— Cardiac
PE 9774	— Nutritional
PE 9774	— Skeletal
PE 9774	Myopathies ; Toxic
PE 7808	Myopathy associated congenital articular rigidity
PE 9774	Myopathy calves ; Nutritional
PE 9774	Myopathy ; Degenerative
PE 7826	Myopathy ; Fibrotic
PE 9774	Myopathy ; Ischaemic
PE 9774	Myopathy lambs ; Nutritional
PE 7826	Myopathy ; Ossifying
	Myopathy [Transport...]
PE 9774	— Transport
PE 9774	— turkeys ; Deep pectoral
PE 9774	— turkeys ; Transport
PE 9774	Myopathy wild animals ; Capture
PF 7771	Myopia
PD 6963	Myopia ; Political
PF 7283	Myopic advice leadership
PF 2394	Myopic decision-making methods
PE 9774	Myositis ; Eosinophilic
PF 2590	Mysticism ; Dependence
PA 6393	Mystification Miseducation
PA 5502	Mystification Reason
PA 7411	Mystification Uncommunicativeness
PA 7367	Mystification Unintelligibility
PF 6224	Mystique attitudes problems disabled people
PC 3458	Myth classlessness
PF 4162	Myth deterrence
PC 1258	Myth national security increase military capacity
PF 1138	Myth ; Oligarchy effectiveness
PF 8887	Myths ; Disintegration accepted
PE 0652	Myxoedema
PD 3480	Myxomatosis
PE 0652	Myxooedema

N

PD 7424	Naalehu disease
PE 2455	Nacardiosis
PC 8534	Nail ; Diseases
PE 7826	Nail prick
PE 6268	Nairobi sheep disease
PF 1870	Naive approach marriage covenant
PF 7116	Naive ; Assessment international nongovernmental organizations
PE 7002	Naive optimism
PA 5568	Naïvety Ignorance
PA 7131	Naïvety Oldness
	Naivety [Unbelief...]
PA 7392	— Unbelief
PA 7371	— Unintelligence
PA 7232	— Unskillfulness
PE 7233	Naked ; Interrogation while
PF 8241	Nakedness developing countries
PF 3350	Nakedness developing countries ; Ill considered pressure eliminate
PE 1796	Name brand food sector developing countries ; Domination transnational corporations domestic
PE 4187	Name-calling
	Name [Denial...]
PE 4624	— Denial right
PF 5223	— dropping
PE 5755	— drugs ; Excessive pricing brand
PE 4624	Name married women ; Imposition husband's
PD 7981	Name merchandise ; Forgery brand
PF 8511	Names geographical features ; Political changes
PF 8511	Names ; Non standardization geographical
PF 8511	Names ; Non standardization topographic
PF 7728	Names people convicted crimes which they are innocent ; Failure clear
PD 3492	Napalm
PF 7248	Narcissism
PF 7248	Narcissitic behaviour disorders
PF 7248	Narcissitic personality disorders
PE 4888	Narco analysis ; Involuntary
PE 9210	Narco-terrorism
PE 5037	Narcotic investigations ; Police crimes during
PF 6787	Narcotics legislation ; Inadequate

PD 0991	Narcotics trade ; Illicit
PF 0823	Narrow context counsel
	Narrow [economic...]
PF 8602	— economic foresight
PF 3871	— egalitarian application trade measures
PF 2949	— employment base
PF 2949	— employment opportunities
PD 2968	— export base developing countries
PF 2998	Narrow job context
PD 1501	Narrow legal definition family developing countries
PF 0076	Narrow media options
PF 3379	Narrow-mindedness
	Narrow [paths...]
PF 7580	— paths ; Steep
PF 2182	— political interests small groups
PE 7313	— profit margin
PF 6548	— programme offerings
	Narrow range
PE 6554	— business skills
PF 3628	— cultural exposure
PD 4100	— food crops
PD 6075	— goods
PF 2477	— practical skills
PE 6526	Narrow road ways
PF 3552	Narrow scope education
PF 4644	Narrow time span poor
PF 7988	Narrow visioning leadership
PF 7967	Narrow world perspective
PF 2444	Narrowing tax base
PA 7306	Narrowmindedness
PA 7256	Narrowness
PE 8187	Narwhal ; Endangered species white whale
PE 2727	Nasal acariasis animals
PG 9838	Nasal catarrh
PJ 3620	Nasal infections
PE 2727	Nasal mites
PD 9667	Nasal solar dermatitis
PD 9667	Nasi dogs ; Eczema
PG 9838	Naso-pharyngeal catarrh
PE 7733	Nasopharyngitis
PC 0799	Natal depression ; Post
PE 3659	Natal infections ; Pre
PF 0090	Nation boundaries ; Artificial
	Nation states
PF 5691	— Crazy
PF 0292	— disregard world opinion
PD 0055	— obstacles world order
	National [accounts...]
PF 7842	— accounts developing countries ; Inadequate systems
PF 0023	— accounts ; Non valuation housework
PD 0351	— advantage ; Restrictions international freedom movement
PE 5132	— affairs ; Concentration national governments activity
	National [basic...]
PF 4763	— basic food stocks ; Secrecy
PF 4216	— bias judges international athletic competitions
PF 8235	— boundaries
PF 0090	— boundaries ; Arbitrary
	National [capitals...]
PF 3455	— capitals ; Unnecessary relocation
PF 5452	— character
PB 0534	— chauvinism
PE 8309	— commodity price support ; Government protection
PF 3130	— communism
PF 4210	— concepts corrupt practices ; Divergent
	National [data...]
PF 0510	— data sources ; Weaknesses
PC 2546	— debt burden
PF 5472	— decision making developing industrialized countries ; Displacement
PE 4097	— defence industries
PE 4097	— defence procurement procedures
PF 4842	— development ; Obstacles
PB 3384	— disintegration
PF 3249	— divorce laws ; Conflicting
	National economic
PJ 4512	— goals ; Contradictory solutions
PD 9436	— recession
PE 0042	— sectors transnational enterprises ; Control
	National economies
PE 7413	— food subsidies ; Distortion
PE 1922	— limited number corporations ; Disproportionate influence
PF 2344	— limited number individuals ; Undisclosed control
PF 9697	— vagaries external markets goods services ; Vulnerability
PC 7873	— vagaries international financial system ; Vulnerability
	National [economy...]
PD 4800	— economy ; Government interference
PC 1524	— educational qualifications ; Non equivalence
PD 0428	— empathy external hardship ; Inadequate
PD 0626	National federalism
PF 2574	— forms currency ; Variations
PE 5132	National governments activity national affairs ; Concentration
PD 0055	National governments obstacle representative democratic political organization
PF 8946	National hegemony over United Nations agencies
PE 4647	National holidays ; Inefficiency due mismatch religious
	National Identity
PE 8513	— due tribalism ; Weak
PF 1934	— Lack
PE 8610	— post colonial countries ; Weak
	National insecurity
PD 3835	— developing countries
PD 8030	— due excessive foreign investment

Needs developing

National insecurity cont'd
PB 1149 — vulnerability
National [instability...]
PC 1910 — *instability due tribalism*
PE 4994 — insurance markets offshore insurers ; Destabilization
PF 2600 — interests ; Parochial
PF 2723 — international anniversaries years ; Proliferation
PF 2141 — isolation
PF 2141 — isolationism
PF 1373 *National language proficiency ; Inadequate*
National law
PE 4768 — enforcement ; Inadequate
PE 4339 — regulations ; Evasion
PF 9571 — relation international transactions ; Conflicts
National laws
PF 2645 — [*National laws*]
PF 0216 — *Conflict*
PE 6679 — governing cheques ; Conflicting
PF 9571 — relating international transactions ; Divergence different
National leaders
PD 1893 — Misrepresentation facts
PD 2710 — Psychopathic
PD 2710 — Self aggrandizing
National [leadership...]
PD 2710 — leadership ; Abusive
PF 9385 — leadership ; Criticism
PF 7034 — legal provision international nongovernmental organizations ; Lack
PE 1375 — level ; Inadequate coordination action intergovernmental programmes
PF 4851 — local legal systems ; Deficiencies
PF 5000 National macroeconomic policies industrialized countries ; Mismatch
PF 7081 National nongovernmental organizations developing countries ; Absence
PE 1922 National oligopolistic trading systems
National [policy...]
PF 2344 — policy making ; Disproportionate influence some individuals
PE 1922 — policy making ; Undisclosed control major corporations
PF 1452 — political dependence
PF 6441 — population ; Declining
PA 2173 — *prejudice*
PF 8434 — prestige ; Pursuit
National [regulations...]
PG 5429 — regulations banking security firms
PE 2200 — regulations transnational corporations ; Burden conflicting
PC 8999 — religious minorities ; Inadequate protection linguistic,
PF 2174 — resources ; Profit over emphasis use
PC 0114 — rivalry
PE 7020 National sections international nongovernmental organizations ; Incompatible equivalent
National security
PD 8153 — Armed crimes against
PC 0554 — Crimes against
PD 7407 — criminals ; Aiding
PC 4341 — Environmental threats
PC 1258 — increase military capacity ; Myth
National security information
PE 3997 — Crimes related
PE 4343 — foreign power ; Revealing
PE 3749 — Mishandling
PE 0484 National security offenders ; Harbouring
National self determination
PE 3123 — capitalist exploitation ; Denial effective
PC 3177 — communist systems ; Denial right
PC 1450 — Denial right
PF 1804 — Habitual overemphasis
National [self...]
PF 5466 — self-indulgence
PD 3546 — self interest ; Natural resources used
PC 1518 — setting ; Multiplicity languages
PE 2488 — settings regional linguistic controversies ; Bilingualism
PF 2636 — socialism
National sovereignty
PE 7906 — Denial right
PD 3496 — foreign military presence ; Erosion
PE 5015 — Infringement
PE 7906 — Loss
PD 0055 — obstacle peaceful socio-economic development
PE 1539 — transnational enterprises ; Erosion
National [statistics...]
PF 0510 — statistics ; Deficiencies
PF 8434 — status race
PE 8280 — supervision adherence international law ; Inadequate
PD 1791 National tax systems ; Disparity
PD 3765 National trade war ; Intra
PC 3293 *National unity*
PF 8107 National unity ; Lack
PB 0534 Nationalism
PB 0534 Nationalism ; Arab
PB 0534 Nationalism ; Dependence
PC 5842 Nationalism ; Economic
PF 3352 Nationalism ; Race reinforcement
PF 3351 Nationalism ; Religion reinforcement
PD 2490 Nationalist agitation
PF 4761 Nationalist exploitation sporting events
PF 6094 Nationalistic attitudes currency
Nationalistic images
PF 2600 — citizenship
PF 2600 — Defensive
PF 2600 — Overruling global interdependence

PD 5475 Nationalistic media
PF 0817 Nationalistic policy responses world economic crisis
PF 0817 Nationalistic response global issues
PD 3546 Nationalistically determined development natural resources
PE 6677 Nationalities ; Conflicting multiple
PF 0699 National acquisition ; Commercialization
Nationality [children...]
PE 6676 — children ; Involuntary loss
PE 8953 — Conflict laws over
PE 6678 — Conflicting military obligations persons multiple
Nationality [Denial...]
PE 4912 — Denial right
PE 1736 — Denial right change
PD 3225 — Deprivation
PE 4016 — Discrimination against women right
PE 6677 — Dual
PE 2485 Nationality ; Involuntary loss
PE 2657 Nationality ; Refusal grant
PE 4912 Nationality ; Restrictions recognition
PE 4016 Nationality women ; Denial right
PD 1994 Nationalization domestic enterprises
PC 2172 Nationalization foreign investments
PC 2172 Nationalization overseas investments ; Risk
PC 3240 *Nationalized industries*
PD 2579 Nationally oriented language systems ; Exclusive
PD 0973 Nationals ; Discrimination against non
PE 6564 Nationals ; Disregard international law capture foreign
PE 0812 Nationals ; Errant
PE 6424 Nationals ; Failure notify imprisonment death prison foreign
PE 6564 Nationals foreign countries ; Government seizure foreign
PE 8414 Nationals leaving their own country ; Restrictions
PE 6422 Nationals military service ; Discrimination against foreign
PE 8817 Nationals returning their country ; Restrictions
PE 1911 Nationals states ; Financial economic disputes states
PE 8746 Nationals who have lived abroad ; Denial social security
PC 0720 Native peoples ; Abuse
PD 4252 Native trade patterns ; Disruption
PF 2186 Nativism
PF 5375 Nativistic movements
PC 4764 Natural areas ; Vulnerability protected
PC 1247 Natural barriers ; Destruction
Natural [dams...]
PC 0715 — dams ; Disastrous failure
PE 8339 — death ; Denial animals right
PF 6026 — disaster ; Inadequate insurance against damages arising
PB 1151 — disasters
PE 0299 — disasters least developed countries
PD 0408 Natural energy resources ; Aggressive uses
PF 7517 Natural energy sources ; Aggression against
Natural environment
PB 5250 — degradation
PC 4777 — Hazards human health
PE 2443 — mankind ; Inevitable destruction
PE 6920 Natural environmental degradation recreation tourism
PD 1472 Natural food poisons
PD 4238 Natural foodstuffs ; Harmful
Natural gas
PE 8464 — Instability trade mineral tar crude chemicals derived coal, petroleum
PD 2596 — *production environmental hazards*
PE 8955 — production instability ; Petroleum
PE 9137 — trade infrastructure ; High cost
Natural [habitats...]
PC 3152 — habitats flora fauna ; Degradation semi natural
PD 8328 — hazards air traffic
PD 0173 — human abortion
PE 1914 Natural killer cell syndrome ; Low
PF 6198 Natural light buildings ; Displacement
PF 8045 Natural manufactured gas ; Long term shortage
PC 3152 Natural natural habitats flora fauna ; Degradation semi
PF 3710 Natural objects ; Inadequate legal rights
PF 1676 Natural origin ; Radio noise
Natural [parks...]
PD 7585 — parks ; Excessive use
PC 2416 — pollutants
PD 3155 — prey ; Extermination wild animal
PC 1299 Natural radiation ; Contamination
Natural resource
PD 4002 — depletion due high-level consumption
PF 3043 — endowments ; Disparity countries
PD 3546 — extraction ; Parochial planning
PF 2466 — product expansion ; Neglect research
PF 2882 — use decisions ; Limited access
Natural resources
PD 2338 — *Absentee owned*
PF 3043 — countries ; Inequality distribution
PC 4824 — Decreasing
PD 3109 — Denial sovereignty over
PB 5250 — Degradation
PD 3922 — developing countries ; Depletion
PD 3625 — developing countries ; Lack
PD 4007 — due population growth ; Depletion
PF 2855 — Emphasis mass extraction
PC 4824 — Excessive demands
PD 3109 — Foreign control
PC 4020 — *Inaccessible*
PF 2204 — Inefficient extraction utilization
PC 7928 — Lack
PE 0273 — least developed countries ; Insufficient use

Natural resources cont'd
PC 4824 — Long term shortage
PD 3546 — Nationalistically determined development
PF 2855 — Unconstrained exploitation
PF 1459 — Underutilization
PD 3546 — used national self-interest
Natural [scientists...]
PD 9182 — scientists ; Deception
PC 2122 — *seascape ; Degradation*
PF 5997 — selection
Natural [water...]
PD 2586 — water cycles ; Underreporting disruptions
PD 9670 — water systems ; Disruption
PE 6955 — wealth ; Denial right people freely dispose
PC 2507 Naturalization
PE 3889 Naturalization passports ; Crimes related immigration,
PD 1472 Naturally occurring poisonous substances food-stuffs
PF 8264 Nature ; Belief humanity's dominance over
PF 7917 Nature ; Corruption good human
PC 4584 Nature ; Crimes against
Nature [Dependence...]
PF 6803 — Dependence fear
PF 4175 — development issues ; Political
PF 0379 — Disconnection people rhythms
PF 9582 Nature ; End
PF 9582 Nature ; False assumptions concerning recuperative power
PF 6803 Nature ; Fear
PE 5754 Nature inherited titles ; Fraudulent
PE 8216 Nature problems ; Inadequate education concerning
PF 0379 Nature ; Separation
PF 2231 *Nature ; Unrealized potential*
PE 4513 Nausea
PD 8982 Naval accidents
PD 8412 Naval arms race
PD 3704 Naval manoeuvres sensitive areas
PE 3580 *Naval warfare*
PD 2731 Navel ill
PE 7826 Navicular disease
PF 6749 Navigable waters ; Non uniformity marking
PE 3868 Navigation ; Hazards
PE 8899 Navigation ; Violation right international freedom
PF 2636 Nazism
PF 2636 Nazism ; Neo
PE 2300 Neapolitan disease
PD 1582 Near collisions ; Aircraft
PE 6685 Near sewage discharge ; Bathing
PF 2813 Nebulous city plans
PF 2631 Necessary information ; Minimal access
PA 0831 Necessities ; Deprivation basic
PF 1238 Necessities ; Inflated cost
PF 1373 *Necessities ; Neglect basic*
PF 2385 Necessities rural communities ; Prohibitive cost
PA 6387 Necessity
PD 1147 *Necessity business payoffs*
PD 0164 Necessity child labour
PE 3948 Necessity creative establishment disestablishment tensions ; Failure understand
PF 6908 Necessity second jobs
PF 9461 Neck ; Pain
PE 4161 Necrobacillosis ; Interdigital
PF 8042 Necromancy
PF 6957 Necrophilia
PE 2684 Necrosing arteritis
PD 3978 Necrosis animals ; Acute pancreatic
PD 9774 Necrosis ; Back muscle
PD 7841 Necrosis ; Cerebrocortical
PD 5535 Necrosis ; Ear
PE 5617 Necrosis ; Infective bulbar
PD 3978 Necrosis ; Peritoneal fat
PD 5535 Necrotic auricular dermatitis
PD 9667 Necrotic dermatitis chicken
PD 5535 Necrotic ear syndrome swine
PD 3978 Necrotic enteritis poultry
PE 7769 Necrotic hepatitis ; Infectious
PE 4161 Necrotic pododermatitis ; Chronic
PE 7553 Necrotic rhinitis
PE 2684 *Necrotizing arteritis ; Generalized*
PD 2565 *Necrotizing arthritis ; Generalized*
PE 3508 Nectoriasis ; Ancylostomises
PC 7597 Need agricultural land
PE 7530 Need availability viable organs transplantations ; Imbalance
Need [consensus...]
PD 2438 — consensus ; Insufficient
PF 0021 — continuing education ; Unsatisfied
PJ 0646 — credit ; Individual
PF 2709 Need ; Decisional paralysis specialized services relation world's
PF 2995 Need functional skills ; Unrecognized
PE 4508 Need increased cooperation central banks
PJ 0646 Need recognition ; Personal
PF 4984 *Need ; Unperceived preschool*
PF 3844 Need ; Vocational obsolescence face overwhelming
PF 3387 Need world religion
PE 4346 Needle casts conifers
PE 0616 *Needle ; Common use hypodermic*
PE 6255 Needle rusts
PE 5112 Needless incarceration
PF 1321 Needs demands ; Gap material technological
PE 8902 Needs ; Devaluation education survival
Needs developing countries
PE 1011 — Insensitivity transnational corporations consumer
PC 5180 — Lack political will respond
PE 4120 — Non responsiveness transnational corporations pharmaceutical

-993-

Needs

PE 0893 Needs ; Disrelationship production work force
PF 1825 Needs ; Elders' concealed
PF 1578 Needs ; Failure adapt general initiatives specific
Needs [Inadaptation...]
PF 5145 — Inadaptation work family
PF 2878 — Inadequate information funding
PD 1080 — Inadequate response societal
PE 3939 — Industrial processes geared reduced social
PE 5397 Needs ; Legal contract system reduced individual
PF 3243 Needs mentality ; Immediate
PF 6552 Needs ; *Obscured intergenerational*
PF 3243 Needs ; Overemphasis immediate superficial
PF 2639 Needs ; Production serving false consumption
PF 1025 Needs promotion marketing ; Disrelated
PF 3608 Needs they address ; Gap function social techniques
PF 2477 Needs ; *Unarticulated training*
PF 0963 Needs ; *Unassessed corporate*
Needs Unclear
PD 4790 — *care*
PF 2477 — *vocational*
PD 1544 — *youth*
Needs [Unconsensed...]
PD 4790 — *Unconsensed health*
PF 1781 — *Uncoordinated expression*
PF 2477 — *Uncorrelated training*
PD 1544 — *Undefined youth*
PF 1169 — Unmet
PF 1169 — Unsatisfied
PF 6471 — *Unsurveyed consumer*
PD 4790 — *Unsurveyed nutritive*
PD 8638 Needs women ; Architecture insensitive
PF 2152 Nefarious benefits ; Misuse plausible arguments conceal
PE 1488 Negated oligarchic control decision making ; Social service quality
PF 1651 *Negated paramedic role*
Negation [Destruction...]
PA 6542 — *Destruction*
PA 5982 — *Disagreement*
PA 6838 — *Dissent*
PA 6979 Negation *Opposition*
Negative [aspects...]
PF 7727 — aspects euphenics
PC 1455 — attitude people faiths
PE 4838 — attitudes towards menstruation
Negative [capability...]
PF 7109 — capability ; Inadequate
PF 2575 — community reputation
PF 6575 — community story
PF 7583 — consequences remedial action ; Disguised
PE 2305 — consequences shifting ecology coastal communities
Negative [ecological...]
PD 4007 — ecological impact overpopulation
PF 0589 — economic repercussions disarmament
PE 4225 — economic social effects advertising
PF 7990 — effect glamour
Negative effects
PF 3376 — claims religious infallibility
PD 1562 — factory farming
PF 0107 — family allowances
PE 0341 — family planning education children
PF 8120 — foreign aid
PD 2761 — *gastronomic fashions*
PF 0129 — nuclear family
PF 3850 — over-crowding mental health
PF 4351 — rejection
PJ 8374 — relative wage rates youth employment
Negative [election...]
PD 9384 — election campaigning
PA 7090 — emotions attitudes
PA 7090 — emotions attitudes ; Dependence
PF 5311 Negative feedback ; Avoidance
Negative [identifying...]
PF 1735 — identifying slogan
PF 4133 — images opponents ; Promotion
PF 9003 — influence peer groups
PE 0598 — information ; Delay disseminating
PE 1478 — information media ; Excessive portrayal
PD 4840 Negative life events
PC 3134 Negative net resource transfers developing countries
PF 9843 Negative personal relationships leaders countries
Negative [self...]
PE 6784 — self-image ageing women
PF 9003 — social context
PE 5134 — social effects introduction new technologies
PF 4133 — socialization towards opponents
PF 0441 Negative thoughts
PF 2575 Negative village self-identity
PF 5950 Negativism
PA 5982 Negativity *Disagreement*
PA 6838 Negativity *Dissent*
PA 6099 Negativity *Hopelessness*
PA 6979 Negativity *Opposition*
PA 7321 Negativity *Refusal*
PA 5438 Neglect
Neglect [adolescent...]
PF 6061 — adolescent health care
PD 8945 — aged
PF 7047 — agricultural rural life developing countries
PF 1373 *Neglect basic necessities*
PF 2346 Neglect community space
Neglect [Dependence...]
PB 0883 — Dependence social
PD 2092 — dependents war victims
PA 6822 — *Disrespect*

Neglect [education...]
PD 0190 — education women
PE 5288 — elderly institutional care
PF 6570 — *elders ; Anticipated*
PF 1664 — elders' skills
PE 9295 — environmental consequences government policies
PF 7820 — expert advice
PE 9295 Neglect future ecological problems ; Government
PE 9295 Neglect future problems social environment arising current policies ; Government
PE 0437 Neglect ; Homicide
PD 8114 Neglect human resource development debtor developing countries
Neglect [individual...]
PF 1403 — *individual goals*
PE 4245 — Infant mortality due parental abuse chronic
PF 7057 — international nongovernmental organization network
Neglect [Neglect...]
PA 5438 — *Neglect*
PF 7069 — non-Western structures international nongovernmental organizations
PA 6662 — Nonaccomplishment
Neglect [peasant...]
PF 7047 — peasant farmers
PD 8607 — personal health
PD 8894 — property maintenance
PF 8072 — public authorities
PF 8072 — Public authorities
Neglect [remote...]
PE 5760 — remote regions islands
PF 2466 — research natural resource product expansion
PF 4959 — role women rural development
Neglect [school...]
PD 8894 — school buildings
PF 5147 — sexual health women
PF 7047 — small agricultural producer
PB 0883 — Social
Neglect [Unskillfulness...]
PA 7232 — *Unskillfulness*
PC 2616 — Urban
PC 2616 — urban housing
Neglect victims
PD 4823 — crime
PE 8824 — motor accidents
PE 6936 — torture
Neglect [war...]
PD 2077 — war veterans ; Social
PD 2092 — war widows orphans
PF 6024 — women farmers
Neglected [children...]
PD 4522 — children
PF 6512 — community education
PF 6575 — *community uniqueness*
PF 6514 — cooperative action
PF 1781 — *council representation*
PF 7808 Neglected food resources
PD 8607 Neglected health practices
PD 0645 Neglected maintenance local infrastructure
PE 5689 Neglected marine space island developing states
PD 3789 *Neglected organization training*
PD 8894 Neglected personal property
PF 6578 Neglected public space
PF 9686 Neglected rural populations
PF 8085 Neglected social skills training
PD 3722 Neglected teachers ; Children
PE 8924 Neglected welfare handicapped persons
PE 4245 Neglected young children
PA 2658 Negligence
PD 2623 Negligence anthropologists
PA 6379 Negligence *Avoidance*
PD 7731 Negligence biologists
Negligence [chemists...]
PD 4265 — chemists
PC 2563 — commerce
PA 2658 — Criminal
PA 2658 Negligence ; Dependence
PA 7361 Negligence *Disorder*
Negligence [employers...]
PD 2879 — employers
PA 6180 — *Error*
PF 0221 — experts
PD 1045 Negligence food industry
PE 0561 Negligence ; Food poisoning
PD 0708 Negligence geologists
PE 7169 Negligence housing tenants
PD 2586 Negligence hydrologists
Negligence [Inattention...]
PA 6247 — *Inattention*
PE 4329 — Increase insurance claims medical
PD 2916 — industry
PD 5380 Negligence lawyers
PD 5948 Negligence local government
Negligence [maintenance...]
PD 7964 — maintenance
PE 0437 — manslaughter
PC 0981 — Maternal
PD 5251 — media
PD 5770 — Medical
PD 4182 — meteorologists
PA 5438 Negligence *Neglect*
Negligence [oceanographers...]
PD 4277 — oceanographers
PF 8046 — official inspectors
PE 1369 — opticians
Negligence [Paternal...]
PD 7297 — Paternal

Negligence [pedologists...] cont'd
PD 1110 — pedologists
PE 5538 — performing socialist responsibilities ; Criminal
PE 3540 — pharmacists
PD 1710 — physicists
PD 9193 — Police
PC 5517 — politics
PC 8019 — Professional
PD 8290 Negligence radiologists
PD 5422 Negligence real estate agents
PD 6626 Negligence social scientists
PD 1012 Negligence transport industry
PA 7341 Negligence *Unpreparedness*
PA 7232 Negligence *Unskillfulness*
PD 7726 Negligence veterinarians
PD 4721 Negligence zoologists
PD 4334 Negligent employees
PD 8114 Negligent implementation austerity measures
PE 6740 Negligent interpretation
PE 0437 Negligent murder
PD 5501 Negligent old age care
PF 1815 Negligent owner improvements
PA 2658 Negligent people
PE 6702 Negligent public service employees
PF 9431 Negotiable facts
PF 5287 Negotiate effective multilateral safeguard systems ; Inability
PE 3970 Negotiate freely employers ; Violation right trade unions
PF 4787 Negotiated treaties ; Unfairly
PE 9646 Negotiating capacity developing countries ; Inadequate
PD 1958 Negotiation employment reward ; Inadequate access
PE 0853 Negotiation entrance terms transnational corporations developing countries ; Inadequate
PE 7004 Negotiation failure ; ASAT
PF 1786 Negotiation procedures ; Outdated labour
PC 8694 Negotiations ; Limited progress North South
PE 4355 Negotiations transnational banks ; Domination loan
PF 8695 Neighbour contacts ; Infrequent
PE 5504 Neighbourhood disputes
PF 6556 Neighbourhood leaders ; Insufficient
PE 2274 Neighbourhood level ; Fragmented forms care
PE 5443 Neighbourhood noise
PF 2575 Neighbourhood relations ; Estranged
PF 2575 Neighbourhood residents ; Limited involvement
PF 6147 Neighbourhood signs ; Insufficient
PE 4780 Neighbourhood technology ; Lack
PC 2616 Neighbourhoods ; Deteriorating inner city
PF 6147 Neighbourhoods ; Unidentifiable urban
PD 0664 Neighbouring developing countries ; Lack integration transport systems
PF 9023 Neighbouring villages ; Social isolation
PE 7903 Neighbours ; Boundary disputes
PE 5504 Neighbours ; Disputes
PE 5443 Neighbours ; Noisy
PD 0120 *Nematicides pollutants*
PE 6461 *Nematode infections*
PD 2228 *Nematodes*
PE 9854 *Nematodes animals*
PD 2228 Nematoid plant diseases
PC 1876 Neo-colonialism
PF 2636 Neo-fascism
PF 2636 Neo-nazism
PD 1763 *Neofelis ; Endangered species*
PD 3978 *Neonatal diarrhoea ruminants*
PD 7841 *Neonatal maladjustment syndrome*
PD 9750 Neonatal mortality
PE 5050 *Neonatal septic polyarthritis*
Neoplasm [bone...]
PD 8347 — *bone cartilage ; Benign*
PE 9229 — bone ; Malignant
PD 8347 — *breast ; Benign*
PE 1175 — breast ; Malignant
PE 9229 Neoplasm connective tissue ; Malignant
PE 4303 Neoplasm digestive organs ; Malignant
PD 8347 *Neoplasm digestive system ; Benign*
PD 8347 *Neoplasm endocrine glands ; Benign*
PE 9705 Neoplasm genital organs ; Benign
PE 5100 Neoplasm genito urinary organs ; Malignant
PE 9086 Neoplasm kidney urinary organs ; Benign
PE 7085 Neoplasm lung ; Malignant
Neoplasm [male...]
PE 8372 — *male genital organs ; Malignant*
PE 9819 — mouth throat ; Malignant
PD 8347 — *musculoskeletal system ; Benign*
PD 8347 *Neoplasm nervous system ; Benign*
PE 0368 Neoplasm ovary ; Benign
PE 7837 *Neoplasm respiratory system ; Benign*
PE 7572 Neoplasm respiratory system ; Malignant
PD 8347 *Neoplasm skin ; Benign*
PE 5016 Neoplasm skin ; Malignant
PD 8347 *Neoplasm throat mouth ; Benign*
PE 7922 Neoplasm uterus ; Benign
PC 3853 Neoplasms
PD 7799 *Neoplasms animals ; Prostatic*
PD 8347 Neoplasms ; Benign
PE 1905 *Neoplasms female genital organs ; Malignant*
PD 2283 *Neoplasms joints*
PE 4637 Neoplasms lymphatic haematopoietic tissue
PC 0092 Neoplasms ; Malignant
PC 3853 Neoplastic diseases
PE 2272 *Nephritis*
PE 2053 *Nephroptosis*
PE 2053 *Nephrosis*
PD 7704 *Nepotism*
PD 1672 Nepotism developing countries
PD 7704 Nepotism public office

-994-

Non ferrous

PF 6013 Nepotism socialist countries
PF 8611 Nepotism ; Tolerance
PJ 5588 Nerd
PE 4584 Nerve ; Diseases optic
PC 0872 Nerve gases
PD 8786 Nerve ; Inflammation optic
PE 7808 Nerve paralysis ; Femoral
PE 4161 Nerve paralysis ; Peroneal
PF 8026 Nerves ; Damage small blood vessels kidneys, eyes
PE 8932 Nerves ; Diseases cranial
PE 8932 Nerves peripheral ganglia ; Diseases
PE 5758 Nervosa ; Anorexia
PE 4538 Nervosa ; Bulimia
PE 6322 Nervous breakdown
PC 8756 Nervous disorders
PE 4724 Nervous dyspepsia
Nervous system
PE 0965 — animals ; Congenital diseases
PD 7841 — animals ; Diseases
PD 8347 — *Benign neoplasm*
PE 9296 — Congenital anomalies
PE 1041 — Demyelinating diseases central
PC 8756 — Diseases
PE 4683 — diseases caused parasites
Nervous system Diseases
PE 9037 — central
PE 8932 — peripheral
PE 8932 — peripheral autonomic
Nervous system [disorders...]
PE 9037 — disorders ; Central
PE 0965 — *dogs ; Disorders*
PE 7915 — Familial disease
PE 7915 — Hereditary disorders central
PE 9468 — Symptoms referable
PE 2300 — *Syphilis central*
PE 2280 — *Tuberculosis central*
PE 4171 Nervousness
PA 5838 Nervousness *Agitation*
PA 6030 Nervousness *Fear*
PA 5467 Nervousness *Inexcitability*
PD 9082 Nets ; Deployment excessively efficient fishing
PD 9082 Nets ; Excessive use drift
PE 5735 Netting ; Bird
PD 9667 Nettle rash
PE 1017 Nettle rash
PE 7044 Network ; Lack identity international nongovernmental organization
PF 7057 Network ; Neglect international nongovernmental organization
PE 1055 Network unconnected major roads ; Road
PE 1055 Network ; Underdeveloped road
PF 2350 *Networks ; Chaotic communication*
Networks [Inadequate...]
PD 2213 — Inadequate political
PE 0174 — Inflexible intermediary political implementation
PF 1681 — *Insufficient community*
PF 4549 Networks ; Old boy
PF 6470 Networks rural areas ; Poor communications
PF 1959 Networks urban ghettos ; Ineffective operation community
PE 8932 Neuralgia
PE 8932 Neuralgia ; Trigeminal
PE 3520 Neurasthenia
PE 8932 Neuritis
PE 8932 Neuritis ; Brachial
PD 8786 Neuritis ; Optic
PC 0699 Neuro-muscular disabilities
PE 4814 Neurofibromatosis
PE 0873 Neurogenic arthritis
PD 4755 Neurological effects torture
PC 8756 Neurological illness
PE 0429 Neurological rage
PE 7915 Neuromuscular disorders ; Hereditary
PD 7424 *Neuromuscular paresis cows following parturient paresis*
PE 8813 Neurone disease ; Motor
PD 5453 *Neuropenia grey Collie dogs ; Cyclic*
PE 4696 Neuropharmacological torture
PF 7490 Neurophysiological compulsion violence
PC 0799 Neuroses ; Depressive
PD 0270 Neuroses ; Psycho
PD 0270 Neurosis
PA 1635 Neurosis ; Anxiety
PE 3717 Neurosis ; Childhood
PD 0270 Neurosis ; Depressive
PF 3904 Neurosis food
PD 0270 Neurosis ; Hysterical
PE 3571 Neurosis ; Infantile
PA 6739 Neurosis *Maladjustment*
PD 0270 Neurosis ; Occupational
PD 0270 Neurosis ; Success
PD 4571 Neurosis ; Traumatic
PE 7912 Neurosis ; War
PE 8349 Neutral countries ; Capitalist subversion communist
PD 3180 Neutral countries ; Communist subversion capitalist
PF 0801 Neutralism
PF 0473 Neutrality
PE 4031 Neutrality civil servants ; Lack political
PF 3076 Neutrality intergovernmental official information ; Excessive
PJ 1743 Neutrality science
PC 2659 Neutrality ; Violation
PD 7501 New animal forms ; Monopolistic control
PF 8767 New ; Bias favour
PE 2399 New born ; Haemolytic disease
PD 2732 *New duck disease*

PF 4668 New initiatives ; Restricted government ability undertake
PF 4306 New International Economic Order ; Lack progress establishing
PD 7840 New life forms ; Monopolistic control
PF 2813 *New machinery ; Unscheduled introduction*
PE 1444 New plant species ; Irresponsible introduction
New [scientific...]
PD 1709 — scientific theories ; Non acceptance
PF 3699 — social problems ; Invention
PD 0822 — social structures developing countries ; Inadequate development
PF 2212 — societal service systems ; Incomplete understanding
New species
PD 1290 — animals ; Irresponsible introduction
PD 3602 — fish ; Irresponsible introduction
PE 3592 — insect pests ; Introduction
PC 1617 — introduction ; Environmental hazards
PC 0776 — Irresponsible creation
PC 0776 — Uncontrollable
PE 1793 New states ; Administrative difficulties
New technologies
PE 4391 — Elimination jobs developing countries due introduction
PE 8758 — Hindrances international spread
PJ 9034 — Inadequate assessment
PE 3337 — Incompatibility traditional
PE 5134 — Negative social effects introduction
PD 0528 — Unemployment introduction
New [technology...]
PF 9127 — technology ; Fear
PF 1789 — technology ; Pollution constraints testing
PF 3699 — threats ; Deliberate invention
PF 1503 New urban families ; Adjustment difficulties
New [weapons...]
PC 0012 — weapons ; Competitive development
PE 1570 — *world monkeys ; Endangered species*
PE 8847 — world porcupines ; Endangered species
PE 3604 New Zealand short tailed bat ; *Endangered species*
PE 7423 Newborn ; Enteric diseases
PD 5228 *Newborn pigs ; Iron dextran toxicity*
PE 1400 Newcastle disease
PE 1400 Newcastle disease poultry
PF 8743 Newcomers ; Mistrust
PF 6091 Newly founded churches ; Non recognition
PD 6081 News ; Biased media coverage
PD 1718 News ; Biased presentation
PD 3030 News censorship
News coverage
PE 5610 — capitalist countries ; Political interference television
PF 3073 — legal affairs ; Restrictions
PF 3072 — parliamentary affairs ; Restrictions direct
PD 0917 — Parochial
PF 3081 News ; Denial access
PF 3082 News distribution media ; Restriction access
PE 1478 News ; Getting use violent
News [Inadequate...]
PF 6541 — *Inadequate schedule*
PE 1750 — inflation
PD 4982 — information ; Deliberate distortion corporate
PD 1893 — information ; Deliberate distortion official
PE 0598 — items ; Calculated delays releasing controversial
PE 1478 — *items ; Reduced impact impersonal repetitive*
PE 1750 News journalism ; Junk
PF 6573 *News ; Limited access*
News [media...]
PD 6081 — media ; Confiscation
PE 1643 — media ; Conflict government
PE 1478 — Media emphasis bad
PE 0598 News ; Postponement bad
PF 6541 *News resources ; Unidentified*
PF 7998 News ; Unpublicized community
PF 7998 News ; Unpublished local
PD 0184 Newspaper journal propaganda
PD 3043 Newspaper journal propaganda ; Foreign controls
PE 0246 Newspaper monopoly
PD 3027 Newspaper periodical censorship
PD 1152 Newspaper waste
PE 2287 Niacin deficiency
PC 3152 Niches ; Destruction biological
PC 5773 Niches ; Vulnerability ecosystem
PE 1315 Nickel pollutant
PE 0824 *Nickel trade ; Instability*
PD 0713 Nicotine abuse
PE 9253 Nicotine withdrawal
PD 2542 *Night blindness*
PE 6126 Night life entertainment facilities ; Inadequate
PE 7589 Night work
PE 6958 Nightmares
PA 7090 Nihilism
PD 2730 *Nile fever ; West*
PE 1514 Nitrate fertilizers ; Overuse
Nitrate [poisoning...]
PD 5228 — *poisoning animals*
PD 2503 — pollution aquifers
PD 2503 — pollution groundwater
PE 1956 Nitrates ; Peroxyacetyl
PE 1956 Nitrates pollutants
PD 5228 *Nitrite poisoning animals*
PE 6087 Nitrites pollutants
PD 2965 Nitrogen compounds pollutants
PE 4328 Nitrogen overdosage plants
PD 2965 Nitrogen oxides pollutants
PE 9443 Nitrous oxide atmosphere ; Increase
PD 4194 Nobbling ; Jury
PF 5778 Nobility ; Insufficient

PD 2728 *Nocardiosis animals*
PE 2197 *Nocturnus ; Pavor*
PF 1790 Nodes ; Insufficient cultural
PF 2822 Nodes ; Lack central
PE 6886 Nodes ; Milker's
PE 9509 *Nodosa animals ; Polyarteritis*
PD 2226 *Nodules legumes ; Root*
PE 5799 Noise ; Aircraft
PE 5799 Noise ; Airport
PC 0268 *Noise ; Building construction*
PE 4109 Noise damage ear
PD 4172 Noise ; Electronic
PD 2567 *Noise ; Head*
PC 0268 Noise ; Health hazards exposure
PE 2473 Noise industrial origin ; Radio
PE 5799 Noise ; Jet
PC 0268 Noise levels ; High
PC 0268 Noise levels ; Lack legislation restricting
PD 3664 Noise ; Motor vehicle
PF 1676 Noise natural origin ; Radio
PE 5443 Noise ; Neighbourhood
PC 0268 Noise pollution
PE 5691 Noise ; Torture continuous
PD 3664 Noise ; Traffic
PD 2831 Noise working environment
PA 5454 Noisome *Badness*
Noisome [Unhealthfulness...]
PA 7226 — *Unhealthfulness*
PA 7107 — *Unpleasantness*
PA 7204 — *Unsavouriness*
PE 5443 Noisy neighbours
PE 4415 Noisy sleeping habits
PE 6206 Nomadic children ; Inadequate education
PF 3700 Nomadism
PD 2520 Nomadism ; Desert
PE 8077 Nomads pastoralists ; Endangered lifestyles
PF 4849 Nomenklatura
Non acceptance
PE 6039 — embryo transfer technology
PF 2499 — *government legislation*
PC 1523 — mixed races
PD 1709 — new scientific theories
PF 1079 — reality
Non accountability
PE 3780 — control production processes ; Public
PE 4997 — government action
PE 4032 — organizations developing technology ; Public
PC 1522 — Social
PF 6574 — state-controlled enterprises
PF 1072 — transnational enterprises
Non [adaptive...]
PE 5246 — adaptive local structure social care
PE 3035 — alcoholic beverages ; Environmental hazards
PE 8748 — alcoholic beverages ; Instability trade
PF 0801 — alignment
PE 4768 — application existing laws
PE 2595 — articulated educational goals
PE 1407 — assistance endangered peoples ; Government
PE 0139 Non barbiturate hypnotics ; Abuse barbiturates
PD 1180 Non-biodegradable plastic waste
Non [citizens...]
PD 0973 — citizens ; Discrimination against
PF 8703 — collective decision-making
PE 8361 — combatants war zones ; Inhumane treatment
PF 1187 — communication ; Parent teacher
Non competitive
PD 8916 — business wages
PF 2411 — *consumer services*
PF 7313 — *products*
PE 8645 — teachers' salaries
Non [compliance...]
PF 0423 — *compliance law*
PD 1133 — comprehensive wage scales
PF 2158 — concerned attitudes
PE 3812 — concessional lending developing countries ; Erratic flows concessional
PF 3069 — convertibility currencies
PF 8195 — cooperation
PF 2788 — cooperation ; Non violent
Non cooperative
PF 7118 — community groups
PC 0025 — economic relations
PF 9474 — governments ; Internationally
PJ 2493 Non-corporate operations existing groups
PD 0634 *Non criminal reasons ; Confinement*
Non [destructible...]
PD 1754 — destructible packaging containers
PF 4854 — disclosure threats public safety
PF 9187 — diversification market crops
PF 2135 — diversification subsistence fishing economies
PE 2471 — durability bridges
Non [electrical...]
PJ 2425 — electrical machinery ; Environmental hazards
PE 8828 — electrical machinery ; Instability trade
PF 3895 — epidemic typhus
PC 1524 — equivalence national educational qualifications
PE 2462 — *ergonomic design*
PC 3651 — essentials ; Production
PD 4191 — existent commodities ; Sale
PE 4878 — Non-fatal deliberate self-harm
PE 8131 Non ferrous base metals ; Instability trade ores concentrates
Non ferrous metal
PE 2601 — *basic industries instability*
PE 0824 — ores ; Long term shortage
PE 0353 — *scrap ; Shortage*

-995-

Non-ferrous metals

Non [ferrous...]
PE 8248 — ferrous metals basic industries environmental hazards
PE 1406 — ferrous metals ; Instability trade
PD 7565 — food crops tropical economies ; Preponderance
Non fuel materials
PE 0546 — Environmental hazards inedible crude
PD 0280 — Instability trade inedible crude
PE 0461 — Long term shortage inedible crude
PE 8351 — Restrictive practices trade inedible crude
PE 8106 Non fur skins ; Instability trade undressed hides
PF 2835 Non-globalized citizenship
PF 6359 Non human life ; Killing
Non human primates
PE 1570 — Endangered species
PE 1621 — research ; Inhumane use
PF 5073 — Shortage experimental
Non [immediate...]
PF 9119 — immediate hazards society ; Insensitivity
PF 2497 — implementation international treaty provisions
PE 4003 — implementation workers wage increases provided legislation collective agreements
PF 2754 — inclusive management decisions
PF 1772 — integrating images personal freedom
PF 1187 — involvement ; Parent school
PE 7651 — ionizing radiation ; Environmental hazards
PF 3855 Non-juridical fault
Non local
PF 6531 — adult training
PS 8148 — *educational planning*
PF 2827 — *employment patterns*
Non [manual...]
PD 9628 — manual workers ; Discrimination against women
PF 3278 — marital cohabitation
PE 3828 — medical purposes ; Administering medical drugs
PD 8629 — members professional bodies ; Discrimination against
Non metallic mineral
PE 0890 — products industries ; Environmental hazards
PE 1851 — products industries ; Underdevelopment
PE 2599 — products industry ; Instability
PF 3100 Non military conflict ; International
PE 0117 Non monetary foreign exchange ; Underutilization
Non [nationals...]
PD 0973 — nationals ; Discrimination against
PD 1521 — nuclear weapon states ; Insecurity vulnerability
PE 5187 — *nutritive substance ; Eating*
Non [parental...]
PF 3253 — parental custody children
PD 1564 — participating elders urban environments ; Isolated
PC 0588 — participation
PF 6023 — participation youth decision-making bodies
PF 8703 — participative decision-making
PF 1959 — *participatory financial priorities*
PF 8650 — payment agreed contributions international organizations ; Government
PF 1382 — *payment bills ; Habitual*
Non payment compensation
PD 7179 — damages
PE 0290 — damages consumers
PE 8898 — forced relocation
PE 5229 — victims catastrophes
PE 3913 — victims crime
PE 0811 — victims malpractice
PE 8824 — victims motor accidents
Non [payment...]
PE 4446 — payment reparations government
PF 8650 — payment United Nations membership dues
PF 6218 — phonetic spelling
PF 1978 — physical psychological violence ; Covert
PD 1211 — political market allocation
PD 5540 — prescription drugs ; Phoney
PE 6526 — *priority road surfacing*
Non productive
PF 4202 — athletic activities
PE 3816 — capitalist elites
PF 4000 — members society
PD 1802 — use cattle livestock
PF 0422 — vocational training
Non [profit...]
PF 3049 — profit organizations ; Inappropriate taxation international
PF 6197 — proximate offices ; Inhibition communication
PC 0300 — psychotic mental disorders
PE 7300 Non-ratification human rights treaties
PF 0977 Non-ratification international treaties
Non recognition
PF 8040 — foreign governments
PE 6418 — foreign powers attorney
PF 9081 — international law
PF 6091 — *newly-founded churches*
PF 8112 — problems
PF 6091 — *religions*
PD 5001 Non-reinforced safety factors
Non renewable resources
PC 4412 — Competition
PD 1018 — Dependence government revenues exploitation
PC 8642 — Unproductive use
PC 8642 — Waste
Non [repatriation...]
PD 3132 — *repatriation export proceeds*
PE 0948 — repatriation prisoners war
PC 9585 — representational government
Non resident
PD 2338 — home ownership
PF 2261 — local authorities
Non resident cont'd
PF 2421 — *school personnel*
Non [residents...]
PD 3048 — residents country ; Tax discrimination against
PF 7754 — resistance evil
PE 2215 — response governments international surveys
PE 4120 — responsiveness transnational corporations pharmaceutical needs developing countries
Non restitution
PE 6074 — cultural property
PC 7859 — monetary gold
PC 7859 — property
PD 1291 Non-return developing country students studying foreign countries
PC 7859 Non-return property
Non [self...]
PF 2943 — self governing status disputed administering government ; Territories accorded United Nations
PC 0519 — settled refugees
PE 1508 — settled refugees living outside camps
PE 8002 — socialist countries ; Lack economic adaptation indigenous society
PF 8511 — standardization geographical names
PF 8511 — standardization topographic names
PF 2974 — standardized social services
PF 4861 — surveillance medical high risk persons
Non [tariff...]
PC 2725 — tariff barriers international trade
PC 0933 — *tariff trade preferences developing countries ; Non uniformity tariff*
PF 2859 — termination pregnancy
PC 1888 — *terrestrial territory ; Conflicting claims*
PE 4717 — toxic chemicals ; Health hazards long term, low level exposure toxic mixtures
PF 1664 — transferring wisdom aged
Non [uniformity...]
PE 6749 — uniformity marking navigable waters
PC 0933 — *uniformity tariff non-tariff trade preferences developing countries*
PD 6019 — union workers ; Discrimination against
PF 9683 — use available health facilities
Non [validity...]
PF 3250 — validity divorce
PF 3283 — validity marriage
PF 0023 — valuation housework national accounts
PF 6310 — *verifiability*
PF 4460 — verifiability compliance nuclear arms treaties
PF 4455 — verifiability compliance nuclear power safeguards
PF 6310 — verification compliance
Non viability
PF 8663 — cold countries
PF 1389 — *local electrical generating capacity*
PD 0068 — small developing countries ; Economic
PD 0441 — small states territories
PF 3826 Non-viable alternative modes social organization
Non violent
PF 2788 — non-cooperation
PD 3228 — political revolution
PF 9327 — weapons
PF 7069 Non Western structures international nongovernmental organizations ; Neglect
PD 4365 Non worshippers ; Prejudice against
PF 2499 *Nonacceptance government legislation*
PA 6387 Nonacceptance *Necessity*
PA 7321 Nonacceptance *Refusal*
PC 1523 *Nonacceptance social groups*
PA 6662 Nonaccomplishment
PA 6662 Nonaccomplishment *Nonaccomplishment*
PA 6858 Nonadherence *Disintegration*
PA 7250 Nonadherence *Disobedience*
PE 0873 *Nonarticular rheumatism*
PF 2313 *Nonavailability technical training*
PA 5878 Nonconformity
PA 6485 Nonconformity *Counteraction*
Nonconformity [Disagreement...]
PA 5982 — *Disagreement*
PA 7250 — *Disobedience*
PA 6838 — *Dissent*
PA 5869 Nonconformity *Exclusion*
PA 5878 Nonconformity *Nonconformity*
PB 1125 Nonconformity social norms
PF 6575 *Nonevident monetary rewards*
PA 5870 Nonexistence
PF 2477 *Nonexistence career counselling*
PF 6556 *Nonexistent development groups*
PC 2163 *Nonexistent high school transport*
PF 7239 Nonexistent people ; Officially
Nonexistent [salary...]
PE 6726 — salary base
PF 4375 — school funds
PF 8652 — security patrol
PF 2444 *Nonexpandable tax base*
PE 9220 Nonfactor services ; Distorting effects commodity taxes transaction goods
PA 5438 Nonfeasance *Neglect*
PA 6662 Nonfeasance *Nonaccomplishment*
PA 7232 Nonfeasance *Unskillfulness*
PA 5644 Nonfeasance *Vice*
PA 5473 Nonfulfilment *Insufficiency*
PA 6662 Nonfulfilment *Nonaccomplishment*
PE 7104 Nongovernmental activities ; Duplication international
Nongovernmental organization
PF 2016 — action ; Inadequate facilities international
PE 7080 — actions ; Divisive responses international
PE 0179 — coordination bodies ; Proliferation duplication international
Nongovernmental organization cont'd
PE 7032 — identity ; Lack international
PE 0362 — information systems ; Proliferation duplication international
PE 7092 — membership apathy ; International
PE 7044 — network ; Lack identity international
PE 7057 — network ; Neglect international
PE 7010 — personnel ; Disadvantaged status international
PE 7140 — response intergovernmental calls action ; Inadequate international
Nongovernmental organizations
PF 7081 — developing countries ; Absence national
PE 7128 — Establishment orientation international
PE 7045 — Excess western based secretariats international
PF 7117 — Inadequate statistical data
PE 7020 — Incompatible equivalent national sections international
PE 0064 — Jurisdictional conflict antagonism international
PE 1169 — Jurisdictional conflict antagonism international
Nongovernmental organizations Lack
PE 7009 — financial information systems international
PF 7058 — international legal provision
PE 0069 — legal provision international
PF 7034 — national legal provision international
Nongovernmental organizations [Limited...]
PF 7093 — Limited recognition international
PD 6800 — Manipulation
PE 7129 — Multinationalism international
PF 7116 — naive ; Assessment international
PF 7069 — Neglect non Western structures international
PF 7082 — Obstacles effective international
PE 7056 — Operational ineffectiveness international
PE 7008 — Political ineffectiveness international
Nongovernmental organizations programmes
PE 1209 — Inadequate coordination international
PE 0741 — Inadequate funding international
PE 1973 — Inadequate relationship international governmental
PF 1595 — Ineffectiveness international
Nongovernmental organizations [regional...]
PF 7033 — regional development ; Absence policies associate international
PE 0259 — Rivalry international
PE 0259 — scarce resources ; Competition international
PE 0777 — specialized agencies United Nations ; Inadequate relationship international
PE 7105 — Uncontrolled structures international
PE 7021 — Unrepresentative international
PF 4095 Nongovernmental resources ; Ineffective governmental use
PA 6926 Nonhumanity
PJ 8129 Nonidentification intended beneficiaries projects
PF 1781 *Noninclusive council representation*
PF 1548 *Nonintellectual home environment*
PF 3278 Nonmarital unions
PA 7362 Nonobservance
PA 7250 Nonobservance *Disobedience*
PA 6247 Nonobservance *Inattention*
PA 7321 Nonobservance *Refusal*
PE 6909 Nonorganic failure thrive
PD 2224 Nonparasitic plant diseases
PE 0436 Nonprofit associations front organizations government ; Misuse
PE 9438 *Nonprogressive pneumonia sheep*
PF 8156 Nonrecognition ethnic differences ; Bureaucratic
PD 4403 Nonrenewable fossil water reserves ; Misuse
PF 6991 Nonsense
PA 6977 Nonsense *Meaninglessness*
PA 7371 Nonsense *Unintelligence*
PD 2731 *Nonsuppurative polyarthritis lambs*
PA 6890 Nonuniformity
PA 5490 Nonuniformity *Changeableness*
PA 5828 Nonuniformity *Discontinuity*
PA 7361 Nonuniformity *Disorder*
PA 6695 Nonuniformity *Inequality*
PA 6774 Nonuniformity *Irregularity*
PA 6890 Nonuniformity *Nonuniformity*
PB 0731 *Nonvenomous insect bite*
PE 0533 Nonverbal communication deficiency
PE 0533 Nonverbal communication skills ; Ignorance
PE 0533 Nonverbal messages ; Insensitivity
PF 3758 Normalism
PF 3976 Norms ; Deviation behavioural
PB 1125 Norms ; Nonconformity social
PF 5301 Norms ; Systematic human behaviour according cultural
PD 2728 *North American blastomycosis*
North South
PC 8694 — Failure restructure economic relations
PC 8694 — gap
PC 8694 — Inequality
PC 8694 — negotiations ; Limited progress
PE 8332 — spheres influence UN related agencies
PE 3639 *Northern fowl mite*
Nose [bleed...]
PE 7864 — *bleed*
PE 9438 — *bot ; Sheep*
PE 7553 — *Bull*
PD 9667 *Nose ; Collie*
Nose [Diseases...]
PD 7924 — Diseases
PE 5122 — Diseases
PE 2412 — Dripping
PE 3635 *Nose fly*
PE 3635 *Nose fly ; Deer*
PE 7864 *Nose ; Haemorrhage*
PD 0659 *Nose inflammation*

PD 3672 Nose irritation
PE 3604 Nosed bats ; Endangered species leaf
PE 3604 Nosed bats ; Endangered species spear
PE 9091 Nosophobia
PJ 5282 Nostalgia
PE 6680 Notes ; Conflicting laws concerning bills exchange promissory
PE 5903 Notes transnational corporations ; Destabilization monetary systems exchange
PA 7270 Nothingness Absence
PA 5870 Nothingness Nonexistence
PD 0982 Notification ; Failure disease
PE 6424 Notify imprisonment death prison foreign nationals ; Failure
PA 2173 Notions ; Preconceived
PA 5454 Notoriety Badness
PA 6839 Notoriety Disrepute
PA 7363 Notoriety Improbity
PA 5612 Notoriety Unchastity
PB 1498 Nourished people ; Badly
PE 4793 Nourishment rest ; Denial working animals restorative
PF 3563 Novae
PF 8767 Novelty ; Obsession
PE 2501 Noxious algae
PC 0119 Noxious fumes
PE 8154 Noxious substances food chains ; Concentration

Nuclear accidents
PD 0771 — [Nuclear accidents]
PE 5324 — Panic result
PE 5080 — space

Nuclear arms
PD 0014 — Proliferation strategic
PD 5076 — race
PF 4460 — treaties ; Non verifiability compliance
PE 3968 Nuclear bomb making materials ; Covert trade
PF 3494 Nuclear bombs ; Ease manufacture
PF 0466 Nuclear conflict ; Erroneous precipitation

Nuclear [decapitation...]
PD 4049 — decapitation
PD 4049 — defence control systems ; Vulnerability
PF 7052 — disarmament ; Obstacles unilateral
PC 2201 Nuclear explosions peacetime
PE 2095 Nuclear explosions underground
PF 0129 Nuclear family ; Negative effects
PF 0679 Nuclear freeze
PF 6611 Nuclear irresponsibility

Nuclear materials
PD 0837 — Diversion
PE 3968 — Illegal exports
PE 7174 — irresponsible countries ; Trade
PD 3495 — Theft
PF 6611 Nuclear mentality
PD 3493 Nuclear missiles ; Unsafe
PD 7663 Nuclear power generation ; Inadequate infrastructure
PF 8531 Nuclear power ; Irrational rejection

Nuclear power plants
PE 5260 — Deterioration
PE 7539 — Environmental hazards decommissioned
PD 2718 — Hazardous locations
PD 7663 — Poor viability
PF 7543 — Prohibitive cost
PE 7539 — Waste resources obsolete

Nuclear power [production...]
PD 4977 — production ; Environmental hazards
PD 4977 — Proliferation
PF 8654 — Restrictive regulation
PF 4455 — safeguards ; Non verifiability compliance
PE 0403 — sources ; Aggression against
PD 0365 — sources ; Vulnerability
PD 7663 — stations ; Insufficient
PD 1741 — techniques ; Monopoly
PE 5080 Nuclear powered reactors ; Radioactive fallout accidents orbiting
PF 4352 Nuclear proliferation ; Risk unintentional global nuclear war due

Nuclear reactor
PD 7579 — accidents
PF 1441 — emissions satellites
PF 6084 — safeguards ; Inadequate

Nuclear reactors
PD 1584 — Environmental pollution
PE 7539 — Excessive cost decommissioning
PD 4977 — Unsafe
PD 5076 Nuclear rivalry countries
PE 3968 Nuclear smuggling
PF 4162 Nuclear strategy ; Illusion
PD 8124 Nuclear trade ; Discriminatory
PC 0842 Nuclear war

Nuclear war due
PF 4346 — accidents ; Risk unintentional
PF 4302 — international crises ; Risk unintentional
PF 4352 — nuclear proliferation ; Risk unintentional global

Nuclear war [Effects...]
PC 0842 — Effects
PF 4156 — generated developments strategic doctrine ; Risk unintentional
PF 4162 — generated strategy deterrence ; Risk unintentional
PD 6265 — Inadequate health services following
PE 6350 — induced winter chill
PC 3541 — Pervasive fear
PF 4435 — Risk intentional
PF 0466 — Risk unintentional
PF 3807 — Theological justification

Nuclear warfare
PF 1236 — alerts ; False
PC 0311 — material ; Stockpiles

Nuclear warfare cont'd
PE 6350 — soot ; Sunlight inhibition
PD 4445 — technology ; Super power monopoly advanced
PE 4199 Nuclear warheads ; Waste
PD 4396 Nuclear waste

Nuclear weapon
PD 0837 — free zones ; Insufficient
PC 4440 — states ; Insecurity vulnerability
PD 1521 — states ; Insecurity vulnerability non

Nuclear weapons
PD 3493 — Accidental loss
PE 3968 — Clandestine trade
PE 9052 — developing countries ; Proliferation
PC 0311 — Excessive number
PF 4727 — Illegality
PE 5698 — industry ; Environmental hazards
PE 5698 — industry ; Violation safety regulations
PE 7476 — Lack appreciation
PE 5698 — manufacture ; Pollution
PC 0012 — research
PD 3493 — systems ; Accidents
PD 8412 — systems ; Maritime
PD 0837 — technology ; Proliferation
PE 3769 — Terrorists armed
PC 2201 — testing
PF 4450 — testing ; Government secrecy concerning
PE 4051 — Unilateral structural disarmament
PE 7476 — Unrecognized merits
PE 5324 Nuclear winter ; Fear
PF 6611 Nuclearism
PF 2660 Nudism
PD 2547 Nudity advertising
PE 5042 Nudity art
PF 2660 Nudity ; Compulsory
PF 2660 Nudity ; Public
PF 2660 Nudity ; Ritual
PF 2660 Nudity ; Shameless
PD 9022 Nuisance
PA 7365 Nuisance Boredom
PD 1175 Nuisance organisms drinking water
PD 9022 Nuisance ; Private
PD 9022 Nuisance ; Public
PA 7107 Nuisance Unpleasantness
PA 6977 Nullity Meaninglessness
PA 5870 Nullity Nonexistence
PA 5942 Nullity Unimportance
PC 0143 Number-blindness
PA 7412 Number∗complex
PF 5670 Numbering systems ; Terminological confusion weights, measures
PF 4660 Numbers ; Unlucky
PF 1216 Numbing feelings
PF 1216 Numbness ; Emotional
PF 1216 Numbness towards others
PD 5707 Nuremberg obligation ; Citizen resistance
PE 3044 Nurse infants during working hours ; Denial right periods
PE 1085 Nursing ; Protein energy malnutrition women during pregnancy
PE 2587 Nut peanut trade ; Instability ground
PE 2587 Nut trade ; Instability fresh fruit edible
PD 0077 Nutrients ; Depleted soil
PD 9227 Nutrients ; Depletion soil
PE 0265 Nutrition education least developed countries ; Inadequate

Nutrition [Ignorance...]
PD 9021 — Ignorance women concerning infant
PC 0382 — Inadequate
PC 2765 — Inadequate animal
PE 1085 Nutrition women ; Denial sufficient

Nutritional anaemia
PD 0321 — [Nutritional anaemia]
PE 9554 — animals
PE 9185 — women developing countries
PC 0382 Nutritional deficiencies
PD 0287 Nutritional diseases
PF 3310 Nutritional food ; Uneaten

Nutritional [habits...]
PF 6078 — habits food aid ; Modification environmentally adapted
PD 0800 — habits ; Poor
PD 7424 — hyperparathyroidism animals
PE 5773 Nutritional ignorance

Nutritional [metabolic...]
PD 8301 — metabolic diseases endocrine
PE 9774 — myopathies swine
PE 9774 — myopathy calves
PE 9774 — myopathy lambs
PE 9774 Nutritional panniculitis
PC 0382 Nutritional practices ; Deficient
PD 1189 Nutritional quackery
PE 8938 Nutritional quality food ; Decline
PF 3838 Nutritional resources ; Religious opposition
PE 9774 Nutritional steatitis
PF 3310 Nutritional taboos
PD 2868 Nutritional taboos ; Violation
PC 0300 Nutritionally induced mental illness
PF 5050 Nutritionally induced skeletal abnormalities swine
PF 1345 Nutritious food ; Export
PF 1212 Nutritious food ; Prohibitive cost
PF 4790 Nutritive needs ; Unsurveyed
PE 5187 Nutritive substance ; Eating non
PD 0386 Nuts oil kernels ; Instability trade oil seeds, oil
PE 8250 Nuts, oil kernels oil seeds ; Environmental hazards oil
PD 2542 Nyctalopia
PE 8213 Nymphomania

PD 7799 Nymphomania
PC 2514 Nyong ; O'nyong

O

PC 2514 O'nyong-nyong
PD 5228 Oak acorns ; Poisoning
PD 5228 Oak buds ; Poisoning
PE 2984 Oak ; Pests diseases
PD 2285 Oases ; Environmental degradation desert
PD 2285 Oasis ecosystems ; Destruction
PF 2418 Obedience ; Blind
PF 2418 Obedience ; Irresponsible
PE 1177 Obesity
PD 7420 Obesity animals
PE 1814 Obesity ; Childhood
PE 4463 Obesity due low metabolism
PE 1177 Obesity ; Overweight
PF 7149 Obey elders ; Failure
PF 7467 Obey religious advisers ; Failure
PF 2152 Obfuscatory arguments
PF 4843 Object disorders ; Self
PD 4738 Objection ; Conscientious
PE 7007 Objection factory ; Conscientious
PD 1800 Objection military service ; Denial right conscientious
PE 8207 Objectives ; Military expeditions against distant
PF 7053 Objectivism
PF 9693 Objectivity ; Erosion scientific
PF 6691 Objectivity institutions ; Inadequate
PF 2827 Objectivity ; Lack
PJ 5626 Objectivity ; Lack religious
PD 4738 Objectors ; Conscientiousness
PD 1800 Objectors ; Internment conscientious
PB 0731 Objects ; Accidents falling
PF 3710 Objects ; Denial rights inanimate
PD 8041 Objects ; Misappropriation sacred
PB 0731 Objects ; Suffocation ingested
PD 5707 Obligation ; Citizen resistance Nuremberg
PA 8658 Obligations ; Avoidance
PD 5941 Obligations ; Crimes related military service
PD 3051 Obligations developing countries ; Inconsiderate insistence creditors fulfilment debt service
PA 8658 Obligations ; Evasion
PE 2215 Obligations ; Failure governments fulfil international reporting
PF 1509 Obligations ; Family crediting
PE 2215 Obligations ; Governmental delays fulfilling international reporting
PD 1157 Obligations ; Irregular payments international financial
PE 6678 Obligations persons multiple nationality ; Conflicting military
PD 4556 Obligations politicians ; Avoidance legal
PF 3838 Obligations ; Religious
PF 1509 Obligations ; Rigid family
PF 1491 Obligations ; Self determination unrelated global

Obligations [Undesired...]
PF 4948 — Undesired sexual
PF 2497 — Unfulfilled treaty
PE 9061 — Unreported tax
PD 4287 Obligatory social kissing
PE 3874 Obliteration footpaths development
PA 6651 Oblivion Forgetfulness
PA 5438 Oblivion Neglect
PA 6940 Oblivion Thoughtlessness
PA 7364 Oblivion Unfeelingness
PE 5042 Obscene art
PF 7427 Obscene humour
PF 1384 Obscene literature
PF 7807 Obscene songs
PE 5757 Obscene telephone calls
PF 7427 Obscene wit
PF 2634 Obscenity
PD 0132 Obscenity ; Exposure children
PF 7427 Obscenity ; Verbal
PF 1357 Obscurantism
PF 1357 Obscurantism ; Dependence
PF 6703 Obscurantist diplomatic language
PF 6574 Obscure accountability public services
PF 6556 Obscure community leadership
PF 6550 Obscure decision mechanisms
PD 2338 Obscure land ownership
PF 2346 Obscure ownership records
PF 6494 Obscured community history
PF 6552 Obscured intergenerational needs
PF 2365 Obscured land rights
PA 6900 Obscurity Formlessness
PA 6978 Obscurity Invisibility

Obscurity [Uncertainty...]
PA 7309 — Uncertainty
PA 5942 — Unimportance
PA 7367 — Unintelligibility
PF 3607 Observance ; Absence holy day
PE 7735 Observance ; Discriminatory effect holy day
PF 7838 Observance multilaterally agreed trade commitments ; Conditional
PE 7244 Observation environmental pollution ; Obstruction astronomical
PF 7838 Observe trade commitments ; Failure political will
PA 6448 Obsessed people
PA 6448 Obsession
PF 6713 Obsession books
PF 9617 Obsession celebrities
PF 1288 Obsession ; Computer
PF 1706 Obsession efficiency
PF 9119 Obsession immediate problems

–997–

Obsession

PA 7157 Obsession *Insanity*
PA 6739 Obsession *Maladjustment*
PF 8767 Obsession novelty
Obsession [people...]
PA 6448 — *people things*
PF 6389 — personal health
PF 6389 — personal safety
PE 7031 Obsession ; *Sexual*
PA 6448 Obsessive-compulsive disorder
PE 7632 Obsessive compulsive personality disorder
PF 7106 Obsolescence building structures ; *Architectural*
PF 3844 Obsolescence face overwhelming need ; *Vocational*
PE 6852 Obsolescence *Inappropriateness*
PE 2862 Obsolescence modern industrialised societies ; Institutional
PA 7131 Obsolescence *Oldness*
PC 2008 Obsolescence ; Planned
PF 1309 Obsolescence rituals customs
PE 6704 Obsolescence roads ; Functional
PD 6150 Obsolescence suburban mode human settlement
PE 9346 Obsolete weapons
PE 5961 Obsolete agricultural cooperative structures
PE 9346 Obsolete armaments ; Waste resources invested
PF 0836 Obsolete basis cultural identity
PF 6578 *Obsolete buildings*
Obsolete [decision...]
PF 5009 — decision-making
PF 5009 — *defence planning*
PD 0975 — deliberative systems
Obsolete [educational...]
PF 8161 — educational values
PD 3548 — employment training
PD 6520 — *equipment*
PD 3998 *Obsolete goods ex socialist countries ; Dumping*
PF 3713 *Obsolete group methods*
PF 5435 Obsolete laws
PF 5435 Obsolete legislation
Obsolete [machinery...]
PD 6520 — *machinery*
PF 3713 — methods
PF 1822 — methods agricultural production
PC 8390 — *military bases*
PE 7539 Obsolete nuclear power plants ; Waste resources
Obsolete [planning...]
PF 5009 — planning
PF 5009 — policy-making
PF 8846 — *problems ; Institutional preoccupation*
PF 1572 — programmes against social problems
Obsolete [security...]
PF 5009 — *security planning*
PF 6573 — *self-sufficiency patterns*
PC 8390 — systems ; Preservation
PD 3548 Obsolete vocational skills
PD 7574 Obsolete weapons ; Marine disposal
PF 1776 *Obsolete zoning restrictions*
PE 8246 Obstacle acquiring skills ; Illiteracy
PA 7391 Obstacle *Closure*
PB 1997 Obstacle development ; Social stratification
PA 5497 Obstacle *Difficulty*
PD 0055 Obstacle peaceful socio economic development ; National sovereignty
PD 0055 Obstacle representative democratic political organization ; National governments
PF 5496 Obstacles aquaculture
PF 7130 Obstacles availability community space
PF 7106 Obstacles building construction
Obstacles [commodity...]
PF 4870 — commodity futures trading
PF 7118 — community achievement
PF 7014 — conceptual integration societal learning
PE 5909 Obstacles developing countries ocean shipping
PF 7923 Obstacles development multidisciplinary approaches
Obstacles [economic...]
PF 3689 — economic reform socialist countries
PF 4852 — education
PF 7082 — effective international nongovernmental organizations
PE 5801 — effective use oceans waterways
PE 5921 — efficient port utilization operation
PF 7022 — efficient utilization time
PF 3127 — extended families
PF 7094 Obstacles family life
PJ 0230 Obstacles industrialization developing countries
Obstacles international
PF 4857 — cultural exchange
PD 0673 — investment ; Tax
PD 5885 — ocean shipping
PD 6223 — trade services
PD 2784 — *transfer corpses*
Obstacles [leadership...]
PF 7011 — leadership
PE 8442 — learning due hunger
PF 4886 — legal relations socialist countries
PF 4865 Obstacles medical experimentation
PF 4842 Obstacles national development
Obstacles [patent...]
PD 0455 — patent protection ; Distortion international trade
PE 8965 — peaceful development space ; Military
PE 4627 — political union island developing countries
PF 0631 — problem alleviation ; General
PC 1153 — proper use science technology
PJ 8292 — proper use technology
PF 4785 Obstacles redeployment military resources peaceful applications
PE 9055 Obstacles restructuring production industrialised countries

PF 6072 Obstacles satellite communications developing countries
PE 4525 Obstacles trade fertilizer products
Obstacles [Underground...]
PE 1944 — *Underground construction*
PF 7052 — *unilateral nuclear disarmament*
PF 5717 — use bicycles
PF 4767 — utilization coastal deep sea water resources
Obstacles world
PD 0055 — order ; Nation states
PC 4890 — trade
PE 9107 — trade ; Informational procedural
PA 7250 Obstinacy *Disobedience*
PA 7325 Obstinacy *Irresolution*
PA 6979 Obstinacy *Opposition*
PA 6509 Obstinacy *Unwillingness*
PE 8829 Obstruct administration justice ; Use undue influence
PE 7423 Obstruction animals ; *Small intestinal*
PE 7244 Obstruction astronomical observation environmental pollution
PE 9829 *Obstruction ; Bile duct*
PE 6553 Obstruction bowel
PF 5660 Obstruction change crisis emergency responses
PF 7391 Obstruction *Closure*
PA 5497 Obstruction *Difficulty*
PD 3982 Obstruction elections
PF 6710 Obstruction government function
PF 7277 Obstruction international criminal investigations
PD 9045 *Obstruction ; Intestinal*
PD 5515 Obstruction justice
PD 5515 Obstruction law enforcement ; Forcible
PF 9756 Obstruction ; *Legal*
PF 5983 Obstruction ; *Medical*
PE 9829 *Obstruction outflow bile*
PD 0790 Obstruction proceedings ; Soliciting
PE 6553 Obstruction pylorus
PE 3912 Obstruction recruiting induction armed forces
PF 9625 Obstruction research
PE 6553 Obstruction stricture bowel
PA 7006 Obstruction *Untimeliness*
PD 3978 *Obstructions animals ; Acute intestinal*
PE 4809 Obstructions international athletic exchange
PE 4785 Obstructions international personnel exchanges cultural cooperation
PD 0637 Obstructive pulmonary disease ; Chronic
PE 5182 *Obstructive pulmonary disease ; Chronic*
PD 9293 *Obstructive uropathy animals*
PE 6544 Obtaining available local funding ; Lack channels
PE 5076 Obtaining property false pretences
PE 3005 *Obtaining property titles ; Delay*
PE 7780 Obtaining travel visas ; Administrative impediments
PE 4161 *Obturator paralysis*
PE 2575 *Occasions ; Unorganized family*
PF 2099 *Occult rites*
PF 7066 Occultations ; Fear eclipses
PE 3312 Occultism
PF 1681 *Occupants ; Irresponsible transient*
PE 8021 Occupation ; Belligerent
PE 6778 Occupation handicap
PC 6716 Occupation ; Hazardous
PE 8021 Occupation ; Military
PE 5462 Occupation trade union premises public authorities
PD 0820 Occupation unoccupied property ; Illegal
PD 1078 Occupation ; Vulnerability women during foreign military
PC 0646 Occupational accidents
PE 6806 Occupational blood diseases
Occupational [cancer...]
PE 3509 — cancer
PD 1239 — *carcinogens ; Medicinal use*
PE 6817 — cataract
PE 5606 — cervicobrachial disorders
PD 0637 — chronic pulmonary diseases
Occupational [dangers...]
PD 6885 — dangers developing countries
PD 1361 — deafness
PE 5684 — dermatoses
PD 3218 — discrimination politics ; Property
PD 0215 — diseases
PE 6866 — diseases voice
PE 4961 — domestic accidents
PD 1500 Occupational exposure radiation ; Excessive
Occupational hazard
PE 3001 — Asbestos
PE 1657 — Carbon monoxide
PF 5744 — Cold
PE 5767 — Dust
PE 5720 — Heat
PE 5780 — Improper lighting
PE 5650 — Lead
PE 5768 — Shift work stress
PE 5708 — Solvents
PE 1210 — Sulphur dioxide
Occupational hazards
PC 6716 — [Occupational hazards]
PE 1849 — benzene
PE 5696 — Biological agents
PE 5656 — Combined stresses
PE 3957 — commerce offices
PE 5651 — Conflicting standards protection against chemical
PE 3760 — farming
PE 6902 — female workers
PE 6904 — male workers
PE 8428 — mining industry
PE 9746 — painters
PE 5684 — Skin irritants
PE 7528 — workers small industries

Occupational hazards cont'd
PE 6868 — youth workers
PC 0865 Occupational health risks
PE 8305 Occupational health safety ; Inadequate teaching
PD 8841 Occupational illness developing countries
PC 0646 Occupational injuries
PD 1493 Occupational mobility developing countries ; Low
PD 0270 *Occupational neurosis*
PD 9219 *Occupational performance ; Passive resistance demands adequate*
PD 6880 Occupational psychopathology
Occupational [rheumatism...]
PE 0502 — rheumatism
PC 0865 — risk health
PE 5355 — risks hazards medical profession
PJ 3578 Occupational stigmata
PE 6937 Occupational stress
Occupations [Dangerous...]
PC 1640 — Dangerous
PJ 3707 — Dirty
PF 2845 — *Disoriented traditional*
PC 1788 Occupied nations
PD 8021 Occupied territories
PE 6074 Occupied territories ; War time exportation cultural property
PD 8825 Occupying soldiers ; Children engendered
PC 2087 Ocean characteristics ; Destructive changes
PC 2087 Ocean current shift
PD 3666 Ocean disposal waste
PE 8971 *Ocean energy ; Underutilization*
PE 5850 Ocean freight rates ; Instability
PF 5922 *Ocean freight ; Unfair surcharges*
Ocean [salinity...]
PC 2087 — salinity change
PD 1241 — sea bed ; Militarization deep
PE 5886 — shipment delays
Ocean shipping
PE 5909 — Obstacles developing countries
PD 5885 — Obstacles international
PE 5898 — services ; Reduction
PD 9082 Ocean stripping
PC 2087 Ocean temperature change
PD 4277 Oceanographers ; Corruption
PD 4277 Oceanographers ; Irresponsible
PD 4277 Oceanographers ; Negligence
PD 4277 Oceanography ; Malpractice
PE 8971 *Oceans ; Underutilization temperature differentials*
PE 5801 Oceans waterways ; Obstacles effective use
PD 3478 *Octopus ; Endangered species squid*
PD 5535 *Ocular squamous cell carcinoma ; Bovine*
PE 2665 *Oculomotor disturbance*
PC 1326 *Odd toed ungulates ; Endangered species*
PE 8288 *Odobenus ; Endangered species*
PE 4481 Odour ; Disagreeable human body
PE 6820 *Odours ; Unpleasant*
PE 5524 Oedema ; Acute bovine pulmonary
PE 7769 *Oedema animals ; Malignant*
PE 3830 Oedema ; Cerebral
PD 7841 *Oedema disease swine*
Oedema [Laryngeal...]
PD 7307 — *Laryngeal*
PD 7307 — Laryngeal
PD 5453 — *Lung*
PD 5453 — *Lung*
PE 7769 *Oedema ; Malignant*
PE 3830 Oedema ; Pulmonary
PE 5524 Oedema syndrome feeder cattle ; Tracheal
PE 9774 *Oedema syndrome ; Leg*
PE 4630 *Oesophageal dilatation*
PE 4630 *Oesophageal diverticulum*
PE 8961 *Oesophageal hiatus hernia*
PD 3978 *Oesophageal spasm animals*
PE 4630 *Oesophageal stenosis animals*
PE 4630 *Oesophagitis animals*
PD 3978 *Oesophagogastric ulcers swine*
PE 4630 *Oesophagus animals ; Diseases*
PE 4630 *Oesophagus animals ; Foreign bodies*
PE 9819 *Oesophagus ; Cancer*
PE 3711 *Oesophagus ; Dilatation*
PE 8636 *Oesophagus ; Diseases*
PE 8624 Oesophagus, stomach duodenum ; Diseases
PE 4787 Off disease plants ; Damping
PD 1285 *Off-farm employment*
PF 6570 *Off fragmented families ; Cut*
PE 1720 Off road terrain vehicles ; Environmental degradation
PC 1108 Off season jobs ; Limited
Off shore
PE 4315 — borrowing domestic monetary policies ; Interference transnational banks'
PF 1066 — broadcasting vessels
PD 1628 — territorial waters ; Conflicting claims concerning
PD 1628 — territorial waters ; Unilateral claims
Offence [Defence...]
PA 5445 — *Defence*
PA 7250 — *Disobedience*
PA 6822 — *Disrespect*
PA 7253 Offence *Envy*
PA 5952 Offence *Illegality*
PA 5644 Offence *Vice*
PD 0124 Offences aboard aircraft
Offences against
PD 5515 — administration justice
PC 4584 — environment ; Criminal
PC 0554 — government
PC 6239 — peace security mankind
PD 7520 — public order

Ophthalmic practice

PD 4158 Offences general applicability
PA 3369 *Offences heterosexual pairing ; Moral*
PD 6632 Offences ; Ideological
PD 5300 Offences involving danger person
PD 9394 Offences juveniles ; Sexual
PC 0742 Offences ; Military
PD 9179 Offences ; Moral
PE 0930 Offences ; Road traffic
PD 4082 Offences ; Sexual
PD 8910 Offences ; Vice sex traffic
PD 0520 *Offenders after completion sentence ; Detention*
PE 9545 Offenders after conviction ; Arbitrary reduction sentences
PE 6929 Offenders ; Discrimination against ex
PE 5043 Offenders ; Discrimination against families
Offenders [effectively...]
PE 9545 — effectively ; Government failure prosecute
PE 6059 — Excessive cost effective prosecution
PF 4723 — Excessive leniency sentencing
PF 0399 — *Execution young*
PD 2484 Offenders ; Flogging
PF 5947 *Offenders ; Fugitive*
PE 9545 Offenders ; Government pardoning convicted human rights
PE 0484 Offenders ; Harbouring national security
Offenders Inadequate
PF 8661 — evidence convict known
PE 8803 — rehabilitation juvenile
PE 8215 — segregation different categories juvenile
PF 5356 Offenders ; Inconsistent sentencing
PF 8661 Offenders ; Law enforcement failure against known
PD 9316 Offenders ; Legal discrimination favour
PC 7373 *Offenders ; Preponderance male criminal*
PE 4907 Offenders ; Racial bias sentencing
PD 0520 *Offenders ; Repeated arrest released*
PE 1837 Offenders ; Women
PF 2826 Offensive corporate logos
PF 7807 Offensive lyrics
PF 2826 Offensive symbols ; Use
PF 6548 *Offerings ; Narrow programme*
PD 3501 *Offerings ; Religious*
PE 5608 Office ; Denial right hold public
PD 7001 Office machine industries ; Protectionism computer
PD 8227 Office ; Misconduct public
PD 7704 Office ; Nepotism public
PF 2827 *Office positions ; Unappealing public*
PE 9857 Office ; Secret intelligence agents public
PE 6660 Office space cities ; Excessive
PE 6948 Office ; Trading public
PF 1106 Office work-place thefts
PE 0998 Officers ; Maltreatment prisoners prison
PE 8251 Offices ; Inadequate working conditions employees commerce
PF 6197 Offices ; Inhibition communication non proximate
PF 1653 *Offices ; Lack extension*
PE 3957 Offices ; Occupational hazards commerce
PF 8072 Offices ; Unresponsive public
Official [abuse...]
PF 4564 — abuse statistics
PF 6574 — accountability ; Inhibited
PF 9459 — apathy
PE 1728 — appointments ; Political bias
PE 5724 — assistance ; Dependence island developing countries
PD 9842 Official blackmail
Official [communications...]
PF 8382 — communications ; Misunderstanding
PF 9288 — connivance human rights abuses
PF 8650 — contributions ; Imbalance disbursements multilateral agencies
PC 9533 — corruption
Official cover
PF 3819 — up government harassment political activists
PF 3819 — up government sanctioned assassinations
PC 1812 — ups
PF 0152 Official cultural pluralism ; Divisive effects
Official [decisions...]
PF 8649 — decisions ; Imposed
PE 6260 — detention ; Escape
PF 0392 — development assistance developing countries ; Decline
PF 6492 — *directives ; Unclear*
PD 8017 — disregard individuals
PF 9699 — documents public scrutiny ; Withdrawal
PF 4699 — documents ; Tampering
PF 9157 Official evasion complaints
PD 8716 Official fabrication evidence
PF 3707 Official harassment ; Fear
Official [inaction...]
PF 6707 — inaction response potential problems
PF 7979 — infighting
PF 3076 — information ; Excessive neutrality intergovernmental
PF 7649 — information ; Increasing scepticism accuracy
PE 4146 — inspection regulated activities ; Ineffective
PF 8046 — inspectors ; Corruption
PF 8046 — inspectors ; Negligence
PF 9385 — institutions ; Criticism
PF 6027 Official languages ; Multiplicity
PF 0963 *Official liaison ; Ineffective*
PD 8716 Official misinterpretation evidence
PD 1893 Official news information ; Deliberate distortion
PD 1893 Official over-reporting under-reporting
PD 5725 Official privilege
PD 8392 Official profiteering
Official [religion...]
PF 6091 — religion
Official [reprisals...] cont'd
PF 3707 — reprisals ; Threat
PF 3060 — reserves developing countries ; Depletion
Official [secrecy...]
PC 1812 — secrecy
PD 5183 — secrets ; Leaking
PF 7702 — self-deception
PC 1812 — silence
PF 4118 — statistics ; Discrepancies
PF 4564 — statistics government ; Biased adjustment
PF 5421 Official texts ; Inconsistent
PD 9498 Officialdom ; Fear
PF 8500 Officialdom ; Lack cooperation
PE 4299 Officially endorsed homophobia
PF 7239 Officially nonexistent people
PE 2719 Officials ; Antagonism government agencies
Officials [Blackmail...]
PD 9842 — Blackmail government
PE 4033 — Bribery customs
PD 4541 — Bribery government
Officials Corruption
PC 9533 — [Officials ; Corruption]
PE 4033 — customs excise
PD 2918 — law enforcement
Officials [Deliberate...]
PD 4982 — Deliberate lying corporation
PD 1893 — Deliberate lying government
PD 8696 — Drug abuse government
Officials [Excessive...]
PF 3434 — Excessive career mobility aid
PE 8881 — Excessive use force law enforcement
PD 9842 — Extortion public
Officials [Harassment...]
PD 4915 — Harassment public
PE 7780 — Harassment travellers immigration
PE 9339 — Homosexual
PE 7687 Officials ; Impersonating
PD 4734 Officials ; Intimidation public
PD 8422 Officials ; Underpayment government
PD 9653 Offshore assets heads state ; Corrupt acquisition
PE 4994 Offshore insurers ; Destabilization national insurance markets
PE 1839 Offshore oil disasters
PC 0065 *Oil ; Dependence*
PE 1839 Oil disasters ; Offshore
PE 0386 Oil kernels ; Instability trade oil seeds, oil nuts
PE 8250 Oil kernels oil seeds ; Environmental hazards oil nuts,
PE 0386 Oil nuts oil kernels ; Instability trade oil seeds,
PE 8250 Oil nuts, oil kernels oil seeds ; Environmental hazards
Oil [petroleum...]
PE 8080 — petroleum crises ; World tension
PE 2134 — pollutant
PE 1839 — pollution
PD 0909 — price fluctuations
PC 4824 *Oil resources ; Scarcity*
Oil [seeds...]
PE 8250 — seeds ; Environmental hazards oil nuts, oil kernels
PE 0386 — seeds, oil nuts oil kernels ; Instability trade
PF 0445 — shale energy source ; Underutilization
PD 1245 — *smoke*
PE 1839 — spillage
PE 1839 — spills
PE 1839 Oil tanker disasters
PE 0313 Oil trade ; Instability olive
PE 9056 Oils ; Environmental hazards essential
PE 8136 Oils fats environmental hazard ; Vegetable
PE 8135 Oils fats ; Environmental hazards animal
Oils fats Instability
PE 8896 — trade animal
PE 0735 — trade animal vegetable
PE 0861 — trade fixed vegetable
Oils fats [Long...]
PE 1188 — Long term shortage animal vegetable
PE 8880 — Restrictive practices trade animal vegetable
PE 8277 — Shortage fixed vegetable
PE 8498 — waxes ; Shortage animal vegetable
PE 8232 Oils perfume materials ; Instability trade essential
PE 8159 Oils trade instability ; Rape mustard
Old age
PB 0477 — [Old age]
PD 0162 — *benefits ; Denial right*
PD 5501 — care ; Negligent
PC 1966 — Inadequate income
PD 0633 — Loneliness
PD 0276 — pensioners' homes
PE 7942 — pensions men women ; Unequal distribution
PE 5956 — pensions ; Progressive erosion
PC 1966 — Reduced income
PF 4549 Old boy networks
PF 1822 Old farming methods
PF 6570 *Old ; Fear growing*
Old [people...]
PF 2875 — *people ; Insufficient transportation*
PF 2830 — people significance past ; Denial
PD 1043 — physical ill health ; Susceptibility
PE 8991 Old world monkeys ; Endangered species
PE 8845 Old world porcupine ; Endangered species
PE 5050 *Older animals ; Osteomyelitis*
PE 5050 *Older animals ; Suppurative arthritis*
PF 1664 Older generation ; Limited social guidance
PD 0512 Older homes ; Unserviced
PD 5350 Older industries
Older people
PD 2318 — Constraints against employment
PF 1825 — due overemphasis economic productivity ; Diminishing role
Older people cont'd
PF 2012 — Lack facilities educating
PE 5951 — Unemployment
PE 9071 — Unemployment educated
PA 7131 Oldness
PE 1488 Oligarchic control decision making ; Social service quality negated
PD 2931 Oligarchic political control elections
PF 1138 Oligarchy effectiveness myth
PD 5193 Oligarchy ; Financial industrial
PD 3238 Oligarchy ; Political
PC 3825 Oligopolies
PE 1922 Oligopolistic trading systems ; National
PE 0135 Oligopolistic world trading system ; Emerging
PE 0313 Olive oil trade ; Instability
PE 2978 Olives ; Pests diseases
PF 4761 Olympic Games ; Political exploitation
PF 8577 Omened events ; Ill
PF 8577 Omens ; Bad
PF 8577 Ominous events
PD 6845 *Omission act ; Criminal*
PA 6180 Omission *Error*
PA 6652 Omission *Incompleteness*
PA 5473 Omission *Insufficiency*
PA 5438 Omission *Neglect*
PA 6662 Omission *Nonaccomplishment*
PA 7232 Omission *Unskillfulness*
PA 5644 Omission *Vice*
PD 9022 Omissions ; Wrongful
PD 2731 *Omphalitis poultry*
PE 2388 Onchocerciasis
PD 2732 *Ondiri disease*
PD 2681 One-parent families
PF 4787 One-sided agreements
PE 6633 Oneself ; Denial right freedom testifying against
PF 4436 Oneupmanship ; Sexual
PF 3575 Ongoing education rural communities ; Limited opportunity
PD 7841 *Ontario encephalitis*
PE 5465 Oomycetes
PD 2728 *Oomycosis*
PD 9742 *Oophoritis*
PD 8786 *Opacity ; Corneal*
PA 7371 Opacity *Unintelligence*
PA 7367 Opacity *Unintelligibility*
PF 5374 Opaque budgetary procedures public sector
PD 1637 Open cast mining ; Landscape disfigurement
PE 8233 Open irrigation wastage
PD 2596 Open pit mines quarries ; Environmental destruction
PE 6163 Open space urban environments ; Wastage
PF 1815 Open spaces urban communities ; Poor condition
PE 5194 Openings various professions ; Imbalance training existence
Operating [capital...]
PF 2042 — *capital ; Insufficient*
PF 2575 — *community cliques*
PF 1062 — context ; Loss past
PF 7932 — costs ; Increase
PF 5354 Operating images ; Self depreciatory
PD 2294 Operating structure marriage ; Individually defined
PE 4736 Operating theatre mistakes
PF 6492 Operation ; Absence support systems effective community
PF 1959 Operation community networks urban ghettos ; Ineffective
PF 7556 Operation ; Imperfect market
PE 9440 Operation international commodity exchanges ; Discriminatory
PE 5921 Operation ; Obstacles efficient port utilization
PE 3988 Operation ; Police
PD 1415 *Operation ; Single product*
PE 2240 Operation unsafe motor vehicles ; Continued
PE 7056 Operational ineffectiveness international nongovernmental organizations
Operations [Clandestine...]
PD 4501 — Clandestine intelligence
PF 7669 — Clandestine military
PF 1781 — *Closed federation*
PF 2439 — Confined scope business
Operations Covert
PD 4501 — intelligence agency
PF 7669 — military
PC 2175 — psychological warfare
Operations [Crimes...]
PD 1163 — Crimes against integrity effectiveness government
PF 5660 — Crisis oriented
PD 5272 — Cross border military
PJ 2493 Operations existing groups ; Non corporate
PF 1205 *Operations ; Family focused*
PF 1205 Operations ; Fragmented conduct community
PF 2457 *Operations ; Independent farming*
PE 5827 Operations ; Political interference port
PF 7669 Operations ; Secret military
Operations [Uncontrolled...]
PC 1142 — Uncontrolled police
PE 6331 — Undercover police
PF 6559 — Unfocused style community
PE 9271 — Unnecessary surgical
PF 1933 — Unprofitable scope industrial
PF 3702 Operator error
PD 3158 *Ophiaroidea ; Endangered species*
PF 7974 Ophthalmia
PD 5535 *Ophthalmia ; Infectious*
PD 5535 *Ophthalmia ; Periodic*
PD 8786 Ophthalmic diseases
PE 1369 Ophthalmic practice ; Unethical

PF 9350 Opiate ; Religion
PE 1329 Opiates ; Abuse
PD 1140 Opinion ; Blocked minority
PD 7219 Opinion ; Denial freedom
PF 9704 Opinion development ; Inadequate mobilization public
PD 2108 Opinion ; Innate expectation suppression minority
PE 7448 Opinion ; Manipulation public
PF 0292 Opinion ; Nation states' disregard world
PF 2909 *Opinion polls ; Misleading public*
PF 3340 Opinions cultures ; Unfavourable
PF 3217 Opinions ; Denial freedom hold
PF 7137 Opinions ; Lengthy legal
PD 4027 *Opioid intoxication*
PE 1329 *Opioid withdrawal*
PE 1329 *Opium abuse*
PD 1762 *Opossums ; Endangered species*
PD 1762 *Opossums ; Endangered species rat*
PF 3388 Opponents ; Ideological
PF 4133 Opponents ; Negative socialization towards
PF 4133 Opponents ; Promotion negative images
PF 6353 Opportunism
PC 1897 *Opportunism ; Political*
PF 3846 Opportunist bias public discussion research development
PF 6531 Opportunities adult training ; Minimal
PF 1352 *Opportunities apprenticeship ; Restricted*
Opportunities [capitalist...]
PD 3122 — capitalist systems ; Restriction educational
PF 2316 — community activities ; Insufficient
PF 6512 — community education ; Unexplored
PF 2316 — corporate activities ; Minimal
Opportunities [Denial...]
PD 9628 — Denial right equal promotion
PD 0176 — developing countries ; Insufficient employment
PE 0706 — disabled persons ; Unequal
PE 0783 — disabled persons ; Unequal employment
PE 4916 Opportunities employment elderly ; Denial rights equal
PE 7726 Opportunities foreign students ; Unequal
Opportunities [Ignored...]
PF 3539 — *Ignored business*
PF 1658 — Inadequate job
PF 1352 — Inadequate training
PF 9008 — income available ; Imbalances people's aspirations structure
PF 0963 — *Infrequent discussion*
PF 3037 — *Infrequent travel*
Opportunities Insufficient
PF 1352 — *apprenticeship*
PF 1658 — job
PF 3184 — learning
PF 1352 — *training*
PF 3037 — *travel*
PE 1753 Opportunities ; Lack skills
Opportunities Limited
PF 3184 — availability learning
PF 3628 — *awareness child care*
PF 2316 — family
PF 3550 — *financing*
PF 1944 Opportunities local communities ; Lack relevant training
PD 3039 Opportunities media reception ; Unequal
PF 2949 *Opportunities ; Narrow employment*
PF 2837 Opportunities practical training communities ; Lack
PF 0581 Opportunities ; Restricted market
PF 3575 Opportunities rural areas ; Limited educational
Opportunities [significant...]
PF 1403 — significant work ; Limited
PF 2949 — small rural communities ; Limited employment
PD 2972 — some sectors society ; Lack job
PC 1131 *Opportunities ; Training employment*
Opportunities [Unattractive...]
PF 3454 — *Unattractive social*
PF 6552 — *Uncommunicated resource*
PC 5576 — unemployed ; Rejection job
PF 3539 — *Unexplored business*
PF 9004 — Unexplored career
PF 2453 — *Unknown training*
PF 9004 — Unperceived career
PF 2453 — *Unperceived educational*
PF 4852 — *Unpublished training*
PF 6925 — Unrecognized
PF 4852 — *Unused training*
PF 0581 — Unutilized marketing
PD 5115 Opportunities women ; Unequal employment
PC 3435 Opportunity ; Dependence inequality
PD 1847 Opportunity developing countries ; Inequality employment
PC 3435 Opportunity ; Inequality
PF 1658 Opportunity ; Limited job
PF 3575 Opportunity ongoing education rural communities ; Limited
PF 5124 Opposed ideologies ; Cover up convergence practice apparently
PF 1746 Opposing groups appropriate wisdom past ; Failure
PF 2199 Opposing policies ; Government action against regimes
PF 3720 Opposing views ; Exclusion
PF 6910 Opposite sex ; Fear
PA 6979 Opposition
Opposition [abortion...]
PF 1022 — abortion ; Religious
PF 2628 — administrative action ; Political
PF 7088 — arms race ; Lack business
PD 7966 Opposition ; Bureaucratic
PF 1022 Opposition contraception ; Religious
Opposition [Difference...]
PA 6698 — *Difference*
PA 5497 — *Difficulty*

Opposition [Disagreement...] cont'd
PA 5982 — *Disagreement*
PA 6191 — *Disapproval*
PF 3821 — *Disorganized political*
PA 6838 — *Dissent*
PF 3826 Opposition existing social order ; Ineffective
PF 1594 Opposition ; Frustration role
Opposition groups
PF 5401 — Idealism
PC 7662 — individuals ; Suppression
PF 3821 — Structural failure
Opposition [Ideological...]
PF 3388 — Ideological
PF 9077 — information disciplines
PF 3162 — international organizations ; Communist
PD 0361 — irradiation technology
PE 9114 Opposition manual intellectual labour ; Artificial
PF 3838 Opposition nutritional resources ; Religious
PA 6979 *Opposition Opposition*
PE 3797 Opposition out personal responses ; Speaking
PD 3015 *Opposition ; Political*
Opposition population control
PF 1021 — family planning
PF 1023 — Government
PF 1022 — Religious
Opposition [private...]
PF 4095 — private initiative ; Government
PE 6722 — profit sharing ; Trade union
PF 3838 — public health practices ; Religious
PF 7890 Opposition ; Unreported
PA 6509 Opposition *Unwillingness*
PE 4634 Oppositional defiant disorder
PB 8656 Oppression
PE 6949 Oppression ; Denial right resist
PA 6852 Oppression *Inappropriateness*
PC 1919 Oppression ; Political
PA 6731 Oppression *Solemnity*
PA 7107 Oppression *Unpleasantness*
PF 3553 Oppression police questioning
PF 1365 Oppressive prevalent images
PF 7053 Oppressive reality
PA 6191 Opprobrium *Disapproval*
PA 6839 Opprobrium *Disrepute*
Optic [nerve...]
PE 4584 — nerve ; Diseases
PD 8786 — nerve ; Inflammation
PD 8786 — neuritis
PE 8470 Optical goods trade instability ; Photographic
PD 2389 *Opticatrophy ; Hereditary*
PE 1369 Opticians ; Corruption
PE 1369 Opticians ; Irresponsible
PE 1369 Opticians ; Malpractice
PE 1369 Opticians ; Negligence
PE 7002 Optimism ; Ingrained
PE 7002 Optimism ; Misplaced
PE 7002 Optimism ; Naive
PE 7002 Optimism ; Unrealistic
PE 7002 Optimism ; Unwarranted
PF 7186 Optimistic assessment problems ; Over
PE 7002 Optimistic bias ; Over
PF 1267 *Options communicating ; Indefinite*
PF 2949 *Options ; Insufficient job*
PF 6535 Options involvement culture ; Scarce
PF 1658 Options ; Limited employment
PF 0076 *Options ; Narrow media*
PF 0718 *Options ; Restricted banking*
PF 2949 Options rural employment ; Minimal
PE 9819 Oral cancer
PE 8505 Oral injury deficiency ; Facial
PF 3756 Oratory ; Empty meeting
PD 0089 Orbit junk
PF 0545 Orbit ; Saturation geosynchronous
PD 0089 Orbital space ; Pollution
PE 5080 Orbiting nuclear powered reactors ; Radioactive fallout accidents
PD 0089 Orbits ; Accumulation aerospace hardware earth
PF 0545 Orbits ; Limited number geostationary satellite
PF 0545 Orbits ; Maldistribution geosynchronous
PF 2231 *Orchard potential ; Unexplored*
PD 9154 *Orchitis*
PD 7799 *Orchitis animals*
PF 0866 Order ; Antiquated world socio economic
PA 7199 Order∗complex
Order [Dependence...]
PB 6021 — Dependence discriminatory international
PB 6021 — Discriminatory international
PD 3879 — Disobedience judicial
PF 4891 Order ; Excessive socio economic
PF 2136 — Feudal society
PF 3826 — Ineffective opposition existing social
PF 4306 Order ; Lack progress establishing New International Economic
PD 0055 Order ; Nation states obstacles world
PD 7520 Order ; Offences against public
PF 3826 Order ; Prevalence psychological conditions unfavourable transition post capitalist social
PF 8753 Order problem relevant information ; Failure international information systems
PC 9112 Order ; Unjust international economic
PA 7361 Orderlessness *Disorder*
PA 6794 Orderlessness *Formlessness*
PA 7309 Orderlessness *Uncertainty*
PE 0310 Orderly marketing arrangements ; Discriminatory
PF 9542 Orders armed forces ; Disregard government
Orders [endangering...]
PF 2418 — endangering environment ; Execution

Orders [Execution...] cont'd
PF 2418 — Execution illegal
PF 2418 — Execution inappropriate
PE 4768 Orders ; Failure enforce judgements
PJ 3258 Orders ; Irregular international money
PE 6323 Orders ; Liability crimes committed disregard
PF 2418 Orders violation human rights ; Execution
PF 3526 *Ordinances ; Restrictive land*
PE 4768 *Ordinances ; Unenforced civic*
PD 5001 Ordinances ; Unenforced safety
PE 6326 Ordination women ; Bias against
PE 0353 *Ore concentrates shortage ; Iron*
PE 8772 Ore concentrates trade instability ; Iron
Ores concentrates
PE 8131 — non ferrous base metals ; Instability trade
PE 0353 — *shortage ; Thorium*
PE 0353 — *shortage ; Uranium*
PE 8212 — thorium trade instability ; Uranium
PE 0824 Ores ; Long term shortage non ferrous metal
Ores metal scrap
PE 8906 — Environmental hazards metalliferous
PE 0553 — Instability trade metalliferous
PE 0353 — Long term shortage metalliferous
PE 0353 *Ores shortage ; Platinum*
PE 0353 *Ores shortage ; Silver*
PE 8556 Ores trade instability ; Silver platinum
PD 9667 *Orf*
PE 7530 Organ donors ; Shortage
PE 9038 Organ failure ; Human
Organ [transplantations...]
PE 7530 — transplantations ; Unauthorized
PE 7530 — transplants humans ; Animal
PD 4271 — transplants ; Traffic children source
PD 1271 *Organic compounds ; Gaseous*
PE 0617 Organic compounds water pollutants ; Toxic
PD 0992 Organic disease brain
PE 0524 *Organic dusts*
PE 1785 Organic farming ; Inadequate subsidies
PE 1256 *Organic matter ; Damage degradable*
PD 5719 Organic proteins ; Loss micro
PF 1212 Organically grown foodstuffs ; Prohibitive cost
PB 5647 Organism ; Vulnerability human
PC 1617 *Organisms ; Disruption ecosystems exotic*
PD 1175 *Organisms drinking water ; Nuisance*
PD 2226 *Organisms ; Mycoplasma like (MLOs)*
Organisms [Pathogenic...]
PG 2302 — *Pathogenic*
PD 3574 — Pesticide destruction soil fauna micro
PE 2280 — *Pyogenic*
PD 7183 Organisms ; Release genetically engineered micro
PC 1617 *Organisms ; Translocation living*
PB 5658 *Organisms ; Vulnerability*
Organization [action...]
PD 0929 — action ; Inadequate facilities international
PF 2016 — action ; Inadequate facilities international nongovernmental
PE 7080 — actions ; Divisive responses international nongovernmental
Organization community
PF 6501 — action ; Ineffective
PF 1790 — environment ; Poor
PF 2822 — space ; Haphazard
PE 0179 Organization coordination bodies ; Proliferation duplication international nongovernmental
PC 1015 Organization ; Dependence undemocratic political
PF 5153 Organization favouring political incumbents ; Electoral
PF 6525 Organization ; Fragmented pattern community
PF 5301 Organization human thought
Organization [identity...]
PE 7032 — identity ; Lack international nongovernmental
PF 4240 — Ineffective anti terrorist
PF 6501 — Ineffective community
PE 0362 — information systems ; Proliferation duplication international nongovernmental
PF 0944 Organization knowledge ; Inadequate
PF 5409 Organization ; Lack recreational
Organization leadership
PE 6947 — Corruption intergovernmental
PD 7471 — Harassment workers'
PE 6947 — Malpractice intergovernmental
PE 6947 — Mismanagement intergovernmental
Organization [membership...]
PE 7092 — membership apathy ; International nongovernmental
PD 4098 — membership ; Compulsory
PD 8832 — Minimal community
PD 6628 — mismanagement ; Inter Governmental
Organization [National...]
PD 0055 — National governments obstacle representative democratic political
PE 7044 — network ; Lack identity international nongovernmental
PF 7057 — network ; Neglect international nongovernmental
PF 3826 — Non viable alternative modes social
PF 4891 Organization ; Over
Organization [personnel...]
PE 7010 — personnel ; Disadvantaged status international nongovernmental
PE 9124 — programme action country level ; Ineffectiveness international
PF 0074 — programmes ; Ineffectiveness intergovernmental
PE 7140 Organization response intergovernmental calls action ; Inadequate international nongovernmental
PF 1460 Organization ; Shortage financial resources United Nations system
PD 3789 *Organization training ; Neglected*
PC 1015 Organization ; Undemocratic political

Outdated property

PF 2810 Organization ; Unformed structures community
PD 0457 Organizational debate ; Politicization intergovernmental
PD 0460 Organizational disease ; Bureaucracy
PF 1232 Organizational empire-building
PF 2825 Organizational empire-building
PC 2921 Organizational intelligence
PE 8776 Organizational mechanism international action ; Inadequate
PF 2431 Organizational mechanisms act against problems ; Inadequate
PF 2810 *Organizational participation ; Restrictive*
PF 2825 Organizational systems ; Institutional domination
PE 1579 Organizational units coordinating bodies United Nations system ; Proliferation duplication
PE 4071 Organizations affiliate internationally ; Violation right workers'
PE 3970 Organizations bargain collectively ; Violation right workers
Organizations [common...]
PE 1184 — common membership ; Inadequate coordination regional intergovernmental
PE 1583 — common membership ; Jurisdictional conflict antagonism regional intergovernmental
PF 3162 — Communist opposition international
PE 3832 — Conditional membership international
PC 3166 — *Control over international*
PE 1029 — coordinating bodies ; Proliferation duplication international
PE 2417 — coordination bodies ; Proliferation duplication intergovernmental
PE 8742 — Corruption charitable
PE 7973 — country level ; Jurisdictional conflict antagonism international
Organizations [developing...]
PF 7081 — developing countries ; Absence national nongovernmental
PE 4032 — developing technology ; Public non accountability
PE 5179 — Distortion democratic procedures international
Organizations [establish...]
PE 4071 — establish confederations ; Violation right workers'
PE 7128 — Establishment orientation international nongovernmental
PE 7045 — Excess western based secretariats international nongovernmental
PF 2806 — Excessive complexity intergovernmental
PE 9405 Organizations ; Failure economic micro
PE 4358 Organizations ; Front
Organizations [government...]
PE 1651 — government ; Misuse business enterprises front
PE 0436 — government ; Misuse nonprofit associations front
PF 8650 — Government non payment agreed contributions international
PE 6388 Organizations ; High costs staff intergovernmental
PE 5090 Organizations ; Inadequacy staff qualifications intergovernmental
Organizations Inadequate
PE 4344 — coordination governmental representation intergovernmental
PE 0730 — coordination intergovernmental system
PF 9175 — power intergovernmental
PF 7117 — statistical data nongovernmental
Organizations [Inappropriate...]
PF 3049 — Inappropriate taxation international non profit
PF 3049 — Inappropriate taxation not profit philanthropic charitable
PE 7020 — Incompatible equivalent national sections international nongovernmental
PF 1262 — Ineffective worker
PF 1451 — Ineffectiveness United Nations system
PE 3538 — Inefficient location facilities international
PE 5222 — Infringement functioning legitimate
PF 6550 — *Insufficient civic*
PE 3538 — Inter state rivalry secretariats intergovernmental
PE 8384 — Interference employers affairs workers'
PE 1758 — Interference public authorities functioning workers
PF 9175 — Intrinsic weakness international
PE 8742 — Irresponsible voluntary
Organizations Jurisdictional conflict
PE 7901 — antagonism intergovernmental
PE 9011 — antagonism intergovernmental
PD 0047 — antagonism international
PD 0138 — antagonism international
PE 0064 — antagonism international nongovernmental
PE 1169 — antagonism international nongovernmental
Organizations Lack
PF 3434 — continuity amongst personnel international
PF 9175 — enforcement power international
PE 7009 — financial information systems international nongovernmental
PF 7058 — international legal provision nongovernmental
PE 0069 — legal provision international nongovernmental
PF 7034 — national legal provision international nongovernmental
PF 7493 — realism intergovernmental
PF 2575 — *youth*
PF 7093 Organizations ; Limited recognition international nongovernmental
PF 2575 *Organizations ; Limited youth*
Organizations [Manipulation...]
PD 6800 — Manipulation nongovernmental
PD 2239 — members ; Parochial attitudes
PD 2942 — Micro state participation international
PF 4876 — mini state membership ; Distortion intergovernmental
PF 2856 — Monopolization information
PE 7129 — Multinationalism international nongovernmental
PF 7116 Organizations naive ; Assessment international nongovernmental
PF 7069 Organizations ; Neglect non Western structures international nongovernmental
Organizations [Obstacles...]
PF 7082 — Obstacles effective international nongovernmental
PE 7056 — Operational ineffectiveness international nongovernmental
PF 8650 — Overdue payments governments international
Organizations [Political...]
PE 7008 — Political ineffectiveness international nongovernmental
PF 9175 — Powerlessness international
PF 8356 — Preponderance Western style
PE 1553 — programmes ; Alienation skilled committed personnel international
PD 1809 — programmes ; Alienation support international
Organizations programmes Inadequate
PD 0285 — coordination international
PE 1209 — coordination international nongovernmental
PF 0498 — funding international
PE 0741 — funding international nongovernmental
PE 1973 — relationship international governmental nongovernmental
Organizations programmes [Ineffectiveness...]
PF 1074 — Ineffectiveness international
PF 1595 — Ineffectiveness international nongovernmental
PE 8064 — Low credibility international
Organizations [Proliferation...]
PF 5992 — Proliferation documents international
PE 3077 — Propaganda intergovernmental
PE 5793 — protection against suspension ; Violation right workers
Organizations [Rapid...]
PF 3434 — Rapid personnel turnover international
PF 7033 — regional development ; Absence policies associate international nongovernmental
PE 3970 — Restrictions collective bargaining workers
Organizations Rivalry
PE 0063 — intergovernmental
PC 1463 — international
PE 0259 — international nongovernmental
Organizations scarce resources
PE 0063 — Competition intergovernmental
PC 1463 — Competition international
PE 0259 — Competition international nongovernmental
PF 0498 Organizations ; Shortfall aid international
PE 0777 Organizations specialized agencies United Nations ; Inadequate relationship international nongovernmental
Organizations [Uncontrolled...]
PE 7105 — Uncontrolled structures international nongovernmental
PC 8676 — Undemocratic
PE 8742 — Unethical practices philanthropic
PD 4873 — Unrepresentative international
PE 7021 — Unrepresentative international nongovernmental
PF 5477 Organizations ; Verbosity intergovernmental
PF 8650 Organizations ; Withholding membership payments intergovernmental
PJ 8271 Organize ; Incapacity developing countries
PE 9110 Organize political parties ; Denial right
Organize [time...]
PF 9980 — time ; Inability
PE 5398 — trade unions ; Denial right
PE 5938 — trade unions government services ; Violation right
PD 2669 Organized action against problems ; Inadequate buildings, services facilities
PE 2309 Organized cooperatives ; Government
Organized crime
PC 2343 — [Organized crime]
PC 2343 — families territories
PE 6699 — Ineffective legislation against
PF 6330 Organized cult machinery
PF 7916 *Organized expertise ; Lack*
PF 6170 Organized fantastic happenings ; Inhibition collectively
PF 8703 Organized participation ; Lack
Organized religion
PC 3359 — Corruption
PD 3423 — Disintegration
PE 0567 — Sexual bigotry
PF 3364 Organized religions ; Fragmentation
Organized [savings...]
PF 1348 — savings funds ; Lack
PD 3267 — sexual abuse children
PD 0055 — sovereign states ; Domination world territorially
PF 5753 — students ; Fragmentation
PD 0610 Organizers ; Denial rights trade unions
PD 0610 Organizers ; Harassment
PD 0983 *Organochlorine compounds pollutants*
PD 0983 *Organochlorine insecticide pollutants*
PD 0983 *Organochlorine pollution*
PD 1694 *Organophosphorous insecticides*
PD 0983 *Organophosphorus insecticides pollutants*
Organs [Trade...]
PE 7530 — Trade human
PE 7530 — transplantation ; Lack human
PE 7530 — transplantations ; Imbalance need availability viable
PD 8016 *Orgasm disorders*
PE 6408 Orgasmic dysfunction
PE 8923 Orgies
PE 0921 *Oriental schistosomiasis*
PF 4644 Orientation ; Degrading people's time
PF 5660 Orientation elected bodies ; Immediacy
PE 6772 Orientation handicaps
PF 3519 Orientation high school ; Overacademic
PE 7128 Orientation international nongovernmental organizations ; Establishment
PF 1382 *Orientation ; Overriding economic*
PF 6530 Orientation ; Subsistence economic
PF 8298 Original sin
PE 2578 Ornithosis
PE 5567 *Ornithosis*
PC 0233 *Orphan care ; Untenable*
PD 7046 Orphan children
PE 5623 Orphaned embryos
PD 7046 Orphans
PD 2092 Orphans ; Neglect war widows
PF 2720 Orthodoxy ; Scientific
PF 0552 Orthographic standards ; Instability
PB 1044 *Orthopaedic disabilities*
PE 3641 Orthoptera insect pests
PE 7826 *Os pedis ; Fracture*
PE 7826 *Osselets*
PE 7826 *Osseous cyst-like lesion*
PD 7424 *Ossificans ; Enteque*
PF 2005 Ossification faith rigid doctrine
PE 7826 *Ossifying myopathy*
PE 7826 *Osslets*
PD 7424 *Osteitis animals*
PE 2298 *Osteitis fibrosa ; Generalized*
PA 6400 Ostentation *Affectation*
PA 7411 Ostentation *Uncommunicativeness*
PD 2565 *Osteo-arthritis*
PE 4161 *Osteoarthritis*
PD 2565 *Osteoarthritis*
PE 5913 *Osteoarthritis animals*
PE 7826 *Osteoarthritis hip*
PE 5913 Osteoarthritis ; Traumatic
PE 3822 *Osteochondritis dissecans*
PD 7424 *Osteochondritis dissecans*
PJ 5172 Osteochondroses
PE 5050 *Osteochondrosis*
PE 3822 *Osteochondrosis*
PD 7424 *Osteodystrophy ; Fibrous*
PE 8589 *Osteogenesis imperfecta*
PG 2295 *Osteomalacia*
PE 8912 Osteomyelitis
PE 5050 *Osteomyelitis older animals*
PE 8912 Osteomyelitis, periostitis infections involving bones acquired deformities bone
PE 2280 *Osteomyelitis ; Pyogenic*
PE 9846 *Osteomyelitis ; Tuberculous*
PD 7424 *Osteopathy ; Hypertrophic*
PE 8589 *Osteopetrosis*
PE 3822 Osteoporosis
PF 1009 Ostracism
PF 2776 Ostracism ; Fear
PF 1009 Ostracized people
PF 7029 Otherness ; Dread
PE 5303 Others ; Exploitation prostitution
PF 1216 Others ; Numbness towards
PF 6382 Others ; Shifting responsibility onto
Otitis externa
PE 5444 — *Acute*
PE 5444 — animals
PE 5444 — *Chronic*
PE 5444 — *Mycotic*
PE 5444 — *Parasitic*
PD 5535 *Otitis interna animals*
PD 2567 *Otitis media*
PD 5535 *Otitis media animals*
PE 5444 *Otoacariasis*
PE 5444 *Otodectic mange*
PE 5444 *Otomycosis*
PE 2746 Otosclerosis
PD 3479 *Otter shrews ; Endangered species*
PE 8936 Otters ; Endangered species weasels, badgers, skunks,
PF 1015 Ousted repressive regime ; Post revolutionary re employment government security services
PF 3226 Out ; Contracting
PE 4421 Out kids ; Stressed
PD 6808 Out ; Lock
PF 1822 Out-of-date methods agricultural production
PF 5435 Out-of-date regulations
PE 0540 Out patient care ; Limited psychiatric
PF 2535 *Out town shopping ; Increase*
Out wedlock
PC 1874 — Children born
PE 9122 — Criminalization sexual relations
PF 5434 — Sexual intercourse
PD 2774 Outbreak ; Inadequate disinfection pastureland after disease
PE 8409 Outbreaks ; Inadequate disinfection humans after animal disease
PE 8409 Outbreaks ; Inadequate disinfection measures humans during animal disease
PC 1968 Outcastes
PD 6017 Outcasts
PD 6017 Outcasts ; Social
PF 6814 Outdated assumptions
PF 1815 *Outdated basic utilities*
Outdated [farm...]
PF 1822 — farm methods
PF 1319 — *farming techniques*
PF 1319 — *farming technology*
PF 1608 — forms community health
PD 3548 Outdated job training
PF 2580 Outdated judicial precedent
PF 1786 Outdated labour negotiation procedures
PF 8793 Outdated procedures
PF 3550 *Outdated property assessments*

Outdated regulations

PF 8793	Outdated regulations						

PF 8793 Outdated regulations
PF 8793 Outdated rules
PD 0530 Outer space benefits limited number developed countries ; Restriction
PF 4444 Outer space ; Invaders
PE 9829 *Outflow bile ; Obstruction*
PE 8957 Outflow developing developed countries ; Foreign private investment income
PD 3132 Outflow developing developed countries ; Private domestic capital
PE 1139 *Outflow ; Excessive capital*
PC 3134 Outflow financial resources developing countries
PE 5338 Outflow funds developing countries ; Aid lenders beneficiaries net
PD 3132 Outflow indigenous capital developing countries
PD 3132 Outflow local investment
PE 4409 Outlaws
PD 8517 Outlets ; Unmarked public water
PF 2428 Outlook civic minded groups ; Limited fragmented
PF 6572 *Outlook ; Frightening investment*
PE 1137 *Outlook ; Inhibiting business*
PE 6578 *Outmoded commercial buildings*
Outmoded concepts
PF 7673 — Institutionalization discriminatory
PF 2580 — legal systems ; Institutionalization discriminatory
PF 2580 — legal systems ; Persistence
PF 7673 — Persistence
PD 0975 Outmoded deliberative systems
PD 0975 Outmoded deliberative systems government
Outmoded [farm...]
PF 1822 — farm techniques
PF 2095 — forms social education
PF 2986 — functional skills rural communities
PF 2580 Outmoded legal systems
PF 8161 Outmoded motivational methods
PF 1262 Outmoded trade unions
PF 8161 Outmoded values ; Imposition young
PF 3548 Outmoded work-study programs
Output [developed...]
PF 1737 — developed market economies ; Slow down growth
PD 5114 — developing countries ; Constraints increased agricultural
PD 5114 — developing countries ; Lack incentives increase agricultural
PA 5454 Outrage *Badness*
PA 6822 Outrage *Disrespect*
PA 7253 Outrage *Envy*
PA 6852 Outrage *Inappropriateness*
PA 6486 Outrage *Injustice*
PA 5644 Outrage *Vice*
PF 0995 Outrageous legal fees
PG 8937 Outreach ; Reduced service
PF 3628 Outside exposure ; Lack
PF 3628 Outside exposure ; Limited
PF 2296 Outside influences rural villages ; Limited exposure
PF 3575 Outside instruction rural communities ; Irregular
PD 2338 Outside land ownership
PE 4780 Outside services ; Excessive dependence local communities
PF 8743 Outsiders ; Fear
PS 6553 *Outward cash flow*
PE 0616 *Ovale malaria*
PE 1905 Ovarian cancer
PD 8301 *Ovarian dysfunction*
PG 7551 *Ovary ; Abscess*
PE 0368 Ovary ; Benign neoplasm
Ovary [disease...]
PD 7799 — *disease ; Follicular cystic*
PD 7799 — *disease ; Luteal cystic*
PD 9742 — Diseases
PF 0296 *Ovens ; Expensive home*
PF 8855 Over-acceptance socio-economic dependency
PE 9580 Over-anxious disorder
Over [centralization...]
PF 2711 — centralization
PF 5472 — centralization global decision-making
PF 5057 — choice ; Consumer
Over commitment
PD 3954 — credit financing
PD 7051 — learning ; Societal
PF 8855 — welfare dependency
Over [competitive...]
PE 4266 — competitive sports
PE 5825 — concentration ownership maritime fleets
PF 0867 — conformity building design
PF 3850 — crowding mental health ; Negative effects
PE 7265 — cultivated gardens
Over [dependence...]
PD 9802 — dependence farm subsidies developed countries
PE 2214 — dependency international financial institutions developing countries
PD 2511 — development bureaucracy ex-colonial countries
PD 4907 — diversification manufactured goods
PD 4907 — diversification services
PE 5722 Over-eating
PC 6262 Over-education
Over emphasis
PF 2997 — economic development
PF 4059 — immediate solutions resource development research
PF 2174 — use national resources ; Profit
Over [emphasizing...]
PD 1827 — emphasizing method rather than content ; Educational curricula
PF 4580 — enthusiastic lending banks
PD 9578 — exploitation biological resources
PD 0826 — exploitation tourism

Over [fastidiousness...]
PF 5735 — fastidiousness
PD 5361 — fishing rights ; Conflicts
PE 3972 — flying foreign aircraft
PE 8525 — foreign investment ; Distortion international trade restrictive controls
PF 1033 — formalised decision-making
PC 0052 Over-intensive soil exploitation
Over [optimistic...]
PF 7186 — optimistic assessment problems
PF 7002 — optimistic bias
PF 4891 — organization
PF 7737 Over-planning
Over population
PB 0035 — [Over-population]
PD 4007 — Environmental pollution due
PE 6024 — *seals*
PE 6024 — wild animals
Over [priced...]
PD 1891 — priced goods services
PE 5523 — pricing
PD 1465 — production commodities
PD 2967 — production primary commodities developing countries
PF 7737 — programmed budgets
PF 7737 — programming
PF 5255 — protectiveness ; Parental
PF 3462 Over-qualification
PD 9628 Over qualification women ; Discriminatory requirement
PB 1845 Over-rapid industrialization
PD 9235 Over-rapid timber exploitation
Over reaction
PF 1083 — against environmental hazards
PF 1622 — past sexual ethics
PF 4110 — Political
Over reliance
PF 9205 — convenient beliefs
PF 7195 — developing countries foreign borrowing
PF 1070 — economic interest groups policy agencies
PF 9560 — government money creation
PD 4982 Over reporting reporting ; Corporate
PD 1893 Over reporting reporting ; Official
PE 1708 Over-spacing suburban housing
Over specialization
PF 0256 — [Over-specialization]
PF 5709 — medical care
PF 7532 — sciences
Over [specialized...]
PF 5712 — specialized study global ecosystem
PF 3588 — specialized supervisory personnel
PD 9802 — subsidised agriculture industrialized countries
Over [United...]
PF 8946 — United Nations agencies ; National hegemony
PF 7585 — use designated wilderness areas
PE 6214 — use formaldehyde building materials personal products
PF 7870 Over-valuation art
PF 3391 Over-warning
PE 9311 Over-watering during irrigation
PA 5473 Overabundance *Insufficiency*
PF 3519 Overacademic orientation high school
PA 5806 Overactivity *Inaction*
PD 8301 Overactivity parathyroid glands
PG 9483 *Overbite*
PG 9483 *Overbite ; Horizontal*
PG 9483 *Overbite ; Vertical*
PE 8667 Overbooking ; Reservation
PF 2797 Overburden youth leadership
PF 1822 Overburdened production plan
PD 0779 Overcapacity
PD 0779 Overcapacity industrial production
PE 6350 Overcast skies
PE 5755 Overcharging proprietary pharmaceutical products
PF 9699 Overclassification embarrassing documents
PA 6739 Overcompensation *Maladjustment*
PC 5075 *Overcomplicated implementation citizen rights*
PA 7115 Overconfidence *Rashness*
PA 5821 Overconscientiousness *Vulgarity*
PF 1815 *Overcrowded building conditions*
PE 5851 Overcrowded port warehousing
PD 4004 *Overcrowded public clinics*
PD 6585 Overcrowded school classes
PF 0734 Overcrowded spectrum radio frequencies
PB 0469 Overcrowding
PD 2760 Overcrowding housed farm animals
PD 0758 Overcrowding housing accommodation
PD 0520 Overcrowding prisons
PE 3757 Overcrowding schools
PF 1257 *Overcrowding schools transients*
PC 3813 Overcrowding ; Urban
PD 9235 Overcutting ; Forest
PF 3448 Overdemand resources
PF 1196 *Overdependence capital*
PE 3959 Overdependence computers universal panacea
PE 8988 Overdependence education
PF 9530 Overdependence government
PD 0678 Overdependence subsidies
PA 5473 Overdeveloped *Insufficiency*
PA 7285 Overdeveloped *Littleness*
PC 0111 Overdevelopment
PF 6475 *Overdivided farmland plots*
PE 4328 Overdosage plants ; Nitrogen
PD 0094 *Overdose*
PF 1179 Overdue government payments
PF 8650 Overdue payments governments international organizations

PE 5722 Overeating
PE 7769 Overeating disease
PF 0443 *Overemphasis academic education*
PF 0440 Overemphasis analysis
PF 7118 Overemphasis community problems
PD 5774 Overemphasis consumer role
PF 6580 Overemphasis discrete problems
Overemphasis economic
PF 1040 — administration ; Ideological
PF 1825 — productivity ; Diminishing role older people due
PF 1545 — progress society
PF 2959 Overemphasis effective use technical resources
PE 8864 Overemphasis governments large firms
Overemphasis [immediate...]
PF 3243 — immediate superficial needs
PF 6580 — Individual problem
PF 2475 — individual rights
PC 1835 — institutional security
PA 5473 — *Insufficiency*
PF 7118 *Overemphasis local problems*
PF 1804 Overemphasis national self determination ; Habitual
PF 7118 *Overemphasis parochial problems*
PF 1275 Overemphasis rapid returns investment
PF 3460 Overemphasis self-sufficiency respect interdependence
PF 6580 Overemphasis symptoms more fundamental problems
PF 2956 Overemphasis ; Technological skill
PF 2956 Overemphasis technological skills
PA 7411 Overemphasis *Uncommunicativeness*
PD 2283 *Overemphasized adult sports*
PF 8181 Overestimated economic benefits projects
PF 7166 Overestimation ability forecast
PA 6607 Overestimation *Misjudgement*
PA 7411 Overestimation *Uncommunicativeness*
PA 5473 Overexpansion *Insufficiency*
PD 4403 Overexploitation aquifers
PD 9082 Overexploitation fish resources
PD 4403 Overexploitation underground water resources
PA 2360 *Overexposure issues*
PE 3883 Overexposure sun
PA 5806 Overextension *Inaction*
PA 5473 Overextension *Insufficiency*
PE 5722 Overfeeding
PD 9082 Overfishing
PD 0876 Overflow ; Contaminated toilet
PF 9100 Overgeneralized policy models
PC 3153 Overgrazing
PD 4812 Overgrazing developing countries
PF 1600 *Overgrown cemeteries*
PE 5050 *Overgrown hooves*
PD 2672 *Overgrown vacant lots*
PG 8903 *Overgrowth service industries sector*
PE 0652 Overgrowth ; Thyroid glands tumorous
PF 9249 *Overhang ; Debt*
PD 9578 Overharvesting
PF 8158 Overhead costs ; Inflated
PF 8158 Overhead costs ; Prohibitive administrative
PF 8158 Overhead erosion funding effectiveness ; Administrative
PE 8063 Overhead ; High transportation
PF 8158 Overhead restrictions efficacy aid
PE 1684 *Overheated rooms*
PC 0254 Overheating economy
PC 0918 Overheating planet
PF 2277 *Overindulgence physical pleasures*
PA 6340 Overload *Adversity*
PF 6411 Overload during control complex equipment ; Information
PE 4630 *Overload ; Grain*
Overload [Immigration...]
PD 4605 — Immigration
PC 1298 — Information
PA 5473 — *Insufficiency*
PA 5491 Overload *Lightness*
PE 1341 Overload ; Power supply system
PF 6411 Overload process controllers ; Information
PE 5259 Overload ; Torture sensory
PD 0689 Overloaded airports
PF 1698 Overloaded party lines
PD 0876 Overloaded sewage systems
PE 4127 Overloaded vehicles
PF 1487 Overloading courts
PF 1176 *Overlooked community services*
PF 4611 Overlooked media channels
PF 6538 Overlooked personal skills
PF 2471 Overlooked potential industrial development rural communities
PF 2346 Overlooked space care
PF 1177 Overnutrition
PF 7578 Overpaid business executives
PF 8763 Overpaid employees
PF 8317 Overpaid experts
PD 7500 Overpayments ; Benefit
PD 4007 Overpopulation ; Negative ecological impact
PF 0905 *Overpowering competitive markets*
PF 6461 *Overpowering traditional habits*
PE 9087 Overprescription drugs
PE 1514 Overprescription fertilizers
PJ 6673 *Overpriced equipment replacement parts*
PE 4162 *Overpriced valuable land*
PC 2049 *Overprivileged regions*
PD 9448 Overproduction ; Agricultural
PD 9448 Overproduction food
PD 2539 Overproduction ; Plutonium
PF 3462 *Overqualification women*
PA 7157 *Overreligiousness Insanity*
PD 1644 Overruns large scale public programmes ; Cost
PF 8079 Overruns ; Unreported cost

PC 2172 Overseas investments ; Risk nationalization
PF 6477 *Overseas work ; Attractiveness*
PA 7253 Oversensitiveness *Envy*
PA 5821 Oversensitiveness *Vulgarity*
PA 6180 Oversight *Error*
PA 5438 Oversight *Neglect*
PF 8455 Oversimplification
PA 6468 Oversimplification *Complexity*
PA 7341 Oversimplification *Unpreparedness*
PF 6580 Oversimplified problem approaches
PF 1107 Overstaffing
PF 8181 Overstated programme advantages
PC 3153 Overstocking
PA 5806 Overstrain *Inaction*
PA 5473 Overstrain *Insufficiency*
PA 5473 Oversupply *Insufficiency*
PA 7193 Overtax *Cheapness*
PA 5473 Overtax *Insufficiency*
PA 5491 Overtax *Lightness*
PF 5922 Overtaxed shipping systems
PD 1964 Overthrow government
PD 1964 Overthrow government ; Advocating
PD 2778 Overtime ; Excessive
PE 5897 Overtonnaging
Overturn [Defeat...]
PA 7289 — *Defeat*
PA 5493 — *Depression*
PA 6542 — *Destruction*
PA 5901 Overturn *Revolution*
PC 0442 Overurbanization
PD 5261 Overuse environmental resources poor
PE 8166 Overuse machinery equipment development projects
PE 1514 Overuse nitrate fertilizers
PE 3959 Overutilization personal computers
PF 6102 Overvalued currencies
PD 2463 Overvalued dollar
PA 5473 Overweight *Insufficiency*
PA 5491 Overweight *Lightness*
PA 7285 Overweight *Littleness*
PE 1177 Overweight obesity
PD 2778 Overwork
PD 2778 Overwork ; *Death*
PA 5473 Overwork *Insufficiency*
PE 4421 Overworked children
PD 7762 Overworked women
PA 6379 Overzealousness *Avoidance*
PA 7157 Overzealousness *Insanity*
PA 7325 Overzealousness *Irresolution*
PA 7115 Overzealousness *Rashness*
PD 7799 Ovine genital campylobacteriosis
PE 5617 Ovine interdigital dermatitis
PD 7420 Ovine ketosis
PF 4875 *Own countries ; Inadequate punishment international criminals their*
PE 8414 Own country ; Restrictions nationals leaving their
PF 2827 Own group values ; Subjective assessment according
Own [language...]
PE 2142 — language ; Denial right indigenous peoples use their
PE 2142 — language ; Denial right peoples use their
PE 0241 — leaders ; Denial right nations select their
PC 2381 — lives ; Denial people control over their
PE 4399 — means subsistence ; Denial right people their
PF 0297 Own political status ; Denial right people determine their
PD 2338 *Owned natural resources ; Absentee*
PE 0988 Owned state controlled enterprises ; Abuse monopoly power state
PD 8056 Owned water ; Privately
PF 1815 *Owner improvements ; Negligent*
PE 5875 Owner registration ; Inequities ship
Ownership Absentee
PD 2338 — [Ownership ; Absentee]
PD 2338 — business
PD 2338 — land
PD 2338 — property
PF 6460 Ownership basis land allocation
PE 0246 Ownership ; Concentration press
PF 2457 *Ownership costs ; Prohibitive*
PE 8411 Ownership ; Denial right
PF 6475 *Ownership ; Diminishing land*
PE 0042 Ownership ; Foreign
PE 4738 Ownership ; Foreign
PB 0750 *Ownership ; Land*
Ownership [maritime...]
PE 5825 — maritime fleets ; Over concentration
PF 2014 — means production ; Unregulated
PF 3005 — *Monopolized land*
PD 2338 Ownership ; Non resident home
PD 2338 Ownership ; Obscure land
PD 2338 Ownership ; Outside land
Ownership [Private...]
PD 4974 — Private
PD 3182 — productive systems ; Limited
PF 4432 — Protection company
PE 2346 *Ownership records ; Obscure*
PE 8580 Ownership ; Risks transfer
PF 3005 *Ownership ; Static land*
PF 3211 *Ownership ; Traditional land*
PF 2365 Ownership tropical villages ; Disorganized approach land
PB 0750 *Ownership ; Unfair distribution land*
PD 3663 Oxidant formation ; Photochemical
PE 9443 Oxide atmosphere ; Increase nitrous
PE 9076 Oxides halogen salts ; Trade instability inorganic chemicals, elements,
PD 2965 Oxides pollutants ; Nitrogen

Oxygen [deficiency...]
PE 2467 — deficiency blood
PE 6289 — deficiency ; Marine
PE 5196 — Destruction environmental
PE 2467 *Oxygen supply foetus ; Defective*
PD 6113 Ozone depletion ; Stratospheric
PE 1359 Ozone pollutant

P

PD 3481 *Pacarana ; Endangered species*
PF 2304 Pace duration work
PF 2304 Pace life ; Increasing
PF 9664 Pace structural change ; Faltering adaptation economic agents accelerating
PE 7898 Pace technological advance ; Fast irregular
PF 7806 Pace technological advancement ; Inability educational systems keep
PF 0010 Pacifism
PD 3521 Packages ; Inadequate labelling
PD 3521 Packages ; Misleading labelling
PF 3143 Packaging agricultural products ; Inadequate
PD 1754 Packaging containers ; Non destructible
PE 0635 Packaging ; Energy intensive
PE 7417 Packaging ; Leakage
PE 0635 Packaging ; Resource intensive
PJ 2270 Pacts ; Abrogation
PF 0419 Pacts ; Secret intergovernmental
PE 2305 Padi fields ; Salty
PD 2728 Paecilomycosis
PF 8068 Paganism
PA 5568 Paganism *Ignorance*
PA 6058 Paganism *Impiety*
PA 0643 Pain
PD 8725 Pain ; *Abdominal*
PE 9604 Pain ; *Abdominal*
PE 1310 Pain ; Back
PE 2694 Pain ; Chronic
Pain [Dependence...]
PA 0643 — Dependence
PA 6799 — *Disease*
PD 0270 — *disorder ; Somatoform*
PF 0916 Pain ; Escaping reality sophisticated shields against life's
PE 1670 Pain ; Experimental exposure animals
PG 3554 Pain ; Facial
Pain [Insensibility...]
PA 5451 — *Insensibility*
PF 1216 — Insensitivity personal
PE 2694 — Intractable
PE 1309 Pain ; Knee
PE 5483 Pain ; Low back
PE 9461 Pain neck
PA 0643 Pain ; People
PA 0643 Pain ; Phantom limb
PE 9461 Pain syndrome ; Vertebrogenic
PA 7107 Pain *Unpleasantness*
PE 3732 Pains due torture ; Joint
PE 3798 Pains due torture ; Residual traumatic
PE 0538 Paint industry ; *Instability*
PE 9746 Painters cancer ; Vulnerability
PE 9746 Painters ; Occupational hazards
PD 2478 Paintings ; Pollution damage
PA 3369 Pairing ; Moral offences heterosexual
PE 9103 Palaearctic mole rats ; Endangered species
PE 5117 Palate ; Cleft
PE 9702 Palatitis
PE 3711 Palatochisis
PD 2730 Pale bird syndrome
PF 1741 Pale skin
PE 9774 Pale soft exudative pork
PE 3247 Palimony ; Inadequate provision
PE 7915 Pallidal system ; Hereditary diseases striato
PE 2981 Palm ; Pests diseases coconut
PE 2981 Palms ; Pests diseases
PE 3933 Palpitation
PD 2632 Palsy
PE 0763 Palsy ; Cerebral
PE 2206 Palsy ; Shaking
PD 2632 *Palsy ; Wasting*
PE 0616 Paludism
PE 3959 Panacea ; Overdependence computers universal
PE 1132 Pancreas ; Diseases
PE 6461 Pancreatic fluke
PD 3978 *Pancreatic necrosis animals ; Acute*
PE 1132 Pancreatitis
PE 2395 Pancreatitis ; Haemorrhagic
PE 7894 Pancytopenia
PE 5303 Pandering
PE 5726 Pandering youth market
PC 1326 *Pangolins ; Endangered species*
PD 9654 *Panhypopituitarism animals*
PE 2633 Panic
PE 2633 Panic consumer buying
PA 7289 Panic *Defeat*
PE 3575 Panic disorder
PA 6030 Panic *Fear*
PA 5467 Panic *Inexcitability*
PE 5324 Panic result nuclear accidents
PF 5937 Panic ; Susceptibility people
PD 2730 *Panleukopenia ; Feline*
PE 9774 *Panniculitis ; Nutritional*
PF 4209 Pantheism
PD 1763 *Panthera ; Endangered species*
PF 0326 Paper companies

PE 1669 Paper deterioration
PD 1152 Paper disposal ; Improper
PE 7914 Paper ; Instability trade pulp waste
PE 1616 Paper ; Long term shortage pulp waste
Paper printing industries
PE 1425 — Environmental hazards
PE 1927 — Instability
PE 1136 — Underdevelopment
PE 7416 Paper products ; Environmental hazards chlorine bleached
PD 4552 Paper proliferation due computerization
PF 4494 Paper qualification syndrome
PF 4494 Paper qualifications ; Reliance
PE 1616 Paper shortage
PD 1152 Paper ; Waste
PF 5856 Paperwork ; Excessive
PD 8786 Papillitis
PD 9667 Papillomatosis animals
PF 8071 Papism
PF 8071 Papist conspiracy
PD 2255 Pappataci fever
PE 2143 Papular stomatitis cattle
PD 0527 Para-military activities
PE 2455 Paracoccidioidomycosis
PE 7562 Paracolon infection
PD 0894 Paradigms ; Lack meaningful personal social
PD 7463 Paradoxes technology transfer programmes
PE 9468 Paraesthesia
PD 9667 Parafilaria infection
PE 0596 Paragonimiasis
PD 9667 Parakeratosis animals
PD 3978 *Parakeratosis ; Ruminal*
PC 6641 Parallel economies ; Development
PF 1781 Parallel social structures
PF 1853 Parallel urban services ; Proliferation
PD 2632 Paralysis
Paralysis [agitans...]
PE 2206 — agitans
PD 0689 — Air traffic
PE 2946 — anatomical extremities
PD 7841 — *animals ; Facial*
Paralysis [Calving...]
PE 4161 — *Calving*
PE 0763 — *Cerebral*
PE 0763 — *Cerebral spastic infantile*
PE 4604 — *Coonhound*
PF 3513 — *Corporate planning*
PD 9449 Paralysis developing countries ; Financial
PA 6799 Paralysis *Disease*
PF 1216 Paralysis ; Emotional
Paralysis [Facial...]
PD 2632 — Facial
PC 2270 — Familial periodic
PE 7808 — Femoral nerve
PD 7841 — forelimb
PD 2632 — Functional
Paralysis [Inaction...]
PA 5806 — *Inaction*
PF 1457 — individual decision-making
PE 0504 — Infantile
PE 9799 — Infectious bulbar
PE 0763 — *Intracranial*
PE 4566 Paralysis limb ; Transient
PD 1140 Paralysis minority voices
PE 4161 Paralysis ; Obturator
PF 3973 Paralysis over future ; Elder
PE 4161 Paralysis ; Peroneal nerve
PD 3978 Paralysis ; Pharyngeal
Paralysis [social...]
PF 2701 — social response
PF 2709 — specialized services relation world's need ; Decisional
PD 2626 — Spinal
Paralysis [throat...]
PE 8173 — throat, eye, limb respiratory muscles
PD 7841 — *Tick*
PD 3978 — *tongue animals*
PE 0965 *Paralytic animal diseases*
PE 9774 *Paralytic myoglobinuria*
PG 4624 Paralytic shellfish poisoning
PF 6575 Paralyzing boom/bust mentality
PF 1776 Paralyzing complexity urban structures
PD 2670 Paralyzing image impossibility change
PF 1389 Paralyzing patterns villages administrative structures
PD 3436 Paralyzing property tax
PF 2444 Paralyzing small communities ; Complex regulations
PE 1651 *Paramedic role ; Negated*
PD 9160 *Paramedic training ; Limited*
PD 9742 Parametrium ; Diseases
PD 0527 Paramilitary gangs
PE 0349 Paramnesia confabulation
PE 6461 Paramphistomes
PE 0435 Paranoia
PE 0435 Paranoia ; *Erotomanic*
PE 0435 Paranoia ; *Jealous*
PE 0435 Paranoia ; *Persecutory*
PE 0435 Paranoia ; *Somatic*
PE 0435 Paranoid personality disorder
Paranoid [schizophrenia...]
PE 0435 — schizophrenia
PE 0435 — state ; Persecutory
PE 0435 — states
PE 0435 Paranoid traits
PD 9154 *Paraphimosis*
PD 7799 *Paraphimosis animals*
PD 0438 *Paraphrenic schizophrenia*

Paraplegia

PE 2945 Paraplegia
PE 0763 *Paraplegia ; Spastic infantile*
PF 4866 Parapsychic warfare
PF 9032 Parascience
PE 3511 *Parasite ; Gid*
PE 3511 Parasite ; Tapeworm
PD 0868 Parasites
PE 2436 Parasites animals ; Gastrointestinal
PE 9427 Parasites cardiovascular system animals
PE 0958 Parasites ; Gastrointestinal infections animals due
PE 0596 Parasites human body
PE 4683 Parasites ; Nervous system diseases caused
PD 4659 Parasites plants
PE 3676 Parasites ; Protozoan
PJ 4169 Parasites ; Social
PD 2735 Parasitic diseases animals
PD 0982 Parasitic diseases ; Infectious
PE 5444 *Parasitic otitis externa*
PD 6284 Parasitic plants
PD 0868 Parasitism
PD 0339 *Parasitosis*
PE 3936 *Parasitosis swine*
PD 5642 Parastatals ; Lack enterprise
PF 4739 Parastatals ; Proliferation
PE 4878 *Parasuicide*
Parathyroid [gland...]
PD 8301 — *gland disorders*
PD 8301 — *gland ; Dysfunction*
PD 8301 — *glands ; Overactivity*
PD 2731 *Paratuberculosis*
PE 4036 Paratyphoid fever
PE 7562 *Paratyphoid infection poultry*
PE 6886 Paravaccinia
PE 6754 *Parcel bombs*
PE 9545 Pardoning convicted human rights offenders ; Government
PF 6186 Parent child relationship ; Excessive intensification
PC 0600 Parent child relationship ; Isolation
Parent families
PD 2681 — Lone
PD 2681 — One
PD 2681 — Single
PF 6461 *Parent input ; Insufficient*
PE 4010 Parent men ; Denial rights
PF 1187 Parent school liaison ; Inadequate
PF 1187 Parent-school non-involvement
PF 1187 Parent-teacher non communication
PF 1187 Parent teacher separation
PE 4019 Parent women ; Denial rights
PE 4245 Parental abuse chronic neglect ; Infant mortality due
PD 3255 Parental affiliation ; Denial
PD 7187 Parental beating children
PF 2060 Parental control children's thoughts reflections
PF 3253 Parental custody children ; Non
Parental [defensiveness...]
PF 2602 — *defensiveness ; Excessive*
PD 5344 — discipline ; Lax
PF 4391 — domination
PE 7700 — drunkenness ; Excessive
PC 3280 Parental fulfilment ; Lack
PF 6461 *Parental influence ; Diminishing*
PF 4391 Parental interference
PE 6075 Parental kidnapping
PD 9145 Parental lying
PF 5255 Parental over-protectiveness
Parental [participation...]
PF 2810 — *participation ; Blocked*
PF 2810 — *participation ; Limited*
PD 5344 — permissiveness
PD 7187 — punishment
Parental [responses...]
PF 6461 — *responses ; Lax*
PE 4010 — rights ; Discrimination against men
PE 4019 — rights ; Discrimination against women
PF 6461 — role ; Surrendered
Parental toleration
PD 5344 — alcoholism children
PD 5344 — drug addiction children
PD 5344 — sexual activity children
PD 0614 Parenthood ; Adolescent
PF 1625 Parenthood ; Prohibitive cost
PD 2681 Parenthood ; Single
PE 0651 Parenticide
PF 2747 Parenting ; Surrogate
PF 6461 *Parenting techniques ; Ineffective*
PE 7700 Parents ; Alcoholic
Parents [children...]
PD 9145 — children ; Lying
PE 4700 — choose their children's education ; Violation freedom
PF 3282 — consent adoption ; Refusal
PF 4205 Parents ; Decreasing number adoptive
PF 3979 Parents ; Disrespect
PE 0651 Parents ; Killing
PF 2093 *Parents ; Lack confidence*
PE 4669 Parents ; Lack freedom movement children separated
PD 1317 Parents ; Possessive attitude
PF 1187 Parents school ; Disrelation
Parents [Undervaluation...]
PF 9306 — Undervaluation education
PF 2810 — *Uninvolved working*
PF 0782 — Unknown
PD 3257 — Unmarried
PE 0763 *Paresis*
PD 7841 *Paresis ; Benign enzootic*
Paresis cows
PD 7424 — *following parturient paresis ; Neuromuscular*

Paresis cows cont'd
PD 7424 — *following parturient paresis ; Skeletal*
PD 7420 — *Parturient*
PD 7420 *Paresis ewes ; Parturient*
PD 7424 *Paresis ; Neuromuscular paresis cows following parturient*
PD 7424 *Paresis ; Skeletal paresis cows following parturient*
PE 4161 *Paresis ; Spastic*
PF 9360 Park lands ; Inaccessible
PF 7130 *Park potential ; Untapped*
PJ 8596 *Park safety ; Diminished*
PD 2025 Park trees ; Endangered
PF 6152 Parking space ; Automobile
PF 7130 *Parking space ; Undesignated*
PE 2206 Parkinson's disease
PE 2206 Parkinsonism
PF 6135 Parkland ; Artificiality
PE 6175 Parkland urban environment ; Inaccessibility green
PE 9282 Parklands ; Endangered
PD 7585 Parks ; Excessive use natural
PF 4550 Parliament ; Abusive language
PF 4550 Parliament ; Indecent conduct
PE 3735 Parliamentarians ; Conflict interest
PE 9759 Parliamentarians ; Violation human rights individual
Parliamentary [affairs...]
PF 3072 — affairs ; Restriction press coverage
PF 3072 — affairs ; Restrictions direct news coverage
PF 4550 — assemblies ; Violence
PD 2167 Parliamentary constituencies ; Unequal
PF 8876 Parliamentary debating time ; Inadequate
PF 0609 Parliamentary immunity ; Threat
PF 8876 Parliamentary inability respond minority concerns
PF 0609 Parliamentary privilege ; Threat
PF 0639 Parliamentary procedures ; Contempt
PF 3214 Parliamentary systems ; Weakness multi party
PF 8876 Parliamentary time approve needed legislation ; Lack
PD 2239 Parochial attitudes organizations' members
PD 0917 Parochial escapist media entertainment
PF 2600 Parochial external relations
PD 1668 Parochial family responsibility
PF 2627 Parochial images society leaders
PD 1231 Parochial interests
PF 2627 Parochial leadership posture
PF 6928 Parochial market research
PD 2469 Parochial monetary agreements
PF 2600 Parochial national interests
PD 0917 Parochial news coverage
PD 3546 Parochial planning natural resource extraction
PE 7118 *Parochial problems ; Overemphasis*
PF 1418 Parochial scientific view
PE 1840 Parochial telecommunications standards
PF 6573 *Parochial value system*
PF 2896 Parochial work images
PF 7593 Parochialism established religions
PF 1728 Parochialism internal values images ; Reinforced
PE 4602 Parole applications ; Delay processing
PE 1121 Parole failure
PE 1121 Parole violation
PE 2356 Parotidea ; Cynanche
PE 2356 Parotitis ; Epidemic
PE 0651 Parricide
PE 2320 Parson's knee
Part time
PB 0750 — employment ; Insufficient
PB 0750 — farm employment
PE 6241 — work ; Discrimination against
PE 6241 — work ; Exploitative
PF 6057 Partial censorship
PE 6512 Partial community education
PE 4196 Partial loss vision
PF 8652 Partial police protection
PF 9409 Partial solutions problems ; Fixation
PF 5978 Partial teacher engagement
PD 7536 *Partially skilled farmers*
PE 6086 Participate conduct public affairs ; Violation right
Participate [globally...]
PF 1006 — globally ; Refusal families
PE 6086 — government ; Denial right
PC 1001 — government women ; Denial right
PE 7312 Participate political processes ; Denial right indigenous peoples
PF 3226 Participate ; Refusal
PF 0749 Participate social processes ; Failure individuals
PD 1564 Participating elders urban environments ; Isolated non
PF 0955 Participating society ; Collapse meaning
PF 3307 Participation ; Absence community
PE 8357 Participation attempting solve intergroup conflicts ; Lack
PF 2810 *Participation ; Blocked parental*
PF 0574 Participation business decision making ; Lack worker
Participation [collective...]
PE 6018 — collective bargaining ; Inadequate employee
PF 8905 — collective religious worship ; Decreasing
PD 3386 — community activity ; Enforced
PF 3560 — community development ; Limits
PE 3869 — control joint venture ; Inadequate
Participation [decision...]
PF 8703 — decision making ; Exclusion
PE 8430 — decision making ; Ineffective means
PC 0588 — Decline civic
PF 5472 — developing countries international decision making ; Restrictions
PE 0402 — developing countries manufacturing ; Unemployment developed countries resulting
PF 3339 — development ; Lack
PE 8690 — disabled persons decision making ; Lack

Participation [Discouraged...] cont'd
PF 3307 — Discouraged community
PD 2029 — Distortion international trade result government
Participation [economies...]
PD 1902 — economies developing countries ; Excessive government
PD 1564 — Elders' self images preclude
PE 5074 — excluded company polity ; Worker
PF 3307 — experience ; Frustrating
PF 3307 Participation ; Factions frustrations community
Participation [image...]
PD 1564 — image ; Preclusion elders'
PC 0352 — *indigenous peoples international decision making ; Lack*
PF 2437 — Ineffective structures community
PD 2942 — international organizations ; Micro state
PE 5761 — island developing countries ; Inadequate international
Participation Lack
PF 3307 — community
PF 8703 — organized
PD 1544 — youth
Participation [large...]
PF 6127 — large communities ; Ineffectiveness individual
PD 1564 — *Limited*
PF 2810 — *Limited parental*
PC 1917 — local politics ; Lack
PE 8503 — local welfare programmes ; Lack
PF 0574 Participation management ; Lack
PE 4015 Participation medical profession torture ; Inhumane
PC 0588 Participation ; Non
PD 2910 Participation politics ; Forced
PC 1917 Participation politics ; Lack
Participation [Reluctant...]
PF 1205 — *Reluctant personal*
PF 3287 — Restricted local
PF 2810 — *Restrictive organizational*
Participation [social...]
PC 5387 — social processes ; Forced
PF 2440 — Sociological ignorance citizen
PF 2347 — Structural failure citizen
Participation [torture...]
PD 4478 — torture
PE 4119 — torture ; Military police personnel
PE 1996 — transnational corporations apartheid system
Participation [Unconfident...]
PF 3307 — Unconfident citizen
PF 5978 — Uninspired teacher
PD 1544 — Unplanned youth
PD 1544 — Unreleased potential youth
PF 4959 Participation women development ; Lack
PE 9009 Participation women ; Lack
PF 6023 Participation youth decision making bodies ; Non
PF 8703 Participative decision making ; Non
PF 1959 *Participatory financial priorities ; Non*
PF 2451 *Participatory methods ; Inhibiting*
PD 2008 Particles ; Airborne
PD 2008 Particulate atmospheric pollution
PD 2906 Parties ; Abuse electoral process political
PD 3536 *Parties ; Banned political*
PC 2494 *Parties ; Clash personal ambitions political*
PC 0116 Parties ; Corruption political
PE 9110 Parties ; Denial right organize political
PD 0548 Parties developing countries ; Inadequate political
PD 0923 Parties ; Fragmentation communist
PC 2494 *Parties ; Inner tensions political*
PE 0752 Parties ; Unjust financing political
PD 4517 Partisan reporting governmental action
PD 1263 Partition
PD 4392 Partner ; Ineffective means searching
PE 4890 Partners ; Deception sexual
PJ 2657 Partners ; Lack marriageable
PF 9818 Partners ; Psychological inconsistency marriage
PE 2203 Partnership model developed countries ; Contradictions growth
PE 7504 Parts agricultural machinery ; Unavailability replacement
PE 5319 Parts ; Counterfeit machine
PJ 4688 Parts ; Overpriced equipment replacement
Parturient paresis
PD 7420 — cows
PD 7420 — ewes
PD 7424 — *Neuromuscular paresis cows following*
PD 7424 — *Skeletal paresis cows following*
PD 2001 Party democracies ; Single
PD 4798 Party entryism ; Political
PF 1698 Party lines ; Overloaded
PD 2906 Party manipulation elections ; Political
PF 3214 Party parliamentary systems ; Weakness multi
PD 2730 *Parvoviral infection ; Canine*
PD 2730 *Parvovirus ; Feline*
PE 9458 Paspalum staggers
PF 1674 Passage ; Absence rites
PD 2948 Passage straits interoceanic canals ; Restrictions
PE 6755 Passage territorial waters ; Denial right innocent
PE 6755 Passage territorial waters ; Unlawful interference rights innocent
PE 5431 Passage urine ; Involuntary
PE 5431 Passage urine ; Unconscious
PD 0791 Passages ; Hazardous road
PF 4480 Passenger cars ; Maldistribution
PF 1495 Passenger service ; Irregular
PF 6382 Passing ; Buck
PC 0300 *Passing faeces ; Involuntary*
PA 7030 Passion
PA 7030 *Passion ; Crimes*
PA 7030 *Passion ; Dependence*

Peculiar gait

PF 5190 Passionate abstract music
PA 6379 Passionlessness *Avoidance*
PA 7364 Passionlessness *Unfeelingness*
PD 9219 *Passive aggressive personality disorder*
PC 4710 Passive atrocities
PF 6765 Passive historical perspective leaders
PF 6846 Passive public support terrorism
PF 2788 Passive resistance
PD 9219 *Passive resistance demands adequate occupational performance*
PE 5146 Passive smoking ; Health hazards
PF 6657 Passive village promotion
PF 6177 Passivity
PA 7250 Passivity *Disobedience*
PA 6200 Passivity *Impatience*
PA 5806 Passivity *Inaction*
PA 6444 Passivity *Quiescence*
PA 7364 Passivity *Unfeelingness*
PE 3889 Passports ; Crimes related immigration, naturalization
PE 4496 Passports ; Fraudulent acquisition use
PF 0699 Passports governments ; Sale
PE 0325 Passports, visas ; Refusal issue travel documents,
PF 1062 Past ; Confusion being detached
PF 9729 Past cooperative cheating
PF 0608 Past ; Denial
PF 2830 Past ; Denial old people significance
Past [efforts...]
PF 5272 — efforts ; Ineffective
PF 2830 — Elders' disapproval
PF 0608 — events ; Evasion
PF 0836 — *experience ; Irrelevancy*
PF 1746 — experience ; Perceived irrelevance
PF 5272 — experiences ; Awkward
PF 1746 Past ; Failure opposing groups appropriate wisdom
PF 5272 Past goals ; Frustrated
PF 1985 Past heritage ; Lack awareness
PF 1062 Past operating context ; Loss
Past oriented
PF 2256 — *concerns*
PF 3007 — educational vision
PF 1365 — *farming images*
PF 1577 Past ; Romantic understanding
PF 1577 Past ; Sentimental attachment
PF 1622 Past sexual ethics ; Over reaction
PD 2732 *Pasteurella antipestifer infection*
PE 7553 *Pasteurellosis*
PE 5524 *Pasteurellosis ; Bovine pneumonic*
PD 2732 *Pasteurellosis sheep ; Septic*
PE 8077 Pastoralists ; Endangered lifestyles nomads
PD 8625 Pasture arable land ; Agricultural mismanagement
PD 2774 Pastureland after disease outbreak ; Inadequate disinfection
PD 3668 *Pastureland ; Contaminated*
PC 2765 *Pastureland ; Insufficient*
PD 7424 *Patellar luxation*
PF 3538 Patent abuse
PF 3538 Patent application ; Difficulty
PF 3538 Patent control ; Inadequate
PF 3538 Patent coverage ; Inadequacy
PF 3538 Patent extension ; Inadequate
PE 4039 Patent pool cross-licensing agreements
PD 0455 Patent protection ; Distortion international trade obstacles
PF 3538 Patent restrictions ; Avoidance
PE 8403 Patent restrictions impeding sustainable development
PD 7501 Patenting genetically transformed animals ; Irresponsible
PE 6788 Patenting genetically transformed plants ; Irresponsible
PD 7840 Patenting life forms ; Irresponsible
PF 7779 Patents ; Delay recognition
PF 4050 Patents ; Government refusal declassify innovative
PF 4050 Patents ; Suppression innovation purchase
PE 0346 Patents trademarks ; Restrictive business practices relation
PD 7297 Paternal deprivation
PD 7297 Paternal negligence
PF 2183 Paternalism
PF 1871 Paternalism ; International
PF 5397 Paternalism ; Medical
PF 3609 Paternalism ; Religious
PF 7635 Paternalistic lies
PF 2183 *Paternalistic punishment*
PD 3255 Paternity ; Confusion
PD 3255 Paternity ; Denial
PE 0155 Pathogenic air pollutants ; Plant
PD 9094 Pathogenic bacteria ; Water borne
PD 2225 *Pathogenic fungi*
PG 2302 *Pathogenic organisms*
PD 0594 Pathogenic viruses
PD 1866 Pathogens ; Plant
PA 6448 *Pathological gambling*
PD 0153 *Pathological intoxication*
PF 1602 Pathological science
PB 3674 Pathologies civilization
PF 7580 *Paths ; Dangerous*
PF 7580 *Paths ; Dangerous foot*
PE 8971 Paths ; Soft energy
PF 7580 *Paths ; Steep narrow*
PF 7580 *Pathways ; Difficult internal*
PE 0540 Patient care ; Limited psychiatric out
PE 9105 Patients ; Carelessness dealing infectious
PD 1662 Patients ; Denial rights medical
PD 1148 Patients ; Denial rights mental
PD 1662 *Patients ; Institutionalized*
PE 5027 Patients ; Medical practitioners refusing treat
PD 0584 Patients psychiatric hospitals ; Abusive treatment
PE 1704 Patients radiation ; Excessive exposure medical
PD 3555 Patriarchal role man ; Family dependence
PE 0651 Patricide
PD 0218 *Patriotism ; Lack*
PF 8652 Patrol ; Nonexistent security
PE 0712 Patrols ; Unidentified submarine
PF 8535 Patronage ; Abusive distribution political
PF 1345 *Patronizing communities*
PD 1415 Pattern business activities small communities ; Restrictive
PF 6525 Pattern community organization ; Fragmented
PC 1716 Pattern ; Continuing dropout
PD 9219 *Pattern deficits interpersonal relatedness ; Pervasive*
PF 0193 Pattern ; Immobility social
PF 2256 *Pattern ; Irritation response*
PF 2256 *Pattern ; Laissez faire response*
PE 7632 Pattern perfectionism inflexibility
PF 6559 *Pattern ; Sporadic spending*
PF 2845 *Pattern ; Unchanging business*
PF 6471 *Pattern ; Unresolved development*
PF 7713 Patterns ; Antiquated thought
Patterns community
PF 6504 — activity ; Fragmented
PF 6545 — groupings ; Divisive
PF 2575 — groupings ; Insular
PF 2845 — identity ; Disjointed
PF 3531 — life ; Absence traditional
PF 0275 Patterns curtail creativity
PE 3337 *Patterns ; Customary working*
Patterns [Difficulties...]
PD 4125 — Difficulties establishing changes rural economic
PC 2279 — Disruption animal migration movement
PD 4252 — Disruption native trade
PF 2303 — Disruptive effect changing employment
PF 2346 — *Divisive road*
PF 6545 — Divisive traditional
PF 1509 Patterns extended family relationships ; Fragmented
Patterns [Family...]
PF 1933 — *Family based production*
PF 2644 — family lifestyle ; Inflexible
PF 0963 — *Fluid community*
Patterns [Haphazard...]
PF 2346 — *Haphazard growth*
PF 1746 — history ; Failure profit
PE 1848 — *Holiday residency*
Patterns [Imbalance...]
PC 8415 — Imbalance international trade
PF 1205 — *Inappropriate study*
PF 2575 — individual involvement ; Constricting
PF 3091 — Inflexible management
PF 2845 — *Ingrained individualistic*
PF 0193 — Inhibiting social
PF 2971 — *Irregular work*
Patterns [Late...]
PF 1382 — *Late bill paying*
PF 7142 — *Limited engagement*
PE 4263 — Long term changes precipitation
PF 2827 *Patterns ; Non local employment*
PF 6573 *Patterns ; Obsolete self sufficiency*
PF 8598 Patterns ; Rigid cultural
PC 1108 Patterns ; Seasonal work
PF 9276 Patterns social choice ; Unknowable future
Patterns Static
PF 4984 — *education*
PF 3091 — management
PF 6545 — *power*
PC 0492 Patterns ; Subsistence farming
Patterns [Temporary...]
PG 8978 — *Temporary resident*
PF 3129 — traditional life ; Restrictive
PF 2644 — traditional way life ; Delimiting family
Patterns [Uncertain...]
PF 5817 — Uncertain economic
PD 2438 — Unclear consensus
PF 2845 — Unconsolidated community
PC 2494 — Unpopular voting
PF 0193 — Unvaried social
PF 2845 Patterns village identity ; Disrupted
PF 1389 Patterns villages administrative structures ; Paralyzing
PF 0836 *Patterns ; We / they language*
PF 2208 Paucity ; Leadership skills
PB 0388 Pauperism
PB 1047 *Pauperization*
PE 4685 Pavements ; Dog faeces
PE 2197 *Pavor nocturnus*
PD 1977 Pay equal work ; Denial right equal
PD 8916 Pay ; Low
PE 6725 Pay system erosion
PE 3872 Pay top managers ; High severance
PD 0309 Pay women ; Denial right equal
PD 0309 Pay women ; Unequal
PF 3448 *Paying jobs ; Scarcity well*
PF 1382 *Paying patterns ; Late bill*
PF 2845 *Paying skills ; Low*
PF 2949 *Paying surcharges ; Excessive bill*
PF 8650 Payment agreed contributions international organizations ; Government non
PF 1382 *Payment bills ; Habitual non*
Payment compensation damages
PE 0290 — consumers ; Non
PD 7179 — Delay
PD 7179 — Non
PE 4446 Payment compensation ; Delay governmental
PE 8898 Payment compensation forced relocation ; Non
Payment compensation victims
PE 5229 — catastrophes ; Non
Payment compensation victims cont'd
PE 3913 — crime ; Delay
PE 3913 — crime ; Non
PE 5229 — disasters ; Delay
PE 5229 — major accidents ; Delay
PE 0811 — malpractice ; Delay
PE 0811 — malpractice ; Non
PE 8824 — motor accidents ; Non
PF 1179 Payment government financial commitments ; Arrears
PE 4446 Payment government reparations ; Delay
PE 4789 Payment housework ; Lack
PF 5514 Payment interest
PD 1157 Payment interest due ; Irregular international
PE 0197 *Payment prizes athletic events ; Discrimination against women*
PE 4446 Payment reparations government ; Non
PE 3926 Payment times creditors ; Inconsiderate choice
PF 8650 Payment United Nations membership dues ; Non
Payments [Abuse...]
PF 6723 — Abuse kind
PF 3062 — adjustment ; Inadequate mechanism balance
PE 1771 — Adverse effect transnational corporations balance
PF 5292 — after expiration industrial property rights
Payments Arrears
PE 4571 — home loan
PD 3053 — international debt
PE 4571 — mortgage
PE 4571 — rental
PC 0998 Payments deficits ; Balance
PC 0998 Payments ; Dependence imbalance
PD 5335 Payments ; Failure make payroll
PE 9700 Payments government agencies ; Arrears financial
PF 8650 Payments governments international organizations ; Overdue
PE 4571 Payments housing ; Arrears
Payments [Imbalance...]
PC 0998 — Imbalance
PD 1229 — *Insufficient health*
PF 8650 — intergovernmental organizations ; Withholding membership
PD 1157 — international financial obligations ; Irregular
PF 1179 Payments ; Overdue government
PE 9567 Payments position developing countries ; Deterioration external
PD 2552 Payments ; Profit oriented interest
PE 9700 Payments state controlled enterprises ; Delays
PD 1147 *Payoffs ; Necessity business*
PD 3736 Payola
PD 5335 Payroll payments ; Failure make
PD 5228 *PBB poisoning animals*
PD 5228 *PCB poisoning animals*
PE 2432 PCBs
PD 1434 *Pea distribution unevenness ; Bean*
PE 2357 Pea picker's disease
PF 9996 Peace
PF 4787 Peace agreements ; Inequitable
PF 5826 Peace ; Collapsed concept
PC 6239 Peace ; Crimes against
PD 5253 Peace ; Denial right people live
PF 6895 Peace forces ; Attacks
PF 6208 Peace ; Inadequate education
PF 9149 Peace proposals ; Rejection
PF 4785 Peace recession
PF 4848 Peace research support ; Inadequate
PC 6239 Peace security mankind ; Offences against
PB 0009 Peace security ; Threats international
Peace [time...]
PE 0736 — time ; Environmental degradation military activity during
PE 2324 — treaty following war ; Failure sign
PF 5826 — Trivialization
Peace [Unauthentic...]
PF 7643 — Unauthentic
PB 7694 — Unjust
PF 7643 — Unmeaningful
PB 0009 Peace ; Violation international
PF 4785 Peaceful applications ; Obstacles redeployment military resources
PE 8965 Peaceful development space ; Military obstacles
PF 2788 Peaceful resistance
PD 0055 Peaceful socio economic development ; National sovereignty obstacle
PE 9294 Peacetime ; Anti personnel use toxic substances
PC 2201 Peacetime ; Nuclear explosions
PE 4586 *Peach leaf curl*
PF 0827 Peaks ; Underutilization facilities due daily seasonal
PE 2587 *Peanut trade ; Instability ground nut*
PE 8259 Peas, lentils leguminous vegetables ; Instability trade beans, dried,
Peasant farmers
PE 4243 — Inadequate working conditions
PD 9155 — Lack social security
PF 7047 — Neglect
PC 0492 Peasant farming
PC 8862 Peasant landholdings ; Dispossession
PC 8862 Peasant title land ; Loss
PC 8862 Peasantry ; Deprivation
PC 8862 Peasants ; Landless
PD 4273 Peasants onto marginal lands ; Forced relocation
PB 1992 *Peasants ; Threats against*
PE 8194 Peat energy source ; Underutilization
PC 3486 Peat land ; Destruction
PD 5110 Peat land ; Inadequate empolderment
PE 9774 *Pectoral myopathy turkeys ; Deep*
PE 8158 *Pectoris ; Angina*
PE 4566 *Peculiar gait*

Peculiar people

PF 4548 Peculiar people
PE 1032 Pecuniary guarantees condition provisional release criminal cases
PE 8152 Pedagogical material ; Limited access culturally adapted
PE 7826 *Pedal bone ; Fracture*
PE 7826 *Pedal bone ; Lameness caused bone cyst*
PA 7170 *Pedantry*
PA 6400 Pedantry *Affectation*
PA 7170 Pedantry *Informality*
PC 6307 Peddling ; Influence
PE 3265 Pederasty
PD 0994 Pedestrian accidents
PE 6151 Pedestrian environments urban areas ; Unattractive
PE 2429 *Pedestrians ; Drunken*
PE 6139 Pedestrians vehicles ; Intimidation
PE 4276 Pediatric AIDS
PD 0982 *Pediculosis*
PD 9667 *Pediculosis animals*
PE 7826 *Pedis ; Fracture os*
PE 2455 *Pedis ; Mycetoma*
PD 1110 Pedologists ; Corruption
PD 1110 Pedologists ; Irresponsible
PD 1110 Pedologists ; Negligence
PD 1110 Pedology ; Malpractice
PE 3265 Pedophilia
PE 3272 Peeping toms
PF 6570 *Peer behaviour expectation*
PF 9003 Peer groups ; Negative influence
PF 9003 Peer influence ; Disruptive
PE 2287 Pellagra
PE 5380 *Pelodera dermatitis*
PD 4575 Pelts ; Slaughter animals
PE 7826 *Pelvic fracture*
PD 8239 *Pelvic inflammatory disease ; Chronic*
PE 7511 *Pelvis ; Fracture*
PC 8534 *Pemphigus*
PE 9509 *Pemphigus vulgaris animals*
PD 0520 Penal systems ; Inadequate prison conditions
PF 0399 Penalty ; Death
PF 0399 Penalty ; Denial rights those punished death
PD 9366 *Pendulous crop poultry*
PJ 8039 Penis envy
PE 2764 Penning conditions factor animal diseases ; Insanitary
PE 2763 Penning domestic animals ; Inadequate housing
PF 5956 Pension costs ; Escalating
PF 5956 Pension funds ; Crisis long term
PC 1966 Pension income ; Inadequate
PD 0162 *Pension rights ; Unequal*
PD 0276 Pensioners' homes ; Old age
PE 7942 Pensions men women ; Unequal distribution old age
PF 5956 Pensions ; Progressive erosion old age
People [abnormal...]
PE 9402 — abnormal height ; Discrimination against
PA 0587 — Aggressive
PD 1611 — Alcohol consumption children young
PA 1635 — Anxious
PA 2360 — Apathetic
PA 7646 — Arrogant
PB 1638 — Authoritarian
PE 0742 — authority ; Alcohol abuse
People [Badly...]
PB 1498 — Badly nourished
PC 6727 — be self determining ; Denial right
PE 1024 — be self governing ; Denial rights indigenous
PC 7652 — Bigoted
PF 0379 — biological rhythms ; Separation
PD 8034 — Boat
PA 7365 — Bored
PA 7365 — Boring
People [Censured...]
PE 0189 — Censured
PA 1742 — Complacent
PD 2318 — Constraints against employment older
PC 2381 — control over their own lives ; Denial
PF 7728 — convicted crimes which they are innocent ; Failure clear names
PA 1986 — Corrupt
PD 5300 — Crimes which endanger
PB 2642 — Cruel
People [Decadent...]
PB 2542 — Decadent
PB 4731 — Deceitful
PB 4731 — Deceptive
PA 1757 — Dehumanized
PC 0352 — Denial rights indigenous
PE 8933 — Dependence excitement danger young
PA 0831 — Deprived
PF 4004 — Despairing
PB 2150 — Destitute
PA 0832 — Destructive
PF 0297 — determine their own political status ; Denial right
PA 0839 — Dominating
PF 1825 — due overemphasis economic productivity ; Diminishing role older
PA 1387 People ; Elitist
PF 3702 People ; Erroneous
People [Failure...]
PE 9071 — Failure employ skills educated elderly
PC 1455 — faiths ; Negative attitude
PF 4357 — Fallacious
PA 6030 — Fearful
PE 6955 — freely dispose natural wealth ; Denial right
PB 5747 — Friendless
PA 2252 — Frustrated
PA 6999 People ; Greedy
People [Homeless...]
PB 2150 — Homeless
PD 8017 — human beings ; Governmental disregard
PF 3856 — Humiliated
PB 0262 — Hungry
People [Ignorant...]
PA 5568 — Ignorant
PF 9536 — Ill starred
PC 0210 — Illiterate
PA 3369 — Immoral
PA 6000 — Improper
PE 8833 — Inadequate recreational facilities disabled
PE 6187 — Inappropriate accommodations single
PA 6416 — Incompetent
PA 7604 — Indifferent
PB 0843 — Inefficient
PB 0855 — Injured
PA 0857 — Insecure
PF 2875 — *Insufficient transportation old*
PB 1992 — Intimidating
PD 8204 — Invisible
PC 6203 — Involuntary movement
PA 8658 — Irresponsible
PF 8508 People ; Labelling
People Lack
PD 5344 — constructive family discipline young
PF 2012 — facilities educating older
PD 0211 — facilities severely deformed
People [lie...]
PE 3909 — lie ; Educating
PD 5253 — live peace ; Denial right
PD 4392 — Loneliness single
PF 2386 — Lonely
People [Malevolent...]
PA 7102 — Malevolent
PA 6359 — Manipulative
PC 0799 — Mentally depressed
PF 6224 — Mystique attitudes problems disabled
PA 2658 People ; Negligent
People [Obsessed...]
PA 6448 — Obsessed
PF 7239 — Officially nonexistent
PF 1009 — Ostracized
People [pain...]
PA 0643 — pain
PF 5937 — panic ; Susceptibility
PF 4548 — Peculiar
PJ 4468 — Pernicious
PB 7709 — Persecuted
PB 0869 — Perverted
PD 7238 — Physically dependent
PB 0388 — Poor
PA 2173 — Prejudice
PE 1536 — pursue development ; Denial right
People [Repressed...]
PB 0871 — Repressed
PF 0379 — rhythms nature ; Disconnection
PD 1291 — rural urban areas ; Migration skilled
PF 9008 People's aspirations structure opportunities income available ; Imbalances
PF 2068 People's lack context future ; Young
PF 5311 *People's representatives process feedback constituents ; Inability*
PF 4644 People's time orientation ; Degrading
People [Sacrilegious...]
PF 0662 — Sacrilegious
PA 0005 — Secretive
PD 6818 — Security risk
PD 5177 — Short
PF 2830 — significance past ; Denial old
PC 0193 — Socially rejected
PB 1875 — Starving
PJ 3525 — Stubborn
PA 0430 — Superstitious
People [their...]
PE 4399 — their own means subsistence ; Denial right
PA 6448 — *things ; Obsession*
PE 1832 — towards family ; Irresponsibility young
PC 0130 — Trafficking
People [Ugly...]
PA 7240 — Ugly
PF 2373 — Unassertive
PA 0911 — Uncentered
PA 5568 — Uneducated
PB 0750 — Unemployed
PE 9071 — Unemployment educated older
PE 5951 — Unemployment older
PC 8247 — Unethical
PA 6486 — Unjust
PF 9536 — Unlucky
PD 8204 — Unrecognized
People [Vain...]
PA 6491 — Vain
PE 9071 — valuable skills ; Premature retirement
PD 5218 — Violation land rights
PA 0429 — Violent
PE 6868 People work ; Hazards young
PB 0855 People ; Wounded
Peoples [Abuse...]
PC 0720 — Abuse native
PE 4054 — Abuse traditional cultural expressions
PE 7242 — Alcoholism amongst indigenous
PC 7384 — Alien domination
PE 8802 Peoples ; Biased media image foreign groups
Peoples Denial right
PE 5484 — adequate standard living indigenous
Peoples Denial right cont'd
PE 4332 — freedom religion indigenous
PE 1506 — social security indigenous
PE 1506 — social welfare services indigenous
Peoples [Denial...]
PE 4972 — Denial rights development indigenous
PE 7765 — Denial social rights indigenous
PC 3320 — Discrimination against indigenous populations housing Denial right adequate housing indigenous
Peoples [education...]
PC 3322 — education ; Denial right indigenous
PC 0720 — Endangered tribes indigenous
PE 4054 — Exploitation artistic expressions
PC 3293 Peoples ; Forced integration
PE 1407 Peoples ; Government non assistance endangered
Peoples [Illiteracy...]
PD 3321 — Illiteracy indigenous
PC 3322 — Inadequate education indigenous
PC 3320 — Inadequate housing indigenous
PC 0352 — *international decision making ; Lack participation indigenous*
PC 0352 — *international decision making processes ; Inadequate access indigenous*
PC 3319 Peoples ; Malnutrition indigenous
PE 7312 Peoples participate political processes ; Denial right indigenous
PF 4548 Peoples perceiving themselves specially chosen
PE 3735 Peoples representatives ; Conflict interest
PD 2490 Peoples ; Separatism indigenous
PE 1506 Peoples ; Social insecurity indigenous
PE 2142 Peoples use their own language ; Denial right
PE 2142 Peoples use their own language ; Denial right indigenous
PC 0720 Peoples ; Victimization indigenous
PE 3000 Pepper trade ; Instability
PE 2308 Peptic ulcers
PD 0556 *Perception disorder ; Post hallucinogen*
PF 8352 Perception ; Extrasensory
PF 2453 Perception facts ; Selective
PE 4589 Perception personal meaning ; Socially reinforced shallow
PF 1682 Perception sexual activity ; Individualistic
PA 6542 Perdition *Destruction*
PA 7148 Perdition *Ungodliness*
PE 7832 Perennial irrigation ; Increase pests diseases
PD 5228 *Perennial ryegrass staggers*
PE 7632 Perfectionism inflexibility ; Pattern
PF 9624 Perfectionists ; Stress
PA 7363 Perfidiousness *Improbity*
PA 7411 Perfidiousness *Uncommunicativeness*
PF 4153 Perfidy ; Government
PE 2308 *Perforation*
PD 9045 *Perforation ; Intestinal*
PE 0891 Performance animals ; Gambling
PD 7257 Performance developing countries ; Differential economic
PF 3845 Performance developing countries ; Inadequate investment
PE 8273 Performance ; Interference trade unions contract
PD 9219 *Performance ; Passive resistance demands adequate occupational*
PF 3498 *Performance requirements ; Low*
PD 3028 *Performances ; Refusal licence theatre*
PD 0188 Performers' rights ; Vulnerability
PE 4810 Performing animals ; Maltreatment
PF 1756 Performing arts
PE 0906 Performing dolphins ; Maltreatment
PE 5538 Performing socialist responsibilities ; Criminal negligence
PE 3936 *Perfringens type C enteritis swine ; Clostridium*
PE 8232 Perfume materials ; Instability trade essential oils
PD 0448 *Pericarditis ; Acute*
PE 0920 *Pericarditis ; Rheumatic*
PE 0471 *Pericardium diseases animals*
PE 3503 Peridontitis
PD 2387 Perinatal morbidity
PD 2387 Perinatal morbidity mortality
PD 3978 *Perineal hernia animals*
PE 0616 Periodic fever
PD 1318 *Periodic insanity*
PF 4149 Periodic mass extinctions species
PD 5535 *Periodic ophthalmia*
PC 2270 *Periodic paralysis ; Familial*
PD 3027 Periodical censorship ; Newspaper
PE 3503 *Periodontal atrophy*
Periodontal [disease...]
PE 3503 — *disease*
PE 9702 — *disease*
PE 3503 — diseases
PE 3503 Periodontitis
PE 9702 *Periodontitis animals*
PE 3503 Periodontosis
PE 9702 *Periostitis animals ; Alveolar*
PE 7826 *Periostitis fetlock joint*
PE 8912 Periostitis infections involving bones acquired deformities bone ; Osteomyelitis,
PE 8932 Peripheral autonomic nervous system ; Diseases
PE 8932 Peripheral ganglia ; Diseases nerves
PE 8932 Peripheral nervous system ; Diseases
PE 0430 *Perishable foodstuffs ; Unsafe transport*
PD 9045 *Peritoneal adhesions*
PD 3978 *Peritoneal fat necrosis*
PE 8228 Peritoneum cancers bronchi ; Pleura
PE 2663 Peritoneum ; Inflammation
PE 2663 Peritonitis
PD 3978 *Peritonitis animals*

PD 2730 Peritonitis ; Feline infectious
PE 7733 Peritonsillar abscess
PD 2630 Perjury
PD 4982 Perjury corporation representatives
PD 1893 Perjury government agents
PD 5725 Perks
PD 1165 Permafrost instability
PA 6802 Permanence
PE 1134 Permanent employment inner cities ; Limited availability
PF 1213 Permanent residency ; Discouragement
PE 7533 Permissible murder
PD 5344 Permissive child-rearing
PF 0593 Permissive education
PF 1252 Permissive society
PF 1252 Permissiveness
PA 5740 Permissiveness Compulsion
PA 5438 Permissiveness Neglect
PD 5344 Permissiveness ; Parental
PA 7321 Permissiveness Refusal
PD 3529 Permits ; Refusal work
PJ 4468 Pernicious people
PJ 4468 Perniciousness
PE 4161 Peroneal nerve paralysis
PE 1956 Peroxyacetyl nitrates
PF 9023 Perpetuated rural isolation
PF 4741 Perpetuating attitudes technological innovation ; Self
PE 8457 Perpetuator failure ; Unemployment
PA 5497 Perplexity Difficulty
PA 6247 Perplexity Inattention
PA 7309 Perplexity Uncertainty
PA 7367 Perplexity Unintelligibility
PB 7709 Persecuted people
PB 7709 Persecution
PD 3453 Persecution animals ; Superstitious
PB 7709 Persecution ; Dependence
PE 1281 Persecution ; Gypsy
PC 5994 Persecution ; Religious
PF 3353 Persecution religious sects
PE 0435 Persecutory paranoia
PE 0435 Persecutory paranoid state
PF 4240 Persistence
PF 4947 Persistence hierarchy
PE 2324 Persistence hostilities following cease-fire agreements
PF 7673 Persistence outmoded concepts
PF 2580 Persistence outmoded concepts legal systems
PE 2324 Persistence technical state war following cease-fire agreements
PF 4240 Persistence terrorist activities
PE 2248 Persistent cough
PE 5569 Persistent ductus arteriosus animals
PE 5569 Persistent right aortic arch animals
PE 4716 Person law ; Denial right recognition
PF 5973 Person ; Uncritical acceptance another
PF 5203 Personal advantage ; Failure sacrifice any
PC 2494 Personal ambitions political parties ; Clash
Personal appearance
PF 4010 — Boring
PF 4010 — Unchanging
PF 4010 — Unfashionable
PF 4010 — Unsatisfactory
PF 5203 Personal austerity ; Resistance
PE 4635 Personal authority political purposes ; Misuse
PD 9376 Personal bankruptcy
Personal [capacity...]
PF 9429 — capacity ; Diminished
PE 6770 — care disabilities
PF 1729 — commitment social vision ; Shallow
PE 3959 — computers ; Overutilization
Personal consumption
PF 5931 — Unnecessary
PF 5931 — Unsustainable
PF 5931 — Wasteful
PD 5770 Personal counsellors ; Bogus psychiatrists
PF 1978 Personal covert himsa
Personal [data...]
PF 4176 — data ; Accumulation
PF 4176 — data ; Errors computerized
PD 3954 — debt
PF 4371 — development ; Lack
PF 0549 — disempowerment
PE 3733 — disfigurement ; Malicious
PF 2437 Personal effectiveness ; Frustrated
PF 8855 Personal entitlement ; Belief
PF 4387 Personal failure
PF 1772 Personal freedom ; Non integrating images
PD 9653 Personal greed government leaders
Personal [habits...]
PD 5494 — habits ; Inappropriate
PD 8607 — health ; Neglect
PF 6389 — health ; Obsession
PD 2569 — honour reputation ; Denial right freedom attacks
PD 2459 — hygiene ; Inadequate
Personal [identity...]
PF 1934 — identity ; Inadequate sense
PF 4389 — illusions
PF 4010 — image ; Dissatisfaction
PD 2769 — income developing countries ; Low level
PD 8568 — income ; Insufficient
PD 0270 — information ; Inability recall important
PD 9376 — insolvency
PA 0911 — integration ; Inadequate
PA 0839 — interest emphasis
PA 0839 — interests ; Dominating
PF 5203 — investment ; Resistance loss any
PD 2495 — isolation communities industrialized countries
PD 4840 Personal life crises

PF 1889 Personal ; Limiting responsibility
Personal [meaning...]
PE 4589 — meaning ; Social reinforcement shallow
PE 4589 — meaning ; Socially reinforced shallow perception
PF 4389 — misperceptions
PC 2089 — mistrust ; Lack international cooperation due
PJ 0646 Personal need recognition
Personal [pain...]
PF 1216 — pain ; Insensitivity
PF 1205 — participation ; Reluctant
PD 8076 — physical disfigurement
Personal physical insecurity
PD 8657 — [Personal physical insecurity]
PE 5391 — developing countries
PE 4639 — industrialized countries
PE 7750 — women
Personal [physical...]
PF 4010 — physical unattractiveness
PA 2173 — prejudices ; Disruptive
PF 8145 — prestige ; Pursuit
PF 6504 — priorities ; Consuming
PF 5203 — privilege ; Unwillingness sacrifice
PE 6214 — products ; Over use formaldehyde building materials
PF 3859 — property ; Irresponsible finders
PD 8894 — property ; Neglected
PF 4153 Personal records ; Government misuse
PF 9843 Personal relations world leaders ; Antipathic
Personal relationships
PF 8759 — Corruption
PF 8759 — Dishonest
PF 8759 — Irresponsible
PF 9843 — leaders countries ; Negative
PF 8759 — Unethical
PF 3797 Personal responses ; Speaking opposition out
PF 6536 Personal responsibility ; Restricted
Personal [safety...]
PF 6389 — safety ; Obsession
PC 0927 — savings ; Uninvested
PF 7142 — schedule ; Self centred
PC 3299 — services ; Exploitative
PF 3777 — shallow meaning ; Institutionally reinforced sense
PF 6538 — skills ; Overlooked
PC 8337 — social maladjustment
PD 0894 — social paradigms ; Lack meaningful
PF 8145 — status ; Pursuit
PF 5013 — success ; Jealousy
PF 6504 — survival ; Consuming
PF 2316 Personal time ; Insufficient
PF 2799 Personal transport rural areas ; Lack
PF 4641 Personal unpopularity
PF 6536 Personal viewpoint ; Rigid
Personal wealth
PC 8222 — [Personal wealth]
PD 9653 — rulers countries ; Hypocritical accumulation
PC 8222 — Unnecessary
PF 2445 Personalistic use training
PA 0911 Personalities ; Unintegrated individual
PD 0919 Personality ; Changes
Personality cults
PC 1123 — [Personality cults]
PC 1123 — Media
PC 1123 — politics
PE 7931 Personality development exiled children ; Inhibition
Personality disorder
PE 3901 — Avoidant
PE 6696 — Dependent
PE 4561 — Hysterical
PE 5048 — Multiple
PE 7632 — Obsessive compulsive
PE 0435 — Paranoid
PD 9219 — Passive aggressive
PD 9219 — Sadistic
PD 9219 — Schizoid
PD 9219 — Schizotypal
PD 4418 — Self defeating
Personality disorders
PD 9219 — [Personality disorders]
PF 1721 — Anti social
PE 4396 — Borderline
PF 8387 — leadership
PE 7248 — Narcissitic
PA 0911 — Personality ; Dissociation human
PA 0911 — Personality ; Fragmentation human
PE 4561 — Personality ; Histrionic
PE 4561 — Personality ; Labile
PC 1777 — Personality ; Repression elements
PE 1275 — Personation ; False
PF 9576 Personification
PF 7018 Personified evil
PF 0559 Personnel act against problems ; Shortage adequately trained
PD 7602 Personnel ; Brutalization military
PD 7602 Personnel ; Bullying amongst military
PF 1098 Personnel convicted atrocities ; Politically motivated amnesty military
Personnel [Denial...]
PE 4758 — Denial rights wounded military
PF 0903 — developing countries ; Lack trained administrative
PD 1291 — developing developed areas ; Emigration trained
PE 7010 — Disadvantaged status international nongovernmental organization
PE 5579 — Drug abuse military
PE 8881 Personnel ; Excessive use force military
PE 4785 Personnel exchanges cultural cooperation ; Obstructions international
PE 9294 Personnel gases ; Health hazards anti

PF 3168 Personnel hiring policies ; Rigid
PF 1098 Personnel homicide ; Absolving military
Personnel [Inadequate...]
PE 8728 — Inadequate conditions work employment public service
PF 2995 — Inadequate maintenance
PE 8093 — industrialized countries ; Emigration trained
Personnel Insufficient
PD 0366 — health
PF 4832 — medical facilities
PE 4920 — military
PE 2274 — supervisory
PF 3434 Personnel international organizations ; Lack continuity amongst
PE 1553 Personnel international organizations programmes ; Alienation skilled committed
PD 0366 Personnel ; Limited health
PF 4126 Personnel ; Maldistribution health
PF 2421 Personnel ; Non resident school
PF 3588 Personnel ; Over specialized supervisory
PE 4119 Personnel participation torture ; Military police
Personnel practices
PD 0862 — Corruption
PD 0862 — Irresponsible
PD 0862 — Unethical
Personnel [Racism...]
PD 3519 — Racism amongst military
PF 3945 — Restrictions use
PF 9786 — reward systems ; Counter productive
PE 1396 — routine maintenance complex systems ; Wastage highly skilled
PE 8495 Personnel troops ; Drunkenness military
PF 3434 Personnel turnover international organizations ; Rapid
Personnel [Untrained...]
PD 8388 — Untrained clerical
PD 8388 — Untrained security
PE 9294 — use asphyxiating gases ; Anti
PE 9294 — use toxic substances peacetime ; Anti
PD 1718 Perspective ; Lack
PF 6765 Perspective leaders ; Passive historical
PF 7967 Perspective ; Narrow world
PF 9428 Perspectives ; Government promotion unrealistic
PE 3831 Perspectives industrialized cultures media ; Excessive portrayal
PE 3831 Perspectives media ; Inadequate coverage developing country
PE 4481 Perspiration
PE 5173 Perthes disease ; Legg
PA 5838 Perturbation Agitation
PA 7361 Perturbation Disorder
PA 6030 Perturbation Fear
PA 6247 Perturbation Inattention
PA 7309 Perturbation Uncertainty
PE 2481 Pertussis
PF 1097 Pervasive distrust government
PC 3541 Pervasive fear nuclear war
PD 9219 Pervasive pattern deficits interpersonal relatedness
PD 7463 Perverse impact technological development capitalist society
PB 0869 Perversion
PB 0869 Perversion ; Dependence
PA 6790 Perversion Distortion
PA 6180 Perversion Error
PA 6088 Perversion Impairment
PA 6852 Perversion Inappropriateness
PF 8479 Perversion justice
Perversion [Miseducation...]
PA 6393 — Miseducation
PA 6741 — Misinterpretability
PA 6644 — Misrepresentation
PA 5502 Perversion Reason
PF 3375 Perversion religion
PD 2198 Perversion ; Sexual
PA 7411 Perversion Uncommunicativeness
PF 5190 Perversions ; Musical
PA 6698 Perversity Difference
PA 7253 Perversity Envy
PA 7325 Perversity Irresolution
PB 0869 Perverted people
PA 6099 Pessimism Hopelessness
PA 6731 Pessimism Solemnity
PF 2818 Pessimism ; Unwarranted
PF 6475 Pessimistic approach agricultural development
Pest control
PD 8426 — Animal
PF 3409 — Insufficient
PF 3409 — Limited
PD 3522 Pest infestations buildings
PD 3696 Pest resistance pesticides
PD 2730 Peste petits ruminants
PF 2899 Pesticide application ; Prohibitive cost
PD 0120 Pesticide contamination food chains
PD 2581 Pesticide damage crops
PD 3574 Pesticide destruction soil fauna micro-organisms
PD 3680 Pesticide hazards wildlife
PE 2349 Pesticide intoxication
PE 2349 Pesticide poisoning
PD 5228 Pesticide poisoning small animals
PE 6480 Pesticide residues food
PD 0120 Pesticide residues pollutants
PD 0120 Pesticides
PD 0120 Pesticides ; Bio accumulation
PD 0120 Pesticides ; Chemical
PE 6480 Pesticides ; Food contamination
PE 1785 Pesticides ; Inappropriate agricultural subsidies sales
PD 1245 Pesticides mist

Pesticides

PD 4629	Pesticides ; Misuse		**Pet animals cont'd**	PF 6221	Philanthropy ; Decline
PD 3696	Pesticides ; Pest resistance	PF 1085	— Disproportionate allocation resources	PF 5039	Philosophical ignorance
PD 0120	Pesticides pollutants	PE 0805	— Kidnapping	PF 2334	Philosophy controlled risk ; Economic
PE 4853	Pesticides ; Shortage	PD 1265	— Maltreatment	PF 3392	Philosophy ; Dependence
	Pesticides Use	PE 0805	— Ransoming	PD 9154	Phimosis
PD 0120	— chemical	PE 4771	— Unethical practices	PD 7799	Phimosis animals
PD 4629	— illegal	PD 2094	— Unwanted	PE 9037	Phlebitis intracranial venous sinuses
PD 4629	— inappropriate	PE 7175	— Violent	PC 9042	Phlebitis ; Puerperal
PE 8729	Pesticides ; Water pollution	PD 2732	Petechial fever ; Bovine	PE 6354	Phobia
PA 6799	Pestiferousness *Disease*	PD 7609	Petition authority ; Denial right		**Phobia [School...]**
PA 7226	Pestiferousness *Unhealthfulness*	PD 2730	*Petits ruminants* ; Peste	PE 4554	— School
PA 7107	Pestiferousness *Unpleasantness*	PE 0538	Petrochemical industry ; Instability chemical	PF 6910	— Sexo
PC 0728	Pests	PC 1192	Petrochemicals industries ; Environmental hazards chemicals	PE 6354	— Simple
	Pests [Animal...]			PE 6374	— Social
PD 8426	— Animal	PE 1483	Petrochemicals industry ; Underdevelopment chemical	PF 9127	Phobia ; Techno
PE 3613	— Aphids	PE 1161	Petrol ; Lead	PF 5758	Phobia ; Weight
PE 3986	— Arachnida		**Petroleum [coal...]**	PE 1656	*Phoca* ; Endangered species
PE 3644	— Army worms	PE 7960	— coal products manufacture underdevelopment	PC 1867	Phone installations ; Unreliable
	Pests [Bee...]	PC 7517	— consumption ; Inability reduce	PD 1632	Phone tapping ; Illegal
PD 3634	— *Bee*	PE 8080	— crises ; World tension oil	PF 6218	Phonetic spelling ; Non
PE 1679	— Beetles insect	PE 1409	Petroleum ; Environmental hazards	PJ 5588	Phoney
PD 1689	— Birds	PE 8464	Petroleum natural gas ; Instability trade mineral tar crude chemicals derived coal,	PD 5550	Phoney non prescription drugs
PE 3646	— Black flies			PE 4555	Phosgene
PE 3627	— Blowflies	PE 8955	Petroleum natural gas production instability	PE 1313	Phosphates pollutants
PE 3635	— Botflies		**Petroleum [petroleum...]**	PD 3663	Photochemical oxidant formation
	Pests [Cockroaches...]	PD 0909	— petroleum products ; Instability trade	PD 3663	Photochemical smog
PE 1633	— Cockroaches	PE 8626	— petroleum products ; Long term shortage	PF 9707	Photographic bias
PE 3648	— Corn borers	PE 0760	— precious stones ; Instability trade crude fertilizers crude minerals, excluding coal,	PD 3086	Photographic propaganda
PE 3640	— Crickets			PE 8470	Photographical optical goods trade instability
PC 0728	— *Crustacean*	PE 1353	— precious stones ; Long term shortage crude fertilizers crude minerals, excluding coal,	PD 2337	Photography ; Censorship
PE 3618	Pests ; Dermaptera insect			PE 2665	Photophobia
	Pests diseases	PD 0909	— prices ; Instability	PD 7420	Photosensitization animals
PE 2983	— chestnut	PD 0909	— products ; Instability trade petroleum	PG 5561	Photosensitization sheep ; Congenital
PE 2976	— citrus fruit	PE 8626	— products ; Long term shortage petroleum	PD 0566	Phthisis
PE 2979	— cocoa	PC 1192	Petroleum refineries ; Environmental hazards	PE 2455	Phycomycosis
PE 2981	— *coconut palm*	PD 2094	Pets ; Abandoned animals	PE 3611	Phylloxera pests
PE 2218	— coffee	PE 4931	Pets ; Bites animal	PE 7826	*Physeal dysplasia*
PE 2220	— cotton	PD 1179	Pets ; Capture use wild animals		**Physical [activity...]**
PE 7783	— Crops	PF 1085	Pets ; Luxury treatment	PC 0716	— activity ; Lack
PE 3591	— deciduous fruit	PD 2689	Pets ; Proliferation	PF 6182	— activity urban environments ; Unhealthy lack daily
PE 2982	— elm	PE 0805	Pets ; Theft		
PD 8567	— fish	PA 6839	Pettiness *Disrepute*	PD 1239	— agents ; Carcinogenic chemical
PE 3590	— grain sorghum	PA 5652	Pettiness *Inferiority*	PA 0587	— *aggression*
PE 7783	— *groundnut*	PA 7306	Pettiness *Narrowmindedness*	PF 1733	— *appearance* ; *Debilitating*
PE 3589	— Maize	PA 7211	Pettiness *Selfishness*	PD 0568	Physical blindness
PE 2984	— oak	PA 5942	Pettiness *Unimportance*	PD 0584	Physical brutality psychiatric hospitals
PE 2978	— olives	PD 5552	*Petty larceny*	PD 1955	Physical condition art objects ; Deterioration
PE 2981	— palms	PD 5552	Petty theft	PE 5763	Physical confinement ; Disease injury
PE 7832	— perennial irrigation ; Increase	PD 2615	Petty treason		**Physical [decline...]**
PE 4413	— poplars	PA 6011	Petulance *Discontentment*	PF 7815	— decline human race
PE 2219	— potato	PA 7253	Petulance *Envy*	PD 3773	— developmental defects
PE 2221	— rice	PA 7325	Petulance *Irresolution*	PD 6020	— disability
PE 2977	— rubber	PA 5479	Petulance *Lamentation*	PE 3733	— disablement ; Malicious
PE 2975	— sugar-beet	PE 5550	Pfeiffer's disease	PD 8076	— disfigurement ; Personal
PE 2217	— sugar cane	PE 8537	Phalangers cuscuses ; Endangered species	PD 0270	— *disorder* ; *Alteration physical functioning suggesting*
PE 2980	— tea		**Phalanx Fracture**		
PD 3585	— trees	PE 7826	— distal	PD 0270	— *disorder* ; *Loss physical functioning suggesting*
PE 2985	— vines	PE 7826	— first		
PE 2222	— wheat	PE 7826	— second	PB 1044	— *disorders*
PD 1207	Pests ; Excessive use chemicals control	PE 7826	— third	PE 3974	— disorders ; Psychogenic
	Pests [Flies...]	PA 0643	*Phantom limb pain*		**Physical [emotional...]**
PE 2254	— Flies insect	PE 4658	Phantom workers	PJ 3545	— emotional strain ; *Mental*,
PD 3585	— Forest	PF 3801	Phantoms	PD 2672	— environment ; Debilitating deterioration
PE 3607	— Fruit flies	PE 2390	Pharmaceutical advertising ; Irresponsible	PF 7291	— evidence ; Tampering
PE 3642	Pests ; Grasshoppers insect	PE 6036	Pharmaceutical drugs developing countries ; Marketing banned	PD 4475	— exertion ; Reduced
	Pests [Hemiptera...]			PD 0134	— expansion cities developing countries ; Uncontrolled
PE 3615	— Hemiptera insect	PE 1278	Pharmaceutical drugs ; Limited access		
PE 3614	— Homoptera insect	PE 6618	Pharmaceutical industry ; Corporate crime		**Physical [feebleness...]**
PE 3609	— Houseflies	PE 4120	Pharmaceutical industry developing countries ; Underdevelopment	PE 2332	— *feebleness*
PD 3522	— Household			PD 0270	— *functioning suggesting physical disorder ; Alteration*
	Pests [industry...]	PE 0564	Pharmaceutical manufacture distribution ; Unauthorized		
PD 1607	— industry ; Microbial	PE 8502	Pharmaceutical medical devices industries ; Ineffective self regulation	PD 0270	— *functioning suggesting physical disorder ; Loss*
PC 1630	— Insect			PD 1618	Physical genetic abnormalities ; Human
PF 3592	— Introduction new species insect	PE 4120	Pharmaceutical needs developing countries ; Non responsiveness transnational corporations	PD 5177	Physical growth ; Inhibited human
	Pests [Lack...]				**Physical [health...]**
PD 1207	— Lack biological control		**Pharmaceutical practices**	PC 0716	— health ; Deterioration
PE 3660	— Leeches	PF 1624	— Inconsistent	PF 1773	— health ; Inadequate maintenance
PE 3649	— Lepidoptera insect	PF 3540	— Irresponsible	PD 5177	— height ; Inadequate human
PE 1439	— Lice insect	PF 3540	— Unethical	PE 9789	— hyperactivity
	Pests [Mites...]		**Pharmaceutical products**		**Physical [ill...]**
PE 3639	— Mites	PE 2390	— Aggressive promotion	PD 1043	— ill health ; Susceptibility old
PE 7832	— modification micro climate ; Increase insect	PE 9611	— Cruel treatment animals testing	PF 0836	— *improvements* ; *Unimportance*
PC 0728	— *Mollusc*	PE 4120	— developing countries ; Inadequate supply	PF 6773	— independence handicap
PE 3641	Pests ; Orthoptera insect	PE 4120	— developing countries ; Inappropriate	PD 4004	— infrastructure health care delivery ; Insufficient
	Pests [Phylloxera...]	PE 1274	— Distortion international trade discriminatory formulation health sanitary regulations agricultural		
PE 3611	— Phylloxera				**Physical insecurity**
PC 1627	— *plants*	PE 5755	— Overcharging proprietary	PE 5391	— developing countries ; Personal
PD 3634	— plants ; Insect	PD 9807	— Side effects	PE 4639	— industrialized countries ; Personal
PE 6741	— Protozoa	PE 7996	— trade instability ; Medicinal	PD 8657	— Personal
PE 3177	Pests ; Rats	PF 3540	Pharmacists ; Negligence	PD 6364	— refugees asylum-seekers
PE 2537	Pests ; Rodents	PE 4696	Pharmacological torture	PE 7370	— university campuses
	Pests [Scale...]	PG 9838	*Pharyngeal catarrh* ; *Naso*	PE 7750	— women ; Personal
PE 3612	— Scale insects	PD 3978	*Pharyngeal paralysis*		**Physical [intimidation...]**
PE 8150	— Screw worm flies	PE 7307	Pharyngitis animals	PC 2934	— intimidation
PE 3643	— Siphonaptera insect	PE 3098	Pharyngitis ; Streptococcal	PE 2876	— intimidation children
	Pests [Teredos...]	PE 9819	Pharynx ; Cancer	PF 9023	— isolation
PE 3624	— Teredos	PE 4651	Pharynx ; Inflammation	PF 6683	Physical locations ; Unpropitious
PE 1747	— Termites	PD 7841	Pheasants ; Eastern encephalitis		**Physical [malformation...]**
PE 3619	— Thysanoptera insect	PD 2730	Pheasants ; Marble spleen disease	PE 2042	— malformation foetus
PE 3620	— Thysanura insect	PF 8352	Phenomena ; Inexplicable	PC 2584	— maltreatment children
PE 1766	— Ticks	PF 2409	*Phenomena* ; *Scientific explanations*	PC 0663	— mental health ; Denial right highest obtainable
PE 1335	— Tsetse flies	PF 8352	Phenomena ; Unexplained		
PE 3645	Pests ; Wireworms	PE 9291	*Phenylketonuria*	PA 0294	— mental ill-health
PD 3586	Pests wood ; Insect	PF 3049	Philanthropic charitable organizations ; Inappropriate taxation not profit		**Physical [pleasures...]**
	Pet animals			PF 2277	— *pleasures* ; *Overindulgence*
PE 7175	— Dangerous	PE 8742	Philanthropic organizations ; Unethical practices	PD 2838	— *posture* ; *Unhealthy*
				PF 1978	— psychological violence ; Covert non
				PF 9023	*Physical remoteness*
				PF 0462	*Physical restraint*
				PF 2971	*Physical stamina* ; *Short term*
				PD 4143	Physical structures ; Collapsing
					Physical suffering
				PB 5646	— Dependence human

Plants animals

Physical suffering cont'd
PB 5646 — Human
PD 7550 — Self inflicted
PC 0300 Physical symptoms ; *Intentional feigning*
PD 8734 Physical torture
PF 4010 Physical unattractiveness ; *Personal*
PD 4475 Physical unfitness
PD 5173 Physical warmth ; *Lack*
Physically [dependent...]
PD 7238 — dependent people
PD 8627 — disabled ; Discrimination against
PD 8314 — disabled ; Lack facilities
Physically handicapped
PD 0196 — children
PE 2087 — Family rejection
PD 6020 — persons
PC 4020 Physically inaccessible resources
PC 7674 Physically inaccessible services
PC 0699 Physically mentally handicapped persons
PD 1405 Physically mentally handicapped ; Stress families
PD 6334 Physician-induced diseases
PE 8303 Physician ; Unaffordable fulltime
PE 5746 Physicians' ability ; Impairment
PE 7568 Physicians ; Smoking
PE 7568 Physicians ; Substance abuse
PE 8303 Physicians ; Unavailability local
PD 5770 Physicians ; Unethical practice
PD 1710 Physicists ; Corruption
PD 1710 Physicists ; Irresponsible
PD 1710 Physicists ; Negligence
PD 1710 Physics ; Malpractice
PD 1710 Physics ; Unethical practice
PD 2224 Physiogenic plant diseases
PE 7161 *Physiological adaptability women ; Poor*
PE 8925 Physiological malnutrition arising mental factors
PD 2224 Physiological plant diseases
PF 4417 Physiological processes ; *Unsociable human*
PE 7826 *Physitis*
PE 5187 *Pica*
PF 6473 Pick up ; *Undependable garbage*
PE 8712 Picket ; Denial right
PA 0429 *Picketing ; Violent*
PE 0559 Pickpocketing
PF 0981 Pictorial arts
PE 2691 Picture story books ; *Harmful effects comic strips*
PF 6580 Piecemeal approach problem-solving
PF 3979 Piety ; *Erosion filial*
PD 9667 *Pig disease ; Greasy*
PD 7420 Piglets ; *Hypoglycaemia*
PE 7808 Piglets ; *Splaylegs*
PD 7841 Piglets ; *Viral encephalomyelitis*
PD 1762 *Pigmy possums ; Endangered species*
PD 7841 Pigs ; *Dancing*
PD 2730 Pigs ; *Encephalomyocarditis virus disease*
PD 5228 Pigs ; *Iron dextran toxicity newborn*
PD 9667 Pigs ; *Pityriasis rosea*
PD 7841 Pigs ; *Shaker*
PD 3480 *Pikas ; Endangered species*
PE 3504 Piles
PE 3504 *Piles ; External*
PE 3504 *Piles ; Internal*
PE 3504 *Piles ; Mixed*
PD 4691 Pilferage
PE 5103 Pilferage theft ; Freight
PF 2346 *Piling removal ; Difficult*
PE 4152 Pillage
PE 1821 Pills ; Abuse sleeping
PD 1038 Pills ; Increased use birth control
PE 3870 Pilot fatigue ; Aircraft
PD 0680 Pilots ; Drug use
PE 5303 Pimps
PE 6255 *Pine blister rust ; White*
PE 5402 Pink messaging services
PD 7420 *Pink tooth*
PD 2730 Pinkeye
PD 5535 Pinkeye
PD 5535 *Pinna ; Dermatitis*
PE 1656 *Pinnipedia*
PE 3254 *Pinta*
PF 1724 Pioneer institutions developing countries ; Unrelated
PF 2358 *Pioneer spirit ; Lack*
PE 6251 Pipelines ; Environmental degradation
PE 3624 Pipeworms
PD 1877 Piracy
PD 0124 Piracy ; Aerial
PD 0188 *Piracy ; Commercial*
PE 6625 *Piracy ; Computer*
PF 8854 *Piracy ; Industrial*
PF 8854 *Piracy ; Intellectual*
PD 0188 *Piracy ; Literary*
PD 7981 *Piracy ; Product*
PD 8438 *Piracy sea*
PE 6625 *Piracy ; Software*
PD 0188 *Pirates ; Copyright*
PD 2596 Pit mines quarries ; Environmental destruction open
PF 8713 Pitch ; *Competitive distortion musical*
PF 8713 Pitch ; *Distortion musical quality escalating*
PE 9458 *Pithomycotoxicosis*
PA 7023 Pitilessness
PA 7156 Pitilessness *Moderation*
PA 7023 *Pitilessness Pitilessness*
PA 5643 Pitilessness *Unkindness*
PE 2286 Pituitary gland disorders
Pituitary [tissue...]
PE 2286 — tissue ; Injury
PE 2286 — tissue ; Tumours

Pituitary [tumours...] cont'd
PD 9654 — *tumours animals*
PE 6325 *Pityriasis rosea*
PD 9667 *Pityriasis rosea pigs*
PD 2545 *Pityriasis versicolour*
PD 7799 *Pizzle rot*
PE 1106 Place thefts ; Office work
Placement burden proof
PF 3918 — abuse vulnerable groups
PF 3918 — disempowered
PF 3918 — victims environment pollution
PD 0606 Placement ; Cinematic product
PD 7799 *Placenta animals ; Retained*
PD 0173 Placental anomalies
PD 3996 Plagiarism
PD 1284 Plagiarism ; Conceptual
PD 3996 *Plagiarizing*
PE 0987 Plague
PE 0987 Plague ; Bubonic
PE 2786 Plague ; Cattle
PD 3978 Plague ; Duck
PE 0987 Plague ; Pneumonic
PE 0987 Plague ; Septicaemic
PE 3643 Plague ; *Sylvatic*
PE 0725 Plagues ; Grasshopper
PE 0725 Plagues ; Locust
PE 0743 Plain settlement ; Flood
PE 0743 Plains ; Building flood
PE 0743 Plains ; Construction flood
PF 2865 Plan ; Absence long range, world wide capital flow
PF 4984 *Plan ; Inadequate pre school*
PF 3448 *Plan ; Inadequate reservoir*
PF 1822 *Plan ; Overburdened production*
Plan [Uncomprehensive...]
PD 0516 — Uncomprehensive river
PE 7695 — Undesigned tree planting
PF 8176 — Undeveloped forward
PF 8176 — Undeveloped future
PF 9478 — *Unimaginative city*
PF 9455 Planet ; Irresponsibility long term future
PC 0918 Planet ; Overheating
PF 1744 Planetary glaciation ; *Cyclical*
PF 7741 Planned degradation product quality
PC 3894 Planned economies ; Failure centrally
PD 0515 Planned economies ; Lack consumer choice centrally
PC 2008 Planned obsolescence
PF 9269 Planned weapons
PF 2895 Planners ; Arrogance
Planning [Absence...]
PF 2804 — Absence accountability construction
PF 1467 — action against problems ; Inadequate
PF 3610 — agencies ; Absence long term economic
PF 3211 — *authorities ; Dependence*
PF 0954 Planning ; Boundary constraints land
PF 2813 Planning community life ; Fragmented
PF 6530 Planning ; Crippling subsistence
Planning [decision...]
PF 5660 — decision making ; Lack reward long range
PE 5226 — Denial right family
PD 1141 — developing countries ; Defective land use
PF 4837 — developing countries ; Imbalance economic social
PF 8931 — Difficulties cooperative
PF 0148 — disinterest ; Family
PE 1980 — disrelationship political industrial sectors
PE 4479 Planning economy ; Static grassroots involvement
Planning education
PE 0341 — children ; Negative effects family
PE 4241 — facilities ; Unequal global distribution family
PD 1039 — Inadequate family
PD 1039 — Lack family
PE 9060 — men ; Inadequate family
Planning [expansion...]
PF 1345 — expansion income ; Absence imaginative
PF 2477 — *experience ; Limited*
PF 1055 — expertise development ; Failure global scale
PF 1036 Planning facilities ; Discrimination family
PF 9166 Planning ; Fragmented resource
PF 1010 Planning implementation ; Uncertain
Planning Inadequate
PD 1038 — health care family
PF 1467 — *manpower*
PF 0963 — procedures community
Planning [Individualistic...]
PF 1205 — *Individualistic agricultural*
PF 0148 — Ineffective family
PD 1050 — information ; Lack family
PF 1959 — *Insufficient industrial*
PF 5409 — Insufficient recreational
PF 2804 — Irresponsible construction
Planning Lack
PF 2605 — community
PF 3731 — demolition
PF 0148 — family
PF 1467 — *manpower*
PF 8185 — social
Planning Limited
PF 3500 — approaches economic
PF 5660 — long range
PF 2813 — scope town
PF 5660 Planning long term development ; Short range
PD 3563 Planning natural resource extraction ; Parochial
PS 8148 *Planning ; Non local educational*
Planning Obsolete
PF 5009 — [Planning ; Obsolete]
PF 5009 — *defence*
PF 5009 — security

PF 1021 Planning ; Opposition population control family
PF 7737 Planning ; Over
Planning [paralysis...]
PF 3513 — paralysis ; Corporate
PF 6550 — *process ; Incomplete*
PF 1740 — product life cycles ; Short term
PD 1039 Planning research ; Inadequate family
Planning [second...]
PF 5958 — second language policy ; Antisocial attitudes
PF 6456 — Self defeating style community
PF 5660 — Short sighted
PF 1740 — Short sighted business
PF 0148 — Sporadic family
PF 2540 — structures small communities ; Lack central
PF 2885 Planning training programmes ; Social bias
Planning Uncoordinated
PF 1205 — *agricultural*
PF 7619 — government
PD 4790 — health care
PF 0963 — *joint*
PC 1217 — production
Planning [Unformed...]
PF 1959 — *Unformed industrial*
PF 9478 — Unimaginative future
PF 1442 — *Uninformed budget*
PF 2477 — *Uninformed career*
PF 0148 — Unmotivated family
PJ 9394 — Unproductive grass roots
PF 2018 — Plans ; Delayed development regional
PF 2813 — Plans ; *Nebulous city*
PF 1010 Plans programmes against problems ; Inadequate implementation
PF 2018 Plans ; Protracted development regional
Plans [Undeveloped...]
PF 5660 — Undeveloped long range
PF 2813 — *Unfunded educational*
PF 6550 — Unstructured beautification
PD 9480 Plant abnormalities
PE 1899 Plant cancer
PE 9773 Plant cultivation ; Regulatory banning
Plant disease
PD 3601 — Bird vectors
PE 8138 — Fungi vectors
PD 3593 — Human vectors
PE 9017 — Mites vectors
PD 3599 — Plant vectors
PD 3596 — vectors
Plant diseases
PC 0555 — [Plant diseases]
PD 3601 — Avian vectors
PD 2226 — Bacterial
PD 2224 — Environmental
PD 2225 — Fungal
PF 7732 — Insect vectors
PD 3593 — man ; Dissemination
PD 2228 — Nematoid
PD 2224 — Nonparasitic
PD 2224 — Physiogenic
PD 2224 — Physiological
PD 2226 — *Rhizobium*
PD 3587 — storage transit
PC 0555 — *Uncontrolled*
PD 2227 — *Viral*
PD 0022 Plant drugs ; Abuse
Plant [genetic...]
PD 9480 — genetic defects ; Increase
PF 3581 — genetic resources conservation ; Inadequate
PF 3146 — growth regulators
Plant life
PD 9480 — Deformation
PD 5157 — due radioactive contamination ; Endangered animal
PF 6337 — *information ; Insufficient*
PD 0779 Plant ; Low capacity utilization manufacturing
Plant [pathogenic...]
PE 0155 — pathogenic air pollutants
PD 1866 — pathogens
PC 0238 — protection ; Lack
PE 0714 Plant quarantine ; Inadequate
PE 3946 Plant remedies ; Ignorance traditional
PE 6788 Plant seeds ; Restricted access
PE 3537 Plant shutdowns
Plant species
PE 9773 — Banned cultivation
PD 4100 — Dependence
PE 1444 — Irresponsible introduction new
PE 6396 — Substitution fast growing
PF 5457 — Plant tumours
PD 3599 — Plant vectors plant disease
PD 7598 — Plantation agriculture
PE 2254 *Plantation flies*
PF 6455 Plantation plots ; Small
PF 6455 Plantation space ; Unproductive utilization
PD 7598 Plantation system ; Capitalist
PE 4243 Plantation workers ; Inadequate working conditions
PE 7695 Plantations ; Inappropriate tree
PC 0238 *Plantations long lifed trees ; Endangered*
PF 9187 Plantations ; Loss land
PD 8716 Planting evidence
PE 7695 Planting plan ; Undesigned tree
Plants [Accumulation...]
PD 0381 — Accumulation contaminant residues terrestrial
PD 0381 — Accumulation pollutants terrestrial
PE 4346 — *Alternaria diseases*
Plants animals
PD 5021 — Accumulation contaminant residues
PD 5021 — Accumulation pollutants

Plants animals cont'd
PC 0776 — Creation transgenic
PB 1395 — Endangered species
PE 8717 — Loss beneficial
PE 1910 Plants ; Anthracnose diseases
Plants [Bacterial...]
PE 4706 — Bacterial soft rot
PE 1607 — Black mildew
PE 4235 — Blotch diseases
PE 4346 — *Botrytis diseases*
PD 5730 Plants crops ; Vulnerability
PD 4148 Plants ; Cruelty
Plants [Damping...]
PE 4787 — Damping off disease
PD 5936 — Declining breeds cultivated
PC 2223 — Decreasing genetic diversity cultivated
PD 3653 — Deficiency diseases
PD 9480 — Deformed
PE 5260 — Deterioration nuclear power
PE 0501 — Downy mildews
Plants Endangered species
PC 0238 — [Plants ; Endangered species]
PE 4314 — flowering
PD 4171 — medicinal
PE 7539 Plants ; Environmental hazards decommissioned nuclear power
PE 8453 Plants ; Environmental impacts coal conversion
PE 4346 *Plants ; Fusarium diseases*
PE 3715 Plants ; Galls
Plants [Harmful...]
PC 0728 — Harmful animals
PD 2718 — Hazardous locations nuclear power
PD 5706 — Hazards
PE 4744 — Health hazards air pollution
Plants [Insect...]
PD 3634 — Insect pests
PD 3600 — Insect vectors viral diseases
PE 6788 — Irresponsible patenting genetically transformed
PD 4217 Plants ; Killing
PE 7519 Plants ; Lack acceptable sites power
PE 4586 *Plants ; Lichen*
PE 8051 Plants ; Moulds
PE 4328 Plants ; Nitrogen overdosage
Plants [Parasites...]
PD 4659 — Parasites
PD 6284 — Parasitic
PC 1627 — Pests
PD 2291 — Poisonous
PD 7663 — Poor viability nuclear power
PE 0529 — Powdery mildews
PE 3865 — Premature drying
PE 2866 — processes ; Ageing industrial
PF 7543 — Prohibitive cost nuclear power
Plants [Radioactive...]
PD 0710 — Radioactive contamination
PE 3363 — Rots
PE 6255 — Rusts
Plants [Scab...]
PE 1040 — Scab diseases
PE 4586 — *Sclerotinia diseases*
PE 1371 — Slime moulds
PE 0857 — Smut diseases
PE 0916 — Sooty mould
PE 4586 — *Spot anthracnose*
PC 7825 — Suffering
PD 4148 Plants ; Torture
PF 2231 *Plants ; Unrealized potential gober gas*
PE 4346 *Plants ; Verticillium wilt*
PD 2227 Plants ; Virus diseases
Plants [Waste...]
PE 7539 — Waste resources obsolete nuclear power
PE 5682 — Weather hazards crops
PE 5465 — *White rusts*
PE 1056 — Wilt diseases
PD 8566 Plastic materials ; Environmental hazards
PE 8651 Plastic products ; Environmental hazards manufacture
PE 4429 Plastic surgery ; Incompetent
PE 4429 Plastic surgery ; Malpractice
PE 3741 *Plastic waste ; Marine pollution*
PD 1180 Plastic waste ; Non biodegradable
PE 5143 Plastics industry ; Instability
PE 7956 Platinum group trade instability ; Silver
PE 0353 *Platinum ores shortage*
PE 8556 Platinum ores trade instability ; Silver
PD 3474 Platyhelminthes ; Endangered species
PF 2152 Plausible arguments conceal nefarious benefits ; Misuse
PD 0549 Play areas ; Unprotected
PF 2017 Play ; Attitude manipulation children
PD 0549 Play ; Inadequate facilities children's
PF 3973 Play role developing future ; Inability elderly
PD 0549 Play ; Unsupervised children's
PD 0549 Play ; Unsupervised road
PD 6851 Playground accidents
PE 7768 Playgrounds ; Harassment
PE 2605 *Playing field ; Unfinished*
PD 3338 *Playing ; Social role*
PD 3028 Plays ; Banned
PF 1216 *Pleasure ; Incapacity*
PA 6731 *Pleasurelessness Solemnity*
PA 7107 *Pleasurelessness Unpleasantness*
PF 2277 *Pleasures ; Overindulgence physical*
PF 4558 Pledges ; Breach administrative
PF 4558 Pledges ; Broken government
PF 4558 Pledges ; Government reneging public
PE 8228 Pleura peritoneum cancers bronchi

PD 0637 *Pleurisy*
PD 0637 *Pleuritis*
PD 2730 Pleuritis ; Feline infectious
PE 1775 Pleuropneumonia ; Contagious bovine
PE 6293 Pleuropneumonia ; Contagious caprine
PE 7553 Pleuropneumonia swine
PB 7125 Plots ; Manipulative international
PF 6475 Plots ; Overdivided farmland
Plots Small
PF 3211 — *farm*
PF 2135 — *farm*
PF 3211 — *land*
PF 6455 — *plantation*
PE 4586 Plum ; Black knot
PE 8272 Plumbing, heating, lighting fixtures fittings ; Instability trade sanitary
PE 7940 Plumbing, heating lighting fixtures fittings ; Shortage sanitary
PF 6527 *Plumbing systems ; Freezing*
PF 2448 Plural society tensions
PF 0152 Pluralism ; Divisive effects official cultural
PF 2182 Pluralism ; Political
PF 3404 Pluralism ; Religious
PF 2440 Pluralistic ignorance
PF 2448 Pluralistic society
PA 1387 *Plutocracy*
PD 2539 Plutonium overproduction
PE 6285 Plutonium pollution
PE 2280 Pneumococcol meningitis
PD 2034 Pneumoconiosis
PE 1400 Pneumoencephalitis ; Avian
PE 2293 Pneumonia
PD 7307 Pneumonia ; Aspiration
PE 5524 Pneumonia ; Bovine atypical interstitial
PE 5524 Pneumonia calves ; Endemic
PE 7553 Pneumonia ; Endemic
PD 7307 Pneumonia ; Foreign body
PD 7307 Pneumonia ; Gangrenous
PD 7307 Pneumonia ; Hypostatic
PD 7307 Pneumonia ; Inhalation
PE 7553 Pneumonia ; Mycoplasmal
PD 7307 Pneumonia ; Mycotic
Pneumonia sheep
PE 9438 — *Atypical*
PE 9438 — *Nonprogressive*
PE 9438 — *Progressive*
PE 9854 Pneumonia small animals
PD 7307 Pneumonia ; Verminous
PE 5524 Pneumonic pasteurellosis ; Bovine
PE 0987 Pneumonic plague
PD 2664 Poaching
PD 2664 Poaching ; Inadequate enforcement against game
PD 3481 Pocket gophers ; Endangered species
PE 8674 Pocket mice kangaroo mice ; Endangered species
PC 2270 *Podagra*
PE 4161 Pododermatitis aseptica diffusa
PE 4161 *Pododermatitis ; Chronic necrotic*
PE 4161 *Pododermatitis circumscripta*
PE 4161 *Pododermatitis ; Infectious*
PE 7826 *Podotrochlitis*
PE 7826 Podotrochlosis
PA 0429 Pogroms
PA 7365 *Pointlessness Boredom*
PA 6852 *Pointlessness Inappropriateness*
PD 5426 Poison food ; Threats
PE 7645 Poisoners ; Psychotic
PD 0105 *Poisoning ; Alcohol*
PE 4969 Poisoning ; Aluminium
Poisoning animals
PD 5228 — [Poisoning animals]
PD 5228 — Acaricide
PD 5228 — *Algal*
PD 5228 — Arsenic
PD 5228 — Blister beetle
PD 5228 — Bracken fern
PD 5228 — Cantharidin
PD 5228 — Coal tar
PD 5228 — Copper
PD 5228 — Cyanide
PD 5228 — Dioxin
PD 5228 — Fluoride
PD 5228 — Herbicide
PD 5228 — Insecticide
PE 9228 — Lead
PE 1155 — Mercury
PD 5228 — Metaldehyde
PD 5228 — Molybdenum
PD 5228 — Nitrate
PD 5228 — Nitrite
PD 5228 — PBB
PD 5228 — PCB
PD 5228 — Polybrominated biphenyls
PD 5228 — Polychlorinated biphenyls
PD 5228 — Rodenticide
PD 5228 — Salt
PD 5228 — Selenium
PD 5228 — Senecio
PD 5228 — Sorghum
PD 5228 — Strychnine
PE 9374 Poisoning ; Bacterial food
PE 9422 Poisoning ; Blood
PE 7555 Poisoning ; Dioxin
PD 2732 Poisoning disease ; Salmon
PD 5228 Poisoning ; Ethylene glycol
PD 0372 Poisoning ; Fish
PG 3050 Poisoning gases vapours

PD 0105 Poisoning ; Human
PE 2349 *Poisoning ; Insecticide*
PE 5650 Poisoning ; Lead
PD 0105 *Poisoning ; Metal*
PE 0561 Poisoning negligence ; Food
PD 5228 *Poisoning oak acorns*
PD 5228 *Poisoning oak buds*
PG 4624 Poisoning ; Paralytic shellfish
PE 2349 Poisoning ; Pesticide
PD 5228 *Poisoning ; Ragweed*
Poisoning [sabotage...]
PD 5426 — sabotage ; Food
PD 5228 — *small animals ; Pesticide*
PB 0731 — solids liquids
PD 5228 — *Sweet clover*
PD 5426 — Poisoning ; Terrorist food
PE 9228 Poisoning wildlife ; Lead
PE 1935 *Poisonings ; Accidental*
PD 2341 *Poisonings ; Intentional fatal*
Poisonous [algae...]
PE 2501 — algae
PE 6867 — allergenic biologically active wood
PE 0175 — animals
PG 3050 *Poisonous gases*
PD 2291 Poisonous plants
PD 1472 Poisonous substances food stuffs ; Naturally occurring
PA 5454 *Poisonousness Badness*
PA 6088 *Poisonousness Impairment*
PA 7226 *Poisonousness Unhealthfulness*
PD 5277 Poisons ; Agricultural
PD 1472 Poisons ; Natural food
PE 1320 Polar bear species ; Endangered
PE 5993 Polar pollution
PF 1333 Polarization local conflicts
PF 1333 Polarization ; Unresolved local
PF 9691 Polarized protest against problems
PF 2152 Polarizing rhetoric
PC 1142 *Police abuse ; Military*
PD 3543 Police brutality
PF 3550 *Police budget ; Inadequate*
Police [control...]
PF 8652 — control ; Inadequate
PD 2918 — corruption
PE 5037 — crimes during narcotic investigations
PE 4183 — criminal suspects ; Withholding information
PE 8881 Police ; Excessive use force
PD 8716 Police falsification evidence
PF 8378 Police ; Fear
Police forces
PE 4847 — Abuses private
PD 3519 — Racism
PD 9193 — Unethical practices
PF 3550 *Police funds ; Inadequate*
PD 0736 Police harassment
Police [immunity...]
PE 5832 — immunity
PF 8125 — indifference community
PC 1142 — *interrogations ; Abusive*
PD 4811 — intervention meetings ; Violent
PE 2007 — intervention workers meetings ; Violent
PD 0736 — intimidation
PF 8559 Police ; Lack confidence
PD 9193 Police malpractice
PF 8559 Police ; Mistrust
PD 9193 Police negligence
Police [operation...]
PE 3988 — operation
PC 1142 — operations ; Uncontrolled
PE 6331 — operations ; Undercover
Police [personnel...]
PE 4119 — personnel participation torture ; Military
PD 3542 — Political
PC 1142 — power ; Abuse
PF 8652 — protection ; Breakdown
PF 8652 — protection ; Partial
PE 7659 — Public assaults
PE 3553 Police questioning ; Improper
PE 3553 Police questioning ; Oppressive
Police [reprisals...]
PD 0736 — reprisals
PD 0736 — reprisals ; Threat
PF 8125 — reputation unhelpful
PF 8652 — response ; Delayed
PD 0736 — retaliation
Police [search...]
PD 3544 — search ; Unauthorized
PE 6331 — Secret
PF 8652 — services ; Distant
PD 7910 — state
PF 8652 — structures ; Absence
PF 3550 *Police ; Underpayment*
PF 3550 *Police wages ; Low*
PE 5606 Policeman's heel
PF 7033 Policies associate international nongovernmental organizations regional development ; Absence
PF 2553 Policies ; Collectivist
PF 6046 Policies concerning employment ; Restrictive trade union
Policies [Deterioration...]
PC 9584 — Deterioration long term trade
PF 2889 — Detrimental international repercussions domestic agricultural
PF 5964 — developing countries ; Unsustainability macroeconomic
PE 8583 — Discriminatory exchange rate

Political monoculture

Policies [Distortion...] cont'd	**Policy making cont'd**	**Political discrimination cont'd**
PE 0347 — Distortion international trade discriminatory government private procurement	PF 5964 — developing countries ; Inadequate economic	PC 3222 — based illiteracy
PF 3458 — Policies ; Environmentally insensitive government	PF 2344 — Disproportionate influence some individuals national	PC 3221 — politics
PF 5658 — Policies ; Failure development	PF 0531 — errors	**Political [disintegration...]**
Policies Government action	PF 9166 — Fragmented	PC 3204 — disintegration
PF 2199 — against regimes alternative	PF 9166 — Inconsistent	PF 5360 — displacement activity
PF 2199 — against regimes hostile	PF 8846 — initiatives ; Irrelevant	PC 2361 — dissent
PF 2199 — against regimes opposing	PF 8703 — Insider	PF 0732 — dissidents ; Inadequate legal counsel
Policies [Government...]	PF 9796 — Messianic	PF 3394 — doctrine ; Inadequacy
PD 7171 — Government dirty tricks	PF 3709 — methods ; Ineffective	PC 8512 — domination
PE 9295 — Government neglect future problems social environment arising current	PF 5009 — Obsolete	PD 2929 — dossiers individuals ; Maintenance
	PF 5061 — Self destructive government	**Political [education...]**
PF 5000 — governments ; Uncoordinated macroeconomic	PF 0317 — Short term	PF 4984 — *education ; Inadequate*
PD 5984 — Policies ; Imprudent budget	PF 0317 — short term considerations ; Domination government	PE 3647 — elitism
PE 7563 — Policies ; Inability developing countries adopt appropriate exchange rate	PF 8176 — Surprise free	PE 6948 — endorsement ; Trading
	PF 9166 — Uncoordinated	PE 4833 — ends ; Exploitation athletic competition commercial
Policies Inadequate	PF 7619 — Uncoordinated government	PC 1868 — espionage ; International
PF 0245 — credit	PF 8703 — Undemocratic	PD 3017 — evidence ; False
PF 4850 — fiscal	PE 1922 — Undisclosed control major corporations national	PC 2367 — exit ; Illegal
PF 0037 — governmental energy conservation	PE 0135 — Undisclosed influence transnational corporations global	PC 7356 — exploitation
PF 0038 — governmental resource conservation		PF 4761 — exploitation Olympic Games
PF 7713 — intellectual cognitive infrastructure formulate effective global	PF 9428 — Unrealistic	PF 2177 — extremism
	PF 9100 Policy models ; Overgeneralized	**Political [favouritism...]**
Policies [Inappropriate...]	**Policy [responses...]**	PF 8535 — favouritism
PF 5645 — Inappropriate	PF 0817 — responses world economic crisis ; Nationalistic	PA 6030 — *fear*
PF 1559 — Incoherent government	PF 7823 — responsibilities ; Irresponsible delimitation	PD 4846 — feuding
PF 2497 — *Inconsistency international commodity*	PE 4965 — restriction trade developing countries ; Environment	PF 5061 — folly
PF 5000 — industrialized countries ; Mismatch national macroeconomic	PF 1559 — reversals ; Expedient	PF 3216 — fragmentation
	PE 1815 Policy smaller banks ; High risk	PC 1833 — fragmentation urban areas
PE 4325 — Interference transnational banks domestic economic	PE 5104 Policy technical advance ; Antagonism employment	PC 2030 *Political gifts*
PE 4315 — Interference transnational banks' off shore borrowing domestic monetary	**Policy [Unactivated...]**	PD 1886 — Political hostage-taking
	PF 4672 — Unactivated	PF 9050 Political hypocrisy
PF 5000 Policies leading industrialized countries ; Divergences macroeconomic	PF 9450 — Uncertain environmental impact current	**Political [ignorance...]**
	PF 2346 — *Unclear rental*	PC 1982 — ignorance
PE 9295 — Policies ; Neglect environmental consequences government	PF 1559 — Unpredictable governmental	PF 8413 — immaturity
	PF 1559 — Unstable government	PC 3193 — imperialism ; Capitalist
Policies [Political...]	PE 0504 — Polio	PC 3164 — imperialism ; Communist
PF 9428 — Political support unsustainable	PD 7841 *Polioencephalomalacia*	PE 0174 — implementation networks ; Inflexible intermediary
PF 9101 — Prohibitive cost inadequate development	PD 7841 *Polioencephalomyelitis ; Porcine*	**Political impotence**
PF 7472 — Public resentment government	PE 0504 — Poliomyelitis	PF 1389 — [*Political impotence*]
Policies [Restrictions...]	PC 0350 Politic ; Crimes against body	PE 5729 — students
PE 4905 — Restrictions industrial economic development due environmental	**Political [abductions...]**	PE 6263 — youth
	PD 5129 — abductions children	**Political [imprisonment...]**
PF 8749 — Restrictive monetary	PE 4745 — action ; Violation right trade unions engage	PC 0562 — imprisonment
PF 8282 — Restrictive social	PF 3819 — activists ; Official cover up government harassment	PF 5153 — incumbents ; Electoral organization favouring
PF 3168 — Rigid personnel hiring	PD 8877 — aggression	PF 0297 — independence ; Lack
Policies transnational corporations	PJ 5138 — agitation	PF 3194 — independence ; Lack internal
PE 6952 — developing countries ; Exploitative financial	PC 3227 — alienation	PD 1624 — indoctrination
PE 1957 — Disruption domestic social	PE 5210 — *annexation*	PE 1980 — industrial sectors ; Planning disrelationship
PE 2396 — Restrictive pricing	PC 0030 — antagonism ; Religious	PE 7008 — ineffectiveness international nongovernmental organizations
Policies [Unacceptability...]	PC 1917 — apathy	
PF 3826 — Unacceptability conserver society	PC 1917 — apathy ; Dependence	PC 3425 — inequality
PF 4114 — Unacknowledged future costs present	PF 2031 — appointees	PC 3425 — inequality ; Dependence
PF 1559 — Unclear government	PE 5614 — assassination	**Political inertia**
PF 1385 — Unjustified military defence	PF 1075 — asylum ; Denial right	PC 1907 — [Political inertia]
PF 9428 — Unrealistic	PF 1075 — asylum ; Lack individual rights	PC 1907 — Dependence
Policy [Abuse...]	PF 0544 — awareness developing countries ; Imbalance distribution	PF 4713 — Pre electoral
PF 8389 — Abuse government		PD 4798 Political infiltration
PF 1070 — agencies ; Over reliance economic interest groups	**Political [barriers...]**	**Political influence**
PC 4667 — Aggressive foreign	PC 3201 — barriers effective legislation	PB 3209 — influencing politics
PF 5958 — Antisocial attitudes planning second language	PE 1728 — bias official appointments	PB 3209 — traditional vested interests ; Excessive
Policy [contrary...]	PJ 3966 — bitterness	PF 9686 — urban elites ; Disproportionate
PF 5061 — contrary self-interest	PD 2912 — blackmail	**Political [injustice...]**
PF 9216 — cross-conditionality restrictions multilateral development aid	PE 4206 — boycott international sports events	PC 2181 — injustice
	PC 2030 — bribery	PC 2181 — injustice ; Dependence
PF 8757 — current scientific knowledge ; Mismatch public	PD 1943 — burglary	PB 1149 — insecurity
Policy [decisions...]	**Political [campaigns...]**	PD 3835 — insecurity developing countries
PF 2447 — decisions ; Ineffective educational	PE 6311 — campaigns ; Avoidance issues	**Political instability**
PF 5964 — developing countries ; Failure public finance	PE 6311 — campaigns ; Vacuous	PC 2677 — [Political instability]
PE 2890 — difficulties developing countries ; Domestic agricultural price	PD 2931 — candidates ; Bias selection	PC 2677 — Dependence
	PF 3087 — *cartoons*	PD 8323 — developing countries
PF 1559 — dilemmas ; Government	PF 8511 — changes names geographical features	**Political [integration...]**
PF 1010 — directives ; Failure implement	PE 4997 — checks balances ; Inadequate system	PF 3215 — integration ; Inadequate
PF 5766 — disagreement	PC 0632 — civil rights ; Denial	PF 3216 — integration ; Lack
PF 7047 — discrimination against rural areas developing countries	PC 9104 — coercion	PF 0796 — integrity ; Lack
	PD 8992 — competition	PF 2182 — interests small groups ; Narrow
PE 8700 Policy educational difficulties ; Inadequate adaptation	PD 0702 — *concentration camps*	PE 5827 — interference port operations
	PE 3016 — confessions ; Forced	PE 5610 — interference television news coverage capitalist countries
Policy [formation...]	PD 3012 — confiscation property	
PF 1076 — formation ; Cumbersome methods	PC 0368 — conflict	PF 4761 — intervention international athletic exchanges ; Excessive
PF 8703 — formulation processes ; Restricted access	PE 6715 — contributions agents foreign principals	
PD 8468 — formulation ; Unscientifically based	PD 2931 — control elections ; Oligarchic	PE 0032 — intervention transnational corporations
PF 1559 — Frequently changing government	PC 0116 — corruption	PC 2938 — intimidation
PF 4753 — Policy ; Immoral public	PC 0647 — corruption judiciary	PF 5298 — involvement ; Insufficient images
Policy Inadequate government	PC 0350 — crime	PC 7569 — isolation
PF 1559 — [Policy ; Inadequate government]	**Political debate**	PC 7569 — isolationism
PF 5645 — [Policy ; Inadequate government]	PF 4600 — Avoidance substantive issues	**Political [kidnapping...]**
PF 2447 — educational	PF 4600 — Impoverishment	PD 1886 — kidnapping
Policy [Inappropriate...]	PF 4600 — Superficiality	PD 7221 — killings
PF 9187 — Inappropriate cash crop	PF 4600 — Trivialization	PE 5614 — killings
PF 8757 — Inappropriate development	**Political [deception...]**	PF 2401 — *know how ; Limited*
PF 9208 — innovation ; Constraint inherited problems	PF 9583 — deception	**Political [leaders...]**
Policy makers	PF 7535 — decision making ; Lack rational global	PD 4749 — leaders ; Self absorption
PF 2895 — Arrogance	PF 1452 — dependence ; National	PF 7455 — leadership ; Loss international
PD 8696 — Drug abuse	PD 2374 — *dependency*	PF 1389 — *leverage ; Lack*
PF 9576 — Static models social change	PD 9792 — destabilization developing countries	PD 8276 — liberty ; Lack
PF 8703 — Unrepresentative	PD 8673 — development ; Lack	PF 9583 — lies
Policy making	PD 8673 — developments ; Limited	**Political [machinery...]**
PF 8703 — conspiracy	PD 2263 — dialogue ; Distrust	PC 1833 — machinery ; Inadequate urban
PE 4439 — contexts ; Employment criminals	PC 1547 — dialogue review ; Inadequate structures	PC 5517 — malpractice
PE 4439 — contexts ; Employment war criminals	PC 0845 — dictatorship	PD 1211 — market allocation ; Non
PF 5660 — Crisis management approach long term development	PF 2013 — *discontent*	PD 5590 — mass murder
	PF 2013 — *discontent*	PD 5207 — media events
PF 8989 — delays	**Political discrimination**	PD 1075 — minority status
PF 9664 — delays response accelerating structural change	PC 0934 — [Political discrimination]	PF 1440 — miscalculation ; International
PF 8278 — Deliberate ignorance during	PE 1828 — administration justice	PF 4405 — monoculture

—1011—

Political murder

Political [murder...] cont'd
PE 5614 — murder
PD 7221 — murders authorized governments
PD 6963 — myopia
Political [nature...]
PF 4175 — nature development issues
PD 2213 — networks ; Inadequate
PE 4031 — neutrality civil servants ; Lack
PD 3238 Political oligarchy
PC 1897 Political opportunism
Political opposition
PD 3015 — [Political opposition]
PF 2628 — administrative action
PF 3821 — Disorganized
PC 1919 Political oppression
Political organization
PC 1015 — Dependence undemocratic
PD 0055 — National governments obstacle representative democratic
PC 1015 — Undemocratic
PF 4110 Political over-reaction
Political parties
PD 2906 — Abuse electoral process
PD 3536 — Banned
PC 2494 — *Clash personal ambitions*
PC 0116 — Corruption
PE 9110 — Denial right organize
PD 0548 — developing countries ; Inadequate
PC 2494 — *Inner tensions*
PE 0752 — Unjust financing
Political [party...]
PD 4798 — party entryism
PD 2906 — party manipulation elections
PF 8535 — patronage ; Abusive distribution
PF 2182 — pluralism
PD 3542 — police
Political power
PC 2510 — Economically controlled
PE 1101 — *Limited*
PC 9104 — Misuse
PF 7163 — Unwillingness sacrifice
Political [prejudice...]
PC 8641 — prejudice
PB 3209 — pressure ; Dependence
PB 3209 — pressure ; Undue
Political prisoners
PC 0562 — [Political prisoners]
PD 3171 — communist systems
PD 3015 — detainees ; Civilian
PD 3014 — detainees ; Military
PD 0702 — Mass detention
PE 4430 — mental institutes
PF 3020 — Mis classification
Political [process...]
PC 1015 — process ; Absence democratic
PE 7312 — processes ; Denial right indigenous peoples participate
PD 5457 — processes ; Violation
PF 4558 — promises ; Extravagant
PE 7680 — proposals ; Deceptive
PC 2933 — purges
PE 5667 — purposes ; Deprivation government benefits
PE 4635 — purposes ; Misuse personal authority
Political [radicalism...]
PF 2177 — radicalism
PD 1624 — re-education
PE 9762 — reasons ; Imposition trade quotas
PD 2549 — refugees
PC 1461 — regimes ; Illegal
PC 1461 — regimes ; Illegitimate
PC 0655 — representation ; Dependence unequal
PC 0655 — representation ; Unequal
PC 1919 — repression
PC 1919 — repression ; Dependence
PF 3707 — reprisals
Political revolution
PF 3237 — [Political revolution]
PD 3228 — Non violent
PD 3230 — Violent
Political [rights...]
PD 8276 — rights ; Denial
PC 1917 — rights ; Underutilization
PF 5932 — risk
PD 8992 — rivalry
Political [scandal...]
PD 4651 — scandal
PC 2361 — schism
PD 9384 — smear campaigns
PC 2494 — stagnation
PJ 4988 — stalemate
PF 0297 — status ; Denial right people determine their own
PC 9058 — structure ; Inadequate
PF 1097 — structures ; Mistrust
PF 9428 — support unsustainable policies
PF 9428 — surrealism
PD 8871 — surveillance
PD 8673 — systems ; Underdevelopment
Political [thinking...]
PF 9821 — thinking concerning global problematique ; Inadequate development
PE 1101 — ties ; *Ineffective*
PD 0181 — torture
PD 7276 — tourism
PD 3013 — trials
Political [uncertainty...]
PF 1559 — uncertainty

Political [union...] cont'd
PE 4627 — union island developing countries ; Obstacles
PC 3204 — unity ; Breakdown
PF 3216 — unity ; Lack
PF 9428 — unreality
PD 8168 — unrest
PC 5180 — unwillingness change
PC 7660 — upheaval ; Risk
PC 7660 — upheavals
PD 9384 Political vilification
PD 4425 Political violence
PF 6202 Political weapon ; Misuse food
Political will
PC 5180 — act problems ; Lack
PC 5180 — cooperate ; Lack
PC 5180 — Lack
PF 7838 — observe trade commitments ; Failure
PC 5180 — respond needs developing countries ; Lack
PE 1101 Politically determined services
PF 3128 Politically emotive words terms
PF 7955 Politically exploitative humanitarian aid
Politically motivated
PF 1098 — amnesty military personnel convicted atrocities
PD 9349 — arrests
PF 9794 — exaggeration human rights infractions
PD 6970 — *mass arrests*
PF 4723 — reduction prison sentences
PF 9558 Politically sensitive incidents ; Fabrication
PF 1214 Politically unrealistic strategic warfare analysis
PD 4556 Politicians ; Avoidance legal obligations
PD 4556 Politicians ; Circumvention law
PF 1097 Politicians ; Citizen mistrust
PE 1101 *Politicians ; Distrusted local*
PF 3962 Politicians' lack character
PE 5614 Politicians ; Terrorism targeted against
PD 6298 Politicization curriculum
PD 8468 Politicization decision-making
PE 6948 Politicization diplomatic positions
PD 6298 Politicization education
PD 4519 Politicization health standards
PF 7925 Politicization human rights issues United Nations
Politicization [intergovernmental...]
PD 0457 — intergovernmental organizational debate
PF 4761 — international sports competitions
PF 4761 — international sports events
PD 2860 — issues
PD 5475 Politicization media ; Excessive
PE 4226 Politicization public service developing countries
PF 7220 Politicization research
PF 7220 Politicization scholarship
PD 2860 Politicization technical debates
PC 0116 Politics ; Corruption
Politics [Dependence...]
PC 0934 — Dependence discrimination
PB 3202 — Dependence power
PC 5517 — Dirty
PC 0934 — Discrimination
PC 1001 — Discrimination against women
PD 7704 — Dynastic
PD 1787 Politics ; Espionage domestic
PD 2910 Politics ; Forced participation
PE 8080 Politics fuel shortages ; Distortion international
PC 3219 Politics ; Ideological discrimination
PC 5517 Politics ; Irresponsible
Politics [Lack...]
PC 1917 — Lack participation
PC 1917 — Lack participation local
PD 3223 — Language discrimination
PF 9631 Politics ; Media theatricalization public life
PD 3385 Politics ; Military influence
PC 5517 Politics ; Negligence
Politics [Personality...]
PC 1123 — Personality cults
PC 3221 — Political discrimination
PB 3209 — Political influence influencing
PF 8535 — Pork barrel
PB 3202 — Power
PD 3218 — Property occupational discrimination
Politics [Racial...]
PD 3329 — Racial
PD 3329 — Racial discrimination
PC 3220 — Religious discrimination
PF 3358 — Religious influence
PC 5517 Politics ; Unethical practices
PE 1101 *Politics ; Unprofitable local*
PF 1651 Polity roles ; Unclear
PE 5074 Polity ; Worker participation excluded company
PE 6555 Poll evil
PE 2069 Pollens
PE 2069 Pollens ; Allergy inducing
PE 4732 Pollens grasses, weeds trees
PE 3880 Pollinating insects ; Destruction crop
PE 3880 Pollinating species ; Destruction crop
PE 6197 Pollinosis
PF 2909 Polls ; Misleading public opinion
Pollutant [Ammonia...]
PD 2965 — *Ammonia*
PE 1732 — *Arsenic*
PE 1127 — *Asbestos*
Pollutant [Cadmium...]
PE 1160 — *Cadmium*
PE 0535 — *Chlorine*
PE 4072 — *Chromium*
PE 2339 — *Cobalt*
PE 0535 — *Copper*
PC 0062 — *Cyanide*

PE 5028 — Pollutant ; DDT
PE 2329 — Pollutant ; Hydrogen sulphide
PE 1161 — Pollutant ; Lead
PE 1155 — Pollutant ; Mercury
PE 1315 — Pollutant ; Nickel
PE 2134 — Pollutant ; Oil
PE 1359 — Pollutant ; Ozone
PD 0120 — Pollutant residues
Pollutant [Selenium...]
PE 1726 — Selenium
PD 1414 — Sewage
PD 2267 — Smoke
PE 1210 — Sulphur dioxide
PE 2964 — Sulphur trioxide
PE 1438 — Pollutant ; Tin
PE 5195 — Pollutant ; Titanium dioxide
PE 2668 — Pollutant ; Vanadium
PE 7229 — Pollutant ; Zinc
PC 5690 Pollutants
Pollutants [Acaricides...]
PD 0120 — *Acaricides*
PC 0119 — Air
PD 0414 — Automobile emissions air
Pollutants Biological
PC 5276 — [Pollutants ; Biological]
PD 0450 — air
PD 1175 — water
Pollutants [Carbamate...]
PE 1282 — Carbamate insecticides
PD 1271 — Chemical air
PE 0535 — Chemical water
PE 1087 — Pollutants ; Detergents
PD 2800 — Pollutants ; Domestic waste water
PD 1670 — Pollutants environment ; Chemical
Pollutants [Fluorides...]
PE 1311 — Fluorides
PD 4925 — freshwater wildlife ; Accumulation
PD 1612 — Fungicides
PE 0524 — *Pollutants ; Gaseous air*
Pollutants [Halogen...]
PE 4483 — Halogen compounds
PD 1143 — Herbicides
PC 0936 — Human exposure hazardous environmental
PE 0754 — Hydrocarbons
Pollutants [Industrial...]
PE 2824 — Industrial domestic heating emissions air
PD 0575 — Industrial waste water
PD 5227 — Inorganic salts
PD 0983 — Insecticides
PE 9813 — Pollutants ; Laboratory waste water
PD 3934 — Pollutants marine wildlife ; Accumulation
Pollutants [Natural...]
PC 2416 — Natural
PD 0120 — *Nematicides*
PE 1956 — Nitrates
PE 6087 — Nitrites
PD 2965 — Nitrogen compounds
PD 2965 — Nitrogen oxides
Pollutants [Organochlorine...]
PD 0983 — *Organochlorine compounds*
PD 0983 — *Organochlorine insecticide*
PD 0983 — *Organophosphorus insecticides*
Pollutants [Pesticide...]
PD 0120 — Pesticide residues
PD 0120 — Pesticides
PE 1313 — Phosphates
PE 0155 — Plant pathogenic air
PD 5021 — plants animals ; Accumulation
PE 3677 — Pollutants ; Rodenticides
Pollutants [Silicates...]
PE 1514 — *Silicates*
PD 0177 — Solid wastes
PD 1670 — *Sulphur compounds*
Pollutants [terrestrial...]
PD 0381 — terrestrial plants ; Accumulation
PD 5278 — terrestrial wildlife ; Accumulation
PC 1609 — Thermal
PE 5436 — Thiols mercaptans
PD 0948 — Toxic metal
PE 0617 — Toxic organic compounds water
PE 7141 — Pollute developing countries ; Licence
PB 6336 Polluters
PD 0563 — Polluters ; Agricultural
PD 4524 — Polluters ; Domestic
PD 5370 — Polluters ; Governmental
PC 8745 — Polluters ; Industrial
PF 8921 — Polluting industries developing countries ; Transfer highly
PB 6336 Pollution
Pollution [Agricultural...]
PD 0563 — Agricultural
PE 4744 — agricultural crops ; Harmful effects air
PC 0119 — Air
PE 4802 — aircraft
PE 3934 — animal production ; Water
PE 9609 — animals ; Health hazards air
PE 5993 — Antarctic
PD 2503 — aquifers ; Nitrate
PE 5993 — Arctic
PD 6283 — Arctic air
PC 0119 — Atmospheric
Pollution [Badness...]
PA 5454 — *Badness*
PE 8175 — bays estuarine waters
PD 1356 — Beach

-1012-

Populations employment

Pollution [children...]
PE 8617 — children ; Health hazards air
PD 1356 — Coastal water
PA 6468 — *Complexity*
PF 1789 — constraints testing new technology
PF 4854 — cover up ; Environmental
PD 2478 Pollution ; Damage cultural artefacts environmental
Pollution damage
PD 2478 — frescoes
PD 2478 — monuments
PD 2478 — paintings
PD 2478 — sculpture
Pollution [Death...]
PF 3960 — Death
PB 1166 — Dependence environmental
PB 6248 — Dependence mental
Pollution developing countries
PC 2023 — [Pollution developing countries]
PE 7141 — Inadequate legislation against environmental
PE 7141 — Lack regulations against
PD 3675 — Water
PD 4007 Pollution due over population ; Environmental
Pollution [Electromagnetic...]
PD 4172 — Electromagnetic
PF 7836 — Employment risk elimination industrial
PB 1166 — Environmental
Pollution [fertilizers...]
PE 8729 — fertilizers ; Water
PE 7584 — Fish kills caused
PD 5605 — Food
PD 1223 — Freshwater
Pollution [groundwater...]
PD 2503 — groundwater
PD 2503 — groundwater ; Nitrate
PE 8476 — growth crops ; Adverse impact
PE 6172 — Pollution ; Havens environmental
PB 6336 *Pollution hazard ; Wildlife*
Pollution hazards
PF 1789 — industrial research development
PD 0708 — Underreporting groundwater
PB 6336 — *wild life*
PB 6336 — *wildlife*
Pollution [Health...]
PC 0936 — Health hazards environmental
PD 1584 — Heat
PF 9150 — Human internal
Pollution [immediate...]
PF 7836 — immediate victims ; Defence industrial
PE 8617 — impact lungs children ; Air
PA 6088 — *Impairment*
PF 9299 — Inadequate legislation against environmental
PA 6852 — *Inappropriateness*
PD 6627 — Indoor air
PE 7584 — induced fish diseases
PE 6650 — information
PE 8476 — Inhibition crop growth
PD 1223 — inland waters
PC 0328 — intensive goods
Pollution [Lack...]
PF 9299 — Lack regulations against environmental
PD 8628 — Lake
PE 7244 — Light
PD 3391 — Long range transboundary air
Pollution [Marine...]
PC 1117 — Marine
PD 1983 — mass media ; Psychological
PB 6248 — Mental
PC 6675 — military activities
PB 2480 — Moral
PE 6256 — mountain environments mountaineers trekkers
Pollution [Noise...]
PC 0268 — Noise
PD 1584 — nuclear reactors ; Environmental
PE 5698 — nuclear weapons manufacture
Pollution [Obstruction...]
PE 7244 — Obstruction astronomical observation environmental
PE 1839 — Oil
PD 0089 — orbital space
PD 0983 — *Organochlorine*
Pollution [Particulate...]
PD 2008 — Particulate atmospheric
PE 8729 — pesticides ; Water
PF 3918 — Placement burden proof victims environment
PE 4744 — plants ; Health hazards air
PE 3741 — plastic waste ; Marine
PE 6285 — Plutonium
PE 5993 — Polar
PE 5385 — Pond
PD 3102 — Program
Pollution [reactors...]
PD 1584 — reactors ; Radiation
PF 3960 — Ritual
PF 7636 — River
Pollution [salinity...]
PE 7837 — salinity effects ; Water
PE 5539 — sediments
PE 8175 — semi-enclosed bodies seawater
PD 9197 — socialist countries ; Environmental
PC 0058 — Soil
PC 0268 — Sonorous
PD 0089 — Space
PF 7636 — Stream
PF 7945 Pollution ; Trans frontier
PD 1096 Pollution ; Transboundary water
Pollution [Uncleanness...]
PA 5459 — *Uncleanness*

Pollution [Underreporting...] cont'd
PD 4265 — Underreporting chemical
PD 4277 — Underreporting marine
PA 7226 — *Unhealthfulness*
PC 8745 — Urban Industrial
PC 0119 — *Urban Industrial air*
PC 0062 Pollution ; Water
Pollution water
PE 3934 — fish farms
PE 8545 — infected faeces
PE 8545 — supplies human excreta
PD 5641 Pollution ; Water surface
PF 5697 Poltergeists
PF 3289 Polyandry
PD 5453 *Polyarteritis animals*
PE 9509 *Polyarteritis nodosa animals*
PE 5913 Polyarthritis
PE 9509 *Polyarthritis animals ; Idiopathic*
PE 5913 *Polyarthritis ; Chlamydial*
PD 2732 *Polyarthritis ; Infectious*
PD 2731 *Polyarthritis lambs ; Nonsuppurative*
PE 5050 *Polyarthritis ; Neonatal septic*
PD 5228 Polybrominated biphenyls poisoning animals
PE 2432 Polychlorinated biphenyls health hazard
PD 5228 *Polychlorinated biphenyls poisoning animals*
PD 2322 Polychemia
PE 4637 *Polycythemia*
PD 5453 *Polycythemia animals*
PE 4637 *Polycythemia vera*
PD 2184 Polygamy
PF 2100 Polygamy ; Serial
PB 0284 *Polygraph tests ; Enforced*
PD 2184 Polygyny
PD 2184 Polygyny ; Sororal
PE 5509 Polymyositis
PE 8932 Polyneuritis
PE 8932 *Polyradiculitis*
PE 4604 *Polyradiculoneuritis*
PD 2732 *Polyserositis ; Haemophilus*
PE 5050 *Polyserositis ; Mycoplasmal*
PD 2732 *Polyserositis ; Porcine*
PD 3825 Polysubstance dependence
PE 4957 *Polytheism*
PE 9369 *Polyuria*
PD 3477 Polyzoa
PA 6400 Pomposity *Affectation*
PE 5385 Pond pollution
PA 7365 Ponderousness *Boredom*
PA 5497 Ponderousness *Difficulty*
PA 7232 Ponderousness *Unskillfulness*
PC 1630 Ponds ; Mosquito breeding
PE 4039 Pool cross licensing agreements ; Patent
PF 7727 Pool ; Deterioration gene
PF 6540 *Pool ; Small employee*
PF 1781 *Pools ; Lack car*
PE 2634 Pools ; Unmanageable stagnant
PF 1055 Poor access roads
PF 1403 Poor adult supervision
PC 0387 Poor climate
PF 1352 *Poor commercial practices*
Poor communications
PF 2350 — [*Poor communications*]
PF 2350 — *methods*
PF 6470 — networks rural areas
Poor [condition...]
PF 1815 — condition open spaces urban communities
PD 1251 — *construction dwellings*
PF 3550 — credit alternatives urban areas
Poor [Deterioration...]
PB 0388 — Deterioration standard living
PC 8866 — digestion
PE 1093 — due legal delays ; Excessive burden
PC 0999 Poor families
PF 3499 Poor geographical location
PA 0294 Poor health ; General
PF 1213 Poor housing construction ; Depressing effect
Poor [Inaccessible...]
PE 8698 — Inaccessible housing
PD 0382 — *Inadequate diet*
PE 1399 — Inequitable justice
PF 2401 — information channels
PS 7068 — *international relations*
PE 1129 Poor judicial sanctions ; Discrimination against
Poor [Lack...]
PF 4662 — Lack commitment protection
PF 8704 — Lack solidarity
PF 5780 — lighting
PD 9156 — living conditions
Poor [managerial...]
PF 1528 — managerial communications
PD 1260 — minority population urban ghettos ; Segregation
PF 0760 — money management
PF 4644 Poor ; Narrow time span
PD 0800 Poor nutritional habits
PF 1790 Poor organization community environment
PD 5261 Poor ; Overuse environmental resources
Poor [people...]
PB 0388 — people
PE 7161 — *physiological adaptability women*
PE 8698 — Prohibitively expensive housing
PF 5703 Poor quality
PD 2743 Poor quality domestic livestock
PF 0905 *Poor-risk commercial farming*
PE 6526 Poor road conditions
Poor [sanitation...]
PD 0876 — sanitation facilities

Poor [service...] cont'd
PF 2583 — service delivery urban environments
PF 2583 — services motivation
PC 4588 — *social environment*
PC 0052 — soil management
PF 6473 — *street lighting*
PF 3498 — *supervision teachers*
Poor [Taxation...]
PE 4601 — Taxation
PF 2721 — teaching methods
PF 1790 — television reception
PF 9980 — time-keeping
PF 1495 — transport systems
PF 5969 Poor vaccination practices
PD 7663 Poor viability nuclear power plants
Poor [water...]
PE 8233 — water management
PD 9170 — work environment
PD 9170 — working conditions
PD 9170 — working environment
PD 1435 — workmanship
PD 9736 Poorly cooled homes
PC 0300 *Poorly developed social responsiveness*
PD 5173 Poorly heated homes
PF 6473 *Poorly lighted roadways*
PF 8071 Popery
PE 4413 Poplars ; Pests diseases
PE 7826 Popped knee
PF 5627 Popular errors
PF 4164 Popular problems
PF 1112 Popular psychological screens ; Escaping reality
PC 4864 *Popular uprisings*
PF 2040 Popular views ; Fear contradicting
PF 2426 Popular wisdom ; Underutilization
PF 2426 Popular wisdom ; Unrecorded
PF 5242 Popularization concepts
PC 0027 *Population ; Ageing rural*
PC 0027 Population ; Ageing world
Population [base...]
PF 6441 — *base ; Diminishing*
PF 4470 — base ; Small
PF 4470 — basic services users ; Small
PE 8768 Population cities ; Migration rural
Population control
PF 1021 — family planning ; Opposition
PF 1023 — Government opposition
PF 0148 — Ineffective
PF 1020 — Ineffective
PF 1022 — Religious opposition
PC 8192 Population countries ; Maldistribution
Population [data...]
PF 0214 — data ; Undetermined
PF 6441 — *Declining community*
PF 6441 — Declining national
PF 6441 — decrease
PB 0469 — density ; Excessively high
PD 6131 — density gradients ; Unbalanced urban
PF 0167 — Dependence maldistribution world
PB 0035 — disease
PD 4007 Population ; Environmental pollution due over
Population growth
PD 4007 — Depletion natural resources due
PE 4124 — developed developing countries ; Imbalance
PB 0035 — Excessive
PF 3743 — Increasing development lag against
PE 5369 — Increasing lag education against
PB 0035 — Rapid
PD 4007 — resource development ; Imbalance
PF 9671 — *Theological justification*
PB 0035 — Unsustainable
PD 8568 Population ; Increasing low income
PB 0035 Population levels ; Unsustainable
PF 0167 Population ; Maldistribution world
PF 3806 Population mobility ; Excessive
Population [Over...]
PB 0035 — Over
PE 8735 — overall growth economy ; Imbalances growth labour force, urban
PE 8437 — overall growth economy labour force ; Urban
PF 6441 *Population ; Reduced village*
PC 6203 Population ; Relocation
Population [seals...]
PE 6024 — *seals ; Over*
PF 3382 — sex ratio ; Deliberate imbalancing
PF 4470 — Socio economically inactive rural
PF 4470 — Static town
PF 1768 *Population ; Unintegrated farming*
PD 1260 Population urban ghettos ; Segregation poor minority
PE 6024 Population wild animals ; Over
PE 7995 Populations ; Abuse control wild animal
PD 8561 Populations ; Ageing
Populations Denial right
PC 3319 — adequate food indigenous
PD 1092 — employment indigenous
PE 4459 — health indigenous
PE 2317 — legal services indigenous
PD 4252 — traditional economies indigenous
Populations [Denial...]
PF 1844 — Denial rights transient
PD 4751 — Difficulty controlling insect
PE 5784 — Disaster hazards island
PC 0352 — Discrimination against indigenous
Populations [education...]
PC 3322 — education ; Discrimination against indigenous
PD 1092 — employment ; Discrimination against indigenous
PD 1092 — employment ; Exploitation indigenous

Populations

Populations [Excessive...] cont'd
- PE 5776 — Excessive stray animal
- PC 3304 — Expropriation land indigenous
- PC 3320 Populations housing Denial right adequate housing indigenous peoples ; Discrimination against indigenous

Populations [imported...]
- PE 3721 — imported diseases ; Massacre indigenous
- PF 1844 — Inadequate community care transient urban
- PE 2541 — Infantilization deprived
- PE 8564 — institutions ; Destruction civilian
- PE 3721 — introduction diseases ; Vulnerability indigenous
- PE 1848 Populations ; Mobility village
- PF 9686 Populations ; Neglected rural
- PD 3305 Populations reservations ; Restriction indigenous
- PB 1395 Populations species ; Vulnerability
- PD 0359 Populations ; Stray dog
- PD 4721 Populations ; Underreporting hazards animal
- PE 7573 Populations ; Violation treaties indigenous
- PF 3410 Populism
- PD 7841 Porcine congenital tremor syndrome
- PD 9293 Porcine cystitis
- PD 7841 Porcine enteroviral encephalomyelitis

Porcine [polioencephalomyelitis...]
- PD 7841 — polioencephalomyelitis
- PD 2732 — polyserositis
- PE 3936 — proliferative enteritis
- PD 2731 — Porcine streptococcal infections
- PE 9774 Porcine stress syndrome
- PE 8845 Porcupine ; Endangered species old world
- PE 8847 Porcupines ; Endangered species new world
- PD 3476 Porifera
- PE 8535 Pork barrel politics
- PE 9774 Pork ; Pale soft exudative
- PE 3511 Pork tapeworm
- PF 1384 Pornographic magazines comics
- PD 0132 Pornography
- PF 1349 Pornography ; Child
- PE 5402 Pornography ; Electronic
- PD 0132 Pornography ; Exposure children
- PF 5190 Pornography sound
- PE 5402 Pornography telephone
- PE 3822 Porous bones
- PD 3476 Porphyria
- PC 2270 Porphyria
- PD 7420 Porphyria erythropoietica ; Congenital
- PD 7420 Porphyriuria
- PE 3806 Porpoises ; Endangered species

Port [cargo...]
- PE 5899 — cargo handling ; Excessive costs inefficient
- PF 5887 — charges ; Inconsistent
- PE 4766 — congestion
- PE 5815 Port delays
- PE 4897 Port facilities ; Unsafe
- PE 5792 Port infrastructure ; Inadequate
- PE 5827 Port operations ; Political interference
- PE 5911 Port services ; Underpricing
- PE 5908 Port storage facilities land locked developing countries ; Inadequate
- PE 4766 Port traffic congestion
- PE 5921 Port utilization operation ; Obstacles efficient
- PE 5851 Port warehousing ; Overcrowded
- PD 0981 Portraiture
- PE 7354 Portrayal crime media ; Excessive
- PE 1478 Portrayal destabilizing information media ; Excessive
- PE 3980 Portrayal drugs, alcohol cigarettes media ; Excessive
- PE 1478 Portrayal negative information media ; Excessive
- PE 3831 Portrayal perspectives industrialized cultures media ; Excessive
- PE 7930 Portrayal sex media ; Excessive
- PE 3980 Portrayal substance abuse media ; Excessive
- PE 6844 Portrayal terrorist activity media ; Excessive
- PD 6279 Portrayal violence mass media ; Excessive

Portrayal [war...]
- PF 9312 — war ; Unrealistic
- PF 7638 — women mass media ; Biased
- PF 7638 — women media ; Exploitative
- PE 4471 Position torture ; Stationary
- PE 5311 Positive bias reporting
- PF 9539 Positive discrimination
- PF 4377 Positive self assessment ; Unrealistically
- PF 3199 Positive spheres influence
- PF 2179 Positivism
- PE 5390 Possessing contraband useful escape ; Introducing
- PA 6686 Possession*complex
- PF 5781 Possession ; Demonic spirit
- PE 5556 Possession drugs
- PE 8364 Possession fencing stolen property
- PE 5283 Possession unlawful distilled spirits
- PD 1317 Possessive attitude parents
- PE 1137 Possibilities ; Debasing retirement
- PE 8985 Possibilities gathering elders ; Insufficient
- PF 8105 Possibilities ; Limited entertainment
- PE 2535 Possibilities local commerce ; Unexploited
- PF 0846 Possibilities ; Refusal family
- PE 2842 Possibilities ; Scarce employment
- PJ 9762 Possibilities training urban areas ; Unrecognized

Possibilities [Unclear...]
- PF 4984 — Unclear preschool
- PD 1240 — Undiversified basis income
- PF 7118 — Unimaginative vision community
- PF 6925 — Unknown
- PE 1301 — Unpublished educational
- PE 5304 Possibility expanding preschool ; Unimagined
- PF 3005 Possible expansion ; Impeded
- PD 1762 Possum ; Endangered species honey

- PD 1762 Possums ; Endangered species pigmy
- PE 5846 Post abortion syndrome
- PF 3826 Post capitalist social order ; Prevalence psychological conditions unfavourable transition
- PE 8610 Post colonial countries ; Weak national identity
- PD 2731 Post-dipping lameness sheep
- PE 0661 Post epileptic automatism
- PD 1162 Post graduate unemployment ; Graduate
- PD 0556 Post-hallucinogen perception disorder
- PC 2027 Post-harvest losses

Post hepatitis
- PE 0517 — disorders
- PE 0517 — hyperbilirubinaemia
- PE 0517 — syndrome
- PC 0799 Post-natal depression
- PF 1015 Post-revolutionary re-employment government security services ousted repressive regime
- PE 0351 Post-traumatic amnesia
- PE 0351 Post-traumatic stress disorder
- PF 7165 Post ; Unattractive teacher
- PF 2717 Postal agents ; Lack
- PD 3033 Postal censorship
- PF 2717 Postal delays

Postal [service...]
- PE 6754 — service ; Irresponsible use
- PE 6754 — service ; Unethical use
- PF 2717 — services ; Inadequacy
- PD 2683 — surveillance governments ; Misuse
- PA 6073 Posterity
- PF 9455 Posterity ; Denial rights
- PF 9455 Posterity ; Irresponsibility towards
- PF 7799 Posthitis animals ; Ulcerative
- PD 7420 Postparturient haemoglobinuria
- PE 0598 Postponement bad news
- PF 3211 Postponement bridge construction
- PD 8557 Postponement street repair
- PF 4498 Posts ; Frustrated talent government
- PD 2838 Posture
- PF 2627 Posture ; Parochial leadership
- PD 2838 Posture ; Unhealthy physical
- PF 3756 Posturing meetings ; Public
- PD 8557 Pot holes ; Roadway
- PE 5465 Potato ; Late blight
- PD 2227 Potato latent mosaic
- PD 2219 Potato ; Pests diseases
- PE 4346 Potato scurf ; Sweet
- PE 8258 Potatoes ; Instability trade fresh
- PE 4346 Potatoes ; Silver scurf
- PF 6532 Potential ; Arrested development labour
- PF 3448 Potential basic resources ; Underdeveloped
- PF 6925 Potential benefits ; Unawareness
- PF 2231 Potential commercial enterprises ; Unrealized
- PF 4871 Potential ; Disregard self healing
- PC 1820 Potential entrepreneurs ; Unconfident
- PF 2969 Potential external relations ; Unrealized
- PF 2231 Potential gober gas plants ; Unrealized
- PF 6707 Potential hazards ; Bureaucratic procrastination response evidence

Potential [industrial...]
- PF 2471 — industrial development rural communities ; Overlooked
- PF 1196 — informal leadership ; Undeveloped
- PF 2042 — investment small, inner city enterprises ; Lack awareness
- PF 6513 Potential local communities ; Underutilization
- PD 8568 Potential ; Low income
- PF 0581 Potential markets ; Unexplored
- PF 0581 Potential markets ; Unidentified
- PF 2231 Potential nature ; Unrealized
- PF 6707 Potential problems ; Official inaction response
- PF 6570 Potential social interaction different age groups ; Stifled

Potential [Uncertain...]
- PF 0581 — Uncertain sales
- PF 2575 — Unchallenged youth
- PF 2989 — Underdeveloped export
- PF 7063 — Underestimation human
- PF 0581 — Undetermined sales
- PF 0581 — Undeveloped market

Potential Unexplored
- PF 2797 — avenues leadership
- PF 9729 — cooperative
- PF 2989 — export
- PF 2231 — orchard

Potential [Unknown...]
- PF 0581 — Unknown market
- PF 8233 — Unmanaged irrigation
- PF 6471 — Unorganized labour
- PF 0422 — Unprepared vocational

Potential Unrealized
- PF 2231 — agronomic
- PF 7063 — human development
- PF 2568 — teaching
- PF 6479 Potential ; Untapped leadership
- PF 7130 Potential ; Untapped park
- PF 1544 Potential youth participation ; Unreleased
- PD 3978 Potomac horse fever
- PE 1314 Potters' asthma

Pottery earthenware
- PE 9019 — industries ; Environmental hazards china,
- PE 8928 — industries instability ; China,
- PE 8427 — manufacture underdevelopment ; China,
- PE 7953 Potto loris ; Endangered species
- PE 5182 Pouches ; Diseases guttural

Poultry [Air...]
- PD 9366 — Air sac mite

Poultry [Ammonia...] cont'd
- PD 9366 — Ammonia burn
- PE 7562 — Arizona infection
- PD 9366 Poultry ; Breast blisters

Poultry [Chlamydiosis...]
- PE 5567 — Chlamydiosis
- PE 2738 — Coccidiosis
- PD 2731 — Colibacillosis
- PE 6461 Poultry ; Digestive tract helminthiasis
- PE 7808 Poultry ; Disorders skeletal system
- PD 7841 Poultry ; Encephalomyelitis
- PD 2732 Poultry ; Erysipelothrix infection
- PD 2735 Poultry ; Histomoniasis

Poultry Infectious
- PE 5567 — bronchitis
- PE 5567 — coryza
- PE 5567 — laryngotracheitis
- PD 7307 Poultry ; Influenza

Poultry [Mansons...]
- PE 2395 — Manson's eyeworm infection
- PD 2732 — Mycoplasma gallisepticum infection
- PD 2732 — Mycoplasma synoviae infection
- PD 2728 — Mycotoxicosis

Poultry [Necrotic...]
- PD 9667 — Necrotic dermatitis
- PD 3978 — Necrotic enteritis
- PE 1400 — Newcastle disease
- PD 2731 Poultry ; Omphalitis

Poultry [Paratyphoid...]
- PE 7562 — Paratyphoid infection
- PD 9366 — Pendulous crop
- PE 7562 — Pullorum disease
- PE 5567 Poultry ; Respiratory diseases
- PD 5453 Poultry ; Round heart disease

Poultry [Spirochetosis...]
- PD 2731 — Spirochetosis
- PD 2731 — Staphylococcosis
- PD 2731 — Streptococcosis
- PD 2868 Poultry ; Taboos against eating
- PD 3978 Poultry ; Trichomoniasis
- PD 3978 Poultry ; Ulcerative enteritis
- PE 5913 Poultry ; Viral arthritis
- PA 6434 Poverty

Poverty [capitalist...]
- PC 3107 — capitalist systems
- PD 4966 — Child
- PD 4966 — Children
- PF 9828 — Concealment information extent
- PE 5252 — consequence war
- PF 2346 — cycle ; Unchecked
- PB 0388 — Poverty ; Dependence
- PC 0444 Poverty developed countries

Poverty developing countries
- PC 0149 — [Poverty developing countries]
- PD 4125 — Economic stagnation due rural
- PD 4125 — Rural
- PD 0723 Poverty disability

Poverty [economy...]
- PD 4125 — economy ; Subsistence
- PF 6530 — economy ; Subsistence
- PC 1966 — elderly
- PD 5261 — Environmental
- PB 0388 — Extreme
- PC 0999 Poverty ; Family

Poverty [image...]
- PF 1365 — image ; Debilitating
- PC 0444 — industrialized countries
- PD 1998 — industrialized countries ; Family
- PF 1258 — Inefficient public spending alleviate
- PC 5052 — Inner city
- PA 5473 — Insufficiency
- PC 4992 Poverty ; Rural
- PB 0388 Poverty ; Socio economic
- PC 5052 Poverty ; Urban
- PD 3489 Poverty ; Urban fringe
- PF 1365 Poverty victimage ; Common
- PE 0529 Powdery mildews plants

Power Abuse
- PB 6918 — [Power ; Abuse]
- PC 6873 — economic
- PC 9104 — government
- PC 1142 — police
- PC 9104 — public
- PF 2692 — scientific
- PF 4174 Power ; Accumulated
- PE 7778 Power arrogance ; Super
- PE 1341 Power blackouts
- PF 5115 Power ; Centralization religious
- PE 7778 Power chauvinism ; Super
- PA 7314 Power*complex
- PC 5323 Power ; Concentration investment
- PE 7665 Power ; Control judiciary executive military

Power [Dependence...]
- PC 6873 — Dependence abuse economic
- PB 1969 — Dependence imbalance
- PE 8707 — developing countries ; Lack purchasing
- PF 0377 — developing countries ; Underproductivity draught animal
- PE 0081 — due advertising ; Monopoly

Power [Economically...]
- PC 2510 — Economically controlled political
- PD 3133 — elderly ; Monopoly
- PC 1002 — Elimination bourgeois
- PE 7778 — exclusiveness ; Super
- PE 1341 Power failure ; Electrical

-1014-

Preparatory education

Power [generation...]
PE 1412 — generation ; Degradation environment electrical
PD 7663 — generation ; Inadequate infrastructure nuclear
PF 5472 — Global centralization
PD 8890 — government bureaucracy ; Unchecked
PF 4668 — government ; Constraints
Power [ideological...]
PF 5124 — ideological diversities ; Maintenance super
PB 1969 — Imbalance
PE 7390 — Inadequate assistance victims abuse
PD 5807 — independence transnational corporations ; Excessive
PF 4135 — industry developing countries ; Underdevelopment
PE 3528 — *Inequality bargaining*
PD 8362 — Insufficient buying
PD 9033 — Insufficient electrical
PF 9175 — intergovernmental organizations ; Inadequate
PF 9175 — international organizations ; Lack enforcement
PC 3185 — intervention countries ; Super
PF 8531 — Irrational rejection nuclear
Power [Lack...]
PE 3528 — *Lack contract*
PF 0223 — Lack enforcement
PF 4668 — Limitations government
Power Limited
PD 8362 — buying
PE 1101 — *political*
PD 8362 — purchasing
Power [Misuse...]
PD 5193 — Misuse financial industrial
PC 9104 — Misuse political
PC 8410 — Monopolization
PC 8410 — Monopoly
PD 4445 — monopoly advanced nuclear warfare technology ; Super
PF 9582 Power nature ; False assumptions concerning recuperative
PF 6545 *Power patterns ; Static*
Power plants
PE 5260 — Deterioration nuclear
PE 7539 — Environmental hazards decommissioned nuclear
PD 2718 — Hazardous locations nuclear
PE 7519 — Lack acceptable sites
PD 7663 — Poor viability nuclear
PF 7543 — Prohibitive cost nuclear
PE 7539 — Waste resources obsolete nuclear
PB 3202 Power politics
PB 3202 Power politics ; Dependence
Power production
PD 4977 — Environmental hazards nuclear
PE 4835 — Risks
PE 9134 — weather ; Adverse effects
PD 4977 Power ; Proliferation nuclear
Power [Regional...]
PC 4291 — Regional imbalance
PF 8654 — Restrictive regulation nuclear
PE 4343 — Revealing national security information foreign
Power [safeguards...]
PF 4455 — safeguards ; Non verifiability compliance nuclear
PC 8270 — Seizure
PE 5825 — shipping industry ; Imbalance
PE 0403 — sources ; Aggression against nuclear
PD 0365 — sources ; Vulnerability nuclear
PE 0988 — state owned state controlled enterprises ; Abuse monopoly
PD 7663 — stations ; Insufficient nuclear
PE 1900 — supply developing countries ; Inadequate electrical
PE 1341 — supply system overload
Power [techniques...]
PD 1741 — techniques ; Monopoly nuclear
PE 9642 — transmission lines ; Environmental hazards electrical
PD 1665 — transmission lines ; Unaesthetic location
Power transnational
PE 0766 — corporations ; Concentration
PE 9646 — corporations developing countries ; Asymmetry bargaining
PE 0207 — enterprises against labour ; Coercive use economic
Power [Uncontrolled...]
PF 2692 — Uncontrolled scientific
PF 2093 — Undeveloped community
PF 6506 — *Uneven concentrations*
PF 2358 — *Unrealized corporate*
PF 7163 — Unwillingness sacrifice political
PJ 3570 Power vacuum
PF 3690 Power ; Wasted woman
PE 5080 Powered reactors ; Radioactive fallout accidents orbiting nuclear
PE 1101 Powerful relationships rural communities ; Unsystematic use
PF 2241 Powerlessness bureaucracy
PF 2347 *Powerlessness ; Citizen*
PF 2093 Powerlessness ; Community image
PJ 5262 Powerlessness ; Economic
Powerlessness [Impotence...]
PA 6876 — *Impotence*
PA 6882 — *Influencelessness*
PF 9175 — international organizations
PF 8618 Powerlessness ; Sense
PD 2670 Powerlessness ; Social change
PA 5558 Powerlessness *Weakness*
PE 3972 Powers ; Aerial surveillance foreign
PE 6418 Powers attorney ; Non recognition foreign
PD 3246 Powers ; Excessive extralegal
PD 7261 Powers ; Military expeditions against friendly
PC 3185 Powers protect investments their citizens foreign countries ; Intervention major
PE 7775 Pox ; Chicken

PE 7304 Pox diseases animals
PD 2730 *Pox ; Fowl*
PE 2300 Pox ; French
PE 2300 Pox ; Great
PE 5530 Pox ; Rickettsial
PE 6272 Pox ; Sheep
PE 7304 *Pox virus infection cats*
Practical education
PF 2721 — Ineffective methods
PF 3498 — Ineffective systems
PF 2840 — Limited
PF 2840 — Limited access
PF 3519 — Unrealized intentions
PE 2810 *Practical engagement ; Limited*
PF 2848 *Practical examples ; Lack*
PF 2477 Practical skills ; Narrow range
PE 5170 Practical skills rural communities ; Limited availability
Practical [technology...]
PF 3409 — technology ; Haphazard transmission
PF 2837 — training communities ; Lack opportunities
PF 6472 — training rural areas ; Inadequate
PF 2426 Practical wisdom ; Unshared
PE 4332 Practice indigenous religion ; Prohibition
Practices Restrictive
PB 9136 — [Practices ; Restrictive]
PC 0073 — business
PD 8831 — health
PD 8614 — legal
PD 8831 — medical
PF 8749 — monetary
PD 8027 — professional
PD 8439 — religious
PD 7875 — scientific
PD 0312 — shipping
PC 5537 — social
PC 0073 — trade
PD 8146 — trade union
PD 0881 — transport insurance
PD 8146 — union
Practices Unethical
PC 8247 — [Practices ; Unethical]
PD 2886 — archival
PC 2563 — business
PE 6615 — catering
PC 2563 — commercial
PD 2625 — consumption
PD 2886 — documentation
PE 0682 — financial
PD 1045 — food
PD 2916 — industrial
PE 1826 — insurance
PC 2915 — intellectual
PE 1826 — life assurance
PD 7964 — maintenance
PD 5251 — media
PD 7360 — military
PD 0862 — personnel
PE 3540 — pharmaceutical
PC 8019 — professional
PD 4341 — trade union
PE 4771 Practices zoo animals ; Unethical
PE 5027 Practitioners refusing treat patients ; Medical
PF 4436 Praise ; Self
PE 2268 Pre adults family decisions ; Exclusion
PF 4713 Pre-electoral political inertia
PF 7134 Pre-logical limitations comprehension international information
PD 5107 Pre-marital sexual intercourse
PF 3659 *Pre-natal infections*
PF 4984 *Pre school plan ; Inadequate*
PF 4887 Pre trial internment ; Excessive length
PE 1692 Pre trial publicity ; Unfair trials due
PD 3092 Pre war propaganda ; War
PF 2005 *Preaching ; Superficial*
PF 1382 Precarious basis family economics
PF 6389 Precautions ; Excessive safety
PD 5001 Precautions ; Inadequate safety
PF 2580 *Precedent ; Outdated judicial*
PF 5295 Precedent undermines accountability ; Unchanging legal
PE 0760 Precious stones ; Instability trade crude fertilizers crude minerals, excluding coal, petroleum
PE 1353 Precious stones ; Long term shortage crude fertilizers crude minerals, excluding coal, petroleum
PE 7705 *Precipitate ill-judged forceps delivery*
PD 4904 Precipitation ; Acidic
PF 0466 Precipitation nuclear conflict ; Erroneous
PE 4263 Precipitation patterns ; Long term changes
PD 1564 Preclusion elders' participation image
PA 2173 *Preconceived notions*
PA 2173 Preconceptions
PF 3354 Predestination heaven
PF 2648 Predetermined employer evaluation
PF 0980 Predict future ; Monopoly knowledge
PF 8352 Prediction
PF 7166 Prediction capacity ; Excessive confidence
PF 4774 Predictions ; Irrelevant
PF 5940 Predominance fast food
PC 0027 Predomination society ; Elders'
PD 0340 Preference agreements ; Distortion international trade discriminatory
PC 0933 Preferences developing countries ; Non uniformity tariff non tariff trade
PD 4064 Preferences ; Racial discrimination sexual
PC 1876 *Preferential tariffs*
PD 0340 Preferential trade arrangements

PF 5968 Pregancies ; Reduced intervals
PE 4808 Pregnancies ; Foetal malformation diabetic
PD 0173 *Pregnancies ; Twin*
PF 2859 Pregnancies ; Unwanted
PD 0614 Pregnancy ; Adolescent
PD 7758 *Pregnancy ; Anaemia*
PE 5026 Pregnancy breast feeding ; Smoking during
PE 3858 Pregnancy breast feeding ; Substance abuse during
Pregnancy [Career...]
PD 7692 — Career interruption due
PE 4894 — childbirth ; Haemorrhage
PD 2289 — Complications
PD 2289 — Pregnancy disorders
PD 7799 *Pregnancy dogs ; False*
PD 2289 *Pregnancy ; Ectopic*
PE 5919 Pregnancy ; Forced
PE 5919 Pregnancy ; Government imposed
PF 2148 Pregnancy ; Incest
PD 0173 *Pregnancy ; Multiple*
PF 2859 Pregnancy ; Non termination
PE 1085 Pregnancy nursing ; Protein energy malnutrition women during
PE 8022 Pregnancy puerperium ; Toxaemias
PD 3266 Pregnancy ; Rape
PC 3258 *Pregnancy ; Social barriers male*
Pregnancy [Teenage...]
PD 0614 — Teenage
PD 7420 — *toxaemia cattle*
PD 7420 — *toxaemia ewes*
PD 0614 Pregnant schoolgirls
Pregnant women
PE 8343 — foetuses ; Medical experimentation
PE 4820 — Inadequate medical care
PE 5316 — Job insecurity
PA 2173 Prejudice
Prejudice against
PC 0507 — animals
PF 0076 — communication visual imagery
PD 8800 — languages
PC 8494 — minorities
PD 4365 — non-worshippers
PC 2541 Prejudice ; Age
PA 5454 Prejudice *Badness*
Prejudice [Caste...]
PC 1968 — Caste
PD 8973 — children
PC 8520 — Cultural
PA 2173 Prejudice ; Dependence
PA 2173 *Prejudice ; Economic*
PD 3469 Prejudice ; Housing
PA 7395 Prejudice *Inexpedience*
PC 3406 Prejudice ; Intellectual
PF 8065 Prejudice ; Legal
PA 6607 Prejudice *Misjudgement*
PA 7306 Prejudice *Narrowmindedness*
PA 2173 *Prejudice ; National*
PA 2173 Prejudice people
PC 8641 Prejudice ; Political
PC 8773 Prejudice ; Racial
PD 4365 Prejudice ; Religious
PC 2022 Prejudice ; Sexual
PC 8774 Prejudice ; Skin colour
PF 4832 Prejudiced medical care
PA 2173 *Prejudices ; Disruptive personal*
PC 0244 Prejudicial employment practices
PF 0756 Prejudicial generation relationships
PF 6472 *Preliteracy ; Inhibiting adult*
PD 1947 Premature birth
PD 5183 Premature disclosure government reports
PE 3865 Premature drying plants
PF 6415 Premature ejaculation
PE 9071 Premature retirement people valuable skills
Premature [school...]
PE 0015 — school leavers developing countries ; Unemployment
PC 1716 — school leaving
PD 1947 — still-birth
PA 7006 Prematurity *Untimeliness*
PD 2341 Premeditated killing
PE 3761 Premenstrual syndrome
PE 3761 Premenstrual tension
PD 6627 *Premises ; Badly laid out work*
PF 2137 *Premises ; Gambling church*
PD 6627 *Premises ; Ill designed*
PE 5462 Premises public authorities ; Occupation trade union
PE 5462 Premises public authorities ; Raids trade union
PD 5859 *Premises ; Risk damage*
PE 5462 Premises ; Seizure trade union
PD 8937 *Premises ; Violation*
PE 8736 Premiums ; High life assurance
PE 4820 Prenatal care ; Lack
PD 6967 Prenatal wrongful death
PA 6448 *Preoccupation*
PF 1706 Preoccupation accelerating time
PF 6957 Preoccupation corpses ; Excessive
PD 5086 Preoccupation death ; Morbid
PD 0270 *Preoccupation imagined defect appearance*
PF 6580 Preoccupation isolated problems
PD 8846 *Preoccupation obsolete problems ; Institutional*
PF 3871 Preoccupation reciprocity trading relations
Preoccupation [sex...]
PE 7031 — sex
PF 9119 — short-term problems
PE 8644 — sustenance urban life style ; Perpetual
PC 2303 *Preoccupations ; Debilitating economic*
PC 6381 Preparatory education ; Lack

Prepare trial

PE 7624 Prepare trial defence ; Denial right time
PC 7373 *Preponderance male criminal offenders*
PF 7565 Preponderance non-food crops tropical economies
PF 8356 Preponderance Western-style organizations
PD 9154 *Prepuce ; Redundant*
PD 8179 Presbyopia
PF 4984 *Preschool advantages ; Undefined*
Preschool [education...]
PF 4984 — *education ; Inadequate*
PF 0905 — *equipment ; Insufficient*
PF 0905 — *equipment ; Limited*
PD 0847 Preschool facilities ; Minimal
PF 4984 *Preschool need ; Unperceived*
PF 4984 *Preschool possibilities ; Unclear*
PE 5304 Preschool ; Unimagined possibility expanding
PF 5188 Prescribed chemical residue levels ; Dangerous allegedly
PE 4946 Prescribed drugs ; Illicit trade
PF 0443 Prescribed irrelevant curriculum
PE 0139 Prescription drugs ; Abuse
PD 5540 Prescription drugs ; Phoney non
PE 3799 Prescription inappropriate drugs
PE 3083 Presenile dementia
PF 4644 Present orientedness ; Extreme
PF 4114 Present policies ; Unacknowledged future costs
PD 1718 Presentation news ; Biased
PD 6081 Presentations ; Distorted media
PF 1677 Presentations ; Incomprehensible
PE 1669 Preservation books ; Inadequate
PF 7542 Preservation cultural property ; Inadequate protection
PD 0361 Preservation ; Irradiation method food
PB 9171 Preservation ; Lack species
PC 8390 Preservation obsolete systems
PF 1688 Preservation status quo ; Uncritical
PD 0487 Preservatives ; Toxic food
PE 8741 Preserved fruit ; Instability trade
PD 9603 Press abuses privacy
Press [Clandestine...]
PD 2366 — Clandestine
PE 7945 — coverage legal affairs ; Restriction
PF 3072 — coverage parliamentary affairs ; Restriction
PE 8951 Press ; Denial right freedom
PD 0160 Press ; Harassment
PE 8951 Press ; Lack freedom
PE 0246 Press ownership ; Concentration
PD 5377 Press standards ; Deterioration
PD 2366 Press ; Underground
PE 0628 Press weakness ; Student
PE 0472 Pressure ; Abnormal blood
PB 3209 Pressure ; Dependence political
PF 3350 Pressure eliminate nakedness developing countries ; Ill considered
PF 2009 *Pressure ; Excessive financial*
PC 3188 Pressure ; Foreign
PE 0472 Pressure ; High blood
PD 0722 Pressure ; Inadequate water
PD 0722 Pressure ; Insufficient water
PE 0472 Pressure ; Low blood
PD 4178 Pressure ; School
PF 7227 Pressure standards quality ; Quantitative
PB 3209 Pressure ; Undue political
PD 9090 Pressure ; War time conditions
PE 7628 Pressured early marriage
PE 6937 Pressured employees
PD 3389 Pressures ; Undemocratic
PF 3455 Prestige projects ; Excessively costly
Prestige Pursuit
PF 7983 — corporate
PF 8434 — national
PF 8145 — personal
PC 5348 Prestige race ; International
PE 7393 Presumed innocent until proven guilty ; Denial right
PA 6862 Presumption *Intrusion*
PA 6607 Presumption *Misjudgement*
PA 7115 Presumption *Rashness*
PA 6921 Presumption *Unduenesss*
PA 6491 Presumption *Vanity*
PF 6814 Presuppositions ; Basic
PE 5076 Pretences ; Obtaining property false
PF 8229 Pretended ignorance events
PF 5943 Pretenders ; Royal
PA 6400 Pretentiousness *Affectation*
PA 6491 Pretentiousness *Vanity*
PD 1285 *Prevailing chicken diseases*
PD 9044 Prevailing community insecurity
PF 7713 Prevailing learning systems ; Inadequacy
PF 7713 Prevailing mental structures challenge human survival ; Inadequacy
PF 3826 Prevalence psychological conditions unfavourable transition post-capitalist social order
PF 6530 Prevalence subsistence earning
PF 1365 Prevalent images ; Oppressive
PF 9756 Prevarication ; Legal
PA 5502 Prevarication *Reason*
PA 7411 Prevarication *Uncommunicativeness*
PC 4710 Preventable deaths
PC 4710 Preventable disabilities
PE 3286 Prevention ; Birth
PE 8542 Prevention computer disasters ; Inadequate
PF 4924 Prevention crime ; Inadequate
PF 0709 Prevention disabilities ; Inadequate
PD 8731 Prevention exchange ideas
PC 0185 Prevention exchange information
PF 9401 Prevention ; Inadequate disease
PD 1631 Prevention ; Inadequate fire
PF 3566 Prevention mitigation ; Inadequate disaster
PF 4240 Prevention terrorism ; Ineffective

PF 6472 *Prevention ; Unpractised dental*
PD 8023 Preventive health care ; Misunderstanding
PE 0751 Preventive medicine ; Insufficient
PF 8515 Preventive sanitation ; Lack
PE 1348 Prevents saving ; Immediacy
PD 3155 Prey ; Extermination wild animal natural
PE 3290 *Price ; Bride*
PE 7313 Price cash crops ; Low
PE 3528 *Price control ; Middle man*
PD 9802 Price difficulties developed countries ; Domestic agricultural
Price [fixing...]
PE 4301 — fixing
PE 4301 — *fixing commodity markets*
PF 8635 — Fluctuation feed
PD 0909 — *fluctuations ; Oil*
PF 8635 — *fluctuations ; Unanticipated*
PE 2890 Price policy difficulties developing countries ; Domestic agricultural
PC 2618 *Price regulation*
PE 4162 Price ; Rising land
PE 8309 Price support ; Government protection national commodity
PD 9802 Price supports subsidies ; Counter productive
PF 8635 Price ; Uncontrolled market
PC 0492 *Price ; Undercutting vendor*
PF 3409 *Price ; Variation fertilizer's*
PF 8635 Price volatility
PC 0840 Price warfare
PF 1267 Priced cooking fuel ; High
PD 1891 Priced goods services ; Over
PD 2968 Prices commodities exported developing countries ; Continuing low level
PD 2968 Prices ; Cyclic swings commodity
Prices [Decline...]
PD 2968 — Decline developing country commodity
PE 4986 — developing countries ; Instability food
PD 2968 — developing country earnings ; Instability commodity
PE 1182 — Distortion international trade minimum pricing regulations measures regulate domestic
PF 1238 Prices ; Excessive retail
PD 1891 Prices goods services ; Inflated
PF 1238 Prices ; High consumer
PF 2385 Prices ; High local
PF 8635 Prices ; Instability
PD 0909 Prices ; Instability petroleum
PF 6980 Prices ; Misleading market
PF 2385 Prices rural communities ; Inflated
PF 2385 Prices rural communities ; Rising food
PE 0245 Prices transnational corporations ; Manipulation transfer
Prices [underlying...]
PE 3827 — *underlying asset values ; Disparity share*
PF 8635 — Unskilled market
PF 8635 — Unstable crop
PE 5755 Pricing brand name drugs ; Excessive
PE 3528 *Pricing ; Exploitation rural*
PE 5523 *Pricing ; Over*
PE 2396 Pricing policies transnational corporations ; Restrictive
PE 1193 Pricing ; Profit repatriation concealed transfer
PE 1182 Pricing regulations measures regulate domestic prices ; Distortion international trade minimum
PE 1193 Pricing ; Transfer
PE 5855 Pricing transnational corporations ; Unfair
PE 7826 *Prick ; Nail*
PE 7826 *Pricked foot*
PE 2398 *Prickly heat*
PA 7599 Pride
PA 7599 Pride ; Dependence
PF 2093 *Pride ; Embarrassed civic*
PF 6570 *Pride ; Misplaced independent*
PF 1390 *Priesthood ; Homosexuality*
PF 6326 Priesthood ; Religious discrimination against women
PF 8889 Priesthood ; Unethical practices
Primary commodities
PD 2898 — because rising living standards ; Inadequate demand
PD 1276 — Competition synthetics
PE 0537 — Dependence industrialized countries imports
Primary commodities developing
PD 2968 — countries ; Excessive income dependence
PD 1554 — countries ; Lack processing industry
PD 2967 — countries ; Over production
PD 3042 — countries ; Underproduction
Primary commodities [due...]
PD 1276 — due technological change ; Reduction demand
PD 2968 — Economic dependence developing countries export
PD 0651 — Hoarding
PD 2968 — Instability export trade developing countries producing
PD 0057 — International trade barriers
PD 1276 — Production synthetic substitutes
PC 0463 — trade ; Instability
PD 1465 Primary commodity surplus
PE 2951 Primary commodity trade amongst developing countries ; Weakness
Primary education
PF 0674 — *Decline government expenditure*
PC 6381 — Denial right free
PC 6381 — Insufficient
PE 2264 *Primary glaucoma*
Primary [health...]
PD 9021 — health care ; Ignorance women concerning
PE 8553 — health care ; Inadequate
PD 7424 — *hyperparathyroidism animals*
PF 0674 — *Primary schools ; Decline public spending*
PE 1570 Primates ; Endangered species non human
PE 1621 Primates research ; Inhumane use non human

PF 5073 Primates ; Shortage experimental non human
PC 8160 Prime land ; Minimal
PC 8160 Prime land ; Unavailability
PD 3501 *Primitive cultures*
PF 2928 *Primitive secret societies*
PE 6715 Principals ; Political contributions agents foreign
PF 3420 Principles ; Compromise betrayal
PB 8031 Principles ; Erosion democratic
Principles [practice...]
PF 4705 — practice ; Discrepancies
PF 2960 — practices developing countries ; Differences trading
PC 2952 — practices different economic systems ; Differences trading
PD 5001 Principles techniques ; Disregard safety
PF 9629 Principles ; Unilateral interpretations multilateral
PF 5520 Principles ; Vulnerability universal
PD 4552 Printed matter ; Proliferation
PF 3628 *Printed media ; Unenticing*
Printing industries
PE 1425 — Environmental hazards paper
PE 8095 — Environmental hazards publishing
PE 1927 — Instability paper
PE 1136 — Underdevelopment paper
PJ 0451 Prior programme approval ; Government requirement that economic benefits be proven
Priorities Conflicting
PF 5766 — [Priorities ; Conflicting]
PF 1776 — *agency*
PF 7118 — *citizen*
PF 7118 — *community*
PF 1740 Priorities construction industry ; Short term
PF 6504 Priorities ; Consuming personal
PF 5766 Priorities ; Disagreement concerning
PF 1382 Priorities ; Family money
PF 6504 Priorities ; Home centred
Priorities [Immediate...]
PF 0317 — Immediate economic
PF 5766 — Incompatible
PF 9276 — Indeterminate future
PF 9068 Priorities ; Low educational
PF 5645 Priorities ; Misguided
PE 1702 Priorities ; Monopolization interest groups development community
PF 1959 *Priorities ; Non participatory financial*
Priorities [Reallocation...]
PF 0648 — Reallocation aid funds alternative
PE 4059 — research ; Emphasis economic
PF 0317 — rule ; Economic
PF 3318 *Priorities ; Sustenance determined*
PD 2884 Prioritized attachment security structures
PF 2261 *Prioritized funding ; Ineffectively*
PF 3523 *Priority farm work*
PF 1030 *Priority ; Individual family*
PF 7274 Priority innovations ; Disputed
PF 8876 Priority legislative reform ; Failure process low
Priority Low
PF 2605 — *athletics*
PF 7118 — *commune*
PF 6472 — *dental*
PF 9068 — *educational*
PF 9068 — *high school*
PF 2605 — *sports*
PF 6526 *Priority road surfacing ; Non*
PF 2969 *Priority status villages ; Low*
PF 2997 Priority system ; Inadequate economic
PF 0131 Prison ; Children reared
PD 0520 Prison conditions penal systems ; Inadequate
PE 6424 Prison foreign nationals ; Failure notify imprisonment death
PD 0165 Prison labour ; Abuse
PE 0998 Prison officers ; Maltreatment prisoners
PE 1675 Prison riots
Prison sentences
PD 0520 — Arbitrary extension
PE 4602 — Disproportionately long
PE 4602 — Excessive length
PF 4723 — Politically motivated reduction
PD 8680 Prison suicides
PD 2484 Prisoners ; Beating
PE 0428 Prisoners ; Bullying amongst
Prisoners [Child...]
PE 6636 — Child
PD 3171 — communist systems ; Political
PC 6935 — conscience
Prisoners [defence...]
PE 8637 — defence ; Lack access
PD 0520 — Denial rights
PD 3015 — detainees ; Civilian political
PD 3014 — detainees ; Military political
PE 6883 — Discriminatory treatment foreign
PE 0978 — Drug abuse
PE 6929 Prisoners ex detainees ; Discrimination against ex
Prisoners [families...]
PE 5043 — families ; Deprivation
PE 5043 — families ; Discrimination against
PE 0428 — fellow inmates ; Maltreatment
PE 0428 Prisoners ; Harassment amongst
PF 0131 Prisoners ; Inadequate care children
PE 0428 Prisoners inmates ; Intimidation
Prisoners [Maltreatment...]
PD 6005 — Maltreatment
PD 0702 — Mass detention political
PD 1480 — medical treatment ; Depriving
PE 4430 — mental institutes ; Political
PF 3020 — Mis classification political
PC 0562 Prisoners ; Political

PE 0998 Prisoners prison officers ; Maltreatment
PF 4875 *Prisoners ; Releasing repatriated*
PD 0520 *Prisoners ; Repeated resentencing released*
PE 4889 Prisoners ; Unethical medical experimentation
PE 0998 Prisoners ; Unsanctioned maltreatment
Prisoners war
PC 8848 — [Prisoners war]
PD 1652 — Brainwashing
PE 1200 — enemy aliens ; Aiding escape
PD 0218 — Forced repatriation
PD 2617 — Ill treatment
PE 0948 — Non repatriation
PD 9414 Prisons ; Corruption
PE 0428 Prisons ; Crime
PE 0978 Prisons ; Drug abuse
PE 1363 Prisons ; Homosexuality
PD 0702 Prisons ; Labour
PD 0520 Prisons ; Overcrowding
PE 1363 Prisons ; Rape
PE 1363 Prisons ; Unreported rape
PB 0284 Privacy ; Arbitrary interference
PE 0223 Privacy compulsory telecommunications ; Invasion
PB 0284 Privacy ; Denial right
PB 0284 Privacy ; Dependence infringement
PB 0284 Privacy ; Infringement
PB 0284 Privacy ; Invasion
PD 9603 Privacy media ; Invasion
PD 9603 Privacy ; Press abuses
PB 0284 *Privacy testing ; Invasion*
PB 0284 Privacy ; Unlawful interference
PF 4176 Privacy use computer databases ; Erosion human rights
PD 0527 Private armies
PF 4480 Private automobiles ; Maldistribution
Private [capital...]
PC 1453 — capital ; Exodus
PF 2573 — control communications mass media ; Dangers
PD 7608 — correspondence ; Denial right
PD 3954 Private debt
PD 3132 Private domestic capital outflow developing developed countries
PE 2700 Private economic expansion ; Restrictions
Private enterprise
PF 1879 — Bias against
PD 2048 — Dependence socialist countries
PD 2048 — socialist countries ; Suppression
PD 3032 *Private film clubs*
PD 3032 *Private ; Film clubs*
PD 4543 Private grants ; Fraudulent procurement use government
Private [hall...]
PC 4867 — *hall management*
PE 6168 — home life ; Denial right
PF 6578 — *homes ; Abandoned*
PF 4095 Private initiative ; Government opposition
Private international
PD 2107 — arms dealers
PJ 8060 — labour management relations ; Inadequate
PE 8547 — trade investment disputes
Private investment
PD 3138 — developing countries ; Disincentives foreign
PE 1911 — disputes ; Transfrontier
PE 8957 — income outflow developing developed countries ; Foreign
PF 6477 — Limited
PF 1653 *Private investors ; Uninterested*
Private [land...]
PF 4670 — *land ; Idle*
PD 5727 — law ; Violations
PD 4840 — life ; Upheavals
PF 8016 Private medical care ; Prohibitive cost
PD 9022 *Private nuisance*
PD 4974 Private ownership
Private [police...]
PE 4847 — police forces ; Abuses
PE 0347 — procurement policies ; Distortion international trade discriminatory government
PC 0939 — profit socialist countries ; Illegal
PA 0832 — *property ; Destruction*
PD 3055 — property ; Government expropriation
PF 3526 Private resources ; Undeveloped channels public
PJ 2918 Private schools ; Proliferation
Private sector
PE 8572 — developing countries ; Inadequate development
PE 7760 — employees ; Disparity remuneration public
PD 4800 — Excessive government intervention
PE 0780 — initiatives government ; Cooptation
PE 8599 — Lack conservation energy
PF 4095 — resources government ; Ineffective use
PD 3032 *Private theatre clubs*
PJ 0914 Private utilization public goods services ; Clandestine
PF 2477 *Private vehicles ; Unmaintained*
PD 8056 Privately owned water
PA 7382 Privation *Loss*
PA 6434 Privation *Poverty*
PF 2139 Privatization intergovernmental monetary system
PE 3391 Privatization public services
PE 3391 Privatization satellite systems
PE 3391 Privatization state monopolies
PF 8025 Privilege
PD 5725 Privilege ; Official
PF 0609 Privilege ; Threat parliamentary
PF 5203 Privilege ; Unwillingness sacrifice personal
PF 4849 Privileged classes citizens
PF 6670 Privileged classes developing countries ; Reinforcement inappropriate development
PE 8261 *Privileged ; Discriminatory use medical resources heal*
PD 5616 Privileged families
PF 8025 Privileges ; Accumulation
PD 5725 Privileges ; Corporate
PD 1231 Privileges ; Entrenched
Privileges immunities
PF 5649 — diplomats ; Abuse
PF 5649 — international civil servants ; Abuse
PE 5488 — international civil servants ; Violation
PE 8766 Prizefighting
PF 9693 Prizes academic integrity ; Corrupting effect valuable
PE 0197 *Prizes athletic events ; Discrimination against women payment*
Problem [alleviation...]
PF 0631 — alleviation ; General obstacles
PF 6580 — approaches ; Oversimplified
PF 5210 — avoidance
PD 0586 Problem children
PA 5497 Problem *Difficulty*
PF 9790 Problem emergence ; Irreversible
PD 6881 Problem families
PA 6997 Problem *Imperfection*
PF 7706 Problem interrelationships ; Assumption lack
PF 6580 Problem overemphasis ; Individual
PF 8753 Problem relevant information ; Failure international information systems order
PF 6580 Problem solving ; Piecemeal approach
Problem [Uncertainty...]
PA 7309 — *Uncertainty*
PA 7367 — *Unintelligibility*
PA 7107 — *Unpleasantness*
PE 1309 Problematic knees
PF 8753 Problematique ; Failure integrate knowledge empower humanity response global
PF 9821 Problematique ; Inadequate development political thinking concerning global
PF 7706 Problems ; Assumption absence causal relationship
PF 5864 Problems ; Belief that wealth eliminates
PF 7745 Problems ; Biased recognition
Problems [Complex...]
PF 0364 — Complex interrelationship world
PF 5210 — Conceptual repression
PE 8307 — countries ; Inadequate inter regional cooperation
Problems [Defeatism...]
PF 8781 — Defeatism face
PF 8112 — Delay recognition
PF 1393 — Delay religions acknowledging social
PF 9328 — Delayed emergence
PF 7706 — Denial causal relationship
PF 1120 — developing countries ; Inadequate research development
PF 6224 — disabled people ; Mystique attitudes
PF 4164 Problems ; Excessive emphasis fashionable
PF 4473 Problems ; Excessive reliance fashionable solutions
PF 2003 Problems facing society ; Multiplicity
Problems Failure
PF 5210 — acknowledge
PF 6580 — focus fundamental systemic
PF 8753 — interrelate wisdom different cultures response social
PF 2256 *Problems ; Fear urban*
PF 9409 Problems ; Fixation partial solutions
Problems Government
PF 6707 — delay remedial response
PF 6707 — delay response symptoms
PF 9704 — failure mobilize support against
PE 9295 — neglect future ecological
PC 3950 Problems ; Governmental resistance inertia response
PF 4473 Problems ; Habitual responses
PB 1423 Problems ; Inaction
Problems Inadequate
PF 8191 — application available knowledge solve
PD 2669 — buildings, services facilities organized action against
PF 1467 — coordination action against
PF 0301 — delivery mechanism response
PE 8216 — education concerning nature
PF 0209 — exchange technical information concerning
PF 0817 — global cooperation solve world
PF 1010 — implementation plans programmes against
PF 1645 — legislation relating action against
PF 2431 — organizational mechanisms act against
PF 1467 — planning action against
PF 5701 — public information concerning
PF 1077 — research
PF 1572 — research proposed solutions
PF 0625 — statistical information data
PF 1981 — technical advice consultation
PF 0863 — technical cooperation
Problems [Insoluble...]
PF 4706 — Insoluble scientific
PF 8846 — *Institutional preoccupation obsolete*
PF 3699 — Invention new social
Problems Lack
PF 1572 — innovative projects against world
PC 5180 — political will act
PF 1572 — well researched projects alleviate
PF 9328 Problems ; Latent
Problems [Marital...]
PD 0518 — Marital
PC 0180 — migrant labour
PF 9409 — Minimally cooperative solutions
PF 7186 — Minimization
PF 0364 — Multi disciplinary characteristics
PF 2003 — Multiplicity
PF 8112 Problems ; Non recognition
Problems [Obsession...]
PF 9119 — Obsession immediate
Problems [Obsolete...] cont'd
PF 1572 — Obsolete programmes against social
PF 6707 — Official inaction response potential
PF 7186 — Over optimistic assessment
Problems Overemphasis
PF 7118 — *community*
PF 6580 — discrete
PF 7118 — *local*
PF 7118 — *parochial*
PF 6580 — symptoms more fundamental
Problems [Polarized...]
PF 9691 — Polarized protest against
PF 9208 — policy innovation ; Constraint inherited
PF 4164 — Popular
PF 6580 — Preoccupation isolated
PF 9119 — Preoccupation short term
PF 9821 — prospects humanity ; Inadequate global consensus concerning
Problems [realistically...]
PF 8435 — realistically ; Inability resolve
PF 7027 — Recurrence misapprehended world
PE 7111 — related human hair ; Cosmetic health
PF 8781 — Resignation
PF 2823 — Restriction funding research social
Problems Shortage
PF 0559 — adequately trained personnel act against
PE 8789 — equipment materials needed act against
PF 0404 — financial resources action against
Problems [Simplistic...]
PF 0799 — Simplistic technical solutions complex environmental
PE 9295 — social environment arising current policies ; Government neglect future
PF 8753 — society ; Failure focus available insight response
PF 9828 — Suppression information concerning social
Problems [Unawareness...]
PD 8023 — *Unawareness health*
PF 1572 — Uncreative strategies against
PF 7706 — Unproven relationships
PF 9148 — Unreported
PE 9107 Procedural obstacles world trade ; Informational
PF 2491 Procedure ; Complex loan
PF 8616 Procedure ; Inadequacy international legal
PF 1442 *Procedure ; Uncoordinated budget*
PD 9641 Procedure ; Unjust voter registration
Procedures [Absence...]
PF 0342 — Absence follow up
PF 2661 — Abuse bureaucratic
PD 8697 — Ambiguous enforcement
Procedures [community...]
PF 0963 — community planning ; Inadequate
PF 8053 — Complex bureaucratic
PF 8519 — Complex legal
PF 8053 — *Complexity land*
PF 8519 — Complicated legal
PF 0639 — Contempt parliamentary
Procedures [delaying...]
PF 5120 — delaying welfare ; Government
PE 2832 — Denial right grievance
PF 2491 — developing countries ; Restrictive loan
PE 2603 — Distortion international trade discriminatory customs administrative entry
PF 5872 — documentation ; Inefficient shipping
PF 1266 Procedures education ; Inappropriate selection examination
PC 0666 Procedures equipment ; Inadequate standardization
PF 0292 Procedures ; Governmental disregard international values
PE 7780 Procedures ; Harassment travellers consular
PF 3702 *Procedures ; Hypersensitive military mobilization*
Procedures [Ignorance...]
PE 1234 — *Ignorance*
PF 1653 — *Ignorance bureaucratic*
PE 3580 — Inadequate abortion
PF 1266 — Inappropriate educational testing
PE 8884 — Ineffectiveness consultation
PE 5179 — international organizations ; Distortion democratic
PE 4097 Procedures ; National defence procurement
PF 8793 Procedures ; Outdated
PF 1786 Procedures ; Outdated labour negotiation
PF 5374 Procedures public sector ; Opaque budgetary
PC 6757 Procedures ; Secret government
PF 0245 Procedures ; Static credit
Procedures [Tedious...]
PF 6505 — *Tedious funding*
PF 7022 — *Time consuming*
PF 6494 — transposing ancient traditions ; Unclarified
PD 2438 Procedures ; Undetermined consensus
Procedures Unfamiliar
PF 2009 — *banking*
PF 1653 — *bureaucratic*
PF 2845 — *government*
PF 1379 *Procedures unsafe products ; Inadequate recall*
PF 7984 Procedures using public health programmes ; Unfamiliar
PE 7780 Procedures ; Victimization immigration control
Proceedings [Discrimination...]
PF 1798 — Discrimination against foreigners legal
PD 7359 — disorderly conduct ; Hindering
PE 1295 — due protective legislation ; Discrimination against juveniles judicial
PE 0405 Proceedings ; Legal profession's monopoly court
PE 1743 Proceedings ; Prohibitive cost linguistic interpretation legal
PD 0790 Proceedings ; Soliciting obstruction
PE 6781 Proceedings ; Tampering witnesses informants
PD 3132 *Proceeds ; Non repatriation export*

Processing capacity

PF 9292 Processing capacity ; Limited human information
PD 0425 Processing developed countries commodities exported developing countries
Processing industries
PE 1280 — Environmental hazards food
PE 8472 — Ineffective self regulation food
PD 0908 — Underdevelopment food
PD 1554 Processing industry primary commodities developing countries ; Lack
PE 4602 Processing parole applications ; Delay
PE 2829 Processing solid wastes ; Inadequate facilities storing, transporting
PE 4428 Processing ; Unreliability computer
PF 5299 Procrastination
PF 8989 Procrastination
PF 6707 Procrastination response evidence potential hazards ; Bureaucratic
PF 3682 Procrastination science face unexplained
PC 6870 Procreate ; Denial right
PE 4544 Procreate severely mentally handicapped ; Denial right
PF 8053 *Procurement ; Complex land*
PE 0347 Procurement policies ; Distortion international trade discriminatory government private
PE 4097 Procurement procedures ; National defence
PD 4543 Procurement use government private grants ; Fraudulent
PE 5303 Procurers
PA 7269 Prodigality
PA 7193 Prodigality *Cheapness*
PF 7318 Prodigality ; Environmental
PA 5473 Prodigality *Insufficiency*
PA 6466 Prodigality *Intemperance*
PA 7269 Prodigality *Prodigality*
PA 7411 Prodigality *Uncommunicativeness*
PA 5644 Prodigality *Vice*
PE 1756 Produce information be sworn ; Failure appear witness
PD 1285 *Produce ; Limited farm*
PE 8063 Produce shipment ; Expensive
PF 2875 *Produce transport ; Difficult*
PF 2875 *Produce transport ; Undependable*
PF 3161 Produced survival living ; Limited horizons
PF 7047 Producer ; Neglect small agricultural
PE 8516 Producers ; Lack credit facilities agricultural
PD 2968 Producing primary commodities ; Instability export trade developing countries
PE 2328 Product differentiation transnational corporations ; Limited market access due
Product [economy...]
PF 6540 — *economy ; Single*
PF 2466 — expansion ; Neglect research natural resource
PD 5426 — extortion
Product [liability...]
PE 4404 — liability protection ; Prohibitive cost
PC 2008 — life cycles ; Manipulated
PF 1740 — life cycles ; Short term planning
Product [manufacture...]
PD 3998 — manufacture ; Defective
PF 8299 — market ; Distant
PF 0581 — markets ; Insufficient
PD 1415 *Product operation ; Single*
PE 0778 Product origin ; Distortion international trade discriminatory requirements respect marks
PD 7981 Product piracy
PD 0606 Product placement ; Cinematic
PD 1435 Product quality ; Deterioration
PF 7741 Product quality ; Planned degradation
PE 1722 Product repair ; Prohibitive cost
Product [shipments...]
PE 7064 — shipments ; Infected animal, meat animal
PE 0083 — standards measures ; Distortion international trade discriminatory requirements respect
PD 0343 — standards ; Restrictive agreements
PD 8804 Product tampering
Production [activities...]
PC 1217 — activities ; Instability economic industrial
PE 7188 — alcoholic beverages ; Illicit
PD 1283 — animal feed ; Use agricultural resources
Production capacity
PD 4219 — developing countries ; Inadequate
PD 0779 — Excess
PD 0779 — utilization ; Decline
PD 1465 Production commodities ; Over
PF 4336 Production countries ; Unequal distribution
PE 2305 *Production ; Decreasing coconut*
PF 9239 Production deliberately distorted maps
Production developing countries
PD 5092 — Deterioration domestic food
PF 4130 — Economic disadvantages excessive food
PE 8188 — Fluctuations food
PF 4385 Production distribution ; Area disparities book
PE 8236 *Production domestic consumers developing countries ; Lack*
Production [Elitist...]
PD 0154 — Elitist control
PD 0376 — Environmental hazards agricultural livestock
PD 6693 — Environmental hazards energy
PD 2596 — *environmental hazards ; Natural gas*
PD 4977 — Environmental hazards nuclear power
PD 8851 Production facilities ; War time disruption economies
PD 2894 Production food live animals ; Instability
Production Health
PD 1714 — hazards drug use meat
PD 1714 — hazards hormone use animal
PE 0524 — risks workers agricultural livestock
PC 0328 Production ; Hidden environmental costs economic
Production [Inappropriate...]
PE 1785 — Inappropriate agricultural subsidies livestock
PE 9055 — industrialized countries ; Obstacles restructuring
PD 5830 — industries ; Protectionism agriculture food
PD 2743 — *Inefficient animal commodity*
PE 8998 — Instability agricultural livestock
PE 8955 — instability ; Petroleum natural gas
PE 7957 — Internationalization capitalist
PE 9115 Production ; Lack meat egg
PE 3547 *Production ; Lack modern*
PC 3651 Production non-essentials
Production [Obsolete...]
PF 1822 — Obsolete methods agricultural
PF 1822 — Out date methods agricultural
PD 0779 — Overcapacity industrial
Production [patterns...]
PF 1933 — *patterns ; Family based*
PF 1822 — *plan ; Overburdened*
PC 1217 — planning ; Uncoordinated
PD 2967 — primary commodities developing countries ; Over
Production processes
PE 8140 — Insufficient knowledge local
PE 3780 — Public non accountability control
PS 2881 — *Unfamiliar*
Production [rejects...]
PD 3998 — rejects
PF 9432 — respond fall demand ; Failure commodity
PE 4835 — Risks power
Production [serving...]
PF 2639 — serving false consumption needs
PC 4815 — Shortage water food
PC 1960 — Slowing growth food
PD 1285 — Stagnated development agricultural
PD 2970 — structures developing countries ; Rigidities
PC 2062 — *Surplus domestic animal*
PD 1276 — synthetic substitutes primary commodities
Production [techniques...]
PD 0528 — techniques ; Unemployment due mass
PE 7313 — *techniques ; Unsophisticated*
PF 7022 — *time ; Extended*
Production [Underdeveloped...]
PF 6493 — Underdeveloped approaches local food
PD 0629 — Underdevelopment agricultural livestock
PF 2821 — Underdevelopment food live animal
Production Unequal distribution
PD 4316 — agricultural
PF 4490 — livestock
PD 4322 — meat
Production [unit...]
PF 8155 — unit ; Growth size
PF 2014 — Unregulated ownership means
PS 2881 — *Unsystematic vegetable*
PD 6103 — Use agricultural land fuel
PD 5114 Production ; Vulnerability agriculture developing countries future declines
Production [Water...]
PE 3934 — Water pollution animal
PE 9134 — weather ; Adverse effects power
PE 0893 — work force needs ; Disrelationship
PE 2858 Production zones export enclaves ; Misuse free
PF 4202 Productive athletic activities ; Non
Productive [capacity...]
PE 9265 — capacity ; Decline capital investment
PC 9785 — capacity ; Maldistribution
PE 3816 — capitalist elites ; Non
PF 8120 Productive effects foreign aid ; Counter
PF 1492 *Productive government constraints programme beneficiaries ; Counter*
PF 4000 Productive members society ; Non
Productive [personnel...]
PF 9786 — personnel reward systems ; Counter
PF 9802 — price supports subsidies ; Counter
PD 4974 — property ; Individualistic disposition
PD 3182 Productive systems ; Limited ownership
Productive [units...]
PE 8787 — units foreign investment ; Dislocation
PD 1802 — use cattle livestock ; Non
PF 6911 — use crime screening ; Counter
PF 0422 Productive vocational training ; Non
PD 7480 Productivity agricultural land ; Declining
PE 5883 Productivity agricultural workers developing countries ; Low
Productivity [Decline...]
PD 1052 — Decline soil
PC 8908 — Declining economic
PC 6031 — Declining worker
PE 8506 — developing countries ; Inadequate incentives increased
PF 1825 — Diminishing role older people due overemphasis economic
PE 8098 — diseased animals ; Economic loss reduced
PE 8098 — diseased animals ; Economic loss reduced
PD 8469 Productivity farm animals ; Lack
PD 5543 Productivity increase ; Decline rate
PD 5543 Productivity industrialized countries ; Declining
Productivity [Lack...]
PC 8908 — Lack
PD 1052 — land ; Diminution biological
PD 8469 — Loss animal
PC 8908 Productivity ; Reduction
PF 7610 Productivity socialist countries ; Declining
PE 4331 Productivity ; Unsustainable short term improvements agricultural
PE 2390 Products ; Aggressive promotion pharmaceutical
PE 5678 Products capitalist socialist countries ; Imbalance trade cultural
PE 3808 Products chemical warfare ; Trade
PA 6058 Profanation *Impiety*
PA 6852 Profanation *Inappropriateness*
PF 7484 Profanation sacred doctrine
PA 7280 Profanation *Wrongness*
PF 7427 Profanity
PD 5380 Profession ; Corruption legal
PD 5770 Profession ; Degradation medical
PE 0803 Profession, judiciary jury membership ; Economic barriers access legal
PE 5355 Profession ; Occupational risks hazards medical
PE 0405 Profession's monopoly court proceedings ; Legal
PE 4015 Profession torture ; Inhumane participation medical
PD 5380 Profession ; Unethical practices legal
PC 8019 Professional activities ; Irresponsible
PF 3294 Professional arrogance
Professional [bodies...]
PD 8629 — bodies ; Discrimination against non members
PJ 7921 — brain drain ; Inter
PF 4833 — burnout
PF 0897 Professional ; Collapse public servant role
PC 2178 Professional discrimination
PE 0810 Professional discrimination educators
PF 0021 Professional fields ; Lack refresher courses
PC 8019 Professional fraud
PF 8675 Professional gain ; Short term
PD 8488 Professional jealousy
PC 8019 Professional malpractice
PC 8019 Professional negligence
Professional practices
PC 8019 — Corruption
PD 8027 — Restrictive
PC 8019 — Unethical
PD 8027 Professional restrictiveness
Professional [salary...]
PF 2042 — *salary losses*
PF 7961 — scientific controlling instruments ; Shortage
PE 8169 — Scientific controlling instruments trade instability
PD 6576 — secrecy
PD 6576 — secretism
PD 0974 — service delivery ; Distrust
PE 6649 — stagnation
PF 3411 — standards ; Lack
PC 2178 Professionalism ; Discriminatory
PJ 2489 Professionalized services ; Exaggerated reliance
PF 1068 Professionals experts ; Impediments internationally mobile
PE 3170 Professionals ; Inadequate working conditions
PF 2110 Professionals ; Individualistic style
PD 9836 Professionals ; Lying medical
PC 0592 *Professionals ; Shortage resident*
PJ 3472 Professionals ; Threats independence
PD 8629 Professions ; Closed
PC 8019 *Professions conceal socially unacceptable initiatives ; Misuse societally endorsed*
PE 5194 Professions ; Imbalance training existence openings various
PF 1373 *Proficiency ; Inadequate national language*
PF 3551 Profit ; Accountability based solely
PE 7313 *Profit base ; Questionable*
PF 9529 Profit ; Cultural traditions blocking business
Profit [making...]
PF 2174 — making ; Business
PF 2174 — making motive ; Distortion
PE 7313 — *margin ; Narrow*
PF 2174 — maximization ; Short term
PF 2464 — motivated utilization construction technology
Profit [organizations...]
PF 3049 — organizations ; Inappropriate taxation international non
PD 2552 — oriented interest payments
PF 2174 — over-emphasis use national resources
PF 1746 Profit patterns history ; Failure
PF 3049 Profit philanthropic charitable organizations ; Inappropriate taxation not
PE 1193 Profit repatriation concealed transfer pricing
Profit [sharing...]
PE 6722 — sharing ; Trade union opposition
PE 3551 — social indicator ; Unreliability
PC 0939 — socialist countries ; Illegal private
PD 8007 Profitability ; Dismissal workers improve
PC 2618 Profiteering
PD 8392 Profiteering ; Official
PE 7313 *Profits ; Decreasing fish*
PF 2457 *Profits ; Delayed*
PC 2618 Profits ; Excess
PC 2618 Profits ; Excessive
PC 3134 Profits foreign investors developing countries ; Excessive repatriation
PE 7313 *Profits ; Low agricultural*
PD 4632 Profits ; Stifling middleman
PF 6530 Profits ; Subsistence level
PC 2618 Profits ; Windfall
PA 7193 Profligacy *Cheapness*
PA 5644 Profligacy *Vice*
PA 5821 Profligacy *Vulgarity*
PE 6413 Profound mental retardation
PD 3102 Program pollution
Programme [action...]
PE 9124 — action country level ; Ineffectiveness international organization
PF 8181 — advantages ; Overstated
PJ 0451 — approval ; Government requirement that economic benefits be proven prior
PF 1492 *Programme beneficiaries ; Counter productive government constraints*

Prolongation life

PF 8499 Programme costs ; Underestimation
PF 3498 *Programme development ; Unskilled*
PF 0342 Programme effort ; Lack consistent
PF 9243 Programme evaluations ; Falsification
Programme [follow...]
PF 0342 — follow up ; Weak
PF 8711 — formulators ; Collusion administrators funding agencies
PF 3409 — *funding ; Insufficient*
PG 7844 *Programme impact ; Limited*
PF 1578 Programme initiatives inappropriate local requirements
PF 2401 *Programme knowledge ; Insufficient*
PF 4984 Programme ; Makeshift educational
PJ 7793 Programme methods ; Lack
PF 6548 Programme offerings ; Narrow
PF 1959 *Programme responsibility ; Undefined*
PF 5969 Programme ; Unorganized immunization
PF 2721 *Programme ; Unscheduled summer*
PF 7737 Programmed budgets ; Over
Programmes [Abuses...]
PD 0136 — Abuses international assistance
PF 1010 — against problems ; Inadequate implementation plans
PF 1572 — against social problems ; Obsolete
PE 1553 — Alienation skilled committed personnel international organizations
PD 1809 — Alienation support international organizations
PD 1644 — Cost overruns large scale public
Programmes [Delays...]
PF 6502 — Delays community building
PE 7195 — Dependence developing countries external financing development
PE 8400 — developing countries ; Failure medical
PF 3339 — *Difficulty implementing development*
PF 4984 — Discontinued educational
PD 0512 — Distant elders'
PD 2046 — *dropouts ; Lack*
PF 4300 — due short term loans ; Uncertainty development
PF 4196 Programmes ; Excessive expense athletic training
PF 8368 Programmes ; Failure development
PF 0170 Programmes ; Fluctuations government social
Programmes [Impractical...]
PF 4944 — Impractical development
PF 3519 — Impractical training
PF 2421 — *Improper implementation*
PE 9643 — Inability generate sufficient culture specific
PF 2583 — *Inaccessible religious*
Programmes Inadequate
PF 4944 — conception development
PE 1209 — coordination international nongovernmental organizations
PD 0285 — coordination international organizations
PF 4984 — educational
PE 0741 — funding international nongovernmental organizations
PF 0498 — funding international organizations
PD 3712 — implementation development
PF 4013 — industrial retraining
PE 1973 — relationship international governmental nongovernmental organizations
PF 4180 — social development
Programmes Ineffectiveness
PD 5429 — conservation
PF 8368 — development
PF 0074 — intergovernmental organization
PF 1595 — international nongovernmental organizations
PF 1074 — international organizations
PF 6377 Programmes ; Inefficient public
PF 7483 Programmes ; Internal inefficiency public
Programmes Lack
PE 8503 — participation local welfare
PF 2810 — *social*
PF 4837 — understanding social economic contexts development
PE 8064 Programmes ; Low credibility international organizations
PE 9643 Programmes media ; Excessive use foreign
PF 9796 Programmes ; Millenarist
PE 1375 Programmes national level ; Inadequate coordination action intergovernmental
PD 3548 Programmes ; Outmoded work study
PD 7463 Programmes ; Paradoxes technology transfer
PF 4739 Programmes ; Proliferation government
Programmes [Rejection...]
PF 9149 — Rejection remedial
PE 7965 — Restrictions eligibility government
PF 4984 — *Rigid school*
PF 2885 Programmes ; Social bias planning training
PE 4360 Programmes transnational banks developing countries ; Socially irresponsible
Programmes [Unconsensed...]
PF 1733 — *Unconsensed environmental*
PE 7984 — Unfamiliar procedures using public health
PF 5409 — Unorganized recreation
PD 0433 Programming ; Advertiser controlled television
PE 4428 Programming defects
PF 7737 Programming ; Over
PD 5377 Programming standards ; Deterioration television
PF 3931 Progress ; Adverse consequences scientific technological
PF 4718 Progress against underdevelopment ; Lack
PF 3931 Progress ; Dogmatic belief benefits technical
PF 1545 Progress equated efficiency
PF 4306 Progress establishing New International Economic Order ; Lack
PF 1502 Progress ; Family structure barrier
PF 8648 Progress ; Inappropriate understanding
PF 1545 Progress ; Lack social
PF 3310 *Progress ; Lack technological*

PC 8694 Progress North South negotiations ; Limited
PF 4718 Progress ; Reversal development
PF 1545 Progress society ; Overemphasis economic
PD 7841 *Progressive degenerative myeloencephalopathy brown Swiss cattle ; Bovine*
PE 5956 Progressive erosion old age pensions
PD 0949 Progressive loss top soil
PE 5188 Progressive muscular atrophy
PF 9438 *Progressive pneumonia sheep*
PF 5502 Progressive reduction government action commitment
PF 3037 Prohibited travel women
PF 4735 Prohibition addictive drugs
PA 5497 Prohibition *Difficulty*
PA 5869 Prohibition *Exclusion*
PE 4332 Prohibition practice indigenous religion
PA 7321 Prohibition *Refusal*
PA 7296 Prohibition *Restraint*
PD 7210 Prohibition trade union meetings
PF 3862 Prohibitions ; Unwarranted
PF 8158 Prohibitive administrative overhead costs
PJ 9259 Prohibitive agricultural diversification costs
PJ 8591 *Prohibitive construction cost rural community buildings*
PF 3550 Prohibitive construction costs
Prohibitive cost
PJ 3244 — *abortion*
PD 1842 — accommodation
PF 1176 — *advertising*
PF 7870 — art
PF 6527 — basic services
PE 8261 — *birth control*
PF 3550 — building materials
PF 2874 — *business security*
PF 2410 — *capital equipment*
PJ 8591 — clinic construction
PF 1653 — connection electricity services
PF 1653 — connection public utilities
PE 8261 — *contraceptives*
PF 2899 — *crop treatment*
PF 2779 — disease control
PF 4375 — education
PD 2931 — *electoral campaigns*
PF 3318 — *electricity*
PE 1722 — equipment maintenance
PF 6527 — essential services rural communities
PE 7548 — eye-glasses developing countries
Prohibitive cost farm
PF 3630 — *equipment*
PF 8763 — *labour*
PF 2457 — *machinery*
Prohibitive cost [fertilizer...]
PF 3409 — *fertilizer*
PF 6478 — *fishing*
PF 6478 — *fishing equipment*
PF 1267 — *fuel*
PD 1891 — *goods services*
PE 8261 — health care
PF 1212 — *healthy foods*
PF 4375 — higher education
PF 0296 — home maintenance
PE 4154 — hospital facilities
PJ 7877 — imported goods
PE 9101 — inadequate development policies
PE 8632 — insurance
PD 4632 — intermediaries
PF 0703 — *knowledge information*
PE 4162 — land
PE 8736 — life assurance
PE 1743 — linguistic interpretation legal proceedings
PF 1238 — *living*
PE 1122 — maintaining comprehensive document collections
PF 0296 — *maintenance*
PF 9014 — *money*
PF 2385 — necessities rural communities
PF 7543 — nuclear power plants
PF 1212 — *nutritious food*
PF 1212 — *organically grown foodstuffs*
PF 1625 — *parenthood*
PF 2899 — *pesticide application*
PF 8016 — *private medical care*
PE 4404 — product liability protection
PE 1722 — *product repair*
PF 2605 — *road construction*
PE 8063 — *transportation*
PF 2899 — *weed control*
PF 5743 Prohibitive costs purification effluents emissions
PF 3526 *Prohibitive daycare requirements*
PF 1345 *Prohibitive drilling cost*
PJ 1768 *Prohibitive equipment costs*
PF 6548 *Prohibitive facility costs*
PF 1345 *Prohibitive irrigation costs*
Prohibitive [labour...]
PF 8763 — labour costs
PF 0995 — legal fees
PF 2378 — *livestock husbandry costs*
Prohibitive [marriage...]
PF 3290 — marriage dowry
PE 8261 — medical expenses
PD 5674 — minimum wages
PF 2457 *Prohibitive ownership costs*
PF 3630 *Prohibitive refrigeration costs*
PF 4375 *Prohibitive school expenses*
PD 4632 *Prohibitive start-up costs businesses*
PF 1653 *Prohibitive telephone costs*
PE 8698 Prohibitively expensive housing poor
PF 9243 Project assessments ; Misrepresentation
PF 9243 Project cost benefit analyses ; Manipulation

PF 8499 Project disadvantages ; Underestimated
PF 2706 Project funding ; Short range
PF 1470 Project implementation ; Delay
PE 4330 Project loans transnational banks developing countries ; Domination restrictive
PE 7613 Project tied migration ; Abuse
PF 4774 *Projections ; Incomplete cost*
Projects [against...]
PF 1572 — against world problems ; Lack innovative
PF 1572 — alleviate problems ; Lack well researched
PE 8687 — available labour ; Unsuitability development
PE 7832 Projects ; Diseases introduced water
PF 3455 Projects ; Excessively costly prestige
PF 8368 Projects ; Failure development
PF 9243 Projects ; False justification
PE 2309 Projects ; Government imposition rural cooperative
PF 4975 Projects ; Implementation development mega
Projects Inadequate
PF 4340 — commercial finance rural development
PF 9244 — control development
PF 9244 — evaluation development
PF 9244 — monitoring development
Projects [Inappropriate...]
PF 4944 — Inappropriate design development
PD 3712 — Inappropriate management development
PD 3712 — Ineffective execution development
PF 4975 Projects ; Misappropriation resources high cost civil engineering
PF 0716 Projects ; Misappropriation resources high cost research
PJ 8129 Projects ; Nonidentification intended beneficiaries
PF 8181 Projects ; Overestimated economic benefits
PF 8166 Projects ; Overuse machinery equipment development
PF 4944 Projects ; Unfeasible development
PF 3455 Projects ; Unnecessary status
Prolapse animals
PD 3978 — *Rectal*
PD 7799 — *Uterine*
PD 7799 — *Vaginal*
PD 7799 Prolapse ; *Cervical*
PE 2481 Prolapse rectum
PD 8775 Prolapse ; *Uterovaginal*
PD 2626 Prolapsed intervertebral disc
PD 3239 Proletariat ; Dictatorship
PD 5034 Proliferation advertising
PD 2072 Proliferation automobiles motor vehicles
PE 9052 Proliferation ballistic missiles developing countries
Proliferation [cash...]
PF 9187 — cash crops
PF 0815 — commercialism
PE 3959 — computers
PJ 0253 — *Conceptual*
PF 5057 — consumer products
PD 8942 — consumer waste
Proliferation [Decrease...]
PA 5589 — *Decrease*
PE 1810 — direct mail advertising
PF 5992 — documents international organizations
PD 4552 — due computerization ; Paper
PE 2417 — duplication intergovernmental organizations coordination bodies
Proliferation duplication international
PE 0458 — information systems
PE 0179 — nongovernmental organization coordination bodies
PE 0362 — nongovernmental organization information systems
PE 1029 — organizations coordinating bodies
PE 1579 Proliferation duplication organizational units coordinating bodies United Nations system
PE 0075 Proliferation duplication United Nations information systems
PF 4739 Proliferation government programmes
PD 4800 Proliferation government regulations
PD 4605 Proliferation immigrants
PC 1298 Proliferation information
PD 5315 Proliferation legislation
PF 0361 Proliferation litigation
PD 0644 Proliferation medical drugs ; Excessive
PD 0071 Proliferation mergers takeovers
PF 2723 Proliferation national international anniversaries years
Proliferation nuclear
PD 4977 — power
PE 9052 — weapons developing countries
PD 0837 — weapons technology
Proliferation [parallel...]
PF 1853 — parallel urban services
PF 4739 — parastatals
PD 2689 — pets
PD 4552 — printed matter
PJ 2918 — private schools
PF 4739 — public sector institutions
PF 4352 Proliferation ; Risk unintentional global nuclear war due nuclear
Proliferation [second...]
PF 1286 — second homes
PF 4739 — state-controlled enterprises
PD 0014 — strategic nuclear arms
PF 2420 Proliferation technology
PF 1065 Proliferation unprocessed scientific data
PE 2449 Proliferation weapons civilian hands
PE 3936 *Proliferative enteritis ; Porcine*
PE 3936 *Proliferative haemorrhagic enteropathy*
PA 7365 *Prolixity Boredom*
PA 5974 *Prolixity Diffuseness*
PA 5473 *Prolixity Insufficiency*
PF 4936 Prolongation dying process ; Excessive
PF 4936 Prolongation life ; Compulsory

Prolongation life

PF 4936 Prolongation life ; Inappropriate
PD 7799 *Prolonged gestation cattle*
PD 7799 *Prolonged gestation sheep*
PF 2471 *Prolonged repair services*
PC 0745 Promiscuity
PA 7361 Promiscuity *Disorder*
PA 6446 Promiscuity *Indiscrimination*
PF 2687 Promiscuity ; Sexual
PA 5612 Promiscuity *Unchastity*
PF 7150 Promise ; Breach
Promises Broken
PF 7150 — [Promises ; Broken]
PF 4558 — election
PF 4558 — government
PF 4558 Promises ; Extravagant political
PF 7150 Promises ; Unfulfilled
PF 4558 Promises ; Unrealistic election
PE 6680 Promissory notes ; Conflicting laws concerning bills exchange
PD 1714 Promoters ; Hormonal growth
PE 9125 Promoting fast livestock growth ; Use artificial methods
PE 5303 Promoting prostitution
PF 6557 Promotion community assets ; Minimum
PE 0523 Promotion developing country small industry products ; Inadequate export
PF 1025 Promotion dominates market research ; Sales
PD 4418 Promotion ; Fear self
PC 4844 Promotion health ; Unequal development
Promotion [Inadequate...]
PF 0963 — *Inadequate industrial*
PD 4418 — *Inhibited self*
PF 6557 — Insufficient community
PF 0963 — *Insufficient industrial*
PF 6471 Promotion ; *Lack industrial*
PF 1025 Promotion marketing ; Disrelated needs
PF 4133 Promotion negative images opponents
PD 9628 Promotion opportunities ; Denial right equal
PF 6657 Promotion ; Passive village
PE 2390 Promotion pharmaceutical products ; Aggressive
PE 1598 Promotion transnational corporations developing countries ; Minimal export
Promotion [Unequal...]
PD 9628 — Unequal
PF 0963 — *Unorganized industrial*
PF 9428 — unrealistic perspectives ; Government
PD 9628 Promotion women ; Unequal
PF 1878 Propaganda
PF 3087 Propaganda ; Art
PD 7171 Propaganda ; Black
PD 3090 Propaganda ; Book
PD 3092 Propaganda ; Denial right freedom war
PF 1878 Propaganda ; Dependence
PD 3089 Propaganda ; Film
PD 3041 Propaganda ; Foreign controls newspaper journal
PC 3074 Propaganda ; Government
PE 3077 Propaganda intergovernmental organizations
PD 0184 Propaganda ; Newspaper journal
PD 3086 Propaganda ; Photographic
Propaganda [Racist...]
PD 3093 — Racist
PD 3085 — Radio television
PD 3094 — Religious
PD 3088 Propaganda ; Theatre
PD 3092 Propaganda ; War pre war
PE 9059 Propagation ; Instability hunting, trapping game
PF 3706 Propagation knowledge ; Authoritarian
PE 8252 Propagation ; Underdevelopment hunting, trapping game
PC 1153 Proper use science technology ; Obstacles
PJ 8292 Proper use technology ; Obstacles
PA 5939 Properties∗complex ; Absolute
PA 6091 Properties∗complex ; Relative
PD 2338 Property ; Absentee landlords urban
PC 6873 *Property ; Abuse individual*
Property Accumulation
PC 8346 — [Property ; Accumulation]
PD 6907 — cultural
PE 3354 — misuse religious
Property [acquisition...]
PD 5552 — acquisition crimes
PB 5486 — *Arbitrary sequestration real*
PF 3550 — *assessments ; Outdated*
PC 1784 — assets ; Insecure
PF 3550 *Property base ; Insufficient*
PE 7903 Property boundaries ; Confusing
Property [Compulsory...]
PJ 0456 — Compulsory seizure
PJ 0456 — Confiscation
PE 7581 — Confiscation trade union
Property Crimes against
PD 5511 — [Property ; Crimes against]
PE 5634 — government public
PE 6486 — intangible
PD 5511 Property ; Criminal harm
Property damage
PD 5859 — [Property damage]
PD 5859 — Criminal
PD 5859 — Uncontrolled
Property Denial right
PD 3055 — freed arbitrary deprivation
PF 0886 — inherit
PF 8854 — intellectual
Property [Deprivation...]
PE 8170 — Deprivation trade union funds
PD 5859 — Destruction
PA 0832 — *destruction*

Property [Destruction...] cont'd
PA 0832 — *Destruction private*
PD 8492 — development ; Exploitative
PF 6537 — *Disrespected public*
PC 3438 — distribution ; Unequal
PD 7298 — during warfare ; Destruction cultural
PF 2346 *Property exchanges ; Unrecorded*
Property [false...]
PE 5076 — false pretences ; Obtaining
PE 8580 — Faulty title
PF 3512 — *Fear loss*
PD 3055 Property ; Government expropriation private
PF 7542 Property ; Gradual destruction cultural
Property [Illegal...]
PD 0820 — Illegal occupation unoccupied
PF 8854 — Inadequate protection intellectual
PF 7542 — Inadequate protection preservation cultural
PD 4974 — Individualistic disposition productive
PC 6873 — *Individualistic use*
PF 0886 — inheritance ; Unequal
PC 1784 — Insecurity
PF 1600 — Insufficient care community
PF 3859 — Irresponsible finders personal
Property Lack
PF 4516 — incentive users care common
PC 1784 — protection
PF 8854 — protection industrial
Property [Larceny...]
PD 4691 — Larceny public
PB 1997 — *laws ; Inflexible*
PE 4545 — loans ; Bad
PE 4545 — loans ; Inadequately secured commercial
PF 2613 — *Loss*
PD 8894 Property maintenance ; Neglect
PE 6074 Property ; Misappropriation cultural
PD 8894 Property ; Neglected personal
Property Non
PC 7859 — restitution
PE 6074 — restitution cultural
PC 7859 — return
Property [occupational...]
PD 3218 — occupational discrimination politics
PE 6074 — occupied territories ; War time exportation cultural
PD 2338 — ownership ; Absentee
Property [Political...]
PD 3012 — Political confiscation
PE 8364 — Possession fencing stolen
PE 7581 — public authorities ; Destruction trade union
PE 7581 — public authorities ; Seizure trade union
PE 8364 Property ; Receiving stolen
Property rights
PA 0833 — *because discrimination ; Denial equal*
PE 8411 — Disregard
PD 9162 — Ill defined
PF 5292 — Payments after expiration industrial
PD 8937 — Restrictions
PD 9162 — Traditional
PD 9162 — Unclear
PA 0833 — *Unequal*
PD 8937 — Violation
PE 4018 — women ; Unequal
PD 8202 Property speculation
Property [tax...]
PD 3436 — tax ; Paralyzing
PF 2444 — *taxes ; Unrelated*
PD 4691 — Theft
PC 1784 — theft damage ; Vulnerability
PF 8854 — Theft intellectual
PD 4691 — *Theft public*
PF 3005 — *titles ; Delay obtaining*
Property [Uncoordinated...]
PF 3005 — Uncoordinated use community
PD 8894 — Unkept rental
PC 1784 — Unlocked
Property [vandalism...]
PD 1350 — vandalism ; Recurring
PD 9343 — Vulnerability elders'
PF 8854 — Vulnerability intellectual
PE 4018 Property women ; Denial right
PF 9694 Prophecies ; Self fulfilling
PF 3495 Prophets ; False
PE 7680 Proposals ; Deceptive political
PF 9149 Proposals ; Rejection peace
PF 9149 Proposals social change ; Rejection
PF 1572 Proposed solutions problems ; Inadequate research
PD 3537 Proprietary information ; Infringement
PE 5755 Proprietary pharmaceutical products ; Overcharging
PF 3862 Proscribed thinking behaviour
PF 2607 Proscriptive controls favouring investor
PD 9545 Prosecute offenders effectively ; Government failure
PE 5515 Prosecution felon ; Hindering
PE 5267 Prosecution giving testimony ; Flight avoid
PE 0621 Prosecution ; Havens
PE 6386 Prosecution ; Indiscriminate anti trust
PD 1464 Prosecution ; Ineffective war crime
PF 8479 Prosecution ; Malicious
PE 6059 Prosecution offenders ; Excessive cost effective
PF 8661 Prosecutions crimes ; Difficulty ensuring
PF 9821 Prospects humanity ; Inadequate global consensus concerning problems
PE 8372 Prostate ; Abscess
PE 8372 Prostate ; Cancer
PE 8372 Prostate diseases
PE 8372 Prostate ; Hyperplasia
PD 7799 *Prostatic abscess*
PD 7799 *Prostatic cysts animals*

PD 7799 *Prostatic neoplasms animals*
PE 8372 Prostatitis
PD 7799 *Prostatitis animals*
PF 0772 Prosthetic devices ; Lack
PF 6231 Prostitutes ; Discrimination against
PE 0392 Prostitutes ; Discrimination against children
PE 0209 Prostitutes ; Violence against
PD 0693 Prostitution
PD 6213 Prostitution ; Adolescent
PE 5376 Prostitution AIDS
PD 3888 Prostitution brothels ; Involuntary
PE 7582 Prostitution ; Child
PF 6231 Prostitution ; Criminalization
PE 5303 Prostitution ; Facilitating
PD 3380 Prostitution ; Female
PA 6088 Prostitution *Impairment*
PA 6852 Prostitution *Inappropriateness*
PD 6213 Prostitution ; Juvenile
PD 3381 Prostitution ; Male
PD 4402 Prostitution ; Male homosexual
PE 5303 Prostitution others ; Exploitation
PE 5303 Prostitution ; Promoting
PD 6213 Prostitution ; Teenage
PD 4525 Prostitution ; Transvestite
PA 5612 Prostitution *Unchastity*
PE 3390 Prostitution women ; Military
PE 0648 Protect aged ; Misallocation resources
PC 3185 Protect investments their citizens foreign countries ; Intervention major powers
PD 2664 Protected endangered species ; Illegal hunting
PC 4764 Protected habitats ; Endangered
PC 4764 Protected natural areas ; Vulnerability
PC 4764 Protected sites ; Endangered
PC 0380 Protected wildlife ; Smuggling
Protection against
PE 5651 — chemical occupational hazards ; Conflicting standards
PE 5793 — dissolution ; Violation right trade unions
PE 5793 — suspension ; Violation right workers organizations
PD 0096 Protection aged during disasters ; Lack
PF 8652 Protection ; Breakdown police
PE 8361 Protection civilians armed conflict ; Inadequate
PF 4432 Protection company ownership
Protection [Distortion...]
PD 0455 — Distortion international trade obstacles patent
PE 4694 — domestic animals during disasters ; Lack
PE 7119 — during maternity leave ; Denial right job
PD 6733 Protection employment ; Inadequate maternity
PE 8240 Protection extortion victims ; Lack legal
Protection [facilities...]
PD 1846 — facilities workers' representatives ; Inadequate
PD 9155 — farmers ; Inadequate social
PE 7793 — following death threats ; Lack
PE 3977 — Fragmented social structures environmental
PF 7556 — free markets ; Lack
Protection [handicapped...]
PD 0098 — handicapped during disasters ; Lack
PE 8497 — humankind ; Denial animals right attention, care
PF 5121 — humankind ; Denial animals right attention, care
PC 0123 Protection ; Inadequate consumer
PD 1631 Protection ; Inadequate fire
Protection individual
PF 0922 — rights due excessive court costs ; Ineffective
PE 1479 — rights ; Psychological barriers judicial
PB 0883 — welfare ; Inadequate
Protection [industrial...]
PF 8854 — industrial property ; Lack
PD 1987 — industries customs duties ; Excessive
PF 8854 — intellectual property ; Inadequate
Protection [Lack...]
PC 0238 — Lack plant
PD 1590 — Lack typhoon
PC 1268 — law ; Denial right equal
PC 8999 — linguistic, national religious minorities ; Inadequate
PD 4469 Protection money ; Extortion
PE 8309 Protection national commodity price support ; Government
Protection [Partial...]
PF 8652 — Partial police
PF 4662 — poor ; Lack commitment
PF 7542 — preservation cultural property ; Inadequate
PE 4404 — Prohibitive cost product liability
PC 1784 — property ; Lack
PF 4095 — public sector agencies competition
PF 7614 Protection ; Reduced images environmental
PF 3021 Protection refugees ; Inadequate
PF 8652 Protection security services ; Remote
PE 8643 Protection their rights ; Denial animals legal
PD 1631 Protection ; Unavailability fire
PD 1846 Protection union representatives ; Denial right
PE 7793 Protection victims intimidation ; Lack
Protection vulnerable
PD 3785 — groups during disasters ; Lack
PF 4662 — groups ; Lack commitment
PB 4353 — Lack
Protection [war...]
PE 3034 — war correspondents ; Inadequate
PE 4694 — wild animals during disasters ; Lack
PD 1078 — women children during disasters ; Lack
PC 4275 Protectionism
Protectionism [advertising...]
PD 7108 — advertising industry
PE 7025 — against imports service-related goods
PD 5830 — agriculture food production industries
PD 7132 — air transportation industry
PE 8134 — alcoholic beverages industry

Public disclosure

Protectionism [automobile...] cont'd
PC 5842 — *automobile industry*
PD 7120 Protectionism banking industry
Protectionism [commodities...]
PC 5842 — *commodities sectors*
PD 7001 — computer office machine industries
PD 7001 — computer services industry
PD 7049 — construction engineering services industries
PE 8795 — consumer products industries
Protectionism [defence...]
PE 8664 — defence arms industries
PE 8321 — developed countries against agricultural products developing countries
PD 3714 — developing countries
PD 2958 — developing countries ; Trade barriers
PD 7060 Protectionism entertainment products film industries
PD 7121 Protectionism franchising services industry
PE 7987 Protectionism free-zone international financial services
PE 8458 Protectionism high-technology industries
Protectionism [insurance...]
PD 7012 — insurance industry
PC 5842 — international trade
PD 9679 — international trade against exports developing countries
PC 5842 Protectionism labour-intensive industries
PD 8614 Protectionism legal services
PC 5842 Protectionism mining industries
PD 7073 Protectionism public accounting auditing services
Protectionism [services...]
PD 7135 — services industries
PE 5888 — shipping industry
PE 5866 — steel basic metal industries
PE 5819 Protectionism textile apparel industries
PC 4275 Protectionism ; Trade
PC 5842 *Protectionism ; Union*
PD 5033 Protectionist measures ; Unpredictable introduction
PE 1295 Protective legislation ; Discrimination against juveniles judicial proceedings due
PC 6848 *Protective legislation ; Lack*
PF 5255 Protectiveness ; Parental over
PD 2374 *Protectorates*
PF 2477 *Protectors ; Insufficient fire*
PD 0339 Protein-calorie malnutrition
PF 1773 *Protein consumption ; Insufficient*
PE 3147 Protein deficiency cereals
PD 0339 Protein ; Deficiency dietary
Protein energy malnutrition
PD 0339 — [Protein-energy malnutrition]
PD 0331 — infants early childhood
PD 0363 — vulnerable groups
PE 1085 — women during pregnancy nursing
PD 7089 Protein ; Excessive consumption
PE 4784 Protein ; Excessive cost animal
PD 0339 Protein intake ; Low
PD 0339 Protein ; Lack
PE 0873 Protein reactions ; Arthritis accompanying foreign
PC 4998 Protein ; Shortage animal
PC 1916 Protein supply ; Inadequate
PF 2758 Proteins factory farming ; Inefficient use
PD 5719 Proteins ; Loss micro organic
PF 8029 Protest
PF 9691 Protest against problems ; Polarized
PD 8522 Protest meetings
PD 5586 Protest workers
PD 1544 Protest ; Youth
PD 9557 Protests ; Governmental disregard legitimate
PC 0690 Protests ; Mass
PJ 1219 *Protocol ; Ignorance*
PD 3475 Protozoa ; Endangered species
PD 0594 *Protozoa infections ; Viral*
PE 6741 Protozoa pests
PD 0594 *Protozoa ; Virus diseases*
PE 3676 Protozoan diseases
PE 3676 Protozoan parasites
PF 2018 Protracted development regional plans
PE 7393 Proven guilty ; Denial right presumed innocent until
PF 8767 Proven methods ; Intolerance
PJ 0451 Proven prior programme approval ; Government requirement that economic benefits be
PF 5653 Proverbial lore errors
PF 5472 *Provincialism ; Western*
PE 3247 Provision alimony ; Inadequate
PF 6473 Provision basic services developing countries ; Underdeveloped
PF 2411 Provision consumer services ; Haphazard
PD 0512 Provision home care services elderly ; Inadequate
PE 0069 Provision international nongovernmental organizations ; Lack legal
PF 7034 Provision international nongovernmental organizations ; Lack national legal
PF 7058 Provision nongovernmental organizations ; Lack international legal
Provision [palimony...]
PE 3247 — palimony ; Inadequate
PF 2874 — public safety ; Inadequate
PF 2694 — public services communication ; Insufficient
PE 1032 Provisional release criminal cases ; Pecuniary guarantees condition
PE 5306 Provisions ; Grant back
PF 2497 Provisions ; Non implementation international treaty
PE 7300 Provisions ratified human rights agreements ; Failure governments implement
PF 6042 Provisions ; Time lag legal
PF 5395 Provisions UN charter ; Antiquated
PA 7107 *Provocation*
PA 7378 Provocation *Aggravation*

PA 7253 Provocation *Envy*
PE 1293 Provocation men ; Sexual
PF 3271 Provocation men ; Sexual
PD 1116 Provocation ; Sexual
PA 7107 Provocation *Unpleasantness*
PE 1293 Provocation women ; Sexual
PC 0937 Provocations capitalist militarism
PE 5607 Provocative dress
PF 9558 Provocative incidents ; Fabrication
PE 3704 Provocative military exercises
PF 2721 Prowess ; Absence social
PE 5050 Proximal femoral epiphysiolysis
PE 7826 Proximal sesamoids ; Fracture
PF 6197 Proximate offices ; Inhibition communication non
PF 8780 Proximity males females ; Unauthorized
PF 3284 *Proxy ; Marriage*
PF 5892 Prudery
PF 5892 Prudishness
PJ 5588 Pseudo
PE 6886 Pseudo-cowpox
PF 5513 Pseudo-culture
PE 2246 Pseudo hermaphroditism ; Human
PF 1626 *Pseudo-leadership*
PF 1626 Pseudo-revolutionaries
PF 1602 Pseudo-science
PF 4778 Pseudo-socialism state socialism
PE 6886 Pseudocowpox
PF 7799 *Pseudocyesis*
PD 9654 *Pseudohyperparathyroidism animals*
PD 2226 Pseudomonas
PF 7799 *Pseudopregnancy*
PE 9799 *Pseudorabies*
PE 9844 *Pseudotuberculosis*
PF 7986 Psionic warfare
PE 2578 Psittacosis
PE 5567 *Psittacosis*
PE 6325 Psoriasis
PE 0873 *Psoriasis ; Arthritis accompanying*
PD 0556 Psychedelic drug abuse
PE 2932 Psychiatric diagnosis ; Misuse
Psychiatric hospitals
PD 0584 — Abusive treatment patients
PD 0584 — Physical brutality
PE 2932 — Unjust commitment
PE 2932 Psychiatric institutions ; Abusive detention
PF 4925 Psychiatric institutions ; Inadequate
PD 5267 Psychiatric malpractice
PE 0540 Psychiatric out patient care ; Limited
PE 6351 Psychiatrism
PE 9172 Psychiatrists ; Inadequacy
PE 9172 Psychiatrists ; Inept testimony psychologists
PE 9172 Psychiatrists ; Lack concensus
PD 5770 *Psychiatrists personal counsellors ; Bogus*
PE 9172 Psychiatry ; Inadequacy
PE 9172 Psychiatry ; Lack diversity
PE 9172 Psychiatry ; Recreational
PD 5267 Psychiatry ; Unethical practices
PF 4968 Psychic conflict
PF 0508 Psychic interference decision-making
PF 4213 Psychic murder
PA 0430 *Psychic surgery*
PE 6968 Psychic traumatization ; Massive
PF 4866 Psychic warfare
PD 0270 Psycho-neuroses
PE 9172 Psychobabble
PD 4248 Psychochemical agents
PF 7076 Psychogenetic constraints behaviour
PD 0270 *Psychogenic amnesia*
PF 7918 Psychogenic death
PE 7451 Psychogenic fugue
PE 3974 Psychogenic physical disorders
PA 0587 *Psychological aggression*
PB 0147 Psychological alienation
PE 1479 Psychological barriers judicial protection individual rights
PF 3826 Psychological conditions unfavourable transition post capitalist social order ; Prevalence
PE 5087 Psychological conflict
Psychological [development...]
PF 6148 — development life cycle ; Inhibition individual
PD 8375 — *disorders*
PE 5087 — distress
PE 5087 — disturbance
PE 4299 Psychological effects AIDS
PC 3144 Psychological environment degradation
PA 6448 Psychological fixation
PA 0911 *Psychological fragmentation*
Psychological [impediments...]
PF 3344 — impediments marriage
PF 9818 — inconsistency marriage partners
PF 0421 — inertia
PF 0421 — inertia ; Dependence
PF 6339 — inhibition
PC 2935 — intimidation
PD 1983 Psychological pollution mass media
PF 1418 Psychological regression ; Human
PF 5950 Psychological resistance ; Chronic
Psychological [screens...]
PF 1112 — screens ; Escaping reality popular
PE 1670 — stress animals ; Experimental induction
PE 6299 — stress urban environment
PC 0300 — *symptoms ; Intentional feigning*
PD 5770 Psychological techniques ; Abuse medical
PD 4559 Psychological torture
PF 1185 Psychological ungroundedness
PA 0587 *Psychological violence*

PF 1978 Psychological violence ; Covert non physical
Psychological warfare
PC 2175 — [Psychological warfare]
PC 2175 — Clandestine
PC 2175 — operations ; Covert
PF 1269 Psychological well being ; Retirement threat
PF 0421 *Psychological withdrawal*
PE 9172 Psychologists psychiatrists ; Inept testimony
PD 1307 Psychomotor development children ; Retardation
PB 1044 *Psychomotor retardation*
PD 0270 Psychoneurosis
PD 2710 Psychopathic national leaders
PJ 6661 Psychopathological psychotherapy
PD 6880 Psychopathology ; Occupational
PF 1721 Psychopaths - Sociopaths
PF 1721 Psychopathy
PD 1722 Psychoses
PD 1318 Psychoses ; Affective
PD 1722 Psychosis
Psychosis [Affective...]
PD 1318 — Affective
PE 9263 — Alcoholic
PD 0919 — *Arteriosclerotic*
PD 0919 — *Atrophic senile*
PD 1318 Psychosis ; Brief reactive
PD 1318 Psychosis ; Circular
PC 0799 Psychosis ; Depressive
PD 1318 Psychosis ; Manic depressive
PD 1722 *Psychosis ; Postencephalitic*
PD 0438 *Psychosis ; Schizo affective*
PE 6965 Psychosis ; Socio
PE 7867 Psychosis ; War
PD 1967 Psychosomatic disorders
PE 5294 Psychosomatic effects torture ; Somatic
PE 1951 Psychosurgery ; Abusive
PD 5267 Psychotherapists ; Irresponsible
PJ 6661 Psychotherapy ; Alienating
PD 5267 Psychotherapy ; Corruption
PJ 6661 Psychotherapy ; Psychopathological
PD 5267 Psychotherapy ; Unethical practices
PE 7645 Psychotic arsonists
PE 0989 Psychotic children
PE 7645 Psychotic mass-murderers
PC 0300 Psychotic mental disorders ; Non
PE 7645 Psychotic poisoners
Psychotic [slashers...]
PE 7645 — slashers
PE 7645 — snipers
PE 7645 — stranglers
PE 7645 Psychotic urban bombers
PE 7645 Psychotic violence
PD 4248 Psychotomimetica
PD 7986 Psychotronic warfare
PD 8786 *Pterygium*
PE 7570 *Ptyalism*
PD 4571 *Puberty*
PD 4571 *Puberty trauma*
PF 6190 Public access ; Lack places urban environments encouraging unstructured
PD 7073 Public accounting auditing services ; Protectionism
Public administration
PF 0903 — developing countries ; Inefficient
PF 2335 — Inefficient
PF 1559 — Unpredictable
Public [affairs...]
PE 6086 — affairs ; Violation right participate conduct
PF 8072 — agencies ; Unresponsive
PD 0172 — archives ; Destruction historic documents
PF 1194 — archives ; Lack access
PE 7659 — assaults police
PF 1001 — assets ; Undervaluation
PE 4796 — assistance developing countries ; Inadequate
Public authorities
PF 6508 — assist financial investment ; Failure
PF 7581 — Destruction trade union property
PE 1758 — functioning workers organizations ; Interference
PF 8072 — neglect
PF 8072 — Neglect
PE 5462 — Occupation trade union premises
PE 5462 — Raids trade union premises
PF 7581 — Seizure trade union property
PE 5070 — Suppression strikes
PF 8072 — Unresponsive
Public [bridges...]
PE 2471 — *bridges ; Unrailed*
PF 1600 — *buildings ; Locked*
PF 1600 — *buildings ; Unavailable*
Public [celebrations...]
PF 5250 — *celebrations ; Unconsensed*
PF 1815 — *cleaning ; Shortage*
PD 4004 — *clinics ; Overcrowded*
PF 5937 — concern ; Undue
PF 4198 — Conspiracy against
PE 6702 — contracts ; Unethical allocation
PF 2437 — *criticism ; Feared*
PC 2162 Public debate ; Suppression
Public debt
PD 3051 — developing countries ; Burden servicing foreign
PD 2133 — developing countries ; Excessive foreign
PC 3056 — Disproportionate external
PC 2546 — Excessive
PD 9168 — industrialized countries ; Excessive foreign
PD 3051 — relief developing countries ; Inadequate
Public [Defecating...]
PE 1602 — Defecating
PA 0005 — disclosure ; Inadequate

Public discussion

Public [discussion...] cont'd
- PF 3846 — discussion research development ; Opportunist bias
- PF 2819 — disillusionment aid developing countries
- PE 4882 — displays sexuality
- PE 2429 — drunkenness

Public [education...]
- PF 3519 — education ; Impractical
- PF 3856 — embarrassment
- PE 6888 — employees ; Denial trade union rights
- PD 5642 — enterprises ; Inefficiency
- PE 5938 — enterprises ; Restrictions trade union rights
- PF 8105 — entertainment ; Inadequate
- PF 6543 — environment ; Ineffective utilization
- PF 0611 — expenditure human development ; Decline
- PD 8114 — expenditure human resources developing countries ; Reduction

Public facilities
- PE 7647 — Deficient
- PD 1566 — *Inadequate*
- PD 1566 — *Unavailable*
- PF 9617 Public figures ; Delusions concerning

Public finance
- PF 7842 — data developing countries ; Inaccurate
- PF 5964 — policy developing countries ; Failure
- PF 7842 — statistics ; Inadequate

Public [finances...]
- PF 5374 — finances ; Lack transparency
- PF 9108 — funding ; Difficult
- PF 9108 — funding ; Elusive

Public funds
- PC 2920 — Embezzlement
- PF 9108 — Limited
- PC 2920 — Misappropriation

Public [gatherings...]
- PF 5250 — gatherings ; Infrequent
- PF 4730 — good ; Lack relationship wealth generation
- PJ 0914 — goods services ; Clandestine private utilization

Public health
- PF 8515 — conditions ; Inadequate
- PF 6234 — expenditures ; Increasing
- PE 3838 — practices ; Religious opposition
- PE 7984 — programmes ; Unfamiliar procedures using
- PE 4586 — subsidies ; Limited
- PF 2241 *Public housing assignments ; Alienating*
- PE 1764 Public ; Human flatulence
- PE 1137 *Public image ; Declining*
- PE 1137 *Public image ; Undeveloped*

Public information
- PF 9699 — Classified
- PF 5701 — concerning problems ; Inadequate
- PF 1267 — Fragmented utilization
- PF 1267 — Inaccessible
- PF 7805 — Inoperative forums
- PF 8074 — Unaccountable management
- PF 7722 *Public injustices ; Riskless responses*

Public interest broadcasting
- PF 5622 — Decline
- PF 5622 — Inadequate support
- PF 5622 — Resistance
- PE 7575 Public interest groups ; Bogus

Public land
- PC 8160 — *Limited*
- PC 8160 — *Unassigned*
- PC 8160 — *Undesignated*

Public life
- PD 3335 — Discrimination against women
- PE 6922 — Discrimination harassment children
- PF 9631 — politics ; Media theatricalization
- PD 2541 Public litter ; Unremoved
- PF 5222 Public meetings ; Unpublicized
- PC 5005 Public morality ; Crimes against

Public [non...]
- PE 3780 — non-accountability control production processes
- PE 4032 — non-accountability organizations developing technology
- PF 2660 — nudity
- PD 9022 — *nuisance*

Public office
- PE 5608 — Denial right hold
- PD 8227 — Misconduct
- PD 7704 — Nepotism
- PF 2827 — *positions ; Unappealing*
- PE 9857 — Secret intelligence agents
- PE 6948 — Trading
- PF 8072 Public offices ; Unresponsive

Public officials
- PD 9842 — Extortion
- PD 4915 — Harassment
- PD 4734 — Intimidation

Public opinion
- PF 9704 — development ; Inadequate mobilization
- PE 7448 — Manipulation
- PF 2909 — *polls ; Misleading*
- PD 7520 Public order ; Offences against

Public [places...]
- PE 1602 — places ; Excreting
- PD 5347 — places ; Spitting
- PF 4558 — pledges ; Government reneging
- PF 8757 — policy current scientific knowledge ; Mismatch
- PF 4753 — policy ; Immoral
- PF 3756 — posturing meetings
- PC 9104 — *power ; Abuse*
- PF 3526 — private resources ; Undeveloped channels
- PE 7760 — private sector employees ; Disparity remuneration

Public programmes
- PD 1644 — Cost overruns large scale

Public programmes cont'd
- PF 6377 — Inefficient
- PF 7483 — Internal inefficiency

Public property
- PE 5634 — Crimes against government
- PF 6537 — *Disrespected*
- PD 4691 — *Larceny*
- PD 4691 — *Theft*

Public records
- PD 4239 — Falsification
- PF 9699 — Secret
- PD 0577 — Theft
- PF 7472 Public resentment government policies
- PF 4197 Public resources ; Ineffective mobilization

Public [safety...]
- PF 2874 — safety ; Inadequate provision
- PF 4854 — safety ; Non disclosure threats
- PD 8422 — salaries ; Limited
- PF 9699 — scrutiny ; Withdrawal official documents

Public sector
- PF 5374 — activity ; Inadequate monitoring
- PF 4095 — agencies competition ; Protection
- PD 7230 — Deficiency innovation
- PD 5984 — deficits
- PF 5374 — Inadequate financial reporting
- PF 4739 — institutions ; Proliferation
- PE 4398 — Lack scope intellectual ability
- PF 6574 — management ; Inadequate accountability
- PD 5642 — Monopolization resources
- PF 5374 — Opaque budgetary procedures
- PF 4574 — savings developing countries ; Decline
- PF 0897 Public servant role professional ; Collapse

Public servants
- PD 4541 — Bribery
- PD 4541 — Corruption
- PD 5399 — Retaliation against
- PD 0540 — Threatening
- PD 4541 — Unlawful rewarding

Public service
- PE 5608 — Denial
- PE 4226 — developing countries ; Politicization
- PE 6702 — employees ; Negligent
- PE 6702 — employees ; Unethical practices
- PE 8509 — Inability communities take advantage training business, industry
- PE 8728 — personnel ; Inadequate conditions work employment
- PD 3335 — women ; Denial right access

Public services
- PF 2694 — communication ; Insufficient provision
- PE 7971 — De facto racial requirement qualifications
- PJ 5608 — Disbarment
- PD 8460 — Discrimination
- PE 8507 — Discrimination against men
- PF 2006 — Institutionalized callousness
- PF 6574 — Limited accountability
- PF 6574 — Obscure accountability
- PF 3391 — Privatization
- PD 3326 — Racial discrimination
- PF 6539 — small towns developed countries ; Limited availability
- PE 4927 Public signs ; Missing

Public space
- PF 1790 — *insufficient*
- PF 1790 — *Limited*
- PF 6578 — Neglected
- PF 1790 — *Undesignated*
- PF 2575 *Public spaces ; Deserted*
- PF 2346 Public spaces ; Inattentive care

Public spending
- PF 1258 — alleviate poverty ; Inefficient
- PF 9108 — Decline
- PF 1447 — defence ; Increasing
- PF 0674 — education ; Decline
- PF 6377 — government ; Inappropriate
- PF 6377 — government ; Misallocation
- PF 4586 — health ; Decline
- PF 6377 — Ineffectiveness
- PF 6377 — Inefficiency
- PF 0674 — *primary schools ; Decline*
- PF 0611 — social sector ; Decline

Public [squares...]
- PF 6165 — squares cities ; Impersonality
- PE 9758 — subsidies ; Inequitable distribution
- PF 6846 — support terrorism ; Passive

Public transportation
- PF 1495 — Distant
- PF 1495 — Inadequate
- PF 1495 — Insufficient
- PF 1495 — Limited
- PF 1495 — Unavailability
- PD 4827 Public trial ; Denial right fair
- PE 3964 Public trial ; Denial right fair

Public [understanding...]
- PD 8003 — understanding science ; Inadequate
- PF 6193 — Unnatural urban environments inhibiting sleeping
- PE 1602 — Urinating
- PE 7647 — utilities ; Dysfunctional
- PF 1653 — *utilities ; Prohibitive cost connection*
- PE 0756 Public ; Verbal sexual harassment women

Public [water...]
- PD 8517 — water outlets ; Unmarked
- PE 1458 — water supplies sabotage ; Contamination
- PD 4143 — *works ; Collapsing*
- PF 3455 — works ; Grandiose
- PF 1538 Publications ; Delays delivery books
- PF 3075 Publications ; Inadequate government

- PF 6541 *Publicity ; Inadequate job*
- PF 1575 *Publicity ; Restrictions*
- PE 1692 Publicity ; Unfair trials due pre trial
- PF 9148 Publicized issues ; Inadequately
- PF 1781 *Publicized services ; Inadequately*
- PE 5312 Publish ; Violation right trade unions
- PE 8095 Publishing printing industries ; Environmental hazards
- PA 7131 Puerility *Oldness*
- PA 7371 Puerility *Unintelligence*
- PC 9042 Puerperal phlebitis
- PC 9042 Puerperal pyrexia
- PC 9042 Puerperal sepsis
- PD 7420 Puerperal tetany
- PC 9042 Puerperal thrombosis
- PC 9042 *Puerperalis ; Mastitis*
- PC 9042 Puerperium ; Complications
- PE 8022 Puerperium ; Toxaemias pregnancy
- PE 1323 Pull down
- PE 7562 Pullorum disease poultry
- PE 9438 Pulmonary adenomatosis
- PE 7085 Pulmonary carcinoma

Pulmonary disease
- PD 0637 — [Pulmonary disease]
- PD 0637 — Chronic obstructive
- PE 5182 — *Chronic obstructive*
- PE 1556 — due torture ; Cardio
- PD 0637 Pulmonary diseases
- PD 0637 Pulmonary diseases ; Occupational chronic
- PD 0637 Pulmonary embolism

Pulmonary emphysema
- PE 2248 — [Pulmonary emphysema]
- PE 5524 — *Acute bovine*
- PD 7307 — *animals*

Pulmonary [haemorrhage...]
- PE 5182 — *haemorrhage ; Exercise induced*
- PE 9415 — heart disease
- PD 9366 — *hypertensive heart disease*
- PE 3830 — Pulmonary oedema
- PE 5524 — Pulmonary oedema ; Acute bovine
- PE 2526 — Pulmonary tuberculosis
- PE 5569 — *Pulmonic stenosis animals*
- PE 7914 — Pulp waste paper ; Instability trade
- PE 1616 — Pulp waste paper ; Long term shortage
- PE 7769 — *Pulpy-kidney disease*
- PE 6360 — Pulses ; Environmental hazards electromagnetic
- PE 3296 — *Pumps ; Unavailability*
- PE 0768 — Punch drunk syndrome
- PE 7826 — *Puncture wounds foot*
- PF 0399 — Punished death penalty ; Denial rights those
- PA 5583 — Punishment

Punishment [Collective...]
- PD 6970 — Collective
- PD 8575 — Corporal
- PE 3488 — criminals mutilation
- PC 3768 — Cruel

Punishment [Degrading...]
- PC 3768 — Degrading
- PC 3768 — Denial right freedom cruel, inhumane degrading
- PF 0399 — deterrent ; Ineffectiveness capital
- PF 7228 Punishment ; Eternal
- PF 0399 Punishment ; Inhumanity capital
- PF 4875 *Punishment international criminals their own countries ; Inadequate*
- PD 7187 Punishment ; Parental
- PF 2183 *Punishment ; Paternalistic*
- PE 0192 Punishment schools ; Corporal
- PF 0399 Punishment torture ; Anticipation capital
- PC 3768 Punishment ; Unusual
- PD 1464 Punishment war criminals ; Inadequate
- PD 4779 Punishments crimes ; Unjust
- PE 4743 Punishments ; Denial right freedom retroactive laws
- PE 8724 Punitive amputations
- PD 5584 Punitive regulations
- PC 3186 Puppet government
- PE 6948 Purchase military commissions
- PF 4050 Purchase patents ; Suppression innovation
- PF 8675 Purchase ; Short term cash
- PF 3211 *Purchasing habits ; Traditional*
- PE 8707 Purchasing power developing countries ; Lack
- PD 8362 Purchasing power ; Limited
- PJ 4787 Purchasing ; Tied
- PF 9729 Purchasing ; Undemonstrated cooperative
- PF 9693 Pure applied science ; Conflict interest
- PD 8402 *Purgatives ; Violent*
- PC 2933 Purges ; Political
- PF 5743 Purification effluents emissions ; Prohibitive costs
- PF 1954 Purism ; Linguistic
- PF 2577 Puritanism
- PA 6977 Purpose ; Lack
- PA 6714 Purposelessness *Chance*
- PA 6852 Purposelessness *Inappropriateness*
- PA 6977 Purposelessness *Meaninglessness*
- PF 8026 *Purpura*
- PG 4410 Purpura animals
- PB 6321 Pusillanimity
- PD 9667 *Pustular dermatitis ; Contagious*
- PE 5524 *Pustular vulvovaginitis ; Infectious*
- PA 6088 Putrefaction *Impairment*
- PE 2053 Pyelitis
- PE 2053 Pyelocystitis
- PE 2053 Pyelonephritis
- PD 9293 *Pyelonephritis ; Bovine*
- PD 9293 *Pyelonephritis ; Contagious bovine*
- PE 6553 Pylorus ; Obstruction
- PC 8534 Pyoderma
- PD 9667 *Pyoderma animals*

PD 9667 Pyoderma ; Secondary
PD 9667 Pyogenic dermatitis
PE 2280 Pyogenic infections
PE 2280 Pyogenic meningitis
PE 2280 Pyogenic organisms
PE 2280 Pyogenic osteomyelitis
PE 9286 Pyometra
PD 7799 Pyometra animals
PE 3503 Pyorrhoea alveolaris
PE 9702 Pyorrhoea alveolaris
PE 7826 Pyramidal disease
PC 9042 Pyrexia ; Puerperal
PA 6448 Pyromania
PE 9177 Pyrotechnic products ; Environmental hazards explosives
PD 5228 Pyrrolizidine alkaloidosis

Q

PE 2534 Q-fever
PE 0912 Qat ; Abuse
PD 1725 Quack doctors
PE 4429 Quackery ; Cosmetic surgery
PD 1725 Quackery ; Medical
PD 1189 Quackery ; Nutritional
PE 2946 Quadriplegia
PE 5567 Quail bronchitis
PD 3978 Quail disease
PF 4828 Qualification earning ; Diversion education
PF 3462 Qualification ; Over
PF 4494 Qualification syndrome ; Paper
PD 9628 Qualification women ; Discriminatory requirement over
PF 2066 Qualifications ; Artificial arbitrary job
PF 0704 Qualifications ; False
PE 5090 Qualifications intergovernmental organizations ; Inadequacy staff
PF 3411 Qualifications ; Lack
PC 1524 Qualifications ; Non equivalence national educational
PE 7971 Qualifications public services ; De facto racial requirement
PF 4494 Qualifications ; Reliance paper
PF 3019 Qualified amnesty
PD 8108 Qualified teachers ; Shortage
PF 5703 Qualitative excellence ; Lack
Quality [Commercial...]
PF 7227 — Commercial erosion standards
PD 7723 — construction work ; Low
PD 1435 — control ; Inadequate
PD 2392 — control ; Inadequate drug
Quality [Decline...]
PF 8466 — Decline educational
PC 8943 — Denial right environmental
PD 1435 — Deterioration product
PD 2743 — domestic livestock ; Poor
PD 1565 Quality equipment maintenance servicing ; Deterioration
PF 8713 Quality escalating pitch ; Distortion musical
PE 8938 Quality food ; Decline nutritional
PE 2770 Quality intensive animal farming units ; Inferior meat
Quality life
PF 7142 — Deteriorating
PB 4993 — International imbalance
PE 7734 — least developed countries ; Deterioration
PF 2971 Quality medicine ; Unavailability
PD 1435 Quality merchandise ; Low
PE 1488 Quality negated oligarchic control decision making ; Social service
PF 7741 Quality ; Planned degradation product
PF 5703 Quality ; Poor
PF 7227 Quality ; Quantitative pressure standards
PF 3409 Quality seeds ; Scarcity high
PC 9714 Quality teaching ; Low
PD 9170 Quality working life ; International imbalance
PF 7098 Quantitative limit societal learning
PF 7227 Quantitative pressure standards quality
PE 9027 Quantitative restrictions ; Distortion international trade
PE 9762 Quantitative trade quotas discretionary character ; Imposition
PF 5587 Quantitative understanding responsibility
PA 6108 Quantity∗complex
PF 1555 Quarantine ; Animal
Quarantine Inadequate
PE 2850 — [Quarantine ; Inadequate]
PE 2756 — animal
PE 0714 — plant
PE 4021 Quarrels ; Domestic
PA 5532 Quarrelsomeness Discaccord
PA 5446 Quarrelsomeness Enmity
PA 7253 Quarrelsomeness Envy
PA 5502 Quarrelsomeness Reason
PD 2596 Quarries ; Environmental destruction open pit mines
Quarrying [industries...]
PE 1858 — industries ; Underdevelopment mining
PE 0993 — industry ; Instability mining
PE 8809 — Instability miscellaneous mining
PE 0616 Quartan
PE 7826 Quarter crack
PE 1429 Quelea
PE 7155 Question context process ; Exclusion decision making processes those who
PF 2135 Questionable economic viability
PF 9431 Questionable facts
PE 7313 Questionable profit base
PF 2135 Questionable trade viability
PE 3553 Questioning ; Improper police

PE 3553 Questioning ; Oppressive police
PA 6444 Quiescence
PF 6160 Quiet zones urban environment ; Inaccessibility
PE 2292 Quinsy
PE 7826 Quittor
PF 2497 Quota cheating ; Trade
PE 9762 Quotas discretionary character ; Imposition quantitative trade
PE 9027 Quotas ; Import
PE 9762 Quotas political reasons ; Imposition trade

R

PE 2254 Rabbit bot fly ; Rodent
PD 3480 Rabbits ; Endangered species
PE 8794 Rabbits hares ; Endangered species
PF 7428 Rabble
PE 1325 Rabies
PE 7004 Race ; Anti satellite arms
PC 1258 Race ; Arms
PE 5183 Race children ; Discrimination against mixed
PC 0006 Race ; Denial right equality because
PC 1258 Race ; Dependence arms
Race [Information...]
PC 0840 — Information technology arms
PF 7815 — Intellectual decline human
PC 5348 — International prestige
PC 5348 — International status
PF 7088 Race ; Lack business opposition arms
Race [National...]
PF 8434 — National status
PD 8412 — Naval arms
PD 5076 — Nuclear arms
PF 7815 Race ; Physical decline human
Race Refusal
PE 8137 — let because applicant's
PE 9054 — loans because borrower's
PE 8823 — sale because buyer's
PF 3352 Race reinforcement nationalism
PF 4152 Race ; Risk unintentional war generated arms
PD 0087 Race ; Space weapons arms
PE 9085 Race ; Uncertainty survival human
PC 6664 Race weapons
PC 1523 Races ; Non acceptance mixed
PE 4907 Racial bias sentencing offenders
PD 4365 Racial bigotry
PC 3684 Racial conflict
PF 7815 Racial degradation
Racial discrimination
PC 0006 — [Racial discrimination]
PE 9054 — according financial loans
PE 4167 — Capital investments supporting
PC 0006 — Dependence
PD 3328 — education
PD 3442 — housing
PE 4907 — judiciary
PC 3683 — Legalized
PD 3329 — politics
PD 3326 — public services
PD 3519 — security forces
PD 4064 — sexual preferences
Racial [elitism...]
PA 1387 — elitism
PC 3334 — exploitation
PC 3334 — exploitation ; Dependence
PF 1199 Racial groups ; Inferior average intelligence certain
PC 0006 Racial groups ; Unequal rights different
Racial [identity...]
PF 0684 — identity ; Lack
PC 1523 — impediments marriage
PF 1199 — inequality
PC 2936 — intimidation
PC 0805 Racial minorities ; Underprivileged
PD 3329 Racial politics
PC 8773 Racial prejudice
Racial [regression...]
PF 0411 — regression ; Human
PD 8762 — repression
PE 7971 — requirement qualifications public services ; De facto
PC 3688 Racial segregation
PF 5452 Racial stereotyping
PC 2936 Racial violence
PD 8718 Racial war
PB 1047 Racialism
PF 1199 Racially-determined differences intelligence
PD 3442 Racially separated residences
PB 1047 Racism
PD 3519 Racism amongst military personnel
PC 4773 Racism child youth literature
PB 8344 Racism ; Cultural
PB 1047 Racism ; Dependence
PC 3686 Racism ; Ethnic
PC 3685 Racism ; Masked
PD 3519 Racism police forces
PC 4513 Racism ; Religious
PD 3328 Racism schools
PF 1199 Racism ; Scientific
PD 3093 Racist propaganda
PC 4513 Racist religion
PE 7430 Rack
PG 3896 Rack Mountain spotted fever
PE 4914 Racketeering
PE 6699 Racketeering ; Ineffective laws against
PE 3482 Racoons ; Endangered species
PC 6114 Radiant energy ; Long term variations solar

PE 9528 Radiant energy ; Medium term cyclic variations solar
Radiation [accidents...]
PD 1949 — accidents
PD 1949 — accidents ; Risk
PE 0962 — aircraft ; Health hazards
PC 0918 Radiation balance ; Deterioration atmospheric
Radiation [carcinogenesis...]
PD 8050 — carcinogenesis humans
PE 1909 — consumer goods electronic devices ; Excessive exposure
PC 1299 — Contamination natural
Radiation [damage...]
PD 8050 — damage human body
PD 1206 — damage materials
PE 2465 — dermatitis
Radiation Environmental hazards
PC 0229 — atomic
PE 7560 — extremely low frequency electromagnetic
PE 7651 — non ionizing
PE 3883 — solar
Radiation [Excessive...]
PE 1704 — Excessive exposure medical patients
PD 1500 — Excessive occupational exposure
PE 0689 — Experimental exposure animals
PC 6912 — Experimental exposure humans
Radiation [Harmful...]
PE 6294 — Harmful biological effects ionizing
PF 5188 — hazard thresholds ; Uncertain
PE 5672 — hazard ; Ultraviolet
PE 4057 — Hazards low level exposure
PE 6056 — health hazard ; Microwave
PD 8050 — Health hazards
PF 1686 — Health hazards exposure cosmic
PD 0748 — human body ; Harmful effects ultrasonic
PF 5069 Radiation ; Increased reflection solar
PE 4057 Radiation ; Long term hazards exposure
PF 6635 Radiation monitoring systems ; Inadequate
PD 1584 Radiation pollution reactors
PD 8290 Radiation risks ; Underreporting
PC 6666 Radiation weapon ; Use
PF 9269 Radiation weapons ; Infrasonic
PF 6015 Radicalism ; Islamic
PF 2177 Radicalism ; Political
PF 8352 Radiesthesia
Radio frequencies
PE 5099 — health hazards
PD 2045 — Jamming
PF 0734 — Limited number available
PF 0734 — Maldistribution
PF 0734 — Overcrowded spectrum
PD 2045 Radio frequency interference
PE 9528 Radio interference associated solar flares
PE 2473 Radio interference industrial origin
PE 2473 Radio noise industrial origin
PF 1676 Radio noise natural origin
Radio television
PD 3029 — censorship
PD 2045 — communications ; Interference
PD 3085 — propaganda
PD 9538 Radio transmission surveillance governments ; Misuse
PE 8843 Radioactive associated materials trade instability
Radioactive contamination
PC 0229 — [Radioactive contamination]
PD 1119 — animals animal products
PD 5157 — Endangered animal plant life due
PE 1431 — marine environment fisheries products
PD 0710 — plants
PE 3383 — soil
PE 2441 — water
PC 0314 Radioactive dust
PC 1242 Radioactive effluents
Radioactive fallout
PC 0314 — [Radioactive fallout]
PE 5080 — accidents orbiting nuclear-powered reactors
PE 5324 — Uncertainty long term health effects
PE 6884 Radioactive material ; Transport storage
PE 7419 Radioactive substances ; Health hazards mining
PC 1242 Radioactive waste disposal
PC 1242 Radioactive wastes
PE 7544 Radiological services ; Inadequate
PC 6666 Radiological warfare
PD 8290 Radiologists ; Corruption
PD 8290 Radiologists ; Irresponsible
PD 8290 Radiologists ; Negligence
PD 8290 Radiology ; Malpractice
PD 8290 Radiology ; Unethical practice
PE 4588 Radiomimetic chemicals
PF 4142 Radios ; Maldistribution
PC 1299 Radon health ; Ill effects
PF 7428 Raff ; Riff
PE 0429 Rage ; Neurological
PE 0429 Rage ; Violent
PD 5228 Ragweed poisoning
PD 5228 Ragwort
PD 5272 Raids ; Cross border anti terrorist
PE 5462 Raids trade union premises public authorities
PD 0078 Rail traffic congestion
Rail transport
PD 0126 — accidents
PE 5848 — developing land locked transit countries ; Inadequate
PD 0496 — facilities ; Inadequate
PD 1367 — practices ; Unfair
PD 0126 Railroad collisions
PD 7420 Railroad disease
PD 7420 Railroad sickness

Railroad usage

PF 6572 *Railroad usage ; Limited*
PD 0126 Railway accidents
PD 4904 Rain ; Acid
PE 0082 Rain ; Deterioration stained glass due acid
PE 5274 Rain forests agricultural development ; Misuse tropical
PD 6204 Rain forests ; Destruction
PE 4103 Rain-storms
PD 4904 Rain ; Toxic
PE 4103 Rainfall ; Excessive
PE 4263 Rainfall ; Increase
PD 0489 Rainfall ; Reduction
PE 4263 Rainfall ; Reduction
PD 0489 Rainfall ; Unreliable
PD 0489 Rainy season ; Failure
PE 0924 Rams ; Epididymitis
PD 8697 Random code enforcement
PE 4619 *Random defecation*
PF 7700 Random events ; Inadequacy methodologies deal discontinuities
PF 1822 *Random land subdivision*
PA 6714 Randomness *Chance*
PA 7361 Randomness *Disorder*
PA 7309 Randomness *Uncertainty*
PE 6554 Range business skills ; Narrow
PD 5133 Rangeland ; Misuse grassland
PD 5133 Rangelands grasslands ; Deterioration
PF 2015 Ranking ; Subjective cultural
PF 7580 *Ransom*
PF 7580 *Ransomed ; Risk being*
PE 0805 Ransoming pet animals
PA 5688 Rapacity *Appropriation*
PA 6379 Rapacity *Avoidance*
PA 6466 Rapacity *Intemperance*
PD 3266 Rape
PE 5621 Rape attempts ; Unreported
PE 6522 Rape children
PD 3266 *Rape ; Gang*
PE 4449 Rape ; Inadequate assistance victims
Rape [Marital...]
PF 4948 — Marital
PF 4948 — marriage
PE 6522 — minor consent
PE 8159 — mustard oils trade instability
Rape [pregnancy...]
PD 3266 — pregnancy
PE 1363 — prisons
PE 1363 — prisons ; Unreported
PE 6522 Rape ; Statutory
PE 5621 Rape ; Unreported
PF 4948 *Rapes ; Date*
PG 2295 *Rapid growth ; Unusually*
PB 1845 Rapid industrialization ; Over
PF 3434 Rapid personnel turnover international organizations
PB 0035 Rapid population growth
PF 1275 Rapid returns investment ; Overemphasis
PF 6712 Rapid social change ; Confusion induced
PD 9235 Rapid timber exploitation ; Over
PF 1844 *Rapid transient flux*
PF 8521 Rapidly changing culture
PF 8521 Rapidly changing cultures
PG 9635 *Rash ; Heat*
PD 9667 *Rash ; Nettle*
PE 1017 *Rash ; Nettle*
PC 8534 *Rash ; Skin*
PE 1017 *Rashes*
PA 7115 Rashness
PD 6363 *Rat-bite fever*
PD 6363 *Rat bite fever*
PD 3481 *Rat chinchillas ; Endangered species*
Rat Endangered species
PE 8684 — African cane
PE 8253 — African mole
PE 9120 — bush rat rock
PE 3177 Rat-infested areas
PD 1762 *Rat opossums ; Endangered species*
PE 9120 Rat rock rat ; Endangered species bush
PE 7300 Ratification human rights treaties ; Delays
PE 7300 Ratification human rights treaties ; Non
PF 0977 Ratification international treaties ; Delays
PF 0977 Ratification international treaties ; Non
PE 7300 Ratified human rights agreements ; Failure governments implement provisions
PF 0473 *Ratify international conventions ; Refusal*
PF 7535 Rational global political decision making ; Lack
PF 3400 Rationalism
PF 9026 Rationing
PE 8261 Rationing health care
PD 7254 Rationing medical treatment
PE 3177 Rats
Rats Endangered species
PD 3481 — bamboo
PE 9103 — palaearctic mole
PD 3481 — spiny
PE 3177 Rats pests
PD 3978 *Rattle belly lambs*
PC 3150 Ravaging wild animals
PA 5688 Ravishment *Appropriation*
PA 5612 Ravishment *Unchastity*
PE 0060 Raw material reserves transnational enterprises ; Excessive exploitation
Raw materials
PD 4270 — [Raw materials]
PE 0669 — components utilization ; Restrictions
PD 2968 — Dependence developing countries export limited range
PD 4270 — Lack

Raw materials cont'd
PE 0194 — markets transnational corporations ; Excessive control
PF 6590 — Under utilized
PD 4270 — Unstable supply
PF 1441 Ray disturbance space ; Gamma
PD 1624 Re education ; Political
PF 1015 Re employment government security services ousted repressive regime ; Post revolutionary
PE 7695 Re forestation ; Inappropriate
PA 6355 Reaction
PC 0300 *Reaction adolescence ; Adjustment*
PF 1083 Reaction against environmental hazards ; Over
PF 1622 Reaction past sexual ethics ; Over
PF 4110 Reaction ; Political over
PB 6332 Reactionary forces
PF 4612 Reactionary methods avoiding danger
PE 1017 Reactions ; Allergic
PE 0873 *Reactions ; Arthritis accompanying foreign protein*
PC 0300 *Reactive attachment disorder infancy*
PD 1318 *Reactive psychosis ; Brief*
PD 7579 Reactor accidents ; Nuclear
PE 2462 Reactor design ; Faults
PF 1441 Reactor emissions satellites ; Nuclear
PF 6084 Reactor safeguards ; Inadequate nuclear
PD 7579 Reactors accidents
PD 1584 Reactors ; Environmental pollution nuclear
PF 7539 Reactors ; Excessive cost decommissioning nuclear
PD 1584 Reactors ; Radiation pollution
PE 5080 Reactors ; Radioactive fallout accidents orbiting nuclear powered
PD 4977 Reactors ; Unsafe nuclear
PD 1950 Reading disabilities
PD 1950 Reading disorders
PS 3268 *Reading levels ; Low*
Real estate
PD 5422 — agents ; Negligence
PD 5422 — corruption
PD 5422 — fraud
PD 5422 — practice ; Unethical
PB 5486 *Real property ; Arbitrary sequestration*
PD 9356 Real value money ; Fluctuations
PD 8916 Real wages ; Decline
PD 2769 Real wages developing countries ; Decline
PF 7493 Realism intergovernmental organizations ; Lack
PF 8435 Realistically ; Inability resolve problems
PF 7414 Reality ; Avoidance
PF 1079 Reality ; Denial
PF 7053 *Reality ; Elusiveness*
PF 1728 Reality ; Lack correspondence basic value images social
PF 1079 Reality ; Non acceptance
PF 7053 Reality ; Oppressive
PF 1112 Reality popular psychological screens ; Escaping
PF 0916 Reality sophisticated shields against life's pain ; Escaping
PF 8922 Reality ; Substitution fantasy
PF 8922 Reality ; Treating fiction
PF 0648 Reallocation aid funds alternative priorities
PF 0648 Reallocation development capital developing countries socialist countries
PF 4707 Reappearance aristocracy
PF 0131 Reared prison ; Children
PD 5344 Rearing ; Permissive child
PD 4633 Rearing ; Subjugation women child
PA 5502 Reason
PF 7521 Reason ; Eclipse
PD 0140 Reasonable work hours ; Denial right
PF 5711 Reasoning ; Defective
PF 2152 Reasoning ; Fatuous
PD 0634 *Reasons ; Confinement non criminal*
PE 9762 Reasons ; Imposition trade quotas political
PD 0917 Reassurance ; Mass media bias towards entertainment
PD 0668 Rebate cartels
PD 1964 Rebellion
PD 1964 Rebellion ; Incitement
PA 5901 Rebellion *Revolution*
PA 7250 Recalcitrance *Disobedience*
PA 7325 Recalcitrance *Irresolution*
PA 6979 Recalcitrance *Opposition*
PA 6509 Recalcitrance *Unwillingness*
PD 0270 *Recall important personal information ; Inability*
PF 1379 *Recall procedures unsafe products ; Inadequate*
PE 4214 Receive information ; Denial right
PE 0261 Receivers ; Foreign controlled direct satellite broadcasting individual
PC 4701 Receiving bribes
PE 8364 Receiving stolen property
PF 1790 *Reception ; Poor television*
PD 3039 Reception ; Unequal opportunities media
PE 9300 Receptive language disorder ; Developmental
PA 6917 Recession
PF 1172 Recession ; Aggravation cyclical
PD 9436 Recession ; Domestic
PF 6071 Recession ; Imbalance recovery
PF 1172 Recession ; International economic
PD 9436 Recession ; National economic
PF 4785 Recession ; Peace
PE 1277 Recessions ; Cyclic business
PF 4785 Rechannelling reduced expenditure defence ; Rigidity
PE 5581 Recidivists
PF 6932 Recipient distortion aid developing countries
PF 2708 Recipients ; Belittling grant
PF 3871 Reciprocity trading relations ; Preoccupation
PD 5349 *Reckless children snowmobile drivers*
PE 9334 Reckless driving

PD 5349 Reckless endangerment ; Criminally
PD 5349 Recklessness
PE 0701 Reclaimed rubber ; Instability trade crude synthetic
PF 2055 Reclamation ; Land
PF 8040 Recognition ; Absence diplomatic
PF 1985 Recognition common heritage ; Fragmented
PF 8040 Recognition foreign governments ; Non
PE 6418 Recognition foreign powers attorney ; Non
Recognition [institutions...]
PF 6173 — institutions transition adolescence ; Inadequate
PF 9081 — international law ; Non
PF 7093 — international nongovernmental organizations ; Limited
PD 0683 *Recognition ; Lack trade union*
PE 4912 Recognition nationality ; Restrictions
PF 6091 *Recognition newly founded churches ; Non*
Recognition [patents...]
PF 7779 — patents ; Delay
PE 4716 — person law ; Denial right
PJ 0646 — Personal need
Recognition problems
PF 7745 — Biased
PF 8112 — Delay
PF 8112 — Non
PF 6091 *Recognition religions ; Non*
PF 9728 Recognize environment resource base economic activity ; Failure
PF 1750 Recognize uniqueness family members ; Failure
PF 7315 Recognized merit ; Accumulation
PF 3701 Reconnaissance satellites
PF 5728 *Record ; Lack historical*
Records [Destruction...]
PF 1836 — Destruction medical
PF 4633 — Destruction scientific
PE 7061 — destruction transnational enterprises
Records Falsification
PF 4932 — government historical
PD 4239 — public
PF 1602 — scientific
PF 0704 *Records ; Falsifying academic*
PF 4153 *Records ; Government misuse personal*
PF 1776 *Records ; Inaccessible landowner*
PF 1836 Records ; Lack medical
PF 4708 *Records ; Misfiled documents*
PF 2346 *Records ; Obscure ownership*
PE 8542 *Records ; Risks computers computer*
Records [Secrecy...]
PF 5983 — Secrecy medical facts
PF 9699 — Secret public
PF 2714 — Suppression safety
Records [Tampering...]
PF 4699 — Tampering scientific
PD 0577 — Theft public
PE 2145 — transnational corporations ; Foreign currency manipulations accounting
PF 6472 *Records ; Unresearched land*
PD 7609 Recourse ; Lack legal
PF 6465 *Recovery ; Inadequate waste*
PF 6071 Recovery recession ; Imbalance
PF 6071 Recovery ; Uneven economic
PE 6276 Recreation areas ; Closure
PF 6503 Recreation areas ; Inaccessible
PF 0202 Recreation facilities elderly ; Inadequate
PF 0202 Recreation funds ; Lack
Recreation [Inaccessibility...]
PF 6138 — Inaccessibility water
PD 0549 — Inadequate spatial facilities youth
PF 0374 — Irresponsible
PD 0549 Recreation ; Lack school
PF 5409 Recreation leaders ; Untrained
PF 0386 Recreation ; Meaningless
PF 5409 Recreation programmes ; Unorganized
Recreation [Social...]
PE 1277 — Social isolation computer based
PF 7130 — *space ; Undeveloped*
PF 7130 — *spaces ; Unused*
PF 9062 Recreation time ; Insufficient
Recreation tourism
PD 0826 — Cultural degradation
PE 6920 — Natural environmental degradation
PD 0826 — Social environmental degradation
PF 5409 Recreation ; Unorganized community
PF 8105 Recreational activities ; Limited
PF 6685 Recreational contact sewage
PF 0374 *Recreational crime games*
Recreational facilities
PE 8833 — disabled people ; Inadequate
PF 0202 — Inadequate
PD 1350 — *Vandalism*
PC 2024 Recreational hunting
PF 5409 Recreational organization ; Lack
PF 5409 Recreational planning ; Insufficient
PE 9172 Recreational psychiatry
PE 9172 Recreational therapy
PE 7403 Recreational use unsurfaced country roads tracks ; Environmental degradation
PF 1406 Recreational war games
PA 7237 Recrimination *Condemnation*
PE 7036 Recruit union members ; Denial right
PE 4484 Recruiting enlistment foreign armed forces ; Unlawful
PE 3912 Recruiting induction armed forces ; Obstruction
PE 4920 Recruitment ; Failure military
PF 3605 Recruitment ; Ineffective leadership
PF 6557 *Recruitment ; Minimal industrial*
PE 9399 Rectal cancer ; Colon
PD 3978 *Rectal prolapse animals*

PE 3711 *Rectovaginal fistula animals*
PE 2481 *Rectum ; Prolapse*
PF 9582 Recuperative power nature ; False assumptions concerning
PF 7027 Recurrence misapprehended world problems
PE 0616 *Recurrent induced malaria*
PD 5535 *Recurrent iridocyclitis*
PD 5535 *Recurrent uveitis ; Equine*
PD 1350 Recurring property vandalism
PJ 0052 Recycled water ; Use
PF 6465 *Recycling materials ; Insufficient*
Red tape
PF 8053 — [Red tape]
PD 0460 — *cooperatives*
PC 2550 — delays
PE 2501 Red tide
PE 7769 *Red water disease*
PA 7259 Redemption*complex
PF 4785 Redeployment military resources peaceful applications ; Obstacles
PD 0050 Redistribute land ; Impossibility
PE 7902 Redlining denies credit
PE 0290 Redress consumers' loss ; Inadequate
PE 4173 Redress rights violations ; Denial right
PC 7517 *Reduce petroleum consumption ; Inability*
PE 4465 Reduced activity malnourished children
PF 1373 *Reduced agricultural reliance*
Reduced [celebrational...]
PF 1560 — celebrational images
PF 0544 — *civic awareness*
PF 2093 — community image
PF 1870 Reduced dimension marriage covenant
PF 4785 Reduced expenditure defence ; Rigidity rechannelling
PF 5079 Reduced family education
PF 1078 Reduced family life styles
PF 9564 Reduced government commitment multilateralism
Reduced images
PF 7614 — environmental protection
PF 6098 — having significant life work
PF 6537 — having significant life work
PF 4323 — humanness
Reduced [impact...]
PE 1478 — *impact impersonal repetitive news items*
PC 1966 — income old age
PE 5397 — individual needs ; Legal contract system
PF 3783 — interior structure families
PF 1000 — interior structures marriage
PF 5968 — intervals pregancies
PE 5397 Reduced legal contract ; Individualistically
Reduced [physical...]
PD 4475 — physical exertion
PE 8098 — productivity diseased animals ; Economic loss
PE 8098 — productivity diseased animals ; Economic loss
PE 0921 *Reduced resistance disease*
Reduced [scope...]
PF 4794 — scope intergovernmental development assistance
PG 8937 — service outreach
PF 1729 — social commitment contemporary life-styles
PE 3939 — social needs ; Industrial processes geared
PF 3605 Reduced training leadership
PF 7071 Reduced understanding globality
PF 6441 *Reduced village population*
PF 4366 Reducing terrorism ; Inadequate international cooperation
PF 0498 Reduction aid expenditure
PF 5069 Reduction albedo Earth's surface
PF 8905 Reduction church attendance
PD 1276 Reduction demand primary commodities due technological change
PD 9082 Reduction fish yields
Reduction [genetic...]
PE 6788 — genetic diversity available seeds
PE 4256 — glacier size
PF 5502 — government action commitment ; Progressive
PD 8945 Reduction motivation care aged
PE 5898 Reduction ocean shipping services
Reduction [prison...]
PF 4723 — prison sentences ; Politically motivated
PC 8908 — productivity
PD 8114 — public expenditure human resources developing countries
PD 0489 Reduction rainfall
PE 4263 Reduction rainfall
Reduction [sentences...]
PE 9545 — sentences offenders after conviction ; Arbitrary
PC 2566 — share developing countries world exports
PE 8782 — soil fertility downstream due impoundment
PF 1560 — symbolic celebrations
PF 4741 Reduction technological diversity
PF 7515 Reduction workforce enterprises
PD 0140 *Reduction working time*
PF 7967 Reductionism
PF 7742 Reductionistic decision making criteria construction industry
PD 8007 Redundancy workers
PD 9154 *Redundant prepuce*
PE 6288 Redwater
PE 2357 *Redwater calves*
PD 5769 Reefs ; Destruction coral
PE 9604 Referable abdomen ; Symptoms
PE 3933 Referable cardiovascular system ; Symptoms
PE 9604 Referable digestive system ; Symptoms
PE 9369 Referable genito urinary system ; Symptoms
PE 4566 Referable joints ; Symptoms
PE 4566 Referable limbs ; Symptoms
PE 3933 Referable lymphatic system ; Symptoms
PE 4566 Referable musculoskeletal system ; Symptoms
PE 9468 Referable nervous system ; Symptoms
PF 7864 Referable respiratory system ; Symptoms
PE 2665 Referable sense organs ; Symptoms
PD 2716 Reference books ; Inadequate inaccurate textbooks
PB 5249 Reference suffering ; Avoidance
PF 2909 Referenda ; Manipulative use
PG 8883 *Referral ; Insufficient financial*
PG 8893 *Referral ; Uncoordinated agency*
PE 0952 Referrals ; Fraudulent medical
PE 0952 Referrals ; Unnecessary health system
PC 1192 *Refineries ; Environmental hazards petroleum*
PF 5069 Reflection solar radiation ; Increased
PF 2060 Reflections ; Parental control children's thoughts
PE 1310 *Reflex ; Loss tendon*
PF 7334 Reform ; Counter
PF 5736 Reform ; Deliberate governmental avoidance legislative
PF 8876 Reform ; Failure process low priority legislative
PD 9189 Reform ; Inadequate land
PF 0677 Reform ; Inadequate social
Reform Lack
PF 0677 — [Reform ; Lack]
PF 2335 — administrative
PD 9189 — agrarian
PF 5736 Reform law ; Reluctance
PF 3689 Reform socialist countries ; Obstacles economic
PD 1624 Reform ; Thought
PD 8179 Refractive disorders eye
PF 0021 Refresher courses professional fields ; Lack
PF 3630 *Refrigeration costs ; Prohibitive*
PD 4790 *Refrigeration facilities ; Lack*
PF 2009 *Refrigeration ; Lack market*
PD 6364 Refugee camps settlements ; Military attacks
PD 5136 Refugee children
PE 6375 Refugee status ; Abuse
PB 0205 Refugees
PE 6376 Refugees asylum seekers ; Detention
PD 6364 Refugees asylum seekers ; Physical insecurity
PD 8034 Refugees boat
Refugees [Denial...]
PE 6375 — Denial economic social rights
PE 3751 — Denial right work
PB 0205 — Dependence
PC 0768 — Disabled
Refugees [Economic...]
PD 4379 — Economic
PE 3728 — Environmental
PD 5025 — Exploitation women
PF 3021 — Expulsion
PE 4868 Refugees ; Handicapped
PF 3021 Refugees ; Inadequate protection
PE 1508 Refugees living outside camps ; Non settled
PC 0519 Refugees ; Non settled
PE 3751 Refugees obtain employment ; Inability
PD 2549 Refugees ; Political
PF 3021 Refugees ; Rejection
PD 1507 Refugees ; Socially handicapped
PC 0519 Refugees transit
PA 7321 Refusal
Refusal [accept...]
PF 7897 — accept international tribunal arbitration ; Government
PF 7897 — accept jurisdiction international courts justice ; Government
PF 5163 — admit error
PF 3282 — adoption
PF 5163 — apologize mistakes
PF 3079 *Refusal broadcasting licence*
PF 4050 Refusal declassify innovative patents ; Government
PF 3248 Refusal divorce ; Religious civil
PE 8423 Refusal entry foreign workers' families
PF 2645 Refusal ; Extradition
PF 1006 Refusal families participate globally
PF 0846 Refusal family possibilities
PF 5296 Refusal government leaders admit mistakes
Refusal grant
PF 0182 — amnesty
PE 5453 — citizenship
PF 3079 — *licences*
PF 3079 — licences media
PF 2657 — nationality
PF 3070 Refusal issue foreign currency use abroad
PE 0325 Refusal issue travel documents, passports, visas
PE 8137 Refusal let because applicant's race
Refusal licence
PF 3079 — media
PF 8964 — renewal
PD 3028 — *theatre performances*
PE 9054 Refusal loans because borrower's race
PF 4244 Refusal medical care
PF 0420 Refusal medical intervention ; Irrational conscientious
PF 3282 Refusal parents' consent adoption
PF 3226 Refusal participate
PF 0473 *Refusal ratify international conventions*
Refusal [sale...]
PE 8823 — sale because buyer's race
PF 0468 — sell
PF 5329 — share knowledge
PE 4349 Refusal testify
PE 0325 Refusal visas
PC 2494 Refusal vote
PD 3529 *Refusal work permits*
PD 0177 Refuse
Refuse [disposal...]
PD 0807 — disposal ; Domestic
PF 3454 — *disposal ; Unsanitary*
PF 3454 — *dumping ; Indiscriminate*
Refuse [dumps...] cont'd
PF 3454 — *dumps ; Uncovered*
PF 1916 Refuse vaccination ; Denial right
PE 4554 Refusers ; School
PE 5027 Refusing treat patients ; Medical practitioners
PE 8769 Regenerated cellulose artificial resins ; Instability trade
PF 8950 Regenerated fibres ; Instability trade synthetic
PF 1015 Regime ; Post revolutionary re employment government security services ousted repressive
PC 1738 *Regime ; Unstable*
PF 2199 Regimes alternative policies ; Government action against
PC 9585 Regimes ; Authoritarian
PD 5393 Regimes ; Environmental destruction communist
Regimes Government support
PF 4821 — corrupt
PF 4821 — repressive
PF 4821 — undemocratic
PF 2199 Regimes hostile policies ; Government action against
PC 1461 Regimes ; Illegal political
PC 1461 Regimes ; Illegitimate political
PC 0698 Regimes ; Military
PE 6052 Regimes ; Military aid repressive
PF 2199 Regimes opposing policies ; Government action against
Regimes [Undemocratic...]
PC 9585 — Undemocratic
PD 2615 — *Unpopular*
PC 1738 — *Unstable*
PE 5141 Regional accents ; Discrimination against
Regional [complementarity...]
PE 7255 — complementarity developing countries economies ; Low
PC 5917 — conflicts
PF 9129 — cooperation ; Fragmented
PE 8307 — cooperation problems countries ; Inadequate inter
Regional [deprivation...]
PD 4424 — deprivation
PF 7033 — development ; Absence policies associate international nongovernmental organizations
PC 2049 — *development countries ; Imbalanced*
PC 2049 — disparities
PC 4312 — distribution deaths ; Unequal
Regional [ecosystems...]
PD 5845 — ecosystems degraded exploitative development
PD 5845 — environmental degradation
PF 3628 — exposure ; Limited
Regional [global...]
PF 2499 — global legislation ; Limited local respect
PE 1604 — groupings developed countries ; Discrimination against developing countries formation
PD 9412 — groupings developing countries ; Inadequate economic integration
Regional [imbalance...]
PD 1494 — imbalance
PC 4291 — imbalance power
PF 9129 — integration ; Inadequate
PE 1184 — intergovernmental organizations common membership ; Inadequate coordination
PE 1583 — intergovernmental organizations common membership ; Jurisdictional conflict antagonism
PE 2488 Regional linguistic controversies ; Bilingualism national settings
PF 8299 Regional markets ; Untapped
PF 2018 Regional plans ; Delayed development
PF 2018 Regional plans ; Protracted development
Regional [tensions...]
PC 5917 — tensions
PD 0169 — trade developing countries ; Weakness
PC 0933 — trade ; Weakness South South inter
PD 4424 Regional underdevelopment
PD 1837 Regional unemployment countries
PC 5917 Regional warfare
Regions [Deforestation...]
PD 6282 — Deforestation mountainous
PD 1908 — Depopulation mountainous
PC 2049 — Deprived
PD 8183 — developed countries ; Depressed
PD 6120 Regions ; Excessive size metropolitan
PC 2049 Regions ; Favoured
PE 5760 Regions islands ; Neglect remote
PC 2049 Regions ; Overprivileged
PD 0110 Regions ; Rivalry disunity developing
PC 2049 Regions ; Underprivileged
PE 5875 Registration ; Inequities ship owner
PD 9641 Registration procedure ; Unjust voter
PE 6853 Registration wills ; Inadequate
PA 6338 Regression
PE 1418 Regression ; Ego
Regression [Hereditary...]
PD 8149 — Hereditary
PE 1418 — Human psychological
PF 0411 — Human racial
PA 6088 Regression *Impairment*
PE 1418 Regression ; Libidinal
Regression [Regression...]
PA 6338 — *Regression*
PA 5619 — *Relapse*
PA 5699 — *Reversion*
PD 8114 Regressive social development
PC 0834 *Regressive welfare system*
PD 5494 *Regular habit ; Lack*
PF 2996 *Regular transportation ; Lack*
PE 1182 Regulate domestic prices ; Distortion international trade minimum pricing regulations measures
PF 0401 Regulate family size ; Inability governments
PE 4146 Regulated activities ; Ineffective official inspection

PE 8067 Regulation automobile manufacturing industry ; Ineffective self
PE 8956 Regulation chemicals manufacturing industries ; Ineffective self
PE 8574 Regulation consumer goods manufacturing industries ; Ineffective self
PE 6226 Regulation electronic messages ; Ineffective
PE 8121 Regulation financial services sectors ; Ineffective self
PE 8472 Regulation food processing industries ; Ineffective self
PE 9048 Regulation health care sector ; Ineffective self
PE 8265 Regulation housing construction industry ; Ineffective self

Regulation [Inadequate...]

PD 1970 — Inadequate firearm
PF 5841 — Ineffective industry self
PF 2782 — Ineffective monitoring restrictive business practices due inadequate
PE 8630 Regulation mining industries ; Ineffective self
PF 8654 Regulation nuclear power ; Restrictive
PE 8502 Regulation pharmaceutical medical devices industries ; Ineffective self
PC 2618 *Regulation ; Price*

Regulation restrictive business

PE 0789 — practices ; Domestic bias
PF 1596 — practices ; Ineffective
PF 0591 — practices service industries ; Inadequate
PE 0225 — practices state enterprises ; Inadequate
PF 5840 Regulation shipping industry ; Ineffective self

Regulation [telecommunications...]

PF 5877 — telecommunications sectors ; Ineffective self
PE 6982 — television ; Excessive
PE 6914 — Torture behavioural
PF 0691 — transnational corporations ; Ineffective international

Regulations [against...]

PF 9299 — against environmental pollution ; Lack
PE 7141 — against pollution developing countries ; Lack
PE 1274 — agricultural pharmaceutical products ; Distortion international trade discriminatory formulation health sanitary
PF 5421 — Ambiguous
PD 8697 — Arbitrary enforcement
PF 4370 Regulations banking industry ; Antiquated
PG 5429 Regulations banking security firms ; National

Regulations Complex

PF 8053 — *business*
PF 2444 — *education*
PF 8053 — government
PF 8443 — housing
PF 2444 — *tax*
PF 4722 — trade
PF 8046 Regulations ; Connivance regulatory inspectors violators

Regulations [Delay...]

PC 1613 — Delay approval urgent
PE 0191 — Distortion international trade discriminatory application antidumping
PE 9073 — Distortion international trade discriminatory formulation equipment safety
PE 4339 Regulations ; Evasion national law
PF 1555 Regulations ; Excessive health

Regulations [Inadequate...]

PD 5001 — Inadequate enforcement safety
PC 9513 — Ineffective
PF 1555 — international travel ; Excessive animal sanitary
PE 1182 Regulations measures regulate domestic prices ; Distortion international trade minimum pricing
PE 5698 Regulations nuclear weapons industry ; Violation safety
PF 5435 Regulations ; Out date
PF 8793 Regulations ; Outdated

Regulations [paralyzing...]

PF 2444 — paralyzing small communities ; Complex
PD 4800 — Proliferation government
PD 5584 — Punitive
PF 3499 *Regulations ; Restrictive logging*
PC 6757 Regulations ; Secret

Regulations [taxes...]

PE 5873 — taxes flags convenience ; Evasion shipping
PF 2444 — *training ; Restrictive*
PE 2200 — transnational corporations ; Burden conflicting national

Regulations [Vague...]

PF 9849 — Vague
PD 7438 — Violations against economic
PE 4006 — Violations health safety
PF 3146 Regulators ; Plant growth

Regulatory [agencies...]

PF 8046 — agencies ; Corruption
PF 4937 — agencies ; Inadequate world
PF 3064 — authority ; Diffuseness
PD 7792 — authority ; Flouting
PE 9773 Regulatory banning plant cultivation

Regulatory codes

PE 0671 — culpability ; Violations
PD 4539 — Violation
PE 3809 — Wilful violation

Regulatory inspectors

PF 2793 — Lack
PF 8046 — Unethical practices
PF 8046 — violators regulations ; Connivance
PC 9513 Regulatory loopholes
PE 4339 Regulatory loopholes countries underdeveloped legislation ; Exploitation
PE 5187 *Regurgitation food*
PF 7728 Rehabilitate victims miscarriage justice ; Failure legally
PE 2451 Rehabilitation ; Demoralizing constraints housing

Rehabilitation facilities

PE 7317 — disabled persons ; Inadequate vocational
PD 1089 — Inadequate
PE 8151 — mentally handicapped ; Inadequate
PE 8803 Rehabilitation juvenile offenders ; Inadequate
PE 8983 Rehabilitation methods blind ; Inadequate
PF 1179 Reimbursements ; Delay government
PF 7752 Reinforce self image being victim circumstances ; Marketing skills which
PD 2019 Reinforced egocentric attitudes ; Educationally
PF 1728 Reinforced parochialism internal values images

Reinforced [safety...]

PD 5001 — safety factors ; Non
PF 3777 — sense personal shallow meaning ; Institutionally
PE 4589 — shallow perception personal meaning ; Socially
PF 1663 Reinforcement delimited self worth ; Social
PE 3995 Reinforcement ; Disparity facilities military mobilization
PF 6670 Reinforcement inappropriate development privileged classes developing countries
PF 4373 Reinforcement male dominance language
PF 1673 Reinforcement materialism ; Media
PF 3352 Reinforcement nationalism ; Race
PF 3351 Reinforcement nationalism ; Religion
PE 4589 Reinforcement shallow personal meaning ; Social
PD 0024 Reinforcement unemployment
PD 7727 Reinforcing contractionary economic spiral ; Self
PD 1290 Reintroduction animals
PC 0193 Rejected people ; Socially
PF 1822 *Rejection agriculture youth*
PF 3826 Rejection alternative images
PC 8127 Rejection children ; Family
PD 9557 Rejection citizen grievances ; Governmental

Rejection [Disapproval...]

PA 6191 — *Disapproval*
PF 3565 — disaster warnings
PA 6838 — *Dissent*
PA 5869 Rejection *Exclusion*
PF 2409 Rejection god
PA 6852 Rejection *Inappropriateness*
PC 5576 Rejection job opportunities unemployed
PC 0981 Rejection ; Maternal

Rejection [Necessity...]

PA 6387 — *Necessity*
PF 4351 — Negative effects
PF 8531 — nuclear power ; Irrational
PA 6979 Rejection *Opposition*

Rejection [peace...]

PF 9149 — peace proposals
PE 2087 — physically handicapped ; Family
PF 9149 — proposals social change

Rejection [refugees...]

PF 3021 — refugees
PA 7321 — *Refusal*
PF 9149 — remedial programmes
PF 0278 — rituals
PF 9127 Rejection technological change ; Irrational
PA 7392 Rejection *Unbelief*
PF 9149 Rejectionism
PD 3998 Rejects ; Production
PA 5619 Relapse
PF 7787 Relapsing fever
PE 7787 Relapsing fevers
PD 9219 *Relatedness ; Pervasive pattern deficits interpersonal*
PD 7522 Relations allies superpowers ; Instability
PF 3192 Relations beyond local environments ; Limited

Relations [developed...]

PC 7459 — developed countries ; Failure restructure economic
PE 7355 — developing countries transnational corporations ; Interference labour
PE 8308 — different parts government ; Unclear fiscal
PF 2575 Relations ; Estranged neighbourhood

Relations [Inadequate...]

PJ 8060 — Inadequate private international labour management
PF 2969 — Incomplete utilization external
PE 0667 — industrialized countries transnational corporations ; Interference labour
PF 2803 — inhibiting vision future ; Static social
PF 0945 — Isolating social forms enhancing ethical
PE 4890 Relations ; Lying concerning sexual
PC 0410 Relations ; Multiplicity languages international
PC 0025 Relations ; Non cooperative economic
PC 8694 Relations North South ; Failure restructure economic
PE 9122 Relations out wedlock ; Criminalization sexual

Relations [Parochial...]

PF 2600 — Parochial external
PS 7068 — *Poor international*
PF 3871 — Preoccupation reciprocity trading
PE 6515 Relations relating sportsmen ; Ineffective use external

Relations [socialist...]

PF 4886 — socialist countries ; Obstacles legal
PF 8040 — states ; Abnormal diplomatic
PF 8040 — states ; Cessation diplomatic
PE 5331 Relations trade ; Crimes related foreign
PE 1187 Relations transnational corporations ; Domination labour
PF 6505 Relations ; Unexercised responsibility external
PF 2969 Relations ; Unrealized potential external
PC 0237 *Relations ; Village leader distant*
PF 9843 Relations world leaders ; Antipathic personal
PD 4233 Relationship ; Breakup affective
PA 6484 Relationship*complex
PF 6186 Relationship ; Excessive intensification parent child

Relationship [international...]

PE 1973 — international governmental nongovernmental organizations programmes ; Inadequate

Relationship [international...] cont'd

PE 0777 — international nongovernmental organizations specialized agencies United Nations ; Inadequate
PC 0600 — Isolation parent child
PD 1941 Relationship ; Limited spheres
PF 7706 Relationship problems ; Assumption absence causal
PF 7706 Relationship problems ; Denial causal
PE 1511 Relationship transnational corporations local industry developing countries ; Inadequate
PE 0106 Relationship transnational corporations specialized agencies United Nations ; Inadequate
PF 2875 *Relationship ; Uncontinuous doctor*
PF 4730 Relationship wealth generation public good ; Lack
PF 8759 Relationships ; Corruption personal
PF 8759 Relationships ; Dishonest personal
PF 4274 Relationships ; Distrust interpersonal

Relationships [Fear...]

PF 8012 — Fear authentic
PD 2495 — *Fragmentation resident*
PF 1509 — Fragmented patterns extended family

Relationships [Indifference...]

PD 9219 — *Indifference social*
PC 7459 — industrialized countries ; Imbalance economic
PF 8759 — Irresponsible personal

Relationships Lack

PF 4416 — intimate
PF 0967 — motivating collegial
PF 2063 — symbolism local
PF 9843 Relationships leaders countries ; Negative personal
PF 0756 Relationships ; Prejudicial generation
PF 7706 Relationships problems ; Unproven
PE 1101 Relationships rural communities ; Unsystematic use powerful
PF 4274 Relationships ; Suspicion adult youth
PF 8759 Relationships ; Unclear
PF 8759 Relationships ; Unethical personal
PE 4203 Relationships work society ; Collapsed
PF 2811 Relative authority leadership
PA 6752 Relative motion*complex
PA 6091 Relative properties*complex
PJ 8374 Relative wage rates youth employment ; Negative effects
PF 9608 Relatives ; Being burden
PB 0034 Relatives ; Estranged
PE 3308 Relatives friends ; Torture
PE 9140 Relatives government leaders ; Corruption amongst
PF 0782 Relatives lost during warfare
PF 0782 Relatives ; Unknown
PB 2480 Relativism ; Moral
PF 7079 Relativism ; Religious
PE 1032 Release criminal cases ; Pecuniary guarantees condition provisional
PE 4577 Release ; Failure appear trial after
PD 7183 Release genetically engineered micro-organisms
PD 0520 *Released offenders ; Repeated arrest*
PD 0520 *Released prisoners ; Repeated resentencing*
PE 0598 Releasing controversial news items ; Calculated delays
PF 4875 *Releasing repatriated prisoners*
PA 5740 Relentlessness *Compulsion*
PA 7325 Relentlessness *Irresolution*
PA 7023 Relentlessness *Pitilessness*
PF 9068 Relevance education ; Unrecognized
PF 1944 Relevance formal education rural communities ; Unperceived
PF 8753 Relevant information ; Failure international information systems order problem
PF 1944 Relevant training opportunities local communities ; Lack
PF 4836 Reliable systems justice developing countries ; Absence

Reliance [canned...]

PD 0800 — *canned food*
PJ 9314 — Cash exchange
PF 9205 — convenient beliefs ; Over
PE 7195 Reliance developing countries external sources development financing ; Excessive
PE 7195 Reliance developing countries foreign borrowing ; Over
PF 0019 Reliance economic indicators human development ; Excessive
PF 1070 Reliance economic interest groups policy agencies ; Over
PF 4473 Reliance fashionable solutions problems ; Excessive
PF 0334 Reliance finite energy resources
PF 9560 Reliance government money creation ; Over
PF 4742 Reliance infallibility equipment ; Excessive
PC 6631 Reliance local governments grants ; Excessive
PE 1514 Reliance mineral fertilizers ; Excessive
PF 4494 Reliance paper qualifications
PJ 2489 Reliance professionalized services ; Exaggerated
PF 1373 *Reliance ; Reduced agricultural*
PF 5107 Relics ; Abuse
PF 8363 Relics ; Cult
PF 8889 Relics ; Fraudulent
PD 3051 Relief developing countries ; Inadequate public debt
PF 9007 Relief facilities ; Inaccessible emergency
PF 0286 Relief ; Inadequate disaster rescue
PF 0727 Relief ; Lack war
PF 6025 Relief ; Uncontrolled costs disaster
PF 6022 Relief ; Uncoordinated disaster
PF 0150 Religion

Religion [capitalist...]

PE 8701 — capitalist countries ; Enforcement
PB 1540 — Civil
PC 3175 — communist systems ; Denial
PC 3359 — Corruption organized
PF 1466 — Culturally biased

Repercussions domestic

PB 1540 Religion ; Decline	**Religious [historical...]**	PF 1205 *Reluctant personal participation*
PF 3863 Religion ; Decline	PE 5051 — *historical forgery*	PE 4725 Remains ; Unsanitary disposal human
Religion Denial right	PF 8889 — hoaxes	PE 4887 Remand ; Excessive imprisonment
PE 6397 — change	PF 3983 — hypocrisy	PF 7583 Remedial action ; Disguised negative consequences
PD 8445 — freedom	PF 6966 — hysteria	PC 1613 Remedial legislation ; Delays elaboration
PF 2850 — manifest	**Religious [images...]**	PF 9149 Remedial programmes ; Rejection
Religion [Dependence...]	PF 1365 — *images ; Incongruous*	PF 6707 Remedial response problems ; Government delay
PF 0150 — Dependence	PD 0355 — impediments marriage	PE 3946 Remedies ; Ignorance traditional herbal
PD 0127 — Discrimination against women	PD 4890 — indoctrination	PE 3946 Remedies ; Ignorance traditional plant
PD 3423 — Disintegration organized	PF 4433 — indulgences ; Commerce	PF 1098 Remission sentences crimes against humanity
PF 3983 — Double standards	PF 3376 — infallibility ; Negative effects claims	PF 2613 Remnants war ; Hazardous
PF 1466 Religion ; Eurocentric	**Religious influence**	PA 5479 Remorselessness *Lamentation*
PF 3367 Religion ; Experimental	PD 3357 — law	PA 7023 Remorselessness *Pitilessness*
PF 9350 Religion ; Harmonial	PD 3357 — law ; Discriminatory	PF 2875 Remote basic services
PF 2005 Religion ; Inadequacy	PF 3358 — politics	PF 2261 Remote borough authorities
PE 4332 Religion indigenous peoples ; Denial right freedom	PF 3358 — secular life ; Undue	PF 4832 Remote clinic service
Religion [Lack...]	PF 3358 — society	PF 1985 Remote cultural heritage
PF 8234 — Lack	PF 3358 — society ; Discriminatory	PF 4984 Remote driver education
PF 3387 — Lack world	**Religious [institutions...]**	PC 7674 Remote facilities ; Geographically
PF 3863 — Loss faith	PF 2137 — *institutions ; Abuse status*	PF 9023 Remote geographical location
PF 3361 Religion masses ; Restrictive influence	PF 5630 — insult	PD 4790 Remote health care services
PF 3387 Religion ; Need world	PC 2937 — intimidation	PE 5664 Remote island developing countries ; Deficient transport
PF 6091 Religion ; Official	PC 1808 — intolerance	PE 8797 Remote junior-high schools countryside
PF 9350 Religion opiate	PC 1808 — intolerance ; Dependence	PE 9135 Remote places ; Unaccompanied foreign travel
PF 3375 *Religion ; Perversion*	PC 2371 Religious justification militarism	PF 8652 Remote protection security services
PE 4332 Religion ; Prohibition practice indigenous	**Religious [leaders...]**	PF 5760 Remote regions islands ; Neglect
PF 4513 Religion ; Racist	PF 9288 — leaders human rights abuses ; Connivance	PC 4020 Remote resources ; Geographically
PF 3351 Religion reinforcement nationalism	PD 8445 — liberty ; Denial	PE 9072 Remote sensing systems ; Inequality inducing effects
Religion [Sexual...]	PF 3495 — lies	PF 8299 Remote supply centers
PE 0567 — Sexual bigotry organized	**Religious minorities**	PD 9051 Remote training resources
PF 6091 — State	PF 3403 — Alienation	PE 1848 *Remote village schools*
PC 0030 — state ; Conflict	PC 2129 — Denial rights	PB 8685 Remoteness
PF 2005 — *Superficial*	PC 8999 — Inadequate protection linguistic, national	PF 5351 Remoteness government
PF 9350 Religion ; Utilitarian	PC 2129 — Underprivileged	PE 9001 Remoteness legal services developing countries
PD 0127 Religion women ; Denial right freedom	PJ 5154 Religious miracles	PF 9023 Remoteness ; Physical
PF 1393 Religions acknowledging social problems ; Delay	PF 1618 Religious monologue	PF 2346 *Removal ; Difficult piling*
PF 3364 Religions ; Fragmentation organized	PE 4647 Religious national holidays ; Inefficiency due mismatch	PF 6527 Removal snow ; Insufficient
PF 5115 Religions ; Hierarchical structures established	**Religious [objectivity...]**	PD 0807 Removal ; Sporadic trash
PF 3364 Religions ; Lack unity	PJ 5626 — objectivity ; Lack	PF 6527 *Removal ; Uneconomic snow*
PF 3404 Religions ; Lack unity	PF 3838 — obligations	PD 8916 Remuneration labour ; Unfair
PF 3404 Religions ; Multiplicity	PD 3501 — *offerings*	PE 7760 Remuneration public private sector employees ; Disparity
PF 6091 *Religions ; Non recognition*	**Religious opposition**	PE 2053 Renal asthma
PF 7593 Religions ; Parochialism established	PF 1022 — abortion	PE 2053 Renal diseases
PF 3403 Religions society ; Inadequate integration	PF 1022 — contraception	**Renal failure**
Religious [activity...]	PF 3838 — nutritional resources	PE 2053 — [Renal failure]
PF 3863 — activity ; Lack	PF 1022 — population control	PD 9293 — *animals ; Acute*
PF 7467 — advisers ; Failure obey	PF 3838 — public health practices	PD 9293 — *animals ; Chronic*
PC 3365 — affiliation ; Segregation based	**Religious [paternalism...]**	PE 2053 Renal glycosuria
PC 3414 — apathy	PF 3609 — paternalism	PE 2053 Renal sclerosis
PC 3414 — apathy ; Dependence	PC 5994 — persecution	PD 7424 *Renal secondary hyperparathyroidism animals*
PF 6826 Religious backsliding	PF 3404 — pluralism	PF 4558 Reneging public pledges ; Government
Religious belief	PC 0030 — political antagonism	PF 0357 Renewable biofuels ; Insufficient utilization
PF 9274 — ecumenical dialogue ; Erosion	PF 5115 — power ; Centralization	PE 8971 Renewable energy resources ; Underutilization
PF 4954 — Excessive integration	PD 8439 — practices ; Restrictive	**Renewable resources**
PF 3404 — Fragmentation	PD 4365 — prejudice	PC 4412 — Competition non
Religious [beliefs...]	PF 2583 — *programmes ; Inaccessible*	PD 1018 — Dependence government revenues exploitation non
PF 6829 — beliefs ; Irrational	PD 3094 — propaganda	PC 8642 — Unproductive use non
PD 4365 — bigotry	PE 3354 — property ; Accumulation misuse	PC 8642 — Waste non
PF 3364 — brethren ; Separated	**Religious [racism...]**	PD 7320 Renewal ; Insensitive urban
PF 1433 — broadcasting ; Decline	PF 4513 — racism	PF 3550 Renewal ; Limited availability investment capital urban
PF 1433 — broadcasting ; Resistance	PF 7079 — relativism	PF 8964 Renewal ; Refusal licence
PD 7278 — buildings ; Desecration	PC 0578 — repression	PF 0296 *Renovation ; Costly home*
Religious [censorship...]	PC 0578 — repression ; Dependence	PF 1238 *Rent increases ; Uncontrolled household*
PD 5998 — censorship	PF 7220 — resolution scientific issues	**Rental [accommodation...]**
PF 3313 — ceremonies ; Burdensome cost	PE 1417 — riots	PD 3465 — accommodation ; Exploitative
PF 3248 — civil refusal divorce	PC 3355 — rivalry	PF 6195 — accommodation ; Socially sterile
PF 1951 — complacency	**Religious [sacrifice...]**	PF 2346 — *agreements ; Risky*
PC 3292 — conflict	PD 3373 — sacrifice	**Rental [payments...]**
PC 3363 — conflict sects	PF 1939 — schism	PE 4571 — payments ; Arrears
PD 6637 — conversion ; Forced	PF 3869 — scriptures ; Inaccessibility	PF 2346 — *policy ; Unclear*
PF 3863 — conviction ; Lack	PF 1106 — secrecy	PD 8894 — property ; Unkept
PE 0397 — courts ; Injustice	PF 3353 — sects ; Banned	PF 2346 Rental returns ; Insufficient
Religious [deception...]	PF 3353 — sects ; Persecution	PE 5913 Reoviral infection
PF 3495 — deception	PD 5534 — smear campaigns	PF 3499 Repair materials ; Inaccessible supply
PD 8807 — denominational education ; Discriminatory effects	PF 1270 — superstition	PD 8557 Repair ; Postponement street
PF 1618 — dialogue ; Ineffective	PF 7079 — syncretism	PE 1722 Repair ; Prohibitive cost product
PF 8010 — discipline ; Lack	**Religious [terrorism...]**	**Repair services**
Religious discrimination	PD 4134 — terrorism	PE 8997 — countryside ; Irregular
PC 1455 — [Religious discrimination]	PF 5694 — theology disrelated life	PF 7959 — developing countries ; Insufficient
PE 0168 — administration justice	PC 7101 — torture	PF 2471 — *Prolonged*
PF 2100 — *against divorcees*	PF 3863 — tradition ; Loss	PD 3056 *Repair ; Uncompensated tenant*
PF 6326 — against women priesthood	PD 4365 Religious unbelievers ; Discrimination against	PE 6574 *Repairs ; Uncompleted sidewalk*
PC 1455 — Dependence	PD 5534 Religious vilification	PE 4446 Reparations ; Delay payment government
PD 8807 — education	**Religious [war...]**	PE 4446 Reparations government ; Non payment
PC 3220 — politics	PC 2371 — war	PF 4875 *Repatriated prisoners ; Releasing*
Religious [dissent...]	PF 3364 — witness ; Lack common	PE 1193 Repatriation concealed transfer pricing ; Profit
PE 4875 — dissent	PE 8905 — worship ; Decreasing participation collective	PF 3132 *Repatriation export proceeds ; Non*
PF 3364 — disunity	PC 6203 Relocation ; Forced	PD 8099 Repatriation ; Forced
PF 3404 — disunity	PE 6687 Relocation hazardous industries developing countries	PE 6854 Repatriation minors ; Unlawful
PF 1115 — doctrine ; Disagreement concerning	PE 8898 Relocation ; Inadequate compensation forced	**Repatriation [prisoners...]**
PF 2005 — doctrine ; Inadequacy	PF 3455 *Relocation national capitals ; Unnecessary*	PD 0218 — prisoners war ; Forced
PF 6988 — dogmatism	PE 8898 Relocation ; Non payment compensation forced	PE 0948 — prisoners war ; Non
Religious [education...]	PE 4273 Relocation peasants onto marginal lands ; Forced	PC 3134 — profits foreign investors developing countries ; Excessive
PE 3802 — education ; Denial right choose moral	PC 6203 Relocation population	PD 1157 Repayment international debts ; Irregular
PA 1387 — *elitism*	PJ 0451 Reluctance government invest high-risk development initiatives	PE 3926 Repayments developing countries ; Excessive unnecessary burden interest
PD 4782 — ethnic divisions ; Internal language,	PF 1735 Reluctance join community action	PD 0520 *Repeated arrest released offenders*
PF 4371 — experience ; Barriers	PF 5736 Reluctance reform law	PF 2100 Repeated marriage divorces
PJ 3532 — experience ; Disparagement	PF 4418 *Reluctance ; Sales call*	PD 0520 *Repeated resentencing released prisoners*
PF 4954 — extremism	PF 6530 *Reluctance towards corporate investment*	PF 3391 Repeated warnings ; Backlash against
Religious [fanaticism...]	PF 0845 Reluctance transform administrative structures ; Government	PF 0589 *Repercussions disarmament ; Negative economic*
PF 4954 — fanaticism	PF 7142 *Reluctancy individual involvement*	PF 2889 Repercussions domestic agricultural policies ; Detrimental international
PF 8889 — fraud	PF 1226 Reluctant claims external resources	
PF 1816 — friction ; Holy places focus	PF 6530 Reluctant investment capital resources	
PF 1338 — fundamentalism	PF 0336 Reluctant local law enforcement	
PF 3858 Religious gerontocracy ; Superannuated		

–1027–

Repetition educational

PC 1716 Repetition educational stages
PD 1893 Repetition untruths ; Susceptibility electorate government
PD 2656 Repetitious work ; Monotonous
PE 1478 *Repetitive news items ; Reduced impact impersonal*
PE 5606 Repetitive strain injuries
PE 7504 Replacement parts agricultural machinery ; Unavailability
PJ 4688 Replacement parts ; Overpriced equipment
PB 5249 Report suffering ; Failure
PD 4982 Reporting ; Corporate over reporting
PD 4517 Reporting governmental action ; Partisan
PF 0203 Reporting ; Inadequate international standards financial accounting
PF 7939 Reporting man made disease ; Inadequate knowledge
Reporting [obligations...]
PE 2215 — obligations ; Failure governments fulfil international
PE 2215 — obligations ; Governmental delays fulfilling international
PD 1893 — Official over reporting
PF 5311 Reporting ; Positive bias
PF 5374 Reporting public sector ; Inadequate financial
PD 4982 Reporting reporting ; Corporate over
PD 1893 Reporting reporting ; Official over
PF 5311 Reporting ; Uncritical upbeat
PF 3096 Reports ; Biased
PD 4982 Reports corporate competitors ; Fabrication
PF 3096 Reports ; Misleading
PF 4564 Reports ; Misleading
PD 5183 Reports ; Premature disclosure government
PE 4581 Representation agreements ; Exclusive sales
PF 3517 Representation ; Denial legal
PC 0655 Representation ; Dependence unequal political
PE 4344 Representation intergovernmental organizations ; Inadequate coordination governmental
PF 3468 Representation ; Lack
PF 3517 Representation ; Lack legal
PF 1781 *Representation ; Neglected council*
PF 1781 *Representation ; Noninclusive council*
PC 0655 Representation ; Unequal political
PC 9585 Representational government ; Non
PF 0981 Representative arts
PD 0055 Representative democratic political organization ; National governments obstacle
PE 0252 Representatives ; Assassination workers
PE 3735 Representatives ; Conflict interest peoples
Representatives [Death...]
PE 4869 — Death threats against workers
PD 1846 — Denial right protection union
PD 7630 — Detention workers
PC 0237 — *Distant*
PE 7620 Representatives following legal strike action ; Dismissal trade union
PE 5882 Representatives ; Forced disappearances workers
Representatives [Inadequate...]
PD 1846 — Inadequate protection facilities workers'
PD 0610 — Intimidation worker's
PD 7471 — Intimidation workers'
PJ 0272 — *Isolation trade union members their*
PD 4982 Representatives ; Perjury corporation
PF 5311 *Representatives process feedback constituents ; Inability people's*
PD 1846 Representatives ; Victimization workers'
PE 1758 Representatives ; Violation right trade unions freely elect their
PF 0889 Represented constituency ; Extreme detachment
PF 0960 Repressed inflation socialist countries
PB 0871 Repressed people
PF 2922 Repressed sexual tendencies
PB 0871 Repression
PF 4821 Repression ; Assisting
PB 8785 Repression ; Communist
PC 8425 Repression ; Cultural
PD 4811 Repression demonstrations ; Violent
Repression Dependence
PB 0871 — [Repression ; Dependence]
PC 1919 — political
PC 0578 — religious
PA 5497 Repression *Difficulty*
PC 8471 Repression ; Economic
PC 1777 Repression elements personality
PF 1216 Repression feelings
PA 6651 Repression *Forgetfulness*
Repression [Ideological...]
PC 8083 — Ideological
PE 4332 — indigenous spirituality
PD 0434 — intellectual dissidents
PA 7382 Repression *Loss*
PC 1919 Repression ; Political
PF 5210 Repression problems ; Conceptual
Repression [Racial...]
PD 8762 — Racial
PA 7321 — *Refusal*
PC 0578 — Religious
PA 7296 — *Restraint*
PC 1777 Repression self-consciousness
PF 2922 Repression ; Sexual
PE 2007 Repression trade union demonstrations ; Violent
PD 0634 Repressive detention juveniles
Repressive [regime...]
PF 1015 — regime ; Post revolutionary re employment government security services ousted
PF 4821 — regimes ; Government support
PE 6052 — regimes ; Military aid
PF 9181 Reprisals
PF 3707 Reprisals against citizens ; Governmental
PF 9078 Reprisals ; Fear
PD 9389 Reprisals ; Fear economic
PF 9181 *Reprisals ; Military*
PD 0736 Reprisals ; Police
PF 3707 Reprisals ; Political
Reprisals Threat
PD 9389 — commercial
PF 3707 — official
PD 0736 — police
PD 9389 Reprisals ; Trade
PA 7237 Reproach *Condemnation*
PA 6191 Reproach *Disapproval*
PA 6839 Reproach *Disrepute*
PA 7237 Reprobation *Condemnation*
PA 7389 Reproduction
PE 5197 Reproduction disabled ; Sexual
PB 1044 *Reproductive organs ; Diseases*
PD 7994 Reproductive processes ; Ignorance
PD 7799 *Reproductive system animals ; Congenital anomalies*
PD 7799 Reproductive system diseases animals
PC 0604 Reptiles ; Endangered species
PC 0604 Reptilia
PD 3053 Repudiation ; Debt
PA 6838 Repudiation *Dissent*
PA 5869 Repudiation *Exclusion*
PA 7325 Repudiation *Irresolution*
PA 6387 Repudiation *Necessity*
PA 7321 Repudiation *Refusal*
PA 6698 Repugnance *Difference*
PA 5982 Repugnance *Disagreement*
PA 5446 Repugnance *Enmity*
PA 7338 Repugnance *Hate*
PA 7107 Repugnance *Unpleasantness*
PA 6509 Repugnance *Unwillingness*
PA 5799 Repulsion
PA 5454 Repulsion *Badness*
PA 6979 Repulsion *Opposition*
PA 5799 Repulsion *Repulsion*
PA 5459 Repulsion *Uncleanness*
PA 7107 Repulsion *Unpleasantness*
PD 2569 Reputation ; Denial right freedom attacks personal honour
PF 2575 Reputation ; Negative community
PF 8125 Reputation unhelpful ; Police
PE 4394 Reputations ; Sullying
PE 8157 Requested services ; Delay delivery
PF 1735 *Requests services ; Disunified*
PF 0905 *Requests ; Unstrategic funding*
PJ 5365 *Required curriculum ; Rigid*
PD 9628 Requirement over qualification women ; Discriminatory
PE 7971 Requirement qualifications public services ; De facto racial
PJ 0451 Requirement that economic benefits be proven prior programme approval ; Government
PD 8561 Requirements aged persons ; Increasing
Requirements [Defeating...]
PE 1437 — Defeating size
PF 2457 — *Demanding*
PC 1131 — Disparity workers skills job
PJ 9248 Requirements ; Irrelevant job
PF 3498 *Requirements ; Low performance*
PE 1843 Requirements ; Market indicators exclusion human
PF 1578 Requirements ; Programme initiatives inappropriate local
PF 3526 *Requirements ; Prohibitive daycare*
PE 0778 Requirements respect marks product origin ; Distortion international trade discriminatory
PE 0083 Requirements respect product standards measures ; Distortion international trade discriminatory
PF 1653 *Requirements ; Unfamiliar bureaucratic*
PF 6471 *Requirements ; Unpolled consumer*
PE 8893 Requirements welfare benefits ; Excessive bureaucratic
PE 4452 Requiring women lift heavy loads ; Work practices
PF 0757 Requisitioning workers prevent strike action
PE 8110 Rescheduling debts developing countries market-related interest rates
PF 0286 Rescue relief ; Inadequate disaster
PF 0198 Research activity ; Imbalanced
PF 9693 Research ; Bias scientific
Research [Clandestine...]
PC 2921 — Clandestine market
PD 7231 — cover illegal activity ; Misuse
PD 0260 — Cruel treatment animals
Research [Deceptive...]
PD 7231 — Deceptive misuse
PF 7634 — Deceptive social science
PF 2112 — Dehumanized individual scientific
PF 9625 — Delays
Research development
PE 4880 — capacity developing countries ; Inadequate
PF 3846 — Opportunist bias public discussion
PF 1789 — Pollution hazards industrial
PF 1120 — problems developing countries ; Inadequate
PD 7231 Research disinformation
Research [efforts...]
PF 2306 — efforts ; Uncoordinated
PE 4059 — Emphasis economic priorities
PF 1789 — Environmental hazards constraining scientific
PF 0059 — Excessive emphasis fashionable areas
PE 4805 Research foetal cadavers foetal tissue
PF 9830 Research ; Fragmentation
PE 4805 Research human embryos
Research [Ideological...]
PF 7220 — Ideological domination
PD 1039 — Inadequate family planning
PC 1449 — Inhumane scientific
Research [Inhumane...] cont'd
PE 1621 — Inhumane use non human primates
PC 6329 — *Insufficient diversification energy*
PF 5419 — Insufficient government funding
PF 0145 — Intrusive social science
Research [Lack...]
PE 0612 — Lack funds medical
PE 8463 — Lack funds veterinary
PF 1086 — Limited market
PF 2529 — Limits areas
Research [methodologies...]
PF 2202 — methodologies ; Top down
PD 7231 — Misrepresentation socially unacceptable activity
PC 9188 — Misuse scientific
PF 2466 Research natural resource product expansion ; Neglect
PC 0012 Research ; Nuclear weapons
PF 9625 Research ; Obstruction
PE 4059 Research ; Over emphasis immediate solutions resource development
Research [Parochial...]
PF 6928 — Parochial market
PF 7220 — Politicization
PF 1077 — problems ; Inadequate
PF 0716 — projects ; Misappropriation resources high cost
PF 1572 — proposed solutions problems ; Inadequate
Research [Restrictions...]
PF 0725 — Restrictions
PD 3154 — Restrictions sharing technical
PF 9693 — Reward biased
PF 9693 — Reward oriented
Research [Sales...]
PF 1025 — Sales dominated market
PF 1025 — Sales promotion dominates market
PF 1430 — Secrecy scientific
PF 5419 — Shortage funds
PF 2823 — social problems ; Restriction funding
PF 9693 — Sponsor biased
PF 9693 — sponsored vested interests
PF 7889 — Superficial
PF 4848 — support ; Inadequate peace
PF 2415 Research total human process ; Superficial
PF 9141 Research ; Unreported
PC 0080 Research using human subjects ; Irresponsible
PC 0012 Research ; Waste resources armaments
PF 1572 Researched projects alleviate problems ; Lack well
PD 0520 *Resentencing released prisoners ; Repeated*
PF 8374 Resentment
PC 3458 Resentment ; Class disparity
PA 6011 Resentment *Discontentment*
PA 7253 Resentment *Envy*
PF 7472 Resentment government policies ; Public
PE 6246 Reservation apartheid system ; Job
PD 3443 *Reservation jobs men*
PD 3443 *Reservation jobs women*
PE 8667 Reservation overbooking
PC 0582 *Reservations*
PD 3305 Reservations ; Restriction indigenous populations
PF 2378 Reserve ; Limited local availability capital
PF 2848 Reserve technical skills rural communities ; Shallow
PF 3060 Reserves developing countries ; Depletion official
PC 5016 Reserves ; Inadequacy emergency food
Reserves Inadequate
PB 4653 — financial
PF 3060 — level developing country monetary
PF 3059 — level world monetary
PF 3061 — mechanism creation liquid
PB 4653 Reserves ; Lack capital
PF 2899 Reserves ; Limited individual capital
Reserves [marine...]
PE 4913 — marine mammals ; Depletion fish
PF 0687 — material ; Unnecessary
PD 4403 — Misuse nonrenewable fossil water
PE 0060 Reserves transnational enterprises ; Excessive exploitation raw material
PD 9515 *Reservoir induced increases seismicity*
PF 3448 *Reservoir plan ; Inadequate*
PF 2848 Reservoir technical skills rural communities ; Limited
PC 3595 *Reservoirs ; Disease*
PE 2572 *Reservoirs disease ; Vertebrate*
PC 2507 *Resettlement countryside ; Forced*
PF 9030 Resettlement ; Fear
PC 6203 Resettlement ; Involuntary mass
PF 4181 Resettlement ; Rural
PD 7776 Resettlement stress
PC 0935 *Residence ; Denial right return country*
PG 8978 *Residence ; Temporary*
PF 2822 *Residences ; Distant farm*
PD 3442 Residences ; Racially separated
PF 1213 *Residency ; Discouragement permanent*
PE 1848 *Residency patterns ; Holiday*
PD 2338 Resident home ownership ; Non
PF 2261 Resident local authorities ; Non
PG 8978 *Resident patterns ; Temporary*
PC 0592 *Resident professionals ; Shortage*
PD 2495 *Resident relationships ; Fragmentation*
PF 2421 *Resident school personnel ; Non*
PF 5354 Resident self image ; Low
PF 6559 Resident unity ; Unfocused
PD 2011 *Residential animal lodging*
PF 7118 *Residential community ; Dispersed*
PF 6511 Residential housing rural communities ; Inadequate
PD 8075 Residential land ; Scarcity
PD 3048 Residents country ; Tax discrimination against non
PF 6132 Residents ; Inaccessibility city centres suburban
PF 2575 Residents ; Limited involvement neighbourhood

Resources

Residents [Unengaged...]
PF 2813 — *Unengaged town*
PF 2813 — *Uninvolved transient*
PJ 3760 — *urban areas ; Disengaged transient*
PD 0438 Residual schizophrenia
PE 3798 Residual traumatic pains due torture
PF 5188 Residue levels ; Dangerous legally prescribed chemical
PE 6480 Residues food ; Pesticide
PD 4925 Residues freshwater animals, fish birds ; Accumulation contaminant
PD 3934 Residues marine animals fish ; Accumulation contaminant
Residues [plants...]
PD 5021 — plants animals ; Accumulation contaminant
PD 0120 — Pollutant
PD 0120 — pollutants ; Pesticide
PD 5278 Residues terrestrial animals ; Accumulation contaminant
PD 0381 Residues terrestrial plants ; Accumulation contaminant
PE 4903 Residues veterinary drugs foods
PF 8781 Resignation problems
PF 8611 Resignation towards bribery
PE 8769 Resins ; Instability trade regenerated cellulose artificial
PA 6448 *Resist impulses gamble ; Failure*
PA 6448 *Resist impulses steal ; Failure*
PE 6949 Resist oppression ; Denial right
PE 6629 Resistance antibiotics ; Bacterial
Resistance [change...]
PF 0557 — change
PF 3010 — changing agricultural methods
PF 2819 — charitable giving ; Increasing
PF 5950 — Chronic psychological
PC 0690 — Civil
Resistance [demands...]
PD 9219 — *demands adequate occupational performance ; Passive*
PE 0921 — *disease ; Reduced*
PF 9659 — Drug
PF 7754 Resistance evil ; Non
PE 4456 Resistance fungicides ; Fungal
PC 0690 Resistance government
PF 5266 Resistance grace
Resistance [inclusive...]
PF 9576 — inclusive environmental models
PE 4901 — incorporating midwives medical care systems
PC 3950 — inertia response problems ; Governmental
Resistance insecticides
PE 3578 — Blackfly
PE 3579 — Cockroach
PE 3572 — Flea
PE 3583 — Housefly
PD 2109 — Insect
PE 3576 — Louse
PE 3582 — Mosquito
PF 0845 Resistance institutional change ; Government
PC 4591 Resistance internationally agreed standards
PF 5203 Resistance loss any personal investment
PC 1869 *Resistance movements*
PD 5707 Resistance Nuremberg obligation ; Citizen
Resistance [Passive...]
PF 2788 — Passive
PF 2788 — Peaceful
PF 5203 — personal austerity
PD 3696 — pesticides ; Pest
PF 5622 — public interest broadcasting
PJ 1433 — Resistance religious broadcasting
PE 3573 Resistance rodenticides ; Rodent
PF 2335 Resistance self review ; Bureaucratic
PF 7134 Resistance ; Societal knowledge
PF 6300 Resistance technology education
PE 6007 Resistant bacteria
PE 6257 Resistant construction ; Inadequate earthquake
PB 6384 *Resistant diseases ; Encouragement drug*
PE 0616 *Resistant malaria ; Drug*
PF 9659 *Resistant viruses ; Drug*
Resolution scientific issues
PF 7220 — Ethical
PF 7220 — Ideological
PF 7220 — Religious
PF 2086 Resolutions ; Irrelevant
PF 2086 Resolutions ; Stock
PF 2086 Resolutions ; Tokenistic meeting
PF 2086 Resolutions ; Vague
PF 8435 Resolve problems realistically ; Inability
PF 7220 Resolve theoretical issues ; Misuse democratic processes
PE 6256 Resorts ; Destruction mountain ecosystems ski
PF 9728 Resource base economic activity ; Failure recognize environment
Resource [channels...]
PF 2517 — channels ; Lack intersocietal
PD 0177 — *conservation ; Lack*
PF 0038 — conservation policies ; Inadequate governmental
Resource depletion
PB 6112 — [Resource depletion]
PB 6112 — Dependence
PD 4002 — due high level consumption ; Natural
Resource development
PD 8114 — debtor developing countries ; Neglect human
PF 2855 — Exploitative
PD 4007 — Imbalance population growth
PE 4059 — research ; Over emphasis immediate solutions
Resource [endowments...]
PF 3043 — endowments ; Disparity countries natural
PB 1016 — exchange ; Breakdown
PD 3546 — extraction ; Parochial planning natural

Resource [import...]
PC 0065 — import dependence
PE 0537 — import dependence developed countries
PE 0635 — intensive packaging
PF 1086 — Resource ; Limited market
PD 0465 Resource mobilization developing countries ; Low level domestic
PF 6552 *Resource opportunities ; Uncommunicated*
PF 9166 Resource planning ; Fragmented
PF 2466 Resource product expansion ; Neglect research natural
PF 2875 Resource services ; Ineffective delivery basic human
PF 5039 *Resource thinking ; Single*
PC 3134 Resource transfers developing countries ; Negative net
Resource [usage...]
PF 2477 — *usage ; Inefficient*
PF 2882 — use decisions ; Limited access natural
PF 0827 — use ; Seasonal
PB 1016 — utilization ; Maldistribution
PF 1316 — utilization ; Unimaginative vision
PF 3994 — Resource ; Violence
PC 3108 Resource wastage capitalist systems
Resources [Absentee...]
PD 2338 — *Absentee owned natural*
PF 0404 — action against problems ; Shortage financial
PD 0408 — Aggressive uses natural energy
PC 0012 — armaments research ; Waste
PE 3812 — available developing countries ; Decline concessional financial
PC 3113 Resources capitalist systems ; Waste human
Resources Competition
PE 0063 — intergovernmental organizations scarce
PE 0259 — international nongovernmental organizations scarce
PC 1463 — international organizations scarce
PC 4412 — non renewable
PC 4412 — scarce
Resources [Conflicting...]
PD 7087 — Conflicting claims over shared inland water
PF 6654 — Conflicting use
PF 3581 — conservation ; Inadequate plant genetic
PD 2625 — Corruption consumption
PF 3043 — countries ; Inequality distribution natural
Resources [Decreasing...]
PC 4824 — Decreasing natural
PC 8419 — Degradation agricultural
PB 5250 — Degradation natural
PC 3304 — Denial indigenous rights lands
PD 3109 — Denial sovereignty over natural
Resources Dependence
PC 0065 — external
PD 1018 — government revenues exploitation non renewable
PE 0537 — industrialized countries import
Resources [Depleted...]
PD 9082 — Depleted fish
PD 9357 — Depletion mineral
PB 5250 — Depletion natural
PD 9118 — *Destruction land*
PD 1052 — Deterioration soil
PE 5551 — developed countries ; Excessive consumption
Resources developing countries
PD 3922 — *Depletion natural*
PF 5964 — Inadequate mobilization financial
PF 5964 — Inefficient management government financial
PF 5964 — Inefficient use domestic
PD 3625 — Lack natural
PC 3134 — Outflow financial
PD 8114 — Reduction public expenditure human
Resources [development...]
PE 9764 — development least developed countries ; Inadequate human
PE 1597 — Discriminatory exploitation sea bed
PF 1979 — Dispersion local capital
PF 7458 — Disregarded financial
PE 8865 — due alcohol ; Draining
PD 4007 — due population growth ; Depletion natural
Resources [Emphasis...]
PF 2855 — Emphasis mass extraction natural
PE 6682 — Environmental hazards mineral exploitation seabed
PF 9612 — Erosion investment human
PC 4824 — Excessive demands natural
PD 9357 — Exhaustion mineral
PE 8261 — *expensive medical techniques ; Waste*
Resources [facilities...]
PD 2762 — facilities ; Disparities distribution communication
PF 4979 — Failure mobilize
PD 3109 — Foreign control natural
PF 0716 — fundamental sciences ; Inappropriate use
PC 4020 — Resources ; Geographically remote
Resources government
PD 0183 — agencies ; Wastage
PD 0183 — agents ; Extravagant waste
PF 4095 — Ineffective use private sector
Resources [heal...]
PE 8261 — *heal privileged ; Discriminatory use medical*
PF 9587 — health ; Inadequate
PF 4975 — high cost civil engineering projects ; Misappropriation
PF 0716 — high cost research projects ; Misappropriation
PD 7669 Resources ; Inaccessible health
PC 4020 *Resources ; Inaccessible natural*
Resources Inadequate
PD 7254 — medical
PF 9587 — medical
PF 4979 — mobilization
Resources [Inappropriate...]
PD 9338 — Inappropriate use financial
PF 2401 — Incomplete access information

Resources [industrialized...] cont'd
PD 1291 — industrialized countries ; Loss developing country human
PE 5551 — industrialized countries ; Unsustainable consumption
PF 4095 — Ineffective governmental use nongovernmental
PF 4197 — Ineffective mobilization public
Resources Inefficient
PF 2204 — extraction utilization natural
PE 5001 — use
PC 7517 — use energy
Resources Inequitable
PD 3210 — access
PE 1597 — allocation rights exploit sea bed marine
PC 4844 — *use medical*
Resources [Insecurity...]
PB 8678 — Insecurity
PB 4653 — Insufficient financial
PB 1016 — International disparity consumption
PE 9346 — invested obsolete armaments ; Waste
PD 2625 — Irresponsible consumption
PC 7928 Resources ; Lack natural
PE 0273 Resources least developed countries ; Insufficient use natural
Resources Limited
PF 1653 — access external
PF 6573 — access society's
PD 7669 — availability health
PF 2009 — availability local commercial
PC 5165 — forest
PF 6477 — local investment
PF 5539 Resources ; Local control
Resources Long term
PB 6112 — shortage
PF 0334 — shortage energy
PD 9357 — shortage mineral
PC 4824 — shortage natural
PB 6112 — shortage uranium
PC 7721 Resources ; Loss human
Resources [Maldistribution...]
PB 1016 — Maldistribution
PD 2705 — Maldistribution medical
PD 2046 — Mass unemployment human
PF 0780 — military activities ; Absorption manpower
PE 6115 — Mismanagement food
PB 5151 — Misuse
PE 6788 — Monopolization agricultural genetic
PD 3546 Resources ; Nationalistically determined development natural
PF 7808 Resources ; Neglected food
Resources [obsolete...]
PE 7539 — obsolete nuclear power plants ; Waste
PF 4767 — Obstacles utilization coastal deep sea water
PD 9578 — Over exploitation biological
PD 3448 — *Overdemand*
PF 2959 — Overemphasis effective use technical
PD 9082 — Overexploitation fish
PD 4403 — Overexploitation underground water
Resources [peaceful...]
PF 4785 — peaceful applications ; Obstacles redeployment military
PF 1085 — pet animals ; Disproportionate allocation
PC 4020 — Physically inaccessible
PD 5261 — poor ; Overuse environmental
PD 1283 — production animal feed ; Use agricultural
PF 2174 — Profit over emphasis use national
PE 0648 — protect aged ; Misallocation
PD 5642 — public sector ; Monopolization
Resources [Reliance...]
PF 0334 — Reliance finite energy
PF 3838 — Religious opposition nutritional
PF 1226 — Reluctant claims external
PF 6530 — Reluctant investment capital
PD 9051 — Remote training
Resources [Scarcity...]
PC 4824 — Scarcity oil
PD 4403 — Shortage groundwater
PF 3665 — small communities ; Strained capital
PA 5984 — Strain world
PF 6530 — Subsistence approach capital
Resources [Unadvertised...]
PF 6573 — *Unadvertised educational*
PF 1345 — Unconnected business
PF 2855 — Unconstrained exploitation natural
Resources Underdeveloped
PF 3448 — *benefits community*
PF 3448 — potential basic
PF 2164 — use agricultural
PF 2842 *Resources ; Underused industrial*
Resources Underutilization
PF 3523 — human
PF 1459 — natural
PE 8971 — renewable energy
PD 7526 Resources ; Underutilized animal genetic
PE 9325 Resources ; Underutilized government
Resources Undeveloped
PF 3526 — channels public private
PF 3523 — human
PC 4815 — water
Resources [Unidentified...]
PF 6541 — *Unidentified news*
PF 1460 — United Nations system organization ; Shortage financial
PC 6031 — Unproductive labour
Resources Unproductive use
PB 8376 — [Resources ; Unproductive use]
PC 8914 — human

-1029-

Resources

Resources Unproductive use cont'd
PC 8642 — non renewable
Resources [Unregulated...]
PF 3183 — Unregulated global
PC 8419 — Unsustainable development agricultural
PD 9082 — Unsustainable exploitation fish
PE 9325 — Untapped government
Resources Unused
PF 2164 — agricultural
PF 1373 — *cultural*
PF 1240 — *fodder*
Resources [Unutilized...]
PF 3523 — Unutilized human
PF 1316 — *Unutilized local*
PE 9133 — urban services ; Insufficient financial
PF 3550 — urban services ; Limited availability funding
PD 3546 — used national self interest ; Natural
PB 4788 Resources ; Vulnerability world genetic
PC 8914 Resources ; Wastage human
Resources Waste
PB 8376 — [Resources ; Waste]
PC 8914 — human
PC 8642 — non renewable
PC 7517 Resources ; Wasted energy
PB 6719 Resources ; Worldwide misallocation
PF 6146 Respect common land ; Inadequate arrangement housing
PA 6822 Respect ; Lack
PF 0879 *Respect ; Loss self*
PF 2499 Respect regional global legislation ; Limited local
PD 2732 *Respiratory disease ; Chronic*
PE 9854 *Respiratory disease complex ; Feline*
Respiratory diseases
PD 0637 — [Respiratory diseases]
PE 5524 — cattle
PE 9438 — goats
PE 5182 — horses
PE 5567 — poultry
PE 9438 — sheep
PE 9854 — small animals
PE 7553 — swine
PE 7159 Respiratory distress syndrome
PD 7924 Respiratory failure
Respiratory [illness...]
PD 7924 — illness
PD 7924 — infections
PE 7591 — infections ; Acute
PE 8173 Respiratory muscles ; Paralysis throat, eye, limb
PE 9151 Respiratory organs ; Inflammatory infections
Respiratory system
PD 7307 — animals ; Diseases
PE 3637 — Benign neoplasm
PD 1618 — *Congenital anomalies*
PD 7924 — Diseases
PE 7572 — Malignant neoplasm
PE 7864 — Symptoms referable
PE 7733 Respiratory tract ; Diseases upper
PE 2526 Respiratory tuberculosis
PD 7924 Respiratory virus diseases
PF 9432 Respond fall demand ; Failure commodity production
PF 8876 Respond minority concerns ; Parliamentary inability
PC 5180 Respond needs developing countries ; Lack political will
PF 9664 Response accelerating structural change ; Policy making delays
PF 2819 Response appeals aid ; Fluctuating
PF 9358 Response creationism ; Censorship textbooks
PF 8652 Response ; Delayed police
PF 3567 Response disasters ; Confused government
PF 6707 Response evidence potential hazards ; Bureaucratic procrastination
Response [global...]
PF 0817 — global issues ; Nationalistic
PF 8753 — global problematique ; Failure integrate knowledge empower humanity
PE 2215 — governments international surveys ; Non
PF 7925 Response human rights ; Bias United Nations
PB 5249 Response injustice ; Indifference
PE 7140 Response intergovernmental calls action ; Inadequate international nongovernmental organization
PD 1432 Response monetary incentives developing countries ; Lack
Response [Paralysis...]
PF 2701 — Paralysis social
PF 2256 — *pattern ; Irritation*
PF 2256 — *pattern ; Laissez faire*
PF 6707 — potential problems ; Official inaction
Response problems
PF 6707 — Government delay remedial
PC 3950 — Governmental resistance inertia
PF 0301 — Inadequate delivery mechanism
PF 8753 — society ; Failure focus available insight
Response [social...]
PF 8753 — social problems ; Failure interrelate wisdom different cultures
PD 1080 — societal needs ; Inadequate
PB 5249 — suffering ; Silence
PF 6707 — symptoms problems ; Government delay
PF 1336 Response technological changes ; Indecisive
PF 0913 Responses atomic energy ; Unhealthy emotional
PA 1999 Responses ; Delayed
PE 7080 Responses international nongovernmental organization actions ; Divisive
PF 6461 *Responses ; Lax parental*
PF 0317 Responses leadership ; Immediate appeasement
PE 4883 Responses malnourished persons ; Inadequate immune
PF 5660 Responses ; Obstruction change crisis emergency
PF 4473 Responses problems ; Habitual
PF 7722 Responses public injustices ; Riskless
PE 3797 Responses ; Speaking opposition out personal
PF 0817 Responses world economic crisis ; Nationalistic policy
PE 5538 Responsibilities ; Criminal negligence performing socialist
PE 7206 Responsibilities ; Discrimination employment women family
PF 7823 Responsibilities ; Irresponsible delimitation policy
Responsibility [Abdication...]
PA 8658 — Abdication
PF 3553 — Absence images social
PF 1731 — adults ; Limited community
PF 5555 Responsibility ; Collapsed tension care
PF 1064 Responsibility ; Deluding familial image social
PF 6505 Responsibility external relations ; Unexercised
PF 2421 Responsibility ; Fragmented citizen
Responsibility [images...]
PF 3553 — images ; Insufficient
PF 2575 — Individual avenues civic
PF 2560 — Individualistic welfare
PE 4225 — Insufficient awareness advertisers social
PD 1544 — Insufficient youth
PD 1544 *Responsibility ; Lack community*
PF 6382 Responsibility onto others ; Shifting
PD 1668 Responsibility ; Parochial family
PF 1889 Responsibility personal ; Limiting
PF 5587 Responsibility ; Quantitative understanding
PF 6536 *Responsibility ; Restricted personal*
PC 3915 Responsibility ; Sectoral fragmentation institutional
Responsibility [Unclaimed...]
PF 1781 — *Unclaimed local*
PD 1544 — *Unclear community*
PF 0963 — *Unconceived corporate*
PF 1959 — *Undefined programme*
PF 6570 — *Unshared mutual*
PF 6536 Responsible involvement community affairs ; Lack
PE 5795 Responsible ; Story that self sufficiency is socially
PF 9209 Responsive behaviour ; Fear emotionally
PC 0300 *Responsiveness ; Poorly developed social*
PE 4120 Responsiveness transnational corporations pharmaceutical needs developing countries ; Non
PA 5500 Rest
PF 9062 Rest ; Denial right leisure
PE 4793 Rest ; Denial working animals restorative nourishment
PE 6615 Restaurant practices ; Irresponsible
PE 1542 Restaurant waste ; Hotel
PE 6074 Restitution cultural property ; Non
PC 7859 Restitution monetary gold ; Non
PC 7859 Restitution property ; Non
PA 5838 Restlessness *Agitation*
PA 5490 Restlessness *Changeableness*
PA 6011 Restlessness *Discontentment*
PE 4074 Restlessness due torture
PA 6200 Restlessness *Impatience*
PE 4793 Restorative nourishment rest ; Denial working animals
PF 2093 Restraining community image
PA 7296 Restraint
PD 4488 Restraint ; Felonious
PF 0462 *Restraint ; Physical*
PE 3296 *Restraints legal land*
PE 0310 Restraints ; Trade restrictions due voluntary export
Restricted access
PF 1471 — developing country products world markets
PF 3550 — municipalities capital funds
PE 6788 — plant seeds
PF 8703 — policy formulation processes
PC 4844 — *sophisticated medical equipment*
PF 2378 Restricted availability capital
PE 5170 Restricted availability urban skills local communities
Restricted [banking...]
PE 0718 — *banking options*
PF 2822 — *beach use*
PF 3499 — *building sites*
PF 6451 *Restricted cash flow*
PF 3005 *Restricted community expansion*
Restricted [daytime...]
PF 6550 — *daytime involvement*
PF 1959 — *decision-making*
PF 1667 — delivery essential services developing country rural communities
PF 2875 — delivery essential services isolated communities
PD 1566 *Restricted extra-curricular activities*
Restricted [family...]
PC 0382 — *family menus*
PD 6571 — *farming alternatives*
PF 6451 — flow local economy
PE 4716 — franchise
PC 3222 — *franchise*
PF 4668 Restricted government ability undertake new initiatives
PF 1471 Restricted growth export markets developing countries
PF 6458 *Restricted higher education*
PF 3552 *Restricted job training*
Restricted [land...]
PF 6475 — *land choices*
PF 6528 — land use
PF 9062 — leisure time
PF 7142 — life-style
PF 3287 — local participation
PF 0581 Restricted market opportunities
PF 1352 *Restricted opportunities apprenticeship*
PF 6536 *Restricted personal responsibility*
Restricted [scope...]
PF 2423 — scope local employment
PE 5376 — sexual liberation AIDS
Restricted [sports...] cont'd
PF 2605 — *sports space*
PE 2425 — *store capital*
PF 2444 *Restricted tax base*
PE 1134 *Restricted union entry*
PF 6098 Restricted vocational images
PF 6537 Restricted vocational images
PC 0268 Restricting noise levels ; Lack legislation
PF 3082 Restriction access news distribution media
PD 1304 Restriction arms supply
PA 5497 Restriction *Difficulty*
Restriction [educational...]
PD 3122 — educational opportunities capitalist systems
PE 8143 — effectiveness ; Individualism
PA 7078 — *Environment*
PA 5869 — *Exclusion*
PF 3467 — *exports ; Government*
Restriction [free...]
PE 0051 — free market competition transnational corporations
PC 2162 — freedom expression
PC 2162 — freedom expression ; Dependence
PF 2823 — funding research social problems
PD 3305 Restriction indigenous populations reservations
PD 3080 Restriction information ; Conflict laws international
PD 0530 Restriction outer space benefits limited number developed countries
PE 7945 Restriction press coverage legal affairs
PF 3072 Restriction press coverage parliamentary affairs
PA 7296 Restriction *Restraint*
PD 6380 Restriction rights during war
PF 4741 Restriction technical basis competition
PE 4965 Restriction trade developing countries ; Environment policy
PC 0475 Restriction wild animal range size
Restrictions [access...]
PE 9440 — access developing countries commodity exchanges
PF 1319 — acquisition knowledge
PF 5248 — adaptations
PD 5584 — against small enterprise
PD 5830 — Agricultural trade
PE 4299 — AIDS infected persons ; Travel
PF 3538 — Avoidance patent
Restrictions [carcass...]
PE 8398 — carcass meat exports ; Inadequate hygiene
PE 3970 — collective bargaining workers organizations
PD 4728 — commercial transactions
PF 3070 — Counterproductive exchange control
PF 3070 — Currency
Restrictions [developing...]
PD 1873 — developing countries ; Domestic market
PF 3310 — *Dietary*
PF 3072 — direct news coverage parliamentary affairs
PE 0522 — Distortion international trade embargoes similar
PE 9027 — Distortion international trade quantitative
PD 2926 — distribution confidential government information
PE 0310 — due intergovernmental arrangements ; Trade
PE 0310 — due voluntary export restraints ; Trade
Restrictions [early...]
PC 2163 — *early apprenticeship*
PF 2798 — effective means transport
PF 8158 — efficacity aid ; Overhead
PE 7965 — eligibility government programmes
PC 3208 — emigration
PC 3208 — Emigration
PD 3529 — employment foreigners
PD 8731 — exchange ideas
PE 0895 — export activity due licensing arrangements
Restrictions [fishing...]
PF 2135 — *fishing ; Seasonal*
PD 3135 — foreign access capital bond markets
PF 3070 — Foreign exchange
PE 8747 — foreigners leaving country sojourn
PE 0897 — free movement commercial vehicles ; Unjustified
Restrictions freedom
PC 5075 — [Restrictions freedom]
PC 0185 — information
PE 8408 — movement countries
PC 0935 — movement countries
PF 3217 — thought
PD 5105 — worship
PE 4335 Restrictions home countries transnational banking activities developing countries
Restrictions [immigration...]
PC 0970 — immigration
PE 8403 — impeding sustainable development ; Patent
PF 1492 — imposed aid developing countries
PE 4905 — industrial economic development due environmental policies
Restrictions international
PC 0931 — freedom information
PD 0351 — freedom movement national advantage
PC 5842 — trade ; Unfair
Restrictions [labour...]
PD 4011 — labour mobility
PJ 5595 — legalizing medications ; Excessive
PF 4681 — liberty disabled
PF 2457 *Restrictions market access*
PF 9216 Restrictions multilateral development aid ; Policy cross conditionality
Restrictions [nationals...]
PE 8414 — nationals leaving their own country
PE 8817 — nationals returning their country
PF 3073 — news coverage legal affairs
PF 1776 *Restrictions ; Obsolete zoning*

Rhythms

Restrictions [participation...]
PF 5472 — participation developing countries international decision-making
PD 2948 — passage straits interoceanic canals
PE 2700 — private economic expansion
PD 8937 — property rights
PF 1575 — publicity
Restrictions [raw...]
PE 0669 — raw materials components utilization
PE 4912 — recognition nationality
PF 0725 — research
PD 6380 — rights martial law
PE 0897 — Road traffic
Restrictions [Shallow...]
PE 2305 — *Shallow river*
PD 3154 — sharing technical research
PE 1659 — socialist citizens working abroad
PE 5938 — Restrictions trade union rights public enterprises
PC 8452 — Restrictions ; Travel
Restrictions [Unfair...]
PC 4275 — Unfair trade
PE 7655 — Unreasonable licensing
PF 3945 — use personnel
PD 8056 — use water ; Abusive
Restrictive [access...]
PE 4704 — access loans
PF 5862 — Accusing channels dialogue being
PD 0343 — agreements product standards
PE 8443 — Restrictive building codes urban areas
PE 0161 — Restrictive business practice ; Direct foreign investment transnational enterprises
Restrictive business practices
PC 0073 — [Restrictive business practices]
PC 0073 — Distortion international trade
PE 0789 — Domestic bias regulation
PF 2782 — due inadequate regulation ; Ineffective monitoring
PE 2999 — Economic inefficiencies developing countries due
PF 1596 — Ineffective regulation
PE 8011 — legislation ; Extraterritorial application
PE 5926 — markets developed countries against exports developing countries
PE 0346 — relation patents trademarks
PE 0591 — service industries ; Inadequate regulation
PE 0225 — state enterprises ; Inadequate regulation
PE 1978 — technology transactions
PE 5915 — transnational corporations
PE 1799 — transnational enterprises ; Consequences
Restrictive [channels...]
PF 3037 — channels cultural interchange
PF 4674 — commercialization definitions
PF 2605 — *community size*
PE 9116 — conditions loans developing countries intergovernmental facilities
PE 8882 — controls movement labour ; Distortion international trade
PE 8525 — controls over foreign investment ; Distortion international trade
PE 0497 — customs valuation practices ; Distortion international trade
Restrictive [early...]
PE 7628 — early marriages
PF 3318 — effects external capital development
PF 3454 — effects traditional community decision-making
PF 3630 — *Restrictive farm practices*
PD 8831 — Restrictive health practices
PF 3361 — Restrictive influence religion masses
Restrictive [land...]
PF 3526 — *land ordinances*
PF 6528 — land-use vision
PD 8614 — legal practices
PF 8065 — *legal structures*
PD 9012 — legislation
PF 2491 — loan procedures developing countries
PF 3499 — *logging regulations*
Restrictive [market...]
PE 3196 — market divisions transnational corporations
PD 8831 — medical practices
PF 8749 — monetary policies
PF 8749 — monetary practices
PF 2810 — *Restrictive organizational participation*
PD 1415 — Restrictive pattern business activities small communities
PF 3129 — Restrictive patterns traditional life
Restrictive practices
PB 9136 — [Restrictive practices]
PE 7899 — beverages tobacco trade
PE 5910 — cargo airline services
PD 8614 — courts
PE 0342 — food live animals trade
PE 0141 — mineral fuels trade
Restrictive practices trade
PE 8880 — animal vegetable oils fats
PE 8600 — chemicals
PE 8351 — inedible crude non-fuel materials
PE 7958 — machinery transport equipment
PD 1797 — manufactured goods
Restrictive [practices...]
PE 9657 — practices transnational banks
PE 2396 — pricing policies transnational banks
PD 8027 — professional practices
PE 4330 — project loans transnational banks developing countries ; Domination
Restrictive [regulation...]
PF 8654 — regulation nuclear power
PF 2444 — *regulations training*
PD 8439 — religious practices

PF 7875 Restrictive scientific practices
PD 0312 Restrictive shipping practices
Restrictive social
PC 5537 — *groups*
PF 8282 — *policies*
PC 5537 — *practices*
PD 3436 Restrictive tax structure
PF 2721 Restrictive teaching methods
Restrictive trade
PC 0073 — practices
PC 0073 — practices ; Dependence
PF 6046 — union policies concerning employment
PD 8146 — union practices
PD 0881 Restrictive transport insurance practices
PD 8146 Restrictive union practices
PF 6528 Restrictive use available land
PD 8027 Restrictiveness ; Professional
PC 7459 Restructure economic relations developed countries ; Failure
PC 8694 Restructure economic relations North South ; Failure
PE 9055 Restructuring production industrialized countries ; Obstacles
PD 3947 Results amniocentesis ; Misuse
PF 1602 Results ; Falsification scientific test
PF 6467 Results formal schooling ; Inadequate
PD 7415 Results ; Misuse scientific medical test
PF 0704 Résumé ; Misleading
PF 6471 *Retail inventory ; Limited*
PF 1238 Retail prices ; Excessive
PF 6572 *Retailed merchandise ; Minimal*
PF 7799 *Retained placenta animals*
PF 1705 Retaining local tradition ; Individualistic
PF 9181 Retaliation
PD 5399 Retaliation against public servants
PF 3707 Retaliation authorities ; Fear
PF 9389 Retaliation ; Economic
PF 9078 Retaliation ; Fear
PF 9181 *Retaliation ; Fear military*
PF 9181 *Retaliation ; Military*
PD 0736 Retaliation ; Police
PF 9078 Retaliation ; Threat
PF 9181 *Retaliation ; Threat military*
PA 6606 Retaliation Vengeance
PE 4791 Retardation ; Borderline mental
PA 5497 Retardation *Difficulty*
PF 9664 Retardation economic structural adjustment
Retardation [Mental...]
PC 1587 — Mental
PE 4821 — Mild mental
PC 1587 — *Moderate mental*
Retardation [Profound...]
PE 6413 — Profound mental
PB 1044 — *Psychomotor*
PD 1307 — psychomotor development children
PA 7296 Retardation *Restraint*
PE 6314 Retardation ; Severe mental
PA 7371 Retardation *Unintelligence*
PA 7006 Retardation *Untimeliness*
PE 3992 Retarded adults ; Inadequate guardianship mentally
PD 0914 Retarded children ; Mentally
PF 6415 Retarded ejaculation
PF 2187 Retarded socialization
PE 0234 Retarding development transnational corporations
PD 2730 *Reticuloendotheliosis*
PD 3978 *Reticuloperitonitis ; Traumatic*
PE 5548 Reticuloses
PE 5548 Reticulum-cell sarcoma
PE 4584 Retina ; Detachment
PE 4584 Retina ; Diseases
PD 8786 *Retina ; Inflammation*
PD 8786 Retinitis
PD 5535 *Retinopathies animals ; Inherited*
PE 6069 Retirement age ; Discrimination against women
PB 0477 *Retirement ; Compulsory*
PD 4458 Retirement ; Denial right
PC 1966 Retirement ; Income maintenance after
PE 9071 Retirement people valuable skills ; Premature
PF 1137 *Retirement possibilities ; Debasing*
PF 1269 Retirement threat psychological well-being
PF 4375 Retraining ; Expensive
PF 8161 *Retraining incentives ; Lack*
PF 4013 Retraining programmes ; Inadequate industrial
PF 8085 *Retraining ; Unavailable skills*
PF 8562 Retribution
PA 7037 *Retribution∗complex*
PA 5583 Retribution *Punishment*
PA 6606 Retribution *Vengeance*
PE 4743 Retroactive laws punishments ; Denial right freedom
PD 0354 Retrograde amnesia
PA 6088 Retrogression *Impairment*
PA 6338 Retrogression *Regression*
PA 5699 Retrogression *Reversion*
PC 0935 *Return country residence ; Denial right*
PD 1291 Return developing country students studying foreign countries ; Non
PE 9811 Return investment developing countries ; Low
PE 3445 Return migration ; Early
PC 7859 Return property ; Non
PE 8817 Returning their country ; Restrictions nationals
PF 2457 *Returns farming ; Insufficient*
PF 2346 *Returns ; Insufficient rental*
PF 1275 Returns investment ; Overemphasis rapid
PF 8562 Revenge
PD 5168 Revenge ; Infectious
PA 6606 Revenge *Vengeance*
PD 2955 Revenue ; Dependence developing countries customs

PF 4197 Revenue ; Inefficient mobilization government
PF 2813 Revenue ; *Insufficient utility*
PF 2174 Revenue maximization ; Short term
PF 7432 Revenue mobilization expenditure allocation ; Imbalance
PD 1018 Revenue substance abuse ; Dependence government
PD 1018 Revenues exploitation environmentally inappropriate products ; Dependence government
PD 1018 Revenues exploitation non renewable resources ; Dependence government
PD 4962 Revenues ; Inequitable distribution
PF 2444 *Revenues ; Limited tax*
PD 4962 Revenues ; Maldistribution
PD 0644 *Reverence drugs medicine*
PC 4321 Reverence life ; Loss
PD 9202 Reversal development process developed countries
PF 4718 Reversal development progress
PF 1588 Reversal Earth's magnetic field
PF 1588 Reversal ; *Geomagnetic*
PF 1559 Reversals ; Expedient policy
PF 9539 Reverse discrimination
PD 1291 Reverse transfer technology
PA 5699 Reversion
Reversion [Regression...]
PA 6338 — *Regression*
PA 5619 — *Relapse*
PA 5699 — *Reversion*
PF 6826 Reversion sin
PF 2335 Review ; Bureaucratic resistance self
PD 5317 Review conviction higher tribunal ; Violation right
PC 1547 Review ; Inadequate structures political dialogue
PA 6191 Revilement *Disapproval*
PE 5546 Revisionism anti-marxist crimes
PF 4932 Revisionism ; Historical
PF 1338 Revivalism
PF 5375 Revivalist movements
PF 8964 Revoking licences
PE 5144 Revolt
PF 0278 Revolt against formalism
PF 0278 Revolt against ritualism
PC 2052 Revolt ; Student
PA 5901 Revolution
PD 3228 Revolution ; Bloodless
Revolution [communism...]
PC 3163 — communism
PF 3232 — Counter
PF 3235 — Cultural
PA 7250 Revolution *Disobedience*
PC 3233 Revolution ; Economic
PF 3146 Revolution ; Green
PC 3231 Revolution ; Ideological
PA 5563 Revolution *Lawlessness*
PD 3228 Revolution ; Non violent political
PF 3237 Revolution ; Political
PA 5901 Revolution *Revolution*
PF 7236 Revolution Separation
PC 3236 Revolution ; Social
PC 3234 Revolution ; Technological
PC 3229 Revolution ; Violent
PD 3230 Revolution ; Violent political
PF 1626 Revolutionaries ; Pseudo
PF 3232 Revolutionary activities ; Counter
PC 3163 Revolutionary communism
PF 7664 Revolutionary courts
PF 1015 Revolutionary re employment government security services ousted repressive regime ; Post
PA 5699 Revulsion
PA 5699 Revulsion *Reversion*
PA 5901 Revulsion *Revolution*
PF 9693 Reward biased research
PD 1958 Reward ; Inadequate access negotiation employment
PF 5660 Reward long range planning decision making ; Lack
PF 9693 Reward oriented research
PF 9786 Reward systems ; Counter productive personnel
PF 9786 Reward systems ; Inappropriate institutionalized
PD 4541 Rewarding public servants ; Unlawful
PF 6575 *Rewards ; Nonevident monetary*
PF 4932 Rewriting history
PE 5380 Rhabditic dermatitis
PF 9774 Rhabdomyolysis dogs ; Exertional
PE 9774 *Rhabdomyolysis ; Exertional*
PF 2152 Rhetoric ; Polarizing
PF 5311 Rhetoric ; Uncritical feel good
PF 3756 Rhetorical inflation meetings
PE 0873 Rheumatic diseases
PE 0920 Rheumatic fever
PE 5081 Rheumatic gout
PE 0920 Rheumatic heart disease
PE 0920 *Rheumatic pericarditis*
PE 0873 Rheumatism ; Nonarticular
PE 0502 Rheumatism ; Occupational
PE 5081 Rheumatoid arthritis
PE 9509 *Rheumatoid arthritis ; Canine*
PE 9854 Rhinitis
PE 9509 Rhinitis animals ; Allergic
PE 7553 *Rhinitis ; Atrophic*
PE 7553 *Rhinitis ; Necrotic*
PE 6010 Rhinoceros ; Endangered black
PE 5182 *Rhinopneumonitis ; Equine viral*
PD 2728 Rhinosporidiosis
PE 5524 *Rhinotracheitis ; Infectious bovine*
PE 2226 *Rhizobium plant diseases*
PE 1904 Rhythm international travel ; Desynchronization bodily
PE 8219 Rhythm ; Irregular heart
PE 1904 Rhythms ; Disruption biological
PF 0379 Rhythms ; Disruption biological

PF 0379 Rhythms nature ; Disconnection people
PF 0379 Rhythms ; Separation people biological
PE 7511 Rib ; Fracture
PE 7511 Ribs ; Fracture
PD 1285 Rice crops ; Low yield
PE 1218 Rice ; Long term shortage
PE 2221 Rice ; Pests diseases
PE 0696 Rice trade ; Instability
PD 5995 Richer nations ; Economic backlash
PC 8222 Richness
PD 0287 Rickets
PD 7424 Rickets animals
PE 2572 Rickettsiae
PE 5530 Rickettsial diseases
PE 5530 *Rickettsial pox*
PE 5530 Rickettsiosis
PE 5530 Rickettsiosis ; Tick borne
PE 5330 Rida
PD 5238 *Ridicule*
PA 6191 Ridicule *Disapproval*
PA 6822 Ridicule *Disrespect*
PA 6491 Ridicule *Vanity*
PJ 4783 Ridigity
PF 7428 Riff-raff
PE 7552 Rift valley fever
PD 5214 Rigging ; Ballot
PE 4301 Rigging ; Bid
PE 3754 Rigging sports contests
PD 5214 Rigging ; Vote
PF 6169 Right abortion ; Denial woman's
Right access
PF 6458 — *higher education ; Denial*
PF 9046 — historical libraries ; Denial
PE 1743 — interpreters judicial hearings ; Denial
PD 3335 — public service women ; Denial
PD 8056 — water ; Inequitable
PE 7345 Right accused segregation convicted criminals ; Denial
Right adequate
PC 3319 — food indigenous populations ; Denial
PE 8924 — health care disabled ; Denial
PD 5254 — housing ; Denial
PC 3320 — housing indigenous peoples ; Discrimination against indigenous populations housing Denial
PD 2028 — medical care ; Denial
PF 0344 — standard living ; Denial
PE 5484 — standard living indigenous peoples ; Denial
PC 0834 — welfare services ; Denial
Right [aortic...]
PE 5569 — *aortic arch animals ; Persistent*
PC 5280 — *appeal ; Abuse*
PD 5317 — appeal ; Denial
PF 9735 — artistic freedom ; Denial
PC 2383 — assembly ; Denial
PD 3224 — association ; Denial
PD 3224 — association ; Denial
PF 3021 — asylum ; Denial
PF 5121 — attention, care protection humankind ; Denial animals
PE 5608 Right be elected government positions ; Violation
PE 7336 Right be informed criminal charges ; Denial
Right benefits
PE 5211 — invalids ; Denial
PF 6077 — science ; Denial
PE 4531 — survivors ; Denial
PE 2700 Right business growth ; Denial
Right [change...]
PE 1736 — change nationality ; Denial
PE 6397 — change religion ; Denial
PD 1915 — choice marriage ; Denial
PE 3802 — choose moral religious education ; Denial
PE 3970 — collective bargaining ; Denial
PC 2162 — communicate ; Denial
PD 7609 — complaint ; Denial
PE 6270 — conditions life liberty proper their species ; Denial animals
PD 6612 — confidentiality ; Denial
PE 7217 — confront accusers ; Denial
PD 1800 — conscientious objection military service ; Denial
PE 7349 — correct misinformation ; Denial
Right cultural
PE 3450 — asylum ; Denial
PF 9005 — identity ; Denial
PE 6561 — life ; Denial
Right [defence...]
PE 4628 — defence against criminal charges ; Violation
PF 4813 — Denial
PF 2364 — develop human beings ; Denial
PE 8911 — dignified birth ; Denial
PE 6623 — dignity ; Denial
PE 9573 — dignity ; Denial animals
PD 6927 — due process law ; Denial
Right economic
PE 6212 — asylum ; Denial
PD 0808 — security ; Denial
PE 5406 — security during periods unemployment ; Denial
Right education
PD 8102 — Denial
PC 3459 — minorities ; Violation
PD 0190 — women ; Denial
Right [educational...]
PE 4700 — educational choice ; Denial
PD 1092 — employment indigenous populations ; Denial
PD 0086 — employment women ; Denial
PE 4507 — enjoyment arts ; Denial
PC 8943 — environmental quality ; Denial

Right equal
PD 1977 — pay equal work ; Denial
PD 0309 — pay women ; Denial
PD 9628 — *promotion opportunities ; Denial*
PC 1268 — protection law ; Denial
PE 1625 — work benefits aged ; Violation
Right equality
PC 0006 — because race ; Denial
PA 0833 — Denial
PE 4712 — states ; Denial
PC 0308 — women ; Denial
Right [euthanasia...]
PF 2643 — euthanasia ; Denial
PE 7489 — examine witnesses ; Denial
PE 5241 — extended family ; Denial
Right [fair...]
PD 4827 — fair public trial ; Denial
PE 3964 — fair public trial ; Denial
PE 7267 — family ; Denial
PE 5226 — family planning ; Denial
PD 9170 — favourable work conditions ; Denial
PC 6870 — found family ; Denial
PE 3963 — free choice work ; Denial
PC 6381 — free primary education ; Denial
PD 3055 — freed arbitrary deprivation property ; Denial
PE 5952 — freedom advocacy discrimination ; Denial
Right freedom arbitrary
PD 1576 — detention ; Denial
PC 2507 — *exile ; Denial*
PC 5313 — expulsion ; Denial
Right freedom [attacks...]
PD 2569 — attacks personal honour reputation ; Denial
PD 8445 — belief ; Denial
PE 6342 — compulsory labour ; Denial
PC 3768 — cruel, inhumane degrading punishment ; Denial
PA 0833 — discrimination ; Denial
PF 6158 — double jeopardy ; Denial
PD 0635 — exploitation children ; Denial
PC 2162 — expression ; Denial
PE 0324 — hunger ; Denial
PE 7235 — imprisonment failure fulfil contractual agreement ; Denial
PC 0185 — information ; Denial
PC 0935 — international movement ; Denial
PE 9650 — mass killing ; Denial animals
PC 8452 — movement persons ; Denial
PE 8408 — movement state ; Denial
PE 8951 — press ; Denial
Right freedom religion
PD 8445 — Denial
PE 4332 — indigenous peoples ; Denial
PD 0127 — women ; Denial
Right freedom [retroactive...]
PE 4743 — retroactive laws punishments ; Denial
PE 4964 — servitude ; Denial
PC 0146 — slavery ; Denial
PE 8024 — suffering ; Denial experimental animals
PE 3899 — suffering ; Denial food animals
PE 6633 — testifying against oneself ; Denial
PF 3217 — thought ; Denial
PC 3429 — torture ; Denial
PD 3092 — war propaganda ; Denial
PE 2832 Right grievance procedures ; Denial
Right [habeas...]
PD 6927 — *habeas corpus ; Denial*
PC 0663 — *health ; Denial*
PE 4459 — health indigenous populations ; Denial
PC 0663 — *highest obtainable physical mental health ; Denial*
PE 5608 — hold public office ; Denial
PD 0520 — humane imprisonment ; Denial
PE 7665 Right independent judges ; Denial
Right indigenous peoples
PC 3322 — education ; Denial
PE 7312 — participate political processes ; Denial
PE 2142 — use their own language ; Denial
Right [inform...]
PE 7337 — inform ; Denial
PF 0886 — inherit property ; Denial
PE 6755 — innocent passage territorial waters ; Denial
PF 8854 — intellectual property ; Denial
PE 8899 — international freedom movement shipping ; Violation
PE 8899 — international freedom navigation ; Violation
PD 4282 — investigate ; Denial
PE 7119 Right job protection during maternity leave ; Denial
PD 8568 Right just income ; Denial
Right justice
PA 6486 — Denial
PC 6162 — Denial
PD 0162 — women ; Denial
PE 7209 Right juvenile criminals segregation adult criminals ; Denial
PE 3463 Right leave any country ; Denial
Right legal
PF 8869 — aid ; Denial
PF 3517 — counsel ; Denial
PE 4628 — defence ; Denial
PE 2317 — services indigenous populations ; Denial
Right [leisure...]
PF 9062 — leisure rest ; Denial
PF 0705 — liberty ; Denial
PD 4234 — life ; Denial
PF 8243 — life ; Denial animals
Right [manifest...]
PF 2850 — manifest religion ; Denial
PF 3343 — marriage ; Denial
PC 2157 — material well being because discrimination ; Denial

Right [maternity...] cont'd
PE 3951 — maternity leave ; Denial
PE 3828 — medical consent ; Denial
PE 6726 — minimum wage ; Denial
PE 1693 — minimum work age ; Denial
PE 4624 Right name ; Denial
Right national
PC 3177 — self determination communist systems ; Denial
PC 1450 — self determination ; Denial
PE 7906 — sovereignty ; Denial
Right nationality
PE 4912 — Denial
PE 4016 — Discrimination against women
PE 4016 — women ; Denial
PE 0241 Right nations select their own leaders ; Denial
PE 8339 Right natural death ; Denial animals
PD 0162 *Right old age benefits ; Denial*
Right organize
PE 9110 — political parties ; Denial
PE 5398 — trade unions ; Denial
PE 5938 — trade unions government services ; Violation
PE 8411 Right ownership ; Denial
PE 8411 Right ownership ; Denial
Right participate
PE 6086 — conduct public affairs ; Violation
PE 6086 — government ; Denial
PC 1001 — government women ; Denial
Right people
PC 6727 — be self determining ; Denial
PF 0297 — determine their own political status ; Denial
PE 6955 — freely dispose natural wealth ; Denial
PD 5253 — live peace ; Denial
PE 1536 — pursue development ; Denial
PE 4399 — their own means subsistence ; Denial
Right [peoples...]
PE 2142 — peoples use their own language ; Denial
PE 3044 — periods nurse infants during working hours ; Denial
PD 7609 — petition authority ; Denial
PE 8712 — picket ; Denial
PF 1075 — political asylum ; Denial
PE 7393 — presumed innocent until proven guilty ; Denial
PB 0284 — privacy ; Denial
PD 7608 — private correspondence ; Denial
PE 6168 — private home life ; Denial
PC 6870 — procreate ; Denial
PE 4544 — procreate severely mentally handicapped ; Denial
PE 4018 — property women ; Denial
PD 1846 — protection union representatives ; Denial
PE 7310 — pursue spiritual well being because discrimination ; Denial
Right [reasonable...]
PD 0140 — reasonable work hours ; Denial
PE 4214 — receive information ; Denial
PE 4716 — recognition person law ; Denial
PE 7036 — recruit union members ; Denial
PE 4173 — redress rights violations ; Denial
PF 1916 — refuse vaccination ; Denial
PE 6949 — resist oppression ; Denial
PD 4458 — retirement ; Denial
PC 0935 — *return country residence ; Denial*
PD 5317 — review conviction higher tribunal ; Violation
PF 6140 — roam countryside ; Denial
Right [security...]
PD 7212 — security ; Denial
PD 2592 — self determination ; Use mercenaries means impeding
PE 9364 — *side abomasal displacement*
PE 4319 — silence ; Denial defendant's
Right social security
PD 3120 — capitalist systems ; Denial
PD 7251 — Denial
PE 1506 — indigenous peoples ; Denial
Right [social...]
PE 1506 — social welfare services indigenous peoples ; Denial
PE 4887 — speedy trial ; Denial
PE 0241 — state succession ; Denial
PE 5070 — strike ; Denial
PE 5070 — strike ; Violation
Right sufficient
PE 7616 — clothing ; Denial
PE 0324 — food ; Denial
PD 5254 — shelter ; Denial
Right [time...]
PE 7624 — time prepare trial defence ; Denial
PD 4695 — trade union activity ; Denial
PD 0683 — trade union association ; Denial
Right trade unions
PE 4745 — engage political action ; Violation
PE 1758 — freely elect their representatives ; Violation
PE 1758 — function freely ; Violation
PE 3970 — negotiate freely employers ; Violation
PE 5793 — protection against dissolution ; Violation
PE 5312 — publish ; Violation
PD 4252 Right traditional economies indigenous populations ; Denial
PE 4737 Right trial court ; Denial
Right [unborn...]
PF 6616 — unborn children ; Denial
PE 3910 — union activity rural workers ; Denial
PE 1355 — union activity special groups ; Denial
PF 2904 Right vote ; Denial
PF 2904 Right vote ; Limitations
PF 6140 Right walk country ; Denial
Right welfare services
PD 0512 — aged ; Denial
PD 0542 — blind ; Denial

Right welfare services cont'd
PD 0601 — deaf ; Denial
PD 1593 *Right whales ; Endangered species*
Right work
PC 3119 — capitalist systems ; Denial
PC 5281 — Denial
PE 3751 — refugees ; Denial
PD 2872 — women socialist countries ; Denial
PD 4695 Right workers freedom association ; Violation
PE 5192 Right workers join trade unions : Violation
Right workers organizations
PE 4071 — affiliate internationally ; Violation
PE 3970 — bargain collectively ; Violation
PE 4071 — establish confederations ; Violation
PE 5793 — protection against suspension ; Violation
PA 6058 Righteousness ; Self
PC 3406 *Righteousness ; Self*
PD 4695 Rights ; Abuse trade union
Rights abuses
PF 9288 — Connivance authorities human
PF 9288 — Connivance religious leaders human
PF 9288 — Government connivance human
PF 9288 — Official connivance human
Rights activists
PE 6934 — Denial rights human
PF 3819 — Government complicity killing human
PE 6934 — Government harassment human
PE 6934 — Vulnerability human
Rights [administration...]
PD 6927 — administration justice ; Denial human
PE 7300 — agreements ; Failure governments implement provisions ratified human
PC 5456 — animals ; Denial
PC 1454 — armed conflicts ; Denial human
PD 4680 — Arrogation
Rights [because...]
PA 0833 — because discrimination ; Denial equal property
PF 7925 — Bias United Nations response human
PD 4728 — businesses ; Denial
Rights [campaigners...]
PE 4438 — campaigners ; Intrusive animal
PC 3124 — capitalist systems ; Denial human
PD 0513 — children youth ; Denial
PC 3178 — communist systems ; Denial human
PF 2365 — *Conflicting land*
PD 5361 — Conflicts over fishing
PF 2583 — *Consumer ignorance*
PB 5405 — Contravention
PE 4183 — criminal suspects ; Abuse
PD 8709 — Criminal violation civil
Rights Denial
PB 5405 — [Rights ; Denial]
PE 8643 — animals legal protection their
PD 5907 — cultural
PD 4150 — economic
PF 6616 — foetal
PB 3121 — human
PD 8276 — political
PC 0632 — political civil
PC 0663 — social
PD 4814 — state's
Rights [Deprivation...]
PB 5405 — Deprivation
PC 7379 — Deprivation human
PD 0520 — detainees ; Denial
PE 4972 — development indigenous peoples ; Denial
PC 0006 — different racial groups ; Unequal
PC 3461 — disabled ; Denial
Rights Discrimination against
PE 4010 — men parental
PE 8879 — women divorce
PE 4019 — women parental
Rights [Disregard...]
PE 8411 — Disregard property
PF 0922 — due excessive court costs ; Ineffective protection individual
PD 6380 — during states emergency ; Suspension
PD 6380 — during war ; Restriction
Rights [elderly...]
PC 2541 — elderly ; Denial
PC 2541 — elders ; Denial
PD 0164 — employed children ; Denial
PD 4916 — equal opportunities employment elderly ; Denial
PC 8999 — ethnic minorities ; Denial
PE 6929 — ex convicts ; Denial
PF 2418 — Execution orders violation human
PE 1597 — exploit sea bed marine resources ; Inequitable allocation
Rights [female...]
PE 5741 — female homosexuals ; Denial
PE 5741 — female homosexuals ; Violation
PC 6870 — Fertility
PE 6369 — foetuses ; Violation
PF 4821 Rights ; Government aid countries violating human
PE 4860 Rights governments ; Denial trade union
Rights [homosexuals...]
PE 1903 — homosexuals ; Denial
PE 1903 — homosexuals ; Violation
PE 6934 — human rights activists ; Denial
Rights [Ignorance...]
PF 2583 — *Ignorance consumer*
PD 9162 — Ill defined property
PD 0943 — illegitimate children ; Denial
PF 3474 — Impunity violators human
PC 4608 — Inadequate enforcement civil
PC 4608 — Inadequate enforcement human

Rights [inanimate...] cont'd
PF 3710 — inanimate objects ; Denial
Rights indigenous
PE 1024 — people be self governing ; Denial
PC 0352 — people ; Denial
PE 7765 — peoples ; Denial social
Rights [individual...]
PE 9759 — individual parliamentarians ; Violation human
PF 9794 — infractions ; Politically motivated exaggeration human
PB 5405 — Infringement
PC 6003 — Infringement human
PE 6755 — innocent passage territorial waters ; Unlawful interference
PF 6365 — instruments ; Inadequacy international human
PF 7925 — Inter Governmental double standard human
PF 7925 — issues international community ; Avoidance human
PF 7925 — issues United Nations ; Politicization human
PD 3035 Rights journalists ; Denial
PC 3304 Rights lands resources ; Denial indigenous
Rights [machine...]
PF 4334 — machine intelligences ; Denial
PE 3882 — male homosexuals ; Denial
PE 3882 — male homosexuals ; Violation
PD 6380 — martial law ; Restrictions
PD 1662 — medical patients ; Denial
PD 1148 — mental patients ; Denial
PD 1148 — mentally ill persons ; Violation
PD 0973 — migrant workers ; Denial
PC 8999 — minorities ; Denial
PF 9794 — Misinformation concerning infringement human
PF 1585 — monitors ; Harassment human
PF 3710 Rights natural objects ; Inadequate legal
Rights [Obscured...]
PF 2365 — *Obscured land*
PE 9545 — offenders ; Government pardoning convicted human
PC 5075 — *Overcomplicated implementation citizen*
PE 2475 — Overemphasis individual
Rights [parent...]
PE 4010 — parent men ; Denial
PE 4019 — parent women ; Denial
PF 5292 — Payments after expiration industrial property
PD 5218 — people ; Violation land
PF 1075 — political asylum ; Lack individual
PF 9455 — posterity ; Denial
PD 0520 — prisoners ; Denial
PF 4176 — privacy use computer databases ; Erosion human
PE 1479 — Psychological barriers judicial protection individual
PE 6888 — public employees ; Denial trade union
PE 5938 — public enterprises ; Restrictions trade union
Rights [refugees...]
PE 6375 — refugees ; Denial economic social
PC 2129 — religious minorities ; Denial
PD 8937 — Restrictions property
PF 4334 — robots ; Denial
Rights [sexual...]
PD 1914 — sexual minorities ; Denial
PD 1914 — sexual minorities ; Violation
PD 4089 — soldiers ; Denial
PD 4814 — sovereign nations ; Denial
PF 6365 — standards ; Inadequacy existing human
PD 6380 — state emergency ; Violation
PD 6346 — students ; Denial
Rights [territorial...]
PC 2945 — territorial integrity ; Denial
PD 6620 — territories ; Denial
PF 0399 — those punished death penalty ; Denial
PD 0610 — trade unions organizers ; Denial
PD 9162 — Traditional property
PF 1844 — transient populations ; Denial
PE 8548 — transsexuals ; Denial
PE 8548 — transsexuals ; Violation
Rights treaties
PE 7300 — Delays ratification human
PE 7300 — Limited acceptance human
PE 7300 — Non ratification human
PF 3710 Rights trees ; Inadequate legal
Rights [Unclear...]
PD 9162 — Unclear property
PF 3464 — Underutilization legal
PC 1917 — Underutilization political
PF 4062 — Undocumented violations human
PD 0162 — *Unequal pension*
PA 0833 — *Unequal property*
PF 3464 — Unfamiliar legal
PE 5152 — Unjust customary
Rights Violation
PB 5405 — [Rights ; Violation]
PC 5285 — civil
PB 3860 — human
PD 8937 — property
PD 6346 — student's
PD 4695 — trade union
Rights violations
PE 4173 — Denial right redress
PF 9794 — False allegations human
PE 1407 — Government inaction alleged human
PF 9474 — Governments unwilling cooperate United Nations eliminating human
PD 5122 — Inadequate assistance victims human
Rights [Vulnerability...]
PD 0188 — Vulnerability performers'
PC 4405 — vulnerable groups ; Denial
PD 3785 — vulnerable groups during states emergency ; Violation
PE 4018 Rights women ; Unequal property

PE 4758 Rights wounded military personnel ; Denial
PF 8598 Rigid cultural patterns
PF 2005 Rigid doctrine ; Ossification faith
PF 4984 *Rigid educational forms*
PF 1509 *Rigid family obligations*
PF 6545 *Rigid local groupings*
PF 6536 *Rigid personal viewpoint*
PF 3168 Rigid personnel hiring policies
PJ 5365 *Rigid required curriculum*
PF 4984 *Rigid school programs*
PD 4011 Rigidities labour markets ; Structural
PD 2970 Rigidities production structures developing countries
PA 5740 Rigidity *Compulsion*
PD 2970 Rigidity developing country economies ; Structural
PD 3515 Rigidity inadaptability aged
PA 7325 Rigidity *Irresolution*
PF 4028 Rigidity labour markets ; Wage
PE 7808 Rigidity ; *Myopathy associated congenital articular*
PA 6802 Rigidity *Permanence*
PF 4785 Rigidity rechannelling reduced expenditure defence
PF 1765 Rigidly entrenched social traditions rural areas
PE 2786 Rinderpest
PE 7826 Ringbone
PE 1323 Ringlet ; Devil's
PC 2343 Rings ; International crime
PD 1762 *Ringtail gliders ; Endangered species*
PD 2545 Ringworm
PD 9667 *Ringworm animals*
PD 2730 *Rio Bravo fever*
PD 2207 Riot control ; Inadequate
PD 1156 Riot control ; Inhumane methods
PD 4091 Riot ; Engaging
PD 6392 Riot ; Inciting
PE 5327 Rioters ; Arming
PC 2551 Riots
PE 1675 Riots ; Prison
PF 1417 Riots ; Religious
PC 2052 Riots ; Student
PD 0516 Riparian states ; Lack cooperation
PF 7911 Rising cost unemployment benefits
PF 2385 Rising food prices rural communities
Rising [land...]
PE 4162 — land price
PD 8888 — level underground water
PD 2898 — living standards ; Inadequate demand primary commodities because
PF 2385 — local expenses
PF 7435 Rising sea level
PD 8888 Rising water level
PF 7580 Risk
Risk [accident...]
PB 0731 — accident
PE 3153 — accidental explosion
PD 0771 — atomic accidents
PF 4612 — aversion strategy
Risk [bad...]
PC 0293 — bad weather
PD 8744 — being kidnapped
PF 7580 — *being ransomed*
PE 5762 — breach contract
Risk capital
PF 3665 — *Insufficient*
PF 6572 — investment
PF 3665 — *Scarcity*
PF 3665 — *Unavailable*
Risk [car...]
PE 4826 — car theft
PD 6033 — Coitus cancer
PF 0905 — *commercial farming ; Poor*
PE 4428 — computer error
PE 4362 — computer fraud
PF 1379 — contracting AIDS blood transfusions
PD 5111 — contracting AIDS kissing
PF 2166 — Credit
Risk [damage...]
PD 5859 — damage premises
PE 4447 — damage stock
PA 6971 — *Danger*
PJ 0451 — development initiatives ; Reluctance government invest high
PD 9108 — during loading unloading
Risk [ecoaccidents...]
PD 2718 — ecoaccidents
PD 2718 — ecocatastrophe
PF 2334 — Economic philosophy controlled
PF 7836 — elimination industrial pollution ; Employment
PD 9397 — employee infidelity
PD 5349 — Endangerment awareness high
PC 1904 — Engineering failure
PF 5482 — evaluation ; Inconsistent
PF 9424 — Excessive desire
PF 6389 — Excessive fear
PD 8054 — Fire
PD 7981 — *Risk forgery shares, stocks, bonds*
PC 0865 — Risk health ; Occupational
PF 6572 — Risk ; High business
Risk [incurring...]
PF 0995 — incurring legal expenses
PF 6572 — Inhibition investment
PD 2591 — insolvency
PF 4435 — intentional nuclear war
PF 9106 — Interruption
PD 3022 Risk libelling
PD 1292 Risk lightning striking
Risk [maritime...]
PD 8982 — maritime accidents

Risk missing

Risk [missing...] cont'd
PF 4690 — missing documents
PE 4107 — multiple birth
PC 2172 Risk nationalization overseas investments
Risk [people...]
PD 6818 — people ; Security
PF 4861 — persons ; Non surveillance medical high
PE 1815 — policy smaller banks ; High
PF 5932 — Political
PC 7660 — political upheaval
PD 1949 Risk radiation accidents
Risk [Shipping...]
PE 0430 — Shipping
PD 3023 — slandering
PE 2435 — sonic bangs
Risk [taking...]
PF 6389 — taking ; Deprivation
PF 3514 — technologies ; High
PD 5552 — theft
PD 0079 — traffic accidents
PD 0971 — transporting dangerous goods
Risk [Uncertainty...]
PA 7309 — *Uncertainty*
PF 6572 — Unfavourable capital
PF 4352 — unintentional global nuclear war due nuclear proliferation
Risk unintentional nuclear
PF 0466 — war
PF 4346 — war due accidents
PF 4302 — war due international crises
PF 4156 — war generated developments strategic doctrine
PF 4162 — war generated strategy deterrence
PF 4152 Risk unintentional war generated arms race
PE 7750 Risk violence women
PF 4215 Risk war
PA 6822 Riskiness *Disrespect*
PA 7309 Riskiness *Uncertainty*
PF 7722 Riskless responses public injustices
PF 7580 Risks
PC 4906 Risks ; Agricultural
PD 8328 Risks ; Aviation
PF 7106 Risks ; Building erection
PE 8542 Risks computers computer records
PF 1824 Risks disease ; Enhanced
PE 3928 Risks due delays delivery goods services
PE 1412 Risks electricity ; Health
PF 3065 Risks ; Export credit
PD 0414 Risks fuel exhaust ; Health
Risks [hazards...]
PE 5355 — hazards medical profession ; Occupational
PF 2166 — High credit
PE 8966 — Holiday
Risks [Impoldering...]
PE 6347 — Impoldering
PD 4519 — Inappropriate legal definition health
PD 2784 — *international transfer corpses* ; *Health*
PF 6572 — Investment blocked high
PF 4841 — Risks medical self experimentation ; Errors
PC 0865 Risks ; Occupational health
PE 4835 Risks power production
PF 6572 Risks ; Speculation
PF 1762 Risks subcontractors ; Economic
Risks [teenage...]
PE 6969 — teenage sex ; Health
PE 8580 — transfer ownership
PD 7716 — Travel
Risks [Underreporting...]
PD 0708 — Underreporting earthquake
PD 8290 — Underreporting radiation
PF 5482 — Unequal evaluation
Risks workers
PE 0524 — agricultural livestock production ; Health
PE 0688 — commerce ; Health
PE 0526 — construction industry ; Health
PE 1159 — electricity, gas, water sanitary services ; Health
PE 1605 — manufacturing industries ; Health
PE 0875 — service industries ; Health
PE 1581 — transport, storage communication industries ; Health
PF 2346 *Risky rental agreements*
PF 7580 Risky situations
PF 1309 Rites customs ; Underutilization historical
PF 9190 Rites dead ; Denial
PF 2099 *Rites ; Occult*
PF 1674 Rites passage ; Absence
PF 8944 Ritual cannibalism
PF 2641 Ritual dismemberment
PF 5392 Ritual hazings
PF 3960 Ritual impurity
PF 2641 Ritual killings
PF 2641 Ritual murder
PF 2660 Ritual nudity
PF 3960 Ritual pollution
Ritual [sexual...]
PD 3267 — sexual abuse children
PF 0319 — slaughter animals
PC 0417 — *suicide*
PF 0278 *Ritualism*
PF 0278 Ritualism ; Anti
PF 0278 Ritualism ; Revolt against
PF 7887 Ritualistic sexual abuse
PF 1309 Rituals customs ; Obsolescence
PF 0278 Rituals ; Rejection
PF 7887 Rituals ; Satanic
PB 0848 Rivalries ; Blocked action due
PF 7979 Rivalries ; Bureaucratic
PB 0848 Rivalry

PD 9252 Rivalry armed forces
Rivalry [Commercial...]
PD 8897 — Commercial
PD 8897 — Corporate
PD 5076 — countries ; Nuclear
PA 5532 Rivalry *Disaccord*
PD 0110 Rivalry disunity developing regions
PD 8897 Rivalry ; Economic
PA 7253 Rivalry *Envy*
Rivalry [Ideological...]
PF 3388 — Ideological
PD 8897 — Industrial
PE 0063 — intergovernmental organizations
PC 0114 — International
Rivalry international
PF 2216 — groups
PD 8897 — markets
PE 0259 — nongovernmental organizations
PC 1463 — organizations
PD 7617 Rivalry ; Interpersonal
PD 9252 Rivalry ; Military
PC 0114 Rivalry ; National
PC 0114 Rivalry nations
PA 6979 Rivalry *Opposition*
PD 8992 Rivalry ; Political
PC 3355 Rivalry ; Religious
Rivalry [Scientific...]
PD 1709 — *Scientific*
PE 3538 — secretariats intergovernmental organizations ; Inter state
PD 7617 — *Sibling*
PD 0014 — Strategic arms
PD 9655 — Superpower
PF 1262 Rivalry ; Trade union
River [banks...]
PD 2290 — banks ; Unlandscaped
PD 0516 — basin development ; Uncoordinated international
PE 2388 — blindness
PD 2154 River courses ; Changing
PD 3673 River dolphins ; Endangered species
PD 2290 River erosion ; Uncontrolled
PD 6571 River experts ; Unresponsive
PD 3142 River ice
PD 0516 River plan ; Uncomprehensive
PD 7636 River pollution
PE 2305 *River restrictions ; Shallow*
PF 3448 *River transport ; Infrequent*
PF 3448 *River transport ; Undependable*
PD 0516 River water disputes
PD 4976 Riverine floods
PD 1990 Rivers ; Desiccation
PD 7636 Rivers ; Discharge dangerous substances
PD 2257 Rivers ; Eutrophication lakes
PD 0767 Rivers ; Impounded
PE 5813 Rivers land locked countries ; Vulnerability lakes
PD 3654 Rivers ; Siltation
PE 2388 Rivers streams ; Fly infested
PD 0516 Rivers ; Upstream diversion international
PE 7808 *Roachback*
PE 1055 Road access village ; Inadequate
Road [conditions...]
PE 6526 — *conditions* ; *Inadequate*
PE 6526 — *conditions* ; *Poor*
PE 2605 — *construction* ; *Prohibitive cost*
PE 1690 Road deaths ; Animal
PD 0791 Road hazards
Road highway
PD 2106 — traffic congestion
PD 0543 — transport facilities developing countries ; Inadequate
PD 0490 — transport facilities ; Inadequate
PF 2605 Road improvements ; Costly
PE 6526 *Road improvements ; Inaccessible*
Road [machinery...]
PF 2410 — *machinery* ; *Unavailable*
PF 2410 — *machines* ; *Unavailability*
PD 8557 — maintenance ; Inadequate
PE 8336 — motor vehicles ; Environmental hazards
PE 1055 Road network unconnected major roads
PE 1055 Road network ; Underdeveloped
Road [passages...]
PD 0791 — passages ; Hazardous
PF 2346 — patterns ; Divisive
PD 0549 — play ; Unsupervised
Road [surface...]
PE 6526 — *surface ; Slippery*
PE 6526 — *surfacing ; Inadequate*
PE 6526 — *surfacing ; Non priority*
PE 1720 Road terrain vehicles ; Environmental degradation off
Road traffic
PD 0079 — accidents
PD 0426 — congestion ; Urban
PE 0930 — offences
PE 0897 — restrictions
PE 0930 — violations
PF 1267 *Road transport fuel ; Expensive*
PD 2072 Road vehicles ; Increase
PE 6526 *Road ways ; Narrow*
PE 9605 Roadblocks ; Illegal
PD 8557 Roads ; Broken surfaces streets
PD 6124 Roads ; Environmental degradation high speed
PE 6704 Roads ; Functional obsolescence
Roads [Impassable...]
PE 6526 — *Impassable*
PE 1055 — Inadequate access
PE 1055 — Insufficient good
PE 1055 — Poor access

PE 1055 ; Road network unconnected major
PF 1249 Roads ; Snowmobiles
PE 7403 Roads tracks ; Environmental degradation recreational use unsurfaced country
PE 5824 Roads transport land locked developing countries ; Inadequate
PE 6526 *Roads ; Unpaved, ungraded*
PD 8557 Roads ; Unrepaired
PE 3547 Roadway adequacy ; Seasonal
PF 2605 *Roadway improvements ; Unfunded*
PD 8557 Roadway pot-holes
PF 1495 *Roadways ; Inadequate commercial*
PF 6473 *Roadways ; Poorly lighted*
PF 6140 Roam countryside ; Denial right
PE 5182 Roaring
PD 5575 Robberies ; Train armoured car
PD 5575 Robbery
PE 4739 Robbery ; Armed
PF 0491 Robbing ; Grave
PF 4334 Robots ; Denial rights
PD 1146 Rock avalanches
PD 0077 Rock infested soils
PG 3896 Rock Mountain spotted fever
PE 9120 Rock rat ; Endangered species bush rat
PD 1146 Rock slides
PG 3896 Rocky Mountain spotted fever
PE 2254 Rodent-or-rabbit bot fly
PE 3573 Rodent resistance rodenticides
PE 9702 *Rodent ulcer*
PE 3629 Rodent vectors disease
PD 3481 Rodentia
PD 5228 Rodenticide poisoning animals
PD 4629 *Rodenticides ; Misuse*
PE 3677 Rodenticides pollutants
PE 3573 Rodenticides ; Rodent resistance
PE 2537 Rodents
PD 3481 Rodents ; Endangered species
PE 2537 Rodents pests
PE 4810 Rodeo animals ; Maltreatment
PA 7363 Roguery *Improbity*
PA 6498 Roguery *Misbehaviour*
PD 1308 Role conditioning ; Sex
PF 3973 Role developing future ; Inability elderly play
PF 8905 Role ; Diminished church
PF 1664 Role elders ; Unrecognized
PF 7456 Role expectations ; Diffusion family
PF 3498 Role learning ; Inappropriate
PD 3555 Role man ; Family dependence patriarchal
Role models
PF 8451 — Inappropriate
PF 8451 — Insufficient
PF 6479 — Lack local leadership
PE 0742 — Substance abuse
PF 1651 *Role ; Negated paramedic*
Role [older...]
PF 1825 — older people due overemphasis economic productivity ; Diminishing
PF 1594 — opposition ; Frustration
PD 5774 — Overemphasis consumer
PD 3338 *Role playing ; Social*
PF 0897 Role professional ; Collapse public servant
Role [society...]
PF 7456 — society ; Lost family
PF 1078 — Subsistence limits
PF 6461 — *Surrendered parental*
PF 1651 Role tasks ; Compartmental
PE 7144 Role United Nations ; Intellectual confusion concerning
PE 7919 Role urban services ; Undefined government
Role [western...]
PE 1046 — western women ; Elimination socio cultural
PF 4959 — women countryside ; Limited
PF 4959 — women rural development ; Neglect
PD 0579 — world affairs ; Government loss leadership
PF 1825 Roles aged ; Unmeaningful social
PF 3115 Roles commodities capitalism ; Conflicting
PF 1651 Roles ; Disrelated images social
PF 6098 Roles ; Disrelated images vocational
PF 1825 Roles elderly ; Inadequate
PD 1544 Roles ; Inadequate youth
PD 1308 Roles ; Inflexible sexual
PE 8806 Roles ; Lack diversity social
PD 1308 Roles marriages ; Inadequate image
PF 3114 Roles money capitalist systems ; Conflicting
PF 1651 Roles ; Static unrelated social
PD 1544 Roles ; Unclarity youth
Roles Unclear
PF 2595 — *educational*
PF 1651 — polity
PF 2595 — *school*
PF 1651 — societal
PD 1544 — youth
Roles [Undefined...]
PF 2595 — *Undefined educational*
PF 1651 — Unexamined community
PF 7456 — Unplanned family
PD 6273 Roles women ; Conflicting
PD 6273 Roles women ; Dual
PD 4233 Romance ; Cessation
PE 1281 Romanies
PF 8071 Romanism
PF 7418 Romantic love ; Dependence
PD 4233 Romantic separation
PF 1577 Romantic understanding past
PE 1684 *Rooms ; Overheated*
PD 2225 *Root cabbage ; Club*
PD 2226 *Root nodules legumes*

Sadness

PE 4453 Root rots ; Cereal
PF 3806 Rootlessness
PF 6143 Roots community initiatives ; Inadequate facilities grass
PJ 9394 Roots planning ; Unproductive grass
PE 8901 Roots tuber trade instability vegetable products
PE 8208 Roots tubers ; Instability trade vegetables,
PD 1593 *Rorquals ; Endangered species*
PE 4586 *Rose blackspot*
PA 5454 Rot *Badness*
PE 5617 *Rot ; Benign foot*
PE 5617 *Rot ; Contagious foot*
PA 6799 Rot *Disease*
Rot Foot
PE 4346 — *[Rot ; Foot]*
PE 5050 — *[Rot ; Foot]*
PE 4161 — *[Rot ; Foot]*
PE 5373 Rot ; Fruit
PE 6269 Rot fungi ; Heart
PE 4586 *Rot grape ; Black*
PE 1314 *Rot ; Grinders'*
PA 6088 Rot *Impairment*
PE 1028 *Rot ; Liver*
PE 5617 *Rot ; Malignant foot*
PA 6977 Rot *Meaninglessness*
PD 7799 *Rot ; Pizzle*
PE 4706 Rot plants ; Bacterial soft
Rot [Sheath...]
PD 7799 — *Sheath*
PE 4161 — *Stable foot*
PE 4586 — *stone fruits ; Brown*
PD 9667 — *Strawberry foot*
PA 5459 Rot *Uncleanness*
PE 5617 *Rot ; Virulent foot*
PE 1606 Rot wood ; Dry
PE 4767 Rot wood ; Wet
PE 3538 Rotation activities intergovernmental facilities ; Costly
PF 3698 Rotation ; Inadequate crop
PE 3936 *Rotaviral enteritis swine*
PD 3978 *Rotaviral infections chickens*
PD 3978 *Rotaviral infections turkeys*
PF 2721 *Rote learning*
PD 3159 *Rotifera ; Endangered species*
PE 4453 Rots ; Cereal root
PE 3363 Rots plants
PE 4455 Rots ; Wood
PD 4769 Rough ; Shortage wood
PE 8302 Rough toothed white dolphins ; Endangered species
PE 4950 Roughage diet ; Inadequate
PA 7156 Roughness *Moderation*
PA 5643 Roughness *Unkindness*
PA 7341 Roughness *Unpreparedness*
PA 5821 Roughness *Vulgarity*
PD 5453 *Round heart disease poultry*
PE 2395 Roundworm infestation
PD 3159 *Roundworms ; Endangered species*
PJ 5068 Rout
PE 1396 Routine maintenance complex systems ; Wastage highly skilled personnel
PF 2256 *Routines ; Habitual lifestyle*
PF 5943 Royal pretenders
PD 3050 Royalties ; International double taxation
PF 5943 Royalty ; Displaced
PE 0701 Rubber ; Instability trade crude synthetic reclaimed
PE 2977 Rubber ; Pests diseases
PD 0807 Rubbish collection ; Infrequent
PD 0807 Rubbish disposal
PA 6852 Rubbish *Inappropriateness*
PA 6977 Rubbish *Meaninglessness*
PE 0785 Rubella
PE 1603 Rubeola
PC 7013 Rudeness
PA 7143 Rudeness *Discourtesy*
PA 7341 Rudeness *Unpreparedness*
PA 6491 Rudeness *Vanity*
PA 5821 Rudeness *Vulgarity*
PF 1822 Rudimentary farming methods
PF 0951 Rugged individualism
Ruin [Defeat...]
PA 7289 — *Defeat*
PA 6542 — *Destruction*
PA 5497 — *Difficulty*
PA 6088 Ruin *Impairment*
PA 7382 Ruin *Loss*
PC 0845 Rule ; Autocratic
PF 0317 Rule ; Economic priorities
PD 2710 Rule fiat
PF 0851 Rule mindset ; Majority
PC 5528 Rule ; Working
PE 7937 Rulers associated sex scandals ; Government
PC 7587 Rulers ; Corruption government
PD 9653 Rulers countries ; Hypocritical accumulation personal wealth
PF 8053 Rules ; Complex government
PF 8793 Rules ; Outdated
Ruling [classes...]
PD 8380 — classes ; Corruption
PF 4849 — classes ; Elitist
PC 9585 — cliques
PE 3004 Rumen ; Digestive disorders
PE 6461 *Rumen flukes*
PE 4630 *Rumen impaction*
PD 3978 *Ruminal parakeratosis*
PD 3978 *Ruminants ; Bloat*
PE 0924 Ruminants ; Infectious abortion
PD 3978 *Ruminants ; Neonatal diarrhoea*
PD 2730 *Ruminants ; Peste petits*
PD 7420 *Ruminants ; Transport tetany*
PE 4656 *Ruminants ; Urolithiasis*
PE 5187 *Rumination disorder infancy*
PF 5596 Rumour
PF 5596 Rumour-mongering
PE 8233 *Run off ; Uncontrollable water*
PD 8340 Runaway adolescents
PD 8340 Runaway children
PD 8340 Runaway youth
PD 4858 Running ; Gun
PD 3142 Runs ; Ice
PD 2730 *Runting syndrome*
PD 7424 *Rupture Achilles tendon*
PD 5453 *Rupture ; Aortic*
PD 7424 *Rupture cruciate ligaments*
PE 4161 *Rupture gastrocnemius muscle*
PE 4161 *Rupture gastrocnemius tendon*
PD 9293 *Ruptured bladder*
PE 1139 Rural areas ; Constricting level capital development
Rural areas developing
PD 1225 — countries ; Lack sanitation
PE 5170 — countries ; Lack skilled manpower
PF 7047 — countries ; Policy discrimination against
PE 8429 — countries ; Unavailability trained teachers
Rural areas [Fragmented...]
PF 1442 — Fragmented forms commercial management
PF 6472 — Inadequate practical training
PF 2799 — Lack personal transport
PF 3575 — Limited availability education
PF 3575 — Limited educational opportunities
PF 4181 — Migration
PF 6470 — Poor communications networks
PF 1765 — Rigidly entrenched social traditions
PF 2875 — Underprovision basic services
PF 8105 Rural attractions ; Uncompetitive
PF 2996 Rural citizens ; Inadequate transportation systems
Rural communities
PE 3296 — Cumulative depletion corporate initiative
PD 2011 — Debilitating conditions health
PD 9504 — Decay
PE 2524 — Deepening debt cycle
PF 2986 — Deskilling
PE 6526 — developing countries ; Inadequate transportation facilities
PD 1204 — developing countries ; Inadequate water supply
Rural communities Inadequate
PE 1139 — *income*
PF 1442 — management skills
PF 6511 — residential housing
Rural communities [Inflated...]
PF 2385 — Inflated prices
PF 4132 — Insufficient telephones
PF 3575 — Irregular outside instruction
PE 8103 — Lack housing teachers
Rural communities Limited
PE 5170 — availability practical skills
PF 2949 — employment opportunities small
PF 3575 — opportunity ongoing education
PF 2848 — reservoir technical skills
Rural communities [Low...]
PF 6478 — *Low fishing income*
PF 2986 — Outmoded functional skills
PF 2471 — Overlooked potential industrial development
PF 6527 — Prohibitive cost essential services
PF 2385 — Prohibitive cost necessities
PF 1667 — Restricted delivery essential services developing country
PF 2385 — Rising food prices
PF 2848 — Shallow reserve technical skills
PE 8171 — Subsistence agricultural income level
PD 3054 — *Uncreditworthiness*
PF 1944 — Unperceived relevance formal education
PE 1101 — Unsystematic use powerful relationships
Rural community
PJ 8591 — *buildings ; Prohibitive construction cost*
PF 3558 — cooperation ; Deteriorating structures
PF 2358 — identity ; Demoralizing images
PF 4470 — market ; Inadequate
PF 7047 — Underemphasized importance
Rural [cooperative...]
PE 2309 — cooperative projects ; Government imposition
PE 2309 — cooperatives ; Failure
PD 1095 — customs traditions developing countries ; Decline
PC 0056 Rural depopulation
Rural development
PC 6242 — Conflicts integrated
PF 4959 — Neglect role women
PF 4340 — projects ; Inadequate commercial finance
PD 4537 — Unsustainable
PF 9023 Rural dispersion
PE 9688 Rural drug abuse
Rural [economic...]
PD 4125 — economic patterns ; Difficulties establishing changes
PF 2949 — employment ; Minimal options
PF 2949 — employment structures ; Inadequate
PF 0393 — energy sources ; Underdeveloped
PF 3630 Rural funding ; Uncertain sources
PF 6511 Rural housing developing countries ; Inadequate
PF 6511 Rural housing ; Insufficient
Rural [importance...]
PF 7047 — importance ; Forgotten
PE 8180 — industrialization developing countries ; Lack
PF 2875 — infrastructure developing countries ; Inadequate
PF 9023 — isolation ; Perpetuated
PF 2973 Rural labour force ; Depleted expertise
PF 7047 Rural life developing countries ; Neglect agricultural
PE 3528 Rural marketing ; Middleman control
Rural population
PC 0027 — *Ageing*
PE 8768 — cities ; Migration
PF 4470 — Socio economically inactive
PF 9686 Rural populations ; Neglected
Rural poverty
PC 4992 — [Rural poverty]
PD 4125 — developing countries
PD 4125 — developing countries ; Economic stagnation due
PE 3528 *Rural pricing ; Exploitation*
PF 4181 Rural resettlement
PC 2237 Rural subsistence economy ; Destruction
PF 2996 Rural transport ; Ineffective
Rural [underdevelopment...]
PC 0306 — underdevelopment
PD 0295 — underemployment developing countries
PF 2949 — unemployment
PD 0295 — unemployment developing countries
Rural urban
PD 1291 — areas ; Migration skilled people
PE 8569 — development ; Accentuated inequality
PE 5022 — income differential
PE 8584 — incomes developing countries ; Imbalance
PE 5022 — incomes ; Imbalance
PF 2648 Rural values urban cultures ; Incompatibility
Rural villages
PF 6496 — Inadequate systems transport communications
PF 2296 — Limited exposure outside influences
PF 2358 — Unformed confidence community corporateness
PD 2011 — Unsanitary environment basic health small
PE 3720 Rural violence
PE 4947 Rural women ; Discrimination against
PE 3910 Rural workers ; Denial right union activity
PF 3575 Rural youth education ; Limited access
PF 6473 *Rush hour transportation ; Crowded*
PE 2348 Russian spring-summer encephalitis
PD 4862 Russification
PE 1945 Rust
PE 6255 *Rust ; Cedar apple*
PE 6255 Rust fungi
PE 6255 *Rust ; Hollyhock*
PD 5803 Rust marine equipment ; Fouling
PE 2233 *Rust wheat ; Stem*
PE 6255 *Rust ; White pine blister*
PE 5465 *Rusts crucifers ; White*
PE 2233 Rusts grasses cereals
PE 6255 *Rusts ; Needle*
PE 6255 Rusts plants
PE 5465 *Rusts plants ; White*
PA 7023 Ruthlessness *Pitilessness*
PA 5643 Ruthlessness *Unkindness*
PD 5228 *Ryegrass staggers ; Annual*
PD 5228 *Ryegrass staggers ; Perennial*

S

PF 7887 Sabbaths ; Witches'
PD 0405 Sabotage
PF 1458 Sabotage ; Contamination public water supplies
PA 6542 Sabotage *Destruction*
PA 5497 Sabotage *Difficulty*
PD 5582 Sabotage ; Ecological
PD 5582 Sabotage environmental activists
PD 5426 Sabotage ; Food poisoning
PE 4454 Sabotage ; Hunt
PE 4454 Sabotage hunting expeditions direct action environmentalists
PA 6088 Sabotage *Impairment*
PA 6876 Sabotage *Impotence*
PD 3102 Sabotage ; Software
PD 9366 *Sac mite poultry ; Air*
PE 3604 *Sac winged bats ; Endangered species*
PF 7484 Sacred doctrine ; Profanation
PF 5495 Sacred documents ; Misreading
PE 5190 Sacred erotic advertising ; Vulgar combination
PD 8041 Sacred objects ; Misappropriation
Sacred [scriptures...]
PF 7129 — scriptures ; Cult
PD 6128 — sites ; Violation
PD 6128 — sites ; Vulnerability
PF 0319 Sacrifice ; Animal
PF 5203 Sacrifice any personal advantage ; Failure
PF 2641 *Sacrifice children*
PF 8118 Sacrifice children holy war
PA 7055 Sacrifice *Death*
PF 2641 Sacrifice ; Human
PA 7382 Sacrifice *Loss*
PF 5203 Sacrifice personal privilege ; Unwillingness
PF 7163 Sacrifice political power ; Unwillingness
PD 3373 Sacrifice ; Religious
PF 0662 Sacrilege
PA 6058 Sacrilege *Impiety*
PA 7280 Sacrilege *Wrongness*
PF 0662 Sacrilegious people
PE 7826 *Sacroiliac desmitis*
PD 9667 *Saddle sores*
PF 3270 Sadism
PE 6748 Sadism ; Sexual
PA 5643 Sadism *Unkindness*
PA 7107 Sadism *Unpleasantness*
PD 9219 *Sadistic personality disorder*
PF 3270 *Sadists ; Difficulty identifying*
PA 6731 Sadness *Solemnity*
PA 5942 Sadness *Unimportance*

PA 7107 Sadness *Unpleasantness*
PE 6137 Sado-masochism
PE 6137 Sadomasochism ; Homosexual
PD 2561 *Safecracking*
PF 5287 Safeguard systems ; Inability negotiate effective multilateral
PE 8542 Safeguards against computer crimes ; Inadequate
PD 1631 Safeguards against fire ; Inadequate
PF 6084 Safeguards ; Inadequate nuclear reactor
PF 4455 Safeguards ; Non verificability compliance nuclear power
PJ 8596 *Safety ; Diminished park*
PD 5001 Safety factors ; Non reinforced
PE 4961 Safety home
PF 2874 Safety ; Inadequate provision public
PE 8305 Safety ; Inadequate teaching occupational health
PC 6848 *Safety legislation ; Inadequate*
PF 4854 Safety ; Non disclosure threats public
PF 6389 Safety ; Obsession personal
PD 5001 Safety ordinances ; Unenforced
Safety [precautions...]
PF 6389 — precautions ; Excessive
PD 5001 — precautions ; Inadequate
PD 5001 — principles techniques ; Disregard
PF 2714 Safety records ; Suppression
Safety regulations
PE 9073 — Distortion international trade discriminatory formulation equipment
PD 5001 — Inadequate enforcement
PE 5698 — nuclear weapons industry ; Violation
PE 4006 — Violations health
PE 6630 Safety standards ; Inadequate implementation maritime
PF 2714 Safety tests ; Fraudulent
PD 1111 Safety ; Threat birds aircraft
PE 2348 Saint Louis encephalitis
PE 5757 Salacious telephone services
PE 9702 *Saláframine toxicosis*
PD 3156 Salamander ; Endangered species
PF 7578 Salaries corporate executives ; Excessive
PF 8763 Salaries ; Disproportionately high
PF 8763 Salaries ; Excessive
PF 8317 Salaries experts ; Excessive
PD 9653 Salaries government leaders ; Excessive
PD 9430 Salaries ; Inequitable range
PE 6388 Salaries international civil servants ; Excessive
PD 8422 Salaries ; Limited public
PE 8645 Salaries ; Non competitive teachers'
PE 6726 Salary base ; Nonexistent
PE 5305 Salary increases ; Inadequate constraints
PF 2042 *Salary losses ; Professional*
PD 9430 Salary scales countries ; Disproportionate
PD 9430 Salary scales ; Unfair
PE 8823 Sale because buyer's race ; Refusal
PD 8405 Sale children
PD 3998 *Sale defective products developing countries*
PE 6369 Sale foetuses
PF 4433 Sale indulgences
PD 4191 Sale non-existent commodities
PF 0699 Sale passports governments
PE 2449 Sale weapons civilians
PC 3298 Sale women
PD 4418 *Sales-call reluctance*
PF 1025 Sales-dominated market research
Sales [pesticides...]
PE 1785 — pesticides ; Inappropriate agricultural subsidies
PF 0581 — potential ; Uncertain
PF 0581 — potential ; Undetermined
PF 1025 — promotion dominates market research
PE 4581 Sales representation agreements ; Exclusive
PE 1727 Saline soils
PE 7837 Saline water ; Adjacency
PC 2087 Salinity change ; Ocean
PE 7837 Salinity effects ; Water pollution
PE 1727 Salinization ; Soil
PE 7837 Salinization ; Water
PE 2353 *Saliva ; Contaminated*
PE 7570 *Saliva ; Diminished secretion*
PE 7570 *Saliva ; Hypersecretion*
PE 7570 *Salivary cyst*
PE 7570 Salivary disorders animals
Salivary gland
PE 7570 *— abscess*
PD 2730 *— fever ; Bat*
PE 7570 *— inflammation*
PD 2732 *Salmon poisoning disease*
PE 7562 Salmonella infections
PE 7562 Salmonellosis
PE 3936 *Salmonellosis swine ; Enteric*
PD 9742 *Salpingitis*
PE 4231 Salt ; Excessive consumption
PD 5228 *Salt poisoning animals*
Salt water
PE 1783 — fish, crustacea molluscs preparations thereof ; Long term shortage
PE 7837 — flooding
PE 7837 — intrusion
PE 8257 Salted smoked meat ; Instability trade dried
PD 5227 Salts pollutants ; Inorganic
PE 9076 Salts ; Trade instability inorganic chemicals, elements, oxides halogen
PE 1727 Salty land
PE 2305 *Salty padi fields*
PD 4502 Salvage archaeology
PE 6254 Salvinia auriculata
PA 7365 Sameness *Boredom*
PD 2730 *San Miguel sea lion virus disease*

PA 6400 Sanctimony *Affectation*
PA 7411 Sanctimony *Uncommunicativeness*
PF 4366 Sanction state supported terrorism ; Government failure
Sanctioned [assassination...]
PE 6356 — assassination ; Corporation
PD 7221 — assassination ; Government
PF 3819 — assassinations ; Official cover up government
PD 4563 Sanctioned cultivation illegal drugs ; Government
PD 1886 Sanctioned hostage taking ; State
PE 6356 Sanctioned killing ; Corporation
PD 7221 Sanctioned killing ; Government
PD 7221 Sanctioned murder ; Government
PD 0181 Sanctioned torture ; Government
PD 0181 Sanctioned torture ; State
PF 4260 Sanctions against governments ; Economic
PD 0610 Sanctions against trade union workers
PF 1976 Sanctions barriers ; Smuggling
PF 1976 Sanctions ; Circumvention international
PE 1129 Sanctions ; Discrimination against poor judicial
PF 1976 Sanctions ; Disregard internationally imposed economic
PF 1870 *Sanctity marriage ; Lessening regard*
PF 1075 Sanctuary ; Denying
Sand [drift...]
PD 0493 — drift
PD 0493 — dune encroachment
PD 0493 — dunes ; Migrating
PE 8612 Sand gravel trade instability stone
PD 3650 Sand storms
PE 7826 *Sandcrack*
PE 2281 *Sandfly*
PE 8370 Sands energy source ; Underutilization tar
PE 3911 Sandy soils
PE 5480 Sanitary facilities ; Torture denial
PE 8272 Sanitary plumbing, heating, lighting fixtures fittings ; Instability trade
PE 7940 Sanitary plumbing, heating lighting fixtures fittings ; Shortage
PE 1274 Sanitary regulations agricultural pharmaceutical products ; Distortion international trade discriminatory formulation health
PF 1555 Sanitary regulations international travel ; Excessive animal
PE 1159 Sanitary services ; Health risks workers electricity, gas, water
PF 2583 Sanitary services ; Underdevelopment electricity, gas, water
PE 8475 Sanitary wastes ; Inadequate facilities transport
PD 5001 *Sanitation codes ; Unenforced*
PD 8023 Sanitation expertise ; Lack
PD 0876 Sanitation facilities ; Poor
PD 0876 Sanitation infrastructure ; Inadequate
PD 8023 Sanitation knowledge ; Lack
PF 8515 Sanitation ; Lack preventive
PF 8515 Sanitation practices ; Unhealthy
PD 1225 Sanitation rural areas developing countries ; Lack
Sanitation [services...]
PF 8515 *— services ; Discrimination access*
PD 0876 — services ; Undeveloped
PE 4570 — system accidents
PD 0876 — systems ; Inappropriate
PD 8837 Sanitation ; Uninformed animal
PF 7428 Sans-culottes
PA 6822 Sarcasm *Disrespect*
PD 7424 *Sarcocystosis*
PE 0264 Sarcoidosis
PC 3853 Sarcoma
PE 5548 Sarcoma ; Reticulum cell
PJ 5177 Sarcomas
PD 7424 *Sarcosporidiosis*
PE 0596 *Sarcosporidiosis*
PF 7018 Satan
PD 3267 Satanic child abuse children
PF 5633 Satanic governments
PF 7887 Satanic rituals
PF 8260 Satanism
PF 7004 Satellite arms race ; Anti
PE 0261 Satellite broadcasting individual receivers ; Foreign controlled direct
PE 0261 Satellite broadcasting ; Uncontrolled
PF 0045 Satellite capacity telecommunications ; Inadequate
Satellite communications
PD 1244 — Deliberate interference
PF 6072 — developing countries ; Obstacles
PD 1244 — Jamming
PF 0545 Satellite orbits ; Limited number geostationary
PF 3701 Satellite surveillance governments ; Misuse
PE 3391 Satellite systems ; Privatization
PF 3703 Satellite transmission ; Allocation television frequency bands
PD 3040 *Satellite transmission monopoly*
PD 0087 Satellite weapons ; Anti
PF 0054 Satellites ; Interference communications
PE 0101 Satellites ; Military expropriation
PF 1441 Satellites ; Nuclear reactor emissions
PF 3701 Satellites ; Reconnaissance
PF 3701 Satellites ; Spy
PF 4819 Sati
PA 6822 *Satire*
PA 6191 Satire *Disapproval*
PA 6822 Satire *Disrespect*
PA 8886 *Satisfaction ; Lack*
PF 0171 Satisfaction ; Lack job
Satisfaction Self
PA 6011 — [Satisfaction ; Self]
PA 1742 — [Satisfaction ; Self]
PA 6491 — [Satisfaction ; Self]

PF 3260 Satisfaction sexual intercourse ; Inadequate
PE 7654 Satisfied style informal leadership advisors ; Self
PE 4261 Saturated fats ; Consumption excessive
PF 0545 Saturation geosynchronous orbit
PE 5650 Saturnism
PE 7826 *Saucer fractures*
PF 1392 Saucers ; Flying (UFOs)
PA 5568 Savagery *Ignorance*
PA 7156 Savagery *Moderation*
PA 5643 Savagery *Unkindness*
PA 5821 Savagery *Vulgarity*
PF 1348 Saving foresight ; Lack
PF 1348 Saving ; Immediacy prevents
PE 1398 Savings banks ; Fraud
PE 4574 Savings capacity developing countries ; Inadequate government
Savings developing countries
PD 0465 — Bad investment
PE 4574 — Decline public sector
PD 0465 — Inadequate domestic
PF 1348 Savings funds ; Lack organized
PC 0927 Savings ; Inadequate
PE 0964 Savings institutions ; Failure
PF 1348 Savings structures ; Lack
PD 0465 Savings ; Uninvested community
PC 0927 Savings ; Uninvested personal
PE 4586 *Scab ; Apple*
PE 1040 Scab diseases plants
PE 3359 Scabies
PE 2727 *Scabies goats*
PE 2727 *Scabies sheep*
PE 5617 *Scald ; Foot*
PE 0394 Scalds ; Burns
PD 9430 Scales countries ; Disproportionate salary
PD 1133 Scales ; Non comprehensive wage
PD 9430 Scales ; Unfair salary
PE 7511 *Scalp wounds*
PE 3639 *Scaly leg mite*
PD 3481 *Scaly tailed squirrels ; Endangered species*
PC 8391 Scandal
Scandal [Defence...]
PD 4361 — Defence
PA 6191 *— Disapproval*
PA 6839 *— Disrepute*
PD 2458 Scandal ; Financial
PD 4651 Scandal ; Implication government leaders
PD 4651 Scandal ; Political
PD 9398 Scandal ; Sex
PD 3841 Scandal ; Stock
PA 5644 Scandal *Vice*
PA 7280 Scandal *Wrongness*
PE 7937 Scandals ; Government leaders associated sex
PE 7937 Scandals ; Government rulers associated sex
PD 9398 Scandals ; Homosexual
PE 5340 Scandals ; Unreported
PF 3448 *Scant animal feed*
PA 5473 Scantiness *Insufficiency*
PA 7285 Scantiness *Littleness*
PA 7408 Scantiness *Smallness*
PF 3332 Scapegoats
PF 2842 *Scarce employment possibilities*
PC 4815 Scarce fresh water
PS 2881 *Scarce local cash*
PF 1658 Scarce local jobs
PF 6535 Scarce options involvement culture
Scarce resources Competition
PC 4412 — [Scarce resources ; Competition]
PE 0063 — intergovernmental organizations
PE 0259 — international nongovernmental organizations
PC 1463 — international organizations
PA 5984 Scarcity
PE 8551 Scarcity appropriate transport
PF 2009 *Scarcity business collaterals*
PC 8160 *Scarcity corporate land*
PF 3002 Scarcity ; False image
PF 3409 *Scarcity high-quality seeds*
PA 5473 Scarcity *Insufficiency*
PC 4824 *Scarcity oil resources*
PD 8075 Scarcity residential land
PF 3665 *Scarcity risk capital*
Scarcity [skilled...]
PE 6243 — skilled manpower developing countries
PF 6477 — start-up funds
PF 3448 *— subsidized fertilizers*
PF 3448 *Scarcity well-paying jobs*
PA 6030 Scare *Fear*
PF 4384 Scaremongering
PD 9094 *Scarlet fever*
PE 3662 *Scarring*
PF 7118 *Scattered housing locations*
PF 1078 *Scavenging survival*
PF 3417 Scepticism
PF 7649 Scepticism accuracy official information ; Increasing
PF 1681 *Scepticism ; Community development*
Schedule [disorder...]
PE 2197 *— disorder ; Sleep wake*
PF 2644 *— disruptions ; Frequent*
PE 5074 — due computerization ; Disruption work
PF 6541 *Schedule news ; Inadequate*
PF 7142 *Schedule ; Self centred personal*
PE 5345 Schedule ; Uncommunicated bus
PF 2316 *Schedule ; Unpredictable work*
PF 2644 *Schedules ; Conflicting time*
Scheduling [Inconvenient...]
PF 6476 *— Inconvenient class*
PF 6476 *— Ineffective class*

Scriptures

	Scheduling [Inflexible...] cont'd		School [pressure...] cont'd		Scientific [dogmatism...] cont'd
PF 6476	— *Inflexible class*	PD 4178	— pressure	PF 6988	— dogmatism
PF 8675	Scheme ; Short range	PD 9068	— priority ; Low high	PC 1937	Scientific elitism
PF 6475	Scheme ; Unorganized land	PF 4984	— *programmes ; Rigid*		**Scientific evidence**
PE 8692	Schemes ; Health hazards water development		**School [recreation...]**	PF 1602	— Faking
PF 0422	Schemes ; Lack vocational training	PD 0549	— recreation ; Lack	PF 4633	— Loss
PE 8233	Schemes ; Mismanagement irrigation	PE 4554	— refusers	PF 4633	— Suppression
PD 1410	Schemes ; Undercapitalized waste use	PF 2595	— *roles ; Unclear*	PF 5055	Scientific expertise ; Inadequacy
PF 3534	Schism		**School [security...]**	PF 2409	*Scientific explanations phenomena*
PF 3181	Schism communism ; Ideological	PF 2874	— *security ; Loose*	PF 1602	Scientific fraud
PF 3534	*Schism ; Economic*	PE 0192	— students ; Beating	PF 1602	Scientific hoaxes
PC 2361	*Schism ; Political*	PF 4375	— supply costs		**Scientific [ignorance...]**
PF 1939	Schism ; Religious	PF 2256	— *systems ; Dual*	PD 8003	— ignorance
PE 0921	Schistosomiasis	PB 2496	— *systems ; Failure*	PD 8003	— illiteracy
PE 0921	*Schistosomiasis ; Haemic*		**School [transport...]**	PF 1615	— information ; Suppression
PE 0921	*Schistosomiasis ; Intestinal*	PC 2163	— transport ; Expensive	PF 2720	— investigation ; Lack
PE 0921	*Schistosomiasis ; Oriental*	PC 2163	— transport ; Nonexistent high		**Scientific issues**
PE 0921	*Schistosomiasis ; Vesical*	PC 2163	— transportation ; Lack	PF 7220	— Ethical resolution
PE 4593	Schizmogenesis	PC 2163	— transportation ; Unavailability	PF 7220	— Ideological resolution
PD 0438	Schizo-affective psychosis	PC 2163	— travel ; Expensive	PF 7220	— Religious resolution
PD 0438	Schizoaffective disorder	PF 0422	School ; Unfacilitated vocational		**Scientific knowledge**
PD 9219	Schizoid personality disorder	PE 8369	School work ; Difficulty transition	PD 8003	— Lack
PD 0438	Schizophrenia	PD 0614	Schoolgirls ; Pregnant	PF 4014	— Limitation current
PD 0438	*Schizophrenia ; Catatonic*	PF 9306	Schooling blocked family	PF 8757	— Mismatch public policy current
PD 0438	*Schizophrenia ; Disorganized type*	PF 3836	Schooling ; Divisive effects formal	PF 1468	— Unapplied
PD 0438	*Schizophrenia ; Hebephrenic*	PE 1848	*Schooling facilities ; Distant*		**Scientific medical**
PE 0435	*Schizophrenia ; Paranoid*	PF 6467	Schooling ; Inadequate results formal	PD 7415	— test results ; Misuse
PD 0438	*Schizophrenia ; Paraphrenic*	PF 6601	Schooling ; Unproductive extension	PD 7415	— testing ; Inappropriate
PD 0438	*Schizophrenia ; Residual*	PE 2876	Schools ; Bullying	PD 7415	— tests ; Inconclusiveness
PD 0438	Schizophreniform disorder		**Schools [Closure...]**	PF 9693	Scientific objectivity ; Erosion
PD 9219	Schizotypal personality disorder	PF 6575	— Closure	PF 2720	Scientific orthodoxy
PE 8883	Scholarship funds developing country students ; Unavailability	PE 0192	— Corporal punishment		**Scientific [power...]**
		PE 8797	— countryside ; Lack secondary	PF 2692	— power ; Abuse
PE 3569	Scholarship funds students ; Unavailability	PE 8797	— countryside ; Remote junior high	PF 2692	— power ; Uncontrolled
PJ 5626	Scholarship ; Ignoring implications biblical	PF 4852	— *Crimes committed urban*	PD 7875	— practices ; Restrictive
PF 7220	Scholarship ; Politicization	PF 0674	*Schools ; Decline public spending primary*	PF 4706	— problems ; Insoluble
PF 9065	Scholastic thinking	PE 9092	Schools ; Disruptive behaviour	PF 1199	Scientific racism
PF 9065	Scholasticism	PF 4375	*Schools ; Expensive distant*		**Scientific records**
	School [Absence...]	PE 0192	Schools ; Flogging	PF 4633	— Destruction
PE 4200	— Absence	PE 9092	Schools ; Indiscipline	PF 1602	— Falsification
PE 1990	— accidents	PE 9092	Schools ; Misbehaviour	PF 4699	— Tampering
PF 3519	— activities ; Impractical	PE 3757	Schools ; Overcrowding		**Scientific research**
PF 3184	— admission ; Limitations	PJ 2918	*Schools ; Proliferation private*	PF 9693	— Bias
PF 0443	— *athletic activities education ; Interference*	PD 3328	Schools ; Racism	PF 2112	— Dehumanized individual
	School attendance	PE 1848	*Schools ; Remote village*	PF 1789	— Environmental hazards constraining
PE 4200	— Inadequate enforcement	PC 2022	*Schools ; Sexually segregated*	PC 1449	— Inhumane
PE 4200	— Irregular	PC 2022	*Schools ; Single sex*	PC 9188	— Misuse
PE 4200	— Low	PE 2062	Schools ; Torture	PF 1430	— Secrecy
PD 8894	*School buildings ; Neglect*	PF 1257	*Schools transients ; Overcrowding*	PD 1709	Scientific rivalry
PE 0242	School buildings ; Underheated	PE 2428	Sciatica	PF 9693	Scientific studies ; Manipulation
	School [classes...]	PC 9188	Science ; Abuse		**Scientific technological**
PD 6585	— classes ; Inappropriate size	PF 2685	Science ; Anti	PC 1153	— activity ; Irresponsible
PD 6585	— classes ; Overcrowded	PD 9848	Science ; Complacency	PC 8885	— capacity ; Gap
PF 6575	— *closure ; Imposed high*	PF 9693	Science ; Conflict interest pure applied	PF 3931	— progress ; Adverse consequences
PF 1187	— communication ; Lost	PF 6077	Science ; Denial right benefits		**Scientific [test...]**
PF 9068	— consequences ; Unrecognized	PF 0770	Science ; Disillusionment	PF 1602	— test results ; Falsification
PJ 8591	— *construction ; Costly*	PF 3682	Science face unexplained ; Procrastination	PC 9188	— theories ; Misuse
	School curriculum		**Science [Imperfections...]**	PD 1709	— theories ; Non acceptance new
PF 0443	— Disrelated	PF 4014	— Imperfections	PF 1418	Scientific view ; Parochial
PF 0443	— Irrelevant	PF 8326	— Inadequacy medical	PE 0439	Scientific whaling
PF 0443	— *Static*	PD 8003	— Inadequate public understanding	PF 3366	Scientism
PF 0443	— *Unsupplemented*	PF 6349	— Inconclusiveness	PD 2091	Scientism ; Sexual
	School [discipline...]	PF 8032	— Irresponsible social	PD 6626	Scientists ; Corruption social
PE 9092	— discipline ; Lack	PE 8671	Science ; Lack diversity medical	PD 9182	Scientists ; Deception natural
PF 1187	— Disrelation parents	PF 7634	Science lying ; Social	PF 4014	Scientists ; Ignorance
PE 1848	— *Distant elementary*	PJ 1743	Science ; Neutrality	PD 6626	Scientists ; Irresponsible social
PE 1848	— *Distant secondary*	PF 1602	Science ; Pathological	PD 9182	Scientists ; Lying
PC 2163	— *distribution ; Unequal*	PF 1602	Science ; Pseudo	PD 6626	Scientists ; Negligence social
PC 1716	— dropouts	PF 7634	Science research ; Deceptive social	PD 2565	*Scleroderma ; Diffuse*
	School enrolment	PF 0145	Science research ; Intrusive social	PE 1041	*Sclerosis ; Multiple*
PC 1716	— *Declining*	PF 6697	Science ; Second class	PE 2053	*Sclerosis ; Renal*
PC 1716	— *Decreasing*		**Science technology**	PE 4586	*Sclerotinia diseases plants*
PC 1716	— *Stagnating*	PE 3105	— capitalism ; Abuse	PF 4113	Scoliosis
	School [equipment...]	PC 8885	— Concentration	PF 2439	Scope business operations ; Confined
PF 0905	— *equipment ; Shortage*	PE 4880	— Inadequate capacity developing countries	PF 3552	Scope education ; Narrow
PC 1716	— examination failures ; Wastage	PF 0770	— Irrelevance		**Scope [industrial...]**
PF 4375	— expenses ; Prohibitive	PC 8885	— Maldistribution	PF 1933	— industrial operations ; Unprofitable
PF 3498	— *experience ; Segmental*	PC 1153	— Obstacles proper use	PE 4398	— intellectual ability public sector ; Lack
	School [facilities...]	PD 0086	— *Underrepresentation women*	PF 4794	— intergovernmental development assistance ; Reduced
PD 0847	— facilities ; Undesignated	PC 1153	— Unethical use		
PC 1716	— Failure	PF 6697	Science ; Unprofessional	PF 2423	Scope local employment ; Restricted
PE 4554	— Fear	PF 7490	Science ; Violence justified	PF 2813	Scope town planning ; Limited
PF 4375	— fees ; Inequitable	PF 7634	Sciences ; Deception social	PE 3272	Scopophilia
PF 4375	— funds ; Nonexistent	PF 8868	Sciences ; Fragmentation	PE 4586	*Scorch ; Leaf*
PF 4984	— *School ; Hopelessness inspired*	PF 0716	Sciences ; Inappropriate use resources fundamental	PD 3473	*Scorpions ; Spiders,*
PF 8905	— School involvement ; Minimal church /	PF 4014	Sciences ; Incomplete state	PE 2665	*Scotoma*
PJ 9371	— *School laboratories ; Hazardous*	PD 6626	Sciences ; Malpractice social	PD 6253	*Scours*
	School leavers	PF 7532	Sciences ; Over specialization	PD 3978	*Scours*
PE 0015	— developing countries ; Unemployment premature		**Sciences Unethical practice**	PD 3978	*Scours ; Winter*
		PD 0708	— earth	PE 8906	*Scrap ; Environmental hazards metalliferous ores metal*
PD 8723	— Mal educated	PA 4277	— marine	PE 0553	*Scrap ; Instability trade metalliferous ores metal*
PD 8723	— Uneducated	PD 1110	— soil	PE 0353	*Scrap ; Long term shortage metalliferous ores metal*
	School [leaving...]	PD 6626	Sciences ; Unethical practices social	PE 0353	*Scrap ; Shortage non ferrous metal*
PC 1716	— leaving ; Premature		**Scientific activity**	PE 8312	Scrap trade instability ; Iron steel
PF 0443	— lessons ; Disrelated	PC 1449	— Dependence inhumane	PE 5330	Scrapie
PF 1187	— liaison ; Inadequate parent	PC 1449	— Inhumane	PE 5339	Scrapped automobiles
	School lunches	PF 1202	— Irrelevant	PB 0731	*Scratch*
PC 0834	— *Inappropriate*	PF 7843	Scientific arrogance	PE 5105	Scratch fever ; Cat
PC 0834	— *Unprovided*		**Scientific [censorship...]**	PE 7826	*Scratches*
PC 0834	— *Unsupplied*	PD 1709	— censorship	PF 6911	Screening ; Counter productive use crime
PE 8733	School ; Maldistribution students enrolled	PE 7961	— controlling instruments ; Shortage professional,	PF 1114	Screens ; Collapse providing ethical value
PF 1187	School non involvement ; Parent	PE 8169	— controlling instruments trade instability professional	PF 1112	Screens ; Escaping reality popular psychological
PF 3519	School ; Overacademic orientation high	PD 7470	— cooperation ; Structurally blocked	PE 8150	Screw-worm flies pests
	School [personnel...]		**Scientific [data...]**	PF 4958	Script ; Indecipherable
PF 2421	— *personnel ; Non resident*	PF 0255	— data ; Forged	PF 7129	Scriptures ; Cult sacred
PE 4554	— phobia	PF 1065	— data ; Proliferation unprocessed	PF 3869	Scriptures ; Inaccessibility religious
PF 4984	— *plan ; Inadequate pre*				

—1037—

PE 0566 Scrofula
PG 3896 Scrub typhus
PF 6404 Scrupulosity
PF 9699 Scrutiny ; Withdrawal official documents public
PF 0981 Sculpture
PD 2478 Sculpture ; Pollution damage
PF 7428 Scum
PE 4346 Scurf potatoes ; Silver
PE 4346 Scurf ; Sweet potato
PE 2380 Scurvy
PE 5812 Sea access land locked developing countries ; Excessive costs
Sea bed
PE 1597 — marine resources ; Inequitable allocation rights exploit
PD 1241 — Militarization deep ocean
PE 1597 — resources ; Discriminatory exploitation
PD 8982 Sea ; Collisions
PD 3158 Sea cucumbers ; Endangered species
PD 8982 Sea disasters
PD 3666 Sea ; Discharge dangerous substances
PE 1289 Sea ice
PE 5923 Sea ; Inadequate laws
PF 7435 Sea level ; Rising
PD 2730 Sea lion virus disease ; San Miguel
PE 2613 Sea mines ; Uncleared
PD 8438 Sea ; Piracy
PE 2501 Sea slime
PD 2788 Sea surges
Sea [traffic...]
PD 1486 — traffic congestion
PJ 3023 — transport facilities ; Inadequate
PE 1839 — transportation ; Environmental hazards
PD 3158 Sea urchins ; Endangered species
PF 4767 Sea water resources ; Obstacles utilization coastal deep
PD 0033 Sea waves ; Seismic
PC 1647 Seabed ecosystems ; Disruption
PC 6682 Seabed resources ; Environmental hazards mineral exploitation
PE 3484 Seals ; Cruel culling
PE 1656 Seals ; Endangered species
PE 6024 Seals ; Over population
PE 4766 Seaport facilities ; Inadequate
PD 3544 Search ; Arbitrary street
PF 6796 Search individualistic meaning
PD 3544 Search ; Unauthorized police
PD 3544 Search warrant
PD 4392 Searching partner ; Ineffective means
PD 1990 Seas ; Desiccation inland
PE 5309 Seas ; Disused machinery accumulation
PC 2122 Seascape ; Degradation natural
PE 2611 Seasickness
PD 0489 Season ; Failure rainy
PC 1108 Season jobs ; Limited off
PD 5212 Season ; Short growing
PF 2135 Season ; Unstable fishing
PF 8163 Seasonal adversity
PE 0258 Seasonal affective disorder
PC 1108 Seasonal employment
PD 0452 Seasonal flooding ; Unchecked
Seasonal fluctuations
PF 8163 — [Seasonal fluctuations]
PD 5212 — agriculture
PC 1108 — work
PC 1108 Seasonal labour
PD 5212 Seasonal malnutrition
PF 2135 Seasonal migration fish
PF 0827 Seasonal peaks ; Underutilization facilities due daily
Seasonal [resource...]
PF 0827 — resource use
PF 2135 — restrictions fishing
PE 3547 — roadway adequacy
PE 3547 Seasonal transportation ; Confined
PC 1108 Seasonal unemployment
PD 5212 Seasonal variability food supplies
PE 3547 Seasonal variations undeveloped transportation ; Isolating effects
PC 1108 Seasonal work patterns
PD 5212 Seasonally determined diets
PE 8175 Seawater ; Pollution semi enclosed bodies
PD 2498 Seaways ; Ice blocked
PD 2490 Secession
PA 7339 Seclusion Duality
PA 5869 Seclusion Exclusion
PA 7296 Seclusion Restraint
PA 6653 Seclusion Unsociability
PF 7424 Seco ; Enteque
Second class
PC 3458 — citizenship
PF 6697 — science
PD 0579 — states
PF 0638 Second coming Christ ; Unreadiness
PF 6908 Second employment
PE 4990 Second generation immigrants ; Marginalization
PD 1484 Second hand equipment ; Underutilization
PE 1286 Second homes ; Proliferation
PF 6908 Second jobs ; Necessity
PF 5958 Second language policy ; Antisocial attitudes planning
PE 6028 Second language teaching ; Limits
PE 7826 Second phalanx ; Fracture
PC 0939 Second, shadow economy socialist countries
PC 0939 Second undeclared employment socialist countries
PD 5345 Secondary education ; Inadequate
PE 2264 Secondary glaucoma
PD 7424 Secondary hyperparathyroidism animals ; Renal

PD 9667 Secondary pyoderma
PE 1848 Secondary school ; Distant
PE 8797 Secondary schools countryside ; Lack
PA 0005 Secrecy
PF 5991 Secrecy ; Abuse banking
PE 6786 Secrecy ; Abuse commercial
Secrecy [concerning...]
PF 4331 — concerning existence extraterrestrials
PF 4450 — concerning nuclear weapons testing ; Government
PE 1571 — Corporation financial
PA 0005 Secrecy ; Dependence
PC 1812 Secrecy ; Excessive government
PF 1136 Secrecy international aid institutions
PF 5983 Secrecy medical facts records
PC 1144 Secrecy ; Military
PF 4763 Secrecy national basic food stocks
PC 1812 Secrecy ; Official
PD 6576 Secrecy ; Professional
PF 1106 Secrecy ; Religious
PF 1430 Secrecy scientific research
PA 7411 Secrecy Uncommunicativeness
PA 6653 Secrecy Unsociability
PE 6786 Secret commercial agreements
PE 6786 Secret deals
PE 8004 Secret detention
Secret government
PF 2990 — agency agreements
PF 2990 — deals
PC 6757 — procedures
PF 9184 — security services
PE 9441 — security vetting job applicants
Secret [imprisonment...]
PD 4259 — imprisonment
PE 9857 — intelligence agents public office
PF 0419 — intergovernmental pacts
PF 0419 — international agreements
PD 8871 — investigation
PC 6757 Secret laws
PF 7669 Secret military operations
PE 6331 Secret police
PF 9699 Secret public records
PC 6757 Secret regulations
Secret [services...]
PF 9184 — services ; Covert cooperation
PF 2508 — societies
PF 2928 — societies ; Primitive
PF 0396 Secret trade cartels
PD 3013 Secret trials
PF 4450 Secret weapons development
PE 3538 Secretariats intergovernmental organizations ; Inter state rivalry
PE 7045 Secretariats international nongovernmental organizations ; Excess western based
PE 7570 Secretion saliva ; Diminished
PD 6576 Secretism ; Professional
PA 0005 Secretive people
PD 3537 Secrets ; Infringement trade
PF 5183 Secrets ; Leaking official
PA 0005 Secrets Secrecy
PF 7780 Sectarian tension
PF 7780 Sectarianism
PE 7020 Sections international nongovernmental organizations ; Incompatible equivalent national
PF 7230 Sector ; Deficiency innovation public
PC 3915 Sectoral fragmentation institutional responsibility
PF 3353 Sects ; Banned religious
PC 2129 Sects ; Discrimination against
PE 6336 Sects ; Exclusive
PC 6642 Sects ; Financial manipulation
PF 3353 Sects ; Persecution religious
PC 3363 Sects ; Threatened
PE 0176 Secular humanism
PF 7735 Secular impact holy days ; Disruptive
PF 3358 Secular life ; Undue religious influence
PB 1540 Secularism
PB 1540 Secularism ; Militant
PB 1540 Secularization
PB 1540 Secularization ; Dependence
PE 4545 Secured commercial property loans ; Inadequately
PD 4401 Secured creditors ; Defrauding
PF 2857 Securing sufficient food supplies ; Inadequate mechanisms
PD 4500 Securities commodities exchange violations
PC 5503 Securities ; Counterfeit money government
PD 3841 Securities fraud
PD 9155 Security agricultural sector ; Insufficient social
PE 8153 Security ; Armed crimes against national
Security [capitalist...]
PD 3120 — capitalist systems ; Denial right social
PF 4050 — classification ; Suppression information
PC 0554 — Crimes against national
PD 7407 — criminals ; Aiding national
Security Denial right
PD 7212 — [Security ; Denial right]
PD 8080 — economic
PD 7251 — social
Security [Dependence...]
PE 8670 — Dependence family social
PD 1229 — Dependence social
PF 1543 — developing countries ; Loss traditional forms social
PE 5406 — during periods unemployment ; Denial right economic
PC 4341 Security ; Environmental threats national
PF 7911 Security ; Erosion social

Security [farmers...]
PD 9155 — farmers ; Lack social
PG 5429 — firms ; National regulations banking
PD 3519 — forces ; Racial discrimination
PD 8859 — fraud ; Social
Security [Inadequate...]
PD 1147 — Inadequate community
PC 1258 — increase military capacity ; Myth national
PF 7911 — Increasing cost social
PE 1506 — indigenous peoples ; Denial right social
PD 1147 — Individualistic images
PF 7911 — Inflated cost social
Security information
PE 3997 — Crimes related national
PE 4343 — foreign power ; Revealing national
PE 3749 — Mishandling national
Security Lack
PE 8231 — airport travel
PD 8262 — emotional
PB 0009 — international
PF 2874 Security ; Loose school
PC 6239 Security mankind ; Offences against peace
PE 7611 Security migrants ; Inadequate social
PE 8746 Security nationals who have lived abroad ; Denial social
PE 0484 Security offenders ; Harbouring national
PC 1835 Security ; Overemphasis institutional
Security [patrol...]
PF 8652 — patrol ; Nonexistent
PD 9155 — peasant farmers ; Lack social
PD 8388 — personnel ; Untrained
PF 5009 — planning ; Obsolete
PF 2874 — Prohibitive cost business
PD 6818 Security risk people
Security services
PF 1015 — ousted repressive regime ; Post revolutionary re employment government
PF 8652 — Remote protection
PF 9184 — Secret government
PD 2884 Security structures ; Prioritized attachment
Security system
PD 6589 — Inadequate
PC 1867 — Inadequate
PF 5137 — Lack world food
PD 1147 Security systems ; Breakdown community
PC 1867 Security systems ; Inadequate
PB 0009 Security ; Threats international peace
PE 7664 Security tribunals
Security [Unequal...]
PF 0852 — Unequal coverage social
PF 0852 — Unequal distribution social
PE 8144 — urban life style ; Struggle financial
PE 9441 Security vetting job applicants ; Secret government
PB 0009 Security ; Violation international
PE 4796 Security welfare services developing countries ; Inadequate social
PD 4027 Sedative intoxication
PE 0139 Sedative withdrawal
PE 1821 Sedatives ; Abuse
PE 0139 Sedatives tranquillizers ; Abuse
PD 5494 Sedentary habits
PD 3654 Sedimentation dams
PE 5539 Sediments ; Pollution
PE 5539 Sediments toxic substances ; Contamination
PC 2414 Sedition
PA 7250 Sedition Disobedience
PA 7363 Sedition Improbity
PC 2414 Seditious writings
PD 4082 Seduction
PD 3265 Seduction children
PA 5612 Seduction Unchastity
PE 4559 Seed-borne diseases
PF 2899 Seed costs ; Excessive hybrid
PF 2899 Seed costs ; Inflated
PF 2899 Seed money ; Lack
PF 3467 Seed supply ; Untimely
PE 8250 Seeds ; Environmental hazards oil nuts, oil kernels oil
PE 6271 Seeds ; Infestation
PF 3409 Seeds ; Low yield
PE 0386 Seeds, oil nuts oil kernels ; Instability trade oil
PE 6788 Seeds ; Reduction genetic diversity available
PE 6788 Seeds ; Restricted access plant
PF 3409 Seeds ; Scarcity high quality
PE 7826 Seedy toe
PF 1113 Seeking ; Attention
PF 5153 Seeking election ; Discrimination against those
PF 6145 Seeking ; Status
PE 9088 Seepage water losses irrigation systems
PF 3498 Segmental school experience
PD 6744 Segmented labour markets
PD 3160 Segmented worms
PC 2022 Segregated schools ; Sexually
PC 0031 Segregation
Segregation [adult...]
PE 7209 — adult criminals ; Denial right juvenile criminals
PD 3444 — Age
PF 6136 — age group ; Household
PE 4299 — AIDS virus victims ; Social
PC 3365 Segregation based religious affiliation
PC 3458 Segregation ; Class
PE 7345 Segregation convicted criminals ; Denial right accused
PE 8215 Segregation different categories juvenile offenders ; Inadequate
Segregation [education...]
PD 3441 — education
PD 3443 — employment

Sepsis

Segregation [Ethnic...] cont'd
PC 3315 — Ethnic
PC 3383 — *eunuchs ; Social*
PA 5869 — *Exclusion*
Segregation [habits...]
PF 2256 — *habits ; Ingrained*
PE 8424 — handicapped children education
PD 3442 — housing
PA 6446 Segregation *Indiscrimination*
PD 4131 Segregation language
PD 3520 Segregation ; Legal
PD 3347 Segregation marriage
PA 7306 Segregation *Narrowmindedness*
PD 1260 Segregation poor minority population urban ghettos
PC 3688 Segregation ; Racial
PA 7296 Segregation *Restraint*
Segregation [Sex...]
PC 3383 — Sex
PD 3440 — social services
PC 0031 — *Spatial*
PA 6653 Segregation *Unsociability*
PE 5152 Seigneur ; Droit
PD 0033 Seismic sea waves
PD 9515 *Seismicity ; Reservoir induced increases*
PE 0125 Seizure cargoes
PE 6564 Seizure foreign nationals foreign countries ; Government
PC 8270 Seizure power
PJ 0456 Seizure property ; Compulsory
PE 5462 Seizure trade union premises
PE 7581 Seizure trade union property public authorities
PE 0241 Select their own leaders ; Denial right nations
PF 1266 Selection examination procedures education ; Inappropriate
PE 4700 *Selection ; Inequitable education*
PF 5997 Selection ; Natural
PD 2931 Selection political candidates ; Bias
PD 8468 Selective avoidance facts decision making
PD 0678 Selective domestic subsidies ; Distortion international trade
PF 1249 Selective ethics
PE 8867 Selective indirect taxes import charges ; Distortion international trade
PF 6057 Selective information
PF 7027 Selective memory ; Societal
PE 8583 Selective monetary controls ; Distortion international trade
PF 2453 Selective perception facts
PD 3483 *Selenartos ; Endangered species*
PD 5228 *Selenium poisoning animals*
PE 1726 Selenium pollutant
PE 3580 Self-abortion
Self absorption
PF 7248 — [Self-absorption]
PA 7363 — *Improbity*
PD 4749 — political leaders
PA 7364 — *Unfeelingness*
Self [accountability...]
PF 2275 — accountability grant funding
PF 0979 — actualizing stance being victim external forces
PA 7211 — admiration *Selfishness*
PA 6491 — admiration *Vanity*
PF 5354 — affirmation ; Absence
PF 4436 — aggrandizement
PD 2710 — aggrandizing national leaders
PF 4377 — assessment ; Unrealistically positive
Self [censorship...]
PF 6080 — censorship
PF 7142 — *centered personal schedule*
PF 1176 — *centred business*
PA 7211 — centredness *Selfishness*
PA 6491 — centredness *Vanity*
PF 1178 — conception ethical void
Self confidence
PF 0879 — Lack
PF 9066 — Limited village
PF 0879 — Low
PF 0879 Self confidences ; Lack
PC 1777 Self consciousness ; Repression
Self contradiction
PA 6698 — *Difference*
PA 5982 — *Disagreement*
PA 6180 — *Error*
PC 1707 Self-created isolationism
PF 6362 Self-deception
PF 7702 Self deception ; Official
Self defeating
PD 4418 — behaviour
PD 4418 — personality disorder
PF 6456 — style community planning
PF 5354 Self-depreciatory operating images
Self destruction
PF 8587 — [Self-destruction]
PA 7055 — *Death*
PF 3849 — Wish
PF 6044 Self-destructive excuses
PF 5061 Self-destructive government policy-making
Self determination
PE 3123 — capitalist exploitation ; Denial effective national
PC 3177 — communist systems ; Denial right national
PC 1450 — Denial right national
PF 1804 — Habitual overemphasis national
PF 1491 — unrelated global obligations
PD 2592 — Use mercenaries means impeding right
Self [determining...]
PC 6727 — determining ; Denial right people be
Self [development...] cont'd
PE 8341 — development family ; Lack
PF 4843 — disorders
PF 5354 — doubt
Self esteem
PA 6659 — *Humility*
PF 5354 — Low
PA 7211 — *Selfishness*
PA 6491 — *Vanity*
PF 4841 Self experimentation ; Errors risks medical
PF 0593 Self expression ; Unconstrained educational
PF 9694 Self-fulfilling prophecies
Self [glorification...]
PF 4436 — glorification
PF 2666 — *glorification*
PE 1024 — governing ; Denial rights indigenous people be
PF 2943 — governing status disputed administering government ; Territories accorded United Nations non
PF 1807 — gratification ; Conditioned
Self [harm...]
PE 4878 — harm ; Non fatal deliberate
PC 3428 — *hatred*
PF 4871 — *healing potential ; Disregard*
PF 2575 Self identity ; Negative village
Self image
PE 6784 — ageing women ; Negative
PE 7752 — being victim circumstances ; Marketing skills which reinforce
PF 2093 — Demeaning community
PF 1529 — Demeaning minority
PF 9098 — due illiteracy ; Low
PF 5354 — Inadequate
PF 5354 — Low
PF 5354 — Low resident
Self [images...]
PF 8781 — images citizens' law enforcement groups ; Defeatist
PD 1564 — images preclude participation ; Elders'
PA 6400 — importance *Affectation*
PA 6491 — importance *Vanity*
PE 2274 — *imposed isolation ; Mothers'*
PE 4703 — imposed social isolation due torture
PE 7252 — incrimination innocent
Self indulgence
PA 6466 — *Intemperance*
PF 5466 — National
PA 7211 — *Selfishness*
Self [indulgent...]
PD 5774 — indulgent consumerism
PF 5466 — indulgent societies
PD 7550 — inflicted physical suffering
Self interest
PA 8760 — [Self-interest]
PC 2576 — driven investment
PD 3546 — Natural resources used national
PF 5061 — Policy contrary
PF 3679 Self-interested industrial vision
PF 5529 Self interested manipulation timing
PF 4741 Self-limiting technical innovation
PF 5354 Self love ; Lack
PF 4843 Self object disorders
Self [perpetuating...]
PF 4741 — perpetuating attitudes technological innovation
PF 4436 — praise
PD 4418 — *promotion ; Fear*
PD 4418 — *promotion ; Inhibited*
Self regulation
PE 8067 — automobile manufacturing industry ; Ineffective
PE 8956 — chemicals manufacturing industries ; Ineffective
PE 8574 — consumer goods manufacturing industries ; Ineffective
PE 8121 — financial services sectors ; Ineffective
PE 8472 — food processing industries ; Ineffective
PE 9048 — health care sector ; Ineffective
PE 8265 — housing construction industry ; Ineffective
PF 5841 — Ineffective industry
PE 8630 — mining industries ; Ineffective
PE 8502 — pharmaceutical medical devices industries ; Ineffective
PF 5840 — shipping industry ; Ineffective
PF 5877 — telecommunications sectors ; Ineffective
Self [reinforcing...]
PD 7727 — reinforcing contractionary economic spiral
PF 0879 — *respect ; Loss*
PF 2335 — review ; Bureaucratic resistance
PC 3406 — *righteousness*
PA 6058 — righteousness *Impiety*
Self satisfaction
PA 1742 — *Complacency*
PA 6011 — *Discontentment*
PA 6491 — *Vanity*
Self [satisfied...]
PF 7654 — satisfied style informal leadership advisors
PF 1681 — story ; Demoralizing community
PF 2013 — story ; Failure
Self sufficiency
PE 6780 — handicap ; Economic
PE 5795 — is socially responsible ; Story that
PF 0841 — Lack economic
PF 6573 — *patterns ; Obsolete*
PF 3460 — respect interdependence ; Overemphasis
PD 7550 Self-torture
PF 3512 Self worth ; Lack community
PF 1663 Self worth ; Social reinforcement delimited
PA 7211 Selfishness
PA 7211 Selfishness *Selfishness*
PA 6491 Selfishness *Vanity*
PF 0468 Sell ; Refusal
PC 3298 Selling ; Bride
PA 0911 Selves ; Fragmented
PF 5985 Semantic confusion
PF 2789 Semilinguism
PD 5228 *Senecio poisoning animals*
PD 5228 *Senecìosis*
PB 0477 Senescence
PJ 1688 Senescent crime
PE 3083 Senile dementia
PD 5501 Senile ; Mistreatment
PD 0919 *Senile psychosis ; Atrophic*
PD 2341 *Senilicide*
PE 6402 Senility
PD 1564 Senior citizens ; Isolated
PD 1718 Sensationalism
PD 1718 *Sensationalism*
PA 7685 Sense abandonment
PF 2575 Sense community ; Declining
Sense community solidarity
PE 4179 — amongst workers ; Inadequate
PD 0110 — developing countries ; Lack
PF 8704 — world level ; Lack
PA 6236 Sense*complex
PF 8781 Sense defeat
Sense organs
PC 9623 — Diseases
PC 9623 — Disorders
PE 2665 — Symptoms referable
Sense [personal...]
PF 1934 — personal identity ; Inadequate
PF 3777 — personal shallow meaning ; Institutionally reinforced
PF 8618 — powerlessness
PF 1246 Sense sexual identity ; Conflicting
PE 1066 Sense smell ; Lack
PF 9980 Sense time ; Inadequate
PF 9980 Sense time ; Loss
PF 2093 *Sensed futility job-hunting*
PF 6991 Senselessness
PA 7361 Senselessness *Disorder*
PA 7157 Senselessness *Insanity*
PA 6977 Senselessness *Meaninglessness*
PA 5502 Senselessness *Reason*
PA 7371 Senselessness *Unintelligence*
PD 5535 Senses animals ; Diseases
PE 9072 Sensing systems ; Inequality inducing effects remote
PE 3704 Sensitive areas ; Naval manoeuvres
PE 3704 Sensitive border areas ; Military manoeuvres
PF 9558 Sensitive incidents ; Fabrication politically
PF 9209 Sensitivity ; Fear emotional
PD 3514 Sensori motor activities aged ; Slowness
PE 4234 Sensory aphasia
Sensory [deprivation...]
PE 5787 — deprivation ; Harmful effects
PE 6797 — deprivation ; Torture
PU 3287 — *disorders*
PE 5259 Sensory overload ; Torture
PA 5612 Sensuality *Unchastity*
PF 1756 Sensuous dance
PD 0520 *Sentence ; Detention offenders after completion*
PF 4723 Sentence squashing
PD 0520 *Sentences ; Arbitrary extension prison*
PF 1098 Sentences crimes against humanity ; Remission
PE 4602 Sentences ; Disproportionately long prison
PE 4602 Sentences ; Excessive length prison
PE 9545 Sentences offenders after conviction ; Arbitrary reduction
PF 4723 Sentences ; Politically motivated reduction prison
Sentencing offenders
PF 4723 — Excessive leniency
PF 5356 — Inconsistent
PE 4907 — Racial bias
PF 3390 Sentiment ; Geo
PF 1073 Sentiment ; Socio
PF 1577 Sentimental attachment past
PF 6961 Sentimentalism
PF 9209 Sentimentality ; Fear
PA 7364 Sentimentality *Unfeelingness*
PA 7371 Sentimentality *Unintelligence*
PF 8416 Separate unequal development different societies
PE 4669 Separated parents ; Lack freedom movement children
PF 3364 Separated religious brethren
PD 3442 Separated residences ; Racially
PA 6858 Separateness *Disintegration*
PA 7339 Separateness *Duality*
PA 7236 Separateness *Separation*
PA 5794 Separateness *Unrelatedness*
PA 7236 Separation
PE 2401 Separation anxiety
PE 2401 Separation anxiety disorder
PE 4959 Separation ; Family
PE 4959 Separation family members
PF 6122 Separation home workplace ; Artificial
PF 4754 Separation individual's social functions
PF 3251 Separation marriage law
PF 0379 Separation nature
PF 1187 Separation ; Parent teacher
PF 0379 Separation people biological rhythms
PD 4233 Separation ; Romantic
PF 6137 Separation urban subcultures ; Insufficient
PD 1544 Separation youth / adult activities
PD 2490 Separatism
PD 2490 Separatism ; Cultural
PD 2490 Separatism indigenous peoples
PD 2490 Separatism ; Minority group
PC 9025 Sepsis

Sepsis

PC 9042 Sepsis ; Puerperal
PE 5569 Septal defects animals
PE 5913 Septic arthritis
PE 5617 Septic laminitis
PD 2732 Septic pasteurellosis sheep
PE 5050 Septic polyarthritis ; Neonatal
PE 9422 Septicaemia
PD 7841 Septicaemia cattle ; Haemophilus
PD 2731 Septicaemia foals
PD 2731 Septicaemia ; Haemorrhagic
PE 0987 Septicaemic plague
PB 5486 Sequestration real property ; Arbitrary
PE 9447 Serial killings
PF 2100 Serial polygamy
PD 2732 Serositis ; Infectious
PE 5913 Serositis ; Transmissible
PE 7826 Serous arthritis fetlock joint
PE 0517 Serum hepatitis
PF 0897 Servant role professional ; Collapse public
PF 5649 Servants ; Abuse privileges immunities international civil
PD 4541 Servants ; Bribery public
         Servants [Conspicuous...]
PE 3457 — Conspicuous consumption international civil
PE 6702 — Corruption civil
PD 4541 — Corruption public
         Servants [Detention...]
PE 5488 — Detention international civil
PE 5278 — developing countries ; Demotivated civil
PE 4964 — Discrimination against domestic
PD 8392 Servants ; Embezzlement civil
PE 6388 Servants ; Excessive salaries international civil
PD 5554 Servants ; Fraud international civil
PE 6702 Servants ; Irresponsible civil
PE 4031 Servants ; Lack political neutrality civil
PD 5399 Servants ; Retaliation against public
PC 0592 Servants ; Shortage domestic
PD 0540 Servants ; Threatening public
PD 4541 Servants ; Unlawful rewarding public
PE 5488 Servants ; Violation privileges immunities international civil
PE 1862 Service agencies ; Cultural diversity ignored social
PD 8017 Service bureaucracy ; Unresponsive social
PF 6051 Service ; Compulsory military
         Service [delivery...]
PD 0974 — delivery ; Distrust professional
PF 2583 — delivery urban environments ; Poor
PE 5608 — Denial public
PD 1800 — Denial right conscientious objection military
PE 4226 — developing countries ; Politicization public
PE 6422 — Discrimination against foreign nationals military
         Service [employees...]
PE 6702 — employees ; Negligent public
PE 6702 — employees ; Unethical practices public
PF 2969 — expectations ; Unfulfilled
PE 5721 Service ; Fragmentation health
         Service [ideologies...]
PD 3190 — ideologies ; Conflicting social
PD 6667 — Impressment military
PE 8509 — Inability communities take advantage training business, industry public
PF 1495 — Inadequate air transport
PE 6754 — Inappropriate use mail
PF 2992 — inconsistencies ; Social
         Service industries
PE 0875 — Health risks workers
PF 0591 — Inadequate regulation restrictive business practices
PG 8903 — sector ; Overgrowth
         Service [Ineffective...]
PF 2335 — Ineffective civil
PE 5345 — Infrequent bus
PF 1495 — Irregular passenger
PE 6754 — Irresponsible use postal
PE 4450 — Irresponsible use telecommunications
PE 5800 Service land locked developing countries ; Inadequate air transport
PE 4768 Service legal process ; Failure
PF 1205 Service ; Misused ambulance
         Service [obligations...]
PD 5941 — obligations ; Crimes related military
PD 3051 — obligations developing countries ; Inconsiderate insistence creditors fulfilment debt
PG 8937 — outreach ; Reduced
PE 8728 Service personnel ; Inadequate conditions work employment public
PE 1488 Service quality negated oligarchic control decision making ; Social
PE 7025 Service related goods ; Protectionism against imports
PF 4832 Service ; Remote clinic
         Service [society...]
PF 6433 — society ; Blocked
PD 4633 — Subjugation women domestic
PF 2212 — systems ; Incomplete understanding new societal
PF 3498 Service training ; Ineffective
         Service [Undependable...]
PE 8254 — Undependable telephone
PE 6754 — Unethical use postal
PE 4450 — Unethical use telephone
PE 6342 Service ; Violation freedom bond
PD 3335 Service women ; Denial right access public
PF 4892 Service workers ; Inadequate utilization volunteer social
         Services [Absence...]
PC 7674 — Absence essential
PD 0512 — aged ; Denial right welfare
PD 0512 — aged ; Inadequate welfare
PD 0542 — Services blind ; Denial right welfare

PD 0542 — Services blind ; Inadequate welfare
         Services [Clandestine...]
PJ 0914 — Clandestine private utilization public goods
PF 2694 — communication ; Insufficient provision public
PF 2451 — community leadership training ; Lack local
PF 4955 — Constraints development mental health
PF 6527 — Costly distribution
PD 7981 — Counterfeit products
PE 8997 — countryside ; Irregular repair
PF 9184 — Covert cooperation secret
         Services [De...]
PE 7971 — De facto racial requirement qualifications public
PD 0601 — deaf ; Denial right welfare
PD 0601 — deaf ; Inadequate welfare
PF 2875 — Declining volunteer
PD 4790 — Deficient health
PF 5120 — Delay access social
PE 8157 — Delay delivery requested
PE 3928 — Delays delivery goods
PC 0834 — Denial right adequate welfare
         Services developing countries
PE 2603 — Absence coordinated customs
PF 4195 — Inadequate industrial
PE 4796 — Inadequate social security welfare
PE 8236 — Insufficient consumer
PE 7959 — Insufficient repair
PE 8986 — Limited mechanical
PE 9001 — Remoteness legal
PF 6473 — Underdeveloped provision basic
         Services [developing...]
PF 1667 — developing country rural communities ; Restricted delivery essential
PJ 5608 — Disbarment public
PF 8515 — Discrimination access sanitation
         Services Discrimination against
PE 8507 — men public
PD 3336 — men social
PD 3691 — women social
PD 8460 Services ; Discrimination public
PC 3433 Services ; Discrimination social
         Services Distant
PE 0718 — banking
PF 9007 — emergency
PF 2477 — fire
PD 4790 — health
PF 4832 — medical
PF 8652 — police
         Services [Distorting...]
PE 9220 — Distorting effects commodity taxes transaction goods nonfactor
PF 1681 — Distrust
PF 1735 — Disunified requests
         Services [elderly...]
PD 0512 — elderly ; Inadequate provision home care
PJ 2489 — Exaggerated reliance professionalized
PC 2518 — Excessive consumption goods
PE 4780 — Excessive dependence local communities outside
PJ 2520 — Exclusive
PF 9007 — Expensive emergency
PC 3299 — Exploitative personal
         Services [facilities...]
PD 2669 — facilities organized action against problems ; Inadequate buildings,
PF 8819 — Failure government intelligence
PD 6265 — following nuclear war ; Inadequate health
PF 0041 — Services ; Governmental inaction concerning trade
PF 2411 — Services ; Haphazard provision consumer
PE 1159 — Services ; Health risks workers electricity, gas, water sanitary
PF 3526 — Services ; Inaccessibility bureaucratic
         Services Inaccessible
PE 0718 — commercial financial
PE 0718 — financial
PD 2926 — government information
PF 2717 — Services ; Inadequacy postal
         Services Inadequate
PE 8319 — conditions work employment utility supply
PD 1428 — emergency medical
PE 0718 — financial
PD 4790 — health
PF 1790 — industrial
PF 4832 — medical
PF 4955 — mental health
PE 7544 — radiological
PC 0834 — social welfare
PF 6492 — support
PF 8517 — water
PE 7718 — working conditions health medical
         Services [Inadequately...]
PF 1781 — Inadequately publicized
PE 4450 — Inappropriate use telecommunications
PF 1506 — indigenous peoples ; Denial right social welfare
PE 2317 — indigenous populations ; Denial right legal
PD 7135 — industries ; Protectionism
PD 7049 — industries ; Protectionism construction engineering
PD 7001 — industry ; Protectionism computer
PD 7121 — industry ; Protectionism franchising
PF 2875 — Ineffective delivery basic human resource
PF 9007 — Ineffective emergency
PC 3433 — Inequality social
PD 1891 — Inflated prices goods
PF 3083 — Inflexible attitudes toward community social
PF 2006 — Institutionalized callousness public
         Services Insufficient
PF 9007 — emergency
PF 3037 — family

         Services Insufficient cont'd
PE 9133 — financial resources urban
         Services [Irregular...]
PE 5345 — Irregular transport
PE 3328 — Irresponsible health care
PF 2875 — isolated communities ; Inaccessibility essential
PF 2875 — isolated communities ; Restricted delivery essential
         Services [Lack...]
PF 2009 — Lack local commercial
PF 6510 — Lack support local commercial
PE 5860 — land locked developing countries ; Unreliable telecommunication
PE 5836 — land locked developing countries ; Unreliable transit
PD 1566 — Lapsed volunteer
PF 4297 — least developed countries ; Inadequate development communication
PE 0289 — least developed countries ; Inadequate infrastructure
         Services Limited access
PF 2875 — available
PF 6577 — health
PF 2583 — urban
         Services Limited [accountability...]
PF 6574 — accountability public
PF 3550 — availability funding resources urban
PF 2652 — distribution basic
PF 2989 — local markets goods
PF 3526 — social
PF 1653 Services logistics ; Complex
         Services [materials...]
PF 2115 — materials needed development ; Lack essential
PE 8151 — mentally handicapped ; Inadequate support
PD 8832 — Minimal community
PF 2583 — motivation ; Poor
PF 2411 Services ; Non competitive consumer
PF 2974 Services ; Non standardized social
         Services [Obscure...]
PF 6574 — Obscure accountability public
PD 6223 — Obstacles international trade
PF 1015 — ousted repressive regime ; Post revolutionary re employment government security
PD 4907 — Over diversification
PD 1891 — Over priced goods
PF 1176 — Overlooked community
         Services [Perpetuated...]
PJ 8892 — Perpetuated unessential
PC 7674 — Physically inaccessible
PE 5402 — Pink messaging
PE 1101 — Politically determined
PE 3391 — Privatization public
         Services Prohibitive cost
PF 6527 — basic
PF 1653 — connection electricity
PD 1891 — goods
PF 1853 Services ; Proliferation parallel urban
PF 2471 Services ; Prolonged repair
         Services Protectionism
PE 7987 — free zone international financial
PD 8614 — legal
PD 7073 — public accounting auditing
         Services [Racial...]
PD 3326 — Racial discrimination public
PE 5898 — Reduction ocean shipping
PF 2709 — relation world's need ; Decisional paralysis specialized
         Services Remote
PF 2875 — basic
PD 4790 — health care
PF 8652 — protection security
         Services [Restrictive...]
PE 5910 — Restrictive practices cargo airline
PE 3928 — Risks due delays delivery goods
PF 2875 — rural areas ; Underprovision basic
PF 6527 — rural communities ; Prohibitive cost essential
         Services [Salacious...]
PE 5757 — Salacious telephone
PF 9184 — Secret government security
PE 8121 — sectors ; Ineffective self regulation financial
PD 4344 — Segregation social
PF 2583 — Slow city
PF 6539 — small towns developed countries ; Limited availability public
PD 4711 Services ; Theft
         Services [Unacknowledged...]
PF 2583 — Unacknowledged availability
PE 0718 — Unavailability banking
PF 9007 — Unavailable emergency
PD 4790 — Uncoordinated health
PE 7919 — Undefined government role urban
PF 2583 — Underdevelopment electricity, gas, water sanitary
PE 5911 — Underpricing port
PF 2583 — Underprovision basic urban
         Services Undeveloped
PD 4790 — health
PF 1790 — industrial
PD 0876 — sanitation
         Services [Undue...]
PE 8444 — Undue taxation certain goods
PE 8603 — Unequal distribution goods
PC 3437 — Unequal distribution social
PE 3328 — Unethical practices health
PD 4790 — Uninitiated health
PF 2583 — Unknown available
PF 2491 — Unknown loan
PF 2845 — Unpublished government
PE 3296 — Unresponsive social
PG 8913 — Unsympathetic token

Sharks

Code	Entry
	Services [Unusable...] cont'd
PG 8866	— Unusable community
PD 4790	— Unused health
PD 4790	— Unutilized health
PE 8959	— urban areas ; Limited transportation
PF 1853	— urban areas ; Uncoordinated social
PF 4470	— users ; Small population basic
PE 5938	Services ; Violation right organize trade unions government
PF 9697	Services ; Vulnerability national economies vagaries external markets goods
PD 1565	Servicing ; Deterioration quality equipment maintenance
PD 3051	Servicing foreign public debt developing countries ; Burden
PF 3467	Servicing ; Inadequate vehicle
PA 7250	Servility *Disobedience*
PA 6822	Servility *Disrespect*
PA 6659	Servility *Humility*
PA 5652	Servility *Inferiority*
PA 7296	Servility *Restraint*
PF 2639	Serving false consumption needs ; Production
PE 7627	Servitude ; Breastfeeding female
PE 4964	Servitude ; Denial right freedom
PE 7826	Sesamoiditis
PE 7826	Sesamoids ; Fracture proximal
PE 2488	Settings regional linguistic controversies ; Bilingualism national
PE 1508	Settled refugees living outside camps ; Non
PC 0519	Settled refugees ; Non
PE 0743	Settlement ; Flood plain
PD 6150	Settlement ; Obsolescence suburban mode human
PF 9841	Settlement systems ; Inadequate financial
PE 3813	Settlement unprotected coastlines
PD 5077	Settlements developing countries ; Inadequacy training human
PD 4368	Settlements ; Makeshift
PD 6364	Settlements ; Military attacks refugee camps
PD 6130	Settlements ; Spatial imbalance human
PE 3813	Settlements tidal flooding ; Exposure
PD 3489	Settlements ; Transitional urban
PE 8020	Settling building foundations ; Uneven
PA 5856	Severance *Exclusion*
PE 3872	Severance pay top managers ; High
PA 7236	Severance *Separation*
PE 6314	Severe mental retardation
PD 0211	Severely deformed people ; Lack facilities
PF 0581	Severely limited markets
PE 4544	Severely mentally handicapped ; Denial right procreate
PA 6400	Severity *Affectation*
PA 5740	Severity *Compulsion*
PA 7143	Severity *Discourtesy*
PA 7156	Severity *Moderation*
PA 5643	Severity *Unkindness*
PE 6685	Sewage contaminated water ; Bathing
PE 6685	Sewage discharge ; Bathing near
PD 0876	Sewage disposal ; Improper
PD 3666	Sewage ; Marine dumping
PD 1414	Sewage pollutant
PE 6685	Sewage ; Recreational contact
	Sewage systems
PD 1414	— Discharge dangerous substances
PD 9196	— Failure maintain
PD 0876	— Overloaded
PD 0876	Sewerage systems ; Inadequate
PF 9109	Sex
	Sex [Abuse...]
PD 2198	— Abuse
PD 2547	— advertising ; Gratuitous
PE 6784	— appeal women ; Loss
PF 3277	Sex ; Change
PD 4082	Sex crimes
	Sex [Dependence...]
PE 7031	— Dependence
PF 9109	— Dependence
PF 3382	— determination birth ; Artificial
	Sex [education...]
PE 0341	— education ; Excessive youth
PD 0759	— education ; Inadequate
PE 5402	— Electronic
PF 6910	Sex ; Fear opposite
PF 3288	Sex ; Group
PD 8016	Sex guilt
PE 6969	Sex ; Health risks teenage
PD 7994	Sex ; Ignorance concerning
PC 1523	Sex ; Interracial
PD 0061	Sex love ; *Dissociation*
PE 7930	Sex media ; Excessive portrayal
PE 7930	Sex media ; Glorification demeaning
PE 7031	Sex ; Preoccupation
	Sex [ratio...]
PF 3382	— ratio ; Deliberate imbalancing population
PF 1128	— ratio ; Imbalance human
PD 3267	— rings ; Child
PD 1308	— role conditioning
	Sex [scandal...]
PD 9398	— scandal
PE 7937	— scandals ; Government leaders associated
PE 7937	— scandals ; Government rulers associated
PC 2022	— schools ; Single
PC 3383	— segregation
PF 4221	— surrogacy ; Therapeutic
PD 7439	Sex ; Teenage
PD 8910	Sex traffic offences ; Vice
PE 9776	Sex ; Unsafe
PC 3432	Sexism
PF 5967	Sexist education children
PF 6014	Sexist language
PD 0247	Sexist violence
PF 6910	Sexo-phobia
PF 6910	Sexophobia
PD 2198	Sexual abnormality
PE 3298	Sexual abstention ; Total
	Sexual abuse children
PE 3265	— [Sexual abuse children]
PD 3267	— Organized
PD 3267	— Ritual
	Sexual abuse [Ritualistic...]
PF 7887	— Ritualistic
PF 7755	— wards
PJ 7167	— *women*
	Sexual activity
PD 5344	— children ; Parental toleration
PF 5868	— Excessive
PF 1682	— Individualistic perception
	Sexual [arousal...]
PD 8016	— *arousal disorders*
PD 3266	— assault
PE 5925	— attraction ; Genetic
	Sexual [behaviour...]
PE 5376	— behaviour ; Effects AIDS
PE 2687	— behaviour ; Individualistic standards
PE 0567	— bigotry organized religion
PF 2687	Sexual conduct ; Collapse standards
PE 7031	Sexual craving
	Sexual [desire...]
PE 9607	— desire disorders
PE 7031	— desire ; Excessive
PF 5868	— development ; Adverse
PD 2198	— deviation
	Sexual discrimination
PC 2022	— [Sexual discrimination]
PF 1035	— contraceptive methods
PD 9563	— developing countries
PD 1468	— education
	Sexual [disorders...]
PD 8016	— disorders ; Human
PD 8016	— dysfunction
PE 3932	— dysfunction torture victims
PF 1622	Sexual ethics ; Over reaction past
	Sexual exhibitionism
PD 1116	— [Sexual exhibitionism]
PE 1293	— Female
PF 3271	— Male
	Sexual exploitation
PC 3261	— [Sexual exploitation]
PD 3267	— children
PD 3263	— men
PE 6613	— Trafficking children
PD 3262	— women
PF 6406	Sexual fetishism
PE 6408	Sexual frigidity
	Sexual harassment
PD 1116	— [Sexual harassment]
PD 7707	— Environmental
PE 1293	— men
PF 3271	— men
PE 8220	— military
PF 3271	— women
PE 0756	— women public ; Verbal
PE 8466	— working place
PF 5147	Sexual health women ; Neglect
	Sexual [identity...]
PF 1246	— identity ; Conflicting sense
PF 1246	— identity ; Indistinct awareness
PF 8413	— *immaturity*
PF 2687	— immorality
PF 6415	— impotence men
PC 1892	— *inadequacy elderly*
PC 1892	— inadequacy ; Human
PG 1957	— innuendo
	Sexual intercourse
PD 7439	— Adolescent
PE 3274	— animals ; Human
PF 6910	— Fear
PF 4948	— Forced
PF 3260	— Inadequate satisfaction
PE 6522	— minors
PF 5434	— out wedlock
PD 5107	— Pre marital
PD 7297	Sexual irresponsibility ; Male
	Sexual [liberation...]
PE 5376	— liberation AIDS ; Restricted
PF 2687	— licence
PE 3265	— love children
PE 3851	Sexual masochism
	Sexual minorities
PD 1914	— Denial rights
PD 1914	— Discrimination against
PD 1914	— Violation rights
	Sexual [morality...]
PF 3259	— morality ; Double standards
PD 5718	— mutilation
PE 6054	— mutilation ; Male
PE 6055	— mutilations ; Female
	Sexual [obligations...]
PF 4948	— obligations ; Undesired
PE 7031	— obsession
PD 4082	— offences
PD 9394	— offences juveniles
PF 4436	— oneupmanship
	Sexual [partners...]
PE 4890	— partners ; Deception
	Sexual [perversion...] cont'd
PD 2198	— perversion
PD 4064	— preferences ; Racial discrimination
PC 2022	— prejudice
PF 2687	— promiscuity
	Sexual provocation
PD 1116	— [Sexual provocation]
PE 1293	— men
PF 3271	— men
PE 1293	— women
	Sexual purposes
PE 4437	— Abuse tourism
PE 1348	— Misuse spiritual authority
PE 4437	— Travel
	Sexual [relations...]
PE 4890	— relations ; Lying concerning
PE 9122	— relations out wedlock ; Criminalization
PF 2922	— repression
PE 5197	— reproduction disabled
PD 1308	— roles ; Inflexible
	Sexual [sadism...]
PE 6748	— sadism
PD 2091	— scientism
PD 5843	— stereotyping
	Sexual [taboos...]
PD 2198	— taboos ; Infringement
PF 2922	— tendencies ; Repressed
PE 5108	— torture
PF 3260	Sexual unfulfilment
PE 5197	Sexual unfulfilment disabled
PD 3276	Sexual violence
PF 3277	Sexualism ; Trans
PF 1764	Sexuality ; Lost covenantal understanding
PE 4882	Sexuality ; Public displays
PE 7930	Sexuality ; Trivialization
PA 5612	Sexuality *Unchastity*
PA 5499	Sexuality *Unsexiness*
PF 6014	Sexually discriminating job terminology
PC 2022	*Sexually segregated schools*
	Sexually transmitted diseases
PD 0061	— [Sexually transmitted diseases]
PD 8821	— Ignorance concerning
PD 5168	— Intentional spread
PC 6641	Shadow economy
PC 0939	Shadow economy socialist countries ; Second,
PE 6419	Shaker disease ; Hairy
PD 7841	Shaker pigs
PE 2206	Shaking palsy
PF 0445	Shale energy source ; Underutilization oil
PD 3647	Shallow alkaline topsoil
PF 1004	Shallow approaches motivational techniques
PF 3777	Shallow meaning ; Institutionally reinforced sense personal
	Shallow [perception...]
PE 4589	— perception personal meaning ; Socially reinforced
PF 1729	— personal commitment social vision
PE 4589	— personal meaning ; Social reinforcement
PF 2848	Shallow reserve technical skills rural communities
PE 2305	*Shallow river restrictions*
PF 2005	*Shallow theology*
PA 6993	Shallowness
PA 5568	Shallowness *Ignorance*
PA 6247	Shallowness *Inattention*
PA 5942	Shallowness *Unimportance*
PA 7371	Shallowness *Unintelligence*
PF 3302	Sham adoption children
PE 4407	Sham executions
PF 9991	Shame
PF 3364	Shame churches
PF 2660	Shameless nudity
PA 6400	Shamelessness *Affectation*
PA 7363	Shamelessness *Improbity*
PA 5479	Shamelessness *Lamentation*
PA 5612	Shamelessness *Unchastity*
PA 6491	Shamelessness *Vanity*
PA 5644	Shamelessness *Vice*
PA 7280	Shamelessness *Wrongness*
PD 3489	Shanty-towns
PE 8450	Shape-shifting
PA 5490	Shapelessness *Changeableness*
PA 7361	Shapelessness *Disorder*
PA 6900	Shapelessness *Formlessness*
PA 5878	Shapelessness *Nonconformity*
	Shapelessness [Ugliness...]
PA 7240	— *Ugliness*
PA 7309	— *Uncertainty*
PA 7367	— *Unintelligibility*
	Share [Declining...]
PD 8994	— Declining international market
PF 1471	— developing countries world commodity exports ; Declining
PC 2566	— developing countries world exports ; Reduction
PF 8675	Share holders ; Short term
PF 5329	Share knowledge ; Refusal
PE 3827	Share prices underlying asset values ; Disparity
PF 5912	Share shipping ; Imbalance developing countries
PD 7087	Shared inland water resources ; Conflicting claims over
PF 1822	*Shared machinery ; Lack*
PF 7022	*Shared time ; Limited*
PD 7981	*Shares bonds ; Forgery*
PD 7981	*Shares, stocks, bonds ; Risk forgery*
PF 3393	Sharing community skills ; Lack
PD 3154	Sharing technical research ; Restrictions
PE 6722	Sharing ; Trade union opposition profit
PE 1181	Sharking ; Loan
PC 2321	*Sharks*

PE 5697	Shaving ; Mandatory hair		**Shock [animals...] cont'd**		**Shortage developing countries cont'd**
PE 5085	Shear hazard ; Wind	PD 9366	— animals	PD 3068	— Foreign exchange
PE 7826	*Sheared heels*	PE 4247	— *animals ; Electric*	PD 3068	— Hard currency
PD 7799	*Sheath-rot*	PC 2673	Shock ; Culture	PD 5045	— Labour
PE 9438	Sheep ; Atypical pneumonia	PA 6799	Shock *Disease*	PF 3060	— Liquidity
PE 7769	*Sheep ; Black disease*	PA 6030	Shock *Fear*	PC 0592	*Shortage domestic servants*
PG 5561	*Sheep ; Congenital photosensitization*	PE 7912	Shock ; Shell	PC 4815	Shortage drinking water
PE 6268	Sheep disease ; Nairobi	PE 5106	Shock syndrome ; Toxic		**Shortage [Educational...]**
PE 6594	Sheep diseases	PA 7107	Shock *Unpleasantness*	PE 8438	— Educational materials
PD 3978	*Sheep ; Enteric diseases*	PE 6603	Shocks body ; Sudden	PE 1216	— electric energy ; Long term
PE 8861	Sheep goats ; Instability trade	PF 6210	Shocks external factors ; Adverse economic	PE 8878	— Electric machinery apparatus appliances
PD 7420	*Sheep ; Hypomagnesemic tetany*	PD 7723	Shoddy construction work	PC 6329	— Energy
PE 0958	*Sheep ; Intestinal trichostrongylosis*	PD 1435	Shoddy workmanship	PF 0334	— energy resources ; Long term
PE 2254	*Sheep ked*	PE 2693	Shooting ; Bird	PE 8789	— equipment materials needed act against problems
PE 5617	Sheep ; Lameness	PE 6339	Shooting ; Torture	PF 1205	— *exciting events*
PE 2727	Sheep ; Mange	PE 1134	*Shop unions ; Closed*	PF 5073	— experimental non-human primates
PE 9438	*Sheep ; Nonprogressive pneumonia*	PD 4695	Shopfloor militancy		**Shortage [financial...]**
PE 9438	*Sheep nose bot*	PE 1113	Shoplifting	PF 0404	— financial resources action against problems
	Sheep [Post...]	PE 8299	Shopping districts ; Inaccessible	PF 1460	— financial resources United Nations system organization
PD 2731	— *Post dipping lameness*	PE 6153	Shopping facilities ; Impersonality mass market	PD 4769	— firewood
PE 6272	— pox	PE 6144	Shopping facilities ; Maldistribution urban	PE 8277	— fixed vegetable oils fats
PE 9438	— *Progressive pneumonia*	PE 2535	Shopping ; Increase out town	PE 0976	— food live animals ; Long term
PD 7799	— *Prolonged gestation*	PD 6019	Shops ; Closed union	PB 2846	— food supplies
PE 9438	Sheep ; Respiratory diseases	PF 6478	*Shops ; Understocked*	PC 8182	— foreign exchange
PE 2727	Sheep ; Scabies	PE 4315	Shore borrowing domestic monetary policies ; Interference transnational banks' off	PC 8182	— foreign exchange ; Chronic
PD 2732	*Sheep ; Septic pasteurellosis*	PF 1066	Shore broadcasting vessels ; Off	PE 8224	— fresh, chilled frozen meat
PE 2254	*Sheep tick*	PD 1628	Shore territorial waters ; Conflicting claims concerning off	PC 4815	— fresh-water sources
PE 7769	*Sheep ; Type D enterotoxaemia*	PD 1628	Shore territorial waters ; Unilateral claims off	PE 1013	— fruit vegetables preparations thereof ; Long term
PD 9667	*Sheep ; Ulcerative dermatosis*	PE 2388	*Shoroiditis*	PB 4653	— funds
PF 0326	Shell corporations	PB 0750	*Short duration athletic careers*	PF 5419	— funds research
PE 7912	Shell shock	PE 8255	Short duration breastfeeding ; Unnaturally	PE 0511	— furniture ; Long term
PG 4624	Shellfish poisoning ; Paralytic	PD 6520	*Short fire hoses*	PE 0427	Shortage girls
PD 5254	Shelter ; Denial right sufficient	PD 5212	Short growing season	PD 4403	Shortage groundwater resources
PD 2011	Shelter ; *In home animal*	PF 9729	Short-lived cooperative commitment		**Shortage [hero...]**
PB 2150	Shelter ; Inadequate	PD 5177	Short people	PF 2834	— hero images
PD 9736	Shelters ; Inadequately cooled		**Short range**	PE 8327	— hides ; Skins fur skins
PD 5173	Shelters ; Inadequately heated	PE 8675	— business ventures	PD 8778	— Housing
PF 0916	Shields against life's pain ; Escaping reality sophisticated	PF 5660	— development strategy	PA 6997	Shortage *Imperfection*
PC 1867	Shift ; Disorienting urban	PE 8675	— financing image	PA 6652	Shortage *Incompleteness*
PE 1137	Shift life styles mining communities ; Traumatic	PF 5660	— planning long-term development		**Shortage industrial**
PC 2087	Shift ; Ocean current	PF 2706	— project funding	PC 1195	— *diamonds*
PE 5768	Shift work ; Inability adapt	PE 8675	— scheme	PE 0802	— goods
PE 5768	Shift work stress occupational hazard	PF 2093	— *Short-shipments*	PC 1820	— leadership entrepreneurial ability developing countries
PD 7516	Shifting agriculture ; Unstable		**Short sighted**	PF 6500	— skills
PD 7516	Shifting cultivation	PF 1740	— business planning	PE 5464	— water
PE 2305	Shifting ecology coastal communities ; Negative consequences	PE 4477	— decisions intersocial interaction		**Shortage [inedible...]**
PF 6382	Shifting responsibility onto others	PF 5660	— planning	PE 0461	— inedible crude non fuel materials ; Long term
PE 8450	Shifting ; Shape	PE 3604	*Short tailed bat ; Endangered species New Zealand*	PA 5473	— *Insufficiency*
PA 5806	Shiftlessness *Inaction*		**Short term**	PE 0353	— *Iron ore concentrates*
PA 7341	Shiftlessness *Unpreparedness*	PF 8675	— cash purchase		**Shortage [Labour...]**
PD 2731	Shigella	PF 1984	— climatic change	PC 0592	— Labour
PB 1151	*Shigellosis*	PF 0317	— considerations ; Domination government policy making	PC 8160	— Land
PE 7826	*Shins ; Bucked*	PF 8675	— gain	PE 9095	— land cemeteries
PE 7826	*Shins ; Sore*	PE 4331	— improvements agricultural productivity ; Unsustainable	PE 9130	— leather, miscellaneous leather manufactures dressed fur skins
PF 7768	Ship blast ; Cover up	PD 9162	— land tenure	PF 3059	— Liquidity
PD 8982	Ship collisions	PF 4300	— loans ; Uncertainty development programmes due	PE 0976	— *livestock ; Long term*
PE 5875	Ship owner registration ; Inequities	PF 2971	— *physical stamina*		**Shortage [machinery...]**
PE 5863	Shipbuilding industry distribution ; Imbalance	PF 1740	— planning product life cycles	PE 1436	— machinery transport equipment ; Long term
PE 5886	Shipment delays ; Ocean	PF 0317	— policy making	PE 9764	— managerial skills least developed countries
PE 8063	Shipment ; Expensive produce	PF 1740	— priorities construction industry	PC 0592	— Manpower
PD 4858	Shipments ; Illegal international arms	PF 9119	— problems ; Preoccupation	PE 8771	— manpower socialist countries
PE 7064	Shipments ; Infected animal, meat animal product	PF 8675	— professional gain	PE 0997	— manufactured fertilizers ; Long term
PF 2093	*Shipments ; Short*	PF 2174	— profit maximization	PE 0802	— manufactured goods ; Long term
PE 3829	Shipping cartels ; Liner	PF 2174	— revenue maximization	PE 9379	— marriageable men
PF 3499	*Shipping distances ; Long*	PF 8675	— share holders	PE 0427	— marriageable women
	Shipping [fever...]	PC 0293	— variations climate	PE 8606	— Meal flour than wheat meslin
PE 5524	— *fever*	PF 2174	Short termism ; Financial	PE 8980	— Meal flour wheat meslin
PD 0700	— Flag discrimination	PD 0140	*Short working time*	PE 8693	— meat meat preparations
PE 5790	— fleets ; Lack investment capital developing countries'	PB 8238	Shortage	PE 1490	— meat meat preparations ; Long term
PE 3868	Shipping hazards	PF 0559	Shortage adequately trained personnel act against problems	PE 1634	— medical drugs
PF 5912	Shipping ; Imbalance developing countries share		**Shortage animal**	PE 3502	— Medicinal products
	Shipping industry	PE 8514	— feedstuffs excluding unmilled cereals ; Long term	PE 0353	— metalliferous ores metal scrap ; Long term
PE 5897	— Excess capacity	PC 4998	— protein	PE 4920	— military manpower
PE 5825	— Imbalance power	PE 1188	— vegetable oils fats ; Long term	PE 1712	— mineral fuels lubricants ; Long term
PE 5839	— Imbalances dry bulk	PE 8498	— vegetable oils fats waxes	PD 9357	— mineral resources ; Long term
PF 5840	— Ineffective self regulation		**Shortage [beverages...]**	PE 0613	— miscellaneous manufactured articles ; Long term
PE 5791	— Instability maritime	PE 1253	— beverages tobacco ; Long term		**Shortage [natural...]**
PE 5888	— Protectionism	PF 5097	— biological specimens medical study	PE 8045	— natural manufactured gas ; Long term
PE 1620	— transnational corporations ; Domination	PF 0118	— books developing countries	PC 4824	— natural resources ; Long term
PE 5909	Shipping ; Obstacles developing countries ocean	PF 0118	— books textbooks developing countries	PE 0824	— non ferrous metal ores ; Long term
PD 5885	Shipping ; Obstacles international ocean		**Shortage [capital...]**	PE 0353	— *non-ferrous metal scrap*
	Shipping practices	PB 4653	— capital	PE 7530	Shortage organ donors
PE 5849	— bulk trades ; Unfair	PE 1218	— cereals cereal preparations ; Long term		**Shortage [Paper...]**
PD 0312	— Restrictive	PD 4769	— charcoal fuel	PE 1616	— Paper
PD 0312	— Unfair	PE 1261	— chemicals ; Long term	PE 4853	— pesticides
PF 5872	Shipping procedures documentation ; Inefficient	PE 0303	— clothing ; Long term	PE 8626	— petroleum petroleum products ; Long term
PE 5873	Shipping regulations taxes flags convenience ; Evasion	PE 8242	— coal-derived tar crude chemicals	PE 0353	— *Platinum ores*
PE 0430	Shipping risk	PE 1054	— coal ; Long term	PE 7961	— professional, scientific controlling instruments
PE 5898	Shipping services ; Reduction ocean	PE 1197	— coffee, tea, cocoa, spices manufactures thereof ; Long term	PF 1815	— *public cleaning*
PF 5922	Shipping systems ; Overtaxed	PC 1195	— commodities ; Long term	PE 1616	— pulp waste paper ; Long term
PE 5804	Shipping ; Transnational corporation control bulk	PE 1353	— crude fertilizers crude minerals, excluding coal, petroleum precious stones ; Long term	PD 8108	Shortage qualified teachers
PE 6630	Shipping vessels ; Substandard	PC 0219	— cultivable land		**Shortage [resident...]**
PE 8899	Shipping ; Violation right international freedom movement		**Shortage [dairy...]**	PC 0592	— *resident professionals*
PE 5874	Ships built ; Imbalances types	PE 8225	— dairy products eggs ; Long term	PB 6112	— resources
PE 5873	Ships ; Convenience flagging	PC 0592	— Daytime manpower	PE 1218	— *rice ; Long term*
PD 5803	Ships ; Corrosion	PB 2846	— Dependence food		**Shortage [salt...]**
PE 3624	Shipworms		**Shortage developing countries**	PE 1783	— salt water fish, crustacea molluscs preparations thereof ; Long term
PD 8982	Shipwrecks	PD 3137	— Capital	PE 7940	— sanitary plumbing, heating lighting fixtures fittings
PC 8245	Shock	PD 3068	— Foreign currency	PF 0905	— *school equipment*
	Shock [Adversity...]			PA 6041	— *Shortcoming*
PA 6340	— *Adversity*			PE 0353	— *Silver ores*
PE 9509	— Anaphylactic				

Skills

Shortage skilled
PD 0044 — labour
PD 0044 — manpower
PD 0044 — workers
PE 4920 Shortage soldiers
PE 1120 Shortage sugar honey preparations thereof ; Long term
Shortage [tanning...]
PE 8501 — tanning dyeing materials
PE 6500 — technical skills
PE 8436 — textile, yarn fabrics, made-up articles
PE 0353 — *Thorium ores concentrates*
PC 4498 — time
PE 7926 — tobacco tobacco manufactures
PD 8108 — trained teachers
Shortage [Uranium...]
PE 0353 — *Uranium ores concentrates*
PB 6112 — *uranium resources ; Long term*
PD 0384 — urban land
Shortage [water...]
PC 4815 — water food production
PC 1173 — water ; Long term
PE 2903 — wheat ; Long term
PC 0592 — *women instructors*
Shortage wood
PD 4769 — fuel
PE 1372 — lumber cork ; Long term
PE 4769 — rough
PE 7920 Shortage ; Wood shaped simply worked
PE 0976 Shortages ; Agricultural
PE 8080 Shortages ; Distortion international politics fuel
PC 8182 Shortages ; Foreign exchange
PB 8891 Shortages supply goods
PA 6041 Shortcoming
PE 9407 Shortening fallow periods agricultural land
PE 1339 Shorter life expectancy men
PF 0498 Shortfall aid international organizations
PD 5174 Shortfalls ; Crop
PE 4915 Shortfalls ; Export earnings
PD 2968 Shortfalls export earnings developing countries ; Commodity related
PE 7771 Shortsightedness
PE 7826 *Shoulder atrophy*
PE 7826 *Shoulder joint ; Arthritis*
PE 7826 *Shoulder ; Slipped*
PE 0906 Show animals ; Maltreatment marine
PD 3013 *Show trials*
PF 0039 Showers ; Mass extinction due comet
Shrews Endangered species
PD 3479 —
PD 3479 — *elephant*
PD 3479 — *otter*
PE 1570 — *tree*
PF 2989 *Shrinking local market*
PF 6683 Shui ; Bad feng
PE 3537 Shutdowns ; Plant
PE 2346 *Shutoff valving ; Inoperative*
PE 1278 Shyness ; Inhibiting
PE 7570 *Sialadenitis*
PE 7570 *Sialocele*
PD 7617 *Sibling rivalry*
PD 2565 *Sicca ; Keratoconjunctivitis*
PD 6627 Sick building syndrome
PF 2634 *Sick jokes*
PE 3724 Sickle cell anaemia
PE 3724 Sickle cell disease
PA 0294 *Sickliness Human illness*
Sickness [African...]
PE 1805 — African horse
PE 2611 — Air
PD 9366 — *animals ; Motion*
PE 2611 Sickness ; Car
PA 6799 Sickness *Disease*
PD 7841 *Sickness ; Equine grass*
PE 0661 Sickness ; Falling
PA 0294 Sickness *Human illness*
PA 7157 Sickness *Insanity*
PE 2611 Sickness ; Motion
PD 2322 Sickness ; Mountain
PD 7420 *Sickness ; Railroad*
Sickness [Simulator...]
PE 4929 — Simulator
PE 1778 — Sleeping
PF 4004 — Spirit
PD 2732 — *Sweating*
PD 2730 *Sickness ; Three day*
PE 2611 Sickness ; Travel
PE 7788 Sickness ; Vibration
PE 0566 Sickness ; Wasting
Side [effect...]
PD 5750 — effect drug trade drug addiction ; Delinquency
PD 9807 — effects ; Medication
PD 9807 — effects pharmaceutical products
PF 3084 Side income jobs ; Insufficient
PE 4274 Siderosis
PF 6574 Sidewalk repairs ; Uncompleted
PC 4311 *Siege*
PE 8189 Sifakas, avahis indri ; Endangered species
PE 2611 Sights ; Unpleasant
PE 2833 Sign languages deaf ; Multiplicity manual
PE 2324 Sign peace treaty following war ; Failure
PD 4172 Signals ; Intermittent electrical
PF 1373 Significance cultural tradition ; Untransposed
PF 2830 Significance past ; Denial old people
PF 3676 Significance work ; Loss
PF 2093 Significant community images ; Lack
PF 1933 *Significant employment ; Lack*

PF 6098 Significant life work ; Reduced images having
PF 6537 Significant life work ; Reduced images having
PF 1403 Significant work ; Limited opportunities
PF 6537 *Signs ; Incomplete identification*
PF 6147 Signs ; Insufficient neighbourhood
PE 4927 Signs ; Missing public
PF 4983 *Silence ; Compromising*
PF 9493 Silence ; Conspiracies
PE 4319 Silence ; Denial defendant's right
PF 0608 Silence historical situations
PC 1812 Silence ; Official
PB 5249 Silence response suffering
PE 1514 *Silicates pollutants*
PE 1314 Silicosis
PE 0566 *Silicotuberculosis*
PE 1323 Silk ; Clover
PE 1550 Silk trade ; Instability
PE 0579 Silt
PD 3654 Siltation dams
PD 3654 Siltation rivers
PD 3654 Silting water systems
PD 5453 *Silver Collie syndrome*
PE 0353 *Silver ores shortage*
PE 7956 Silver platinum group trade instability
PE 8556 Silver platinum ores trade instability
PE 4346 *Silver scurf potatoes*
PE 3620 *Silverfish*
PE 4224 Simian acquired immunodeficiency syndrome
PD 4747 Simian immuno-deficiency virus (SIV)
PE 0616 *Simian malaria*
PE 0522 Similar restrictions ; Distortion international trade embargoes
PE 9840 *Simony*
PE 1144 Simple assault
PG 5976 *Simple goitre*
PD 3978 *Simple indigestion animals*
PE 6354 Simple phobia
PF 2097 *Simplistic family decision-making structures*
PF 0799 Simplistic technical solutions complex environmental problems
PE 7920 Simply worked shortage ; Wood shaped
PE 4929 Simulator sickness
PF 2457 Simultaneous demands machinery
PF 0641 *Sin*
PF 6327 Sin against Holy Spirit
PF 8298 Sin ; Original
PE 6826 Sin ; Reversion
PF 2031 *Sinecures*
PA 6793 *Sinfulness ; Feelings*
Single [cash...]
PF 9187 — cash crop
PF 5382 — combat
PF 6158 — criminal act ; Multiple convictions
PC 3606 — *crop farming*
PE 8516 — cropping credit ; Limits
PF 0443 — *Single language instruction*
PD 2681 Single mothers
Single [parent...]
PD 2681 — parent families
PD 2681 — parenthood
PD 2001 — party democracies
PE 6187 — people ; Inappropriate accommodations
PD 4392 — people ; Loneliness
PA 2103 — *person households ; Increasing number*
PF 6540 — *product economy*
PD 1415 — *product operation*
PF 5039 — Single resource thinking
Single [sex...]
PC 2022 — *sex schools*
PE 4188 — source finance ; Developing country dependence
PF 1240 — source income
PF 4741 Single technology ; Dependence
PF 3552 Single trade training
PE 1215 Sinistrality
PF 0641 Sinners
PE 7826 *Sinus ; Coronary*
PC 9025 *Sinus infection*
PC 9025 *Sinuses ; Infection*
PC 9025 *Sinuses ; Infection*
PE 9037 Sinuses ; Phlebitis intracranial venous
PE 7591 *Sinusitis*
PE 9854 *Sinusitis*
PD 2732 *Sinusitis ; Infectious*
PE 3643 Siphonaptera insect pests
PE 5308 Sita ; Carcinoma
PD 7258 Site desecration ; Burial
PE 7099 Site development ; Land erosion brought
PE 6005 Site ; Unfinished industrial
Sites [dams...]
PD 0253 — *dams ; Submergence historical*
PD 6128 — Desecration ancient burial
PD 4502 — Destruction archaeological
Sites Endangered
PD 0253 — ancient
PD 0253 — monuments historic
PC 4764 — protected
PF 3499 Sites ; Excessive cost construction
PE 4146 Sites ; Ineffective inspection toxic waste
PE 1256 Sites ; Methane gas emissions landfill
PF 3523 Sites ; Migration youth urban
PF 7519 Sites power plants ; Lack acceptable
PF 3499 Sites ; Restricted building
Sites [Unavailability...]
PF 3499 — Unavailability building
PF 7130 — Unavailable commercial
PF 3499 — Undesignated building

Sites [Unprepared...] cont'd
PF 7130 — Unprepared industrial
PD 6128 Sites ; Violation sacred
PD 6128 Sites ; Vulnerability sacred
PE 5021 Situation corporate capital companies ; Fraud
PF 2699 Situation ; Inadequate solution oriented images global economic
PF 2828 Situational demands vocational decisions ; Unrelated ability application
PC 0300 *Situational disturbances ; Transient*
PF 7580 Situations ; Risky
PF 0608 Situations ; Silence historical
PD 4747 SIV Simian immuno-deficiency virus
PF 3560 *Size ; Excessive family*
PJ 8128 *Size farms ; Uneconomic*
PE 8706 Size impersonality firms ; Growing
PF 0401 Size ; Inability governments regulate family
PF 2423 *Size limits industry*
PD 6120 Size metropolitan regions ; Excessive
PF 8155 Size production unit ; Growth
Size [Reduction...]
PE 4256 — Reduction glacier
PE 1437 — requirements ; Defeating
PC 0475 — Restriction wild animal range
PF 2605 — *Restrictive community*
PD 6585 Size school classes ; Inappropriate
PF 8798 Size social institutions ; Excessive
PF 3560 *Size ; Unsupportable family*
PE 5050 *Skeletal abnormalities swine ; Nutritionally induced*
PE 9774 *Skeletal myopathies swine*
PD 7424 *Skeletal paresis cows following parturient paresis*
Skeletal system
PD 2565 — Diseases musculo
PE 2298 — *disorders*
PE 7808 — *poultry ; Disorders*
PF 6575 *Skeptical farming attitude*
PE 6256 Ski resorts ; Destruction mountain ecosystems
PE 6350 *Skies ; Overcast*
PC 0699 *Skill disabilities*
PE 2956 Skill overemphasis ; Technological
PF 8552 *Skill transference ; Decreased*
PE 8552 Skill ; Underdeveloped technological
PE 1553 Skilled committed personnel international organizations programmes ; Alienation
PD 7536 *Skilled farmers ; Partially*
PE 1753 Skilled jobs ; Lack
Skilled labour
PD 0044 — Decrease
PD 0044 — Shortage
PE 1753 — Unemployed
Skilled manpower
PE 6243 — developing countries ; Scarcity
PE 5170 — rural areas developing countries ; Lack
PD 0044 — Shortage
PD 1291 Skilled people rural urban areas ; Migration
PE 1396 Skilled personnel routine maintenance complex systems ; Wastage highly
PD 2479 Skilled specialists ; Inequitable distribution
PD 8108 Skilled teachers ; Insufficient
PJ 8370 *Skilled tradesmen ; Uncertified*
Skilled workers
PE 5170 — countryside ; Absence
PE 4804 — Insufficient manually
PE 5737 — island developing countries ; Lack
PD 0044 — Shortage
PE 5884 — transport sectors land locked developing countries ; Lack
PB 0750 — Underemployment
PD 0044 Skilled workmen ; Insufficient
PF 3523 *Skilled youth ; Departure*
PF 8552 *Skills advancement ; Blocked*
PE 8768 *Skills cities ; Drain*
PF 8085 Skills curriculum ; Inadequate
Skills developing countries
PE 0046 — *Inadequate administrative*
PE 0046 — *Inadequate cartographic*
PE 7313 — Lack local management
PE 0046 — Lack management
PF 5006 — Limited exchange
PE 5170 — Wastage local
PE 7230 Skills disorder ; Motor
PF 3499 *Skills drain ; Business*
PE 9071 Skills educated elderly people ; Failure employ
PD 8872 *Skills ; Elimination traditional*
PF 5006 Skills flow ; Inadequate inter developing country
PF 2477 *Skills gap ; Intra group*
PD 8546 Skills home bound educated women ; Failure employ
Skills [Ignorance...]
PE 0533 — Ignorance nonverbal communication
PE 8246 — Illiteracy obstacle acquiring
PF 1403 — Imbalanced distribution
PF 2208 — Impractical leadership
PF 2401 — *information ; Incomplete*
PF 2401 — *information ; Uncollected*
Skills Insufficient
PC 6445 — [Skills ; Insufficient]
PC 4867 — *commercial*
PC 4867 — *management*
PE 6500 — technical
PC 1131 Skills job requirements ; Disparity workers
Skills [Lack...]
PF 1528 — Lack management communication
PF 3393 — Lack sharing community
PF 8085 — Lagging training social
PE 6554 — Latent business

–1043–

Skills least

	Skills [least...] cont'd
PE 9764	— least developed countries ; Shortage managerial
	Skills Limited
PF 2477	— *basic*
PE 7344	— means marketing employable
PD 9160	— medical
PF 2477	— range functional
PF 8594	— technical
PD 8123	— verbal
	Skills [Little...]
PD 9113	— Little employable
PE 5170	— local communities ; Restricted availability urban
PF 1373	— *Lost farming*
	Skills Low
PC 4867	— level management
PF 6500	— level technical
PF 2845	— *paying*
PF 8085	Skills ; Marginal grooming
PC 1820	Skills ; Missing entrepreneur
	Skills [Narrow...]
PE 6554	— Narrow range business
PF 2477	— Narrow range practical
PF 1664	— Neglect elders'
	Skills [Obsolete...]
PD 3548	— Obsolete vocational
PE 1753	— opportunities ; Lack
PF 2956	— Overemphasis technological
PF 6538	— Overlooked personal
PF 2208	Skills paucity ; Leadership
PE 9071	Skills ; Premature retirement people valuable
PF 8085	Skills retraining ; Unavailable
	Skills rural communities
PF 1442	— Inadequate management
PE 5170	— Limited availability practical
PF 2848	— Limited reservoir technical
PF 2986	— Outmoded functional
PF 2848	— Shallow reserve technical
	Skills [Shortage...]
PF 6500	— Shortage industrial
PF 6500	— Shortage technical
PD 7719	— society ; Loss technical
	Skills [trainers...]
PC 2163	— *trainers ; Unavailability*
PF 8085	— training ; Limited
PF 8085	— training ; Neglected social
	Skills [Unavailability...]
PF 1790	— *Unavailability construction*
PF 6538	— *Uncatalogued community*
PD 7536	— *Underdeveloped farming*
PC 4867	— *Underused management*
PF 6538	— *Underutilization locally available*
PE 1753	— *Underutilized labour*
PC 4867	— *Undiscovered managerial*
PC 2880	— *Unequal global distribution basic*
PF 6538	— *Unknown local*
PA 5568	— *Unlearned fundamental*
PE 7344	— *Unmarketable available*
PC 6445	— *Unorganized transfer*
PF 2845	— *Unprofitable former*
PF 2845	— *Unprofitable traditional*
PF 2995	— *Unrealized marketable*
PF 6538	— *Unrecognized income*
PF 2995	— *Unrecognized need functional*
PF 3605	— *Untransferred leadership*
PF 1373	— *Untransmitted fishing*
PF 6538	— *Unused local*
PE 8048	— urban areas ; Undeveloped business
	Skills [Vanishing...]
PD 8872	— *Vanishing artistic*
PF 2995	— *Victimized market*
PF 2995	— *Victimizing market*
PE 7752	Skills which reinforce self image being victim circumstances ; Marketing
PD 8347	Skin ; Benign neoplasm
PE 5016	Skin cancers
PE 5684	Skin ; Chapped
	Skin colour
PC 8774	— Discrimination based
PF 1741	— Dissatisfaction
PC 8774	— prejudice
PF 1741	Skin colourants ; Abusive use
PD 9667	Skin ; Congenital anomalies
	Skin [Dark...]
PF 1741	— Dark
PE 4966	— Desiccation human
PD 9667	— *disease cattle ; Lumpy*
PC 8534	— diseases
PD 9667	— diseases animals
PC 8534	— disorders
PE 4966	— Dry
PE 5380	Skin ; Helminths
	Skin [infections...]
PC 8534	— infections
PC 8534	— Injury
PE 5684	— irritants occupational hazards
PC 8534	— irritation
	Skin [Leishmaniasis...]
PE 2281	— *Leishmaniasis*
PE 7154	— lesions due torture
PF 1741	— lighteners ; Abusive use
PE 5016	Skin ; Malignant neoplasm
PF 1741	Skin ; Pale
PC 8534	Skin rash
PC 8534	Skin subcutaneous tissue ; Diseases
PF 1741	Skin ; Untanned
PD 0389	Skins endangered species ; Trade furs

PE 0828	Skins ; Environmental hazards undressed hides, skins fur
	Skins fur skins
PE 0828	— Environmental hazards undressed hides,
PE 1235	— Instability trade undressed hides,
PE 8327	— shortage hides
	Skins Instability trade
PE 8740	— dressed fur
PE 8106	— undressed hides non fur
PE 1235	— undressed hides, skins fur
PE 8327	Skins shortage hides ; Skins fur
PE 9130	Skins ; Shortage leather, miscellaneous leather manufactures dressed fur
PE 7511	*Skull fracture*
PE 7511	*Skull fractures*
PE 8936	Skunks, otters ; Endangered species weasels, badgers,
PD 0124	Skyjacking
PD 3023	Slander
PD 4982	Slander ; Corporate
PD 3023	Slandering ; Risk
PF 5213	Slang
PA 6228	Slant *Deviation*
PA 6790	Slant *Distortion*
PA 6644	Slant *Misrepresentation*
PA 7411	Slant *Uncommunicativeness*
PE 6264	Slashburning
PE 7645	Slashers ; Psychotic
	Slaughter animals
PD 4575	— fur industry
PD 8486	— humans
PD 4575	— pelts
PF 0319	— Ritual
PE 8109	Slaughter diseased animals
PE 8109	Slaughter diseased animals ; Economic loss
PE 9767	Slaughter facilities ; Lack
PF 4617	Slaughter ; God
PE 9767	Slaughter houses ; Inadequate hygiene
PE 0358	Slaughterhouse practices ; Cruel inefficient
PD 2772	*Slaughtering meat ; Inhumane livestock*
PC 0130	Slave trade
PD 3303	Slave trade ; White
PD 3978	*Slavers*
PC 0146	Slavery
PD 3888	Slavery ; Brothel
PC 3300	Slavery ; Chattel
	Slavery [Debt...]
PD 3301	— Debt
PC 0146	— Denial right freedom
PC 0146	— Dependence
	Slavery like practices
PC 0146	— [Slavery-like practices]
PE 3681	— apartheid
PC 0798	— colonialism
PD 3978	*Slavery mouth*
	Sleep [deprivation...]
PE 2741	— deprivation
PE 0451	— difficulties due torture
PE 2197	— disorders
PE 2197	— disturbance
PE 2741	Sleep loss
PE 2197	*Sleep terror disorder*
PE 2197	*Sleep-wake schedule disorder*
PE 2197	Sleep walking
PE 2197	*Sleepiness ; Excessive daytime*
PD 8052	Sleepiness work
PD 8052	Sleeping during work
PE 4415	Sleeping habits ; Noisy
PE 1821	Sleeping pills ; Abuse
PF 6193	Sleeping public ; Unnatural urban environments inhibiting
PE 1778	Sleeping sickness
PE 2197	Sleeplessness
PE 2197	*Sleepwalking disorder*
PD 1146	*Slides ; Rock*
PE 1371	Slime moulds plants
PE 2501	Slime ; Sea
PE 7826	*Slipped shoulder*
PE 6526	*Slippery road surface*
PE 3604	*Slit faced bats ; Endangered species*
PE 9702	*Slobber factor*
PF 1735	*Slogan ; Negative identifying*
PE 5952	Slogans hate
PF 3212	Slogans mottoes ; Empty
PD 6282	Slopes ; Inappropriate cultivation steep
PA 3275	Sloth
PA 6379	Slothfulness *Avoidance*
PA 6099	Slothfulness *Hopelessness*
PA 5806	Slothfulness *Inaction*
PA 7364	Slothfulness *Unfeelingness*
PA 5644	Slothfulness *Vice*
PD 3603	*Sloths ; Endangered species*
PF 2583	*Slow city services*
PF 1737	Slow-down growth output developed-market economies
PE 1117	Slow economical growth socialist countries
PF 9007	Slow emergency care
PE 6478	Slow rate income expansion
PC 1960	Slowing growth food production
PA 7365	Slowness *Boredom*
PA 5806	Slowness *Inaction*
PD 3514	Slowness sensori-motor activities aged
PA 6166	Slowness *Slowness*
PA 7371	Slowness *Unintelligence*
PA 7006	Slowness *Untimeliness*
PD 3489	Slums developing countries ; Urban
PE 7877	Slums ; Inadequate health care urban
PE 1887	Slums industrialized countries ; Urban

PD 7473	Slums ; Malnutrition
PD 3139	Slums ; Urban
PE 3934	Slurry ; Contaminated farm
PF 7187	Smacking children
PF 7047	Small agricultural producer ; Neglect
	Small animals
PE 9702	— *Cheilitis*
PD 9667	— *Cuterebra infestation*
PD 5228	— *Pesticide poisoning*
PE 9854	— *Pneumonia*
PE 9854	— Respiratory diseases
PF 6455	— *Uncontrolled*
PE 1815	Small bank failure
PF 8026	Small blood vessels kidneys, eyes nerves ; Damage
	Small business
PD 5603	— Discouraging conditions
PE 9405	— failures
PF 3008	— methods ; Ineffectiveness traditional
PD 1415	*Small businesses ; Insufficient*
PJ 2079	*Small businesses ; Unprofitable*
	Small communities
PF 2444	— Complex regulations paralyzing
PF 6477	— Diminishing capital investment
PF 2540	— Lack central planning structures
PF 6540	— larger towns ; Transfer business
PF 1658	— Limited availability jobs
PD 1415	— Restrictive pattern business activities
PF 3665	— Strained capital resources
PD 0068	Small developing countries ; Economic non viability
PF 6540	*Small employee pool*
PD 5584	Small enterprise ; Restrictions against
	Small [farm...]
PF 3211	— *farm plots*
PF 2135	— *farm plots*
PE 8516	— farmers ; Limited access credit
PD 1007	— farms developing countries ; Uneconomical
PF 1535	Small gathering places
PF 2182	Small groups ; Narrow political interests
PF 6167	Small households ; Inhibition adult life
	Small [industries...]
PE 7528	— industries ; Occupational hazards workers
PE 0523	— industry products ; Inadequate export promotion developing country
PF 2042	— inner city enterprises ; Lack awareness potential investment
PE 7423	— *intestinal obstruction animals*
PE 3936	— *intestine swine ; Mesenteric torsion*
PE 4733	— *jurisdictions ; Bias jury trials*
PF 3211	*Small land plots*
PF 2989	*Small local markets*
PD 2374	Small nations foreign intervention ; Vulnerability
	Small [plantation...]
PF 6455	— plantation plots
PF 4470	— population base
PF 4470	— population basic services users
PF 2949	Small rural communities ; Limited employment opportunities
PD 2011	Small rural villages ; Unsanitary environment basic health
PD 0928	Small size domestic market developing countries
	Small states
PD 2374	— International insecurity
PD 0441	— territories ; Non viability
PD 2374	— Vulnerability
PD 2374	— world system ; Plight
PE 0958	*Small-strongyle infection horses*
PF 1226	*Small student enrolment*
PF 6539	Small towns developed countries ; Limited availability public services
PD 6125	Small towns ; Vulnerability
PE 3988	Small war
PE 1815	Smaller banks ; High risk policy
PA 7408	Smallness
PA 6839	Smallness *Disrepute*
PA 7152	Smallness *Fewness*
PE 9296	Smallness head ; Abnormal
PA 6659	Smallness *Humility*
PA 5652	Smallness *Inferiority*
PA 5473	Smallness *Insufficiency*
PA 7285	Smallness *Littleness*
PA 7306	Smallness *Narrowmindedness*
PA 7211	Smallness *Selfishness*
PA 7408	Smallness *Smallness*
PA 5942	Smallness *Unimportance*
PE 0097	Smallpox
	Smear campaigns
PD 7171	— government ; Covert
PD 9384	— Political
PD 5534	— Religious
PE 1066	Smell ; Lack sense
PE 6820	Smells ; Unpleasant
PD 4172	Smog ; Electronic
PD 3663	Smog ; Photochemical
PD 3663	Smog ; Urban
PD 3663	Smog ; Volcanic
PD 2267	Smoke
PD 1245	*Smoke ; Oil*
PD 2267	Smoke pollutant
PE 8257	Smoked meat ; Instability trade dried salted
PE 5146	Smokers ; Health hazards children
PD 0713	Smoking
PE 6219	Smoking adolescents
PE 6219	Smoking children
PD 0713	Smoking ; Cigarette
PE 4996	Smoking developing countries
PE 5026	Smoking during pregnancy breast-feeding

Social science

PD 0713	Smoking epidemic
PE 5146	Smoking ; Health hazards passive
PE 7568	Smoking physicians
PD 0713	Smoking ; Tobacco
PE 4995	Smoking women ; Health hazards
PD 0713	*Smoking work places*
PE 3604	*Smoky bats ; Endangered species*
PD 2620	Smuggling
PD 4858	Smuggling ; Arms
PE 9004	Smuggling cultural artefacts
PE 1880	Smuggling ; Drug
PE 3968	Smuggling ; Nuclear
PC 0380	Smuggling protected wildlife
PF 1976	Smuggling sanctions barriers
PD 9765	Smuggling ; Toxic waste
PE 0857	Smut diseases plants
PE 2234	Smuts grasses cereals
PE 2747	Snail vectors animal diseases
PD 3478	*Snails ; Endangered species*
PD 5228	Snakebite
	Snatching [Baby...]
PE 6154	— Baby
PD 4691	— *Bag*
PF 0491	— Body
PE 1427	Sniffing ; Glue
PE 7645	Snipers ; Psychotic
PF 2146	Snobbery
PA 6822	Snobbery *Disrespect*
PE 0779	Snobbery ; Educational
PA 5869	Snobbery *Exclusion*
PF 2146	Snobbery ; Inverted
PA 6491	Snobbery *Vanity*
PA 5821	Snobbery *Vulgarity*
PF 2146	Snobbishness
PE 4415	Snoring
PE 6280	Snotsiekte
PC 0293	Snow
PD 4904	Snow ; Acidic
PE 7838	Snow avalanches
PE 6527	Snow ; Insufficient removal
PE 6527	*Snow removal ; Uneconomic*
PD 5349	*Snowmobile drivers ; Reckless children*
PF 1249	*Snowmobiles roads*
PE 4346	*Snowmold lawns*
PE 4346	*Snowmold turf*
PD 1150	*Snowstorms*
PE 2224	Snuff movies
PF 0995	Soaring cost legal aid
PF 0963	*Sociability ; Hindered winter*
PF 2386	Sociability ; Unfulfilled
	Social [accountability...]
PC 1522	— accountability ; Lack
PE 8518	— accounting business community ; Lack
PF 2216	— agency competition
PC 2130	— alienation
PF 1088	— anarchy ; Fear
PC 3412	— apathy
PC 3412	— apathy ; Dependence
PF 8855	— assistance ; Unquestioning dependency
PD 0061	— attitudes ; Inhibiting
	Social [background...]
PF 2241	— *background ; Disrelated*
PC 3458	— barrier ; Class consciousness
PC 3258	— barriers male pregnancy
PD 1566	— *base ; Undetermined*
	Social behaviour
PC 4726	— Anti
PD 0329	— developing countries ; Increase anti
PE 7370	— university students ; Anti
	Social [benefits...]
PF 1303	— benefits ; Limited access
PF 2885	— bias planning training programmes
PF 4436	— boasting
PB 2496	— breakdown
PE 5246	Social care ; Non adaptive local structure
PF 2493	Social care structures developing countries ; Inflexible
	Social change
PF 6712	— Confusion induced rapid
PC 6989	— Delays implementation
PF 9576	— policy makers ; Static models
PD 2670	— powerlessness
PF 9149	— Rejection proposals
	Social [chaos...]
PB 2496	— chaos
PF 9276	— choice ; Unknowable future patterns
PF 1088	— collapse anarchy ; Fear
PF 1729	— commitment contemporary life styles ; Reduced
PF 1729	— commitment ; Lack individual
PF 1723	— compassion ; Unexpressed
PF 5242	— concepts ; Facile
PF 8076	— confidence ; Undermined
PC 0137	— conflict
PF 9144	— conscience ; Lack
PD 3557	— consequences excessive employment married women ; Adverse
PF 4882	— constraints freedom imagination
PF 8695	— contact ; Lack
	Social context
PF 6617	— developing technology ; Limited
PF 0963	— *Entrapping*
PF 9003	— Negative
	Social [control...]
PD 0144	— control developing countries ; Loss traditional forms
PE 3149	— costs companies ; Evasion
PD 8114	— costs structural adjustment debtor developing countries ; Excessive

	Social [cultural...] cont'd
PD 0784	— cultural life ; Exclusion disabled persons
	Social [data...]
PF 0214	— data ; Lack comparable
PJ 1014	— debt
PB 0001	— defences ; Vulnerability
PF 0214	— demographic statistics ; Inadequate
	Social development
PF 5692	— Constraint time individual
PD 0266	— developing countries ; Disparity
PF 9576	— Inappropriate theoretical models
PC 0242	— Lack
PF 4180	— programmes ; Inadequate
PD 8114	— Regressive
	Social [deviation...]
PC 3452	— deviation
PF 7798	— dialogue ; Erosion
PD 3241	— dictatorship
PD 3517	— disadvantage aged
PD 1544	— disaffection young
	Social discipline
PD 0095	— developing countries ; Inadequate
PF 8078	— Lack
PD 6893	— socialist countries ; Inadequate
	Social discrimination
PC 1864	— [Social discrimination]
PF 2100	— *against divorcees*
PC 1864	— Dependence
PF 5929	— foreign language teaching ; Ethnic
	Social [disguise...]
PF 2505	— disguise ambiguity
PC 3309	— disintegration
PC 1707	— distance
PF 3398	— doctrine ; Inadequacy
	Social [economic...]
PF 4837	— economic contexts development programmes ; Lack understanding
PF 2095	— education ; Outmoded forms
PE 4225	— effects advertising ; Negative economic
PE 5134	— effects introduction new technologies ; Negative
PE 9295	— environment arising current policies ; Government neglect future problems
PC 4588	— *environment ; Poor*
PD 0826	— environmental degradation recreation tourism
PF 1249	— ethics ; Haphazard forms
PF 5250	— events ; Infrequent
PF 2196	— evolutionism
PC 0193	— exclusion
PF 0611	— expenditure ; Decline government
PC 6215	— expenditure ; Excessive growth
	Social forms
PD 2050	— Disorientation young due lack
PF 0945	— enhancing ethical relations ; Isolating
PF 1058	— Immediate gratification based
PF 1324	Social fragmentation
PF 4754	Social functions ; Separation individual's
	Social [gap...]
PC 1707	— gap
PF 1073	— group ; Undue attachment
PC 1523	— *groups ; Nonacceptance*
PC 5537	— *groups ; Restrictive*
PB 0035	— growth ; Unconstrained
PF 1664	— guidance older generation ; Limited
PD 6626	Social hazards ; Underreporting
PF 4947	Social hierarchy
	Social [identities...]
PE 1082	— identities developing countries transnational corporations ; Disruption cultural
PF 6516	— identity ; Ambiguous shape
PF 1934	— *identity ; Lack*
PE 5134	— ill-effects automation
PF 3341	— impediments marriage
PC 0237	— inaccessibility
PF 6194	— inadequacy large buildings
PF 3613	— inadequacy men
PF 8085	— Inadequate
PF 3551	— indicator ; Unreliability profit
PF 8078	— indiscipline
PJ 5588	— ineptitude
PE 1339	— inequalities life expectancy
PB 0514	— inequality
PB 0514	— inequality ; Dependence
PF 5737	— infrastructure islands ; Weak
PC 0797	— injustice
PC 0797	— injustice ; Dependence
	Social insecurity
PC 1867	— [Social insecurity]
PE 4796	— developing countries
PE 1506	— indigenous peoples
	Social institutions
PF 3831	— Closure
PF 3831	— Destruction
PF 8798	— Excessive size
PF 2081	— Faceless
PF 1755	— Fostering dependency
PE 4796	Social insurance developing countries ; Inadequate
	Social integration
PC 1532	— foreigners ; Inadequate
PE 6779	— handicap
PF 1324	— Lack
	Social [interaction...]
PF 6570	— interaction different age groups ; Stifled potential
PD 4287	— intimacy ; Forced
PC 2940	— intimidation
PC 2940	— intimidation ; Dependence
PD 8204	— invisibility

	Social [irrelevance...] cont'd
PF 1870	— irrelevance marriage vows
PE 5796	— irresponsibility transnational corporations
	Social isolation
PF 2386	— [Social isolation]
PC 1707	— [Social isolation]
PE 1277	— computer-based recreation
PE 1277	— computer users
PE 4703	— due torture ; Self imposed
PD 1564	— elderly
PE 8681	— housewives
PF 9023	— mountain valley communities
PF 9023	— neighbouring villages
PD 6810	— torture
PE 8681	— women home
PD 4287	Social kissing ; Obligatory
PC 0663	Social liberty ; Lack
	Social life
PE 8806	— forms ; Lack variety
PF 3454	— *Unattractive*
PD 8113	— Unbalanced
PE 4856	Social losses due disability ; Economic
PF 0910	Social loyalties ; Futility
	Social [maladjustment...]
PE 4258	— maladjustment children migrants
PC 8337	— maladjustment ; Personal
PD 0065	— methodologies global ideology ; Absence
PF 3884	— methods ; Stagnating
PF 3884	— methods ; Uninitiated
PF 2195	— mobility ; Lack
PE 3939	Social needs ; Industrial processes geared reduced
	Social neglect
PB 0883	— [Social neglect]
PB 0883	— Dependence
PD 2077	— war veterans
PC 1522	Social non-accountability
PB 1125	Social norms ; Nonconformity
PF 3454	*Social opportunities ; Unattractive*
	Social order
PF 2136	— Feudal
PF 3826	— Ineffective opposition existing
PF 3826	— Prevalence psychological conditions unfavourable transition post capitalist
PF 3826	Social organization ; Non viable alternative modes
PD 6017	Social outcasts
	Social [paradigms...]
PD 0894	— paradigms ; Lack meaningful personal
PJ 4169	— parasites
PF 0193	— pattern ; Immobility
PF 0193	— patterns ; Inhibiting
PF 0193	— patterns ; Unvaried
PF 1721	— personality disorders ; Anti
PE 6374	— phobia
PF 4837	— planning developing countries ; Imbalance economic
PF 8185	— planning ; Lack
PF 8282	— policies ; Restrictive
PE 1957	— policies transnational corporations ; Disruption domestic
PC 5537	— practices ; Restrictive
	Social problems
PF 1393	— Delay religions acknowledging
PF 8753	— Failure interrelate wisdom different cultures response
PF 3699	— Invention new
PF 1572	— Obsolete programmes against
PF 2823	— Restriction funding research
PF 9828	— Suppression information concerning
	Social processes
PC 0588	— Abstention
PC 0588	— Disinterest
PF 0749	— Failure individuals participate
PC 5387	— Forced participation
	Social [programmes...]
PF 0170	— programmes ; Fluctuations government
PC 2810	— *programmes ; Lack*
PF 1545	— progress ; Lack
PD 9155	— protection farmers ; Inadequate
PF 2721	— *prowess ; Absence*
	Social [reality...]
PF 1728	— reality ; Lack correspondence basic value images
PF 0677	— reform ; Inadequate
PF 1663	— reinforcement delimited self-worth
PE 4589	— reinforcement shallow personal meaning
PF 2803	— relations inhibiting vision future ; Static
PD 9219	— *relationships ; Indifference*
PF 2701	— response ; Paralysis
	Social responsibility
PF 3553	— Absence images
PF 1064	— Deluding familial image
PE 4225	— Insufficient awareness advertisers
PC 0300	*Social responsiveness ; Poorly developed*
PC 3236	Social revolution
	Social rights
PC 0663	— Denial
PE 7765	— indigenous peoples ; Denial
PE 6375	— refugees ; Denial economic
PD 3338	*Social role-playing*
	Social roles
PF 1825	— aged ; Unmeaningful
PF 1651	— Disrelated images
PE 8806	— Lack diversity
PF 1651	— Static unrelated
	Social science
PF 8032	— Irresponsible
PF 7634	— lying
PF 7634	— research ; Deceptive

Social science

Social science cont'd
- PF 0145 — research ; Intrusive

Social sciences
- PF 7634 — Deception
- PD 6626 — Malpractice
- PD 6626 — Unethical practices

Social scientists
- PD 6626 — Corruption
- PD 6626 — Irresponsible
- PD 6626 — Negligence
- PF 0611 Social sector ; Decline public spending

Social security
- PD 9155 — agricultural sector ; Insufficient
- PD 3120 — capitalist systems ; Denial right
- PD 7251 — Denial right
- PD 1229 — Dependence
- PE 8670 — Dependence family
- PD 1543 — developing countries ; Loss traditional forms
- PF 7911 — Erosion
- PD 9155 — farmers ; Lack
- PD 8859 — fraud
- PF 7911 — Increasing cost
- PE 1506 — indigenous peoples ; Denial right
- PF 7911 — Inflated cost
- PE 7611 — migrants ; Inadequate
- PE 8746 — nationals who have lived abroad ; Denial
- PD 9155 — peasant farmers ; Lack
- PF 0852 — Unequal coverage
- PF 0852 — Unequal distribution
- PE 4796 — welfare services developing countries ; Inadequate
- PE 4299 Social segregation AIDS virus victims
- PC 3383 *Social segregation eunuchs*

Social service
- PE 1862 — agencies ; Cultural diversity ignored
- PD 8017 — bureaucracy ; Unresponsive
- PD 3190 — ideologies ; Conflicting
- PF 2992 — inconsistencies
- PE 1488 — quality negated oligarchic control decision-making
- PE 4892 — workers ; Inadequate utilization volunteer
- PF 5120 Social services ; Delay access

Social services Discrimination
- PC 3433 — [Social services ; Discrimination]
- PD 3336 — against men
- PD 3691 — against women

Social services [Inequality...]
- PC 3433 — Inequality
- PF 3083 — Inflexible attitudes toward community
- PF 3526 — *Limited*
- PF 2974 — Non standardized
- PD 3440 — Segregation
- PC 3437 — Unequal distribution
- PE 3296 — *Unresponsive*
- PF 1853 — urban areas ; Uncoordinated

Social [skills...]
- PF 8085 — skills ; Lagging training
- PF 8085 — skills training ; Neglected
- PB 5577 — status
- PD 0884 — stigma
- PE 7627 — stigma breast-feeding
- PE 4299 — stigmatization AIDS-infected persons
- PB 5577 — stratification
- PB 1997 — stratification obstacle development
- PB 1997 — structure ; Inflexible

Social structures
- PD 1566 — depressed areas ; Fragmentation
- PD 0822 — developing countries ; Inadequate development new
- PE 3977 — environmental protection ; Fragmented
- PF 6178 — Ideal tension free
- PF 3826 — Low credibility alternative
- PF 1781 — *Parallel*

Social [studies...]
- PF 3824 — studies ; Terminological crisis
- PD 4633 — subjugation women
- PE 6671 — symbols dominated economy

Social systems
- PB 5417 — Destabilization
- PF 1443 — Uncontrollability world
- PB 8031 — Undemocratic
- PB 2853 — Vulnerability

Social [techniques...]
- PF 3608 — techniques needs they address ; Gap function
- PF 3884 — technologies ; Inadequate
- PF 1765 — traditions rural areas ; Rigidly entrenched
- PD 0917 — transformation ; Bias media against development
- PC 1522 Social unaccountability

Social underdevelopment
- PC 0242 — [Social underdevelopment]
- PC 0242 — Dependence
- PB 0539 — Dependence economic
- PB 0539 — Economic

Social values
- PF 1118 — Irrelevant
- PF 5242 — Misappropriation
- PF 5242 — Trivialization
- PF 1729 Social vision ; Shallow personal commitment

Social welfare
- PD 8859 — benefits ; Unethical use
- PD 8859 — corruption
- PD 1229 — Debilitating effects
- PD 1229 — Dependence
- PF 0019 — indicators ; Inadequate
- PC 0834 — services ; Inadequate
- PE 1506 — services indigenous peoples ; Denial right
- PD 3518 Social withdrawal aged
- PC 0115 Socialism
- PD 2993 Socialism ; Bureaucratization

- PF 2636 Socialism ; National
- PF 4778 Socialism ; Pseudo socialism state
- PF 4778 Socialism state socialism ; Pseudo
- PF 9485 Socialism ; Subversion
- PE 3594 Socialism ; Workers alienation

Socialist [capitalist...]
- PE 8093 — capitalist countries ; Brain drain
- PE 1659 — citizens working abroad ; Restrictions
- PD 4862 — colonialism

Socialist countries
- PC 0939 — Black market economies
- PD 2413 — Currency black market
- PF 7610 — Declining productivity
- PD 2872 — Denial right work women
- PD 2872 — Discrimination against working women
- PC 0940 — Disguised unemployment
- PC 0939 — Double employment
- PD 3998 — *Dumping obsolete goods ex*
- PE 5975 — Embourgeoisement
- PD 9197 — Environmental pollution
- PD 1104 — External dependence vulnerability
- PC 0939 — Illegal private profit
- PF 5678 — Imbalance trade cultural products capitalist
- PF 4884 — Inadequate economic integration
- PD 6893 — Inadequate social discipline
- PE 7908 — Inefficient labour use
- PE 4230 — international data systems ; Underparticipation
- PD 0515 — Lack consumer goods
- PE 8002 — Lack economic adaptation indigenous society non
- PF 6013 — Nepotism
- PF 3689 — Obstacles economic reform
- PF 4886 — Obstacles legal relations
- PD 2048 — private enterprise ; Dependence
- PF 0648 — Reallocation development capital developing countries
- PF 0960 — Repressed inflation
- PC 0939 — Second, shadow economy
- PC 0939 — Second undeclared employment
- PE 8771 — Shortage manpower
- PE 1117 — Slow economical growth
- PD 2048 — Suppression private enterprise
- PC 0939 — Undeclared employment
- PD 1882 — Undeclared strikes
- PF 0888 — Unreliability statistics
- PD 2502 — West ; Excessive debt
- PE 2954 Socialist developed market economies ; Weakness trade
- PE 2953 Socialist developing economies ; Weakness trade
- PE 5538 Socialist responsibilities ; Criminal negligence performing
- PD 4952 Socialist states ; War
- PA 6373 Socialization∗complex
- PF 2187 Socialization ; Retarded
- PF 4133 Socialization towards opponents ; Negative
- PF 3560 Socializing ; *Family oriented*
- PF 6552 *Socializing systems ; Absent*
- PF 8780 Socializing unrelated men women
- PF 6345 Socially disruptive effects video games
- PD 1507 Socially handicapped refugees

Socially [inadequate...]
- PE 1925 — inadequate housing estates
- PD 8638 — inappropriate housing
- PD 4572 — ineffective family units
- PE 4360 — irresponsible programmes transnational banks developing countries
- PC 0193 — isolated groups
- PF 1696 — *isolated women*
- PF 3341 *Socially ; Marrying down*
- PF 5430 Socially ; Marrying up

Socially [reinforced...]
- PE 4589 — reinforced shallow perception personal meaning
- PC 0193 — rejected people
- PE 5795 — responsible ; Story that self sufficiency is
- PF 6195 Socially sterile rental accommodation

Socially [unacceptable...]
- PD 7231 — unacceptable activity research ; Misrepresentation
- PC 8019 — *unacceptable initiatives ; Misuse societally endorsed professions conceal*
- PD 2675 — unintegrated expatriates
- PC 0381 — unsustainable development
- PD 6760 — Socially vulnerable groups ; Medical experimentation
- PF 5242 Societal concepts ; False concreteness
- PB 7125 Societal control ; Conspiracies
- PF 2340 Societal engagement ; Collapse
- PE 8318 Societal impact education ; Delay
- PF 8870 Societal impact innovation ; Delay
- PF 7134 Societal knowledge resistance
- PF 7050 Societal learners ; Mis communications

Societal learning
- PF 7003 — Collective attention span limit
- PF 7122 — Disconnectedness
- PD 7086 — Inadequate use visual imagery
- PF 7038 — Limit collective comprehension span
- PF 7074 — Limits
- PF 7014 — Obstacles conceptual integration
- PF 7098 — Quantitative limit
- PF 7110 — Superficiality collective comprehension
- PF 0704 Societal memory ; Erosion
- PD 1080 Societal needs ; Inadequate response
- PF 7051 Societal over-commitment learning
- PF 1651 Societal roles ; Unclear
- PF 7027 Societal selective memory
- PF 2212 Societal service systems ; Incomplete understanding new
- PC 8019 *Societally endorsed professions conceal socially unacceptable initiatives ; Misuse*

- PF 2666 Societies ; Head hunting tribal

Societies [Inadaptation...]
- PE 5023 — Inadaptation technology man industrialized
- PJ 8295 — Inadequate interface fundamentally different
- PE 2862 — Institutional obsolescence modern industrialized
- PF 0245 — Societies ; Lack mutual credit
- PF 2928 — Societies ; Primitive secret

Societies [Secret...]
- PF 2508 — Secret
- PF 5466 — Self indulgent
- PE 8416 — Separate unequal development different
- PE 8139 — Societies ; Traditional values subordination women developing
- PF 3826 — Societies ; Unconvincing alternatives existing
- PF 9608 — Society ; Being burden
- PF 6433 — Society ; Blocked service

Society [Collapse...]
- PF 3750 — Collapse corporate engagement
- PF 0955 — Collapse meaning participating
- PE 4203 — Collapsed relationships work
- PF 0815 — Commercialization
- PD 3169 — Communist closed
- PC 2405 — Society ; Deviant
- PF 3358 — Society ; Discriminatory religious influence
- PC 0027 — Society ; Elders' predomination
- PF 0304 — Society ; Excessive government intervention
- PF 8753 — Society ; Failure focus available insight response problems

Society Inadequate
- PF 3402 — integration ideology
- PF 3403 — integration religions
- PF 9478 — models ideal

Society [Insensitivity...]
- PF 9119 — Insensitivity non immediate hazards
- PB 1151 — *Instability complex*
- PF 4001 — Institutionalized members
- PF 1664 — Isolation wisdom elders

Society [Lack...]
- PF 1324 — Lack integration
- PD 2972 — Lack job opportunities some sectors
- PF 2627 — leaders ; Parochial images
- PD 7719 — Loss technical skills
- PF 7456 — Lost family role
- PF 3368 — Society ; Multidenominational
- PF 2003 — Society ; Multiplicity problems facing
- PF 4000 — Society ; Non productive members
- PE 8002 — Society non socialist countries ; Lack economic adaptation indigenous
- PF 1545 — Society ; Overemphasis economic progress

Society [Permissive...]
- PF 1252 — Permissive
- PD 7463 — Perverse impact technological development capitalist
- PF 2448 — Pluralistic
- PF 3826 — policies ; Unacceptability conserver
- PF 3358 — Society ; Religious influence
- PF 6573 — Society's resources ; Limited access
- PB 2853 — Society ; Structural tensions

Society [tensions...]
- PF 2448 — tensions ; Plural
- PD 8942 — Throwaway
- PF 5937 — truth ; Vulnerability
- PE 5508 — Society ; Underworld
- PE 6149 — Society widows ; Burden
- PF 6570 — *Society ; Youth oriented*
- PC 4588 — Socio-cultural environment degradation
- PF 1046 — Socio cultural role western women ; Elimination
- PF 1042 — Socio dramas ; Failure motivating

Socio economic
- PF 1447 — burden militarization
- PF 7428 — class ; Low
- PF 8855 — dependency ; Over acceptance

Socio economic development
- PF 9576 — Competing models
- PF 9576 — Inadequate models
- PD 0055 — National sovereignty obstacle peaceful

Socio economic [diversity...]
- PF 4405 — diversity ; Erosion
- PC 5879 — growth ; Increasing development lag against
- PB 9015 — growth ; Unconstrained
- PC 1059 — infrastructure ; Weakness
- PA 0857 — insecurity
- PF 2969 — interdependencies ; Unrecognized
- PF 0866 — order ; Antiquated world
- PF 4891 — order ; Excessive
- PB 0388 — poverty

Socio [Economic...]
- PC 6759 — Economic stress
- PF 1245 — economic systems globalization ; Vulnerability
- PF 4470 — economically inactive rural population
- PE 6965 — Socio-psychosis
- PF 1073 — Socio-sentiment
- PF 2924 — Sociological espionage
- PF 2440 — Sociological ignorance citizen participation
- PJ 5254 — Sociologism ; Vulgar
- PF 1721 — Sociopaths ; Psychopaths
- PE 3273 — Sodomy
- PJ 1382 — Soft commissions
- PE 3035 — Soft drinks ; Carcinogenic
- PE 8971 — Soft energy paths
- PE 9774 — *Soft exudative pork ; Pale*
- PE 4706 — Soft rot plants ; Bacterial
- PA 5988 — Softness
- PA 6030 — Softness *Fear*
- PA 6876 — Softness *Impotence*
- PA 7392 — Softness *Unbelief*

PA 7371 Softness *Unintelligence*
PE 0199 Softness ; Water
PA 5558 Softness *Weakness*
PE 6625 Software copies ; Illegal
PD 7450 Software design ; Biased computer
PF 6067 Software developing countries ; Lack specifically designed
PE 4428 Software errors ; Hazardous computer
PE 6625 Software piracy
PD 3102 Software sabotage
PE 4428 Software ; Unreliability computer
Soil [acidification...]
PD 3658 — acidification
PD 1416 — aeration ; Lack
PD 3647 — Alkaline
PE 2739 Soil-borne diseases animals
Soil [compaction...]
PD 1416 — compaction
PD 6482 — conservation developing countries ; Inadequate
PD 0949 — conservation ; Lack
Soil [deficiency...]
PD 0077 — deficiency
PD 1052 — degradation
PD 9227 — Demineralization
PD 0949 — depletion
PD 1052 — deterioration
Soil erosion
PD 0949 — [Soil erosion]
PD 0949 — Top
PD 2290 — water
PE 3656 — wind
PD 0077 Soil exhaustion
PC 0052 Soil exploitation ; Over intensive
Soil [fauna...]
PD 3574 — fauna micro organisms ; Pesticide destruction
PD 0077 — fertility ; Deterioration
PE 8782 — fertility downstream due impoundment ; Reduction
PD 0077 Soil infertility
PC 2696 *Soil ; Laterization*
Soil [management...]
PC 0052 — management ; Poor
PD 3668 — Metal contamination
PC 0052 — mismanagement
PD 6482 — mismanagement developing countries
PD 0077 Soil nutrients ; Depleted
PD 9227 Soil nutrients ; Depletion
Soil [pollution...]
PC 0058 — pollution
PD 1052 — productivity ; Decline
PD 0949 — Progressive loss top
PE 3383 Soil ; Radioactive contamination
PD 1052 Soil resources ; Deterioration
Soil [salinization...]
PE 1727 — salinization
PD 1110 — sciences ; Unethical practice
PD 1052 — structure ; Destruction
PC 0058 Soil toxic substances ; Contamination
PD 3699 Soil-transmitted diseases
PC 0052 Soil ; Unfertilized
PD 3658 Soils ; Acidic
PE 7698 Soils ; Calcareous
PE 5036 Soils ; Expansive
PD 0077 *Soils ; Rock infested*
Soils [Saline...]
PE 1727 — Saline
PE 3911 — Sandy
PD 0077 — Sterile
PD 0077 — *Stone filled*
PE 1936 Soils ; Trace element deficiencies
PD 1110 Soils ; Underreporting hazards
PC 0052 Soils ; Unimproved farm
PE 8747 Sojourn ; Restrictions foreigners leaving country
PD 9667 *Solar dermatitis ; Nasal*
PF 0370 Solar energy ; Underutilization
PE 9528 Solar flares
PE 9528 Solar flares ; Radio interference associated
Solar [radiant...]
PC 6114 — radiant energy ; Long term variations
PE 9528 — radiant energy ; Medium term cyclic variations
PE 3883 — radiation ; Environmental hazards
PF 5069 — radiation ; Increased reflection
PD 8825 Soldiers ; Abandoned children foreign
PE 5986 Soldiers ; Child
PD 8825 Soldiers ; Children engendered occupying
PD 7602 Soldiers ; Dehumanization
PD 4089 Soldiers ; Denial rights
PE 4920 Soldiers ; Shortage
PE 8220 Soldiers ; Women
PE 7826 *Sole ; Bruised*
PE 4161 *Sole ; Ulceration*
PF 3551 Solely profit ; Accountability based
PA 6731 Solemnity
PD 3479 *Solenodons ; Endangered species*
PD 7676 Solicitation ; Criminal
PF 6557 *Solicitation ; Ineffective business*
PD 0790 Soliciting obstruction proceedings
PE 2829 Solid wastes ; Inadequate facilities storing, transporting processing
PD 0177 Solid wastes pollutants
PE 4179 Solidarity amongst workers ; Inadequate sense community
PE 0835 Solidarity developing countries ; Decline communal spirit village
PD 0110 Solidarity developing countries ; Lack sense community
PF 8704 Solidarity poor ; Lack
PF 8704 Solidarity world level ; Lack sense community
PB 0731 *Solids liquids ; Poisoning*
PF 5657 Solipsism
PE 4056 Solitary confinement
PF 2699 Solution oriented images global economic situation ; Inadequate
PF 0799 Solutions complex environmental problems ; Simplistic technical
PD 2888 Solutions ecological issues ; Lack faith
PJ 4512 Solutions national economic goals ; Contradictory
Solutions problems
PF 4473 — Excessive reliance fashionable
PF 9409 — Fixation partial
PF 1572 — Inadequate research proposed
PF 9409 — Minimally cooperative
PE 4059 Solutions resource development research ; Over emphasis immediate
PE 8357 Solve intergroup conflicts ; Lack participation attempting
PF 8191 Solve problems ; Inadequate application available knowledge
PF 0817 Solve world problems ; Inadequate global cooperation
PE 1349 Solvent methylated spirits drinking
PF 1427 Solvents anaesthetic drugs ; Inhaling
PE 5708 Solvents occupational hazard
PF 6580 Solving ; Piecemeal approach problem
PE 0435 *Somatic paranoia*
PE 5294 Somatic psychosomatic effects torture
PD 0270 *Somatization disorder*
PD 0270 *Somatoform pain disorder*
PE 5296 Some developing countries drug trade ; Economic dependence
PF 2344 Some individuals national policy making ; Disproportionate influence
PD 2972 Some sectors society ; Lack job opportunities
PF 7807 Songs ; Obscene
PE 2435 Sonic bangs ; Risk
PE 2435 Sonic boom generated supersonic aircraft
PC 0268 Sonorous pollution
PE 1953 Soot
PE 6350 Soot ; Sunlight inhibition nuclear warfare
PE 0916 Sooty mould plants
PC 4844 *Sophisticated medical equipment ; Restricted access*
PF 0916 Sophisticated shields against life's pain ; Escaping reality
PD 6571 Sophisticated technology development ; Dependence
PF 8008 Sophistry
PA 6393 Sophistry *Miseducation*
PA 5502 Sophistry *Reason*
PA 7411 Sophistry *Uncommunicativeness*
PA 7232 Sophistry *Unskillfulness*
PF 2099 Sorcery
PA 6379 Sordidness *Avoidance*
PA 5454 Sordidness *Badness*
PA 7193 Sordidness *Cheapness*
PA 7361 Sordidness *Disorder*
PA 6839 Sordidness *Disrepute*
PA 5459 Sordidness *Uncleanness*
PE 7826 *Sore knee*
PD 9667 *Sore mouth*
PE 7826 *Sore shins*
PE 4651 Sore throat
PE 5380 *Sorehead*
PA 5446 Soreness *Enmity*
PA 7253 Soreness *Envy*
PA 5451 Soreness *Insensibility*
PE 4651 Soreness throat
PE 2455 *Sores*
PE 8510 Sores infected persons ; Discharge
PE 5380 *Sores ; Jack*
PD 9667 *Sores ; Saddle*
PE 5380 *Sores ; Summer*
PE 3590 Sorghum ; Pests diseases grain
PD 5228 *Sorghum poisoning animals*
PD 2184 Sororal polygyny
PA 5479 Sorrow *Lamentation*
PA 6731 Sorrow *Solemnity*
PA 7107 Sorrow *Unpleasantness*
PF 4213 Soul murder
PJ 5193 *Soulless institutions*
PA 7364 Soullessness *Unfeelingness*
PF 5190 Sound ; Pornography
PE 4497 Sour crop
PE 4188 Source finance ; Developing country dependence single
PF 1240 Source income ; Single
PD 4271 Source organ transplants ; Traffic children
Source Underutilization
PF 0445 — oil shale energy
PE 8194 — peat energy
PE 8370 — tar sands energy
PC 7517 Sources ; Aggression against natural energy
PE 0403 Sources ; Aggression against nuclear power
PE 7195 Sources development financing ; Excessive reliance developing countries external
PD 1204 Sources ; Distant water
PC 4412 Sources energy ; Competition
PD 6923 Sources ; Environmental degradation fresh water
PF 3084 Sources income ; Dependency unpredictable
PF 1345 Sources income expansion ; Underdeveloped
PF 3630 Sources rural funding ; Uncertain
PC 4815 Sources ; Shortage fresh water
PF 0393 Sources ; Underdeveloped rural energy
PE 9325 Sources ; Untapped government
PD 0365 Sources ; Vulnerability nuclear power
PF 0510 Sources ; Weaknesses national data
PA 6011 Sourness *Discontentment*
PA 5446 Sourness *Enmity*
PA 7253 Sourness *Envy*
PE 0653 South American trypanosomiasis
PC 8694 South ; Failure restructure economic relations North
PC 8694 South gap ; North
PC 8694 South ; Inequality North
PC 0933 South inter regional trade ; Weakness South
PC 8694 South negotiations ; Limited progress North
PC 0933 South South inter regional trade ; Weakness
PE 8332 South spheres influence UN related agencies ; North
PD 4814 Sovereign nations ; Denial rights
PD 0055 Sovereign states ; Domination world territorially organized
PF 0478 Sovereign states ; Unbankruptability
PE 7906 Sovereignty ; Denial right national
PE 5015 Sovereignty ; Erosion
PD 3496 Sovereignty foreign military presence ; Erosion national
PE 5015 Sovereignty ; Infringement national
PE 7906 Sovereignty ; Loss national
PD 0055 Sovereignty obstacle peaceful socio economic development ; National
PD 3109 Sovereignty over natural resources ; Denial
PE 0261 Sovereignty trans border broadcasting ; Violation
PE 1539 Sovereignty transnational enterprises ; Erosion national
PD 7799 *Sow ; Lactation failure*
PE 0961 *Soybean trade ; Instability*
PF 6152 Space ; Automobile parking
PD 0530 Space benefits limited number developed countries ; Restriction outer
PF 1695 Space boulders
Space care
PF 2346 — frustration
PF 2346 — Overlooked
PF 2346 — Undemonstrable
PE 6660 Space cities ; Excessive office
PA 6713 Space*complex
PF 1815 Space ; Defaced community
PF 7130 *Space ; Expensive meeting*
PF 2261 *Space ; Externally controlled*
PF 1441 Space ; Gamma ray disturbance
PF 2822 Space ; Haphazard organization community
PF 7130 *Space homes ; Insufficient storage*
Space [Inadequate...]
PF 2346 — Inadequate care community
PF 7130 — *Inadequate industrial*
PF 1815 — *Insufficient business*
PE 1535 — Insufficient meeting
PF 1790 — *insufficient public*
PF 4444 — Invaders outer
PE 5689 — island developing states ; Neglected marine
PD 0089 Space junk
Space [Lack...]
PD 4004 — *Lack clinic*
PC 8160 — Lack land
PF 1790 — *Limited public*
PF 2605 — *Limited training*
Space [Marginal...]
PF 7130 — *Marginal living*
PE 8965 — Military obstacles peaceful development
PD 0087 — Military targets
Space [Neglect...]
PF 2346 — Neglect community
PF 6578 — Neglected public
PE 5080 — Nuclear accidents
Space [objects...]
PE 0250 — objects ; Damage caused
PF 7130 — Obstacles availability community
PF 1681 — *Overwhelming deteriorating*
PD 0089 Space pollution
PD 0089 Space ; Pollution orbital
PF 2605 *Space ; Restricted sports*
Space [Undefined...]
PF 7130 — *Undefined cultural*
PF 7130 — *Underutilized available*
PF 7130 — *Undesignated parking*
PF 1790 — *Undesignated public*
PF 1546 — Undeveloped community
PF 7130 — *Undeveloped recreation*
PF 1546 — Unfocused design community
PF 6521 — Unidentified community
PF 2519 — Unplanned use community
PF 6455 — Unproductive utilization plantation
PF 7130 — *Unused living*
PF 2813 — *upgrading ; Unintentional*
PE 6163 — urban environments ; Wastage open
PE 5458 — usage ; Ineffective
PE 5458 — usage ; Inefficient
Space [warfare...]
PD 6439 — warfare
PD 0087 — weapons arms race
PF 4450 — weapons ; Clandestine
PC 6716 *Spaces ; Confined*
PF 6385 Spaces ; Desecration holy
PF 2575 *Spaces ; Deserted public*
PF 2346 Spaces ; Inattentive care public
PF 7130 *Spaces ; Unused recreation*
PF 1815 Spaces urban communities ; Poor condition open
PF 5968 Spacing families ; Insufficient birth
PE 1708 Spacing suburban housing ; Over
PF 7003 Span limit societal learning ; Collective attention
PF 2384 Span ; Limited individual attention
PF 4644 Span poor ; Narrow time
PF 7038 Span societal learning ; Limit collective comprehension
PE 3511 *Sparganosis*
PA 5473 Sparsity *Insufficiency*
PD 3978 *Spasm animals ; Oesophageal*

Spasmodicness

PA 5838 Spasmodicness *Agitation*
PA 5828 Spasmodicness *Discontinuity*
PA 7361 Spasmodicness *Disorder*
PA 5451 Spasmodicness *Insensibility*
PA 7156 Spasmodicness *Moderation*
PE 0965 *Spastic diseases animals*
PE 0763 Spastic infantile paralysis ; Cerebral
PE 0763 *Spastic infantile paraplegia*
PE 4161 *Spastic paresis*
PE 4161 *Spastic syndrome*
PA 5838 Spasticity *Agitation*
PA 6799 Spasticity *Disease*
PA 7156 Spasticity *Moderation*
PE 0763 Spastics
PD 8179 *Spatial ability ; Impairment visual*
PD 0549 Spatial facilities youth recreation ; Inadequate
PD 6130 Spatial imbalance human settlements
PC 0031 *Spatial segregation*
PE 7826 *Spavin ; Bog*
PE 7826 *Spavin ; Bone*
PJ 4592 Spaying
PF 8565 Speaking foreigners
PE 3797 Speaking opposition out personal responses
PE 3604 *Spear nosed bats ; Endangered species*
PF 0713 *Special care vehicles ; Insufficient*
PE 0088 Special courts ; Injustice
PE 1355 Special groups ; Denial right union activity
PC 6307 Special influence ; Trading
PD 8388 *Special training ; Insufficient*
PF 3498 *Special training ; Uncoordinated*
PE 8303 Specialist doctors developing countries ; Lack
PF 6012 Specialists ; Conflicting viewpoints
PD 2479 Specialists ; Inequitable distribution skilled
PC 0432 Specialization education ; Excessive
PF 0256 Specialization ; Excessive
PF 5709 Specialization medical care ; Over
PF 0256 Specialization ; Over
PF 7532 Specialization sciences ; Over
Specialized agencies United
PE 0777 — Nations ; Inadequate relationship international nongovernmental organizations
PE 0106 — Nations ; Inadequate relationship transnational corporations
PE 2486 — Nations ; Jurisdictional conflict antagonism
PE 8799 — Nations ; Jurisdictional conflict antagonism
PD 6571 *Specialized farming ; Limited*
PF 1748 Specialized jargon
PF 1748 Specialized jargon ; Incomprehensibility
Specialized [services...]
PF 2709 — services relation world's need ; Decisional paralysis
PF 5712 — study global ecosystem ; Over
PF 3588 — supervisory personnel ; Over
PF 1653 *Specialized technology ; Lack*
PF 2875 *Specialized transportation ; Lack*
PF 4548 Specially chosen ; Peoples perceiving themselves
Species [acinonyx...]
PD 1763 — *acinonyx ; Endangered*
PE 8684 — African cane rat ; Endangered
PE 8253 — African mole rat ; Endangered
PE 8738 — agoutis acouchis ; Endangered
PD 3156 — amphibia ; Endangered
PC 1713 — animals ; Endangered
PD 1290 — animals ; Irresponsible introduction new
PD 3160 — annelida ; Endangered
PD 3603 — *anteaters ; Endangered*
PD 3473 — arachnida ; Endangered
PE 1656 — *arctocephalus ; Endangered*
PD 3603 — *armadillos ; Endangered*
PD 3471 — arthropoda ; Endangered
PD 3159 — aschelminthes ; Endangered
PD 3158 — *asteroidea ; Endangered*
PD 0332 — aves ; Endangered
PE 1570 — *aye aye ; Endangered*
Species [bamboo...]
PD 3481 — *bamboo rats ; Endangered*
PD 1762 — *bandicoots ; Endangered*
PE 9773 — Banned cultivation plant
PD 1593 — *beaked whales ; Endangered*
PD 3483 — bear ; Endangered
PD 3481 — *beavers ; Endangered*
PD 0332 — birds ; Endangered
PD 3158 — *brittle stars ; Endangered*
PD 3477 — bryozoa ; Endangered
PE 3604 — *bulldog bats ; Endangered*
PE 9120 — bush rat rock rat ; Endangered
PE 3762 — butterfly ; Endangered
Species [capybara...]
PD 3481 — *capybara ; Endangered*
PD 3482 — carnivores ; Endangered
PD 1763 — cats ; Endangered
PE 8683 — cavies dolichotids ; Endangered
PD 1593 — cetacea ; Endangered
PD 3481 — *chinchillas ; Endangered*
PE 3604 — chiroptera ; Endangered
PE 9417 — Collectors exotic
PE 3604 — *common bats ; Endangered*
PF 9201 — Competition
PF 9275 — Competition
PD 3158 — *crinodiea ; Endangered*
PD 3472 — crustacea ; Endangered
Species [dasyures...]
PE 8527 — dasyures marsupial mice ; Endangered
PD 7302 — Decreasing diversity biological
PE 5064 — Degradation environment destruction
PE 6270 — Denial animals right conditions life liberty proper their
Species [Dependence...] cont'd
PD 4100 — Dependence plant
PE 3880 — Destruction crop pollinating
PE 3604 — *disc winged bats ; Endangered*
PE 8739 — dog wolves foxes ; Endangered
PD 3481 — *dormice ; Endangered*
Species [echinodermata...]
PD 3158 — echinodermata ; Endangered
PD 3158 — *echinoidea ; Endangered*
PD 3603 — edentates ; Endangered
PD 3771 — elephant ; Endangered
PD 3479 — *elephant shrews ; Endangered*
Species Endangered
PE 4938 — migratory bird
PE 1320 — polar bear
PE 4184 — totemic
Species [Even...]
PC 1326 — *Even toed ungulates ; Endangered*
PD 9578 — Excessive economic exploitation animal
PB 9171 — extinction
PB 9171 — Extinction
Species [false...]
PE 3604 — *false vampires ; Endangered*
PD 7506 — farm animals ; Endangered
PD 3158 — *feather stars ; Endangered*
PD 1763 — *felis ; Endangered*
PC 1535 — fish ; Endangered
PF 3602 — fish ; Irresponsible introduction new
PE 4314 — flowering plants ; Endangered
PC 1326 — *Flying lemurs ; Endangered*
PE 3604 — *free tailed bats ; Endangered*
PD 3156 — frog ; Endangered
PE 3604 — *fruit bats ; Endangered*
PE 3604 — *funnel eared bats ; Endangered*
Species [Genetically...]
PC 0776 — Genetically modified feral
PE 9121 — genets, civets mongooses ; Endangered
PE 1570 — *gibbons ; Endangered*
PE 3604 — *golden bat ; Endangered*
PD 3479 — *golden moles ; Endangered*
PE 1570 — *great apes ; Endangered*
PD 1593 — *grey whale ; Endangered*
PD 3481 — *gundis ; Endangered*
Species [hedgehogs...]
PD 3479 — *hedgehogs ; Endangered*
PD 3158 — *holothuroidea ; Endangered*
PD 1762 — *honey possum ; Endangered*
PE 3604 — *horseshoe bats ; Endangered*
PC 1617 — Hostile introduction
PD 2419 — Hybridization wild animal
PD 3482 — *hyenas ; Endangered*
PC 1326 — *Hyracoidea ; Endangered*
Species [Illegal...]
PD 2664 — Illegal hunting protected endangered
PF 3592 — insect pests ; Introduction new
PD 3479 — insectivores ; Endangered
PC 2326 — insects ; Endangered
PC 0380 — International trade endangered
PC 1617 — introduction ; Environmental hazards new
PD 7513 — invertebrates ; Endangered
PC 0776 — Irresponsible creation new
PE 1444 — Irresponsible introduction new plant
PD 3481 *Species jerboas ; Endangered*
PD 3481 *Species jumping mice ; Endangered*
PE 8708 Species kangaroos wallabies ; Endangered
PD 1762 *Species koala ; Endangered*
Species [leaf...]
PE 3604 — *leaf nosed bats ; Endangered*
PE 1570 — *lemurs ; Endangered*
PD 4100 — Limited food grain
PE 0904 — llamas ; Endangered
Species [macropods...]
PD 1762 — macropods ; Endangered
PC 1326 — mammals ; Endangered
PD 3673 — marine mammals ; Endangered
PD 3160 — *marine worms ; Endangered*
PE 3981 — marsupial moles ; Endangered
PD 1762 — marsupials ; Endangered
PD 4171 — medicinal plants ; Endangered
PE 8967 — mice hamsters ; Endangered
PE 8750 — moles desmans ; Endangered
PD 3478 — molluscs ; Endangered
PE 1656 — *monachus ; Endangered*
PC 1326 — *monotremes ; Endangered*
PE 3604 — *mouse tailed bats ; Endangered*
PD 3481 — *murids ; Endangered*
Species [neofelis...]
PD 1763 — *neofelis ; Endangered*
PE 1570 — *new world monkeys ; Endangered*
PE 8847 — new world porcupines ; Endangered
PE 3604 — *New Zealand short tailed bat ; Endangered*
PE 1570 — non human primates ; Endangered
Species [Odd...]
PC 1326 — *Odd toed ungulates ; Endangered*
PE 8288 — odobenus ; Endangered
PE 8991 — old world monkeys ; Endangered
PE 8845 — old world porcupine ; Endangered
PD 3158 — *ophiaroidea ; Endangered*
PD 1762 — *opossums ; Endangered*
PD 3479 — *otter shrews ; Endangered*
Species [pacarana...]
PD 3481 — *pacarana ; Endangered*
PE 9103 — palaearctic mole rats ; Endangered
PC 1326 — *Pangolins ; Endangered*
PD 1763 — *panthera ; Endangered*
PF 4149 — Periodic mass extinctions
Species [phalangers...] cont'd
PE 8537 — phalangers cuscuses ; Endangered
PE 1656 — *phoca ; Endangered*
PD 1762 — *pigmy possums ; Endangered*
PD 3480 — *pikas ; Endangered*
PB 1395 — plants animals ; Endangered
PC 0238 — plants ; Endangered
PD 3474 — platyhelminthes ; Endangered
PD 3481 — *pocket gophers ; Endangered*
PE 8674 — pocket mice kangaroo mice ; Endangered
PE 3806 — porpoises ; Endangered
PE 7953 — potto loris ; Endangered
PB 9171 — preservation ; Lack
PD 3475 — protozoa ; Endangered
Species [rabbits...]
PD 3480 — rabbits ; Endangered
PE 8794 — rabbits hares ; Endangered
PD 3482 — *racoons ; Endangered*
PD 3481 — *rat chinchillas ; Endangered*
PD 1762 — *rat opossums ; Endangered*
PC 0604 — reptiles ; Endangered
PD 1593 — *right whales ; Endangered*
PD 1762 — *ringtail gliders ; Endangered*
PD 3673 — *river dolphins ; Endangered*
PD 3481 — rodents ; Endangered
PD 1593 — *rorquals ; Endangered*
PD 3159 — *Rotifera ; Endangered*
PE 8302 — rough toothed white dolphins ; Endangered
PD 3159 — *Roundworms ; Endangered*
Species [sac...]
PE 3604 — *sac winged bats ; Endangered*
PD 3156 — salamander ; Endangered
PD 3481 — *scaly tailed squirrels ; Endangered*
PD 3158 — *sea cucumbers ; Endangered*
PD 3158 — *sea urchins ; Endangered*
PE 1656 — seals ; Endangered
PD 3483 — *selenartos ; Endangered*
PD 3479 — *shrews ; Endangered*
PE 8189 — sifakas, avahis indri ; Endangered
PE 3604 — *slit faced bats ; Endangered*
PD 3603 — *sloths ; Endangered*
PE 3604 — *smoky bats ; Endangered*
PD 3478 — *snails ; Endangered*
PD 3479 — *solenodons ; Endangered*
PE 3604 — *spear nosed bats ; Endangered*
PD 1593 — *sperm whales ; Endangered*
PD 3481 — *spiny dormice ; Endangered*
PD 3481 — *spiny rats ; Endangered*
PD 3476 — sponge ; Endangered
PD 3481 — *springhaas ; Endangered*
PD 3478 — *squid octopus ; Endangered*
PD 3481 — *squirrels ; Endangered*
PD 3158 — *starfish ; Endangered*
PE 6604 — Striga
PE 6396 — Substitution fast growing plant
Species [tamarins...]
PE 7933 — tamarins marmosets ; Endangered
PE 1570 — *tarsiers ; Endangered*
PD 3479 — *tenrecs ; Endangered*
PD 1762 — *thylacine ; Endangered*
PD 3156 — toad ; Endangered
Species Trade
PD 0389 — animal products endangered
PC 0380 — exotic
PD 0389 — furs skins endangered
Species [tree...]
PE 1570 — *tree shrews ; Endangered*
PD 3483 — *tremarctos ; Endangered*
PE 3604 — *true vampires ; Endangered*
PC 1326 — *Tubulidentata ; Endangered*
PD 3481 — *tucotucas ; Endangered*
PC 0776 Species ; Uncontrollable new
PD 3483 *Species ursus ; Endangered*
Species [vicuna...]
PE 0904 — vicuña ; Endangered
PA 5414 — violence ; Inter
PA 5414 — violence ; Intra
PB 1395 — Vulnerability populations
Species [walrus...]
PE 8288 — walrus ; Endangered
PF 1925 — warfare ; Inter
PE 5067 — water fowl ; Endangered
PE 8936 — weasels, badgers, skunks, otters ; Endangered
PD 1593 — whale ; Endangered
PE 8187 — white whale narwhal ; Endangered
PD 1762 — *wombats ; Endangered*
PE 8935 Species zagoutis coypu hutia ; Endangered
PE 1656 *Species zalophus ; Endangered*
PC 0507 Speciesism
PF 1624 Specific diseases ; Culture
PC 3908 Specific foodstuffs ; Excessive consumption
PF 1578 Specific needs ; Failure adapt general initiatives
PF 9643 Specific programmes ; Inability generate sufficient culture
PF 6067 Specifically designed software developing countries ; Lack
PE 9417 Specimens ; Abusive collection
PF 5097 Specimens medical study ; Shortage biological
PA 6785 Speciousness *Disappearance*
PA 5502 Speciousness *Reason*
PA 7411 Speciousness *Uncommunicativeness*
PE 0891 Spectator sports ; Exploitation animals
PF 0734 Spectrum radio frequencies ; Overcrowded
Speculation [capitalism...]
PC 2194 — capitalism
PC 2194 — Capitalist

Standard living

Speculation [commodities...] cont'd
PD 9637 — commodities futures markets
PD 9637 — Commodity
PD 1614 Speculation developing countries
PD 9489 Speculation ; Exchange rate
PD 9489 Speculation ; Foreign currency
PD 9489 Speculation money markets
PD 8202 Speculation ; Property
PF 6572 Speculation risks
PC 1453 Speculative flight capital
Speech [defects...]
PE 2265 — defects
PC 2162 — Denial freedom
PE 2265 — disorders
PF 7427 Speech ; Indecent
PF 7798 Speech ; Irresponsible expression emotions equated free
PE 2147 Speed motor vehicles ; Excessive
PD 6124 Speed roads ; Environmental degradation high
PE 4887 Speedy trial ; Denial right
PE 4302 Spell ; Inability
PF 3956 Spellbound
PF 6218 Spelling ; Complicated
PE 4302 Spelling ; Incorrect
PF 6218 Spelling ; Non phonetic
PF 0552 Spelling variations
PF 3311 Spells
PF 1258 Spending alleviate poverty ; Inefficient public
PE 5302 Spending cost effective activities ; Insufficient government
Spending [Decline...]
PF 9108 — Decline public
PF 1447 — defence ; Increasing public
PD 5492 — Deficit
PD 5774 — Diverted consumer
PF 0674 Spending education ; Decline public
PD 0183 Spending ; Excessive governmental
PF 6377 Spending government ; Inappropriate public
PF 6377 Spending government ; Misallocation public
PF 4586 Spending health ; Decline public
PF 6377 Spending ; Ineffectiveness public
PF 6377 Spending ; Inefficiency public
PF 6559 Spending pattern ; Sporadic
PF 0674 Spending primary schools ; Decline public
PF 0611 Spending social sector ; Decline public
PF 6559 Spending ; State highway
Spending [Unchecked...]
PF 6559 — Unchecked community
PF 2990 — Underreported government
PF 6559 — Unfocused community
PF 2990 — Unreported government
PC 0927 Spendthrift economies
PE 8659 Spent food ; Excessive proportion income
PD 1593 Sperm whales ; Endangered species
Spheres influence
PF 3199 — Division world
PF 3199 — Positive
PE 8332 — UN related agencies ; North South
PD 1941 Spheres relationship ; Limited
PC 3908 Spices ; Excessive consumption
PE 0915 Spices manufactures thereof ; Instability trade coffee, tea, cocoa,
PE 1197 Spices manufactures thereof ; Long term shortage coffee, tea, cocoa,
PE 0481 Spices their manufacture ; Environmental hazards coffee, tea, cocoa,
PE 1619 Spices trade ; Instability
PD 3473 Spiders, scorpions
PE 1839 Spillage ; Oil
PE 1839 Spills ; Oil
PE 1221 Spina bifida
PE 9037 Spinal abscesses ; Intra
PE 4604 Spinal column animals ; Diseases
Spinal cord
PE 4604 — animals ; Diseases
PE 8813 — Diseases
PE 8813 — injuries
PD 2626 Spinal paralysis
PE 8589 Spine ; Congenital anomalies
PE 4113 Spine ; Curvature
PD 2626 Spine ; Diseases
PE 7511 Spine ; Fracture
PE 7826 Spines syndrome ; Kissing
PE 7826 Spinous processes ; Overriding dorsal
PD 8622 Spinsterhood
PD 3481 Spiny dormice ; Endangered species
PD 3481 Spiny rats ; Endangered species
PD 7727 Spiral ; Self reinforcing contractionary economic
PF 7118 Spirit ; Divided community
PF 7118 Spirit ; Flagging community
PF 2358 Spirit ; Lack pioneer
PF 5781 Spirit possession ; Demonic
PF 4004 Spirit sickness
PE 0835 Spirit village solidarity developing countries ; Decline communal
PF 5191 Spirited animals ; Broken
PF 9195 Spiritism
PA 7365 Spiritlessness *Boredom*
PA 6030 Spiritlessness *Fear*
PA 6731 Spiritlessness *Solemnity*
PA 7364 Spiritlessness *Unfeelingness*
PF 9043 Spirits ; Affliction malevolent
PE 1349 Spirits drinking ; Solvent methylated
PF 6734 Spirits ; Malevolent
PE 5283 Spirits ; Possession unlawful distilled
PE 2585 Spirits trade ; Instability

PE 1348 Spiritual authority sexual purposes ; Misuse
PE 9840 Spiritual benefits ; Commerce
Spiritual [defilement...]
PF 6657 — defilement
PA 8038 — desperation
PF 8010 — discipline ; Lack
PF 7467 — disobedience
Spiritual [emotional...]
PF 1568 — emotional hindrance
PF 1568 — emotional malaise
PF 4371 — experience ; Barriers
PF 7005 Spiritual guidance ; Loss
Spiritual [healing...]
PF 1568 — healing ; Lack understanding
PA 8038 — hunger
PF 3983 — hypocrisy
PF 6657 Spiritual impurity
PF 7467 Spiritual insubordination
PF 6657 Spiritual uncleanliness
PA 6220 Spiritual void
PE 7310 Spiritual well being because discrimination ; Denial right pursue
PF 9195 Spiritualism
PE 4332 Spirituality ; Repression indigenous
PE 3254 Spirochaetal diseases
PE 3254 Spirochaetosis
PD 2731 *Spirochetosis* poultry
PD 5347 Spitting public places
PE 7808 *Splaylegs* piglets
PD 2730 *Spleen disease* pheasants ; Marble
PE 6155 Spleen ; Diseases
PE 1028 Spleen ; Enlargement liver
PE 7826 *Splints*
PE 4111 Split eastern western Europe
PF 6545 Split tribal urban loyalty
PA 6088 *Spoilage*
PC 2027 Spoilage agricultural products
PD 0161 Spoilage ; Food grain
PA 6088 Spoilage *Impairment*
PD 2243 Spoilage storage ; Food
PA 5688 Spoliation *Appropriation*
PA 6542 Spoliation *Destruction*
PA 7382 Spoliation *Loss*
PE 8990 Spondylitis ; Ankylosing
PE 5913 Spondylitis ; Ankylosing
PE 5913 Spondylosis deformans
PD 3476 Sponge ; Endangered species
PE 5191 Spongiform encephalopathy ; Bovine (BSE)
PD 9693 Sponsor biased research
PD 9370 Sponsored education ; Commercially
PD 6008 Sponsored terrorism ; State
PD 9693 Sponsored vested interests ; Research
PD 0173 Spontaneous abortion
PF 4043 Spontaneous human combustion
PE 4732 Spora ; Air
PD 2732 *Sporadic bovine encephalomyelitis*
PF 7118 Sporadic community growth
PE 2421 Sporadic community lobbying
PF 0148 Sporadic family-planning
PF 6559 Sporadic spending pattern
PD 0807 Sporadic trash removal
PD 0722 Sporadic water supply
PE 2455 Sporotrichosis
PE 4284 Sporozoa birds ; Blood
PD 2283 *Sport*
PE 5286 Sport ; Fishing
PF 4761 Sporting events ; Nationalist exploitation
Sports [accidents...]
PE 4262 — accidents
PE 2605 — activities ; Unfinanced
PE 4893 — Animal fighting
Sports athletic competitions
PE 3754 — Bribery connection
PE 3754 — Corruption
PE 4576 — Gambling
PE 4576 Sports betting ; Clandestine
PD 1323 Sports ; Blood
Sports [competitions...]
PF 4761 — competitions ; Politicization international
PE 3754 — contests ; Rigging
PD 1323 — Cruel
Sports [Discrimination...]
PE 4232 — Discrimination against men
PE 0197 — Discrimination against women
PE 4250 — Drug abuse
PD 0094 — Drug use animal
PF 3560 Sports equipment ; Inadequate
PF 3560 Sports equipment ; Insufficient
Sports events
PE 4222 — Commercialization athletic activities
PE 4206 — Political boycott international
PF 4761 — Politicization international
PE 3388 Sports ; Excessive claims human development
PE 0891 Sports ; Exploitation animals spectator
Sports [fans...]
PE 6281 — fans ; Violent
PE 2605 — field ; Unsatisfactory
PE 2605 — funds ; Lack
PE 4262 Sports injuries
PE 5686 Sports ; Long term injuries
PE 4266 Sports ; Over competitive
PD 2283 Sports ; Overemphasized adult
PE 2605 Sports priority ; Low
PE 2605 Sports space ; Restricted
PE 4026 Sports violence
PE 4026 Sports ; Violent

PF 1406 Sports ; War
PE 6515 Sportsmen ; Ineffective use external relations relating
PE 6516 Sportsmen's practices ; Wasteful
PE 4586 Spot anthracnose plants
PE 4586 Spot elm ; Black leaf
PE 4586 Spot maple ; Tar
PE 4346 Spots ; Fruit
PE 1954 Spots ; Leaf
Spotted fever
PG 3896 — Rack Mountain
PG 3896 — Rock Mountain
PG 3896 — Rocky Mountain
PD 6881 Spouse cohabitant abuse
PE 0753 Spouses torture victims
PD 1739 Spouts ; Water
PE 7808 *Spraddled legs*
PB 0731 *Sprains* joints
PD 2752 Spread animal diseases factory farming
PE 2401 Spread ; Inadequate information
PE 8758 Spread new technologies ; Hindrances international
PD 5168 Spread sexually transmitted diseases ; Intentional
PE 2348 Spring summer encephalitis ; Russian
PD 1204 Spring water ; Distant
PD 3481 *Springhaas* ; Endangered species
PE 7826 *Springhalt*
PA 6030 Spunklessness *Fear*
PA 7364 Spunklessness *Unfeelingness*
PA 6400 Spurious *Affectation*
PA 6180 Spurious *Error*
PA 5952 Spurious *Illegality*
PA 7411 Spurious *Uncommunicativeness*
PF 3701 Spy satellites
PC 2140 Spying
PD 0527 Squads ; Death
PD 7221 Squads ; Death
PA 5454 Squalor *Badness*
PA 7361 Squalor *Disorder*
PA 6839 Squalor *Disrepute*
PA 5459 Squalor *Uncleanness*
PD 5535 Squamous cell carcinoma ; Bovine ocular
PE 6165 Squares cities ; Impersonality public
PD 3139 Squatters ; Urban
PD 0820 Squatting
PD 8926 Squealer
PF 5735 Squeamishness
PD 3478 Squid octopus ; Endangered species
PD 8786 Squinting
PD 3481 Squirrels ; Endangered species
PD 3481 Squirrels ; Endangered species scaly tailed
Stable [flies...]
PE 2254 — flies
PE 2254 — flies
PE 4161 — foot rot
PF 4375 Staff ; Costly teaching
PE 6388 Staff intergovernmental organizations ; High costs
PE 5090 Staff qualifications intergovernmental organizations ; Inadequacy
PE 4893 Stag hunting
PD 7221 Staged encounter killings
PC 1716 Stages ; Repetition educational
PC 2536 Stagflation
PD 7420 *Staggers*
PD 5228 *Staggers ; Annual ryegrass*
PD 7420 *Staggers ; Grass*
PD 9458 *Staggers ; Paspalum*
PD 5228 *Staggers ; Perennial ryegrass*
PE 2634 Stagnant pools ; Unmanageable
PE 2634 Stagnant surface water
PE 2634 Stagnant waste water
PD 1285 Stagnated development agricultural production
PF 6537 Stagnated images community identity
PF 5978 Stagnating educators
PC 1716 Stagnating school enrolment
PF 3884 Stagnating social methods
PA 3917 Stagnation
PF 0440 Stagnation ; Analytical
PC 8269 Stagnation ; Cultural
PD 5326 Stagnation developing countries ; Economic
PD 4125 Stagnation due rural poverty developing countries ; Economic
PF 1277 Stagnation ; Economic
PC 0002 Stagnation ; Economic
PF 3949 Stagnation food aid developing countries
PA 5806 Stagnation *Inaction*
PC 2494 Stagnation ; Political
PF 6649 Stagnation ; Professional
PA 6444 Stagnation *Quiescence*
PA 6839 Stain *Disrepute*
PF 6997 Stain *Imperfection*
PA 5459 Stain *Uncleanness*
PE 0082 Stained glass due acid rain ; Deterioration
PA 5497 Stalemate *Difficulty*
PJ 4988 Stalemate ; Political
PF 2189 Stalinism
PA 1999 Stalling *Delay*
PF 2971 Stamina ; Short term physical
PE 2265 Stammering
PF 0979 Stance being victim external forces ; Self actualizing
PF 0979 Stance ; Defensive life
PF 5949 Stance ; Unmotivated youth
PF 9429 Stand trial ; Incompetence
PF 7925 Standard human rights ; Inter Governmental double
Standard living
PD 4037 — Decadent
PF 7142 — Declining
PF 0344 — Denial right adequate

Standard living cont'd
- PE 4052 — developing countries ; Inadequate
- PE 5484 — indigenous peoples ; Denial right adequate
- PE 7734 — least developed countries ; Inadequate
- PB 0388 — poor ; Deterioration
- PF 2271 Standardization
- PF 2271 Standardization ; Excessive
- PF 8511 Standardization geographical names ; Non
- PC 0666 Standardization procedures equipment ; Inadequate
- PF 8511 Standardization topographic names ; Non
- PF 2974 Standardized social services ; Non
- PE 7136 Standards borrower nations ; Enforced curtailment living
- PE 8443 Standards ; Complicated building
- PF 8466 Standards ; Decline educational

Standards Deterioration
- PD 5377 — film
- PD 5377 — media
- PD 5377 — press
- PD 5377 — television programming
- PD 5377 — video

Standards [developing...]
- PD 0142 — developing countries ; Inequitable labour
- PD 5229 — Discriminatory imposition
- PE 5225 — Double
- PE 5409 — dress ; Enforcement

Standards [Failure...]
- PE 5239 — Failure conform international health
- PC 4591 — Failure conform internationally agreed
- PF 0203 — financial accounting reporting ; Inadequate international
- PF 9050 Standards government ; Double moral

Standards [Imposition...]
- PF 1432 — Imposition exogenous time
- PF 7178 — Inability define moral
- PF 6365 — Inadequacy existing human rights
- PF 5072 — Inadequacy international

Standards Inadequate
- PF 8829 — building
- PD 2898 — demand primary commodities because rising living
- PF 6630 — implementation maritime safety
- PC 4591 — international harmonization
- PF 8927 Standards ; Informality international
- PF 0552 Standards ; Instability orthographic

Standards [Lack...]
- PF 1297 — Lack livelihood
- PF 3411 — Lack professional
- PF 8927 — Limited enforceability international
- PF 0344 — living ; Inadequate
- PF 8466 — Low educational

Standards [measures...]
- PE 0083 — measures ; Distortion international trade discriminatory requirements respect product
- PF 3006 — Medical double
- PD 7723 — Minimal construction
- PF 3377 — Moral double

Standards morality
- PF 7178 — Breakdown
- PF 5225 — Double
- PF 5225 — Double
- PF 2621 Standards ; Multiplicity time

Standards [Parochial...]
- PE 1840 — Parochial telecommunications
- PD 4519 — Politicization health
- PE 5651 — protection against chemical occupational hazards ; Conflicting
- PF 5864 — Pursuit affluent living
- PF 7227 Standards quality ; Commercial erosion
- PF 7227 Standards quality ; Quantitative pressure

Standards [religion...]
- PF 3983 — religion ; Double
- PC 4591 — Resistance internationally agreed
- PD 0343 — Restrictive agreements product

Standards sexual
- PF 2687 — behaviour ; Individualistic
- PF 2687 — conduct ; Collapse
- PF 3259 — morality ; Double
- PC 9714 Standards teaching ; Inadequate
- PF 2901 Standards ; Uncritical acceptance dogmas
- PF 9075 Standards workers ; Deterioration living
- PE 1438 Stannosis
- PD 2731 Staphylococcosis poultry
- PE 4101 Staple food supply developing countries ; Inadequate
- PE 1806 Staples developing countries due transnational corporations ; Import dependency food
- PD 6439 Star wars
- PD 3158 Starfish ; Endangered species
- PD 3158 Stars ; Endangered species brittle
- PD 3158 Stars ; Endangered species feather
- PD 4632 Start up costs businesses ; Prohibitive
- PF 6477 Start up funds ; Scarcity
- PB 1875 Starvation
- PB 0315 Starvation despite sufficient world food supply ; Massive
- PE 5758 Starvation ; Wilful
- PF 1113 Starving attention
- PB 1875 Starving people
- PE 9465 State banks ; Failure

State [Conflict...]
- PC 0030 — Conflict church
- PC 0030 — Conflict religion
- PD 4597 — control communications mass media

State controlled enterprises
- PE 0988 — Abuse monopoly power state owned
- PE 9700 — Delays payments
- PE 7474 — Excessive borrowing

State controlled enterprises cont'd
- PD 5642 — Inefficiency
- PF 6574 — Non accountability
- PF 4739 — Proliferation

State [controlled...]
- PF 1262 — *controlled trade unions*
- PD 9653 — Corrupt acquisition offshore assets heads
- PC 0350 — *Crimes*
- PC 0350 — Crimes against
- PD 0550 — custody deprived children
- PC 3437 State ; Deficiencies welfare
- PE 8408 State ; Denial right freedom movement

State [emergency...]
- PD 6380 — emergency ; Violation rights
- PE 0225 — enterprises ; Inadequate regulation restrictive business practices
- PF 1001 — enterprises ; Undervaluation

State [health...]
- PA 0294 — health ; Lowered
- PF 6559 — *highway spending*
- PG 6455 — *Homicidal crimes*

State [immunity...]
- PE 4930 — immunity
- PC 3197 — *imperialism ; Intra*
- PE 3601 — imposed functional breeding humans

State [membership...]
- PF 4876 — membership ; Distortion intergovernmental organizations mini
- PA 9226 — *mind ; Criminal*
- PE 3391 — monopolies ; Privatization
- PC 0521 — *monopoly*
- PF 7947 — monopoly capitalism
- PE 0988 State owned state controlled enterprises ; Abuse monopoly power

State [participation...]
- PD 2942 — participation international organizations ; Micro
- PD 0435 — Persecutory paranoid
- PD 7910 — Police
- PF 6091 State religion
- PE 3538 State rivalry secretariats intergovernmental organizations ; Inter
- PD 4814 State's rights ; Denial

State [sanctioned...]
- PD 1886 — sanctioned hostage-taking
- PD 0181 — sanctioned torture
- PF 4014 — sciences ; Incomplete
- PF 4778 — socialism ; Pseudo socialism
- PF 6008 — sponsored terrorism
- PE 0241 — succession ; Denial right

State supported
- PD 6008 — international terrorism
- PF 4366 — terrorism ; Government failure sanction
- PD 5560 — violence against citizens
- PD 6008 State terrorism
- PE 8267 State trading government monopoly practices ; Distortion international trade
- PE 2324 State war following cease fire agreements ; Persistence technical
- PE 6676 Stateless children
- PE 7906 Stateless nations
- PE 2485 Stateless persons
- PE 2485 Statelessness
- PE 4016 Statelessness women
- PE 2032 Statements ; Distortion corporation financial
- PF 4583 Statements ; False
- PF 5691 *States ; Crazy nation*

States emergency
- PE 4363 — Excessive imposition
- PD 6380 — Suspension rights during
- PE 4363 — Undeclared
- PD 3785 — Violation rights vulnerable groups during

States emergency Vulnerability
- PE 4694 — animals during
- PD 0098 — disabled during
- PD 0096 — elderly

Static [credit...]
- PF 0245 — credit procedures
- PF 0245 — credit uses
- PF 0443 — *curriculum format*
- PF 4984 *Static education patterns*
- PF 1365 *Static farming image*
- PF 4479 Static grassroots involvement planning economy
- PF 2896 Static job image
- PF 3005 *Static land ownership*
- PF 3091 Static management patterns
- PF 9576 Static models social change policy makers
- PF 6545 *Static power patterns*

Static [school...]
- PF 0443 — *school curriculum*
- PF 2803 — social relations inhibiting vision future
- PF 2883 — superficial adult values
- PF 4470 Static town population
- PF 1651 Static unrelated social roles
- PF 4471 Stationary position torture
- PF 7117 Statistical data nongovernmental organizations ; Inadequate
- PF 4118 Statistical errors
- PF 5711 Statistical fallacies
- PF 0625 Statistical information data problems ; Inadequate
- PF 4106 Statistical information tourism ; Inadequate
- PF 4564 Statistical malpractice
- PF 4564 Statistical ; Biased adjustment
- PF 0510 Statistics ; Deficiencies national
- PF 4118 Statistics ; Discrepancies official
- PF 4564 Statistics government ; Biased adjustment official

Statistics cont'd
- PF 0019 Statistics ; Governmental bias

Statistics Inadequate
- PF 0011 — environmental
- PF 7842 — public finance
- PF 0214 — social demographic
- PF 2622 Statistics ; Lack comparability international
- PF 2622 Statistics ; Limitations international

Statistics [Misinterpretation...]
- PF 4564 — Misinterpretation
- PF 4564 — Misreporting
- PF 4564 — Misuse
- PF 4564 Statistics ; Official abuse
- PF 0888 Statistics socialist countries ; Unreliability
- PF 4564 Statistics ; Unethical use
- PE 6375 Status ; Abuse refugee
- PF 2137 *Status ; Church abuse tax exempt*
- PF 0297 Status ; Denial right people determine their own political
- PF 2943 Status disputed administering government ; Territories accorded United Nations non self governing
- PD 8996 Status employment ; Inferior
- PD 3427 Status immigrants industrialized countries ; Menial work
- PE 7010 Status international nongovernmental organization personnel ; Disadvantaged

Status [language...]
- PF 8145 — language
- PF 8145 — life-style
- PB 5577 — Low
- PF 8145 Status mannerisms
- PF 2342 Status monetary gold ; Uncertain

Status [Political...]
- PD 1140 — Political minority
- PF 3455 — projects ; Unnecessary
- PF 7983 — Pursuit corporate
- PF 8145 — Pursuit personal
- PE 5408 Status quo ; Family adaptation community
- PF 1688 Status quo ; Uncritical preservation

Status [race...]
- PC 5348 — race ; International
- PF 8434 — race ; National
- PF 2137 — *religious institutions ; Abuse*

Status [seeking...]
- PF 8145 — seeking
- PB 5577 — Social
- PF 8145 — symbol
- PF 8145 — symbols
- PF 2969 *Status villages ; Low priority*
- PC 7074 Statutory crime
- PE 6522 Statutory rape
- PA 6448 *Steal ; Failure resist impulses*
- PD 4691 Stealing
- PE 7679 Stealth weaponry
- PD 2682 Steaming
- PE 9774 *Steatitis ; Nutritional*

Steel basic industries
- PE 8397 — environmental hazards ; Iron
- PE 8070 — instability ; Iron
- PE 8223 — Underdevelopment iron
- PE 5866 Steel basic metal industries ; Protectionism
- PE 1945 Steel ; Corrosion iron
- PE 8263 Steel industry ; Health hazards
- PE 8969 Steel manufactures trade instability ; Iron
- PE 8312 Steel scrap trade instability ; Iron
- PF 7580 *Steep narrow paths*
- PD 6282 Steep slopes ; Inappropriate cultivation
- PF 3563 Stellar explosions
- PE 2233 Stem rust wheat
- PA 5981 Stench

Stenosis animals
- PE 5569 — Aortic
- PE 4630 — *Oesophageal*
- PE 5569 — *Pulmonic*
- PF 6064 Stepfamilies
- PE 5380 *Stephanofilariasis*
- PF 8508 Stereotypes
- PF 8508 Stereotypes ; Acceptance
- PF 8508 Stereotypes ; Dependence
- PF 1244 Stereotypes ; Inaccurate criminal
- PF 1509 *Stereotypes ; Inaccurate youth*
- PD 3443 *Stereotyping employment ; Gender*
- PF 5452 Stereotyping ; Ethnic
- PD 5843 Stereotyping ; Gender
- PF 5452 Stereotyping ; Racial
- PD 5843 Stereotyping ; Sexual
- PD 6195 Sterile rental accommodation ; Socially
- PD 0077 Sterile soils
- PD 6133 Sterile working environment
- PA 7365 Sterility *Boredom*
- PE 9302 Sterility ; Female
- PC 6037 Sterility ; Human
- PD 9154 *Sterility ; Male*
- PA 7208 *Sterility Unproductiveness*
- PD 3240 Sterilization ; Compulsory
- PD 3240 Sterilization ; Involuntary
- PE 4544 Sterilization mentally disabled
- PF 4250 *Steroids ; Abuse anabolic*
- PF 4646 Stewardship ; Failure
- PD 8786 Sticky eye
- PE 9774 Stiff lamb disease
- PA 7365 Stiffness *Boredom*
- PA 5740 Stiffness *Compulsion*
- PA 7170 Stiffness *Informality*
- PA 7325 Stiffness *Irresolution*
- PD 2283 Stiffness joint
- PA 6976 Stiffness *Toughness*
- PE 7826 Stifle joint ; Inflammation

Structural disarmament

PF 2797 Stifled emerging leadership	PE 2691 Story books ; Harmful effects comic strips picture	**Stress [children...]**
PF 6570 Stifled potential social interaction different age groups	PF 6575 Story community future ; Detrimental	PE 4421 — children
PD 4632 Stifling middleman profits	PF 9729 Story ; Cooperative failure	PE 4421 — children ; Academic
PE 7627 Stigma breast feeding ; Social	**Story [Debilitating...]**	PE 5053 — Computer
PD 0884 Stigma ; Social	PF 2168 — Debilitating content village	**Stress [Dietary...]**
PJ 3578 Stigmata ; Occupational	PF 1681 — Defeating community	PC 1648 — *Dietary energy*
PD 0884 Stigmatization	PF 2093 — Defeating economic	PE 0351 — disorder ; Post traumatic
PE 4299 Stigmatization AIDS infected persons ; Social	PF 1681 — Demoralizing community self	PE 5164 — due automation ; Mental
PD 9757 Stigmatization handicapped	PF 2013 Story ; Failure self	PD 4178 — due examinations
PD 7279 Stigmatized diseases	PF 2845 Story ; Ineffective community	**Stress [Ecological...]**
Still birth	PF 6575 Story ; Negative community	PC 1282 — Ecological
PD 4029 — [Still-birth]	PE 5795 Story that self-sufficiency is socially responsible	PE 1635 — entrepreneurs
PD 4029 — mortality	PF 6528 Story ; Unconvincing land use	PC 1282 — Environmental
PD 1947 — Premature	PE 7904 Stoves ; Inadequate cooking	PF 9624 — environmentalists
PD 4029 Still-borne babies	PB 0205 Stowaway asylum-seekers	PE 1635 — Executive
PE 4874 Stimuli ; Damage infant brains malnutrition insufficient	PB 0205 Stowaways	**Stress [factory...]**
PE 3636 Stings ; Deadly insect bites	PD 0595 Stowing away	PD 2760 — factory farming ; Animal
PE 3636 Stings ; Insect bites	PD 8786 Strabismus	PD 1405 — families physically mentally handicapped
PD 7758 *Stock ; Alimentary anaemia*	PB 9165 Strain	PD 8130 — Family
PC 1408 *Stock ; Decreasing genetic diversity animal breeding*	**Strain [Emotional...]**	**Stress [Heat...]**
PD 4500 Stock-exchange bourse related crimes	PG 2572 — *Emotional*	PE 2398 — Heat
PD 4511 Stock exchanges bourses ; International collapse	PA 5446 — Enmity	PD 2322 — High altitude
PD 4511 Stock market crash	PE 6785 — Eye	PC 1648 — human beings
Stock markets	PA 6030 Strain *Fear*	**Stress [idealists...]**
PD 5676 — Excessive interdependence	**Strain [Impairment...]**	PF 9624 — idealists
PE 4508 — Lack international coordination supervisors financial	PA 6088 — Impairment	PD 7776 — immigrants
PE 5102 — Low confidence investment	PA 5806 — Inaction	PE 6996 — industry
PD 5676 — Vulnerability	PA 5467 — Inexcitability	PE 4953 — inhabitants tall buildings ; Environmental
PJ 5262 — weaker foreign economies ; Effect dominant	PA 5606 — injuries ; Repetitive	PC 0677 Stress ; Juvenile
PF 2086 Stock resolutions	PA 5473 — Insufficiency	PF 6174 Stress large building complexes ; Disorientation
PE 4447 Stock ; Risk damage	PA 7325 — *Irresolution*	**Stress [Managerial...]**
PD 3841 Stock scandal	PJ 3545 Strain ; Mental, physical emotional	PE 1635 — Managerial
PC 0311 Stockpiles nuclear warfare material	PE 6865 Strain modern technology ; Visual	PD 0518 — Marital
PD 7981 *Stocks, bonds ; Risk forgery shares,*	PA 7411 Strain Uncommunicativeness	PD 7216 — materials
PC 5016 Stocks ; Inadequate food	PA 5984 Strain world resources	PD 7216 — Mechanical
PF 4763 Stocks ; Secrecy national basic food	PF 3665 Strained capital resources small communities	PE 6937 Stress ; Occupational
PF 0687 Stocks ; Unnecessary	PB 0731 *Strains joints*	PE 5768 Stress occupational hazard ; Shift work
PD 0577 *Stolen archives*	PB 0731 *Strains muscles*	PF 9624 Stress perfectionists
PE 0323 Stolen artworks	PD 2948 Straits interoceanic canals ; Restrictions passage	PD 7776 — Resettlement
PE 8364 Stolen property ; Possession fencing	PA 7157 Strangeness *Insanity*	PC 6759 Stress ; Socio Economic
PE 8364 Stolen property ; Receiving	PA 5878 Strangeness *Nonconformity*	PE 9774 Stress syndrome ; *Porcine*
PE 7509 Stomach cancer	PE 8743 Strangers ; Fear	PE 0525 Stress trauma context civil violence
PE 4724 Stomach ; Catarrh	PE 8743 Strangers ; Mistrust	PD 4571 Stress ; Traumatic
PE 8624 Stomach duodenum ; Diseases oesophagus,	PE 1323 Strangle weed	PE 6299 Stress urban environment ; Psychological
PE 1599 Stomach upsets	PE 7645 Stranglers ; Psychotic	PE 5083 Stress video display terminal work ; Job
PE 0958 *Stomach worm infections swine*	PE 9844 *Strangles*	PE 6937 Stress work
Stomatitis animals	PD 2341 Strangulation	PE 5720 Stress work ; Heat
PE 2143 — [Stomatitis animals]	PE 6861 Strangulation ; Developing country import	PE 4421 Stressed-out kids
PE 2143 — *Gangrenous*	**Strategic arms**	PE 5656 Stresses occupational hazards ; Combined
PE 2143 — *Mycotic*	PC 1606 — capability ; Unequal	PD 4840 Stressful life experiences
PE 2143 — *Ulcenomembranous*	PD 0014 — competition	PE 4161 Stretches
PE 2143 — *Ulcerative*	PC 1606 — Imbalance	PE 7915 Siriato pallidal system ; Hereditary diseases
PD 2730 — *Vesicular*	PD 0014 — rivalry	PD 2543 Strict army discipline codes
PE 2143 *Stomatitis cattle ; Papular*	PF 4156 Strategic doctrine ; Risk unintentional nuclear war generated developments	PE 6553 Stricture bowel ; Obstruction
PD 0077 *Stone filled soils*	PE 7174 Strategic goods hostile countries ; Trade	PA 7156 Stridency *Moderation*
PE 4586 *Stone fruits ; Brown rot*	PC 1606 Strategic instability ; International	PE 6604 Striga species
PE 8612 Stone ; Sand gravel trade instability	PD 0014 Strategic nuclear arms ; Proliferation	**Strike action Dismissal**
PC 4864 Stone throwing	PF 1214 Strategic warfare analysis ; Politically unrealistic	PE 7620 — trade union representatives following legal
PE 0760 Stones ; Instability trade crude fertilizers crude minerals, excluding coal, petroleum precious	PJ 9080 Strategically distributed products ; Vulnerability abuse	PE 7620 — workers following legal
PE 1353 Stones ; Long term shortage crude fertilizers crude minerals, excluding coal, petroleum precious	PF 1572 Strategies against problems ; Uncreative	PE 7620 — workers prevent legal
PE 5470 Stonework ; Corrosion	PF 5645 Strategies ; Inadequate	PE 0757 Strike action ; Requisitioning workers prevent
PC 4864 *Stoning*	PD 9661 Strategies ; Inappropriate infant feeding	PE 1541 Strike action trade unions ; Transnational
PD 8926 Stool pigeon	PF 4162 Strategy deterrence ; Risk unintentional nuclear war generated	PE 5070 Strike ; Denial right
PD 0694 Stoppages work ; Collective	PF 4162 Strategy ; Illusion nuclear	PD 9667 Strike ; Fly
PF 3467 Storage capacity ; Limited	PE 4612 Strategy ; Risk aversion	PE 5070 Strike ; Violation right
PE 1581 Storage communication industries ; Health risks workers transport,	PF 5660 Strategy ; Short range development	PF 5805 Strikebreaking
Storage [dangerous...]	PB 1997 Stratification obstacle development ; Social	PD 0694 Strikers
PC 6913 — dangerous substances	PB 5577 Stratification ; Social	PD 0694 Strikes
PE 4293 — *disease ; Glycogen*	PD 6113 Stratospheric ozone depletion	PD 1111 Strikes against aircraft ; Bird
PE 4293 — *diseases ; Glycogen*	PD 9667 Strawberry foot rot	PD 5384 Strikes ; Illegal
Storage facilities	PE 5776 Stray animal populations ; Excessive	PE 5070 Strikes public authorities ; Suppression
PF 3467 — *Inadequate*	PE 2759 Stray animals ; Inhumane killing	PD 1882 Strikes socialist countries ; Undeclared
PE 4877 — Inadequate food	PE 0359 Stray dog populations	PD 5384 Strikes ; Undeclared
PF 3467 — Lack	PE 0360 Stray dogs ; Cruel methods destruction	PD 5384 Strikes ; Wildcat
PE 5908 — land locked developing countries ; Inadequate port	PJ 4775 Stray voltage	PD 1292 Striking ; Risk lightning
PD 2243 — Food spoilage	PD 9366 Stray voltage animal housing	PE 7826 *Stringhalt*
PC 6913 Storage hazardous materials	PD 7636 Stream pollution	PD 1637 Strip-mining
PC 6913 Storage hazardous materials ; Underground	PD 1990 Streams ; Desiccation	PD 2198 *Strip tease*
PE 8233 Storage ; Inadequate water	PE 2388 Streams ; Fly infested rivers	PE 9224 Stripping ; Asset
PE 6884 Storage radioactive material ; Transport	PE 5980 Street children	PC 7605 Stripping ; Conceptual asset
PF 7130 *Storage space homes ; Insufficient*	PD 8557 Street dust	PD 9082 Stripping ; Ocean
PD 3587 Storage transit ; Plant diseases	PD 2682 Street gangs	PE 2691 Strips picture story books ; Harmful effects comic
PE 5623 Storage ; Unlimited practice human embryo	**Street [life...]**	PE 1684 Stroke
PE 2857 Storage ; Unsafe firearm	PE 6705 — life ; Decline	PD 9366 Stroke ; Lightning
PE 2425 Store capital ; *Restricted*	PE 6473 — *lighting ; Insufficient*	PD 8402 Strong laxatives
PE 1669 Stored documents archives ; Deterioration	PE 6473 — lighting ; Poor	PD 0122 Strong toxic substances ; Hazards
PD 3657 Stored manufactured goods ; Insect damage	PD 8557 Street repair ; Postponement	PD 0958 *Strongyle infection horses ; Small*
PF 1205 *Storekeepers ; Distrust*	PD 3544 Street search ; Arbitrary	PE 0958 *Strongyle infections horses ; Large*
PF 6504 *Stories ; Unfounded conflicting*	PJ 1150 Street vending	PE 2395 *Strongyloidiasis*
PE 2829 Storing, transporting processing solid wastes ; Inadequate facilities	PD 5980 Street youth ; Unprotected	PE 0958 *Strongyloidosis*
PE 7647 *Storm drains ; Clogged*	PD 8557 Streets roads ; Broken surfaces	PC 1522 Structural accountability ; Lack
PD 2788 Storm surges	PE 2845 Streets ; Unnamed	**Structural adjustment**
PD 2788 Storm tides	PE 2575 Strength elders ; Limited	PD 8114 — debtor developing countries ; Excessive social costs
PD 1150 Storms	PE 4920 Strength military forces ; Under	PF 9664 — Retardation economic
PD 3655 Storms ; Dust	PE 3098 *Streptococcal infection throat ; Haemolytic*	PF 9664 — world economy ; Faltering
PE 3881 *Storms ; Electrical*	PD 2731 *Streptococcal infections ; Porcine*	PF 4165 Structural adjustments ; Inadequate trade related
PD 1661 *Storms ; Geomagnetic*	PE 9844 *Streptococcal lymphadenitis swine*	PF 7745 Structural amnesia institutional systems
PD 0251 *Storms ; Hail*	PE 3098 *Streptococcal pharyngitis*	PE 0707 Structural barriers disabled persons
PD 1661 *Storms ; Magnetic*	PD 2731 *Streptococcosis poultry*	**Structural [change...]**
PE 4103 *Storms ; Rain*	PE 3098 *Streptococcus infection*	PF 9664 — change ; Faltering adaptation economic agents accelerating pace
PD 3650 *Storms ; Sand*	PD 9667 *Streptotrichosis ; Cutaneous*	PF 9664 — change ; Policy making delays response accelerating
PE 2223 *Storms ; Wind*	PB 9165 Stress	PF 8100 — complexity ; Confusing
	PE 4951 Stress addiction	PE 7670 Structural disarmament ; Insecurity unilateral
	PE 1670 Stress animals ; Experimental induction psychological	PE 4051 Structural disarmament nuclear weapons ; Unilateral

Structural failure

Structural failure
PD 1230 — [Structural failure]
PF 2347 — citizen participation
PF 3821 — opposition groups
PF 7745 Structural forgetfulness
Structural [imbalances...]
PC 7459 — *imbalances three largest market economies*
PB 1935 — injustice
PF 4983 — *injustice ; Complicity*
PD 4011 Structural rigidities labour markets
PD 2970 Structural rigidity developing country economies
Structural [technological...]
PE 4596 — technological changes ; Training inappropriate
PE 7185 — technological changes ; Working hours inappropriate
PB 2853 — tensions society
PB 0750 Structural unemployment
PB 1935 Structural violence
PB 1935 Structural violence ; Dependence
PD 7470 Structurally blocked scientific cooperation
PF 1502 Structure barrier progress ; Family
PD 2438 Structure ; Collapsed consensus
PA 6944 Structure∗complex
PD 0276 Structure ; Demise three generation family
PD 1052 Structure ; Destruction soil
PF 1064 Structure ethical decision ; Absence common
PF 3783 Structure families ; Reduced interior
PF 6712 *Structure ; Fraying tribal*
PC 0442 Structure ; Haphazard urban
Structure [Inadequate...]
PD 2438 — Inadequate consensus
PC 9058 — Inadequate political
PF 3091 — Inflexible management
PB 1997 — Inflexible social
PD 2294 Structure marriage ; Individually defined operating
PF 9008 Structure opportunities income available ; Imbalances people's aspirations
PF 5351 Structure ; Removed governmental
PD 3436 Structure ; Restrictive tax
PE 5246 Structure social care ; Non adaptive local
PE 3296 *Structure ; Uncorporate economic*
Structures Absence
PC 1522 — accountability
PF 8652 — police
PF 2575 — *youth*
PD 2802 Structures achieving global unity ; Inadequate
PF 7106 Structures ; Architectural obsolescence building
Structures [challenge...]
PF 7713 — challenge human survival ; Inadequacy prevailing mental
PD 4143 — Collapsing physical
PF 2350 — communication ; Inadequate
Structures community
PF 1781 — decision making ; Ineffective
PF 2810 — organization ; Unformed
PF 2437 — participation ; Ineffective
PF 6476 Structures ; Constricting effect educational
PD 5305 Structures ; Defacement urban
PF 1566 Structures depressed areas ; Fragmentation social
Structures developing countries
PD 0822 — Inadequate development new social
PE 0365 — Inadequate local authority legal
PE 2493 — Inflexible social care
PD 2970 — Rigidities production
PA 8653 Structures ; Distrust loan
Structures [engagement...]
PD 1564 — engagement elders ; Isolation fostering
PE 3977 — environmental protection ; Fragmented social
PF 1301 — essential corporateness ; Deteriorated
PF 5115 — established religions ; Hierarchical
PD 1230 — Failure engineering materials
PF 0845 Structures ; Government reluctance transform administrative
Structures [Ideal...]
PF 6178 — Ideal tension free social
PF 1000 — Inadequate family
PF 2949 — Inadequate rural employment
PF 1861 — Individualistic family
PF 4818 — industrial nations ; Ineffective economic
PF 3448 — *Insufficient enabling*
PF 7069 — international nongovernmental organizations ; Neglect non Western
PE 7105 — international nongovernmental organizations ; Uncontrolled
Structures Lack
PF 3307 — *actuation*
PF 1653 — *funding*
PF 6556 — local leadership decision making
PF 1364 — savings
Structures [Limited...]
PF 3184 — Limited availability learning
PF 6506 — local consensus ; Ineffective
PE 0365 — local government developing countries ; Inadequate institutional
PF 3826 — Low credibility alternative social
PF 1000 Structures marriage ; Reduced interior
PF 1097 Structures ; Mistrust political
PF 5961 Structures ; Obsolete agricultural cooperative
Structures [Parallel...]
PF 1781 — *Parallel social*
PF 1776 — Paralyzing complexity urban
PF 1389 — Paralyzing patterns villages administrative
PC 1547 — political dialogue review ; Inadequate
PD 2884 — Prioritized attachment security
PF 8065 *Structures ; Restrictive legal*
PF 3558 Structures rural community cooperation ; Deteriorating
PF 2097 Structures ; Simplistic family decision making

PF 2540 Structures small communities ; Lack central planning
Structures [unemployed...]
PE 7326 — unemployed ; Deficient local
PE 2845 — *Unfamiliar government*
PF 2699 — Unimaginative vision existing international economic
PE 2575 — *Unplanned youth*
PE 2568 — Unrealized use education
PS 3268 — *Unsupportive home*
PF 6450 — Untransposed community
PE 1334 Structures ; Wind damage
PB 4411 Struggle existence
PE 8144 Struggle financial security urban life style
PF 1078 Struggle mindset ; Subsistence
PD 5228 *Strychnine poisoning animals*
PJ 3525 Stubborn people
PA 5740 Stubbornness *Compulsion*
PA 7250 Stubbornness *Disobedience*
PA 7325 Stubbornness *Irresolution*
PA 5988 Stubbornness *Softness*
PA 6509 Stubbornness *Unwillingness*
Student [absenteeism...]
PE 4200 — absenteeism
PC 1522 — *accountability ; Ineffective*
PE 4200 — attendance ; Erratic
PC 2052 Student dissent
PF 1226 *Student enrolment ; Small*
PF 7577 Student immobility
PE 0628 Student press weakness
PC 2052 Student revolt
PD 6346 Student's rights ; Violation
PF 2602 *Student teacher disinterest ; Mutual*
PF 6467 *Student teacher ratio ; High*
PC 2052 Student unrest
PE 7370 Student violence campus
PD 6346 Students ; Abuse
PE 7370 Students ; Anti social behaviour university
PE 0192 Students ; Beating school
PD 9370 Students ; Commercial exploitation
PD 4178 Students ; Competitive burden
Students [Denial...]
PD 6346 — Denial rights
PE 6258 — Discrimination against adult
PF 5507 — Drug abuse
PD 2624 Students ; Economically disadvantaged
PE 8733 Students enrolled school ; Maldistribution
PF 1205 *Students' families ; Indifference*
PF 5753 Students ; Fragmentation organized
Students Lack
PB 4653 — *fees*
PF 7577 — foreign
PF 2168 — *incentives*
PD 8723 Students ; Learning disabled college
PF 5777 Students ; Manipulation
PE 5729 Students ; Political impotence
PD 1291 Students studying foreign countries ; Non return developing country
Students [Unavailability...]
PE 3569 — Unavailability scholarship funds
PE 8883 — Unavailability scholarship funds developing country
PE 7726 — Unequal opportunities foreign
PF 9693 Studies ; Manipulation scientific
PF 3824 Studies ; Terminological crisis social
PF 1205 *Study environment ; Inadequate*
PF 5712 Study global ecosystem ; Over specialized
PF 1205 *Study patterns ; Inappropriate*
PD 3548 Study programmes ; Outmoded work
PF 5097 Study ; Shortage biological specimens medical
PF 3560 *Study time ; Limited*
PD 1291 Studying foreign countries ; Non return developing country students
PA 6400 Stuffiness *Affectation*
PA 7365 Stuffiness *Boredom*
PA 7325 Stuffiness *Irresolution*
PA 7306 Stuffiness *Narrowmindedness*
PA 6444 Stuffiness *Quiescence*
PA 5981 Stuffiness *Stench*
PA 6738 Stuffiness *Unimaginativeness*
PA 7371 Stuffiness *Unintelligence*
PD 1452 Stuffs ; Naturally occurring poisonous substances food
PF 6155 Stultifying homogeneity modern cities
PE 4921 Stunting growth children
PJ 8777 Stupidity
PA 6180 Stupidity *Error*
PA 7371 Stupidity *Unintelligence*
PA 5806 Stupor *Inaction*
PA 6739 Stupor *Maladjustment*
PA 7364 Stupor *Unfeelingness*
PE 2265 Stuttering
PE 2357 *Stuttgart disease*
PD 8786 *Stye*
PF 2874 *Style ; Bravado youth*
Style [City...]
PF 1437 — City life
PF 6559 — community operations ; Unfocused
PF 6456 — community planning ; Self defeating
PF 6514 — cooperative action ; Unformed
PD 4368 Style ; Dislocated life
PD 4037 Style ; Extravagant life
Style [Independent...]
PF 7142 — *Individualistic life*
PF 7654 — informal leadership advisors ; Self satisfied
PF 1078 Style ; Lack variety life
PE 8356 Style organizations ; Preponderance Western
PE 8644 Style ; Perpetual preoccupation sustenance urban life

PF 2110 Style professionals ; Individualistic
PF 7142 Style ; Restricted life
Style [Status...]
PF 8145 — Status life
PE 8144 — Struggle financial security urban life
PF 1078 — Subsistence life
PE 8477 Styles ; Apathy toward improvement urban life
PD 4069 Styles ; Escapist family life
PF 3211 Styles ; Inhibiting effects traditional life
PE 8755 Styles ; Lack variety life
PE 8296 Styles ; Minimal family interest urban life
PE 1137 Styles mining communities ; Traumatic shift life
PF 1078 Styles ; Reduced family life
PF 1729 Styles ; Reduced social commitment contemporary life
PF 8663 Sub-arctic countries
PF 5508 Sub culture ; Criminal
PE 7826 *Subchondral bone cyst*
PF 1762 Subcontracting
PF 1762 Subcontractors ; Economic risks
PE 5508 Subculture ; Criminal
PF 6137 Subcultures ; Insufficient separation urban
PE 3639 *Subcutaneous mite*
PC 8534 Subcutaneous tissue ; Diseases skin
PF 1822 *Subdivision ; Random land*
PA 7250 Subjection *Disobedience*
PA 5652 Subjection *Inferiority*
PA 7296 Subjection *Restraint*
PF 2827 Subjective assessment according own group values
PF 2015 Subjective cultural ranking
PD 8697 Subjective law enforcement
PF 8015 Subjectivism
PF 2827 Subjectivity
PA 5688 Subjugation *Appropriation*
PA 7289 Subjugation *Defeat*
PA 7296 Subjugation *Restraint*
Subjugation women
PD 4633 — Biological
PD 4633 — child-rearing
PD 4633 — domestic service
PD 4633 — Social
PD 0606 Subliminal advertising
PE 5309 Submarine junk
PE 0712 Submarine objects ; Unidentified
PE 0712 Submarine patrols ; Unidentified
PD 0253 *Submergence historical sites dams*
PE 9643 Submission media cultural imperialism
PD 0914 Subnormal children ; Mentally
PF 1199 Subordination ; Economic
PA 5652 Subordination *Inferiority*
PA 7296 Subordination *Restraint*
PE 8139 Subordination women developing societies ; Traditional values
PE 8792 Subsequent economic growth ; Import substitution barrier
PD 5156 Subsidence ; Land
PE 4393 Subsidence mining
PD 0669 Subsidiaries transnational enterprises ; Tying supplies
Subsidies [chemicalized...]
PE 1785 — chemicalized farming ; Inappropriate agricultural
PA 4532 — Concealed government
PD 9802 — Counter productive price supports
PE 1961 — countervailing duties ; Distortion international trade export
Subsidies [Dependence...]
PE 7413 — Dependence economies food
PD 9802 — developed countries ; Over dependence farm
PE 1785 — developing countries ; Inappropriate irrigation
PD 0678 — Discriminatory
PD 0678 — Distortion international trade selective domestic
PE 7413 — Distortion national economies food
PD 9802 Subsidies farmers developed countries ; Excessive
PD 4543 Subsidies fraud
PE 9758 Subsidies ; Inequitable distribution public
PF 4586 Subsidies ; Limited public health
PE 1785 Subsidies livestock production ; Inappropriate agricultural
PE 1785 Subsidies organic farming ; Inadequate
PD 0678 Subsidies ; Overdependence
PE 1785 Subsidies sales pesticides ; Inappropriate agricultural
PF 1653 *Subsidies ; Unknown availability*
PD 3775 Subsiding coastal areas
PD 9802 Subsidized agriculture industrialized countries ; Over
PF 3448 *Subsidized fertilizers ; Scarcity*
PF 2489 Subsidy ; Unavailability income
Subsistence [agricultural...]
PE 8171 — agricultural income level rural communities
PC 0492 — agriculture ; Unproductive
PC 6530 — approach capital resources
PE 4399 Subsistence ; Denial right people their own means
PF 3008 *Subsistence discourages vision*
Subsistence [earning...]
PF 6530 — earning ; Prevalence
PF 6530 — economic orientation
PC 2237 — *economy ; Destruction rural*
PF 3008 — *employment ; Unmotivating*
Subsistence [farmer...]
PF 1078 — farmer hero
PC 0492 — farming patterns
PF 2135 — fishing economies ; Non diversification
Subsistence [level...]
PB 1078 — level malnutrition
PF 6530 — level profits
PF 1078 — life style
PF 1078 — limits role
PD 4125 — living ; Continued
PF 1078 — living cycle

Support services

Subsistence [planning...]
PF 6530 — planning ; Crippling
PD 4125 — poverty economy
PF 6530 — poverty economy
PF 1078 Subsistence struggle mindset
PF 1078 Subsistence style ; Independent
PE 7826 Subsolar abscess
Substance abuse
PC 5536 — [Substance abuse]
PD 1018 — Dependence government revenue
PE 0680 — during control complex equipment
PE 3858 — during pregnancy breast-feeding
PE 3980 — media ; Excessive portrayal
PE 3980 — media ; Glorification
PE 7568 — physicians
PE 0742 — role models
PE 1427 — Volatile
PD 9805 — work
PD 3825 Substance dependence
PE 5187 Substance ; Eating non nutritive
PD 4027 Substance intoxication
PE 9267 Substance ; Symbolic international agreements
PE 5058 Substandard hospital buildings
PD 1251 Substandard housing accommodation
PD 1435 Substandard products ; Manufacture
PE 6630 Substandard shipping vessels
PD 6795 Substandard waste treatment
PF 4600 Substantive issues political debate ; Avoidance
PD 1276 Substitutes primary commodities ; Production synthetic
PD 7682 Substituting commodities exported developing countries ; Development industrialized countries products
PE 8792 Substitution barrier subsequent economic growth ; Import
PD 7682 Substitution developed countries ; Import
PF 8922 Substitution fantasy reality
PE 6396 Substitution fast growing plant species
PE 8255 Substitution inappropriate foodstuffs breast feeding
PA 5502 Subterfuge *Reason*
PA 7411 Subterfuge *Uncommunicativeness*
PA 7232 Subterfuge *Unskillfulness*
PE 1708 Suburban housing ; Over spacing
PD 6150 Suburban mode human settlement ; Obsolescence
PF 6132 Suburban residents ; Inaccessibility city centres
PD 2345 Suburbia ; Environmental degradation
PD 2345 Suburbs ; Destruction green inner
Subversion [capitalist...]
PD 3180 — capitalist neutral countries ; Communist
PF 9485 — communism
PE 8349 — communist neutral countries ; Capitalist
Subversion [democracy...]
PD 3180 — democracy
PA 6542 — *Destruction*
PE 9210 — Drug financed
PD 5876 Subversion international agreements
PA 5699 Subversion *Reversion*
PA 5901 Subversion *Revolution*
PF 9485 Subversion socialism
PD 0557 Subversive activities
PD 0557 Subversive activities ; Dependence
PD 7276 Subversive tourism
PE 3909 Subverting minds young
PF 5390 Success ; Fear
PF 5013 Success ; Jealousy personal
PD 0270 Success neurosis
PF 9729 Success ; Uncertain cooperative
PF 6540 *Successes ; Infrequent business*
PF 9729 Successes unknown ; Co
PF 6572 *Successful businesses ; Undiscovered*
PE 0241 Succession ; Denial valid state
PD 5453 *Sudden death syndrome chickens*
PD 7420 *Sudden death syndrome feeder cattle*
PD 1885 Sudden infant death syndrome
PE 6603 Sudden shocks body
PD 1885 Sudden unexpected infant death
PA 5806 Suddenness *Inaction*
PA 7325 Suddenness *Irresolution*
PA 5451 Sufferance *Insensibility*
PA 7107 Sufferance *Unpleasantness*
PA 7690 Suffering
PD 8812 Suffering ; Animal
PB 5249 Suffering ; Avoidance reference
Suffering [Denial...]
PE 8024 — Denial experimental animals right freedom
PE 3899 — Denial food animals right freedom
PB 5646 — Dependence human physical
PB 5680 — Dependence mental
PB 5249 Suffering ; Failure report
PB 5955 Suffering ; Human
PB 5646 Suffering ; Human physical
PB 5249 Suffering ; Indifference
PB 5680 Suffering ; Mental
PC 7825 Suffering plants
PD 7550 Suffering ; Self inflicted physical
PB 5249 Suffering ; Silence response
PE 2224 Suffering videos ; Trivialization death
PE 7616 Sufficient clothing ; Denial right
PE 9643 Sufficient culture specific programmes ; Inability generate
PE C324 Sufficient food ; Denial right
PF 2857 Sufficient food supplies ; Inadequate mechanisms securing
PE 1085 Sufficient nutrition women ; Denial
PD 5254 Sufficient shelter ; Denial right
PB 0315 Sufficient world food supply ; Massive starvation despite
PA 7055 *Suffocation*

PA 7055 Suffocation *Death*
PA 6542 Suffocation *Destruction*
PB 0731 *Suffocation ingested objects*
PB 0731 *Suffocation ; Mechanical*
PE 2975 Sugar beet ; Pests diseases
Sugar [cane...]
PE 2217 — cane ; Pests diseases
PE 8958 — confectionery preparations trade instability
PE 1894 — craving
PE 1894 Sugar ; Excessive consumption
PE 1894 Sugar honey preparations thereof ; Environmental hazards
PE 1120 Sugar honey preparations thereof ; Long term shortage
PE 1926 Sugar ; Low blood
PE 0383 Sugar preparations honey ; Instability trade sugar,
PE 0383 Sugar, sugar preparations honey ; Instability trade
PE 0383 Sugar trade ; Instability
PD 5574 Suggestibility ; Interrogative
PD 5574 Suicidal terrorists
PC 0417 Suicide
PE 5771 Suicide ; Adolescent
PE 4878 Suicide ; Attempted
PF 5240 Suicide ; Cognitive
PF 5957 Suicide ; Cultural
PE 2643 Suicide ; Illegality doctor assisted
PE 5771 Suicide ; Juvenile
PC 0417 *Suicide ; Ritual*
PE 5771 Suicide ; Teenage
PD 5574 *Suicide terrorists*
PD 8680 Suicides ; Gaol
PD 8680 Suicides ; Jail
PD 8680 Suicides ; Prison
PE 4394 Sullying reputations
PE 2329 Sulphide health hazard ; Hydrogen
PE 2329 Sulphide pollutant ; Hydrogen
PD 1670 *Sulphur compounds pollutants*
PE 1210 Sulphur dioxide occupational hazard
PE 1210 Sulphur dioxide pollutant
PE 0760 *Sulphur ; Instability trade*
PE 2964 Sulphur trioxide pollutant
PE 4825 Sulphuric acid mist
PE 6366 Summary executions
PE 6197 Summer catarrh
PE 2348 Summer encephalitis ; Russian spring
PE 9458 *Summer fescue toxicosis*
PF 2721 *Summer programme ; Unscheduled*
PE 5380 *Summer sores*
PA 6400 Sumptuousness *Affectation*
PE 3883 Sun ; Overexposure
PE 3883 Sunburn
PE 6350 *Sunlight ; Absence direct*
PE 6350 Sunlight inhibition nuclear warfare soot
PE 9528 Sunspot cycle ; Epidemics associated
Sunstroke
PE 2398 — [Sunstroke]
PE 3883 — [Sunstroke]
PD 9366 — [Sunstroke]
Super power
PE 7778 — arrogance
PE 7778 — chauvinism
PE 7778 — exclusiveness
PF 5124 — ideological diversities ; Maintenance
PC 3185 — intervention countries
PD 4445 — monopoly advanced nuclear warfare technology
PF 3858 Superannuated religious gerontocracy
PE 6007 Superbug
PA 6822 Superciliousness *Disrespect*
PA 6491 Superciliousness *Vanity*
PF 2883 Superficial adult values ; Static
PF 7493 Superficial consensus intergovernmental bodies
PF 7889 Superficial consultation
PB 0731 *Superficial injury*
PF 7889 Superficial investigations
PE 6789 Superficial keratitis
PF 3243 Superficial needs ; Overemphasis immediate
PF 2005 *Superficial preaching*
Superficial [religion...]
PF 2005 — religion
PF 7889 — research
PF 2415 — research total human process
PF 7889 Superficial surveys
PF 3638 Superficial symbols economy
PA 7365 Superficiality *Boredom*
PF 7110 Superficiality collective comprehension societal learning
PA 6785 Superficiality *Disappearance*
PA 5568 Superficiality *Ignorance*
PA 6247 Superficiality *Inattention*
PF 4600 Superficiality political debate
PA 5942 Superficiality *Unimportance*
PA 7371 Superficiality *Unintelligence*
PC 1259 Superiority ; Bureaucratic
PF 8433 Supernatural ; Belief
PF 3437 Supernatural entities ; Disruption human activities
PF 8433 Supernaturalism
PF 3563 Supernovae
PD 9655 Superpower rivalry
PJ 3570 Superpower vacuum
PD 7522 Superpowers ; Instability relations allies
PE 2435 Supersonic aircraft ; Sonic boom generated
PA 0430 Superstition
PF 0565 Superstition ; Astrology
PA 0430 Superstition ; Dependence
PF 5673 Superstition ; Exorcism
PF 0565 Superstition ; Horological
PF 1270 Superstition ; Religious
PA 7392 Superstition *Unbelief*

PA 0430 Superstitious people
PD 3453 Superstitious persecution animals
PF 8754 Superstitious symbolic acts
PE 8280 Supervision adherence international law ; Inadequate national
PF 2793 Supervision ; Inadequate
PF 1403 Supervision ; Poor adult
PF 3498 Supervision teachers ; Poor
PF 2477 Supervision ; Vacated youth
PE 4508 Supervisors financial stock markets ; Lack international coordination
PE 2274 Supervisory personnel ; Insufficient
PF 3588 Supervisory personnel ; Over specialized
PE 8545 Supplies human excreta ; Pollution water
Supplies Inadequate
PC 5016 — emergency food
PD 1294 — facilities transport water
PF 2857 — mechanisms securing sufficient food
PD 1607 Supplies ; Infection industrial water
PE 1634 Supplies ; Insufficient medical
PE 7954 Supplies least developed countries ; Inadequate food
PD 4403 Supplies ; Misuse artesian water
Supplies [sabotage...]
PE 1458 — sabotage ; Contamination public water
PD 5212 — Seasonal variability food
PB 2846 — Shortage food
PE 0669 — subsidiaries transnational enterprises ; Tying
PF 3913 Supply agreed multilateral assistance ; Delay
PE 6243 Supply appropriate trained manpower developing countries ; Inadequate
Supply [centres...]
PF 8299 — centres ; Inaccessible market
PF 8299 — centres ; Remote
PF 4375 — costs ; School
Supply [developing...]
PE 1900 — developing countries ; Inadequate electrical power
PD 4101 — developing countries ; Inadequate staple food
PF 6495 — distribution ; Ineffective means goods
PE 8979 — due military activities ; Disruption food
PE 8496 — dwarfs, midgets giants ; Lack adequate clothing
PE 2467 Supply foetus ; Defective oxygen
PB 8891 Supply goods ; Shortages
Supply Inadequate
PF 2115 — electricity
PE 0366 — emergency blood
PC 1916 — protein
Supply [Inconvenient...]
PD 1204 — Inconvenient water
PF 9432 — Inflexibility commodity
PD 0722 — Instability water
Supply Insufficient
PD 0335 — *diversification urban industrial energy*
PF 2115 — *domestic electricity*
PB 2846 — food
PC 4815 — fresh water
PC 4815 Supply ; Limited water
PB 0315 Supply ; Massive starvation despite sufficient world food
PE 4120 Supply pharmaceutical products developing countries ; Inadequate
Supply [raw...]
PD 4270 — raw materials ; Unstable
PF 3499 — *repair materials ; Inaccessible*
PD 1304 — Restriction arms
PD 1204 — rural communities developing countries ; Inadequate water
Supply [services...]
PE 8319 — services ; Inadequate conditions work employment utility
PD 0722 — Sporadic water
PE 1341 — system overload ; Power
Supply systems
PD 9196 — Deterioration water
PD 0482 — *Fouling water*
PD 9196 — Leakage water
Supply [Unpredictable...]
PD 0722 — Unpredictable water
PD 8122 — Unsanitary water
PF 3467 — *Untimely seed*
PD 8122 — Untreated water
PD 0722 Supply ; Variable water
PE 2449 Supplying firearms criminal activity
PF 9704 Support against problems ; Government failure mobilize
PF 4821 Support corrupt regimes ; Government
PF 2947 Support ; Disorganized liaison formal
PA 8653 Support ; Distrust blocking
PF 6076 Support enforcement ; Inadequate system child
PE 7944 Support families ; Lack interim
PF 4821 Support foreign dictatorships ; Government
PE 8309 Support ; Government protection national commodity price
Support [Inadequate...]
PF 4848 — Inadequate peace research
PF 1653 — *Insufficient councilman*
PF 5659 — Insufficient minority culture
PD 1809 — international organizations programmes ; Alienation
Support [Lack...]
PF 8960 — Lack governmental
PF 8960 — Limited government
PF 6510 — local commercial services ; Lack
PF 8960 Support ; Marginal government
PF 5622 Support public interest broadcasting ; Inadequate
PF 4821 Support repressive regimes ; Government
Support [services...]
PF 6492 — services ; Inadequate
PE 8151 — services mentally handicapped ; Inadequate

-1053-

Support systems

Support [systems...] cont'd
PF 6492 — systems effective community operation ; Absence
PF 4936 — systems ; Inhumane use medical life
PE 6846 Support terrorism ; Passive public
Support [undemocratic...]
PF 4821 — undemocratic regimes ; Government
PF 2199 — unfriendly governments ; Eroding
PF 9428 — unsustainable policies ; Political
PE 4531 *Support widowers ; Inefficient*
PE 4167 Supporting racial discrimination ; Capital investments
PD 9802 Supports subsidies ; Counter productive price
PF 0275 Suppression creativity innovation
PA 6542 Suppression *Destruction*
PA 5497 Suppression *Difficulty*
PD 7385 Suppression evidence
PF 1615 Suppression experimental data
PF 4932 Suppression historical information
PD 9146 Suppression information
Suppression information concerning
PF 4854 — environmental hazards
PF 4854 — health hazards
PF 9828 — social problems
Suppression information [corporations...]
PD 9146 — corporations
PD 9146 — government
PF 4050 — security classification
PF 4050 Suppression innovation purchase patents
PC 5018 Suppression intellectual freedom
PA 7382 Suppression *Loss*
PD 2108 Suppression minority opinion ; Innate expectation
PC 7662 Suppression opposition groups individuals
PD 2048 Suppression private enterprise socialist countries
PC 2162 Suppression public debate
PA 7321 Suppression *Refusal*
PA 7296 Suppression *Restraint*
Suppression [safety...]
PF 2714 — safety records
PF 4633 — scientific evidence
PF 1615 — scientific information
PE 5070 — strikes public authorities
PE 2053 *Suppuration kidney*
PE 5617 *Suppuration ; Lamellar*
PE 5913 *Suppurative arthritis animals*
PE 5050 *Suppurative arthritis older animals*
PF 3354 Supralapsarianism
PC 3024 Supremacy ; Male
PF 2949 *Surcharges ; Excessive bill paying*
PF 5922 Surcharges ocean freight ; Unfair
PE 5931 Surface devoted urbanization ; Increasing proportion land
PE 5096 Surface faulting ; Earth
PE 4515 Surface missiles ; Surface
PD 5641 Surface pollution ; Water
PF 5069 — Reduction albedo Earth's
PE 6526 *Surface ; Slippery road*
PE 4515 Surface surface missiles
PE 2634 Surface water ; Stagnant
PD 8557 Surfaces streets roads ; Broken
PE 6526 *Surfacing ; Inadequate road*
PE 6526 *Surfacing ; Non priority road*
PA 5473 Surfeit *Insufficiency*
PE 4429 Surgeons ; Untrained cosmetic
PE 2663 *Surgery ; Abdominal*
PE 0412 Surgery animals ; Experimental
PD 5119 Surgery ; Delayed
PE 4429 Surgery ; Incompetent plastic
PE 4429 Surgery ; Malpractice plastic
PA 0430 *Surgery ; Psychic*
PE 4429 Surgery quackery ; Cosmetic
PE 9271 Surgery ; Unnecessary
PD 2788 *Surges ; Sea*
PD 2788 *Surges ; Storm*
PE 2863 Surgical complications
PE 4736 Surgical errors
PE 4736 Surgical malpractice
PE 1951 Surgical manipulation brain
PE 9271 Surgical operations ; Unnecessary
PF 7547 Surgical torture
PA 7143 Surliness *Discourtesy*
PA 7253 Surliness *Envy*
PF 4752 Surplus
PC 1258 Surplus armaments
PC 2062 *Surplus domestic animal production*
PA 5473 Surplus *Insufficiency*
Surplus [labour...]
PE 3105 — labour
PF 3630 — Lack financial
PF 5044 — leisure time ; Unpreparedness
PD 1465 Surplus ; Primary commodity
PC 3134 Surplus wealth developing industrialized countries ; Transfer
PC 2062 Surpluses ; Agricultural
PC 2062 *Surpluses animal products*
PD 0156 Surpluses developing countries ; Labour
PE 2902 Surpluses ; Excessive wheat
PC 4879 *Surpluses ; Large trade*
PE 3705 Surprise attack
Surprise free
PF 4774 — forecasts
PF 7700 — methodologies ; Limitations
PF 8176 — policy-making
PF 7700 — thinking ; Limitations
PF 9769 Surprises ; Environmental
PF 9428 Surrealism ; Political
PF 6533 Surrendered control marketing systems
PF 6461 *Surrendered parental role*

PE 3973 Surreptitious entry during surveillance
PA 7411 Surreptitiousness *Uncommunicativeness*
PF 4221 Surrogacy ; Therapeutic sex
PF 2747 Surrogate mothers ; Commercial
PF 2747 Surrogate parenting
PF 6573 *Surrounding territory ; Unexplored*
PD 9538 Surveillance communications ; Misuse government
PE 3972 Surveillance foreign powers ; Aerial
Surveillance governments Misuse
PD 2930 — electronic
PD 2683 — postal
PD 9538 — radio transmission
PF 3701 — satellite
PD 1632 — telephone
PD 0366 Surveillance ; Insufficient health
PF 8871 Surveillance ; Internal
PF 4861 Surveillance medical high risk persons ; Non
PD 8871 Surveillance ; Political
PE 3973 Surveillance ; Surreptitious entry during
PE 2215 Surveys ; Non response governments international
PF 7889 Surveys ; Superficial
PF 6504 *Survival ; Consuming personal*
PF 9187 Survival ; Discouragement food crops guaranteeing
PE 9085 Survival human race ; Uncertainty
Survival [Inadequacy...]
PF 7713 — Inadequacy prevailing mental structures challenge human
PF 4946 — Individual unfitness
PF 3161 — induced individualism
PF 3161 Survival living ; Limited horizons produced
PF 1870 Survival marriage ; Threat
PF 3161 Survival mindset ; Individualist
PE 8902 Survival needs ; Devaluation education
PF 1078 *Survival ; Scavenging*
PF 2602 Survivalism ; Limiting effect individual
PE 4531 Survivors ; Denial right benefits
PA 6971 Susceptibility *Danger*
PD 2389 *Susceptibility disease ; Genetic*
PD 1893 Susceptibility electorate government repetition untruths
PC 9025 *Susceptibility infection*
PA 6882 *Susceptibility Influencelessness*
PD 1043 Susceptibility old physical ill-health
PF 5937 Susceptibility people panic
PE 7161 Susceptibility women alcohol
PE 4183 Suspects ; Abuse rights criminal
PE 4183 Suspects ; Withholding information police criminal
PE 0579 Suspended matter water
PA 6030 Suspense *Fear*
PA 6444 Suspense *Quiescence*
PA 7309 Suspense *Uncertainty*
PE 5869 Suspension *Exclusion*
PD 6927 *Suspension habeas corpus*
PA 5806 Suspension *Inaction*
PA 6852 Suspension *Inappropriateness*
PD 6380 Suspension rights during states emergency
PA 7006 Suspension *Untimeliness*
PE 5793 Suspension ; Violation right workers organizations protection against
PF 4274 Suspicion adult-youth relationships
PF 8335 Suspicion bureaucracy
PF 9729 Suspicion ; Co ops
PA 8653 Suspicion *Distrust*
PA 8653 *Suspicion imposed change*
PC 2089 Suspicion ; Inter Governmental
PA 7392 Suspicion *Unbelief*
PA 8653 *Suspicions ; Misinformed growth*
PE 6367 Suspicious deaths during detention
PF 2528 Sustainable development ; False assumptions
PE 8403 Sustainable development ; Patent restrictions impeding
PF 1870 Sustaining symbols marriages ; Lack
PF 3318 *Sustenance determined priorities*
PE 8644 *Sustenance urban life style ; Perpetual preoccupation*
PE 4819 Suttee
PE 9604 *Swallowing ; Difficulty*
PE 0616 *Swamps*
PD 4751 Swarms ; Insect
PE 7808 Swayback
PF 7427 Swearing
PD 2630 Swearing ; False
PC 8534 *Sweat glands ; Diseases*
PD 2732 *Sweating sickness*
PE 7826 Sweeney
PD 5228 *Sweet clover poisoning*
PD 9667 *Sweet itch horses*
PE 4346 *Sweet potato scurf*
PC 2558 Sweeteners
PE 6390 Sweeteners ; Unsafe artificial
PG 6761 Swelling mucous membrane larynx windpipe
PD 0486 Swindling
Swine [Cardiac...]
PE 9774 — *Cardiac myopathies*
PE 3936 — *Clostridium perfringens type C enteritis*
PD 7841 — *Coronaviral encephalomyelitis*
PE 3936 — *Swine dysentery*
Swine Enteric
PE 3936 — *colibacillosis*
PE 3936 — *diseases*
PE 3936 — *salmonellosis*
PD 2731 *Swine erysipelas*
Swine fever
PE 6266 — [Swine fever]
PE 5207 — African
PE 6266 — Classical
PE 7553 Swine influenza
PD 9293 Swine kidney worm infection
PE 5050 Swine ; Lameness

PE 2727 *Swine ; Mange*
PE 3936 *Swine ; Mesenteric torsion small intestine*
Swine [Necrotic...]
PD 5535 — *Necrotic ear syndrome*
PE 9774 — *Nutritional myopathies*
PE 5050 — *Nutritionally induced skeletal abnormalities*
PD 7841 *Swine ; Oedema disease*
PE 3978 *Swine ; Oesophagogastric ulcers*
PE 3936 *Swine ; Parasitosis*
PE 7553 *Swine ; Pleuropneumonia*
PE 7553 *Swine ; Respiratory diseases*
PE 3936 *Swine ; Rotaviral enteritis*
Swine [Skeletal...]
PE 9774 — *Skeletal myopathies*
PE 0958 — *Stomach worm infections*
PE 9844 — *Streptococcal lymphadenitis*
PE 3936 *Swine ; Transmissible gastroenteritis*
PD 2730 *Swine vesicular disease*
PD 2730 *Swine ; Vesicular exanthema*
PE 2357 Swineherds disease
PE 7304 *Swinepox*
PD 2968 Swings commodity prices ; Cyclic
PD 7841 Swiss cattle ; Bovine progressive degenerative myeloencephalopathy brown
PE 5567 Swollen head syndrome
PF 2314 *Swopping ; Wife*
PE 1756 Sworn ; Failure appear witness produce information be
PE 4586 *Sycamore anthracnose*
PE 3643 *Sylvatic plague*
PE 8814 Symbol ; Lack common
Symbol [Status...]
PF 8145 — Status
PF 3715 — system failure
PF 6461 — systems ; Modern disruption traditional
PF 8754 Symbolic acts ; Superstitious
PF 1560 Symbolic celebrations ; Reduction
PE 9267 Symbolic international agreements substance
PF 2870 Symbolic wealth ; Leadership
PF 2063 Symbolism local relationships ; Lack
PF 1081 Symbols connecting individual's life cosmos ; Absence convincing
Symbols [Desecration...]
PF 9138 — Desecration
PF 1365 — despair
PE 6671 — dominated economy ; Social
PF 3638 Symbols economy ; Superficial
PF 1681 *Symbols ; Faded community*
PE 8814 Symbols ; Lack commitment common
PF 1870 Symbols marriages ; Lack sustaining
PC 7605 Symbols ; Misappropriation cultural
PF 8145 Symbols ; Status
PF 9070 Symbols unrelated human experience
PF 2826 Symbols ; Use offensive
PE 4208 Symptomatic heart disease
PF 2780 Symptoms animal diseases ; Confusion
PC 9067 Symptoms disease ; Ill defined
Symptoms [illness...]
PD 8023 — illness ; Unawareness
PC 0300 — *Intentional feigning physical*
PC 0300 — *Intentional feigning psychological*
PD 8775 Symptoms ; Menopausal
PE 6580 Symptoms more fundamental problems ; Overemphasis
PF 6707 Symptoms problems ; Government delay response
Symptoms referable
PE 9604 — abdomen
PE 3933 — cardiovascular system
PE 9604 — digestive system
PE 9369 — genito-urinary system
PE 4566 — joints
PE 4566 — limbs
PE 3933 — lymphatic system
PE 4566 — musculoskeletal system
PE 9468 — nervous system
PE 7864 — respiratory system
PE 2665 — sense organs
PF 0248 Synarchism
PE 3933 *Syncope*
PD 5453 *Syncope ; Fatal*
PF 7079 *Syncretism ; Religious*
PE 7808 *Syndactyly animals*
PF 8493 Syndicalism
PF 8493 Syndicalism ; Criminal
PE 9324 Syndromes affecting multiple systems ; Congenital
PD 3978 *Syndromes animals ; Malabsorption*
PE 5606 Syndromes ; Cumulative trauma
PC 1306 Synergistic interaction biological agents ; Harmful
PD 2732 *Synoviae infection poultry ; Mycoplasma*
PD 2283 *Synovitis*
PD 2732 *Synovitis ; Infectious*
PE 7826 *Synovitis ; Villonodular*
PE 6390 Synthetic food products
PE 0701 Synthetic reclaimed rubber ; Instability trade crude
PE 8950 Synthetic regenerated fibres ; Instability trade
PD 1276 Synthetic substitutes primary commodities ; Production
PD 1276 Synthetics primary commodities ; Competition
PE 2300 Syphilis
Syphilis [Cardiovascular...]
PE 2300 — *Cardiovascular*
PE 2300 — *central nervous system*
PE 2300 — *Congenital*
PE 2300 *Syphilis ; Early*
PE 2280 *Syphilitic meningitis*
PE 9037 *Syringomyelia*
PE 9291 *Syrup urine disease ; Maple*
PD 8984 System ; Absence maintenance
PE 4570 System accidents ; Sanitation

Tactlessness

System [child...]
PF 6076 — child support enforcement ; Inadequate
PD 8122 — contamination ; Water
PF 0048 — Crisis international monetary financial
System [decision...]
PF 2843 — decision making ; Inadequacy committee
PC 9584 — Declining credibility international trading
PF 4875 — Deficiencies criminal justice
PD 2886 — design ; Irresponsible information
System [Emerging...]
PE 0135 — Emerging oligopolistic world trading
PE 6725 — erosion ; Pay
PD 0722 — Erratic water
PF 2070 — ethical decision making ; Lack
System [failure...]
PF 3715 — failure ; Symbol
PE 0296 — Fragmentation complexity United Nations
PC 9584 — Fragmentation international trading
PD 7374 — System ; Imperialistic distribution
PF 0048 — System ; Inadequacies international monetary
System Inadequate
PE 0296 — coordination United Nations
PD 2269 — drainage
PF 2997 — economic priority
PF 2113 — international judicial
PC 9584 — international trading
PD 8839 — irrigation
PD 6589 — security
PC 1867 — security
PC 9112 — world economic
System [Inappropriate...]
PF 0245 — Inappropriate credit
PF 4516 — incentives ; Asymmetry economic
PD 8839 — Incomplete irrigation
PF 1205 — *Individualized farming*
PF 2261 — *Ineffective care taking*
PF 4984 — *Inflexible educational*
PD 8517 — infrastructure ; Inadequate water
PF 0944 — integration knowledge ; Lack
PE 0117 — international trade ; Inadequate barter
PE 6246 System ; Job reservation apartheid
PD 8217 System justice ; Mistrust
System Lack
PF 3058 — confidence international monetary
PF 0048 — coordination international monetary
PF 5137 — world food security
PF 6539 *System ; Lenient judicial*
PF 2350 System ; Minimal communication
PF 1874 System ; Mismanagement exchange rate
System [organization...]
PF 1460 — organization ; Shortage financial resources United Nations
PE 0730 — organizations ; Inadequate coordination intergovernmental
PF 1451 — organizations ; Ineffectiveness United Nations
PE 1341 — overload ; Power supply
System [Parochial...]
PF 6573 — *Parochial value*
PE 1996 — Participation transnational corporations apartheid
PD 2374 — Plight small states world
PE 4997 — political checks balances ; Inadequate
PF 2139 — Privatization intergovernmental monetary
PE 1579 — Proliferation duplication organizational units coordinating bodies United Nations
System [reduced...]
PE 5397 — reduced individual needs ; Legal contract
PE 0952 — referrals ; Unnecessary health
PC 0834 — *Regressive welfare*
PC 3458 System ; Two tier economic
System [Uncertainty...]
PC 9584 — Uncertainty international trading
PD 8839 — Undeveloped irrigation
PC 9112 — Unjust global economic
PD 3778 — Unrepresentative control international monetary
PD 8122 — Unsanitary water
PC 2677 — Unstable electoral
PC 7873 System ; Vulnerability national economies vagaries international financial
PE 8254 System ; Vulnerability telephone
PF 5301 Systematic human behaviour according cultural norms
PD 2565 *Systemic lupus erythematosus*
PE 9509 *Systemic lupus erythematosus animals*
PF 2071 *Systemic movements ; Growth anti*
PF 6580 Systemic problems ; Failure focus fundamental
Systems [Absent...]
PF 6552 — *Absent socializing*
PD 9544 — Abuse computer
PD 9544 — Abuse information
PD 3493 — Accidents nuclear weapons
PF 6506 — achieving grassroots consensus ; Lack
PD 3112 — Alienation capitalist
Systems [Bias...]
PF 6743 — Bias document classification
PD 9330 — Bias information
PF 8065 — Biased legal
PD 9641 — Biased voting
PD 1147 — Breakdown community security
Systems [Censorship...]
PD 3172 — Censorship communist
PC 3125 — Competition capitalist
PF 3114 — Conflicting roles money capitalist
PE 9324 — Congenital syndromes affecting multiple
PD 3118 — Contradictions capitalist
PD 3179 — Contradictions communist
PF 4197 — Costly taxation
PF 9786 — Counter productive personnel reward

Systems [Cover...] cont'd
PF 4854 — Cover up unsafe transportation
PE 6529 — Culturally biased legal
PF 4899 Systems ; Deficiencies civil justice
PF 4851 Systems ; Deficiencies national local legal
Systems Denial
PC 3176 — democracy communist
PC 3174 — freedom expression thought communist
PC 3173 — freedom movement communist
PC 3124 — human rights capitalist
PC 3178 — human rights communist
PC 3175 — religion communist
Systems Denial right
PC 3177 — national self determination communist
PD 3120 — social security capitalist
PC 3119 — work capitalist
Systems [Destabilization...]
PB 5417 — Destabilization social
PE 8233 — Deterioration irrigation
PD 9196 — Deterioration water supply
Systems developing countries
PF 4132 — Inadequate telephone
PF 2124 — Ineffective tax
PD 3046 — Inequitable tax
PF 4136 — Insufficient television
Systems [Differences...]
PC 2952 — Differences trading principles practices different economic
PD 1414 — Discharge dangerous substances sewage
PD 7450 — Discriminatory design information
PD 1791 — Disparity national tax
PD 9670 — Disruption natural water
PD 3436 — Distortionary tax
PE 4476 — Dual exchange rate
PF 2256 — *Dual school*
Systems [Economic...]
PC 3167 — Economic competition communist
PF 6492 — effective community operation ; Absence support
PC 3170 — Elitism communist
Systems Excessive
PE 1122 — cost comprehensive information
PC 3116 — demand goods capitalist
PE 2525 — land usage transportation
Systems [exchange...]
PE 5903 — exchange notes transnational corporations ; Destabilization monetary
PD 2579 — Exclusive nationally oriented language
PC 3117 — Exploitation capitalist
Systems [Failure...]
PD 9196 — Failure maintain sewage
PB 2496 — *Failure school*
PD 0482 — *Fouling water supply*
PF 6527 — Freezing plumbing
Systems [General...]
PF 3103 — General unproductivity capitalist
PF 1245 — globalization ; Vulnerability socio economic
PF 7059 — government ; Inflexible deliberative
PD 0975 — government ; Outmoded deliberative
PF 5287 Systems ; Inability negotiate effective multilateral safeguard
PF 7713 Systems ; Inadequacy prevailing learning
Systems Inadequate
PE 1728 — appointment
PE 2820 — budgetary coordination United Nations
PF 5172 — correctional
PF 9841 — financial clearing
PF 9841 — financial settlement
PE 8066 — integration international information
PF 6157 — integration transport
PF 4816 — intergovernmental legal
PD 9196 — maintenance water
PD 0520 — prison conditions penal
PF 6635 — radiation monitoring
PC 1867 — security
PD 0876 — sewerage
PF 9716 — use modern information
Systems [Inappropriate...]
PF 9786 — Inappropriate institutionalized reward
PD 0876 — Inappropriate sanitation
PE 7753 — Incompatibility document classification
PF 2212 — Incomplete understanding new societal service
PF 3671 — Ineffective data
PF 1462 — Ineffective tax
PF 4197 — Inefficient taxation
PE 9072 — Inequality inducing effects remote sensing
PD 3436 — Inequitable tax
PF 5514 — Inflation based monetary
PF 4936 — Inhumane use medical life support
PF 2825 — Institutional domination organizational
PF 2580 — Institutionalization discriminatory outmoded concepts legal
PF 2350 — Insufficient communications
PF 1739 — Interaction deficiencies world economical
PD 0104 — international governmental decision making processes ; Inadequate information
PE 7009 — international nongovernmental organizations ; Lack financial information
PF 4836 Systems justice developing countries ; Absence reliable
PF 7806 Systems keep pace technological advancement ; Inability educational
Systems Lack
PE 2820 — centralized financial control United Nations
PE 0075 — coordination United Nations information
PE 8002 — economic adaptation indigenous
PD 3106 — individualism capitalist
PF 6541 — local information

Systems Lack cont'd
PD 8517 — water
Systems [Leakage...]
PD 9196 — Leakage water supply
PE 0366 — Limited blood transfusion
PD 3182 — Limited ownership productive
Systems [Maldistribution...]
PD 0813 — Maldistribution land customary tenure
PD 8412 — Maritime nuclear weapons
PC 3191 — Military industrial complex capitalist
PD 1330 — Military industrial governmental complex communist
PF 6461 — Modern disruption traditional symbol
PF 2905 — monitoring industrial growth ; Inadequate
Systems [national...]
PF 7842 — national accounts developing countries ; Inadequate
PE 1922 — National oligopolistic trading
PD 0664 — neighbouring developing countries ; Lack integration transport
Systems [Obsolete...]
PD 0975 — Obsolete deliberative
PF 8753 — order problem relevant information ; Failure international information
PD 0975 — Outmoded deliberative
PF 2580 — Outmoded legal
PD 0876 — Overloaded sewage
PF 5922 — Overtaxed shipping
Systems [Persistence...]
PF 2580 — Persistence outmoded concepts legal
PD 3171 — Political prisoners communist
PF 1495 — Poor transport
PC 3107 — Poverty capitalist
PF 3498 — practical education ; Ineffective
PC 8390 — Preservation obsolete
PE 3391 — Privatization satellite
Systems Proliferation duplication
PE 0458 — international information
PE 0362 — international nongovernmental organization information
PE 0075 — United Nations information
Systems [Resistance...]
PE 4901 — Resistance incorporating midwives medical care
PC 3108 — Resource wastage capitalist
PD 3122 — Restriction educational opportunities capitalist
PF 2996 — rural citizens ; Inadequate transportation
Systems [Seepage...]
PE 9088 — Seepage water losses irrigation
PD 3654 — Silting water
PF 7745 — Structural amnesia institutional
PF 6533 — Surrendered control marketing
PF 5670 Systems ; Terminological confusion weights, measures numbering
PF 6496 Systems transport communications rural villages ; Inadequate
PE 9831 Systems ; Unauthorized access computer information
PF 1443 Systems ; Uncontrollability world social
Systems Undemocratic
PB 8031 — *economic*
PB 8031 — *social*
PD 9641 — *voting*
Systems [Underdevelopment...]
PD 8673 — Underdevelopment political
PE 4230 — Underparticipation socialist countries international data
PD 7731 — Underreporting hazards biological
PD 4790 — Undeveloped health care
PF 7801 — Unreliability weapons
PD 9641 — Unrepresentative electoral
PF 6471 — *Unutilized wholesale*
PF 7798 Systems verbal exchange ; Collapse
Systems Vulnerability
PE 8542 — computer
PD 4049 — nuclear defence control
PB 2853 — social
Systems [Wastage...]
PE 1396 — Wastage highly skilled personnel routine maintenance complex
PC 3113 — Waste human resources capitalist
PE 8796 — Water losses irrigation
PF 3214 — Weakness multi party parliamentary
PC 2724 — Weakness trade different economic
PD 0356 — Weaknesses military conscription

T

PD 4403 Table ; Lowering water
PF 3310 Taboo
PF 4166 Taboo ; Mother law
PE 7017 Taboos
PD 2868 *Taboos against eating poultry*
PF 3310 *Taboos ; Food*
Taboos Infringement
PF 3976 — [Taboos ; Infringement]
PD 2868 — dietary
PD 2198 — sexual
PF 3310 *Taboos ; Nutritional*
PE 1296 Taboos ; Torture violation
Taboos [Violating...]
PF 3976 — Violating
PD 2868 — Violation food
PD 2868 — Violation nutritional
PF 0327 Tactical methodologies ; Total absence
PF 0327 Tactical methods ; Absence
PF 6119 Tactics ; Government delaying
PA 7143 Tactlessness *Discourtesy*
PA 6446 Tactlessness *Indiscrimination*

Tactlessness

PA 5438 Tactlessness *Neglect*
PE 3511 Taeniasis
PF 4432 Takeover bids abroad ; Blocks hostile
PD 0071 Takeovers ; Proliferation mergers
PE 4398 Talent government posts ; Frustrated
PF 2792 Talent jobs ; Limited ways matching
PA 0005 *Talents ; Hidden individual*
PS 1567 *Talents ; Unused female*
PD 7841 *Talfan disease*
PF 8363 Talismans ; Dependence amulets, charms
PE 4953 Tall buildings ; Environmental stress inhabitants
PE 7933 Tamarins marmosets ; Endangered species
PD 1179 Taming wild animals
PF 4699 Tampering evidence
PF 4699 Tampering official documents
PF 7291 Tampering physical evidence
PD 8804 Tampering ; Product
PF 4699 Tampering scientific records
PE 6781 Tampering witnesses informants proceedings
PF 1214 Tank analysis ; Think
PE 1839 Tanker disasters ; Oil
PE 8233 *Tanks ; Lack water*
Tanning dyeing
PE 8571 — industries ; Environmental hazards
PE 8501 — materials ; Shortage
PE 9089 — trade ; Instability
PE 6586 Tapeworm ; Beef
PE 3511 *Tapeworm ; Dog*
PE 3511 *Tapeworm ; Dwarf*
PE 3511 Tapeworm parasite
PE 3511 *Tapeworm ; Pork*
PD 1632 Tapping ; Illegal phone
PE 8464 Tar crude chemicals derived coal, petroleum natural gas ; Instability trade mineral
PE 8242 Tar crude chemicals ; Shortage coal derived
PD 5228 *Tar poisoning animals ; Coal*
PE 8370 Tar sands energy source ; Underutilization
PE 4586 *Tar spot maple*
PA 7006 Tardiness *Untimeliness*
PD 0087 Targets space ; Military
Tariff barriers
PC 0569 — international trade
PC 2725 — international trade ; Non
PC 2369 — trade developing developed countries
PC 0933 *Tariff non tariff trade preferences developing countries ; Non uniformity*
PC 0933 *Tariff trade preferences developing countries ; Non uniformity tariff non*
PC 1876 *Tariffs ; Preferential*
PE 7826 *Tarsal hydrarthrosis*
PE 7808 *Tarsal valgus*
PE 1570 *Tarsiers ; Endangered species*
PC 1522 Task accountability ; Irregular
PF 1651 Tasks ; Compartmental role
PA 7365 Tastelessness *Boredom*
PE 2770 *Tastelessness ; Food*
PA 5821 Tastelessness *Vulgarity*
PA 5558 Tastelessness *Weakness*
PD 8926 Tattletales
PD 2559 *Tattooing*
PE 4187 Taunting
Tax [administration...]
PE 5009 — administration developing countries ; Inefficient
PD 3436 — assessment ; Inappropriate
PD 4882 — avoidance
PF 1462 — avoidance
PD 3050 Tax barriers dissemination technical knowledge
Tax base
PF 2444 — *Low*
PF 2444 — *Narrowing*
PF 2444 — *Nonexpandable*
PF 2444 — *Restricted*
PF 2444 — *Undiversified*
PD 3047 Tax discrimination against investment foreign country
PD 3048 Tax discrimination against non-residents country
Tax [evasion...]
PD 1466 — evasion
PD 1466 — evasion ; Customs
PF 2137 — *exempt status ; Church abuse*
PE 2370 Tax havens ; Abuse
PE 4290 Tax holidays
PE 6681 Tax impediments international motor traffic
PF 3342 *Tax impediments marriage*
PE 9061 Tax liability ; Understatement
PF 1462 Tax loopholes
PJ 3491 Tax manipulations ; International
PE 9061 Tax obligations ; Unreported
PD 0673 Tax obstacles international investment
PD 3436 Tax ; Paralyzing property
Tax [raise...]
PF 2949 — *raise ; Fear*
PF 2444 — *regulations ; Complex*
PF 2444 — *revenues ; Limited*
PD 3436 Tax structure ; Restrictive
Tax systems
PF 2124 — developing countries ; Ineffective
PD 3046 — developing countries ; Inequitable
PD 1791 — Disparity national
PD 3436 — Distortionary
PF 1462 — Ineffective
PD 3436 — Inequitable
PD 1477 Tax treaties developed developing countries ; Inequitable
PD 3436 Tax ; Unequal graduation income
PD 4221 Taxable objects ; Unlawful trafficking
PE 8444 Taxation certain goods services ; Undue
PD 3436 Taxation ; Excessive
Taxation [Increased...]
PF 2949 — *Increased fear*
PD 0858 — International double
PF 3049 — international non profit organizations ; Inappropriate
PF 3049 Taxation not profit philanthropic charitable organizations ; Inappropriate
PE 4601 Taxation poor
PD 3050 Taxation royalties ; International double
PF 4197 Taxation systems ; Costly
PF 4197 Taxation systems ; Inefficient
PC 8452 *Taxation travel*
PE 5873 Taxes flags convenience ; Evasion shipping regulations
PE 8867 Taxes import charges ; Distortion international trade selective indirect
PE 9220 Taxes transaction goods nonfactor services ; Distorting effects commodity
Taxes [Underpayment...]
PD 4882 — Underpayment duties
PD 3436 — Unfair
PF 2444 — *Unrelated property*
Tea cocoa spices
PE 0915 — manufactures thereof ; Instability trade coffee,
PE 1197 — manufactures thereof ; Long term shortage coffee,
PE 0481 — their manufacture ; Environmental hazards coffee,
PE 0915 *Tea mate ; Instability trade*
PE 2980 Tea ; Pests diseases
PE 2054 Tea trade ; Instability
PF 3470 *Teach unwritten language ; Inability*
PF 2602 *Teacher disinterest ; Mutual student*
PF 5978 Teacher engagement ; Partial
PE 7165 *Teacher facilities ; Limited*
PF 1187 Teacher non communication ; Parent
PF 5978 Teacher participation ; Uninspired
PE 7165 Teacher post ; Unattractive
PF 6467 *Teacher ratio ; High student*
PF 1187 Teacher separation ; Parent
PF 4852 *Teacher training ; Inadequate*
PE 8221 Teacher turnover ; High
PD 3722 Teachers ; Children neglected
PF 5978 *Teachers ; Detached local*
PE 8103 Teachers' houses ; Unconstructed
Teachers [Inadequate...]
PE 7165 — Inadequate working conditions
PE 7165 — *Insufficient incentives*
PD 8108 — Insufficient skilled
PC 2163 *Teachers ; Lack vocational*
PE 6183 Teachers ; Maldistribution
PC 2163 *Teachers ; Minimal technical*
PF 3498 *Teachers ; Poor supervision*
PE 8429 Teachers rural areas developing countries ; Unavailability trained
PE 8103 Teachers rural communities ; Lack housing
Teachers [salaries...]
PE 8645 — salaries ; Non competitive
PD 8108 — Shortage qualified
PD 8108 — Shortage trained
Teachers [Underpaid...]
PE 8645 — Underpaid
PE 8645 — Underpayment
PF 5978 — Unmotivated
PE 3909 Teaching children lie
PF 0905 *Teaching equipment ; Inadequate*
PF 5929 Teaching ; Ethnic social discrimination foreign language
Teaching [Inadequate...]
PC 9714 — Inadequate
PC 9714 — Inadequate standards
PF 0956 — intellectual methods ; Faulty approaches
PF 6028 Teaching ; Limits second language
PC 9714 Teaching ; Low quality
Teaching methods
PF 2721 — Limited
PF 2721 — Poor
PF 2721 — Restrictive
PE 8305 Teaching occupational health safety ; Inadequate
PF 2568 Teaching potential ; Unrealized
PF 4375 *Teaching staff ; Costly*
PE 4640 Tear gas ; Indiscriminate use
PB 1701 Tear ; Wear
PE 4187 Teasing
Technical [advance...]
PE 5104 — advance ; Antagonism employment policy
PE 1981 — advice consultation problems ; Inadequate
PF 2698 — agricultural business training ; Limited availability
PF 8120 — aid ; Unproductive
PF 0772 — aids disabled persons ; Insufficient
PE 4866 — assistance ; Imbalance capital
PF 2813 — *assistance ; Inadequate*
PC 4382 Technical barriers trade
PF 4741 Technical basis competition ; Restriction
PF 0863 Technical cooperation problems ; Inadequate
Technical [data...]
PF 0625 — data ; Inadequate economic
PD 2860 — debates ; Politicization
PE 4933 — development excess manpower developing countries ; Lack
PE 8190 — development ; Lack economic
PE 2462 Technical faults ; Unexplained
Technical [information...]
PF 0209 — information concerning problems ; Inadequate exchange
PF 3397 — information gap
PE 5814 — infrastructure maritime commerce developing countries ; Lack
PF 4741 — innovation ; Self limiting
PD 7719 Technical know how ; Loss
PD 3050 Technical knowledge ; Tax barriers dissemination
PF 3931 Technical progress ; Dogmatic belief benefits
PD 3154 Technical research ; Restrictions sharing
PF 2959 Technical resources ; Overemphasis effective use
Technical skills
PF 6500 — Insufficient
PF 8594 — Limited
PF 6500 — Low level
PF 2848 — rural communities ; Limited reservoir
PF 2848 — rural communities ; Shallow reserve
PF 6500 — Shortage
PD 7719 — society ; Loss
PF 0799 Technical solutions complex environmental problems ; Simplistic
PE 2324 Technical state war following cease fire agreements ; Persistence
PC 2163 *Technical teachers ; Minimal*
Technical training
PF 4375 — Costly
PE 8716 — Inadequate
PE 8716 — Ineffective
PF 2313 — *Nonavailability*
PD 0366 Technicians ; Insufficient health
PF 3523 *Technicians ; Unemployed trained*
PF 6330 Technicism
PF 9127 Techno-fear
PE 7153 Techno-mercenaries
PF 9127 Techno-phobia
PF 4989 Technocentrism
PF 6330 Technocracy
PF 6330 Technocracy ; Machine
PF 6330 *Technocratic futurism*
PF 6330 Technocratic leadership
Technological [activities...]
PF 9328 — activities ; Delayed environmental impact
PC 1153 — activity ; Irresponsible scientific
PF 3184 — adaptation ; Insufficient
PE 7898 — advance ; Fast irregular pace
PE 7943 — advance ; Functional changes due
PF 7806 — advancement ; Inability educational systems keep pace
Technological [capacity...]
PD 7719 — capacity ; Disintegration
PC 8885 — capacity ; Gap scientific
PF 9127 — change ; Irrational rejection
PD 1276 — change ; Reduction demand primary commodities due
Technological changes
PF 1336 — Indecisive response
PE 4596 — Training inappropriate structural
PE 7185 — Working hours inappropriate structural
PF 0953 Technological communications ; Thwarted
PF 2401 *Technological contact ; Insufficient*
PF 4636 Technological decisions ; Limited access
Technological development
PD 7463 — capitalism ; Abusive
PD 7463 — capitalist society ; Perverse impact
PC 1227 — Fragmentation
PE 7095 — Intellectual methods which disenfranchise grassroots
PF 6454 — Lack means local
PF 4741 Technological diversity ; Reduction
PF 2410 Technological equipment ; Inappropriate level
PC 1730 Technological expansion ; Undirected
PD 4426 Technological failures ; Disastrous
PE 3078 Technological growth ; Increasing development lag against
PE 3745 Technological household gimmicks
Technological innovation
PF 2707 — Centralized decisions local
PD 0528 — Endangered jobs
PF 9127 — Fear
PF 4741 — Self perpetuating attitudes
PF 3409 *Technological know how ; Insufficient*
PF 9345 Technological leadership ; Loss international
PF 4741 Technological monoculture
PF 1321 Technological needs demands ; Gap material
Technological [process...]
PF 5438 — process ; Dehumanization man
PF 3931 — progress ; Adverse consequences scientific
PF 3310 — *progress ; Lack*
PC 3234 Technological revolution
Technological [skill...]
PF 2956 — skill overemphasis
PF 8552 — skill ; Underdeveloped
PF 2956 — skills ; Overemphasis
PF 8835 Technologically untrained farmers
PE 5336 Technologies ; Decreasing rate development major agriculture
PF 0878 Technologies developing countries ; Use inappropriate
PE 4391 Technologies ; Elimination jobs developing countries due introduction new
PF 3514 Technologies ; High risk
PE 8758 Technologies ; Hindrances international spread new
Technologies [Inadequate...]
PJ 9034 — Inadequate assessment new
PF 3884 — Inadequate social
PE 3337 — Incompatibility traditional new
PC 1227 — Incompatible
PE 5134 Technologies ; Negative social effects introduction new
PD 0528 Technologies ; Unemployment introduction new
Technology [agricultural...]
PF 3467 — agricultural upgrading ; Insufficient access
PC 0840 — *arms race ; Information*
PJ 9034 — assessment ; Inadequate capacity
PF 4989 Technology ; Blind faith

Territories

PC 0418 Technology ; Bulldozer
Technology [capitalism...]
PE 3105 — capitalism ; Abuse science
PC 8885 — Concentration science
PB 0001 — *criminal use*
Technology Dependence
PD 2420 — [Technology ; Dependence]
PF 1489 — developing countries imported
PF 4741 — dominant
PF 4741 — single
PE 0046 *Technology developing countries ; Inadequate cartographic*
PD 6571 Technology development ; Dependence sophisticated
PF 6300 Technology education ; Resistance
PD 2712 Technology ; Elitist control construction
PF 9127 Technology ; Fear new
PD 0338 Technology gap developed countries
PE 7985 Technology gap developed developing countries
PF 3409 Technology ; Haphazard transmission practical
PE 7174 Technology hostile countries ; Diversion high
Technology [Inadequate...]
PE 4880 — Inadequate capacity developing countries science
PE 5820 — Inappropriate transfer
PE 7637 — Inconsistent application telecommunications
PC 0418 — Indiscriminate application distribution
PE 8458 — industries ; Protectionism high
PF 2848 — *Ineffective agricultural*
PF 6524 — *Insufficient modern*
PE 5820 — Introduction inappropriate
PF 0770 — Irrelevance science
PE 7174 — irresponsible groups ; Diversion high
Technology [Lack...]
PE 4780 — Lack neighbourhood
PF 1653 — *Lack specialized*
PF 6617 — Limited social context developing
Technology [Maldistribution...]
PC 8885 — Maldistribution science
PE 5023 — man industrialized societies ; Inadaptation
PE 8261 — *medical cures ; Inaccessibility high*
PF 3554 — Minimal access appropriate
PE 6039 Technology ; Non acceptance embryo transfer
Technology [Obstacles...]
PJ 8292 — Obstacles proper use
PC 1153 — Obstacles proper use science
PD 0361 — Opposition irradiation
PF 1319 — *Outdated farming*
Technology [Pollution...]
PF 1789 — Pollution constraints testing new
PF 2464 — Profit motivated utilization construction
PD 2420 — Proliferation
PD 0837 — Proliferation nuclear weapons
PE 4032 — Public non accountability organizations developing
PD 1291 Technology ; Reverse transfer
PD 4445 Technology ; Super power monopoly advanced nuclear warfare
Technology [Trade...]
PD 8621 — Trade biological warfare
PD 9692 — Trade chemical warfare
PE 1978 — transactions ; Restrictive business practices
Technology transfer
PE 4922 — developing countries ; Decline
PE 4922 — developing countries ; Impasse
PE 4922 — Disadvantageous terms
PD 7463 — programmes ; Paradoxes
PE 5820 — transnational corporations ; Adverse impact
PE 1918 Technology transnational corporations ; Monopolization
Technology [Unbalanced...]
PE 7637 — Unbalanced application communications
PC 0418 — Uncontrolled application
PC 1174 — Uncontrolled environmental impact
PD 0086 — *Underrepresentation women science*
PC 1153 — Unethical use science
PF 3409 — *Unexposed farming*
PE 6865 Technology ; Visual strain modern
PC 7041 Technology ; Worker maladjustment
PA 7090 *Technophobia*
PE 5053 Technostress
PC 6684 Tectonic disasters ; Telluric
PF 6505 *Tedious funding procedures*
PA 7365 Tedium *Boredom*
PA 6997 Tedium *Imperfection*
PA 6890 Tedium *Nonuniformity*
PA 7107 Tedium *Unpleasantness*
PD 1302 Teenage abortion
PD 1302 Teenage abortions
PF 6570 *Teenage language idiom*
PD 0614 Teenage motherhood ; Early
PD 0614 Teenage pregnancy
PD 6213 Teenage prostitution
Teenage [sex...]
PD 7439 — sex
PE 6969 — sex ; Health risks
PE 5771 — suicide
PF 7498 Teenage violence
PF 6061 Teenagers ; Inaccessibility health care
PE 3711 *Teeth ; Abnormalities*
PD 1185 Teeth disorders
PF 4832 *Teeth ; Inadequate care*
PD 1185 *Teeth ; Irregularity*
PD 1185 *Teeth ; Loosening*
PE 0004 Telecommunication facilities developing countries ; Inadequacy
PE 5860 Telecommunication services land locked developing countries ; Unreliable
PD 5701 Telecommunications capabilities ; Disparity world
PE 7637 Telecommunications capability ; Gaps
PD 0187 Telecommunications ; Destabilizing international
PE 3657 Telecommunications facilities ; Delay connection
PF 4132 Telecommunications facilities ; Maldistribution
Telecommunications [Inadequate...]
PF 0045 — Inadequate satellite capacity
PD 2930 — Interception
PE 0223 — Invasion privacy compulsory
PE 5655 — island developing countries ; Inadequate
Telecommunications [sectors...]
PF 5877 — sectors ; Ineffective self regulation
PE 4450 — service ; Irresponsible use
PE 4450 — services ; Inappropriate use
PE 1840 — standards ; Parochial
PE 7637 Telecommunications technology ; Inconsistent application
PF 8352 Telepathy
PD 1632 Telephone bugging
Telephone [calls...]
PE 4367 — calls ; Illegal long distance
PE 5757 — calls ; Obscene
PF 1653 — *costs ; High*
PF 1653 — *costs ; Prohibitive*
PF 1698 Telephone delays
PF 1653 *Telephone lines ; Costly*
PE 5402 Telephone ; Pornography
Telephone [service...]
PE 8254 — service ; Undependable
PE 4450 — service ; Unethical use
PE 5757 — services ; Salacious
PD 1632 — surveillance governments ; Misuse
PE 8254 — system ; Vulnerability
PF 4132 — systems developing countries ; Inadequate
PF 1205 *Telephone use ; Inconsiderate*
PE 3657 Telephones ; Delay connection
PF 4132 Telephones rural communities ; Insufficient
Television [censorship...]
PD 3029 — censorship ; Radio
PD 2045 — communications ; Interference radio
PE 6982 — Confusion values monitoring
PE 6982 Television ; Excessive regulation
PD 0433 Television ; Exploitative commercial
Television [film...]
PE 4260 — film violence ; Effect
PE 4260 — films ; Effect violence
PF 3703 — frequency bands satellite transmission ; Allocation
PD 1533 Television ; Increase antisocial behaviour due
PD 5833 Television ; Inequality inducing effects
PD 3040 *Television monopoly*
PE 5610 Television news coverage capitalist countries ; Political interference
Television [programming...]
PD 0433 — programming ; Advertiser controlled
PD 5377 — programming standards ; Deterioration
PD 3085 — propaganda ; Radio
PF 1790 *Television reception ; Poor*
PF 4136 Television sets ; Maldistribution
PF 4136 Television systems developing countries ; Insufficient
Television [viewer...]
PE 4395 — viewer fatigue
PD 1533 — viewing ; Excessive
PE 4260 — violence
PD 1533 Television watching ; Habitual
PD 1533 Television watching ; Late
PC 6684 Telluric tectonic disasters
PA 7253 Temper *Envy*
PC 2087 Temperature change ; Ocean
PE 8971 *Temperature differentials oceans ; Underutilization*
PC 0918 Temperature ; Increasing air
PC 0119 *Temperature inversion*
PE 3734 Temperatures ; Torture exposing freezing
PA 7156 Tempestuousness *Moderation*
Temporal [deprivation...]
PF 4644 — deprivation
PF 4644 — discrimination
PF 4644 — disorientation
PF 4644 Temporal ghettos
PF 9980 Temporal illiteracy
PF 1432 Temporal imperialism
PD 2269 Temporary drainage ; Ineffective
PG 8978 *Temporary residence*
PG 8978 *Temporary resident patterns*
PF 1623 Temporizing
PA 7736 Temptation
PA 7736 Temptation ; Dependence
PA 7343 Temptation *Dissuasion*
PA 7736 Tempters *Temptation*
PD 3056 *Tenant repair ; Uncompensated*
PE 3528 Tenant's dependence middlemen ; Farm
PD 8916 Tenant wages ; Low
PD 3465 Tenants ; Exploitation housing
PE 8633 Tenants ; Inadequate working conditions land
PE 7169 Tenants ; Irresponsible housing
PE 7169 Tenants ; Negligence housing
PE 7169 Tenants ; Unethical practices housing
PF 2922 Tendencies ; Repressed sexual
PE 4301 Tendering ; Collusive
PE 7072 Tendering international trade ; Collusive
PF 9209 Tenderness ; Fear
PE 5913 *Tendinitis animals*
PE 5913 *Tendon ; Bowed*
Tendon [reflex...]
PE 1310 — *reflex ; Loss*
PD 7424 — *Rupture Achilles*
PE 4161 — *Rupture gastrocnemius*
PE 7826 *Tendons ; Contracted*
PE 7808 *Tendons ; Contracted flexor*
PD 2647 *Tendons ; Inflammation*
PE 9604 *Tenesmus*
PE 5606 Tenosynovitis
PE 5913 *Tenosynovitis*
PE 5913 *Tenosynovitis animals*
PD 3479 *Tenrecs ; Endangered species*
PB 6370 Tension
PF 5555 Tension care responsibility ; Collapsed
PA 5532 Tension *Disaccord*
PA 5446 Tension *Enmity*
PA 6030 Tension *Fear*
PF 6178 Tension free social structures ; Ideal
PA 5473 Tension *Insufficiency*
PB 8287 Tension ; International
PB 6302 Tension ; Mental
PE 8080 Tension oil petroleum crises ; World
PE 3761 Tension ; Premenstrual
PF 7780 Tension ; Sectarian
PJ 0095 Tension ; Undervalued creative establishment/dis establishment
PF 1675 Tensionless image free choice
PC 5917 Tensions ; Area
PC 3685 Tensions ; Ethnic
PE 3948 Tensions ; Failure understand necessity creative establishment disestablishment
PE 5927 Tensions involving transnationals ; Labour
PF 2241 *Tensions ; Lifestyle*
PF 2448 Tensions ; Plural society
PC 2494 *Tensions political parties ; Inner*
PC 5917 Tensions ; Regional
PB 2853 Tensions society ; Structural
PA 6858 Tenuousness *Disintegration*
PA 7285 Tenuousness *Littleness*
PD 0406 Tenure ; Feudalistic land
PD 9162 Tenure ; Insecure land
PE 2451 *Tenure ; Insecure lease*
PD 9162 Tenure ; Short term land
PD 0813 Tenure systems ; Maldistribution land customary
PE 0697 Teratogens
PE 3624 Teredos pests
Terminal [ileitis...]
PE 3936 — *ileitis*
PE 4906 — illness ; Chronic
PE 1277 — isolation
PE 5083 Terminal work ; Job stress video display
PF 2859 Termination pregnancy ; Non
PF 2152 *Terminological ambiguities*
PF 5670 Terminological confusion weights, measures numbering systems
PF 3824 Terminological crisis social studies
PF 5383 Terminological deception
PF 0091 Terminological equivalents languages ; Lack
PF 6014 Terminology ; Sexually discriminating job
PF 2174 Termism ; Financial short
PE 1747 Termites pests
PE 4603 Terms financial loans developing countries ; Deteriorating
PF 3128 Terms ; Politically emotive words
Terms [technology...]
PE 4922 — technology transfer ; Disadvantageous
PD 2897 — trade developing countries ; Deterioration
PE 0853 — transnational corporations developing countries ; Inadequate negotiation entrance
PB 5250 Terracide
PE 1720 Terrain vehicles ; Environmental degradation off road
PD 5278 Terrestrial animals ; Accumulation contaminant residues
PD 0381 Terrestrial plants ; Accumulation contaminant residues
PD 0381 Terrestrial plants ; Accumulation pollutants
PC 1888 *Terrestrial territory ; Conflicting claims non*
PD 5278 Terrestrial wildlife ; Accumulation pollutants
PE 6958 Terrifying dreams
PF 1066 Territorial bases ; Unilateral declarations independence extra
PF 0090 Territorial boundaries ; Artificial
PC 1888 Territorial claims ; Unsettled
PC 1888 Territorial disputes states
PC 9547 Territorial expansionism
PF 0090 Territorial fragmentation
PC 2944 Territorial fragmentation
PC 2945 Territorial integrity ; Denial rights
PC 2945 Territorial integrity ; Disruption
PD 3496 Territorial military bases intelligence centres ; Extra
Territorial waters
PD 1628 — Conflicting claims concerning off shore
PE 6755 — Denial right innocent passage
PD 1628 — Unilateral claims off shore
PE 6755 — Unlawful interference rights innocent passage
PD 0055 Territorially organized sovereign states ; Domination world
PF 2943 Territories accorded United Nations non-self-governing status disputed administering government
PC 2362 Territories ; Conflicting claims states
Territories [Denial...]
PD 6620 — Denial rights
PF 1452 — *Dependent*
PC 2362 — Disputed
PF 1066 Territories ; Extra jurisdictional
PF 2943 Territories ; Lack United Nations jurisdiction administered
PE 5785 Territories ; Military threats island countries
PD 0441 Territories ; Non viability small states
PD 8021 Territories ; Occupied
PC 2343 Territories ; Organized crime families
PE 5700 Territories ; Vulnerability island developing countries

PE 6074 Territories ; War time exportation cultural property occupied
PC 1888 *Territory ; Conflicting claims concerning Antarctic*
PC 1888 *Territory ; Conflicting claims non terrestrial*
PF 7435 Territory ; Decreasing
PF 3390 Territory ; Undue attachment
PF 6573 *Territory ; Unexplored surrounding*
PF 8483 Terror
PE 2197 *Terror disorder ; Sleep*
PA 6030 Terror *Fear*
PA 7156 Terror *Moderation*
PD 5574 Terrorism
PD 6656 Terrorism against idea
PE 4089 Terrorism ; Air
PD 5582 Terrorism ; Environmental
PE 9210 Terrorism financed trade drugs
PF 4366 Terrorism ; Government failure sanction state supported
PD 6008 Terrorism ; Government incitement
Terrorism [Inadequate...]
PF 4366 — Inadequate international cooperation reducing
PF 4240 — Ineffective prevention
PD 6656 — Intellectual
PE 9210 Terrorism ; Narco
PF 6846 Terrorism ; Passive public support
PD 4134 Terrorism ; Religious
Terrorism State
PD 6008 — [Terrorism ; State]
PD 6008 — sponsored
PD 6008 — supported international
Terrorism targeted against
PE 5944 — business corporations
PE 5944 — corporate executives
PE 5614 — politicians
PD 9997 Terrorism ; Urban
PF 4240 Terrorist activities ; Persistence
PE 6844 Terrorist activity media ; Excessive portrayal
PE 2368 Terrorist bombing
PD 5426 Terrorist food poisoning
PD 5541 Terrorist havens
PF 4240 Terrorist organization ; Ineffective anti
PD 5272 Terrorist raids ; Cross border anti
PE 9207 Terrorists armed biochemical weapons
PE 3769 Terrorists armed nuclear weapons
PD 5574 *Terrorists ; Suicidal*
PD 5574 *Terrorists ; Suicide*
PE 4466 Terrorizing
PE 0616 *Tertian ; Benign*
PD 7841 *Teschen disease*
PF 1266 Test-led education
PF 1602 Test results ; Falsification scientific
PD 7415 Test results ; Misuse scientific medical
PE 8634 Test tube conception
PE 6750 Testamentary dispositions ; Ineffective
PD 2557 *Testaments ; Forgery wills*
PD 8301 *Testicular dysfunction*
PE 4349 Testify ; Refusal
PE 6633 Testifying against oneself ; Denial right freedom
PE 5267 Testimony ; Flight avoid prosecution giving
PE 9172 Testimony psychologists psychiatrists ; Inept
PD 1190 Testing drugs ; Inadequate
PB 0284 *Testing ; Enforced drug*
PF 5304 Testing errors ; Laboratory
PF 4450 Testing ; Government secrecy concerning nuclear weapons
Testing [Inadequacy...]
PD 1975 — Inadequacy insensitivity intelligence
PD 7415 — Inappropriate scientific medical
PB 0284 — *Invasion privacy*
PB 0284 *Testing ; Mandatory AIDS*
PF 1789 Testing new technology ; Pollution constraints
PC 2201 Testing ; Nuclear weapons
PE 9611 Testing pharmaceutical products ; Cruel treatment animals
PF 1266 Testing procedures ; Inappropriate educational
PC 2201 Testing ; Thermonuclear weapons
PE 9611 Testing toiletries cosmetics ; Abusive treatment animals
PE 8372 *Testis ; Cancer*
PF 5679 Tests ; Dubious medical
PB 0284 *Tests ; Enforced lie detector*
PB 0284 *Tests ; Enforced polygraph*
PF 2714 Tests ; Fraudulent safety
PD 7415 Tests ; Inconclusiveness scientific medical
PF 5679 Tests ; Unnecessary health
PE 2530 Tetanus
PD 7420 *Tetany cattle ; Hypomagnesemic*
PD 7420 *Tetany ; Grass*
PD 7420 *Tetany mares ; Lactation*
PD 7420 *Tetany ; Puerperal*
PD 7420 *Tetany ruminants ; Transport*
PD 7420 *Tetany sheep ; Hypomagnesemic*
PD 7420 *Tetany ; Transit*
PE 5569 *Tetralogy fallot animals*
PE 2946 Tetraplegia
PD 0752 *Texas cattle fever*
Textbooks Biased inaccurate
PF 9358 — biology
PF 1780 — geography
PD 2082 — history
PF 0118 Textbooks developing countries ; Shortage books
PD 2716 Textbooks reference books ; Inadequate inaccurate
PF 9358 Textbooks response creationism ; Censorship
PE 5819 Textile apparel industries ; Protectionism
Textile clothing industries
PE 1103 — Environmental hazards
PE 1008 — Instability

Textile clothing industries cont'd
PE 0453 — Underdevelopment
PE 1550 Textile fibres their waste ; Instability trade unprocessed
PE 9074 Textile fibres waste ; Environmental hazards
PE 5823 Textile industry ; Inadequate conditions work
PE 8970 Textile yarn fabrics, made up articles ; Instability trade
PE 8436 Textile, yarn fabrics, made up articles ; Shortage
PF 5421 Texts ; Inconsistent official
PJ 3021 Thalassaemia
PD 1827 Than content ; Educational curricula over emphasizing method rather
PF 3549 Than method ; Educational curricula based content rather
PE 8606 Than wheat meslin shortage ; Meal flour
PD 5086 Thanatomania
PF 3849 Thanatophilia
PF 0462 Thanatophobia
PF 3849 Thanos
PE 4786 Theatre acts ; Maltreatment animals used
PD 3028 Theatre censorship
PD 3032 *Theatre clubs ; Private*
PE 4736 Theatre mistakes ; Operating
PD 3028 *Theatre performances ; Refusal licence*
PD 3088 Theatre propaganda
PF 9631 Theatricalization public life politics ; Media
PD 5552 Theft
PD 0577 Theft archives
PE 4826 Theft ; Automobile
PD 0577 Theft books
Theft [damage...]
PC 1784 — damage ; Vulnerability property
PD 2957 — data
PD 0577 — documents
PE 3684 Theft ; Employee
PE 5103 Theft ; Freight pilferage
Theft [ideas...]
PD 1284 — ideas
PF 8854 — ideas
PF 8854 — intellectual property
PD 3495 Theft nuclear materials
Theft [pets...]
PE 0805 — pets
PD 5552 — *Petty*
PD 4691 — property
PD 4691 — *public property*
PD 0577 — public records
PD 5552 Theft ; Risk
PE 4826 Theft ; Risk car
PD 4711 Theft services
PE 4826 Theft vehicles
PE 0323 Theft works art
PE 1106 Thefts ; Office work place
PE 3996 Theileriases
PE 3996 Theileriasis
PD 2735 *Theileriosis*
PF 3422 Theism
PD 5535 *Thelaziasis*
PF 4548 Themselves specially chosen ; Peoples perceiving
PD 3357 *Theocracy*
PF 6358 Theological collapse
PF 3807 Theological justification nuclear war
PF 9671 Theological justification population growth
PF 5694 Theology disrelated life ; Religious
PE 0567 Theology ; Gender bias
PF 2005 *Theology ; Shallow*
PF 6966 Theomania
PF 7220 Theoretical issues ; Misuse democratic processes resolve
PF 9576 Theoretical models social development ; Inappropriate
PF 3348 Theories ; Misuse evolutionary
PC 9188 Theories ; Misuse scientific
PD 1709 Theories ; Non acceptance new scientific
PF 9728 Theory ; Inappropriate economic
PF 4221 Therapeutic sex surrogacy
PF 6751 Therapeutic substances human origin ; Limited availability
PE 9172 Therapy ; Recreational
PC 1609 Thermal pollutants
PD 1165 Thermokarst
PC 2201 Thermonuclear weapons testing
PF 7728 They are innocent ; Failure clear names people convicted crimes which
PF 0836 *They language patterns ; We /*
PE 2185 Thiamine deficiency
PE 5567 *Thick head*
PE 4161 *Thimbling*
PF 1214 Think-tank analysis
PF 3862 Thinking behaviour ; Proscribed
PF 9821 Thinking concerning global problematique ; Inadequate development political
PF 4510 Thinking ; Fuzzy environmental
PE 7040 Thinking industry ; Inflexible military
PF 7700 Thinking ; Limitations surprise free
PF 9065 Thinking ; Scholastic
PF 5039 *Thinking ; Single resource*
PF 5039 Thinking ; Uncritical
PF 5039 Thinking ; Unreflective
PF 8922 Thinking ; Wishful
PA 7152 Thinness *Fewness*
PA 5473 Thinness *Insufficiency*
PA 7371 Thinness *Unintelligence*
PA 7204 Thinness *Unsavouriness*
PE 6290 Thinning ; Eggshell
PE 5436 Thiols mercaptans pollutants
PE 7826 *Third phalanx ; Fracture*
PE 3818 Thirst

PE 4247 *Thoracic trauma animals*
PE 0353 *Thorium ores concentrates shortage*
PE 8212 Thorium trade instability ; Uranium ores concentrates
PC 3174 Thought communist systems ; Denial freedom expression
Thought [Denial...]
PF 3217 — Denial freedom
PF 3217 — Denial right freedom
PF 3217 — Dependence denial freedom
PF 0441 Thought ; Harmful
PF 5301 Thought ; Organization human
PF 7713 Thought patterns ; Antiquated
PD 1624 Thought reform
PF 3217 Thought ; Restrictions freedom
PF 9205 Thoughtless acceptance convenient beliefs
PA 6940 Thoughtlessness
PA 6247 Thoughtlessness *Inattention*
PA 5438 Thoughtlessness *Neglect*
Thoughtlessness [Unintelligence...]
PA 7371 — *Unintelligence*
PA 5643 — *Unkindness*
PA 7232 — *Unskillfulness*
PF 5205 Thoughts ; Dirty
PF 5205 Thoughts ; Impure
PF 0441 Thoughts ; Negative
PF 2060 Thoughts reflections ; Parental control children's
PE 1323 Thread ; Gold
PB 1992 Threat
PE 5505 Threat arson attacks
PD 1111 Threat birds aircraft safety
PD 9389 Threat commercial reprisals
PE 4661 Threat ; Criminal
Threat [Danger...]
PA 6971 — *Danger*
PE 5488 — diplomatic immunity
PA 6191 — *Disapproval*
PE 8014 Threat industrial combines farming communities
PF 8874 Threat ; Military
PF 9181 *Threat military retaliation*
PF 3707 Threat official reprisals
Threat [parliamentary...]
PF 0609 — parliamentary immunity
PF 0609 — parliamentary privilege
PD 0736 — police reprisals
PF 1269 — psychological well being ; Retirement
PF 9078 Threat retaliation
PF 1870 Threat survival marriage
PF 8874 Threat war
PE 4609 Threat young ; Drug
PC 3295 Threatened minorities
PC 1995 Threatened sects
PC 3295 Threatened vulnerable minorities
PE 8174 Threatened warfare ; Children
PD 0540 Threatening public servants
Threats against
PE 3308 — family friends
PB 1992 — *peasants*
PD 7471 — trade union leaders
PE 4869 — trade union leaders ; Death
PD 2044 — voters
PE 4869 — workers representatives ; Death
PD 0337 Threats ; Death
PF 3699 Threats ; Deliberate invention new
PE 1622 Threats governments against governments
Threats [ideological...]
PC 3362 — ideological movements minorities
PJ 3472 — independence professionals
PB 0009 — international peace security
PE 5785 — island countries territories ; Military
PE 7793 Threats ; Lack protection following death
PE 5837 Threats land locked countries ; Military
PC 4341 Threats national security ; Environmental
PD 5426 Threats poison food
PF 4854 Threats public safety ; Non disclosure
PF 4729 Threats ; Unreported
PE 2260 Three-day fever
PD 2730 *Three-day sickness*
PD 0276 *Three generation family structure ; Demise*
PD 4519 Thresholds ; Commercial distortion toxicity
PE 5239 Thresholds ; Failure respect internationally agreed toxicity
PF 5188 Thresholds ; Uncertain radiation hazard
PF 5188 Thresholds ; Uncertain toxicity
PA 7341 Thriftlessness *Unpreparedness*
PE 3619 Thrips
PE 6909 Thrive ; Nonorganic failure
PD 7924 Throat ; Diseases
PE 7733 Throat ; Diseases
PE 8173 Throat, eye, limb respiratory muscles ; Paralysis
PE 3098 *Throat ; Haemolytic streptococcal infection*
Throat [Infection...]
PE 7733 — *Infection*
PE 4651 — inflammation
PE 7733 — *irritation*
PE 9819 Throat ; Malignant neoplasm mouth
PD 8347 *Throat mouth ; Benign neoplasm*
PE 4651 Throat ; Sore
PE 4651 Throat ; Soreness
PG 2234 *Thrombocytopenia*
PE 9509 *Thrombocytopenia*
PE 9509 *Thrombocytopenia animals ; Autoimmune*
PD 2732 *Thrombocytopenia ; Canine infectious cyclic*
PD 7841 *Thromboembolic meningitis ; Infectious*
PE 9037 *Thrombophlebitis*
PE 5783 Thrombosis
PD 5453 *Thrombosis animals*

PE 8456 Thrombosis ; Coronary
PC 9042 *Thrombosis ; Puerperal*
PD 8942 Throwaway society
PC 4864 *Throwing ; Stone*
PG 5292 *Thrush*
PE 4497 Thrush chickens
PE 4497 Thrush turkeys
PE 3881 Thunderstorms
PF 2575 *Thwarted community enthusiasm*
PF 0953 Thwarted technological communications
PD 1762 *Thylacine ; Endangered species*
PE 7708 Thymic tumours
PE 7708 Thymus gland ; Diseases
Thyroid [gland...]
PE 0652 — gland disorders
PE 0652 — glands inflammatory lesions
PE 0652 — glands tumorous overgrowth
PG 3408 *Thyroidism ; Hyper*
PG 3408 *Thyroidism ; Hypo*
PE 0652 *Thyroiditis*
PE 9509 *Thyroiditis animals ; Autoimmune*
PE 0652 *Thyrotoxicosis*
PE 3619 Thysanoptera insect pests
PE 3620 Thysanura insect pests
Tic disorder
PE 9344 — *Chronic motor*
PE 9344 — *Chronic vocal*
PE 9344 — *Transient*
PE 9344 Tic disorders
Tick borne
PE 3897 — diseases
PE 2348 — encephalitis
PE 5530 — rickettsiosis
PD 7841 *Tick paralysis*
PE 2254 *Tick ; Sheep*
PE 1766 *Ticks ; Fowl*
PE 1766 Ticks pests
PE 9344 Tics
PE 8971 *Tidal energy ; Underutilization*
PE 3813 Tidal flooding ; Exposure settlements
PE 5006 Tidal floods
PE 2305 *Tidal water damage*
PD 0033 Tidal waves
PE 2305 *Tide determination transportation*
PE 2501 Tide ; Red
PD 2788 Tides ; Storm
PF 1492 Tied aid
PE 7613 Tied migration ; Abuse project
PJ 4787 Tied purchasing
PF 2358 *Ties ; Divisive family*
PE 1101 *Ties ; Ineffective political*
PD 1380 *Ties ; Weakening family*
PA 7193 Tightness *Cheapness*
PA 5869 Tightness *Exclusion*
PA 7152 Tightness *Fewness*
PD 9235 Timber exploitation ; Over rapid
PF 3448 *Timber ; Unavailability*
PF 8876 Time approve needed legislation ; Lack parliamentary
PF 5044 Time arising automation ; Excess leisure
PE 6974 Time communications enemy ; War
PA 7222 Time*complex
Time [conditions...]
PD 9090 — conditions pressure ; War
PF 7022 — *consuming jobs*
PF 7022 — *consuming procedures*
Time [design...]
PF 9980 — design ; Erratic
PF 6665 — Differing conceptions
PD 8851 — disruption economies production facilities ; War
Time [Environmental...]
PE 0736 — Environmental degradation military activity during peace
PF 1706 — Excessively efficient use
PE 6074 — exportation cultural property occupied territories ; War
PF 7022 — *Extended production*
PF 2795 Time ; Fatalistic attitudes use
PE 8283 Time flexibility labour market ; Lack
Time [imprisonment...]
PF 0726 — imprisonment deportation ; Delayed consequences war
PF 5044 — Inability make use leisure
PF 9980 — Inability organize
PF 8876 — Inadequate parliamentary debating
PF 9980 — Inadequate sense
PF 5692 — individual social development ; Constraint
PE 4781 — Inhumane medical experimentation during war
PF 2316 — *Insufficient personal*
PF 9062 — Insufficient recreation
PF 9980 Time keeping ; Poor
PC 4498 Time ; Lack
PF 6042 Time lag legal provisions
Time Limited
PF 9062 — leisure
PF 7022 — *shared*
PF 3560 — *study*
PF 9980 Time ; Loss sense
PF 7022 Time ; Obstacles efficient utilization
PF 4644 Time orientation ; Degrading people's
PF 1706 Time ; Preoccupation accelerating
PE 7624 Time prepare trial defence ; Denial right
PD 0140 *Time ; Reduction working*
PF 9062 Time ; Restricted leisure
Time [schedules...]
PF 2644 — *schedules ; Conflicting*
PD 0140 — *Short working*

Time [Shortage...] cont'd
PC 4498 — Shortage
PF 4644 — span poor ; Narrow
PF 1432 — standards ; Imposition exogenous
PF 2621 — standards ; Multiplicity
Time [Underutilized...]
PF 6476 — *Underutilized class*
PF 5044 — Uninventive leisure
PF 5044 — Unpreparedness surplus leisure
PF 8939 — Unproductive meeting
PF 8993 — use ; Unresourceful
Time [Wasted...]
PF 8993 — Wasted
PF 1761 — Wasted waiting
PE 8907 — women ; Insufficient leisure
PJ 7586 Timidity
PA 6030 Timidity *Fear*
PF 5529 Timing ; Self interested manipulation
PD 2907 Timing ; Unjust election
PF 6683 Timing ; Unpropitious
PE 1438 Tin pollutant
PE 0824 *Tin trade ; Instability*
PD 2567 *Tinnitus*
PA 7365 Tiresomeness *Boredom*
PA 5806 Tiresomeness *Inaction*
PA 7107 Tiresomeness *Unpleasantness*
PE 4637 Tissue ; Cancer lymphatic
PD 2565 Tissue ; Diseases connective
PC 8534 Tissue ; Diseases skin subcutaneous
PE 2286 Tissue ; Injury pituitary
PE 9229 Tissue ; Malignant neoplasm connective
PE 4637 Tissue ; Neoplasms lymphatic haematopoietic
PE 4805 Tissue ; Research foetal cadavers foetal
PE 2286 Tissue ; Tumours pituitary
PD 2565 Tissues ; Diseases muscoskeletal connective
PD 2565 Tissues ; Diseases musculoskeletal system connective
PE 5195 Titanium dioxide pollutant
PD 1718 *Titillation*
PC 8862 Title land ; Loss peasant
PE 8580 Title property ; Faulty
PF 7198 Titles ; Accumulation
PF 3005 *Titles ; Delay obtaining property*
PE 5754 Titles ; Fraudulent nature inherited
PD 3156 Toad ; Endangered species
PD 0713 Tobacco ; Chewing
PE 9093 Tobacco cigarette advertising ; Irresponsible
PD 0713 Tobacco habit
PE 1641 Tobacco ; Instability trade beverages
PE 1253 Tobacco ; Long term shortage beverages
Tobacco manufactures
PE 0483 — Environmental hazards tobacco
PE 0572 — instability
PE 0572 — Instability trade tobacco
PE 7926 — Shortage tobacco
PD 2227 *Tobacco mosaic*
PD 0713 Tobacco smoking
Tobacco tobacco manufactures
PE 0483 — Environmental hazards
PE 0572 — Instability trade
PE 7926 — Shortage
PE 7899 Tobacco trade ; Restrictive practices beverages
PE 5617 *Toe abscess*
PE 7826 *Toe crack*
PE 7826 *Toe ; Seedy*
PC 1326 *Toed ungulates ; Endangered species Even*
PC 1326 *Toed ungulates ; Endangered species Odd*
PD 0876 *Toilet facilities ; Inadequate*
PD 0876 Toilet facilities ; Unsanitary
PD 0876 *Toilet overflow ; Contaminated*
PE 9611 Toiletries cosmetics ; Abusive treatment animals testing
PA 5497 Toilsomeness *Difficulty*
PA 5806 Toilsomeness *Inaction*
PE 9267 Token intergovernmental treaties
PG 8913 Token services ; Unsympathetic
PF 2086 Tokenistic meeting resolutions
PF 8611 Tolerance corruption
PB 2542 *Tolerance ; Exaggerated*
PF 8611 Tolerance nepotism
PC 4710 Tolerated atrocities
PD 5344 Toleration alcoholism children ; Parental
PD 5344 Toleration drug addiction children ; Parental
PD 5344 Toleration sexual activity children ; Parental
PG 5461 *Tomato ; Fusarium wilt*
PG 5461 *Tomato wilt*
PE 8688 Tomatoes ; Instability trade fresh
PD 3978 *Tongue animals ; Paralysis*
PD 1185 *Tongue ; Diseases*
PE 9702 *Tongue ; Inflammation*
PG 5613 *Tongue ; Inflammation*
PG 5613 *Tongue ; Ulcer*
PE 2292 Tonsillitis
PC 9025 *Tonsils ; Infection*
PB 0001 *Tools criminal use*
PD 6520 Tools ; Lack maintenance
PD 1185 Tooth decay
PD 1185 *Tooth fang ; Inflammation*
PD 7420 *Tooth ; Pink*
PE 1974 *Toothache*
PE 8302 Toothed white dolphins ; Endangered species rough
PF 2202 Top down research methodologies
PE 3872 Top managers ; High severance pay
PD 0949 Top soil erosion
PD 0949 Top soil ; Progressive loss
PF 8511 Topographic names ; Non standardization
PC 5010 Topological disaster
PD 3647 Topsoil ; Shallow alkaline

PE 0560 *Tor ; Cholera El*
PA 5454 Torment *Badness*
PA 5497 Torment *Difficulty*
PA 6030 Torment *Fear*
PA 6852 Torment *Inappropriateness*
PA 5451 Torment *Insensibility*
PA 7107 Torment *Unpleasantness*
PD 1739 Tornadoes
PA 5806 Torpor *Inaction*
PA 6444 Torpor *Quiescence*
PA 7364 Torpor *Unfeelingness*
PE 7423 *Torsion ; Gastric*
PE 3936 *Torsion small intestine swine ; Mesenteric*
PD 9022 Torts
PA 5451 Tortuousness *Insensibility*
PA 7107 Tortuousness *Unpleasantness*
PB 3430 Torture
Torture [animals...]
PC 3532 — animals
PF 0399 — Anticipation capital punishment
PE 0969 — Anxiety resulting
Torture [Badness...]
PA 5454 — *Badness*
PD 2484 — beating
PE 6914 — behavioural regulation
PE 7015 — breaking bones
PE 4078 — bruising
PE 2678 — burning
Torture [Cardio...]
PE 1556 — Cardio pulmonary disease due
PD 5204 — Chemical
PD 2851 — children
PE 3734 — Cold
PD 4590 — confinement
PE 5691 — continuous noise
PE 3719 — crushing
Torture [Denial...]
PC 3429 — Denial right freedom
PE 5480 — denial sanitary facilities
PE 7201 — Dental
PB 3430 — Dependence
PE 0885 — Depression due
PD 3763 — deprivation
PE 5279 — destroying cherished things
PE 4371 — Dietary
PD 0764 — Disabled victims
PA 6790 — *Distortion*
Torture [Electrical...]
PE 9000 — Electrical
PE 4687 — Emotional instability due
PE 3734 — exposing freezing temperatures
PE 7327 — exposure animals
PE 3915 — exposure weather
Torture [False...]
PF 9794 — False evidence
PE 4229 — Fatigue due
PE 1255 — Forced witness
Torture [Gait...]
PE 6791 — Gait disturbances due
PE 6724 — Gastro intestinal disorders due
PD 0181 — Government sanctioned
PC 3429 Torture ; Human
Torture [Inability...]
PE 3716 — Inability concentrate due
PE 6936 — Inadequate assistance victims
PE 4015 — Inhumane participation medical profession
PA 5451 — *Insensibility*
PD 6145 — Institutionalized
PE 4687 — Introversion due
PE 1520 — Irritability due
PE 3732 Torture ; Joint pains due
PE 6593 Torture ; Loss memory due
Torture [methods...]
PD 0181 — *methods ; Inter Governmental exchange*
PE 4119 — Military police personnel participation
PD 7576 — mutilation
Torture [Neglect...]
PE 6936 — Neglect victims
PD 4755 — Neurological effects
PE 4696 — Neuropharmacological
Torture [Participation...]
PD 4478 — Participation
PE 4696 — Pharmacological
PD 8734 — Physical
PD 4148 — plants
PD 0181 — Political
PD 4559 — Psychological
Torture [relatives...]
PE 3308 — relatives friends
PC 7101 — Religious
PE 3798 — Residual traumatic pains due
PE 4074 — Restlessness due
Torture [schools...]
PE 2062 — schools
PD 7550 — Self
PE 4703 — Self imposed social isolation due
PE 6797 — sensory deprivation
PE 5259 — sensory overload
PE 5108 — Sexual
PE 6339 — shooting
PE 7154 — Skin lesions due
PE 0451 — Sleep difficulties due
PD 6810 — Social isolation
PE 5294 — Somatic psychosomatic effects
PD 0181 — State sanctioned
PE 4471 — Stationary position

Torture

Torture [Surgical...] cont'd
- PE 7547 — Surgical
- PF 4062 Torture ; Undocumented
- PA 7107 Torture *Unpleasantness*

Torture victims
- PE 0969 — fear
- PE 4579 — Children
- PE 0885 — Depression
- PE 2119 — Family
- PD 4755 — Mental anguish
- PE 3932 — Sexual dysfunction
- PE 0753 — Spouses
- PE 1296 Torture violation taboos

Torture [waiting...]
- PE 3927 — waiting
- PE 4792 — Water
- PE 4078 — wounding
- PE 4792 Tortures using water
- PF 0327 Total absence tactical methodologies
- PC 8974 Total depravity
- PF 8686 Total disarmament
- PF 2415 Total human process ; Superficial research
- PE 3298 Total sexual abstention
- PD 3213 Totalitarian democracy
- PF 2190 Totalitarianism
- PE 4184 Totemic species ; Endangered
- PF 3421 Totemism
- PA 6976 Toughness
- PE 9344 *Tourette's disorder*
- PD 0826 Tourism ; Cultural degradation recreation
- PF 4115 Tourism ; Degradation developing countries
- PE 9825 Tourism ; Deterioration cultural artefacts
- PD 0826 Tourism ; Excessive
- PF 7276 Tourism ; Foreign intervention
- PE 3008 Tourism ; Hunting
- PF 4106 Tourism ; Inadequate statistical information
- PF 0905 *Tourism ; Lack incentives*
- PB 5249 Tourism ; Moral
- PE 6920 Tourism ; Natural environmental degradation recreation
- PD 0826 Tourism ; Over exploitation
- PD 7276 Tourism ; Political

Tourism [sexual...]
- PE 4437 — sexual purposes ; Abuse
- PD 0826 — Social environmental degradation recreation
- PF 7276 — Subversive
- PF 1735 *Tourist attraction ; Insufficient*
- PF 4112 Tourist dependent economies ; Instability
- PE 8966 Tourist hazards
- PF 6572 *Tourist trade ; Unexplored*
- PE 7538 Tourists developing countries ; Health hazards
- PF 3099 *Tours ; Biased cultural*
- PF 6575 *Town brochure ; Undesigned*
- PE 2813 *Town centre ; Unfocused*
- PF 2813 *Town planning ; Limited scope*
- PF 4470 *Town population ; Static*
- PE 2813 *Town residents ; Unengaged*
- PF 2535 *Town shopping ; Increase out*
- PF 6539 Towns developed countries ; Limited availability public services small
- PD 3489 Towns ; Shanty
- PF 6540 Towns ; Transfer business small communities larger
- PD 6125 Towns ; Vulnerability small
- PE 5463 Toxaemia
- PD 7420 Toxaemia cattle ; Pregnancy
- PD 7420 Toxaemia ewes ; Pregnancy
- PE 8022 Toxaemias pregnancy puerperium
- PE 2731 *Toxaemic animal diseases*
- PE 2501 Toxic algae
- PE 4717 Toxic chemicals ; Health hazards long term, low level exposure toxic mixtures non

Toxic food
- PD 0487 — additives
- PD 0487 — colourings
- PD 0487 — preservatives
- PE 4588 Toxic genetic chemicals
- PC 0119 Toxic harmful agents ; Airborne
- PE 6840 Toxic hazards cassava

Toxic [metal...]
- PD 0948 — metal pollutants
- PD 0948 — metals
- PD 5256 — mixtures
- PE 4717 — mixtures non toxic chemicals ; Health hazards long term, low level exposure
- PE 4675 — modern furniture materials
- PE 9774 — *myopathies*
- PE 0617 Toxic organic compounds water pollutants
- PD 9765 Toxic products ; Illicit movement
- PD 4904 Toxic rain
- PE 5106 Toxic shock syndrome

Toxic substances
- PD 1115 — [Toxic substances]
- PD 7205 — Bioaccumulation
- PE 5539 — Contamination sediments
- PC 0058 — Contamination soil
- PE 2061 — developing countries ; Trading products containing
- PD 0122 — Hazards strong
- PF 5188 — Ignorance hazardous levels
- PE 9294 — peacetime ; Anti personnel use

Toxic waste
- PD 1398 — dumping
- PE 4146 — sites ; Ineffective inspection
- PD 9765 — smuggling
- PE 2061 Toxic wastes developing countries ; Dumping
- PA 5454 Toxicity *Badness*
- PD 5228 *Toxicity newborn pigs ; Iron dextran*

Toxicity thresholds
- PD 4519 — Commercial distortion
- PE 5239 — Failure respect internationally agreed
- PF 5188 — Uncertain
- PA 7226 Toxicity *Unhealthfulness*
- PE 9611 Toxicological experiments ; Use animals
- PE 3506 Toxicosis
- PE 9702 *Toxicosis ; Salaframine*
- PE 9458 *Toxicosis ; Summer fescue*
- PD 1115 Toxins
- PE 3659 Toxoplasmosis
- PE 3659 *Toxoplasmosis ; Acquired*
- PE 3659 *Toxoplasmosis ; Congenital*
- PE 1158 Toys ; Dangerous
- PE 4664 Toys ; Gender differentiated
- PE 1158 Toys ; Undesirable influence
- PE 4297 Toys ; Violent interactive

Trace element
- PE 1936 — deficiencies soils
- PE 5328 — deficiency
- PE 5328 — imbalance human body
- PE 5328 Trace elements human body ; Excessive
- PD 1354 Trace gases atmosphere ; Increase
- PE 5524 *Tracheal oedema syndrome feeder cattle*
- PE 9854 Tracheobronchitis
- PE 9854 *Tracheobronchitis dogs ; Infectious*
- PE 1946 Trachoma
- PE 7403 Tracks ; Environmental degradation recreational use unsurfaced country roads
- PD 9293 *Tract animals ; Tumours lower urinary*

Trade [Abuse...]
- PD 6002 — Abuse dominant market position international
- PD 9679 — against exports developing countries ; Protectionism international
- PE 4523 — agricultural commodities developing countries ; Inadequate
- PD 8286 — alcoholic beverages
- PE 1406 — *aluminium ; Instability*
- PE 2951 — amongst developing countries ; Weakness primary commodity

Trade animal
- PE 8816 — feedstuffs, excluding unmilled cereals ; Instability
- PE 8896 — oils fats ; Instability
- PD 0389 — products endangered species
- PE 0735 — vegetable oils fats ; Instability
- PE 8880 — vegetable oils fats ; Restrictive practices
- PS 6553 *Trade areas ; Limited*

Trade arrangements
- PF 7642 — Bilateralism
- PE 0396 — Collusive international
- PD 0340 — Preferential
- PE 8593 Trade asses, mules hinnies ; Instability
- PE 4991 Trade ; Banning ivory

Trade barriers
- PC 4275 — [Trade barriers]
- PD 9651 — Inter cultural
- PE 4170 — manufactured goods developing countries
- PD 0057 — primary commodities ; International
- PD 2958 — protectionism developing countries

Trade [beans...]
- PE 8259 — beans, dried, peas, lentils leguminous vegetables ; Instability
- PE 1680 — beverages ; Instability
- PE 1641 — beverages tobacco ; Instability
- PD 8621 — biological warfare technology
- PD 8621 — biological weapons ; International
- PE 0396 Trade cartels ; Secret
- PE 1769 Trade cereals cereal preparations ; Instability

Trade chemical
- PE 0500 — elements compounds ; Instability
- PD 9692 — warfare technology
- PD 9692 — weapons ; International

Trade [chemicals...]
- PE 8600 — chemicals ; Restrictive practices
- PE 8206 — coal, coke briquettes ; Instability
- PE 0915 — coffee, tea, cocoa, spices manufactures thereof ; Instability
- PE 7072 — Collusive tendering international
- PF 7838 — commitments ; Conditional observance multilaterally agreed
- PF 7838 — commitments ; Failure political will observe
- PD 7981 — counterfeit goods
- PC 5624 — *Crimes against*
- PE 5331 — Crimes related foreign relations

Trade crude
- PE 0760 — fertilizers crude minerals, excluding coal, petroleum precious stones ; Instability
- PE 9049 — fertilizers ; Instability
- PE 0701 — synthetic reclaimed rubber ; Instability
- PE 5678 Trade cultural products capitalist socialist countries ; Imbalance
- PE 5702 Trade cultural products developed developing countries ; Imbalance

Trade [dairy...]
- PE 0576 — dairy products eggs ; Instability
- PE 1496 — deficits developing countries ; Excessive external
- PC 1100 — deficits ; Excessive external
- PE 4190 — developed developing countries ; Bipolarization

Trade developing countries
- PD 3497 — Arms
- PC 0933 — Decline mutual
- PD 2897 — Deterioration terms
- PD 2961 — Developed country limiting
- PE 4965 — Environment policy restriction
- PD 5176 — Inadequate
- PF 4160 — Inadequate industrial

Trade developing countries cont'd
- PD 2968 — producing primary commodities ; Instability export
- PC 0933 — Weakness
- PD 0169 — Weakness regional
- PC 2369 Trade developing developed countries ; Tariff barriers
- PC 2724 Trade different economic systems ; Weakness

Trade discriminatory
- PE 0191 — application antidumping regulations ; Distortion international
- PE 2603 — customs administrative entry procedures ; Distortion international
- PE 9073 — formulation equipment safety regulations ; Distortion international
- PE 1274 — formulation health sanitary regulations agricultural pharmaceutical products ; Distortion international
- PE 0347 — government private procurement policies ; Distortion international
- PD 8124 Trade ; Discriminatory nuclear

Trade discriminatory
- PD 0340 — preference agreements ; Distortion international
- PE 0778 — requirements respect marks product origin ; Distortion international
- PE 0083 — requirements respect product standards measures ; Distortion international

Trade [Distortion...]
- PC 6761 — Distortion international
- PE 8740 — dressed fur skins ; Instability
- PE 8257 — dried salted smoked meat ; Instability
- PD 0991 — Drug
- PD 5750 — drug addiction ; Delinquency side effect drug
- PE 9210 — drugs ; Terrorism financed
- PD 2144 — dumping ; Distortion international

Trade [Economic...]
- PE 5296 — Economic dependence some developing countries drug
- PE 7907 — electric energy ; Instability
- PE 0522 — embargoes similar restrictions ; Distortion international
- PC 0380 — endangered species ; International
- PE 8232 — essential oils perfume materials ; Instability
- PE 1513 — excluding cotton jute ; Instability vegetable fibre
- PC 0380 — exotic species
- PE 1961 — export subsidies countervailing duties ; Distortion international

Trade [fairs...]
- PF 3099 — *fairs ; International*
- PE 4525 — fertilizer products ; Obstacles
- PC 7873 — Financial destabilization world
- PE 0972 — fish, crustacea molluscs preparations thereof ; Instability
- PE 0861 — fixed vegetable oils fats ; Instability
- PD 1434 — food live animals ; Instability
- PE 2603 — formalities ; Excessive customs

Trade fresh
- PE 8591 — frozen chilled meat ; Instability
- PE 8258 — potatoes ; Instability
- PE 8688 — tomatoes ; Instability
- PE 0961 Trade fruit vegetables preparations thereof ; Instability
- PD 0389 Trade furs skins endangered species
- PD 0877 *Trade gas ; Instability*
- PF 7730 Trade ; Government complicity illegal arms

Trade [harassment...]
- PD 7441 — harassment
- PE 1513 — *hard fibres ; Instability*
- PE 7530 — human organs

Trade Illegal
- PC 4645 — [Trade] ; Illegal]
- PD 0991 — drug
- PE 4991 — ivory

Trade [Illicit...]
- PD 0991 — Illicit narcotics
- PC 4879 — imbalances ; Fiscal
- PE 0117 — Inadequate barter system international
- PE 1139 — *income village communities ; Insufficient*
- PD 0280 — inedible crude non fuel materials ; Instability
- PE 8351 — inedible crude non fuel materials ; Restrictive practices
- PE 9107 — information ; Lack transparency international
- PE 9107 — Informational procedural obstacles world
- PE 9137 — infrastructure ; High cost natural gas
- PD 3841 — Insider

Trade Instability
- PE 2585 — alcoholic beverages
- PE 1774 — banana
- PE 2585 — beer
- PE 0619 — chemicals
- PE 1549 — cocoa
- PE 0950 — coffee
- PE 0824 — *copper*
- PE 1510 — cotton
- PE 1552 — date

Trade [instability...]
- PE 8875 — instability electrical machinery, apparatus appliances
- PE 2587 — Instability fresh fruit edible nut
- PE 8119 — instability ; Fuel wood charcoal
- PE 1474 — Instability fur
- PE 2587 — *Instability ground nut peanut*
- PE 9076 — instability inorganic chemicals, elements, oxides halogen salts

Trade instability Iron
- PE 8772 — ore concentrates
- PE 8969 — steel manufactures
- PE 8312 — steel scrap

Trade [Instability...]
- PE 1794 — Instability jute
- PE 0824 — *Instability lead*

-1060-

Trade [instability...] cont'd
PE 8320 — instability ; Leather leather manufactures
PE 1683 — *Instability margarine*
PE 8489 — instability ; Meal flour wheat meslin
PE 7996 — instability ; Medicinal pharmaceutical products
Trade Instability
PE 0824 — nickel
PE 0313 — olive oil
PE 3000 — pepper
PE 8470 Trade instability ; Photographical optical goods
PC 0463 Trade ; Instability primary commodities
Trade instability
PE 8169 — professional ; Scientific controlling instruments
PE 8843 — Radioactive associated materials
PE 8159 — Rape mustard oils
Trade [Instability...]
PE 0696 — Instability rice
PE 1550 — *Instability silk*
PE 7956 — instability ; Silver platinum group
PE 8556 — instability ; Silver platinum ores
Trade Instability
PE 0961 — *soybean*
PE 1619 — *spices*
PE 2585 — *spirits*
Trade [instability...]
PE 8612 — instability stone ; Sand gravel
PE 0383 — Instability sugar
PE 8958 — instability ; Sugar confectionery preparations
Trade Instability
PE 9089 — tanning dyeing
PE 2054 — tea
PE 0824 — *tin*
Trade [instability...]
PE 8212 — instability ; Uranium ores concentrates thorium
PE 1711 — Instability vegetable
PE 8901 — instability vegetable products ; Roots tuber
Trade Instability
PE 0385 — wheat
PE 2522 — wine
PE 2056 — wool
PE 0824 — *zinc*
Trade [International...]
PC 1358 — International arms
PE 1880 — International drug
PE 8547 — investment disputes ; Private international
PC 8930 — Irresponsible international
PD 1434 *Trade legumes ; Instability*
PD 1376 Trade live animals ; Instability
Trade [machinery...]
PD 0620 — machinery transport equipment ; Instability
PE 7958 — machinery transport equipment ; Restrictive practices
PE 8682 — manila fibre ; Instability
PE 0806 — manufactured fertilizers ; Instability
Trade manufactured goods
PE 2966 — developing countries ; Weakness
PE 0882 — Instability
PD 1797 — Restrictive practices
Trade [measures...]
PF 3871 — measures ; Narrow egalitarian application
PE 0755 — meat meat preparations ; Instability
PE 9762 — mechanisms ; Discriminatory managed
PD 0553 — metalliferous ores metal scrap ; Instability
PD 0877 — mineral fuels, lubricants related materials ; Instability
PE 8464 — mineral tar crude chemicals derived coal, petroleum natural gas ; Instability
PE 1182 — minimum pricing regulations measures regulate domestic prices ; Distortion international
PE 1683 — miscellaneous food preparations ; Instability
PE 0814 — miscellaneous manufactured articles ; Instability
Trade non
PE 8748 — alcoholic beverages ; Instability
PE 8828 — electrical machinery ; Instability
PE 1406 — ferrous metals ; Instability
PC 2725 Trade ; Non tariff barriers international
Trade nuclear
PE 3968 — bomb making materials ; Covert
PE 7174 — materials irresponsible countries
PE 3968 — weapons ; Clandestine
Trade [obstacles...]
PD 0455 — obstacles patent protection ; Distortion international
PC 4890 — Obstacles world
PE 0386 — oil seeds, oil nuts oil kernels ; Instability
PE 8131 — ores concentrates non ferrous base metals ; Instability
Trade [patterns...]
PD 4252 — patterns ; Disruption native
PC 8415 — patterns ; Imbalance international
PD 0909 — petroleum petroleum products ; Instability
PC 9584 — policies ; Deterioration long term
PC 0073 — practices ; Dependence restrictive
PC 0073 — practices ; Restrictive
PC 0933 — preferences developing countries ; Non uniformity tariff non tariff
PE 4946 — prescribed drugs ; Illicit
PE 8741 — preserved fruit ; Instability
PE 3808 — products chemical warfare
PC 4275 — protectionism
PC 5842 — Protectionism international
PE 7914 — pulp waste paper ; Instability
Trade [quantitative...]
PE 9027 — quantitative restrictions ; Distortion international
PF 2497 — quota cheating
PE 9762 — quotas discretionary character ; Imposition quantitative
PE 9762 — quotas political reasons ; Imposition

Trade [regenerated...]
PE 8769 — regenerated cellulose artificial resins ; Instability
PF 4722 — regulations ; Complex
PF 4165 — related structural adjustments ; Inadequate
PD 9389 — reprisals
Trade restrictions
PD 5830 — Agricultural
PE 0310 — due intergovernmental arrangements
PE 0310 — due voluntary export restraints
PC 4275 — Unfair
Trade restrictive
PC 0073 — business practices ; Distortion international
PE 8882 — controls movement labour ; Distortion international
PE 8525 — controls over foreign investment ; Distortion international
PE 0497 — customs valuation practices ; Distortion international
Trade Restrictive practices
PE 7899 — beverages tobacco
PE 0342 — food live animals
PE 0141 — mineral fuels
PD 2029 Trade result government participation ; Distortion international
PE 8272 Trade sanitary plumbing, heating, lighting fixtures fittings ; Instability
PD 3537 Trade secrets ; Infringement
Trade selective
PD 0678 — domestic subsidies ; Distortion international
PE 8867 — indirect taxes import charges ; Distortion international
PE 8583 — monetary controls ; Distortion international
Trade [services...]
PF 0041 — services ; Governmental inaction concerning
PD 6223 — services ; Obstacles international
PE 8861 — sheep goats ; Instability
PC 0130 — Slave
PE 2954 — socialist developed market economies ; Weakness
PE 2953 — socialist developing economies ; Weakness
PE 8267 — state trading government monopoly practices ; Distortion international
PE 7174 — strategic goods hostile countries
PE 0383 — sugar, sugar preparations honey ; Instability
PE 0760 — *sulphur ; Instability*
PC 4879 — *surpluses ; Large*
PE 8950 — synthetic regenerated fibres ; Instability
Trade [Tariff...]
PC 0569 — Tariff barriers international
PE 0915 — *tea mate ; Instability*
PC 4382 — Technical barriers
PE 8970 — textile yarn fabrics, made up articles ; Instability
PE 0572 — tobacco tobacco manufactures ; Instability
PF 3552 — *training ; Single*
PE 0135 — transnational corporations ; Escalating control international
PE 8920 — travel goods ; Instability
PE 7621 — tungsten ; Instability
Trade [U235...]
PE 8560 — U235 depleted uranium ; Instability
PE 8106 — undressed hides non fur skins ; Instability
PE 1235 — undressed hides, skins fur skins ; Instability
PD 2897 — Unequal exchange
PF 6572 — *Unexplored tourist*
PC 5842 — Unfair restrictions international
Trade union
PD 4695 — activity ; Denial right
PD 0683 — association ; Denial right
PE 2007 — demonstrations ; Violent repression
PD 4341 — fraud
PE 8170 — funds property ; Deprivation
Trade union leaders
PD 7630 — Arrest
PE 0252 — Assassination
PE 4869 — Death threats against
PE 8367 — employers government ; Collusion
PE 5882 — Forced disappearances
PD 7630 — Ill treatment
PD 7471 — Threats against
Trade union [meetings...]
PD 7210 — meetings ; Prohibition
PJ 0272 — *members their representatives ; Isolation*
PE 6722 — opposition profit-sharing
PF 6046 — policies concerning employment ; Restrictive
PD 8146 — practices ; Restrictive
PD 4341 — practices ; Unethical
Trade union premises
PE 5462 — public authorities ; Occupation
PE 5462 — public authorities ; Raids
PE 5462 — Seizure
Trade union property
PE 7581 — Confiscation
PE 7581 — public authorities ; Destruction
PE 7581 — public authorities ; Seizure
PD 0683 *Trade union recognition ; Lack*
PE 7620 Trade union representatives following legal strike action ; Dismissal
Trade union rights
PD 4695 — Abuse
PE 4860 — governments ; Denial
PE 6888 — public employees ; Denial
PE 5938 — public enterprises ; Restrictions
PD 4695 — Violation
Trade [union...]
PF 1262 — union rivalry
PD 0610 — union workers ; Sanctions against
PF 8493 — unionism
Trade unions
PD 3535 — Banned

Trade unions cont'd
PE 8273 — contract performance ; Interference
PD 4341 — Corruption
PE 5398 — Denial right organize
PC 4613 — Discrimination against
PE 4745 — engage political action ; Violation right
PE 1758 — freely elect their representatives ; Violation right
PE 1758 — function freely ; Violation right
PE 4860 — Government discrimination against
PE 5938 — government services ; Violation right organize
PF 1262 — Inadequate international
PD 4341 — Irresponsible
PE 3970 — negotiate freely employers ; Violation right
PD 0610 — organizers ; Denial rights
PF 1262 — Outmoded
PE 5793 — protection against dissolution ; Violation right
PE 5312 — publish ; Violation right
PF 1262 — *State controlled*
PE 1541 — Transnational strike action
PE 6018 — Unrepresentativity
PE 5192 — Violation right workers join
Trade [unmilled...]
PE 1769 — unmilled cereals ; Instability
PD 5033 — Unpredictable barriers
PE 1550 — unprocessed textile fibres their waste ; Instability
Trade [vegetable...]
PE 9164 — vegetable based foodstuffs ; Instability
PE 8208 — vegetables, roots tubers ; Instability
PF 2135 — *viability ; Questionable*
PF 9697 — Vulnerability adverse conditions foreign
Trade war
PC 0840 — [Trade war]
PC 0840 — *Computer development*
PD 3765 — Intra national
Trade [Weakness...]
PC 0933 — Weakness South South inter regional
PD 3303 — White slave
PE 2521 — wood, lumber cork ; Instability
PC 6873 *Trademarks ; Abusive use*
PE 0346 — Trademarks ; Restrictive business practices relation patents
PF 3558 *Traders ; Disunity*
PE 5849 — Trades ; Unfair shipping practices bulk
PJ 8370 *Tradesmen ; Uncertified skilled*
PC 9584 — Trading agreements ; Circumvention international
PC 6641 — Trading ; Black market
PD 3917 — Trading commodities ; Insider
PC 9584 — Trading discipline ; Breakdown international
PJ 7177 *Trading enemy*
PD 3917 — Trading fraud ; Commodities
PE 8267 — Trading government monopoly practices ; Distortion international trade state
PD 3841 — Trading ; Illicit
PD 3841 — Trading ; Insider
PF 4870 — Trading ; Obstacles commodity futures
Trading [political...]
PE 6948 — political endorsement
PD 2897 — position developing countries ; Decline
PF 2960 — principles practices developing countries ; Differences
PC 2952 — principles practices different economic systems ; Differences
PE 2061 — products containing toxic substances developing countries
PE 6948 — public office
PF 3871 — Trading relations ; Preoccupation reciprocity
PC 6307 — Trading special influence
Trading system
PC 9584 — Declining credibility international
PE 0135 — Emerging oligopolistic world
PC 9584 — Fragmentation international
PC 9584 — Inadequate international
PC 9584 — Uncertainty international
PE 1922 Trading systems ; National oligopolistic
PJ 6424 *Trading ; Unfair*
PF 2493 *Tradition-bound childcare*
PF 1705 Tradition ; Individualistic retaining local
PC 3293 *Tradition ; Loss linguistic*
PF 3863 Tradition ; Loss religious
PF 1373 Tradition ; Untransposed significance cultural
PC 0492 Traditional agricultural methodology
PF 7808 Traditional agricultural products ; Ignorance
Traditional [community...]
PF 3454 — community decision making ; Restrictive effects
PF 6106 — *crop attachment*
PE 4054 — cultural expressions peoples ; Abuse
PD 0826 — *customs ; Abuse*
Traditional economies
PD 4252 — Devalorization
PD 4252 — Discrimination against
PD 4252 — indigenous populations ; Denial right
PD 4252 — Marginalization
PF 7960 — Traditional energy alternatives ; Disuse
PD 0406 — Traditional estates ; Maldistribution land associated large
Traditional [farm...]
PE 7313 — *farm marketing*
PF 1822 — farming methods
PF 3448 — *farming ; Unreliable*
PD 0144 — forms social control developing countries ; Loss
PD 1543 — forms social security developing countries ; Loss
PF 1705 *Traditional gatherings ; Insufficient*
PF 2703 *Traditional gifts ; Untransposed creativity*
Traditional [habits...]
PF 6461 — *habits ; Overpowering*
PE 3946 — herbal remedies ; Ignorance

Traditional hiring

Traditional [hiring...] cont'd
- PF 2256 — *hiring practices*
- PD 0765 Traditional industries developing countries ; Excessive foreign investment

Traditional [land...]
- PF 3211 — *land distribution*
- PF 3211 — *land ownership*
- PF 3129 — *life* ; Restrictive patterns
- PF 3211 — *life styles* ; Inhibiting effects

Traditional [male...]
- PC 3024 — *male dominance*
- PE 7313 — *market practices*
- PD 1415 — *markets* ; Limited
- PE 3946 — *medical practices* ; Ignorance
- PC 4321 — *modes behaviour* ; Contempt
- PF 6592 — *modes life developing countries* ; Westernization
- PE 3337 Traditional new technologies ; Incompatibility
- PF 2845 *Traditional occupations* ; *Disoriented*

Traditional [patterns...]
- PF 3531 — *patterns community life* ; Absence
- PF 6545 — *patterns* ; Divisive
- PE 3946 — *plant remedies* ; Ignorance
- PD 9162 — *property rights*
- PF 3211 — *purchasing habits*

Traditional [skills...]
- PD 8872 — *skills* ; Elimination
- PF 2845 — *skills* ; *Unprofitable*
- PF 3008 — *small business methods* ; Ineffectiveness
- PF 6461 — *symbol systems* ; Modern disruption

Traditional values
- PA 1782 — Decay
- PF 2256 — Inappropriate application
- PE 8139 — subordination women developing societies
- PB 3209 Traditional vested interests ; Excessive political influence
- PF 2644 Traditional way life ; Delimiting family patterns
- PF 4871 Traditional Western medicine ; Lack integration
- PF 2676 Traditionalism
- PF 6461 *Traditionally determined housing*
- PF 9529 Traditions blocking business profit ; Cultural
- PF 1696 Traditions cultural isolation ; Local
- PD 1095 Traditions developing countries ; Decline rural customs
- PJ 3532 Traditions ; Disparagement transcendent experiences
- PF 8156 Traditions ; Insensitivity diversity cultural
- PF 1765 Traditions rural areas ; Rigidly entrenched social
- PF 6494 Traditions ; Unclarified procedures transposing ancient
- PD 0079 Traffic accidents ; Risk
- PD 0079 Traffic accidents ; Road
- PD 4271 Traffic children medical experiments
- PD 4271 Traffic children source organ transplants

Traffic congestion
- PD 0078 — [Traffic congestion]
- PD 0689 — Air
- PE 4766 — Port
- PD 0078 — *Rail*
- PD 2106 — Road highway
- PD 1486 — Sea
- PD 0426 — Urban road
- PE 8266 Traffic control ; Inadequate
- PF 9464 Traffic delays ; Air
- PG 7741 Traffic ; Hazardous commercial
- PD 2722 Traffic immigrant workers
- PD 2722 Traffic immigrant workers ; Abusive
- PD 8328 Traffic ; Natural hazards air
- PD 3664 Traffic noise
- PE 0930 Traffic offences ; Road
- PD 8910 Traffic offences ; Vice sex
- PD 0689 Traffic paralysis ; Air
- PC 4442 Traffic persons
- PE 0897 Traffic restrictions ; Road
- PE 6681 Traffic ; Tax impediments international motor
- PE 0930 Traffic violations ; Road
- PC 0380 Traffic ; Wildlife

Trafficking children
- PD 8405 — [Trafficking children]
- PF 3302 — adoption
- PD 7266 — economic exploitation
- PD 4271 — medical exploitation
- PE 6613 — sexual exploitation
- PE 7271 Trafficking government benefit coupons
- PF 7730 Trafficking ; Government complicity drug

Trafficking [illegal...]
- PE 7711 — illegal firearms
- PD 0991 — Illicit drug
- PD 2722 — Illicit labour
- PC 0130 Trafficking people
- PD 4221 Trafficking taxable objects ; Unlawful
- PC 3298 Trafficking women
- PF 4516 Tragedy commons
- PE 3874 *Trails* ; *Disappearance walking*
- PD 0126 Train accidents
- PD 5575 *Train armoured car robberies*
- PF 0903 Trained administrative personnel developing countries ; Lack
- PF 2477 Trained administrators ; Improperly
- PF 4984 Trained drivers ; Insufficient
- PF 2477 Trained firefighters ; Lack

Trained [labour...]
- PD 9113 — *labour* ; Insufficient
- PF 1332 — *leaders* ; Inadequate
- PF 1332 — *leaders* ; Insufficient
- PC 4867 Trained managers ; Insufficient
- PE 6243 Trained manpower developing countries ; Inadequate supply appropriate

Trained personnel
- PF 0559 — act against problems ; Shortage adequately

Trained personnel cont'd
- PD 1291 — developing developed areas ; Emigration
- PE 8093 — industrialized countries ; Emigration

Trained [teachers...]
- PE 8429 — teachers rural areas developing countries ; Unavailability
- PD 8108 — teachers ; Shortage
- PF 3523 — technicians ; Unemployed
- PC 1131 Trainee employment ; Unavailability
- PF 4852 Trainers ; Inadequate preparation
- PC 2163 Trainers ; Unavailability skills
- PD 3789 Training ; Absence management
- PC 4867 Training ; Absence management
- PE 6650 Training ; Break down communications due difference
- PE 8509 Training business, industry public service ; Inability communities take advantage

Training [communities...]
- PF 2837 — communities ; Lack opportunities practical
- PD 4684 — Corruption
- PF 4375 — cost ; High
- PF 4375 — Costly technical
- PF 4375 — costs ; Unavailability

Training [decision...]
- PD 2036 — decision making ; Inadequate
- PE 4246 — Discrimination against women athletic
- PS 9612 — *Distant commercial*
- PC 1131 Training employment opportunities
- PE 5194 Training existence openings various professions ; Imbalance

Training [facilities...]
- PD 0847 — facilities ; Expensive
- PD 0847 — facilities ; Insufficient
- PD 0847 — facility ; Advanced
- PD 5077 Training human settlements developing countries ; Inadequacy
- PC 2163 *Training ; Inaccessible vocational*

Training Inadequate
- PF 4984 — *driver*
- PF 9007 — *first aid*
- PD 9160 — *medical*
- PF 4852 — *teacher*
- PE 8716 — *technical*
- PE 4596 Training inappropriate structural technological changes

Training Ineffective
- PF 1352 — mechanisms functional
- PF 3498 — *service*
- PE 8716 — *technical*

Training [information...]
- PF 1301 — *information* ; Unpublicized
- PF 3605 — Insufficient leadership
- PD 8388 — *Insufficient special*
- PF 6032 Training judges ; Inadequate

Training Lack
- PD 8388 — [Training ; Lack]
- PD 8388 — *agro urban*
- PF 1352 — *apprentice*
- PF 0551 — *house*
- PF 2451 — *local services community leadership*
- PF 0422 — *vocational*
- PF 3605 Training leadership ; Reduced

Training Limited
- PF 2698 — availability technical agricultural business
- PC 4867 — *finance*
- PD 9160 — *paramedic*
- PF 8085 — *skills*

Training [Minimal...]
- PC 4867 — Minimal business
- PF 6531 — Minimal opportunities adult
- PF 0422 — Misdirected vocational

Training [needs...]
- PF 2477 — *needs* ; Unarticulated
- PF 2477 — *needs* ; Uncorrelated
- PD 3789 — *Neglected organization*
- PF 8085 — Neglected social skills
- PF 6531 — Non local adult
- PF 0422 — Non productive vocational
- PF 2313 — *Nonavailability technical*
- PD 3548 Training ; Obsolete employment

Training opportunities
- PF 1352 — *Inadequate*
- PF 1352 — Insufficient
- PF 1944 — local communities ; Lack relevant
- PF 2453 — *Unknown*
- PF 4852 — *Unpublished*
- PF 4852 — *Unused*
- PD 3548 Training ; Outdated job
- PF 2445 Training ; Personalistic use

Training programmes
- PF 4196 — Excessive expense athletic
- PF 3519 — Impractical
- PF 2885 — Social bias planning

Training [resources...]
- PD 9051 — *resources* ; Remote
- PF 3552 — *Restricted job*
- PF 2444 — *Restrictive regulations*
- PF 6472 — rural areas ; Inadequate practical

Training [schemes...]
- PF 0422 — schemes ; Lack vocational
- PF 3552 — *Single trade*
- PF 8085 — social skills ; Lagging
- PE 2605 — space ; Limited

Training [Unavailability...]
- PF 8835 — Unavailability agricultural
- PC 2163 — Unavailable vocational
- PF 3498 — *Uncoordinated special*
- PF 3552 — *Undeveloped job*

Training [Unexplored...] cont'd
- PC 4867 — Unexplored business
- PF 4013 — Unexplored industrial
- PF 8835 — Unmotivated agricultural
- PC 4867 — Unstructured business
- PF 4852 — *Unused methods*
- PJ 9762 — urban areas ; Unrecognized possibilities
- PF 3007 Training vision ; Unimaginative long term
- PF 6024 Training women farmers ; Unadapted vocational
- PD 0190 Training women ; Lack
- PA 7250 Traitorousness *Disobedience*
- PA 7363 Traitorousness *Improbity*
- PA 7325 Traitorousness *Irresolution*
- PA 5699 Traitorousness *Reversion*
- PD 2615 Traitors
- PE 0435 Traits ; Paranoid
- PE 5460 Tramps
- PE 0139 Tranquillizers ; Abuse sedatives
- PE 1821 Tranquillizers ; Dependence minor
- PE 0261 Trans border broadcasting ; Violation sovereignty
- PF 6855 Trans frontier cooperation ; Lack
- PF 7945 Trans-frontier pollution
- PF 3277 Trans-sexualism
- PE 9220 Transaction goods nonfactor services ; Distorting effects commodity taxes
- PF 9571 Transactions ; Conflicts national law relation international
- PE 0740 Transactions developing countries ; International indebtedness arising insurance
- PF 9571 Transactions ; Divergence different national laws relating international
- PE 7968 Transactions ; Illiteracy inhibitor business
- PD 4728 Transactions ; Restrictions commercial
- PE 1978 Transactions ; Restrictive business practices technology

Transactions [Undocumented...]
- PF 6594 — Undocumented international financial
- PC 4645 — Unlawful business
- PF 5728 — *Unreported financial*
- PC 6641 — Unreported financial
- PD 3391 Transboundary air pollution ; Long range
- PD 1096 Transboundary water pollution
- PF 4371 Transcendent experience ; Barriers
- PJ 3532 Transcendent experiences traditions ; Disparagement
- PF 6540 Transfer business small communities larger towns
- PD 2784 *Transfer corpses* ; *Health risks international*
- PD 2784 *Transfer corpses* ; *Obstacles international*

Transfer [developing...]
- PE 4922 — developing countries ; Decline technology
- PE 4922 — developing countries ; Impasse technology
- PE 4922 — Disadvantageous terms technology
- PD 0084 Transfer educational facilities ; Barriers
- PE 4332 Transfer forms worship ; Inappropriate
- PF 8921 Transfer highly polluting industries developing countries
- PE 8403 Transfer knowledge ; Copyright barriers
- PF 6035 Transfer knowledge ; Language barriers
- PE 8580 Transfer ownership ; Risks

Transfer [prices...]
- PE 0245 — prices transnational corporations ; Manipulation
- PE 1193 — pricing
- PE 1193 — pricing ; Profit repatriation concealed
- PD 7463 — programmes ; Paradoxes technology
- PC 6445 Transfer skills ; Unorganized
- PC 3134 Transfer surplus wealth developing industrialized countries

Transfer technology
- PE 5820 — Inappropriate
- PE 6039 — Non acceptance embryo
- PD 1291 — Reverse
- PE 5820 Transfer transnational corporations ; Adverse impact technology
- PF 8552 *Transference* ; *Decreased skill*
- PF 1664 Transferring wisdom aged ; Non
- PC 3134 Transfers developing countries ; Negative net resource
- PE 7803 Transfers ; Legitimizing illegal fund
- PF 0845 Transform administrative structures ; Government reluctance
- PD 0946 Transformation agriculture developing countries ; Lagging
- PD 0917 Transformation ; Bias media against development social
- PF 9281 Transformation international division labour ; Exploitative
- PD 7501 Transformed animals ; Irresponsible patenting genetically
- PE 6788 Transformed plants ; Irresponsible patenting genetically
- PE 1911 Transfrontier private investment disputes
- PD 0366 Transfusion systems ; Limited blood
- PF 1379 Transfusions ; Risk contracting AIDS blood
- PC 0776 Transgenic plants animals ; Creation
- PA 7250 Transgression *Disobedience*
- PA 5952 Transgression *Illegality*
- PA 2497 Transgression international agreements
- PA 5644 Transgression *Vice*
- PD 7527 Transgressions ; Civil law
- PA 6425 Transience
- PA 5490 Transience *Changeableness*
- PA 6425 Transience *Transience*
- PF 1844 *Transient flux* ; *Rapid*
- PE 0355 Transient global amnesia
- PF 1257 *Transient mobility* ; *Disturbing*
- PE 4566 Transient monoplegia
- PF 1681 *Transient occupants* ; *Irresponsible*
- PE 4566 Transient paralysis limb
- PF 1844 *Transient populations* ; Denial rights
- PF 2813 *Transient residents* ; *Uninvolved*

PJ 3760 *Transient residents urban areas ; Disengaged*
PC 0300 *Transient situational disturbances*
PE 9344 *Transient tic disorder*
PF 1844 *Transient urban populations ; Inadequate community care*
PF 1257 *Transients ; Overcrowding schools*
Transit [conventions...]
PJ 8840 — conventions ; Lack adherence international
PE 5789 — conventions land locked countries ; Lack adherence international
PE 5848 — countries ; Inadequate rail transport developing land locked
PE 5524 *Transit fever*
PD 3587 Transit ; Plant diseases storage
PC 0519 Transit ; Refugees
PE 5836 Transit services land locked developing countries ; Unreliable
PD 7420 *Transit tetany*
PE 6006 Transiting aircraft ; Interference
PF 6173 Transition adolescence ; Inadequate recognition institutions
PF 3826 Transition post capitalist social order ; Prevalence psychological conditions unfavourable
PE 8369 Transition school work ; Difficulty
PD 3489 Transitional urban settlements
PE 5959 Translation ; Cultural dominance
PF 6916 Translation errors ; Interpretation
PD 0825 Translation minority languages ; Insufficient
PC 1617 Translocation living organisms
PC 6203 Transmigration
PD 3978 *Transmissable enteritis*
PD 7799 *Transmissible canine venereal tumour*
PD 3936 *Transmissible gastroenteritis swine*
PE 5913 *Transmissible serositis*
PF 3703 Transmission ; Allocation television frequency bands satellite
PD 8360 *Transmission disease ; Faecal*
Transmission [Ignorance...]
PD 8821 — Ignorance concerning disease
PF 1985 — Insufficient cultural heritage
PD 6421 — international travel ; Disease
Transmission lines
PE 9642 — Environmental hazards electrical power
PE 9642 — Health hazards high voltage
PD 1665 — Unaesthetic location power
PD 3040 *Transmission monopoly ; Satellite*
PF 3409 Transmission practical technology ; Haphazard
PD 9538 *Transmission surveillance governments ; Misuse radio*
Transmitted diseases
PD 8821 — Ignorance concerning sexually
PD 5168 — Intentional spread sexually
PD 0061 — Sexually
PD 3699 — Soil
PD 8360 Transmitters disease ; Animals
PE 4335 Transnational banking activities developing countries ; Restrictions home countries
Transnational banks developing
PE 4320 — countries ; Discrimination against
PE 4310 — countries ; Discrimination lending
PE 4330 — countries ; Domination restrictive project loans
PE 4360 — countries ; Socially irresponsible programmes
Transnational banks [domestic...]
PE 4325 — domestic economic policies ; Interference
PE 4355 — Domination loan negotiations
PF 4350 — Inadequate information concerning
PE 4315 — off shore borrowing domestic monetary policies ; Interference
PE 9657 — Restrictive practices
PE 5804 Transnational corporation control bulk shipping
PE 5891 Transnational corporation imperialism
Transnational corporations
PE 5820 — Adverse impact technology transfer
PE 0980 — Aggravation instability exchange rates
PE 1996 — apartheid system ; Participation
PE 1771 — balance payments ; Adverse effect
PE 2200 — Burden conflicting national regulations
PE 3781 — coffee industry ; Domination
PD 0766 — Concentration power
PE 1011 — consumer needs developing countries ; Insensitivity
PE 5831 — Control industries sectors
PE 2397 — Control marketing distribution channels
PE 5903 — Destabilization monetary systems exchange notes
Transnational corporations developing
PE 9646 — countries ; Asymmetry bargaining power
PE 6952 — countries ; Exploitative financial policies
PE 2004 — countries ; Harmful effects advertising
PE 0853 — countries ; Inadequate negotiation entrance terms
PE 1598 — countries ; Minimal export promotion
Transnational corporations [Disregard...]
PE 1940 — Disregard consumer interests
PE 1082 — Disruption cultural social identities developing countries
PE 1957 — Disruption domestic social policies
PE 1796 — domestic name brand food sector developing countries ; Domination
Transnational corporations Domination
PE 2193 — advertising
PE 1448 — agricultural equipment industry
PE 1469 — automobile industry
PE 2084 — copper industry
PE 0163 — developing countries
PE 1322 — economic integration
PE 1187 — labour relations
PE 1620 — shipping industry
Transnational corporations [Escalating...]
PE 0135 — Escalating control international trade

Transnational corporations [Excessive...] cont'd
PE 0194 — Excessive control raw materials markets
PD 5807 — Excessive power independence
PE 2145 — Foreign currency manipulations accounting records
PE 0135 — global policy making ; Undisclosed influence
PE 1806 — Import dependency food staples developing countries due
PE 1660 — Increasing income disparity developing countries due
PF 0691 — Ineffective international regulation
PE 7355 — Interference labour relations developing countries
PE 0667 — Interference labour relations industrialized countries
PE 0163 — Intimidation developing countries
PE 7892 — least developed countries ; Inadequate investment
PE 2328 — Limited market access due product differentiation
PE 1511 — local industry developing countries ; Inadequate relationship
PE 0245 — Manipulation transfer prices
PE 7061 — Missing documents data
PE 1918 — Monopolization technology
PE 4120 — pharmaceutical needs developing countries ; Non responsiveness
PE 0032 — Political intervention
PE 0051 — Restriction free market competition
Transnational corporations Restrictive
PE 5915 — business practices
PE 3196 — market divisions
PE 2396 — pricing policies
Transnational corporations [Retarding...]
PE 0234 — Retarding development
PE 5796 — Social irresponsibility
PE 0106 — specialized agencies United Nations ; Inadequate relationship
PE 5855 — Unfair pricing
PJ 2555 *Transnational cultural penetration*
Transnational enterprises
PD 5891 — [Transnational enterprises]
PE 0207 — against labour ; Coercive use economic power
PE 1799 — Consequences restrictive business practices
PE 0042 — Control national economic sectors
PE 0226 — Destabilizing financial action
PE 0322 — developing countries ; Bribery
PE 0766 — Economic imperialism
PE 1539 — Erosion national sovereignty
PE 0060 — Excessive exploitation raw material reserves
PE 0109 — Monopolistic activity
PF 1072 — Non accountability
PE 7061 — Records destruction
PE 0161 — restrictive business practice ; Direct foreign investment
PE 0669 — Tying supplies subsidiaries
PE 5751 Transnational monopolies developing countries ; Control
PE 1541 Transnational strike action trade unions
PE 8640 Transnationals global communications industry ; Control
PE 5927 Transnationals ; Labour tensions involving
PE 9107 Transparency international trade information ; Lack
PF 5374 Transparency public finances ; Lack
PD 2289 *Transplacental disorders*
PE 1337 Transplantation industrialized country methods developing countries ; Inappropriate
PE 7530 Transplantation ; Lack human organs
PE 7530 Transplantations ; Imbalance need availability viable organs
PE 7530 Transplantations ; Unauthorized organ
PE 7530 Transplants humans ; Animal organ
PD 4271 Transplants ; Traffic children source organ
Transport [accidents...]
PC 8478 — accidents
PD 0126 — accidents ; Rail
PE 3394 — animals ; Maltreatment
PF 6527 *Transport business ; Unprofitable*
PD 9163 Transport cartels ; Air
PF 6496 Transport communications rural villages ; Inadequate systems
Transport [dangerous...]
PD 0971 — dangerous goods
PE 5848 — developing land locked transit countries ; Inadequate rail
PF 2875 *Difficult freight*
PF 2875 *Difficult produce*
PF 9007 — Distant emergency
PE 1048 Transport electrical energy ; Inadequate facilities
Transport equipment
PE 0738 — Environmental hazards
PD 0620 — Instability trade machinery
PF 1436 — Long term shortage machinery
PE 7958 — Restrictive practices trade machinery
PC 2163 *Transport ; Expensive school*
PD 0543 Transport facilities developing countries ; Inadequate road highway
Transport facilities Inadequate
PD 2487 — inland waterway
PD 0496 — rail
PD 0490 — road highway
PJ 3023 — sea
PF 1495 — urban
PF 1267 *Transport fuel ; Expensive road*
PE 8858 Transport hindrance economy
Transport [industry...]
PD 1012 — industry ; Corruption
PD 1012 — industry ; Negligence
PE 6348 — Ineffective rural
PF 3448 — *Infrequent river*
PD 0881 — insurance practices ; Restrictive

PE 5824 Transport land locked developing countries ; Inadequate roads
Transport [modes...]
PF 2403 — modes ; Incompatibility
PD 9163 — monopolies ; Air
PE 3829 — monopoly ; Multimodal
PE 9774 — *myopathy*
PE 9774 — *myopathy turkeys*
PC 2163 *Transport ; Nonexistent high school*
PE 0430 *Transport perishable foodstuffs ; Unsafe*
PD 1012 Transport practices ; Irresponsible
Transport practices Unfair
PD 1367 — [Transport practices ; Unfair]
PD 9163 — air
PD 1367 — rail
Transport [remote...]
PE 5664 — remote island developing countries ; Deficient
PF 2798 — Restrictions effective means
PF 2799 — rural areas ; Lack personal
Transport [sanitary...]
PE 8475 — sanitary wastes ; Inadequate facilities
PE 8551 — Scarcity appropriate
PE 5884 — sectors land locked developing countries ; Lack skilled workers
PF 1495 — *service ; Inadequate air*
PE 5800 — service land locked developing countries ; Inadequate air
PE 5345 — services ; Irregular
PE 1581 — storage communication industries ; Health risks workers
PE 6884 — storage radioactive material
Transport systems
PF 6157 — Inadequate integration
PD 0664 — neighbouring developing countries ; Lack integration
PF 1495 — Poor
PD 7420 *Transport tetany ruminants*
Transport [Undependable...]
PF 2875 — *Undependable produce*
PF 3448 — *Undependable river*
PF 6495 — *Unfeasible goods*
PF 2875 — *Unreliable freight*
PU 4476 — *Unsafe*
Transport vehicles
PE 8063 — Expensive
PF 0713 — Inadequate
PF 0713 — Lack
PD 1294 Transport water supplies ; Inadequate facilities
PF 2996 *Transportation access ; Limited*
PF 2996 *Transportation availability ; Limited*
Transportation [calamities...]
PC 8478 — calamities
PE 3547 — Confined seasonal
PE 8063 — cost ; High
PF 6473 — *Crowded rush hour*
PD 0390 — Cruel animal
PF 1495 Transportation ; Distant public
PE 1839 Transportation ; Environmental hazards sea
Transportation [facilities...]
PD 1388 — facilities developing countries ; Inadequate
PE 6526 — facilities rural communities developing countries ; Inadequate
PE 8063 — fares ; High
PE 1839 *Transportation hazardous cargoes ; Marine*
Transportation [Inadequate...]
PE 0430 — Inadequate cargo
PF 1495 — Inadequate public
PF 2949 — *Incomplete direct*
PD 7132 — industry ; Protectionism air
PF 1495 — infrastructure ; Insufficient
PF 1495 — Insufficient public
PD 4790 — *Irregular hospital*
PE 3547 — Isolating effects seasonal variations undeveloped
Transportation Lack
PF 2996 — regular
PC 2163 — school
PF 2875 — specialized
PF 1495 Transportation ; Limited public
PF 2875 Transportation old people ; Insufficient
PE 8063 Transportation overhead ; High
PE 8063 Transportation ; Prohibitive cost
PE 8959 Transportation services urban areas ; Limited
Transportation systems
PF 4854 — Cover up unsafe
PE 2525 — Excessive land usage
PF 2996 — rural citizens ; Inadequate
PE 2305 Transportation ; Tide determination
Transportation [Unavailability...]
PF 1495 — Unavailability public
PC 2163 — *Unavailability school*
PF 1495 — Underdevelopment
PD 1012 — Unethical practices
PD 0971 — Unhealthy dangerous goods ; Risk
PE 2829 — Transporting processing solid wastes ; Inadequate facilities storing,
PF 6494 Transposing ancient traditions ; Unclarified procedures
PF 3277 *Transsexualism*
PE 8548 Transsexuals ; Denial rights
PE 8548 Transsexuals ; Discrimination against
PE 8548 Transsexuals ; Violation rights
PF 9269 *Transuranic weapons*
PD 2198 *Transvestic fetishism*
PE 6348 *Transvestism*
PD 4525 *Transvestite prostitution*
PE 5735 Trapping animals
PE 9059 Trapping game propagation ; Instability hunting,

Trapping game

PE 8252 Trapping game propagation ; Underdevelopment hunting,
PE 5735 Traps ; Cruel animal
PD 0807 Trash collection ; Unorganized
PD 0807 Trash disposal
PD 0807 Trash removal ; Sporadic
PD 0807 Trash ; Unremoved heavy
PD 4571 Trauma
PE 4109 Trauma ; Acoustic
PE 4247 *Trauma animals ; Thoracic*
PE 8911 Trauma ; Birth
Trauma [Childhood...]
PD 5597 — Childhood
PE 7912 — Combat
PE 0525 — context civil violence ; Stress
PE 1670 Trauma-inducing experiments animals
PE 4571 *Trauma ; Puberty*
PE 5606 Trauma syndromes ; Cumulative
Traumatic [amnesia...]
PE 0357 — amnesia
PE 0351 — amnesia ; Post
PE 0873 — *arthritis*
PD 3978 *Traumatic gastritis*
PE 6874 Traumatic injuries
PE 4571 Traumatic neurosis
PE 5913 Traumatic osteoarthritis
PE 3798 Traumatic pains due torture ; Residual
PD 3978 *Traumatic reticuloperitonitis*
Traumatic [shift...]
PE 1137 — shift life-styles mining communities
PD 4571 — stress
PE 0351 — stress disorder ; Post
PE 3503 *Traumatism*
PE 6968 Traumatization ; Massive psychic
Travel [delays...]
PE 1977 — delays
PE 1904 — Desynchronization bodily rhythm international
PE 6421 — Disease transmission international
Travel documents
PE 9123 — Delay issue
PE 4496 — Forged
PE 0325 — passports, visas ; Refusal issue
Travel [Excessive...]
PF 1555 — Excessive animal sanitary regulations international
PE 0208 — Excessive frontier formalities international
PE 1538 — Exhausting commuter
PC 2163 — *Expensive school*
PE 8920 Travel goods ; Instability trade
PF 3037 *Travel opportunities ; Infrequent*
PF 3037 *Travel opportunities ; Insufficient*
Travel [remote...]
PE 9135 — remote places ; Unaccompanied foreign
PC 8452 — restrictions
PE 4299 — restrictions AIDS-infected persons
PD 7716 — risks
Travel [security...]
PE 8231 — security ; Lack airport
PE 4437 — sexual purposes
PE 2611 — sickness
PC 8452 *Travel ; Taxation*
PE 7780 Travel visas ; Administrative impediments obtaining
PF 3037 *Travel women ; Prohibited*
PD 7716 Traveller ; Vulnerability
PE 7780 Travellers consular procedures ; Harassment
PD 7716 Travellers ; Dangers
PE 7780 Travellers immigration officials ; Harassment
PF 6196 Travellers ; Unconvivial hotel environments
PA 6822 Travesty *Disrespect*
PA 6644 Travesty *Misrepresentation*
PA 7411 Travesty *Uncommunicativeness*
PE 1578 Treachery double agents
PF 4153 Treachery ; Government
PA 7363 Treachery *Improbity*
PA 7309 Treachery *Uncertainty*
PA 7411 Treachery *Uncommunicativeness*
PD 2615 Treason
PD 2615 Treason felony
PD 2615 Treason ; High
PA 7363 Treason *Improbity*
PA 7325 Treason *Irresolution*
PD 2615 Treason ; Petty
PA 5699 Treason *Reversion*
PE 5027 Treat patients ; Medical practitioners refusing
PE 7064 Treated meat infected animals ; Insufficiently
PF 5421 Treaties ; Ambiguous
PF 2497 Treaties ; Breaching international
PD 8465 Treaties ; Covert violation international
Treaties [Delays...]
PE 7300 — Delays ratification human rights
PF 0977 — Delays ratification international
PE 6992 — Deliberately weakened international
PD 1477 — developed developing countries ; Inequitable tax
PF 4787 Treaties ; Humiliating
PE 7573 Treaties indigenous populations ; Violation
PF 4787 Treaties ; Inequitable
PE 7300 Treaties ; Limited acceptance human rights
PF 0977 Treaties ; Limited acceptance international
Treaties Non
PE 7300 — ratification human rights
PF 0977 — ratification international
PE 4460 — verifiability compliance nuclear arms
PE 9267 Treaties ; Token intergovernmental
PF 4787 Treaties ; Unfairly negotiated
PF 8922 Treating fiction reality
PF 7727 Treating genetic diseases ; Compounding effect
PD 5501 Treatment aged ; Abusive

PD 0153 Treatment alcoholism ; Ineffective
Treatment animals
PD 0260 — research ; Cruel
PE 9611 — testing pharmaceutical products ; Cruel
PE 9611 — testing toiletries cosmetics ; Abusive
PE 5857 Treatment corpses ; Undignified
PD 7821 Treatment ; Degrading medical
PD 1480 Treatment ; Depriving prisoners medical
PF 9794 Treatment ; False declarations ill
PE 6883 Treatment foreign prisoners ; Discriminatory
PE 6795 Treatment ; Inadequate waste
PE 8361 Treatment non combatants war zones ; Inhumane
Treatment [patients...]
PD 0584 — patients psychiatric hospitals ; Abusive
PF 1085 — pets ; Luxury
PE 2617 — prisoners war ; Ill
PF 2899 — *Prohibitive cost crop*
PD 7254 Treatment ; Rationing medical
PE 6795 Treatment ; Substandard waste
PD 7630 Treatment trade union leaders ; Ill
PD 7821 Treatment ; Undignified medical
PD 7821 Treatment women labour ; Abusive
PD 2831 Treatment workplaces ; Insufficient acoustical
PE 2324 Treaty following war ; Failure sign peace
PF 2497 Treaty obligations ; Unfulfilled
PF 2497 Treaty provisions ; Non implementation international
PF 2497 Treaty violations
PC 7896 Tree blight
PC 7896 Tree deaths ; Mass
PE 7695 Tree plantations ; Inappropriate
PE 7695 Tree planting plan ; Undesigned
PE 1570 *Tree shrews ; Endangered species*
PC 0238 Trees beyond their commercial life ; Unproductive use
PE 7695 Trees ; Degradation environment
PD 1642 Trees ; Destruction hedges hedgerow
Trees Endangered
PD 2025 — park
PC 0238 — *plantations long lifed*
PD 2025 — urban
PE 4586 *Trees ; Hypoxylon canker*
PF 3710 Trees ; Inadequate legal rights
PE 4139 Trees ; Mistletoe
PD 3585 Trees ; Pests diseases
PE 4732 Trees ; Pollens grasses, weeds
PC 0238 *Trees ; Unsustainable cultivation long lifed*
PD 2025 Trees ; Urban hazards
PE 6256 Trekkers ; Pollution mountain environments mountaineers
PD 3483 *Tremarctos ; Endangered species*
PE 6461 Trematode diseases
PE 6461 Trematode infection
PE 5330 Tremblante mouton
PD 7841 Trembles ; *Congenital*
PE 9263 Tremens ; *Delirium*
PD 7841 Tremor syndrome ; *Porcine congenital*
PA 6030 Tremulousness *Fear*
PD 2255 Trench fever
PE 3254 Trench mouth
PE 2143 Trenchmouth
PF 3523 Trend ; *Youth exodus*
PF 9790 Trends ; Irreversible environmental
PA 5838 Trepidation *Agitation*
PA 6030 Trepidation *Fear*
PA 5467 Trepidation *Inexcitability*
PD 3794 Trespass
PE 9363 Trespass ; Chemical
PD 3794 Trespass ; Criminal
PA 7250 Trespass *Disobedience*
PA 5952 Trespass *Illegality*
PA 6862 Trespass *Intrusion*
PA 6921 Trespass *Undueness*
PA 5644 Trespass *Vice*
PD 3794 Trespassers
PE 4898 Trespassing cattle
PE 4898 Trespassing livestock
PC 2343 Triads
PD 7254 Triage
PE 4577 Trial after release ; Failure appear
PE 4737 Trial court ; Denial right
PE 7624 Trial defence ; Denial right time prepare
Trial Denial right
PD 4827 — fair public
PE 3964 — fair public
PE 4887 — speedy
Trial [Imprisonment...]
PE 4887 — Imprisonment delayed
PF 9429 — Incompetence stand
PD 1576 — Internment
PE 4887 — internment ; Excessive length pre
PF 3278 Trial marriage
PE 1692 Trial publicity ; Unfair trials due pre
PE 0424 Trials absentia ; Injustice
PE 3916 Trials ; Complex
PE 1692 Trials due pre trial publicity ; Unfair
PE 3916 Trials exceeding competence juries
PE 0597 Trials ; Injustice mass
PE 3916 Trials ; Lengthy
PD 3013 Trials ; Political
Trials [Secret...]
PD 3013 — *Secret*
PD 3013 — *Show*
PE 4733 — small jurisdictions ; Bias jury
PD 4827 Trials ; Unjust
PE 3616 Triatominae
PA 0298 *Tribal conflict states*
PD 2191 Tribal conflicts

PE 6545 Tribal loyalties ; Divisive
PE 2666 Tribal societies ; Head hunting
PF 6712 *Tribal structure ; Fraying*
PE 6545 Tribal urban loyalty ; Split
PD 2191 Tribal warfare ; Disruption development
PC 1910 Tribalism
PC 1910 *Tribalism ; Internal conflict due*
PC 1910 *Tribalism ; National instability due*
PE 8513 *Tribalism ; Weak national identity due*
PC 0720 Tribes indigenous peoples ; Endangered
PA 6340 Tribulation *Adversity*
PA 7107 Tribulation *Unpleasantness*
PF 7897 Tribunal arbitration ; Government refusal accept international
PD 5317 Tribunal ; Violation right review conviction higher
PE 7664 Tribunals ; Extrajudicial courts
PE 0494 Tribunals ; Injustice military
PE 0494 Tribunals ; Military
PE 7664 Tribunals ; Security
PE 2311 Trichiniasis
PE 2311 Trichinosis
PE 3508 *Trichocephaliasis*
PE 2310 Trichomoniasis
PE 7799 *Trichomoniasis ; Bovine*
PD 3978 *Trichomoniasis poultry*
PE 0958 *Trichostrongylosis sheep ; Intestinal*
PE 0958 *Trichuriasis ; Canine*
PA 7411 Trickery *Uncommunicativeness*
PA 7232 Trickery *Unskillfulness*
PD 7171 Tricks policies ; Government dirty
PF 6158 Tried twice same charges ; Violation freedom being
PE 8932 *Trigeminal neuralgia*
PE 2964 Trioxide pollutant ; Sulphur
PF 0919 Tripartite cooperation ; Limited
PA 6448 *Trips*
PE 6354 Triskaidekaphobia
PF 8203 Triumphalism
PF 7414 Trivia ; Cultivation
PA 6852 Triviality *Inappropriateness*
PA 5652 Triviality *Inferiority*
PA 5942 Triviality *Unimportance*
PA 7371 Triviality *Unintelligence*
PE 2224 Trivialization death suffering videos
PF 5155 Trivialization equality
PF 7497 Trivialization liberty
PF 0959 Trivialization love
PF 5826 Trivialization peace
PF 4600 Trivialization political debate
PF 7930 Trivialization sexuality
PF 5242 Trivialization social values
PE 7826 *Trochanteric bursitis*
PE 8495 Troops ; Drunkenness military personnel
PD 2592 Troops ; Mercenary
PD 2289 *Trophoblastic disease*
PE 5644 Trophy hunting ; Animal
Tropical [climates...]
PE 1811 — climates ; Corrosion
PE 1265 — climates ; Destructive action mould
PD 1590 — cyclones
PD 1739 — cyclones
PD 6204 Tropical deforestation
PD 4775 Tropical diseases ; Uncontrolled
PF 7565 Tropical economies ; Preponderance non food crops
PE 3639 *Tropical fowl mite*
PE 5808 Tropical island developing countries ; Inviability
PE 5274 Tropical rain forests agricultural development ; Misuse
PE 2235 Tropical villages ; Detrimental effect jungle environment
PF 2365 Tropical villages ; Disorganized approach land ownership
PA 6340 Trouble *Adversity*
PA 5497 Trouble *Difficulty*
PA 6030 Trouble *Fear*
PA 7395 Trouble *Inexpedience*
PA 6921 Trouble *Undueness*
PA 7107 Trouble *Unpleasantness*
PC 8866 Troubles ; Digestive
PE 4200 Truancy
PE 4200 Truancy ; Unaccounted
PA 2996 Truck availability ; Erratic
PA 5532 Truculence *Disaccord*
PA 7143 Truculence *Discourtesy*
PA 5643 Truculence *Unkindness*
PE 3604 *True vampires ; Endangered species*
PC 4422 Trust ; Abuse
PF 7150 Trust ; Breach
PC 4422 Trust ; Exploitation
PD 4691 *Trust funds ; Misuse*
PF 6386 Trust prosecution ; Indiscriminate anti
PA 6164 Truth∗complex
PF 8203 Truth ; Excessive equation convictions
PF 9595 Truth ; Fear
PF 5937 Truth ; Vulnerability society
PA 7411 Truthlessness *Uncommunicativeness*
PE 5725 Trypanosomiasis
PE 1778 Trypanosomiasis ; African
PE 0653 Trypanosomiasis ; American
PE 0653 Trypanosomiasis ; South American
PE 1335 Tsetse flies pests
PF 5781 Tsukimono
PD 0033 Tsunamis
PE 3895 *Tsutsugamushi disease*
PE 8901 Tuber trade instability vegetable products ; Roots
PE 0566 Tuberculosis
PG 9737 Tuberculosis ; Acute miliary
PE 9846 Tuberculosis bones joints
PE 2280 *Tuberculosis central nervous system*

Unbearableness

PE 0566 Tuberculosis ; Disseminated
PE 0566 Tuberculosis ; Fibro caseus
PE 0566 Tuberculosis ; Fibroid
PE 2053 Tuberculosis genito-urinary system
PE 0566 Tuberculosis intestines
PE 2053 Tuberculosis kidney
PF 2212 Tuberculosis materials ; Lack
PE 2526 Tuberculosis ; Pulmonary
PE 2526 Tuberculosis ; Respiratory
PE 2653 Tuberculous laryngitis
PD 2654 Tuberculous lymphadenitis
PE 2280 Tuberculous meningitis
PD 2283 Tuberculous milk
PE 9846 Tuberculous osteomyelitis
PE 8208 Tubers ; Instability trade vegetables, roots
PC 1326 Tubulidentata ; Endangered species
PD 3481 Tucotucas ; Endangered species
PE 6872 Tularaemia
PE 0652 Tumorous overgrowth ; Thyroid glands
PD 0992 Tumour brain
PD 0992 Tumour ; Brain
PD 2567 Tumour ; Ear
PD 7799 Tumour ; Transmissible canine venereal
PC 3853 Tumours
PD 9654 Tumours animals ; Functional islet cell
PD 9654 Tumours animals ; Pituitary
PD 8347 Tumours ; Benign
PD 5535 Tumours external ear animals
PE 9229 Tumours ; Intracranial
PD 2283 Tumours joints
PE 2053 Tumours ; Kidney
PD 9293 Tumours kidney animals
PD 9293 Tumours lower urinary tract animals
PD 7799 Tumours ; Mammary
PE 2286 Tumours pituitary tissue
PE 5457 Tumours ; Plant
PE 7708 Tumours ; Thymic
PD 1165 Tundra ecosystem fragility
PE 7621 Tungsten ; Instability trade
PA 5838 Turbidity Agitation
PA 5838 Turbulence Agitation
PD 2127 Turbulence ; Air
PD 2127 Turbulence ; Clear air
PA 7361 Turbulence Disorder
PA 7156 Turbulence Moderation
PE 4346 Turf ; Snowmold
PE 3639 Turkey chigger
PE 5567 Turkey coryza
PD 2728 Turkey X disease
PD 3978 Turkeys ; Coronaviral enteritis
PE 9774 Turkeys ; Deep pectoral myopathy
PD 5453 Turkeys ; Dissecting aneurysm
PD 3978 Turkeys ; Haemorrhagic enteritis
PD 7799 Turkeys ; Low fertility
PD 2730 Turkeys ; Lymphoproliferative disease
PD 2732 Turkeys ; Mycoplasma meleagridis infection
PD 3978 Turkeys ; Rotaviral infections
PE 4497 Turkeys ; Thrush
PE 9774 Turkeys ; Transport myopathy
PD 2730 Turkeys ; Viral hepatitis
PC 3685 Turmoil ; Minority
PD 3518 Turned-in elders' mindset
PE 9324 Turner's syndrome
PF 0907 Turnover developing countries ; High labour
PE 8221 Turnover ; High teacher
PF 3434 Turnover international organizations ; Rapid personnel
PA 7363 Turpitude Improbity
PA 5644 Turpitude Vice
PE 2265 Twangs
PD 0173 Twin pregnancies
PD 1739 Twisters
PC 3458 Two-tier economic system
PF 1492 Tying ; Aid
PE 0669 Tying supplies subsidiaries transnational enterprises
PE 9774 Tying-up syndrome horses
PE 3936 Type C enteritis swine ; Clostridium perfringens
PE 7769 Type D enterotoxaemia sheep
PD 4158 Types crimes ; Generalized
PE 5874 Types ships built ; Imbalances
PD 1753 Typhoid fever
PE 7562 Typhoid ; Fowl
PD 1590 Typhoon protection ; Lack
PD 1590 Typhoons
PE 2357 Typhus ; Canine
PD 1753 Typhus ; Classical
PE 3895 Typhus ; Endemic
PD 1753 Typhus ; Epidemic
PD 0640 Typhus fever
PE 3895 Typhus ; Flea borne
PD 1753 Typhus ; Louse borne
          Typhus [Marine...]
PG 3896 — Marine
PE 3895 — Mite borne
PG 3896 — Murine
PE 3895 Typhus ; Non epidemic
PG 3896 Typhus ; Scrub
PD 2710 Tyrannical dictatorship
PC 0845 Tyranny
PA 7296 Tyranny Restraint

U

PE 8560 U235 depleted uranium ; Instability trade
PD 7799 Udder acne dairy cows
PF 1392 UFOs Flying saucers

PF 1392 UFOs Unidentified flying objects
PA 7240 Ugliness
PA 7240 Ugliness ; Dependence
PA 7253 Ugliness Envy
PA 7240 Ugliness Ugliness
PA 7107 Ugliness Unpleasantness
PA 7240 Ugly people
PE 2143 Ulcenomembranous stomatitis animals
PE 2308 Ulcer ; Gastrojejunal
PE 9702 Ulcer ; Rodent
PG 5613 Ulcer tongue
PE 6789 Ulceration cornea
PD 9667 Ulceration ; Leg
PD 9667 Ulceration ; Lip
PE 4161 Ulceration sole
PD 9045 Ulcerative colitis
PE 0873 Ulcerative colitis ; Arthritis accompanying
PD 9667 Ulcerative dermatosis sheep
PD 9045 Ulcerative enteritis
PD 3978 Ulcerative enteritis poultry
PE 6789 Ulcerative keratitis
PD 9844 Ulcerative lymphangitis ; Equine
PD 7799 Ulcerative mammillitis ; Bovine
PD 7799 Ulcerative posthitis animals
PE 2143 Ulcerative stomatitis animals
PD 7799 Ulcerative vulvitis
PE 2308 Ulcers
PE 9364 Ulcers ; Abomasal
PE 2308 Ulcers ; Duodenal
PE 2308 Ulcers ; Gastric
PE 2308 Ulcers ; Peptic
PD 3978 Ulcers swine ; Oesophagogastric
PE 1154 Ulmi ; Ceratcystis
PD 0748 Ultrasonic radiation human body ; Harmful effects
PE 5672 Ultraviolet radiation hazard
PE 8961 Umbilical hernia
PF 5395 UN charter ; Antiquated provisions
PE 8332 UN related agencies ; North South spheres influence
PF 3826 Unacceptability conserver society policies
PD 7231 Unacceptable activity research ; Misrepresentation socially
PD 0520 Unacceptable conditions imprisonment
PC 8019 Unacceptable initiatives ; Misuse societally endorsed professions conceal socially
PE 9135 Unaccompanied foreign travel remote places
PE 0988 Unaccountability decentralized government agencies
PE 4997 Unaccountability government
PF 3458 Unaccountability government agencies ; Environmental
          Unaccountability [institutions...]
PF 3458 — institutions degrading environment
PF 1136 — international aid institutions
PF 1136 — international financial institutions
PC 1522 Unaccountability ; Social
PF 9184 Unaccountable government intelligence agencies
PF 8074 Unaccountable management public information
PE 4200 Unaccounted truancy
PF 2583 Unacknowledged availability services
PD 1284 Unacknowledged copying ideas
PD 4259 Unacknowledged detention
PF 4114 Unacknowledged future costs present policies
PF 2877 Unacknowledged global interdependence
PF 6559 Unacknowledged highway danger
PF 4672 Unactivated policy
PF 6024 Unadapted vocational training women farmers
PF 6573 Unadvertised educational resources
PA 7325 Unadvisedness Irresolution
PA 7371 Unadvisedness Unintelligence
PD 1185 Unaesthetic appearance
PF 0867 Unaesthetic architecture design ; Monotonous
PD 1126 Unaesthetic foodstuffs
PE 2156 Unaesthetic location advertising hoardings billboards
PD 1665 Unaesthetic location power transmission lines
PE 8303 Unaffordable fulltime physician
PF 6005 Unallocated industrial land
PF 6471 Unanalysed available capital
PF 5222 Unannounced community meetings
PF 2535 Unanticipated market variance
PF 8635 Unanticipated price fluctuations
PD 6627 Unappealing business buildings
PF 0836 Unappealing farm labour
PD 3054 Unappealing loan collateral
PF 2827 Unappealing public office positions
PF 1468 Unapplied scientific knowledge
PD 2338 Unapproachable absentee landlords
PF 2400 Unarticulated goal educational methods
PF 2477 Unarticulated training needs
PF 2373 Unassertive people
PF 0963 Unassessed corporate needs
PC 8160 Unassigned public land
PA 6090 Unastonishment
PF 1733 Unattractive appearance deteriorating buildings
PF 1815 Unattractive business facilities
PF 2093 Unattractive community image
PF 6478 Unattractive fishing business
PF 6478 Unattractive fishing employment
PD 8916 Unattractive industrial wages
PF 2439 Unattractive local businesses
PF 3499 Unattractive locale economic development
PF 6578 Unattractive merchandise display
PE 6151 Unattractive pedestrian environments urban areas
PF 3454 Unattractive social life
PF 3454 Unattractive social opportunities
PE 7165 Unattractive teacher post
PF 4010 Unattractiveness ; Personal physical
PD 4287 Unauthentic intimacy
PF 7643 Unauthentic peace

PA 6180 Unauthenticity Error
PA 5502 Unauthenticity Reason
PA 7309 Unauthenticity Uncertainty
PA 7411 Unauthenticity Uncommunicativeness
PE 9831 Unauthorized access computer information systems
PB 0284 Unauthorized circulation confidential information
PF 4176 Unauthorized disclosure computerized information
PE 7188 Unauthorized liquor manufacturing distribution
PE 5637 Unauthorized medical practice
PE 7530 Unauthorized organ transplantations
          Unauthorized [pharmaceutical...]
PE 0564 — pharmaceutical manufacture distribution
PD 3544 — police search
PF 8780 — proximity males females
          Unavailability [agricultural...]
PC 7597 — agricultural land
PF 8835 — agricultural training
PF 1352 — apprenticeship positions
PF 7916 — appropriate expertise
PE 0718 Unavailability banking services
          Unavailability building
PF 3499 — land
PF 3554 — materials
PF 3499 — sites
          Unavailability [community...]
PF 7118 — community centres
PE 8303 — community doctors
PF 6337 — comprehensive information
PF 1790 — construction skills
PF 4832 Unavailability dietary care
PF 1301 Unavailability educational data
PF 2702 Unavailability effective motivational techniques ; General
          Unavailability [farming...]
PF 3630 — farming capital
PC 7597 — farming land
PD 1631 — fire protection
PF 9007 — first aid
PF 6573 — funding information
          Unavailability [impartial...]
PF 6395 — impartial expertise
PF 2489 — income subsidy
PD 8839 — irrigation water
          Unavailability [land...]
PE 5024 — land agricultural purposes developing countries
PF 6477 — land funds
PF 7277 — legal information
PF 6479 — library facilities
          Unavailability local
PF 6477 — capital
PF 3448 — dentists
PF 6574 — directories
PF 1658 — jobs
PE 8303 — physicians
PE 1278 Unavailability medicines
PF 6524 Unavailability modern equipment
          Unavailability [prime...]
PC 8160 — prime land
PF 1495 — public transportation
PE 3296 — pumps
PF 2971 Unavailability quality medicine
PE 7504 Unavailability replacement parts agricultural machinery
PF 2410 Unavailability road machines
          Unavailability [scholarship...]
PE 8883 — scholarship funds developing country students
PE 3569 — scholarship funds students
PC 2163 — school transportation
PC 2163 — skills trainers
          Unavailability [timber...]
PF 3448 — timber
PF 6337 — timely data
PE 8429 — trained teachers rural areas developing countries
PC 1131 — trainee employment
PF 4375 — training costs
PC 2035 Unavailability youth employment
PF 7130 Unavailable commercial sites
PF 2313 Unavailable education effective living
PF 9007 Unavailable emergency services
PF 2378 Unavailable machinery capital
PF 1600 Unavailable public buildings
PD 1566 Unavailable public facilities
PF 3665 Unavailable risk capital
PF 2410 Unavailable road machinery
PF 8085 Unavailable skills retraining
PC 2163 Unavailable vocational training
PD 1942 Unawareness actuality ; Lack essential freedom due
PF 3211 Unawareness health benefits
PD 8023 Unawareness health problems
PF 6925 Unawareness potential benefits
PD 8023 Unawareness symptoms illness
PF 8205 Unawareness welfare benefits
PE 7637 Unbalanced application communications technology
PD 5492 Unbalanced budgets
PC 0382 Unbalanced diet
PF 0204 Unbalanced distribution knowledge
PC 0382 Unbalanced family diets
PF 2813 Unbalanced food usage
PB 0479 Unbalanced growth
PB 0479 Unbalanced growth ; Dependence
PF 3843 Unbalanced industrial distribution
PE 0691 Unbalanced infant diets
PD 8113 Unbalanced social life
PD 6131 Unbalanced urban population density gradients
PJ 2664 Unbanked liquidity
PF 0478 Unbankruptability sovereign states
PA 7107 Unbearableness Unpleasantness

Unbelief

PA 7392 Unbelief	PF 1176 Uncohesive business community	PF 6196 Unconvivial hotel environments travellers
PF 8068 Unbelievers	PF 2401 Uncollected skills information	PE 0596 Uncooked inadequately cooked flesh
PD 4365 Unbelievers ; Discrimination against religious	PA 7107 Uncomfortableness Unpleasantness	PG 8893 Uncoordinated agency referral
PF 6616 Unborn children ; Denial right	PF 2842 Uncommitted local industry	PF 1205 Uncoordinated agricultural planning
PF 2813 Unbudgeted child education	PF 6506 Uncommunicated annexation authority	PF 1442 Uncoordinated budget procedure
PF 6483 Uncatalogued ancient manuscripts	PE 5345 Uncommunicated bus schedule	PF 6022 Uncoordinated disaster relief
PF 6538 Uncatalogued community skills	PF 1301 Uncommunicated educational data	**Uncoordinated [economic...]**
PF 4077 Uncatalogued documents	PF 6552 Uncommunicated resource opportunities	PF 1345 — economic enterprises
PF 6483 Uncatalogued historical documents	PF 1664 Uncommunicated wisdom elders	PF 1768 — efforts agricultural development
PF 8623 Uncelebrated heroes	PA 7411 Uncommunicativeness	PF 1781 — expression needs
PA 0911 Uncentered people	PA 7115 Uncommunicativeness Rashness	**Uncoordinated government**
PF 9729 Uncertain cooperative success	PA 7411 Uncommunicativeness Uncommunicativeness	PF 7619 — decision-making
PF 5817 Uncertain economic patterns	PA 6653 Uncommunicativeness Unsociability	PF 7619 — planning
PF 9450 Uncertain environmental impact current policy	PD 7179 Uncompensated damages	PF 7619 — policy-making
PF 2535 Uncertain market access	PE 4789 Uncompensated housewives' labour	PD 4790 Uncoordinated health care planning
PF 2535 Uncertain market volume	PD 3056 Uncompensated tenant repair	PD 4790 Uncoordinated health services
PF 1010 Uncertain planning implementation	PF 2439 Uncompetitive local merchandising	**Uncoordinated [intergovernmental...]**
PF 5188 Uncertain radiation hazard thresholds	PF 8105 Uncompetitive rural attractions	PF 7133 — intergovernmental information
Uncertain [sales...]	PF 6574 Uncompleted sidewalk repairs	PD 0516 — international river basin development
PF 0581 — sales potential	PE 0375 Uncomplicated alcohol withdrawal	PE 8233 — irrigation control
PF 3630 — sources rural funding	PF 1442 Uncomprehensive budget methods	PF 0963 Uncoordinated joint planning
PF 2342 — status monetary gold	PF 6512 Uncomprehensive community education	**Uncoordinated [macroeconomic...]**
PF 5188 Uncertain toxicity thresholds	PD 0516 Uncomprehensive river plan	PF 5000 — macroeconomic policies governments
PF 4770 Uncertainties animal experimentation	PF 0963 Unconceived corporate responsibility	PF 9576 — multiplicity development models
PA 6438 Uncertainty	PF 3307 Unconfident citizen participation	PF 9576 — multiplicity global financial models
PA 7309 Uncertainty	PC 1820 Unconfident potential entrepreneurs	PF 9166 Uncoordinated policy-making
PA 5490 Uncertainty Changeableness	PF 1345 Unconnected business resources	PC 1217 Uncoordinated production planning
Uncertainty [Danger...]	PE 1055 Unconnected major roads ; Road network	PF 2306 Uncoordinated research efforts
PA 6971 — Danger	PA 7193 Unconscionableness Cheapness	PF 1853 Uncoordinated social services urban areas
PF 0431 — death missing persons	**Unconscionableness [Improbity...]**	PF 3498 Uncoordinated special training
PE 7679 — Defence information	PA 7363 — Improbity	PF 3005 Uncoordinated use community property
PA 6438 — Dependence	PF 6486 — Injustice	PE 8721 Uncoordinated use computers automation
PF 4295 — development expenditures due floating-rate loans	PA 5473 — Insufficiency	PE 3296 Uncorporate economic structure
PF 4300 — development programmes due short-term loans	PA 7156 Unconscionableness Moderation	PF 9323 Uncorporate family decisions
PF 5817 Uncertainty ; Economic	PE 5431 Unconscious passage urine	PF 2477 Uncorrelated training needs
Uncertainty [Insecurity...]	PF 2595 Unconsensed educational goals	PF 3454 Uncovered refuse dumps
PF 8354 — Insecurity future crop	PF 1733 Unconsensed environmental programmes	PF 1572 Uncreative strategies against problems
PC 9584 — international trading system	PD 4790 Unconsensed health needs	PD 3054 Uncreditworthiness developing countries
PF 7109 — Intolerance	PF 0963 Unconsensed institution desirability	PD 3054 Uncreditworthiness rural communities
PA 6978 — Invisibility	PF 6528 Unconsensed land usage	**Uncritical acceptance**
PA 7325 — Irresolution	PF 5250 Unconsensed public celebrations	PF 5973 — another person
PA 6438 Uncertainty land zoning	PF 1262 Unconsensed union farmers	PF 8596 — authority
PE 5324 Uncertainty long-term health effects radioactive fallout	PF 6479 Unconsensed village library	PF 2901 — dogmas standards
PE 1137 Uncertainty miners' future	PF 2845 Unconsolidated community patterns	PF 5311 Uncritical feel-good rhetoric
PF 1559 Uncertainty ; Political	PF 0593 Unconstrained educational self-expression	PF 1688 Uncritical preservation status quo
PE 9085 Uncertainty survival human race	PF 2855 Unconstrained exploitation natural resources	PF 5039 Uncritical thinking
Uncertainty [Unbelief...]	PF 5365 Unconstrained lobbying	PF 5311 Uncritical upbeat reporting
PA 7392 — Unbelief	PB 0035 Unconstrained social growth	PA 6191 Unctuousness Disapproval
PA 7309 — Uncertainty	PB 9015 Unconstrained socio-economic growth	PA 6058 Unctuousness Impiety
PA 7367 — Unintelligibility	PE 8103 Unconstructed teachers' houses	PA 7411 Unctuousness Uncommunicativeness
PC 1524 — university degree equivalencies	PF 2875 Unconvincing doctor relationship	PD 0153 Uncurbed alcohol use
PJ 8370 Uncertified skilled tradesmen	PF 1443 Uncontrollability world social systems	PF 6573 Uncurious attitudes ; Unquestioning,
PF 2575 Unchallenged youth potential	PC 0776 Uncontrollable new species	PF 6559 Undecided bank use
PF 9205 Unchallenging beliefs	PE 8233 Uncontrollable water run-off	PC 0939 Undeclared employment socialist countries
PF 9478 Unchallenging world vision	PC 0418 Uncontrolled application technology	PC 0939 Undeclared employment socialist countries ; Second
PF 2845 Unchanging business pattern	PF 6025 Uncontrolled costs disaster relief	**Undeclared [states...]**
PF 5295 Unchanging legal precedent undermines accountability	PF 6514 Uncontrolled dogs	PE 4363 — states emergency
PF 4010 Unchanging personal appearance	**Uncontrolled [economic...]**	PD 5384 — strikes
PF 5612 Unchastity	PC 0495 — economic growth	PD 1882 — strikes socialist countries
PA 7338 Unchastity Hate	PC 0495 — economic growth ; Dependence	PB 0593 — Undeclared war
PF 5612 Unchastity Unchastity	PC 1174 — environmental impact technology	PF 6531 Undefined adult education
PA 5644 Unchastity Vice	PE 4997 — executive branch action	**Undefined [community...]**
PF 6559 Unchecked community spending	PD 9534 Uncontrolled growth arable land	PF 6556 — community leadership
PD 8544 Unchecked destructive behaviour	PC 8316 Uncontrolled growth debt	PF 6521 — community limits ; Geographically
PD 1350 Unchecked minor vandalism	PF 1238 Uncontrolled household rent increases	PF 7130 — cultural space
PF 2346 Unchecked poverty cycle	PC 2024 Uncontrolled hunting	PF 2595 Undefined educational roles
PD 8890 Unchecked power government bureaucracy	PB 1845 Uncontrolled industrialization	PF 7919 Undefined government role urban services
PD 0452 Unchecked seasonal flooding	PB 1845 Uncontrolled industrialization ; Dependence	PF 1815 Undefined lot usage
PF 1781 Unclaimed local responsibility	**Uncontrolled [land...]**	PF 4984 Undefined preschool advantages
PF 6494 Unclarified procedures transposing ancient traditions	PE 1925 — land markets	PF 1959 Undefined programme responsibility
PF 2303 Unclarity dock use	PF 6528 — land use	PD 1544 Undefined youth needs
PD 1544 Unclarity youth roles	PD 1350 — local vandalism	PB 8031 Undemocratic economic systems
PF 8515 Unclean food	**Uncontrolled market**	PF 8703 Undemocratic government decision-making
PA 5454 Uncleanliness Badness	PF 7880 — economy	PD 2710 Undemocratic leadership
PD 2459 Uncleanliness ; Bodily	PF 7880 — forces	PC 8676 Undemocratic organizations
PF 6657 Uncleanliness ; Spiritual	PF 8635 — price	**Undemocratic [policy...]**
PF 5612 Uncleanliness Unchastity	**Uncontrolled [markets...]**	PF 8703 — policy-making
PA 5459 Uncleanliness Uncleanness	PF 7880 — markets	PC 1015 — political organization
PA 5644 Uncleanliness Vice	PD 0040 — media	PC 1015 — political organization ; Dependence
PA 5459 Uncleanness	PD 2229 — migration	PD 3389 — pressures
PF 3960 Uncleanness ; Ceremonial	**Uncontrolled [physical...]**	PC 9585 Undemocratic regimes
Unclear [care...]	PD 0134 — physical expansion cities developing countries	PF 4821 Undemocratic regimes ; Government support
PD 4790 — care needs	PC 0555 — plant diseases	PB 8031 Undemocratic social systems
PD 1544 — community responsibility	PC 1142 — police operations	PD 9641 Undemocratic voting systems
PD 2438 — consensus patterns	PD 5859 — property damage	PF 2346 Undemonstrable space care
PF 2595 Unclear educational roles	PD 2290 Uncontrolled river erosion	PF 9729 Undemonstrated cooperative benefits
PE 8308 Unclear fiscal relations different parts government	**Uncontrolled [satellite...]**	PF 9729 Undemonstrated cooperative purchasing
PF 1559 Unclear government policies	PE 0261 — satellite broadcasting	PA 5490 Undependability Changeableness
PF 0336 Unclear local enforcement	PF 2692 — scientific power	PA 6971 Undependability Danger
PF 6492 Unclear official directives	PF 6455 — small animals	PA 7363 Undependability Improbity
Unclear [polity...]	PE 7105 — structures international nongovernmental organizations	PA 7325 Undependability Irresolution
PF 1651 — polity roles	PD 4775 Uncontrolled tropical diseases	PA 7309 Undependability Uncertainty
PF 4984 — preschool possibilities	**Uncontrolled [urban...]**	PF 6473 Undependable garbage pick-up
PD 9162 — property rights	PC 0442 — urban development	PF 2875 Undependable produce transport
PF 8759 Unclear relationships	PD 3488 — urbanization developed countries	PF 3448 Undependable river transport
PF 2346 Unclear rental policy	PD 0134 — urbanization developing countries	PE 8254 Undependable telephone service
PF 2595 Unclear school roles	PF 4176 — use computer data	PD 1611 Under-age drinking
PF 1651 Unclear societal roles	PB 8376 Uncontrolled waste	PF 0021 Under-developed continuing education
Unclear [village...]	PF 3499 Uncontrolled waterfront damage	PF 8827 Under-insured
PF 6506 — village consensus	PC 8836 Unconventional war	PE 4920 Under-strength military forces
PF 1365 — vision farmers	PA 7273 Unconventionality	PF 6590 Under-utilized raw materials
PF 2477 — vocational needs	PF 1985 Unconveyed cultural heritage	PC 1716 Underachievement ; Educational
PD 1544 Unclear youth needs	PD 4067 Unconvicted war criminals	PD 1410 Undercapitalized waste use schemes
PD 1544 Unclear youth roles	PF 3826 Unconvincing alternatives existing societies	PD 6017 Underclass
PF 2613 Uncleared land mines	PF 6528 Unconvincing land-use story	PC 5052 Underclass ; Urban
PF 2613 Uncleared sea mines		PE 0596 Undercooked meat

Undisposed dead

PE 6331 Undercover police operations
PC 0492 *Undercutting vendor price*
PE 1234 *Underdeveloped administrative abilities*
PF 6493 Underdeveloped approaches local food production
PF 3448 *Underdeveloped benefits community resources*
PF 1240 Underdeveloped capacity income farming
PF 6479 Underdeveloped community leadership
PE 8190 Underdeveloped economy
PF 2989 *Underdeveloped export potential*
PE 7313 *Underdeveloped farm marketing*
PD 7536 *Underdeveloped farming skills*
PE 4339 Underdeveloped legislation ; Exploitation regulatory loopholes countries
PF 0202 Underdeveloped leisure facilities
PF 3448 Underdeveloped potential basic resources
PF 6473 Underdeveloped provision basic services developing countries
PE 1055 Underdeveloped road network
PF 0393 Underdeveloped rural energy sources
PF 1345 Underdeveloped sources income expansion
PF 8552 Underdeveloped technological skill
PF 2164 Underdeveloped use agricultural resources
PE 8233 Underdeveloped water management
PB 0206 Underdevelopment
PD 0629 Underdevelopment agricultural livestock production
PF 1374 Underdevelopment basic metal industries
PE 1483 Underdevelopment chemical petrochemicals industry
PE 8427 Underdevelopment ; China, pottery earthenware manufacture
Underdevelopment Dependence
PB 0206 — [Underdevelopment ; Dependence]
PB 0539 — economic social
PC 0242 — social
PD 8183 Underdevelopment developed countries
Underdevelopment [Economic...]
PC 0281 — Economic
PB 0539 — Economic social
PF 2583 — electricity, gas, water sanitary services
Underdevelopment [fishing...]
PE 2138 — fishing industry
PF 2821 — food live animal production
PD 0908 — food processing industries
PC 0880 — *forestry industry*
PF 2604 — *Furniture fixtures manufacture*
PE 8715 Underdevelopment ; Glass glass products manufacture
PE 8252 Underdevelopment hunting, trapping game propagation
Underdevelopment [Incompleteness...]
PA 6652 — *Incompleteness*
PC 0880 — industrial economic activities
PE 8223 — iron steel basic industries
PF 4718 Underdevelopment ; Lack progress against
PF 4836 Underdevelopment legal infrastructure
Underdevelopment [Machinery...]
PE 1852 — *Machinery equipment industries*
PF 0854 — manufacturing industries
PF 0942 — metal products, machinery equipment industries
PE 1858 — mining quarrying industries
PE 1851 Underdevelopment non-metallic mineral products industries
Underdevelopment [paper...]
PE 1136 — paper printing industries
PE 7960 — Petroleum coal products manufacture
PE 4120 — pharmaceutical industry developing countries
PD 8673 — political systems
PF 4135 — power industry developing countries
PD 4424 Underdevelopment ; Regional
PC 0306 Underdevelopment ; Rural
PC 0242 Underdevelopment ; Social
PE 0453 Underdevelopment textile clothing industries
PF 1495 Underdevelopment transportation
PA 7341 Underdevelopment *Unpreparedness*
PF 2604 Underdevelopment woodworking industries
PF 7047 Underemphasized importance rural community
PB 1860 Underemployment
Underemployment [Dependence...]
PB 1860 — Dependence
PD 8141 — developing countries
PD 0295 — developing countries ; Rural
PB 0750 *Underemployment skilled workers*
PC 3490 Underemployment ; Urban
PF 8499 Underestimated project disadvantages
PF 7063 Underestimation human potential
PA 6607 Underestimation *Misjudgement*
PF 8499 Underestimation programme costs
PE 1944 Underground construction obstacles
PC 6641 Underground economy
PC 6641 Underground illicit business
PE 2095 Underground ; Nuclear explosions
PD 2366 Underground press
PC 6913 *Underground storage hazardous materials*
PD 4403 Underground water resources ; Overexploitation
PD 8888 Underground water ; Rising level
PE 0242 Underheated school buildings
PE 3827 Underlying asset values ; Disparity share prices
PD 9306 Undermanning critical equipment
PF 2949 *Undermined job motivity*
PF 8076 Undermined social confidence
PF 5295 Undermines accountability ; Unchanging legal precedent
PE 4289 Undermining multilateral forums industrialized countries
PC 0382 Undernourishment
PB 1498 Undernutrition
PC 0382 Undernutrition
PD 8422 Underpaid bureaucrats
PD 8916 Underpaid employees
PE 8645 Underpaid teachers
PE 4145 Underparticipation developing countries airline industry
PE 4230 Underparticipation socialist countries international data systems
PD 4882 Underpayment duties taxes
PD 8422 Underpayment government officials
PD 8422 Underpayment judiciary
PF 3550 *Underpayment police*
PE 8645 Underpayment teachers
PD 8916 Underpayment work
PE 9199 Underpayment work developing countries
PC 1716 Underperformance education
PD 5432 Underpopulation
PE 5911 Underpricing port services
PF 6469 *Underprioritized human values*
PC 0582 Underprivileged ethnic minorities
PE 5199 Underprivileged home environment
PC 3325 Underprivileged ideological minorities
PE 8261 Underprivileged ; Inaccessibility health care
PC 3324 Underprivileged linguistic minorities
PC 3424 Underprivileged minorities
PC 3424 Underprivileged minorities ; Dependence
Underprivileged [racial...]
PC 0805 — racial minorities
PC 2049 — regions
PC 2129 — religious minorities
PD 3042 Underproduction primary commodities developing countries
PF 6524 Underproductive methods agricultural management
PF 1107 Underproductivity
PF 0377 Underproductivity draught animal power developing countries
PF 2875 Underprovision basic services rural areas
PF 2583 Underprovision basic urban services
PF 2990 Underreported government spending
PF 9148 Underreported issues
PD 7726 Underreporting animal diseases
PD 4265 Underreporting chemical pollution
PD 2586 Underreporting disruptions natural water cycles
PD 0708 Underreporting earthquake risks
PD 0708 Underreporting groundwater pollution hazards
Underreporting hazards
PD 4721 — animal populations
PD 7731 — biological systems
PD 2623 — minority cultures
PD 1110 — soils
PD 4277 Underreporting marine pollution
PD 8290 Underreporting radiation risks
PD 6626 Underreporting social hazards
PD 4182 Underreporting weather hazards
PD 0086 *Underrepresentation women science technology*
PE 2635 Undersized persons ; Unequal facilities
PD 0366 *Understaffed health clinics*
PD 9306 Understaffing basic facilities
PE 3948 Understand necessity creative establishment disestablishment tensions ; Failure
PF 7071 Understanding globality ; Reduced
PD 8023 Understanding hygiene ; Inadequate
PF 7050 Understanding information ; Cultural lag
PF 0944 Understanding knowledge ; Lack comprehensive framework
PF 2434 *Understanding ; Lack*
PF 5106 Understanding ; Lack international
PF 3554 *Understanding ; Minimal financial*
PF 2212 Understanding new societal service systems ; Incomplete
PF 1577 Understanding past ; Romantic
PF 8648 Understanding progress ; Inappropriate
PF 5587 Understanding responsibility ; Quantitative
Understanding [science...]
PD 8003 — science ; Inadequate public
PF 1764 — sexuality ; Lost covenantal
PF 4837 — social economic contexts development programmes ; Lack
PF 0761 — spiritual healing ; Lack
PE 1729 Understated debts
PE 9061 Understatement tax liability
PF 6478 *Understocked shops*
PF 4668 Undertake new initiatives ; Restricted government ability
PF 2477 *Undertrained fire fighters*
PD 9113 Undertrained labour force
PF 1664 Underused elders' wisdom
PF 2842 *Underused industrial resources*
PC 4867 Underused management skills
Underutilization [biocontrol...]
PF 6229 — biocontrol
PF 0357 — biogas energy
PF 0357 — biomass energy
PF 0827 Underutilization facilities due daily seasonal peaks
PF 0357 *Underutilization fuelwood energy*
PE 8971 *Underutilization geothermal energy*
Underutilization [historical...]
PF 1309 — historical rites customs
PF 3523 — human resources
PF 0345 — hydropower
PF 0100 Underutilization intellectual ability
PF 4155 Underutilization international data bases developing countries
Underutilization [labour...]
PF 6293 — labour force
PF 6293 — Labour force
PF 4670 — land
PF 3464 — legal rights
PF 8595 — livestock least developed countries
PF 6538 — locally available skills
PF 1459 Underutilization natural resources
PE 0117 Underutilization non-monetary foreign exchange
PE 8971 *Underutilization ocean energy*
PF 0445 Underutilization oil shale energy source
Underutilization [peat...]
PE 8194 — peat energy source
PC 1917 — political rights
PF 2426 — popular wisdom
PF 6513 — potential local communities
PE 8971 Underutilization renewable energy resources
PD 1484 Underutilization second-hand equipment
PF 0370 Underutilization solar energy
Underutilization [tar...]
PF 8370 — tar sands energy source
PE 8971 — *temperature differentials oceans*
PE 8971 — *tidal energy*
Underutilization [waste...]
PF 6465 — *waste*
PE 8971 — *wave energy*
PF 0373 — wind energy
PF 0048 — *world banking facilities*
PD 7526 Underutilized animal genetic resources
PF 7130 *Underutilized available space*
PF 6476 *Underutilized class time*
PD 2672 *Underutilized dilapidated buildings*
PE 9325 Underutilized government resources
PE 1753 Underutilized labour skills
PU 1623 *Underutilized veterinary aid*
PF 9306 Undervaluation education parents
PF 1001 Undervaluation public assets
PF 1001 Undervaluation state enterprises
PJ 0095 Undervalued creative pro-establishment/dis-establishment tension
PF 6102 Undervalued currencies
PC 7373 *Underworld milieu*
PE 5508 Underworld society
PF 3499 *Undesignated building sites*
PF 2575 *Undesignated gathering places*
Undesignated [parking...]
PF 7130 — *parking space*
PC 8160 — *public land*
PF 1790 — *public space*
PD 0847 Undesignated school facilities
PF 4670 *Undesignated vacant land*
PF 6575 *Undesigned town brochure*
PF 7695 Undesigned tree-planting plan
PC 3207 *Undesirable aliens*
PD 1714 Undesirable effects animal feed additives
PC 0018 Undesirable effects migration
PE 1158 Undesirable influence toys
PE 1134 *Undesirable manual labour*
PA 6011 Undesirableness *Discontentment*
PA 7395 Undesirableness *Inexpedience*
PA 5458 Undesirableness *Inhospitality*
PA 7107 Undesirableness *Unpleasantness*
PF 4948 Undesired sexual obligations
PD 2438 Undetermined consensus procedures
PF 0214 Undetermined population data
PF 0581 Undetermined sales potential
PD 1566 *Undetermined social base*
PE 8048 Undeveloped business skills urban areas
PF 6471 Undeveloped channels commercial initiative
PF 3526 Undeveloped channels public private resources
Undeveloped community
PF 3558 — cooperation
PF 2093 — power
PF 1546 — space
PF 2489 Undeveloped credit lines
Undeveloped [financial...]
PE 5902 — financial markets developing countries
PF 8176 — forward plan
PF 8176 — future plan
Undeveloped [health...]
PD 4790 — health care systems
PD 4790 — health services
PF 3523 — human resources
PF 1790 *Undeveloped industrial services*
PD 8839 Undeveloped irrigation system
PF 3552 *Undeveloped job training*
PF 6556 Undeveloped local leadership
PF 5660 Undeveloped long range plans
PF 0581 Undeveloped market potential
PF 1196 Undeveloped potential informal leadership
PF 1137 *Undeveloped public image*
PF 7130 *Undeveloped recreation space*
PD 0876 Undeveloped sanitation services
PE 3547 Undeveloped transportation ; Isolating effects seasonal variations
PC 4815 Undeveloped water resources
PD 8047 Undeveloped work ethic developing countries
PF 2797 *Undeveloped youth leadership*
PD 7821 Undignified medical treatment
PF 5857 Undignified treatment corpses
PF 0905 Undirected expansion economic base
PC 1730 Undirected technological expansion
PF 2575 *Undirected youth activities*
PE 1922 Undisclosed control major corporations national policy-making
PF 2344 Undisclosed control national economies limited number individuals
PE 0135 Undisclosed influence transnational corporations global policy-making
PF 6548 *Undiscovered business contacts*
PC 4867 Undiscovered managerial skills
PF 6572 *Undiscovered successful businesses*
PF 1985 Undisplayed heritage gifts
PE 2778 Undisposed dead animals

PF 1240 Undiversified basis income possibilities
PD 2892 Undiversified economies developing countries
PD 4100 Undiversified feed crops
PE 4188 Undiversified forms credit used developing countries
PF 3499 *Undiversified labour force*
PF 2444 *Undiversified tax base*
PF 5728 Undocumented information
PF 6594 Undocumented international financial transactions
PF 4062 Undocumented massacres
PE 6951 Undocumented migrants
PF 4062 Undocumented torture
PF 4062 Undocumented violations human rights
PF 4062 Undocumented war crimes
PF 0551 Undomesticated men
Undressed hides
PE 8106 — non fur skins ; Instability trade
PE 0828 — skins fur skins ; Environmental hazards
PE 1235 — skins fur skins ; Instability trade
PF 1073 Undue attachment social group
PF 3390 Undue attachment territory
PD 9316 Undue consideration criminals
PE 8829 Undue influence obstruct administration justice ; Use
PB 3209 Undue political pressure
PF 5937 Undue public concern
PF 3358 Undue religious influence secular life
PE 8444 Undue taxation certain goods services
PA 6921 Undueness
PA 6011 Uneasiness *Discontentment*
PA 6030 Uneasiness *Fear*
PA 6200 Uneasiness *Impatience*
PA 5467 Uneasiness *Inexcitability*
PA 7107 Uneasiness *Unpleasantness*
PF 3310 *Uneaten nutritional food*
PJ 8128 *Uneconomic size farms*
PF 6527 *Uneconomic snow removal*
PD 1007 Uneconomical small farms developing countries
PA 5568 Uneducated people
PD 8723 Uneducated school leavers
PE 7326 Unemployed ; Deficient local structures
Unemployed [Educated...]
PD 8550 — Educated
PE 1379 — educated youth
PD 9347 — Exploitation
PD 1162 Unemployed graduates ; Chronically
PD 1273 Unemployed intellectuals developing countries
PB 0750 Unemployed people
PC 5576 Unemployed ; Rejection job opportunities
PE 1753 Unemployed skilled labour
PF 3523 *Unemployed trained technicians*
PB 0750 Unemployment
Unemployment benefits
PD 8859 — fraud
PF 7911 — *Inflated cost*
PF 7911 — *Rising cost*
Unemployment [caused...]
PD 0467 — caused environmental conservation
PF 9828 — Concealment information extent
PD 1837 — countries ; Disparities
PD 1837 — countries ; Regional
PE 4294 — *Crimes committed during high*
Unemployment [Denial...]
PE 5406 — Denial right economic security during periods
PB 0750 — Dependence
PC 9718 — developed countries
PE 0402 — developed countries resulting participation developing countries manufacturing
Unemployment developing countries
PD 0176 — [Unemployment developing countries]
PD 0295 — Rural
PD 1551 — Urban
PD 0528 Unemployment due mass production techniques
PE 9071 Unemployment educated older people
PD 7466 Unemployment ; Export
PC 5916 Unemployment ; Female
PD 1162 Unemployment ; Graduate post graduate
Unemployment [High...]
PC 2035 — High cost youth
PD 1162 — highschool graduates
PD 2046 — human resources ; Mass
PC 9718 Unemployment industrialized countries
PD 0528 Unemployment introduction new technologies
PE 9476 Unemployment least developed countries
PF 2303 *Unemployment married women*
PE 5951 Unemployment older people
PE 8457 Unemployment perpetuator failure
PE 0015 Unemployment premature school leavers developing countries
Unemployment [rate...]
PB 0750 — rate ; High
PD 0024 — Reinforcement
PF 2949 — Rural
Unemployment [Seasonal...]
PC 1108 — Seasonal
PC 0940 — socialist countries ; Disguised
PB 0750 — *Structural*
PC 6720 Unemployment ; Voluntary
PE 5362 Unemployment wood industry
PC 2035 Unemployment ; Youth
PF 6479 Unencouraged local leadership
PE 4768 *Unenforced animal control*
PE 4768 *Unenforced civic ordinances*
PE 4768 Unenforced laws
PE 4768 *Unenforced littering laws*
PD 5001 Unenforced safety ordinances
PD 5001 *Unenforced sanitation codes*
PD 1544 Unengaged community youth
PF 2813 *Unengaged town residents*
PF 2644 *Unengaging family activities*
PF 3628 *Unenticing printed media*
PF 3554 *Unenvisioned building goods*
Unequal access
PC 2163 — education
PC 2801 — food
PD 0190 — women education
PF 0852 Unequal coverage social security
PE 8416 Unequal development different societies ; Separate
PC 4844 Unequal development promotion health
Unequal distribution
PF 4108 — agricultural machinery
PD 4316 — agricultural production
PF 3439 — fame honours
PE 3569 — *fellowships*
PF 4495 — fish catches
PF 6836 — *forces*
PE 8603 — goods services
PC 3438 — land assets
PF 4490 — livestock production
PD 4322 — meat production
PE 7942 — old age pensions men women
PF 4336 — production countries
PF 0852 — social security
PC 3437 — social services
PB 7666 — wealth
Unequal [employment...]
PE 0783 — employment opportunities disabled persons
PD 5115 — employment opportunities women
PF 5482 — evaluation risks
PD 2897 — exchange trade
PE 2635 Unequal facilities undersized persons
Unequal global distribution
PF 2880 — basic skills
PC 5601 — economic growth
PE 4241 — family planning education facilities
PD 3436 Unequal graduation income tax
PE 6835 Unequal health benefits women
Unequal income distribution
PD 4962 — [Unequal income distribution]
PC 2815 — countries
PC 2815 — countries ; Dependence
PD 7615 — developing countries
PE 6891 — industrialized countries
PC 4312 Unequal life expectancy
Unequal [morbidity...]
PC 6869 — morbidity mortality countries
PE 4354 — mortality elderly countries
PC 9586 — mortality rates countries
Unequal opportunities
PE 0706 — disabled persons
PE 7726 — foreign students
PD 3039 — media reception
Unequal [parliamentary...]
PD 2167 — parliamentary constituencies
PD 0309 — pay women
PD 0162 — *pension rights*
PC 0655 — political representation
PC 0655 — political representation ; Dependence
PD 9628 — *promotion*
PD 9628 — promotion women
Unequal property
PC 3438 — distribution
PF 0886 — inheritance
PA 0833 — *rights*
PE 4018 — rights women
PC 4312 Unequal regional distribution deaths
PC 0006 Unequal rights different racial groups
PC 2163 *Unequal school distribution*
PC 1606 Unequal strategic arms capability
PJ 8892 *Unessential services ; Perpetuated*
PE 6702 Unethical allocation public contracts
PD 2886 Unethical archival practices
PD 8227 Unethical behaviour leaders
PC 2563 Unethical business practices
Unethical [catering...]
PE 6615 — catering practices
PC 2563 — commercial practices
PD 2625 — consumption practices
PF 0374 — corporate entertainment
PD 2886 Unethical documentation practices
Unethical [entertainment...]
PF 0374 — entertainment
PE 4805 — experimentation using aborted foetuses
PD 2697 — experiments drugs medical devices
PE 0682 Unethical financial practices
PD 1045 Unethical food practices
Unethical [industrial...]
PD 2916 — industrial practices
PE 1826 — insurance practices
PC 2915 — intellectual practices
PE 3553 — interrogation methods
PE 1826 Unethical life assurance practices
Unethical [maintenance...]
PD 7964 — maintenance practices
PD 5251 — media practices
PE 4889 — medical experimentation prisoners
PD 5770 — medical practice
PD 7360 — military practices
PE 1369 Unethical ophthalmic practice
Unethical [people...]
PC 8247 — people
PF 8759 — personal relationships
PD 0862 — personnel practices
PE 3540 — pharmaceutical practices
Unethical practice
PD 2623 — anthropology
PD 7731 — biosciences
PD 4265 — chemistry
PD 0708 — earth sciences
PD 2586 — hydrology
PD 4277 — marine sciences
PD 4182 — meteorology
PD 5770 — physicians
PD 1710 — physics
PD 8290 — radiology
PD 1110 — soil sciences
PD 4721 — zoosciences
Unethical practices
PC 8247 — [Unethical practices]
PD 8001 — apparel industry
PE 4771 — circus animals
PD 8001 — clothing industry
PF 0221 — consultants
PE 4771 — domesticated animals
PD 4334 — employees
PD 2879 — employers
PE 4771 — farm animals
PD 6701 — forestry
PD 0814 — government
PE 3328 — health services
PE 7169 — housing tenants
PE 6740 — interpreters
PD 5380 — legal profession
PD 5948 — local government
PE 4771 — pet animals
PE 8742 — philanthropic organizations
PD 9193 — police forces
PC 5517 — politics
PF 8889 — priesthood
PD 5267 — psychiatry
PD 5267 — psychotherapy
PE 6702 — public service employees
PF 8046 — regulatory inspectors
PD 6626 — social sciences
PD 1012 — transportation
PE 4771 — zoo animals
PC 8019 Unethical professional practices
PD 5422 Unethical real estate practice
PD 4341 Unethical trade union practices
Unethical use
PE 6754 — postal service
PC 1153 — science technology
PD 8859 — social welfare benefits
PF 4564 — statistics
PE 4450 — telephone service
PD 7726 Unethical veterinary practice
PF 6506 *Uneven concentrations power*
PB 6207 Uneven development
PD 8056 Uneven distribution water
PF 6071 Uneven economic recovery
PE 8020 Uneven settling building foundations
PD 1434 *Unevenness ; Bean pea distribution*
PA 6997 Unevenness *Imperfection*
PA 6695 Unevenness *Inequality*
PF 1651 Unexamined community roles
PF 6505 Unexercised responsibility external relations
PD 1885 Unexpected infant death ; Sudden
PF 0296 Unexpected maintenance costs
PF 2437 *Unexperienced individual effectiveness*
PE 8631 Unexplained appearances disappearances persons objects
PF 8352 Unexplained phenomena
PF 3682 Unexplained ; Procrastination science face
PE 2462 Unexplained technical faults
PF 2613 Unexploded bombs
PF 2535 Unexploited possibilities local commerce
Unexplored [alternate...]
PF 7960 — alternate energy
PF 6548 — alternatives commercial development
PF 2797 — avenues leadership potential
PF 3539 *Unexplored business opportunities*
PC 4867 Unexplored business training
PF 9004 Unexplored career opportunities
PF 9729 Unexplored cooperative potential
Unexplored [energy...]
PF 7960 — energy alternatives
PE 8540 — entertainment alternatives countryside
PF 2989 — *export potential*
PF 6471 *Unexplored industrial inducements*
PF 4013 Unexplored industrial training
PF 1240 Unexplored livestock income
PF 6512 Unexplored opportunities community education
PF 2231 *Unexplored orchard potential*
PF 0581 Unexplored potential markets
PF 6573 *Unexplored surrounding territory*
PF 6572 *Unexplored tourist trade*
PF 3409 *Unexposed farming technology*
PF 1723 Unexpressed social compassion
PF 0422 Unfacilitated vocational school
PD 9163 Unfair air transport practices
Unfair competition
PC 0099 — [Unfair competition]
PE 5506 — convict-made goods
PE 5506 — workers employed forced labour
PB 0750 *Unfair distribution land ownership*
PC 2649 Unfair elections
PD 4695 Unfair labour practices
PD 0312 Unfair practices maritime commerce
PE 5855 Unfair pricing transnational corporations

Unfair [rail...]
PD 1367 — *rail transport practices*
PD 8916 — *remuneration labour*
PC 5842 — *restrictions international trade*
Unfair [salary...]
PD 9430 — *salary scales*
PD 0312 — *shipping practices*
PE 5849 — *shipping practices bulk trades*
PF 5922 — *surcharges ocean freight*
Unfair [taxes...]
PD 3436 — *taxes*
PC 4275 — *trade restrictions*
PJ 6424 — *trading*
PD 1367 — *transport practices*
PE 1692 — *trials due pre-trial publicity*
PF 4787 Unfairly negotiated treaties
PA 7363 Unfaithfulness *Improbity*
Unfamiliar [banking...]
PF 2009 — *banking procedures*
PF 1653 — *bureaucratic procedures*
PF 1653 — *bureaucratic requirements*
PF 7029 Unfamiliar ; Fear
PF 2845 Unfamiliar *government procedures*
PF 2845 Unfamiliar *government structures*
PF 3464 Unfamiliar legal rights
PF 6479 Unfamiliar *library use*
PE 7984 Unfamiliar procedures using public health programmes
PS 2881 Unfamiliar *production processes*
PF 2813 Unfamiliar *use equipment*
PA 5568 Unfamiliarity *Ignorance*
PA 7232 Unfamiliarity *Unskillfulness*
PF 4010 Unfashionable personal appearance
PF 6572 Unfavourable capital risk
PC 0387 Unfavourable climatic conditions
PC 3340 Unfavourable opinions cultures
PF 3826 Unfavourable transition post capitalist social order ; Prevalence psychological conditions
PC 0480 Unfavourable wildlife decisions
PF 4944 Unfeasible development projects
PE 6495 Unfeasible goods transport
PE 8061 Unfeasible housing alternatives urban areas
PF 1790 Unfeasible *industrial images*
PF 6527 Unfeasible *infrastructure costs*
PF 6577 Unfeasible *medical expectations*
PA 7364 Unfeelingness
PC 0052 Unfertilized soil
PF 2605 Unfinanced *sports activities*
PF 5716 Unfinished imperfect universe
PE 6005 Unfinished industrial site
PF 2605 Unfinished *playing field*
PE 4863 Unfit legal defendants
PD 4475 Unfitness ; Physical
PF 4946 Unfitness survival ; Individual
PF 6531 Unfocused adult education
PF 6559 Unfocused *community spending*
PF 1546 Unfocused design community space
PF 6528 Unfocused land usage
PF 6559 Unfocused resident unity
PF 6559 Unfocused style community operations
PF 2813 Unfocused *town centre*
PJ 7929 Unforeseen capitalization expenses
PF 9769 Unforeseen environmental crises
PF 8374 Unforgiving
PF 6506 Unformed community consensus
PF 2358 Unformed confidence community corporateness rural villages
PF 1959 Unformed *industrial planning*
PF 2810 Unformed structures community organization
PF 6514 Unformed style cooperative action
PF 6504 Unfounded conflicting stories
PA 5532 Unfriendliness *Disaccord*
PA 7143 Unfriendliness *Discourtesy*
PA 5446 Unfriendliness *Enmity*
PA 6653 Unfriendliness *Unsociability*
PD 9310 Unfriendly consumer products ; Environmentally
PF 2199 Unfriendly governments ; Eroding support
PC 0328 Unfriendly products ; Ecologically
PF 2842 Unfulfilled aspirations economic life
PF 7150 Unfulfilled promises
PF 2969 Unfulfilled *service expectations*
PF 2386 Unfulfilled sociability
PF 2497 Unfulfilled treaty obligations
PE 5197 Unfulfilment disabled ; Sexual
PF 3260 Unfulfilment ; Sexual
PF 6531 Unfunded adult education
PF 2813 Unfunded *educational plans*
PD 6520 Unfunded fire department
PF 2605 Unfunded roadway improvements
PA 7148 Ungodliness
PA 5644 Ungodliness *Vice*
PA 7143 Ungraciousness *Discourtesy*
PA 5458 Ungraciousness *Inhospitality*
PA 5643 Ungraciousness *Unkindness*
PE 6526 Ungraded roads ; Unpaved,
PF 1185 Ungroundedness ; Psychological
PF 2477 Unguided vocational decision
PE 7826 Ungulae ; *Dystrophia*
PE 4161 Ungulae horizontalis ; *Fissura*
PE 4161 Ungulae longitudinalis ; *Fissura*
PC 1326 Ungulates ; *Endangered species Even toed*
PC 1326 Ungulates ; *Endangered species Odd toed*
PS 9191 Unhappiness
PA 6191 Unhappiness *Disapproval*
PA 6011 Unhappiness *Discontentment*
PA 6731 Unhappiness *Solemnity*
PA 7107 Unhappiness *Unpleasantness*

PD 0941 Unhappy marriages
PA 7226 Unhealthfulness
PD 6627 Unhealthy buildings
PF 0913 Unhealthy emotional responses atomic energy
PB 6384 Unhealthy environment
PF 7896 Unhealthy forests
PF 6182 Unhealthy lack daily physical activity urban environments
PD 2838 Unhealthy physical posture
PF 8515 Unhealthy sanitation practices
PF 8125 Unhelpful ; Police reputation
PF 2358 Unhelpful *vandalism image*
PE 8255 Unhygienic bottle feeding
PF 8515 Unhygienic conditions
PF 1773 Unhygienic *dirt floors houses*
PF 6147 Unidentifiable urban neighbourhoods
PF 6521 Unidentified community space
PF 1392 Unidentified flying objects (UFOs)
PF 6541 Unidentified *news resources*
PF 0581 Unidentified potential markets
PE 0712 Unidentified submarine objects
PE 0712 Unidentified submarine patrols
PF 6235 Unification ; Ethnic
PF 6749 Uniform marking navigable waters ; Non
PC 0933 *Uniformity tariff non tariff trade preferences developing countries ; Non*
PF 5250 Unifying events ; Infrequent
PD 1628 Unilateral claims off-shore territorial waters
PF 1066 Unilateral declarations independence extra-territorial bases
PF 9629 Unilateral interpretations multilateral principles
PF 7052 Unilateral nuclear disarmament ; Obstacles
PF 7670 Unilateral structural disarmament ; Insecurity
PF 4051 Unilateral structural disarmament nuclear weapons
PF 3317 Unilingualism ; Diverse
PF 9478 Unimaginative *city plan*
PF 3007 Unimaginative educational vision
PF 9478 Unimaginative *facility use*
PF 9478 Unimaginative *future planning*
PF 0867 Unimaginative housing design
PF 3007 Unimaginative long-term training vision
Unimaginative vision
PF 7118 — *community possibilities*
PF 2699 — *existing international economic structures*
PF 1316 — *resource utilization*
PF 9478 — *world's future*
PA 6738 Unimaginativeness
PA 7365 Unimaginativeness *Boredom*
PA 6738 Unimaginativeness *Unimaginativeness*
PF 5304 Unimagined possibility expanding preschool
PF 4672 Unimplemented decisions
PA 5942 Unimportance
PA 6659 Unimportance *Humility*
PF 0836 Unimportance physical improvements
PF 7408 Unimportance *Smallness*
PA 5942 Unimportance *Unimportance*
PC 0052 Unimproved farm soils
PD 8837 Uninformed animal sanitation
PF 1442 Uninformed budget planning
PC 0834 Uninformed care techniques
PF 2477 Uninformed career planning
PF 6531 Uninitiated adult education
PD 4790 Uninitiated health services
PF 3884 Uninitiated social methods
PF 1705 Uninitiated *women's groups*
PF 5978 Uninspired teacher participation
PE 0242 Uninsulated buildings
PF 2042 Uninsurable business investments
PF 1768 Unintegrated agricultural development
PC 4824 Unintegrated biosphere ecosystem management
PF 4837 Unintegrated development
PD 2675 Unintegrated expatriates ; Socially
PF 1768 *Unintegrated farming population*
PA 0911 Unintegrated individual personalities
PF 7371 Unintelligence
PA 5568 Unintelligence *Ignorance*
PA 6940 Unintelligence *Thoughtlessness*
PF 7371 Unintelligence *Unintelligence*
PA 7232 Unintelligence *Unskillfulness*
PA 7367 Unintelligibility
PF 4352 Unintentional global nuclear war due nuclear proliferation ; Risk
PD 3022 Unintentional *libel*
Unintentional nuclear war
PF 4346 — due accidents ; Risk
PF 4302 — due international crises ; Risk
PF 4156 — generated developments strategic doctrine ; Risk
PF 4162 — generated strategy deterrence ; Risk
PF 0466 — Risk
PF 2813 Unintentional space upgrading
PF 4152 Unintentional war generated arms race ; Risk
PF 1653 Uninterested private investors
PF 2987 Uninterested workers
PF 5044 Uninventive leisure time
PD 0465 Uninvested community savings
PC 0927 Uninvested personal savings
PE 8061 Uninvestigated building alternatives
PE 6911 Uninvestigated crimes
PF 2575 Uninviting *youth activities*
PD 2338 Uninvolved absentee landlords
PF 2813 Uninvolved transient residents
PF 2810 Uninvolved working parents
Union activity
PD 4695 — Denial right trade
PE 3910 — rural workers ; Denial right
PE 1355 — special groups ; Denial right

PD 0683 Union association ; Denial right trade
PE 5070 *Union busters*
PE 5192 Union busting
PE 0063 Union competition ; Inter
PE 2007 Union demonstrations ; Violent repression trade
PE 1134 *Union entry ; Restricted*
Union [farmers...]
PE 1262 — farmers ; Unconsensed
PD 4341 — fraud ; Trade
PE 8170 — funds property ; Deprivation trade
PE 4627 Union island developing countries ; Obstacles political
Union leaders
PD 7630 — Arrest trade
PD 0252 — Assassination trade
PE 4869 — Death threats against trade
PE 8367 — employers government ; Collusion trade
PE 5882 — Forced disappearances trade
PD 7630 — Ill treatment trade
PD 7471 — Threats against trade
Union [meetings...]
PE 7210 — meetings ; Prohibition trade
PE 7036 — members ; Denial right recruit
PJ 0272 — *members their representatives ; Isolation trade*
PA 0429 — *monopoly violence*
PE 6722 Union opposition profit sharing ; Trade
PF 6046 Union policies concerning employment ; Restrictive trade
Union practices
PD 8146 — Restrictive
PD 8146 — Restrictive trade
PD 4341 — Unethical trade
Union premises
PE 5462 — public authorities ; Occupation trade
PE 5462 — public authorities ; Raids trade
PE 5462 — Seizure trade
Union property
PE 7581 — Confiscation trade
PE 7581 — public authorities ; Destruction trade
PE 7581 — public authorities ; Seizure trade
PC 5842 *Union protectionism*
Union [recognition...]
PD 0683 — recognition ; Lack trade
PD 1846 — representatives ; Denial right protection
PE 7620 — representatives following legal strike action ; Dismissal trade
Union rights
PD 4695 — Abuse trade
PE 4860 — governments ; Denial trade
PE 6888 — public employees ; Denial trade
PE 5938 — public enterprises ; Restrictions trade
PD 4695 — Violation trade
PF 1262 Union rivalry ; Trade
PD 6019 Union shops ; Closed
PD 6019 Union workers ; Discrimination against non
PD 0610 Union workers ; Sanctions against trade
PF 8493 Unionism ; Trade
PE 8345 Unionization working women ; Lack
PD 3535 Unions ; Banned trade
Unions [Closed...]
PE 1134 — *Closed shop*
PE 8273 — contract performance ; Interference trade
PD 4341 — Corruption trade
Unions [Denial...]
PE 5398 — Denial right organize trade
PC 4613 — Discrimination against labour
PC 4613 — Discrimination against trade
PE 4745 Unions engage political action ; Violation right trade
PE 1758 Unions freely elect their representatives ; Violation right trade
PE 1758 Unions function freely ; Violation right trade
PE 4860 Unions ; Government discrimination against trade
PE 5938 Unions government services ; Violation right organize trade
PF 1262 Unions ; Inadequate international trade
PD 4341 Unions ; Irresponsible trade
PF 3278 Unions ; Marriage like
PE 3970 Unions negotiate freely employers ; Violation right trade
PF 3278 Unions ; Nonmarital
PD 0610 Unions organizers ; Denial rights trade
PF 1262 Unions ; Outmoded trade
PE 5793 Unions protection against dissolution ; Violation right trade
PE 5312 Unions publish ; Violation right trade
PF 1262 *Unions ; State controlled trade*
PE 1541 Unions ; Transnational strike action trade
PE 6018 Unions ; Unrepresentativity trade
PE 5192 Unions ; Violation right workers join trade
PF 1750 Uniqueness family members ; Failure recognize
PF 6575 Uniqueness ; *Neglected community*
PF 3448 Unirrigated islands
PF 1781 United action ; Lack
United Nations
PF 8946 — agencies ; National hegemony over
PF 5337 — developing country membership ; Irresponsibility
PF 5934 — Disenchantment
PF 9474 — eliminating human rights violations ; Governments unwilling cooperate
PE 0777 — Inadequate relationship international nongovernmental organizations specialized agencies
PE 0106 — Inadequate relationship transnational corporations specialized agencies
PE 0075 — information systems ; Lack coordination
PE 0075 — information systems ; Proliferation duplication
PE 7144 — Intellectual confusion concerning role
PF 5337 — Irresponsibility member governments

United Nations

United Nations cont'd
PF 2943 — jurisdiction administered territories ; Lack
PE 2486 — Jurisdictional conflict antagonism specialized agencies
PE 8799 — Jurisdictional conflict antagonism specialized agencies
PF 5934 — Loss credibility
PF 5934 — Marginalization
PF 8650 — membership dues ; Non payment
PF 2943 — non self governing status disputed administering government ; Territories accorded
PF 5934 — Perceived irrelevance
PF 7925 — Politicization human rights issues
PF 7925 — response human rights ; Bias

United Nations system
PE 0296 — Fragmentation complexity
PE 0296 — Inadequate coordination
PF 1460 — organization ; Shortage financial resources
PF 1451 — organizations ; Ineffectiveness
PE 1579 — Proliferation duplication organizational units coordinating bodies
PE 2820 United Nations systems ; Inadequate budgetary coordination
PE 2820 United Nations systems ; Lack centralized financial control
PA 6858 Unity ; Abandonment
PC 3204 Unity ; Breakdown political
PD 2802 Unity ; Inadequate structures achieving global

Unity Lack
PA 6233 — [Unity ; Lack]
PF 2434 — human
PF 3388 — ideological
PF 8107 — national
PF 3216 — political
PF 2434 — world
PC 3293 *Unity ; National*
PF 3364 Unity religions ; Lack
PF 3404 Unity religions ; Lack
PF 6559 Unity ; Unfocused resident
PE 3959 Universal panacea ; Overdependence computers
PF 5520 Universal principles ; Vulnerability
PF 5520 Universality ; Jeopardization
PF 5716 Universe ; Unfinished imperfect
PE 8873 Universities ; Gap industry
PD 4282 *University autonomy ; Erosion*
PF 7370 University campuses ; Physical insecurity
PC 1524 University degree equivalencies ; Uncertainty
PE 7370 University students ; Anti social behaviour
PF 2911 Unjust allocation government contracts
PE 2932 Unjust commitment psychiatric hospitals
PE 5152 Unjust customary rights
PD 5965 Unjust dismissal workers

Unjust [election...]
PC 2649 — election administration
PD 2907 — election timing
PD 2919 — electoral campaigns
PE 0752 Unjust financing political parties
PC 9112 Unjust global economic system
PC 9112 Unjust international economic order
PC 7112 Unjust laws
PC 7112 Unjust legislation

Unjust [peace...]
PB 7694 — peace
PA 6486 — people
PD 4779 — punishments crimes
PD 4827 Unjust trials
PD 9641 Unjust voter registration procedure
PE 1327 Unjustified game cropping
PF 1385 Unjustified military defence policies
PE 0897 Unjustified restrictions free movement commercial vehicles
PJ 0255 Unjustified urban conservation
PF 4010 Unkempt appearance
PD 8894 Unkept rental property
PA 5643 Unkindness
PA 5643 *Unkindness Unkindness*
PF 9276 Unknowable future patterns social choice
PD 8837 Unknowledgeable animal care

Unknown [aetiology...]
PD 8588 — aetiology ; Diseases
PF 1653 — *availability subsidies*
PE 2583 — *available services*
PF 6506 *Unknown building codes*

Unknown [cause...]
PD 8588 — cause ; Diseases
PE 5463 — causes mortality
PF 0782 — children
PF 9729 — Co successes
PF 5222 — community meetings
PD 5730 — crop dangers
PF 1985 — cultural heritage
PF 6188 Unknown ; Fear
PF 2491 Unknown loan services
PF 6538 Unknown local skills
PF 0581 Unknown market potential
PF 9167 Unknown origin ; Aerial explosions
PD 8588 Unknown origin ; Diseases
PF 0782 Unknown parents
PF 6925 Unknown possibilities
PF 0782 Unknown relatives
PF 2453 *Unknown training opportunities*
PD 2290 *Unlandscaped river banks*
PF 7520 *Unlawful assembly*
PC 4645 Unlawful business transactions
PC 4645 Unlawful commerce
PE 7176 Unlawful compensation assistance government matters

PE 5283 Unlawful distilled spirits ; Possession
PD 1928 Unlawful entry
PE 5267 Unlawful flight
PE 4997 Unlawful government
PD 5332 Unlawful government action
PD 4489 Unlawful imprisonment

Unlawful interference
PE 4058 — family home
PE 4058 — family life
PJ 9820 — marine activities
PB 0284 — privacy
PE 6755 — rights innocent passage territorial waters

Unlawful [recruiting...]
PE 4484 — recruiting enlistment foreign armed forces
PE 6854 — repatriation minors
PD 4541 — rewarding public servants
PD 4221 Unlawful trafficking taxable objects
PA 5952 Unlawfulness *Illegality*
PE 6486 Unlawfulness *Injustice*
PA 7280 Unlawfulness *Wrongness*
PA 5568 *Unlearned fundamental skills*
PE 5623 Unlimited practice human embryo storage
PA 5559 Unlimitedness
PD 9108 Unloading ; Risk during loading
PF 6559 *Unlocated community centre*
PF 6548 *Unlocated finance counselling*
PC 1784 Unlocked property
PF 9536 Unluckiness
PF 4660 Unlucky numbers
PF 9536 Unlucky people
PE 2471 Unmaintained bridges
PF 2477 *Unmaintained private vehicles*
PA 5497 Unmanageability *Difficulty*
PA 7325 Unmanageability *Irresolution*
PA 7232 Unmanageability *Unskillfulness*
PF 2346 *Unmanageable land maintenance*
PE 2634 Unmanageable stagnant pools
PE 8233 Unmanaged irrigation potential
PD 3488 Unmarked public water outlets
PF 7344 Unmarketable available skills
PF 3278 Unmarried couple
PD 3256 Unmarried fathers ; Discrimination against
PD 0902 Unmarried mothers
PD 3257 Unmarried parents
PD 8622 Unmarried women
PD 8622 Unmarried women ; Discrimination against
PF 7643 *Unmeaningful peace*
PE 1825 Unmeaningful social roles aged
PF 1169 *Unmet needs*

Unmilled cereals
PE 1331 — Environmental hazards animal feedstuffs, excluding
PE 1769 — Instability trade
PE 8816 — Instability trade animal feedstuffs, excluding
PE 1514 — Long term shortage animal feedstuffs excluding
PF 4646 Unmindfulness
PF 8835 Unmotivated agricultural training
PF 0148 Unmotivated family planning
PF 5978 Unmotivated teachers
PF 5949 Unmotivated youth stance
PD 8916 Unmotivating low wages
PF 3008 *Unmotivating subsistence employment*
PD 1229 *Unmotivating welfare dependence*
PF 2845 *Unnamed streets*
PB 0869 Unnatural acts
PD 2544 Unnatural boundaries developing countries
PF 0090 Unnatural boundaries states
PF 6193 Unnatural urban environments inhibiting sleeping public
PE 8255 Unnaturally short duration breastfeeding
PA 3981 Unnaturalness
PA 6400 Unnaturalness *Affectation*
PA 6312 Unnaturalness *Inelegance*
PA 7157 Unnaturalness *Insanity*
PA 5878 Unnaturalness *Nonconformity*
PA 7411 Unnaturalness *Uncommunicativeness*
PA 5643 Unnaturalness *Unkindness*
PE 3926 Unnecessary burden interest repayments developing countries ; Excessive
PF 1345 *Unnecessary business closings*
PD 4125 *Unnecessary education expenditure*
PE 3745 Unnecessary gadgets
PD 0952 Unnecessary health system referrals
PF 5679 Unnecessary health tests
PF 5931 Unnecessary luxury
PD 4632 Unnecessary market middlemen
PE 8261 Unnecessary medical expenses
PF 5931 Unnecessary personal consumption
PC 8222 Unnecessary personal wealth
PF 3455 Unnecessary relocation national capitals
PF 0687 Unnecessary reserves material

Unnecessary [status...]
PF 3455 — status projects
PF 0687 — stocks
PE 9271 — surgery
PE 9271 — surgical operations
PF 7137 Unnecessary verbosity legal documents
PA 5458 *Unneighbourliness Inhospitality*
PA 6436 Unnumbered
PD 0800 Unnutritious eating habits
PC 0382 Unnutritious eating habits
PD 0800 Unnutritious food consumption
PC 0382 Unnutritious food consumption
PD 1611 Unobserved drinking age laws
PD 0820 Unoccupied housing ; Appropriation
PD 0820 Unoccupied property ; Illegal occupation
PF 5409 Unorganized community recreation
PF 2128 Unorganized development work forces

PD 0512 Unorganized elderly aid
PF 2575 *Unorganized family occasions*
PF 4832 Unorganized health facilities
PF 5969 Unorganized immunization programme
PF 0963 *Unorganized industrial promotion*
PC 7997 Unorganized job market
PF 6471 *Unorganized labour potential*
PF 6475 Unorganized land scheme
PF 5409 Unorganized recreation programs
PC 6445 *Unorganized transfer skills*
PD 0807 Unorganized trash collection
PF 2575 *Unorganized welcoming process*
PA 5564 Unorthodoxy
PE 4451 Unpaid female labour ; Dependence developing countries
PD 9437 Unpaid government loans
PE 4789 Unpaid household work
PE 4789 Unpaid household work ; Dependence
PD 3056 Unpaid labour
PD 5335 Unpaid wages
PF 4550 Unparliamentary behaviour
PE 6526 *Unpaved, ungraded roads*
PA 5838 Unpeacefulness *Agitation*
PA 5532 Unpeacefulness *Discaccord*
PF 9004 Unperceived career opportunities
PF 2453 *Unperceived educational opportunities*
PF 1750 *Unperceived interests children*
PF 4984 *Unperceived preschool need*
PF 1944 Unperceived relevance formal education rural communities
PF 6531 Unplanned adult education
PF 7456 Unplanned family roles
PF 6475 Unplanned land development
PF 6475 Unplanned land use
PD 8894 Unplanned maintenance margin

Unplanned urbanization
PC 0442 — [Unplanned urbanization]
PD 0134 — developing countries
PD 3488 — industrialized countries
PF 2519 Unplanned use community space
PF 4670 *Unplanned vacant land*
PF 1546 *Unplanned village design*
PD 1544 Unplanned youth participation
PF 2575 *Unplanned youth structures*
PE 6820 Unpleasant odours
PE 2611 *Unpleasant sights*
PE 6820 *Unpleasant smells*
PA 7107 Unpleasantness
PA 5454 Unpleasantness *Badness*
PA 5532 Unpleasantness *Discaccord*
PA 7107 Unpleasantness *Unpleasantness*
PF 6471 *Unpolled consumer requirements*
PD 2615 *Unpopular regimes*
PC 2494 Unpopular voting patterns
PF 4641 Unpopularity childhood
PF 4641 *Unpopularity ; Personal*
PF 1442 *Unpopulated fishing grounds*
PF 6472 *Unpractised dental prevention*
PF 5490 Unpredictability *Changeableness*
PA 6971 Unpredictability *Danger*
PF 4928 Unpredictability earthquakes
PA 7325 Unpredictability *Irresolution*
PF 5356 Unpredictability judicial decisions
PF 5817 Unpredictability key economic variables
PA 7309 Unpredictability *Uncertainty*
PF 5118 Unpredictability weather
PF 5033 Unpredictable barriers trade
PF 9769 Unpredictable ecological disasters
PF 8171 Unpredictable farming income
PF 2135 *Unpredictable fish yield*
PF 1559 Unpredictable governmental policy
PF 5033 Unpredictable introduction protectionist measures
PF 1559 Unpredictable public administration
PF 3084 Unpredictable sources income ; Dependency
PD 0722 Unpredictable water supply
PF 2316 *Unpredictable work schedule*
PF 6462 *Unprepared adult leadership*
PF 7130 *Unprepared industrial sites*
PF 9113 Unprepared local labour
PF 0422 Unprepared vocational potential
PA 7341 Unpreparedness
PF 8176 Unpreparedness
PF 3567 Unpreparedness ; Disaster
PF 0638 Unpreparedness ; Eschatological
PF 5016 Unpreparedness food emergencies
PF 5933 Unpreparedness ; Military
PF 5438 Unpreparedness *Neglect*
PF 5044 Unpreparedness surplus leisure time
PA 7341 Unpreparedness *Unpreparedness*
PA 7006 Unpreparedness *Untimeliness*
PF 1065 Unprocessed scientific data ; Proliferation
PE 1550 Unprocessed textile fibres their waste ; Instability trade
PC 1420 Unproductive dependents
PF 6601 Unproductive extension schooling
PJ 9394 Unproductive grass-roots planning

Unproductive [labour...]
PC 6031 — labour resources
PF 6528 — land use
PF 6477 — local investments
PF 8939 Unproductive meeting time
PC 0492 Unproductive subsistence agriculture
PF 8120 Unproductive technical aid

Unproductive use
PC 8914 — human resources
PC 8642 — non-renewable resources
PB 8376 — resources

Unskilled programme

Unproductive use cont'd
PC 0238 — *trees beyond their commercial life*
PF 6455 Unproductive utilization plantation space
PC 6031 Unproductive workers
PA 7208 Unproductiveness
PA 6852 Unproductiveness *Inappropriateness*
PA 7208 Unproductiveness *Unproductiveness*
PF 3103 *Unproductivity capitalist systems ; General*
PC 4867 *Unprofessional building management*
PF 6697 Unprofessional science
PF 2411 *Unprofitable entertainment market*
PF 0905 *Unprofitable farm marketing*
PF 2845 *Unprofitable former skills*
PF 1442 *Unprofitable home industries*
PF 6528 Unprofitable land use practices
PE 1101 *Unprofitable local politics*
PF 1086 Unprofitable market practices
PF 1933 Unprofitable scope industrial operations
PJ 2079 Unprofitable small businesses
PF 2845 *Unprofitable traditional skills*
PF 6527 *Unprofitable transport business*
PF 6540 *Unpromoted business choices*
PF 7118 *Unpromoted community business*
PF 6525 *Unpromoted industrial development*
PF 7118 *Unpromoted village businesses*
PF 6683 Unpropitious physical locations
PF 6683 Unpropitious timing
Unprotected [coastlines...]
PE 3813 — *coastlines ; Settlement*
PF 2813 — *commercial establishments*
PC 0123 — *consumers*
PD 0549 Unprotected play areas
PD 5980 Unprotected street youth
PA 7070 Unprovability
PF 6310 Unprovable
PF 7706 Unproven relationships problems
PF 7998 Unpublicized community news
PF 1267 *Unpublicized meeting agendas*
PF 5222 Unpublicized public meetings
PF 1301 *Unpublicized training information*
PF 1301 Unpublished educational possibilities
PF 2845 *Unpublished government services*
PF 7998 Unpublished local news
PF 4852 *Unpublished training opportunities*
PJ 7316 Unpunctuality
PD 1350 *Unpunished garden vandalism*
PF 7847 Unquestionable beliefs
PF 7847 Unquestionable certitude
PF 2312 Unquestioned control economic forces
PF 8855 Unquestioning dependency social assistance
PF 6573 Unquestioning, uncurious attitudes
PE 2471 *Unrailed public bridges*
PF 8176 Unreadiness
PF 5933 Unreadiness ; Military
PF 0638 Unreadiness second coming Christ
PF 1944 *Unrealistic agricultural vision*
PF 2814 *Unrealistic attempts globality*
PF 8922 *Unrealistic beliefs*
Unrealistic [election...]
PF 4558 — election promises
PF 4510 — environmentalism
PF 7002 — expectations
PF 4774 Unrealistic forecasts
PE 7002 Unrealistic optimism
Unrealistic [perspectives...]
PF 9428 — perspectives ; Government promotion
PF 9428 — policies
PF 9428 — policy-making
PF 9312 — portrayal war
PF 1214 Unrealistic strategic warfare analysis ; Politically
PE 5305 Unrealistic wage claims
PF 4377 Unrealistically positive self-assessment
PA 6180 Unreality *Error*
PA 6959 Unreality *Insubstantiality*
PA 5870 Unreality *Nonexistence*
PF 9428 Unreality ; Political
PA 6738 Unreality *Unimaginativeness*
PF 2231 *Unrealized agronomic potential*
PF 2358 *Unrealized corporate power*
PF 7063 *Unrealized human development potential*
PF 3519 *Unrealized intentions practical education*
PF 2989 *Unrealized local market*
PF 2995 *Unrealized marketable skills*
Unrealized potential
PF 2231 — commercial enterprises
PF 2969 — external relations
PF 2231 — *gober-gas plants*
PF 2231 — nature
PF 2568 Unrealized teaching potential
PF 2568 Unrealized use education structures
PE 7655 Unreasonable licensing restrictions
PA 7193 Unreasonableness *Cheapness*
PA 5473 Unreasonableness *Insufficiency*
PA 5502 Unreasonableness *Reason*
PA 7371 Unreasonableness *Unintelligence*
PF 6096 Unreciprocated love
PD 2781 *Unrecognized animal diseases*
PF 9729 Unrecognized benefits cooperatives
PF 1781 *Unrecognized benefits corporate action*
PF 9729 Unrecognized co-op benefits
PF 4114 Unrecognized future financial commitments
PF 7239 Unrecognized government ; Minorities
PE 6538 Unrecognized income skills
PE 7476 Unrecognized merits nuclear weapons
PF 2995 *Unrecognized need functional skills*
PF 6925 Unrecognized opportunities

PD 8204 Unrecognized people
PJ 9762 *Unrecognized possibilities training urban areas*
PF 9068 Unrecognized relevance education
PF 1664 Unrecognized role elders
PF 9068 Unrecognized school consequences
PF 2969 Unrecognized socio-economic interdependencies
PF 9068 Unrecognized values education
PF 5728 Unrecorded insights
PF 5728 Unrecorded knowledge
PF 2426 Unrecorded popular wisdom
PF 2346 *Unrecorded property exchanges*
PF 5728 Unrecorded wisdom
PD 9310 Unrecyclable disposable products
PD 8942 Unrecycled consumer waste
PB 8376 Unrecycled waste
PF 5039 Unreflective thinking
PF 5381 Unregistered births
PE 8233 Unregulated crop water
PF 3183 Unregulated global resources
PD 9578 Unregulated harvesting
PF 6475 Unregulated land development
PD 2697 Unregulated medical experiments humans
PF 2014 *Unregulated ownership means production*
PE 6226 Unregulated videotext
PF 1540 Unrehabilitated historical figures
PF 2828 Unrelated ability application situational demands vocational decisions
PF 6199 Unrelated buildings urban environments
PD 1703 Unrelated daily life ; Human wisdom
PC 1131 *Unrelated education ; Available jobs*
PF 1491 Unrelated global obligations ; Self determination
PF 9070 Unrelated human experience ; Symbols
PF 8780 Unrelated men women ; Socializing
PF 1724 Unrelated pioneer institutions developing countries
PF 2444 *Unrelated property taxes*
PF 1651 Unrelated social roles ; Static
PA 5794 *Unrelatedness*
PF 1681 *Unreleased creativity elderly*
PD 1544 Unrewarded potential youth participation
Unreliability [Changeableness...]
PA 5490 — *Changeableness*
PE 4428 — computer processing
PE 4428 — computer software
PA 6971 Unreliability *Danger*
PC 2297 Unreliability equipment machinery
PF 7363 Unreliability *Impropriety*
PA 7325 Unreliability *Irresolution*
PF 3551 Unreliability profit social indicator
PF 0888 Unreliability statistics socialist countries
PA 7392 Unreliability *Unbelief*
PA 7309 Unreliability *Uncertainty*
PF 7801 Unreliability weapons systems
PF 6832 Unreliable data
PF 2875 *Unreliable freight transport*
PC 1867 *Unreliable phone installations*
PD 3998 Unreliable products
PD 0489 Unreliable rainfall
Unreliable [telecommunication...]
PE 5860 — telecommunication services land-locked developing countries
PF 3448 — *traditional farming*
PE 5836 — transit services land-locked developing countries
PF 2346 *Unrentable vacant housing*
PD 8557 Unrepaired roads
PF 3523 *Unreplenished urban migration*
PF 2887 Unreported accidents
Unreported [births...]
PF 5381 — births
PF 4729 — bullying
PF 5728 — *businesses*
Unreported [conflicts...]
PF 4967 — conflicts
PF 9184 — cooperation intelligence agencies
PF 8079 — cost overruns
PF 1456 — crimes
Unreported [disagreement...]
PF 7890 — disagreement
PF 7768 — disasters
PF 8090 — diseases
PF 7890 — disputes
PF 5728 — Unreported events
Unreported financial
PF 8079 — losses
PF 5728 — transactions
PC 6641 — transactions
PF 5340 Unreported fraud
PF 2990 Unreported government spending
PF 4729 Unreported harassment
PF 7890 Unreported hostility
Unreported [illness...]
PF 8090 — illness
PC 6641 — income business activities
PF 6594 — international movements funds
PF 4729 — intimidation
PF 7890 — opposition
PF 9148 — Unreported problems
Unreported rape
PE 5621 — [Unreported rape]
PE 5621 — attempts
PE 1363 — prisons
PF 9141 Unreported research
PF 5340 Unreported scandals
PE 9061 Unreported tax obligations
PF 4729 Unreported threats
PF 4967 Unreported violence
PF 8090 Unreported wounds

PC 3778 Unrepresentative control international monetary system
PD 9641 Unrepresentative electoral systems
PC 0237 *Unrepresentative formal leaders*
PE 7021 Unrepresentative international nongovernmental organizations
PD 4873 Unrepresentative international organizations
PF 8703 *Unrepresentative policy makers*
PC 0237 *Unrepresentative village leaders*
PE 6018 *Unrepresentativity trade unions*
PF 6096 Unrequited love
PF 6570 *Unresearched adaptable products*
PF 6512 *Unresearched educational choices*
PF 6472 *Unresearched land records*
PF 6528 Unreserved grazing lands
PE 7903 Unresolved community boundaries
PF 6471 *Unresolved development pattern*
PF 1316 *Unresolved legal issues*
PF 1333 Unresolved local polarization
PF 8993 Unresourceful time use
Unresponsive public
PF 8072 — agencies
PF 8072 — authorities
PF 8072 — offices
PD 6571 Unresponsive river experts
PD 8017 Unresponsive social service bureaucracy
PF 3296 *Unresponsive social services*
PE 4294 Unrest ; Crimes committed during civil
PD 4012 Unrest ; Economic
PC 3685 Unrest ; Ethnic
PD 4012 Unrest ; Industrial
PD 8168 Unrest ; Political
PC 2052 Unrest ; Student
PD 1544 Unrest ; Youth
PC 0952 *Unrestrained animal damage*
PE 8881 Unrestrained use force administration justice
PE 5305 Unrestrained wage increases
PF 4690 *Unretrievable documents*
PF 1735 *Unrewarded initiatives ; Apparently*
PF 1249 *Unrewarded volunteer initiatives*
PA 7250 Unruliness *Disobedience*
PA 7325 Unruliness *Irresolution*
PA 5563 Unruliness *Lawlessness*
PA 7156 Unruliness *Moderation*
PA 7296 Unruliness *Restraint*
Unsafe [abortion...]
PE 3580 — abortion
PE 1575 — aircraft
PE 6390 — artificial sweeteners
PE 2462 — automobile design construction
PF 1379 *Unsafe blood-related products*
PF 4854 Unsafe buildings ; Cover up
PE 5170 *Unsafe construction techniques*
PD 4278 *Unsafe countries*
PF 1379 Unsafe design consumer products
PF 4854 Unsafe equipment ; Cover up
PE 2857 Unsafe firearm storage
PD 8894 Unsafe home maintenance
PE 5058 Unsafe hospitals
PF 4854 Unsafe industrial installations ; Cover up
PE 4859 Unsafe industrial, laboratory medical equipment
PE 2240 Unsafe motor vehicles ; Continued operation
PD 3493 Unsafe nuclear missiles
PD 4977 Unsafe nuclear reactors
PE 4897 Unsafe port facilities
PF 1379 Unsafe products ; Inadequate recall procedures
PE 9776 Unsafe sex
Unsafe [transport...]
PU 4476 — transport
PE 0430 — *transport perishable foodstuffs*
PF 4854 — transportation systems ; Cover up
PD 8122 Unsafe water
PE 0998 Unsanctioned maltreatment prisoners
PA 6995 Unsanctity
PE 3580 Unsanitary abortion practices
PE 8515 Unsanitary conditions
PE 4725 Unsanitary disposal human remains
PD 2011 Unsanitary environment basic health small rural villages
PE 0395 Unsanitary inhumane urban food animal conditions
PF 1844 *Unsanitary markets*
PF 3454 *Unsanitary refuse disposal*
PD 0876 Unsanitary toilet facilities
PD 8122 Unsanitary water supply
PD 8122 Unsanitary water system
PE 7903 Unsatisfactory delimitation boundaries
PF 3260 Unsatisfactory love-making
PF 4010 Unsatisfactory personal appearance
PE 2605 *Unsatisfactory sports field*
PF 0021 Unsatisfied need continuing education
PF 1169 Unsatisfied needs
PD 1544 Unsatisfied youth ambitions
PA 7204 Unsavouriness
PE 2813 *Unscheduled introduction new machinery*
PF 2721 *Unscheduled summer programme*
PD 8468 Unscientifically based policy formulation
PC 1784 Unsecured goods
PE 2813 *Unsecured vacated buildings*
PD 0512 Unserviced older homes
PC 3304 Unsettled indigenous land claims
PC 1888 Unsettled territorial claims
PA 5499 Unsexiness
PF 3526 *Unskilled bureaucratic contacts*
PF 3260 Unskilled coitus
PE 8092 *Unskilled labour industrialized countries ; Demand*
PF 8635 *Unskilled market prices*
PF 2995 *Unskilled mechanical maintenance*
PF 3498 *Unskilled programme development*

Unskilled work

PD 9113 Unskilled work force
PA 7232 Unskillfulness
PA 6653 Unsociability
PF 4417 Unsociable human physiological processes
PF 6911 Unsolved crimes
PF 3526 Unsophisticated bureaucratic methods
PF 7313 Unsophisticated production techniques
PA 7392 Unsophistication *Unbelief*
PA 7232 Unsophistication *Unskillfulness*
PF 3661 Unsound intra-corporate communications
PF 8635 Unstable crop prices
Unstable [economic...]
PF 2042 — *economic base*
PC 0002 — *economic growth*
PC 2677 — *electoral system*
PD 7929 Unstable family life
PF 2135 Unstable fishing season
Unstable [government...]
PF 3214 — *government*
PF 1559 — *government policy*
PD 9117 — *growth manufacturing developing countries*
PC 1738 Unstable regime
PC 1738 Unstable regimes
PD 7516 Unstable shifting agriculture
PD 4270 Unstable supply raw materials
PF 8105 Unstimulating entertainment
PF 0905 Unstrategic funding requests
PD 1544 Unstructured afterschool engagement
PF 6550 Unstructured beautification plans
PC 4867 Unstructured business training
PF 7118 Unstructured community gatherings
PD 4368 Unstructured home life
PF 1781 Unstructured improvement efforts
PF 4768 Unstructured law enforcement
PF 6550 Unstructured local decision-making
PF 6190 Unstructured public access ; Lack places urban environments encouraging
PF 0581 Unstudied export market
PF 6493 Unstudied gardening capacity
PE 3296 Unsuccessful economic ventures
Unsuitability [development...]
PE 8687 — *development projects available labour*
PA 5982 — *Disagreement*
PA 6011 — *Discontentment*
Unsuitability [Inappropriateness...]
PA 6852 — *Inappropriateness*
PA 7395 — *Inexpedience*
PA 5473 — *Insufficiency*
PE 5896 — *insurance land locked developing countries ; Excessive costs*
PA 7341 Unsuitability *Unpreparedness*
PA 7006 Unsuitability *Untimeliness*
PA 5821 Unsuitability *Vulgarity*
PA 7280 Unsuitability *Wrongness*
PF 3448 Unsuitable commercial facilities
PD 0549 Unsupervised children's play
PC 9042 Unsupervised home births
PD 0549 Unsupervised road play
PF 0443 Unsupplemented school curriculum
PC 0834 Unsupplied school lunches
PF 3560 Unsupportable family size
PF 2575 Unsupported youth activities
PF 2575 Unsupported youth centre
PE 1478 Unsupportive criticism media ; Excessive
PS 3268 Unsupportive home structures
PE 7403 Unsurfaced country roads tracks ; Environmental degradation recreational use
PF 6471 Unsurveyed consumer needs
PD 4790 Unsurveyed nutritive needs
PC 5829 Unsustainability development ; Increasing
PF 5964 Unsustainability macroeconomic policies developing countries
PC 8419 Unsustainable agricultural development
PE 5551 Unsustainable consumption resources industrialized countries
PC 0238 Unsustainable cultivation long-lifed trees
Unsustainable development
PB 9419 — [Unsustainable development]
PC 8419 — agricultural resources
PD 4671 — coast zones
PC 0381 — Culturally
PC 0111 — Ecologically
PC 0495 — energy industry
PC 7517 — energy use
PC 8419 — farmlands
PD 4900 — forest lands
PD 6923 — fresh waters
PC 0381 — Socially
PC 0495 Unsustainable economic development
PD 9082 Unsustainable exploitation fish resources
PD 5984 Unsustainable fiscal deficits
PD 9578 Unsustainable harvesting rates
Unsustainable [personal...]
PF 5931 — *personal consumption*
PF 9428 — *policies ; Political support*
PB 0035 — *population growth*
PB 0035 — *population levels*
PD 4537 Unsustainable rural development
PE 4331 Unsustainable short-term improvements agricultural productivity
PC 0442 Unsustainable urban development
PF 8623 Unsymbolized village heroes
PG 8913 Unsympathetic token services
PC 5601 Unsynchronized economic growth
PD 3507 Unsystematic allocation market facilities

PE 1101 Unsystematic use powerful relationships rural communities
PS 2881 Unsystematic vegetable production
PF 1741 Untanned skin
PF 2426 Untapped community wisdom
Untapped government
PE 9325 — grants
PE 9325 — resources
PE 9325 — sources
PF 6479 Untapped leadership potential
PF 7130 Untapped park potential
PF 8299 Untapped regional markets
PF 2595 Untargeted educational goals
PF 2009 Untenable investment incentives
PC 0233 Untenable orphan care
PC 2122 Untended landscape areas
PF 6814 Untested assumptions
PF 6475 Untested industrial expansion
PF 7325 Unthoughtfulness *Irresolution*
PA 7371 Unthoughtfulness *Unintelligence*
PA 5643 Unthoughtfulness *Unkindness*
PE 7393 Until proven guilty ; Denial right presumed innocent
PA 7006 Untimeliness
PF 3467 Untimely seed supply
PD 8388 Untrained clerical personnel
PF 4429 Untrained cosmetic surgeons
PF 6462 Untrained elders' leadership
PF 8835 Untrained farmers ; Technologically
PF 5409 Untrained recreation leaders
PD 8388 Untrained security personnel
PF 2477 Untrained youth advisors
PF 2126 Untransferability books countries cultures
PF 2126 Untransferability images countries cultures
PF 3605 Untransferred leadership skills
PF 1373 Untransmitted fishing skills
PF 6494 Untransmitted local history
PF 6450 Untransposed community structures
PF 2703 Untransposed creativity traditional gifts
PF 1373 Untransposed significance cultural tradition
PD 1285 Untreated chicken diseases
PD 8122 Untreated water supply
PA 6971 Untrustworthiness *Danger*
PA 7363 Untrustworthiness *Improbity*
PA 7309 Untrustworthiness *Uncertainty*
PD 1893 Untruths ; Susceptibility electorate government repetition
PG 8866 Unusable community services
PF 3539 Unusable knowledge
PF 2164 Unused agricultural resources
PF 1373 Unused cultural resources
PS 1567 Unused female talents
PF 1240 Unused fodder resources
PF 2274 Unused gathering places
PD 4790 Unused health services
Unused [land...]
PF 4670 — land
PF 4670 — land ; Deliberately
PF 2205 — livestock waste
PF 7130 — living space
PF 6477 — local capital
PF 6538 — local skills
PF 0581 Unused market channels
PF 4852 Unused methods training
PF 7130 Unused recreation spaces
PF 4852 Unused training opportunities
PC 3768 Unusual punishment
PG 2295 Unusually rapid growth
PD 1415 Unutilized fish by-products
PE 9325 Unutilized government funds
PD 4790 Unutilized health services
PF 3523 Unutilized human resources
PF 1316 Unutilized local resources
PF 0581 Unutilized marketing opportunities
PF 6471 Unutilized wholesale systems
PD 2672 Unvalued diplomas
PD 0193 Unvaried social patterns
PE 1139 Unviable commercial ventures
PE 1907 Unwanted children
PB 9015 Unwanted excessive growth
PE 5580 Unwanted facial hair
PF 3382 Unwanted female babies
PF 2859 Unwanted high fertility
PD 2094 Unwanted pet animals
PF 2859 Unwanted pregnancies
PE 7002 Unwarranted optimism
PF 2818 Unwarranted pessimism
PF 3862 Unwarranted prohibitions
PD 0902 Unwed mothers ; Discrimination against
PA 6799 Unwholesomeness *Disease*
PA 7226 Unwholesomeness *Unhealthfulness*
PF 9474 Unwilling cooperate United Nations eliminating human rights violations ; Governments
PA 6509 Unwillingness
PC 5180 Unwillingness change ; Political
PF 8396 Unwillingness contemplate criticism
PE 8230 Unwillingness divulge information industrial concerns
PF 4204 Unwillingness make difficult decisions
PF 5203 Unwillingness sacrifice personal privilege
PF 7163 Unwillingness sacrifice political power
PD 2790 Unwillingness work
PE 7017 Unwritten codes behaviour ; Discriminatory
Unwritten language
PF 3470 — [Unwritten language]
PF 3470 — Inability teach
PF 3470 — Localization
PF 5311 Upbeat reporting ; Uncritical

PF 6458 Upgraded employment ; Disorganized attempts
PF 3467 Upgrading ; Insufficient access technology agricultural
PF 2813 Upgrading ; Unintentional space
PA 6542 Upheaval *Destruction*
PA 5467 Upheaval *Inexcitability*
PA 7156 Upheaval *Moderation*
PC 7660 Upheaval ; Risk political
PC 7660 Upheavals ; Political
PD 4840 Upheavals private life
PF 1817 Upholding global concern ; Inadequate means
PA 2346 Upkeep ; Insufficient land
PD 6282 Upland watersheds ; Deforestation
PE 7511 Upper limb ; Fracture
PF 7733 Upper respiratory tract ; Diseases
PC 4864 Uprising ; General
PC 4864 Uprisings ; Popular
PF 7361 Uproar *Disorder*
PA 7156 Uproar *Moderation*
PD 3139 Uprootedness
PA 5838 Upset *Agitation*
Upset [Destruction...]
PA 6542 — *Destruction*
PA 5497 — *Difficulty*
PA 7361 — *Disorder*
PA 6030 Upset *Fear*
PA 6247 Upset *Inattention*
PA 7156 Upset *Moderation*
PA 5901 Upset *Revolution*
PA 7309 Upset *Uncertainty*
PF 1599 Upsets ; Stomach
PD 0516 Upstream diversion international rivers
PC 9067 Uraemia
PE 8560 Uranium ; Instability trade U235 depleted
PE 0353 Uranium ores concentrates shortage
PE 8212 Uranium ores concentrates thorium trade instability
PB 6112 Uranium resources ; Long term shortage
Urban areas
PJ 3760 — *Disengaged transient residents*
PE 8959 — *Limited transportation services*
PD 1291 — *Migration skilled people rural*
PC 1833 — *Political fragmentation*
PC 3550 — *Poor credit alternatives*
PE 8443 — *Restrictive building codes*
PE 6151 — *Unattractive pedestrian environments*
PE 1853 — *Uncoordinated social services*
PE 8048 — *Undeveloped business skills*
PE 8061 — *Unfeasible housing alternatives*
PJ 9762 — *Unrecognized possibilities training*
PF 9686 Urban bias
PE 7645 Urban bombers ; Psychotic
Urban [communities...]
PF 1815 — *communities ; Poor condition open spaces*
PF 1681 — *community identity ; Demoralizing image*
PJ 0255 — *conservation ; Unjustified*
PD 5948 — *corruption*
PD 7399 — *crime*
PF 2648 — *cultures ; Incompatibility rural values*
PC 2616 Urban decay
PC 1867 Urban dependence mind-set
Urban development
PE 8569 — Accentuated inequality rural
PC 0442 — Uncontrolled
PC 0442 — Unsustainable
PE 6299 Urban distress
Urban [elites...]
PF 9686 — *elites ; Disproportionate political influence*
PF 9686 — *elitism*
PE 1848 — *emigration ; Massive*
Urban environment
PE 6175 — Inaccessibility green parkland
PE 6160 — Inaccessibility quiet zones
PE 6185 — Inadequate interaction humans animals
PF 6159 — Inhibition exploration children
PE 6299 — Psychological stress
Urban environments
PF 6190 — encouraging unstructured public access ; Lack places
PF 6193 — inhibiting sleeping public ; Unnatural
PE 4685 — Insalubrity animal excrement
PE 6171 — Insufficient common land
PD 1564 — Isolated non participating elders
PE 2583 — Poor service delivery
PE 6182 — Unhealthy lack daily physical activity
PE 6199 — Unrelated buildings
PE 6163 — Wastage open space
Urban [families...]
PF 1503 — families ; Adjustment difficulties new
PD 2211 — fires
PE 0395 — food animal conditions ; Unsanitary inhumane
PD 3489 — fringe poverty
Urban [ghettos...]
PF 1959 — ghettos ; Ineffective operation community networks
PC 1260 — ghettos ; Segregation poor minority population
PC 0442 — growth
PD 1988 — guerrillas
PD 2025 Urban hazards trees
PC 2616 Urban housing ; Neglect
Urban [income...]
PE 5022 — income differential ; Rural
PE 8584 — incomes developing countries ; Imbalance rural
PE 5022 — incomes ; Imbalance rural
PC 0119 — industrial air pollution
PD 0335 — industrial energy supply ; Insufficient diversification
PC 8745 — Industrial pollution
PC 6641 — informal sector
PE 6299 — insecurity

-1072-

Vanity

PD 0008 Urban labour market developing countries ; Flooding
PD 0384 Urban land ; Shortage
Urban life
PE 8644 — style ; Perpetual preoccupation sustenance
PE 8144 — style ; Struggle financial security
PE 8477 — styles ; Apathy toward improvement
PE 8296 — styles ; Minimal family interest
PF 6545 Urban loyalty ; Split tribal
PD 7473 Urban malnutrition
PF 3523 *Urban migration ; Unreplenished*
PC 2616 Urban neglect
PF 6147 Urban neighbourhoods ; Unidentifiable
PC 3813 Urban overcrowding
PC 1833 Urban political machinery ; Inadequate
Urban population
PD 6131 — density gradients ; Unbalanced
PE 8735 — overall growth economy ; Imbalances growth labour force,
PE 8437 — overall growth economy labour force
Urban [populations...]
PF 1844 — populations ; Inadequate community care transient
PC 5052 — poverty
PF 2256 — *problems ; Fear*
PD 2338 — property ; Absentee landlords
Urban [renewal...]
PD 7320 — renewal ; Insensitive
PF 3550 — renewal ; Limited availability investment capital
PD 0426 — road traffic congestion
PF 4852 *Urban schools ; Crimes committed*
Urban services
PE 9133 — Insufficient financial resources
PF 2583 — Limited access
PF 3550 — Limited availability funding resources
PF 1853 — Proliferation parallel
PE 7919 — Undefined government role
PF 2583 — Underprovision basic
Urban [settlements...]
PD 3489 — settlements ; Transitional
PC 1867 — *shift ; Disorienting*
PE 6144 — shopping facilities ; Maldistribution
PF 3523 — *sites ; Migration youth*
PE 5170 — skills local communities ; Restricted availability
Urban slums
PD 3139 — [Urban slums]
PD 3489 — developing countries
PE 7877 — Inadequate health care
PE 1887 — industrialized countries
Urban [smog...]
PD 3663 — smog
PD 3139 — squatters
PC 0442 — structure ; Haphazard
PD 5305 — structures ; Defacement
PF 1776 — structures ; Paralyzing complexity
PF 6137 — subcultures ; Insufficient separation
Urban [terrorism...]
PD 9997 — terrorism
PD 8388 — *training ; Lack agro*
PF 1495 — *transport facilities ; Inadequate*
PD 2025 — trees ; Endangered
Urban [underclass...]
PC 5052 — underclass
PC 3490 — underemployment
PD 1551 — unemployment developing countries
PF 1257 Urban villages ; Disunity
PA 0429 *Urban violence*
PF 1437 Urbanism
PC 0442 Urbanization
PE 9020 Urbanization climate ; Adverse effects
Urbanization [developed...]
PD 3488 — developed countries ; Uncontrolled
PD 0134 — developing countries ; Uncontrolled
PD 0134 — developing countries ; Unplanned
Urbanization [Increasing...]
PE 5931 — Increasing proportion land surface devoted
PC 1563 — industrialization developing countries ; Imbalance
PD 3488 — industrialized countries ; Unplanned
PE 5931 Urbanization ; Loss agricultural land
PC 0442 Urbanization ; Unplanned
PD 3158 *Urchins ; Endangered species sea*
PE 2307 *Urethritis*
PD 0061 *Urethritis ; Gonococcal*
PA 6387 Urgency *Necessity*
PC 1613 Urgent regulations ; Delay approval
PE 2307 Urinary bladder disorders
PE 7934 Urinary calculus
PE 2307 Urinary infection ; Chronic
PE 9086 Urinary organs ; Benign neoplasm kidney
PE 5100 Urinary organs ; Malignant neoplasm genito
Urinary system
PD 9293 — *animals ; Congenital anomalies*
PD 9293 — diseases animals
PC 4575 — Diseases genito
PE 9369 — Symptoms referable genito
PE 2053 — *Tuberculosis genito*
PD 9293 *Urinary tract animals ; Tumours lower*
PE 1602 Urinating public
PE 9291 *Urine disease ; Maple syrup*
PE 9369 *Urine ; Excessive discharge*
PE 5431 Urine ; Involuntary passage
PE 5431 Urine ; Unconscious passage
PE 4656 Urolithiasis
PE 4656 *Urolithiasis ; Canine*
PE 4656 *Urolithiasis ; Feline*
PE 4656 *Urolithiasis horses*
PE 4656 *Urolithiasis ruminants*
PE 4656 *Urological syndrome ; Feline*
PD 9293 *Uropathy animals ; Obstructive*
PD 9293 *Uroperitoneum foals*
PJ 3814 Urostomy
PD 3483 Ursidae
PD 3483 *Ursus ; Endangered species*
PE 1017 *Urticaria*
PD 9667 *Urticaria animals*
PD 2463 US dollar dominance world economy
PF 1825 Uselessness ; Elders' feeling
Uselessness [Impotence...]
PA 6876 — *Impotence*
PA 6852 — *Inappropriateness*
PA 7395 — *Inexpedience*
PA 6662 Uselessness *Nonaccomplishment*
PA 5688 Usurpation *Appropriation*
PA 6921 Usurpation *Undueness*
PF 9014 Usury
PE 1181 Usury ; Criminal
PE 2524 Usury developing countries
PE 1905 Uterine cancer
PD 7799 *Uterine eversion*
PD 7799 *Uterine prolapse animals*
PD 8775 *Uterovaginal prolapse*
PE 7922 Uterus ; Benign neoplasm
PD 8775 Uterus ; Diseases
PE 9286 Uterus ; Infective diseases
PD 8775 *Uterus ; Malposition*
PF 9350 Utilitarian religion
PF 7647 Utilities ; Dysfunctional public
PF 1815 *Utilities ; Outdated basic*
PF 1653 *Utilities ; Prohibitive cost connection public*
PF 2813 *Utility revenue ; Insufficient*
PE 8319 Utility supply services ; Inadequate conditions work employment
PF 4767 Utilization coastal deep sea water resources ; Obstacles
PF 2464 Utilization construction technology ; Profit motivated
PD 0779 Utilization ; Decline production capacity
PF 5639 Utilization expertise ; Individualistic
PF 2969 Utilization external relations ; Incomplete
PF 7808 Utilization foodstuffs ; Limited
PB 1016 Utilization ; Maldistribution resource
PD 0779 Utilization manufacturing plant ; Low capacity
PF 2204 Utilization natural resources ; Inefficient extraction
PE 5921 Utilization operation ; Obstacles efficient port
PF 6455 Utilization plantation space ; Unproductive
Utilization public
PF 6543 — environment ; Ineffective
PJ 0914 — goods services ; Clandestine private
PF 1267 — information ; Fragmented
PF 0357 Utilization renewable biofuels ; Insufficient
PE 0669 Utilization ; Restrictions raw materials components
PF 7022 Utilization time ; Obstacles efficient
PF 1316 Utilization ; Unimaginative vision resource
PF 4892 Utilization volunteer social service workers ; Inadequate
PF 6590 Utilized raw materials ; Under
PF 2192 Utopianism
PD 5535 *Uvea animals ; Diseases anterior*
PD 8786 *Uveal tract ; Inflammation*
PD 8786 *Uveitis*
PD 5535 *Uveitis ; Equine recurrent*

V

PF 1213 *Vacant houses ; Deteriorated*
PF 2346 *Vacant housing ; Unrentable*
Vacant land
PF 4670 — *Clustered*
PF 4670 — *Undesignated*
PF 4670 — *Unplanned*
PD 2672 *Vacant lots ; Overgrown*
PF 2813 *Vacated buildings ; Unsecured*
PF 2813 *Vacated business buildings*
PF 2477 *Vacated youth supervision*
PF 1916 Vaccination ; Compulsory
PF 1916 Vaccination ; Denial right refuse
PE 2786 *Vaccination domestic animals ; Inadequate*
PF 1916 Vaccination ; Enforced
PF 5969 Vaccination practices ; Poor
PE 8383 Vaccines factory farming ; Abuse antibiotics
PE 4657 Vaccines ; Lack
PA 5490 Vacillation *Changeableness*
PA 7325 Vacillation *Irresolution*
PA 7309 Vacillation *Uncertainty*
PE 6311 Vacuous political campaigns
PA 6220 Vacuum ; Existential
PF 2258 Vacuum ; International cultural
PJ 3570 Vacuum ; Power
PJ 3570 Vacuum ; Superpower
PC 2035 Vacuum ; Youth job
PF 9697 Vagaries external markets goods services ; Vulnerability national economies
PC 7873 Vagaries international financial system ; Vulnerability national economies
PE 9286 Vagina ; Infective diseases
PD 0061 *Vaginal candiasis*
PD 7799 *Vaginal hyperplasia animals*
PD 7799 *Vaginal prolapse animals*
PD 8016 *Vaginismus*
PE 9286 *Vaginitis*
PD 7799 *Vaginitis animals*
PE 5460 Vagrancy
PA 5806 Vagrancy *Inaction*
PF 6570 *Vague community memory*
PF 9849 Vague laws
PF 9849 Vague regulations
PF 2086 Vague resolutions
PF 2401 Vagueness civic information
PA 6900 Vagueness *Formlessness*
PA 6959 Vagueness *Insubstantiality*
PA 6978 Vagueness *Invisibility*
PF 9849 Vagueness laws
PA 7309 Vagueness *Uncertainty*
PA 7367 Vagueness *Unintelligibility*
PD 3978 *Vagus indigestion*
PA 6491 Vain people
PF 4436 Vainglory
PA 6491 Vainglory *Vanity*
PA 6099 Vainness *Hopelessness*
PA 6876 Vainness *Impotence*
PA 6852 Vainness *Inappropriateness*
PA 5502 Vainness *Reason*
PA 5942 Vainness *Unimportance*
PA 6491 Vainness *Vanity*
PE 7808 *Valgus ; Carpal*
PE 7808 *Valgus ; Tarsal*
PF 1200 Validity ; Challenges
PF 3250 Validity divorce ; Non
PF 3283 Validity marriage ; Non
PF 9023 Valley communities ; Social isolation mountain
PE 2348 Valley encephalitis ; Murray
PE 7552 Valley fever ; Rift
PE 4162 Valuable land ; Overpriced
PF 9693 Valuable prizes academic integrity ; Corrupting effect
PE 9071 Valuable skills ; Premature retirement people
PF 7870 Valuation art ; Over
PF 0023 Valuation housework national accounts ; Non
PE 0497 Valuation practices ; Distortion international trade restrictive customs
PJ 5990 Value asset ; Volatile
PA 1782 Value erosion
PF 1728 Value images social reality ; Lack correspondence basic
PF 8763 Value labour ; Increasing economic
PD 9356 Value money ; Fluctuations real
PF 1114 Value screens ; Collapse prcviding ethical
PF 6573 *Value system ; Parochial*
PA 6852 Valueless *Inappropriateness*
PA 5942 Valueless *Unimportance*
PF 1118 Values ; Collapse common
Values [Decay...]
PA 1782 — Decay traditional
PC 2102 — Declining family
PE 3827 — Disparity share prices underlying asset
PF 9068 Values education ; Unrecognized
Values [Images...]
PF 1728 — images ; Reinforced parochialism internal
PF 8161 — Imposition young outmoded
PF 2256 — Inappropriate application traditional
PF 8161 — Inappropriate educational
PF 9276 — Indeterminate future human
Values Inflated
PF 7870 — art
PD 1842 — house
PE 4162 — land
PF 1118 Values ; Irrelevant social
PF 5242 Values ; Misappropriation social
PE 6982 Values monitoring television ; Confusion
PF 8161 Values ; Obsolete educational
PF 0292 Values procedures ; Governmental disregard international
Values [Static...]
PF 2883 — Static superficial adult
PF 2827 — Subjective assessment according own group
PE 8139 — subordination women developing societies ; Traditional
PF 5242 Values ; Trivialization social
PF 6469 *Values ; Underprioritized human*
PF 2648 Values urban cultures ; Incompatibility rural
PF 2346 *Valving ; Inoperative shutoff*
PE 0920 *Valvular diseases heart*
PE 0920 *Valvular insufficiency ; Chronic*
PE 1890 Vampire bats
PE 3604 *Vampires ; Endangered species false*
PE 3604 *Vampires ; Endangered species true*
PF 6432 Vampirism
PD 0448 *Vanadium deficiency*
PE 2668 Vanadium pollutant
PD 1350 Vandalism
PE 5171 Vandalism ; Art
PA 6542 Vandalism *Destruction*
PF 1249 *Vandalism ; Fear*
PF 2358 *Vandalism image ; Unhelpful*
PA 6498 Vandalism *Misbehaviour*
PA 7156 Vandalism *Moderation*
Vandalism [rates...]
PD 1350 — rates ; High
PD 1350 — *recreational facilities*
PD 1350 — Recurring property
Vandalism [Unchecked...]
PD 1350 — *Unchecked minor*
PD 1350 — *Uncontrolled local*
PA 5643 — *Unkindness*
PD 1350 — *Unpunished garden*
PD 1350 *Vandalized family gardens*
PD 8872 Vanishing artistic skills
PA 6491 Vanity
PA 6491 Vanity ; Dependence
PA 6659 Vanity *Humility*
PA 6852 Vanity *Inappropriateness*
PA 5942 Vanity *Unimportance*

Vanity

PA 6491 Vanity *Vanity*
PA 7289 Vanquishment *Defeat*
PA 7365 Vapidity *Boredom*
PA 6997 Vapidity *Imperfection*
Vapidity [Unimportance...]
PA 5942 — *Unimportance*
PA 7371 — *Unintelligence*
PA 7204 — *Unsavouriness*
PA 5558 Vapidity *Weakness*
PE 9446 Vapour ; Increase concentration atmospheric water
PE 6820 Vapours ; Irritant gases
PG 3050 *Vapours ; Poisoning gases*
PF 1479 Variability crop yields
PD 5212 Variability food supplies ; Seasonal
PF 1479 Variability grain yields
PD 0159 Variable criteria legal induced abortion
PD 0722 Variable water supply
PF 5817 Variables ; Unpredictability key economic
Variance [Disaccord...]
PA 5532 — *Disaccord*
PA 5982 — *Disagreement*
PA 6838 — *Dissent*
PF 2535 Variance ; Unanticipated market
PA 5490 Variation *Changeableness*
PA 6228 Variation *Deviation*
PA 6698 Variation *Difference*
PF 3409 *Variation fertilizer's price*
PA 6890 Variation *Nonuniformity*
PC 0293 Variations climate ; Short term
PF 2574 Variations national forms currency
Variations [solar...]
PC 6114 — solar radiant energy ; Long term
PE 9528 — solar radiant energy ; Medium term cyclic
PF 0552 — Spelling
PE 3547 Variations undeveloped transportation ; Isolating effects seasonal
PE 6864 Varicose veins
PF 3145 Varieties ; Excessive number crop
PF 3146 Varieties ; Introduction high yield crop
Variety [life...]
PB 9748 — life forms ; Decreasing
PE 8806 — life style ; Lack
PC 2223 — Limited crop
PF 0479 — Limited food
PE 8806 Variety social life forms ; Lack
PE 0097 *Variola major*
PE 0097 *Variola minor*
PC 2270 *Vascular disease ; Arteriosclerotic*
PD 5453 *Vasculitis animals*
PE 3933 *Vasovagal attack*
PD 8385 Vector-borne diseases
PF 2775 *Vector immunity animal diseases*
PD 2746 Vectors animal disease ; Domestic animals
Vectors animal diseases
PD 2751 — [Vectors animal diseases]
PD 2784 — Human
PD 2748 — Insect
PE 2747 — Snail
PE 2749 — Wild birds
PD 2750 — Worms
PC 3595 Vectors ; Disease
Vectors disease
PE 3621 — Aedes mosquitoes
PD 8360 — Animal
PD 8360 — Animals
PE 3622 — Anopheline mosquitoes
PE 6659 — Birds
PE 3623 — Culicine mosquitoes
PD 2746 — Domestic animals
PE 4514 — Flies
PC 3597 — Insect
PE 3616 — Kissing bugs
PD 8371 — Man
PE 1923 — Mosquitoes
PE 3629 — Rodent
PE 4514 Vectors diseases ; Flies
Vectors [Human...]
PD 6651 — Human
PD 6651 — human disease
PD 6651 — Human disease
PE 3632 — human disease ; Insect
PE 1923 Vectors ; Mosquito
PE 1923 Vectors ; Mosquitoes
PD 3596 Vectors ; Plant disease
Vectors plant disease
PD 3601 — Bird
PE 8138 — Fungi
PD 3593 — Human
PE 9017 — Mites
PD 3599 — Plant
PD 3601 Vectors plant diseases ; Avian
PD 7732 Vectors plant diseases ; Insect
PD 2562 Vectors viral diseases ; Bacteria
PD 3600 Vectors viral diseases plants ; Insect
PD 4217 Vegecide
PE 9164 Vegetable based foodstuffs ; Instability trade
PE 1513 Vegetable fibre trade, excluding cotton jute ; Instability
Vegetable oils fats
PE 8136 — environmental hazard
PE 0735 — Instability trade animal
PE 0861 — Instability trade fixed
PE 1188 — Long term shortage animal
PE 8880 — Restrictive practices trade animal
PE 8277 — Shortage fixed
PE 8498 — waxes ; Shortage animal
PS 2881 *Vegetable production ; Unsystematic*
PE 8901 Vegetable products ; Roots tuber trade instability
PF 1711 Vegetable trade ; Instability
PE 9029 Vegetables ; Environmental hazards fruit
PE 8259 Vegetables ; Instability trade beans, dried, peas, lentils leguminous
PE 0961 Vegetables preparations thereof ; Instability trade fruit
PE 1013 Vegetables preparations thereof ; Long term shortage fruit
PE 8208 Vegetables, roots tubers ; Instability trade
PD 4217 Vegetation ; Instruction
PE 4161 *Vegetative interdigital dermatitis*
PA 7253 Vehemence *Envy*
PA 7156 Vehemence *Moderation*
PF 2996 *Vehicle availability ; Limited*
PC 8478 Vehicle collisions
PE 5572 *Vehicle driver fatigue ; Motor*
PD 0414 Vehicle emissions ; Motor
PD 3664 Vehicle noise ; Motor
PF 3467 *Vehicle servicing ; Inadequate*
PE 2240 Vehicles ; Continued operation unsafe motor
Vehicles [Environmental...]
PE 1720 — Environmental degradation off road terrain
PE 8336 — Environmental hazards road motor
PE 2147 — Excessive speed motor
PE 8063 — Expensive transport
Vehicles [Inadequate...]
PF 0713 — Inadequate transport
PD 2072 — Increase road
PF 0713 — *Insufficient special care*
PE 6139 — Intimidation pedestrians
PF 0713 Vehicles ; Lack transport
PF 4485 Vehicles ; Maldistribution commercial
PE 4127 Vehicles ; Overloaded
PD 2072 Vehicles ; Proliferation automobiles motor
PE 4826 Vehicles ; Theft
PE 0897 Vehicles ; Unjustified restrictions free movement commercial
PF 2477 *Vehicles ; Unmaintained private*
PE 5607 Veiled women ; Improperly
PE 2684 Veins ; Diseases
PE 6864 Veins ; Varicose
PA 7363 Venality *Improbity*
PE 8210 Vendettas
PJ 1150 Vending ; Street
PC 0492 *Vendor price ; Undercutting*
PE 2465 Venenanta ; Dermatitis
PD 9667 *Veneral balanoposthitis*
PD 9667 *Veneral vulvitis*
PD 0061 Venereal diseases
PD 7799 *Venereal tumour ; Transmissible canine*
PA 6606 Vengeance
PF 8562 Vengeance
PF 7653 Vengeance ; Blood
PA 7237 Venial *Condemnation*
PA 5454 Venom *Badness*
PA 5446 Venom *Enmity*
PA 7156 Venom *Moderation*
PA 7226 Venom *Unhealthfulness*
PA 5643 Venom *Unkindness*
PD 6823 Venomous animals
PE 9037 *Venous sinuses ; Phlebitis intracranial*
PE 2280 *Ventilation ; Bad*
PC 3320 *Ventilation ; Inadequate*
PD 0122 *Ventilation ; Lack*
PB 4653 *Venture capital ; Lack*
PF 7016 Venture capital ; Misapplication
PE 3869 Venture ; Inadequate participation control joint
PF 8675 Ventures ; Short range business
PE 3296 *Ventures ; Unsuccessful economic*
PE 1139 *Ventures ; Unviable commercial*
PE 4637 *Vera ; Polycythemia*
PD 5238 Verbal abuse
PD 5238 Verbal attack
PF 7358 Verbal conflict envisaged destructive
PF 7798 Verbal exchange ; Collapse systems
PF 7427 Verbal obscenity
PE 0756 Verbal sexual harassment women public
PD 8123 Verbal skills ; Limited
PB 2619 Verbicide
PA 5473 Verbosity *Insufficiency*
PF 5477 Verbosity intergovernmental organizations
PF 7137 Verbosity legal documents ; Unnecessary
PF 4460 Verifiability compliance nuclear arms treaties ; Non
PF 6310 *Verifiability ; Non*
PF 4455 Verificability compliance nuclear power safeguards ; Non
PF 6310 Verification compliance ; Non
PD 7307 *Verminous bronchitis*
PD 7307 *Verminous pneumonia*
PE 4161 *Verrucosa animals ; Dermatitis*
PE 7826 *Verrucosa ; Dermatitis*
PE 4161 *Verrucosa granulosa*
PD 2545 *Versicolour ; Pityriasis*
PE 2572 *Vertebrate reservoirs disease*
PE 3511 *Vertebrates intermediate hosts*
PE 9461 Vertebrogenic pain syndrome
PD 1554 Vertical diversification developing countries ; Lack
PE 4161 *Vertical hoof wall fissures*
PG 9483 *Vertical overbite*
PE 4346 *Verticillium wilt plants*
PE 5101 Vertigo
PE 0921 *Vesical schistosomiasis*
PD 2730 *Vesicular disease ; Swine*
PD 2730 *Vesicular exanthema swine*
PD 2730 *Vesicular stomatitis animals*
PF 8026 Vessels kidneys, eyes nerves ; Damage small blood
PF 1495 *Vessels ; Maldistribution merchant*
PF 1066 Vessels ; Off shore broadcasting
PE 6630 Vessels ; Substandard shipping
Vested interests
PF 3918 — Benefit doubt given
PD 1231 — Entrenchment
PB 3209 — Excessive political influence traditional
PF 9693 — Research sponsored
PD 2077 Veterans ; Social neglect war
PD 7726 Veterinarians ; Corruption
PD 7726 Veterinarians ; Irresponsible
PD 7726 Veterinarians ; Negligence
PU 1623 *Veterinary aid ; Underutilized*
PA 8197 *Veterinary approach ; Misunderstanding*
PE 4903 Veterinary drugs foods ; Residues
PD 7726 Veterinary malpractice
PD 7726 Veterinary practice ; Unethical
PE 8463 Veterinary research ; Lack funds
PF 3162 *Veto*
PF 9441 Vetting job applicants ; Secret government security
PA 5454 Vexation *Badness*
PA 7253 Vexation *Envy*
PA 6030 Vexation *Fear*
PA 7107 Vexation *Unpleasantness*
PF 8663 Viability cold countries ; Non
PF 1389 *Viability local electrical generating capacity ; Non*
PD 7663 Viability nuclear power plants ; Poor
PF 2135 *Viability ; Questionable economic*
PF 2135 *Viability ; Questionable trade*
PD 0068 Viability small developing countries ; Economic non
PD 0441 Viability small states territories ; Non
PF 3826 Viable alternative modes social organization ; Non
PE 7530 Viable organs transplantations ; Imbalance need availability
PE 8162 Vibration damage buildings
PE 8162 Vibration damage cultural artefacts
PC 0268 *Vibration ; Environmental hazards*
PE 7788 Vibration sickness
PE 1145 Vibration working environment
PE 1145 Vibrations health hazard
PE 0560 *Vibrio cholerae*
PD 2731 *Vibrionic hepatitis ; Avian*
PD 6363 *Vibriosis*
PE 6323 Vicarious criminal liability
PA 5644 Vice
PA 5644 Vice ; Dependence
PD 8910 Vice sex traffic offences
PA 5454 Viciousness *Badness*
PA 7156 Viciousness *Moderation*
PA 5643 Viciousness *Unkindness*
PA 5644 Viciousness *Vice*
PA 6340 Vicissitude *Adversity*
PA 5490 Vicissitude *Changeableness*
PE 7752 Victim circumstances ; Marketing skills which reinforce self image being
PB 3561 Victim ; Disaster
PF 0979 Victim external forces ; Self actualizing stance being
Victim image
PF 2093 — Community
PF 0979 — Economic
PF 2093 — Economic
PF 0979 — Individual
PF 1529 — Minority
PF 2093 — Villagers
PE 0969 Victim's fear ; Torture
PF 1365 *Victimage ; Common poverty*
PF 6987 Victimization
PF 6987 Victimization archetypes
PC 5512 Victimization children
Victimization [immigration...]
PE 7780 — immigration control procedures
PA 6852 — *Inappropriateness*
PC 0720 — indigenous peoples
PF 6987 Victimization process
PA 7411 Victimization *Uncommunicativeness*
PD 1846 Victimization workers' representatives
PF 2995 *Victimized market skills*
PF 2995 *Victimizing market skills*
PC 5005 Victimless crime
Victims [abuse...]
PE 7390 — abuse power ; Inadequate assistance
PE 4086 — Accident
PE 4086 — accidents ; Inadequate assistance
PE 5229 Victims catastrophes ; Non payment compensation
PE 4579 Victims ; Children torture
Victims crime
PE 3913 — Delay payment compensation
PD 4823 — Inadequate assistance
PD 9316 — Legal bias against
PD 9316 — Legal indifference
PD 4823 — Neglect
PE 3913 — Non payment compensation
Victims [Crimes...]
PC 5005 — Crimes
PD 0762 — crimes ; Disabled
PB 1992 — *crimes ; Intimidation*
Victims [Defence...]
PF 7836 — Defence industrial pollution immediate
PE 0885 — Depression torture
PE 5229 — disasters ; Delay payment compensation
PF 3918 Victims environment pollution ; Placement burden proof
PE 8240 Victims extortion ; Inadequate assistance
PE 2119 Victims ; Family torture
PD 5122 Victims human rights violations ; Inadequate assistance
PE 7793 Victims intimidation ; Lack protection
PE 8240 Victims ; Lack legal protection extortion

Violence

Victims [major...]
- PE 5229 — major accidents ; Delay payment compensation
- PE 0811 — malpractice ; Delay payment compensation
- PE 0811 — malpractice ; Non payment compensation
- PD 4755 — Mental anguish torture
- PF 7728 — miscarriage justice ; Failure legally rehabilitate

Victims motor accidents
- PE 8824 — Delay compensation
- PE 8824 — Neglect
- PE 8824 — Non payment compensation
- PD 2092 Victims ; Neglect dependents war
- PE 4449 Victims rape ; Inadequate assistance

Victims [Sexual...]
- PE 3932 — Sexual dysfunction torture
- PE 4299 — Social segregation AIDS virus
- PE 0753 — Spouses torture

Victims torture
- PD 0764 — Disabled
- PE 6936 — Inadequate assistance
- PE 6936 — Neglect
- PE 0904 Vicuña ; Endangered species
- PE 5966 Video censorship
- PE 5083 Video display terminal work ; Job stress
- PE 6345 Video game addiction
- PE 6345 Video games ; Socially disruptive effects
- PD 5377 Video standards ; Deterioration
- PE 2224 Video violence
- PE 6345 Video virus
- PE 2224 Videos ; Trivialization death suffering
- PE 6226 Videotext ; Unregulated
- PE 4395 Viewer fatigue ; Television
- PD 0433 Viewers ; Exploitation
- PD 1533 Viewing ; Excessive television
- PF 6536 Viewpoint ; Rigid personal
- PF 7011 Viewpoints ; Diverse leadership
- PF 6012 Viewpoints specialists ; Conflicting
- PF 3720 Views ; Exclusion opposing
- PF 2040 Views ; Fear contradicting popular
- PF 6863 Vigilance ; Impaired
- PD 0527 Vigilantism
- PD 5582 Vigilantism ; Environmental
- PA 5454 Vileness *Badness*
- PA 6191 Vileness *Disapproval*
- PA 6839 Vileness *Disrepute*
- PA 7363 Vileness *Improbity*

Vileness [Unchastity...]
- PA 5612 — *Unchastity*
- PA 5459 — *Uncleanness*
- PA 5942 — *Unimportance*
- PA 7107 — *Unpleasantness*
- PA 5644 Vileness *Vice*
- PA 5821 Vileness *Vulgarity*
- PF 7299 Vilification
- PF 7171 Vilification government
- PD 9384 Vilification ; Political
- PD 5534 Vilification ; Religious
- PF 7118 *Village businesses* ; *Unpromoted*

Village [capital...]
- PF 1979 — *capital* ; *Limited*
- PE 1139 — *communities* ; *Insufficient trade income*
- PF 6506 — consensus ; Difficult
- PF 6506 — consensus ; Unclear

Village cooperation
- PF 3558 — Inadequate inter
- PF 3558 — Insufficient
- PF 3558 — Lack
- PD 9504 Village decay
- PF 1546 Village design ; Unplanned

Village [economic...]
- PC 2791 — *economic isolation*
- PC 2791 — *economy* ; *Contained*
- PC 1820 — entrepreneurship ; Inexperienced
- PF 8623 Village heroes ; Unsymbolized

Village [identity...]
- PF 2845 — *identity* ; *Disparate*
- PF 2845 — *identity* ; *Disrupted patterns*
- PE 1055 — Inadequate road access
- PE 1658 — Insufficient jobs

Village [leader...]
- PC 0237 — *leader distant relations*
- PC 0237 — *leaders* ; *Unrepresentative*
- PE 6556 — leadership ; Ill equipped
- PE 6556 — leadership ; Lack
- PF 6479 — *library* ; *Unconsensed*
- PF 9023 — location ; Isolated

Village [management...]
- PF 1442 — management ; Limited
- PF 5250 — meetings ; Irregular
- PE 3296 — *mentality* ; *Isolated*

Village [population...]
- PF 6441 — *population* ; *Reduced*
- PE 1848 — populations ; Mobility
- PF 6657 — promotion ; Passive

Village [schools...]
- PE 1848 — *schools* ; *Remote*
- PF 9066 — self confidence ; Limited
- PE 2575 — self identity ; Negative
- PE 0835 — solidarity developing countries ; Decline communal spirit
- PF 2168 — story ; Debilitating content
- PA 2093 Villagers victim image
- PF 1389 Villages administrative structures ; Paralyzing patterns

Villages [Deportation...]
- PC 6203 — Deportation
- PE 2235 — Detrimental effect jungle environment tropical
- PF 2365 — Disorganized approach land ownership tropical

Villages [Disunity...] cont'd
- PF 1257 — Disunity urban
- PF 6496 Villages ; Inadequate systems transport communications rural
- PF 2296 Villages ; Limited exposure outside influences rural
- PF 2969 *Villages* ; *Low priority status*
- PF 9023 Villages ; Social isolation neighbouring
- PF 2358 Villages ; Uniformed confidence community corporateness rural
- PD 2011 Villages ; Unsanitary environment basic health small rural
- PA 7363 Villainy *Improbity*
- PA 5644 Villainy *Vice*
- PE 7826 Villonodular synovitis
- PE 2143 Vincent's angina
- PE 3254 Vincent's angina
- PE 2143 Vincent's infection
- PF 1542 Vindictive litigation ; Frivolous
- PA 6606 Vindictiveness *Vengeance*
- PE 1323 Vine ; Love
- PE 2985 Vines ; Pests diseases
- PF 4821 Violating human rights ; Government aid countries
- PF 3976 Violating taboos
- PF 5762 Violation agreements
- PD 3018 Violation amnesty
- PC 5285 Violation civil rights
- PD 709 Violation civil rights ; Criminal
- PA 7250 Violation *Disobedience*
- PD 4728 violation entrepreneurial activities
- PD 2868 Violation food taboos

Violation freedom
- PF 6158 — being tried twice same charges
- PE 6342 — bond service
- PE 4700 — parents choose their children's education
- PD 1251 Violation house codes
- PC 7112 Violation human decency ; Laws

Violation human rights
- PB 3860 — [Violation human rights]
- PF 2418 — Execution orders
- PE 9759 — individual parliamentarians

Violation [Illegality...]
- PA 5952 — *Illegality*
- PA 6852 — *Inappropriateness*
- PF 5148 — integrity creation

Violation international
- PF 2497 — agreements
- PF 2497 — conventions
- PB 0009 — peace
- PB 0009 — security
- PD 8465 — treaties ; Covert
- PD 5218 Violation land rights people
- PE 6726 Violation minimum wage laws
- PA 7156 Violation *Moderation*
- PC 2659 Violation neutrality
- PD 2868 Violation nutritional taboos

Violation [Parole...]
- PE 1121 — Parole
- PD 5457 — political processes
- PD 8937 — premises
- PE 5488 — privileges immunities international civil servants
- PD 8937 — property rights
- PD 4539 Violation regulatory codes
- PE 3809 Violation regulatory codes ; Wilful

Violation right
- PE 5608 — be elected government positions
- PE 4628 — defence against criminal charges
- PC 3459 — education minorities
- PE 1625 — equal work benefits aged
- PE 8899 — international freedom movement shipping
- PE 8899 — international freedom navigation
- PE 5938 — organize trade unions government services
- PE 6086 — participate conduct public affairs
- PD 5317 — review conviction higher tribunal
- PE 5070 — strike

Violation right trade
- PE 4745 — unions engage political action
- PE 1758 — unions freely elect their representatives
- PE 1758 — unions function freely
- PE 3970 — unions negotiate freely employers
- PE 5793 — unions protection against dissolution
- PE 5312 — unions publish

Violation right workers
- PD 4695 — freedom association
- PE 5192 — join trade unions
- PE 4071 — organizations affiliate internationally
- PE 3970 — organizations bargain collectively
- PE 4071 — organizations establish confederations
- PE 5793 — organizations protection against suspension

Violation rights
- PB 5405 — [Violation rights]
- PE 5741 — female homosexuals
- PE 6369 — foetuses
- PE 1903 — homosexuals
- PE 3882 — male homosexuals
- PD 1148 — mentally-ill persons
- PD 1914 — sexual minorities
- PD 6380 — state emergency
- PE 8548 — transsexuals
- PD 3785 — vulnerable groups during states emergency

Violation [sacred...]
- PD 6128 — sacred sites
- PE 5698 — safety regulations nuclear weapons industry
- PE 0261 — sovereignty trans-border broadcasting
- PD 6346 — student's rights

Violation [taboos...]
- PE 1296 — taboos ; Torture

Violation [trade...] cont'd
- PD 4695 — trade union rights
- PE 7573 — treaties indigenous populations
- PA 7280 Violation *Wrongness*

Violations [against...]
- PD 7438 — against economic regulations
- PE 1108 — against firearms explosives laws
- PF 0013 — Arms
- PE 1208 Violations ; Banking law
- PE 2324 Violations ; Cease fire
- PE 5786 Violations contract law
- PE 4173 Violations ; Denial right redress rights
- PE 0930 Violations ; Driving law
- PE 9794 Violations ; False allegations human rights
- PE 1407 Violations ; Government inaction alleged human rights
- PE 9474 Violations ; Governments unwilling cooperate United Nations eliminating human rights
- PE 4006 Violations health safety regulations
- PE 4062 Violations human rights ; Undocumented
- PD 5122 Violations ; Inadequate assistance victims human rights
- PD 5727 Violations private law
- PE 0671 Violations regulatory codes culpability
- PE 0930 Violations ; Road traffic
- PA 4500 Violations ; Securities commodities exchange
- PF 2497 Violations ; Treaty
- PF 3474 Violators human rights ; Impunity
- PF 8046 Violators regulations ; Connivance regulatory inspectors
- PA 5414 Violence

Violence against
- PD 5560 — citizens ; State supported
- PC 0720 — indigenous minorities
- PE 6209 — prostitutes
- PD 0247 — women
- PE 7672 — women ; Alcohol related
- PE 7084 Violence ; Alcohol related
- PD 4782 Violence along internal borders
- PD 2950 Violence ; Border incidents

Violence [Campus...]
- PE 7370 — Campus
- PE 7370 — campus ; Student
- PC 4864 — Civil
- PA 0429 — *comic books*
- PA 5740 — Compulsion
- PD 6881 — Conjugal
- PF 1978 — Covert non physical psychological
- PD 4752 — Crimes
- PD 6279 — Culture
- PF 7498 — Culture youth

Violence [Dependence...]
- PB 1935 — Dependence structural
- PF 7490 — determined evolution
- PC 7373 — disorders ; Criminal
- PD 6881 — Domestic

Violence [Effect...]
- PE 4260 — Effect television film
- PD 2044 — Electoral
- PD 5081 — entertainment
- PA 7253 — Envy
- PD 6881 Violence ; Family
- PD 5582 Violence fanatical environmentalists
- PA 0429 *Violence* ; *Gang*
- PF 7490 Violence ; Genetically determined
- PA 0429 Violence ; Human
- PA 0429 Violence ; Human dependence

Violence [Inappropriateness...]
- PA 6852 — *Inappropriateness*
- PD 7520 — Incitement
- PA 0429 — Indiscriminate
- PF 7490 — Instinctual
- PB 1935 — Institutional
- PC 3685 — Inter communal ethnic
- PA 5414 — Inter species
- PA 5414 — Intra species
- PF 7490 Violence justified science

Violence [mass...]
- PD 6279 — mass media ; Excessive portrayal
- PD 6279 — Media
- PD 6279 — media ; Glorification
- PA 0429 — Mob
- PA 7156 — Moderation
- PF 7490 Violence ; Neurophysiological compulsion

Violence [parliamentary...]
- PF 4550 — parliamentary assemblies
- PD 4425 — Political
- PA 0587 — *Psychological*
- PE 7645 — Psychotic

Violence [Racial...]
- PC 2936 — Racial
- PF 3994 — resource
- PE 3720 — Rural

Violence [Sexist...]
- PD 0247 — Sexist
- PD 3276 — Sexual
- PE 4026 — Sports
- PE 0525 — Stress trauma context civil
- PB 1935 — Structural

Violence [Teenage...]
- PF 7498 — Teenage
- PE 4260 — Television
- PE 4260 — television films ; Effect

Violence [Union...]
- PA 0429 — *Union monopoly*
- PA 5643 — *Unkindness*
- PF 4967 — Unreported
- PA 0429 — *Urban*
- PE 2224 Violence ; Video

Violence women

PE 7750 Violence women ; Fear
PE 7750 Violence women ; Risk
PF 7498 Violence ; Youth
PC 2321 Violent animals
PD 3907 Violent children
PD 4752 Violent crime
PD 4666 Violent deaths
PA 0429 *Violent demonstrations*
PE 3407 Violent emotions
PE 4297 Violent interactive toys
PE 1478 *Violent news ; Getting use*
PF 2788 Violent non cooperation ; Non
Violent [people...]
PA 0429 — people
PE 7175 — pet animals
PA 0429 — *picketing*
PD 4811 — police intervention meetings
PE 2007 — police intervention workers meetings
PD 3230 — political revolution
PD 3228 — political revolution ; Non
PD 8402 — *purgatives*
Violent [rage...]
PE 0429 — rage
PD 4811 — repression demonstrations
PE 2007 — repression trade union demonstrations
PC 3229 — revolution
PE 4026 Violent sports
PE 6281 Violent sports fans
PF 9327 Violent weapons ; Non
PC 3151 Violent wildlife
Viral [animal...]
PD 2741 — animal diseases ; Airborne
PD 2730 — animal infections
PD 2730 — *arteritis ; Equine*
PE 5913 — *arthritis poultry*
PE 6403 Viral combinations ; Disease causing
PD 3978 *Viral diarrhoea ; Bovine*
PD 1175 *Viral disease ; Water borne*
Viral diseases
PD 0594 — [Viral diseases]
PD 2730 — animals
PD 2562 — Bacteria vectors
PF 1916 — *Inoculation*
PD 3600 — plants ; Insect vectors
PE 2348 Viral encephalitis
PD 7841 *Viral encephalomyelitis piglets*
Viral hepatitis
PE 0517 — A
PE 0517 — B
PE 0517 — *Canine*
PD 2730 — *ducks*
PD 2730 — *geese*
PD 2730 — *turkeys*
PD 0594 Viral infections
PD 0594 *Viral infections*
PD 2227 Viral plant diseases
PD 0594 *Viral protozoa infections*
PE 5182 *Viral rhinopneumonitis ; Equine*
PF 7127 Virtue ; Excessive
PA 5454 Virulence *Badness*
PA 7055 Virulence *Death*
PA 5446 Virulence *Enmity*
PA 7253 Virulence *Envy*
PA 7156 Virulence *Moderation*
PA 7226 Virulence *Unhealthfulness*
PA 5643 Virulence *Unkindness*
PE 5617 *Virulent foot rot*
PE 5182 *Virus 1 infection ; Equine herpes*
PD 4747 Virus ; Bovine immuno deficiency (BIV)
PE 3852 Virus ; Coxsackie
Virus disease
PD 2730 — *Congo*
PD 2730 — *pigs ; Encephalomyocarditis*
PD 2730 — *San Miguel sea lion*
Virus diseases
PD 0594 — [Virus diseases]
PD 2730 — animals
PD 2562 — bacteria
PD 2227 — plants
PD 0594 — *protozoa*
PD 7924 — Respiratory
PD 4747 Virus ; Feline immuno deficiency (FIV)
PE 7848 Virus ; Feline immuno deficiency (FIV)
PD 5111 Virus ; HIV human immunodeficiency
PD 4747 Virus ; Immuno deficiency
PE 7304 *Virus infection cats ; Pox*
PD 4747 Virus ; Simian immuno deficiency (SIV)
PE 4299 Virus victims ; Social segregation AIDS
PF 6345 Virus ; Video
PD 3102 Virus warfare ; Computer
PE 3680 Viruses
PD 3102 Viruses ; Computer
PF 9659 *Viruses ; Drug resistant*
PD 0594 Viruses ; Pathogenic
PE 7780 Visas ; Administrative impediments obtaining travel
PE 0325 Visas ; Refusal
PE 0325 Visas ; Refusal issue travel documents, passports,
PE 6777 Visceral impairments
Visceral [larva...]
PE 6278 — *larva migrans*
PE 2281 — *leishmaniasis*
PE 3676 — *leishmaniasis animals*
PD 5453 — *lymphoma ; Feline*
PD 5453 — *lymphosarcoma ; Feline*
PE 2593 Visibility degradation
PE 2593 Visibility ; Deterioration atmospheric
PF 2822 *Visibility facilities ; Inadequate*
Visibility [Lack...]
PE 2593 — Lack
PF 2822 — *Low business*
PF 7118 — *Low community*
PD 5510 Visible accumulated junk
PF 7118 *Vision community possibilities ; Unimaginative*
PF 6575 Vision community's future ; Frozen
PF 2699 Vision existing international economic structures ; Unimaginative
PF 1365 *Vision farmers ; Unclear*
PF 2803 Vision future ; Static social relations inhibiting
PD 8179 Vision impairment
PE 3377 Vision ; Inflected loss
PF 9478 Vision ; Limited development
PE 4196 Vision ; Partial loss
PF 3007 Vision ; Past oriented educational
PF 1316 Vision resource utilization ; Unimaginative
PF 6528 Vision ; Restrictive land use
Vision [Self...]
PF 3679 — Self interested industrial
PF 1729 — Shallow personal commitment social
PF 3008 — *Subsistence discourages*
Vision [Unchallenging...]
PF 9478 — Unchallenging world
PF 3007 — Unimaginative educational
PF 3007 — Unimaginative long term training
PF 1944 — *Unrealistic agricultural*
PF 9478 Vision world's future ; Unimaginative
PF 7988 Visioning leadership ; Narrow
PF 2352 Visions ; Dystopian
PF 5691 Visions ; Fanatical
PF 5691 Visions ; Malignant
PF 4832 Visits ; Infrequent medical
PE 9438 *Visna ; Maedi*
PD 8179 Visual deficiencies
PE 5083 Visual display units ; Health hazards computer
PF 0076 Visual imagery ; Prejudice against communication
PD 7086 Visual imagery societal learning ; Inadequate use
PF 0076 Visual media ; Absence audio
PD 8179 *Visual spatial ability ; Impairment*
PE 6865 Visual strain modern technology
PF 2352 Visualize creative future ; Inhibited capacity
PD 2542 Visually handicapped persons
PF 0704 Vitae ; False curriculum
PF 1985 Vitality ; Lost heritage
PE 2538 Vitamin A deficiency
PD 0715 *Vitamin B deficiency*
PD 0715 *Vitamin B1 ; Lack*
PE 2380 Vitamin C deficiency
PD 0287 *Vitamin D deficiency*
PD 0715 Vitamin deficiencies diet
PE 4760 Vitamin E deficiency domestic animals
PE 4665 Vitamins ; Excessive consumption
PA 6088 Vitiation *Impairment*
PA 5644 Vitiation *Vice*
PD 5238 Vituperation
PA 6191 Vituperation *Disapproval*
PE 0616 *Vivax malaria*
PD 0260 Vivisection
PF 3824 Vocabulary ; Lack
PF 1254 Vocal leaders ; Lack
PE 9344 *Vocal tic disorder ; Chronic*
PF 6098 Vocation ; Collapsed images
PF 1765 *Vocational change ; Fear*
PD 2790 Vocational commitment ; Minimal
PF 2477 *Vocational decision ; Unguided*
PF 2828 Vocational decisions ; Unrelated ability application situational demands
Vocational education
PF 0422 — Circumscribed
PF 0422 — Inadequate
PF 0422 — void
Vocational images
PF 6098 — Breakdown
PF 6098 — Restricted
PF 6537 — Restricted
PF 2477 *Vocational misguidance*
PF 2477 *Vocational needs ; Unclear*
PF 3844 Vocational obsolescence face overwhelming need
PF 0422 Vocational potential ; Unprepared
PE 7317 Vocational rehabilitation facilities disabled persons ; Inadequate
PF 6098 Vocational roles ; Disrelated images
PF 0422 Vocational school ; Unfacilitated
PD 3548 Vocational skills ; Obsolete
PC 2163 *Vocational teachers ; Lack*
Vocational training
PC 2163 — *Inaccessible*
PF 0422 — Lack
PF 0422 — Misdirected
PF 0422 — Non productive
PF 0422 — schemes ; Lack
PC 2163 — *Unavailable*
PF 6024 — women farmers ; Unadapted
PF 0422 Vocationally irrelevant ; Education
PD 3663 Vog
PF 2261 *Voice ; Lack community*
PE 6866 Voice ; Occupational diseases
PD 1140 Voices ; Paralysis minority
PF 1254 Void ; Crippling leadership
PF 1178 Void ; Self conception ethical
PA 6220 Void ; Spiritual
PF 0422 Void ; Vocational education
PC 1453 Volatile capital
PE 1427 Volatile substance abuse
PJ 5990 Volatile value asset
PE 5930 Volatility ; Exchange rate
PF 3141 Volatility interest rates
PA 7325 Volatility *Irresolution*
PF 8635 Volatility ; Price
PA 6425 Volatility *Transience*
PA 7371 Volatility *Unintelligence*
PE 5109 Volcanic dust
PD 3568 Volcanic eruptions
PD 3663 Volcanic smog
PD 3568 Volcanoes
PD 9366 *Voltage animal housing ; Stray*
PJ 4775 Voltage ; Stray
PE 9642 Voltage transmission lines ; Health hazards high
PF 2535 Volume ; Uncertain market
PF 4930 Voluntary dissolution family
PE 0310 Voluntary export restraints ; Trade restrictions due
PE 8742 Voluntary organizations ; Irresponsible
PC 6720 Voluntary unemployment
PF 1249 *Volunteer initiatives ; Unrewarded*
Volunteer [services...]
PF 2875 — *services ; Declining*
PD 1566 — *services ; Lapsed*
PF 4892 — social service workers ; Inadequate utilization
PF 2477 *Volunteers ; Dispersed fire*
PE 7423 *Volvulus animals ; Gastric dilatation*
PE 4513 Vomiting
PD 7841 *Vomiting wasting disease*
PE 0985 Vomito amarilli
PE 2348 *Von Economo's disease*
PF 9006 Voodoo
PA 6379 Voracity *Avoidance*
PA 6466 Voracity *Intemperance*
PD 5214 Vote buying
PF 2904 Vote ; Denial right
PF 2904 Vote ; Limitations right
PC 2494 Vote ; Refusal
PD 5214 Vote rigging
PC 2494 Voter apathy
PC 2494 Voter fatigue
PD 9641 Voter registration procedure ; Unjust
PC 2494 Voters ; Decreasing number election
PC 2494 Voters ; Inadequate choice
PD 2044 Voters ; Threats against
PC 2494 *Voting ; Compulsory*
PC 2494 Voting patterns ; Unpopular
PD 9641 Voting systems ; Biased
PD 9641 Voting systems ; Undemocratic
PF 1870 Vows ; Social irrelevance marriage
PE 3272 Voyeurism
PE 5190 Vulgar combination sacred erotic advertising
PF 7427 Vulgar language
PJ 5254 Vulgar sociologism
PE 9509 *Vulgaris animals ; Pemphigus*
PE 0566 *Vulgaris ; Lupus*
PF 5213 Vulgarism
PA 5821 Vulgarity
PA 7143 Vulgarity *Discourtesy*
PA 5652 Vulgarity *Inferiority*
PA 5612 Vulgarity *Unchastity*
PA 5821 Vulgarity *Vulgarity*
PA 1766 Vulnerability
Vulnerability [abuse...]
PJ 9080 — abuse strategically distributed products
PF 9697 — adverse conditions foreign trade
PD 5114 — agriculture developing countries future declines production
PE 4694 — animals during states emergency
Vulnerability [children...]
PE 4276 — children AIDS
PE 8174 — children during armed conflict
PE 8542 — computer systems
PC 0123 — Consumer
PD 8030 — countries destabilization foreign investment
PD 0660 — Crop
PE 5682 — crops weather
Vulnerability [developing...]
PC 6189 — developing countries
PD 0367 — developing countries inflation
PE 5683 — diplomatic agents
PD 0098 — disabled during states emergency
PC 2430 — drought
Vulnerability [economic...]
PF 1172 — economic cycles
PD 7486 — economies import penetration
PC 5773 — ecosystem niches
PD 0096 — elderly states emergency
PD 9343 — elders' property
PC 6329 — *Energy dependence*
PD 1863 — European insecurity
Vulnerability [farming...]
PC 4906 — farming
PB 2253 — food chains
PC 5165 — Forest
PE 6833 — frontier workers
PF 5749 Vulnerability ; Geopolitical
PF 5365 Vulnerability government lobbying
PB 5647 Vulnerability human organism
PE 6934 Vulnerability human rights activists
Vulnerability [indigenous...]
PE 3721 — indigenous populations introduction diseases
PF 8854 — intellectual property
PE 5798 — intravenous drug users AIDS
PE 5700 — island developing countries territories
PE 5813 Vulnerability lakes rivers land-locked countries
PD 5788 Vulnerability land-locked developing countries

Warnings

	Vulnerability marine
PF 7554	— animal communication
PC 1647	— ecosystems
PD 5178	— environment catastrophic warfare damage
	Vulnerability [marriage...]
PF 1870	— marriage institution
PC 1002	— middle-class
PC 0541	— Military insecurity
PD 6282	— mountainous environments
	Vulnerability [national...]
PF 9697	— national economies vagaries external markets goods services
PC 7873	— national economies vagaries international financial system
PB 1149	— National insecurity
PD 1521	— non nuclear weapon states ; Insecurity
	Vulnerability nuclear
PD 4049	— defence control systems
PD 0365	— power sources
PC 4440	— weapon states ; Insecurity
PB 5658	Vulnerability organisms
	Vulnerability [painters...]
PE 9746	— painters cancer
PD 0188	— performers' rights
PD 5730	— plants crops
PB 1395	— populations species
PC 1784	— property theft damage
PC 4764	— protected natural areas
PD 6128	Vulnerability sacred sites
	Vulnerability small
PD 2374	— nations foreign intervention
PD 2374	— states
PD 6125	— towns
	Vulnerability [social...]
PB 0001	— social defences
PB 2853	— social systems
PD 1104	— socialist countries ; External dependence
PF 5937	— society truth
PF 1245	— socio-economic systems globalization
PD 5676	— stock markets
	Vulnerability [telephone...]
PE 8254	— telephone system
PA 6976	— *Toughness*
PD 7716	— traveller
PF 5520	Vulnerability universal principles
PA 7341	Vulnerability *Unpreparedness*
PC 3486	Vulnerability wetlands
	Vulnerability women
PD 1078	— children emergencies
PD 1078	— during armed conflicts
PD 1078	— during foreign military occupation
PD 0407	Vulnerability world cable communications
PB 4788	Vulnerability world genetic resources
PB 4353	Vulnerable ; Exploitation
	Vulnerable groups
PC 4405	— Denial rights
PB 4353	— Dependency
PB 4353	— Disempowerment
PD 3785	— during disasters ; Lack protection
PD 3785	— during states emergency ; Violation rights
PF 8209	— Excessive institutionalization
PF 4662	— Lack commitment protection
PD 6760	— Medical experimentation socially
PF 3918	— Placement burden proof abuse
PD 0363	— Protein energy malnutrition
PB 4353	Vulnerable ; Lack protection
PC 3295	Vulnerable minorities ; Threatened
PD 7799	*Vulvitis animals*
PD 7799	*Vulvitis ; Ulcerative*
PD 9667	*Vulvitis ; Veneral*
PE 5524	*Vulvovaginitis ; Infectious pustular*

W

PF 6745	Wage bargaining ; Government bias
PE 5305	Wage claims ; Unrealistic
PE 6726	Wage ; Denial right minimum
PE 6726	Wage fixing ; Lack minimum
PE 4003	Wage increases provided legislation collective agreements ; Non implementation workers
PE 5305	Wage increases ; Unrestrained
PE 6726	Wage laws ; Violation minimum
PJ 8374	Wage rates youth employment ; Negative effects relative
PF 4028	Wage rigidity labour markets
PE 6726	Wage scale ; Low
PD 1133	Wage scales ; Non comprehensive
PF 2137	Wagering ; Gambling
	Wages [Decline...]
PD 8916	— Decline real
PD 2769	— developing countries ; Decline real
PE 9199	— developing countries ; Low
PD 8916	Wages ; Excessively low
PD 5674	Wages ; High minimum
PD 8916	Wages ; Inadequate
PF 3550	*Wages ; Low police*
PD 8916	Wages ; Low tenant
PD 8916	Wages ; Non competitive business
PD 5674	Wages ; Prohibitive minimum
	Wages [Unattractive...]
PD 8916	— Unattractive industrial
PD 8916	— Unmotivating low
PD 5335	— Unpaid
PF 6192	Waiting environments ; Alienating
PD 5119	Waiting lists ; Hospital
	Waiting [time...]
PF 1761	— time ; Wasted
PF 5120	— times government facilities ; Excessive
PE 3927	— Torture
PE 2197	*Wake schedule disorder ; Sleep*
PC 7896	Waldsterben
PF 6140	Walk country ; Denial right
PD 7585	Walkers ; Degradation countryside careless
PE 2197	*Walking ; Sleep*
PE 3874	Walking trails ; Disappearance
PE 4161	*Wall fissures ; Horizontal hoof*
PE 4161	*Wall fissures ; Vertical hoof*
PE 7826	*Wall ; Hollow*
PE 8708	Wallabies ; Endangered species kangaroos
PE 8288	Walrus ; Endangered species
PD 7841	Wanderers
PA 7325	Wantonness *Irresolution*
PA 5612	Wantonness *Unchastity*
PA 5644	Wantonness *Vice*
PB 0593	War
PD 1652	War ; Brainwashing prisoners
	War [casualties...]
PD 4189	— casualties
PC 1869	— Civil
PE 4294	— *Civil crimes committed during*
PC 1573	— Class
PF 3100	— Cold
PC 0840	— *Computer development trade*
PE 3034	— correspondents ; Inadequate protection
PD 1464	— crime prosecution ; Ineffective
	War crimes
PC 0747	— [War crimes]
PD 1464	— Ineffective deterrent against
PF 4062	— Undocumented
	War criminals
PE 4697	— Government approved employment
PD 5541	— Havens
PD 1464	— Inadequate punishment
PE 4439	— policy making contexts ; Employment
PD 4067	— Unconvicted
	War [damage...]
PD 8719	— damage civilian areas
PD 3057	— debt
PF 2636	— *Defeat*
PB 0593	— Dependence
PC 3431	— Dependence ideological
PE 8915	— Destruction economy due
PD 0874	— disabled ; Ageing
PA 5532	— *Disaccord*
PC 4257	— Disastrous consequences
	War due
PF 4346	— accidents ; Risk unintentional nuclear
PF 4302	— international crises ; Risk unintentional nuclear
PF 4352	— nuclear proliferation ; Risk unintentional global nuclear
	War [Economic...]
PD 3765	— Economic civil
PE 5252	— Economic decline caused
PC 0842	— Effects nuclear
PE 1200	— enemy aliens ; Aiding escape prisoners
PF 4034	— Enjoyment
PC 6675	— Environmental consequences
PD 4714	— Excitement
	War [Failure...]
PE 2324	— Failure sign peace treaty following
PE 2324	— following cease fire agreements ; Persistence technical state
PD 0218	— Forced repatriation prisoners
	War [games...]
PE 3704	— games
PF 1406	— games ; Recreational
PD 4843	— Gang
	War generated
PF 4152	— arms race ; Risk unintentional
PF 4156	— developments strategic doctrine ; Risk unintentional nuclear
PF 4162	— strategy deterrence ; Risk unintentional nuclear
PF 9312	War ; Glorification
PC 1738	War ; Guerrilla
	War [Hazardous...]
PF 2613	— Hazardous remnants
PF 5681	— Holy
PE 2592	— Housing destruction
	War [Ideological...]
PC 3431	— Ideological
PD 2617	— Ill treatment prisoners
PD 6265	— Inadequate health services following nuclear
PD 4714	— Incitement
PE 8979	— induced famine
PE 6350	— induced winter chill ; Nuclear
PD 8359	— Industrial destruction
PD 3765	— Intra national trade
PE 3988	— Limited
PD 7918	War ; Long term effects
	War [neurosis...]
PE 7912	— neurosis
PE 0948	— Non repatriation prisoners
PC 0842	— Nuclear
PF 1447	War-oriented economies
	War [Pervasive...]
PC 3541	— Pervasive fear nuclear
PE 5252	— Poverty consequence
PD 3092	— pre-war propaganda
PC 8848	— Prisoners
PD 3092	— propaganda ; Denial right freedom
PD 3092	— propaganda ; War pre
	War [psychosis...] cont'd
PE 7867	— psychosis
	War [Racial...]
PD 8718	— Racial
PF 0727	— relief ; Lack
PC 2371	— Religious
PD 6380	— Restriction rights during
	War Risk
PF 4215	— [War ; Risk]
PF 4435	— intentional nuclear
PF 0466	— unintentional nuclear
	War [Sacrifice...]
PF 8118	— Sacrifice children holy
PE 3988	— Small
PD 4952	— socialist states
PF 1406	— sports
PF 3807	War ; Theological justification nuclear
PF 8874	War ; Threat
	War time
PE 6974	— communications enemy
PD 9090	— conditions pressure
PD 8851	— disruption economies production facilities
PE 6074	— exportation cultural property occupied territories
PF 0726	— imprisonment deportation ; Delayed consequences
PE 4781	— Inhumane medical experimentation during
PC 0840	War ; Trade
	War [Unconventional...]
PC 8836	— Unconventional
PB 0593	— *Undeclared*
PF 9312	— Unrealistic portrayal
PD 2077	War veterans ; Social neglect
PD 2092	War victims ; Neglect dependents
PD 2092	War widows orphans ; Neglect
PE 8361	War zones ; Inhumane treatment non combatants
PE 3635	Warble fly
PE 5288	Wards ; Inhumane geriatric
PF 7755	Wards ; Sexual abuse
PF 2842	*Warehousing jobs ; Automated*
PE 5851	Warehousing ; Overcrowded port
PB 0593	Warfare
PF 1236	Warfare alerts ; False nuclear
PF 1214	Warfare analysis ; Politically unrealistic strategic
	Warfare [Bacteriological...]
PC 0195	— Bacteriological
PC 1164	— Biochemical
PC 0195	— Biological
	Warfare [Chemical...]
PC 0872	— Chemical
PC 1164	— Chemical biological
PE 8174	— Children threatened
PC 2175	— Clandestine psychological
PD 3102	— Computer virus
PC 4311	— Conventional
PC 1738	— *Counter insurgency*
	Warfare [damage...]
PD 5178	— damage ; Vulnerability marine environment catastrophic
PC 1164	— Dependence biochemical
PD 7298	— Destruction cultural property during
PD 2191	— Disruption development tribal
	Warfare [Economic...]
PC 0840	— Economic
PE 8915	— Economic burden
PC 2696	— Environmental
PC 0195	— Epidemic
PF 4866	— Extrasensory
	Warfare [Gang...]
PD 4843	— Gang
PC 0195	— Germ
PC 1738	— Guerrilla
PF 7669	Warfare ; Illegal
PF 1925	Warfare ; Inter species
PC 0311	Warfare material ; Stockpiles nuclear
PE 3580	*Warfare ; Naval*
PC 2175	Warfare operations ; Covert psychological
	Warfare [Parapsychic...]
PF 4866	— Parapsychic
PC 0840	— *Price*
PD 7986	— Psionic
PF 4866	— Psychic
PC 2175	— Psychological
PD 7986	— Psychotronic
	Warfare [Radiological...]
PC 6666	— Radiological
PC 5917	— Regional
PF 0782	— Relatives lost during
PE 6350	Warfare soot ; Sunlight inhibition nuclear
PD 6439	Warfare ; Space
	Warfare technology
PD 4445	— Super power monopoly advanced nuclear
PD 8621	— Trade biological
PD 9692	— Trade chemical
PE 3808	Warfare ; Trade products chemical
PE 6443	Warfare ; Use animals
PE 4199	Warheads ; Waste nuclear
PA 5532	Warlike *Disaccord*
PC 0918	Warming ; Global
PD 4714	Warmongers
PD 5173	Warmth ; Lack physical
PF 5744	Warmth work ; Lack
PF 3565	Warning disasters ; Inadequate
PF 4298	*Warning hoaxes ; Bomb*
PF 3391	Warning ; Over
PF 1391	Warnings ; Backlash against repeated
PF 4298	*Warnings ; False bomb*
PF 4298	Warnings ; False disaster

Warnings

PF 3565	Warnings ; Ignored disaster	
PF 3565	Warnings ; Rejection disaster	
PA 6790	Warpedness *Disease*	
PA 6997	Warpedness *Imperfection*	
PA 6486	Warpedness *Injustice*	
PA 7306	Warpedness *Narrowmindedness*	
PA 7411	Warpedness *Uncommunicativeness*	
PA 5644	Warpedness *Vice*	
PD 3544	*Warrant ; Search*	
PJ 1730	Wars ; Farm	
PD 5361	Wars ; Fish	
PD 6439	Wars ; Star	
PD 9667	*Warts*	
PJ 5108	Warts	
PC 3108	Wastage capitalist systems ; Resource	
PC 1716	Wastage ; Educational	
PD 8844	Wastage ; Food	
PD 0183	Wastage governmental budgets appropriations	
PE 1396	Wastage highly skilled personnel routine maintenance complex systems	
PC 8914	Wastage human resources	
PE 5170	Wastage local skills developing countries	
PE 8233	Wastage ; Open irrigation	
PE 6163	Wastage open space urban environments	
PD 0183	Wastage resources government agencies	
PC 1716	Wastage school examination failures	
PD 0183	Waste ; Bureaucratic	
PA 7193	Waste *Cheapness*	

Waste [Destruction...]
- PA 6542 — *Destruction*
- PA 6799 — *Disease*
- PD 0807 — disposal ; Consumer
- PD 0201 — *disposal deep wells*
- PD 1398 — Disposal hazardous
- PD 6795 — disposal ; Inadequate
- PC 1242 — disposal ; Radioactive
- PD 8844 — Dumping food products
- PD 1398 — dumping ; Hazardous
- PD 1398 — dumping ; Toxic
- PE 0984 — Waste electricity
- PE 9074 — Waste ; Environmental hazards textile fibres
- PD 8844 — Waste ; Food

Waste [Hospital...]
- PE 4725 — Hospital
- PE 1542 — Hotel restaurant
- PC 8914 — human resources
- PC 3113 — human resources capitalist systems

Waste [Impairment...]
- PA 6088 — *Impairment*
- PA 6852 — *Inappropriateness*
- PE 1550 — Instability trade unprocessed textile fibres their
- PA 7382 — Waste *Loss*
- PE 3741 — Waste ; Marine pollution plastic
- PE 4725 — Waste ; Medical

Waste [Newspaper...]
- PD 1152 — Newspaper
- PD 1180 — Non biodegradable plastic
- PC 8642 — non-renewable resources
- PD 4396 — Nuclear
- PE 4199 — nuclear warheads
- PD 3666 — Waste ; Ocean disposal

Waste paper
- PD 1152 — [Waste paper]
- PE 7914 — Instability trade pulp
- PE 1616 — Long term shortage pulp
- PD 8942 — Waste products ; Dumping consumer
- PD 8942 — Waste ; Proliferation consumer
- PF 6465 — Waste recovery ; Inadequate

Waste resources
- PB 8376 — [Waste resources]
- PC 0012 — armaments research
- PE 8261 — *expensive medical techniques*
- PD 0183 — government agents ; Extravagant
- PE 9346 — invested obsolete armaments
- PE 7539 — obsolete nuclear power plants
- PE 4146 — Waste sites ; Ineffective inspection toxic
- PD 9765 — Waste smuggling ; Toxic
- PD 6795 — Waste treatment ; Inadequate
- PD 6795 — Waste treatment ; Substandard

Waste [Uncontrolled...]
- PB 8376 — Uncontrolled
- PF 6465 — *Underutilization*
- PB 8376 — Unrecycled
- PD 8942 — Unrecycled consumer
- PC 2205 — *Unused livestock*
- PD 1410 — use schemes ; Undercapitalized
- PD 0482 — Waste water contamination

Waste water Discharge
- PD 2800 — dangerous substances domestic
- PD 0575 — dangerous substances industrial
- PE 9813 — dangerous substances laboratory
- PD 0482 — Waste water ; Impurities

Waste water pollutants
- PD 2800 — Domestic
- PD 0575 — Industrial
- PE 9813 — Laboratory
- PE 2634 — Waste water ; Stagnant
- PC 7517 — Wasted energy resources
- PD 0136 — Wasted foreign aid
- PF 8993 — Wasted time

Wasted [waiting...]
- PF 1761 — waiting time
- PD 3669 — water
- PF 3690 — woman power
- PF 5931 — Wasteful personal consumption
- PF 6516 — *Wasteful sportsmen's practices*

PA 7193	Wastefulness *Cheapness*	
PA 6542	Wastefulness *Destruction*	
PE 6005	Wastelands ; Derelict industrial	
PC 2205	Wastes ; Agricultural	
PD 0575	Wastes ; Chemical industry	
PE 2061	Wastes developing countries ; Dumping toxic	
PE 8702	Wastes ; Food manufacturing industry	
PC 9053	Wastes ; Hazardous	
PE 2829	Wastes ; Inadequate facilities storing, transporting processing solid	
PE 8475	Wastes ; Inadequate facilities transport sanitary	
PE 8504	Wastes ; Liquid agricultural	
PD 3666	Wastes ; Marine dumping	
PD 0575	Wastes ; Mine	
PD 0177	Wastes pollutants ; Solid	
PC 1242	Wastes ; Radioactive	
PD 7424	*Wasting disease ; Manchester*	
PD 7841	*Wasting disease ; Vomiting*	
PE 4921	Wasting growth children	
PE 4921	*Wasting ; Muscular*	
PD 2632	*Wasting palsy*	
PD 0566	Wasting sickness	
PD 1533	Watching ; Habitual television	
PD 1533	Watching ; Late television	

Water [Abusive...]
- PD 8056 — Abusive restrictions use
- PD 1204 — access ; Difficult
- PD 1204 — access ; Limited
- PE 7837 — Adjacency saline
- PE 6685 — Water ; Bathing sewage contaminated
- PD 1175 — Water ; Biological contamination

Water borne
- PE 2787 — animal diseases
- PE 3401 — disease
- PE 3401 — diseases
- PD 9094 — pathogenic bacteria
- PD 1175 — *viral disease*

Water [catchments...]
- PE 8233 — *catchments ; Inadequate*
- PE 0535 — Chemical contamination
- PD 1204 — collection ; Arduous
- PE 7433 — Competition industry agriculture
- PF 1653 — *connection ; Expensive*
- PE 8233 — *conservation ; Lack*
- PD 2503 — Contaminated well
- PD 0235 — Contamination drinking
- PD 0482 — contamination ; Waste
- PD 8517 — control ; Inadequate
- PD 2586 — cycles ; Underreporting disruptions natural

Water [damage...]
- PE 2305 — *damage ; Tidal*
- PD 8517 — delivery ; Insufficient
- PE 8692 — development schemes ; Health hazards
- PD 8122 — Dirty

Water Discharge dangerous
- PD 2800 — substances domestic waste
- PD 0575 — substances industrial waste
- PE 9813 — substances laboratory waste

Water [Disease...]
- PD 1175 — *Disease causing microbes drinking*
- PE 7769 — *disease ; Red*
- PE 0516 — disputes ; River
- PD 1204 — Distant spring
- PF 8352 — divining
- PE 7433 — domestic uses ; Loss
- PF 1345 — *Water equipment ; Expensive*

Water [fish...]
- PE 1783 — fish, crustacea molluscs preparations thereof ; Long term shortage salt
- PE 3934 — fish farms ; Pollution
- PE 3634 — *flea*
- PE 7837 — flooding ; Salt
- PE 2871 — Fluoridation drinking
- PC 4815 — food production ; Shortage
- PE 5067 — fowl ; Endangered species
- PE 3815 — Water hyacinth

Water [Impurities...]
- PD 0482 — Impurities waste
- PE 7433 — industrial uses ; Loss
- PD 8056 — Inequitable right access
- PE 8545 — infected faeces ; Pollution
- PC 4341 — insecurity
- PE 7837 — intrusion ; Salt

Water [Lack...]
- PD 8122 — Lack clean
- PE 5650 — Lead contamination drinking
- PD 4403 — level ; Lowering
- PD 8888 — level ; Rising
- PD 9196 — lines ; Deficient
- PC 1173 — Long term shortage
- PE 8796 — looses irrigation canals

Water losses
- PE 8796 — dams
- PE 8796 — irrigation systems
- PE 9088 — irrigation systems ; Seepage

Water [Maldistribution...]
- PD 8056 — Maldistribution
- PE 8233 — management ; Poor
- PE 8233 — management ; Underdeveloped
- PD 1175 — *Water ; Nuisance organisms drinking*
- PD 8517 — Water outlets ; Unmarked public

Water pollutants
- PD 1175 — Biological
- PE 0535 — Chemical
- PD 2800 — Domestic waste
- PD 0575 — Industrial waste

	Water pollutants cont'd	
PE 9813	— Laboratory waste	
PE 0617	— Toxic organic compounds	
	Water pollution	
PC 0062	— [Water pollution]	
PE 3934	— animal production	
PD 1356	— Coastal	
PD 3675	— developing countries	
PE 8729	— fertilizers	
PE 8729	— pesticides	
PE 7837	— salinity effects	
PD 1096	— Transboundary	

Water [pressure...]
- PD 0722 — pressure ; Inadequate
- PD 0722 — pressure ; Insufficient
- PD 8056 — Privately owned
- PE 7832 — projects ; Diseases introduced

Water [Radioactive...]
- PE 2441 — Radioactive contamination
- PF 6138 — recreation ; Inaccessibility
- PD 4403 — reserves ; Misuse nonrenewable fossil

Water resources
- PD 7087 — Conflicting claims over shared inland
- PF 4767 — Obstacles utilization coastal deep sea
- PD 4403 — Overexploitation underground
- PC 4815 — Undeveloped
- PD 8888 — Water ; Rising level underground
- PE 8233 — *Water run off ; Uncontrollable*

Water [salinization...]
- PE 7837 — salinization
- PE 1159 — sanitary services ; Health risks workers electricity, gas,
- PF 2583 — sanitary services ; Underdevelopment electricity, gas,
- PC 4815 — Scarce fresh
- PD 8517 — services ; Inadequate
- PC 4815 — Shortage drinking
- PE 5464 — Shortage industrial
- PE 0199 — softness
- PD 2290 — Soil erosion

Water sources
- PD 1204 — Distant
- PE 6923 — Environmental degradation fresh
- PC 4815 — Shortage fresh

Water [spouts...]
- PD 1739 — spouts
- PE 2634 — Stagnant surface
- PE 2634 — Stagnant waste
- PE 8233 — *storage ; Inadequate*

Water supplies
- PE 8545 — human excreta ; Pollution
- PE 1294 — Inadequate facilities transport
- PD 1607 — Infection industrial
- PD 4403 — Misuse artesian
- PE 1458 — sabotage ; Contamination public

Water supply
- PD 1204 — Inconvenient
- PD 0722 — Instability
- PC 4815 — Insufficient fresh
- PC 4815 — Limited
- PD 1204 — rural communities developing countries ; Inadequate
- PD 0722 — Sporadic

Water supply systems
- PD 9196 — Deterioration
- PD 0482 — Fouling
- PD 9196 — Leakage

Water supply [Unpredictable...]
- PD 0722 — Unpredictable
- PD 8122 — Unsanitary
- PD 8122 — Untreated
- PD 0722 — Variable
- PD 5641 — Water surface pollution
- PE 0579 — Water ; Suspended matter

Water system
- PD 8122 — contamination
- PD 0722 — Erratic
- PD 8517 — infrastructure ; Inadequate
- PD 8122 — Unsanitary

Water systems
- PD 9670 — Disruption natural
- PD 9196 — Inadequate maintenance
- PD 8517 — Lack
- PD 3654 — Silting

Water [table...]
- PD 4403 — table ; Lowering
- PE 8233 — *tanks ; Lack*
- PE 4792 — torture
- PE 4792 — Tortures using

Water [Unavailability...]
- PD 8839 — Unavailability irrigation
- PD 8056 — Uneven distribution
- PE 8233 — Unregulated crop
- PD 8122 — Unsafe
- PE 8233 — *usage ; Ineffective*
- PF 4839 — use ; Lack accord
- PJ 0052 — Use recycled
- PE 9446 — Water vapour ; Increase concentration atmospheric

Water [Wasted...]
- PD 3669 — Wasted
- PD 2232 — weeds
- PE 4740 — wells ; Corrosion encrustation
- PE 3401 — Waterborne diseases
- PD 3499 — *Waterfront damage ; Uncontrolled*
- PE 9311 — Watering during irrigation ; Over
- PD 8888 — Waterlogging land
- PD 8517 — Watermains locations ; Inadequate

Whale narwhal

PJ 5279 Waterproofing ; Inadequate
PE 3035 Waters ; Cancer causing additives mineral
PD 1628 Waters ; Conflicting claims concerning off shore territorial
PE 6755 Waters ; Denial right innocent passage territorial
PD 3669 Waters ; Heated effluent
PE 6749 Waters ; Non uniformity marking navigable
PE 8175 Waters ; Pollution bays estuarine
PD 1223 Waters ; Pollution inland
Waters [Unilateral...]
PD 1628 — Unilateral claims off shore territorial
PE 6755 — Unlawful interference rights innocent passage territorial
PD 6923 — Unsustainable development fresh
PD 6282 Watershed deterioration
PD 6282 Watersheds ; Deforestation upland
PD 6282 Watersheds ; Destruction
PD 2487 Waterway transport facilities ; Inadequate inland
PB 0593 *Waterways ; Mined international*
PE 5801 Waterways ; Obstacles effective use oceans
PD 3978 *Watery mouth*
PE 8971 *Wave energy ; Underutilization*
PE 7837 *Waves ; Flood*
PD 0033 *Waves ; Seismic sea*
PD 0033 *Waves ; Tidal*
PE 8498 Waxes ; Shortage animal vegetable oils fats
PD 1283 *Way life ; Cosmically inharmonious*
PF 2644 Way life ; Delimiting family patterns traditional
PF 0836 *We / they language patterns*
PF 1737 Weak economic growth capitalist countries
PF 3394 *Weak government*
PF 2212 *Weak local initiatives*
PA 7325 Weak-mindedness *Irresolution*
PA 7371 Weak-mindedness *Unintelligence*
PE 8513 Weak national identity due tribalism
PE 8610 Weak national identity post-colonial countries
PF 0342 Weak programme follow-up
PF 5737 Weak social infrastructure islands
PF 2418 *Weakened heart*
PD 1229 Weakened initiative ; Welfare
PF 6992 Weakened international treaties ; Deliberately
PD 1380 *Weakening family ties*
PJ 5262 Weaker foreign economies ; Effect dominant stock markets
PA 5558 Weakness
PE 6816 Weakness ; Cardiac
Weakness [Danger...]
PA 6971 — *Danger*
PE 0269 — developing countries ; Housing infrastructural
PD 2463 — *dollar*
PA 6030 Weakness *Fear*
PF 3394 *Weakness governance*
Weakness [Imperfection...]
PA 6997 — *Imperfection*
PA 6876 — *Impotence*
PA 5806 — *Inaction*
PA 6882 — *Influencelessness*
Weakness infrastructure
PC 1228 — developing countries
PE 5772 — island developing countries
PE 7000 — land-locked developing countries
Weakness [international...]
PF 9175 — international organizations ; Intrinsic
PD 0169 — intra-regional trade developing countries
PA 7325 — *Irresolution*
PF 3214 Weakness multi-party parliamentary systems
PG 2295 Weakness ; *Muscular*
PE 2951 Weakness primary commodity trade amongst developing countries
Weakness [socio...]
PC 1059 — socio-economic infrastructure
PC 0933 — South - South inter-regional trade
PE 0628 — Student press
Weakness trade
PC 0933 — developing countries
PC 2724 — different economic systems
PE 2966 — manufactured goods developing countries
PE 2954 — socialist developed market economies
PE 2953 — socialist developing economies
Weakness [Unbelief...]
PA 7392 — *Unbelief*
PA 7371 — *Unintelligence*
PA 7204 — *Unsavouriness*
PA 5644 Weakness *Vice*
PA 5558 Weakness *Weakness*
PE 3831 Weaknesses media ; Excessive emphasis developing country
PD 0356 Weaknesses military conscription systems
PF 0510 Weaknesses national data sources
PC 5225 Wealth ; Accumulation
PE 6955 Wealth ; Denial right people freely dispose natural
PB 7666 Wealth ; Dependence inequitable distribution
Wealth developing
PD 5258 — countries ; Disparity distribution
PD 5258 — countries ; Maldistribution
PC 3134 — industrialized countries ; Transfer surplus
Wealth [eliminates...]
PF 5864 — eliminates problems ; Belief that
PB 8222 — Excessive
PD 4037 — Extravagant use
PF 4730 Wealth generation public good ; Lack relationship
PD 9653 Wealth government leaders ; Excessive accumulation
PB 7666 Wealth ; Inequitable distribution
PE 0503 Wealth ; Lack accountability disposal
PF 2870 Wealth ; Leadership symbolic
PB 7666 Wealth ; Maldistribution

PC 8222 Wealth ; Personal
PD 9653 Wealth rulers countries ; Hypocritical accumulation personal
PB 7666 Wealth ; Unequal distribution
PC 8222 Wealth ; Unnecessary personal
PF 6670 Wealthy elites developing countries ; Disproportionately
PE 0691 Weaning diet ; Inadequate
PD 9661 Weaning ; Improper infant
PD 9021 Weaning infants ; Ignorance women concerning
PC 0012 Weapon experimentation
PD 0837 Weapon free zones ; Insufficient nuclear
PF 6202 Weapon ; Misuse food political
PD 1521 Weapon states ; Insecurity vulnerability non nuclear
PC 4440 Weapon states ; Insecurity vulnerability nuclear
PC 6666 Weapon ; Use radiation
PF 7678 Weaponry ; Creeping modernization military
PE 7679 Weaponry ; Stealth
PD 0658 Weapons
Weapons [Accidental...]
PD 3493 — Accidental loss nuclear
PD 0087 — Anti satellite
PD 0087 — arms race ; Space
PC 0872 Weapons ; Binary chemical
Weapons [Chemical...]
PC 0872 — Chemical
PE 2449 — civilian hands ; Proliferation
PE 2449 — civilians ; Sale
PF 4450 — Clandestine space
PE 3968 — Clandestine trade nuclear
PC 0012 — Competitive development new
Weapons [Dependence...]
PC 6664 — Dependence genetic ethnic
PD 1519 — Dependence inhumane indiscriminate
PE 9052 — developing countries ; Proliferation nuclear
PF 4450 — development ; Secret
PF 9269 — Weapons ; Envisaged
PC 0311 Weapons ; Excessive number nuclear
Weapons [Genetic...]
PF 9269 — Genetic
PC 6664 — Genetic ethnic
PC 2696 — Geophysical
Weapons [Illegality...]
PF 4727 — Illegality nuclear
PC 5230 — Imbalance conventional
PE 5698 — industry ; Environmental hazards nuclear
PE 5698 — industry ; Violation safety regulations nuclear
PF 9269 — Infrasonic radiation
PD 1519 — Inhumane indiscriminate
PD 8621 — International trade biological
PD 9692 — International trade chemical
PE 7476 Weapons ; Lack appreciation nuclear
Weapons [manufacture...]
PE 5698 — manufacture ; Pollution nuclear
PD 7574 — Marine disposal obsolete
PD 3492 — massive destructiveness ; Incendiary
PF 9327 Weapons ; Non violent
PE 9346 Weapons ; Obsolescent
PF 9269 Weapons ; Planned
PC 6664 Weapons ; Race
PC 0012 Weapons research ; Nuclear
Weapons systems
PD 3493 — Accidents nuclear
PD 8412 — Maritime nuclear
PF 7801 — Unreliability
Weapons [technology...]
PD 0837 — technology ; Proliferation nuclear
PE 9207 — Terrorists armed biochemical
PE 3769 — Terrorists armed nuclear
Weapons testing
PF 4450 — Government secrecy concerning nuclear
PC 2201 — Nuclear
PC 2201 — Thermonuclear
PF 9269 — Weapons ; Transuranic
PE 4051 Weapons ; Unilateral structural disarmament nuclear
PE 7476 Weapons ; Unrecognized merits nuclear
PB 1701 Wear
PA 7365 Wear *Boredom*
PA 6858 Wear *Disintegration*
PA 6088 Wear *Impairment*
PA 7382 Wear *Loss*
PB 1701 Wear tear
PA 7365 Weariness *Boredom*
PA 5806 Weariness *Inaction*
PA 5558 Weariness *Weakness*
PE 1008 Wearing apparel industries ; Instability
PE 1103 Wearing apparel manufacture ; Environmental hazards
PA 7365 Wearisomeness *Boredom*
PA 5806 Wearisomeness *Inaction*
PA 6731 Wearisomeness *Solemnity*
PA 7107 Wearisomeness *Unpleasantness*
PE 8936 Weasels, badgers, skunks, otters ; Endangered species
PE 9134 Weather ; Adverse effects power production
PC 4987 Weather anomalies ; Large scale
PC 0293 Weather ; Bad
PC 0293 Weather conditions ; Adverse
PD 2740 Weather factor animal disease
PF 5118 Weather forecasting ; Inaccurate
PE 5682 Weather hazards crops plants
PD 4182 Weather hazards ; Underreporting
PE 5212 Weather-induced fluctuations food
PD 7941 Weather modification ; Hostile
PC 0293 Weather ; Risk bad
PE 3915 Weather ; Torture exposure
PF 5118 Weather ; Unpredictability
PE 5682 Weather ; Vulnerability crops

PC 0293 Weather ; Winter
PB 2253 Webs ; Food
PC 1874 Wedlock ; Children born out
PE 9122 Wedlock ; Criminalization sexual relations out
PF 5434 Wedlock ; Sexual intercourse out
Weed control
PD 3598 — developing countries ; Inadequate
PD 1574 — Inadequate
PF 2899 — Prohibitive cost
PE 5404 Weed creation genetic engineering ; Accidental
PE 1323 Weed ; Strangle
PD 1574 Weeds
PD 2232 Weeds ; Aquatic
PE 3987 Weeds ; Destruction
PE 4732 Weeds trees ; Pollens grasses,
PD 2232 Weeds ; Water
PD 7424 *Weidektankheit*
PF 5758 Weight phobia
PF 5970 Weights ; Low birth
PF 5670 Weights, measures numbering systems ; Terminological confusion
PE 2357 Weil's disease
PF 2575 *Welcoming process ; Unorganized*
Welfare benefits
PD 1229 — Dependence government
PE 8893 — Excessive bureaucratic requirements
PD 8859 — fraud
PF 8205 — Lack awareness available
PF 8205 — Unawareness
PD 8859 — Unethical use social
PD 8859 Welfare corruption ; Social
Welfare [Debilitating...]
PD 1229 — Debilitating effects social
PD 1229 — Dependence social
PD 1229 — dependence ; Unmotivating
PF 8855 — dependency ; Over commitment
PF 5120 Welfare ; Government procedures delaying
PE 8924 Welfare handicapped persons ; Neglected
Welfare Inadequate
PC 1167 — animal
PC 0233 — child
PE 4245 — infant
PE 5794 — legislation animal
PB 0883 — *protection individual*
PF 0019 Welfare indicators ; Inadequate social
PC 0233 Welfare institutions ; *Lack child*
Welfare Lack
PD 8837 — concern animal
PE 8341 — family
PE 5794 — legislation animal
PE 5794 Welfare legislation ; Lack animal
PE 8503 Welfare programmes ; Lack participation local
PF 2560 Welfare responsibility ; Individualistic
Welfare services
PD 0512 — aged ; Denial right
PD 0512 — aged ; Inadequate
PD 0542 — blind ; Denial right
PD 0542 — blind ; Inadequate
PD 0601 — deaf ; Denial right
PD 0601 — deaf ; Inadequate
PC 0834 — Denial right adequate
PE 4796 — developing countries ; Inadequate social security
PC 0834 — Inadequate social
PC 1506 — indigenous peoples ; Denial right social
PC 3437 Welfare state ; Deficiencies
PC 0834 Welfare system ; Regressive
PD 1229 Welfare weakened initiative
PD 1229 Welfarism ; Destructive
Well being
PC 2157 — because discrimination ; Denial right material
PE 7310 — because discrimination ; Denial right pursue spiritual
PF 1269 — Retirement threat psychological
PF 3448 Well paying jobs ; Scarcity
PF 1572 Well researched projects alleviate problems ; Lack
PE 2503 Well water ; Contaminated
PE 4740 Wells ; Corrosion encrustation water
PF 6527 *Wells ; Lack nearby*
PD 0201 Wells ; Waste disposal deep
PE 8450 Were wolves
PE 8450 Werewolves
PD 2730 Wesselsbron disease
PE 4111 West division Europe ; east
PE 2502 West ; Excessive debt socialist countries
PD 2730 West Nile fever
PE 7045 Western based secretariats international nongovernmental organizations ; Excess
PF 5472 *Western culture ; Excessive bias favour European*
PE 9374 *Western duck disease*
PE 4111 Western Europe ; split eastern
PE 4985 Western marriages ; Insecurity
PF 4871 Western medicine ; Lack integration traditional
PF 5472 *Western provincialism*
PF 7069 Western structures international nongovernmental organizations ; Neglect non
PF 8356 Western style organizations ; Preponderance
PE 1046 Western women ; Elimination socio cultural role
PF 6592 Westernization traditional modes life developing countries
PE 4767 Wet rot wood
PC 3486 Wetland environments ; Destruction
PE 5110 Wetlands ; Inadequate empolderment
PC 3486 Wetlands ; Vulnerability
PE 5431 Wetting ; Bed
PD 1593 Whale ; Endangered species
PD 1593 *Whale ; Endangered species grey*
PE 8187 Whale narwhal ; Endangered species white

Whales Endangered species
- PD 1593 — *beaked*
- PD 1593 — *right*
- PD 1593 — *sperm*
- PE 0439 Whaling
- PE 0439 Whaling ; Scientific
- PE 2903 Wheat ; Long term shortage

Wheat meslin
- PE 8980 — shortage ; Meal flour
- PE 8606 — shortage ; Meal flour than
- PE 8489 — trade instability ; Meal flour
- PE 2222 Wheat ; Pests diseases
- PE 2233 *Wheat ; Stem rust*
- PE 2902 Wheat surpluses ; Excessive
- PE 0385 Wheat trade ; Instability
- PJ 5588 Whimp
- PD 2626 *Whiplash injury*
- PE 7826 *Whirlbone lameness*
- PD 8926 Whistle-blowers
- PF 1585 Whistle blowers ; Harassment
- PE 5516 White-collar crime
- PE 8302 White dolphins ; Endangered species rough toothed
- PF 3455 White elephantiasis
- PF 7631 White lies
- PE 9774 *White muscle disease*
- PE 6255 *White pine blister rust*
- PE 5465 *White rusts crucifers*
- PE 5465 *White rusts plants*
- PD 3303 White slave trade
- PE 8187 White whale narwhal ; Endangered species
- PF 6471 *Wholesale systems ; Unutilized*
- PE 2481 Whooping cough
- PF 2099 Wicca craft
- PE 4531 *Widowers ; Inefficient support*
- PD 0488 Widowhood
- PF 6149 Widows ; Burden society
- PF 4819 Widows ; Burning
- PD 2092 Widows orphans ; Neglect war
- PD 6758 Wife abuse
- PD 6758 Wife beating
- PD 0518 Wife ; Enmity husband
- PD 0518 Wife ; Fighting husband
- PD 3555 Wife husband ; Dependence
- PF 2314 *Wife-swopping*

Wild animal
- PD 3155 — natural prey ; Extermination
- PE 7995 — populations ; Abuse control
- PC 0475 — range size ; Restriction
- PD 2419 — species ; Hybridization

Wild animals
- PC 3151 — Aggressive
- PE 9774 — *Capture myopathy*
- PD 2729 — carriers animal diseases
- PD 1481 — Commercial exploitation
- PC 3151 — Dangerous
- PD 2776 — Difficulty controlling disease
- PD 2776 — Diseases
- PD 1179 — Domestication
- PE 4694 — during disasters ; Lack protection
- PC 1445 — Extermination
- PD 8904 — Misuse
- PE 6024 — Over population
- PD 1179 — pets ; Capture use
- PC 3150 — Ravaging
- PD 1179 — Taming
- PE 2749 Wild birds vectors animal diseases
- PB 6336 *Wild life ; Pollution hazards*
- PD 5384 Wildcat strikes

Wilderness areas
- PD 7585 — campers ; Degradation
- PB 5250 — Destruction
- PF 9360 — Inaccessible
- PD 7585 — Over use designated
- PD 0739 Wildfires
- PD 2682 Wilding
- PD 0739 Wildland fires

Wildlife Accumulation pollutants
- PD 4925 — freshwater
- PD 3934 — marine
- PD 5278 — terrestrial
- PE 3008 *Wildlife ; Buzzing*
- PC 3150 Wildlife ; Crop damage
- PC 0480 Wildlife decisions ; Unfavourable
- PE 7794 Wildlife due artificial flooding ; Loss
- PC 1445 Wildlife extinction
- PC 1445 Wildlife extinction ; Dependence
- PC 0475 Wildlife ; Forcible displacement
- PD 0500 Wildlife ; Forest damage

Wildlife habitats
- PE 7794 — dams ; Inundation
- PC 0480 — Dependence destruction
- PC 0480 — Destruction
- PC 3151 Wildlife ; Harmful
- PE 9228 Wildlife ; Lead poisoning

Wildlife [Pesticide...]
- PD 3680 — Pesticide hazards
- PB 6336 — *pollution hazard*
- PB 6336 — *Pollution hazards*
- PC 0380 Wildlife ; Smuggling protected
- PC 0380 Wildlife traffic
- PC 3151 Wildlife ; Violent
- PF 8278 Wilful ignorance government
- PE 5758 Wilful starvation
- PE 3809 Wilful violation regulatory codes
- PC 5180 Will act problems ; Lack political
- PC 5180 Will cooperate ; Lack political
- PA 7102 Will ; Ill
- PC 5180 Will ; Lack political
- PF 7838 Will observe trade commitments ; Failure political
- PC 5180 Will respond needs developing countries ; Lack political
- PA 7250 Willfulness *Disobedience*
- PA 7325 Willfulness *Irresolution*
- PA 5563 Willfulness *Lawlessness*
- PE 6853 Wills ; Inadequate registration
- PD 2557 *Wills testaments ; Forgery*
- PC 2270 *Wilson's disease*
- PD 2226 Wilt cucumber ; Bacterial
- PD 2226 Wilt cucurbits ; Bacterial
- PE 1056 Wilt diseases plants
- PE 4346 Wilt plants ; Verticillium
- PG 5461 *Wilt ; Tomato*
- PG 5461 *Wilt tomato ; Fusarium*
- PE 2223 Wind
- PD 1245 Wind blown dirt ; Excessive
- PE 2223 Wind-borne animal diseases
- PE 1334 Wind damage structures
- PF 0373 Wind energy ; Underutilization
- PE 6354 *Wind ; Irrational fear*

Wind [shear...]
- PE 5085 — shear hazard
- PE 3656 — Soil erosion
- PE 2223 — storms
- PC 2618 Windfall profits
- PE 7826 Windgalls
- PD 1350 Window-breaking
- PG 6761 Windpipe ; Swelling mucous membrane larynx
- PE 7826 Windpuffs
- PE 2522 Wine trade ; Instability
- PE 3604 Winged bats ; Endangered species disc
- PE 3604 Winged bats ; Endangered species sac
- PE 6350 Winter chill ; Nuclear war induced
- PE 0258 Winter depression
- PD 3978 *Winter dysentery ; Bovine*
- PE 5324 Winter ; Fear nuclear
- PD 3978 *Winter scours*
- PF 0963 *Winter sociability ; Hindered*
- PC 0293 Winter weather
- PD 1632 Wiretapping
- PF 3645 Wireworms pests
- PF 1664 Wisdom aged ; Non transferring
- PF 8753 Wisdom different cultures response social problems ; Failure interrelate

Wisdom [elders...]
- PF 1664 — elders society ; Isolation
- PF 1664 — elders ; Uncommunicated
- PF 1664 — Erosion elders'
- PF 1746 Wisdom past ; Failure opposing groups appropriate

Wisdom [Underused...]
- PF 1664 — Underused elders'
- PF 2426 — Underutilization popular
- PF 5728 — Unrecorded
- PF 2426 — Unrecorded popular
- PD 1703 — unrelated daily life ; Human
- PF 2426 — Unshared practical
- PF 2426 — Untapped community
- PF 8922 Wishful thinking
- PF 7427 Wit ; Obscene
- PF 2099 Witch doctors
- PD 7885 Witch hunt
- PE 1046 Witch hunting ; Long term historical effect
- PF 2099 Witchcraft
- PF 7887 Witchcraft ceremonies
- PF 7887 Witches' sabbaths
- PE 6604 Witchweed
- PF 4402 Withdrawal

Withdrawal [aged...]
- PD 3518 — aged ; Social
- PE 1558 — *Amphetamine*
- PE 0139 — *Anxiolytic*
- PA 6379 — *Avoidance*
- PD 2363 Withdrawal ; Cocaine
- PF 5502 Withdrawal commitment action ; Governmental delayed

Withdrawal [delirium...]
- PE 9263 — *delirium ; Alcohol*
- PA 6838 — *Dissent*
- PA 7339 — *Duality*
- PA 5869 Withdrawal *Exclusion*
- PE 0139 Withdrawal ; *Hypnotic*
- PA 6598 Withdrawal *Incuriosity*
- PA 7325 Withdrawal *Irresolution*
- PE 9253 Withdrawal ; *Nicotine*
- PF 9699 Withdrawal official documents public scrutiny
- PE 1329 Withdrawal ; *Opioid*
- PF 0421 Withdrawal ; *Psychological*
- PA 6338 Withdrawal *Regression*
- PE 0139 Withdrawal ; *Sedative*
- PF 0375 Withdrawal ; Uncomplicated alcohol
- PA 7364 Withdrawal *Unfeelingness*
- PA 6542 Withering *Destruction*
- PA 6822 Withering *Disrespect*
- PA 6088 Withering *Impairment*
- PA 5643 Withering *Unkindness*
- PA 6555 *Withers ; Fistulous*
- PF 9239 Withholding geographical information

Withholding information
- PF 8536 — [Withholding information]
- PF 8536 — government
- PE 4183 — police criminal suspects
- PD 9836 Withholding medical information
- PF 8650 Withholding membership payments intergovernmental organizations

Witlessness [Ignorance...]
- PA 5568 — *Ignorance*
- PA 6247 — *Inattention*
- PA 7157 — *Insanity*
- PA 7371 Witlessness *Unintelligence*
- PF 5127 Witness ; False
- PE 1618 Witness ; Inability bear
- PE 3364 Witness ; Lack common religious
- PE 1756 Witness produce information be sworn ; Failure appear
- PE 1255 Witness torture ; Forced
- PE 7489 Witnesses ; Denial right examine
- PF 6012 Witnesses ; Disagreement expert
- PE 6781 Witnesses informants proceedings ; Tampering
- PB 1992 Witnesses ; Intimidation
- PD 1030 Wives ; Abandoned
- PD 1030 Wives ; Deserted
- PF 4764 Wives ; Disobedient
- PE 9140 Wives government leaders ; Corruption
- PA 5454 Woe *Badness*
- PA 6731 Woe *Solemnity*
- PA 7107 Woe *Unpleasantness*
- PE 0185 Wolf children
- PE 8739 Wolves foxes ; Endangered species dog
- PE 8450 Wolves ; Were
- PE 7062 Woman-hating
- PF 3690 Woman power ; Wasted
- PF 6169 Woman's right abortion ; Denial
- PE 9286 *Womb ; Inflammation*
- PD 1762 *Wombats ; Endangered species*

Women [Abduction...]
- PC 3298 — Abduction
- PD 3557 — Adverse social consequences excessive employment married
- PE 6784 — Ageing
- PE 7672 — Alcohol related violence against
- PE 7161 — alcohol ; Susceptibility
- PD 8638 — Architecture insensitive needs
- PE 4242 — athletic competition ; Exclusion
- PE 4246 — athletic training ; Discrimination against

Women [Battered...]
- PD 6758 — Battered
- PF 8751 — Belief emotional instability
- PF 6326 — Bias against ordination
- PD 4633 — Biological subjugation
- PD 9628 — business ; Discrimination against
- PD 4633 — Women child rearing ; Subjugation

Women children
- PE 8788 — Discrimination against
- PD 1078 — during disasters ; Lack protection
- PD 1078 — emergencies ; Vulnerability
- PE 8220 Women combatants

Women concerning
- PD 9021 — appropriate child care ; Ignorance
- PD 9021 — child bearing ; Ignorance
- PD 9021 — infant nutrition ; Ignorance
- PD 9021 — primary health care ; Ignorance
- PD 9021 — weaning infants ; Ignorance

Women [Conflicting...]
- PD 6273 — Conflicting roles
- PF 1257 — *cooperatives ; Insufficient*
- PF 4959 — countryside ; Limited role
- PD 0247 — Cruelty
- PE 9009 Women decision making ; Exclusion

Women Denial right
- PD 3335 — access public service
- PD 0190 — education
- PD 0086 — employment
- PD 0309 — equal pay
- PC 0308 — equality
- PD 0127 — freedom religion
- PD 0162 — justice
- PE 4016 — nationality
- PC 1001 — participate government
- PE 4018 — property

Women [Denial...]
- PE 4019 — Denial rights parent
- PE 1085 — Denial sufficient nutrition
- PC 3426 — Dependency

Women developing countries
- PC 4898 — Discrimination against
- PE 8660 — Illiteracy
- PE 9185 — Nutritional anaemia

Women [developing...]
- PE 8139 — developing societies ; Traditional values subordination
- PF 4959 — development ; Lack participation
- PC 0729 — Disabled

Women Discrimination against
- PC 0308 — [Women ; Discrimination against]
- PE 6245 — black working
- PE 4947 — rural
- PD 8622 — unmarried

Women [Discriminatory...]
- PD 9628 — Discriminatory requirement over qualification
- PE 8879 — divorce rights ; Discrimination against
- PD 8546 — Domestic deskilling
- PD 4633 — domestic service ; Subjugation
- PD 6273 — Dual roles

Women during
- PD 1078 — armed conflicts ; Vulnerability
- PD 1078 — foreign military occupation ; Vulnerability
- PE 1085 — pregnancy nursing ; Protein energy malnutrition

Women [education...]
- PD 0190 — education ; Discrimination against
- PD 0190 — education ; Unequal access
- PE 1046 — Elimination socio cultural role western

Workers' organization

Women [employment...] cont'd
PD 0086 — employment ; Discrimination against
PD 3557 — Excessive employment married
PD 9628 — executives ; Discrimination against
PC 9733 — Exploitation
Women [Failure...]
PD 8546 — Failure employ skills home bound educated
PE 7206 — family responsibilities ; Discrimination employment
PF 6024 — farmers ; Neglect
PF 6024 — farmers ; Unadapted vocational training
PE 7750 — Fear violence
PE 8343 — foetuses ; Medical experimentation pregnant
Women [Health...]
PE 4995 — Health hazards smoking
PE 8681 — home ; Social isolation
PF 3211 — *House bound*
Women [Illiteracy...]
PE 4380 — Illiteracy
PE 4380 — Illiteracy
PE 8699 — immigrants ; Lack education
PE 4624 — Imposition husband's name married
PE 5607 — Improperly veiled
PE 4820 — Inadequate medical care pregnant
PD 3197 — Inadequate working conditions
PC 0592 — *instructors ; Shortage*
PE 8907 — Insufficient leisure time
PE 5316 Women ; Job insecurity pregnant
PD 7821 Women labour ; Abusive treatment
Women Lack
PE 9009 — participation
PD 0190 — training
PE 8345 — unionization working
Women [law...]
PD 0162 — law ; Discrimination against
PE 4452 — lift heavy loads ; Work practices requiring
PE 8720 — Loss civil capacity married
PE 6784 — Loss sex appeal
Women [marriage...]
PD 3694 — marriage ; Dependency
PE 7638 — mass media ; Biased portrayal
PE 7638 — media ; Exploitative portrayal
PE 8220 — military forces ; Discrimination against
PE 3390 — Military prostitution
Women [Negative...]
PE 6784 — Negative self image ageing
PD 0190 — Neglect education
PF 5147 — Neglect sexual health
PD 9628 — non manual workers ; Discrimination against
Women [offenders...]
PE 1837 — offenders
PF 3462 — *Overqualification*
PD 7762 — Overworked
Women [parental...]
PE 4019 — parental rights ; Discrimination against
PE 0197 — payment prizes athletic events ; Discrimination against
PE 7750 — Personal physical insecurity
PC 1001 — politics ; Discrimination against
PE 7161 — *Poor physiological adaptability*
PF 6326 — priesthood ; Religious discrimination against
PF 3037 — *Prohibited travel*
PD 3335 — public life ; Discrimination against
PE 0756 — public ; Verbal sexual harassment
Women [refugees...]
PD 5025 — refugees ; Exploitation
PD 0127 — religion ; Discrimination against
PD 3443 — *Reservation jobs*
PE 6069 — retirement age ; Discrimination against
PE 4016 — right nationality ; Discrimination against
PE 7750 — Risk violence
PF 4959 — rural development ; Neglect role
PE 1198 Women's access judicial process ; Economic barriers
PE 3296 Women's behaviour ; Dominated
PF 1705 Women's groups ; Insufficient
PF 1705 Women's groups ; Uninitiated
PF 3025 Women's liberation
PC 3298 Women ; Sale
PD 0086 Women science technology ; Underrepresentation
Women Sexual
PJ 7167 — *abuse*
PD 3262 — exploitation
PF 3271 — harassment
PE 1293 — provocation
Women [Shortage...]
PE 0427 — Shortage marriageable
PD 3691 — social services ; Discrimination against
PD 4633 — Social subjugation
PD 2872 — socialist countries ; Denial right work
PD 2872 — socialist countries ; Discrimination against working
PF 8780 — Socializing unrelated men
PF 1696 — *Socially isolated*
PE 8220 — soldiers
PE 0197 — sports ; Discrimination against
PE 4016 — Statelessness
PC 3298 Women ; Trafficking
PF 2303 *Women ; Unemployment married*
Women Unequal
PE 7942 — distribution old age pensions men
PD 5115 — employment opportunities
PE 6835 — health benefits
PD 0309 — pay
PD 9628 — promotion
PE 4018 — property rights
PD 8622 Women ; Unmarried
PD 0247 Women ; Violence against

PE 4102 Women workers multinational enterprises developing countries ; Discrimination against
PE 0886 Wood-boring beetles
PE 8119 Wood charcoal trade instability ; Fuel
PE 8091 Wood cork manufactures ; Environmental hazards
PD 2301 Wood deterioration decay
PE 1606 Wood ; Dry rot
PE 4769 Wood fuel ; Shortage
PE 5362 Wood industry ; Unemployment
PD 3586 Wood ; Insect pests
PE 2521 Wood, lumber cork ; Instability trade
PE 1372 Wood, lumber cork ; Long term shortage
PE 6867 Wood ; Poisonous, allergenic biologically active
PE 4455 Wood rots
PD 4769 Wood rough ; Shortage
PE 7920 Wood shaped simply worked shortage
PE 4767 Wood ; Wet rot
PC 1366 Woodlands ; Destruction
Woodworking industries
PE 0864 — Environmental hazards
PE 0681 — Instability
PF 2604 — Underdevelopment
PE 0886 Woodworm
PD 9667 *Wool ; Lumpy*
PD 9667 *Wool maggots*
PE 2056 Wool trade ; Instability
PF 6389 Wooling ; Cotton
PE 2293 *Woolsorters' disease*
PF 5247 Words ; Misappropriation
PF 3128 Words terms ; Politically emotive
Work [Accidents...]
PC 0646 — Accidents
PE 1693 — age ; Denial right minimum
PE 4243 — agricultural workers ; Inadequate conditions
PE 2033 — Alcoholic intoxication
PD 3076 — Alienation
PE 1277 — Atomization computer based
PF 6477 — *Attractiveness overseas*
PC 5528 — Avoidance
PE 5132 — awareness international affairs ; Insufficient
PE 1625 Work benefits aged ; Violation right equal
PA 0657 *Work ; Boring*
Work [capitalist...]
PC 3119 — capitalist systems ; Denial right
PF 2303 — *changes ; Insecurity*
PD 0694 — Collective stoppages
PD 8047 — commitment developing countries ; Lack
PD 2790 — commitment ; Lack
PD 9170 — conditions ; Denial right favourable
PE 6841 — construction industry ; Inadequate conditions
PD 2656 — Dehumanization
PF 2644 *Work demands ; Exhausting*
Work Denial right
PC 5281 — [Work ; Denial right]
PE 1977 — equal pay equal
PE 3963 — free choice
Work [Dependence...]
PE 4789 — Dependence unpaid household
PE 9199 — developing countries ; Underpayment
PE 8369 — Difficulty transition school
PE 0783 — Discrimination against handicapped
PE 6241 — Discrimination against part time
PD 4514 — Drug abuse
Work [employment...]
PE 8728 — employment public service personnel ; Inadequate conditions
PE 8319 — employment utility supply services ; Inadequate conditions
PE 6811 — environment ; Cardiac conditions
PD 9170 — environment ; Poor
PD 8047 — ethic developing countries ; Undeveloped
PD 2790 — ethic ; Erosion
PC 5576 — Evasion
PD 0140 — Excessive hours
PE 6241 — Exploitative part time
PE 5145 Work family needs ; Inadaptation
Work force
PD 0164 — Child
PE 0893 — needs ; Disrelationship production
PD 9113 — Unskilled
PF 2128 Work forces ; Unorganized development
Work [Hazards...]
PE 6868 — Hazards young people
PE 5720 — Heat stress
PE 4493 — hotel catering industries ; Inadequate conditions
PD 0140 — hours ; Denial right reasonable
Work [Images...]
PF 2896 — images ; Parochial
PE 5768 — Inability adapt shift
PE 2308 — *Incapacity*
PD 0024 — incentives ; Inadequate
PC 0646 — injuries
PE 5083 — Work ; Job stress video display terminal
Work [Lack...]
PD 7650 — Lack discipline
PF 5744 — Lack warmth
PF 1403 — Limited opportunities significant
PF 3676 — Loss significance
PF 7723 — Low quality construction
PD 2656 Work ; Monotonous repetitious
PE 7589 Work ; Night
Work [Pace...]
PF 2304 — Pace duration
PF 2971 — *patterns ; Irregular*
PC 1108 — patterns ; Seasonal
PD 3529 — *permits ; Refusal*

Work [place...] cont'd
PE 8972 — place ; Hazardous combination effects
PE 1106 — place thefts ; Office
PE 8916 — places ; Distant
PD 0713 — *places ; Smoking*
PE 4452 — practices requiring women lift heavy loads
PD 6627 — *premises ; Badly laid out*
PF 3523 — *Priority farm*
Work [Reduced...]
PF 6098 — Reduced images having significant life
PF 6537 — Reduced images having significant life
PE 3751 — refugees ; Denial right
Work [schedule...]
PE 5074 — schedule due computerization ; Disruption
PF 2316 — *schedule ; Unpredictable*
PC 1108 — Seasonal fluctuations
PD 7723 — Shoddy construction
PD 8052 — Sleepiness
PD 8052 — Sleeping during
PE 4203 — society ; Collapsed relationships
PD 3427 — status immigrants industrialized countries ; Menial
PE 6937 — Stress
PE 5768 — stress occupational hazard ; Shift
PD 3548 — study programmes ; Outmoded
PD 9805 — Substance abuse
PE 5823 Work textile industry ; Inadequate conditions
Work [Underpayment...]
PD 8916 — Underpayment
PE 4789 — Unpaid household
PD 2790 — Unwillingness
PD 2872 Work women socialist countries ; Denial right
PE 7920 Worked shortage ; Wood shaped simply
PD 3245 Worker benefits ; Economic bias
PD 5244 Worker disobedience
PD 2790 Worker loyalty corporations ; Lack
PC 7041 Worker maladjustment technology
PF 1262 Worker organizations ; Ineffective
Worker [participation...]
PF 0574 — participation business decision making ; Lack
PE 5074 — participation excluded company polity
PC 6031 — productivity ; Declining
PD 0610 — Worker's representatives ; Intimidation
Workers [Abusive...]
PD 2722 — Abusive traffic immigrant
PD 5265 — Accidents agricultural
PE 0524 — agricultural livestock production ; Health risks
PE 3594 — alienation socialism
PD 0568 *Workers ; Blind*
Workers [commerce...]
PE 0688 — commerce ; Health risks
PD 5586 — Complaints
PE 0526 — construction industry ; Health risks
PD 4334 — Corruption
PE 5170 — countryside ; Absence skilled
PD 9397 — Criminal
Workers Denial
PE 1625 — equal benefits elderly
PE 3910 — right union activity rural
PD 0973 — rights migrant
Workers [Deterioration...]
PE 9075 — Deterioration living standards
PE 6843 — developed countries ; Displacement
PE 5583 — developing countries ; Low productivity agricultural
PD 4673 — Disabled
PE 0769 — Disabled migrant
PD 2790 — Discouraged
Workers Discrimination
PD 6019 — against non union
PD 9628 — against women non manual
PE 4934 — employment against immigrant
Workers [electricity...]
PE 1159 — electricity, gas, water sanitary services ; Health risks
PE 5506 — employed forced labour ; Unfair competition
PD 6930 — Exploitation casual
Workers [families...]
PE 8423 — families ; Refusal entry foreign
PF 9234 — Fear humiliation co
PE 7620 — following legal strike action ; Dismissal
PD 4695 — freedom association ; Violation right
PE 4658 — Workers ; Ghost
PE 6240 — Workers home ; Disadvantages
PD 4506 — Workers ; Ignorance
PD 8007 — Workers improve profitability ; Dismissal
Workers Inadequate
PE 4243 — conditions work agricultural
PE 4179 — sense community solidarity amongst
PE 4892 — utilization volunteer social service
PE 4243 — working conditions agricultural
PE 4243 — working conditions plantation
Workers [Incompetent...]
PD 4535 — Incompetent
PE 4804 — Insufficient manually skilled
PD 4334 — Irresponsible
PF 5737 — island developing countries ; Lack skilled
PE 5192 — Workers join trade unions ; Violation right
Workers [manufacturing...]
PE 1605 — manufacturing industries ; Health risks
PE 2007 — meetings ; Military intervention
PE 2007 — meetings ; Violent police intervention
PE 4102 — multinational enterprises developing countries ; Discrimination against women
Workers Occupational hazards
PE 6902 — female
PE 6904 — male
PE 6868 — youth
PD 7471 Workers' organization leadership ; Harassment

Workers organizations
PE 4071 — affiliate internationally ; Violation right
PE 3970 — bargain collectively ; Violation right
PE 4071 — establish confederations ; Violation right
PE 8384 — Interference employers affairs
PE 1758 — Interference public authorities functioning
PE 5793 — protection against suspension ; Violation right
PE 3970 — Restrictions collective bargaining
Workers [Phantom...]
PE 4658 — Phantom
PE 7620 — prevent legal strike action ; Dismissal
PE 0757 — prevent strike action ; Requisitioning
PD 5586 — Protest
PD 8007 Workers ; Redundancy
Workers representatives
PE 0252 — Assassination
PE 4869 — Death threats against
PD 7630 — Detention
PE 5882 — Forced disappearances
PD 1846 — Inadequate protection facilities
PD 7471 — Intimidation
PD 1846 — Victimization
Workers [Sanctions...]
PD 0610 — Sanctions against trade union
PE 0875 — service industries ; Health risks
PD 0044 — Shortage skilled
PC 1131 — skills job requirements ; Disparity
PE 7528 — small industries ; Occupational hazards
Workers [their...]
PD 0973 — their families ; Discrimination against migrant
PD 2722 — Traffic immigrant
PE 5884 — transport sectors land locked developing countries ; Lack skilled
PE 1581 — transport, storage communication industries ; Health risks
Workers [Underemployment...]
PB 0750 — *Underemployment skilled*
PE 2987 — Uninterested
PD 5965 — Unjust dismissal
PC 6031 — Unproductive
PE 6833 Workers ; Vulnerability frontier
PE 4003 Workers wage increases provided legislation collective agreements ; Non implementation
PC 5017 Workers ; Xenophobia regard migrant
PE 7515 Workforce enterprises ; Reduction
Working [abroad...]
PE 1659 — abroad ; Restrictions socialist citizens
PE 6427 — animals limitation working hours ; Denial
PE 4793 — animals restorative nourishment rest ; Denial
PE 0653 Working capacity ; *Debilitated*
Working conditions
PD 9170 — Adverse
PE 4243 — agricultural workers ; Inadequate
PE 3969 — construction industry ; Inadequate
PD 1476 — developing countries ; Inadequate
PE 8251 — employees commerce offices ; Inadequate
PE 7718 — health medical services ; Inadequate
PD 3427 — immigrant labourers industrialized countries ; Inadequate living
PE 8633 — land tenants ; Inadequate
PE 4243 — peasant farmers ; Inadequate
PE 4243 — plantation workers ; Inadequate
PD 9170 — Poor
PE 3170 — professionals ; Inadequate
PE 7165 — teachers ; Inadequate
PD 3197 — women ; Inadequate
Working environment
PD 2831 — Noise
PD 9170 — Poor
PD 6133 — Sterile
PE 1145 — Vibration
Working hours
PE 3044 — Denial right periods nurse infants during
PE 6427 — Denial working animals limitation
PE 7185 — inappropriate structural technological changes
PF 2971 — *Irregular*
PD 9170 Working life ; International imbalance quality
PD 6812 Working mothers ; Discrimination against
Working [parents...]
PF 2810 — *parents ; Uninvolved*
PE 3337 — *patterns ; Customary*
PE 8466 — *place ; Sexual harassment*
PC 5528 Working rule
PD 0140 *Working time ; Reduction*
PD 0140 *Working time ; Short*
Working women
PE 6245 — Discrimination against black
PE 8345 — Lack unionization
PD 2872 — socialist countries ; Discrimination against
PD 1435 — Workmanship ; Poor
PD 1435 — Workmanship ; Shoddy
PD 0044 — Workmen ; Insufficient skilled
PF 6122 — Workplace ; Artificial separation home
PD 2831 — Workplaces ; Insufficient acoustical treatment
Works art
PE 0558 — Debasement
PE 6252 — Dismembered
PE 9004 — Illicit export
PE 8088 — Inadequate documentation
PE 0323 — Looting
PE 0323 — Theft
PD 4143 *Works ; Collapsing public*
PA 3455 *Works ; Grandiose public*
PD 0579 World affairs ; Government loss leadership role
PF 2071 World anarchy
PF 0048 *World banking facilities ; Underutilization*

World [cable...]
PD 0407 — cable communications ; Vulnerability
PD 2043 — calendar ; Inadequate
PF 1471 — commodity exports ; Declining share developing countries
World [economic...]
PF 0817 — economic crisis ; Nationalistic policy responses
PC 9112 — economic system ; Inadequate
PF 1739 — economical systems ; Interaction deficiencies
World economy
PC 0002 — Declining growth
PD 7727 — Deflation
PC 8073 — Economic financial instability
PF 9664 — Faltering structural adjustment
PD 2463 — US dollar dominance
World [End...]
PF 4528 — End
PC 2566 — exports ; Reduction share developing countries
PF 3628 — exposure ; Limited
PF 2088 World federalism
World food
PD 5046 — crisis
PD 5046 — economy ; Imbalance
PF 5137 — security system ; Lack
PB 0315 — supply ; Massive starvation despite sufficient
World [genetic...]
PB 4788 — genetic resources ; Vulnerability
PD 3984 — geography ; Ignorance
PF 4937 — government ; Lack
World [Illiteracy...]
PD 6645 — Illiteracy fourth
PD 0398 — Inadequate international map
PF 0053 — interests ; Lack autonomous world level actor identify clarify
World [leaders...]
PF 9843 — leaders ; Antipathic personal relations
PF 0053 — level actor identify clarify world interests ; Lack autonomous
PF 8704 — level ; Lack sense community solidarity
PF 2342 — liquidity ; Inadequate
World [maritime...]
PE 5801 — maritime integration ; Lack
PF 1471 — markets ; Restricted access developing country products
PC 7873 — monetary financial conditions ; Fluctuations
PF 3059 — monetary reserves ; Inadequate level
PE 1570 — monkeys ; *Endangered species new*
PE 8991 — monkeys ; *Endangered species old*
PF 7967 — Mono factor explanations
PF 0292 World opinion ; Nation states' disregard
PD 0055 World order ; Nation states obstacles
PF 7967 — World perspective ; Narrow
World population
PC 0027 — Ageing
PF 0167 — Dependence maldistribution
PF 0167 — Maldistribution
PE 8845 — World porcupine ; *Endangered species old*
PE 8847 — World porcupines ; *Endangered species new*
World problems
PF 0364 — Complex interrelationship
PF 0817 — Inadequate global cooperation solve
PF 1572 — Lack innovative projects against
PF 7027 — Recurrence misapprehended
World [regulatory...]
PF 4937 — regulatory agencies ; Inadequate
PF 3387 — religion ; Lack
PF 3387 — religion ; Need
PA 5984 — resources ; Strain
PF 9478 — World's future ; Unimaginative vision
PF 2709 — World's need ; Decisional paralysis specialized services relation
World [social...]
PF 1443 — social systems ; Uncontrollability
PF 0866 — socio economic order ; Antiquated
PF 3199 — spheres influence ; Division
PD 2374 — system ; Plight small states
World [telecommunications...]
PD 5701 — telecommunications capabilities ; Disparity
PE 8080 — tension oil petroleum crises
PD 0055 — territorially organized sovereign states ; Domination
World trade
PC 7873 — Financial destabilization
PE 9107 — Informational procedural obstacles
PC 4890 — Obstacles
PE 0135 World trading system ; Emerging oligopolistic
PF 2434 World unity ; Lack
PF 9478 World vision ; Lack
PF 2865 World wide capital flow plan ; Absence long range,
PF 8922 Worldview ; Imposing fictional
PB 6719 Worldwide misallocation resources
PE 8150 Worm flies pests ; Screw
PE 3510 Worm ; Guinea
Worm infection
PD 9293 — dog ; Giant kidney
PD 9293 — mink ; Giant kidney
PD 9293 — Swine kidney
PE 0958 Worm infections swine ; *Stomach*
PE 3160 Worms ; *Endangered species marine*
PD 9667 Worms ; *Fleece*
PE 3644 Worms pests ; Army
PD 3160 Worms ; Segmented
PD 2750 Worms vectors animal diseases
PA 7499 Worry *Difficulty*
PA 6030 Worry *Fear*
PA 7107 Worry *Unpleasantness*
PD 2315 Worship ; Ancestor

PD 2330 Worship barrier development ; Animal
PE 6795 Worship ; Closure places
PF 8905 Worship ; Decreasing participation collective religious
PF 8260 Worship ; Devil
PF 2650 Worship ; Hero
Worship [Idolatrous...]
PF 3374 — Idolatrous
PE 6795 — Inaccessible places
PE 4332 — Inappropriate transfer forms
PF 1115 — Incompatible forms
PD 5105 Worship ; Restrictions freedom
PD 4365 Worshippers ; Prejudice against non
PF 3512 Worth ; Lack community self
PF 1663 Worth ; Social reinforcement delimited self
PA 5454 Worthlessness *Badness*
PA 6852 Worthlessness *Inappropriateness*
PA 7395 Worthlessness *Inexpedience*
PA 5942 Worthlessness *Unimportance*
PA 5454 Wound *Badness*
PA 6799 Wound *Disease*
PA 7253 Wound *Envy*
PA 6088 Wound *Impairment*
PA 5451 Wound *Insensibility*
PA 7107 Wound *Unpleasantness*
PE 4758 Wounded military personnel ; Denial rights
PB 0855 Wounded people
PA 4078 Wounding ; Torture
PB 0855 Wounds
PA 4247 Wounds animals
PE 7826 Wounds foot ; Puncture
PE 9111 Wounds ; Gunshot
PF 7511 Wounds ; Scalp
PF 8090 Wounds ; Unreported
PA 5532 Wrangle *Disaccord*
PA 5502 Wrangle *Reason*
PF 8563 Wrath God
PE 5340 Wrecks derelicts hazards
PB 0477 Wrinkles ; Facial
PD 0270 Writer's cramp
PA 2837 Writing ; Complex curriculum
PA 0330 Writing disorder ; Developmental expressive
PC 2414 Writings ; *Seditious*
PA 0911 Wrong belief
PD 9022 Wrongful acts
PD 6967 Wrongful death ; Prenatal
PD 6062 Wrongful detention
PD 9022 Wrongful omissions
PA 7280 Wrongness
PA 5454 Wrongness *Badness*
PA 6180 Wrongness *Error*
PA 5952 Wrongness *Illegality*
PA 6486 Wrongness *Injustice*
PA 7006 Wrongness *Untimeliness*
PA 5644 Wrongness *Vice*
PA 7280 Wrongness *Wrongness*

X

PD 2728 X disease ; Turkey
PD 3978 X horses ; Colitis
PE 9816 Xanthomatosis
PD 2226 Xanthomonas
PE 7808 Xanthosis
PD 4957 Xenophobia
PD 4957 Xenophobia ; *Dependence*
PA 5869 Xenophobia *Exclusion*
PA 6030 Xenophobia *Fear*
PA 7338 Xenophobia *Hate*
PA 7306 Xenophobia *Narrowmindedness*
PC 5017 Xenophobia regard migrant workers
PE 2538 Xerophthalmia
PE 7570 *Xerostomia*

Y

PC 2343 Yakuza
PE 8970 Yarn fabrics, made up articles ; Instability trade textile
PE 8436 Yarn fabrics, made up articles ; Shortage textile,
PE 6857 Yaws
PF 2723 Years ; Proliferation national international anniversaries
PE 1028 *Yellow atrophy ; Acute*
Yellow [fat...]
PE 9774 — *fat disease*
PE 0985 — *fever*
PE 3621 — *fever mosquitoes*
PE 0985 Yellow Jack
PE 2227 Yellows ; *Aster*
PD 6363 Yersiniosis
PF 3146 Yield crop varieties ; Introduction high
PE 6524 Yield farm methods ; Low
PE 2890 Yield grain ; Adverse effects high
PD 1285 *Yield rice crops ; Low*
PF 3409 Yield seeds ; Low
PF 2135 Yield ; *Unpredictable fish*
PD 7480 Yields ; Low crop
PD 9082 Yields ; Reduction fish
PF 1479 Yields ; Variability crop
PF 1479 Yields ; Variability grain
PD 1109 Yobbery
PE 4245 Young children ; Neglected
PC 0212 Young criminals
Young [driver...]
PE 6119 — driver accidents
PD 0094 — *drivers ; Drug abuse*
PE 4609 — Drug threat

Young [due...] cont'd
PD 2050 — due lack social forms ; Disorientation
PE 1848 *Young families ; Insufficient*
PF 0399 *Young offenders ; Execution*
PF 8161 Young outmoded values ; Imposition
Young people
PD 1611 — Alcohol consumption children
PE 8933 — Dependence excitement danger
PD 5344 — Lack constructive family discipline
Young [peoples...]
PF 2068 — people's lack context future
PE 1832 — people towards family ; Irresponsibility
PE 6868 — people work ; Hazards
PD 1544 *Young ; Social disaffection*
PE 3909 *Young ; Subverting minds*
PD 1544 *Youth accountability ; Limited*
Youth activities
PF 2575 — *Boring*
PF 5675 — *Irregular*
PF 2575 — *Lack*
PF 2575 — *Limited*
PF 2575 — *Undirected*
PF 2575 — *Uninviting*
PF 2575 — *Unsupported*
Youth [adult...]
PD 1544 — adult activities ; Separation
PF 2477 — *advisors ; Untrained*
PD 1611 — alcoholism
PD 1544 — alienation
PD 1544 — ambitions ; Unsatisfied
PC 0877 — anxiety
PF 5949 — Apathy
PF 2575 *Youth centre ; Unsupported*
PF 6766 Youth ; Cult
Youth [decision...]
PF 6023 — decision making bodies ; Non participation
PD 0513 — Denial rights children
PF 3523 — *Departure skilled*
PD 1544 — displacement ; Community
PD 5987 — drug abuse
Youth [Economic...]
PD 0164 — Economic exploitation
PF 3575 — education ; Limited access rural
PF 6570 — *emphasis*
Youth employment
PD 2318 — Discrimination favour
PJ 8374 — Negative effects relative wage rates
PC 2035 — Unavailability
Youth engagement
PF 5949 — Limited
PD 1544 — Lost direction
PF 5949 — Misdirected

PF 3523 *Youth exodus trend*
Youth [facilities...]
PE 8084 — facilities countryside ; Meagre
PF 2575 — *facilities ; Insufficient*
PD 1544 — Frustration
PD 2682 Youth gangs
PD 1544 Youth ; Grievances
Youth [image...]
PF 1509 — *image ; Destructive*
PF 1509 — incompetence ; Assumed
PD 1544 — Isolated community
PC 2035 Youth job vacuum
Youth [leadership...]
PF 2797 — *leadership ; Overburden*
PF 2797 — *leadership ; Undeveloped*
PF 2874 — *lifestyle ; Idle*
PD 4773 — literature ; Racism child
Youth [market...]
PD 5750 — market ; Criminal investment
PF 5726 — market ; Pandering
PF 3523 — *migration cities*
PD 1544 Youth needs ; Unclear
PD 1544 Youth needs ; Undefined
Youth [organizations...]
PF 2575 — organizations ; Lack
PF 2575 — organizations ; Limited
PF 6570 — oriented society
Youth participation
PD 1544 — Lack
PD 1544 — Unplanned
PD 1544 — Unreleased potential
Youth [Political...]
PE 6263 — Political impotence
PF 2575 — *potential ; Unchallenged*
PD 1544 — protest
Youth [recreation...]
PD 0549 — recreation ; Inadequate spatial facilities
PF 1822 — *Rejection agriculture*
PF 4274 — relationships ; Suspicion adult
PD 1544 — responsibility ; Insufficient
Youth roles
PD 1544 — Inadequate
PD 1544 — Unclarity
PD 1544 — Unclear
PD 8340 Youth ; Runaway
Youth [sex...]
PE 0341 — sex education ; Excessive
PF 5949 — stance ; Unmotivated
PF 1509 — *stereotypes ; Inaccurate*
PF 2575 — *structures ; Absence*
PF 2575 — *structures ; Unplanned*
PF 2874 — *style ; Bravado*

Youth [supervision...] cont'd
PF 2477 — supervision ; Vacated
Youth [Unemployed...]
PE 1379 — Unemployed educated
PC 2035 — unemployment
PC 2035 — unemployment ; High cost
PD 1544 — Unengaged community
PD 5980 — Unprotected street
PD 1544 — unrest
PF 3523 — *urban sites ; Migration*
PF 7498 Youth violence
PF 7498 Youth violence ; Culture
PE 6868 Youth workers ; Occupational hazards

Z

PE 8935 Zagoutis coypu hutia ; Endangered species
PE 1656 *Zalophus ; Endangered species*
PE 3604 *Zealand short tailed bat ; Endangered species New*
PB 3415 Zealotry
PA 6379 Zealotry *Avoidance*
PA 7157 Zealotry *Insanity*
PE 5439 Zinc dust
PE 7229 Zinc pollutant
PE 0824 *Zinc trade ; Instability*
PF 0200 Zionism
PF 0200 Zionist conspiracy
PF 9006 Zombie
PF 2575 *Zoning disputes ; Divisive*
PF 1776 *Zoning restrictions ; Obsolete*
PA 6438 *Zoning ; Uncertainty land*
Zoo animals
PE 8098 — Loss disease game
PE 4834 — Maltreatment
PE 4771 — Unethical practices
PE 4834 Zoo facilities ; Inadequate
PE 4834 Zoo mismanagement
PD 1770 Zooanthroponoses
PD 4721 Zoologists ; Corruption
PD 4721 Zoologists ; Irresponsible
PD 4721 Zoologists ; Negligence
PD 4721 Zoology ; Malpractice
PD 1770 Zoonoses
PD 1770 Zoonoses ; Inadequate control
PD 6363 Zoonotic bacterial diseases
PC 0066 *Zoosadism*
PD 4721 Zoosciences ; Unethical practice
PE 9438 *Zwoegersiekte*
PE 0614 Zygomycetes

References PY

World Problems

Scope of bibliography

The following section constitutes a collection of bibliographical references relevant to the World Problems (Section P).

The entries appear here for any of the following reasons:

(a) Reference is made to them in the individual entries of Section P (where they appear in abridged form, namely with author, title and year of publication).

(b) Reference is made to them in the commentary on Section P (which appears in Section PZ).

(c) The publications are relevant to Section P, even though they have not (yet) been cross-referenced in either of the two ways indicated above.

Bibliographical entries

The entries appear by order of author and, within author, by year.

In certain cases the *International Standard Book Number (ISBN)* is given to facilitate access.

Source of bibliography

The bibliography was derived from a wide variety of sources. These include:

- Information from international organizations, including catalogues of publications and accessions lists

- Information from commercial publishers, especially sales catalogues

- Citations and bibliographies in other publications and reference books

- Systematic compilations of books in print (notably *Books in Print* and *International Books in Print*)

- Book reviews in specialized journals

Abaya, Hernando J Making of a Subversive (1984)
Quezon City, New Day Publishers, 244 p.
ISBN 971-10-0154-3.

Abbot, George C International Indebtedness and the Developing Countries (1980)
New Delhi, Vikas Publishing House, 308 p.
ISBN 0-7069-1064-8.

Abel, Ernest L Fetal Alcohol Syndrome: an annotated and comprehensive bibliography (1981)
Boca Raton FL, CRC Press, 144 p. bibl. Vol 1.
ISBN 0-8493-6192-3.

Abel, Ernest L Fetal Alcohol Syndrome and Fetal Alcohol Effects (1984)
New York, Plenum Publishing, 256 p.
ISBN 0-306-41427-9.

Abel, Ernest L Behavioral Teratology: a bibliography to the study of birth defects of the mind (1985)
New York, Greenwood Press, 206 p.
ISBN 0-313-25066-9.

Ablon, Joan Little People in America: the social dimensions of dwarfism (1984)
New York, Praeger Publishers, 224 p.
ISBN 0-275-91109-8.

Abraham, H J World Problems in the Classroom: a teacher's guide to some United Nations tasks (1973)
Paris, UNESCO, 223 p.

Abraham, Herbert J World Problems in the Classroom (1973)
Paris, UNESCO, 223 p. ISBN 92-3-101048-4.

Abrams, et al Biology of Lung Cancer: diagnosis and treatment (1988)
New York, Dekker Marcel, 384 p. ISBN 0-8247-7642-9.

Abramson, D I and Miller, D S Vascular Problems in Musculoskeletal Disorders of the Limbs (1981)
Berlin, Springer-Verlag, 404 p. illus. ISBN 0-387-90524-3.

Abse, D Wilfred Hysteria and Related Mental Disorders (1987)
Littleton MA, PSG Publishing, 560 p.ISBN 0-7236-0811-3.

Academy of Comparative Philosophy and Religion Validity and Value of Religious Experience: seminar proceedings (1968)
Belgaum, Academy of Comparitive Philosophy and Religion.

ACI Committee Shrinkage and creep in concrete (1972)
Detroit MI, American Concrete Institute. Bibliography No 10.
ISBN 0-685-85151-6.

Acid Rain Foundation Air Pollutants: effects on forest ecosystems; proceedings of a symposium, May 8-9, 1985, St Paul (1985)
Acid Rain Foundation, 439 p. illus. ISBN 0-935577-01-7.

Acker, Robert F et al (Eds) Proceedings of the third international congress on marine corrosion and fouling (1974)
Evanston IL, Northwestern University Press.
ISBN 0-8101-0445-8.

Ackerman, A Bernard and Maize, John C (Eds) Malignant Melanoma and Other Melanocytic Neoplasms (1985)
New York, Raven Press, 275 p. illus.
ISBN 0-88167-184-3.

Ackerman, Robert J Children of Alcoholics: a guidebook for parents, educators and therapists (1987)
S and S, 224 p. 2nd ed.
ISBN 0-671-64527-7.

Ackoff, Russell L Management in Small Doses (1986)
New York, Wiley.

Acquaviva, S S The Decline of the Sacred in Industrial Society (1979)
Oxford, Blackwell Scientific Publications.

Adachi, Masazumi Neuromuscular Diseases (1989)
New York, Igaku-Shoin Medical Publishers, 350 p.
ISBN 0-89640-150-2.

Adam, Carol J The Sexual Politics of Meat: a feminist-vegetarian critical theory (1990)
New York, Continuum Publishing, 256 p.

Adam, Carol J The Sexual Politics of Meat: a feminist-vegetarian critical theory (1990)
New York, Continuum Publishing Group, 256 p.

Adams, Donald D and Page, Walter P Acid Deposition: environmental, economic and policy issues (1985)
New York, Plenum Publishing, 572 p.
ISBN 0-306-42062-7.

Adams, James The Financing of Terror (1986)
New York, Simon and Schuster International, 352 p.
ISBN 0-671-49700-6.

Adams, Patricia and Solomon, Lawrence In the Name of Progress: the underside of foreign aid (1985)
Toronto, Energy Probe Research Foundation.

Adams, R M (Ed) Occupational Skin Diseases (1986)
Oxford, Blackwell Scientific Publications, 176 p.
ISBN 0-932883-27-3.

Adams, Ruth and Cullen, Susan (Eds) The Final Epidemic: physicians and scientists on nuclear war
Wilmington DE, Peace Resource Center, 254 p.

Addo, Herb Beyond Eurocentricity: transformation and transformational responsibility (1985)
In: *Development as Social Transformation; reflections on the global problematique* London, Hodder and Stoughton. in association with the United Nations University.

Addo, Herb Imperialism: the permanent stage of capitalism (1986)
Tokyo, United Nations University, 182 p.
ISBN 92-808-0484-7

Addo, Herb, et al Development as Social Transformation: reflections on the global problematique (1985)
London, Hodder and Stoughteon in association with the United Nations University, 281 p. ISBN 0-340-35634-0.

Adhinarayan, S P Case for Colour (1964)
Bombay, Asia Publishing House, 208 p.

Adjangba, M Inequality and a New Maritime Order: theories and issues of dependent development (1985)
Tampere, Tampereeen yliopisto, 434 p.
ISBN 951-44-1740-2.

Adkins, Virgil R The Static Position of Classifying Alcoholism and Drug Addiction As Identical Illnesses (1986)
Independence MO, International University Press, 91 p.
ISBN 0-89697-276-3.

Adler, Oral Communication Problems in Children and Adolescents (1987)
Orlando FL, Grune and Stratton. ISBN 0-8089-1887-7.

Adler, Patricia A Wheeling and Dealing (1985)
New York, Columbia University Press.
ISBN 0-231-06060-2.

Adriano, D C Trace Elements in the Terrestrial Environment (1985)
Berlin, Springer-Verlag, 545 p. ISBN 3-540-96158-5.

Advisory Committee for the Coordination of Information Systems Directory of United Nations Databases and Information Systems (1985)
New York, UN. A repertory of the particulars of over 600 information Systems in 36 UN affiliated organizations.
ISBN 92-9048-295-8.

Agar, Michael Ripping and Running: a formal ethnograph of urban heroin addicts (1973)
San Diego CA, Academic Press. ISBN 0-12-785020-1.

Agarwal, A N Problem of Management Grasp in Underdeveloped Countries (1970)
Allahabad, Allahabad Univ Publications, 42 p.

Agesta Group Environment-International: twenty years after Stockholm 1972-1992 (report on the implementation of the Stockholm Action Plan and on priorities and institutional arrangements for the 1980s) and a parliamentary view of the state of the world environment (results of a survey carried out by the Secretariat of the International Parliamentary Conference on the Environment) (1982)
Berlin, Erich Schmidt Verlag.

Aggarwal, Vinod K International Debt Threat: bargaining among creditors and debtors in the 1980s (1987)
Berkeley CA, University of California Press, 72 p. illus. Policy Papers in International Affairs: No 29.
ISBN 0-87725-529-6.

Aggarwala, Om Prakash Misconduct and Disciplinary Action Against Workmen (1972)
Delhi, Metropolitan Book Company, 296 p.

Aggerholm, Paula N Headache: health and medical subject analysis (1987)
Annandale VA, ABBE Publishers Association of Washington, 150 p. ISBN 0-88164-484-6.

Aggleton, Peter Rebels Without a Cause?: middle class youth and the transition from school to work (1987)
Hants, Taylor and Francis, 160 p.
ISBN 1-85000-224-X. Issues in Education and Training Series: No 8.

Agranowitz, Aleen and McKeown, Milfred R Alphasia Handbook: for adults and children (1975)
Springfield IL, Thomas Charles C, 336 p. illus.
ISBN 0-398-00017-4.

Agress, Wilhelm Towards an Intelligence Revolution: comments on the future role of social intelligence in international relations (1980)
Lund, Research Policy Institute, 49 p.

Ahmadjian, Vernon and Hale, Mason E The Lichens (1974)
San Diego CA, Academic Press. ISBN 0-12-044950-1.

Ahmed, Paul I and Ahmed, N (Eds) Coping with Juvenile Diabetes (1985)
Springfield IL, Thomas Charles C, 420 p. illus.
ISBN 0-398-05073-2.

Ahne, W (Ed) Cooperative Programme of Research on Aquaculture (Fish Diseases) - 3rd Session, Munich (Germany FR), 23 Oct 1979, fish diseases: third COPRAQ (Cooperative Programme of Research on Aquaculture) Session, (Munich, Federal Republic of Germany, 23-26 October 1979, Proceedings) (1980)
Rome, FAO, 252 p. ISBN 3-540-10406-2.

Ahuja, M M S (Ed) Epidemiology of Diabetes in Developing Countries (1979)
New Delhi, Interprint Mehta House, 124 p.

Aicardi, Jean Epilepsy in Children (1986)
New York, Raven Press, 428 p. International Review of Child Neurology Ser. ISBN 0-88167-182-7.

Ailor, William H (Eds) Atmospheric Corrosion (1982)
New York, Wiley Interscience, 1056 p. Corrosion Monograph.
ISBN 0-471-86558-3.

Ainlay, Stephen Day Brought Back My Night: aging and new vision loss (1989)
London, Routledge, 176 p. ISBN 0-415-00764-X.

Ainlay, Stephen C, et al The Dilemma of Difference: a multidisciplinary view of stigma (1986)
New York, Plenum Publishing, 286 p.
ISBN 0-306-42304-9.

Ainsworth, G C; Hawksworth, D L and Sutton, B C Ainsworth and Bisby's Dictionary of the Fungi (1983)
Kew, CAB International Mycological Institute, 457 p. 7th ed.
ISBN 0-85198-515-7.

Airaksinen, Timo Ethics of Coercion and Authority: a philosophical study of social life (1988)
Pittsburgh PA, University of Pittsburgh Press, 256 p. illus.
ISBN 0-8229-3583-X.

Aita, John A Neurologic Manifestations of General Diseases (1975)
Springfield IL, Thomas Charles C, 936 p.
ISBN 0-398-02675-0.

Aitken, I D Chlamydial Diseases of Ruminants (1986)
Luxembourg, CEE, 162 p. ISBN 92-825-6654-4.

Akindele, R A and Vogt, M A (Eds) Smuggling and Coastal Piracy in Nigeria: proceedings of a workshop (1983)
Lagos, Nigerian Institute of International Affairs, 93 p.

Akinyemi, A B et al Disarmament and Development: utilization of resources for military purposes in Black Africa
Lagos, Nigerian Institute of International Affairs.

Alatas, Syed H Corruption (1986)
Brookfield VT, Gower Publishing, 320 p.
ISBN 0-566-05107-9.

Alatas, Syed Hussein Problem of Corruption
Singapore, Times Books International, 168 p.
ISBN 9971-65-346-X.

Albas, Daniel C and Albas, Cheryl M Student Life and Exams: stresses and coping strategies (1984)
Dubuque IA, Kendall/Hunt Publishing, 192 p.
ISBN 0-8403-3362-5.

Albert, Daniel M et al Herpesvirus: recent studies (1974)
New York, Irvington Publishers. 33 vols.

Albus, James S Peoples' Capitalism: the economics of the robot revolution (1986)
New York, Gordon Press Publishers.ISBN 0-8490-3619-4.

Alcorn, Randy and Alcorn, Nancy Women under Stress (1986)
Portland OR, Multnomah Press. ISBN 0-88070-157-9.

Alderfer, Hannah, et al Caught Looking: feminism, pornography and censorship (1987)
New York, Caught Looking, 96 p. illus.
ISBN 0-9617884-0-2.

Aleksakhin, R M Radioactive Contamination of Soil and Plants (1965)
Jerusalem, Keter Publishing House, 112 p.
ISBN 0-7065-0400-3.

Aleksander, I, et al Robotics Development and Future Applications (1985)
Brussels, Commission of the European Communities, 100 p. 2nd ed. ISBN 92-825-4903-8.

Alexander, Nancy J, et al (Eds) Heterosexual Transmission of AIDS (1990)
Chichester, John Wiley and Sons, 466 p.
ISBN 0-471-56208-4.

Alexander, Shana The Pizza Connection: lawyers, drugs, money, mafia (1988)
New York, Weidenfeld and Nicolson.
ISBN 1-55584-027-2.

Alexander-Williams, J and Irving, M Intestinal Fistulas (1982)
Littleton MA, PSG Publishing, 240 p.ISBN 0-7236-0555-6.

Alexandre, A et al Road Traffic Noise (1975)
New York, Elsevier Science Publishing.
ISBN 0-85334-628-3.

Alford, David V A Colour Atlas of Fruit Pests: their recognition, biology and control (1984)
Wolfe Medical England/Sheridan, 310 p. illus.
ISBN 0-7234-0816-5.

Allaby, Michael (Ed) Macmillan Dictionary of the Environment (1988)
Basingstoke, Macmillan, 448 p. 3rd ed.
ISBN 0-333-45561-4.

Allan, W H; Lancaster, J E and Toth, B Newcastle Disease Vaccines: their production and use (1978)
Rome, FAO, 163 p.
FAO Animal Production and Health Series: No 10.
ISBN 92-5-100484-6.

Allen, Clifford The Sexual Perversions and Abnormalities: a study in the psychology of paraphilia (1979)
Westport CT, Greenwood Press. ISBN 0-313-20627-9.

Allen, David F The Cocaine Crisis (1987)
New York, Plenum Publishing, 270 p.
ISBN 0-306-42482-7.

Allen, Hannah Don't Get Stuck : the case against vaccinations and injections (1985)
Tampa FL, Natural Hygiene Press. ISBN 0-914532-33-2.

Allen, Nancy Homicide: perspectives on prevention (1980)
New York, Human Sciences Press, 192 p.
ISBN 0-87705-382-0.

Allin, Craig W The Politics of Wilderness Preservation (1982)
Westport CT, Greenwood Press, 344 p.
ISBN 0-313-21458-1.

Almli, C Robert and Finger, Stanley Early Brain Damage (1984)
San Diego CA, Academic Press. 2 vols. Vol 1: Research orientation and clinical observations, Vol 2: neurobiology and behavior.

Altbach, Philip G The Foreign Student Delimma
In: *Educational Documentation and Information: Bulletin of the IBE* 1985, pp. 236-237.

Alting von Geusau, F A Uncertain Detente (1979)
Dordrecht, Kluwer Academic Publishers Group, 310 p.
ISBN 90-286-0818-4.

Altman, Arnold and Schwartz, Allen D Malignant Diseases of Infancy, Childhood and Adolescence (1983)
Philadelphia PA, WB Saunders. ISBN 0-7216-1211-3.

Altman, H J Alzheimer's Disease: problems, prospects, and perspectives (1987)
New York, Plenum Publishing, 412 p.
ISBN 0-306-42662-5.

Altura, B M, et al (Eds) Magnesium in Cellular Processes and Medicine (1987)
New York, Karger S AG, 230 p. ISBN 3-8055-4369-7.

Altura, Burton M et al Handbook of Shock and Trauma: basic science (1983)
New York, Raven Press, 484 p. Vol 1.
ISBN 0-89004-583-6.

Alvord, Elssworth C Jr, et al (Eds) Experimental Allergic Encephalomyelitis: a useful model for multiple sclerosis (1984)
New York, Alan R Liss, 582 p. Progress in Clinical and Biological Research Ser: Vol 146. ISBN 0–8451–0146–3.

Aman, Reinhold (Eds) Maledicta 1980 (1980)
Waukesha WI, Maledicta Pr, 320 p. illus. Maledicta: International Journal of Verbal Aggression Ser: Vol 4, No 1 and 2. ISBN 0–916500–55–1.

Amara, Ait H and Founou-Tchuigoua, B (Eds) Africa: the roots of the agrarian crisis (1989)
London, Zed Books.

Amaura, Edward G (Eds) Blindness: medical subject analysis with bibliography (1987)
Washington DC, ABBE Publications Association, 160 p. bibl. ISBN 0–88164–544–3.

American Association for Counseling Substance Abuse: bibliography
Alexandria VA, American Association for Counseling. ISBN 0–317–59914–3.

American Association of School Administrators Student Discipline: problems and solutions
Arlington VA, American Association of School Administrators. ISBN 0–686–36518–6.

American Bar Association National Security Leaks: is there a legal solution? (1986)
American Bar Association, 52 p.
ISBN 0–89707–224–3.

American Cancer Society American Cancer Society's Complete Book of Cancer: prevention, detection, diagnosis, treatment rehabilitation, cure (1986)
New York, Doubleday, 672 p. illus. ISBN 0–385–17847–6.

American Health Research Institute Depression: medical subject analysis and research directory with bibliography (1982)
Washington DC, ABBE Publications Association, 133 p. bibl. ISBN 0–941864–30–8.

American Health Research Institute Medical Subject Research Index of International Bibliography Concerning Cocaine (1982)
Washington DC, ABBE Publications Association, 198 p. bibl. ISBN 0–941864–16–2.

American Health Research Institute Disasters and Disaster Planning: medical analysis index with research (1987)
Annandale VA, ABBE Publishers Association of Washington, 150 p. ISBN 0–88164–390–4.

American Health Research Institute Noise and Adverse Effects on Health: medical subject analysis with reference (1987)
Annandale VA, ABBE Publishers Association of Washington, 150 p. ISBN 0–88164–450–1.

American Health Research Institute Nuclear Warfare: science and medical subject analysis with (1987)
Annandale VA, ABBE Publishers Association of Washington, 150 p. ISBN 0–88164–462–5.

American Health Research Institute and Bartone, John C Neurotic Disorders: medical subject analysis and research guide with (1984)
Annandale VA, ABBE Publishers Association of Washington, 149 p. ISBN 0–88164–118–9.

American Institut Psychology The psychology of the master counterfeiters (1986)
Albuquerque NM, American Institute for Psychological Research, 237 p. 2 Volls. Illus.

American Society for Testing and Materials Symposium on Radiation Effects on Metals and Neutron Dosimetry
Ann Arbor MI, Books on Demand. American Society for Testing and Materials Series: Special Technical Publication: 341. ISBN 0–317–10870–0.

American Society for Testing and Materials Effects of Environment and Complex Load History on Fatigue Life – STP 462 (1970)
Philadelphia PA, American Society for Testing and Materials, 332 p. ISBN 0–8031–0032–9.

American Society for Testing and Materials Corrosion in natural environment (1974)
Philadelphia PA, American Society for Testing and Materials, 352 p. ISBN 0–8031–0315–8.

American Society for Testing and Materials Atmospheric Corrosion Investigation of Aluminum–Coated, Zinc–Coated and Copper–Bearing Steel Wire and Wire Products (1975)
Philadelphia PA, American Society for Testing and Materials, 90 p. STP 585. ISBN 0–8031–0285–2.

American Society for Testing and Materials Pesticides, Resource Recovery (1986)
Philadelphia PA, American Society for Testing and Materials, 1028 p. Water and Environmental Technology: 1104. ISBN 0–8031–0905–9.

American Society of Civil Engineers Structural Failures: modes, causes, responsibilities
Ann Arbor MI, Books on Demand. ISBN 0–317–08323–6.

American Society of Civil Engineers Advisory Notes on Lifeline Earthquake Engineering (1983)
New York, American Society of Civil Engineers, 242 p. ISBN 0–87262–377–7.

Amerio, Alberto, et al Acute Renal Failure: clinical and experimental (1987)
New York, Plenum Publishing, 336 p.
ISBN 0–306–42556–4.

Ames, Frederick C, et al (Eds) Current Controversies in Breast Cancer (1984)
University of Tex Press, 671 p. illus.
Annual Clinical Conference on Cancer Ser: No 26.
ISBN 0–292–71093–3.

Ames, Jessie D Changing Character of Lynching: review of lynching 1931–1941, with discussion of recent developments in this field
New York, AMS Press. ISBN 0–404–00134–3.

Amey, Lorne J Visual Literacy: implications for the production of children's television programs (1976)
Halifax, D U Library School. ISBN 0–7703–0148–7.

Amicarelli, V, et al Identification of Air Quality and Environmental Problems in the European Community (1987)
Luxembourg, CEE, 136 p. Heavy Metals: 2.
ISBN 92–825–6837–7.

Amin, G A Modernization of Poverty (1980)
Leiden, Brill. ISBN 90–04–06193–2.

Amin, Samir Unequal Development: an essay on the social formations of peripheral capitalism (1977)
New York, Monthly Review Press. Transl from French.
ISBN 0–85345–433–7.

Amnesty International Amnesty International Report
New York, Amnesty International. Annual Report.

Amnesty International The Death Penalty (1979)
London, Amnesty International, 209 p. illus.
ISBN 0–900058–88–9.

Amnesty International Political Killings by Governments (1983)
USA, Amnesty International USA, 144 p.
ISBN 0–86210–051–8.

Amnesty International Torture in the Eighties (1984)
London, Amnesty International.

Amnesty International When the State Kills: the death penalty v human rights (1989)
London, Amnesty International.

Anand, R P (Ed) Cultural Factors in International relations (1981)
New Delhi, Abhinav Publications, 115 p.
ISBN 81–7017–134–2.

Anand, V K Insurgency and Counter–Insurgency: a study of modern guerilla warfare (1981)
New Delhi, Deep and Deep Publications, 263 p.

Anchell, Melvin Sex and Insanity (1983)
Portland OR, National Book, 169 p. ISBN 0–89420–238–3.

Andel, J van, et al (Eds) Disturbance in Grasslands: causes, effects and processes (1987)
Dordrecht, Kluwer Academic Publishers Group, 328 p.
ISBN 90–6193–640–3.

Andelman, Julian B and Underhill, Dwight W (Eds) Health Effects from Hazardous Wastes Sites (1987)
Chelsea MI, Lewis Publishers, 294 p. illus.

Andersen, Alfred F Liberating the Early American Dream (1985)
Ukiah CA, Tom Paine Institute, 272 p.
ISBN 0–931803–02–0.

Andersen, Arnold E Practical Comprehensive Treatment of Anorexia Nervosa and Bulimia (1985)
Baltimore MD, Johns Hopkins University Press, 224 p. Series in Contemporary Medicine and Public Health.
ISBN 0–8018–2442–7.

Anderson, and Becker, Pathology of Congenital Heart Disease (1982)
Stoneham MA, Butterworth Publishers.
ISBN 0–407–00137–9.

Anderson, C K, et al Germ Cell Tumours
New York, Taylor and Francis, 450 p.
ISBN 0–85066–223–0.

Anderson, David C Crimes of Justice: improving the police, the courts, the prisons (1988)
New York, Times Books, 336 p. ISBN 0–8129–1607–7.

Anderson, Dennis The Economics of Afforestation: a case study in Africa
Washington DC, International Bank for Reconstruction and Development.

Anderson, F K and Treshow, M Plant Stress from Air Pollution (1989)
Chichester, John Wiley and Sons, 350 p.
ISBN 0–471–92374–5.

Anderson, Jock R and Hazell, Peter B R (Eds) Variability in Grain Yields: implications for agricultural research and policy in developing countries (1989)
Baltimore MD, Johns Hopkins University Press.

Anderson, Kenneth Symptoms after Forty (1988)
New York, William Morrow, 384 p. ISBN 0–688–08245–9.

Anderson, Louis E and Baker, Paul J Social Problems: a critical thinking approach (1987)
Belmont CA, Wadsworth Publishing, 414 p.
ISBN 0–534–07428–6.

Anderson, Martin Conscription: a select and annotated bibliography (1976)
Stanford CA, Hoover Institution Press, 472 p.
ISBN 0–8179–2571–6.

Anderson, Olive Suicide in Victorian and Edwardian England (1987)
New York, Oxford University Press, 488 p.
ISBN 0–19–820101–X.

Andreani, D et al Hypoglycemia (1987)
New York, Raven Press, 328 p. ISBN 0–88167–321–8.

Andreasen, J O Traumatic Injuries of the Teeth (1981)
Philadelphia PA, WB Saunders, 462 p.
ISBN 0–7216–1249–0.

Andree, Michel Down with Stereotypes: eliminating sexism from children's literature (1987)
Lanham MD, UNIPUB, 105 p. ISBN 92–3–102380–2.

Andreoli, M; Monaco, F and Robbins, J Advances in Thyroid Neoplasia: proceedings of the international colloquium on thyroid neoplasia, Rome, Sept 1981
Acta Medica.

Andrewes, Christopher In Pursuit of the Common Cold (1973)
London, Heinemann.

Andrews, J A Keyguide to Information Sources on the International Protection of Human Rights (1986)
New York, Facts on File, 224 p. ISBN 0–8160–1822–7.

Andrews, Theodora A Bibliography of Drug Abuse: supplement 1977–1980 (1981)
Englewood CO, Libraries Unlimited, 312 p. bibl.
ISBN 0–87287–252–1.

Anetzberger, Georgia J The Etiology of Elder Abuse by Adult Offspring (1987)
Springfield IL, Thomas Charles C, 144 p.
ISBN 0–398–05297–2.

Angelli, Susanna Street Children: a growing urban tragedy (1986)
London, Independent Commission on International Humanitarian Issues and Weidenfield and Nicholson, 123 p.

Anglin, Lise Cocaine: a selection of annotated papers from 1880–1984 concerning health effects (1985)
Toronto ON, Addiction Research Foundation, 223 p.
ISBN 0–88868–114–3.

Anker, Richard and Hein, Catherine Sex Inequalities in Urban Employment in the Third World (1986)
New York, Saint Martin's Press, 304 p.
ISBN 0–312–71341–X.

Annas, Julia and Barnes, Jonathan The Modes of Scepticism: ancient texts and modern interpretations (1985)
New York, Cambridge University Press, 216 p.
ISBN 0–521–25682–8.

Annecke, D P and Moran, V C Insects and Mites of Cultivated Plants in South Africa (1983)
Kent, Butterworth, 400 p. illus. ISBN 0–409–08398–4.

Annest, Joseph L; Mahaffey, Kathryn and Cox, Kaludia Blood Lead Levels for Persons Ages 6 Months to 74 Years: United States, 1976–1980 (1984)
Hyattsville MD, National Center for Health Statistics, 50 p.
ISBN 0–8406–0291–X.

Anon Threats and Opportunities of Global Countertrade: marketing, financing and organizational implications
Hong Kong, Business International Asia/Pacific.

Anon Fluctuations in Income and Employment: with special reference to recent American experience and post–war prospects
Millwood NY, Kraus Reprint and Periodicals. 3rd ed. London School of Economic and Political Science Studies in Economics and Commerce: Vol 8.

Anon Population Decline in Europe
London, Edward Arnold.

Anon Insider Trading: coping with use and abuse of market sensitive
San Diego CA, Harcourt Brace Jovanovich.
ISBN 0–317–29525–X.

Anon Psychology of Dictatorships: sketches on the influence of dictatorships on their subjects and on the rest of the humanity
Barcelona, Salador Raich Ullán. ISBN 84–85176–19–7.

Anon The Promise of World Peace: the universal house of justice
London, Oneworld Publications, 192 p. illus.
ISBN 1–85168–002–0.

Anon

Anon

Anon Index Fraud: memoranda submitted to the experts committee on consumer price index number (1963)
New Delhi, All India Trade Union Congress, 54 p.

Anon Report from Iron Mountain on the Possibility and Desirability of Peace: with introductory material by Leonard C Lewin (1967)
New York, Dell Publishing, 109 p.

Anon Student Revolution: a global analysis (1968)
Bombay, Lalvani Publishing House, 408 p.

Anon Conscientious Objector (1968)
New York, Greenwood Press. 8 Vols, No 6. Repr of 1946 ed.

Anon Phallicism in Ancient Worships (1970)
New Delhi, Kumar Brothers Publications, 95 p. Introduction, Additional Notes and an Appendix by Alexander Wilder.

Anon Phallic Tree Worship or Cultus Arborum: with illustrative legends, superstitions, usages (1971)
Varanasi, Bharat Bharati, 111 p.

Anon Frost, Moisture and Erosion: nine reports (1975)
Washington DC, Transportation Research Board, 105 p. Transportation Research Report Ser.
ISBN 0–309–02380–7.

Anon Infectious Mononucleosis (1978)
Lorain OH, Dayton Laboratories. ISBN 0–916750–30–2.

Anon Measuring the Condition of the World's Poor: the physical quality of life index published (1979)
Elmsford NY, Pergamon Books. ISBN 0–08–023890–4.

Anon Preventing Hospital Infections: guidelines proposed by the Health Council of the Netherlands (1980)
's Gravenhage, Netherlands Government Publishing Office, 510 p. ISBN 90–12–02863–9.

Anon Warfare in a Fragile World: military impact on the human environment (1980)
Stockholm, Almqvist and Wiksell, 250 p.
ISBN 0–85066–187–0.

Anon The Problem of Drop–outs: interpretive bibliography (1980)
Teheran, IIALM, 123 p.

Anon An Annotated Bibliography of Works on Juvenile Delinquency in America and Britain in the Nineteenth Century (1980)
Folcroft PA, Folcroft Library Editions.
ISBN 0–8414–7288–2.

Anon Against Concessions to Bourgeois Nationalism (1981)
Moscow, Agentstvo Pecati Novosti, 225 p.

Anon Biologic Effects of Environmental Electromagnetism (1981)
Berlin, Springer–Verlag, 332 p.

ISBN 3-540-90512-X. In: Topics in Environmental Physiology and Medicine.

Anon Environmental and Human Population Problems at High Altitute (1982)
Paris, CNRS. ISBN 2-222-02851-5.

Anon The Paper Qualification Syndrome (PQS) and Unemployment of School Leavers: a comparative subregional study of four East and four West African Countries (1982)
Geneva, JASPA, 227 p. ISBN 92-2-102970-0.

Anon Unilateral Measurements to Prevent Double Taxation (1982)
Norwell MA, Kluwer Academic Publishers, 550 p. ISBN 90-6544-007-0.

Anon Trichomoniasis: scientific papers of the symposium on trichomoniasis, Basel, October 1981 (1983)
New York, Karger S AG, 92 p. ISBN 3-8055-3751-4.

Anon Journal of Social Issues: cumulative index 1945-1975 (1983)
Medford NJ, Learned Information. Vols 1-31. ISBN 0-317-01049-2.

Anon International Symposium on Vaginal Mycoses Vienna, September 1981 (1983)
New York, Karger S AG, 112 p. ISBN 3-8055-3638-0.

Anon Evasion Fiscale, Fraude Fiscale: tax avoidance, tax evasion (1983)
Norwell MA, Kluwer Academic Publishers, 644 p. ISBN 9-06-544123-9.

Anon Nazi Concentration Camps: structure and aims, the image of the prisoner, the Jews in the camps; Proceedings of the 4th Yad Vasham International Conference (1984)
Jerusalem, Yad Vashem, 750 p.

Anon Report of the Committee on Sexual Offences Against Children and Youths (1984)
Washington DC, Government Publication Centre, 1314 p. ISBN 0-660-11639-1.

Anon International Symposium on Inborn Errors of Metabolism in Humans, Third, Munich, March 1984: abstracts (1984)
Basel, S Karger AG, 126 p. ISBN 3-8055-3894-4.

Anon RICO (Racketeer Influenced and the Corrupt Organizations): business disputes and the "racketeering" (1984)
Chicago IL, Commerce Clearing House, 224 p. ISBN 0-317-19245-0.

Anon Readings in Culturally Disadvantaged Children (1984)
Guilford CT, Special Learning Corporation. ISBN 0-89568-449-7.

Anon The Environmental Impacts of Exploitation of Oil Shales and Tar Sands (1985)
Energy Report Series, ERS 13-85.

Anon Advances in Immunity and Cancer Therapy (1985)
Berlin, Springer-Verlag. Vol 1. ISBN 0-387-96083-X.

Anon Health Consequences of Smoking: cancer and chronic lung disease in the workplace, (1985)
Washington DC, United States Government Printing Office, 564 p. ISBN 0-318-21368-0.

Anon Mental Disorders, Alcohol and Drug-Related Problems (1985)
New York, Elsevier Science Publishing, 600 p. ISBN 0-444-80686-5.

Anon The Encyclopedia of Unbelief (1985)
Buffalo NY, Prometheus Books, 819 p. 2 vols. ISBN 0-87975-307-2.

Anon Repetition Strain Injury (1986)
Washington DC, Government Publication Centre, 117 p. ISBN 0-644-05104-3.

Anon Domestic Violence (1986)
Washington DC, Government Publication Centre, 98 p. ISBN 0-644-01339-7.

Anon Criminal Use of False ID (1986)
New York, Gordon Press Publishers. ISBN 0-8490-3653-4.

Anon Abstracts on Crime and Juvenile Delinquency: cumulative index 1968-1982 (1986)
Buffalo NY, William S Hein, 382 p. ISBN 0-89941-465-6.

Anon The History of Prostitution: its extent, causes and effects throughout the world (1986)
New York, Apt Books, 708 p. repr. ISBN 81-210-0053-X.

Anon Toxic Substances Control Act (TSCA) Chemical Substance Inventory (1986)
Washington DC, US Government Printing Office, 3968 p. 5 Vols. ISBN 0-318-20382-0.

Anon The Control of Bureaucratic Corruption: case studies in Asia (1986)
Columbia MO, South Asia Books, 275 p. ISBN 0-8364-1855-7.

Anon Fourteenth International Cancer Congress, Budapest, August 1986: abstracts of lectures, symposia and free communication (1986)
Basel, S Karger AG, 1350 p.

Anon Edge: organized crime, business and labor unions: report to the president and the attorney general (1986)
Washington DC, United States Government Printing Office, 415 p. ISBN 0-318-20370-7.

Anon CIA Improvised Sabotage Techniques (1986)
New York, Gordon Press Publishers. ISBN 0-8490-3543-0.

Anon OSS Sabotage and Demolition Manual (1986)
New York, Gordon Press Publishers. ISBN 0-8490-3550-3.

Anon Failure Analysis and Prevention: metals handbook (1986)
Metals Park OH, ASM International, 850 p. Vol. 11 ISBN 0-87170-017-4.

Anon Black Bag Owner's Manual: international espionage overt and convert operations and methodology (1986)
New York, Gordon Press Publishers. 2 Vols. Criminology Ser. ISBN 0-8490-3695-X.

Anon Foreign Debts in the Present and a New International Economic Order (1987)
Freiburg, Universitätsverlag Freiburg, 364 p. 2nd ed. ISBN 2-8271-0340-0.

Anon Isolated Communities: a major report on the needs of inland Australia (1987)
Armidale NSW, University of New England, 280 p.

Anon International Index on Training in Conservation of Cultural Property (1987)
Malibu CA, J Paul Getty Museum, 140 p. 4th ed. ISBN 0-89236-127-1.

Anon Ice Navigation in Canadian Waters (1987)
Washington DC, Government Publication Centre, 100 p. ISBN 0-660-53754-0. Text in English and French.

Anon Insect Allergy: allergy and toxic reactions to insects and other anthropods (1987)
Saint Louis MO, Warren H Green, 480 p. illus. 2nd ed. ISBN 0-87527-324-6.

Anon Destruction by Demolition-Incendiaries and Sabotage (1987)
New York, Gordon Press Publishers. ISBN 0-8490-3951-7.

Anon Pregnancy and Employment (1987)
Washington DC, BNA Books, 210 p. ISBN 0-87179-936-7.

Anon CIBA Foundation Symposium: autoimmunity and autoimmune disease (1987)
New York, John Wiley and Sons. CIBA Foundation Symposium Ser: 129. ISBN 0-471-91095-3.

Anon Development Aid: a guide to national and international agencies (1988)
London, Butterworth, 560 p. ISBN 0-408-00991-8.

Anon The Place of Endocrine Therapy in Breast Cancer Disease: proceedings from a conference (1988)
Wolfeboro NH, Longwood Publishing Group. No 111. ISBN 0-905958-46-2.

Anon Colloque International sur la Militarisation de l'Espace Extra-atmosphérique (Bruxelles, 28-29 juin 1986) – International Colloquium on the Militarisation of Outer Space (Brussels, June 28-29 1986) (1988)
Brussels, Bruylant, 340 p.

Anon Phantom Encounters (1988)
Alexandria VA, Time-Life Books, 144 p. ISBN 0-8094-6328-8.

Anon Marketsearch: directory of 18,000 published market research studies on worldwide markets (1988)
London, Arlington Management Publications. 12th ed.

Ansari, N (Ed) Epidemiology and Control of Schistosomiasis (1973)
Basel, S Karger AG, 752 p. ISBN 3-8055-1340-2.

Ansell, Barbara M Rheumatic Disorders in Children (1980)
Stoneham MA, Butterworth Publishers, 344 p. ISBN 0-407-00186-7.

Anshen, Ruth N Anatomy of Evil (1985)
Mount Kisco NY, Moyer Bell Limited, 224 p. illus. Orig Title: The Reality of the Devil. ISBN 0-918825-15-6.

Ansley, Robert E Discrimination in Housing (1979)
Chicago IL, CPL Bibliographies, 75 p. ISBN 0-86602-013-6.

Anthony, Ian The Naval Arms Trade (1990)
Oxford, Stockholm International Peace Research Institute, 208 p. SIPRI Strategic Issue Papers. ISBN 0-19-829137-X.

Appel, Max J (Ed) Virus Infections of Carnivores (1987)
Amsterdam, Elsevier Science Publishing, 500 p. ISBN 0-444-42709-0.

Appelman, Henry D Pathology of the Esophagus, Stomach and Duodenum (1984)
New York, Churchill Livingstone, 287 p. ISBN 0-443-08219-7.

Appley, Mortimer H and Trumbull, Richard A Dynamics of Stress: physiological, psychological and social (1986)
New York, Plenum Publishing, 360 p. ISBN 0-306-42252-2.

Appleyard, Reginald and Stahl, Charles (Eds) International Migration Today (1988)
Paris, UNESCO. illus. 2 vols. Vol 1: Trend and Prospects, Vol 2: Emerging Issues.

Apter, Steven J and Goldstein, Arnold P Youth Violence: programs and prospects (1986)
Elmsford NY, Pergamon Books, 304 p. ISBN 0-08-031922-X.

Archer, Dane and Gartner, Rosemary Violence and Crime in Cross-National Perspectives (1987)
New Haven CT, Yale University Press, 351 p. ISBN 0-300-04023-7.

Archer, John Animals under Stress (1980)
New Delhi, Arnold-Heinemann, 64 p.

Archer, Leonie Slavery: and other forms of unfree labour (1988)
New York, Routledge Chapman and Hall, 288 p. ISBN 0-415-00203-6.

Archer, S A, et al (Ed) European Handbook of Plant Diseases (1988)
Oxford, Blackwell Scientific Publications, 598 p. illus. ISBN 0-632-01222-6.

Arendell, Terry Mothers and Divorce: legal, economic, and social dilemmas (1986)
Berkeley CA, University of California Press, 320 p. ISBN 0-520-05708-2.

Arendt, Hannah The Origins of Totalitarianism
San Diego CA, Harcourt Brace Jovanovich, 527 p. ISBN 0-15-670153-7.

Arendt, Hannah Eichmann in Jerusalem: a report of the banality of evil (1977)
New York, Penguin Books. ISBN 0-14-004450-7.

Argy, Victor and Nevile, John (Eds) Inflation and Unemployment: theory, experience and policy-making (1985)
London, Unwin Hyman, 320 p. illus.

Ariel, Irving M and Cleary, Joseph Breast Cancer: diagnosis and treatment (1987)
New York, McGraw-Hill Book Company, 577 p. illus. ISBN 0-07-002190-2.

Armour, Audrey (Ed) Not-in-My-Backyard Syndrome: proceedings of a two-day symposium on public involvement in siting waste management facilities, May 13-14, 1983 (1984)
North York ON, York University, 296 p. ISBN 0-919762-23-9.

Armstrong, Warwick Development as a Smokescreen: the worth of a United Nations Project (1978)
Montreal, McGill-Queen's University Press, 18 p.

Armyr, Gunno; Elmér, ke and Herz, Ulrich Alcohol in the World of the 80s (1984)
Stockholm, Sober Vörlags AB. 2nd ed. ISBN 91-7296-138-4.

Arner, George B Consanguineous marriages in the American population
New York, AMS Press. Columbia University, Studies in the Social Sciences Series No 83, Repr of 1908. ISBN 0-404-51083-3.

Arngrim, Torben Attempted Suicide: etiology and long term prognosis; a persons follow up study (1975)
Odense, Odense University Press, 192 p. ISBN 87-7492-121-5.

Arnold, Terrell E The Violence Formula: why people lend sympathy and support to terrorism (1988)
Lexington MA, Lexington Books, 224 p. ISBN 0-669-13153-9.

Arrighi, G A Crisis of Hegemony
In: *S Amin (Ed) Dynamics of Global Crisis* New York, Monthly Review, 1982.

Arrow, Kenneth J and Boskin, Michael J (Eds) The Economics of Public Debt: proceedings of a conference held by the International Economic Association at Stanford, California (1988)
St Martin, 400 p. ISBN 0-312-01871-1.

Arruga, A (Ed) International Strabismus Symposium (1966, Giessen): an evaluation of the present status of orthopedics, pleotics and related diagnostic treatment regimes: held in connection with the 20th international congress of ophthalmology, Munich 1966 (1968)
Basel, S Karger AG, 478 p. ISBN 3-8055-1036-5.

Art, Robert J and Waltz, Kenneth N The Use of Force: military power and international politics (1988)
Lanham MD, University Press of America, 740 p. ISBN 0-8191-7002-X.

Artsybashev, E S Forest Fires and Their Control (1984)
Rotterdam, AA Balkema, 168 p. ISBN 90-6191-430-2.

Asboth, Tibor World Problems and their Perceptions: a detailed outline of the state of the art (1984)
Laxenburg, IIASA. Prepared for submission to UNESCO by IIASA.

Asch, Peter Economic Theory and the Antitrust Dilemma (1984)
Melbourne FL, Robert E Krieger, 426 p. ISBN 0-89874-699-X.

Ascher, François Tourism: transnational corporations and cultural identities (1986)
Paris, UNESCO, 103 p. ISBN 92-3-102095-1.

Asher, Geoffrey Custody and Control: a social world with imprisoned youth (1986)
London, Allen and Unwin, 180 p. ISBN 0-86861-850-0.

Ashton, John The Devil in Britain and America (1980)
San Bernardino CA, Borgo Press, 363 p. ISBN 0-89370-608-6.

Asimov, Isaac, et al (Eds) Computer Crimes and Capers (1983)
Chicago IL, Academy Chicago Publishers, 242 p.

Asiwaju, A I (Ed) Partitioned Africans (1985)
Yaba, Lagos University Press, 350 p. ISBN 978-2264-54-7.

Aspen Strategy Group and European Strategy Group Chemical Weapons and Western Security Policy (1987)
Aspen Strategy Group/University Pr of Amer, 80 p. ISBN 0-8191-6169-1.

Assink, J W and Brink, W J van den (Eds) Contaminated Soil (1985)
Dordrecht, Kluwer Academic Publishers Group, 948 p. ISBN 90-247-3267-0.

Assmann, G and Lewis, B (Eds) Current Views on the Prevention, Diagnosis and Treatment of Hyperlipaemia: six monographs from a conference, London, November 14, 1986 (1988)
Royal Society of Medicine Services Ltd. No 122. ISBN 0-905958-54-3.

Association of Bay Area Governments The Disposal of Hazardous Waste by Small Quality Generators – Magnitude of the Problem (1985)
Oakland CA, Association of Bay Area Governments, 166 p. ISBN 0-318-22707-X.

Asterita, Mary F, et al Physiology of Stress (1985)
Merrillville IN, Aster Publishing. ISBN 0-933019-00-9.

Asthma and Allergy Foundation of America and Norback, Craig (Eds) The Allergy Encyclopedia (1981)
New York, New American Library, 336 p. illus. Mosby Medical Library.

Atal, Y and Dall'Oglio, L (Eds) Migration of Talent: causes and consequences of brain drain; Three studies form Asia (1987)
Lanham MD, UNIPUB. ISBN 0-317-67231-2.

Atkinson, Christine Step-Parenting: understanding the emotional problems and stresses (1986)
New York, Sterling Publishing, 128 p. ISBN 0-7225-1264-3.

Atkinson, Gary M and Moraczewski, Albert S A Moral Evaluation of Contraception and Sterilization: a dialogical study (1979)
Braintree MA, Pope John Center, 115 p. ISBN 0-935372-05-9.

Atkinson, Roland M (Ed) Alcohol and Drug Abuse in Old Age (1987)
Cambridge MA, Cambridge University Press, 81 p.
ISBN 0–521–34743–2.

Atkinson, Sam F et al Saltwater Intrusion (1986)
Chelsea MI, Lewis Publishers, 433 p.
ISBN 0–87371–054–1.

Auld, John Marijuana Use: a social control (1981)
San Diego CA, Academic Press. ISBN 0–12–068280–X.

Austern, David The Crime Victim's Handbook (1987)
New York, Viking Penguin, 224 p. ISBN 0–670–80475–4.

Austin, Gary Bibliography on Deafness (1976)
Silver Spring MD, National Association of the Deaf.
ISBN 0–913072–20–6.

Austin, Gregory A, et al Drug and Abuse: a guide to research findings (1984)
Santa Barbara CA, ABC Clio.
ISBN 0–87436–413–2. Vol 1: Adults, 443; Vol 2: Adolescents, 510 p.

Australian Academy of Science Ecology of Biological Invasions
Canberra City ACT, Australian Academy of Science.
ISBN 0–85847–128–0. Published jointly with Cambridge University Press.

Averill, James R Anger and Aggression: an essay on emotion (1982)
Berlin, Springer–Verlag, 402 p. illus. Springer Series in Social Psychology. ISBN 0–387–90719–X.

Avery, John Health Effects of War and the Threat of War: an introduction to the literature (1988)
Copenhagen, WHO, 68 p.

Avineri, Shlomo (Ed) Varieties of Marxism (1977)
Jerusalem, The Van Leer Jerusalem Foundation, 404 p.

Ayliffe, G A J and Collins, B Hospital Acquired Infection (1989)
Sevenoaks, Butterworths, 144 p. illus. 2nd ed.
ISBN 0–7236–1259–5.

Ayoob, Mohammed Conflict and Intervention in the Third World (1980)
New York, Saint Martin's Press. ISBN 0–312–16228–6.

Ayres, P G Effects of Disease on the Physiology of the Growing Plant (1982)
Cambridge MA, Cambridge University Press, 200 p. illus.
ISBN 0–521–23306–2. Society for Experimental Biology Symposium Ser: No 11.

Ayres, Robert L Banking on the Poor: the world bank and world poverty (1983)
Cambridge MA, MIT Press, 296 p. ISBN 0–262–51028–6.

Azar, Henry A and Potter, Michael Multiple Myeloma and Related Disorders
Ann Arbor MI, Books on Demand. Vol. 1
ISBN 0–8357–9426–1.

Aziz, Sataj Agricultural Policies for the 1990s (1990)
Paris, OECD, 136 p. ISBN 92–64–13350–X.

Badcock, Blair Unfairly Structured Cities (1984)
New York, Basil Blackwell, 448 p. ISBN 0–631–13395–X.

Badham, Paul and Badham, Linda Death and Immortality in the Religions of the World (1987)
New Era Books/Paragon Hse, 238 p.
ISBN 0–913757–67–5.

Baer, George M The Natural History of Rabies (1975)
San Diego CA, Academic Press. ISBN 0–12–072401–4.

Bagdonas, A; Georg, J C and Gerber, J F Techniques of Frost Prediction and Methods of Frost and Cold Protection (1978)
Geneva, WHO, 101 p. ISBN 92–63–10487–5.

Bagh, Hazari Challenge of Watersheds: proceedings of the Development Centre on Watershed Management for Asia and the Far East
Bihar, Soil Conservation Society, 167 p.

Bagnara, S, et al Interaction of Workers and Machinery: physical and psychological stress (1987)
Brussels, Commission of the European Communities, 120 p.
ISBN 92–825–6485–1.

Bailey, Alice A Glamour: a world problem (1973)
New York, Lucis Publishing. ISBN 0–85330–009–7.

Bailey, Alice A Problems of Humanity (1983)
New York, Lucis Publishing. ISBN 0–85330–113–1.

Bailey, F Lee and Rothblatt, Henry B Crimes of Violence: homicide and assault (1973)
Rochester NY, Lawyers Co–Operative Publishing, 543 p.
ISBN 0–686–05455–5.

Bailey, F Lee and Rothblatt, Henry B Crimes of Violence: rape and other sex crimes (1973)
Rochester NY, Lawyers Co–Operative Publishing.
ISBN 0–686–14500–3.

Bailey, Joe Pessimism (1988)
New York, Routledge Chapman and Hall, 200 p.
ISBN 0–415–00247–8.

Bailey, M R, et al (Eds) The Global Impact of AIDS: proceedings of the First International Conference on the global impact of AIDS, co–sponsored by the World Health Organization and the London School of Hygiene and Tropical Medicine, held in London, March 1988 (1989)
New York, Alan R Liss, 460 p. ISBN 0–8451–4270–4.

Bairoch, Paul Urban Unemployment in Developing Countries. The Nature of the Problem and Proposals for its Solution (1976)
Geneva, ILO, 99 p. ×3 ISBN 92–2–100998–X.

Baker Explosion Hazards and Evaluation (1983)
Amsterdam, Elsevier Science Publishing, 808 p. Fundamental Studies in Engineering: Vol 5.

Baker, Lawrence, et al Biology and Therapy of Acute Leukemia (1985)
Dordrecht, Martinus Nijhoff.

Baker, Paul J and Anderson, Louis E Social Problems: a critical thinking approach (1987)
Belmont CA, Wadsworth Publishing, 414 p.
ISBN 0–534–07428–6.

Baker, R E D and Holliday, P Witches' Broom Disease of Cacao (1957)
Kew, CAB International Mycological Institute, 42 p. illus. Phytopathological Papers: No 2.

Baker, Scott R and Rogul, Marvin (Eds) Environmental Toxicity and the Aging Process (1987)
New York, Alan R Liss, 162 p. Progress in Clinical and Biological Research Ser: Vol 228. ISBN 0–8451–5078–2.

Baker, Timothy B and Cannon, Dale S Assessment and Treatment of Addictive Disorders (1988)
New York, Praeger Publishers, 320 p.
ISBN 0–275–92388–6.

Bakker, D J and Vlught, H van der (Eds) Neuropsychological Correlates and Treatment (1989)
Lisse, Swets en Zeitlinger, 224 p. Learning Disabilities: 1.
ISBN 90–265–0983–9.

Bakker, Dirk J; Satz, Paul and De Wit, Jan Specific Reading Disability: advances in theory and method (1970)
Lewiston NY, Hogrefe International, 166 p.
ISBN 90–237–4103–X.

Bakshi, Rajinder Singh Politicians, Bureaucrats and the Development Process (1986)
New Delhi, Radiant Publishers, 246 p.
ISBN 81–7027–094–4.

Balderston, Judith, et al Malnourished Children of the Rural Poor: the web of food, health, education, fertility (1981)
Dover MA, Auburn House Publishing, 204 p.
ISBN 0–86569–071–5.

Baldry, P E The Battle Against Bacteria: a fresh look (1976)
New York, Cambridge University Press, 140 p.
ISBN 0–521–21268–5.

Balfour, D J Nicotine and the Tobacco Smoking Habit (1984)
Elmsford NY, Pergamon Books, 220 p.
ISBN 0–08–030779–5.

Ball, Jean A Reactions to Motherhood: the role of post–natal care (1987)
New York, Cambridge University Press, 168 p.
ISBN 0–521–30331–1.

Ball, Madeleine and Mann, Jim Lipids and Heart Disease: a practical approach (1989)
New York, Oxford University Press, 120 p.
ISBN 0–19–261701–X.

Ballenger, J C Neurobiology of Panic Disorder (1990)
Chichester, John Wiley and Sons, 410 p.
ISBN 0–471–56210–6.

Ballenger, John J (Ed) Diseases of the Nose, Throat, Ear, Head and Neck (1985)
Philadelphia PA, Lea and Febiger, 1432 p. illus, 13th ed.

Balter, Harry G Tax Fraud and Evasion (1983)
New York, Warren Gorham and Lamont.
ISBN 0–88262–796–1.

Baltruch, H J and Waltz, Millard Cancer and Stress: international psychontology project (1987)
New York, AMS Press. ISBN 0–404–63256–4.

Bambawale, Usha Inter–religious Marriages (1982)
Poona, Dastane Ramchandra, 250 p.

Bamber, J H The Fears of Adolescents (1979)
San Diego CA, Academic Press. ISBN 0–12–077550–6.

Bammer, Kurt and Newberry, Benjamin H (Eds) Stress and Cancer (1981)
Lewiston NY, Hogrefe International, 264 p.
ISBN 0–88937–003–6.

Ban, T A and Lehmann, H E (Eds) Diagnosis and Treatment of Old Age Dementias (1989)
Basel, S Karger AG, 120 p. Available in the UK and Africa only.

Bandman, Elsie L and Bandman, Bertram Bioethics and Human Rights: a reader for health professionals (1986)
Lanham MD, University Press of America, 408 p.
ISBN 0–8191–5257–9.

Bandyopadhyaya, J Climate and World Order: an inquiry into the natural cause of underdevelopment (1983)
Columbia MO, South Asia Books, 178 p.

Banghawe, A F; Mngola, E N and Maina, G (Eds) Use and Abuse of Drugs and Chemicals in Tropical Africa: proceedings of the 1973 Annual Scientific Conference of the East African Medical Research Council Nairobi (1975)
Nairobi, Kenya Literature Bureau, 693 p.

Bank, Arthur, et al (Eds) Fifth Cooley's Anemia Symposium (1985)
New York, New York Academy of Science, 471 p. Annals of the New York Academy of Sciences: Vol 445.
ISBN 0–89766–284–9.

Bankowski, Z and Bryant, J H (Eds) Health Policy, Ethics and Human Values: European and North American perspectives: conference highlights, papers and conclusions, XXIst CIOMS Conference (1988)
Geneva, WHO, 223 p. ISBN 92–9036–034–8.

Bankowski, Z and Carballo, M (Eds) Battered Children and Child Abuse: proceedings of the CIOMS Round Table Conference, XIXth (1986)
Geneva, WHO, 187 p. ISBN 92–9036–026–7.

Bankowski, Z and Mejia, A Health Manpower Out of Balance Conflicts and Prospects: proceedings of the XXth CIOMS Round Table Conference, Acapulco, Mexico, 7–12 Sept 1987 (1987)
Geneva, WHO, 210 p. ISBN 9–2903–6030–5.

Banks, Moira A Stroke (1986)
New York, Churchill Livingstone, 225 p.
ISBN 0–443–02923–7.

Banner, Hubert S and Tremayne, Errol E Calamities of the World (1971)
Detroit MI, Gale Research, 288 p. ISBN 0–8103–3918–8.

Baran, and Dawber, Diseases of the Nails and Their Management (1984)
Saint Louis MO, CV Mosby. ISBN 0–632–01058–4.

Baran, P and Sweezy, P Monopoly Capital (1966)
New York, Monthly Review Press.

Baranov, V I and Khitrov, L M (Eds) Radioactive Contamination of the Sea (1966)
Jerusalem, Keter Publishing House, 200 p.
ISBN 0–7065–0425–9.

Barber, Sotirios A, et al Introduction to Problem Solving in Political Science (1971)
Columbus OH, Charles E Merrill, 109 p. Merrill Political Science Series. ISBN 0–675–09207–8.

Bardsley, Barney Flowers in Hell: an investigation into women and crime (1988)
Pandora Pr/Routledge Chapman and Hall.
ISBN 0–86358–065–3.

Barefoot, J K Employee Theft Investigation (1990)
Sevenoaks, Butterworths, 224 p. 2nd ed.
ISBN 0–409–90211–X.

Barfield, Owen Saving the Appearances: a study in idolatry (1965)
San Diego CA, Harcourt Brace Jovanovich, 190 p.
ISBN 0–15–679490–X.

Baring–Gould, Sabine Book of Werewolves: being an account of terrible superstition
Detroit MI, Gale Research, 266 p. ISBN 0–8103–4241–3.

Barksdale, Byron L Investment Broker Malpractice (1987)
Galveston TX, Yellow Rose Financial Corporation.
ISBN 0–930631–01–3.

Barlow, Charles F Headaches and Migraine in Childhood (1985)
Philadelphia PA, Lippincott JB, 288 p.
ISBN 0–632–01326–5.

Barlow, Geoffrey and Hill, Alison Video Violence and Children (1986)
New York, Saint Martin's Press, 192 p.
ISBN 0–312–84571–5.

Barlow, Robin, et al Economic Behavior of the Affluent (1978)
Westport CT, Greenwood Press. ISBN 0–8371–9454–7.

Barnaby, Frank The Gaia Peace Atlas: survival into the Third Millennium (1988)
London, Pan Books, 271 p. illus. ISBN 0–330–30151–9.

Barnard, George W The Child Molester: an integrated approach to evaluation and treatment (1989)
New York, Brunner/Mazel, 300 p. Clinical Psychiatry Ser.
ISBN 0–87630–526–5.

Barnes, Harry E The Chickens of the Interventionist Liberals Have Come to Roost
Brooklyn NY, Revisionist Press. ISBN 0–87700–194–4.

Barnes, LeRoy W Social Problems 1989–1990 (1989)
Guilford CT, Dushkin Publishing Group, 224 p. illus. 17th ed.
ISBN 0–87967–783–X.

Barnett, A and Bell, R M Rural Energy and the Third World: a review of social science research and technology policy problems (1982)
New York, Pergamon Books, 302 p. illus.
ISBN 0–08–028954–1.

Barney, Gerald O The Global 2000 Report to the President: entering the twenty–first century (1980)
Washington DC, United States Government Printing Office.

Barney, Gerald O The Global 2000 Report to the President: entering the twenty–first century (1980)
Washington DC, GPO.

Baron, R A Human Aggression (1977)
New York, Plenum Publishing, 316 p. illus. Perspectives in Social Psychology Ser. ISBN 0–306–31050–3.

Barrat, John and Louw, M H H International Aspects of Over–Population: proceedings of a conference at Jan Smuts House, Johannesburg 1970 (1972)
S.A.I.I.A., 334 p.

Barrett, James and Rose, Robert M Mental Disorders in the Community: findings from psychiatric epidemiology: annual (1986)
New York, Guilford Press, 377 p. ISBN 0–89862–376–6.

Barrett, Rowland P, et al Advances in Developmental Disorders (1987)
Greenwich CT, Jai Press. vol 1. ISBN 0–89232–841–X.

Barry, Kathleen Female Sexual Slavery (1986)
New York, Avon Books, 336 p. ISBN 0–380–54213–7.

Bartels, Dianne, et al Beyond Baby M: ethical issues in new reproductive techniques (1990)
Clifton NJ, Humana Press, 288 p. ISBN 0–89603–166–7.

Bartknecht, W Explosions: course, prevention, protection (1980)
Berlin, Springer–Verlag, 251 p. ISBN 3–540–10216–7.

Bartlett, Bruce Tailoring Taxes to Promote International Growth

Bartoli, Jill and Botel, Morton Reading – Learning Disability: an ecological approach (1988)
New York, Columbia University Press, 280 p.
ISBN 0–8077–2905–1.

Barton, Len and Walker, Stephen Race, Class and Education (1983)
New York, Routledge Chapman and Hall, 235 p.
ISBN 0–7099–0684–6.

Barton, Russell Institutional Neuroses (1959)
Bristol, John Wright.

Bartone, John C Occupational Diseases: international survey with medical research (1983)
Annandale VA, ABBE Publishers Association of Washington, 152 p. ISBN 0–941864–50–2.

Bartone, John C War: a medical, psychological and scientific subject (1984)
Annandale VA, ABBE Publishers Association of Washington, 160 p. ISBN 0–941864–91–X.

Bartosova, Ludmila, et al Diseases of the Hair and the Scalp (1984)
Basel, S Karger AG, 252 p. illus. Current Problems in Dermatology: Vol 12. ISBN 3–8055–3783–2.

Baskin, V Aid Offered by the West: myths and reality
Los Angeles CA, Progress Publishing, 200 p.

Baskin, Wade and Wedeck, Harry E Dictionary of Pagan Religion (1973)
Secaucus NJ, Lyle Stuart, 324 p. ISBN 0–8065–0386–6.

Bass, Lewis J D Product Liability: design and manufacturing defects (1986)
New York, McGraw–Hill Book Company, 572 p. ISBN 0–07–004036–2.

Bassis, Michael S, et al Social Problems (1982)
San Diego CA, Harcourt Brace Jovanovich, 586 p. ISBN 0–15–581430–3.

Bastenie, P A and Bonny (Eds) Recent Progress in the Diagnosis and Treatment of Hypothyroid Conditions (1980)
Amsterdam, Elsevier Science Publishing, 158 p. International Congress Ser: Vol 529. ISBN 0–444–90161–2.

Batchelor, Edwardd Jr (Ed) Abortion: the moral issues (1982)
New York, The United Christian Pilgrim Press, 256 p. ISBN 0–8298–0612–1.

Batesman, Nils and Peterson, David Social Issues (1989)
Englewood Cliffs NJ, Prentice Hall, 416 p. ISBN 0–13–815994–7.

Batshaw, Mark L; Perret, Yvonne M and Kasmer, Elaine Children with Handicaps: a medical primer (1986)
Baltimore MD, Brookes Paul H Publishing, 413 p. ISBN 0–933716–64–8.

Batten, Peter Living Trophies: a shocking look at the conditions of America's zoos (1976)
New York, Harper and Row.

Baum, Michael Breast Cancer: the facts (1988)
Oxford NY, Oxford University Press, 128 p. illus. 2nd ed. ISBN 0–19–261728–1.

Baumann, B Imaginative Participation: the career of an organizing concept in a multidisciplinary context (1975)
Dordrecht, Kluwer Academic Publishers Group, 198 p. ISBN 90–247–1693–4.

Baumer, Jean–Max and Gleich, Albrecht von Transnational Corporations in Latin America (1982)
Grüsch, Rüegger Verlag, 175 p. ISBN 3–7253–0157–3.

Baumgart, Winfried and Mast, Ben V Imperialism: the idea and reality of British and French colonial (1982)
New York, Oxford University Press. ISBN 0–19–873040–3.

Baur, Susan Hypochondria: woeful imaginings (1988)
Berkeley CA, University of California Press, 260 p. ISBN 0–520–06107–1.

Baxi, Uprendra and Paul, Thomas Mass Disasters and Multinational Liability: the Bhopal case (1986)
Columbia MO, South Asia Books, 230 p. ISBN 0–317–56366–1.

Baxter, Ralph H Sexual Harassment in the Workplace (1986)
New York, Executive Enterprises Publications, 100 p. ISBN 0–88057–407–0.

Bayer, Edward J Rape Within Marriage: a moral analysis delayed (1985)
Lanham MD, University Press of America, 160 p. ISBN 0–8191–4613–7.

Beales, Philip Otosclerosis (1981)
Sevenoaks, Butterworths, 216 p. illus. ISBN 0–7236–0598–X.

Beard, George M and Rosenberg, Charles E American Nervousness, Its Causes and Consequences: a supplement to nervous exhaustion
Salem NH, Ayer Company Publishers, 382 p. ISBN 0–405–03932–8.

Beaud, Michel History of Capitalism Fifteen Hundred to Nineteen Eighty
New York, Monthly Review Press. ISBN 0–317–61677–3.

Beaudry, Micheline Battered Women (1985)
Montreal, Black Rose Books, 125 p. ISBN 0–920057–47–0.

Beazley, Kim and Clark, Ian Politics of Intrusion: the super powers and the Indian Ocean (1979)
Chippendale, Alternative Publishing Cooperative, 148 p. ISBN 0–909188–22–X.

Beck, L, et al (Eds) The Cancer Patient: illness and recovery (1986)
New York, VCH Publications, 171 p. illus. Cancer Campaign Ser: Vol 9. ISBN 0–89574–193–8.

Beck, William A and Avioli, Louis V Osteoporosis (1988)
Washington DC, American Association of Retired Persons. ISBN 0–318–23730–X.

Becker, Ernest The Denial of Death (1973)
New York, Free Press. ISBN 0–02–902150–2.

Becker, Howard S (Ed) Social Problems: a modern approach (1966)
New York, John Wiley and Sons, 31 p.

Becker, W, et al Ear, Nose and Throat Diseases (1988)
New York, Thieme Medical Publishers, 255 p. ISBN 0–86577–226–6.

Becker, Yechiel (Ed) African Swine Fever (1986)
Dordrecht, Kluwer Academic Publishers Group, 176 p. ISBN 0–89838–848–1.

Beckerman, Wilfred, et al Poverty and the Impact of Income Maintenance Programmes in Four Developed Countries: case studies of Australia, Belgium, Norway and Great Britain (1979)
Washington DC, ILO, 90 p. ISBN 9–22–102063–0.

Beckwith, Burnham P Government by Experts: the next stage in political evolution (1972)
Palo Alto CA, Beckwith Burnham Putman. ISBN 0–682–47539–4.

Beevers, Gareth P and Cruickshank, Kennedy J (Eds) Ethnic Factors in Health and Disease (1989)
Sevenoaks, Butterworths, 400 p. illus. ISBN 0–7236–0916–0.

Behrman, S J, et al Progress in Infertility (1987)
Boston MA, Brown Little, 880 p. 3rd ed.

Beilenson, Lawrence W Power Through Subversion (1972)
Washington DC, Public Affairs Press. ISBN 0–8183–0195–3.

Beilke, S and Elshout, A J (Eds) Acid Deposition (1983)
Dordrecht, Kluwer Academic Publishers Group. ISBN 90–2771–588–2.

Bell, Coral Crises and Policy–Makers (1976)
Canberra, Australian National University, 128 p.

Bell, J Bowyer On Revolt: strategies of national liberation (1976)
Cambridge MA, Harvard University Press, 368 p. ISBN 0–674–63655–4.

Bellack, Alan S, et al (Eds) International Handbook of Behavior Modification and Therapy (1982)
New York, Plenum Publishing, 1052 p.

Bellak, Leopold Overload: the new human condition (1975)
New York, Human Sciences Press, 223 p. ISBN 0–87705–245–X.

Belli, Melvin and Krantzler, Mel Divorcing (1988)
New York, Saint Martin's Press, 528 p. ISBN 0–312–01760–X.

Bellisario, R and Mizejewski, G J (Eds) Transplacental Disorders: perinatal detection, treatment and management (including pediatric AIDS) (1990)
New York, Alan R Liss, 288 p. ISBN 0–471–56677–2.

Belveal, L Dee Speculation in Commodity Contracts and Options (1985)
Homewood IL, Dow Jones–Irwin, 250 p. ISBN 0–87094–672–2.

Beman, Lamar T Selected Articles on Capital Punishment (1983)
New York, AMS Press. Repr of 1925. Capital Punishment Ser. ISBN 0–404–62401–4.

Ben–Rafael, E and Lissak, M Social Aspects of Guerilla and Anti–Guerilla Warfare
Jerusalem, Magnes Press, 96 p. ISBN 965–223–314–5.

Ben–Yehuda, Nachman Deviance and Moral Boundaries: witchcraft, the occult, science fiction (1987)
Chicago IL, University of Chicago Press, 260 p. ISBN 0–226–04336–3.

Benda, Julien Treason of the Intellectuals (1969)
New York, WW Norton. ISBN 0–393–00470–8.

Bender, H G and Beck, L Cancer of the Uterine Cervix (1985)
New York, VCH Publishers, 231 p. ISBN 0–89574–184–9.

Bender, Lynn–Darrell Politics of Hostility (1975)
San Germán, Inter American University Press, 168 p. 5th ed. ISBN 0–913480–24–X.

Bendinelli, M and Friedman, H Coxsackie Viruses: a general update (1988)
New York, Plenum Publishing, 450 p. ISBN 0–306–42725–7.

Benegar, John and Johnson, Jacquelyn Global Issues in the Intermediate Classroom (1989)
Boulder CO, Social Science Education Consortium, 186 p. illus. rev ed. ISBN 0–89994–323–3.

Benegar, John and Johnson, Jacquelyn Global Issues in the Intermedia Classroom (1989)
Boulder CO, Social Science Education Consortium, 186 p. illus. rev ed. ISBN 0–89994–323–3.

Beniwal, S P S, et al An Annotated Bibliography of Pigeonpea Diseases 1906–81 (1985)
Patancheru A P, ICRISAT, 114 p.

Benneth, F J; Nsanzumuhire, H and Nhonoli, A M (Eds) Degenerative Disorders in the African Environment: epidemiology and consequences (1976)
Nairobi, Kenya Literature Bureau, 390 p.

Bennett, David Cardiac Arrhythmias (1989)
Sevenoaks, Butterworths, 208 p. illus. ISBN 0–7236–1595–0.

Bennett, James T and Williams, Walter E Strategic Minerals: the economic impact of supply disruptions (1981)
Washington DC, Heritage Foundation, 59 p. ISBN 0–317–47060–4.

Bennett, John; Cuschieri, A and Hennessy, T P J Reflux Oesophagitis (1989)
Sevenoaks, Butterworths, 208 p. illus. ISBN 0–407–01445–4.

Bennett, Wayne W and Hess, Karen M Investigating Arson (1984)
Springfield IL, Thomas Charles C, 422 p. illus. ISBN 0–398–04934–3.

Benson, D Frank Aphasia, Alexia and Agraphia (1979)
New York, Churchill Livingstone. illus. Clinical Neurology and Neurosurgery Monographs: Vol 1.

Benson, Hazel B Behavior Modification and the Child: an annotated bibliography (1979)
Westport CT, Greenwood Press. ISBN 0–313–21489–1.

Bentovin, Arnon, et al Child Sexual Abuse Within the Family: assessment and treatment (1988)
Littleton MA, PSG Publishing, 320 p. illus. ISBN 0–7236–0634–X.

Benyon, John and Solomos, John The Roots of Urban Unrest (1987)
Elmsford NY, Pergamon Books, 200 p. ISBN 0–08–035840–3.

Bequele, A and Boyden, J Combating Child Labour (1988)
Geneva, ILO, 226 p. ISBN 92–2–106388–7.

Berberoglu, Berch The Internationalization of Capital: imperialism and capitalist development on a world (1987)
New York, Praeger Publishers, 245 p. ISBN 0–275–92169–7.

Beres, Louis R Apocalypse: nuclear catastrophe in world politics (1980)
Chicago IL, University of Chicago Press. illus. ISBN 0–226–04360–6.

Beres, Louis R Terrorism and Global Security: the nuclear threat (1987)
Boulder CO, Westview Press, 161 p. ISBN 0–8133–0411–3.

Berg, Leo The Superman in Modern Literature: flaubert, carlyle, emerson, nietzsche
Philadelphia PA, Richard West. ISBN 0–8274–3555–X.

Bergan, John J and Yao, James (Eds) Aneurysms: diagnosis and treatment (1981)
New York, Grune and Stratton, 704 p. ISBN 0–8089–1440–5.

Bergauist, James and Manickam, P Kambar Crisis of Dependency in Third World Ministries (1976)
Blantyre, Christian Literature Association, 160 p.

Berger, Peter L (Ed) Capitalism and Equality in the Third World: modern capitalism, vol II (1987)
London, Hamish Hamilton, 375 p. illus. ISBN 0–8191–5574–8.

Berglas, Steven The Success Syndrome: hitting bottom when you reach the top (1986)
New York, Plenum Publishing, 300 p. ISBN 0–306–42349–9.

Bergmann, Werner (Ed) Without Trial: current research on antisemitism (1987)
Berlin, Walter De Gruyter, 546 p. Psychological Research on Antisemitism Ser: Vol 2. ISBN 0–89925–355–5.

Bergquist, Charles W Alternative Approaches to the Problem of Development: a selected and annotated bibliography (1979)
Durham NC, Carolina Academic Press, 264 p. ISBN 0–89089–081–1.

Bergsma, Daniel X–Linked Mental Retardation and Verbal Disability
White Plains NY, March of Dimes Birth Defects Foundation. ISBN 0–686–10022–0.

Bergsma, Daniel (Ed) Limb Malformations
White Plains NY, March of Dimes Birth Defects Foundation. Symposia Ser: Vol 10, No 5.

Bergsma, Daniel (Ed) Cytogenetics, Environmental Malformation Syndromes (1976)
White Plains NY, March of Dimes Birth Defects Foundation. Alan R Liss, Inc Ser: Vol 12, No 5. ISBN 0–686–18079–8.

Berkman, Alexander ABC of Anarchism
New York, Free Press, 86 p. ISBN 0–686–46369–2.

Berlage, Gai and Egelman, William Understanding Social Issues: sociological fact finding (1989)
Needham Heights MA, Allyn and Bacon, 150 p. 2nd ed. ISBN 0–205–12255–8.

Berlin, G Lennis Earthquakes and the Urban Environment (1980)
Boca Raton FL, CRC Press. 3 Vols. ISBN 0–8493–5173–1.

Berme, Necip (Ed) Biomechanics of Normal and Pathological Human Articulating Joints (1985)
Dordrecht, Kluwer Academic Publishers Group, 384 p. ISBN 90–247–3164–X.

Bernard, F Paedophilia: a factual report (1985)
Rotterdam, Enclave.

Bernard, Jessie Social Problems at Midcentury: role, status and stress in a context of abundance (1957)
New York, Holt, Rinehart and Windston.

Bernardoni, Claudia and Werner, Vera (Eds) Wasted Wealth: the Participation of women in public life (1985)
Munich, K G Saur Verlag, 238 p. ISBN 3–598–10603–3.

Bernhardt, R and Jolowicz, J–A (Eds) International Enforcement of Human Rights: reports submitted to the colloquium of the International Association of Legal Science, Heidelberg 1985 (1987)
Berlin, Springer–Verlag, 170 p. ISBN 3–540–17574–1.

Bernholz, Peter International Game of Power: past, present and future (1985)
Berlin, Mouton de Gruyter, 216 p. ISBN 3–11–009784–2.

Berns, Kenneth I The Parvoviruses (1984)
New York, Plenum Publishing, 424 p. ISBN 0–306–41412–0.

Bernstein, Arthur H Avoiding Medical Malpractice (1987)
Chicago IL, Pluribus Press, 150 p. ISBN 0–931028–91–4.

Bernstein, J; et al Low Back Pain: an historical and contemporary overview of the occupational, medical and psychosocial issues of chronic back pain (1989)
Thorofare NJ, Slack, 250 p.

Bernstein, Penny L Theoretical Approaches in Dance–Movement Therapy (1984)
Dubuque IA, Kendall Hunt Publishing, 224 p. vol II. ISBN 0–8403–3463–X.

Bernthal, John E and Bankson, Nicholas W Articulation and Phonological Disorders (1988)
Englewood Cliffs NJ, Prentice Hall, 352 p. ISBN 0–13–048943–3.

Bernzweig, Eli The Nurse's Liability for Malpractice
New York, McGraw–Hill Book Company, 416 p. 4th ed. ISBN 0–07–005065–1.

Berry, Wendell The Futility of Global Thinking
In: *Resurgence* Bideford UK, March/April 1990, 139, pp. 4–6.

Bertell, Rosalie No Immediate Danger?: prognosis for a radioactive earth
Toronto ON, The Women's Press, 400 p. ISBN 0–88961–092–4.

Berth, Verstappen (Ed) Human Rights Reports: an annotated bibliography of fact–finding missions (1987)
Munich, K G Saur Verlag, 393 p. ISBN 3–598–10745–5.

Berthold, and Schwarz, UFO–Dynamics: psychiatric and psychic aspects of the ufo (1983)
Moore Haven FL, Rainbow Books, 260 p. Bk. 2 ISBN 0–935834–13–3.

Berthoud, Richard Disadvantages of Inequality: a study of social deprivation (1976)
Woodstock NY, Beekman Publications. illus.
ISBN 0-8464-0338-2.

Beschner, and Friedman, Youth Drug Abuse: problems, issues, and treatment (1979)
Lexington MA, Lexington Books. ISBN 0-669-02804-5.

Best, Joel Images of Issues: typifying contemporary social problems (1989)
Hawthorne NY, Aldine de Gruyter, 280 p. Social Problems and Social Issues Series. ISBN 0-202-30352-7.

Béteille, André (Ed) Equality and Inequality: theory and practice
Oxford NY, Oxford University Press, 300 p.

Béteille, André Idea of Natural Inequality: and other essays
Oxford NY, Oxford University Press, 240 p.

Betsworth, Roger G The Radical Movement of the Nineteen Sixties (1980)
Metuchen NJ, Scarecrow Press, 363 p.
ISBN 0-8108-1307-6.

Betts, Raymond F and Shafer, Boyd C The False Dawn: european imperialism in the nineteenth century (1975)
Minneapolis MN, University of Minnesota Press.
ISBN 0-8166-0762-1.

Betts, Richard K Nuclear Blackmail and Nuclear Balance (1987)
Washington DC, Brookings Institution, 240 p.
ISBN 0-8157-0936-6.

Beumont, P J V and Touyz, S W Anorexia Nervosa and Bulimia: a guide for clinicians (1985)
San Diego CA, Academic Press.

Bevington, Ch F P Identification and Quantification of Atmospheric Emission Sources of Heavy Metals and Dust from Metallurgical Processes and Waste Incineration (1987)
Luxembourg, CEE, 106 p. ISBN 92-825-7478-4.

Beyer, G Brain Drain: a Selected bibliography on temporary and permanent migration of skilled workers and high level manpower 1965-1972 (1972)
Dordrecht, Kluwer Academic Publishers Group, 78 p.
ISBN 90-247-1453-2.

Bezanson, Randall P, et al Libel Law and the Press: myth and reality (1987)
New York, Free Press. ISBN 0-02-905870-8.

Bhagavan, M R Critique of Appropriate Technology for Underdeveloped Countries (1979)
Uppsala, Scandinavian Institute of African Studies, 56 p.
ISBN 91-7106-150-9.

Bhalla, A S (Ed) Technology and Employment in Industry (1985)
Geneva, ILO, 436 p. 3rd rev ed.
ISBN 92-2-103969-2. Foreword by Amartya Sen.

Bhalla, V K and Sarma, N A World Monetary and Financial System: issues and reforms (1987)
Columbia MO, South Asia Books. ISBN 81-7017-227-6.

Bhattacharya, Mohit Bureaucracy and Development Administration (1985)
New Delhi, Uppal Publishing House.

Bhumralker, Williams Atmospheric Effects of Waste Heat Discharges: energy, power and environment (1982)
New York, Dekker Marcel, 203 p. Vol. 13
ISBN 0-8247-1653-1.

Bianchi, G; Prolow, K and Oledzki, A (Eds) Man under Vibration: suffering and protection (1981)
Amsterdam, Elsevier Science Publishing, 200 p.

Bickner, Mei L and Shaughnessey, Marlene Women at Work: annotated bibliography (1977)
Los Angeles CA, University of California Press, 420 p.
ISBN 0-89215-064-5.

Bier, William C (Eds) Alienation: plight of modern man? (1972)
Bronx NY, Fordham University Press, 271 p. Pastoral psychology ser: No 7. ISBN 0-8232-0950-4.

Bignon, J and Scarpa, G L (Eds) Biochemistry, Pathology and Genetics of Pulmonary Emphysema: proceedings of a Meeting on Emphysema helt at Porto Conte, Sassari (Sardinia), April 27-30, 1980 (1981)
New York, Pergamon Books, 430 p. illus.
ISBN 0-08-027379-3.

Bilal, J and Kuehnhold, W W Marine Oil Pollution in Southeast Asia, Regional (1980)
Rome, FAO, 128 p.

Bilquees, Faid and Hamid, Shahnaz Impact of International Migration on Women and Children Left behind: a Case study of Punjabi village (1981)
Islamabad, Pakistan Institute of Development Economics.

Bilson, Beth (Ed) Wage Restraint and the Control of Inflation: an international survey (1987)
St Martin, 208 p.
ISBN 0-312-00374-9.

Bingham, Richard D, et al The Homeless in Contemporary Society (1987)
Newbury Park CA, Sage Publications, 320 p.
ISBN 0-8039-2888-2.

Bint AIDS and AIDS-related Infections: current strategies for prevention and therapy
San Diego CA, Academic Press. ISBN 0-12-099200-0.

Binyon, Pamela M Concepts of "Spirit" and "Demon": a study in the use of different languages describing the same phenomena (1977)
New York, Peter Lang Publishing, 132 p.
ISBN 3-261-01787-2.

Bird, Eric C Coastline Changes: a global review (1985)
New York, John Wiley and Sons, 219 p.
ISBN 0-471-90646-8.

Bird, Julio and Maramorosch, Karl Tropical Diseases of Legumes: papers presented at the Rio Piedras (1975)
San Diego CA, Academic Press. ISBN 0-12-099950-1.

Birren, J E and Danon, D Aging: a challenge to science and society
New York, Oxford University Press. 3 vols.

Bishop, Beverly Spasticity: its physiology and management (1977)
Alexandria VA, American Physical Therapy Association.
ISBN 0-912452-20-X.

Bishop, David H and Compans, Richard W Nonsegmented Negative Strand Viruses: paramyxonviruses and rhabodoviruses (symposium) (1984)
San Diego CA, Academic Press. ISBN 0-12-102480-6.

Bishop, Duane S Behavioral Problems and the Disabled (1984)
Melbourne FL, Robert E Krieger, 494 p.
ISBN 0-89874-726-0.

Bisley, Geoffrey G A Handbook of Ophthalmology for Developing Countries (1980)
New York, Oxford University Press. ISBN 0-19-261244-1.

Biswas, Asit K and El-Hinnawi, Essam (Eds) Third World and the Environment (1987)
Nairobi, UNEP, 256 p. Natural Resources and Environment Series: No 11. ISBN 1-85148-011-0.

Black, Bertram J Work and Mental Illness: transitions to employment (1988)
Baltimore MD, Johns Hopkins University Press, 288 p.
ISBN 0-8018-3565-8.

Black, H C Black's Law Dictionary (1979)
St Paul, West Publishing Company. ISBN 0-8299-2041-2.

Black, Perry (Ed) Brain Dysfunction in Children: ethiology, diagnosis and management (1981)
New York, Raven Press, 320 p. ISBN 0-89004-022-2.

Blackman, Julie Intimate Violence: a study of injustice (1989)
Irvington NY, Columbia University Press, 205 p.
ISBN 0-231-05094-1.

Blackstock, Paul W and Schaf, Frank Intelligence, Espionage, Counterespionage and Covert Operations: a guide to information sources (1978)
Detroit MI, Gale Research, 272 p. International Relations Information Guide Ser: Vol 2. ISBN 0-8103-1323-5.

Blackstone, William T and Heslep, Robert D (Eds) Social Justice and Preferential Treatment: women and racial minorities in education and business (1977)
Athens GA, University of Georgia Press, 228 p.
ISBN 0-8203-0434-4.

Blagg, Nigel School Phobia and Its Treatment (1987)
New York, Routledge Chapman and Hall, 240 p.
ISBN 0-7099-3938-8.

Blaikie, Piers The Political Economy of Soil Erosion in Developing Countries (1986)
New York, John Wiley and Sons, 184 p.
ISBN 0-470-20419-2.

Blake, W O History of Slavery and the Slave Trade (1970)
Brooklyn NY, Haskell Booksellers. ISBN 0-8383-1105-9.

Blakeslee, Berton The Limb-Deficient Child (1963)
Berkeley CA, University of California Press.
ISBN 0-520-00125-7.

Bland, John H Disorders of the Cervical Spine: diagnosis and medical management (1987)
Philadelphia PA, WB Saunders, 400 p.
ISBN 0-7216-1187-7.

Blaskow, Ned Steamroller and the Pebbles: militant atheism in the Soviet Union
Cronulla, Fact Research, 144 p. ISBN 90-277-1536-X.

Blau, Sheldon P The Body Against Itself (1984)
New York, Doubleday, 144 p. ISBN 0-385-18800-5.

Blaug, Mark Education and the Employment Problem in Developing Countries (1981)
Lanham MD, UNIPUB, 89 p. ISBN 92-2-101005-8.

Blaustein, Albert Human Rights Sourcebook (1987)
New York, Paragon House Publishers, 970 p.
ISBN 0-88702-202-2.

Blauth, W and Schneider-Sickert, F Congenital Deformities of the Hand: an atlas of their surgical treatment (1980)
Berlin, Springer-Verlag, 394 p. ISBN 3-540-10084-9.

Blennerhassett, Evelyn and Gorman, Patricia Absenteeism in the Public Service: information systems and control strategies (1986)
Dublin, Institute of Public Administration, 164 p.

Bles, W and Brandt, T Disorders of Posture and Gait (1986)
New York, Elsevier Science Publishing, 358 p.
ISBN 0-444-80756-X.

Bleuler, Manfred and Clemens, Siegfried M The Schizophrenic Disorders: long-term patient and family studies (1978)
New Haven CT, Yale University Press.
ISBN 0-300-01663-8.

Bliss, Eugene L Multiple Personality, Allied Disorders and Hypnosis (1986)
New York, Oxford University Press, 300 p.
ISBN 0-19-503658-1.

Bliss, William D and Binder, Rudolph M New Encyclopedia of Social Reform
Salem NH, Ayer Company Publishers.
ISBN 0-405-02436-3.

Bloch, Howard R and Ferguson, Frances (Eds) Misogyny, Misandry, and Misanthropy
Berkeley CA, University of California Press, 242 p. illus.
ISBN 0-520-06544-1.

Block, J Bradford The Signs and Symptoms of Chemical Exposure (1980)
Springfield IL, Thomas Charles C, 164 p.
ISBN 0-398-03958-5.

Blong, R J Volcanic Hazards: a sourcebook on the effects of eruptions (1984)
San Diego CA, Academic Press, 440 p.
ISBN 0-12-107180-4.

BloomBecker, Jay (Ed) Computer Crime, Computer Security, Computer Ethics (1986)
National Center Computer Crime, 32 p.
ISBN 0-933561-02-4.

Bloomfield, Horace R Female Executives and the Degeneration of Management (1983)
Albuquerque NM, Institute for Economic and Financial Research, 129 p. ISBN 0-86654-063-6.

Bloor, Colin M and Liebow, Averill A The Pulmonary and Bronchial Circulations In Congenital Heart Disease (1980)
New York, Plenum Publishing, 294 p.
ISBN 0-306-40383-8.

Bluestone, Barry, et al Low Wages and the Working Poor (1973)
Ann Arbor MI, University of Michigan, 215 p.
ISBN 0-87736-126-6.

Blum, et al (Eds) Pharmaceuticals and Health Policy: international perspectives on provision and control of medicines (1981)
Holmes and Meier, 387 p. ISBN 0-8419-0682-3.

Blumenfeld, Samuel L The New Illiterates (1988)
Boise ID, Paradigm Company, 358 p. 2nd ed.
ISBN 0-941995-05-4.

Blumenthal S J and Osofsky, H J Premenstrual Syndrome: current findings and future directions (1986)
Cambridge MA, Cambridge University Press, 110 p.
ISBN 0-521-33276-1.

Boalt, Gunnar The Sociology of Research (1969)
Carbondale IL, Southern Illinois University Press.

Boalt, Gunnar Competing Belief Systems (1984)
Stockholm, Almqvist and Wiksell, 164 p.
ISBN 91-22-00678-8.

Bobath, Berta Abnormal Postural Reflex Activity Caused by Brain Lesions (1985)
Oxford, Heineman Professional Publishing, 128 p. 3rd ed.
ISBN 0-433-03300-2.

Bobbitt, Philip Democracy and Deterrence: the history and future of nuclear strategy (1987)
New York, Saint Martin's Press, 334 p.
ISBN 0-312-00522-9.

Bock, S Allan Food Allergies: a primer for people (1988)
New York, Vantage Press. ISBN 0-533-07562-9.

Bodansky, David et al Indoor Radon and Its Hazards (1987)
Seattle WA, University of Washington Press, 192 p.
ISBN 0-295-96516-9.

Bodey, Gerald P and Fainstein, Victor (Eds) Candidiasis (1985)
New York, Raven Press, 294 p. illus.
ISBN 0-88167-046-4.

Bodley, John H Tribal People and Development Issues: a global overview (1988)
Mountain View CA, Mayfield Publishing Company.

Bodrova, Valentina and Anker, Richard Working Women in Socialist Countries: the fertility connection (1985)
Lanham MD, UNIPUB, 234 p. ISBN 92-2-103910-2.

Boels, D, et al Soil Degradation: proceedings of the land use seminar on soil (1982)
Brookfield VT, Brookfield Publishing, 286 p.
ISBN 90-6191-220-2.

Boerstler, Richard W Letting Go: a holistic and meditative approach to living and dying (1982)
Assocs Thanatology, 112 p. illus. ISBN 0-9607928-0-5.

Bogdanov, A Triangle of Rivalry: USA, Western Europe, Japan
Los Angeles CA, Progress Publishing, 160 p.

Boggs, Carl Social Movements and Political Power: emerging forms of radicalism in the west (1987)
Philadelphia PA, Temple University Press, 304 p.
ISBN 0-87722-447-1.

Boguslaw, Robert Systems Analysis and Social Planning: human problems of post industrial society (1986)
New York, Irvington Publishers. Repr of 1982.
ISBN 0-8290-2011-X.

Bohigian, George M Handbook of External Diseases of the Eye (1987)
Thorofare NJ, Slack, 220 p. ISBN 1-55642-007-2.

Bohrer, Stanley P and Alavi, Abass Bone Ischaemia and Infarction in Sickle Cell Disease (1981)
Saint Louis MO, Warren H Green, 347 p.
ISBN 0-87527-188-X.

Bok, Sissela Lying: moral choice in public and private life (1979)
New York, Random House. ISBN 0-394-72804-1.

Bok, Sissela Secrets: on the ethics of concealment and revelation (1984)
New York, Random House. ISBN 0-394-72142-X.

Bolin, B Climate Changes and their Effect on the Biosphere (1980)
Geneva, WHO, 49 p. ISBN 92-63-10542-1.

Bolle, Kees W (Ed) Secrecy in Religions (1987)
Leiden, Brill, 164 p. ISBN 90-04-8342-1.

Bollens, John C and Schmandt, Henry J Political Corruption: power, money, and sex (1979)
Pacific Palisades CA, Palisades Publishers.
ISBN 0-913530-18-2.

Boller, Paul F Jr and George, John They Never Said It: a book of fake quotes, misquotes and curious citations (1989)
Oxford NY, Oxford University Press, 160 p.
ISBN 0-19-505541-1.

Bollet, Alfred J Plagues and Poxes: the rise and fall of epidemic disease (1987)
New York, Demos Publications. ISBN 0-939957-06-X.

Bolognesi Human Retroviruses, Cancer and AIDS: approaches to prevention and therapy (1988)
New York, John Wiley and Sons. ISBN 0-471-60912-9.

Bolt, Bruce A Nuclear Explosions and Earthquakes: the parted veil (1976)
New York, Freeman WH, 309 p. ISBN 0-7167-0276-2.

Bolton, Frank G and Bolton, Susan R Working with Violent Families (1987)
London, Sage, 400 p. ISBN 0-8039-2586-7.

Bonadio, George R Fires: index of modern information (1988)
Washington DC, ABBE Publications Association, 160 p. ISBN 0-88164-788-8.

Bonatti, Luigi Uncertainty: studies among philosophy, economics and sociopolitical theory (1983)
Amsterdam, BR Grüner, 132 p. ISBN 90-6032-230-4.

Bond, D and Chandley, A C Aneuploidy (1983)
Oxford NY, Oxford University Press. illus. OMMG Ser. ISBN 0-19-261376-6.

Bondi, A et al Urogenital Infections (1988)
New York, Plenum Publishing, 146 p. ISBN 0-306-42799-0.

Bonhoeffer, Dietrich Creation and Fall (1965)
New York, Macmillan. ISBN 0-02-083890-5.

Bonner, R et al The Visually Limited Child (1970)
New York, Irvington Publishers. ISBN 0-8422-0061-4.

Bonnicksen, Andrea In Vitro Fertilization: building policy from laboratories to legislators (1989)
Irvington NY, Columbia University Press, 208 p. ISBN 0-231-06904-9.

Bonomo, Luca and Higginson, A E (Eds) International Overview on Solid Waste Management: a report from the International Solid Wastes and Public Cleansing Association (ISWA) (1988)
San Diego CA, Academic Press, 450 p. ISBN 0-12-114975-7.

Bontrager, G Edwin Divorce and the Faithful Church (1978)
Scottdale PA, Herald Press, 224 p. ISBN 0-8361-1850-2.

Booth, B, et al Source-Book for Volcanic-Hazards Zonation (1984)
Paris, UNESCO, 97 p. illus. ISBN 92-3-102111-7. Natural Hazards Series: No 4.

Bordsky, A and Edelwich, J Burn-Out: stages of disillusionment in the helping professions (1980)
New York, Human Sciences Press.

Borgstrom, Georg Too Many: an ecological overview of earth's limitations (1969)
New York, Macmillan, 400 p. illus.

Bories, J (Ed) Cerebral Ischaemia (1985)
Berlin, Springer-Verlag, 155 p. illus. ISBN 0-387-16158-9.

Bornschier, V and Chase-Dunn, C Transnational Corporations and Under Development (1985)
New York, Praeger.

Borriello, S P (Ed) Antibiotic Associated Diarrhoea and Colitis (1984)
Amsterdam, Kluwer Academic Publishers Group, 188 p. Developments in Gastroenterology Ser.ISBN 0-89838-623-3.

Borschberg, E Prevalence of Varicose Veins in the Lower Extremities (1967)
New York, Karger S AG, 140 p. ISBN 3-8055-0763-1.

Bortner, M A Delinquency and Justice: an age of crisis (1988)
New York, McGraw-Hill Book Company, 397 p. ISBN 0-07-006561-6.

Bos, L Symptoms of Virus Diseases in Plants: with indexes of names of symptoms in English, Dutch, French and Spanish (1978)
Wageningen, Centre for Agricultural Publishing and Documentation, 225 p. 3rd rev ed. ISBN 90-220-0658-1.

Bose, Christine, et al Hidden Aspects of Women's Work (1987)
New York, Praeger Publishers, 384 p. ISBN 0-275-92415-7.

Boskind-White, Marlene and White, William C Jr Bulimarexia: the binge-purge cycle (1987)
New York, WW Norton. 2nd ed.

Bossche, H Vanden et al Aspergillus and Aspergillosis (1988)
New York, Plenum Publishing, 342 p. ISBN 0-306-42828-8.

Bosson, Rex and Varon, Bension The Mining Industry and the Developing Countries (1977)
New York, Oxford University Press. ISBN 0-19-920096-3.

Boström, Harry and Ljungstedt, Nils (Eds) Detection and Prevention of Adverse Drug Reactions (1984)
Stockholm, Almqvist and Wiksell, 294 p. ISBN 91-22-00680-X.

Boswell, David M and Wingrove, Janet M The Handicapped Person in the Community (1974)
New York, Routledge Chapman and Hall. ISBN 0-422-74760-2.

Boswell, John The Abandonment of Children in Western Europe from Late Antiquity to the Renaissance
New York, Pantheon Books, 488 p. illus.

Bottino, Joseph C, et al Liver Cancer (1985)
Norwell MA, Kluwer Academic Publishers. ISBN 0-89838-713-2.

Bottom, Norman R and Gallati, Robert R Industrial Espionage: intelligence techniques and countermeasures (1984)
Stoneham MA, Butterworth Publishers, 352 p. ISBN 0-409-95108-0.

Bottoms, Anthony E and Light, Roy Problems of Long-Term Imprisonment (1987)
Brookfield VT, Gower Publishing, 250 p. Cambridge Criminology Ser: No 58. ISBN 0-566-05427-2.

Boüart, J C Tax Problems of Cultural Foundations and of Patronage in the European Community (1976)
Dordrecht, Kluwer Academic Publishers Group, 128 p. ISBN 90-200-0474-3.

Boud, D J (Ed) Problem-Based Learning in Education for the Professions (1985)
Kensington, Higher Education Research and Development Society, 180 p. ISBN 0-909528-92-6.

Boulding, Elise The Underside of History: a view of women through time (1976)
Boulder CO, Westview, 829 p. illus.

Bourdillon, M F C and Fortes, M Sacrifices (1980)
San Diego CA, Academic Press. ISBN 0-12-119040-4.

Bourgoignie, Thierry, et al (Eds) Unfair Terms in Consumer Contracts: legal treatment, effective implementation and final impact on the consumer: proceedings of the second European Workshop on Consumer Law (1983)
Brussels, Bruylant, 268 p. ISBN 2-87077-199-1.

Bourguignon, Erika Religion, Altered States of Consciousness, and Social Change
Columbus OH, Ohio State University Press, 399 p. ISBN 0-8142-0167-9.

Bourne, Richard and Levin, Jack Social Problems: causes, consequences, interventions (1983)
Saint Paul MN, West Publishing College and School, 422 p. ISBN 0-314-69661-X.

Bouscaren, R; Frank, R and Veldt, C Hydrocarbons: identification of air quality in member states of the European Communities (1987)
Luxembourg, CEE, 326 p. ISBN 92-825-6942-X.

Bousset, Wilhelm The Antichrist legend
London.

Bowker Prison Victimization: a gruesome catalog of unintended punishment (1980)
Amsterdam, Elsevier Science Publishing, 232 p. ISBN 0-444-99077-1.

Bowman, James S, et al Professional Ethics: dissent in organizations, an annotated bibliography and resource guide (1984)
New York, Garland Publishing, 325 p. bibl. Public affairs and administration ser: Vol 2. ISBN 0-8240-9217-1.

Boyadjian, Haig and Warren, James F Risks (1988)
New York, John Wiley and Sons, 392 p. ISBN 0-471-91207-7.

Boyd, Gavin and Pentland, Charles (Eds) Issues in Global Politics (1981)
New York, Free Press. ISBN 0-02-904470-7.

Boyer, J L, et al (Eds) Liver Cirrhosis (1987)
Dordrecht, Kluwer Academic Publishers Group. ISBN 0-85200-993-3. Falk Ser: No 44.

Boyle, Francis A Defending Civil Resistance under International Law (1987)
Ardsley-on-Hudson NJ, Transnational Publications, 240 p. ISBN 0-941320-43-X.

Boyle, Kevin (Ed) Article Nineteen World Report 1988: information, freedom and censorship (1988)
New York, Times Books, 352 p.

Bracewell-Milnes, B, et al International Tax Avoidance (1979)
Dordrecht, Kluwer Academic Publishers Group, 368 p. 2 vols.

Bracho, Frank Utopia and Reality of South-South Economic Cooperation (1986)
In: *Development and Socio-Economic Progress* 35, pp. 16-23.

Bracken, Jeanne M Children with Cancer: a comprehensive reference guide for parents (1986)
Oxford NY, Oxford University Press, 288 p. illus. ISBN 0-19-503482-1.

Bradford, David S, et al Moe's Textbook of Scoliosis and Other Spinal Deformities (1987)
Philadelphia PA, WB Saunders, 672 p. ISBN 0-7216-6428-8.

Bradford, Leland and Bradford, Martha Retirement: coping with emotional upheavals (1979)
Chicago IL, Nelson-Hall. ISBN 0-88229-564-0.

Bradshaw, A D and Chadwick, M J The Restoration of Land: the ecology and reclamation of derelict and (1980)
Berkeley CA, University of California Press. ISBN 0-520-03961-0.

Bradshaw, A D and Chadwick, M J The Restoration of the Land: the ecology and reclamation of derelict and degraded land (1980)
California CA, University of California Press.

Brailsford, Henry N and Strauss, S Why Capitalism Means War (1972)
New York, Garland Publishing. ISBN 0-8240-0287-3.

Brain, Walter R Speech Disorders: aphasia, apraxie and agnosia
Ann Arbor MI, Books on Demand. ISBN 0-317-41727-4.

Brainard, Willard T Women and Spouse Abuse: index of modern information (1988)
Washington DC, ABBE Publications Association, 150 p. ISBN 0-88164-784-5.

Brainard, Williard T Injuries of the Spinal Cord: medical subject analysis and research guide (1985)
Annandale VA, ABBE Publishers Association of Washington, 150 p. ISBN 0-88164-106-5.

Braithwaite, John Corporate Crime in the Pharmaceutical Industry (1984)
New York, Routledge Chapman and Hall, 500 p. ISBN 0-7102-0049-8.

Braithwaite, John Crime, Shame and Reintegration (Date not set)
New York, Cambridge University Press, 240 p. ISBN 0-521-35567-2.

Bramwell, Anna C (Ed) Refugees in the Age of Total War (1988)
London, Unwin Hyman, 384 p. ISBN 0-04-445194-6.

Branch, Turner W Construction Accident Pleading and Practice (1988)
New York, John Wiley and Sons, 426 p. ISBN 0-471-83996-5.

Brandt Commission North-South: a programme for survival (1980)
London, Pan.

Brandt Commission Common Crisis; North-South: cooperation for world recovery (1983)
London, Pan Books, 174 p. ISBN 0-330-28130-5.

Brandt, Lilian Five Hundred Seventy Four Deserters and Their Families: a descriptive study of their characteristics and circumstances (1972)
Salem NH, Ayer Company Publishers, 210 p. 2nd ed. Family in America Ser. ISBN 0-405-03850-X.

Branson, D R and Dickson, K L Aquatic Toxicology and Hazard Assessment (Fourth Conference)- STP 737 (1981)
Philadelphia PA, American Society for Testing and Materials, 466 p. ISBN 0-8031-0799-4.

Brasser, L J and Colley, J R T Chronic Respiratory Diseases in Children in Relation to Air Pollution: report on a WHO study (1980)
Geneva, WHO, 89 p. ISBN 92-9020-167-3. Euro Reports and Series: No 28.

Braun, Ernst Wayward Technology (1984)
Westport CT, Greenwood Press, 224 p. ISBN 0-313-24398-0.

Braunwald, Eugene Heart Disease: a textbook of cardiovascular medicine (1988)
Philadelphia PA, WB Saunders, 2016 p. ISBN 0-7216-1953-3.

Brawner, Carroll O First International Conference on Uranium Mine Waste Disposal (1980)
Littleton CO, Society of Mining Engineers, 626 p. ISBN 0-89520-279-4.

Bray, Mark Are Small Schools the Answer?: cost-effective strategies for rural school provision
London, ComSec, 89 p. ISBN 0-85092-302-6.

Breakwell, Glynis M Threatened Identities (1983)
New York, John Wiley and Sons, 269 p. ISBN 0-471-10233-4.

Brecher, Michael, et al Crises in the Twentieth Century (1988)
Elmsford NY, Pergamon Books, 900 p. 2 vols. ISBN 0-08-034981-1.

Breer, William The Adolescent Molester (1987)
Springfield IL, Thomas Charles C, 240 p. illus.

Brennan, William The Abortion Holocaust: today's final solution (1983)
Saint Louis MO, Landmark Press, 230 p. illus. ISBN 0-911439-01-3.

Brenner, Barry M and Lazarus, J Michael Acute Renal Failure (1983)
Philadelphia PA, WB Saunders, 837 p. ISBN 0-7216-1964-9.

Brenner, M Harvey Mental Illness and the Economy (1973)
Cambridge MA, Harvard University Press, 320 p. ISBN 0-674-56875-3.

Brenner, Reuven Rivalry: in business, science, among nations (1990)
Cambridge, Cambridge University Press, 244 p. ISBN 0-521-38584-9.

Bresciani, Francesco, et al (Eds) Hormones and Cancer Two: proceedings of the second international congress on hormones and cancer (1984)
New York, Raven Press, 736 p. 3 Vols. Progress in Cancer Research and Therapy Ser: Vol 31,35.

Breuer, Georg and Fabian, P Air in Danger: ecological perspectives of the atmosphere (1980)
New York, Cambridge University Press, 180 p. ISBN 0-521-22417-9.

Brewer, Earl J Juvenile Rheumatoid Arthritis (1982)
Philadelphia PA, WB Saunders. ISBN 0-7216-1986-X.

Briant, C L Metallurgical Aspects of Environmental Failures (1985)
Amsterdam, Elsevier Science Publishing, 238 p. ISBN 0-444-42491-1.

Brink, Carla J (Ed) Cocaine: a symposium (1985)
Madison WI, Wisconsin Institute Drug Abuse, 69 p. illus. ISBN 0-9615363-0-6.

Brinton, Maurice Irrational in Politics
Montreal, Black Rose Books, 76 p. ISBN 0-919618-50-2.

Bristow, M R Drug Induced Heart Disease (1981)
New York, Elsevier Science Publishing, 476 p. ISBN 0-444-80206-1.

Britt, Beverley A (Ed) Malignant Hyperthermia (1987)
Dordrecht, Kluwer Academic Publishers Group, 448 p. ISBN 0-89838-960-7.

Brock, D J Early Diagnosis of Fetal Defects (1983)
New York, Churchill Livingstone, 165 p. illus. Current Reviews in Obstetrics and Gynaecology Ser: No 2. ISBN 0-443-02302-6.

Brock, T D Eutrophic Lake (1985)
Berlin, Springer-Verlag, 308 p. ISBN 3-540-96184-4.

Brockington, I F and Kumar, R Motherhood and Mental Illness (1982)
Orlando FL, Grune and Stratton, 288 p. ISBN 0-8089-1481-2.

Brockner, J and Rubin, J Z Entrapment in Escalating Conflicts: a Social psychological analysis (1985)
Berlin, Springer-Verlag, 335 p. ISBN 3-540-96089-9.

Brod, Craig and St John, Wes Technostress: the human cost of the computer revolution
Reading MA, Addison-Wesley, 288 p. ISBN 0-201-11211-6.

Brody, Baruch A (Ed) Suicide and Euthanasia: historical and contemporary themes (1989)
Dordrecht, Kluwer Academic Publishers Group, 250 p. ISBN 0-7923-0106-4.

Brogan, J C (Ed) Nitrogen Losses and Surface Run-Off from Landspreading of Manures: proceedings of a workshop in the EEC Programme of co-ordination and research on effluents from livestock (1981)

Dordrecht, Kluwer Academic Publishers Group, 487 p. ISBN 90–247–2471–6.

Bromley, David G (Ed) Falling from the Faith: causes and consequences of religious apostasy (1988)
London, Sage, 248 p. Sage Focus Editions: Vol 95. ISBN 0–8039–3188–3.

Bromley, Ida Tetraplegia and Paraplegia (1985)
New York, Churchill Livingstone, 261 p. ISBN 0–443–03233–5.

Bromley, J, et al (Eds) Chemical Waste: handling and treatment (1986)
Berlin, Springer–Verlag, 370 p. ISBN 3–540–13246–5.

Brongersma, Edward Loving Boys: a multidisciplinary study of sexual relations between adult and minor males (1988)
Amsterdam, Global Academic Publishers. 0

Bronsen, Hugo H Sports and Athletic Injuries: medical subject analysis and research index with (1985)
Annandale VA, ABBE Publishers Association of Washington, 120 p. ISBN 0–88164–052–2.

Brookes, Vincent J Poisons: properties, chemical identification, symptoms (1975)
Melbourne FL, Robert E Krieger, 318 p. ISBN 0–88275–148–4.

Brookfield, Stephen D Developing Critical Thinkers (1987)
San Francisco CA, Jossey–Bass, 293 p.

Brookins, Douglas G The Indoor Radon Problem (1990)
Irvington NY, Columbia University Press, 228 p. ISBN 0–231–06748–8.

Brooksby, J B (Ed) The Aerial Transmission of Disease (1984)
Port Washington NY, Scholium International, 166 p. illus. Philosophical Transactions of the Royal Society: Ser B: Vol 302. ISBN 0–85403–214–2.

Brown, A J The Great Inflation Nineteen Thirty–Nine to Nineteen Fifty–One (1983)
New York, Garland Publishing, 333 p. Gold, Money, Inflation and Deflation Ser. ISBN 0–8240–5226–9.

Brown, A J World Inflation since Nineteen Fifty: an international comparative Study (1985)
Cambridge MA, Cambridge University Press, 350 p. National institute of Economic and Social Research Economic and Social Studies: No 13. ISBN 0–521–30351–6.

Brown, Burnell R Jr Anesthesia in Hepatic and Biliary Tract Disease (1988)
Philadelphia PA, Davis FA Co, 299 p. illus. ISBN 0–8036–1253–2.

Brown, Darrell, et al Reclamation and Vegetative Restoration of Problem Soils and Disturbed Lands (1987)
Park Ridge NJ, Noyes Publications, 560 p. ISBN 0–8155–1102–7.

Brown, David E and Morehouse, Bonnie S Arizona Wetlands and Waterfowl (1985)
Tucson AZ, University of Arizona Press, 169 p. ISBN 0–8165–0904–2.

Brown, George W and Harris, Tirril Social Origins of Depression: a study of psychiatric disorder in women (1978)
New York, Free Press. ISBN 0–02–904890–7.

Brown, Hanbury The Wisdom of Science: its relevance to culture and religion (1986)
Cambridge, Cambridge University Press, 194 p. illus. ISBN 0–521–31448–8.

Brown, Harold O Heresies (1988)
Grand Rapids MI, Baker Book House, 503 p. ISBN 0–8010–0953–7.

Brown, Henry P Egalitarianism and the Generation of Inequality (1988)
New York, Oxford University Press, 560 p. ISBN 0–19–828648–1.

Brown, J P The Economic Effects of Floods (1972)
Berlin, Springer–Verlag, 87 p. Lecture Notes in Economics and Mathematical Systems: Vol 70. ISBN 0–387–05925–3.

Brown, Lester Soil Erosion: quiet crisis in the world economy (1984)
Washington DC, Worldwatch Institute. Worldwatch Papers. ISBN 0–916468–60–7.

Brown, Lester R Resource Trends and Population Policy: a time for reassessment (1979)
Washington DC, Worldwatch Institute. Worldwatch Papers. ISBN 0–916468–28–3.

Brown, Lester R Food or Fuels: new competition for the world's cropland (1980)
Washington DC, Worldwatch Institute. Worldwatch Papers. ISBN 0–916468–34–8.

Brown, Lester R The Changing World Food Prospect: the nineties and beyond (1988)
Washington DC, Worldwatch Institute, 60 p. Worldwatch Papers. ISBN 0–916468–86–0.

Brown, Lester R and Jacobson, Jodi L Our Demographically Divided World (1986)
Washington DC, Worldwatch Institute, 64 p. Worldwatch Papers Series: 74. ISBN 0–916468–75–5.

Brown, Lester R and Jacobson, Jodi L The Future of Urbanizations: facing the ecological and economic constraints (1987)
Washington DC, Worldwatch Institute, 56 p. illus. Worldwatch Papers. ISBN 0–916468–78–X.

Brown, Lester R, et al State of the World 1990 (1990)
New York, W W Norton, 253 p.

Brown, Thomas M and Scammell, Henry Rheumatoid Arthritis: its cause and its treatment (1988)
New York, Evans M, 256 p. ISBN 0–87131–543–2.

Brown, William J et al Syphilis and Other Venereal Diseases
Ann Arbor MI, Books on Demand. ISBN 0–317–55361–5.

Browne, Ray B Objects of Special Devotion: fetishes and fetishism in popular culture (1982)
Bowling Green OH, Bowling Green University Popular Press. ISBN 0–87972–191–X.

Browne, Ray B Forbidden Fruits: taboos and tabooism in culture (1984)
Bowling Green OH, Bowling Green University Popular Press, 192 p. ISBN 0–317–14769–2.

Brozek, Josef Malnutrition and Human Behavior (1985)
New York, Reinhold Van Nostrand, 432 p. ISBN 0–442–21108–2.

Bruce–Chatt, L J and DeZulueta, Julian The Rise and Fall of Malaria in Europe: a historico–epidemiological study (1980)
Oxford NY, Oxford University Press. illus. ISBN 0–19–858168–8.

Bruckmann, Gerard; Meadows, Donella and Richardson, John Groping in the Dark: the first decade of global modelling (1982)
New York, Wiley.

Brumberg, Joan J Fasting Girls: the emergence of anorexia nervosa (1988)
Cambridge MA, Harvard University Press, 368 p. ISBN 0–674–29501–3.

Brundage, James A Law, Sex and Christian Society in Medieval Europe (1987)
Chicago IL, University of Chicago Press, 646 p. ISBN 0–226–07783–7.

Brunner, and Gravas, Clinical Hypertension and Hypotension (1982)
New York, Dekker Marcel, 498 p. ISBN 0–8247–1279–X.

Brunner, R and Crecine, J P A Fragmented Society: hard to govern democratically
In: *Information Technology, Some Critical Implications for Decision Makers* New York, Conference Board, 1972.

Brunner, S and Langfeldt, B (Eds) Breast Cancer (1987)
Berlin, Springer–Verlag, 145 p. Recent Results in Cancer Research Ser: Vol 105. ISBN 0–387–17301–3.

Brushwood, David B Medical Malpractice: pharmacy law (1986)
New York, McGraw–Hill Book Company, 500 p. ISBN 0–07–008576–5.

Brusten, M, et al (Eds) Youth Crime, Social Control and Prevention: theoretical perspectives and policy implications (1986)
ISBN 3–89085–140–1.

Bruun, Kettil, et al The Gentlemen's Club: international control of drugs and alcohol
Chicago IL, University of Chicago Press, 338 p. ISBN 0–226–07777–2.

Bryans, J T and Gerber, H (Eds) Equine Infectious Diseases 3: 3rd international conference on equine infectious diseases, Paris 1972 (1973)
New York, Karger S AG, 560 p. ISBN 3–8055–1392–5.

Bryant, Clifton D Khaki–Collar Crime: deviant behavior in the military context (1979)
New York, Free Press. ISBN 0–02–904930–X.

Bryant, N J Disputed Paternity: the value and application of blood tests (1980)
New York, Thieme Medical Publishers, 185 p. ISBN 3–13–599001–X.

Brzezinski, Z J; Heikkinen, E and Waters, W E (Eds) The Elderly in Eleven Countries (1983)
Geneva, WHO, 231 p. ISBN 92–890–1157–2.

Brzoska, Michael and Ohlson, Thomas Arms Transfers to the Third World, 1971–1985 (1987)
Oxford, Stockholm International Peace Research Institute, 400 p. illus. SIPRI Publication Series. ISBN 0–19–829116–7.

Buchanan, Allen E Marx and Justice: the radical critique of liberalism (1982)
Totowa NJ, Rowman and Littlefield Publishers, 220 p. ISBN 0–8476–7356–1.

Bucherl, Wolfgang et al Venomous Animals and Their Venoms (1971)
San Diego CA, Academic Press. Vol. 1 ISBN 0–686–66781–6.

Buchheim, Hans, et al Totalitarian Rule: its nature and characteristics (1987)
Middletown CT, Wesleyan University Press, 112 p. ISBN 0–8195–6021–9.

Büchner, T, et al (Ed) Acute Leukemias (1987)
Berlin, Springer–Verlag, 740 p. ISBN 3–540–16556–8.

Buckholdt, David R and Gubrium, Jaber F Caretakers: treating emotionally disturbed children (1985)
Washington DC, University Press of America, 268 p. ISBN 0–8191–4265–4.

Buckle, Peter Musculo–Skeletal Disorders at Work: proceedings of the conference of university of (1987)
New York, Taylor and Francis, 250 p. ISBN 0–85066–381–4.

Buckley, Michael J Modern Atheism (1987)
New Haven CT, Yale University Press, 432 p. ISBN 0–300–03719–8.

Buckley, Peter Essential Papers on Psychosis (1988)
New York, New York University Press, 384 p. ISBN 0–8147–1096–4.

Budenz, Louis F The Techniques of Communism (1977)
Salem NH, Ayer Company Publishers. ISBN 0–405–09942–8.

Buford, Bill Among the Thugs (1990)
London, Martin Secker and Warburg, 224 p. ISBN 0–436–07526–1.

Buikhuizen, Wouter and Mednick, Sarnoff A (Eds) Explaining Criminal Behaviour: interdisciplinary approaches (1988)
Leiden, Brill, 260 p. ISBN 90–04–08514–9.

Bukharin, Nikolai and Lenin, V I Imperialism and World Economy
New York, Monthly Review Press, 176 p. ISBN 0–85345–290–3.

Bukharin, Nikolai and Preobrazhensky, Evgeny The ABC of Communism (1988)
Ann Arbor MI, University of Michigan, 432 p. Ann Arbor Paperbacks Ser. ISBN 0–472–09112–3.

Bull, David A Growing Problem: pesticides and the third world poor
San Francisco CA, Institute for Food and Development Policy. ISBN 0–85598–064–8.

Bull, Hedley Intervention in World Politics (1984)
New York, Oxford University Press. ISBN 0–19–827467–X.

Bullough, Vern L and Elcano, Barrett W A Bibliography of Prostitution (1977)
New York, Garland Publishing, 430 p. ISBN 0–8240–9947–8.

Buranelli, Vincent and Buranelli, Nan Spy–Counterspy: an encyclopedia of espionage (1982)
New York, McGraw Hill Book Company, 352 p. ISBN 0–07–008915–9.

Burchell, Robert W and Sternleib, George Plant Closings in the New Industrial Revolution (1988)
New Brunswick NJ, Center for Urban Policy Research, 412 p. ISBN 0–317–64577–3.

Burdon, Jeremy J Diseases and Plant Population Biology (1987)
New York, Cambridge University Press, 224 p. ISBN 0–521–30283–8.

Burgdorfer, Willy and Anacker, Robert Rickettsiae and Rickettsial Diseases (1981)
San Diego CA, Academic Press. ISBN 0–12–143150–9.

Burger, Julian Report from the Frontier: the state of the world's indigenous people (1987)
Cambridge MA, Cultural Survival, 320 p. ISBN 0–939521–41–5.

Burgess, Ann W and Hartman, Carol R Sexual Exploitation of Patients by Health Professionals (1986)
New York, Praeger Publishers, 207 p. ISBN 0–275–92171–9.

Burish, Thomas G and Bradley, Laurence A Coping with Chronic Disease: research and applications (1983)
San Diego CA, Academic Press. ISBN 0–12–144450–3.

Burka, Jane and Yuen, Lenora Procrastination: why you do it, what to do about it (1983)
Reading MA, Addison–Wesley Publishing, 256 p. illus. ISBN 0–201–10191–2.

Burkholder, H C High Level Nuclear Waste Disposal (1986)
Columbus OH, Battelle Press, 936 p. ISBN 0–935470–29–8.

Burley, Jeffrey Obstacles to Tree Planting in Arid and Semi–Arid Lands: comparative case studies from India and Kenya (1983)
Lanham MD, UNIPUB, 52 p. ISBN 92–808–0391–3.

Burnell, Peter Economic Nationalism in the Third World (1985)
Boulder CO, Westview Press, 288 p. ISBN 0–8133–0338–9.

Burns, Gerald P and Bank, Simmy Disorders of the Pancreas (1989)
New York, Macmillan, 512 p. ISBN 0–02–317280–0.

Burrows, B Respiratory Disorders (1983)
Chicago IL, Year Book Medical Publishers,. ISBN 0–8151–1351–X.

Burtch, Brian E and Ericson, Richard V Silent System: an inquiry into prisoners who suicide and annotated bibliography (1979)
UT Criminology, 113 p. ISBN 0–919584–43–8.

Burton, John Deviance Territorism and War: process of solving unsolved social and political problems (1979)
Potts Point NSW, Australian National University Press, 255 p. ISBN 0–08–032987–X.

Burton, John Deviance, Terrorism and War (1979)
New York, Saint Martin's Press. ISBN 0–312–19753–5.

Burton, Robert Anatomy of Melancholy
AMS Press. repr of 1893. ISBN 0–404–07822–2.

Burton, Theodore E Financial Crises and Periods of Industrial and Commercial Depression
Salem NH, Ayer Company Publishers. ISBN 0–8369–5925–6.

Busch, Briton Cooper War Against the Seals: a history of the North American seal fishery (1984)
Montreal, McGill–Queen's University Press, 432 p. ISBN 0–7735–0578–4.

Bussman, W D and Beisel, A Acute and Chronic Heart Failure (1986)
New York, Springer Publishing, 290 p. ISBN 0–387–15905–3.

Butler, and Lewis, Aging and Mental Health (1983)
New York, New American Library. ISBN 0–452–25405–1.

Butler, E J Fungi and Disease in Plants (1981)
Dehra Dun, International Book Distributors.

Butler, Smedley D; Crozier, Frank P and Thomson, Christopher B Three Generals on War: war is a racket (1973)
New York, Garland Publishing. ISBN 0–8240–0423–X.

Butler, Thomas C Plague and Other Yersinia Infections (1983)
New York, Plenum Publishing, 232 p. ISBN 0–306–41414–7.

Butters, Nelson and Cermak, Laird S Alcoholic Korsakoff's Syndrome: an information processing approach to amnesia (1980)
San Diego CA, Academic Press. ISBN 0–12–148380–0.

Button, John A Dictionary of Green Ideas Vocabulary for a Sane and Sustainable Future (1988)
London, Routledge, 524 p. Introduced by Chomsky, Noam. "Colourless Ideas Sleep Furiously". ISBN 0–415–00231–1.

Buttram, Harold E Dangers of Immunization (1983)
Quakertown PA, Humanitarian Publishing, 72 p. ISBN 0–916285–27–8.

Buzan, Barry G (Ed) The International politics of Deterrence (1987)
St Martin, 231 p. ISBN 0–312–00917–8.

Bychowski, Gustav Evil in Man: anatomy of hate and violence (1968)
New York, Grune and Stratton, 104 p.
Byers, R B (Ed) Deterrence in the 1980s: crisis and dilemma (1984)
London, C Helm, 235 p. illus. ISBN 0-7099-3288-X.
Byers, R B (Ed) Deterrence in the 1980's: crisis and dilemma (1985)
Toronto ON, Canadian Institute of Strategic Studies, 256 p. ISBN 0-7099-3288-X.
Byers, R B (Ed) De-Nuclearization of the Oceans (1986)
Toronto ON, Canadian Institute of Strategic Studies, 288 p. ISBN 0-7099-3936-1.
Byers, R B and Leyton-Brown, D (Eds) Superpower Intervention in the Pesian Gulf (1982)
Toronto ON, Canadian Institute of Strategic Studies. ISBN 0-919769-11-X.
Bylund, Erik; Linderholm, Håkan and Rune, Olof (Eds) Ecological Problems of the Circumpolar Area: papers from the International Symposium at Luleå, Sweden, June 28-30, 1971 (1974)
Luleå, Norrbottens museum, 340 p.
Bynum, Jack E and Thompson, William E Juvenile Delinquency: a sociological approach (1989)
Hemel Hempstead, Simon and Schuster International, 500 p. ISBN 0-205-11774-0.
Bynyan, John Fear of God
Sterling VA, Reiner Publications. ISBN 0-685-19828-6.
Byrne, Elizabeth; Cunningham, Cliff and Sloper, Patricia Families and Their Children with Down's Syndrome: one feature in common
London, Routledge, 160 p. ISBN 0-415-00607-4.
Byron, William The Causes of World Hunger (1983)
Mahwah NJ, Paulist Press. ISBN 0-8091-2483-1.
Byron, William J On the Protection and Promotion of the Right to Food: an ethical reflection (1988)
Washington DC, Smithsonian Institution Press, pp. 14-36. in: Science, Ethics, and Food: Papers of a colloquium organized by the Smithsonian Institution/ ed by LeMAy Brian WJ. ISBN 0-87474-605-1.
C D Howe Research Institute Anticipating the Unexpected: policy review and outlook, 1979 (1979)
Toronto ON, CD Howe Institute. ISBN 0-88806-049-1.
CAB International Mycological Institute Diseases of Tropical Forage Legumes and Grassess
Kew, CAB International Mycological Institute. Annotated Bibliographies Series, Plant Series: M2.
CAB International Mycological Institute Diseases of Rape
Kew, CAB International Mycological Institute. Annotated Bibliographies Series, Plant Series: M3.
CAB International Mycological Institute Diseases of Coconut
Kew, CAB International Mycological Institute. Annotated Bibliographies Series, Plant Series: M4.
CAB International Mycological Institute Diseases of Banana
Kew, CAB International Mycological Institute. Annotated Bibliographies Series, Plant Series: M8.
CAB International Mycological Institute Diseases of Sorghum
Kew, CAB International Mycological Institute. Annotated Bibliographies Series, Plant Series: M9.
CAB International Mycological Institute Diseases of Oil Palm
Kew, CAB International Mycological Institute. Annotated Bibliographies Series, Plant Series: M10.
CAB International Mycological Institute Diseases of Tropical Root Crops
Kew, CAB International Mycological Institute. Annotated Bibliographies Series, Plant Series: M11, M12, M13.
CAB International Mycological Institute Diseases and Cultivation of Clove
Kew, CAB International Mycological Institute. Annotated Bibliographies Series, Plant Series: M15.
CAB International Mycological Institute Post-Harvest Diseases and Disorders of Tropical Root Crops
Kew, CAB International Mycological Institute. Annotated Bibliographies Series, Plant Series: M16.
CAB International Mycological Institute Coffee Rust (Hemileia Vastatrix and H Coffeicola): update 1984
Kew, CAB International Mycological Institute. Annotated Bibliographies Series, Plant Series: M22.
CAB International Mycological Institute Irrigation and Plant Disease
Kew, CAB International Mycological Institute. Annotated Bibliographies Series, Plant Series: M23.
CAB International Mycological Institute Mycoses of the Heart
Kew, CAB International Mycological Institute. Annotated Bibliographies Series, Animal/Human Series: L1.
CAB International Mycological Institute Mycoses of the Eye and Orbit
Kew, CAB International Mycological Institute. Annotated Bibliographies Series, Animal/Human Series: L3.
CAB International Mycological Institute Cryptococcosis: immunology
Kew, CAB International Mycological Institute. Annotated Bibliographies Series, Animal/Human Series: L5.
CAB International Mycological Institute Histoplasmosis: immunology
Kew, CAB International Mycological Institute. Annotated Bibliographies Series, Animal/Human Series: L6.
CAB International Mycological Institute Blastomycosis: immunology
Kew, CAB International Mycological Institute. Annotated Bibliographies Series, Animal/Human Series: L8.
CAB International Mycological Institute Aspergillosis: immunology: I: laboratory studies
Kew, CAB International Mycological Institute. Annotated Bibliographies Series, Animal/Human Series: L9.
CAB International Mycological Institute Mycotoxins of Maize: update 1983
Kew, CAB International Mycological Institute. Annotated Bibliographies Series, Animal/Human Series: L13.
Cailliet, Rene Hand Pain and Impairment (1982)
Philadelphia PA, Davis FA, 232 p. ISBN 0-8036-1618-X.
Cailliet, Rene Knee Pain and Disability (1983)
Philadelphia PA, Davis FA, 177 p. ISBN 0-8036-1621-X.
Caird, F I and Williamson, J Eye and Its Disorders in the Elderly (1986)
Littleton MA, PSG Publishing, 175 p. ISBN 0-7236-0706-0.
Cairns, John Ecoaccidents (1985)
New York, Plenum Publishing, 172 p. ISBN 0-306-42223-9.
Calabrese, E J, et al (Eds) Inorganics in Drinking Water and Cardiovascular Disease (1985)
Princeton NJ, Princeton University Press, 340 p. illus. Advances in Modern Environmental Toxicology Ser. ISBN 0-911131-10-8.
Calabrese, Edward J Pollutants and High Risk Groups: the biological basis of increased human susceptibility to environmental and occupational pollutants (1978)
New York, John Wiley and Sons, 266 p. Environmental Science and Technology: Wiley-Interscience Series of Texts and Monographs.
Calabresi, Guido Costs of Accidents: legal and economic analysis (1970)
New Haven CT, Yale University Press. ISBN 0-300-01115-6.
Caldwell, John C (Ed) Persistence of High Fertility: population prospect in the Third World
Canberra, Australian National University. 2 Vols.
California University Committee on Problems of War and Peace in the Society of Nations (1937)
Salem NH, Ayer Company Publishers. ISBN 0-8369-0270-X.
Calimari, D Selected Bibliography on Studies and Research Relevant to Pollution in the Mediterranean (1977)
Lanham MD, UNIPUB, 100 p. ISBN 92-5-000253-X.
Callagher, R P (Ed) Epidemiology of Malignant Melanoma (1986)
Berlin, Springer-Verlag, 210 p. ISBN 3-540-16020-5.
Calvert, E Roy Capital Punishment in the Twentieth Century: with intro and index added (1973)
Montclair NJ, Smith and Patterson Publishing. 5th rev ed. Criminology Law Enforcement and Social Problems Ser: No 153. ISBN 0-87585-153-3.
Cameron, D and Frazer, E The Lust to Kill: a feminist investigation of sexual murder (1987)
Polity Press, 207 p. ISBN 0-7456-0336-X.
Cameron, David M Regionalism and Supranationalism (1981)
Brookfield VT, Brookfield Publishing, 136 p. ISBN 0-920380-74-3.
Camille, Paglia Sexual Personae: art and decadence from Nefertiti to Emily Dickinson (1990)
London, Yale University Press, 700 p. ISBN 0-300-04396-1.
Campbell, Barry R and Herzstein, Robert E The International Debt Problem and Its Impact on Finance and Trade (1984)
New York, Practising Law Institute, 248 p. illus. bibl. Commercial Law and Practice: Course Handbook Series: 318.
Campbell, Duncan That Was Business, This is Personal: the changing faces of professional crime (1990)
London, Martin Secker and Warburg, 256 p. ISBN 0-436-19990-4.
Campbell, Elaine The Childless Marriage: an exploratory study of couples who do not want children (1986)
London, Tavistock Publications, 200 p. ISBN 0-422-60070-9.
Campbell, William C Trichinella and Trichinosis (1983)
New York, Plenum Publishing, 606 p. ISBN 0-306-41140-7.
Campion, M G Worry: a maieutic analysis (1986)
Brookfield VT, Gower Publishing, 350 p. ISBN 0-566-05118-4.
Camporesi, Piero and Murray, Tania C The Incorruptible Flesh (1988)
New York, Cambridge University Press, 272 p. ISBN 0-521-32003-8.
Canadian Institute of Chartered Accountants Fraud and Error (1982)
Toronto ON, Canadian Institute of Chartered Accountants.
Canadian Institute of Child Health Effects of Alcohol, Tobacco and Caffeine on the Fetus: a bibliography of the recent human literature (1979)
Ottawa ON, Canadian Institute of Child Health, 41 p.
Canavaggio, Pierre Dictionnaire Raisonne Des Superstitions et Des Croyances Populaires (1977)
New York, French and European Publications, 247 p. ISBN 0-686-56937-7.
Cantor, Sheila Childhood Schizophrenia (1988)
New York, Guilford Press, 193 p. ISBN 0-89862-713-3.
Cantril, Hadley The Pattern of Human Concerns (1965)
New Brunswick NJ, Rutgers University Press, 427 p.
Cantwell, D and Carlson, G (Eds) Affective Disorders in Childhood and Adolescence: an update (1983)
Great Neck NY, PMA Publishing, 480 p. Child Behavior and Development ser: Vol 5. ISBN 0-89335-189-X.
Cantwell, Dennis and Baker, Lorian Developmental Speech and Language Disorders (1987)
New York, Guilford Press, 214 p. ISBN 0-89862-400-2.

Cardoso, Joel A Metabolism with Inborn Errors: medical subject analysis with bibliography (1987)
Annandale VA, ABBE Publishers Association of Washington, 150 p. ISBN 0-88164-635-0.
Carelli, Anne O Sex Equity in Education: readings and strategies (1988)
Springfield IL, Thomas Charles C, 412 p. ISBN 0-398-05415-0.
Carey, John UN Protection of Civil and Political Rights (1970)
Charlottesville VA, University Press of Virginia, 205 p. Vol. 8 ISBN 0-8156-2146-9.
Carlson, Sven H Trade and Dependency: studies in the expansion of Europe (1984)
Stockholm, Almqvist and Wiksell, 188 p. ISBN 91-554-1628-4.
Carman, John B and Marglin, Frédérique Apffel (Eds) Purity and Auspiciousness in Indian Society (1985)
Leiden, Brill, 129 p. ISBN 90-04-077789-9.
Carmichael, Edward A Confronting Global Challenges (1986)
Toronto ON, CD Howe Institute, 74 p. ISBN 0-88806-141-2.
Carnes, Patrick Out of the Shadows: understanding sexual addiction (1985)
Irvine CA, CompCare Publishers, 185 p. ISBN 0-89638-086-6.
Carnoy, Martin Education and Employment: a critical appraisal (1977)
Paris, UNESCO, 91 p. bibl. ISBN 92-803-1078-X. Fundamentals of Educational Planning: No 26.
Carone, Pasquale A, et al (Eds) Mental Health Needs of Workers and Management: expectations and reality (1988)
Huntington NY, Maple Hill Press, 170 p. Problems of Industrial Psychiatric Medicine Ser: Vol XII. ISBN 0-930545-05-2.
Carone, Pasquale et al The Emotionally Troubled Employee: a challenge to industry (1976)
Albany NY, State University of New York Press, 99 p. ISBN 0-87395-801-2.
Carone, Pasquale, et al Addictive Disorders Update: alcoholism, drug abuse, gambling (1982)
New York, Human Sciences Press, 192 p. ISBN 0-89885-034-7.
Carone, Pasquale, et al Mental Health Problems of Workers and Their Families (1985)
New York, Human Sciences Press, 160 p. ISBN 0-89885-227-7.
Caroscio, James T Amyotrophic Lateral Sclerosis: a guide to patient care (1986)
New York, Thieme Medical Publishers, 360 p. ISBN 3-13-691701-4.
Carp, E (Comp) Directory of Western Palearctic Wetlands (1980)
Gland, IUCN, 506 p. illus. ISBN 2-88032-300-2.
Carpenter, Kenneth J The History of Scurvy and Vitamin C (1988)
New York, Cambridge University Press, 304 p. ISBN 0-521-34773-4.
Carpentier, J and Cazamian, P Night Work: its effect on the health and welfare of workers (1988)
Geneva, ILO, 85 p. 3rd ed. ISBN 92-2-101676-5.
Carpentier, James and Cazamian, Pierre Noise and Vibration in the Working Environment (1978)
Geneva, ILO, 131 p. Occupational Safety and Health Series: 33. ISBN 92-2-101676-5.
Carr, Arthur C, et al (Eds) Grief: selected readings (1974)
Center Thanatology, 155 p. ISBN 0-930194-76-4.
Carsten, F L War Against War: British and German radical movements in the First World War (1982)
Berkeley CA, University of California Press, 300 p. ISBN 0-520-04581-5.
Carter, April Success and Failure in Arms Control Negotiations (1989)
Oxford, Stockholm International Peace Research Institute, 320 p. illus. SIPRI Publication Series. ISBN 0-19-829128-0.
Carter, John M Rape in Medieval England: an historical and sociological study (1985)
Lanham MD, University Press of America, 196 p. ISBN 0-8191-4503-3.
Carter, Vernon G and Dale, Tom Topsoil and Civilization (1981)
Norman OK, University of Oklahoma Press, 240 p. illus. rev ed. ISBN 0-8061-1107-0.
Cartwright, Frederick Disease and History (1972)
London, Hart-Davis.
Cartwright, Robert E and Phillips, Jerry J Products Liability (1986)
New York, Kluwer Law Book, 1690 p. 3 Vols. Kluwer Products Liability Library. ISBN 0-930273-41-9.
Cartwright, Robert E, et al Products Liability (1986)
New York, Kluwer Law Book, 1690 p. 3 vols. ISBN 0-930273-41-9.
Cary, J W and Weston, R E Social Stress in Agriculture: the implications of rapid economic change (1978)
Parkville VIC, University of Melbourne, 64 p. ISBN 0-909467-03-X.
Casey, D E, et al (Eds) Dyskinesia: research and treatment (1985)
Berlin, Springer-Verlag, 235 p. illus. Psychopharmacology Supplementum: No 2. ISBN 0-387-15009-9.
Casselman, Jo and Moorthamer, Lut Violent Social Behaviour and Alcohol Use: review of the literature (1988)
Copenhagen, WHO, 82 p.
Cassels, Alan Fascism (1975)
Arlington Heights IL, Harlan Davidson. illus. ISBN 0-88295-718-X.

Cassidy, E L Political Kidnapping (1986)
New York, Gordon Press Publishers. ISBN 0–8490–3852–9.

Cassiers, Juan The Hazards of Peace: a european view of detente (1984)
Washington DC, University Press of America, 94 p. Harvard Studies in International Affairs: No 34.
ISBN 0–8191–4019–8.

Catovsky, D and Foa, R The Lymphoid Leukaemias (1989)
Sevenoaks, Butterworths, 320 p. illus.
ISBN 0–407–00259–6.

Catrina, Christian and Frei, Daniel Risks of Unintentional Nuclear War
New York, UN.

Catton, William R and Udall, Stewart L Overshoot: the ecological basis of revolutionary change (1980)
Champaign IL, University of Illinois Press, 320 p.
ISBN 0–252–00818–9.

Cavalloro, R (Ed) Integrated Tse–tse Fly Control, Methods and Strategies: proceedings of the CEC International Symposium, Ispra 4–6 March 1988 (1987)
Rotterdam, AA Balkema, 213 p. ISBN 90–6191–702–6.

Cavin, Susan, et al Lesbian Origins (1985)
San Francisco CA, Ism Press, 263 p.
ISBN 0–910383–16–2.

Cebula, Richard The Deficit Problem in Perspective (1987)
Lexington MA, Lexington Books, 128 p.
ISBN 0–669–14303–0.

Cellarius, R A and Platt, John Councils of Urgent Studies
In: *Science* 25 Aug 1972, pp. 670–676.

Cépède, M and Abensour, E S Rural Problems in the Alpine Region: an international study (1961)
Rome, FAO, 201 p. ISBN 92–5–101667–4.

Cervos–Navarro, J and Sarkander, H I (Eds) Brain Aging: neuropathology and neuropharmacology (1983)
New York, Raven Press, 454 p. illus. Aging Ser: Vol 21.
ISBN 0–89004–739–1.

Chaffee, Judy K Leprosy: medical subject analysis and research guide (1987)
Washington DC, ABBE Publications Association, 150 p.
ISBN 0–88164–552–4.

Chaffee, S and Petrick, M Using the Mass Media: communication problems in American society (1975)
New York, McGraw–Hill Book Company, 384 p.
ISBN 0–07–010375–5.

Chakotin, Serge Rape of the Masses: the psychology of totalitarian political
Brooklyn NY, Haskell Booksellers. ISBN 0–8383–1264–0.

Chalude, M The Re–insertion of Women in Working Life: initiatives and problems (1987)
Luxembourg, CEE, 232 p. ISBN 92–825–7165–3.

Chamberlain, Neil W and Schilling, Jane M Impact of Strikes: their social and economic costs (1973)
Westport CT, Greenwood Press, 257 p.
ISBN 0–8371–7066–4.

Chambers, Robert; Longhurst, Richard and Pacey, Arnold (Eds) Seasonal Dimensions to Rural Poverty (1983)
Brighton, Institute of Development Studies, 259 p.
ISBN 0–861–87334–3.

Chambers, William and Pilowsky, Daniel (Eds) Hallucinations, in Children (1987)
Cambridge MA, Cambridge University Press, 139 p.
ISBN 0–521–34898–6.

Champ, B R and Dyte, C E Report of the FAO Global Survey of Pesticide Susceptibility of Stored Grain Pests (1976)
Rome, FAO, 297 p. FAO Plant Protection and Protection Series: No 5.

Champ, Michael A and Park, P K Global Marine Pollution Bibliography: ocean dumping of municipal and industrial wastes (1982)
New York, Plenum Publishing, 424 p.
ISBN 0–306–65205–6.

Chand, Attar Terrorism, Political Violence and Security of Nations: a global survey of socio–political, legal and economic aspects of national, regional and international terrorism (1986)
Delhi, Modern Publishers, 620 p.

Chandler, Craig C, et al Fire in Forestry: forest fire behavior and effects (1983)
New York, John Wiley and Sons, 450 p. Vol. I
ISBN 0–471–87442–6.

Chandler, William U Banishing Tobacco
Washington DC, Worldwatch Institute. Worldwatch Papers: 68.

Chandler, William U Energy Productivity: key to environmental protection and economic progress (1985)
Washington DC, Worldwatch Institute. Worldwatch Papers.
ISBN 0–916468–63–1.

Chandler, William U Banishing Tobacco (1986)
Washington DC, Worldwatch Institute. Worldwatch Papers.
ISBN 0–916468–68–2.

Chandra, Mahesh Socio–Economic Crimes (1979)
Bombay, N R Tripathi, 176 p.

Chang, Dae H and Blazicek, Donald L An Introduction to Comparative and International Criminology (1986)
Acorn NC, 110 p. ISBN 0–89386–017–4.

Channa, V C Caste: indentity and continuity (1979)
Delhi, BR Publication Corporation, 180 p.
ISBN 81–7018–042–2.

Chantagul Instability of Export Earning of LDC Primary Products and the Role of International Commodity Agreement
Bangkok, Chualongkorn University Press.

Chapman, Antony J and Gale, Anthony (Eds) Psychology and Social Problems: an introduction to applied psychology (1985)
New York, John Wiley and Sons, 374 p.
ISBN 0–471–90313–2.

Chapman, C P and Morrison, D Cosmic Catastrophes (1989)
New York, Plenum Publishing, 316 p.
ISBN 0–306–43163–7.

Chapman, M and Dixon, R A (Eds) Meaning and the Growth of Understanding (1987)
Berlin, Springer–Verlag, 290 p. ISBN 3–540–17516–4.

Chapman, Simon Pushing Smoke: tobacco advertising and promotion (1988)
Copenhagen, WHO, 68 p. Smoke–free Europe: 8.

Char, Devron H Thyroid Eye Disease (1985)
Baltimore MD, Williams and Wilkins, 225 p.
ISBN 0–683–01519–2.

Charles, Sara C and Kennedy, Eugene Defendant: a psychiatrist on trial for medical malpractice (1985)
New York, Free Press, 236 p. ISBN 0–02–905910–0.

Charnaz, Kathlene C The Social Reality of Death (1980)
New York, Random House, 384 p. ISBN 0–394–34832–X.

Charny, Israel W Genocide: the critical bibliographic review (1988)
New York, Facts on File, 288 p. ISBN 0–8160–1903–7.

Charny, Israel W (Ed) Genocide: a critical bibliographic review (1988)
London, Mansell, 288 p. ISBN 0–7201–1876–X.

Charters, D and Tugwell, M Armies in Low Intensity Conflict: a comparative study of institutional adaptation (1988)
Elmsford NY, Pergamon Books, 270 p.
ISBN 0–08–036253–2.

Chase, Salmon P Reclamation of Fugitives from Service (1847)
Salem NH, Ayer Company Publishers. Black Heritage Library Collection Ser. ISBN 0–8369–8726–8.

Chasle (Ed) World Problems by the Year 2000: report of the Scientific Workshop held at UNESCO, Paris, 22 and 23 June 1987
In: *Bureau of Studies and Programming – The World by the Year 2000* 1987, Paris, UNESCO, BEP/GPI/1, pp. 5–15.

Chatterjee, L and Nijkamp, P (Eds) Urban Problems and Economic Development: proceedings of ASI, Amsterdam, The Netherlands, 1980 (1981)
Dordrecht, Kluwer Academic Publishers Group, 359 p.
ISBN 90–286–2661–1.

Chatterjee, S K Drug Abuse and Drug–Related Crimes (1988)
Amsterdam, Kluwer Academic Publishers Group.

Chattopadhyaya, S B Diseases of Plants, Yieldings, Drugs, Dyes and Spices (1969)
New Delhi, Indian Council of Agricultural Research, 107 p.

Chaturvedi, M K Rural Middlemen: network of patronage (1986)
New Delhi, Ashish Publishing House, 136 p.

Châu, Ta Ngoc and Carron, Gabriel (Eds) Regional Disparities in Educational Development: a controversial issue (1980)
Paris, International Institute for Educational Planning, 257 p.
ISBN 92–803–1085–2.

Chaudhurry, Muzaffar Ahmed Examination of the Criticism Against Bureaucracy (1965)
Dhaka, National Institute of Public Administration, 69 p.

Chavkin, Wendy Double Exposure: women's health hazards on the job and at home (1984)
New York, Monthly Review Press, 288 p.
ISBN 0–85345–632–1.

Chazov, E I and Smirnov, V N Thrombosis and Thrombolysis (1986)
New York, Plenum Publishing, 444 p.
ISBN 0–306–10989–1.

Chen, Lincoln C and Scrimshaw, Nevin S Diarrhea and Malnutrition: interactions, mechanisms, and interventions (1983)
New York, Plenum Publishing, 334 p.
ISBN 0–306–41046–X.

Cheng, Thomas C Parasitic and Related Diseases: basic mechanisms, manifestations, and control (1986)
New York, Plenum Publishing, 176 p.
ISBN 0–306–42119–4.

Chénier, Nancy Miller Reproductive Hazards at Work (1983)
Washington DC, Government Publication Centre, 105 p.

Chervenak, Frank A, et al Anomalies of the Fetal Head, Neck and Neural Axis: ultrasound diagnosis and management (1988)
Philadelphia PA, WB Saunders, 192 p. illus.
ISBN 0–7216–1957–6.

Cheshire, Paul C and Hay, Dennis G Urban Problems in Western Europe (1988)
UK, Unwin Hyman, 256 p.

Cheslock, Charles J Understanding the Common Cold (1987)
New York, Vantage Press, 142 p. ISBN 0–533–07066–X.

Chester, R Divorce in Europe (1977)
Dordrecht, Kluwer Academic Publishers Group, 316 p.
ISBN 90–207–0652–7.

Chester, Ronald Inheritance, Wealth, and Society (1982)
Bloomington IN, Indiana University Press, 256 p.
ISBN 0–253–33009–2.

Chesterman, John and Lipman, Andy The Electronic Pirates: DIY crime of the century
London, Routledge, 272 p. ISBN 0–415–00738–0.

Chetley, Andrew and Gilbert, David Problem Drugs (1986)
Den Haag, IOCU.

Chey, William Y Functional Disorders of the Digestive Tract (1983)
New York, Raven Press, 368 p. ISBN 0–89004–859–2.

Chigier, E (Ed) Youth and Disability (1986)
Tel Aviv, Freund Publishing House, 430 p.
ISBN 965–294–018–6.

Chiles, John Teenage Depression and Suicide (1986)
New York, Chelsea House Publishers.
ISBN 0–87754–771–8.

Chilman, Catherine S, et al Employment and Economic Problems (1988)
Newbury Park CA, Sage Publications, 272 p.
ISBN 0–8039–2707–X.

Chilman, Catherine S, et al Variant Family Forms (1988)
Newbury Park CA, Sage Publications, 320 p.
ISBN 0–8039–2709–6.

Chilman, Catherine S, et al (Eds) Chronic Illness and Disability (1988)
London, Sage, 300 p. Families in Trouble Ser: Vol 2.
ISBN 0–8039–2703–7.

Chinkin, Davidson and Ricquier (Eds) Current Problems of International Trade Financing (1983)
Singapore, Malaya Law Review. Dist Butterworths.

Chirban, John T Thalassemia: an interdisciplinary approach (1987)
Lanham MD, University Press of America, 106 p.
ISBN 0–8191–5675–2.

Chisci, G and Morgan, R P Soil Erosion in European Community—Impact of Changing Agriculture: proceedings of a seminar on land degradation (1986)
Brookfield VT, Brookfield Publishing, 248 p.
ISBN 90–6191–657–7.

Chisholm, Anthony and Dumsday, Robert Land Degradation: problems and policies (1988)
New York, Cambridge University Press, 412 p.
ISBN 0–521–34079–9.

Chisholm, J Julian and O'Hara, David M (Eds) Lead Absorption in Children: management, clinical and environmental aspects (1982)
Munich, Urban and Schwarzenberg, 229 p.
ISBN 3–541–70331–8.

Chivian, Eric, et al Last Aid: the medical dimensions of nuclear war (1982)
New York, Freeman WH, 338 p. ISBN 0–7167–1434–5.

Choldin, Marianne T A Fence Around the Empire: Russian Censorship of Western Ideas under the Tsars (1985)
Durham NC, Duke University Press, 282 p. illus. Policy Studies Ser. ISBN 0–8223–0625–5.

Chomsky, Noam The Political Economy of Human Rights (1979)
Boston, South End. Vol 1: The Washington Connection and Third World Fascism.

Chomsky, Noam Towards a New Wold War (1982)
London, Sinclair Brown.

Chomsky, Noam Pirates and Emperors: international terrorism in the real world (1987)
Montreal, Black Rose Books, 215 p. 0–920057–93–4.

Chopra, O P Tax Ethics: unaccounted income or black money (1985)
New Delhi, Inter–India Publications, 119 p.
ISBN 81–210–0092–0.

Chopra, S K (Ed) Brain Drain: causes, consequences and solutions (1986)
New Delhi, Lancer's International.

Chopra, Sanjiv Disorders of the Liver (1988)
Philadelphia PA, Lea and Febiger, 249 p.
ISBN 0–8121–1101–X.

Choucri, Nazli Multidisciplinary Perspectives on Population and Conflict (1984)
Syracuse NY, Syracuse University Press, 240 p.
ISBN 0–8156–2314–3.

Chretien, et al Current Therapy of Head and Neck Cancer: proceedings of the first international conference of the Society of Head and Neck Surgeons (1985)
Saint Louis MO, CV Mosby, 600 p. ISBN 0–941158–40–3.

Christie Infectious Diseases (1987)
New York, Churchill Livingstone. 4th ed. 2 vols.
ISBN 0–443–03585–7.

Christodoulou, Demetrios The Unpromised Land: agrarian reform and conflict worldwide (1989)
London, Zed Books. ISBN 0–86232–778–4.

Church, Diana et al Gambling: crime or recreation (1988)
Plano TX, Information Plus, 88 p. ISBN 0–936474–80–7.

Chuta, Enyinna and Sethuraman, S V Rural Small–Scale Industrie and Employment in Africa and Asia: a review of problems and policies (1983)
Geneva, ILO, 160 p. ISBN 92–2–103514–X.

Chvála, Milan, et al Horse Flies of Europe (Diptera, Tabanidae) (1972)
Leiden, Brill, 500 p. ISBN 0–900848–57–X.

Ciba Foundation Congenital Disorders of Erythropoiesis
Ann Arbor MI, Books on Demand. Ciba Foundation Foundation – New Ser: No 37. ISBN 0–317–29170–X.

CIBA Foundation Symposium Fibrosis (1986)
New York, John Wiley and Sons, 320 p. Symposium: 114.
ISBN 0–471–91083–X.

Cihal, V Intergranular Corrosion of Steels and Alloys (1984)
New York, Elsevier Science Publishing.
ISBN 0–444–99644–3.

Cizik, Richard The High Cost of Indifference (1984)
Ventura CA, Regal Books.

Claridge, Gordon Origins of Mental Illness: temperament, deviance and disorder (1985)
New York, Basil Blackwell, 240 p. ISBN 0–631–14198–7.

Clark, Champ Flood (1982)
Alexandria VA, Time–Life Books. Planet Earth Ser.
ISBN 0–8094–4308–2.

Clark, David D, et al Computer Security: risks, technologies and policies

Clark, David, et al Corrosion of glass (1979)
Ashlee Pub Co, 75 p. ISBN 0–911993–18–5.

Clark, Eric The Want Makers: lifting the lid of the world advertising industry: how they make you buy (1988)
Geneva, UN, 416 p. ISBN 0–340–32028–1.

Clark, Lorenne and Lewis, Debra Rape: the price of coercive sexuality
Toronto ON, The Women's Press, 224 p.
ISBN 0-88961-033-9.

Clark, Lorenne M G and Lange, Lynda (Eds) Sexism of Social and Political Theory: women and reproduction from Plato to Nietzsche (1979)
Toronto ON, University of Toronto Press.
ISBN 0-8020-6375-6.

Clark, O and Roeher, H D Thyroid Tumors (1988)
New York, Karger S AG, 228 p. ISBN 3-8055-4713-7.

Clark, R B The Long-Term Effects of Oil Pollution on Marine Populations, Communities and Ecosystems: proceedings (1982)
Port Washington NY, Scholium International, 260 p.
ISBN 0-85403-188-X.

Clark, W C and Munn, R E (Eds) Sustainable Development of the Biosphere (1988)
Cambridge MA, Cambridge University Press.

Clarke, B and Show, S B Forest Fire Control (1978)
Rome, FAO, 100 p. 3rd ed.
FAO Forestry Series: No 6. ISBN 92-5-100488-9.

Clarke, Michael (Eds) Corruption: causes, consequences and control (1984)
St Martin, 250 p.
4 ISBN 0-312-17007-6.

Clarke, R V G and Hope, Tim Coping with Burglary: research perspectives on policy (1984)
Dordrecht, Kluwer Academic Publishers Group, 272 p.
ISBN 0-89838-151-7.

Clayton, R M, et al Problems of Normal and Genetically Abnormal Retinas (1983)
San Diego CA, Academic Press. ISBN 0-12-176180-0.

Clignet, Remi Many Wives, Many Powers: authority and power in polygynous families
Ann Arbor MI, Books on Demand. ISBN 0-8357-9464-4.

Cline, Ray S; Alexander, Yonah and Denton, Jermiah Terrorism As State Sponsored Covert Warfare (1986)
Fairfax VA, Hero Books, 128 p. ISBN 0-915979-19-5.

Cline, William R, et al World Inflation and the Developing Countries (1981)
Washington DC, Brookings Institution, 266 p.
ISBN 0-8157-1468-8.

Clygout, Sanivar H Homosexuality (1985)
Annandale VA, ABBE Publishers Association of Washington, 156 p. ISBN 0-88164-090-5.

Coate, Roger A Global Issue Regimes (1982)
New York, Praeger, 218 p. bibl. ISBN 0-03-059276-3.

Coccheri, S and Donati, Maria B (Eds) International Congress on Fibrinolysis (1984)
Basel, S Karger AG, 156 p. Journal: Haemostasis 14 (1984) 1.

Cockburn, Forrester and Gitzelmann, Richard Inborn Errors of Metabolism in Humans (1982)
New York, Alan R Liss, 308 p. ISBN 0-8451-3008-0.

Code, Chris (Ed) Characteristics of Aphasia (1989)
Basingstoke, Taylor and Francis, 220 p.
ISBN 0-85066-470-5.

Code, Chris and Muller, Dave J (Eds) Second International Aphasia Rehabilitation Congress: proceedings of the conference, Göteborg, Sweden, June 1986 (1987)
New York, Taylor and Francis, 116 p. Aphasiology Special Issue Ser: Vol 1, No 3, May-June 1987.
ISBN 0-85066-913-8.

Coffey, D S et al Multidisciplinary Analysis of Controversies in the Management of Prostate Cancer (1988)
New York, Plenum Publishing, 344 p.
ISBN 0-306-42927-6.

Coffman, Derek A, et al Gastrointestinal Disorders (1987)
New York, Churchill Livingstone, 207 p. illus. Library of General Practice: Vol 13. ISBN 0-443-02976-8.

Cohen, Betsy The Snow White Syndrome: all about envy (1987)
New York, Macmillan, 256 p. ISBN 0-02-526970-4.

Cohen, Daniel The Encyclopedia of Ghosts (1987)
New York, Dodd-Mead. illus. ISBN 0-396-09050-8.

Cohen, Eliot A Citizens and Soldiers: the dilemmas of military service (1985)
Ithaca NY, Cornell University Press, 227 p.
ISBN 0-8014-1581-0. Studies in Security Affairs.

Cohen, Raymond Threat Perception in International Crisis (1979)
Madison WI, University of Wisconsin Press, 214 p.
ISBN 0-299-08000-5.

Cohen, Sidney The Substance Abuse Problems: new issues for the 1980's (1985)
New York, Haworth Press, 323 p. Vol 2.
ISBN 0-86656-368-7.

Cohen, Steven M Interethnic Marriage and Friendship (1980)
Salem NH, Ayer Company Publishers. Dissertations on Sociology Ser.

Cohn, D V and Martin, T J Osteoporosis (1987)
New York, Elsevier Science Publishing, 52 p.
ISBN 0-444-80910-4.

Cohn, Lawrence H and Ionescu, Marian I (Eds) Mitral Valve Disease: diagnosis and treatment (1985)
Sevenoaks, Butterworths, 384 p. illus.
ISBN 0-407-00267-7.

Cole, George F and Frankowski, Stanislaw (Eds) Abortion and Protection of the Human Fetus: legal problems in a cross-cultural perspective (1988)
Dordrecht, Kluwer Academic Publishers Group, 352 p.
ISBN 0-89838-922-4.

Cole, H S, et al Models of Doom: a critique of the limits to growth (1973)
New York, Universe Books, 252 p. ISBN 0-87663-184-7.

Cole, Sam and Miles, Ian Worlds Apart (1984)
Totowa NJ, Rowman and Littlefield Publishers, 256 p.
ISBN 0-8476-7374-X.

Colheart, M; Marshall, J C and Patterson, K (Eds) Deep Dyslexia (1987)
London, Routledge, 472 p. rev paperback ed.
ISBN 0-7102-1235-6.

Colhoun, J Club Root Disease of Crucifers Caused by Plasmodiophora Brassicae Woron (1958)
Kew, CAB International Mycological Institute, 108 p. illus. Phytopathological Papers: No 3.

Collar, N J; Stuart, S N and Arlott, Norman Threatened Birds of Africa and Related Islands (Date not set)
Princeton NJ, Princeton University Press, 797 p.
ISBN 2-88032-604-4.

Colledge, M Migration and Health: towards an understanding of the health care (1986)
Geneva, WHO, 210 p. ISBN 92-890-1045-2.

Collingridge, David The Social Control of Technology (1981)
New York, Saint Martin's Press. ISBN 0-312-73168-X.

Collins, Joseph and Lappe, Frances M World Hunger: twelve myths (1986)
San Francisco CA, Institute Food and Development Policy, 208 p. ISBN 0-317-66242-2.

Collins, N M and Morris, M G IUCN Swallowtail Red Data Book (Lepidoptera: papilionidae) (1984)
Gland, IUCN. illus. ISBN 2-88032-603-6.

Collins, N M; Pyle, R M and Wells, S M IUCN Invertebrate Red Data Book (1983)
Gland, IUCN, 632 p. illus. ISBN 2-88032-602-8.

Colman, Robert W Disorders of Thrombin Formation (1983)
New York, Churchill Livingstone, 161 p.
ISBN 0-443-08184-0.

Combs, J Who's Who in the World Zionist Conspiracy (1982)
Brooklyn NY, Revisionist Press. ISBN 0-87700-327-0.

Comer, Michael J, et al Bad Lies in Business: the commonsense guide to detecting deceit in negotiations, interviews and investigations (1988)
New York, McGraw Hill Book Company, 256 p.
ISBN 0-07-707073-9.

Commission of the European Communities and Environmental Resources Ltd The Law and Practice Relating to Pollution Control in the Member States of the European Communities (1983)
Norwell MA, Graham and Trotman. Vol. 110 vols.
ISBN 0-86010-806-6.

Commission on Health Research for Development Health Research: essential link to equity in development (1990)
Oxford, Oxford University Press, 136 p. illus.
ISBN 0-19-520838-2.

Committee on Criteria and Guidelines for the Evaluation of Projects Designed to Protect or Enhance Biodiversity, Board on Biology Evaluation of Biodiversity Projects (1989)
Commission on Life Sciences, 50 p.

Committee on Irrigation-Induced Water Quality Problems (Eds) Irrigation-Induced Water Quality Problems (1989)
Washington DC, National Academy Press, 104 p.
ISBN 0-3090-3991-6.

Committee on Review of Switching, Synchronization and Network Control in National Security Telecommunications Growing Vulnerability of the Public Switched Networks: implications for national security emergency preparedness (1989)
Commission on Engineering and Technical Systems, 140 p.

Commonwealth Towards a Commonwealth of Learning: a proposal to create the University of the Commonwealth for cooperation in distance education
London, ComSec, 83 p. ISBN 0-85092-311-5.

Commonwealth Jobs for Young People: a way to a better future
London, ComSec, 152 p. ISBN 0-85092-313-1.

Commonwealth Secretariat Resources for the Development of Entrepreneurs: a guided reading list and annotated bibliography
London, ComSec, 115 p. ISBN 0-85092-305-0.

Commonwealth Secretariat (Ed) Confronting Violence: a manual for Commonwealth action
London, ComSec, 192 p. bibl.

Commonwealth Secretariat Protectionism: threat to international order (1982)
London, ComSec, 152 p.

Commonwealth Secretariat The Copyright System: practice and problems in developing countries (1983)
London, ComSec, 87 p.

Commonwealth Secretariat Vulnerability: small states in the global society (1985)
London, ComSec, 126 p.

Compans, Richard W and Bishop, David H Segmented Negative Strand Viruses: arenaviruses, bunyaviruses and orthomyxoviruses (1984)
San Diego CA, Academic Press. Vol. 1
ISBN 0-12-183501-4.

Comstock, Anthony Frauds Exposed: or, How People Are Deceived and Robbed, and Youth Corrupted
Montclair NJ, Smith and Patterson Publishing.
ISBN 0-87585-079-0.

Condie, L W Water Chlorination: chemistry, environment impact and health effects (1989)
Chelsea MA, Lewis Publishers, 1150 p.
ISBN 0-87371-167-X.

Confavreux, C, et al Trends in European Multiple Sclerosis Research: proceedings of the european committee for (1988)
New York, Elsevier Science Publishing, 428 p.
ISBN 0-444-80978-3.

Conklin, Agnes M Failure of Highly Intelligent Pupils
New York, AMS Press. ISBN 0-404-55792-9.

Conners, C Keith; Wells, Karen C and Kazdin, Alan E Hyperkinetic Children (1986)
Newbury Park CA, Sage Publications, 160 p.
ISBN 0-8039-2278-7.

Connors, S and Campbell, D On the Record: surveillance, computers and privacy: The inside story
UK, Michael Joseph.

Conquest, Robert H The Great Terror: a reassessment
New York, Oxford University Press, 560 p.

Conservation Foundation State of the Environment: a view toward the nineties (1987)
Washington DC, Conservation Foundation, 614 p.

Consumers' Association of Pinang Fighting Tobacco in the Third World
Pinang, Consumers' Association of Pinang, 52 p.
ISBN 967-9950-12-3.

Conver, Christopher C and Conver, Leigh E Self-Defeating Life Styles (1988)
Nashville TN, Broadman Press. ISBN 0-8054-5441-1.

Conway, Donald J Human Response to Tall Buildings (1982)
New York, Reinhold Van Nostrand ISBN 0-87933-268-9.

Conway, Douglas R Human Rights: index of modern information with bibliography (1988)
Annandale VA, ABBE Publishers Association of Washington, 150 p. ISBN 0-88164-874-4.

Conybeare, Irene Civilization or Chaos: a study of the present world crisis in the light of Eastern metaphysics (1959)
Bombay, Chetna Private, 247 p. 2nd rev ed.

Conybeare, John A C Trade Wars: the theory and practice of international commercial rivalry (1987)
New York, Columbia University Press, 319 p. bibl. The Political Economy of International Change.

Cook, Alice H The Working Mother: a survey of problems and programs in nine (1978)
Ithaca NY, I L R Press, 84 p. ISBN 0-87546-067-4.

Cook, G C Communicable and Tropical Diseases (1988)
Oxford, Heineman Professional Publishing, 336 p.
ISBN 0-433-00029-5.

Cook, James Remedies and Rackets: the truth about patent medicines today (1976)
Salem NH, Ayer Company Publishers.
ISBN 0-405-08059-X.

Cook, Lauren; Osterholt, Jack B and Riley, Edward C Jr Anticipating Tomorrow's Issues: a handbook for policymakers (1988)
Council of State Policy and Planning Agencies, 77 p.

Cook, Mark and Howells, Kevin Adult Sexual Interest in Children (1981)
San Diego CA, Academic Press. ISBN 0-12-187250-5.

Cook, Mark and Wilson, Glenn Love and Attraction: an international conference (1979)
Elmsford NY, Pergamon Books. ISBN 0-08-022234-X.

Cook, T Vagrancy: some new perspectives (1979)
San Diego CA, Academic Press. ISBN 0-12-187550-4.

Coombs, P H World Educational Crisis: a system analysis (1970)
Allahabad, A H Wheeler, 115 p.

Cooper, Arlin J Computer and Communications Security: strategies for the 1990s (1989)
New York, McGraw-Hill Book Company.
ISBN 0-07-012926-6.

Cooper, C L and Payne, R Causes, Coping and Consequences of Stress at Work (1988)
New York, John Wiley and Sons, 375 p.
ISBN 0-471-91879-2.

Cooper, Carey L Psychosocial Stress and Cancer (1985)
New York, John Wiley and Sons, 263 p.
ISBN 0-471-90477-5.

Cooper, J E and Jackson, O F Diseases of the Reptilia (1982)
San Diego CA, Academic Press. 2 vols.
ISBN 0-12-187901-1.

Cooper, M A The Uncertainty of the Signs of Death (1902)
London.

Cooper, M G (Ed) Risk: man-made hazards to man (1985)
Oxford NY, Oxford University Press. illus.

Cooper, Martha and Chalfant, Henry Subway Art (1984)
New York, H Holt and Company, 104 p. illus.
ISBN 0-03-071963-1.

Coote, Belinda The Hunger Crop: poverty and the sugar industry (1987)
Oxford, Oxfam Publications, 160 p. illus.
ISBN 0-85598-081-8.

Copely, Ursula E Directory of Homosexual Organizations and Publications: 1985-1986
Bossier City LA, Homosexual Information Center, 100 p.
ISBN 0-686-26160-7.

Coppock, Rob Social Constraints on Technological Progress (1984)
Brookfield VT, Gower Publishing, 291 p.
ISBN 0-566-00754-1.

Corbett, Margaret-Ann and Meyer, Jerrilyn H The Adolescent and Pregnancy (1987)
Oxford, Blackwell Scientific Publications, 284 p. illus.

Corbett, Robin Guerilla Warfare: from nineteen thirty-nine to the present day (1986)
Philadelphia PA, Trans-Atlantic Publications, 224 p.
ISBN 0-85613-469-4.

Cornblath, Marvin and Schwartz, Robert Disorders of Carbohydrate Metabolism in Infancy
Ann Arbor MI, Books on Demand. ISBN 0–317–26428–1.
Cornish, Edward (Ed) Global Solutions: innovative approaches to world problems (1984)
Bethesda MD, World Future Society, 160 p.
ISBN 0–930242–22–X.
Cornwall, Hugo Datatheft: computer fraud, industrial espionage and information crime (1987)
Oxford, Heineman Professional Publishing.
ISBN 0–434–90265–9.
Correa, P and Haenzel, W (Eds) Epidemiology of Cancer of the Digestive Tract (1982)
Dordrecht, Kluwer Academic Publishers Group, 264 p.
ISBN 90–247–2601–8.
Corrigan, James J Hemorrhagic and Thrombotic Diseases in Childhood and Adolescence (1985)
New York, Churchill Livingstone, 216 p.
ISBN 0–443–08425–4.
Cortes, Carlos E (Ed) The Latin American Brain Drain to the United States: an original anthology (1981)
Salem NH, Ayer Company Publishers. illus. Hispanics in the United States Ser. ISBN 0–405–13176–3.
Corvalan–Vasquez, Oscar Youth Employment and Training in Developing Countries: an annotated bibliography (1984)
Lanham MD, UNIPUB, 172 p. ISBN 92–2–103420–8.
Costa, Joseph J Abuse of the Elderly (1984)
Lexington MA, Lexington Books, 320 p.
ISBN 0–669–06142–5.
Costlow, J D and Tipper, R C (Eds) Marine Biodeterioration: an interdisciplinary study (1983)
Annapolis MD, Naval Institute Press, 512 p. illus.
ISBN 0–87021–530–2.
Cotes, et al Occupational Lung Disorders (1986)
Saint Louis MO, CV Mosby. ISBN 0–8016–1403–1.
Couchiching Institute on Public Affairs World Challenge: international development at a time of East-West tension (1981)
Toronto ON, Couchiching Institute on Public Affairs.
Coulmas, Florian Linguistic Minorities and Literacy: language policy issues in developing countries (1984)
Hawthorne NY, Mouton de Gruyter, 133 p.
ISBN 3–11–009867–9.
Coulson, R I and Coulson, G M Effect of Forestry Practices on Bird Breeding in Open Forest (1980)
Hobart, University of Tasmania, 40 p.
Council of Europe Convention on the Reduction of Cases of Multiple Nationality and Military Obligations in Cases of Multiple Nationality: protocol amending the convention and additional protocol to the convention (1978)
Strasbourg, Council of Europe. ISBN 92–871–0095–0.
Council of Europe Lung Cancer in Western Europe: statistical report (1978)
Strasbourg, Council of Europe. 3rd ed.
ISBN 92–871–0725–4.
Council of Europe Economic Crime (1981)
Strasbourg, Council of Europe. ISBN 92–871–0588–X.
Council of Europe Night Work: comparative study of legislation and regulations; problems and social repercussions (1981)
Strasbourg, Council of Europe. ISBN 92–871–0678–9.
Council of Europe Alcohol and Crime (1984)
Strasbourg, Council of Europe. ISBN 92–871–0310–0.
Council of Europe Economic Crises and Crime (1985)
Strasbourg, Council of Europe. ISBN 92–871–0780–7.
Council of Europe Temporary Employment Businesses: general problems: specific problems relating to legal or illegal hiring out of workers across borders (1985)
Strasbourg, Council of Europe. ISBN 92–871–0396–8.
Council of Europe Protection of Personal Data Used for the Purposes of Direct Marketing (1986)
Strasbourg, Council of Europe. Recommandation: R (85) 20.
ISBN 92–871–0876–5.
Council of Europe Protection of Personal Data Used for Social Security Purposes (1986)
Strasbourg, Council of Europe. ISBN 92–871–0924–9.
Council of Europe Extortions under Terrorist Threats (1986)
Strasbourg, Council of Europe. ISBN 92–871–0841–2.
Council of Europe The European Convention on Human Rights (1986)
Strasbourg, Council of Europe, 54 p.
Council of Europe Human Rights in Prisons (1987)
Strasbourg, Council of Europe. ISBN 92–871–0926–5.
Council of Europe The Psychological and Social Consequences of Unemployment (1987)
Strasbourg, Council of Europe. ISBN 92–871–1026–3.
Cousteau, Jacques–Yves and Cousteau Society The Cousteau Almanac: an inventory of life on our watery planet (1979)
New York, Doubleday, 838 p. illus.
Covelli, Pasquale and Wiedman, Melvin Diabetes: current research and future directions in (1988)
Jefferson NC, McFarland Publishers, 128 p.
ISBN 0–89950–361–6.
Covitz, Joel D Emotional Child Abuse: the family curse (1986)
Boston MA, Sigo Press, 162 p. ISBN 0–938434–22–5.
Cox, David and Goldblat, Jozef (Eds) Nuclear Weapon Tests: prohibition or limitation? (1988)
Solna, Stockholm International Peace Research Institute, 448 p. illus. SIPRI Publication Series.
ISBN 0–19–829120–5.
Cox, Del Corruption and Cover-Up (1988)
Vicksburg MS, Liame Press, 200 p. ISBN 0–9620425–0–1.
CPPS/UNEP Sources, Levels and Effects of Marine Pollution in the South-East Pacific (1983)
Nairobi, UNEP, 354 p. UNEP Regional Seas Reports and Studies: No 7.
Craig, H L Stress Corrosion – New Approaches (1976)
Philadelphia PA, American Society for Testing and Materials, 429 p. STP 610. ISBN 0–8031–0580–0.
Craven, Gregory Secession: the ultimate states rights (1986)
Portland OR, International Specialized Book Services, 250 p.
ISBN 0–522–84317–4.
Cravioto, Joaquin; Hambraeus, Leif and Vahlquist, Bo (Eds) Early Malnutrition and Mental Development: proceedings of a symposium jointly sponsored by the National Institute of Child Health and Human Development, Bethesda, Maryland, the Swedish Nutrition Foundation and the World Health Organization (1974)
Stockholm, Almqvist and Wiksell, 244 p.
Crepaldi, Gaetano, et al Diabetes, Obesity and Hyperlipidemias (1983)
San Diego CA, Academic Press. ISBN 0–12–195480–3.
Crescenzi, G Serlupi A Multidisciplinary Approach to Myelin Diseases (1987)
New York, Plenum Publishing, 416 p.
ISBN 0–306–42776–1.
Crew, Jennifer and Wright, D Revolutions in the Modern World
University N E, 336 p. ISBN 0–85834–251–0.
Critchley, John Feudalism (1977)
London, Unwin Hyman.
Croft, B A and Hoyt, S C (Eds) Integrated Management of Insect Pests of Pome and Stone Fruit (1983)
New York, Wiley Interscience, 454 p. Environmental Science and Technology Texts and Monographs.
ISBN 0–471–05334–1.
Crohn, Joel Ethnic Identity and Marital Conflict: Jews, Italians and Wasps (1986)
New York, American Jewish Committee, 44 p.
ISBN 0–87495–078–3.
Croll, Neil A and Cross, John H Human Ecology and Infectious Diseases (1983)
San Diego CA, Academic Press. ISBN 0–12–196880–4.
Crompton, D W, et al Ascariasis and Its Prevention and Control (1988)
New York, Taylor and Francis, 300 p.
ISBN 0–85066–424–1.
Crompton, Rosemary White–Collar Proletariat: deskilling and gender in clerical work (1984)
Basingstoke, Macmillan, 282 p. ISBN 0–333–32752–7.
Crooke, Stanley T and Prestayko, Archie W Introduction to Clinical Oncology (1981)
San Diego CA, Academic Press. Cancer and Chemotherapy: 2. ISBN 0–12–197802–8.
Crooker, T W and Leis, B N (Eds) Corrosion Fatigue: mechanics, metallurgy, electrochemistry and engineering (1983)
Philadelphia PA, American Society for Testing and Materials, 522 p. ISBN 0–8031–0245–3.
Crosby, John F Illusion and Disillusion: the self in love and marriage (1985)
Belmont CA, Wadsworth Publishing, 323 p.
ISBN 0–534–04470–0.
Cross, Jean and Farrer, Donald Dust Explosions (1982)
New York, Plenum Publishing, 260 p.
ISBN 0–306–40871–6.
Cross, John and Guyer, Mel Social Traps (1980)
Ann Arbor MI, University of Michigan Press. illus.
ISBN 0–472–06315–4.
Crosson, Pierre R The Cropland Crisis: myth or reality? (1982)
Washington DC, Resources for the Future, 205 p.
ISBN 0–8018–2816–3.
Crouch, Colin Class Conflict and the Industrial Relations Crisis (1977)
Brookfield VT, Gower Publishing. ISBN 0–435–82250–0.
Crow, Iain, et al Unemployment, Crime and Offenders (1989)
London, Routledge, 216 p. ISBN 0–415–01834–X.
Crowe, Catherine The Night-Side of Nature: of ghosts and ghost-seers (1988)
Aquarian Pr England/Sterling, 452 p. illus.
ISBN 0–85030–519–5.
Crowell, David H, et al Childhood Aggression and Violence: sources of influence, prevention, and control (1987)
New York, Plenum Publishing, 316 p.
ISBN 0–306–42355–3.
Crozier, Michael J Landslides: causes, consequences and environment (1986)
New York, Routledge Chapman and Hall, 272 p.
ISBN 0–7099–0790–7.
Cruetz, W New Light on the Protocols of Zion (1982)
Brooklyn NY, Revisionist Press. ISBN 0–87700–366–1.
Cruickshank, William M and Hallahan, Daniel P Perceptual and Learning Disabilities in Children: psychoeducational practices
Syracuse NY, Syracuse University Press, 498 p. Vol. 1
ISBN 0–8156–2165–5.
Cullen, S I (Ed) Focus on Acne Vulgaris: proceedings of a symposium sponsored by Lederle International, Held in Athens, March 28–29, 1985 (1985)
Wolfeboro NH, Longwood Publishing Group, 176 p. International Congress and Symposium Ser: No 95.
ISBN 1–85315–074–6.
Culligan, Matthew J, et al Back to Basics Management: the lost craft of leadership (1983)
New York, Facts on File, 168 p. ISBN 0–87196–755–3.
Cullinan, Tim Visual Disability in the Elderly (1986)
New York, Routledge Chapman and Hall, 128 p.
ISBN 0–7099–3409–2.
Cumming, G and Bonsignore, G Smoking and the Lung (1985)
New York, Plenum Publishing, 514 p.
ISBN 0–306–41828–2.
Cumming, Gordon and Bonsignore, Giovanni Drugs and the Lung (1984)
New York, Plenum Publishing, 294 p.
ISBN 0–306–41600–X.
Cummings, Nancy B and Klahr, Saulo Chronic Renal Disease: causes, complications and treatment (1985)
New York, Plenum Publishing, 624 p.
ISBN 0–306–41764–2.
Cunha, Burke A Infectious Diseases in the Elderly (1988)
Littleton MA, PSG Publishing, 336 p. illus.
ISBN 0–88416–475–6.
Curb, Rosemary and Manahan, Nancy Lesbian Nuns: breaking silence (1986)
New York, Warner Books, 400 p. ISBN 0–446–32659–3.
Curran, and Renzetti, Social Problems: society in crisis (1987)
Needham Heights MA, Allyn and Bacon.
ISBN 0–205–10482–7.
Curry, Richard O (Ed) Freedom at Risk: secrecy, censorship and repression in the 1980s (1988)
Philadelphia PA, Temple University Press.
ISBN 0–87722–543–5.
Curtin, Brian J The Myopias: basic science and clinical management (1985)
New York, Lippincott, 495 p. illus. ISBN 0–06–140672–4.
Curtis Inequality in American Communities (1977)
San Diego CA, Academic Press. ISBN 0–12–200250–4.
Curtis Poultry Diseases: short notes containing strategic information for veterinary students (1989)
Liverpool, Liverpool University Press, 64 p. 3rd ed.
ISBN 0–85323–356–X.
Curtis, Michael Totalitarianism
Ann Arbor MI, Books on Demand. ISBN 0–317–27282–9.
Curtis, Michael (Ed) Antisemitism in the Contemporary World (1985)
Boulder CO, Westview Press, 200 p. ISBN 0–8133–0157–2.
Curtis, W S and Donlon, E T Observational Evaluation of Severely Multi–Handicapped Children (1985)
Lisse, Swets en Zeitzinger, 204 p. ISBN 90–265–0595–7.
Cuvillier, Rolande The Reduction of Working Time: scope and implications in industrialised market (1984)
Washington DC, ILO, 150 p. ISBN 92–2–103817–3.
Cyriax, James The Slipped Disc (1980)
Hampshire, Gower Publications Group, 256 p.
ISBN 0–566–02218–4.
Czerski, P, et al (Eds) Biological Effects and Health Hazards of Microwave Radiation: proceedings of an international symposium (Warsaw 1973) (1974)
Copenhagen, WHO, 350 p.
D'Arcy, P F and Green, J P Iatrogenic Disease (1986)
New York, Oxford University Press, 950 p.
ISBN 0–19–261440–1.
D'itri, P A and D'itri, F M Mercury Contamination: a human tragedy (1977)
Melbourne FL, Robert E Krieger, 311 p.
ISBN 0–471–02654–9.
Dabberdt, Walter F Atmospheric Dispersion of Hazardous–Toxic Materials from Transport Accidents: proceedings of a course, international center (1985)
New York, Elsevier Science Publishing, 200 p.
ISBN 0–444–87518–2.
Daems, H and Wee, M van der (Eds) Rise of Managerial Capitalism (1974)
Dordrecht, Kluwer Academic Publishers Group, 235 p.
ISBN 90–618–6015–6.
Dahlberg, Ingetraut ICC – Information Coding Classification: principles, structure and application possibilities
In: *International Classification* 9, 1982, 2, pp. 87–93.
Dalakas, Marinos Polymyositis and Dermatomyositis (1988)
Stoneham MA, Butterworth Publishers, 352 p.
ISBN 0–409–95191–9.
Dallin, Alexander (Ed) Diversity in International Communism: a documentary record 1961–1963 (1963)
New York, Columbia University Press, 867 p.
ISBN 0–231–08611–3.
Dalton, Katharina The Premenstrual Syndrome and Progesterone Theory (1983)
Oxford, Heineman Professional Publishing, 300 p. illus. 2nd ed. ISBN 0–433–07092–7.
Dalton, Thomas F The Effects of Heat and Stress on Cleanup Personnel Working with Hazardous Materials (1984)
Detroit MI, Spill Control Association of America.
ISBN 0–318–01766–0.
Daly, Martin and Wilson, Margo Homicide (1988)
Berlin, Walter de Gruyter, 328 p. ISBN 3–11–011725–8.
Dangerous Goods Panel of Air Navigations Technical Instructions for the Safe Transport of Dangerous Goods by Air 1989–90 (1988)
Chicago IL, International Regulations Publishing, 535 p.
ISBN 0–940394–28–6.
Dangerous Goods Panel of Air Navigations and Commission of ICAO Technical Instructions for the Safe Transport of Dangerous Goods by Air, 1987–1988 (1986)
Intereg, 535 p. ISBN 0–940394–21–9.
Daniel, J W, et al Mercury Poisoning, No 2 (1972)
New York, Irvington Publishers, 220 p. illus.
ISBN 0–8422–7073–6.
Daniels, Cora L and Stevans, C M Encyclopedia of Superstitions, Folklore and the Occult Sciences of the World (1971)
Detroit MI, Gale Research, 1885 p. 3 vols.
ISBN 0–8103–3286–8.
Daniels, Robert V Documentary History of Communism (1984)
New Delhi, DK Agencies. REPR 1986,
ISBN 81–85007–12–8.
Danmole, Masood Heritage of Imperialism (1974)
Bombay, Asia Publishing House, 330 p.

Darby, Padraig L, et al Anorexia Nervosa: recent developments in research (1983)
New York, Alan R Liss, 472 p. Neurology and Neurobiology: Vol 3.

Dareer, Asma El Woman, Why Do You Weep?: circumcision and its consequences (1983)
Atlantic Highlands NJ, Humanities Press International, 144 p. ISBN 0-86232-098-4.

Darnbrough, Ann and Kinrade, Derek Directory for Disabled People (1985)
Wolfeboro NH, Longwood Publishing Group, 358 p. ISBN 0-85941-255-5.

Darrow, Clarence Resist Not Evil (1972)
Montclair NJ, Patterson Smith Publishing, 200 p. Intro and index added. Criminology, Law Enforcement and Social Problems Series: 148. ISBN 0-87585-147-7.

Das, A K Unemployment of Educated Youth in Asia: a comparative analysis of the situation in India, Bangladesh and the Philippines (1981)
Paris, International Institute for Educational Planning, 49 p.

Das, Man Singh and Harry, Joseph Homosexuality in International Perspective (1980)
New Delhi, Vikas Publishing House, 132 p.

Das, P K Monsoons (1986)
Geneva, WHO, 155 p. ISBN 92-63-10613-4.

Dasberg, H, et al Society and Trauma of War (1987)
Assen, Van Gorcum, 80 p. ISBN 90-232-2275-X.

Dascal, M The Controversy over Ideas and the Idea of Controversy
In: *F Gil (Ed) Controvérsias Científicas e filosoficas – Proceedings of the Evora University Colloquium 1986* Evora, 1989.

Dasmann, R F Planet in Peril?: man and the biosphere today (1972)
Paris, UNESCO, 125 p. ISBN 92-3-100947-8.

Dässler, H G and Böritz, S (Eds) Air Pollution and Its Influence on Vegetation: causes – effects – prophylaxis and therapy (1987)
Dordrecht, Kluwer Academic Publishers Group, 244 p. ISBN 90-6193-619-5.

Daunton, N G, et al Mechanisms of Motion-Induced Vomiting (1983)
New York, Karger S AG, 80 p. ISBN 3-8055-3790-5.

Davenport, Charles B and Rosenberg, Charles The Feebly Inhibited (1984)
New York, Garland Publishing, 156 p. ISBN 0-8240-5804-6.

David, M The Complications of Modern Medicine: a treatise on iatrogenic diseases (1963)
New York.

Davies, J C When Men Revolt and Why (1971)
New York, Free Press. ISBN 0-02-907310-3.

Davies, Nigel Human Sacrifice in History and Today (1988)
New York, Hippocrene Books, 320 p. ISBN 0-88029-211-3.

Davies, P Alcohol-related Problems in the European Community (1985)
Luxembourg, CEE, 131 p. ISBN 92-825-4918-6.

Davis, C J, et al Nausea and Vomiting: mechanisms and treatment (1986)
New York, Springer Publishing, 200 p. ISBN 0-387-15436-1.

Davis, David B The Fear of Conspiracy: images of un-american subversion from the (1972)
Ithaca NY, Cornell University Press, 396 p. ISBN 0-8014-0598-X.

Davis, John M and Maas, James W (Eds) The Affective Disorders (1985)
Washington DC, American Psychiatric Press, 448 p. illus. ISBN 0-88048-214-1.

Davis, John W, et al Infectious Diseases of Wild Mammals (1981)
Ames IA, Iowa State University Press, 446 p. ISBN 0-8138-0445-0.

Davis, Lenwood G Ecology of Blacks in the Inner City: an exploratory bibliography (1975)
Chicago IL, CPL Bibliographies. Nos. 785-786 ISBN 0-686-20352-6.

Davis, Peter R (Ed) Industrial Back Pain in Europe (1985)
New York, Taylor and Francis, 416 p. Ergonomics Special Issue Ser: Vol 28, No1. ISBN 0-85066-985-5.

Davis, R D; Hucker, G and L'Hermite, P (Eds) Environmental Effects of Organic and Inorganic Contaminants in Sewage Sludge: proceedings of a workshop held at Stevenage, 1982 (1983)
Dordrecht, Kluwer Academic Publishers Group, 272 p. ISBN 90-277-1586-6.

Dawber, R and Rook, A Diseases of the Hair and Scalp (1982)
Oxford, Blackwell Scientific Publications, 582 p. ISBN 0-632-00822-9.

Dawson, John and Phillips, Melanie Doctor's Dilemmas (1985)
Hemel Hempstead, Simon and Schuster International, 240 p. ISBN 0-7108-0983-2.

Day, Alan J (Eds) Border and Territorial Disputes (1987)
Detroit MI, Gale Research, 500 p. 2nd ed. ISBN 0-8103-2543-8.

Day, Alan J and East, Roger Government Economic Agencies of the World: an international directory of governmental organizations concerned with economic development and planning (1985)
Detroit MI, Gale Research. A Keesing's Reference Publication Ser. ISBN 0-8103-2104-1.

Day, Stacey B Cancer, Stress, and Death (1986)
New York, Plenum Publishing, 392 p. ISBN 0-306-42187-9.

De Baets, M, et al (Eds) Myasthenia Gravis: European conference on myasthenia gravis, Maastricht, 1st June 1987 (1988)
Basel, S Karger AG, 160 p. illus. ISBN 3-8055-4736-6. Monographs in Allergy: Vol 25.

De Boismont, A Brierre A Treatise on Magnetism and Hallucinations As an Expression of Nervous Disorders (1988)
Albuquerque NM, American Institute for Psychological Research, 137 p. illus. ISBN 0-89920-180-6.

De Catanzaro, Denys Suicide and Self-Damaging Behavior: a sociobiological perspective (1981)
San Diego CA, Academic Press. ISBN 0-12-163880-4.

De Gier, J J and O'Hanlon, J F Drugs and Driving (1986)
New York, Taylor and Francis, 300 p. ISBN 0-85066-290-7.

de Grazia, Raffaele Clandestine Employments. The situation in industrialised economy countries (1984)
Geneva, ILO, 118 p. ISBN 92-2-103355-4.

de Grolier, Eric From Theories to Concepts and From Facts to Words
In: *ISSI* 1990, 124.

De Huszar, George B Persistent International Issues
Salem NH, Ayer Company Publishers. ISBN 0-8369-2772-9.

De Mause, Lloyd (Ed) The History of Childhood: the untold story of child abuse (1988)
New York, Peter Bedrick Books, 458 p. ISBN 0-87226-181-6.

De Soto, Hernando The Other Path: the invisible revolution in the Third World (1989)
New York, Harper and Row, 271 p. Foreword by Mario Vargas Llosa. ISBN 0-06-016020-9.

De Vries, Jan Cancer and Leukaemia: an alternative approach (1988)
Mnstream Scotland/David and Charles, 144 p.
By Appointment Only Ser. ISBN 1-85158-136-7.

De Vries, Jan Stress and Nervous Disorders (1988)
North Pomfret VT, David and Charles, 126 p. ISBN 0-906391-81-4.

De Young, Mary Incest: an annotated bibliography (1985)
Jefferson NC, McFarland Publishers, 171 p. bibl. ISBN 0-89950-142-7.

De Zayas, Alfred The Wehrmacht War Crimes Bureau

Deacon, Richard The Truth Twisters: disinformation; the making and spreading of official distortions, half-truths and lies (1988)
London, Futura Publications, 240 p. ISBN 0-7088-3644-5.

Deble, Isabell The School Education of Girls: an international comparative study on school (1980)
Lanham MD, UNIPUB, 180 p. ISBN 92-3-101782-9.

Dechesne, B H, et al (Eds) Sexuality and Handicap: problems of motor handicapped people (1986)
Springfield IL, Thomas Charles C, 240 p. illus. ISBN 0-398-04746-4.

Decker, David L, et al Urban Structure and Victimization (1982)
Lexington MA, Lexington Books, 128 p. ISBN 0-669-02951-3.

Dede, Spiro Counter-revolution in the Counter-revolution
Tirana, Book Distribution Enterprise, 303 p.

DeDombal, F T (Ed) Inflammatory Bowel Disease: some international data and reflections (1986)
Oxford NY, Oxford University Press, 608 p. illus. ISBN 0-19-261354-5.

Deer, John Polygamy and Polyandry: a comprehensive bibliography (1986)
Gibson LA, Research and Discovery Publications, 108 p. bibl. Orig. ISBN 0-940519-08-9.

Deere, Derek H Corrosion in Marine Environment International Sourcebook I: ship painting and corrosion (1977)
New York, Halsted Press, 259 p. ISBN 0-470-15203-6.

DeFrain, John, et al Stillborn: the invisible death (1986)
Lexington MA, Lexington Books, 247 p. ISBN 0-669-11352-2.

DeGaay Fortman, B (Ed) Overdevelopment: a series of public lectures (1979)
's Gravenhage, Institue of Social Studies, 164 p.

Degen, R and Niedermeyer, E (Eds) The Lennox-Gastaut Syndrome (1989)
New York, Alan R Liss, 500 p. ISBN 0-8451-2749-7.

Degenhardt, Henry W Political Dissent: an international guide to dissident, extra-parliamentary, guerrilla and illegal political movements (1983)
Harlow, Longman, 592 p. bibl. ISBN 0-582-90255-X.

Deger, Saadet and Sen, Somnath Military Expenditure: the political economy of international security (1990)
Oxford, International Peace Research Institute, 186 p. Strategic Issue Papers. ISBN 0-19-829141-8.

DeGrazia, Edward and Newman, Roger K Banned Films: movies, censors, and the first amendment (1982)
Munich, K G Saur Verlag, 455 p. ISBN 0-8352-1509-1.

Deits, Bob Life after Loss: a personal guide to dealing with death, divorce, job change and relocation (1988)
Stuttgart, Fischer, 204 p.

Deitz, Samuel M and Hummel, John H Discipline in the Schools: a guide to reducing misbehavior (1978)
Englewood Cliffs NJ, Educational Technology Publications, 280 p. ISBN 0-87778-127-3.

Dekker, J, et al Fungicide Resistance in Crop Protection (1982)
Wageningen, Centre for Agricultural Publishing and Documentation. ISBN 90-220-0797-9.

Dekker, Lies, et al Management of Toxic Materials in an International Setting: a case study of cadmium in the North Sea (1987)
Rotterdam, AA Balkema, 140 p. ISBN 90-6191-795-6.

Delacoste, Frederique and Alexander, Priscilla Sex Work: writings by women in the sex industry (1987)
Pittsburgh PA, Cleis Press, 360 p. ISBN 0-939416-10-7.

Delamaide, Darrell Debt Shock: the full story of the world credit crisis (1984)
Garden City NY, Doubleday, 280 p. ISBN 0-385-18899-4.

Delamotte, Yves and Takezawa, S Quality of Working Life in International Perspective (1984)
Washington DC, ILO, 89 p. ISBN 92-2-103402-X.

Delarue, et al Esophageal cancer (1988)
Saint Louis MO, CV Mosby, 496 p. illus. International Trends in General Thoracic Surgery: Vol 4. ISBN 0-8016-2048-1.

Deleuze, Gilles and Guattari, Felix A Thousand Plateaus: capitalism and schizophrenia (1987)
Minneapolis MN, University of Minnesota Press, 629 p. ISBN 0-8166-1401-6.

Delmont, J (Ed) Milk Intolerance and Rejection (1983)
Basel, S Karger AG, 170 p. ISBN 3-8055-3546-5.

Deloitte; Haskins and Sells Taxation of International Executives (1985)
Dordrecht, Kluwer Academic Publishers Group, 370 p. ISBN 90-6544-227-8.

Delp, Jeanne L and Martinson, Ruth A A Handbook for Parents of Gifted and Talented (Also Helpful for Educators)
Los Angeles CA, National State Leadership Training Institute on Gifted and Talented, 202 p. ISBN 0-318-02117-X.

Dembroski, T M, et al Biobehavioral Bases of Coronary Heart Disease (1983)
New York, Karger S AG, 482 p. ISBN 3-8055-3629-1.

Denno, Deborah W and Schwarz, Ruth M Biological, Psychological, and Environmental Factors in Delinquency and Mental Disorder: an interdisciplinary bibliography (1985)
Westport CT, Greenwood Press, 222 p. ISBN 0-313-24939-3.

Denton, R M and Pogson, C I Metabolic Regulation (1976)
New York, Routledge Chapman and Hall. ISBN 0-412-13150-1.

DePury, J M S Crop Pests of East Africa (1968)
Oxford NY, Oxford University Press, 244 p. ISBN 0-19-644052-1.

Desai, I P Caste, Caste-Conflict and Reservation (1985)
New Delhi, Ajanta Books, 220 p. ISBN 81-202-0143-4.

Dessypris, Emmanuel N Pure Red Cell Aplasia (1988)
Baltimore MD, Johns Hopkins University Press, 176 p. ISBN 0-8018-3572-0.

Dev, Som Multinational Corporations and the Third World (1986)
New Delhi, Ashish Publishing House, 224 p.

Devereux, D and Greco, G Color Atlas on the Treatment of Breast Diseases (1989)
Saint Louis MO, Ishiyaku Euro-America, 300 p. illus. ISBN 0-912791-65-9.

Devlin, H B Management of Abdominal Hernias (1988)
Stoneham MA, Butterworth Publishers, 208 p. ISBN 0-407-00348-7.

Dewart, Joanne M Death and Resurrection (1986)
Wilmington DE, Glazier Michael. ISBN 0-89453-362-2.

Dews, P B (Ed) Caffeine: perspectives from recent research (1984)
Berlin, Springer-Verlag, 300 p. ISBN 3-540-13532-4.

Deyo, R A (Ed) Back Pain in Workers (1988)
Oxford, Blackwell Scientific Publications, 176 p. illus. ISBN 0-932883-63-X.

Dhanda, M R and Sen, G P Haemorrhagic Septicaemia and Allied Conditions in Sheep, Goats, Pigs and Poultry (1972)
New Delhi, Indian Council of Agricultural Research, 40 p.

Diamant, Louise (Ed) Male and Female Homosexuality: psychological perspectives (1987)
Hants, Taylor and Francis, 292 p. ISBN 0-89116-449-9. The Series in Clinical and Community Psychology.

Diamond, Jed Looking for Love in All the Wrong Places: overcoming romantic and sexual addictions (1988)
New York, Putnam Publishing Group, 224 p. ISBN 0-399-13372-0.

Diamond, Susan A Helping Children of Divorce: a handbook for parents and teachers (1986)
New York, Schocken Books, 130 p. ISBN 0-8052-0821-6.

Dick, James C Violence and Oppression (1979)
Athens GA, University of Georgia Press, 224 p. ISBN 0-8203-0446-8.

Dick, W C Immunological Aspects of Rheumotology (1981)
New York, Elsevier Science Publishing, 262 p. ISBN 0-444-19474-6.

Dickson, James G Diseases of Field Crops (1971)
New Delhi, Oxford and IBH Publishing, 517 p. 2nd ed.

Diekstra, René F W and Hawton, Keith E (Eds) Suicide in Adolescence (1986)
Dordrecht, Kluwer Academic Publishers Group, 196 p. ISBN 0-89838-780-9.

Dierkes, Meinolf, et al Technological Risk (1980)
Boston MA, Oelgeschlager Gunn and Hain, 160 p. ISBN 0-89946-059-3.

Dietz, O and Wiesner, E (Eds) Diseases of the Horse (1984)
Basel, S Karger AG, 196 p. ISBN 3-8055-3497-3.

Dijk, Tean A van Prejudice in Discourse: an Analysis of ethnic prejudice in cognition and conversation (1984)
Amsterdam, Benjamins Publishing Company, 170 p. ISBN 0-915027-43-7.

Dilman, Ilham Love and Human Separateness
New York, Basil Blackwell, 169 p. ISBN 0-317-58084-1.

DiMaio, V J M Gunshot Wounds: practical aspects of criminal and forensic investigations (1987)
Amsterdam, Elsevier Science Publishing. 2nd repr of 1985. ISBN 0-444-00928-0.

Dimitrov, Georgi Against Fascism and War
Sofija, Sofia Press Agency, 132 p.

Dion, Robert Crimes of the Secret Police (1982)
Montreal, Black Rose Books, 228 p. ISBN 0-919619-57-6.

Dixon, Norman On the Psychology of Military Incompetence

Dmitriev, Val and Oelwein, Pat Advances in Down Syndrome (1988)
Seattle WA, Special Child Publications, 336 p.
ISBN 0-87562-092-2.

Do It Now Foundation Drugs, Alcohol and Pregnancy (1988)
Phoenix AZ, Do It Now Foundation. ISBN 0-89230-190-2.

Dobkowski, Michael and Willimann, Isidor (Eds) Research in Inequality and Social Conflict (1988)
Greenwich CT, Jai Press. Vol 1. ISBN 0-89232-745-6.

Dobson, Christopher and Payne, Ronald Counterattack: the West's battle against the terrorists (1984)
New York, Facts on File, 222 p. ISBN 0-87196-878-9.

Docter, R F Transvestites and Transsexuals: toward a theory of cross-gender behavior (1988)
New York, Plenum Publishing, 244 p.
ISBN 0-306-42878-4.

DOE Technical Information Center Acid Precipitation: a compilation of worldwide literature – A bibliography (1983)
Oak Ridge TN, United States Department of Energy, 732 p. bibl. ISBN 0-87079-500-7.

DOE Technical Information Center Radioactive Waste Management: low-level radioactive waste: a bibliography (1984)
Oak Ridge TN, US Department of Energy, 183 p.
ISBN 0-87079-524-4.

DOE Technical Information Center and McLaren, Lynda H Radioactive Waste Management: high-level radioactive wastes: a bibliography (1984)
Oak Ridge TN, US Department of Energy, 349 p. Supplement 1 ISBN 0-87079-528-7.

Doehring, Donald et al Reading Disabilities: the interaction of reading, leading and (1981)
San Diego CA, Academic Press. ISBN 0-12-219180-3.

Doh, Joon-Chien Eastern Intellectuals and Western Solutions: follower syndrome in India (1980)
New Delhi, Vikas Publishing House, 160 p.
ISBN 0-7069-0968-2.

Dolitsky, Marlene Under the Tumtum Tree: From Nonsense to Sense: a study in non-automatic comprehension (1984)
Amsterdam, Benjamins Publishing Company, 118 p.
ISBN 0-915027-39-9.

Dolman, Antiony (Ed) RIO: Reshaping the International Order: a report to the Club of Rome (1977)
New York, Signet, 432 p. Coordinated by Jan Tinbergen.

Domhoff, G William and Dye, Thomas R (Eds) Power Elites and Organizations (1987)
London, Sage, 320 p. Focus Editions Ser: Vol 82.
ISBN 0-8039-2680-4.

Dong, Pham Van Some Cultural Problems
Hanoi, Xunhasaba, 220 p.

Donham, K Occupational Disease of Agricultural Workers (1985)
New York, Praeger Publishers. ISBN 0-275-91302-3.

Donnelly, Jack and Howard, Rhoda E International Handbook on Human Rights (1987)
Westport CT, Greenwood Press. ISBN 0-313-24788-9.

Donnison, David and Middleton, Alan Regenerating the Inner City (1987)
New York, Routledge Chapman and Hall, 304 p.
ISBN 0-7102-1116-3.

Dooley, Tricia C Insomnia: index of modern information (1988)
Washington DC, ABBE Publications Association, 150 p.
ISBN 0-88164-844-2.

Dosa, Marta L and Froehlich, Thomas J (Eds) Curriculum Development in a Changing World (1985)
's Gravenhage, International Federation for Information and Documentation, 240 p. ISBN 92-66-00645-9.

Dosman, James A and Cotton, David J Occupational Pulmonary Disease: focus on grain dust and health (1980)
San Diego CA, Academic Press. ISBN 0-12-221240-1.

Dost, H and VanBreemen, M Bangkok Symposium on Acid Sulphate Soils: proceedings of the 2nd international symposium on Acid Sulphate Soils, Bangkok, 1981 (1982)
Wageningen, International Institute for Land Reclamation and Improvement. ISBN 90-70260-71-9.

Douglas, C H The Land for the Chosen People Racket (1982)
Brooklyn NY, Revisionist Press. ISBN 0-87700-415-3.

Douglas, Mary Purity and Danger: an analysis of the concepts of pollution and taboo (1984)
New York, Routledge Chapman and Hall, 196 p.
ISBN 0-7448-0011-0.

Doury, P et al Algodystrophy (1981)
Berlin, Springer-Verlag, 190 p. illus. ISBN 0-387-10624-3.

Dovidio, John D and Gaertner, Samuel L Prejudice, Discrimination, and Racism (1986)
San Diego CA, Academic Press. ISBN 0-12-221425-0.

Dowling, Colette Perfect Women: hidden fears of inadequacy (1988)
New York, Summit Books, 272 p. ISBN 0-671-54747-X.

Downing, A B (Ed) Euthanasia and the Right to Death (1969)
London, Peter Owen.

Downing, Theodore; Kruijt, Dirk and Ufford, Philip Quarles van (Eds) The Hidden Crisis in Development: development bureaucracies
Tokyo, United Nations University. ISBN 90-6256-641-3.

Dowty, Alan Closed Borders (1987)
New Haven CT, Yale University Press, 272 p.
ISBN 0-300-03824-0.

Doyle, D (Ed) International Symposium on Pain Control, 1986: proceedings of a symposium, Cannes, France, October 17-19, 1986, No 123 (1988)
Wolfeboro NH, Longwood Publishing Group.
ISBN 0-905958-55-1.

Doyle, Robert J Readings in Wealth Accumulation Planning (1987)
Bryn Mawr PA, American College. ISBN 0-943590-10-8.

Drake, Marie R Rape: index of modern information with bibliography (1988)
Annandale VA, ABBE Publishers Association of Washington, 150 p. ISBN 0-88164-880-9.

Drèze, Jean and Sen, Amartya Hunger and Public Action
Clarendon Press, 373 p.

Drummond, Harold P Schizophrenia: medical and psychological subject index (1987)
Annandale VA, ABBE Publishers Association of Washington, 150 p. ISBN 0-88164-496-X.

Drummond, Harold P Sex Offenses: medical and psychological subject analysis (1987)
Annandale VA, ABBE Publishers Association of Washington, 150 p. ISBN 0-88164-310-6.

Drummond, R O and Kunz, S E (Eds) Arthropod Pests of Livestock: a review of technology (1988)
Boca Raton FL, CRC Press, 272 p. ISBN 0-8493-6860-X.

Dubashi, Jay Snakes and Ladders: the development game (1985)
New Delhi, Allied Publishers, 296 p.

Dube, S C and Basilov, Vladimir N Secularization in Multi-religious Societies: Indo-Soviet perspectives (1983)
New Delhi, Concept Publishing, 322 p.

Duberman, Lucile Reconstituted Family: a study of remarried couples and their children (1975)
Chicago IL, Nelson-Hall, 200 p. ISBN 0-88229-168-8.

Duby, Georges; Goldhammer, Arthur and Bisson, Thomas N The Three Orders: feudal society imagined (1980)
Chicago IL, University of Chicago Press, 432 p.
ISBN 0-226-16771-2.

Duchaine, Nina The Literature of Police Corruption: a selected, annotated bibliography (1979)
New York, John Jay Press. Vol. II ISBN 0-89444-008-X.

Dudley, Walter C and Lee, Min Tsunami (1988)
Honolulu, University of Hawaii Press.

Dudley, Walter C and Lee, Min Tsunami (1988)
University of Hawaii Press.

Duke, Simon United States Military Forces and Installations in Europe (1989)
Oxford, Stockholm International Peace Research Institute, 464 p. SIPRI Publication Series. ISBN 0-19-829132-9.

Duker, Marilyn and Slade, Roger Anorexia and Bulimia Nervosa: how to help (1988)
New York, Taylor and Francis, 256 p.
ISBN 0-335-09836-3.

Dukes, M Side Effects of Drugs
New York, Elsevier Science Publishing, 476 p. Annual: 1985-1988.

Dumon, Wilfried and Paepa, Chrisiane de (Eds) International CFR Seminar on Divorce and Remarriage (19,1981, Leuven): papers (1981)
Leuven, Katholieke Universiteit Leuven, 269 p.

Dumont, J J and Nakken, H (Eds) Cognitive, Social and Remedial Aspects (1989)
Lisse, Swets en Zeitlinger, 224 p. Learning Disabilities: 2.
ISBN 90-265-0984-7.

Duncan, R and Weston-Smith, M (Eds) Encyclopedia of Ignorance (1977)
Oxford, Pergamon Press.

Dunkle, Ruth E and Schmidley, James Stroke in the Elderly (1987)
New York, Springer Publishing, 224 p.
ISBN 0-8261-5430-1.

Dunnigan, James F and Bay, Austin A Quick and Dirty Guide to War: briefings on present and potential wars (1985)
New York, William Morrow, 384 p. ISBN 0-688-04199-X.

Dunning, A J and Wils, W I M (Eds) Heart of the Future/Future of the Heart: scenarios on cardiovascular diseases 1985-2010, vol 2: background and approach 1986 (1986)
Dordrecht, Kluwer Academic Publishers Group, 176.
ISBN 0-89838-868-6.

Dunning, A J and Wils, W I M (Eds) Heart and the Future/Future of the Heart: scenarios on cardiovascular diseases 1985-2010, vol 1: scenario report 1986 (1987)
Dordrecht, Kluwer Academic Publishers Group, 109 p.
ISBN 0-89838-877-5.

DuPerron, William Annotated Research Bibliography on the Female Offender (1978)
Ottawa ON, Canadian Criminal Justice Association, 157 p.

Dupraz, J Probability, Signal, Noise (1986)
New York, McGraw Hill Book Company, 334 p.
ISBN 0-07-018330-9.

Dupuis, H and Zerlett, G Effects on Whole-Body Vibration (1986)
Berlin, Springer-Verlag, 162 p. ISBN 3-540-16584-3.

Durbin, R D Toxins in Plant Disease (1981)
San Diego CA, Academic Press. ISBN 0-12-225050-8.

Durning, Alan Poverty and the Environment: reversing the ownward spiral (1989)
Washington DC, Worldwatch Institute. Worldwatch Papers.
ISBN 0-916468-93-3.

Durning, Alan B Mobilizing at the Grassroots: local attention on poverty and the environment (1989)
Washington DC, Worldwatch Institute. Worldwatch Papers.
ISBN 0-916468-89-5.

Durrington, P N Diagnosis and Management of Hyper-lipidaemias (1989)
Sevenoaks, Butterworths, 250 p. illus.
ISBN 0-7236-0915-2.

Durron, Daskin R Heart Injuries: medical and scientific subject analysis with (1987)
Annandale VA, ABBE Publishers Association of Washington, 150 p. ISBN 0-88164-422-6.

Durron, Daskin Rice Injuries and Wounds I: medical subject analysis and research guide (1985)
Annandale VA, ABBE Publishers Association of Washington, 150 p. ISBN 0-88164-098-0.

DuToit, Brian M Cannabis in Africa (1980)
Rotterdam, AA Balkema, 512 p. ISBN 90-6191-030-7.

Dwivedi, S N Political Corruption in India (1967)
New Delhi, Popular Book Services, 180 p.

Dworkin, Andrea Woman Hating: a radical look at sexuality
New York, EP Dutton. ISBN 0-525-48397-7.

Dy, Fe Josefina F Visual Display Units: job content and stress in office work (1985)
Geneva, ILO, 138 p. ISBN 92-2-105084-X.

Dyer, Frederick C and Dyer, John M Bureaucracy vs Creativity: the dilemma of modern leadership (1965)
Baltimore MD, University of Miami Press.
ISBN 0-87024-134-6.

Dykes, P W and Keighley, M R Gastrointestinal Haemorrhage (1981)
Littleton MA, PSG Publishing, 448 p.
ISBN 0-7236-0584-X.

Eadington, William R The Gambling Papers (1982)
Reno NV, University of Nevada, 2000 p. 13 Vols.
ISBN 0-942828-17-8.

Eadington, William R (Ed) Gambling Papers: proceedings of the Sixth National Conference on Gambling and Risk Taking (1985)
Reno NV, University of Nevada, 1776 p. 5 Vols.
ISBN 0-317-20704-0.

Eagleton, Clyde (Ed) Analysis of the Problem of War (1972)
Salem NH, Ayer Company Publishers. repr of 1937.
ISBN 0-8369-9961-4. Select Bibliographies Reprint Ser.

Eaker, Elaine D, et al Coronary Heart Disease in Women: proceedings of an N I H workshop (1987)
New York, Haymarket-Doyma, 207 p.
ISBN 0-937716-26-X.

Ebel, Karl-H and Ulrich, Erhard The Computer in Design and Manufacturing – Servant or Master?: social and labour effects of computer-aided design/computer-aided manufacturing (CAD/CAM)
Geneva, ILO. ISBN 92-2-106072-1.

Eberhart, George M A Geo-Bibliography of Anomalies: primary access to observations of UFOs, ghosts, and other mysterious phenomena (1980)
New York, Greenwood Press, 1114 p.
ISBN 0-313-21337-2.

Eberhart, George M and Hynek, J Allen UFO's and the Extraterrestrial Contact Movement: a bibliography (1986)
New York, Garland Publishing, 600 p.
ISBN 0-8240-8755-0.

Eckenfelder, W Wesley Industrial Water Pollution Control (1988)
New York, McGraw Hill Book Company, 352 p. 2nd ed. Water Resources and Environmental Engineering Ser.

Eckhoff, Rolf Kristian Dust Explosions in the Process Industry (1990)
Sevenoaks, Butterworths, 250 p. illus.
ISBN 0-408-04803-4.

Eckholm, Erik The Dispossessed of the Earth: land reform and sustainable development (1979)
Washington DC, Worldwatch Institute. Worldwatch Papers.
ISBN 0-916468-29-1.

Eckholm, Erik P The Other Energy Crisis: firewood (1975)
Washington DC, Worldwatch Institute.
ISBN 0-916468-00-3.

Ecobichon, Donald J and Joy, Robert M Pesticides and Neurological Diseases (1982)
Boca Raton FL, CRC Press, 296 p. ISBN 0-8493-5571-0.

Economopoulos, A P (Ed) Fruit Flies (1987)
Amsterdam, Elsevier Science Publishing, 590 p.
ISBN 0-444-98946-3.

ECOSOC Statistical Commission System of Social and Demographic Statistics: potential uses and usefulness (1974)
New York, United Nations.

Edelman, marian W Families in Peril: an agenda for social change (1987)
Cambridge MA, Harvard University Press, 152 p. illus. W E B Du Bois Lectures Ser. ISBN 0-674-29228-6.

Edelstein, Michael R and Levine, Adeline Contaminated Communities (1988)
Boulder CO, Westview Press, 215 p.ISBN 0-8133-7447-2.

Edgerton, Robert B and Meyers, C Edward Lives in Process: mildly retarded adults in a large city (1984)
Washington DC, American Association on Mental Retardation, 192 p. ISBN 0-940898-13-6.

Educational Policy Research Center Contemporary Societal Problems (1971)
Menlo Park, Stanford Research Institute, 134 p.

Edwards, Chris The Fragmented World: competing perspectives on trade, money and crisis (1985)
New York, Routledge Chapman and Hall. Development and Underdevelopment Ser. ISBN 0-416-73390-5.

Edwards, Corwin, et al Economic and Political Aspects of International Cartels (1976)
Salem NH, Ayer Company Publishers.
ISBN 0-405-09275-X.

Edwards, Felicity, et al (Eds) Epilepsy and Employment – A Medical Symposium on the Current Law: proceedings of a symposium sponsored by Labaz Sanofi UK Ltd, held at the Royal College of Physicians, London, June 3, 1985 (1986)
Wolfeboro NH, Longwood Publishing Group, 114 p. International Congress and Symposium Ser: No 86.

Edwards, John Positive Discrimination and Social Justice (1987)
New York, Routledge Chapman and Hall, 220 p.
ISBN 0–422–78990–9.

Edwards, M Disorders of Articulation: aspects of dysarthria and verbal dyspraxia (1984)
New York, Springer Publishing, 130 p.
ISBN 0–387–81787–5.

Edwards, R G and Purdy, J M Human Conception In Vitro (1982)
San Diego CA, Academic Press. ISBN 0–12–232740–3.

Edwards, R G, et al Implantation of the Human Embryo (1985)
San Diego CA, Academic Press. ISBN 0–12–232455–2.

Egami, Nobuo (Ed) Radiation Effects on Aguatic Organisms (1980)
Tokyo, Japan Scientific Societies Press, 310 p.

Egenter, Richard and Matussek, Paul Moral Problems and Mental Health (1967)
Alba. ISBN 0–8189–0095–4.

Egli, H and Inwood, M J The Haemophiliac in the Eighties (1981)
New York, Karger S AG, 310 p. ISBN 3–8055–2885–X.

Ehrenberg, B Sleep and Sleep Disorders (1989)
Oxford, Blackwell Scientific Publications, 350 p.
ISBN 0–86542–057–2.

Ehrenfeld, David The Arrogance of Humanism (1978)
Oxford, Oxford University Press, 5 p.

Ehrlich, Paul and Ehrlich, Anne Extinction: the causes and consequences of the disappearance (1981)
New York, Random House. ISBN 0–394–51312–6.

Ehrlich, Paul R and Holdren, John P (Ed) The Cassandra Conference: resources and the human predicament (1988)
A University Press, 330 p.

Eide, Asbjørn and Mubanga–Chipoya, Chama Conscientious Objection to Military Service
Oslo, International Peace Research Institute, 32 p.

Eigen, Manfred How Does Information Originate?: principles of biological self–organization (1978)
Advances in Chemical Physics, 38, p.211–262.

Einstein and Catanzaro (Eds) Proceedings of the Fourth International Conference on Coccidioidomycosis (1985)
Washington DC, National Foundation for Infectious Diseases, 532 p. ISBN 0–9614520–0–5.

Einstein, Stanley Drug and Alcohol Use: issues and factors (Date not set)
New York, Plenum Publishing, 480 p.
ISBN 0–306–41378–7.

Eisenberg, Mickey, et al Sudden Cardiac Death in the Community (1984)
New York, Praeger Publishers, 163 p.
ISBN 0–275–91428–3.

Eisenberg, Myron G et al Disabled People As Second Class Citizens (1982)
New York, Springer Publishing, 320 p.
ISBN 0–8261–3220–0.

Eisler, Robert and Lathrop, Donald Man into Wolf: an anthropological interpretation of sadism (1978)
Santa Barbara CA, Ross–Erikson, 264 p.
ISBN 0–915520–16–8.

Eitzen, Social Problems (1985)
Needham Heights MA, Allyn and Bacon.
ISBN 0–205–08584–9.

Eitzen, D Stanley Society's Problems: sources and consequences (1989)
Needham Heights MA, Allyn and Bacon, 456 p.
ISBN 0–205–11979–4.

Eitzen, Stanley D and Timmer, Doug A (Eds) Crime in the Streets and Crime in the Suites: perspectives on crime and criminal justice (1989)
Hemel Hempstead, Simon and Schuster International, 320 p.
ISBN 0–205–11977–8.

Ekstrand, L H (Ed) Ethnic Minorities and Immigrants in a Cross–Cultural Perspective (1986)
Lisse, Swets en Zeitlinger, 256 p. ISBN 90–265–0725–9.

El–Ayouty, Yassin The Dissemination, Use and Impact of Knowledge Relevant to UNITAR: a program for research and action
In: *Social Science Information* Oct 1971, 10, pp. 55–72.

El–Hinnawi, Essam Environmental Refugees (1985)
Nairobi, UNEP, 41 p.

El–Swaify, S A, et al Soil Erosion and Conservation
Ankeny IA, Soil and Water Conservation Society, 793 p.
ISBN 0–935734–11–2.

Elias, N; Martins, H and Whitley, R (Eds) Scientific Establishment and Hierarchies (1982)
Dordrecht, Kluwer Academic Publishers Group, 384 p.
ISBN 90–277–1322–7.

Eliasson, Gunnar; Sharefkin, Mark and Ysander, Bengt–Christer (Eds) Policy Making in a Disorderly World Economy
Stockholm, Almqvist and Wiksell, 417 p.
ISBN 91–7204–166–8.

Elisens, Wayne J; Anderson, Christiane and Angell, Bobbi Monograph of the Maurandyinae (Scrophulariaceae–Antirrhinae) (1985)
Ann Arbor MI, American Society of Plant Taxonomists, 97 p.
ISBN 0–912861–05–3.

Elkind, Jerome B Non–Appearance Before the International Court of Justice: functional and comparative analysis (1984)
Dordrecht, Martinus Nijhoff, 233 p. ISBN 90–247–2921–1.

Elliot, Jeffrey M Annual Editions: third world 1988–89 (1988)
Guilford CT, Dushkin Publishing Group, 256 p.
ISBN 0–87967–707–4.

Elliott, E N Cotton Is King, and Pro–Slavery Arguments
New York, Johnson Reprint Corporation.
ISBN 0–384–14175–7.

Elliott, Kimberly A and Williamson, John (Eds) World Econic Problems (1988)
Washington DC, Institute for International Economics, 293 p.
Special Reports Series: 7. ISBN 0–88132–055–2.

Ellis, Albert and Sagarin, Edward Nymphomania (1968)
London.

Ellis, K V Surface Water Pollution and Its Control (1989)
Basingstoke, Macmillan, 350 p. ISBN 0–333–42764–5.

Ellis, Lee Theories of Rape: inquiries into the cause of sexual aggression (1989)
Hants, Taylor and Francis, 193 p. ISBN 0–89116–172–4.

Ellis, W A and Little, T W (Eds) The Present State of Leptospirosis Diagnosis and Control (1986)
Dordrecht, Martinus Nijhoff. Current Topics in Veterinary Medicine and Animal Science Ser. ISBN 0–89838–777–9.

Ellul, Jacques The Technological System (1980)
New York, Continuum Publishing, 384 p.
ISBN 0–8264–0002–7.

Eltis, David and Walvin, James The Abolition of the Atlantic Slave Trade (1981)
Madison WI, University of Wisconsin Press, 328 p.
ISBN 0–299–08490–6.

Elworthy, Frederick T The Evil Eye (1986)
New York, Crown Publishers. ISBN 0–517–55971–4.

Emmelkamp, P M G; Everaerd, W T A M and Kraaimaat, F (Eds) Fresh Perspectives on Anxiety Disorders (1989)
Lisse, Swets en Zeitlinger, 380 p.

Emmelkamp, Paul M G Phobic and Obsessive–Compulsive Disorders: theory, research, and practice (1982)
New York, Plenum Publishing, 368 p.
ISBN 0–306–41044–3.

Energy Probe Research Foundation Fuelwood: a global crisis (1983)
Toronto, Energy Probe Research Foundation, 64 p.

Engelbert, Ernest A Water Scarcity: impacts on western agriculture (1985)
Berkeley CA, University of California Press, 550 p.
ISBN 0–520–05300–1.

Engels, Friedrich and Marx On Colonialism (1980)
Moscow, Izdatel'stvo Progress, 383 p.

English, E Philip Great Escape?: an examination of north–south tourism (1986)
Ottawa ON, North–South Institute, 89 p.
ISBN 0–920494–59–5.

English, O Spurgeon and Pearson, Gerald H J Emotional Problems of Living
New York, WW Norton, 640 p. ISBN 0–393–01078–3.

Englund, Tomas Curriculum as a Political Problem: changing educational conceptions, with special reference to citizenship education (1986)
Stockholm, Almqvist and Wiksell, 384 p.
ISBN 91–22–00807–1.

Engstrom, Paul F, et al (Eds) Advances in Cancer Control: the war on cancer – 15 years of progress (1987)
New York, Alan R Liss, 346 p. Progress in Clinical and Biological Research Ser: Vol 248. ISBN 0–8451–5098–7.

Enna, C D Peripheral Denervation of the Hand (1989)
New York, Alan R Liss, 196 p. ISBN 0–8451–4252–6.

Enna, C D Peripheral Denervation of the Foot (1989)
New York, Alan R Liss, 250 p. ISBN 0–8451–4253–4.

Ennew, Judith The Sexual Exploitation of Children (1986)
New York, Saint Martin's Press, 200 p.
ISBN 0–312–71353–3.

Enzi, G et al Obesity: pathogenesis and treatment (1981)
San Diego CA, Academic Press. ISBN 0–12–240150–6.

EPA Pesticides: drinking water health advisory (1990)
Chelsea MI, Lewis Publishers, 900 p.
ISBN 0–87371–235–8.

Epanchin, Betty C and Paul, James L Emotional Problems of Childhood and Adolescence: a multi–disciplinary approach (1987)
Columbus ON, Merrill Publishing, 448 p.
ISBN 0–675–20566–2.

Epstein, M A The Epstein–Barr Virus: recent advances (1986)
Oxford, Heineman Professional Publishing, 298 p.
ISBN 0–433–09450–8.

Epstein, Simon Cry of Cassandra: the resurgence of European anti–semitism (1986)
Royal Oak MI, Zenith Press, 256 p. Trans. from Fr by Norman S Posel. ISBN 0–915765–13–6.

Eriksson, A W et al Population Structure and Genetic Disorders: proceedings of the 7th Sigfred Juselius (1981)
San Diego CA, Academic Press. ISBN 0–12–241450–0.

Erlandsen, Stanley L and Meyer, Ernest A Giardia and Giardiasis: biology, pathogenesis and epidemiology (1984)
New York, Plenum Publishing, 432 p.
ISBN 0–306–41539–9.

Erlanger, Ellen Eating Disorders: a question and answer book about anorexia nervosa (1987)
Minneapolis MN, Lerner Publications.
ISBN 0–8225–0038–8.

Escalante, E (Eds) Underground corrosion (1981)
Philadelphia PA, American Society for Testing and Materials, 210 p. ISBN 0–8031–0703–X.

Escalona, Sibylle Critical Issues in the Early Development of Premature Infants (1988)
New Haven CT, Yale University Press, 256 p.
ISBN 0–300–03516–0.

Eth, Spencer and Pynoos, Robert S Post–Traumatic Stress Disorder in Children (1985)
Washington DC, American Psychiatric Press, 208 p.
ISBN 0–88048–067–X.

Ettorre, Betty Women and Substance Abuse (1989)
Basingstoke, Macmillan, 208 p. ISBN 0–333–48310–3.

Etzioni, Amitai Social Problems (1976)
Englewood Cliffs NJ, Prentice Hall, 192 p.
ISBN 0–13–817403–2.

Etzioni–Halevy, Eva The Knowledge Elite and the Failure of Prophecy (1985)
London, Unwin Hyman, 120 p. Controversies in Sociology Ser: No 18. ISBN 0–04–301192–6.

Etzkowitz, Henry Is America Possible? (1980)
Saint Paul MN, West Publishing College and School, 407 p.
ISBN 0–8299–0329–1.

Europa Institute, Leiden (Ed) European Competition Policy (1973)
Dordrecht, Kluwer Academic Publishers Group, 265 p.
ISBN 90–286–0363–8.

European Communities International Symposium on Bovine Leukosis (1985)
Luxembourg, European Communities, 657 p.
ISBN 92–825–4559–8.

European Communities Safety Aspects of Hazardous Wastes: proceedings of a round table, Dublin, November 27–29, 1985 (1986)
Luxembourg, European Communities, 475 p.
ISBN 92–825–6075–9.

European Foundation for the Improvement of Living and Working Conditions The Effects of Shiftwork on Health, Social and Family Life
Dublin, EFILWC.

European Foundation for the Improvement of Living and Working conditions Shiftwork: quantity and quality of sleep
Dublin, EFILWC.

European Foundation for the Improvement of Living and Working conditions Shiftwork and Accidents
Dublin, EFILWC.

European Foundation for the Improvement of Living and Working conditions The Working Environment at Visual Display Units: a literature study
Dublin, EFILWC.

European Foundation for the Improvement of Living and Working conditions Effect of Introduction of a Visual Display Unit in a Computerised Office on the Health of Operators: a multidisciplinary research design
Dublin, EFILWC.

European Foundation for the Improvement of Living and Working conditions A European Study of Commuting and Its Consequences
Dublin, EFILWC.

European Foundation for the Improvement of Living and Working conditions The Journey Between Home and Work: the effects of commuting on health and safety of commuters/workers
Dublin, EFILWC.

European Foundation for the Improvement of Living and Working conditions Noise, Stress and Work
Dublin, EFILWC.

European Foundation for the Improvement of Living and Working conditions Physical and Psychological Stress at Work: state of the art study
Dublin, EFILWC.

European Organization for Quality Control Quality in a World of Limited Resources: proceedings of EOQC conference in Madrid, 1983
Bern, European Organization for Quality Control.

Evan, Harry Z Employers and the Environmental Challenge (1986)
Washington DC, ILO, 101 p. ISBN 92–2–105647–3.

Evans, A S Viral Infections of Humans: epidemiology and control (Date not set)
New York, Plenum Publishing, 790 p.
ISBN 0–306–42731–1.

Evans, Alfred S and Feldman, Harry A Bacterial Infections of Humans (1982)
New York, Plenum Publishing, 744 p.
ISBN 0–306–40967–4.

Evans, D Morier Facts, Failures, and Frauds Revelations, Financial, Mercantile, Criminal (1968)
New York, Kelley Augustus M. ISBN 0–678–00394–7.

Evans, Glen and Farberow, Norman L The Encyclopedia of Suicide (1988)
New York, Facts on File, 368 p. ISBN 0–8160–1397–7.

Evans, H J and Lloyd, D Mutagen–Induced Chromosome Damage in Man (1979)
New Haven CT, Yale University Press.
ISBN 0–300–02315–4.

Evans, Martha M (Ed) Dyslexia: an annotated bibliography (1982)
New York, Greenwood Press, 644 p. illus. bibl. Contemporary Problems of Childhood Ser: No 5. ISBN 0–313–21344–5.

Evernden, Neil Natural Alien: humankind and the environment (1985)
Toronto ON, University of Toronto Press.
ISBN 0–8020–6639–9.

Everyman, Ron False Consciousness and Ideology in Marxist Theory (1981)
Stockholm, Almqvist and Wiksell, 320 p.
ISBN 91–22–00468–8.

Evron, Y (Ed) International Violence: terrorism, surprise and control
Jerusalem, Magnes Press, 300 p. ISBN 965–223–485–0.

Ewen, Cecil H Witchcraft and Demonianism
New York, AMS Press. ISBN 0–404–18410–3.

Exton–Smith, A N and Caird, F Metabolic and Nutritional Disorders in the Elderly (1980)
Littleton MA, PSG Publishing, 238 p. ISBN 0–7236–0537–8.

Eze, Osita C Human Rights in Africa: some selected problems (1987)
New York, Saint Martin's Press, 310 p.
ISBN 0–312–39962–6.

Fabbri, Paolo (Ed) Recreational Uses of Coastal Areas: a research project of the Commission on the Coastal Environment, International Geographical Union (1989)
Dordrecht, Kluwer Academic Publishers Group, 308 p.
ISBN 0–7923–0279–6.

Fahy, P C and Persley, G J (Eds) Plant Bacterial Diseases: a diagnostic guide (1983)
San Diego CA, Academic Press, 416 p.
ISBN 0–12–247660–3.

Fair, D E and Bertrand, R International Lending in a Fragile World Economy (1983)
Dordrecht, Kluwer Academic Publishers Group, 433 p.
ISBN 90–247–2809–6.

Fairlie, Henry and Lawrence, Vint The Seven Deadly Sins Today (1979)
Notre Dame IN, University of Notre Dame Press.
ISBN 0–268–01698–4.

Falk, Peter A Law, Morality, and War in the Contemporary World (1984)
Westport CT, Greenwood Press, 120 p.
ISBN 0–313–24682–3.

Fanon, Frantz The Wretched of the Earth (1966)
USA, 252 p.
Trans. from French: C Farington. Foreword by: J–P Sartre.

Fanon, Frantz; Farrington, Constance and Sartre, Jean-Paul Wretched of the Earth (1965)
New York, Grove Press. ISBN 0–394–17327–9.

Fantechi, R and Ghazi, A (Eds) Carbon Dioxide and Other Greenhouse Gases: climatic and associated impacts (1989)
Dordrecht, Kluwer Academic Publishers Group, 292 p.
ISBN 0–7923–0191–9.

FAO Pesticide Residues in Food: joint report of the FAO Working Party on Pesticide Residues and the WHO Expert Committee on Pesticide Residues
Rome, FAO. FAO Agricultural Studies: 73, 78, 87, 88, 90, 92, 97. FAO Plant Protection and Protection Series: 1, 8.

FAO Animal Health Yearbooks
Rome, FAO.
FAO Animal Production and Health Series: No 20, 24, 25.

FAO Mammals of the Seas
Rome, FAO.
FAO Fisheries Series: No 5. Vol 1: Report of the FAO Advisory Committee on Marine Resources Research: Working Party on Marine Mammals with the Cooperation of UNEP (1978, 531 p). Vol 2: Pinneped Species Summaries and Report on Sirenians (1979, 151 p). Vol 3: General Papers and Large Cetaceans (19814, 504 p). Vol 4: Small Cetaceans, Seals, Sirenians and Otters (1982, 531 p).

FAO Catches and Landings 1983–1984
Rome, FAO.
Yearbooks of Fishery Statistics: Vol 56, 58.

FAO Fishery Commodities 1983–1984
Rome, FAO.
Yearbooks of Fishery Statistics: Vol 57, 59.

FAO FAO Nutrition Meetings Report Series
Rome, FAO.
No 1 (1976), 29 (1974), 31 (1974), 48 (1974), 50 (1972), 51 (1972), 55 (197 5).

FAO Fish Diseases: technical notes submitted to EIFAC 3rd session (1965)
Rome, FAO, 35 p. out of print. ISBN 92–5–102052–3.

FAO Calcium Requirements: report of an FAO/WHO Expert Group, Rome, May 1961 (1968)
Rome, FAO, 54 p. 2nd ed.
FAO Nutrition Meetings Report Series: No 30.
ISBN 92–5–100466–8.

FAO Vicuña Conservation Legislation (1971)
Rome, FAO, 8 p. out of print.
ISBN 92–5–101976–2.

FAO Symposium on the Nature and Extent of Water Pollution Problems Affecting Inland Fisheries in Europe: synthesis on national reports (1972)
Rome, FAO, 20 p. in Eng/French. out of print.
ISBN 92–5–002066–X.

FAO Summary of the Organized Discussion on the Economic Evaluation of Sport Fishing, Dublin, 1967 (1973)
Rome, FAO, 27 p. 2nd ed.
ISBN 92–5–102040–X.

FAO Potential Uses of Waste Waters and Heated Effluents (1973)
Rome, FAO, 23 p. rev ed. out of print.
ISBN 92–5–102044–2.

FAO Specifications for the Identity and Purity of Food Additives and their Toxicological Evaluation; Emulsifiers, Stabilizers, Bleaching and Maturing Agents: seventh report of the joint FAO/WHO Expert Committee on Food Additives, Rome, 1963 (1974)
Rome, FAO, 189 p. 2nd ed. FAO Nutrition Meetings Report Series: No 35. ISBN 92–5–101830–8.

FAO Report of the Symposium on the Major Communicable Fish Diseases in Europe and Their Control (1974)
Rome, FAO, 43 p. 2nd ed. out of print. Suppl 1: The Major Communicable Fish Diseases of Europe and North America: A review of national and international measures and their control (1973, 48 p. Out of print). Suppl 2: Symposium on the Major Communicable Fish Diseases in Europe and Their Control: Panel reviews and relevant papers (1973, 257 p. 2nd ed. In Engl/French).

FAO Water and the Environment (1975)
Rome, FAO, 67 p. 6th ed. out of print.
ISBN 92–5–100364–5.

FAO Incentives and Disincentives for Farmers in Developing Countries (1976)
Rome, FAO, 47p. 2nd ed.
ISBN 92–5–100608–3.

FAO Eradication of Hog Cholera and African Swine Fever (1976)
Rome, FAO, 25 p. out of print.

FAO Man's Influence on the Hydrological Cycle (1976)
Rome, FAO, 78 p. 3rd ed.
ISBN 92–5–101999–1.

FAO Pest Resistance to Pesticides and Crop Loss Assessment: report of the first session of the FAO Panel of Experts, Washington DC, 1976 (1977)
Rome, FAO, 48 p. out of print. ISBN 92–5–100222–3.

FAO Water Quality Criteria for European Freshwater Fish: report on cadmium and freshwater fish (1977)
Rome, FAO, 31 p. out of print. ISBN 92–5–100297–5.

FAO Rodent Pest: biology and control (Bibliography 1970–1974) (1977)
Rome, FAO, 836 p. bibl.
ISBN 92–5–100435–8.

FAO Insecticides and Application Equipment for Tsetse Control (1977)
Rome, FAO, 80 p.
ISBN 92–5–100183–9.

FAO Impact of Oil and the Marine Environment (1977)
Rome, FAO, 250 p.
ISBN 92–5–100219–3. GESAMP Reports and Studies: No 10.

FAO Wholesomeness of Irradiated Food: report of the joint FAO/WHO Expert Committee, Geneva, 1976 (1977)
Rome, FAO, 44 p.
FAO Food and Nutrition Series: No 6. ISBN 92–5–100282–7.

FAO Second European Consultation on the Economic Evaluation of Sport and Commercial Fisheries: report and technical papers, Göthenborg, Sweden, 1975 (1977)
Rome, FAO, 190 p. in Engl/French. ISBN 92–5–000256–4.

FAO Land Degradation (1977)
Rome, FAO, 117 p. 2nd ed.
ISBN 92–5–100106–5.

FAO Calcareous Soils (1977)
Rome, FAO, 276 p. 2nd ed.
ISBN 92–5–100276–2.

FAO Assessing Soil Degradation (1977)
Rome, FAO, 87 p. out of print.
ISBN 92–5–100410–2.

FAO Mycotoxins: report on the joint FAO/WHO/UNEP conference, Nairobi, 1977 (1977)
Rome, FAO, 115 p.
ISBN 92–5–100489–7.

FAO Pesticide Residues in Food: index and summary of report of joint meetings of FAO and WHO Expert Bodies on pesticide residues 1965–1978. (1978)
Rome, FAO, 41 p.

FAO Declining Breeds of Mediterranean Sheep (1978)
Rome, FAO, 68 p. illus. ISBN 92–5–100543–5.

FAO Pesticide Residues in Food: evaluations (1978)
Rome, FAO, 468 p.
Suppliment.

FAO Pollution: an international problem for fisheries (1978)
Rome, FAO, 85 p. 2nd ed.
FAO Fisheries Series: No 4. ISBN 92–5–100376–9.

FAO Soil Erosion by Water: some measures for its control on cultivated lands (1978)
Rome, FAO, 284 p. 2nd ed. FAO Land and Water Development Series: 71. ISBN 92–5–100474–9.

FAO Soil Erosion by Wind and Measures for its Control on Agricultural Lands (1978)
Rome, FAO, 88 p. 4th ed.
FAFO Land and Water Development Series: No 6.

FAO Improving Soil Fertility in Africa (1978)
Rome, FAO, 150 p. 2nd ed. out of print.
ISBN 92–5–100426–9.

FAO Effects of Intensive Fertilizer Use on the Human Environment (1978)
Rome, FAO, 368 p. 3rd ed. ISBN 92–5–100657–1.

FAO Shifting Cultivation and Soil Conservation in Africa (1978)
Rome, FAO, 254 p. 2nd ed.
ISBN 92–5–100393–9.

FAO Systematic Index of International Water Resources Treaties, Declarations, Acts and Cases by Basin (1978)
Rome, FAO, 511 p. in Engl/French/Span.

FAO Guidelines for Integrated Control of Rice Insect Pests (1979)
Rome, FAO, 123 p. ISBN 92–5–100705–5.

FAO Rodenticides: analyses, specifications, formulations for use in public health and agriculture (1979)
Rome, FAO, 81 p.
ISBN 92–5–100798–5.

FAO The African Trypanosomiasis (1979)
Rome, FAO, 105 p.
ISBN 92–5–100240–1.

FAO Groundwater Pollution: technology, economics and management (1979)
Rome, FAO, 149 p.
ISBN 92–5–100699–7.

FAO Guidelines for Integrated Control of Maize Pests (1979)
Rome, FAO, 98 p.
ISBN 92–5–100875–2.

FAO Elements of Integrated Control of Sorghum Pests (1979)
Rome, FAO, 167 p.
ISBN 92–5–100884–1.

FAO Pesticide Residues in Food (1977): report of the joint meeting FAO/WHO, Geneva, 1977 (1979)
Rome, FAO, 88 p. 3rd ed.
ISBN 92–5–100578–8.

FAO Water Quality Criteria for European Freshwater Fish: report on chlorine and freshwater fish (1979)
Rome, FAO, 16 p. 2nd ed. out of print.
ISBN 92–5–102072–8.

FAO Water Quality Criteria for European Freshwater Fish: report on zinc and freshwater fish (1979)
Rome, FAO, 30 p. 2nd ed. out of print.
ISBN 92–5–102073–6.

FAO Water Quality Criteria for European Freshwater Fish (1979)
Rome, FAO, 32 p. 2nd ed. out of print.
ISBN 92–5–100122–7.

FAO Trace Elements in Soils and Agriculture (1979)
Rome, FAO, 70 p. 3rd ed.
ISBN 92–5–100485–4.

FAO Sandy Soils (1979)
Rome, FAO, 251 p. 2nd ed.
ISBN 92–5–100613–X.

FAO Arsenic and Tin in Food: reviews and commonly used methods of analysis (1979)
Rome, FAO, 114 p. ISBN 92–5–100727–6.

FAO Prevention of Mycotoxins: recommended practices for the prevention of mycotoxins in food, feed and their products (1979)
Rome, FAO, 71 p.
ISBN 92–5–100703–9.

FAO The Economic Value of Breast–Feeding (1979)
Rome, FAO, 97 p.
ISBN 92–5–100797–7.

FAO Perspectives on Mycotoxins (1979)
Rome, FAO, 171 p.
ISBN 92–5–100870–1.

FAO The Environmental Impact of Tsetse Control Operations (1980)
Rome, FAO, 74 p.
ISBN 92–5–101001–3.

FAO Recommended Methods for Measurement of Pest Resistance to Pesticides (1980)
Rome, FAO, 136 p.
ISBN 92–5–100883–3.

FAO East Coast Fever and Related Tick–Borne Diseases (1980)
Rome, FAO, 996 p.

FAO Pesticide Residues in Food (1979): report of the Joint Meeting FAO/WHO, Geneva, 1979 (1980)
Rome, FAO, 97 p. out of print. ISBN 92–5–100922–8.

FAO Pesticide Residues in Food (1979): evaluations; data and recommendations of the joint meeting FAO/WHO, Geneva, 1979 (1980)
Rome, FAO, 569 p. out of print. Suppliment.
ISBN 92–5–100958–9.

FAO Parasites, Infections and Diseases of Fish in Africa (1980)
Rome, FAO, 224 p. out of print. ISBN 92–5–100982–1.

FAO Impact on Soils of Fast–Growing Species in Lowland Humid Tropics (1980)
Rome, FAO, 121 p. out of print. ISBN 92–5–100972–4.

FAO Dietary Fats and Oils in Human Nutrition (1980)
Rome, FAO, 102 p.
FAO Food and Nutrition Series: No 20. ISBN 92–5–100802–7.

FAO Water Quality Criteria for European Freshwater Fish: report on combined effects on freshwater fish and other aquatic life of mixture of toxicants in water (1980)
Rome, FAO, 57 p. out of print.

FAO Corrosion and Encrustation in Water Wells (1980)
Rome, FAO, 108 p.
ISBN 92–5–100933–3.

FAO Water Law in Selected African Countries: Benin, Burundi, Ethiopia, Gabon, Kenya, Mauritius, Sierra Leone, Swaziland, Upper Volta and Zambia (1980)
Rome, FAO, 273 p. 2nd ed.
ISBN 92–5–100748–9.

FAO Legislation on Wildlife, Hunting and Protected Areas in Some European Countries (1980)
Rome, FAO, 54 p. ISBN 92–5–100878–7.

FAO Weeds in Tropical Crops: selected abstracts on constraints on production caused by weeds in maize, rice, sorghum, millet, groundnuts and cassava, 1952–1980 (1981)
Rome, FAO, 94 p. ISBN 92–5–101146–X.

FAO Animal Genetic Resources Conservation and Management (1981)
Rome, FAO, 399 p.
ISBN 92–5–101118–4.

FAO Pesticide Residues in Food (1980): report of the joint meeting FAO/WHO, Rome, 1980 (1981)
Rome, FAO, 88 p. out of print. ISBN 92–5–101058–7.

FAO Pesticide Residues in Food: evaluations; data and recommendations of the joint meeting FAO/WHO, Rome, 1980 (1981)
Rome, FAO, 468 p.
Suppliment. ISBN 92–5–101148–6.

FAO Conservation of the Genetic Resources of Fish: problems and recommendations report of the expert consultation on genetic resources of fish, Rome, 1980 (1981)
Rome, FAO, 51 p.
ISBN 92–5–101173–7.

FAO Agriculture: toward 2000 (1981)
Rome, FAO, 134 p. ISBN 92–5–101080–3.

FAO Handbook on Human Nutritional Requirements (1981)
Rome, FAO, 66 p. 3rd ed.
FAO Food and Nutrition Series: No 4. ISBN 92–5–100129–4.

FAO Report of the Symposium on New Developments in the Utilization of Heated Effluents and of Recirculation System for Intensive Aquaculture, Stavanger, 1980 (1981)

Rome, FAO, 44 p.
ISBN 92-5-101059-5.
FAO Soil Conservation for Developing Countries (1981)
Rome, FAO, 104 p. 3rd ed.
ISBN 92-5-100101-4.
FAO Rodent Control in Agriculture: a handbook on the biology and control of commensal rodents as agricultural pests (1982)
Rome, FAO, 95 p.   ISBN 92-5-101295-4.
FAO Echinococcosis – Hydatidosis Surveillance, Prevention and Control: FAO–UNEP–WHO guidelines (1982)
Rome, FAO, 157 p.
ISBN 92-5-101205-9.
FAO Hormones in Animal Production (1982)
Rome, FAO, 62 p.   ISBN 92-5-101213-X.
FAO Pesticide Residues in Food: report on the joint meeting FAO/WHO, Geneva, 1981 (1982)
Rome, FAO, 75 p. Out of print.   ISBN 92-5-101202-4.
FAO Pesticide Residues in Food: evaluations of the joint meeting FAO/WHO, Geneva 1981 (1982)
Rome, FAO, 576 p. Supplement.   ISBN 92-5-101306-3.
FAO The Prevention of Losses in Cured Fish (1982)
Rome, FAO, 98 p. 2nd ed.
FAO Food Loss Prevention in Perishable Crops (1983)
Rome, FAO, 91 p. 2nd ed.
ISBN 92-5-101028-5.
FAO Selected Bibliography on Major African Reservoirs (1983)
Rome, FAO, 58 p. in Engl/French.   ISBN 92-5-001349-3.
FAO Guidelines for the Control of Soil Degradation FAO/UNEP (1983)
Rome, FAO, 43 p.
ISBN 92-5-101404-3.
FAO Ticks and Tick–Borne Diseases: selected articles from the World Animal Review (1983)
Rome, FAO, 81 p.
ISBN 92-5-101289-X.
FAO African Animal Trypanosomiasis: selected articles from the World Animal Review (1983)
Rome, FAO, 86 p.
ISBN 92-5-101288-1.
FAO Guidelines for Watershed Management (1983)
Rome, FAO, 306 p. 3rd ed.   ISBN 92-5-100242-8.
FAO Pesticide Residues in Food: report of the Joint Meeting FAO/WHO, Rome, 1982 (1983)
Rome, FAO, 79 p.
ISBN 92-5-101360-8.
FAO Pesticide Residues in Food (1982): evaluations, monographs (1983)
Rome, FAO, 434 p.
Supplement.   ISBN 92-5-101432-9.
FAO Management of Upland Watersheds: participation of the mountain communities (1983)
Rome, FAO, 216 p.   ISBN 92-5-101337-3.
FAO Fuelwood Supplies in the Developing Countries (1983)
Rome, FAO, 134 p.
FAO Water Quality Criteria for European Freshwater Fish: report on chromium and freshwater fish (1983)
Rome, FAO, 37 p. out of print.
FAO Keeping the Land Alive: soil erosion: Its causes and cures (1983)
Rome, FAO, 95 p.
ISBN 92-5-101342-X.
FAO Assessment and Collection of Data on Pre–Harvest Foodgrain Losses (1983)
Rome, FAO, 130 p.   ISBN 92-5-101314-4.
FAO Approaches to World Food Security (1983)
Rome, FAO, 185 p.
FAO Institutional Aspects of Shifting Cultivation in Africa (1984)
Rome, FAO, 177 p.
ISBN 92-5-101498-1.
FAO Improving Weed Management: proceedings of the FAO/IWSS Expert Consultation, Rome, 1982 (1984)
Rome, FAO, 190 p. 2nd ed.
ISBN 92-5-101335-7.
FAO Economic Guidelines for Crop Pest Control (1984)
Rome, FAO, 102 p.
FAO Changes in Shifting Cultivation in Africa (1984)
Rome, FAO, 67 p.
ISBN 92-5-102151-1.
FAO Guidelines for Integrated Control of Cotton Pests (1984)
Rome, FAO, 199 p.   ISBN 92-5-101430-2.
FAO Pesticide Residues in Food: report of the joint meeting FAO/WHO, Geneva, 1983 (1984)
Rome, FAO, 75 p.
ISBN 92-5-102094-9.
FAO Proceedings of the Joint FAO/UNEP Expert Panel Meeting, 1983 (1984)
Rome, FAO. out of print. 2 Vols.
FAO Approaches to the Regulation of Fishing Effort (1984)
Rome, FAO, 45 p.
ISBN 92-5-101492-2.
FAO World Review of Interactions Between Marine Mammals and Fisheries (1984)
Rome, FAO, 197 p. illus.   ISBN 92-5-102145-7.
FAO Water Quality Criteria for European Freshwater Fish: report on nickel and freshwater fish (1984)
Rome, FAO, 29 p.
ISBN 92-5-102176-7.

FAO Water Quality Criteria for European Freshwater Fish: report on nitrite and freshwater fish (1984)
Rome, FAO, 29 p.
ISBN 92-5-102177-5.
FAO Report of the Symposium on Habitat Modification and Freshwater Fisheries, Aarhus, Denmark, 1984 (1984)
Rome, FAO, 43 p. in Engl/French.   ISBN 92-5-002178-X.
FAO Promoting Agricultural Trade Among Developing Countries (1984)
Rome, FAO, 188 p.
ISBN 92-5-101501-5.
FAO Development Strategies for the Rural Poor (1984)
Rome, FAO, 117 p.
ISBN 92-5-102122-8.
FAO Slaughterhouse Cleaning and Sanitation (1985)
Rome, FAO, 52 p.
ISBN 92-5-102296-8.
FAO Breeding Poplars for Disease Resistance (1985)
Rome, FAO, 72 p.
ISBN 92-5-102200-3.
FAO Report of the FAO/UNEP Meeting in the Toxicity and Bioaccumulation of Selected Substances in Marine Organism, Rovinj (Yugoslavia), 1984 (1985)
Rome, FAO, 26 p.
ISBN 92-5-102278-X.
FAO Soil Conservation and Management in Developing Countries (1985)
Rome, FAO, 216 p. 3rd ed.   ISBN 92-5-100430-7.
FAO Watershed Development: with special reference to soil and water conservation (1985)
Rome, FAO, 266 p.
FAO The Future of Shifting Cultivation in Africa and the Task of Universities: proceedings of the international workshop on shifting cultivation: teaching and research at university level, Ibadan (Nigeria), 1982 (1985)
Rome, FAO, 198 p.
ISBN 92-5-102092-9.
FAO Prevention of Post–Harvest Food Losses: a training manual (1985)
Rome, FAO, 120 p.
FAO Training Series: No 10.   ISBN 92-5-102209-7.
FAO Manual of Pest Control for Food Security Reserve Grain Stocks (1985)
Rome, FAO, 208 p.   ISBN 92-5-102235-6.
FAO Pesticide Residues in Food (1983): evaluations, monographs (1985)
Rome, FAO, 515 p.
Supplement.   ISBN 92-5-102222-4.
FAO Pesticide Residues in Food (1984): report of the joint meeting FAO/WHO, Rome, 1984 (1985)
Rome, FAO, 108 p.
ISBN 92-5-102216-X.
FAO Pesticide Residues in Food (1984): the monographs (1985)
Rome, FAO, 739 p.
Supplement.
FAO Avalanche control (1985)
Rome, FAO, 248 p.
ISBN 92-5-100736-5.
FAO Sand Dune Stabilization, Shelterbelts and Afforestation in Dry Zones (1985)
Rome, FAO, 247 p.
ISBN 92-5-102261-5.
FAO World Food Security: selected themes and issues (1985)
Rome, FAO, 108 p.   ISBN 92-5-102165-1.
FAO Food Aid and Food Security: past performance and future potential (1985)
Rome, FAO, 60 p.   ISBN 92-5-102331-X.
FAO Residues of Veterinary Drugs in Foods: report of a joint FAO/WHO expert consultation, Rome, 1984 (1985)
Rome, FAO, 66 p. in Engl/French/Spanish.
ISBN 92-5-002210-7.
FAO Nutritional Implications of Food Aid: an annotated bibliography (1985)
Rome, FAO, 119 p. bibl.
FAO Pesticide Residues in Food (1985): report of the joint meeting FAO/WHO, Geneva, 1985 (1986)
Rome, FAO, 87 p.
ISBN 92-5-102365-4.
FAO Residues (1986)
Rome, FAO, 384 p. Pesticide Residues in Food, Evaluations: 1.   ISBN 92-5-102402-2.
FAO Wildland Fire Management Terminology (1986)
Rome, FAO, 298 p. in Engl/French/Germ/Ital.
ISBN 92-5-100874-4.
FAO/UNEP Marine Mammals: global plan of action (1985)
Nairobi, UNEP, 112 p. UNEP Regional Seas Reports and Studies: No 55.
FAO–WHO Experts on Pesticide Residues Pesticide Residues in Food: report of the FAO–WHO experts on pesticide
Geneva, WHO. Annual Reports: 1969, 1970, 1971, 1972, 1973, 1974, 1975.
Farber, E M, et al Psoriasis: proceedings of the fourth international (1987)
New York, Elsevier Science Publishing, 600 p.
ISBN 0–444–01212–5.
Farberow The Many Faces of Suicide: indirect self–destructive behavior
New York, McGraw-Hill Book Company, 448 p.
ISBN 0–07–019944–2.

Farmer, P Lead Pollution from Motor Vehicles, 1974–86: a select bibliography (1987)
New York, Elsevier Science Publishing, 95 p.
ISBN 1–85166–066–6.
Farnworth, E G and Golley, F B (Eds) Fragile Ecosystems: evaluation of research and applications in the neotropics (1974)
Berlin, Springer-Verlag, 258 p.   ISBN 3–540–06695–0.
Faro, Sebastian and Gilstrap III, Larry C (Eds) Infections in Pregnancy (1990)
Chichester, John Wiley and Sons, 290 p.
ISBN 0–471–56221–1.
Farr, M J Long term retention of Knowledge and Skills (1987)
Berlin, Springer-Verlag, 185 p.   ISBN 3–540–96531–9.
Farrell, Alexander D Anxiety Disorders: medical subject analysis with bibliography (1987)
Washington DC, ABBE Publications Association, 150 p.
ISBN 0–88164–612–1.
Farrell, James L and Fuller, Russell Modern Investments and Security Analysis (1987)
New York, McGraw-Hill Book Company, 640 p. illus.
ISBN 0–07–022621–0.
Faulstich, H, et al Amanita Toxins and Poisoning: international amanita symposium Heidelberg, November 1–3, 1978 (1980)
Königstein im Taunus, Sven Koeltz Scientific Books, 236 p.
Faunce, William A and Munson, Eric M Problems of an Industrial Society (1981)
New York, McGraw Hill Book Company, 256 p.
ISBN 0–07–020105–6.
Fauriol, Georges (Ed) Latin American Insurgencies (1985)
Washington DC, US Government Printing Office, 227 p. illus.
ISBN 0–318–18780–9.
Favazza, Armando R Bodies under Siege: self–mutilation in culture and psychiatry (1987)
Baltimore MD, Johns Hopkins University Press, 304 p.
ISBN 0–8018–3453–8.
Favre, David S International Trade in Endangered Species: a guide to CITES (1989)
Dordrecht, Martinus Nijhoff, 424 p.   ISBN 0–7923–0114–5.
Fayerweather, John Host National Attitudes Toward Multinational Corporations (1982)
New York, Praeger Publishers, 366 p.
ISBN 0–275–90789–9.
Feagin, Clairece and Feagin, Joe R Social Problems: a critical power–conflict perspective (1989)
Englewood Cliffs NJ, Prentice Hall, 496 p. 3rd ed.
ISBN 0–13–817552–7.
Fehr, Kevin and Kalant, Harold (Eds) Cannabis and Health Hazards (1983)
Toronto ON, Addiction Research Foundation, 843 p.
ISBN 0–88868–084–8.
Fein, Helen (Ed) Current Research on Antisemitism, Vol 1: the persisting question: sociological perspectives and social context of modern antisemitism (1987)
Berlin, Walter de Gruyter, 430 p.   ISBN 0–89925–320–2.
Feinberg, Renee Women, Education, and Employment: a bibliography of periodical citations (1982)
Hamden CT, Shoe String Press, 274 p.
ISBN 0–208–01967–7.
Feinberg, Richard E and Kallab, Valerina Adjustment Crisis in the Third World (1984)
New Brunswick NJ, Transaction Books, 200 p.
ISBN 0–88738–040–9.
Fellendorf, George W (Ed) Bibliography on Deafness (1977)
Alexander Graham, bibl.
ISBN 0–88200–1116–.
Feller, Irving International Bibliography on Burns
Ann Arbor MI, National Institute for Burn Medicine. 1950–1984.
Felsenfeld, Oscar Borrelia, Borreliosis and Relapsing Fever: strains, vectors, human and animal borreliosis (1971)
Saint Louis MO, Warren H Green, 192 p. illus.
ISBN 0–87527–032–8.
Fenner, F, et al Smallpox and Its Eradication
Geneva, WHO, 1500 p.   ISBN 92–4–156110–6.
Fenner, Frank, et al The Orthopoxviruses (1988)
San Diego CA, Academic Press, 525 p.
ISBN 0–12–253045–4.
Fenton, Thomas P and Heffron, Mary J Third World Resource Directory: a guide to organizations and publications (1984)
Maryknoll NY, Orbis Books, 304 p.   ISBN 0–88344–509–3.
Ferber, Marianne A Women and Work, Paid and Unpaid: a selected, annotated bibliography (1987)
New York, Garland Publishing, 408 p.
ISBN 0–8240–8690–2.
Ferencz, Benjamin B Defining International Aggression: the search for world peace (1975)
New York, Oceana. 2 vols.   ISBN 0–379–00271–X.
Ferguson Wood, E J and Johannes, R E (Eds) Tropical Marine Pollution (1975)
Amsterdam, Elsevier Science Publishing, 192 p.
ISBN 0–444–41298–0.
Fernandez, Clara M Edema Research: subject analysis with bibliography (1987)
Washington DC, ABBE Publications Association, 150 p. bibl.
ISBN 0–88164–631–8.
Fernando, R Traditional and Non-Traditional Foods (1981)
Rome, FAO, 156 p.
FAO Food and Nutrition Series: No 2.   ISBN 92–5–100167–7.
Ferranti, M P and Ferrero, G L (Eds) Sorting of Household Waste and Thermal Treatment of Waste (1985)
Amsterdam, Elsevier Science Publishing, 521 p. illus.
ISBN 0–85334–382–9.
Ferris, Brian and Toyne, Peter World Problems (1979)
Chester Springs CA, Dufour Editions, 236 p.
ISBN 0–7175–0509–X.

Feshbach, Murray (Ed) National Security: proceedings of the workshop held after the 27th party congress of the USSR, NATO headquarters, Brussels, Belgium, November 6–7, 1986 (1987)
Dordrecht, Kluwer Academic Publishers Group, 322 p.
ISBN 90–247–3553–X.

Fetherolf, Loufti Martha Rural Development, Taking into Account the Problems of the Indigenous Populations as well as the Drift of the Rural Population to the Cities and its Integration in the Urban Informal Sector (1986)
Geneva, ILO, 91 p. ISBN 92–2–105335–0.

Fetherolf, Loufti Martha Rural Women. Unequal Partners in Development (1987)
Geneva, ILO, 80 p.
ISBN 92–2–102389–3.

Fichter, Manfred M (Ed) Bulimia Nervosa: basic research, diagnosis and therapy (1990)
Chichester, John Wiley and Sons, 368 p.
ISBN 0–471–92405–9.

Fidel, Kenneth Militarism in Developing Countries (1975)
New Brunswick NJ, Transaction Books, 300 p.
ISBN 0–87855–092–5.

Fieldhouse, Richard (Ed) Security at Sea: naval forces and arms control (1990)
Oxford, Stockholm International Peace Research Institute, 320 p. SIPRI Publication Series. ISBN 0–19–829130–2.

Fieldhouse, Richard and Taoka, Shunji Superpowers at Sea: an assessment of the naval arms race (1989)
Oxford, Stockholm International Peace Research Institute, 208 p. illus. SIPRI Strategic Issue Papers.
ISBN 0–19–829135–3.

Fielding, Henry An Enquiry into the Causes of the Late Increase of Robbers
Montclair NJ, Smith and Patterson Publishing. 2nd ed. Criminology, Law Enforcement and Social Problems Ser.
ISBN 0–87585–210–6.

Fiennes, Richard N Zoonoses and the Origins and Ecology of Human Disease (1979)
San Diego CA, Academic Press. ISBN 0–12–256050–7.

Fiennes, T W Infectious Cancers of Animals and Man (1982)
San Diego CA, Academic Press. ISBN 0–12–256040–X.

Figley, Charles R Trauma and Its Wake
New York, Brunner/Mazel. 2 vols. Vol 1: The Study and Treatment of Post–Traumatic Stress (1985, 475 p.), Vol 2: Traumatic Stress Theory, Research, and Intervention (1986, 368 p.).

Figley, Charles R and McCubbin, Hamilton I Stress and the Family: vol II coping with catastrophe (1983)
New York, Brunner/Mazel, 272 p. ISBN 0–87630–332–7.

Finch, R G, et al (Eds) Infective Endocarditis (1988)
San Diego CA, Academic Press, 192 p.

Finch, R G, et al Infective Endocarditis (1988)
San Diego CA, Academic Press, 192 p.
ISBN 0–12–256357–3.

Finch, Ron Exporting Danger: a history of the Canadian nuclear energy export programme (1986)
Montreal, Black Rose Books, 250 p. ISBN 0–920057–74–8.

Finegold Anaerobic Bacteria in Human disease (1977)
San Diego CA, Academic Press. ISBN 0–12–256750–1.

Finegold, Sydney M and George, W Lance Anaerobic Infections in Humans (1989)
San Diego CA, Academic Press, 875 p.
ISBN 0–12–256745–5.

Finkelhor, David, et al The Dark Side of Families: current family violence research (1983)
Newbury Park CA, Sage Publications, 384 p.
ISBN 0–8039–1934–4.

Finn, M C Complete Book of International Smuggling (1986)
New York, Gordon Press Publishers.ISBN 0–8490–3562–7.

Finnes, R N (Ed) Pathology of Simian Primates (1972)
New York, Karger S AG. 2 Vols.

Fioretti, P The Menopause: clinical and pathophysiological aspects (1982)
San Diego CA, Academic Press. ISBN 0–12–256080–9.

Firman, I D and Waller, J M Coffee Berry Disease and Other Colletotrichum Diseases of Coffee (1977)
Kew, CAB International Mycological Institute, 53 p. illus. ISBN 0–85198–367–7. Phytopathological Papers: No 20.

Fischer, David; Goldblat, Jozef and Szasz, Paul (Eds) Safeguarding the Atom: a critical appraisal (1985)
Solna, Stockholm International Peace Research Institute.

Fisher, H W The Pre-Menstrual Syndrome: proceedings of a round–table discussion, (1988)
Wolfeboro NH, Longwood Publishing Group. No. 119
ISBN 0–905958–51–9.

Fishman, Alfred P Pulmonary Diseases and Disorders (1988)
New York, McGraw Hill Book Company, 1680 p. 3 vols.
ISBN 0–07–079982–2.

Fitter, M The Impact of New Technology on Workers and Patients in the Health Services: physical and psychological stress (1987)
Luxembourg, CEE, 166 p. ISBN 92–825–6797–4.

Fitter, Richard and Fitter, Maisie The Road to Extinction: problems of categorizing the status of taxa (1987)
New York, Columbia University Press, 132 p.
ISBN 2–88032–929–9.

Flaherty, David, et al (Eds) Privacy and Access to Government Data for Research: an international bibliography (1979)
Bronx NY, Mansell, 208 p. ISBN 0–7201–0920–5.

Flaherty, David H Privacy and Government Data Banks: an international perspective (1979)
Bronx NY, Mansell, 352 p. ISBN 0–7201–0930–2.

Flatters, Frank and Olewiler, Nancy D Dominant Government Firms in an Oligopolistic Industry (1984)
Albuguerque NM, Inter–Hemispheric Educational Resource Centre, 43 p.

Flavin, Christopher World Oil: coping with the dangers of success (1985)
Washington DC, Worldwatch Institute. Worldwatch Papers.
ISBN 0–916468–66–6.

Flavin, Christopher Electricity for a Developing World: new directions (1986)
Washington DC, Worldwatch Institute, 68 p. Worldwatch Papers. ISBN 0–916468–71–2.

Flavin, Christopher Reassessing Nuclear Power: the fallout from Chernobyl (1987)
Washington DC, Worldwatch Institute, 92 p. Worldwatch Papers. ISBN 0–916468–76–3.

Flavin, Christopher Slowing Global Warming: a worldwide strategy (1989)
Washington DC, Worldwatch Institute. Worldwatch Papers.
ISBN 0–916468–92–5.

Fledelius, H C; Alsbirk, P H and Goldschmidt, E (Eds) International Conference on Myopia (3, 1980, Copenhagen): proceedings (1981)
Dordrecht, Kluwer Academic Publishers Group, 253 p.
ISBN 90–6193–725–6.

Fleischmajer, Raul, et al (Eds) Biology, Chemistry and Pathology of Collagen (1986)
New York, New York Academic of Science, 537 p. Annals of the New York Academic of Sciences: Vol 460.
ISBN 0–89766–315–2.

Fletcher, Joseph and Menninger, Karl Morals and Medicine (1979)
Princeton NJ, Princeton University Press.
ISBN 0–691–07234–5.

Flisser, Ana et al Cysticercosis: symposium (1982)
San Diego CA, Academic Press. ISBN 0–12–260740–6.

Florman, Samuel C Blaming Technology: the irrational search for scapegoats (1982)
New York, Saint Martin's Press, 224 p.
ISBN 0–312–08363–7.

Flynn, Charles P Insult and Society: patterns of comparative interaction (1976)
New York, Associated Faculty Press.
ISBN 0–8046–9152–5.

Flynn, Robert J Parasites of Laboratory Animals
Ann Arbor MI, Books on Demand. ISBN 0–317–58224–0.

Fogarty, Michael P, et al Sex, Career and Family
Ann Arbor MI, Books on Demand. ISBN 0–317–29679–5.

Fogelson, Robert M Violence As Protest: a study of riots and ghettos (1980)
Westport CT, Greenwood Press, 265 p.
ISBN 0–313–22642–3.

Foley, C W, et al Abnormalities of Companion Animals: analysis of heritability (1979)
Ames IA, Iowa State University Press, 270 p.
ISBN 0–8138–0940–1.

Foley K, Sue The Political Blacklist in the Broadcasting Industry (1979)
Salem NH, Ayer Company Publishers. Dissertations in broadcasting Ser. ISBN 0–405–11757–4.

Folmer, H and Oosterhaven, J (Ed) Spatial Inequalities and Regional Development (1979)
Dordrecht, Kluwer Academic Publishers Group, 258 p.
ISBN 0–89838–006–5.

Fooner, M Interpol: issues in world crime and international justice (1989)
New York, Plenum Publishing, 238 p. illus. Criminal Justice and Public Safety Series. ISBN 0–306–43135–1.

Ford, C E and Snyder, C R Coping With Negative Life Events: clinical and social psychological perspectives (1987)
New York, Plenum Publishing, 436 p.
ISBN 0–306–42432–0.

Ford, Franklin L Political Murder: from tyrannicide to terrorism (1987)
Cambridge MA, Harvard University Press, 456 p.
ISBN 0–674–68636–5.

Ford, J Massingberd Trilogy on Wisdom and Celibacy
Notre Dame IN, University of Notre Dame Press.
ISBN 0–268–00285–1.

Forer, Lois G A Chilling Effect: the growing threat of libel and invasion of (1987)
New York, WW Norton. ISBN 0–393–02396–6.

Forest, Marsha Education – Integration: a collection of readings on the integration of children with mental handicaps into regular school systems (1984)
Downsview ON, G Allan Roeher Institute, 87 p.

Forester, William S and Skinner, John International Perspectives on Hazardous Waste Management: report from twelve ISWA countries (1987)
San Diego CA, Academic Press, 350 p.
ISBN 0–12–262165–4.

Forrest, Gary G Alcoholism and Human Sexuality (1983)
Springfield IL, Thomas Charles C, 408 p.
ISBN 0–398–04691–3.

Forsdyke, A G Meteorological Factors of Air Pollution (1970)
Geneva, WMO, 32 p.

Forssman, S and Coppee, G H Occupational Health Problems of Young Workers (1975)
Washington DC, ILO. ISBN 92–2–101051–1.

Foucault, Michel Madness and Civilization: a history of insanity in the age of reason (1988)
Vin/Random, 2nd.
ISBN 0–679–72110–X.

Fourlanos, Gerassimos Sovereignty and the Ingress of Aliens: with special focus on family unity and refugee law (1986)
Philadelphia PA, Coronet Books, 186 p.
ISBN 91–22–00855–1.

Fourquin, G The Anatomy of Popular Rebellion in the Middle Ages (1978)
New York, Elsevier Science Publishing, 182 p.
ISBN 0–444–85006–6.

Fowden, L, et al Trace Element Deficiency: metabolic and physiological consequences (1982)
Port Washington NY, Scholium International, 213 p.
ISBN 0–85403–171–5.

Fowler, B A (Ed) Biological and Environmental Effects of Arsenic (1984)
Amsterdam, Elsevier Science Publishing, 288 p. Topics in Environmental Health: Vol 6. ISBN 0–444–80513–3.

Fowler, Murray E Zoo and Wild Animal Medicine (1986)
Philadelphia PA, WB Saunders, 1127 p.
ISBN 0–7216–1013–7.

Fox, J L Intracranial Aneurysms (1983)
Berlin, Springer–Verlag, 676 p. 3 Vols.

Fox, Michael Allen and Groarke, Leo (Eds) Nuclear War: philosophical perspectives (1987)
Bern, Verlag Peter Lang AG, 278 p. ISBN 0–8204–0686–4.

Fox, Richard G (Ed) Extra Y Chromosome and Deviant Behaviour: a bibliography (1970)
Toronto ON, University of Toronto Press, 21 p.
ISBN 0–919584–15–2.

Fracs, Brian North and Fracs, Peter Reilly Raised Intracranial Pressure: A clinical guide (1990)
Oxford, Heinemann Professional Publishing, 128 p.

Francavilla, A et al Liver and Hormones (1987)
New York, Raven Press, 372 p. ISBN 0–88167–371–4.

Francis, Dorothy B Shoplifting: the crime everybody pays for (1980)
New York, Lodestar Books. ISBN 0–525–66658–3.

Francki, R I, et al The Plant Viruses
New York, Plenum Publishing. 4 vols. Vol 1: Polyhedral Virions With Tripartite Gnomes (1985, 324 p.), Vol 2: Rod–Shaped Plant Viruses (1986, 420 p.), Vol 3: Polyhedral Virions With Monopartite Rna Genomes (1987, 308 p.), Vol 4: The Filamentous Plant Viruses (1988, 408 p.)
ISBN 0–306–41958–0.

Frandsen, Public and Health Aspects of Periodontal Disease (1984)
Lombard IL, Quintessence Publishing, 220 p.
ISBN 0–86715–153–6.

Frank, Andre F On Capitalist Underdevelopment (1975)
New York, Oxford University Press. ISBN 0–19–560475–X.

Frank, André G Dependent Accumulation and Underdevelopment (1979)
New York, Monthly Review Press, 226 p.
ISBN 0–85345–468–X.

Frank, Andrew (Ed) Disabling Diseases: physical, environmental and psychological management (1989)
Oxford, Heineman Professional Publishing, 320 p.
ISBN 0–433–00057–0.

Frank, H (Ed) Man and Wolf: advances, issues and problems in captive wolf research (1986)
Dordrecht, Kluwer Academic Publishers Group, 448 p.
ISBN 90–6193–614–4.

Franklin, Michael Rich Man's Farming: the crisis in agriculture
London, Routledge, 122 p. ISBN 0–415–01061–6.

Franklyn, J Death by Enchantment (1971)
London, Hamish Hamilton.

Fraser, et al Diagnosis of Diseases of the Chest, Vol II (1988)
Philadelphia PA, WB Saunders, 900 p. 3rd ed..
ISBN 0–7216–3871–6.

Fraser, George R The Causes of Profound Deafness in Childhood: a study of 3,535 individuals with severe hearing loss present at birth or of childhood onset
Ann Arbor MI, Books on Demand.

Fraser, T G Partition in Ireland, India, and Palestine: theory and practice (1984)
New York, Saint Martin's Press, 256 p.
ISBN 0–312–59752–5.

Fraser, T Morris Human Stress Work and Job Satisfaction (1987)
Geneva, ILO, 72 p. ISBN 92–2–103042–3.

Frederickson, George M The Arrogance of Race: historical perspectives on slavery, racism (1988)
Middletown CT, Wesleyan University Press, 298 p.
ISBN 0–8195–5177–5.

Frederiksen, R A; Girard, J C and Williams, R J Sorghum and Pearl Millet Disease Identification Handbook (1978)
Patancheru A P, ICRISAT, 88 p. Information Bulletin: No 2.

Freedman, Warren International Products Liability (1986)
New York, Kluwer Law Book, 946 p. 2 vols.
ISBN 0–930273–12–5.

Freedman, Warren The Right of Privacy in the Computer Age (1987)
Westport CT, Greenwood Press, 173 p.
ISBN 0–89930–187–8.

Freeman, M D The Rights and the Wrongs of Children (1983)
Wolfeboro NH, Longwood Publishing Group, 295 p.
ISBN 0–903804–20–4.

Freeman, Peggy MBE The Deaf/Blind Baby: a programma for care (1986)
Oxford, Heineman Professional Publishing, 160 p.
ISBN 0–433–10906–8.

Freeman, R M and Malvern, John The Unstable Bladder (1988)
Sevenoaks, Butterworths, 192 p. illus.
ISBN 0–7236–1439–3.

Fregert, Sigfrid Manual of contact dermatitis (1981)
Chicago IL, Year Book Medical Publishers.
ISBN 0–8151–3282–4.

Fregly, Melvin and Kare, Morley The Role of Salt in Cardiovascular Hypertension (1982)
San Diego CA, Academic Press, 473 p.
ISBN 0–12–267280–1.

Freiberg, Marcos A and Walls, Jerry G The World of Venomous Animals (1984)
Neptune NJ, TFH Publications, 192 p. ISBN 0–87666–567–9.

French, Alfred and Berlin, Irving Depression in Children and Adolescents (1979)
New York, Human Sciences Press, 298 p. ISBN 0–87705–390–1.

French, Barbara Coping with Bulimia: the binge–purge syndrome (1988)
San Bernardino CA, Borgo Press, 160 p. ISBN 0–8095–7055–6.

French, S The Big Brother Game: an encyclopedia of bugging, wiretapping, tailing, optical and electronic surveillance and surreptitious entry (1986)
New York, Gordon Press Publishers. ISBN 0–8490–3674–7.

Freudenberg, Nicholas Not in our Backyards: community action for health and environment (1984)
New York, Monthly Review Press, 320 p. ISBN 0–85345–653–4.

Freund, Paul (Ed) Experimentation with Human Subjects (1972)
London, Allen and Unwin.

Friday, Paul C and Stewart, V Lorne Youth Crime and Juvenile Justice: international perspectives (1977)
New York, Praeger Publishers, 200 p. ISBN 0–275–90265–X.

Fried, Robert The Hyperventilation Syndrome: research and clinical treatment (1987)
Baltimore MD, Johns Hopkins University Press, 192 p. Contemporary Medicine and Public Health Ser. ISBN 0–8018–3394–9.

Friedman, Julian R and Sherman, Marc I Human Rights: an international and comparative law bibliography (1985)
Westport CT, Greenwood Press, 868 p. ISBN 0–313–24767–6.

Friedman, Mendel Nutritional and Toxicological Aspects of Food Safety (1984)
New York, Plenum Publishing, 596 p. ISBN 0–306–41708–1.

Friedman, Richard Male Homosexuality: a contemporary psychoanalytic perspective (1988)
New Haven CT, Yale University Press, 320 p. ISBN 0–300–03963–8.

Friedman, Richard C (Ed) Behavior and the Menstrual Cycle (1982)
New York, Dekker Marcel, 480 p. illus. Sexual Behavior Ser. ISBN 0–8247–1852–6.

Friedrich, C J von and Brazezinski, Z K Totalitarian Dictatorship and Autocracy (1969)
Bombay, Times of India Press, 156 p.

Frisch, Jean–Romain (Ed) Future Stresses for Energy Resourses (1986)
Dordrecht, Kluwer Academic Publishers Group, 226 p. ISBN 0–86010–824–4.

Frolov, I Global Problems and the Future of Mankind (1982)
Chicago IL, Imported Publications, 311 p. ISBN 0–8285–2479–3.

Fromm, Erich Anatomy of Human Destructiveness (1981)
New York, Fawcett Book Group, 576 p. ISBN 0–449–24021–5.

Frost, N E, et al Metal Fatigue
Ann Arbor MI, Books on Demand. Oxford Engineering science Ser. ISBN 0–317–08550–6.

Fuchs–Bruninghoff, E, et al Functional Illiteracy and Literacy Provision in Developed Countries: the case of the Federal Republic of Germany (1987)
Lanham MD, UNIPUB, 90 p. illus. orig. ISBN 92–820–1046–5. UIE Case Studies: No 7.

Fuller, R Inflation (1980)
Worldwatch Paper: 34.

Fulton, Robert Death, Grief and Bereavement: a bibliography 1845–1975 (1976)
Salem NH, Ayer Company Publishers. bibl. Death and Dying Serv. ISBN 0–405–09570–8.

Furet, François (Ed) Unanswered Questions: Nazi Germany and the genocide of the Jews (1989)
New York, Schocken Books, 384 p. ISBN 0–8052–0908–5.

Furman, Erna and Freud, Anna A Child's Parent Dies: studies in childhood bereavement
New Haven CT, Yale University Press. ISBN 0–300–01719–7.

Furstenberg, Frank F Jr Unplanned Parenthood: the social consequences of teenage childbearing (1976)
New York, Free Press. ISBN 0–02–911010–6.

Futuribles International The world by the Year 2000: international future survey, synoptic report
In: *Bureau of Studies and Programming – The World by the Year 2000* 1987, Paris, UNESCO, BEP/GPI/1, pp. 17–62.

Gaafar, S M, et al (Eds) Parasites, Pests and Predators (1985)
Amsterdam, Elsevier Science Publishing, 576 p. ISBN 0–444–42175–0.

Gabor, Thomas, et al Armed Robbery: cops, robbers and victims (1987)
Springfield IL, Thomas Charles C, 244 p. illus. ISBN 0–398–05328–6.

Gabriel, Marcel The Decline of Wisdom (1955)
New York, Allied Publications.

Gabus, A Les Problèmes Mondiaux: analyse de perception (1972)
Paper presented to the Third World Futures Research Conference, Bucharest, 1972.

Gabus, A The Purpose of Contemporary Society and the DEMATEL Research Project (1973)
Extracts of a presentation to the Japan section of the Club of Rome.

Gadbaw, R Michael and Richards, Timothy J (Eds) Intellectual Property Rights: global consensus, global conflict? (1987)
Boulder CO, Westview Press, 400 p. ISBN 0–8133–7550–9.

Gadow, Kenneth D and Swanson, H Lee Memory and Learning Disabilities (1987)
Greenwich CT, Jai Press. Advances in Learning and Behavioral Disabilities Ser: Suppl 2. ISBN 0–89232–836–3.

Gagne, Eve E School Behavior and School Discipline: coping with deviant behavior in the schools (1983)
Lanham MD, University Press of America, 176 p. ISBN 0–8191–2748–5.

Gaitz, C M and Samorajski, T (Eds) Biomedical Issues (1985)
Berlin, Springer–Verlag, 545 p. Aging 2000: our health care desting: 1. ISBN 3–540–96057–0.

Gal, Allon Socialist Zionism: theory and issues in contemporary jewish (1973)
New Brunswick NJ, Transaction Books, 225 p. ISBN 0–87073–669–8.

Galasko, C S Skeletal Metastases (1986)
Stoneham MA, Butterworth Publishers, 180 p. ISBN 0–407–00409–2.

Gale, R P and Hoelzer, D (Eds) Acute Lymphoblastic Leukemia (1990)
New York, Alan R Liss, 342 p. ISBN 0–471–56719–1.

Gale, Robert and Golde, David Leukemia: recent advances in biology and treatment (1985)
New York, Alan R Liss, 764 p. UCLA Ser: Vol 28. ISBN 0–8451–2627–X.

Gallery, Shari Computer Security: readings from "security management magazine" (1986)
Stoneham MA, Butterworth Publishers, 300 p. ISBN 0–409–90084–2.

Galloway, David M and Goodwin, Carole Educating Slow–Learning and Maladjusted Children: integration or segregation? (1979)
White Plains NY, Longman. ISBN 0–582–48914–8.

Galski, Thomas and Nora, Rena The Handbook of Pathological Gambling (1987)
Springfield IL, Thomas Charles C, 228 p. ISBN 0–398–05268–9.

Gamstorp, Ingrid and Sarnat, Harvey B (Eds) Progressive Spinal Muscular Atrophies (1984)
New York, Raven Press, 256 p. illus. International Review of Child Neurology Ser. ISBN 0–89004–952–1.

Gander, T Nuclear, Biological and Chemical Warfare (1987)
New York, Hippocrene Books, 128 p. illus. ISBN 0–87052–451–8.

Gandert, Slade R Protecting Your Collection (1982)
New York, Haworth Press, 144 p. ISBN 0–917724–78–X.

Gangrade, K D Crisis of Values: a study in generation gap (1975)
New Delhi, Chetana Publications, 295 p.

Gapany–Gapanavicius, B Otosclerosis: genetics and surgical rehabilitation (1975)
Jerusalem, Keter Publishing House, 260 p. ISBN 0–7065–1452–1.

Gara, Larra and Chatfield, Charles International War Resistance Through World War II
New York, Garland Publishing. ISBN 0–8240–0449–3.

Garbarino, James, et al Troubled Youth, Troubled Families: understanding families at–risk for adolescent maltreatment (1986)
Berlin, Walter de Gruyter, 356 p. ISBN 3–11–010818–6.

Garbarino, James, et al (Eds) Special Children – Special Risks: the maltreatment of children with disabilities (1987)
New York, Aldine, 311 p. ISBN 3–11–011422–4.

Gardiner, Muriel and Spender, Stephen The Deadly Innocents: portraits of children who kill (1985)
New Haven CT, Yale University Press, 216 p. ISBN 0–300–03306–0.

Gardner, Richard A Separation Anxiety Disorder: psychodynamics and psychotherapy (1985)
Cresskill NJ, Creative Therapeutics, 180 p. ISBN 0–933812–10–8.

Gardos, George and Casey, Daniel E Tardive Dyskinesia and Affective Disorders (1984)
Washington DC, American Psychiatric Press, 96 p. ISBN 0–88048–060–2.

Garg, Usha and Vibhakar, Jagdish Poverty of Nations and New Economic Order (1985)
Delhi, UDH Publishers and Distributors, 275 p.

Gargas, Eivind, et al Lectures in Coastal Pollution: WHO training course (1976)
Water Quality Inst, 72 p.

Garling, Tommy and Valsiner, Jaan (Eds) Children within Environments: toward a psychology of accident prevention (1985)
New York, Plenum Publishing, 262 p. ISBN 0–306–42116–X.

Garnick, Marc and Richie, Jerome Urologic Cancer: a multidisciplinary approach (1983)
New York, Plenum Publishing, 288 p. ISBN 0–306–41473–2.

Garnsey, P D and Whittaker, C R Imperialism in the Ancient World (1979)
New York, Cambridge University Press. ISBN 0–521–21882–9.

Garratty, George (Ed) Hemolytic Disease of the Newborn (1984)
American Association Blood. illus. ISBN 0–9155355–05–1.

Gartner, Leslie P Alcohol and Pregnancy (1984)
Reisterstown MD, Jen House Publishing, 80 p. ISBN 0–910841–03–9.

Gaskell, Martin Slums (1989)
London, Pinter Publishers, 300 p. ISBN 0–7185–1293–6.

Gath, Ann Down's Syndrome and the Family: the early years (1978)
San Diego CA, Academic Press. ISBN 0–12–277450–7.

Gattis, Louie S Living in The Shadows: families of the accused (1988)
Altamore Springs FL, Cheetah Publishing, 225 p. illus. ISBN 0–936241–01–2.

Gaudier, Maryse Workers' Participation in Management: selected bibliography, 1984–1988 (1988)
Geneva, ILO, 204 p. ISBN 92–9014–445–9. Bibliography No 13.

Gavras, H and Gavras, I Hypertension in the Elderly (1983)
Littleton MA, PSG Publishing, 256 p. ISBN 0–7236–7047–1.

Gearheart, Bill R, et al The Exceptional Student in the Regular Classroom (1988)
Columbus ON, Merrill Publishing, 480 p. 4th ed.

Gebreah, S Ahmed M and Tay, A Qirbi A K B Study on the Problems of the Financing of the Education in the LDCs (1985)
Paris, UNESCO.

Geissler, Erhard (Ed) Strenghtening the Biological Weapons Convention by Confidence–Building Measures (1990)
Oxford, Stockholm International Peace Research Institute, 206 p. SIPRI Chemical and Biological Warfare Studies: 10. ISBN 0–19–829139–6.

Gelder, Michael G; Johnston, Derek W and Matthews, Andrew M Agoraphobia: nature and treatment (1981)
London, Routledge, 248 p. ISBN 0–422–78060–X.

Gelles, Richard J and Cornell, Claire P International Perspectives on Family Violence (1983)
Lexington MA, Lexington Books, 176 p. ISBN 0–669–06199–9.

Gelles, Richard J and Lancaster, Jane B (Eds) Child Abuse and Neglect: biosocial dimensions (1988)
Berlin, Walter de Gruyter, 334 p. ISBN 3–11–011552–2.

Gelstein, Sylvia S Juvenile Delinquency (1985)
Annandale VA, ABBE Publishers Association of Washington, 154 p. ISBN 0–88164–002–6.

Genest, Jacques, et al Hypertension: physiopathology and treatment (1983)
New York, McGraw Hill Book Company, 1344 p. ISBN 0–07–023061–7.

Geneva Institute of International Relations Problems of Peace: lectures (1968)
Salem NH, Ayer Publishers. repr of 1934. Essay Index Reprint Series. ISBN 0–8369–0470–2.

George, F H Machine Takeover (1977)
Elmsford NY, Pergamon Books, 208 p. ISBN 0–08–021229–8.

George, Henry Social Problems (1981)
New York, Robert Schalkenbach Foundation, 310 p. ISBN 0–911312–17–X.

George, Susan How the Other Half Dies: the real reasons for world hunger (1977)
Totowa NJ, Rowman and Littlefield Publishers, 328 p. ISBN 0–916672–07–7.

George, Vic Wealth, Poverty and Starvation: an international perspective (1988)
Hemel Hempstead, Simon and Schuster International, 224 p. ISBN 0–7450–0070–3.

Georgii, H W and Pankrath, J (Eds) Deposition of Atmospheric Pollutants: proceedings of a colloquium held at Oberursel/Taunus, West Germany, 1981 (1982)
Dordrecht, Kluwer Academic Publishers Group, 200 p. ISBN 90–277–1438–X.

Gerber, David A (Ed) Anti–Semitism in American History (1986)
Champaign IL, University of Illinois Press, 440 p. ISBN 0–252–01477–4.

Gerber, Lynn (Ed) Psoriatic Arthritis (1985)
New York, Grune and Stratton, 224 p. ISBN 0–8089–1709–9.

Gerber, William American Liberalism: laudable end, controversial means (1987)
Lanham MD, University Press of America, 338 p. ISBN 0–8191–6266–3.

Gereffi, Gary The Pharmaceutical Industry and Dependency in the Third World (1983)
Princeton NJ, Princeton University Press, 232 p. ISBN 0–691–09401–2.

Gerety, R J Non–A, Non–B Hepatitis (1981)
San Diego CA, Academic Press. ISBN 0–12–280680–8.

Gerety, Robert J Hepatitis A (1984)
San Diego CA, Academic Press. ISBN 0–12–280670–0.

Gerety, Robert J Hepatitis B (1985)
San Diego CA, Academic Press. ISBN 0–12–280672–7.

Gerhardt, Uta E and Wadsworth, Michael E J (Eds) Stress and Stigma: explanation and evidence in the sociology of crime and illness (1984)
Ann Arbor MI, Campus Publishers, 208 p. ISBN 3–593–33381–3.

Gerhold, H D et al The Breeding of Pest Resistant Trees (1966)
Elmsford NY, Pergamon Books. ISBN 0–08–011764–3.

Gerlach, S A Marine Pollution: diagnosis and therapy (1981)
Berlin, Springer–Verlag, 218 p. ISBN 3–540–10940–4.

Geronimo, Roger J, et al Liberalism Exposed: pick pocket economics (1987)
Columbus GA, Brentwood Communications Group, 119 p. ISBN 1–55630–037–9.

Gerosa, M, et al (Eds) Brain Tumors – Biopathology and Therapy: proceedings of the brain tumor workshop – Verona 1 held in Verona, Italy, 13–14 June 1985 (1986)
New York, Pergamon Books, 254 p. illus. Advances in the Biosciences Ser: No 58. ISBN 0–08–032013–9.

Geroski, Paul A, et al (Eds) Oligopoly, Competition and Welfare (1985)
New York, Basil Blackwell, 272 p. ISBN 0–631–14479–X.

Gerrick, David J Sore Throat-The Danger Within (1978)
Lorain OH, Dayton Laboratories. ISBN 0-685-89681-1.

Gerrick, David J The Rh Problem (1979)
Lorain OH, Dayton Laboratories. ISBN 0-916750-49-3.

Gerver, Israel; Howton, William and Rosenberg, Bernard (Eds) Mass Society in Crisis: social problems ans social pathology (1971)
New York, Macmillan. 2nd ed.

GESAMP Atmospheric Transport on Contaminants Into the Mediterranean Region (1985)
Nairobi, UNEP, 53 p. UNEP Regional Seas Reports and Studies: No 68.

Geyer, Felix and Schweitzer, David (Eds) Alienation: problems of meaning, theory and method (1981)
New York, Routledge Chapman and Hall, 288 p.
ISBN 0-7100-0835-X.

Geyer, R A Marine Environmental Pollution (1980)
Amsterdam, Elsevier Science Publishing. 2 Vols.

Geyer, R F and Schweitzer, D R (Eds) Theories of Alienation: critical perspectives in philosophy and the social sciences (1976)
Dordrecht, Kluwer Academic Publishers Group, 306 p.
ISBN 90-207-0630-6.

Geyer, R F and Zouwen, J van der Dependence and Inequality (1982)
Elmsford NY, Pergamon Books, 336 p.
ISBN 0-08-027952-X.

Ghai Dharam P, et al Overcoming Rural Underdevelopment (1983)
Geneva, ILO, 98 p. ISBN 92-2-102149-1.

Ghosh, B N Economics of Underemployment (1983)
New Delhi, Deep and Deep Publications, 450 p.

Ghosh, Pradip K Energy Policy and Third World Development (1984)
Westport CT, Greenwood Press, 392 p.
ISBN 0-313-24140-6.

Gibbons, Alice M Divorce and Divorce Factors: subject analysis and research index with (1987)
Annandale VA, ABBE Publishers Association of Washington, 150 p. ISBN 0-88164-682-2.

Gibbons, Kenneth and Rowat, Donald C (Eds) Political Corruption in Canada: cases, causes, cures
Ottawa ON, Carleton University Press.
ISBN 0-7710-9795-6.

Gibson, D Down's Syndrome: the psychology of mongolism (1979)
New York, Cambridge University Press.
ISBN 0-521-21914-0.

Gibson, Diane The Evaluation and Treatment of Eating Disorders (1986)
New York, Haworth Press, 160 p. ISBN 0-86656-541-8.

Gibson, J A Diseases of Forest Trees Widely Planted As Exotics in the Tropics and Southern Hemisphere
New York, State Mutual Book and Periodical Service. 2 Vols.

Gibson, T E (Ed) Weather and Parasitic Animal Disease (1978)
Geneva, WHO, 174 p. ISBN 92-63-10497-2.

Gibson, Terry Rheumatic Diseases (1986)
Stoneham MA, Butterworth Publishers, 216 p.
ISBN 0-407-00315-0.

Giersch, Herbert (Ed) Capital Shortage and Unemployment in the World Economy: symposium 1977 (1978)
Tübingen, JCB Mohr, 348 p. ISBN 3-16-341191-6.

Giersch, Herbert (Ed) International Debt Problem: lessons for the future (1986)
Tübingen, JCB Mohr, 200 p. ISBN 3-16-345085-7.

Gifkins, R C (Ed) Strength of Metals and Alloys (ICSMA 6) Proceedings of the 6th International Conference, Melbourne, Australia, August 16-20, 1982 (1982)
New York, Pergamon Books, 1200 p. 3 Vols. International Series on the Strength and Fracture of Materials and Structures. ISBN 0-08-029325-5.

Gilbert, Michael Inflation and Social Conflict (1986)
Hemel Hempstead, Simon and Schuster International, 256 p.
ISBN 0-7108-0238-2.

Gilbert, Neil Capitalism and the Welfare State: dilemmas of social benevolence (1985)
New Haven CT, Yale University Press.
ISBN 0-300-03477-6.

Gillian, J L Social Problems (1965)
Bombay, Times of India Press, 496 p.

Gilliatt, R W and Asbury, A K Peripheral Nerve Disorders (1984)
Stoneham MA, Butterworth Publishers, 320 p.
ISBN 0-407-02297-X.

Gilligan, Stephen G Therapeutic Trances: the cooperation principle in Ericksonian psychotherapy (1986)
New York, Brunner/Mazel, 384 p. ISBN 0-87630-442-0.

Gillmore, Robert Liberalism and the Politics of Plunder (1987)
Dublin NH, Bauhan, William L. ISBN 0-87233-087-7.

Gilman, Sander L Difference and Pathology: stereotypes of sexuality, race, and madness (1985)
Ithaca NY, Cornell University Press, 304 p.
ISBN 0-8014-1785-6.

Gilmore, Thomas N Making a Leadership Change: how organizations and leaders can handle (1988)
San Francisco CA, Jossey-Bass Publishers.
ISBN 1-55542-114-8.

Gilroy, E and Miles, T R Dyslexia at College (1986)
London, Routledge, 160 p.

Ginneken, Wouter Van and Garzuel, Michel Unemployment in France, the Federal Republic of Germany and the Netherlands. A Survey of Trends, Causes and Policy Options (1985)
Geneva, ILO, 137 p. ISBN 92-1-103032-7.

Ginneken, Wouter Van and Park, Jong-goo Generating Internationally Comparable Income Distribution Estimates (1984)
Geneva, ILO, 176 p. ISBN 92-2-10366-9.

Ginsberg, Harold S The Adenoviruses (1984)
New York, Plenum Publishing, 622 p.
ISBN 0-306-41592-5.

Ginsburg, Benson E and Carter, Bonnie F Premenstrual Syndrome (1987)
New York, Plenum Publishing, 452 p.
ISBN 0-306-42498-3.

Giovannelli, G et al Hypertension in Children and Adolescents (1981)
New York, Raven Press, 364 p. ISBN 0-89004-523-2.

Gise Premenstrual Syndromes: new findings and controversies (1987)
New York, Churchill Livingstone. Vol 2.
ISBN 0-443-08537-4.

Gist, Psychosocial Aspects of Disaster (1988)
New York, John Wiley and Sons. ISBN 0-471-84894-8.

Gittinger, J Price Compounding and Discounting Tables for Project Analysis: with a guide to their applications (1984)
Baltimore MD, Johns Hopkins University Press.
ISBN 0-8018-2409-5.

Glassner, M I Access to the Sea for Developing Landlocked States (1970)
Dordrecht, Kluwer Academic Publishers Group, 298 p.
ISBN 90-247-5022-9.

Glassner, M J Bibliography on Land-locked States (1986)
Dordrecht, Kluwer Academic Publishers Group, 224 p. 2nd ed. ISBN 90-247-3261-1.

Gleditsch, Nils Petter, et al Economic Incentives to Arm?: effects of military spending on industrialized markets economics
Oslo, International Peace Research Institute, 72 p.

Glendon, A, et al Bibliographical Review of Data Sources on Occupational Accidents and Diseases: consolidated report (1986)
Luxembourg, European Communities, 116 p.
ISBN 92-825-6426-6.

Gleser, Goldine C et al Prolonged Psychological Effects of a Disaster: a study of buffalo creek (1981)
San Diego CA, Academic Press. ISBN 0-12-286260-0.

Glossop, Ronald J Confronting War (1987)
Jefferson NC, McFarland Publishers, 385 p.
ISBN 0-89950-273-3.

Godfried, Robert S The Black Death: natural and human disaster in Medieval Europe (1983)
New York, Free Press, 240 p. ISBN 0-02-912630-4.

Godish, T Indoor Air Pollution Control (1989)
Chelsea MA, Lewis Publishers, 400 p.
ISBN 0-87371-098-3.

Godschalk, David R, et al Catastrophic Coastal Storms: hazard mitigation and development management (1989)
Durham NC, Duke University Press, 344 p.
ISBN 0-8223-0855-X.

Goff, Kenneth (Ed) Brain-Washing (1983)
Clackamas OR, Emissary Publications, 64 p.
ISBN 0-941380-03-3.

Goffman, Erving Asylums: essays on the social situation of mental patients and other inmates (1961)
New York, Aldine, 400 p. ISBN 0-202-30000-5.

Goffman, Erving Asylums (1970)
Harmondsworth, Penguin Books. Penguin Books.

Goffman, Irving Stigma: notes on the management of spoiled identity (1963)
Englewood Cliffs NJ, Prentice Hall. ISBN 0-13-846626-2.

Gohain, Bikash Human Sacrifice and Head Hunting in North Eastern India (1977)
Assam, Lawyer's Book Stall, 102 p.

Gold, M S Marijuana (Date not set)
New York, Plenum Publishing, 180 p.
ISBN 0-306-43062-2.

Gold, Mark Assault and Battery (1983)
London, Pluto Press.

Goldanskii, V I and Rotblat, J (Eds) Global Problems and Common Security (1989)
New York, Springer Verlag, 300 p. illus.
ISBN 0-387-51699-9.

Goldberg, Leonard (Ed) Alcohol, Drugs and Traffic Safety: proceedings of the 8th international conference on alcohol, drugs and traffic safety held in Stockholm 1980 (1981)
Stockholm, Almqvist and Wiksell, 1489 p. 3 Vols.

Goldberg, Samuel Army Training of Illiterates in World War Two
New York, AMS Press. ISBN 0-404-55966-2.

Goldblat, Jozef Non-Profileration: the why and the wherefore (1985)
Solna, Stockholm International Peace Research Institute.

Goldenberg, Naomi Ruth End of God: important directions for a feminist critique of religion in the works of Siegmund Freud and Carl Jung (1981)
Ottawa ON, University of Ottawa Press, 144 p.
ISBN 2-7603-0043-9.

Goldhaber, Samuel Z Pulmonary Embolism and Deep Venous Thrombosis (1985)
Philadelphia PA, WB Saunders, 295 p.
ISBN 0-7216-4151-2.

Golding, D N and Barrett, J The Practical Treatment of Backache and Sciatica (1984)
Norwell MA, Kluwer Academic Publishers.
ISBN 0-85200-773-6.

Golding, Jean, et al Sudden Infant Death: patterns, puzzles, and problems (1985)
Seattle WA, University of Washington Press, 272 p.
ISBN 0-295-96302-6.

Goldman, Marvin and Bustad, Leo K Biomedical Implications of Radiostrontium Exposure: proceedings (1972)
Oak Ridge TN, US Department of Energy, 411 p.
ISBN 0-87079-152-4.

Goldman, S J, et al Erosion and Sediment Control Handbook (1986)
New York, McGraw Hill Book Company.
ISBN 0-07-023655-0.

Goldschmidt-Clermont, Luisella Economic Evaluations of Unpaid Household Work. Africa, Asia, Latin America and Oceania (1987)
Geneva, ILO, 213 p. ISBN 92-2-105827-1.

Goldschmidt-Clermont, Luisella Unpaid Work in the Household: a review of economic evaluation methods (1989)
Geneva, ILO, 137 p. 4th impression.
ISBN 92-2-103085-7. Women, Work and Development: No 1.

Goldsmith, Edward and Hildyard, Nicholas (Eds) The Social and Environmental Effects of Large Dams
Cornwall, The Ecologist. bibl. 3 Vols.

Goldsmith, Edward and Hildyard, Nicholas (Eds) The Earth Report: monitoring the battle for our environment (1988)
London, Mitchell Beazley International, 240 p. illus.
ISBN 0-85533-695-1.

Goldstein, Eleanor C Pollution
Boca Raton FL, Social Issues Resources Series. 3 Vols. Vol 1: incl. 1972-1974 Supplements, Vol 2: incl. 1975-1979 Supplements, Vol 3: incl. 1980-1984 Supplements.

Goldstein, Eleanor C Pollution: incl 1985-1987 supplement (1987)
Boca Raton FL. Social Issues Resources Series: 4.
ISBN 0-89777-118-4.

Goldstein, Eleanor C (Ed) Consumerism (1987)
Boca Raton FL, Social Issues Resources Series. Vol 2 (including 1982-1986 supplements). ISBN 0-89777-057-9.

Goldstein, Herbert and Goldstein, Marjorie T The Reasoning Ability of Mildly Retarded Learners (1980)
Reston VA, Council for Exceptional Children, 80 p.
ISBN 0-86586-102-1.

Goldstein, J H (Ed) Sports Violence (1983)
Berlin, Springer-Verlag, 370 p. ISBN 3-540-90828-5.

Goldstein, Michael J, et al Pornography and Sexual Deviance (1973)
Berkeley CA, University of California Press.
ISBN 0-520-02406-0.

Goldstein, Phillip Neurological Disorders of Pregnancy (1986)
Mount Kisco NY, Futura Publishing, 296 p.
ISBN 0-87993-272-4.

Goldstein, Robert D Mother-Love and Abortion: a legal interpretation (1988)
Berkeley CA, University of California Press, 250 p.
ISBN 0-520-06084-9.

Goldstein, S Sexual Exploitation of Children (1986)
Amsterdam, Elsevier Science Publishing, 440 p.
ISBN 0-444-01117-X.

Golembiewski, Robert T, et al Stress in Organizations: toward a phase model of burnout (1985)
New York, Praeger Publishers, 286 p.
ISBN 0-275-90024-X.

Golikere, R K Vegetarian vs Non-Vegetarian (1960)
Bombay, Popular Book Depot, 80 p.

Golman and Coghill Disruptive Behaviour in Schools: causes, treatment and prevention (1987)
Lund, Utbildningshuset Studentlitteratur, 303 p.
ISBN 91-44-26411-9.

Golub, Sharon (Ed) Lifting the Curse of Menstruation: a feminist appraisal of the influence of menstruation on women's lifes (1983)
New York, Haworth Press, 156 p. Women and Health Ser: Vol 8, Nos 2/3. ISBN 0-86656-242-7.

Golubev, G N (Ed) Environmental Management of Agricultural Watersheds (1983)
Laxenburg, International Institute for Applied Systems Analysis, 289 p. ISBN 3-7045-0059-3.

Gomez Buendia, Hernando (Ed) Urban Crime: global trends and policies (1989)
Tokyo, United Nations University. ISBN 92-808-0679-3.

Gomez, E D South China Sea Fisheries Development and Coordinating Programme (1980)
Rome, FAO, 128 p.

Gondolf, Edward W and Roy, Maria Men Who Batter: an integrated approach for stopping wife abuse (1985)
Holmes Beach FL, Learning Publications, 224 p.
ISBN 0-918452-56-2.

Gontzea, I and Sutzescu, P Natural Antinutritive Substances in Foodstuffs and Forages (1968)
Basel, S Karger AG, 184 p. ISBN 3-8055-0856-5.

Goode, Elizabeth W (Comp) Drug Abuse Bibliography for 1980 (1984)
Troy NY, Whitston Publishing, 689 p. no 10.
ISBN 0-87875-285-4.

Goode, Polly T (Comp) Drug Abuse Bibliography
Troy NY, Whitston Publishing. Annual.

Goodman, David (Ed) The International Farm Crisis (1989)
Hampshire, Macmillan, 400 p. illus. ISBN 0-333-46947-X.

Goodman, Felicitas D How About Demons? Possession and Exorcism in the Modern World (1988)
Bloomington IN, Indiana University Press, 160 p.
ISBN 0-253-32856-X.

Goodman, G T and Chadwick, M J (Eds) Environmental Management of Mineral Wastes (1978)
Dordrecht, Kluwer Academic Publishers Group, 367 p.
ISBN 90-286-0054-X.

Goodman, Paul Moral Ambiguity of America
Toronto ON, Canadian Broadcasting Coorporation, 100 p.

Goodman, Paul Compulsory Miseducation (1971)
London, Penguin Books, 9 p.

Goodman, Paul S, et al Absenteeism: new approaches to understanding, measuring and managing employee absence (1984)
San Francisco CA, Jossey-Bass Publishers. Management Ser.

Goodmann, Richard M and Gorlin, Robert J The Malformed Infant and Child: an illustrated guide (1983)
Oxford NY, Oxford University Press. illus.
ISBN 0-19-503255-1.

Goodspeed, D J Conspirators (1984)
Agincourt ON, Macmillan of Canada, 274 p.
ISBN 0-7715-9856-4.

Goodwin-Gill, Guy S The Refugee in International Law (1984)
New York, Oxford University Press. ISBN 0-19-825372-9.

Goodwin, Jean M Sexual Abuse: incest victims and their families (1988)
Littleton MA, PSG Publishing, 288 p. illus. 2nd ed.
ISBN 0-88416-588-4.

Gopalan, C Nutrition - Problems and Programmes in South-East Asia (1987)
Geneva, WHO, 164 p.
ISBN 92-9022-184-4. SEARCO Regional Health Papers: No 15.

Gorbach, Sherwood L (Ed) Infectious Diarrhea (1986)
Oxford, Blackwell Scientific Publications, 328 p. illus.

Gordon, David M Theories of Poverty and Underemployment (1973)
Lexington MA, Heath DC. ISBN 0-669-89268-8.

Gordon, David R and Dangerfield, Royden The Hidden Weapon: the story of economic warfare (1976)
Jersey City NJ, Da Capo Press. ISBN 0-306-70769-1.

Gordon, Richard L An Economic Analysis of World Energy Problems (1981)
Cambridge MA, MIT Press, 320 p. illus.
ISBN 0-262-07080-4.

Gordon, Sheila World Problems (1971)
London, Batsford, 96 p.

Gorizontov, B Capitalism and the Ecological Crisis
Los Angeles CA, Progress Publishing, 160 p.

Gorlenko, M V Bacterial Diseases of Plants (1961)
Jerusalem, Keter Publishing House, 184 p.

Gosavi, M S Management Gap in a Developing Economy (1971)
Bombay, Progressive Corporation, 156 p.

Gosney, E S; Popenoe, Paul and Grob, Gerald N Sterilization for Human Betterment (1979)
Salem NH, Ayer Company Publishers.
ISBN 0-405-11915-1.

Goss, Michael Poltergeists: an annotated bibliography of works in english, (1979)
Metuchen NJ, Scarecrow Press, 389 p.
ISBN 0-8108-1181-2.

Gotoh, F and Letchner, H (Eds) Clinical Haemorheology - A New Approach to Cerebrovascular Diseases: proceedings of a satellite symposium sponsored by Hoechst, in conjunction with the XVII international congress of internal medicine held in Kyote, October 8, 1984 (1986)
Wolfeboro NH, Longwood Publishing Group, 98 p. International Congress and Symposium Ser: No 100.
ISBN 0-905958-28-4.

Gottesman, Irving I and Shields, James Schizophrenia and Genetics (1972)
San Diego CA, Academic Press. ISBN 0-12-293450-4.

Gottesman, Irving I and Shields, James Schizophrenia: the epigenetic puzzle (1982)
New York, Cambridge University Press, 275 p.
ISBN 0-521-22573-6.

Gottheil, Edward, et al Etiologic Aspects of Alcohol and Drug Abuse (1983)
Springfield IL, Thomas Charles C, 360 p.
ISBN 0-398-04732-4.

Gottheil, Edward, et al The Combined Problems of Alcoholism, Drug Addiction and Aging (1985)
Springfield IL, Thomas Charles C, 386 p.
ISBN 0-398-05046-5.

Gottlieb, Geoffrey J and Ackerman, A Bernard Kaposi's Sarcoma: a text and atlas (1988)
Philadelphia PA, Lea and Febiger, 330 p.
ISBN 0-8121-1041-2.

Gottschalk, G; Pfenning, W and Werner, H (Eds) Anaerobes and Anaerobic Infections: symposia held at the 12th international congress of microbiology in Munich 1978 (1980)
Stuttgart, Fischer, 170 p. ISBN 3-437-30310-4.

Goudzwaard, Bob Aid for the Overdeveloped West (1975)
Jordan Station ON, Wedge Publishing Foundation.
ISBN 0-88906-100-9.

Gould, David J and Amaro-Reyes, Jose A The Effects of Corruption on Administrative Performance: illustrations from developing countries (1983)
Washington DC, International Bank for Reconstruction and Development. Working Papers: 580.

Gowda, K G V Eurodollar Flows and International Monetary Stability (1978)
Mysore, Geetha Book House, 539 p.

Graaff, W de and Reijnen, G C M Pollution of Outer Space, in Particular of the Geostationary Orbit: scientific, policy and legal aspects (1989)
Dordrecht, Martinus Nijhoff, 188 p. ISBN 90-247-3750-8.

Grainger, Alan The Threatening Desert: controlling desertification
London, Earthscan, 288 p. ISBN 1-85383-041-0.

Grainger, Alan Desertification: how people make deserts, how people can stop them and why they don't (1986)
Wolfeboro NH, Longwood Publishing Group, 96 p. illus.
ISBN 0-905347-37-4.

Grais, Bernard Lay-Offs and Short-Time Working in Selected OECD Countries (1983)
Brussels, Organization for Economic Cooperation and Development.

Grathoff, R H Structure of Social Inconsistencies (1978)
Dordrecht, Kluwer Academic Publishers Group, 186 p.
ISBN 90-247-5006-7.

Graubard, Mark Witchcraft and the Nature of Man (1985)
Lanham MD, University Press of America, 326 p.
ISBN 0-8191-4313-8.

Graumann C F and Moscovici, S (Eds) Changing conceptions of conspiracy (1987)
Berlin, Springer-Verlag, 280 p. Springer Series in Social Psychology. Illus. ISBN 0-387-96223-9.

Gray, Frank D Pulmonary Embolism
Ann Arbor MI, Books on Demand. ISBN 0-317-07818-6.

Gray, H Peter International Economic Problems and Policies (1987)
London, St Martin's Press. ISBN 0-312-42088-9.

Gray, Jeffrey A The Psychology of Fear and Stress (1988)
New York, Cambridge University Press, 432 p.
ISBN 0-521-24958-9.

Grayson, Merrill Diseases of the Cornea (1983)
Saint Louis MO, CV Mosby, 668 p. 2nd ed. Illus.
ISBN 0-8016-1973-4.

Green, Arnold W Social Problems: arena of conflict (1975)
New York, McGraw-Hill Publishing.

Green, G Dorsey and Clunis, D Merilee Lesbian Couples (1988)
Seattle WA, Seal Pr Feminist, 260 p.
ISBN 0-931188-59-8.

Green, Richard The "Sissy Boy Syndrome" and the Development of Homosexuality (1987)
New Haven CT, Yale University Press, 432 p.
ISBN 0-300-03696-5.

Green, Richard, et al (Eds) Transsexualism and Sex Reassignment
Ann Arbor MI, Books on Demand. illus.
ISBN 0-317-07865-8.

Greenawalt, R Kent Discrimination and Reverse Discrimination (1983)
New York, Random House, 260 p. ISBN 0-394-33577-5.

Greenberg, Bernard Flies and Diseases (1970)
Princeton NJ, Princeton University Press. 2 Vols: 1 - Ecology, Classification, and Biotic Associations; 2 - Biology and Disease Transmission. ISBN 0-691-08093-3.

Greenberg, Martin H and Waugh, Charles G (Eds) Cults: an anthology of secret societies, sects, and the supernatural (1983)
New York, Beaufort Books Publishers, 368 p.
ISBN 0-8253-0159-9.

Greenfield, Jeanette The Return of Cultural Treasures

Greenland, Cyril Preventing CAN Deaths: an international study of deaths due to child abuse and neglect (1987)
London, Routledge, 250 p. ISBN 0-422-61210-3.

Greenough, P R; MacCallum, F J and Weaver, A D Lameness in Cattle (1981)
Sevenoaks, Butterworths, 496 p. illus. 2nd ed.
ISBN 0-85608-030-6.

Greenway, James C Extinct and Vanishing Birds of the World
New York, Dover Publications. ISBN 0-486-21869-4.

Greenwood, M Epidemics and Crowd Diseases (1935)
New York, Macmillan.

Greer-Wootten, Bryn, et al Waste Management: a contextual bibliography based on periodical (1986)
Monticello IL, Vance Bibliographies, 53 p.
ISBN 0-89028-789-9.

Gregor, A James The Fascist Persuasion in Radical Politics (1974)
Princeton NJ, Princeton University Press, 424 p.
ISBN 0-691-07556-5.

Greve, E L (Ed) Glaucoma and Cataract (1986)
Amstelveen, Kugler Publications, 120 p.

Grey, Margot Return from Death: an exploration of the near-death experience (1985)
New York, Routledge Chapman and Hall, 224 p.
ISBN 1-85063-019-4.

Gribbin, J Climatic Change (1978)
New York, Cambridge University Press.
ISBN 0-521-21594-3.

Griffin, Charles Ethesey Appearance and Eventuality: or parasitism, rejuvenescing and the visual lie
Woolloomooloo NSW, Legal Books, 621 p.
ISBN 0-9588948-0-9.

Griffith, Linda L Rattle Fatigue: and you thought you were busy before you had (1986)
San Luis Obispo CA, Impact Publishers, 176 p.
ISBN 0-915166-44-5.

Griffiths, Franklyn and Polanyi, John C (Ed) Dangers of Nuclear War (1979)
Toronto ON, University of Toronto Press.
ISBN 0-8020-2356-8.

Grigulevich, Iosif (Ed) Global Problems of Mankind and the State
Moscow, Izdatel'stvo Nauka.

Grinspoon, Lester (Ed) The Long Darkness: psychological and moral perspectives on nuclear winter (1986)
New Haven CT, Yale University Press, 224 p.
ISBN 0-300-03663-9.

Grizzle, Anne F and Proctor, William Mother Love, Mother Hate: breaking dependent love patterns in family (1988)
New York, Ballantine Books, 304 p. ISBN 0-449-90246-3.

Grob, Gerald N Social Problems and Social Policy Series (1975)
Salem NH, Ayer Company Publishers. 51 vols.
ISBN 0-405-07474-3.

Groher, Michael E Dysphagia: diagnosis and management (1984)
Stoneham MA, Butterworth Publishers, 256 p.
ISBN 0-409-95112-9.

Groom, A J and Taylor, Paul Global Issues in the United Nations' Framework (1989)
New York, Saint Martin's Press, 387 p.
ISBN 0-312-02846-6.

Gross, F A Crime: a universal problem (1977)
Kenwin, Juta, 187 p. ISBN 0-7021-0871-5.

Gross, H (Eds) Dictionary of Corrosion and corrosion control (1985)
Amsterdam, Elsevier Science Publishing. Eng and Ger.
ISBN 0-317-47233-X.

Groth, A N Men Who Rape: the psychology of the offender (1979)
New York, Plenum Publishing, 246 p.
ISBN 0-306-40268-8.

Groupement International des Associations Nationales de Fabricants de Produits Agrochimiques Coping with Resistance to Fungicides: GIFAP seminar, 1981 (1981)
Brussels, Groupement International des Associations Nationales de Fabricants de Pesticides, 52 p.

Grove, D I (Ed) Strongyloidiasis: a major roundworm infection of man (1989)
Basingstoke, Taylor and Francis, 220 p.
ISBN 0-85066-732-1.

Gruber, R T, et al The Pathophysiology of Combined Injury and Trauma (1987)
San Diego CA, Academic Press. ISBN 0-12-304755-2.

Grundfest-Broniatowski and Esselstyn Controversies in Breast Disease: diagnosis and management (1988)
New York, Dekker Marcel, 544 p. ISBN 0-8247-7880-4.

Grunebaum, Henry, et al Mentally Ill Mothers and Their Children (1982)
Chicago IL, University of Chicago Press, 378 p.
ISBN 0-226-31029-9.

Grunwell, P The Nature of Phonological Disability in Children (1981)
San Diego CA, Academic Press, 256 p.
ISBN 0-12-305250-5.

Grusec, J E and Lytton, H Social Development: history, theory and research (1988)
Berlin, Springer-Verlag, 500 p. ISBN 3-540-96591-2.

Gryboski, Joyce Gastrointestinal Problems in the Infant
Ann Arbor MI, Books on Demand. ISBN 0-317-08715-0.

Guérin, Laurent and Leblond, Bernard Soil Conservation: project design and implementation using labour-intensive techniques (1988)
Geneva, ILO, 206 p. 2nd impression.
ISBN 92-2-103395-3.

Gugler, Josef The Urbanization of the Third World (1988)
New York, Oxford University Press, 448 p.
ISBN 0-19-823260-8.

Guha, Bimalendu Study on the Language Barrier in the Production, Dissemination and Use of the Scientific and Technical Information with Special Reference to the Problems of the Developing Countries (1985)
Paris, UNESCO, 66 p. bibl.

Guilleminault, Christian Sleeping and Waking Disorders: indications and techniques (1982)
Stoneham MA, Butterworth Publishers, 435 p.
ISBN 0-409-90051-6.

Guinagh, B J Catharsis and Cognition in Psychotherapy (1987)
Berlin, Springer-Verlag, 135 p. illus.

Gunderson, John G Borderline Personality Disorder (1987)
Cambridge MA, Cambridge University Press, 216 p.
ISBN 0-521-34494-8.

Gunn, John C (Ed) The Mentally Disordered Offender (1990)
Oxford, Heineman Professional Publishing, 288 p.
ISBN 0-433-01468-7.

Gunter, Barrie Dimensions of Television Violence (1985)
New York, Saint Martin's Press, 282 p.
ISBN 0-312-21077-9.

Guppy, N Tropical Deforestation: a global view (1984)
In: Foreign Affairs 4, Spring, pp. 928–966.

Gupta, Joyeeta Toxic Terrorism: dumping hazardous wastes (1990)
London, Earthscan, 160 p. ISBN 1-85383-061-5.

Gupta, Krishna K Pickpockets: the mysterious species (1987)
New York, Apt Books, 106 p. ISBN 81-7018-403-7.

Gurney, T R Fatigue of Welded Structures (1980)
New York, Cambridge University Press.
ISBN 0-521-22558-2.

Guroff, Gordon Growth and Maturation Factors
New York, John Wiley and Sons. 2 Vols.

Gutek, Barbara A Sex and the Workplace (1985)
San Francisco CA, Jossey-Bass Publishers.
ISBN 0-87589-656-1.

Gutkind, Peter C W Bibliography on Unemployment (1977)
Montreal, McGill-Queen's University Press, 76 p.
ISBN 0-88819-030-1.

Guttentag The Mental Health of Women (1980)
San Diego CA, Academic Press. ISBN 0-12-310850-0.

Guttman, L Spinal Cord Injuries: comprehensive management and research (1976)
Oxford, Blackwell Scientific Publications, 768 p.
ISBN 0-632-00079-1.

Guttman, V and Merz, M Corrosion and Mechanical Stress at High Temperatures (1981)
New York, Elsevier Science Publishing, 477 p.
ISBN 0-85334-956-8.

Gyöngyössy, István International Money Flows and Currency Crises (1984)
Dordrecht, Kluwer Academic Publishers Group, 160 p.
ISBN 90-247-2647-6.

Haagensen, C D Diseases of the Breast (1986)
Philadelphia PA, WB Saunders, 1050 p. illus. 3rd ed. ISBN 0-7216-4442-2.

Habermas, Jurgen and McCarthy, Thomas Legitimation Crisis
Boston MA, Beacon Press, 192 p. ISBN 0-8070-1521-0.

Habermehl, G G Venomous Animals and Their Toxins (1981)
Berlin, Springer-Verlag, 195 p. ISBN 3-540-10780-0.

Hadenius, Axel Crisis of the Welfare State?: opinions about taxes and public expenditure in Sweden (1986)
Stockholm, Almqvist and Wiksell, 152 p. ISBN 91-22-00821-7.

Hafer, Keith W Coping with Bereavement from Death or Diverce (1988)
Melbourne FL, Robert E Krieger, 108 p. ISBN 0-89464-317-7.

Hafez, E S (Ed) Spontaneous Abortion (1984)
Dordrecht, Kluwer Academic Publishers Group, 500 p. ISBN 0-318-01662-1.

Hagerman, Randi J, et al The Fragile X Syndrome (1983)
Dillon CO, Spectra Publishing. ISBN 0-915667-00-2.

Hägglund, Solveig Sex-Typing and Development in an Ecological Perspective (1986)
Göteborg, Kungliga Vetenskaps- och Vitterhets-samhället. ISBN 91-7346-167-9.

Hails, Michael R Plant Poisoning in Animals: a bibliography of world literature, 2 (1986)
Forestburgh NY, Lubrecht and Cramer, 92 p. ISBN 0-85198-578-5.

Hajela, P D Problems of Monetary Policy in Underdeveloped Countries (1973)
Bombay, Lalvani Publishing House, 306 p.

Haken, H (Ed) Synergetics: cooperative phenomena in multicomponent systems: proceedings of the International Symposium on Synergetics 1972 (1973)
Stuttgart, Teubner GmbH, 279 p. ISBN 3-519-03011-X.

Hakovirta, H Third World Conflicts and Refugeeism: dimensions, dynamics and trends of the world refugee problem (1986)
Helsinki, Finska Vetenskaps-Societeten, 160 p. ISBN 951-653-140-7.

Hale, Andrew R and Glendon, A Ian Individual Behaviour in the Control of Danger (1987)
Amsterdam, Elsevier Science Publishing, 464 p. ISBN 0-444-42838-0.

Hales, J R S and Richards, D A B (Eds) Heat Stress (1987)
Amsterdam, Elsevier Science Publishing, 540 p. ISBN 0-444-80932-5.

Halkes, R and Olson, J K (Eds) Teacher Thinking: a new perspective on persisting problems in education, Proceedings of the first symposium of the International Study Association on Teacher Thinking (1984)
Lisse, Swets en Zeitlinger, 232 p. ISBN 90-265-0558-2.

Hall, A J Flash Flood Forecasting (1981)
Geneva, WHO, 38 p. ISBN 92-63-10577-4.

Hall, D W Handling and Storage of Food Grains in Tropical and Subtropical Areas (1980)
Rome, FAO, 350 p. 3rd ed. FAO Agricultural Development Paper: No 90. ISBN 92-5-100854-X.

Hall, Lawrence B and Shilling, Charles W Planetary Quarantine: principles, methods and problems (1971)
New York, Gordon and Breach, 184 p. ISBN 0-677-15100-4.

Hall, Richard Dirorganized Crime
St Lucia, University of Queensland Press, 256 p. ISBN 0-7022-1955-X.

Hall, Richard Uninvited Guests: a documented history of UFO sightings (1988)
Santa Fe NM, Aurora Press, 384 p. ISBN 0-943358-32-9.

Halstead, Bruce W and Saunders, Paul Poisonous and Venomous Marine Animals of the World (1988)
Princeton NJ, Darwin Press, 1500 p. ISBN 0-87850-050-2.

Halsted, Charles H and Rucker, Robert B Nutrition and the Origin of Disease (1988)
San Diego CA, Academic Press, 692 p. ISBN 0-12-319640-X.

Hamberg, Daniel Economic Growth and Instability: a study in the problem of capital accumulation, employment and the business cycle (1978)
Westport CT, Greenwood Press. illus. Repr of 1956. ISBN 0-313-20215-X.

Hamilton, Lyle H, et al High Frequency Ventilation (1986)
Boca Raton FL, CRC Press, 192 p. ISBN 0-8493-6739-5.

Hamilton, M B and Hirszowicz, Maria Class and Inequality: in pre-industrial, capitalist and communist (1987)
New York, Saint Martin's Press, 320 p. ISBN 0-312-01222-5.

Hamline University, Advanced Legal Education Staff Issues of Illegality (1985)
Saint Paul MN, Hamline University School of Law, 162 p. ISBN 0-317-42984-1.

Hámori, D Constitutional Disorders and Hereditary Diseases in Domestic Animals (1983)
Amsterdam, Elsevier Science Publishing, 728 p. ISBN 0-444-99683-4.

Han, Henry H Terrorism, Political Violence and World Order (1984)
Lanham MD, University Press of America, 790 p. ISBN 0-8191-3739-1.

Hand, I and Wittchen, H U (Eds) Panic and Phobias Two (1988)
Berlin, Springer-Verlag, 280 p. illus. ISBN 0-387-19088-0.

Handel, Michael Military Deception in Peace and War
Jerusalem, Magnes Press, 64 p.

Hanen, Marsha P; Osler, Margaret J and Weyant, Robert G (Eds) Science, Pseudo-Science and Society (1980)
Waterloo ON, Wilfrid Laurier University Press, 303 p. ISBN 0-88920-100-5.

Haner, F T and Ewing, John S Country Risk Assessment: theory and worldwide practice (1985)
New York, Praeger Publishers, 336 p. ISBN 0-275-90208-0.

Hanington, Edda Migraine (1974)
London, Priory Press Publications.

Hankins, W David and Puett, David (Eds) Hormones, Call Biology and Cancer: perspectives and potentials (1988)
New York, Alan R Liss, 316 p. Progress in Clinical and Biological Research Ser: Vol 262. ISBN 0-8451-5112-6.

Hanmer, Jalna, et al Women, Policing and Male Violence: an internation perspective (1988)
New York, Routledge Chapman and Hall, 240 p. ISBN 0-415-00692-9.

Hanscombe, Gillian E and Forster, Jackie Rocking the Cradle: lesbian mothers, a challenge in family living (1982)
Boston MA, Alyson Publications,, 160 p. ISBN 0-932870-17-1.

Hansen, Art and Oliver-Smith, Anthony Involuntary Migration and Resettlement: the problems and responses of dislocated people (1982)
Boulder CO, Westview Press. ISBN 0-89158-976-7.

Hanshaw, J B; Plowright, W and Weiss, K E Cytomegloviruses (1968)
New York, Springer Publishing, 131 p. ISBN 0-387-80891-4.

Hanson, C William Toxic Emergencies (1984)
New York, Churchill Livingstone, 336 p. ISBN 0-443-08196-4.

Haque, Serajul and Mohammed, Nur Poverty, Inequality and Income Distribution: a select bibliography (1977)
Inst Development Studies.

Hardert, Ronald A; Gordon, Leonard and Laner, Mary R Confronting Social Problems (1984)
Saint Paul MN, West Publishing Company, 489 p. illus. ISBN 0-314-78013-0.

Hardin, Garrett J Naked Emperors: essays of a taboo-stalker (1982)
Los Altos CA, Kaufmann William, 300 p. ISBN 0-86576-033-0.

Hardine, Rosetta R Psychology of Hypertension: medical analysis index with research (1985)
Annandale VA, ABBE Publishers Association of Washington, 150 p. ISBN 0-88164-340-8.

Harding, Geoffrey Opiate Addiction, Morality and Medicine: from moral illness to pathological disease (1987)
New York, Saint Martin's Press, 224 p. ISBN 0-312-01201-2.

Hardison, O B Disappearing Through the Skylight: culture and technology in the twentieth century
New York, Viking, 389 p.

Hardoy, Jorge E and Satterthwaite, David Squatter Citizen: life in the urban Third World (1989)
London, Earthscan Publications.

Hardy, G Doom of the Welfare Society (1975)
Carnegie VIC, Reform Publishing, 100 p.

Hardy, Phil The Encyclopedia of Horror Movies (1987)
New York, Harper and Row, 440 p. ISBN 0-06-055050-3.

Hardy, Russell W Lumbar Disc Disease (1982)
New York, Raven Press, 344 p. ISBN 0-89004-616-6.

Hare, F K Climatic Variations, Drought and Desertification (1985)
Geneva, WHO, 35 p. ISBN 92-63-10653-3.

Hare, J W (Ed) Diabetes Complicating Pregnancy: the Joslin Clinic method (1989)
New York, Alan R Liss, 206 p. ISBN 0-8451-4268-2.

Harkavy, Robert E Bases Abroad: the global foreign military presense (1989)
Oxford, Stockholm International Peace Research Institute, 418 p. illus. SIPRI Publication Series. ISBN 0-19-829131-0.

Harney, David M Medical Malpractice (1987)
Charlottesville VA, Michie Company, 700 p. ISBN 0-87473-279-4.

Harrington, Michael The Politics of God's Funeral: the spiritual crisis of Western civilization (1985)
New York, Penguin, 320 p. Penguin Nonfiction Series. ISBN 0-14-007689-1.

Harris, John E and Morris, Peter E Everyday Memory and Action and Absent Mindedness (1984)
San Diego CA, Academic Press. ISBN 0-12-327640-3.

Harris, M L Cultural Materialism (1979)
New York, Random House.

Harris, Stuart The Permafrost Environment (1985)
London, Routledge, 288 p. ISBN 0-7099-3713-X.

Harrison, Deborah Limits of Liberalism (1981)
Montreal, Black Rose Books. ISBN 0-919619-21-5.

Hart, David M Banditry in Islam: studies in Morocco, Algeria and the Pakistan North West Frontier (1987)
Boulder CP, Lynne Rennier Pubs, 80 p. ISBN 0-906559-22-7.

Hart, Julian T Hypertension (1987)
New York, Churchill Livingstone, 296 p. ISBN 0-443-01665-8.

Hart, William C, et al Lightning and Lightning Protection (1979)
Gainesville VA, White Consultants, 181 p. ISBN 0-932263-14-3.

Hartjen, Clayton A A Possible Trouble: an analysis of social problems (1977)
New York, Praeger.

Hartman, Sven S (Ed) Syncretism: symposium on cultural contact, meeting of religions, syncretism held at Abo 1966 (1969)
Stockholm, Almqvist and Wiksell, 294 p.

Hartman, Sven S; Böcher, Otto and Benrath, G A Antichrist
In: Theologische Realenzyklopädie (1978, Berlin).

Hartmann, Ernest The Sleeping Pill (1978)
New Haven CT, Yale University Press. ISBN 0-300-02248-4.

Hartshorne, Hugh and May, Mark Studies in the Nature of Character: studies in deceit (1975)
Salem NH, Ayer Company Publishers, 440 p. Classics in Child Development Serv: Vol 1. ISBN 0-405-06465-9.

Harwell, M A Nuclear Winter: the human and environmental consequences of nuclear war (1984)
Berlin, Springer-Verlag, 179 p. ISBN 3-540-96093-7.

Harwell, M A Nuclear Winter (1984)
New York, Springer Publishing, 179 p. ISBN 0-387-96093-7.

Harwell, Mark A and Hutchinson, Thomas C The Environmental Consequences of Nuclear War (SCOPE 28): ecological, agricultural, and human effects (1986)
New York, John Wiley and Sons. Vol. 2 ISBN 0-471-90898-3.

Hasluck, A Unwilling emigrants (1979)
Melbourne VIC, Oxford University Press, 175 p. ISBN 0-19-550574-3.

Hasselman, D P and Heller, R A Thermal Stresses in Severe Environments (1980)
New York, Plenum Publishing, 750 p. ISBN 0-306-40544-X.

Hastings, D W Impotence and Frigidity (1963)
Boston MA, Brown Little.

Hastings, Robert C (Ed) Leprosy (1986)
New York, Churchill Livingstone, 331 p. illus. Medicine in the Tropics Ser. ISBN 0-443-02893-1.

Hatano S, et al (Eds) Hypertension and Stroke Control in the Community: proceedings, WHO Meeting Tokyo, Mar 11-13, 1974 (1976)
Geneva, WHO. illus. ISBN 92-4-156052-5.

Hatfield, Agnes B and Lefley, Harriet P Families of the Mentally Ill: coping and adaptation (1987)
New York, Guilford Press, 340 p. ISBN 0-89862-683-8.

Hauerwas, Stanley Responsibility for Devalued Persons: ethical interaction between society (1982)
Springfield IL, Thomas Charles C, 122 p. ISBN 0-398-04705-7.

Haug, Wolfang F; Bock, Robert and Hall, Stuart Critique of Commodity Aesthetics (1986)
Minneapolis MN, University of Minnesota Press, 192 p. ISBN 0-8166-1531-4.

Haugrud, Wilderness Preservation (1985)
Stanford CA, Stanford Environmental Law Society, 123 p. ISBN 0-318-04413-7.

Hauk, Warren C (Ed) Motivating People to Work (1984)
Norcross GA, Institute of Industrial Engineers. ISBN 0-89806-057-5.

Hausen, Björn M Woods Injurious to Human Health: a manual (1981)
Berlin, Walter de Gruyter, 189 p. ISBN 3-11-008485-6.

Hausfater, Glenn and Hrdy, Sarah B (Eds) Infanticide: comparative and evolutionary perspectives (1984)
Berlin, Aldine de Gruyter, 630 p. Biological Foundations of Human Behavior Ser. ISBN 0-202-02022-3.

Haworth, Lawrence Decadence and Objectivity: ideals for work in the post-consumer society (1977)
Toronto ON, University of Toronto Press. ISBN 0-8020-6398-5.

Hay, Donald A and Vickers, John S The Economics of Market Dominance (1987)
New York, Basil Blackwell, 208 p. ISBN 0-631-14784-5.

Hayes, M Horace Veterinary Notes for Horse Owners: an illustrated manual (1988)
New York, Prentice Hall, 688 p. ISBN 0-13-941956-X.

Hayter, T The Creation of World Poverty: an alternative view to the Brandt Report (1981)
London, Pluto Press.

Hayward, J L Hormones and Human Breast Cancer: an account of fifteen years study (1970)
Berlin, Springer-Verlag. illus. Recent Results in Cancer Research: Vol 24. ISBN 0-387-04989-4.

Hazardous Materials Control Research Institute Management of Hazardous Wastes and Environmental Emergencies (1985)
Silver Spring MD, Hazardous Materials Control Research Institute. illus. ISBN 0-318-02812-3.

Hearse, David J et al (Eds) Life-Threatening Arrhythmias during Ishemia and Infarction (1987)
New York, Raven Press, 256 p. illus. ISBN 0-88167-266-1.

Heath, Donald (Ed) Aspects of Hypoxia (1986)
Liverpool, Liverpool University Press, 264 p. ISBN 0-85323-295-4.

Heathcote, J G and Hibbert, J R Aflatoxins: chemical and biological aspects (1978)
Amsterdam, Elsevier Science Publishing, 212 p. Developments in Food Science Ser: Vol 1. ISBN 0-444-41686-2.

Heathcote, R L Perception of Desertification (1980)
Lanham MD, UNIPUB, 134 p. ISBN 92-808-0190-2.

Hecker, Justus F Dancing Mania of the Middle Ages (1970)
New York, B Franklin. repr of 1837. Research and Source Works: No 540. ISBN 0-8337-1637-9.

Heckscher, E Mercantilism (1955)
London, Allen and Unwin. 2 Vols.

Hefner, Philip and Schroeder, W Widick (Eds) Belonging and Alienation: religious foundations for the human future (1976)
Chicago IL, Centre for the Scientific Study of Religion, 200 p. Studies in religion and society. ISBN 0-913348-07-4.

Heginbotham, Chris The Rights of Mentally Ill People (1987)
Cambridge MA, Cultural Survival, 12 p.
ISBN 0–946690–49–9.

Heidenheimer, Arnold J et al (Eds) Political Corruption: a handbook (1988)
New Brunswick NJ, Transaction Books, 1010 p. 2nd rev ed.
ISBN 0–88738–163–4.

Heidensohn, Frances Women and Crime (1985)
Basingstoke, Macmillan, 228 p. ISBN 0–333–36216–0.

Heilbrunn, Otto The Soviet Secret Services (1981)
Westport CT, Greenwood Press, 216 p.
ISBN 0–313–22892–2.

Heimann, H (Ed) Alcohol Abuse in Indians of North and South America: cultural difference in the conflict with Western civilization (1978)
New York, Karger S AG, 72 p. ISBN 3–8055–3043–9.

Heinze, Ruth–Inge Trance and Healing in Southeast Asia Today (1988)
Bangkok, White Lotus, 400 p. ISBN 0–9592092–0–4.

Helleiner, Gerald K International Economic Disorder: essays in North–South relations (1981)
Toronto ON, University of Toronto Press.
ISBN 0–8020–6534–1.

Heller, Kurt A, et al Identifying and Nurturing the Gifted: an international perspective (1986)
Lewiston NY, Hogrefe International, 250 p.
ISBN 0–920887–11–2.

Helpman, Elhanan and Leiderman, Leonardo Stabilization in High Inflation Countries: analytical foundations and recent experience (1987)
Tel Aviv, David Horowitz Institute for the Research of Developing Countries.

Hemert, P A van; Ramhorst, J D van and Regamey, H (Eds) International Symposium on Pertussis: proceedings of the 25th symposium organized by the Permanent Section of Microbiological Standardization, Bilthoven, 1969 (1970)
Basel, S Karger AG, 294 p. ISBN 3–8055–0636–8.

Hemminki, Kari, et al (Eds) Occupational Hazards and Reproduction (1985)
New York, Hemisphere Publishing, 333 p. illus.

Henderson, M Allen How Con Games Work (1986)
Secaucus NJ, Stuart Lyle, 256 p. ISBN 0–8065–1014–5.

Hendre, Sudhir Laxman Poverty, Corruption and Inflation (1974)
Bombay, Supraja Prakashan, 224 p.

Henly, Arthur Phobias: the crippling fears (1988)
New York, Avon Books, 224 p. ISBN 0–380–70659–8.

Hennelly, Alfred and Langan, John Human Rights in the Americas: the struggle for consensus (1982)
Washington DC, Georgetown University Press, 304 p.
ISBN 0–87840–401–5.

Hennssey, T C, et al Stress Physiology and Forest Productivity (1986)
Norwell MA, Kluwer Academic Publishers.
ISBN 90–247–3359–6.

Henschen, Folke The History and Geography of Disease (1966)
New York, Delacorte Press.

Henslin, James M and Light, Donald Social Problems (1983)
New York, McGraw Hill Book Company, 656 p.
ISBN 0–07–037836–3.

Heny, F (Ed) Ambiguities in Intentional Contexts (1980)
Dordrecht, Kluwer Academic Publishers Group, 285 p.
ISBN 90–277–1167–4.

Herbert, Carrie M H Talking in Silence: the sexual harassment of schoolgirls (1989)
Hants, Taylor and Francis, 208 p. ISBN 1–85000–585–0.

Herman, Edward S Real Terror Network: terrorism in fact and propaganda (1984)
250 p. ISBN 0–920057–25–X.

Herman, Judith Father–Daughter Incest (1982)
Cambridge MA, Harvard University Press, 296 p.
ISBN 0–674–29506–4.

Hermann, A H Conflicts of National Laws with International Business Activity: issues of extraterritoriality (1982)
Toronto, CD Howe Institute, 96 p. ISBN 0–902594–41–9.

Hermann, Charles F and Coate, Roger A Substantive Problem Areas
In: *Linda P Brady; Patrick Callahan and Margaret G Herman (Eds) Describing Foreign Policy Behavior* 1982, Berverly Hills CA, Sage, pp. 7–114.

Herner, Torsten Challenge of Schizophrenia (1982)
Stockholm, Almqvist and Wiksell, 170 p.

Hershman, Jablow and Lieb, Julian The Key to Genius: manic–depression and the creative life (1988)
Buffalo NY, Prometheus Books, 220 p.
ISBN 0–87975–437–0.

Herzog, Peter E Harmonization of Laws in the European Communities: products liability, conflict of laws, and (1983)
Charlottesville VA, University Press of Virginia, 164 p.
ISBN 0–8139–0985–6.

Hess, Albert G and Clement, Priscilla F (Eds) History of Juvenile Delinquency: a collection of essays (1988)
Aalen, Scientia Verlag und Antiquariat Kurt Schilling. 2 vols.

Heston, Stanley M Bed Wetting: cause and correction (1987)
Dexter OR, Medical Information International, 146 p.
ISBN 0–9620473–0–9.

Hethcote, H W and Yorke, J A Gonorrhea Transmission Dynamics and Control (1984)
Berlin, Springer–Verlag, 105 p. Lecture Notes in Biomathematics Ser: Vol 56. ISBN 0–387–13870–6.

Hetzel, B S and Smith, R M (Eds) Fetal Brain Disorders: recent approaches to the problem of mental deficiency (1981)
Amsterdam, Elsevier Science Publishing, 490 p.
ISBN 0–444–80321–1.

Heurlin, Bertel Threat as a Concept in International Politics (1977)
Copenhagen, Forsvarets Oplysnings– og Velfaerdstjeneste, 32 p.

Hewitt, Patricia and Rose–Neil, Wendy Your Second Baby
Winchester MA, Unwin Hyman

Heyzer, Noeleen (Ed) Missing Women: development planning in Asia and the Pacific (1985)
Kuala Lumpur, Asian and Pacific Development Centre, 452 p.
ISBN 967–99954–3–7.

Hibino, Susumu; Takaku, Fumimaro and Shahidi, N T (Eds) Aplastic Anemia: proceedings of the first international symposium on aplastic anemia (1978)
Tokyo, University of Tokyo Press, 450 p.

Hickin, N Pest Animals in Buildings: a world review (1986)
New York, John Wiley and Sons, 398 p.
ISBN 0–470–20636–5.

Hickman, Bert G (Ed) Global International Economic Models: selected papers from an IIASA conference (1983)
Amsterdam, Elsevier Science Publishing, 316 p. 1st ed.
ISBN 0–444–86718–X.

Higgins, Ronald Plotting Peace: the owl's reply to hawks and doves (1990)
Riverside NJ, Macmillan Publishing. ISBN 0–08–033618–3.

Hightower, J Howard and Baker, Eugene A New Type of Phobia is Isolated: famanotophobia (1987)
New York, Vantage Press, 93 p. ISBN 0–533–06280–2.

Higuchi PCB Poisoning and Pollution (1976)
San Diego CA, Academic Press. ISBN 0–12–347850–2.

Higuchi, Kentaro (Ed) P C B Poisoning and Pollution (1976)
Amsterdam, Elsevier Science Publishing, 192 p.
ISBN 4–06–129884–4.

Hilgard, Ernest R The Experience of Hypnosis (1968)
San Diego CA, Harcourt Brace Jovanovich, 353 p.

Hill, D S Agricultural Insect Pests of the Tropics and their Control (1987)
Cambridge MA, Cambridge University Press, 758 p. 2nd ed.
ISBN 0–521–28867–3.

Hill, D S Agricultural Insect Pests of Temperate Regions and Their Control (1987)
New York, Cambridge University Press, 600 p.
ISBN 0–521–24013–1.

Hill, O Psychosomatic Medicine (1970)
London, Butterworth.

Hillary, Irene B and Hennessen, W (Eds) Enteric Infections in Man and Animals: standardization of immunological procedures (1983)
Basel, S Karger AG, 356 p. ISBN 3–8055–3714–X.

Hindmarsh, Albert E Force in Peace: force short of war in international relations (1973)
New York, Associated Faculty Press, 264 p.
ISBN 0–8046–1757–0.

Hippchen, Leonard J and Yim, Yong S Terrorism, International Crime, and Arms Control (1982)
Springfield IL, Thomas Charles C, 306 p.
ISBN 0–398–04626–3.

Hirono, I Naturally Occurring Carcinogens of Plant Origin: toxicology, pathology and biochemistry (1987)
New York, Elsevier Science Publishing, 240 p.
ISBN 0–444–98972–2.

Hirsch, E W Sexual Fear (1950)
New York.

Hirsch, F Social Limits to Growth (1977)
London, Routledge and Kegan Paul.

Hiruki, Chuji (Ed) Tree Mycoplasma and Mycoplasma Diseases (1987)
Edmonton, University of Alberta Press, 220 p.
ISBN 0–88864–126–5.

Hobson, John A The Origins, Growth and Potential Survival of Modern Capitalism (1985)
Albuquerque NM, Institute for Economic and Financial Research, 197 p.

Hobson, John A Cartels, Trusts and the Economic Power of Bankers, Financiers and Money–Moguls (1985)
Albuquerque NM, Institute for Economic and Financial Research, 135 p. ISBN 0–86654–145–4.

Hochschild, Arlie R The Managed Heart: commercialization of human feeling (1983)
Berkeley CA, University of California Press, 307 p.
ISBN 0–520–04800–8.

Hochschild, Jennifer L The New American Dilemma: liberal democracy and school desegregation (1984)
New Haven CT, Yale University Press, 279 p.
ISBN 0–300–03113–0.

Hodge, John L et al Cultural Bases of Racism and Group Oppression: an examination of traditional "western"
Chestnut Hill MA, Two Riders Press, 275 p.
ISBN 0–915860–02–3.

Hodgetts, Richard M; Luthans, Fred and Thompson, Kenneth R Social Issues in Business: strategic and public policy perspectives (1987)
New York, Macmillan Publishing Group, 694 p.
ISBN 0–02–372900–7.

Hodgkinson, Liz Bodyshock: the truth about changing sex (1987)
Harrap Ltd England/State Mutual Book.
ISBN 0–86287–317–7.

Hodson, John D Ethics of Legal Coercion (1983)
Dordrecht, Kluwer Academic Publishers Group, 168 p.
ISBN 90–277–1494–0.

Hoff, Gerald L and Davis, John W Noninfectious Diseases of Wildlife (1982)
Ames IA, Iowa State University Press, 174 p.
ISBN 0–8138–0990–8.

Hoff–Jørgensen, R and Pétursson, G (Eds) Maedi–Visna and Related Diseases (1989)
Dordrecht, Kluwer Academic Publishers Group, 208 p.
ISBN 0–7923–0481–0.

Hoffer, Abram and Osmond, H Hallucinogens (1967)
San Diego CA, Academic Press. ISBN 0–12–351850–4.

Hoffman, Donald B and Warren, Kenneth S Schistosomiasis IV: condensations of the selected literature (1978)
New York, Hemisphere Publishing, 538 p. 2 vols.
ISBN 0–89116–164–3.

Hoffman, Lance J Computers and Privacy in the Next Decade (1980)
San Diego CA, Academic Press. ISBN 0–12–352060–6.

Hofstad, M S et al Diseases of Poultry (1984)
Ames IA, Iowa State University Press, 832 p.
ISBN 0–8138–0430–2.

Hofstede, Geert Cultures Consequences: international differences in work–related values (1980)
London, Sage.

Hoksbergen, R A C Adoption in Worldwide Perspective: a Review of programs, policies and legislation in 14 countries (1986)
Lisse, Swets en Zeitlinger, 200 p. ISBN 90–265–0738–0.

Holden, Peter; Pfäfflin, Georg Fredrich and Horlemann, Jügen (Eds) Third World People and Tourism: approaches to a dialogue (1986)
Bangkok, The Ecumenical Coalition on Third World Tourism, 156 p.

HÖll, Otmar Austria's Technological Dependence: basic dimensions and current trends (1980)
Vienna, Braumüller Universitäts–Verlachsbuchhandlung.
ISBN 3–7003–0460–9.

Holland, Joe The Spiritual Crisis of Modern Culture (1983)
Washington DC, Center of Concern, 13 p.

Hollender, L F, et al Acute Pancreatitis: an interdisciplinary synopsis (1983)
Munich, Urban and Schwarzenberg, 175 p.
ISBN 3–541–70841–7.

Holliday, Paul Fungus Diseases of Tropical Crops (1980)
New York, Cambridge University Press, 500 p.
ISBN 0–521–22529–9.

Hollingsworth, J Rogers and Hanneman, Robert Centralization and Power in Social Service Delivery Systems (1984)
Dordrecht, Kluwer Academic Publishers Group, 224 p.
ISBN 0–89838–142–8.

Hollon, W Eugene Frontier Violence: another look (1974)
New York, Oxford University Press. ISBN 0–19–502098–7.

Holmes, John R (Ed) Managing Solid Wastes in Developing Countries (1984)
New York, Wiley Interscience, 320 p.
ISBN 0–471–90234–9.

Holmes, King K, et al Sexually Transmitted Diseases (1984)
New York, McGraw Hill Book Company, 1108 p.
ISBN 0–07–029675–8.

Holton, Inherited Metabolic Disorders (1987)
New York, Churchill Livingstone. ISBN 0–443–03195–9.

Holy, M Erosion and Environment (1980)
New York, Pergamon Books, 266 p. illus. Environmental Sciences and Applications: Vol 9. ISBN 0–08–024466–1.

Holzworth, Jean Diseases of the Cat: medicine and surgery (1987)
Philadelphia PA, WB Saunders, 960 p. illus. Vol 1.
ISBN 0–7216–4763–4.

Homburger, F Safety Evaluation and Regulation of Chemicals (1983)
New York, Karger S AG, 294 p. No. 1
ISBN 3–8055–3578–3.

Honey Does Accent Matter?: the pygmalion factor (1989)
Winchester MA, Faber and Faber.

Hong, Evelyne See the Third World While It Lasts: the social and environmental impact of tourism with special reference to Malaysia
Pinang, Consumers' Association of Pinang, 98 p.
ISBN 967–9950–10–7.

Honig, Alice S Risk Factors in Infancy (1986)
New York, Gordon and Breach, 186 p.
ISBN 0–677–21420–0.

Honrubia, Vicente and Brazier, Mary (Eds) Nystagmus and Vertigo: clinical approach to the patient with dizziness (1982)
San Diego CA, Academic Press, 320 p.
ISBN 0–12–355080–7. UCLA Forum in Medical Sciences Ser: No 24.

Hood, D E and Tarrant, P V (Eds) Problem of Dark Cutting in Beef: seminar in the EEC Programme of Coordination of Research on Animal Welfare held in Brussels, Belgium, 1980 (1981)
Dordrecht, Kluwer Academic Publishers Group, 504 p.
ISBN 90–247–2522–4.

Hopkins, Anthony (Ed) Epilepsy (1987)
New York, Demos Publications, 650 p. illus.
ISBN 0–939957–04–3.

Hopkins, R H Forecasting Techniques of Clear Air Turbulence Including that Associated with Mountain Waves (1977)
Geneva, WHO, 31 p. ISBN 92–63–10482–4.

Hopkins, Sidney J Suicide: medical subject analysis (1987)
Annandale VA, ABBE Publishers Association of Washington, 160 p. ISBN 0–88164–588–5.

Hörburger, H Job–Sharing: probleme und möglichkeiten (1986)
Luxembourg, CEE, 117 p. ISBN 92–825–5863–0.

Horobin, Gordon and May, David (Eds) Living with Handicap: transitions in the lives of people with mental handicaps (1987)
St Martin, 140 p. Research Highlights in Social Worl Ser.

Horosz, William The Promise and Peril of Human Purpose (1970)
Saint Louis MO, Fireside Books, 350 p.
ISBN 0–87527–018–2.

Horowitz, Donald Ethnic Groups in Conflict (1985)
Berkeley CA, University of California Press.
ISBN 0-520-05385-0.

Horton, Paul B, et al The Sociology of Social Problems (1988)
Englewood Cliffs NJ, Prentice Hall, 480 p.
ISBN 0-13-821687-8.

Horton, Paul B; Larson, Richard F and Leslie, Gerald R The Sociology of Social Problems (1988)
Englewood Cliffs NJ, Prentice Hall, 480 p. illus. 9th ed.
ISBN 0-13-821687-8.

Horwitz, Orville, et al (Eds) Diseases of Blood Vessels (1985)
Philadelphia PA, Lea and Febiger, 429 p. illus.

Horzinck, Marian C Non-Arthropod Borne Togaviruses (1981)
San Diego CA, Academic Press, 216 p.
ISBN 0-12-356550-2.

Hotaling, Gerald T, et al (Eds) Coping with Family Violence: research and policy perspectives (1988)
London, Sage, 336 p. ISBN 0-8039-2722-3.

Hotaling, Gerald T, et al Family Abuse and Its Consequences: new directions in research (1988)
London, Sage, 336 p. ISBN 0-8039-2720-7.

Hotchin, J Persistent and Slow Virus Infections (1971)
Basel, S Karger AG, 212 p.
ISBN 3-8055-1176-0. Monographs in Virology: Vol 3.

Houdek, Frank G Protection of Cultural Property and Archaeological Resources: a comprehensive bibliography of law-related materials (1988)
New York, Oceana, 122 p. ISBN 0-379-20911-X.

Houlton, John E and Taylor, Polly Trauma Management in the Dog and Cat (1987)
Littleton MA, PSG Publishing, 172 p.
ISBN 0-7236-0696-X.

Houtman, J P W and Hamer, C J A van den (Eds) Physiological and Biochemical Aspects of Heavy Elements in Our Environment (1975)
Dordrecht, Kluwer Academic Publishers Group, 128 p.
ISBN 90-298-0700-8.

Howard, Michael Carlton The Impact of the International Mining Industry on Native Peoples (1988)
Geneva, UN, 258 p. illus. ISBN 0-908470-75-4.

Howe, Charles W Managing Renewable Natural Resources in Developing Countries (1982)
Boulder CO, Westview Press, 212 p.
ISBN 0-86531-313-X.

Howe, G Melvyn A World Geography of Human Diseases (1978)
San Diego CA, Academic Press. ISBN 0-12-357150-2.

Howe, Melvyn G (Ed) Global Geocancerology: a world geography of human cancers (1986)
New York, Churchill Livingstone, 350 p. illus.
ISBN 0-443-02765-X.

Howells, John G Modern Perspectives in the Psychiatry of the Neuroses (1988)
New York, Brunner/Mazel. ISBN 0-87630-515-X.

Hoxha, Enver Superpowers
Tirana, Book Distribution Enterprise, 678 p.

Hoyle, Fred, et al Viruses from Space (1986)
Wolfeboro NH, Longwood Publishing Group, 119 p.
ISBN 0-906449-93-6.

Hsu, Teng H Stress and Strain Data Handbook (1986)
Houston TX, Gulf Publishing, 350 p. ISBN 0-87201-159-3.

Hubbert, William T, et al Diseases Transmitted from Animals to Man (1975)
Springfield IL, Thomas Charles C, 1236 p.
ISBN 0-398-03056-1.

Hudson, J R Contagious Bovine Pleuropneumonia (1971)
Rome, FAO, 120 p.
FAO Agricultural Studies: No 86.

Hudson, Jame I and Pope, Harrison G (Eds) The Psychology of Bulimia
Cambridge MA, Cambridge University Press, 267 p.
ISBN 0-521-35412-9.

Hueck, Gerda and Finser, R The Problem of Lefthandedness (1986)
Spring Valley NY, Saint George Book Service.
ISBN 0-916786-83-8.

Hueper, Wilhelm C Occupational and Environmental Cancers of the Urinary System
New Haven CT, Yale University Press.
ISBN 0-300-01126-1.

Huggings, Martha K From Slavery to Vagrancy in Brazil: crime and social control in the Third World (1985)
New Brunswick NJ, Rutgers University Press, 190 p. Crime, Law and Deviance Ser. ISBN 0-8135-1044-9.

Huhner, M The Diagnosis and Treatment of Sexual Disorders in the Male and Female, Including Sterility and Impotence (1946)
Philadelphia PA, FA Davis.

Huismans, J W and Suess, M J (Eds) Management of Hazardous Waste: policy guidelines and code of practice (1983)
Geneva, WHO, 100 p. ISBN 92-890-1105-X.

Hukins, D W and Mulholland, R C (Eds) Back Pain: new methods for clinical investigation and assessment (1986)
Manchester Univ Pr/St Martin, 176 p.
ISBN 0-7190-2311-4.

Hula, Erich and Thompson, Kenneth W Nationalism and Internationalism: european and american perspectives (1984)
Lanham MD, University Press of America, 326 p.
ISBN 0-8191-3704-9.

Hulby, David and Chappell, Willard et al Risk Assessment of Wastewater Disinfection (1985)
Lancaster PA, Technomic Publishing, 175 p.
ISBN 0-87762-517-4.

Hulett, Louisa S Decade of Detente: shifting definitions and denouement (1982)
Washington DC, University Press of America, 200 p.

Huls, Mary E Partitions: a bibliography of recent periodical literature (1986)
Monticello IL, Vance Bibliographies, 6 p.
ISBN 0-89028-908-5.

Hume, Michael et al Venous Thrombosis and Pulmonary Embolism (1970)
Cambridge MA, Harvard University Press.
ISBN 0-674-93320-6.

Humphrey, James H (Ed) Stress in Childhood (1984)
New York, AMS Press. Studies in Modern Society: Political and Social Issues: No 17.

Humphreys, S C and King, H Mortality and Immortality: the anthropology and archaeology of death (1982)
San Diego CA, Academic Press. ISBN 0-12-361550-X.

Hurley, Patricia S Female Genital Diseases: subject analysis and research index with (1985)
Annandale VA, ABBE Publishers Association of Washington, 150 p. ISBN 0-88164-059-X.

Hursch, Carolyn J The Trouble with Rape (1977)
Chicago IL, Nelson-Hall, 144 p. ISBN 0-88229-323-0.

Hurst, G W Meteorology and the Colorado Potato Beetle (1975)
Geneva, WHO, 52 p. ISBN 92-63-10391-7.

Hurst, G W and Rumney, R P Protection of Plants Against Adverse Weather (1971)
Geneva, WMO, 64 p.

Hurt, R Douglas The Dust Bowl: an agricultural and social history (1981)
Chicago IL, Nelson-Hall, 240 p. ISBN 0-88229-541-1.

Husain, Syed A and Vandiver, Trish Suicide in Children and Adolescents (1984)
Great Neck NY, PMA Publishing, 302 p.
ISBN 0-89335-190-3.

Huskisson, E C and Hart, F Dudley Joint Disease (1987)
Littleton MA, PSG Publishing, 208 p. ISBN 0-7236-0571-8.

Hussain, S Jaffer Marriage Breakdown and Divorce Law Reform in Contemporary Society: a Comparative Study of USA, UK and India (1983)
New Delhi, Concept Publishing, 240 p.

Hutchings, M and Carver, M Man's Dominion: our violation of the animal world (1970)
London, Hart-Davis.

Hutchinson, T C and Meema, K M Effects of Atmospheric Pollutants on Forests, Wetlands, and Agricultural Ecosystems (1987)
New York, Springer Publishing, 652 p.
ISBN 0-387-16084-1.

Hutnik, Russell and Davis, Grant Ecology and Reclamation of Devastated Land (1973)
New York, Gordon and Breach, 1070 p. 2 Vols
ISBN 0-677-15580-8.

Hutson, D H and Roberts, T R (Eds) Insecticides (1986)
New York, John Wiley and Sons, 400 p. Progress in Pesticide Biochemistry Ser: 5. ISBN 0-471-90758-8.

Hutton, Gary W Welfare Fraud Investigation (1985)
Springfield IL, Thomas Charles C, 330 p.
ISBN 0-398-05140-2.

Hyman, Richard Strikes (1989)
Basingstoke, Macmillan, 200 p. 4th ed. 1st, 2nd and 3rd ed published by Fontana 1972, 1977, 1984.
ISBN 0-333-47360-4.

Ichinoe, Kihyoe Trophoblastic Diseases (1986)
New York, Igaku-Shoin Medical Publishers, 194 p.
ISBN 0-89640-122-7.

ICIHI Indigenous Peoples: a global quest for justice
Atlantic Highlands NJ, Humanities Press International, 208 p.
ISBN 0-86232-758-X.

ICIHI Refugees: the dynamics of displacement (1986)
Atlantic Highlands NJ, Humanities Press International, 180 p.
ISBN 0-86232-696-6.

Ickes, W (Ed) Compatible and Incompatible Relationships
Berlin, Springer-Verlag, 615 p. ISBN 3-540-96024-4.

ICRISAT Sorghum Diseases, a World review: proceedings of the International Workshop on Sorghum Diseases, 11-15 Dec 1978, ICRISAT, Hyderabad, India (1980)
Patancheru A P, ICRISAT, 478 p.

ICRISAT Proceedings of the Consultants' Group Discussion on the Resistance to Soilborne Diseases of Legumes, 8-11 Jan 1979, ICRISAT Center, India (1980)
Patancheru A P, ICRISAT, 176 p.

ICRISAT Proceedings of the Second International Workshop on Striga, 5-8 Oct 1981, Ouagadougou, Burkina Faso (1983)
Patancheru A P, ICRISAT, 136 p.

ICRISAT Groundnut Rust Disease: proceedings of a Discussion Group Meeting, 24-28 Sep 1984, ICRISAT Center, India (1987)
Patancheru A P, ICRISAT, 212 p.

ICRISAT Adaptation of Chickpea and Pigeonpea to Abiotic Stresses: proceedings of the Consultants' Workshop, 19-21 Dec 1984, ICRISAT Center, India (1987)
Patancheru A P, ICRISAT, 184 p.

ICRISAT Summary and Recommandation of the International Workshop on Aflatoxin Contamination of Groundnut, 6-9 Oct 1987, ICRISAT Center, India (1988)
Patancheru A P, ICRISAT, 44 p.

ICRISAT Coordination of Research on Peanut Stripe Virus: summary proceedings of the First Meeting to Coordinate Research on Peanut Stripe Virus Disease of Groundnut, 9-12 June 1987, Malang, Indonesia (1988)
Patancheru A P, ICRISAT, 32 p.

Ihde, D and Zaner, R M Interdisciplinary Phenomenlogy (1975)
Dordrecht, Kluwer Academic Publishers Group, 187 p. Selected Studies in Phenomenology and Existential Philosophy: 6. ISBN 90-247-1922-4.

Ikeda, Daisaku and Gage, Richard The Human Revolution (1976)
New York, John Weatherhill. Vol. 3 ISBN 0-8348-0118-3.

Ikwue, K Inconvenient Marriage (1982)
Ibadan, Olaiya Fagbamigbe Publishers, 303 p.

Illich, Ivan Medical Nemesis: the expropriation of health (1976)
New York, Pantheon Books.

ILO Economic and Social Effects of Multinational Enterprises in Export Processing Zones: a joint ILO/UNCTC project
Geneva, ILO. ISBN 92-2-106194-9.

ILO From Pyramid to Pillar: population change and social security in Europe
Geneva, ILO. ISBN 92-2-106497-2.

ILO Collective Bargaining Problems and Practices on Plantations and the Exercise of Trade Union Rights: report II, 7th Session of the Committee on Work and Plantations, Geneva, 1976
Geneva, ILO, 101 p. ISBN 92-2-101601-3.

ILO Conditions of Work in the Textile Industry, Including Problems Related to Organization of Work: report III: 10th Session of the Textiles Committee, Geneva, 1978
Geneva, ILO, 87 p. ISBN 92-2-101830-X.

ILO Maintenance of Rights in Social Security: report V, 69th Session of the International Labour Conference, Geneva, 1983
Geneva, ILO, 101 p. ISBN 92-2-103131-4.

ILO Ad Hoc Meeting on Civil Aviation, Geneva, 1977
Geneva, ILO, 108 p. Occupational Health and Safety in Civil Aviation: II. ISBN 92-2-101786-9.

ILO Problems Specific to Employees in Commerce and Offices: report II: 9th session of the Advisory Committee on Salaried Employees and Professional Workers, 1985
Geneva, ILO, 115 p. ISBN 92-2-103933-1.

ILO Social Security Protection for Seafarers Including Those Serving in Ships Flying Flags Other Than Those of Their Own Country: report III, Part 2, 74th (Maritime) Session of the International Labour Conference, Geneva, 1987
Geneva, ILO, 47 p.
ISBN 92-2-105790-9. Proceedings Series.

ILO Major Hazard Control: a practical manual
Geneva, ILO. ISBN 92-2-106432-8.

ILO Improvements of the Conditions of Life and Work of Peasants, Agricultural Workers and Other Comparable Groups
Geneva, ILO, 80 p. ISBN 92-2-101188.

ILO Prevention of Accidents Due to Electricity Underground in Coal Mines (1959)
Geneva, ILO.
ISBN 92-2-100148-2.

ILO Prevention of Accidents Due to Fire Underground in Coal Mines (1959)
Geneva, ILO.
ISBN 92-2-100149-0.

ILO Guide to the Prevention and Suppression of Dust in Mining, Tunelling and Quarrying (1965)
Geneva, ILO, 421 p. ISBN 92-2-100954-8.

ILO Discrimination in Employment and Occupations. Standards and Policy Statements Adopted Under the Auspices of the ILO (1967)
Geneva, ILO, 56 p. ISBN 92-2-100916-9.

ILO Guide to Safety in Agriculture (1969)
Geneva, ILO, 247 p. ISBN 92-2-100030-3.

ILO Pneumoconiosis Conference (1971)
Geneva, ILO, 808 p. illus. ISBN 92-2-000463-1.

ILO Automation in Developing Countries (1974)
Geneva, ILO, 246 p. Round-table discussion on the manpower problems assioted with the introduction of automation and advanced technology in developing countries, Geneva, 1-3 July 1970. ISBN 92-2-100158-X.

ILO Conditions of Work and Life of Migrant and Seasonal Workers Employed in Hotels, Restaurants and Similar Establishments. Report II. 2nd Tripartite Technical Meeting fro Hotels, Restaurants and Similar Establishments (1974)
Geneva, ILO, 54 p. ISBN 92-2-101211-5.

ILO Control and Prevention of Occupational Hazards Caused by Cancerogenic Substances and Agents (1974)
Geneva, ILO, 45 p. ISBN 92-2-101104-6.

ILO Social Problems of Contract, Sub-contract and Casual Labour in the Petroleum Industry (1974)
Geneva, ILO, 95 p.
ISBN 92-2-101103-8.

ILO Problems and Opportunities of Employment and Re-employment of Older Workers in Commerce and Offices (1974)
Geneva, ILO, 43 p. ISBN 92-2-101142-9.

ILO Prevention of Accidents Due to Explosions Underground in Coal Mines (1974)
Geneva, ILO.
ISBN 92-2-101062-7.

ILO Equality of Opportunity and Treatment for Women Workers (1975)
Geneva, ILO, 124 p. ISBN 92-2-101234-4.

ILO Minimum Wage-fixing and Economic Development (1975)
Geneva, ILO, 217 p. ISBN 92-2-100003-6.

ILO Substandard Vessels, Particularly those Registered under Flags of Convenience (1976)
Geneva, ILO. 2 Vols.

ILO Conditions of Work and Employment of Public Service Personnel of Local, Regional or Provincial Authorities. Report II. 2nd Session of the Joint Committee on the Public Service

(1976)
Geneva, ILO, 61 p. ISBN 92–2–101410–X.

ILO Continuity of Employment of Seafarers (1976)
Geneva, ILO, 29 p. ISBN 92–2–105904–9.

ILO Employment and Conditions of Work and Life of Nursing Personnel (1976)
Geneva, ILO, 108 p. Prelimanary report and questionnaire. ISBN 92–2–101368–5.

ILO Radiation Protection in Mining and Milling of Uranium and Thorium (1976)
Geneva, ILO, 346 p. ISBN 92–2–101504–1.

ILO Protection Against Anti–Union Discrimination (1976)
Geneva, ILO, 123 p. ISBN 92–2–101348–0.

ILO Migrant Workers – Occupational Safety and Health: joint ILO/WHO Committee on Occupational Health, Seventh Session, Geneva, 5–11 August 1975 (1977)
Geneva, ILO, 82 p.
ISBN 92–2–101643–9. Occupational Safety and Health Series: No 34.

ILO The Right to Organize (1977)
Geneva, ILO.
ISBN 92–2101789–3 (Hard cover); ISBN 92–2–101790–7 (Limp cover).

ILO Improvement and Harmonisation of Social Security Systems in Africa (1977)
Geneva, ILO, 67 p. ISBN 92–2–101798–2.

ILO Employment, Growth and Basic Needs. A One–world Problem (1978)
Geneva, ILO, 211 p. ISBN 92–2100903–3.

ILO Employment of Women with Families Responsabilities: summary of Reports on Recommendation N 123 (article 19 of the Constitution) (1978)
Geneva, ILO, 89 p. ISBN 92–2–101748–6.

ILO Labour and Social Problems Arising out of Seasonal Fluctuations of the Food Products and Drink Industries (1978)
Geneva, ILO, 51 p. ISBN 92–2–101879–2.

ILO Night Work. Its Effects on the Health and Welfare of the Worker (1978)
Geneva, ILO, 82 p. ISBN 92–2–101676–5.

ILO Guide to Health and Hygiene in Agricultural Work (1979)
Geneva, ILO, 309 p. ISBN 92–2–101974–8.

ILO Conditions of Work, Vocational Training and Employment of Women (1979)
Geneva, ILO, 92 p ISBN 92–2–102011–8.

ILO Pensions and Inflation: an international discussion (1980)
Geneva, ILO, 136 p. 2nd impression.
ISBN 92–2–101684–6.

ILO Fifth International Report on the Prevention and Suppression of Dust in Mining, Tunelling and Quarrying, 1968–1972 (1980)
Geneva, ILO, 106 p. ISBN 92–2–101899–7.

ILO Pensions in Inflation (1980)
Geneva, ILO, 136 p. ISBN 92–2–101684–6.

ILO The World of Work and the Protection of the Child: African symposium on the world of work and the protection of the child (Yaoundé, December 1979) (1981)
Geneva, ILO, 143 p. ISBN 92–9014–170–0.

ILO Problems of Women Non–Manual Workers: work organization, vocational training, equality of treatment at the work–place, job opportunities: Report III, 8th Session of the Advisory Committee on Salaried Employees and Professional Workers, Geneva, 1981 (1981)
Geneva, ILO, 92 p. ISBN 92–2–102556–X.

ILO Equal opportunities and Equal Treatment for Men and Women Workers. Workers with family responsabilities (1981)
Geneva, ILO, 88 p. ISBN 92–2–102406–7.

ILO Industries in Trouble (1981)
Geneva, ILO, 178 p.
ISBN 92–2–102678–7 (Hard Cover); ISBN 92–2102679–5 (Limp Cover).

ILO Vocational Rehabilitation of Leprosy Patients Report on the ILO–DANIDA Asian Seminar Bombay, India (26 October–6 November, 1981) (1982)
Washington DC, ILO, 126 p. ISBN 92–2–103047–4.

ILO Conditions of Work and Employment in Water, Gas and Electricity Supply Services (1982)
Geneva, ILO, 62 p. ISBN 92–2–102912–3.

ILO Occupational Injuries (1982)
Geneva, ILO. ISBN 92–2–102949–2.

ILO Labour Force, Employment, Unemployment and Underemployment (1982)
Geneva, ILO, 84 p. ISBN 92–2–102948–4.

ILO Labour Force Participation and Development (1982)
Geneva, ILO, 267 p. ISBN 92–2–102762–7.

ILO Prevention of Occupational Cancer – International Symposium (1982)
Geneva, ILO, 658 p. ISBN 92–2–002907–3.

ILO The Adaptation of the Training of Managerial Staff and Employees to Structural and Technological Changes in Hotels, Restaurants and Similar Establishments (1983)
Geneva, ILO, 93 p. ISBN 92–2–105060–2.

ILO Bibliography of Periodicals on the Quality of Working Life (1983)
Geneva, ILO, 88 p. ISBN 92–2–103475–5.

ILO The Improvement of Working Conditions and of the Working Environment in the Construction Industry (1983)
Geneva, ILO, 96 p. ISBN 92–2–13051–2. **Pro** #E6841.

ILO Position of African Countries Regarding the Ratification and Implementation of International Labour Standards (1983)
Geneva, ILO, 46 p. ISBN 92–2–103527–1.

ILO Social Problems and Employment in Hotels, Restaurants and Similar Establishments in Developing Countries: report ii–third tripartite technical meeting (1983)
Washington DC, ILO, 107 p. ISBN 92–2–103379–1.

ILO Introduction of Social Security (1984)
Geneva, ILO, 184 p. ISBN 92–2–105623–6.

ILO Occupational Hazards from Non–Ionising Electromagnetic Radiation (1984)
Geneva, ILO, 133 p.
ISBN 92–2–103540–9. Occupational Safety and Health Series: No 53.

ILO Adaptation of New Jobs and the Employment of the Disabled (1984)
Geneva, ILO, 112 p. ISBN 92–2–103826–2.

ILO Automation, Work Organisation and Occupational Stress (1984)
Geneva, ILO, 188 p. Report, conclusions and working papers of Meetings of Experton Automation, Work Organisation, Work Intensity and Occupational Stress, Geneva, 28 November–7 December 1983. ISBN 92–2–103866–1.

ILO Measures to Overcome Obstacles to the Observance in the Construction Industry of ILO standards (1984)
Geneva, ILO, 89 p. ISBN 92–2–105710–0.

ILO Financing Social Security The Options (1984)
Geneva, ILO, 144 p. ISBN 92–2–103634–0.

ILO Employment of Disabled Persons. Manual on Selective Placement (1984)
Geneva, ILO, 119 p. ISBN 92–2–103709–6.

ILO Group–based Savings and Credit for the Rural Poor (1984)
Geneva, ILO, 126 p. ISBN 92–2–103891–2.

ILO Improving Working Conditions and Environment: an international programme (PIACT) (1984)
Geneva, ILO, 129 p. ISBN 92–2–103804–1.

ILO Protection of Workers Against Noise and Vibration in the Working Environment (1984)
Geneva, ILO, 74 p. ISBN 92–2–101709–5.

ILO Collective Bargaining: a response to the recession in industrialised (1984)
Washington DC, ILO, 275 p. ISBN 92–2–103628–6.

ILO Safety in the Use of Asbestos: an ILO code of practice (1984)
Washington DC, ILO, 116 p. ISBN 92–2–103872–6.

ILO Conciliation in Industrial Disputes: a practical guide (1985)
Geneva, ILO, 133 p. 5th impression. ISBN 92–2–101007–4.

ILO Access to decision–makers in multinational and multi–plant enterprises (1985)
Geneva, ILO, 69 p.
ISBN 92–2–105120–X.

ILO Blending of New Technologies with Traditional Activities (1985)
Geneva, ILO, 24 p.
ISBN 92–2–105065–3.

ILO The Achievement of Full Employment in the Wood Industries (1985)
Geneva, ILO, 82 p. ISBN 92–2–105060–2.

ILO Basic Principles of Vocational Rehabilitation of the Disabled (1985)
Geneva, ILO, 59 p. 3rd ed. ISBN 92–2–105130–7.

ILO Grievance Arbitration. A Practical Guide (1985)
Geneva, ILO, 71 p. ISBN 92–2–101722–2.

ILO The Cost of Social Security (1985)
Geneva, ILO, 113 p. ISBN 92–2–003873–0.

ILO Employment and Poverty in a Troubled World (1985)
Geneva, ILO, 55 p.
ISBN 92–2–100528–3.

ILO Employment and Conditions of Work and Life in Health and Medical Services (1985)
Geneva, ILO, 142 p. ISBN 92–2–105525–7.

ILO Employment Effects of Multinational Enterprises in Industrialised Countries (1985)
Geneva, ILO, 100 p. ISBN 92–2–102741–4.

ILO Occupational Exposure to Airborne Substances Harmful to Health (1985)
Geneva, ILO, 44 p. ISBN 92–2–102442–3.

ILO Occupational Hazards and Diseases in Commerce and Offices (1985)
Geneva, ILO, 54 p. ISBN 92–2–103934–X.

ILO Prevention and Settlement of Labour Disputes in ASEAN (1985)
Geneva, ILO, 339 p. ISBN 92–2–105158–7.

ILO World Labour Report (1985)
Washington DC, ILO, 245 p. Vol. 2 ISBN 92–2–103848–3.

ILO Into the Twenty–First Century. The Development of Social Security (1986)
Geneva, ILO, 115 p. ISBN 92–2–103631–6.

ILO Annotated Bibliography on Child Labour (1986)
Geneva, ILO, 69 p. ISBN 92–2–105465–9.

ILO Bibliography of Published Research of the World Employment Programme (1986)
Geneva, ILO, 177 p. 6th ed. ISBN 92–2–105464–0.

ILO The Cost of Occupational Accidents and Diseases (1986)
Geneva, ILO, 142 p.
ISBN 92–2–103758–4. Occupational Safety and Health Series: No 54.

ILO Accident Prevention: a workers' education manual (1986)
Geneva, ILO, 175 p. 2nd Ed. ISBN 92–2–103392–9.

ILO Accident Prevention on Board Ship at Sea and in Port (1986)
Geneva, ILO, 188 p. ISBN 92–2–101837–7.

ILO Child Labour. A Briefing Manual (1986)
Geneva, ILO, 82 p.
ISBN 92–2–105639–2.

ILO Protection of Workers Against Radio–Frequency and Microwave Radiation. A technical review (1986)
Geneva, ILO, 72 p. ISBN 92–2–105604–X.

ILO Personnel Decision–making in Wholly Owned Foreign Subsidiaries and in International Joint Ventures (1986)
Geneva, ILO, 43 p. ISBN 92–2–105691–0.

ILO Employment and Conditions of Work in Water, Gas and Electricity Supply Services (1987)
Geneva, ILO, 153 p. ISBN 92–2–105872–7.

ILO Employment Promotion and Vocational and Social Security (1987)
Geneva, ILO, 156 p.
ISBN 92–2–105574–4. Preliminary report and questionnaire.

ILO Industrial Disputes (1987)
Geneva, ILO.
ISBN 92–2–105905–7.

ILO Annotated Bibliography on Clandestine Employment (1987)
Washington DC, ILO, 132 p. ISBN 92–2–105726–7.

ILO Incomes from Work: between equity and efficiency (1987)
Washington DC, ILO, 169 p. World Labour Report: 3.
ISBN 92–2–105951–0.

ILO Women in Rural Development: critical issues (1988)
Geneva, ILO, 51 p. ISBN 92–2–102388–5. A WEP Study.

ILO Work and Family: the child care challenge – conditions of work digest (1988)
Geneva, ILO, 303 p.
ISBN 92–2–106498–0. Vol 7: No 2, 1988.

ILO Year Book of Labour Statistics, 1988 (1988)
Geneva, ILO, 1148 p. 48th issue. ISBN 92–2–006424–3.

ILO Meeting the Social Debt (1988)
Geneva, ILO, 124 p. ISBN 92–2–106805–6.

ILO Occupational Cancer: prevention and control (1988)
Geneva, ILO, 122 p. 2nd rev ed.
ISBN 92–2–106454–9. Occupational Safety and Health Series: No 39.

ILO The Emerging Response to Child Labour: conditions of work digest (1988)
Geneva, ILO, 250 p. ISBN 92–2–106391–7. Vol 7: No 1.

ILO World Employment Review (1988)
Geneva, ILO, 66 p. ISBN 92–2–106408–5.

ILO Part–time Work: conditions of work digest (1989)
Geneva, ILO, 311 p.
ISBN 92–2–106508–1. Vol 8: No 1 1989.

ILO Safety in the Use of Industrial Robots (1989)
Geneva, ILO, 80 p.
ISBN 92–2–106434–4. Occupational Safety and Health Series: No 60.

ILO World Labour Report 1989: employment and labour incomes, government and its employees, statistical appendix (1989)
Geneva, ILO, 159 p. ISBN 92–2–106444–1.

ILO Night Work (1989)
Geneva, ILO, 140 p.
ISBN 92–2–106664–9. International Labour Conference: 75th Session, 1988, Report V: Part 2.

ILO and United Centre on Transnational Corporations Women Workers in Multinational Enterprises in Developing Countries (1985)
Geneva, ILO, 119 p. ×3
ISBN 92–2–105827–1. ORG #B2183; #E6753.

IMCO/UNEP The Status of Oil Pollution and Oil Pollution Control in the West and Central African Region (1982)
Nairobi, UNEP, 187 p. UNEP Regional Seas Reports and Studies: No 4.

Imhauser, G Club Foot (1986)
New York, Thieme Medical Publishers, 128 p. illus.
ISBN 0–86577–169–3.

Imura, H and Kuzuya, H (Eds) Hormone Receptors and Receptor Diseases: proceedings of the Int Symposium on Hormone Receptors and Receptor Diseases, Kyoto 1982 (1983)
Amsterdam, Elsevier Science Publishing, 220 p.
ISBN 0–444–90298–8.

Imwinkelried, Edward J The Methods of Attacking Scientific Evidence (1982)
Charlottesville VA, Michie Company, 547 p.
ISBN 0–87215–559–5.

Inaba, Y, et al Brain Edema (1986)
New York, Springer Publishing, 690 p.
ISBN 0–387–15780–8.

Indian Institute of Foreign Trade Study on Freight Tariffs and Practices of Shipping Conferences (1979)
New Delhi, Indian Institute of Foreign Trade, 118 p.

Indian Institute of Foreign Trade Growing Protectionism in Developed Countries: implications for India (1980)
New Delhi, Indian Institute of Foreign Trade, 287 p.

Indian Society of Agricultural Economics Seminar on Problems of Small Farmers (1967, Bombay) (1968)
Bombay, Indian Society of Agricultural Economics, 133 p.

Industrial Health Foundation Industrial Noise: a selective bibliography, 1963–1973 (1973)
Pittsburg PA, Industrial Health Foundation, 111 p. bibl.

Ingraham, Barton Political Crime in Europe (1979)
Berkeley CA, University of California Press.
ISBN 0–520–03562–3.

Inkley, F A Oil Loss Control in the Petroleum Industry: proceedings of an international meeting (1986)
New York, John Wiley and Sons, 345 p.
ISBN 0–471–90813–4.

Inlander, Charles, et al Medicine on Trial: medical mistakes and incompetence in the practice (1988)
New York, Prentice Hall, 288 p. ISBN 0–13–573544–0.

Inokuchi, Kiyoshi et al Digestive Tract Tumors (1986)
New York, Plenum Publishing, 288 p.
ISBN 0–306–42297–2.

Inozemtsev, N N Global Problems of Our Age (1985)
Los Angeles CA, Progress Publishing, 328 p.
ISBN 0–8285–2944–2.

Institut de Philosophie de l'Académie des Sciences de l'URSS and Institut de Recherches Scientifiques sur les Systèmes Analyse Systématique des Problèmes Mondiaux (1988)
Paris, UNESCO, 163 p. bibl. Grand Programme I: Réflection sur les problèmes mondiaux et études prospectives: BEP/GPI/24.

Institut de Recherches Scientifiques sur les Systèmes/Institut de Philosophie de l'Académie des Sciences de l'URSS Analyse Systématique des Problèmes Mondiaux (1986)
Paris, UNESCO, 163 p. Grand Programme I, Réflections sur les Problèmes Mondiaux et Etudes Prospectives: BEP/GPI/24.

Institut Français d'Histoire Sociale Anarchism: catalogue of 19th and 20th century books and pamphlets from different countries (1982)
Munich, K G Saur Verlag, 170 p. ISBN 3-598-10442-1.

Institut Soviétique de Recherche Scientifique sur les Systèmes Les Problèmes Mondiaux et les Relations internationales - Globalnye Problemy i Mejdounarodnye Otnochenia (1981)
Moscow, Institut Soviétique de Recherche Scientifique sur les Systèmes.

Institute for Futures Research and Education Directory of Social and Human Forecasting (1975)
Roma, Edizioni Previsionali, 633 p.

Institute for Propaganda Analysis Propaganda Analysis (1977)
New York, Gordon Press Publishers. 5 vols.
ISBN 0-8490-2486-2.

Institute of Paper Chemistry Forest Fires: methods, meteorological aspects, and statistics, Vol 2 (1974)
Appleton WI, Institute of Paper Chemistry, 225 p. Bibliographic Ser: No 262. ISBN 0-317-34388-2.

Inter-Parliamentary Union Participating of Women in Political Life and the Decision-making Process (1988)
Geneva, Inter-Parliamentary Union, 53 p.

InterAction Council and InterAction Policy Board Reports and Recommandations of the High-Level Expert Groups (1990)
New York, InterAction Council, 126 p.

International Air Transport Association Airline Guide to Human Factors (1981)
Geneva, IATA.

International Astronomical Union and Kresak, L and Millman, P M Physics and Dynamics of Meteors (1968)
Norwell MA, Kluwer Academic Publishers, 525 p. Proceedings of the IAU Symposium: 33. ISBN 90-277-0127-X.

International Atomic Energy Agency Severe Accidents in Nuclear Power Plants: proceedings of an International Symposium on Severe Accidents in Nuclear Power Plants jointly organized by the International Atomic Energy Agency and the Nuclear Energy Agency of the OECD and held in Sorrento, Italy, 21-25 March 1988
Vienna, IAEA, 1277 p. illus. 2 Vols.

International Atomic Energy Agency Assessing the Impact of Deep Sea Disposal of Low Level Radioactive Waste of Living Marine Resources (1988)
Vienna, IAEA, 127 p. illus.
ISBN 92-0-125488-1. Technical Reports Series: No 288.

International Atomic Energy Agency and NEH Optimization of Radiation Protection (1987)
Lanham MD, UNIPUB, 605 p. ISBN 92-0-020386-8.

International Bank for Reconstruction and Development World Debt Tables, 1987-1988 Edition
Washington DC, International Bank for Reconstruction and Development. 2 Vols.

International Bank for Reconstruction and Development Preventing the Tragedy of Maternal Deaths: report of the International Safe Motherhood Conference, Nairobi, Kenya, February 1987
Washington DC, International Bank for Reconstruction and Development.

International Bank for Reconstruction and Development Road Deterioration in Developing Countries: causes and remedies
Washington DC, International Bank for Reconstruction and Development, 72 p.

International Bank for Reconstruction and Development World Debt Tables, 1989-1990 Edition (1990)
Washington DC, International Bank for Reconstruction and Development. 2 Vols. Vol 1: Analysis and Summary Tables, 100 p. Vol 2: Country Tables, 550 p.ISBN 0-8213-1408-4.

International Bureau of Fiscal Documentation International Tax Avoidance and Evasion: colloquy of the Council of Europe, Strasbourg, 1980 (1981)
Amsterdam, International Bureau of Fiscal Documentation, 184 p. ISBN 90-70125-19-6.

International Center for Mechanical Sciences Introduction to Gasdynamics of Explosions (1972)
Berlin, Springer-Verlag, 220 p. illus. CISM Pubns Ser: No 48.
ISBN 0-387-81083-8.

International Chamber of Commerce Piracy at Sea
Paris, ICC.

International Civil Aviation Organization Flight Crew Fatique and Flight Time Limitations (1984)
Montreal, ICAO, 312 p. 6th ed. Circ 52.

International Civil Aviation Organization Wind Shear (1987)
Montreal, ICAO, 189 p. Circ 186.

International Commission for the Study of Communication Problems Many Voices One World: communication and society, today and tomorrow: towards a new more just and more efficient world information and communication order (1980)
Paris, UNESCO, 312 p. ISBN 0-92-3-101802-7.

International Commission on Radiological and Sowby, F D Limits for Inhalation of Radon Daughters by Workers (1981)
Elmsford NY, Pergamon Books, 32 p. ICRP Publication: 32.
ISBN 0-08-028864-2.

International Commission on Radiological Protection (Ed) Lung Cancer Risk From Indoor Exposures to Radon Daughters (1987)
New York, Pergamon Books, 60 p. illus.
ISBN 0-08-035579-X. International Commission of Radiological Protection Ser: No 50.

International Commission on Snow and Ice of the International Association of Hydrological Sciences Avalanche Atlas: illustrated international avalanche classification (1981)
Lanham MD, UNIPUB, 265 p. illus. (In Eng, Fr, Span, Ger and Rus). Natural Hazards Ser. ISBN 92-3-001696-9.

International Confederation of Free Trade Unions Annual Survey of Violations of Trade Union Rights
Brussels, ICFTU.

International Confederation of Free Trade Unions Breaking Down the Wall of Silence: how to combat child labour (1986)
Brussels, ICFTU.

International Confederation of Free Trade Unions Trade Union Rights: survey of violations 1985/86 (1986)
Brussels, ICFTU.

International Confederation of Free Trade Unions Sexual Harassment at Work (1987)
Brussels, ICFTU.

International Council of Shopping Centers Store Security: how to reduce thefts and avoid legal problems (1994)
New York, International Council of Shopping Centers.
ISBN 0-317-16408-2.

International Council of Social Welfare The Struggle for Equal Opportunity (1977)
New York, Columbia University Press.
ISBN 0-231-04346-5.

International Environment Liason Centre International Environment-Development Facts March 1989 (1989)
Nairobi, International Environment Liason Centre, 138 p. illus. Newspaper clippings.

International Fiscal Association Cahiers 1981
Norwell MA, Kluwer Academic Publishers, 550 p. Vol LXVIB.
ISBN 90-654-4007-0.

International Fiscal Association (Ed) International Double Taxation of Inheritances and Gifts (1985)
Dordrecht, Kluwer Academic Publishers Group, 576 p.
ISBN 90-6544-219-7.

International Fiscal Association (Ed) Currency Fluctuations and International Double Taxation (1986)
Dordrecht, Kluwer Academic Publishers Group, 576 p.
ISBN 90-6544-259-6.

International Fiscal Association (Ed) International Tax Problems of Charities and Other Private Institutions with Similar Tax Treatment (1986)
Dordrecht, Kluwer Academic Publishers Group, 72 p.
ISBN 90-6544-270-7.

International Fiscal Association (Ed) Transfer of Assets Into and Out of a Taxing Jurisdiction (1986)
Dordrecht, Kluwer Academic Publishers Group, 552 p.
ISBN 90-6544-258-8.

International Institute for Environment and Wetland Drainage in Europe (1984)
Washington DC, IIED. ISBN 0-905347-52-8.

International Institute for Environment and Development Acid Earth: the global threat of acid pollution (1985)
Washington DC, IIED. ISBN 0-905347-61-7.

International Institute of Refrigeration Status of CFCs: refrigeration systems and refrigerant properties (1988)
Paris, International Institute of Refrigeration, 437 p.
ISBN 2-903-63342-8.

International Labour Conference, 67th Session, Safety and Health and the Working Environment, Report VI (1981)
Washington DC, ILO, 79 p. Pt. 2 ISBN 92-2-102408-3.

International Labour Organization Conditions of Work and Employment of Professional Workers. Tripartite Meeting on Conditions of Work and Employment of Professional Workers (1977)
Geneva, ILO, 111 p. ISBN 92-2-101836-9.

International Maritime Organization Inter-Governmental Conference on the Convention on the Dumping of Wastes at Sea
London, IMO.

International Maritime Organization Manual on Chemical Pollution: section I: problem assessment and response arrangements
London, IMO.

International Maritime Organization Oily-Water Separators and Monitoring Equipment (1987)
London, IMO.

International Monetary Fund Economic Adjustment: policies and problems (1987)
Washington DC, IMF. ISBN 0-939934-98-1.

International Organization for the Elimination Zionism and Racism (1979)
New Brunswick NJ, North American Publications.
ISBN 0-930244-03-6.

International Tanker Owners Pollution Federation (Ed) Aerial Observation of Oil at Sea (1981)
London, International Tanker Owners Pollution Federation, 6 p. ill.
Technical Information Paper: Number 1.

International Tanker Owners Pollution Federation (Ed) Use of Booms in Combating Oil Pollution (1981)
London, International Tanker Owners Pollution Federation, 8 p. ill.
Technical Information Paper: Number 2.

International Tanker Owners Pollution Federation (Ed) Aerial Application of Oil Spill Dispersants (1982)
London, International Tanker Owners Pollution Federation, 8 p. illus.
Technical Information Paper: Number 3.

International Tanker Owners Pollution Federation (Ed) Use of Oil Spill Dispersants (1982)
London, International Tanker Owners Pollution Federation, 8 p. illus.
Technical Information Paper: Number 4.

International Tanker Owners Pollution Federation (Ed) Recognition of Oil Shorelines (1983)
London, International Tanker Owners Pollution Federation, 7 p. ill. Technical Information Paper: Number 6.

International Tanker Owners Pollution Federation (Ed) Contingency Planning for Oil Spills (1985)
London, International Tanker Owners Pollution Federation, 8 p. ill. Technical Information Paper: Number 9.

International Tanker Owners Pollution Federation (Ed) Effects of Marine Oil Spills (1985)
London, International Tanker Owners Pollution Federation, 8 p. ill. Technical Information Paper: Number 10.

International Tanker Owners Pollution Federation (Ed) Fate of Marine Oil Spills (1986)
London, International Tanker Owners Pollution Federation, 8 p. ill. Technical Information Paper: Number 11.

International Tanker Owners Pollution Federation (Ed) Action: oil spill (1986)
London, International Tanker Owners Pollution Federation, 8 p. ill. Technical Information Paper: Number 12.

International Technical Information Institute (Ed) Toxic and Hazardous: for handling and disposal with toxicity and hazard data
Tokyo, International Technical Information Institute, 650 p.

Irwin, Edna M Growing Pains: a study of teenage distress (1978)
Totowa NJ, Biblio Distribution Centre, 310 p.
ISBN 0-7130-0166-6.

Ishinishi, Noburu, et al (Eds) Carcinogenic and Mutagenic Effects of Diesel Engine Exhaust (1986)
Amsterdam, Elsevier Science Publishing, 540 p.
ISBN 0-444-80854-X.

Italiano, Michael L Liability for Underground Storage Tanks (1987)
New York, Practising Law Institute, 384 p.
ISBN 0-317-66634-7.

Itoh, Makoto The Basic Theory of Capitalism: the forms and substance of the capitalist economy (1987)
Totowa NJ, Barnes and Noble Books-Imports, 256 p.
ISBN 0-389-20729-2.

IUCN (Eds) Convention on International Trade in Endangered Species of Wild Fauna and Flora - CITES: proceedings of the meeting of the Conference of the Parties
Gland, IUCN.

IUCN Conservation Monitoring Center IUCN Red List of Threatened Animals, 1988 (1988)
New York, Columbia University Press, 172 p.
ISBN 2-88032-935-3.

IUCN; UNEP and WWF (Eds) Caring for the World: a strategy for sustainability (1990)
Gland, IUCN, 135 p. 2nd draft.

Ivanyi, L Immunological Aspects of Oral Diseases (1986)
Norwell MA, Kluwer Academic Publishers.
ISBN 0-85200-961-5.

Iversen, L L The Biology of Nicotine Dependence (1990)
Chichester, John Wiley and Sons, 280 p. CIBA Foundation Symposium: 152. ISBN 0-471-92688-4.

Iversen, Leslie L et al Neuroleptics and Schizophrenia (1978)
New York, Plenum Publishing, 262 p. Handbook of Psychopharmacology: 10. ISBN 0-306-38930-4.

Iwai, Junichi Salt and Hypertension: proceedings of the Lewis K Dahl symposium (1982)
New York, Igaku-Shoin Medical Publishers, 320 p.
ISBN 0-89640-072-7.

Izrael, Y A and Stukin, E D The Gamma Emission of Radioactive Fallout (1970)
Bainbridge, Coronet Books, 154 p.

Jacks, Graham V and Whyte, Robert O Vanishing Lands: a world survey of soil erosion
Salem NH, Ayer Company Publishers, 384 p.
ISBN 0-405-04573-5.

Jackson, Jerome A (Eds) Bird Conservation (1987)
Madison WI, University of Wisconsin Press, 156 p. No 3.
ISBN 0-299-11124-5.

Jackson, M; Ford-Lloyd, B V and Parry, M I (Eds) Climate Changes and Plant Genetic Resources (1990)
London, Pinter Publishers, 242 p. illus.

Jackson, Michael (Ed) Prisoners of Isolation: solitary confinement in Canada (1983)
Toronto ON, University of Toronto Press.
ISBN 0-8020-5620-2.

Jackson, Michael P Strikes: industrial conflict in Britain, USA and Australia (1987)
Hemel Hempstead, Simon and Schuster International, 272 p.
ISBN 0-7450-0209-9.

Jackson, P (Ed) Elephants and Rhinos in Africa: a time for decision (1982)
Gland, IUCN, 36 p. ISBN 2-88032-208-1.

Jackson, Robert (Ed) Global Issues: 1989-1990 (1989)
Guilford CT, Dushkin Publishing Group, 256 p. illus. 5th ed.
ISBN 0-87967-797-X.

Jackson, Robert H and Rosberg, Carl G Personal Rule in Black Africa: prince, autocrat, prophet, tyrant (1982)
Berkeley CA, University of California Press, 350 p.
ISBN 0-520-04185-2.

Jackson, Stanley W Melancholia and Depression: from Hippocratic times to modern times (1987)
New Haven CT, Yale University Press, 480 p.
ISBN 0-300-03700-7.

Jacobson, Jodi L Environmental Refugees: a yardstick of habitability
Washington DC, Worldwatch Institute. Worldwatch Papers: 86.

Jacobson, Jodi L Environmental Refugees: a yardstick of habitability (1988)
Washington DC, Worldwatch Institute, 43 p. Worldwatch Papers.
ISBN 0-916468-87-9.

Jacobsson, Staffan and Sigurdson, Jon Technological Trends and Challenges in Electronics (1983)
New York, Learning Resources in International Studies, 314 p.
ISBN 91-86002-27-9.

Jacoby, Neil H, et al Bribery and Extortion in World Business (1977)
New York, Free Press. Studies in modern corporation.
ISBN 0-02-916000-6.

Jadwani, Hassanand T and Bruchey, Stuart Some Aspects of the Multinational Corporation's Exposure to the Exchange Rate Risk (1980)
Salem NH, Ayer Company Publishers.
ISBN 0-405-13369-3.

Jahn, Egbert and Sakomoto, Yoshikazu (Ed) Elements of World Instability: armaments, communication, food, international division of labour (1981)
Frankfurt, Campus Verlag, 392 p. ISBN 3-593-32851-8.

Jahoda, Gustav The Psychology of Superstition (1970)
New York, Penguin Books.

Jameelah, Maryam Western Civilization Condemned by Itself: a comprehensive study of moral retrogression and its consequences (1979)
Lahore, Mohammed Yusuf Khan and Sons. 2 vols.

James, Adrian L and Wilson, Kate Couples, Conflict and Change: social work with marital relationships (1986)
London, Routledge, 256 p. ISBN 0-422-79900-9.

James, D Geraint and Williams, W Jones Sarcoidosis and Other Granulomatous Disorders (1985)
Philadelphia PA, WB Saunders, 256 p.
ISBN 0-7216-1044-7.

James, Lynn F, et al The Ecology and Economic Impact of Poisonous Plants on Livestock Production (1987)
Boulder CO, Westview Press, 400 p.ISBN 0-8133-7453-7.

Jamieson, Katherine and Flanagan, Timothy J (Eds) Sourcebook of Criminal Justice Statistics, 1986. (1987)
Washington DC, US Government Printing Office, 534 p.
ISBN 0-318-23844-6.

Janke, Peter and Sim, Richard Guerrilla and Terrorist Organizations: a world directory and bibliography (1983)
New York, Macmillan. ISBN 0-02-916150-9.

Janke, Peter and Sim, Richard Guerilla and Terrorist Organizations: a world directory and bibliography (1983)
New York, Macmillan. ISBN 0-02-916150-9.

Jankovic, Joseph and Tolosa, Eduardo Parkinson's Disease and Movement Disorders (1988)
Baltimore MD, Urban and Schwarzenberg, 464 p.
ISBN 0-8067-0971-5.

Janos, Andrew C (Ed) Authoritarian Politics in Communist Europe: uniformity and diversity in one-party states (1976)
Berkeley CA, University of California Press, 220 p. Research Ser: No 28. ISBN 0-87725-128-2.

Janowitz, Morris Professional Soldier: a social and political portrait (1964)
New York, Free Press. ISBN 0-02-916170-3.

Jansen, Willy Women without Men: gender and marginality in an Algerian town (1986)
Leiden, Brill, 303 p. ISBN 90-04-08345-6.

Jansson, Mats and Persson, Gunnar (Eds) Phosphorus in Freshwater Ecosystems: proceedings of a symposium held in Uppsala, Sweden, September 25-28, 1985 (1988)
Dordrecht, Kluwer Academic Publishers Group, 352 p.
ISBN 90-6193-657-8.

Janus, Mark D, et al Adolescent Runaways Causes and Consequences (1987)
Lexington MA, Lexington Books, 176 p.
ISBN 0-669-13047-8.

Janz, Dieter, et al (Eds) Epilepsy, Pregnancy and the Child (1982)
New York, Raven Press, 576 p. ISBN 0-89004-654-9.

Japan Container Association (Ed) Index of Dangerous Substances Classified in Various Regulations (1975)
Tokyo, Kaibundo Publishing.

Jarvis, William T, et al Quackery and You (1983)
Hagerstown MD, Review and Herald Publishing Association.
ISBN 0-8280-0148-0.

Jasani, Bhupendra Space Weapons and International Security (1987)
New York, Oxford University Press, 366 p.
ISBN 0-19-829102-7.

Jay, Martin Marxism and Totality (1984)
Berkeley CA, University of California Press, 576 p.
ISBN 0-520-05096-7.

Jayaraman, M Stagnation and Wastage in Primary Schools
New Delhi, National Council of Educational Research and Training, 33 p.

Jeanmart, L Tumors of the Spine (1985)
New York, Springer Publishing, 130 p.
ISBN 0-387-15326-8.

Jefferies, Richard The Gamekeeper at Home and the Amateur Poacher (1978)
New York, Oxford University Press. Oxford Paperback Series. ISBN 0-19-281240-8.

Jelliffe, D B and Stanfield, J P (Eds) Diseases of Children in the Subtropics and Tropics (1978)
Baltimore MD, Williams and Wilkins, 1070 p. 3rd ed.
ISBN 0-7131-4277-4.

Jenkins, R Exploitations (1970)
London, Paladin.

Jennings, Charles The Confidence Trick (1988)
London, Hamilton. ISBN 0-241-12522-7.

Jéquier, Nicholas (Ed) Appropriate Technology, Problems and Promises (1976)
Paris, OECD, 344 p. ISBN 92-64-11492-0.

Jerath, Bal K Homicide - A Bibliography: (Supplement 1984) (1984)
Augusta GA, Pine Tree Publications. ISBN 0-943974-01-1.

Jerath, Bal K, et al Homicide - A Bibliography of over 4,500 Items (1982)
Augusta GA, Pine Tree Publications,.
ISBN 0-943974-00-3.

Jewell, Peter A Problems in Management of Locally Abundant Wild Mammals (1982)
San Diego CA, Academic Press. ISBN 0-12-385280-3.

Jirasek, Jan E and Cohen, M Michael Development of the Genital System and the Male Pseudohermaphroditism
Ann Arbor MI, Books on Demand. ISBN 0-317-19868-8.

Joardar, Biswanath Prostitution in Historical and Modern Perspectives (1984)
New Delhi, Inter-India Publications, 300 p.

John, J, et al (Ed) Influence of Economic Instability on Health: proceedings of a symposium organized by the Gesellschaft für Strahlen- und Umweltforschung München, Federal Republic of Germany, 9-11 September 1981 (1983)
Berlin, Springer-Verlag, 528 p. ISBN 3-540-12274-5.

John, M and Katz, Stanford N Marriage and Cohabitation in Contemporary Societies: ares of legal, social and ethical change (1980)
Scarborough, Butterworths, 454 p.

Johnson, Branden, B and Covello, Vincent, T (Eds) Social and Cultural Construction of Risk: essays on risk selection and perception (1987)
Dordrecht, Kluwer Academic Publishers Group, 424 p.
ISBN 1-55608-033-6.

Johnson, Elmer H (Ed) International Handbook of Contemporary Developments in Criminology (1983)
Westport CT, Greenwood Press. Illus. 2 Vols. Vol I: General Issues and the Americas. Vol II: Europe, Africa, the Middle East and Asia.

Johnson, Kendall Trauma in the Lives of Children (1989)
Basingstoke, Macmillan, 160 p. ISBN 0-333-51094-1.

Johnson, Loch K America's Secret Power: the CIA in a democratic society

Johnson, Wayne G and Tracy, Martin B (Eds) Social Development Issues: alternative approaches to global human needs (1986)
Iowa City IA, University of Iowa. Repr of 1977 ed. Vol 1, No 2. ISBN 0-317-58804-4.

Johnstone, Jay and Talley, Rick Temporary Insanity (1986)
New York, Bantam Books, 208 p. ISBN 0-553-26167-3.

Joint Group of Exports on the Scientific Aspects of Marine Pollution Cadmium, Lead and Tin in the Marine Environment (1985)
Nairobi, UNEP, 85 p. UNEP Regional Seas Reports and Studies: No 56.

Joint Meeting on Conditions of Work Employment and Conditions of Work of Teachers (1981)
Washington DC, ILO, 158 p. ISBN 92-2-102777-5.

Joint WHO/FAO Expert Committee on Food Additives (Eds) Residues of Some Veterinary Drugs in Animals and Food (1988)
Rome, FAO.

Jones, David Lloyd Three International Conventions on Hijacking and Offences on Board Aircraft: explanatory documentation for Commonwealth jurisdictions (1985)
London, ComSec, 146 p.

Jones, Elise F, et al Teenage Pregnancy in Industrialized Countries (1987)
New Haven CT, Yale University Press, 304 p.
ISBN 0-300-03705-8.

Jones, Frances M Defusing Censorship: the librarian's guide to handling censorship (1983)
Phoenix AZ, Oryx Press, 240 p. ISBN 0-89774-027-0.

Jones, G E Contagious Agalactia and Other Mycoplasmal Diseases of Small Ruminants (1987)
Luxembourg, European Communities, 118 p.
ISBN 92-825-7008-8.

Jones, Howard M Revolution and Romanticism (1974)
Cambridge MA, Harvard University Press, 490 p.
ISBN 0-674-76710-1.

Jones, Robert S Noise and Vibration Control in Buildings (1984)
New York, McGraw-Hill Book Company, 384 p. illus.
ISBN 0-07-006431-8.

Jones, Russell R and Wigley, T (Eds) Ozone Depletion: health and environmental consequences (1990)
Chichester, John Wiley and Sons, 264 p.
ISBN 0-471-92316-8.

Jones, Warren H, et al Shyness: perspectives on research and treatment (1986)
New York, Plenum Publishing, 410 p.
ISBN 0-306-42033-3.

Jordan, David S and Marchand, C R War and Waste: a series of discussions of war and war accessories (1972)
New York, Garland Publishing. ISBN 0-8240-0265-2.

Jordan, Shannon M Decision Making for Incompetent Persons: the law and morality of who shall decide (1985)
Springfield IL, Thomas Charles C, 170 p.

Joshi, Nandini Umashankar Challenge of Poverty: the developing countries in the New International Order (1978)
New Delhi, Arnold-Heinemann, 101 p.

Josso, Nathalie The Intersex Child (1981)
New York, Karger S AG, 272 p. ISBN 3-8055-0909-X.

Jovanovic, L Controversies in Diabetes and Pregnancy (1988)
New York, Springer Publishing, 260 p.
ISBN 0-387-96622-6.

Joyner, Nancy D Aerial Hijacking As an International Crime (1974)
New York, Oceana, 352 p. ISBN 0-379-00004-0.

Judge, A J N Atlas of International Relationship Network: reviews of mapping networks of organization/problem relationship in order to produce an "Atlas" as a complement to the Yearbook of International Organizations and to the Encyclopedia of World Problems and Human Potential (1987)
Brussels, Union of International Associations.

Julian, D G and Wenger, N K Cardiac Problems of the Adolescent and Young Adult (1985)
Stoneham MA, Butterworth Publishers, 320 p.
ISBN 0-407-02268-6.

Julian, Desmond G Angina Pectoris (1985)
New York, Churchill Livingstone, 233 p. illus. 2nd ed.

Julian, Joseph and Kornblum, William Social Problems (1986)
Englewood Cliffs NJ, Prentice Hall, 624 p.
ISBN 0-13-816851-2.

Jurji (Ed) Religious Pluralism and World Community (1969)
Leiden, Brill. ISBN 90-04-06-1616-3.

Kaayk, Jan Education Estrangement and Adjustment: a study among pupils and school leavers in Bukumbi: a rural community in Tanzania (1976)
Berlin, Mouton de Gruyter, 268 p. ISBN 90-279-7702-X.

Kabata, Z Parasites and Diseases of Fish Cultures in the Tropics (1985)
New York, Taylor and Francis, 310 p.
ISBN 0-85066-285-0.

Kahler, Miles The Politics of International Debt (1986)
Ithaca NY, Cornell University Press, 272 p.
ISBN 0-8014-1911-5.

Kahnert, Friedrich Improving Urban Employment and Labor Productivity (1987)
Washington DC, International Bank for Reconstruction and Development, 80 p. ISBN 0-8213-0908-0.

Kahrs, Robert F10 Viral Diseases of Cattle (1981)
Ames IA, Iowa State University Press, 224 p. illus.
ISBN 0-8138-0860-X.

Kaiser, Ronald A Liability and Law in Recreation, Parks and Sports (1986)
Englewood Cliffs NJ, Prentice Hall, 368 p.
ISBN 0-13-535089-1.

Kalant, O J, et al (Eds) Cannabis: health risks (1983)
Toronto ON, Addiction Research Foundation, 1983.
ISBN 0-88868-081-3.

Kaldor, Mary The Baroque Arsenal (1983)
London, Sphere Books, 239 p.

Kalter, Harold (Ed) Issues and Reviews in Teratology (1984)
New York, Plenum Publishing, 532 p. Vol 2.
ISBN 0-306-41652-2.

Kamble, N D Deprived Castes and Their Struggle for Equality (1983)
New Delhi, Ashish Publishing House, 400 p.

Kameir, El-Wahtig and Kursany, Ibrahim Corruption as the Fifth Factor of Production in the Sudan (1985)
Uppsala, Scandinavian Institute of African Studies, 33 p.
ISBN 91-7106-223-8.

Kamerman, Jack B Death in the Midst of Life (1988)
Englewood Cliffs NJ, Prentice Hall, 192 p.
ISBN 0-13-197708-3.

Kameyama, Masakuni and Tomonaga, Masanori Cerebrovascular Disease (1988)
New York, Igaku-Shoin Medical Publishers, 240 p.
ISBN 0-89640-133-2.

Kandiyoti, Denis The Reproduction of Subordination: rural women in the Third World (1984)
Paris, UNESCO, 225 p.

Kandiyoti, Deniz Women in Rural Production Systems: problems and policies (1986)
Lanham MD, UNIPUB, 119 p. ISBN 92-3-102296-2.

Kaniki, M Religious Conflict and Cultural Accomodation: the Impact of Islam on some aspects of the African societies (1976)
Dar es Salaam, University of Dar es Salaam.

Kano, Eiichi (Ed) Current Researches in Hyperthermia Oncology (1988)
San Diego CA, Academic Press, 400 p.
ISBN 0-12-396430-X.

Kanungo, Rabindra N Alienation: an integrative approach (1982)
New York, Praeger Publishers, 220 p.
ISBN 0-275-90832-1.

Kao, Charles H Brain Drain (1980)
Mei Ya China/Int Spec Book, 178 p.
ISBN 0-89955-157-2.

Kaplan, Abraham The Conduct of Inquiry (1964)
San Francisco, Chandler, p. 70-71.

Kaplan, D W Diabetes in Childhood and Adolescence (1986)
New York, Karger S AG, 72 p. ISBN 3-8055-4400-6.

Kaplan, David W Violent Deaths in Childhood and Adolescence (1985)
New York, Karger S AG, 80 p. ISBN 3-8055-4231-3.

Kaplan, Henry S Hodgkin's Disease (1980)
Cambridge MA, Harvard University Press.
ISBN 0-674-40485-8.

Kaplan, Katheen R Hostility, Characteristics and Behavior: medical subject analysis with reference bibliography (1988)
Washington DC, ABBE Publications Association, 150 p.
ISBN 0-88164-507-9.

Kaplan, Morton A Isolation or Interdependence: today's Choice for tomorrow's world (1977)
Agincourt ON, Macmillan of Canada.

Kaplan, Norman M et al The Kidney in Hypertension (1987)
New York, Raven Press, 288 p. ISBN 0–88167–298–X.
Kaplinsky, Raphael Micro-Electronics and Employment Revisited: a review (1987)
Washington DC, ILO, 181 p. ISBN 92–2–105610–4.
Kapoor, A S (Ed) Cancer and the Heart (1986)
Berlin, Springer-Verlag, 318 p. illus. ISBN 0–387–96245–X.
Kapp, K William Social Costs, Economic Development and Environmental Disruption (1983)
Pr of Amer, 222 p. Intro by John Ullmann.
ISBN 0–8191–3208–X.
Kapur, Basant K Studies in Inflationary Dynamics: financial repression and financial liberalization in less developed countries (1986)
Singapore, Singapore University Press, 164 p.
ISBN 9971–69–099–3.
Kapur, Rajiv A Sikh Separatism: the politics of faith (1986)
Winchester MA, Unwin Hyman, 240 p.
ISBN 0–04–320179–2.
Kardelj, Edvard Contradictions of Social Property in a Socialist Society (1981)
Beograd, Aktuelna Pitanja Socijalizma.
Karmen, Andrew Crime Victims: an introduction to victimology (1984)
Pacific Grove CA, Brooks/Cole Publishing.
ISBN 0–534–02997–3.
Karnes, Elizabeth L, et al Discipline in Our Schools: an annotated bibliography (1983)
Westport CT, Greenwood Press, 700 p.
ISBN 0–313–23521–X.
Karp, Aaron (Ed) Shades of Grey: the hidden arms trade today (1989)
Solna, Stockholm International Peace Research Institute, 175 p. ISBN 0–19–829134–5.
Karpenko, G V Stress-Corrosion Cracking of Steels (1979)
Brookfield VT, Trans Tech Publications, 200 p.
ISBN 0–317–54813–1.
Karstad, L, et al (Ed) Wildlife Disease Research and Economic Development (1981)
Ottawa ON, International Development Research Centre, 80 p. ISBN 0–88936–307–2.
Kastenbaum, Robert (Ed) Death and Dying (1977)
Salem NH, Ayer Company Publishers. illus. 40 Vols.
ISBN 0–405–09550–3.
Katz, Jacob Exclusiveness and Tolerance (1983)
West Orange NJ, Behrman House, 208 p.
ISBN 0–87441–365–6.
Katz, Jacob F Legg-Calve-Perthes Disease (1984)
New York, Praeger Publishers, 252 p.
ISBN 0–275–91437–2.
Katz, Jay Experimentation with Human Beings (1972)
New York, Russell Sage Foundation, 1160 p.
ISBN 0–87154–438–5.
Katz, Jonathan Government Versus Homosexuals: an original anthology (1975)
Salem NH, Ayer Company Publishers.
ISBN 0–405–07350–X.
Katzman, Robert (Ed) Biological Aspects of Alzheimer's Disease (1983)
Cold Spring Harbor, 495 p. Banbury Report Ser: 15.
ISBN 0–87969–213–8.
Kavka, Gregory Moral Paradoxes of Nuclear Deterrence (1987)
New York, Cambridge University Press, 230 p.
ISBN 0–521–33043–2.
Kavolis, Vytautas (Ed) Comparative Perspectives on Social Problems (1969)
Boston MA, Little, Brown and Company.
Kay, A B Allergy and Inflammation (1987)
San Diego CA, Academic Press. ISBN 0–12–402745–8.
Kayyali, Abdul W (Ed) Zionism, Imperialism and Racism (1979)
New York, Routledge Chapman and Hall, 304 p.
ISBN 0–85664–761–6.
Keeler, Leo W The Problem of Error from Plato to Kant (1977)
New York, Gordon Press Publishers.
ISBN 0–8490–2482–X.
Keeler, Richard F, et al Effects of Poisonous Plants on Livestock (1978)
San Diego CA, Academic Press. ISBN 0–12–403250–8.
Keen, Sam Faces of the Enemy: reflections of the hostile imagination (1986)
New York, Harper and Row. ISBN 0–06–250471–1.
Kegley, Charles W and Wittkopf, Eugene R The Global Agenda: issues and perspectives (1988)
New York, McGraw-Hill, 408 p. ISBN 0–07–553724–9.
Kegley, Charles W and Wittkopf, Eugene R The Global Agenda: issues and perspectives (1988)
New York, McGraw-Hill Publishing, 408 p. 2nd ed.
ISBN 0–07–553724–9.
Keith, Arthur The Place of Prejudice in Modern Civilization (1982)
Brooklyn NY, Revisionist Press. ISBN 0–87700–336–X.
Keith, Louis (Ed) Common Infections (1985)
Dordrecht, Kluwer Academic Publishers Group.
ISBN 0–85200–733–7.
Keith, Louis (Ed) Uncommon Infections and Special Topics (1985)
Dordrecht, Kluwer Academic Publishers Group.
ISBN 0–85200–733–7.
Keith, R W (Ed) Central Auditory and Language Disorders in Children (1985)
Basingstoke, Taylor and Francis, 210 p.
ISBN 0–85066–533–7.
Keller, Mark, et al International Bibliography of Studies on Alcohol: references, 1901–1950 (1981)
New Brunswick NJ, Rutgers Center of Alcohol Studies Publications. Vol. 13 vols. ISBN 0–911290–07–9.
Keller, P J (Ed) Female Infertility (1978)
Basel, S Karger AG. illus. Contributions to Gynecology and Obstetrics: Vol 4. ISBN 3–8055–2791–8.
Keller, Paul J Hormonal Disorders in Gynecology (1981)
Berlin, Springer-Verlag, 113 p. illus. ISBN 0–387–10341–4.
Kellhammer, U and Überla, K (Eds) Long-Term Studies on Side-Effects of Contraception: state and planning: symposium of the study group 'Side-Effects of oral contraceptives, pilot phase', Munich 1977 (1978)
Berlin, Springer-Verlag, 240 p. ISBN 3–540–09093–2.
Kelly, David H and Kelly, Gail P Education of Women in Developing Countries
In: *Educational Documentation and Information: Bulletin of the IBE* 222. ISSN 0303–3899.
Kelly, James E Decayed, Missing and Filled Teeth Among Youths 12–17 Years (1975)
Hyattsville MD, National Center for Health Statistics, 50 p.
ISBN 0–8406–0022–4.
Kelly, Robert J Organized Crime: a global perspective (1986)
Totowa NJ, Rowman and Littlefield Publishers, 312 p.
ISBN 0–86598–085–3.
Kelterborn, E Catalogue of Salmonella First Isolations 1965–1984 (1987)
Norwell MA, Kluwer Academic Publishers.
ISBN 0–89838–832–5.
Kelvin, Peter and Jarrett, Joanna Unemployment: its social psychological effects (1985)
New York, Cambridge University Press, 180 p.
ISBN 0–521–30481–4.
Kemm, J R Vitamin Deficiency in the Elderly: prevalence, clinical significance and effects on brain function (1985)
Oxford, Blackwell Scientific Publications, 224 p.
ISBN 0–632–01365–6.
Kempe, C Henry and Kempe, Ruth The Common Secret: sexual abuse of children and adolescents (1984)
New York, Freeman WH, 284 p. ISBN 0–7167–1624–0.
Kendall Anxiety and Depression: distinctive and overlapping features
San Diego CA, Academic Press. ISBN 0–12–404170–1.
Kendig, Edwin L and Chernick, Victor Disorders of the Respiratory Tract in Children (1983)
Philadelphia PA, WB Saunders. ISBN 0–7216–5379–0.
Kendrick, Walter The Secret Museum: pornography in modern culture (1988)
New York, Penguin Books, 288 p. ISBN 0–14–010947–1.
Kennedy, Donald and Bates, Richard R (Eds) Air Pollution, the Automobile and Public Health (1988)
Washington DC, National Academy Press, 650 p.
ISBN 0–309–03726–3.
Keppe, Norberto R The Decay of the American People (and of the United States) (1985)
Sao Paulo, Proton Editora, 263 p.
Keppel, David and Keppel, John Uncertainty: the ground for life (1982)
Essex. First part of a manuscript in progress.
Kerkar, A V Obstetrics and Gynaecology in Developing Countries (1985)
New Delhi, Allied Publishers, 400 p.
Kernan, R P, et al (Eds) Introduction of Exotic Species, Advantages and Problems: proceedings of a symposium, 1979 (1979)
Dublin, Royal Irish Academy, 111 p. ISBN 0–901714–14–3.
Kerr, Donna H Barriers to Integrity: modern modes of knowledge utilization (1983)
Boulder CO, Westview Press, 130 p.ISBN 0–86531–661–9.
Kesse-Adu, K Politics of Political Detention
Accra, Ghana Publishing Corporation, 226 p.
Kessler, Edwin The Thunderstorm in Human Affairs (1988)
Norman OK, University of Oklahoma Press, 200 p.
ISBN 0–8061–2153–X.
Kesteloot, H and Joossens, J V (Ed) Epidemiology of Arterial Blood Pressure (1980)
Dordrecht, Kluwer Academic Publishers Group, 515 p.
ISBN 90–247–2386–8.
Khairallah, David Insurrection Under International Law (1973)
International Book Ctr.
ISBN 0–86685–176–3.
Khan, A M and Hoelzl, A Evolution of Future Energy Demands Till 2030 in Different World Regions: an assessment made for the two IIASA scenarios (1982)
Laxenburg, International Institute for Applied Systems Analysis, 137 p. ISBN 3–7045–0034–8.
Khan, Mazahar-ul Hag Purdah and Polygamy: a Social pathology of Muslim society (1982)
New Delhi, Amar Prakashan, 204 p.
Khanna, P N Problems of Love and Sex in Marriage (1970)
New Delhi, Social Publications, 400 p.
Khera, S S and Sharma, G L Important Exotic Diseases of Live-Stock Including Poultry (1967)
New Delhi, Indian Coucil of Agricultural Research, 185 p.
Khogali, Mustafa M, et al Heat Stroke and Temperature Regulation (1984)
San Diego CA, Academic Press. ISBN 0–12–406180–X.
Khushalani, Y Dignity and Honour of Women as Basic and Fundamental Human Rights (1982)
Amsterdam, Kluwer Academic Publishers Group.
Kierzkowski, Henryk Monopolistic Competition and International Trade (1984)
New York, Oxford University Press. ISBN 0–19–828467–5.
Kigin, Denis J Teacher Liability in School-Shop Accidents (1983)
Ann Arbor MI, Prakken Publications. ISBN 0–911168–51–6.
Kihl, Young W and Lutz, James M World Trade Issues: regime, structure and policy (1985)
New York, Praeger Publishers, 288 p.
ISBN 0–275–90127–0.
Kihlstrom, J F and Evans, F J (Eds) Functional Disorders of Memory (1979)
Hillsdale NJ, Lawrence Erlbaum Assocs, 432 p.
ISBN 0–89859–471–5.
Kilbourne, Edwin D Influenza (1987)
New York, Plenum Publishing, 382 p.
ISBN 0–306–42456–8.
Killick-Kendrick, R and Peters, W Rodent Malaria (1978)
San Diego CA, Academic Press. ISBN 0–12–407150–3.
Kim, Charles W Trichinellosis (1985)
Albany NY, State University of New York Press, 343 p.
ISBN 0–88706–091–9.
Kim, Linda Women Workers in Multinational Enterprises in Developing Countries (1985)
Lanham MD, UNIPUB, 119 p. ISBN 92–2–100532–1.
Kimmerling, Baruch Zionism and Territory (1983)
Berkeley CA, University of California Press, 288 p.
ISBN 0–87725–151–7.
Kincaid-Smith, Priscilla, et al Progress in Glomerulonephritis (1979)
New York, Wiley Interscience, 458 p.
ISBN 0–471–04424–5.
King, H C and Marchan (Eds) Otolaryngolic Allergy (1981)
Amsterdam, Elsevier Science Publishing, 508 p.
ISBN 0–444–00592–7.
King, Warren B Endangered Birds of the World: the ICBP bird red data book (1981)
Washington DC, Smithsonian Institution Press, 624 p.
ISBN 0–87474–584–5.
Kinsman, Gary Regulation of Desire: sexuality in Canada (1987)
Montreal, Black Rose Books, 200 p. ISBN 0–920057–81–0.
Kirby, Stephen and Robson, Gordon (Ed) The Militarisation of Space (1987)
Hemel Hempstead, Harvester Wheatsheaf, 272 p.
ISBN 0–7450–0346–X.
Kirkbride, C A Control of Livestock Diseases (1986)
Springfield IL, Thomas Charles C, 160 p.
ISBN 0–398–05181–X.
Kirkby, M J and Morgan, R P Soil Erosion: landscape systems (1981)
New York, John Wiley and Sons, 312 p.
ISBN 0–471–27802–5.
Kirkendall, Lester A Premarital Intercourse and Interpersonal Relationships: a research study of interpersonal relationships based on case histories of 668 premarital intercourse experiences reported by 200 college level males (1984)
New York, Greenwood Press, 302 p.
ISBN 0–313–24293–3.
Kirsner and Shorter Diseases of the Colon, Rectum, and Anal Canal (1988)
Baltimore MD, Williams and Wilkins, 858 p.
ISBN 0–683–04623–3.
Kiss, I (Ed) Microbial Association and Interactions in Food: proceedings of the 12th International IUMS-ICFMH Symposium, Budapest, Hungary 12–15 July, 1983
Dordrecht, Kluwer Academic Publishers Group, 470 p.
ISBN 90–277–1802–4.
Kitch, Edmund, et al Selected Statutes and International Agreements on Unfair Competition, Trademarks, Copyrights and Patents (1986)
Mineola NY, Foundation Press, 322 p. University casebook Ser. ISBN 0–88277–331–3.
Kitching, G Development and Underdevelopment in Historical Perspective: population nationalism and industrialization (1982)
London, Methuen.
Kitching, R L and Jones, R E Ecology of Pests (1981)
Dickson, CSIRO, 264 p. ISBN 0–643–00408–4.
Klass, Morton Caste: the emergence of the South Asian social system (1980)
ISHI PA, 224 p.
ISBN 0–915980–97–5.
Klass, Philip J UFO-Abductions: a dangerous game (1988)
Buffalo NY, Prometheus Books, 200 p.
ISBN 0–87975–430–3.
Klastersky, J (Ed) Infections in Cancer Patients (1982)
New York, Raven Press, 232 p. European Organization for Research on Treatment of Cancer (EORTC) Monographs: Vol 10. ISBN 0–89004–627–1.
Klein, Fritz and Wolf, Timothy (Eds) Two Lives to Lead: bisexuality in men and women (1985)
New York, Harrington Park Press, 255 p.
ISBN 0–918393–22–1.
Klein, Gillian Reading into Racism: bias in children's literature and learning (1986)
New York, Routledge Chapman and Hall, 192 p.
ISBN 0–7102–0160–5.
Klein, H Zionism Rules the World (1982)
Brooklyn NY, Revisionist Press. ISBN 0–87700–424–2.
Klein, Susan S Handbook for Achieving Sex Equity Through Education (1985)
Baltimore MD, Johns Hopkins University Press, 560 p.
ISBN 0–8018–3172–5.
Kleinman, Arthur Social Origins of Distress and Disease: depression, neurasthenia, and pain in modern china (1986)
New Haven CT, Yale University Press, 256 p.
ISBN 0–300–03541–1.
Kleinman, Arthur and Good, Byron Culture and Depression (1985)
Berkeley CA, University of California Press, 500 p.
ISBN 0–520–05493–8.

Klement, Alfred W Radioactive Fallout from Nuclear Weapons Tests: proceedings (1965)
Oak Ridge TN, US Department of Energy, 965 p. ISBN 0–87079–323–3.

Klenerman, L The Foot and its Disorders (1982)
Oxford, Blackwell Scientific Publications, 472 p. illus. ISBN 0–632–00863–6.

Kletz, T Learning from Accidents in Industry (1988)
Stoneham MA, Butterworth Publishers, 168 p. ISBN 0–408–02696–0.

Klinowska, Margaret Dolphins, Porpoises, and Whales of the World: the IUCN red data book (1988)
New York, Columbia University Press, 550 p. ISBN 2–88032–936–1.

Klitgaard, Robert Elitism and Meritocracy in Developing Countries: selection policies for higher education (1986)
Baltimore MD, Johns Hopkins University Press, 246 p. ISBN 0–8018–3269–1.

Kluger, Matthew J Fever: its biology, evolution and function (1979)
Princeton NJ, Princeton University Press. illus.

Knight, Edwin W Emotional Illness (1979)
Melbourne VIC, Inter–Church Trade and Industry Mission. ISBN 0–909917–23–X.

Knight, Stephen The Brotherhood: the explosive exposé of the secret world of the Freemasons (1985)
London, Panther Books, 326 p. bibl. ISBN 0–586–05983–0.

Knutsen, Torbjorn, L Hegemony in the Modern International System
Oslo, International Peace Research Institute, 25 p.

Kobayashi, Michele Y Birth Injuries: medical subject analysis with research bibliography (1987)
Washington DC, ABBE Publications Association, 150 p. bibl. ISBN 0–88164–237–1.

Kocanda, S Fatigue Failure of Metals (1978)
Dordrecht, Kluwer Academic Publishers Group, 367 p. ISBN 90–286–0025–6.

Koelega, Harry S (Ed) Environmental Annoyance: characterization, measurement and control (1987)
Amsterdam, Elsevier Science Publishing, 424 p. ISBN 0–444–80865–5.

Kohen–Raz, R Learning Disabilities and Postural Control (1986)
Tel Aviv, Freund Publishing House, 250 p. ISBN 965–294–017–8.

Kohn, George Encyclopedia of American scandal (1989)
New York, Facts on file, 480 p. Illus.ISBN 0–8160–1313–6.

Kokoski, Richard and Müller, Harald The Non–Proliferation Treaty: political and technological prospects and dangers in 1990; a SIPRI research report (1990)
Solna, Stockholm International Peace Research Institute, 138 p. ISBN 91–85114–51–0.

Kokot, Waltraud Perceived Control and the Origins of Misfortune: a case study in cognitive anthropology (1982)
Berlin, Dietrich Reimer Verlag, 155 p. ISBN 3–496–00714–1.

Kolar, F Export of Counter–Revolution: past and present
Moscow, Izdatel'stvo Progress, 254 p.

Komuli, Suresh (Ed) Corruption in India: the Growing evil (1975)
New Delhi, Chetana Publications, 128 p.

Koo, Anthony Y Land Market Distortion and Tenure Reform (1982)
Ames IA, Iowa State University Press, 138 p. ISBN 0–8138–1078–7.

Koobatian, James (Comp) Faking It: an international bibliography of art and literary forgeries 1949–1986 (1987)
Washington DC, Special Libraries Association, 240 p. bibl. ISBN 0–87111–320–1.

Koocher, Gerald P and O'Malley, John E The Damocles Syndrome: psychological consequences of surviving childhood cancer (1981)
New York, McGraw–Hill Book Company, 218 p. ISBN 0–07–035340–9.

Koranyi, Erwin K Transsexuality in the Male: the spectrum of gender dysphoria (1980)
Springfield IL, Thomas Charles C, 192 p. ISBN 0–398–03924–0.

Korbin, Jill E, et al Child Abuse and Neglect: cross–cultural perspectives (1981)
Berkeley CA, University of California Press, 214 p. ISBN 0–520–04432–0.

Kornai, J Economics of Shortage (1980)
Amsterdam, Elsevier Science Publishing. 2 Vols. Vol A: 296 p, Vol B: 336 p.

Kornai, János Contradictions and Dilemmas: studies on the socialist economy and society
Budapest, Corvina Kiadó V, 166 p. Coed with MIT Press.

Korner, Peter, et al The IMF and the Debt Crisis: a guide to the third world's dilemma (1986)
Atlantic Highlands NJ, Humanities Press International, 240 p. ISBN 0–86232–487–4.

Kornhauser, Arthur, et al Industrial Conflict (1977)
Salem NH, Ayer Company Publishers. ISBN 0–405–10179–1.

Korpas, J and Tomori, Z Cough and other respiratory reflexes (1979)
Basel, S Karger AG. Progress in respiration research: Vol 12. Illus. ISBN 3–8055–3007–2.

Korten, David C Getting to the 21st Century: voluntary action and the global agenda (1990)
West Hartford CT, Kumarian Press, 253 p.

Koso–Thomas, Olayinka The Circumcision of Women: a strategy for eradication (1987)
Atlantic Highlands NJ, Humanities Press International, 128 p. ISBN 0–86232–700–8.

Kosterlitz, H W and Terenius, L Y Pain and Society (1980)
New York, VCH Publishers, 523 p. ISBN 0–89573–099–5.

Kostopoulos, Tryphon Decline of the Market: the ruin of capitalism and anti–capitalism (1987)
Stockholm, Almqvist and Wiksell, 258 p. ISBN 91–22–00833–0.

Kostopoulos, Tryphon The Decline of the Market: the ruin of capitalism and anti–capitalism (1987)
Stockholm, Almqvist and Wiksell, 258 p. ISBN 91–22–00833–0.

Kovacevic, Z and Gruder, W G Molecular Nephrology: biochemical aspects of kidney functions (1987)
Hawthorne NY, Walter de Gruyter, 424 p. ISBN 3–11–011121–7.

Kovarsky, Joel Arthritis: a guide for patient and family (1987)
New York, Raven Press, 150 p.

Kowalski, Charlotte J Obesity: medical subject analysis with bibliography (1987)
Annandale VA, ABBE Publishers Association of Washington, 160 p. ISBN 0–88164–522–2.

Krant, Melvin J Dying and Dignity: the meaning and control of a personal death (1974)
Springfield IL, Thomas Charles C, 164 p. ISBN 0–398–02995–4.

Kratochwil, Friedrich, et al Peace and Disputed Sovereignty: reflections on conflict over territory (1985)
Washington DC, University Press of America, 172 p. ISBN 0–8191–4954–3.

Kratochwill, Thomas R Selective Mutism: implications for research and treatment (1981)
Hillsdale NJ, Erlbaum Lawrence Associates, 208 p. ISBN 0–89859–064–7.

Kraus, Hans Backache Stress and Tension (1984)
New York, Pocket Books. ISBN 0–671–50850–4.

Kravchenko, M Social Problems: Soviet approach (1981)
Moscow, Agentstvo Pecati Novosti, 150 p.

Kreier Parasitic Protozoa (1977)
San Diego CA, Academic Press. ISBN 0–12–426002–0.

Kreier Malaria (1980)
San Diego CA, Academic Press. Vol 3. ISBN 0–12–426103–5.

Kreier, Julius P Malaria: epidemiology, chemotherapy, morphology and (1980)
San Diego CA, Academic Press. Vol. 1 ISBN 0–12–426101–9.

Kreier, Julius P Malaria: pathology, vector studies and culture (1980)
San Diego CA, Academic Press. Vol. 2 ISBN 0–12–426102–7.

Kreir, Julius P Parasitic Protozoa (1977)
San Diego CA, Academic Press. Vol. 3 ISBN 0–12–426003–9.

Kretschmer, V Leukozytenseparation und Transfusion (1981)
Basel, S Karger AG, 188 p. illus. Beitraege zu Infusionstherapie und Klinische Ernaehrung: Vol 6. ISBN 3–8055–1946–X.

Kriegel, H, et al Developmental Effects of Prenatal Irradiation (1982)
Stuttgart, Fischer, 387 p. ISBN 3–437–30375–9.

Kriesberg, Louis Social Conflicts (1982)
Englewood Cliffs NJ, Prentice Hall, 352 p. ISBN 0–13–815589–5.

Krishna Iyer, V R Law, Society and Collective Consciousness (1982)
New Delhi, Allied Publishers, 213 p.

Krishnamurty, S Dowry Problem (1981)
Bombay, Popular Prakashan.

Kritsberg, Wayne The Adult Children of Alcoholics Syndrome (1988)
New York, Bantam Books, 176 p. ISBN 0–553–27279–9.

Kritzinger, Erna E and Taylor, Kenneth G Diabetic Eye Disease (1984)
Norwell MA, Kluwer Academic Publishers. ISBN 0–85200–736–1.

Kronick, Doreen Social Development of Learning Disabled Persons (1981)
San Francisco CA, Jossey–Bass Publishers. ISBN 0–87589–499–2.

Kronk, Gary W Meteor Showers: a descriptive catalog (1988)
Hillside NJ, Enslow Publishers, 320 p. ISBN 0–89490–072–2.

Krugman, et al Infectious Diseases of Children (1985)
Saint Louis MO, CV Mosby. 8th ed. ISBN 0–8016–2795–8.

Krugman, Paul Exchange Rate Instability (1988)
Cambridge MA, MIT Press, 130 p. Lionel Robbins Lecture Ser.

Kruskemper, Hans L and Doering, Charles H Anabolic Steroids (1968)
San Diego CA, Academic Press. ISBN 0–12–426950–8.

Kruus, P and Valeriote, I M Controversial Chemicals: a citizen's guide (1984)
Niles IL, Polyscience, 243 p. ISBN 0–919868–22–3.

Kryter, Karl D The Effects of Noise on Man (1985)
San Diego CA, Academic Press. ISBN 0–12–427460–9.

Kubler–Ross, Elisabeth (Ed) Death: the final stage of growth (1975)
Englewood Cliffs NJ, Prentice Hall, 192 p. illus. Human Development Ser. ISBN 0–13–196998–6.

Kuiper, J and Brink, W J van den (Eds) Fate and Effects of Oil in Marine Ecosystems (1987)
Dordrecht, Kluwer Academic Publishers Group, 352 p. ISBN 90–247–3489–4.

Kumar, Mahendra Violence and Non–Violence in International Relations (1974)
Conway NH, Thompson Press, 260 p.

Kumarappa, Jagadisan Mohandas Our Begger Problems: how to tackle it (1945)
Bombay, Padma Publications, 294 p.

Kunio, Yoshihara Rise of Ersatz Capitalism in South–East Asia (1988)
Melbourne VIC, Oxford University Press, 320 p. ISBN 0–19–588885–5.

Kuo–Nan Liou, An Introduction to Atmospheric Radiation (1980)
San Diego CA, Academic Press. ISBN 0–12–451450–2.

Kuper, Jessica Social Problems and Mental Health (1987)
New York, Routledge Chapman and Hall, 176 p. ISBN 0–7102–1170–8.

Kuper, Leo Genocide: its political use in the twentieth century (1982)
New Haven CT, Yale University Press, 256 p. ISBN 0–300–02795–8.

Kupperman, Robert H Technological Advances and Consequent Dangers: growing threats to civilization (1984)
Washington DC, Center for Strategic and International Studies, 11 p. ISBN 0–89206–053–0.

Kurata, Hiroshi and Ueno, Yoshio (Eds) Toxigenic Fungi: their toxins and health hazard (1984)
Amsterdam, Elsevier Science Publishing, 364 p. ISBN 0–444–99630–3.

Kurata, Hiroshi and Ueno, Yoshio (Eds) Toxigenic Fungi – Their Toxins and Health Hazard: proceedings of the mycotoxin symposia held in the 3rd International Mycological Congress, Tokyo, August 30 – Sept 3, 1983 (1984)
Amsterdam, Elsevier Science Publishing, 384 p. ISBN 4–06–200880–7.

Kurian, George and Ghosh, Ratna Women in the Family and the Economy: an international comparative survey (1981)
Westport CT, Greenwood Press, 448 p. ISBN 0–313–22275–4.

Kurjak, A (Ed) The Fetus As a Patient: proceedings of the first international symposium held in Sveti Stefan, Yugoslavia, 4–7 June, 1984 (1985)
Amsterdam, Elsevier Science Publishing, 320 p. International Congress Ser: No 665. ISBN 0–444–80663–6.

Kuroiwa, Yoshigoro (Ed) Multiple Sclerosis in Asia (1976)
Toronto ON, University of Toronto Press, 284 p.

Kurstak, E, et al Viruses, Immunity and Mental Disorders (1987)
New York, Plenum Publishing, 425 p. ISBN 0–306–42337–5.

Kurtz, Norman R, et al (Eds) Occupational Alcoholism: an annotated bibliography (1984)
Toronto ON, Addiction Research Foundation, 218 p. ISBN 0–88868–101–1.

Kurzman, Dan A Killing Wind: inside the Bhopal catastrophe (1987)
New York, McGraw Hill Book Company, 320 p. ISBN 0–07–035687–4.

Kuschinsky, K Opiate Dependence (1977)
Stuttgart, Fischer, 39 p. ISBN 3–437–10479–9.

Kushi, Michio and Mann, John D Diabetes and Hypoglycemia (1985)
Briarcliff Manor NY, Japan Publications, 128 p. ISBN 0–87040–615–9.

Kydes, A S and Geraghty, D M Energy Markets in the Longer Term: planning under uncertainty (1985)
New York, Elsevier Science Publishing, 430 p. ISBN 0–444–87750–9.

Lacoius–Petruccelli, Alberto Perinatal Asphyxia (1987)
New York, Plenum Publishing, 188 p. ISBN 0–306–42358–8.

Lagadec, P Major Technological Risk: an assessment of industrial disasters (1982)
Elmsford NY, Pergamon Books, 536 p. ISBN 0–08–028913–4.

Lakos, Amos International Terrorism: a bibliography (1986)
Boulder CO, Westview Press, 481 p.ISBN 0–8133–7157–0.

Lakshman, T R and Johansson, B (Eds) Large–Scale Energy Projects: assessment of regional consequences (1985)
Amsterdam, Elsevier Science Publishing, 330 p. ISBN 0–444–87724–X.

Lal, Deepak The Poverty of Development Economics (1985)
Cambridge MA, Harvard University Press, 144 p. illus. ISBN 0–674–69470–8.

Lal, H Corruption: complex social phenomenon (1970)
New Delhi, Department of Personnel and Administrative Reforms, 112 p.

Lal, S N Problems of Public Borrowing in Under–Developed Countries (1978)
Allahabad, Chugh Publications, 320 p.

Lambert, H P and Caldwell, A D Pneumonia and Pneumococcal Infections (1980)
Wolfeboro NH, Longwood Publishing Group, 117 p. ISBN 1–85315–031–2.

Lampman, Robert J The Share of Top Wealth–Holders in National Wealth, 1922–56 (1984)
Westport CT, Greenwood Press, 286 p. ISBN 0–313–24425–1.

Lan, Ngo Manh (Ed) Unreal Growth: critical studies on Asian development (1985)
Delhi, Hindustan Publication Corporation, 900 p. 2 Vols.

Lancaster, Jane B and Hamburg, Beatrix A (Eds) School–Age Pregnancy and Parenthood: biosocial dimensions (1986)
Berlin, Walter de Gruyter, 403 p.

Landner, Lars and Wahlgren, Ulf Eutrophication of Lakes and Reservoirs in Warm Climates (1988)
Copenhagen, WHO, 170 p. Environmental Health Series: No 30.

Langen, Dietrich Speaking of Sleeping Problems: learning to sleep well again (1983)
New Delhi, Sterling Publishers, 127 p. ISBN 81–207–0262–X.

Laqueur, Walter (Ed) Fascism: a readers' guide – analysis, interpretations and bibliography (1977)
Berkeley CA, University of California Press. bibl. ISBN 0–520–03033–8.

Larson, John A, et al Lying and Its Detection: a study of deception and deception tests
Montclair NJ, Smith and Patterson Publishing.
ISBN 0–87585–078–2.

LaRue, Robert D Jr and Parisi, Lynn Global–International Issues and Problems: a resources book for secondary schools (1989)
Santa Barbara CA, ABC–Clio, 250 p. Social Studies Resources for Secondary School Librarian, Teachers and Students. ISBN 0–87436–536–8.

LaRue, Robert Jr and Parisi, Lynn Global–International Issues and Problems: a resources book for secondary schools (1989)
Santa Barbara CA, ABC–Clio, 250 p. Social Studies Resources for Secondary School Librarian, Teachers and Students. ISBN 0–87436–536–8.

Lasch, Christopher The Culture of Narcissism (1978)
New York, Norton.

Lash, Scott and Urry, John The End of Organized Capitalism (1987)
Madison WI, University of Wisconsin Press, 360 p.
ISBN 0–299–11670–0.

Lasker, Bruno Human Bondage in Southeast Asia (1972)
New York, Greenwood Press, 406 p. 2nd ed.
ISBN 0–8371–5612–2.

Laslett, Peter, et al (Eds) Bastardy and Its Comparative History: studies in the history of illegitimacy and martial nonconformism (1980)
Cambridge MA, Harvard University Press, 446 p. illus. Studies in Social and Demographic History.
ISBN 0–674–06338–4.

Lasne, Sophie and Gaultier, Andre P A Dictionary of Superstitions (1984)
Englewood Cliffs NJ, Prentice Hall, 304 p.
ISBN 0–13–210881–X.

Lasswell, Harold, et al (Eds) Propaganda and Communication in World History: the symbolic instrument in early times (1979)
Honolulu HI, University of Hawaii Press. Propaganda and Communication in World History Series: 1.
ISBN 0–8248–0496–1.

Last, Cynthia and Hersen, Michel (Eds) Handbook of Anxiety Disorders (1988)
New York, Pergamon Books, 600 p. General Psychology Series: 151. ISBN 0–08–032766–4.

Laszlo, Ervin The Inner Limits of Mankind: heretical reflections on today's values, culture and politics (1989)
London, Oneworld Publications, 143 p.
ISBN 1–85168–015–2.

Lathrop, J W Planning for Rare Events: nuclear accident preparedness (1981)
Elmsford NY, Pergamon Books, 280 p.
ISBN 0–08–028703–4.

Laubenfels, Jean M The Gifted Student: an annotated bibliography (1977)
Westport CT, Greenwood Press. ISBN 0–8371–9760–0.

Lauer, Hans E; Castelliz, K and Davies, Saunders Aggression and Repression in the Individual and Society (1981)
Hudson NY, Anthroposophic Press, 111 p.
ISBN 0–85440–359–0.

Lauer, Robert Social Problems and the Quality of Life (1988)
Dubuque IA, William C Brown. 4th ed.
ISBN 0–697–03178–0.

Lauer, Robert H Social Problems and the Quality of Life (1986)
Dubuque IA, William C Brown, 592 p.
ISBN 0–697–00434–1.

Laurence, John The History of Capital Punishment (1983)
New York, Citadel Press, 230 p. illus.
ISBN 0–8065–0840–X.

Laurie, G and Raymond, J Proceedings of Rehabilitation Gazette's Second International Post–Polio Conference and Symposium on Living Independently with Severe Disability (1984)
Saint Louis MO, Gazette International Networking Institute, 74 p. ISBN 0–931301–01–7.

Laurie, G and Raymond, J Proceedings of GINI's Third International Polio and Independent Living Conference, May 10–12, 1985, St Louis, Missouri (1986)
Saint Louis MO, Gazette International Networking Institute, 68 p. ISBN 0–931301–02–5.

Lauterpacht, Hersh Recognition in International Law
New York, AMS Press. ISBN 0–404–15338–0.

Lawlor, William R (Ed) Cross–Border Transactions Between Related Companies (1985)
Dordrecht, Kluwer Academic Publishers Group, 312 p.
ISBN 90–6544–232–4.

Lawrence, Peter (Ed) World Recession and the Food Crisis in Africa (1986)
Boulder CO, Westview Press, 320 p.
ISBN 0–8133–0511–X.

Lawson, Annette Adultry: an analysis of love and betrayal (1988)
New York, Basic Books. ISBN 0–465–00075–4.

Lawson, Edward H Encyclopedia of Human Rights (1990)
New York, Taylor and Francis, 800 p.
ISBN 0–8002–8003–2.

Le Magnen, Jacques Hunger (1986)
New York, Cambridge University Press, 176 p.
ISBN 0–521–26450–2.

Lea, David A M and Chaudhri, D P (Eds) Rural Development and the State: contradictions and dilemmas in developing countries (1983)
London, Methuen, 351 p. illus. bibl.

Lea, Henry C and Burr, George L Materials Toward a History of Witchcraft
New York, AMS Press. 3 vols. ISBN 0–404–18420–0.

Lea, John Tourism and Development in the Third World (1988)
New York, Routledge Chapman and Hall, 80 p.
ISBN 0–415–00671–6.

Leahy, Robert L The Child's Construction of Social Inequality (1983)
San Diego CA, Academic Press. ISBN 0–12–439880–4.

Leatherman, K D and Dickson, R A The Management of Spinal Deformities (1988)
Littleton MA, PSG Publishing, 672 p. ISBN 0–7236–0740–0.

Leavitt, Jerome E (Ed) Child Abuse and Neglect: research and innovation: proceedings of the NATO Advanced Study Institute on Research and Innovations in Child Abuse and Neglect, Les Arcs, France, 28 June, 10 July, 1982 (1983)
Dordrecht, Kluwer Academic Publishers Group, 352 p.
ISBN 90–247–2862–2.

Leblanc, The OAS and the Promotion and Protection of Human Rights (1977)
Norwell MA, Kluwer Academic Publishers.
ISBN 90–247–1943–7.

Lebow, Richard N Nuclear Crisis Management: a dangerous illusion (1988)
Ithaca NY, Cornell University Press, 232 p. Cornell Studies in Security Affairs. ISBN 0–8014–9531–8.

Lecaillon Jacques, et al Income Distribution and Economic Development (1986)
Geneva, ILO, 211 p. ISBN 92–2–103559–X.

Lee, C W and Maurice, J M African Trypanosomiasis (1983)
Washington DC, International Bank for Reconstruction and Development, 107 p. ISBN 0–8213–0191–8.

Lee, Chin–Chuan and Katz, Elihu Media Imperialism Reconsidered: the homogenizing of television culture (1980)
Newbury Park CA, Sage Publications, 276 p.
ISBN 0–8039–1495–4.

Lee, Douglas H Environmental Factors in Respiratory Disease (1972)
San Diego CA, Academic Press. ISBN 0–12–440655–6.

Lee, Rance P L Corruption and Its Control in Hong Kong (1981)
Bangkok, Chualongkorn University Press, 221 p.
ISBN 962–201–251–5.

Lee–Wright, Lee Child Slaves (1990)
London, Earthscan, 288 p. ISBN 1–85383–044–5.

Lee, Yong Leng Razor's Edge: boundaries and boundary disputes in Southeast Asia (1980)
Singapore, Institute of Southeast Asian Studies, 29 p.

Leeuwenberg, A J Medical and Poisonous Plants of the Tropics (1987)
Lanham MD, UNIPUB, 152 p. ISBN 90–220–0921–1.

LeGrand, Julian and Robinson, Ray The Economics of Social Problems (1980)
San Diego CA, Harcourt Brace Jovanovich, 200 p.
ISBN 0–15–518910–7.

Lehner, T and Cimasoni, G The Borderland Between Caries and Periodontal Disease II (1980)
Orlando FL, Grune and Stratton, 288 p.
ISBN 0–8089–1313–1.

Leibee, Howard C Tort Liability for Injuries to Pupils (1965)
Ann Arbor MI, Campus Publishers, 104 p.
ISBN 0–87506–009–9.

Leiblum, Sandra R and Rosen, Raymond C Sexual Desire Disorders (1988)
New York, Guilford Press, 470 p. ISBN 0–89862–714–1.

Leibowitz, Howard M Corneal Disorders: clinical diagnosis and management (1984)
Philadelphia PA, WB Saunders, 550 p. Illus.
ISBN 0–7216–5727–3.

Leitch, D A Environmental Impact and Regulation of Recreational All–Terrain Vehicles in Manitoba (1975)
Winnipeg, Natural Resources Institute, 79 p.

Leitner, Michael J and Leitner, Sara F Leisure in Later Life: a sourcebook for the provision of recreation services for elders (1986)
New York, Haworth Press, 340 p. Activities, Adaptation and Aging Ser: Vol 7 Nos 3–4. ISBN 0–86656–452–7.

Lella, Joseph W Perils of Patient Government: professionals and patients in a chronic–care hospital (1986)
Waterloo ON, Wilfrid Laurier University Press, 232 p.
ISBN 0–88920–197–8.

Lenin, V I Imperialism: the highest stage of capitalism
Beijing, Foreign Languages Press, 188 p.

Lenin, V I Marxism and Revisionism
Tirana, Book Distribution Enterprise, 160 p.

Lenin, V I Against Dogmatism and Sectarianism (1978)
Los Angeles CA, Progress Publishing, 215 p.
ISBN 0–8285–0066–5.

Lenin, V I Against Liquidationism (1980)
Moscow, Izdatel'stvo Progress, 320 p.

Lenin, V I Against Right–Wing and Left–Wing Opportunism, Against Trotskyism (1983)
Moscow, Izdatel'stvo Progress, 608 p.

Lentz, Harris M Assassinations and Executions: an encyclopedia of political violence, 1865–1986 (1988)
Jefferson NC, McFarland Publishers, 296 p.
ISBN 0–89950–312–8.

Lenz, George, et al Opiates (1986)
San Diego CA, Academic Press. ISBN 0–12–443830–X.

Leong, C K Children With Specific Reading Disabilities (1987)
Lisse, Swetz en Zeitlinger, 374 p.
ISBN 90–265–0291–5. Modern Approaches to the Diagnostic and Instruction of Multi–Handicapped Children: Vol 19.

Lessof, M H Allergy: an international text book (1988)
Baltimore MD, Williams and Wilkins, 655 p.
ISBN 0–683–04951–8.

Lethbridge, H J Hard Graft in Hong Kong: scandal, corruption and the ICAC (1985)
Melbourne VIC, Oxford University Press, 256 p.
ISBN 0–19–583896–3.

Lever, R J A W Pests of the Coconut Palm (1980)
Rome, FAO, 190 p. 2nd ed.
FAO Plant Protection and Protection Series: No 18.
ISBN 92–5–100857–4.

Levi, L Stress in Industry: causes, effects and prevention (1984)
Geneva, ILO, 59 p. ISBN 92–2–103826–2.

Levi, L and International Labour Office Stress in Industry: cause, effects, and prevention (1984)
Washington DC, ILO, 70 p. ISBN 92–2–103539–5.

Levi, Lennart Society, Stress, and Disease: the productive and reproductive age, male–female (1978)
New York, Oxford University Press. Vol. 3
ISBN 0–19–261306–5.

Levie, Howard S Law of War and Neutrality: a selective english language bibliography (1988)
Dobbs Ferry NY, Oceana Publications.
ISBN 0–379–20914–4.

Levin, Bernard (Ed) Gastrointestinal Cancer: current approaches to diagnosis and treatment (1987)
University of Texas Press, 439 p.
M D Anderson Hospital and Tumor Institute Clinical Conference on Cancer Ser: No 30. ISBN 0–292–72735–6.

Levin, D M Pathologies of the Modern Self (1987)
New York, University Press.

Levin, Harvey J The Invisible Resource: use and regulation of the radio spectrum
Baltimore MD, Johns Hopkins University Press, 432 p.
ISBN 0–8018–1316–6.

LeVine, Victor Political Corruption and the Informal Policy (1971)
Accra, Ghana Universities Press, 21 p.
ISBN 9964–3–0026–3.

Lévy–Leboyer, Claude (Ed) Vandalism: behaviour and motivations (1984)
Amsterdam, Elsevier Science Publishing, 364 p.
ISBN 0–444–86775–9.

Levy, Oscar The Revival of Aristocracy
New York, Gordon Press Publishers. ISBN 0–8490–0952–9.

Lewin, B Sex and Family Planning: how we teach the young: report on a study (1984)
Geneva, WHO, 170 p. ISBN 92–890–1159–9.

Lewin, Keith Educational Finance in Recession
In: *Prospects: quarterly review of education* 1986, XVI, 2.

Lewis Anesthesiology Malpractice (1989)
New York, John Wiley and Sons.
ISBN 0–471–60954–4. Medico–Legal Library.

Lewis, C S The Problem of Pain (1978)
New York, Macmillan. ISBN 0–02–086850–2.

Lewis, Dan A and Salem, Greta W Fear of Crime: incivility and the production of a social problem (1986)
New Brunswick NJ, Transaction Books, 145 p.
ISBN 0–88738–086–7.

Lewis, Michael (Ed) Research in Social Problems and Public Policy
Greenwich CT, Jai Press. 4 Vols.

Lewis, Michael Research in Social Problems and Public Policy (1979)
Greenwich CT, Jai Press. Vol. 1 ISBN 0–89232–068–0.

Lewis, Michael and Beck, Bernard The Culture of Inequality (1978)
Amherst MA, University of Massachusetts Press, 224 p.
ISBN 0–87023–247–9.

Lewis, Michael and Rosenblum, Leonard A The Origins of Fear
Ann Arbor MI, Books on Demand. ISBN 0–317–08216–7.

Lewis, P, et al Visual Blight in America (1973)
Washington DC, Association of American Geographers.
ISBN 0–89291–070–4.

Lewis, Scott M and McCutchen, Jeffrey R Emergency Medical Malpractice (Date not set)
New York, John Wiley and Sons. ISBN 0–471–85110–8.

Leyton–Brown, David The Utility of International Economic Sanctions (1987)
New York, Saint Martin's Press, 336 p.
ISBN 0–312–00369–2.

Leyton, Elliott Myth of Delinquency: an anatomy of juvenile nihilism
Toronto ON, McClelland and Stewart.

Leyton, Elliott Hunting Humans: the rise of the modern multiple murderer (1986)
Toronto ON, McClelland and Stewart.
ISBN 0–7710–5308–8.

Liben, Lynn S Deaf Children: developmental perspectives (1979)
San Diego CA, Academic Press. ISBN 0–12–447950–2.

Licata, Salvatore J and Petersen, Robert P Historical Perspectives on Homosexuality (1982)
New York, Haworth Press, 224 p. ISBN 0–917724–27–5.

Lichtenstein, Michael J Vitamin Deficiencies: index of modern information with bibliography (1988)
Washington DC, ABBE Publications Association, 150 p. bibl.
ISBN 0–88164–888–4.

Lidin, G D Air Pollution in Mines: theory, hazards and controls (1966)
Jerusalem, Keter Publishing House, 304 p.
ISBN 0–7065–0411–9.

Liebich, André (Ed) Future of Socialism in Europe: proceedings of ICES third international colloquium (1979)
Toronto ON, Institute for Christian Studies, 328 p.

Liebl, Hildegard and Liebl, Karlhans International Bibliography of Economic Crime (1985)
Pfaffenweiler, Centaurus-Verlagsgesellschaft, 660 p. ISBN 3-89085-012-X.

Liechti, R Hip Arthrodesis and its Problems (1978)
Berlin, Springer-Verlag. illus. Trans. from Ger: P Casey. ISBN 0-387-08614-5.

Liegeois, P La Scolarisation des Enfants Tziganes et Voyageurs: rapport de synthèse (1986)
Luxembourg, CEE, 286 p. ISBN 92-825-6429-0.

Liehaus, Jacob L Premenstrual Syndrome: index of modern information (1988)
Annandale VA, ABBE Publishers Association of Washington, 150 p. ISBN 0-88164-868-X.

Lier, Irene H van Acid Rain and International Law (1981)
Dordrecht, Kluwer Academic Publishers Group, 220 p. ISBN 90-286-2231-4.

Lifton, Robert Jay and Falk, Richard Indefensible Weapons (1984)
Toronto ON, Canadian Broadcasting Corporation, 192 p. ISBN 0-88794-108-7.

Lightner, Otto C History of Business Depressions: a vivid portrayal of periods of economic (1970)
New York, Franklin Burt. ISBN 0-8337-2061-9.

Lincoln, Alan J Crime in the Library: a study of patterns, impact, and security (1984)
New York, Bowker RR, 179 p. ISBN 0-8352-1863-5.

Lindberg, L N, et al (Eds) Stress and Contradiction in Modern Capitalism (1975)
Lexington MA, D C Heath.

Lindenfeld, Frank Radical Perspectives on Social Problems: readings in critical sociology (1987)
Dix Hills NY, General Hall, 414 p. ISBN 0-930390-74-1.

Lindsay, Jeanne W and Monserrat, Catherine Teenage Marriage: coping with reality (1988)
Buena Park CA, Morning Glory Press, 208 p. ISBN 0-930934-31-8.

Linear, Marcus Zapping the Third World: the disaster of development aid (1985)
London, Pluto Press, 242 p. ISBN 0-7453-0013-8.

Linebarger, Paul M Psychological Warfare
Salem NH, Ayer Company Publishers, 318 p. ISBN 0-405-04755-X.

Lingeman, Richard R Drugs from A to Z: a dictionary (1974)
New York, McGraw Hill Book Company, 320 p. 2nd. ISBN 0-07-037912-2.

Linton, Ralph Culture and Mental Disorders (1956)
Springfield IL, Thomas Charles C.

Lippman, Marc E, et al Diagnosis and Management of Breast Cancer (1988)
Philadelphia PA, WB Saunders, 480 p. illus. ISBN 0-7216-1958-4.

Lipshulz, Larry I and Howards, Stuart S (Eds) Infertility in the Male (1983)
New York, Churchill Livingstone, 409 p. illus. ISBN 0-443-08159-X.

Lipton, Merle Sanctions and South Africa: the dynamics of economic isolation (1988)
London, The Economist Intelligence Unit, 169 p. bibl. ISBN 0-85058-240-7. Special Report: The Economist Intelligence Unit: No 119.

Litan, Robert E and Winston, Clifford Liability: perspectives and policy (1988)
Washington DC, Brookings Institution, 248 p. ISBN 0-8157-5272-5.

Littrell, W Boyd, et al Bureaucracy As a Social Problem (1983)
Greenwich CT, Jai Press. ISBN 0-89232-368-X.

Lloyd, Evan L Hypothermia – Cold Stress (1986)
Aspen Pub, 400 p.

Lloyd, Peter J International Trade Problems of Small Nations
Ann Arbor MI, Books on Demand. ISBN 0-317-20457-2.

Locker, David Disability and Disadvantage: the consequences of chronic illness (1984)
New York, Routledge Chapman and Hall, 220 p. ISBN 0-422-78740-X.

Lodge Drug and Alcohol Abuse in the Workplace: an assessment of economic and productivity losses (1987)
New York, Dekker Marcel, 448 p. ISBN 0-8247-7769-7.

Lodge, Juliet The Threat of Terrorism: combating political violence in Europe (1987)
Hemel Hempstead, Harvester Wheatsheaf, 288 p. ISBN 0-7450-0329-X.

Loeb, Edwin M Blood Sacrifice Complex (1924)
Millwood NY, Kraus Reprint and Periodicals. ISBN 0-527-00529-0.

Logemann, J A Evaluation and Treatment of Swallowing Disorders (1983)
Basingstoke, Taylor and Francis, 262 p. ISBN 0-85066-543-4.

Loiy, Paul and Daisey, Joan M Toxic Air Pollution (1987)
Chelsea MI, Lewis Publishers, 300 p. illus. ISBN 0-87371-057-6.

Lonetto, Richard and Templer, Donald E Death Anxiety (1986)
New York, Hemisphere Publishing, 155 p. Series in Health Psychology and Behavariol Medicine. ISBN 0-89116-554-1.

Long, Duncan Surviving Major Chemical Accidents and Chemical-Biological Warfare (1986)
Port Townsend WA, Loompanics Unlimited, 152 p. ISBN 0-915179-38-5.

Long, Frank Restrictive Business Practices, Transnational Corporations and Development (1981)
Norwell MA, Kluwer Academic Publishers, 192 p. ISBN 0-89838-057-X.

Long, M and Roberson, B S World Problems: a topic geography (1973)
London, English Universities Press, 216 p.

Loomis, Mildred Major Universal Problems (1978)
Spring Grove PA, School of Living Adult Education, 83 p.

Looney, John G (Ed) Mental Illness in Children and Adolescents
Cambridge MA, Cambridge University Press, 284 p. ISBN 0-521-35907-4.

Lopez, Alan D and Ruzicka, Lado T (Eds) Sex Differentials in Mortality: trends, determinants and consequences; selection of the papers presented at the ANU/UN/WHO Meeting held in Canberra (Australia) 1-7 December 1981
Canberra, Australian National University.

Lopez de la Osa, L (Ed) Aspects and Treatment of Vulvar Cancer: 1st international symposium on vulvar cancer, Madrid 1979
New York, Karger S AG, 206 p. ISBN 3-8055-1399-2.

Lopez, George A, et al Testing Theories of State Violence, State Terror, and Repression (1988)
Boulder CO, Westview Press, 265 p. ISBN 0-8133-7525-8.

Lopez-Rey, Manual Guide to United Nations Criminal Policy (1985)
Brookfield VT, Gower Publishing, 200 p. ISBN 0-566-05070-6.

Lorenz, Konrad The Waning of Humaneness (1988)
London, Unwin Paperbacks, 250 p. First published in German, 1983. Trans. by Robert Warren Kickert. ISBN 0-04-440442-5.

Losse, H, et al Pyelonephritis: urinary tract infections (1984)
New York, Thieme Medical Publishers, 179 p. Vol. 5 ISBN 0-86577-145-6.

Lotz, Robert Inter-Noise 86
Poughkeepsie NY, Noise Control Foundation, 1472 p. 2 Vols. ISBN 0-317-56939-2.

Louis Harris and Associates Victims of Crime: a research report of experiencing victimization (1983)
New York, Garland Publishing, 64 p. ISBN 0-8240-9145-0.

Lowder, Stella Inside Third World Cities (1986)
London, Routledge, 304 p. ISBN 0-7099-1647-7.

Lowry, Ritchie P Social Problems: a critical analysis of theories and public policy (1974)
Lexington MA, D C Heath.

Lozano, J Carlos (Ed) Fifth International Conference on Plant Pathogenic Bacteria (1981)
Cali, Centro Internacional de Agricultura Tropical, 640 p. ISBN 84-89206-21-X.

Lozoya, Jorge A (Ed) The Social and Cultural Issues of the New International Economic Order (1981)
Elmsford NY, Pergamon Books. ISBN 0-08-025123-4.

Lucas, G and Synge, H IUCN Plant Red Data Book (1978)
Gland, IUCN, 540 p. ISBN 2-88032-202-2.

Ludlow, Christy L and Cooper, Judith A Genetic Aspects of Speech and Language Disorders (1983)
San Diego CA, Academic Press. ISBN 0-12-459350-X.

Ludz, Peter C Alienation as a Concept in the Social Sciences: a trend report and bibliography prepared for the International Sociological Association under the auspices of the International Committee for Social Sciences Information and Documentation (1975)
Berlin, Mouton de Gruyter. Current sociology – La sociologie contemporaine: Vol 21, No 1. ISBN 90-2797-841-7.

Luisada, Aldo A Pulmonary Edema in Man and Animals (1970)
Saint Louis MO, Warren H Green, 168 p. illus. ISBN 0-87527-050-6.

Lukasiewicz, J Railway Game: a study in socio-technological obsolescence
Bangkok, Chualongkorn University Press. ISBN 0-7710-9905-3.

Lundberg, Erik Instability and Economic Growth
Ann Arbor MI, Books on Demand. illus. Studies in Comparative Economics: No 8. ISBN 0-317-09709-1.

Lundborg, Per The Economics of Export Embargoes (1987)
Croom Helm, Routledge Chapman and Hall, 128 p. ISBN 0-7099-4151-X.

Lundin, S J (Ed) Non-Production by Industry of Chemical-Warfare Agents: technical verification under a chemical weapons convention (1989)
Oxford, Stockholm International Peace Research Institute, 280 p. illus. SIPRI Chemical and Biological Warfare Studies Series: 9. ISBN 0-19-829129-9.

Luntz, Maurice H (Ed) Glaucoma: international viewpoints (1989)
Great Neck NY, PMA Publishing, 250 p. illus. ISBN 0-89335-312-4.

Luria, A R Traumatic Aphasia: its syndromes, psychology and treatment (1970)
Berlin, Mouton de Gruyter, 480 p. ISBN 90-279-0717-X.

Lynch, J The Broken Heart: the medical consequences of loneliness (1977)
New York, Basic Books.

Lynn, Robert E Chicorel Abstracts to Reading and Learning Disabilities (1984)
New York, American Library Publishing, 325 p. ISBN 0-934598-85-1.

Lyon, Bryce D From Fief to Indenture: the transition from feudal to non-feudal contract in Western Europe (1971)
New York, Hippocrene Books. ISBN 0-374-95211-6.

Lyubashenko, S Y (Ed) Diseases of Fur-Bearing Animals (1983)
New Delhi, Amerind Publishing Company, 402 p.

Mabbutt, J and Wilson, A (Eds) Social and Environmental Aspects of Desertification (1980)
Tokyo, United Nations University, 40 p. ISBN 92-808-0127-9.

Mabro, Robert (Ed) The 1986 Oil Price Crisis: economic effects and policy responses: proceedings of the Eight Oxford Energy Seminar, September 1986 (1988)
Oxford NY, Oxford University Press, 286 p. illus. bibl. ISBN 0-19-730007-3.

MacBean, Alasdair I Export Instability and Economic Development (1966)
Cambridge MA, Harvard University Press. Center for International Affairs Ser. ISBN 0-674-28600-6.

Macdermott, R P (Ed) Inflammatory Bowel Disease – Current Status and Future Approach: proceedings of the international symposium on future research approaches – mechanisms of chronic infection and inflammation (1988)
Amsterdam, Elsevier Science Publishing, 748 p. International Congress Ser: No 775.

Macdonald, John M Indecent Exposure (1973)
Springfield IL, Thomas Charles C, 180 p. ISBN 0-398-02598-3.

MacDonald, Scott B Dancing on a Volcano: the Latin American drug trade (1988)
New York, Praeger Publishers, 176 p. ISBN 0-275-93105-6.

Mace, Gillian S, et al The Bereaved Child: analysis, education, and treatment (1981)
New York, Plenum Publishing, 292 p. ISBN 0-306-65197-1.

Mace, M E and Bell, A A Fungal Wilt Diseases of Plants (1981)
San Diego CA, Academic Press. ISBN 0-12-464450-3.

Macesich, George The International Monetary Economy and the Third World (1981)
New York, Praeger Publishers, 314 p. ISBN 0-275-90674-4.

MacFarlane, S N Superpower Rivalry and Third World Radicalism (1985)
London, Croom Helm.

MacFarlane, T W and Samaranayake, L P Oral Candidosis (1989)
Sevenoaks, Butterworths, 256 p. ISBN 0-7236-0983-7.

Machado, Manuel A An Industry in Crisis: Mexican-United States cooperation in the control of foot and mouth disease
Ann Arbor MI, Books on Demand. University of California Publications in History: Vol 80.

Mackay, Charles and Templeton, John M Extraordinary Popular Delusions and the Madness of Crowds (1985)
Fort Lauderdale FL, Templeton Publications, 730 p. ISBN 0-934405-00-X.

MacKenzie, Andrew Hauntings and Apparitions: an investigation of the evidence
Chicago IL, Academy Chicago Publishers, 288 p. ISBN 0-586-08430-4.

Mackenzie, F Child Health and Mortality in Sub-Saharan Africa-Annotated Bibl. of 1975-86 Lit. – Sante et Mortalite Infantiles: en afrique subsaharienne-bibliographie annotee (1987)
Lanham MD, UNIPUB, 223 p. ISBN 0-88936-499-0.

Mackenzie, J S Viral Diseases in South East Asia and the Western Pacific (1983)
San Diego CA, Academic Press, 672 p. ISBN 0-12-484820-6.

Mackenzie, R Alec The Time Trap (1975)
New York, McGraw Hill Book Company, 208 p. ISBN 0-07-044650-4.

MacKie, Rona M Eczema and Dermatitis (1983)
New York, Arco Pub, 112 p. illus. Positive Health Guides Ser.

MacKinnon, Catharine A and Emerson, Thomas I Sexual Harassment of Working Women: a case of sex discrimination (1979)
New Haven CT, Yale University Press. ISBN 0-300-02298-0.

Macklin, R and Gaylin, W Mental Retardation and Sterilization: a problem of competency and paternalism (1981)
New York, Plenum Publishing, 274 p. ISBN 0-306-40689-6.

MacMillan, Donald Mental Retardation in School and Society (1982)
Glenview IL, Scott Foresman. ISBN 0-673-39164-7.

MacMunn, George F Leadership Through the Ages (1935)
Salem NH, Ayer Company Publishers. ISBN 0-8369-0657-8.

Macnicol, M F The Problem Knee: diagnosis and management of the younger patient (1986)
Oxford, Heineman Professional Publishing, 220 p. illus. ISBN 0-433-20130-4.

MacPherson, C B The Political Theory of Possessive Individualism (1962)
New York, Oxford University Press.

MacSweeney, D The Crazy Ape: sanity, madness, your brain, and you (1982)
Atlantic Highlands NJ, Humanities Press International, 244 p. ISBN 0-7206-0547-4.

Madkour, Monir M (Ed) Brucellosis (1989)
Sevenoaks, Butterworths, 336 p. illus. ISBN 0-7236-0941-1.

Maestri, William Choose Life and Not Death: a primer on abortion, euthanasia, and suicide
Staten Island NY, Alba House. ISBN 0-8189-0490-9.

Magel, Charles R A Bibliography on Animal Rights and Related Matters (1981)
Washington DC, University Press of America.

Maggied, H S Transportation for the Poor: research in rural mobility (1982)
Dordrecht, Kluwer Academic Publishers Group, 224 p. ISBN 0-89838-081-2.

Maggiore, Dolores J Lesbianism: an annotated bibliography (1987)
Metuchen NJ, Scarecrow Press, 156 p.
ISBN 0-8108-2048-X.

Mahaffey, Kathryn (Ed) Dietary and Environmental Lead: human health effects (1985)
Amsterdam, Elsevier Science Publishing, 460 p.
ISBN 0-444-80609-1.

Mahendra, B Depression (1987)
Dordrecht, Kluwer Academic Publishers Group.
ISBN 0-85200-983-6.

Maher, Peter Child Abuse (1987)
New York, Basil Blackwell, 200 p. ISBN 0-631-15071-4.

Maier, Norman R Frustration: the study of behavior without a goal (1982)
New York, Greenwood Press, 264 p. illus. 2nd ed.
ISBN 0-313-23340-3.

Malamuth, Neil M and Donnerstein, Edward Pornography and Sexual Aggression (1984)
San Diego CA, Academic Press. ISBN 0-12-466280-3.

Malatesha, R N and Aaron, P G Reading Disorders: varieties and treatment (1982)
San Diego CA, Academic Press. ISBN 0-12-466320-6.

Malatesha R N and Whitaker, Harry A (Eds) Dyslexia: a global issue (1984)
Dordrecht, Martinus Nijhoff. NATO Series D: Advanced Science Institutes Series; Behavioral and Social Sciences.
ISBN 90-247-2909-2.

Malcolmson, Robert W Nuclear Fallacies: how we have been misguided since Hiroshima (1985)
Montreal, McGill-Queen's University Press, 128 p.
ISBN 0-7735-0585-7.

Malik, S Surendra Supreme Court on Preventive Detention from 1950 to Present (1985)
New York, State Mutual Book and Periodical Service, 478 p.
ISBN 0-317-54842-5.

Malinvaud, Edmond Mass Unemployment (1985)
New York, Basil Blackwell, 140 p. ISBN 0-631-13704-1.

Mallen, Ronald E and Smith, Jeffrey M Legal Malpractice (1988)
Saint Paul MN, West Publishing College and School, 2000 p. 2 Vols. ISBN 0-314-06927-5.

Mallon, Thomas Stolen Words: forays into the origins and ravages of plagiarism
New York, Ticknor and Fields, 300 p.

Malmberg, Torsten Human Territoriality: survey on the behavioural territories of Man with preliminary discussion of meaning (1980)
Berlin, Mouton de Gruyter, 346 p. ISBN 90-279-7948-0.

Maloney, Clarence (Ed) The Evil Eye (1976)
New York, Columbia University Press, 334 p. illus.
ISBN 0-231-05825-X.

Malony, H Newton Clergy Malpractice (1986)
St Louisville KY, John Knox Westminster, 192 p.
ISBN 0-664-24591-9.

Maltsev, G Illusion of Equal Rights: legal inequality in the capitalist world (1984)
Moscow, Izdatel'stvo Progress, 256 p.

Maltz, Michael D Recidivism (1984)
San Diego CA, Academic Press. ISBN 0-12-468980-9.

Mandela, Nelson L'Apartheid: précédé d'une lettre de Breyten Breytenbach (1985)
Paris, Les Editions de Minuit, 111 p. ISBN 2-7073-0549-9.

Manis, Jerome G Serious Social Problems (1983)
Needham Heights MA, Allyn and Bacon.
ISBN 0-205-08044-8.

Manis, Jerome G Serious Social Problems (1983)
Needham Heights MA, Allyn and Bacon.
ISBN 0-205-08044-8.

Mankabady, Samir The International Maritime Organisation: accidents at sea (1986)
Croom Helm, Routledge Chapman and Hall, 350 p. Vol 2.
ISBN 0-7099-4640-6.

Manning, N P Social Problems and Welfare Ideology (1985)
Brookfield VT, Gower Publishing, 238 p.
ISBN 0-566-00938-2.

Manocha, B L Marriage Conflicts: causes and cures (1983)
New Delhi, Sterling Publishers, 288 p.
ISBN 81-207-0478-9.

Manocha, Sohan L Malnutrition and Retarded Human Development (1972)
Springfield IL, Thomas Charles C, 400 p. illus.

Manshard, Walther and Ruddle, Kenneth Renewable Natural Resources and the Environment: pressing problems in the developing world (1981)
London, Tycooly International, 396 p.
ISBN 0-907567-01-0.

Mansukhani, H L Jungle of Customs Law and Procedures (1974)
New Delhi, Vikas Publishing House, 752 p.

Mansukhani, H L Corruption and Public Servants (1979)
New Delhi, Vikas Publishing House, 346 p.
ISBN 0-7069-0730-2.

Mao Zedong On the Correct Handling of Contradictions among the People (1966)
Beijing, Foreign Languages Press, 64 p.

Maragos, G D Seminar on Pediatric Allergy (1977)
New York, Karger S AG, 80 p. ISBN 3-8055-2648-2.

Maramorosch, K and Harris, K F Leafhopper Vectors and Plant Disease Agents (1979)
San Diego CA, Academic Press. ISBN 0-12-470280-5.

Maramorosch, K and Raychaudhuri, S P (Eds) Mycoplasma Diseases of Crops: basic and applied aspects (1988)
Berlin, Springer-Verlag, 450 p. ISBN 3-540-96646-3.

Maramorosch, Karl and Harris, Kerry Plant Diseases and Vectors: ecology and epidemiology (1981)
San Diego CA, Academic Press. ISBN 0-12-470240-6.

Maramorosch, Karl and Raychaudhuri, S P Mycoplasma Diseases of Trees and Shrubs (1981)
San Diego CA, Academic Press. ISBN 0-12-470220-1.

Maramorsoch, Karl and McKelvey, John J Subviral Pathogens of Plants and Animals: viroids and prions (1985)
San Diego CA, Academic Press. ISBN 0-12-470230-9.

Maran, A G Logan Turner's Diseases of the Nose, Throat and Ear (1988)
Littleton MA, PSG Publishing, 464 p. ISBN 0-7236-0945-4.

Marberger, H, et al Prostatic Disease: proceedings of the american - european (1976)
New York, Alan R Liss, 432 p. ISBN 0-8451-0006-8.

March, Audrey and Markel, Lester Global Challenge to the United States: a study of the problems, the perils and the proposed solutions involved in Washington's search for a new rol (1975)
Cranbury NJ, Fairleigh Dickinson University Press, 241 p.
ISBN 0-8386-1822-7.

Marcu, E D Nationalism in the Sixteenth Century (1975)
Pleasantville NY, Abaris Books. ISBN 0-913870-08-0.

Mardh, P A, et al Chlamydia (Date not set)
New York, Plenum Publishing, 325 p.
ISBN 0-306-42965-9.

Margolis, Stephen Sexually Transmitted Diseases: an annotated selective bibliography (1984)
New York, Garland Publishing, 200 p.
ISBN 0-8240-9092-6.

Maric, D Adaptating Working Hours to Modern Needs: the time factor in the new approach to working conditions (1980)
Geneva, ILO, 50 p. ISBN 92-2-101659-5 (Hard Cover).

Marien, Michael A Future Survey Guide to 50 Overviews and Agendas
In: *Futures Research Quarterly* Bethesda MD, World Future Society, Special Issue, Spring 1990, 6, 1, pp. 103-112.

Marien, Michael (Ed) Future Survey: a guide to recent books an articles concerning trends, forecasts and policy proposals
Bethesda MD, World Future Society. Annual.

Marien, Michael (Comp) Societal Directions and Alternatives: a critical guide to the literature (1975)
LaFayette NY, Information for Policy Design.

Mark, Jeffrey An Analysis of Usury (1980)
New York, Gordon Press Publishers. ISBN 0-8490-3085-4.

Markovits, Andrei S and Deutsch, Karl W Fear of Science-Trust in Science: conditions for change in the climate of opinion (1980)
Boston MA, Oelgeschlager Gunn and Hain, 288 p.
ISBN 0-89946-038-0.

Markowitz, Irving L Power and Class in Africa: an introduction to change and conflict in African politics (1977)
Englewood Cliffs NJ, Prentice Hall. ISBN 0-13-686642-5.

Markowitz, Milton and Gordis, Leon Rheumatic Fever
Ann Arbor MI, Books on Demand. ISBN 0-317-26114-2.

Marks, Joel (Ed) The Ways of Desire: new essays in philosophical psychology on the concept of wanting (1986)
Chicago IL, Precedent Publishing. ISBN 0-913750-44-1.

Marongiu, Pietro and Newman, Graeme Vengeance: the fight against injustice (1987)
Totowa NJ, Rowman and Littlefield Publishers, 176 p.
ISBN 0-317-56047-6.

Marquardt, Thomas P Acquired Neurogenic Disorders (1982)
Englewood Cliffs NJ, Prentice Hall, 208 p.
ISBN 0-13-003814-8.

Marris, Peter and Rein, Martin Dilemmas of Social Reform: poverty and community action in the US (1982)
Chicago IL, University of Chicago Press.
ISBN 0-226-50657-6.

Marrus, Michael R The Unwanted: european refugees in the twentieth century (1987)
New York, Oxford University Press, 432 p.
ISBN 0-19-505186-6.

Mars, Gerald Cheats at Work: an anthology of workplace crime (1983)
London, Unwin Hyman, 242 p. Counterpoint Ser.
ISBN 0-04-301166-7.

Marsh, Peter and Campbell, Anne Aggression and Violence (1982)
St Martin, 256 p.
ISBN 0-312-01402-3.

Marshall, Victor C Major Chemical Hazards (1987)
New York, Halsted Press, 587 p. ISBN 0-470-20813-9.

Martens, Richard A and Martens, Sherlyn Milk Sugar Dilemma: living with lactose intolerance (1987)
East Lansing MI, Medi-Ed Press, 260 p.
ISBN 0-936741-01-5.

Martin, A de and Lemkowitz, Saul Environmentally Acceptable Incineration of Chlorinated Chemical Waste: review of theory and practice
Delft, Delftse Universitaire Pers. ISBN 90-6275-229-2.

Martin, Bradley E The International Trade in Rhinoceros Products (1980)
Gland, IUCN, 83 p. illus. ISBN 2-88032-203-0.

Martin, Esmond Bradley International Trade in Rhinoceros Products (1980)
Gland, IUCN, 83 p.

Martin, Frederick N Hearing Disorders in Children: pediatric audiology (1987)
Austin TX, Pro-Ed, 502 p. ISBN 0-89079-144-9.

Martin, Gray and Pear, Joseph Behavior Modification and What it is and How to do it (1988)
Englewood Cliffs NJ, Prentice Hall, 576 p. illus. 3rd ed.

Martin, W B (Ed) Respiratory Diseases in Cattle: a seminar of the EEC Programme of Coordination of Research on Beef Production (1977, Edingburgh) (1978)
Dordrecht, Kluwer Academic Publishers Group, 574 p.
ISBN 90-247-2134-2.

Martin, W B Diseases of Sheep (1983)
Oxford, Blackwell Scientific Publications, 294 p. illus.
ISBN 0-632-01008-8.

Martineau, R Insects Harmful to Forest Trees (1984)
Niles IL, Polyscience, 288 p. ISBN 0-919868-21-5.

Martinez-Palomo, A (Ed) Amebiasis (1986)
Amsterdam, Elsevier Science Publishing, 270 p.
ISBN 0-444-80728-4.

Martinez-Palomo, A (Ed) Ambebiasis (1986)
Amsterdam, Elsevier Science Publishing, 270 p. Human Parasitic Diseases: 2. ISBN 0-444-80728-4.

Martini, M Tuberculosis of the Bones and Joints (1988)
New York, Springer Publishing, 230 p.
ISBN 0-387-18166-0.

Martinsons, Janis School of Wisdom: a project for reforms of schools
Silverdale NSW, Janis Martinsons.

Martyn, E B (Ed) Plant Virus Names: an annotated list of names and synonyms of plant viruses and diseases (1968)
Kew, CAB International Mycological Institute, 240 p. Phytopathological Papers: No 9.

Martyn, E B (Ed) Additions and corrections to phytopathological paper no 9, 1968, and newly recorded plant virus names (1971)
Kew, CAB International Mycological Institute, 41 p. Supplement: 1.

Marwick, Max G Witchcraft and Sorcery (1987)
New York, Penguin Books, 494 p. ISBN 0-14-022678-8.

Mashbits, Ya G and Utkin, G N (Comp) Problems of Economic Regionalization in the Developing Countries (1984)
Moscow, Institute of Geography, 167 p. illus. bibl.

Mason, Abelle Ports of Entry: social concerns (1985)
San Diego CA, Harcourt Brace Jovanovich, 186 p.
ISBN 0-15-570749-3.

Mason, Edward S Controlling World Trade: cartels and commodity agreements
Salem NH, Ayer Company Publishers, 308 p.
ISBN 0-405-04574-3.

Mason, Henry L Mass Demonstrations Against Foreign Regimes (1966)
New Orleans LA, Tulane University. Vol. 10
ISBN 0-930598-09-1.

Mason, Jim and Singer, Peter Animal Factories (1982)
New York, Crown Publishers.

Massie, Henry N and Rosenthal, Judith Childhood Psychosis in the First Four Years of Life
New York, McGraw-Hill Book Company, 256 p.
ISBN 0-07-040765-7.

Masters, N C and Shapiro, H A Medical Secrecy and the Doctor-Patient Relationship (1966)
Rotterdam, AA Balkema, 99 p.

Masters, R E Eros and Evil: the sexual psychopathology of witchcraft
New York, AMS Press. ISBN 0-404-18427-8.

Masters, R E L The Hidden World of Erotica: an objective re-examination of perverse sex practices in different cultures (1973)
Lyrebird Press.

Masters, William H and Johnson, Virginia E Human Sexual Inadequacy (1981)
New York, Bantam Books. ISBN 0-553-26317-X.

MATCOM Shoplifting (1980)
Vienna, MATCOM, 35 p.
ISBN 92-2-10248-4.

MATCOM Food Spoilage and Preservation. A Learning Element for Staff of Consumer Cooperatives (1984)
Vienna, MATCOM, 36 p. ISBN 92-2-103693-6.

MATCOM Leakage. A Learning Element for Staff of Consumer Cooperatives (1986)
Vienna, MATCOM, 28 p.
ISBN 92-2-102049-5.

Mather, T H Environmental Management for Vector Control in Rice Fields (1985)
Lanham MD, UNIPUB, 152 p. ISBN 92-5-102104-X.

Mathews, Andrew M, et al Agoraphobia and Treatment (1986)
New York, New York University Press, 233 p.

Mathews, G Harvest from Weather (1967)
Geneva, WMO, 48 p.

Mathews, M R Disclosure of Leases: a survey of the literature with particular reference to lessees (1980)
Kitchener ON, McBain Publications.

Mathis, James L Clear Thinking about Sexual Deviations: a new look at an old problem (1972)
Chicago IL, Nelson-Hall, 232 p. ISBN 0-911012-40-0.

Matjushkin, E N; Smirnov, E N and Zhivotchenko, V I The Amur Tiger in the USSR (1980)
Gland, IUCN, 50 p. illus. ISBN 2-88032-204-9.

Matlby, Edward Waterlogged Earth: why waste the world's wet places? (1986)
Wolfeboro NH, Longwood Publishing Group, 200 p.
ISBN 0-905347-63-3.

Matsumura, Fumio and Murti, Krishna C (Eds) Biodegradation of Pesticides (1982)
New York, Plenum Publishing, 326 p.

Matthews, Catherine J (Ed) Police Stress: a selected bibliography concerning police stress and the psychological evaluation and counseling of police (1979)
Toronto ON, University of Toronto Press, 43 p.
ISBN 0-919584-44-6.

Matthews, Catherine J and Chunn, Dorothy E Congestion and Delay in the Criminal Courts: a selected bibliography (1979)
Toronto ON, University of Toronto Press, 66 p.
ISBN 0-919584-45-4.

Matthews, Gwyneth Ferguson Voices from the Shadows: women with disabilities speak out
Toronto ON, The Women's Press, 200 p.
ISBN 0–88961–080–0.

Maul–Mellott, Susan K and Adams, Jeanette, N Childhood Cancer: a nursing overview (1987)
Boston MA, Jones and Bartlett.

Maunder, J W Human Ectoparasites (1988)
New York, Taylor and Francis, 350 p.
ISBN 0–85066–420–9.

Maurer, K and Wurtman, R J Organic Brain Disorders (1990)
Germany FR, F Vieweg, 64 p.

Maurice–Williams, R S Spinal Degenerative Diseases (1981)
Littleton MA, PSG Publishing, 356 p. ISBN 0–7236–0583–1.

Mauricev, Francis and Chamberlain, Hugh The Diseases of Women with Child and in Childbed (1985)
New York, Garland Publishing, 440 p.
ISBN 0–8240–5925–5.

Mauss, Armand L Social Problems as Social Movements (1975)
Philadelphia PA, J B Lippincott.

Mauvais–Jarvis, P et al Hirsutism (1981)
New York, Springer Publishing, 110 p.
ISBN 0–387–10509–3.

Mawdesley–Thomas, Lionel E, et al Diseases of Fish (1974)
New York, Irvington Publishers, 277 p.
ISBN 0–8422–7178–3.

Mawke, Michael and Jahn, Anthony F Diseases of the Ear: clinical and pathologic aspects (1987)
Philadelphia PA, Lea and Febiger, 324 p.

Maximova, M Global Problems and Peace Among Nations (1982)
New York, State Mutual Book, 80 p. ISBN 0–317–53759–8.

Maxwell, Nicholas From Knowledge to Wisdom: a revolution in the aims and methods of science (1987)
New York, Basil Blackwell, 298 p. ISBN 0–631–13602–9.

Maxwell, Ph D and Milton, A The Alcoholics Anonymous Experience (1984)
New York, McGraw–Hill Book Company, 192 p.
ISBN 0–07–040996–X.

Mayer, E International Congress on Diseases of Cattle (1980) (1981)
Hannover, Schlütersche Verlagsanstalt und Druckerei, 1560 p. 2 Vols.

Mayer, Egon Children of Intermarriage: a study in pattern of identification and family (1983)
New York, American Jewish Committee, 56 p.
ISBN 0–87495–055–4.

Mayes, Andrew R Human Organic Memory Disorders (Date not set)
New York, Cambridge University Press, 324 p.
ISBN 0–521–34418–2.

Mayo, E The Human Problems of an Industrial Civilization (1933)
New York, Macmillan.

Mayo, Elton and Stein, Leon The Social Problems of an Industrial Civilization (1977)
Salem NH, Ayer Company Publishers.
ISBN 0–405–10185–6.

McAllister, James and Wilson, R Glaucoma (1986)
London, Butterworth, 320 p. illus. Butterworths International Medical Reviews Ophamology Ser: Vol 4.
ISBN 0–407–02343–7.

McBrien, David C and Slater, Trevor F Biomedical and Clinical Aspects of Cancer of the Uterine Cervix (1984)
San Diego CA, Academic Press. ISBN 0–12–481760–2.

McCaffrey, Stephan C Private Remedies for Transfrontier Environmental Disturbance (1975)
Gland, IUCN, 156 p.

McCaffrey, Stephan C and Lutz, R E (Eds) Environmental Pollution and Individual Rights: an international symposium (1978)
Dordrecht, Kluwer Academic Publishers Group, 238 p.
ISBN 90–268–1021–0.

McCarthy, Gillian T The Physically Handicapped Child: an interdisciplinary approach to management (1984)
Winchester MA, Faber and Faber, 384 p.
ISBN 0–571–13263–4.

McCarthy, Wendy and Fegan, Lydia Sex Education and the Intellectually Handicapped (1984)
Baltimore MD, Williams and Wilkins, 96 p.
ISBN 0–86792–031–9.

McClellan, George B The Armies of Europe (1976)
New York, Gordon Press Publishers. ISBN 0–8490–1450–6.

McCormick, John Acid Earth: the global threat of acid pollution (1990)
London, Earthscan. ISBN 1–85383–033–X.

McCormick, Richard A and Ramsey, Paul (Eds) Doing Evil to Achieve Good: moral choice in conflict situations (1985)
Washington DC, University Press of America, 274 p.
ISBN 0–8191–4586–6.

McCuen, Gary E Pornography and Sexual Violence (1985)
Hudson WI, McCuen Gary E Publications, 123 p.
ISBN 0–86596–053–4.

McCuen, Gary E Our Endangered Atmosphere: global warming and the ozone layer (1987)
Hudson WI, McCuen Gary E Publications, 133 p.
ISBN 0–86596–063–1.

McCulloch, J Wallace and Prins, Herschel A Signs of Stress: the social problems of psychiatric illness (1978)
Totowa NJ, Biblio Distribution Centre, 207 p.
ISBN 0–7130–0165–8.

McCulloch, Jock Asbestos: its human cost
St Lucia, University of Queensland Press, 320 p.
ISBN 0–7022–2001–9.

McDermott, Jeanne The Killing Winds: the menace of biological warfare (1987)
New York, William Morrow, 230 p. ISBN 0–87795–896–3.

McDonnell, Kathleen (Ed) Adverse Effects: women and the pharmaceutical industry (1986)
Penang, IOCU, 217 p. ISBN 967–9973–17–4.

McDowell, Elizabeth V Educational and Emotional Adjustments of Stuttering Children
New York, AMS Press. ISBN 0–404–55314–1.

McDowell, Gary L Curbing the Courts: the constitution and the limits of judicial power (1988)
Baton Rouge LA, Louisiana State University Press, 248 p.
ISBN 0–8071–1339–5.

McElwain, T J and Selby, P Hodgkin's Disease (1987)
Oxford, Blackwell Scientific Publications, 426 p. illus.
ISBN 0–632–01335–4.

McFarland, Barbara and Baumann, Tyeis B Feeding the Empty Heart: adult children and compulsive eating (1988)
New York, Harper and Row, 128 p. ISBN 0–06–255483–2.

McFarland, Barbara and Baumann, Tyeis B Feeding the Empty Heart: adult children and compulsive eating (1988)
New York, Harper and Row, 128 p. ISBN 0–06–255483–2.

McFerran, J B and McNulty, M S (Eds) Acute Virus Infections of Poultry (1986)
Dordrecht, Kluwer Academic Publishers Group, 256 p.
ISBN 0–89838–809–0.

McGrath, James J and Barnes, Charles D Air Pollution–Physiological Effects (1982)
San Diego CA, Academic Press. ISBN 0–12–483880–4.

McGrath, William J Freud's Discovery of Psychoanalysis: the politics of hysteria (1987)
Ithaca NY, Cornell University Press, 336 p. Paperback Ser.
ISBN 0–8014–9411–7.

McGraw–Hill (Eds) McGraw–Hill Encyclopedia of Energy (1981)
New York, McGraw–Hill Book Company, 840 p. 2nd ed.
ISBN 0–07–045268–7.

McGuffin, John Internment
Elanora Heights, Anvil Press, 228 p. ISBN 0–900068–19–1.

McHale, John World Design Science Decade 1965–1975: five two years phases of a world retooling design proposed to the International Union of Architects for adoption by world architectural schools
Carbondale IL, World Resources Inventory. 2 Vols. Phase I (1965), Document 4: The Ten Year Program, 115 p, illus. Phase II (1967), Document 6: The Ecological Context: energy and materials, 135 p, illus.

McIlroy, Ken School Failure and What to do About It (1979)
Palmerstone North, Dunmore Press, 156 p.
ISBN 0–908564–34–1.

McIlroy, Ken Casebook of Learning Problems: a book for parents and teachers (1986)
Palmerstone North, Dunmore Press, 126 p.
ISBN 0–86469–051–7.

McInnes, John M and Treffry, Jacqueline Deaf–Blind Infants and Children: a developmental guide (1982)
University of Toronto Press, 264 p.
ISBN 0–8020–2415–7.

McIntosh, John L Research on Suicide: a bibliography (1985)
Westport CT, Greenwood Press, 323 p.
ISBN 0–313–23992–4.

McIvor, D Overuse Injuries: the RSI phenomenon (1986)
Carlton South VIC, Pitman Publishing, 96 p.
ISBN 0–85896–374–4.

McKelvey, John J, et al Vectors of Disease Agents: interactions with plants, animals, and men (1981)
New York, Praeger Publishers, 256 p.
ISBN 0–275–90521–7.

McKenna, C J The Economics of Uncertainty (1986)
New York, Oxford University Press, 256 p.
ISBN 0–19–520525–1.

McKenna, Virgina, et al Beyond the Bars: the zoo dilemma (1987)
Wellingborough, Thorsons.

McKenzie, John R (Ed) Uncivilized Races of Men: in All countries of the World
Delhi, Daya Publishing House, 1530 p.
ISBN 81–7035–016–6.

McKern, R B and Lowenthal, G C (Eds) Limits to Prediction
Mosman, Australian Professional Publications.
ISBN 0–949416–02–9.

McKnight, Thomas L Feral Livestock in Anglo–America
Ann Arbor MI, Books on Demand. University of California Publications in Geography: 16. ISBN 0–317–29511–X.

McLaren, D J and Skinner, B J (Eds) Resources and World Development: report of the Dahlem workshop on resources and world development (1987)
New York, John Wiley and Sons, 600 p. Physical, Chemical, Earth Science Ser. ISBN 0–471–91568–8.

McLatchie, Greg Combat Sports Injuries (1982)
New York, State Mutual Book and Periodical Service, 192 p.
ISBN 0–9506989–2–X.

McLaurin, Robert L, et al Spina Bifida: a multidisciplinary approach (1986)
New York, Praeger Publishers, 509 p.
ISBN 0–275–92100–X.

McLennan, Barbara The Impact of Transnational Corporations on Developing Countries: a selected review of the literature and annotated bibliography in the areas of education, science and culture and a report on the science and technology research and training policies of transnational and multinational corporations (1977)
Paris, UNESCO, 58 p.

McMillan, Richard C et al (Ed) Euthanasia and the Newborn: conflicts regarding saving lives (1987)
Dordrecht, Kluwer Academic Publishers Group, 340 p.
ISBN 90–277–2299–4.

McNally, Ruth and Wheale, Peter Genetic Engineering: catastrophe or utopia? (1988)
Hemel Hempstead, Simon and Schuster International, 288 p.
ISBN 0–7450–0010–X.

McNown, John S Staff Development for Institutions Educating and Training Engineers and Technicians: a study dealing with the special problems of developing countries (1977)
Paris, UNESCO, 127 p. bibl.
ISBN 92–3–101444–7. Studies in Engineering Education: No 4.

McRoy, Ruth G, et al Emotional Disturbance in Adopted Adolescents: origins and development (1988)
New York, Praeger Publishers, 229 p.
ISBN 0–275–92913–2.

McWhinney, E Conflicts and Compromise: international law and world order in a revolutionary age (1981)
Dordrecht, Kluwer Academic Publishers Group, 160 p.
ISBN 90–286–2671–9.

McWhinney, Edward Aerial Piracy and International Terrorism: the Illegal diversion of aircraft and international law (1987)
Dordrecht, Kluwer Academic Publishers Group, 64 p.
ISBN 0–89838–919–4.

Mechoulam, Raphael Marijuana (1973)
San Diego CA, Academic Press. ISBN 0–12–487450–9.

Medawar, Charles Drugs and World Health: an international consumer perspective (1984)
Den Haag, IOCU, 64 p.

Meer, Jeff Drugs and Sports (1987)
New York, Chelsea House Publishers, 104 p. illus. Encyclopedia of Psychoactive Drugs Ser: No 2.
ISBN 1–55546–226–X.

Mehl, Lewis E Mind and Matter: healing approaches to chronic illness (1986)
San Francisco CA, Mindbody Communications, 372 p.
ISBN 0–939508–14–1.

Meier, Gerald M Problems of a World Monetary Order (1982)
Oxford, Oxford University Press. illus. 2nd ed.
ISBN 0–19–503010–9.

Meier, Robert F Crime and Society (1989)
Hemel Hempstead, Simon and Schuster International, 600 p.
ISBN 0–205–11770–8.

Meighan, Thomas An Investigation of the Self–Concept of Blind and Visually Handicapped Adolescents (1971)
New York, American Foundation for the Blind, 49 p.
ISBN 0–89128–045–6.

Meister, Robert Hypochondria: toward a better understanding (1980)
New York, Taplinger Publishing. ISBN 0–8008–4044–5.

Mellanby, K Scabies (1972)
New York, State Mutual Book and Periodical Service, 87 p.
ISBN 0–317–07173–4.

Mello, Nancy K Advances in Substance Abuse (1988)
Greenwich CT, Jai Press. Vol. 4 ISBN 0–89232–349–3.

Melnick, J L and Maupas, P Hepatitis B Virus and Primary Liver Cancer (1981)
New York, Karger S AG, 212 p. ISBN 3–8055–1784–X.

Melton, J Gordon Magic, Witchcraft and Paganism in America: a bibliography (1982)
New York, Garland Publishing, 300 p.
ISBN 0–8240–9377–1.

Melvern, Linda, et al Techno–Bandits (1984)
Boston MA, Houghton Mifflin, 305 p. ISBN 0–395–36066–8.

Memmi, Albert The Colonizer and The Colonized: introduction by Jean–Paul Sartre with a new introduction by Liam O'Dowd (1990)
London, Earthscan, 192 p. ISBN 1–85383–070–4.

Mendelievich, Elias (Ed) Child Labour. A Briefing Manual (1980)
Geneva, ILO, 176 p. ISBN 92–2–102165–3.

Mendelsohn, Geoffrey, et al Diagnosis and Pathology of Endocrine Diseases (1988)
New York, Lippincott, 624 p. illus. ISBN 0–397–50731–3.

Mendelsohn, J The Final Solution in the Extermination Camps and the Aftermath (1982)
New York, Garland Publishing, 250 p. The Holocaust Ser.
ISBN 0–8240–4886–5.

Mendelsohn, John Medical Experiments on Jewish Inmates of Concentration Camps (1982)
New York, Garland Publishing, 282 p. The Holocaust Ser.

Mendelson, George and Slovenko, Ralph Psychiatric Aspects of Personal Injury Claims (1988)
Springfield IL, Thomas Charles C, 296 p.
ISBN 0–398–05411–8.

Mendelson, Wallace B The Use and Misuse of Sleeping Pills: a clinical guide (1980)
New York, Plenum Publishing, 232 p.
ISBN 0–306–40370–6.

Mendelson, Wallace B, et al Human Sleep and Its Disorders (1977)
New York, Plenum Publishing, 274 p.
ISBN 0–306–30966–1.

Mendglievich, Elias Children at Work (1979)
Lanham MD, UNIPUB, 176 p. ISBN 92–2–102165–3.

Mendlovitz, Saul H (Ed) Preferred Worlds for the 1990s
New York, Free Press. 6 vols (1972–79).

Mengistu, A, et al An Annotated Bibliography of Chickpea Diseases 1915–1976 (1978)
Patancheru A P, ICRISAT, 50 p. Information Bulletin: No 1.

Menninger, W Walter and Hannah, Gerald (Eds) The Chronic Mental Patient–II (1987)
New York, American Psychiatric Association, 224 p.
ISBN 0–88048–278–8.

Merchant, I A and Barner, Ralph D Outline of Infectious Diseases of Domestic Animals (1964)
Ames IA, Iowa State University Press, 478 p. 3rd ed.
ISBN 0–8138–0831–6.

Merchant, James A Occupational Respiratory Diseases (1986)
Washington DC, United States Government Printing Office, 851 p. ISBN 0-318-22599-9.
Meredith, D S Banana Leaf Spot Disease (Sigatoka) Caused by Mycosphaerella Musicola Leach (1970)
Kew, CAB International Mycological Institute, 147 p. illus. Phytopathological Papers: No 11.
Merino-Rodriguez, Manuel (Ed) Lexicon of Plant Pests and Diseases (1966)
Amsterdam, Elsevier Science Publishing, 351 p. In Engl, Lat, Span, Ital and Ger. Elsevier Lexica Ser: Vol 7.
Merke, F (Ed) History and Iconography of Endemic Goitre and Cretinism (1984)
Dordrecht, Kluwer Academic Publishers Group, 339 p. ISBN 0-85200-646-2.
Merkl, Peter H Political Violence and Terror: motifs and motivations (1986)
Berkeley CA, University of California Press, 400 p. ISBN 0-520-05605-1.
Meron, Theodor Human Rights in International Law: legal and policy issues (1986)
New York, Oxford University Press, 608 p. 2 vols. in 1 ISBN 0-19-825540-3.
Merrick, Lamar The Mysterious, Conflicting Operations Carried Out in Both Europe and the United States by Spies and Secret Service (1986)
Albuquerque NM, Institute for Economic and Political World, 375 p. 2 vols. ISBN 0-86722-134-8.
Mersky, Roy M (Ed) Conference on Transnational Economic Boycotts and Coercion (1978)
New York, Oceana. 2 vols. ISBN 0-379-20335-9.
Merté, H-J (Ed) Genesis of Glaucoma: contributions of the participants of the Wessely symposium in Munich with final considerations by H Goldmann (1978)
Dordrecht, Kluwer Academic Publishers Group, 320 p. ISBN 90-6193-156-8.
Merton, Robert K Social Problems and Sociological Theory
New York, Harcourt Brace Jovanovich, pp. 793-845.
Merton, Robert K, et al (Eds) Contemporary Social Problems (1976)
San Diego CA, Harcourt Brace Jovanovich, 782 p. 4th ed. ISBN 0-15-513793-X.
Mesarovic, Mihajlo D and Pestel, Eduard Mankind at the Turning Point: the second report of the Club of Rome (1974)
New York, E P Dutton, 210 p. illus. ISBN 0-525-15230-X.
Messerli, Franz H (Ed) The Heart of Hypertension (1987)
Sevenoaks, Butterworths, 510 p. illus. ISBN 0-914-31645-1.
Messerli, Franz H The Heart and Hypertension (1987)
Stoneham MA, Butterworth Publishers, 482 p. ISBN 0-914316-45-1.
Messerschmidt, James Capitalism, Patriarchy and Crime (1986)
Totowa NJ, Rowman and Littlefield Publishers, 224 p. ISBN 0-8476-7496-7.
Metcalf, et al Destructive and Useful Insects (1962)
New York, McGraw-Hill Book Company, 1087 p. illus. 4th ed. ISBN 0-07-099445-5.
Mettrick, D F and Desser, S S (Eds) Parasites: their World and Ours: Proceedings of the 5th International Congress of Parasitology, Toronto, Canada, 1982 (1982)
Amsterdam, Elsevier Science Publishing, 466 p. ISBN 0-444-80433-1.
Meulen, V Ter and Katz, M Slow Virus Infections of the Central Nervous System (1977)
New York, Springer Publishing. ISBN 0-387-90188-4.
Mew, James Traditional Aspects of Hell (1971)
Detroit MI, Gale Research, 474 p. repr of 1903 ed. ISBN 0-8103-3693-6.
Meyer, Secondary and Functional Rhinoplasty: the difficult nose (1988)
Orlando FL, Grune and Stratton. ISBN 0-8089-1879-6.
Meyer, Marshall W Limits to Bureaucratic Growth (1985)
Berlin, Walter De Gruyter, 259 p. ISBN 3-11-009865-2.
Meyer, Mary C, et al Sexual Harassment at Work (1981)
Princeton NJ, Petrocelli Books, 256 p. ISBN 0-89433-156-6.
Meyers, C Edward Quality of Life in Severely and Profoundly Mentally Retarded People: research foundations for improvement (1978)
Washington DC, American Association on Mental Retardation. ISBN 0-686-23879-6.
Meyler, L and Peck, H M Drug Induces Disease (1962)
Assen, Van Gorcum.
Michelet, Jules Satanism and Witchcraft (1983)
Secaucus NJ, Stuart Lyle, 352 p. ISBN 0-8065-0059-X.
Mickan, B Parameters Characterizing Toxic and Hazardous Waste Disposal Sites: management and monitoring (1987)
Luxembourg, European Communities, 234 p. ISBN 92-825-7157-2.
Micklous, Edward F and Flemming, Peter A Terrorism Nineteen Eighty to Nineteen Eighty-Seven: a selectively annotated bibliography (1988)
Westport CT, Greenwood Press, 328 p. ISBN 0-313-26248-9.
Midlarsky, Manus I Inequality and Contemporary Revolutions (1986)
Denver CO, University of Denver, 184 p. ISBN 0-87940-081-1.
Mies, Maria Patriarchy and Accumulation on a World Scale: women in the international division of labour (1986)
Atlantic Highlands NJ, Humanities Press International, 260 p. ISBN 0-86232-341-X.
Milavsky, J Ronald et al Television and Aggression: results of a panel study (1982)
San Diego CA, Academic Press, 493 p. ISBN 0-12-495980-6.
Milburn, Thomas W and Watman, Kenneth H On the Nature of Threat: a social psychological analysis (1981)
New York, Praeger Publishers, 160 p. ISBN 0-275-90683-3.
Miles, The Official History of the Underground Press (Date not set)
New York, Boyars Marion Publishers, 160 p. ISBN 0-906890-40-3.
Miles, Agnes Women and Mental Illness: the social context of female neurosis (1988)
Herts, Simon and Schuster International, 224 p.
Miles, Robert Capitalism and Unfree Labour: anomaly or necessity? (1987)
New York, Routledge Chapman and Hall, 256 p.
Milikan, Clark H, et al Stroke (1987)
Philadelphia PA, Lea and Febiger, 341 p. illus. ISBN 0-8121-1016-1.
Milkman, Harvey and Shaffer, Howard J (Eds) The Addictions: multidisciplinary perspectives and treatments (1984)
Lexington MA, Lexington Books, 224 p. ISBN 0-669-08739-4.
Miller, Albert J Confrontation, Conflict and Dissent: a bibliography of a decade of controversy (1972)
Metuchen NJ, Scarecrow Press, 567 p. ISBN 0-8108-0490-5.
Miller, C J Contempt of court (1988)
Oxford NY, Oxford University Press, 550 p. 2nd ed. ISBN 0-19-825478-4.
Miller, David Anarchism (1984)
Evman England/Biblio Dist, 224 p.
Modern Ideologies Ser. ISBN 0-460-10093-9.
Miller, Donald S and Brighouse, B A Thermal Discharges: a guide to power and process plant cooling water (1984)
B H R A Fluid Engineering, 221 p. ISBN 0-906085-93-4.
Miller, E Willard and Miller, Ruby M Environmental Hazards-Radioactive Materials and Wastes: a bibliography (1985)
Monticello IL, Vance Bibliographies, 67 p. ISBN 0-89028-266-8.
Miller, E William and Miller, Ruby M Environmental Hazards: air pollution (1988)
Santa Barbara CA, ABC Clio, 250 p. Contemporary World Issues Ser. ISBN 0-87436-528-7.
Miller, E Williard and Miller, Ruby M Environmental Hazards-Industrial and Toxic Wastes: a bibliography (1985)
Monticello IL, Vance Bibliographies, 99 p. bibl. Public Administration Ser: Bibliography P-1615. ISBN 0-89028-265-X.
Miller, Joseph C Slavery: a worldwide bibliography (1985)
White Plains NY, Kraus International Publications, 480 p. ISBN 0-527-63659-2.
Miller, K Toxicological Aspects of Food (1987)
New York, Elsevier Science Publishing, 458 p. ISBN 1-85166-080-1.
Miller, L L Marijuana: effects on human behavior (1974)
San Diego CA, Academic Press. ISBN 0-12-497050-8.
Miller, Morris Coping is Not Enough The international debt crisis and the roles of the World Bank and the International Monetary Fund (1986)
Homewood IL, Dow Jones-Irwin, 200 p. ISBN 0-87094-933-0.
Miller, Nancy E and Cohen, Gene D Schizophrenia and Aging: schizophrenia, paranoia, and schizophreniform (1987)
New York, Guilford Press, 367 p. ISBN 0-89862-228-X.
Millman, Irving, et al Hepatitis B: the virus, the disease, and the vaccine (1984)
New York, Plenum Publishing, 264 p. ISBN 0-306-41723-5.
Mills, David Overcoming Religion (1980)
New York, Citadel Press. ISBN 0-8065-0742-X.
Mills, John F and Mansfield, John M The Genuine Article: the making and unmasking of fakes and forgeries (1982)
New York, Universe Books, 240 p. ISBN 0-87663-401-3.
Milne, Teddy The Unseen Holocaust (1987)
Northampton MA, Pittenbruach Press, 24 p. ISBN 0-938875-09-4.
Milunsky, Aubrey Genetic Disorders and the Fetus: diagnosis, prevention and treatment (1986)
New York, Plenum Publishing, 924 p. ISBN 0-306-42301-4.
Milunsky, Aubrey, et al (Eds) Advances in Perinatal Medicine (1981)
New York, Plenum Publishing, 456 p. Vol 1. ISBN 0-306-40482-6.
Mims, Cedric, et al Viruses and Demyelinating Diseases (1984)
San Diego CA, Academic Press. ISBN 0-12-498280-8.
Minority Rights Group Ltd World Minorities
London, Minority Rights Group. 3 Vols.
Mirron, Murray S and Goldstein, Arnold P Hostage (1979)
New York, Pergamon Books, 170 p. rev ed. ISBN 0-08-023875-0. Pergamon General Psychology Ser: No 79.
Miserez, D (Ed) Refugees - The Trauma of Exile: based on a workshop on psycho-social and medical needs of refugees/asylum seekers and how to meet them (1989)
Dordrecht, Martinus Nijhoff, 360 p. ISBN 0-7923-0112-9.
Mislin, Hans and Ravera, Oscar (Eds) Cadmium in the Environment (1986)
Basel, Birkhäuser Verlag, 156 p. ISBN 3-7643-1760-4.
Misra, A, et al Plant Tumors (1983)
New Delhi, Today and Tomorrow's Printers and Publishers, 200 p. ISBN 1-55528-045-5.
Misra, Girishwar and Tripathi, Lal Baccan Psychological Consequences of Prolonged Deprivation (1980)
Agra, National Psychological Corporation, 251 p.
Misra, R P Development Issues of Our Times (1985)
New Delhi, Concept Publishing, 368 p.
Misra, R P Third World Peasantry: a continuing saga of deprivation (1986)
New Delhi, Sterling Publishers, 624 p. ISBN 81-207-0158-5.
Mitchell, Edward Porpoise, Dolphin and Small Whale Fisheries of the World: status and problems (1975)
Gland, IUCN, 129 p.
Mitchell, J Urinary Tract Trauma (1984)
Littleton MA, PSG Publishing, 288 p. ISBN 0-7236-0816-4.
Mitsuhashi, Susumu (Ed) Drug Resistance in Bacteria: genetics, biochemistry and molecular biology (1983)
Stuttgart, George Thieme Verlag, 429 p. ISBN 3-13-641401-2. Dist all countries except Japan by publisher.
Mizejewski, Gerald and Porter, Ian S Alpha-Fetoprotein and Congenital Disorders (1985)
San Diego CA, Academic Press. ISBN 0-12-501630-1.
Mngola, E N (Ed) Diabetes Nineteen Eighty-Two: proceedings of the 11th Congress of the International Diabetes Federation, Nairobi, Nov 10-17, 1982 (1983)
Amsterdam, Elsevier Science Publishing. ISBN 0-444-90291-0. International Congress Ser: Vol 600.
Moavenzadeh, F and Geltner, D Transportation, Energy and Economic Development: a dilemma in the developing world (1984)
New York, Elsevier Science Publishing, 530 p. ISBN 0-444-42338-9.
Mockler, Anthony The New Mercenaries: the history of the hired soldier from the Congo to the Seychelles (1987)
New York, Paragon House Publishers, 374 p. illus. ISBN 0-913729-72-8.
Moll, H Atlas of Pediatric Diseases (1976)
Stuttgart, George Thieme Verlag, 275 p. ISBN 3-13-521601-2.
Mollat, Michel and Goldhammer, Arthur The Poor in the Middle Ages: an essay in social history (1986)
New Haven CT, Yale University Press, 336 p. ISBN 0-300-02789-3.
Molnar, Alex (Ed) Social Issues and Education: challenge and responsibility (1987)
Alexandria VA, Association for Supervision and Curriculum Development, 138 p. ISBN 0-87120-141-0.
Molnar, Thomas Theists and Atheists: a Typology of non-belief (1980)
Berlin, Mouton de Gruyter, 220 p. ISBN 90-279-7788-7.
Molyneux, John World Prospects: a contemporary study (1979)
New York, Prentice Hall, 400 p. ISBN 0-13-968826-9. Ed by Greenaway, Rob.
Monahan, John Food Poisoning (1987)
Lackawaxen PA, Medical-Info Books, 125 p. ISBN 0-317-67410-2.
Monahan, John and Steadman, Henry J Mentally Disordered Offenders (1983)
New York, Plenum Publishing, 296 p. ISBN 0-306-41151-2.
Monat, Rosalyn K Sexuality and the Mentally Retarded: a clinical and therapeutic guidebook (1982)
Boston MA, College-Hill Press, 150 p. ISBN 0-316-57820-7.
Money, John Sex Errors of the Body: dilemmas, education, counseling
Baltimore MD, Johns Hopkins University Press, 145 p. ISBN 0-8018-0467-1.
Mönnig, H O and Veldman, F J Handbook on Stock Diseases
Cape Town, Tafelberg Publishers, 392 p. ISBN 0-624-01666-8.
Monsen, R J and Walters, K D Nationalized Companies: a threat to american business (1983)
New York, McGraw Hill Book Company. ISBN 0-07-071569-6.
Montague, Meg Ageing and Autonomy: who makes decisions for older people in supported accomodation? (1982)
Saint Laurence, 198 p.
Montero, McDowell Social Problems (1986)
New York, Macmillan, 525 p. ISBN 0-02-382250-3.
Montgomery, John D International Dimensions of Land Reform (1984)
Boulder CO, Westview Press, 300 p. ISBN 0-86531-781-X.
Montgomery, John W Demon Possession (1976)
St Louis, Bethany Press. IBSN 0-87123-102-6.
Moody, Roger Indigenous Peoples: visions and realities (1987)
Atlantic Highlands NJ, Humanities Press International. 2 vols. ISBN 0-86232-305-3.
Mooney, Thomas O, et al Sexual Options for Paraplegics and Quadriplegics (1975)
Boston MA, Brown Little, 150 p. ISBN 0-316-57937-8.
Moonman, Eric (Ed) The Violent Society
London, Frank Cass, 168 p. ISBN 0-7146-3309-7.
Moore, Joan W and Moore, Burton M Social Problems (1982)
Englewood Cliffs NJ, Prentice Hall, 464 p. illus. ISBN 0-13-817387-7.
Moore, M R, et al Disorders of Porphyrin Metabolism (1987)
New York, Plenum Publishing, 396 p. ISBN 0-306-42625-0.
Moore, R I The Origins of European Dissent (1985)
New York, Basil Blackwell, 338 p. ISBN 0-631-14404-8.
Moorehead, Caroline (Ed) Betrayal: child exploitation in today's world (1989)
London, Barrie and Jenkins, 192 p. ISBN 0-7126-2170-9.
Moos, Rudolf H (Ed) Coping with Life Crises: an integrated approach (1986)
New York, Plenum Publishing, 444 p.

ISBN 0-306-42133-X. Plenum Series in Stress and Coping.

Moran, W B Covert Surveillance and Electronic Penetration (1986)
New York, Gordon Press Publishers. ISBN 0-8490-3714-X.

Morehouse, Ward Third World Panacea or Global Boondoggle?: the UN Conference on Science and Technology for Development Revisited (1984)
Lund, Research Policy Institute, 89 p.

Morgan, Lanier Understanding and Modification of Delinquent Behavior (1984)
Libra.

Morgan, R P Soil Erosion and Its Control (1985)
New York, Reinhold Van Nostrand, 336 p. ISBN 0-442-26441-0.

Morgan, Royston P Soil Erosion and Conservation (1986)
New York, John Wiley and Sons. ISBN 0-470-20671-3.

Morgenstern, Felice International Conflicts of Labour Law. A Survey of the Law Applicable to the International Employment Relation (1986)
Geneva, ILO, 129 p. ISBN 92-2-103593-X.

Morici, Peter Global Competitive Struggle: challenges to the United States and Canada (1984)
Toronto ON, CD Howe Institute, 124 p. ISBN 0-89068-076-0.

Morio, Simone and Zoctizum, Yarrise Two Studies on Unemployment Among Educated Young People (1980)
Lanham MD, UNIPUB. 2 Pts. ISBN 92-3-101618-0.

Morone, Joseph G and Woodhouse, Edward J Averting Catastrophe: strategies for regulating risky technologies (1988)
Berkeley CA, University of California Press, 225 p. ISBN 0-520-05754-6.

Morris, Dirk Government Debt in International Financial Markets (1988)
London, Pinter Publishers, 220 p. ISBN 0-86187-994-5.

Morris, J C H The Conflict of Laws (1984)
UK, Sweet and Maxwell. 3rd ed.

Morris, Roger Riots and Disturbances in Correctional Institutions (1982)
College Park MD, American Correctional Association. rev ed. ISBN 0-942974-07-7.

Morris, Samuel M Coronary Disease: medical subject analysis with reference bibliography (1987)
Washington DC, ABBE Publications Association, 150 p. bibl. ISBN 0-88164-514-1.

Morris, Samuel M Phobias and Disorders: index of modern information (1988)
Annandale VA, ABBE Publishers Association of Washington, 150 p. ISBN 0-88164-834-5.

Morse, Bradford Aboriginal Self-Government in Australia and Canada
Kingston ON, Queen's University, 130 p. ISBN 0-88911-424-2.

Morton, Andrew B Food Contamination: medical subject analysis with bibliography (1987)
Annandale VA, ABBE Publishers Association of Washington, 160 p. ISBN 0-88164-536-2.

Moscovitch, Allen and Drover, Glenn (Ed) Inequality: essays on the political economy of social welfare (1981)
Toronto ON, University of Toronto Press. ISBN 0-8020-2403-3.

Moser, Robert H Diseases of Medical Progress: a contemporary analysis of illness produced by drugs and other therapeutic agents (1969)
Springfield IL, Thomas Charles C. 2nd ed.

Mould, R F and Bose, Aranbinda International Strategies for the Eradication of Carcinoma of the Cervix in Developing Areas: proceedings of the meeting of the international (1987)
New York, Taylor and Francis, 350 p. ISBN 0-85274-473-0.

Mountfort, Paul Exposure on Hypothermia
Washington DC, Government Publication Centre, 48 p.

Moussaieff Masson, J Oceanic Feeling: the Origins of religious sentiment in ancient India (1980)
Dordrecht, Kluwer Academic Publishers Group, 213 p. ISBN 90-277-1050-3.

Mouton, Pierre Social Security in Africa: trends, problems and prospects (1976)
Geneva, ILO, 166 p. ISBN 92-2-101414-2.

Mowat, Alex P Liver Disorders in Childhood (1987)
Stoneham MA, Butterworth Publishers, 320 p. ISBN 0-407-00480-7.

Mowday, Richard et al Employee-Organization Linkages (1981)
San Diego CA, Academic Press. ISBN 0-12-509370-5.

Mowlana, Hamid International Flow of News: an annotated bibliography (1985)
Paris, UNESCO, 272 p.

Moyle, Alen Conquering Constipation
Vancouver BC, Forbez Enterprises, 128 p.

Mudd Responses of Plants to Air Pollution (1975)
San Diego CA, Academic Press. ISBN 0-12-509450-7.

Mufti, El M Hydatid Disease (1989)
Sevenoaks, Butterworths, 160 p. illus. ISBN 0-407-01329-6.

Muggler, Dixie M Poisoning and Medicine: guidebook for reference and research (1985)
Annandale VA, ABBE Publishers Association of Washington, 150 p. ISBN 0-88164-152-9.

Muir, I F, et al Burns and their Treatment (1987)
London, Butterworth, 192 p. illus. 3rd ed. ISBN 0-407-00333-9.

Mukherji, S K Prostitution in India (1986)
New Delhi, Inter-India Publications, 528 p. ISBN 81-210-0054-8.

Mulder, D C (Ed) Secularization in global Perspective (1981)
Amsterdam, Vrije Universiteit Boekhandel, 195 p.

Mullen, Barbara D and McGinn, Kerry A The Ostomy Book: living comfortably with colostomies, Ileostomies, and urostomies (1980)
Menlo Park CA, Bull Publishing, illus. ISBN 0-915950-41-3.

Muller Associates and SYSCON Corporation and Brookhaven National Laboratory, et al Handbook of Radon in Buildings: detection, safety, and control (1988)
New York, Hemisphere Publishing, 286 p. ISBN 0-89116-823-0.

Müller, Helga Hydrocarbons in the Freshwater Environment: a literature review (1987)
Stuttgart, E Schweizerbart'sche Verlagsbuchhandlung, 69 p. ISBN 3-510-47022-2.

Mulloy, Paul; Smith, Karen and Switzer, Kenneth A Global Issues: activities and resources for the high school teacher (1987)
Boulder CO, Social Science Education Consortium, 120 p. illus. 2ne rev ed. ISBN 0-89994-312-8.

Mulloy, Paul; Smith, Karen and Switzer, Kenneth A Global Issues: activities and resources for the high school teacher (1987)
Boulder CO, Social Science Education Consortium, 120 p. illus. 2nd rev ed. ISBN 0-89994-312-8.

Munday, Brian (Ed) The Crisis in Welfare: an international perspective on social services and social work (1989)
Hemel Hempstead, Simon and Schuster International, 224 p. ISBN 0-7450-0163-7.

Munn, Robert F Strip Mining: an annotated bibliography (1973)
Morgantown WV, West Virginia University Press, 110 p. ISBN 0-937058-09-2.

Munnichs, J M A and Heuwel, J A van den (Eds) Dependency or Interdependency in Old Age (1976)
Dordrecht, Kluwer Academic Publishers Group, 175 p. ISBN 90-247-1895-3.

Munsat, Theodore L Post-polio Syndrome (1989)
Sevenoaks, Butterworths, 224 p. ISBN 0-409-90153-9.

Munthe, E and Roland, E (Eds) Bullying: an international perspective (1989)
London, David Fulton Publishers, 144 p.

Murakami, Y Stress Intensity Factors Handbook (1987)
Elmsford NY, Pergamon Books, 1566 p. 2 vols. ISBN 0-08-034809-2.

Muravchik, Joshua and Kirkpatrick, Jeane The Uncertain Crusade: Jimmy Carter and the dilemmas of human rights (1986)
Lanham MD, University Press of America, 265 p. ISBN 0-8191-5108-4.

Murdock, Steve H, et al Nuclear Waste: socioeconomic dimensions of long-term storage (1983)
Boulder CO, Westview Press, 343 p. ISBN 0-86531-447-0.

Murphy, John W, et al The Underside of High-Tech (1986)
Westport CT, Greenwood Press, 226 p. ISBN 0-313-24612-2.

Murray, Allan P Depreciation (1971)
Cambridge MA, Harvard Law International Tax, 138 p. illus. Tax Technique Handbook Ser. ISBN 0-915506-12-2.

Mushkat, Marion The Third World and Peace (1983)
New York, Saint Martin's Press, 356 p. ISBN 0-312-80039-8.

Muslehuddin, Muhammad Crime and the Islamic Doctrine of Preventive Measures (1985)
Islamabad, Islamic Research Institute, 76 p. ISBN 969-408-058-4.

Mussell, Harry and Staples, Richard C Stress Physiology in Crop Plants (1979)
New York, John Wiley and Sons, 510 p. ISBN 0-471-03809-1.

Mutharika, A Peter The Regulation of Statelessness Under International and National Law: text and documents (1977)
Dobbs Ferry NY, Oceana Publications. 2 bdrs. ISBN 0-379-10040-1.

Myers, Edward When Parents Die: a guide for adults (1987)
New York, Penguin Books, 224 p.

Myers, N The Cheetah Acinonyx Jubatus in Africa (1975)
Gland, IUCN, 90 p. illus. ISBN 2-88032-015-1.

Myers, N The Leopard Panthera Pardus in Africa (1976)
Gland, IUCN, 79 p. illus. ISBN 2-88032-017-8.

Myers, Norman (Ed) The Gaia Atlas of Planet Management: for today's caretakers of tomorrow's world (1985)
London, Pan.

Nadelmann, K H Conflicts of Law: international and interstate: selected essays (1972)
Dordrecht, Kluwer Academic Publishers Group, 401 p. ISBN 90-247-1212-2.

Nagai, Michio Education and Indoctrination: social and philosophical bases (1976)
Toronto ON, University of Toronto Press, 128 p. ISBN 0-86008-148-6.

Nagy, and Verakis, Development and Control of Dust Explosions (1983)
New York, Dekker Marcel, 352 p. ISBN 0-8247-7004-8.

Nakamura, Susumu and Dykstra, Andrew (Eds) Language, Thought and Culture Symposium: quest for peace: international understanding and language barriers (1979)
Hirakatashi, Kansai University of Foreign Studies, 182 p.

Nakanishi, T (Ed) Long Term Clinical Care of Parkinson's Disease, Tokyo, April 1987: second symposium (1988)
Basel, S Karger AG, 42 p. ISBN 3-8055-4833-8. Journal: European Neurology: Vol 28, Suppl 1, 1988.

Nalivkin, D and Bhattacharya, B B Hurricaines, Storms and Tornadoes: geographic characteristics and geological activity (1983)
Brookfield VT, Brookfield Publishing, 609 p. ISBN 0-317-65012-2.

Nance, Walter, E et al (Eds) International Congress on Twin Studies, 2nd: twin research: proceedings (1987)
New York, Alan R Liss. Progress in Clinical and Biological Research Ser: Vol 24A: Psychology and Methodoloty. Vol 24B: Biology and Epidemiology. Vol24C: Clinical Studies.

Narayan, Raj Falling Education Standards (1970)
Agra, Lakshmi Narain Agarwal, 134 p.

Narveson, Jan The Libertarian Idea (1988)
Philadelphia PA, Temple University Press, 416 p. ISBN 0-87722-569-9.

Nash, Jay R Darkest Hours: a narrative encyclopedia of worldwide disasters (1976)
Chicago IL, Nelson-Hall, 826 p. ISBN 0-88229-140-8.

Nath, R Environmental Pollution of Cadmium: biological, physiological and health effects (1986)
New Delhi, Interprint Mehta House.

National Academy Press Drug and Alcohol Problems: exploring biology, behavior, and public health (1988)
Washington DC, National Academy Press, 150 p. ISBN 0-309-03931-2.

National Center for State Courts Alternatives to Incarceration: an annotated bibliography 1978-1980 (1981)
New Brunswick NJ, Center for Urban Policy Research, 97 p. bibl.

National Corrosion Engl Corrosion Basics - An introduction (1984)
Houston TX, National Association of Corrosion Engineers, 364 p. ISBN 0-915567-02-4.

National Corrosion England Automotive corrosion by de-icing salts (1981)
Houston TX, National Association of Corrosion Engineers, 426 p. Illus. ISBN 0-317-36473-1.

National Electrical Manufacturers Association PCB Health Effects
Washington DC, National Electrical Manufacturers Association. 2 Vols. Vol 1: Potential Health Effects in the Human from Exposure to Polychlorinated Biphenyls (PCBs) and Related Impurities. Vol 2: Literature Update on Potential Health Effects From Exposure to Polychlorinated Biphenyls and Related Chlorinated Heterocycles. ISBN 0-318-18042-1.

National Electrical Manufacturers Association PCB Health Effects
Washington DC, National Electrical Manufacturers Association. Vol. 12 vols. ISBN 0-318-18042-1.

National Institute for Burn Medicine International Bibliography on Burns: supplement Feller, I
Ann Arbor MI, National Institute for Burn Medicine. bibl. 1985-1986.

National Research Council Toxic Shock Syndrome (1982)
Washington DC, National Academy Press. ISBN 0-309-03286-5.

National Research Council Women's Work, Men's Work: sex segregation on the job (1985)
Washington DC, National Academy Press, 173 p. ISBN 0-309-03429-9.

National Research Council Safety of Dams: flood and earthquake criteria (1985)
Washington DC, National Academy Press, 320 p. ISBN 0-309-03532-5.

National Research Council Guide to Infectious Diseases of Mice and Rats (1988)
Washington DC, National Academy Press, 370 p. ISBN 0-309-03794-8.

National Research Council Health Risks of Radon and Other Internally Deposited Alpha-Emitters: beir iv (1988)
Washington DC, National Academy Press, 600 p. ISBN 0-309-03797-2.

National Research Council Contaminated Marine Sediments: assessment and remediation (1990)
Washington DC, National Academy Press, 506 p. ISBN 0-309-04095-7.

National Research Council, Division of Medical Ozone and Other Photochemical Oxidants (1977)
Washington DC, National Academy Press, 719 p. ISBN 0-309-02531-1.

National Water Well Association Radon in Ground Water: hydrogeologic impact and application to indoor (1987)
Chelsea MI, Lewis Publishers, 550 p. ISBN 0-87371-117-3.

NATO Advanced Study Institute and Mansinha, L and Smylie, D E and Beck, A E et al Earthquake Displacement Fields and the Rotation of the Earth: proceedings of the nato advanced study (1970)
Norwell MA, Kluwer Academic Publishers, 308 p. ISBN 90-277-0159-8.

Nauman, Elmo Jr Exorcism Through the Ages (1974)
New York, Citadel Press, 256 p. illus. ISBN 0-8065-0450-1.

Naviaux, James L Horses in Health and Disease (1985)
Philadelphia PA, Lea and Febiger, 300 p. ISBN 0-8121-0935-X.

Naylor, R T Dominion of Debt: centre, periphery and the international economic order (1985)
Montreal, Black Rose Books, 200 p. ISBN 0-920057-50-0.

Neiderhaus, Lee B Radon: index of modern information (1988)
Annandale VA, ABBE Publishers Association of Washington, 150 p. ISBN 0-88164-766-7.

Nene, Yeshwant Laxman and Aggarwal, V K Some Seed Borne Diseases and their Control (1978)
New Delhi, Ministry of Information and Broadcasting, 44 p.

Nerozzi, Dina, et al (Eds) Hypothalamic Dysfunction in Neuropsychiatric Disorders (1987)
New York, Raven Press, 382 p. illus.
ISBN 0–88167–304–8.

Neth, R and Gallo, R C (Eds) Modern Trends in Human Leukemia VII (1987)
Berlin, Springer–Verlag, 595 p. illus. Haematology and Blood Transfusion Ser: Vol 31.

Nettler, Gwynn Explaining Crime (1984)
New York, McGraw Hill Book Company, 416 p.
ISBN 0–07–046313–1.

Nettleship, Martin A; Givens, R Dale and Nettleship, Anderson (Eds) War, Its Causes and Correlates (1975)
Berlin, Mouton de Gruyter, 814 p. ISBN 90–279–7659–7.

Neu, H C and Williams, J D New Trends in Urinary Tract Infections (1988)
New York, Karger S AG, 358 p. ISBN 3–8055–4637–8.

Neufeld, Maurice F Poor Countries and Authoritarian Rule (1965)
Ithaca NY, ILR Press, 256 p. International Report Ser: No 6.

Nevile, John W Root of All Evil: essays on economics, ethics and capitalism
Sydney NSW, Australian Council of Churches.
ISBN 0–85821–024–X.

Newcomb, Michael D Drug Use in the Workplace (1988)
Dover MA, Auburn House Publishing, 180 p.
ISBN 0–86569–182–7.

Newland, Kathleen Women and Population Growth: choice beyond childbearing (1977)
Washington DC, Worldwatch Institute. Worldwatch Papers.
ISBN 0–916468–15–1.

Newland, Kathleen Global Employment and Economic Justice: the policy challenge (1979)
Washington DC, Worldwatch Institute. Worldwatch Papers.
ISBN 0–916468–27–5.

Newland, Kathleen International Migration: the search for work (1979)
Washington DC, Worldwatch Institute. Worldwatch Papers.

Newland, Kathleen Refugees: the new international politics of displacement (1981)
Washington DC, Worldwatch Institute. Worldwatch Papers.
ISBN 0–916468–42–9.

Newman, Charles M and Czechowicz, James International Risk Management (1983)
Morristown NJ, Financial Executives Research Foundation, 189 p. ISBN 0–910586–51–9.

Nicholas, Darrel D Wood Deterioration and Its Prevention by Preservative Treatments: degradation and protection of wood (1984)
Syracuse NY, Syracuse University Press, 448 p. Vol. 1
ISBN 0–8156–2303–8.

Nicoletti, Paul Diagnosis and Vaccination for the Control of Brucellosis in the Near East (1982)
Lanham MD, UNIPUB, 45 p. Animal production and health papers: No 38. ISBN 92–5–101278–4.

Nieveen, J (Ed) Arrhythmias of the Heart (1981)
Amsterdam, Elsevier Science Publishing, 256 p.
ISBN 0–444–90206–6.

Nincic, Miroslav and Wallensteen, Peter (Eds) Dilemmas of Economic Coercion: sanctions in world politics (1983)
New York, Praeger Publishers, 250 p. illus. bibl.
ISBN 0–03–064236–1. Praeger Special Studies, Praeger Scientific.

Nisbet, Robert Prejudices: a philosophical dictionary (1982)
Cambridge, Harvard University Press, 318 p.
ISBN 0–674–70065–1.

Nistico, Giuseppe et al Neurotransmitters, Seizures, and Epilepsy III (1986)
New York, Raven Press, 528 p. ISBN 0–88167–229–7.

Noam, Eli M and Dennis, Everette The Cost of Libel: economic and policy implications (1989)
New York, Columbia University Press, 320 p.
ISBN 0–231–06692–9.

Nobe, Kenneth C and Sampath, Rajan K (Eds) Issues in Third World Development (1984)
Boulder CO, Westview Press, 500 p. Special Studies in Social, Political and Economic Development.
ISBN 0–86531–686–4.

Nobe, Kenneth C and Sampath, Rajan K (Eds) Issues in Third World Development (1984)
Boulder CO, Westview Press, 500 p. Special Studies in Social, Political and Economic Development.
ISBN 0–86531–686–4.

Noll, Kenneth E, et al Recovery, Recycle and Reuse of Industrial Waste (1985)
Chelsea MI, Lewis Publishers, 204 p. illus. Industrial Waste Management Ser. ISBN 0–87371–002–9.

Noonan, John T Bribes (1987)
Berkeley CA, University of California Press, 839 p.
ISBN 0–520–06154–3.

Noorani, A G Ministerial Misconduct (1972)
New Delhi, Vikas Publishing House.

Nordau, Max Egomania and the Psychology of Contemporary Man (1986)
Albuquerque NM, American Institute for Psychological Research, 225 p. illus. 2 Vols. ISBN 0–89920–129–6.

Nordentoft, E L; Wallin, Johan A and Nielsen, Hans Victor (Eds) Traffic Speed and Casualties: epidemiological effects of traffic speed and speed limitations: Proceedings of an International Interdisciplinary Symposium, GI Avernas, Funen April 1975 (1975)
Melbourne VIC, Oxford University Press, 239 p.
ISBN 87–7492–145–2.

Nordquist, Joan (Ed) Domestic Violence: a bibliography (1986)
Santa Cruz CA, Reference and Research Services, 50 p. bibl. Contemporary Social Issues: A bibliographic Ser: No 4.
ISBN 0–937855–07–3.

Normandeau, André International Bibliography on Criminal Statistics: 1945–1968
Ottawa ON, Canadian Criminal Justice Association.

Norris, Mirriam and Spaulding, Patricia J Blindness in Children
Ann Arbor MI, Books on Demand. ISBN 0–317–09463–7.

Norris, Ruth, et al Pills, Pesticides and Profits: the international trade in toxic substances (1982)
Croton–on–Hudson NY, North River Press, 182 p.
ISBN 0–88427–050–5.

Northern, Jerry L and Downs, Marion P Hearing in Children (1984)
Baltimore MD, Williams and Wilkins, 408 p.
ISBN 0–683–06573–4.

Norton, Bryan G The Preservation of Species: the value of biological diversity (1986)
Princeton NJ, Princeton University Press, 272 p.
ISBN 0–691–08389–4.

Nourse, Alan E Fractures, Dislocations and Sprains (1978)
New York, Franklin Watts. ISBN 0–531–01494–0.

Novack, George and Mandel, Ernest Marxist Theory of Alienation (1973)
Haymarket NSW, Pathfinder Press, 96 p.
ISBN 0–87348–229–8.

Novak, Daniel A The Wheel of Servitude: black forced labor after slavery (1978)
University Press of Kentucky, 144 p.
ISBN 0–8131–1371–7.

Novicow, Jacques and Cooper, S E War and Its Alleged Benefits (1972)
New York, Garland Publishing. ISBN 0–8240–0274–1.

Novotny, Vladimir and Chesters, Gordon Handbook of Nonpoint Pollution Sources and Management (1981)
New York, Reinhold Van Nostrand, 528 p.
ISBN 0–442–22563–6.

Nriagu, J O (Ed) Biochemistry of Lead in the Environment; part A: ecological cycles (1978)
In: *Topics in Environmental Health: Vol 1.*

Nriagu, J O (Ed) Biogeochemistry of Lead in the Environment: part B: biological effects (1978)
Amsterdam, Elsevier Science Publishing, 398 p.
ISBN 0–444–80110–3. In: Topics in Environmental Health: Vol 1.

Nriagu, J O (Ed) Biogeochemistry of Mercury in the Environment (1979)
Amsterdam, Elsevier Science Publishing, 696 p.
ISBN 0–444–80110–3. In: Topics in Environmental Health: Vol 3.

Nriagu, J O (Ed) Changing Mental Cycles and Human Health: report of the Dahlem Workshop on changing mental cycles and human health, Berlin 1983, March 20–25 (1984)
Berlin, Springer–Verlag, 450 p. illus. Dahlem Workshop Reports, Life Sciences Research Report: Vol 28.
ISBN 0–387–12748–8.

Nugent, Ward J Prejudice: index of modern information with bibliography (1988)
Annandale VA, ABBE Publishers Association of Washington, 150 p. ISBN 0–88164–902–3.

Nukhovich, E International Monopolies and the Developing Countries (1980)
Chicago IL, Imported Publications. ISBN 0–8285–1801–7.

Nunnally, Elam W, et al Mental Illness, Delinquency, Addictions, and Neglect (1988)
Newbury Park CA, Sage Publications, 264 p.
ISBN 0–8039–2705–3.

Nunnally, Elam W, et al Troubled Relationships (1988)
Newbury Park CA, Sage Publications, 304 p.
ISBN 0–8039–2701–0.

Nunnenkamp, Peter The International Debt Crisis of the Third World: causes and consequences for the world economy (1986)
New York, Saint Martin's Press, 208 p.
ISBN 0–312–42003–X.

Nurkse, Ragnar Problems of Capital Formation in Underdeveloped Countries
Melbourne VIC, Oxford University Press.

Nussenblatt, Uvetis: fundamental and practice (1988)
Chicago IL, Year Book Medical Publishers,.
ISBN 0–8151–6457–2.

Nygård, Fredrik and Sandström, Arne Measuring Income Inequality (1982)
Stockholm, Almqvist and Wiksell, 436 p.
ISBN 91–22–00439–4.

O'Brien, David W Misconduct Cases Book (1985)
Covina CA, Winterbrook Publications, 501 p.
ISBN 0–9602204–1–0.

O'Brien, P M S The Premenstrual Syndrome (1987)
Oxford, Blackwell Scientific Publications, 232 p. illus.
ISBN 0–632–01343–5.

O'Brien, William V The Conduct of a Just and Limited War (1981)
New York, Praeger Publishers, 512 p.
ISBN 0–275–90693–0.

O'Connell, Jeffrey and Kelly, C Brian The Blame Game: injury, insurance and injustice (1986)
Lexington MA, Lexington Books, 176 p.
ISBN 0–669–11129–5.

O'Flynn, Gráinne World Survival: the Third World struggle
Portland OR, O'Brien FM Bookseller, 128 p.
ISBN 0–86278–040–3.

O'Neill, Brad E, et al (Eds) Insurgency in the Modern World (1980)
Boulder CO, Westview Press, 320 p. illus. Westview Special Study Ser. ISBN 0–8133–0056–8.

O'Neill, Michael J Terrorist Spectaculars: should TV coverage be curbed ? (1986)
New York, Priority Press Publications, 109 p.
ISBN 0–87078–202–9.

O'Neill, Sherry B Starving for Attention (1982)
New York, Continuum Publishing, 212 p. illus.
ISBN 0–8264–0209–7.

O'Sullivan, Noel Terrorism, Ideology and Revolution: the origins of modern political violence (1986)
Boulder CO, Westview Press, 256 p.ISBN 0–8133–0345–1.

O'Toole, Roger Precipitous Path: studies in political sects; a study of the analogy between the political sects and religious groups on the fringes of Canadian society (1981)
Richmond Hill ON, Irwin Publishing, 160 p.

Oakland, Thomas P and Terry, Edwin J Divorced Fathers: reconstructing a quality life (1983)
New York, Human Sciences Press, 201 p.
ISBN 0–89885–101–7.

Obe, G (Ed) Mutations in Man (1984)
Berlin, Springer–Verlag, 350 p. ISBN 3–540–13113–2.

Obeng, L E Man-made Lakes and Their Problems (1979)
Rome, FAO, 138–142 p.

Oberai, A S Migration, Urbanisation and Development (1987)
Washington DC, ILO, 108 p. No. 5 ISBN 92–2–106129–9.

Ochberg, Frank Victims of Terrorism (1981)
Boulder CO, Westview Press, 201 p.ISBN 0–89158–463–3.

Ochberg, Frank M and Hamburg, David A Post–Traumatic Therapy and Victims of Violence (1988)
New York, Brunner/Mazel, 384 p. ISBN 0–87630–490–0.

Odekunle, Femi (Ed) Nigeria: corruption in development (1986)
Ile–Ife, University of Ife Press, 306 p.

Odera, H O Punishment and Terrorism in Africa: problems in the philosophy and practice of punishment (1976)
Nairobi, Kenya Literature Bureau, 102 p.

OECD Science, Growth and Society: report of the Secretary-General's Ad Hoc Group on New Concepts of Science Policy (1971)
Paris, OECD, p. 59–63.

OECD List of Social Concerns Common to OECD Countries (1973)
Paris, OECD.

OECD Young Driver Accidents (1975)
Paris, OECD, 194 p. ISBN 92–64–11341–X.

OECD Unemployment Compensation and Related Employment Policy Measures: general report and county studies (1979)
Washington DC, OECD, 286 p. ISBN 92–64–11909–4.

OECD Equal Opportunities for Women (1979)
Paris, OECD, 216 p. ISBN 92–64–11904–3.

OECD The Impact of Tourism on the Environment: general report (1980)
Washington DC, OECD, 150 p. ISBN 92–64–12060–2.

OECD Youth Unemployment: the causes and consequences (1980)
Washington DC, OECD, 134 p. ISBN 92–64–12137–4.

OECD Buying Power: the exercise of market power by dominant buyers (1981)
Washington DC, OECD, 178 p. ISBN 92–64–12168–4.

OECD Transfrontier Pollution and the Role of States (1981)
Washington DC, OECD, 202 p. ISBN 92–64–12197–8.

OECD The Welfare State in Crisis (1981)
Paris, OECD, 274 p. ISBN 92–64–12192–7.

OECD International Aspects of Inflation – The Hidden Economy (1982)
Paris, OECD. OECD Occasional Studies.
ISBN 92–64–12330–X.

OECD The Cost of Oil Spills: expert studies presented to OECD seminar (1982)
Washington DC, OECD, 252 p. ISBN 92–64–12339–3.

OECD Photochemical Smog: contribution of volatile organic compounds (1982)
Washington DC, OECD, 98 p. ISBN 92–64–12297–4.

OECD Measuring the Effects of Inflation on Income, Saving and Wealth (1983)
Paris, OECD, 170 p. ISBN 92–64–12449–7.

OECD Coal: environmental issues and remedies (1983)
Washington DC, OECD, 88 p. ISBN 92–64–12512–4.

OECD Economic Aspects of International Chemicals Control (1983)
Washington DC, OECD, 96 p. ISBN 92–64–12508–6.

OECD Impacts of Heavy Freight Vehicles (1983)
Washington DC, OECD, 170 p. ISBN 92–64–12423–3.

OECD Job Losses in Major Industries: manpower strategy responses (1983)
Washington DC, OECD, 125 p. ISBN 92–64–12408–X.

OECD Microelectronic, Robotics and Jobs (1983)
Paris, OECD, 72 p. ICCP Series: no 10.
ISBN 92–64–12852–2.

OECD International Comparison Study on Reactor Accident Consequence Modeling (1984)
Paris, OECD, 110 p. ISBN 92–64–12554–X.

OECD Shorter Working Time: a dilemma for collective bargaining (1984)
Washington DC, OECD, 92 p. ISBN 92–64–12640–6.

OECD Metrology and Monitoring of Radon, Thoron and Their Daughter Products (1985)
Washington DC, OECD, 148 p. ISBN 0–318–18487–7.

OECD Transfrontier Movements of Hazardous Wastes: legal and institutional aspects (1985)
Washington DC, OECD, 304 p. ISBN 92–64–12694–5.

OECD Environmental Effects of Electricity Generation (1985)
Paris, OECD, 154 p. ISBN 92–64–12697–X.

OECD Problems of Trade in Fishery Products (1985)
Paris, OECD, 384 p. ISBN 92–64–12775–5.

OECD Nuclear Reactor Accident Source Terms: report by an NEA Group of Experts, March 1986 (1986)
Paris, OECD, 50 p. ISBN 92–64–12808–5.

OECD Fighting Noise, Strengthening Noise Abatement Policies (1986)
Washington DC, OECD, 145 p. ISBN 92–64–12827–1.

OECD OECD Guidelines for Testing of Chemicals (1987)
Paris, OECD, 114 p. ISBN 92-64-12900-6.
OECD Financing and External Debt of Developing Countries - 1986 Survey (1987)
Paris, OECD, 228 p. ISBN 92-64-12853-0.
OECD Adolescents and Comprehensive Schooling (1987)
Washington DC, OECD, 148 p. ISBN 92-64-13035-7.
OECD Newly Industrialising Countries: challenge and opportunity for OECD industries (1988)
Paris, OECD, 150 p. ISBN 92-64-13041-1.
OECD External Debt: definition, statistical coverage and methodology (1988)
Paris, OECD, 178 p. ISBN 92-64-13039-X.
OECD Demographic Change and Public Policy: ageing populations: the social policy implications (1988)
Paris, OECD, 90 p. ISBN 92-64-13113-2.
OECD Disabled Youth: the right to adult status (CERI) (1988)
Paris, OECD, 60 p. ISBN 92-64-13132-9.
OECD Heavy Trucks, Climate Change and Pavement Damage: report prepared by and OECD Scientific Experts Group (1988)
Paris, OECD, 174 p. ISBN 92-64-13150-7.
OECD The Costs of Restricting Imports: the automobile industry (1988)
Washington DC, OECD, 174 p. ISBN 92-64-13037-3.
OECD Substitute Fuels for Road Transport: a technology assessment (1990)
Paris, OECD, 114 p. ISBN 92-64-13324-0.
OECD Transport for Disabled People: a review of provisions and standards for journey planning and pedestrian access (1990)
Paris, OECD, 114 p. ISBN 92-821-1145-8.
OECD and NEA Severe Accidents in Nuclear Power Plants (1986)
Washington DC, OECD, 32 p. ISBN 9-2641-2821-2.
OECD and NEA Near-Field Assessment of Repositories for Low and Medium Level Radioactive Waste (1988)
Paris, OECD, 314 p. ISBN 92-64-03060-3.
OECD and NEA The Radiological Impact of the Chernobyl Accident in OECD Countries (1988)
Washington DC, OECD, 184 p. ISBN 92-64-13043-8.
Oesterreich, Traugott Konstatin Obsession and Possession by Spirits Both Good and Evil (1930)
London, Kegan Paul.
Offiong, Daniel O Imperialism and Dependency: obstacles to african development (1982)
Washington DC, Howard University Press, 304 p.
ISBN 0-88258-126-0.
Oh, Tai K The Asian Brain Drain: a factual and casual analysis (1977)
R and E Publications.
ISBN 0-88247-425-1.
Okerman, Lieve Diseases of Domestic Rabbits (1988)
Oxford, Blackwell Scientific Publications, 140 p. illus.
ISBN 0-632-02254-X.
Okidi, C O Regional Control of Ocean Pollution: legal and institutional problems and prospects (1978)
Dordrecht, Kluwer Academic Publishers Group, 283 p.
ISBN 90-286-0367-0.
Okken, P A; Swart, R J and Zwerver, S (Eds) Climate and Energy: the feasibility of controlling Co2 emissions (1989)
Dordrecht, Kluwer Academic Publishers Group, 267 p.
ISBN 0-7923-0519-1.
Okolicsanyi, L, et al (Ed) Assessment and Management of Hepatobiliary Disease (1987)
Berlin, Springer-Verlag, 440 p. ISBN 3-540-17760-4.
Okun, Lewis Woman Abuse: facts replacing myths (1985)
Albany NY, State University of New York Press, 298 p.
ISBN 0-88706-077-3.
Olivier, Christiane Jocasta's Children
London, Routledge.
Olsen, Steen Tumors of the Kidney and Urinary Tract (1985)
Philadelphia PA, WB Saunders, 291 p.
ISBN 0-7216-1588-0.
Olton, R and Plano, J C The International Relations Dictionary (1982)
Oxford, Clio Press. ISBN 0-87436-332-2.
Olusoga, S O Management of Corruption
Ikeja, Literamed Publications.
Olweus, Dan Aggression in the Schools: bullies and whipping boys (1978)
Basingstoke, Taylor and Francis, 218 p.
ISBN 0-470-99361-8.
Olweus, Dan, et al Development of Antisocial and Prosocial Behavior: research, theories and issues (1985)
San Diego CA, Academic Press. ISBN 0-12-525880-1.
Onesti, Gaddo and Klimt, Christian R Hypertension: determinants, complications and intervention (1978)
Orlando FL, Grune and Stratton, 480 p.
ISBN 0-8089-1108-2.
Ontario, Ministry of Education Corporal Punishment in the Schools (1981)
Ontario ON, Ministry of Education, 257 p.
ISBN 0-7743-6192-1.
Ontiveros, Suzanne R Global Terrorism: a historical bibliography (1986)
Santa Barbara CA, ABC Clio, 168 p. ISBN 0-87436-453-1.
Ophuls, William Ecology and the Politics of Scarcity (1977)
Salt Lake City UT, WH Freeman.
Orenstein, Neil S and Bingham, Sarah L Food Allergies (1988)
New York, Putnam Publishing Group, 160 p.
ISBN 0-317-59799-X.
Organization of American States (Ed) Basic Documents Pertaining to Human Rights in the Inter-American System: (Updated to 1 March 1988) (1988)
Washington DC, OAS.

Orie, N G and Lende, R van der (Eds) Bronchitis III: proceedings of the third international symposium on bronchitis, the Netherlands, 1969 (1970)
Assen, Van Gorcum, 448 p. ISBN 90-232-0704-1.
Oriel, D et al Chlamydial Infections (1986)
New York, Cambridge University Press, 595 p.
ISBN 0-521-32453-X.
Orlandi, Mario, et al Encyclopedia of Health: stress and mental health (1988)
New York, Facts on File, 128 p. ISBN 0-8160-1668-2.
Orloff, M J and Stipa, S Medical and Surgical Problems of Portal Hypertension (1981)
San Diego CA, Academic Press. ISBN 0-12-528380-6.
Ornstein, Robert and Ehrlich, Paul New World, New Mind: changing the way we think to save our future (1989)
London, Metheun, 302 p. ISBN 0-413-61680-0.
Osahon, Naiwu (Ed) Oil Conspiracy
Apapa, Heritage Books.
Osborne, William The Rape of the Powerless (1971)
New York, Gordon and Breach, 212 p.
ISBN 0-677-14720-1.
Osgood, Nancy and McIntosh, John L Suicide and the Elderly: an annotated bibliography and review (1986)
Westport CT, Greenwood Press, 206 p.
ISBN 0-313-24786-2.
Oshimata, Tana Therapy of Mental Disorders: medical analysis index with research (1987)
Annandale VA, ABBE Publishers Association of Washington, 150 p. ISBN 0-88164-410-2.
Oski, Frank A and Naiman, J Lawrence Hematologic Problems in the Newborn (1982)
Philadelphia PA, WB Saunders, 376 p. illus. 3rd ed. Major Problems in Clinical Pediatrics: Vol 4.
Ostrow, David G, et al Sexually Transmitted Diseases in Homosexual Men: diagnosis, treatment, and research (1983)
New York, Plenum Publishing, 292 p.
ISBN 0-306-41337-X.
Otlet, Paul An Introduction to the Study of International Problems: opening lecture at the Interallied School for Advanced Education, Paris, 18 December 1919
In: *Revue Internationale de Sociologie* Paris, UNESCO, 35.
Otlet, Paul Les Problèmes Internationaux (1916)
Geneva, Librairie Kundig, 503 p.
Otlet, Paul Monde: essai d'universalisme (1935)
Brussels, Editions Mundaneum, 467 p.
Ottens, Allen J Coping with Romantic Breakup (1987)
New York, Rosen Publishing Group, 147 p.
ISBN 0-8239-0649-3.
Ou, S H Rice Diseases (1985)
Kew, CAB International Mycological Institute, 391 p. 2nd ed.
ISBN 0-85198-545-9.
Ouchterlony, O and Holmgren, J Cholera and Related Diarrheas: molecular aspects of a global health problem (1980)
New York, Karger S AG. ISBN 3-8055-3060-9.
Overholt, William H Political Risk (1982)
Woodhead-Faulkner, 152 p.
Euromoney Ser. ISBN 0-903121-33-6.
Owen, Wilfred Transportation and World Development (1987)
Baltimore MD, Johns Hopkins University Press, 176 p.
ISBN 0-8018-3495-3.
Padover, Saul K Aggression Without Weapons (1968)
New York, Ethical Cultural Publications.
Page, Robert M Stigma (1984)
New York, Routledge Chapman and Hall, 156 p.
ISBN 0-7100-9786-7.
Pakrasi, Kanthi Bhusan Female Infanticide in India (1976)
Calcutta, Editions India, 419 p.
Palazzini, Pietro Sin: its reality and nature (1964)
New Rochelle NY, Scepter Publishers, 238 p.
ISBN 0-933932-25-1.
Palen, J John Social Problems (1979)
New York, McGraw Hill Book Company. illus.
ISBN 0-07-048103-2.
Palme Commission Common Security: a programme for disarmament: the report of the Independent Commission on Disarmament and Security Issues under the Chairmanship of Olof Palme (1982)
London, Pan. ISBN 0-330-26846-5.
Palmer, Edward L Children and the Faces of Television: teaching, violence, selling (1980)
San Diego CA, Academic Press. ISBN 0-12-544480-X.
Palmier, Leslie Control of Bureaucratic Corruption: case studies in Asia (1985)
New Delhi, Allied Publishers, 292 p.
Pandey, R (Ed) Nononcogenic Avian Viruses (1989)
Basel, S Karger AG, 144 p.
Paquet, K J, et al Portale Hypertension (1982)
New York, Karger S AG, 282 p. ISBN 3-8055-3480-9.
Paranjape, H K Flight of Technical Personnel in Public Undertakings: a study report for the government of India on behalf of the Indian Institute of Public Administration (1964)
New Delhi, 191 p.
Parcel, Toby L and Mueller, Charles W Ascription and Labor Markets: race and sex difference in earnings (1983)
San Diego CA, Academic Press. ISBN 0-12-545020-6.
Pareto, Vilfredo The Theory of the Economic and Political Elites in the Historical Scenario of the 20th Century (1984)
Find Class Reprints, 181 p.
ISBN 0-89901-220-5.
Parillo, Vincent N, et al Contemporary Social Problems (1989)
New York, Macmillan, 660 p. ISBN 0-02-391731-8.
Parillo, Vincent N; Stimson, Ardyth and Stimson, John Contemporary Social Problems (1989)
New York, Macmillan Publishing, 660 p. 2nd ed.
ISBN 0-02-391731-8.

Park, James Absurdity, Insecurity and Despair (1975)
Minneapolis MN, Existential Books. Existential Freedom Ser: No 8. ISBN 0-89231-008-1.
Park, Kilho P, et al Radioactive Wastes and the Ocean (1983)
New York, John Wiley and Sons, 522 p. Wastes in the Ocean: 3. ISBN 0-471-09770-5.
Parkes, Colin M Bereavement (1987)
International Universal Press. 2nd ed.
Parkin, Alan J Memory and Amnesia: an introduction (1987)
New York, Basil Blackwell, 240 p. ISBN 0-631-14868-X.
Parkin, M D, et al (Eds) International Incidence of Childhood Cancer (1988)
Oxford NY, Oxford University Press, 300 p. illus. IARC Scientific Publications: No 87. ISBN 92-832-1187-1.
Parmeggiani, Luigi (Ed) Encyclopedia of Occupational Health and Safety (1983)
Lanham MD, UNIPUB, 2538 p. 2 vols. 3rd rev ed.
Parrish, Ruth G Eugenics: index of modern information with bibliography (1988)
Washington DC, ABBE Publications Association, 150 p. bibl.
ISBN 0-88164-898-1.
Parsons, Howard L Self, Global Issues and Ethics (1980)
Amsterdam, BR Grüner, 210 p. ISBN 90-6032-178-2.
Parsons, Tony Men: the darker continent (1990)
New York, James H Heineman.
Passchier, W F and Bosnjakovic, B F M (Eds) Human Exposures to Ultraviolet Radiation: risks and regulations (1987)
Jonesboro AR, ESP, 580 p. ISBN 0-444-80914-7.
Passmore, John Man's Responsibility for Nature: ecological problems and Western traditions (1980)
London, Duckworth, 228 p. 2nd ed. ISBN 0-7156-0819-3.
Patnaik, Prabhat (Ed) Lenin and Imperialism: an appraisal of theories and contemporary reality
Hyderabad, Orient Longman. ISBN 0-86131-612-6.
Patosaari, P Chemicals in Forestry: health hazards and protection (1987)
Helsinki.
Pattemore, W J Effects of the Pulpwood Industry on Wildlife in Tasmania 1: design for long term studies
National Parks Sany Bay.
Patterson, K David Pandemic Influenza Seventeen Hundred to Nineteen Hundred: a study in historical epidemiology (1987)
Totowa NJ, Rowman and Littlefield Publishers, 128 p.
ISBN 0-8476-7512-2.
Pattillo, Ronald A and Hussa, Robert O Human Trophoblast Neoplasms (1984)
New York, Plenum Publishing, 516 p.
ISBN 0-306-41735-9.
Paukert, Liba The Employment and Unemployment of Women in OECD Countries (1984)
Paris, OECD, 88 p. ISBN 92-64-12570-1.
Paulus, P B Prison Crowding: a psychological perspective (1988)
Berlin, Springer-Verlag, 105 p. ISBN 3-540-96650-1.
Paustenbach, Dennis J The Risk Assessment of Environmental Hazards (1988)
New York, John Wiley and Sons, 1600 p.
ISBN 0-471-84998-7.
Pavalko, Ronald Social Problems (1986)
Itasca IL, Peacock FE, 520 p. ISBN 0-87581-280-5.
Pavey, Roger V The Kindest Cut: a study of circumcision
New York, Associated Faculty Press, 200 p. 2nd ed. Judaic Studies. ISBN 0-8046-9385-4.
Pavone-Macaluso, M, et al Testicular Cancer and Other Tumors of the Genitourinary Tract (1985)
New York, Plenum Publishing, 550 p.
ISBN 0-306-41906-8.
Pawlick, Thomas Killing Rain: the global threat of acid precipitation
Vancouver, Douglas and McIntyre, 192 p.
ISBN 0-88894-442-X.
Pawlowski, A; Seminara, D and Watson, R R (Eds) Alcohol, Immunomodulation and AIDS (1990)
New York, Alan R Liss, 476 p. ISBN 0-471-56215-7.
Payne, L N (Ed) Marek's Disease: scientific basis and methods of control (1985)
Dordrecht, Kluwer Academic Publishers Group, 384 p.
ISBN 0-89838-730-2.
Peach, Ceri, et al Ethnic Segregation in Cities (1982)
Athens GA, University of Georgia Press, 220 p.
ISBN 0-8203-0599-5.
Peach, Richard K Readings in Agnosia (1986)
White Plains NY, Longman, 125 p. Special Education Ser.
ISBN 0-582-28615-8.
Peacock, Thomas B On Malformations of the Human Heart (1977)
Wolfeboro NH, Longwood Publishing Group.
ISBN 0-89341-506-5.
Pearse, Andrew Seeds of Plenty, Seeds of Want: social and economic implications of the green revolution (1980)
Oxford NY, Oxford University Press, 262 p.
Pearson, et al Aquatic Toxicology and Hazard Assessment (Fifth Conference)- STP 766 (1982)
Philadelphia PA, American Society for Testing and Materials, 414 p. ISBN 0-8031-0796-X.
Pearson, Geoffrey Hooligan: a history of respectable fears (1983)
Basingstoke, Macmillan, 296 p. ISBN 0-333-23400-6.
Pearson, Lester Partners in Development: report of the Commission on International Development (1969)
New York, Praeger.
Pena, A S Recent Advances in Crohn's Disease: developments in gastroenterology one (1981)
Norwell MA, Kluwer Academic Publishers, 549 p.
ISBN 90-247-2475-9.

Penfield, Joyce Women and Language in Transition (1987)
Albany NY, State University of New York Press, 224 p.
ISBN 0–88706–485–X.

Penman, A D Deterioration of Dams and Reservoirs: examples and their analysis (1984)
Rotterdam, AA Balkema, 368 p. Trans. from French.
ISBN 90–6191–546–5.

Pennington, James E Respiratory Infections: diagnosis and management (1988)
New York, Raven Press, 700 p. ISBN 0–88167–460–5.

Penny, Malcolm; Quinn, David and Werikhe, Micheal Rhinos: an endangered species (1988)
New York, Facts on File, 128 p. ISBN 0–8160–1882–0.

Perk, Kalman, et al Immunodeficiency disorders and retroviruses (1988)
San Diego CA, Academic Press, 240 p. Advances in Veterinary Science and Comparative Medicine: 33.
ISBN 0–12–039232–1.

Perkins, H C Air Pollution (1974)
New York, McGraw Hill Book Company, 448 p.
ISBN 0–07–049302–2.

Perlmutter, Amos Modern Authoritarianism (1984)
New Haven CT, Yale University Press, 192 p.
ISBN 0–300–03178–5.

Perpiñan, Mary Soledad International Meeting of Experts on the Social and Cultural Causes of Prostitution and Strategies Against Procuring and Sexual Exploitation of Women, Madrid, 1985 (1985)
Paris, UNESCO, 65 p. bibl.

Perring, F H and Mellanby, K Ecological Effects of Pesticide (1978)
San Diego CA, Academic Press. ISBN 0–12–551350–X.

Perron, Charles Normal Accidents: living with high–risk technologies (1985)
New York, Basic Books, 386 p. ISBN 0–465–05142–1.

Perrucci, Robert, et al Plant Closings: international context and social costs (1988)
Hawthorne NY, Aldine de Gruyter, 256 p.
ISBN 0–202–30338–1.

Perry, C M and Pfaltzgraff, R L Selling the Rope to Hang Capitalism?: the debate on west–east trade and technology (1987)
Elmsford NY, Pergamon Books, 216 p.
ISBN 0–08–034959–5.

Persley, G J (Ed) Bacterial Wilt Disease in Asia and the South Pacific (1986)
Inkata Pr Australia/International Spec Book, 146 p.
ISBN 0–949511–20–X.

Persson, B (Ed) Surviving Failures: the psychology and politics of project mismanagement: Patterns and case of project mismanagement (1979)
Stockholm, Almqvist and Wiksell, 257 p.

Peschel, Richard E and Peschel, Enid R When a Doctor Hates a Patient: chapters from a young physicians life (1986)
Berkeley CA, University of California Press.
ISBN 0–520–05755–4.

Peters, Edward Heresy and Authority in Medieval Europe (1980)
Philadelphia PA, University of Pennsylvania Press, 384 p.
ISBN 0–8122–7779–1.

Peters, Edward Torture (1985)
New York, Basil Blackwell, 160 p. ISBN 0–631–13164–7.

Peters, George A and Peters, Barbara J Sourcebook on Asbest Diseases, Vol II: medical, legal and engineering aspects (1986)
New York, Garland Publishing.

Petersilia, Joan Racial Disparities in the Criminal Justice System (1983)
Chicago IL, Rand Corporation, 128 p.
ISBN 0–8330–0506–5.

Peterson, H and Marquardt, J Appraisal and Diagnosis of Speech and Language Disorders (1981)
Englewood Cliffs NJ, Prentice Hall. ISBN 0–13–043505–8.

Peterson, Martin International Interest Organizations and the Transmutation of Postwar Society (1979)
Stockholm, Almqvist and Wiksell, 720 p.
ISBN 91–22–00255–3.

Peterson, Richard R Women, Work, and Divorce (1988)
Albany NY, State University of New York Press, 192 p.
ISBN 0–88706–858–8.

Petitpierre, Dom Robert Exorcising Devils (1976)
London, Robert Hale.

Petrie, Asenath Individuality in Pain and Suffering (1978)
Chicago IL, University of Chicago Press.
ISBN 0–226–66347–7.

Petterson, Olaf and Akerberg, H Interpreting Religious Phenomena: studies with reference to the phenomenology of religion (1981)
Stockholm, Almqvist and Wiksell, 202 p.
ISBN 91–22–00405–X.

Petti, Theodore A (Ed) Childhood Depression (1983)
New York, Haworth Press, 95 p. Journal of Children in Contemporary Society Ser: Vol 15, No 2.
ISBN 0–917724–95–X.

Pettman, Ralph (Ed) Moral Claims in World Affairs (1979)
Potts Point NSW, Australian National University, 199 p.
ISBN 0–08–033016–9.

Petursson, Hannes and Lader, Malcolm Dependence on Tranquilizers (1984)
New York, Oxford University Press. ISBN 0–19–712152–7.

Pfaltz, C R (Ed) Controversial Aspects of Meniere's Disease (1986)
New York, Thieme Medical Publishers, 164 p. illus.
ISBN 0–317–53436–X.

Phelan, Richard J and Ross, Kenneth Product Liability: warnings, instructions, and recalls 1984 (1984)
New York, Practising Law Institute, 300 p.
ISBN 0–317–27552–6.

Phelps Brown, E H The Inequality of Pay (1978)
Berkeley CA, University of California Press.
ISBN 0–520–03380–9.

Philips, L Predatory Pricing (1987)
Luxembourg, European Communities, 73 p.
ISBN 92–825–6715–X.

Phillips, E Lakin Stress, Health and Psychological Problems in the Major Professions (1983)
Lanham MD, University Press of America, 478 p.
ISBN 0–8191–2773–6.

Phillips, Paul Regional Disparities: new updated edition
Toronto ON, James Lorimer, 176 p. ISBN 0–88862–206–6.

Phillips, Roderick Putting Asunder: a history of divorce in western society (Date not set)
New York, Cambridge University Press, 816 p.
ISBN 0–521–32434–3.

Phillipson, Chris, et al (Eds) Dependency and Interdependency in Old Age (1986)
New York, Routledge Chapman and Hall, 300 p.
ISBN 0–7099–3987–6.

Piazza, Robert Readings in Hyperactivity (1979)
Guilford CT, Special Learning Corporation.
ISBN 0–89568–107–2.

Pickett, E E (Ed) Atmospheric Pollution (1987)
Berlin, Springer–Verlag, 290 p. Dist USA, Canada, Mexico: Hemisphere Publ Corp. ISBN 3–540–17997–6.

Picquet, D Cheryn and Best, Reba A Post–Traumatic Stress Disorder, Rape Trauma, Delayed Stress and Related Conditions: a bibliography, with a directory of veterans (1986)
Jefferson NC, McFarland Publishers, 208 p.
ISBN 0–89950–213–X.

Pierce, Grant N, et al Heart Dysfunction in Diabetes (1988)
Boca Raton FL, CRC Press, 256 p. ISBN 0–8493–6887–1.

Pimlott, D H (Ed) Wolves (1975)
Gland, IUCN, 144 p. illus. ISBN 2–88032–019–4.

Pimpley and Sharma Struggle for Status (1985)
Delhi, BR Publishing Corporation, 232 p.
ISBN 81–7018–265–4.

Pincher, Chapman Traitors: the anatomy of treason (1987)
New York, Saint Martin's Press, 640 p.
ISBN 0–312–00696–9.

Pindberg, J J Oral Cancer and Precancer (1980)
Littleton MA, PSG Publishing, 192 p. ISBN 0–7236–0529–7.

Pinkert, Theodore Current Research on the Consequences of Maternal Drug Abuse (1985)
Washington DC, United States Government Printing Office, 118 p. ISBN 0–318–18741–8.

Pires, Manuel International Juridical Double Taxation of Income (1989)
Deventer, Kluwer Law and Taxation, 336 p. Series on International Taxation: No 11. ISBN 90–6544–426–2.

Pitt, Brice (Ed) Dementia (1987)
New York, Churchill Livingstone, 246 p. illus. Medicine in Old Age Ser. ISBN 0–443–03264–5.

Pledge, P J and Tonkin, C B One World: sources and study guide (1971)
Cassell, 281 p.

Plekhanov, G Fundamental Problems of Marxism (1980)
Moscow, Izdatel'stvo Progress, 118 p.

Plewig, G and Kligman, A M Acne: morphogenesis and treatment (1975)
Berlin, Springer–Verlag, 333 p. ISBN 3–540–07212–8.

Plum, F Brain Dysfunction in Metabolic Disorders (1974)
New York, Elsevier Science Publishing.
ISBN 0–7204–7521–X.

Plum, Fred and Posner, Jerome The Diagnosis of Stupor and Coma (1980)
Philadelphia PA, Davis FA, 373 p. ISBN 0–8036–6992–5.

Poleszynski, Dag Food, Social Cosmology and Mental Health: the case of suger (1982)
Lanham MD, UNIPUB, 52 p. ISBN 92–808–0324–7.

Pollitt, Ernesto Poverty and Malnutrition in Latin America: early childhood intervention programs (1980)
New York, Praeger Publishers, 150 p.
ISBN 0–275–90538–1.

Pollitt, Ernesto and Leibel, Rudolph Iron Deficiency: brain biochemistry and behavior (1982)
New York, Raven Press, 231 p. ISBN 0–89004–690–5.

Pollitzer, R Cholera (1959)
Geneva, WHO, 1019 p. ISBN 92–4–140043–9.

Pollock, Cynthia Decommissioning: nuclear power's missing link
Washington DC, Worldwatch Institute. Worldwatch Papers: 69.

Pollock, Cynthia Decommissioning: nuclear power's missing link (1986)
Washington DC, Worldwatch Institute, 56 p. Worldwatch Papers. ISBN 0–916468–70–4.

Pontificia Accademia delle Scienze Immunology, Epidemiology and Social Aspects of Leprosy (1984)
Rome, Pontificia Accademia delle Scienze.
ISBN 88–7761–022–0.

Pontificia Accademia delle Scienze The Artificial Prolongation of Life and The Determination of The Exact Moment of Death (1985)
Rome, Pontificia Accademia delle Scienze.
ISBN 88–7761–001–8.

Pontificia Accademia delle Scienze The Interaction of Parasitic Diseases and Nutrition (1985)
Rome, Pontificia Accademia delle Scienze.
ISBN 88–7761–002–6.

Pontificia Accademia delle Scienze Remote Sensing and Its Impact on Developing Countries (1986)
Rome, Pontificia Accademia delle Scienze.
ISBN 88–7761–007–7.

Ponting, Clive Breach of Promise
H Hamilton.

Pope, Harrison G Jr and Hudson, James I New Hope for Binge Eaters: advances in the understanding and treatment for bulimia (1985)
New York, Harper and Row, 256 p. ISBN 0–06–091239–1.

Popkin, Richard H The History of Scepticism from Erasmus to Spinoza (1979)
Berkeley CA, University of California Press.
ISBN 0–520–03827–4.

Popper, Karl The Open Society and its Enemies

Porter, Ian H and Hook, Ernest B Human Embryonic and Fetal Death (1980)
San Diego CA, Academic Press. ISBN 0–12–562860–9.

Porter, Roy, et al (Eds) The Anatomy of Madness (1988)
New York, Routledge Chapman and Hall. 3 Vols.

Posner, Gerald L Warlords of Crime: the new mafia (1988)
New York, McGraw Hill Book Company, 288 p.
ISBN 0–07–050600–0.

Postal, Sandra Altering the Earth's Chemistry: assessing the risks
Washington DC, Worldwatch Institute. Worldwatch Papers: 71.

Postal, Sandra Conserving Water: the untapped alternative (1985)
Washington DC, Worldwatch Institute. Worldwatch Papers.
ISBN 0–916468–67–4.

Postel, Sandra Water for Agriculture: facing the limits
Washington DC, Worldwatch Institute. Worldwatch Papers: 93.

Postel, Sandra Air Pollution, Acid Rain, and the Future of Forests (1984)
Washington DC, Worldwatch Institute. Worldwatch Papers.
ISBN 0–916468–57–7.

Postel, Sandra Altering the Earth's Chemistry: assessing the risks (1986)
Washington DC, Worldwatch Institute, 68 p. Worldwatch Papers. ISBN 0–916468–72–0.

Postel, Sandra Defusing the Toxics Threat: controlling pesticides and industrial waste (1987)
Washington DC, Worldwatch Institute, 70 p. Worldwatch Papers. ISBN 0–916468–80–1.

Postman, Neil Conscientious Objections: stirring up trouble about language, technology, (1988)
New York, Alfred A Knopf, 208 p. ISBN 0–394–57270–X.

Pothen, S Divorce: its causes and consequences in hindu society (1986)
New York, Advent Books, 320 p. ISBN 0–7069–2932–2.

Potter, Carole A Knock on Wood: an encyclopedia of talismans, charms, (1983)
New York, Beaufort Books Publishers, 272 p.
ISBN 0–8253–0138–6.

Pottier, P E and Glasser (Ed) Characterization of Low and Medium–Level Radioactive Waste Forms: final report, 2nd programme, 1980–1984 (1986)
Luxembourg, European Communities, 278 p.
ISBN 92–825–6582–3.

Potts, M, et al Childbirth in Developing Countries (1983)
Norwell MA, Kluwer Academic Publishers, 300 p.
ISBN 0–85200–493–1.

Powell, Barbara Overcoming Shyness (1981)
New York, McGraw Hill Book Company.
ISBN 0–07–050570–5.

Powell, Gloria J and Inouye, Daniel K The Psychosocial Development of Minority Group Children (1983)
New York, Brunner/Mazel, 600 p. ISBN 0–87630–277–0.

Powell, Robert; Nicholson, Shirley and Ellwood, Robert The Great Awakening (1983)
Wheaton IL, Theosophical Publishing House, 179 p.
ISBN 0–8356–0577–9.

Powelson, John P The Story of Land: a world history of land tenure and agrarian reform (1987)
Boston MA, Oelgeschlager Gunn and Hain, 350 p.
ISBN 0–89946–218–9.

Powelson, John P and Stock, Richard The Peasant Betrayed: agriculture and land reform in the third world (1987)
Boston MA, Oelgeschlager Gunn and Hain, 302 p.
ISBN 0–89946–216–2.

Powers, Gene R Cleft Palate (1986)
Austin TX, Pro–Ed, 52 p. illus. ISBN 0–89079–094–9.

Powills, Leo I Dyslexia: subject, reference and research guidebook (1987)
Washington DC, ABBE Publications Association, 160 p.
ISBN 0–88164–602–4.

Practising Law Institute Product Counterfeiting: remedies (1984)
New York, Practising Law Institute, 332 p. Patents, Copyrights, Trademarks and Literacy Property Course Handbook: Vol 180. ISBN 0–317–11486–7.

Practising Law Institute Trading on Inside Information (1984)
New York, Practising Law Institute, 712 p. Vol. 270
ISBN 0–317–27560–7.

Practising Law Institute Protecting Trade Secrets 1986 (1986)
New York, Practising Law Institute, 969 p.
ISBN 0–317–27497–X.

Prada, Manuel Ganzalez Grafitos
New York, Hispanic Institute in the United States, 254 p.
ISBN 0–318–14270–8.

Prasad, Ananda S Essential and Toxic Trace Elements in Human Health and Disease (1988)
New York, Alan R Liss, 704 p. ISBN 0–8451–1617–7.

Prasad, N Rural Violence in India (1985)
Allahabad, Vohra Publishers and Distributors, 120 p.

Pratt, Dallas Painful Experiments on Animals (1976)
New York, Argus Archives.

Preece, Paul, et al (Eds) Cancer of the Stomach (1986)
New York, Grune and Stratton, 320 p.
ISBN 0–8089–1835–4.

Premary, Y S Polyandry in the Himalayas (1975)
New Delhi, Vikas Publishing House, 200 p.

Preobrazhensky, E A and Day, Richard B The Decline of Capitalism (1985)
Armonk NY, ME Sharpe, 200 p. ISBN 0–87332–295–9.

Preston, S H Mortality Patterns in National Populations (1976)
San Diego CA, Academic Press. ISBN 0–12–564450–7.

Preus, Alf Edward Identifying Subgroups of Stutterers (1981)
Oslo, Norwegian University Press, 230 p. ISBN 82–00–05801–8.

Pribram, E Deidre Female Spectators: looking at film and television (1988)
New York, Routledge Chapman and Hall, 224 p. ISBN 0–86091–204–3.

Prince, R (Ed) Trance and Possession States (1966)
Montreal, Bucke Memorial Society.

Prinzing, Friedrich Epidemics Resulting from Wars (1977)
New York, Gordon Press Publishers. ISBN 0–8490–1781–5.

Prior, Margot R; Griffin, Michael W and Rutter, Michael Hyperactivity: diagnosis and management (1985)
New York, Heineman James H, 282 p. ISBN 0–433–26350–4.

Proal, Louis Political Crime (1973)
Montclair NJ, Smith and Patterson Publishing. repr of 1898. ISBN 0–87585–146–0. Criminology, Law Enforcement and Social Problems Ser: No 146. With intro and index added.

Proctor, R A Clinical Aspects of Endotoxin Shock (1985)
New York, Elsevier Science Publishing, 250 p. Vol. 4 ISBN 0–444–90390–9.

Pueschel, and Steinberg, Down Syndrome: a comprehensive bibliography (1980)
Cambridge MA, Academic Guild Publishers. ISBN 0–8240–7158–1.

Pullman, Maynard E, et al Retroviruses and Disease (1989)
San Diego CA, Academic Press, 270 p. ISBN 0–12–322570–1.

Pulzer, Peter The Rise of Political Anti–Semitism in Germany and Austria (1988)
Cambridge MA, Harvard University Press, 384 p. rev ed. ISBN 0–674–77166–4.

Purves, David Trace–Element Contamination of the Environment: fundamental aspects of pollution control and environmental science, 7 (1985)
Amsterdam, Elsevier Science Publishing, 244 p. 2nd rev ed. ISBN 0–444–42503–9.

Purvis, R N Corporate Crime (1979)
Sevenoaks, Butterworths, 806 p. ISBN 0–409–46340–X.

Puskár, Anton and Golovin, Stanislav A Fatigue in Materials: cumulative damage processes (1984)
Amsterdam, Elsevier Science Publishing, 316 p. ISBN 0–444–99597–8.

Putman, Robert D Comparative Study of Political Elites (1976)
Englewood Cliffs NJ, Prentice Hall, 256 p. illus. ISBN 0–13–154195–1.

Putnam, George H Censorship of the Church of Rome and Its Influence upon the Production and Distribution of Literature (1967)
Salem NH, Ayer Company Publishers. 2nd ed. ISBN 0–405–08871–X.

Puttman, George The Censorship of the Church of Rome
New York, Gordon Press Publishers. 2 Vols. ISBN 0–87968–826–2.

Putz–Anderson, Vern Cumulative Trauma Disorders (1988)
New York, Taylor and Francis, 112 p. ISBN 0–85066–405–5.

Pyatt, Sherman E Apartheid: a selective annotated bibliography, 1979–1987 (1989)
New York, Garland Publishing. Garland Reference Library of Social Science: Vol 587. ISBN 0–824–07637–0.

Pye, Kenneth Aeolian Dust and Dust Deposits (1987)
San Diego CA, Academic Press. ISBN 0–12–568690–0.

Qing–Nan, Meng Land–Based Marine Pollution (1987)
Amsterdam, Kluwer Academic Publishers Group.

Quay, Richard H In Pursuit of Equality of Educational Opportunity: a selective bibliography and guide to the (1977)
New York, Garland Publishing, 200 p. ISBN 0–8240–9872–2.

Quick, James C and Quick, Jonathan D Organizational Stress and Preventive Management (1984)
New York, McGraw Hill Book Company, 336 p. ISBN 0–07–051070–9.

Quigley, Gary H Homeless and Street People: index of modern information (1988)
Annandale VA, ABBE Publishers Association of Washington, 150 p. ISBN 0–88164–954–6.

Quinn, Mary J and Tomita, Susan K Elder Abuse and Neglect: causes, diagnosis and intervention strategies (1986)
New York, Springer Publishing, 336 p. ISBN 0–8261–5120–5.

Quinn, Michael A, et al Lead Paint Poisoning in Urban Children: an annotated bibliography (1976)
Chicago IL, CPL Bibliographies. No. 1130 ISBN 0–686–20406–9.

Qureshi, Muhammed Siddiq Zionism and Racism
Lahore, Islamic Publications.

Rabald, E Corrosion Guide (1978)
Amsterdam, Elsevier Science Publishing, 900 p. 2nd rev ed. ISBN 0–444–40465–1.

Radha, Sinha Landlessness: a growing problem (1984)
Rome, FAO, 112 p. FAO Economic and Social Development Series: No 28. ISBN 92–5–101372–1.

Radioactive Waste Campaign Deadly Defense: military radioactive landfills (1988)
Wilmington DE, Peace Resource Center, 170 p.

Raghaviah, V Nomadism: its cause and cure (1968)
Hyderabad, Tribal Cultural Research and Training Institute, 48 p.

Raivio, Kari O et al Respiratory Distress Syndrome (1984)
San Diego CA, Academic Press. ISBN 0–12–576180–5.

Ralovich, B Listeriosis Research: present situation and perspective (1984)
Budapest, Akadémiai Kiadó és Nyomada, 222 p. ISBN 963–05–3657–9.

Ram, N V R Games Bureaucrats Play (1978)
New Delhi, Vikas Publishing House, 77 p. ISBN 0–7069–0631–4.

Ramadhan, S Islam and Nationalism
Chicago IL, Kazi Publications. ISBN 0–686–18586–2.

Rambo, A Terry Primitive Polluters: Semang impact on the Malaysian tropical rain (1986)
Ann Arbor MI, University of Michigan, 104 p. ISBN 0–915703–04–1.

Ramesh, Jairam and Weiss, Charles Jr (Eds) Mobilizing Technology for World Development (1979)
Washington DC, Overseas Development Council, 240 p. ISBN 0–03–055451–9.

Ramsey, Paul Fabricated Man: the ethics of genetic control (1970)
New Haven CT, Yale University Press.

Ramström, Lars M (Ed) Smoking Epidemic: a matter of worldwide concern: Proceedings of the Fourth World Conference on Smoking and Health, Stockholm 1979 (1980)
Stockholm, Almqvist and Wiksell, 352 p. ISBN 91–22–00389–4.

Randall, Vicky Women in Politics: an international perspective (1987)
Basingstoke, Macmillan, 362 p. illus. bibl. ISBN 0–333–44897–9.

Randle, P J et al Carbohydrate Metabolism and Its Disorders (1981)
San Diego CA, Academic Press. Vol. 3 ISBN 0–12–579703–6.

Randles, Jenny The UFO Conspiracy: the first forty years (1988)
New York, Sterling Publishing, 224 p. ISBN 0–7137–1867–6.

Ranke–Heinemann, Uta Eunuchs for Heaven: the Catholic Church and sexuality (1990)
Andre Deutsch.

Ransom, W H Building Failures (1987)
New York, Routledge Chapman and Hall, 200 p. 2nd ed. ISBN 0–419–14260–6.

Rao, Rama K Classified and Annotated Bibliography on Child Abuse and Neglects (1986)
Monticello IL, Vance Bibliographies, 24 p. Public Administration Ser: P 1993. ISBN 0–89028–973–5.

Rao, T V S Ramamohan and Sriraman, S Disequilibrium in the Rail Freight Services (1985)
New Delhi, Ajanta Books, 242 p. ISBN 81–202–0122–1.

Rao, V Venkata Century of Tribal Politics (1977)
New Delhi, Chand S, 556 p.

Rapaport, Jacques, et al Small States and Territories: status and problems
Salem NH, Ayer Company Publishers. ISBN 0–405–02237–9.

Rapoport, David C Assassination and Terrorism
Toronto ON, Canadian Broadcasting Corporation, 96 p.

Rapoport, Judith L Obsessive–Compulsive Disorder in Children and Adolescents (1988)
Washington DC, American Psychiatric Press, 300 p. ISBN 0–88048–282–6.

Rappaport, Julian and Seidman, Edward (Eds) Redefining Social Problems (1986)
New York, Plenum Publishing, 334 p. Perspectives in Social Psychology Series. ISBN 0–306–42052–X.

Rasmussen, Richard M The UFO Literature: a comprehensive annotated bibliography of works (1985)
Jefferson NC, McFarland Publishers, 271 p. ISBN 0–89950–136–2.

Rattray, Jamie and Howells, Bill and Siegler, Irv et al Kids and Smoking (1983)
Deerfield Beach FL, Health Communications, 199 p. ISBN 0–932194–14–1.

Read, Clark P Parasitism and Symbiology: an introductory text
Ann Arbor MI, Books on Demand. repr of 1970. ISBN 0–8357–9947–6.

Read, Stanley E and Zabriskie, John B Streptococcal Diseases and the Immune Response (1980)
San Diego CA, Academic Press. ISBN 0–12–583880–8.

Reardon, Betty Discrimination (1977)
New York, Decade Media Books and Communications, 111 p. Vol. 2: No. 2 ISBN 0–910365–03–2.

Reason, J T and Brand, J J Motion Sickness (1976)
San Diego CA, Academic Press. ISBN 0–12–584050–0.

Rechcigl, M (Ed) Nutrition and the World Food Problem (1979)
Basel, S Karger AG, 374 p. 2nd unchanged ed. ISBN 3–8055–2779–9.

Rediger, G Lloyd Coping with Clergy Burnout (1982)
Valley Forge PA, Judson Press, 112 p. ISBN 0–8170–0956–6.

Rees, A R and Purcell, H J (Eds) Society for the Environmental Therapy, Inaugural Conference, 1981: proceedings of the inaugural conference of the society for environmental therapy, held in Oxford, March 21–23, 1981
Ann Arbor MI, Books on Demand. ISBN 0–317–58693–9.

Regamey, R H, et al (Eds) International Symposium on Foot–and–Mouth Disease: proceedings of the symposium organized by the Permanent Section of Microbiological Standardization, Lyon 1967 (1968)
Basel, S Karger AG, 184 p. ISBN 3–8055–0630–9.

Regamey, R H, et al (Eds) International Symposium on Rabies, 2: proceedings of the 40th symposium organized by the International Association of Biological Standardization, Lyon 1972 (1974)
Basel, S Karger AG, 392 p. ISBN 3–8055–1649–5.

Regamey, R H, et al International Symposium on Viral Hepatitis: proceedings (1975)
New York, Karger S AG, 492 p. ISBN 3–8055–2312–2.

Regan, Tom and Bowker, John Animal Sacrifices (1988)
Philadelphia PA, Temple University Press, 288 p. ISBN 0–87722–511–7.

Reginald, Jorge S Smoking: medical subject analysis and research guide (1987)
Annandale VA, ABBE Publishers Association of Washington, 150 p. ISBN 0–88164–270–3.

Regional Employment Programme for Latin America and the Caribbean Adjustment and Social Debt: structuralist approach
Santiago, PRELAC. ISBN 92–2–105895–6.

Regnier Victor A, Pynoos Jon (Eds) Housing the Aged: design directives and policy considerations (1987)
Amsterdam, Elsevier Science Publishing, 500 p. ISBN 0–444–01012–2.

Rehabilitation International The Economics of Disability: international perspectives (1981)
New York, Rehabilitation International, 238 p.

Reich, Michael Racial Inequality: a political–economic analysis (1981)
Princeton NJ, Princeton University Press, 345 p. ISBN 0–691–04227–6.

Reid, Daniel, et al Infections in Current Medical Practice (1986)
London, Butterworth, 160 p. ISBN 0–407–00355–X.

Reid, Ken Disaffection from School (1986)
New York, Routledge Chapman and Hall, 230 p. ISBN 0–423–51540–3.

Reid, Stephen E and Reid, Stephen E Head and Neck Injuries in Sports (1984)
Springfield IL, Thomas Charles C, 212 p. ISBN 0–398–04974–2.

Reid, Sue T Crime and Criminology (1988)
New York, Holt, Rinehart and Winston, 672 p. illus. 5th ed.

Reiffel, James et al Psychosocial Aspects of Cardiovascular Disease (1980)
Brooklyn NY, Center for Thanatology Research and Education, 365 p. ISBN 0–930194–32–2.

Réiffers, Jean–Louis and Cartapanis, André Transnational Corporations and Endogenous Development (1982)
ISBN 92–3–101853–1.

Rein, Martin Social Policy: issues of choice and change (1983)
Armonk NY, M E Sharpe, 384 p. Repr. of 1970 ed. ISBN 0–87332–235–5.

Reisen, Helmut and Van Trotsenburg, Axel Developing Countries Debt: the budgetary and transfer problem (1988)
Paris, OECD, 194 p. ISBN 92–64–13053–5.

Reissner, Will (Ed) Dynamics of World Revolution Today (1975)
Haymarket NSW, Pathfinder Press, 188 p. ISBN 0–87348–374–X.

Rementsova, M M Brucellosis in Wild Animals (1987)
Oxford, IBH Publishing and South Asia Books. ISBN 81–7087–005–4.

Remer, C F Study of Chinese Boycotts: with special reference to their economic effectiveness (1966)
Taipei, Chen Wen Publishing Company, 306 p. ISBN 0–89644–076–1. Repr of 1933.

Rendle–Short, et al Synopsis of Children's Diseases (1985)
Littleton MA, PSG Publishing, 568 p. illus. 6th ed. ISBN 0–7236–0743–5.

Rengert, George and Wasilchick, John Suburban Burglary: a time and a place for everything (1985)
Springfield IL, Thomas Charles C, 136 p. ISBN 0–398–05142–9.

Renner, M J and Rosenzweig, M R Enriched and Impoverished Environments: effects on brain and behaviour (1987)
Berlin, Springer–Verlag, 150 p. ISBN 3–540–96523–8.

Renner, Michael Rethinking the Role of the Automobile (1988)
Washington DC, Worldwatch Institute, 72 p. Worldwatch Papers. ISBN 0–916468–85–2.

Renner, Michael National Security: the economic and environmental dimensions (1989)
Washington DC, Worldwatch Institute. Worldwatch Papers. ISBN 0–916468–91–7.

Renshaw, Geoffrey Employment, Trade and North–South Co–Operation (1981)
Washington DC, ILO, 263 p. ISBN 92–2–102530–6.

Renvoize, Jean Going Solo: single mothers by choice (1985)
New York, Routledge Chapman and Hall, 280 p. ISBN 0–7102–0065–X.

Renvoize, Jean Incest: a family pattern (1985)
New York, Routledge Chapman and Hall, 224 p. ISBN 0–7100–9073–0.

Repetto, Robert (Ed) The Global Possible: resources, development and the new century (1985)
New Haven, Yale University Press.

Repetto, Robert and Gillis, Malcolm (Eds) Public Policy and the Misuse of Forest Resources
Cambridge MA, Cambridge University Press, 200 p. ISBN 0–521–34022–5.

Res, Zannis and Motamen, Sima International Debt and Central Banking in the 1980s (1987)
New York, Saint Martin's Press, 256 p. ISBN 0–312–00530–X.

Research Group on Living and Surviving Inhabiting the Earth as a Finite World: an examination of the prospects of housing in a finite world in which is fairly shared, natural resources are not depleted and the environment is protected
Dordrecht, Kluwer Academic Publishers Group.
ISBN 0–89838–018–9.

Reutlinger, Shlomo and Selowsky, Marcelo Malnutrition and Poverty: magnitude and policy options (1976)
Baltimore MD, Johns Hopkins University Press, 106 p.
ISBN 0–8018–1868–0.

Reynolds, Cecil R and Mann, Lester (Eds) Encyclopedia of Special Education: reference for the education of the handicapped and other exceptional children and adults (1987)
New York, John Wiley and Sons, 1793 p.

Reynolds, Vernon, et al (Eds) The Sociobiology of Ethnocentrism: evolutionary dimensions and xenophobia, discrimination, racism and nationalism (1987)
Athens GA, University of Georgia Press, 357 p.
ISBN 0–8203–0915–X.

Rhee, Kyu H Struggle for National Identity in the Third World (1983)
Elizabeth NJ, Hollym International, 233 p.
ISBN 0–930878–31–0.

Ribes, Bruno Biology and Ethics: reflections inspired by a UNESCO symposium (1978)
Lanham MD, UNIPUB, 202 p. ISBN 92–3–101568–0.

Ribton–Turner, C J A History of Vagrants and Vagrancy, and Beggars and Begging (1972)
Montclair NJ, Smith and Patterson Publishing, 780 p.
ISBN 0–87585–138–X.

Riccardi, Vincent M and Eichner, June E Neurofibromatosis: phenotype, natural history, and pathogenesis (1986)
Baltimore MD, Johns Hopkins University Press, 352 p.
ISBN 0–8018–2987–9.

Riccardo, Martin V Vampires Unearthed (1983)
New York, Garland Publishing, 150 p.
ISBN 0–8240–9128–0.

Rice, Edward E Wars of the Third Kind: conflict in underdeveloped countries (1988)
Berkeley CA, University of California Press, 180 p.
ISBN 0–520–06236–1.

Rich, Avery E Potato Diseases (1983)
San Diego CA, Academic Press. ISBN 0–12–587420–0.

Richards, David A Sex, Drugs, Death and the Law: an essay on human rights and overcriminalization (1982)
Totowa NJ, Rowman and Littlefield Publishers, 328 p.
ISBN 0–8476–7063–5.

Richards, John F and Tucker, Richard P (Eds) World Deforestation in the Twentieth Century (1988)
Durham NC, Duke University Press, 304 p.
ISBN 0–8223–0784–7.

Richardson, M J An Annotated List of Seedborne Diseases (1979)
Kew, CAB International Mycological Institute, 320 p. 3rd ed. ISBN 0–85198–429–0. Phytopathological Papers: No 23. Supplements 1 (1981) and 2 (1983) published by: International Seed Testing Association, Zürich, Switzerland.

Richardson, Ruth Death, Dissection and the Destitute (1987)
New York, Routledge Chapman and Hall, 288 p. illus.
ISBN 0–7102–0919–3.

Richelson, Jeffrey The US Intelligence Community (1988)
Cambridge MA, Ballinger Publishing, 392 p.
ISBN 0–88730–245–9.

Richter, L Upgrading Labour Market Information in Developing Countries: problems, progress and prospects (1989)
Geneva, ILO, 62 p. ISBN 92–2–106453–0.

Rickel, Annette U Teenage Pregnancy and Parenting (1989)
Basingstoke, Taylor and Francis, 225 p.
ISBN 0–89116–808–7.

Riddell, Barry Economic Nationalism (1972)
Toronto ON, Maclean–Hunter. 2nd ed. 6 Vols.
ISBN 0–919290–08–6.

Riddell, Roger C Foreign Aid Reconsidered (1987)
Baltimore MD, Johns Hopkins University Press, 320 p.
ISBN 0–8018–3546–1.

Ridker, Ronald G Changing Resource Problems of the Fourth World
Baltimore MD, Johns Hopkins University Press, 162 p.
ISBN 0–8018–1847–8.

Rieber, Robert W Communication Disorders (1981)
New York, Plenum Publishing, 366 p.
ISBN 0–306–40527–X.

Riedijk, W (Ed) Appropriate Technology for Developing Countries (1987)
Delft, Delftse Universitaire Pers. ISBN 90–6275–284–5.

Riemann, H P and Burridge, M J (Eds) Impact of Diseases on Livestock Production in the Tropics (1984)
Amsterdam, Elsevier Science Publishing, 632 p.
ISBN 0–444–42326–5.

Riemann, Hans and Bryan, Frank L Food–Borne Infections and Intoxication (1979)
San Diego CA, Academic Press. ISBN 0–12–588360–9.

Rifaat, Ahmed M International Aggression: a study of the legal concept (1979)
Stockholm, Almqvist and Wiksell, 359 p. Its Development and Definition in International Law Ser.
ISBN 91–22–00296–7.

Riffer, Jeffey K Sports and Recreational Injuries (1986)
Colorado Springs CO, Shepard's/McGraw–Hill, 623 p.
ISBN 0–07–052828–4.

Rifkin, Jeremy Time Wars: the primary conflict in human history (1987)
New York, Henry Holt, 263 p. bibl. ISBN 0–8050–0377–0.

Riley, Tom Proving Punitive Damages: the complete handbook (1981)
Englewood Cliffs NJ, Prentice Hall, 347 p.
ISBN 0–13–731778–6.

Riordan, Michael (Ed) The Day After Midnight: the effects of nuclear war
Wilmington DE, Peace Resource Center, 143 p.

Ristic, Miodrag and Kreier, Julius P Babesiosis (1981)
San Diego CA, Academic Press. ISBN 0–12–588950–X.

Ristic, Miodrag and McIntyre, Ian (Eds) Diseases of Cattle in the Tropics: economic and zoonotic relevance (1981)
Dordrecht, Kluwer Academic Publishers Group, 662 p.
ISBN 90–247–2399–X.

Ritch, et al The Glaucomas (1988)
Saint Louis MO, CV Mosby, 1472 p. illus.

Ritzer, George Social Problems (1986)
New York, Alfred A Knopf, 608 p. illus. 2nd ed.
ISBN 0–394–35427–3.

Rivkin, Robert S and Stichman, Barton F The Rights of Military Personnel (1981)
New York, Avon Books. ISBN 0–380–01668–0.

Rivlin, Harry R, et al Advantage – Disadvantaged Gifted: national conference on disadvantaged gifted
Los Angeles CA, National State Leadership Training Institute on, 100 p. ISBN 0–318–15992–9.

Roberts Microbial Diseases of Fish (1983)
San Diego CA, Academic Press. ISBN 0–12–589660–3.

Roberts, Brad Slow and Uneven Global Economic Growth: implications and perscriptions (1986)
CSIS/University Press of America, 60 p. Significant Issues Ser: Vol VIII, No 5. ISBN 0–8191–6071–7.

Roberts, Catharine Women and Rape (1989)
Hemel Hempstead, Simon and Schuster International, 224 p.
ISBN 0–7108–1214–0.

Roberts, Kenneth School Leavers and Their Prospects: youth and the labor market in the 1980's (1984)
New York, Taylor and Francis, 169 p.
ISBN 0–335–10418–5.

Roberts, Paula Women, Poverty, and Child Support (1986)
Chicago IL, National Clearinghouse for Legal Services, 166 p. ISBN 0–941077–15–2.

Robertson, Ian and McKee, Michael Social Problems (1980)
New York, Random House, 494 p. ISBN 0–394–32025–5.

Robertson, Jeffrey D Psychiatric Malpractice: liability of mental health professionals (1988)
New York, John Wiley and Sons, 575 p.
ISBN 0–471–84098–X.

Robineault, Manfred J Communicable Diseases: medical subject analysis and research bibliography (1985)
Washington DC, ABBE Publications Association, 150 p.
ISBN 0–88164–195–2.

Robinson, Halbert B and Robinson, Nancy M The Mentally Retarded Child (1976)
New York, McGraw Hill Book Company, 672 p.
ISBN 0–07–053202–8.

Robinson, J P Chemical and Biological Warfare Developments 1986–1987 (1990)
Oxford, Oxford University Press, 160 p. SIPRI Chemical and Biological Warfare Studies: 11. ISBN 0–19–829140–X.

Robinson, Julian (Ed) The Chemical Industry and the Projected Chemical Weapons Convention: proceedings of a SIPRI–Pugwash conference (1986)
Oxford, Stockholm International Peace Research Institute. 2 vols. SIPRI Chemical and Biological Warfare Studies: 5, 6.

Robinson, Thomas A The Bauer Thesis Examined: the geography of heresy in the early christian church (1988)
Lewistone NY, Edwin Mellen Press, 240 p.
ISBN 0–88946–611–4. Studies in the Bible and Early Christianity: Vol 11.

Robson, Ivan Bernard Ostrich Syndrome: coping in the eighties and nineties (1984)
Winnipeg, Alpha Publications.

Robson, Kenneth S The Borderline Child: approaches to etiology, diagnosis and treatment (1982)
New York, McGraw Hill Book Company, 320 p.
ISBN 0–07–053346–6.

Roche, M and Rodier, J A World Catalogue of Maximum Observed Floods
Washington DC, IAHS. Follow–up to the 'World Catalogue of Very Large Floods, UNESCO, 1976. ISBN 0–947571–00–0.

Rockford, Doris E Drug Effects on the Fetus: medical research subject analysis with bibliography (1987)
Washington DC, ABBE Publications Association, 150 p. bibl.
ISBN 0–88164–248–7.

Rodgers, Gerry Poverty and Population. Approaches and Evidence (1984)
Geneva, ILO, 213 p. ISBN 92–2–103803–3.

Rodgers, Gerry and Standing, Guy Child Work, Poverty and Underdevelopment: issues for research in low–income countries (1982)
Lanham MD, UNIPUB, 310 p. ISBN 92–2–102813–5.

Roe, J C The Chemical Industry and the Health of the Community: proceedings of an anglo–american conference (1985)
Wolfeboro NH, Longwood Publishing Group, 177 p.
ISBN 1–85315–009–6.

Rogers, Michael A Living with Paraplegia (1986)
Winchester MA, Faber and Faber, 200 p.
ISBN 0–571–13951–5.

Rogge, John Refugees: a third world dilemma (1987)
Totowa NJ, Rowman and Littlefield Publishers, 384 p.
ISBN 0–8476–7557–2.

Rogow, Arnold The Psychiatrists (1971)
London, Allen and Unwin.

Roizman, Bernard The Herpesviruses
New York, Plenum Publishing. 3 Vols.

Roizman, Bernard and Lopez, Carlos Immunobiology and Prophylaxis of Human Virus (1985)
New York, Plenum Publishing, 458 p. The Herpesviruses: 4.
ISBN 0–306–41793–6.

Rolfe, Rial D and Finegold, Sydney M Clostridium Difficile (1988)
San Diego CA, Academic Press, 384 p.
ISBN 0–12–593410–6.

Rollins, Leighton and Corrigan, Daniel Disasters of War (1981)
Tiburon CA, Cadmus Editions, 48 p. ISBN 0–932274–16–1.

Rolls, B J and Rolls, E T Thirst (1982)
New York, Cambridge University Press, 192 p.
ISBN 0–521–22918–9.

Roman, Stephen and Eugen, Loebl Alternative to Communism and Capitalism: the Responsible society (1979)
New Delhi, Abhinav Publications, 215 p.
ISBN 81–7017–087–7.

Rondia, D; Cooke, M and Haroz, R K (Eds) Mobile Source Emissions Including Polycyclic Organic Species: proceedings of the NATO Advanced Research Workshop on Mobile Source Emissions including polycyclic organic species, Liège, Belgium, August 30 – September 2, 1982 (1982)
Dordrecht, Kluwer Academic Publishers Group, 404 p.
ISBN 90–277–1633–1.

Rood, Lois S and Faison, Karen Beyond Severe Disability: a functional bibliography (1985)
Omaha NE, University of Nebraska at Omaha, 32 p.
ISBN 1–55719–014–3.

Roonwal, M L Termite Life and Termite Control in Tropical South Asia (1979)
Jodhpur, Scientific Publications, 177 p.
ISBN 81–85046–02–6.

Ropke, John C Concrete Problems: causes and cures (1982)
New York, McGraw Hill Book Company, 192 p.
ISBN 0–07–053609–0.

Rose–Ackerman, Susan Corruption: a study in political economy (1978)
San Diego CA, Academic Press. ISBN 0–12–596350–5.

Rose, F Clifford Advances in Migraine Research and Therapy (1982)
New York, Raven Press, 248 p. ISBN 0–89004–848–7.

Rose, I Nelson Gambling and the Law (1985)
Hollywood CA, Gambling Times, 308 p.
ISBN 0–89746–066–9.

Rose, Noel and Mackay, Ian R (Eds) The Autoimmune Disease (1985)
San Diego CA, Academic Press. ISBN 0–12–596920–1.

Rosen, Ellen I Bitter Choices: blue collar women in and out of work (1987)
Chicago IL, University of Chicago Press, 232 p.
ISBN 0–226–72644–4.

Rosen, Ismond Sexual Deviation (1979)
New York, Oxford University Press. ISBN 0–19–263208–6.

Rosenberg, Charles Problems in Eugenics: papers communicated to the first international congress (1985)
New York, Garland Publishing, 679 p. The History of Hereditarian Thought Ser. ISBN 0–8240–5806–2.

Rosenberg, H S and Bernstein, J (Eds) Central Nervous System Diseases (1987)
Basel, S Karger AG, 1987 p. Perspectives in Pediatric Pathology Ser: Vol 10.

Rosenberg, Roy A, et al Happily intermarried: authoritative advice for a joyous Jewish–Christian marriage (1988)
New York, Macmillan. ISBN 0–02–604870–1.

Rosenblum, Simon Misguided Missiles: Canada, the cruise and star wars (1985)
Toronto ON, James Lorimer, 200 p. ISBN 0–88862–698–3.

Rosenthal, J (Ed) Arterial Hypertension: pathogenesis, diagnosis and therapy (1982)
Berlin, Springer–Verlag, 529 p. ISBN 3–540–90611–8.

Rosenwasser, Harvey M Malpractice and Contact Lenses: a guide to limiting liability in contact lens (1988)
Philadelphia PA, Gillman–Marcuse, 96 p.
ISBN 0–9620349–0–8.

Rosnow, R L and Fine, G A Rumor and Gossip (1976)
New York, Elsevier Science Publishing, 166 p.
ISBN 0–444–99035–6.

Ross, Alan O Psychological Disorders of Children: a behavioral approach to theory, research and (1979)
New York, McGraw Hill Book Company.
ISBN 0–07–053883–2.

Ross, R R Reading Disability and Crime: link and remediation: An annotated bibliography (1977)
Ottawa ON, Canadian Criminal Justice Association, 37 p.

Roth, Alexander (Ed) Allergy in the World: a guide for physicians and travelers (1978)
Honolulu HI, University of Hawaii Press, 185 p.
ISBN 0–8248–0521–6.

Roth, June Living Better with a Special Diet (1983)
New York, Arco Pub, 276 p. ISBN 0–668–05718–1.

Rothermund, Dietmar and Simon, John Education and the Integration of Ethnic Minorities (1986)
New York, Saint Martin's Press, 233 p.
ISBN 0–312–23725–1.

Rothman, Milton A A Physicist's Guide to Skepticism (1988)
Buffalo NY, Prometheus Books, 200 p.
ISBN 0–87975–440–0.

Rothman, Sheila M and Rothman, David J (Eds) Women in Prison (1986)
New York, Garland Publishing, 500 p. Women and Children First Ser. ISBN 0–8240–7692–3.

Rothstein, Stanley W The Power to Punish: a social inquiry into coercion and control (1984)
Lanham MD, University Press of America, 188 p.
ISBN 0–8191–3731–6.

Rotunda, Ronald D Professional Responsibility (1988)
Saint Paul MN, West Publishing College and School, 414 p.
ISBN 0–314–73052–4.

Rowe, M Laurens Backache at Work (1983)
Fairport NY, Perinton Press, 122 p. illus.
ISBN 0-931157-00-5.

Rowe, Patrick J and Vikhlyaeva, E M (Eds) Diagnosis and Treatment of Infertility (1988)
Lewiston NY, Hogrefe International, 250 p.
ISBN 0-920887-14-7.

Roy, E A, et al (Ed) Neuropsychological Studies of Apraxia and Related Disorders (1985)
Amsterdam, Elsevier Science Publishing, 414 p. Advances in Psychology Ser: Vol 1. ISBN 0-444-87669-3.

Roy, Ramashray Bureaucracy and Development: the case of Indian agriculture
New Delhi, Satvahan Publications.

Roy, Sarojendra Nath Radicalism: philosophy of democratic revolution (1946)
Calcutta, Renaissance Publishers, 123 p.

Rubin, I Leslie and Crocker, Allen C (Eds) Developmental Disabilities: delivery of medical care for children and adults (1989)
Philadelphia PA, Lea and Febiger, 512 p. illus.
ISBN 0-8121-1082-X.

Rubin, Lewis J Pulmonary Heart Disease (1984)
Norwell MA, Kluwer Academic Publishers.
ISBN 0-89838-632-2.

Rubington, Earl and Weinberg, Martin The Study of Social Problems (1989)
New York, Oxford University Press, 304 p.
ISBN 0-19-505723-6.

Rubington, Earl and Weinberg, Martin The Study of Social Problems: six perspectives (1989)
Oxford, Oxford University Press, 322 p. illus. 4th ed.
ISBN 0-19-505723-6.

Rudenstine, David The Rights of Ex-Offenders (1981)
New York, Avon Books.

Ruebsaat, Helmut J and Hull, Raymond The Male Climacteric (1975)
New York, Hawthorn Books.

Rüedi, L (Ed) Deaf Mutism (1959)
New York, Karger S AG, 170 p. ISBN 3-8055-0221-4.

Ruesch, Hans Slaughter of the Innocent (1978)
New York, Bantam Books.

Ruggiero, Roberta The Do's and Dont's of Low Blood Sugar: an everyday guide to hypoglycemia (1988)
Hollywood FL, Fell Frederick Pubs, 104 p. illus.
ISBN 0-8119-0745-7.

Rusch, Wilbert H The Argument: creationism vs evolutionism (1984)
Norcross GA, Creation Research Society Books, 86 p. illus. Creation Research Society Monograph Ser: No 3.
ISBN 0-940384-04-3.

Rushbrook, Frank Fire Aboard (1979)
Dobbs Ferry NY, Sheridan House, 638 p.
ISBN 0-85174-331-5.

Russell, D E Rebellion, Revolution, and Armed Force: a comparative study of fifteen countries with special emphasis on Cuba and South Africa (1974)
San Diego CA, Academic Press. ISBN 0-12-785745-1.

Russell, Diane E Sexual Exploitation (1984)
Newbury Park CA, Sage Publications, 319 p.
ISBN 0-8039-2354-6.

Russell, Jeffrey Burton The Prince of Darkness: radical evil and the power of good in history
Ithaca NY, Cornell University Press, 288 p.

Russo-Marie, Francoise, et al (Eds) Advances in Inflammation Research (1986)
New York, Raven Press, 462 p. Vol 10.
ISBN 0-88167-123-1.

Rutherford, Robert B, et al Severe Behavior Disorders of Children and Youth (1987)
Boston MA, College-Hill Press, 280 p. Vol. 9
ISBN 0-316-76363-2.

Ryan, Stephen J, et al Retinal Disease (1985)
Orlando FL, Grune and Stratton, 304 p.
ISBN 0-8089-1677-7.

Ryan, William Blaming the Victim (1976)
New York, Random House.

Ryder, Richard Victims of Science (1984)
London, Davis-Poyntner.

Ryder, Richard D Animal Revolution (1990)
Basil Blackwell.

Sabine, Lorenzo Notes on Duels and Duelling: alphabetically arranged with a preliminary historical essay
Salem NH, Ayer Co Publications. ISBN 0-8369-8191-X.

Sable, Martin H Industrial Espionage and Trade Secrets: an international bibliography (1985)
New York, Haworth Press, 93 p. ISBN 0-86656-417-9.

Sabljak, Mark and Greenberg, Martin H Sports Babylon: sex, drugs, and other dirty dealings in the world (1988)
New York, Crown Publishers, 308 p. ISBN 0-517-66717-7.

Sachdev, Paul (Ed) Adoption: current issues and trends (1983)
Sevenoaks, Butterworths, 240 p. ISBN 0-409-86516-8.

Sachdev, Paul (Ed) International Handbook on Abortion (1988)
New York, Greenwood Press, 544 p.
ISBN 0-313-23463-9.

Sachrieder, Jürgen Animal Diseases in Tropical and Subtropical Regions (1970)
Bombay, Asia Publishing House, 178 p.

Sachs, Chritina Child Abduction (1987)
New York, State Mutual Book and Periodical Service, 150 p.
ISBN 0-85308-104-2.

Sack, Robert D Libel, Slander, and Related Problems (1980)
New York, Practising Law Institute, 697 p.
ISBN 0-686-68826-0.

Sackman, Harold (Ed) Computers and International Socio-Economic Problems (1987)
Amsterdam, Elsevier Science Publishing, 608 p.
ISBN 0-444-70229-6.

Sadoux, Jean-Jacques Racism in the Western Film from D W Griffith to John Ford: Indians and blacks (1980)
Brooklyn NY, Revisionist Press. ISBN 0-87700-272-X.

Sahabat Alam Malaysia and Asia Pacific Peoples' Environment Network Forest Resources Crisis in the Third World
Penang, Sahabat Alam Malaysia and Asia Pacific Peoples' Environment Network, 516 p.

Saint Gompert, Eva Nutrition Disorders: subject, reference and research guidebook (1987)
Annandale VA, ABBE Publishers Association of Washington, 160 p. ISBN 0-88164-594-X.

Saito, H et al Neurotransmitters as Modulators of Blood (1987)
Philadelphia PA, Coronet Books, 380 p. Progress in Hypertension: 1. ISBN 90-6764-100-6.

Sajhau, J P and von Muralt, J Plantations and Plantation Workers (1987)
Geneva, ILO, 207 p. ISBN 92-2-105651-1.

Sakai, A and Larcher, W Frost Survival of Plants (1987)
Berlin, Springer-Verlag, 340 p. illus. Ecological Studies: Vol 62. ISBN 0-387-17332-3.

Sakai, Toshiaki and Tsuboi, Takayuka Genetic Aspects of Human Behavior (1985)
New York, Igaku-Shoin Medical Publishers, 275 p.
ISBN 0-89640-115-4.

Salánki, J (Ed) Heavy Metals in Water Organisms: proceedings of a symposium held at the Balaton Limnological Research Institute, Tihany 1984 (1985)
Budapest, Akadémiai Kiadó és Nyomada, 442 p.
ISBN 963-05-4195-5.

Saldeen, Tom (Ed) The Microembolism Syndrome (1979)
Philadelphia PA, Coronet Books, 240 p. illus.
ISBN 0-317-46479-5.

Salin, Pascal (Ed) Currency Competition and Monetary Union (1984)
Dordrecht, Kluwer Academic Publishers Group, 308 p.
ISBN 90-247-2817-7.

Salk, Darrell, et al The Werner's Syndrome and Human Aging (1985)
New York, Plenum Publishing, 672 p.
ISBN 0-306-42101-1.

Sallantin, Xavier L'invariant des jeux militaires, économiques et politiques (1976)
Paris, Laboratoire BENA de Logique Générale.

Salthouse, Timothy Theory of Cognitive Aging (1985)
Amsterdam, Elsevier Science Publishing, 456 p.
ISBN 0-444-87827-0.

Salvayre, R et al Lipid Storage Disorders: biological and medical aspects (1988)
New York, Plenum Publishing, 826 p.
ISBN 0-306-42928-4.

Samir Amin, et al La Crise, Quelle Crise? (1982)
Paris, Maspero.

Sammen, Peter D The Nails in Disease (1986)
Oxford, Heineman Professional Publishing, 224 p.
ISBN 0-433-2915-. 0.

Sanchez, Jose M Anticlericalisme: a brief history (1973)
Notre Dame IN, University of Notre Dame Press, 256 p.
ISBN 0-268-00471-4.

Sanday, Peggy R Divine Hunger: cannibalism as a cultural system (1986)
Cambridge MA, Cambridge University Press, 304 p. illus.
ISBN 0-521-32226-X.

Sandbacka, Carola (Ed) Cultural Imperialism and Cultural Identity: proceedings of the 8th conference of Nordic ethnographers and anthropologists (1977)
Helsinki, Suomen Antropologinen Seura, 262 p.
ISBN 951-95433-1-7.

Sandercock, Leonie Land Racket: the real costs of property speculation (1979)
Petersham NSW, Hale and Iremonger, 128 p.
ISBN 0-908094-42-6.

Sanders, Douglas E Aboriginal Self-Government in the United States
Kingston ON, Queen's University, 69 p.
ISBN 0-88911-430-7.

Sandford-Smith, J Eye Disease in Hot Climates (1986)
Littleton MA, PSG Publishing, 282 p. ISBN 0-7236-0750-8.

Sandizzo, Pasquale L and Diakosawas, Dimitris Instability in the Terms of Trade of Primary Commodities 1900-1982 (1987)
Lanham MD, UNIPUB, 224 p. ISBN 92-5-102531-2.

Sandler, G (Eds) Coronary Heart Disease (1987)
Berlin, Springer-Verlag. Practical Clinical Medicine Ser.
ISBN 0-85200-692-6.

Santos, Adolfo L Liver Diseases: medical subject analysis with bibliography (1987)
Annandale VA, ABBE Publishers Association of Washington, 160 p. ISBN 0-88164-554-0.

Sargent, J R Europe and the Dollar in the World-Wide Disequilibrium (1981)
Dordrecht, Kluwer Academic Publishers Group, 348 p.
ISBN 90-286-0700-5.

Sarno, Martha Taylor and Höök, Olle (Eds) Aphasia: assessment and treatment (1980)
Stockholm, Almqvist and Wiksell, 234 p.
ISBN 91-22-00388-6.

Sarrazin, R; Vincent, F and Vrousos, C (Eds) Thymic Tumors (1989)
Basel, S Karger AG, 180 p. Available in the UK and Africa only.

Sathyamurthy, T V Nationalism in the Contemporary World: political and sociological perspectives (1983)
Totowa NJ, Rowman and Littlefield Publishers, 258 p.
ISBN 0-86598-117-5.

Sau, Ranjit Unequal Exchange, Imperialism and Underdevelopment: an essay on the political economy of world capitalism
Melbourne VIC, Oxford University Press.

Sawicka, E and Branthwaite, M Respiratory Emergencies (1988)
Stoneham MA, Butterworth Publishers, 96 p.
ISBN 0-407-00861-6.

Sawyer, Roger Slavery in the Twentieth Century (1986)
New York, Routledge Chapman and Hall, 400 p.
ISBN 0-7102-0475-2.

Sawyer, Roger Children Enslaved (1988)
New York, Routledge Chapman and Hall, 240 p.
ISBN 0-415-00273-7.

Sax, N Irving Cancer Causing Chemicals (1981)
New York, Reinhold Van Nostrand, 400 p.
ISBN 0-442-21919-9.

Sax, N Irving and Lewis, Richard J Hazardous Chemicals Desk Reference (1987)
New York, Reinhold Van Nostrand, 1000 p.
ISBN 0-442-28208-7.

Saxena, R K Social Reforms: infanticide and Sati (1975)
New Delhi, Trimurti Publications, 155 p.

Scanlan, John A, et al The Global Refugee Problem: US and world response (1983)
Newbury Park CA, Sage Publications, 256 p. Vol. 467
ISBN 0-8039-2014-8.

Scarff, R W and Torloni, H Historical Typing of Breast Tumours (1977)
Geneva, WHO. illus and slides. 2nd ed. World Health Organization: International Historical Classification of Tumours Ser: No 2.

Scarlato, G and Matthews, W B Multiple Sclerosis: present and future (1984)
New York, Plenum Publishing, 274 p.
ISBN 0-306-41823-1.

Schad, G A and Warren, K S (Ed) Hookworm Disease: current status and new directions (1989)
Basingstoke, Taylor and Francis, 300 p.
ISBN 0-85066-762-3.

Schaff, Adam Alienation as a Social Phenomenon (1980)
New York, Pergamon Books. ISBN 0-08-021807-5.

Schechter, Roger E Unfair Trade Practices and Intellectual Property (1986)
Saint Paul MN, West Publishing Co College, 272 p. Black Letter Ser. ISBN 0-314-98619-7.

Scheinberg, I Herbert and Sternlieb, Irmin Wilson's Disease (1984)
Philadelphia PA, WB Saunders, 192 p.
ISBN 0-7216-7953-6.

Schell, Jonathan The Fate of the Earth (1982)
Bungay, The Chauser Press, 244 p.

Schell, Musher Pathogenesis and Immunology of Treponemal Infections (1983)
New York, Dekker Marcel, 424 p. ISBN 0-8247-1384-2.

Schenken, Robert S Endometriosis (1988)
New York, Lippincott, 320 p. illus. ISBN 0-397-50866-2.

Scheper-Hughes, Nancy (Ed) Child Survival: anthropological perspective on treatment and maltreatment of children (1987)
Dordrecht, Kluwer Academic Publishers Group, 412 p.
ISBN 1-55608-028-X.

Schervish, Paul G The Structural Determinants of Unemployment: vulnerability and power in market relations (1983)
San Diego CA, Academic Press. ISBN 0-12-623950-9.

Schevill, William E The Whale Problem: a status report (1974)
Cambridge MA, Harvard University Press, 384 p.
ISBN 0-674-95075-5.

Schinzel, Albert Catalogue of Unbalanced Chromosome Aberrations in Man (1983)
Berlin, Walter de Gruyter, 886 p. ISBN 3-11-008370-1.

Schlesinger, Ben and Schlesinger, Rachel Abuse of the Elderly: issues and annotated bibliography (1988)
Toronto ON, University of Toronto Press, 188 p.

Schlesinger, Benjamin Sexual Abuse of Children: a resource guide and annotated bibliography (1982)
Cheektowaga NY, University of Toronto Press, 212 p.
ISBN 0-8020-6481-7.

Schlesinger, R Walter The Togaviruses (1980)
San Diego CA, Academic Press. ISBN 0-12-625380-3.

Schlossberg, D Tuberculosis (1988)
New York, Springer Publishing, 225 p.
ISBN 0-387-96552-1.

Schmalzl, F, et al (Eds) Preleukemia International Workshop (1979)
Berlin, Springer-Verlag, 200 p. illus. ISBN 0-387-09698-1.

Schmicker, R, et al Metabolic Disturbances in the Predialytic Phase of Chronic Renal Failure (1988)
New York, Karger S AG, 140 p. ISBN 3-8055-4739-0.

Schmitt, Hans A Neutral Europe Between War and Revolution, 1917-23 (1988)
Charlottesville VA, University Press of Virginia, 325 p.
ISBN 0-8139-1153-2.

Schmitz, Charles A Disaster: the United Nations and international relief management (1987)
New York, Council on Foreign Relations, 98 p. bibl.

Schnaars, Steven Megamistakes (1989)
New York, Free Press, 202 p.

Schneider, Hans Joachim The Victim in International Perspective (1982)
Hawthorne NY, Walter de Gruyter, 513 p.
ISBN 3-11-007510-5.

Schneider, T (Ed) Acidification and Its Policy Implications: proceedings of an international conference, Amsterdam 1986 (1986)
Amsterdam, Elsevier Science Publishing, 514 p.
ISBN 0-444-42725-2.

Schneider, T and Grant, L (Eds) Air Pollution by Nitrogen Oxides: proc of the US–Dutch Int Symp, Maastricht, Netherlands, 1982 (1982)
Amsterdam, Elsevier Science Publishing, 1100 p.
ISBN 0–444–42127–0.

Schneller, Donald P The Prisoner's Family: a study of the effects of imprisonment on the families of prisoners (1976)
R and E Publications. ISBN 0–88247–407–3.

Schniderman, Harry L and Leverich, Bingham B Price Discrimination in Perspective (1987)
American Law Institute, 314 p. 2nd ed.
ISBN 0–8318–0534–X.

Scholl, Erwin, et al Diseases of Swine (1986)
Ames IA, Iowa State University Press, 930 p.
ISBN 0–8138–0441–8.

Scholl, Geraldine (Ed) Foundations of Education for Blind and Visually Handicapped Children and Youth: theory and practice (1986)
New York, American Foundation for the Blind, 525 p. illus.
ISBN 0–89128–124–X.

Schoning, U Complexity and Structure (1986)
Berlin, Springer–Verlag. ISBN 3–540–16079–5.

Schopenhauer, Arthur Pantheism and the Christian System (1987)
Albuquerque NM, American Classical College Press, 119 p.
ISBN 0–89266–588–2.

Schopflin, George Censorship and Political Communication in Eastern Europe: a collection of documents (1983)
St Martin, 250 p.
ISBN 0–312–12728–6.

Schopler, Eric and Mesibov, Gary Autism in Adolescents and Adults (1983)
New York, Plenum Publishing, 456 p.
ISBN 0–306–41057–5.

Schopler, Eric and Mesibov, Gary B Communication Problems in Autism (1985)
New York, Plenum Publishing, 350 p.
ISBN 0–306–41859–2.

Schott, Kerry Policy, Power and Order (1984)
New Haven CT, Yale University Press, 288 p.
ISBN 0–300–03237–4.

Schove, D Justin Sunspot Cycles (1983)
New York, Reinhold Van Nostrand, 416 p.
ISBN 0–87933–424–X.

Schramm, Wilbur and Atwood, Erwin Circulation of News in the Third World: a study of Asia (1981)
Hong Kong, Chinese University Press, 352 p.
ISBN 962–201–238–8.

Schrier, Robert W and Gottschalk, Carl W Diseases of the Kidney (1988)
Boston MA, Brown Little, 1200 p. 3 vols.
ISBN 0–316–77480–4.

Schuck, Peter H Agent Orange on Trial: mass toxic disasters in the courts (1986)
Cambridge MA, Harvard University Press, 352 p.
ISBN 0–674–01025–6.

Schuknecht, Harold F Pathology of the Ear (1974)
Cambridge MA, Harvard University Press, 448 p. illus. Commonwealth Fund Publications Ser. ISBN 0–674–65786–1.

Schulman, Joseph D and Simpson, Joe L Genetic Diseases in Pregnancy: maternal effects and fetal outcome (1981)
San Diego CA, Academic Press. ISBN 0–12–630940–X.

Schuster, Robert L (Ed) Landslide Dams: processes, risk and mitigation (1986)
New York, American Society of Civil Engineers, 164 p. Sessions Proceedings, Geotechnical Special Publication: No 3. ISBN 0–87262–524–9.

Schwartz, Barry Queuing and Waiting (1975)
Chicago IL, University of Chicago Press, 217 p.
ISBN 0–226–74210–5.

Schwartz–Nobel, L Starving in the Shadow of Plenty (1982)
New York, McGraw Hill Book Company, 240 p.
ISBN 0–07–055776–4.

Schwartz–Salant, Nathan Narcissism and Character transformation: the Psychology of narcissistic character disorders (1982)
Toronto ON, Inner City Books, 192 p.
ISBN 0–919123–08–2.

Schweitzer Corrosion and corrosion protection handbook (1988)
New York, Dekker Marcel, 680 p. 2nd rev and enl ed.

Scitovsky, Tibor The Joyless Economy (1976)
New York, Oxford University Press. ISBN 0–19–502183–5.

Scott, Hilda Working Your Way to the Bottom: the feminization of poverty (1985)
New York, Routledge Chapman and Hall, 180 p.
ISBN 0–86358–011–4.

Seager, Joni (Ed) The State of the Earth (1990)
New York, Simon and Schuster, 234 p. illus.

Seaquist, Edgar O Diagnosing and Repairing House Structure Problems (1980)
New York, McGraw Hill Book Company.
ISBN 0–07–056013–7.

Seeley, Robert A A handbook for conscientious objectors (1981)
Philadelphia PA, Central Committee for Conscientious Objectors, 240 p. Rev 13th ed. Illus. ISBN 0–933368–03–8.

Seeliger, Heinz P Listeriosis (1961)
New York, Hafner Publications. 2nd ed.
ISBN 0–02–852020–3.

Seligman, Martin E Helplessness: on depression, development and death (1975)
New York, Freeman WH, 250 p. ISBN 0–7167–0752–7.

Selikoff, Irving and Lee, Douglas H Asbestos and Disease (1978)
San Diego CA, Academic Press. ISBN 0–12–636050–2.

Sellen, Betty–Carol and Young, Patricia A Feminists, Pornography, and the Law: an annotated bibliography of conflict, 1970–1986 (1987)
Hamden CT, Shoe String Press, 204 p.
ISBN 0–208–02124–8.

Selth, Andrew Against Every Human Law: the terrorist threat to diplomacy (1988)
Elmsford NY, Pergamon Books, 208 p.
ISBN 0–08–034404–6.

Semmler, W (Ed) Competition, Instability, and Nonlinear Cycles (1986)
Berlin, Springer–Verlag, 340 p. ISBN 3–540–16794–3.

Sen, Chanakya Against the Cold War: a Study of Asian–African policies since World War II (1962)
Bombay, Asia Publishing House, 288 p.

Sen, Dhirani Problem of Minorities (1940)
Allahabad, Allahabad Publishing House, 793 p.

Sendlein, L V, et al Surface Mining Environmental Monitoring and Reclamation Handbook (1983)
New York, Elsevier Science Publishing, 750 p.
ISBN 0–444–00791–1.

Serafim, J Laginha (Ed) Safety of Dams: proceedings of an international conference, Coimbra, 23–28th April 1984 (1984)
Rotterdam, AA Balkema, 599 p. 2 Vols.
ISBN 90–6191–521–X.

Sereda, P J and Litvan, G G (Eds) Durability of Building Materials and Components (1980)
Philadelphia PA, American Society for Testing and Materials, 1034 p. ISBN 0–8031–0325–5.

Seren, E and Mattiolo, M Definition of the Summer Infertility Problem of the Pig (1987)
Luxembourg, European Communities, 162 p.
ISBN 92–825–7222–6.

Sereny, Gitta and Wilson, Victoria The Invisible Children: child prostitution in America, West Germany and Great Britain (1985)
New York, Alfred A Knopf, 254 p. ISBN 0–394–53389–5.

Serjeant, Graham R Sickle Cell Disease (1988)
New York, Oxford University Press, 496 p.
ISBN 0–19–261753–2.

Servais, Jean–Michel Inviolability of Trade Union Premises and Communications (1980)
Geneva, ILO, 43 p. ISBN 92–2–102178–5.

Sethi, Amarjit S et al (Ed) Strategic Management of Technostress in an Information Society (1987)
Göttingen, Verlag für Psychologie, 399 p.

Sethi, S P Railway Accidents: their causes and remedies (1965)
New Delhi, Tech India Publications, 224 p.

Sethuraman, S V The Urban Informal Sector in Developing Countries: employment, poverty, and environment (1981)
Lanham MD, UNIPUB, 225 p. ISBN 92–2–102591–8.

Seuss, Michael J Nonionizing Radiation Protection (1982)
Geneva, WHO, 267 p. ISBN 92–890–1101–7.

Seward, Jack Hara–Kiri: Japanese ritual suicide (1968)
Rutland VT, Charles E Tuttle. ISBN 0–8048–0231–9.

Shah, Nandkumar S and Donald, Alexander G Psychoneuroendocrine Dysfunction (1984)
New York, Plenum Publishing, 660 p.
ISBN 0–306–41320–5.

Shah, Nandkumar S and Donald, Alexander G Movement Disorders (1986)
New York, Plenum Publishing, 424 p.
ISBN 0–306–42135–6.

Shaked, Ami Human Sexuality in Physical and Mental Illnesses and Disabilities: an annotated bibliography
Ann Arbor MI, Books on Demand. bibl.

Shames, George H and Wiig, Elisabeth H Human Communication Disorders (1982)
Columbus OH, Merrill Publishing, 544 p.
ISBN 0–675–09837–8.

Shannon, Thomas A Surrogate Motherhood: the ethics of using human beings (1988)
New York, Crossroad Publishing, 212 p.
ISBN 0–8245–0899–8.

Shapiro, Harry L Race Mixture (1953)
Lanham MD, UNIPUB, 58 p. ISBN 92–3–100416–6.

Shapiro, Sam, et al Infant, Perinatal, Maternal, and Childhood Mortality in the United States
Cambridge MA, Harvard University Press.
ISBN 0–674–45301–8.

Shapiro, Samuel et al Sudden Death: medical subject analysis and research directory (1983)
Annandale VA, ABBE Publishers Association of Washington, 134 p. ISBN 0–941864–34–0.

Shariff, M Crime and Punishment in Islam
Chicago IL, Kazi Publications. ISBN 0–686–18560–9.

Sharma, S P International Boundary Disputes and International Law (1977)
Bombay, NM Tripathi, 340 p.

Sharman, Julian Cursory History of Swearing (1968)
New York, Franklin Burt. ISBN 0–8337–3240–4.

Sharp, F and Symonds, M Hypertension in Pregnancy (1987)
Ithaca NY, Perinatology Press. ISBN 0–916859–28–2.

Sharp, Jane M O (Ed) Europe After an American Withdrawal: economic and military issues (1990)
Oxford, Stockholm International Peace Research Institute, 501 p. ISBN 0–19–827836–5.

Shaver, K G Attribution of Blame (1985)
Berlin, Springer–Verlag, 210 p. ISBN 3–540–96120–8.

Shaw, Peter The War Against the Intellect: episodes in the decline of discourse
Iowa City, University of Iowa Press, 181 p.

Shchetinin, V D U S Monopolies and Developing Countries (1985)
Los Angeles CA, Progress Publishing, 178 p.
ISBN 0–8285–3400–4.

Shea, Cynthia P A Vanishing Shield: protecting the ozone layer (1988)
Washington DC, Worldwatch Institute, 60 p. Worldwatch Papers. ISBN 0–916468–88–7.

Shearing, Clifford D (Ed) Organizational Police Deviance (1981)
Sevenoaks, Butterworths, 244 p. ISBN 0–409–84880–8.

Shedd, William G Doctrine of Endless Punishment (1986)
Carlisle PA, Banner of Truth, 201 p. repr of 1885 ed.
ISBN 0–85151–491–X.

Sheils, W J (Ed) Persecution and Toleration (1984)
New York, Basil Blackwell, 500 p.
ISBN 0–631–13601–0. Studies in Church History: Vol 21.

Sheleff, Leon S Ultimate Penalties (1987)
Columbus OH, Ohio State University Press, 492 p.
ISBN 0–8142–0436–8.

Shepher, Joseph Incest: a biosocial view (1983)
San Diego CA, Academic Press, 184 p.
ISBN 0–12–639460–1.

Sheppard, C R C and Wells, S M Coral Reef Directory (1984)
Gland, IUCN. ISBN 2–88032–708–3.

Sheridan, Lilian B Death: index of modern information (1988)
Washington DC, ABBE Publications Association, 150 p.
ISBN 0–88164–864–7.

Sherman, Lawrence W Scandal and Reform: controlling police corruption (1978)
Berkeley CA, University of California Press.
ISBN 0–520–03523–2.

Sherman, William L Forced Native Labor in Sixteenth Century Central America (1979)
Omaha NE, University of Nebraska Press, 496 p. illus.
ISBN 0–8032–4100–3.

Shibles, Warren and Falkenberg, Gabriel Lying: a critical analysis (1985)
Whitewater WI, Language Press, 242 p.
ISBN 0–912386–20–7.

Shields, William M Philopatry, Inbreeding and the Evolution of Sex (1983)
Albany NY, State University of New York Press, 245 p.
ISBN 0–87395–617–6.

Shihata, Ibrahim Case for the Arab Oil Embargo (1975)
Beirut, Institute for Palestine Studies, 114 p.

Shipley, Elizabeth H Bacterial Infections: medical subject analysis with bibliography (1987)
Washington DC, ABBE Publications Association, 160 p. bibl.
ISBN 0–88164–524–9.

Shiraki, Kiezo; Yousef, Mohamed and Wilber, Charles G Man in Stressful Environments: diving, hyper and hyperbaric physiology (1988)
Springfield IL, Thomas Charles C, 286 p.
ISBN 0–398–05361–8.

Shirley, Andrew Plots and conspiracies (1975)
New York, Greenwood Press, 142 p. Repr of 1957. Illus.
ISBN 0–8371–8459–2.

Shklar, Judith N Ordinary Vices (1985)
Cambridge MA, Harvard University Press, 280 p.
ISBN 0–674–64176–0.

Shoham, S Giora and Rahau, Giora The Mark of Cain: the stigma theory of crime and social deviance (1982)
New York, Saint Martin's Press, 240 p.
ISBN 0–312–51446–8.

Shoman, S Giora The Violence of Silence: the impossibility of dialogue (1982)
New Brunswick NJ, Transaction Books, 300 p.
ISBN 0–905927–06–0.

Shore, Susan A Migraine: a medical subject analysis with research (1988)
Annandale VA, ABBE Publishers Association of Washington, 150 p. ISBN 0–88164–654–7.

Short, Marjorie Essential Oral Anatomies: teeth, head and neck (1987)
Albany NY, Delmar Publishers, 256 p.
ISBN 0–8273–2742–0.

Shortridge, K F (Ed) Newcastle Disease and its Control in Southeast Asia: proceedings of a UNESCO–sponsored training course and workshop on viral vaccines poultry (1982)
Hong Kong, Hong Kong University Press, 107 p.
ISBN 962–209–038–9.

Shover, Neal Aging Criminals (1985)
Newbury Park CA, Sage Publications.
ISBN 0–8039–2528–X.

Shrader–Frechette, Kristin Nuclear Power and Public Policy: the social and ethical problems of fission technology (1980)
Dordrecht, Kluwer Academic Publishers Group, 176 p.
ISBN 90–277–1054–6.

Shreeve, Caroline Cystitis (1987)
Rochester VT, Inner Traditions International, 128 p.

Shukla, K S Adolescent Thieves: a study in socio–cultural dynamics (1979)
New Delhi, Inter–India Publications, 223 p.

Shuler, Alexanderina V Malaria: meeting the global challenge (1985)
Boston MA, Oelgeschlager Gunn and Hain, 128 p.
ISBN 0–89946–193–X.

Shuval, H I Thalassogenic Diseases (1986)
Nairobi, UNEP, 44 p.

Sicuteri, F L, et al Trends in Cluster Headache (1987)
New York, Elsevier Science Publishing, 396 p.
ISBN 0–444–80871–X.

Siddiqui, M I Animal Sacrifice in Islam
Chicago IL, Kazi Publications. ISBN 0–686–63893–X.

Sieber, Ulrich The International Handbook on Computer–Related Crime and the Infringements of Privacy (1986)
New York, John Wiley and Sons. ISBN 0–471–91224–7.

Siebert, Horst (Ed) Global Environmental Resources: the ozone problem (1981)
New York, Peter Lang Publishing, 104 p.
ISBN 3–8204–5999–5.

Siedek, Shelby V Jurisprudence and Iatrogenic Problems: index and bibliography (1988)
Annandale VA, ABBE Publishers Association of Washington, 150 p. ISBN 0–88164–720–9.

Siegal, Mary-Ellen The Cancer Patient's Handbook: everything you need to know about today's care and treatment (1986)
New York, Walker Publishing Company.
ISBN 0–8027–0898–6.

Siegfried, J and Zimmermann, M Phantom and Stump Pain (1982)
New York, Springer Publishing, 192 p.
ISBN 0–387–11041–0.

Siegmann, Heinrich World Modelling (1987)
Paris, UNESCO, 175 p. Major Programme I, Reflection on World Problems and Future–Oriented Studies: BEP/GPI/2.

Siegmann, Heinrich World Modeling (1987)
Paris, UNESCO, 175 p. Major Programme I, Reflection on World Problems and Future–Oriented Studies: BEP/GPI/2.

Siemens, Heide H Infection and Infections: medical subject research analysis with bibliography (1985)
Washington DC, ABBE Publications Association, 150 p.
ISBN 0–88164–256–8.

Siewert J R and Holscher, A H (Eds) Diseases of the Esophagus (1987)
Berlin, Springer Verlag, 1400 p. illus. ISBN 0–387–17697–7.

Sifakis, Carl The Mafia Encyclopedia (1988)
New York, Facts on File, 384 p. ISBN 0–8160–1856–1.

Signorielli, Nancy and Gerbner, George Violence and Terror in the Mass Media: an annotated bibliography (1988)
Westport CT, Greenwood Press, 264 p.
ISBN 0–313–26120–2.

Signorielli, Nancy; Milke, Elizabeth and Katzman, Carol Role Portrayal and Stereotyping on Television (1985)
Westport CT, Greenwood Press, 214 p.
ISBN 0–313–24855–9.

Silverman, H, et al Early Identification and Intervention: selected proceedings from the Fourth International Symposium on Learning Problems (1979)
Ontario ON, Ministry of Education, 257 p.
ISBN 0–7743–6842–X.

Silverman, Maylin and Salisbury, Richard House Divided?: anthropological studies of factionalism (1976)
St John's, Memorial University of Newfoundland.
ISBN 0–919666–13–2.

Silverman, Milton, et al Prescription for Death: the drugging of the third world (1982)
Berkeley CA, University of California Press, 208 p.
ISBN 0–520–04721–4.

Sim, Foo Gaik The Pesticide Poisoning Report: a survey of some Asian countries (1985)
Penang, IOCU, 88 p. ISBN 967–9973–06–9.

Simai, M Interdependence and Conflicts in the World Economy (1981)
Budapest, Akadémiai Kiadó és Nyomada, 219 p. Coedition with Sijthoff and Nordhoff International Publishers, Alphen a/d Rijn. ISBN 963–05–2470–8.

Simpson, George E and Yinger, Milton Racial and Cultural Minorities: an analysis of prejudice and discrimination
New York, Plenum Publishing, 506 p.
ISBN 0–306–41777–4.

Simpson, John and Bennett, Jana The Disappeared and the Mothers of the Plaza (1985)
New York, Saint Martin's Press, 416 p.
ISBN 0–312–21229–1.

Simri, M Inderdependence and Conflict in the World Economy (1981)
Dordrecht, Kluwer Academic Publishers Group, 219 p.
ISBN 90–286–2241–1.

Sinclair, Wayne A, et al Diseases of Trees and Shrubs (1987)
Ithaca NY, Cornell University Press, 576 p.
ISBN 0–8014–1517–9.

Singer, Frederick Paget's Disease of Bone (1977)
New York, Plenum Publishing, 172 p.
ISBN 0–306–30996–3.

Singer, Peter Animal Liberation (1977)
New York, Avon Books.

Singer, Peter and Kuhse, Helga Should the Baby Live?: the problem of handicapped infants
Melbourne VIC, Oxford University Press, 280 p.
ISBN 0–19–286062–3.

Singer, S F (Ed) Global Effects of Environmental Pollution: a symposium organized by the American Association for the Advancement of Science held in Dallas, 1968 (1970)
Dordrecht, Kluwer Academic Publishers Group, 218 p.
ISBN 90–277–0151–2.

Singh Diseases of Vegetable Crops
New Delhi, Oxford and IBH Publishing.

Singh, Charlene P Inflammation and Health Sciences: medical analysis index with research (1987)
Annandale VA, ABBE Publishers Association of Washington, 150 p. ISBN 0–88164–348–3.

Sinha, Durganand et al (Eds) Deprivation: its social roots and psychological consequences (1982)
New Delhi, Concept Publishing, 269 p.

Sinha, M R (Ed) Problem of World Liquidity and Other Essays (1970)
Bombay, Asian Studies Press, 122 p.

Siriwardena, R (Ed) Equality and the Religious Traditions of Asia (1987)
St Martin, 300 p.
ISBN 0–312–00401–X.

Sitwell, Sacheverell Poltergeists – Fact or Fiction? (1988)
New York, Dorset Press, 256 p. ISBN 0–88029–165–6.

Sivanesan, A and Waller, J M Sugercane Diseases (1986)
Kew, CAB International Mycological Institute, 88 p.
ISBN 0–85198–564–5. Phytopathological Papers: No 29.

Sixel, Friedrich W Crisis and Critique: on the logic of late capitalism (1987)
Leiden, Brill, 156 p. ISBN 90–04–08284–0.

Skidmore, Gail and Spahn, Theodore J From Radical Left to Extreme Right: a bibliography of current periodicals (1987)
Metuchen NJ, Scarecrow Press, 503 p.
ISBN 0–8108–1967–8.

Skinner, F A and Quesnel, L B Streptococci (1978)
San Diego CA, Academic Press. ISBN 0–12–648035–4.

Sklar, Holly (Ed) Trilateralism: the Trilateral Commission and elite planning for world management
Montreal, Black Rose Books, 604 p. ISBN 0–919618–43–X.

Slater, Philip Wealth Addiction (1980)
New York, EP Dutton.

Slatoff, Walter J The Look of Distance (1985)
Columbus OH, Ohio State University Press, 309 p.
ISBN 0–8142–0385–X.

Slatta, Richard W (Ed) Bandidos: the varieties of Latin American banditry (1987)
New York, Greenwood Press, 229 p. Contributions to criminology and penology ser: No 14. ISBN 0–313–25301–3.

Sloan, Irving J The Right to Die: legal and ethical problems (1988)
Dobbs Ferry NY, Oceana Publications, 160 p.
ISBN 0–379–11167–5.

Sloan, S J and Cooper, C L Pilots under Stress (1987)
New York, Routledge Chapman and Hall, 240 p.
ISBN 0–7102–0479–5.

Smil, V and Knowland, W E Energy in the Developing World: the real energy crisis (1980)
New York, Oxford University Press. ISBN 0–19–854425–1.

Smith, A D Concepts of Labour Force Underutilisation (1972)
Geneva, ILO, 88 p.
ISBN 92–2–100108–3.

Smith, Alrick L Reliability of Engineering Materials (1985)
Stoneham MA, Butterworth Publishers, 160 p.
ISBN 0–408–01507–1.

Smith, Ann Grandchildren of Alcoholics (1988)
Deerfield Beach FL, Health Communications.
ISBN 0–932194–55–9.

Smith, Barbara E Digging Our Own Graves: coal miners and the struggle over black lung (1987)
Philadelphia PA, Temple University Press, 240 p.
ISBN 0–87722–451–X.

Smith, C V Meteorological Observations in Animal Experiments (1970)
Geneva, WMO, 37 p.

Smith, Christopher Public Problems: the management of urban distress (1988)
New York, Guilford Press, 276 p. ISBN 0–89862–782–6.

Smith, Christopher Public Problems: the management of urban distress (1988)
New York, Guilford Press, 276 p. ISBN 0–89862–782–6.

Smith, I M CO2 and Climate Changes (1988)
Paris, OECD, 52 p. ISBN 92–9029–157–5.

Smith, Joan, et al Racism, Sexism, and the World–System (1988)
Westport CT, Greenwood Press. ISBN 0–313–26331–0.

Smith, K R Biofuels, Air Pollution, and Health: a global review (1987)
New York, Plenum Publishing, 476 p.
ISBN 0–306–42519–X.

Smith, L P Weather and Food (1962)
Geneva, WMO, 80 p. Freedom from Hunger Campaign, Basic Study 1.

Smith, Louis Botulisme: the organism, its toxins, the disease (1988)
Springfield IL, Thomas Charles C, 204 p. 2nd ed.

Smith, M A Contaminated Land: reclamation and treatment (1985)
New York, Plenum Publishing, 456 p.
ISBN 0–306–41928–9.

Smith, M J Ulcer and Non-Ulcer Dyspepsias (1987)
Norwell MA, Kluwer Academic Publishers.
ISBN 0–85200–970–4.

Smith, Mickey C Small Comfort: a history of the minor tranquilizers (1985)
New York, Praeger Publishers, 272 p.
ISBN 0–275–91325–2.

Smith, Phillip H and Prout, George R Bladder Cancer (BIMR Urology) (1983)
Stoneham MA, Butterworth Publishers. Vol. 1
ISBN 0–407–02358–5.

Smith, Robert J The Psychopath in Society (1978)
San Diego CA, Academic Press. ISBN 0–12–652550–1.

Smith, Roberts A HIV and Other Highly Pathogenic Viruses (1988)
San Diego CA, Academic Press, 135 p.
ISBN 0–12–652245–6.

Smith, Wrynn Diabetes, Liver and Digestive Diseases (1988)
New York, Facts on File, 160 p. ISBN 0–8160–1459–0.

Snyder, Solomon H Biological Aspects of Mental Disorder (1980)
New York, Oxford University Press. ISBN 0–19–502715–9.

Snyder, Stanley G Embezzler's Dirty Tricks: and how to spot them (1986)
WestView Pub, 336 p. illus.
ISBN 0–937535–00–1.

Sobhan, Rehman Crisis of External Dependence
New York, University Press.

Soble, Alan Pornography: marxism, feminism and the future of sexuality (1986)
New Haven CT, Yale University Press, 224 p.
ISBN 0–300–03524–1.

Socarides, Charles Homosexuality (1978)
Northvale NJ, Aronson Jason, 664 p.
ISBN 0–87668–355–3.

Society of Automobile Engineers Crash Injury Impairment and Disability: long term affects (1986)
Warrendale PA, Society of Automotive Engineers.
ISBN 0–89883–932–7.

Society of Automotive Engineers Automotive Crash Avoidance Research (1987)
Warrendale PA, Society of Automotive Engineers.
ISBN 0–89883–970–X.

Sodeman, W A (Ed) Acute Gastrointestinal Problems (1986)
Oxford, Blackwell Scientific Publications, 184 p.
ISBN 0–932883–32–X.

Soedjatmoko Policymaking for Long-Term Global Issues (1988)
Washington DC, Georgetown University, 24 p. illus. Occasional Papers. ISBN 0–934742–49–9.

Soelle, Dorothee and Kalin, Everett R Suffering (1975)
Philadelphia PA, Fortress Press, 192 p.
ISBN 0–8006–1813–0.

Solov'eva, G I Parasitic Nematodes of Woody and Herbaceous Plants (1975)
New Delhi, Amerind Publishing Company, 134 p.

Sommer, Alfred Nutritional Blindness: xerophthalmia and keratomalacia (1982)
Oxford NY, Oxford University Press. illus.
ISBN 0–19–502977–1.

Somogyi, J C (Ed) World-Wide Problems of Nutrition Research and Nutrition Education: symposium der Internationalen Stiftung zur Förderung der Ernährungsforschung und Ernährungsaufklärung, Luzern, July 1981 (1982)
Basel, S Karger AG, 76 p. ISBN 3–8055–3586–4.

Somogyi, J C Malnutrition – A Problem of Industrial Societies? (1988)
New York, Karger S AG, 128 p. ISBN 3–8055–4811–7.

Somogyi, J C and Varela, G (Eds) Nutritional Deficiencies in Industrialized Countries: 17th symposium of the Group of European Nutritionist, Santiago de Compostela 1979 (1981)
Basel, S Karger AG, 176 p. ISBN 3–8055–1994–X.

Sondhi, Krishan Problems of Communication in Developing Countries (1980)
Coos Bay OR, Vision Books, 272 p.

Soni, R Control of Marine Pollution in International Law (1985)
Kenwin, Juta, 328 p. ISBN 0–7021–1660–2.

Sorenson, Joyce and Murray, Nancy Digestive Disorders (1983)
West Hartford CT, Witkower Press, 36 p.
ISBN 0–911638–08–3.

Sorokin, P A SOS: the meaning of our crisis
Millwood NY, Kraus Reprint and Periodicals.
ISBN 0–527–84832–8.

Sors, Andrew I and Coleman, David Pollution Research Index
Detroit MI, Gale Research, 555 p. ISBN 0–582–90006–9.

Soubrier, J P and Vedrinne, J Depression and Suicide (1983)
Elmsford NY, Pergamon Books, 912 p.
ISBN 0–08–027080–8.

Soulshy, E J Parasitic Zoonoses: clinical and experimental studies (1974)
San Diego CA, Academic Press. ISBN 0–12–655360–2.

South Commission The Challenge to the South: the report of the South Commission (1990)
New York, Oxford University Press, 325 p.
ISBN 0–19–877311–0.

Spark, R F The Infertile Male: the clinician's guide to diagnosis and treatment (1988)
New York, Plenum Publishing, 358 p.
ISBN 0–306–42859–8.

Sparks, S N Birth Defects and Speech-Language Disorders (1984)
Basingstoke, Taylor and Francis, 224 p.
ISBN 0–85066–561–2.

SPC/SPEC/ESCAP/UNEP Radioactivity in the South Pacific (1984)
Nairobi, UNEP, 211 p. UNEP Regional Seas Reports and Studies: No 40.

Special Learning Corporation Exceptional Children: a reference book (1984)
Guilford CT, Special Learning Corporation.

Spector, Malcolm Constructing Social Problems (1987)
Hawthorne NY, Aldine de Gruyter, 192 p. Repr of 1977 ed.
ISBN 0–202–30337–3.

Spector, Malcolm and Kitsuse, John I Constructing Social Problems (1987)
Berlin, Aldine de Gruyter, 184 p. ISBN 3–11–011555–7.

Spencer, Christopher and Navaratham, V Drug Abuse in East Asia (1981)
New York, Oxford University Press. ISBN 0–19–580476–7.

Spender, Dale Man Made Language (1985)
London, Routledge and Kegan Paul, 250 p.
2nd ed. ISBN 0–7102–0315–2.

Sperling, Susan Animal Liberators: research and morality (1988)
Berkeley CA, University of California Press, 220 p.
ISBN 0–520–06198–5.

Spicker, S F, et al (Eds) Contraceptive Ethos: reproductive rights and responsibilities (1987)
Dordrecht, Kluwer Academic Publishers Group, 276 p.
ISBN 1–55608–035–2.

Spicker, Stuart F, et al The Use of Human Beings in Research (1988)
Norwell MA, Kluwer Academic Publishers. ISBN 1–55608–043–3.

Spielberger, Chas D, et al Stress and Anxiety (1988)
New York, Hemisphere Publishing, 258 p. Vol. 11 ISBN 0–89116–312–3.

Spoerri, T Nekrophilie (1959)
Basle.

Spooner, Henry J and Lord Leverhulme, Wealth from Waste: elimination of waste a world problem (1973)
Easton PA, Hive Publishing, 332 p. ISBN 0–87960–063–2.

Spooner, Philip, et al Slurry Trench Construction for Pollution Migration Control (1985)
Park Ridge NJ, Noyes Data Corp, 237 p. ISBN 0–8155–1020–9. Pollution Technology Review Ser: No 118.

Spriestersbach, D C Psychosocial Aspects of the "Cleft Palate Problem" (1973)
Iowa City IA, University of Iowa Press, 560 p. 2 vols. ISBN 0–87745–044–7.

Springer, Allen L The International Law of Pollution: protecting the global environment in a world of (1983)
Westport CT, Greenwood Press, 218 p. ISBN 0–89930–052–9.

Sreenivas, V Acute Disorders of the Abdomen: diagnosis and treatment (1980)
Berlin, Springer–Verlag, 200 p. ISBN 3–540–90483–2.

Sreenivasan, K Anatomy of Wealth
New Delhi, Somaiya Publications, 140 p.

Srinivasan, T N and Bardhan, Pranab K Rural Poverty in South Asia (1988)
New York, Columbia University Press, 608 p. ISBN 0–231–06224–9.

St Louis, Kenneth O The Atypical Stutterer: principles and practices of rehabilitation (1985)
San Diego CA, Academic Press. ISBN 0–12–661620–5.

St Onge, K R The Melancholy Anatomy of Plagiarism (1988)
Lanham MD, University Press of America, 118 p. ISBN 0–8191–6859–9.

Stack, John F The Primordial Challenge: ethnicity in the contemporary world (1986)
Westport CT, Greenwood Press, 242 p. ISBN 0–313–24759–5.

Stagnara, Pierre and Dove, John Spinal Deformity (1988)
Stoneham MA, Butterworth Publishers. ISBN 0–407–00427–0.

Stajner, Rikard Crisis: anatomy of contemporary crises and theory of crises in the neo–imperial stage of capitalism (1976)
Beograd, Aktuelna Pitanja Socijalizma.

Stangvik, Gunnar Self–Concept and School Segregation (1979)
Stockholm, Almqvist and Wiksell, 528 p. ISBN 91–7346–064–8.

Stanley, C Maxwell Managing Global Problems: a guide to survival (1979)
Muscatine IA, Stanley Foundation, 286 p. ISBN 0–9603112–1–1.

Stanley, Charles Temptation (1988)
Nashville TN, Oliver–Nelson. ISBN 0–8407–9036–8.

Stannic, Dominic L Pneumonia: medical subject analysis, reference and research (1987)
Annandale VA, ABBE Publishers Association of Washington, 160 p. ISBN 0–88164–592–3.

Stark, James and Goldstein, Howard The Rights of Crime Victims (1985)
Carbondale IL, World Resources Inventory, 448 p. ISBN 0–8093–9952–0.

Starr, Gerald Minimum Wage Fixing: an international review of practices and problems (1981)
Geneva, ILO, 203 p. ISBN 92–2–102510–1.

Starr, Gerald Minimum Wage–fixing. An International Review of Practices and Problems (1981)
Geneva, ILO, 203 p. ISBN 92–2–102511–X.

Starr, Jerold M Cultural Politics: radical movements in modern history (1984)
New York, Praeger Publishers, 368 p. ISBN 0–275–91276–0.

Stavenhagen, Rodolfo Between Underdevelopment and Revolution: a Latin American perspective (1981)
New Delhi, Abhinav Publications, 220 p. ISBN 81–7017–139–3.

Stedman, Ray C Spiritual Warfare: winning the daily battle with satan (1985)
Portland OR, Multnomah Press, 145 p. ISBN 0–88070–094–7.

Steel, E W and McGhee, Terence Water Supply and Sewerage (1979)
New York, McGraw Hill Book Company. ISBN 0–07–060929–2.

Steffen, Lloyd H Self–Deception and the Common Life (1986)
Bern, Verlag Peter Lang AG, 415 p. ISBN 0–8204–0243–5.

Steffensmeier, Darrell J The Fence: in the shadow of two worlds (1986)
Totowa NJ, Rowman and Littlefield Publishers, 304 p. ISBN 0–8476–7494–0.

Steidlmeier, Paul Paradox of Poverty: a reappraisal of economic development policy (1987)
Cambridge MA, Ballinger Publishing, 344 p. ISBN 0–88730–184–3.

Steihm, Immunology Disorders in Infants and Children (1988)
Philadelphia PA, WB Saunders, 8996 p. ISBN 0–7216–1242–3.

Stein, Ruth Care of Children with Chronic Illness: issues and strategies (1988)
New York, Springer Publishing. ISBN 0–8261–5900–1.

Steinbrugge, Karl V and Busch, Charles U Earthquakes, Volcanoes, and Tsunamis: an anatomy of hazards (1982)
New York, Skandia America Group. ISBN 0–9609050–0–6.

Stekel, Wilhelm Patterns of Psychosexual Infantilism (1952)
New York, Liveright Publishing. ISBN 0–87140–840–6.

Stephan, Walter G and Feagin, Joe R School Desegregation: past, present, and future (1980)
New York, Plenum Publishing, 370 p. ISBN 0–306–40378–1.

Steptoe, Andrew Psychological Factors in Cardiovascular Disorders (1981)
San Diego CA, Academic Press. ISBN 0–12–666450–1.

Sterling, Claire The Mafia (1990)
New York, Hamish Hamilton, 384 p.

Stern, Arthur Air Pollution
San Diego CA, Academic Press. 8 vols. Environmental Sciences Ser: Vol 1: Air Pollutants, Their Transformation and Transport (1976, 715 p.). Vol 2 (3rd ed, 1977). Vol 3: Measuring, Monitoring and Surveillance (1976, 799 p.). Vol 4: Engineering Control of Air Pollution (1977, 946 p.). Vol 5 (3rd ed, 1977). Vol VI: Supplement Part A–Air Pollutants, their Transformation, Transport and Effects (3rd ed, 1986). Vol 7: Supplement Part B: Measurement, Monitoring, Surveillance and Engineering Control of Air Pollution (3rd ed, 1986). Vol 8: Supplement Part C: Management of Air Quality (3rd ed, 1986).

Stern, Arthur Air Pollution (1977)
San Diego CA, Academic Press. 8 Vols.

Stern, Arthur C Air Pollution: supplement part a–air pollutants, their (1986)
San Diego CA, Academic Press. Vol. VI ISBN 0–12–666606–7.

Stern, Marvin Death, Grief and Friendship in the Eighteenth Century: Edward Gibbon and Lord Sheffield (1985)
Center Thanatology. ISBN 0–930194–35–7.

Stern, R M, et al Health Hazards and Biological Effects of Welding Fumes and Gases (1986)
New York, Elsevier Science Publishing, 612 p. ISBN 0–444–80784–5.

Sternbach, Richard A Pain Patients: traits and treatments (1974)
San Diego CA, Academic Press. ISBN 0–12–667235–0.

Sternlicht, Manny and Sternlicht, Manny Social Behavior of the Mentally Retarded: an annotated bibliography (1983)
New York, Garland Publishing, 400 p. ISBN 0–8240–9137–X.

Stevens, David A Coccidioidomycosis: a text (1980)
New York, Plenum Publishing, 296 p. ISBN 0–306–40410–9.

Stevens, Neil E and Stevens, Russell B Disease in Plants: an introduction to agricultural phytopathology
New Delhi, International Books and Periodicals Supply Service.

Stevenson, Colin Challenging Adult Illiteracy: reading and writing disability in the British army (1986)
New York, Columbia University Press, 224 p. ISBN 0–8077–2737–7.

Stevenson, Michael (Ed) Readings in RSI: the ergonomics approach to repetition strain injury
Kensington NSW, New South Wales University Press. ISBN 0–86840–162–5.

Steward, D E Contact Inhibition (1985)
Avant Books/Slawson Comm, 144 p. ISBN 0–932238–29–7.

Stewart, Alva W Wilderness Protection: a bibliographic review (1985)
Monticello IL, Vance Bibliographies. ISBN 0–89028–332–X.

Stierlin, Helm and Weber, Gunthard Unlocking the Family Door: a systematic approach to the understanding and treatment of anorexia nervosa (1989)
New York, Brunner/Mazel, 240 p. ISBN 0–87630–541–9.

Stilwell, F J B Economic Crisis–Cities and Regions
New York, Pergamon Books. ISBN 0–08–024810–1.

Stimson, Gerry V and Oppenheimer, Edna Heroin Addiction (1982)
New York, Routledge Chapman and Hall, 300 p. ISBN 0–422–77890–7.

Stipa, S and Cavallaro, A Peripheral Arterial Diseases: medical and surgical problems (1982)
San Diego CA, Academic Press. ISBN 0–12–671460–6.

Stjernquist, Per Poverty on the Outskirts: on cultural impoverishment and cultural integration (1987)
Stockholm, Almqvist and Wiksell, 36 p. ISBN 91–22–01045–9.

Stock, Thomas and Sutherland, Ronald (Eds) National Implementation of the Future Chemical Weapons Convention (1990)
Oxford, Stockholm International Peace Research Institute, 200 p. SIPRI Chemical and Biological Warfare Studies: 11. ISBN 0–19–827837–3.

Stockholm International Peace Research Institute Incendiary Weapons (1975)
Stockholm, Almqvist and Wiksell, 255 p. ISBN 0–262–19139–3.

Stockholm International Peace Research Institute The Law of War and Dubious Weapons (1976)
Solna, SIPRI, 78 p. ISBN 91–85114–31–6.

Stockholm International Peace Research Institute Weapons of Mass Destruction and the Environment (1977)
Solna, SIPRI, 95 p. ISBN 0–85066–132–3.

Stockholm International Peace Research Institute Outer Space: battlefield of the future? (1978)
London, Taylor and Francis, 202 p. illus. ISBN 0–85066–130–7.

Stockholm International Peace Research Institute Ecological Consequences of the Second Indochina War (1979)
Solna, SIPRI, 119 p. ISBN 91–22000–62–3.

Stockholm International Peace Research Institute Warfare in a Fragile World: military impact on the human environment (1980)
Solna, SIPRI, 249 p. ISBN 0–85066–187–0.

Stockholm International Peace Research Institute Nuclear Radiation in Warfare (1981)
London, Taylor and Francis, 149 p. illus. ISBN 0–85066–217–6.

Stockholm International Peace Research Institute Arms Race and Arms Control: facts and figures on the arms race and arms control efforts (1982)
London, Taylor and Francis, 242 p. ISBN 0–85066–232–X.

Stockholm International Peace Research Institute Success and Failure in Arms Control Negotiations (1989)
Solna, Stockholm International Peace Research Institute.

Stockholm International Peace Research Institute Yearbook 1990: world armaments and disarmament (1990)
Oxford, Oxford University Press, 576 p. illus. ISBN 0–19–827862–4.

Stocking, George W and Watkins, Myron W Cartels or Competition? (1986)
Buffalo NY, Hein William S, 516 p. ISBN 0–89941–478–8.

Stohl, Leslee L Cat Diseases and Veterinary Medicine: subject analysis index with research bibliography (1987)
Washington DC, ABBE Publications Association, 150 p. bibl. ISBN 0–88164–352–1.

Stoller, Robert J and Stoller, Robert J Perversion: the erotic form of hatred (1986)
Washington DC, American Psychiatric Press, 256 p. ISBN 0–88048–262–1.

Stolz, Barbara A Still Struggling: a portrait of low–income women in the 1980's (1985)
Lexington MA, Lexington Books, 224 p. ISBN 0–669–10930–4.

Stone, Christopher D Should Trees Have Standing: towards legal rights for natural objects

Stonehouse, A and Sutherland, B After a Miscarriage (1986)
Carlton South VIC, Pitman Publishing, 72 p. ISBN 0–85896–235–7.

Storey, Moorfield Problems of To–Day (1920)
Salem NH, Ayer Company Publishers. ISBN 0–8369–0908–9.

Stormon, E J and Stransky, Thomas F Towards the Healing of Schism (1987)
Mahwah NJ, Paulist Press, 576 p. ISBN 0–8091–2910–8.

Stormorken, Bjørn and Zwaak, Leo Human Rights Terminology in International Law: a thesaurus (1988)
Strasbourg, Council of Europe, 234 p. ISBN 90–247–3643–9.

Stover, R H Fusarial Wilt (Panama Disease) of Bananas and Other Musa Species (1962)
Kew, CAB International Mycological Institute, 128 p. illus. Phytopathological Papers: No 4.

Stover, R H Banana, Plantain and Abaca Diseases (1972)
Kew, CAB International Mycological Institute, 316 p. illus. ISBN 0–85198–088–0.

Strahler, Arthur N Science and Earth History: the evolution–creation controversy (1988)
Buffalo NY, Prometheus Books, 552 p. ISBN 0–87975–414–1.

Strano, A Thrombosis and Cardiovascular Disease (1984)
New York, Plenum Publishing, 438 p. ISBN 0–306–41261–6.

Strauss, Herbert A and Bergmann, Werner (Ed) Current Research on Antisemitism (1987)
Berlin, Walter de Gruyter. 2 Vols.

Strauss, John S and Carpenter, William T Schizophrenia (1981)
New York, Plenum Publishing, 232 p. ISBN 0–306–40704–3.

Strauss, José (Ed) Acute Renal Disorders and Renal Emergencies (1984)
Dordrecht, Kluwer Academic Publishers Group, 448 p. ISBN 0–89838–663–2.

Streit, Gary Psychology of a Broken Heart: an essay on romantic love (1987)
Medford OR, Golden Blossom Publishing. ISBN 0–9618180–0–X.

Stringer, C Wildlife of Tropical Australia (1969)
Princeton NJ, Darwin Press, 112 p.

Stringer, P Confronting Social Issues: applications of social psychology (1982)
San Diego CA, Academic Press. Vol 2. ISBN 0–12–673802–5.

Strober, Warren, et al (Eds) Mucosal Immunity and Infections at Mucosal Surfaces (1988)
Oxford NY, Oxford University Press, 464 p. illus. ISBN 0–19–504329–4.

Stroebe, Wolfgang and Stroebe, Margaret S Bereavement and Health (1987)
New York, Cambridge University Press, 288 p. ISBN 0–521–24470–6.

Strom, Margaret A Societal Issues – Scientific Viewpoints (1987)
New York, American Institute of Physics, 256 p. ISBN 0–88318–537–7.

Stromsta, Courtney Elements of Stuttering (1986)
Oshtemo MI, Atsmorts Publishing, 256 p. ISBN 0–9620355–4–8.

Stroud, Robert Bird Disease by Stroud
Neptune NJ, TFH Publications. ISBN 0–87666–435–4.

Stulman, Julius Evolving Mankind's Future; The World Institute: a problem–solving methodology (1967)
Philadelphia PA, Lippincott, 95 p.

Stumm, Werner (Ed) Global Chemical Cycles and their Alternations by Man (1977)
New York, VCH Publications, 347 p. illus. Physical and Chemical Sci Rsch Rapt Ser: No 2. ISBN 0–89573–084–7.

Stur, O (Ed) Mental Retardation, 2nd congress, Vienna, 1961: part 2: psychological and sociological problems in imbecility (1963)
New York, Karger S AG, 259 p. ISBN 3–8055–1568–5.

Sturgeon, Susan and Rans, Laurel The Woman Offender: a bibliographic sourcebook (1975)
Lincoln MA, Entropy Ltd, 63 p. ISBN 0–938876–02–3.

Sturzo, Luigi and Gooch, G P The International Community and the Right of War (1971)
New York, Fertig Howard. ISBN 0–86527–104–6.

Subrahmanyam, K (Ed) Insecurity of Developing Nations (1986)
New Delhi, Lancer's International.

Suckling, K E and Groot, P H Hyperlipidaemia and Atherosclerosis (1988)
San Diego CA, Academic Press, 240 p. ISBN 0–12–675645–7.

Sudhir, Chandra (Ed) Social Transformation and Creative Imagination (1984)
New Delhi, Allied Publishers, 376 p.

Suess, M J, et al (Eds) Ambient Air Pollutants from Industrial Sources: a reference handbook (1985)
Amsterdam, Elsevier Science Publishing, 843 p. ISBN 0–444–80605–9.

Sugimura, Takashi, et al (Eds) Environmental Mutagens and Carcinogens (1981)
Toronto ON, University of Toronto Press, 794 p. Dist outside Asia and Australasia us Liss.

Sujan, M A and Trivadi, V D Smuggling: the inside story (1976)
Bombay, Jaico Publishing House, 200 p.

Sullivan, Charles A and Zimmer, Michael J Employment Discrimination (1988)
Boston MA, Brown Little, 2000 p. 2nd ed. 3 Vols. ISBN 0–316–82192–6.

Sullivan, Eleanor, et al Chemical Dependency in Nursing (1987)
Reading MA, Addison–Wesley, 150 p. ISBN 0–201–07581–4.

Sullivan–Fowler, Micaela Alternative Therapies, Unproven Methods, and Health Fraud (1988)
Chicago IL, American Medical Association. ISBN 0–89970–319–4.

Sullivan, H S Schizophrenia as a Human Process (1962)
New York, WW Norton.

Sullivan, John J and Foster, Joyce C Stress and Pregnancy (1987)
New York, AMS Press. ISBN 0–404–63261–0.

Sullivan, Thomas J and Thompson, Kenrick S Introduction to Social Problems (1991)
New York, Macmillan Publishing, 1156 p. 2nd ed. ISBN 0–02–418481–0.

Sullivan, Thomas J, et al Social Problems: divergent perspectives (1980)
New York, Macmillan, 728 p. ISBN 0–02–418430–6.

Sundberg–Weitman, Brita Discrimination on Grounds of Nationality: free movement of workers and freedom of establishment under the EEC Treaty (1977)
Amsterdam, Elsevier Science Publishing, 248 p. ISBN 0–7204–0477–0.

Sundmacher, R Herpetic Eye Disease: international symposium (1981)
New York, Springer Publishing, 560 p. ISBN 0–387–00324–X.

Sutcliffe, Charles The Dangers of Low Level Radiation (1987)
Brookfield VT, Gower Publishing, 284 p. ISBN 0–566–05482–5.

Suter, Keith Alternative to War: the peaceful settlement of international disputes
Hornsby NSW, Transpareon Press, 100 p.

Sutherland, Edwin; Geis, Gilbert and Goff, Colin White Collar Crime: the uncut version (1983)
New Haven CT, Yale University Press, 320 p. ISBN 0–300–02921–7.

Suttie, Ian The Origins of Love and Hate
Free Association Books.

Sutton, John Stubborn Children (1988)
Berkeley CA, University of California Press. ISBN 0–520–06093–8.

Sutton, R A L (Ed) Urolithiasis (1987)
Basel, S Karger AG, 96 p. ISBN 3–8055–4567–3.

Svalastoga, Kaare On Deadly Violence (1982)
Oslo, Norwegian University Press, 160 p. ISBN 82–00–06071–3.

Swaab, D F, et al (Eds) Aging of the Brain and Alzheimer's Disease (1986)
Amsterdam, Elsevier Science Publishing, 526 p. ISBN 0–444–80793–4.

Swarup, Ram Facts and Fallacies About Selfrelief Wrongly Termed as 'Selfabuse' Solitary Vice, Sin of Youth and 'Onanism' (1961)
Uttar Pradesh, Madhuri Publishers, 104 p.

Swerdlow, M and Ventafridda, V Cancer Pain (1986)
Norwell MA, Kluwer Academic Publishers. ISBN 0–85200–858–9.

Swidler, Leonard Religious Liberty and Human Rights in Nations and in Religions (1986)
Philadelphia PA, Ecumenical Press, 255 p. ISBN 0–931214–06–8.

Swineford, Oscar Asthma and Hay Fever and Other Allergic Diseases for Victims and Their Families (1973)
Springfield IL, Thomas Charles C, 186 p. ISBN 0–398–02767–6.

Swisher Biodegradation (1986)
New York, Dekker Marcel, 1176 p. 2nd rev enl ed. Surfactant Science Ser. ISBN 0–8247–6938–4.

Swoboda, A K (Ed) Capital Movements and Their Control: conference of the International Center for Monetary and Banking Studies (2)
Dordrecht, Kluwer Academic Publishers Group. ISBN 90–286–0295–X.

Symons, C, et al Specific Heart Muscle Disease (1983)
Littleton MA, PSG Publishing, 160 p. ISBN 0–7236–0641–2.

Symons, Ronald C and Hughes, Charles C (Eds) Culture–Bound Syndromes: folk illnesses of psychiatric and anthropological interest (1985)
Dordrecht, Kluwer Academic Publishers Group, 536 p. ISBN 90–277–1858–X.

Symposium on Comparative Leukemia Research, 7th International Copenhagen, October 1975 Comparative Leukemia Research 1975: proceedings (1976)
Basel, S Karger AG, 600 p. illus. Biblbiotheca Haemalologica: No 43. ISBN 3–8055–2316–5.

Systems Research Institute Dictionary of Administration and Management (1986)
Los Angeles CA, Systems Research Center, 1340 p. ISBN 0–912352–08–6.

Szabo, Kalman T Congenital Malformations in Laboratory and Farm Animals (1988)
San Diego CA, Academic Press, 1000 p. ISBN 0–12–680130–4.

Szabo, L Cultural Rights (1974)
Dordrecht, Kluwer Academic Publishers Group, 115 p. ISBN 90–286–0364–6.

Szabolcs, I Review of Research on Salt–Affected Soils (1979)
Lanham MD, UNIPUB, 137 p. ISBN 92–3–101613–X.

Szasz, Kathleen Petishism: pets and their people in the Western World (1969)
New York, Holt, Rinehart and Winston.

Szasz, Thomas The Myth of Mental Illness (1967)
New York, Dell Books.

Szasz, Thomas S The Manufacture of Madness (1971)
London, Routledge and Kegan Paul.

Szumski, Bonnie Terrorism: opposing viewpoints ser (1986)
Saint Paul MN, Greenhaven Press, 200 p. ISBN 0–89908–389–7.

Taeni, Rainer Latent Anxiety (1978)
Chippendale, Alternative Publishing Cooperative, 207 p. ISBN 0–909188–14–9.

Taheri, Amir Holy Terror: inside the world of islamic terrorism (1987)
Bethesda MD, Adler and Adler Publishers, 332 p. ISBN 0–917561–45–7.

Talbot, Eugene S Degeneracy: its causes, signs and results (1984)
New York, Garland Publishing, 372 p. The History of Hereditarian Thought Ser. ISBN 0–8240–5830–5.

Talbott, J H and Yü, T–F Gout and Uric Acid Metabolism (1976)
Stuttgart, George Thieme Verlag, 303 p. ISBN 3–13–544301–9.

Tanabe, T, et al (Eds) Nephrotoxicity of Antibiotics and Immunosuppressants: proceedings of the Satellite Symposium of the Fourth International Congress, Sapporo, Japan, July 26–28, 1986 (1987)
Developments in Toxicology and Environmental Sciences Ser: 14. ISBN 0–444–80858–2.

Tanaka, T M and Komai, K (Eds) Current Research on Fatigue Cracks (1987)
Amsterdam, Elsevier Science Publishing, 318 p. ISBN 1–85166–091–7.

Tanimoto, Helene S and Inaba, Gail T Women's Work, Collective Bargaining, Comparable Worth–Pay Equity, Job Evaluation – and All That: an annotated bibliography (1987)
Honolulu HI, University of Hawaii Press, 76 p. ISBN 0–318–23504–8.

Taperell, G Q; Vermeesch, R B and Harland, D J Trade Practices and Consumer Protection (1983)
Sevenoaks, Butterworths, 975 p. 3rd ed. ISBN 0–409–49111–X.

Tarizzo, M L Field Methods for the Control of Trachoma (1973)
Geneva, WHO. ISBN 92–4–154033–8.

Tarshis, Lorie World Economy in Crisis: unemployment, inflation and international debt
Toronto ON, James Lorimer. ISBN 0–88862–626–6.

Tarter, Ralph E and Van Thiel, David H Alcohol and the Brain: chronic effects (1985)
New York, Plenum Publishing, 366 p. ISBN 0–306–41998–X.

Taske, C E, et al Thermal and Environmental Effects in Fatigue: research–design interface (1983)
New York, American Society of Mechanical Engineers, 256 p. ISBN 0–317–02651–8.

Tataryn, L Dying for a Living: the politics of industrial death (1979)
Ottowa, Deneau and Breenberg.

Taussig, Helen B Congenital Malformations of the Heart: general considerations (1960)
Cambridge MA, Harvard University Press. Vol. 12 vols. ISBN 0–674–16150–5.

Taylor, David Pig Diseases (1983)
Alexandria Bay NY, Diamond Farm Book Publishers, 250 p. ISBN 0–9506932–2–7.

Taylor, I G and Markides, Andreas Disorders of Auditory Function, III (1981)
San Diego CA, Academic Press. ISBN 0–12–684780–0.

Taylor, James B et al Tornado: a community responds to disaster (1970)
Seattle WA, University of Washington Press, 205 p. ISBN 0–295–95088–9.

Taylor, Joseph C and Mittman, Charles Pulmonary Emphysema and Proteolysis, 1986 (1987)
San Diego CA, Academic Press, 568 p. ISBN 0–12–684570–0.

Taylor, T Ajibola Crop Pests and Diseases
New York, University Press. ISBN 0–19–575390–9.

Teichman, Jenny Meaning of Illegitimacy
South Edmonton, Academic Printing and Publishing.

Teitelbaum, Michael S and Winter, Jay M The Fear of Population Decline (1985)
San Diego CA, Academic Press. ISBN 0–12–685190–5.

Templin, Linda A Mental Health and Criminal Justice (1984)
London, Sage, 320 p. Criminal Justice System Annuals Ser: Vol 20. ISBN 0–8039–2084–9.

Ten Horn, G H M M, et al (Eds) Psychiatric Case Registers in Public Health: a worldwide inventory 1960–1985 (1986)
Amsterdam, Elsevier Science Publishing, 444 p. ISBN 0–444–90441–7.

Tenney, Louise Modern Day Plagues (1987)
Provo UT, Woodland Books. ISBN 0–913923–59–1.

Teoh, Eng–Soon, et al (Eds) Endometriosis: and other disorders and infections (1987)
Park Ridge, Parthenon Publishing Group, 250 p. illus. Advances in Fertility and Sterility Series. ISBN 0–940813–20–3.

Terry, Robert D (Eds) Aging and the Brain (1988)
New York, Raven Press, 320 p. illus. Aging ser: Vol 32. ISBN 0–88167–329–3.

Teubner, Gunther (Ed) Juridification of Social Spheres: a Comparative analysis in the areas of labor, corporate, antitrust and social welfare law (1987)
Berlin, Walter de Gruyter, 446 p.

Textile Industrial Hygiene Roundtable Cotton Dust Exposures (1987)
Akron OH, American Industrial Hygiene Association. Vol. II ISBN 0–932627–26–9.

The Documentation of the Council of Europe (Eds) The Education of Migrant Workers' Children: report of the European Contact Workgroup, Dillingen, Federal Republic of Germany, 1980 (1981)
Lisse, Swets en Zeitlinger, 186 p.

The Documentation of the Council of Europe (Eds) Sex Stereotyping in Schools: report of the Educational Research Workshop, Honefoss, Norway, 1981 (1982)
Lisse, Swets en Zeitlinger, 216 p.

Théberge, R L (Ed) Common African Pests and Diseases of Cassava, Yam, Sweet, Potato and Cocoyam (1985)
Ibadan, International Institute of Tropical Agriculture, 108 p.

Thibault, Gisele Marie Dissenting Feminist Academy: a history of the barriers to feminist scholarship (1987)
New York, Peter Lang Publishing, 270 p. ISBN 0–8204–0396–2.

Thio, Alex Deviant Behavior (1987)
New York, Harper and Row, 538 p. 3rd ed. ISBN 0–06–046683–9.

Third World Publications Third World Guide: nineteen eighty–six to nineteen eighty–seven (1986)
New York, Grove Press, 632 p. ISBN 0–394–62330–4.

Thirlway HWA Non–Appearance Before the International Court of Justice (1984)
Cambridge, Cambridge University Press, 184 p. ISBN 0–521–26594–0.

Thomas, Alan J Acquired Hearing Loss: psychological and psychosocial implications (1985)
San Diego CA, Academic Press. ISBN 0–12–687920–6.

Thomas, Alexander and Chess, Stella Temperament and Development (1977)
New York, Brunner/Mazel. ISBN 0–87630–139–1.

Thomas, Caroline In Search of Security: the Third World in international relations (1987)
Hemel Hempstead, Harvester Wheatsheaf, 256 p. ISBN 0–7450–0394–X.

Thomas, G, et al (Eds) Summer Mastitis (1987)
Dordrecht, Kluwer Academic Publishers Group, 226 p. ISBN 0–89838–982–8.

Thompson, J M and Hunt, G W Elastic Instability Phenomena (1984)
New York, John Wiley and Sons, 209 p. ISBN 0–471–90279–9.

Thompson, Robert J Jr Behavior Problems in Children with Development and Learning Disabilities (1986)
Ann Arbor MI, University of Michigan, 250 p. International Academy for Research in Learning Disabilities Monograph Ser: No 3.

Thorkelson, Lori Emotional Dependency: a threat to close friendships (1984)
San Rafael CA, Exodus International North America. illus. ISBN 0–931593–00–X.

Thronton, W T Overpopulation and its remedy (1971)
I.A.P., 446 p. ISBN 0–7165–1751–5.

Thung, Mady A (Ed) Exploring the new Religious Consciousness: an Investigation of religious change by a Dutch working group (1985)
Amsterdam, Vrije Universiteit Boekhandel, 209 p.

Thyistrup, A (Ed) Dentine and Dentine Reactions in the Oral Cavity: proceedings of the meeting of the Council of Europe Research Group on Surface and Colloid Phenomena in the Oral Cavity held in Copenhagen, Denmark, February 24–28, 1987 (1987)
IRL Pr US, 250 p. ISBN 1–85221–050–8.

Tibbits, Donald F Language Disorders in Adolescents (1982)
Lincoln NE, Cliffs Notes, 120 p. ISBN 0–8220–1832–2.

Tibbs, David J Venous Disorders of the Lower Limbs (1989)
Sevenoaks, Butterworths, 240 p. illus.
ISBN 0-407-00393-2.

Tietze, Christopher and Henshaw, Stanley K Induced Abortion: a world review (1986)
New York, Allan Guttmacher Institute. 6th ed.

Tildon, Tyson, et al Sudden Infant Death Syndrome (1983)
San Diego CA, Academic Press. ISBN 0-12-691050-2.

Timagenis, G J International Control of Marine Pollution (1980)
Dordrecht, Kluwer Academic Publishers Group, 1338 p.
ISBN 90-286-0560-6.

Tiryakian, Edward A (Ed) Global Crisis: sociological analysis and responses (1984)
Leiden, Brill, 135 p. ISBN 90-04-07284-5.

Tisdell, C On the Economics of Saving Wildlife from Extinction (1979)
New York, UN, 29 p. ISBN 0-7259-0329-5.

Tjomas, Edwin J (Ed) Behavior Modification Procedure (1974)
Berlin, Aldine de Gruyter, 368 p. ISBN 0-202-36019-9.

Tobias, Andrew Treating Malpractice: report of the twentieth century fund task force on medical malpractice insurance (1986)
New York, Priority Press Publications, 70 p.
ISBN 0-87078-173-1.

Tolley, Howard B Children and War (1973)
New York, Columbia University Press, 274 p.
ISBN 0-8077-2280-4.

Tomasic, Roman Failure of Imprisonment (1979)
London, Allen and Unwin, 166 p. ISBN 0-86861-097-6.

Tomlinson, John Is Band-Aid Social Work Enough? (1979)
East Victoria Park, Wobbly Press, 270 p.
ISBN 0-909-38803-2.

Tonini, J Corrosion of Reinforcing Steel in Concrete (1980)
Philadelphia PA, American Society for Testing and Materials, 224 p. STP 713. ISBN 0-8031-0316-6.

Toribara, T Y et al Polluted Rain (1980)
New York, Plenum Publishing, 514 p.
ISBN 0-306-40353-6.

Törnqvist, G et al (Ed) Division of Labour, Specialization and Technical Change: global, regional and workplace level (1987)
Stockholm, Almqvist and Wiksell, 278 p.
ISBN 91-22-00894-2.

Torrey, E Fuller Nowhere to Go: the tragic odyssey of the homeless mentally ill (1988)
New York, Harper and Row, 256 p. ISBN 0-06-015993-6.

Torrie, Jill (Ed) Banking on Poverty: the global impact of the IMF and World Bank (1983)
Toronto ON, Between the Lines, 336 p.
ISBN 0-919946-39-9.

Trachtenberg, Peter The Casanova Complex: compulsive lovers and their women (1989)
London, Angus and Robertson, 288 p.
ISBN 0-207-16171-2.

Trad, Paul V Infant and Childhood Depression: developmental factors (1987)
New York, John Wiley and Sons, 463 p.
ISBN 0-471-85230-9.

Trager, Oliver C (Ed) Fighting Terrorism: negotiation or retaliation? (1986)
New York, Facts on File, 224 p. ISBN 0-8160-1395-0.

Train, John Famous Financial Fiascos (1984)
New York, Crown Publishers, 128 p. ISBN 0-517-54583-7.

Traub, R and Starcke, H (Eds) Fleas: proceedings of the International Conference on Fleas, Ashton World, 1977 (1980)
Rotterdam, AA Balkema, 450 p. ISBN 90-6191-018-8.

Trepelkov, V General Crisis of Capitalism (1984)
Chicago IL, Imported Publications, 168 p.

Tresemer, David Fear of Success (1977)
New York, Plenum Publishing, 258 p.
ISBN 0-306-31012-0.

Treverton, Gregory F Covert Action: the limits of intervention in the post-war world (1989)
New York, Basic Books, 304 p. ISBN 0-465-01440-2.

Tribe, Laurence Abortion: the clash of absolutes (1990)
New York, Norton, 270 p.

Tribe, Laurence Abortion: the clash of absolutes (1990)
New York, Norton, 270 p.

Troiden, Richard R Gay and Lesbian Identity: a sociological analysis (1988)
Dix Hills NY, General Hall, 176 p. ISBN 0-930390-80-6.

Trowell, H C and Burkitt, D P Western Diseases: their emergence and prevention (1981)
Cambridge MA, Harvard University Press, 480 p.
ISBN 0-674-95020-8.

Trowell, H C, et al Dietary Fibre, Fibre-Depleted Foods and Disease (1985)
San Diego CA, Academic Press. ISBN 0-12-701160-9.

Tse Tung, Mao On Contradiction (1937)
Peking, Foreign Language Press.

Tsuang, Ming T and VanderMey, Randall Genes and the Mind: inheritance of mental illness (1980)
New York, Oxford University Press. ISBN 0-19-261268-9.

Tsubaki, T and Yase, Y (Eds) Amyotrophic Lateral Sclerosis: recent advances in research and treatment (1988)
Amsterdam, Elsevier Science Publishing, 394 p. International Congress Ser: No 769. ISBN 0-444-80979-1.

Tsubaki, Tadao and Irukayama, Katsuro (Eds) Minamata Disease: methylmercury poisoning in Minamata and Niigata, Japan (1977)
Amsterdam, Elsevier Science Publishing, 328 p.
ISBN 4-06-129951-4.

Tuchman, Barbara W The March of Folly from Troy to Vietnam

Tulenko, Thomas and Cox, Robert H Recent Advances in Arterial Diseases: atherosclerosis advances in arterial diseases: atherosclerosis, hypertension and vasospasm (1986)
New York, Alan R Liss, 404 p. Progress in Clinical and Biological Research Ser: Vol 219. ISBN 0-8451-5069-3.

Tummala, K K Ambiguity of Ideology and Adminstrative Reform (1979)
New Delhi, Allied Publishers, 368 p.

Tuomi, Helena and Värynen, Raimo Transnational Corporation, Armaments and Development: a study of transnational military production, international transfer of military technology and their impact on development (1980)
Tampere, Peace Research Institute, 312 p.
ISBN 951-706-046-7.

Turner, P D Oil Palm Diseases and Disorders
Kuala Lumpur, Incorporated Society of Planters, 279 p.

Turner, Roger N The Hay Fever Handbook: a self-help program that works (1988)
Thorsons, Sterling, 192 p. illus. ISBN 0-7225-1592-8.

Turnham, David; Salomé, Bernard and Schwarz, Antoine (Eds) Development Centre Seminars: the informal sector revisited (1990)
Paris, OECD, 224 p. ISBN 92-64-13328-3.

Turok, Ben Revolutionary Thought in the Twentieth Century (1980)
Atlantic Highlands NJ, Humanities Press International, 360 p.
ISBN 0-905762-42-8.

Tutorow, Norman E and Winnovich, Karen War Crimes, War Criminals, and War Crimes Trials: an annotated bibliography and source book (1986)
Westport CT, Greenwood Press, 568 p.
ISBN 0-313-24412-X.

Tweedie, Joyce Children's Hearing Problems (1987)
Littleton MA, PSG Publishing, 116 p.ISBN 0-7236-0910-1.

Twitchell, James B Dreadful Pleasures: an anatomy of modern horror (1985)
New York, Oxford University Press, 353 p.
ISBN 0-19-503566-6.

Twitchell, James B Forbidden Partners: the incest taboo in modern culture (1986)
New York, Columbia University Press, 288 p.
ISBN 0-231-06412-8.

Tyler, Leona E Individual Differences: abilities and motivational directions (1974)
Englewood Cliffs NJ, Prentice Hall. illus.
ISBN 0-13-458042-7.

Tyrell, D A J (Ed) Aspects of Slow and Persistant Virus Infections (1979)
Dordrecht, Kluwer Academic Publishers Group, 296 p.
ISBN 90-247-2281-0.

Udupa, K N, et al Disorders of the Thyroid Gland in Tropics (1983)
New York, Advent Books, 320 p. ISBN 0-7069-1173-3.

Uhlig, Mark (Ed) Apartheid in Crisis (1986)
New York, UN, 334 p.

Ulmsten, U (Ed) Female Stress Incontinence (1983)
Basel, S Karger AG, 72 p. ISBN 3-8055-3665-8.

Ulrich, B and Pankrath, J (Eds) Effects of Accumulation of Air Pollutants in Forest Ecosystems: proceedings of a workshop held at Göttingen, West Germany, 1982 (1983)
Dordrecht, Kluwer Academic Publishers Group, 408 p.
ISBN 90-277-1476-2.

Uma Ram Nath Smoking: Third World alert (1986)
New York, Oxford University Press, 200 p.
ISBN 0-19-261402-9.

UNCTAD Protectionism and Structural Adjustment in the World Economy (1982)
New York, UNCTAD. ISBN 92-1-112139-6

UNCTAD Control of Restrictive Practices in Transfer of Technology Transactions: selected principal regulations, policy guidelines and case law at the national and regional levels (1982)
New York, UNCTAD. ISBN 92-1-112143-4

UNCTAD The International Monetary System and Financial Markets: recent developments and the policy challenge (1985)
New York, UNCTAD. ISBN 92-1-112187-6

UNCTAD Compensatory Financing of Export Earnings Shortfalls (1985)
New York, UNCTAD. This treatise discusses the issue of compensatory finance as well as the continued severity of the problem of instability in the export trade of developing primary producing countries. ISBN 92-1-112184-1

UNCTAD Collusive Tendering (1985)
New York, UNCTAD. ISBN 92-1-112197-3

UNCTAD The Least Developed Countries: 1985 Report (1986)
New York, UNCTAD. ISBN 92-1-112216-3

UNCTAD Current Problems of Economic Integration: the problems of promoting and financing integration projects (1986)
New York, UNCTAD, 115 p. ISBN 92-1-112216-3

UNEP/ASEAMS Oil Pollution and Its Control in the East Asian Seas Region (1988)
Nairobi, UNEP, 290 p. UNEP Regional Seas Reports and Studies: No 96.

UNEP/FAO/UNESCO/WHO/WMO/IOC/IAEA Selected Bibliography on the Pollution of the Mediterranean Sea (1981)
Geneva, UNEP, 135 p. UNEP Regional Seas Directories and Bibliographies.

UNEP/UNESCO/UN Coastal Erosion in West and Central Africa (1985)
Nairobi, UNEP, 237 p. UNEP Regional Seas Reports and Studies: No 67.

UNEP/UNESCO/UN-DIESA Bibliography on Coastal Erosion in West and Central Africa (1985)
Rome, FAO, 92 p. UNEP Regional Seas Directories and Bibliographies.

UNESCO Effects of Urbanization and Industrialization on the Hydrological Regime and on Water Quality: proceedings (1977)
Paris, UNESCO, 572 p. illus. bibl.
ISBN 92-3-001537-7. Studies and Reports in Hydrology: No 24.

UNESCO Statistics of Educational Attainment and Illiteracy, 1945-1974 (1977)
Paris, UNESCO, 233 p. bibl.
ISBN 92-3-001506-7. Statistical Reports and Studies: No 22.

UNESCO Interim Report on Communication Problems in Modern Society (1978)
Paris, UNESCO, 79 p.

UNESCO Select Bibliography of Unesco Publications, Reports and Documents Relating to Natural Hazards, 1961-1978 (1979)
Paris, UNESCO, 10 p.

UNESCO Wastage in Primary and General Secondary Education: a statistical study of trends and patterns in repetition and drop-out (1980)
Paris, UNESCO, 157 p. bibl.

UNESCO World Problems in the Classroom (1981)
Paris, UNESCO, 59 p. illus. Educational Studies and Documents: 4. ISBN 92-3-101817-5.

UNESCO Colloque International sur la Création d'un Réseau Décentralisé d'Analyses et de Recherches sur la Problématique Mondiale (1984)
Paris, UNESCO. BEP/802/4 Prov.

UNESCO Unesco Yearbook on Peace and Conflict Studies 1984 (1986)
Paris, UNESCO, 229 p. ISBN 92-3-102398-5.

UNESCO Evaluating Long-Term Developments by Using Global Models (1987)
Paris, UNESCO, 41 p. bibl. Major Programme I, Reflection on World Problems and Future-Oriented Studies: BEP/GPI/4.

UNIDO Industry and Development: global report 1986 (1986)
New York, UNIDO. ISBN 92-1-106217-9.

Orgs #E3386.

Union des Associations Internationales Annuaire de la Vie Internationale
Bruxelles, Union des Associations Internationales, 2 vols. Vol 1: 1908-1909, 1370p; Vol II: 1910-1911, 2652p.

Union des Associations Internationales Code des Voeux Internationaux: codification générale des voeux et résolutions des organismes internationaux (1923)
Bruxelles, Union des Associations Internationales, 940p.

Union of International Associations Global Action Networks (1985/86)
Munich, K G Saur Verlag. (Yearbook of International Organizations, vol. 3).

Union of International Associations Encyclopedia of World Problems and Human Potential (1986)
München, K G Saur, 1440p, 2nd ed. Previous edition entitled: Yearbook of World Problems and Human Potential.

Union of International Associations Yearbook of International Organizations (1990)
München, K G Saur, 27th ed, 3 vols.

Union of International Associations / Mankind 2000 Yearbook of World Problems and Human Potential (1976)
Brussels, K G Saur.

United Nations Thirty-Ninth Session supplement: report of the ad hoc committee on the drafting of an international convention against the recruitment, use, financing and training of mercenaries
New York, UN. Official Record Ser: 43.

United Nations Satellite Warfare: a challenge for the international community
Geneva, UN. ISBN 92-9045-018-5.

United Nations Industrialization of Developing Countries: problems and prospects
New York, UN.

United Nations First World-wide Study on the Food Processing Industrie
New York, UN.

United Nations Levels and Trends of Fertility Throughout the World, 1950-1970
New York, UN.

United Nations Transnational Corporations in South Africa and Namibia: United Nations public hearings
New York, UN, 242 p. bibl.

United Nations Estimated World Requirements of Narcotic Drugs
New York, UN.

United Nations Banned Products: consolidated list of products whose consumption and/or sale have been banned, withdrawn, severely restricted and not approved by governments
New York, UN, 466 p. ISBN 92-1-130120-3.

United Nations Airborne Sulphur Pollution Effects and Control
New York, UN.

United Nations Air Pollution Across Boundaries
New York, UN, 157 p.

United Nations Disarmement: problems related to outer space
Geneva, UN. ISBN 92-9045-023-1.

United Nations World Population Trends and Policies - 1987 Monitoring Report - Special Topics: fertility and women's life cycle and socio-economic differentials in mortality
Geneva, UN.
ISBN 92-1-151168-2. Population Studies: No 13.

United Nations Sources, Effects and Risks of Ionizing Radiation: United Nations Scientific Committee on the Effects of Atomic Radiation, 1988 report to the General Assembly, with annexes
Geneva, UN. ISBN 92-1-142143-8.

United Nations Materials on Jurisdictional Immunities of States and Their Property
New York, UN. ISBN 92-1-033005-6.

United Nations Proceedings of the ESCAP-FAO-UNEP Expert Group on Fuelwood and Charcoal
New York, UN, 120 p. ISBN 92-1-119178-5.

United Nations Groundwater Problems in Coastal Areas
Lanham MD, UNIPUB, 596 p. ISBN 92-3-102415-9.

United Nations Emerging Diseases of Animals (1968)
Geneva, UN. Agricultural Planning Studies: No 61.
ISBN 0-685-09378-6.

United Nations Environmental Contamination by Radioactive Materials (1969)
Lanham MD, UNIPUB, 736 p. ISBN 92-0-020169-5.

United Nations The Capacity Study of the United Nations Development System (1969)
New York, UN.

United Nations Cultural Rights as Human Rights (1970)
Lanham MD, UNIPUB, 125 p. 3rd ed 1977.
ISBN 92-3-100846-3. Studies and Documents on Cultural Policies.

United Nations Disposal of Radioactive Wastes into Rivers, Lakes and Estuaries (1971)
Lanham MD, UNIPUB, 77 p. ISBN 92-0-123171-7.

United Nations Management of Low and Intermediate Level Radioactive Wastes (1971)
Lanham MD, UNIPUB, 814 p. ISBN 92-0-020170-9.

United Nations Mercury Contamination in Man and His Environment (1973)
Lanham MD, UNIPUB, 181 p. ISBN 92-0-115172-1.

United Nations Aspects of the Introduction of Food Irradiation in Developing Countries (1974)
Geneva, UN, 113 p. illus. ISBN 92-0-111673-X.

United Nations Management of International Water Resources (1975)
New York, UN. ISBN 92-1-104078-7.

United Nations Environmental Effects of Cooling Systems at Nuclear Power Plants: proceedings (1975)
Lanham MD, UNIPUB, 832 p. ISBN 92-0-020075-3.

United Nations International Conventions on Civil Liability for Nuclear Damage (1976)
Geneva, UN, 261 p. Legal Ser: No 4.
ISBN 92-0-076076-7.

United Nations Biological and Environmental Effects of Low-Level Radiation: proceedings of a symposium on biological (1976)
Lanham MD, UNIPUB, 370 p. 2 Vols.
ISBN 92-0-010076-7.

United Nations Impact of Nuclear Releases into the Aquatic Environment (1976)
Lanham MD, UNIPUB, 521 p. ISBN 92-0-020375-2.

United Nations Radiation for a Clean Environment: proceedings (1976)
Lanham MD, UNIPUB, 672 p. ISBN 92-0-060075-1.

United Nations The Aging in Slums and Uncontrolled Settlements (1977)
New York, UN.
ISBN 92-1-130028-2.

United Nations Collective Bargaining in Industrialized Countries: recent trends and problems (1978)
Lanham MD, UNIPUB, 113 p. ISBN 92-2-101943-8.

United Nations The International Bill of Human Rights (1978)
New York, UN, 42 p.

United Nations International Acceptance of Irradiated Food: legal aspects (1979)
Geneva, UN, 70 p. Legal Ser: No 11.
ISBN 92-0-176079-5.

United Nations Behaviour of Tritium in the Environment (1979)
Lanham MD, UNIPUB, 711 p. ISBN 92-0-020079-6.

United Nations Problems Associated with Export of Nuclear Power Plants (1979)
Lanham MD, UNIPUB, 484 p. ISBN 92-0-020178-4.

United Nations Protein-Energy Requirements Under Conditions Prevailing in Developing Countries: current knowledge and research needs (1979)
Lanham MD, UNIPUB, 73 p. ISBN 92-808-0018-3.

United Nations Efficiency and Distributional Equity in the Use and Treatment of Water: guidelines for pricing and regulations (1980)
New York, UN. ISBN 92-1-104084-1.

United Nations Aquifer Contamination and Protection (1980)
Lanham MD, UNIPUB, 442 p. ISBN 92-3-101886-8.

United Nations Fish Feed Technology: lectures presented at the fao-undp training (1980)
Lanham MD, UNIPUB, 400 p. ISBN 92-5-100901-5.

United Nations Trypanotolerant Livestock in West and Central Africa (1980)
Lanham MD, UNIPUB. ISBN 92-5-100978-3.

United Nations The Hunger Problematique and a Critique of Research (1981)
Lanham MD, UNIPUB, 72 p. ISBN 92-808-0163-5.

United Nations Integration of Disabled Persons into Community (1981)
New York, UN.
ISBN 92-1-130075-4.

United Nations Current Nuclear Power Plant Safety Issues (1981)
Lanham MD, UNIPUB, 518 p. Vol. 13 Vols.
ISBN 92-0-020181-4.

United Nations Management of Gaseous Wastes from Nuclear Facilities (1981)
Lanham MD, UNIPUB, 701 p. ISBN 92-0-020380-9.

United Nations Report of the World Assembly on Aging (1982)
New York, UN. ISBN 92-1-100200-1.

United Nations Cooperation Admits Uncertainty: priorities for international and south-southlines for pricing and regulations action views and recommandations of the Committee for Development Planning (1982)
New York, UN. ISBN 92-1-104131-7.

United Nations Main Issues in Transport for Developing Countries During the Third United Nations Development Decade (1981-1990) (1982)
New York, UN. ISBN 92-1-104126-0.

United Nations Technologies for Coastal Erosion Control (1982)
New York, UN. ISBN 92-1-104130-9.

United Nations Ionizing Radiation: sources and biological effects (1982)
New York, UN, 773 p.

United Nations Exploitation of Child Labour (1982)
New York, UN.
ISBN 92-1-154023-2.

United Nations From Peasant Girls to Bangkok Masseuses (1982)
Lanham MD, UNIPUB. ISBN 92-2-103013-X.

United Nations Apartheid, Poverty and Malnutrition (1982)
Lanham MD, UNIPUB, 106 p. ISBN 92-5-101203-2.

United Nations Conservation and Development of Tropical Forest Resources: based on an fao-unep-unesco expert meeting on (1982)
Lanham MD, UNIPUB, 131 p. ISBN 92-5-101267-9.

United Nations Cultural Industries: a challenge for the future of culture (1982)
Lanham MD, UNIPUB, 236 p. ISBN 92-3-102003-X.

United Nations Issues and Priorities in Public Administration and Finance in the Third United Nations Development Decade (1983)
New York, UN. ISBN 92-1-104131-7.

United Nations Overcoming Economic Disorder: international action for recovery and development (1983)
New York, UN. ISBN 92-1-109026-1.

United Nations The Least Developed Countries and Action in their Favour by the International Community (1983)
New York, UN. ISBN 92-1-100195-1.

United Nations Economic and Social Consequences of the Arms Race and of Expenditures (1983)
New York, UN.
ISBN 92-1-142040-7.

United Nations National Laws and Regulations Relating to the Control of Narcotic Drugs Psychotropic Substances (1983)
New York, UN. ISBN 92-1-148058-2.

United Nations Cassava Toxicity and Thyroid: research and public health issues (1983)
Lanham MD, UNIPUB, 147 p. ISBN 0-88936-368-4.

United Nations Guidelines on Technology Issues in the Pharmaceutical Sector in the Developing Countries (1983)
New York, UN, 68 p. ISBN 92-1-112138-8.

United Nations Occupational Safety and Health in the Iron and Steel Industry (1983)
Lanham MD, UNIPUB, 341 p. ISBN 92-2-103471-2.

United Nations Human Rights in Urban Areas (1983)
Lanham MD, UNIPUB, 169 p. ISBN 92-3-101983-X.

United Nations Racism, Science and Pseudo-Science (1983)
Lanham MD, UNIPUB, 158 p. ISBN 92-3-101993-7.

United Nations Crisis of Reform: breaking the barriers of development (1984)
New York, UN. ISBN 92-1-109042-3.

United Nations The United Nations and Human Rights (1984)
New York, UN. ISBN 92-1-100256-7.

United Nations The New World Information and Communication Order: a selective bibliography (1984)
New York, UN. ISBN 92-1-100035-8.

United Nations World Recovery and Monetary Reform (1984)
New York, UN. ISBN 92-1-104151-1

United Nations A Bibliographic Guide to Studies on the Status of Women: development and population trends (1984)
Lanham MD, UNIPUB, 292 p. ISBN 92-3-102122-2.

United Nations World Cartography (1985)
New York, UN. Issued irregularly since 1951.
ISBN 92-1-1002284-2.

United Nations Challenge to Multilateralism: a time for renewal (1985)
New York, UN. ISBN 92-1-109109-8.

United Nations The Debt Problem: acute and chronic aspects (1985)
New York, UNCTC. ISBN 92-1-104199-1.

United Nations Savings for Development: report of the Third Symposium on the Mobilization of Personal Savings in Developing Countries (1985)
New York, UN. ISBN 92-1-104172-4.

United Nations Report of the FAO-UNEP Meeting on Toxity and Bioaccumulation of Selected Substances in Marine Organisms: Rovinj, Yugoslavia, 5-9 November 1984 (1985)
Lanham MD, UNIPUB, 22 p.
ISBN 92-5-102278-X. Fisheries Report: No 334.

United Nations Conscientious objection to military service (1985)
New York, UN, 32 p. ISBN 92-1-154053-4.

United Nations Report on the World Social Situation (1985)
New York, UN. Issued quadrenially since 1961.
ISBN 92-1-130097-5.

United Nations The World Aging Situation: strategies and policies (1985)
New York, UN.
ISBN 92-1-130100-9.

United Nations The Law of the Sea: pollution by dumping (1985)
New York, UN.
ISBN 92-1-133274-5.

United Nations Socio-Economic Differentials in Child Mortality in Developing Countries (1985)
New York, UN.
ISBN 92-1-151154-2.

United Nations Transport of Dangerous Goods/ Recommandations on the Transport of Dangerous Goods: tests and criteria (1985)
New York, UN, 189 p. This Vol contains specific technical guidelines on the transport of dangerous explosive substances and organic peroxides. ISBN 92-1-139021-4.

United Nations Extradition for Drug-Related Offences (1985)
New York, UN.
ISBN 92-1-148069-8.

United Nations Principles, Guidelines and Guarantees for the Protection of Persons Detained on Grounds of Mental Ill-Health or Suffering from Mental Disorder (1985)
New York, UN, 38 p.
ISBN 92-1-154056-9.

United Nations European Agreement Concerning the International Carriage of Dangerous Goods by Road (ADR) and Protocol of Signature: 1985 (1985)
New York, UN, 210 p. Vol. I3 vols. ISBN 92-1-139017-6.

United Nations Risks and Benefits of Energy Systems: proceedings of a symposium, Julrich, 9-13 april (1985)
Lanham MD, UNIPUB, 671 p. ISBN 92-0-020784-7.

United Nations World Labour Report: labour relations, international labour (1985)
Lanham MD, UNIPUB, 245 p. No. 2 ISBN 92-2-103848-3.

United Nations The Determination of Polychlorinated Biphenyls in Open Ocean Waters (1985)
Lanham MD, UNIPUB, 48 p. ISBN 92-3-102262-8.

United Nations Land, Food and People (1985)
Lanham MD, UNIPUB, 96 p. ISBN 92-5-102079-5.

United Nations World Concerns and The United Nations (1986)
New York, UN. Revised ed. This book provides model teaching units for primary, secondary and teacher education that are directly usable in the classroom. ISBN 92-1-100292-3.

United Nations Doubling Development Finance: meeting a global challenge (1986)
New York, UN. 47 p. ISBN 92-1-104185-6.

United Nations United Nations Conferences on the Standardization of Geographical Names (1986)
New York, UN. Includes reports and technical papers submitted to the conferences, covering areas such as national standardization, training courses, terminilogy, cooperative arrangements for the naming of features beyond a single sovereignty, and international cooperation measures.
ISBN 92-1-100304-0.

United Nations Microbial Technologies to Overcome Environmental Problems of Persistent Pollutants (1986)
New York, UN, 132 p. ISBN 92-807-1110-5.

United Nations Assessment of Radioactive Contamination in Man 1984: proceedings of an international symposium on the assessment of radioactive contamination in man organized by the International Atomic Energy Agency in Co-operation with the World Health Organization and held in Paris, 19-23 November 1984 (1986)
Lanham MD, UNIPUB, 583 p.
ISBN 92-0-020085-0. Proceedings Ser.

United Nations Guidelines for Consumer Protection (1986)
New York, UN. ISBN 92-1-030035-1.

United Nations Economic Recession and Specific Population Groups (1986)
New York, UN, 99 p.
ISBN 92-1-130110-6.

United Nations Developmental Social Welfare: a global survey of issues and priorities since 1968 (1986)
New York, UN, 57 p.
ISBN 92-1-130112-2.

United Nations Disability: Situation, Strategies and Policies: United Nations Decade of Disabled Persons, 1983-1992 (1986)
New York, UN, 57 p. ISBN 92-1-130116-1.

United Nations The Social Situation of Migrant Workers and Their Families (1986)
New York, UN, 63 p.
ISBN 92-1-130118-1.

United Nations Classification by Broad Economic Categories: defined in terms of Standard International Trade Classification (SITC) (1986)
New York, UN, 77 p. ISBN 92-1-161282-9.

United Nations The Naval Arms Race (1986)
New York, UN.
ISBN 92-1-142116-0.

United Nations Genetic and Somatic Effects of Ionizing Radiation (1986)
New York, UN, 366 p.
ISBN 92-1-142123-3.

United Nations Manual for the Development of Criminal Justice Statistics (1986)
New York, UN, 99 p.
ISBN 92-1-161269-1.

United Nations Standard International Trade Classification (1986)
New York, UN, 169 p.
ISBN 92-1-161265-9.

United Nations Exploitation of Labour Through Illicit and Clandestine Trafficking (1986)
New York, UN, 192 p.
ISBN 92-1-154058-5.

United Nations An Oceanographic Model for the Dispersion of Wastes Disposed of in the Deep Sea (1986)
Lanham MD, UNIPUB, 166 p. ISBN 92-0-125186-6.

United Nations Report of the International Narcotics Control Board for 1985: demand and supply of opiates for medical and (1986)
New York, UN, 40 p. ISBN 92-1-148072-8.

United Nations Women, Work and Demographic Issues: report of an ILO-UNITAR Seminar, Tashkent, UK (1986)
Lanham MD, UNIPUB, 159 p. ISBN 92-2-103886-6.

United Nations Femmes Au Pays (1986)
Lanham MD, UNIPUB, 187 p. ISBN 92-3-202177-3.

United Nations Radiation Doses, Effects, Risks (1986)
New York, UN, 64 p.

United Nations The United Nations and Drug Control: active cooperation in the struggle against drug abuse and illicit trafficking (1987)
New York, UN. ISBN 92-1-100314-8.

United Nations Study on Deterrence: its implications for disarmament and the arms race, negotiated arms reductions and international security and other related matters (1987)
New York, UN, 142 p. Study Ser: No 17.
ISBN 92-1-142127-6.

United Nations Advisory Material for the IAEA Regulations for the Safe Transport of Radioactive Material (1987)
Geneva, UN, 204 p. 3rd ed. ISBN 92-0-123487-2.

United Nations Coping with Drop-Out: a handbook (1987)
Lanham MD, UNIPUB, 34 p. ISBN 0-317-59390-0.

United Nations Consolidated List of Products Whose Consumption and/or Sale Have Been Banned, Withdrawn, Severely Restricted or Not Approved by Governments (1987)
New York, UN. Pharmaceuticals, agricultural chemicals, industrial chemicals and consumer products – what is one country's poison may be for sale in an other country.
ISBN 92-1-130120-3.

United Nations Land-Locked States (1987)
New York, UN.
ISBN 92-1-133292-3.

United Nations International Cooperation in Tax Matters: guidelines for International Cooperation Against the Evasion and Avoidance to Taxes (with Special Reference to Taxes on Income, Profits, Capital and Capital Gains) (1987)
New York, UN. ISBN 92-1-159036-1.

United Nations Demographic Development and Social Security (1987)
Lanham MD, UNIPUB, 100 p. ISBN 92-2-105992-8.

United Nations Effects and Control of Transboundary Air Pollution (1987)
New York, UN, 133 p. ISBN 92-1-116410-9.

United Nations Packaging and Transportation of Radioactive Materials-Patram 1986 (1987)
Lanham MD, UNIPUB, 634 p. Vol 1. ISBN 92-0-020087-7.

United Nations Radiotherapy in Developing Countries (1987)
Lanham MD, UNIPUB, 455 p. ISBN 92-0-010087-2.

United Nations Recommendations on the Transport of Dangerous Goods (1987)
New York, UN, 504 p. ISBN 92-1-139023-0.

United Nations Report of the International Conference on Drug Abuse and Illicit Trafficking, Vienna, 17-26 June 1987 (1987)
New York, UN, 143 p. ISBN 92-1-100320-2.

United Nations Climate Crisis (1987)
New York, UN, 105 p. ISBN 92-807-1169-5.

United Nations IRPTC Legal File, 1986 (1987)
New York, UN. Vols I and II. ISBN 92-807-1163-6.

United Nations Sanctions Against South Africa: the peaceful alternative to violent change (1988)
New York, UN, 201 p. bibl.

United Nations Transnational Corporations – A Selective Bibliography 1983-1987 (1988)
Geneva, UN. 2 Vols.

United Nations Emergency Response Planning and Preparedness for Transport Accidents Involving Radioactive Material (1988)
Lanham MD, UNIPUB, 103 p. ISBN 92-0-123088-5.

United Nations Pesticides: food and environmental implications (1988)
Lanham MD, UNIPUB, 331 p. ISBN 92-0-010288-3.

United Nations UNCTC Bibliography 1974-1987 (1988)
New York, UN, 83 p. ISBN 92-1-104218-6.

United Nations Mortality of Children Under Age 5: world estimates and projections, 1950-2025 (1989)
Geneva, UN.
ISBN 92-1-151169-0. Population Studies: No 105.

United Nations Centre on Transnational Corporations Bibliography on Transnational Corporations (1978)
New York, UNCTC. ISBN 92-1-004000-7.

United Nations Centre on Transnational Corporations Transnational Corporations in Advertising: a technical paper (1979)
New York, UNCTC. ISBN 92-1-104019-1.

United Nations Centre on Transnational Corporations Measures Strenghtening the Negociating Capacity of Governments in their Relations with Transnational Corporations: technology Transfer through Transnational Corporations (1979)
New York, UNCTC. ISBN 92-1-104158-9.

United Nations Centre on Transnational Corporations Transnational Reinsurance Operations: a technical paper (1980)
New York, UNCTC. ISBN 92-1-104034-5.

United Nations Centre on Transnational Corporations Transnational Corporations in the Copper Industry (1981)
New York, UNCTC. ISBN 92-1-104033-7.

United Nations Centre on Transnational Corporations Transnational Corporations Linkages in Developing Countries The case of Backward Linkages via Subcontracting (1981)
New York, UNCTC. ISBN 92-1-104032-9.

United Nations Centre on Transnational Corporations Transnational Corporations Linkages in the Bauxite/Aluminium Industry (1981)
New York, UNCTC. ISBN 92-1-104028-0.

United Nations Centre on Transnational Corporations Transnational Corporations in Food and Beverage Processing (1981)
New York, UNCTC. ISBN 92-1-104029-9.

United Nations Centre on Transnational Corporations Towards International Standardization of Corporate Accounting and Reporting (1982)
New York, UNCTC. ISBN 92-1-104115-5.

United Nations Centre on Transnational Corporations Transnational Corporations and Transborder Data Flows: a technical paper (1982)
New York, UNCTC. ISBN 92-1-104030-2.

United Nations Centre on Transnational Corporations Measures Strengthening the Negociation Capacity of Governments in their Relations with Transnational Corporations: regional Integration Cum/Versus Corporations Integration: A Technical Paper (1982)
New York, UNCTC. ISBN 92-1-104224-0.

United Nations Centre on Transnational Corporations Transnational Corporations in International Tourism (1982)
New York, UNCTC. ISBN 92-1-104041-8.

United Nations Centre on Transnational Corporations Transnational Corporations in the Fertilizer Industry (1982)
New York, UNCTC. ISBN 92-1-104047-1.

United Nations Centre on Transnational Corporations Transnational Corporations in the Power Equipment Industry (1982)
New York, UNCTC. ISBN 92-1-104046-9.

United Nations Centre on Transnational Corporations Management Contracts in Developing Countries: an Analysis of their Substantive Provisions (1982)
New York, UNCTC. ISBN 92-1-104125-2.

United Nations Centre on Transnational Corporations Transborder Data Flows: access to the International On-Line Data-Base Market Reviews the on-line database industry in terms of costs, products and relevance for developing countries (1983)
New York, UNCTC. ISBN 92-1-104043-6.

United Nations Centre on Transnational Corporations Transnational Corporations in the Agricultural Machinery and Equipment Industry (1983)
New York, UNCTC. ISBN 92-1-104039-6.

United Nations Centre on Transnational Corporations Transnational Corporations in the International Auto Industry (1983)
New York, UNCTC. ISBN 92-1-104038-8.

United Nations Centre on Transnational Corporations Salient Features and Trends in Foreign Direct Investment (1983)
New York, UNCTC. ISBN 92-1-104058-2.

United Nations Centre on Transnational Corporations Main Features and Trends in Petroleum and Mining Agreements (1983)
New York, UNCTC. ISBN 92-1-104057-4

United Nations Centre on Transnational Corporations Transnational Corporations and Contractual Relations in the World Uranium Industry: a technical paper (1983)
New York, UNCTC. ISBN 92-1-104049-5.

United Nations Centre on Transnational Corporations Issues in Negociating International Loan Agreements with Transnational Banks (1983)
New York, UNCTC. ISBN 92-1-104048-5.

United Nations Centre on Transnational Corporations International Standards of Accounting and Reporting: report of the Ad-Hoc Intergovernmental Working Group of Experts on International Standards of Accounting and Reporting (1984)
New York, UNCTC. ISBN 92-1-104228-3.

United Nations Centre on Transnational Corporations Transnational Corporations in the Pharmaceutical Industry of Developing Countries (1984)
New York, UNCTC. ISBN 92-1-104155-4.

United Nations Centre on Transnational Corporations Transborder Data Flows: Transnational Corporations and Remote Sensing Data: A Technical Paper (1984)
New York, UNCTC. ISBN 92-1-104056-2.

United Nations Centre on Transnational Corporations Transnational Corporations and International Trade: selected Issues (1985)
New York, UNCTC, 93 p. ISBN 92-1-104232-1.

United Nations Centre on Transnational Corporations Environmental Aspects of the Activities of Transnational Corporations (1985)
New York, UNCTC, 114 p. ISBN 92-1-104165-1.

United Nations Centre on Transnational Corporations Trends and Issues in Foreign Direct Investment and Related Flows (1985)
New York, UNCTC. ISBN 92-1-104170-8.

United Nations Centre on Transnational Corporations Transnational Corporations in the International Semiconductor Industry (1986)
New York, UNCTC, 471 p. ISBN 92-1-104174-0.

United Nations Centre on Transnational Corporations Analysis of Engineering and Technical Assistance Consultancy Contracts (1986)
New York, UNCTC. ISBN 92-1-104177-5.

United Nations Centre on Transnational Corporations International Accounting and Reporting Issues: 1986 Review (1986)
New York, UNCTC. ISBN 92-1-104192-9.

United Nations Centre on Transnational Corporations Transnational Corporations and Technology Transfer: effects and Policy Issues (1987)
New York, UNCTC. ISBN 92-1-104199-1.

United Nations Conference on Trade and Development Commodity Export Earnings Shortfalls, Existing Financial Mechanisms and the Effects of Shortfalls on the Economic Development of Developing Countries (1987)
Geneva, UNCTAD.

United Nations Conference on Trade and Development Characterization and In-dept Analysis of the Origins of the Current Crisis of the Market and of the Primary Tungsten and Intermediate Products Industry (1987)
Geneva, UNCTAD.

United Nations Conference on Trade and Development Problems of Protectionism and Structural Adjustment, Introduction and Part 1: restrictions on trade (1987)
Geneva, UNCTAD.

United Nations Conference on Trade and Development Ocean Freight Rates and Their Effects on Exports of Developing Countries (1987)
Geneva, UNCTAD.

United Nations Economic Commission for Europe Regulations and Legislation on Food Additives and Chemicals for Food Packaging (1984)
Geneva, ECE. ISBN 92-1-116326-9

United Nations Economic Commission for Europe Impact of Air-pollution Damage to Forests for Roundwood Supply and Forest Products Markets (1987)
Geneva, UN.

United Nations Economic Commission for Europe New and Renewable Sources of Energy: the course not yet taken (1987)
Geneva, UN.

United Nations Economic Commission for Latin External Debt in Latin America: adjustment policies and renegotiation (1985)
Boulder CO, Lynne Rienner Publishers, 125 p.
ISBN 0-931477-40-9.

United Nations Educational, Scientific and Cultural Organization Violence and Its Causes: methodological and theoretical aspects of recent research on violence (1980)
Paris, UNESCO, 269 p. ISBN 92-3-101809-4.

United Nations Environment Programme Environmental Aspects of the Aluminium Smelting: a technical review (1981)
Nairobi, UNEP, 166 p.

United Nations Environment Programme Environmental Management in the Pulp and Paper Industry (1981)
Nairobi, UNEP, 1202 p. Industry and Environment Manual Series: No 1, Vols 1, 2.

United Nations Environment Programme Environmental Aspects of the Iron and Steel Industry: workshop proceedings, Geneva, 17-20 October 1978 (1983)
Nairobi, UNEP. 465 p. Vol 1, Part 1 and 2.

United Nations Environment Programme Comparative Data on the Emissions, Residuals and Health Hazards of Energy Sources: phase I (1985)
Nairobi, UNEP, 149 p. Energy Report Series, ERS 14-85.

United Nations Environment Programme Environmental Refugees (1985)
New York, UNEP.
ISBN 92-807-1103-2.

United Nations Environment Programme Radiation: doses, effects, risks (1986)
New York, UNEP.
ISBN 92-807-1104-0.

United Nations Environment Programme The Ozone Layer (1987)
Nairobi, UNEP, 335 p.

United Nations Environment Programme The Greenhouse Gases (1987)
Nairobi, UNEP, 40 p.

United Nations Environment Programme Environmental Data Report (1989)
Cambridge, Basil Blackwell, 547 p. illus. 2nd ed. Prepared for UNEP by the GEMS Monitoring and Assessment Research Centre, London, in cooperation with the World Resources Institute, Washington DC and the UK Department of the Environment, London. ISBN 0956-9324.

United Nations, Geneva Water Hyacinth. Proceedings of the International Conference on Water Hyacinth (1984)
Geneva, UN, 1026 p. ISBN 92-807-1090-6

United Nations, Geneva Is universality in Jeopardy ? (1986)
Geneva, UN. ISBN 92-1-100334-2

United Nations Industrial Development Organization The Growth of the Leather Industry in Developing Countries: problems and prospects
Vienna, UNIDO.

United Nations Industrial Development Organization The Growth of the Pharmaceutical Industry in Developing Countries: problems and prospects
Vienna, UNIDO.

United Nations Institute for Disarmament Research Disarmament (1984)
New York, UN, 124 p. A short guide to United Nations and other sources of informations. ISBN 92-9045-008-8

United Nations Institute for Disarmament Research Prevention of the Arms Race in Outer Space: international law aspect (1986)
New York, UN, 26 p. ISBN 92-9045-012-6

United Nations Institute for Training and Research Relations Between the UN and Non-UN Regional Inter-Governmental Organizations (1975)
New York, UNITAR. ISBN 92-1-157016-6.

United Nations Institute for Training and Research Towards Greater Order, Coherence and Coordination in the United Nations System (1975)
New York, UNITAR. ISBN 92-1-157043-3.

United Nations Institute for Training and Research Crowded Agendas, Crowded Rooms, Institutional Arrangements at UNCLOS III: some lessons in global negociations (1981)
New York, UNITAR. ISBN 92-1-157033-6.

United Nations Institute for Training and Research Racism and Its Elimination (1981)
New York, UNITAR.
ISBN 92-1-157057-3.

United Nations Institute for Training and Research Assessing the UN Scale of Assessments: is it fair? is it aquitable? (1982)
New York, UNITAR.
ISBN 92-1-157017-4.

United Nations Institute for Training and Research Prevention of Nuclear War: a United Nations perspective (1984)
New York, UNITAR.
ISBN 0-89946-184-0.

United Nations Institute for Training and Research Cooperation in the 1980s: principles and prospects (1986)
New York, UNITAR.
ISBN 0-907567-73-8.

United Nations Institute for Training and Research The United Nations and Collective Management of International Conflict (1986)
New York, UNITAR, 73 p. ISBN 92-1-157092-1.

United Nations Statistical Office A Global Review of Human Settlements and Statistical Annex (1976)
Elmsford NY, Pergamon Books. 2 vols.
ISBN 0-08-021045-7.

Vaeth, J M and Meyer, J (Eds) Cancer and the Elderly (1986)
Basel, S Karger AG, 202 p. Frontiers of Radiation Therapy and Oncology Ser: Vol 20. ISBN 3-8055-4145-7.

Vahrmeijer, J Poisonous Plants of Southern Africa that Cause Stock Losses
Cape Town, Tafelberg Publishers, 168 p.
ISBN 0-624-01459-2.

Vaillant, George E Empirical Studies of Ego Mechanisms of Defense (1986)
New York, American Psychiatric Association, 176 p. Clinical Insights Monograph. ISBN 0-88048-131-5.

Vaina, Lucia and Hintikka, J (Eds) Cognitive Constraints on Communication: representations and processes (1983)
Dordrecht, Kluwer Academic Publishers Group, 456 p.
ISBN 90-277-1456-8.

Valentine, G H The Chromosomes and Their Disorders: an introduction for clinicians (1986)
Dobbs Ferry NY, Sheridan Medical Books, 228 p.
ISBN 0-433-33603-X.

Vallee, Jacques UFO's in Space: anatomy of a phenomenon (1987)
New York, Ballantine Books, 304 p. ISBN 0-345-34437-5.

Vallee, Jacques Dimensions: a casebook of alien contact (1988)
London, Sphere Books Ltd, 304 p. ISBN 0-7474-0529-8.

Valorian Society Human History Viewed As Sovereign Individuals vs Manipulated Masses (1986)
Rochester WA, Sovereign Press, 112 p.
ISBN 0-914752-23-5.

Van Cleve, John V Gallaudet Encyclopedia of Deaf People and Deafness (1987)
New York, McGraw Hill Book Company, 1440 p. 3 vols.
ISBN 0-07-079229-1.

Van de Vate, Dwight Romantic Love: a philosophical inquiry (1981)
University Park PA, Pennsylvania State University Press, 176 p. ISBN 0-271-00288-3.

Van den Haag, Ernest and Conrad, John P The Death Penalty: a debate (1983)
New York, Plenum Publishing, 320 p.
ISBN 0-306-41416-3.

Van der Hoeven R, Richards Peter J (Eds) World Recession and Global Interdependence: effects on employment, poverty and policy formation in developing countries (1987)
Geneva, ILO, 139 p. ISBN 92-2-105609-0.

Van Der Plank, J E Plant Diseases: epidemics and control (1964)
San Diego CA, Academic Press. ISBN 0-12-711450-5.

Van Dijk, J Rubella Handicapped Children: the effects of bi-lateral cataract and or hearing (1982)
Lewiston NY, Hogrefe International, 254 p.
ISBN 90-265-0432-2.

Van Dijk, Teun A Communicating Racism: ethnic prejudice in thought and talk (1987)
Newbury Park CA, Sage Publications, 376 p.
ISBN 0-8039-2674-X.

Van Driel, G J; Hartog, J A and Van Ravenzwaaij, C Limits to the Welfare State (1980)
Dordrecht, Kluwer Academic Publishers Group, 192 p.
ISBN 0-89838-026-X.

Van Keep, Pieter A and Utian, Wulf H (Eds) The Controversial Climateric (1982)
Dordrecht, Kluwer Academic Publishers Group, 200 p. illus.
ISBN 0-85200-410-9.

Van Rens, and Kayser, F H Local Antibiotic Treatment in Osteomyelitis and Soft Tissue Infections (1981)
New York, Elsevier Science Publishing, 196 p.
ISBN 0-444-90222-8.

Vandenberg, Steven G, et al The Heredity of Behavior Disorders in Adults and Children (1986)
New York, Plenum Publishing, 316 p.
ISBN 0-306-42191-7.

Varady, David P Neighborhood Upgrading: a realistic assessment (1986)
Albany NY, State University of New York Press, 184 p.
ISBN 0-88706-299-7.

Varma, Ved P Anxiety in Children (1984)
New York, Routledge Chapman and Hall, 228 p.
ISBN 0-416-01031-8.

Vaubel, Roland Strategies for Currency Unification: the economics of currency competition and the case for a european parallel currency (1978)
Tübingen, JCB Mohr, 471 p. ISBN 3-16-340571-1.

Vaughn, Trudy W Leg Injuries: medical subject analysis and research guidebook with bibliography (1987)
Washington DC, ABBE Publications Association, 150 p. bibl.
ISBN 0-88164-292-4.

Vedros, Neylan A (Ed) Evolution of Meningococcal Disease (1987)
Boca Raton FL, CRC Press. 2 Vols. ISBN 0-8493-4643-6.

Veith, Ilza Hysteria: the history of a disease (1970)
Chicago IL, University of Chicago Press. illus.
ISBN 0-226-85253-9.

Velimirovic, B Infectious Diseases in Europe: a fresh look (1984)
Geneva, WHO, 330 p. ISBN 92-890-1015-0.

Venkatasubramanian, K Wastage in Primary Education (1978)
Hyderabad, Orient Longman, 112 p. ISBN 0-86131-098-5.

Vergers, Charles A Handbook of Electrical Noise: measurement and technology (1987)
Blue Ridge Summit PA, TAB Books, 496 p. illus. 2nd ed.
ISBN 0-8306-2802-9.

Veronesi, Tetanus: important new concepts (1981)
New York, Elsevier Science Publishing, 284 p.
ISBN 0-444-90203-1.

Veronesi, Umberto, et al Cutaneous Melanoma: status of knowledge and future perspectives (1987)
San Diego CA, Academic Press, 709 p.
ISBN 0-12-718855-X.

Verstraete, M and Vermylen, J Thrombosis (1984)
Elmsford NY, Pergamon Books, 328 p.
ISBN 0-08-030799-X.

Vesenjak-Hirjan, J; Portersfield, J S and Arslanagic, E (Eds) Arboviruses in the Mediterranean Countries (1980)
Stuttgart, Fischer, 300 p. ISBN 3-437-10622-8.

Vickers, John and Yarrow, George Privatization: economic analysis (1988)
Cambridge MA, MIT Press.

Vierdag, E W Concept of Discrimination in International Law (1973)
Dordrecht, Kluwer Academic Publishers Group, 176 p.
ISBN 90-247-1525-3.

Vig, Baldev K and Sandberg, Avery A (Eds) Aneuploidy, Part A (1987)
New York, Alan R Liss, 448 p. Progress and Topics in Cytogenetics Ser: Vol 7A.

Vikler, Henry Assertiveness (1988)
Grand Rapids MI, Zondervan Publishing House, 368 p.
ISBN 0-310-39471-6.

Vincent-Daviss, Diana (Ed) Bibliography of Human Rights: a collection of bibliographies and research resources, 2 binders
New York, Oceana. bibl. ISBN 0-379-20909-8.

Vingerhoets, A Psychosocial Stress (1985)
Lisse, Swets en Zeitlinger, 160 p.

Visher, Emily B and Visher, John S Stepfamilies: myths and realities (1980)
Secaucus NJ, Stuart Lyle. ISBN 0-8065-0743-8.

Viski, L Road Traffic Offenders and Crime Policy (1982)
Budapest, Akadémiai Kiadó és Nyomada, 204 p.
ISBN 963-05-2841-X.

Visser, J and Chapman, D S Snakes and Snakebites: venomous snakes and management of snakes in Southern Africa
Cape Town, C Struik Publishers, 152 p.
ISBN 0-86977-218-X.

Vladutiu, Adrian O Pleural Effusion (1986)
Mount Kisco NY, Futura Publishing, 440 p.
ISBN 0-87993-285-6.

Volk, Bruno W and Arquilla, Edward R The Diabetic Pancreas (1985)
New York, Plenum Publishing, 652 p.
ISBN 0-306-41781-2.

Von Furer-Haimendorf, Christoph (Ed) Caste and Kin in Nepal, India and Ceylon: anthropological studies in Hindu-Buddhist contact zones (1982)
Calcutta, Asia Publishing Company, 364 p.
ISBN 0-85692-019-3.

Von Hirsch, Andrew Past or Future Crimes: deservedness and dangerousness in the sentencing (1985)
New Brunswick NJ, Rutgers University Press, 250 p.
ISBN 0-8135-1115-1.

Von Rosenstein, Nicholas R The Diseases of Children and their Remedies
Washington DC, Nutrition Foundation, 364 p. Trans. from Swed.

Voort, T H A van der Televison Violence: a Child's-Eye View (1986)
Amsterdam, Elsevier Science Publishing.
ISBN 0-444-87978-1.

Vyvyan, John The Dark Face of Sciences (1971)
London, Michael Joseph.

Wade, Carlson Great Hoaxes and Famous Imposters
Flushing NY, Jonathan David Publishers.
ISBN 0-8246-0200-5.

Wade, Nicholas The Art and Science of Visual Illusions (1983)
New York, Routledge Chapman and Hall, 224 p. illus. International Library of Psychology. ISBN 0-7100-0868-6.

Wadel, Cato Now Whose Fault is That?: the struggle for self-esteem in the face of chronic unemployment (1973)
St John's, Memorial University of Newfoundland.
ISBN 0-919666-05-1.

Wadsworth, G R The Diet and Health of Isolated Populations (1984)
Boca Raton FL, CRC Press, 232 p. ISBN 0-8493-6101-X.

Wagner, G, et al (Eds) Cancer of the Liver, Esophagus and Nasopharynx (1987)
Berlin, Springer-Verlag, 220 p. illus. ISBN 0-387-16967-9.

Wagner, Gorm and Green, Richard Impotence: physiological, psychological, and surgical (1981)
New York, Plenum Publishing, 192 p.
ISBN 0-306-40719-1.

Wahlqvist, Mark, et al Use and Abuse of Vitamins: food versus pills
New York, Macmillan, 120 p. ISBN 0-7251-0540-2.

Waitakin, Howard and Waterman, Barbara The Exploitation of Illness in Capitalist Society (1974)
New York, Macmillan. ISBN 0-02-424540-2.

Walker, Charles A, et al Too Hot to Handle? (1983)
New Haven CT, Yale University Press, 240 p.
ISBN 0-300-02899-7.

Walker, K R (Ed) The Evolution-Creation Controversy Perspectives on Religion, Philosophy, Science and Education: a handbook
University of Tenn Geo, 155 p. illus.
Paleontological Society Special Publications Ser.
ISBN 0-931377-00-5.

Walker, Mabel L Urban Blight and Slums (1971)
New York, Russell and Russell Publishers.
ISBN 0-8462-1546-2.

Walker, R and Knowles, M E International Symposium on Food Toxicology: a special issue of food additives (1984)
New York, Taylor and Francis, 164 p. Vol. 1
ISBN 0-85066-996-0.

Walker-Smith, J A and McNeish, A S Diarrhea and Malnutrition in Childhood (1986)
Stoneham MA, Butterworth Publishers, 320 p.
ISBN 0-407-00401-7.

Walker-Smith, John A Diseases of the Small Intestine in Childhood (1988)
London, Butterworth, 256 p. illus. 3rd ed.
ISBN 0-407-01320-2.

Wallace, James D Virtues and Vices (1986)
Ithaca NY, Cornell University Press, 208 p.
ISBN 0-8014-9372-2.

Wallace, Joseph A Crack busters workbook: how to recognize and resist the pressure to use crack (1988)
New Brunswick NJ, North American Publications, 96 p.
ISBN 0-929105-01-X.

Wallach, Gerldine P and Butler, Katherine G Language Learning Disabilities in School-Age Children (1983)
Baltimore MD, Williams and Wilkins, 436 p.
ISBN 0-683-08707-X.

Wallerstein, I The Capitalist World Economy (1979)
London, Cambridge University Press.

Wallerstein, Immanuel (Ed) World Inequality
Montreal, Black Rose Books, 225 p. ISBN 0-919618-66-9.

Wallett, Tim Shark Attack: in Southern African waters and treatment of victims
Cape Town, C Struik Publishers, 192 p.

Wallich, Henry C The Cost of Freedom: a new look at capitalism (1979)
Westport CT, Greenwood Press. ISBN 0-313-20935-9.

Walsh, Barent W and Rosen, Paul M Self-Mutilation: theory, research, and treatment (1988)
New York, Guilford Press, 270 p. ISBN 0-89862-731-1.

Walsh, Dermot Heavy Business: commercial burglary and robbery (1986)
New York, Routledge Chapman and Hall, 240 p.
ISBN 0-7102-0668-2.

Walsh, Kenneth Strikes in Europe and the United States: measurement and incidence (1984)
New York, Saint Martin's Press, 242 p.
ISBN 0-312-76641-6.

Walsh, Marilyn E The Fence: a new look at the world of property theft (1976)
Westport CT, Greenwood Press, 215 p.
ISBN 0-8371-8910-1.

Walsh, Mary R Doctors Wanted - No Women Need Apply: sexual barriers in the medical profession (1977)
New Haven CT, Yale University Press.
ISBN 0-300-02024-4.

Walter, Ingo Secret Money: the world of international financial secrecy (1985)
Lexington MA, Lexington Books, 224 p.
ISBN 0-669-11563-0.

Walters, William and Ross, William Transsexualism and Reassignment (1986)
Oxford NY, Oxford University Press, 232 p.
ISBN 0-19-554462-5.

Walton A Handbook of Pig Diseases (1987)
Liverpool, Liverpool University Press, 172 p. 2nd ed.
ISBN 0-85323-156-7.

Walton, John Elites and Economic Development: comparative studies on the political economy of Latin American cities (1977)
University of Tex Press, 269 p.
Latin American Monographs: No 41. ISBN 0–292–72018–1.

Wanatabe, Shaw et al Digestive Disease Pathology (1988)
New York, Macmillan, 275 p. ISBN 0–02–424570–4.

Wanatabe, Shaw, et al Pathology (1988)
New York, Macmillan, 275 p. Digestive Disease: 1.
ISBN 0–02–424570–4.

Wang, Charleston C How to Manage Workplace Derived Hazards and Avoid Liability (1988)
Park Ridge NJ, Noyes Publications, 335 p.
ISBN 0–8155–1134–5.

Wang, Guichen, et al Smashing the Communal Pot: formulation and development of rural responsibility system (1985)
New York, New World Press, 250 p.

Ward, Robert Debugging (1986)
Carmel IN, Que Corp, 350 p. ISBN 0–88022–261–1.

Ward, S Alexander How to Overcome Laziness and Achieve Your Goals (1986)
North Miami Beach FL, Ward S Alexander, 151 p.
ISBN 0–939189–00–3.

Wardlaw, Grant Political Terrorism: theory, tactics and counter-measures
Cambridge MA, Cambridge University Press, 250 p. 2nd ed.
ISBN 0–521–36296–2.

Ware, G W (Ed) Reviews of Environmental Contamination and Toxicology (1987)
Berlin, Springer-Verlag, 200 p. Vol 101.
ISBN 0–387–96593–9.

Warr, Peter Work, Unemployment and Mental Health (1987)
Oxford NY, Oxford University Press, 384 p.
ISBN 0–19–852159–6.

Warren, Bennis G The Unconscious Conspiracy: why leaders can't lead
Ann Arbor MI, Books on Demand. ISBN 0–317–42056–9.

Warren, Bill Imperialism: pioneer of capitalism (1980)
London, New Left Books.

Warren, Carol Madwives: schizophrenic women in the nineteen fifties (1987)
New Brunswick NJ, Rutgers University Press, 300 p.
ISBN 0–8135–1225–5.

Washton, Arnold M Cocaine Addiction: treatment, recovery, and relapse prevention (1989)
New York, WW Norton. ISBN 0–393–70069–0.

Waterhouse, J, et al (Eds) Cancer Incidence in Five Continents (1986)
Oxford NY, Oxford University Press, 811 p. illus. IARC Scientific Publications: Vol 4, No 36. ISBN 0–19–723042–3.

Waters, W G; Heaver, Trevor D and Verrier, T Oil Pollution from Tanker Operations: causes, costs, controls
Vancouver BC, University of British Colombia.
ISBN 0–919804–17–9.

Watkin, J E (Ed) Mercury in Man's Environment (1971)
Ottawa ON, Royal Society of Canada, 210 p.

Watson, D H (Ed) Natural Toxicants in Food: modell systems, progress and prospects (1987)
New York, VCH Publications, 254 p. ISBN 3–527–26439–6.

Watson, Douglas, et al Mental Health, Substance Abuse and Deafness (1983)
Little Rock AR, American Deafness and Rehabilitation Association, 61 p. ISBN 0–914494–08–2.

Watson, Joyce M Solvent Abuse: the adolescent epidemic? (1986)
New York, Routledge Chapman and Hall, 208 p.
ISBN 0–7099–3683–4.

Watt, Kenneth E, et al The Unsteady State: environmental problems, growth, and culture (1977)
Honolulu HI, University of Hawaii Press.
ISBN 0–8248–0480–5.

Watts, Michael R The Dissenters: from the reformation to the french revolution (1986)
New York, Oxford University Press, 568 p.
ISBN 0–19–822956–9.

Watts, R W and Gibbs, D A Lysosome Storage Diseases: biochemical and clinical aspects (1986)
New York, Taylor and Francis, 284 p.
ISBN 0–85066–326–1.

Watts, Tim J Clergy Malpractice: a bibliography (1988)
Monticello IL, Vance Bibliographies, 8 p. bibl.
ISBN 1–55590–609–5.

Weart, Spencer R Nuclear Fear: a history of images (1988)
Cambridge MA, Harvard University Press, 544 p.
ISBN 0–674–62835–7.

Webb, R H and Wilshire, H G (Eds) Environmental Effects of Off-Road Vehicles: impacts and management in arid regions (1983)
Berlin, Springer-Verlag, 560 p.
ISBN 3–540–90737–8. In: Springer Series on Environmental Management.

Webber, Kathleen C Thrombosis: medical subject analysis with bibliography (1987)
Annandale VA, ABBE Publishers Association of Washington.
ISBN 0–88164–574–5.

Webster, Christopher, et al Constructing Dangerousness: scientific legal and policy implications (1985)
Toronto ON, University of Toronto Press, 161 p.
ISBN 0–919584–62–4.

Webster, Nesta H Secret Societies and Subversive Movements (1972)
Hollywood CA, Angriff Press, 419 p. ISBN 0–913022–05–5.

Webster, Noah Brief History of Epidemics and Pestilential Diseases (1970)
New York, B Franklin. repr of 1799. 2 Vols. Research and Source Works Ser: No 539. ISBN 0–8337–3710–4.

Wedeen, Richard P Poison in the Pot: the legacy of lead (1984)
Carbondale IL, World Resources Inventory.
ISBN 0–8093–1156–9.

Weglyn, Michi Years of Infamy: the untold story of America's concentration camps (1976)
New York, William Morrow. illus. ISBN 0–688–07996–2.

Weiner, Betty Abuse in Sports: an annotated bibliography (1985)
New York, Vantage Press, 34 p. bibl. CompuBibs Ser: No 10.
ISBN 0–914791–09–5.

Weiner, H (Ed) Duodenal Ulcer (1971)
New York, Karger S AG, 200 p. Advances in Psychosomatic Medicine: Vol 6. ISBN 3–8055–1159–0.

Weininger, Otto P Motherhood and Prostitution (1983)
Albuquerque NM, American Institute for Psychological Research,, 121 p. ISBN 0–89920–062–1.

Weiser, J Atlas of Insect Diseases (1977)
Dordrecht, Kluwer Academic Publishers Group, 240 p.
ISBN 90–6193–549–0.

Weisman, Charlotte Lender Liability (1988)
Philadelphia PA, Morris Robert Associates, 96 p.
ISBN 0–936742–49–6.

Weiss, Roger D and Mirin, Steven M Cocaine (1986)
New York, American Psychiatric Association, 178 p.
ISBN 0–88048–216–8.

Weissmann, Gerald Advances in Inflammation Research
New York, Raven Press. Annual.

Welles, James F Understanding Stupidity (1987)
Mt Pleasant Press, 266 p. 2nd.
ISBN 0–9617729–0–5.

Wesson, Donald R and Adams, Kenneth Polydrug Abuse: the results of a national collaborative study (1978)
San Diego CA, Academic Press. ISBN 0–12–745250–8.

West, D J Sexual Victimisation: two recent researches into sex problems (1985)
Brookfield VT, Gower Publishing, 194 p.
ISBN 0–566–00832–7.

West, L H T and Hore, T (Eds) Analysis of Drunk Driving Research (1980)
Clayton VIC, Monash University, 393 p.
ISBN 0–86746–022–9.

Westhuizen, J van der and Oosthuizen, H Prediction of Parole Failure and Maladjustment
Pretoria, University of South Africa, 340 p.
ISBN 0–86981–280–7.

Westing, Arthur H (Ed) Environmental Warfare: a technical, legal and policy appraisal (1984)
Stockholm, Almqvist and Wiksell, 108 p.
ISBN 0–85066–278–8.

Westing, Arthur H (Ed) Herbicides in War: the long-term ecological and human consequences (1984)
Stockholm, Almqvist and Wiksell, 210 p.
ISBN 0–85066–265–6.

Westing, Arthur H (Ed) Explosive Remnants of War: mitigating the environmental effects (1985)
London, Taylor and Francis, 141 p.

Westing, Arthur H (Ed) Global Resources and International Conflict: environmental factors in strategic policy and action (1986)
Oxford, Stockholm International Peace Research Institute, 292 p. SIPRI-UNEP Series. ISBN 0–19–829104–3.

Westing, Arthur H (Ed) Cultural Norms, War and the Environment (1988)
Oxford, Stockholm International Peace Research Institute, 192 p. illus. SIPRI Publication Series.
ISBN 0–19–829125–6.

Wheeler, Gerald R Counterdeterrence: a report on juvenile sentencing and effects of prisonization (1978)
Chicago IL, Nelson-Hall, 208 p. ISBN 0–88229–315–X.

White, Anne and White, Nelson Spiritual Intimidation (1984)
Pasadema CA, The Technology Group, 65 p.
ISBN 0–939856–39–5.

White, Michael Working Hours: assessing the potential for reduction (1987)
Geneva, ILO, 104 p. bibl.
ISBN 92–2–106151–5.

White, W R (Ed) Sedimentation Problems in River Basins (1982)
Paris, UNESCO, 152 p. illus.
ISBN 92–3–102014–5. Studies and Reports in Hydrology: No 35.

White, William L and Hagen, Russell J Incest in the Organizational Family: the ecology of burnout in closed systems (1986)
Bloomington IL, Lighthouse Training Institute, 284 p.
ISBN 0–938475–00–2.

Whitehead, William E and Schuster, Marvin M Psychophysiological Gastrointestinal Disorders (1985)
San Diego CA, Academic Press. ISBN 0–12–747030–1.

Whitlock, F A Symptomatic Affective Disorders (1983)
San Diego CA, Academic Press. ISBN 0–12–747580–X.

Whitman, Frederick L and Mathy, Robin M Male Homosexuality in Four Societies: Brazil, Guatemala, The Philippines (1985)
New York, Praeger Publishers, 240 p.
ISBN 0–275–90037–1.

Whitman, Thomas L, et al Behavior Modification with the Severely and Profoundly Retarded: research and application (1983)
San Diego CA, Academic Press. ISBN 0–12–747280–0.

WHO Mosquito-Borne Haemorrhagic Fevers of South-East Asia and the Western Pacific (1966)
Geneva, WHO, 104 p. Bulletin of WHO 35, 1.
ISBN 0–686–09224–4.

WHO Echinococcosis (Hydatidosis) (1968)
Geneva, WHO, 136 p. Bulletin of WHO 39 (1968) 1.
ISBN 0–686–09223–6.

WHO Dysentery: a report (1969)
Geneva, WHO. World Health Statistics Ser: Vol 22, No 4.
ISBN 0–686–09184–1.

WHO Typhoid and Paratyphoid Fevers: a report (1969)
Geneva, WHO. ISBN 0–686–09179–5.

WHO Health Aspects of Chemical and Biological Weapons: report of a WHO group of consultants (1970)
Geneva, WHO, 132 p. ISBN 92–4–156034–7.

WHO International Conference on Alternative Insecticides for Vector Control, Atlanta, Feb 1971: proceedings (1971)
Geneva, WHO, 470 p. Bulletin of WHO: Vol 44, Nos 1–3.
ISBN 0–686–09008–X.

WHO Vector Control in International Health (1972)
Geneva, WHO. ISBN 92–4–154016–8.

WHO Drought and Agriculture (1975)
Geneva, WHO, 128 p. ISBN 92–63–10392–5.

WHO Drought: lectures presented at the twenty-fifth session of the WMO Executive Committee (1975)
Geneva, WHO. Ser: No 5. ISBN 92–63–00403–X.

WHO Proceedings of the WMO/IAMAP Symposium on Long-term Climate Fluctuations (Norwich, 18–23 August 1975) (1975)
Geneva, WHO, 503 p. ISBN 92–63–10421–2.

WHO Review of the Present Knowledge of Plant Injury by Air Pollution (1976)
Geneva, WHO, 27 p. TN: No 147. ISBN 92–63–10431–X.

WHO The Quantitative Evaluation of the Risk of Disaster from Tropical Cyclones (1976)
Geneva, WHO, 143 p. Ser: No 8. ISBN 92–63–10455–7.

WHO Systems for Evaluating and Predicting the Effects of Weather and Climate on Wildland Fires (1978)
Geneva, WHO, 40 p. ISBN 92–63–10496–4.

WHO Present Techniques of Tropical Storm Surge Prediction (1978)
Geneva, WHO, 87 p. Report on Marine Science Affairs: 13.
ISBN 92–63–10500–6.

WHO Schizophrenia: an international follow-up study (1979)
New York, John Wiley and Sons, 438 p.
ISBN 0–471–99623–8.

WHO Problems Related to Alcohol Consumption: report on a WHO Export Committee (Geneva 1979) (1980)
Geneva, WHO, 72 p.
ISBN 92–4–120650–0. Technical Report Series: No 650.

WHO Manganese (1981)
Geneva, WHO, 110p.
ISBN 92–4–154077–X. Environmental Health Criteria: No 17.

WHO Arsenic (1981)
Geneva, WHO, 173 p.
ISBN 92–4–154078–8. Environmental Health Criteria: No 18.

WHO Sea-ice Information Services in the World (1981)
Geneva, WHO. loose-leaf. updated by supplements.
ISBN 92–63–10574–X.

WHO Nuclear Power: health implications of transuranium elements: Report on a working group (Brussels 1979) (1982)
Geneva, WHO, 88 p.
ISBN 92–890–1102–5. WHO Regional Publications, European Series: No 11.

WHO Treponemal Infections: report of a WHO Scientific Group (Geneva, 1980) (1982)
Geneva, WHO, 75 p.
ISBN 92–4–120674–8. Technical Report Series: No 674.

WHO Bacterial and Viral Zoonoses: report of a WHO Expert Committee with the participation of FAO (Geneva, 1981) (1982)
Geneva, WHO, 146 p.
ISBN 92–4–120682–9. Technical Report Series: No 682.

WHO Chlorine and Hydrogen Chloride (1982)
Geneva, WHO, 95 p.
ISBN 92–4–154081–8. Environmental Health Criteria: No 21.

WHO Health Aspects of Chemical Safety: health effects of combined exposures to chemicals in work and community environments: Proceedings of a course (Lodz 1982) (1983)
Copenhagen, WHO, 453 p. Interim Document: No 11.

WHO Health Aspects of Chemical Safety: allergic responses and hypersensitivities induced by chemicals: Proceedings of a joint WHO/CEC workshop (Frankfurt am Main 1982) (1983)
Copenhagen, WHO, 288 p. Interim Document: No 12.

WHO Gestational Trophoblastic Diseases: report on a WHO Scientific Group (Geneva, 1982) (1983)
Geneva, WHO, 81 p.
ISBN 92–4–120692–6. Technical Report Series: No 692.

WHO Selected Radionuclides (1983)
Geneva, WHO, 237 p.
ISBN 92–4–154085–0. Environmental Health Criteria: No 25.

WHO Styrene (1983)
Geneva, WHO, 123 p.
ISBN 92–4–154086–9. Environmental Health Criteria: No 26.

WHO Acrylonitrile (1983)
Geneva, WHO, 125 p.
ISBN 92–4–154088–5. Environmental Health Criteria: No 28.

WHO Meteorological Aspects of Certain Processes Affecting Soil Degradation – Especially Erosion (1983)
Geneva, WHO, 149 p. ISBN 92–63–10591–X.

WHO Health Aspects of Chemical Safety: studies in epidemiology – part I: exposure of elderly to cadmium – lead neurotoxicity in children – Welder's exposure to chromium and nickel (1984)
Copenhagen, WHO, 205 p. Interim Document: No 15.

WHO Health Aspects of Chemical Safety: pesticide residue analysis: proceedings of a joint FAO/WHO course (Eger, Hungary, 1983) (1984)
Copenhagen, WHO, 333 p. Interim Document: No 14.

WHO Lymphatic Filariasis: fourth Report on the Expert Committee on Filariasis (Geneva, 1983) (1984)
Geneva, WHO, 112 p.
ISBN 92–4–120702–7. Technical Report Series: No 702.

WHO The Leishmaniases: report of a WHO Expert Committee (Geneva, 1982) (1984)
Geneva, WHO, 140 p.
ISBN 92–4–120701–9. Technical Report Series: No 701.

WHO Extremely Low Frequency (ELF) Fields (1984)
Geneva, WHO, 131 p.
ISBN 92–4–154095–8. Environmental Health Criteria: No 35.

WHO Fluorine and Fluorides (1984)
Geneva, WHO, 136 p.
ISBN 92–4–154096–6. Environmental Health Criteria: No 36.

WHO Aquatic (marine and freshwater) Biotoxins (1984)
Geneva, WHO, 95 p.
ISBN 92–4–154097–4. Environmental Health Criteria: No 37.

WHO Health Hazards from Nitrates in Drinking Water: report on a WHO meeting (Copenhagen 1984) (1985)
Copenhagen, WHO, 102 p. Environmental Health Series: No 1.

WHO Health Policy Implications of Unemployment (1985)
Geneva, WHO, 409 p. ISBN 92–890–1029–0.

WHO Prenatal and Perinatal Infections (1985)
Geneva, WHO, 147 p. ISBN 92–890–1259–5.

WHO Papers Presented at the WMO Technical Conference on Observations and Measurements of Atmospheric Contaminants (TECOMAC) (Vienna, Austria 1983) (1985)
Geneva, WHO, 660 p. ISBN 92–63–10647–9.

WHO Joint FAO/WHO Expert Committee on Brucellosis (1986)
Geneva, WHO, 133 p. Technical report ser: No 740.
ISBN 92–4–120740–X.

WHO WHO Expert Committee on Malaria: eighteenth Report (1986)
Geneva, WHO, 104 p.
ISBN 92–4–120735–3. Technical Report: No 735.

WHO Acute Respiratory Infections in South-East Asia (1986)
Geneva, WHO, 58 p. ISBN 92–9022–147–X.

WHO Health Impact of Different Energy Source: a challenge for the end of the century (1986)
Geneva, WHO, 71 p. ISBN 92–890–1110–6.

WHO Tick-Born Encephalitis and Haemorrhagic Fever with Renal Syndrome in Europe (1986)
Geneva, WHO, 79 p. ISBN 92–890–1270–6.

WHO Tobacco Smoking (1986)
Geneva, WHO, 421 p. ISBN 92–832–1538–9.

WHO Land Use and Agrosystem Management under Severe Climatic Conditions (1986)
Geneva, WHO, 133 p. ISBN 92–63–10633–9.

WHO El Niño Phenomenon and Fluctuations of Climate: lectures presented at the thirty-sixth session of the WMO Executive Council (1984) (1986)
Geneva, WHO, 46 p. ISBN 92–63–10649–5.

WHO Report of the International Conference on the Assessment of the Role of Carbon Dioxide and of Other Greenhouse Gases in Climate Variations and Associated Impacts (1985) (1986)
Geneva, WHO, 78 p. ISBN 92–63–10661–4.

WHO Tick-Borne Encephalitis and Haemorrhagic: fever with renal syndrome in Europe (1986)
Geneva, WHO, 79 p. EURO Reports and Studies: No 104.
ISBN 9–2890–1270–6.

WHO Ammonia (1986)
Geneva, WHO, 210 p. ISBN 92–4–154194–6.

WHO WHO Expert Committee on Venereal Diseases and Treponematoses: sixth report (1986)
Geneva, WHO, 141 p. ISBN 92–4–120736–1.

WHO Sexuality and People with Physical Disabilities: report on a study by Mary Porter (1987)
Copenhagen, WHO, 69 p.

WHO Sexuality, Family Planning and Migrant Populations (1987)
Copenhagen, WHO, 177 p. bibl.

WHO Experience on Water Fluoridation in Europe (1987)
Copenhagen, WHO, 117 p. English only.

WHO PCDD and PCDF Emissions from Incinerators for Municipal Sewage Sludge and Solid Waste: evaluation of human exposure: Report on a WHO meeting (Naples 1986) (1987)
Copenhagen, WHO, 56 p. Environmental Health Series: No 17. English only.

WHO Health Effects of Methylmercury in the Mediterranean Sea: report on a joint WHO/FAO/UNEP meeting (Athens, 1986) (1987)
Copenhagen, WHO, 51 p. English and French.

WHO Pentachlorophenol (1987)
Geneva, WHO, 236 p.
ISBN 92–4–154271–3. Environmental Health Criteria: No 71.

WHO Silica and Some Silicates (1987)
Geneva, WHO, 289 p.
ISBN 92–832–1242–8. IARC Monographs on the Evaluation of the Carcinogenic Risk of Chemicals to Humans: Vol 42.

WHO Control of Lymphatic Filariasis: a manual for health personnel (1987)
Geneva, WHO, 89 p. ISBN 92–4–154217–9.

WHO The Hypertensive Disorders of Pregnancy: report of a WHO Study Group (1987)
Geneva, WHO, 114 p.
ISBN 92–4–120758–2. Technical Report Series: No 758.

WHO Overall Evaluations of Carcinogenicity: an updating of IARC Monographs (1987)
Geneva, WHO, 440 p. Vols 1–42.
ISBN 92–832–1411–0. IARC Monographs on the Evaluation of Carcinogenic Risks to Humans: 7.

WHO Mycotic Diseases in Europe: report on a WHO Meeting (Hamburg 1983) (1987)
Geneva, WHO, 84 p. ISBN 92–890–1271–4.

WHO WHO Expert Committee on Onchocerciasis Third Report (1987)
Geneva, WHO, 167 p. ISBN 9–2412–0752–3.

WHO Assessment of Health Risks in Infants Associated with Exposure to PCBs, PCDDs and PCDFs in Breast Milk: report on a WHO working group (Abano Terme/Padua 1987) (1988)
Copenhagen, WHO, 116 p. Environmental Health Series: No 29. English only.

WHO Chromium (1988)
Geneva, WHO, 197 p.
ISBN 92–4–154261–6. Environmental Health Criteria: No 61.

WHO Phosphine and Selected Metal Phosphides (1988)
Geneva, WHO, 100 p.
ISBN 92–4–154273–X. Environmental Health Criteria: No 73.

WHO Urban Vector and Pest Control: eleventh report of the WHO Expert Committee on Vector Biology and Control (1988)
Geneva, WHO, 77 p.
ISBN 92–4–120767–1. Technical Report Series: No 767.

WHO Collaborating Centre on Air Pollution Control and WHO Collaborating Centre on Clinical and Epidemiological Aspects of Air Pollution Selected Methods of Measuring Air Pollutants (1976)
Geneva, WHO. Offset Pub: No 24. ISBN 92–4–170024–6.

WHO Expert Committee, Athens WHO Expert Committee on Filariasis: report (1973)
Geneva, WHO. 3rd ed. Technical Report Ser: No 542.
ISBN 92–4–120542–3.

WHO Expert Committee, Geneva Prevention of Perinatal Mortality and Morbidity: a report (1970)
Geneva, WHO, 60 p. Technical Report Ser: No 457.
ISBN 92–4–120457–5.

WHO Expert Committee Geneva, 1969, 4th WHO Expert Committee on Plague: report (1970)
Geneva, WHO. ISBN 92–4–120447–8.

WHO Expert Committee on Insecticides, Geneva, 1970 Application and Dispersal of Pesticides: a report (1971)
Geneva, WHO, 66 p. Technical Report Ser: No 465.
ISBN 92–4–120465–6.

WHO Meeting Geneva, 1974 Viral Hepatitis: report (1975)
Geneva, WHO. ISBN 92–4–120570–9.

WHO Scientific Group – Geneva 1971 Inherited Blood Clotting Disorders: report (1972)
Geneva, WHO. Technical Report Ser.
ISBN 0–686–16785–6.

WHO Scientific Group, Geneva 1972 Vector Ecology: report (1972)
Geneva, WHO. Technical Report Ser: No 501.
ISBN 92–4–120501–6.

WHO Scientific Group Geneva, 1972 Viral Hepatitis: report (1973)
Geneva, WHO. ISBN 92–4–120512–1.

WHO Scientific Group, Geneva, 1975 Epidemiology of Infertility: report (1975)
Geneva, WHO. Technical Report Ser: No 582.
ISBN 92–4–120582–2.

WHO Seminar, Tours Prevention of Perinatal Mortality and Morbidity: a report (1972)
Geneva, WHO, 97 p. Public Health Papers Ser: No 42.
ISBN 92–4–130042–6.

WHO Study Group Cerebrospinal Meningitis Control: report (1976)
Geneva, WHO. Technical Report Ser: No 588.
ISBN 92–4–120588–1.

Whybrow, Peter C, et al Mood Disorders: toward a new psychobiology (1984)
New York, Plenum Publishing, 244 p.
ISBN 0–306–41568–2.

Wiegele, Thomas C, et al Leaders under Stress: a psychophysiological analysis (1985)
Durham NC, Duke University Press, 229 p.
ISBN 0–8223–0641–7.

Wiersma, G E Cohabitation, an Alternative to Marriage ?: a Cross-national study (1983)
Dordrecht, Kluwer Academic Publishers Group, 225 p.
ISBN 90–247–2845–2.

Wight, Nancy (Ed) Law and Current World Issues, Columbia Law School Symposium 1986 (1987)
Dobbs Ferry NY, Oceana Publications, 174 p.

Wigoder, Geoffrey (Ed) Anti-Semitism
Jerusalem, Keter Publishing House, 232 p.

Wiking, Staffan Military Coups in Sub-Saharan Africa: how to justify illegal assumptions of power (1983)
Stockholm, Almqvist and Wiksell. ISBN 91–7106–214–9.

Wilberger, James E Spinal Cord Injuries in Children (1986)
Mount Kisco NY, Futura Publishing, 304 p.
ISBN 0–87993–264–3.

Wilcox, Laird Terrorism, Assassination, Espionage and Propaganda: a master bibliography (1988)
Olathe KS, Laird Wilcox Editorial Research.
ISBN 0–933592–50–7.

Wild, Victor The Science of Revolution: fundamentals of Marxism-Leninism, Mao Tse Tung thought and the line of the Revolutionary Communist Party, U S A (1980)
Chicago IL, RCP Publications. ISBN 0–89851–038–4.

Wilfred, Lucia M Septicemia: subject, reference and research guidebook (1987)
Washington DC, ABBE Publications Association, 160 p.
ISBN 0–88164–578–8.

Wilgus, Neal and Wilson, Robert A The Illuminoids: secret societies and political paranoia (1978)
Santa Fe NM, Sun Publishing, 262 p.
ISBN 0–89540–045–6.

Wilkinson, Edward J Pathology of the Vulva and Vagina (1986)
New York, Churchill Livingstone, 340 p.
ISBN 0–443–08514–5.

Willerson, James T et al Ischemic Heart Disease: clinical and pathophysiological aspects (1982)
New York, Raven Press, 384 p. ISBN 0–89004–563–1.

Williams, A Olufemi Virus-Associated cancers in Africa (1984)
Oxford NY, Oxford University Press, 806 p. LARC Ser.
ISBN 0–19–723063–6.

Williams, Carolyn L and Westermeyer, Joseph Refugee Mental Health in Resettlement Countries (1986)
New York, Hemisphere Publishing, 267 p.
ISBN 0–89116–445–6.

Williams, Christine L Gender Differences at Work: women and men in nontraditional occupations
Berkeley CA, University of California Press, 210 p.
ISBN 0–520–06373–2.

Williams, J D and Burnie, J Bacterial Meningitis (1987)
San Diego CA, Academic Press, 248 p.
ISBN 0–12–755155–7.

Williams, J H Chromium in Sewage Sludge Applied to Agricultural Land: final report (1988)
Brussels, Commission of the European Communities, 63 p.
ISBN 92–825–8645–6.

Williams, John J Electromagnetic Brainblaster: brainwashing through electromagnetics (1985)
Alamogordo NM, Consumertronics, 52 p. illus.
ISBN 0–934274–05–3.

Williams, John S Jr, et al Environmental Pollution and Mental Health (1973)
Arlington VA, Information Resources Press, 136 p.

Williams, Joyce E and Holmes, Karen A The Second Assault: rape and public attitudes (1981)
Westport CT, Greenwood Press, 256 p.
ISBN 0–313–22542–7.

Williams, Norman P The Ideas of the Fall and of Original Sin: a historical and critical study
New York, AMS Press. repr of 1927.
ISBN 0–404–18439–1.

Williams, Robert Political curruption in Africa (1987)
Brookfield VT, Gower Publishing, 251 p. Pub by Gower Pub England.

Williams, Robert B and Venturini, Joseph L School Vandalism: cause and cure (1981)
Saratoga CA, R and E Publishers, 130 p.
ISBN 0–86548–060–5.

Williams, Robert R Toward the Conquest of Beriberi (1961)
Cambridge MA, Harvard University Press. illus.

Williams, Roger Liver Failure (1986)
New York, Churchill Livingstone, 230 p.
ISBN 0–443–03109–6.

Willich, Ray Troubled Ones: sexually and emotionally abused children (1979)
Melbourne VIC, Hill of Content Publishing, 152 p.
ISBN 0–85572–104–9.

Willie, Charles V and Beker, Jerome Race Mixing in Public Schools (1973)
New York, Irvington Publishers. ISBN 0–275–28812–9.

Willing, Jules Z The Reality of Retirement: the inner experience of becoming a retired (1981)
Chapel Hill NC, Lively Mind Books. ISBN 0–688–00298–6.

Willis, Evan Medical Dominance: division of labour in Australian health care (1983)
London, Allen and Unwin, 200 p. ISBN 0–86861–254–5.

Willoughby, John Capitalist Imperialism, Crisis and the State (1986)
New York, Harwood Academic Publishers, 101 p.
ISBN 3–7186–0322–5.

Wills, Geoff and Cooper, Cary L Pressure Sensitive: popular musicians under stress (1988)
Newbury Park CA, Sage Publications, 160 p.
ISBN 0–8039–8141–4.

Wilsnack, Sharon C and Beckman, Linda J Alcohol Problems in Women: antecedents, consequences and intervention (1984)
New York, Guilford Press, 480 p. ISBN 0–89862–164–X.

Wilson, Anne Mixed Race Children: a study of identity (1987)
London, Unwin Hyman, 172 p. ISBN 0–04–370168–X.

Wilson, Glenn D and Cox, David N Child-Lovers: a study of paedophiles in society (1983)
Chester Springs PA, Dufour Editions, 132 p.
ISBN 0–7206–0603–9.

Wilson, J P, et al Human Adaptation to Extreme Stress: from the holocaust to Vietnam (1988)
New York, Plenum Publishing, 382 p.
ISBN 0–306–42873–3.

Wilson, J Walter and Plunkett, Orda A The Fungus Diseases of Man (1965)
Berkeley CA, University of California Press.
ISBN 0–520–01344–1.

Wilson, James Q and Herrnstein, Richard J Crime and Human Nature (1986)
S and S, 640 p.

Wilson, John and International Agency for the Prevention of Blindness World Blindness and its Prevention
Oxford NY, Oxford University Press. 3 vols.

Wilson, Robert A The New Inquisition (1987)
Phoenix AZ, Falcon Press, 240 p. ISBN 0–941404–49–8.

Winchell, Carol A The Hyperkinetic Child: an annotated bibliography, 1974 to 1979 (1981)
Westport CT, Greenwood Press, 488 p. ISBN 0–313–21452–2.

Winder, Christopher The Development Neurotoxicity of Lead (1984)
Dordrecht, Kluwer Academic Publishers Group.

Winick, Myron Malnutrition and Brain Development (1976)
New York, Oxford University Press. ISBN 0–19–501983–0.

Winn, Kenneth The Manipulated Mind: brainwashing, conditioning and indoctrination (1983)
London, Octagon Press.

Winograd, Jesse S Glaucoma: subject, reference and research guide (1987)
Washington DC, ABBE Publications Association, 160 p. ISBN 0–88164–538–9.

Winslow, Robert M and C, Carlos Monge Hypoxia, Polycythemia, and Chronic Mountain Sickness (1987)
Baltimore MD, Johns Hopkins University Press, 288 p. ISBN 0–8018–3448–1.

Winteringham, F P Soil and Fertilizer Nitrogen (1985)
Lanham MD, UNIPUB, 107 p. ISBN 92–0–115184–5.

Wintrobe, Maxwell M Blood, Pure and Eloquent: a story of discovery, of people, and of ideas (1980)
New York, McGraw Hill Book Company, 352 p. ISBN 0–07–071135–6.

Wishnie, Howard A The Impulsive Personality (1977)
New York, Plenum Publishing, 226 p. ISBN 0–306–30973–4.

Withers, John Major Industrial Hazards: their appraisal and control (1988)
Brookfield VT, Gower Publishing, 240 p. ISBN 0–291–39725–5.

Witt, James G and Fischer, William E Deadly Deceptions (1987)
Milwaukee WI, WELS Board for Parish Education, 44 p. ISBN 0–938272–32–2.

Wittfogel, Karl A Oriental Despotism: a comparative study of total power (1981)
New York, Random House, 556 p. ISBN 0–394–74701–1.

Wittke, Carl F Refugees of Revolution
Westport CT, Greenwood Press. ISBN 0–8371–2988–5.

Wittmann, G (Ed) Herpesvirus Diseases of Cattle, Horses, and Pigs (1989)
Boston MA, Kluwer Academic Publishers Group, 192 p. ISBN 0–7923–0118–8.

Wittmann, G and Hall, S A (Eds) Aujeszkiy's Disease: a seminar in the Animal Pathology Series of the CEC Programme of Coordination of Agricultural Research (1981, Tübingen) (1982)
Dordrecht, Kluwer Academic Publishers Group, 295 p. ISBN 90–247–2638–7.

Wober, Mallory and Gunter, Barrie Television and Social Control (1988)
New York, Saint Martin's Press, 250 p. ISBN 0–312–01305–1.

Wolchik, Sharlene A and Karoly, Paul (Eds) Children of Divorce: empirical perspectives on adjustment (1988)
New York, Gardner Press, 356 p. ISBN 0–89876–120–4.

Wolf, Edward Soil Erosion: quiet crisis in the world economy
Washington DC, Worldwatch Institute. Worldwatch Papers: 60.

Wolf, Edward C On the Brink of Extinction: conserving biological diversity
Washington DC, Worldwatch Institute. Worldwatch Papers: 78.

Wolf, Edward C Beyond the Green Revolution: new approaches for Third World agriculture
Washington DC, Worldwatch Institute. Worldwatch Papers: 73.

Wolf, Edward C Beyond the Green Revolution: new approaches for Third World agriculture (1986)
Washington DC, Worldwatch Institute, 48 p. Worldwatch Papers. ISBN 0–916468–74–7.

Wolf, Edward C On the Brink of Extinction: conserving the diversity of life (1987)
Washington DC, Worldwatch Institute, 52 p. Worldwatch Papers. ISBN 0–916468–79–8.

Wolf, John B Fear of Fear: a survey of terrorist operations (1981)
New York, Plenum Publishing, 256 p. ISBN 0–306–40766–3.

Wolf, Peter, et al (Eds) Advances in Epileptology: the 16th epilepsy international symposium (1987)
New York, Raven Press, 816 p. illus. ISBN 0–88167–222–X.

Wolfe, Alan The Limits of Legitimacy (1980)
New York, Free Press. ISBN 0–02–934860–9.

Wolfe, Douglas A and O'Connor, Thomas P Urban Wastes in Coastal Marine Environments (1988)
Melbourne FL, Robert E Krieger. Oceanic Processes in Marine Pollution: 5. ISBN 0–89874–963–8.

Wolff, Harold G Wolff's Headache and Other Head Pain (1987)
Oxford NY, Oxford University Press, 448 p. 5th ed. ISBN 0–19–504356–1.

Wolff, Robert P Poverty of Liberalism (1969)
Boston MA, Beacon Press.

Wolff, Robert P, et al Critique of Pure Tolerance
Boston MA, Beacon Press. ISBN 0–8070–1559–8.

Wolffsohn, A Fire Control in Tropical Pine Forests (1981)
New York, State Mutual Book and Periodical Service. ISBN 0–85074–056–8.

Wolfgang, Charles H and Glickman, Carl D Solving Discipline Problems: strategies for classroom teachers (1986)
Needham Heights MA, Allyn and Bacon, 348 p. ISBN 0–205–08630–6.

Wolman, M Healing and Scarring of Atheroma (1984)
New York, Plenum Publishing, 164 p. ISBN 0–306–41514–3.

Wolpin, Miles D Military Radicalism in the Middle–East and in the Mediterranean
Oslo, International Peace Research Institute, 59 p.

Wolpin, Miles D Militarization, Internal Repression and Social Welfare in the Third World
Oslo, International Peace Research Institute, 221 p.

Wolpin, Miles D State Terrorism and Repression in the Third World: parameters and prospects
Oslo, International Peace Research Institute, 115 p.

Wood, Barbara L Children of Alcoholism: struggle for self and intimacy in adult life (1987)
New York, New York University Press, 166 p. ISBN 0–8147–9219–7.

Wood, Clive Haemorrhoids: current concepts on causation and management (1979)
Wolfeboro NH, Longwood Publishing Group, 36 p. ISBN 1–85315–055–X.

Wood, Edward J Giants and Dwarfs (1976)
Folcroft PA, Folcroft Library Editions. IBSN 0–8414–9609–9.

Wood, S (Ed) The Degradation of Work? (1982)
London, Hutchinson.

Woodcock, George Civil Disobedience
Toronto ON, Canadian Broadcasting Corporation, 80 p.

Wooden, Wayne S and Parker, Jay Men Behind Bars: sexual exploitation in prison (1982)
New York, Plenum Publishing, 264 p. ISBN 0–306–41074–5.

Woods, John Engineered Death: abortion, suicide, euthanasia and senecide (1978)
Ottawa ON, University of Ottawa Press, 180 p. ISBN 0–7766–1020–1.

Woodward, John C and Queen, Janel The Solitude of Loneliness (1988)
Lexington MA, Lexington Books, 128 p. ISBN 0–669–14505–X.

World Commission on Environment and Development Our Common Future (1987)
Oxford, Oxford University Press, 383 p. ISBN 0–19–282080–X.

World Congress of Rehabilitation International Participation of People with Disabilities: international perspectives (1981)
New York, Rehabilitation International. ISBN 0–686–94877–7.

World Conservation Union Proceedings of the World Lagomorph Conference
Gland, IUCN, 983 p.

World Conservation Union Threatened Deer (1978)
Gland, IUCN, 434 p.

World Conservation Union Feral Mammals: problems and potential (1984)
Gland, IUCN, 152 p. illus. ISBN 2–88032–902–7.

World Conservation Union Plants in Danger: what do we know? (1986)
Gland, IUCN, 488 p. ISBN 2–88032–707–5.

World Intellectual Property Organization Licensing Guide for Developing Countries: a guide on the legal aspects of the negotiation and preparation of industrial property licenses and technology transfer agreements appropriate to the needs of developing countries (1977)
Geneva, WIPO, 184 p.

World Intellectual Property Organization Symposium on the Effective Protection of Industrial Property Rights, Geneva 1987 (1987)
Geneva, WIPO, 299 p.

World Intellectual Property Organization Background Reading Material on Intellectual Property (1988)
Geneva, WIPO, 400 p.

World Meteorological Organization The Effect of Weather and Climate Upon the Keeping Quality of Fruit (1963)
Geneva, WMO, 180 p.

World Meteorological Organization Air Pollutants, Meteorology and Plant Injury (1969)
Geneva, WMO, 73 p.

World Meteorological Organization Weather and Animal Diseases (1970)
Geneva, WMO, 49 p.

World Meteorological Organization Dispersion and Forecasting of Air Pollution (1972)
Geneva, WMO, 116 p.

World Rehabilitation Fund Transitions and Adults with Learning Disabilities: an international perspective (1985)
New York, World Rehabilitation Fund, 120 p.

World Resources Institute World Resources 1990–1991: a guide to the global environment (1990)
Oxford, Oxford University Press, 383 p.

World Tourism Organization WTO/UNEP Workshop on Environmental Aspects of Tourism, 1983
Madrid, WTO, 160 p.

Wortis, J Mental Retardation and Development Disabilities (1986)
New York, Elsevier Science Publishing, 400 p. Vol. XIV ISBN 0–444–00990–6.

Wötzel, Horst Trip into Illusion: misuse of drugs by adolescents (1975)
Stuttgart, Verlag der Stiftung Gralsbotschaft, 68 p. ISBN 3–87860–073–9.

Woube, Mengistu The Geography of Hunger: some aspects of the causes and impacts of hunger (1987)
Philadelphia PA, Coronet Books, 146 p. ISBN 91–506–0656–5.

Wright, Anthony Dizziness: a guide to disorders of balance (1987)
Dobbs Ferry NY, Sheridan Medical Books, 230 p. ISBN 0–7099–3659–1.

Wright, Dudley Vampires and Vampirism
New York, Gordon Press Publishers. ISBN 0–87968–093–8.

Wright, James D and Rossi, Peter H Armed and Considered Dangerous: a survey of felons and their firearms (1986)
Berlin, Walter de Gruyter, 247 p. ISBN 3–11–010966–2.

Wright, James D; Rossi, Peter H and Daly, Katleen Under the Gun: weapons, crime and violence in America (1983)
New York, Aldine, 360 p. ISBN 0–202–30305–5.

Wurman, Richard Saul Information Anxiety
New York, Doubleday, 356 p. illus.

Wyke, Maria A Developmental Dysphasia (1979)
San Diego CA, Academic Press. ISBN 0–12–766950–7.

Wynne, B Risk Management and Hazardous Wastes: implementation and the dialectics of credibility (1987)
Berlin, Springer–Verlag, 459 p. ISBN 3–540–18243–8.

Xanthou, M New Aspects of Nutrition in Pregnancy, Infancy and Prematurity (1987)
New York, Elsevier Science Publishing, 208 p. ISBN 0–444–80936–8.

Yablonsky, Lewis Juvenile Delinquency (1988)
New York, Harper and Row, 510 p. ISBN 0–06–047291–X.

Yachir, Fayçal The Conflict over Mineral Resources in Africa
London, Zed Books.

Yaffe, Maurice and Nelson, Edward The Influence of Pornography on Behaviour (1983)
San Diego CA, Academic Press. ISBN 0–12–767850–6.

Yaffey, M J H False Import Valuation (1967)
Dar es Salaam, ERB Publications Committee.

Yang, S J and Ellison, A J Machinery Noise Measurement (1985)
Oxford NY, Oxford University Press. illus. Monographs in Electrical and Electronic Engineering. ISBN 0–19–859333–3.

Yardley, Stella S Fear and Panic: index of modern information (1988)
Annandale VA, ABBE Pubishers Association of Washington, 150 p. ISBN 0–88164–848–5.

Yari, Labo Climate of Corruption (1978)
Enugu, Fourth Dimension Publishing, 176 p. ISBN 978–156–014–2.

Yemin, Edward Workforce reductions in undertakings Policies and Measures for the Protection of Redundant Workers in Seven Industrialised Economy Countries (1982)
Geneva, ILO, 214 p. ISBN 92–2–102910–7.

Yin, Peter Victimization and the Aged (1985)
Springfield IL, Thomas Charles C, 222 p. ISBN 0–398–05079–1.

Young, Mary de Child Molestation: an annotated bibliography (1987)
Jefferson NC, McFarland Publishers, 190 p. ISBN 0–89950–243–1.

Young, Peter and Tyre, Colin Dyslexia or Illiteracy?: realizing the right to read (1983)
New York, Taylor and Francis, 179 p. ISBN 0–335–10192–5.

Youngs, Bettie, B Stress in Children: how to recognise, avoid and overcome it (1986)
Melbourne VIC, Nelson Publishers, 176 p. ISBN 0–17–006777–7.

Youssef, Michael Revolt Against Modernity: Muslim zealots and the West (1985)
Leiden, Brill, 189 p. ISBN 90–04–07559–3.

Yunis, G J New Chromosomal Syndromes (1977)
San Diego CA, Academic Press. ISBN 0–12–775165–3.

Zachar, D Soil Erosion (1982)
New York, Elsevier Science Publishing, 548 p. ISBN 0–444–99725–3.

Zadoks, J C and Rijsdijk, F H Atlas of Cereal Growing in Europe (1985)
Geneva, UN, 169 p. illus.

Zakladnoi, G A and Ratanova, V F Stored–Grain Pests and Their Control (1987)
Rotterdam, AA Balkema, 279 p. ISBN 90–6191–494–9.

Zastrow, Charles Social Problems: issues and solutions (1988)
Chicago IL, Nelson–Hall, 656 p. 2nd ed. ISBN 0–8304–1111–9.

Zastrow, Charles and Bowker, Lee Social Problems: issues and solutions (1984)
Chicago IL, Nelson–Hall, 618 p. ISBN 0–8304–1051–1.

Zenzinov, Vladimir M Deserted: the story of the children abandoned in Soviet Russia (1975)
Westport Ct, Hyperion Press, 216 p. Repr of 1931. Trans. from Russia. ISBN 0–88355–190–X.

Zetterberg, J Peter Evolution Versus Creationism: the public education controversy (1983)
Phoenix AZ, Oryx Press, 528 p.

Zeylstra, W G Aid or Development: the relevance of development aid to problems of developing countries (1977)
Dordrecht, Kluwer Academic Publishers Group, 269 p. 2nd ed. ISBN 90–286–0115–5.

Zhuang, Weisong (Ed) Equity and Urban Environment in the Third World: with special reference to Asean countries and Singapore (1975)
Bombay, Architects Publishing of India, 197 p. ISBN 0–88911–424–2.

Zieve, Philip D and Levin, Jack Disorders of Hemostasis
Ann Arbor MI, Books on Demand. ISBN 0–8357–9542–X.

Zimbardo, Philip Shyness (1987)
New York, Jove Publications. ISBN 0–515–08919–2.

Zimmerman, Ekkart Political Violence, Crisis, and Revolution: theories and research (1983)
Rochester VT, Schenkman Books, 792 p. ISBN 0–87073–894–1.

Zimmerman, Roy R Malpractice I: medical subject analysis and research guide (1984)
Annandale VA, ABBE Publishers Association of Washington, 140 p. ISBN 0-88164-108-1.

Zimmerman, Roy R Malpractice II (1985)
Annandale VA, ABBE Publishers Association of Washington, 150 p. ISBN 0-88164-242-8.

Ziswiler, V Extinct and Vanishing Animals: a biology of extinction and survival (1967)
Berlin, Springer-Verlag, 133 p. rev ed. ISBN 3-540-90003-9.

Zorab, P A and Siegler, David Scoliosis, 1979 (1980)
San Diego CA, Academic Press. ISBN 0-12-781860-X.

Zuckerman, Arie J Viral Hepatitis and Liver Disease (1988)
New York, Alan R Liss, 1160 p. ISBN 0-8451-4247-X.

Zumpt, F Stomoxyine Biting Flies of the World: diptera, muscidae: taxonomy, biology, economic importance and control (1973)
Stuttgart, Fischer, 175 p. ISBN 3-437-30146-2.

Zurawicki, L Multinational Enterprises in the West and East (1979)
Dordrecht, Kluwer Academic Publishers Group, 207 p. ISBN 90-286-0419-7.

Zwirnmann, K H (Ed) Nonpoint Nitrate Pollution of Municipal Water Supply Sources: issues of analysis and control (1982)
Laxenburg, International Institute for Applied Systems Analysis, 302 p. ISBN 3-7045-0032-1.

Notes   PZ

World Problems

Significance

Acknowledgement of the universe of problems	1143
Constraints on a problem-focused approach	1147
Framework for interelating incompatible perspectives	1150
Unique features	1151
Prededents and parallels	1152
Precedents in history and tradition	1156

Criteria

Assumptions	1157
Problem disguises and problem evasion	1159
Definitions	1161
Problem inclusion	1163
Problem exclusion	1164
Problem importance	1166

Method

Identification procedure	1167
Problem naming	1168
Document control and problem description	1169
Inter-problem relationships	1171
Conceptual processes summarized	1172

Patterning problems

Concept refinement process	1173
Classification and section attribution	1176
Language games	1177
Patterning the problematique	1178
Computer representation of problem networks	1180

Comments

General	1181
Approaches to problems	1182
Beyond the problem-lobby mindset	1184
Problem metaphors	1185
Future possibilities	1187

***   Bibliographical references identified in abridged form in the following section refer to publications detailed, by author, in Section PY, which is the bibliography for Section P.

Significance: acknowledgement of the universe of problems

The flood of documents produced by international organizations contains a very large number of facts, preoccupations, statements of belief, programme proposals and criticisms of other initiatives. Faced with this flood, most bodies survive by ignoring all but a small fraction of it. They endeavour to carve out a small niche, cultivating a support network of similarly minded bodies and formulating the most powerful strategy possible for them in order to act on the problems they perceive. This includes undermining the initiatives of those whom they perceive to be causing or sustaining such problems.

Many coalitions of organizations have "answers" to the current crisis, however they choose to perceive it. The proponents of each such answer naturally attach special importance to their own as being of crucial relevance at this time, whether in the short-term or for tactical reasons, or in the long-term as being the only appropriate basis for a viable world society in the future. However this widespread focus on "answer production", a vital moving force in society, obscures both the significance of the lack of fruitful integration between existing answers and the manner in which such answers undermine each other's significance. The mind-set also fails to recognize the positive significance of the continuing disruptive emergence of new "alternative" answers.

Amongst this multitude of answers, explanations put forward as objective, rational and factually-based by scientific and government authorities are increasingly questionable because of peer group, political, security, religious and commercial pressures guiding evaluation and reporting. The many exercises in producing global strategies based on an overview of extensive ranges of problems are themselves far from free from such influences. They tend to appear successful when they succeed in reducing the complexity of the problematique. There is considerable confusion about the nature of integration whether amongst the disciplines or especially in relation to policy initiatives.

The communication space of the international community is thus characterized by claims and counter-claims attesting to or denying the importance of particular problems, or questioning the manner in which they are defined. The challenge is to determine what new kind of information tool could usefully reflect this communication condition, offering integrative insights, but without simply adding to the existing confusion. Adding to this challenge is the fact that any such attempt is in many respects totally presumptuous -- particularly when undertaken with limited resources.

In the paragraphs which follow, and with the aid of quotations from a variety of authors, an attempt is made to show the significance of a problem-focused approach. By this is meant an approach which focuses on problems in all their negativity rather than on solutions to problems. The basic point being that **only by knowing more about the nature of problems and how they combine together will it be possible to conceive of more adequate solutions which have any hope of widespread support.**

1. Multiplicity and gravity of problems
There is widespread recognition of the number and seriousness of the problems faced by mankind, as acknowledged in texts such as the following:

"It is unforgivable that so many problems from the past are still with us, absorbing vast energies and resources desperately needed for nobler purposes: a horrid and futile armaments race instead of world development; remnants of colonialism, racism and violations of human rights instead of freedom and brotherhood; dreams of power and domination instead of fraternal coexistence; exclusion of great human communities from world co- operation instead of universality; extension of ideological domains instead of mutual enrichment in the art of governing men to make the world safe for diversity; local conflicts instead of neighbourly co-operation. While these antiquated concepts and attitudes persist, the rapid pace of change around us breeds new problems which cry for the world's collective attention and care: the increasing discrepancy between rich and poor nations; the scientific and technological gap; the population explosion; the deterioration of the environment; the urban proliferation; the drug problem; the alienation of youth; the excessive consumption of resources by insatiable societies and institutions. The very survival of a civilized and humane society seems to be at stake." (U. Thant, Secretary-General of the United Nations on the occasion of United Nations Day, 1970).

Although there is agreement that there are many problems and that many are serious, little concerted effort has been made to determine how many problems there are. Such efforts as have been made have generally been limited to identifying **major** or **critical** problems, usually guided either by political expediency or by the particular objective of a major agency. For example, one study for the President of the USA, resulting in 6 problems analyzed in detail, was based on a procedure whereby 1000 problems were deliberately filtered through a succession of phases down to 100, to 50, to 20 before the *"final sort and aggregation"*. (Assessment of Future National and International Problems. US National Science Foundation, 1977, NSF/STP 76- 02573). Only the final 6 were submitted to the President. No further mention is made of the 994, whatever their importance to particular constituencies. At that time UNESCO engaged in an exercise to identify the "major world problems" with which it was concerned and identified 12 (Medium-Term Plan 1977-1982. UNESCO, 19 C/4).

2. Interrelationships between problems
It is becoming increasingly evident, and increasingly accepted, that problems interact with one another. This situation is illustrated by the following:

"In spite of much publicity, the complexity and magnitude of the problems faced by man if he is to survive as a social animal is still only adequately conceived by specialists, and it derives not so much from the mere multiplicity and gravity of problems awaiting a solution in our technological society, or in what the Battelle Institute has described as the "frightening series of problems now appearing over the horizon", as from the fact that between these multiple problems there exists an incalculable number of inter-relationships which, whether ascertained or not, greatly restrict the range of action open to the policy-maker. It is this situation which has brought about the tendency for the solution of one problem to create a number of new ones, often in fields only distantly related at first sight to the original matter. In particular, this not being fully understood, there is a general disposition to envisage and treat the symptoms of trouble, particularly the more obvious ones such as the various forms of pollution of the environment, rather than to deal with the root cause which is to be found in the inadequacy of the decision-making machinery of human society under any form of government at present known. Serious errors in decision-making with regard to the Tennessee Valley or the rivers feeding the Caspian Sea or the application of DDT have produced disastrous consequences which cannot be remedied by going back to the starting point. It is necessary to start from the position as it now exists, and necessary to fully understand it, for which purpose full and processed information is required." (Sir Peter Smithers. Governmental Control; a prerequisite for effective relations between the United Nations and non-UN regional organizations. New York, United Nations Institute for Training and Research, 1972, p.45-46).

"The systems of international trade, payments and finance are component parts of an interdependent world economy. The functioning of each is intimately related to that of the others; and present or prospective arrangements in the three spheres must be viewed in terms of the requirements for economic expansion and structural change in the world as a whole. The interrelationships have many facets and may take a number of forms. Examples are not hard to find: inadequacies in the flow of finance, long-term or short-term, may obstruct a mutually advantageous international division of labour; an improperly working adjustment process may exert deflationary or inflationary pressures, and encourage restrictions on flows of goods and finance; rigidities in trade patterns may generate chronic instability in currency markets; the capacity to service accumulated

debt may be impaired by an inadequate rate of growth in the export markets of debtor countries. Any tension between the established international economic mechanism and the dynamics of economic growth will be reflected in difficulties in the monetary, financial and commercial spheres. A malfunctioning in any one of these spheres will generally produce stresses in the others also. Acute problems, when they arise, may emerge in the form of commercial, financial or monetary imbalances that appear to be localized in particular countries or groups of countries. Deeper analysis will, however, often show that the problems of one country or group of countries in one sphere are intimately related to concomitant problems in other countries and in other spheres and that adequate overall solutions depend on parallel and consistent measures in several different fields, having regard to the interests of all countries. What may appear to be a problem unique to one sphere may be symptomatic of wider and more far-reaching tensions in the international economic system as a whole." (United Nations Conference on Trade and Development. Interdependence of problems of trade, development, finance and the international monetary system; report by the Secretary-General. Geneva, UNCTAD, 6 July 1973, TD/B/459, para. 1- 3).

3. Problem generation

Not only are there many interrelationships between problems, but in some cases the combined presence or interaction of two or more problems can lead to the emergence of new problems. Thus workers in factories are often exposed simultaneously to different physical or chemical agents which interact and which have a combined effect significantly different to the sum of the effects of the various agents encountered independently. Another example is the synergistic relationship between malnutrition and infection which in its combined form constitutes a major problem in developing areas.

"There is, then, no such thing as the food crisis. Similarly, there is no such thing, in isolation, as the population crisis, the urbanisation crisis, the pollution crisis, the armaments crisis, the oil crisis, the energy crisis, the fertiliser crisis, the resources crisis, the water crisis, the soil crisis, the fish crisis, the technology crisis or the trade crisis. Each of these crises acts on the others, and while it may be useful to focus attention on them one at a time, none of them can be solved unless the others are taken into account. This hydra-headed world crisis is difficult to comprehend... The dilemma at Rome, as at Stockholm, Caracas, Bucharest and elsewhere, is that the poor and hungry nations sense that the isolated crisis on the agenda is but a part of a wider population- resources-development crisis which unless resolved in toto will condemn them for good to the status of second-class citizens on their own planet...the present series of international conferences suffers from a universal catch-22, which states that any problem we can solve is part of a larger problem which we cannot." (Jon Tinker. The Green Revolution is over. *New Scientist*, 7 November 1974, p.388-393).

Although there is agreement that there are interrelationships between problems, little concerted effort has been made to identify how many there are and between which problems. Such efforts as have been made have generally been limited to determining adequate descriptions (in mathematical terms) for the nature of the relationships between a handful of **major** or **critical** problems. The relationships between other problems have only been explored within the various specialized domains, irrespective of any wider significance. Communication between such domains is generally agreed to be poor or non-existent.

4. Complexity of the inter-problem network

By the manner in which the simple interactions between the problems combine together, a new condition, namely a problem system or problem network is identifiable, as illustrated by the following:

"Many of the problems we experience today have been with us for a long time and those of recent vintage do not seem insurmountable, of themselves. The feature that is wholly new in the problematic aspects of our situation is rather a frightening growth in the size of the issues and a tendency toward congealment whose dynamics appears to be irreversible. The congruence of events appears suddenly possessed of a direction and a total meaning which emphasizes the insufficiency of all the proposed solutions increasingly and reveals rigidities that are not stable or set, that do not confine the problems but enlarge them, while also deepening them. This suggests that our situation has an inner momentum we are unable fully to comprehend; or, rather, that we are trying to cope with it by means of concepts and languages that were never meant to penetrate complexities of this kind; or, again, that we are trying to contain it with institutions which were never intended for such use. Therefore, even to be able to talk meaningfully about these problems (or, is it a single problem that is facing us?) we need first to develop a conceptual approach and a language we can use, which correspond better than what we now have to the essence of the situation." (Hasan Ozbekhan. Toward a general theory of planning. In: Eric Jantsch (Ed). Perspectives of Planning. Paris, OECD, 1969, p.144).

"Problems misbehave. Instead of neatly slipping into clean-cut categories that correspond with the names of ministries, scientific disciplines, and problem-solving programs, they tend to fuse with each other and become a tangled web. Thus, as a society becomes more complex, analysis of the housing problem leads one into industrial location, transportation, technological development, fiscal policy and intergovernmental relations. Any serious analysis of the population problem leads one into the consideration of the resource base for supporting any given population level, appropriate technologies in the use of such resources as well as in birth control, social security, opportunities for female education and employment, and a variety of cultural and motivational questions. Any problem of ethnic or geographic imbalance within a country cuts across all problems and programs that affect any ethnic or regional subdivision of the country." (Bertram M Gross. Strategy for economic and social development. *Policy Sciences*, 2, 1971, p.353).

The Club of Rome introduced the term "world problematique" to denote the current situation in which mankind is no longer confronted by identifiable, discrete problems, each one amenable to being dealt with on its own terms, but by an intricate and dynamic maze of situations, mechanisms, phenomena, and dysfunctions, which, even when they are apparently disjointed, interfere and interact with one another, creating a veritable problem system. *"Our present situation is so complex and is so much a reflection of man's multiple activities, however, that no combination of purely technical, economic, or legal measures and devices can bring substantial improvement. Entirely new approaches are required to redirect society towards goals of equilibrium rather than growth. Such a reorganization will involve a supreme effort of understanding, imagination, and political and moral resolve."* (Commentary by The Club of Rome Executive Committee on The Limits to Growth. New York, Universe Books, 1973, p.193).

Although there is agreement that interrelationships between problems are so numerous as to constitute a complex network or system, little concerted effort has been made to map this complexity. Such tentative efforts as have been made have generally been limited to the production of simple maps of the relationships between **major** or **critical** problems, or (in a few cases) the production of more detailed maps for some particular problem area.

5. Increasing inadequacy in response to the problem network

The traditional and planned approaches to problems are recognized as increasingly incapable of containing the problem complex as it is now emerging. This situation is illustrated by the following:

"Evidence is mounting that the environment which managers seek to control - or, at least, to guide or restrain - is increasing in turbulence and complexity at a rate that far exceeds the capacity of management researchers to provide new and improved methodologies to affect management's intentions. Faced with the consequences of force-fed technological change, and the concomitant changes in the social, political, psychological, and theological spheres, there is real danger that the process by which new concepts of management control are invented and developed may itself be out of control relative to the demands that are likely to be imposed upon it." (Introduction to a 1968 management conference session of the College of Management Control Systems, The Institute of Management Sciences)

"While the difficulties and dangers of problems tend to increase at a geometric rate, the knowledge and manpower qualified to deal with those problems tend to increase at an arithmetical rate." (Yehezkel Dror. Prolegomenon to policy sciences, AAAS symposium, Boston, 1969)

"Social institutions face growing difficulties as a result of an ever increasing complexity which arises directly and indirectly from the development and assimilation of technology. Many of the most serious conflicts facing mankind result from the interaction of social, economic, technological, political and psychological forces and can no longer be solved by fractional approaches from individual disciplines." (Bellagio Declaration on Planning. In: Erich Jantsch (Ed) Perspectives on Planning. Paris, OECD, 1969).

"What finally makes all of our crises still more dangerous is that they are now coming on top of each other. Most administrations...are not prepared to deal with...multiple crises, a crisis of crises all at one time...Every problem may escalate because those involved no longer have time to think straight." (John Platt. What we must do. Science, 28 November 1969, p.1115-1121).

"Scientists and business and political leaders in virtually every country are becoming increasingly aware that the human race is facing more crises than its social and political institutions can handle adequately....Many important steps are now being taken to meet these problems. These steps, however, are often shaped to fit existing institutional patterns or to be politically or commercially expedient, while other measures of perhaps equal or greater importance have not yet been started. Moreover, the multitude of crises and their complexity and interactions so overburden the mechanisms that have been designed to handle them that there is a valid fear that these mechanisms will break down at the critical moment and make the disasters worse." (R A Cellarius and John Platt. Councils of Urgent Studies. Science, 25 August 1972, p.670-676).

"...the world is becoming so complex and changing so rapidly and dangerously and the need for anticipating problems is so great, that we may be tempted to sacrifice (or may not be able to afford) democratic political processes." (H Kahn and J Wiener. Faustian powers and human choices. In: W R Ewald, Jr (Ed). Environment and Change. Bloomington, Indiana University Press, 1968).

6. Inadequacy of institutional response to problems

The weaknesses of the organized response to problems are best illustrated by the following:

"The map of organizations or agencies that make up the society is, as it were, a sort of clear overlay against a page underneath it which represents the reality of the society. And the overlay is always out of phase in relation to what's underneath: at any given time there's always a mismatch between the organizational map and the reality of the problems that people think are worth solving...There's basically no social problem such that one can identify and control within a single system all the elements required in order to attack that problem. The result is that one is thrown back on the knitting together of elements in networks which are not controlled and where network functions and the network roles become critical." (Donald Schon. Beyond the Stable State; public and private learning in a changing society. London. Temple Smith, 1971)

"Since problems were for so long deemed to be immutable, functions already assumed became more important than aims. Thus the attainment of major national goals, such as the elimination of illiteracy or the improvement of agricultural yields, called for the development of the relevant government functions, such as education and agricultural policy. In the sequel, within each of these functions, new goals were inferred from extrapolations of goals already achieved; the functions defined the problems to be met, and reassessment of the problems at hand did not lead to the redefinition of the function...The rigidity, fragmentation, and institutional competitiveness of bureaucratic practices are obviously both causes and consequences of this state of affairs. Bureaucratic development is partly a result of the vagueness of aims pursued. The determination of new aims is often not sufficient, however, to overcome these weaknesses, which also stem from the inclination of bureaucracies to resist innovation. For these reasons, contemporary societies are called upon to challenge certain forms of organisation that can no longer render the services they require, because in these societies, change and uncertainty have become the constant companions of prosperity. Thus, it has become a commonplace that many new problems, over the last quarter of a century, have been recognized too late by the government machine, which has often been moved to action only by the advent of a crisis...Any attempt to assess dissatisfactions, define opportunities, and formulate new goals inevitably runs counter to established policies that have been instrumental in the emergence of new problems. It will therefore always be difficult to look to operational agencies and policies for an objective effort to redefine aims that may involve agonizing reappraisals, challenge existing interests, or simply call for a sense of perspective incompatible with the responsibilities of day-to-day action. For this reason the identification of emerging problems is a function that tends to be overlooked by traditional public administration and therefore cannot be wholly integrated with it..." (Organisation for Economic Cooperation and Development. Science, Growth and Society. Paris, OECD, 1971, p.60-61)

"Consider the problem of poverty among minority groups. Our nation is committed to reducing poverty. We do not know how to approach solving the problem without creating other undesirable conditions in the process. Our government comes at a problem, like minority group poverty, from many directions: some officials are convinced that all that is necessary is to stimulate economic growth, others call for better education, still others advocate a direct transfer of income, and of welfare. This is much like many blind men feeling parts of an elephant and then being asked to describe it. The man who describes the trunk is as right as the man who describes a leg; both are partially right. Division of problems into subproblems without knowing their overall dimensions hardly ever contributes to a situation." (John Crecine and R D Brunner. A fragmented society; hard to govern democratically. In: Information Technology; some critical implications for decision makers. New York. The Conference Board, 1972, p.178)

"Institutions, firms and (thanks to television) private citizens today receive critical information very quickly indeed; the aggregate picture at federal level is slow by comparison to materialize. To put the point the other way round, then, the body politic has wildly overactive reflexes. In the body physiologic this is the condition of clonus - it is a symptom of spasticity. If we live, as I suspect, in a spastic society it is because of clonic response. And by the expectations of these arguments, the clonus will get worse." (Stafford Beer. Managing modern complexity. In: Committee on Science and Astronautics, US House of Representatives. The Management of Information and Knowledge. Washington, US Government Printing Office, 1970, p.45).

"...increasing specialization makes all problems more difficult. With more economic and social development, the subdivision of labour is carried to extremes never dreamt of in previous historic periods. The more effective and efficient organizations and planning bodies are those that operate for narrow and segmental purposes, thereby rendering much more difficult any effort to achieve mutual adjustment or coordination. The more able, honoured and highly valued expert is the one who works within an increasingly narrow sphere and who has great difficulty in communicating with other experts as well as laymen." (Bertram M Gross. Strategy for economic and social development. *Policy Science*, 2, 1971, p.353)

7. Competing "key problems"

The relationships and significance of each of the lesser-known problems may well be recognized in the appropriate sectors of the available scientific literature, but may thus only influence a limited sector of society. This means that information systems, organizations and programmes often recognise only one particular set of problems and over-identify with them. This results in a multiplicity of candidates for "the key problem" requiring maximum allocation of resources to bodies,which may not intercommunicate even though each may stress the importance of defining its own problem in relationship to other problems. Hasan Ozbekhan makes the point: "This almost subconsciously motivated attempt, that of a sector to expand over the whole space of the system in its own particular terms and in accordance with its own particular outlooks and traditions, compounds the problem by further fragmenting the wholeness of the system. For sectors cannot become systems, they can only dominate them; and when they do they warp them. Hence this tendency toward the spreading of sectoral primacies over the full social space must be viewed with alarm. It is a portent, and an ominous one, of the conflicts and dislocations that await us unless a system-wide integrative approach is worked out..." (Hasan Ozbekhan. Toward a general theory of planning. In: Eric Jantsch (Ed). Perspectives of Planning. Paris, OECD, 1969, p.83-84). There is also the suspicion that the network of problems may be better integrated than the networks of organizational and conceptual resources which could be brought to bear upon them.

8. Institutional difficulties in identifying problems

The "Bertrand Report", a recent major internal review of the difficulties afflicting the United Nations system (Maurice Bertrand. Some Reflections on Reform of the United Nations. Geneva, UN Joint Inspection Unit, 1985, JIU/REP/85/9) notes: *"In short, it is the sectoralized, decentralized and fragmented structures of the System that are the reason for its failure to adapt to the solution of development problems."* (para 104) *"The countries concerned need a World Organization capable of facilitating syntheses, organizing co-ordination, helping to find long-term financial arrangements, and granting many-sided aid to solve the most urgent problems. What the United Nations System offers them is a series of divergent and contradictory recommendations, some 30 bodies whose action has to be coordinated with that of some 20 sources of bilateral aid, but it does not help them to solve their medium and long-term financial problems."* (para 106)

"In other words, since the Organization here is confronting the essential mission it should fulfil, we have to ask ourselves whether it is properly equipped to do so; whether the results obtained so far are satisfactory or negligible; and whether the Organization really does possess the organs capable of reflecting upon and identifying the problems and the framework of negotiations which the modern world needs. The replies to these questions are inevitably negative; the machinery of negotiation is not easily identifiable and separable from the rest of the activities under the various sectoral programmes and does not constitute a coherent system. The results achieved relate only to a few limited fields and do not represent solid progress in the direction of changing world consensus. This situation has its political reasons, which are well known, but they do not explain everything. Actually, it is the structure of negotiations offered by the World Organization that is ill-adapted to solving the problems of the modern world." (paras 107-8)

"They call for considerable preliminary efforts to identify the problems which are susceptible to negotiation before any negotiations can begin. This work of identification is complex, and it comes up against difficulties of a cultural, technical, ideological and semantic kind; it can often only be concluded when a preliminary agreement is beginning to take shape on a given concept; so that it is no longer surprising that it implies attempt after attempt at formulation, often clumsily done, and that it is a source of endless talk. Negotiation among 160 parties presents specific technical difficulties other than those of the size of the meeting chamber or the organization of simultaneous interpretation. It involves the definition of interest groups whose composition and dimensions vary according to the subjects dealt with, and the method of representation of these groups." (para 109)

9. Absence of consensus concerning problem priorities

In 1974 Jan Tinbergen noted that only two years after the (Pearson) report of the International Commission of Development suggested accelerated growth for the developing world, the results of the Club of Rome study indicated the necessity for decelerating world growth. He suggests that these two objectives are not necessarily irreconcilable, but are very close to being incompatible. The two sets of recommendations clearly emerged from studies which detected different problems as being of major importance. Robin Clarke, in demonstrating the pressing need for alternative technology, examined 9 problems (from pollution to alienation) and showed how five different functionally significant constituencies perceived the problems and the necessary solutions. Consensus appeared to be minimal.

It is a fact of political and social life that there is no general consensus on the relative priority of problems. As noted in the first report of the Social Indicator Development Programme of the Organisation for Economic Cooperation and Development: *"Commonality of social concerns among Member countries tends to be greatest at the highest level of generality, diminishing as the definition becomes more specific."* The degree of consensus also increases when the problem is perceived as being so extensive that it can only be solved by some improbable combination of institutions or "everyone acting together".

Significance: constraints on a problem-focused approach

A number of arguments against a problem-focused approach have been encountered during the course of this project. Although the arguments overlap, in that they are based on common conceptions, they are examined separately below.

1. Major problems versus minor problems
It may be argued that the major problems are well-known and have been adequately described and that all other problems are either components of the major problems or relatively unimportant. This raises the question as to how the importance of the major problems was determined. Was the problem of the environment important before the United Nations conference in Stockholm in 1972? Some international organizations have been working on this problem since the 1950's, but a book produced in 1967 by the well-respected Hudson Institute (Herman Kahn and A J Wiener. The Year 2000; a framework for speculation on the next thirty-three years) makes no mention of either pollution or environment. Importance in this sense means simply as a political issue, since all the information concerning the problem was available whilst the problem was still unimportant.

There are however other ways in which a problem can be important. A problem may not be of importance in its own right but primarily by virtue of its relationship to other problems in the problem complex or network. Consider the case where no significant impact has been achieved by the allocation of considerable resources to the mutually reinforcing problems A, B, D and E, considered to be of greatest importance because of their immediate tangible effects. If it can be shown that A, B, D, and E are all dependent on reinforcement from the seemingly insignificant and little-known problems C and F, then C and F may acquire considerable importance in any policy relating to the problem complex. Their relationships to the other problems, and the possibility that they may lend themselves more easily to available remedies, makes them of vital importance in any general strategy, since any positive results will have beneficial multiplier effects which may alleviate the more tangible problems.

Furthermore, if it can be shown that action on problems C and F is impeded by problems G, M and Q, then the latter may acquire even greater significance because of they way in which they obscure critical leverage points in the problem network at which action and research may be most beneficial with a minimum of resources. The difficulty at this time is that it is apparently not possible to determine which problems are like C and F, and which are like G, M and Q, since all attention is devoted to A, B, D and E, except in the plaintive reports from those attempting to implement the solutions to the latter.

Only by exploring the networks of interrelationships between problems of all degrees of importance and visibility will it be possible to locate the critical leverage points, as opposed to those action areas which can continue to absorb resources without any significant result.

2. Major problems versus subproblems
It may be argued that once the major problems have been identified it is unnecessary to attempt to identify the component subproblems with any degree of precision, whether because the precision is illusory or because the subproblems are merely aspects of the major problem without any significant degree of autonomy. In contrast to this view, the OECD Social Indicator Development Programme (1973) in identifying 24 fundamental social concerns stated that each of these "...*may be viewed as the summit of a vertically linked hierarchy of an indefinite number of subconcerns representing the important aspects and means of influencing the fundamental concern. At the same time, there are various kinds of horizontal linkages or relationships among these hierarchies; a particular concern or subconcern may have simultaneous effects on a number of other social concerns.*" (3)

On this point John Crecine and R D Brunner (1972) note: "*Division of problems into subproblems without knowing their overall dimensions hardly ever contributes to a solution. But, it is precisely this division into subproblems that must be achieved, however badly, if an organization is to effectively pursue an objective or execute a program. Without knowing the structure of a problem, it is difficult, if not impossible, to efficiently design solutions or government organization.*"

Also: "*The sad fact of the matter is that we know very little about dividing the social problems with which government must deal into component subproblems. Without effective division of overall problems and subsequent assignment of the parts to specific units, government is likely to remain the blunt instrument it now is. All the information, communication, computer capability, all the coordination in the world is useless if not properly mobilized.*"

The difficulty in identifying subproblems is to determine down to what level of detail it is useful to go in different problem areas. This project explores this difficulty in a number of different problem areas where many levels of subproblem exist (*eg* commodity problems, endangered species).

3. Irresponsibility of drawing attention away from major problems
It may be argued that drawing attention away from the 5 to 10 problems currently in favour as "major", and giving a comparatively greater amount of attention to seemingly "minor" problems, serves to dilute the already inadequate effort to solve the major problems. In order to understand the major problems better it is however necessary to focus on the minor problems through which they may be connected in unforeseen ways. It is by analyzing the network of all problems that it becomes possible to determine what the major problems are under any particular set of conditions.

But perhaps of greatest importance, **people may identify more easily with non-major problems** and unless the interrelationship of all problems can be demonstrated such people cannot be convinced of the merit of allocating resources to the major problems. It could also be argued that programmes to mobilize public opinion in support of the major problems have been in operation for sufficient time to have been able to make any significant impact possible. In the report of the United Nations Secretary-General reviewing the *Dissemination of Information and Mobilization of Public Opinion Relative to Problems of Development* (E/5358, 21 May 1973) it is noted that: "*...it is difficult to escape the conclusion that... the state of public opinion on matters of development, particularly in the industrialized countries, is generally less favourable today than it has been in the past.*" It also notes: "*It would probably be unfair to conclude that a sudden callousness had overcome public opinion in the developed countries. It is more like a closing of the gates to a pattern of generalizations perceived as outworn by over-use.*"

Since a high proportion of available resources will continue to be allocated to the major problems, experiment with alternative approaches is justified to see whether it is possible to break out of the pattern of out-worn generalization. The greatest danger lies in the probability that the United Nations system's public relations and public information programmes will lead the informed public and many decision-makers to believe that the U.N. is doing all that can or need be done and has the attack on every world problem well-coordinated. This automatically devalues the activities of other bodies, reduces the allocation of resources and support to them, dampens initiative from the local and national level which is not channelled through governmental and U.N. channels against U.N.-perceived problems, and effectively nullifies the type of constructive criticism which can lead to renewal of effort, new approaches, and galvanization of the political will necessary to the accomplishment of all international programme objectives.

4. Problems versus values
It may be argued that it is a mistake to focus on the negative features of society, namely problems, rather than on the positive features, namely values or goals, which are a basis for consensus formation and the coherence of society. And yet it is the irony of the times that problems have greater currency than values and would often appear to be the focal point for greater consensus. People can agree about problems and they lend themselves to action-oriented debate. To an

important degree, with the loss of common positive symbols and the absence of a universal ethic, common problems perform a unifying function. In addition they are easier to identify with precision.

The Organisation for Economic Cooperation and Development, through its Secretary-General's Ad Hoc Group on New Concepts of Science Policy (1971), in discussing the formulation of problems, notes that: "*The systematic identification and formulation of new problems are the more necessary because the distinguishing characteristic of many of the present social demands is that they are defined more by the dissatisfactions they engender than by a precise formulation of the satisfactions looked for: existence of dissatisfaction, in other words, does not automatically imply a recognition of preferable alternatives. The complexity of society and the limitations of knowledge make it difficult or impossible to envisage realistic alternatives. This is one of the frustrations of modern society: today's "hungers" are not easily defined. Thus, environmental pollution, the chaos of city life, and the inadequacies of the universities arouse discontent that is not expressed in precise alternative concepts of the types of environment, city, or university desired. Although in many cases these discontents may be based on misperceptions of the objective situation, we must recognise that the perception is itself part of reality. Thus the discontent cannot be alleviated by physical measures alone: it requires an understanding of the total situation.*"

The emergence of problems may therefore be considered as the actualization of hitherto unrecognized values. A problem is in some ways a value in disguise and may signal the presence of new values. In the DEMATEL Project of the Battelle Institute, one element of the definition of a problem was that it related implicitly to a value system (A Gabus, 1972). A problem is an instance of value-dissonance.

5. Problems versus solutions
It may be argued that at a time when everyone is aware of the problems, and many are suffering from excessive awareness, any further emphasis on problems rather than solutions is unconstructive. From this perspective, what is lacking at this time are collections of information on solutions, not collections of information on problems. As will be seen below, however, most of the available information tends to be either on the major problems or on conventional solutions to existing problems. Unless a clear picture of the range of problems is available, and it is not, the solutions proposed may either be solutions to non-critical problems or solutions which will simply aggravate other problems as a result of their successful implementation. There is also the point that solutions envisaged for today's problems are already inadequate by the time they are implemented because of the evolution of the problem environment. A focus must therefore be maintained on tomorrow's problems in the light of current predictions. This approach does not preclude cross-referencing the problems identified here to a parallel collection of information on solutions.

6. Unmanageable number of problems
It may be argued that once any attempt is made to look beyond the 5 to 10 currently favoured major problems there is no limit to the number of problems which can be identified and described. Any problem area can be broken down into subproblem areas which can in turn be broken down further. The exercise then becomes impossible because of the amount of information, and is of questionable value for the same reason. This argument could, however, also be applied to the activities of the botanist and zoologist who now recognize some million species of plants and animals respectively. But zoologists, for example, have found ways of handling this degree of diversity without needing to limit themselves to such basic categories as mammals, reptiles, birds, fish, and insects. The question is whether some similar approach can be made to the range of problems. Only a deliberate attempt to collect such information can provide a basis for any response. This section is itself a demonstration that it is possible to collect information on more than 10 problems without the data becoming uncontrollable.

7. Multiplicity of problem interrelationships
It may be argued that any attempt to record the potential interrelationships between a large number of problems leads to such a large number of interrelationships as to be unmanageable. Thus 1,000 problems could give rise to over 990,000 interrelationships. If however the information collection is limited to those relationships which have been recognized, the number of actual interrelationships is much more limited and therefore quite manageable. Again only a deliberate attempt to collect such information can prove whether such an approach is impractical.

8. General, unstructured approach versus particular, structured approach
It may be argued that any such project is only manageable and of significance if it is conducted in terms of some particular viewpoint such as the policy requirements of a given organization. Or else, it may be argued that a particular classification scheme or model must first be developed to guide the subsequent collection and presentation of information. These are however precisely the difficulties at this time. There is a multiplicity of oriented projects and models with no effort at interrelating them or suggesting that they should be interrelated. And yet it is the disagreement amongst the advocates of different approaches which hinders the formation of any consensus or general strategy and the mobilization of adequate resources. The challenge is to develop a project which is as general and minimally structured in its approach as is feasible, but without losing coherence and utility. This project is an experiment in that direction.

9. Inadequate theoretical framework
It can be argued that a project of this kind cannot be fruitfully undertaken without a well-articulated theoretical framework. This can be viewed as essential in defining what are to be considered problems. The theory of problems is however poorly developed to the point of being non-existent. There is even the suspicion that problems are in an important respect undefinable. A special weakness of relying on any particular theory, is that only those phenomena which fit the theory are then considered worthy of collection. But such inadequacies should not prevent the collection of information which may assist in articulating more appropriate theoretical frameworks in the future.

10. Erroneous conception of a problem as a well-defined entity
It may be argued that problems, by their very nature, are nebulous and poorly defined, and that therefore a numeric identifier cannot be usefully and meaningfully allocated to a problem. Any such treatment of the problem in fact distorts the nature of the problem and gives it a precision which it lacks in reality and implies that it possesses characteristics which it may not have. It is therefore impossible to make a list of world problems because what is perceived as a problem is in fact a cultural variable. Any such attempt therefore forces all problems into the same mould and implies that they can all be conceived as having common features particularly when embedded in a network of problems. The notion of a relationship as a simple link between two problems may also be considered unsatisfactory for similar reasons.

This project is however not so much concerned with what a problem is as a problem but rather with how a problem is perceived and discussed in terms of the labels given to it. It is in denoting the variety of phenomena as "problems" that the above errors may be encountered, but once this has been done and has achieved the present acceptability it then becomes permissible to identify the semantic domain in question by a numeric identifier and to attempt a summary description of the processes believed to be associated with that domain.

11. Sufficiency of information on problems
It may be argued that there is already a very large amount of information available on most problems. Some problems have one or more books describing them; some have whole specialized libraries devoted to them; many are covered by specialized periodicals and abstracting systems. Under such circumstances a summary description could not do justice to the complexity of the subject matter and the available knowledge.

Against this it must be said that only specialists can afford the time to scan such quantities of information, and only well-endowed institutes can afford to obtain even a small proportion of the available material. In addition, as was discovered during this project, the information is rarely structured in such a way as to make evident the nature of the problem, let alone the relationships between one problem and another. Such information is scattered through a multitude of documents, except in a few isolated cases. Whilst many documents exist, they may well be effectively unobtainable during the time they are needed.

Current international information systems do not facilitate access to many vital documents. Such documents may be quickly out of print, and normal booktrade delays may be up to two months between Europe and America and up to six months in the case of some developing countries. But whether available or not, the widening gap between the exponentially increasing quantity of data available for consumption and man's very limited capacity for acquiring and processing useful information needs to be bridged by new methods of presenting information. The attempt in this project to hold problem information in networks of relationships which can be plotted on maps or displayed on computer graphics devices is an experiment at reducing the current difficulty.

12. Theoretical superficiality
This project emerged partly in response to the initial Club of Rome exercise on "Limits to Growth" which promised hopes of understanding the complex dynamics of the world system through computer models. It was believed that authenticated facts could be married with theory to provide an integrated, in-depth framework through which the difficulties and opportunities of the world system could be explored. Collection of problem perceptions, in which all "facts" are a matter of interpretation, can be viewed as totally superficial by comparison. Since that time many competing world models have been produced, each in response to perceived inadequacies in the others. The models have been unable to reflect the full range of problems in the real world, especially those which are less readily quantifiable, even though they may have considerable impact on world dynamics (eg "alienation", "blasphemy", "corruption"). With the failure of planned economies, the limited value of such models for elitist policy-making in the real-world has become even more apparent. They deny the impact of images of problems, however misguided. This project is an effort to register the expressed concerns of different segments of the population with a view to using a different class of mathematical tool to approach such complexity and render it comprehensible.

13. Inappropriateness of "problem-solving"
The response to problems, especially in the international community, has come to be dominated by a "problem-solving" perspective which is increasingly questionable. Donald Schön (1979) in commenting on this notes: *"The problem-solving perspective contains three central components. It directs our attention, first of all, to the search for solutions. The problems themselves are generally assumed to be given...If problems are assumed to be given, this is in part because they are taken always to have the same form."* He sees this as based on an instrumentalist assumption. *"Problem solving consists in the effort to find means for the achievement of our objectives in the face of the constraints that make such achievement difficult...The problem solver...is always engaged in searching some problem-space in order to find means well-suited, in the face of constraints, to the achievement of some objective function."* He also notes however that there have been increasing difficulties with this perspective and that a sense of inadequacy has begun to spread among practitioners of social policy and among the public at large. The social situations have turned out to be more complex than was supposed. According to Schön, it becomes increasingly doubtful in the case of social policy that we can make accurate temporal predictions or design models which converge upon a true description of reality. *"Moreover, the unexpected problems created by our search for acceptable means to the ends we have chosen reveals...a stubborn conflict of ends traceable to the problem setting itself."* This project might therefore be considered a response to his concluding point that, in the domain of social policy, there is a need to understand better how problems are set.

14. Project approach as instance of the key problem
It may be argued that the allocation of resources to the collection of information on problems is in itself an example of the general tendency to substitute action about a problem for action on the problem. The problems are denatured by the process and lose the potency that they have in the real world. Worse still, any attempt to draw attention away from the key problem (such as capitalism or communism) to a multiplicity of pseudo-problems, which at best are symptoms of the key problem, merely serves to aggravate current difficulties, whilst profiting from them. All action can however be criticized in this way, particularly when there is disagreement on what **the** key problem is or what problem components should be tackled in what order in any strategy. This project is an experiment in alleviating both the difficulties from which such disagreement arises and those to which it gives rise.

15. Problems as human constructs
There is increasing recognition that problems are human constructs, namely artefacts of concerned minds. In the words of Donald Schön in discussing problem-setting in social policy (1979): *"Problems are not given. They are constructed by human beings in their attempts to make sense of complex and troubling situations. Ways of describing problems move into and out of good currency."* From this perspective it might be argued that any attempt to document problems is quiet unrealistic, resembling in some respect efforts to document dreams and to attribute some reality to them. But it may also be argued that if problems are treated as realities, with some persistence over time, then it is important to understand how people are influenced and moved by such realities.

16. Problem perception as cultural bias
In the light of the previous point, if problems are artefacts of concerned human minds, it has also been suggested by Kuang-ming Wu (1982) that they are the artefacts of Western minds. For him the supposition of "problematicalness", with its attendant implications for reason, for principles, and for history, is so deeply ingrained in Western consciousness that its denial seems absurd. But, in the light of his interpretation of Chuang Tzu, to conceive of life as presenting problems to be solved is a misconception of life. If there are indeed major problems of culture, and cultural attempts to respond to them, then history is not merely a chronicle of episodes but allows of interpretation as a form of drama. With a problem-oriented vision it is possible to speak of the rise and fall of civilizations, of a dialectic of progress or devolution, and of the importance of roles in history in relation to problems. But if it is not necessary to see life as presenting problems or to understand life in relation to problems, then these features of historical consciousness are not as important as they presently seem. Alternative views are then also possible and may even prove more appropriate.

Significance: framework for interrelating incompatible perspectives

1. Need for a common frame
Before achieving consensus for purposes of action, some framework needs to be developed within which the different problems can be interrelated **prior** to the determination of their relative importance. Geoffrey Vickers argues that: *"The changes that will flow from all of these impacts are unpredictable and perhaps unimaginable, but we can prepare to recognize and understand them more quickly as they emerge, by finding some common frame within which to comprehend them."*

2. Myth of consensus
Consensus does not have to be total for effective action to take place. Different constituencies can pursue different problems provided that there is some general understanding of how the different problems being tackled by different groups are interrelated, at least in the terms of each perspective.

3. Possible criterion for a framework
From the previous paragraphs, such a framework should be able to contain:

(a) problems which are incompatible in the light of the conclusions of different kinds of scientific analysis;

(b) problems which may be perceived by one group to be irrelevant or trivial, and by another to be of major importance;

(c) problems which are normally unmentionable in intergovernmental circles for political reasons, namely wholesale massacres, torture, political imprisonment, and other sensitive problems, whether current or in recent history;

(d) problems which are potentially, but not currently, important as political issues (such as environmental pollution prior to the 1972 UN Conference in Stockholm);

(e) problems, recognized as such by the United Nations, but which catch many others unprepared because of the strength of the counter-claim (*eg* the 1975 UN vote to recognize zionism as a form of racism);

(f) problems, recognized as fundamental as a result of very sophisticated analysis, which are extremely subtle and essentially beyond the capability of existing institutions (*eg* Kenneth Boulding's identification of the reduction of psychosocial variety as being a major threat to society's ability to respond successfully to future crises).

4. Recognizing the dynamic between incompatible perspectives
As things stand no existing framework even attempts to reflect such incompatible perspectives. And yet the dynamics of their interaction are the reality of social life. Just as in the case of the arms race, it is the action-reaction phenomenon between the protagonists which contributes directly to its continuation. In this connection it is valuable to recall the technique used, in very difficult times, by Diderot and d'Alembert, the editors of The Encyclopaedia (1751-1772). *"The editors of The Encyclopaedia were well aware of the dangers they faced, and so they cleverly maintained an air of innocence throughout. By a brilliant device of cross-reference, however, they were able to annihilate the effect of an orthodox view in one article with the arguments expressed in another article to which the reader was referred."*

In contrast to those times, the right view cannot be simply brought to light by a brilliant argument (or other device) cross-referencing the outmoded incorrect view. Nowadays, there are many intellectual and other authorities, each with its own set of arguments. It is no longer easy to determine which set of arguments protects an outmoded view, or by which view it should be replaced, since all the functionally significant groups (even amongst the sciences) compete in advocating their own perspectives and in criticizing every other perspective. The arguments of many of the groups may be well-documented, although the absence of evidence in the case of the others does not curb their advocacy or the sincerity of belief in their particular perspective.

5. Registering disagreement
It is however possible to envisage a framework in which the problems perceived by each group could be combined, or registered separately if there is disagreement, accompanied by their supporting arguments and the relationships perceived to other problems. The problems emerging from each perspective can be handled in this way. So can the focal points of disagreement. If the claims by one group for the existence or importance of a problem are contested by another, then the arguments of the counter-claim can be recorded with the claim.

6. Refining opposing perspectives
In contrast to the example of The Encyclopaedia, in such a case each group supplies its own brilliant arguments, annihilating or ignoring the competing groups. The functions of any editorial group are then limited to locating the best formulated argument for each position and for the problem inter-relationships which they consider significant. This task can best be performed with the collaboration of the interested groups, preferably through their representatives at the international level, whether inter-governmental, nongovernmental, or informal bodies.

7. Tolerance of ambiguity
Clearly the results of such an exercise would not satisfy those with a thirst for the immediate and final answer on any particular problem, because when any such final answer is contested, the aim would be to reflect the dissent, even of a minority group. As Abraham Kaplan (1964) has explained in discussing methodology in the behavioral sciences:

"The demand for exactness of meaning and for precise definition of terms can easily have a pernicious effect, as I believe it often has had in behavioral science. It results in what has been aptly named the premature closure of our ideas. That the progress of science is marked by successive closures can be stipulated; but it is just the function of inquiry to instruct us how and where closure can best be achieved.... That a cognitive situation is not as well structured as we would like does not imply that no inquiry made in that situation is really scientific... Tolerance of ambiguity is as important for creativity in science as it is anywhere else."

8. Providing contextual understanding
But irrespective of the scientific value of such a framework, it is a necessity to policy formulation. In discussing the problems of developing contextual knowledge John P Crecine and R D Brunner (1972) note: *"It is not enough for the masses and the government to understand one another and to be able to communicate effectively. Knowing what the problems are, in and of themselves, seldom proves sufficient to improve situations. A different kind of knowledge and ability is required concerning the context of public-sector decisions and the workings of those societal mechanisms the public sector attempts to alter. Uncovering necessary contextual knowledge to support public policy moves is difficult at present and likely to becomes more so.....Little effort is made to determine the content or the timing of research to maximize its contribution to the solving of social problems. The means of achieving full employment of minorities in an urban ghetto, for example, is not a problem which an economist, a political scientist, or a sociologist alone is likely to solve. To the extent that scholars focus their attention on increasingly narrow details without relating the results to a more comprehensive map of society, they are not likely to provide public officials with the knowledge necessary to grapple effectively with the problems of society."*

Significance: unique features

1. Phenomenological approach
The "world problems" incorporated in this section are those identified by specific constituencies and groups of experts in the light of their own criteria and world views. Where information from different constituencies appears to be identifying the same problem, it is integrated into a single description. The editorial research endeavoured to honour the language and emphases of the original constituencies rather than to impose particular interpretations upon the information so as to fit the problem into the editors' own classificatory framework.

2. Scan-based problem search
Although the better-known problems can be identified in information systems by their names ("inflation", "torture", etc), most problems cannot be readily detected in this way. Many problems do not have well-defined names and may only be identified by phrases of some length. Several variants may then be possible when synonyms are used. Interrogating an information system for "problems" is not productive since that term, or its synonyms, is seldom a descriptor for the document containing information on a specific problem. Furthermore, although many specific problems are the subject of documents which are identified in their titles, the majority are only mentioned as chapter or section headings, or in the body of the text.

Detecting problems for addition to this section is therefore most effectively done by extensive scanning of a wide variety of documents. This procedure extends the catchment area much beyond that based on any predetermined classificatory framework and ensures the incorporation, rather than the exclusion, of unforeseen problems described in the literature.

3. International and multi-cultural sources
Information is included on problems recognized in many countries and cultures. Deliberate efforts have been made to include problems perceived in non-western cultures, even when they do not accord with the views prevailing in the international community.

4. Multi-disciplinary sources
Efforts have been made to include problems recognized by a wide variety of disciplines, whether those of the natural or of the social sciences. Some problems may be acknowledged by single disciplines, others by clusters of disciplines, and others by no currently existing discipline.

5. Sensitivity to undervalued constituencies
The search for problem perceptions has been extended to include some problems recognized by no international or academic authority, especially those only recognized by practitioners in the field, or by those actually experiencing the problem. Whilst there is widespread consensus about the better-known problems, there are constituencies around the world which are sensitive to other kinds of problems. Such constituencies may be either numerically small, disempowered, misinformed, relatively disorganized, or simply unskilled at gaining recognition for their problems by the international community. This section aims deliberately to incorporate such problems where they can be detected. The intent is to identify what different constituencies experience as painful and important, whatever the views of others may be. Such problems as "apathy", "sin", "witchcraft", "abduction by extra-terrestrials", "blasphemy" and "wife-beating" are therefore also included, even though they may be of little interest to many conventional approaches to "world problems". A major reason for this approach is that people are moved by what touches them, however irrational this may appear from another perspective. Unless there is greater recognition of the relative importance currently attached to problems in a democratic society, there will be little understanding of the steps required to reallocate resources to "refugees" from the considerable amount people currently choose to allocate to "wrinkles".

6. Collection of biases
The problems included reflect many biases, whether political, ideological, cultural or religious. The particular bias of this project is that recognition of the spectrum of such biases is considered as basic to the formulation of more appropriate strategies. Since initiatives are engendered and sustained by such biases, it is assumed that they will continue to contribute directly to the dynamics of society and must therefore be woven into any more general strategy, however incompatible they appear to be.

7. Inter-problem relationships
There is a widespread tendency to treat problems in isolation and to design organizations insensitive to the relationships between problems. In this section much effort has been devoted to indicating relationships between problems. One set of relationships indicates the general/specific (broader/narrower) links between them. Another indicates the functional (aggravating, reducing) links between them.

8. Integration of detail within overview
Through the indication of an extensive pattern of relationships between problems, it becomes possible to integrate very specific problems, with which people may identify more closely, into more general (or fundamental) problems on which policy-making and organizational strategies tend to focus.

9. Open categorization
The problems included are of many types and reflect many levels of expertise and degrees of sensitivity. This section does not establish definitive categories to cluster such problems. The pattern of relationships links problems of quite different types. How such categories should be established in the light of the priorities of any particular user, thus becomes a decision to be made by the user. Users benefit by being confronted with the need to make that decision because it offers them the opportunity to review the assumptions that it is so easy to make when excluding certain types of problem as irrelevant to some current concern. This is one reason for randomly distributing the problems within any section. Ideally, on any given page, any user should find a third of the problems to be obviously relevant, a third as raising challenging questions as to whether they might be relevant, and a third that could only be considered as irrelevant (if not ridiculous). Such is the diversity of preoccupations that it is to be expected that different users would select and reject different problems. The presentation helps to encourage reflection on which problems are important or irrelevant to whom under what conditions.

10. Counter-claims
In order to reflect the questionable status of many problems in the eyes of particular schools of expertise, the descriptions of problems include a "counter-claim" where possible. This presents any arguments against the existence of the problem as formulated. Such counter-claims help to demonstrate that the problem domain is a highly turbulent one in which many so-called facts are treated as totally questionable from other perspectives.

11. Establishment of a framework
This section constitutes a framework within which problems can be "registered" and located whatever their status in the eyes of authorities or experts of any persuasion. Other attempts to document ranges of problems tend to exclude those which do not correspond to the current fashions and priorities of the international community. This section establishes a framework within which degrees of relevance can be explored and then revised, without endangering the pattern of perceptions.

12. Longer-term perspective
Although much effort has been devoted to collecting information on problems currently perceived as important, whether fashionable or not, the framework is designed to permit the inclusion of problems of less immediate concern. Potential problems of the medium and long-term future, especially those for future generations, are also included in order to provide a "foresight" dimension. The framework also includes what might be considered problems of the past (or of the ignorant and superstitious), such as "eclipses" and "demons".

Significance: precedents and parallels

1. Previous UIA initiatives

The Union of International Associations, an international non-governmental organization founded in Brussels in 1907 partly on the initiative of two Nobel Peace Prize laureates (Henri La Fontaine, 1913; Auguste Beernaert, 1909), had activities prior to 1939 which are of historical interest in relation to the current project. These include:

(a) *Annuaire de la Vie Internationale* (Vol I: 1908-1909, 1370 pages; Vol II: 1910-1911, 2652 pages) which included information on problems with which international organisations were concerned at that time;

(b) *Code des Voeux Internationaux: codification générale des voeux et résolutions des organismes internationaux* (1923, 940 pages, under the auspices of the League of Nations), which listed those portions of the texts of international organisation resolutions which covered substantive matters, including what are now regarded as world problems. It covered 1216 resolutions adopted at 151 international meetings. The subject index lists some 1200 items.

(c) Paul Otlet, co-founder of the UIA, produced in 1916 a book entitled *Les Problèmes Internationaux et la Guerre* which identified many problems giving rise to and caused by war, and proposing the creation of a League of Nations. In 1935 he attempted a synthesis which touched upon many problems and their solution within a society in transformation. The preface bore the title "*The Problem of Problems*". He also dealt with this question in 1918.

The different series of publications of the UIA since 1949 constitute a useful source of information on problems recognized by international organizations, especially the *Yearbook of International Organizations*. The programme to produce this Encyclopedia was initiated in 1972 with the support of Mankind 2000. The first edition was produced in 1976 (under the title *Yearbook of World Problems and Human Potential*). Work on the second was initiated in 1982, leading to publication in 1986.

2. Reference books and surveys

(a) **By intergovernmental bodies**: Several of the major bodies within the United Nations system publish reference books which include descriptions of a broad range of many world problems. The World Health Organization has published *Health Hazards of the Human Environment*. The International Labour Organisation publishes *Encyclopaedia of Occupational Health and Safety*. The World Bank has published *Environmental, Health and Human Ecological Considerations in Economic Development Projects*. It continues to publish annually the *World Development Report*. The United Nations Environment Programme (in cooperation with the World Resources Institute) publishes the *Environmental Data Report*.

Individual divisions within the United Nations system produce a very large range of document series which present summaries of the current state of a particular world problem area, *eg* the periodic *Report on the World Social Situation* produced by the UN Department of Economic and Social Affairs. Statistical yearbooks or reviews are produced by the major agencies.

Apart from the United Nations system, many of the 300 other conventional intergovernmental organizations (and the 1500 of other categories of intergovernmental body) produce detailed analyses, summaries, or statistical surveys relating to the world problems in their domain. For example, the Environment Directorate of the Organisation for Economic Cooperation and Development produces a series of reports on individual environmental problems. The Council for Mutual Economic Assistance has created an International Institute for the Study of Economic Problems in the Worldwide Socialist System.

Some major international organizations periodically attempt to review the range of world problems with which they are concerned, in an effort to redefine their priorities for the future. Thus, for example, UNESCO has produced an *Analysis of problems and table of objectives to be used as a basis for medium-term planning (1977-1982)*. This exercise resulted in the identification of 12 major world problems which were linked to 59 objectives. The Organisation for Economic Cooperation and Development produced in 1973 a *List of Social Concerns Common to Most OECD Countries*.

Such bodies hold a multiplicity of international meetings, frequently on specific world problems, which give rise to meeting reports which may identify new world problems and contain valuable material on them. Overviews of this activity may be obtained from publications of the Union of International Associations such as the 3-volume *Yearbook of International Organizations* (1990).

(b) **By international nongovernmental bodies:** As with the intergovernmental bodies, many of the 4600 conventional international nongovernmental organizations (as well as the 4000 of other categories of international body) undertake equivalent studies.

The annual report of Amnesty International is a well-publicised example. The International Institute for Environment and Development and the World Resources Institute collaborate in the production of *World Resources*, initially intended as an annual publication, with each volume complementing previous ones rather than being updates of them. The volumes offer reviews of issues together with data tables.

Again, overviews of this activity may be obtained from the 3-volume *Yearbook of International Organizations* (1990).

(c) **By individuals and other bodies:** Encyclopaedias and similar general reference works contain descriptive information concerning a wide range of problems, although the problem is generally not recognized as a problem but rather as a phenomenon. Important problems may be omitted. Thus although Diderot's Encyclopaedia in the 18th century includes an entry on torture, the 1975 edition of the *Encyclopaedia Britannica* does not, nor does the *International Encylopaedia of the Social Sciences*.

A World Design Science Decade (1965-1975) was proposed by R Buckminster Fuller to the International Union of Architects at their 6th World Congress in 1961. This proposal called for the initiation, by schools of architectural and environmental planning around the world, of a continuing survey of the total chemical and energy resources available to man on a global scale, and of human trends and needs in relation to these resources, and of how the use of these resources may be redesigned to serve all humanity. This proposal led to the creation of the World Resources Inventory at Southern Illinois University (Carbondale) and to the production of a series of documents by R Buckminster Fuller and John McHale relating to each phase of the programme. Phase 1 was entitled *World Literacy re World Problems*, for which one of the documents produced in 1963 was *Inventory of World Resources, Human Trends and Needs*.

In the period 1970-72, the Institute of Cultural Affairs and the associated Ecumenical Institute (Chicago) undertook an extremely comprehensive survey of the range of contradictions with which society was confronted. This material was ordered in various ways in a series of unpublished studies one of which identified 385 contradictions grouped into 77 categories. These contradictions were perceived as underlying problems in many sectors (economic, cultural, social, *etc*). From 1974-78 this material was used to guide 50 community dialogues in some 30 countries. Each of these gave rise to further sets of contradictions described in a series of internal reports.

Lester Brown, with colleagues, has produced an annual *State of the World* report, since 1984, which is widely distributed. The English-language editions are published by the Worldwatch Institute (Washington DC) of which he is director. The report reviews current problems in fields such as ecology, resources, hunger, population and energy.

In an effort to reach a wider audience, two overviews of the planetary situation have been provided in similar formats. The *Gaia Atlas of Planetary Management* (1985) is edited by Norman Myers and reviews and illustrates problems and possibilities in relation to land, ocean, elements, evolution, humankind, civilization and management. The *Gaia Peace Atlas; survival into the third millenium* (1988) is edited by Frank Barnaby.

There are an increasing number of reports by individuals and organizations focusing on environment-related problems. For example, Jacques Cousteau and the Cousteau Society have produced *The Cousteau Almanac; an inventory of life on our water planet* (1979) which reviews many problems.

A number of school textbooks and teachers guides have been produced on world problems: *World Problems in the Classroom; a teacher's guide to some United Nations tasks*, which gives information on 12 problems (1973); *One World; sources and study guide*, which gives information on 5 problems (1971); *World Problems*, which gives information on 6 problems (1971); and *World Problems; a topic geography* (1973), which gives information on 36 problems in 8 groups. Many others have been produced in more recent years.

3. Independent commission reports
A series of "independent commissions", loosely related to the United Nations system, have produced influential, authoritative reports on broad ranges of issues. These have provided a vital focus for reviews of problems and strategies free from many of the constraints of particular institutional frameworks. It is unfortunate that these reports each fail to integrate the findings of those that preceded them, often failing even to mention them:

- Capacity Study of the United Nations Development System (Jackson Report, 1969)
- Commission on International Development (Pearson Report, 1969).
- Reshaping the International Order, meeting from 1974-76 (Tinbergen RIO Commission, 1976)
- Independent Commission on International Development (Brandt Commission), meeting from 1977-82 (1980, 1983).
- International Commission for the Study of Communication Problems (MacBride Commission), reporting in 1980.
- Independent Commission on Disarmament and Security Issues (Palme Commission), meeting from 1980 (1982).
- Independent Commission on International Humanitarian Issues, meeting from 1983.
- World Commission on Environment and Development (Brundtland Commission), meeting from 1983 (1987).
- Commission on Health Research for Development (Evans Commission), meeting from 1987-90 (1990).
- South Commission (Nyerere Commission), meeting from 1987-90 (1990).

These reports derive much of their importance from their role in articulating possible global strategies. Intergovernmental and nongovernmental bodies may collaborate in this process, as is especially noteworthy in the case of IUCN, WWF and UNEP in producing a **** strategy document on the environment (1990).

4. Research and modelling
(a) **By intergovernmental bodies:** The United Nations University is chartered to devote its work to research into the pressing global problems of human survival, development and welfare that are the concern of the United Nations and its agencies. This is done through a network of research and post-graduate training centres and programmes located around the world and coordinated by a central body. Its project on Goals, Processes and Indicators of Development (1978-82) strongly influenced the content of a number of sections of this publication.

The United Nations Institute for Training and Research directs research into problems which are of interest to the Secretariat and the Assembly of the United Nations and which is primarily of interest to national officials and diplomats. UNITAR has undertaken a future studies programme, particularly in terms of impact on the United Nations. The United Nations Research Institute for Social Development conducts research into problems and policies of social development during different phases of economic growth. The United Nations Social Defence Research Institute undertakes research into the field of prevention and control of juvenile delinquency and adult criminality. Many other such specialized international research units exist.

The International Institute for Applied Systems Analysis (whose members are the principal scientific academies in each country) initiates and supports collaborative and individual research in relation to problems of modern societies arising from scientific and technological development.

The General Conference of UNESCO adopted a resolution at its 23rd Session (1983), creating a major programme concerned with reflection on world problems and future-oriented studies. This was reconfirmed at its 24th Session (1985). During the first two-year period a symposium was held on the creation of a decentralized network for analysis and research on world problems (1984). During the second two-year period, with a budget of $1.8 million, it was proposed to track the evolution of the global problematique and its perception by different schools of thought, encourage research on it and promote exchanges of information and ideas on world problems through the network. The programme, severely threatened by internal and budgetary problems, has given rise to a series of over 50 studies. The first of these contains the report of a Scientific Workshop in 1987 on world problems in the year 2000 (UNESCO, 1987). Another study, jointly by the Institute for Scientific Research on Systems and the Institute of Philosphy of the USSR Academy of Sciences, deals with the systems analysis of world problems (Institut de recherches scientifiques sur les systèmes, 1988)

Within the United Nations system efforts are being made by the various statistical units to move towards the implementation of a System of Social and Demographic Statistics which could serve as the principal data base on many social problems, but particularly for the preparation of social indicators by which many problems are identified and tracked (1974).

The Organisation for Economic Cooperation and Development, as the result of the first phase of its Social Indicator Development Programme, has produced a *List of Social Concerns Common to Most OECD Countries* (1973) with the object of identifying the social demands, aspirations and problems which are or will likely be major concerns of socio-economic planning processes. The social concerns identified are those *"which are of sufficient importance, present or potential, to the governments of those countries for them to want to have indicators available on a comparable basis."* A social concern is defined as *"an identifiable aspiration or concern of fundamental and direct importance to human well-being as opposed to a matter of instrumental or indirect importance to well-being."* Social concerns involving means rather than ends are excluded. The list identifies 24 concerns in 8 groups; 14 of the concerns also have a total of 56 sub-concerns indicated against them. Each of the 24 fundamental social concerns *"may be viewed as the summit of a vertically linked hierarchy of an indefinite number of sub-concerns representing the important aspects and means of influencing the fundamental concern. At the same time, there are various kinds of horizontal linkages or relationships among these hierarchies; a particular concern or sub-concern may have simultaneous effects on a number of other social concerns...It will remain with the planners for specific sectors to extend the hierarchy further downwards to suit their more detailed sector planning, evaluation and programme needs and to establish horizontal relationships between the diverse components of the hierarchies."* The document notes that *"Commonality of social concerns among Member countries tends to be greatest at the highest level of generality, diminishing as the definition becomes more specific."*

(b) **By international nongovernmental bodies:** The Club of Rome (created in 1968 by a group of 30 individuals and limited in membership to 100) initiated in 1970 a Project on the Predicament of Mankind. This had as its objective the examination of the complex of problems in the world, conceived as a world problematique in that: the problems occur to some degree in all societies; they contain technical, social, economic and political elements; and that they all interact. The project has been conducted in phases:

- The first phase led in 1972 to the very well-publicised study under Dennis Meadows entitled *The Limits to Growth*. This examined the interaction of five basic factors (or problem areas) that determine and limit growth on the planet.
-

The second phase resulted in 1974 in the production of a report *Mankind at the Turning Point* by M D Mesarovic and E Pestel in which the global system outlined in the previous phase was disaggregated into ten major interacting geographical regions and analyzed with new methods.

- The third phase in 1976 led to the production of a report on *Goals for Global Society* (under the direction of Ervin Laszlo) which identified sociological, psychological and cultural inner limits which could give positive direction to human aspirations. The Club of Rome world system modelling exercise has stimulated many emulators and rectifiers. A survey of these has recently been produced (1982). The Club was also responsible for initiating in 1974 the project on Reshaping International Order, mentioned above.

The International Federation of Institutes for Advanced Study is a mechanism for transdisciplinary and transnational initiatives to assist society to cope with an increasingly complex, rapidly changing and interdependent world. The Institute sponsors global modelling activities. Since 1987 it has been active in establishing a Human Dimensions of Global Change Programme (also known as Human Responses to Global Change Programme) to complement the UNESCO-ICSU International Geosphere-Biosphere Programme.

The Battelle Memorial Institute, through its Geneva Research Centre, conducted the DEMATEL project (namely DEcision-MAking Trial and Evaluation Laboratory). The objectives were to help find better solutions to world and generalized problems based on a better understanding of the problem structure or so-called world problematique, in order to avoid the selection of solutions which are in fact problem-generating. A survey of problems perceived by about 100 responsible and knowledgeable persons was prepared in 1972, and led to the production of a list of 48 problems in 14 groups. The initial questionnaire and proposed follow-up questionnaire were designed to determine the perceived relative importance of problems and their affect on each other. Mathematical techniques for the analysis of these systems of interrelationships were then developed. An objective was to produce a map of the world problematique (1973).

The organization Futuribles International undertook in 1986-87, under UNESCO auspices, a world-wide survey among eminent persons from varying geopolitical and ideological backgrounds selected for the quality of their thinking on major world problems and the prospects for change (Futuribles International, 1987).

The experience of elder statesmen has been harnessed through the InterAction Council of Former Heads of Government. The reports of its meetings endeavour to provide a synthesis of perspectives on the longer-term issues to which society is vulnerable (InterAction Council, 1990).

(c) By individuals and other bodies: A large number of research-oriented institutes have programmes which attempt to identify and focus on one or more of the most critical world problems. Such institutes are usually related to some aspect of planning, forecasting, futures, technology assessment, or policy sciences. An overview of the range of this activity may be obtained from Michael Marien's *Future Survey Guide to 50 Overviews and Agendas* (1990). This is essentially an update of his *Societal Directions and Alternatives; a critical guide to the literature* (1976) and to the regular *Future Survey Annual; a guide to recent books and articles concerning trends, forecasts and policy proposals,* produced by him for the World Future Society which publishes it.

A number of institutes maintain the results of extensive surveys of current activities around the world in their own data banks. Such a survey, in the field of future studies, has been conducted for the United Nations Institute for Training and Research (UNITAR) by the Center for Integrative Studies. The World Future Society produces a directory of future-related resources and a periodical surveying them.

Most institutes are primarily concerned with a limited range of major problems, such as population, resources, or environment (within which are of course grouped many other problems although usually not distinguished as such). An exception is the Hudson Institute which has identified 78 technological crises in 7 groups. Many institutes necessarily conduct such research to identify the problems which will affect the body or area from which their funding is derived, eg Europe 2000, Hawaii Commission on the Year 2000, or individual corporations interested in predicting the environment within which their products must be profitable. There is a well-recognized tendency for institutes to switch programmes from year to year as new problems appear on the horizon of funding bodies.

There is a tendency for special institutes to be created in each country for the comprehensive analysis of policy alternatives, national goals, and national priorities. These necessarily involve a focus on the world problem context. An example is the Institute for the Analysis of Public Choices established by the Aspen Institute for Humanistic Studies.

There is of course an unknown amount of government-sponsored classified research as well as corporation-sponsored proprietary research. This may well be superior to anything that is publicly available, although it is likely to suffer from the disadvantage of being oriented in terms of the sponsoring body.

Numerous books and articles by individual researchers identify specific world problems or groups of problems and propose taxonomies for them. However the number of problems taken into consideration is usually less than 10. An exception is Hasan Ozbekhan's series of 28 Continuous Critical Problems (1969). During his period as Executive Director of the Club of Rome (prior to the Limits to Growth exercise) this was extended in an internal document to approximately 50 problems. These problems are system-wide and are characterized by the fact that they cannot be solved independently of the rest of the set.

There is a very extensive literature on social problems and social issues. An *Encyclopedia of Social Reform* was even produced at the end of the 19th Century. There appears however to be an important difference between what are currently included under the term world problems and what is currently meant by a social problem, although even amongst sociologists there is disagreement as to the definition of a social problem. Thus in *Contemporary Social Problems* edited by R K Merton and R Nisbet, 15 major social problems are identified. The exclusion of other possible problems is justified by the statement: *"Sociology is a special science characterized by concepts and conclusions, which are based on analysis and research, yielding in turn perspectives on society and its central problems. For many decades now, sociologists have worked carefully and patiently on these problems."* (1971) Social problems would therefore appear to be those problems perceived by sociologists as being the central problems of society.

A major study was commissioned by President Carter and resulted in the production of the *Global Report 2000; a report to the President* by the US Council on Environmental Quality.

Numerous studies of world order have been made in which the focus is placed on the political-social-legal forms, organizations and institutions envisaged as being relevant to the solution of world problems, especially those connected with organized violence. The most extensive of these is the World Order Models Project (sponsored by the Institute for World Order), which has given rise to a series of publications (1972-1979).

Numerous surveys have been conducted of community attitudes towards local problems and problem-solving. An example is the Benchmark programme of the Academy of Contemporary Problems (Ohio State University). Other surveys have been made of some special-interest membership organization concerning the relative importance of current problems or those that they perceive as emerging in the foreseeable future. A few nation-wide surveys of this type have been conducted. Thus, for example, in 1968 the *Sunday Times* in the United Kingdom requested that readers write in to draw attention to problems or suggested remedies, and then published a compilation of the results.

In 1984-85 the BBC sponsored a *Domesday Project* in which 10,000 British schools participated. The results have been made available to the schools on laser disk. An international equivalent is envisaged. The Institut Français d'Opinion Publique conducted a survey in France concerning 40 problems to determine their relative probability, gravity, and ability to stimulate individuals to activity. (The results were reported at a Colloque International sur la Perception

Nouvelle des Menaces in 1973). The Center for Integrative Studies, on behalf of the World Academy of Art and Science, questioned 3000 international organizations concerning the relative importance of 25 problem areas in an effort to identify world priorities; the survey respondents added 196 other items. (The results were reported at the second Conference on Environment and Society in Transition in 1974).

The Educational Policy Research Center of the Stanford Research Institute, produced a study in 1971 on *Contemporary Societal Problems*. This attempted "*to identify and to interrelate the driving problems of our time, both national and international, to develop a useful perspective from which to better understand these problems, and to thereby identify crucial dilemmas whose understanding seems necessary if societal continuity is to be ensured.*" The report explored the use of resource allocation analysis as a tool for the identification of neglected societal problems and presented it as part of a more general problem analysis procedure. The study made "*a comprehensive attempt to list all relevant societal problems.*" Three overlapping procedures were used: (a) a selection of prominent (mainly American) persons of known divergence in both ideology and professional background were asked to nominate other persons whom they regarded as having the best grasp of current problems, to identify key materials on current problems, and to identify they key problems they saw as being most crucial at that time and in the future; (b) published results of previous systematic attempts to identify, categorize, or list societal problems were collected; (c) using the information collected a core sample of texts was collected for detailed analysis. The body of the report (27 pages) distinguishes between substantive, process, normative, and conceptual problems, and then compared the conventional and a proposed transformational view of societal problems. The appendix (46 pages) listing societal problem descriptions and taxonomies, consists of six items: Ralph Borsodi's *Seventeen Problems of Man and Society;* the US National Industrial Conference Board's *Perspectives for the 70's and 80's.* Karl Deutsch's *Issues which the proposed center for national goals and alternatives should address;* the Institute for the Future's *Future Opportunities for Foundation Support;* and John Platt's *What we must do.* These identify 17, 118, 35, 64, and 8 problem areas, respectively. The sixth item, resulting from the literature search and the leading thinker survey, lists 46 problem areas in seven groups.

A proposal was made in 1972, by Richard Cellarius and John Platt for the creation of International Councils of Urgent Studies to seek out and support the kind of research effort on world problems that would be inappropriate (or suspect) if sponsored by national governments. They identified some 210 areas of urgent research under 25 headings within 6 main groups.

A number of universities have courses on problem-solving. For example, the Mershon Center Program of Transnational Intellectual Cooperation in the Policy Sciences (directed by Chadwick Alger) at Ohio State University has a graduate course in problem-solving in international organizations. In addition to identifying and comparing the various problem networks and their interdependencies, a focus is placed upon the networks of organizations concerned with the networks of problems. At Swarthmore College there is a programme on problem complexes in public technology. Southern Illinois University, through a programme originally directed by R Buckminster Fuller, operates a World Game which introduces students to interactions between problems and resources.

5. Historical and traditional initiatives
See following note

Significance: precedents in history and tradition

It might easily be assumed that social problems exist since the dawn of history. However this does not appear to be the case, especially if it is assumed that for human suffering to qualify as a problem there should be a recognition that something should be done about it. It has been argued by Arnold Green (*Social Problems: Arena of Conflict*, 1975) that a consciousness of social problems did not arise until the latter part of the eighteenth century with the emergence of the notions of equality, humanitarianism, the goodness of human nature, and the modifiability of social conditions. It may also be argued that the religions of the world have responded to the condition of personal and collective suffering since their origin (Bowker***. *Problems of Suffering in the Religions of the World*).

Depending on what is meant by a social problem, the following may be explored as early examples of recognition of problematic social conditions.

1. Arthasastra

Seemingly, the first deliberate attempt to document problems appears in the Kautilya's *Arthasastra*. This classical Indian text on statecraft was written sometime in the period 321-300 BC. Many of the chapters deal explicitly with the nature of particular problems, including various forms of corruption and subversion, robbery, assault, defamation, juvenile delinquency, sexual intercourse with immature girls, the "calamities" of sovereignty, the "troubles" of men, etc. The commentary on "national calamities" covers fire, floods, pestilences, famines, rats, snakes, tigers and demons.

2. Japanese "tsumi"

Traditionally Japanese statecraft and government regulations enumerated what were termed "tsumi". In modern Japanese this is equivalent to notions of sin, offence and crime. In much earlier times it was a broad term applied to actions or conditions causing the degeneration of, or hindering, the proper growth and development of the life-force. As such it was related to the notion of ritual impurity. The oldest enumeration, dating from the 10th century is that of the *Oharae no Kotoba* in the *Engi Shiki*. This divided the tsumi into the heavenly tsumi and the earthly tsumi.

Heavenly tsumi included: destroying ridges between fields, burying irrigation ditches, destroying aqueducts, double planting of seeds, driving stakes in mud to cause harm, skinning animals alive, skinning animals backwards, polluting a pure place with excrement.

Earthly tsumi included: injuring the skin and causing blood to run, desecrating a corpse, irregularities in skin pigmentation, skin eruptions such as warts or tumours, incest, bestiality, calamities due to noxious pests, celestial calamities such as lightning and eclipses, calamities caused by birds, harming draught animals with curses, and placing curses on people.

3. Himsa in the Jain tradition

Within the Jain tradition in India, the concept of "ahimsa" was articulated in the period 599 to 527 BC and through subsequent development. The term is subject to a variety of interpretations but includes notions of non-violence, non-resistance to evil and passive resistance. The converse notion of "himsa" denotes a wide range of forms of violence of which some 432 have been distinguished and documented by scholars of that tradition.

4. Afflictions and hindrances (Buddhism)

In various Buddhist traditions considerable importance is attached to fundamental afflictions as the cause of suffering (and as responsible for maintaining the cycle of rebirth). All other problems are seen as engendered by them. In the *Visuddhimagga* by Bhadantacariya Buddhagosa, prepared in the 5th century AD, the following detailed checklist is given (followed there by indications of which forms of knowledge ensure release from them in each case). The seeming duplication is due to the emphasis on the different ways a limited set of "problems" act, as indicated by the often metaphoric categories:

(a) *Fetters*: greed for material benefits, greed for non-material benefits, conceit/pride, excitement/agitation, ignorance, delusion of selfhood (false view of individuality), doubt, susceptibility to rites and rituals, greed for sense desires, and resentment.

(b) *Corruptions/Defilements*: greed, hatred, delusion, conceit/pride, false view, uncertainty, mental sloth, excitement/agitation, consciencelessness, shamelessness.

(c) *Wrongnesses*: wrong view, wrong thinking, wrong speech (falsehood), wrong action, wrong livelihood, wrong effort, wrong mindfulness, wrong concentration, possibly together with wrong understanding of deliverance and wrong knowledge.

(d) *Worldly conditions (despondency/servitude to states)*: gain, loss, fame, disgrace, pleasure, pain, blame, praise.

(e) *Meannesses* (kinds of avarice): avarice about dwellings, families, gain, dhamma, praise.

(f) *Perversions* (Reversals): perversion of perception, of consciousness, and of view (whereby, in each case, the inappropriate is misapprehended as appropriate).

(g) *Ties*: covetousness, ill will, susceptibility to rites and rituals, dogmatic misinterpretation of truth.

(h) *Tendencies to inappropriate action*: partiality (desire/zeal), hatred, delusion, fear.

(i) *Hindrances*: sensuous desire, ill-will, sloth/torpor, distraction (agitation/worry), doubt.

(j) *Misapprehension/Wrong views*: ignoring essentials in favour of non-essentials.

(k) *Graspings/Clingings:* clinging to views, susceptibility to rites and rituals, clinging to selfhood, desire.

(l) *Inherent tendencies/Biases*: sensuous passion, resentment, conceit/pride, false view, doubt, craving for existence, ignorance.

(m) *Courses of immoral action*: life-taking, theft, sexual misconduct, lying, slanderous speech, harsh speech, gossip, covetousness, ill-will, wrong view.

(n) *Immoral states of consciousness*: eight rooted in greed, two rooted in hate, two rooted in delusion.

5. Sins and vices (Christianity)

In the Chrisitian approach to sins, some of the earliest listings dating back to the 4th century, gave eight sins. Eventually the church settled on seven sins classified in the order: pride, envy, anger, sloth, greed, gluttony, and lust. Occasionally an eighth sin, melancholy, was added. They have been both personal faults and great social evils, being at the origin of the multiplicity of other problems, notably the ordinary vices, such as creulety, hypocrisy, snobbery, betrayal, and misanthropy..

6. Crises and opportunities (Taoism)

The Chinese Book of Changes (I Ching) is a sophisticated effort to map out the pattern of changes which occur in any psycho-social system. Implicit in this complex pattern is a recognition that change occurs once some condition of imbalance or excess has been reached. In these terms a problem may be seen as a phase in the transition from one condition to another. A problem is thus the accumulation of imbalance which necessarily triggers the transition to a new phase within the pattern of possible changes. The I Ching marries a rigidly ordered binary system with an extremely metaphoric interpretation of its significance. The ordering could supposedly lead to categories of problems of different degrees of articulation, from a set of 2 fundamental categories, through 4, 8, 16, 32 and 64. Amongst the 64 categories there are 384 transformational pathways. Each of these could be interpreted as a particular kind of crisis or opportunity and could thus be used as a way of ordering problems.

Criteria: assumptions

It is a basic mistake to assume that the concept of a problem is held in the same way, whether between cultures or within any culture. The question as to whether problems "exist", and the nature of that existence is not understood in the same way in different contexts. At one extreme, as noted above, problems may be considered as artefacts of concerned minds, and possibly only of western minds. In the West the supposition of "problematicalness", with its attendant implications for reason, for principles, and for history, is so deeply ingrained in western consciousness that its denial can only seem absurd.

In the light of other cultures or philosophical systems, notably that of Chuang Tzu (Kuang-min Wu, 1982), to conceive of life as presenting problems to be solved is a misconception of life. With a problem-oriented vision it is possible to speak of the rise and fall of civilizations, of a dialectic of progress or devolution, and of the importance of roles in history in relation to problems. But if it is not necessary to understand life as presenting problems, or to understand life in relation to problems, then these features of historical consciousness are not as important as they presently seem. Alternative views are then also possible and may be more appropriate.

It is useful to attempt to identify alternative ways in which problems can be perceived, as a means of increasing understanding of the constraints on providing any satisfactory definition. This will also make evident the difficulty of attracting any consensus on the global problematique. Whilst it is possible to discuss these perceptual modes as models, a broader and more insightful discussion results from treating such models as part of a set of metaphors.

The following are therefore discussed as metaphors of the problematique:

1. Ordered array
Problems can be viewed as constituting an ordered array, like atoms in a complex molecule, or like an opposing array of military units. This view would tend to be favoured by those who are used to defining their environment in an orderly manner (cf Descartes, Hegel, Hume, Toynbee, Spengler, Marx), in terms which favour management and control, whatever the degree of simplification necessary. To deal with obstacles they must be named and placed, preferably so that the hierarchies of importance are evident. *(To be contrasted with...)*

2. Disorder and chaos
Problems can be viewed as synonymous with chaos and disorder. This view would tend to be favoured by those who have lost control over their environment, realize that they are subject to more forces than they originally assumed, or simply prefer the challenge of the disorderly (cf William James, Bergson, Schopenhauer, Rousseau). Problems are then too confusing to present any stable or orderly features.

3. Static structure
Problems can be viewed as forming a static, semi-permanent configuration of elements, especially by those opposed to political change. This view would tend to be favoured by government agencies mandated to respond to particular problems over an extended period of time. The view is reinforced by legislation and regulatory procedures. The problems are seen to be unchanging or to change quite slowly. *(To be contrasted with...)*

4. Dynamic structure
Problems can be viewed as constituting a dynamic, in which the problems arise in the dynamic relations between non-problematic, static elements. As such the problems cannot be readily located and named. They only exist as dynamic relationships changing continuously (cf Comte, Hegel, Marx, Whitehead, Bergson). This view would tend to be favoured by those whose survival depends on very short-term considerations, such as in politics, public relations and certain forms of commercial trading.

5. Discrete phenomena
Problems can be viewed as distinct phenomena with some form of boundaries separating them (cf Aristotle, Aquinas, Leibniz, Burke, Malinowski). This view would tend to be favoured by those who need to distinguish or allocate mandates, and divisions of responsibility, as well as by those in bureaucracies that resist any attempt to establish any continuity between the problem they are concerned with and those of other departments or agencies. *(To be contrasted with...)*

6. Continuous phenomena
Problems can be viewed as forming a continuous, possibly "seamless", field of tensions (cf Plotinus, Augustine, Spinoza, Locke, Durkheim). This view might be held by those favouring single-factor explanations in terms of pervasive conspiracy, subversion or evil forces. It would also be held by those favouring field theories in which problems might emerge as interference effects.

7. External relationship to phenomena
Problems can be viewed as externalities, as objects of experience to be perceived from without (cf Descartes, F W Taylor). As such they have an existence independent of any particular observer. This view would be favoured by those with either a rationalist or an empiricist orientation. It is basic to the strategic assumptions in many international programmes designed to "deal with" problems. *(To be contrasted with...)*

8. Identification with phenomena
Problems can be held to be only genuinely comprehensible through an intuitive identification with the experience they constitute, experienced by the observer as he experiences himself (cf Bergson, Hegel, Beatrice Webb, Mayo) This view would be favoured by those whose views have been strongly formed by the personal experience of suffering in themselves or in others, and who identify strongly with others in a condition of suffering.

9. Sharply defined phenomena
Problems can be viewed as being directly experienceable (cf Descartes, Hume, Russell). This view would tend to be favoured by those concerned with the concrete reality of such problems as destitution, torture and disease. For them, any other kinds of problem are unreal abstractions of no significance, other than as distractions from the concrete reality of human suffering. *(To be contrasted with...)*

10. Implicitly defined phenomena
Problems can be viewed as implying levels of significance greater than that immediately present (cf Hegel, Whitehead, Niebuhr, Proust). This view would tend to be favoured by those who detect more fundamental problems in conditions which may not themselves be experienced as problematic. This might include the catastrophic long-term implications of seemingly innocent phenomena.

11. Inherently comprehensible phenomena
Problems can be viewed as comprehensible in terms of existing paradigms or through their natural evolution (cf Hobbes, Machiavelli, Gibbon). This view would tend to be favoured by pragmatists and those with a scientific orientation for whom a satisfactory explanation in terms of known factors must eventually be possible. *(To be contrasted with...)*

12. Inherently incomprehensible phenomena
Problems can be viewed as calling for explanation in terms of other frames of reference, which may not necessarily be accessible to man (cf Plato, Schopenhauer, Hegel, Plotinus, Niebuhr, Toynbee). This view would tend to be favoured by certain religious groups and in cultures sympathetic to belief in other levels of being or realms of existence.

13. Phenomena in a context of due process
Problems can be viewed as subject to known laws as a part of definable processes (cf Marx). This view would tend to be favoured

by those endeavouring to model such processes as in econometrics and related disciplines. *(To be contrasted with...)*

14. Spontaneous phenomena
Problems can be viewed as totally spontaneous events, happenings or catastrophes unconnected to each other (cf H A L Fisher). This view could tend to be favoured by those who perceive chance and accident to be prime explanatory factors, as in the insurance industry, or important to the way they work, as with the media. It is also natural to those in the political arena for whom events may be of more significance than the multitude of interpretations placed upon them.

Comment
Clearly these different views are not mutually exclusive and overlap in complex ways in the case of any group or discipline. The 14 views have in fact been elaborated on the basis of work by W T Jones (Jones, 1961), who developed 7 axes of bias by which many academic debates could be characterized.

The 14 views above form 7 pairs of extremes corresponding to the extreme positions on such axes. Jones showed how any individual had a profile of pre-logical preferences based on the degree of inclination towards one or other extreme of each pair. The scholars named in each case are those given by Jones as examples.

In this project, although the information may derive from individuals or groups holding any combination of the above biases, the assumption made is that there is value in collecting, ordering and presenting the information as though problems did take the form of an ordered array (even if it is only "partially ordered"). The bias, in terms of the above checklist, is therefore towards understanding problems as: an ordered array (a), essentially static (c), discrete (e), experienced as externalities (g), sharply-defined (i), inherently comprehensible (k), and as part of a due process (m).

This is not to deny that a radically different set of biases does offer valuable insights and is more appropriate under certain circumstances. In fact many of the problems are only articulated by people having those other biases. It is quite probable that the design of appropriate, sustainable solutions also calls for another pattern of biases. But those emphasized are valuable in creating a framework to permit insights arising from those other biases to be bought more effectively into play on the problematique as a whole.

Criteria: problem disguises and problem evasion

Considerable difficulty was experienced because the available material, from whatever source, rarely provides a comprehensive and succinct description of a problem as a problem. There seem to be a variety of ways by which societal problems are transformed and diluted by processes in society with different perspectives. Such processes may also be seen as ways of evading or avoiding recognition of problems.

1. Assemblies, conferences (agenda items)

Such occasions are usually highly structured by agenda item. If societal problems are to be discussed they are reconceived as items in the conference process. As such it is their procedural features and disturbance to the current activities of existing bodies which come to the fore. In this context problems are distinguished with difficulty from routine meeting agenda items. This is especially so when the main function of the assembly is to review the work of other bodies which implement its directives.

Agenda items may give rise to resolutions. Again these may concern societal problems, but it is only by careful examination that problem-oriented resolutions are distinguished from other types of resolution. For example, research on UN ECOSOC resolutions by UNITAR categorized resolutions and their paragraphs according to 10 categories: recognition of issues ("*identifying, defining, assessing importance of, and commenting upon substantive problems, facts, conditions, events and causal connections external to the UN*"), delineating potential UN participation in world problems, setting standards and goals, creating or modifying UN organization, establishing programmes and strategies, detailed implementation, information transfer and coordination, monitoring and evaluation, exhorting governments, and internal administration.

Only 5% to 7% (depending on the level of analysis) were concerned with recognition of issues, and even this percentage included "*restating, reiterating or making reference to information on substantive problems, needs, facts, states, and conditions.*"

2. Administrative reports on substantive problems

The report of the Director-General of UNESCO (mentioned above), identifying world problems and supplying each "*with a brief and general description*", typifies the confused nature of problem descriptions currently available. Thus with respect to the first problem, human rights, the nearest equivalent to a description is the statement that:

"*The Organization's constitutional responsibilities with regard to human rights may be summed up as follows (a) to assist in combatting all forms of discrimination; (b) to promote certain fundamental rights, such as the right to education; (c) to extend the opportunities for leading a more satisfactory life, at the individual and community levels, through participation in scientific advancement and in cultural life and access to full and objective information.*"

The societal problem under (a) is "discrimination", but it is embedded in a concern with UNESCO's own constitution, which surely is irrelevant to any description of the problem. In addition the problem is described in terms of combatting such discrimination. Again what UNESCO does about the problem is surely irrelevant to any description of it, unless the problem is in fact that of "combatting discrimination", namely the strategic, tactical, and logistical problem of combatting discrimination. This is not the external problem of discrimination but a problem internal to the organizational system in some way related to the undefined external problem.

Similar difficulties could be brought to light with respect to the eleven other problems. The descriptions are all embedded in preoccupations with organizational and program goals (or, in some cases, with the theoretical preoccupations of the predominant discipline, or of the department responsible for formulating the description).

3. Political arena, government (issues)

In the political arena societal problems are merged into the maze of issues which galvanize the political process. Issues, as with news, may be very short-term, highly personalized or concerned with threats to the credibility or image of some establishment unit. Problems only become identified as issues when they excite a significantly powerful pressure group. The extent to which issues become issues, or get lost in limbo, is to a large extent fortuitous. Many issues are deliberately projected as problems when in fact they are only pseudo- problems, which may nevertheless be sufficiently magnetic to attract short- term electoral support. Power groups appropriate issues as a means of establishing relevance to a constituency. Once the dramatic appeal and novelty is lost relative to other issues, a problem issue is discarded.

4. Administrations, agencies, secretariats (programmes)

Many organizations hold that too much effort is put into recognizing problems, whereas the real need is for solutions. The documentation from such bodies tends to recognize problems in passing or by implication only. Their material reports on the range of programmes they are implementing (with emphasis on their success), whether or not such programmes can be related to specific problems or not. In a number of cases, especially with bureaucracies, it is legitimate to ask whether the programmes are simply memorials to problems that have long disappeared or have completely changed their form.

Administrative bodies and agencies tend to work in terms of programme and budget items. The problems, supposedly defined at a plenary or planning meeting, are here disguised and defined by the action programmes agreed upon. Just as intelligence has been defined as "what is measured by an IQ test", the problem becomes "that which the programme is designed to combat". A secretariat official of one major intergovernmental agency, questioned about material on illiteracy, put the point very simply by stating: "*Illiteracy is not our business; we are concerned with literacy programmes.*" At any stage up to or following its full recognition, the problem may be absorbed into some section of the administrative apparatus. It is internalized so that it is almost impossible to distinguish (from the organization's perspective) between action to solve the problem and the routine activity of the administrative section, or even between the external problem and the internal administrative or political difficulties in solving it.

5. Public relations, public information (symbols)

A problem has to be transmuted by a public relations operation into a symbol in order for it to permeate the world of images. There are many symbols which do not represent problems. The process of conversion into a symbol involves a simplification, a dramatization and a humanization. This may strip the problem of subtle cross-linking relationships to other problems, introduce ambiguity, and may even distort it beyond recognition by those who originally defined it. The symbol of the problem is designed to incite to specific action, not to facilitate new thinking about the nature of the problem or whether or how to act against it. Where the public information is disseminated by an organization or agency with programmes designed to reduce or eliminate the problem, it is in the agency's interest to concentrate its information on the success (however partial) of its programmes, rather than the gravity (however great) of the problem. This is best demonstrated by an examination of the catalogues of photographs available to the press from intergovernmental agencies. Only a very small percentage attempt to illustrate the problems, most illustrate actions to solve the problems.

6. Journalism, newsmedia (events)

Journalism tends to focus on events, news items and stories, possibly illustrative of an underlying social problem. But more often than not, the problem is interpreted to give meaning to a personalized event rather than vice versa. Nevertheless this sector is possibly least reluctant to record, if in over-dramatized form, the announcement of an unforeseen problem.

7. Legislation and treaties (crimes)

Legislation is concerned to proscribe certain activities (abuses, offenses, *etc*) which create or constitute societal problems. A body of legislation may be conceived as a set of contained problems -

problems "behind bars". All crimes may be considered problems. The societal problems of interest are those that escape from these constraints to a significant degree - beyond the threshold level up to which the legislation may be considered adequate. An international agreement may signal the presence of a world problem, and may of course contain it, if properly implemented. The difficulty is to determine when legislation disguises the presence of uncontained problems. Much legislation is a compromise and has often been deliberately designed to focus on the more obvious manifestations of a problem. Other dimensions of a problem, including various sub-problems, may not be recognized by the legislation. Legal loopholes ensure that certain problems remain unrecognized and uncontained.

8. Insurance (risks)
The insurance sector of the economy is not concerned with problems as such but is concerned with risks. Risks may however be considered as potential problems. In this sense the insurance sector is the most explicitly concerned with the definition of problems. These problems are in most cases defined in terms of the financial interests of the insurers. The insurance sector may however prove to be a rich source of information on the incidence of many problems.

9. Business (markets)
Business, especially the service sector, depends on the ability of entrepreneurs to detect problems for which people are prepared to pay for some form of remedy or preventive facility. People vulnerable to such problems constitute well-defined markets. Advertising addressed to specific markets is often structured to identify the problem explicitly and may even use the term "problem". The question is how to distinguish markets based on problems of interest to this section from other markets. The scope of this approach is indicated by the 18,000 market research studies published annually (Marketsearch, 1988). And is there a point at which the commercialization of a problem remedy can be said to effectively contain or tame the problem so that it is no longer problematic in any sense of interest to this section ?

10. Celebration of values (visions)
Some individuals and groups consider that it is unhelpful to devote any time to recognizing problems. All effort should be devoted to emphasizing positive values, appropriate visions of the future, and the necessary actions to give form to such visions. The documentation from such bodies tends to be "problem-free", except for a marked tendency to identify some other bodies as unconstructive (or even evil) in promoting opposing initiatives. There is little effort to face up to the problems opposing implementation of the visions, or to recognize the problems that such implementation will cause for others.

11. Religion (sins)
Historically religions have played a major role in clarifying the values in the light of which social problems such as poverty or injustice are perceived. But such problems tend to be perceived by religions as being a consequence of sins and vices (Judeo-Christianity) or afflictions (Buddhism), which are not usually considered as problems in their own right. There is a distinction between what a religion perceives as a sin and what secular society chooses to perceive as a problem, especially in the case of sins or afflictions of the mind having little recognizable impact on society.

12. Conflicts
Conflicts, whether violent or not, may be considered as a definite manifestation of a problem. The turbulence of the conflict, in its very concreteness with all the visible side effects, tends to obscure the underlying problem. Those involved in a particular border conflict naturally tend to resist interpretation of their conflict as an instance of the general problem of border conflicts.

13. Documentation (subjects)
Clearly all problems which form the subject of an article or book should be detected by documentation, library and abstracting systems. This is so, but only as "subjects" completely merged into the multitude of other subjects which are the preoccupation of classification systems. Unfortunately, subject headings and descriptions fail to detect problems which are not yet labelled by an accepted descriptor - namely those at present defined by a phrase or a mathematical relationship (eg between resource flows). Nor do the documentation systems detect problems noted in the body of a text.

14. Research disciplines (conceptual puzzles)
Groups of academic orientation, are primarily interested in new theories suggested by the phenomena associated with a problem. Academic literature of any quality can only refer to problems in passing, as an illustration of the steps in a theoretical argument. The situation is somewhat different in the case of the applied sciences explicitly concerned with bringing academic knowledge to bear on a problem. Here however the concern tends to be solution oriented, namely how to remedy the problem, rather than documenting its nature and extent.

The problems detected by disciplines are normally intimately bound up with the characteristics of the theory or model used to research them. T S Kuhn clarifies the relationship between research problems and societal problems in the following quotation (1962): "*Bringing the normal research problem to a conclusion is achieving the anticipated in a new way, and it requires the solution of all sorts of complex instrumental, conceptual, and mathematical puzzles... It is no criterion of goodness in a puzzle that its outcome be interesting or important. On the contrary, the really pressing problems, eg a cure for cancer or the design of a lasting peace, are often not puzzles at all, largely because they may not have any solution... One of the things that a research community acquires with a paradigm is a criteria for choosing problems that, while the paradigm is taken for granted, can be assumed to have solutions. To a great extent these are the only problems that the community will admit as scientific or will permit its members to undertake. Other problems, including many that had previously been standard are rejected as metaphysical, as the concern of another discipline, or sometimes as just too problematic to be worth the time. A paradigm can, for that matter, even insulate the community from those socially important problems that are not reducible to puzzle form, because they cannot be stated in terms of the conceptual and instrumental tool the paradigm supplies... One of the reasons why normal science seems to progress so rapidly is that its practitioners concentrate on problems that only their own lack of ingenuity should keep them from solving.*"

An external societal problem may be internalized by the discipline, as is the case with administrative agencies, such that it is impossible to distinguish (from within the discipline) between action to solve the problem and the normal advance of theoretical knowledge within the discipline, or even between the external problem and the internal theoretical or practical difficulties in solving it.

The situation is further complicated by the relationship between the problem as researched and the problem as perceived by the body concerned with the formulation of policy. Yassin El-Ayouty (1971) makes the point: "*There are certain misconceptions held by the operational official as regards what research should do for him. An important misconception is the operator's assumption that the solution of his problem would be primarily advanced through the mere collection of facts. In this regard, the operator may erroneously conceive of the research process as an exercise aimed at providing him with specific replies or answers to questions or problems which he has selected for research. As a result of this misconception, the operator finds the problems, as researched, appear different from those in which he, the action man, is interested. The disappointment of the operator does not stop only at finding that he is no nearer to the answers he is seeking through research than when he began. It is compounded by the fact that the whole research process may appear to be a complicated way of saying the obvious. As to the researcher, he may have his own frustrations in responding to the demands of policy through research. As his research proceeds, his conviction may grow that the action official has asked the wrong questions, and that the concepts and categories in which the policy problem has been posed are neither meaningful nor useful. If he reformulates the problem or restates the questions, the result may be that his customer, the action official, makes little or no use of his investigations.*"

Criteria: definitions

World problems may be readily defined for particular purposes, but such definitions do not exhaust the complex significance of problems. There is a need to continue exploring how such problems should be thought about.

1. Nebulous characteristic of problems
Problems are strange nebulous entities having a shadowy existence. They may be described or bounded by negatives - as *"the substantial, unwanted discrepancies between what exists in a society and what a functionally significant collectivity within that society seriously (rather than in fantasy) wants to exist in it."* (R K Merton, 1976).

(a) Subjectivity: The shadowy nature of problems derives from the fact that they represent in part an objective state of affairs and in part a subjective state of mind. Thus a UNESCO expert meeting on violence reporting on its definition notes: *"What do we mean by violence. That depends on who "we" are."* But even this objective quality may be questioned.

"Problems and solutions are, however, based on the perceptions of individuals. They are not objective conditions of the real world. They are subjective constructions - what Kenneth Boulding would call "images" of the real world - although such perceptions may be and often are shared in roughly the same form by many people. Nevertheless, problems may appear in different forms to different people. What is a critical problem to one person may appear unimportant, or even not a problem at all, to another person. To paraphrase Boulding, a problem is what somebody perceives as a problem; and, without somebody or something to perceive it, a problem is an absurdity." (T J Cartwright, 1973). In the words of Donald Schön in discussing problem-setting in social policy (1979): *"Problems are not given. They are constructed by human beings in their attempts to make sense of complex and troubling situations. Ways of describing problems move into and out of good currency."*

The emergence or disappearance of the discrepancies noted above may be affected by raising or lowering standards or by the improvement or deterioration of social conditions. *"We must therefore be prepared to find that the same social conditions and behaviours will be defined by some as a social problem and by others as an agreeable and fitting state of affairs. For the latter, indeed, the situation may begin to become a problem only when the presumed remedy is introduced by the former.....There is no paradox then in finding that some complex, industrial societies, having a comparatively high plane of material life and rapid advancement of cultural values, may nevertheless be regarded by their members as more problem-ridden than other societies with substantially less material wealth and cultural achievement. Nor is there any longer a paradox in finding that as conditions improve in a society (as gauged by widespread values), popular satisfaction may nevertheless decline."* (R K Merton, 1976)

(b) Fuzzily defined problems
Statements supposedly defining "world problems" may lack any precision permitting a specific problem to be identified. This is illustrated by the answer of Margaret Thatcher to a question to identify "the most pressing international problem". She responded: *"The great unknown is whether Gorbachev's bold reforms of the Soviet system can be taken to completion. To us, this would mean a freer society with a proper rule of law and a genuine respect for human rights. Such a change would have an effect on som many other nations and on how people see socialism -- which is not about human beings at all. Socialism is about economic plans and people having to conform to them, not about government serving the fundamental dignity and freedom of the individual;. The fact that we are even considering the possibility of such a change in the Soviet Union is an enormous step forward."* (International Herald Tribune, April 1989).

(c) Problems as boundary phenomena
The domains, noted earlier, through which problems are disguised and evaded, each have problems of vital *internal* concern. The problems of interest in this section are however those which appear to have some existence "out beyond" the various conceptual frameworks which society has evolved to respond to unforeseen social change of an unexpected nature. Such problems overflow and are not contained by such frameworks.

Problems then indicate the presence of an "outside" with respect to society - uncontained processes. It is almost as though the layers of problems and matching procedures internal to organizations, disciplines, legal systems, politics, *etc*, constituted a distorting factor hindering and even blocking the perception of a problem. Every attempt is made within such different domains to perceive the problem within some familiar framework, if it is not possible to deny its existence.

There is therefore a parallel between the following statements about Problem "X", where "X":

(a) "has no theoretical significance";

(b) "is not on the current agenda of our general assembly";

(c) "is not the subject of any existing legislation";

(d) "is not an issue of political importance";

(e) "is not a matter of concern within the current two-year programme of our organization;"

(f) "is not of interest to our readership (or viewers)".

Each such sector experiences great difficulty and reluctance in grasping the problem as a negative condition in its own right. Each sector rapidly separates its attentions from the social and human impacts of the problem, reinterpreting it and transforming it. This reduces the significance of its particular content and diverts attention to the various formalistic features of the manner by which the original problem is contained and encapsulated. The problem is converted into: a story, an issue, a case, a programme focus, an agenda item, *etc*. This is accompanied by an effort to concentrate more upon what is being done to remedy the problem situation than to clarify the nature of that problem.

(c) Solutions as problems: Even the distinction between problems and solutions is blurred and confused. A supposedly less desirable state of affairs is conveniently called a problem situation and the more desirable situation is termed the solution situation. But as Bertram Gross (1971) notes: *"..all solutions create problems. Adequate solutions lead to large problems. Good solutions create fantastic problems."* He cites the consequences of successful agricultural development in developing countries. Frank Trippett (1972) notes: *"The Politician can appeal solely to the boundless and inextinguishable nostalgia of the human race. So he talks about "problems" for which he proposes "solutions"... But he does not solve these problems, simply because from the folkloric world he can scarcely see, let alone touch, the actual world. His is a phantom reality. The very things he calls problems are, in fact, solutions in the real world."* He cites the unemployment problem as a conventional solution to economic problems, and the urban problem of overcrowding as a solution to the problems of housing increasing numbers of people.

Donald Schön (1976) notes however that there have been increasing difficulties with the "problem-solving" perspective and that a sense of inadequacy has begun to spread among practitioners of social policy and among the public at large. The social situations have turned out to be more complex than was supposed. According to Schön, it becomes increasingly doubtful in the case of social policy that we can make accurate temporal predictions or design models which converge upon a true description of reality. *"Moreover, the unexpected problems created by our search for acceptable means to the ends we have chosen reveals...a stubborn conflict of ends traceable to the problem setting itself."*

2. Problem definability

It can be easily assumed that what is meant by a "world problem" can be readily defined. Undoubtedly this is so in some cases and for some constituencies. The special challenge is to respond to the worldview of those labelling as a "problem" a perception which is totally without meaning in another framework. From this second framework, in turn, other sorts of "problem" may be perceived, possibly with quite different characteristics.

(a) Problems as undefinable phenomena: Whilst it is possible to produce conceptually neat definitions of what constitutes a problem, it could be argued that in its most genuine sense a problem is essentially, and paradoxically, an undefinable phenomena. Definitions can be projected onto the perceived phenomena. But what characterizes a problem is the inability to encapsulate it within an appropriate definition. This is not to deny that a definition cannot be provided for "poverty", for example. It is rather that, to the extent that the definition meets the formal requirements of a particular discipline or school of thought, the theoretical refinement required by any methodology (economics, for example) will effectively deprive it of the meanings it has to the poor who experience it existentially as a problem. In a sense a problem is that which does not lend itself to being encompassed conceptually. In effect it is an "anti-concept" network of such problems might even be viewed as an "anti-theory".

(b) Absence of accepted definition of a problem: Possibly for such reasons, there is no generally accepted definition of a "world problem" and there is considerable debate about the nature of a "social problem". No attempt is therefore made at a final definition of a world problem at this stage. The distinction between "world" problem (as "global" problems) and "world-wide" problems (as present in many countries, but without significant transboundary effects) is occasionally stressed. In this sense, world problems are those of the world as a whole, whereas world-wide problems are those of parts of the world or of particular groups around the world. Although there may be circumstances under which some such distinction is appropriate, the approach here is to establish the universe of problems - the problem context - within which users may determine those problems which merit identification as global as opposed to world-wide.

(c) Documenting perceptions of problems: In order to build up as comprehensive a data base as possible, the criteria for problem inclusion and exclusion were initially kept to a minimum. The emphasis during the selection procedure was not on whether adequate proof existed that a problem was a valid and significant one according to some objective standard. Rather, the emphasis was placed on including those "problems" which well-established constituencies indicated as significant in terms of their own frame of reference - even when the validity and existence of the problem is challenged by the perspective from some other frame of reference. The emphasis is thus on documenting what people believe to be factual, irrespective of whether that belief is challenged by others as being totally subjective and ill-founded. The problems documented are those which preoccupy people and move them to act, individually or collectively, whether or not such concerns are considered ridiculous from some other perspective. In effect, all problems are sought which are identified as being of importance by some functionally significant collectivity that manifests itself in some way at the international level (whether as an organization or through self-selected groups of spokespersons).

(d) Problem existence engendered by belief systems: This open-ended approach permits the registration of all the problems perceived as real, whether or not, as Stafford Beer suggests (6), most of the problems with which society believes it is faced, are bogus problems generated by theories about social progress and the way society works. The existence of information questioning the validity of a perceived problem is treated as information about that problem. Each perceived problem is envisaged as having a certain probability of existence for some groups in society and is therefore treated like a proposition carrying annotations commenting on its validity - but it is included.

3. Some definitions

(a) Social problem: Sociologists usually consider a social problem to be an *"alleged situation that is incompatible with the values of a significant number of people who agree that action is needed to alter the situation"* (E Rubington et al. *The Study of Social Problems*, 1981). In this same study five more specific definitions are noted:

- Social pathology perspective: A social problem is a violation of moral expectations. Desirable social conditions and arrangements are considered health, while persons or situations that diverge from moral expectations are regarded as "sick".

- Social disorganization perspective: A social problem is a failure of rules. Three major types of social disorganization are normlessness, culture conflict, and breakdown (in which conformity to rules is counter-productive).

-- Value conflict perspective: Social problems are conditions that are incompatible with the values of some group whose members succeed in publicizing a call for action.

-- Deviant behaviour perspective: Social problems reflect violations of normative expectations; behviour or situations that depart from norms are deviant.

-- Labelling perspective: A social problem or social deviant is defined by social reactions to an alleged violation of rules or expectations. This persepctive focuses on the conditions under which behaviors or situations come to be defined as deviant or problematic.

(b) Global problem: Tibor Asboth in a report for UNESCO defined a "global problem" as having the following attributes: long-term, persistent, pervasive, affecting many people, the "ownership" of the problem being difficult to establish, the characteristics of the "solution" being unknown, and with proposed solutions requiring new styles of cooperation for implementation (*World Problems and their Perception*, 1984).

(c) World-wide problem: A distinction is occasionally made between a global problem, affecting the world in its entirety, and a world-wide problem as existing in many countries but without special significance at the global level.

4. Use of pragmatic guidelines

The approach taken has been to avoid any well-formed definition. Instead a set of guidelines is used to include or exclude particular types of "problem" arising in different source materials. But the guidelines are treated as flexible and open to challenge as new information is received. Two basic techniques were used:

(a) Recognition in published documents: Problems registered had to be based on published documents. The documents preferred were those arising from the work of international organizations, which cover most matters which have emerged as being of more than national significance. However, use was also made of material from other publications. Individual responses to a questionnaire sent mainly to international organizations were used only as an indication of the existence of a problem for which published documents were required.

(b) Determination of cut-off points for exclusion of detail: Criteria were progressively elaborated to reduce the inclusion, in this first series, of very detailed problems which were nested within other problems. In other words, when a distinct hierarchy of problems was encountered (*eg* problems relating to commodities, or to the extinction of species) suitable cut-off points were selected within the hierarchy below which more detailed problems were not considered (*eg* a commodity class level within a classification of commodities).

This approach led to the elaboration of:

(a) a list of tentative positive definitions as a guideline for problem identification;

(b) a list of general criteria for inclusion of problems identified;

(c) a more specific set of criteria for the exclusion of certain kinds of problem.

These are discussed in subsequent Notes (see also Notes on Notes on *Problem disguises and problem evasion* and on *Problem naming*.

Criteria: problem inclusion

1. Tentative positive definitions
(a) Any condition believed to threaten the balanced physical and psycho- social development of the individual in society, whether the threat is directly to his personal well-being, to the values which he upholds, or to features of his environment on which he is dependent.

(b) Any condition believed to cause or constitute social regression or degradation.

(c) Any condition before which society is currently believed to be in some way helpless, because resources cannot be brought to bear upon the problem.

(d) Any condition believed to render social change uncontrollable or discontinuous, or which so increases the complexity of society that it becomes incomprehensible in its totality and consequently unmanageable as a whole.
The following guidelines for problem inclusion have been used. As guidelines, exceptions are made whenever this appears appropriate.

2. Geographical spread
The problem should be recognized in at least three countries or considered to exist in at least three countries. Resources should preferably be allocated to its solution in at least three countries. The problems included relating specifically to one country only are those which are the subject of a United Nations resolution (eg apartheid, zionism). Problems can be considered as "world" problems, either because they require solutions on a global scale (eg the international monetary crisis), or because they are present in a number of different countries, even if only local solutions are required (eg urban problems). Regional problems are included for such geopolitical regions as "developing countries", "tropical zones", but not for specific geographic regions (eg Africa, Caribbean).

3. Disciplinary spread
The problem should be common to, or with implications for, more than one discipline and should preferably have implications for different classes of discipline (eg natural and social science disciplines). This excludes problems internal to a discipline.

4. Expert recognition
The problem should be recognized by more than one expert, and preferably by experts in different countries, and if possible by an international governmental or nongovernmental body. In other words, the problem should have an adequate "constituency".

5. Expert documentation
The problem should be the subject of serious articles, scholarly studies, official reports, or reports of international meetings. The problem must be adequately documented, or its recognition must be adequately argued. This does not however imply the need for any check on the validity of the argument.

6. Time period
The problem should have been the subject of documents dating from 1970. Problems no longer considered to be active are not included (although more exceptions to this have been accepted than for the previous edition).

7. Duration
Short-term calamities, natural disasters, man-made disasters, massacres, wars, or calamitous events, are not be treated as individual problems, although appropriate groups of such disasters (eg earthquakes in general, as contrasted with an individual earthquake disaster) are so treated.

8. Potential problems
The problem can be a potential or future problem, even a "vulnerability", namely a problem which does not currently exist, because some threshold has not yet been passed, but whose emergence is predicted for some foreseeable future time and for which preventive action is advocated now.

9. Autonomous problems
The problem should preferably be in some way distinct and clearly possible to isolate. But where the relation between a sub-problem and the problem of which it is a part is not immediately apparent, or the dependence of one on the other is questionable or ambiguous, sub- problems should be treated as problems in their own right, particularly where the sub-problem is perceived as having distinct relationships with other problems. (The nature of the problem-subproblem relationship is indicated by cross-references within each entry.)

10. Seriousness
There should be some indication that the problem, if not solved, will aggravate or cause some social tension, or alternatively is a key factor in preventing the solution to other problems which result in such tensions. This means that seemingly trivial problems may be included if relatively large amounts of resources are allocated to their solution rather than to the solution of what others may consider to be more serious problems.

11. Secret problems
The documentation available which legitimates concern with the problem should not be classified or secret material, for obvious reasons. (Clearly, however, such secret problems may exist and, for that very reason, be of special importance.) For those problems for which secrecy and cover-up policies are believed by significant constituencies to be in operation, isolated examples of problems may be considered sufficient evidence for the existence of the problem as a world problem. (Counter- arguments refuting the claims for the existence of the problem would then be sought from the published documents of the institutions held to be responsible for such policies.)

12. Natural environment problems
Problems of pollution, resources, population, and the reduction of environmental variety have been considered in detail with an effort to locate suitable cut-off points for nested problems.

13. Emotions as problems
Emotions such as anger, hate, jealousy, fear, and anxiety have been considered for inclusion as general problems.

14. Structural violence problems
Problems relating to any forms of discrimination, imbalance in resource usage, social injustice, or unparticipative decision-making have been considered in detail.

15. Moral and ethical problems
In contrast to the previous edition, clearly defined problems of this type have been included. The emphasis has been on problems experienced in practice, not on those which emerge as distinctions in philosophical or theological debate.

Criteria: problem exclusion

The following guidelines for problem exclusion have been used. As guidelines, exceptions are made whenever this appears appropriate. In particular, excluded classes of problems may be signalled by a single general entry, with, or without description.

1. General problems
Very general problems, such as the inadaptability of man to change, have been considered for inclusion if they have been precisely formulated. A number of widespread problems common to any form of organized action are also considered as general problems, as will be clear from the points below. These include problems of inadequate finance, training, knowledge, *etc*.

2. Operational problems (routine)
The normal operational problems of a problem-solving organizational system have not been included. This means that no problems which arise (are encountered and solved) as part of normal technical, academic, research, legal, administrative, or political activity (namely "contained" problems) have been considered. This includes: institutional development, technology development, programme or process implementation, and operation and maintenance of equipment and services.

However, whilst the problem of obtaining spare parts for maintaining a particular machine or group of machines (*eg* agricultural machinery) would normally not be considered, it may be considered if, as in the developing countries, this is recognized as an important obstacle to industrialization. In that case it would be considered as a general problem for developing countries. Normal operational constraints, such as political problems, human resource problems, resource availability problems, and problems of public acceptance, have also been excluded, except as general problems.

In the case of problems of the supply of resources, these have only been considered if the resource in question was a basic item (*eg* cereals, construction materials, *etc*) and not a luxury item.

3. Operational problems (insoluble)
Insoluble operational problems, whether institution, technology, or concept dependent have not been considered as such.

4. Institutional change problems
Problems of institution building, management (including intra-organizational coordination), financing, and adaptation have not been considered as such. However, some consideration has been given to problems relating to inadaptability of institutions, or the inadequacy of financing, or the lack of management skills, but only as general problems. Although problems of building specific institutions have not been considered, some consideration has been given to the problems of specific institutions of world importance (such as the United Nations) or to types of institutions (such as transnational corporations).

5. Structural modification problems
Problems of changing attitudes, technology, institutions, or legislation have been considered only as general problems.

6. Coordination problems
Problems of coordination between institutions, between disciplines, between regions, or between sectors have been considered only as general problems and not in relation to specific disciplines, institutions, regions, or sectors. The exception made for institutions concerned the major international systems (*eg* United Nations, intergovernmental organizations in general, and international nongovernmental organizations in general). The exceptions made for regions concerned the relation between developed and developing countries, and between the major power blocs.

7. Planning problems
Problems of evaluation of objectives, strategy formulation, and resource allocation have been considered only as general problems. Problems of choosing between alternative courses of action have not been considered.

8. Compatibility problems
Problems of equipment standardization, compatibility of procedures and legislation, or problems of language in this context, have not been considered except as general problems.

9. Interaction problems
Problems of an inter-cultural, inter-ethnic, inter-faith, inter-ideology, or similar kind, have only been considered as general problems.

10. Consensus-formation problems
Problems such as that of mobilizing opinion, freedom of information, freedom of association, and over-simplification of issues, have only been considered as general problems.

11. Knowledge storage problems
Problems of language, technology, semantics, and cost, in relation to knowledge storage, retrieval and dissemination have only been considered as general problems.

12. Communication problems
Problems of language, technology, semantics, and cost, in relation to communication in general and public information in particular have only been considered as general problems.

13. Data usage problems
Problems of undemocratic control of data, invasion of privacy, and commercial abuse of collected data have only been considered as general problems.

14. Operational side effects
Problems arising from the deterioration of the natural environment and the normal operations of institutions and industries have been considered in detail with an effort to locate suitable cut-off points for nested problems.

15. Professional problems
The problems internal to a profession, as perceived by its members, have not been considered. However, those problems created by its activities, as perceived by outsiders, have.

16. Controversies
Controversies may be treated as resulting from the conflict between different schools of thought, applying their respective disciplines, over some unresolved and challenging issue (M Dascal, 1989). The existence of such a controversy may signal the presence of a problem of interest here, but this is not necessarily the case, especially when the controversy is primarily over methodological rather than substantive issues. In the sense of interest here a controversy marks the defiance of a threatening real world issue in response to attempts to encompass it with the available disciplines.

17. Problems of belief
Problems of belief as such have not been considered although general problems such as evil, superstition and animism have been included, as well as extremes of belief such as fundamentalism and fanaticism. Problems of protecting or promoting a particular belief were not considered unless the belief related to human rights or other beliefs relating directly to societal problems.

18. Moral dilemmas
In moral discourse this may refer to several distinct situations of moral conflict and preplexity: (a) when there is some evidence that an act is morally right or obligatory and some evidence that it is wrong; (b) there is conclusive evidence that an act is both morally right or obligatory and morally wrong; and (c) the moral reasons for (or against) an act are in conflict with nonmoral reasons against (or for) an act. Problems are not included as dilemmas, but dilemmas may be associated with problems that are included for other reasons.

19. Institutional protection problems
Problems of protecting existing procedures or institutions (other than the general problem of security) have not been considered unless such procedures were designed to protect human rights or other matters relating directly to societal problems. (Thus school

absenteeism was not considered as a problem for schools but rather as a problem for the child or for society.)

20. Conflict problems
Territorial, political, and industrial disputes (including war) have not been considered individually (eg civil war has been considered as a problem but not individual civil wars).

21. Anti-group problems
Problems documented by one group of bodies as being caused by the dangerous activity of another group have not been considered unless recognized by the United Nations. Thus the problem of the existence and activity of a particular named capitalist or communist institution, for example, have not been considered, although the general problem of capitalism and communism to which they relate have been considered.

22. For-profit problems
Technical problems defined as an open challenge, with prize money offered to the solver (as in the case of man-powered flight, for example) have not been included. However "sponsored" problems, deliberately created in secret by a group in order to derive commercial or political profit from "solving" them would be included (eg bugging and bomb attacks by security firms, new diseases by pharmaceutical corporations, semi-addictive additives by food and beverage corporation, vulnerable varieties of seed by seed corporations).

23. Action obstacles arising from specific objectives
When a specific programme or objective is defined, problems are perceived as obstacles to its achievement. Such problems have not been included, unless they are common to many programmes or objectives. Examples of those excluded are:

(a) Problems affecting the progress of tourism in the developing countries are not included if they only relate to tourism.

(b) Problems of river development, such as improving hydrological services, improving analytical tools in water resource utilization, and encouraging scientific and technical investigation are not included.

The following quotations illustrate some problems of this kind that were not, for the purposes of this volume, considered for inclusion:

- An immediate problem was the most-favoured-nation principle...

- The prevention of human rights violations was therefore a world-wide problem

- One basic problem is to break with the traditional belief that a national policy for children should be confined to dealing with the under-privileged and handicapped

- EEC spending and expansion problems - The importance of the problem of the relationship between over-all and industrial programming derives from certain essential considerations

- The problem of intermittence (of power demand) may, however, be over-stressed and with it the storage of energy...

- If you have money to invest: you have a problem

24. Technological problems
Problems of application of research to solve some urgent technical problem have not been considered. For example:

(a) the problem of economic desalination of sea-water;

(b) the problems of adapting man to space travel;

(c) the problem of machine translation of texts;

(d) the problem of extraction of energy from the fusion process.

25. Measurement problems
Problems of quantifiability, data collection and comparability have not been considered. This means that problems such as the following have been excluded:

(a) the problem of measuring poverty;

(b) problems of quantifying the effects of special preferences;

(c) problems of the lack of suitable monitoring instruments for environmental pollutants.

The only exception is the general problem of comparability of statistics.

26. Research or scientific problems
Problems of research, methodology, and analysis have not been considered, as with theoretical problems in general (eg Hilbert's list of 23 outstanding unsolved problems for mathematicians in 1900), and problems of paradigm change. However some attention has been given to the general problems of inadequate concepts, logical or semantic fallacies, irresponsible research, or research which legitimates some abuse of human rights.

Different kinds of research have in fact been distinguished by Gunnar Boalt (7) on the basis of the relationship between the problems considered and the theory in question:

(a) the problem independent, not associated with the theory;

(b) the problem independent, with a secondary association with theory, which is of lesser importance;

(c) the problem is of about equal importance with the theory used;

(d) the problem is of some interest, but the theory is of more interest;

(e) the problem consists of the testing of a theory.

Only the first two categories would merit consideration for inclusion here as problems of possible world importance.

A useful distinction between in-house research problems and social systemic problem is illustrated by the remarks of T S Kuhn (as quoted in a later Note on *Problem disguises and problem naming*).

27. Regional problems
Problems specific to particular regions (eg "drought in the Sahel") are excluded or incorporated into more general problems (eg "drought") or possibly into more general sub-problems eg ("drought in developing countries" or "drought in arid zones"), if this seems appropriate.

Criteria: problem importance

No effort has been made to determine the relative importance of problems for which entries have been included. In this preliminary exercise, effort has been limited to locating problem descriptions and relationships between problems, which would then permit further attention to be given to the question of the relative importance of the problems.

1. Varying importance attached to the same problem
Different minority groups and interest groups approach the universe of problems from different perspectives and with different value preferences. Such differences are reflected in the very wide variety of international organizations representing such views. The resulting differences in the weighting of the relative importance of problems leads to different: priorities for action; time-scales within which action must take place; relative amounts of resources to be allocated; and hence to different perceived critical paths through the problem network. The evaluation of the relative importance of problems is itself a major task beyond the scope or immediate intentions of this section, particularly since a large number of reputable authors and organizations have devoted effort to isolating the 5 to 10 key problems which merit immediate attention. It is the fact that these authors and organizations are not in agreement which is of interest here.

2. Varying importance over time
The importance attached to particular problems varies greatly over time, especially over extended periods of time. But as an example, even over a three year period, The New York Times/CBS News Poll polled Americans in January 1985 and September 1989 concerning what they perceived to be the most important problems. In January 1985, 23 percent noted war, nuclear war and defence, in contrast with only 1 percent in September 1989. In the case of unemployment it was 16 percent and 4 percent, whereas in the case of drugs it was 1 percent and 54 percent.

3. Objective importance vs. Subjective importance
In deciding what is a valid problem for inclusion, the "objective importance" according to experienced analysts was considered not so significant as the "subjective importance" to those who perceive a particular problem as of major importance from their place in the social system. The social response to eclipses is interesting in this respect. Thus a relatively "trivial" or "irrelevant" problem (according to some "objective" analysis) which looms large in the daily preoccupations of an individual or an organization may from that perspective appear to be of much greater importance than some "major" problem. Whether a "major" problem is held to be major because of the results of some specialized, up-to-the-minute method of analysis, or because it looms large in the daily preoccupations of some other segment of society, is immaterial.

3. Importance and resource allocation
Any segment of society may legitimately attempt to convince other segments of the importance of those problems to which it is sensitive. Whether or not it succeeds, if it is free to do so, it will allocate resources to remedy those problems which it considers to be of relatively greater importance, in the light of its own standards of objectivity, social justice, *etc*. Such resources are not then available to allocate to the solution of other problems, judged by other segments of society to be the major problem

The question raised is where the line should be drawn. Does the allocation of hundreds of million dollars annually to the reduction of personal facial and physical defects (to take an extreme example) justify the inclusion of "unmentionable" but widespread conditions such as ugliness, halitosis, obesity, excess body hair, and the like, as world problems? The point being that such funds are currently not available for the better legitimated problems such as underdevelopment.

4. Determining the relative importance of problems
The relative importance of problems is therefore not clear. Any attempt to clarify the matter could proceed in one of two ways:

(a) Exclusion of "irrelevant" problems: In order to isolate those of "major" importance which merit further analysis, the "irrelevant" problems may be excluded by some set of criteria. This immediately alienates all those individuals and groups whose problems are not admitted as being of major importance by the body responsible for the selection procedure. Being alienated and excluded, and perceived as misinformed or motivated by self-interest, they will continue to allocate the resources over which they have control to the problems which they perceive to be of importance. This is one reality behind the current shortage of funds for "worthy" problem areas.

(b) Clarifying the problem context: By first collecting information on as comprehensive a range of problems as is feasible as a basis for clarifying the debate as to the relative merit of the problems. This alternative opens up the possibility of demonstrating the interrelationships between the problems (including the direct or indirect relationships between the problems perceived by the different groups to be either of "major" or of "trivial" importance). A clearer understanding of the merit of the opposing viewpoints may then be achieved by all concerned. The consequence is that the psychosocial needs acknowledged by policy-makers may then become more subtle. The value of allocating resources to less popular problems may then be recognized.

5. Dangerous depersonalization of problems
Major problems are in danger of acquiring the same status in people's minds as governmental agencies. They are perceived as being too vast and impersonal to be related to in any meaningful way. But even though a problem may only be a symptom (according to some method of analysis), if a significant group believes it to be a problem, and relates to it as such, then it should be registered as a problem. The same people many not identify with or understand the nature of the underlying or causative problem which is more important in the eyes of experts. The relationship between the symptomatic problem and the underlying problem can be identified and registered as a particular kind of relationship appropriately labelled.

It is important to include problems with which people identify.

Method: identification procedure

The following procedures for identifying problems and locating material were used in parallel, although some of them have been used to a greater extent in building up the database either for the 1976 or for the 1986 edition of the Encyclopedia:

1. Requests to international organizations
Requests for documents on specific problems were sent to selected secretariats amongst the 20,000 international organizations selected from the companion *Yearbook of International Organizations*. They requested either that specific problems be identified and described within a questionnaire framework, or that the organizations send any documents or other material from which the required information could be obtained. With the questionnaire was sent a preliminary list of criteria by which suitable problems could be identified, with the request that additional or alternative criteria be supplied. Where appropriate a proof of an entry from the previous edition was included for amendment or comment. This mainly served to increase the flow of problem-oriented documents already received from international organizations in connection with other information processing activities of the Union of International Associations (see below). The value of the replies in both cases lay mainly in (a) their identification of new problems for which documentation was either supplied or had to be obtained, or in (b) their identification of problem categories which it was not useful to include.

2. Documents from international organizations
International organizations send a stream of documents to the Union of International Associations to facilitate the production of reference books on their activities. The relevant publications are the 3-volume *Yearbook of International Organizations*, and the quarterly *International Congress Calendar*, for both of which supplements are included in the bi-monthly periodical *Transnational Associations*. This incoming stream and the documents already filed were scanned for problem descriptions.

3. Intergovernmental organizations
The United Nations and its Specialized Agencies produce considerable quantities of material about world problems. Where feasible, the relevant documents and publications were obtained or photocopied, partly as a result of research in the appropriate libraries. A similar approach was used with bodies such as OECD, the Council of Europe and the Commonwealth Secretariat. The main library research work was done at the United Nations Geneva Library where the publications on public access shelves were scanned. Given the nature of the search it was not possible to make effective use of the UNIDOC series. The principal difficulty is that the presence of much valuable problem information is not signalled by the procedural and administrative titles used for many of the documents. And where a problem-oriented descriptor is used, especially for common problems, it is seldom possible to distinguish a document containing mainly substantive information from one containing mainly procedural information.

4. Reference works
Extensive use was made of a wide variety of encyclopedias and standard reference works to locate and build up files on problems. These were important in ensuring comprehensive coverage of ranges of problems which could not have been foreseen otherwise.

5. Periodicals
Other than periodicals received directly from international organizations, only limited use was made of specialized journals. After a number of trials, use of this source did not prove to be cost effective. Extensive use was however made of press cuttings derived from a number of English-language newspapers. These proved very useful in locating new problems, identifying name variants for them, and ensuring that the database reflected current sensitivity to the importance of certain problems, from whatever constituencies they derived.

6. Market survey literature
Reference material on market surveys was used to identify those problems which were considered commercially important because of the products or services that could be provided to alleviate them. This source was especially interesting in that it pointed to seemingly trivial problems (such as "dry skin", "wood rots") to which large amounts of money were allocated. This source is difficult to use effectively since many of the indexed reports can only be obtained at great cost because of the commercial value of the information.

7. Publication catalogues
As part of the effort to increase sensitivity to unforeseen problems, two of the main sources on current English-language publications in print were scanned. These were *Books in Print* and *International Books in Print*. The scan was also used as part of the process of identifying books for inclusion in the bibliography. Special requests were also addressed to the principal English-language book publishers for sales catalogues which could be scanned for problem-oriented books.

8. Key resource people
In an effort to increase sensitivity to poorly-documented fundamental problems, a special request was sent to several hundred people recognized for their continuing insights into the crises of the times. The request was for photocopies of any documents which could assist the documentation process, rather than for specially written contributions to the database.

9. Multilateral treaties, resolutions and recommendations
In principle those problems recognized by the international community, for which a collective response is possible, should be reflected in the texts of multilateral treaties, resolutions and recommendations. Some use was made of this material. But despite its potential and the great value of documenting the link between problems and the legal instruments responding to them, systematic work on this source did not prove cost effective. Much more remains to be done. The principal difficulty is that the problem is often heavily disguised by the legal text concerned primarily with the response to it.

Whilst many techniques were used to collect information from a wide variety of sources, it would be totally inappropriate to imply that this procedure is capable of gathering all the "available" relevant information. Neither time, financial resources, nor personnel make this feasible, even if the information is not subject to restricted access or stringent copyright protection. For example, resources are not available for on-line searches of external databases. What is achieved is achieved within quite definite constraints. In a specific instance a book may be available on a problem in a distant library or for a certain price. The cost of obtaining access to this information, which may be the latest and best, may be too great. Qualitatively inferior information may have to be used. This of course corresponds to the real world situation in which few have the resources to operate with the best information. Much may however be accomplished by juxtaposing items of "low grade" information. The art has been to compensate for any inadequacies by presenting information so that users are oriented toward the more appropriate source, even if not precisely to it.

Method: problem naming

1. Misnamed problems
There seems to be a general lack of precision in thinking about and naming problems. A recent major internal review of the United Nations notes, for example, that: *"The United Nations System does not possess precise criteria for defining problems which have some chance of being taken seriously by the international community as a whole. The identification of problems which should be the subject of "major conferences", for example, is done mainly on the basis of the existing sectoral schedule of problems: industrialization, science and technology, agrarian reform, population, women, environment, water resources, etc. But frequently the subjects chosen do not represent really new problems, or they are only repetitive devices for driving home the claims of the Group of 77 (increased aid from the industrialized countries, etc). Hence major conferences of this type often culminate in "action programmes" which in spite of their title do not embody anything concrete and do not contribute to any change in the respective attitudes of the participants...."* (para 127).

Elsewhere the same report states: *"The task of identifying world problems occurs when the time has not yet come for negotiating but only for recognizing as a whole the existence of elements of a "problematique" common to all countries, but without any suggestion of going beyond the analysis stage. These problems are dealt with at all levels. In virtually all programmes of all organizations there is a research and identification component of this type. However, only a few problems gradually emerge from all this corpus...and they are gradually identified as suitable for possible discussion on the convergence of national policies or the negotiation of common standards. Thus the questions of the environment, population, certain social problems, economic and monetary problems are at various stages of identification within the world forum."* (para 112). (Maurice Bertrand. Some Reflections on the Reform of the United Nations. Geneva, UN Joint Inspection Unit, 1985, JIU/REP/85/9).

2. Shorthand denotation of problems
The lack of precision in thinking about problems arises in part from shorthand usages by which issues are identified in the media and in political debate. As an example, the Director-General of UNESCO produced a report to the 18th General Conference (Paris 1974) concerning the "Analysis of problems and table of objectives to be used as a basis for medium-term planning (1977-1982)". It was specified that this should include *"all major world problems...coming within UNESCO's purview and relevant to its goals."* The major world problems identified as such were: 1. Human rights; 2. Peace; 3. The advance of knowledge - scientific and artistic creativity; 4. Exchange of information; 5. Communication between persons and between peoples; 6. Concepts and methodologies of development; 7. Policies and strategies for development; 8. Infrastructures and training for development; 9. Greater participation by certain groups in development; 10. Man's natural environment and its resources; 11. Man in his environment, 12. Population.

In these examples the problems are not named in such a way as to be recognizable as problems. "Peace" and "youth" are not problems as such. Both are values to many. Peace is a major value and goal of UNESCO. The use of such words on their own to name problems is therefore quite unhelpful. Peace and disarmament are only problems in a very special and cynical sense explored by the Iron Mountain Report (Lewin, 1967). Use of such words to denote problems therefore has to be questioned. Quite different words may be called for in such cases to name meaningfully the problem implied by such shorthand usage (for example, in the above case: conflict and alienated youth). "Peace" as a positive and desirable condition cannot also be the name for a problem. "War", "conflict" and other tensions are what is presumably meant. The only problem which could justifiably bear the name "peace" is that arising from any negative features of peace as a condition (*eg* lack of stimulus, *etc*), and even then some negative qualifier should be supplied for clarity (*eg* "unjust peace").

For an organizational system to consider the problem and the objective as identical can only lead to considerable confusion. It is even counter-productive because the organization is then motivated to perpetuate problem-solving activity irrespective of whether or not the problem persists as originally perceived. It is presumably for such reasons that the Batelle Institute's DEMATEL Project required that problems had to be stated not as goals to be attained but as unacceptable situations for which there are numerous perceived solutions.

It may be argued that the UNESCO document does not suggest that "human rights" is a problem, but rather that the problem is "the problem of human rights". In other cases this technique of adding "problem" to the descriptor is widely used when no adequate term is available (*eg* the urban problem, the youth problem, the drug problem). This technique has been avoided in this section because it tends to blur (and even discourage) any focus on component problems. As the report of the UNESCO Executive Board (93 EX/4, 31 July 1973, para 51) notes: *"There are no youth problems as such, but only problems that affect youth."* What problems make up the issue areas known as "the youth problem" and "the urban problem"?

3. Negative-naming of problems
None of these names given by UNESCO to the problems it is facing would be considered acceptable as problem names for entries in this section. It is the "lack of human rights", or their infringement, which constitutes the problem. Similarly, it is the presence of conflict, or the instability of any period of peace, which are the problems. (Human rights and peace, as such, are goals or values) Exchange of information is the name of a process, which if it operated inadequately, as it does, would constitute a problem. "Man in his environment" does not denote a problem but a subject of study or debate. "Population" denotes the (number of) people living in a place, country, *etc*, or a special group of people. Only when this number is too high, too low, or increasing too rapidly, *etc*, can problems be considered, to exist. An effort has therefore been made in this section to locate an appropriately negative name to clarify and make evident the supposedly negative nature of the societal problems for which entries are included. With respect to the population issue therefore, it may be made up of "overpopulation", "underpopulation", "inadequate birth control", and similar problems.

Only by requiring that a negative name or phrase be found for the problem, and by avoiding the use of "problem", could problems be satisfactorily isolated from issue areas and programme objectives (see also *Language-determined distinctions* in the Introduction).

4. Ambiguity
As noted earlier, problems are not necessarily named in an unambiguous manner. There is no standard problem terminology. As a result the same problem may be named in a variety of ways. But it is also the case that different problems may be referred to by the same name. A more specific problem may in one context be given the name of its broader problem in another context. Some problems are more effectively named, or are more widely acknowledged, under a metaphoric name (e.g. the greenhouse effect).

5. Degrees of distinction
Because of the variety of ways in which a given problem may be named, especially when different constituencies use different names for the same problem, there is some difficulty in locating and eliminating duplicates in the database. The question that must then be asked is whether very different names are referring to the same problem, to different aspects of the same problem, or to different problems. And even if they are referring to distinct problems, is it appropriate to reflect this distinction by attaching such names to different filing numbers.

6. Combining problems
To restrain premature proliferation of problems in the database, especially in the absence of adequate information, closely related problems may be held as a single problem but with a string of (indexed) names for these various different problems. These can later be split off into separate problems when this is justified.

Method: document control and problem description

1. Administration vs. conceptual preoccupations
A fundamental distinction is made between the administrative concern with controlling the flow of documents relating to problems and the conceptual concern with naming such problems (which includes differentiating them from other problems, or merging ill-formed problems with others) and interrelating them.

2. Arbitrary problem numbering
Control is maintained by allocating an arbitrary filing number to each "problem" as it is encountered. This number identifies the problem in the database and identifies the physical file of material on it. In the database one or more problem "names" (incorporating useful descriptors and common names) are associated with that number in the database. The descriptors are immediately indexed so that the numbered problem is accessible through any of the descriptors (or through the categories to which such descriptors have been allocated). Any material subsequently collected on that problem is channelled into individual physical files bearing the corresponding number.

The use of arbitrary numbers as filing points has a number of advantages:
- simplification of administrative operations, movement of documents between files, etc
- simplification of computer database operations
- separation of administrative concerns from conceptual concerns about the scope of the problem denoted by that number

This last point is of considerable importance in an evolving system. As indicated by the sequence of points in Figure 1 (see following note), information accumulates around a filing number which may have a variety of words associated with it as partial, or even tentative, descriptors. When a problem file number is first "opened" as a "new" problem, the information associated with it and the descriptors used may be quite tentative, especially in the case of complex problems. Thus whilst some problems can be clearly labelled with unambiguous descriptors (e.g. malaria, loneliness), others may eventually have a string of synonymous descriptors associated with them. In the most complex cases, requiring a string of descriptors, several variants of such strings, with different combinations of synonyms, may become associated with the file number.

3. Redistribution of physical documents
It should be emphasized that problems treated as "new" may in fact be subsequently merged with other problems or split into distinct problems. The physical documents are then moved into the files corresponding to the numbers of the destination problems. The information in the database is also transferred accordingly.

4. Convergence towards permanent numbers
Over time, especially given that the programme has been maintained since 1972, the majority of problems become clearly established under their fixed numbers. However new problems entering the database may at any time lead to a conceptual re-evaluation of the status of the older problems, although the probability of any major shift becomes increasingly remote. But although the number becomes increasingly permanent, the name (or names) given to the problem may continue to be refined by the process described below.

5. Computer environment
Some details of the computer software used to facilitate and control the editorial research on a network of conceptual entities is given in Section Z.

6. "Definition"
Problem descriptions are based on the information which accumulates in the physical file bearing the same number as the problem in the database. Editorial work on the descriptions usually takes place after extensive work on the relations between the problems in that domain (see following note). This means that when an editor examines the file and compares the contents with the computer record, it becomes apparent whether items in the physical file need to be moved to other locations because they are more appropriate there, or whether photocopies of certain items need to be made and transferred because they contain information relevant in several places.

The editorial intent is not to provide a final "definition" of the problem but to indicate its "nature" and to clarify the preoccupations of the constituency concerned by the problem. The process resembles the procedures of a prosecutor preparing a brief to present the defendants' case in the manner most likely to ensure prosecution. The actual text may therefore be either very precise, amounting to a definition, or very loose, depending on the kind of problem and the information available. The text may be revised on a number of occasions, possibly as a result of being sent to an international organization in proof form for comment. Paragraphs may be moved into the description from other problems as a result of the processes described above.

7. Editorial intervention
The preparation of a description is very much an editorial process, ideally with minimal intervention by the editors. The intention is to allow the arguments of different constituencies (as "witnesses for the prosecution") to speak for themselves. Where information is available from many sources, this may involve gleaning material from different documents and combining the elements in a suitable manner. The quality of the description thus depends above all on the availability of appropriate texts and the copyright constraints surrounding them. One of the great merits of working with the documents of international organizations is that much of their material is either in the public domain or that they welcome any use of it.

8. Language bias
Material prepared on problems by international organizations has the additional merit that it has already had excessive national and cultural biases removed or at least attenuated. This is of considerable importance because of a major resource constraint, namely the question of language. Although the Union of International Associations receives information in a variety of international languages, its publications are normally in English only. And in the case of the Encyclopedia, non-English material is rarely used in order to avoid translation costs. This inherent bias is partially corrected by the use of international organization material which is designed for publication in several languages and may indeed have been translated into English from one of those languages.

9. Organization of description
In addition to the "Nature" of the problem, other possible headings under which descriptive information may be provided include: Background, Incidence, Claim, Counter-claim. Background is used when some historical context is required for an understanding of the problem. Incidence is used if there is some statistical or other information indicative of the dimensions of the problem. Claim is used to present examples of strong statements from bodies advocating priority attention to the problem, especially when the statements succinctly dramatize the overriding importance of the problem. Counter-claim is used for examples of statements from bodies who consider the problem non-existent, totally misrepresented, or who deny its importance as a problem and may even consider the "problem" to be a solution.

10. Distribution of material between descriptive paragraphs
Whilst the material available on some problems can be clearly distributed amongst the descriptive paragraphs, for others this is not the case. Especially when information is inadequate or of low quality, or when the problem is anyway difficult to articulate, one or other paragraph may be used to carry the available material. In some cases it may seem more appropriate to use the Claim paragraph only, accepting the bias. In other cases, all that may be available is information on the situation in a particular country, which is then given as an example under Incidence. Statistical data indicative of the importance of a problem is often only available for those countries that can allocate resources to the collection of such information and which are motivated to do so. For this reason use is often made of data from the USA or Western European countries on

the assumption that this at least indicates what the situation might be like in other countries if the data were available for them.

11. Claims and counter-claims
It is important to stress that the editors are not attempting to present "the objective truth" by making editorial judgements on what is factual and what is not. The editors endeavour to present problems as they are each perceived from the framework within which each is experienced as significant, using whatever "facts" are considered appropriate by bodies working within that framework. This is especially the case with the Claim and Counter-claim paragraphs. When the information is available, these provide a means of reflecting more explicitly the dynamics within the international community between advocates and detractors of particular problem conceptions. The existence of such dynamics is of course implicit in the juxtaposition of problems which may easily be seen to be mutually exclusive.

12. Problems without descriptions: description by relationship
Problems may appear without descriptions for several reasons. The information received may not readily lend itself to the preparation of a description, typically because of the non-problematic or solution-oriented style. Even when adequate information has been received, editorial resources may not justify preparation of a description at this stage. The problem may be considered too specific to warrant a description, as is typical of those allocated to Section PG. The problem may be so general or fundamental that it is not useful to attempt the challenge of producing an adequate description at this time, as is typical with those allocated to Section PA. Such problems may also be effectively "described" by their function in grouping more specific problems within the logical context of relationships.

13. Comprehensiveness of description
It is important to remember that the number of the problem serves as an arbitrary marker denoting its "existence". To this may be attached one or more "names" which give some indication as to the scope of the problem, especially when they are not synonyms. The set of names may imply the presence of other name variants which are not present. But in many cases the set of names will not exhaustively indicate the scope of the problem. The information available for the description may stress only an aspect of the problem and thus be much less comprehensive, at this stage, than the set of names imply. The reverse may also occur in that the description may be somewhat broader than that implied by the names, especially when the description must provide contextual material in order to be meaningful.

14. Quality of information and description
Because of the logistical problems of assembling and editing appropriate material, as well as the lack of adequate information on many of the subtler problems, the quality of descriptions varies. Some entries reflect an understanding of a problem carefully articulated by an international organization. Others are based on information assembled from a variety of sources. Since the editorial bias is towards inclusion of entries in order at least to acknowledge the sensitivity of some constituency, some entries are based on what in intelligence circles is described as "low grade information". Although once established in this way, and appropriately indexed, higher quality of information may become available to improve the description.

15. Source citation
Although statements used in building up problem descriptions are, in almost every case, very closely based on existing published documents, no explicit link is established between statement and source documents. This has been avoided for several reasons. Resources only permitted problem statements to be located and did not necessarily permit the location of the best document(s) devoted to that problem. A problem description may be built up from fragments of many documents making source indication impracticable and of limited value. Any principal document used may be more appropriately associated with a broader problem, especially when the information on a specific problem is only a small part of it. The editorial process of selecting and restructuring texts from different sources may unintentionally distort the meaning in the original contexts (particularly when the original statements did not constitute clear descriptions). Non-citation of sources is particularly regrettable in the case of little known documents from international organizations. In this edition a major step towards remedying this deficiency has been taken by indicating references to books on the problem, where these could be traced. In some cases such references have been used as a substitute for any further description.

16. International organization citation
It has been the continuing ambition of this programme to include in descriptions specific references to international organizations concerned with a problem. This continues to be beyond the available editorial resources. A principal reason is that many interesting problems are only briefly mentioned in passing in documents of international bodies and it cannot be said that they are actively "concerned" with them. On the other hand, many international organizations claim to be concerned with the more fashionable problems. The current remedy for this deficiency is to group organizations and problems together under some 3,000 subject categories in the companion Volume 3 of the *Yearbook of International Organizations*

Method: inter-problem relationships

As indicated above, two main groups of cross-references are provided between problems. These are the conventional broader/narrower group and a group of "functional" cross-references. The process of indicating the initial relationships between problems is in effect an extension, if not an integral part, of the naming process. The hierarchical relationships, even if only first approximations, position the problem with respect to others and confirm the distinction from them. Particular attention was therefore given to the relationships between problems.

A problem may have any number of cross-references, but the maximum number of any one type seldom exceeds 20. The 13,167 problems in the database currently have some 80,394 cross-references between them.

1. Hierarchical relationships

In the case of the broader/narrower group, there are three well-established types: broader problems, narrower problems, and "related" problems. These have the usual meanings, with the related category being used as a catch-all in those exceptional cases when the relationship cannot be more appropriately expressed through any of the other cross-reference types. In contrast to conventional use however, a problem may have several "broader" problems.

Various series of problems necessitated some regrouping of problems into problem groups to avoid inclusion of too many problems at too great a level of detail. Different methods of handling this matter and establishing cut-off points were used on an experimental basis. Constructing such problem hierarchies was considerably facilitated when the available documents had attempted some such categorization of the problems.

2. Functional relationships

In the four types of functional cross-reference, the described problem: aggravates (cited problem), is aggravated by (cited problem), alleviates (cited problem), or is alleviated by (cited problem). Clearly this group forms two complementary pairs. In certain cases a problem may both aggravate and be aggravated by the same cited problem.

Relationships between problems, other than hierarchical ones, were included either where they were specifically mentioned in the available documents or where they could be reasonably inferred from such material. It is rare for documents to be systematic in their description of the relationships between problems. Relationship networks have to be built up from several different sources. Often it was not clear whether the relationship applied for the whole of a problem hierarchy or for only some component part. Some effort was made to "tidy up" such networks, but in general the practice adopted was to include relationships at this time in order that the networks could be more thoroughly criticized with a view to improvement. It is general easier to criticize errors of commission than to undertake the extra effort to remedy errors of omission.

3. Tentative relationships

It must be strongly emphasized that no cross-reference can be considered "permanent". Cross-references are treated more like pointers. During the editorial process pointers may be modified into a more appropriate configuration. Typically a pointer from Problem A to Problem B may be replaced by one from Problem A to Problem C, plus another from Problem C to Problem B -- if Problem C appears to be an appropriate intermediary. Some pointers may be more obvious and permanent than others which are tentative or only approximate.

Clearly all the different forms of cross-reference interweave to form a very complex network. When indicating functional relationships between problems, the information available may not be sufficiently unambiguous as to whether the pointer should be made to a broader problem or to a narrower problem. Or the information may only mention the relationship to the narrower problem, when the context suggests that it could be more appropriately made to the broader. In this sense whatever indication is given can only be considered tentative, subject to modification later in the editorial process for a future edition.

4. Computer checking

The editorial process of checking relationships is both extremely time consuming and subject to error. Much use is therefore made of software routines both to ensure reciprocity of relationships and to check for various forms of redundancy. Examples of such redundant patterns of linkages include:

- both Problem A and a narrower problem of Problem A indicated as aggravated by Problem B.

- more subject to query, several narrower problems of Problem A aggravated by Problem B, which could possibly be replaced by Problem A aggravated by Problem B.

5. "Unrelated" problems

As part of the editorial process, it is useful to ask such questions as:

- which problems are not part of another problem ?

- which problems are not aggravated by other problems ?

- which problems do not aggravate other problems ?

- which problems do not alleviate some other problem ?

Whilst it is probable that there are only a limited number of problems which are at the top of problem trees, it is less clear whether all problems are parts of such trees. It is probable that for a problem to be a problem it should cluster or aggravate some other problem, but whether problems linked in this way form "islands" separate from other similarly linked problems remains to be explored.

From an editorial point of view, it is clearly important to focus on the first question, if only as a means of detecting duplicate narrower problems. This question is also important in the case of highly specific problems (such as some rare disease), which it may not be useful to represent in the system at all, but for which some broader problem can be usefully included, even if no information on it exists (the class of such rare diseases, for example).

In the final published product the distinction between problem perceptions substantiated by information received and those based on interpolations of this kind is quite evident. The interpolated problems are present in skeletal form, with name(s) and cross-references, but without any other descriptive text.

Method: conceptual processes summarized

1.a Detecting/Finding/Scanning -- accumulating variants -- symptoms/pre-problems	1.b Eliminating/Rejecting/Filtering -- nonproblems (cases, solutions, theories, events, projects)
2.a Clustering/Grouping -- combining synonyms, aspects, duplicates -- tidying-up variants	2.b Splitting/Distinguishing -- combining synonyms, aspects, duplicates -- tidying-up variants
3.a Responding to sources -- sensitivity to where people are -- what people identify with -- accept unforeseen categories	3.b Detachment from fashionable -- reservations concerning solution/value hype -- articulating value dimension -- clarifying value qualifiers as ordering principle
4.a Sharpening names/keywords -- renaming / negativizing -- de-hyping/de-solutionizing -- what is the problem the solution is designed to remedy -- alternative names to facilitate location	4.b Broadening / Balancing -- inserting anthropocentric terms where implicity -- opening possibility of non-anthropocentric equivalent. -- opening up sets /series -- is there a complementary problem
5.a Naming more general problems -- naming clusters -- creating intermediary problems to group/label sub-clusters	5.b Determining level of specificity -- appropriate cut-off points at bottom of hierarchies -- what should only be covered at a more general level
6.a Elaborating description -- sharp, not waffly -- appropriate amount -- transfer aspects of mega-problem texts to sub-problems	6.b Caring for poorly articulated and inadequately documented problems

Figure 1. Problem entries: conceptual processes

1.a Detecting/Finding/Scanning -- recognized relationships	1.b Eliminating/Rejecting/Filtering -- misconceived relationships -- too vague, too specific
2.a Forming hierarchy -- tidying up	2.b Distinguishing hierarchies -- splitting off branches
3.a Shifting level to more general relationships	3.b Shifting level to more specific relationships
4.a Inserting implicit relationships	4.b Eliminating erroneous relationships
5.a Responding to necessity of ordering clusters -- appropriate cut-off points -- adjusting from minimal to optimal ordering	5.b Avoiding excessive imposition of simplistic ordering
6.a Cross-referencing responsible organizations	6.b Cross-referencing bibliographic sources

Figure 2. Problem relationships: conceptual processes

Patterning problems: concept refinement process

The above procedure was initiated in 1972 and has increasingly become a purely "administrative" matter, especially since the proportion of new problems or new classes of problems continues to diminish. The ongoing concern is therefore much more with the conceptual processes whereby the "problem" associated with any given number is clarified through the naming (descriptor allocation) process. This may involve grouping different problems under one number or splitting one problem into several different problems. An important question is the clarification of relationships to more general and to more specific problems.

There are two major constraints to be borne in mind in the following discussion:

(a) The logistical constraint of ensuring that any changes, whether inspired by information gathered or by the concept refinement process, should not jeopardize maintenance of documentary control over the range of problems at this stage of the programme. This especially governs decisions on the degree of specificity explicitly permissable within any group of problems -- namely to what level of detail are numbers to be allocated to problems. In other words, the need for appropriate cut-off points.

(b) The information gathered also imposes constraints on what degree of "concept refinement" is appropriate without distortion. This is a major issue given the differences in quality of information and the differences in interpretation of "facts" in different documents, even those supposedly of equivalent quality. It is especially difficult when the original information is vague and defines the problem using a loose terminology or when the terminology used is precise but applies to a more general or a more specific problem. It is obviously also difficult in the absence of adequate information on a problem whose existence is only suggested by the pattern of information in the database.

In both cases the decision is ultimately a matter of editorial judgement, relying on a multiplicity of documents to suggest the most appropriate compromise until further information becomes available.

An overview of the concept refinement processes is presented in Figures 1 and 2 in the previous note. It is important to understand that many of these processes occur in parallel or are undertaken simultaneously by editors working on the network of problems from different computer workstations. Some are discussed below.

1. Problem identification

Once the bulk of the material has been filed by problem (as described above), the filed documents are then individually re-examined to locate descriptions of other problems which are then transferred to appropriate existing files or made the subject of new files. This is an integral part of the process of identifying relationships between problems. Scanning the material brings out possible variants on the name of the problem, which have to be distinguished from names which are more appropriately associated with other problems (which may need to be entered into the database). Variant names, even colloquialisms, may be vital as providing alternative means for locating the problem through the index. Also to be distinguished are various kinds of non-problem, according to the criteria guidelines, which appear in the documentation as problems or are identified by problem names. Such poorly-formed problems are coded as low-priority items in the database, not to be included in the published volume unless there is later reason to change their status.

2. Problem clustering

Because of the fuzziness of problem names, some of which take the form of phrases of some length, detection of duplicates is a major concern. The challenge is that similar names may have been associated with significantly different problem descriptions. The question then becomes whether to merge the two problems or to associate both names with one description and rename the other problem to distinguish it appropriately. Another challenge is the question of whether variants or aspects of a root problem should be distinguished as separate problems or merely as alternative names of that root problem. Here the decision may be strongly influenced by the amount of material available on a "variant", suggesting that it merits separate treatment. This tendency may be restrained to some degree by the process of systematization discussed below.

3. Problem (re-)naming

A high proportion of the documentation on problems makes use of loose terminology, especially in the case of the less tangible problems. Problems are often identified by positive terms more closely associated with the values attached to their solution. Implicit anthropocentric qualifiers may be omitted. An effort is therefore made to sharpen the negative aspects of the name to give greater focus to it. Clearer descriptor phrases may be discovered in the literature. Colloquial descriptors are added when they facilitate access to the problem. The ability to add alternative names is also used as a way of grouping into one problem the names of problem variants when it does not (yet) seem appropriate to treat these as separate problems, although it is useful to maintain an index trace on them. The significance of a collection of names attached to a problem thus varies depending on whether it can be considered well-defined or simply as a loose cluster of concerns around an ill-defined "proto-problem".

4. Adequacy of problem names

In the case of complex problems, the varying adequacy of sets of problem names may be illustrated by Figure 3. There the problem to which reference is being made, to which the number refers, and which the descriptive text is intended to encompass, is indicated by a circle. The names associated with that problem are indicated by the lines, each line representing one name.

Fig. 3A: In this ideal case in which the set of problem names encompasses and contains the content, although each name is in itself inadequate.

Fig. 3B: In this case the content implied is larger than that effectively encompassed by the set of names. The set does however share a common centre of reference with the content.

Fig. 3C: In this case the set of names implies more than the content. The problem is of lesser scope than the set of names implies. The names and content still share a common centre however.

Fig. 3D: In this development of the ideal case (Fig. 3A), implications of two of the names, combined with a third, serve to delineate an additional problem, distinct from the main problem. This suggests the need to split off the subsidiary problem, removing from the main problem the name which contributes to its enclosure.

Fig. 3E: In this case the set of names no longer shares a common centre with the content, and as such is no longer adequate. There are aspects of the problem which are not captured by the names, and the names imply a problem which does not exist.

Fig. 3F: In this case the set of names is inadequate to the content, even less so than in the case of Fig. 3B.

Using the convention of Figure 3, other cases could be indicated, notably:

- Case of a circle without lines, which would indicate a problem for which no names had been found (a "nameless problem");

- Case of set of names without a circle, which would indicate names referring to a non-existent problem ("problem-less names");

Other possibilities include cases of single or parallel lines in various positions in relation to a circle. Of special interest is a line which extends much beyond the circle which it intersects.

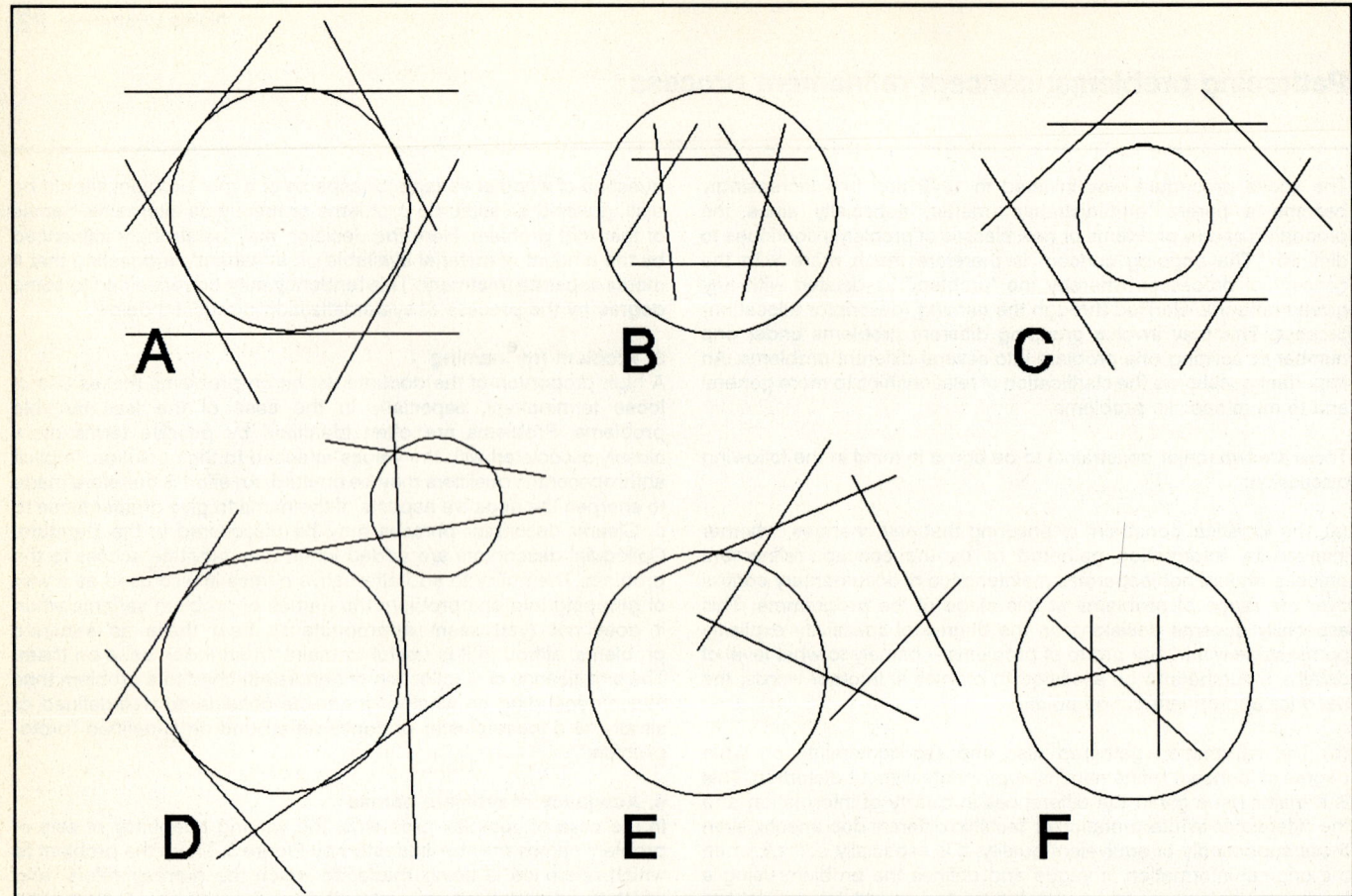

Figure 3. Illustration of varying adequacy of sets of problem names (lines) in relation to substantive content (circles)

whole forming a partially ordered pattern. Where several lines enclose an area, there is the possibility that a circle may be detectable.

There may be a case for using computer graphics to explore such patterning, with the object of detrmining whether there are ways of massaging the relationship between names and content into more ordered patterns. Of even more interest is the possibility of using a three-dimensional representation to extend the power of the geometric metaphor.

The geometrical convention may also be used to explore the ways in which underlying problems are conceptually contained. It provides a visual illustration of how problems may escape containment. As such it might be used as a way of mapping "problem-space" and the conceptual response to it.

5. Responding appropriately to source material
Much material available on problems makes what can easily be treated as exaggerated claims for particular problems whilst disparaging others. A careful balance must therefore be sought between sensitivity to unusual (even seemingly ridiculous) problems and detachment from the seemingly evident importance of currently fashionable problems (however eminent the legitimating authorities). This calls for openness to unforeseen perspectives and reservations concerning institutional and ideological hyperbole. Given the disproportionate amount of information on fashionable problems, this balance must be sought so as to give useful amounts of information on problems which may be of quite different significance.

6. Elaboration of logical context
Few, if any, problems exist in isolation. Many problems are defined as much by the context of other perceived problems as by any particular textual description that can be given to them. Most problems can be considered as "part of" some broader problem. Many problems therefore group together more specific problems. These relationships form problem hierarchies. Because of the variety of ways in which problems can be so grouped, any one problem may be part of several different broader problems. Effort must therefore be made to position a given problem within a context of broader problems, if necessary using it to group more specific problems. The problem needs to be named so as to distinguish it effectively from both the broader and the narrower problems. The process of "tidying up" such relationships is a continuing one, which may be dependent on the availability of some new document clarifying the pattern. Difficulties may be created because of conflicting perspectives in the literature on how certain problems should be grouped. At any particular point in time, there is always more that could be done which would take more time than resources permit. Some hierarchies must necessarily be left in a crude state. The logical framework around any problem may be more or less subject to further change.

7. Problem classification and section attribution
Because of the potential complexity of a problem's logical context, any effort to classify problems neatly by subject according to conventional principles has been resisted. An alternative approach has been developed which is used annually to group problems and organizations in Volume 3 of the *Yearbook of International Organizations* in a pattern of some 3,000 categories. In large part this classification is done automatically on the basis of descriptors in the problem title and constitutes one of the online editorial tools for exploring the database. For this edition the process of editing was further facilitated by attributing a section code to each problem. This is a pragmatic effort to group problems so as to distinguish those when tend to be higher in any logical hierarchy from those lower in the hierarchy (using the letter codes A through E). Other codes, notably F, are used for exceptional cases. During the editorial process the code attributed to a problem may be modified to position it more usefully within this pragmatic system, as well as to give it a lower editorial priority. (The significance of these codes is explained in more detail in the following note on *Classification and section attribution*).

8. Identification of functional relationships
Just as few problems exist in logical isolation, so few are independent of each other. Most problems aggravate some others directly, and in their turn are aggravated by others. Some problems reduce others, or are reduced by them.

This information may be present in the material collected on one problem enabling the relationship to be inserted. It may however only be present in the affected problem, so that the relationship gets automatically inserted in the former (given the relational nature of the database), even when there is no confirming documentation there.

Just as with the logical relationships, the documentation may suggest functional links to other problems at a level which is too general or too specific because the database structure is already more finely articulated. Again it is a matter of editorial judgement how this is to be handled. No relationships may be given in the documentation, in which case it may be appropriate to insert those that are obvious. In many cases the insertion of obvious functional links, is more appropriate than the use of "see also references" (which is effectively the significance of the "Related problems" relationship).

A major difficulty in inserting functional relationships is that of ensuring that they are made between the appropriate levels of problem hierarchies. Thus it is not useful to show all the narrower problems of Problem A as being aggravated by Problem B, when it suffices to show Problem A as being aggravated by B.

Whereas in some cases the logical and functional relationships between problems are reasonably clear and distinct, in others this is far less so. Problem A may both aggravate B and be aggravated by it. It may be far from clear of what problem Problem A is a part. These difficulties emerge particularly in the case of those allocated to Section PF, and may be one reason why they are so allocated.

9. Confrontation with new documents
Since so many problems and relationships are now present in the database, an important process is that of challenging the existing database with new problem-focused material. Each such document may raise questions as to whether it:

- can be directly associated with an existing number and its set of descriptors;

- necessitates additions to the descriptors, namely does it suggest additional alternative names for the problem, or a correction to the existing name, or revision of the names attached to a whole set of problems;

- calls for the creation of a more general or a more specific problem, possibly requiring that some existing problem should have its scope and descriptors modified to make it more general or more specific;

- implies the existence of a whole new category of problems previously unrepresented in the database.

10. Systematization of problems
Although the prime emphasis is on registering those problems detected in available documentation, as these build up in the database certain patterns are suggested. The extent to which this patterning can be allowed to influence the inclusion or exclusion of problems is discussed is later notes.

The question is then to what extent the problem names should be rationalized. Again these are matters for editorial judgement, where the editorial resources are available. Some tentative steps have been taken towards such rationalization, but much more is called for as the opportunities become clearer. This process is important in preparing sets of problems to be checked from a particular perspective by external authorities.

11. Systematization of problem relationships
As noted above, when entering problem relationships into the database it is quite difficult to avoid redundancy. Computer programmes can be used to highlight certain redundancies, and even to eliminate some of them. But it is a matter of judgement as to how far this process can be allowed to go. For example, if 75 per cent of the narrower problems of Problem A are aggravated by Problem B, can these explicit relationships be replaced by a single relationship between A and B ? But what if the percentage indicating such aggravation is only 50 or 25 ?

12. "Massaging" the database
The point to be made is that editors are attempting to identify a pattern of significance in relation to each number, but the question always remains as to whether that pattern is stable and well-formed or whether some portion of that significance should not be moved to a distinct number. This depends not only on editorial judgement but on whether some constituency believes that a problem is distinct from the problem where it would otherwise be filed.

It should be stressed again that the database is designed to reflect not "facts" but perceptions of facts, however questionable they may appear in the eyes of others. The editors do not attempt to determine what is the most "authoritative" view, but rather what views are representative of significant constituencies (which normally would include those that are widely considered as authoritative). Decisions as to what are representative views, even if conflicting, are assisted by the international context within which the information is obtained.

13. Constraining the anticipatory function
In the light of the above points and the discussion of patterning in the following notes), to what extent should the database perform an anticipatory function ? It is clear that seemingly credible problem names can easily be "generated" by appropriate combinations of descriptors from descriptor sets (in a manner reminiscent of Ramon Llull's category generator).

The availability of such descriptor sets encourages the recognition by the international community of problems which may or may not exist. Just as there are "letterhead organizations", there may be descriptor-generated problems (for which project funding may even by sought). Such problems may emerge into the literature as a result of efforts by delegations or the media to appear innovative. Given the institutional resistance to the recognition of new problems however, systematic explorations of such a procedure suggests an approach to identifying potential problems rather than allowing them only to be registered in a haphazard manner as information is found on them.

Perhaps category generators could usefully be developed to draw attention to categories of problem which may be neglected in conventional decision-making. In this project the procedure has only been systematically adopted in the case of economic sectors and endangered species. It is as a means of ordering problems at different levels of specificity, particularly in order to avoid isolated excessively-publicised entries (eg endangered panda, or endangered monkey-eating eagle) where more general entries would be more appropriate (eg endangered bears, or endangered birds of prey, respectively).

The question remains however as to whether the distinctions between problems, generated using different verbal "operators", correspond to valuable nuances with which people identify, or whether such distinctions are purely contrived in order to arouse emotive effects through their novelty

Patterning problems: classification and section attribution

1. Complexity of problem classification

The approach to problem classification is treated quite separately from the administrative question of providing a filing point for information (whether physically or electronically). In the first edition (UIA, 1976), problems were quite deliberately not classified in any way -- other than under the arbitrary filing number. The principal reason for this approach is that it was considered desirable to separate the logistical issues of managing the information from the highly controversial issues of how problems should be grouped. As has been argued elsewhere (Judge, 1981), classification is a highly political act - especially, when dealing with "world problems". These points are discussed further in the Introduction.

Problems may be classified by the object susceptible to the disease (problems of plants or animals), by the age of the object (problems of youth or elderly), by part of the object (problems of the eye or foot), by type of symptom (inflammation of tissue or dry rot), by location of occurrence (problems in transit or in storage), by geopolitical region (problems of industrialized or developing countries), or by causal agent (deficiency problems or problems of pests). Other bases for classification might also be envisaged.

2. Subject classification of problems

With the development of the *Yearbook of International Organizations* into a 3-volume publication in 1983, the third volume (UIA, 1989c) entitled *Global Action Networks* (classified directory by subject and region) has been used to group together both international organizations described in the first volume and the world problems from this Encyclopedia.

In a research oriented system it has been considered desirable to create an information processing context in which the manner in which the problems were grouped could be continually reviewed. This is the approach taken with successive annual editions of *Global Action Networks*. For each edition efforts are made to fine-tune the thesaurus structure currently numbering some 3,000 categories. New categories are added and the attribution of organizations and problems to categories and category combinations is modified. The 1990/91 edition incorporates world problems from this volume, but according to the information available in early 1990.

The system of classification was developed after examining the possibility of using other international systems (UIA, 1989c, Appendix). It was partly inspired by the system developed by Ingetraut Dahlberg (Dahlberg, 1982) and partly by structural features of the periodic table of chemical elements (van Spronsen, 1969). It was deliberately designed to highlight integrative or interdisciplinary relations between categories. The thesaurus is continually redesigned as a system of categories to reflect the systemic relation between the preoccupations of international organizations.

A computer programme is used to reallocate problems to categories whenever a significant number of thesaurus modifications have been made. This is usually done annually. Interim changes are however relatively easily made. During the editorial process, any change made to indexed names results in the problem being reindexed and allocated to any relevant categories associated with the new words indexed. At any time therefore problems can be accessed via word, via specific subject category, via subject group, or via various Boolean combinations of these elements.

3. Section attribution of problems

The policy of resistance to subject classification has also been applied to the international organizations in the *Yearbook of International Organizations*. But in the latter volume the organizations are grouped into sections (A through U) which are based partially on degree of internationality. Although having little theoretical justification, the system has proved to be useful in practice in dealing with many exceptional forms of international body. Experience with this system led to the development of an equivalent one for world problems. This was tentatively presented in the 1986 edition as a special index (see Section PX in 1986 edition).

For the current volume, this system has been refined and used to order the actual descriptions in Sections PA through PG (which absorbed entries originally allocated to Sections PP and PQ of the 1986 edition). Within the sections, problems are filed by their arbitrary number as in the *Yearbook of International Organizations*. Part of the editorial process, from edition to edition involves decisions on the appropriateness of any reassignment between the A through G sections. The resulting sections are as follows:

Section PA: Used for ubiquitous, fundamental, "abstract" problems, which tend not to be considered sufficiently tangible to appear on the agendas of international organizations (e.g. apathy, corruption, greed, etc)

Section PB: Used for major cross-category world-wide problems (or problem complexes), which tend to be prominent on the agendas of international organizations (e.g. war, environmental degradation, etc) and to group many subsidiary problems.

Section PC: Used for major cross-sectoral problems of a specific nature.

Section PD: Used for detailed problems.

Section PE: Used for emanations (combinations) of other problems.

Section PF: Used for exceptional, "fuzzy", potential problems.

Section PG: Used for very specific problems on which it is not considered appropriate to provide descriptions. (Entries are not printed, although problems are indexed to any parent problem in earlier sections).

Section PJ: Used for editorial purposes as a temporary section for new, unconfirmed or inadequately named problems. (Entries are not printed, although problems are index as for Section PG)

Section PS: Used for editorial purposes as a provisional reject section for "problems" which are considered inappropriate for this volume. (Not included in this volume).

Section PU: Used for editorial purposes as a semi-permanent reject section for "problems" which are considered inappropriate for this volume, or very low priority. (Not included in this volume).

4. Derivation of section attributions

The following overlapping guidelines were used in allocating the section code letter during the editorial process:

General to specific: Classes of phenomena, such as living species, are allocated "B" at the kingdom level (*eg* plants) down to "G" for a specific species (*eg* bald eagle), or a specific disease or commodity.

Universality: All classes ("A") of living beings, for example, through to particular classes or specific types ("E"), such as tropical islanders.

Fundamental to dependent: ("A" through "E")

Hierarchical level: Top of several hierarchies ("A") to a specific feature of a single hierarchy ("A" or "G")

Discipline specificity: Transcending any group of disciplines ("A"), through major conjunction of disciplines ("B") to single discipline ("D") and sub-discipline ("E").

Geographical/cultural locus: From global ("A"), without any specified region or division, through intercontinental ("C"), such as developing, industrialized or socialist, to specific ("D"), such as mountain or tropical regions, through multiple qualifiers ("E") such as "disabled women in developing countries", to problems of one single country which are of wider significance ("G"), such as apartheid.

Set membership: Where the problem name suggests the possibility that the problem is a member of a set of similar problems, they are coded "C" or lower. For example "blindness in children" suggests "blindness in the elderly".

Fashionable problems: Care has been taken not to give exaggerated prominence to highly-publicized specific problems (*eg* endangered species of whale, AIDS).

Exceptions: The code "F" has been used for potential problems, dormant problems, extra-terrestrial problems and various subtle or intangible problems.

Attribution to a particular section is not considered definitive. It is a pragmatic convenience. Some problems coded "F" could well have been coded otherwise, especially "A". Problems which otherwise would have been coded "G", but have nevertheless acquired a description, are coded "E" so that they can be printed in this volume.

Patterning problems: languages games

This project is in many ways an exploration of the use of language and the effects of its use on the distinctions which are accepted in international discourse. The question is to what extent use of legitimate-sounding problem names in practice signals the distinct existence of the named problems. Furthermore, to what extent problems can be effectively named without their having meaningful existence. This project is on the receiving end of documents naming problems. The names attributed may be intended to signal the existence of new problems. But the names used may effectively imply the existence of other problems to which analogous names have not yet been attributed (possibly because the relevant documents have not been received). These questions constitute a real challenge in any attempt to impose a greater degree of order on the rich use of language to evoke a response to concrete problems.

1. Emerging distinctions
In a project which is designed to be responsive to the problem perceptions of different constituencies and to the distinctions that they choose to make, the words used to denote problems acquire a special importance. In an international context, in which many problems have been identified, such constituencies may carefully choose an unusual combination of words to give greater precision to a problem, or problem variant, which they perceive others to have neglected when it has been associated with some better known problem. In order to sharpen perceptions new variant names may be formulated, especially (and most creatively) by journalists. The question then becomes when is the word combination to be considered as naming a new and distinct problem and when is it to be considered as a valuable synonym for an existing problem name.

2. Artificial distinctions
Journalists provide a good example, since good journalism is in the business of reporting emergent issues and of naming them in a meaningful way. On the other hand journalists are under pressure to make recurring topics interesting by providing a new slant or angle. But the same may be said of politicians who are also in the business of naming problems which they believe may arouse the interest of their constituencies. A seemingly "new" problem may thus emerge into the literature purely because of skills with language and not because a real distinction has been made from some existing problem in the same domain.

3. Completing problem sets
A related concern results from any recognition of a problem series or set. If a set of problems has been recognized, such as: medical malpractice, legal malpractice, insurance malpractice, etc. These may be grouped under a broader problem such as professional malpractice (thus naming the set). It may become apparent that there is no problem on malpractice in some other domain such as "architectural malpractice" , raising the question of whether it should be included in this set in the absence of information indicating that the problem "exists". A search of the database may reveal that such material has accumulated under "irresponsibility", "unethical practices" or "negligence" in that domain. This may lead to redistributing material on all problems using those qualifiers for every domain. Such techniques compensate for the logistical difficulties in gathering and processing information for individual problems. There may indeed be information on architectural malpractice, but it may be quite impractical to allocate resources to obtaining it.

4. Naming an unnamed problem set
A similar concern results from the editorial recognition of a set of problems in the absence of any information on the broader problem implied by the existence of the elements of the set. Thus in the example above, if information was available on the problems in the set but not on "professional malpractice", is it appropriate to create the latter problem in order to group the others ? This raises the very interesting question as to whether an unnamed set of problems constitutes a new problem in its own right, and the further question of the status of the set of sets of problems. The unnamed set may indeed be a meaningful problem even though at that level of generality it may not be currently recognized. It may be more appropriate to point functional cross-references to such a broader problem rather than to duplicate such references to a number of its sub-problems.

5. Problem generation using sets of categories
By taking a set of subject categories or even some set of problems, quite legitimate-seeming problem names can be "generated". Using the above example, "malpractice" could be combined with the complete set of professions (such as engineering, surveying, etc). In the case of "corruption" as a problem, how many narrower problems is it useful to open up if only information on "political corruption" is readily available ? Is it appropriate to combine "corruption" with the names of the major classes of activities: science, religion, culture, education, military, etc ? In such cases it is extremely unlikely that the combination is meaningless, and quite probable that such problems do "exist".

6. Problem generation using regional and group qualifiers
The previous point illustrates the possibility of identifying more specific problems by the addition of qualifiers:
- Racial (Economic/Social/...)
- discrimination (exploitation/...) in the
- construction industry (agricultural industry/...) against
- disabled (impoverished/...)
- women (youth/...) in
- tropical (semi-arid/...)
- developing (industrialized/mountainous/...)
- countries (regions/...)

7. Problem generation using negative "operators"
The editorial process of ensuring that a problem has an appropriately negative name leads to the accumulation in the database of problem names with a fairly limited set of negative qualifiers. These are used in the literature to denote problems when applied to members of the set of subject categories. For example:
- "shortage of" food (funds, personnel, equipment, etc)
- "misuse of" power (funds, equipment, science, etc)
- "hazards of" nuclear power (toxic waste, computers, sugar consumption, etc)
- "maldistribution of" food (electricity, health facilities, wealth, etc)

These "negative operators" can also be used to generate problems. Some related issues are explored in the discussion of the link between such negative values and problems (see *****). Care must be taken to avoid an accumulation of distinct problems in the same domain using negative qualifiers which are to varying degrees synonymous. This increases pressure for a more systematic approach.

8. "Nameless problems" versus "Problem-less names"
Whether in tidying up the database or in opening it to new problems suggested by the pattern of existing problems, there is clearly a fundamental issue which needs to be further explored as the database is developed. The techniques described above provide a way of broadening the scope of the database and introducing levels of order where sets of problems can be recognized. Given the desire to orient users to potential problems, "opening up" problems, in order to complete a set, increases sensitivity to information which can be usefully collected under the number prior to any editorial description of the problem. But these techniques can only be used with considerable caution. Sensitivity is indeed required to the existence of "nameless problems" implied by the relational context in the database. Providing such problems with a name so that they can be used to group already named problems introduces order into the database. It can create a focus for real problems on which information is scarce to non-existent. At the same time care has to be taken to guard against the creation of "problem-less names" and against shifting the centre of gravity of the project towards what amounts to language and word games.

Patterning problems: patterning the problematique

The number of problems and their degree of interrelationship are a continuing challenge to comprehension. Without any patterning, the amount of information is overwhelming. The simplistic patternings characteristic of conventional practice in documentation systems are however part of the conceptual problem rather than the solution. They disguise complexity and create deceptive impressions of order where order is lacking, or rather where higher forms of order are implicit. The very simple ordering used in this Section P is designed to keep this challenge to the forefront. A more complex regrouping of the problems by subject, inspired by the periodic classification of chemical elements, is available in a companion volume (*Global Action Networks*).

This note reports on the possible value of a patterning of the whole set of problems based on three possible elements in the name of a problem. The elements, which in each case constitute a relatively limited set, are: negative operators; subject categories; and specific qualifiers.

1. Negative operators
As discussed earlier in connection with *Language Games*, the editorial process of ensuring that a problem has an appropriately negative name leads to the accumulation in the database of problem names with a fairly limited set of negative qualifiers. These are used in the literature to denote problems when applied to members of the set of subject categories. For example:
- "shortage of" food (funds, personnel, equipment, etc)
- "misuse of" power (funds, equipment, science, etc)
- "hazards of" nuclear power (toxic waste, computers, sugar consumption, etc)
- "maldistribution of" food (electricity, health facilities, wealth, etc)

These "negative operators" can also be used to generate problems. Some related issues are explored in the discussion of the link between such negative values and problems (see Section VZ).

In each of the above cases, the "negative operator" could be replaced by synonymous terms, like "insufficiency of ", "irresponsible use of ", "dangers of" and "imbalance of". There could be a question as to whether the substitutes were true synonyms or whether they signalled real alternatives. More interesting is the possibility that the range of negative operators indeed constitutes a limited set, such as might be ordered as in Figure 4 (based on value sets in Section VT).

2. Subject categories
The history of classification is largely the history of ways of patterning subjects. Unfortunately, few of these initiatives are of much use in patterning problems. It was for this reason that an alternative approach was used with the companion volume (*Global Action Networks*). This allows for continuing refinement and experiment between annual editions. The approach is explained in detail in that volume. The aim was to interweave the pattern of categories to facilitate recognition of interdisciplinary patterns of a higher order. Within any cell of the matrix (identified by a 2-digit number). More detailed categories are available since a 4-digit numeric code is used to identify subjects. Some 3,000 categories are currently used. The categories are specifically designed to be responsive to the range of activities of some 25,000 international organizations and the problems they encounter.

3. Specific qualifiers
Problems may share the same negative operator and subject category but be distinguished by specific qualifiers. For example:
- Maldistribution of resources in industrialized countries
- Maldistribution of resources in developing countries
- Maldistribution of resources in arid zones
- Maldistribution of resources in mountainous regions
- Violation of rights of disabled elderly
- Violation of rights of disabled children
- Violation of rights of disabled war veterans
- Violation of rights of disabled workers
- Violation of rights of disabled women in developing countries
- Violation of rights of disabled women in industrialized countries

In the first case, the specific qualifier is a geopolitical term. In the second, it is a category of people, usually an exploited group. In the third case, both types of qualifier are used in combination.

4. Patterning
A systematic pattern of problems can be generated by combining these three sets of elements. Those currently in the database are an approximation to this comprehensive pattern of problems. It should be quickly noted that the pattern is itself simplistic, although much more complex than existing patterns. But paradoxically it is perhaps just this quality that might make it satisfactory to the Cartesian mindset that so dominates the international community. Its main merit would be in providing a focus for criticism, since its weaknesses are so evident. It corresponds to efforts to achieve satisfactory urban planning through use of the grid system. The advantage is that problems are effectively "zoned". The weakness is that the zoning is an imposition on a higher order of complexity which cannot be satisfactorily represented through the grid system.

The pattern of negative operators can be explored further. In particular there is merit in exploring the relationship to the pattern of values clustered in Section VT. The link lies in the basic recognition that a problem can only exist in the light of a value. It is the negative operators that provide the problematic dimension and thus there should be some kind of mapping of them onto the set of values.

For any patterning to be meaningful, it should aim for more than can be achieved by a simple grid. This is the objective of continuing work on the pattern of subjects (in *Global Action Networks*), which can be represented on a grid although the relationship between the columns and rows is based on an analogy to the much more complex periodic table of chemical elements. Attempts have been made to represent this in a variety of forms, including circles and spirals. The intent is to render explicit integrative dimensions which are lost, or implicit, in the grid pattern.

In the discussion of the pattern of values, the suggestion has been made that the value polarities might usefully be mapped onto the surface of a sphere, interwoven in a "tensegrity" structure to reflect the integrity of their tensional interdependence. Imposing spherical curvature on the grid renders the pattern of values finite but unbounded. In this form it constitutes a whole which may prove psychologically more meaningful, for the same reasons that mandalas are used to render complexes of psychic functions meaningful in an integrative manner. The set of negative operators might then lend themselves to similar treatment. Just as the value polarities indicate extremes of imbalance, the negative operators might be configured to indicate the same.

In effect the set of negative operators provides a pattern of ways in which "things can go wrong". Further confrontation between the value polarities and the negative operators could increase the sophistication of the latter set without jeopardizing its comprehensibility. In this form it could constitute a valuable checklist for policy weaknesses in any domain.

5. Future "massaging" of the database
The value of any pattern of problems obviously needs to be tested by attempting to present all problems in the database in terms of that pattern. As discussed earlier, one merit of this is to challenge names currently given to problems and to identify problems which are in effect duplicates. This approach would also expose the possible existence of problems not currently reflected in the database. On the other hand, the exercise would also highlight problems to which the current set of negative operators is poorly adapted, thus suggesting lines for further improvement. Of special interest is the way in which problems, which are unusual in terms of the implicit operator (not the subject category or qualifier), raise questions about values which have not been rendered sufficiently explicit. They may even point to the emergence of new values.

In future editions there may even be merit in presenting the problems in terms of the pattern of negative operators.

	Existence	Quality	Evolution	

NATURE:	+ emergence of -
	   presence of -
	
	- failure of +
	   death of +
	   absence of +
		+ excess quality
		   overdesigned
		
		- poor quality
		   defective
			+ overdeveloped
			   overevolved
			
			- underdeveloped
			   undeveloped
			   deterioration
			   degradation
				(Pride)

	Quantity	Distribution	Accessibility/Exposure	

SUFFICIENCY:
(accumulation)
	+ excess, surplus
	   oversufficiency
	
	- shortage, lack of
		+ pervasive -
		   carriers of -
		   vectors of -
		
		- maldistribution
		   concentration
		   imbalance
		   monopoly of
			+ forced exposure
			   overexposure
			   conspicuousness of -
			
			- inaccessible +
			   unavailable +
			   underexposure
				(Aquisitiveness)

	Action quantity	Rate of change/growth	Rate of transformation	

DYNAMICS:
(motivation)
	+ hyperactive
	   rapid pace
	
	- inactive
	   inertia
	   paralysis
	   slowness
	   slow
		+ proliferation
		   too rapid change
		
		- too slow change
		   unchanging
			+ too rapid transformation
			   chaotic revolution
			
			- too slow transformation
			   untransformative
				(Inordinate desire)

	Comprehensiveness	Veracity	Certitude	

REPRESENTATION:
(assessment)
	+ overcomplex
	   incomprehensible
	   excess variety
	
	- superficial
	   trivial
	   shallow
	   unrequisite variety
		+ secrecy, cover-up
		   obscurantism
		   suppression of info.
		
		- disinformation
		   hype, bias
		   misleading
		   fallacy
			+ dogmatic clarity
			   determinism
			   fatalism
			   unquestionable
			
			- unclarity, doubt
			   confusion
			   limited vision
			   too many questions
			   too many options
			   uncertainty
			   non-verifiability
				(Envy)

	Structure	Regulation	Regularity/Phasing	

ORGANIZATION:
(governance)
	+ overorganized
	   overstructured
	
	- fragmented
	   unintegrated
	   uncoordinated
	   disunited
	   unstructured
		+ overregulated
		   restrictive
		   barriers
		   overstandardized
		   overconventional
		   homogenization
		
		- unregulated,
		   uncontrolled
		   unstandardized
		   unconventional
		   disorders, diseases
		   undisciplined
			+ overpredictable
			   routine, habitual
			   short-term, too regular
			   overprogrammed
			   monotonous
			
			- unreliable
			   unpredictable
			   unforeseen
			   unprogrammed
			   delay
			   out-of-phase
				(Insatiability)

	Orientation	Utilization	Appropriateness	

ACTION:
(reaction)
	+ indoctrination
	   forced conversion
	   propaganda
	   conditioning
	
	- discrimination
	   denial of right
	   disparagement
	   demeaning
		+ abuse, overuse
		   manipulative use
		   profligate use
		   unethical use
		   avoidance of rules
		   overefficient
		
		- malpractice
		   negligence
		   evasion of rules
		   inefficient
			+ slickness
			   overreadiness
			   overreaction
			   opportunistic
			   overeffective
			
			- inappropriate
			   unsuitable
			   unsustainable
			   unpreparedness
			   ineffective
			   inequitable
				(Indignation/
				Anger)

	Exposure to environment	Impact on environment	Insight/Learning	

CONSEQUENCE:
(of inaction)
	+ dangerous
	   hazardous, risky
	   overstimulating
	   violence from
	   disruption by
	   destruction by
	   hazards
	
	- unstimulating
	   unchallenging
	   perversion by
	   corruption by
		+ endangering
		   violence to
		   violation of
		   harassment of
		   disruption of
		   destruction of
		
		- endangered
		   unviable
		   corruption of
		   perversion of
		   pollution
			+ overconfidence
			   self-righteousness
			   self-appreciation
			
			- forgetfulness
			   unlearning
			   corruption of insight
			   loss of integrity/will
			   self-depreciation
			   dependency
				(Stagnation)

Figure 4. Possible set of "negative value qualifiers" organized in terms of "principles of integrity"

Patterning problems: computer representation of problem networks

1. Background
Since the inception of this project in 1972, efforts have been made to bring together the resources, hardware and expertise to permit networks of problems to be portrayed as networks in graphic form. The data is organized to that end. Ideally this technique would be used to communicate with international organizations concerning the networks of problems within their area of concern. By representing networks to organizations, they would be encouraged to identify weaknesses in the pattern of relationships indicated and would be able to make suggestions for specific additions or deletions. Similarly the editorial work on networks of problems would be considerably facilitated. And of course a set of network maps would be an extremely valuable tool as a complement to the text information displayed in this volume.

Despite efforts following the 1976 edition, it was not possible to achieve this for the 1986 edition. And despite efforts following the latter edition, no graphic representations have been possible for this edition or as a product of it. The approach is discussed in Section Z, with an indication of progress to date.

Aside from the technical issues discussed in Section Z, the concern here is with new ways of exploring information on complex networks of problems. The point has been repeatedly made that information on problems is now overwhelming. Conventional ways of exploring that information are totally inadequate to any insightful overview. Some possibilities are outlined below.

It should be noted that the emphasis here is not on the dynamic approach favoured by "global modellers" in order to represent complex systems. Extensive resources have been allocated to this approach for nearly 20 years without any major breakthroughs of policy significance. Such modelling totally ignores the question of the comprehensibility of any output to those who have to make decisions, even though it concentrates on a small sub-set of quantifiable problems.

The concern here is with the larger challenge of representing the extensive network of problems, whether tangible or "fuzzy", in ways which give a sense of context. The intent is therefore both less ambitious from a theoretical point of view and more ambitious in terms of encompassing the full range of issues with which people are concerned -- whether or not they can be reflected in any systems model. From a mathematical point of view, in contrast with conventional modelling the tools used are those of graph theory, topology and work on partially ordered sets. But the intent is to use these to design appropriate algorithms to simplify the user's understanding of the patterns of problems.

The following facilities acquire greater significance with the increasing use of CD-ROM to hold information on large databases and to facilitate access to them.

2. Representation of problem "hierarchies"
There is a need to be able to portray a problem hierarchy fully for inspection. Whilst this can be done in a non-graphic mode (as is currently the case in editorial work), the approach is cumbersome. Ideally users should be able to view the whole hierarchy or zoom into parts of it, accessing the descriptive data when required. This facility is of special interest given that problem hierarchies are not simple. A single problem may be part of several tree structures.

3. Representation of sets of problems
Although relationships between problems may not be known, it should be possible to view those in related subject domains. Again this can be done in a non-graphic mode, but users need the facility of looking at the problems in any domain at different levels of detail, whether the tops of hierarchies only or at any of a succession of levels below. The need for this facility is obvious in the case of geographic information, where the need may be for more or less detail (eg including or excluding towns of smaller size). Increasing detail would thus include in the graphic display the hierarchical links to more specific problems, just as minor roads to smaller towns are included when a more detail geographical map is used.

4. Functional networks linking problems
There are two patterns of functional relationships between problems in the database at present: aggravated by/aggravating and reduced by/reducing. Again these can be explored in non-graphic mode but this becomes especially difficult when combined with the question of hierarchical level. Specifically the difficulty lies in determining to which hierarchical level a functional relationship should point. Much greater flexibility is required by users in navigating through the networks in order to get a sense of the pattern of relationships at the appropriate level of detail.

5. Configuring patterns of problems
In the features discussed above the emphasis is placed on holding the data in a form which respects its inherent complexity. The user is obliged to navigate through it using the new manoeuvring possibilities. This only responds partially to the user's needs to obtain an integrative overview of the data. Other approaches are possible if the user is allowed to experiment with various distorted presentations of the data (analogous to the need of map-makers to use various "projections" when presenting data on the globe on the surface of a sheet of paper). Some possibilities include:

(a) Grid representation: Forcing the problems onto a 2-dimensional grid, possibly zoned by subject, and allowing problem hierarchies to project vertically according to the number of levels required. Many-levelled hierarchies in a particular zone would then be represented like sky-scrapers compared to "single-story" problems elsewhere. This would trigger an understanding of the "urban" organization of the set of problems, distinguishing between the "built-up, down-town" area and "suburbia". A modified version of this approach could also be used by treating different subject areas as distinct "towns" and providing maps for them, each with its own "down-town" area. In both versions the grid lines can be treated as "roads" with "traffic" of a certain frequency corresponding to the intensity of the relationship between problems. Use of the visual "town" metaphor may evoke new approaches to problems in the light of the well-developed skills of "town management" and "traffic management".

(b) Other two-dimensional projections: As the map-making analogy suggests, there are many other "projections" in two dimensions onto which problem network information might usefully be projected. In each case the emphasis is on using the geometry of the projection to provide an artificially integrated overview and a way of articulating the detail of the networks.

(c) Spherical projections: The complete pattern of problem networks could be projected onto the surface of a sphere as a way of emphasizing the bounded nature of the system within which the problems occur. This may have advantages in rendering explicit symmetry and other features which might render the whole more comprehensible. Approximations to such a spherical representation may be more practical, along the lines suggested by Buckminster Fuller.

(d) Unconventional suraces: Computer environments increasingly permit the projection of data onto surfaces which do not need to correspond to conventional rules of geometry. This may be especially fruitful to encompass contradictory patterns of information which can best be embedded on dynamic surfaces, or shifting patterns, where some of the integrative power comes from the periodicity or rhythm with which data is portrayed in a particular way.

6. Experimental emphasis
Given the present conceptual bankruptcy in the face of the complexities of the problematique, there is a strong case for experimenting with unusual ways of portraying networks of problems. There is little hope that major breakthroughs can emerge from agendas formulated in terms of a linear sequence of points, as remains standard practice in policy-making discussions. The hardware and software skills to engage in such experiments are readily available and widely used outside the policy environment

Comments: general

1. Degree of disagreement
There is a high level of disagreement about the "facts" relating to any so-called problem. Institutions are resistant to treating as "factual" that with which they are not yet organized to deal. This is also true of academic disciplines, when the social reality of a problem is in advance of the theory required to recognize its existence and structure any informed response. In principle there is merit in such disagreement because it allows for alternative frameworks through which a problem may be viewed through the creative juxtaposition of different perspectives -- thus avoiding the trap of being inappropriately "locked into" a particular framework. In practice the level of disagreement amongst "competent authorities" leaves much room for manoeuvre by those who wish to manipulate the ambiguity to their own advantage.

2. Solutions as problems
Increasingly responses to problems are perceived as problems in their own right. Where this is not immediately the case in the short-term, the longer-term effects of any solution often prove to be problematic. As in medicine, for most drugs there are "side effects". This reflects the insight that "for every problem there is a solution that is immediate, simple and wrong".

3. Problems as solutions
Many problems are in effect "solutions" to other problems, however much the nature of the solution may be regretted. Clearly any problems resulting in the death of individuals contribute to easing the problem of "overpopulation". There is merit in reflecting on the possibility that the human race engenders appropriate problems to constrain its own excesses, where it fails to deal adequately with those excesses by some consciously organized means. It may be more a question of the human race becoming increasingly vulnerable to those problems by which its excesses can be constrained.

4. Peter Principle of problem perception
There is a tendency for world problems to be most clearly defined at a level just beyond that at which individuals or institutions are competent to deal with them. By analogy with the Peter Principle for the promotion of individuals within organizations, problems get aggregated or analyzed into a form which "disenables" those who might be expected to act upon them -- and who are thus provided with a convenient excuse for their limited impact on them. At the more limited level at which action could be effectively taken the problem definition is then not considered appropriate (thus failing to discourage the inappropriate action at that level based on inadequate insight). The problem at the more limited level is considered to be "part of", or a "symptom of" some larger, or more fundamental, problem on which it is believed action should be focused. It is precisely at this level that the effectiveness of action is severely inhibited. Action at both levels tends to be inhibited or inappropriate.

5. Short-termism
The general response to problems by institutions is governed by short-termism. Institutions endeavour to devise solutions which can be made to appear as though effective action is being taken, whether or not that action is effective either in the short-term or, more significantly, the long term. Institutional action is evaluated through reporting systems designed to emphasize positive achievements by each hierarchical level reporting to that above it. Such systems encourage suppression of information on inadequacies, or at least delay its emergence until some later reporting cycle. Often organizational units cannot be usefully held accountable for failures made during a previous programme/budget cycle. The focus is on institutional survival through the current cycle. Long-term problems, and the need for longer-term programmes of action, tend to have a very low priority, except for public relations purposes.

6. Suppression of information and "cover-ups"
Large amounts of information are "classified", primarily because of its value for "national security" and "competitive advantage", or to avoid "embarrassment" to interested parties. It is in the interest of institutions and individuals to suppress information that might endanger their own professional survival and advancement. The practice of "cover-up" is widespread. It ensures wide opportunities for "official denials" that there is any matter for concern.

Increasingly information of any value is somebody's "intellectual property" and is withheld unless an appropriate, and often costly, payment is made. Intergovernmental organizations produce documents which are distributed on a "restricted" basis, only partly for cost reasons. There is therefore merit in querying whether the information in the public domain adequately describes the network of problems with which humanity is faced. If information is so important that it needs to be withheld or suppressed, then presumably a proportion of it concerns problems on which it is judged inappropriate to inform the public because of the seriousness of the implications. To this extent, studies of the operations of the international community can only be superficial and unrealistic if they are solely based on documents in the public domain, as are most studies at this time.

7. Corruption
The problem least frequently mentioned in connection with any studies of the crises facing humanity is that of corruption in whatever form. This is especially true of studies by intergovernmental organizations which are anxious to avoid embarrassing their member states.

Informally, industrialized countries have drawn attention to the level of corruption in developing countries -- as one of the principal excuses for reduced commitment to development assistance. Despite daily press reports, the prevalence of corruption, of many types and at all levels in industrialized countries (even involving presidents of countries), remains unexamined as a systemic ill.

Corruption is frequently mentioned in investigative reports by journalists (quoting appropriate sources) on the failure of action against problems. The press, whether in industrialized or developing countries, tends to be the only place where reports of corruption appear, since they tend not to be the subject of any further investigation, except when they take the form of isolated scandalous incidents. Policies in response to problems tend to be designed as though systematic corruption was not a significant factor in undermining their effectiveness, whether partially or completely.

8. Inability to handle unwelcome information
This may be seen in governmental response to such problems as acid precipitation, CFCs, food poisoning and smoking. In each case there is a phase in which experts are found to argue that there is "no proven link" justifying rapid action. Delaying tactics are readily used. Governments renege on commitments or scale down any attempts to implement remedial action.

9. Need for "bad" problems
There would seem to be a fundamental need for the presence of a special class of problems which the majority of society can agree are totally repugnant, at least in any formal communications. Problems which have served that function include: nazism, torture, and apartheid. The prime characteristic of such problems is that they should be prevalent in a distant society, in the past, or in secret locations (whose existence can be readily denied in one's own society). Such problems have a global consensus building function across boundaries that usually fragment society. No "reasonable" person would be associated with them. The cessation of the Cold War and the reconfiguration of the apartheid situation in South Africa may lead to situations in which other problems have to rise to prominence to fulfil their functions in this respect. Examples of such problems might be the increasing concern with sexual abuse of children (paedophilia) on the one hand and the totally repugnant behaviour of Iraq on the other. If people need the existence of such problems in order for their own behaviour to shine by contrast, it would seem that care needs to be taken in eliminating them.

Comments: approaches to problems

It is useful to review some of the ways in which individuals and organizations tend to approach problems. Russell Ackoff (1986), for example, suggests that there are four ways of treating problems: absolution, resolution, solution, and dissolution. To absolve a problem is to ignore it and hope it will go away or solve itself. To resolve a problem is to do something that yields an outcome that is good enough, that satisfies. To solve a problem is to do something that yields the best possible outcome, that optimizes. To dissolve a problem is to eliminate it by redesigning the system that has it. This note develops points raised earlier concerning *Problem disguises and problem avoidance*.

1. "Key" or "Log-jam" approach: With this approach, the emphasis is on finding the "key" problem or focusing on it (once the assumption has been made that it has been found). The implication is that through the "key" problem a way forward can be opened. This conceptual dependence on the existence of a single factor or "handle", necessarily ignores other factors as irrelevant. It can perhaps be better illustrated by a "log-jam", as a complex of problems, in which appropriate strategy is the search by experts for the key logs which need to be moved in order to release the whole pile of logs to flow down-river. This mindset implies reliance on expert analysis and application of optimal force at a precise location. Any action on logs viewed as non-key is disparaged as irrelevant, if not counter-productive and dangerous.

2. Problem avoidance approach: With this approach, the way forward is not to focus on problems but on next steps, whether in terms of solutions, strategies or visions. Any focus on problems is considered disempowering. The assumption here is that the problems will tend to resolve themselves once appropriate actions are undertaken to implement visions of a desirable future. Focusing on problems exaggerates their importance and distorts thinking counter-productively.

3. Saviour approach: Some individuals or groups put themselves forward, or are put forward by their disciples, as providing the appropriate approach to problems. Whether religious or ideological, the emphasis is placed on associating with their belief system through which appropriate responses to problems emerge. Large advertisements may be placed in the international press encouraging people to subscribe to the approach, or appeals may be launched through other media. This process is favoured by certain religious groups and sects who see little hope through other methods.

4. "Concrete" approach: Especially following the failure of other kinds of initiatives, many individuals and groups favour dealing with problems on a case-by-case basis, avoiding general explanations, models and solutions. Conceiving of problems divorced from the concrete setting is considered inappropriate, especially when a focus on "poverty" in the abstract distracts attention from poverty in a particular community. There is even some question as to whether problems can be effectively approached by any other strategy.

5. Prerequisite explanation approach: Some individuals and groups require a satisfactory explanation, usually scientific, of the nature of a problem, before taking steps to deal with it. The assumption is that a problem must be fully understood before any useful action can be taken. Some bodies use this prerequisite, by the selection of appropriate experts, as a means of postponing action when "no proven link" has been demonstrated (eg air pollution and acid rain).

6. "Macro-mega" approach: Problems may be viewed as so vast and all-embracing that they are beyond the capacity for effective action by individuals or groups. The prerequisite for action thus becomes some form of consensus, usually unobtainable, amongst the disparate forces that govern society. This may also be used as an effective device for avoiding individual responsibility and as a way of "sweeping smaller problems under large carpets."

7. Mandate approach: Any individual or group with a specified responsibility can reject consideration of a new problem as not falling within a pre-defined mandate. Here the prerequisite is some form of mandate, which may be quite unobtainable in the case of a new, cross-sectoral problem that may fall within the concerns of all or none. Declaring that a problem is somebody else's responsibility is frequently used as a means for avoiding it.

8. "Allopathic physician" approach: Problems may be approached like diseases of the body. Symptoms then need to be noted in order to diagnose the underlying disease. Encouraged by the mindset of allopathic medicine and its many miraculous cures, specific remedies may then be envisaged, whether in the form of "drugs" or "surgery". Just as the physician has available hundreds (if not thousands) of drugs, an array of specific remedies can be envisaged to the many problems which may emerge. But just as with physicians, the number of such "drugs" is now so great that it is difficult, even for specialists, to keep up to date with the latest and best techniques, since many remedies of the past go out of fashion. The situation is exacerbated by consultants associated with particular schools of thought who, like pharmaceutical companies, distinguish themselves by advocating new remedies (if necessary misrepresenting their advantages and concealing their weaknesses) in order to outmanoeuvre their competitors. This approach favours the idea that to every specific problem there is a specific solution. It plays down the complementary idea that problems emerge within a context on which attention might more usefully be focused, as in homeopathic medicine. The allopathic approach may be used to avoid dealing with systemic problems.

9. Image approach: Where individuals and groups are primarily concerned with their own competitive advantage in a complex society, problems may be approached in terms of their image-building effect. In this situation the problem itself is only of importance as a vehicle which may advance or undermine the cause of those mandated to respond to it. Especially for politicians, problems may be acknowledged as important for only as long as they are a current political issue, particularly one receiving media coverage. In this case a positive image is a prerequisite for action.

10. Profiteering approach: Problems may approached only in terms of their profit-making potential. A problem becomes significant to the extent that goods or services can be sold in the process of remedying it. The stress is not on whether such products are an appropriate response to the problem but whether a market can be defined and developed that perceives this to be the case. In the worst cases this approach leads to sale of totally inappropriate, and even dangerous, goods -- or, worse still, to the provocation of problems for which solutions can be sold.

11. Doom-monger approach: Some groups and individuals benefit from viewing the accumulation of problems as leading to a catastrophic situation in the not too distant future. For some this is viewed as Nature's judgement on humanity, for others it is God's -- especially in the light of apocalyptic visions. According to the latter, such final catastrophes may be viewed quite optimistically as presaging the arrival of God's Kingdom on Earth.

12. Grass-roots approach: For some, problems can only be effectively understood at grass-roots level by the people. Grass-roots insight and resources are what is required to respond to the problem. The way forward is through trusting in and mobilizing the people -- or, better still, enabling them to organize themselves in response to the issues they consider relevant.

13. Technocratic expertise approach: Technocrats tend to hold the view that they have all the systems modelling skills necessary to identify and deal with problems. In this view, all that is hindering an appropriate response is the complex of irrelevant political issues which prevent the technocratic elites from marshalling and deploying the necessary resources as required. This approach is favoured by those who believe that there is a technical "fix" for every problem. It is also favoured by those scientists who hold that only the objectivity, rationality and intelligence of science is adequate to the task.

14. Elitist diplomacy approach: Those in power, or with access to the powerful, tend to hold the view that problems can best be approached by suitable negotiation amongst the power elites. Only they can decide what problems merit consideration and with what degree of urgency. Only they are in a position to design appropriate strategies to deal with them. Once an appropriate strategy has been determined, preferably behind closed doors, it can then be suitably packaged for media presentation to ensure political acceptability.

15. **Multinational enterprise approach:** Some within the multinational business community hold the view that only they have the necessary combination of resources to be able to manage the problems of the planet in all their complexity. Here the stress is placed on managerial expertise in dealing with a complex environment and forming the necessary coalitions of resources in response to a highly dynamic situation. In this view, the realism and experience of the business community in ensuring a viable economic and social environment is what is required.

16. **Conspirator approach:** Given the complexity of society and the inadequacies of most formal modes of action, some hold that the only effective approach to problems is through "behind the scenes" action by the like-minded. This allows for much more sophisticated and far-reaching strategies, especially in the light of esoteric or new paradigms. Problems can then be redefined within more creative contexts allowing for approaches which would otherwise be impossible, especially if they had to be explained to non-initiates.

17. **Conspiracy approach:** Problems may be viewed as resulting from the activities of secret coalitions of forces, which may or may not be acting in the interests of the planet as a whole, however they define their objectives. These may include cartels, secret societies, intelligence agencies, religious groups, financiers, or even extra-terrestrials or the agents of Satan. From this perspective, problems are merely symptoms of the actions of some conspiracy, and are deliberately provoked to achieve certain designed effects.

18. **Empathy approach:** Problems may be viewed according to whether they evoke an empathetic response or not, namely whether the individual or group can identify with those exposed to a problem. Problems which do not arouse empathy are considered meaningless or the responsibility of others.

19. **Deviation-from-orthodoxy approach:** Problems may be viewed as symptoms of deviation from some mode of behaviour or thinking prescribed by doctrine or dogma, whether ideological or religious. As such they constitute a form of infringement of taboos, which may indeed have been established by custom. In this approach the emphasis is on reforming the offender rather than dealing with the consequences of his action.

20. **Divine intercession approach:** The problems of the world may be viewed as quite beyond the capacity of humanity. The only viable approach may be considered some form of divine intervention to assist individuals or groups in their efforts. For those holding this view, prayer invoking such intercession may be perceived as the most appropriate response to problems.

21. **Reorganization approach:** Problems may be approached as resulting from defects in the organization of the bodies or programmes within whose mandates they ought to fall and which are expected to be able to contain or regulate any imbalances in the social or natural environments. Typically this results in shifts from centralization to decentralization, or from privatization to nationalization, in order to respond to problems more effectively. *We trained hard, but it seemed that every time we were beginning to form up in teams we would be reorganized. I was to learn later in life that we tend to meet any new situation by reorganizing and a wonderful method it can be for creating an illusion of progress while producing confusion, inefficiency and demoralization.* (Petronius Arbiter, Roman Governor of Bithynia, who committed suicide in A.D. 65).

22. **Problems as enemies:** Whereas enemies have traditionally taken the form of opposing groups or individuals, increasingly humanity's enemies take the form of problems. There has been much learning concerning responses to enemies of the traditional kind. The question is whether this suggests any insights into understanding the challenge of problems and options for dealing with them, especially given the tendency to confuse the problems with those perceived as representing them. Traditional responses include:

(a) Assassination: This has already been used against those deemed to be opposing reform as well as against activists promoting reform. However, it seems difficult to eliminate problems themselves by such "termination with prejudice", although the use of a tactical strike force to eliminate a factory producing biochemical weapons may constitute an example. Where a particular incident, case or condition provides a focus for awareness of a more general problem, resolving that specific case may (be used to) destroy any momentum in the development of awareness of the more general problem.

(b) Imprisonment: This is widely used against those disrupting the peace and contravening regulations as well as against activists promoting reform. This may be done on a large scale as in concentration camps and prison colonies. This strategy of "containment" has been extensively used to restrict the spread of a problem, especially through the use of quarantine measures in the case of epidemics. It is also used to prevent the spread of information about an emerging problem.

(c) Sabotage: Use of this approach against those deemed to be exacerbating problems is developing, as in eco-tage, notably in the case of animal rights activists. It is also used to undermine reform movements by bombing their facilities and destroying evidence collected by them. Again it seems difficult to employ this strategy against problems, unless tactical strikes against factories are an example, or the introduction of species to counteract a pest.

(d) Subversion: This has long been used to penetrate and undermine both reform movements and groups opposing reform. It is a major tool of covert operations by security agencies. It is used by activists to ensure the presence of friends in the enemy camp. Against problems as such it is unclear what scope there may be for turning problems against themselves, although this would be a very elegant option. The example of introduction of species to counteract the spread of a pest is also a form of subversion.

(e) Harassment: This is widely used against reform activists (by police, tax authorities, employers) and by them against those opposing reform (in the form of demonstrations, mailings). Greenpeace has developed this technique in responding to problems, although it is not clear that problems as such can be subject to the strategy.

(f) Disinformation and propaganda: This is very widely employed by those opposing reform through the use of fabricated evidence, planted stories or rumours, and smear campaigns. It has also been used by reform activists, making exaggerated claims concerning certain problems. It is possible that this strategy could be used in dealing with certain problems of belief systems and information handling, as is the case in de-programming of cult members. It is of course used by those of different beliefs in protecting themselves from the pernicious influences of competing systems.

(g) Conventional warfare: This strategy is the standard for dealing with reform activists or with those opposing reforms. It is also applied to dealing with problems. Extensive use of military terminology is made, even by international agencies: mobilization, marshalling, targeting, ammunition, army (*cf* Earth Army, Peace Corps, Salvation Army), and campaign (*cf* illiteracy, hunger). In the case of the "drug war", attempts have even been made to deploy military forces.

(h) High-tech warfare: Although especially characterized by nuclear warfare, it is above all characterized by the use of sophisticated electronic detection and tracking equipment, together with associated computer-aided decision-making ("situation rooms"). Such facilities are increasingly used to monitor the activities of reform activists (*cf* electronic surveillance, credit card trails) but are also being used more modestly to track those opposing reform. Limited examples of the extension of this approach to problems themselves include various forms of environmental monitoring and remote sensing.

Comments: beyond the problem-lobby mindset

1. Competition for problem recognition
Many of the problems of society are believed to call for initiative by local, national or international government bodies. Such bodies are held to be the focus for action decisions within society, especially given their central role in any political process. To a lesser degree, an equivalent role is played by non-governmental bodies that are a focus for deciding priorities concerning action on problems and the allocation of resources to them. Foundations are one example.

Advocates of action against particular problems, whether official or private bodies (or even individuals), are expected to present their cases to focal points. Such focal points are naturally restricted in their resources. The restrictions generally cover availability of funds, time available to process any petition, and the expertise which can be devoted to evaluating any proposal. Consequently, advocates of action on problems are expected to compete with one another, both for the attention of the focal points and for any resources which are available through them. The problem is assessed in terms of the effectiveness of the advocates in this process and not in terms of the seriousness of the problems in their own right.

2. Unaccountability and irresponsibility
The focal points are generally in a position to claim that they have appropriately performed their task if they are able to play off the different petitioners so that the most influential petitioners receive preferential treatment and the less influential are content with whatever recognition is given to the problems for which they are pleading. In this resource allocation process the focal points are not necessarily required to respect other criteria than those to which they are subject by the competing petitioners. In fact the focal points need have no substantive concerns other than those to which they choose, or are obliged, to respond from amongst the advocated problems. Those functioning within the focal points may also respond to pressures from those mandating them to ensure that further resources are allocated to the geographical area they represent, irrespective of the substantive justification for that decision. Resources may therefore be allocated for reasons which have little to do with the urgency of problems and may have more to do with responding to a traditional clientele. Decisions may also be taken for unstated reasons, including various forms of bribery or exchanges of favours, or even because of threats, whether implicit or explicit.

3. Sustaining the importance of spurious problems
As a result of this process a problem acquires the status of "important" for two reasons. Firstly because its advocates manoeuvred more effectively with the resources they were able to invest in the lobbying process, and secondly because the preferential treatment their problem receives as a result of a favourable decision is perceived as a guarantee of that importance. Throughout this process, the question of whether a favoured problem merits more attention than others is not of concern to either the focal points or to those competing for their attention. Neither group has any need to demonstrate effective concern for the complex of problems as a whole, other than as pointing to the critical role of the problems selected for attention. Those worthy of attention are those that received attention. It is then in the interests of both the focal points and the successful advocates, once a programme is underway, to reinforce appreciation for the importance of the selected problems through effective use of the media and appropriate public relations exercises (awards, etc.). This ensures appreciation for the discrimination exercised in the decision to act on the problems selected. In effect the approach to problems develops into a closed system increasingly divorced from the full range of problems to which people are sensitive.

4. Manipulating expertise
It may be argued that the most effective advocates are those that are able to present a strong case. However, experts of necessary eminence, or authorities of necessary weight, can be found to support any case. Ultimately it is increasingly a question of the impression created, especially through the media. A weak case can succeed where a good case fails if appropriate lobbying tactics are used, especially when few holds are barred. In this context high quality expertise may be out-manoeuvred by low quality expertise if the latter is appropriately supported. In addition, if concern for a problem is not out-manoeuvred in the short-term, this may be achieved in the longer-term. Thus focal points may commit to a programme in the short-term only to progressively downgrade their commitment in later periods.

5. Courts, courtiers and petitioners
There is little that is new or innovative in this lobbying process. Most aspects of it would be familiar in the courts of emperors and kings through extensive historical periods. Lobbying around the European Commission and the associated bodies can thus be seen as repeating a well-tried pattern, as has been the case around the United Nations and its Specialized Agencies. Replete as it is with arrogance and sycophancy, the question to be asked is how well this pattern is adapted to the complexities of the problematique as opposed to the needs of a focal point to filter petitioners to ensure the continued survival of that focal point.

6. Neglected problems
The difficulty in a lobbying context is that relevance is defined competitively. But some problems do not have advocates, and some problems worthy of attention do not have strong advocates. This is especially true of new and emerging problems. They may be deprived of resources without jeopardy to the focal point in the short-term, but this may prove to be exceptionally unwise in the longer-term. Of course the focal points do not have a genuine concern for the longer-term because their policy cycles are determined by preoccupation with relatively short-term electoral cycles. Those functioning in relation to them are necessarily preoccupied by short-term career opportunities.

7. Mapping the problem context
It would be naive to expect any significant change to this situation, which tends to be viewed quite uncritically as appropriate by those who engage in it. It is however legitimate to envisage how the set of problems might be viewed in the future in relation to the resource allocation process.

What seems to be missing is a database or registry of problems, irrespective of currently fashionable criteria of relevance and irrelevance which can distort the design of information systems into ad hoc exercises of little long term significance. Such a database provides a context in which information on problems and their relationships may be held. Information in this sense also includes claims and arguments in favour and against particular perceptions of problems. This database may be interrogated in a variety of ways by advocates of particular sets of problems. But the lobbying process can then be seen as selecting sub-sets of problems for action. In this sense a case needs also to be made, explicitly or implicitly, for the problems to which resources are not allocated by that process. Of special interest is the manner in which the selected problems are impacted upon by those to which resources are not allocated.

Comments: problem metaphors

The nebulous, shadowy nature of problems discussed earlier suggests the value of trying to understand them through metaphors. Not only can valuable insights be obtained, but this helps to comprehend how people acknowledging problems as understood through one metaphor have difficulty in attaching significance to problems as understood through another metaphor. The favoured metaphor may render some problems even more evanescent. There is also the possibility of considering problems as metaphors in their own right (see Section MZ).

1. Atoms, molecules
Problems may be considered as discrete entities, like atoms, having relationships to one another, like molecular bonds. This perspective corresponds to the simplest ways of portraying atoms and molecules (the solar system and billiard ball models) and is still in use for teaching and in graphic displays for research (see Appendix YF***). The structure of the information in this volume is effectively based on this metaphor. But although such models are useful, it is the implied discreteness of atoms which obscures their other properties, such as field effects (which are less easy to visualize) that may lead to alternative descriptions. Problems also have non-discrete characteristics which can better be understood as field effects.

In the light of this metaphor, the editorial process can be thought of as an attempt to discover the detailed structure of the macro-molecular complex which so strongly orders the life processes of society. It thus bears some resemblance to efforts to discover the structure of DNA or to the current Human Genome Project.

2. States of matter
It is useful to compare problems using the metaphor of the different states of matter. **Solid:** problems can be described as solid barriers, and rock-like obstacles. **Liquid:** the fluid, shifting, interconnected nature of problems can be understood using liquid metaphors. (In Section MZ, a chemical metaphor of problems dissolving into solutions and being precipitated out as solids is mentioned). **Air:** the manner in which problems "clamour for attention" or people are "bombarded by problems" can be understood in terms of the pressure exerted by molecules in a gaseous state.

In the light of this metaphor, the editorial process can be thought of as an attempt to collect and relate information concerning the different states in order to be able to plot out a phase diagram showing their relationship.

3. Topography and surveying
Problems are often described in terms of topographical features or geological processes. They may be compared to mountain barriers, earthquakes, volcanoes, landslides, or avalanches.

In the light of the topographical metaphor, the editorial process may be compared to a form of geological survey. An important concern is fixing positions of features relative to one another by the process of triangulation. Articulating the pattern of relationships between problems can be viewed in this light.

4. Weather and meteorology
Problems are often described in terms of weather effects. Crises can also be described in terms of storms and hurricanes, whether of wind or accompanied by waves (as in the phrase "making waves"), or in terms of severe extremes of temperature. A variant is the use of "high" to denote a positively valued personal condition and "low" to denote a negative, problematique condition.

In the light of the meteorological metaphor, the editorial process bears some resemblance to efforts to map and interrelate weather patterns. This metaphor might be used to suggest the futility of implying any permanence to a particular and temporary "low". But it might also be used to show that "lows" tend to arise in particular places and under certain conditions, however they shift thereafter. The high-low polarity has the great advantage of being able to encode the value-problem polarity and to suggest the level of complexity of the resulting system.

5. Ecology
Problems can be described in terms of ecosystems (jungle, desert, ice-field), wild beasts, insect pests, plagues or viruses.

In the light of this ecological metaphor, the editorial process can be viewed as an attempt to identify species and to map out food chains between them.

6. Disorders and diseases
Problems may be considered to be in some way the social equivalent of foreign bodies circulating in the human bloodstream (requiring the action of antibodies), or of different diseases affecting the different structures and processes of the human body.

In the light of this metaphor the editorial process can be viewed as an exercise in social pathology, an effort to identify the range of ills to which the human environment is subject. To some extent the product therefore has a function analogous to that of the WHO *International Classification of Diseases* or the *Diagnostic and Statistical Manual of Mental Disorders* (of the American Psychiatric Association).

7. Stars and astronomy
Problems may be considered to be like stellar objects spread throughout the universe as "gravity-wells" and localized distortions of space-time. It is then as distortions that their problematic nature becomes manifest and that they exert effects on the social continuum. The metaphor may be pursued further by comparing the different kinds of problems to the different kinds of stellar objects, and even comparing their evolution to that of stellar evolution (as charted on the Herzsprung-Russell diagram). There is also the implication that under certain conditions problems can themselves become life-supporting.

In the light of this metaphor the editorial process can be compared to a star charting astronomical project. Of special interest are then the implications of the constraints in endeavouring to generate such charts from a single, eccentrically located position, when observations are further distorted by a variety of interference effects. The constraints are counteracted by establishing long base-lines and arrays, as in radio-telescopy. Networks of international organizations can be viewed as arrays of detectors from which problem information is derived by bringing that information to an appropriate focus.

8. Conceptual "anti-matter"
Problems bear a resemblance to "negative theories" (or "anti-theories"), namely they that exist in the same way that theories exist (bearing the same relationship to data and values), but instead of providing explanatory and predictive power to link related phenomena within a coherent framework, they mark the presence of confusion and unpredictable relationships between seemingly unrelated phenomena. They challenge and disrupt the conceptual frameworks which claim to be able to handle them. To the extent that they involve a comprehension vortex, with some form of "event horizon", they resemble "black holes" in the universe of information. Note that the supply of an adequate number and variety of problems may be necessary as a structuring device for a complex society since it provides a sufficient number of "sinks" (perhaps the psycho-social equivalent to the astronomers' "black holes") to absorb the excess energy generated by social processes.

9. Photography
Information on problems can be considered as detected by a process similar to that by which light or radiation is captured on photographic film. Whereas the normal investigatory procedure depends on the development of such film into "positives", problems are more appropriately viewed on "negatives".

10. Discontinuities
Problems can be understood as discontinuities in the seamless pattern of the known and predictable. As such they lend themselves to exploration as "catastrophes" using the mathematics of

catastrophe theory, or as "strange attractors" using the mathematics of chaos theory.

11. Sins
Many problems can be viewed as sins of omission or commission on the part of humanity, especially since many of them can be held to derive from the fundamental ("deadly") sins, whether singly or in combination.

In the light of this metaphor the editorial process is equivalent to studying and cataloguing sins (in the Christian tradition) or afflictions (in the Buddhist tradition).

12. Demons and demonology
Traditionally, and from some psychoanalytic perspectives, there is merit in viewing problems as demons or gods of wrath ****. Different demons then manifest to humans through problems of different types. Classes of problems are then the work of classes of demons.

In the light of this metaphor, the editorial process then becomes an exercise in demonology. The product is thus equivalent to a Dictionary of Demons.

13. Monsters and bestiaries
Traditionally the unknown is indicated by the risk of encountering monsters and legendary beasts, to which problems can usefully be compared in the light of the devastation which both effect on societies. Ancient maps of the known world were traditionally bordered by regions inhabited by monsters. This is equivalent to the known and predictable world of routine from which problematic monsters must be kept at bay.

In the light of this metaphor, the editorial process can be viewed as an effort to document such beasts and monsters and to produce a "bestiary". As with such supposedly factual bestiaries, the information is based largely on travellers tales - the travellers being either ill-equipped to register unambiguously the nature of the beasts encountered or otherwise unable to confirm their experiences to communities insensitive to the rich variety of life. International organizations may be considered such travellers to distant and specialized domains.

14. Topography of hell
Traditionally Hell is the realm in which problems may be viewed as reigning explicitly supreme. In some cultures, this realm may be subdivided into many hells of different nature. Some forms of the board game "snakes and ladders", especially in South East Asia, are also used to indicate the transformational pathways to these distinct hells.

In the light of this metaphor, the editorial process might be viewed as an effort to provide a topographical map of Hell -- mapping the wrinkles on Satan's face. The hierarchies of problems detected might then be viewed as the skyscrapers in downtown hell. The challenge is to get a clearer sense of the relationship between such urban structures and to map the pathways on which resources flow to and from them.

15. Evils from Pandora's box
Problems can be viewed as the evils which, in Greek mythology, escaped from Pandora's box as a result of her curiosity and carelessness. They might be understood as having escaped because of the inadequacy of the mental framework with which she approached the original container. Hers lacked the requisite variety to contain the variety to which she exposed herself.

In the light of this metaphorical context, the editorial process might be viewed as an effort to create a framework or "box" by which the evils could once again be contained. At this point in time no such conceptual container exists. Different conceptual frameworks endeavour to contain different problems, but many escape any such efforts. (The challenge is reminiscent of the current technical challenge of designing an adequate container for plasma to derive energy from nuclear fusion reactions.)

16. Shape-changing monsters and protective devices
Greek mythology offers examples of monsters that shift their form once effectively attacked, or sprout additional heads if one is cut off. Problem complexes often appear to be of a similar nature, shifting their point of impact, disappearing to manifest in some other form, or multiplying in vigour in response to efforts to contain them. Greek mythology also offers examples of monsters, such as Medusa, which destroy or transform those who view them directly (eg into stone). The petrifying potential of such monsters seems to carry over into the superstitious fear that some have in exposing themselves too closely to problems in all their negativity. Such proximity is believed to undermine the positive energy necessary for social advance, or even to bring bad luck (possibly attracting the unwelcome attention of supernatural powers).

In the light of this metaphorical context, the editorial process may be viewed as calling for appropriate protective conceptual devices (analogous to the shield of Perseus) to avoid being transformed by the horrific nature of some of the problems. As in laboratories and surgeries, a certain form of conceptual hygiene is required to avoid being transfixed and traumatized by exposure to the problems.

17. Hallucinations and delusions
It may also be the case that the greater recognition of problems in an increasingly sophisticated society based on communications media, is to some extent a social equivalent of individual hallucination under conditions of prolonged sensory deprivation. The increasing proportion of the population living and interacting with, and through, a world of images reduces the collective daily necessity to relate directly to traditional grounded realities. In so doing it creates a generalized sense of eventlessness which provokes the emergence of compensatory collective hallucinations to which the collectivity can respond with positive activity.

From this perspective the editorial process is that of the psychoanalyst faithfully recording the patient's delusions and accepting them as realities as a prerequisite for fruitful and transformative dialogue.

Comments: future possibilities

1. Possible future improvements

(a) Inclusion of new problem entries and revision of information included in each entry, with addition of information where appropriate to produce a more complete description.

(b) Revision and extension of the system of relationships between problems to include such features as (a) relationships arising from a situation in which one problem is perceived as having displaced another, as a result of new understanding of the nature of the problem (whether or not this understanding is widespread); (b) relationships arising from recognition, as a result of analysis, that a problem is a symptom of a more fundamental problem (as distinct from cause-effect relationships between problems of equivalent level); (c) relationships arising from educational considerations, namely indicating the next more complex problem, in which the nature of the problem is reformulated in more sophisticated terms; and (d) relationships arising from the historical order in which problems were perceived and displaced by other later problem perceptions.

(c) Extension of the system of relationships between problems to other series to include such features as: (a) international organization sub-units specifically concerned with the problem; (b) resolutions of major United Nations bodies dealing specifically with the problem; (c) qualification of relationships, such as those with international bodies, to specify whether they are concerned with policy, research, management, public information, or information exchange.

(d) Inclusion in entry descriptions, where appropriate and where such information is available, of statements criticizing the existence of the problem as described (namely counter-arguments or counter-claims).

(e) Inclusion in the entries, where appropriate, of other subheadings such as (a) details of how the problem has developed over time and how it is expected to develop in the future; (b) list of countries in which the problem is known to occur; (c) information centres which keep track of the problem (other than international organizations); (d) standard reference books dealing with the problem; (e) international meetings dealing with the problem.

(f) Development of alternative classification systems for the problems.

(g) Identification of key people who are closely associated with action against particular problems by functioning as catalysts for the generation of new organizations, programmes, or other initiatives. A separate section listing such people could be cross-referenced to the problems series.

(h) Development of computer programmes to draw attention to errors in the ways in which the hierarchies of cross-references for particular problem-areas have been indicated.

(i) Development of computer programmes to plot out onto "maps" certain problem networks around core problems. Such maps could be included as illustrations accompanying the descriptions of such problems in future editions. More complicated maps could also be constructed showing how the network of organizations matches, or fails to match, the network of problems. Collections of such problem-based maps could be published in a form of atlas accompanying future editions of this volume (see discussion in Section Z).

2. Questions for the future

Interesting questions that emerged during the course of work on this project include:

(a) How can networks of relationships be analyzed systematically as networks to determine what are the most important focal points for action, and what different meanings could then be attached to "importance"?

(b) How can comprehension of complexity be improved without artificially forcing relationships into (definitive) hierarchical groupings thus doing violence to any inter-hierarchical linkages?

(c) Might it not be useful to investigate the result of using the mathematical technique to convert relationships between points into points in a network? Useful insights may then emerge from being able to switch between the perception of problems as linked in a network of relationships and the perception of problems as relationships which intersect at certain points.

(d) Given that the number, variety and relationships of human diseases, and the nature of their effects on the individual are now well understood, do they not suggest ways for organizing thought about the range and variety of psycho-social problems and their impact on the psycho-social system?

(e) Is it as ecologically inappropriate to ask the question "What are the five most important problems (organizations, etc) in the social system" as it is to ask the question "What are the five most important animals (plants, etc) in the natural environment"?

(f) Can the relationships between problems (or between organizations) be usefully conceived as analogous to the food webs and trophic levels within which animals are embedded? Does this help to suggest why different kinds of problems emerge as being of major importance at different times? How might the evolution of problems and problem systems be conceived in this light?

(g) From what is the stability of a "problem ecosystem" (as it might emerge from the previous point) derived? Is it useful to distinguish between degrees of (negative) maturity of problem ecosystems and to attempt to determine the amount of energy required to maintain them? Is anything suggested for better understanding of problem systems by the fact that a highly diversified ecosystem has the capacity for carrying a high amount of organization and information and requires relatively little energy to maintain it, whereas, conversely, the lower the maturity of the system, the less the energy required to disrupt it (as emphasized by R Margalef)? Thus anything that keeps an ecosystem oscillating (or "spastic"), retains it in a state of low maturity, whence the possible danger of simplistic reorganization of organizational, conceptual, or value systems. Is the problem of understanding and organizing the maturation of natural ecosystems of a similar form to that of understanding and organizing the disruption of problem ecosystems?

(h) Given the absence of sufficient comparable information to produce sensitive, widely-acceptable, quantitative world models covering all aspects of the psycho-social system, to what extent can increasing the number and variety of non-quantitative relationships and entities documented lead to valuable insights of greater acceptability? In other words, to what extent can knowing less about more (and organizing that knowledge) compensate for not being able to know more about less? Can any relationships be established between the amount of information, the type (quantitative, structured or unstructured qualitative), the manner of representation, and its degree of acceptability?

(i) To what extent is the complexity of the problem system with which humanity is faced greater than that which its organizational and intellectual resources are capable of handling? Worse, is there a widespread unacknowledged preference for simplifying the representation of complex problem (and other) systems down to less than 10 elements so that they lend themselves to easy debate in public and in a policy-making environment? Are organizational and conceptual resources then marshalled and structured to match the problem system as simplified rather than to handle it in its more dangerous complexity, thus running the (unacknowledged) risk of leaving the problems uncontained and uncontainable by the resources available? Does this suggest a corollary to Ashby's Law of Requisite Variety which might read: That any attempt to control a psycho-social system with a control system of less complexity (ie of less variety) than that of the psycho-social system itself can only be made to succeed by suppressing or ignoring the variety (ie reducing the diversity) in the psycho-social system so that it is less than the relative simplicity of the control system? Such suppression tends to breed violence, however.

UIA Current publications

Yearbook of International Organizations / ed. by UAI. - München, New York, London, Paris : Saur, 1990. - 27th ed. 1990/1991. - 30 cm. - ISSN 0084-3814.

Vol.1 Organization Descriptions and Index. 1776 p. + Appendices (14). - ISBN 3-598-22205-X.

Organization descriptions (24,209 entries)
The non-profit organizations included may be intergovernmental, non-governmental, or mixed in character. They cover every field of human activity. Descriptions, varying in length from several lines to several pages, are grouped into the following section
- Federations of international organizations
- Universal membership organizations
- Inter-continental organizations
- Regional membership organizations
- Semi-autonomous bodies
- Organizations of special form
- Internationally-oriented national bodies
- Inactive and dissolved bodies
- Religious orders and fraternities

Contents of descriptions
The descriptions, based almost entirely on data by the organizations themselves, include:
- Organization names in all relevant languages
- Principal and secondary addresses
- Main activites and programmes
- Personnel and finances
- Technical and regional commissions
- History, goals, structure
- Inter-organizational links
- Languages used
- Membership by country

Multilingual index
The computer-generated index provides the most detailed available means of identifying international bodies. Access is possible via:
- Organization names in English, French, and other working languages
- Former names in various languages
- Name initials or abbreviations in various languages
- Organization subject categories in English, French, German, Rusian and Spanish
- Personal names of principal executive officers

Checklist of publication titles
Periodical and non-periodical publications of international organizations

Vol.2 International Organization Participation : Country Directory of Secretariats and Membership (Geographic Volume). - 8th ed. 1990/91. - 1760 p. - ISBN 3-598-22206-1.

Secretariat countries (Section S) This part lists by country the international organizations which maintain headquarters or other offices in that country. Address are given in each case.
Membership countries (Section M) This part lists, for each country, the international organizations which have members in that country. For each organization listed, the international headquarters address is given, in whatever country that is located.

Vol.3 Global Action Networks : Classified Directory by Subject and Region (Subject Volume). - 8th ed. 1990/91. - 1684 p. - ISBN 3-598-22203-3.

Classified by subject (Sections W, X) These parts list over 23,000 international organizations by subject according to their principal preoccupations. Subjects are grouped into both general and detailed categories, as well as on the basis of interdisciplinary subject combinations. The classification scheme highlights functional relationships between distinct preoccupations.
Classified by region (Section Y). This part lists international organizations by subject according to the region with which they are particularly concerned.

International Congress Calendar/ ed. by UAI. - Brussels: UAI, 1991. 4 vol .- 331st ed. 1989, 30 cm. - ISSN 0538-6349.

The International Congress Calendar is intended as a convenient reference work for anyone seeking information on international events. From the 23rd edition (1983) the Calendar appears quarterly. Each of the four volumes is self-contained including indexes. Amendments and additions occuring between volumes are specially indicated so that every issue contains the most up-to-date information on international events. Again, this year events listed in the Calendar have increased considerably. All the information on these events is derived from primary sources, i.e. the organizations themselves through regular questionnaire mailings. The proven structure of the Calendar remains unchanged, ensuring convenient access to all events by means of: a geographical section, a chronological section and a subject/organizations index.

Encyclopaedia of World Problems and Human Potential / ed.by UAI. - München, New York, London, Paris : Saur, 1991. - 3rd ed., 2 vols, 2140 p., - ISBN 3-598-10842-7.

World Problems and Human Potential, now in its third edition, is a comprehensive source of information on world problems that have been been recognized, on how they are perceived to be interrelated, and on the human resources available to challenge them. Detailed sections draw attention to a variety of alternative insights into the ways in which human development and the world problematique mutually inhibit, enable, and provoke each other.

International Association Statutes Series / ed. by UAI. - München, New York, London, Paris : Saur, 1988. . - 30 cm - ISSN 0933-2588. Vol.1. 1 ed. - 600 p, 30 cm. - ISBN 3-598-21671-8 (Saur München).

The first volume includes the official texts of 393 statutes of international nongovernmental organizations described in Sections A, B, C of the Yearbook of International Organizations, namely bodies with membership in countries in at least two continents. Future volumes will include statutes of organizations from other sections, namely regional bodies and those of a less conventional structure. They may also include statutes of lesser known intergovermental bodies or those of a hybrid governmental/nongovernmental nature. In contrast to the Yearbook series, each volume of the Statute series will only include information not published in previous volumes of the series.

Who's Who in International Organizations ? / ed by UAI. - München, New York, London, Paris: Saur 1991. - ISBN 3-598-10908-3.

This new Who's Who in International Organizations ? is an indispensable reference work for all international non-governmental organizations, intergovernmental organizations, journalists, libraries, universities and research institutes. This 3 volume set contains approximatively 12,000 biographies of eminent individuals from organizations in every field of human endeavour. Intergovernmental organizations; international non-governmental, non-profit bodies; international committees, centers, institutes; information systems, conference series and informal networks; and national non-profit groups concerned with international issues will be represented. The biographies contain: full name, organization, position, nationality, profession, date and place of birth, family, detailed biography, own publications, memberships, and honours. Three indexes list entries by nationality, by field of work and by organization name. The set is scheduled for publication in Summer 1991.